DETROIT PUBLIC LIBRARY
3 5674 04903550 5

P9-EMG-873

CHILTON®

ASIAN SERVICE MANUAL 2008 EDITION VOLUME III

Lexus
Scion
Toyota

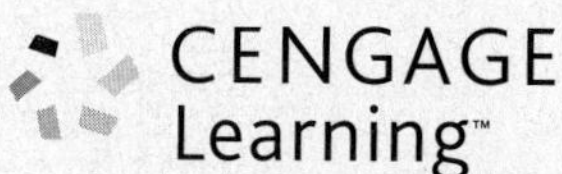

CENGAGE Learning™

Australia • Brazil • Japan • Korea • Mexico • Singapore • Spain • United Kingdom • United States

CHILTON®
Asian Service Manual
2008 Edition
Volume III
Lexus, Scion, Toyota

Vice President, Technology & Trades Professional Business Unit:
Gregory L. Clayton

Publisher, Technology & Trades Professional Business Unit:
David Koontz

Director of Marketing:
Beth A. Lutz

Marketing Manager:
Jennifer Stall

Marketing Assistant:
Rachael Conover

Production Director:
Carolyn Miller

Editorial Assistant:
Jason Yager

Production Manager:
Andrew Crouth

Publishing Coordinator:
Paula Baillie

Sr. Content Project Manager:
Elizabeth C. Hough

Managing Editor:
Terry L. Blomquist

Editors:
Jim Bailey
Ken Burdette
Sherry Burdette
Tim Crain
Scott Critchfield
Nick D'Andrea
Eugene F. Hannon, Jr.
James R. Marotta
Kyla Nyjordet
Christine Sheeky
Lance Williams

Graphical Designer:
Melinda Possinger

For more information contact:
Cengage Learning
Executive Woods
5 Maxwell Drive, PO Box 8007,
Clifton Park, NY 12065-8007
Visit us at **www.chilton.cengage.com**
Visit our corporate website at **www.cengage.com**
For permission to use material from the text or product, contact us by
Tel. (800) 730-2214
Fax (800) 730-2215
www.cengage.com/permissions

Cengage Learning products are represented in Canada by Nelson Education, Ltd.

ISBN 10: 1-4283-2217-5
ISBN 13: 978-1-4283-2217-2
ISSN: 1939-621X

NOTICE TO THE READER

Publisher does not warrant or guarantee any of the products described herein or perform any independent analysis in connection with any of the product information contained herein. Publisher does not assume, and expressly disclaims, any obligation to obtain and include information other than that provided to it by the manufacturer.

The reader is expressly warned to consider and adopt all safety precautions that might be indicated by the activities herein and to avoid all potential hazards. By following the instructions contained herein, the reader willingly assumes all risks in connection with such instructions.

The publisher makes no representation or warranties of any kind, including but not limited to, the warranties of fitness for particular purpose or merchantability, nor are any such representations implied with respect to the material set forth herein, and the publisher takes no responsibility with respect to such material. The publisher shall not be liable for any special, consequential, or exemplary damages resulting, in whole or part, from the readers' use of, or reliance upon, this material.

Printed in the United States of America
1 2 3 4 5 xx 13 12 11 10 09 08

Table of Contents

Sections

Model Index

USING THIS INFORMATION

Organization

To find where a particular model section or procedure is located, look in the Table of Contents. Main topics are listed with the page number on which they may be found. Following the main topics is an alphabetical listing of all of the procedures within the section and their page numbers.

Manufacturer and Model Coverage

This product covers 2007–2008 Asian models that are produced in sufficient quantities to warrant coverage, and which have technical content available from the vehicle manufacturers before our publication date. Although this information is as complete as possible at the time of publication, some manufacturers may make changes which cannot be included here. While striving for total accuracy, the publisher cannot assume responsibility for any errors, changes, or omissions that may occur in the compilation of this data.

Part Numbers & Special Tools

Part numbers and special tools are recommended by the publisher and vehicle manufacturer to perform specific jobs. Before substituting any part or tool for the one recommended, you must be completely satisfied that neither your personal safety, nor the performance of the vehicle will be endangered.

ACKNOWLEDGEMENT

The publisher would like to express appreciation to the following vehicle manufacturers for their assistance in producing this manual. No further reproduction or distribution of the material in this manual is allowed without the expressed written permission of the vehicle manufacturers and the publisher. Toyota Motor Sales USA including Lexus, Scion and Toyota.

PRECAUTIONS

Before servicing any vehicle, please be sure to read all of the following precautions, which deal with personal safety, prevention of component damage, and important points to take into consideration when servicing a motor vehicle:

- Always wear safety glasses or goggles when drilling, cutting, grinding or prying.
- Steel-toed work shoes should be worn when working with heavy parts. Pockets should not be used for carrying tools. A slip or fall can drive a screwdriver into your body.
- Work surfaces, including tools and the floor should be kept clean of grease, oil or other slippery material.
- When working around moving parts, don't wear loose clothing. Long hair should be tied back under a hat or cap, or in a hair net.
- Always use tools only for the purpose for which they were designed. Never pry with a screwdriver.
- Keep a fire extinguisher and first aid kit handy.
- Always properly support the vehicle with approved stands or lift.
- Always have adequate ventilation when working with chemicals or hazardous material.
- Carbon monoxide is colorless, odorless and dangerous. If it is necessary to operate the engine with vehicle in a closed area such as a garage, always use an exhaust collector to vent the exhaust gases outside the closed area.
- When draining coolant, keep in mind that small children and some pets are attracted by ethylene glycol antifreeze, and are quite likely to drink any left in an open container, or in puddles on the ground. This will prove fatal in sufficient quantity. Always drain the coolant into a sealable container.
- To avoid personal injury, do not remove the coolant pressure relief cap while the engine is operating or hot. The cooling system is under pressure; steam and hot liquid can come out forcefully when the cap is loosened slightly. Failure to follow these instructions may result in personal injury. The coolant must be recovered in a suitable, clean container for reuse. If the coolant is contaminated it must be recycled or disposed of correctly.
- When carrying out maintenance on the starting system be aware that heavy gauge leads are connected directly to the battery. Make sure the protective caps are in place when maintenance is completed. Failure to follow these instructions may result in personal injury.
- Do not remove any part of the engine emission control system. Operating the engine without the engine emission control system will reduce fuel economy and engine ventilation. This will weaken engine performance and shorten engine life. It is also a violation of Federal law.
- Due to environmental concerns, when the air conditioning system is drained, the refrigerant must be collected using refrigerant recovery/recycling equipment. Federal law requires that refrigerant be recovered into appropriate recovery equipment and the process be conducted by qualified technicians who have been certified by an approved organization, such as MACS, ASI, etc. Use of a recovery machine dedicated to the appropriate refrigerant is necessary to reduce the possibility of oil and refrigerant incompatibility concerns. Refer to the instructions provided by the equipment manufacturer when removing refrigerant from or charging the air conditioning system.
- Always disconnect the battery ground when working on or around the electrical system.
- Batteries contain sulfuric acid. Avoid contact with skin, eyes, or clothing. Also, shield your eyes when working near batteries to protect against possible splashing of the acid solution. In case of acid contact with skin or eyes, flush immediately with water for a minimum of 15 minutes and get prompt medical attention. If acid is swallowed, call a physician immediately. Failure to follow these instructions may result in personal injury.
- Batteries normally produce explosive gases. Therefore, do not allow flames, sparks or lighted substances to come near the battery. When charging or working near a battery, always shield your face and protect your eyes. Always provide ventilation. Failure to follow these instructions may result in personal injury.
- When lifting a battery, excessive pressure on the end walls could cause acid to spew through the vent caps, resulting in personal injury, damage to the vehicle or battery. Lift with a battery carrier or with your hands on opposite corners. Failure to follow these instructions may result in personal injury.
- Observe all applicable safety precautions when working around fuel. Whenever

servicing the fuel system, always work in a well-ventilated area. Do not allow fuel spray or vapors to come in contact with a spark, open flame, or excessive heat (a hot drop light, for example). Keep a dry chemical fire extinguisher near the work area. Always keep fuel in a container specifically designed for fuel storage; also, always properly seal fuel containers to avoid the possibility of fire or explosion. Do not smoke or carry lighted tobacco or open flame of any type when working on or near any fuel-related components.

• Fuel injection systems often remain pressurized, even after the engine has been turned OFF. The fuel system pressure must be relieved before disconnecting any fuel lines. Failure to do so may result in fire and/or personal injury.

• The evaporative emissions system contains fuel vapor and condensed fuel vapor. Although not present in large quantities, it still presents the danger of explosion or fire. Disconnect the battery ground cable from the battery to minimize the possibility of an electrical spark occurring, possibly causing a fire or explosion if fuel vapor or liquid fuel is present in the area. Failure to follow these instructions can result in personal injury.

• The EPA warns that prolonged contact with used engine oil may cause a number of skin disorders, including cancer! You should make every effort to minimize your exposure to used engine oil. Protective gloves should be worn when changing oil. Wash your hands and any other exposed skin areas as soon as possible after exposure to used engine oil. Soap and water, or waterless hand cleaner should be used.

• Some vehicles are equipped with an air bag system, often referred to as a Supplemental Restraint System (SRS) or Supplemental Inflatable Restraint (SIR) system. The system must be disabled before performing service on or around system components, steering column, instrument panel components, wiring and sensors. Failure to follow safety and disabling procedures could result in accidental air bag deployment, possible personal injury and unnecessary system repairs.

• Always wear safety goggles when working with, or around, the air bag system. When carrying a non-deployed air bag, be sure the bag and trim cover are pointed away from your body. When placing a non-deployed air bag on a work surface, always face the bag and trim cover upward, away from the surface. This will reduce the motion of the module if it is accidentally deployed.

• Electronic modules are sensitive to electrical charges. The ABS module can be damaged if exposed to these charges.

• Brake pads and shoes may contain asbestos, which has been determined to be a cancer-causing agent. Never clean brake surfaces with compressed air. Avoid inhaling brake dust. Clean all brake surfaces with a commercially available brake cleaning fluid.

• When replacing brake pads, shoes, discs or drums, replace them as complete axle sets.

• When servicing drum brakes, disassemble and assemble one side at a time, leaving the remaining side intact for reference.

• Brake fluid often contains polyglycol ethers and polyglycols. Avoid contact with the eyes and wash your hands thoroughly after handling brake fluid. If you do get brake fluid in your eyes, flush your eyes with clean, running water for 15 minutes. If eye irritation persists, or if you have taken brake fluid internally, immediately seek medical assistance.

• Clean, high quality brake fluid from a sealed container is essential to the safe and proper operation of the brake system. You should always buy the correct type of brake fluid for your vehicle. If the brake fluid becomes contaminated, completely flush the system with new fluid. Never reuse any brake fluid. Any brake fluid that is removed from the system should be discarded. Also, do not allow any brake fluid to come in contact with a painted or plastic surface; it will damage the paint.

• Never operate the engine without the proper amount and type of engine oil; doing so will result in severe engine damage.

• Timing belt maintenance is extremely important! Many models utilize an interference-type, non-freewheeling engine. If the timing belt breaks, the valves in the cylinder head may strike the pistons, causing potentially serious (also time-consuming and expensive) engine damage.

• Disconnecting the negative battery cable on some vehicles may interfere with the functions of the on-board computer system (s) and may require the computer to undergo a relearning process once the negative battery cable is reconnected.

• Steering and suspension fasteners are critical parts because they affect performance of vital components and systems and their failure can result in major service expense. They must be replaced with the same grade or part number or an equivalent part if replacement is necessary. Do not use a replacement part of lesser quality or substitute design. Torque values must be used as specified during reassembly to ensure proper retention of these parts.

LEXUS

ES350

1

SPECIFICATIONS AND MAINTENANCE CHARTS

ENGINE AND VEHICLE IDENTIFICATION

Engine						
Code ①	Liters (cc)	Cu. In.	Cyl.	Fuel Sys.	Engine Type	Eng. Mfg.
2GR-FE	3.5 (3456)	211	6	SFI	DOHC	Toyota

Model Year	
Code ②	Year
7	2007
8	2008

SFI: Sequential Fuel Injection

DOHC: Double Overhead Camshaft

NA: Information not available

① Stamped on the left side of the engine block

② 10th digit of the Vehicle Identification Number (VIN)

22140_ES35_C0001

GENERAL ENGINE SPECIFICATIONS

Year	Model	Engine Displacement Liters	Engine Series ID	Net Horsepower @ rpm	Net Torque @ rpm (ft. lbs.)	Bore x Stroke (in.)	Com-pression Ratio	Oil Pressure @ rpm
2007	ES350	3.5	2GR-FE	272@6200	254@4700	3.70x3.27	10.8:1	36-78@3000
2008	ES350	3.5	2GR-FE	272@6200	254@4700	3.70x3.27	10.8:1	36-78@3000

22140_ES35_C0002

ENGINE TUNE-UP SPECIFICATIONS

Year	Engine Displacement Liters	Engine ID	Spark Plug Gap (in.)	Ignition Timing (deg.)*	Fuel Pump (psi)	Idle Speed (rpm)	Valve Clearance	
							Intake	Exhaust
2007	3.5	2GR-FE	0.043	NA	44-50	650-750	NA	NA
2008	3.5	2GR-FE	0.043	NA	44-50	650-750	NA	NA

NOTE: The Vehicle Emission Control Information label often reflects specification changes made during production. The label figures must be used if they differ from those in this chart.

NA: Not available

22140_ES35_C0003

CAPACITIES

Year	Model	Engine Displacement Liters	Engine ID	Engine Oil with Filter (qts.)	Transmission (qts.) 5-Spd	Transmission (qts.) Auto.*	Transfer Case (pts.)	Drive Axle Front (pts.)	Drive Axle Rear (pts.)	Fuel Tank (gal.)	Cooling System (qts.)
2007	ES350	3.5	2GR-FE	6.4	—	6.8	—	—	—	18.5	9.5
2008	ES350	3.5	2GR-FE	6.4	—	6.8	—	—	—	18.5	9.5

22140_ES35_C0004

FLUID SPECIFICATIONS

Year	Model	Engine Displacement Liters (cc)	Engine ID/VIN	Engine Oil	Auto. Trans.	Drive Axle	Power Steering Fluid	Brake Master Cylinder
2007	ES350	3.5 (3456)	2GR-FE	5W-20	—	—	ATF Dexron II Or III	DOT 3
2008	ES350	3.5 (3456)	2GR-FE	5W-20	—	—	ATF Dexron II Or III	DOT 3

DOT: Department Of Transpotation

22140_ES35_C0005

VALVE SPECIFICATIONS

Year	Engine Displacement Liters	Engine ID	Seat Angle (deg.)	Face Angle (deg.)	Spring Test Pressure (lbs. @ in.)	Spring Installed Height (in.)	Stem-to-Guide Clearance (in.) Intake	Stem-to-Guide Clearance (in.) Exhaust	Stem Diameter (in.) Intake	Stem Diameter (in.) Exhaust
2007	3.5	2GR-FE	45	44.5	NA	NA	0.0010-0.0024	0.0012-0.0026	0.2154-0.2159	0.2151-0.2157
2008	3.5	2GR-FE	45	44.5	NA	NA	0.0010-0.0024	0.0012-0.0026	0.2154-0.2159	0.2151-0.2157

NA: Information not available

22140_ES35_C0006

CRANKSHAFT AND CONNECTING ROD SPECIFICATIONS

All measurements are given in inches.

Year	Engine Displacement Liters	Engine ID	Crankshaft Main Brg. Journal Dia.	Crankshaft Main Brg. Oil Clearance	Crankshaft Shaft End-play	Crankshaft Thrust on No.	Connecting Rod Journal Diameter	Connecting Rod Oil Clearance	Connecting Rod Side Clearance
2007	3.5	2GR-FE	2.4011-2.4016	0.0010-0.0019	0.0016-0.0095	2	2.0863-2.0866	0.0018-0.0026	0.0059-0.0157
2008	3.5	2GR-FE	2.4011-2.4016	0.0010-0.0019	0.0016-0.0095	2	2.0863-2.0866	0.0018-0.0026	0.0059-0.0157

22140_ES35_C0007

PISTON AND RING SPECIFICATIONS

All measurements are given in inches.

Year	Engine Displ. Liters	Engine ID	Piston Clearance	Ring Gap Top Comp.	Ring Gap Bottom Comp.	Ring Gap Oil Control	Ring Side Clearance Top Comp.	Ring Side Clearance Bottom Comp.	Ring Side Clearance Oil Control
2007	3.5	2GR-FE	0.0018-0.0020	0.0098-0.0138	0.0197-0.0236	0.0039-0.0157	0.0008-0.0028	0.0008-0.0024	0.0028-0.0059
2008	3.5	2GR-FE	0.0018-0.0020	0.0098-0.0138	0.0197-0.0236	0.0039-0.0157	0.0008-0.0028	0.0008-0.0024	0.0028-0.0059

22140_ES35_C0008

TORQUE SPECIFICATIONS

All readings in ft. lbs.

Year	Engine Displacement Liters	Engine ID	Cylinder Head Bolts	Main Bearing Bolts	Rod Bearing Bolts	Crankshaft Damper Bolts	Flywheel Bolts	Manifold Intake	Manifold Exhaust	Spark Plugs	Oil Pan Drain Plug
2007	3.5	2GR-FE	①	②	③	184	61	15	15	13	30
2008	3.5	2GR-FE	①	②	③	184	61	15	15	13	30

① Step 1: 10mm bolts to 27 ft. lbs.
Step 2: 10mm point cap bolts plus 90 degrees
Step 3: 10mm point cap bolts plus 90 degrees

② Step 1: 16 cap bolts to 45 ft. lbs.
Step 2: 16 cap bolts plus 90 degrees
Step 3: 8 side bolts to 38 ft. lbs.
Step 4: Front bolts to 22 ft. lbs.

③ Step 1: 18 ft. lbs.
Step 2: Plus 90 degrees

22140_ES35_C0009

WHEEL ALIGNMENT

Year	Model		Caster Range (+/-Deg.)	Caster Preferred Setting (Deg.)	Camber Range (+/-Deg.)	Camber Preferred Setting (Deg.)	Toe-in (in.)	Steering Axis Inclination (Deg.)
2007	ES350	Front	0.75	2.95	0.75	-0.72	0+/-0.08	12.33
		Rear	—	—	0.75	-1.35	0.16+/-0.08	—
2008	ES350	Front	0.75	2.95	0.75	-0.72	0+/-0.08	12.33
		Rear	—	—	0.75	-1.35	0.16+/-0.08	—

22140_ES35_C0010

TIRE, WHEEL AND BALL JOINT SPECIFICATIONS

Year	Model	OEM Tires Standard	OEM Tires Optional	Tire Pressures (psi) Front	Tire Pressures (psi) Rear	Wheel Size	Ball Joint Inspection	Lug Nut Torque (ft. lbs.)
2007	ES350	P215/55R17	—	32	32	7-J	①	76
2008	ES350	P215/55R17	—	32	32	7-J	①	76

OEM: Original Equipment Manufacturer

PSI: Pounds Per Square Inch

STD: Standard

OPT: Optional

① Replace if any measurable movement is found.

22140_ES35_C0011

BRAKE SPECIFICATIONS

All measurements in inches unless noted

Year	Model		Brake Disc Original Thickness	Brake Disc Minimum Thickness	Brake Disc Maximum Runout	Minimum Lining Thickness	Brake Caliper Bracket Bolts (ft. lbs.)	Brake Caliper Mounting Bolts (ft. lbs.)
2007	ES350	F	1.102	0.983	0.0020	0.039	79	25
		R	0.390	0.334	0.0059	0.039	46	20
2008	ES350	F	1.102	0.983	0.0020	0.039	79	25
		R	0.390	0.334	0.0059	0.039	46	20

F: Front

R: Rear

22140_ES35_C0012

SCHEDULED MAINTENANCE INTERVALS

LEXUS—ES350

TO BE SERVICED	TYPE OF SERVICE	VEHICLE MILEAGE INTERVAL (x1000)												
		5	10	15	20	25	30	35	40	45	50	55	60	90
Engine oil & filter	R	✓	✓	✓	✓	✓	✓	✓	✓	✓	✓	✓	✓	✓
Automatic transmission fluid	S/I			✓			✓			✓			✓	✓
Ball joints & dust covers	S/I			✓			✓			✓			✓	✓
Bolts & nuts on chassis & body	S/I			✓			✓			✓			✓	✓
Brake linings & drums	S/I	✓	✓	✓	✓	✓	✓	✓	✓	✓	✓	✓	✓	✓
Brake line pipes & hoses	S/I			✓			✓			✓			✓	✓
Brake pads & discs (front & rear)	S/I	✓	✓	✓	✓	✓	✓	✓	✓	✓	✓	✓	✓	✓
Brake fluid	R						✓						✓	✓
Rack and pinion assembly	S/I			✓			✓			✓			✓	✓
Steering linkage & boots	S/I			✓			✓			✓			✓	✓
Air cleaner filter	R						✓						✓	✓
Spark plugs ①	R													
Drive belts	S/I												✓	✓
Exhaust pipes & mountings	S/I			✓			✓			✓			✓	✓
Fuel lines & connections	S/I						✓						✓	✓
Engine coolant ②	S/I			✓			✓			✓			✓	✓
Fuel tank cap gasket	S/I						✓						✓	✓
Rotate tires	S/I			✓			✓			✓			✓	
Clean air conditioning filter ③	S/I			✓			✓			✓			✓	
Axle shaft bolts	S/I			✓			✓			✓			✓	✓
Brake pad thickness and rotor runout	S/I						✓						✓	✓

R: Replace S/I: Service or Inspect

① Spark plugs are replaced at 120,000 miles

② Replace engine coolant at 100,000 miles and then inspect every 15,000 miles

③ Replace air conditioning filter every 30,000 miles

FREQUENT OPERATION MAINTENANCE (SEVERE SERVICE)

If a vehicle is operated under any of the following conditions it is considered severe service:

- Extremely dusty areas.

- 50% or more of the vehicle operation is in 32°C (90°F) or higher temperatures, or constant temperatures below 0°C (32°F).

- Prolonged idling (vehicle operation in stop and go traffic).

- Frequent short running periods (engine does not warm to normal operating temperatures).

- Police, taxi, delivery usage or trailer towing usage.

Air cleaner filter: service or inspect every 5000 miles

Ball joints & dust covers: service or inspect every 5000 miles.

Bolts & nuts on chassis & body: service or inspect every 5000 miles.

Axle shaft bolts: service or inspect every 5000 miles.

Steering linkage: service or inspect every 5000 miles.

22140_ES35_C0013

PRECAUTIONS

Before servicing any vehicle, please be sure to read all of the following precautions, which deal with personal safety, prevention of component damage, and important points to take into consideration when servicing a motor vehicle:

- Never open, service or drain the radiator or cooling system when the engine is hot; serious burns can occur from the steam and hot coolant.
- Observe all applicable safety precautions when working around fuel. Whenever servicing the fuel system, always work in a well-ventilated area. Do not allow fuel spray or vapors to come in contact with a spark, open flame, or excessive heat (a hot drop light, for example). Keep a dry chemical fire extinguisher near the work area. Always keep fuel in a container specifically designed for fuel storage; also, always properly seal fuel containers to avoid the possibility of fire or explosion. Refer to the additional fuel system precautions later in this section.
- Fuel injection systems often remain pressurized, even after the engine has been turned **OFF**. The fuel system pressure must be relieved before disconnecting any fuel lines. Failure to do so may result in fire and/or personal injury.
- Brake fluid often contains polyglycol ethers and polyglycols. Avoid contact with the eyes and wash your hands thoroughly after handling brake fluid. If you do get brake fluid in your eyes, flush your eyes with clean, running water for 15 minutes. If eye irritation persists, or if you have taken brake fluid internally, IMMEDIATELY seek medical assistance.
- The EPA warns that prolonged contact with used engine oil may cause a number of skin disorders, including cancer. You should make every effort to minimize your exposure to used engine oil. Protective gloves should be worn when changing oil. Wash your hands and any other exposed skin areas as soon as possible after exposure to used engine oil. Soap and water, or waterless hand cleaner should be used.
- All new vehicles are now equipped with an air bag system, often referred to as a Supplemental Restraint System (SRS) or Supplemental Inflatable Restraint (SIR) system. The system must be disabled before performing service on or around system components, steering column, instrument panel components, wiring and sensors. Failure to follow safety and disabling procedures could result in accidental air bag deployment, possible personal injury and unnecessary system repairs.
- Always wear safety goggles when working with, or around, the air bag system. When carrying a non-deployed air bag, be sure the bag and trim cover are pointed away from your body. When placing a non-deployed air bag on a work surface, always face the bag and trim cover upward, away from the surface. This will reduce the motion of the module if it is accidentally deployed. Refer to the additional air bag system precautions later in this section.
- Clean, high quality brake fluid from a sealed container is essential to the safe and proper operation of the brake system. You should always buy the correct type of brake fluid for your vehicle. If the brake fluid becomes contaminated, completely flush the system with new fluid. Never reuse any brake fluid. Any brake fluid that is removed from the system should be discarded. Also, do not allow any brake fluid to come in contact with a painted surface; it will damage the paint.
- Never operate the engine without the proper amount and type of engine oil; doing so WILL result in severe engine damage.
- Timing belt maintenance is extremely important. Many models utilize an interference-type, non-freewheeling engine. If the timing belt breaks, the valves in the cylinder head may strike the pistons, causing potentially serious (also time-consuming and expensive) engine damage. Refer to the maintenance interval charts for the recommended replacement interval for the timing belt, and to the timing belt section for belt replacement and inspection.
- Disconnecting the negative battery cable on some vehicles may interfere with the functions of the on-board computer system(s) and may require the computer to undergo a relearning process once the negative battery cable is reconnected.
- When servicing drum brakes, only disassemble and assemble one side at a time, leaving the remaining side intact for reference.
- Only an MVAC-trained, EPA-certified automotive technician should service the air conditioning system or its components.

BRAKES

ANTI-LOCK BRAKE SYSTEM (ABS)

GENERAL INFORMATION

PRECAUTIONS

- Certain components within the ABS system are not intended to be serviced or repaired individually.
- Do not use rubber hoses or other parts not specifically specified for and ABS system. When using repair kits, replace all parts included in the kit. Partial or incorrect repair may lead to functional problems and require the replacement of components.
- Lubricate rubber parts with clean, fresh brake fluid to ease assembly. Do not use shop air to clean parts; damage to rubber components may result.
- Use only DOT 3 brake fluid from an unopened container.
- If any hydraulic component or line is removed or replaced, it may be necessary to bleed the entire system.
- A clean repair area is essential. Always clean the reservoir and cap thoroughly before removing the cap. The slightest amount of dirt in the fluid may plug an orifice and impair the system function. Perform repairs after components have been thoroughly cleaned; use only denatured alcohol to clean components. Do not allow ABS components to come into contact with any substance containing mineral oil; this includes used shop rags.
- The Anti-Lock control unit is a microprocessor similar to other computer units in the vehicle. Ensure that the ignition switch is **OFF** before removing or installing controller harnesses. Avoid static electricity discharge at or near the controller.
- If any arc welding is to be done on the vehicle, the control unit should be unplugged before welding operations begin.

BRAKES — BLEEDING THE BRAKE SYSTEM

BLEEDING PROCEDURE

MASTER CYLINDER BLEEDING

If the master cylinder is reinstalled or if the reservoir becomes empty, bleed the air from the master cylinder. To prevent brake fluid from adhering, cover nearly painted surfaces with a shop rag or a piece of cloth.

1. Using a union nut wrench (10 mm), disconnect the 2 brake lines from the master cylinder, using a suitable brake line wrench.
2. Have an assistant slowly depress the brake pedal and hold it.
3. Cover the 2 outer holes with your fingers, and have your assistant release the brake pedal.
4. Repeat the previous 2 steps 3 or 4 times.
5. Using a union nut wrench (10 mm), connect the 2 the brake lines to the master cylinder and tighten to 11 ft. lbs. (15 Nm).

➡Use a torque wrench with a fulcrum length of 250 mm (9.84 in.).

➡This torque value is effective when the union nut wrench is parallel to the torque wrench.

BRAKE LINE BLEEDING

➡Bleed air from the brake line of the wheel farthest from the master cylinder.

1. Raise and safely support the vehicle.
2. Connect a vinyl tube to the bleeder plug.
3. Have an assistant depress the brake pedal several times, then loosen the bleeder plug while the pedal is depressed.
4. When fluid stops coming out, tighten the bleeder plug, then release the brake pedal.
5. Repeat steps 3 and 4 until all the air in the fluid has been bled out.
6. Tighten the brake bleeder plug to 73 inch lbs. (8.3 Nm).
7. Repeat the above steps to bleed the air out of the brake line for each wheel.

BRAKES — FRONT DISC BRAKES

CAUTION

Dust and dirt accumulating on brake parts during normal use may contain asbestos fibers from production or aftermarket brake linings. Breathing excessive concentrations of asbestos fibers can cause serious bodily harm. Exercise care when servicing brake parts. Do not sand or grind brake lining unless equipment used is designed to contain the dust residue. Do not clean brake parts with compressed air or by dry brushing. Cleaning should be done by dampening the brake components with a fine mist of water, then wiping the brake components clean with a dampened cloth. Dispose of cloth and all residue containing asbestos fibers in an impermeable container with the appropriate label. Follow practices prescribed by the Occupational Safety and Health Administration (OSHA) and the Environmental Protection Agency (EPA) for the handling, processing, and disposing of dust or debris that may contain asbestos fibers.

BRAKE CALIPER

REMOVAL & INSTALLATION

See Figure 1.

1. Before servicing the vehicle, refer to the Precautions Section.
2. Remove the front wheel.

➡Do not let brake fluid sit on painted surfaces, as it will eat through the paint. Wash it off immediately.

3. Drain brake fluid.
4. Remove the union bolt and gasket from the disc brake cylinder assembly, then disconnect the flexible hose.

➡Remove the disc brake cylinder assembly while holding both of the brake pads or the anti-squeal springs may fall off the brake pads.

5. Hold the front disc brake cylinder slide pin and remove the 2 bolts and disc brake cylinder assembly.

To install:

6. Install the disc brake cylinder assembly with the 2 bolts and tighten to 25 ft. lbs. (34 Nm).
7. Check the installation of the anti-squeal springs. Visually check for any clearance between the brake pad and front disc brake pad support plates.

➡ If the anti-squeal springs are installed correctly, there will be no clearance between the brake pad and the front disc brake pad support plates. If there is a clearance, the anti-squeal springs may not be installed properly.

➡ Check all 4 contact surfaces between the brake pad and the front disc brake pad support plates.

8. Connect the flexible hose with the union bolt and a new gasket and tighten to 21 ft. lbs. (29 Nm).

➡Install the front brake flexible hose lock securely in the lock hole in the disc brake cylinder.

9. Fill reservoir with brake fluid.
10. Bleed master cylinder.
11. Bleed brake line.
12. Bleed brake actuator assembly.
13. Inspect for brake fluid leak.
14. Inspect brake fluid level in reservoir.
15. Install the front wheel and tighten the lug nuts to 76 ft. lbs. (103 Nm).

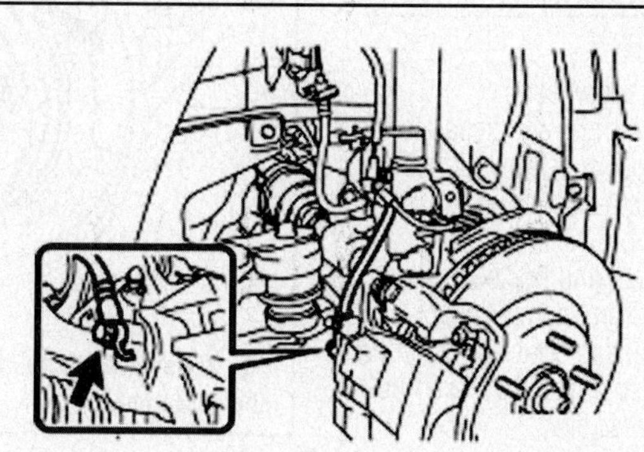

Fig. 1 Removing the union bolt and gasket

DISC BRAKE PADS

REMOVAL & INSTALLATION

1. Before servicing the vehicle, refer to the Precautions Section.
2. Remove the 2 front disc brake cylinder slide pins (upper and lower) from the front disc brake cylinder mounting.
3. Remove brake cylinder.
4. Remove the 2 anti-squeal springs.
5. Remove the 2 brake pads from the front disc brake cylinder mounting.

To install:

6. Install the 2 brake pads with front anti-squeal shims to the front disc brake cylinder mounting.

➡When replacing worn pads, the front anti-squeal springs must be replaced at the same time.

➡Be sure to install the anti-squeal springs into the front disc brake pad installation holes as far as they will go.

7. Install the 2 front disc brake cylinder slide pins (upper and lower) from the front disc brake cylinder mounting.
8. Install the brake cylinder.

BRAKES — REAR DISC BRAKES

⁂ CAUTION

Dust and dirt accumulating on brake parts during normal use may contain asbestos fibers from production or aftermarket brake linings. Breathing excessive concentrations of asbestos fibers can cause serious bodily harm. Exercise care when servicing brake parts. Do not sand or grind brake lining unless equipment used is designed to contain the dust residue. Do not clean brake parts with compressed air or by dry brushing. Cleaning should be done by dampening the brake components with a fine mist of water, then wiping the brake components clean with a dampened cloth. Dispose of cloth and all residue containing asbestos fibers in an impermeable container with the appropriate label. Follow practices prescribed by the Occupational Safety and Health Administration (OSHA) and the Environmental Protection Agency (EPA) for the handling, processing, and disposing of dust or debris that may contain asbestos fibers.

BRAKE CALIPER

REMOVAL & INSTALLATION

1. Before servicing the vehicle, refer to the Precautions Section.
2. Remove the rear wheel.
3. Drain brake fluid.

⁂ WARNING

Do not let brake fluid sit on painted surfaces, as it will eat through the paint. Wash it off immediately.

4. Remove the union bolt and the gasket from the rear disc brake cylinder assembly, then disconnect the rear brake flexible hose.
5. Hold the 2 rear disc brake cylinder slide pins and remove the 2 bolts and rear disc brake cylinder assembly.

To install:

6. Install the rear disc brake cylinder assembly with the 2 bolts and tighten to 20 ft. lbs. (77 Nm).
7. Connect the rear brake flexible hose with the union bolt and a new gasket and tighten to 24 ft. lbs. (33 Nm).
8. Fill reservoir with brake fluid.
9. Bleed brake line.
10. Inspect for brake fluid leak.
11. Inspect brake fluid level in reservoir.
12. Install the rear wheel and tighten the lug nuts to 76 ft. lbs. (103 Nm).

DISC BRAKE PADS

REMOVAL & INSTALLATION

1. Before servicing the vehicle, refer to the Precautions Section.
2. Remove the 2 rear disc brake cylinder slide pins (upper and lower) from the rear disc brake cylinder mounting.
3. Remove brake cylinder.
4. Remove the 2 brake pads with the rear anti-squeal shims.

To install:

5. Installation is the reverse of removal procedure.

BRAKES — PARKING BRAKE

PARKING BRAKE SHOES

REMOVAL & INSTALLATION

See Figures 2 through 4.

1. Remove the rear wheel.
2. Remove the 2 bolts and separate the rear disc brake caliper assembly. Do not disconnect the flexible hose from the disc brake caliper assembly.
3. Remove the parking brake shoe adjusting hole plug from the rear disc.
4. Release the parking brake and place the matchmarks on the rear disc and the axle hub.
5. Remove the rear disc.

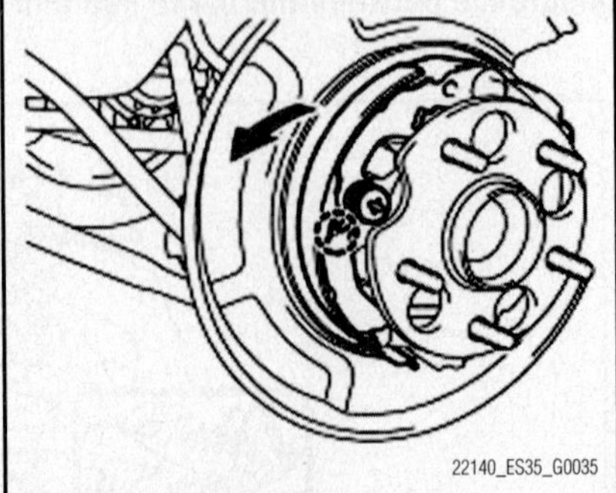

Fig. 2 Remove the No. 1 parking brake shoe assembly

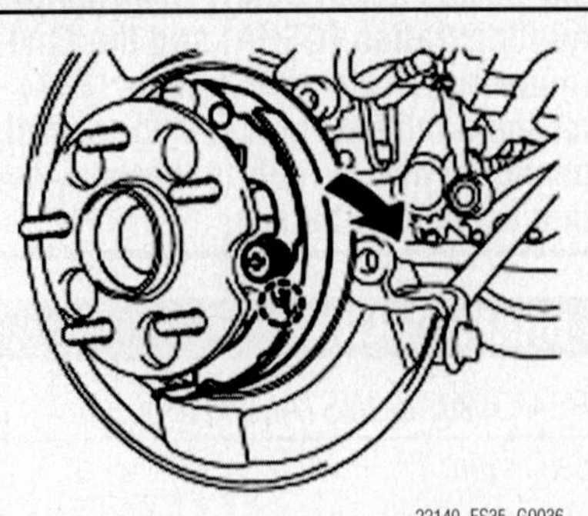

Fig. 3 Remove the No. 2 parking brake shoe assembly

➡**If the disc cannot be removed easily, turn the shoe adjuster until the disc turns freely.**

6. Using needle-nose pliers, remove the 2 parking brake shoe return tension No. 1 springs.
7. Remove the parking brake shoe strut and the parking brake shoe strut compression spring.
8. Remove the No. 1 parking brake shoe assembly, as follows:
   a. Release the claw of the parking brake shoe hold down spring No. 2 cup.
   b. Remove the No. 1 parking brake shoe assembly as shown in the illustration.
   c. Remove the parking brake shoe hold down spring No. 1 cup, the parking brake shoe hold down spring, the parking brake shoe hold down spring No. 2 cup, and the parking brake shoe hold down spring No. 1 pin.
9. Remove the parking brake shoe adjusting screw set.
10. Remove the parking brake shoe return tension No. 2 spring.
11. Remove the No. 2 parking brake shoe assembly, as follows:
   a. Release the claw of the parking brake shoe hold down spring No. 2 cup.
   b. Remove the No. 2 parking brake shoe assembly as shown in the illustration.
   c. Remove the parking brake shoe hold down spring No. 1 cup, the parking brake shoe hold down spring, the parking brake shoe hold down spring No. 2 cup, and the parking brake shoe hold down spring No. 2 pin.
   d. Using needle-nose pliers, disconnect the No. 3 parking brake cable assembly from the parking brake shoe lever.

➡**Be careful not to damage the No. 3 parking brake cable assembly.**

12. Using a screwdriver, remove the C-washer, shim and the parking brake shoe lever.
13. Remove the parking brake shoe guide plate set bolt and the parking brake shoe guide plate.

To install:

14. Apply high temperature grease to the backing plate where it contacts the shoe.
15. Apply adhesive (Toyota Genuine Adhesive 1344, Three Bond 1344 or equivalent) to the threads of the parking brake shoe guide plate set bolt.
16. Install the parking brake shoe guide plate with the parking brake shoe guide plate set bolt and tighten to 13 ft. lbs. (18 Nm).
17. Install the parking brake shoe lever and shim to the No. 2 parking brake shoe assembly with a new C-washer.
18. Using a feeler gauge, measure the clearance between the No. 2 parking brake shoe assembly and parking brake shoe lever. Standard clearance: Less than 0.35 mm (0.014 in.).
19. If the clearance is not as specified, replace the shim with one of the correct size.
20. Install the No. 2 parking brake shoe assembly as follows:
   a. Using needle-nose pliers, connect the No. 3 parking brake cable assembly to the parking brake shoe lever.
   b. Install the No. 2 parking brake shoe assembly with the parking brake shoe hold down spring No. 2 pin, the parking brake shoe hold down spring No. 2 cup, the parking brake shoe hold down spring and the parking brake shoe hold down spring No. 1 cup.
   c. Engage the claw of the parking brake shoe hold down spring No. 2 cup to the No. 2 parking brake shoe assembly.
21. Install the parking brake shoe adjusting screw set, as follows:
   a. Apply high temperature grease to the parking brake shoe adjusting screw set as shown in the illustration.
   b. Install the parking brake shoe return tension No. 2 spring to the No. 1 parking brake shoe assembly and the No. 2 parking brake shoe assembly.
   c. Install the parking brake shoe adjusting screw set to the No. 1 parking brake shoe assembly and the No. 2 parking brake shoe assembly.
22. Install the No. 1 parking brake shoe assembly as follows:
   a. Install the No. 1 parking brake shoe assembly with the parking brake shoe hold down spring No. 1 pin, parking brake shoe hold down spring No. 2 cup, parking brake shoe hold down spring and parking brake shoe hold down spring No. 1 cup.
   b. Engage the claw of the parking brake shoe hold down spring No. 2 cup to the No. 1 parking brake shoe assembly.
23. Attach the parking brake shoe strut and the parking brake shoe strut compression spring to the No. 1 parking brake shoe assembly and No. 2 parking brake shoe assembly.
24. Using needle-nose pliers, install the 2 parking brake shoe return tension No. 1 springs. First install the front side spring and then the rear side spring.
25. Inspect parking brake installation and check that each part is installed properly.

➡**There should be no oil or grease on the friction surfaces of the shoe linings and discs.**

26. Install the rear disc.
27. Install the parking brake shoe adjusting hole plug.
28. Adjust parking brake shoe clearance.
29. Install the rear disc brake caliper assembly with the 2 bolts and tighten to 46 ft. lbs. (62 Nm).

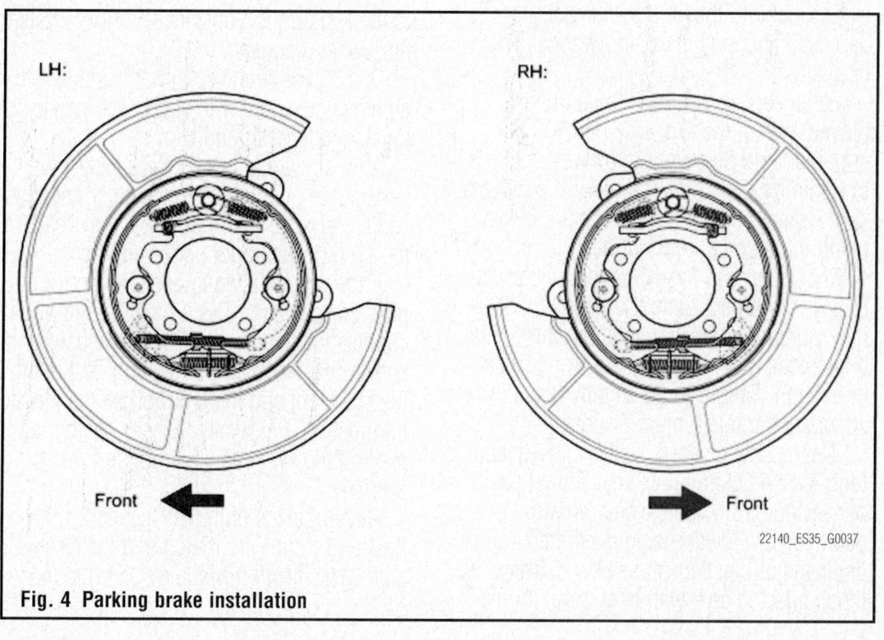

Fig. 4 Parking brake installation

30. Install the rear wheel.
31. Adjust the parking brake pedal travel.
32. Bed in parking brake shoes to discs, as follows:
    a. Drive the vehicle at about 31 mph (50 km/h) on a safe, level and dry road.
    b. Depress the parking brake pedal with 34 lbs. (150 N) of force.
33. Drive the vehicle about 0.25 miles (400 m) in this condition.
    a. Repeat this procedure 3 times using 5-minute intervals between each procedure to prevent the parking brake assembly from overheating.
34. Remove the rear wheel.
35. Adjust parking brake shoe clearance.
36. For A/T vehicles, adjust the parking brake pedal travel.
37. For M/T vehicles, adjust the parking brake pedal travel.
38. Install the rear wheel and tighten the lug nuts to 76 ft. lbs. (103 Nm).

CHASSIS ELECTRICAL

AIR BAG (SUPPLEMENTAL RESTRAINT SYSTEM)

GENERAL INFORMATION

** CAUTION

These vehicles are equipped with an air bag system. The system must be disarmed before performing service on, or around, system components, the steering column, instrument panel components, wiring and sensors. Failure to follow the safety precautions and the disarming procedure could result in accidental air bag deployment, possible injury and unnecessary system repairs.

SERVICE PRECAUTIONS

Disconnect and isolate the battery negative cable before beginning any airbag system component diagnosis, testing, removal, or installation procedures. Allow system capacitor to discharge for two minutes before beginning any component service. This will disable the airbag system. Failure to disable the airbag system may result in accidental airbag deployment, personal injury, or death.

Do not place an intact undeployed airbag face down on a solid surface. The airbag will propel into the air if accidentally deployed and may result in personal injury or death.

When carrying or handling an undeployed airbag, the trim side (face) of the airbag should be pointing towards the body to minimize possibility of injury if accidental deployment occurs. Failure to do this may result in personal injury or death.

Replace airbag system components with OEM replacement parts. Substitute parts may appear interchangeable, but internal differences may result in inferior occupant protection. Failure to do so may result in occupant personal injury or death.

Wear safety glasses, rubber gloves, and long sleeved clothing when cleaning powder residue from vehicle after an airbag deployment. Powder residue emitted from a deployed airbag can cause skin irritation. Flush affected area with cool water if irritation is experienced. If nasal or throat irritation is experienced, exit the vehicle for fresh air until the irritation ceases. If irritation continues, see a physician.

Do not use a replacement airbag that is not in the original packaging. This may result in improper deployment, personal injury, or death.

The factory installed fasteners, screws and bolts used to fasten airbag components have a special coating and are specifically designed for the airbag system. Do not use substitute fasteners. Use only original equipment fasteners listed in the parts catalog when fastener replacement is required.

During, and following, any child restraint anchor service, due to impact event or vehicle repair, carefully inspect all mounting hardware, tether straps, and anchors for proper installation, operation, or damage. If a child restraint anchor is found damaged in any way, the anchor must be replaced. Failure to do this may result in personal injury or death.

Deployed and non-deployed airbags may or may not have live pyrotechnic material within the airbag inflator.

Do not dispose of driver/passenger/curtain airbags or seat belt tensioners unless you are sure of complete deployment. Refer to the Hazardous Substance Control System for proper disposal.

Dispose of deployed airbags and tensioners consistent with state, provincial, local, and federal regulations.

After any airbag component testing or service, do not connect the battery negative cable. Personal injury or death may result if the system test is not performed first.

If the vehicle is equipped with the Occupant Classification System (OCS), do not connect the battery negative cable before performing the OCS Verification Test using the scan tool and the appropriate diagnostic information. Personal injury or death may result if the system test is not performed properly.

Never replace both the Occupant Restraint Controller (ORC) and the Occupant Classification Module (OCM) at the same time. If both require replacement, replace one, then perform the Airbag System test before replacing the other.

Both the ORC and the OCM store Occupant Classification System (OCS) calibration data, which they transfer to one another when one of them is replaced. If both are replaced at the same time, an irreversible fault will be set in both modules and the OCS may malfunction and cause personal injury or death.

If equipped with OCS, the Seat Weight Sensor is a sensitive, calibrated unit and must be handled carefully. Do not drop or handle roughly. If dropped or damaged, replace with another sensor. Failure to do so may result in occupant injury or death.

If equipped with OCS, the front passenger seat must be handled carefully as well. When removing the seat, be careful when setting on floor not to drop. If dropped, the sensor may be inoperative, could result in occupant injury, or possibly death.

If equipped with OCS, when the passenger front seat is on the floor, no one should sit in the front passenger seat. This uneven force may damage the sensing ability of the seat weight sensors. If sat on and damaged, the sensor may be inoperative, could result in occupant injury, or possibly death.

DISARMING THE SYSTEM

To avoid personal injury when working on vehicles equipped with an air bag, the negative battery cable must be disconnected and at least 90 seconds must elapse before working on the system. Failure to do so may result in deployment of the air bag.

ARMING THE SYSTEM

To arm the system after service is finished, connect the negative battery cable.

CLOCKSPRING CENTERING

See Figures 5 and 6.

1. Before servicing the vehicle, refer to the Precautions Section.
2. Check that the ignition switch is **OFF**.
3. Check that the battery negative (-) terminal is disconnected.

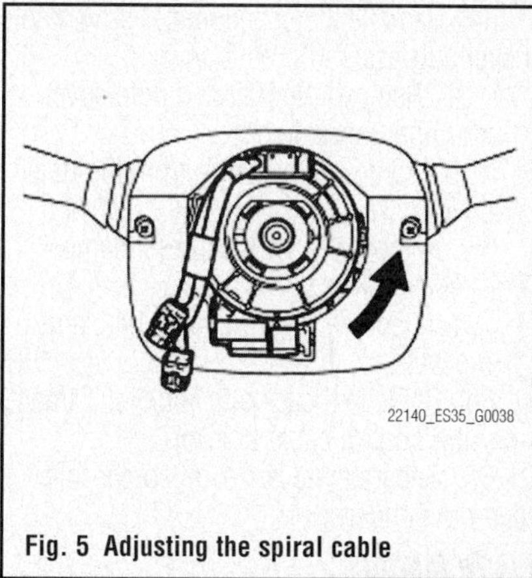

Fig. 5 Adjusting the spiral cable

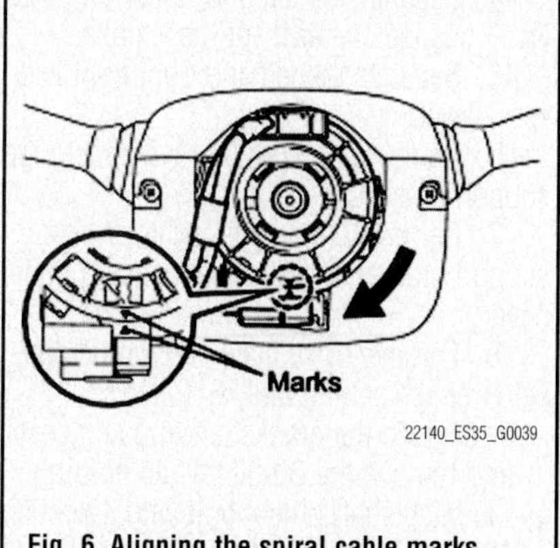

Fig. 6 Aligning the spiral cable marks

CAUTION

After removing the terminal, wait for at least 90 seconds before starting the operation.

4. Rotate the spiral cable counterclockwise slowly by hand until it feels firm.
5. Rotate the spiral cable clockwise approximately 2.5 turns to align the marks.

➡Do not turn the spiral cable by the airbag wire harness.

➡The spiral cable will rotate approximately 2.5 turns to both the left and right from the center.

DRIVETRAIN

AUTOMATIC TRANSAXLE ASSEMBLY

REMOVAL & INSTALLATION

See Figures 7 through 14.

➡When the automatic transaxle is replaced, the transaxle's compensation code must be input into the TCM. When the TCM is replaced, the existing transaxle compensation codes must be input into the new TCM.

➡If the valve body assembly has been replaced, initialize the TCM and perform a road test to allow the TCM to learn.

➡If the shift solenoid SL1 has been replaced, perform a road test to allow the TCM to learn.

➡If the TCM or transaxle has been replaced, register the transmission compensation code in the TCM.

➡Perform RESET MEMORY (A/T initialization) when replacing the automatic transmission assembly, engine assembly or ECM.

1. Before servicing the vehicle, refer to the Precautions Section.

CAUTION

Observe all applicable safety precautions when working around fuel. Whenever servicing the fuel system, always work in a well ventilated area. Do not allow fuel spray or vapors to come in contact with a spark or open flame. Keep a dry chemical fire extinguisher near the work area. Always keep fuel in a container specifically designed for fuel storage; also, always properly seal fuel containers to avoid the possibility of fire or explosion.

2. Discharge the fuel system pressure.
3. Disconnect cable from negative battery terminal.
4. Place front wheels facing straight ahead.
5. Remove front wheels.
6. Remove the right and left engine under covers.
7. Remove the right front fender apron seal.
8. Drain engine coolant.
9. Drain automatic transaxle fluid, as follows:
   a. Remove the refill plug and gasket.
   b. Using a 6 mm socket hexagon wrench, remove the overflow plug and gasket.
   c. Using a 6 mm socket hexagon wrench, remove the No. 1 transmission oil filler tube.
   d. Drain the automatic transaxle fluid.
   e. Using a 6 mm socket hexagon wrench, install the No. 1 transmission oil filler tube and tighten to 15 inch lbs. (1.7 Nm).
   f. Using a 6 mm socket hexagon wrench, install the overflow plug with a new gasket and tighten to 30 ft. lbs. (40 Nm).
   g. Install the refill plug with a new gasket and tighten to 36 ft. lbs. (49 Nm).
10. Remove windshield wiper link assembly.
11. Remove the cowl top panel outer sub-assembly.
12. Remove the cool air intake duct seal.
13. Remove the V-bank cover sub-assembly.
14. Remove the V-ribbed belt.
15. Remove air cleaner inlet assembly.
16. Remove air cleaner cap sub-assembly.
17. Remove air cleaner case sub-assembly.
18. Remove the No. 1 air cleaner inlet.
19. Remove the battery.
20. Remove the intake air resonator sub-assembly.

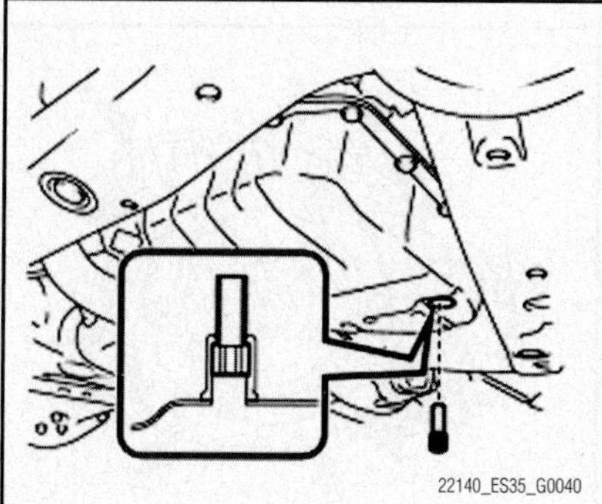

Fig. 7 Remove the No. 1 transmission oil filler tube

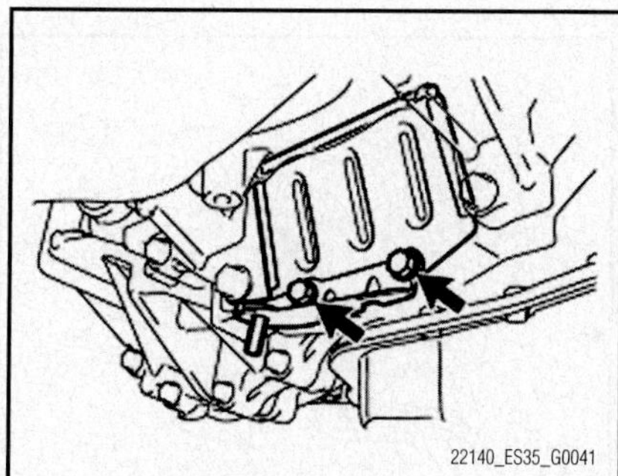

Fig. 8 Removing the 2 bolts, exhaust pipe support bracket and flywheel housing under cover

21. Remove the right No. 2 engine mounting stay.
22. Remove the engine moving control rod sub-assembly.
23. Disconnect the No. 1 fuel vapor feed hose.
24. Disconnect check valve to brake booster hose.
25. Disconnect the radiator inlet and outlet hoses.
26. Disconnect the No. 1 oil cooler inlet and outlet hoses.
27. Disconnect the heater water inlet hose.
28. Disconnect the heater water outlet hose.
29. Remove the ECM.
30. Remove the upper relay block cover.
31. Disconnect the engine wire.
32. Remove the bolt and disconnect the wire harness.
33. Disconnect the park/neutral position switch connector.
34. Disconnect the transmission control cable assembly, as follows:
    a. Remove the nut from the control shaft lever.
    b. Disconnect the transmission control cable assembly from the control shaft lever.
    c. Remove the clip and disconnect the transmission control cable assembly from the No. 1 transmission control cable bracket.
35. Separate the fuel tube sub-assembly.
36. Disconnect the No. 1 oil reservoir to pump hose.
37. Disconnect return tube sub-assembly.
38. Remove the No. 1 exhaust pipe support bracket.
39. Remove front exhaust pipe assembly.
40. Remove both front axle hub nuts.
41. Disconnect both front stabilizer link assemblies.
42. Disconnect the left and right front speed sensors.
43. Separate the left and right tie rod end sub-assemblies.
44. Separate the left and right front lower No. 1 suspension arm sub-assemblies.
45. Separate the left and right front axle assemblies.
46. Remove the front No. 1 exhaust pipe support bracket.
47. Remove the 2 bolts, exhaust pipe support bracket and flywheel housing under cover.
48. Remove drive plate and torque converter clutch setting bolt, as follows:
    a. Turn the crankshaft to gain access and remove the 6 bolts while holding the crankshaft pulley bolt with a wrench.
49. Disconnect steering sliding yoke.
50. Remove generator assembly.
51. Separate cooler compressor assembly.
52. Remove engine assembly with transaxle.
53. Remove vane pump assembly.
54. Install engine hangers.
55. Remove front frame assembly.
56. Remove both front drive shaft assemblies.
57. Remove engine wire.
58. Remove starter assembly.
59. Remove the automatic transaxle assembly, as follows:
    a. Remove the bolt and breather bracket from the left camshaft housing sub-assembly.
    b. Remove the breather hose from the breather bracket.
    c. Remove the 11 bolts.
    d. Separate and remove the automatic transaxle.
60. Put matchmarks on the transaxle housing and torque converter clutch assembly.
61. Remove the torque converter clutch assembly from the automatic transaxle assembly.
62. Remove the 3 bolts and engine mounting front bracket.
63. Remove the 2 bolts and TCM from the transaxle.
64. Remove the speedometer driven hole cover sub-assembly, as follows:
    a. Remove the bolt and hole cover from the transaxle case.
    b. Remove the O-ring from the hole cover.
65. Remove the wire harness clamp bracket, as follows:
    a. Remove the 4 bolts and 4 clamp brackets.
66. Remove the 2 bolts and No. 1 transmission control cable bracket.
67. Separate the No. 1 oil cooler inlet and outlet hoses.

To install:

68. Inspect the torque converter clutch assembly.
69. Install the torque converter clutch assembly, as follows:
    a. Engage the splines of the input shaft and turbine runner.
    b. Engage the splines of the stator shaft and the stator while turning the torque converter clutch assembly.

➡If the stator shaft splines are difficult to engage with the stator splines, move the torque converter back approximately 10 mm (0.40 in.) and engage the splines while rotating the torque converter.

    c. Turn the torque converter clutch assembly to engage the key of the oil pump drive gear into the slot on the torque converter clutch assembly.
    d. Using a Vernier caliper and a straightedge, measure dimension "A" between the transaxle fitting part of the engine and the converter fitting part of the drive plate.
    e. Using a Vernier caliper and a straightedge, measure dimension "B"

22140_ES35_G0042

Fig. 9 Removing drive plate and torque converter clutch setting bolt

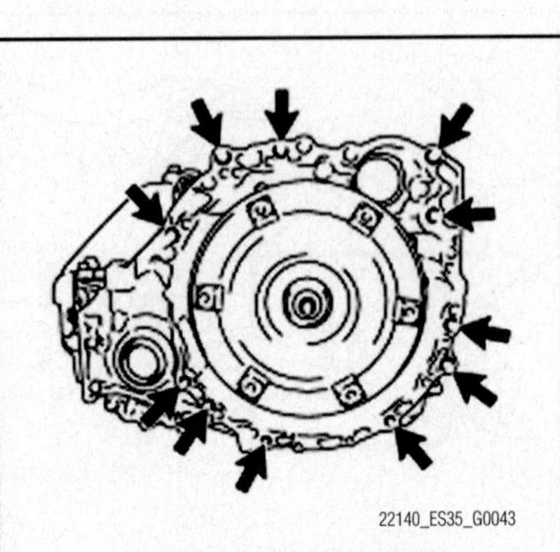

Fig. 10 Removing the 11 bolts

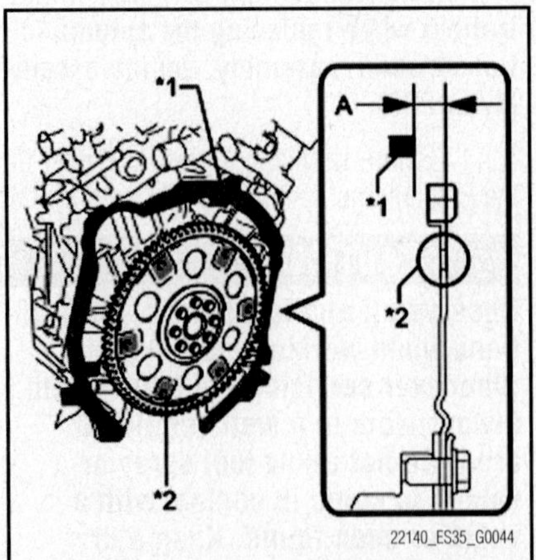

Fig. 11 Measure dimension "A" between the transaxle fitting part of the engine and the converter fitting part of the drive plate

shown in the illustration and check that "B" is greater than "A" (measured in previous step). Standard: A + 1 mm (0.04 in.) or more.

➡Make sure to deduct the thickness of the straightedge.

➡If the transaxle is installed to the engine with the torque converter not sufficiently inserted, the torque converter may be damaged.

➡In the case of where a U660E is used with a 2GR-FE, the standard installation depth is 0.55 inches (14 mm) or more.

70. Install the No. 1 oil cooler inlet and outlet hoses.
71. Install the No. 1 transmission control cable bracket with the 2 bolts and tighten to 9 ft. lbs. (12 Nm).
72. Install the 4 clamp brackets and 4 bolts and tighten to 74 inch lbs. (8.4 Nm).
73. Install the speedometer driven hole cover sub-assembly, as follows:
    a. Coat a new O-ring with ATF, and install it to the hole cover.
    b. Install the hole cover to the transaxle case with the bolt and tighten to 49 inch lbs. (5.5 Nm).
74. Install the TCM, as follows:
    a. Install the TCM to the automatic transaxle assembly.
    b. Install and tighten the 2 bolts in the order shown in the illustration to 9 ft. lbs. (11 Nm).
75. Install the engine mounting front bracket to the automatic transaxle with the 3 bolts and tighten to 47 ft. lbs. (64 Nm).
76. Install the automatic transaxle assembly, as follows:
    a. Install the automatic transaxle to the engine with the 11 bolts and tighten to:
    - A: 47 ft. lbs. (64 Nm)
    - B: 47 ft. lbs. (64 Nm)
    - C: 34 ft. lbs. (46 Nm)
    - D: 32 ft. lbs. (43 Nm)

➡Make sure that the 2 knock pins are installed on the engine before installing the transaxle.

    b. Install the breather bracket to the camshaft housing sub-assembly with the bolt and tighten to 49 inch lbs. (5.5 Nm).
    c. Install the breather hose to the breather bracket.
77. Install starter assembly.
78. Connect the park/neutral position switch connector.
79. Connect the wire harness with the bolt and tighten to 9 ft. lbs. (12 Nm).
80. Install the engine wire.
81. Install the left and right front drive shaft assemblies.
82. Install the front frame assembly.
83. Install the vane pump assembly.
84. Install the engine assembly with transaxle.
85. Install the cooler compressor assembly.
86. Install the generator assembly.
87. Install the steering sliding yoke.
88. Install the drive plate and torque converter clutch setting bolt, as follows:
    a. Apply a few drops of adhesive (Toyota Genuine Adhesive 1324, Three Bond 1324 or equivalent) to 2 threads on the tip of the 6 torque converter clutch mounting bolts.
    b. Install the 6 torque converter clutch mounting bolts and tighten to 30 ft. lbs. (41 Nm). First install the black colored bolt, and then the remaining 5 bolts.
89. Install the flywheel housing under cover and exhaust pipe support bracket to the automatic transaxle with the 2 bolts tighten to 69 inch lbs. (7.8 Nm).
90. Install the No. 1 exhaust pipe support bracket.
91. Install the left and right front axle assemblies.
92. Install the left and right front lower No. 1 suspension arm sub-assemblies.
93. Install the left and right tie rod end sub-assemblies.
94. Install the left and right front speed sensors.
95. Install the left and right front stabilizer link assemblies.
96. Install the left and right front axle shaft nuts.
97. Install the front exhaust pipe assembly.
98. Install the No. 1 exhaust pipe support bracket.
99. Connect the steering gear outlet return tube.
100. Connect the No. 1 oil reservoir to pump hose.
101. Connect the transmission control cable assembly, as follows:
    a. Install the transmission control cable assembly to the control shaft lever with the nut and tighten to 10 ft. lbs. (13 Nm).
    b. Connect the transmission control cable assembly with a new clip to the No. 1 transmission control cable bracket.
102. To complete installation, reverse remaining removal procedure.
103. Connect the negative battery cable.
104. Add coolant and power steering fluid.
105. Bleed the power steering system.
106. Add automatic transaxle fluid.
107. Inspect and adjust shift lever position.
108. Inspect for coolant leak, fuel leak, and exhaust gas leak.
109. Inspect ignition timing.
110. Inspect the engine idle speed.
111. Check the function of throttle body.
112. Install the V-bank cover sub-assembly.
113. Install the cool air intake duct seal.
114. Inspect CO/HC.
115. Inspect the ABS speed sensor signal.

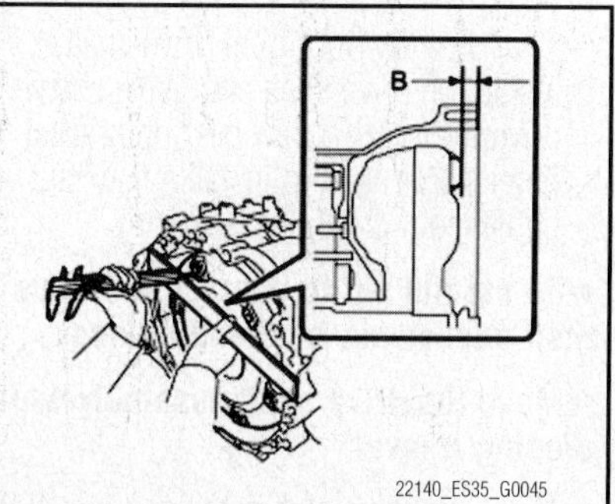

Fig. 12 Measure dimension "B" shown in the illustration and check that "B" is greater than "A"

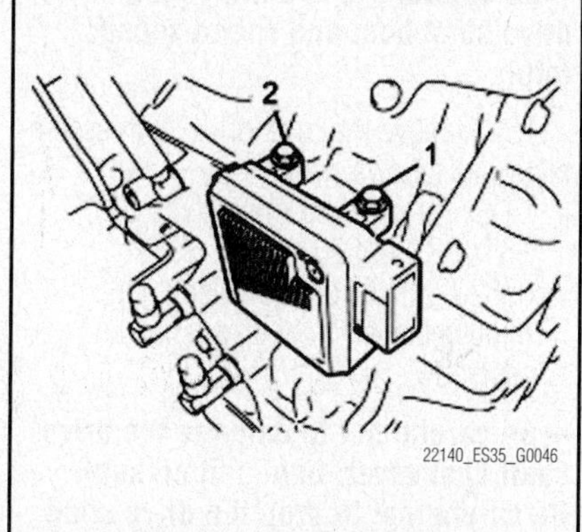

Fig. 13 TCM bolt tightening sequence

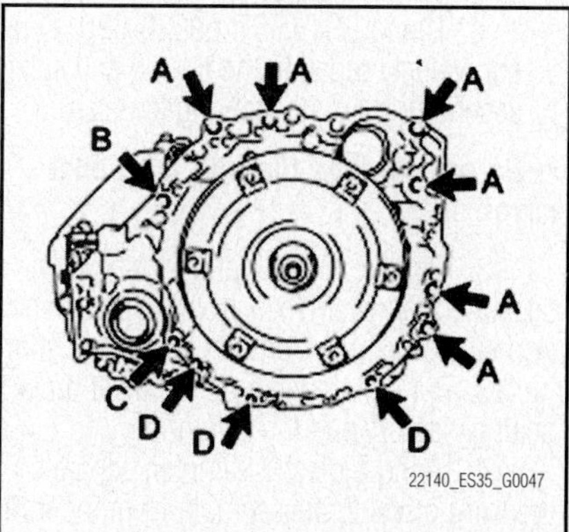

Fig. 14 Automatic transaxle bolt sequence

116. Inspect and adjust front wheel alignment.

117. Install the right front fender apron seal.

118. Install the right and left engine under covers.

FRONT HALFSHAFT

REMOVAL & INSTALLATION

See Figures 15 through 20.

1. Before servicing the vehicle, refer to the Precautions Section.
2. Remove the engine under cover.
3. Remove the drain plug and gasket, and then drain the automatic transaxle fluid.
4. Install a new gasket and drain plug and tighten to 36 ft. lbs. (49 Nm).
5. Remove front wheel.
6. Using SST (SST: 09930-00010) and hammer, release the staked part of the front axle hub nut.

➡Loosen the staked part of the nut completely, otherwise the screw of the drive shaft may be damaged.

7. While applying the brakes, remove the front axle hub nut.
8. Remove the nut and separate the front stabilizer link assembly.

➡If the ball joint turns together with the nut, use a hexagon wrench (6mm) to hold the stud.

9. Remove the bolt and clip, and separate the speed sensor wire and flexible hose from the shock absorber.
10. Remove the bolt and separate the front speed sensor from the steering knuckle.

➡Do not allow foreign matter to adhere to the speed sensor. Be careful not to damage the speed sensor.

11. Separate tie rod end sub-assembly, as follows:
    a. Remove the cotter pin and nut.
    b. Using SST (SST: 09628-62011) or equivalent, separate the tie rod end sub-assembly from the steering knuckle.

➡Do not damage the ball joint dust cover.

12. Remove the bolt and 2 nuts, and separate the lower No. 1 front suspension arm sub-assembly from the lower ball joint.
13. Put matchmarks on the front drive shaft assembly and the axle hub.
14. Using a plastic hammer, separate the front drive shaft assembly from the front axle hub sub-assembly.

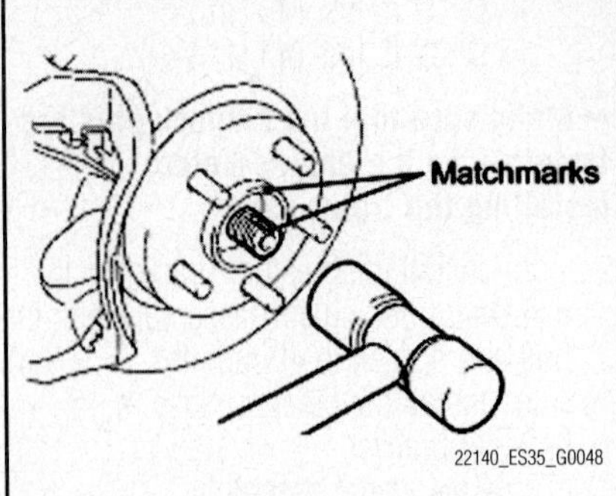

Fig. 15 Put matchmarks on the front drive shaft assembly and the axle hub

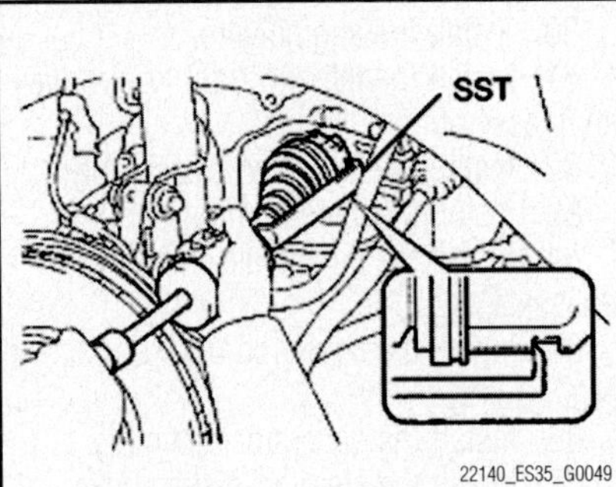

Fig. 16 Remove the left front drive shaft assembly

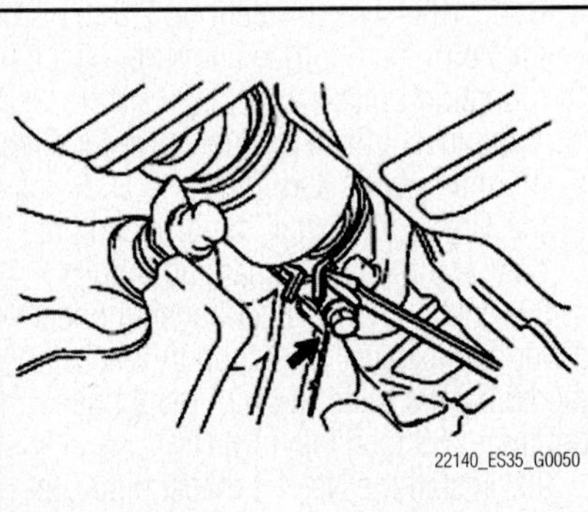

Fig. 17 Remove the right front drive shaft assembly

➡Be careful not to damage the drive shaft boot and speed sensor rotor.

15. Remove the front drive shaft assembly(s), as follows:
    a. For left front drive shaft, use SST (SST: 09520-01010, SST: 09520-24010) or equivalent, and remove the left front drive shaft assembly.

➡Be careful not to damage the drive shaft dust cover, boot and oil seal. Be careful not to drop the drive shaft assembly.

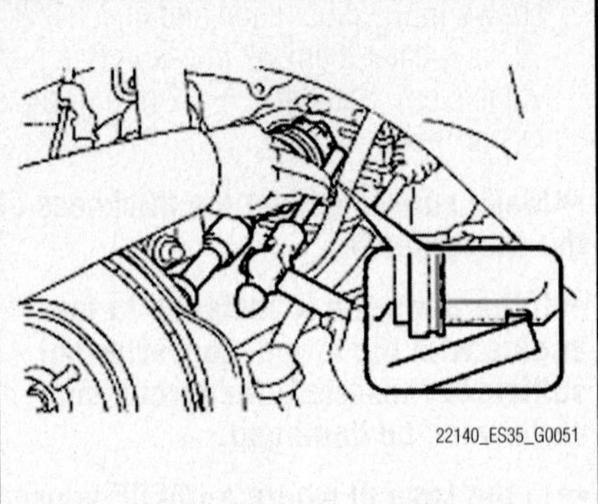

Fig. 18 Install the left front drive shaft assembly

    b. For right front drive shaft, use a screwdriver and remove the bearing bracket hole snap ring.
    c. Remove the bolt and right front drive shaft assembly from the drive shaft bearing bracket.

➡Do not damage the boot and oil seal.

16. Fix front axle hub bearing. The hub bearing could be damaged if it is subjected to the vehicle's full weight, such as moving the vehicle with the drive shaft removed. If it is necessary to place the vehicle's weight on the hub bearing, first support it with SST (SST: 09608-16042).

To install:

17. Install the front drive shaft assembly(s), as follows:
    a. Coat the spline of the inboard joint shaft assembly with automatic transaxle fluid.
    b. For the left front drive shaft, align the shaft splines and install the drive shaft assembly with a brass bar and hammer.

➡Set the shaft snap ring with the opening side facing down. Be careful not to damage the drive shaft dust cover, boot, and oil seal. Move the drive shaft assembly while keeping it level.

    c. For the right front drive shaft, install the drive shaft and use a screwdriver to install a new bearing bracket hole snap ring, and install a new bolt tightened to 24 ft. lbs. (32 Nm).

➡Be careful not to damage the drive shaft dust cover, boot and oil seal.

➡Move the drive shaft assembly while keeping it level.

18. Align the matchmarks and install the front drive shaft assembly to the front axle hub sub-assembly.

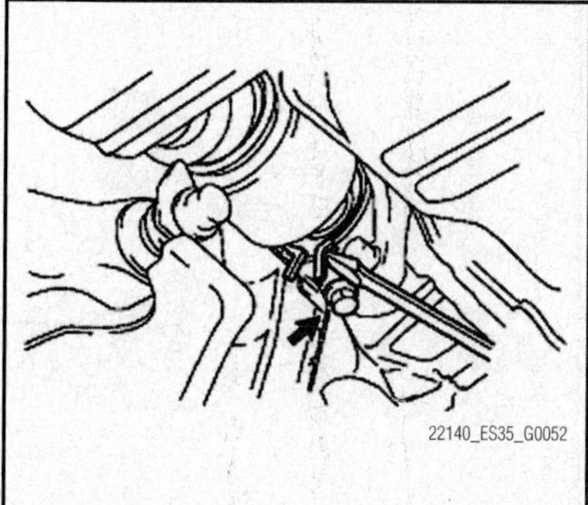

Fig. 19 Install the right front drive shaft assembly

➡**Be careful not to damage the drive shaft boot and speed sensor rotor.**

19. Install the lower ball joint to the lower No. 1 front suspension arm sub-assembly with the bolt and 2 nuts and tighten to 55 ft. lbs. (75 Nm).
20. Install the tie rod end sub-assembly to the steering knuckle with the nut and tighten to 36 ft. lbs. (49 Nm).
21. Install a new cotter pin. If the holes for the cotter pin are not aligned, tighten the nut up to 60° further.

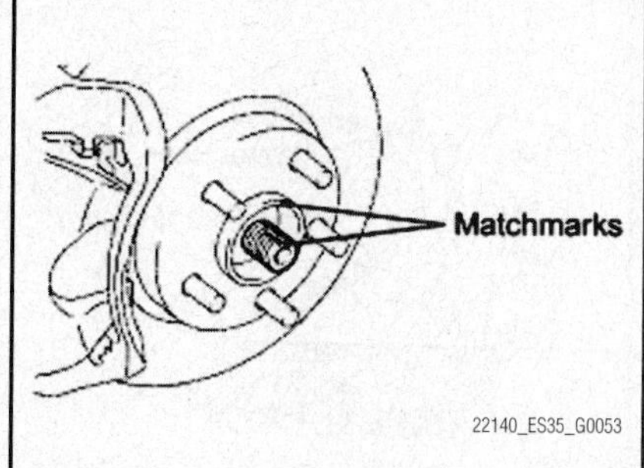

Fig. 20 Align the matchmarks and install the front drive shaft assembly

22. Install the front speed sensor to the steering knuckle with the bolt and tighten to 71 inch lbs. (8 Nm).
23. Install the flexible hose and the speed sensor to the shock absorber with the bolt and set the sensor clip on the knuckle and tighten to 14 ft. lbs. (19 Nm).

➡**Be careful not to damage the speed sensor. Do not allow foreign matter to adhere to the speed sensor. Do not twist the sensor wire when installing the speed sensor.**

24. Install the stabilizer link assembly with the nut and tighten to 55 ft. lbs. (74 Nm).

➡**If the ball joint turns together with the nut, use a hexagon wrench (6 mm) to hold the stud.**

25. Clean the threaded parts on the drive shaft and axle hub nut using a non-residue solvent.

➡**Be sure to perform this work for a new drive shaft. Keep the threaded parts free of oil and foreign objects.**

26. Using a socket wrench (30 mm), install a new axle hub nut and tighten to 217 ft. lbs. (294 Nm).
27. Using a chisel and hammer, stake the front axle hub nut.
28. Install front wheel.
29. Add automatic transaxle fluid.
30. Inspect automatic transaxle fluid.
31. Inspect and adjust front wheel alignment.
32. Install the engine under cover.
33. Check the ABS speed sensor signal.

ENGINE COOLING

THERMOSTAT

REMOVAL & INSTALLATION

See Figures 21 and 22.

1. Before servicing the vehicle, refer to the Precautions Section.
2. Drain engine coolant.
3. Remove the V-bank cover sub-assembly.
4. Remove the RH front fender apron seal.
5. Remove the RH No. 2 engine mounting stay.
6. Remove the drive belt.
7. Remove No. 2 idler pulley sub-assembly.
8. Separate the radiator hose outlet.
9. Remove the 2 nuts and disconnect the water inlet from the cylinder block.
10. Remove the thermostat and the gasket from the thermostat.

To install:

11. Install a new gasket to the thermostat.
12. Install a new gasket to the thermostat.

➡**The jiggle valve may be set within 10° of either side of the prescribed position.**

13. Install the thermostat with the jiggle valve facing up.
14. Install the water inlet and tighten to 7 ft. lbs. (10 Nm).
15. The remainder of installation is the reverse of removal.
16. After installation, inspect for coolant leak.

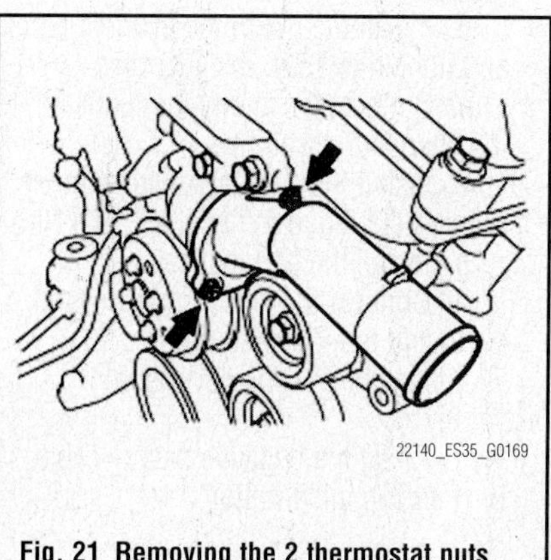

Fig. 21 Removing the 2 thermostat nuts

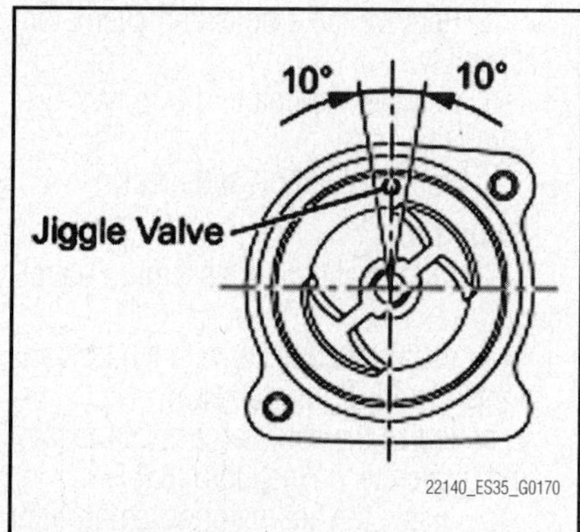

Fig. 22 Radiator jiggle valve

WATER PUMP

REMOVAL & INSTALLATION

See Figures 23 through 26.

1. Before servicing the vehicle, refer to the Precautions Section.
2. Remove engine assembly and transaxle. secure engine stand.
3. Remove RH front No. 1 engine mounting bracket.
4. Remove the No. 2 idler pulley sub-assembly, as follows:
   a. Remove the 2 bolts, 2 idler pulley cover plates and 2 idler pulley sub-assemblies.
5. Remove the 5 bolts and V-ribbed belt tensioner assembly.
6. Using SST: 09960-10010, hold the water pump pulley. Remove the 4 bolts and water pump pulley.

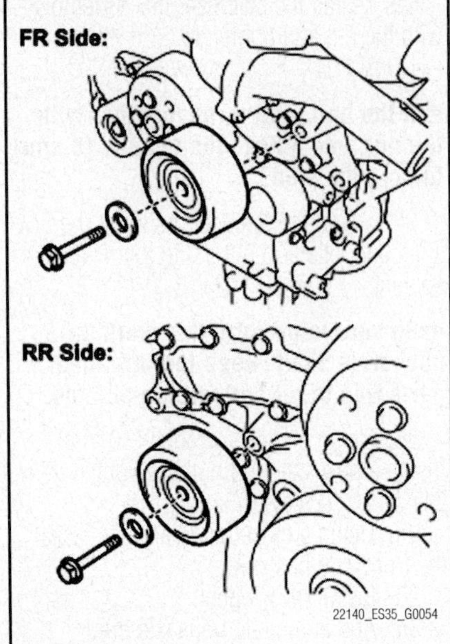

Fig. 23 Remove the bolts and idler pulley sub-assemblies

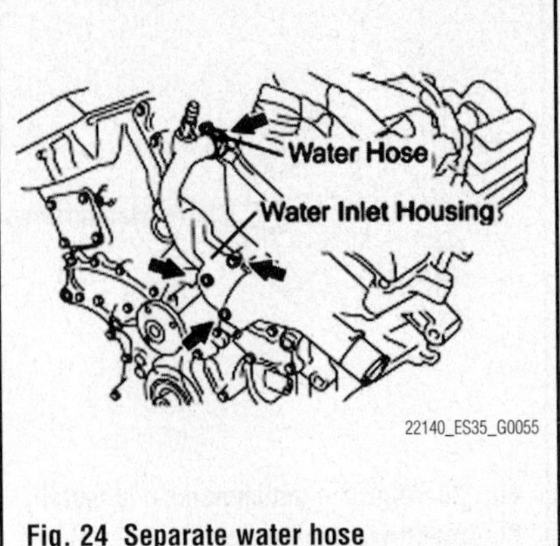

Fig. 24 Separate water hose

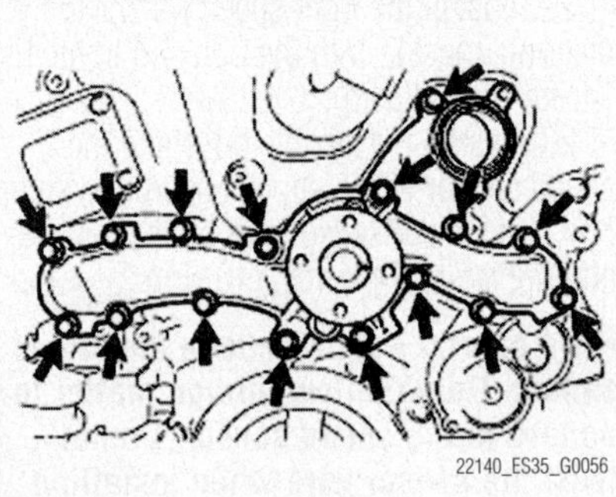

Fig. 25 Remove the 16 bolts, water pump assembly

Fig. 26 Water pump tightening sequence

7. Remove water inlet housing, as follows:
   a. Separate the water hose.
   b. Remove the 2 bolts, nut and water inlet housing.
   c. Remove the water inlet housing gasket and water outlet pipe O-ring.
8. Remove the 16 bolts, water pump assembly and water pump gasket.

To install:

➡Make sure that there is no oil on the threads of the A bolts.

➡Be sure to replace the 2 C bolts with new ones or reuse them after applying adhesive (Part No. 08833-00080, three bond 1344 or equivalent).

9. Install a new water pump gasket and the water pump assembly with the 16 bolts and tighten to:
   a. Bolt A: 15 ft. lbs. (21 Nm)
   b. Bolt B: 81 inch lbs. (9.1 Nm).
   c. Bolt C: 81 inch lbs. (9.1 Nm).
10. Install a new water inlet housing No. 1 gasket and water outlet pipe O-ring..

➡Be careful not to allow the O-ring to get caught between the parts.

11. Install water inlet housing, as follows:
   a. Install a new No. 1 water inlet housing gasket and water outlet pipe O-ring.
   b. Install the water inlet with the 2 bolts and nut and tighten to 7 ft. lbs. (10 Nm).
12. Temporarily install the water pump pulley with the 4 bolts.
   a. Using SST (SST: 09960-10010) or equivalent, hold the water pump pulley.
   b. Tighten the 4 bolts to 15 ft. lbs. (21 Nm).
13. Install the V-ribbed belt tensioner assembly with the 5 bolts and tighten to 32 ft. lbs. (43 Nm).
14. Install the 2 idler pulley cover plates and idler pulley sub-assemblies with the 2 bolts and tighten to 32 ft. lbs. (43 Nm).
15. To complete installation, reverse remaining removal procedure.
16. Add engine coolant.
17. Inspect for coolant leaks.

ENGINE ELECTRICAL

CHARGING SYSTEM

ALTERNATOR

REMOVAL & INSTALLATION

1. Disconnect the negative battery cable.
2. Remove the V-bank cover sub-assembly.
3. Remove the V-ribbed belt.
4. Remove the alternator assembly, as follows:
   a. Disconnect the wire harness clamp.
   b. Remove the terminal cap.
   c. Remove the nut and disconnect the wire harness from terminal B.
   d. Disconnect the alternator connector from the alternator assembly.
   e. Remove the nut from the cylinder block.
   f. Remove the 2 bolts and alternator assembly.
   g. Remove the bolt and wire harness clamp stay.
   h. Remove the bolt and bracket.

To install:

5. Install the alternator assembly, as follows:
   a. Install the bracket with the bolt and tighten to 15 ft. lbs. (20 Nm).
   b. Install the wire harness clamp stay and tighten to 74 inch lbs. (8.4 Nm).
   c. Install the alternator assembly with the 2 bolts and tighten to 32 ft. lbs. (43 Nm).
   d. Install the nut to the cylinder block and tighten to 15 ft. lbs. (20 Nm).
   e. Connect the alternator connector to the alternator assembly.
   f. Install the alternator wire with the nut and tighten to 87 inch lbs. (9.8 Nm).
   g. Install the terminal cap.
   h. Connect the wire harness clamp.
6. Install the V-ribbed belt.
7. Install the V-bank cover sub-assembly.
8. Connect the negative battery cable.
9. Perform initialization.

ENGINE ELECTRICAL

IGNITION SYSTEM

FIRING ORDER

Firing order for 3.5L engine: 1–2–3–4–5–6

IGNITION COIL

REMOVAL & INSTALLATION

See Figure 27.

1. Before servicing the vehicle, refer to the Precautions Section.

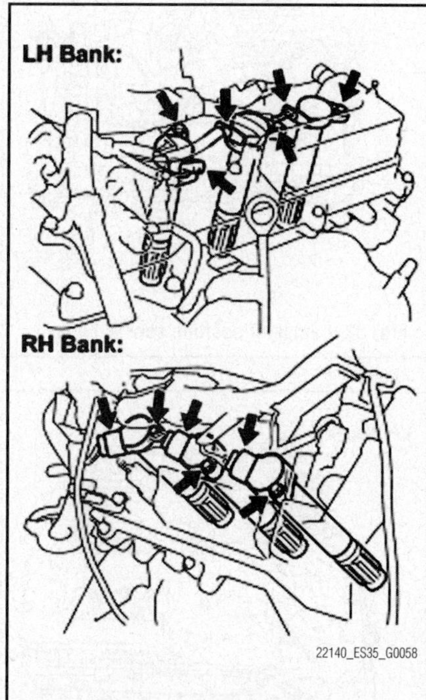

Fig. 27 Removing ignition coils

2. Disconnect the negative battery cable.
3. Drain and recycle the engine coolant.
4. Remove windshield wiper link assembly.
5. Remove cowl top panel outer sub-assembly.
6. Remove v-bank cover sub-assembly.
7. Remove air cleaner cap sub-assembly.
8. Remove intake air surge tank assembly.
9. Remove No. 1 surge tank stay by performing the following:
   a. Remove the bolt and disconnect the harness clamp.
   b. Remove the bolt and No. 1 surge tank stay.
10. Disconnect the 6 ignition coil connectors.
11. Remove the 6 bolts and 6 ignition coils.
12. To install, reverse removal procedure.
13. Tighten the following to specification:
   a. 6 ignition coil bolt: 10 ft. lbs. (10 Nm).
   b. No. 1 surge tank stay bolt: 15 ft. lbs. (21 Nm).
   c. No. 1 surge tank stay bolt and clamp: 62 inch lbs. (7 Nm).

IGNITION TIMING

ADJUSTMENT

All engines are equipped with a Distributorless Ignition System (DIS). No timing adjustment is possible.

SPARK PLUGS

REMOVAL & INSTALLATION

See Figure 28.

1. Before servicing the vehicle, refer to the Precautions Section.
2. Remove the V-bank cover.
3. Remove the intake air surge tank.
4. Disconnect the 6 ignition coil connectors.
5. Remove the 6 bolts and 6 ignition coils.
6. Using a 16 mm (0.63 in.) plug wrench, remove the spark plugs.
7. Installation is the reverse of removal, noting the following:
   a. Torque the ignition coils to 66 inch lbs. (7.5 Nm) and the spark plugs to 13 ft. lbs (18 Nm).

Fig. 28 Removing spark plugs

ENGINE ELECTRICAL

STARTING SYSTEM

STARTER

REMOVAL & INSTALLATION

See Figure 29.

1. Before servicing the vehicle, refer to the Precautions Section.
2. Disconnect the negative battery cable.
3. Remove cool air intake duct seal.
4. Remove v-bank cover sub-assembly.
5. Remove air cleaner inlet assembly.
6. Remove air cleaner cap sub-assembly.
7. Remove air cleaner case sub-assembly.
8. Remove No. 1 air cleaner inlet.

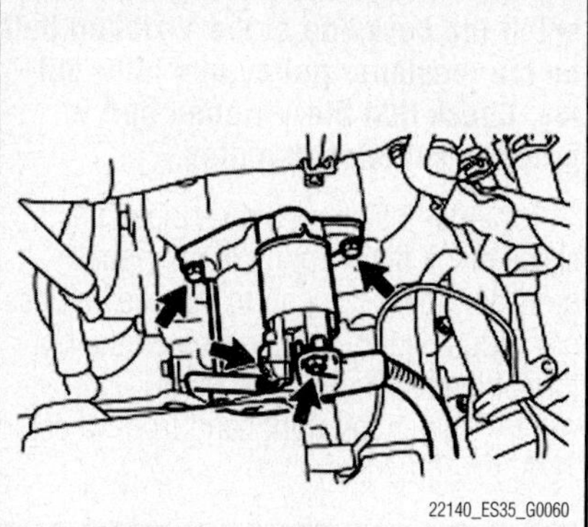

Fig. 29 Removing starter assembly

9. Disconnect the terminal 50 connector from the starter assembly.
10. Remove the nut and disconnect the wire harness from terminal 30.
11. Remove the 2 bolts and starter assembly.

To install:

12. Install the starter assembly with the 2 bolts and tighten to 26 ft. lbs. (37 Nm).
13. Connect the wire harness to terminal 30 and install the nut and tighten to 87 inch lbs. (9.8 Nm).
14. Cover the nut with the cap.
15. Connect terminal 50 to the starter assembly.
16. To complete installation, reverse removal procedure.

ENGINE MECHANICAL

➡**Disconnecting the negative battery cable may interfere with the functions of the on board computer systems and may require the computer to undergo a relearning process, once the negative battery cable is reconnected.**

ACCESSORY DRIVE BELTS

ACCESSORY BELT ROUTING

See Figure 30.

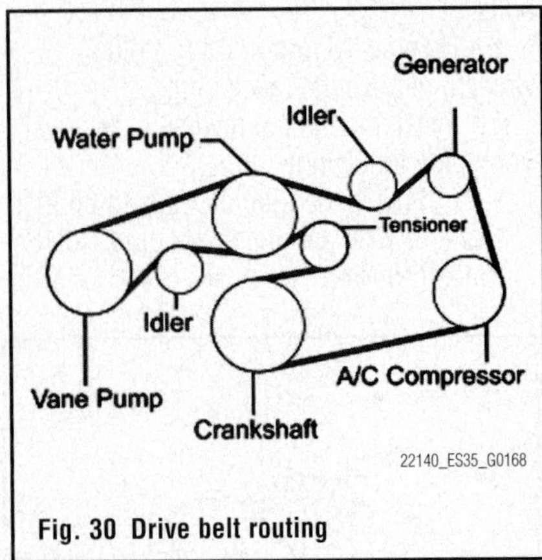

Fig. 30 Drive belt routing

INSPECTION

See Figure 31.

Visually check the V-ribbed belt for excessive wear, frayed cords, etc. If any defect has been found, replace the V-ribbed belt.

- Cracks on the rib side of a belt are considered acceptable. If the belt has chunks missing from the ribs, it should be replaced

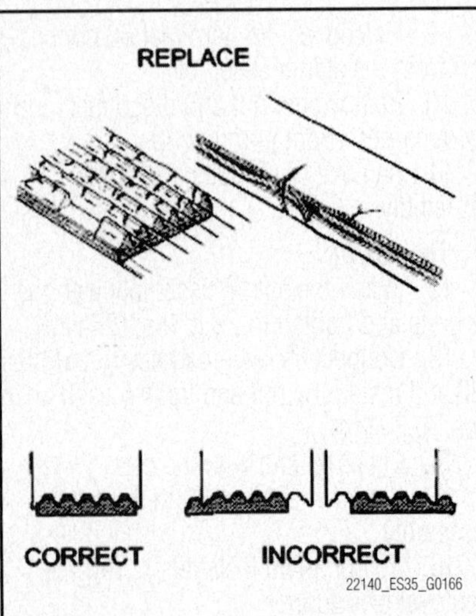

Fig. 31 Inspecting the drive belt

- A "new belt" is a belt which has been used for less than 5 minutes with the engine running
- A "used belt" is a belt which has been used for 5 minutes or more with the engine running

ADJUSTMENT

This vehicle is equipped with an auto-tensioner and cannot be adjusted.

REMOVAL & INSTALLATION

1. Before servicing the vehicle, refer to the Precautions Section.
2. Remove the right hand front wheel.
3. Remove the right hand front fender apron seal.
4. Remove the V-bank cover sub-assembly.
5. Using Special Tool: 09249—63010, release the belt tension by turning the belt tensioner counterclockwise, and remove the V-ribbed belt from the belt tensioner.
6. While turning the belt tensioner counterclockwise, align with its holes and then insert the 5 mm bi-hexagon wrench into the holes to fix the V-ribbed belt tensioner.
7. Remove the v-ribbed belt.

To install:

8. To install, reverse removal procedure.
9. If it is difficult to install the V-ribbed belt, perform the following procedure:
   a. Put the V-ribbed belt on every pulley except the tensioner pulley.
   b. While releasing the belt tension by turning the belt tensioner counterclockwise, put the V-ribbed belt on the tensioner pulley.

➡**Put the backside of the V-ribbed belt on the tensioner pulley and idler pulley. Check that the V-ribbed belt is properly set to each pulley.**

10. After installing the V-ribbed belt, check that it fits properly in the ribbed grooves. Check to confirm that the belt has not slipped out of the grooves on the bottom of the crank pulley by hand.
11. Tighten the right hand front wheel to: 76 ft. lbs. (103 Nm).

CAMSHAFT AND VALVE LIFTERS

REMOVAL & INSTALLATION

See Figures 32 through 55.

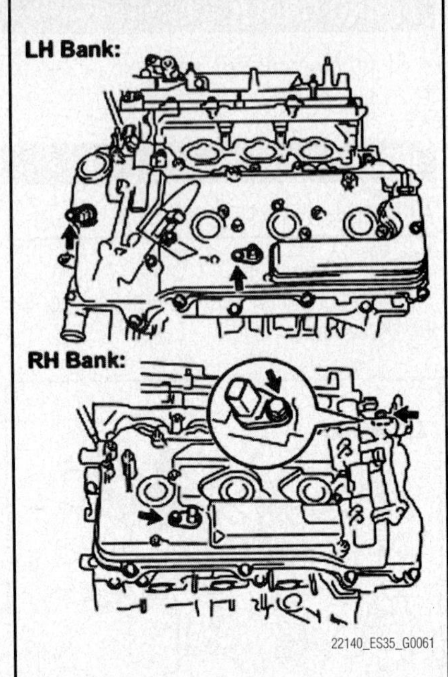

Fig. 32 Camshaft position sensor bolts

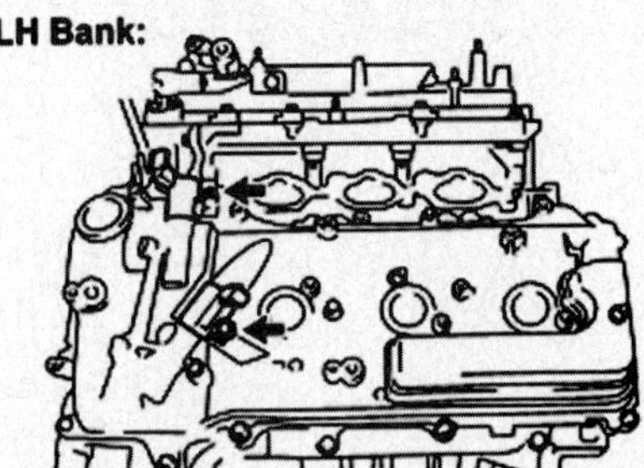

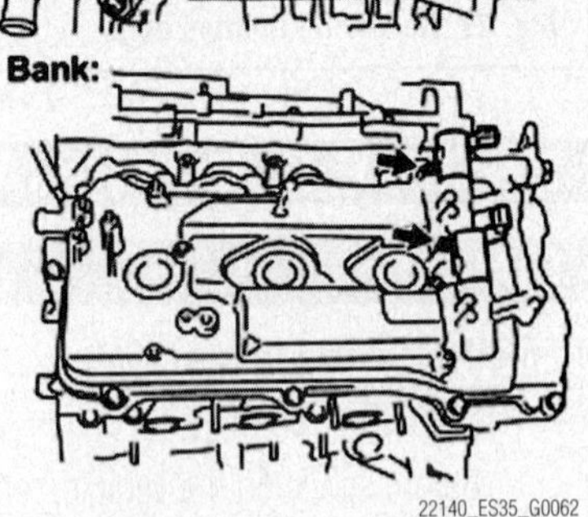

Fig. 33 Camshaft timing oil control valve assembly bolts

1. Before servicing the vehicle, refer to the Precautions Section.
2. Remove the engine assembly.
3. Install on engine stand.
4. Remove the oil filler cap and gasket.
5. Remove the spark plugs and ignition coil assembly.
6. Remove the drain plug and gasket.
7. Remove the ventilation valve.

8. Remove the 4 bolts and 4 camshaft position sensors.

9. Remove the 4 bolts and 4 camshaft timing oil control valves.

10. Remove the bolt and crankshaft position sensor.

11. Remove the 2 oil pipe unions and oil pipe. Remove the LH oil control valve filter and gaskets.

12. Remove the oil pipe bolt. Remove the 2 oil pipe unions and oil pipe. Remove the RH oil control valve filter and gaskets.

13. Remove the cylinder block water drain cock sub-assembly, as follows:

a. Remove the water drain cocks from the cylinder block.

b. Remove the water drain cock plugs from the water drain cocks.

14. Remove the oil filter.

15. Remove the crankshaft pulley, as follows:

a. Using SST (SST: 09213-70011, SST: 09330-00021) or equivalent, loosen the crankshaft pulley bolt.

b. Using SST (SST: 09950-50013) or equivalent, remove the crankshaft pulley bolt and crankshaft pulley.

16. Remove the 6 bolts and the left hand No. 1 front engine mounting bracket. Using "Torx" socket wrench E8, remove the 2 stud bolts.

17. Remove the water inlet housing, as follows:

a. Remove the 2 nuts, water inlet and thermostat.

b. Remove the gasket.

c. Remove the drain cock plug.

d. Remove the drain cock.

e. Remove the 2 stud bolts.

f. Remove the 2 bolts, nut, and water inlet housing.

g. Remove the 2 O-rings.

18. Remove the water outlet, as follows:

a. Remove the 2 bolts, 4 nuts and water outlet.

b. Remove the 2 gaskets and O-ring.

19. Remove the 12 bolts, valve cover (for Bank 1) and gasket, then remove the 3 gaskets.

20. Remove the 12 bolts, valve cover (for Bank 2) and gasket, then remove the 3 gaskets.

21. Remove the No. 2 oil pan sub-assembly.

22. Remove the oil strainer sub-assembly.

23. Remove the oil pan sub-assembly.

24. Remove the No. 1 oil pan baffle plate.

25. Remove the engine rear oil seal, as follows:

a. Remove the 6 bolts.

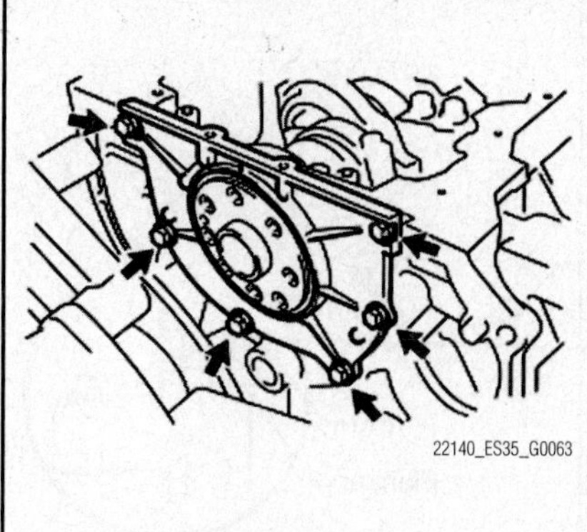

Fig. 34 Engine rear oil seal bolts

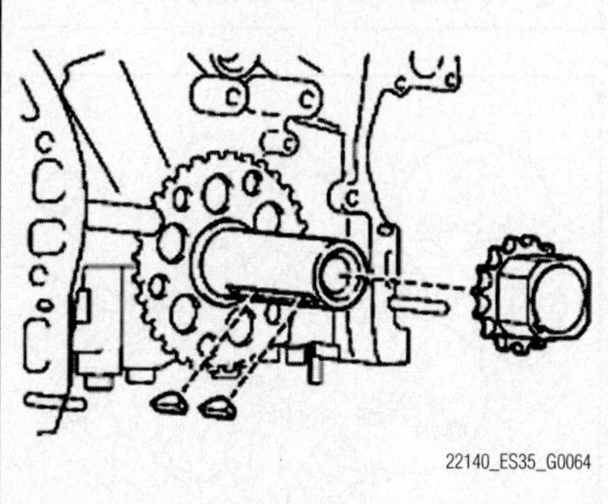

Fig. 35 Removing 2 pulley set keys from the crankshaft

b. Using a screwdriver with the tip taped, pry out the oil seal retainer. Be careful not to damage the oil seal retainer.

➡**Be careful not to damage the engine rear oil seal retainer.**

➡**Tape the screwdriver tip before use.**

26. Place the oil seal retainer on wooden blocks. Using a screwdriver and a hammer, tap out the oil seal.

27. Remove the water pump assembly.

28. Remove the timing chain cover sub-assembly.

29. Remove timing chain case oil seal.

30. Set the No. 1 cylinder to TDC/compression.

31. Remove the No. 1 chain tensioner assembly.

32. Remove the chain tensioner slipper.

33. Remove the chain sub-assembly.

34. Remove the idle sprocket assembly.

35. Remove the No. 1 chain vibration damper.

36. Remove crankshaft timing sprocket. Remove the 2 pulley set keys from the crankshaft.

37. Remove camshaft timing gears and No. 2 chain (for Bank 1), as follows:

a. While raising the No. 2 chain tensioner, insert a pin of 0.039 in (1.0 mm) into the hole to fix the No. 2 chain tensioner.

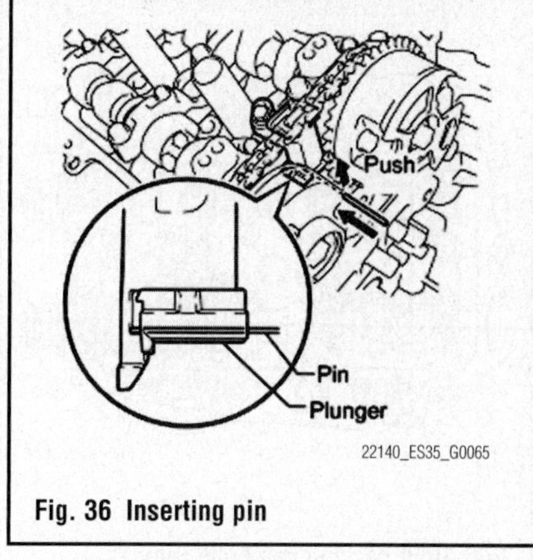

Fig. 36 Inserting pin

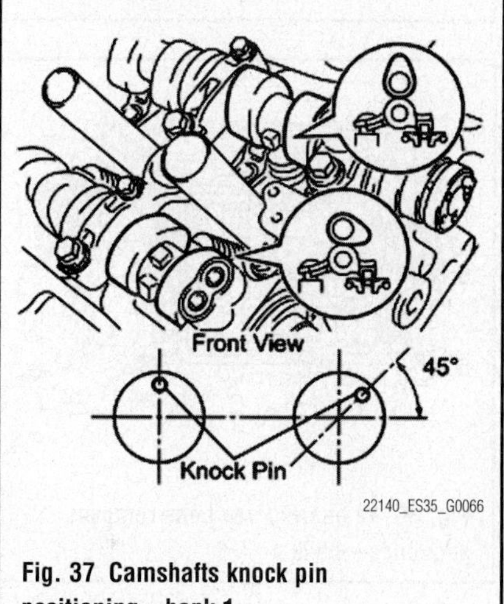

Fig. 37 Camshafts knock pin positioning—bank 1

➡**Be careful not to damage the cylinder head with the wrench.**

➡**Do not disassemble the camshaft timing gear assemblies.**

b. Hold the hexagonal portion of the camshaft with a wrench, and remove the 2 bolts and 2 camshaft timing gears.

c. Remove the No. 2 chain.

38. Remove the bolt and No. 2 chain tensioner assembly.

39. Remove camshaft bearing cap (for Bank 1), as follows:

a. Check that the camshafts are positioned as shown in the illustration.

b. Uniformly loosen and remove the 8 bearing cap bolts in the sequence shown in the illustration.

c. Uniformly loosen and remove the 12 bearing cap bolts in the sequence shown in the illustration. Uniformly loosen the bolts while keeping the camshaft level.

d. Remove the 5 bearing caps.

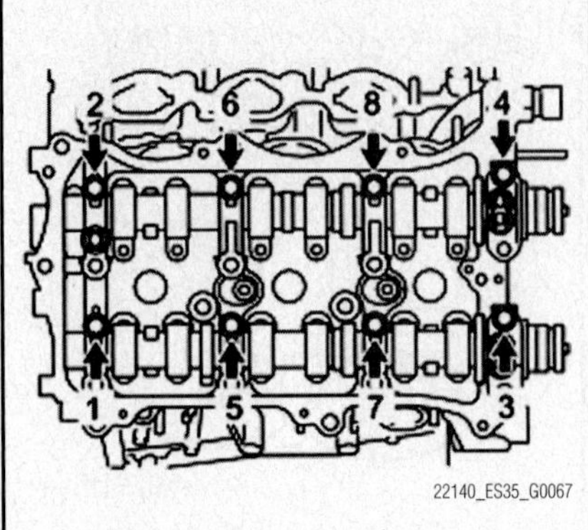

Fig. 38 8 bearing cap bolts removal sequence—bank 1

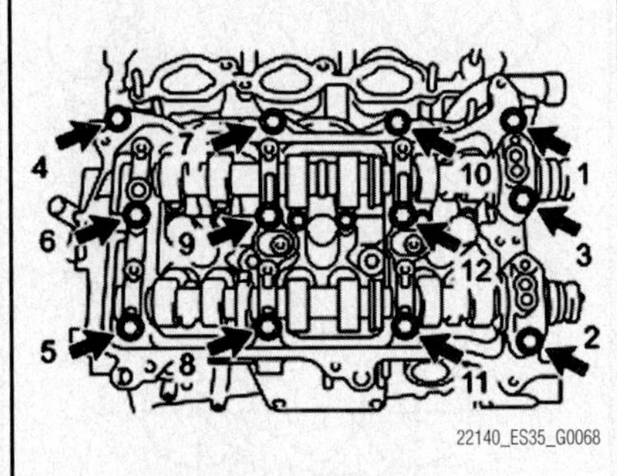

Fig. 39 12 bearing cap bolts removal sequence—bank 1

40. Remove the camshaft.

41. Remove the No. 2 camshaft.

42. If necessary, remove the right hand camshaft housing sub-assembly by prying between the cylinder head and the camshaft housing with a screwdriver with the tip taped.

➡Be careful not to damage the contact surfaces of the cylinder head and the camshaft housing.

43. Remove the camshaft timing gears and No. 2 chain (for Left-hand Bank), as follows:

a. While pushing down the No. 3 chain tensioner, insert a pin of 1.0 mm (0.039 in.) into the hole to fix the No. 3 chain tensioner.

b. Hold the hexagonal portion of the camshaft with a wrench, and remove the 2 bolts and 2 camshaft timing gears.

➡Be careful not to damage the cylinder head with the wrench.

➡Do not disassemble the camshaft timing gear assemblies.

c. Remove the No. 2 chain.

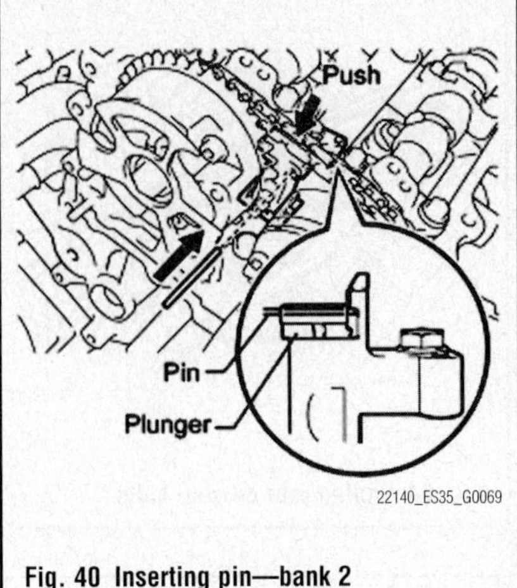

Fig. 40 Inserting pin—bank 2

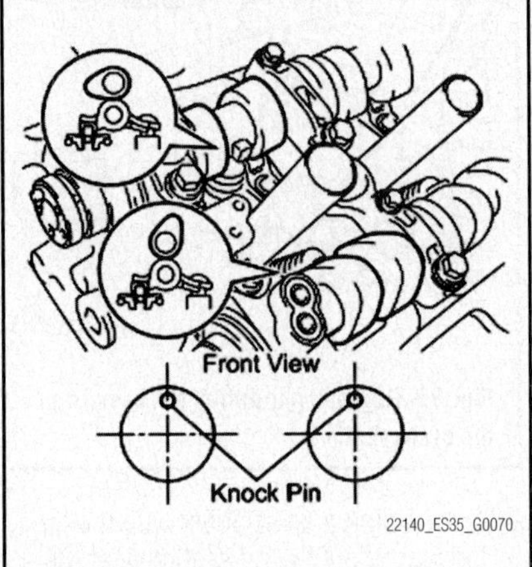

Fig. 41 Camshafts knock pin positioning—bank 2

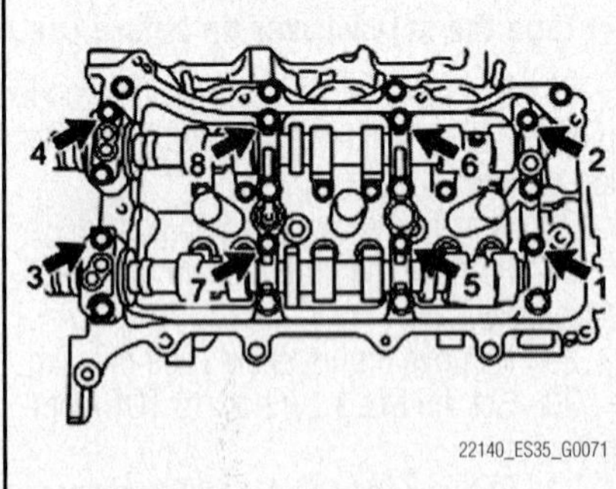

Fig. 42 8 bearing cap bolts removal sequence—bank 1

44. Remove the bolt and No. 3 chain tensioner.

45. Remove the camshaft bearing cap (bank 2), as follows:

a. Make sure that the knock pin of the camshaft is positioned as shown in the illustration.

b. Uniformly loosen and remove the 8 bearing cap bolts in the sequence shown in the illustration.

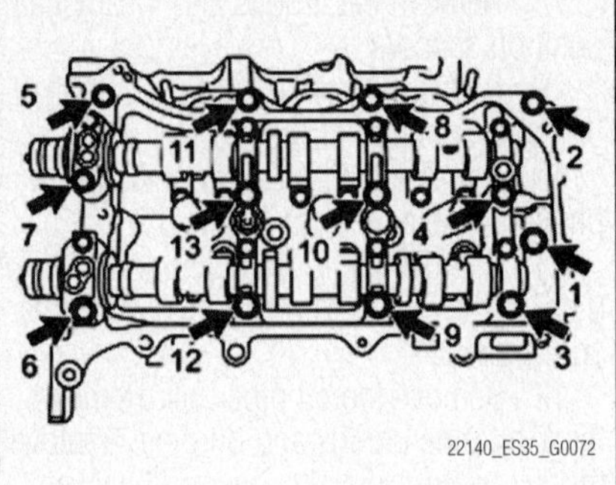

Fig. 43 13 bearing cap bolts removal sequence—bank 1

c. Uniformly loosen and remove the 13 bearing cap bolts in the sequence shown in the illustration. Loosen the bolts while keeping the camshaft level.

d. Remove the 5 camshaft bearing caps.

46. Remove the No. 3 camshaft.

47. Remove the No. 4 camshaft.

48. If necessary, remove the left hand camshaft housing sub-assembly by prying between the cylinder head and the camshaft housing with a screwdriver with the tip taped.

➡Be careful not to damage the contact surfaces of the cylinder head and the camshaft housing.

49. Remove the No. 1 valve rocker arm sub-assembly, as follows:

a. Remove the 24 valve rocker arms.

➡Arrange the removed parts in the correct order.

50. Remove the valve lash adjuster assembly, as follows:

a. Remove the 24 valve lash adjusters from the cylinder head.

➡Arrange the removed parts in the correct order.

To install:

➡Keep the lash adjuster free of dirt and foreign objects. Only use clean engine oil.

51. Install the valve lash adjuster assembly, as follows:

a. Place the lash adjuster into a container filled with engine oil. Insert the SST 09276-75010 tip into the lash adjuster's plunger and use the tip to press down on the check ball inside the plunger.

b. Squeeze the SST and lash adjuster together to move the plunger up and down 5 to 6 times.

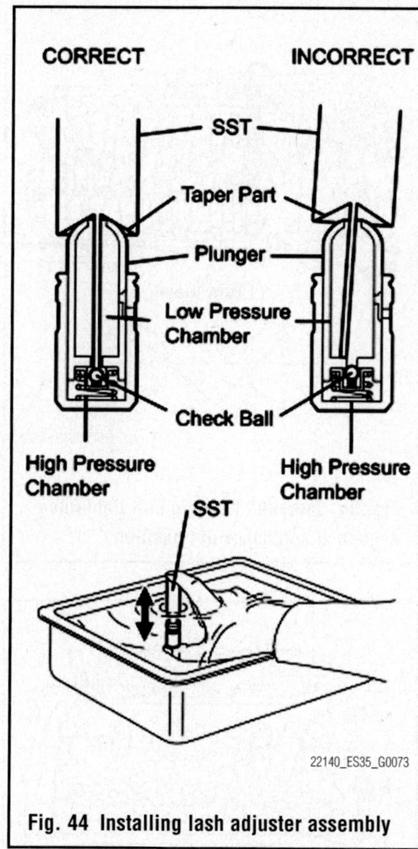

Fig. 44 Installing lash adjuster assembly

c. Check the movement of the plunger and bleed the air. Make sure the plunger moves up and down.

➡**When bleeding air from the high-pressure chamber, make sure that the tip of the SST is actually pressing the check ball as shown in the illustration. If the check ball is not pressed, air will not bleed.**

d. After bleeding the air, remove the SST. Then, try to press the plunger quickly and firmly with a finger. Make sure the plunger is very difficult to move. If the result is not as specified, replace the lash adjuster.

➡**Install the lash adjuster to the same place it was removed from.**

e. Install the lash adjusters.

52. Install the No. 1 valve rocker arm sub-assembly, as follows:

a. Apply engine oil to the lash adjuster tips and valve stem cap ends.

b. Make sure that the valve rocker arms are installed as shown in the illustration.

53. Install the right hand camshaft bearing cap, as follows:

a. Apply engine oil to the camshaft journals, camshaft housing and bearing caps.

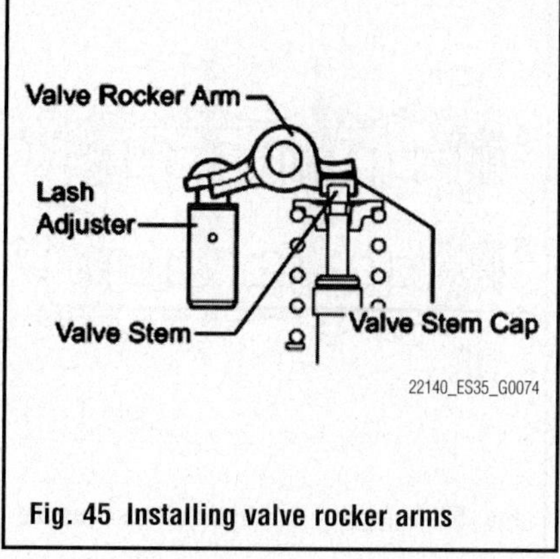

Fig. 45 Installing valve rocker arms

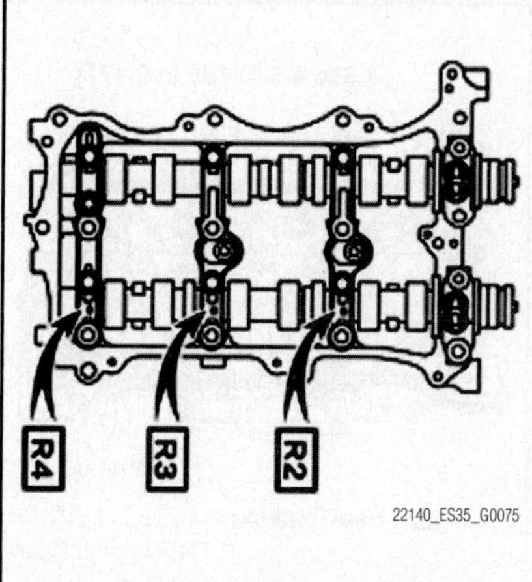

Fig. 46 Camshaft bearing caps placement

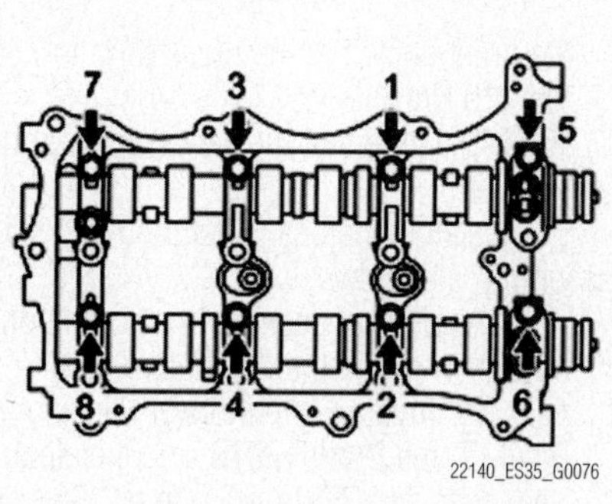

Fig. 47 Camshafts bearing cap tightening sequence

b. Install the camshaft and No. 2 camshaft to the right camshaft housing.

c. Make sure of the marks and numbers on the camshaft bearing caps and place them in each proper position and direction.

d. Temporarily tighten the 8 bearing cap bolts to 7 ft. lbs. (10 Nm) in the order shown in the illustration.

54. Install the right hand camshaft housing sub-assembly, as follows:

a. Apply seal packing in a continuous line as shown in the illustration. Seal packing: Toyota Genuine Seal Packing Black, Three Bond 1207B or equivalent. Seal diameter: 3.5 to 4.5 mm (0.138 to 0.177 in.).

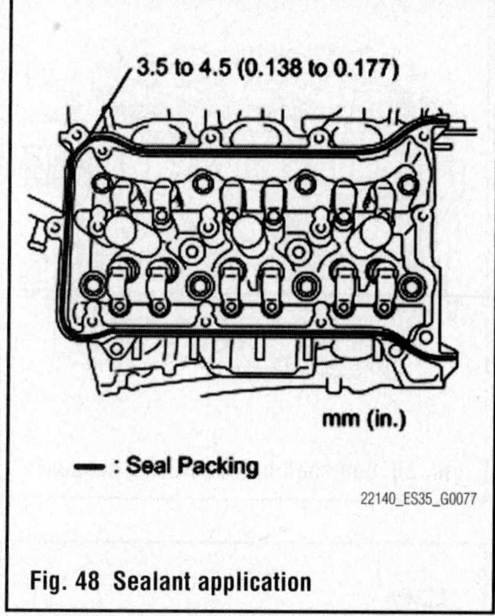

Fig. 48 Sealant application

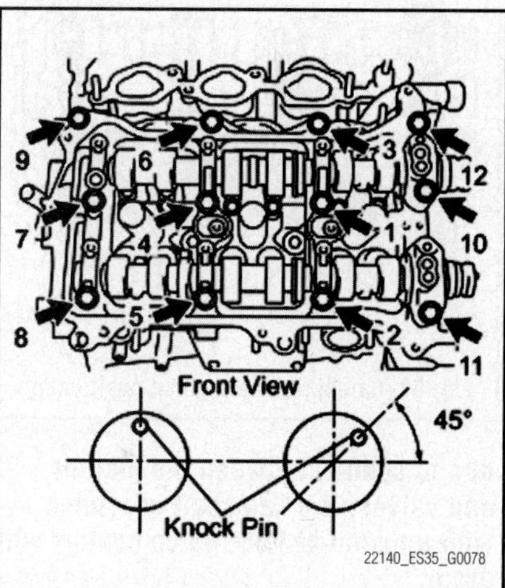

Fig. 49 Camshaft bolt tightening sequence and knock pin position

➡**Remove any oil from the contact surface. Install the camshaft housing sub-assembly within 3 minutes and tighten the bolts within 15 minutes after applying sealant. Do not start the engine for at least 2 hours after installing.**

b. Make sure that the knock pins of the camshafts are positioned as shown. Install the right hand camshaft housing and tighten the 12 bolts in the order shown in the illustration to 21 ft. lbs. (28 Nm).

➡**When installing the camshaft housing RH, it is necessary to correctly position the camshafts as shown in the illustration. Failure to correctly position these parts may result in damage**

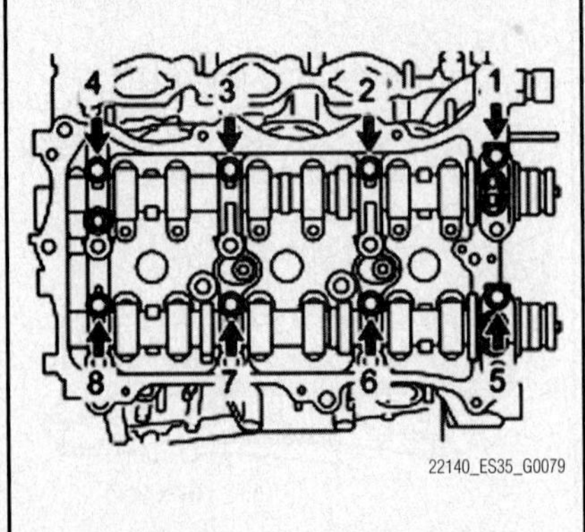

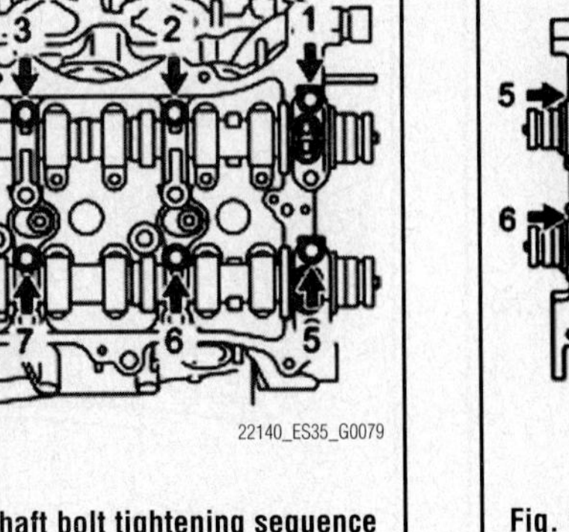

Fig. 50 Camshaft bolt tightening sequence

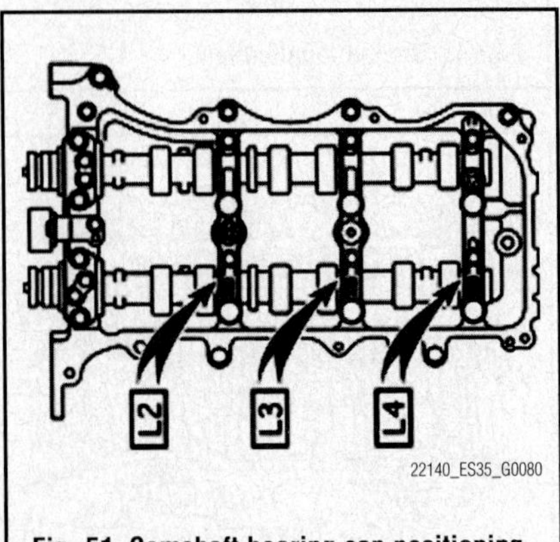

Fig. 51 Camshaft bearing cap positioning

due to contact between the pistons and valves. If a camshaft is rotated with a piston at TDC, valve contact will occur.

➡If any of the bolts are loosened during installation, remove the camshaft housing, clean the installation surfaces, and reapply seal packing.

➡If the camshaft housing is removed because any of the bolts are loosened during installation, make sure that the previously applied seal packing does not enter any oil passages.

c. Tighten the 8 bolts to 12 ft. lbs. (16 Nm) in the order shown in the illustration.

55. Install the camshaft bearing cap (bank 2), as follows:

a. Apply engine oil to the camshaft journals, camshaft housing and bearing caps.

b. Install the No. 3 camshaft and No. 4 camshaft to the left hand camshaft housing.

c. Make sure of the marks and numbers on the camshaft bearing caps and place them in each proper position and direction.

d. Temporarily tighten the 8 bolts in the order shown in the illustration to 7 ft. lbs. (10 Nm).

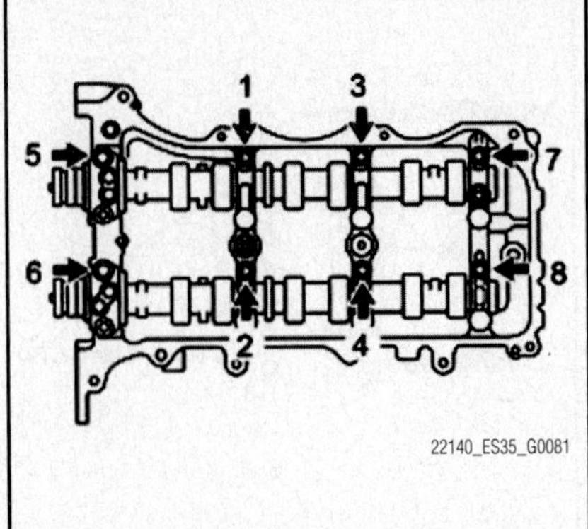

Fig. 52 Camshaft bolt tightening sequence

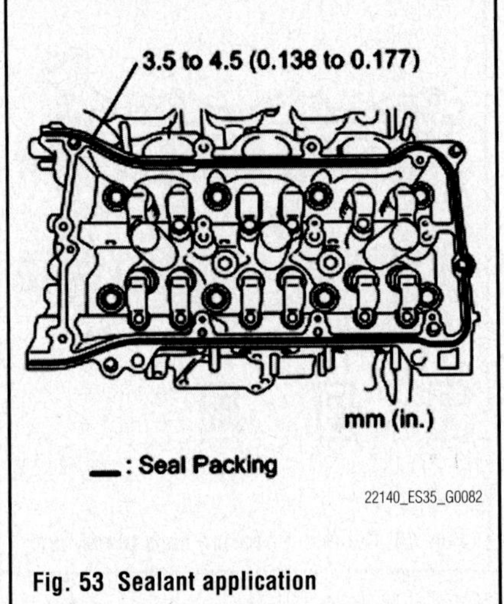

Fig. 53 Sealant application

56. Make sure that the valve rocker arm is installed.

57. Install the left camshaft housing subassembly, as follows:

a. Apply seal packing in a continuous line as shown in the illustration. Seal packing: Toyota Genuine Seal Packing Black, Three Bond 1207B or equivalent. Seal diameter: 3.5 to 4.5 mm (0.138 to 0.177 in.).

➡Remove any oil from the contact surface. Install the camshaft housing subassembly within 3 minutes and tighten the bolts within 15 minutes after applying sealant. Do not start the engine for at least 2 hours after installing.

b. Make sure that the knock pins of the camshafts are positioned as shown. Install the left hand camshaft housing and tighten the 13 bolts in the order shown in the illustration to 21 ft. lbs. (28 Nm).

➡When installing the camshaft housing LH, it is necessary to correctly position the camshafts as shown in the illustration. Failure to correctly position these parts may result in damage due to contact between the pistons and valves. If a camshaft is rotated with a piston at TDC, valve contact will occur.

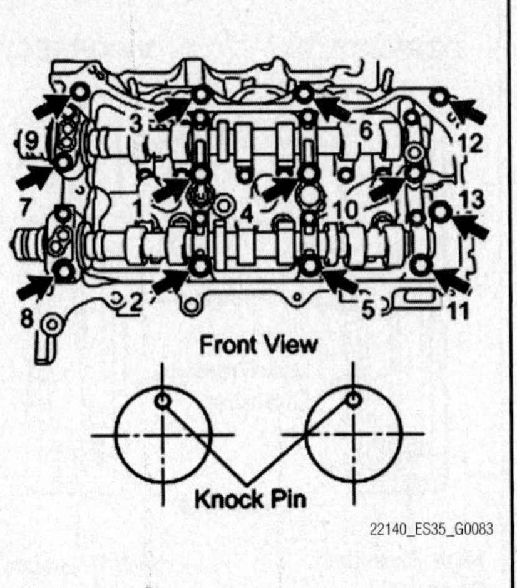

Fig. 54 Camshaft housing bolt tightening sequence and knock pin position

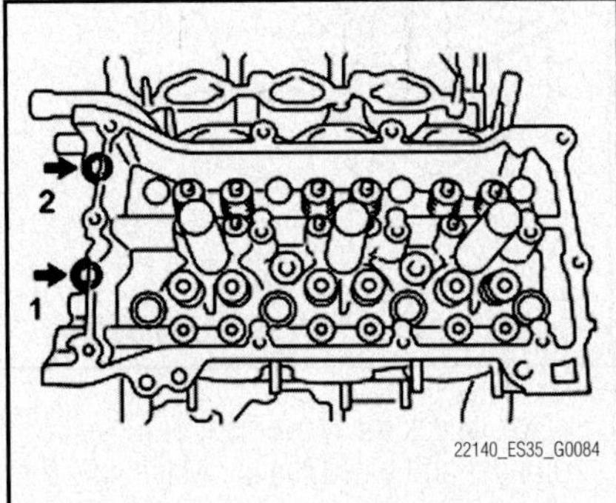

Fig. 55 Camshaft housing bolt tightening sequence

➡If any of the bolts are loosened during installation, remove the camshaft housing, clean the installation surfaces, and reapply seal packing.

➡If the camshaft housing is removed because any of the bolts are loosened during installation, make sure that the previously applied seal packing does not enter any oil passages.

c. Tighten the 8 bolts to 12 ft. lbs. (16 Nm) in the order shown in the illustration.

58. Install the No. 2 chain tensioner assembly with the bolt and tighten to 15 ft. lbs. (21 Nm).

59. While pushing in the tensioner, insert a pin of 1.0 mm (0.039 in.) diameter into the hole to hold it.

60. Install the camshaft timing gears and No. 2 chain (for Right-hand Bank).

61. Install the No. 3 chain tensioner assembly with the bolt and tighten to 15 ft. lbs. (21 Nm).

62. While pushing in the tensioner, insert a pin of 1.0 mm (0.039 in.) diameter into the hole to hold it.

63. Install the camshaft timing gears and No. 2 chain (for Left-hand Bank).

64. Install the No 1 chain vibration damper with the 2 bolts and tighten to 17 ft. lbs. (23 Nm).

65. Install the No 2 chain vibration damper.

66. Install the timing gear set keys and timing gear as shown in the illustration.

67. Install the idle sprocket assembly.

68. Install the chain sub-assembly.

69. Install the chain tensioner slipper.

70. Install the No. 1 chain tensioner assembly.

71. Install the water pump assembly.

72. Install the timing chain cover sub-assembly.

73. Install the water inlet housing.

74. Install the No. 1 left front engine mounting bracket, as follows:

a. Install the No. 1 left front engine mounting bracket with the 6 bolts and tighten to 40 ft. lbs. (54 Nm).

➡Install the water inlet and mounting bracket within 15 minutes after installing the chain cover. Do not start the engine for at least 2 hours after installation.

75. Install the No. 1 oil pan baffle plate with the 7 bolts and tighten to 7 ft. lbs. (10 Nm).

76. Install the oil pan sub-assembly.

77. Install the oil strainer sub-assembly.

78. Install the No. 2 oil pan sub-assembly.

79. Install a new gasket and oil pan drain plug and tighten to 30 ft. lbs. (40 Nm).

80. Install the cylinder head cover sub-assembly, as follows:

a. Apply seal packing (Toyota Genuine Seal Packing Black, Three Bond 1207B or equivalent).

➡Remove any oil from the contact surface. Install the crankcase within 3 minutes after applying seal packing. Do not start the engine for at least 2 hours after installation.

b. Install the gasket to the head cover.

c. Install the head cover with the 12 bolts. Tighten bolt A to 15 ft. lbs. (21 Nm), and other bolts to 7 ft. lbs. (10 Nm). Be certain to tighten bolt 1.

81. Install the left-hand cylinder head cover sub-assembly, as follows:

a. Apply seal packing (Toyota Genuine Seal Packing Black, Three Bond 1207B or equivalent).

➡Remove any oil from the contact surface. Install the crankcase within 3 minutes after applying seal packing. Do not start the engine for at least 2 hours after installation.

b. Install the gasket to the head cover.

c. Install the head cover with the 14 bolts. Tighten bolt A to 15 ft. lbs. (21 Nm), and other bolts to 7 ft. lbs. (10 Nm). Be certain to tighten bolts 1 and 10.

82. Install water outlet.

83. Install the crankshaft pulley.

84. Install the oil filter element.

85. Install the cylinder block water drain cock sub-assembly, as follows:

a. Apply adhesive around the drain cocks. Adhesive: Toyota Genuine Adhesive 1324, Three Bond 1324 or Equivalent.

b. Install the water drain cocks and tighten to 18 ft. lbs. (25 Nm). Do not rotate the drain cocks more than 1 revolution (360°) after tightening the drain cocks with the specified torque. Do not loosen after setting correctly.

c. Install the water drain cock plug to the water drain cocks and tighten to 9 ft. lbs. (13 Nm).

86. Install the No. 1 oil pipe.

87. Install the oil pipe.

88. Install the crankshaft position sensor with the bolt and tighten to 7 ft. lbs. (10 Nm).

89. Install the 4 camshaft timing oil control valves with the 4 bolts and tighten to 7 ft. lbs. (10 Nm).

90. Install the 4 camshaft position sensors with the 4 bolts and tighten to 7 ft. lbs. (10 Nm).

91. Install the ventilation valve sub-assembly, as follows:

a. Apply adhesive (Toyota Genuine Adhesive 1324, Three Bond 1324 or equivalent) around the ventilation valve.

b. Install the ventilation valve and tighten to 20 ft. lbs. (27 Nm).

92. Install the 6 spark plugs and the ignition coil assembly.

93. Install the oil filler cap sub-assembly.

94. Remove the engine stand.

95. Install the engine assembly.

CRANKSHAFT FRONT SEAL

REMOVAL & INSTALLATION

See Timing Chain Cover and Seal.

CYLINDER HEAD

REMOVAL & INSTALLATION

See Figures 56 through 62.

1. Before servicing the vehicle, refer to the Precautions Section.

2. Remove the engine assembly with transaxle.

3. Secure engine.

4. Remove the oil filler cap sub-assembly.

5. Remove the spark plugs and ignition coil assembly.

6. Remove the oil pan drain plug and gasket.

7. Remove the ventilation valve sub-assembly.

8. Remove the camshaft position sensor.

9. Remove the camshaft timing oil control valve assembly.

10. Remove crankshaft position sensor.

11. Remove the No. 1 oil pipe.

12. Remove the oil pipe.

13. Remove the cylinder block water drain cock sub-assembly.

14. Remove the oil filter.

15. Remove the crankshaft pulley.

16. Remove the left hand No. 1 front engine mounting bracket.

17. Remove the water inlet housing.

18. Remove the water outlet.

19. Remove the cylinder head covers and gaskets.

20. Remove the No. 2 oil pan sub-assembly.

21. Remove the oil strainer sub-assembly.

22. Remove the oil pan sub-assembly.

23. Remove the No. 1 oil pan baffle plate.

24. Remove the engine rear oil seal.

25. Remove the water pump assembly.

26. Remove the timing chain cover.

27. Set the No. 1 cylinder to TDC/compression.

28. Remove the No. 1 chain tensioner assembly.

29. Remove the chain tensioner slipper.

30. Remove the chain sub-assembly.

31. Remove the idle sprocket assembly.

32. Remove the No. 1 and 2 chain vibration damper.

➡Be careful not to drop washers into the cylinder head.

➡Cylinder head warpage or cracking could result from removing bolts in an incorrect order.

➡Be sure to keep separate the removed parts for each installation position.

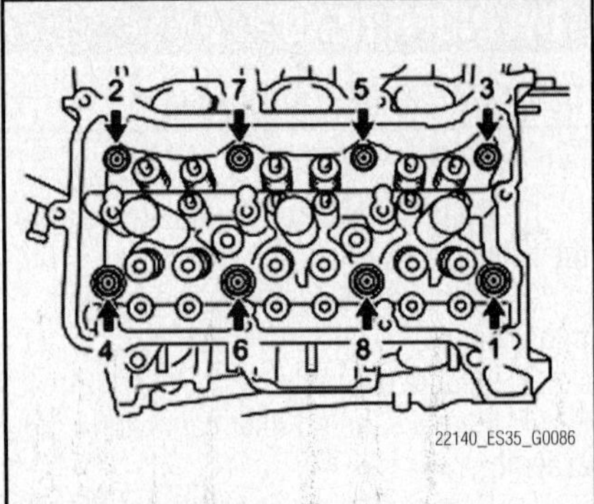

Fig. 56 Right hand cylinder head bolt removal sequence

c. Remove the cylinder head and gasket.

33. Remove the right hand cylinder head sub-assembly, as follows:

a. Using a 10 mm bi-hexagon wrench, uniformly loosen the 8 bolts in the sequence shown in the illustration. Remove the 8 cylinder head bolts and plate washers.

WARNING

Be careful not to drop washers into the cylinder head.

WARNING

Cylinder head warpage or cracking could result from removing bolts in an incorrect order.

➡Be sure to keep separate the removed parts for each installation position.

b. Remove the cylinder head and gasket.

34. Remove the left hand cylinder head sub-assembly, as follows:

a. Uniformly loosen and remove the 2 bolts in the sequence shown in the illustration.

➡Be careful not to drop washers into the cylinder head.

➡Cylinder head warpage or cracking could result from removing bolts in an incorrect order.

b. Using a 10 mm bi-hexagon wrench, uniformly loosen the 8 bolts in the sequence shown in the illustration. Remove the 8 cylinder head bolts and plate washers.

c. Remove the cylinder head and gasket

To install:

35. Place the right hand cylinder head

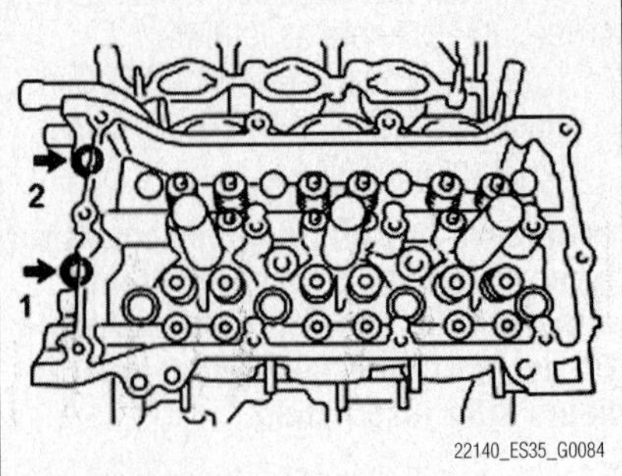

Fig. 57 Left hand cylinder head sub-assembly bolt removal sequence

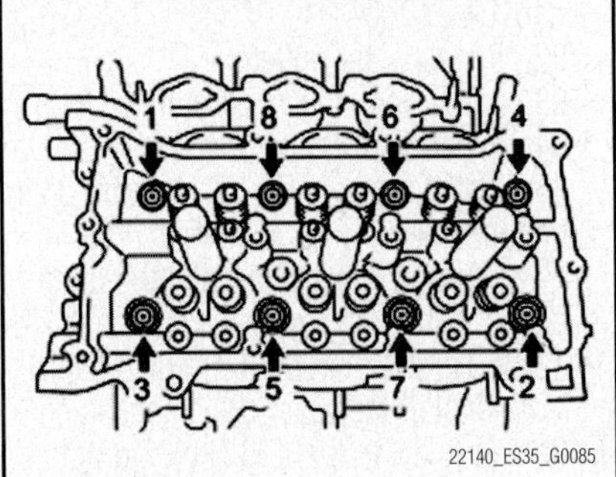

Fig. 58 Left hand cylinder head bolt removal sequence

gasket on the cylinder block surface with the front face of the Lot No. stamp upward.

➡Be careful of the installation direction.

➡Gently place the cylinder head in order not to damage the gasket with the bottom part of the head.

36. Place the cylinder head on the cylinder block.

CAUTION

Do not allow oil to adhere to the mounting surface of the cylinder head.

37. Apply a light coat of engine oil to the threads and under the heads of the cylinder head bolts.

38. The cylinder head bolts are tightened in 3 progressive steps:

a. Step 1: Using a 10 mm bi-hexagon wrench, install and uniformly tighten the 8 cylinder head bolts with the plate washers in several steps and in the sequence shown in the illustration. Tighten to 27 ft. lbs. (36 Nm).

b. Step 2: Mark the cylinder head bolt head with paint as shown in the

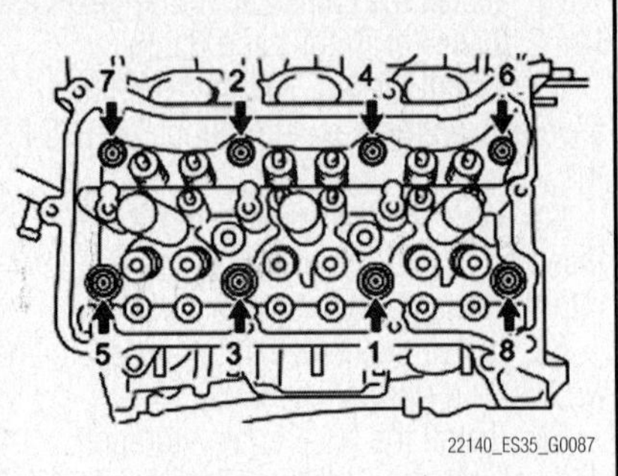

Fig. 59 Right hand cylinder head bolt tightening sequence

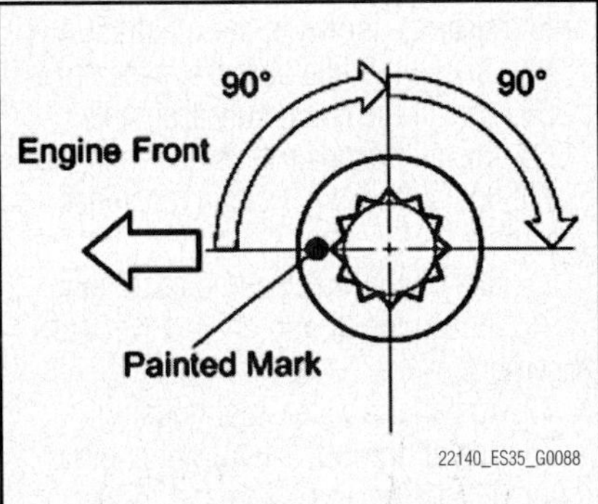

Fig. 60 Mark the cylinder head bolt and tighten another 90°

illustration. Tighten the cylinder head bolts another 90°.

c. Step 3: Tighten the cylinder head bolts an additional 90°. Check that the painted mark is now facing rearward.

d. Seal packing will seep out on the engine's front side. Thoroughly wipe clean any seal packing.

39. Place the left hand cylinder head gasket on the cylinder block surface with the front face of the Lot No. stamp upward.

➡Be careful of the installation direction.

➡Gently place the cylinder head in order not to damage the gasket with the bottom part of the head.

40. Place the cylinder head on the cylinder block.

➡Do not allow oil to adhere to the mounting surface of the cylinder head.

41. Apply a light coat of engine oil to the threads and under the heads of the cylinder head bolts.

42. The cylinder head bolts are tightened in 3 progressive steps:

a. Step 1: Using a 10 mm bi-hexagon wrench, install and uniformly tighten the 8 cylinder head bolts with the plate

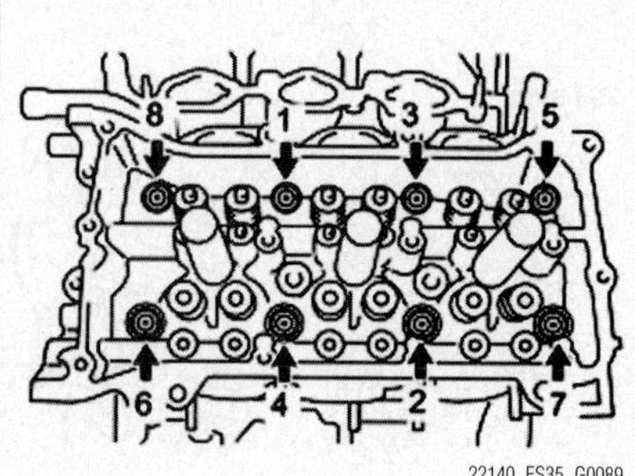

Fig. 61 Left hand cylinder head bolt tightening sequence

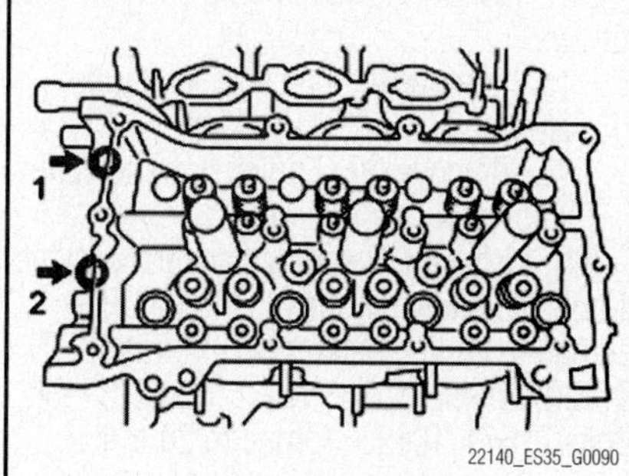

Fig. 62 Left hand cylinder head sub-assembly bolt tightening sequence

washers in several steps and in the sequence shown in the illustration. Tighten to 27 ft. lbs. (36 Nm).

b. Step 2: Mark the cylinder head bolt head with paint as shown in the illustration. Tighten the cylinder head bolts another 90°.

c. Step 3: Tighten the cylinder head bolts an additional 90°. Check that the painted mark is now facing rearward.

d. Tighten the 2 bolts in the order shown in the illustration to 22 ft. lbs. (30 Nm). Only use the specifications stated above when tightening the bolts 1 and 2 shown in the illustration.

e. Seal packing will seep out on the engine's front side. Thoroughly wipe clean any seal packing.

43. Install the No. 2 chain tensioner assembly.
44. Install the No. 3 chain tensioner.
45. Install the No. 1 and 2 chain vibration damper.
46. Install the idle sprocket assembly.
47. Install the chain sub-assembly.
48. Install the chain tensioner slipper.
49. Install the No. 1 chain tensioner assembly.
50. Install the water pump assembly.
51. Install the timing chain cover.
52. Install the water inlet housing.
53. Install the left hand No. 1 front engine mounting bracket.
54. Install the No. 1 oil pan baffle plate.
55. Install the oil pan sub-assembly.
56. Install the oil strainer sub-assembly.
57. Install the No. 2 oil pan sub-assembly.
58. Install the oil pan drain plug and gasket.
59. Install the cylinder head covers and gaskets
60. Install the water outlet.
61. Install the crankshaft pulley.
62. Install the oil filter.
63. Install the cylinder block water drain cock sub-assembly.
64. Install the No. 1 oil pipe.
65. Install the oil pipe.
66. Install the crankshaft position sensor.
67. Install the camshaft timing oil control valve assembly.
68. Install the camshaft position sensor.
69. Install the ventilation valve sub-assembly.
70. Install the spark plugs and ignition coil assembly.
71. Install the oil filler cap sub-assembly.
72. Install the engine assembly with transaxle.

ENGINE ASSEMBLY

REMOVAL & INSTALLATION

See Figures 63 through 69.

1. Before servicing the vehicle, refer to the Precautions Section.
2. Prevent gasoline from spilling out.
3. Remove front wheels.
4. Remove the engine under covers.
5. Remove the right front fender apron seal.
6. Drain engine oil.
7. Drain engine coolant.
8. Drain automatic transaxle fluid.
9. Remove the windshield wiper link assembly.
10. Remove the cowl top panel outer sub-assembly.
11. Remove the 7 clips and intake duct seal.
12. Remove the V-bank cover sub-assembly.
13. Remove the drive belt.
14. Remove the 2 bolts, clamp and air cleaner inlet.
15. Remove air cleaner cap sub-assembly.
16. Remove the air cleaner assembly with hose, as follows:

a. Disconnect the vacuum hoses and hose clamp.

Fig. 63 Removing the 4 bolts and engine moving control rod

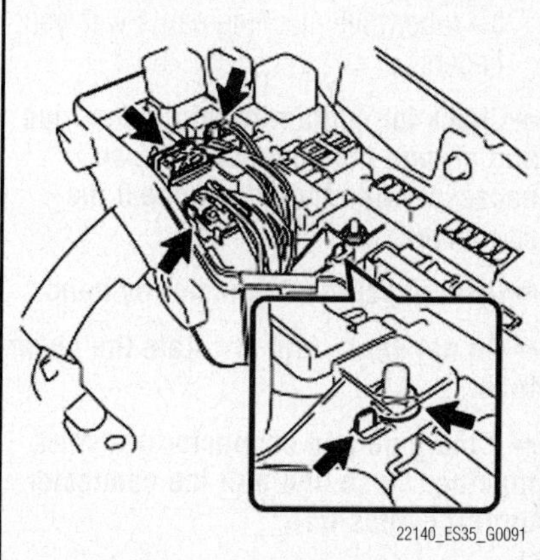

Fig. 64 Disconnecting the engine wire from the engine room junction block

b. Remove the 3 bolts and air cleaner case.

17. Remove the bolt and No. 1 air cleaner inlet.
18. Loosen the bolt and nut, and remove the battery clamp.
19. Remove the clip, bolt and intake air resonator.
20. Remove the bolt, 2 nuts, and No. 2 RH mounting stay.
21. Remove the 4 bolts and engine moving control rod
22. Remove the clamp and disconnect the No. 1 fuel vapor feed hose.
23. Remove the clamp and disconnect the check valve to brake booster hose.
24. Remove the clamp and disconnect the radiator inlet and outlet hose.
25. Remove the oil cooler inlet and outlet hoses.
26. Disconnect the heater water inlet and outlet hose.
27. Remove the upper relay block cover.
28. Disconnect the engine wire from the engine room junction block, as follows:

a. Remove the nut and separate the wire harness.

b. Using a screwdriver, unlock the engine room junction block. Pull the engine room junction block upward.

c. Disconnect the engine wire connectors.

d. Remove the 2 bolts and 2 clamps from the body.

e. Remove the bolt and clamp from the bracket.

29. Disconnect the transmission control cable assembly. Remove the clip and nut, and separate the cable from the transaxle.

30. Separate the fuel tube sub-assembly, as follows:

a. Remove the No. 1 fuel pipe clamp.

b. Disconnect the connector from the tube while pinching part A with your fingers.

➡Check for contamination in the pipe and around the connector. Clean if necessary and then disconnect the connector.

➡Disconnect the connector by hand.

➡Do not bend, fold or rotate the nylon tube.

➡If the pipe and connector are stuck together, push and pull the connector until it comes free.

➡Put the pipe and connector ends in vinyl bags to prevent damage and contamination.

31. Disconnect the oil reservoir to No. 1 pump hose.

32. Disconnect the return tube sub-assembly.

33. Remove the No. 1 exhaust pipe support bracket.

34. Remove the front exhaust pipe assembly.

35. Remove the left front axle shaft nut, as follows:

36. Using SST (SST: 09930-00010) or equivalent and a hammer, strike the lock nut covering to remove it.

a. Set the drive shaft's groove so that it faces up. Then use the SST and a hammer.

b. Remove the covering from the lock nut completely. Otherwise the screw of the drive shaft may be damaged.

c. Do not sharpen the tip of the SST.

d. Using a 30 mm socket wrench, remove the lock nut.

37. Remove the right front axle shaft nut.

38. Disconnect the left front stabilizer link assembly, as follows:

a. Using a 6 mm socket hexagon wrench, hold the ball stud.

b. Remove the nut and disconnect the stabilizer link.

39. Disconnect the right front stabilizer link assembly.

40. Remove the bolts and disconnect the right and left front speed sensors from the steering knuckle.

41. Disconnect the left and right tie rod assemblies, as follows:

a. Remove the cotter pin and nut.

b. Using SST (SST: 09628-62011) or equivalent, disconnect the tie rod ends from the steering knuckle.

➡Do not damage the dust cover of the ball joint.

42. Disconnect the lower No. 1 left and right front suspension arm sub-assemblies, as follows:

a. Remove the bolt and 2 nuts, and separate the front suspension arm from the lower ball joint.

b. Using a plastic hammer, disconnect the drive shaft from the axle hub.

43. Separate left and right front axle assembly.

44. Remove the No. 1 exhaust pipe support bracket with the 2 bolts.

45. Remove the drive plate and torque converter clutch setting bolt, as follows:

a. Using SST (SST: 09213-70011, SST: 09330-00021) or equivalent, hold the crankshaft.

b. Remove the 2 bolts and flywheel housing under cover.

c. Remove the 6 torque converter clutch setting bolts.

46. Disconnect steering sliding yoke.

47. Remove generator assembly.

48. Separate the cooler compressor assembly, as follows:

a. Remove the 2 connector clamps.

b. Remove the 4 bolts and separate the compressor.

➡Hang up the hoses instead of detaching them.

49. Remove the engine assembly with transaxle, as follows:

a. Set the engine lifter.

b. Remove the 4 bolts, 2 nuts, and right and left frame side rail plates.

c. Remove the 4 bolts, 2 nuts, front suspension member right and left rear brace.

d. Operate the engine lifter, then remove the engine assembly from the vehicle.

➡Make sure the engine is clear of all wiring and hoses.

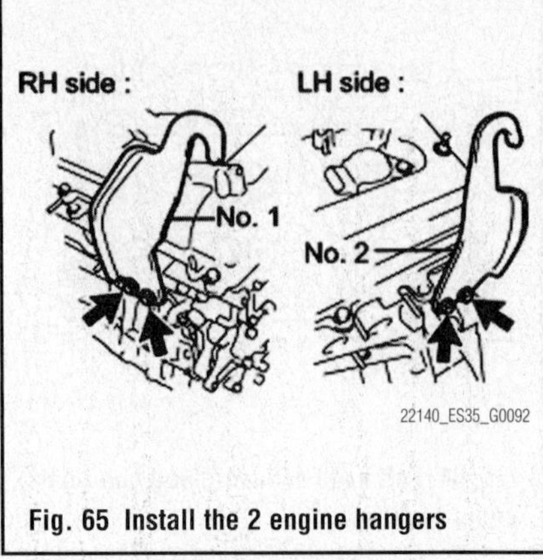

Fig. 65 Install the 2 engine hangers

50. Remove the vane pump assembly, as follows:

a. Remove the pressure feed tube clamp bolt.

b. Remove the 2 bolts, nut and vane pump.

51. Install the 2 engine hangers with the 4 bolts as shown in the illustration. Part numbers: Engine hanger No. 1 12281-31120, Engine hanger No. 2 12282-31100, Bolts 91671-10825. Torque to 24 ft. lbs. (33 Nm).

a. Attach the engine sling device and hang the engine with the chain block.

52. Remove the front frame assembly, as follows:

a. Disconnect the connector and clamp.

b. Disconnect the 2 clamps.

c. Remove the 2 nuts and disconnect the right and left engine mounting insulators.

d. Remove the bolt and disconnect the engine mounting insulator FR.

53. Remove the right and left front drive shaft assemblies.

54. Remove the engine wire.

55. Remove the starter assembly.

56. Remove the automatic transaxle assembly.

57. Remove the drive plate and ring gear sub-assembly.

58. Install engine stand.

59. Remove intake air surge tank assembly, as follows:

a. Disconnect the 2 water by-pass hoses from the throttle with motor body assembly

b. Disconnect the vapor feed hose.

c. Disconnect the throttle with motor body assembly connector and clamp.

d. Disconnect the No. 1 ventilation hose.

e. Remove the bolt and vacuum hose clamp.

f. Disconnect the connector.

g. Remove the 4 bolts, No. 1 surge tank stay and throttle body bracket.

h. Using a 5 mm socket hexagon wrench, remove the 4 bolts. Remove the 2 nuts and intake air surge tank. Remove the gasket from the intake air surge tank.

60. Remove the ignition coil assembly.

61. Remove the bolt and the No. 2 RH engine mounting stay.

62. Remove the intake manifold, as follows:

a. Uniformly loosen and remove the 6 bolts and 4 nuts.

b. Remove the intake manifold and 2 gaskets.

63. Remove the right exhaust manifold sub-assembly, as follows:

a. Disconnect the A/F sensor connector clamp.

b. Uniformly loosen and remove the 6 nuts.

c. Remove the manifold and gasket.

64. Remove the oil level gauge guide sub-assembly, as follows:

a. Remove the oil level gauge.

b. Remove the 2 bolts, and No. 1 and No. 2 oil level gauge guides.

c. Remove the O-rings from the oil level gauge guide.

65. Remove the 2 bolts and the No. 2 manifold stay.

66. Remove the 3 bolts and the No. 2 exhaust manifold heat insulator.

67. Remove the left exhaust manifold sub-assembly, as follows:

a. Uniformly loosen and remove the 6 nuts.

b. Remove the manifold and gasket.

68. Remove the 3 bolts and the right engine mounting bracket.

69. Remove the 3 bolts and drive shaft bearing bracket.

70. Remove the 5 bolts and V-ribbed belt tensioner assembly.

71. Remove the 2 bolts and the No. 2 timing gear cover.

72. Remove the bolt, plate and No. 2 idler pulley sub-assembly.

73. Remove the water pump pulley.

74. Remove the 6 bolts and No. 1 left front engine mounting bracket.

75. Remove the 2 bolts and 2 radio setting condensers.

76. Remove the bolt and No. 1 vacuum switching valve.

77. Using a 24 mm deep socket wrench, remove the engine oil pressure switch assembly.

78. Remove 2 bolts and 2 knock control sensors.

79. Using a 19 mm deep socket wrench, remove the EFI engine coolant temperature sensor and gasket.

80. Replace partial engine assembly as necessary.

To install:

81. Using a 19 mm deep socket wrench, install the EFI engine coolant temperature sensor and a new gasket and tighten to 14 ft. lbs. (20 Nm).

82. Install the knock control sensor.

83. Install the engine oil pressure switch assembly, as follows:

a. Clean the threads of the oil pressure switch. Apply adhesive (Toyota Genuine Adhesive 1344, Three Bond 1344 or equivalent) to 2 or 3 threads of the oil pressure switch.

b. Using a 24 mm deep socket wrench install the oil pressure switch and tighten to 15 ft. lbs. (21 Nm).

Install the left front No. 1 engine mounting bracket with the 6 bolts and tighten to 40 ft. lbs. (54 Nm).

84. Install the bolt and No. 1 vacuum switching valve and tighten to 7 ft. lbs. (10 Nm).

85. Install the 2 bolts and 2 radio setting condensers and tighten to 7 ft. lbs. (10 Nm).

86. Install the No. 1 engine mounting bracket front LH with the 6 bolts and tighten to 40 ft. lbs. (54 Nm).

87. Install the water pump pulley.

88. Install the No. 2 idler pulley sub-assembly and cover plate with the bolt and tighten to 32 ft. lbs. (43 Nm).

89. Install the No. 2 timing gear cover with the 2 bolts and tighten to 53 inch lbs. (6 Nm).

90. Install the V-ribbed belt tensioner assembly, as follows:

a. Temporarily install the V-ribbed belt tensioner with the 5 bolts.

b. Install the V-ribbed belt tensioner by tightening the bolt 1 and bolt 2 in the order shown in the illustration to 32 ft. lbs. (43 Nm).

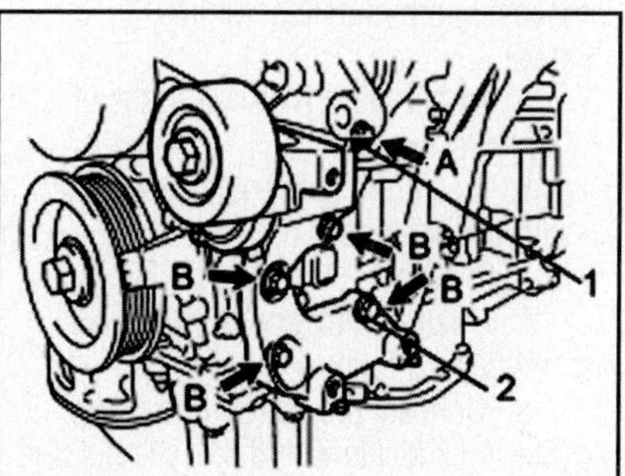

Fig. 66 V-ribbed belt tensioner bolt tightening sequence

c. Tighten the other bolts to 32 ft. lbs. (43 Nm). Bolt length: A: 70 mm (2.76 in.), B: 33 mm (1.30 in.)

91. Install the drive shaft bearing bracket with the 3 bolts and tighten to 47 ft. lbs. (64 Nm).

92. Install the engine mounting bracket RH with the 3 bolts and tighten to 40 ft. lbs. (54 Nm).

93. Install the left exhaust manifold sub-assembly, as follows:

a. Install a new gasket.

b. Install the exhaust manifold sub-assembly with the 6 nuts and tighten to 15 ft. lbs. (21 Nm).

94. Install the No. 2 exhaust manifold heat insulator with the 3 bolts and tighten to 75 inch lbs. (8.5 Nm).

95. Install the No. 2 manifold stay with the 2 bolts and tighten to 25 ft. lbs. (34 Nm).

96. Install the oil level gauge guide sub-assembly, as follows:

a. Install 2 new O-rings to the oil level gauge guide.

b. Apply a coat of engine oil lightly to the O-rings.

c. Push in the oil level gauge guide end into the guide hole.

d. Install the No. 1 oil level gauge guide with the bolt and tighten to 15 ft. lbs. (21 Nm).

e. Install the No. 2 oil level gauge guide with the bolt and tighten to 15 ft. lbs. (21 Nm).

f. Install the oil level gauge.

97. Install the right exhaust manifold sub-assembly, as follows:

a. Install a new gasket.

b. Install the exhaust manifold sub-assembly with the 6 nuts and tighten to 15 ft. lbs. (21 Nm).

98. Install the intake manifold.

99. Install the right No. 2 mounting stay with the bolt and tighten to 15 ft. lbs. (21 Nm).

100. Install the 6 ignition coil assemblies with the 6 bolts and tighten to 7 ft. lbs. (10 Nm).

⁂ CAUTION

DO NOT apply oil to the following bolts: surge tank and intake manifold, No. 1 surge tank stay and cylinder head cover, No. 1 surge tank stay and surge tank, throttle body bracket and cylinder head cover, throttle body bracket and surge tank.

a. Install a new gasket to the intake air surge tank.

b. Using a 5 mm hexagon socket wrench, install the 4 bolts and 2 nuts. Tighten the bolts to 13 ft. lbs. (18 Nm). Tighten the nuts to 12 ft. lbs. (16 Nm).

c. Install the throttle body bracket, No. 1 surge tank stay and 4 bolts and tighten to 15 ft. lbs. (21 Nm).

d. Connect the connector.

e. Install the vacuum hose clamp with the bolt and tighten to 48 inch lbs. (5.4 Nm).

f. Connect the No. 1 ventilation hose.

g. Install the clamp and connect the throttle with motor body assembly connector.

h. Connect the vapor feed hose.

i. Connect the 2 water by-pass hoses to the throttle with motor body assembly.

101. Remove the engine stand.

102. Install the drive plate and ring gear sub-assembly.

103. Install the automatic transaxle assembly.

104. Install the starter assembly.

105. Install the engine wire.

106. Install the right and left front drive shaft assemblies.

107. Install the front frame assembly, as follows:

a. Install the right and left engine mounting insulators and with the 2 nuts and tighten 70 ft. lbs. (95 Nm).

b. Install the engine mounting insulator FR with the bolt and tighten 64 ft. lbs. (87 Nm).

c. Connect the connector and clamp.

d. Connect the 2 clamps.

108. Install the vane pump assembly, as follows:

a. Install the vane pump with the 2 bolts and nut and tighten 32 ft. lbs. (43 Nm).

b. Install the pressure feed tube clamp bolt and tighten 87 inch lbs. (9.8 Nm).

c. Connect the power steering oil pressure switch connector.

109. Install the engine assembly with transaxle, as follows:

a. Set the engine assembly with transaxle on the engine lifter.

b. Install the engine assembly to the vehicle.

c. Install the right and left frame side rail plates and with the 4 bolts and 2 nuts. Tighten A to 63 ft. lbs. (85 Nm) and B to 24 ft. lbs. (32 Nm).

d. Install the right and left front suspension member rear braces with the 4 bolts and 2 nuts. Tighten C to 63 ft. lbs. (85 Nm) and D to 24 ft. lbs. (32 Nm).

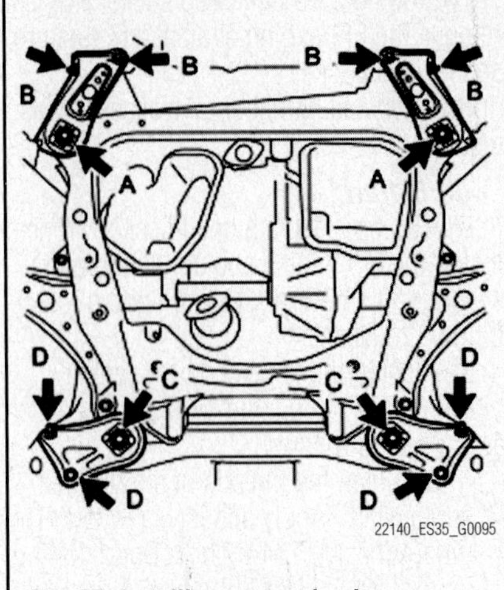

Fig. 67 Installing suspension braces

110. Install the cooler compressor assembly.

111. Install generator assembly.

112. Install the steering intermediate shaft assembly.

113. Install drive plate and torque converter clutch setting bolt, as follows:

a. Using SST (SST: 09213-70011, SST: 09330-00021) or equivalent, hold the crankshaft.

b. Using kerosene or gasoline, clean the bolts thoroughly.

c. Apply adhesive (Toyota Genuine Adhesive 1324, Three Bond 1324 or equivalent) to 2 or 3 threads of the bolt end.

d. Install the 6 torque converter set bolts and tighten 30 ft. lbs. (41 Nm). First tighten the green colored bolt, then the other 5 bolts.

e. Install the flywheel housing under cover with the 2 bolts and tighten 69 inch lbs. (7.8 Nm).

114. Install the No. 1 exhaust pipe support bracket with the 2 bolts and tighten 15 ft. lbs. (21 Nm).

115. Install both front axle assemblies.

116. Install the lower right and left No. 1 suspension arm sub-assemblies, as follows:

a. Install the drive shaft to the steering knuckle.

b. Install the suspension lower arm with the bolt and 2 nuts and tighten 55 ft. lbs. (75 Nm).

117. Install the left and right tie rod assemblies, as follows:

a. Connect the tie rod end to the steering knuckle with the castle nut and tighten 36 ft. lbs. (49 Nm).

➡Do not allow lubricants to contact the thread and taper portions.

➡After tightening the castle nut, tighten additional 60° so that a cotter pin can be inserted.

b. Insert a new cotter pin.

118. Install the right and left front speed sensors.

119. Install the right and left front stabilizer link assemblies, as follows:

a. Using a 6 mm socket hexagon wrench, hold the ball stud and install the nut and tighten 55 ft. lbs. (74 Nm).

120. Install the right and left front axle shaft hub nuts, as follows:

a. Using a 30 mm socket wrench, install a new hub nut and tighten 217 ft. lbs. (294 Nm).

b. Using a chisel and hammer, tap the hub nut.

121. Install the front exhaust pipe assembly.

122. Install the No. 1 exhaust pipe support bracket.

123. Connect the oil reservoir to the No. 1 pump hose.

124. Connect the steering gear outlet return tube.

125. Connect the fuel tube sub-assembly, as follows:

a. Push in the fuel tube connector to the fuel pipe until connector makes a "click" sound.

➡Check for damage or contamination on the connected part of the pipe. Check if the pipe and the connector are securely connected by trying to pull them apart.

b. Install the No. 1 fuel pipe clamp .

126.Connect the transmission control cable assembly. Install the clip and nut, and connect the cable to the transaxle. Tighten the nut to 9 ft. lbs. (13 Nm).

127. Connect engine wire, as follows:

a. Install the bolt and clamp to the bracket. Tighten 74 inch lbs. (8.4 Nm).

b. Install the 2 bolts and 2 clamps to the body.

c. Connect the wire to the engine room junction block. Then, install it with the nut and 3 connectors. Tighten to 74 inch lbs. (8.4 Nm).

128. Install upper relay block cover upper.

129. Connect the heater water outlet and inlet hose.

130. Connect the oil cooler inlet and outlet hoses.

131. Install the radiator inlet and outlet hoses.

132. Install the clamp and connect the check valve to brake booster hose.

Fig. 68 Engine moving control rod bolt tightening sequence

133. Install the clamp and connect the No. 1 fuel vapor feed hose.
134. Install the engine moving control rod, as follows:
  a. Temporarily install the moving control rod with the 4 bolts.
  b. Tighten the bolt A to 28 ft. lbs. (38 Nm).
  c. Tighten the bolt B in the order shown in the illustration to 28 ft. lbs. (38 Nm).
135. Temporarily install the right No. 2 mounting stay with the bolts and 2 nuts. Tighten to 17 ft. lbs. (23 Nm).
136. Install the intake air resonator with the bolt and clip and tighten to 44 inch lbs. (5 Nm).
137. Install the battery and battery tray.
138. Install the No. 1 air cleaner inlet with the bolt and tighten 44 inch lbs. (5 Nm).
139. Install the air cleaner case sub-assembly, as follows:
  a. Install the air cleaner case with the 3 bolts and tighten 44 inch lbs. (5 Nm).
  b. Connect the vacuum hose and hose clamp.
140. Install the air cleaner cap sub-assembly.
141. Install the air cleaner inlet with the clamp and 2 bolts and tighten to 44 inch. lbs. (5 Nm).
142. Connect vacuum hoses.
143. Install drive belt.
144. Install the outer cowl top panel.
145. Install the windshield wiper link assembly.
146. Install the front wheels.
147. Add automatic transaxle fluid.
148. Add engine oil.
149. Add engine coolant.
150. Add power steering fluid.
151. Bleed power steering system.
152. Inspect for oil leak, coolant leak, fuel leak, and exhaust gas leak.
153. Inspect and adjust front wheel alignment.
154. Inspect ignition timing and engine idle speed.

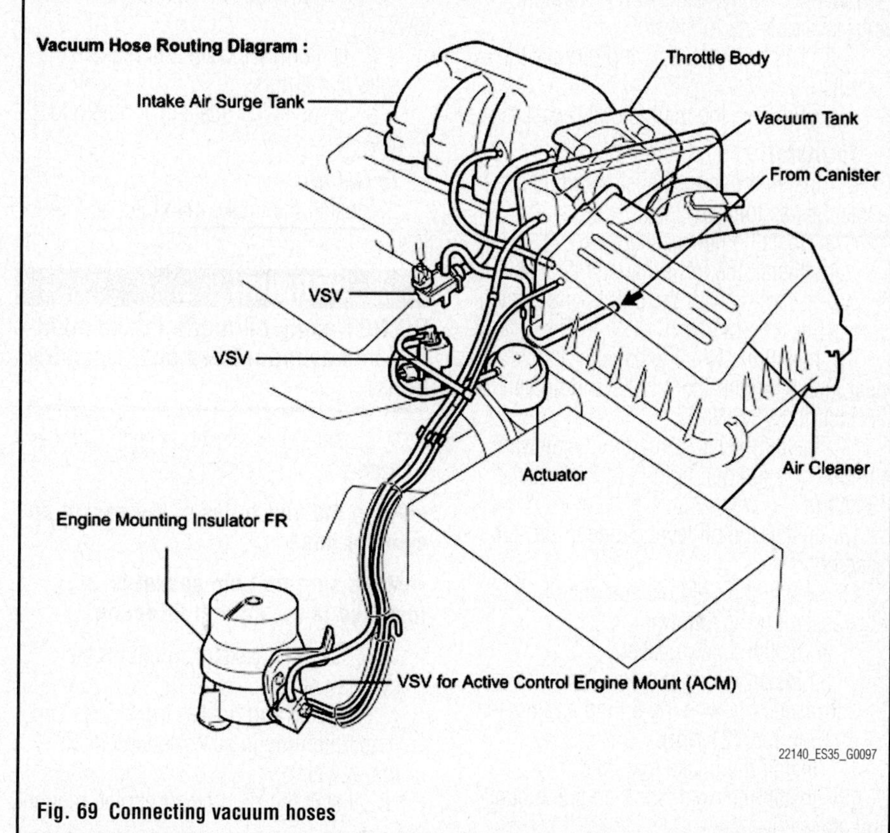

Fig. 69 Connecting vacuum hoses

155. Check function of throttle body.
156. Install the V-bank cover sub-assembly.
157. Inspect CO/HC.
158. Check ABS speed sensor signal.
159. Perform initialization.

EXHAUST MANIFOLD

REMOVAL & INSTALLATION

See Figures 70 and 71.

1. Before servicing the vehicle, refer to the Precautions Section.
2. Remove the engine assembly with transaxle.
3. Secure the engine.
4. Remove the ignition coil assembly.
5. Remove the right No. 2 engine mounting stay.
6. Remove the intake manifold.
7. Remove the right exhaust manifold sub-assembly, as follows:
  a. Uniformly loosen and remove the 6 nuts.
  b. Remove the manifold and gasket.
8. Remove the oil level gauge guide sub-assembly.
9. Remove the bolt, nut and No. 2 manifold stay.
10. Remove the 3 bolts and No. 2 exhaust manifold heat insulator.

Fig. 70 Removing the exhaust manifold 6 nuts—Right

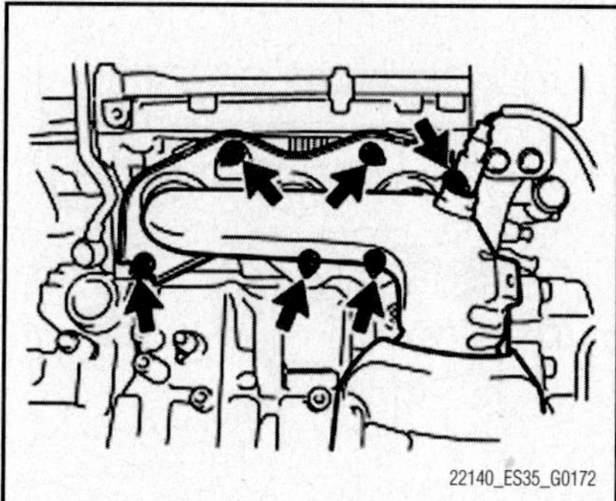

Fig. 71 Removing the exhaust manifold 6 nuts—Left

11. Remove the left exhaust manifold sub-assembly, as follows:
a. Uniformly loosen and remove the 6 nuts.
b. Remove the manifold and gasket.

To install:

12. Install the left exhaust manifold sub-assembly, as follows:
a. Install a new gasket.
b. Install the left exhaust manifold sub-assembly with the 6 nuts and tighten to 15 ft. lbs. (21 Nm).
13. Install the No. 2 exhaust manifold heat insulator with the 3 bolts and tighten to 75 inch lbs. (8.5 Nm).
14. Install the No. 2 manifold stay with the bolt and nut and tighten to 25 ft. lbs. (34 Nm).
15. Install the oil level gauge guide sub-assembly.
16. Install the right exhaust manifold sub-assembly, as follows:
a. Install a new gasket.
b. Install the right exhaust manifold sub-assembly with the 6 nuts and tighten to 15 ft. lbs. (21 Nm).
17. Install the intake manifold.
18. Install the right No. 2 engine mounting stay.
19. Install the ignition coil assembly.
20. Install the engine assembly with transaxle.

INTAKE MANIFOLD

REMOVAL & INSTALLATION

See Figure 72.

1. Before servicing the vehicle, refer to the Precautions Section.
2. Remove the engine assembly with transaxle.
3. Secure the engine.
4. Remove the ignition coil assembly.
5. Remove the right No. 2 engine mounting stay.

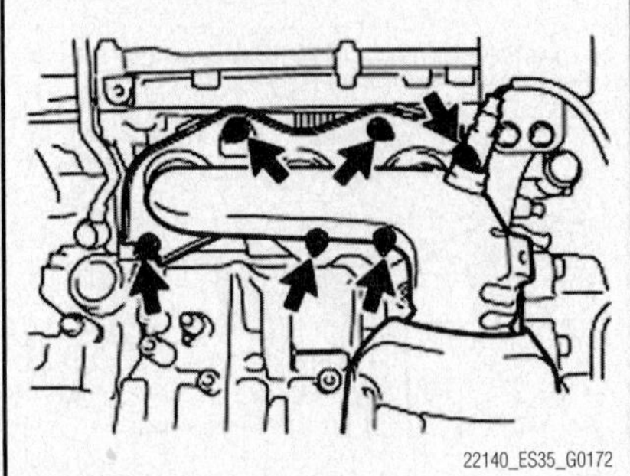

Fig. 72 Locating intake manifold bolts and nuts

6. Remove the intake manifold, as follows:
a. Uniformly loosen and remove the 6 bolts and 4 nuts.
b. Remove the intake manifold and 2 gaskets.

To install:

7. Install the intake manifold, as follows:

⁂ WARNING

DO NOT apply oil to the intake manifold and cylinder head sub-assembly bolts.

a. Set a new gasket on each cylinder head.

➡Align the port holes of the gasket and cylinder head.

➡Make sure that the gasket is installed in the correct direction.

b. Set the intake manifold on the cylinder heads.
c. Install and tighten the 6 bolts and 4 nuts uniformly in several steps to 15 ft. lbs. (21 Nm).
8. Install the right No. 2 engine mounting stay.
9. Install the ignition coil assembly.
10. Install the engine assembly with transaxle.

OIL PAN

REMOVAL & INSTALLATION

See Figures 73 through 77.

1. Before servicing the vehicle, refer to the Precautions Section.
2. Drain the engine oil.
3. Remove the engine assembly with transaxle.
4. Secure the engine.
5. Remove the oil filler cap and gasket.
6. Remove the oil pan drain plug and gasket.
7. Remove the oil pan drain plug and gasket.
8. Remove the No. 1 oil pipe, as follows:
a. Remove the 2 oil pipe unions and oil pipe.
b. Remove the left hand oil control valve filter and gaskets.
9. Remove the oil pipe, as follows:
a. Remove the bolt.
b. Remove the 2 oil pipe unions and oil pipe.
c. Remove the right oil control valve filter and gaskets.

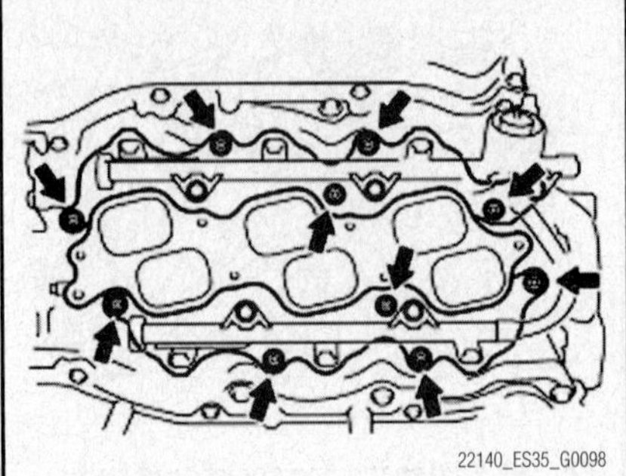

Fig. 73 No. 2 oil pan sub-assembly removal

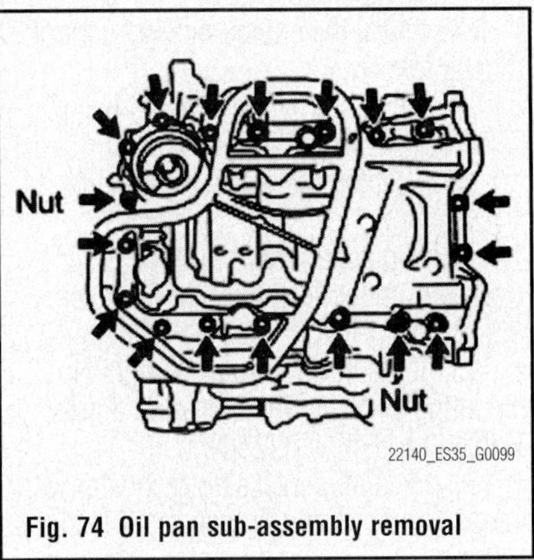

Fig. 74 Oil pan sub-assembly removal

10. Remove the oil filter element, as follows:
a. Remove the drain plug. Do not remove the O-ring.
b. Connect the hose to the pipe.
c. Insert the pipe with the hose into the oil filter cap.
d. Make sure that the oil is completely drained and remove the pipe and O-ring.
e. Using SST (SST: 09228-06501) or equivalent, remove the oil filter cap.
f. Remove the oil filter element and O-ring from the oil filter cap. Do not use any tools when removing the O-ring to prevent the O-ring groove from being damaged.
11. Remove the No. 2 oil pan sub-assembly, as follows:
a. Remove the 16 bolts and 2 nuts.
b. Insert the blade of SST (SST: 09032-00100) or equivalent tool between the oil pans. Cut through the applied sealer and remove the No. 2 oil pan sub-assembly.

➡Be careful not to damage the contact surfaces of the oil pans.

12. Remove the oil pan sub-assembly, as follows:
a. Remove the 16 bolts and 2 nuts.

Timing Chain Cover:

Lower Cylinder Block:

16 mm
(0.63 in.)

34 mm
(1.34 in.)

12 mm
(0.47 in.)

22140_ES35_G0100

Fig. 75 Locating stud bolts

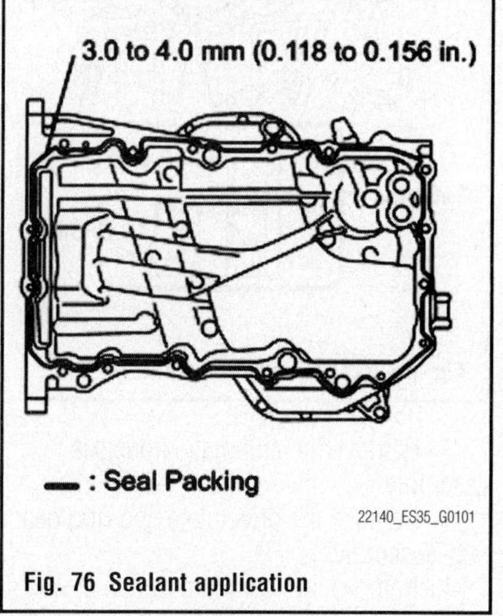

Fig. 76 Sealant application

➡**Be sure to clean the bolts and stud bolts and check the threads for cracks or other damage.**

b. Remove the oil pan by prying between the oil pan and cylinder block with a taped screwdriver.

➡**Be careful not to damage the contact surfaces of the cylinder block and oil pan.**

c. Remove the 2 O-rings.

To install:

13. Install the oil pan sub-assembly, as follows:

a. Using an E8 "TORX" socket wrench, install the stud bolts as shown in the illustration. Tighten to 7 ft. lbs (10 Nm).

b. Install 2 new O-rings.

➡**Remove any oil from the contact surface.**

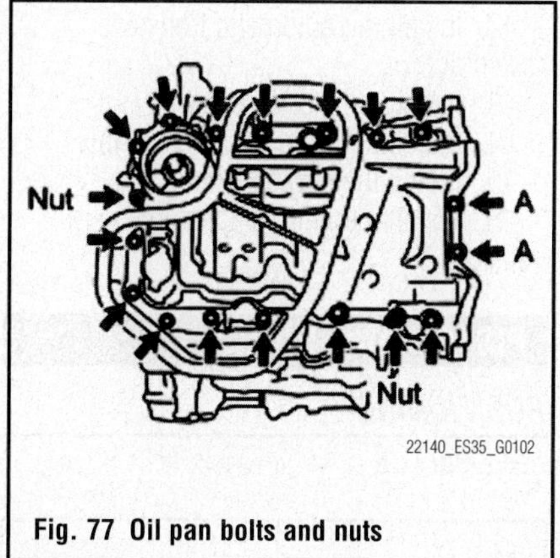

Fig. 77 Oil pan bolts and nuts

➡**Install the oil pan within 3 minutes after applying seal packing.**

➡**Do not start the engine for at least 2 hours after installing.**

c. Apply seal packing (Toyota Genuine Seal Packing Black, Three Bond 1207B or equivalent) in a continuous line as shown in the illustration. Seal diameter: 3.0 to 4.0 mm (0.118 to 0.156 in.).

d. Install the oil pan with the 16 bolts and 2 nuts and tighten to 7 ft. lbs (10 Nm), and 15 ft. lbs (21 Nm).

14. Install the No. 2 oil pan sub-assembly, as follows:

➡**Remove any oil from the contact surface.**

➡**Install the No. 2 oil pan within 3 minutes after applying seal packing.**

➡**Do not start the engine for at least 2 hours after installing.**

a. Using an E6 "TORX" socket wrench, install the stud bolts as shown in the illustration and tighten to 35 inch lbs (4 Nm).

b. Apply seal packing (Toyota Genuine Seal Packing Black, Three Bond 1207B or equivalent) in a continuous line as shown in the illustration. Seal diameter: 3.0 to 4.0 mm (0.118 to 0.156 in.).

c. Install the No. 2 oil pan with the 16 bolts and 2 nuts and tighten to 7 ft. lbs (10 Nm).

15. Install the oil pan drain plug and a new gasket. Tighten to 30 ft. lbs (40 Nm).

16. Install the oil filter element, as follows:

a. Clean the inside of the oil filter cap, the threads and O-ring groove.

b. Apply a small amount of engine oil to a new O-ring and install it to the oil filter cap.

c. Set a new oil filter element to the oil filter cap.

d. Remove dirt or foreign matter from the installation surface and inside of the engine.

e. Apply a small amount of engine oil to the O-ring again and install the oil filter cap.

➡**Be careful that the O-ring does not get caught between the parts. The O-ring must not be twisted on the groove.**

f. Using SST (SST: 09228-06501) or equivalent, install the oil filter cap and tighten to 18 ft. lbs (25 Nm). Make sure that the oil filter is installed securely as shown in the illustration.

17. Install the oil filler cap sub-assembly.

18. Install the engine assembly with transaxle.

19. Check for oil leaks.

OIL PUMP

REMOVAL & INSTALLATION

1. Before servicing the vehicle, refer to the Precautions Section.
2. Remove the engine assembly with transaxle.
3. Secure engine.
4. Remove the engine wire.
5. Remove the front frame assembly.
6. Remove the starter assembly.
7. Remove the automatic transaxle assembly.
8. Remove the oil level gauge guide sub-assembly.
9. Remove the right and left exhaust manifold sub-assemblies.
10. Remove the drive plate and ring gear sub-assembly.
11. Remove the No. 2 idler pulley sub-assembly.
12. Remove the V-ribbed belt tensioner assembly.
13. Remove the water pump pulley.
14. Remove the water inlet housing.
15. Remove the crankshaft pulley.
16. Remove the No. 2 oil pan sub-assembly.
17. Remove the oil strainer sub-assembly.
18. Remove the oil pan sub-assembly.
19. Remove the intake air surge tank assembly.
20. Remove the ignition coil assembly.
21. Remove the No. 1 and 2 oil pipes.
22. Remove the right and left cylinder head cover sub-assemblies.
23. Remove the timing chain or belt cover sub-assembly.
24. Remove the timing gear case or timing chain case oil seal, as follows:
    a. Using a screwdriver with the tip taped, pry out the oil seal.

To install:

25. Install timing gear case or timing chain case oil seal, as follows:
    a. Using SST (SST: 09316-60011) or equivalent tool, tap in a new oil seal until its surface is flush with the timing chain case edge.

➡Keep the lip free from foreign matter.

➡Do not tap on the oil seal at an angle.

➡Make sure that the oil seal edge does not stick out of the timing chain case.

    b. Apply MP grease to the oil seal lip.

26. Install timing chain or belt cover sub-assembly.
27. Install the right and left cylinder head cover sub-assemblies.
28. Install the No. 1 and 2 oil pipes.
29. Install the ignition coil assembly.
30. Install the intake air surge tank assembly.
31. Install the oil pan sub-assembly.
32. Install the oil strainer sub-assembly.
33. Install the No. 2 oil pan sub-assembly.
34. Install the crankshaft pulley.
35. Install the water inlet housing.
36. Install the water pump pulley.
37. Install the V-ribbed belt tensioner assembly.
38. Install the No. 2 idler pulley sub-assembly.
39. Install the drive plate and ring gear sub-assembly.
40. Install the right and left exhaust manifold sub-assemblies.
41. Install the oil level gauge guide sub-assembly.
42. Install the automatic transaxle assembly.
43. Install the starter assembly.
44. Install the front frame assembly.
45. Install the engine wire.
46. Install the engine assembly with transaxle.

PISTON AND RING

POSITIONING

See Figure 78.

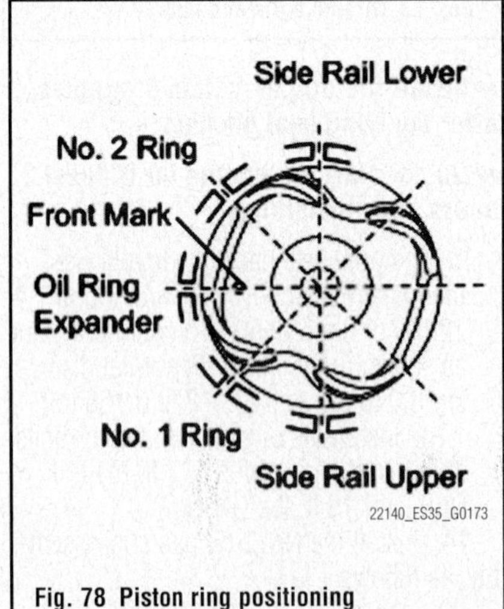

Fig. 78 Piston ring positioning

REAR MAIN SEAL

REMOVAL & INSTALLATION

See Figures 79 and 80.

1. Before servicing the vehicle, refer to the Precautions Section.

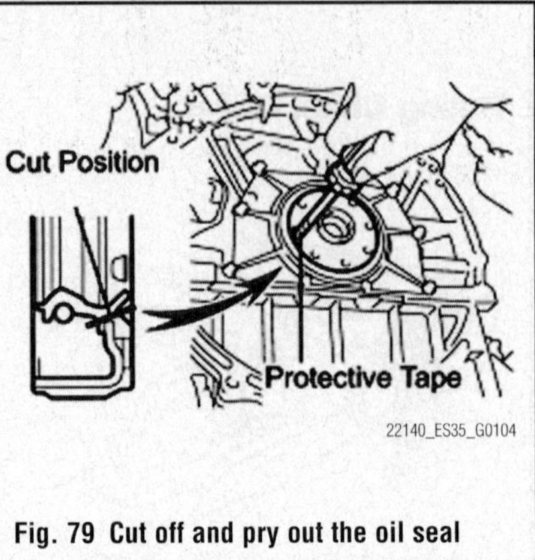

Fig. 79 Cut off and pry out the oil seal

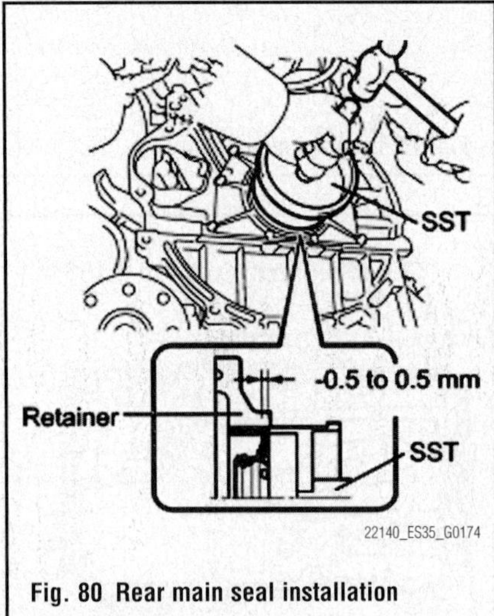

Fig. 80 Rear main seal installation

2. Remove the automatic transaxle assembly.
3. Remove the drive plate and ring gear sub-assembly.
4. Remove the rear main seal, as follows:
    a. Using a knife, cut off the oil seal lip.

To install:

5. Apply MP grease to a new oil seal lip.
6. Using SST (SST: 09223-15030, SST: 09950-70010) or equivalent and a hammer, tap in the oil seal. Oil seal tap in depth: -0.020 to 0.020 in. (-0.5 to 0.5 mm)
7. Install the drive plate and ring gear sub-assembly.
8. Install automatic transaxle assembly.

TIMING CHAIN, SPROCKETS, FRONT COVER AND SEAL

REMOVAL & INSTALLATION

Timing Chain Cover and Seal

See Figures 81 through 86.

1. Before servicing the vehicle, refer to the Precautions Section.

2. Remove the engine assembly with transaxle.
3. Secure engine.
4. Remove the oil filler cap sub-assembly.
5. Remove the spark plugs and ignition coil assembly.
6. Remove the oil pan drain plug and gasket.
7. Remove the ventilation valve sub-assembly.
8. Remove the camshaft position sensor.
9. Remove the camshaft timing oil control valve assembly.
10. Remove crankshaft position sensor.

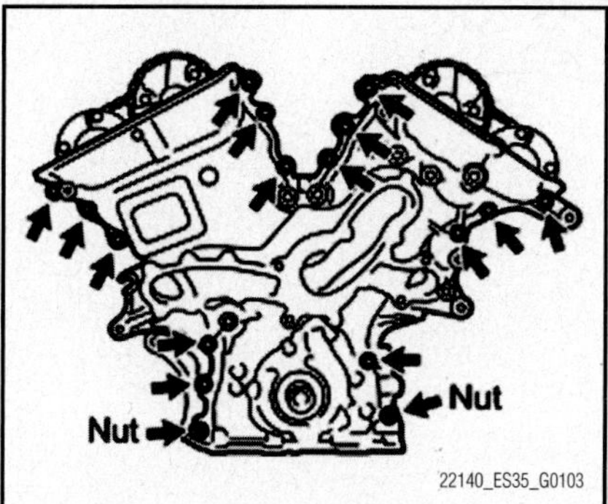

Fig. 81 Locating timing chain cover sub-assembly bolts and nuts

11. Remove the No. 1 oil pipe.
12. Remove the oil pipe.
13. Remove the cylinder block water drain cock sub-assembly.
14. Remove the oil filter.
15. Remove the crankshaft pulley.
16. Remove the left hand No. 1 front engine mounting bracket.

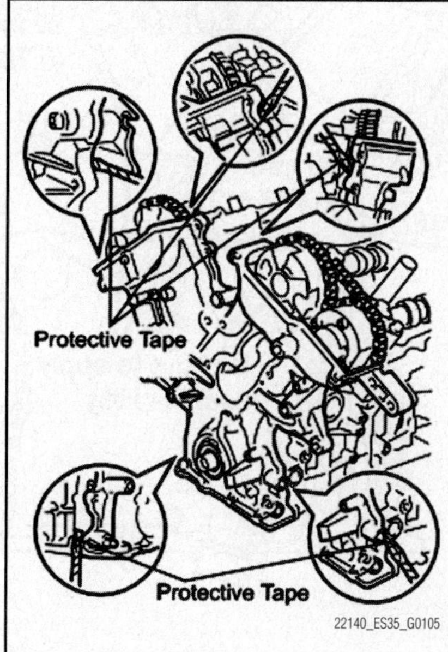

Fig. 82 Timing chain cover removal

17. Remove the water inlet housing.
18. Remove the water outlet.
19. Remove the left-hand cylinder head cover sub-assembly and gasket.
20. Remove the cylinder head cover sub-assembly and gasket.
21. Remove the No. 2 oil pan sub-assembly.
22. Remove the oil strainer sub-assembly.
23. Remove the oil pan sub-assembly.
24. Remove the water pump assembly.
25. Remove the timing chain cover sub-assembly, as follows:
   a. Remove the 15 bolts and 2 nuts as shown in the illustration.

➡Be careful not to damage the contact surfaces of the cylinder head, cylinder block and chain cover.

   b. Remove the timing chain cover by prying between the timing chain cover and cylinder head or cylinder block with a screwdriver with the tip taped.
   c. Remove the 4 bolts, chain cover plate and gasket.
   d. Remove the gasket.

To install:

26. Install timing gear case or timing chain cover oil seal, as follows:

➡Keep the lip free from foreign matter.

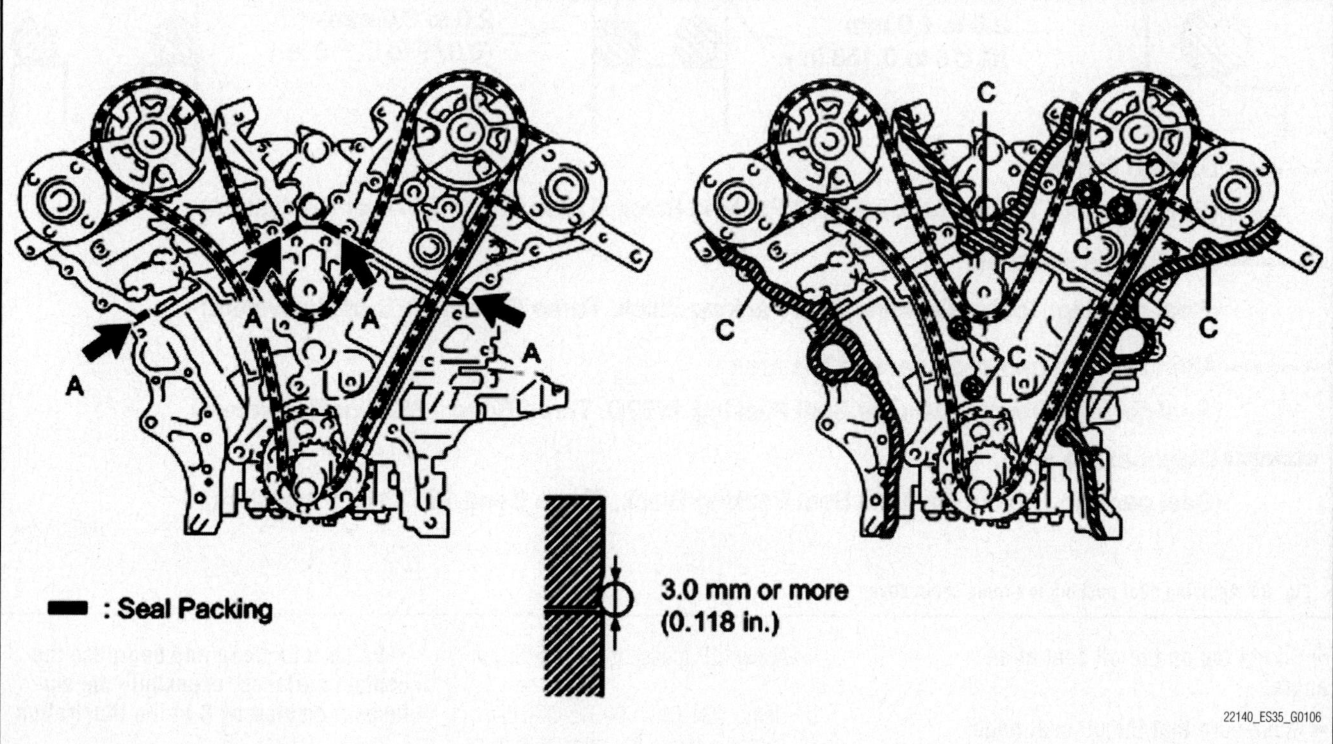

Fig. 83 Applying seal packing to timing chain cover sub-assembly

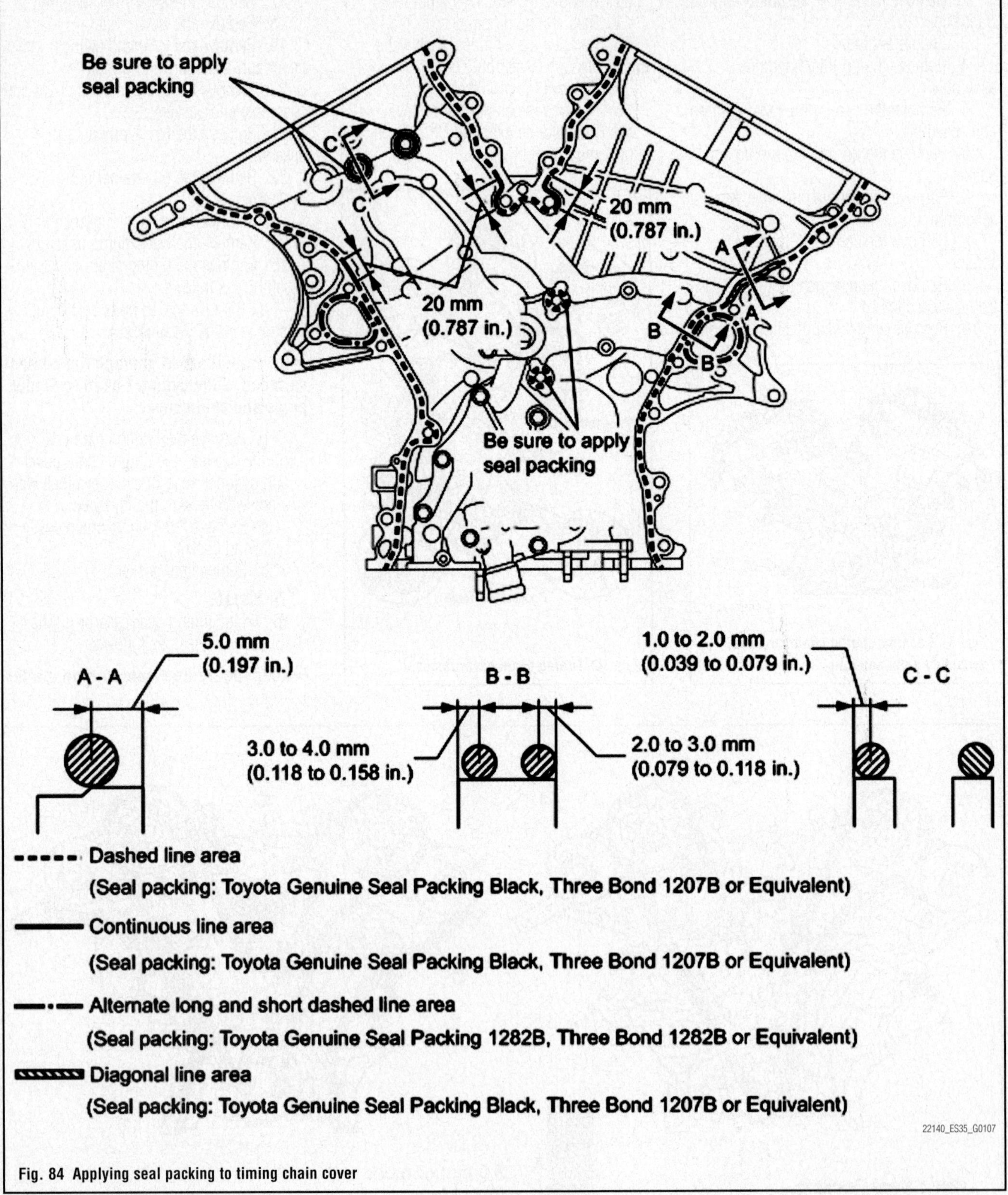

Fig. 84 Applying seal packing to timing chain cover

➡Do not tap on the oil seal at an angle.

➡Make sure that the oil seal edge does not stick out of the timing chain cover.

a. Apply MP grease to a new oil seal lip.

b. Using SST (SST: 09316-60011) and a hammer, tap in the oil seal until its surface is flush with the timing chain cover edge.

➡Be sure to clean and degrease the contact surfaces, especially the surfaces indicated by C in the illustration.

➡When the contact surfaces are wet, wipe them with an oil-free cloth before applying seal packing.

➡**Install the chain cover within 3 minutes after applying seal packing.**

➡**Do not start the engine for at least 2 hours after installing.**

27. Install the timing chain cover sub-assembly, as follows:
   a. Apply seal packing (Toyota Genuine Seal Packing Black, Three Bond 1207B or equivalent) in a continuous line to the engine unit as shown in the illustration. Seal diameter: 3.0 mm (0.118 in.).

➡**When the contact surfaces are wet, wipe them with an oil-free cloth before applying seal packing.**

➡**Install the crankcase within 3 minutes and tighten the bolts within 15 minutes after applying seal packing.**

➡**Do not start the engine for at least 2 hours after installing.**

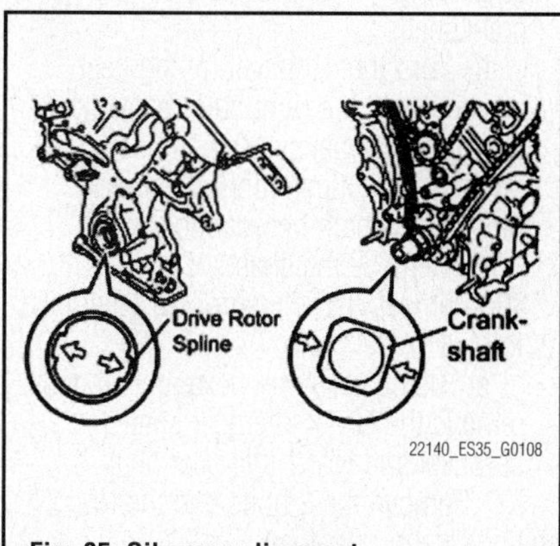

Fig. 85 Oil pump alignment

   b. Apply seal packing in a continuous line to the timing chain cover as shown in the following illustration. Seal packing: Toyota Genuine Seal Packing Black, Three Bond 1207B or equivalent, Toyota Genuine Seal Packing Black, Three Bond 1282B, Three Bond 1282B or equivalent.
   c. Install a new gasket.
   d. Align the oil pump's drive rotor spline and the crankshaft as shown in the illustration. Install the spline and chain cover to the crankshaft.
   e. Loosely install the timing chain cover with the 23 bolts and 2 nuts, but do not tighten the bolts and 2 nuts yet.

CAUTION

Make sure that there is no oil on the bolt and nut threads.

   f. Fully tighten the bolts in this order: Area 1 and Area 2, tighten to 15 ft. lbs. (21 Nm).
   g. Fully tighten the bolts in Area 3 to 15 ft. lbs. (21 Nm). Tighten the bolts and nuts in the order of upper to lower as shown in the illustration.
   h. Fully tighten the bolts in Area 4 to 32 ft. lbs. (43 Nm), and to 15 ft. lbs. (21 Nm). Tighten the bolts and nuts in the order of lower to upper as shown in the illustration.
   - Bolt A: 1.57 inches (40 mm)
   - Bolt B: 2.17 inches (55 mm)
   - Bolt C: 0.98 inches (25 mm)

28. Install the water pump assembly.
29. Install the water inlet housing.
30. Install the left hand No. 1 front engine mounting bracket.
31. Install the oil pan sub-assembly.
32. Install the oil strainer sub-assembly.
33. Install the No. 2 oil pan sub-assembly.
34. Install the oil pan drain plug and gasket.
35. Install the cylinder head cover sub-assembly.
36. Install the left-hand cylinder head cover sub-assembly.
37. Install the water outlet.
38. Install the crankshaft pulley.
39. Install the oil filter.
40. Install the cylinder block water drain cock sub-assembly.
41. Install the No. 1 oil pipe.
42. Install the oil pipe.
43. Install the crankshaft position sensor.
44. Install the camshaft timing oil control valve assembly.
45. Install the camshaft position sensor.
46. Install the ventilation valve sub-assembly.
47. Install the spark plugs and ignition coil assembly.
48. Install the oil filler cap sub-assembly.
49. Install the water pump assembly.
50. Install the engine assembly with transaxle.

Timing Chain and Sprockets

See Figures 87 through 98.

1. Before servicing the vehicle, refer to the Precautions Section.
2. Remove the engine assembly with transaxle.

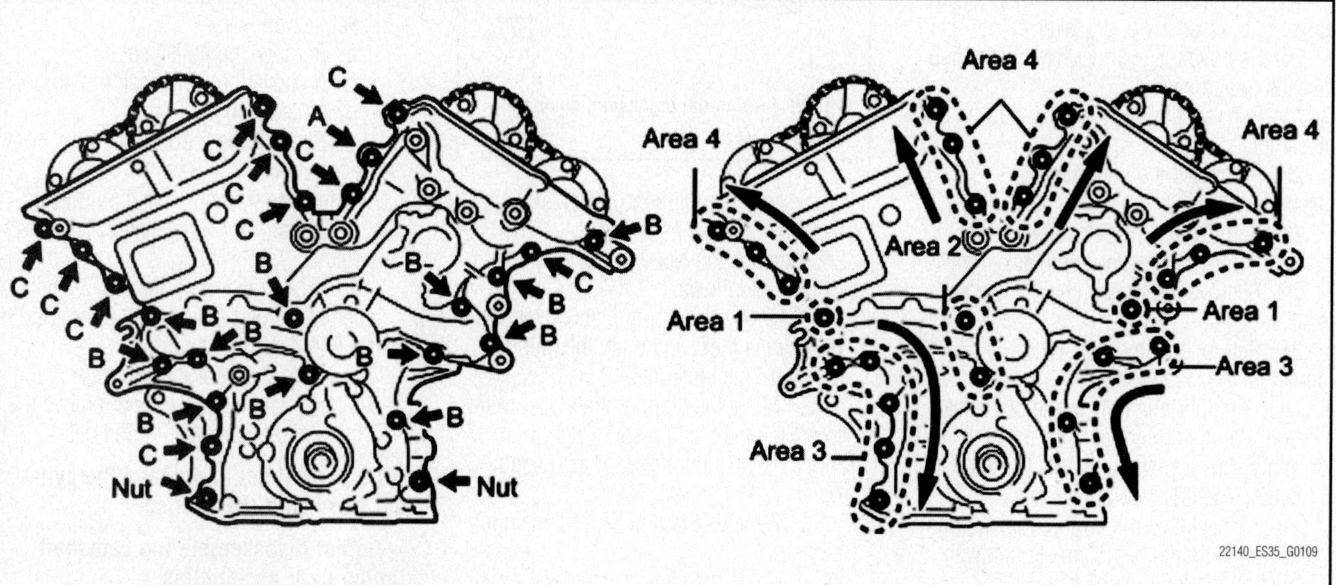

Fig. 86 Timing chain cover bolts and nuts tightening sequence

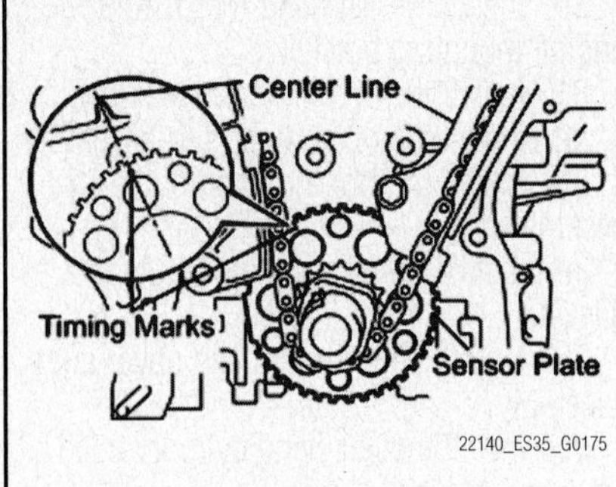

Fig. 87 Set the timing mark on the crank angle sensor plate

3. Secure engine.
4. Remove the oil filler cap sub-assembly.
5. Remove the spark plugs and ignition coil assembly.
6. Remove the oil pan drain plug and gasket.
7. Remove the ventilation valve sub-assembly.
8. Remove the camshaft position sensor.
9. Remove the camshaft timing oil control valve assembly.
10. Remove crankshaft position sensor.
11. Remove the No. 1 oil pipe.
12. Remove the oil pipe.
13. Remove the cylinder block water drain cock sub-assembly.
14. Remove the oil filter.
15. Remove the crankshaft pulley.
16. Remove the left hand No. 1 front engine mounting bracket.
17. Remove the water inlet housing.
18. Remove the water outlet.
19. Remove the left-hand cylinder head cover sub-assembly and gasket.
20. Remove the cylinder head cover sub-assembly and gasket.
21. Remove the No. 2 oil pan sub-assembly.
22. Remove the oil strainer sub-assembly.
23. Remove the oil pan sub-assembly.
24. Remove the water pump assembly.
25. Remove the timing chain cover and seal.
26. Set no. 1 cylinder to TDC/compression, as follows:
a. Temporarily tighten the pulley set bolt. Set the timing mark on the crank angle sensor plate to the RH block bore center line (TDC/compression).
b. Check that the timing marks of the camshaft timing gears are aligned with the timing marks of the bearing cap as shown in the illustration. If not, turn the crankshaft 1 revolution (360°) and align the timing marks as above.

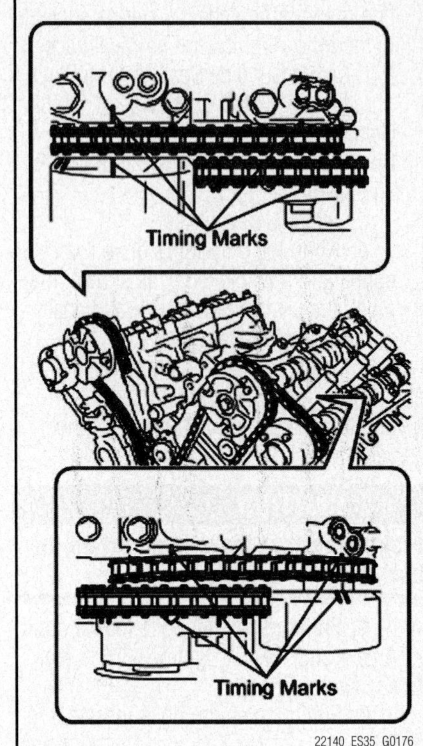

Fig. 88 Check that the timing marks of the camshaft timing gears

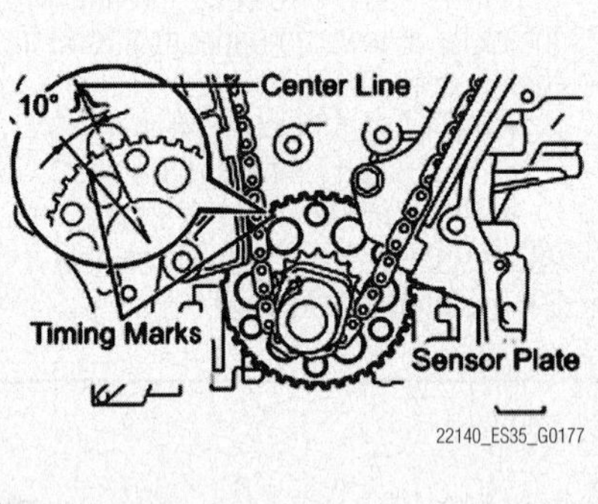

Fig. 89 Turning the crankshaft counter-clockwise 10°

27. Remove the No. 1 chain tensioner assembly, as follows:
a. Move the stopper plate upward to release the lock, and push the plunger deep into the tensioner.
b. Move the stopper plate downward to set the lock, and insert a pin of φ1.27 mm (0.05 in.) into the stopper plate's hole.
c. Remove the 2 bolts and chain tensioner.
28. Remove the chain tensioner slipper.

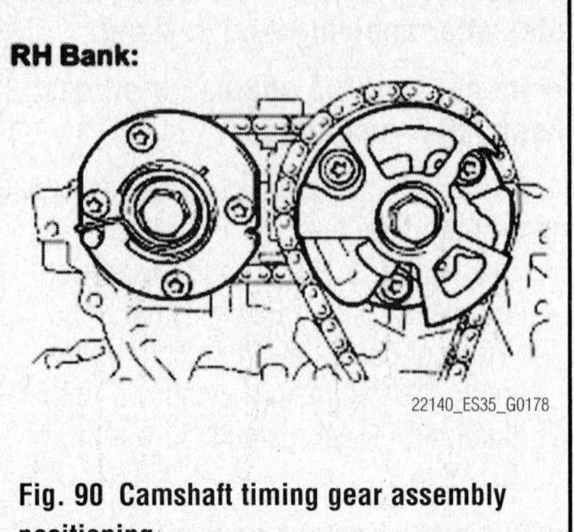

Fig. 90 Camshaft timing gear assembly positioning

29. Remove the chain sub-assembly, as follows:
a. Turn the crankshaft counterclockwise 10° to loosen the chain of the crankshaft timing sprocket.
b. Remove the pulley set bolt.
c. Remove the chain from the crankshaft timing sprocket and place it on the crankshaft.
d. Turn the camshaft timing gear assembly on the right hand bank clockwise (approximately 60°) and set it as shown in the illustration. Be sure to loosen the chain between the banks.
e. Remove the chain.
30. Remove the idle sprocket assembly, as follows:
a. Using a 10 mm hexagon wrench, remove the No. 2 idle gear shaft, sprocket and No. 1 idle gear shaft.
31. Remove the 2 bolts and the No. 1 chain vibration damper.
32. Remove the No. 2 chain vibration damper.
33. Remove the crankshaft timing sprocket, as follows:
a. Remove the pulley set bolt.
b. Remove the crankshaft timing gear from the crankshaft.
c. Remove the 2 pulley set keys from the crankshaft.
34. Remove the camshaft timing gears and No. 2 chain (for Right-hand Bank), as follows:
a. While raising up the No. 2 chain tensioner, insert a pin of φ1.0 mm (0.039 in.) into the hole to hold it.
b. Hold the hexagonal portion of the camshaft with a wrench, and remove the 2 bolts and 2 camshaft timing gears.

➡Be careful not to damage the cylinder head with the wrench.

➡Do not disassemble the camshaft timing gear assemblies.

c. Remove the No. 2 chain.

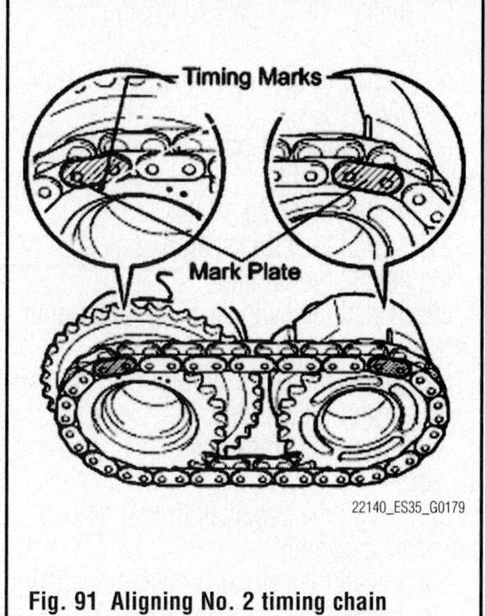

Fig. 91 Aligning No. 2 timing chain

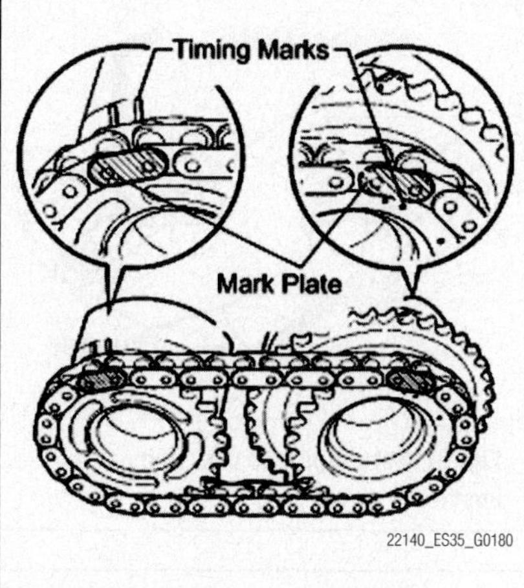

Fig. 92 Aligning No. 2 timing chain (bank 2)

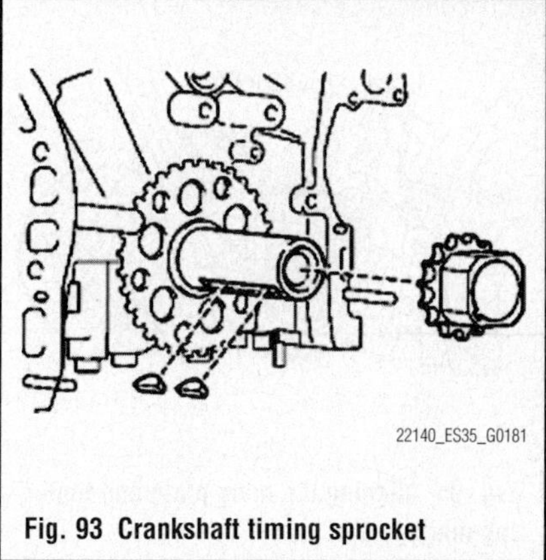

Fig. 93 Crankshaft timing sprocket

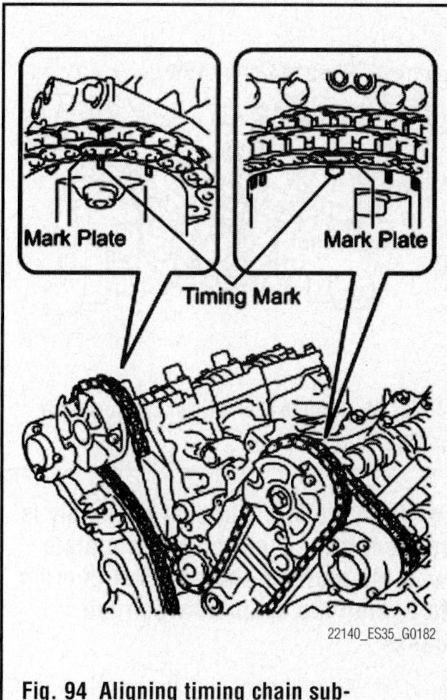

Fig. 94 Aligning timing chain sub-assembly

35. Remove the bolt and No. 2 chain tensioner.

36. Remove the camshaft timing gears and No. 2 chain (for Left-hand Bank), as follows:

a. While pushing down on the No. 3 chain tensioner, insert a pin of φ1.0 mm (0.039 in.) into the hole to hold it.

b. Hold the hexagonal portion of the camshaft with a wrench, and remove the 2 bolts and 2 camshaft timing gears.

➡Be careful not to damage the cylinder head with the wrench.

➡Do not disassemble the camshaft timing gear assemblies.

c. Remove the No. 2 chain.

37. Remove the bolt and the No. 3 chain tensioner.

To install:

38. Install the No. 2 chain tensioner assembly with the bolt and tighten to 15 ft. lbs. (21 Nm).

39. While pushing in the tensioner, insert a pin of 1.0 mm (0.039 in.) into the hole to hold it.

40. Install the camshaft timing gears and No. 2 chain (for Right-hand Bank), as follows:

a. Align the mark plate with the timing marks (1-dot mark) of the camshaft timing gears as shown.

b. Apply a light coat of engine oil to the bolt threads and bolt-seating surface.

c. Align the knock pin of the camshaft with pin hole of the camshaft timing gear. Install the camshaft timing gear and the right camshaft timing exhaust gear with the No. 2 chain installed.

d. Hold the hexagonal portion of the camshaft with the wrench and tighten the two bolts to 74 ft. lbs. (100 Nm).

e. Remove the pin from the No. 2 chain tensioner.

41. Install the No. 3 chain tensioner assembly with the bolt and tighten to 15 ft. lbs. (21 Nm).

42. While pushing in the tensioner, insert a pin of 1.0 mm (0.039 in.) into the hole to hold it.

43. Install the camshaft timing gears and No. 2 chain (for Left-hand Bank), as follows:

a. Align the mark plate (yellow) with the timing marks (2-dot mark) of the camshaft timing gears as shown.

b. Apply a light coat of engine oil to the bolt threads and bolts seating surface.

c. Align the knock pin of the camshaft with pin hole of the camshaft timing gear. Install the camshaft timing gear and the left camshaft timing exhaust gear with the No. 2 chain installed.

d. Hold the hexagonal portion of the camshaft with the wrench and tighten the two bolts to 74 ft. lbs. (100 Nm).

e. Remove the pin from the No 2 chain tensioner.

44. Install the No. 1 and 2 chain vibration dampers.

45. Install the timing gear set keys and crankshaft timing sprocket as shown in the illustration.

46. Install the idle sprocket assembly, as follows:

a. Apply a light coat of engine oil to the rotating surface of the No. 1 idle gear shaft.

b. Temporarily install the No. 1 idle gear shaft and idle sprocket with the No. 2 idle gear shaft while aligning the knock pin of the No. 1 idle gear with the knock pin groove of the cylinder block. Be careful of the idle gear direction.

c. Using a 10 mm hexagon wrench, tighten the No. 2 idle gear shaft to 44 ft. lbs. (60 Nm).

d. After installing the idle sprocket assembly, check that the idle sprocket turns smoothly.

47. Install the chain sub-assembly, as follows:

a. Align the mark plate and timing marks as shown in the illustration and install the chain. The camshaft mark plate is orange.

b. Do not pass the chain over the crankshaft, just put it on.

c. Turn the camshaft timing gear assembly on the right bank counterclockwise to tighten the chain between the banks.

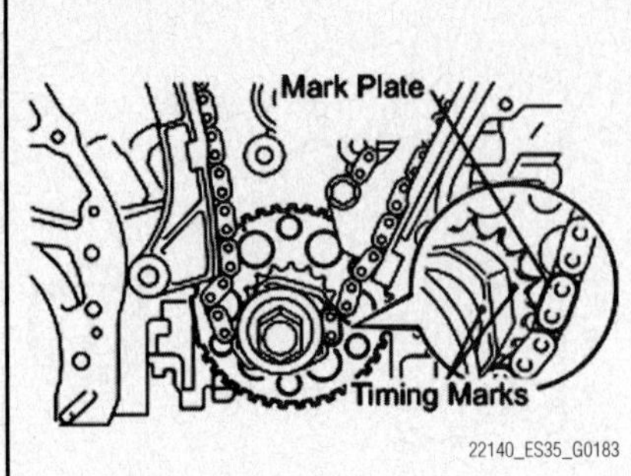

Fig. 95 Aligning the mark plate and timing mark

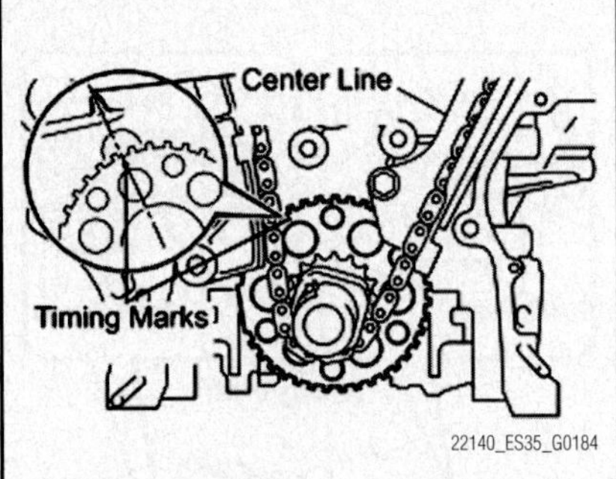

Fig. 96 Turning the crankshaft clockwise (TDC/compression)

➡**When the idle sprocket assembly is reused, align the timing chain plate with the mark on the sprocket in order to tighten the chain between the banks.**

d. Align the mark plate and timing marks as shown in the illustration and install the chain onto the crankshaft timing sprocket. The crankshaft to mark plate is yellow.

e. Temporarily tighten the pulley set bolt.

f. Turn the crankshaft clockwise to set it to the right-hand block bore more centerline. (TDC/compression).

48. Install the chain tensioner slipper.

49. Install the No. 1 chain tensioner assembly, as follows:

a. Move the stopper plate upward to release the lock, and push the plunger deep into the tensioner.

b. Move the stopper plate downward to set the lock, and insert a hexagon wrench into the hole of the stopper plate.

c. Install the No. 1 chain tensioner with the 2 bolts and tighten to 7 ft. lbs. (10 Nm).

d. Remove the lock pin of the No. 1 chain tensioner. Check that each timing

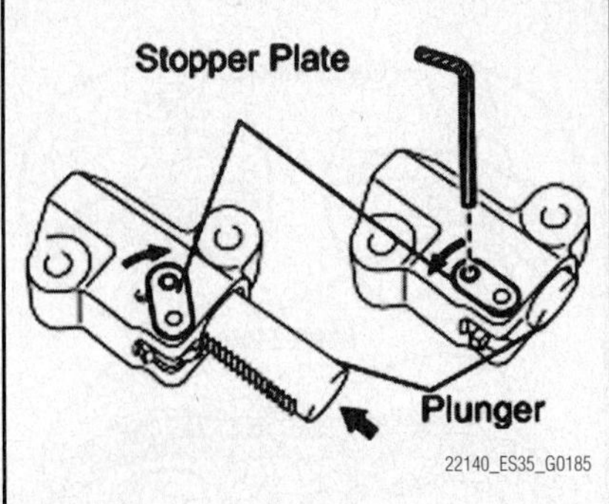

Fig. 97 Set chain tensioner plunger position

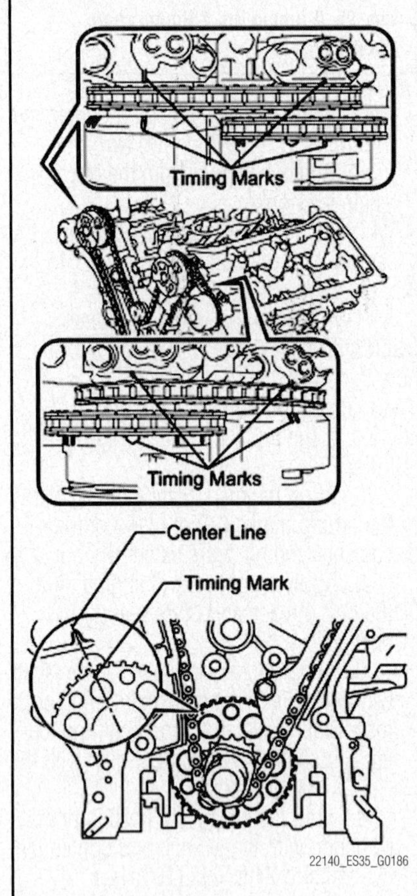

Fig. 98 Aligning timing marks

mark is aligned with the crankshaft at TDC/compression.

e. Remove the pulley set bolt.

50. Install timing chain cover and seal.

51. Install the water pump assembly.

52. Install the water inlet housing.

53. Install the left hand No. 1 front engine mounting bracket.

54. Install the oil pan sub-assembly.

55. Install the oil strainer sub-assembly.

56. Install the No. 2 oil pan sub-assembly.

57. Install the oil pan drain plug and gasket.

58. Install the cylinder head cover sub-assembly.

59. Install the left-hand cylinder head cover sub-assembly.

60. Install the water outlet.

61. Install the crankshaft pulley.

62. Install the oil filter.

63. Install the cylinder block water drain cock sub-assembly.

64. Install the No. 1 oil pipe.

65. Install the oil pipe.

66. Install the crankshaft position sensor.

67. Install the camshaft timing oil control valve assembly.

68. Install the camshaft position sensor.

69. Install the ventilation valve sub-assembly.

70. Install the spark plugs and ignition coil assembly.

71. Install the oil filler cap sub-assembly.

72. Install the engine assembly with transaxle.

VALVE LASH

ADJUSTMENT

➡**Keep the lash adjuster free of dirt and foreign objects.**

➡**Only use clean engine oil.**

1. Place the lash adjuster into a container filled with engine oil.

2. Insert the SST's (SST: 09276-75010) tip into the lash adjuster's plunger and use the tip to press down on the check ball inside the plunger.

3. Squeeze the SST and lash adjuster together to move the plunger up and down 5 to 6 times.

4. Check the movement of the plunger and bleed the air. OK: Plunger moves up and down.

➡**When bleeding air from the high-pressure chamber, make sure that the tip of the SST is actually pressing the check ball as shown in the illustration. If the check ball is not pressed, air will not bleed.**

5. After bleeding the air, remove the SST. Then, try to press the plunger quickly and firmly with a finger. OK: Plunger is very difficult to move. If the result is not as specified, replace the lash adjuster.

6. Install the lash adjusters.

➡**Install the lash adjuster to the same place where it was removed from.**

ENGINE PERFORMANCE & EMISSION CONTROL

COMPONENT LOCATIONS

See Figures 99 through 106.

ACCELERATOR PEDAL POSITION (APP) SENSOR

LOCATION

See Figure 107.

REMOVAL & INSTALLATION

See Figure 108.

1. Before servicing the vehicle, refer to the Precautions Section.
2. Remove left center floor carpet cover.
3. Disconnect the accelerator pedal connector.
4. Remove the 2 nuts and accelerator pedal assembly.

➡Avoid physical shock to the accelerator pedal assembly.

➡Do not disassemble the accelerator pedal assembly.

To install:

5. Installation is the reverse of the removal procedure. Tighten the accelerator pedal rod nuts to 43 inch lbs. (5.4 Nm).

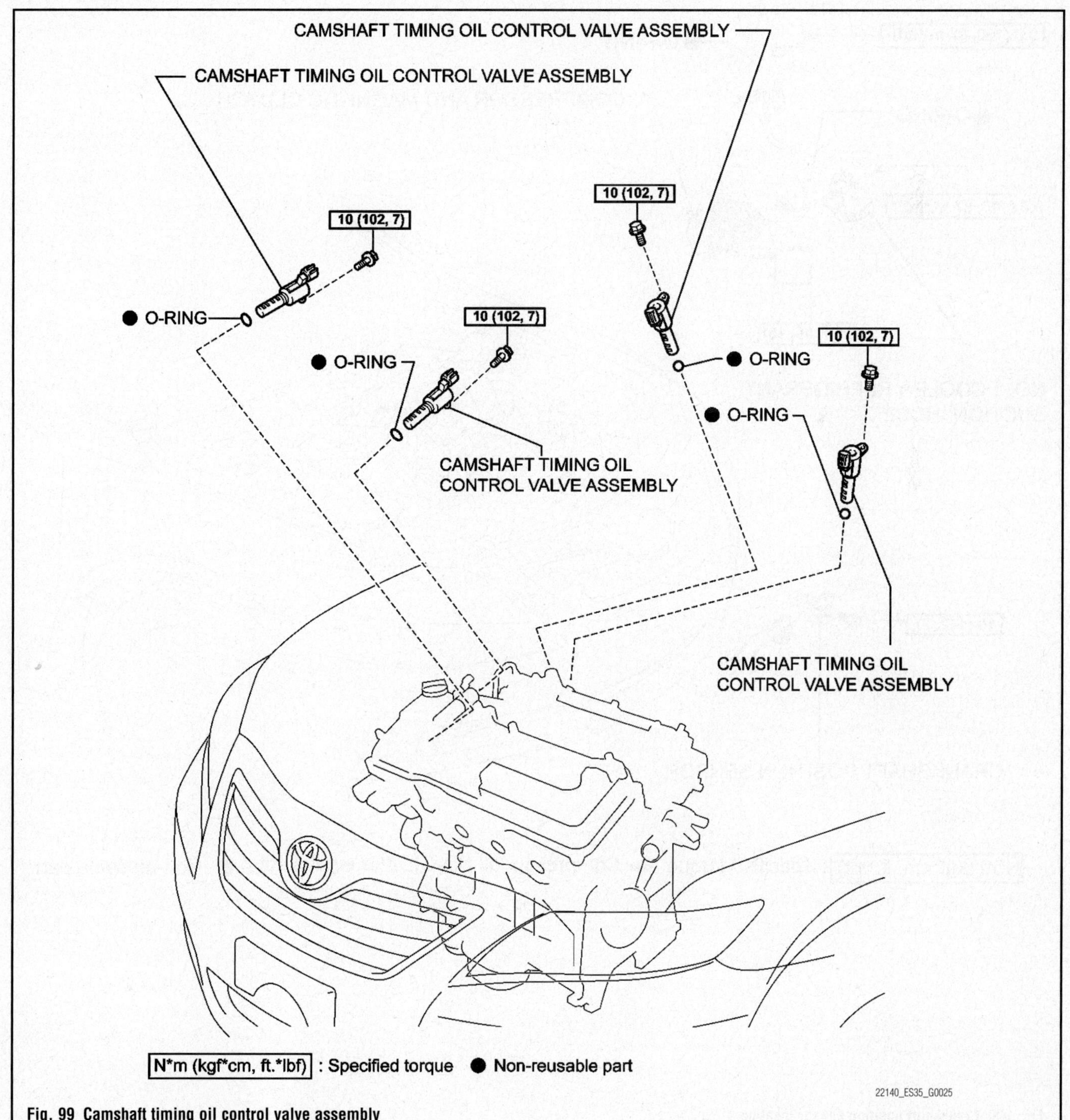

Fig. 99 Camshaft timing oil control valve assembly

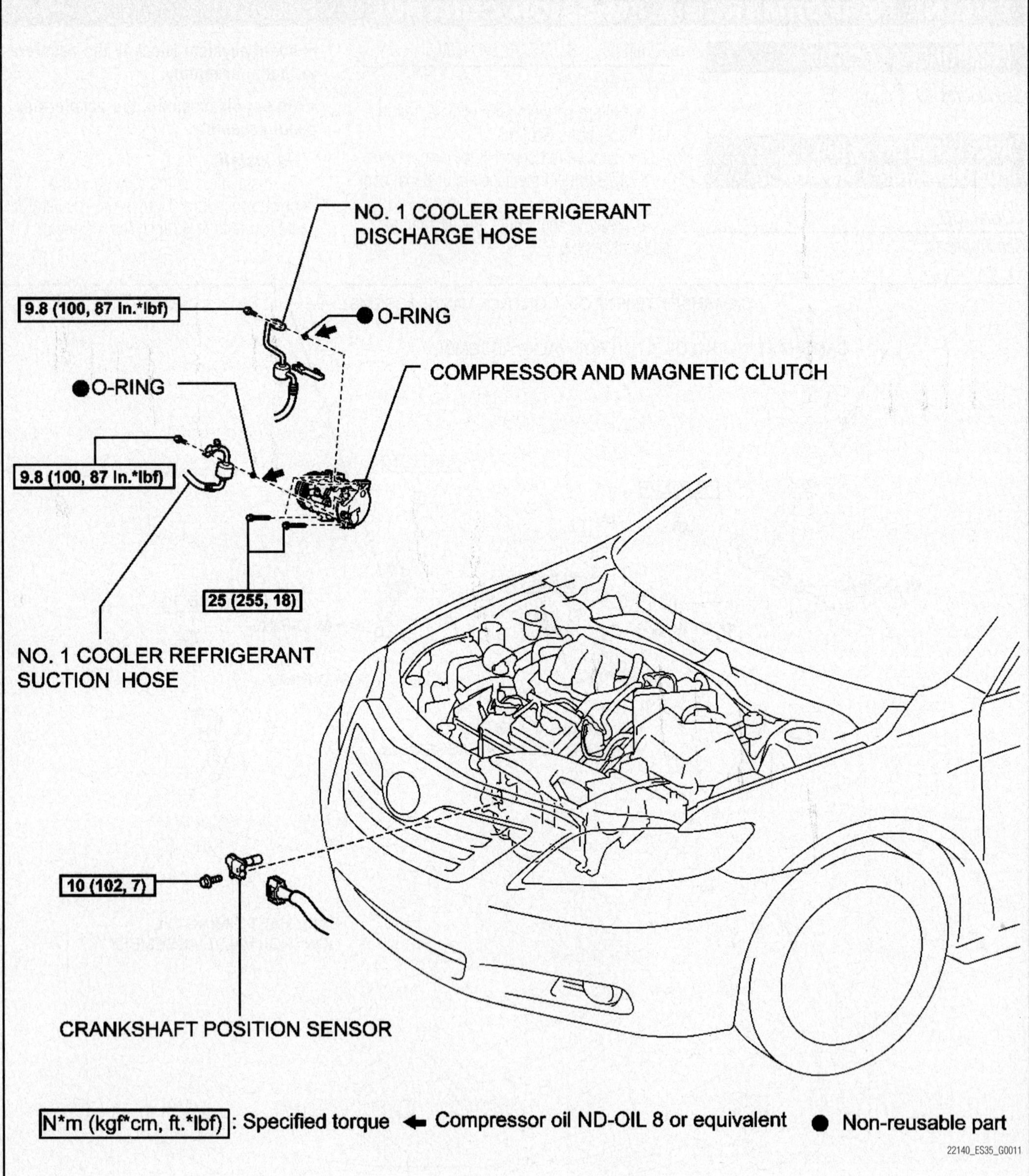

Fig. 100 Crankshaft position sensor location

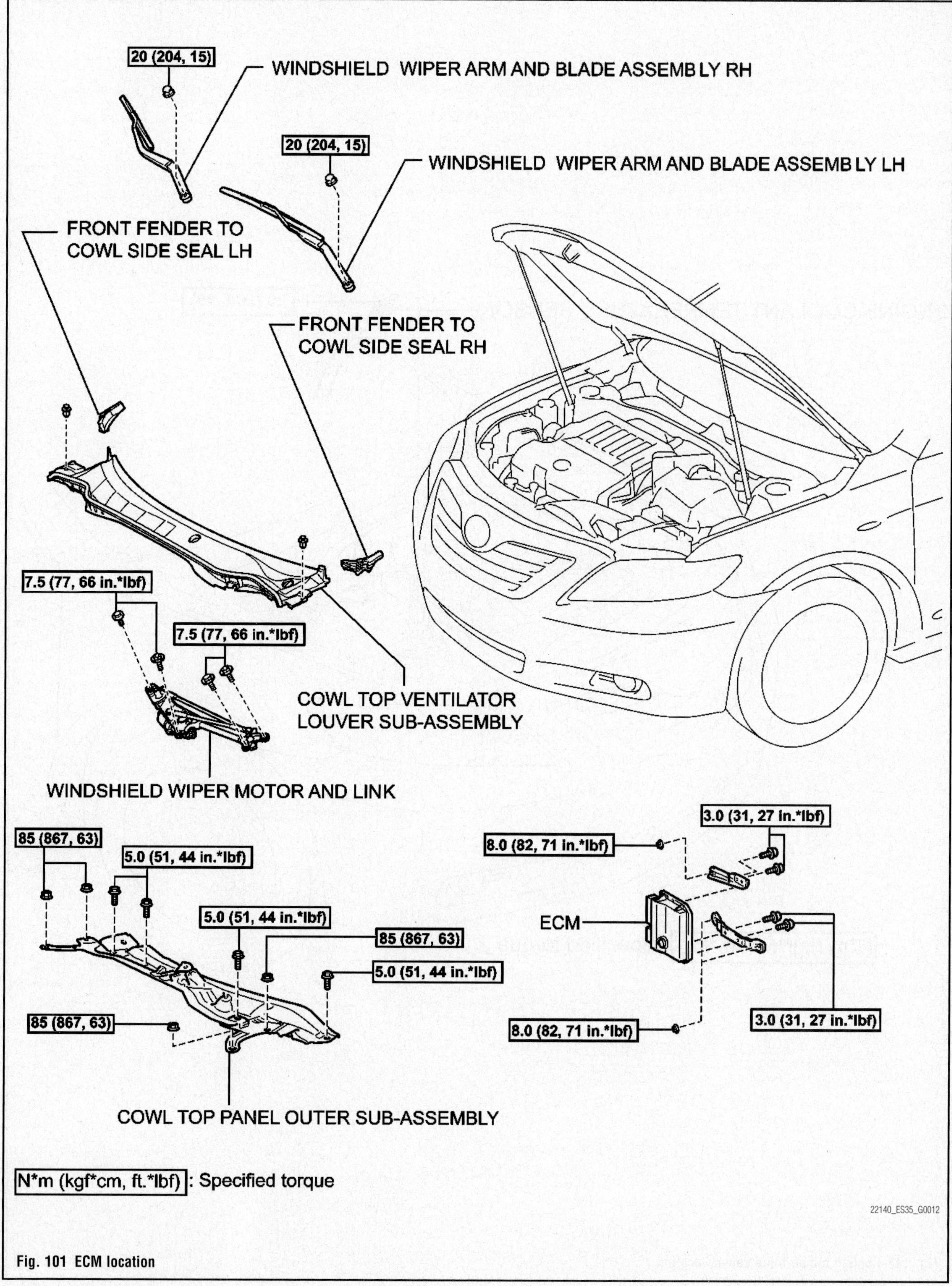

Fig. 101 ECM location

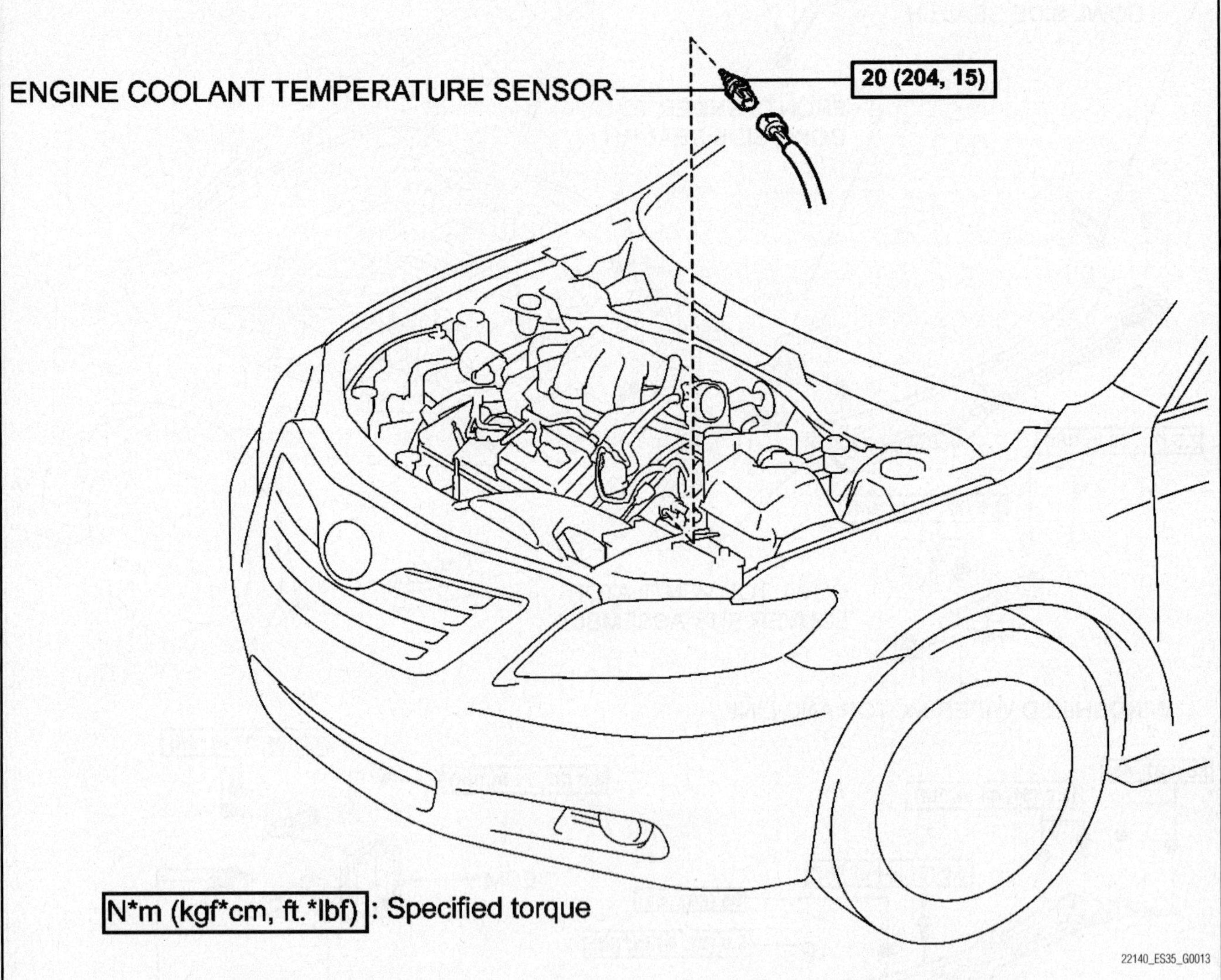

Fig. 102 Coolant temperature sensor location

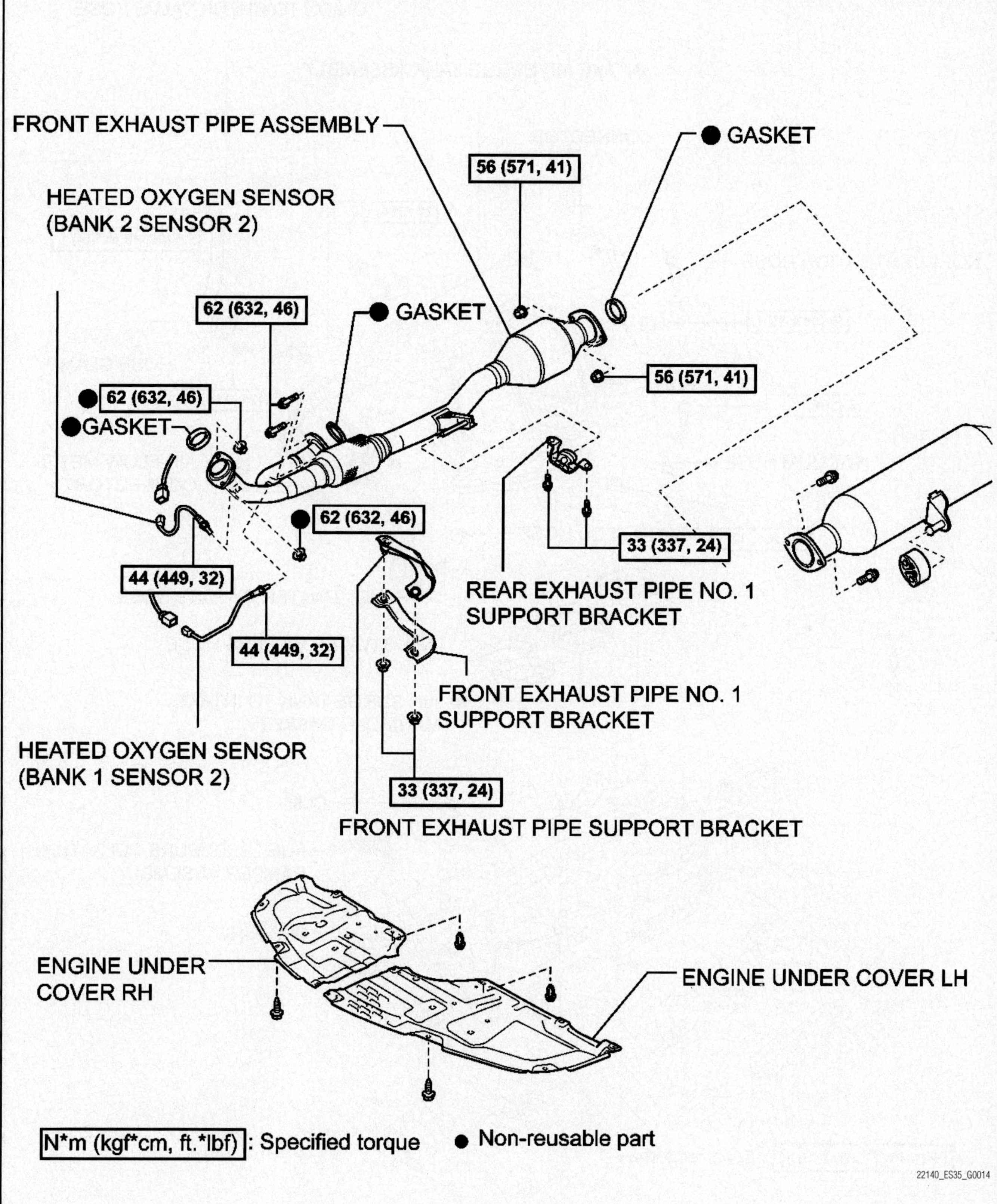

Fig. 103 Oxygen Sensor location

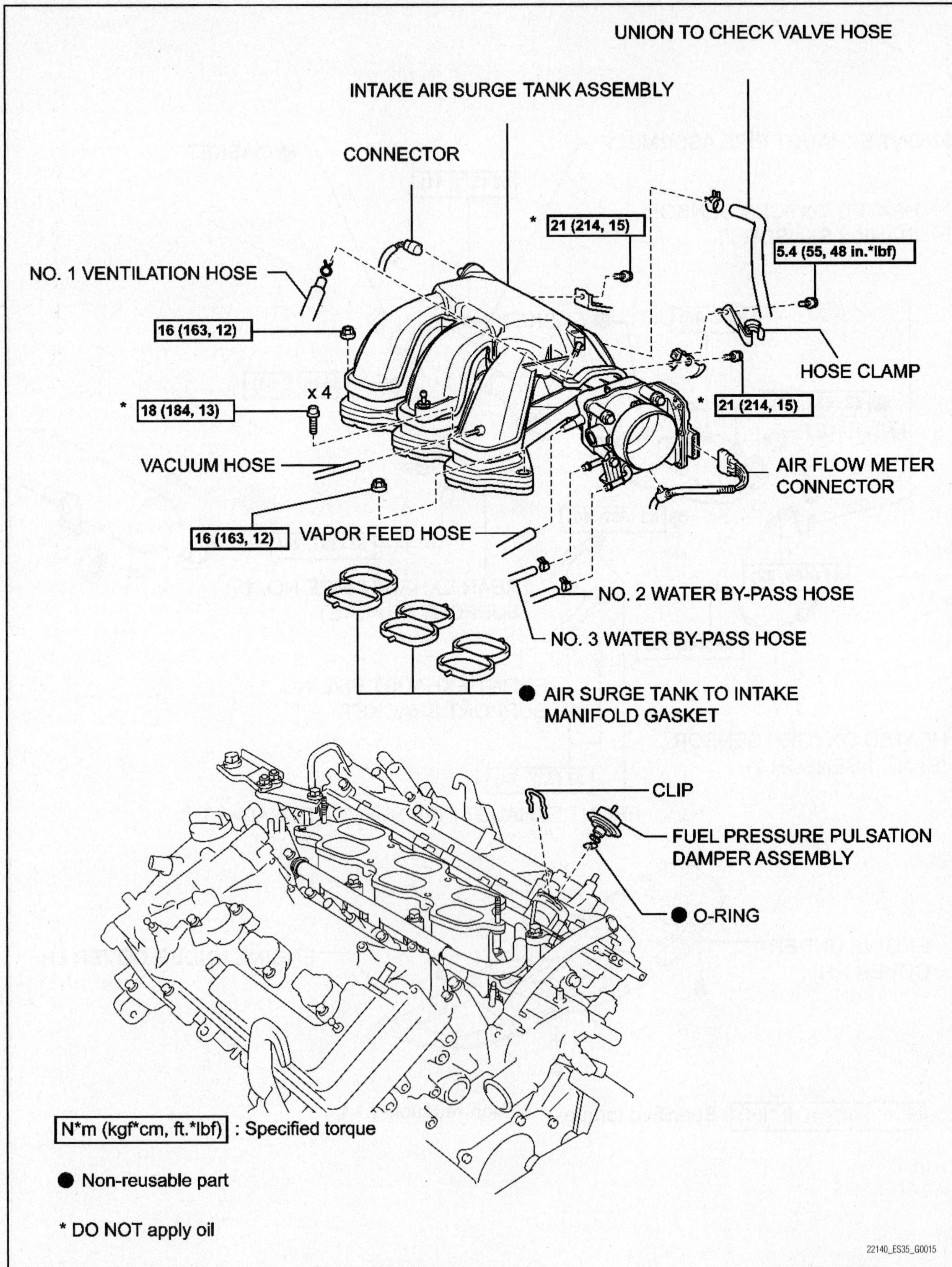

Fig. 104 Fuel pressure pulsation damper (injection pressure regulator) location

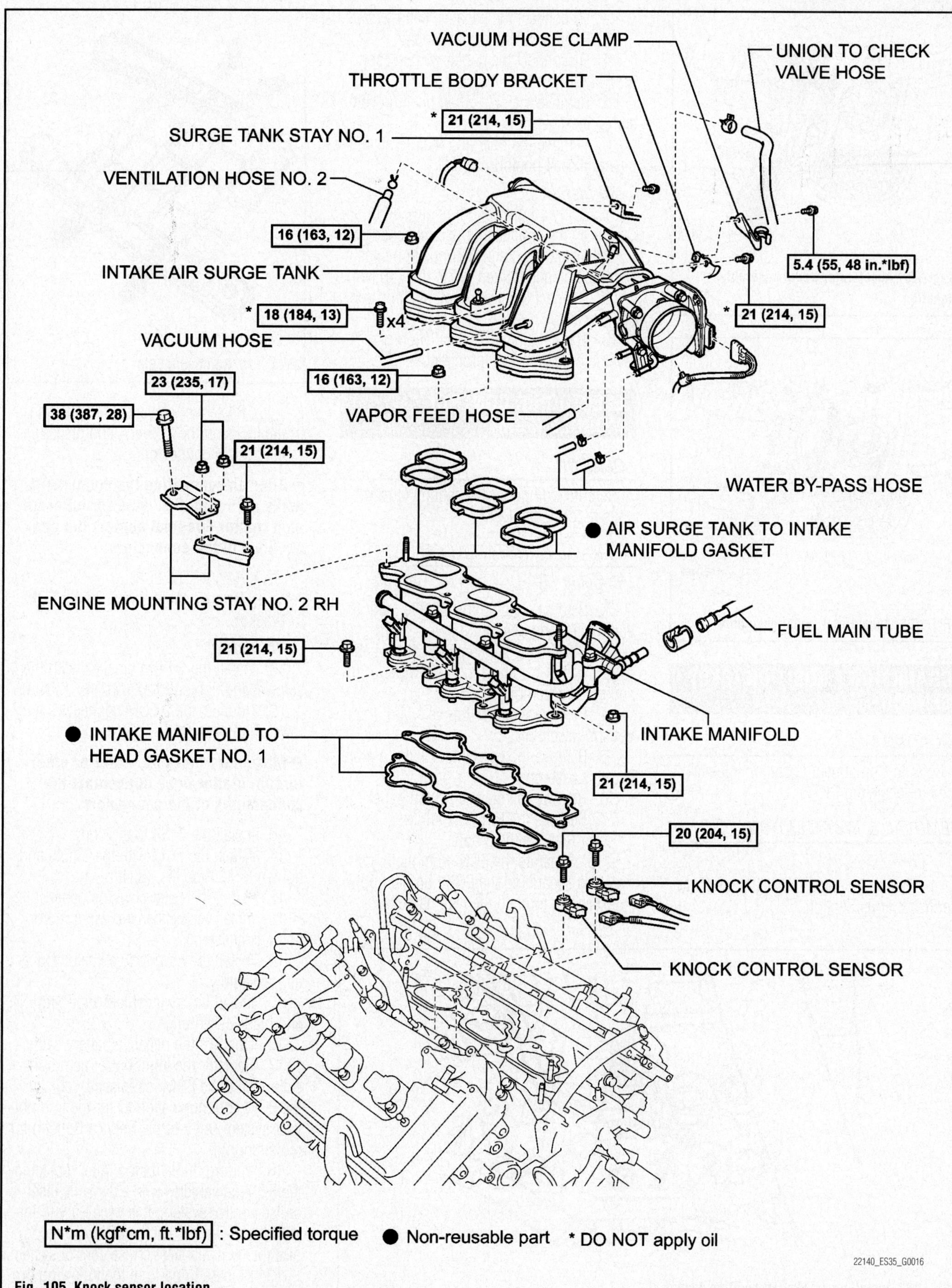

Fig. 105 Knock sensor location

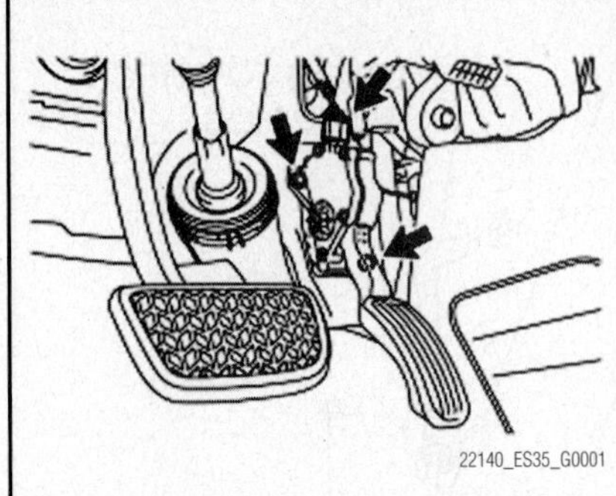

Fig. 107 Accelerator pedal assembly location

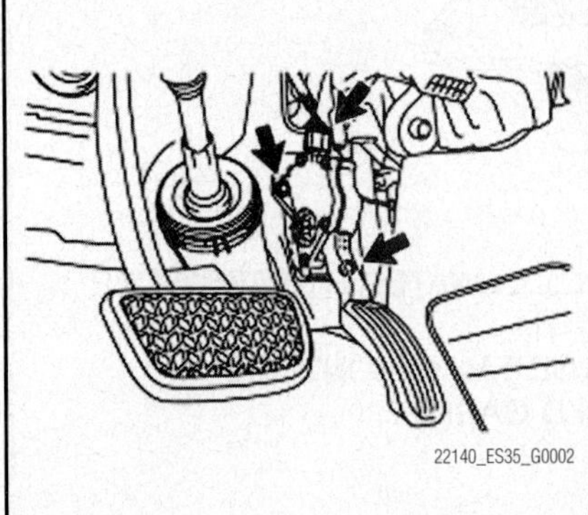

Fig. 108 Accelerator pedal assembly

CRANKSHAFT POSITION (CKP) SENSOR

LOCATION

Refer to Component Locations illustrations to locate the CKP.

REMOVAL & INSTALLATION

See Figure 109.

1. Before servicing the vehicle, refer to the Precautions Section.
2. Disconnect the negative battery cable.
3. Remove alternator assembly.
4. Disconnect the cooler compressor assembly.
5. Remove the crankshaft position sensor connector.
6. Remove the bolt, and then remove the crankshaft position sensor.

To install:

7. Apply a light coat of engine oil to the O-ring on the crankshaft position sensor.
8. Install the crankshaft position sensor with the bolt and tighten to 7ft. lbs. (10 Nm).
9. Connect the crankshaft position sensor connector.
10. The remainder of installation is the reverse of the removal procedure.

ELECTRONIC CONTROL MODULE (ECM)

LOCATION

Refer to Component Locations illustrations to locate the ECM.

REMOVAL & INSTALLATION

See Figure 110.

1. Before servicing the vehicle, refer to the Precautions Section.
2. Disconnect the negative battery cable.
3. Remove both windshield wiper arm and blade assemblies.
4. Remove the cowl top ventilator louver sub-assembly.
5. Remove the windshield wiper motor and link assembly.
6. Remove the outer cowl top panel.
7. Remove the ECM, as follows:
   a. Remove the 3 nuts.
   b. Separate the ECM from the body. When separating the ECM, do not apply excessive force to the wire harness.

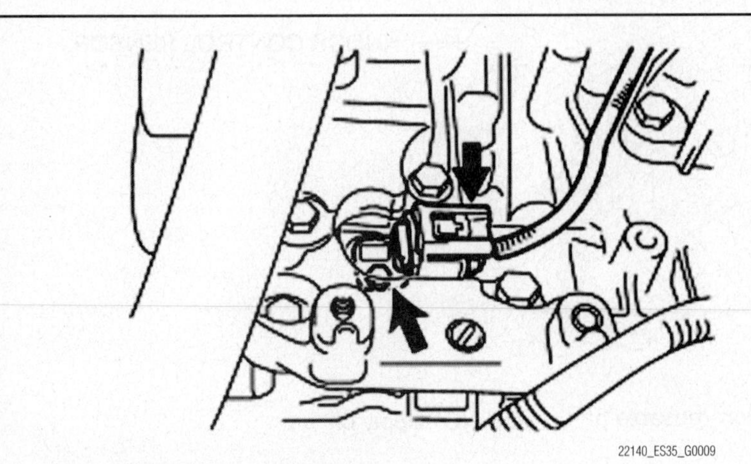

Fig. 109 Installing crankshaft position sensor

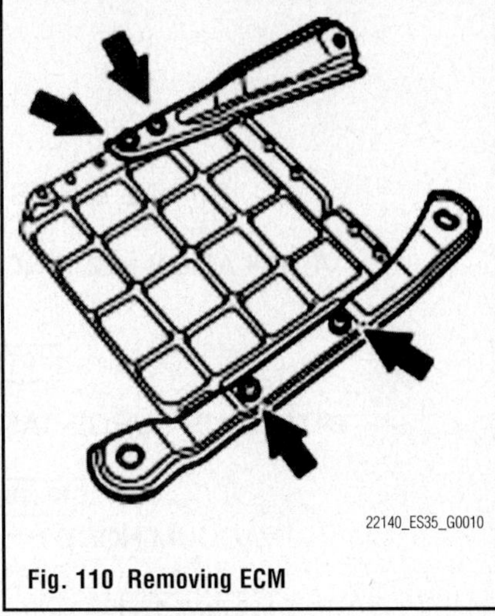

Fig. 110 Removing ECM

   c. Raise the 2 levers while pushing the locks on the 2 levers, and disconnect the 2 ECM connectors.

➡After disconnecting the connectors, make sure that dirt, water or other foreign matter does not contact the connections of the connectors.

   d. Remove the ECM.
   e. Remove the 4 screws and 2 ECM brackets.

To install:

8. Install the 2 ECM brackets with the 4 screws, and tighten to 27 inch lbs. (3 Nm).
9. Connect the 2 ECM connectors and lower the 2 levers.

➡Make sure that dirt, water or other foreign matter does not contact the connections of the connectors.

10. Install the ECM to the body.
11. Attach the ECM with the 3 nuts and tighten to 71 inch lbs. (8 Nm).
12. Install the outer cowl top panel.
13. Install the windshield wiper motor and link assembly.
14. Install the cowl top ventilator louver sub-assembly.
15. Install both windshield wiper arm and blade assemblies.
16. Connect the negative battery cable.
17. Register the immobilizer communication ID. If the ECM is replaced, register the ECM communication ID for the immobilizer system (refer to the Service Bulletin for registration).
18. Perform initialization. After replacing the ECM on vehicles with a dynamic laser cruise control system, it is necessary to initialize the ECM so that the ECM can recognize the dynamic laser cruise control system.
19. Be sure to perform the following procedure after replacing the ECM:

a. Turn the ignition switch on (IG).

b. Turn the cruise control main switch on.

c. With the brake pedal depressed, push the cruise control main switch to RES/ACC 3 times within 3 seconds. Check that the buzzer sounds at this time.

➡Do not turn the headlight dimmer switch on at this time because the optical axis automatic adjustment mode has already started, which may lead to an incorrect optical axis setting. If the headlight dimmer switch is turned on by mistake, readjust the optical axis.

ENGINE COOLANT TEMPERATURE (ECT) SENSOR

LOCATION

Refer to Component Locations illustrations to locate the ECT sensor.

REMOVAL & INSTALLATION

See Figure 111.

1. Before servicing the vehicle, refer to the Precautions Section.
2. Drain engine coolant.
3. Remove V-bank cover sub-assembly.
4. Remove the air cleaner inlet assembly.
5. Remove the air cleaner cap sub-assembly.
6. Remove the air cleaner case sub-assembly.
7. Remove No. 1 air cleaner inlet.
8. Remove the engine coolant temperature sensor connector.
9. Using a19 mm deep socket wrench, remove the engine coolant temperature sensor and gasket.

To install:

10. Installation is the reverse of the removal procedure. Torque the engine coolant temperature sensor to 15 ft. lbs. (20 Nm).

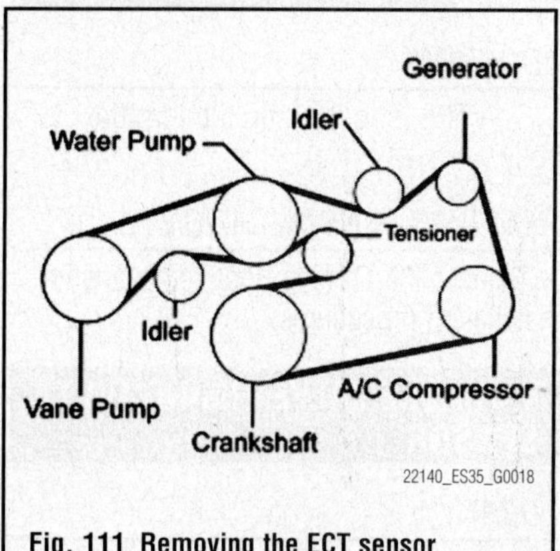

Fig. 111 Removing the ECT sensor

HEATED OXYGEN (HO2S) SENSOR

LOCATION

Refer to Component Locations illustrations to locate the Oxygen Sensor.

REMOVAL & INSTALLATION

See Figures 112 and 113.

1. Before servicing the vehicle, refer to the Precautions Section.
2. Remove the front exhaust pipe assembly.
3. Disconnect the 2 oxygen sensor connectors.
4. Using Special Tool: 09224-00010 or equivalent, remove the 2 oxygen sensors from the front pipe assembly.

To install:

5. Install the 2 oxygen sensors to the front pipe assembly. Tighten to 32 ft. lbs. (44 Nm) and 30 ft. lbs. (40 Nm). Use a torque wrench with a fulcrum length of 300 mm (11.81 in.).
6. Connect the 2 oxygen sensor connectors.
7. Remove the front exhaust pipe assembly.

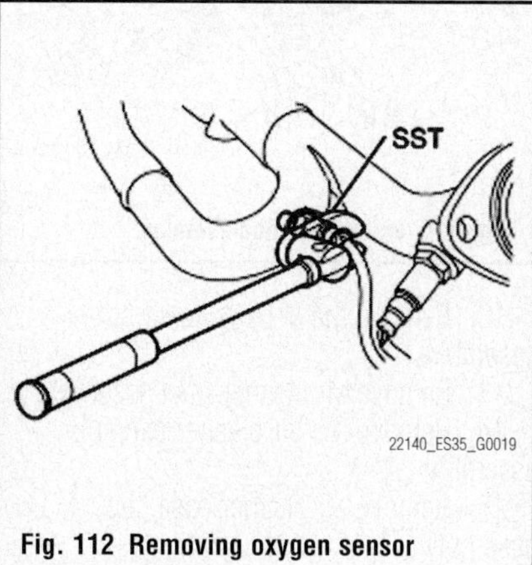

Fig. 112 Removing oxygen sensor

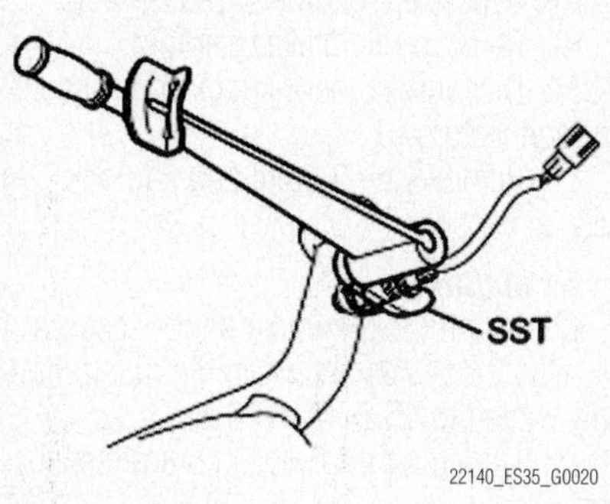

Fig. 113 Installing oxygen sensor

INJECTION PRESSURE REGULATOR (IPR)

LOCATION

Refer to Component Locations illustrations to locate the fuel pressure pulsation damper (injection pressure regulator).

REMOVAL & INSTALLATION

See Figure 114.

1. Before servicing the vehicle, refer to the Precautions Section.

✲✲ CAUTION

Observe all applicable safety precautions when working around fuel. Whenever servicing the fuel system, always work in a well ventilated area. Do not allow fuel spray or vapors to come in contact with a spark or open flame. Keep a dry chemical fire extinguisher near the work area. Always keep fuel in a container specifically designed for fuel storage; also, always properly seal fuel containers to avoid the possibility of fire or explosion.

2. Discharge fuel system pressure.
3. Disconnect the negative battery cable.
4. Drain engine coolant.
5. Remove both windshield wiper arm and blade assemblies.
6. Remove the right cowl top ventilator louver.
7. Remove the windshield wiper motor and link assembly.
8. Remove the outer cowl top panel.
9. Remove the V-bank cover sub-assembly.
10. Remove the air cleaner cap air cleaner with hose.
11. Remove the intake air surge tank.
12. Remove the fuel pressure pulsation damper assembly, as follows:

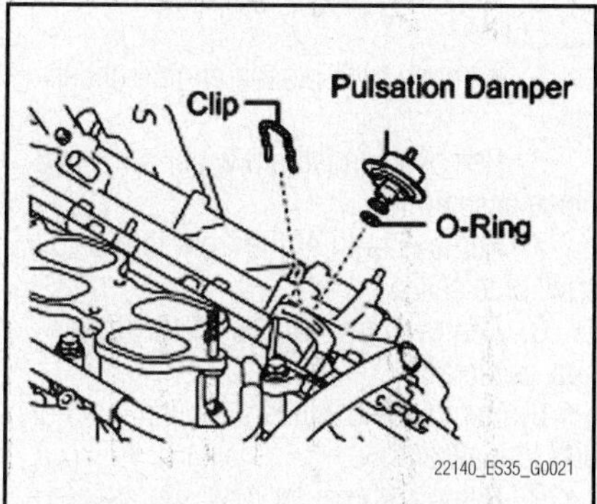

Fig. 114 Removing fuel pressure pulsation damper

a. Remove the clip.

b. Pull out the fuel pressure pulsation damper from the fuel delivery pipe.

c. Remove the O-ring from the fuel pressure pulsation damper.

To install:

13. Apply a light coat of spindle oil or gasoline to a new O-ring and install it to the fuel pressure pulsation damper.
14. Install the fuel pressure pulsation damper to the fuel delivery pipe.
15. Install the clip.
16. The remainder of installation is the reverse of the removal procedure.
17. Check for coolant and fuel leaks.
18. Perform initialization procedure.

KNOCK SENSOR (KS)

LOCATION

Refer to Component Locations illustrations to locate the knock sensor.

REMOVAL & INSTALLATION

See Figures 115 and 116.

CAUTION

Observe all applicable safety precautions when working around fuel. Whenever servicing the fuel system, always work in a well ventilated area. Do not allow fuel spray or vapors to come in contact with a spark or open flame. Keep a dry chemical fire extinguisher near the work area. Always keep fuel in a container specifically designed for fuel storage; also, always properly seal fuel containers to avoid the possibility of fire or explosion.

1. Before servicing the vehicle, refer to the Precautions Section.
2. Properly discharge the fuel system pressure.
3. Disconnect battery negative cable.
4. Drain and recycle the engine coolant.
5. Remove both plastic engine under covers.
6. Remove windshield wiper arms and blade assemblies.
7. Remove both of the front fender to cowl side seals.
8. Remove cowl top ventilator louver sub-assembly.
9. Remove the windshield wiper motor and link assembly.
10. Remove the outer cowl top panel.
11. Remove the cool air intake duct seal.

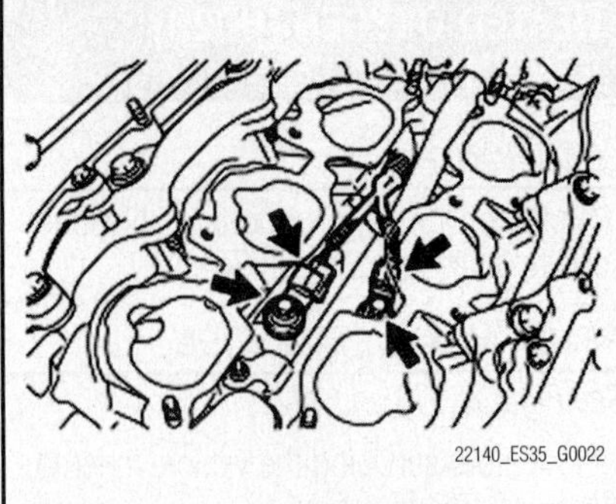

Fig. 115 Removing knock sensor

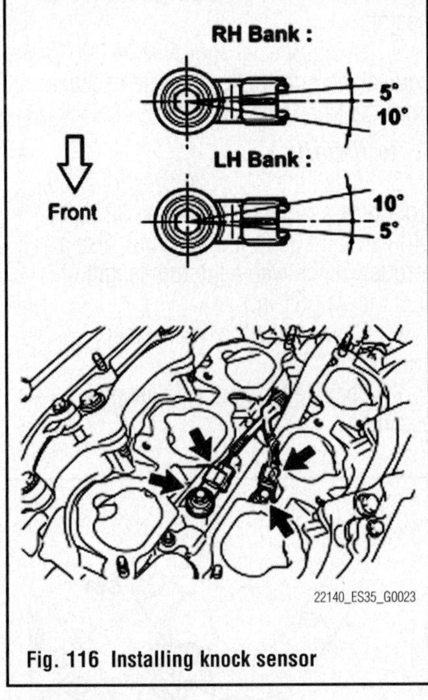

Fig. 116 Installing knock sensor

12. Remove the V-bank cover sub-assembly.
13. Remove air cleaner inlet assembly.
14. Remove the air cleaner cap sub-assembly.
15. Remove air cleaner case sub-assembly.
16. Remove the intake air surge tank.
17. Remove no. 1 air cleaner inlet.
18. Separate fuel tube sub-assembly.
19. Remove the intake manifold.
20. Disconnect the 2 knock control sensor connectors.
21. Remove the 2 bolts and 2 knock control sensors.

To install:

22. Install the 2 knock control sensors with the 2 bolts as shown in the illustration and tighten to 15 ft. lbs. (20 Nm).
23. Connect the 2 knock control sensor connectors.
24. The remainder of installation is the reverse of the removal procedure.
25. Inspect for fuel leak and check the function of throttle body.

MASS AIR FLOW (MAF) SENSOR

LOCATION

The MAF sensor is between the throttle body and air cleaner housing. Refer to Component Locations illustrations at the top of this section to locate the MAF sensor.

REMOVAL & INSTALLATION

See Figure 117.

1. Before servicing the vehicle, refer to the Precautions Section.
2. Disconnect the mass air flow meter connector.
3. Remove the 2 screws and mass air flow meter.

Fig. 117 Mass air flow meter

To install:

4. Installation is the reverse of the removal procedure.

THROTTLE POSITION SENSOR (TPS)

LOCATION

The TPS is located on the throttle body.

REMOVAL & INSTALLATION

Refer to the Throttle Body removal and installation procedures.

VARIABLE VALVE TIMING (VVT) SENSOR

LOCATION

See Figure 118.

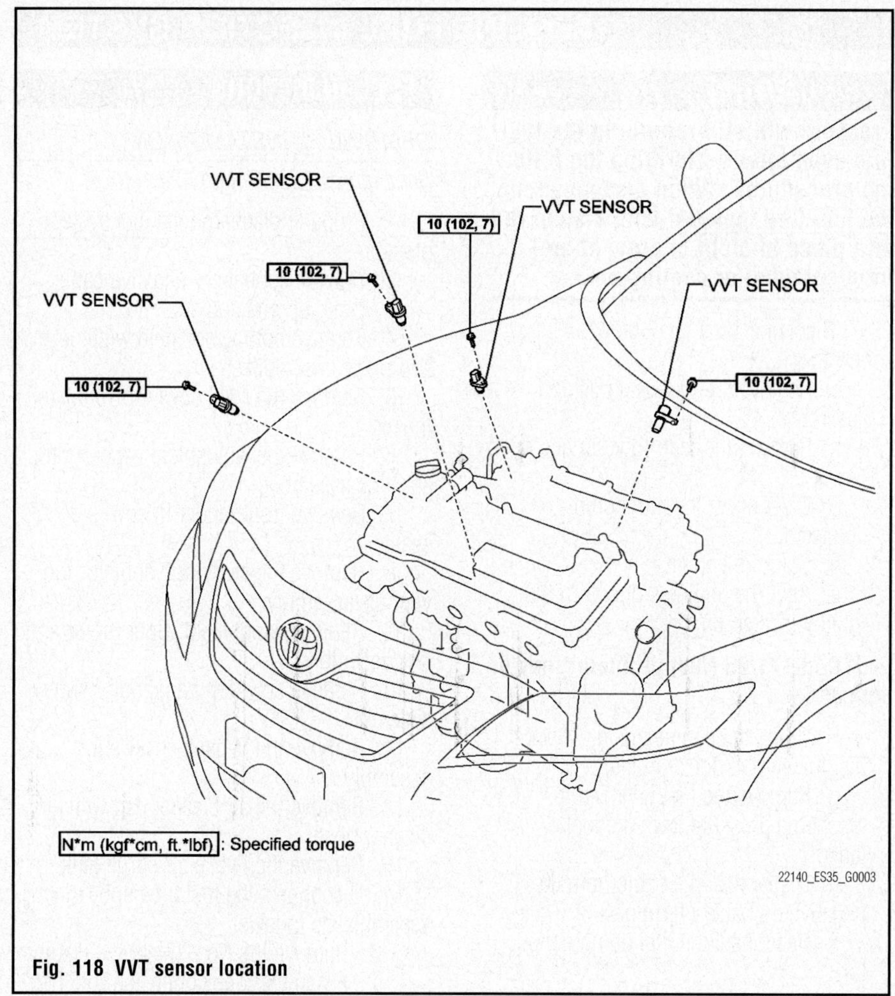

Fig. 118 VVT sensor location

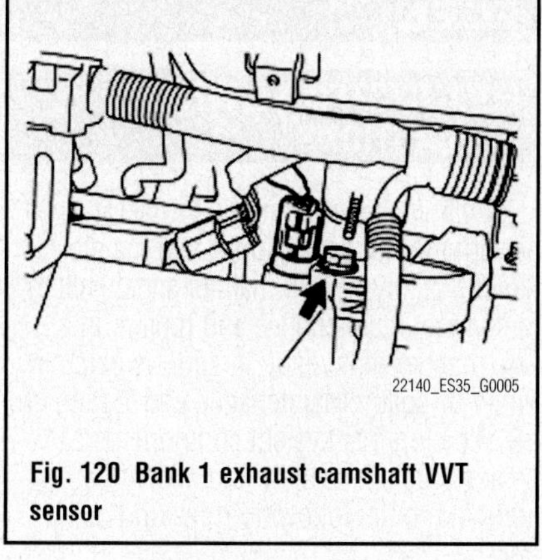

Fig. 120 Bank 1 exhaust camshaft VVT sensor

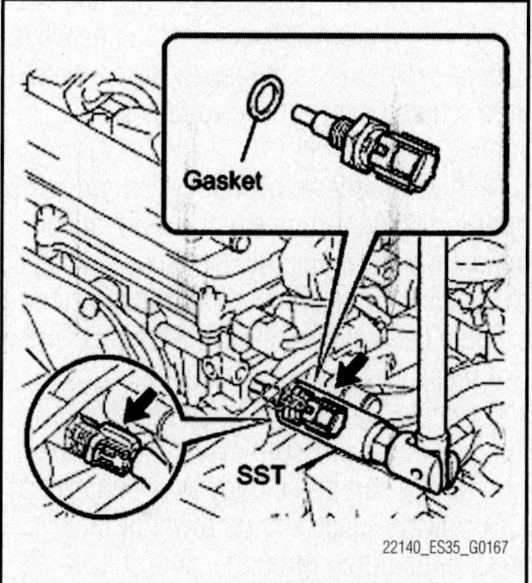

Fig. 121 Bank 2 exhaust camshaft VVT sensor

REMOVAL & INSTALLATION

See Figures 119 through 121.

1. Before servicing the vehicle, refer to the Precautions Section.
2. Drain and recycle the engine coolant.
3. Disconnect the negative battery cable.
4. Remove the V-bank cover sub-assembly.
5. Remove the windshield wiper link assembly.
6. Remove the front cowl top outside panel.
7. Remove the Intake camshaft VVT sensor (Bank 1), as follows:
   a. Disconnect the VVT sensor connector.

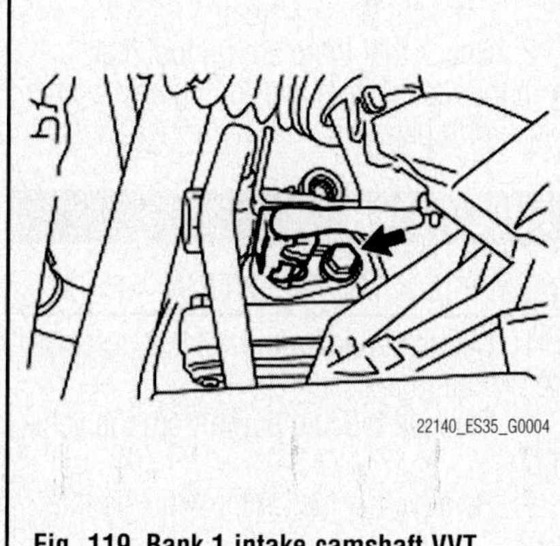

Fig. 119 Bank 1 intake camshaft VVT sensor

   b. Remove the bolt and VVT sensor.
8. Remove the Exhaust camshaft VVT sensor (Bank 1), as follows:
   a. Disconnect the VVT sensor connector.
   b. Remove the bolt and VVT sensor.
9. Remove the Exhaust camshaft VVT sensor (Bank 2), as follows:
   a. Disconnect the VVT sensor connector.
   b. Remove the bolt and VVT sensor.
10. Remove the Intake camshaft VVT sensor (Bank 2), as follows:
   a. Disconnect the VVT sensor connector.
   b. Remove the bolt and VVT sensor.

FUEL SYSTEM SERVICE PRECAUTIONS

Safety is the most important factor when performing not only fuel system maintenance but any type of maintenance. Failure to conduct maintenance and repairs in a safe manner may result in serious personal injury or death. Maintenance and testing of the vehicle's fuel system components can be accomplished safely and effectively by adhering to the following rules and guidelines.

- To avoid the possibility of fire and personal injury, always disconnect the negative battery cable unless the repair or test procedure requires that battery voltage be applied.
- Always relieve the fuel system pressure prior to disconnecting any fuel system component (injector, fuel rail, pressure regulator, etc.), fitting or fuel line connection. Exercise extreme caution whenever relieving fuel system pressure to avoid exposing skin, face and eyes to fuel spray. Please be advised that fuel under pressure may penetrate the skin or any part of the body that it contacts.
- Always place a shop towel or cloth around the fitting or connection prior to loosening to absorb any excess fuel due to spillage. Ensure that all fuel spillage (should it occur) is quickly removed from engine surfaces. Ensure that all fuel soaked cloths or towels are deposited into a suitable waste container.
- Always keep a dry chemical (Class B) fire extinguisher near the work area.
- Do not allow fuel spray or fuel vapors to come into contact with a spark or open flame.
- Always use a back-up wrench when loosening and tightening fuel line connection fittings. This will prevent unnecessary stress and torsion to fuel line piping.
- Always replace worn fuel fitting O-rings with new Do not substitute fuel hose or equivalent where fuel pipe is installed.

Before servicing the vehicle, make sure to also refer to the precautions in the beginning of this section as well.

RELIEVING FUEL SYSTEM PRESSURE

CAUTION

Perform the following procedures to prevent fuel from spilling out before removing any fuel system parts.

CAUTION

Pressure will still remain in the fuel line even after performing the following procedures. When disconnecting the fuel line, cover it with a shop rag or a piece of cloth to prevent fuel from spraying or coming out.

1. Disconnect the fuel pump connector:
   a. Remove the rear seat cushion assembly.
   b. Remove the rear floor service hole cover.
   c. Disconnect the fuel pump connector.
   d. Start the engine.
   e. After the engine stops, turn the ignition switch off.

➡DTC P0171/25 (fuel problem) may be detected.

   f. Crank the engine again. Check that the engine does not start.
   g. Remove the fuel tank cap to discharge pressure from the fuel tank.
   h. Disconnect the cable from the negative (-) battery terminal.
   i. Reconnect the fuel pump connector.
   j. Install the rear floor service hole cover.
   k. Install the rear seat.
2. Check that there are no fuel leaks from the fuel system after doing any maintenance or repairs.

FUEL FILTER

REMOVAL & INSTALLATION

1. Before servicing the vehicle, refer to the precautions section.
2. Remove the fuel pump from the vehicle.
3. Remove the fuel pump filter, as follows:

➡Do not damage the fuel pump filter. Do not remove the suction filter.

   a. Using a screwdriver, pry out the clips.
   b. Pull out the fuel pump filter from the fuel pump.

To install:

4. Install the fuel pump filter with a new clip.
5. Install the fuel pump.

FUEL INJECTORS

REMOVAL & INSTALLATION

See Figures 122 through 124.

1. Properly discharge the fuel system pressure.
2. Disconnect battery negative cable.
3. Drain engine coolant.
4. Remove both windshield wiper arm and blade assemblies.
5. Remove the right cowl top ventilator louver.
6. Remove the windshield wiper motor and link assembly.
7. Remove front fender to cowl side seal.
8. Remove the cowl top ventilator louver sub-assembly.
9. Remove windshield wiper motor and link assembly.
10. Remove cowl top panel outer sub-assembly.
11. Remove the V-bank cover sub-assembly.
12. Remove the air cleaner cap with air cleaner hose.
13. Remove the intake air surge tank.
14. Disconnect the fuel tube sub-assembly, as follows:
    a. Remove the No. 2 fuel pipe clamp.
    b. Pinch the tube connector and then pull out the fuel pipe.

➡Check that there is no dirt or other foreign objects around the connector

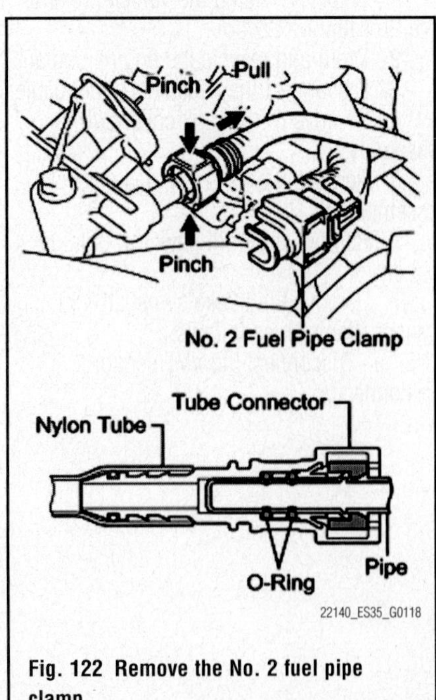

Fig. 122 Remove the No. 2 fuel pipe clamp

before removing fuel tube, and clean the connector as necessary.

➡It is necessary to prevent mud or dirt from entering the connector. If mud or dirt gets in the connector, the O-rings may not seal properly.

➡Do not use any tools in this operation.

➡Do not bend, kink or twist the nylon tube. Protect the connector by covering it with a plastic bag.

➡When the pipe and connector are stuck, push and pull the connector to release and pull the connector out carefully.

15. Remove the fuel injector assembly, as follows:

a. Disconnect the 6 fuel injector connectors.

b. Remove the 5 bolts and fuel delivery pipe together with the 6 fuel injectors.

➡Be careful not to drop the fuel injectors when removing the fuel delivery pipe.

c. Remove the 6 insulators from the intake manifold.

d. Pull out the fuel injector from the fuel delivery pipe.

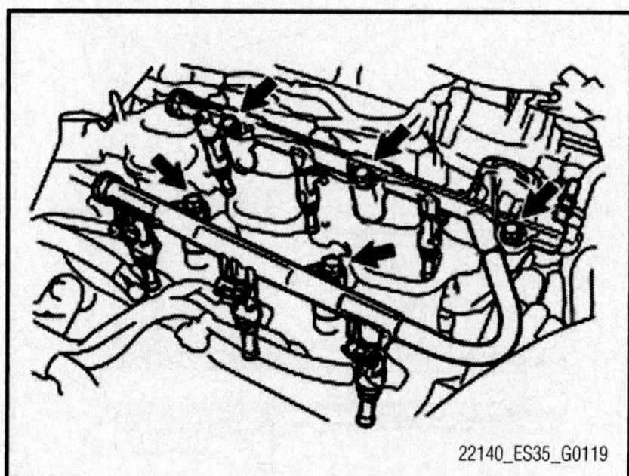

22140_ES35_G0119

Fig. 123 Removing the 5 bolts and fuel delivery pipe together with the 6 fuel injectors

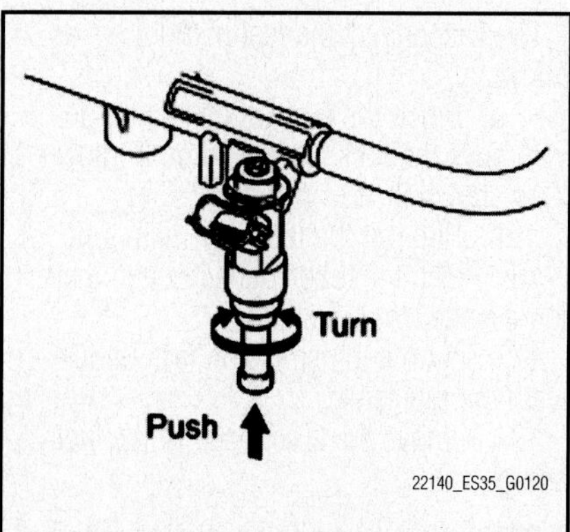

22140_ES35_G0120

Fig. 124 Installing fuel injector to fuel rail

e. Remove the 6 O-rings from the injectors.

To install:

16. Install the fuel injector assembly, as follows:

a. Apply a light coat of spindle oil or gasoline to new O-rings, and install one to each injector.

b. Apply a light coat of spindle oil or gasoline where the fuel delivery pipe contacts the O-ring.

➡Be careful not to twist the O-ring.

➡After installing the fuel injector, check that it turns smoothly. If not, reinstall it with a new O-ring.

c. Push the fuel injector while twisting it back and forth to install it in the fuel delivery pipe.

d. Position the fuel injector connector outward.

e. Install 6 new insulators to the intake manifold.

f. Place the fuel delivery pipe and the 6 fuel injectors together to the intake manifold.

➡Be careful not to drop the fuel injectors when installing the fuel delivery pipe.

g. Temporarily install the 6 bolts which are used to hold the fuel delivery pipe to the intake manifold.

➡After installing the fuel injector, check that it turns smoothly. If not, reinstall it with a new O-ring.

h. Tighten the 5 bolts which are used to hold the fuel delivery pipe to the intake manifold to 15 ft. lbs. (21 Nm).

17. Push in the tube connector to the pipe until the tube connector makes a "click" sound.

➡Before connecting the tube, make sure that it is not damaged. Make sure that there is no dirt present on the connecting surfaces.

➡After connecting, check if the fuel tube connector and the pipe are securely connected by pulling on them.

18. Install the No. 2 fuel pipe clamp.

19. Te remainder of installation is the reverse of the removal procedure.

20. Check for coolant leak and fuel leak.

FUEL PUMP

REMOVAL & INSTALLATION

See Figures 125 and 126.

1. Before servicing the vehicle, refer to the precautions section.

2. Discharge fuel system pressure.

3. Disconnect battery negative cable.

4. Remove the rear seat cushion assembly.

5. Remove the rear floor service hole cover.

6. Disconnect the fuel pump connector.

7. Separate the fuel pump tube sub-assembly, as follows:

➡Check if there is any dirt or mud around the connector before this operation and remove the dirt as necessary.

➡Be careful of mud because the quick connector has an O-ring which seals the pipe and connector that can be contaminated.

➡Do not use any tools in this operation.

➡Do not bend or twist the nylon tube. Cover the fuel tube joint with a plastic bag.

➡When the fuel tube joint and fuel suction plate are stuck, pinch the fuel tank tube between fingers, and turn it carefully to release it. Disconnect the fuel tank tube.

a. Remove the tube joint clip, and pull out the fuel pump tube.

8. Remove fuel tank vent tube set plate, as follows:

a. Remove the 8 bolts and set plate.

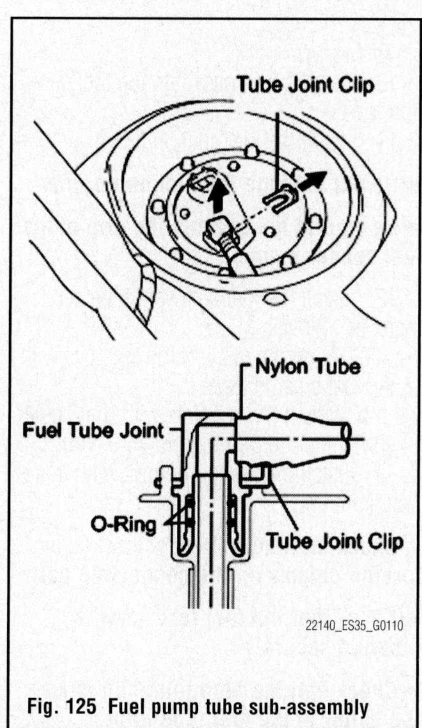

22140_ES35_G0110

Fig. 125 Fuel pump tube sub-assembly

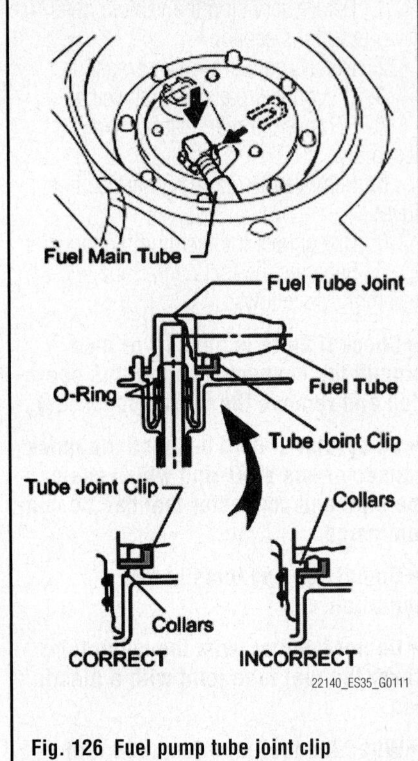

Fig. 126 Fuel pump tube joint clip

9. Remove fuel suction tube assembly with pump and gauge, as follows:
   a. Pull out the fuel suction tube from the fuel tank.

➡Do not damage the fuel pump filter.

➡Be careful not to bend the arm of the fuel sender gauge.

   b. Remove the gasket from the fuel suction tube.

To install:

10. Install a new gasket to the fuel suction tube.
11. Install the fuel suction tube.

➡Do not damage the fuel pump filter.

➡Be careful not to bend the arm of the fuel sender gauge.

12. Install the fuel tank vent tube set plate, as follows:
    a. Align the mark of the set plate with the fuel suction tube.
    b. Install the set plate with the 8 bolts and tighten to 52 inch lbs. (5.9 Nm).
13. Install the fuel pump tube with the tube joint clip.

➡Check that there is no scratches or foreign objects on the connecting part.

➡Check that the fuel tube joint is inserted securely.

➡Check that the tube joint clip is on the collar of the fuel tube joint.

➡After installing the tube joint clip, check that the fuel tube joint is pulled off.

14. Connect battery negative cable.
15. Inspect for fuel leak.
16. Install the rear floor service hole cover.
17. Install the rear seat cushion assembly.

FUEL TANK

REMOVAL & INSTALLATION

See Figures 127 through 132.

1. Before servicing the vehicle, refer to the precautions section.
2. Discharge fuel system pressure.
3. Disconnect battery negative cable.
4. Remove rear seat cushion assembly.
5. Remove rear floor service hole cover.
6. Separate fuel pump tube sub-assembly.
7. Remove fuel tank vent tube set plate.
8. Remove fuel suction tube assembly with pump and gauge.
9. Drain fuel.
10. Remove center exhaust pipe assembly.
11. Disconnect no. 2 parking brake cable assembly.
12. Disconnect no. 3 parking brake cable assembly.
13. Remove rear stabilizer BAR No. 1 bracket.
14. Remove lower center fuel tank protector, as follows:
    a. Remove the 4 bolts and the (Except SE grade).
    b. Remove the 4 bolts and 2 clips (for SE Grade).
    c. Remove the fuel tank protector (for SE Grade).

➡Check that there is no dirt or other foreign objects around the connector before removing fuel tubes, and clean the connector as necessary.

➡It is necessary to prevent mud or dirt from entering the connector. If mud or dirt gets in the connector, the O-rings may not seal properly.

➡Do not use any tools in these operations.

➡Do not bend, kink or twist the nylon tubes. Protect the connector by covering it with a plastic bag.

➡When the pipe and connector are stuck, push and pull the connector to release and pull the connector out carefully.

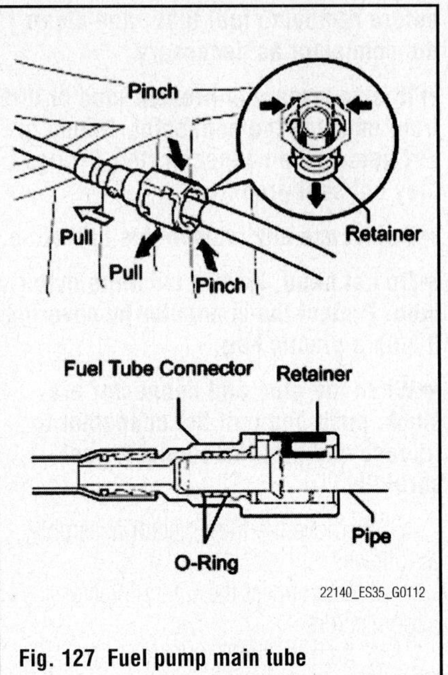

Fig. 127 Fuel pump main tube

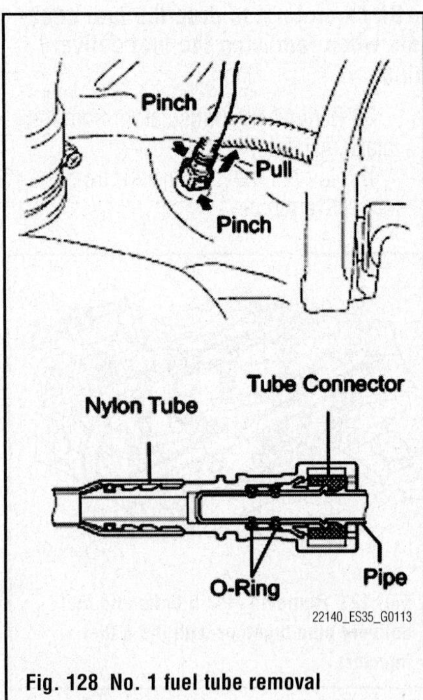

Fig. 128 No. 1 fuel tube removal

15. Disconnect the fuel pump tube, as follows:
    a. Pinch the tabs of the retainer to remove the lock claws and pull it down as shown in the illustration.
    b. Pull out the fuel tank main tube.
16. Pinch the tube connector and then pull out the No. 1 fuel tube.
17. Set up a transmission jack underneath the fuel tank.
18. Remove the 2 set bolts of the fuel tank bands.

➡Check that there is no dirt or other foreign objects around the connector

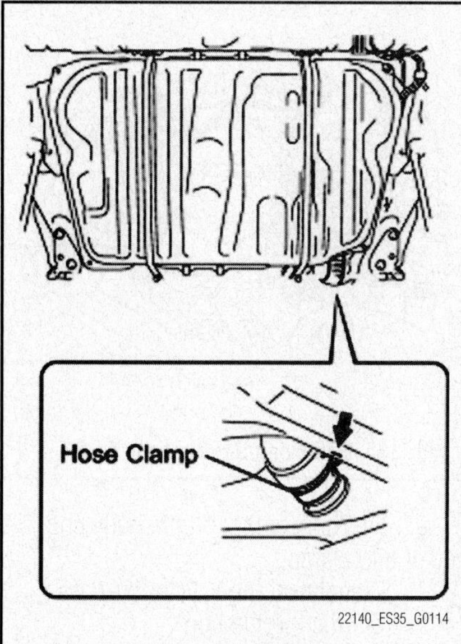

Fig. 129 Disconnecting the fuel tank to filter pipe hose—except PZEV

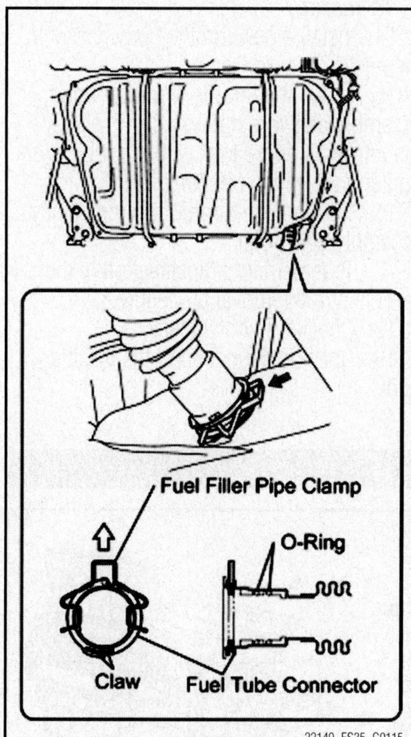

Fig. 130 Disconnecting the fuel tank to filter pipe hose—PZEV

before removing fuel tubes, and clean the connector as necessary.

➡It is necessary to prevent mud or dirt from entering the connector. If mud or dirt gets in the connector, the O-rings may not seal properly.

➡Do not use any tools in these operations.

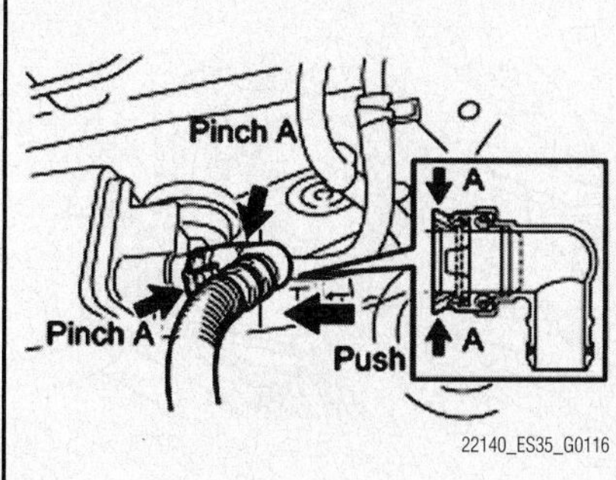

Fig. 131 Disconnect the fuel tank vent hose from the charcoal canister

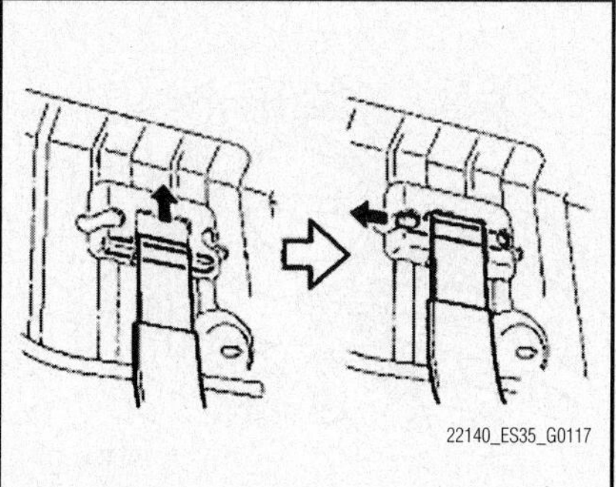

Fig. 132 Removing the 2 pins and 2 fuel tank bands

➡Do not bend, kink or twist the nylon tubes. Protect the connector by covering it with a plastic bag.

➡When the pipe and connector are stuck, push and pull the connector to release and pull the connector out carefully.

19. Remove the hose clamp and disconnect the fuel tank to filter pipe hose (except PZEV).
20. Remove the clamp and disconnect the fuel tank to filter pipe hose (for PZEV).
21. Slightly lower the transmission jack.
22. Disconnect the fuel tank vent hose from the charcoal canister, as follows:
    a. Push the connector deep into the charcoal canister to release the locking pin.
    b. Pinch portion A.
    c. Pull out the connector.
23. Remove the 2 pins and 2 fuel tank bands as shown in the illustration.
24. Remove the 4 clip nuts.

To install:

25. Install the 4 clip nuts.
26. Install the 2 fuel tank bands with the 2 pins.
27. Connect the fuel tank vent hose.
28. Connect the fuel tank inlet pipe with the fuel filter pipe clamp.
29. Tighten the 2 set bolts of the fuel tank bands to 29 ft. lbs. (39 Nm).
30. Connect the No. 1 fuel tube, as follows:
    a. Push the fuel tube connector into the pipe until the fuel tube connector makes a "click" sound.

➡Check that there is no damage or foreign objects on the connected part.

➡After connecting, check if the fuel tube connector and the pipe are securely connected by trying to pull them apart.

31. Connect the fuel pump tube, as follows:
    a. Push in the fuel pump tube connector to the pipe and push up the retainer so that the claws engage.

➡Check that there is no damage or foreign objects on the connected part.

➡After connecting, check if the fuel tube connector and the pipe are securely connected by trying to pull them apart.

32. Install the lower center fuel tank protector and tighten to 48 inch lbs. (5.4 Nm).
33. Install the No. 3 parking brake cable assembly with the bolt and nut and tighten to 53 inch lbs. (6 Nm), and 75 inch lbs. (8.5 Nm).
34. Install the No. 2 parking brake cable assembly with the bolt and nut and tighten to 53 inch lbs. (6 Nm), and 75 inch lbs. (8.5 Nm).
35. Install the center exhaust pipe assembly.
36. Install the fuel suction tube assembly with pump and gauge.
37. Install the fuel tank vent tube set plate.
38. Connect the fuel pump tube sub-assembly.
39. Add fuel.
40. Connect battery negative cable.
41. Inspect for fuel leak and exhaust gas leak.
42. Install the rear floor service hole cover.
43. Install the rear seat cushion assembly.

IDLE SPEED

ADJUSTMENT

Idle speed is maintained by the ECM. No adjustment is necessary or possible.

THROTTLE BODY

REMOVAL & INSTALLATION

See Figures 133 through 136.

1. Before servicing the vehicle, refer to the precautions section.
2. Disconnect battery negative cable.
3. Drain engine coolant.
4. Remove cool air intake duct seal.
5. Remove the V-bank cover sub-assembly.
6. Remove air cleaner inlet assembly.
7. Remove air cleaner cap sub-assembly, as follows:
   a. Disconnect the 3 vacuum hoses.
   b. Disconnect the mass air flow meter connector (1).
   c. Disconnect the No. 2 ventilation hose (2).
   d. Disconnect the hose band (3).

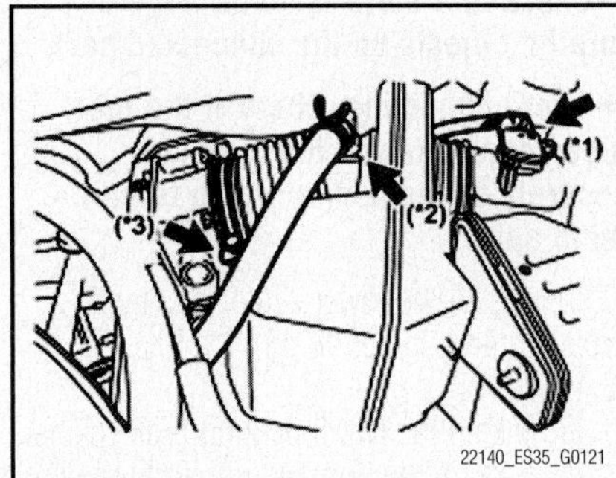

Fig. 133 Removing air cleaner cap sub-assembly

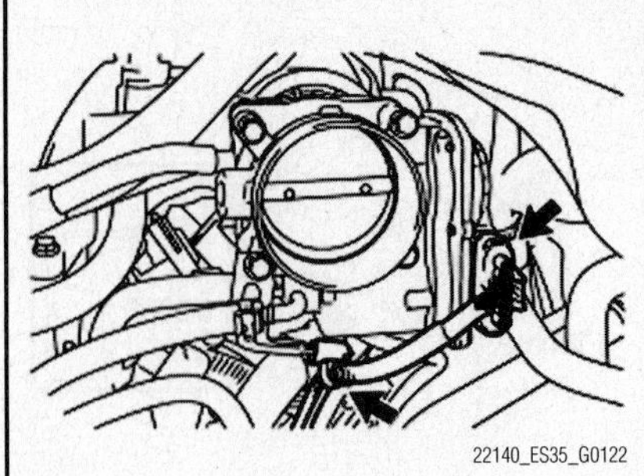

Fig. 134 Disconnecting the throttle body connector

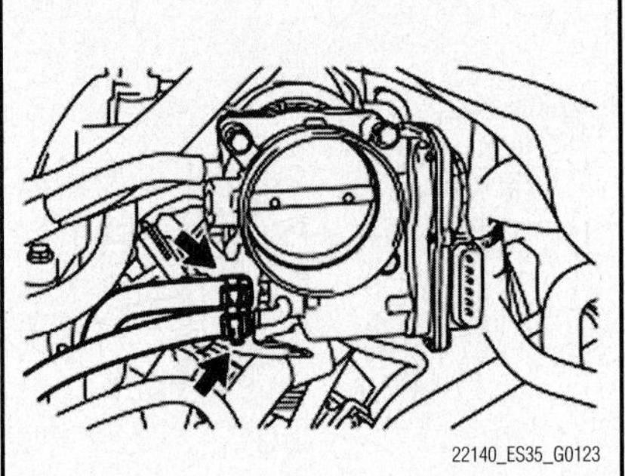

Fig. 135 Disconnecting the 2 water by-pass hoses

   e. Disconnect the 3 bands, and remove the air cleaner cap sub-assembly.
8. Remove air cleaner case sub-assembly.
9. Remove No. 1 air cleaner inlet.

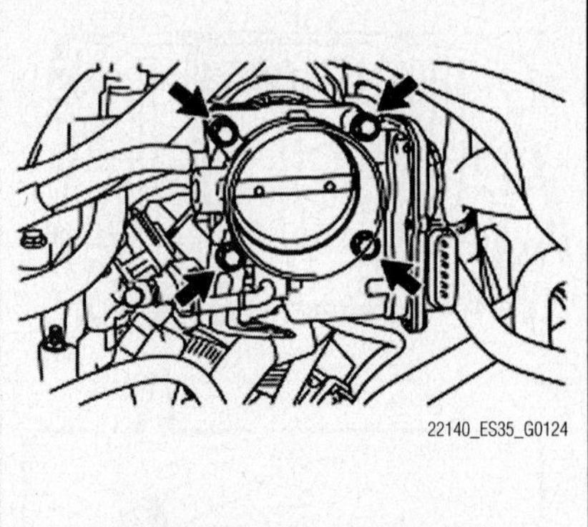

Fig. 136 Removing throttle body bolts

10. Disconnect the throttle body connector and clamp.
11. Disconnect the 2 water by-pass hoses from the throttle body.
12. Remove the 4 bolts and throttle body.
13. Remove the throttle body gasket from the intake air surge tank.

To install:

14. Install a new throttle body gasket to the intake air surge tank.
15. Install the throttle w/ motor body assembly and wire harness clamp stay to the intake air surge tank with the 4 bolts and tighten to 7 ft. lbs. (10 Nm).
16. Connect the throttle w/ motor body assembly connector.
17. Te remainder of installation is the reverse of the removal procedure.
18. Check for coolant leak.
19. Check the function of the throttle body.

HEATING & AIR CONDITIONING SYSTEM

BLOWER MOTOR

REMOVAL & INSTALLATION

See Figures 137 and 138.

1. Drain and recycle the engine coolant.
2. Disconnect the negative battery cable.
3. Remove instrument panel.
4. For TMC made:
   a. Disconnect the connector.
   b. Remove the 2 screws and blower assembly.
   c. Remove the 3 screws and blower with fan motor sub-assembly.
5. For TMMK made:
   a. Remove cooler expansion valve.
   b. Remove the connector and clamp, and disconnect the wire harness.
   c. Remove the 6 screws and then the blower assembly with the cooler evaporator sub-assembly.

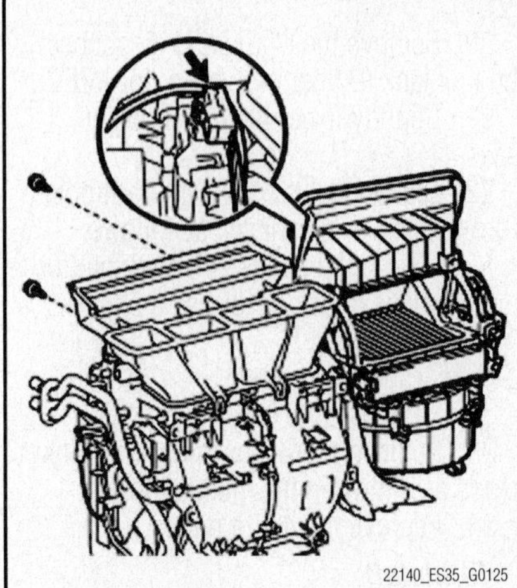

Fig. 137 Removing the connector, clamp and wire harness—TMMK

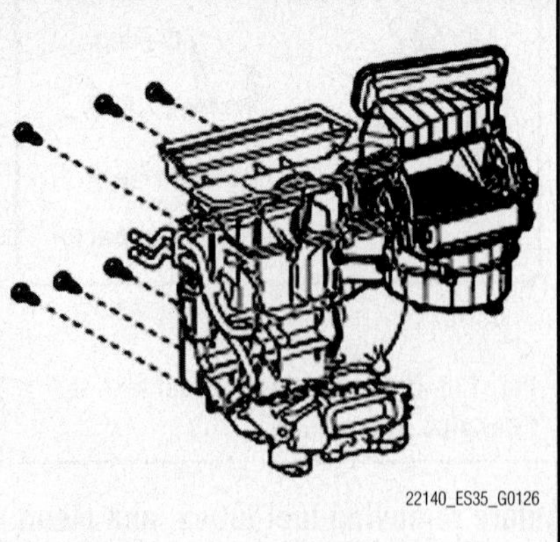

Fig. 138 Removing the 6 screws and blower assembly—TMMK

   d. Remove the 3 screws and blower with fan motor sub-assembly.
6. To install, reverse removal procedure.

HEATER CORE

⁂ CAUTION

Some models covered by this manual may be equipped with a Supplemental Restraint System (SRS), which uses an air bag. Whenever working near any of the SRS components, such as the impact sensors, the air bag module, steering column and instrument panel, disable the SRS, as described in Section 6.

REMOVAL & INSTALLATION

See Figures 139 through 146.

1. Before servicing the vehicle, refer to the Precautions Section.

⁂ CAUTION

Wait for 90 seconds after disconnecting the cable to prevent airbag deployment.

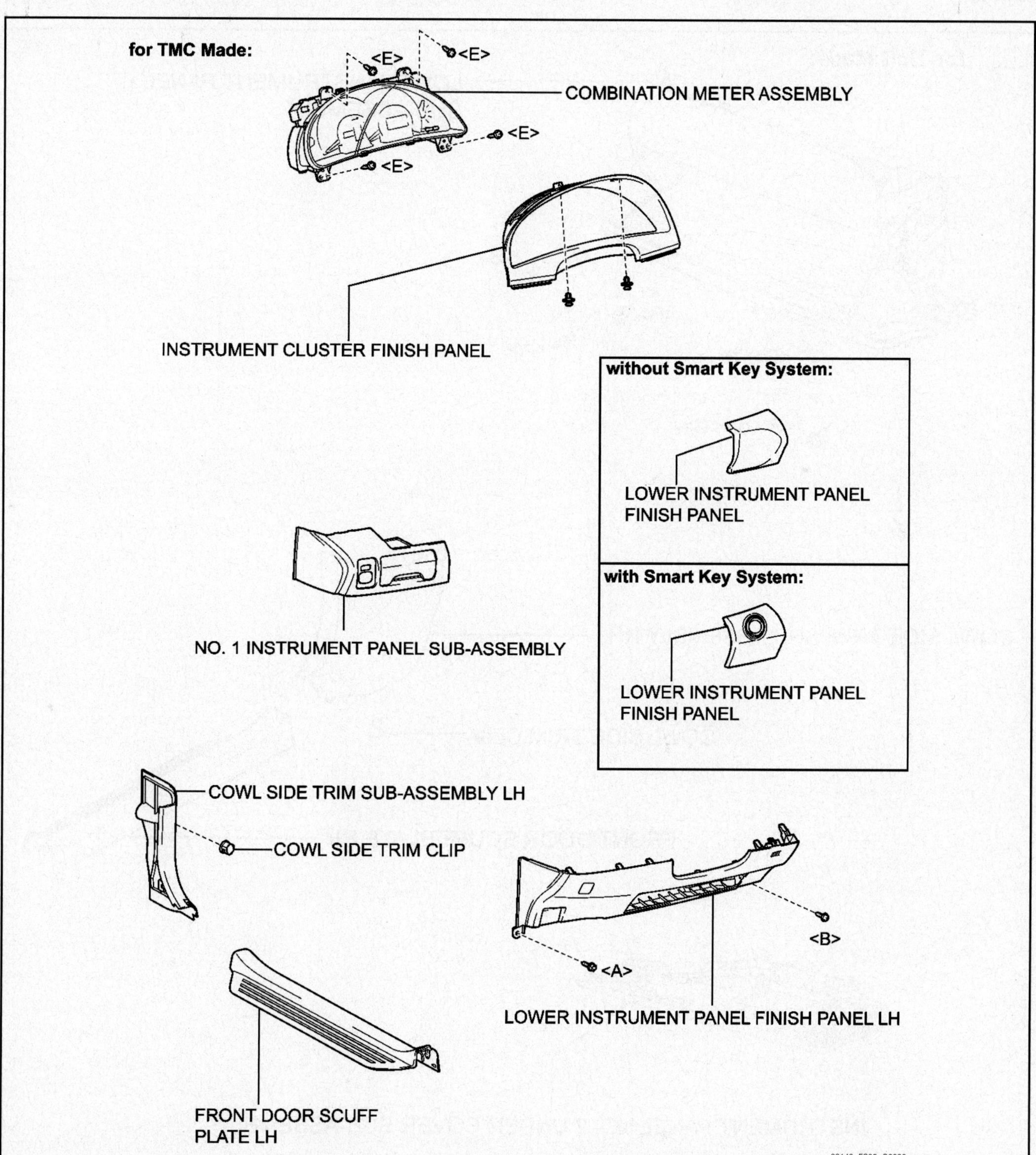

Fig. 139 Instrument cluster, combination meter assembly, instrument panel sub-assembly and lower trip sub-assembly

2. Disconnect battery negative terminal.
3. Remove the lower No. 2 and No. 3 steering wheel covers.
4. Remove the steering pad.
5. Remove the steering wheel assembly.
6. Remove the LH front door scuff plate.
7. Remove the LH cowl side trim sub-assembly.
8. Remove steering column cover.
9. Remove the turn signal switch assembly.
10. For vehicles without Smart Key System, disengage the 2 claws and 2 clips and then remove the lower instrument panel finish panel.
11. For vehicles with Smart Key System, disengage the 2 claws and 2 clips. Disconnect the connector and remove the lower instrument panel finish panel.
12. Using a molding remover, disengage the 2 clips. Disengage the guide and 4 claws, and then remove the No. 1 instrument cluster finish panel.
13. Remove the 4 screws. Disconnect each connector and remove the combination meter assembly.

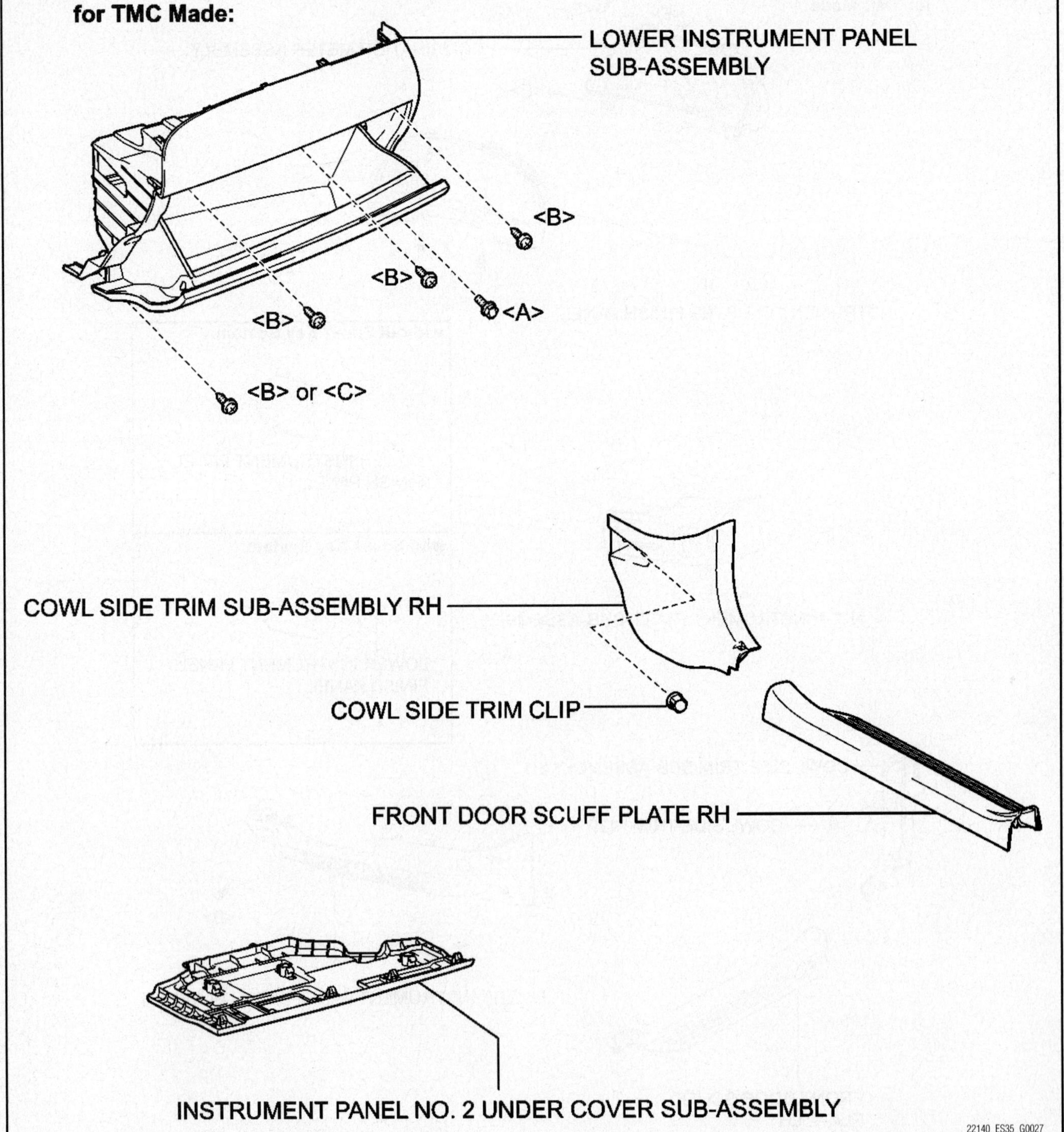

Fig. 140 Lower instrument panel sub-assembly and No. 2 under cover sub-assembly

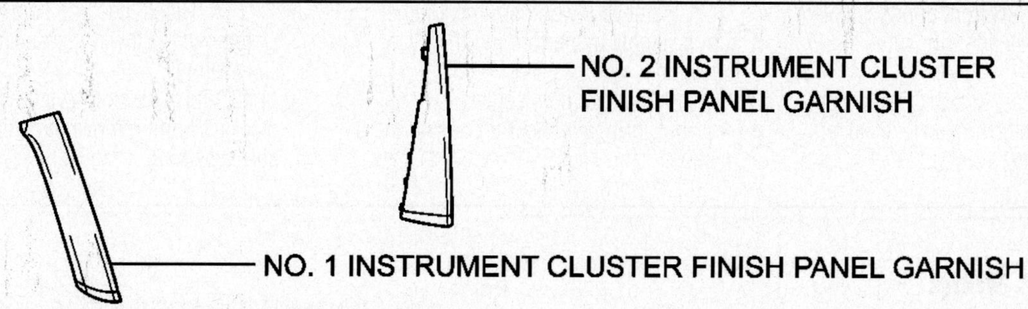

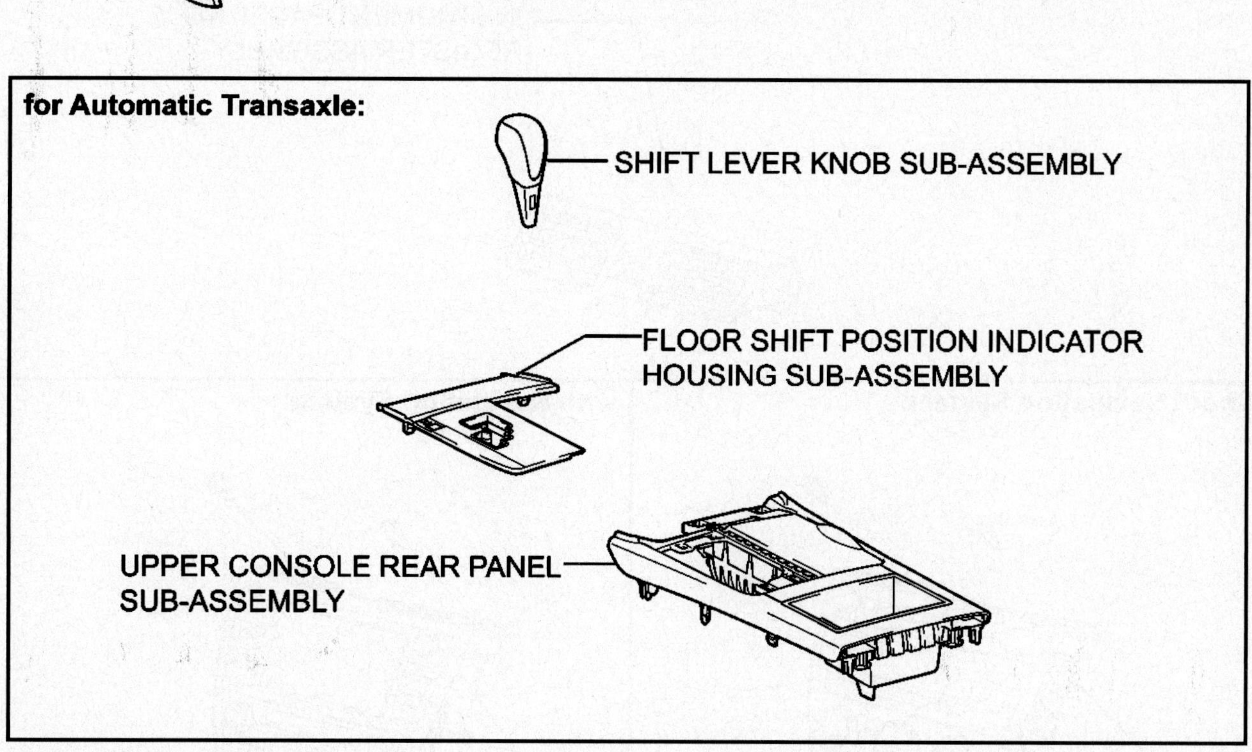

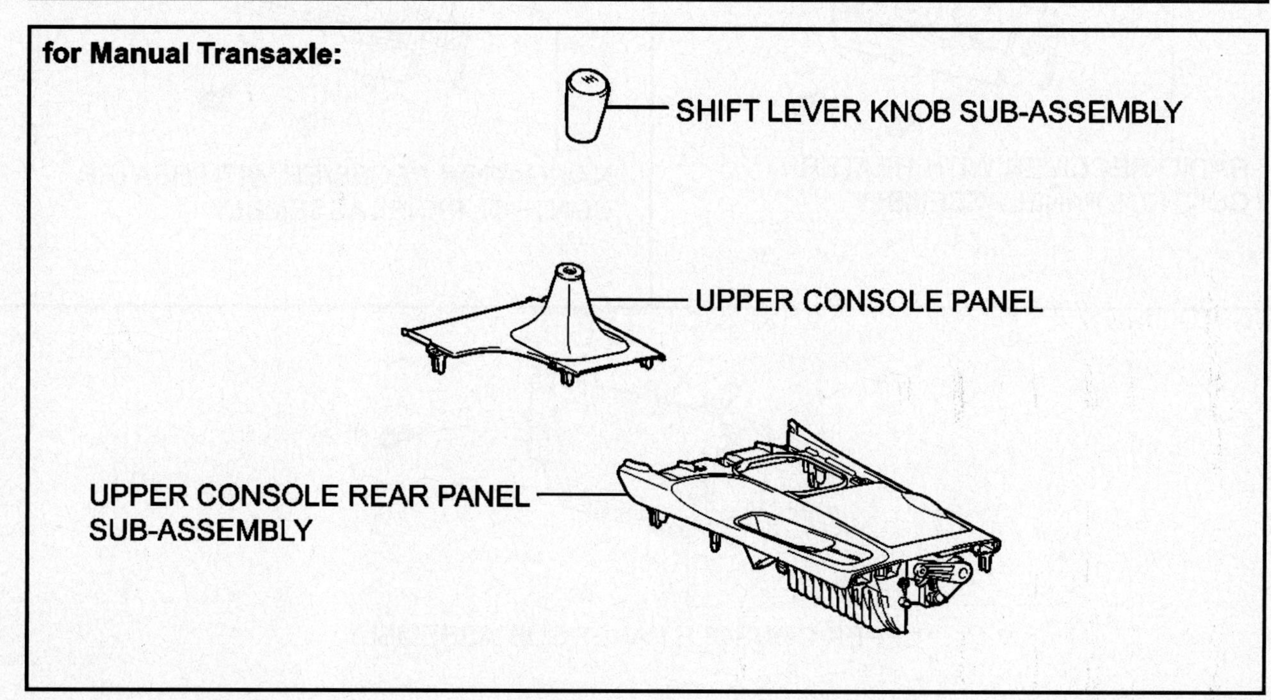

Fig. 141 Instrument cluster panel garnish, floor housing sub-assembly and upper console rear panel sub-assembly

14. Remove the RH front door scuff plate.

15. Remove cowl side trim sub-assembly.

16. Disengage the 4 claws. Disengage the 2 guides and remove the No. 2 under cover sub-assembly.

17. Remove lower instrument panel sub-assembly by performing the following:

a. Remove the 4 screws.

b. Disengage the 3 claws and the 3 clips.

c. Disconnect the connector and remove the lower instrument panel sub-assembly.

18. Turn the shift lever knob counter-clockwise and remove the shift lever knob sub-assembly.

for TMC Made:

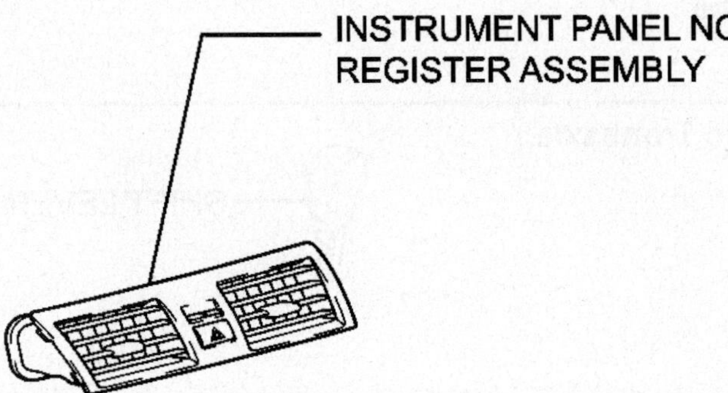

without Navigation System:

with Navigation System:

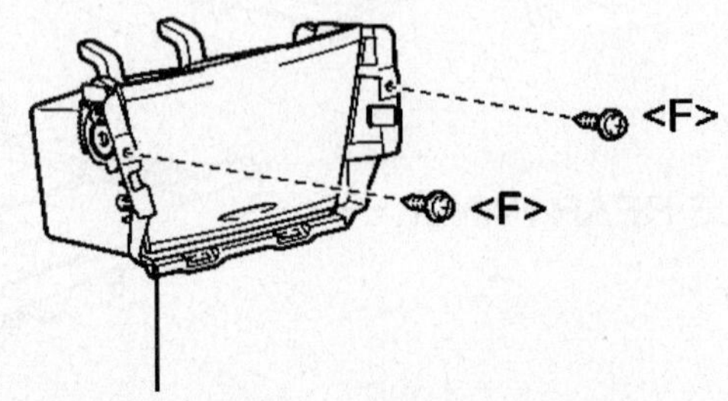

Fig. 142 Instrument panel no. 2 register assembly, control panel assembly and upper console sub-assembly

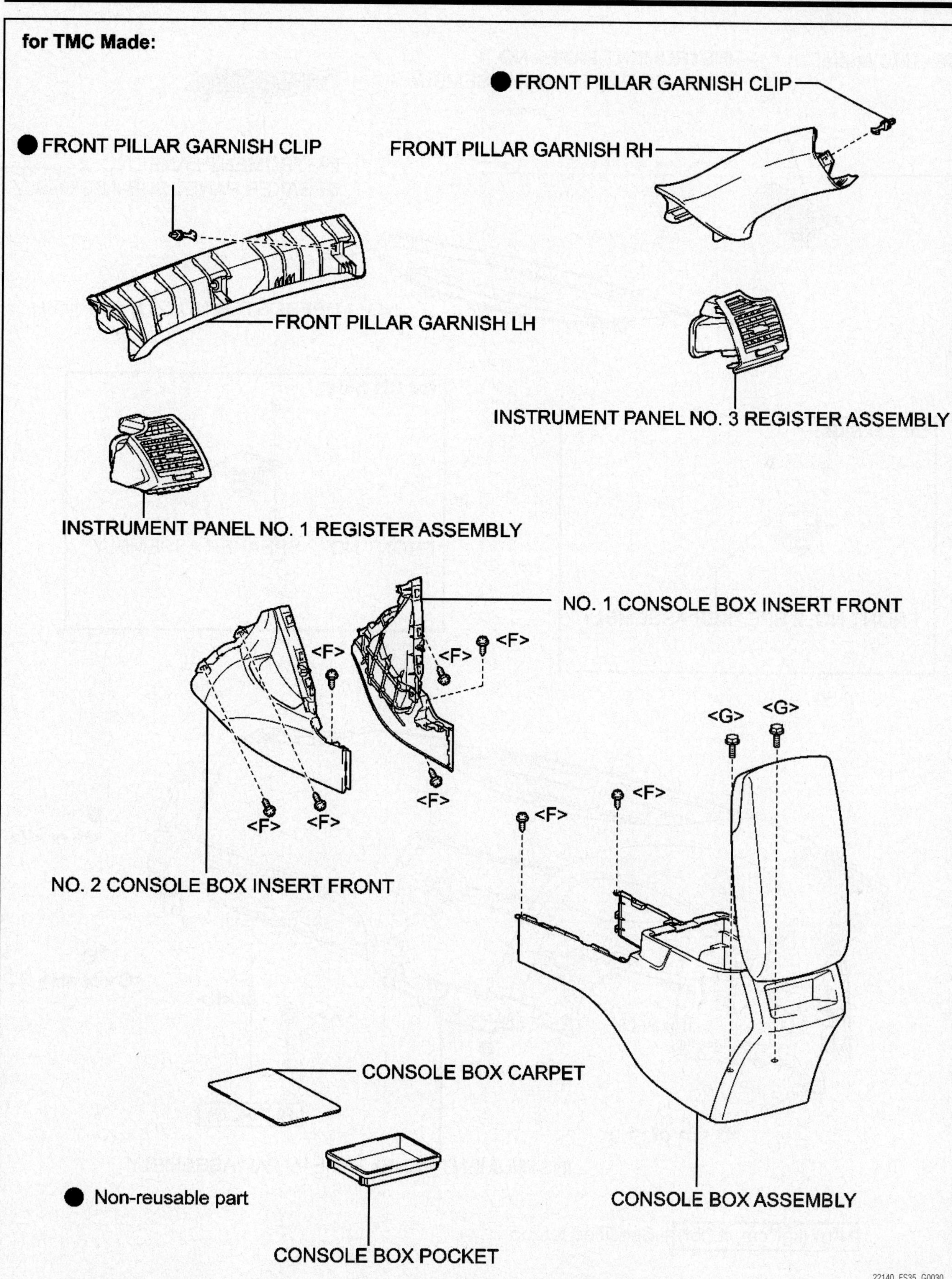

Fig. 143 Console box assembly, front console box inserts, LH front pillar garnish and instrument panel No. 1 and No. 3 register assembly

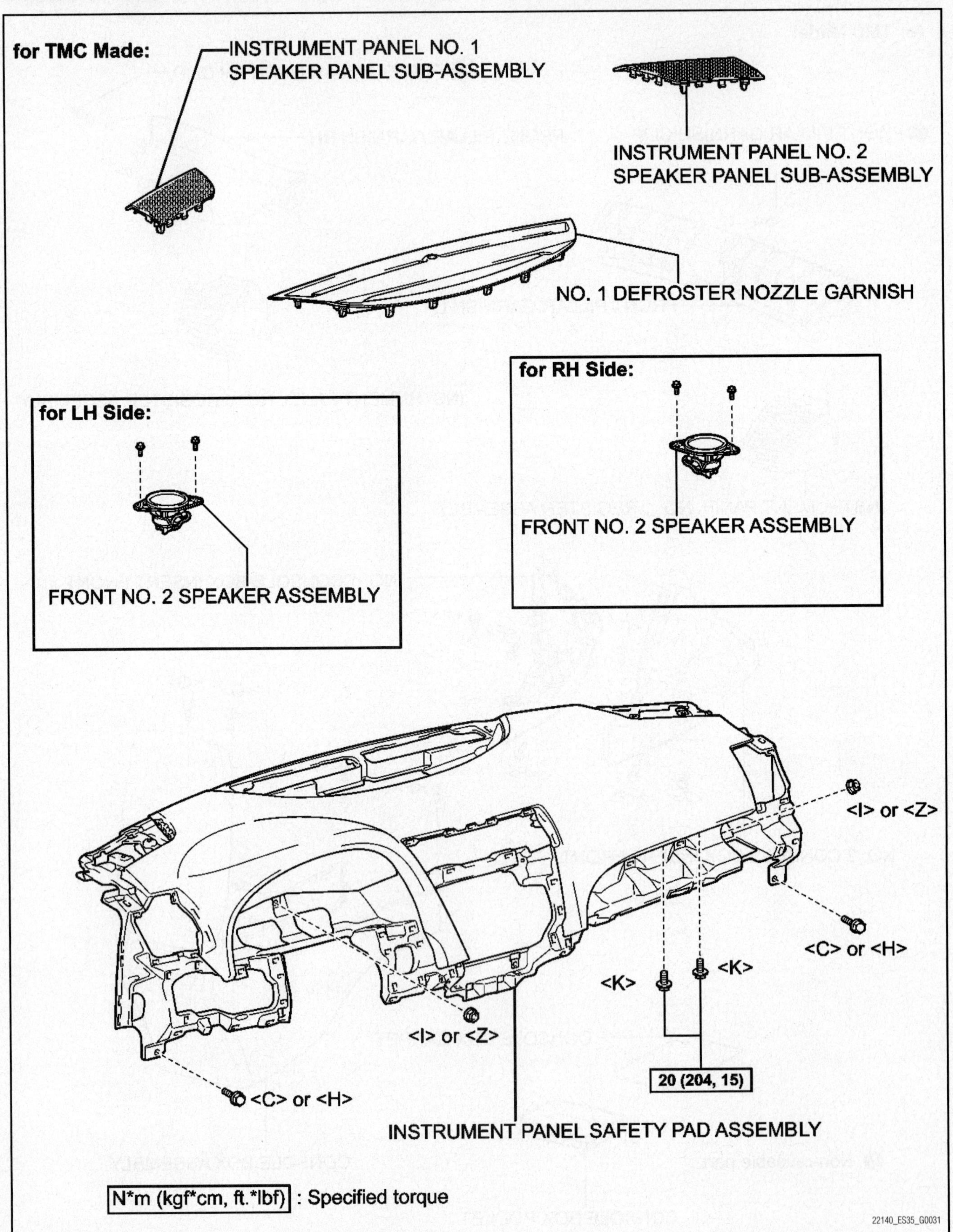

Fig. 144 Speakers, speaker panel sub-assembly and No. 1 defroster nozzle garnish

19. Disengage the 2 clips and remove the No. 1 instrument cluster finish panel garnish.

20. Disengage the 2 clips and remove the No. 2 instrument cluster finish panel garnish.

21. For A/T vehicles, Disengage the 6 claws and the 3 clips, and then remove the floor shift position indicator housing sub-assembly. If equipped with Seat Heater System, disconnect each connector.

22. For M/T vehicles, open the lid of the upper console panel. Apply protective tape to the area. Using a moulding remover, disengage the 2 claws and the 5 clips, and then remove the upper console panel.

23. Disengage the 3 claws and the 5 clips. Disconnect the connector and remove the upper console rear panel sub-assembly.

24. Remove instrument panel no. 2 register assembly by performing the following:

a. Apply protective tape to the areas.

b. Using a moulding remover, disengage the 3 clips.

c. Using a moulding remover, disengage the 4 clips.

d. Disconnect the connector and remove the instrument panel No. 2 register assembly.

25. Remove radio receiver with heater control panel assembly.

26. Remove the console box pocket

27. Remove the console box pocket.

28. Remove the console box assembly by performing the following:

a. Remove the 2 screws.

b. Disengage the clamp.

c. Remove the 2 bolts and the console box assembly.

29. Remove both of the front console box inserts by performing the following:

a. Remove the 3 screws.

b. Disengage the clip and remove the front console box insert.

30. Remove the LH front pillar garnish.

31. Disengage the 4 clips and remove the instrument panel No. 1 register assembly.

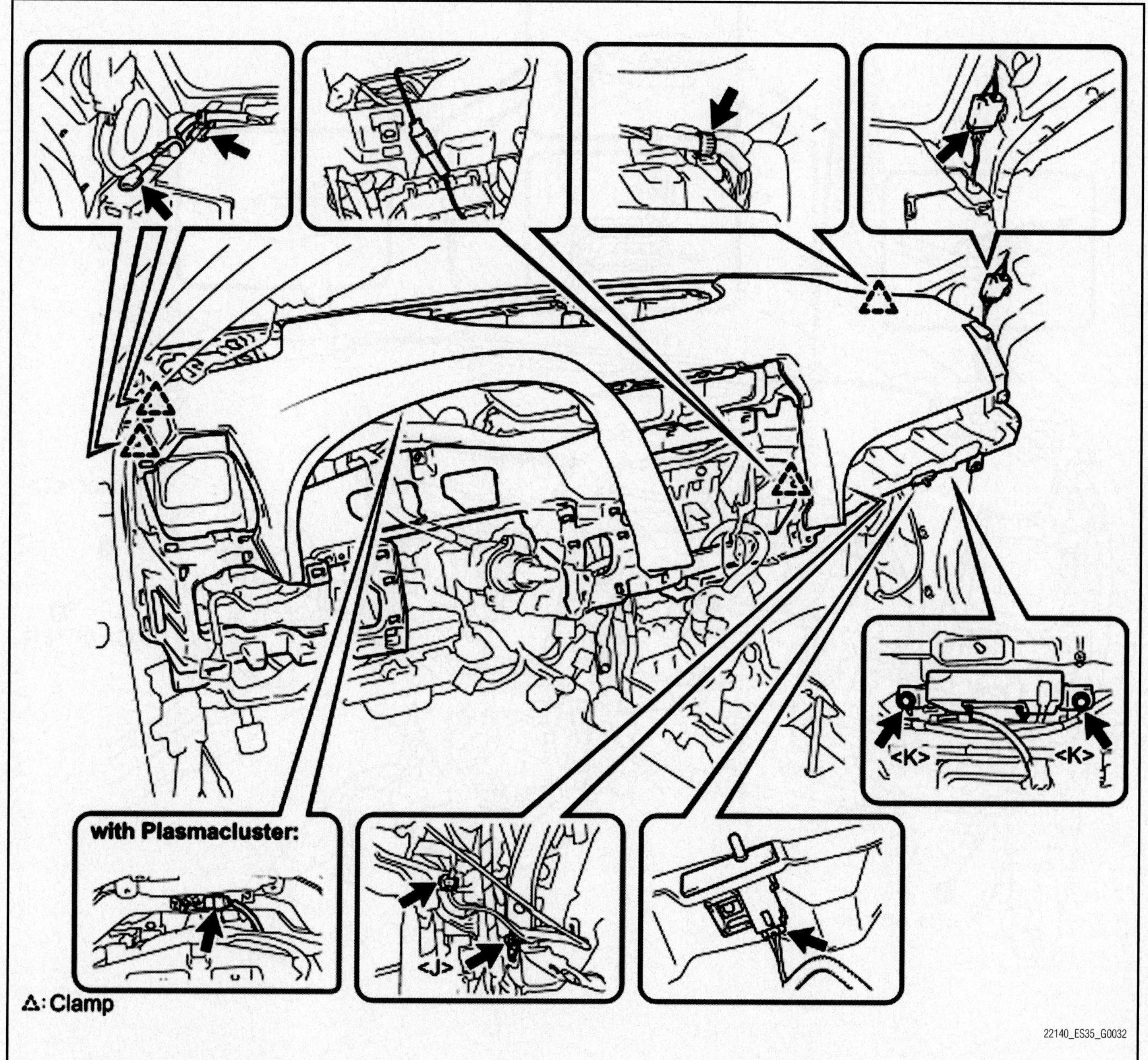

Fig. 145 Instrument panel safety pad assembly (1 of 2)

32. Remove instrument panel no. 1 speaker panel sub-assembly by performing the following:
   a. Disengage the 6 claws and the 2 clips.
   b. Disengage the 2 guides and remove the instrument panel No. 1 speaker panel sub-assembly.
33. Remove RH front no. 2 speaker assembly.
34. Remove the RH front pillar garnish.
35. Disengage the 4 clips and remove the instrument panel No. 3 register assembly.
36. Remove instrument panel no. 2 speaker panel sub-assembly by performing the following:
   a. Disengage the 6 claws and the 2 clips.
   b. Disengage the 2 guides and remove the instrument panel No. 2 speaker panel sub-assembly.
37. Remove LH and RH front no. 2 speaker assemblies.
38. Remove no. 1 defroster nozzle garnish by performing the following:
   a. Disengage the 8 clips and the 4 guides.
   b. Disconnect each connector and remove the No. 1 defroster nozzle garnish.
39. Disconnect instrument panel wire assembly.
40. Remove instrument panel safety pad assembly by performing the following:
   a. Disengage each clamp.

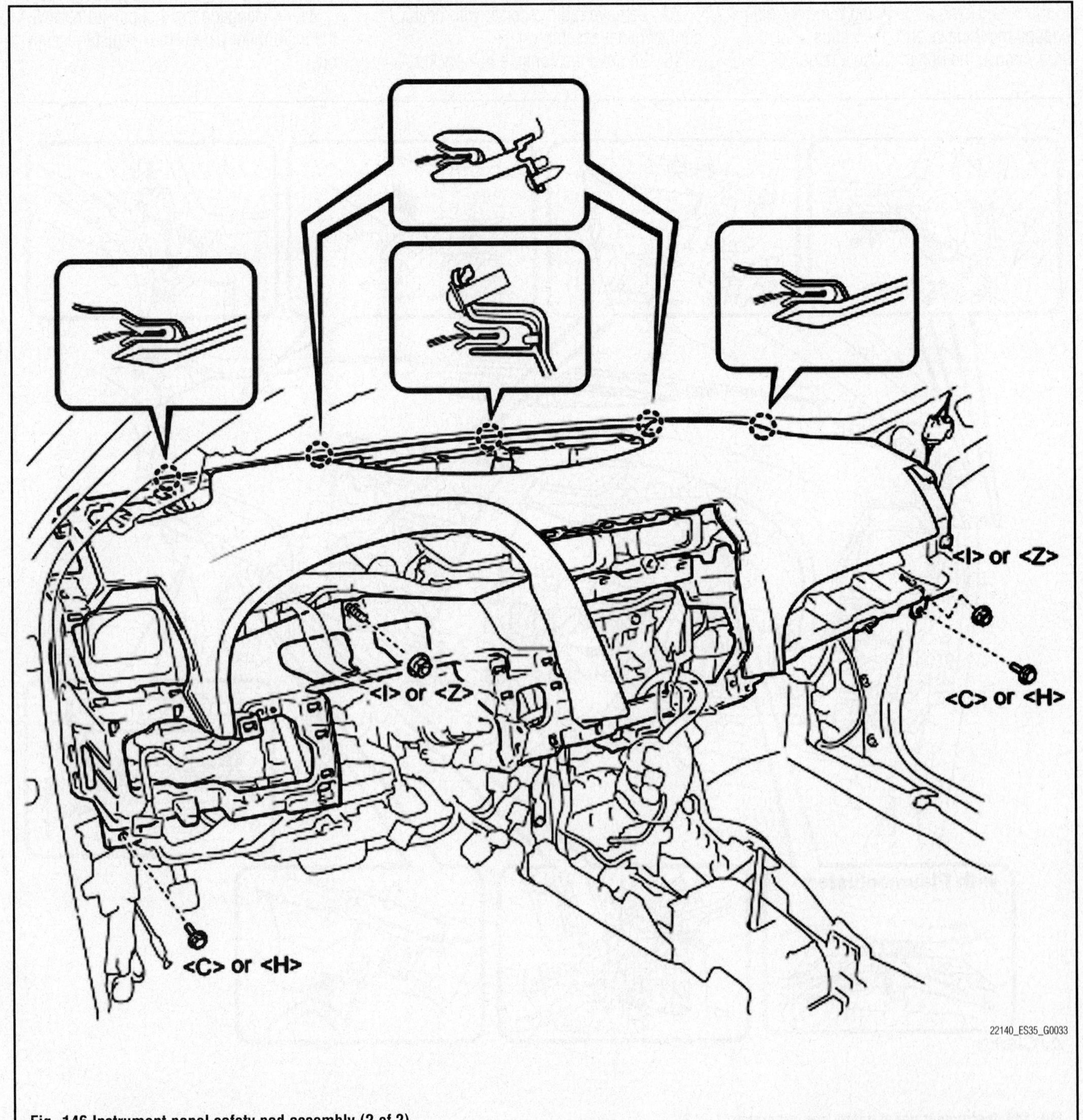

Fig. 146 Instrument panel safety pad assembly (2 of 2)

b. Disconnect each connector.

c. Remove the bolt (J).

CAUTION

Some models covered by this manual may be equipped with a Supplemental Restraint System (SRS), which uses an air bag. Whenever working near any of the SRS components, such as the impact sensors, the air bag module, steering column and instrument panel, disable the SRS, as described in Section 6.

d. Remove the 2 passenger airbag bolts (K).

e. If equipped with Plasmacluster, disconnect the connector

f. Disconnect the connector

g. Remove the 2 bolts (C) or (H).

h. Disengage the 5 claws and remove the instrument panel safety pad assembly.

i. Disengage the claw and remove the 5 instrument panel stays.

41. Remove heater core as necessary.

To install:

42. Installation is the reverse of the removal procedure.

43. Perform initialization.

44. Inspect the steering pad.

45. Inspect the SRS warning light.

STEERING

POWER STEERING GEAR

REMOVAL & INSTALLATION

See Figures 147 and 148.

1. Drain and recycle the engine coolant.
2. Disconnect the negative battery cable.
3. Remove instrument panel.

22140_ES35_G0125

Fig. 147 Removing the connector, clamp and wire harness—TMMK

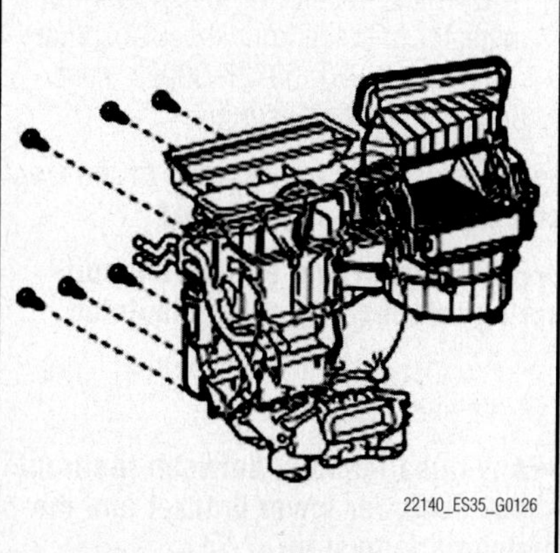

Fig. 148 Removing the 6 screws and blower assembly—TMMK

4. For TMC made:
   a. Disconnect the connector.
   b. Remove the 2 screws and blower assembly.
   c. Remove the 3 screws and blower with fan motor sub-assembly.
5. For TMMK made:
   a. Remove cooler expansion valve.
   b. Remove the connector and clamp, and disconnect the wire harness.
   c. Remove the 6 screws and then the blower assembly with the cooler evaporator sub-assembly.
   d. Remove the 3 screws and blower with fan motor sub-assembly.
6. To install, reverse removal procedure.

POWER STEERING PUMP

REMOVAL & INSTALLATION

See Figures 149 through 152.

1. Before servicing the vehicle, refer to the precautions section.
2. Drain power steering fluid.
3. Remove RH engine under cover.
4. Remove RH front fender apron seal.
5. Remove v-bank cover sub-assembly.
6. Remove fan and generator v belt.

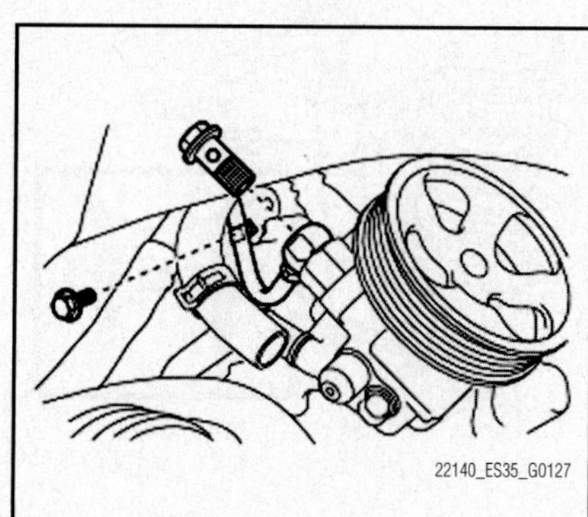

Fig. 149 Removing the union bolt

7. Slide the clip and disconnect the No. 1 fluid reservoir to pump hose from the vane pump assembly.
8. Disconnect pressure feed tube assembly, as follows:
   a. Remove the union bolt and disconnect the pressure feed tube assembly from the vane pump assembly.

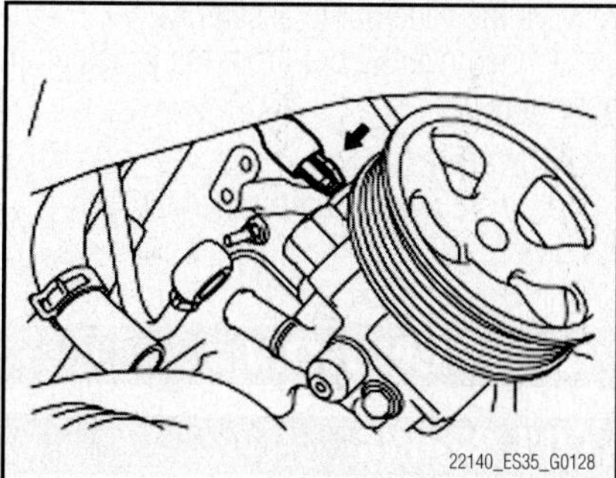

Fig. 150 Disconnecting the power steering fluid pressure switch connector

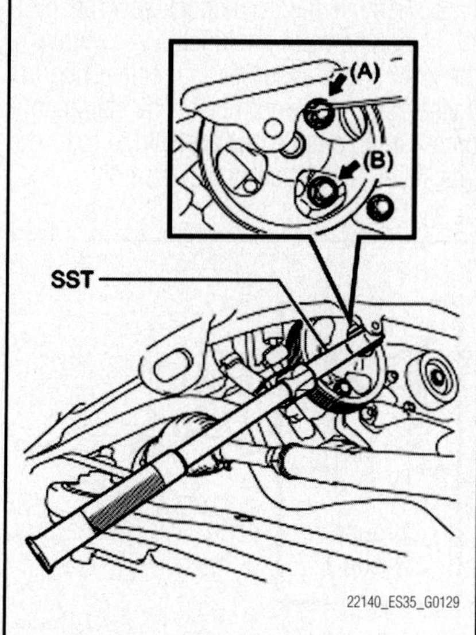

Fig. 151 Removing the vane pump assembly

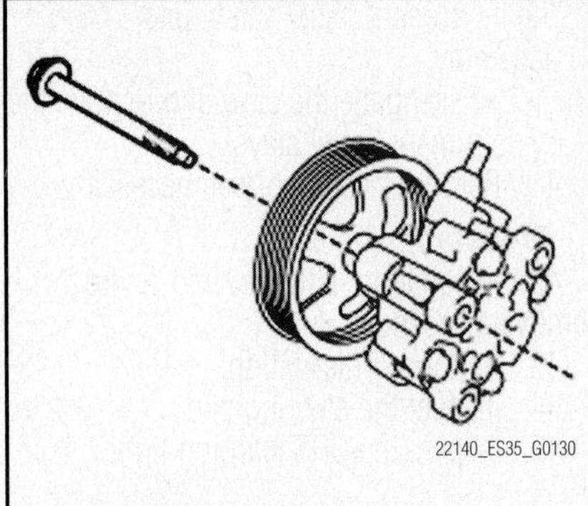

Fig. 152 Removing the bolt from the vane pump assembly

b. Remove the bolt and separate the pressure feed tube clamp.

c. Remove the gasket from the pressure feed tube assembly.

9. Disconnect the power steering fluid pressure switch connector.

10. Using SST: 09249-63010, loosen bolt (A) and remove bolt (B), and then remove the vane pump assembly.

11. Remove the bolt from the vane pump assembly.

To install:

12. Install vane pump assembly, as follows:

a. Temporarily install the bolt to the vane pump assembly.

b. Install the vane pump assembly.

➡ Use a torque wrench with a fulcrum length of 11.81 inches (300 mm).

➡This torque value is effective when SST is parallel to the torque wrench.

c. Using SST: 09249-63010, tighten the 2 bolts to 32 ft. lbs. (43 Nm).

13. Connect the connector to the power steering fluid pressure switch.

14. Connect pressure feed tube assembly, as follows:

a. Install a new gasket to the pressure feed tube assembly.

b. Temporarily connect the pressure feed tube assembly to the vane pump assembly with the union bolt.

c. Install the pressure feed tube assembly clamp with the bolt. Tighten to 87 ft. lbs. (10 Nm).

d. Fully tighten the union bolt and tighten to 37 ft. lbs. (50 Nm).

➡Make sure that the stopper of the pressure feed tube assembly contacts the vane pump assembly securely.

15. Connect No. 1 fluid reservoir to pump hose, as follows:

➡Connect the No. 1 oil reservoir to pump hose with the paint mark facing toward the rear of the vehicle.

➡Push the No. 1 oil reservoir to pump hose as far as it will go as shown in the illustration.

➡Install the clip at the position specified in the illustration.

a. Connect the No. 1 fluid reservoir to pump hose to the vane pump assembly with the clip.

16. To complete installation, reverse removal procedure.

BLEEDING

1. Before servicing the vehicle, refer to the precautions section.
2. Check the fluid level.
3. Jack up the front of the vehicle and support it with stands.
4. With the engine stopped, turn the wheel slowly from lock to lock several times.
5. Lower the vehicle.
6. Start the engine.

SUSPENSION

FRONT SUSPENSION

COIL SPRING

REMOVAL & INSTALLATION

See Figures 153 through 158.

1. Before servicing the vehicle, refer to the precautions section.
2. Remove the front shock absorber.
3. As shown in the illustration, secure the front shock absorber with coil spring in a vise using aluminum plates by clamping onto a double nutted bolt affixed to the bracket at the bottom of the absorber.

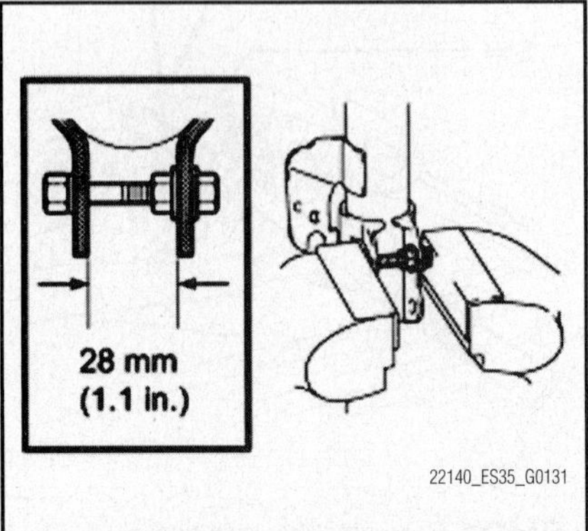

Fig. 153 Secure the front shock absorber

➡ Do not use an impact wrench.

➡ If the front coil spring is compressed at an angle, using 2 SST will make the work easier.

4. Using SST: 09727-30021, compress the front coil spring.

5. Remove the front suspension support sub-assembly, front suspension support bearing, front coil spring upper seat, front coil spring upper insulator, front coil spring, front spring bumper, and front coil spring lower insulator from the front shock absorber.

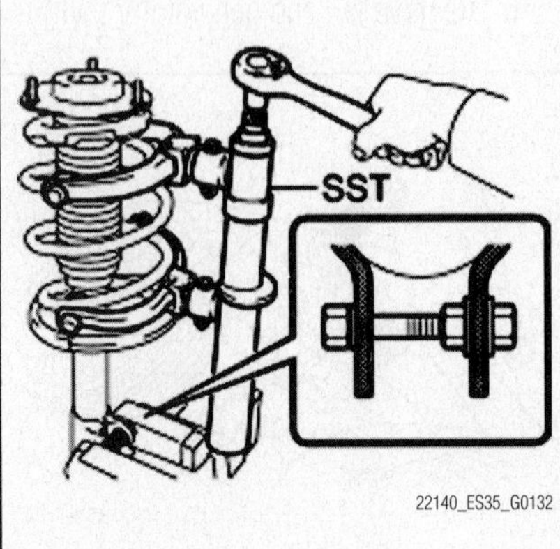

Fig. 154 Removing coil spring components

To install:

6. Install front coil spring as follows:

a. Install the front spring bumper to the piston rod.

➡Align the 2 protrusions of the front coil spring lower insulator and the 2 holes in the front shock absorber.

➡Do not use an impact wrench.

b. Install the front coil spring lower insulator onto the front shock absorber.

c. Using SST: 09727-30021, compress the front coil spring.

➡The smaller diameter end of the front coil spring must face upward.

➡Fit the lower end of the front coil spring into the gap of the insulator.

d. Install the front coil spring to the front shock absorber.

➡Any misalignment between the front shock absorber lower bracket and the matchmark must be +/-5°.

e. Install the front coil spring upper insulator as shown in the illustration.

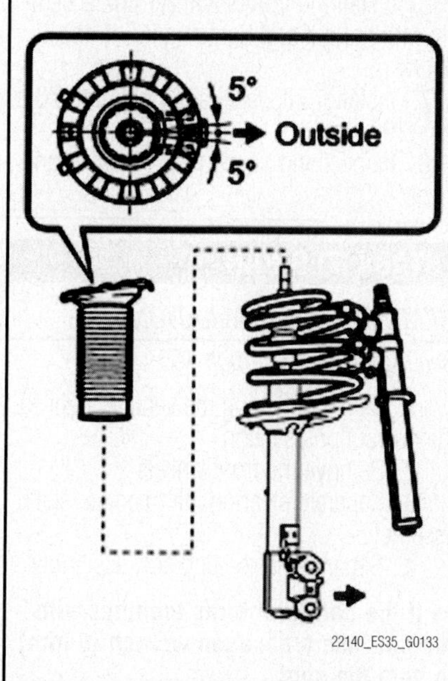

Fig. 155 Installing the front coil spring upper insulator

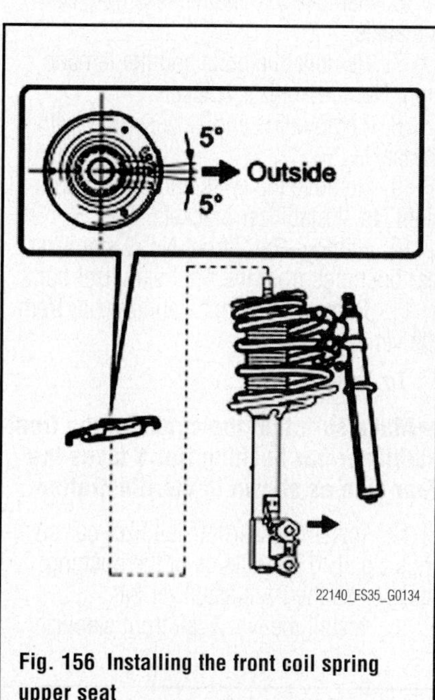

Fig. 156 Installing the front coil spring upper seat

➡Any misalignment between the front shock absorber lower bracket and the matchmark must be +/-5°.

7. Install the front coil spring upper seat with the mark facing to the outside of the vehicle.

➡ If there is foreign matter inside the front suspension support bearing, replace it with a new one.

a. Install a new front suspension support bearing.

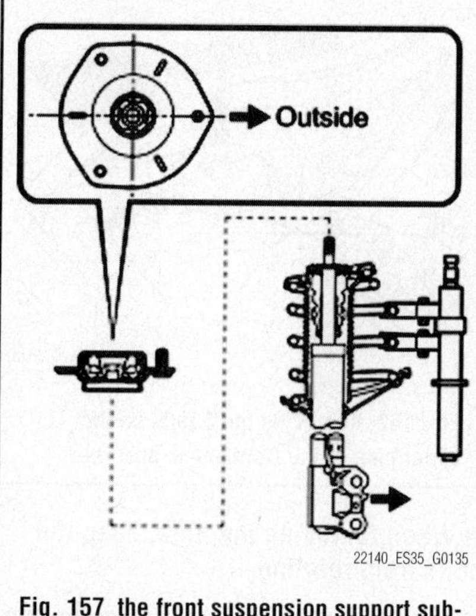

Fig. 157 the front suspension support sub-assembly

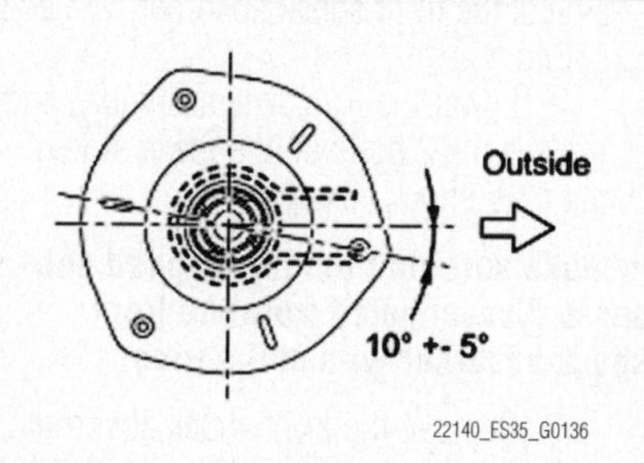

Fig. 158 Aligning the front shock absorber lower bracket and arrows

➡ Check that the flats on the piston rod and the flats on the front suspension support sub-assembly are aligned.

b. Install the front suspension support sub-assembly. Temporarily tighten a new lock nut.

➡ Do not use an impact wrench.

➡Any misalignment between the front shock absorber lower bracket and the matchmark must be +/-5°.

c. Remove the SST slowly in order to release the coil spring.

LOWER BALL JOINT

REMOVAL & INSTALLATION

See Figure 159.

1. Before servicing the vehicle, refer to the precautions section.
2. Remove the front wheel.
3. Remove the front axle hub nut.
4. Separate the front speed sensor.
5. Separate the front disc the brake caliper assembly.

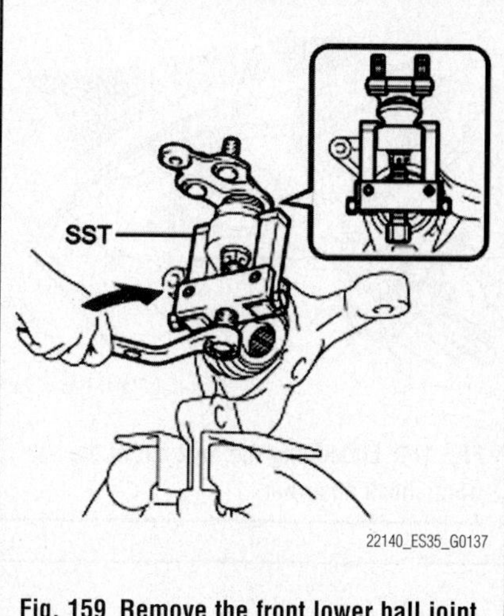

Fig. 159 Remove the front lower ball joint assembly

6. Remove front disc.
7. Separate the tie rod assembly.
8. Separate the No. 1 front lower suspension arm.
9. Remove the front axle assembly.
10. Remove front wheel No. 1 bearing dust deflector.
11. Remove front axle hub hole snap ring.
12. Remove front axle hub.
13. Remove front disc brake dust cover.
14. Remove the front lower ball joint assembly, as follows:
    a. Secure the steering knuckle in a vise using aluminum plates.
    b. Remove the cotter pin and castle nut.

➡Do not damage the dust cover of the ball joint.

➡Do not damage the steering knuckle.

c. Using SST (SST: 09628-62011) or equivalent, remove the front lower ball joint assembly.

To install:

15. Installation is the reverse of the removal procedure, noting the following:
    a. Install the front lower ball joint assembly to the steering knuckle with the castle nut and tighten to 91 ft. lbs. (123 Nm). Further tighten the nut up to 60° if the holes for the cotter pin are not aligned.
    b. Inspect and adjust the front wheel alignment.
    c. Inspect the ABS speed sensor signal.

SHOCK ABSORBERS

REMOVAL & INSTALLATION

See Figures 160 through 162.

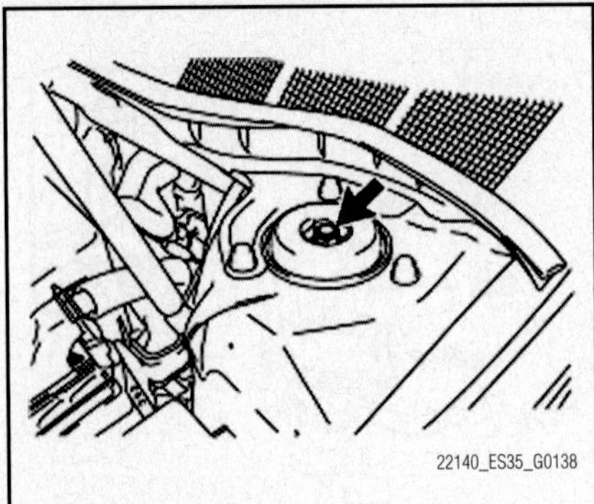

Fig. 160 Loosening the lock nut of the front shock absorber

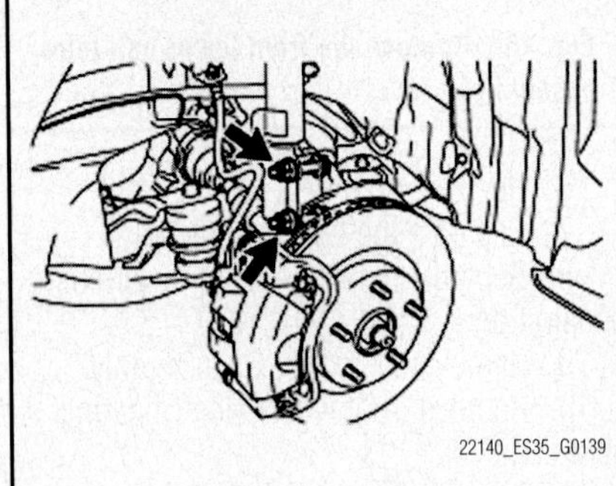

Fig. 161 Remove the 2 nuts on the lower side of the front shock absorber

➡Use the same procedures for the RH side and the LH side. The procedures listed below are for the LH side.

1. Before servicing the vehicle, refer to the precautions section.
2. Remove the front wheel.
3. Remove the nut and disconnect the front stabilizer link assembly from the front shock absorber assembly.
4. Remove front shock absorber with coil spring, as follows:
   a. Loosen the lock nut of the front shock absorber with coil spring.

➡Do not remove the lock nut.

➡Only loosen the nut when disassembling the front shock absorber with coil spring.

   b. Remove the bolt and disconnect the front flexible hose and front speed sensor wire harness from the front shock absorber with coil spring.

➡Be sure to remove the front speed sensor from the front shock absorber with coil spring.

   c. Remove the 2 nuts on the lower side of the front shock absorber with coil spring.

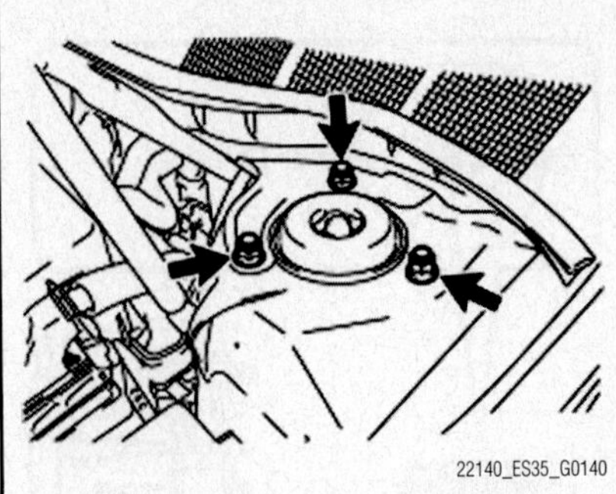

Fig. 162 Removing the 3 nuts on the upper side of the front shock absorber

➡When removing the nuts, keep the bolts from rotating.

➡Keep the bolts inserted to secure the front axle assembly.

   d. Remove the 3 nuts on the upper side of the front shock absorber with coil spring.
   e. Lower the front axle assembly, and remove the 2 bolts on the lower side of the front shock absorber.

➡Make sure that the front speed sensor is disconnected from the front shock absorber with coil spring.

   f. Remove the front shock absorber with coil spring.

To install:

5. Install front shock absorber with coil spring, as follows:
   a. Install the front shock absorber with coil spring to the front axle assembly and insert the 2 bolts from the front side of the vehicle.
   b. Slowly jack up the vehicle using a wooden block and install the front shock absorber with coil spring (upper side) to the vehicle.
   c. Install the 3 nuts to the upper side of the front shock absorber with coil spring and tighten to 63 ft. lbs. (85 Nm).

➡When installing the nuts, keep the bolts from rotating.

   d. Install the 2 nuts to the lower side of the front shock absorber with coil spring and tighten to 155 ft. lbs. (210 Nm).
   e. Install the front flexible hose and front speed sensor wire harness with the bolt and tighten to 14 ft. lbs. (19 Nm).
   f. Fully tighten the lock nut and tighten to 52 ft. lbs. (70 Nm).

➡If the ball joint turns together with the nut, use a hexagon wrench (6 mm) to hold the stud.

6. Install the front stabilizer link assembly with the nut and tighten to 55 ft. lbs. (74 Nm).
7. Install front wheel and tighten to 76 ft. lbs. (103 Nm).
8. Inspect and adjust front wheel alignment.

STABILIZER BAR

REMOVAL & INSTALLATION

See Figures 163 through 165.

1. Before servicing the vehicle, refer to the precautions section.
2. Remove the front wheels.
3. Separate steering intermediate shaft assembly.
4. Separate tie rod end sub-assembly.

➡ If the ball joint turns together with the nut, use a hexagon wrench (6 mm) to hold the stud.

5. Remove the 2 nuts and the front stabilizer link assembly
6. Remove the engine assembly with transaxle.
7. Remove the bolts and the left and right No. 1 stabilizer brackets.
8. Remove the engine assembly with transaxle.
9. Remove the bolts and the left and right No. 1 stabilizer brackets.
10. Remove the 2 front No. 1 stabilizer bar bushings from the front stabilizer bar.
11. Remove the front stabilizer bar from the vehicle.

To install:

➡Make sure that the cutout of the front stabilizer bar bushing No. 1 faces the rear side as shown in the illustration.

12. Install the 2 front stabilizer bar bushings No. 1 to the outside of the bushing stopper on the front stabilizer bar.
13. Install the No. 1 left front stabilizer

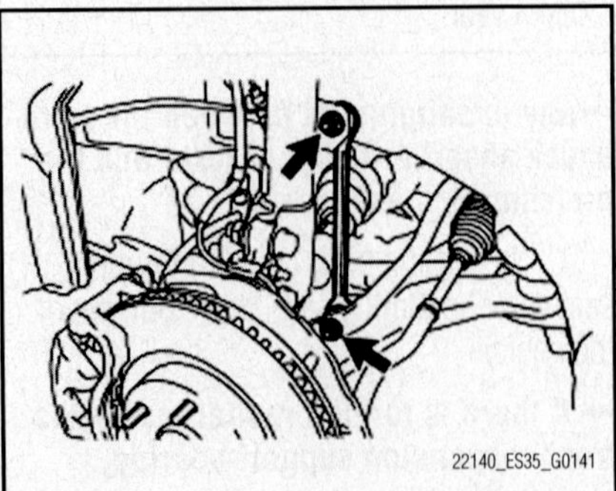

Fig. 163 Remove the 2 nuts and front stabilizer link assembly (left hand shown)

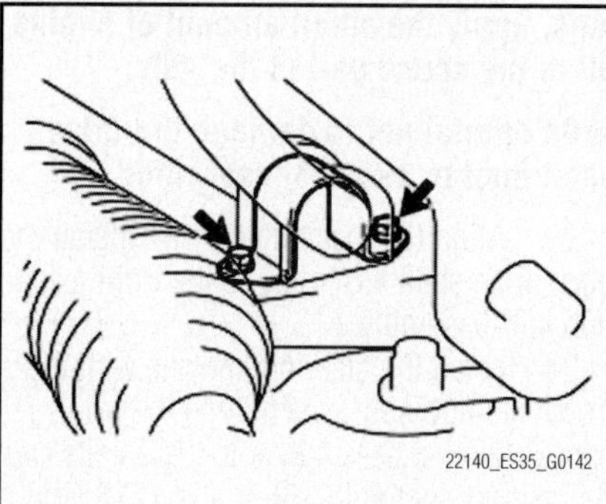

Fig. 164 Remove the 2 bolts and No. 1 stabilizer bracket (left hand shown)

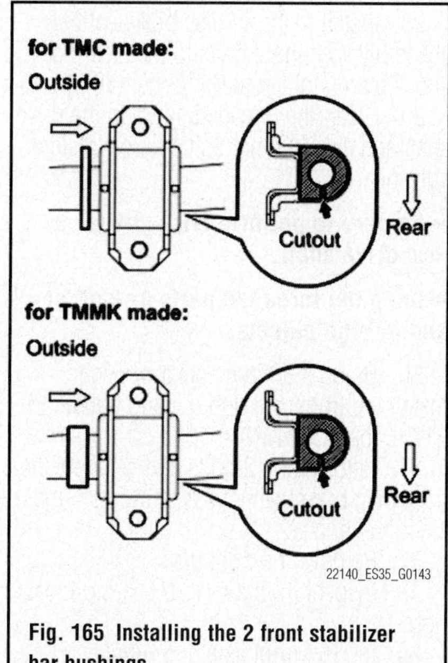

Fig. 165 Installing the 2 front stabilizer bar bushings

bracket with the 2 bolts and tighten to 20 ft. lbs. (27 Nm).

14. Install the No. 1 right front stabilizer bracket with the 2 bolts and tighten to 20 ft. lbs. (27 Nm).
15. Install the engine assembly with transaxle.
16. Install the left front stabilizer link assembly with the 2 nuts and tighten to 55 ft. lbs. (74 Nm).
17. Install the right front stabilizer link assembly with the 2 nuts and tighten to 55 ft. lbs. (74 Nm).
18. To complete installation, reverse removal procedure.
19. Inspect and adjust the front wheel alignment.

STEERING KNUCKLE

REMOVAL & INSTALLATION

See Wheel Bearings.

WHEEL BEARINGS

REMOVAL & INSTALLATION

See Figures 166 through 172.

1. Before servicing the vehicle, refer to the Precautions Section.
2. Remove front wheel.
3. Remove front axle hub nut.
4. Separate front speed sensor.
5. Remove the 2 bolts and separate the front disc brake caliper assembly from the steering knuckle. Use wire or an equivalent tool to keep the brake caliper from hanging down by the flexible hose.
6. Remove front disc.
7. Separate tie rod end sub-assembly.
8. Separate front suspension lower no. 1 arm.
9. Remove front axle assembly.
10. Using a screwdriver with its tip wrapped with vinyl tape, remove the No. 1 front wheel bearing dust deflector. Be careful not to damage the steering knuckle.
11. Using snap ring pliers, remove the front axle hub hole snap ring.

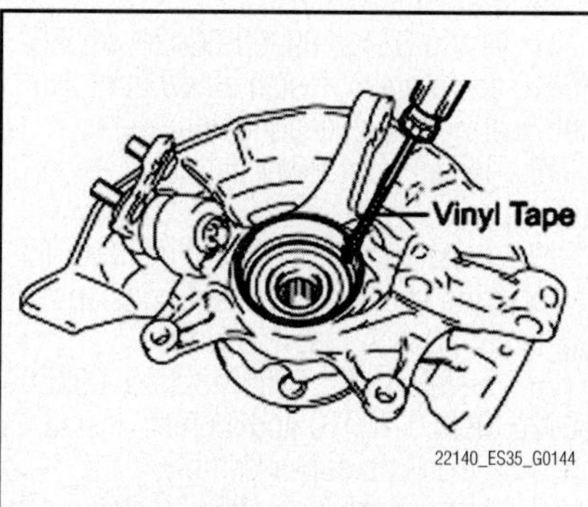

Fig. 166 remove the No. 1 front wheel bearing dust deflector

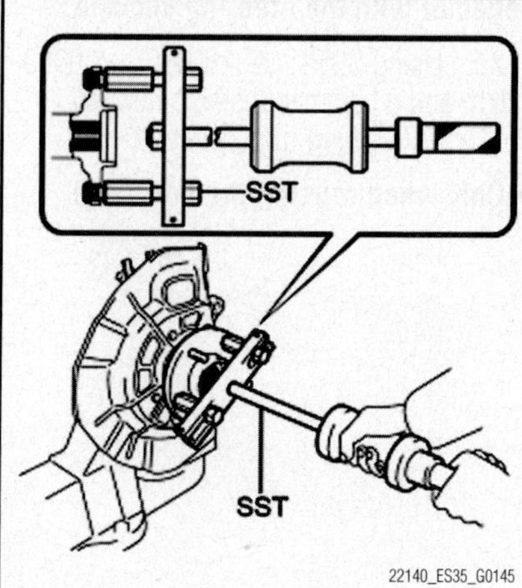

Fig. 167 Remove the front axle hub sub-assembly

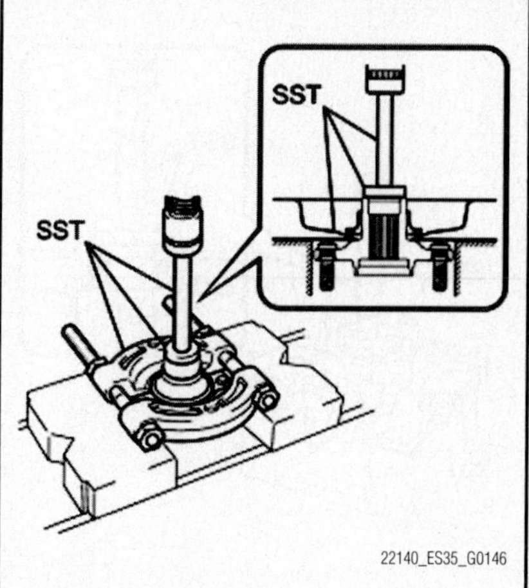

Fig. 168 Remove the bearing inner race (outside) from the front axle hub sub-assembly

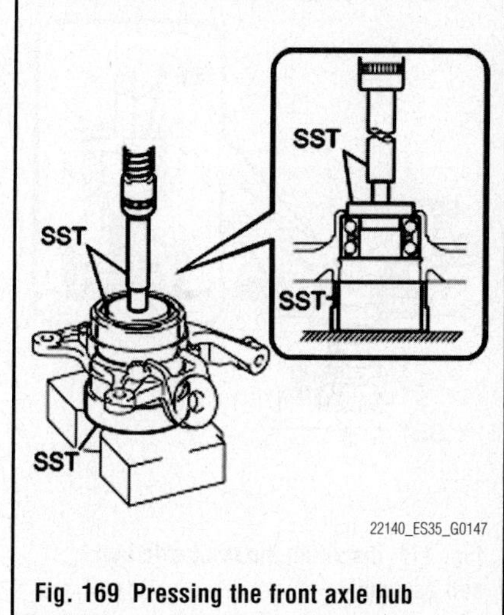

Fig. 169 Pressing the front axle hub bearing

12. Remove front axle hub sub-assembly by performing the following:
    a. Hold the front axle assembly between aluminum plates in a vise.

➡Do not overtighten the vise.

    b. Using SST 09520-00031, remove the front axle hub sub-assembly.

➡Be careful not to drop the front axle hub sub-assembly.

    c. Using SST 09555-55010, SST: 09950-60010 and SST: 09950-70010 and a press, remove the bearing inner race (outside) from the front axle hub sub-assembly.

13. Remove the 4 bolts and disc brake dust cover from the steering knuckle.
14. Remove front lower ball joint assembly.

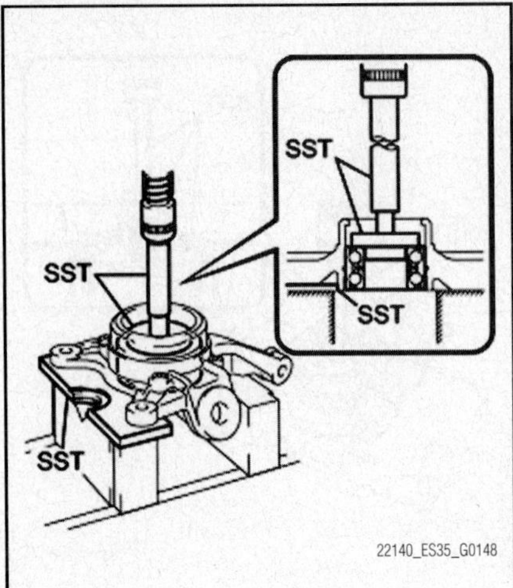

Fig. 170 Removing the front axle hub bearing

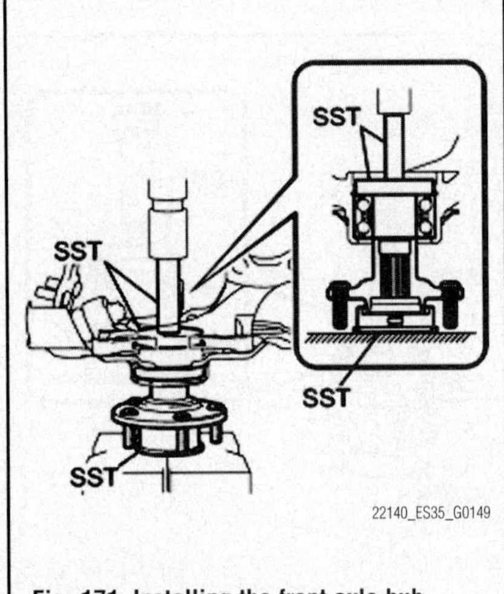

Fig. 171 Installing the front axle hub sub-assembly

15. Remove front axle hub bearing by performing the following:

a. Place the bearing inner race (outside) on the front axle hub bearing.

b. Using SST 09527-17011, SST: 09950-60010 and a press, press the front axle hub bearing until it contacts the SST: 09950-70010.

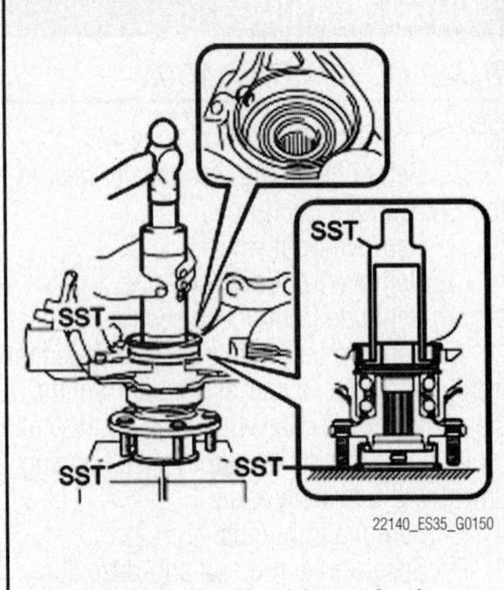

Fig. 172 Installing No. 1 front wheel bearing dust deflector

c. Using SST: 09527-20011, SST: 09950-60010 to make the steering knuckle horizontal, fix it to the V-block.

16. Using SST: 09950-70010 and a press, remove the front axle hub bearing from the steering knuckle.

To install:

17. Using SST's: 09950-60020, 09950-70010 and a press, install a new front axle hub bearing to the steering knuckle.

18. Install front lower ball joint assembly.

19. Install the disc brake dust cover to the steering knuckle with the 4 bolts and tighten to 73 inch lbs. (8.3 Nm).

20. Using SST's: 09608-32010, 09950-60020, 09950-70010 and a press, install the front axle hub sub-assembly.

21. Using snap ring pliers, install a new front axle hub hole snap ring.

➡ Align the hole for the speed sensor in the No. 1 front wheel bearing dust deflector with the steering knuckle.

22. Using SST's: 09316-60011, 09608-32010 and a hammer, install a new No. 1 front wheel bearing dust deflector.

➡Only when reusing the bolts and nuts, apply the small amount of engine oil to the screw part of the nuts.

➡Be careful not to damage the drive shaft boot or speed sensor rotor.

23. Align the matchmarks and install the front drive shaft assembly to the front axle hub sub-assembly.

24. Install the steering knuckle with the front axle hub sub-assembly to the front shock absorber assembly with the 2 bolts and 2 nuts and tighten to 155 ft. lbs. (210 Nm).

25. Install the lower No. 1 front suspension arm sub-assembly.

26. Install the tie rod end sub-assembly.

27. Install the front disc.

28. Install the front disc brake caliper assembly with the 2 bolts to the steering knuckle and tighten to 79 ft. lbs. (107 Nm).

29. Clean the threaded parts on the drive shaft and axle hub nut using a non-residue solvent.

➡Be sure to perform this work for a new drive shaft.

➡Keep the threaded parts free of oil and foreign objects.

30. Using a 30 mm socket wrench, install the front axle hub nut and tighten to 217 ft. lbs. (294 Nm).

31. Remove the 2 bolts and separate the front disc brake caliper assembly from the steering knuckle.

32. Remove the front disc.

33. Inspect front axle hub bearing looseness.

34. Inspect front axle hub runout.

35. Install the front disc.

36. Install the front disc brake caliper assembly with the 2 bolts to the steering knuckle and tighten to 79 ft. lbs. (107 Nm).

37. Install the front speed sensor.

38. Using a chisel and hammer, stake the axle hub nut.

39. Install the front wheel.

40. Inspect and adjust front wheel alignment.

41. Check ABS speed sensor signal.

COIL SPRING

REMOVAL & INSTALLATION

See Figures 173 through 175.

1. Before servicing the vehicle, refer to the precautions section.
2. Secure the rear shock absorber with coil spring in a vise using aluminum plates by closing the vise onto the double nutted bolt affixed to the bracket at the bottom of the absorber.

➡Do not use an impact wrench.

➡If the rear coil spring is compressed at an angle, using 2 SST will make the work easier.

3. Using SST: 09727-30021, compress the rear coil spring
4. Remove the nut, rear shock absorber collar and rear suspension support assembly.
5. Remove the rear coil spring, rear No. 1 spring bumper, and rear coil spring lower insulator.

To install:

6. Install the rear No. 1 spring bumper to the piston rod.
7. Install the rear coil spring lower insulator onto the rear shock absorber.

➡Do not use an impact wrench.

8. Using SST: 09727-30021, compress the rear coil spring.

➡The smaller diameter end must face upward.

➡Fit the lower end of the rear coil spring into the gap of the lower seat.

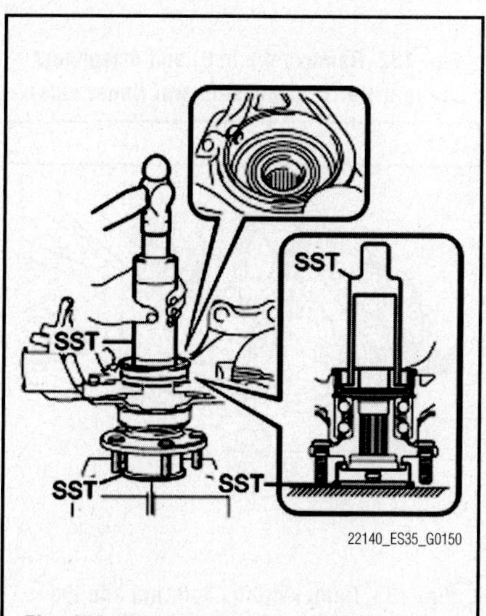

Fig. 173 Remove the rear coil spring

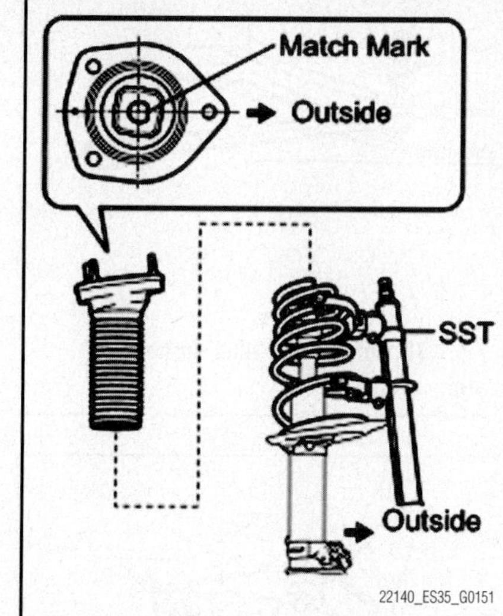

Fig. 174 Align the notches of the shock absorber

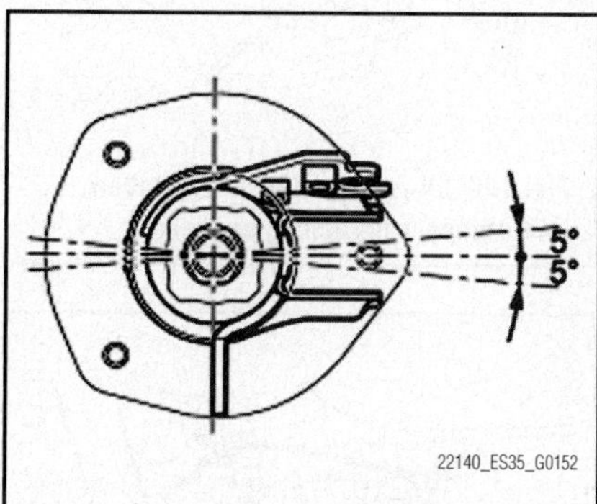

Fig. 175 Lining up the rear suspension support assembly's stud bolts

➡If the front coil spring is compressed at an angle, using 2 SST will make the work easier.

9. Install the rear coil spring to the rear shock absorber.

➡Align the notches of the piston rod and the rear suspension support assembly as shown in the illustration before installing the rear suspension support assembly.

10. Install the rear suspension support assembly.
11. Align the notches of the shock absorber with the notch of the rear suspension support assembly so that the notches face the outside of the vehicle.
12. Install the rear shock absorber collar.
13. Loosely tighten a new lock nut to the rear suspension piston rod.

➡ Do not use an impact wrench.

➡ When lining up the rear suspension support assembly's stud bolts at the middle point between the two sides of the bracket, the maximum permissible degree of error is plus or minus 5°.

14. Release the spring while adjusting the rear suspension support assembly to the position shown in the illustration, and remove the SST from the rear coil spring.

CONTROL ARMS/LINKS

REMOVAL & INSTALLATION

No. 1 Suspension Arm

See Figures 176 through 181.

1. Before servicing the vehicle, refer to the precautions section.

➡Check if an old gasket still remains on the pipe. If so, remove it. Also, check if any bolts or nuts are rusted. If so, replace them.

2. Remove rear wheel.
3. Remove center exhaust pipe assembly.
4. Remove tail exhaust pipe assembly.
5. Separate both rear stabilizer link assemblies.
6. Remove rear stabilizer bar no. 2 and no. 1 bracket.
7. Remove rear stabilizer bar.
8. Remove rear stabilizer bushing.
9. Separate rear strut rod.

➡When removing the bolt, keep the nut from rotating.

10. Remove the bolt, nut and separate the rear suspension No. 2 arm (outer side) from the rear axle carrier.

➡ When removing the bolt, keep the nut from rotating.

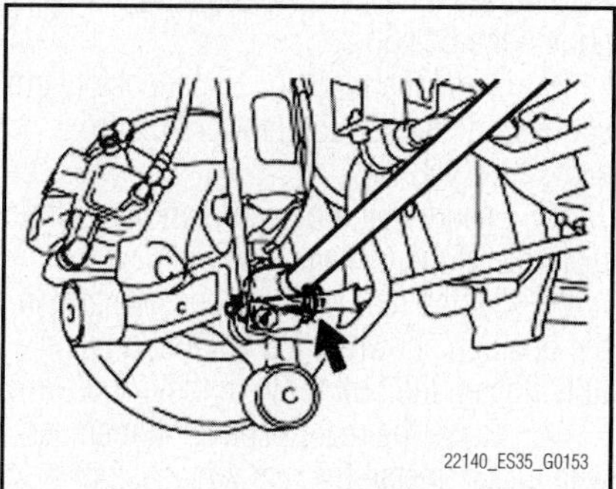

Fig. 176 Removing the bolt, nut and the rear No. 1 suspension arm—LH shown

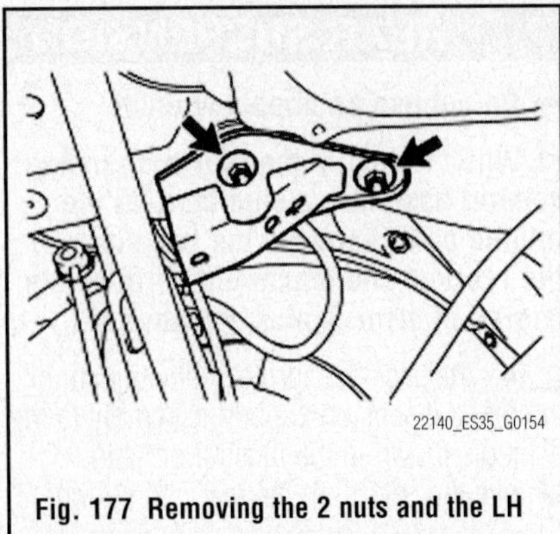

Fig. 177 Removing the 2 nuts and the LH rear suspension member lower stopper

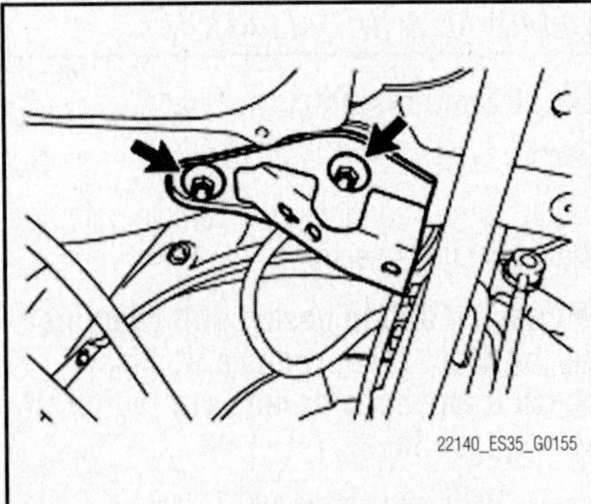

Fig. 178 Removing the 2 nuts and the RH rear suspension member lower stopper

11. Remove the bolt, nut and the rear No. 1 suspension arm (outer side) from the rear axle carrier.
12. Remove the 2 nuts and the LH rear suspension member lower stopper.
13. Remove the 2 nuts and the RH rear suspension member lower stopper.
14. Support the rear suspension member with a jack.
15. Remove the 2 bolts, and the rear suspension member sub-assembly.
16. Remove the bolt and rear No. 1 suspension arm assembly.

To install:

17. Install the No. 1 rear suspension arm (inner side) with the bolt, and temporarily tighten the bolt.
18. Install the rear No. 1 suspension arm so that the bracket leans toward the front side of the vehicle.
19. Ensure that the paint mark faces the rear side of the vehicle.
20. Set the rear No.1 suspension arm in the position shown in the illustration, and fully tighten the bolt to 74 ft. lbs. (100 Nm).
21. Raise the rear suspension member with a jack. Install the rear suspension member with the 2 bolts and tighten to 41 ft. lbs. (56 Nm).
22. Install both the rear suspension member lower stoppers with the 2 nuts and tighten to:
    a. Nut A: 41 ft. lbs. (55 Nm).
    b. Nut B: 28 ft. lbs. (38 Nm).

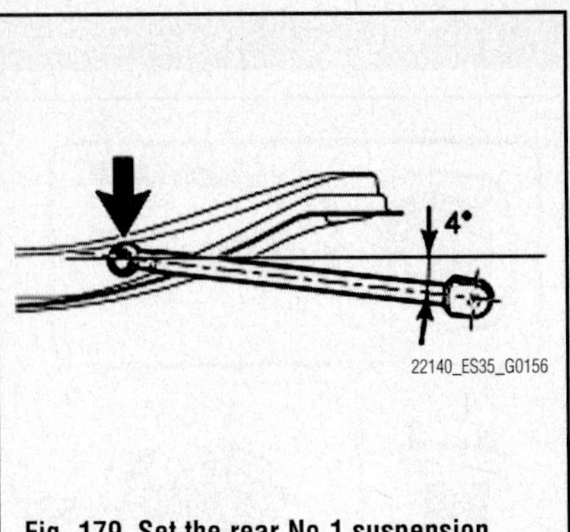

Fig. 179 Set the rear No.1 suspension arm

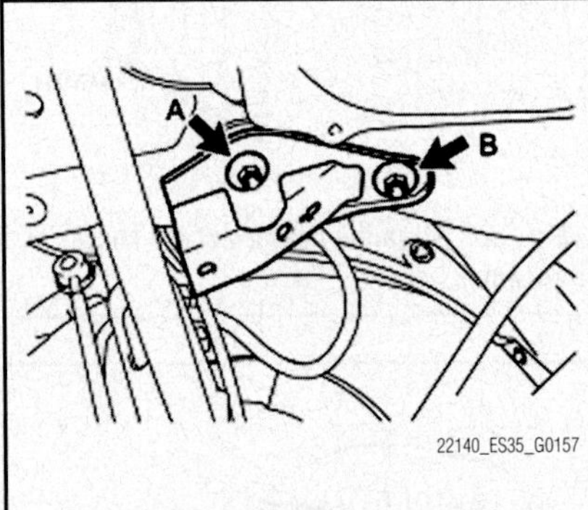

Fig. 180 LH rear suspension member lower stopper tightening sequence

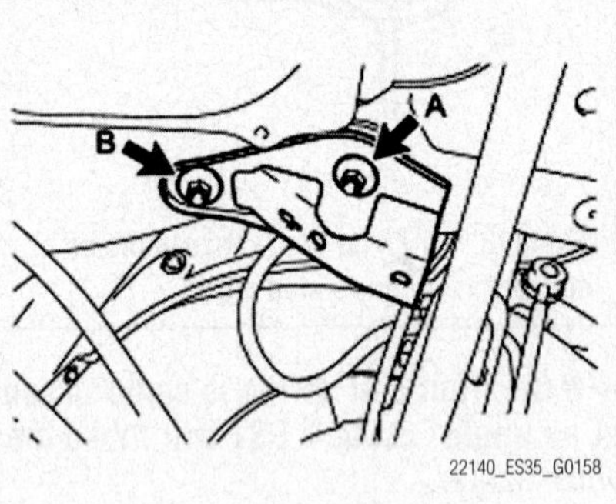

Fig. 181 LH rear suspension member lower stopper tightening sequence

➡Insert the bolt from the front of the vehicle and temporarily install the bolt.

23. Connect the rear No.1 suspension arm (outer side) to the rear axle carrier with the bolt and nut and temporarily tighten the bolt and nut. When temporarily tightening the bolt, keep the nut from rotating.

➡Insert the bolt from the inside of the vehicle and temporarily install the bolt.

24. Connect the strut rod assembly rear to the axle carrier with the bolt and nut and temporarily tighten the bolt. When temporarily tightening the bolt, keep the nut from rotating.
25. Jack up the rear axle carrier, placing a wooden block to avoid damage. Apply load to the suspension so that the installed bolt of the rear No. 1 suspension arm (inner side) is horizontally aligned with the center of the rear axle hub.
26. Fully tighten rear No. 1 suspension arm and tighten the bolt to 74 ft. lbs. (100 Nm).
27. Fully tighten rear No. 2 suspension arm and tighten the bolt to 74 ft. lbs. (100 Nm).
28. To complete installation, reverse removal procedure/

No. 2 Suspension Arm

See Figures 182 and 183.

1. Before servicing the vehicle, refer to the precautions section.
2. Remove the rear wheel.
3. Remove the bolt, and disconnect the rear No. 2 suspension arm (inner side).

➡ When removing the bolt, keep the nut from rotating.

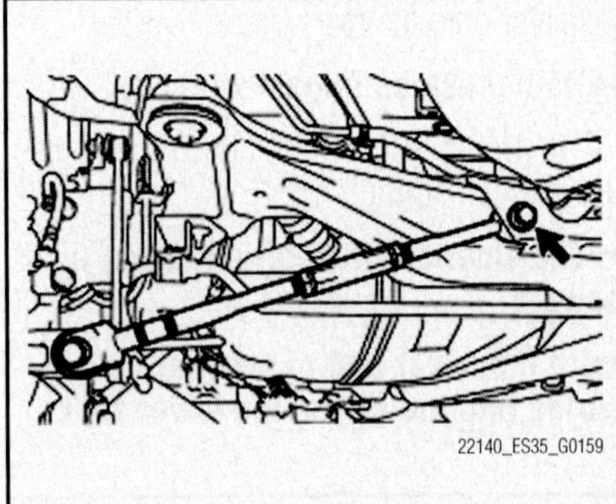

Fig. 182 Remove the bolt, and disconnect the rear No. 2 suspension arm (inner side)

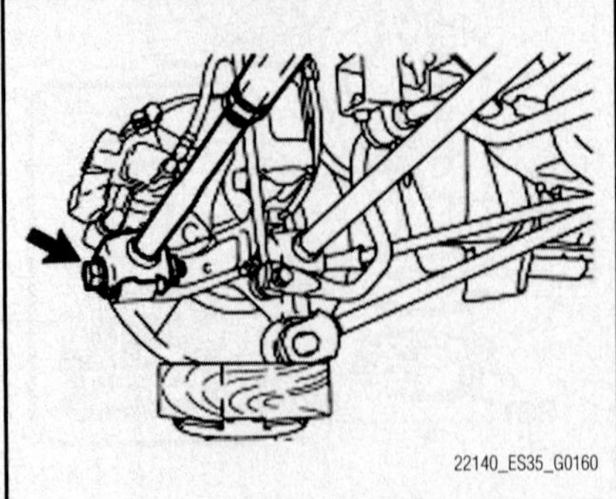

Fig. 183 Removing the bolt, nut and the rear No. 2 suspension arm (outer side)

4. Remove the bolt, nut and the rear No. 2 suspension arm (outer side) from the rear axle carrier.

To install:

➡Ensure that the paint mark faces to the rear of the vehicle.

5. Install the rear No. 2 suspension arm (inner side) with the bolt, and temporarily tighten the bolt.

➡When temporarily tightening the bolt, keep the nut from rotating.

6. Connect the rear No. 2 suspension arm (outer side) to the rear axle carrier with the bolt and nut, and temporarily tighten the bolt.
7. Stabilize suspension.
8. Fully tighten the rear No. 2 suspension arm bolt (inner side) to 74 ft. lbs. (100 Nm).
9. Fully tighten the rear No. 2 suspension arm bolt (outer side) to 74 ft. lbs. (100 Nm).
10. Install the rear wheel.
11. Inspect and adjust the rear wheel alignment.

SHOCK ABSORBER

REMOVAL & INSTALLATION

See Figures 184 and 185.

1. Before servicing the vehicle, refer to the precautions section.
2. Remove the rear seat cushion assembly.
3. Remove rear seat headrest plate cover.
4. Remove rear seat headrest assembly.
5. Remove the rear seatback assembly.
6. Remove the rear wheel.
7. Separate LH rear stabilizer link assembly.
8. Remove the 2 bolts, and disconnect the rear brake flexible hose and rear speed sensor from the rear shock absorber with coil spring and rear axle carrier.
9. Remove the 4 claws and the rear suspension support No. 1 cover.
10. Remove the rear shock absorber with coil spring, as follows:

➡Do not remove the lock nut.

➡Only loosen the nut when disassembling the rear shock absorber with coil spring.

a. Loosen the lock nut of the rear shock absorber with coil spring.

➡When removing the nuts, keep the bolts from rotating.

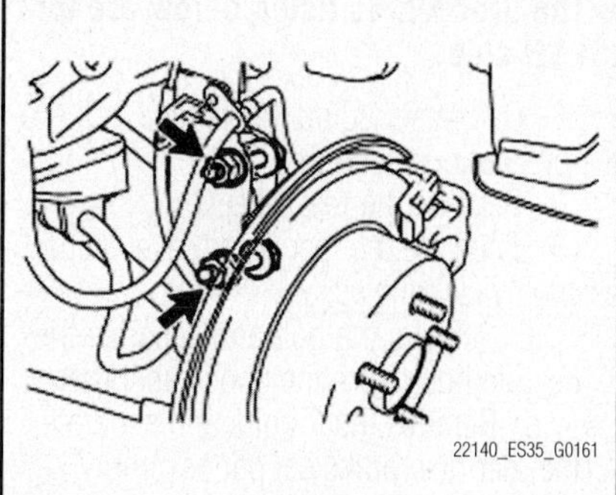

Fig. 184 Loosen the 2 nuts on the lower side of the shock absorber

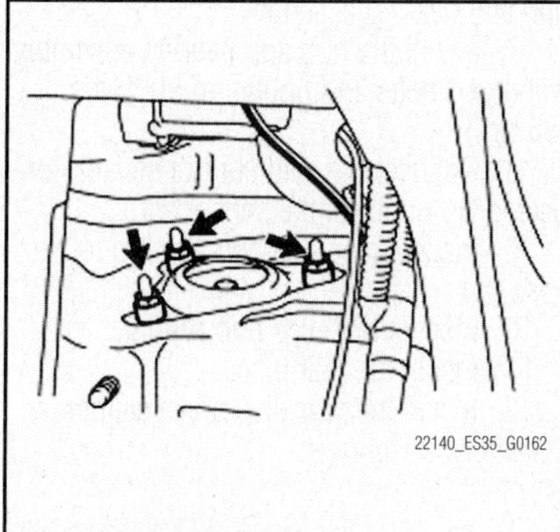

Fig. 185 Remove the 3 nuts

➡Keep one bolt inserted to secure the hub and disc rotor.

b. Remove the 2 nuts and 2 bolts on the lower side of the rear shock absorber with coil spring.

c. Remove the 3 nuts on the upper side of the rear shock absorber with coil spring.

➡Make sure that the rear speed sensor is disconnected from the rear shock absorber with coil spring.

d. Lower the rear axle carrier, and remove the 2 bolts on the lower side of the rear shock absorber with coil spring.

To install:

11. Install the rear shock absorber with coil spring to the rear axle carrier assembly and insert the 2 bolts from the rear of the vehicle.
12. Slowly jack up the vehicle using a wooden block and install the rear shock absorber with coil spring (upper side) to the vehicle.
13. Install the 3 nuts to the upper side of the rear shock absorber with coil spring and tighten to 29 ft. lbs. (39 Nm).

➡When installing the nuts, keep the bolts from rotating.

14. Install the 2 nuts and 2 bolts to the lower side of the rear shock absorber with coil spring and tighten 133 ft. lbs. (180 Nm).
15. Fully tighten the lock nut to 41 ft. lbs. (55 Nm).
16. Connect rear speed sensor.
17. Install the LH rear stabilizer link assembly.
18. Engage the 4 claws and install the rear suspension support No. 1 cover.
19. To complete installation, reverse remaining removal.
20. Check abs speed sensor signal.
21. Inspect and adjust rear wheel alignment.

STABILIZER BAR

REMOVAL & INSTALLATION

See Figures 186 and 187.

1. Before servicing the vehicle, refer to the precautions section.
2. Remove rear wheels.
3. Remove tail exhaust pipe assembly.
4. Center exhaust pipe assembly.
5. Remove rear stabilizer link assembly.
6. Remove the 2 bolts and rear stabilizer bar No. 2 bracket.

Fig. 186 Removing the 2 bolts and No. 2 bracket

Fig. 187 Removing the 2 bolts and No. 1 bracket

7. Remove the 2 bolts and rear stabilizer bar No. 1 bracket.
8. Remove the 2 rear stabilizer bushings from the rear stabilizer bar.
9. Remove rear stabilizer bar.

To install:

10. Install the 2 rear stabilizer bushings to the outside of the stopper ring on the stabilizer bar.
11. Install the rear stabilizer bar No. 2 and No. 1 bracket.
12. Install the rear stabilizer bar with the 2 bolts and tighten to 23 ft. lbs. (31 Nm).
13. Install rear stabilizer link assembly.
14. To complete installation, reverse remaining removal.
15. Check abs speed sensor signal.
16. Inspect and adjust rear wheel alignment.

WHEEL BEARINGS

REMOVAL & INSTALLATION

See Figure 188.

➡**Use the same procedures for the RH side and LH side.**

➡**The procedures listed below are for the LH side.**

1. Before servicing the vehicle, refer to the precautions section.
2. Remove the rear wheel.
3. Separate the rear disc brake caliper assembly, as follows:
   a. Remove the bolt and separate the flexible hose from the shock absorber.
   b. Remove the 2 bolts and separate the rear disc brake caliper assembly.
4. Remove the rear disc.
5. Disconnect the skid control sensor connector.
6. Remove the 4 bolts and the rear axle hub and bearing assembly.
7. Install the hub and bearing assembly with the 4 bolts and tighten to 59 ft. lbs. (80 Nm).
8. Connect the skid control sensor connector. Do not twist the sensor wire.
9. Inspect rear axle hub bearing looseness.
10. Inspect rear axle hub runout.
11. Install the rear disc.
12. Install the rear disc brake caliper assembly, as follows:

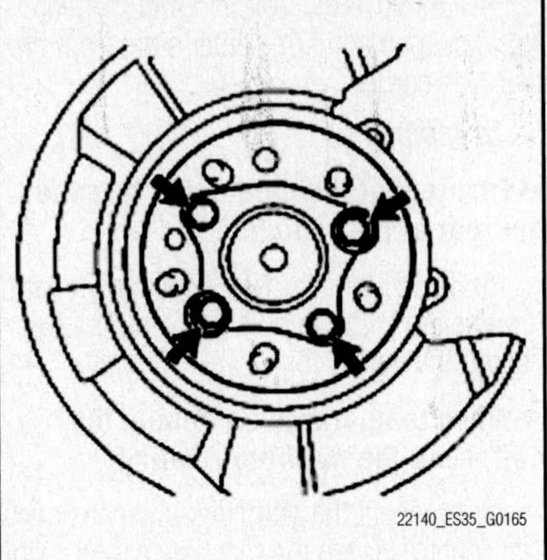

Fig. 188 Remove the 4 bolts and the rear axle hub and bearing assembly

   a. Install the rear disc brake caliper with the 2 bolts and tighten to 46 ft. lbs. (62 Nm).
   b. Install the flexible hose with the bolt and tighten to 14 ft. lbs. (19 Nm).
13. Install the rear wheel.
14. Inspect and adjust the rear wheel alignment.
15. Check ABS speed sensor signal.

LEXUS

GS350 • GS430 • GS460

2

SPECIFICATIONS AND MAINTENANCE CHARTS

ENGINE AND VEHICLE IDENTIFICATION

Engine						
Code ①	Liters (cc)	Cu. In.	Cyl.	Fuel Sys.	Engine Type	Eng. Mfg.
2GR-FSE	3.5 (3456)	211	V6	SFI D4S	DOHC	Toyota
3UZ-FE	4.3 (4293)	262	V8	SFI	DOHC	Toyota
1UR-FSE	4.6 (4608)	282	V8	SFI D4S	DOHC	Toyota

Model Year	
Code ②	Year
7	2007
8	2008

SFI: Sequential Multi-port Fuel Injection

DOHC: Double Overhead Camshaft

① Located on the timing belt cover

② 10th digit of the VIN

22140_GS35_C0001

GENERAL ENGINE SPECIFICATIONS

All measurements are given in inches.

Year	Model	Engine Displacement Liters (cc)	Engine ID/VIN	Fuel System Type	Net Horsepower @ rpm	Net Torque @ rpm (ft. lbs.)	Bore x Stroke (in.)	Com-pression Ratio	Oil Pressure @ rpm
2007	GS 350	3.5 (3456)	2GR-FSE	SFID4S	303@6400	274@4800	3.70x3.27	11.8:1	36@2500
	GS 430	4.3 (4293)	3UZ-FE	SFI	290@5600	319@3400	3.58x3.25	10.5:1	43-85@3000
2008	GS 350	3.5 (3456)	2GR-FSE	SFID4S	303@6400	274@4800	3.70x3.27	11.8:1	36@2500
	GS 460	4.6 (4608)	1UR-FSE	SFID4S	N/A	N/A	3.70x3.27	N/A	31@2500

SFI : Sequential Multi-port Fuel Injection

22140_GS35_C0002

ENGINE TUNE-UP SPECIFICATIONS

Year	Engine Displacement Liters	Engine ID/VIN	Spark Plug Gap (in.)	Ignition Timing (deg.)	Fuel Pump (psi)	Idle Speed (rpm)	Valve Clearance	
							Intake	Exhaust
2007	3.5	2GR-FSE	0.039-0.043	8-18B	①	600-700	HYD	HYD
	4.3	3UZ-FE	0.039-0.043	8-12B	44-50	700-800	0.006-0.010	0.010-0.014
2008	3.5	2GR-FSE	0.039-0.043	8-18B	①	600-700	HYD	HYD
	4.6	1UR-FSE	0.039-0.043	8-12B	①	700-800	HYD	HYD

NOTE: The Vehicle Emission Control Information label often reflects specification changes made during production. The label figures must be used if they differ from those in this chart.

B: Before top dead center

HYD: Hydraulic Valve Lifters

① Low pressure system 28-85 psi. High pressure system 508-653 psi.

22140_GS35_C0003

CAPACITIES

Year	Model	Engine Displacement Liters	Engine ID/VIN	Engine Oil with Filter	Transmission (qts.) Auto.	Transmission (qts.) Manual	Drive Axle (pts.)	Fuel Tank (gal.)	Cooling System (qts.)
2007	GS 350	3.5	2GR-FSE	①	②	—	③	18.7	9.6
	GS 430	4.3	3UZ-FE	5.4	3.6	—	2.8	18.7	10.7
2008	GS 350	6.5	2GR-FSE	①	②	—	③	18.7	9.6
	GS 460	4.6	1UR-FSE	9.1	11.1	—	2.8	18.7	11.7

NOTE: All capacities are approximate. Add fluid gradually and check to be sure a proper fluid level is obtained.

① 2WD models: 6.6 qts.
AWD models: 6.7 qts.

② 2WD models: 8.4 qts. Overhaul
AWD models: 10.6 qts. Overhaul

③ 2WD models: 2.84 pts.
AWD models front: 1.4 pts.
AWD models rear: 2.8 pts.

22140_GS35_C0004

VALVE SPECIFICATIONS

Year	Engine Displacement Liters	Engine ID/VIN	Seat Angle (deg.)	Face Angle (deg.)	Spring Test Pressure (lbs. @ in.)	Spring Free-Length (in.)	Stem-to-Guide Clearance (in.) Intake	Stem-to-Guide Clearance (in.) Exhaust	Stem Diameter (in.) Intake	Stem Diameter (in.) Exhaust
2007	3.5	2GR-FSE	NA	NA	N/A	2.035	0.0010-0.0024	0.0012-0.0026	0.2154-0.2159	0.2151-0.2157
	4.3	3UZ-FE	45	44.5	45.9-50.7@ 1.3795	2.130	0.0010-0.0024	0.0012-0.0026	0.2154-0.2159	0.2152-0.2157
2008	3.5	2GR-FSE	NA	NA	N/A	2.035	0.0010-0.0024	0.0012-0.0026	0.2154-0.2159	0.2151-0.2157
	4.6	1UR-FSE	45	44.5	N/A	2.035	0.0010-0.0024	0.0012-0.0026	0.2154-0.2159	0.2152-0.2157

NA: Not Available

22140_GS35_C0005

CRANKSHAFT AND CONNECTING ROD SPECIFICATIONS

All measurements are given in inches.

Year	Engine Displacement Liters	Engine ID/VIN	Crankshaft Main Brg. Journal Dia.	Crankshaft Main Brg. Oil Clearance	Crankshaft Shaft End-play	Crankshaft Thrust on No.	Connecting Rod Journal Diameter	Connecting Rod Oil Clearance	Connecting Rod Side Clearance
2007	3.5	2GR-FSE	2.4011-2.4016	0.0018-0.0026	0.0016-0.0094	2	2.0863-2.0866	0.0018-0.0026	0.0059-0.0157
	4.3	3UZ-FE	2.6373-2.6378	①	0.0008-0.0087	3	2.0465-2.0472	0.0008-0.0019	0.0063-0.0138
2008	3.5	2GR-FSE	2.4011-2.4016	0.0018-0.0026	0.0016-0.0094	2	2.0863-2.0866	0.0018-0.0026	0.0059-0.0157
	4.6	1UR-FSE	2.6373-2.6378	①	0.0008-0.0087	3	2.0859-2.0866	0.0010-0.0020	0.0059-0.0217

NA: Not Available

① Journal No. 1 and 5: 0.0007 - 0.0012 inch
Remaining journals: 0.0009 - 0.0015 inch

22140_GS35_C0008

PISTON AND RING SPECIFICATIONS

All measurements are given in inches.

Year	Engine Displacement Liters	Engine ID/VIN	Piston Clearance	Ring Gap: Top Compression	Ring Gap: Bottom Compression	Ring Gap: Oil Control	Ring Side Clearance: Top Compression	Ring Side Clearance: Bottom Compression	Ring Side Clearance: Oil Control
2007	3.5	2GR-FSE	0.0008-0.0020	0.0091-0.0130	0.0138-0.0177	0.0039-0.0157	0.0008-0.0028	0.0008-0.0024	0.0008-0.0028
	4.6	3UZ-FE	0.0023-0.0040	0.0118-0.0197	0.0157-0.0236	0.0059-0.0197	0.0012-0.0031	0.0008-0.0024	SNUG
2008	3.5	2GR-FSE	0.0008-0.0020	0.0091-0.0130	0.0138-0.0177	0.0039-0.0157	0.0008-0.0028	0.0008-0.0024	0.0008-0.0028
	4.6	1UR-FSE	0.0014-0.0022	0.0091-0.0130	0.0138-0.0177	0.0039-0.0157	0.0008-0.0028	0.0008-0.0024	0.0008-0.0028

22140_GS35_C0007

TORQUE SPECIFICATIONS

All readings in ft. lbs.

Year	Engine Displacement Liters	Engine ID/VIN	Cylinder Head Bolts	Main Bearing Bolts	Rod Bearing Bolts	Crankshaft Damper Bolts	Flywheel Bolts	Manifold: Intake	Manifold: Exhaust	Spark Plugs	Lug Nuts
2007	3.5	2GR-FSE	①	②	③	184	61	15	30	13	76
	4.3	3UZ-FE	④	⑤	⑥	181	22	13	32	13	76
2008	3.5	2GR-FSE	①	②	③	184	61	15	30	13	76
	4.6	1UR-FSE	⑦	⑧	③	221	22	15	15	13	76

① Step 1: 27 ft. lbs.
Step 2: Tighten an additional 90 degrees
Step 3: Tighten an additional 90 degrees
14 mm bolt: 22 ft. lbs.

⑤ Inside position: 45 ft. lbs.
Inside position Step 2: Tighten an additional 90 degrees
Cylinder block Side postion: 19 ft. lbs.

③ Step 1: 30 ft. lbs.
Step 2: Plus 90 degrees

④ Step 1: 44 ft. lbs.
Step 2: Plus 90 degrees

⑤ Inside position: 20 ft. lbs.
Inside position Step 2: Tighten an additional 90 degrees
Cylinder block Side postion: 36 ft. lbs.

⑥ Step 1: 18 ft. lbs.
Step 2: Plus 90 degrees

⑦ Step 1: 27 ft. lbs.
Step 2: Tighten an additic
Step 3: Tighten an additional 90 degrees
12 mm bolt: 15 ft. lbs.

⑧ Inside position: 45 ft. lbs.
Outside position Step 1: 20 ft. lbs.
Outside position Step 2: Tighten an additional 90 degrees
Cylinder block Side postion: 33 ft. lbs.

22140_GS35_C0006

WHEEL ALIGNMENT

Year	Model		Caster Range (+/-Deg.)	Caster Preferred Setting (Deg.)	Camber Range (+/-Deg.)	Camber Preferred Setting (Deg.)	Toe-in (in.)	Steering Axis Inclination (Deg.)
2007	GS 350 ①	F	0.75	①	0.75	②	0 +/- 0.04	③
		R	—	—	0.75	④	0.12 +/- 0.08	—
	GS 430	F	0.75	+7.43	0.75	-0.43	0 +/- 0.04	9.43 +/- 0.075
		R	—	—	0.75	-1.32	0.12 +/- 0.08	—
2008	GS 350 ①	F	0.75	①	0.75	②	0 +/- 0.04	③
		R	—	—	0.75	④	0.12 +/- 0.08	—
	GS 460	F	0.75	+7.43	0.75	-0.43	0 +/- 0.04	9.43 +/- 0.075
		R	—	—	0.75	-1.32	0.12 +/- 0.08	—

① 2WD: +7.38
AWD: +4.88

② 2WD: -0.38
AWD: -0.42

③ 2WD: 9.38
AWD: 11.18

④ 2WD: -1.17
AWD: -1.05

22140_GS35_C0009

TIRE, WHEEL AND BALL JOINT SPECIFICATIONS

Year	Model	OEM Tires Standard	OEM Tires Optional	Tire Pressures (psi) Front	Tire Pressures (psi) Rear	Wheel Size	Ball Joint Inspection
2007	GS 350/430	225/50R17 94W	245/40R18 93V	Std: 33 Opt: 33	Std: 33 Opt: 33	Std:17x 7.5-JJ Opt:18x8-JJ	U: 9-30 in. ①
2008	GS 350/460	225/50R17 94W	245/40R18 93V	Std: 33 Opt: 33	Std: 33 Opt: 33	Std:17x 7.5-JJ Opt:18x8-JJ	U: 9-30 in. ①

OEM: Original Equipment Manufacturer
PSI: Pounds Per Square Inch
STD: Standard
OPT: Optional
U: Upper
① Torque required in inch lbs. to rotate ball joint when removed from the knuckle

22140_GS35_C0010

BRAKE SPECIFICATIONS

All measurements in inches unless noted

Year	Model	Front Brake Disc Original Thickness	Front Brake Disc Minimum Thickness	Front Brake Disc Maximum Run-out	Rear Brake Disc Original Thickness	Rear Brake Disc Minimum Thickness	Rear Brake Disc Maximum Run-out	Minimum Lining Thickness	Brake Caliper Bracket Bolts (ft. lbs.)	Brake Caliper Mounting Bolts (ft. lbs.)
2007	GS 350	1.181	1.063	0.0020	0.709	0.650	0.0020	0.0390	①	②
	GS 430	1.181	1.063	0.0020	0.709	0.650	0.0020	0.0390	①	②
2008	GS 350	1.181	1.063	0.0020	0.709	0.650	0.0020	0.0390	①	②
	GS 460	1.181	1.063	0.0020	0.709	0.650	0.0020	0.0390	①	②

① Front: 58 ft. lbs.
Rear: 40 ft. lbs.
② Front: 58 ft. lbs.
Rear: 18 ft. lbs.

22140_GS35_C0011

SCHEDULED MAINTENANCE INTERVALS
Lexus—GS350, GS430, GS460

TO BE SERVICED	TYPE OF SERVICE	VEHICLE MILEAGE INTERVAL (x1000)												
		7.5	15	22.5	30	37.5	45	52.5	60	67.5	75	82.5	90	97.5
Engine oil & filter	R	✓	✓	✓	✓	✓	✓	✓	✓	✓	✓	✓	✓	✓
A/C filter (if equipped) ①	S/I	✓	✓	✓	✓	✓	✓	✓	✓	✓	✓	✓	✓	✓
Automatic transaxle fluid & filter	S/I		✓		✓		✓		✓		✓		✓	
Ball joints & dust covers	S/I		✓		✓		✓		✓		✓		✓	
Bolts & nuts on chassis & body	S/I		✓		✓		✓		✓		✓		✓	
Brake fluid ②	S/I		✓		✓		✓		✓		✓		✓	
Brake line pipes & hoses	S/I		✓		✓		✓		✓		✓		✓	
Brake pads & discs (front & rear)	S/I		✓		✓		✓		✓		✓		✓	
Differential oil	S/I		✓		✓		✓		✓		✓		✓	
Driveshaft boots (if equipped)	S/I		✓		✓		✓		✓		✓		✓	
Steering gear housing oil	S/I		✓		✓		✓		✓		✓		✓	
Steering linkage	S/I		✓		✓		✓		✓		✓		✓	
Air filter	R				✓				✓				✓	
Exhaust pipes & mountings	S/I				✓				✓				✓	
Fuel lines & connections	S/I				✓				✓				✓	
Engine coolant	R						✓					✓		
Fuel tank cap gasket	R								✓					
Spark plugs	R								✓					
Charcoal canister	S/I								✓					
Drive belts	S/I								✓					
Valve clearance	S/I								✓					

R: Replace S/I: Service or Inspect

① Replace at 15,000 miles.

② Replace at 30,000 miles (unless previously replaced).

FREQUENT OPERATION MAINTENANCE (SEVERE SERVICE)

If a vehicle is operated under any of the following conditions it is considered severe service

- Extremely dusty areas.

- 50% or more of the vehicle operation is in 32°C (90°F) or higher temperatures, or constant operation in temperatures below 0°C (32°F).

- Prolonged idling (vehicle operation in stop and go traffic).

- Frequent short running periods (engine does not warm to normal operating temperatures).

- Police, taxi, delivery usage or trailer towing usage.

Oil & oil filter: change every 3750 miles.

Ball joints & dust covers: service or inspect every 7500 miles.

Bolts & nuts on chassis & body: service or inspect every 7500 miles.

Brake pads & discs (front & rear): service or inspect every 7500 miles.

Driveshaft boots (if equipped): service or inspect every 7500 miles.

Steering linkage: service or inspect every 7500 miles.

Air filter: service or inspect every 15,000 miles.

Automatic transmission fluid & filter: replace every 15,000 miles.

Differential oil: replace every 15,000 miles.

Exhaust pipes & mountings: service or inspect every 15,000 miles.

Drive belts: service or inspect at 60,000 miles & every 7500 miles thereafter.

Timing belts: replace every 60,000 miles.

22140_GS35_C0012

PRECAUTIONS

Before servicing any vehicle, please be sure to read all of the following precautions, which deal with personal safety, prevention of component damage, and important points to take into consideration when servicing a motor vehicle:

- Never open, service or drain the radiator or cooling system when the engine is hot; serious burns can occur from the steam and hot coolant.
- Observe all applicable safety precautions when working around fuel. Whenever servicing the fuel system, always work in a well-ventilated area. Do not allow fuel spray or vapors to come in contact with a spark, open flame, or excessive heat (a hot drop light, for example). Keep a dry chemical fire extinguisher near the work area. Always keep fuel in a container specifically designed for fuel storage; also, always properly seal fuel containers to avoid the possibility of fire or explosion. Refer to the additional fuel system precautions later in this section.
- Fuel injection systems often remain pressurized, even after the engine has been turned **OFF**. The fuel system pressure must be relieved before disconnecting any fuel lines. Failure to do so may result in fire and/or personal injury.
- Brake fluid often contains polyglycol ethers and polyglycols. Avoid contact with the eyes and wash your hands thoroughly after handling brake fluid. If you do get brake fluid in your eyes, flush your eyes with clean, running water for 15 minutes. If eye irritation persists, or if you have taken brake fluid internally, IMMEDIATELY seek medical assistance.
- The EPA warns that prolonged contact with used engine oil may cause a number of skin disorders, including cancer. You should make every effort to minimize your exposure to used engine oil. Protective gloves should be worn when changing oil. Wash your hands and any other exposed skin areas as soon as possible after exposure to used engine oil. Soap and water, or waterless hand cleaner should be used.
- All new vehicles are now equipped with an air bag system, often referred to as a Supplemental Restraint System (SRS) or Supplemental Inflatable Restraint (SIR) system. The system must be disabled before performing service on or around system components, steering column, instrument panel components, wiring and sensors. Failure to follow safety and disabling procedures could result in accidental air bag deployment, possible personal injury and unnecessary system repairs.
- Always wear safety goggles when working with, or around, the air bag system. When carrying a non-deployed air bag, be sure the bag and trim cover are pointed away from your body. When placing a non-deployed air bag on a work surface, always face the bag and trim cover upward, away from the surface. This will reduce the motion of the module if it is accidentally deployed. Refer to the additional air bag system precautions later in this section.
- Clean, high quality brake fluid from a sealed container is essential to the safe and proper operation of the brake system. You should always buy the correct type of brake fluid for your vehicle. If the brake fluid becomes contaminated, completely flush the system with new fluid. Never reuse any brake fluid. Any brake fluid that is removed from the system should be discarded. Also, do not allow any brake fluid to come in contact with a painted surface; it will damage the paint.
- Never operate the engine without the proper amount and type of engine oil; doing so WILL result in severe engine damage.
- Timing belt maintenance is extremely important. Many models utilize an interference-type, non-freewheeling engine. If the timing belt breaks, the valves in the cylinder head may strike the pistons, causing potentially serious (also time-consuming and expensive) engine damage. Refer to the maintenance interval charts for the recommended replacement interval for the timing belt, and to the timing belt section for belt replacement and inspection.
- Disconnecting the negative battery cable on some vehicles may interfere with the functions of the on-board computer system(s) and may require the computer to undergo a relearning process once the negative battery cable is reconnected.
- When servicing drum brakes, only disassemble and assemble one side at a time, leaving the remaining side intact for reference.
- Only an MVAC-trained, EPA-certified automotive technician should service the air conditioning system or its components.

BRAKES

ANTI-LOCK BRAKE SYSTEM (ABS)

GENERAL INFORMATION

PRECAUTIONS

- Certain components within the ABS system are not intended to be serviced or repaired individually.
- Do not use rubber hoses or other parts not specifically specified for and ABS system. When using repair kits, replace all parts included in the kit. Partial or incorrect repair may lead to functional problems and require the replacement of components.
- Lubricate rubber parts with clean, fresh brake fluid to ease assembly. Do not use shop air to clean parts; damage to rubber components may result.
- Use only DOT 3 brake fluid from an unopened container.
- If any hydraulic component or line is removed or replaced, it may be necessary to bleed the entire system.
- A clean repair area is essential. Always clean the reservoir and cap thoroughly before removing the cap. The slightest amount of dirt in the fluid may plug an orifice and impair the system function. Perform repairs after components have been thoroughly cleaned; use only denatured alcohol to clean components. Do not allow ABS components to come into contact with any substance containing mineral oil; this includes used shop rags.
- The Anti-Lock control unit is a microprocessor similar to other computer units in the vehicle. Ensure that the ignition switch is **OFF** before removing or installing controller harnesses. Avoid static electricity discharge at or near the controller.
- If any arc welding is to be done on the vehicle, the control unit should be unplugged before welding operations begin.

BLEEDING PROCEDURE

BLEEDING PROCEDURE

See Figure 1.

If any work is done on the brake system or if air in the brake lines is suspected, bleed the system of air.

➡Do not let brake fluid remain on painted surfaces. Wash it off immediately.

1. Fill the reservoir with brake fluid.

➡If the master cylinder has been disassembled or if the reservoir becomes empty, bleed the air from the master cylinder.

2. Bleed the brake master cylinder as follows:
   a. Disconnect the brake lines from the master cylinder.
   b. Slowly depress the brake pedal and hold it.
   c. Block off the outer holes with your fingers, and release the brake pedal.
   d. Repeat the previous 2 steps 3 or 4 times.
3. Bleed the brake line as follows:
   a. Connect the vinyl tube to the brake caliper.
   b. Depress the brake pedal several times, then loosen the bleeder plug with the pedal held down.
   c. At the point when fluid stops coming out, tighten the bleeder plug, then release the brake pedal.
   d. Repeat the previous 2 steps until all the air in the fluid has been bled out.
   e. Repeat the above procedure to bleed the air out of the brake line for each wheel.
   f. Tighten the bleeder plug to 8 ft. lbs. (11 Nm).
4. Bleed the brake actuator as follows:
   a. Remove the reservoir cap.
   b. Install the SST 09992-00242, 09992-00350 to the reservoir.
   c. Connect the vinyl tube to the bleeder plug of the brake actuator.
   d. Using SST, apply the 14.2 psi (98.1kpa) of pressure to the reservoir.
   e. Loosen the bleeder plug.
   f. Bleed the air out of the brake actuator, tighten the bleeder plug to 74 inch lbs. (8.3 Nm).
5. Check the fluid level and add fluid if necessary.

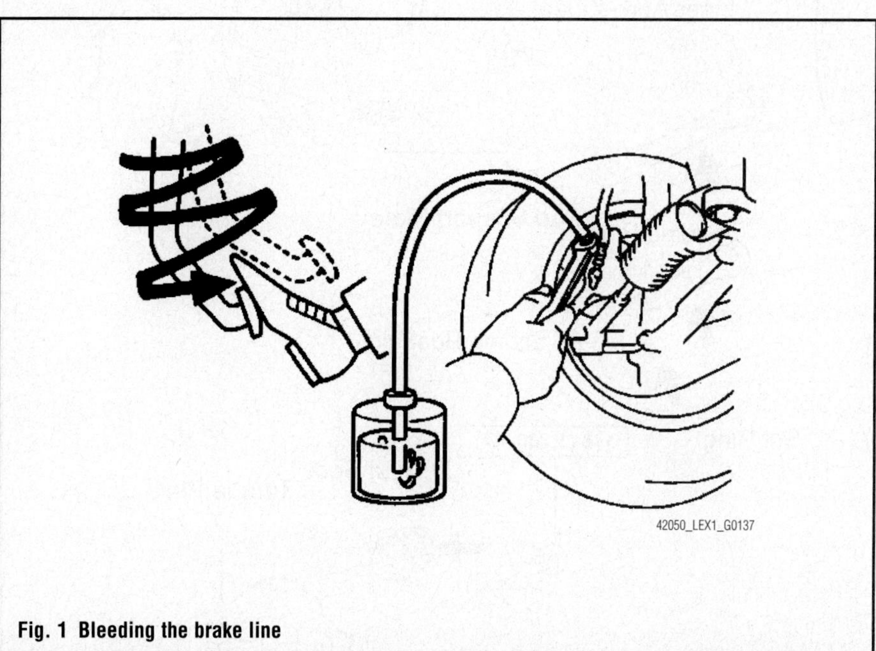

Fig. 1 Bleeding the brake line

⁂ CAUTION

Dust and dirt accumulating on brake parts during normal use may contain asbestos fibers from production or aftermarket brake linings. Breathing excessive concentrations of asbestos fibers can cause serious bodily harm. Exercise care when servicing brake parts. Do not sand or grind brake lining unless equipment used is designed to contain the dust residue. Do not clean brake parts with compressed air or by dry brushing. Cleaning should be done by dampening the brake components with a fine mist of water, then wiping the brake components clean with a dampened cloth. Dispose of cloth and all residue containing asbestos fibers in an impermeable container with the appropriate label. Follow practices prescribed by the Occupational Safety and Health Administration (OSHA) and the Environmental Protection Agency (EPA) for the handling, processing, and disposing of dust or debris that may contain asbestos fibers.

BRAKE CALIPER

REMOVAL & INSTALLATION

See Figure 2.

1. Before servicing the vehicle, refer to the precautions in the beginning of this section.
2. Remove or disconnect the following:
   - Wheels
   - Brake line at the caliper
   - Anti-squeal springs
   - Mounting bolts, while holding the sliding pin with a wrench
   - Caliper assembly

To install:

3. Install or connect the following:
   - Caliper. Hold the sliding pin and tighten the mounting bolts to 58 ft. lbs. (78 Nm).
   - Anti-squeal springs
   - Brake line with 2 new gaskets and tighten the union bolt to 29 ft. lbs. (39 Nm)
4. Bleed the brake system.
   - Wheels
5. Check and if necessary fill the master cylinder reservoir.

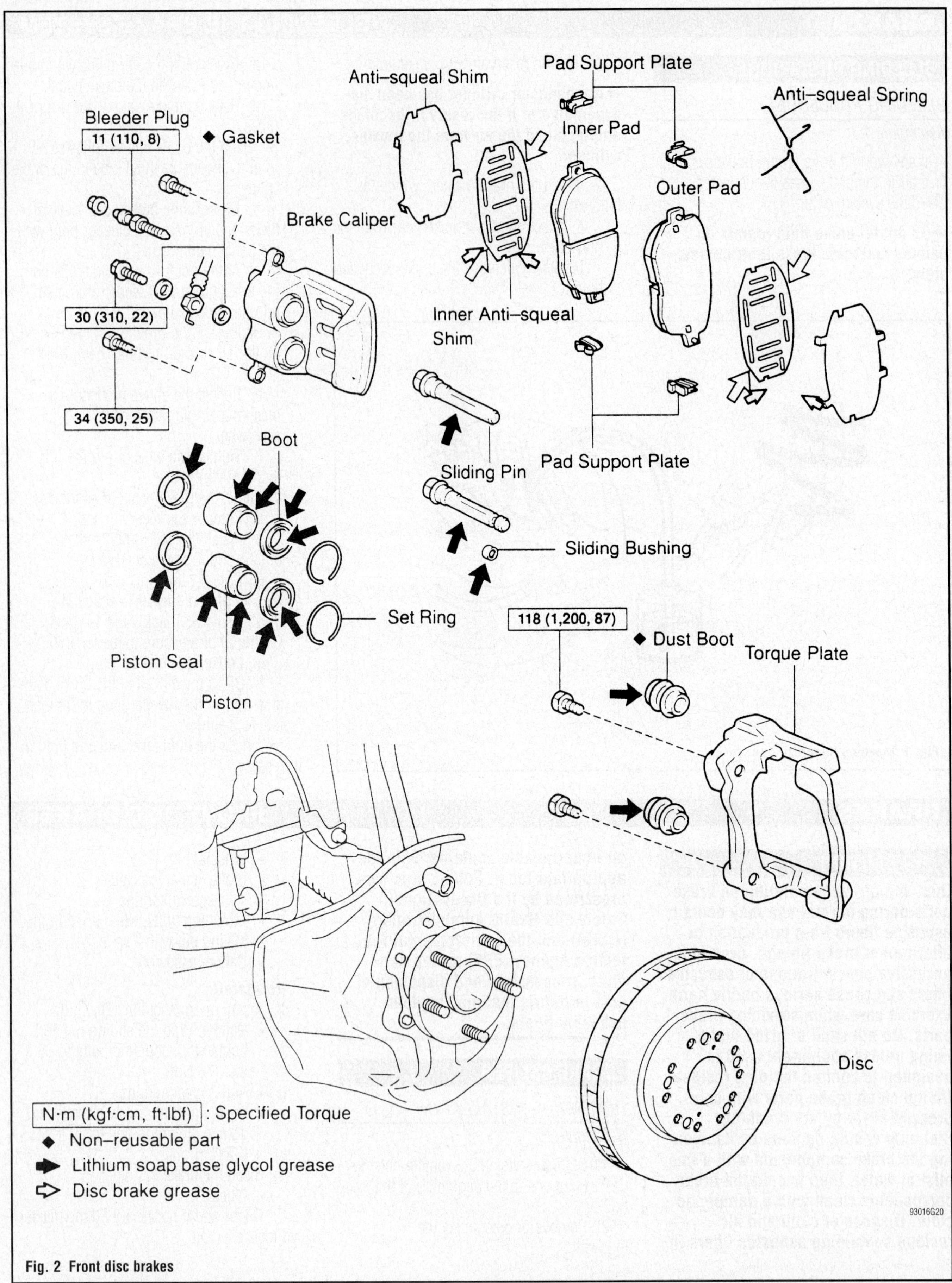

Fig. 2 Front disc brakes

DISC BRAKE PADS

REMOVAL & INSTALLATION

1. Before servicing the vehicle, refer to the precautions in the beginning of this section.
2. Remove or disconnect the following:
   - Wheels
3. Hold the sliding pin on the lower mounting bolt and remove the bolt. Swivel the caliper upward and out of the way.
   - Anti-squeal springs
   - Brake pads, retainers and anti-squeal shims

To install:

4. Install or connect the following:
   - Pad support plates and the pad wear indicator plate on the inside pad
   - Both pads (and anti-squeal shims) with the wear indicator plates facing downward
5. Compress the caliper pistons and install the caliper.
   - Anti-squeal springs
   - Hold the sliding pin and tighten the mounting bolts to 58 ft. lbs. (78 Nm)
   - Wheels
6. Check the brake fluid level in the reservoir.

BRAKES

REAR DISC BRAKES

** CAUTION

Dust and dirt accumulating on brake parts during normal use may contain asbestos fibers from production or aftermarket brake linings. Breathing excessive concentrations of asbestos fibers can cause serious bodily harm. Exercise care when servicing brake parts. Do not sand or grind brake lining unless equipment used is designed to contain the dust residue. Do not clean brake parts with compressed air or by dry brushing. Cleaning should be done by dampening the brake components with a fine mist of water, then wiping the brake components clean with a dampened cloth. Dispose of cloth and all residue containing asbestos fibers in an impermeable container with the appropriate label. Follow practices prescribed by the Occupational Safety and Health Administration (OSHA) and the Environmental Protection Agency (EPA) for the handling, processing, and disposing of dust or debris that may contain asbestos fibers.

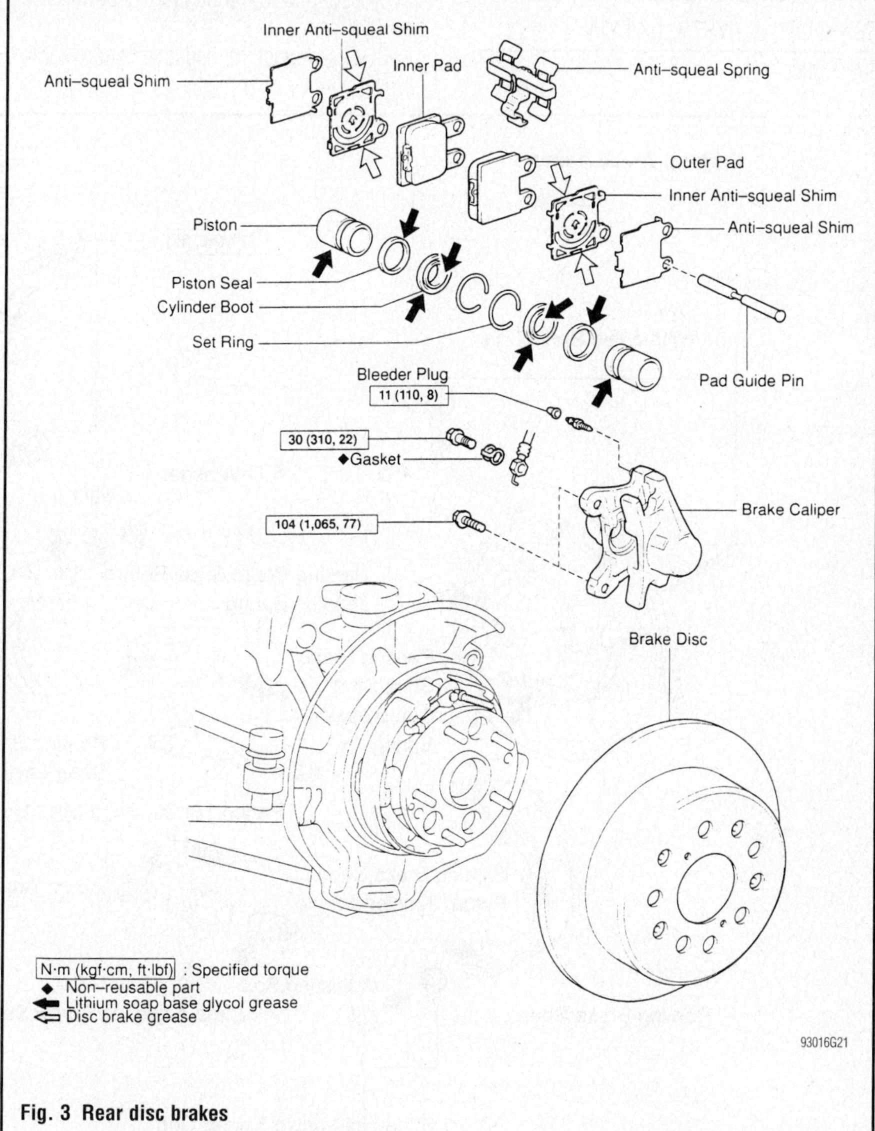

Fig. 3 Rear disc brakes

BRAKE CALIPER

REMOVAL & INSTALLATION

See Figure 3.

1. Before servicing the vehicle, refer to the precautions in the beginning of this section.
2. Remove or disconnect the following:
   - Wheels
   - Brake line at the caliper
   - Anti-squeal springs
   - Mounting bolts, while holding the sliding pin with a wrench
   - Caliper assembly

To install:

3. Install or connect the following:
   - Caliper. Tighten the mounting bolts to 40 ft. lbs. (54 Nm)
   - Slide pins. Tighten to 18 ft. lbs. (25 Nm)
   - Anti-squeal springs
   - Connect the brake line with 2 new gaskets and tighten the union bolt to 22 ft. lbs. (30 Nm)
4. Bleed the brake system.
   - Wheels
5. Check and if necessary fill the master cylinder reservoir.

DISC BRAKE PADS

REMOVAL & INSTALLATION

1. Before servicing the vehicle, refer to the precautions in the beginning of this section.
2. Remove or disconnect the following:
   - Wheels
3. Hold the sliding pin on the lower mounting bolt and remove the bolt. Swivel the caliper upward and out of the way.
   - Anti-squeal springs
   - Brake pads, retainers and anti-squeal shims

To install:

4. Install or connect the following:
   - Pad support plates and the pad wear indicator plate on the inside pad
   - Both pads (and anti-squeal shims) with the wear indicator plates facing downward
5. Compress the caliper pistons and install the caliper.
   - Anti-squeal springs
   - Hold the sliding pin and tighten the mounting bolts to 40 ft. lbs. (54 Nm).
   - Wheels
6. Check the brake fluid level in the reservoir.

BRAKES — PARKING BRAKE

PARKING BRAKE SHOES

REMOVAL & INSTALLATION

See Figure 4.

1. Before servicing the vehicle, refer to the precautions in the beginning of this section.
2. Disconnect the negative battery cable.
3. Remove the rear wheel.
4. Remove the 2 bolts and separate the rear disc brake caliper assy. Do not disconnect the flexible hose from the brake caliper.

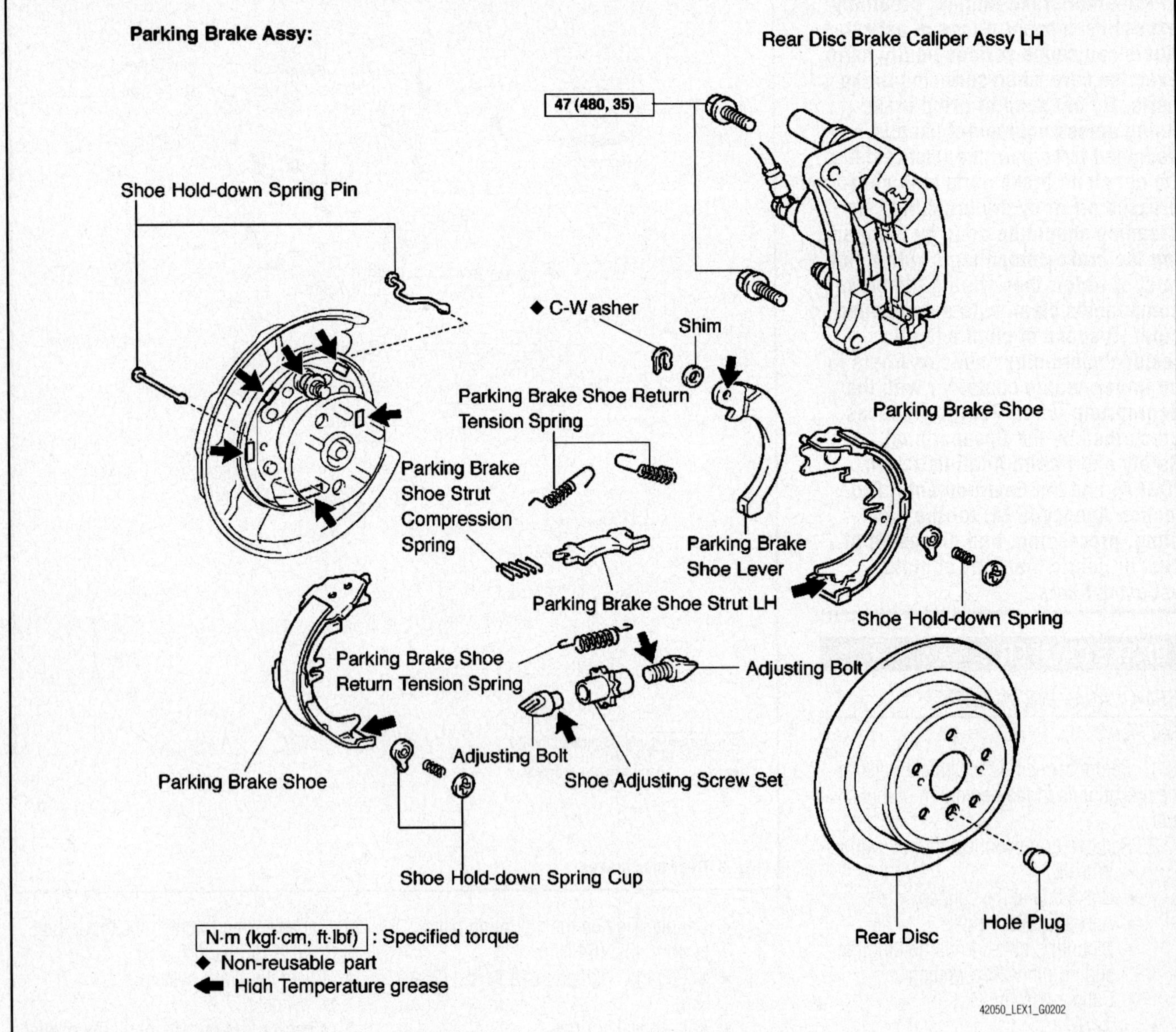

Fig. 4 Exploded view of typical parking brake shoes assembly

5. Release the parking brake, and remove the rear disc.

➡**Put matchmarks on the disc and the axle hub. If the disc cannot be removed easily, turn the shoe adjuster until the wheel turns freely.**

6. Using a needle-nose pliers, remove the 2 return tension springs.
7. Slide out the front shoe and remove the shoe strut compression spring.
8. Remove the parking brake shoe strut.
9. Remove the parking brake shoe as follows:
   a. Release the cup claw and remove the front and rear parking brake shoe.
   b. Disconnect the parking brake cable from the shoe lever.
   c. Remove the tension spring and shoe adjuster screw set from the front and rear shoe.
   d. Remove the 2 shoe hold-down springs, 4 cups and 2 pins.
   e. Using a screwdriver, remove the C-washer.
   f. Remove the shim and shoe lever from the parking brake shoe.
10. Apply chalk to the inside surface of the disc, then grind down the brake shoe lining to fit. If the contact between the brake disc and the shoe lining is improper, repair it using a brake shoe grinder or replace the brake shoe assembly.

To install:

11. Apply the high temperature grease to the shoe attached surface of backing plate.
12. Install the parking brake shoe as follows:
   a. Install the shoe lever and shim to the rear shoe with a new C-washer.
   b. Using a feeler gauge, measure the clearance. Standard clearance: Less than 0.35mm (0.0138 in.). If the clearance is not within the specification, replace the shim with one of the correct size.
   c. Apply the high temperature grease to the adjusting bolt.
   d. Install the shoe adjusting screw set and tension spring to the front and rear shoe.
   e. Install the 2 pins, 4 cups and 2 shoe hold-down springs.
   f. Connect the parking brake cable to the shoe lever.
   g. Install the front and rear parking brake shoe.
13. Install the parking brake shoe strut.
14. Install the parking brake shoe strut compression spring.
15. Using a needle nose pliers, install the 2 return tension springs.
16. Check that each part is installed properly.

➡**There should be no oil or grease adhering to the friction surface of the shoe lining and disc.**

17. Install the rear disc.
18. Adjust the parking brake shoe clearance.
19. Install the rear disc brake caliper assy.
20. Install the rear wheel.
21. Inspect and adjust parking brake pedal or lever travel.

CHASSIS ELECTRICAL

AIR BAG (SUPPLEMENTAL RESTRAINT SYSTEM)

GENERAL INFORMATION

⁂ CAUTION

These vehicles are equipped with an air bag system. The system must be disarmed before performing service on, or around, system components, the steering column, instrument panel components, wiring and sensors. Failure to follow the safety precautions and the disarming procedure could result in accidental air bag deployment, possible injury and unnecessary system repairs.

SERVICE PRECAUTIONS

Disconnect and isolate the battery negative cable before beginning any airbag system component diagnosis, testing, removal, or installation procedures. Allow system capacitor to discharge for two minutes before beginning any component service. This will disable the airbag system. Failure to disable the airbag system may result in accidental airbag deployment, personal injury, or death.

Do not place an intact undeployed airbag face down on a solid surface. The airbag will propel into the air if accidentally deployed and may result in personal injury or death.

When carrying or handling an undeployed airbag, the trim side (face) of the airbag should be pointing towards the body to minimize possibility of injury if accidental deployment occurs. Failure to do this may result in personal injury or death.

Replace airbag system components with OEM replacement parts. Substitute parts may appear interchangeable, but internal differences may result in inferior occupant protection. Failure to do so may result in occupant personal injury or death.

Wear safety glasses, rubber gloves, and long sleeved clothing when cleaning powder residue from vehicle after an airbag deployment. Powder residue emitted from a deployed airbag can cause skin irritation. Flush affected area with cool water if irritation is experienced. If nasal or throat irritation is experienced, exit the vehicle for fresh air until the irritation ceases. If irritation continues, see a physician.

Do not use a replacement airbag that is not in the original packaging. This may result in improper deployment, personal injury, or death.

The factory installed fasteners, screws and bolts used to fasten airbag components have a special coating and are specifically designed for the airbag system. Do not use substitute fasteners. Use only original equipment fasteners listed in the parts catalog when fastener replacement is required.

During, and following, any child restraint anchor service, due to impact event or vehicle repair, carefully inspect all mounting hardware, tether straps, and anchors for proper installation, operation, or damage. If a child restraint anchor is found damaged in any way, the anchor must be replaced. Failure to do this may result in personal injury or death.

Deployed and non-deployed airbags may or may not have live pyrotechnic material within the airbag inflator.

Do not dispose of driver/passenger/curtain airbags or seat belt tensioners unless you are sure of complete deployment. Refer to the Hazardous Substance Control System for proper disposal.

Dispose of deployed airbags and tensioners consistent with state, provincial, local, and federal regulations.

After any airbag component testing or service, do not connect the battery negative cable. Personal injury or death may result if the system test is not performed first.

If the vehicle is equipped with the Occupant Classification System (OCS), do not connect the battery negative cable before performing the OCS Verification Test using the scan tool and the appropriate diagnostic information. Personal injury or death may result if the system test is not performed properly.

Never replace both the Occupant Restraint Controller (ORC) and the Occupant Classification Module (OCM) at the same time. If both require replacement, replace one, then perform the Airbag System test before replacing the other.

Both the ORC and the OCM store Occupant Classification System (OCS) calibration data, which they transfer to one another when one of them is replaced. If both are replaced at the same time, an irreversible fault will be set in both modules and the OCS may malfunction and cause personal injury or death.

If equipped with OCS, the Seat Weight Sensor is a sensitive, calibrated unit and must be handled carefully. Do not drop or handle roughly. If dropped or damaged, replace with another sensor. Failure to do so may result in occupant injury or death.

If equipped with OCS, the front passenger seat must be handled carefully as well. When removing the seat, be careful when setting on floor not to drop. If dropped, the sensor may be inoperative, could result in occupant injury, or possibly death.

If equipped with OCS, when the passenger front seat is on the floor, no one should sit in the front passenger seat. This uneven force may damage the sensing ability of the seat weight sensors. If sat on and damaged, the sensor may be inoperative, could result in occupant injury, or possibly death.

DISARMING THE SYSTEM

To avoid personal injury when working on vehicles equipped with an air bag, the negative battery cable must be disconnected and at least 90 seconds must elapse before working on the system. Failure to do so may result in deployment of the air bag.

ARMING THE SYSTEM

To rearm the air bag system, simply reconnect the battery cable(s).

CLOCKSPRING CENTERING

1. Center the spiral cable as follows:
   a. Check that the ignition switch is OFF.
   b. Check that the battery negative terminal is disconnected. Do not start the operation for 90 seconds after removing the terminal.
   c. Turn the cable counterclockwise by hand until it becomes harder to turn.
   d. Then rotate the cable clockwise about 2.5 turns to align the marks. The cable will rotate about 2.5 turns to both left and right of the center.

DRIVETRAIN

AUTOMATIC TRANSMISSION ASSEMBLY

REMOVAL & INSTALLATION

GS 350—2WD Models

See Figure 5.

1. Disconnect cable from negative battery terminal.
2. Remove engine under cover.
3. Drain coolant.
4. Drain automatic transmission fluid.
5. Remove the propeller shaft with center bearing.
6. Remove cool air intake duct seal.
7. Remove engine room side covers.
8. Remove V-bank cover.
9. Remove exhaust manifold sub-assembly LH.
10. Remove starter assembly.
11. Remove the flywheel housing side cover from the engine assembly.
12. Remove the 6 torque converter clutch setting bolts.
13. Remove the nut, and disconnect the floor shift gear shifting rod.
14. Support the automatic transmission assembly with a transmission jack.
15. Remove the 4 bolts, and disconnect the engine rear mounting member from the body.
16. Remove the 4 nuts and engine rear mounting member from the automatic transmission assembly.
17. Disconnect wire harness and connector.
18. Disconnect the park/neutral position switch connector, transmission wire connector, and 2 transmission revolution sensor connectors.
19. Disconnect the 2 water by-pass hoses from the transmission oil cooler.
20. Remove the 9 bolts and automatic transmission.

To install:

21. Install the automatic transmission assembly to the engine with the 9 bolts and tighten to 52 ft. lbs. (71 Nm) and 27 ft. lbs. (37Nm).

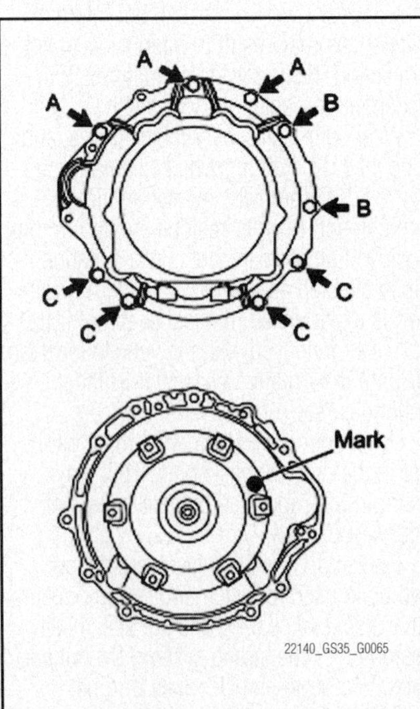

Fig. 5 Transmission bolt locations

➡Make sure that the mark is positioned as shown in the illustration.

22. Connect the 2 water by-pass hoses to the transmission oil cooler.
23. Install the wire harness clamps to the automatic transmission assembly with the 2 bolts.
24. Connect the park/neutral start switch connector, transmission wire connector, and 2 transmission revolution sensor connectors.
25. Install the engine rear mounting member to the automatic transmission assembly with the 4 nuts and tighten to 10 ft. lbs. (13 Nm).
26. Install the engine rear mounting member to the body with the 4 bolts and tighten to 19 ft. lbs (26 Nm).
27. Temporarily tighten the floor shift gear shifting rod with the nut.
28. Install the 6 torque converter clutch setting bolts and tighten to 35 ft. lbs. (48 Nm).
29. Install the flywheel housing side cover to the engine.
30. Install starter assembly.
31. Install exhaust manifold sub-assembly LH.
32. Install the propeller shaft with center bearing.
33. Connect cable to negative battery terminal.
34. Add engine coolant.
35. Add automatic transmission fluid.
36. Check for engine coolant leaks.
37. Adjust shift lever position and tighten.
38. Install engine under covers.

39. Install V-bank cover.
40. Install engine room side covers.
41. Install cool air intake duct seal.
42. Perform initialization, if necessary.

GS 350—AWD Models

See Figure 6.

1. Disconnect cable from negative battery terminal.
2. Remove engine under cover.
3. Drain coolant.
4. Drain automatic transmission fluid.
5. Remove propeller shaft assembly front.
6. Remove the propeller shaft with center bearing.
7. Remove cool air intake duct seal.
8. Remove engine room side covers.
9. Remove V-bank cover.
10. Remove exhaust manifold sub-assembly LH.
11. Remove starter assembly.
12. Remove the flywheel housing side cover from the engine assembly.
13. Remove the 6 torque converter clutch setting bolts.
14. Remove the nut, and disconnect the floor shift gear shifting rod.
15. Support the automatic transmission assembly with a transmission jack.
16. Remove the 4 bolts, and disconnect the engine rear mounting member from the body.
17. Remove the 4 nuts and engine rear mounting member from the automatic transmission assembly.
18. Disconnect wire harness and connector.
19. Disconnect the park/neutral position switch connector, transmission wire connector, and 2 transmission revolution sensor connectors.
20. Disconnect the 2 water by-pass hoses from the transmission oil cooler.
21. Remove the 9 bolts and automatic transmission.

To install:

22. Install the automatic transmission assembly to the engine with the 9 bolts and tighten to 52 ft. lbs. (71 Nm) and 27 ft. lbs. (37Nm).

➡Make sure that the mark is positioned as shown in the illustration.

23. Connect the 2 water by-pass hoses to the transmission oil cooler.
24. Install the wire harness clamps to the automatic transmission assembly with the 2 bolts.
25. Connect the park/neutral start switch connector, transmission wire connector, and 2 transmission revolution sensor connectors.
26. Install the engine rear mounting member to the automatic transmission assembly with the 4 nuts and tighten to 10 ft. lbs. (13 Nm).
27. Install the engine rear mounting member to the body with the 4 bolts and tighten to 19 ft. lbs (26 Nm).
28. Temporarily tighten the floor shift gear shifting rod with the nut.
29. Install the 6 torque converter clutch setting bolts and tighten to 35 ft. lbs. (48 Nm).
30. Install the flywheel housing side cover to the engine.
31. Install starter assembly.
32. Install exhaust manifold sub-assembly LH.
33. Install the propeller shaft with center bearing.
34. Install propeller shaft assembly front.
35. Connect cable to negative battery terminal.
36. Add engine coolant.
37. Add automatic transmission fluid.
38. Check for engine coolant leaks.
39. Adjust shift lever position and tighten.
40. Install engine under covers
41. Install V-bank cover.
42. Install engine room side covers.
43. Install cool air intake duct seal.
44. Perform initialization, if necessary.

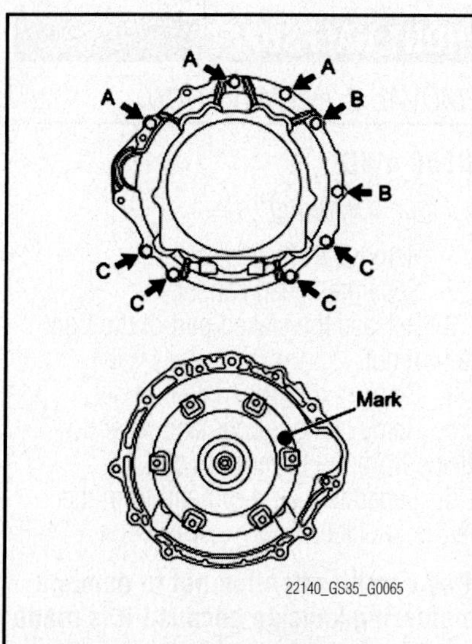

Fig. 6 Transmission bolt locations

GS 430

See Figure 7.

1. Disconnect cable from negative battery terminal.
2. Remove engine under cover.
3. Drain automatic transmission fluid.
4. Remove the propeller shaft with center bearing.
5. Remove cool air intake duct seal.
6. Remove engine room side covers.
7. Remove V-bank cover.
8. Remove exhaust manifold sub-assembly LH.
9. Remove starter assembly.
10. Remove the flywheel housing side cover from the engine assembly.
11. Remove the 6 torque converter clutch setting bolts.
12. Remove the nut, and disconnect the floor shift gear shifting rod.
13. Support the automatic transmission assembly with a transmission jack.
14. Remove the 4 bolts, and disconnect the engine rear mounting member from the body.
15. Remove the 4 nuts and engine rear mounting member from the automatic transmission assembly.
16. Disconnect wire harness and connector.
17. Disconnect the park/neutral position switch connector, transmission wire connector, and 2 transmission revolution sensor connectors.
18. Disconnect the oil cooler outlet tubes.
19. Remove the 9 bolts and automatic transmission.

To install:

20. Install the automatic transmission assembly to the engine with the 9 bolts and tighten to 52 ft. lbs. (71 Nm) and 27 ft. lbs. (37Nm).

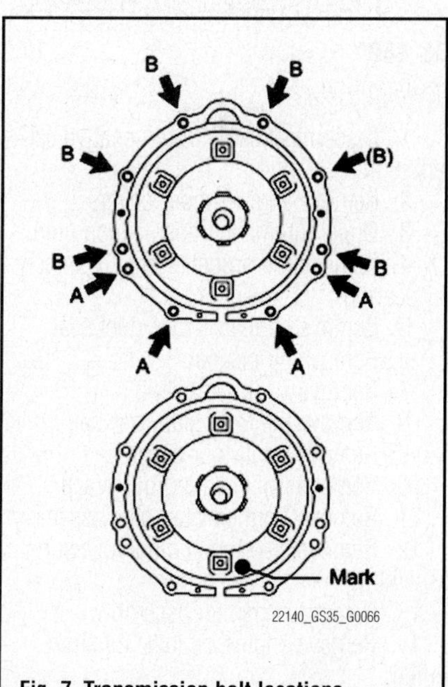

Fig. 7 Transmission bolt locations

➡Make sure that the mark is positioned as shown in the illustration.

21. Connect the oil cooler tubes.
22. Install the wire harness clamps to the automatic transmission assembly with the 2 bolts.
23. Connect the park/neutral start switch connector, transmission wire connector, and 2 transmission revolution sensor connectors.
24. Install the engine rear mounting member to the automatic transmission assembly with the 4 nuts and tighten to 10 ft. lbs. (13 Nm).
25. Install the engine rear mounting member to the body with the 4 bolts and tighten to 19 ft. lbs (26 Nm).
26. Temporarily tighten the floor shift gear shifting rod with the nut.
27. Install the 6 torque converter clutch setting bolts and tighten to 35 ft. lbs. (48 Nm).
28. Install the flywheel housing side cover to the engine.
29. Install starter assembly.
30. Install exhaust manifold sub-assembly LH.
31. Install the propeller shaft with center bearing.
32. Connect cable to negative battery terminal.
33. Add automatic transmission fluid.
34. Check for engine coolant leaks.
35. Adjust shift lever position and tighten.
36. Install engine under covers.
37. Install V-bank cover.
38. Install engine room side covers.
39. Install cool air intake duct seal.
40. Perform initialization, if necessary.

GS 460

See Figure 8.

1. Disconnect cable from negative battery terminal.
2. Remove engine under covers.
3. Drain automatic transmission fluid.
4. Remove the propeller shaft with center bearing.
5. Remove cool air intake duct seal.
6. Remove air cleaner.
7. Remove V-ribbed belt.
8. Remove the front stabilizer bar.
9. Remove front floor cover.
10. Remove front center floor brace.
11. Remove front exhaust pipe assembly.
12. Remove steering sliding yoke sub-assembly.
13. Remove generator assembly.
14. Remove engine oil level dipstick guide.
15. Disconnect compressor and pulley.
16. Remove engine room side covers.
17. Remove V-bank cover.
18. Remove exhaust manifold sub-assemblies.
19. Remove starter assembly.
20. Remove the flywheel housing side cover from the engine assembly.
21. Remove the 6 torque converter clutch setting bolts.
22. Remove the nut, and disconnect the floor shift gear shifting rod.
23. Support the automatic transmission assembly with a transmission jack.
24. Remove the 4 bolts, and disconnect the engine rear mounting member from the body.
25. Remove the 4 nuts and engine rear mounting member from the automatic transmission assembly.
26. Disconnect wire harness and connector.
27. Disconnect the park/neutral position switch connector, transmission wire connector, and 2 transmission revolution sensor connectors.
28. Disconnect the oil cooler outlet tubes.
29. Remove the 10 bolts and automatic transmission.

To install:

30. Install the automatic transmission assembly to the engine with the 10 bolts and tighten to 52 ft. lbs. (71 Nm) and 27 ft. lbs. (37Nm).

➡Make sure that the mark is positioned as shown in the illustration.

31. Connect the oil cooler tubes.
32. Install the wire harness clamps to the automatic transmission assembly with the 2 bolts.
33. Connect the park/neutral start switch connector, transmission wire connector, and 2 transmission revolution sensor connectors.
34. Install the engine rear mounting member to the automatic transmission assembly with the 4 nuts and tighten to 10 ft. lbs. (13 Nm).

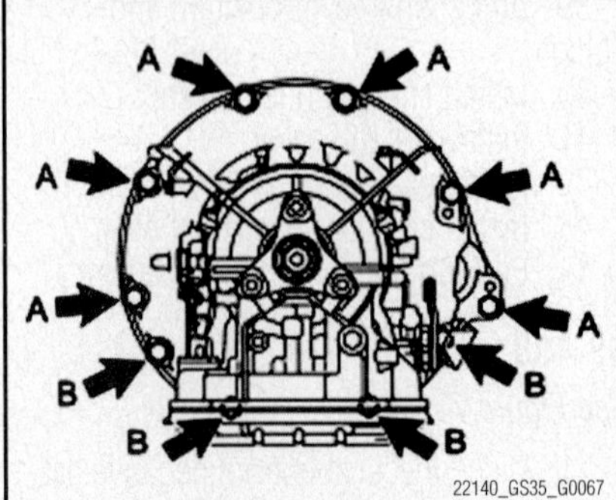

Fig. 8 Transmission bolt locations

35. Install the engine rear mounting member to the body with the 4 bolts and tighten to 19 ft. lbs (26 Nm).
36. Temporarily tighten the floor shift gear shifting rod with the nut.
37. Install the 6 torque converter clutch setting bolts and tighten to 35 ft. lbs. (48 Nm).
38. Install the flywheel housing side cover to the engine.
39. Install starter assembly.
40. Install exhaust manifold sub-assembly LH.
41. Install the propeller shaft with center bearing.
42. Connect compressor and pulley.
43. Install engine oil level dipstick guide.
44. Install generator assembly.
45. Install steering sliding yoke sub-assembly.
46. Install front exhaust pipe assembly.
47. Install front floor covers.
48. Install front center floor brace.
49. Install the front stabilizer bar.
50. Install V-ribbed belt.
51. Install air cleaner.
52. Connect cable to negative battery terminal.
53. Add automatic transmission fluid.
54. Check for engine coolant leaks.
55. Adjust shift lever position and tighten.
56. Install engine under covers.
57. Install V-bank cover.
58. Install engine room side covers.
59. Install cool air intake duct seal.
60. Perform initialization, if necessary.

TRANSFER CASE ASSEMBLY

REMOVAL & INSTALLATION

The transfer case is removed with the transmission assembly.

FRONT HALFSHAFT

REMOVAL & INSTALLATION

GS 350 4WD

See Figures 9 through 12.

1. Remove front wheels.
2. Separate speed sensor.
3. Release the staked part of the front axle hub nut.
4. Remove the front axle hub nut.
5. Remove the 2 bolts and steering knuckle from the lower ball joint.
6. Separate the tie rod end from the steering knuckle.

➡Pay careful attention not to damage the steering knuckle because it is made of aluminum and may be damaged eas-

ily. If the steering knuckle spacer has come off, replace the steering knuckle with a new one.

7. Using a plastic hammer, separate the front axle assembly from the drive shaft assembly.

8. Remove the front drive shaft assemblies.

To install:

9. Coat the spline of the inboard joint shaft assembly with gear oil.

10. Set the shaft snap ring with the opening side facing down.

11. Align the shaft splines and install the drive shaft assembly with a brass bar and hammer.

12. Install the front drive shaft assembly to the front axle assembly.

13. Install the tie rod end to the steering knuckle with the nut and tighten to 48 ft. lbs. (65 Nm). Install a new clip.

14. Install the steering knuckle to the lower ball joint with the 2 bolts and tighten to 89 ft. lbs. (120 Nm).

15. Clean the threaded parts on the drive shaft and front axle hub nut using a non-residue solvent.

16. Install a new axle hub nut and tighten to 217 ft. lbs. (295 Nm).

17. Using a chisel and hammer, stake the front axle hub nut.

18. Install speed sensor front.

19. Install front wheel.

20. Inspect alignment.

REAR HALFSHAFT

REMOVAL & INSTALLATION

See Figures 9 through 12.

1. Remove rear tire.

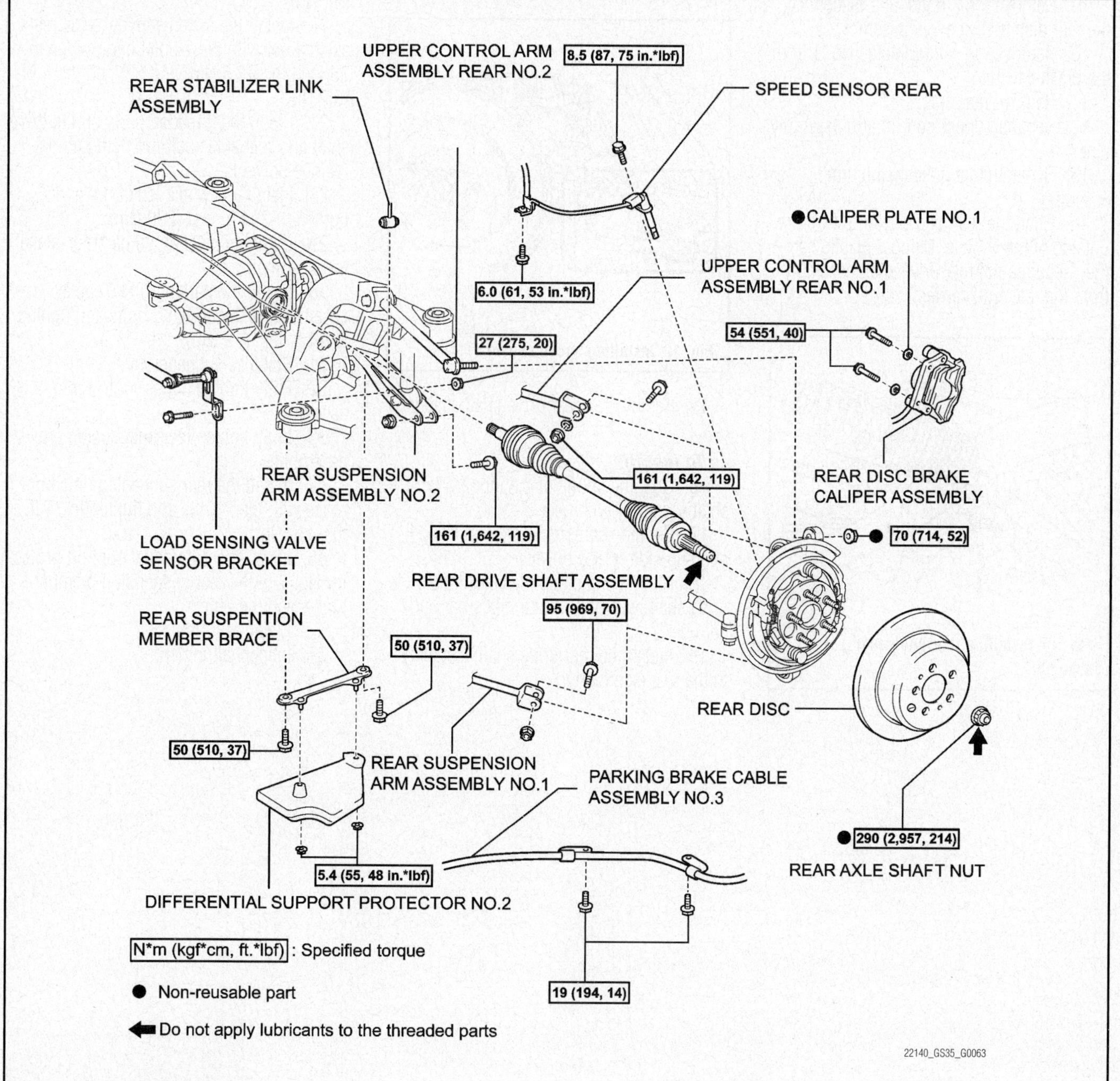

Fig. 9 Rear driveshaft locations

2. Remove the bolt and nut, and separate the load sensing valve sensor bracket and stabilizer link assembly.

3. Remove the 2 nuts and differential support protector No. 2 from the suspension member brace.

4. Remove the 2 bolts and suspension member brace.

5. Remove the 2 bolts, and separate the parking brake cable No. 3.

6. Release the staked part of the axle shaft nut.

7. Remove the axle shaft nut.

8. Remove the 2 bolts, and separate the speed sensor from the axle carrier.

9. Remove the 2 bolts, and disconnect the rear disc brake caliper assembly.

10. Remove the caliper plates No. 1 from the brake caliper.

11. Remove rear disc.

12. Separate upper control arm assembly rear.

13. Separate rear suspension arm assembly.

14. Push the rear axle carrier toward the outside of the vehicle. Using a plastic hammer, separate the rear drive shaft assembly from the rear axle carrier.

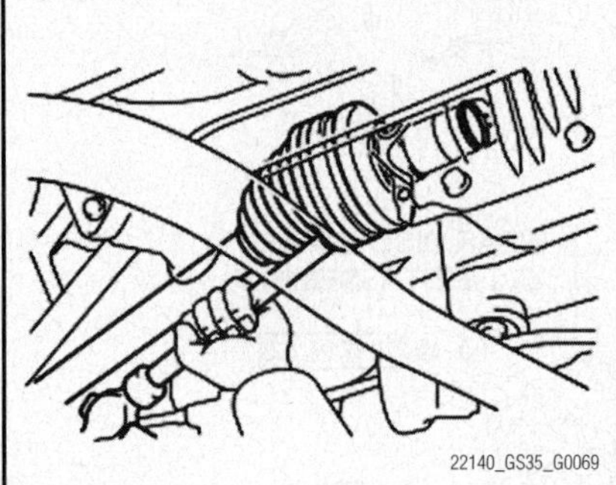

Fig. 10 Installing rear driveshaft assembly

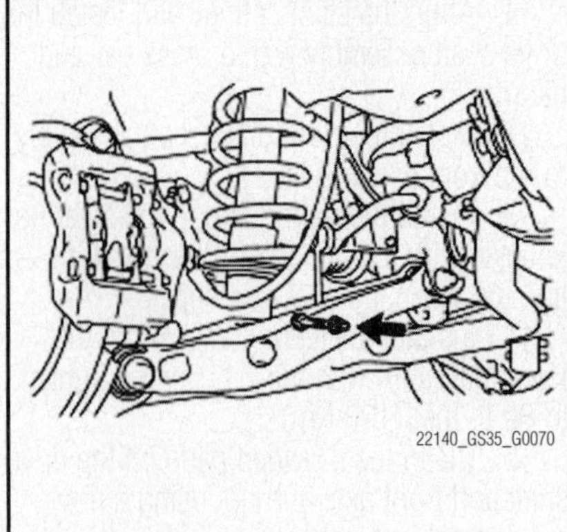

Fig. 11 Installing load sensor bracket

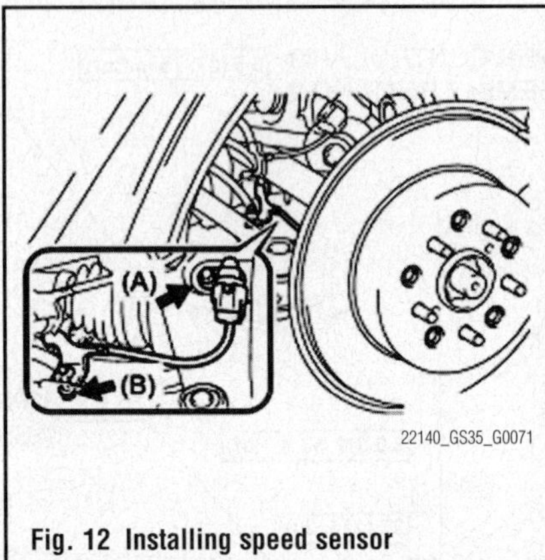

Fig. 12 Installing speed sensor

15. Remove the rear drive shaft assembly.

To install:

16. Coat the spline of the inboard joint shaft assembly with gear oil.

17. Set the shaft snap ring with the opening side facing down.

18. Align the shaft splines and install the drive shaft assembly with a brass bar and hammer.

19. Install the rear drive shaft assembly to the rear axle carrier.

20. Install upper control arm assembly.

21. Temporarily tighten upper control arm assembly.

22. Temporarily tighten rear suspension arm assemblies.

23. Install the stabilizer link assembly and the load sensing valve sensor bracket to the rear suspension arm assembly No. 2 with the bolt and nut and tighten to 20 ft. lbs. (27 Nm).

24. Install rear disc.

25. Install the rear disc brake caliper assembly and caliper plates No. 1 with the 2 bolts and tighten to 40 ft. lbs. (54 Nm).

26. Install the speed sensor to the rear axle carrier with the 2 bolts and tighten to 75 inch lbs. (8.5 Nm) and 53 inch lbs. (6.0 Nm).

27. Clean the threaded parts on the drive shaft and rear axle shaft nut using a non-residue solvent.

28. Install a new axle shaft nut and tighten to 214 ft. lbs. (290 Nm).

29. Using a chisel and a hammer, stake the axle shaft nut.

30. Install the parking brake cable assembly No.3 with the 2 bolts and tighten to 14 ft. lbs. (19 Nm).

31. Stabilize suspension.

32. Fully tighten upper control arm assembly.

33. Fully tighten rear suspension arm assemblies.

34. Install the rear suspension member brace with the 2 bolts and tighten to 37 ft. lbs. (50 Nm).

35. Install the differential support protector No. 2 to the rear suspension member brace with the 2 nuts.

36. Install rear tire.

37. Inspect alignment.

ENGINE COOLING

THERMOSTAT

REMOVAL & INSTALLATION

3.5L (2GR-FSE) Engines

See Figure 13.

1. Before servicing the vehicle, refer to the precautions in the beginning of this section.
2. Disconnect the negative battery cable. Wait at least 90 seconds before performing any other work.
3. Remove engine under cover.
4. Remove the cool air intake duct seal.
5. Remove the bolt and No. 1 air cleaner inlet.
6. Remove the 2 nuts and the V-bank cover sub-assembly.
7. Drain the engine coolant.
8. Disconnect the hose from the water inlet with thermostat.
9. Remove the 3 nuts, water inlet with thermostat and gasket.

To install:

10. Install a new gasket and the water inlet with thermostat with the 3 nuts. Tighten the 3 nuts to 7 ft. lbs. (10 Nm).
11. Connect the radiator outlet hose to the radiator assembly and secure it with the clip. Make sure that the claws on the clip are positioned outward with space between the hose end and the clip end should be less than 1–5mm. Hose contact with the stopper should be less than 2.5mm.
12. Refill the engine coolant.
13. Connect the negative battery cable.
14. Start the engine and check cooling system for leaks.
15. Install the V-bank cover sub-assembly and tighten the 2 nuts to 44 inch lbs. (5 Nm).
16. Install the No. 1 air cleaner inlet. Tighten the bolt to 44 inch lbs. (5 Nm).
17. Install the cool air intake duct seal.
18. Install engine under cover.
19. Perform system initialization procedure.

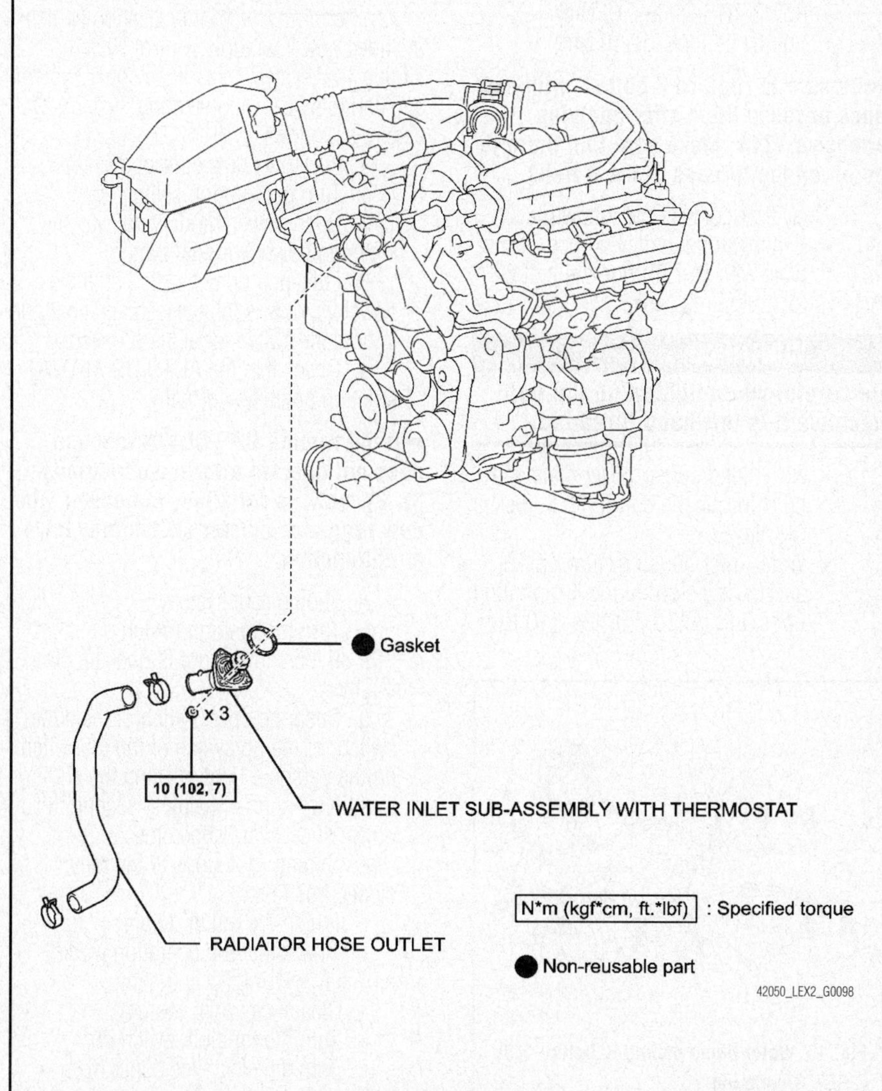

Fig. 13 Exploded view of thermostat assembly and related components—2GR-FSE engine

4.3L (3UZ-FE) Engine

See Figure 14.

1. Before servicing the vehicle, refer to the precautions in the beginning of this section.
2. Disconnect the negative battery cable. Wait at least 90 seconds before performing any other work.
3. Remove the air cleaner inlet No. 1.
4. Drain the engine coolant.
5. Remove the 3 nuts and disconnect the water inlet from the water inlet housing.
6. Remove the thermostat.
7. Remove the gasket from the thermostat.

To install:

8. Install a new gasket to the thermostat.
9. Insert the thermostat into the water inlet housing with the jiggle valve facing straight upward. The jiggle valve may be set within 30° of either side of the prescribed position.
10. Install the water inlet with the 3 nuts and tighten to 14 ft. lbs. (19 Nm).
11. Refill the engine coolant.
12. Connect the negative battery cable.
13. Start the engine and check cooling system for leaks.
14. Install the air cleaner inlet No. 1.
15. Check the coolant level and add if necessary.

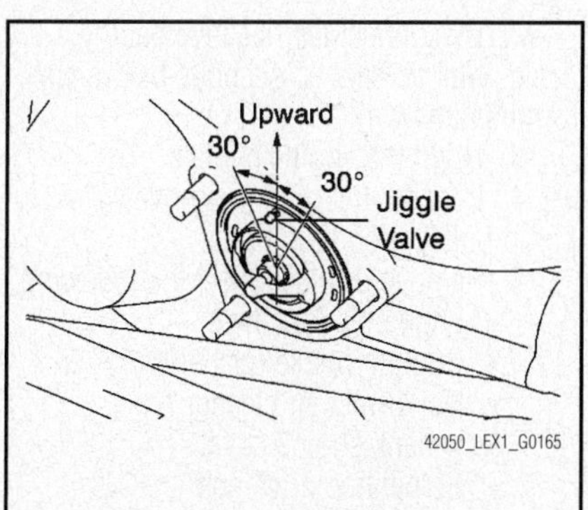

Fig. 14 Correct alignment of the thermostat jiggle valve—3UZ-FE engine

4.6L (1UR-FSE) Engine

1. Before servicing the vehicle, refer to the precautions in the beginning of this section.
2. Disconnect the negative battery cable. Wait at least 90 seconds before performing any other work.
3. Drain the engine coolant.
4. Remove air cleaner inlet.
5. Remove the 3 nuts holding the water inlet to the inlet housing, and disconnect the water inlet together with the radiator hose from the water inlet housing.
6. Remove the thermostat.
7. Remove the gasket from the thermostat.

To install:

8. Install a new gasket to the thermostat.
9. Insert the thermostat into the water inlet housing with the jiggle valve facing straight upward. The jiggle valve may be set within 30° of either side of the prescribed position.
10. Install the water inlet with thermostat with the 3 nuts. Tighten the 3 nuts to 13 ft. lbs. (18 Nm).
11. Install air cleaner inlet.
12. Refill the engine coolant.
13. Connect the negative battery cable.
14. Start the engine and check cooling system for leaks.
15. Check the coolant level and add if necessary.
16. Install the battery clamp cover and air cleaner inlet.

WATER PUMP

REMOVAL & INSTALLATION

3.5L (2GR-FSE) Engine

See Figure 15.

1. Before servicing the vehicle, refer to the precautions in the beginning of this section.
2. Disconnect the negative battery cable. Wait at least 90 seconds before performing any work.
3. Drain the engine coolant.
4. Loosen, but do not remove water pump pulley mounting bolts.
5. Remove or disconnect the following:
   - Engine under cover
   - Cool air intake duct seal
   - No. 1 inlet air cleaner
   - V-bank engine cover
   - Serpentine drive belt
   - Injector driver
   - Water pump pulley
   - No. 2 engine cover
   - No. 1 engine cover
   - Mounting fasteners and injector driver unit
   - Radiator inlet and outlet hoses
   - 5 hoses from the water inlet housing
   - Water inlet housing, gasket and O-ring
   - Bolt, cover plate and No. 2 idler pulley
   - Bolt, cover plate and belt tensioner pulley (marking: L)

➡Do not turn the bolt "L" counterclockwise.

   - 16 bolts, water pump assembly and water pump gasket

To install:

6. Install a new water pump gasket and water pump assembly with the 16 mounting bolts and tighten as follows:
   - Bolt A: 15 ft. lbs. (21 Nm)
   - Bolt B: 81 inch lbs. (9 Nm)
   - Bolt C: 81 inch lbs. (9 Nm)

➡Be sure to replace 2 bolts C with new ones or reuse them after applying adhesive 1344. Make sure that there is no oil on the threads of the A bolts.

7. Install or connect the following:
   - Belt tensioner pulley and cover plate with bolt. Torque the bolt to 32 ft. lbs. (43 Nm).

⁂ WARNING

Be careful when tightening the bolt because it is left-hand threaded.

   - No. 2 idler pulley, cover plate and bolt. Torque the bolt to 32 ft. lbs. (43 Nm).
   - Water inlet housing, new gasket and O-ring. Torque the 4 mounting bolts and nut to 7 ft. lbs. (10 Nm).

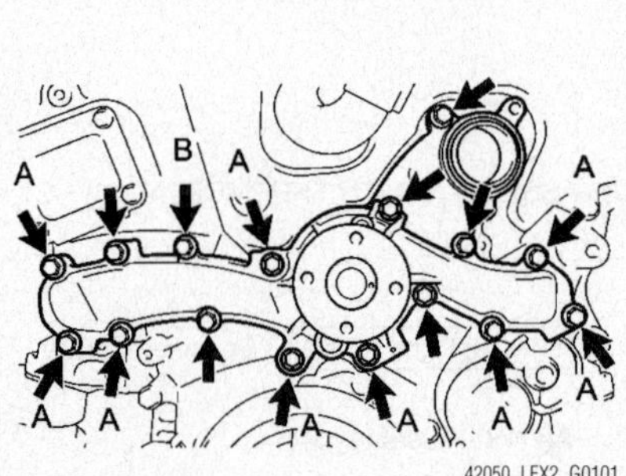

Fig. 15 Water pump mounting bolts—3.5L (2GR-FSE) engine

➡Be careful not to allow the O-ring to get caught between parts.

   - 5 hoses to the water inlet housing
   - Inlet and outlet radiator hoses and new clamps
   - Water pump pulley with 4 bolts finger tight only
   - Injector driver unit
   - No. 1 engine cover
   - No. 2 engine cover
   - Serpentine drive belt
   - V-bank engine cover
   - No. 1 inlet air cleaner
   - Cool air intake duct seal
   - Engine under cover
   - Negative battery cable

8. Torque the 4 water pump pulley bolts to 15 ft. lbs. (21 Nm).
9. Add engine coolant.
10. Start the engine, check for leaks and bleed the cooling system.
11. Recheck all fluid levels and add if necessary.
12. Perform system initialization (which includes power window control system, sliding roof system, clearance sonar system and variable gear ratio steering system) procedure as follows:
   - Power window control system

   a. Turn the ignition switch on.
   b. Open power window halfway by pressing power window switch.
   c. Fully pull up the switch until the power window is fully closed and continue to hold the switch for at least 1 second.
   d. Check that the AUTO UP / DOWN function operates normally.

➡If the remote UP / DOWN function does not operate after the conditions 1), 2), or 3) is satisfied, the power window regulator master switch may have a malfunction.

   - Sliding roof system

   e. Turn the ignition switch on.
   f. If the sliding roof is opened, close it fully.
   g. Push the open switch of the slide switch, or the up switch of the tilt switch on the personal light, making the sliding roof tilt up approximately 1 second, tilt down, slide open, slide close.
   h. Sliding roof stops at the fully closed position.
   i. Finish the initialization.
   j. Check that the operation works normally with AUTO operation.

   - Clearance sonar system

   k. Turn the ignition switch on.
   l. Turn the clearance sonar main switch ON.

m. Turn the steering wheel to the full left and right lock position.

➡Make sure to completely turn the steering wheel to the left and right full lock position.

n. Confirm that the learning operation has been completed by checking the multi-information display.

o. At an area with few turns and curves, and minimal traffic, drive at 20 km/h or more for 5 minutes or more.

- Variable gear ratio steering system

p. Turn the ignition switch on, and check that the master warning light and VSC/ABS warning lights illuminate for a few seconds.

➡If the warning lights remain on or blink, repair the applicable system.

q. Drive the vehicle on a straight road at 35 km/h (22 mph) or more for 5 seconds or longer.

r. Confirm that steering angle sensor initialization is completed by doing the following:

- Drive the vehicle on a straight road at 60 km/h (37 mph) or more for 30 seconds or longer.
- Stop the vehicle (engine running).
- Slowly turn the steering wheel from lock to lock.
- If it turns approximately 2.7 turns, steering angle sensor initialization is completed. If it turns approximately 3.2 turns, steering angle sensor initialization is not completed.

13. Road test the vehicle.

4.3L (3UZ-FE) Engine

See Figures 16 and 17.

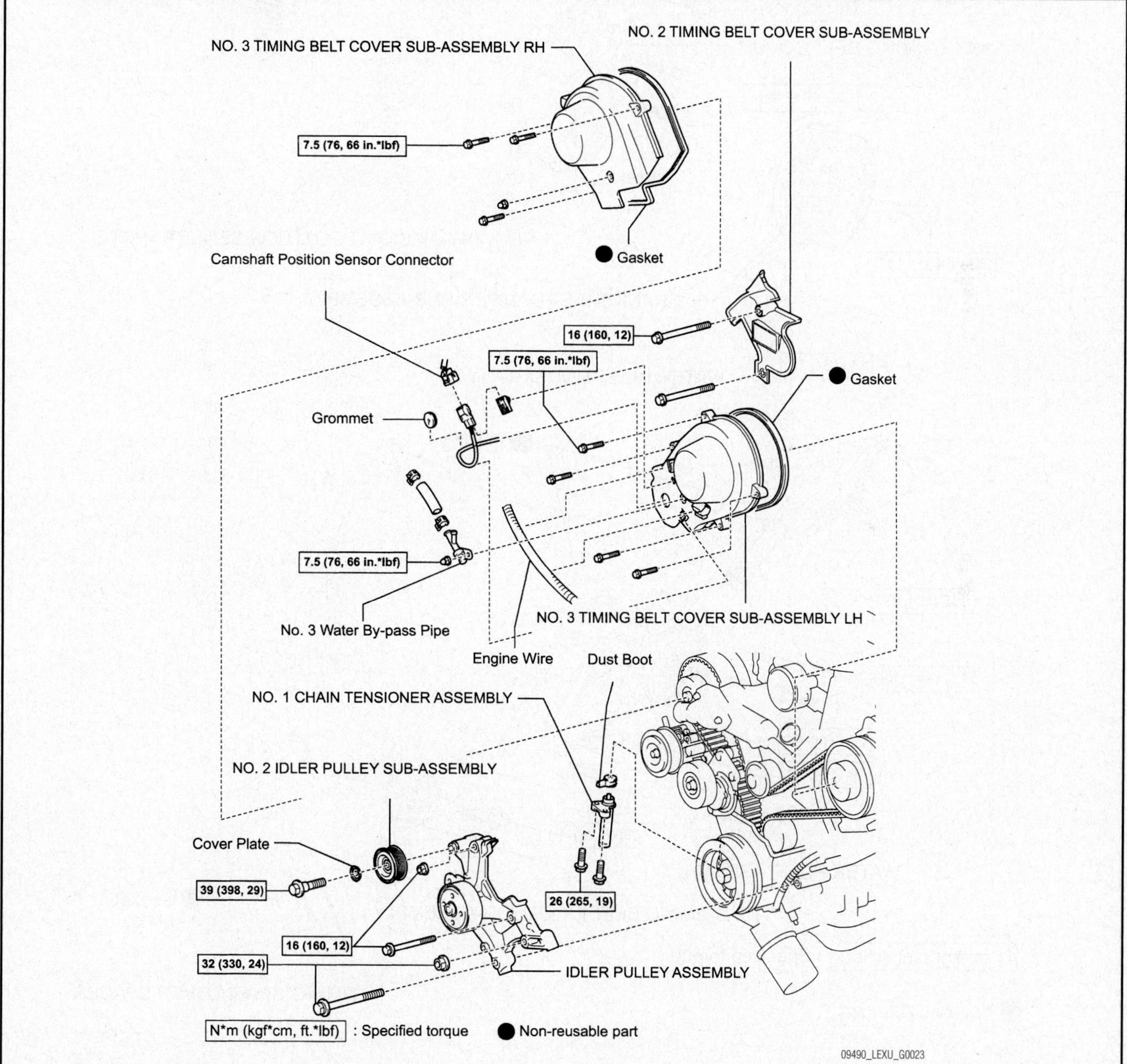

Fig. 16 Exploded view of water pump assembly—3UZ-FE Engine

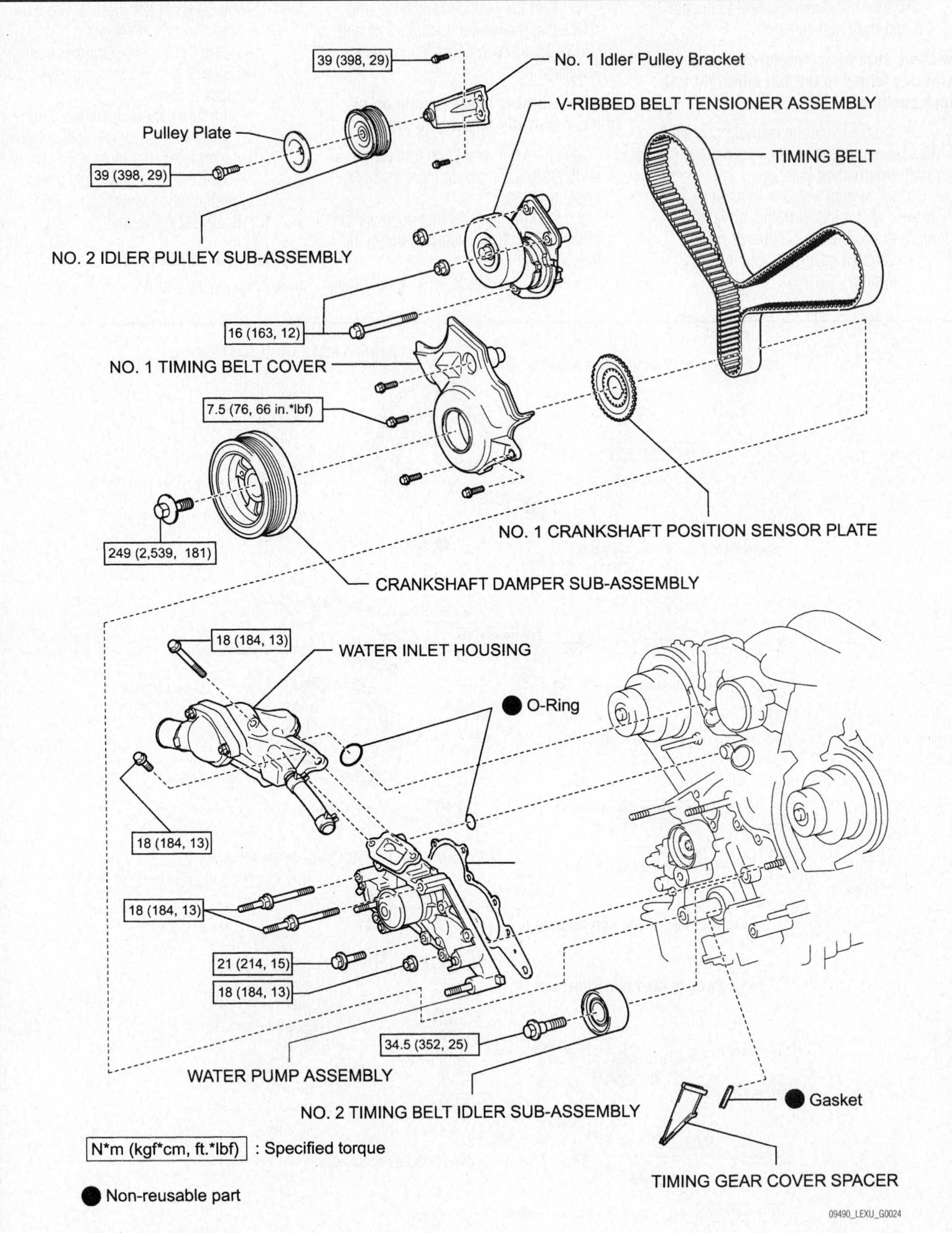

Fig. 17 Exploded view of water pump assembly—GS430

1. Before servicing the vehicle, refer to the precautions in the beginning of this section.
2. Disconnect the negative battery cable.
3. Drain the cooling system.
4. Remove or disconnect the following:
   - Engine under cover
   - Timing belt
   - Water inlet housing
   - Timing belt idler sub assembly
   - Water pump

To install:

5. Install or connect the following:
   - New gasket, water pump with 5 bolts and 2 stud bolts and nuts. Tighten bolt to 15 ft lbs (21 Nm) and stud nuts to 13 ft. lbs (18 Nm)
   - Water inlet housing with new O-ring and seal packing

➡If O-ring contacts engine oil it must be replaced

   - Timing belt idler sub assembly
   - Timing belt assembly
   - Radiator assembly
   - Engine coolant
   - Negative battery cable
6. Check for coolant leaks.
7. Initialize the power windows and seat control systems.
   - Engine under cover

4.6L (1UR-FSE) Engine

See Figure 18.

1. Remove engine under cover.
2. Drain engine coolant.

Fig. 18 Water pump bolt locations

3. Remove V-bank cover sub-assembly.
4. Remove cool air intake duct seal.
5. Remove air cleaner assembly.
6. Remove engine room side cover.
7. Remove radiator reserve tank assembly.
8. Remove V-ribbed belt.
9. Disconnect the No. 2 radiator hose from the water inlet sub-assembly with thermostat.
10. Disconnect the No. 5 water by-pass hose from the water inlet housing.
11. Disconnect the water inlet hose from the water inlet housing.
12. Disconnect the No. 3 water by-pass hose from the water inlet housing.
13. Remove the 3 bolts, water inlet housing and gasket.
14. Using a screwdriver or an equivalent, hold the water pump pulley.
15. Remove the 9 bolts, water pump and gasket.

To install:

16. Install the water pump and a new gasket with the 9 bolts as shown in the illustration. Tighten bolts to 15 ft. lbs. (20 Nm), 17 ft. lbs. (23 Nm), 35 ft. lbs. (47 Nm) in order.
17. Temporarily install the water pump pulley with the 4 bolts.
18. Using a screwdriver or an equivalent, hold the pulley and tighten the 4 bolts to 15 ft. lbs. (21 Nm).
19. Install the water inlet housing with the 3 bolts and tighten to 15 ft. lbs. (21 Nm).
20. Connect the No. 3 water by-pass hose to the water inlet housing.
21. Connect the water inlet hose to the water inlet housing.
22. Connect the No. 5 water by-pass hose to the water inlet housing.
23. Connect the No. 2 radiator hose to the water inlet sub-assembly with thermostat.
24. Install V-ribbed belt.
25. Install radiator reserve tank assembly.
26. Install air cleaner assembly.
27. Add engine coolant.
28. Inspect for coolant leak.
29. Install engine room side cover.
30. Install cool air intake.
31. Install V-bank cover sub-assembly.
32. Install engine under cover.

ENGINE ELECTRICAL

CHARGING SYSTEM

ALTERNATOR

REMOVAL & INSTALLATION

3.5L (2GR-FSE) Engine

1. Before servicing the vehicle, refer to the precautions in the beginning of this section.
2. Remove or disconnect the following:
   - Negative battery cable. Wait at least 90 seconds before performing any other work.
   - V-bank cover sub-assembly
   - Cool air intake duct seal
   - Left engine room side cover
   - Engine under cover
   - Serpentine drive belt
3. Remove or disconnect the with pulley compressor by performing the following:
   - Bolt, nut and bracket (AWD)
   - Magnetic clutch connector
   - Nut and 3 bolts (2WD); or 2 bolts (AWD)
   - Stud bolt using an E8 Torx® socket and with pulley compressor

➡It is not necessary to completely remove the compressor. With the hoses connected to the compressor, hang the compressor on the vehicle body with a rope.

   - No. 2 idler pulley sub-assembly
   - Clamp and bolt from alternator
   - Alternator connector
   - Rubber cap
   - Nut and battery cable.
   - Nut, bolt and alternator bracket
   - 2 mounting bolts and alternator

To install:

4. Install or connect the following:
   - Alternator and 2 mounting bolts. Torque to 32 ft. lbs. (43 Nm).
   - Alternator bracket with bolt and nut at the engine. Torque to 15 ft. lbs. (20 Nm).
   - Battery cable. Tighten the nut to 87 inch lbs. (10 Nm).
   - Wire harness bracket and clamp to the alternator
   - Alternator wire to terminal.
   - Terminal cap
   - No. 2 idler pulley and cover plate with the bolt. Tighten bolt to 32 ft. lbs. (43 Nm).
   - With pulley compressor assembly
   - Serpentine drive belt
   - Engine under cover
   - Left engine room side cover
   - Cool air intake duct seal
   - V-bank cover sub-assembly
   - Negative battery cable

4.3L (3UZ-FE) and 4.6L (1UR-FSE) Engines

1. Before servicing the vehicle, refer to the precautions in the beginning of this section.
2. Remove or disconnect the following:
   - Negative battery cable. Wait at least 90 seconds before performing any other work.
   - Air cleaner inlet
   - Accessory drive belt
   - Oil pan protector
   - Engine under cover
   - Power steering pump
   - Alternator harness connectors
   - Heated Oxygen (HO2S) sensor wiring
   - Alternator

To install:

3. Install or connect the following:
   - Alternator. Tighten the fasteners to 29 ft. lbs. (39 Nm).
   - HO2S sensor wiring
   - Alternator harness connectors
   - Power steering pump
   - Engine under cover
   - Oil pan protector
   - Accessory drive belt
   - Air cleaner inlet
   - Negative battery cable

ENGINE ELECTRICAL

IGNITION SYSTEM

FIRING ORDER

See Figures 19 and 20.

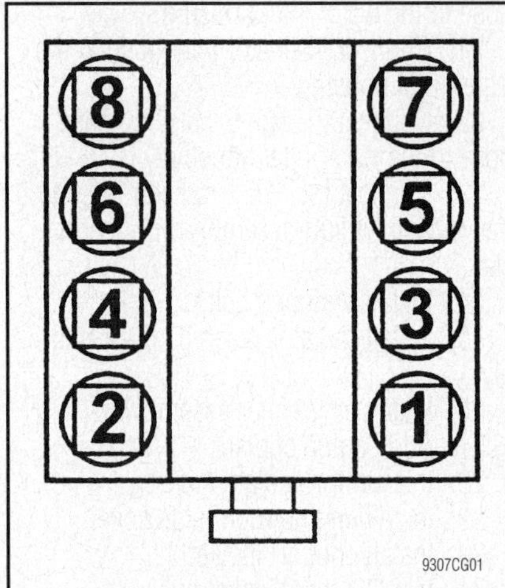

**Fig. 19 4.6L (1UR-FSE) and 4.3L (3UZ-FE) Engines
Firing order: 1–8–4–3–6–5–7–2
Distributorless ignition system
(one coil on each cylinder)**

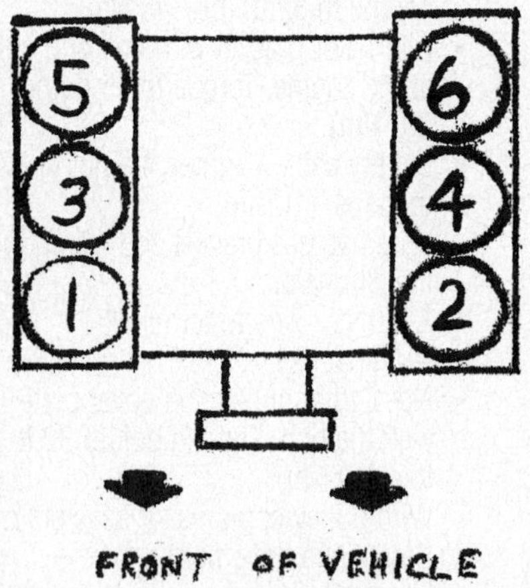

**Fig. 20 3.5L (2GR-FSE) Engines
Firing order: 1–2–3–4–5–6
Distributorless ignition system
(one coil per cylinder)**

IGNITION COIL PACK

REMOVAL & INSTALLATION

GS 350

See Figures 21 through 24.

1. Before servicing the vehicle, refer to the precautions in the beginning of this section.
2. Disconnect the negative battery cable. Wait at least 90 seconds before performing any other work.

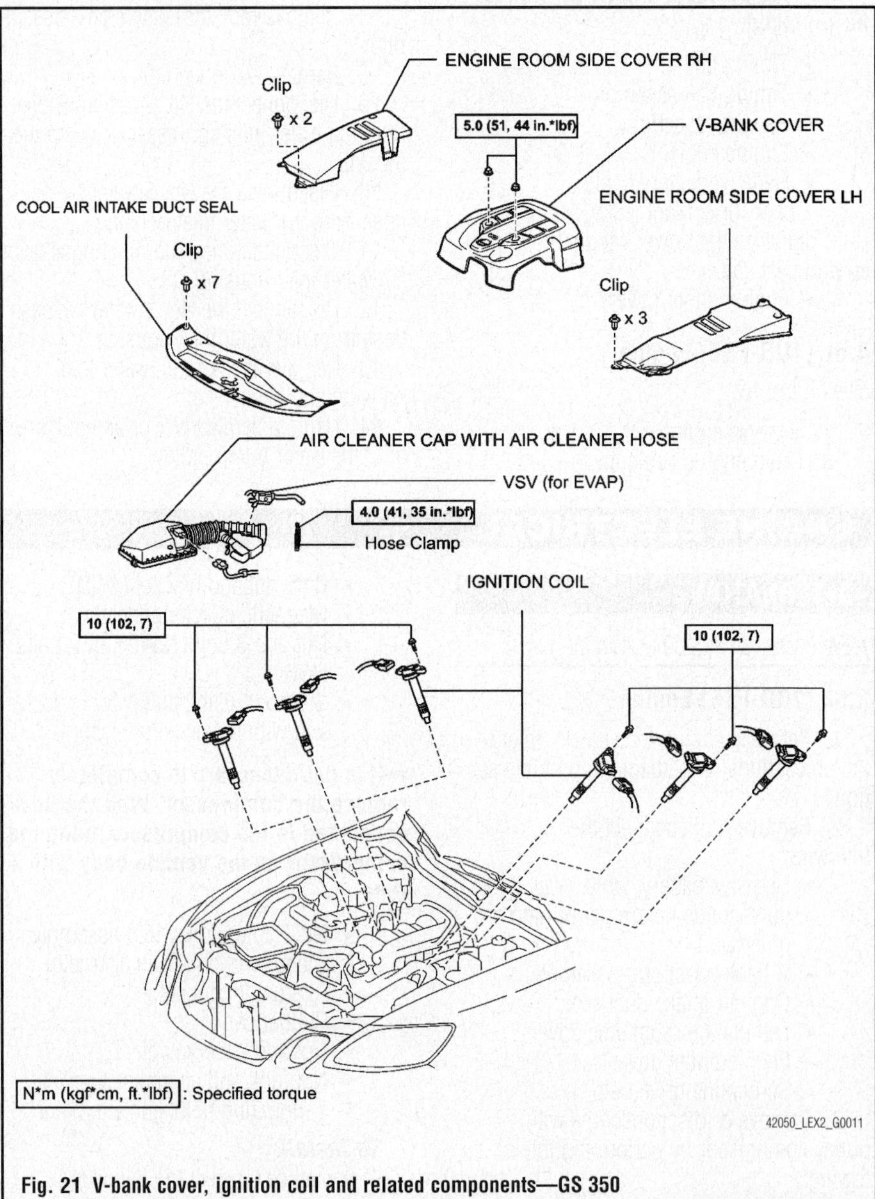

Fig. 21 V-bank cover, ignition coil and related components—GS 350

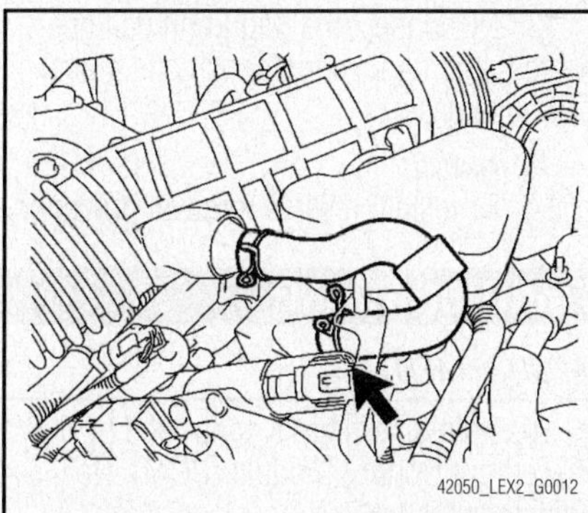

Fig. 22 Disconnect the ventilation hose at the cylinder head—GS 350

Fig. 23 Disconnect the MAF meter connector and clamp from the air cleaner—GS 350

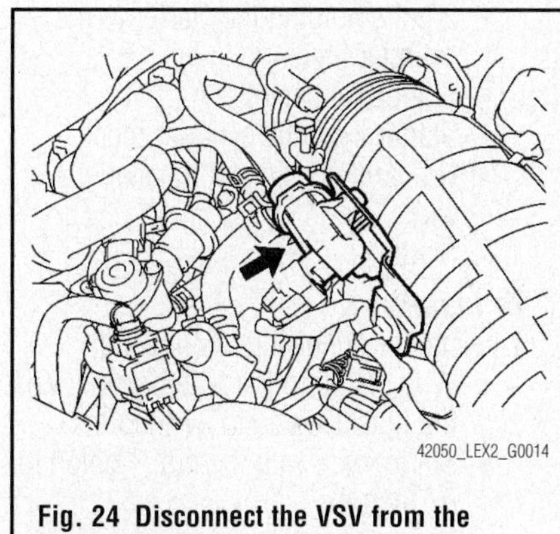

Fig. 24 Disconnect the VSV from the EVAP—GS 350

3. Remove cool air intake duct seal.
4. Remove the right and left engine room side covers.
5. Remove V-bank cover.
6. Disconnect the ventilation hose from the cylinder head.
7. Disconnect the Mass Air Flow (MAF) meter connector.
8. Disconnect the clamp from the air cleaner.
9. Disconnect the VSV from the EVAP.
10. Disconnect the 4 clamps.
11. Remove the hose clamp and air cleaner cap with air cleaner hose.
12. Disconnect the 2 wire harness clamps.
13. Disconnect the 6 ignition coil connectors.
14. Remove the 6 bolts and 6 ignition coils.

To install:

15. Install the 6 ignition coils with the 6 bolts and torque to 7 ft. lbs. (10 Nm).
16. Connect the 6 ignition coil connectors.
17. Connect the 2 wire harness clamps.
18. Install the air cleaner cap with the air cleaner hose assembly with the 4 clamps and hose clamp.

➡Be sure to install the air cleaner assembly so that the screw part of the hose clamp is as shown in the illustration.

19. Install the VSV for EVAP to the air cleaner hose.
20. Connect the MAF meter connector and clamp to the air cleaner.
21. Connect the ventilation hose to the cylinder head cover with the clamp.
22. Install V-bank cover.
23. Install the right and left engine room side covers.
24. Install cool air intake duct seal.
25. Connect the negative battery cable.
26. Perform system initialization procedure.

GS 430

See Figure 25.

1. Before servicing the vehicle, refer to the precautions in the beginning of this section.
2. Disconnect the negative battery cable. Wait at least 90 seconds before performing any other work.
3. Remove cool air intake duct seal.

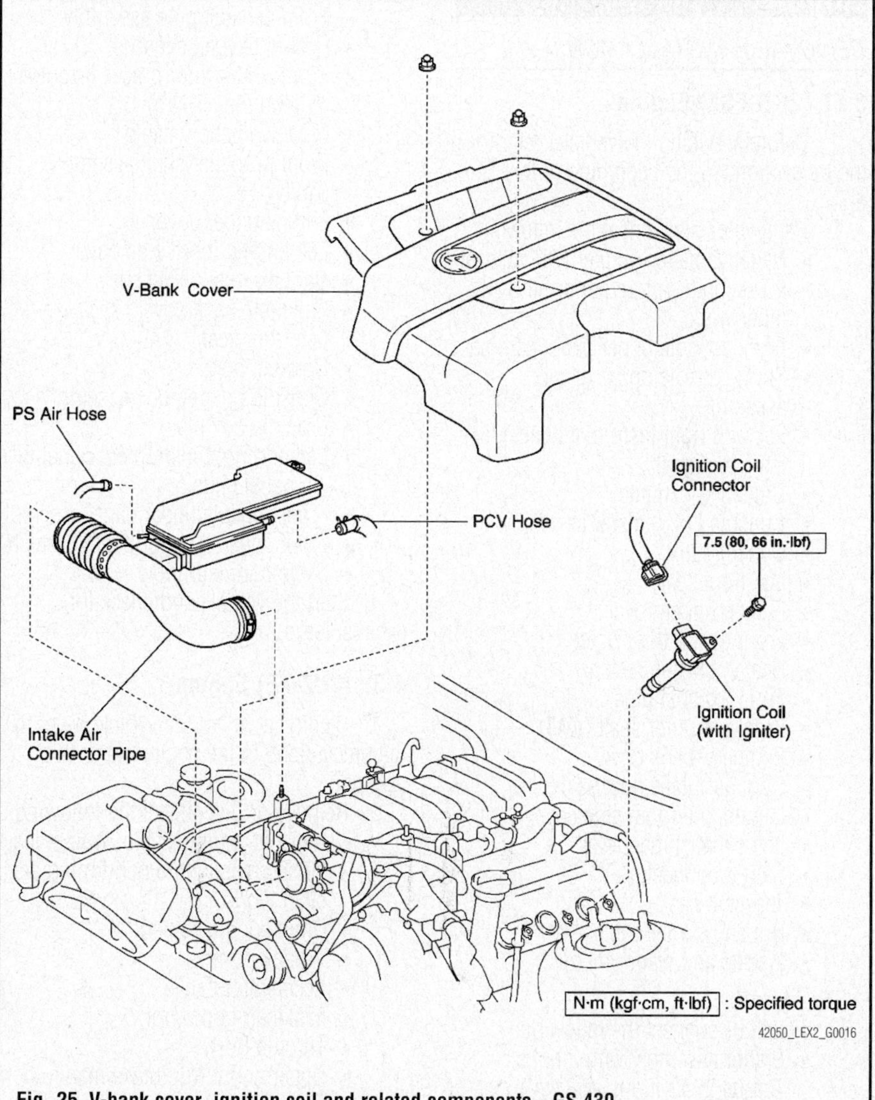

Fig. 25 V-bank cover, ignition coil and related components—GS 430

4. Remove the right and left engine room side covers.
5. Remove the No.1 air cleaner inlet.
6. Remove the 2 nuts and V-bank cover.
7. Disconnect the ignition coil connector.
8. Remove the bolt and pull out the ignition coil.

To install:

9. Install the 8 ignition coils with the 8 bolts and torque to 66 inch lbs. (7.5 Nm).
10. Connect the ignition coil connectors.
11. Install V-bank cover.
12. Install the No.1 air cleaner inlet.
13. Install the right and left engine room side covers.
14. Install cool air intake duct seal.
15. Connect the negative battery cable.
16. Perform system initialization procedure.

GS460

1. Disconnect cable from negative battery terminal.
2. Remove V-bank cover sub-assembly.
3. Remove cool air intake duct seal.
4. Remove engine room side covers.
5. Remove air cleaner assembly.
6. Remove skid control ECU.
7. Remove battery.
8. Remove no. 1 battery tray support.
9. Remove power steering ECU assembly.
10. Remove the 2 bolts and battery tray.
11. Separate No. 1 engine room relay block assembly.
12. Remove the 2 wire harness clamps.
13. Disconnect the 8 ignition coil connectors and remove ignitions coils along with tube gaskets.

To install:

14. To install, reverse removal procedure

IGNITION TIMING

ADJUSTMENT

The ignition timing is controlled by the Powertrain Control Module (PCM). No adjustment is necessary or possible.

SPARK PLUGS

REMOVAL & INSTALLATION

See ignition Coil Pack Removal and Installation.

ENGINE ELECTRICAL

STARTING SYSTEM

STARTER

REMOVAL & INSTALLATION

3.5L (2GR-FSE) Engine

1. Before servicing the vehicle, refer to the precautions in the beginning of this section.
2. Remove or disconnect the following:
   - Negative battery cable. Wait at least 90 seconds before performing any other work.
   - Front console upper panel garnish
   - Front console upper panel assembly
   - Left and right instrument panel finish panel ends
   - Console box plate
   - Console box register assembly
   - Console box
   - Left front seat
   - Engine under cover
   - Front center floor brace
   - Heated oxygen sensor
   - Front exhaust pipe
   - Front propeller shaft (AWD)
   - Engine V-bank cover
   - Cool air intake duct seal
   - Left engine room side cover
   - Left exhaust manifold
   - Starter connector
   - Terminal cap
   - Nut and starter cable
   - 2 bolts and starter motor

To install:

3. Install or connect the following:
   - Starter assembly with 2 bolts. Torque to 43 ft. lbs. (58 Nm).
   - Starter wires
   - Left exhaust manifold
   - Front exhaust pipe assembly
   - Heated oxygen sensor
   - Front center floor brace. Torque to 65 inch lbs. (7 Nm).
   - Negative battery cable
   - Front propeller shaft assembly (AWD)
   - Engine under cover
   - Left engine room side cover
   - Cool air intake duct seal
   - Engine V-bank cover
   - Left front seat
   - Console box
   - Console box register assembly
   - Console box plate
   - Left and right instrument panel finish panel ends
   - Front console upper panel assembly
   - Front console upper panel garnish
   - System initialization
4. Start the vehicle and check for exhaust leaks.

4.3L (3UZ-FE) Engine

1. Before servicing the vehicle, refer to the precautions in the beginning of this section.
2. Remove or disconnect the following:
   - Negative battery cable. Wait at least 90 seconds before performing any other work
   - Drain engine coolant
   - V-bank cover
   - Accelerator cable
   - Intake air connector
   - Throttle Body
   - Upper and lower intake manifold assembly
   - Rear water bypass joint
   - Water bypass pipe
   - Water bypass pipe from the water pump
   - Wire clamp from the bracket on the water bypass pipe
   - O-ring from the water bypass pipe
   - Water bypass pipe bracket from the water bypass pipe
   - 2 bolts holding the starter to the cylinder block
   - Starter connector
   - Starter from the cylinder block
   - Nut, and disconnect the starter wire
   - Starter

To install:

3. Install or connect the following:
   - Wire clamp to the wire bracket with the bolt. Tighten to 87 inch lbs.
   - Starter wire with the nut. Tighten to 87 inch lbs.
   - Starter connector
   - Starter with the 2 bolts. Torque the bolts to 29 ft. lbs. (39 Nm).
   - Water bypass pipe bracket to the water bypass pipe
   - O-ring to the water bypass pipe
   - Water bypass pipe
   - Wire clamp to the bracket on the water bypass pipe
   - Water bypass pipe bolts. Torque the bolts to 13 ft. lbs. (18 Nm).
   - Rear water bypass joint
   - Intake manifold assembly
   - Throttle body
   - Intake air connector
   - Accelerator cable
   - V-bank cover

4.6L (1UR-FSE) Engine

1. Disconnect cable from negative battery terminal.
2. Remove engine under covers.
3. Remove cool air intake duct seal.
4. Remove air cleaner.
5. Remove V-ribbed belt.
6. Remove the front stabilizer bar.
7. Remove front floor cover.
8. Remove front center floor brace.
9. Remove front exhaust pipe assembly.
10. Remove steering sliding yoke sub-assembly.
11. Remove generator assembly.
12. Remove engine oil level dipstick guide.
13. Disconnect compressor and pulley.
14. Remove engine room side covers.
15. Remove V-bank cover.
16. Remove exhaust manifold sub-assemblies.
17. Remove starter assembly.

To install:

18. Install starter assembly.
19. Install exhaust manifold sub-assembly LH.
20. Connect compressor and pulley.
21. Install engine oil level dipstick guide.
22. Install generator assembly.
23. Install steering sliding yoke sub-assembly.
24. Install front exhaust pipe assembly.
25. Install front floor covers.
26. Install front center floor brace.
27. Install the front stabilizer bar.
28. Install V-ribbed belt.
29. Install air cleaner.
30. Connect cable to negative battery terminal.
31. Install engine under covers.
32. Install V-bank cover.
33. Install engine room side covers.
34. Install cool air intake duct seal.

ENGINE MECHANICAL

ACCESSORY DRIVE BELTS

ACCESSORY BELT ROUTING

See Figures 26 through 28.

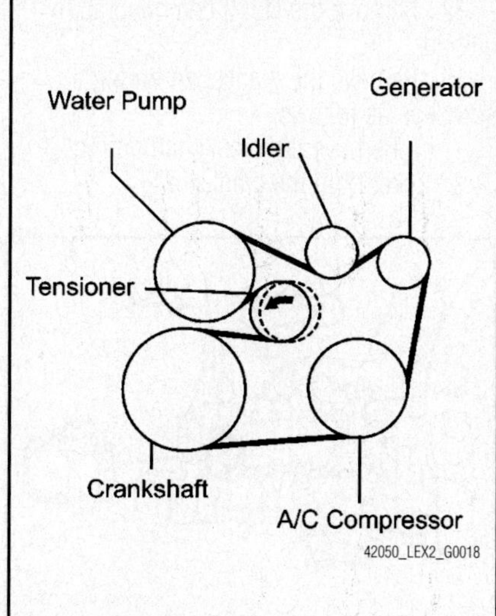

Fig. 26 Serpentine drive belt routing—3.5L (2GR-FSE) engine

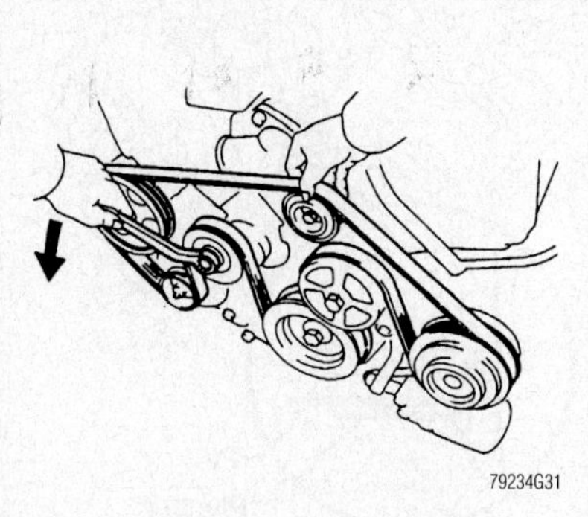

Fig. 27 Serpentine drive belt routing—4.3L (3UZ-FE) engine

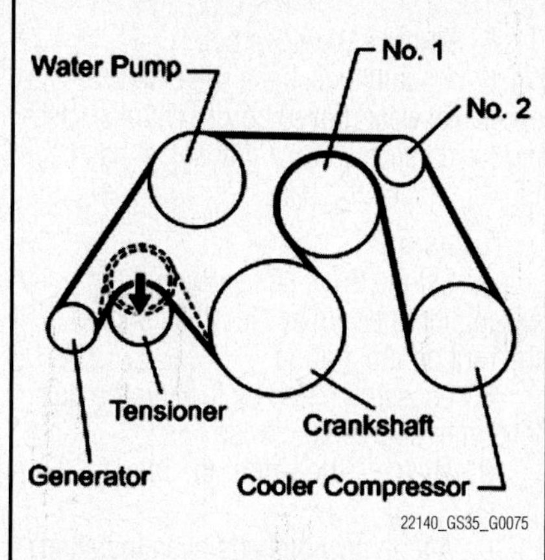

Fig. 28 Serpentine drive belt routing—4.6L (1UR-FSE) engine

INSPECTION

Inspect the drive belt for signs of glazing or cracking. A glazed belt will be perfectly smooth from slippage, while a good belt will have a slight texture of fabric visible. Cracks will usually start at the inner edge of the belt and run outward. All worn or damaged drive belts should be replaced immediately.

ADJUSTMENT

These engines are equipped with automatic belt tensioners. Adjusting the belt tension is not possible or necessary.

REMOVAL & INSTALLATION

3.5L (2GR-FSE) Engine

See Figures 29 and 30.

1. Before servicing the vehicle, refer to the precautions in the beginning of this section.

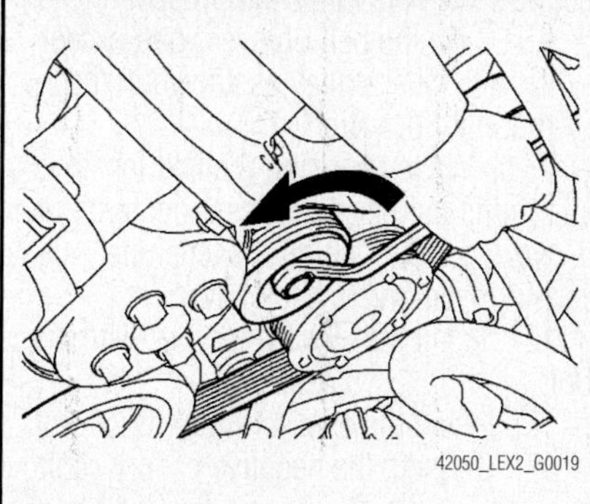

Fig. 29 Loosening the drive belt tension—3.5L (2GR-FSE) engine

Fig. 30 Fix the belt tensioner in place by inserting a 5mm bi-hexagon wrench into the holes—3.5L (2GR-FSE) engine

2. Disconnect the negative battery cable. Wait at least 90 seconds before performing any other work.
3. Remove the cool air intake duct seal.
4. Remove the air cleaner inlet.
5. Remove the V-bank cover.
6. While releasing the belt tension by turning the belt tensioner counterclockwise, and remove the drive belt from the belt tensioner.
7. While turning the belt tensioner counterclockwise, align with its holes, and

then insert a 5mm bi-hexagon wrench into the holes to fix the belt tensioner.

8. Visually check the drive belt for excessive wear, frayed cords, chunks missing from its ribs, etc. If any defect has been found, replace the belt.

9. Check that nothing gets caught in the tensioner by turning it clockwise and counterclockwise. If a malfunction exists, replace the tensioner.

To install:

10. Install the drive belt.

11. While turning the belt tensioner counterclockwise, remove the bar.

12. Put the backside of the drive belt on the tensioner pulley and idler pulley. Check that the belt is properly set to each pulley.

13. If it is difficult to install the drive belt, perform the following procedure:

a. Put the belt on every part except the tensioner pulley as shown in the routing illustration.

b. While releasing the belt tension by turning the belt tensioner counterclockwise, put the belt on the tensioner pulley.

14. Install the V-bank cover.

15. Install the air cleaner inlet with the bolt.

16. Install the cool air intake duct seal.

17. Connect the negative battery cable.

4.3L (3UZ-FE) and 4.6L (1UR-FSE) Engines

See Figures 31 and 32.

1. Before servicing the vehicle, refer to the precautions in the beginning of this section.

2. Disconnect the negative battery cable. Wait at least 90 seconds before performing any other work.

3. Remove the air cleaner inlet.

4. Loosen the belt tension by turning the belt tensioner counterclockwise.

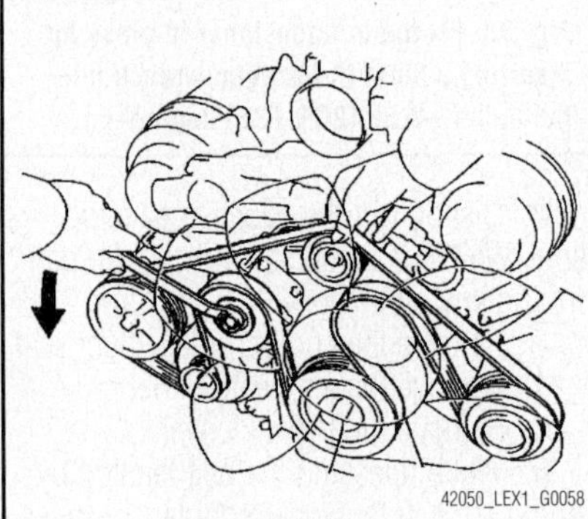

42050_LEX1_G0058

Fig. 31 Loosening the drive belt tension—4.3L (1UZ-FE and 3UZ-FE) engines

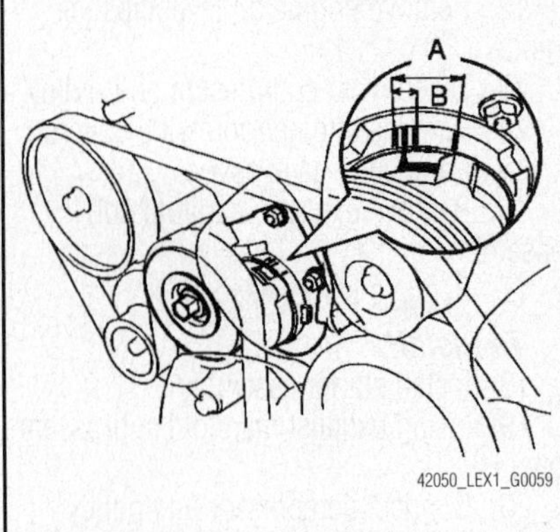

42050_LEX1_G0059

Fig. 32 Check that the mark is within range B as shown on the tensioner pulley—1UR-FSE and 3UZ-FE engines

➡The pulley bolt for the belt tensioner has a left-hand thread.

5. Remove the drive belt.

6. Visually check the drive belt for excessive wear, frayed cords, chunks missing from its ribs, etc. If any defect has been found, replace the belt.

To install:

7. Set the drive belt to everything except the idler pulley No. 2, as shown in the routing illustration.

8. Loosen the belt by turning the belt tensioner counterclockwise.

9. Then set the belt to the idler pulley.

10. After a new belt has been installed, check that the mark is within range B as shown in the illustration.

11. Install the air cleaner inlet.

12. Connect the negative battery cable.

CAMSHAFT AND VALVE LIFTERS

REMOVAL & INSTALLATION

3.5L (2GR-FSE) Engine

See Figures 33 through 57.

1. Before servicing the vehicle, refer to the precautions in the beginning of this section.

2. Drain the cooling system.

3. Drain the engine oil.

4. Relieve the fuel system pressure.

5. Remove or disconnect the following:

- Negative battery cable. Wait at least 90 seconds before performing any other work
- Oil filler cap and gasket
- Radiator cap
- No. 1 and 2 engine hangers
- Spark plugs
- Ventilation valve
- 4 camshaft position sensors
- 4 camshaft timing oil control valves
- Crankshaft position sensor
- Left and right side oil check valve bolt, oil pipe union and oil pipe
- Left and right side oil control valve filter and gaskets
- Cylinder block water drain cocks
- Oil filter element
- Water inlet and thermostat assembly
- Rear water by-pass joint
- Cylinder head cover and gaskets

6. Set the No. 1 cylinder to TDC/compression.

7. Remove the No. 1 chain tensioner assembly.

8. Remove the chain tensioner slipper.

9. Remove the chain sub-assembly.

10. Remove the idle sprocket assembly.

11. Remove the No. 1 chain vibration damper.

12. Remove the No. 2 chain vibration damper.

13. Remove the crankshaft timing sprocket, as follows:

a. Remove the crankshaft timing sprocket from the crankshaft.

22140_RX35_G0068

Fig. 33 Crankshaft timing sprocket

Push
Pin
Plunger

22140_RX35_G0069

Fig. 34 Pinning tensioner

Hold
Turn
22140_RX35_G0070

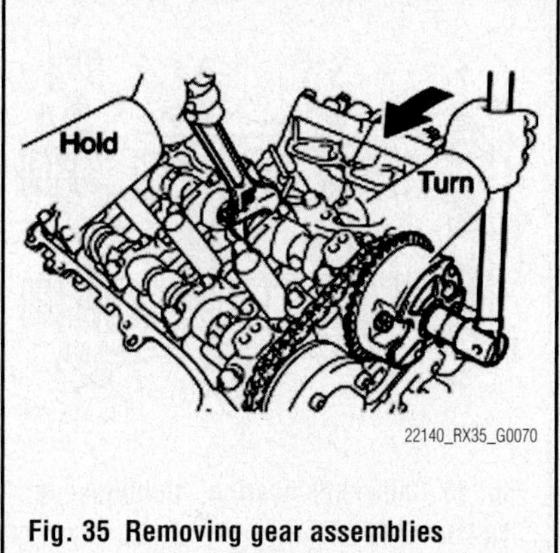

Fig. 35 Removing gear assemblies

Front View
45°
Knock Pin
22140_RX35_G0071

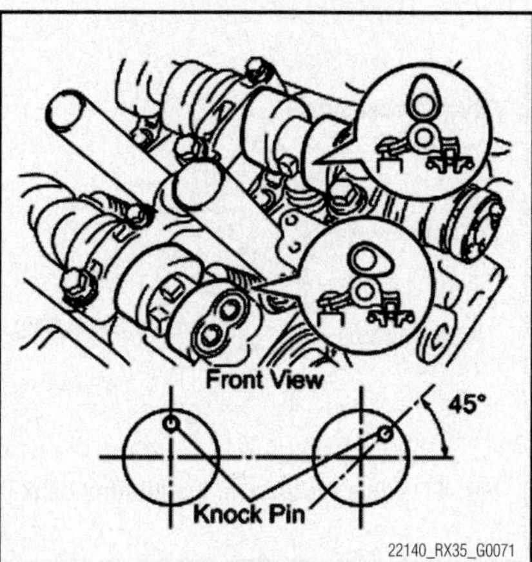

Fig. 36 Positioning camshafts for bearing cap removal bank 1

b. Remove the 2 pulley set keys from the crankshaft.

14. Remove the camshaft timing gears and No. 2 chain (for Bank 1), as follows:

a. While raising the No. 2 chain tensioner, insert a pin of 1.0 mm (0.039 in.) into the hole to fix the No. 2 chain tensioner.

b. Hold the hexagonal portion of the camshaft with a wrench, and remove the 2 bolts and 2 camshaft timing gear assemblies.

➡Be careful not to damage the cylinder head with the wrench.

➡Do not disassemble the camshaft timing gear assemblies.

c. Remove the No. 2 chain.

15. Remove the bolt and No. 2 chain tensioner assembly.

16. Remove the camshaft bearing cap (for Bank 1), as follows:

a. Check that the camshafts are positioned as shown in the illustration.

b. Uniformly loosen and remove the 8 bearing cap bolts in several steps and in the sequence shown in the illustration.

2
6
8
4
1
5
7
3
22140_RX35_G0072

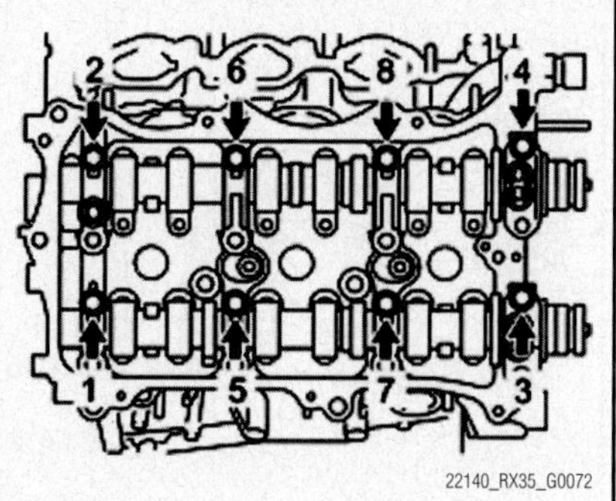

Fig. 37 Camshafts bearing cap removal sequence bank 1

7
4
10
1
6
9
12
3
5
8
2
11
22140_RX35_G0073

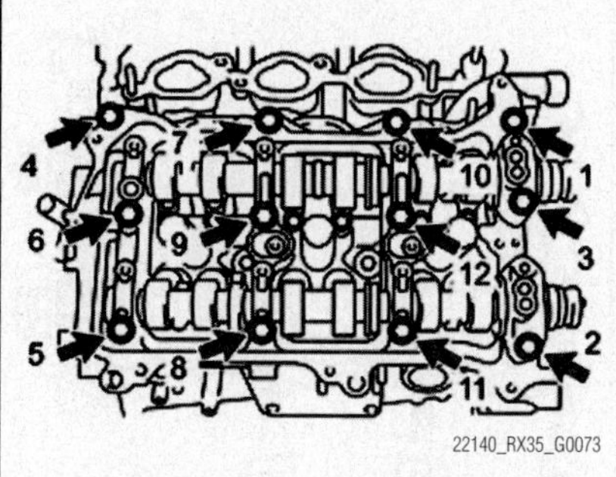

Fig. 38 Bearing cap loosening sequence bank 1

c. Uniformly loosen and remove the 12 bearing cap bolts in several steps and in the sequence shown in the illustration.

➡Uniformly loosen the bolts while keeping the camshaft level.

d. Remove the 5 camshaft bearing caps.

17. Remove the camshaft.

18. Remove the No. 2 camshaft.

19. Remove the right hand camshaft housing sub-assembly by prying between the cylinder head and the camshaft housing with a screwdriver with the tip taped.

➡Be careful not to damage the contact surfaces of the cylinder head and the camshaft housing.

20. Remove the camshaft timing gears and No. 2 chain (for Bank 2), as follows:

a. While pushing down the No. 3 chain tensioner, insert a pin of 1.0 mm (0.039 in.) into the hole to fix the No. 3 chain tensioner.

b. Hold the hexagonal portion of the camshaft with a wrench, and remove the 2 bolts and 2 camshaft timing gear assemblies.

➡Be careful not to damage the cylinder head with the wrench.

Push
Pin
Plunger
22140_RX35_G0074

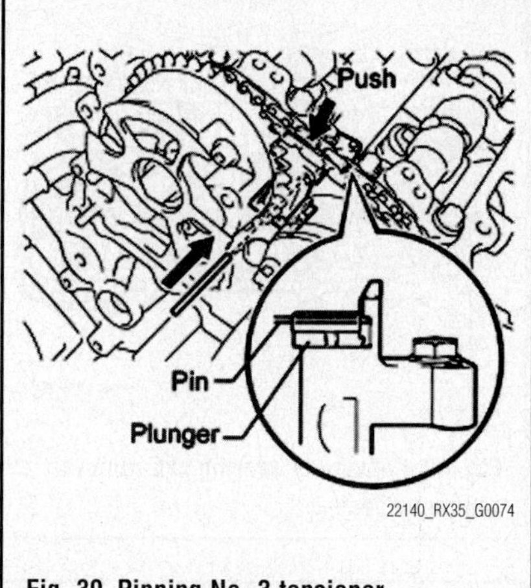

Fig. 39 Pinning No. 3 tensioner

Turn
Hold
22140_RX35_G0075

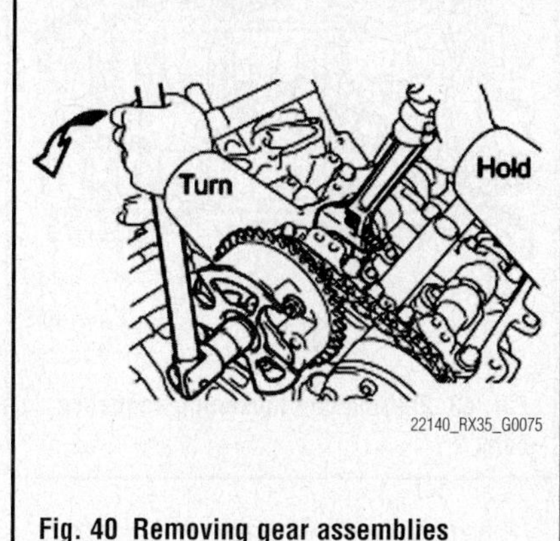

Fig. 40 Removing gear assemblies

Front View
Knock Pin
22140_RX35_G0076

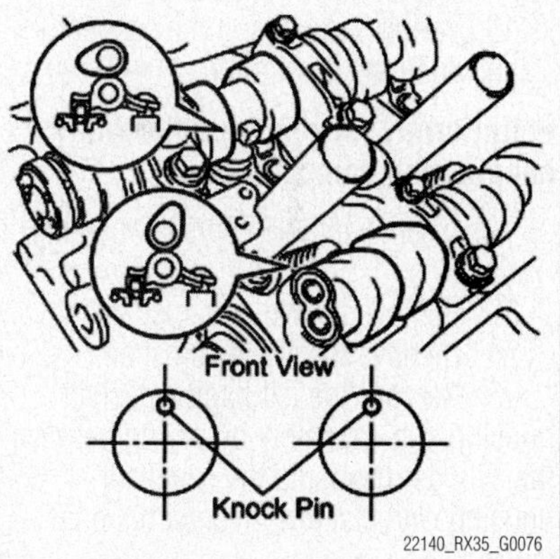

Fig. 41 Positioning camshafts for bearing cap removal bank 2

➡Do not disassemble the camshaft timing gear assemblies.

c. Remove the No. 2 chain.

21. Remove the bolt and No. 3 chain tensioner.

22. Remove the camshaft bearing cap (for Bank 2), as follows:

a. Check that the camshafts are positioned as shown in the illustration.

b. Uniformly loosen and remove the

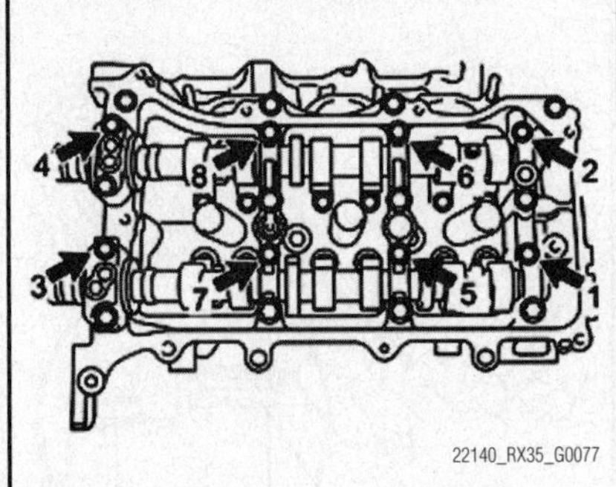

Fig. 42 Camshafts bearing cap removal sequence bank 2

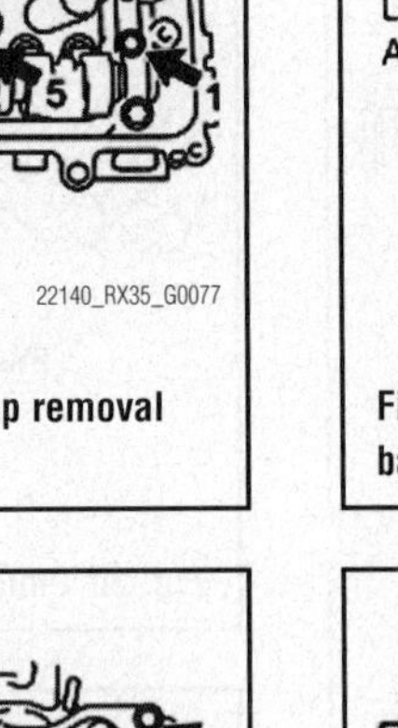

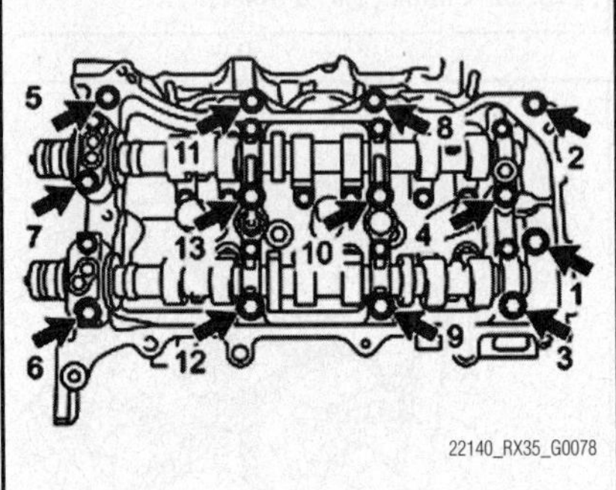

Fig. 43 Bearing cap loosening sequence bank 2

8 bearing cap bolts in the sequence shown in the illustration.

c. Uniformly loosen and remove the 13 bearing cap bolts in the sequence shown in the illustration.

➡Uniformly loosen the bolts while keeping the camshaft level.

d. Remove the 5 camshaft bearing caps.

23. Remove the No. 3 camshaft.
24. Remove the No. 4 camshaft.
25. Remove the left hand camshaft housing sub-assembly by prying between the cylinder head and the camshaft housing with a screwdriver with the tip taped.

➡Be careful not to damage the contact surfaces of the cylinder head and the camshaft housing.

26. Remove the No. 1 valve rocker arm sub-assembly, as follows:

a. Remove the 24 valve rocker arms.

➡Arrange the removed parts in the correct order.

27. Remove the valve lash adjuster assembly, as follows:

a. Remove the 24 valve lash adjusters from the cylinder head.

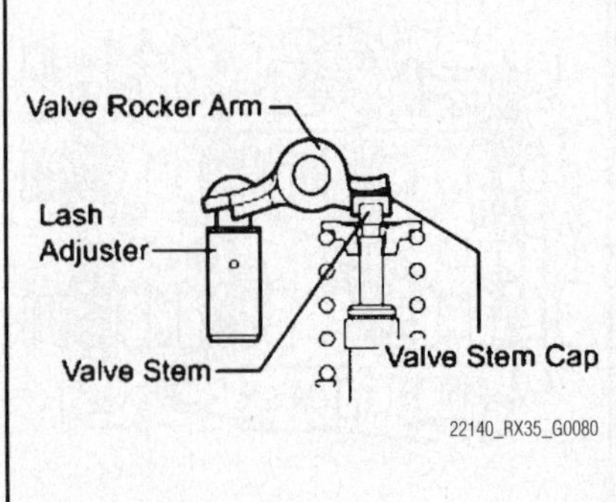

Fig. 44 Installing valve rocker arms bank 1

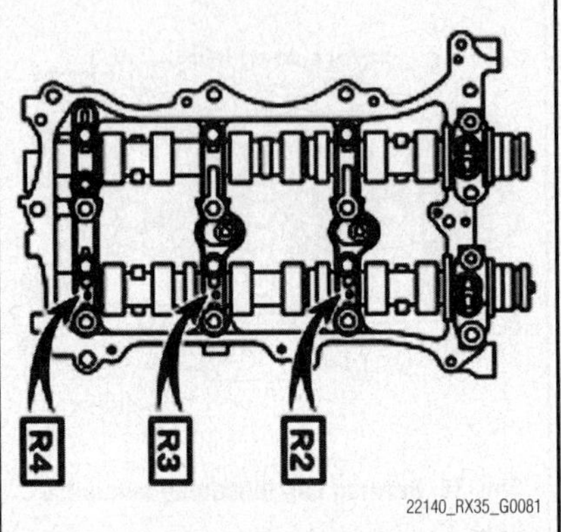

Fig. 45 Camshaft bearing caps placement bank 1

➡Arrange the removed parts in the correct order.

To install:

28. Install the valve lash adjuster assembly.
29. Install the No. 1 valve rocker arm sub-assembly, as follows:

a. Apply engine oil to the lash adjuster tip and valve stem cap end.

b. Make sure that the valve rocker arms are installed as shown in the illustration.

30. Install the camshaft bearing cap (for Bank 1), as follows:

a. Apply engine oil to the camshaft journals, camshaft housing and bearing caps.

b. Install the camshaft and No. 2 camshaft to the right camshaft housing.

c. Make sure of the marks and numbers on the camshaft bearing caps and place them in each proper position and direction.

d. Temporarily tighten the 8 bearing cap bolts to 7 ft. lbs. (10 Nm) in the order shown in the illustration.

31. Install the right camshaft housing sub-assembly, as follows:

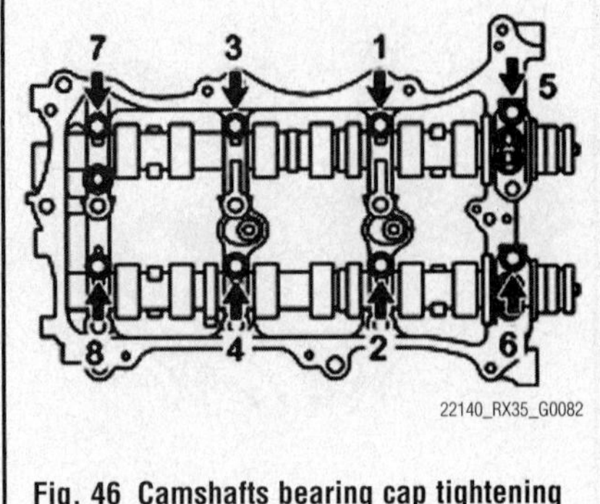

Fig. 46 Camshafts bearing cap tightening sequence bank 1

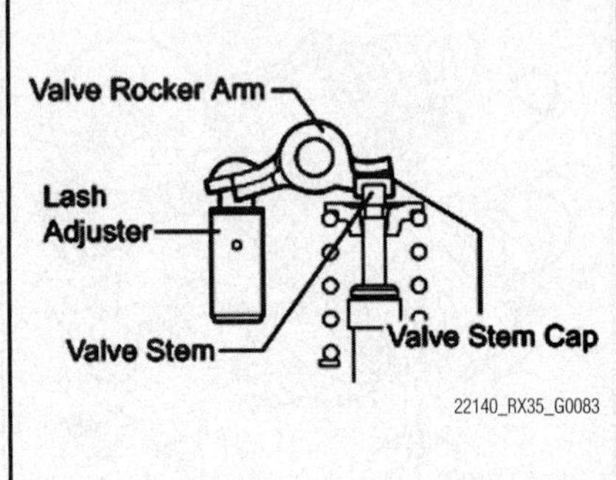

Fig. 47 Valve rocker arm installation bank 1

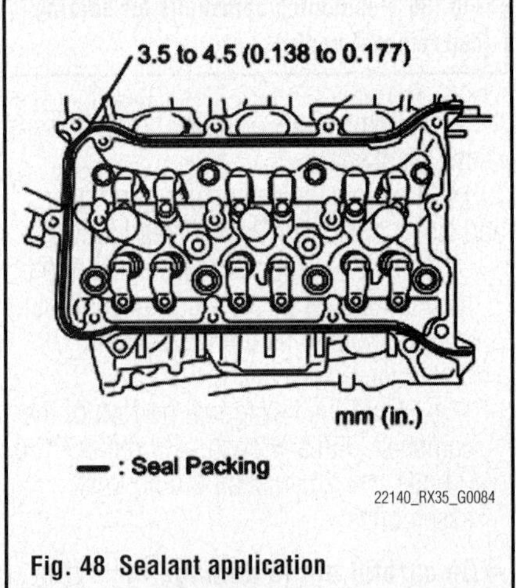

Fig. 48 Sealant application

a. Make sure that the valve rocker arm is installed as shown in the illustration.

b. Apply seal packing in a continuous line as shown in the illustration. Seal packing: Toyota Genuine Seal Packing Black, Three Bond 1207B or equivalent. Seal diameter: 3.5 to 4.5 mm (0.138 to 0.177 in.).

➡Remove any oil from the contact surface. Install the camshaft housing sub-assembly within 3 minutes. Do not

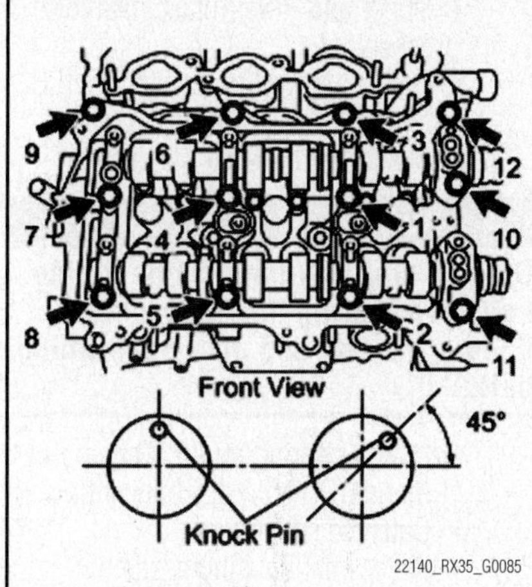

Fig. 49 Camshaft sub-assembly tightening sequence 12 bolt bank 1

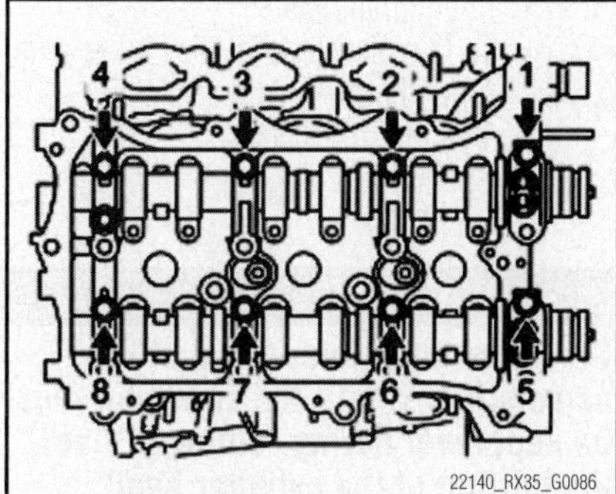

Fig. 50 Camshaft sub-assembly tightening sequence 8 bolt bank 1

start the engine for at least 2 hours after installing.

c. Install the camshaft housing and tighten the 12 bolts in the order shown in the illustration to 21 ft. lbs. (28 Nm).

➡When installing the camshaft housing, it is necessary to correctly position the camshafts as shown in the removal illustration. Failure to correctly position these parts may result in damage due to contact between the pistons and valves. If a camshaft is rotated with a piston at TDC, valve contact will occur. If any of the bolts are loosened during installation, remove the camshaft housing, clean the installation surfaces, and reapply seal packing. If the camshaft housing is removed because any of the bolts are loosened during installation, make sure that the previously applied seal packing does not enter any oil passages.

d. Complete the tightening of the 8 bolts to 12 ft. lbs. (16 Nm) in the order shown in the illustration.

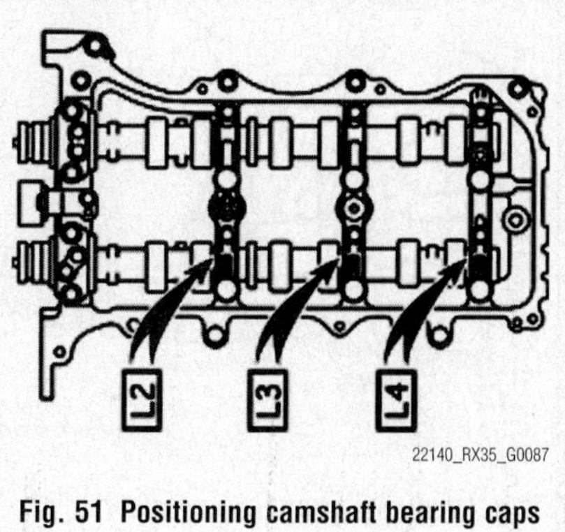

Fig. 51 Positioning camshaft bearing caps bank 1

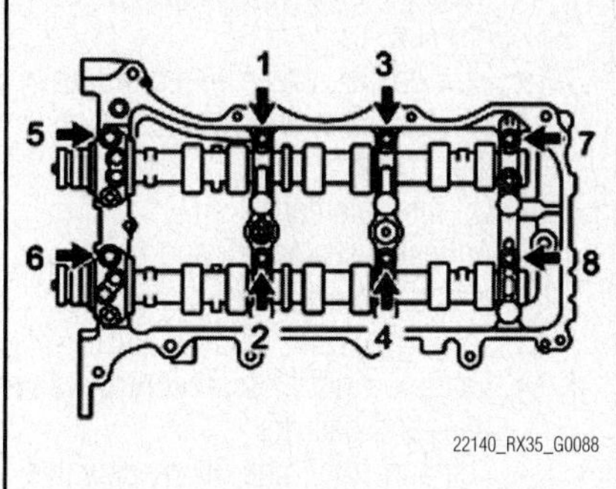

Fig. 52 Camshaft bolt tightening sequence bank 1

32. Install the camshaft bearing cap (for Bank 2), as follows:

a. Apply engine oil to the camshaft journals, camshaft housing and bearing caps.

b. Install the No. 3 camshaft and No. 4 camshaft to the left camshaft housing.

c. Make sure of the marks and numbers on the camshaft bearing caps and place them in each proper position and direction.

d. Temporarily tighten the 8 bolts in the order shown in the illustration to 7 ft. lbs. (10 Nm).

33. Install the left camshaft housing sub-assembly, as follows:

a. Make sure that the valve rocker arm is installed as shown in the illustration.

b. Apply seal packing in a continuous line as shown in the illustration. Seal packing: Toyota Genuine Seal Packing Black, Three Bond 1207B or equivalent. Seal diameter: 3.5 to 4.5 mm (0.138 to 0.177 in.).

➡Remove any oil from the contact surface. Install the camshaft housing sub-assembly within 3 minutes. Do not start the engine for at least 2 hours after installing.

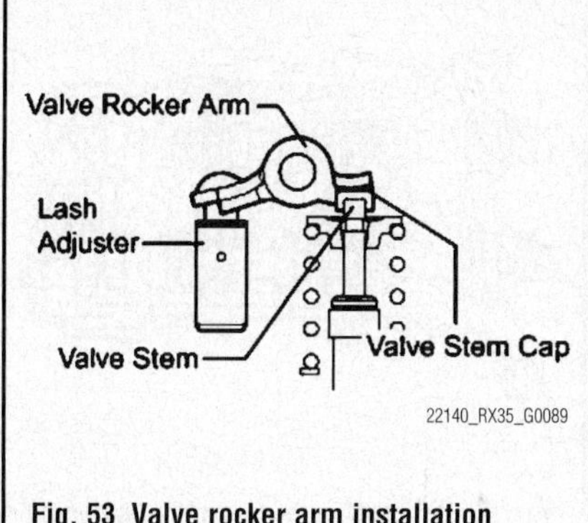

Fig. 53 Valve rocker arm installation bank 2

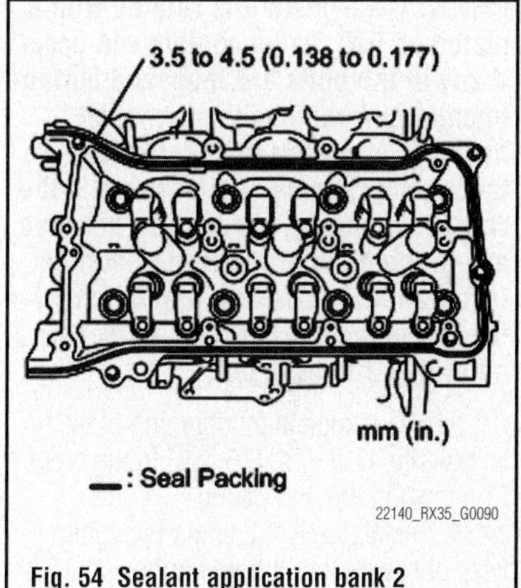

Fig. 54 Sealant application bank 2

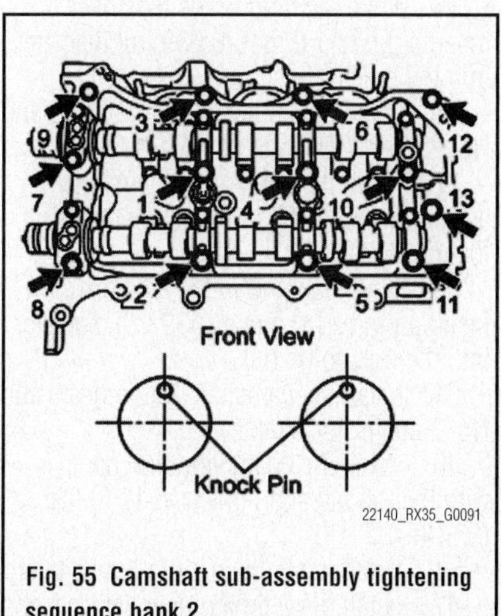

Fig. 55 Camshaft sub-assembly tightening sequence bank 2

c. Install the camshaft housing and tighten the 13 bolts in the order shown in the illustration to 21 ft. lbs. (28 Nm).

➡When installing the camshaft housing, it is necessary to correctly position the camshafts as shown in the removal illustration. Failure to correctly position these parts may result in damage due to contact between the pistons and

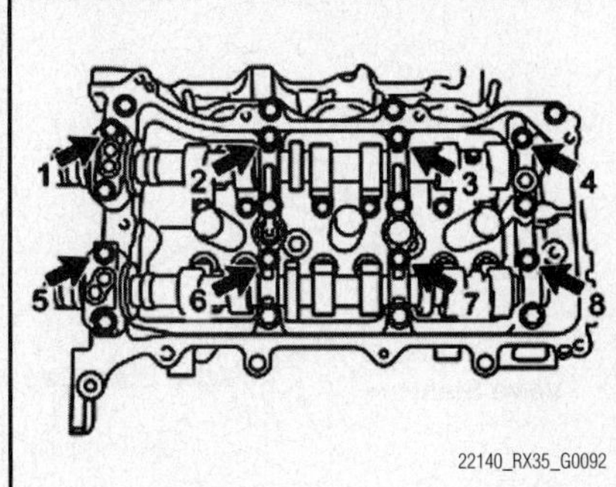

Fig. 56 Camshaft sub-assembly tightening sequence bank 2

Fig. 57 Crankshaft timing sprocket bank 2

valves. If a camshaft is rotated with a piston at TDC, valve contact will occur. If any of the bolts are loosened during installation, remove the camshaft housing, clean the installation surfaces, and reapply seal packing. If the camshaft housing is removed because any of the bolts are loosened during installation, make sure that the previously applied seal packing does not enter any oil passages.

d. Complete the tightening of the 8 bolts to 12 ft. lbs. (16 Nm) in the order shown in the illustration.

34. Install the No. 2 chain tensioner assembly with the bolt and tighten to 15 ft. lbs. (21 Nm).

35. While pushing in the tensioner, insert a pin of 1.0 mm (0.039 in.) diameter into the hole to fix it.

36. Install the camshaft timing gears and No. 2 chain (for Bank 1).

37. Install the No. 3 chain tensioner assembly with the bolt and tighten to 15 ft. lbs. (21 Nm).

38. While pushing in the tensioner, insert a pin of 1.0 mm (0.039 in.) diameter into the hole to hold it.

39. Install the camshaft timing gears and No. 2 chain (for Bank 2).

40. Install the chain vibration damper with the 2 bolts and tighten to 17 ft. lbs. (23 Nm).

41. Install the 2 chain vibration dampers.

42. Install the 2 timing gear set keys and crankshaft timing sprocket as shown in the illustration.

43. Install the idle sprocket assembly.

44. Install chain sub-assembly.

45. Install the chain tensioner slipper.

46. Install the No. 1 chain tensioner assembly.

47. Install the following:

- Timing chain cover,
- Cylinder head cover
- New gaskets, O-ring and rear water by-pass joint. Torque to 7 ft. lbs. (10 Nm).
- New gasket, water inlet and thermostat assembly. Torque to 7 ft. lbs. (10 Nm).
- Oil filter element
- Adhesive sealer, left and right side cylinder block water drain cocks. Torque to 22 ft. lbs. (30 Nm).
- Left and right side oil control valve filter and gaskets
- Left and right side oil check valve bolt, oil pipe union and oil pipe. Torque to 44 ft. lbs. (60 Nm).
- Crankshaft position sensor. Torque to 7 ft. lbs. (10 Nm).
- 4 camshaft timing oil control valves
- 4 camshaft position sensors
- Adhesive and ventilation valve
- Spark plugs. Torque to 13 ft. lbs. (18 Nm).
- Engine hangers. Torque to 24 ft. lbs. (33 Nm).
- Radiator cap
- Oil filler cap and gasket
- Negative battery cable

48. Refill the coolant and engine oil. Start the engine and check for leaks or abnormal conditions. Perform and road test. Then, recheck for leaks and recheck fluid levels.

4.3L (3UZ-FE) Engine

See Figure 58.

1. Before servicing the vehicle, refer to the precautions in the beginning of this section.
2. Relieve the fuel system pressure.
3. Remove or disconnect the following, as applicable to each engine:

- Engine under cover
- Drain engine coolant
- V-bank cover
- Air cleaner assembly
- Intake air pipe
- Radiator (if necessary)
- Throttle body
- Upper and lower intake manifold assembly
- Camshaft position sensor and LH timing belt rear plates
- RH timing belt rear plates

⁂ WARNING

Do NOT drop anything inside timing belt cover during this procedure. Keep oil, water and dust from timing belt.

- Power steering pump from engine mount (Do NOT disconnect hoses)
- Catalytic converters
- Water inlet housing assembly
- Water bypass pipe, front bypass joint, and rear bypass joint
- Ignition coils
- Variable valve timing (VVT) sensors
- Engine hangers
- Oil dipsticks and guides for engine oil and transmission fluid
- Cylinder head covers
- Spark plugs
- Camshafts

⁂ WARNING

Since the thrust clearance of the camshaft is small, the camshaft must be kept level during removal. If not, the portion of the cylinder head receiving the camshaft thrust may crack or be otherwise damaged, causing the camshaft to later seize or break. Follow the camshaft removal procedure carefully as given in this section.

To install:

4. Install or reconnect the following:

- Engine wiring and ground straps
- Oxygen sensor wire bracket on LH cylinder head
- Spark plugs
- Camshafts

5. Inspect and adjust the valve lash.
6. Install or reconnect the following:

- Cylinder head covers, with new gaskets
- Engine hangers
- VVT sensors
- Engine and transmission dipsticks and tubes
- Ignition coils
- Water bypass joints and pipe; torque nuts and bolt to 13 ft. lbs. (18 Nm)
- Water inlet housing assembly
- Catalytic converters, with new gaskets and new nuts; torque nuts to 46 ft. lbs. (62 Nm)

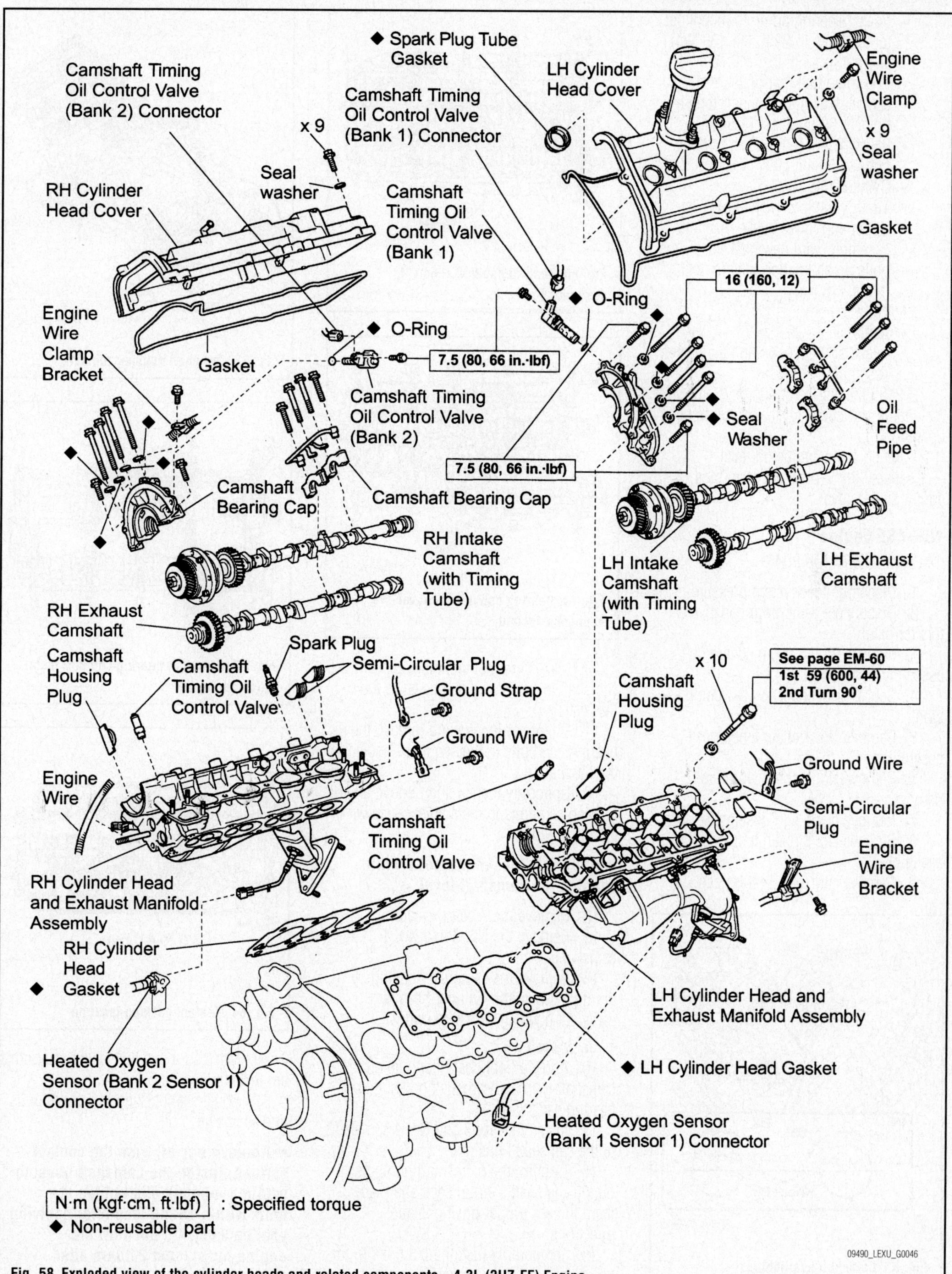

Fig. 58 Exploded view of the cylinder heads and related components—4.3L (3UZ-FE) Engine

- Power steering pump to mounting; torque bolts to 29 ft. lbs. (39 Nm) and nut to 32 ft. lbs. (43 Nm)
- LH timing belt rear plates and camshaft position sensor; torque bolts to 66 inch lbs. (7.5 Nm)
- RH right belt rear plates; torque bolts to 66 inch lbs. (7.5 Nm)
- Camshaft pulleys
- Timing belt to camshaft pulleys
- Upper and lower intake manifold assembly, with new gaskets; torque bolts, in alternating pattern, to 13 ft. lbs. (18 Nm)
- Throttle body
- Radiator
- Intake air connector and air cleaner assembly

7. Refill the engine cooling system. Start the engine and check for leaks and proper operation.
8. Recheck the engine oil level.
9. Install the V-bank cover and the engine under cover.

1UR-FSE Engine

See Figures 59 through 68.

1. Discharge fuel system pressure.
2. Disconnect cable from negative battery terminal.
3. Remove timing chain cover sub-assembly.
4. Set No. 1 cylinder to TDC/compression.
5. Remove No. 1 chain tensioner assemblies.
6. Remove No. 1 chain vibration damper.
7. Remove chain sub-assembly.
8. Remove No. 3 chain tensioner assembly.
9. Remove camshaft bearing cap.

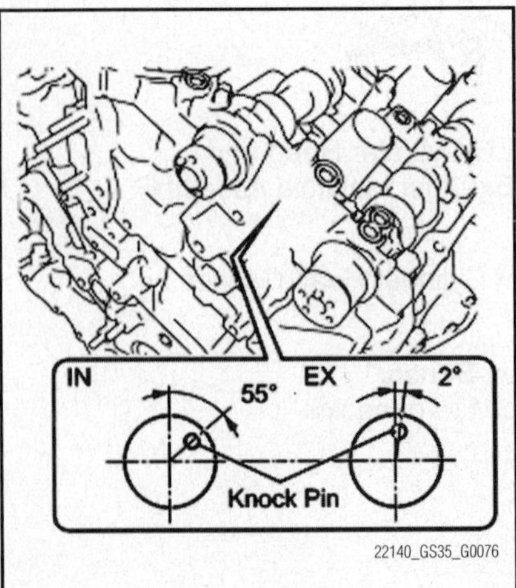

Fig. 59 Camshaft positioning

Fig. 60 Bearing cap bolt removal sequence first

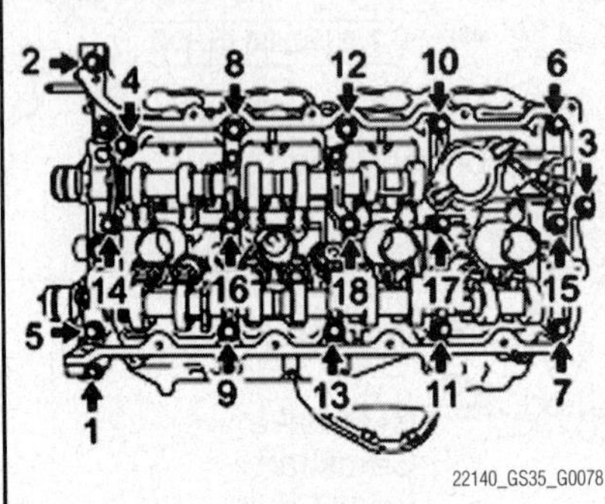

Fig. 61 Bearing cap bolt removal sequence second

10. Make sure that the knock pin of the camshaft is positioned as shown in the illustration.
11. Uniformly loosen and remove the 8 bearing cap bolts in the sequence shown in the illustration
12. Uniformly loosen and remove the 18 bearing cap bolts in the sequence shown in the illustration

➡Uniformly loosen the bolts while keeping the camshaft level

13. Remove the 7 bearing caps.
14. Remove the No. 3 and No. 4 camshafts.
15. Remove the camshaft housing by prying between the cylinder head and camshaft housing with a screwdriver.

To install:

16. Apply a light coat of engine oil to the camshaft journals, camshaft housing and bearing caps.
17. Install the No. 3 and No. 4 camshafts to the camshaft housing.
18. Confirm the marks and numbers on the camshaft bearing caps and place them in their proper positions and directions.
19. Temporarily install the 8 bolts in the order shown in the illustration.

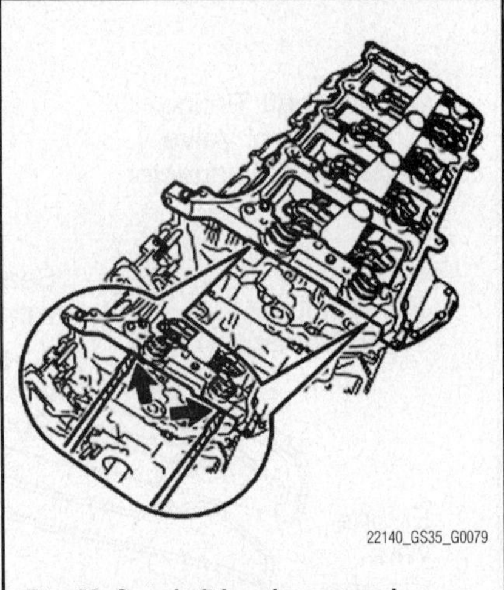

Fig. 62 Camshaft housing removal

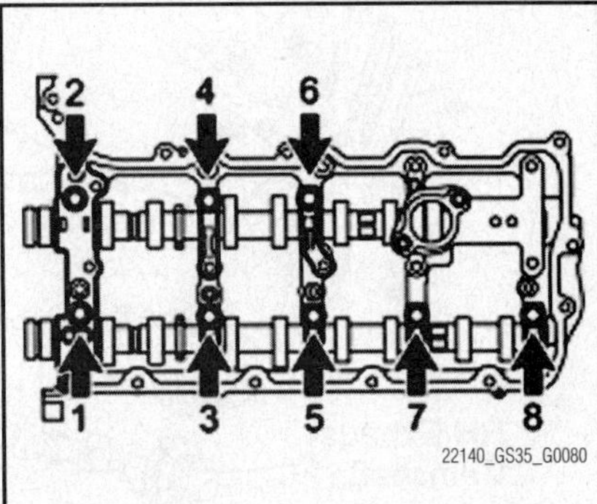

Fig. 63 Camshaft bearing cap bolt installation sequence

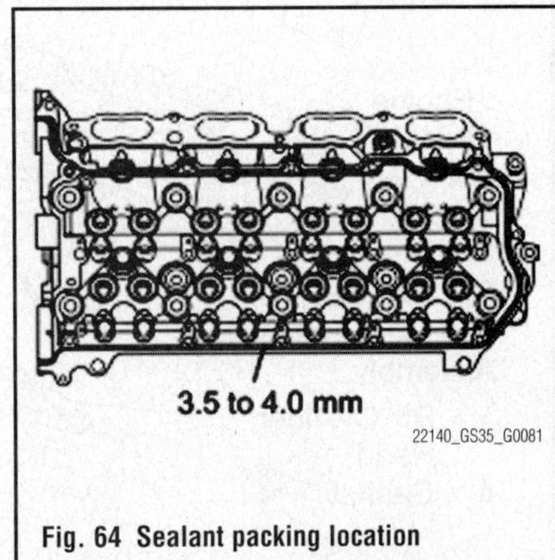

Fig. 64 Sealant packing location

20. Make sure that the valve rocker arms are installed correctly
21. Apply seal packing in a continuous line as shown

➡Remove any oil from the contact surface. Install the camshaft housing within 3 minutes and tighten the bolts within 15 minutes after applying seal packing. Do not start the engine for at least 2 hours after the installation.

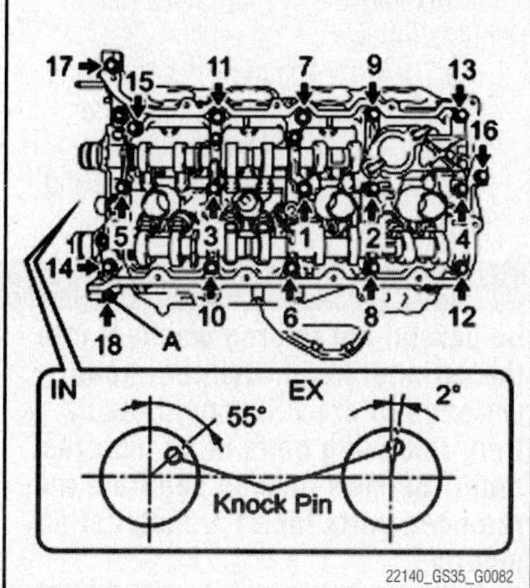

Fig. 65 Camshaft housing bolt tightening sequence

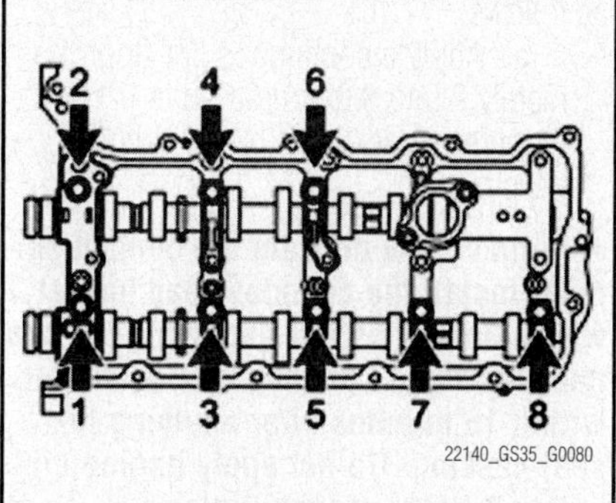

Fig. 66 Camshaft bearing cap bolt tightening sequence

22. Install the camshaft housing, and install the 12 bolts and tighten to 7 ft. lbs. (10 nm) and 22 Ft. lbs. (30 Nm), in the order shown.

➡**Make sure that each knock pin of the camshafts is positioned as shown in the illustration before installing the camshaft housing.**

23. Tighten the 8 bolts to 12 ft. lbs. (16 Nm) in the order shown in the illustration
24. Install No. 3 chain tensioner assembly.
25. Install chain sub-assembly.
26. Install chain tensioner slipper.
27. Install No. 1 chain tensioner assembly.
28. Install No. 1 chain vibration damper.
29. Using a wrench, hold the hexagonal portion of the No. 3 camshaft.
30. Using a 12 mm socket hexagon wrench, tighten the camshaft timing gear assembly with a new bolt to 58 ft. lbs. (79 Nm).

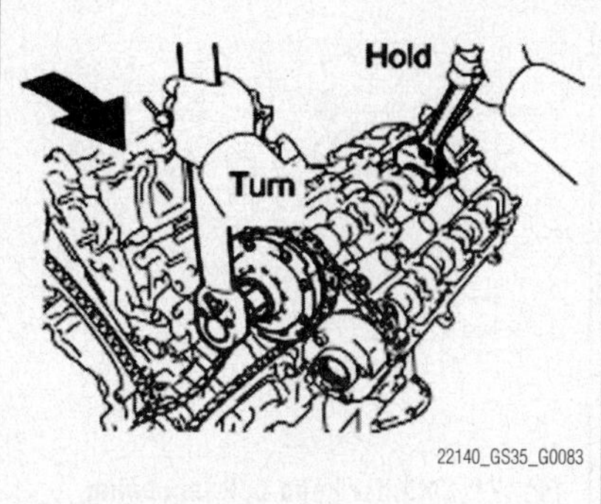

Fig. 67 No. 3 camshaft timing gear tightening

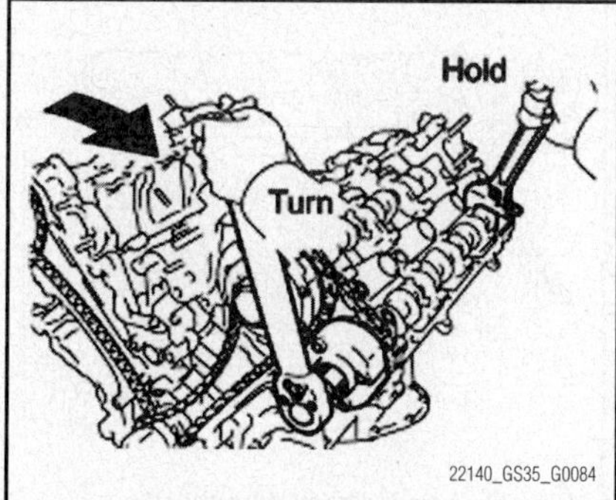

Fig. 68 No. 4 camshaft timing gear tightening

31. Using a wrench to hold the hexagonal portion of the No. 4 camshaft, tighten the camshaft timing exhaust gear assembly with the bolt to 74 ft. lbs. (100 Nm).
32. Check No. 1 cylinder to TDC/compression.
33. Install timing chain and gears.
34. Install timing chain cover sub-assembly.
35. Connect cable from negative battery terminal.
36. Perform similar procedure for opposite bank.

Perform initialization, if necessary.

CRANKSHAFT FRONT SEAL

REMOVAL & INSTALLATION

3.5L (2GR-FSE) and 4.6L (1UR-FSE) Engines

See Figures 69 and 70.

1. Before servicing the vehicle, refer to the precautions in the beginning of this section.
2. Disconnect the negative battery cable. Wait at least 90 seconds before performing any other work.
3. Air cleaner assembly.
4. Remove the radiator reserve assembly.

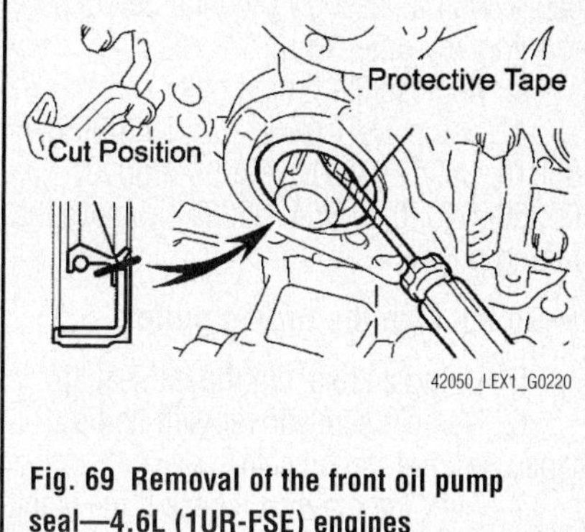

Fig. 69 Removal of the front oil pump seal—4.6L (1UR-FSE) engines

Fig. 70 Installation of the front oil pump seal—4.6L (1UR-FSE) engines

5. Remove the accessory drive belt.
6. Remove the crankshaft pulley.
7. Using a screwdriver with its tip taped, pry out the timing chain case oil seal.

➡**After the removal, check the crankshaft for damage. If it is damaged, smooth the surface with 400-grit sandpaper.**

To install:

8. Using SST 09223-22010, 09506-35010 and a hammer, tap in the oil seal until its surface is flush with the rear oil seal retainer edge. Keep the lip free of foreign matter. Do not tap the oil seal at an angle.
9. Install the crankshaft pulley.
10. Install the accessory drive belt.
11. Install the radiator reserve assembly.
12. Air cleaner assembly.
13. Inspect for oil leaks.
14. Connect the negative battery cable.

4.3L (3UZ-FE) Engine

1. Before servicing the vehicle, refer to the precautions in the beginning of this section.

2. Disconnect the negative battery cable. Wait at least 90 seconds before performing any other work.
3. Remove the timing belt.
4. Using SST 09950–50013 (09951–05010, 09952–05010, 09953–05010, 09953–05020, 09954–05021), remove the timing pulley.

➡**Do not turn the timing pulley.**

5. Using a knife, cut the oil seal lip.
6. Using a screwdriver with its tip taped, pry out the oil seal.
7. After the removal, check if the crankshaft is not damaged. If it is damaged, smooth the surface with 400–grit sandpaper.

To install:

8. Apply MP grease to a new oil seal lip. Keep the lip free from foreign materials.
9. Using SST 09316–60011 (09316–00011) and a hammer, tap in the oil seal until its surface is flush with the oil pump edge. Wipe off any extra grease on the crankshaft. Be careful not to tap the oil seal at an angle.
10. Align the timing pulley set key with the key groove of the pulley.
11. Face the timing pulley's flange side inward. Using SST 09223–46011 and a hammer, tap in the timing pulley.
12. Install the timing belt.
13. Connect the negative battery cable.

CYLINDER HEAD

REMOVAL & INSTALLATION

3.5L (2GR-FSE) Engine

See Figures 71 through 76.

1. Before servicing the vehicle, refer to the precautions in the beginning of this section.
2. Drain the cooling system.
3. Drain the engine oil.
4. Relieve the fuel system pressure.
5. Remove or disconnect the following:
   - Negative battery cable. Wait at least 90 seconds before performing any other work
   - Oil filler cap and gasket
   - Radiator cap
   - No. 1 and 2 engine hangers
   - Spark plugs
   - Ventilation valve
   - 4 camshaft position sensors
   - 4 camshaft timing oil control valves
   - Crankshaft position sensor
   - Left and right side oil check valve bolt, oil pipe union and oil pipe
   - Left and right side oil control valve filter and gaskets

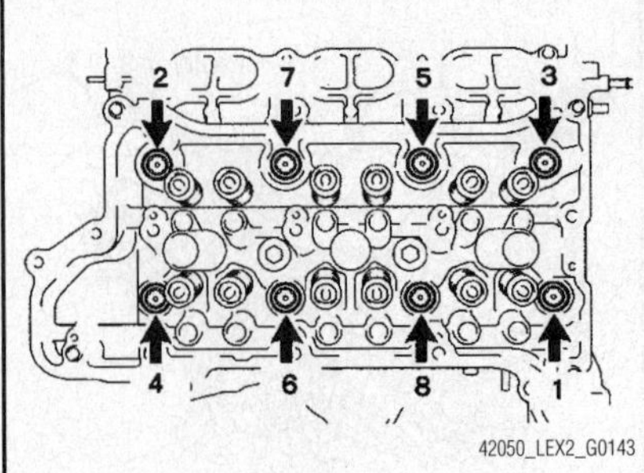

Fig. 71 Cylinder head bolt loosening sequence (right side)—3.5L (2GR-FSE) engine

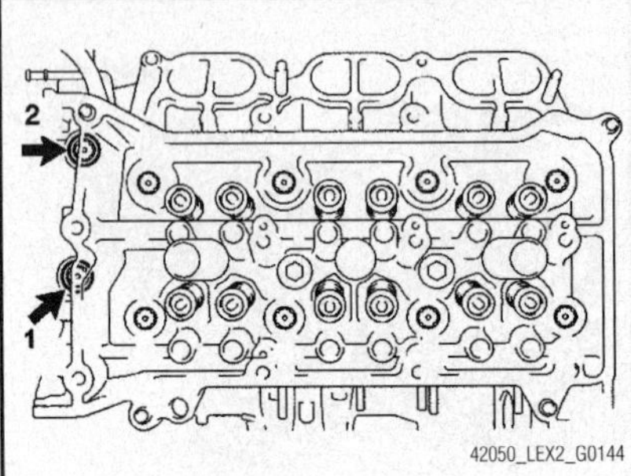

Fig. 72 Cylinder head 14mm bolt loosening sequence (left side)—3.5L (2GR-FSE) engine

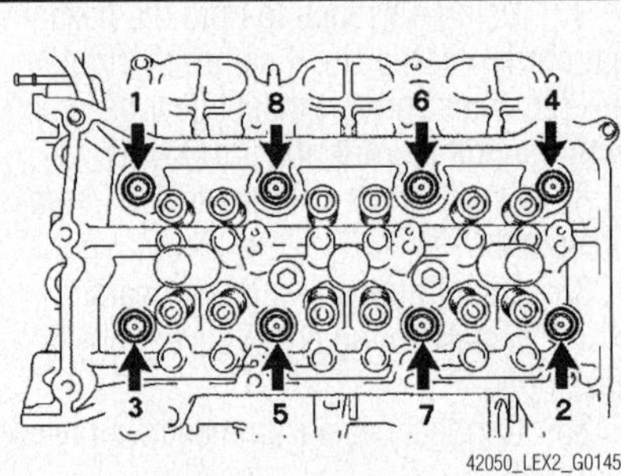

Fig. 73 Cylinder head bolt loosening sequence (left side)—3.5L (2GR-FSE) engine

   - Cylinder block water drain cocks
   - Oil filter element
   - Water inlet and thermostat assembly
   - Rear water by-pass joint
   - Cylinder head cover and gaskets
   - Timing chain cover, timing chain and timing chain sprockets
   - Camshaft and camshaft housing assembly
   - Valve rocker arms. Arrange the removed rocker arms in the correct order.
   - Valve lash adjusters. Arrange the removed valve lash adjusters in the correct order.

6. Remove the cylinder head (left or right) as follows:
   a. Using a 10mm bi-hexagon wrench, uniformly loosen the 8 bolts in the sequence shown in the illustration. Remove the 8 cylinder head bolts and plate washers.

⁂ WARNING

Be careful not to drop washers into the cylinder head. Cylinder head warpage or cracking could result from removing bolts in an incorrect order. Be sure to keep separate the removed parts for each installation position.

   b. Remove the cylinder head and gasket.

To install:

7. Install the cylinder head to the engine as follows:
   a. Apply a continuous line approximately 2.5 to 3.0mm (0.098 to 0.118 in.) of the seal packing to a new cylinder head gasket.

➡**Remove any oil from the contact surface. Install the cylinder head gasket within 3 minutes after applying the seal packing. Install the cylinder head bolt within 15 minutes after applying the seal packing. Do not apply engine oil within 2 hours of installation.**

   b. Place the cylinder head gasket on the cylinder block surface with the Lot No. stamp upward.

⁂ WARNING

Be careful of the installation direction. Gently place the cylinder head in order not to damage the gasket with the bottom part of the head.

   c. Place the cylinder head on the cylinder block.

⁂ WARNING

Be careful not to allow oil to adhere to the bottom part of the cylinder head.

   d. Apply a light coat of engine oil to the threads and under the heads of the cylinder head bolts.
   e. Using a 10mm bi-hexagon wrench, install and uniformly tighten the 8 cylinder head bolts with the plate washers to 27 ft. lbs. (36 Nm) in the sequence shown in the illustration. If any of the bolts does not meet the torque, replace it.

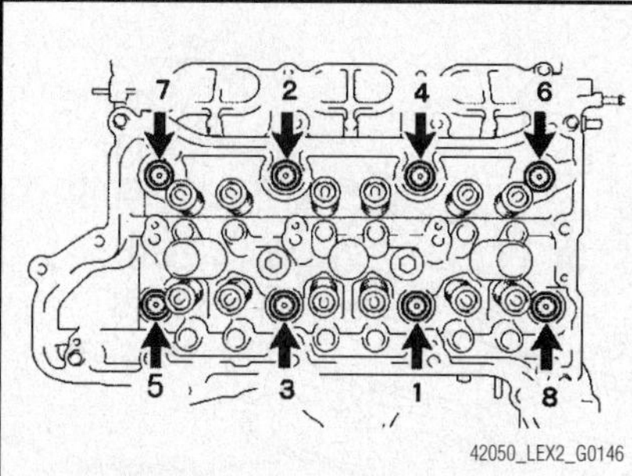

Fig. 74 Cylinder head bolt tightening sequence (right side)—3.5L (2GR-FSE) engine

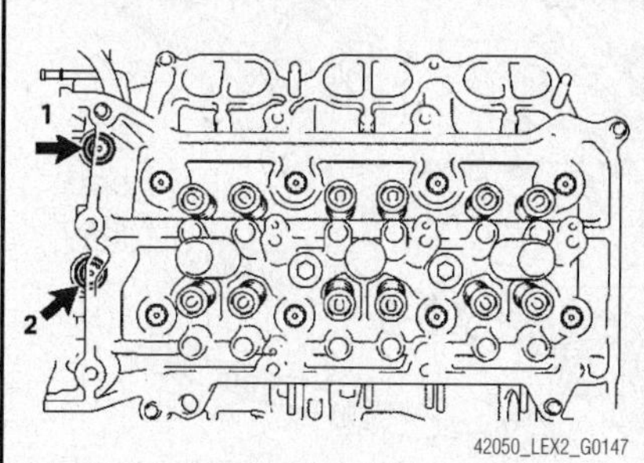

Fig. 75 Cylinder head 14mm bolt tightening sequence (left side)—3.5L (2GR-FSE) engine

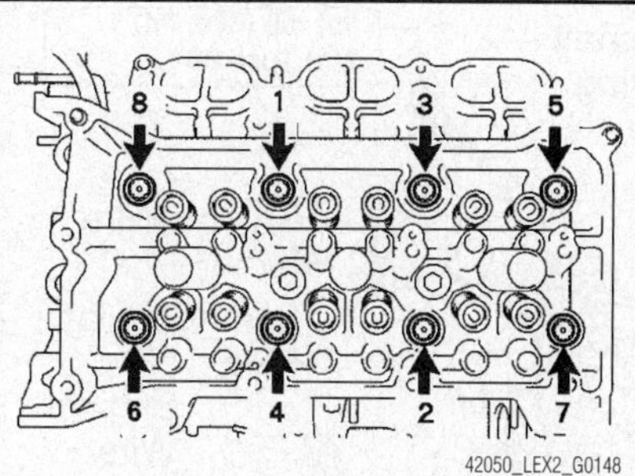

Fig. 76 Cylinder head bolt tightening sequence (left side)—3.5L (2GR-FSE) engine

f. Mark the forward edge of each bolt with paint, then retighten each bolt, in proper sequence, an additional 90 degrees. Check that each painted mark is now at a 90 degrees angle to the front. The paint mark should have been applied to the bolt in the 9 o'clock position and should now be in the 12 o'clock position.

g. Tighten each bolt again, in proper sequence, an additional 90 degrees. Check that each painted mark is now facing rearward.

h. Tighten the 2 bolts on the left cylinder head in the order shown in the illustration. Torque to 22 ft. lbs. (30 Nm).

➡Do not use the tightening procedure for a plastic region bolt (if equipped) when tightening bolts 1 and 2 shown in the illustration.

i. Seal packing will seep out on the engine's front side. Thoroughly wipe off seeped out seal packing.

8. Install or connect the following:
- Valve lash adjusters
- Valve rocker arms
- Camshaft and camshaft housing assembly
- Timing chain cover, timing chain and timing chain sprockets
- Cylinder head cover
- New gaskets, O-ring and rear water by-pass joint. Torque to 7 ft. lbs. (10 Nm).
- New gasket, water inlet and thermostat assembly. Torque to 7 ft. lbs. (10 Nm).
- Oil filter element
- Adhesive sealer, left and right side cylinder block water drain cocks. Torque to 22 ft. lbs. (30 Nm).
- Left and right side oil control valve filter and gaskets
- Left and right side oil check valve bolt, oil pipe union and oil pipe. Torque to 44 ft. lbs. (60 Nm).
- Crankshaft position sensor. Torque to 7 ft. lbs. (10 Nm).
- 4 camshaft timing oil control valves
- 4 camshaft position sensors
- Adhesive and ventilation valve
- Spark plugs. Torque to 13 ft. lbs. (18 Nm).
- Engine hangers. Torque to 24 ft. lbs. (33 Nm).
- Radiator cap
- Oil filler cap and gasket
- Negative battery cable

9. Refill the coolant and engine oil. Start the engine and check for leaks or abnormal conditions. Perform and road test. Then, recheck for leaks and recheck fluid levels.

4.3L (3UZ-FE) Engine

See Figures 77 through 79.

1. Before servicing the vehicle, refer to the precautions in the beginning of this section.
2. Relieve the fuel system pressure.
3. Remove or disconnect the following, as applicable to each engine:
- Engine under cover
- Drain engine coolant
- V-bank cover
- Air cleaner assembly
- Intake air pipe
- Radiator (if necessary)
- Throttle body
- Upper and lower intake manifold assembly
- Camshaft position sensor and LH timing belt rear plates
- RH timing belt rear plates

⁂ WARNING

Do NOT drop anything inside timing belt cover during this procedure. Keep oil, water and dust from timing belt.

- Power steering pump from engine mount (Do NOT disconnect hoses)
- Catalytic converters
- Water inlet housing assembly
- Water bypass pipe, front bypass joint, and rear bypass joint
- Ignition coils
- Variable valve timing (VVT) sensors
- Engine hangers
- Oil dipsticks and guides for engine oil and transmission fluid
- Cylinder head covers
- Spark plugs
- Camshafts

⁂ WARNING

Since the thrust clearance of the camshaft is small, the camshaft must be kept level during removal. If not, the portion of the cylinder head receiving the camshaft thrust may crack or be otherwise damaged, causing the camshaft to later seize or break. Follow the camshaft removal procedure carefully as given in this section.

- Both oxygen sensor connectors
- Ground wire from LH cylinder head
- Engine wire bracket for oxygen sensor on LH cylinder head

4. Uniformly loosen the 10 cylinder head bolts on each cylinder head, in several passes, following loosening sequence as shown.

⁂ WARNING

Use care so that no bolts or washers are dropped into the recesses or enclosed portions of the cylinder head or block.

5. Carefully lift the cylinder head from the locating dowels on the engine block. Place cylinder heads on wooden blocks on the workbench.

6. If necessary, exhaust manifolds may be removed from the cylinder heads at this time.

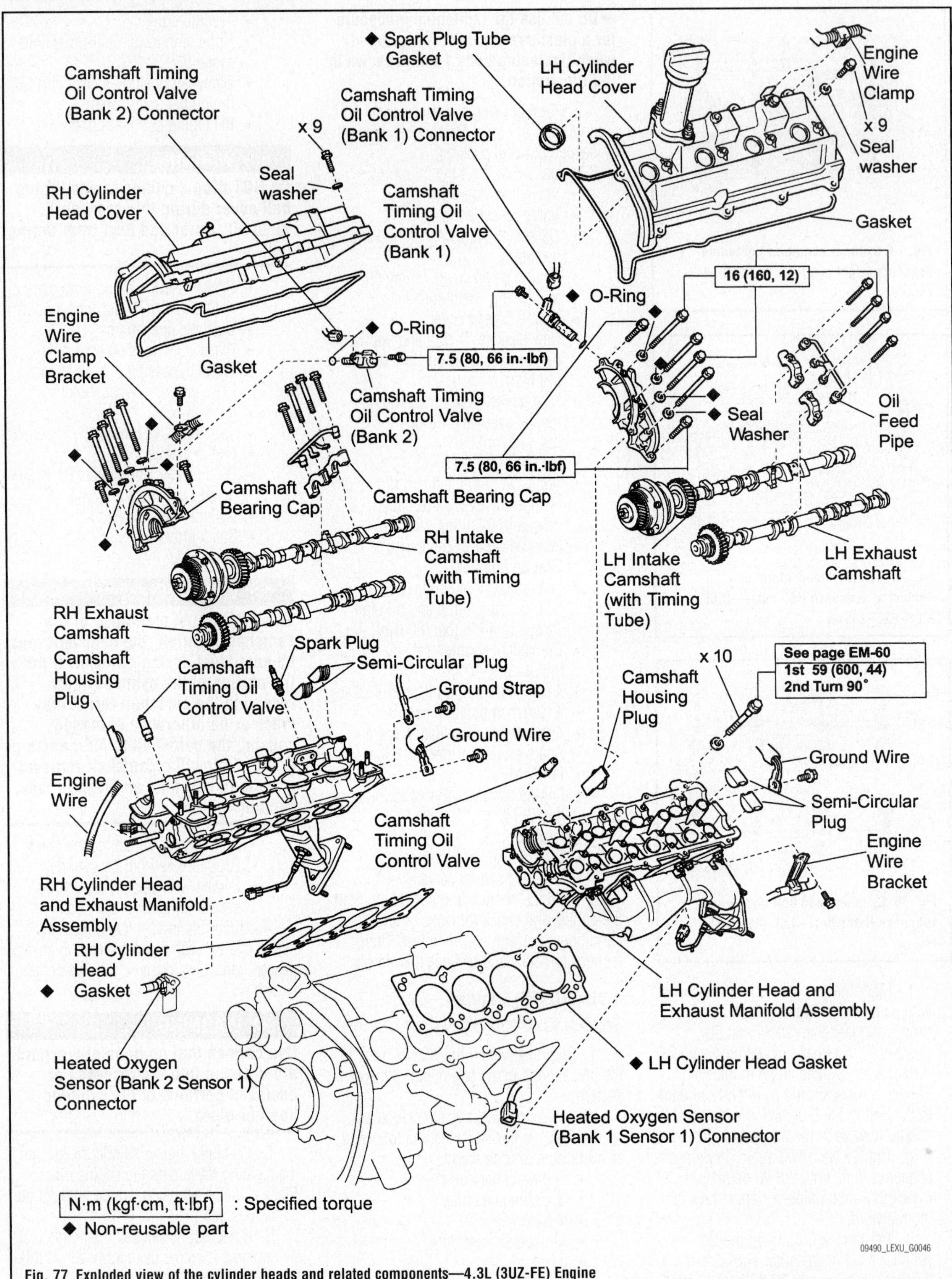

Fig. 77 Exploded view of the cylinder heads and related components—4.3L (3UZ-FE) Engine

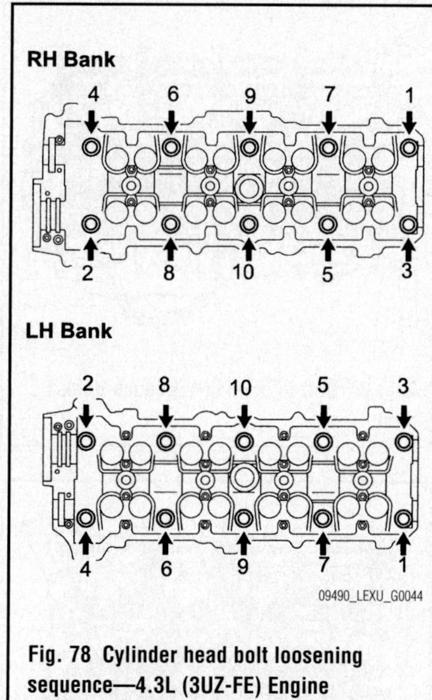

Fig. 78 Cylinder head bolt loosening sequence—4.3L (3UZ-FE) Engine

To install:

7. If removed, install exhaust manifolds to cylinder heads, with new gaskets. Ensure the white mark on the gasket is facing the manifold side.

8. Install and tighten the exhaust manifold bolts, in an alternating pattern, to a final torque of 32 ft. lbs. (44 Nm).

9. Install the heat shields.

10. With a new cylinder head gasket in place, carefully position the cylinder head onto the engine block locating dowels.

➡The cylinder head gaskets have a "3R" marks for the RIGHT cylinder head, and a "3L" mark for the LEFT cylinder head.

➡If any cylinder head bolt appears stretched or damaged, replace it. If a bolt will not reach final torque setting, replace it.

11. Apply a light coat of oil to the cylinder head bolt threads. Install the washers and insert the cylinder head bolts into position.

12. In several passes, following the tightening sequence shown, tighten the cylinder head bolts to a final torque of 44 ft. lbs. (59 Nm).

13. Once the bolts reach this setting, then place a white paint mark on the front of each bolt head. Using the torque wrench, turn each bolt, in the sequence shown, an additional 90 degrees, using the paint mark as a reference.

14. Install or reconnect the following:
- Engine wiring and ground straps

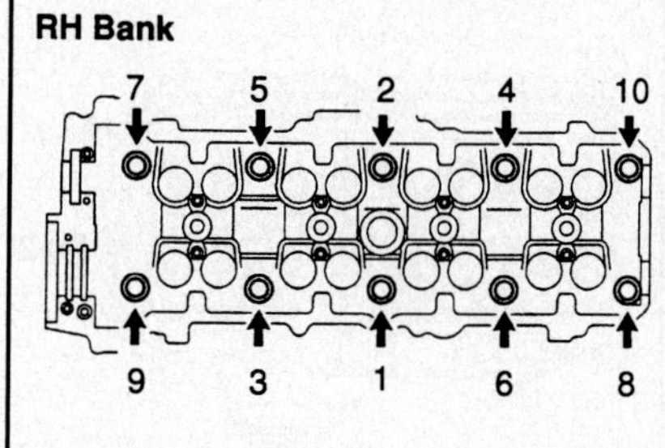

Fig. 79 Cylinder head torque sequence—4.3L (3UZ-FE) engine

- Oxygen sensor wire bracket on LH cylinder head
- Spark plugs
- Camshafts

15. Inspect and adjust the valve lash.

16. Install or reconnect the following:
- Cylinder head covers, with new gaskets
- Engine hangers
- VVT sensors
- Engine and transmission dipsticks and tubes
- Ignition coils
- Water bypass joints and pipe; torque nuts and bolt to 13 ft. lbs. (18 Nm)
- Water inlet housing assembly
- Catalytic converters, with new gaskets and new nuts; torque nuts to 46 ft. lbs. (62 Nm)
- Power steering pump to mounting; torque bolts to 29 ft. lbs. (39 Nm) and nut to 32 ft. lbs. (43 Nm)
- LH timing belt rear plates and camshaft position sensor; torque bolts to 66 inch lbs. (7.5 Nm)
- RH right belt rear plates; torque bolts to 66 inch lbs. (7.5 Nm)
- Camshaft pulleys
- Timing belt to camshaft pulleys
- Upper and lower intake manifold assembly, with new gaskets; torque bolts, in alternating pattern, to 13 ft. lbs. (18 Nm)
- Throttle body
- Radiator
- Intake air connector and air cleaner assembly

17. Refill the engine cooling system. Start the engine and check for leaks and proper operation.

18. Recheck the engine oil level.

19. Install the V-bank cover and the engine under cover.

1UR-FSE Engine

See Figures 80 through 96.

1. Discharge fuel system pressure.
2. Disconnect cable from negative battery terminal.
3. Remove timing chain cover sub-assembly.
4. Set No. 1 cylinder to TDC/compression.
5. Remove No. 1 chain tensioner assemblies.
6. Remove No. 1 chain vibration damper.
7. Remove chain sub-assembly.
8. Remove No. 3 chain tensioner assembly.
9. Remove camshaft bearing cap.
10. Make sure that the knock pin of the camshaft is positioned as shown in the illustration.

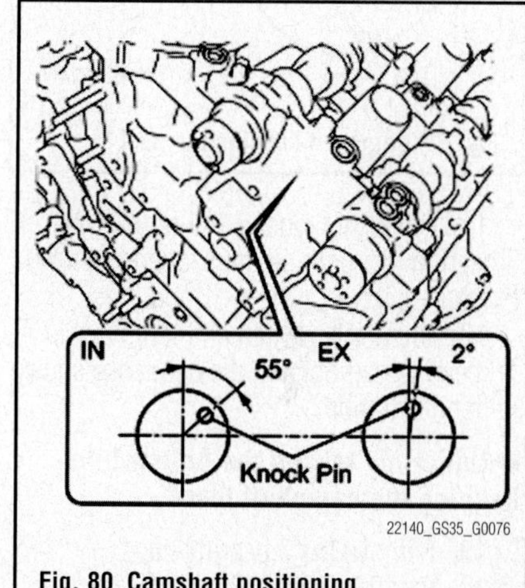

Fig. 80 Camshaft positioning

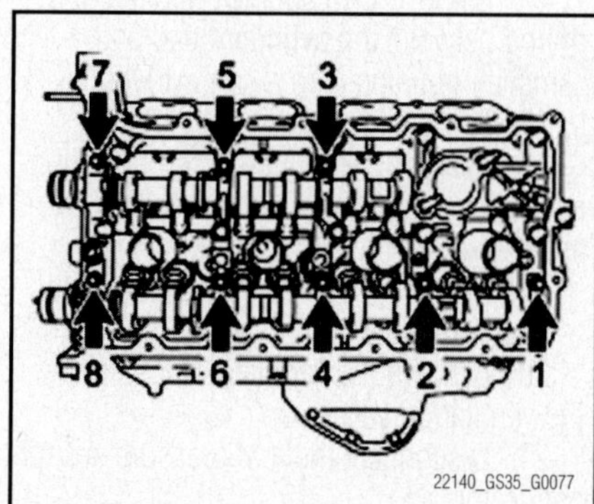

Fig. 81 Bearing cap bolt removal sequence first

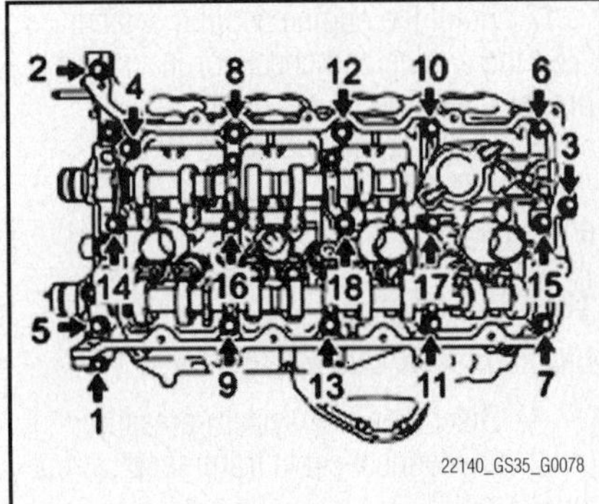

Fig. 82 Bearing cap bolt removal sequence second

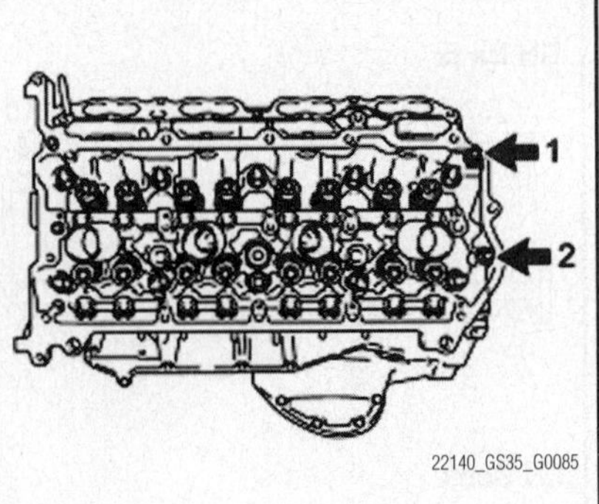

Fig. 84 Bolt loosening sequence Bank 1

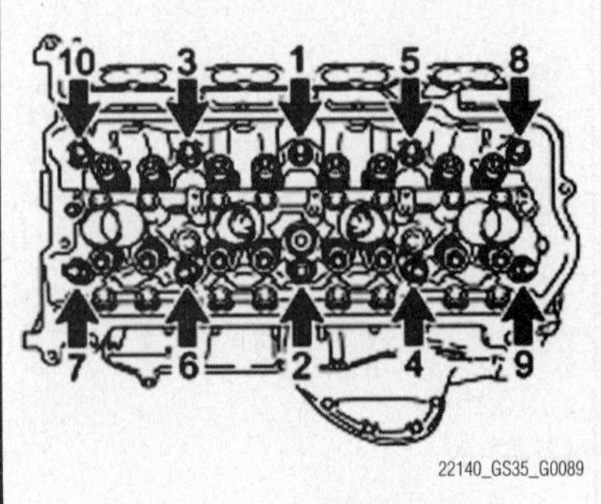

Fig. 86 Bolt tightening sequence Bank 1 first

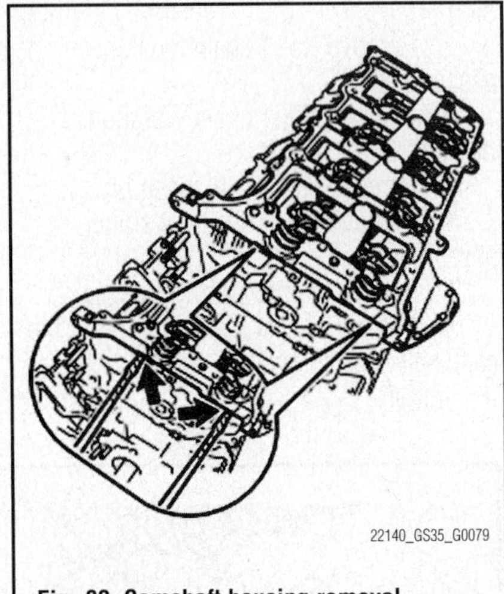

Fig. 83 Camshaft housing removal

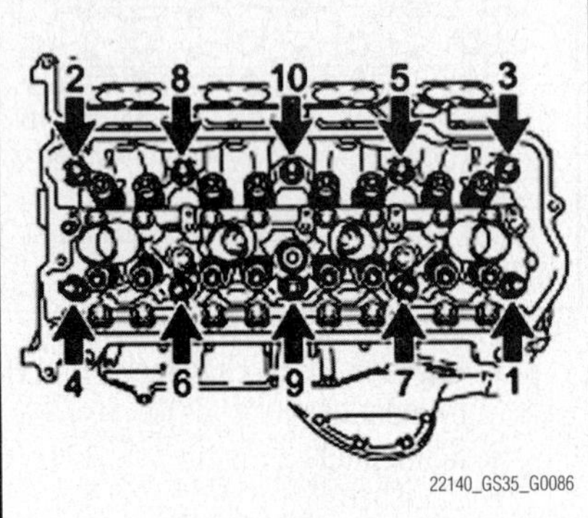
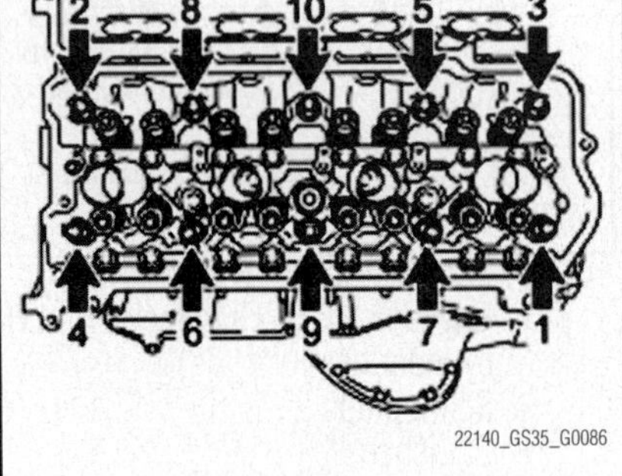

Fig. 85 Bolt loosening sequence Bank 2

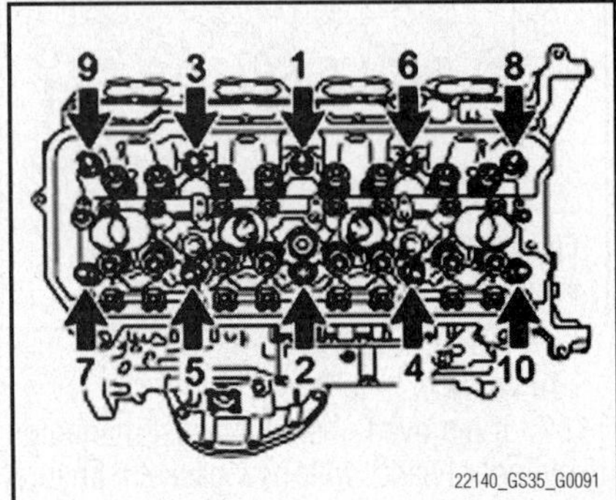

Fig. 87 Bolt tightening sequence Bank 2 first

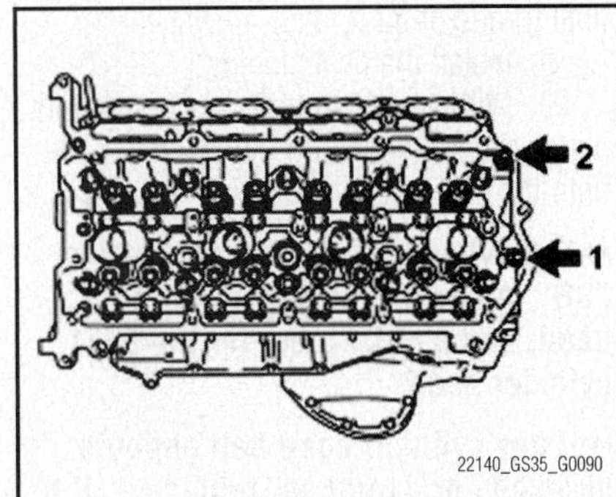

Fig. 88 Bolt tightening sequence Bank 1 second

11. Uniformly loosen and remove the 8 bearing cap bolts in the sequence shown in the illustration

12. Uniformly loosen and remove the 18 bearing cap bolts in the sequence shown in the illustration

➡Uniformly loosen the bolts while keeping the camshaft level

13. Remove the 7 bearing caps

14. Remove the No. 3 and No. 4 camshafts.

15. Remove the camshaft housing by prying between the cylinder head and camshaft housing with a screwdriver.

16. Remove the intake manifold.

17. Remove the fuel pressure pulsation damper assembly and No. 1 fuel pipe sub-assembly.

18. Remove fuel pipe sub-assemblies.

19. Remove engine covers.

20. Remove the ventilation hose from the ventilation valve.

21. Disconnect the fuel pressure sensor connector.

22. Disconnect the injector connectors.

23. Remove the bolts and delivery pipe from the cylinder head.

24. Remove the injector vibration insulators from the cylinder head.

25. Remove the valve rocker arms from the cylinder head.

26. Remove the valve lash adjusters from the cylinder head.

27. Remove the valve stem caps from the cylinder head.

28. Uniformly loosen and remove the 2 bolts in the sequence shown in the illustration.

29. Using a 10 mm bi-hexagon wrench, uniformly loosen the 10 bolts in the sequence shown in the illustration. Remove the 10 cylinder head bolts and plate washers

30. Remove the cylinder heads and gaskets

To install:

31. Place the cylinder head gasket on the cylinder block surface with the Lot No. stamp facing upward.

32. Place the cylinder head on the cylinder block.

➡Ensure that no oil is on the mounting surface of the cylinder head.

33. Apply a light coat of engine oil to the threads and under the heads of the cylinder head bolts.

34. Using a 10 mm bi-hexagon wrench, install and uniformly tighten the 10 cylinder head bolts to 27 ft. lbs. (36 Nm) with the plate washers in several steps, in the sequence shown in the illustration.

35. Mark each cylinder head bolt head with paint as shown in the illustration.

36. Tighten the cylinder head bolts another 90° in the sequence shown in step 1.

37. Tighten the cylinder head bolts by an additional 90° in the sequence shown in step 1.

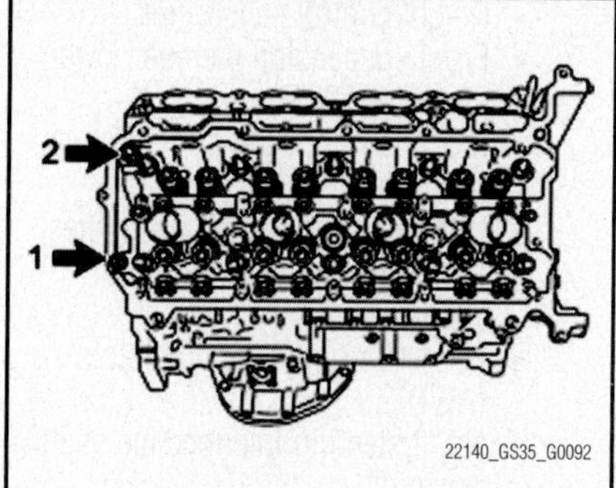

Fig. 89 Bolt tightening sequence Bank 2 second

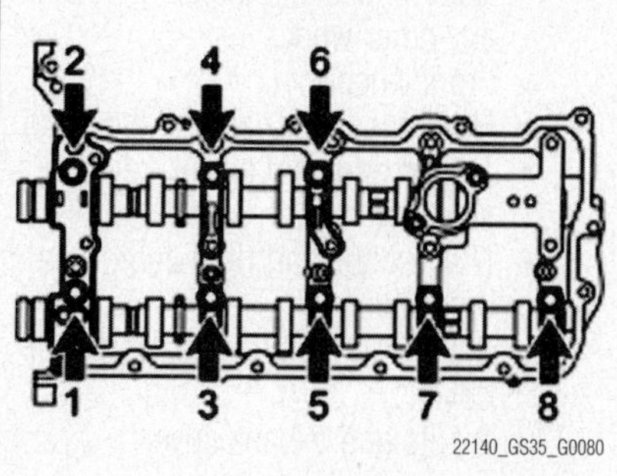

Fig. 91 Camshaft bearing cap bolt installation sequence

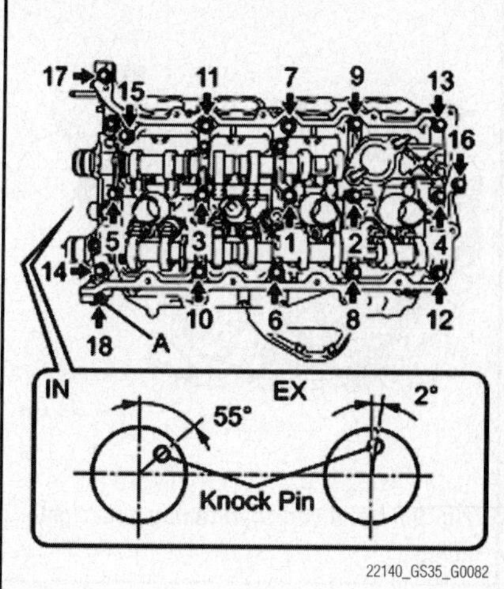

Fig. 93 Camshaft housing bolt tightening sequence

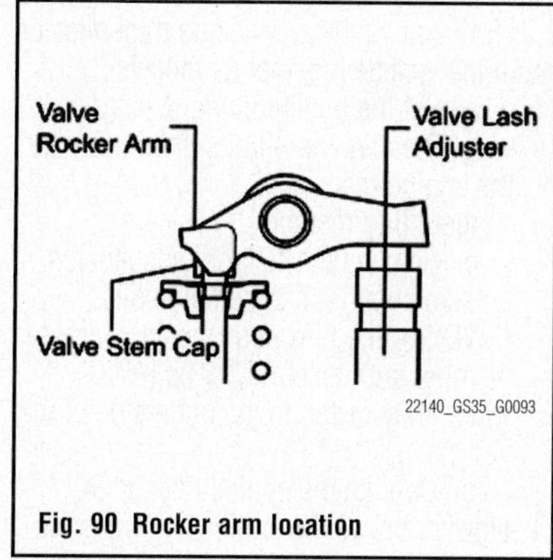

Fig. 90 Rocker arm location

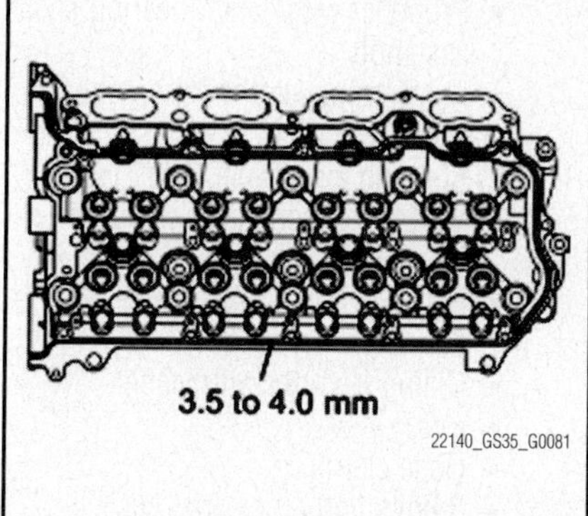

Fig. 92 Sealant packing location

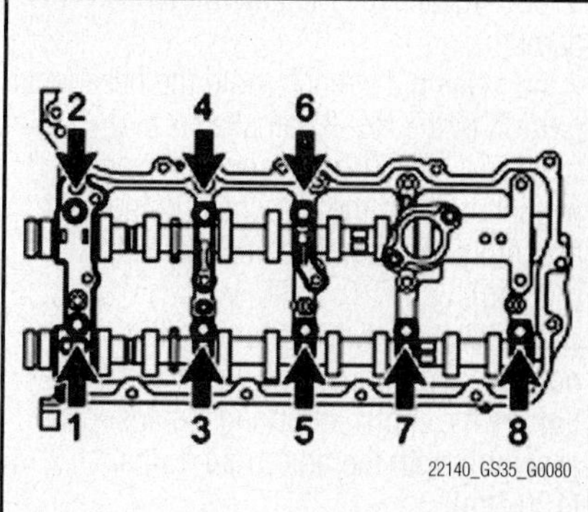

Fig. 94 Camshaft bearing cap bolt tightening sequence

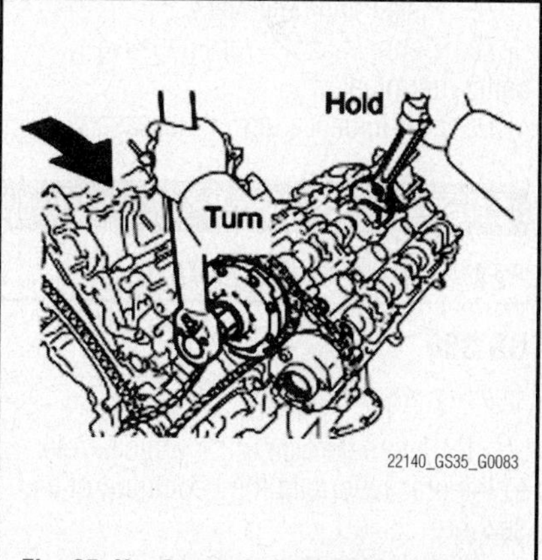

Fig. 95 No. 3 camshaft timing gear tightening

38. Check that the painted marks are now facing rearward.
39. Uniformly install the 2 cylinder head bolts to 15 ft. lbs. (21 Nm).
40. Apply a light coat of engine oil to the valve stem caps.
41. Install the valve stem caps to the cylinder head.
42. Install the valve lash adjusters to the cylinder head.
43. Install the ventilation hose from the ventilation valve.
44. Install fuel pipe sub-assemblies.
45. Install the fuel pressure pulsation damper assembly and No. 1 fuel pipe sub-assembly.
46. Install the intake manifold.
47. Apply engine oil to the lash adjuster tips and valve stem cap ends.
48. Make sure that the valve rocker arms are installed as shown in the illustration.
49. Install the bolts and delivery pipe to the cylinder head.
50. Connect the injector connectors.
51. Connect the fuel pressure sensor connector.
52. Install the injector vibration insulators to the cylinder head.
53. Apply a light coat of engine oil to the camshaft journals, camshaft housing and bearing caps.
54. Install the No. 3 and No. 4 camshafts to the camshaft housing.
55. Confirm the marks and numbers on the camshaft bearing caps and place them in their proper positions and directions.
56. Temporarily install the 8 bolts in the order shown in the illustration.
57. Make sure that the valve rocker arms are installed correctly
58. Apply seal packing in a continuous line as shown

➡Remove any oil from the contact surface. Install the camshaft housing within 3 minutes and tighten the bolts within 15 minutes after applying seal packing. Do not start the engine for at least 2 hours after the installation.

59. Install the camshaft housing, and install the 12 bolts and tighten to 7 ft. lbs. (10 nm) and 22 Ft. lbs. (30 Nm), in the order shown.

➡Make sure that each knock pin of the camshafts is positioned as shown in the illustration before installing the camshaft housing.

60. Tighten the 8 bolts to 12 ft. lbs. (16 Nm) in the order shown in the illustration
61. Install No. 3 chain tensioner assembly.

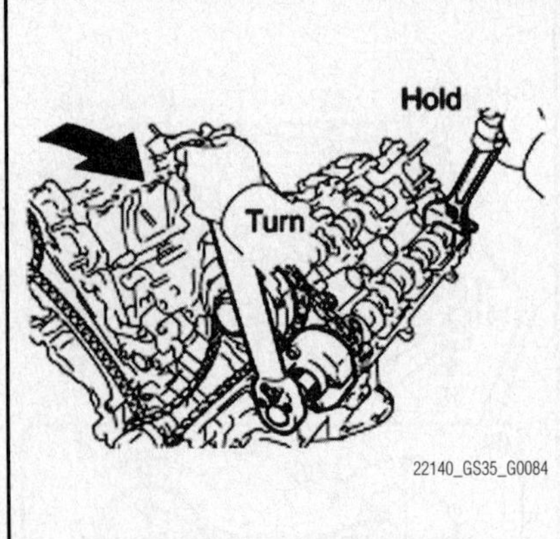

Fig. 96 No. 4 camshaft timing gear tightening

62. Install chain sub-assembly.
63. Install chain tensioner slipper.
64. Install No. 1 chain tensioner assembly.
65. Install No. 1 chain vibration damper.
66. Using a wrench, hold the hexagonal portion of the No. 3 camshaft.
67. Using a 12 mm socket hexagon wrench, tighten the camshaft timing gear assembly with a new bolt to 58 ft. lbs. (79 Nm).
68. Using a wrench to hold the hexagonal portion of the No. 4 camshaft, tighten the camshaft timing exhaust gear assembly with the bolt to 74 ft. lbs. (100 Nm).
69. Check No. 1 cylinder to TDC/compression.
70. Install timing chain cover sub-assembly.
71. Install engine covers.
72. Connect cable from negative battery terminal.

Perform initialization , if necessary.

ENGINE ASSEMBLY

REMOVAL & INSTALLATION

GS 350

See Figures 97 through 101.

1. Before servicing the vehicle, refer to the precautions in the beginning of this section.
2. Release the fuel pressure.
3. Drain the engine coolant and engine oil.
4. Drain transmission fluid.
5. Drain differential fluid (AWD).
6. Remove or disconnect the following:
    - Negative battery cable. Wait at least 90 seconds before performing any other work.
    - Front wheels
    - Hood insulator pad and the hood
    - 7 clips and cool air intake duct seal
    - Nut, 2 clips and right engine room side cover
    - 3 clips and left engine room side cover
    - 2 nuts and V-bank cover
    - Engine under cover
    - 4 screws, 2 grommets, 2 spacers and No. 2 engine under cover (2WD)
    - Front exhaust pipe assembly
    - Propeller with center bearing shaft assembly
    - Front propeller shaft assembly (AWD)
    - Bolt and air cleaner inlet
    - Ventilation hose from the cylinder head
    - Mass Airflow (MAF) meter connector
    - Clamp from the air cleaner
    - EVAP VSV
    - Hose clamp
    - 3 bolts and air cleaner case
    - Radiator inlet, outlet and reserve tank hoses
    - Clamp and union-to-check valve hose
    - Heater water inlet and outlet hoses
    - Fuel main tube
    - Engine room ECM cover
    - ECM connectors and connector holder
    - Positive battery cable
    - Engine room No. 1 relay block, junction block cover
    - Wire from engine room No. 1 relay block, junction block
    - Ground cable
    - Serpentine drive belt
7. Remove or disconnect the with pulley compressor by performing the following:
    - Bolt, nut and bracket (AWD)
    - Magnetic clutch connector
    - Nut and 3 bolts (2WD); or 2 bolts (AWD)
    - Stud bolt using an E8 Torx® socket and with pulley compressor

➡It is not necessary to completely remove the compressor. With the hoses connected to the compressor, hang the compressor on the vehicle body with a rope.

    - Height control sensor link
    - Front suspension member lower protector (2WD)
    - Left and right front axle hub nuts (AWD)
    - Left and right tie rod assemblies
    - Left and right front shock absorbers
    - Left and right lower ball joints (AWD)
    - No. 2 steering intermediate shaft assembly
    - Power steering link wire harness (2WD)
    - Floor shift gear shifting rod sub-assembly
8. Remove the engine and transmission from the vehicle together as follows:
    a. Set the engine lifting device.
    b. Remove the 4 bolts, then separate the engine rear mounting member.
    c. Remove the 12 front suspension crossmember sub-assembly bolts (2WD); or the 8 front suspension crossmember sub-assembly bolts (AWD) where they fasten to the underside of the vehicle.
    d. Operate the engine lifter, then slowly remove the engine and transmission with crossmember sub-assembly from the vehicle.

➡Make sure the engine is clear of all wiring and hoses.

    e. Install engine hangers on each side of the engine using 4 bolts and tighten to 24 ft. lbs. (33 Nm).
    f. Attach an engine sling device and hang the engine with a chain block.
9. Remove right and left side front drive shaft assemblies (AWD).
10. Separate the transmission assembly from the engine.
11. While holding the crankshaft, remove the 8 bolts, front spacer, drive plate and ring gear sub-assembly and rear spacer.
12. Remove the 2 bolts, then remove the front suspension crossmember sub-assembly from the engine.
13. Fix the engine onto engine stand with the bolts.

To install:

14. Remove the bolts and engine from the engine stand.
15. Install the front suspension crossmember sub-assembly with the 2 bolts and torque to 26 ft. lbs. (35 Nm).
16. Install the drive plate and ring gear

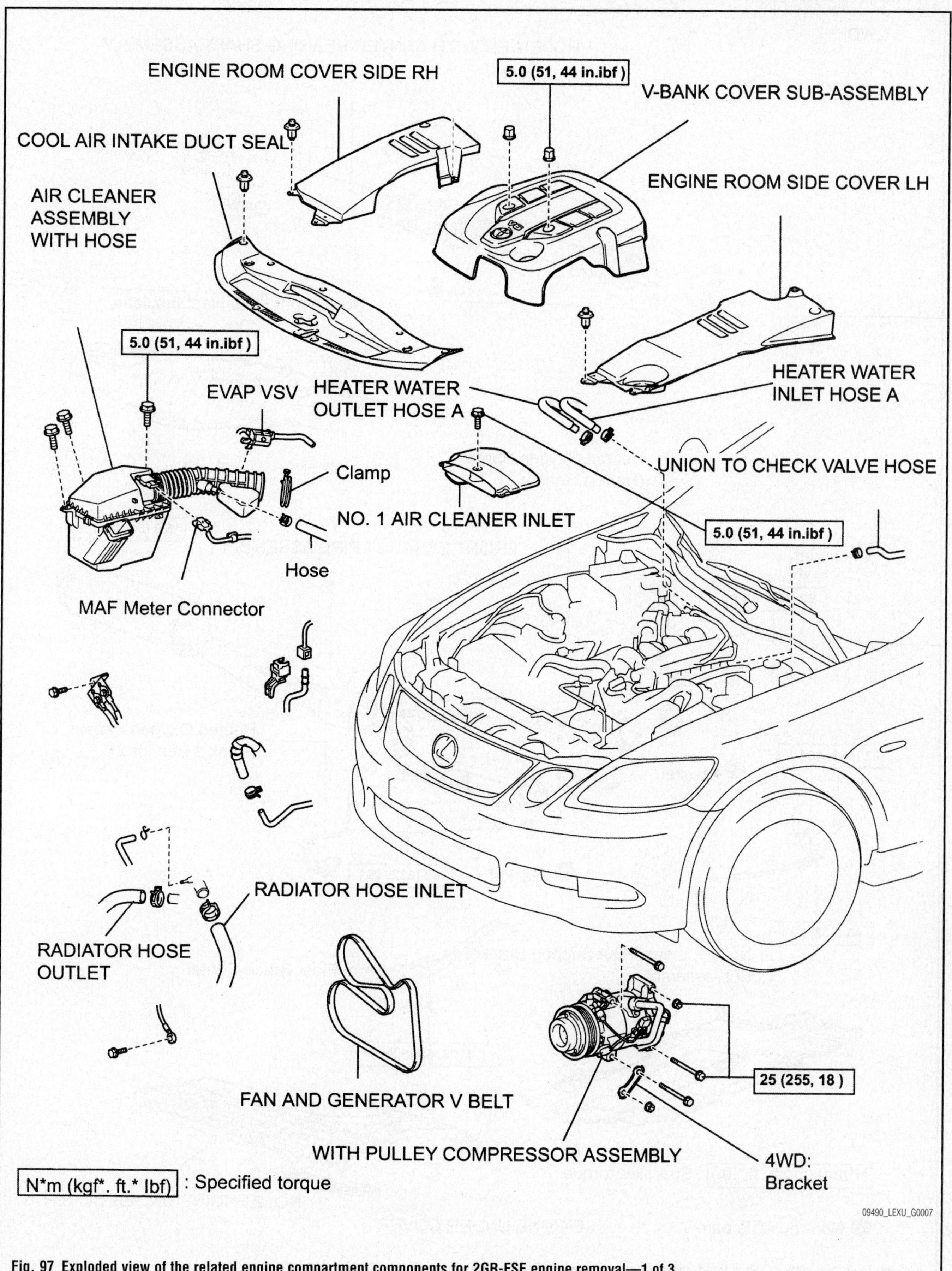

Fig. 97 Exploded view of the related engine compartment components for 2GR-FSE engine removal—1 of 3

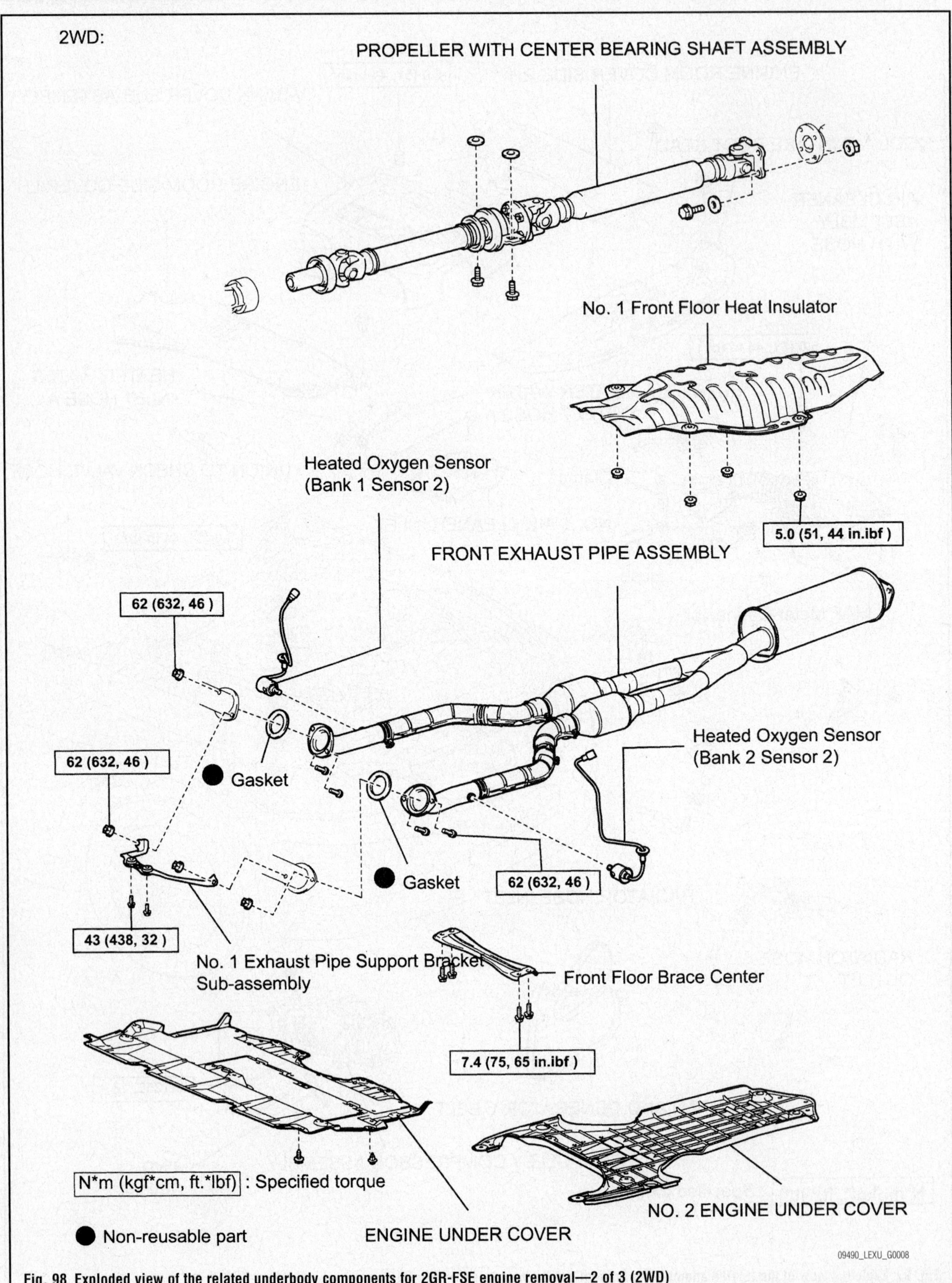

Fig. 98 Exploded view of the related underbody components for 2GR-FSE engine removal—2 of 3 (2WD)

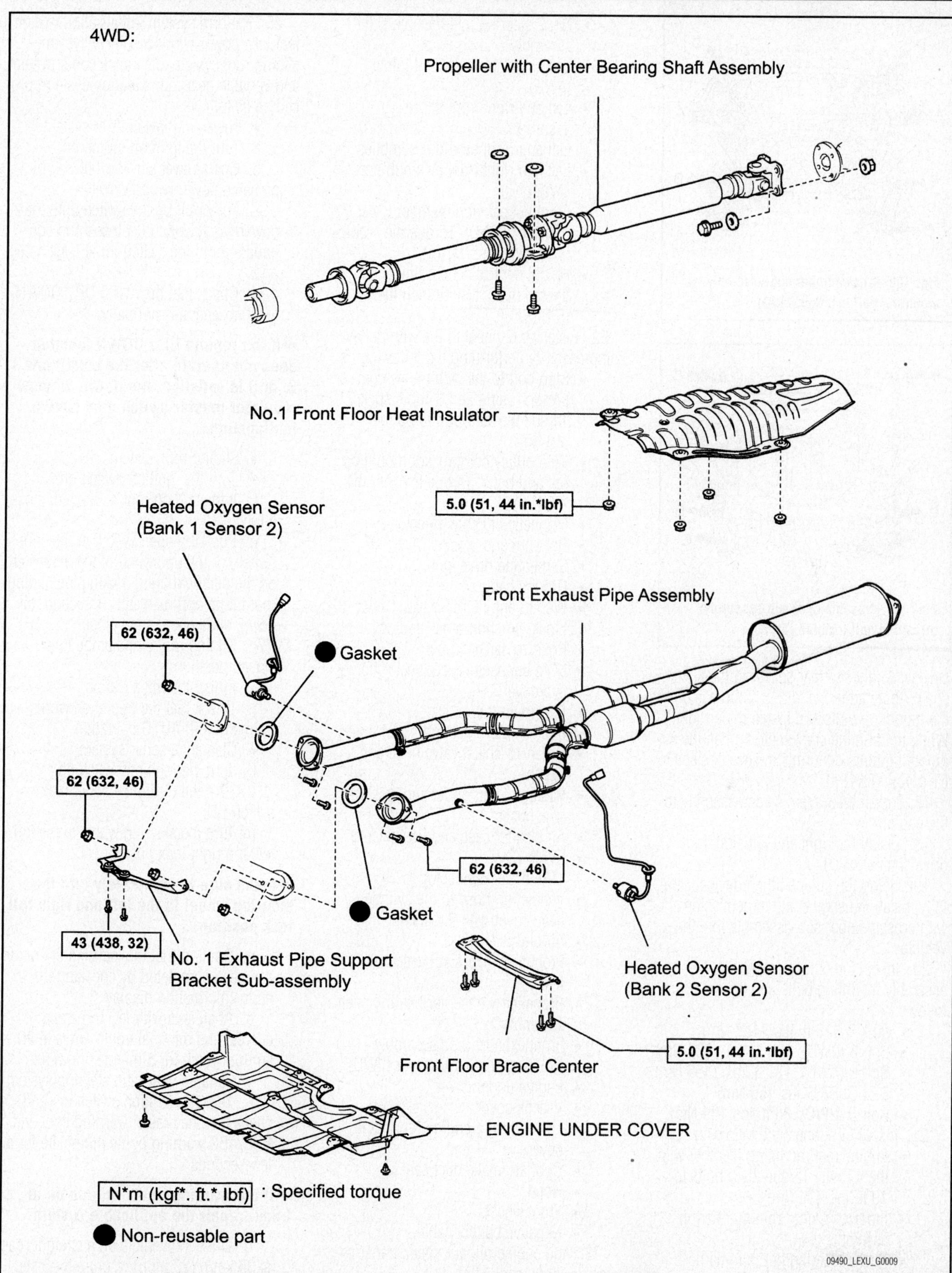

Fig. 99 Exploded view of the related underbody components for 2GR-FSE engine removal—3 of 3 (AWD)

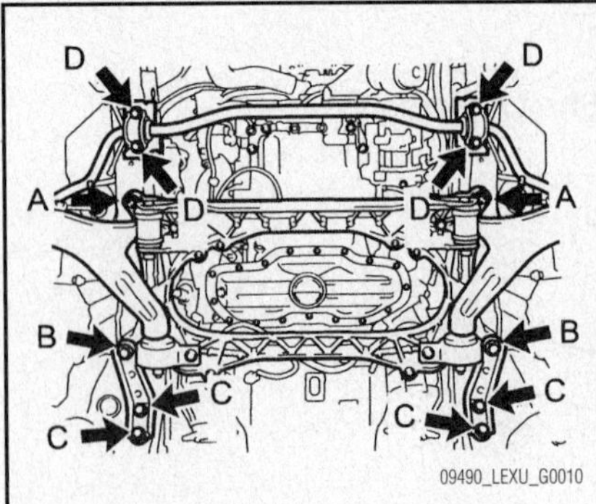

Fig. 100 Crossmember sub-assembly mounting bolt location (2WD)

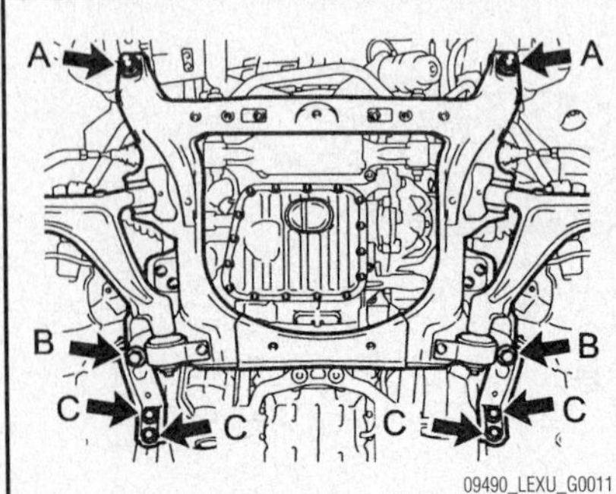

Fig. 101 Crossmember sub-assembly mounting bolt location (AWD)

sub-assembly and rear spacer to the crankshaft. Apply adhesive to 2 or 3 threads of the mounting bolts and install finger tight. While holding the crankshaft, tighten the 8 mounting bolts uniformly in several steps to 61 ft. lbs. (83 Nm).

17. Install the transmission assembly to the engine.

18. Install the right and left side front drive shafts (AWD).

19. Using the engine lifting device, carefully install the engine and transmission with crossmember sub-assembly into the vehicle.

20. Install the crossmember sub-assembly mounting bolts and tighten as follows:

- Bolt A: 123 ft. lbs. (167 Nm)
- Bolt B (2WD): 150 ft. lbs. (204 Nm)
- Bolt B (AWD): 121 ft. lbs. (165 Nm)
- Bolt C: 37 ft. lbs. (50 Nm)
- Bolt D (2WD): 36 ft. lbs. (49 Nm)

21. Install or connect the following:

- Engine rear mounting member with the 4 bolts. Torque the 4 bolts to 19 ft. lbs. (26 Nm).
- Floor shift gear shifting rod sub-assembly
- Power steering link wire harness (2WD)
- No. 2 steering intermediate shaft assembly
- Left and right lower ball joints (AWD)
- Left and right front shock absorbers
- Left and right tie rod assemblies
- Left and right front axle hub nuts (AWD)
- Front suspension member lower protector (2WD). Torque the 4 bolts to 71 inch lbs. (8 Nm).
- Height control sensor link and tighten the nut to 48 inch lbs. (5 Nm).

22. Install or connect to the with pulley compressor by performing the following:

- Stud bolt to the with pulley compressor using an E8 Torx® socket. Torque the stud bolt to 89 inch lbs. (10 Nm).
- With pulley compressor mounting nut and bolts. Torque to 18 ft. lbs. (25 Nm).
- Magnetic clutch connector
- Bolt, nut and bracket (AWD)
- Serpentine drive belt
- Ground cable
- Wire to engine room No. 1 relay block, junction block and cover
- Positive battery cable
- ECM connectors, connector holder and cover
- Fuel main tube
- Heater water inlet and outlet hoses
- Union-to-check valve hose and clamp
- Radiator inlet, outlet and reserve tank hoses
- Air cleaner case and hose clamp
- EVAP VSV
- MAF meter connector
- Ventilation hose to the cylinder head cover with the clamp
- Air cleaner inlet
- Front propeller shaft assembly (AWD)
- Propeller with center bearing shaft assembly
- Front exhaust pipe assembly
- No. 2 engine under cover (2WD)
- Engine under cover
- V-bank cover
- Left and right engine room side covers
- Cool air intake duct seal
- Hood
- Front wheels
- Negative battery cable

23. Add engine oil, automatic transmission fluid and coolant.

24. Add front differential oil (AWD).

25. Perform system initialization (which includes power window control system, sliding roof system, clearance sonar system and variable gear ratio steering system) procedure as follows:

- Power window control system

a. Turn the ignition switch on.

b. Open power window halfway by pressing power window switch.

c. Fully pull up the switch until the power window is fully closed and continue to hold the switch for at least 1 second.

d. Check that the AUTO UP / DOWN function operates normally.

➡If the remote UP / DOWN function does not operate after the conditions 1, 2, or 3 is satisfied, the power window regulator master switch may have a malfunction.

- Sliding roof system

e. Turn the ignition switch on.

f. If the sliding roof is opened, close it fully.

g. Push the open switch of the slide switch, or the up switch of the tilt switch on the personal light, making the sliding roof tilt up approximately 1 second, tilt down, slide open, slide close.

h. Sliding roof stops at the fully closed position.

i. Finish the initialization.

j. Check that the operation works normally with AUTO operation.

- Clearance sonar system

k. Turn the ignition switch on.

l. Turn the clearance sonar main switch ON.

m. Turn the steering wheel to the full left and right lock position.

➡Make sure to completely turn the steering wheel to the left and right full lock position.

n. Confirm that the learning operation has been completed by checking the multi-information display.

o. At an area with few turns and curves, and minimal traffic, drive at 20 km/h or more for 5 minutes or more.

- Variable gear ratio steering system

p. Turn the ignition switch on, and check that the master warning light and VSC/ABS warning lights illuminate for a few seconds.

➡If the warning lights remain on or blink, repair the applicable system.

q. Drive the vehicle on a straight road at 35 km/h (22 mph) or more for 5 seconds or longer.

r. Confirm that steering angle sensor initialization is completed by doing the following:

- Drive the vehicle on a straight road at 60 km/h (37 mph) or more for 30 seconds or longer.
- Stop the vehicle (engine running).
- Slowly turn the steering wheel from lock to lock.
- If it turns approximately 2.7 turns, steering angle sensor initialization is completed. If it turns approximately 3.2 turns, steering angle sensor initialization is not completed.

26. Check shift lever position as follows:
    - Remove the nut and disconnect the shifting rod.
    - Turn the control shaft lever of the neutral start switch counterclockwise until it stops, and turn it clockwise 2 notches to set it to the N position.
    - Move the shift lever to the N position and tighten the nut while lightly pushing the lever toward the R position.
    - After adjustment, check that the shift lever moves smoothly and the shift lever and gear operate correctly.
27. Inspect and adjust front end alignment.
28. Check idle speed and ignition timing.

➡Do not start the engine for at least 1 hour after installing.

29. Recheck all fluid levels and add if necessary.
30. Start engine and check for fluid, fuel and exhaust leaks.
31. Road test the vehicle.

GS 430

See Figure 102.

1. Before servicing the vehicle, refer to the precautions in the beginning of this section.
2. Disconnect the negative battery cable. Wait at least 90 seconds before proceeding with any other work. This provides time to disarm the airbag system.
3. Remove the hood.
4. Relieve the fuel pressure from the fuel lines.
5. Remove the engine under cover.
6. Drain the engine coolant and engine oil.
7. Remove or disconnect the following:
    - V-bank cover, if equipped
    - Air cleaner inlet, air cleaner assembly, and intake air connector pipe
    - Radiator
    - Serpentine drive belt
    - Front suspension member brace.
8. Before disconnecting any wiring, connectors, cables or hoses, be sure they are properly marked for proper

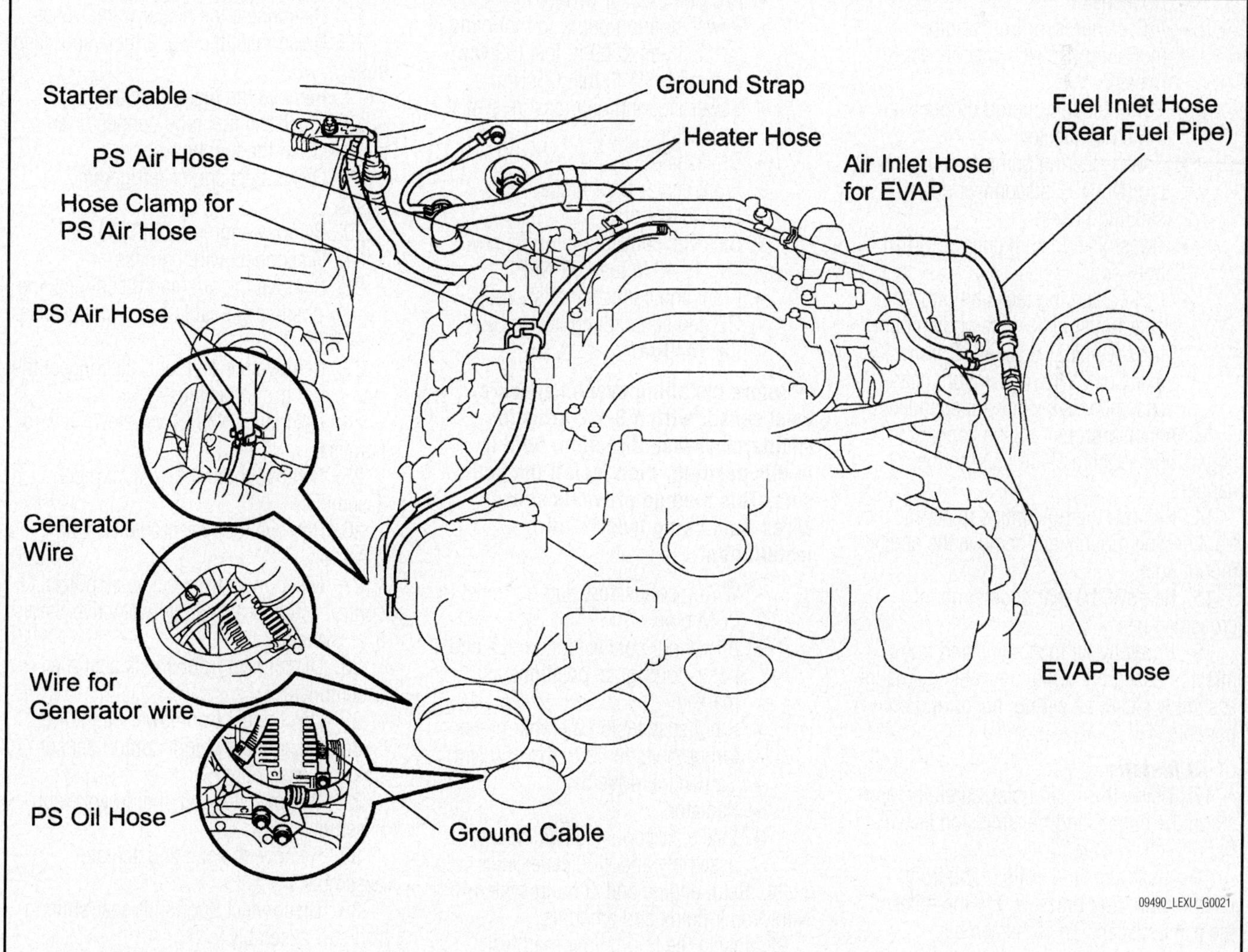

Fig. 102 Showing the locations of wiring and other components to disconnect for engine removal—GS 430

reinstallation. As items are marked, remove the following:

- Alternator wiring
- Power steering oil hose from oil pan
- Power steering hoses and clamp from timing belt cover
- Starter cable from battery
- Ground strap from body
- Both heater hoses
- Fuel inlet hose (rear fuel pipe) from fuel main tube

9. Disconnect all wiring connectors and grommet from ECM box.
10. Disconnect the Heated Oxygen Sensors (HEGO) from the exhaust pipe, after disconnecting the wiring grommets from the floor panel.
11. Remove the front and center exhaust pipes and then remove the catalytic converters.
12. Remove or disconnect the following:
    - Center front floor brace
    - Heat insulators
    - Driveshaft
    - A/C compressor from engine mounting (Do NOT disconnect A/C hoses)
    - Power steering pump oil pressure switch connector
    - Power steering pump from mounting (Do NOT disconnect power steering hoses)
    - Transmission shift control rod at both ends
    - Power steering rack and pinion gear housing from its body mountings (Do NOT disconnect from axles or CV joints); ensure gear assembly is securely suspended
13. Attach a proper engine hangers to engine and attach a chain hoist to the hangers.
14. Remove the hole plugs to access the front engine mounts and remove the engine mount nuts.
15. Remove the rear engine mount crossmember.
16. Carefully lift the engine and transmission assembly out of the vehicle. Ensure assembly clears all wiring, hoses and components.

To install:

17. Using the chain hoist, carefully reposition the engine and transmission assembly into the vehicle.
18. Insert the stud bolts of the front engine mounting brackets into the holes of the front suspension crossmember.

➡Ensure the engine is kept level during reattachment.

19. Install the rear engine mount crossmember (with "V8" mark facing forward) and torque the bolts to 19 ft. lbs. (26 Nm) and the nuts to 10 ft. lbs. (14 Nm).
20. Install the 4 nuts on the front engine mount brackets and tighten to 50 ft. lbs. (68 Nm). Install the hole plugs.
21. Remove the chain hoist.
22. Reinstall the V-bank cover bracket to the engine hanger, with a nut.
23. Install or connect the following:
    - Heat insulators to front side of front exhaust pipe
    - Power steering gear housing to mounting; torque nuts to 48 ft. lbs. (65 Nm)
    - Power steering gear oil tube
    - Steering column yoke bolt to power steering gear; torque to 26 ft. lbs. (25 Nm)
    - Transmission control rod
    - A/C compressor to mounting; torque bolt to 36 ft. lbs. (49 Nm) and nut to 21 ft. lbs. (29 Nm)
    - A/C compressor wiring
    - Power steering pump to mounting; torque bolts to 29 ft. lbs. (39 Nm) and nut to 21 ft. lbs. (29 Nm)
    - Power steering oil pressure switch connector
    - Driveshaft
    - Front center floor brace
    - Heat insulators
    - Catalytic converters; torque new nuts to 46 ft. lbs. (62 Nm)
    - Front and center exhaust pipes
    - Oxygen sensors; tighten to 32 ft. lbs. (44 Nm)

➡Before installing oxygen sensors, twist sensor wiring 3-1/2 turns in counterclockwise direction; hold it in this position, then install the sensors. This method prevents sensor wires from being twisted after installation.

- Wiring connectors and grommet to ECM box
- All wiring, connectors, straps and hoses to original positions as marked
- Front suspension member brace; torque bolts to 43 ft. lbs. (58 Nm)
- Serpentine drive belt
- Radiator
- Intake air connector, air cleaner assembly and air cleaner inlet

24. Refill engine and cooling system with proper fluids and amounts.
25. Install the hood, then road test thc vehicle. Recheck fluid levels after road test.

GS 460

See Figures 103 and 104.

1. Place front wheels facing straight ahead.
2. Discharge refrigerant from refrigeration system.
3. Discharge fuel system pressure.
4. Disconnect cable from negative battery terminal.
5. Disconnect cable from negative battery terminal.
6. Disconnect cable from negative battery terminal.
7. Remove the 7 clips and cool air intake duct seal.
8. Remove the clips and engine room side covers.
9. Remove engine under covers.
10. Drain engine oil.
11. Drain engine coolant.
12. Drain automatic transmission fluid.
13. Remove air cleaner assembly.
14. Remove radiator hoses and oil cooler lines.
15. Disconnect the heater water hoses.
16. Disconnect the No. 2 fuel vapor feed hose.
17. Remove the fuel pipe clamps.
18. Pinch the fuel tube connector and then pull out the fuel hoses.
19. Disconnect cooler refrigerant hoses.
20. Remove engine room ECU cover.
21. Disconnect wire harness.
22. Remove the nut, and disconnect the battery positive (+) cable and clamp from the battery.
23. Remove the bolt, and disconnect the clamp and ground cables.
24. Disconnect the power steering wire harness.
25. Remove the front exhaust pipe assembly.
26. Remove the rear propeller shaft assembly.
27. Disconnect front active stabilizer control actuator assembly (w/ Active Stabilizer System).
28. Disconnect front shock absorber assemblies.
29. Disconnect front lower ball joints.
30. Disconnect height control sensor link sub-assembly.
31. Remove steering sliding yoke sub-assembly.
32. Remove the steering actuator assembly.
33. Disconnect floor shift gear shifting rod sub-assembly.
34. Remove the 4 bolts and disconnect the rear engine mounting member.

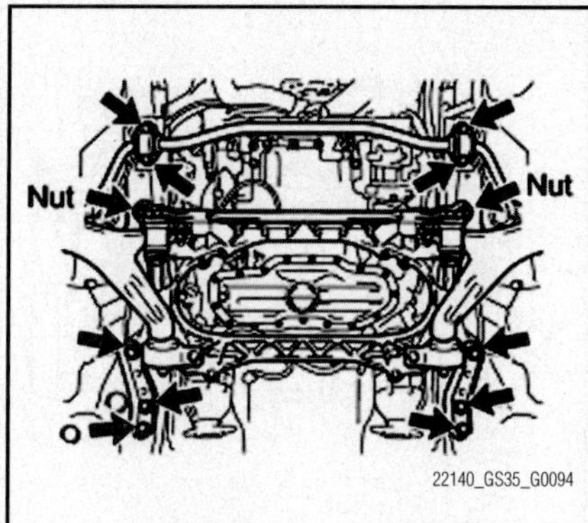

Fig. 103 Engine mounting bolt locations

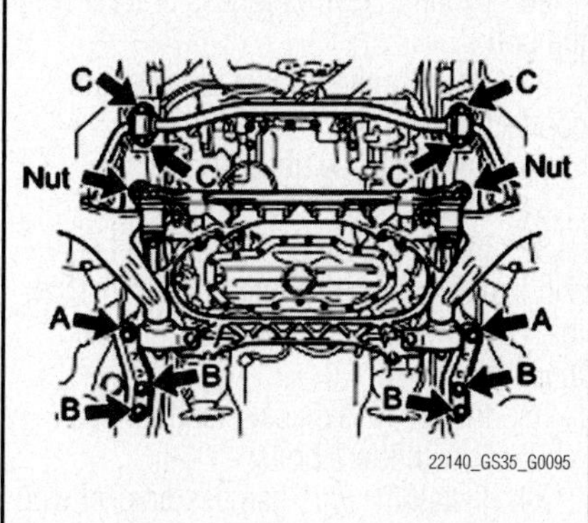

Fig. 104 Engine mounting bolt locations

35. Remove the 10 bolts and 2 nuts shown in the illustration.

36. Operate the engine lifter, then slowly remove the engine from the vehicle

To install:

37. Place the engine on an engine lifter.

38. Operate the engine lifter, then install the engine to the vehicle.

39. Install the 10 bolts and 2 nuts and tighten to the following:.

a. Bolt A–Torque: 150 ft. lbs. (204 Nm).

b. Bolt B–Torque: 37 ft. lbs. (50 Nm).

c. Bolt C–Torque: 36 ft. lbs. (49 Nm).

d. Nut–Torque: 123 ft. lbs. (167 Nm).

40. Install the engine rear mounting member with the 4 bolts and tighten to 19 ft. lbs. (26 Nm).

41. Connect floor shift gear shifting rod sub-assembly.

42. Install the steering actuator assembly.

43. Install steering sliding yoke sub-assembly.

44. Connect height control sensor link sub-assembly.

45. Connect front lower ball joint assemblies.

46. Connect front shock absorber assembly.

47. Connect front active stabilizer control actuator assembly.

48. Install the rear propeller shaft assembly.

49. Install the front exhaust pipe assembly.

50. Connect the power steering wire harness.

51. Connect the ground cables.

52. Connect the positive (+) battery cable with the nut.

53. Connect wiring harness.

54. Install engine room ECU cover.

55. Install cooler refrigerant hoses.

56. Install the fuel pipes.

57. Install the fuel pipe clamps.

58. Connect the No. 2 fuel vapor feed hose.

59. Connect the heater water hoses.

60. Connect the radiator hoses and oil cooler lines with the clamps.

61. Install air cleaner sub-assembly.

62. Add engine oil.

63. Add engine coolant.

64. Add automatic transmission fluid.

65. Connect cable to negative battery terminal.

66. Inspect for leaks.

67. Install front wheel.

68. Install engine under covers.

69. Install engine room side covers.

70. Install V-bank cover sub-assembly.

71. Perform initialization, if necessary.

EXHAUST MANIFOLD

REMOVAL & INSTALLATION

3.5L (2GR-FSE) Engine

See Figure 105.

1. Before servicing the vehicle, refer to the precautions in the beginning of this section.

2. Remove or disconnect the following:

- Negative battery cable. Wait at least 90 seconds before performing any other work
- Cool air intake duct seal
- Left and right engine room side covers
- V-bank cover
- Engine under cover
- No. 2 engine under cover (2WD)
- Left and right rear engine under cover (2WD)
- Front exhaust pipe assembly
- Front lower suspension member protector (2WD)

3. Remove the oil dipstick guide assembly as follows:

a. Remove the oil level gauge.

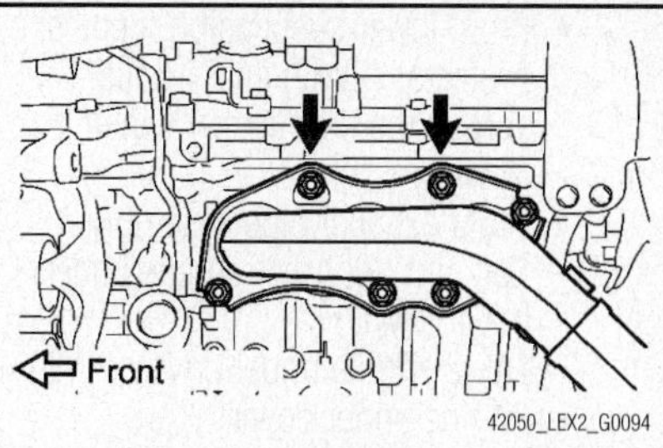

Fig. 105 When tightening the exhaust manifold nuts, start with the nuts depicted by the arrows as shown (left side illustrated, right side the same)—3.0L (3GR-FSE) engine

b. Remove the bolt, then remove the No. 2 oil dipstick guide.

c. Remove the O-ring from the No. 2 oil dipstick guide.

d. Remove the bolt and clamp (2WD), then remove the No. 1 oil dipstick guide.

e. Remove the O-ring from the No. 1 oil dipstick guide.

- Oxygen sensor wiring connectors
- 6 nuts, exhaust manifold and gasket (left and right sides)

To install:

4. Install the exhaust manifold(s) as follows:

a. Install a new exhaust manifold gasket.

b. Install the exhaust manifold with 6 new nuts and torque to 15 ft. lbs. (21 Nm).

✻✻ WARNING

Do not damage the stud bolt when installing the exhaust manifold. Be sure to tighten either of nuts A first as shown in the illustration.

5. Install the oil dipstick guide assembly as follows:

a. Install a new O-ring to the oil dipstick guide.

b. Apply a light coat of engine oil to the O-ring.

c. Push in the oil dipstick guide end into the guide hole.

d. Install the No. 1 oil dipstick guide with the bolt and torque to 7 ft. lbs. (10 Nm).

e. Connect the clamp (2WD).

f. Install the No. 2 oil dipstick guide with the bolt and torque to 15 ft. lbs. (21 Nm).

g. Install the oil dipstick.

6. Install or connect the following:

- Oxygen sensor wiring connectors

- Front lower suspension member protector (2WD) and 4 bolts. Torque the 4 bolts to 71 inch lbs. (8 Nm).
- Front exhaust pipe assembly
- Left and right rear engine under cover (2WD)
- No. 2 engine under cover (2WD)
- Engine under cover
- V-bank cover
- Left and right engine room side covers
- Cool air intake duct seal
- Negative battery cable

7. Start the engine and check for exhaust leaks and proper operation.
8. System initialization

4.3L (3UZ-FE) Engine

See Figure 106.

1. Before servicing the vehicle, refer to the precautions in the beginning of this section.
2. Remove or disconnect the following:
   - Negative battery cable
   - Engine under cover
   - Coolant
   - V-bank cover
   - Air cleaner assembly (if needed for access to exhaust manifold)
3. See CYLINDER HEADS in this section.
4. Remove 4 bolts and heat shield from exhaust manifold.
5. Remove 8 nuts and remove exhaust manifold and gasket.

To install:

6. To install, reverse the removal procedure. Install new manifold gasket and new retaining nuts. Torque the exhaust manifold nuts to 32 ft. lbs. (44 Nm).
7. Install the manifold heat shields.
8. Refill the engine cooling system. Start the engine and check for leaks.

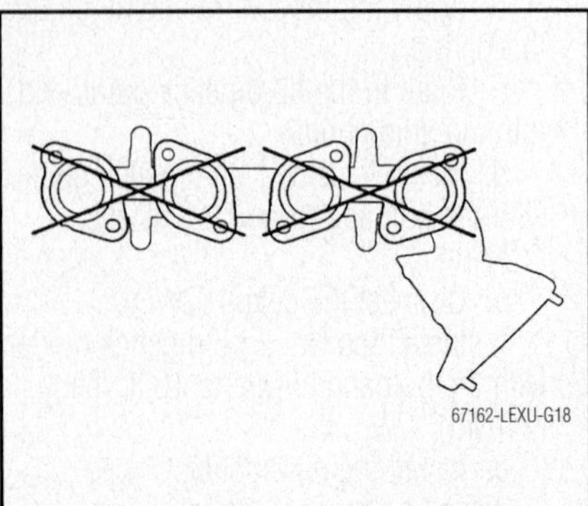

Fig. 106 Measuring exhaust manifold warpage

4.6L (1UR-FSE) Engine

See Figures 107 and 108.

1. Disconnect cable from negative battery terminal.
2. Remove engine under covers.
3. Remove cool air intake duct seal.
4. Remove air cleaner.
5. Remove V-ribbed belt.
6. Remove the front stabilizer bar.
7. Remove front floor cover.
8. Remove front center floor brace.
9. Remove front exhaust pipe assembly.
10. Remove steering sliding yoke sub-assembly.
11. Remove generator assembly.
12. Remove engine oil level dipstick guide.
13. Disconnect compressor and pulley.
14. Remove engine room side covers.
15. Remove V-bank cover.
16. Disconnect the air fuel ratio sensor connector RH.
17. Remove the 3 bolts and No. 1 exhaust manifold heat insulator.
18. Remove the 8 nuts and exhaust manifold RH.
19. Remove the gasket.
20. Disconnect the air fuel ratio sensor connector LH.
21. Remove the bolt and wire harness bracket LH.
22. Remove the 3 bolts and No. 2 exhaust manifold heat insulator.
23. Remove the 8 nuts and exhaust manifold LH.
24. Remove the gasket.

To install:

25. Install a new gasket LH.
26. Install the exhaust manifold LH, and install 8 new nuts and tighten to 15 ft. lbs. (21 Nm) in the order shown in the illustration.
27. Install the exhaust manifold heat insulator with the 3 bolts.

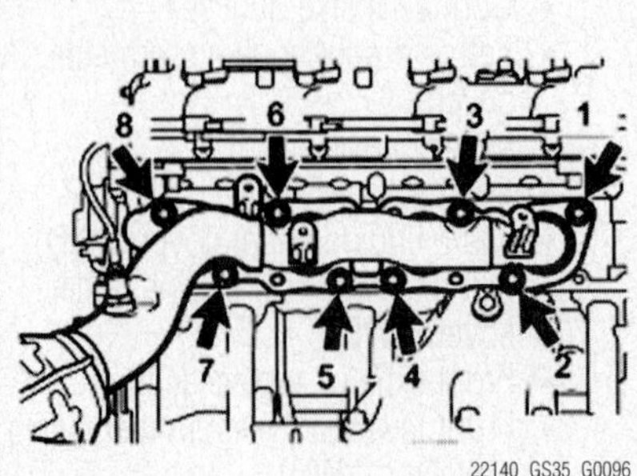

Fig. 107 Exhaust manifold tightening sequence LH

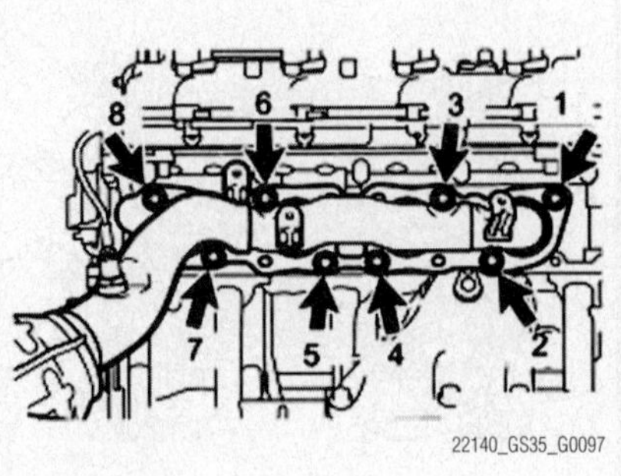

Fig. 108 Exhaust manifold tightening sequence RH

28. Install the wire harness bracket with the bolt.
29. Connect the air fuel ratio sensor connector.
30. Install air fuel ratio sensor.
31. Install a new gasket RH.
32. Install the exhaust manifold LH, and install 8 new nuts and tighten to 15 ft. lbs. (21 Nm) in the order shown in the illustration.
33. Install the exhaust manifold heat insulator with the 3 bolts.
34. Install the wire harness bracket with the bolt.
35. Connect the air fuel ratio sensor connector.
36. Install air fuel ratio sensor.
37. Connect compressor and pulley.
38. Install engine oil level dipstick guide.
39. Install generator assembly.
40. Install steering sliding yoke sub-assembly.
41. Install front exhaust pipe assembly.
42. Install front floor covers.
43. Install front center floor brace.
44. Install the front stabilizer bar.
45. Install V-ribbed belt.
46. Install air cleaner.
47. Connect cable to negative battery terminal.
48. Install engine under covers.
49. Install V-bank cover.
50. Install engine room side covers.
51. Install cool air intake duct seal.

INTAKE MANIFOLD

REMOVAL & INSTALLATION

3.5L (2GR-FSE) Engine

See Figure 109.

1. Before servicing the vehicle, refer to the precautions in the beginning of this section.
2. Relieve fuel system pressure.

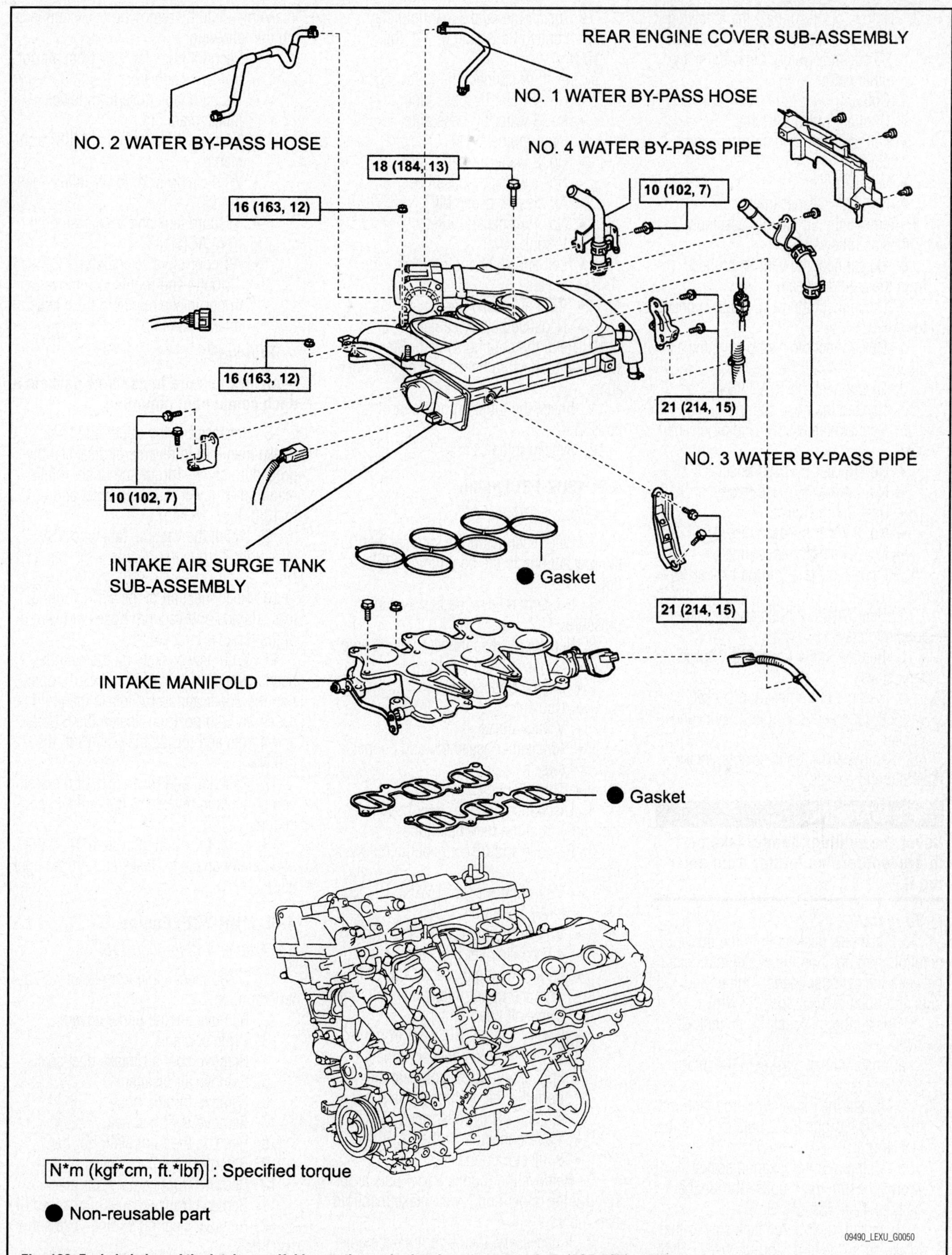

Fig. 109 Exploded view of the intake manifold mounting and related components—3.5L (2GR-FSE) engine

3. Remove or disconnect the following:
   - Negative battery cable. Wait at least 90 seconds before performing any other work.
   - Coolant
   - Cool air intake duct seal
   - Left and right engine room side covers
   - V-bank cover
   - No. 1 air cleaner inlet
4. Remove the air cleaner assembly with hose as follows:
   a. Disconnect the ventilation hose from the cylinder head.
   b. Disconnect the MAF meter connector.
   c. Disconnect the clamp from the air cleaner.
   d. Disconnect the EVAP VSV.
   e. Loosen the hose clamp.
   f. Remove the 3 bolts and air cleaner case.
   - No. 1 water by-pass hose
   - No. 2 water by-pass hose
   - Rear engine cover
   - No. 3 water by-pass pipe
   - No. 4 water by-pass pipe
5. Remove intake air surge tank as follows:
   a. Remove the 2 bolts and intake manifold stay.
   b. Remove the 4 bolts and 2 surge tank stays.
   c. Using a 5mm hexagon socket wrench, remove the 7 bolts, 2 nuts and gasket.
6. Remove the 4 bolts, 2 nuts, intake manifold and gasket.

WARNING

Cover the cylinder head intake port to prevent foreign matter from entering it.

To install:

7. Place new gaskets onto the intake manifold and position the intake manifold between the cylinder heads. Tighten the nuts and bolts to 15 ft. lbs. (21 Nm).
8. Install the intake air surge tank as follows:
   a. Install a new gasket to the intake air surge tank.
   b. Install the intake air surge tank with the 2 nuts. Torque the 2 nuts to 13 ft. lbs. (18 Nm).
   c. Using a 5mm hexagon socket wrench, install the 7 bolts. Torque the bolts to 12 ft. lbs. (16 Nm).
   d. Install the 2 surge tank stay to the intake air surge tank. Torque the 4 bolt to 15 ft. lbs. (21 Nm).
   e. Install the intake manifold stay with 2 bolts and torque to 7 ft. lbs. (10 Nm).
9. Install or connect the following:
   - No. 4 water by-pass pipe
   - No. 3 water by-pass pipe
   - Rear engine cover
   - No. 2 water by-pass hose
   - No. 1 water by-pass hose
   - Air cleaner assembly
   - No. 1 air cleaner inlet
   - V-bank cover
   - Left and right engine room side covers
   - Cool air intake duct seal
   - Negative battery cable
10. Refill the cooling system. Start the engine and check for leaks and proper operation.
11. Check the function of the throttle body unit.
12. System initialization.

4.3L (3UZ-FE) Engine

See Figures 110 and 111.

1. Before servicing the vehicle, refer to the precautions in the beginning of this section.
2. Properly relieve the fuel system pressure.
3. Remove or disconnect the following:
   - Negative battery cable
   - Engine under cover
   - Coolant
   - V-bank cover
   - Air cleaner assembly and connectors
   - Throttle body assembly
4. Disconnect the fuel inlet hose (rear fuel pipe) from the fuel main tube.
5. Remove and disconnect the following:
   - VSV connector for EVAP
   - EVAP hose from VSV
   - VSV from upper intake manifold
   - 4 V-bank cover brackets
   - Engine wiring protector (LH side) from the upper intake manifold and camshaft bearing cap
   - 2 wire clamps (RH side) from the brackets on the delivery pipe
   - Engine wire protector (rear side) from the rear water bypass joint and the RH cylinder head
   - VSV connector for the ACIS
   - 8 injector connectors
6. Remove the 6 bolts and 4 nuts and remove the upper and lower intake manifold assembly.
7. If necessary, the upper intake manifold can be disassembled from the lower intake manifold by removing or disconnecting the following:
   - Vacuum hose for VSV from air control valve actuator
   - Vacuum tank hose from lower intake manifold
   - Vacuum hose (VSV for ACIS) from clamp
   - Wire clamp from lower intake manifold
   - Vacuum tank and VSV assembly from ACIS
   - Air control valve actuator
   - 15 bolts and 5 nuts to remove upper intake manifold from lower intake manifold

To install:

➡Always be sure to use new gaskets at each component mounting.

8. Reassemble the upper and lower intake manifold in reverse of disassembly procedure given. Torque the upper intake manifold-to-lower manifold bolts and nuts to 13 ft. lbs. (18 Nm).
9. Install the vacuum tank and VSV assembly. Torque the nuts to 13 ft. lbs. (18 Nm).
10. Reconnect all of the wiring connectors, clamps and vacuum hoses in reverse of the removal procedure.
11. With new gaskets on the cylinder heads (white marks facing outward), position the upper and lower intake manifold assembly into position. Install the 6 bolts and 4 nuts and torque them to 13 ft. lbs. (18 Nm).
12. Reinstall and reconnect all remaining components in reverse of the removal procedure.
13. Refill the cooling system. Start the engine and check for leaks and proper operation.

4.6L (1UR-FSE) Engine

See Figures 112 through 116.

1. Disconnect cable from negative battery terminal.
2. Remove engine under covers.
3. Drain coolant.
4. Remove cool air intake duct seal.
5. Remove air cleaner.
6. Remove throttle body.
7. Remove V-ribbed belt.
8. Remove the front stabilizer bar.
9. Remove front floor cover.
10. Remove front center floor brace.
11. Remove front exhaust pipe assembly.
12. Remove steering sliding yoke sub-assembly.
13. Remove generator assembly.

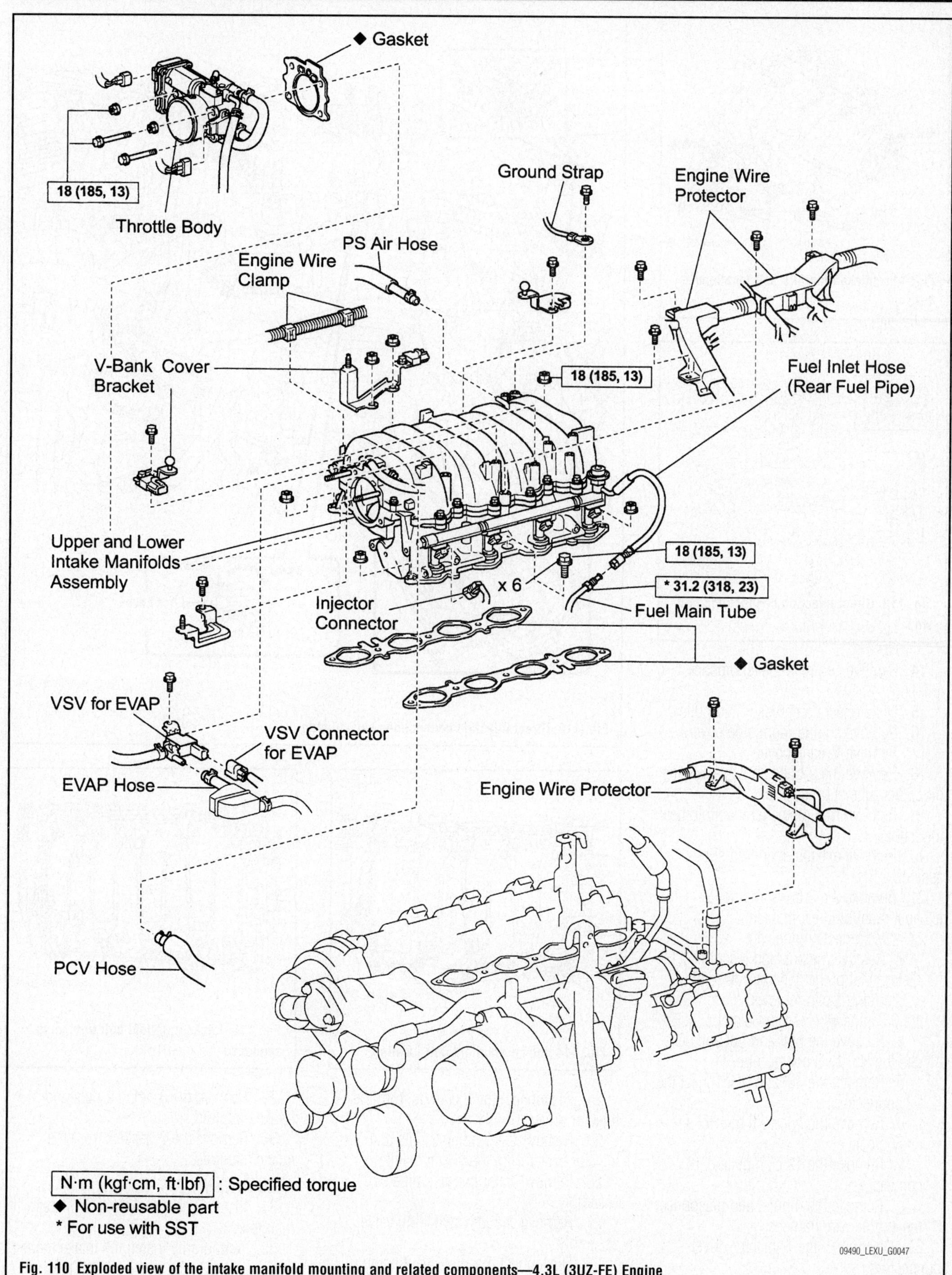

Fig. 110 Exploded view of the intake manifold mounting and related components—4.3L (3UZ-FE) Engine

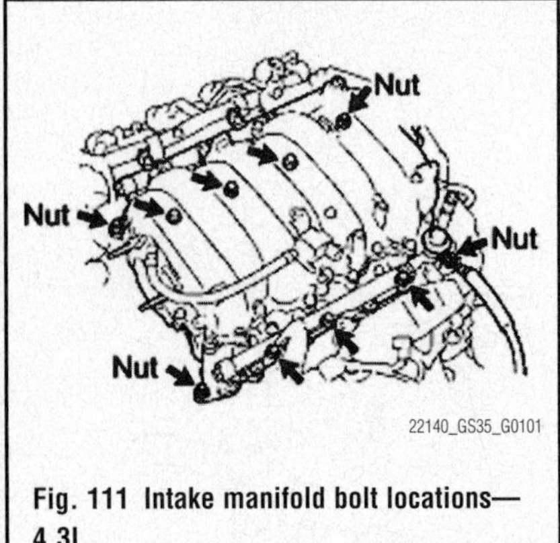

Fig. 111 Intake manifold bolt locations—4.3L

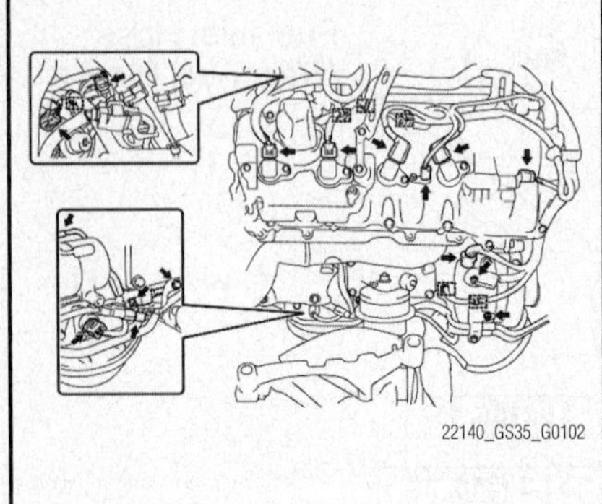

Fig. 112 Direct injection connections first step

14. Remove engine oil level dipstick guide.
15. Disconnect compressor and pulley.
16. Remove engine room side covers.
17. Remove V-bank cover.
18. Remove the 2 bolts and separate the oil cooler tube sub-assembly.
19. Remove No. 1 exhaust manifold heat insulator.
20. Remove exhaust manifold sub-assembly RH.
21. Remove the 3 bolts and no. 3 exhaust manifold heat insulator.
22. Disconnect engine wire.
    a. Remove the nut, and disconnect the +B terminal of the generator assembly.
    b. Using a clip remover, disconnect the 6 engine wire harness clamps.
    c. Remove the bolt and nut and separate the No. 1 oil cooler tube.
    d. Remove the nut and disconnect the 12 connectors.
    e. Remove the nut from the No. 1 relay block.
    f. Remove the 12 clamps and 10 connectors.
    g. Remove the 4 nuts, and disconnect the engine wire harness.
23. Disconnect the 4 injector driver connectors

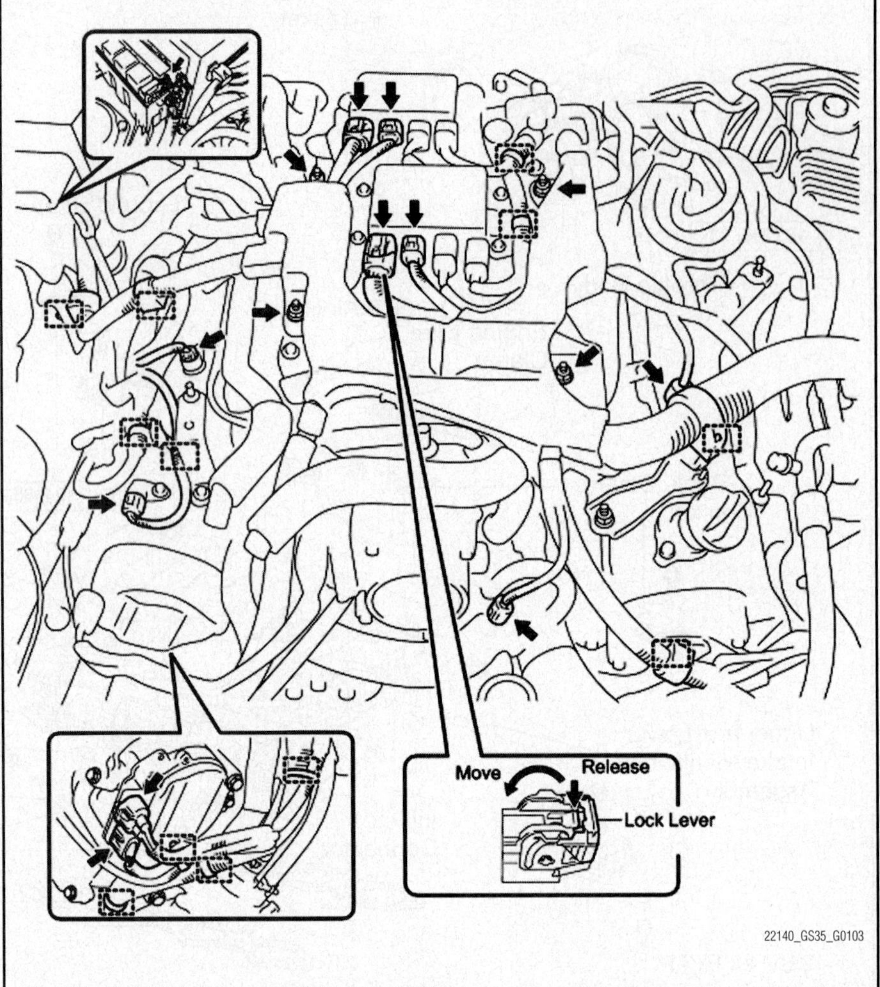

Fig. 113 Direct injection connections second step

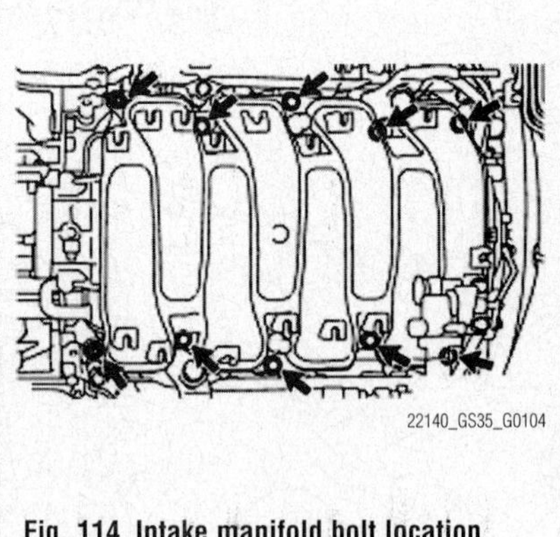

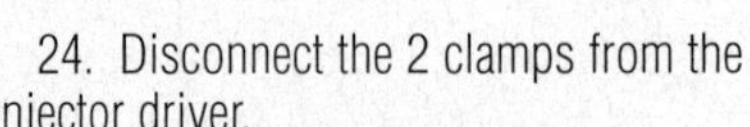
Fig. 114 Intake manifold bolt location

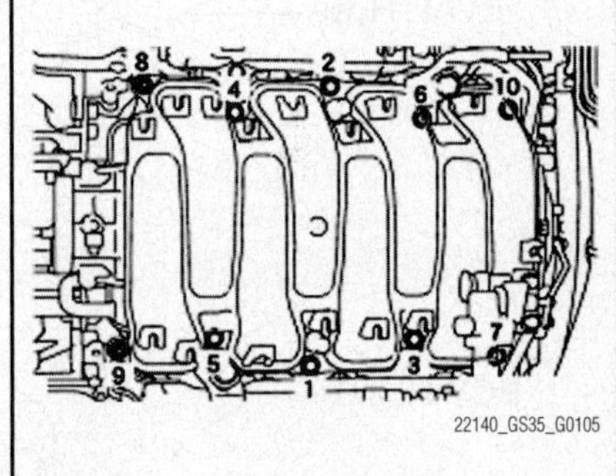

Fig. 115 Intake manifold bolt tightening sequence

24. Disconnect the 2 clamps from the injector driver.
25. Remove the 2 bolts, 2 nuts and injector driver from the intake manifold.
26. Remove water by-pass pipe sub-assembly.
27. Remove vacuum switching valve assembly.
28. Disconnect the No. 1 ventilation hose from the intake manifold.
29. Remove the 8 bolts, 2 nuts and intake manifold.
30. Remove the 2 gaskets from the intake manifold.

To install:

31. Install 2 new gaskets to the intake manifold.
32. Temporarily install the intake manifold with the 2 nuts and 8 bolts. Then

Fig. 116 Injector driver installation

tighten the 2 nuts and 8 bolts uniformly in the order shown in the illustration

33. Connect the No. 1 ventilation hose to the intake manifold.
34. Install vacuum switching valve assembly.
35. Install the water by-pass pipe to the intake manifold with the 2 bolts.
36. Connect the heater inlet water hose, heater outlet water hose, water inlet hose, and No. 3 water by-pass hose to the water by-pass pipe with the 4 clamps.
37. Install the injector driver to the intake manifold by installing the 2 bolts and 2 nuts and tightening to 7 ft. lbs. (10 Nm) in the order shown in the illustration.
38. Connect the 4 wire harness connectors to the injector driver. Then move the lock lever as shown in the illustration to lock the connectors.
39. Connect the 2 clamps to the injector driver.
40. Connect the 4 injector driver connectors.
41. Connect the engine wire harness connector.
    a. Connect the engine wire harness with the 4 nuts.
    b. Connect the 12 clamps and 10 connectors.
    c. Connect the engine wire harness to the No. 1 relay block with the nut.
    d. Connect the 6 clamps and 12 connectors.
    e. Install the engine wire harness clamp bracket with the nut.
    f. Connect the +B terminal of the generator assembly with the nut.
42. Install the exhaust manifold heat insulator with the 3 bolts.
43. Install exhaust manifold sub-assembly RH.
44. Install the wire harness bracket with the bolt.
45. Connect the air fuel ratio sensor connector.
46. Install air fuel ratio sensor.
47. Connect compressor and pulley.
48. Install engine oil level dipstick guide.
49. Install generator assembly.
50. Install the oil cooler tube sub-assembly with the 2 bolts.
51. Install steering sliding yoke sub-assembly.
52. Install front exhaust pipe assembly.
53. Install front floor covers.
54. Install front center floor brace.
55. Install the front stabilizer bar.
56. Install V-ribbed belt.
57. Install throttle body.
58. Install air cleaner.
59. Connect cable to negative battery terminal.
60. Fill coolant.
61. Install engine under covers.
62. Install V-bank cover.
63. Install engine room side covers.
64. Install cool air intake duct seal.
65. Connect cable to negative battery terminal.
66. Check for leaks.
67. Perform initialization, if necessary.

OIL PAN

REMOVAL & INSTALLATION

3.5L (2GR-FSE) Engine

➡The No. 1 oil pan cannot be removed with the engine in the vehicle. The engine and transmission must be removed as a unit, then separated. See ENGINE ASSEMBLY section. It may be possible to remove the No. 2 oil pan from the vehicle while the engine is still in the vehicle.

1. Before servicing the vehicle, refer to the precautions in the beginning of this section.
2. Drain the engine oil.
3. Remove or disconnect the following:
    - Negative battery cable from the battery
    - Engine/transmission assembly
    - Oil filter element
    - Oil filter bracket (AWD)
4. Remove the No. 2 oil pan sub-assembly as follows:
    a. Remove the 16 bolts and 2 nuts (2WD).
    b. Remove the 14 bolts and 2 nuts (AWD).
    c. Insert the blade of a prying tool between the oil pans. Cut through the applied sealer and remove the No. 2 oil pan sub-assembly.

➡Be careful not to damage the contact surfaces of the oil pans.

5. Remove the oil with strainer pipe sub-assembly (AWD) as follows:
    a. Remove the 3 mounting nuts.
    b. Remove the oil with strainer pipe and gasket.
6. Remove the oil pan sub-assembly as follows:
    a. Remove the 16 bolts and 2 nuts.

➡Be sure to clean the bolts and stud bolts and check the threads for cracks or other damage.

    b. Remove the oil pan by prying between the oil pan and cylinder block with a screwdriver.

⁂ WARNING

Be careful not to damage the contact surfaces of the cylinder block and oil pan.

    c. Remove the 2 O-rings.

To install:

7. Install the oil pan sub-assembly as follows:
    a. When replacing a stud bolt, install it by using an E6 "Torx" socket wrench. Torque the stud bolt to 35 inch lbs. (4 Nm).
    b. Apply seal packing in a continuous line of 0.118–0.156 inches (3.0–4.0mm) in diameter.

➡Remove any oil from the contact surface. Install the oil pan within 3 minutes after applying seal packing. Do not start the engine for at least 2 hours after installing.

    c. Install the oil pan with the 16 bolts and 2 nuts. Torque the A bolts to 7 ft. lbs. (10 Nm). Torque the remaining bolts to 15 ft. lbs. (21 Nm).
8. Install the oil with strainer pipe sub-assembly (AWD) as follows:
    a. Install a new gasket.
    b. Install the oil with strainer pipe with the 3 nuts. Torque to 7 ft. lbs. (10 Nm).
9. Install the No. 2 oil pan sub-assembly as follows:
    a. Apply seal packing in a continuous line of 0.118–0.156 inches (3.0–4.0mm) in diameter.

➡Remove any oil from the contact surface. Install the oil pan No. 2 within 3 minutes after applying seal packing. Do not start the engine for at least 2 hours after installing.

b. Install the oil pan with the 16 bolts and 2 nuts (2WD). Torque the bolts to 7 ft. lbs. (10 Nm).

c. Install the oil pan with the 14 bolts and 2 nuts (AWD). Torque the bolts to 7 ft. lbs. (10 Nm).

10. Install or connect the following:
    - New gasket and oil filter bracket (AWD). Torque the bolt and 2 nuts to 15 ft. lbs. (21 Nm).
    - Oil filter element
    - Engine/transmission assembly
    - Negative battery cable from the battery
    - Engine with oil

4.3L (3UZ-FE) Engines

See Figures 117 and 118.

➡The oil pan cannot be removed with the engine in the vehicle. The engine and transmission must be removed as a unit, then separated. It may be possible to remove the oil pan from the vehicle while the engine is still in the vehicle.

1. Before servicing the vehicle, refer to the precautions in the beginning of this section.
2. Remove or disconnect the following:
    - Engine/transmission assembly
3. Remove the 18 bolts and 2 nuts.
4. Using a screwdriver, remove the oil pan by prying the areas between the cylinder block and oil pan.

To install:

5. Apply the seal packing to oil pan No. 1.
6. Temporarily install the oil pan with the 18 bolts, stud bolt and 2 nuts.
7. Set the oil pan as shown.

➡Make sure the clearance between the rear end of the oil pan and cylinder block is 0.2 mm (0.008 in.) or less. If the clearance is more than 0.2 mm (0.008 in.), the oil pan will deform when tightening the bolts and nuts.

8. Uniformly tighten the bolts and nuts in several passes to the following:

a. Bolts A and D to 66 inch lbs. (7.5 Nm).

b. Bolts B and C to 21 ft. lbs, (28 Nm).

c. Nut E to 21 ft. lbs, (28 Nm).

9. Apply seal packing to oil pan No. 2

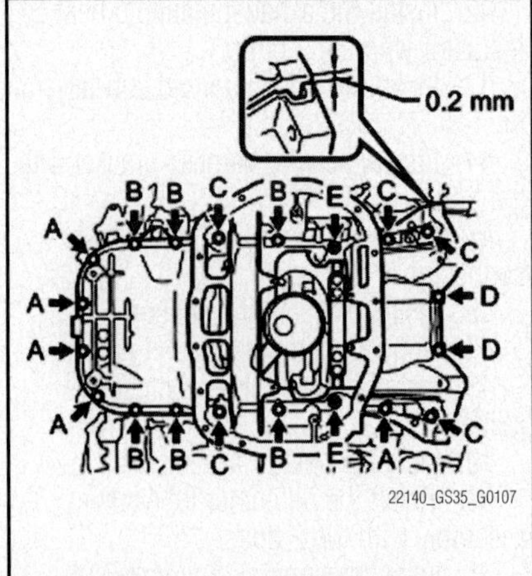

Fig. 117 Oil pan No. 1 bolt location first

Bolt	Bolt Head Size	Length
A	10 mm	22 mm (0.87 in.)
B	12 mm	25 mm (0.99 in.)
C	12 mm	60 mm (2.38 in.)
D	10 mm	35 mm (1.38 in.)

22140_GS35_G0108

Fig. 118 Oil pan No. 1 bolt legend second

10. Install the oil pan with the 15 bolts and 2 nuts to 66 inch lbs. (7.5 Nm). Uniformly tighten the bolts and nuts in several passes.
11. Install or connect the following:
    - Engine/transaxle assembly
    - All fluids

4.6L (1UR-FSE) Engines

See Figure 119.

➡The oil pan cannot be removed with the engine in the vehicle. The engine and transmission must be removed as a unit, then separated. It may be possible to remove the oil pan from the vehicle while the engine is still in the vehicle.

1. Before servicing the vehicle, refer to the precautions in the beginning of this section.
2. Remove or disconnect the following:
    - Engine/transmission assembly
3. Remove the 15 bolts and 2 nuts and remove the No. 2 oil pan.
4. Insert the blade of an oil pan cutter between the oil pans. Cut through the

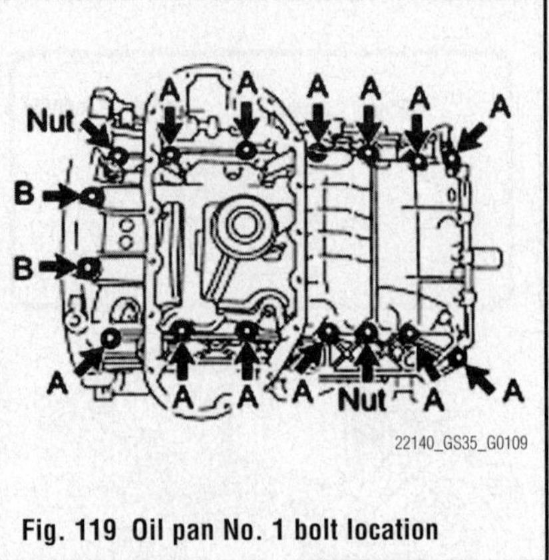

Fig. 119 Oil pan No. 1 bolt location

applied sealer and remove the No. 2 oil pan sub-assembly.

5. Remove the 14 bolts and 2 nuts and remove the No. 1 oil pan.
6. Remove the oil pan sub-assembly by prying between the oil pan and cylinder block with a screwdriver.

To install:

7. Apply the seal packing to oil pan No. 1.
8. Install the oil pan sub-assembly with the 14 bolts and 2 nuts.
9. Uniformly tighten the bolts and nuts in several passes to the following:

a. Bolts A to 7 ft. lbs. (10 Nm).

b. Bolts B 26 ft. lbs, (35 Nm).

c. Nut to 26 ft. lbs, (35 Nm).

10. Apply seal packing to oil pan No. 2.
11. Install the oil pan with the 15 bolts and 2 nuts to 7 ft. lbs. (10 Nm). Uniformly tighten the bolts and nuts in several passes.
12. Install or connect the following:
    - Engine/transaxle assembly
    - All fluids

OIL PUMP

REMOVAL & INSTALLATION

3.5L (2GR-FSE) Engine

See Figures 120 through 125.

➡The oil pump cannot be removed with the engine in the vehicle. The engine and transmission must be removed as a unit, then separated.

1. Before servicing the vehicle, refer to the precautions in the beginning of this section.
2. Remove or disconnect the following:
    - Engine/transmission assembly
    - Front differential assembly (AWD)
    - Serpentine drive belt
    - No. 2 idler pulley
    - Alternator
    - A/C compressor unit, if necessary

Fig. 120 Location of mounting bolts for oil pump cover behind the timing chain cover—3.5L (2GR-FSE) engine

- Left and right engine mounting brackets
- Serpentine belt tensioner
- Water pump pulley
- Fuel injector driver
- Intake air surge tank assembly and No. 2 surge tank stay
- Water hose joint
- Crankshaft pulley
- Water inlet
- Oil pan assembly
- Oil strainer
- No. 1 and 2 fuel pipes
- High pressure side fuel pump
- Ignition coil assembly
- No. 1 and 2 oil pipes
- Left and right cylinder head covers

3. Remove the timing chain cover assembly as follows:

a. Remove bolt and wiring harness clamp bracket.

b. Remove 25 mounting bolts and 2 mounting nuts.

c. Remove the timing chain cover by prying between the timing chain cover and cylinder head or cylinder block with a screwdriver.

d. Remove the gasket.

➡The oil pump assembly is incorporated into the back of the timing chain cover. The oil pump assembly can be disassembled from the back of the timing chain cover for inspection purposes.

To install:

4. Install the timing chain cover assembly as follows:

a. Apply seal packing in a continuous line of 0.197–0.217 inches (5.0–5.5mm) in diameter to the engine at the seam where the cylinder head meets the camshaft bearing cap assembly and the cylinder head meets the cylinder block.

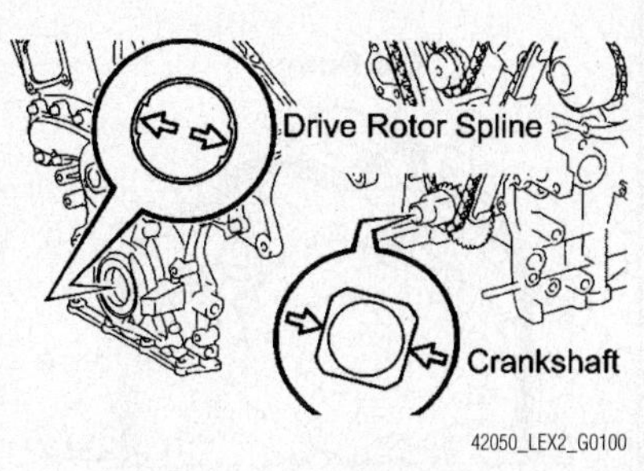

Fig. 121 Align the oil pump drive rotor spline and the crankshaft—3.5L (2GR-FSE) engine

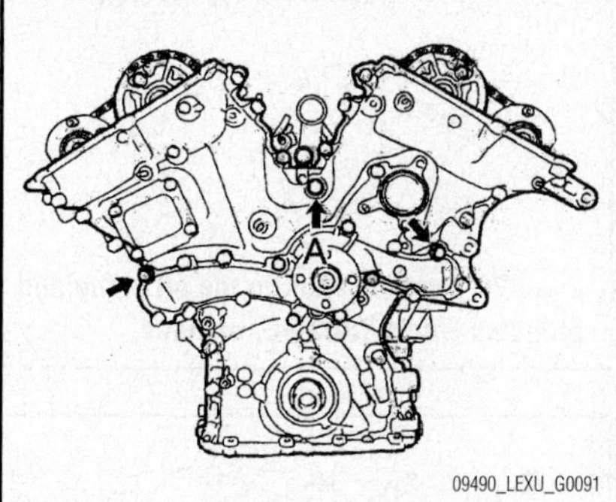

Fig. 122 Location of 3 bolts to be tightened first (location of bolt A shown)—3.5L (2GR-FSE) engine

➡Be sure to clean, degrease and dry the contact surfaces before applying the seal packing. Install the component within 3 minutes after applying seal packing. Do not start the engine for at least 2 hours after installing.

b. Apply seal packing in a continuous line of 0.138–0.158 inches (3.5–4.0mm) in diameter to the timing chain cover.

c. Install a new gasket.

d. Align the oil pump drive rotor spline and the crankshaft. Install the spline and chain cover to the crankshaft.

e. Temporarily tighten the timing chain cover with the 25 bolts and nuts.

f. Fully tighten the 3 bolts shown in the illustration. Torque bolt A to 32 ft. lbs. (43 Nm). Torque the 2 remaining bolts to 15 ft. lbs (21 Nm).

g. Fully tighten the 3 bolts shown in the illustration. Torque the bolts to 15 ft. lbs (21 Nm).

h. Fully tighten the 7 bolts and 2 nuts shown in the illustration. Torque the bolts to 15 ft. lbs (21 Nm).

➡Be sure to tighten the bolts and nuts in order of upper to lower.

i. Fully tighten the 12 bolts shown in the illustration. Torque the bolts to 15 ft. lbs (21 Nm).

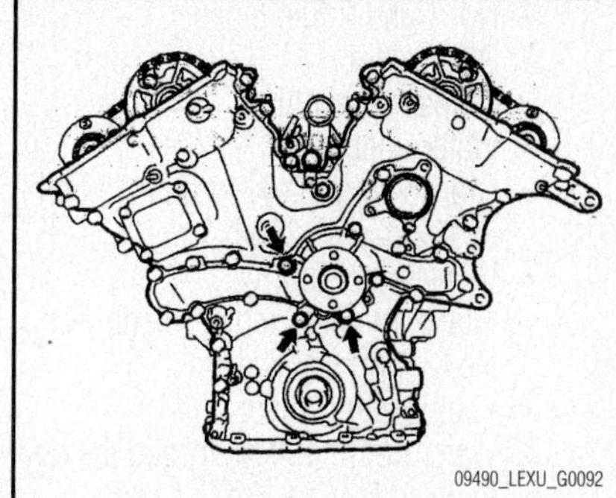

Fig. 123 Location of 3 bolts to be tightened second—3.5L (2GR-FSE) engine

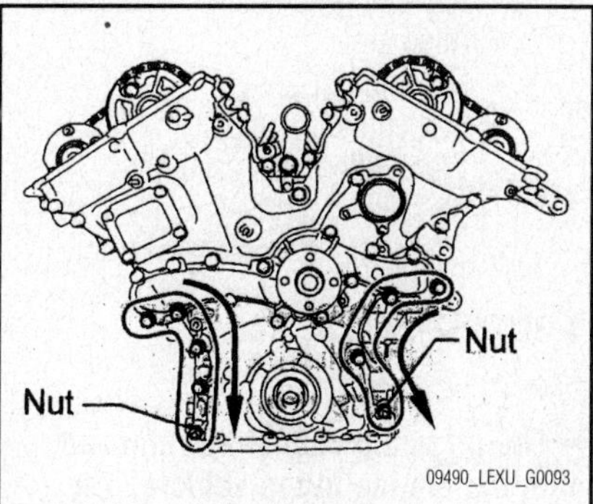

Fig. 124 Location of 7 bolts and 2 nuts to be tightened third—3.5L (2GR-FSE) engine

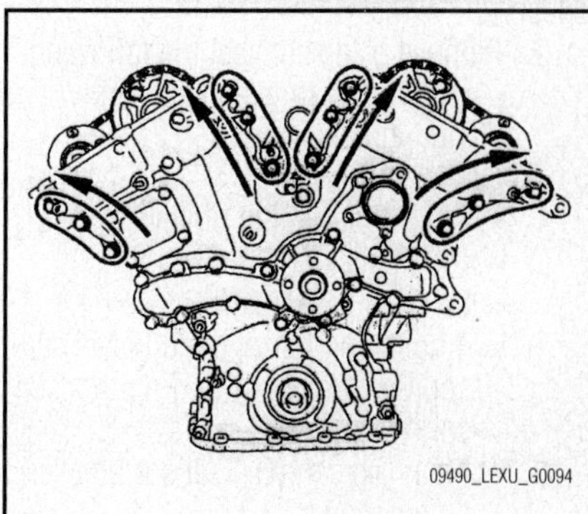

Fig. 125 Location of 12 bolts to be tightened fourth—3.5L (2GR-FSE) engine

➡Be sure to tighten the bolts in order of lower to upper.

j. Install the bolt and wiring harness bracket. Torque the bolts to 7 ft. lbs (10 Nm).

5. Install or connect the following:

- Left and right cylinder head covers
- No. 1 and 2 oil pipes
- Ignition coil assembly
- High pressure side fuel pump

- No. 1 and 2 fuel pipes
- Oil strainer
- Oil pan assembly
- Water inlet
- Crankshaft pulley. Torque the bolt to 192 ft. lbs (260 Nm).
- Water hose joint
- Intake air surge tank assembly and No. 2 surge tank stay
- Fuel injector driver
- Water pump pulley. Torque the bolt to 15 ft. lbs (21 Nm).
- Serpentine belt tensioner
- Left and right engine mounting brackets
- A/C compressor unit, if necessary
- Alternator
- No. 2 idler pulley
- Serpentine drive belt
- Front differential assembly (AWD)
- Engine/transmission assembly

4.3L (3UZ-FE) Engines

See Figures 126 through 128.

➡The oil pump cannot be removed with the engine in the vehicle. The engine and transmission must be removed as a unit, then separated.

1. Before servicing the vehicle, refer to the precautions in the beginning of this section.
2. Remove or disconnect the following:
   - Engine/transmission assembly
   - Timing belt
   - Idler pulleys
   - Crankshaft timing pulley
   - Oil dipstick and guide
   - Oil level sensor lead
   - 4 bolts and lift off the oil level sensor. Be careful not to drop this sensor.
   - Main Oxygen (O_2) sensor bracket, if necessary
   - Oil filter and filter bracket assembly by removing the stud bolt and 2 nuts
   - Engine Crankshaft Position (CKP) sensor. Remove the sensor by removing the bolt.
   - 12 bolts and 2 nuts from the No. 2 oil pan
   - No. 2 (lower) oil pan. Use a gasket-cutting tool
   - 2 bolts and 3 nuts and drop down the baffle plate
   - Oil strainer
   - No. 1 oil pan. There are slots for inserting the prybar.
   - 8 bolts holding the oil pump to the engine

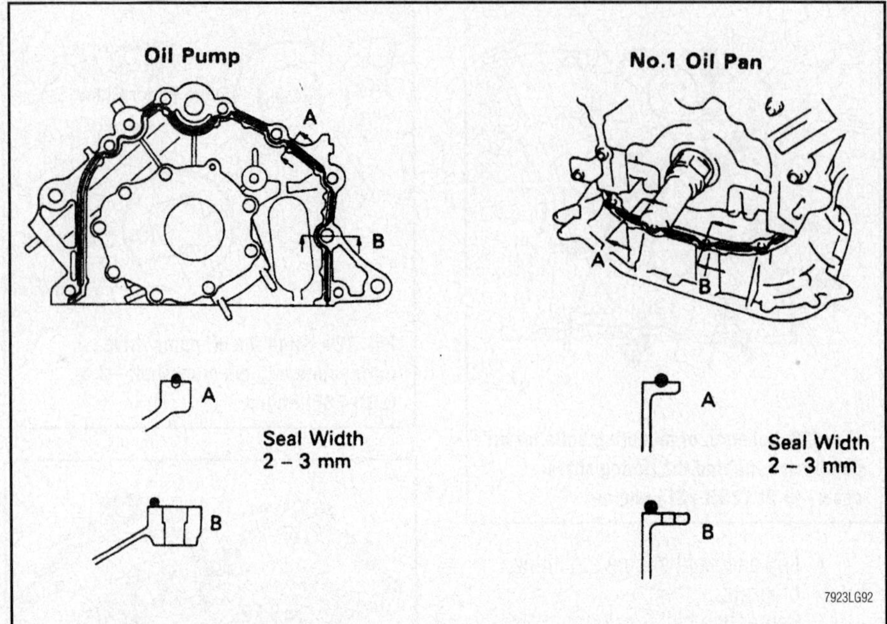

Fig. 126 Apply sealant to the oil pump and the No. 1 oil pan, as shown, before installing the oil pump—4.3L (3UZ-FE) engines

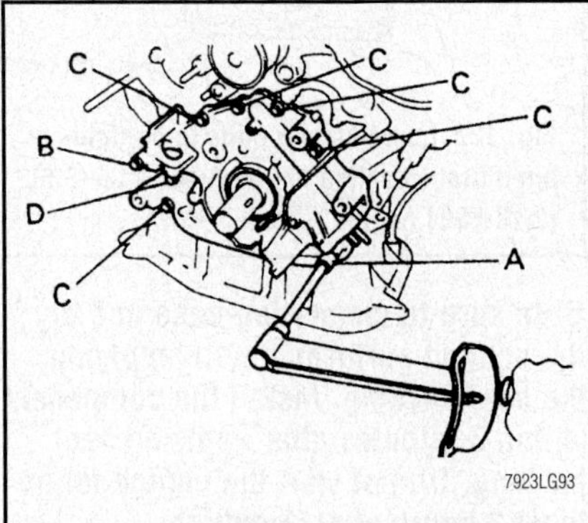

Fig. 127 4.0L engine oil pump mounting bolt locations, according to bolt lengths—(A) 1.97 in. (50mm), (B) 4.17 in. (106mm), (C) 1.18 in. (30mm) and (D) 1.57 in. (40mm)

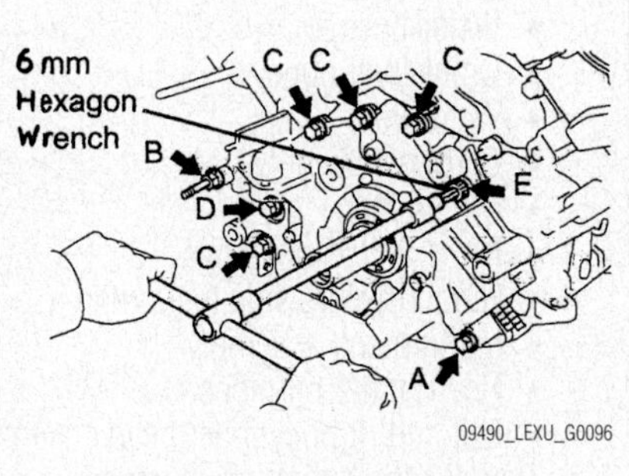

Fig. 128 4.3L engine oil pump mounting bolt locations, according to bolt lengths—(A) 1.97 in. (50mm), (B) 4.17 in. (106mm), (C) 1.18 in. (30mm), (D) 1.73 in. (44mm) and (E) 1.10 in. (28mm)

➡Make certain to observe bolt position during removal. The bolts are different lengths and sizes. Record their position for proper reassembly.

   - Oil pump from the engine block
   - O-ring from the block

To install:

➡Prior to installing the oil pump, lubricate the gears with clean engine oil.

3. Install or connect the following:
   - A 2–3mm wide (0.08–0.12 in.) bead of RTV sealant to the oil pump
   - New O-ring in position on the block
   - Oil pump on the engine
   - The 8 bolts in their correct locations. Tighten the bolts with 12mm or 6mm heads to 12 ft. lbs. (16 Nm) and the bolts with 14mm heads to 22 ft. lbs. (30 Nm).
   - A ⅛ inch (3–4mm) bead of RTV sealant to the pan mating surface.
   - No. 1 pan. Bolts–10mm: 66 inch lbs. (8 Nm); 12mm: 21 ft. lbs. (28 Nm)
   - Oil strainer and tighten the bolts to 66 inch lbs. (8 Nm)
   - Baffle plate and tighten the bolts and nuts to 66 inch lbs. (8 Nm)
   - Remaining components
   - Engine/transaxle

4.6L (1UR-FSE) Engine

See Figures 129 through 134.

1. Discharge fuel system pressure.
2. Disconnect cable from negative battery terminal.
3. Remove intake manifold.
4. Remove battery.
5. Remove battery tray.
6. Remove power steering ECU assembly.
7. Remove wire harness clamp bracket.
8. Drain engine oil.
9. Remove front bumper assembly.
10. Remove radiator support opening cover.
11. Remove millimeter wave radar sensor assembly.
12. Remove hood lock assembly.
13. Remove engine room ECU outlet duct.
14. Remove upper radiator support sub-assembly.
15. Disconnect oil cooler hose.
16. Disconnect radiator hoses and radiator assembly.
17. Separate compressor and pulley.
18. Remove exhaust manifold sub-assembly lh.
19. Disconnect wire harness.
20. Remove the bolt, and disconnect the clamp and ground cables.
21. Remove engine cover sub-assemblies.
22. Remove fuel pressure pulsation damper assembly.
23. Remove fuel pipe sub-assemblies.
24. Remove fuel pump assembly.
25. Remove inlet water housing.
26. Remove water pump pulley.
27. Remove idler pulley sub-assemblies.
28. Remove V-ribbed belt tensioner assembly.
29. Remove front water by-pass joint.
30. Remove camshaft timing control motor assembly.
31. Remove oil filter assembly.
32. Remove crankshaft pulley.
33. Remove ignition coil assembly.
34. Remove V-bank covers.
35. Remove cylinder head cover sub-assemblies.
36. Remove timing chain cover sub-assembly.
37. Remove the oil pump gasket from the cylinder block.
38. Remove the O-ring from the cylinder block.
39. Remove the inlet water pipe.
40. Remove the 2 O-rings from the inlet water pipe.

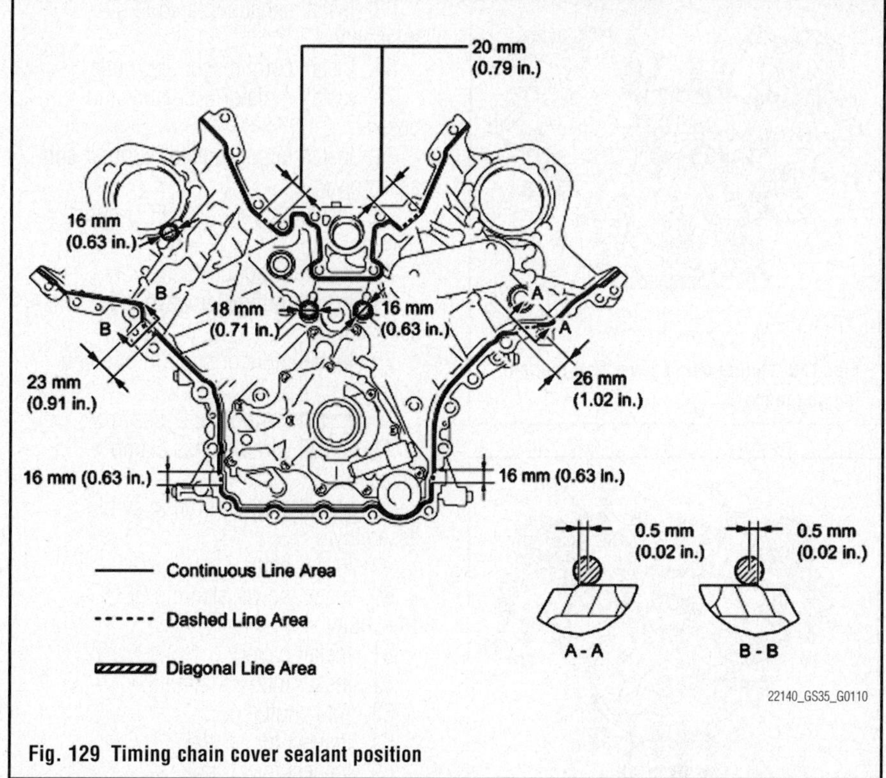

Fig. 129 Timing chain cover sealant position

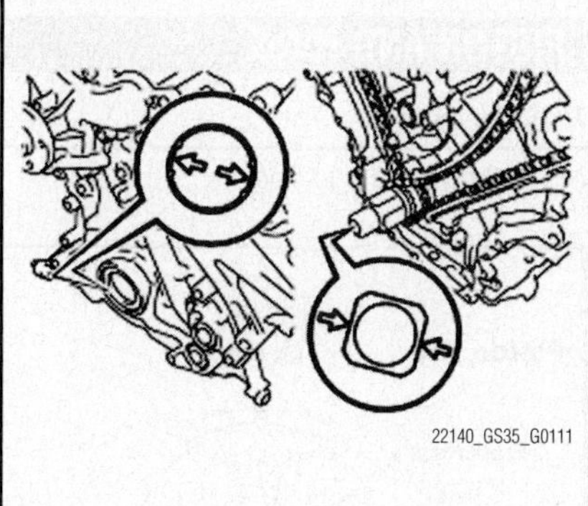

Fig. 130 Timing chain cover oil pump alignment

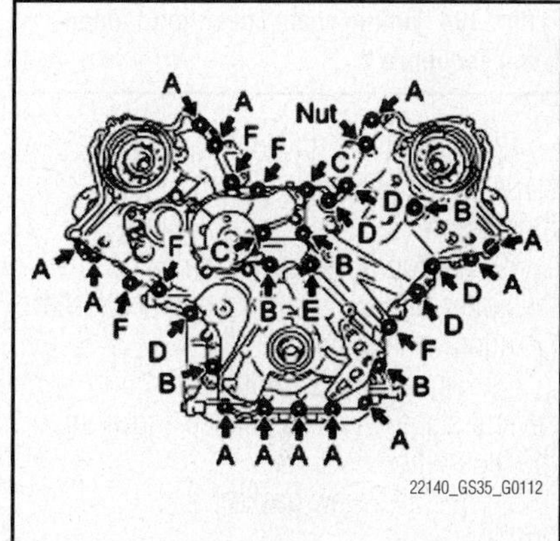

Fig. 131 Timing chain cover bolt position

To install:

41. Apply soapy water to 2 new O-rings and install them to the inlet water pipe.
42. Install the inlet pipe to the No. 1 heat exchanger cover.
43. Install a new oil pump gasket.
44. Install a new O-ring.
45. Apply seal packing in a continuous line to the timing chain cover as shown in the illustration.
46. Align the oil pump's drive rotor spline and the crankshaft as shown in the illustration. Install the spline and chain cover to the crankshaft.
47. Temporarily tighten the timing chain cover with the 30 bolts and nut.

Item	Length	Thread diameter
Bolt A	25 mm (0.984 in.)	8 mm (0.315 in.)
Bolt B	55 mm (2.165 in.)	8 mm (0.315 in.)
Bolt C	70 mm (2.756 in.)	8 mm (0.315 in.)
Bolt D	35 mm (1.378 in.)	10 mm (0.394 in.)
Bolt E	55 mm (2.165 in.)	10 mm (0.394 in.)
Bolt F	80 mm (3.150 in.)	10 mm (0.394 in.)

22140_GS35_G0113

Fig. 132 Timing chain cover bolt legend

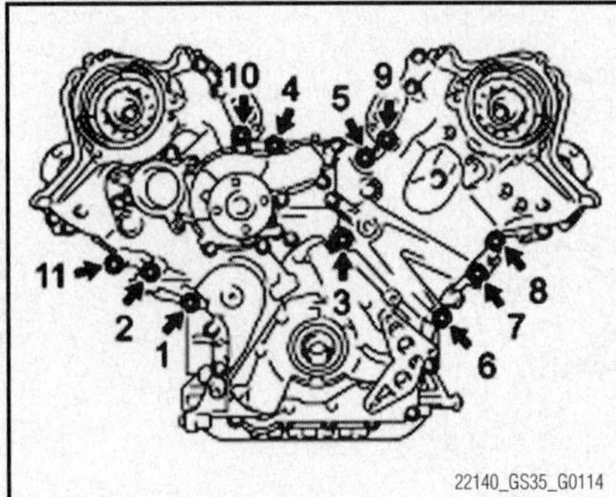

Fig. 133 Timing chain cover bolt tightening sequence 1

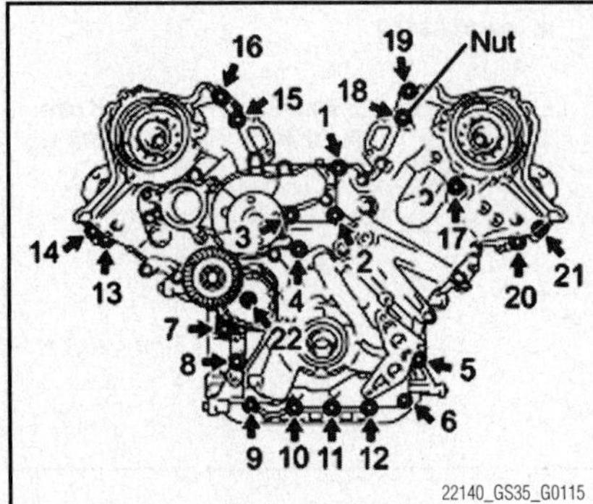

Fig. 134 Timing chain cover bolt tightening sequence 2

48. Tighten the 11 bolts in several steps, in the sequence shown in the illustration.
49. Temporarily tighten the belt tensioner with the standard bolt and 6 mm hexagon wrench bolt.
50. Tighten the 21 bolts and nut in several steps, in the sequence shown in the illustration.
51. Install 2 new gaskets and 2 plugs.
52. Install cylinder head cover sub-assemblies.
53. Install engine covers.
54. Install ignition coil assembly.
55. Install crankshaft pulley.
56. Install oil filter assembly.
57. Install camshaft timing control motor assembly.
58. Install front water by-pass joint.
59. Install idler pulleys.
60. Install water pump pulley.
61. Install inlet water housing.
62. Install fuel pump assemblies.
63. Install fuel pipes.
64. Install fuel pressure pulsation damper assembly.
65. Install engine covers.
66. Connect wire harness.
67. Connect ground wire.
68. Install exhaust manifold sub-assembly lh.
69. Install compressor and pulley.
70. Install radiator assembly and hoses.
71. Install upper radiator support sub-assembly.
72. Install engine room ECU outlet duct.
73. Install hood lock assembly.
74. Install millimeter wave radar sensor assembly.
75. Install radiator support opening cover.
76. Install front bumper assembly.
77. Install wire harness clamp bracket.
78. Install engine room relay block assembly.
79. Install battery tray.
80. Install power steering ECU assembly.
81. Install battery.
82. Install intake manifold.
83. Add engine oil.
84. Inspect for oil leak.
85. Inspect for oil leak.
86. Perform initialization.

PISTON AND RING

POSITIONING

See Figures 135 and 136.

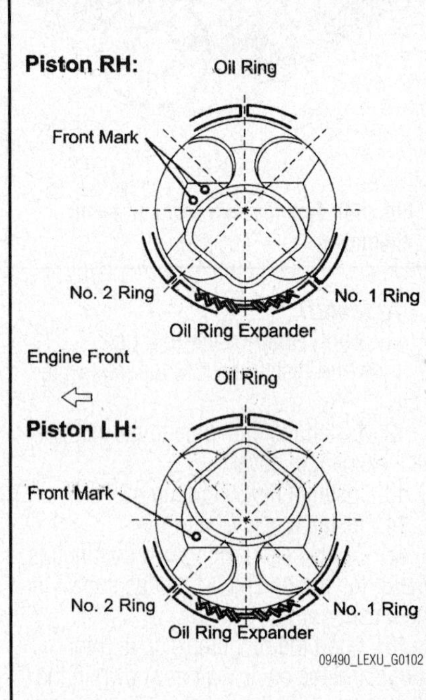

Fig. 135 Piston ring positioning—3.5L (2GR-FSE) Engines

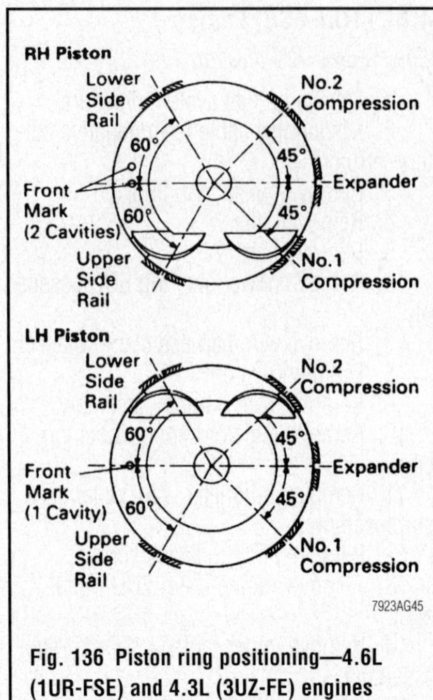

Fig. 136 Piston ring positioning—4.6L (1UR-FSE) and 4.3L (3UZ-FE) engines

REAR MAIN SEAL

REMOVAL & INSTALLATION

1. If the rear oil seal retainer is removed from the cylinder block, perform the following:
   a. Using a screwdriver and hammer, tap out the oil seal.
   b. Using SST 09223-15030, 09950-70010 (09951-07100) and a hammer, tap in a new oil seal until its surface is flush with the rear oil seal retainer edge.
   c. Apply MP grease to the oil seal lip.
2. If the rear seal retainer is installed on the cylinder block, perform the following:
   a. Using a knife, cut off the oil seal lip.
   b. Using a screwdriver, pry out the oil seal.

**** CAUTION**

Be careful not to damage the crankshaft. Tape the screwdriver tip.

   c. Apply MP grease to a new oil seal lip.
   d. Using SST 09223-15030, 09950-70010 (09951-07100) and a hammer, tap in the oil seal until its surface is flush with the rear oil seal retainer edge.

TIMING BELT FRONT COVER

REMOVAL & INSTALLATION

4.3L (3UZ-FE) Engine

See Figures 137 through 141.

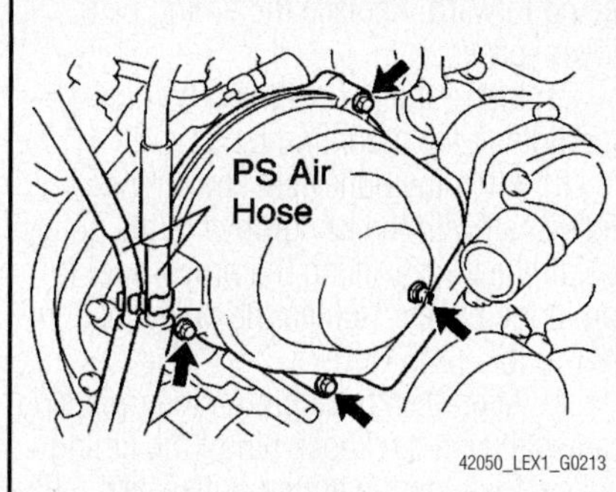

Fig. 137 Removal of timing belt cover No. 2—4.3L (3UZ-FE) engine

Fig. 138 Disconnect the 2 water by-pass hoses as shown—4.3L (3UZ-FE) engine

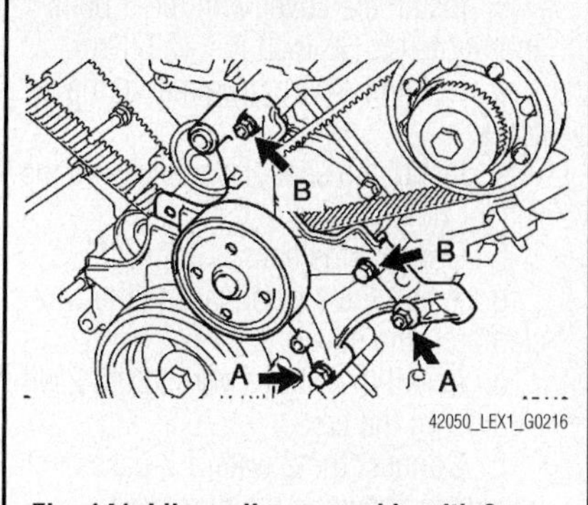

Fig. 141 Idler pulley assembly with 2 bolts and 2 nuts—4.3L (3UZ-FE) engine

1. Before servicing the vehicle, refer to the precautions in the beginning of this section.

2. Disconnect the negative battery cable. Wait at least 90 seconds before performing any other work.

3. Remove air cleaner inlet No. 1.

4. Drain engine coolant.

5. Remove the V-bank cover.

6. Remove the intake air connector pipe.

7. Remove the engine under cover No. 1.

8. Disconnect radiator hose No. 1.

9. Disconnect radiator hose No. 2.

10. Disconnect oil cooler inlet tube No. 1.

11. Disconnect oil cooler outlet tube No. 1.

12. Remove the air cleaner assembly.

13. Remove the radiator assembly.

14. Remove the accessory drive belt

15. Remove the 2 bolts, nut and vane pump assembly. Pump should be removed with the hoses connected and then hang with a rope or wire on the body's side.

16. Remove the alternator.

17. Remove the A/C compressor assembly. The cooler compressor with the magnetic clutch should be removed with the low-pressure and high-pressure hoses connected and then hang with a rope or wire on the body's side.

18. Disconnect the 2 PS air hoses from the clamp on the timing belt cover No. 2.

19. Remove the cap nut, 3 bolts, timing belt cover No. 2 and gasket.

20. Remove the left timing belt cover sub-assembly No. 3 as follows:

a. Remove the cap nut, and disconnect the No.3 water bypass pipe from the cover.

b. Disconnect the 2 water by-pass hoses as shown in the illustration.

c. Disconnect the engine wire from the 2 wire clamps.

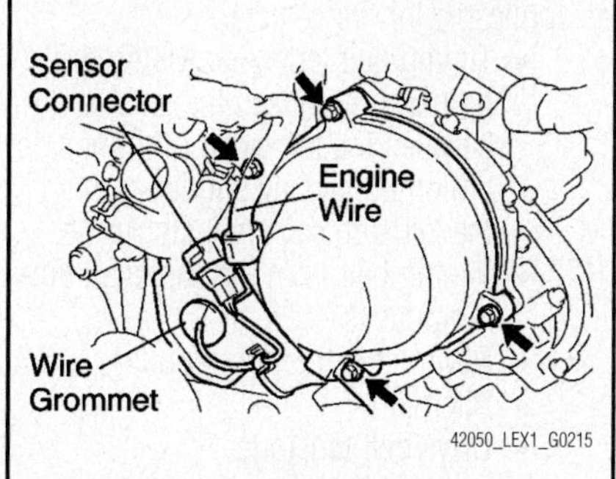

Fig. 139 Left timing belt cover sub-assembly No. 3—4.3L (3UZ-FE) engine

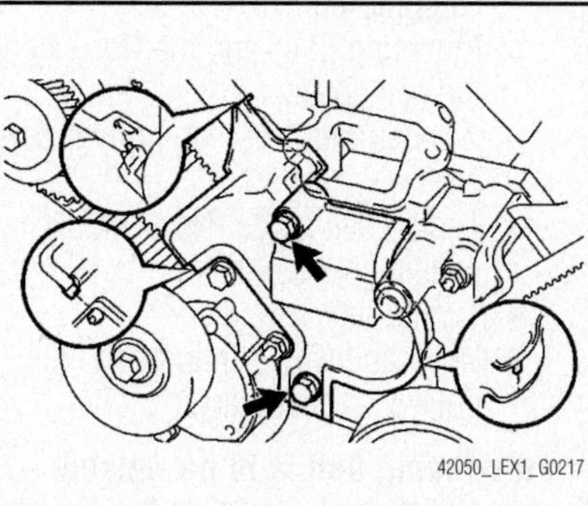

Fig. 140 Timing belt cover sub-assembly No. 2—4.3L (3UZ-FE) engine

d. Disconnect the camshaft position sensor connector.

e. Disconnect the camshaft position sensor wire from the wire clamp on the cover.

f. Remove the wire grommet from the cover.

g. Remove the 4 bolts.

h. Disconnect the cover from the timing plate and camshaft bearing cap.

i. Disconnect the wire clamp for the sensor from the cover.

j. Remove the connector holder from the sensor connector.

k. Remove the left timing belt cover No. 3 and gasket.

21. Remove the 2 bolts and timing belt cover sub-assembly No. 2.

22. Remove the bolt, 2 nuts and drive belt tensioner.

23. Remove the 2 bolts, 2 nuts and idler pulley assembly.

24. Remove the crankshaft pulley (damper).

25. Remove the 4 bolts and timing belt No. 1 cover.

To install:

26. Install the timing belt No. 1 cover with the 4 bolts and tighten to 66 inch lbs. (7.5 Nm).

27. Install the crankshaft pulley (damper).

28. Install the idler pulley assembly with the 2 bolts and 2 nuts. Tighten the bolts and nuts as follows:

- 14mm head bolt A: 24 ft. lbs. (32 Nm)
- 12mm head bolt B: 12 ft. lbs. (16 Nm)

a. Each bolt length is indicated below. Bolt Length:

- 4.49 inches (114mm) for 14mm head (A)
- 4.17 inches (106mm) for 12mm head (B)

29. Install the drive belt tensioner with the bolt and 2 nuts. Tighten the bolt and nuts to 12 ft. lbs. (16 Nm). Use a bolt that is 4.18 inches (106mm) in length.

30. Fit the timing belt cover sub-assembly No. 2, matching the claws and pin with each part. Install the timing belt cover with the 2 bolts and tighten to 12 ft. lbs. (16 Nm). Use bolts that are 4.17 inches (106mm) in length.

31. Install the left timing belt cover sub-assembly No. 3 as follows:

a. Install the gasket to the cover.

b. Run the camshaft position sensor wire through the cover hole.

c. Install the cover with the 4 bolts and tighten to 66 inch lbs. (7.5 Nm).

d. Install the wire grommet to the cover.

e. Install the sensor connector to the sensor holder.

f. Connect the sensor connector.

g. Install the sensor wire to the wire clamp on the cover.

h. Install the engine wire to the 2 wire clamps on the cover.

i. Connect the 2 water by-pass hoses, as shown in the illustration.

j. Install the No. 3 water by-pass pipe to the cover with the cap nut and tighten to 66 inch lbs. (7.5 Nm).

32. Install gasket to timing belt cover No. 2. Install the cover with the cap nut and 3 bolts and tighten to 66 inch lbs. (7.5 Nm).

33. Install the 2 PS air hoses to the clamp on the timing belt cover No. 2.

34. Install the A/C compressor, stay and wire bracket with the 3 bolts and nut. Tighten the bolts to 36 ft. lbs. (49 Nm). Tighten the nut to 21 ft. lbs. (29 Nm).

35. Remove the alternator.

36. Install the vane pump assembly with the 2 bolts and nuts. Alternately tighten the bolts and nut. Tighten the bolts to 29 ft. lbs. (39.2 Nm). Tighten the nut to 32 ft. lbs. (43 Nm).

37. Install the accessory drive belt

38. Install the radiator assembly.

39. Install the air cleaner assembly and tighten the fasteners to 44 inch lbs. (5 Nm).

40. Connect oil cooler outlet tube No. 1.

41. Connect oil cooler inlet tube No. 1.

42. Connect radiator hose No. 2.

43. Connect radiator hose No. 1.

44. Install air cleaner inlet No. 1 and tighten the fasteners to 44 inch lbs. (5 Nm).

45. Install the intake air connector pipe and tighten the fasteners to 44 inch lbs. (5 Nm).

46. Install the V-bank cover and tighten the fasteners to 44 inch lbs. (5 Nm).

47. Refill the engine cooling system and check for leaks.

48. Install the engine under cover No. 1.

49. Connect the negative battery cable.

TIMING BELT AND SPROCKETS

REMOVAL & INSTALLATION

4.3L (3UZ-FE) Engines

See Figures 142 and 143.

1. Remove all necessary components for access to the right-hand side No. 3 and No. 2, and left-hand side No. 2 timing belt covers, then remove the covers.

2. Turn the crankshaft pulley and align its groove with the timing mark **0** of the No. 1 timing cover. Check that the timing marks of the camshaft timing pulleys and timing belt rear plates are aligned. If not, turn the crankshaft 1 full revolution (360°).

3. Loosen crankshaft pulley bolt, then set the No. 1 cylinder to about 50° ATDC on the compression stroke. With the crankshaft pulley notch aligned with the **0** mark on the timing belt cover, turn the crankshaft pulley about 50°clockwise and put timing mark of the crank pulley in line with the centers of the pulley bolt and the No. 2 timing belt idler pulley bolt.

4. Remove or disconnect the following, as applicable for the vehicle:

- Timing belt tensioner. Using the proper tool, loosen the tension between the left side and right side timing pulleys by slightly turning the left side camshaft clockwise.
- Timing belt from the camshaft timing pulleys
- Power steering pump pulley
- Alternator
- Drive belt tensioner
- Bolt and timing pulleys, using the proper tool
- Bolt and the crankshaft pulley with the proper tool.
- Fan bracket
- Mounting bolts and the No. 1 timing belt cover
- 2 upper and lower timing belt covers
- Timing belt guide (No. 1 crank position sensor plate)
- Timing belt
- No. 1 and No. 2 timing belt idler pulleys, if necessary.

➡If the timing belt is to be reused, draw a directional arrow on the timing belt in the direction of engine rotation (clockwise) and place matchmarks on the timing belt and crankshaft gear to match the drilled mark on the pulley.

To install:

5. Align the installation mark on the timing belt with the drilled mark of the crankshaft timing pulley. Install the timing belt on the crankshaft timing pulley, No. 1 idler pulley and the No. 2 idler pulley.

➡If the old timing belt is being reinstalled, be sure the directional arrow is facing in the original direction and that the belt and crankshaft gear matchmarks are properly aligned.

6. Install the timing belt guide (No. 1 crank angle sensor plate) with the cup side facing forward. Replace the timing belt cover spacer.

7. Install the No. 1 timing belt cover and tighten the mounting bolts.

8. Align the pulley set key on the crankshaft with the key groove of the pulley. Install the pulley, using the proper tool to tap in the pulley. Tighten the pulley bolt to 181 ft. lbs. (245 Nm).

9. Align the knock pin on the right side camshaft with the knock pin of the timing pulley. Slide on the timing pulley with the right side mark facing forward. Tighten the bolt to 80 ft. lbs. (108 Nm).

10. Align the knock pin on the left side camshaft with the knock pin of the timing pulley. Slide on the timing pulley with the left side mark facing forward. Tighten the bolt to 80 ft. lbs. (108 Nm).

11. Turn the crankshaft pulley and align its groove with the **0** timing mark on the No. 1 timing belt cover. Using the proper tool, turn the crankshaft timing pulley and align the timing marks of the camshaft timing pulley and the timing belt rear plate.

12. Install the timing belt to the left side camshaft timing pulley by:

a. Using the proper tool, slightly turn the left side timing pulley clockwise. Align the installation mark of the timing belt with the timing mark of the camshaft timing pulley and hang the timing belt on the left side camshaft pulley.

b. Using the proper tool, align the timing marks of the left side camshaft pulley and the timing belt rear plate.

c. Check that the timing belt has tension between crankshaft timing pulley and the left side camshaft pulley.

13. Install the timing belt to the right side camshaft timing pulley by:

a. Using the proper tool, slightly turn the right side timing pulley clockwise. Align the installation mark of the timing belt with the timing mark of the camshaft timing pulley and hang the timing belt on the right side camshaft pulley.

b. Using the proper tool, align the timing marks of the right side camshaft pulley and the timing belt rear plate.

c. Check that the timing belt has tension between the crankshaft timing pulley and the right side camshaft pulley.

14. The timing belt tensioner must be set prior to installation. The tensioner can be set as follows:

a. Place a plate washer between the tensioner and a block. Using a suitable

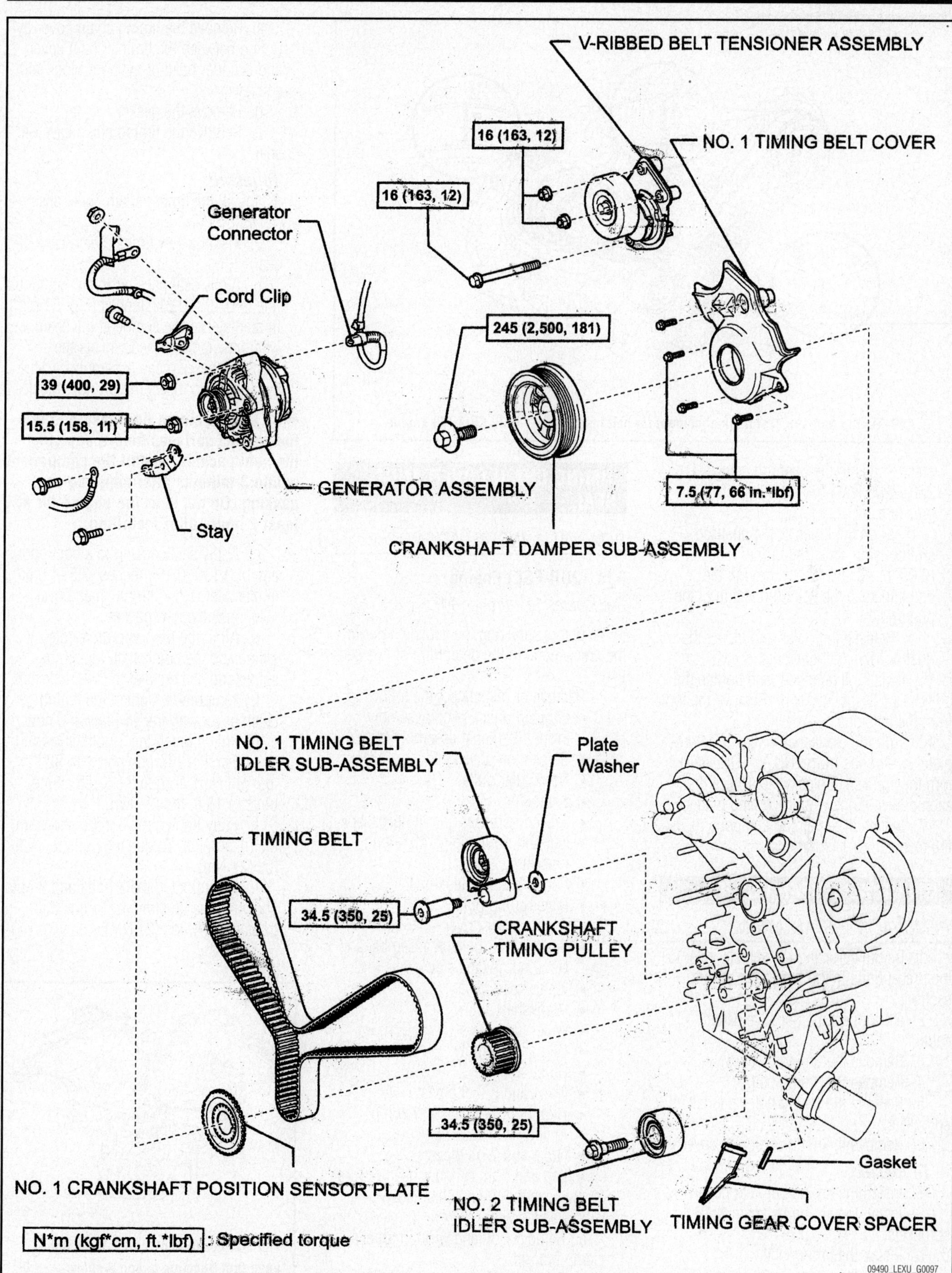

Fig. 142 Exploded view of the timing belt and cover assembly and related components—4.3L (3UZ-FE)

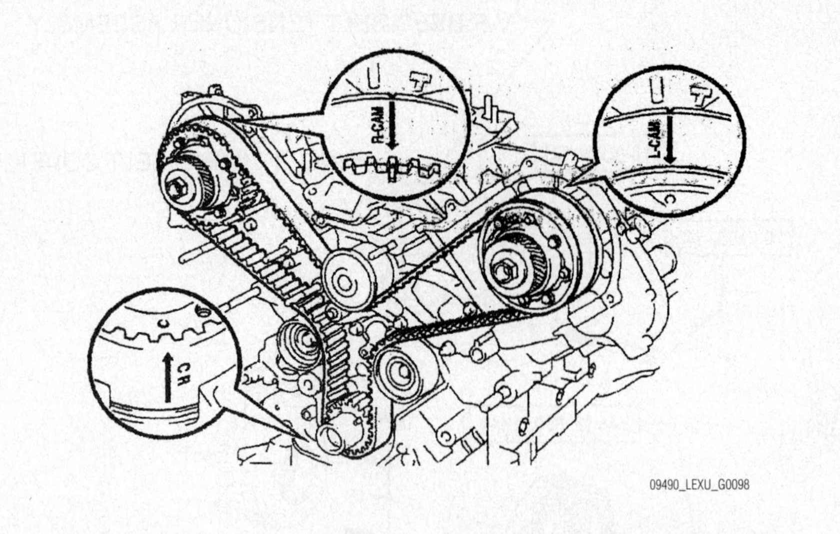

Fig. 143 Timing belt sprocket mark alignment for belt installation—4.3L (3UZ-FE) Engine

press, press in the pushrod using 220–2205 lbs. (100–1000kg) of pressure.

b. Align the holes of the pushrod and housing, pass the proper tool (0.05 in. Allen wrench) through the holes to keep the setting position of the pushrod.

c. Release the press and install the dust boot on the tensioner.

15. Install the tensioner and tighten the bolts to 20 ft. lbs. (26 Nm). Remove the tool from the tensioner.

16. Turn the crankshaft pulley two complete revolutions from TDC-to-TDC. Always turn the crankshaft clockwise. Check that each pulley aligns with the timing marks.

17. Install all remaining components in the reverse order of removal.

TIMING BELT REAR COVER

REMOVAL & INSTALLATION

1. Before servicing the vehicle, refer to the precautions in the beginning of this section.
2. Disconnect the negative battery cable.
3. Remove the timing belt covers.
4. Remove the timing belt.
5. Remove the timing belt camshaft sprockets.
6. Remove the timing belt rear cover.

To install:

7. Install the timing belt rear cover.
8. Install the timing belt camshaft sprockets.
9. Install the timing belt.
10. Install the timing belt covers.
11. Connect the negative battery cable

TIMING CHAIN COVER AND SEAL

REMOVAL & INSTALLATION

3.5L (2GR-FSE) Engine

See Figures 144 through 147.

1. Before servicing the vehicle, refer to the precautions in the beginning of this section.
2. Remove or disconnect the following:
   - Engine/transmission assembly
   - Front differential assembly (AWD)
   - Serpentine drive belt
   - No. 2 idler pulley
   - Alternator
   - A/C compressor unit, if necessary
   - Left and right engine mounting brackets
   - Serpentine belt tensioner
   - Water pump pulley
   - Fuel injector driver
   - Intake air surge tank assembly and No. 2 surge tank stay
   - Water hose joint
   - Crankshaft pulley
   - Water inlet
   - Oil pan assembly
   - Oil strainer
   - No. 1 and 2 fuel pipes
   - High pressure side fuel pump
   - Ignition coil assembly
   - No. 1 and 2 oil pipes
   - Left and right cylinder head covers
3. Remove the timing chain cover assembly as follows:

a. Remove bolt and wiring harness clamp bracket.

b. Remove 25 mounting bolts and 2 mounting nuts.

c. Remove the timing chain cover by prying between the timing chain cover and cylinder head or cylinder block with a screwdriver.

d. Remove the gasket.

e. Remove the timing chain case oil seal.

To install:

4. Install the timing chain cover assembly as follows:

a. Install a new timing chain case oil seal.

b. Apply seal packing in a continuous line of 0.197–0.217 inches (5.0–5.5mm) in diameter to the engine at the seam where the cylinder head meets the camshaft bearing cap assembly and the cylinder head meets the cylinder block

➡Be sure to clean, degrease and dry the contact surfaces before applying the seal packing. Install the component within 3 minutes after applying seal packing. Do not start the engine for at least 2 hours after installing.

c. Apply seal packing in a continuous line of 0.138–0.158 inches (3.5–4.0mm) in diameter to the timing chain cover

d. Install a new gasket

e. Align the oil pump drive rotor spline and the crankshaft. Install the spline and chain cover to the crankshaft.

f. Temporarily tighten the timing chain cover with the 25 bolts and nuts.

g. Fully tighten the 3 bolts shown in the illustration. Torque bolt A to 32 ft. lbs. (43 Nm). Torque the 2 remaining bolts to 15 ft. lbs (21 Nm).

h. Fully tighten the 3 bolts shown in the illustration. Torque the bolts to 15 ft. lbs (21 Nm).

i. Fully tighten the 7 bolts and 2 nuts shown in the illustration. Torque the bolts to 15 ft. lbs (21 Nm).

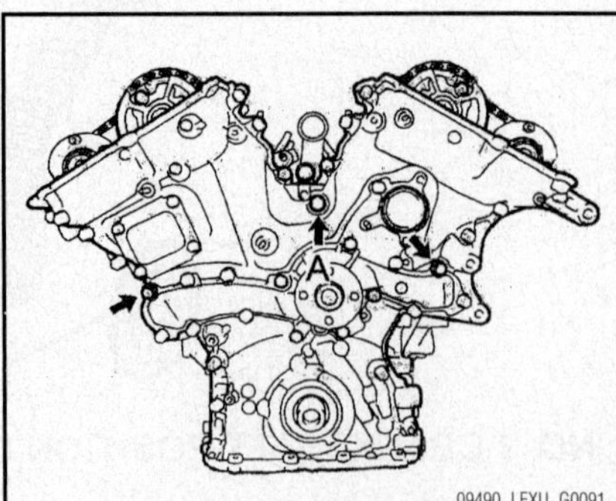

Fig. 144 Location of 3 bolts to be tightened first (location of bolt A shown)—3.5L (2GR-FSE) engines

➡**Be sure to tighten the bolts and nuts in order of upper to lower.**

j. Fully tighten the 12 bolts shown in the illustration. Torque the bolts to 15 ft. lbs (21 Nm).

➡**Be sure to tighten the bolts in order of lower to upper.**

k. Install the bolt and wiring harness bracket. Torque the bolts to 7 ft. lbs (10 Nm).

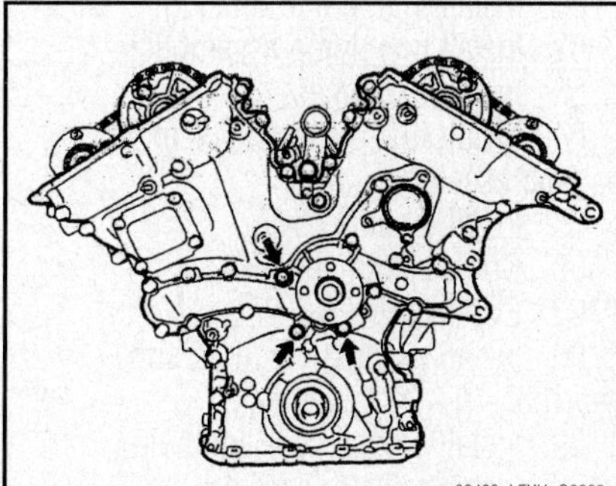

Fig. 145 Location of 3 bolts to be tightened second—3.5L (2GR-FSE) engines

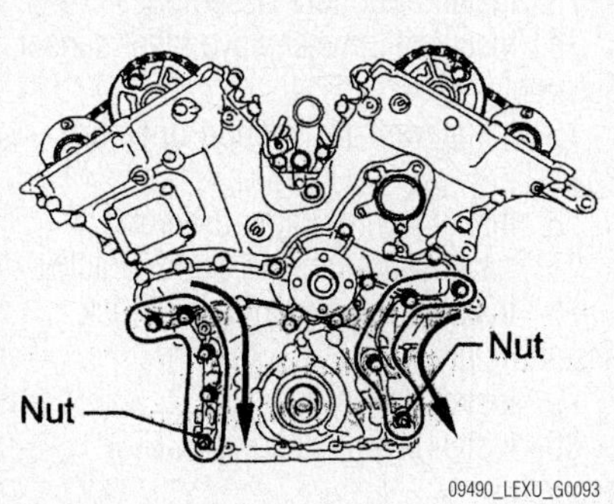

Fig. 146 Location of 7 bolts and 2 nuts to be tightened third—3.5L (2GR-FSE) engines

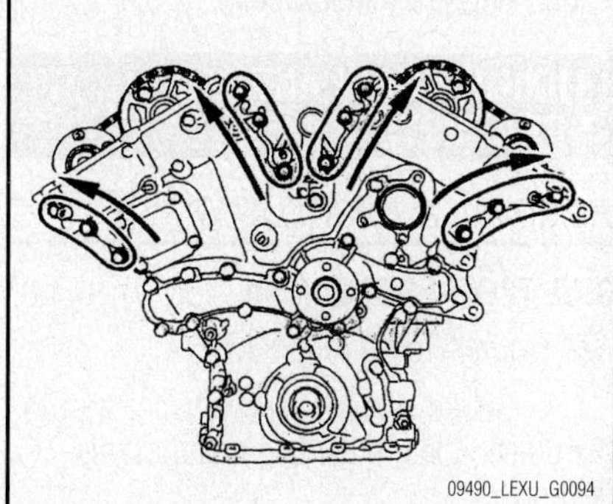

Fig. 147 Location of 12 bolts to be tightened fourth—3.5L (2GR-FSE) engine

5. Install or connect the following:
   - Left and right cylinder head covers
   - No. 1 and 2 oil pipes
   - Ignition coil assembly
   - High pressure side fuel pump
   - No. 1 and 2 fuel pipes
   - Oil strainer
   - Oil pan assembly
   - Water inlet
   - Crankshaft pulley. Torque the bolt to 192 ft. lbs (260 Nm).
   - Water hose joint
   - Intake air surge tank assembly and No. 2 surge tank stay
   - Fuel injector driver
   - Water pump pulley. Torque the bolt to 15 ft. lbs (21 Nm).
   - Serpentine belt tensioner
   - Left and right engine mounting brackets
   - A/C compressor unit, if necessary
   - Alternator
   - No. 2 idler pulley
   - Serpentine drive belt
   - Front differential assembly (AWD)
   - Engine/transmission assembly

4.6L (1UR-FSE) Engine

See Figures 148 through 153.

1. Discharge fuel system pressure.
2. Disconnect cable from negative battery terminal.
3. Remove intake manifold.
4. Remove battery.
5. Remove battery tray.
6. Remove power steering ECU assembly.
7. Remove wire harness clamp bracket.
8. Drain engine oil.
9. Remove front bumper assembly.
10. Remove radiator support opening cover.
11. Remove millimeter wave radar sensor assembly.
12. Remove hood lock assembly.
13. Remove engine room ECU outlet duct.
14. Remove upper radiator support subassembly.
15. Disconnect oil cooler hose.
16. Disconnect radiator hoses and radiator assembly.
17. Separate compressor and pulley.
18. Remove exhaust manifold subassembly lh.
19. Disconnect wire harness.
20. Remove the bolt, and disconnect the clamp and ground cables.
21. Remove engine cover sub-assemblies.
22. Remove fuel pressure pulsation damper assembly.
23. Remove fuel pipe sub-assemblies.
24. Remove fuel pump assembly.
25. Remove inlet water housing.
26. Remove water pump pulley.
27. Remove idler pulley sub-assemblies.

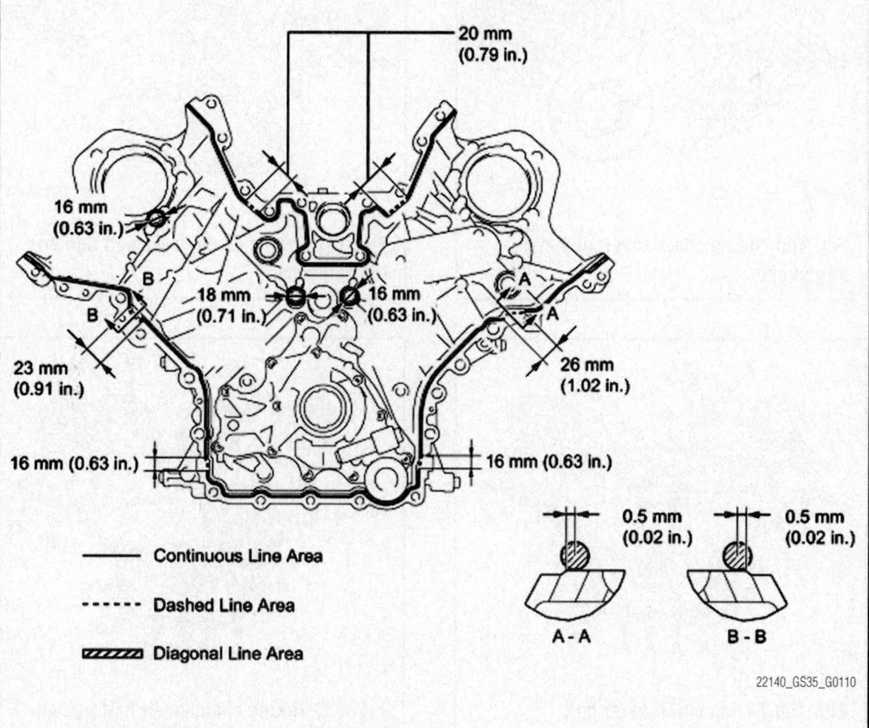

Fig. 148 Timing chain cover sealant position

28. Remove V-ribbed belt tensioner assembly.
29. Remove front water by-pass joint.
30. Remove camshaft timing control motor assembly.
31. Remove oil filter assembly.
32. Remove crankshaft pulley.
33. Remove ignition coil assembly.
34. Remove V-bank covers.
35. Remove cylinder head cover sub-assemblies.
36. Remove timing chain cover sub-assembly.
37. Remove the oil pump gasket from the cylinder block.
38. Remove the O-ring from the cylinder block.
39. Remove the inlet water pipe.
40. Remove the 2 O-rings from the inlet water pipe.

To install:

41. Apply soapy water to 2 new O-rings and install them to the inlet water pipe.
42. Install the inlet pipe to the No. 1 heat exchanger cover.
43. Install a new oil pump gasket.
44. Install a new O-ring.
45. Apply seal packing in a continuous line to the timing chain cover as shown in the illustration.

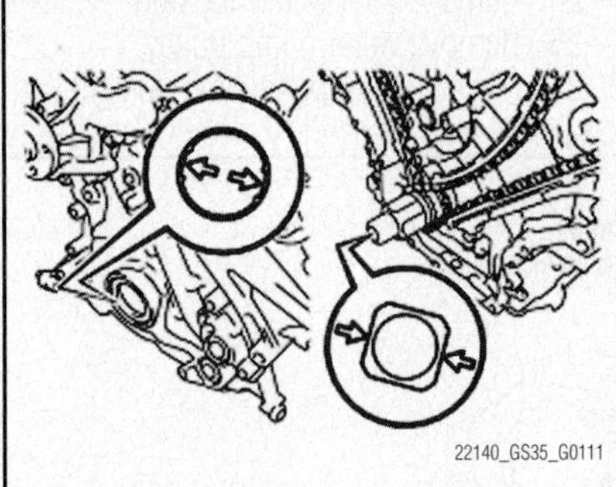

22140_GS35_G0111

Fig. 149 Timing chain cover oil pump alignment

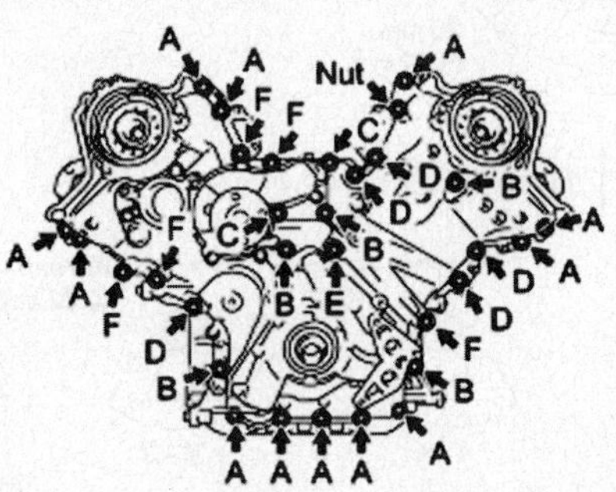

22140_GS35_G0112

Fig. 150 Timing chain cover bolt position

46. Align the oil pump's drive rotor spline and the crankshaft as shown in the illustration. Install the spline and chain cover to the crankshaft
47. Temporarily tighten the timing chain cover with the 30 bolts and nut
48. Tighten the 11 bolts in several steps, in the sequence shown in the illustration
49. Temporarily tighten the belt tensioner with the standard bolt and 6 mm hexagon wrench bolt.
50. Tighten the 21 bolts and nut in several steps, in the sequence shown in the illustration.

Item	Length	Thread diameter
Bolt A	25 mm (0.984 in.)	8 mm (0.315 in.)
Bolt B	55 mm (2.165 in.)	8 mm (0.315 in.)
Bolt C	70 mm (2.756 in.)	8 mm (0.315 in.)
Bolt D	35 mm (1.378 in.)	10 mm (0.394 in.)
Bolt E	55 mm (2.165 in.)	10 mm (0.394 in.)
Bolt F	80 mm (3.150 in.)	10 mm (0.394 in.)

22140_GS35_G0113

Fig. 151 Timing chain cover bolt legend

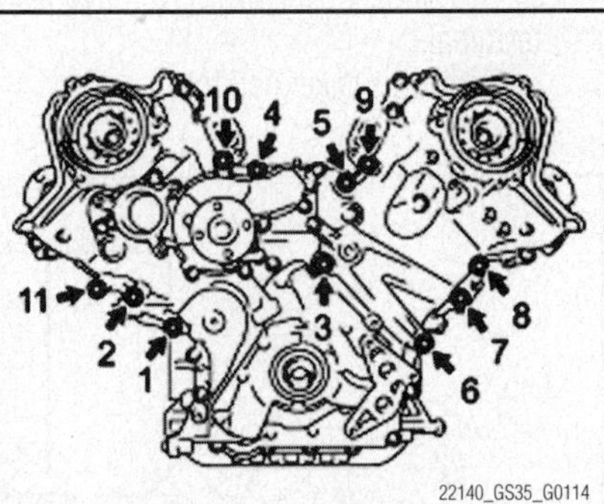

22140_GS35_G0114

Fig. 152 Timing chain cover bolt tightening sequence 1

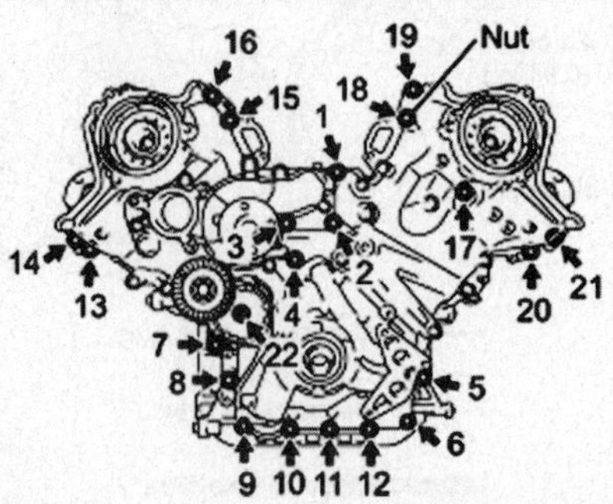

22140_GS35_G0115

Fig. 153 Timing chain cover bolt tightening sequence 2

51. Install 2 new gaskets and 2 plugs.
52. Install cylinder head cover sub-assemblies.
53. Install engine covers.
54. Install ignition coil assembly.
55. Install crankshaft pulley.
56. Install oil filter assembly.
57. Install camshaft timing control motor assembly.
58. Install front water by-pass joint.
59. Install idler pulleys.
60. Install water pump pulley.
61. Install inlet water housing.
62. Install fuel pump assemblies.
63. Install fuel pipes.
64. Install fuel pressure pulsation damper assembly.
65. Install engine covers.
66. Connect wire harness.
67. Connect ground wire.
68. Install exhaust manifold sub-assembly lh.
69. Install compressor and pulley.
70. Install radiator assembly and hoses.
71. Install upper radiator support sub-assembly.
72. Install engine room ECU outlet duct.
73. Install hood lock assembly.
74. Install millimeter wave radar sensor assembly.
75. Install radiator support opening cover.
76. Install front bumper assembly.
77. Install wire harness clamp bracket.
78. Install engine room relay block assembly LH.
79. Install battery tray.
80. Install power steering ECU assembly.
81. Install battery.
82. Install intake manifold.
83. Add engine oil.
84. Inspect for oil leak.
85. Inspect for oil leak.
86. Perform initialization.

TIMING CHAIN AND SPROCKETS

REMOVAL & INSTALLATION

3.5L (2GR-FSE) Engine

See Figures 154 through 170.

1. Before servicing the vehicle, refer to the precautions in the beginning of this section.
2. Drain the cooling system.
3. Drain the engine oil.
4. Relieve the fuel system pressure.

5. Remove or disconnect the following:
 - Negative battery cable. Wait at least 90 seconds before performing any other work
 - Oil filler cap and gasket
 - Radiator cap
 - No. 1 and 2 engine hangers
 - Spark plugs
 - Ventilation valve
 - 4 camshaft position sensors
 - 4 camshaft timing oil control valves
 - Crankshaft position sensor
 - Left and right side oil check valve bolt, oil pipe union and oil pipe
 - Left and right side oil control valve filter and gaskets
 - Cylinder block water drain cocks
 - Oil filter element
 - Water inlet and thermostat assembly
 - Rear water by-pass joint
 - Cylinder head cover and gaskets
 - Timing chain cover, timing chain and timing chain sprockets
 - No. 2 chain tensioner assembly

6. Remove the camshaft timing gears and No. 2 chain (right) as follows:

a. While raising the No. 2 chain tensioner, insert a pin of 1.0mm (0.039 in.) into the hole to fix the No. 2 chain tensioner.

b. Hold the hexagonal portion of the camshaft with a wrench, and remove the 2 bolts and 2 camshaft timing gear assemblies.

c. Remove the No. 2 chain.
 - No. 2 chain tensioner

7. Remove the camshaft bearing cap (right) as follows:

a. Check that the camshafts are positioned as shown in the illustration.

b. Uniformly loosen and remove the 9 bearing cap bolts in the sequence shown in the illustration.

c. Uniformly loosen and remove the 14 bearing cap bolts in the sequence shown in the illustration.

d. Remove the 6 bearing caps.
 - Camshaft
 - No. 2 camshaft
 - Camshaft housing sub-assembly (right) by prying between the cylinder head and camshaft housing sub-assembly (right) with a screwdriver.

WARNING

Be careful not to damage the contact surfaces of the cylinder head and camshaft housing.

8. Remove the camshaft timing gears and No. 2 chain (left) as follows:

a. While pushing down the No. 3 chain tensioner, insert a pin of 1.0mm (0.039 in.) into the hole to fix the No. 3 chain tensioner.

b. Hold the hexagonal portion of the camshaft with a wrench, and remove the 2 bolts and 2 camshaft timing gear assemblies.

c. Remove the No. 2 chain.
 - No. 3 chain tensioner

9. Remove the camshaft bearing cap (left) as follows:

a. Check that the camshafts are positioned as shown in the illustration.

b. Uniformly loosen and remove the 8 bearing cap bolts in the sequence shown in the illustration.

c. Uniformly loosen and remove the 13 bearing cap bolts in the sequence shown in the illustration.

d. Remove the 5 bearing caps.
 - No. 3 camshaft
 - No. 4 camshaft
 - Camshaft housing sub-assembly (left) by prying between the cylinder head and camshaft housing sub-assembly (left) with a screwdriver.

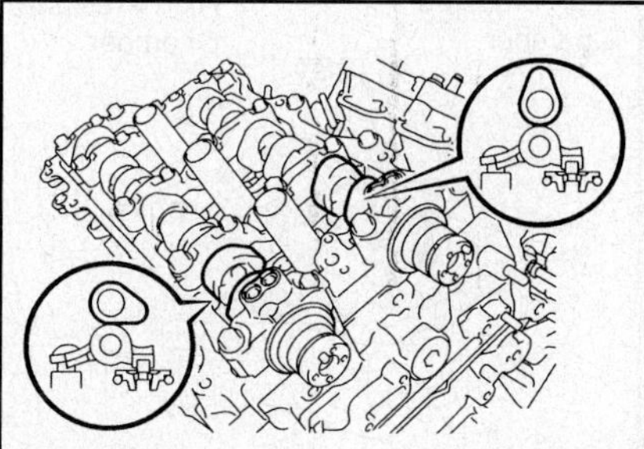

Fig. 154 Check that the camshafts are positioned as shown (right side)—3.5L (2GR-FSE) engines

Fig. 157 Check that the camshafts are positioned as shown (left side)—3.5L (2GR-FSE) Engines

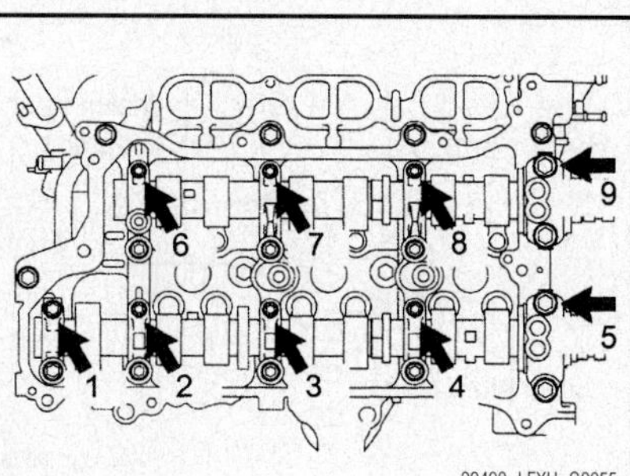

Fig. 155 Right side camshaft bearing inner bolt loosening sequence—3.5L (2GR-FSE) engines

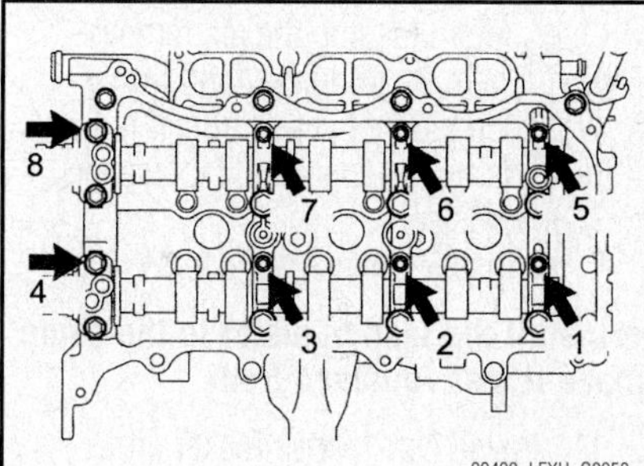

Fig. 158 Left side camshaft bearing inner bolt loosening sequence—3.5L (2GR-FSE) Engines

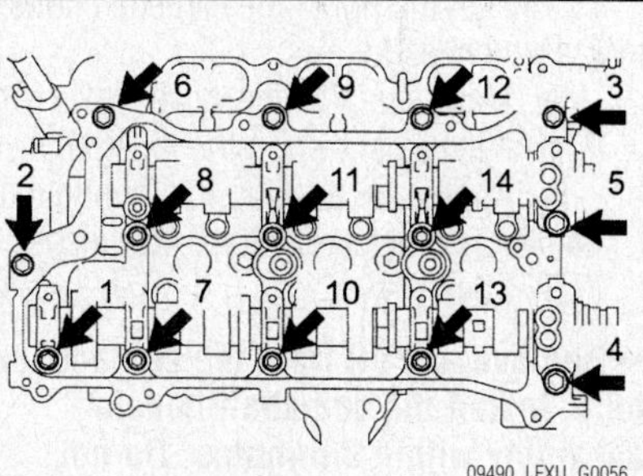

Fig. 156 Right side camshaft bearing outer bolt loosening sequence—3.5L (2GR-FSE) engines

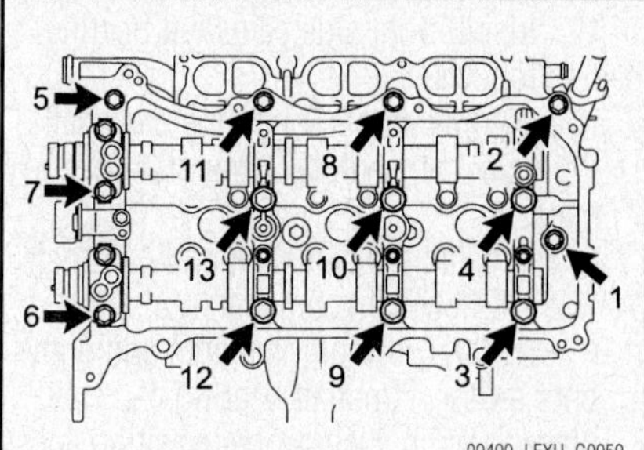

Fig. 159 Left side camshaft bearing outer bolt loosening sequence—3.5L (2GR-FSE) Engines

WARNING

Be careful not to damage the contact surfaces of the cylinder head and camshaft housing.

- Valve rocker arms. Arrange the removed rocker arms in the correct order.
- Valve lash adjusters. Arrange the removed valve lash adjusters in the correct order.

To install:

10. Install valve lash adjusters as follows:

WARNING

Keep the lash adjuster free of dirt and foreign objects. Only use clean engine oil.

a. Place the lash adjuster into a container filled with engine oil.

b. Insert the SST's tip (09276-75010) into the lash adjuster's plunger and use the tip to press down on the check ball inside the plunger as is shown.

c. Squeeze the SST and lash adjuster together to move the plunger up and down 5 to 6 times.

d. Check the movement of the plunger and bleed the air.

e. After bleeding the air, remove SST. Then, try to quickly and firmly press the plunger with a finger. If the result is not as specified, replace the lash adjuster.

f. Install the lash adjusters.

➡Install the lash adjuster to the same place it was removed from.

11. Install No. 1 valve rocker arm assembly as follows:

a. Apply engine oil to the lash adjuster tip and valve stem cap end.

b. Make sure that the valve rocker arms are installed as shown in the illustration.

12. Install right side camshaft bearing cap as follows:

a. Apply engine oil to the camshaft journals, camshaft housing and bearing caps.

b. Install the camshaft and camshaft No. 2 to the right camshaft housing.

c. Make sure of the marks and numbers on the camshaft bearing caps and place them in each proper position and direction.

d. Temporarily tighten the 9 bolts to 7 ft. lbs. (10 Nm) in the order shown in the illustration.

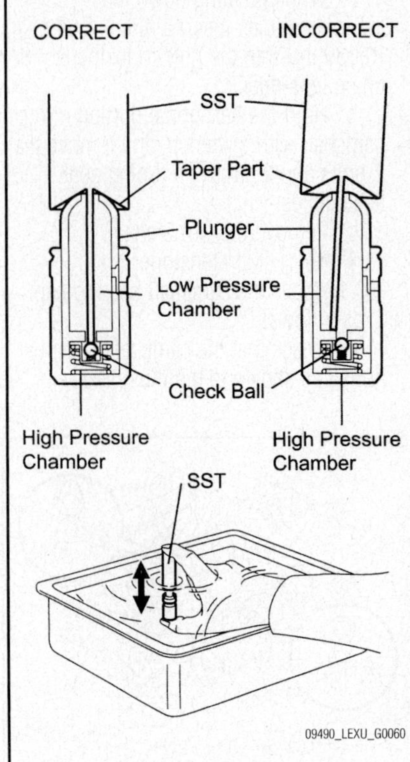

Fig. 160 Bleeding air from the valve lash adjuster assembly—3.5L (2GR-FSE) Engines

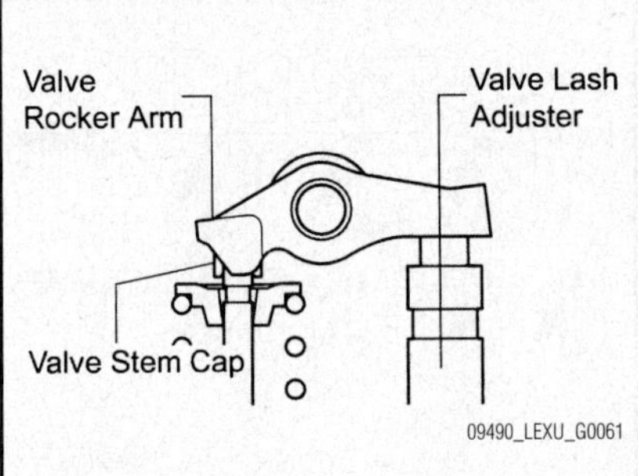

Fig. 161 Correct installation of the valve rocker arm assembly—3.5L (2GR-FSE) Engines

13. Install the right side camshaft housing assembly as follows:

a. Make sure that the valve rocker arm is installed as shown in the illustration.

b. Apply seal packing in a continuous line approximately 0.138–0.158 inches (3.5–4.0mm) wide.

➡Remove any oil from the contact surface. Install the camshaft housing assembly within 3 minutes. Do not start the engine for at least 2 hours after installing.

c. Install the right camshaft housing and tighten the 14 bolts to 21 ft. lbs. (28 Nm) in the order shown in the illustration.

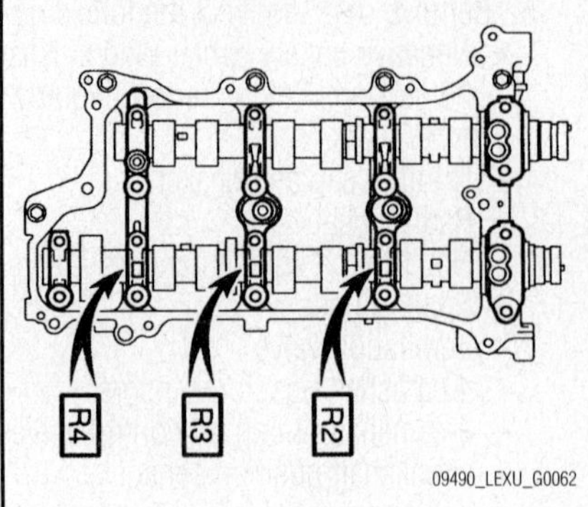

Fig. 162 Make sure of the marks and numbers on the right camshaft bearing caps and place them in each proper position and direction—3.5L (2GR-FSE) Engines

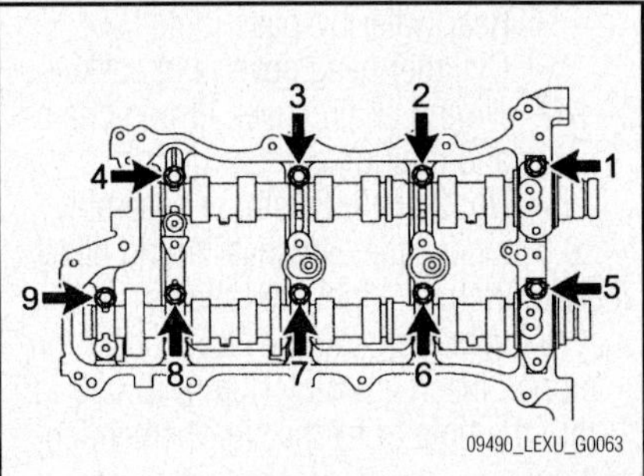

Fig. 163 Temporarily tighten the 9 bolts in the order shown (right side)—3.5L (2GR-FSE) Engines

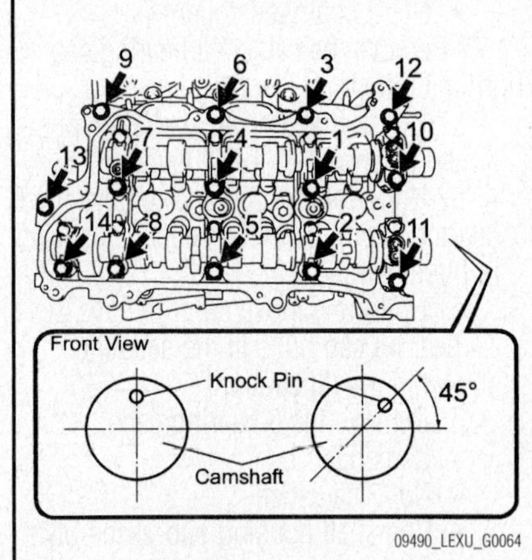

Fig. 164 Right camshaft housing assembly bolt tightening sequence and camshaft positioning—3.5L (2GR-FSE) Engines

WARNING

When installing the camshaft housing, it is necessary to correctly

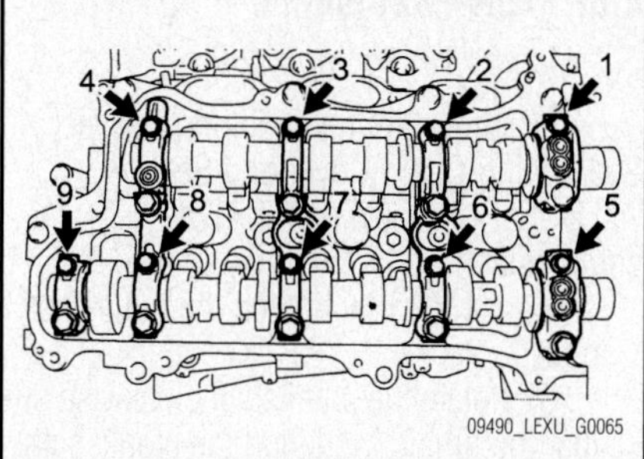

Fig. 165 Right side camshaft bearing inner bolt tightening sequence—3.5L (2GR-FSE) Engines

position the camshafts as shown in the illustration. Failure to correctly position these parts may result in damage due to contact between the pistons and valves. If a camshaft is rotated with a piston at TDC, valve contact will occur. If any of the bolts are loosened during installation, remove the camshaft housing, clean the installation surfaces, and reapply seal packing. If the camshaft housing is removed because any of the bolts are loosened during installation, make sure that the previously applied seal packing does not enter any oil passages.

d. Tighten the 9 bolts to 12 ft. lbs. (16 Nm) in the order shown in the illustration.

14. Install left side camshaft bearing cap as follows:

a. Apply engine oil to the camshaft journals, camshaft housing and bearing caps.

b. Install camshaft No. 3 and camshaft No. 4 to the left camshaft housing.

c. Make sure the marks and numbers on the camshaft bearing caps and place them in each proper position and direction.

d. Temporarily tighten the 8 bolts to 7 ft. lbs. (10 Nm) in the order shown in the illustration.

15. Install the left side camshaft housing assembly as follows:

a. Make sure that the valve rocker arm is installed as shown in the illustration.

b. Apply seal packing in a continuous line approximately 0.197–0.217 inches (5.0–5.5mm) wide.

➡Remove any oil from the contact surface. Install the camshaft housing assembly within 3 minutes. Do not start the engine for at least 2 hours after installing.

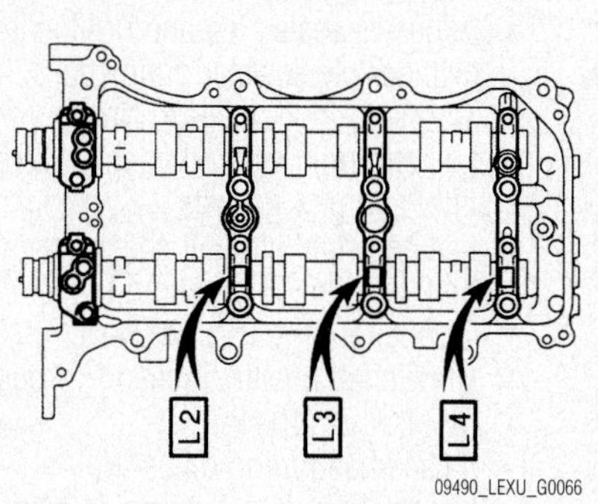

Fig. 166 Make sure of the marks and numbers on the left camshaft bearing caps and place them in each proper position and direction—3.5L (2GR-FSE) Engines

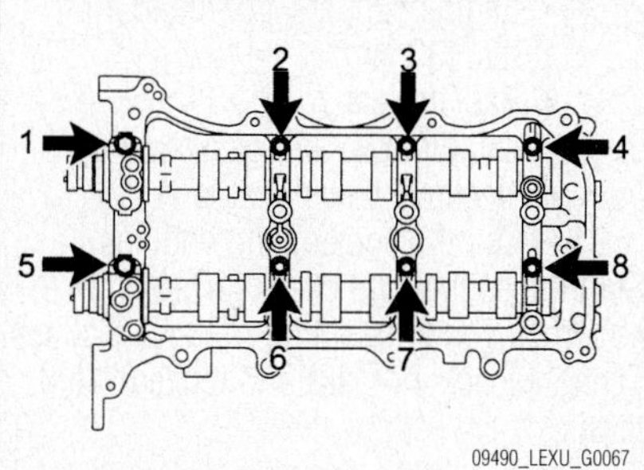

Fig. 167 Temporarily tighten the 8 bolts in the order shown (left side)—3.5L (2GR-FSE) Engines

c. Install the left camshaft housing and tighten the 13 bolts to 21 ft. lbs. (28 Nm) in the order shown in the illustration.

⁂ WARNING

When installing the camshaft housing, it is necessary to correctly position the camshafts as shown in the illustration. Failure to correctly position these parts may result in damage due to contact between the pistons and valves. If a camshaft is rotated with a piston at TDC, valve contact will occur. If any of the bolts are loosened during installation, remove the camshaft housing, clean the installation surfaces, and reapply seal packing. If the camshaft housing is removed because any of the bolts are loosened during installation, make sure that the previously applied seal packing does not enter any oil passages.

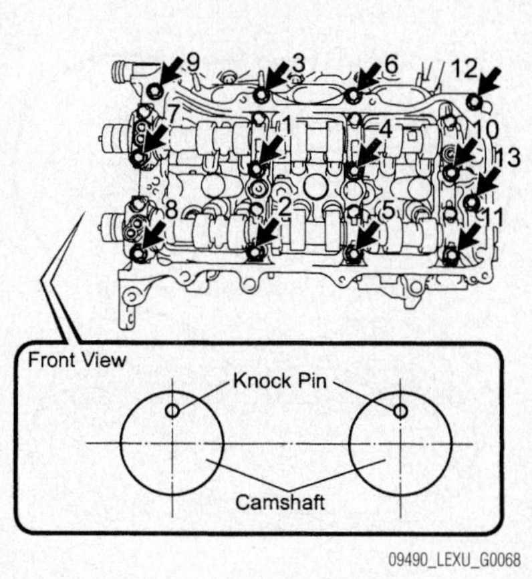

Fig. 168 Left camshaft housing assembly bolt tightening sequence and camshaft positioning—3.5L (2GR-FSE) Engines

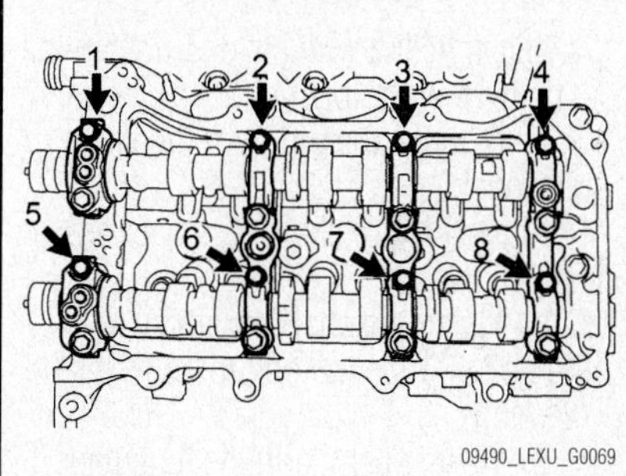

Fig. 169 Left side camshaft bearing inner bolt tightening sequence—3.5L (2GR-FSE) Engines

d. Tighten the 8 bolts to 12 ft. lbs. (16 Nm) in the order shown in the illustration.

16. Install No. 2 chain tensioner assembly as follows:

a. Install the No. 2 chain tensioner with the bolt. Torque the bolt to 15 ft. lbs. (21 Nm).

b. While pushing in the tensioner, insert a pin of 0.039 in. (1.0mm) into the hole to fix it.

17. Install right side camshaft timing gears and No. 2 chain.

a. Align the mark plate (yellow) with the timing marks (1-dot mark) of the camshaft timing gears as shown in the illustration.

b. Apply a light coat of engine oil to the bolt threads and bolt-seating surface.

c. Align the knock pin of the camshaft with the pin hole of the camshaft timing gear. Install the camshaft timing gear and camshaft timing exhaust gear (right) with the No. 2 chain installed.

d. Hold the hexagonal portion of the

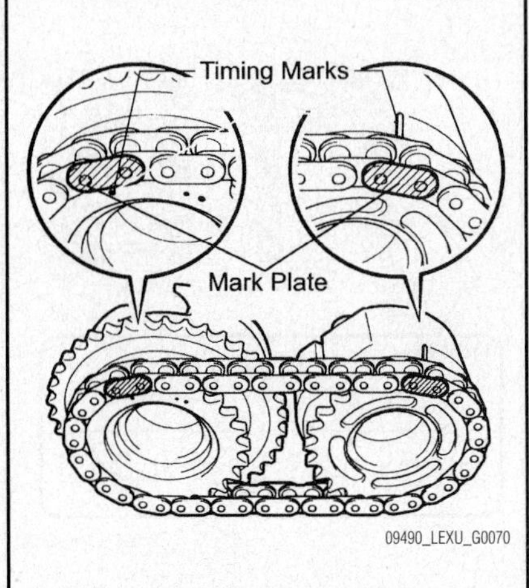

Fig. 170 Align the mark plate with the timing marks of the camshaft timing gears (right side shown, left side similar)—3.5L (2GR-FSE) Engines

camshaft with a wrench, and tighten the 2 bolts to 74 ft. lbs. (100 Nm)

e. Remove the pin from the chain tensioner.

18. Install No. 3 chain tensioner assembly as follows:

a. Install the chain tensioner with the bolt. Torque the bolt to 15 ft. lbs. (21 Nm).

b. While pushing in the tensioner, insert a pin of 0.039 in. (1.0mm) into the hole to hold it.

19. Install left side camshaft timing gears and No. 2 chain.

a. Align the mark plate (yellow) with the timing marks (2-dot mark) of the camshaft timing gears as shown in the illustration.

b. Apply a light coat of engine oil to the bolt threads and bolt-seating surface.

c. Align the knock pin of the camshaft with the pin hole of the camshaft timing gear. Install the camshaft timing gear and camshaft timing exhaust gear (left) with the No. 2 chain installed.

d. Hold the hexagonal portion of the camshaft with a wrench, and tighten the 2 bolts to 74 ft. lbs. (100 Nm)

e. Remove the pin from the chain tensioner.

20. Install or connect the following:

- Timing chain cover, timing chain and timing chain sprockets
- Cylinder head cover and gaskets
- New gaskets, O-ring and rear water by-pass joint. Torque to 7 ft. lbs. (10 Nm).
- New gasket, water inlet and thermostat assembly. Torque to 7 ft. lbs. (10 Nm).
- Oil filter element
- Adhesive sealer, left and right side cylinder block water drain cocks. Torque to 22 ft. lbs. (30 Nm).
- Left and right side oil control valve filter and new gaskets
- Left and right side oil check valve bolt, oil pipe union and oil pipe. Torque to 44 ft. lbs. (60 Nm).
- Crankshaft position sensor. Torque to 7 ft. lbs. (10 Nm).
- 4 camshaft timing oil control valves. Torque to 7 ft. lbs. (10 Nm).
- 4 camshaft position sensors. Torque to 7 ft. lbs. (10 Nm).
- Adhesive and ventilation valve. Torque to 20 ft. lbs. (27 Nm).
- Spark plugs. Torque to 13 ft. lbs. (18 Nm).
- Engine hangers. Torque to 24 ft. lbs. (33 Nm).
- Radiator cap
- Oil filler cap and gasket
- Negative battery cable

21. Refill the coolant and engine oil. Start the engine and check for leaks or abnormal conditions. Perform and road test. Then, recheck for leaks and recheck fluid levels.

4.6L (1UR-FSE) Engine

See Figures 171 through 180.

1. Remove timing chain front cover. See Timing Chain Cover and Seal.

2. Set No. 1 cylinder to TDC/ compression.

a. Temporarily tighten the pulley set bolt.

b. Rotate the crankshaft clockwise so that the timing marks on the crankshaft timing gear and camshaft timing gears are as shown in the illustration.

3. Remove No. 1 chain tensioner assembly (for Bank 1).

a. Move the stopper plate upward to release the lock, and push the plunger deep into the tensioner.

b. Move the stopper plate downward to set the lock, and insert a hexagon wrench into the stopper plate hole.

c. Remove the 2 bolts and chain tensioner.

4. Remove chain tensioner slipper (for Bank 1).

5. Remove the 2 bolts and chain vibration damper.

6. Remove chain sub-assembly (for Bank 1).

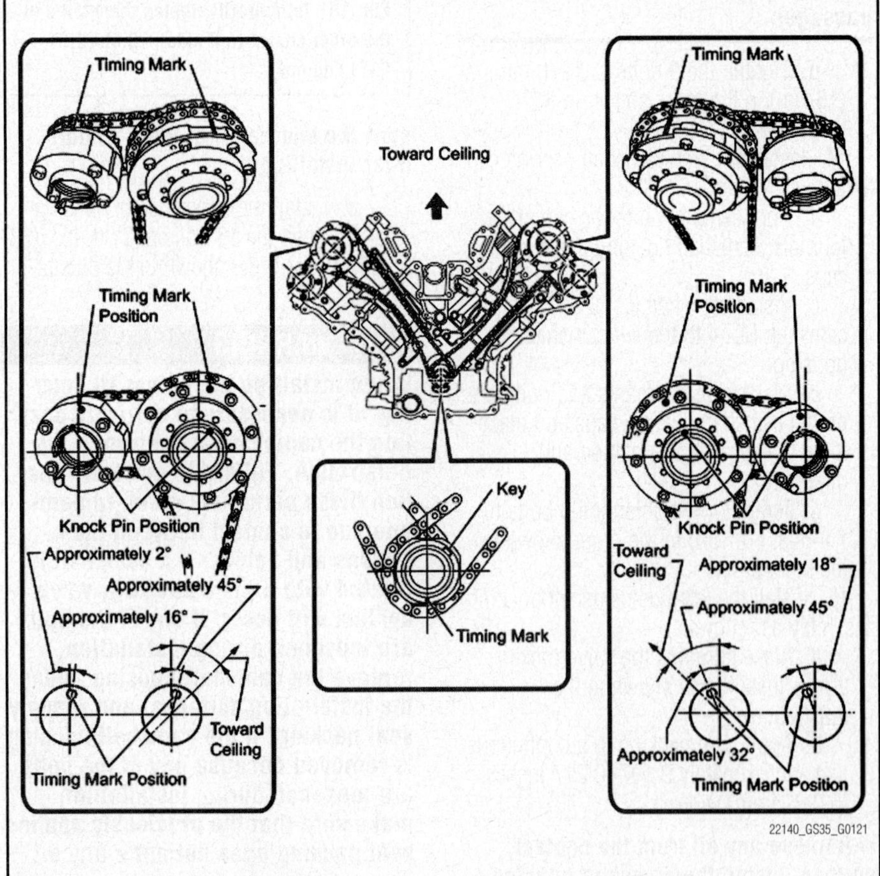

Fig. 171 Timing chain alignment

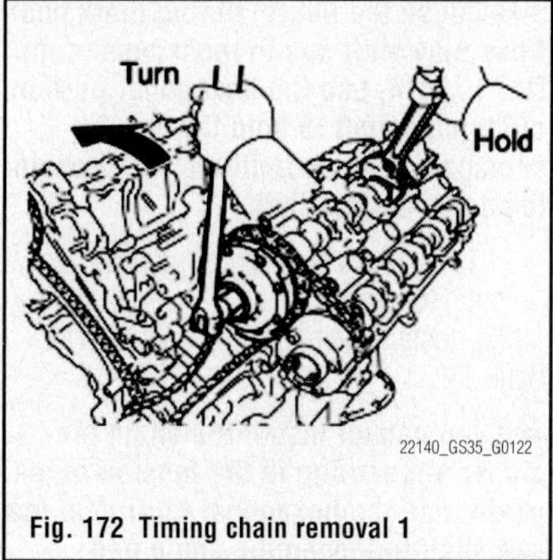

Fig. 172 Timing chain removal 1

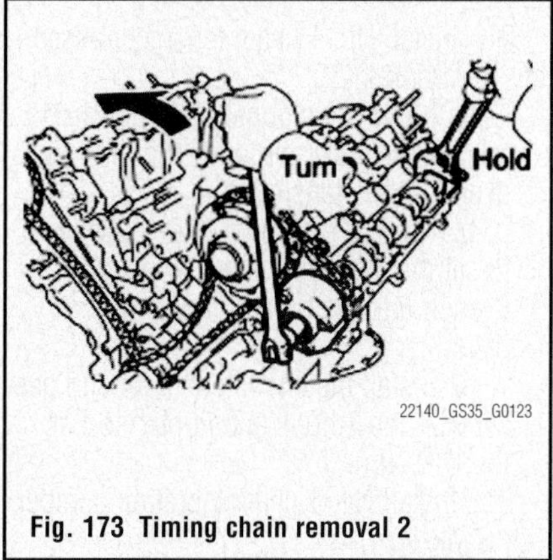

Fig. 173 Timing chain removal 2

a. While pushing down the No. 3 chain tensioner, insert a pin of φ1.0 mm (0.039 in.) into the hole to fix it in place.

b. Hold the hexagonal portion of the camshaft with a wrench and loosen the bolt with a 12mm hexagon wrench.

c. Hold the hexagonal portion of the camshaft with a wrench and loosen the bolt.

d. Remove the 2 bolts. Then with the No. 1 and No. 2 chains still connected to the gears, remove the camshaft timing gear assembly, camshaft timing exhaust gear assembly and crankshaft timing sprocket.

e. Remove the No. 1 and No. 2 chains from the gears.

7. Remove No. 3 chain tensioner assembly.

a. Remove the 2 bolts and chain tensioner.

8. Remove No. 1 chain tensioner assembly (for Bank 2).

a. Move the stopper plate upward to release the lock, and push the plunger deep into the tensioner.

b. Move the stopper plate downward to set the lock, and insert a hexagon wrench into the stopper plate hole.

c. Remove the 2 bolts and chain tensioner.

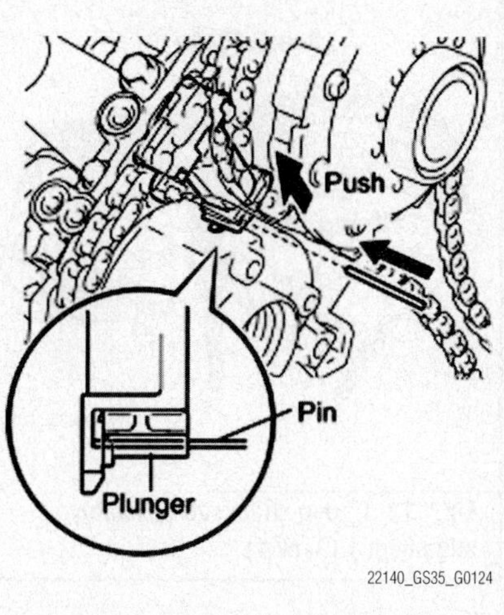

Fig. 174 Timing chain tensioner removal

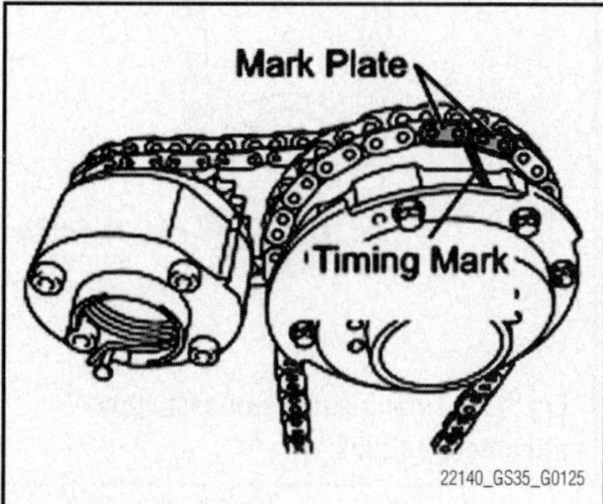

Fig. 175 Timing chain sub-assembly alignment 1 (Bank 2)

9. Remove chain tensioner slipper (for Bank 2).

10. Remove No. 1 chain vibration damper (for Bank 2).

a. Remove the 2 bolts and vibration damper.

11. Remove chain sub-assembly (for Bank 2).

a. While raising up the No. 2 chain tensioner, insert a pin of φ1.0 mm (0.039 in.) into the hole to fix it in place.

b. Hold the hexagonal portion of the camshaft with a wrench and loosen the bolt with a 12 mm hexagon wrench.

c. Hold the hexagonal portion of the camshaft with a wrench and loosen the bolt.

d. Remove the 2 bolts. Then with the No. 1 and No. 2 chains still connected to the gears, remove the camshaft timing gear assembly, camshaft timing exhaust gear assembly and crankshaft timing sprocket.

e. Remove the No. 1 and No. 2 chains from the gears.

Fig. 176 Timing chain sub-assembly alignment 2 (Bank 2)

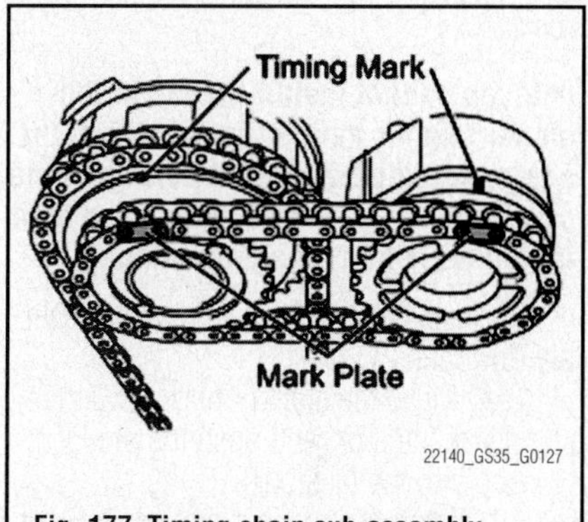

Fig. 177 Timing chain sub-assembly alignment 3 (Bank 2)

12. Remove No. 2 chain tensioner assembly.

To install:

13. Install No. 2 chain tensioner assembly

a. While raising up the No. 2 chain tensioner, insert a pin of φ1.0 mm (0.039 in.) into the hole to fix it in place.

14. Install chain sub-assembly (for Bank 2).

a. Align the No. 1 chain's orange mark plates with the camshaft timing gear's timing mark, and attach the chain to the gear as shown in the illustration.

b. Align the No. 1 chain's orange mark plate with the crankshaft timing gear's timing mark, and attach the chain to the gear as shown in the illustration

c. Align the No. 2 chain's mark plates (yellow) with the timing marks of the camshaft timing gear assembly and camshaft timing exhaust gear assembly, and attach the No. 2 chain to the gears as shown in the illustration.

➡The crankshaft timing gear and camshaft exhaust gear assembly will be installed with the No. 1 and No. 2 chains connected to the gears.

d. Install the crankshaft timing gear to the crankshaft.

e. Align and attach the knock pin of the No. 1 camshaft with the pin hole of the camshaft timing gear assembly.

f. Using the hexagonal portion of the No. 2 camshaft, align and attach the knock pin of the No. 2 camshaft with the pin hole of the camshaft timing exhaust gear assembly.

g. Remove the pin from the No. 2 chain tensioner.

15. Install No. 1 chain vibration damper (for Bank 2).

a. Install the vibration damper with the 2 bolts and tighten to 15 ft. lbs. (21 Nm).

16. Install chain tensioner slipper (for Bank 2).

➡If you cannot install the chain tensioner slipper due to the tension of the chain, use the hexagonal portion of the camshaft to loosen the chain, and then install the chain tensioner slipper.

17. Install No. 1 chain tensioner assembly (for Bank 2).

a. Move the stopper plate upward to release the lock, and push the plunger deep into the tensioner.

b. Move the stopper plate downward to set the lock, and insert a hexagon wrench into the hole of the stopper plate.

c. Install the chain tensioner with the 2 bolts and tighten to 7 ft. lbs. (10 Nm).

d. Remove the hexagon wrench from the chain tensioner.

18. Install No. 3 chain tensioner assembly.

a. Install the chain tensioner with the 2 bolts and tighten to 7 ft. lbs. (10 Nm) .

b. While pushing down the No. 2 chain tensioner, insert a pin of φ1.0 mm (0.039 in.) into the hole to fix it in place.

19. Install chain sub-assembly (for Bank 1).

a. Align the No. 1 chain's orange mark plates with the camshaft timing gear's timing mark, and attach the chain to the gear as shown in the illustration.

b. Align the No. 2 chain's mark plates (yellow) with the timing marks of the camshaft timing gear assembly and camshaft timing exhaust gear assembly, and attach the No. 2 chain to the gears as shown in the illustration

➡The crankshaft timing gear and camshaft exhaust gear assembly will be installed with the No. 1 and No. 2 chains connected to the gears

Fig. 178 Timing chain sub-assembly alignment 1 (Bank 1)

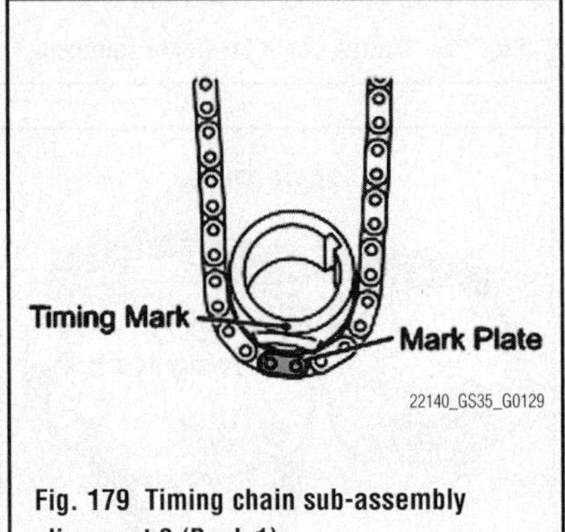

Fig. 179 Timing chain sub-assembly alignment 2 (Bank 1)

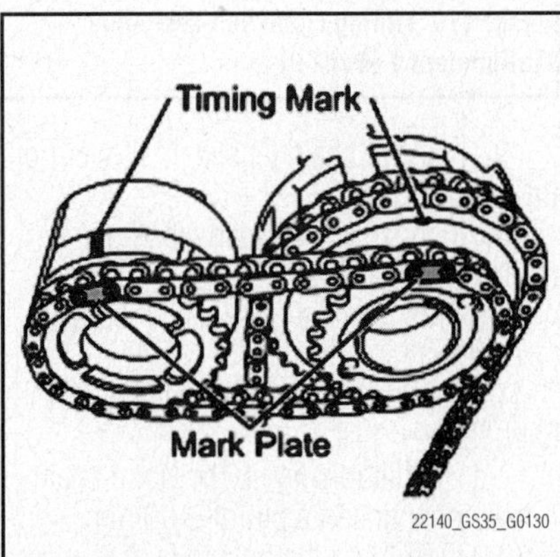

Fig. 180 Timing chain sub-assembly alignment 3 (Bank 1)

c. Install the crankshaft timing gear to the crankshaft.

d. Align and attach the knock pin of the No. 3 camshaft with the pin hole of the camshaft timing gear assembly.

e. Using the hexagonal portion of the No. 4 camshaft, align and attach the knock pin of the No. 4 camshaft with the pin hole of the camshaft timing exhaust gear assembly.

➡Because the gears' timing mark positions may shift due to looseness of the No. 1 chain, use the hexagonal portion of the camshaft to hold the No. 3 camshaft in place until the No. 1 chain tensioner is installed.

f. Remove the pin from the No. 2 chain tensioner.

20. Install chain tensioner slipper (for Bank 1).

➡If you cannot install the chain tensioner slipper due to the tension of the chain, use the hexagonal portion of the camshaft to loosen the chain and install the chain tensioner.

21. Install No. 1 chain tensioner assembly (for Bank 1).

a. Move the stopper plate upward to release the lock, and push the plunger deep into the tensioner.

b. Move the stopper plate downward to set the lock, and insert a hexagon wrench into the hole of the stopper plate.

c. Install the chain tensioner and gasket with the 2 bolts and tighten to 7 ft. lbs. (10 Nm).

22. Install No. 1 chain vibration damper (for Bank 1).

a. Install the vibration damper with the 2 bolts.

b. Remove the hexagon wrench from the No. 1 chain tensioner.

23. Tighten camshaft timing gear assembly for Bank 1.

a. Using a wrench, hold the hexagonal portion of the No. 3 camshaft.

b. Using a 12 mm socket hexagon wrench, tighten the camshaft timing gear assembly with a new bolt and tighten to 58 ft. lbs. (79 Nm).

c. Using a wrench to hold the hexagonal portion of the No. 4 camshaft, tighten the camshaft timing exhaust gear assembly with the bolt and tighten to 74 ft. lbs. (100 Nm).

24. Tighten camshaft timing gear assembly for Bank 2.

a. Using a wrench, hold the hexagonal portion of the No. 1 camshaft.

b. Using a 12 mm socket hexagon wrench, tighten the camshaft timing gear assembly with a new bolt and tighten to 58 ft. lbs. (79 Nm).

c. Using a wrench to hold the hexagonal portion of the No. 2 camshaft, tighten the camshaft timing exhaust gear assembly with the bolt and tighten to 74 ft. lbs. (100 Nm).

25. Check No. 1 cylinder to TDC/compression.
a. Temporarily tighten the pulley set bolt.
b. Rotate the crankshaft clockwise so that the timing marks on the crankshaft timing gear and camshaft timing gears are as shown in the illustration.
26. Install timing chain cover.

VALVE LASH

ADJUSTMENT

No adjustment is necessary.

ENGINE PERFORMANCE & EMISSION CONTROL

ACCELERATOR PEDAL POSITION (APP) SENSOR

LOCATION

See Figure 181.

REMOVAL & INSTALLATION

1. Disconnect negative battery cable.
2. Disconnect the accelerator pedal position sensor connector.
3. Remove the 2 bolts and accelerator pedal.

➡Do not disassemble the accelerator pedal assembly.

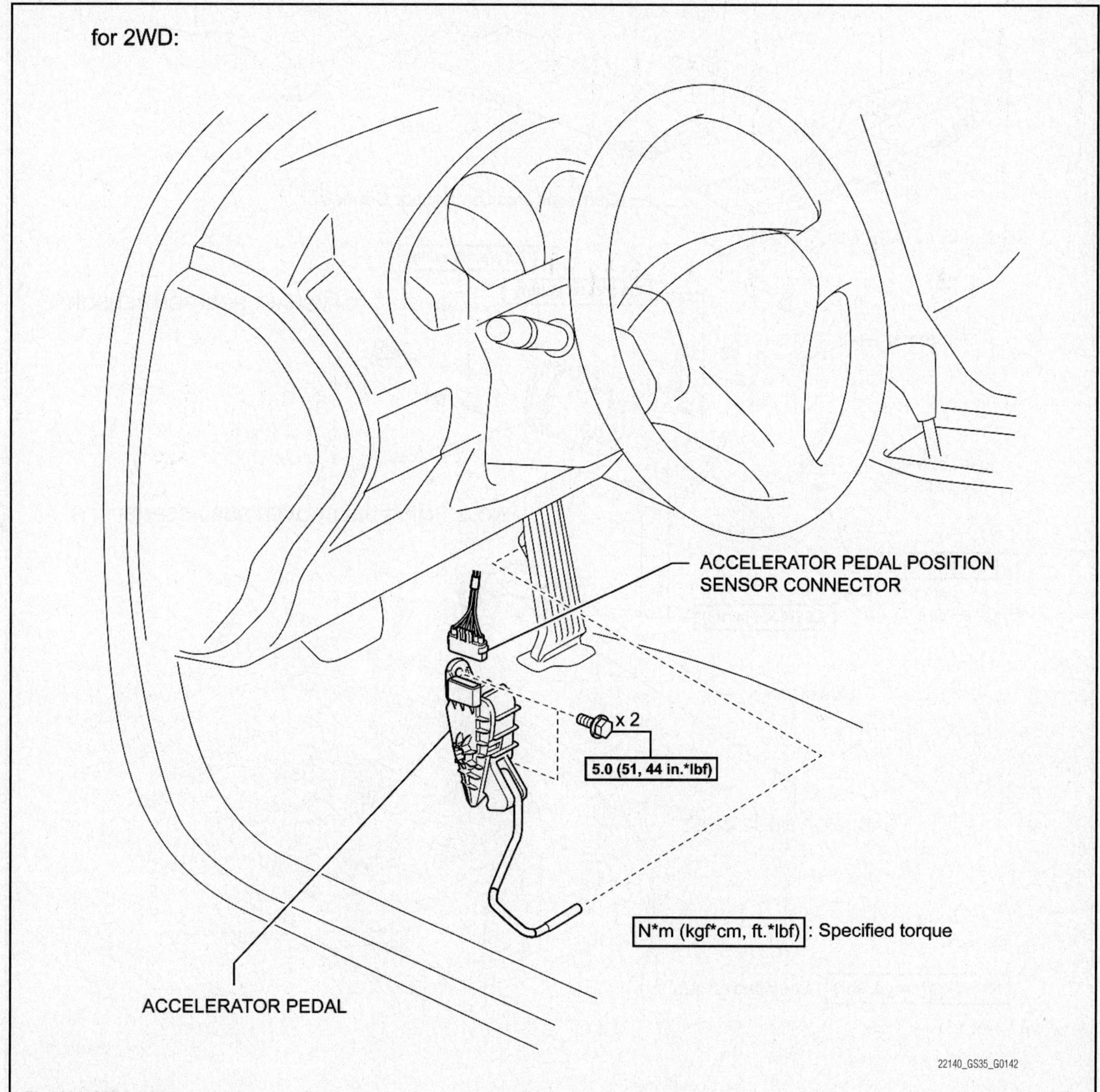

Fig. 181 APP location

CAMSHAFT POSITION (CMP) SENSOR

LOCATION

See Figures 182 and 183.

REMOVAL & INSTALLATION

4.3L (3UZ-FE)

1. Disconnect cable from negative battery terminal.
2. Drain engine coolant.
3. Remove cool air intake duct seal.
4. Remove engine room side cover LH.
5. Remove the 2 nuts and V-bank cover.
6. Remove radiator hose inlet.

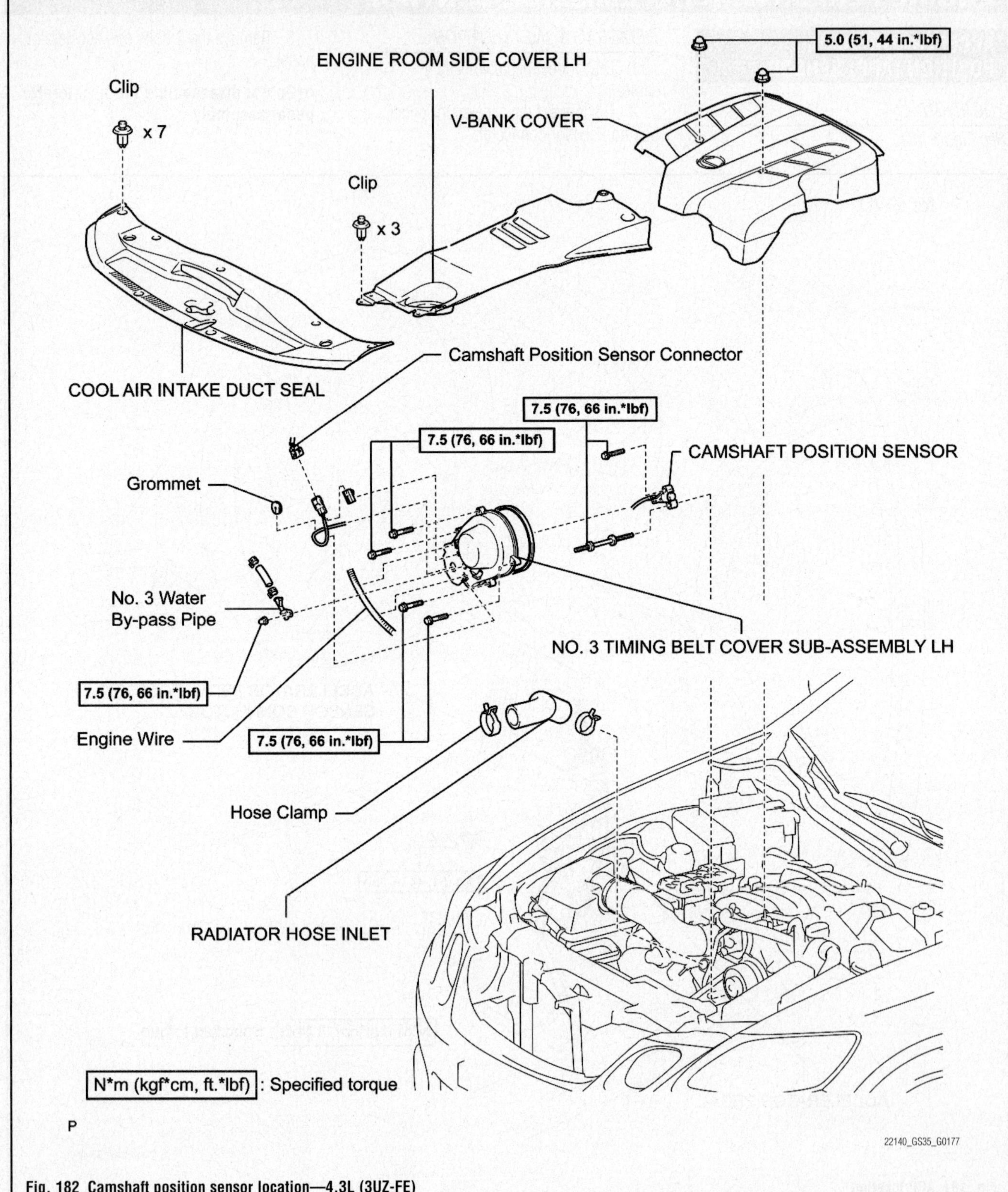

Fig. 182 Camshaft position sensor location—4.3L (3UZ-FE)

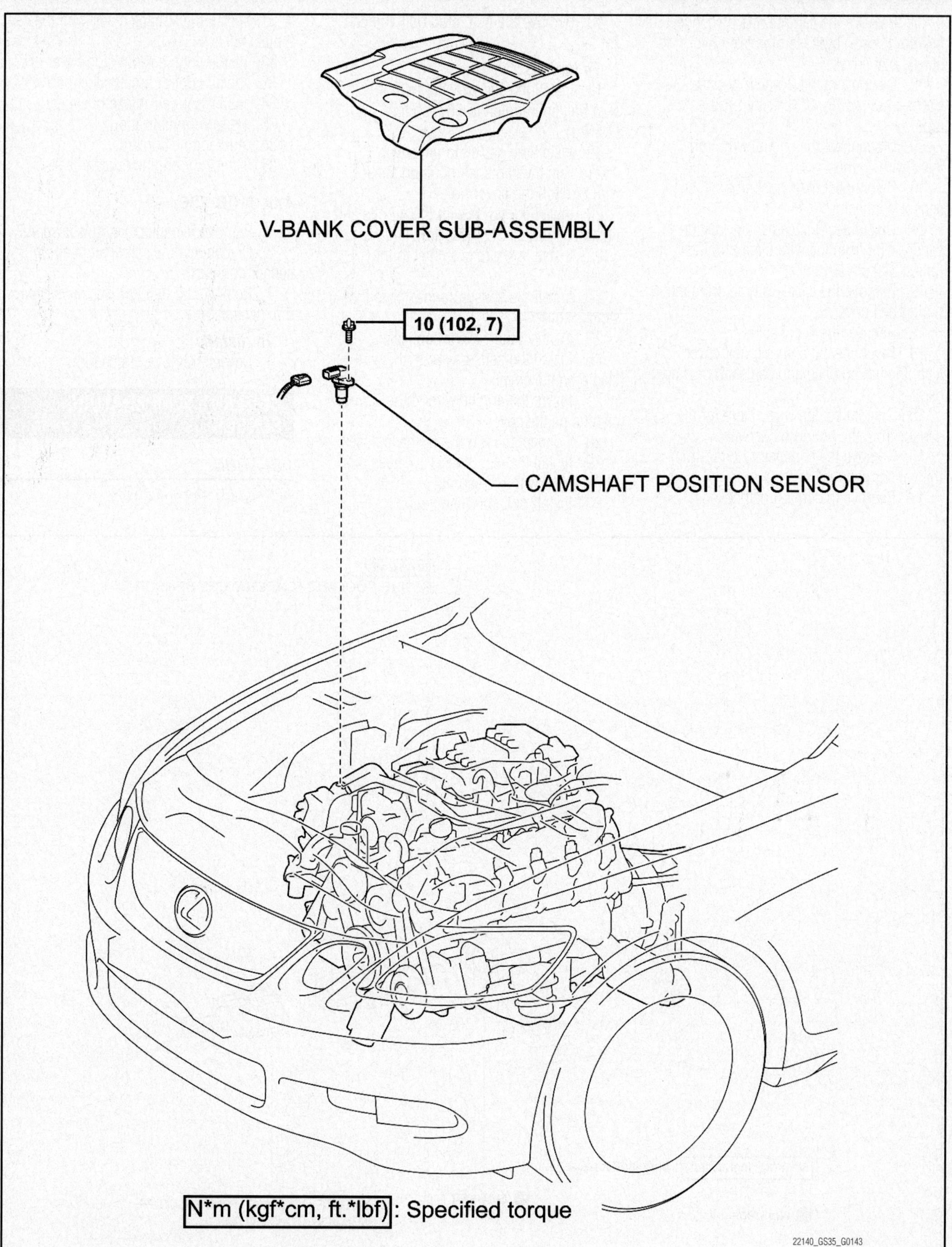

Fig. 183 Camshaft position sensor location—4.6L (1UR-FSE)

7. Remove the cap nut and disconnect the No. 3 water by-pass pipe from the timing belt cover.
8. Disconnect the 2 water by-pass hoses from the No. 3 water by-pass pipe.
9. Disconnect the engine wire from the 2 wire clamps.
10. Disconnect the camshaft position sensor connector.
11. Disconnect the camshaft position sensor wire from the wire clamp on the timing belt cover.
12. Remove the wire grommet from the timing belt cover.
13. Remove the 4 bolts.
14. Disconnect the timing belt cover from the timing plate and camshaft bearing cap.
15. Disconnect the wire clamp for the sensor from the timing belt cover.
16. Remove the connector holder from the sensor connector.
17. Remove the timing belt cover.
18. Remove the bolt, stud bolt and sensor.

To install:

19. Install the sensor with the bolt and stud bolt and tighten to 66 inch lbs. (7.5Nm).
20. Install the gasket to the cover.
21. Run the camshaft position sensor wire through the cover hole.
22. Install the cover with the 4 bolts and tighten to 66 inch lbs. (7.5Nm).
23. Install the wire grommet to the cover.
24. Install the sensor connector to the sensor holder.
25. Connect the sensor connector.
26. Install the sensor wire to the wire clamp on the cover.
27. Install the engine wire to the 2 wire clamps on the cover.
28. Connect the 2 water by-pass hoses.
29. Install the No. 3 water by-pass pipe to the cover with the cap nut.
30. Install radiator hose inlet.
31. Install the V-bank cover with the 2 nuts.
32. Install engine room side cover LH.
33. Install cool air intake duct seal.
34. Install cool air intake duct seal.
35. Perform initialization.
36. Add engine coolant.
37. Check for engine coolant leaks.

4.6L (1UR-FSE)

1. Remove V-bank cover sub-assembly.
2. Disconnect the camshaft position sensor connector.
3. Remove the bolt and camshaft position sensor.

To install:

4. Reverse removal procedure

COOLANT TEMPERATURE SENSOR

LOCATION

See Figures 184 and 185.

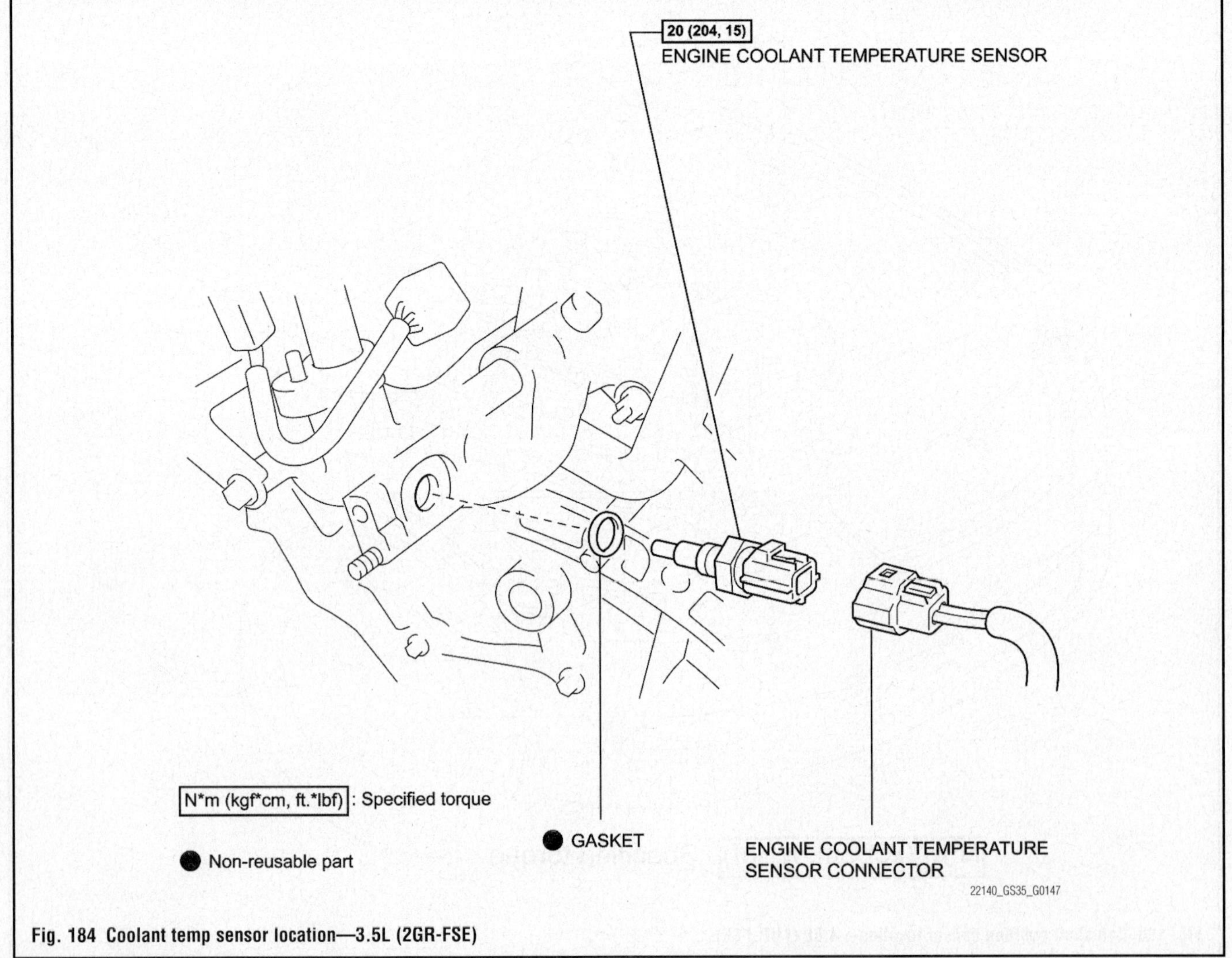

Fig. 184 Coolant temp sensor location—3.5L (2GR-FSE)

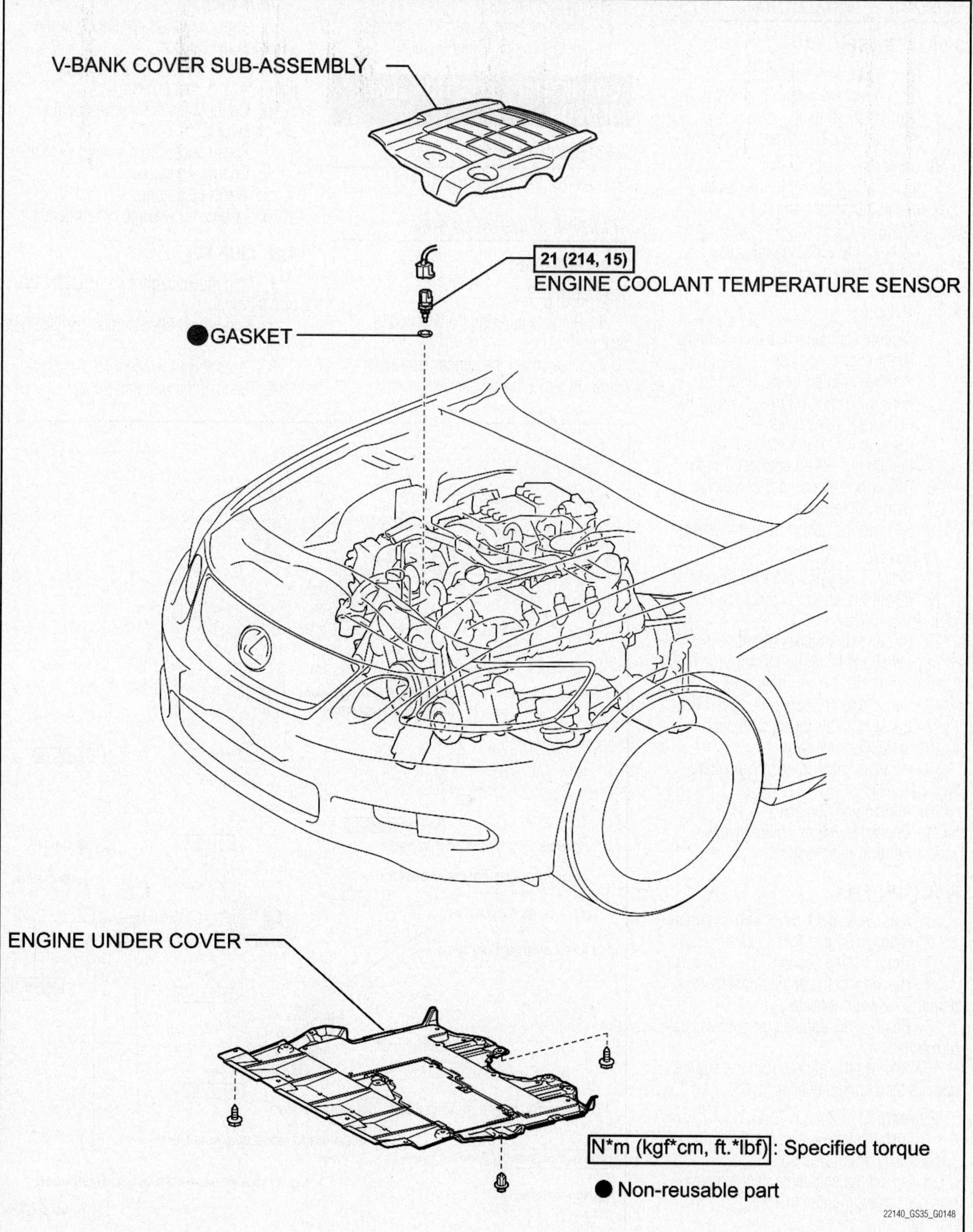

Fig. 185 Coolant temp sensor location—4.6L (1UR-FSE)

REMOVAL & INSTALLATION

3.5L (2GR-FSE)

1. Remove intake manifold.
2. Disconnect the sensor connector.
3. Remove the sensor.
4. Remove the gasket from the sensor.

To install:

5. Install a new gasket to the sensor.
6. Install the sensor and tighten to 15 ft. lbs. (20 Nm).
7. Connect the sensor connector.
8. Install intake manifold.

4.3L (3UZ-FE)

1. Disconnect negative battery terminal.
2. Drain engine coolant.
3. Remove cool air intake duct seal.
4. Remove engine room side cover RH.
5. Remove V-bank cover.
6. Remove No. 1 air cleaner inlet.
7. Remove intake air connector pipe.
8. Disconnect the sensor connector.
9. Remove the sensor.
10. Remove the gasket from the sensor.

To install:

11. Install a new gasket to the sensor.
12. Install the sensor and tighten to 15 ft. lbs. (20 Nm).
13. Install a new gasket to the sensor.
14. Install intake air connector pipe.
15. Install No. 1 air cleaner inlet.
16. Install engine room side cover RH.
17. Install cool air intake duct seal.
18. Install V-bank cover.
19. Connect cable to negative battery terminal.
20. Add engine coolant.
21. Check for engine coolant leaks.
22. Perform initialization.

4.6L (1UR-FSE)

1. Remove V-bank cover sub-assembly.
2. Remove engine under cover.
3. Drain engine coolant.
4. Disconnect the engine coolant temperature sensor connector.
5. Remove the engine coolant temperature sensor.
6. Remove the gasket from the engine coolant temperature sensor.

To install:

7. Install a new gasket to the engine coolant temperature sensor.
8. Install the engine coolant temperature sensor and tighten to 15 ft. lbs. (20 Nm).
9. Connect the engine coolant temperature sensor connector.
10. Add engine coolant.
11. Inspect for coolant leak.
12. Install V-bank cover sub-assembly.
13. Install engine under cover.

CRANKSHAFT POSITION (CKP) SENSOR

LOCATION

See Figures 186 through 188.

REMOVAL & INSTALLATION

3.5L (2GR-FSE)

See Figure 189.

1. Remove air conditioning cooler assembly.
2. Disconnect the sensor connector.
3. Remove the bolt and sensor.

To install:

4. Apply a coat of engine oil to the O-ring of the sensor.
5. Install the sensor with the bolt and tighten to 7 ft. lbs. (10 Nm).
6. Connect the crankshaft position sensor connector.
7. Install cooler compressor assembly.
8. Charge refrigerant.
9. Warm up engine.
10. Inspect for leakage of refrigerant.

4.3L (3UZ-FE)

1. Disconnect cable from negative battery terminal.
2. Remove oil filter element service hole cover.
3. Disconnect the sensor connector.
4. Remove the bolt and sensor.

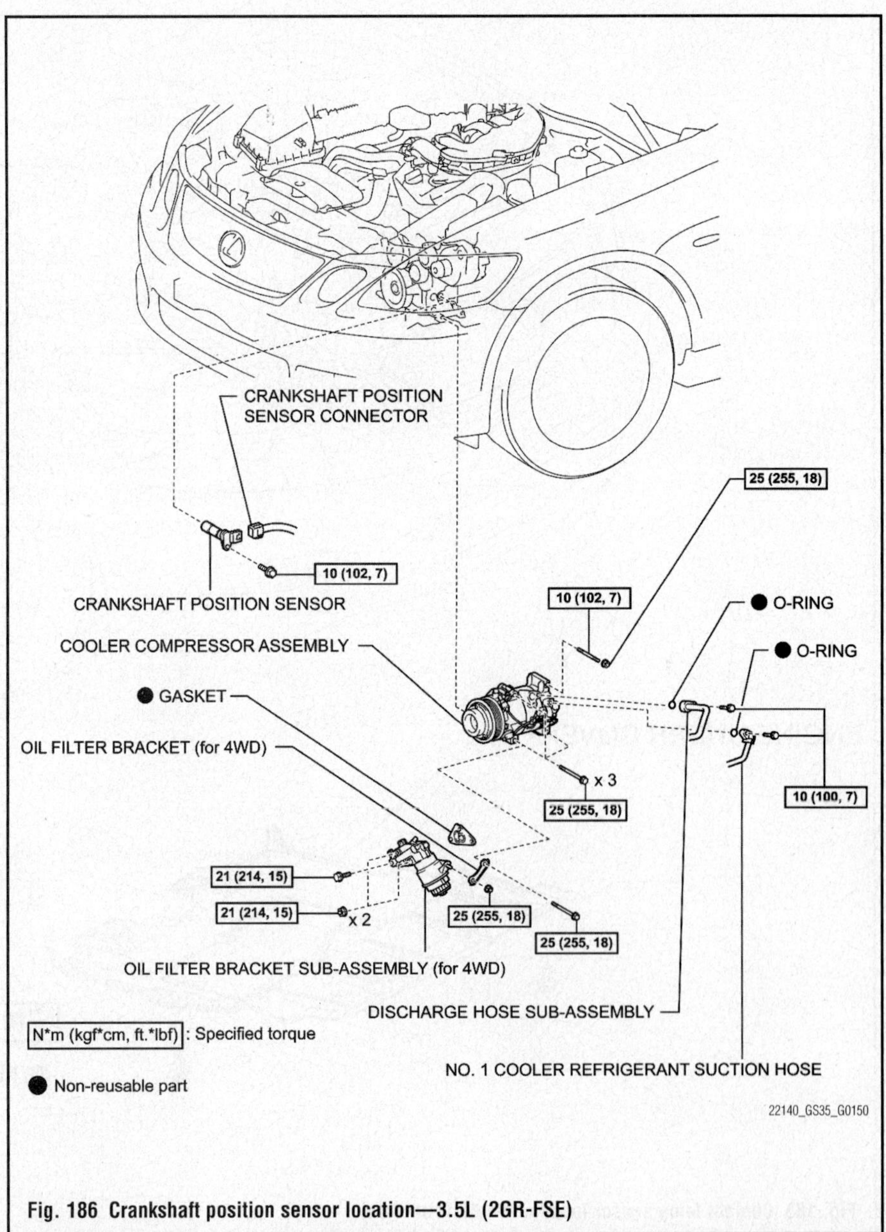

Fig. 186 Crankshaft position sensor location—3.5L (2GR-FSE)

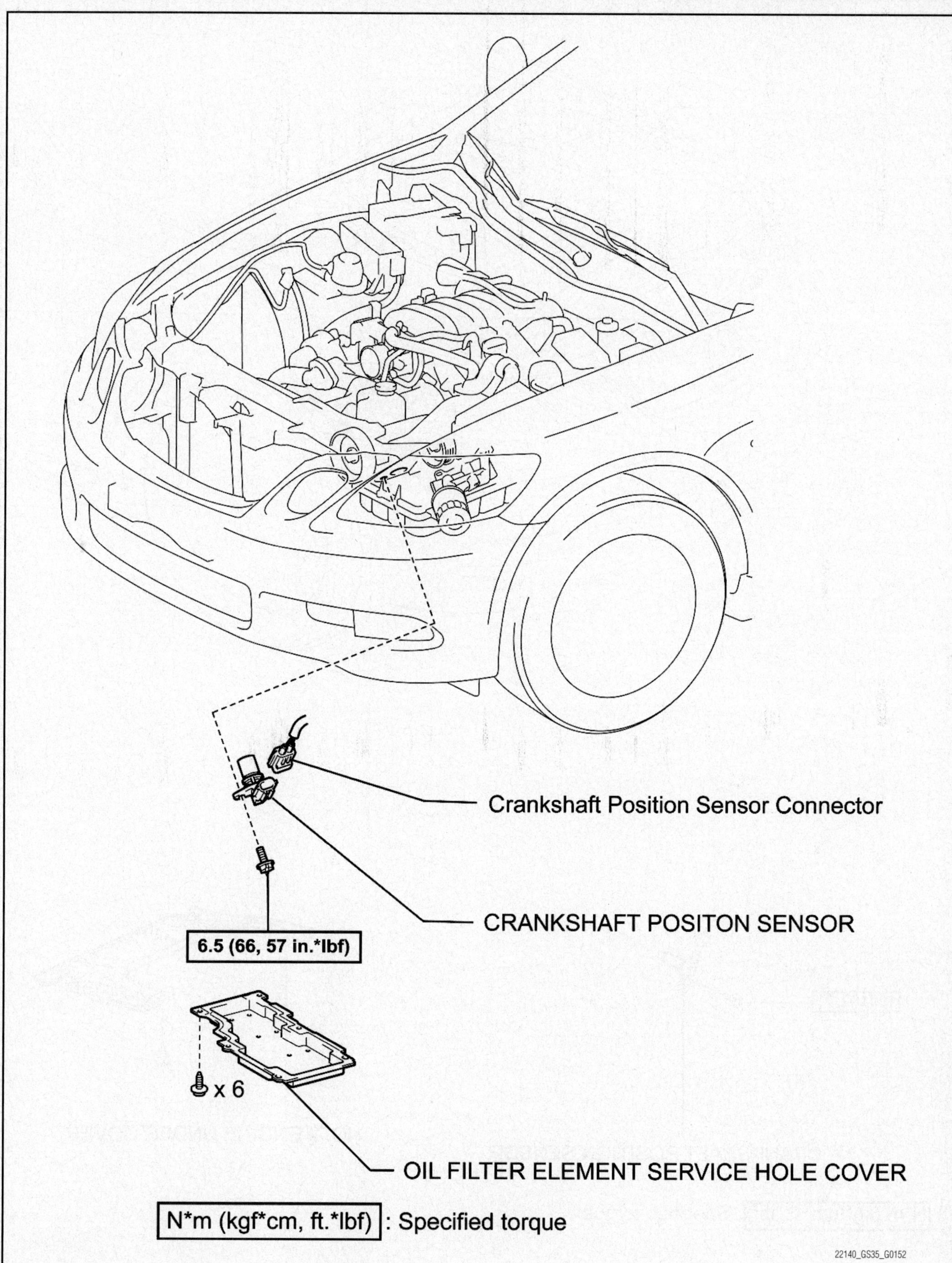

Fig. 187 Crankshaft position sensor location—4.3L (3UZ-FE)

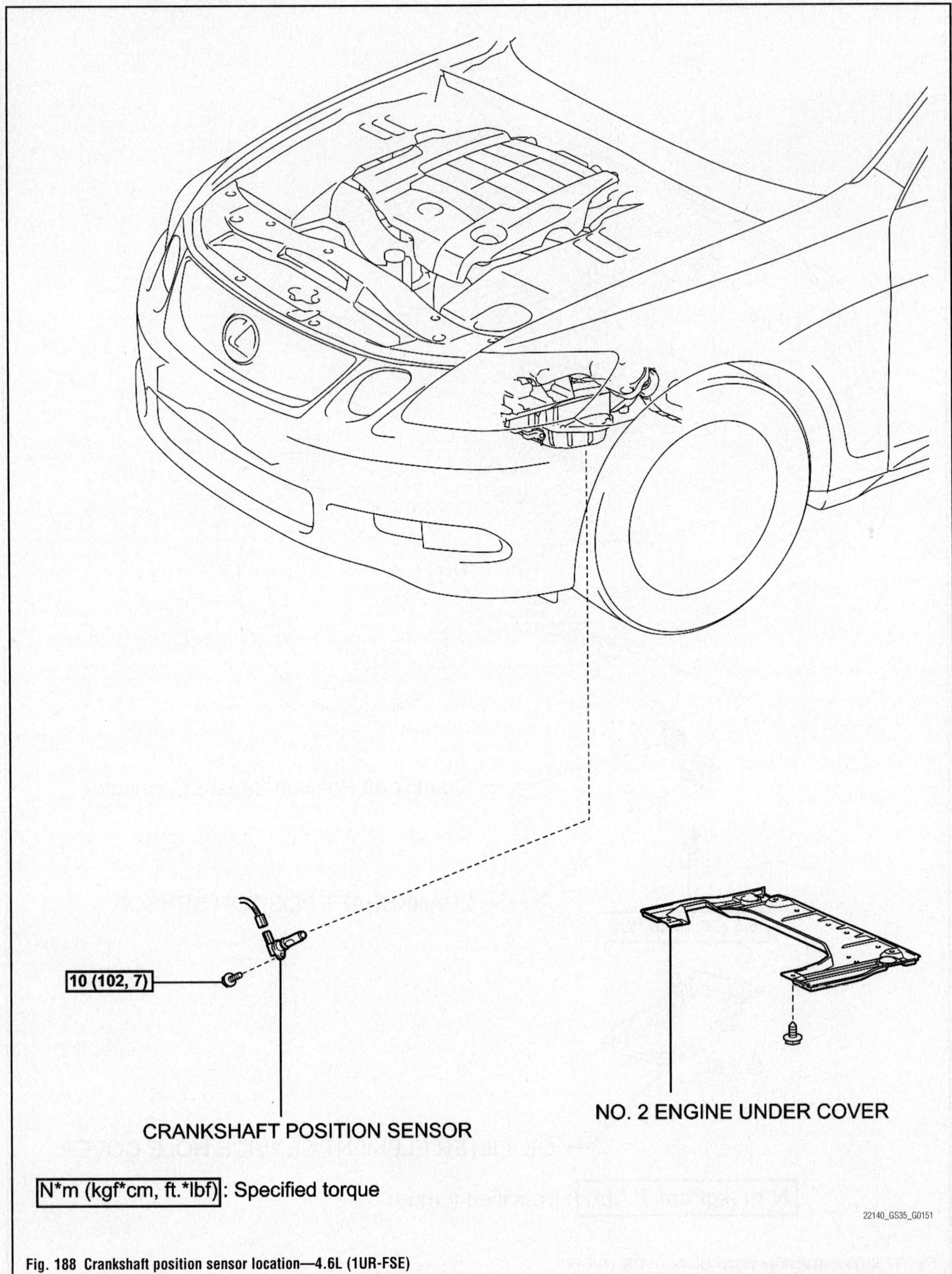

Fig. 188 Crankshaft position sensor location—4.6L (1UR-FSE)

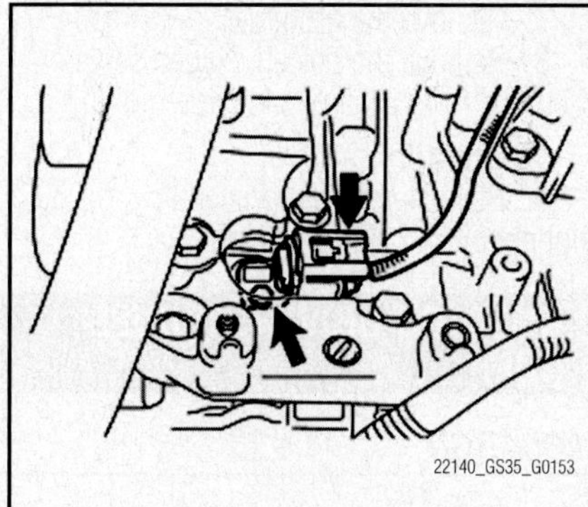

Fig. 189 Crankshaft position sensor removal

To install:

5. Install the sensor with the bolt and tighten to 57 inch lbs. (6.5 Nm).
6. Connect the sensor connector.
7. Install oil filter element service hole cover.
8. Connect cable to negative battery terminal.
9. Perform initialization.

4.6L (1UR-FSE)

1. Remove No. 2 engine under cover.
2. Disconnect the crankshaft position sensor connector.
3. Remove the bolt and crankshaft position sensor.

To install:

4. Install the crankshaft position sensor with the bolt and tighten to 7 ft. lbs. (10 Nm).
5. Connect the crankshaft position sensor connector.
6. Inspect for oil leak.
7. Install No. 2 engine under cover.

ELECTRONIC CONTROL MODULE (ECM)

LOCATION

See Figure 190.

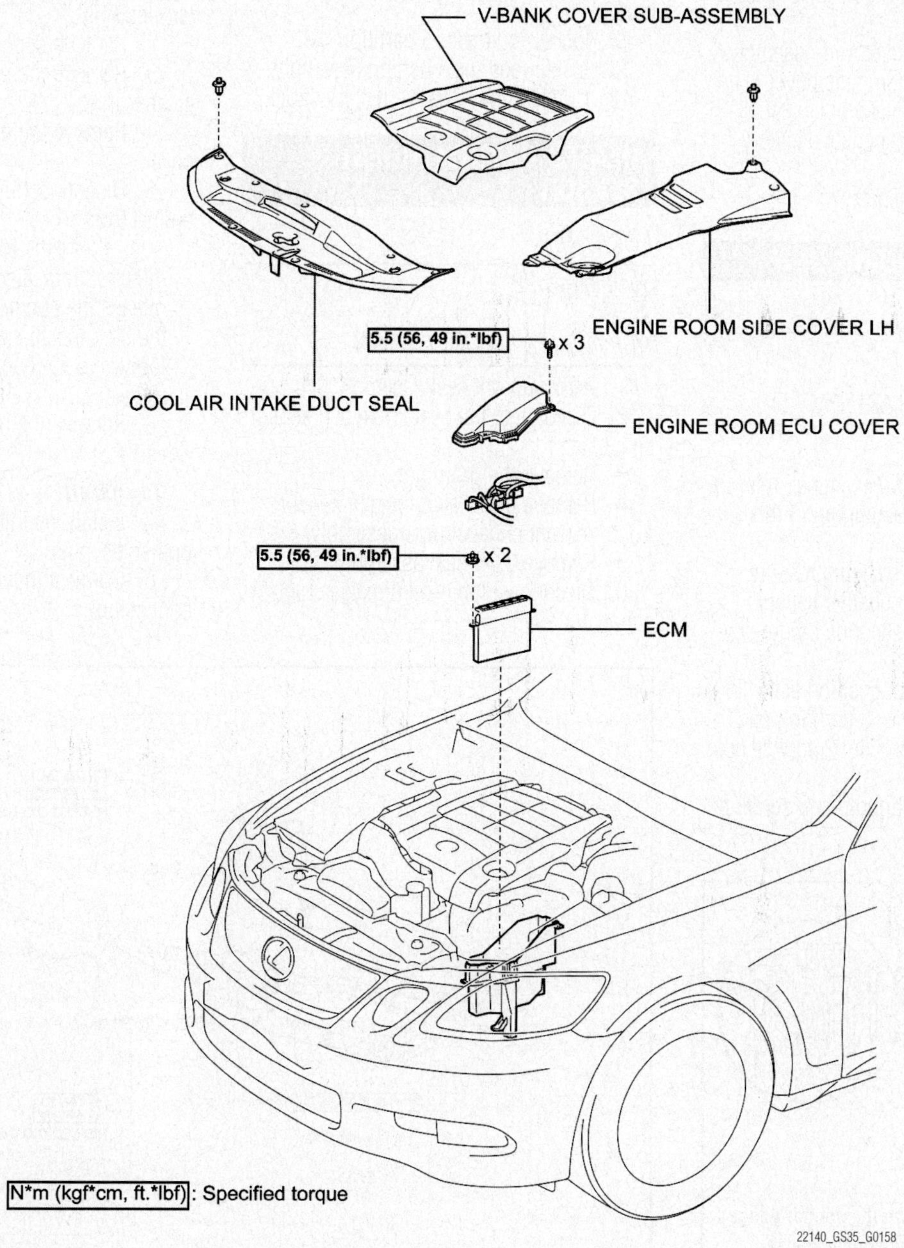

Fig. 190 ECM location

REMOVAL & INSTALLATION

1. Disconnect cable from negative battery terminal.
2. Remove cool air intake duct seal.
3. Remove engine room side cover LH.
4. Remove V-bank cover sub-assembly.
5. Remove the 3 bolts and engine room ECU cover.
6. Detach the claw and disconnect the No. 4 connector holder.
7. Disconnect the 6 ECM connectors.
8. Remove the 2 nuts and ECM.

To install:

9. Install the ECM with the 2 nuts.
10. Connect the 6 ECM connectors.
11. Connect the No. 4 connector holder.
12. Install the engine room ECU cover with the 3 bolts.
13. Install V-bank cover sub-assembly.
14. Install engine room side cover LH.
15. Install cool air intake duct seal.
16. Connect cable to negative battery terminal.
17. Perform reset memory.

FUEL LEVEL SENDING UNIT

LOCATION

In fuel tank.

REMOVAL & INSTALLATION

See Figure 191.

1. Before servicing the vehicle, refer to the precautions in the beginning of this section.
2. Relieve the fuel system pressure.
3. Disconnect the negative battery cable. Wait at least 90 seconds before performing any other work.
4. Remove rear seat cushion assembly.
5. Remove room No. 3 partition pad.
6. Remove rear floor No. 2 service hole cover.
7. Remove the fuel pump and sending gauge assembly.
8. Remove the fuel sending gauge from the fuel pump assembly as follows:
   a. Disconnect the fuel sender gauge connector.
   b. Press down on the sender gauge claw. Then slide the sender gauge upward.

To install:

9. Install the fuel sending gauge to the fuel pump assembly as follows:
   a. Set the fuel sender gauge to the No. 1 fuel sub-tank. Then slide the sender gauge downward to install.
   b. Connect the fuel sender gauge connector.
10. Install the fuel pump and sending gauge assembly.
11. Install rear floor No. 2 service hole cover.
12. Install room No. 3 partition pad.
13. Install rear seat cushion assembly.
14. Connect the negative battery cable.

HEATED OXYGEN (HO2S) SENSOR

LOCATION

See Figure 192.

REMOVAL & INSTALLATION

1. Remove front seat assembly RH.
2. Disconnect the heated oxygen sensor connector.
3. Remove the grommet.
4. Remove the heated oxygen sensor from the front exhaust pipe assembly.
5. Remove front seat assembly LH.
6. Disconnect the heated oxygen sensor connector.
7. Remove the grommet.
8. Remove the heated oxygen sensor from the front exhaust pipe assembly.

To install:

9. Reverse removal procedure, and tighten to 32 ft. lbs. (44 Nm).

INJECTOR DRIVER MODULE (IDM)

LOCATION

See Figure 193.

REMOVAL & INSTALLATION

See Figures 194 and 195.

1. Disconnect cable from negative battery terminal.
2. Remove cool air intake duct seal.
3. Remove the bolt and No. 1 air cleaner inlet.
4. Remove the clip, 2 cap nuts and cover.
5. Disconnect the 4 injector driver connectors.
   a. Move the lock lever in the direction indicated by the arrow to release the connector lock. Disconnect the 3 connectors with wire harness locks and the connector from the injector driver.
6. Remove the bolt, 2 nuts and injector driver

To install:

7. Install the injector driver with the bolt and 2 nuts.
8. Connect the 4 injector driver connectors.

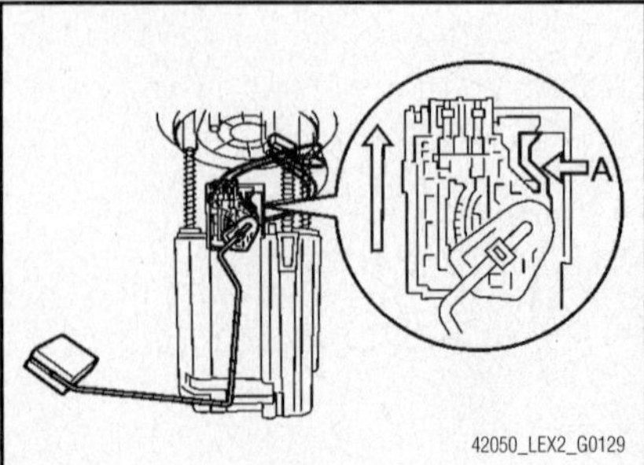

Fig. 191 Press down on the sender gauge claw (A), then slide the sender gauge upward.

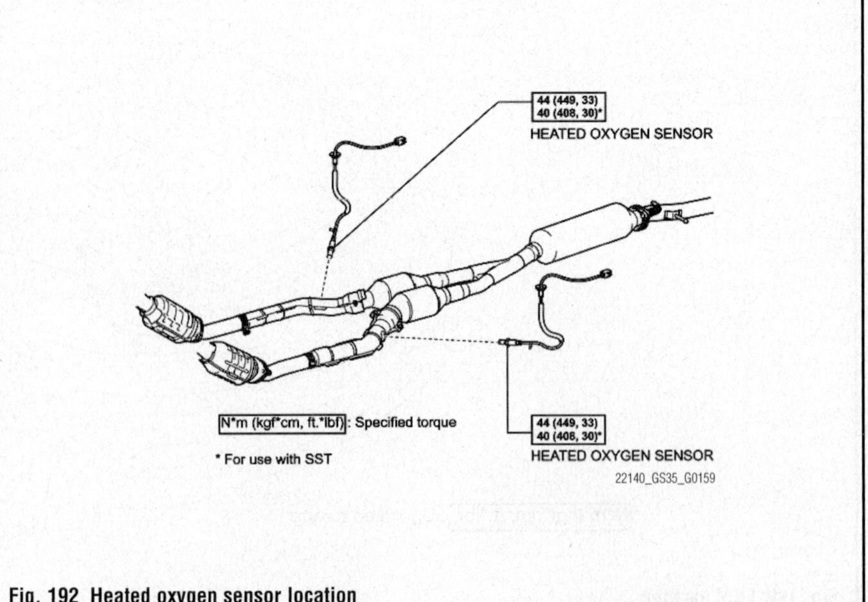

Fig. 192 Heated oxygen sensor location

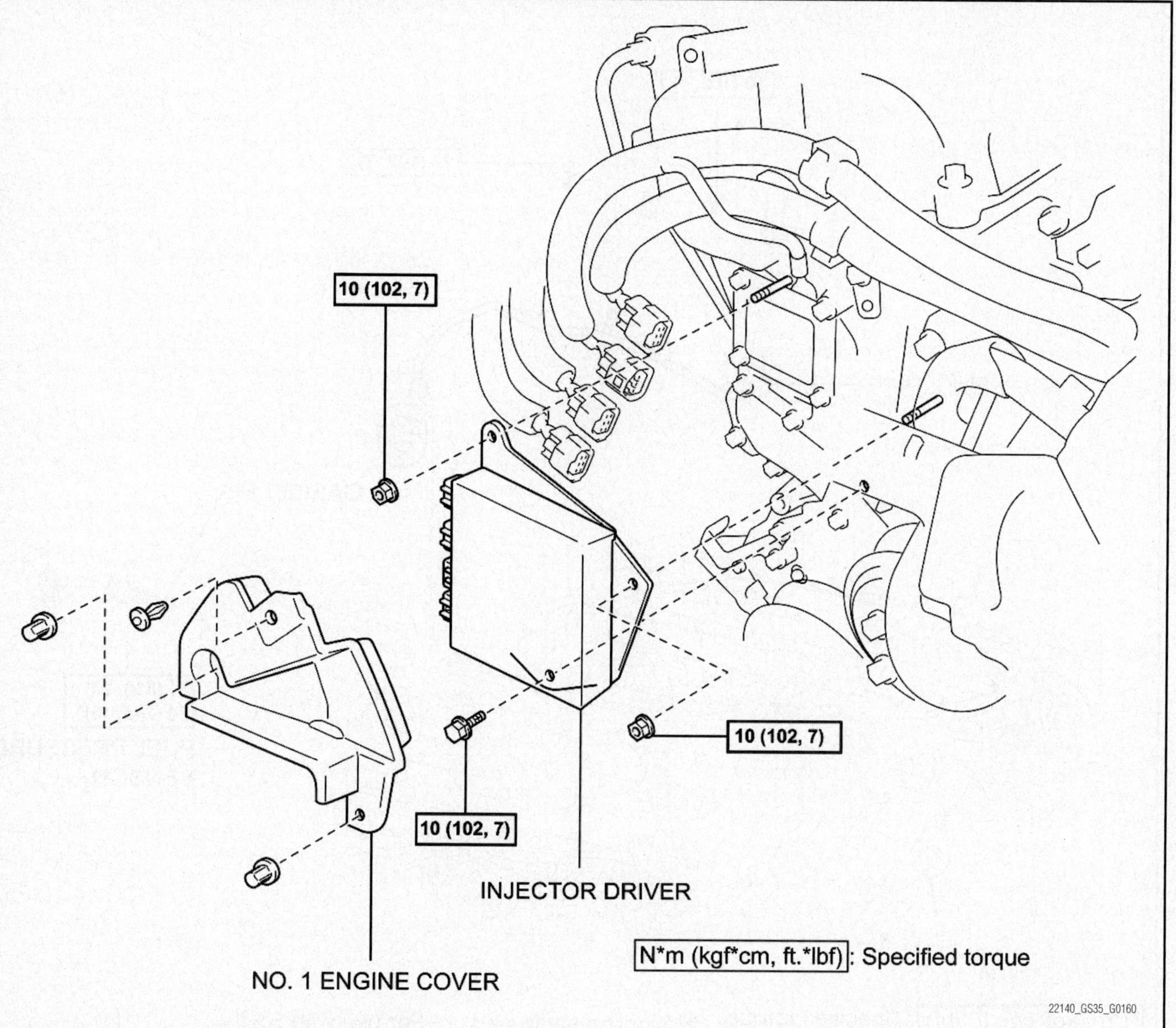

Fig. 193 Injector driver module location—3.5L (2GR-FSE)

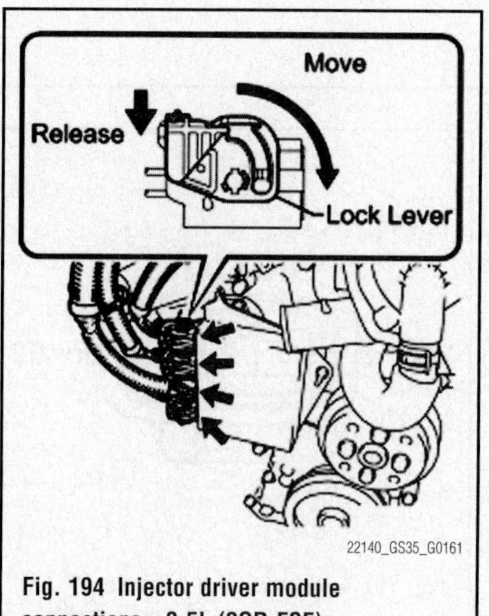

Fig. 194 Injector driver module connections—3.5L (2GR-FSE)

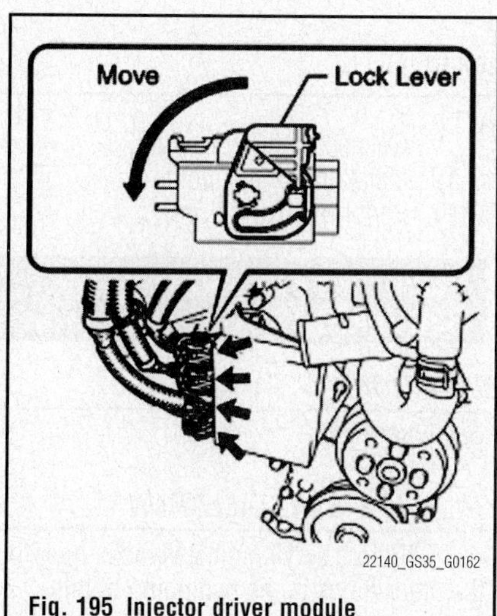

Fig. 195 Injector driver module connections—3.5L (2GR-FSE)

9. Connect the 3 connectors with wire harness locks and the connector to the injector driver. Move the lock lever in the direction indicated by the arrow to lock the connector.
10. Install the cover with the 2 cap nuts and clip.
11. Install the No. 1 air cleaner inlet with the bolt.
12. Install cool air intake duct seal.
13. Connect cable to negative battery terminal.
14. Perform initialization.

INJECTION CONTROL PRESSURE (ICP) SENSOR

LOCATION

See Figure 196.

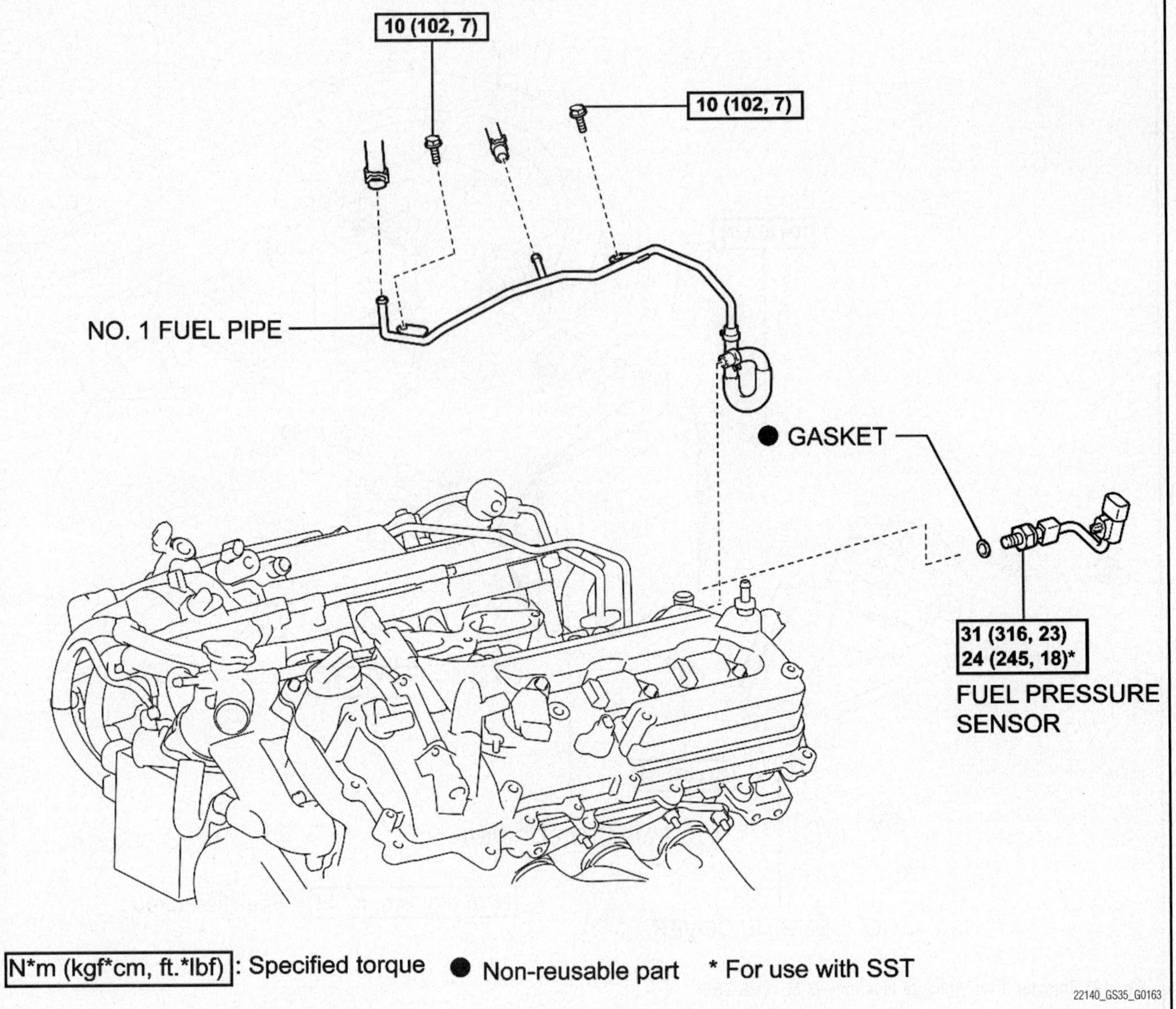

10 (102, 7)
10 (102, 7)
NO. 1 FUEL PIPE
● GASKET
31 (316, 23)
24 (245, 18)*
FUEL PRESSURE
SENSOR
N*m (kgf*cm, ft.*lbf) : Specified torque
● Non-reusable part
* For use with SST
22140_GS35_G0163

Fig. 196 Injector control pressure sensor—3.5L (2GR-FSE)

REMOVAL & INSTALLATION

1. Remove intake manifold.
2. Disconnect water hose joint.
3. Remove No. 1 fuel pipe.
4. Disconnect the connector and clamp.
5. Remove the fuel pressure sensor and gasket.

To install:

6. Install a new gasket to the fuel pressure sensor.

install the fuel pressure sensor to the No. 1 delivery pipe and tighten to 18 ft. lbs. (24 Nm) and 23 ft. lbs. (31 Nm).

7. Connect the connector with the clamp.
8. Install No. 1 fuel pipe.
9. Connect water hose joint.
10. Install intake manifold.
11. Perform initialization.

INJECTION PRESSURE REGULATOR (IPR)

LOCATION

See Figure 197.

REMOVAL & INSTALLATION

1. Before servicing the vehicle, refer to the precautions in the beginning of this section.
2. Relieve the fuel system pressure.

PR
VC
3
2
1
E2
22140_GS35_G0164

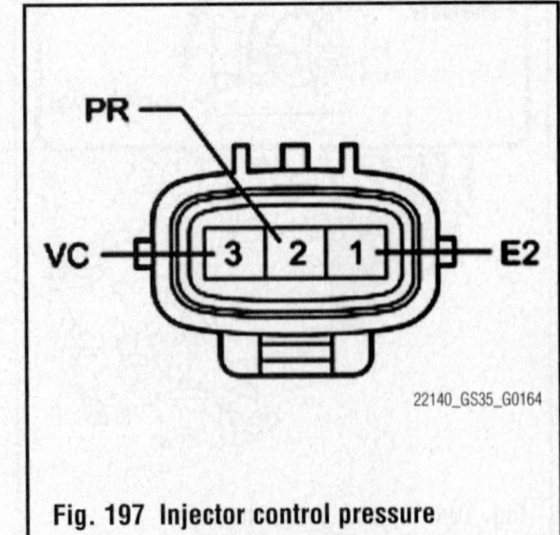

Fig. 197 Injector control pressure sensor pinout—3.5L (2GR-FSE)

3. Disconnect the negative battery cable. Wait at least 90 seconds before performing any other work.
4. Remove rear seat cushion assembly.
5. Remove room No. 3 partition pad.
6. Remove rear floor No. 2 service hole cover.
7. Remove the fuel pump and sending gauge assembly.
8. Disconnect the fuel pump connector from the fuel suction plate.
9. Using needle nose pliers, remove the E-ring.
10. Separate the fuel suction plate and remove the spring from the sub-tank.
11. Remove fuel filter.
12. Remove jet pump.
13. Remove fuel pump harness.
14. Remove No. 1 fuel suction support.
15. Remove fuel pump assembly.
16. Remove the fuel pressure regulator from the fuel filter.
17. Remove the 2 O-rings from the pressure regulator

To install:

18. Apply a light coat of gasoline to 2 new O-rings, and install them onto the fuel pressure regulator.
19. Install the fuel pressure regulator to the fuel filter.
20. Attach the 2 claws of the fuel filter and install the cap.
21. Install fuel pump assembly.
22. Install No. 1 fuel suction support.
23. Install fuel pump harness.
24. Install jet pump.
25. Install fuel filter.
26. Install fuel suction plate sub-assembly.
27. Install the fuel sending gauge to the fuel pump assembly as follows:
    a. Set the fuel sender gauge to the No. 1 fuel sub-tank. Then slide the sender gauge downward to install it.
    b. Connect the fuel sender gauge connector.
28. Install the fuel pump and sending gauge assembly.
29. Install rear floor No. 2 service hole cover.
30. Install room No. 3 partition pad.
31. Install rear seat cushion assembly.
32. Connect the negative battery cable.

INTAKE AIR TEMPERATURE (IAT) SENSOR

LOCATION

Intake Air Temperature (IAT) sensor is mounted on the Mass Air Flow (MAF) meter.

REMOVAL & INSTALLATION

1. Remove air flow meter.
2. Remove intake air temperature sensor.

To install:

3. Reverse removal procedure.

KNOCK SENSOR (KS)

LOCATION

See Figures 198 through 200.

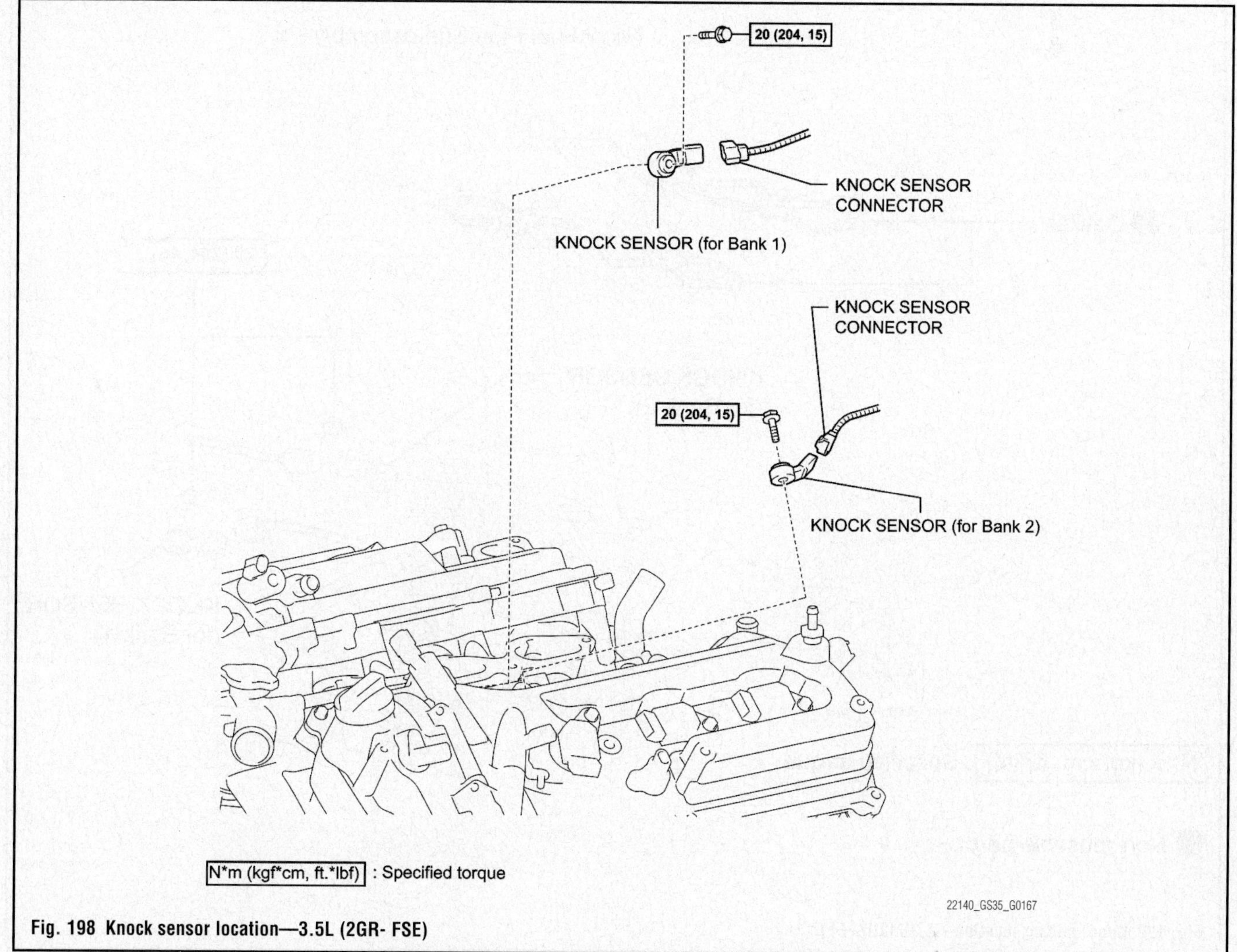

Fig. 198 Knock sensor location—3.5L (2GR- FSE)

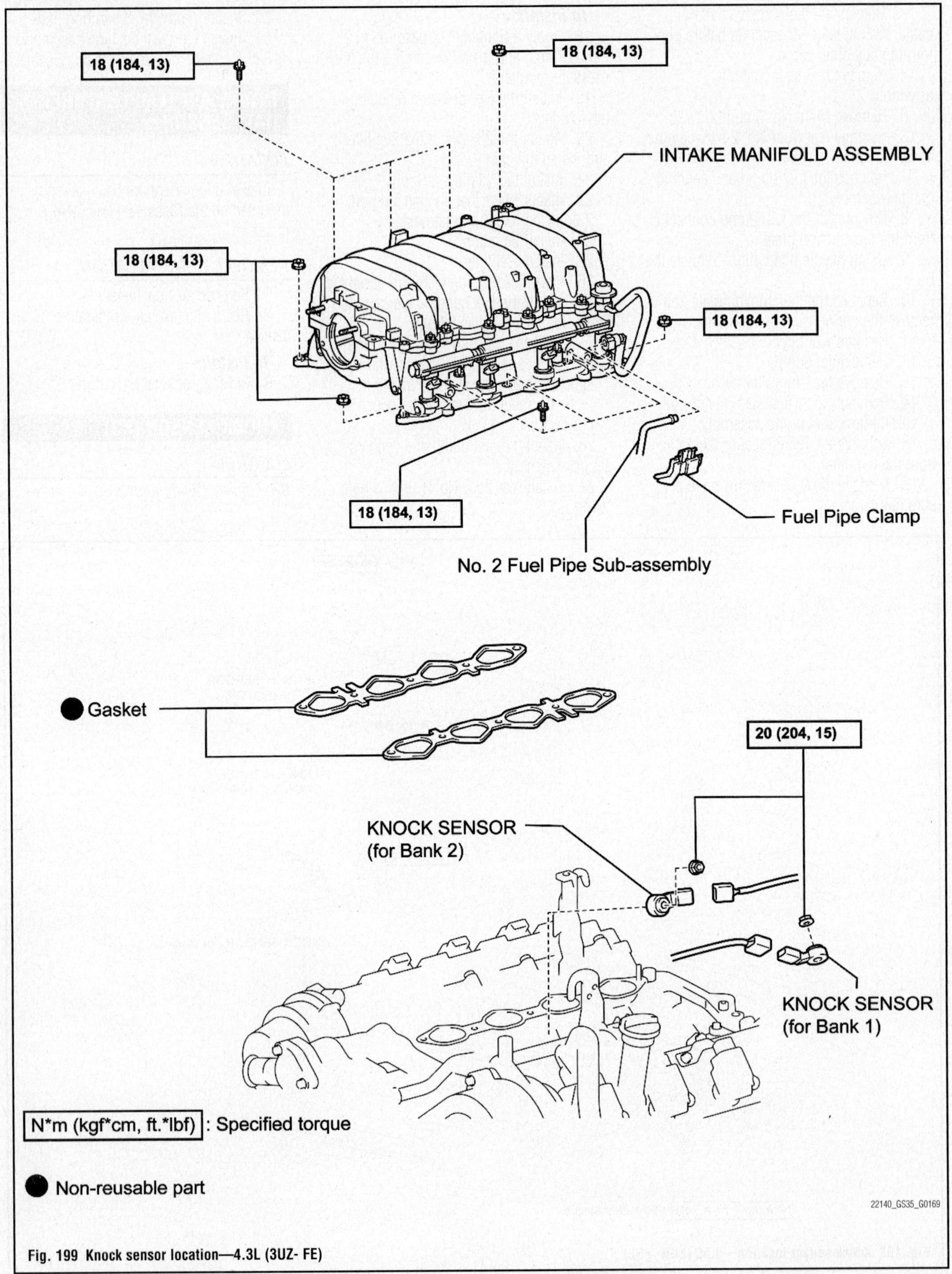

Fig. 199 Knock sensor location—4.3L (3UZ- FE)

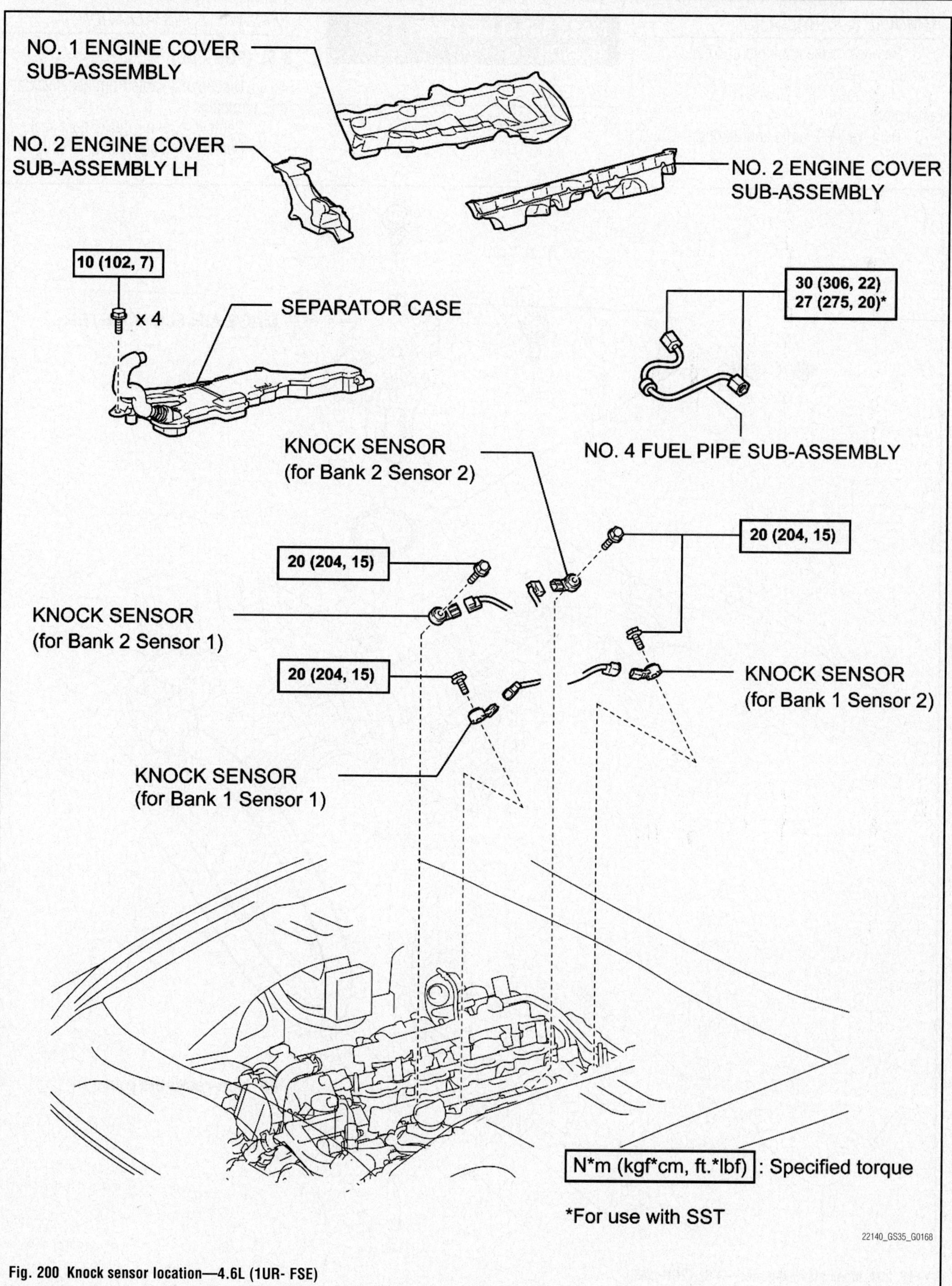

Fig. 200 Knock sensor location—4.6L (1UR- FSE)

REMOVAL & INSTALLATION

1. Remove intake manifold and fuel injection assembly.
2. Disconnect the 2 knock sensor connectors.
3. Remove the 2 bolts and 2 knock sensors.

MASS AIR FLOW (MAF) SENSOR

LOCATION

See Figures 201 through 203.

REMOVAL & INSTALLATION

3.5L (2GR-FSE)

1. Disconnect cable from negative battery terminal.
2. Remove cool air intake duct seal.
3. Remove engine room side cover RH.

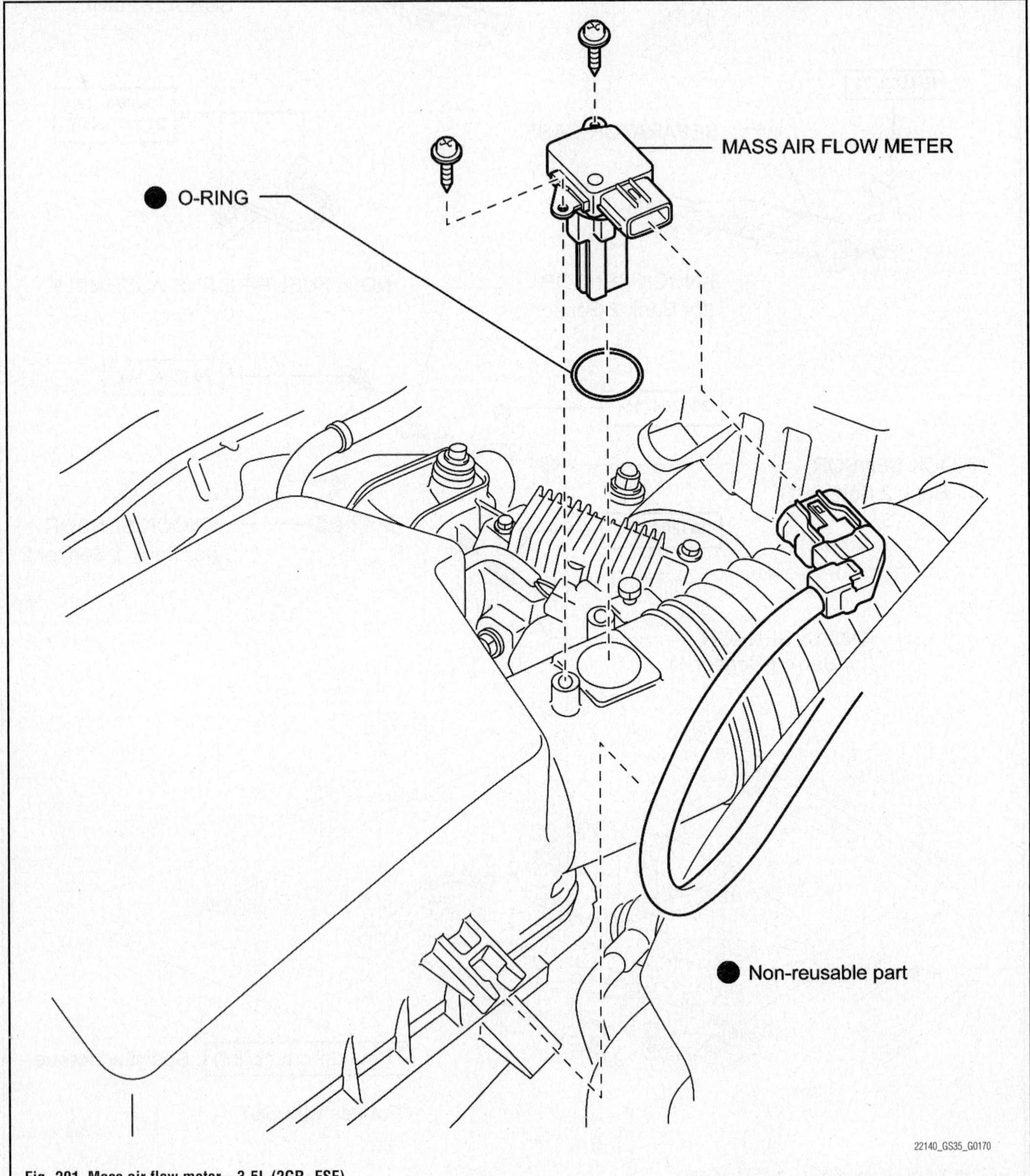

Fig. 201 Mass air flow meter—3.5L (2GR- FSE)

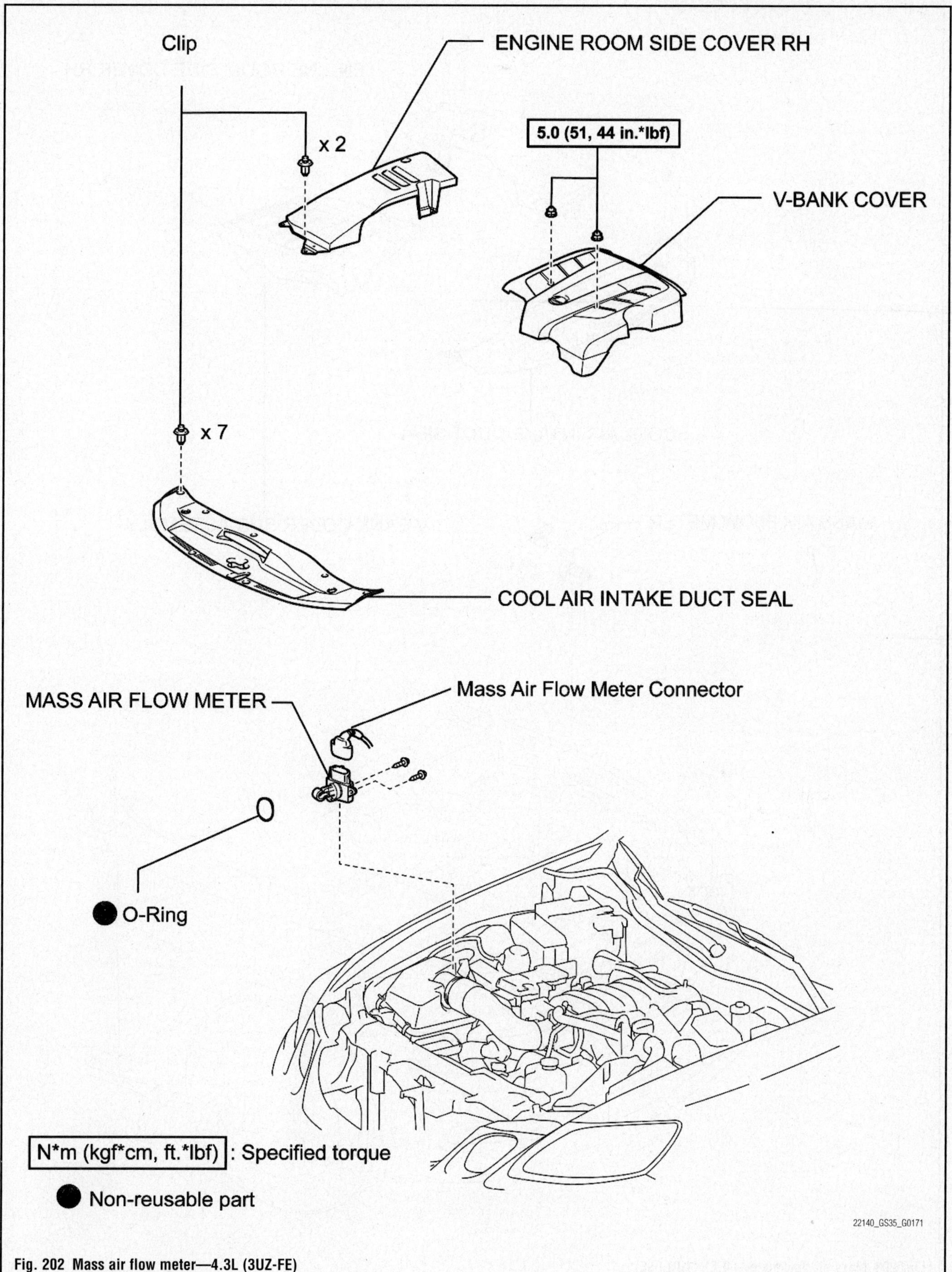

Fig. 202 Mass air flow meter—4.3L (3UZ-FE)

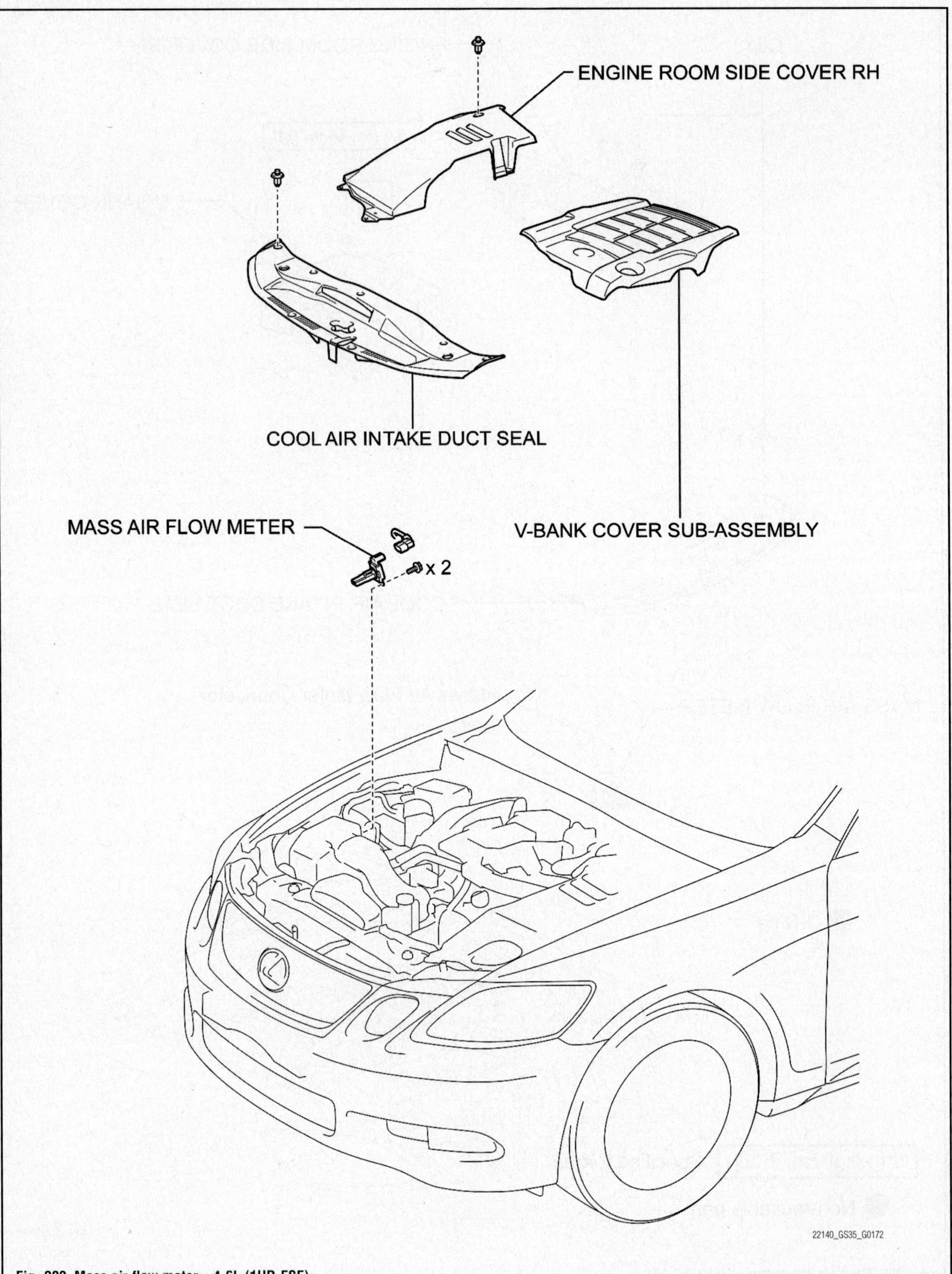

Fig. 203 Mass air flow meter—4.6L (1UR-FSE)

4. Disconnect the MAF meter connector.
5. Remove the 2 screws and MAF meter.
6. Remove the O-ring from the MAF meter.

To install:

7. Install a new O-ring to the MAF meter.
8. Install the MAF meter with the 2 screws.
9. Connect the MAF meter connector.
10. Install engine room side cover RH.
11. Install cool air intake duct seal.
12. Connect cable to negative battery terminal.
13. Perform initialization.

3UZ-FE

1. Disconnect cable from negative battery terminal.
2. Remove cool air intake duct seal.
3. Remove engine room side cover RH.
4. Remove V-bank cover.
5. Disconnect the MAF meter connector.
6. Remove the 2 screws and MAF meter.
7. Remove the O-ring from the MAF meter.

To install:

8. Install a new O-ring to the MAF meter.
9. Install the MAF meter with the 2 screws.
10. Connect the MAF meter connector.
11. Install engine room side cover RH.
12. Install V-bank cover.
13. Install cool air intake duct seal.
14. Connect cable to negative battery terminal.
15. Perform initialization.

4.3L (1UR-FSE)

1. Disconnect cable from negative battery terminal.
2. Remove cool air intake duct seal.
3. Remove engine room side cover RH.
4. Remove V-bank cover.
5. Disconnect the MAF meter connector.
6. Remove the 2 screws and MAF meter.
7. Remove the O-ring from the MAF meter.

To install:

8. Install a new O-ring to the MAF meter.
9. Install the MAF meter with the 2 screws.
10. Connect the MAF meter connector.
11. Install engine room side cover RH.
12. Install V-bank cover.
13. Install cool air intake duct seal.
14. Connect cable to negative battery terminal.
15. Perform initialization.

OIL PRESSURE SENSOR

LOCATION

See Figures 204 through 206.

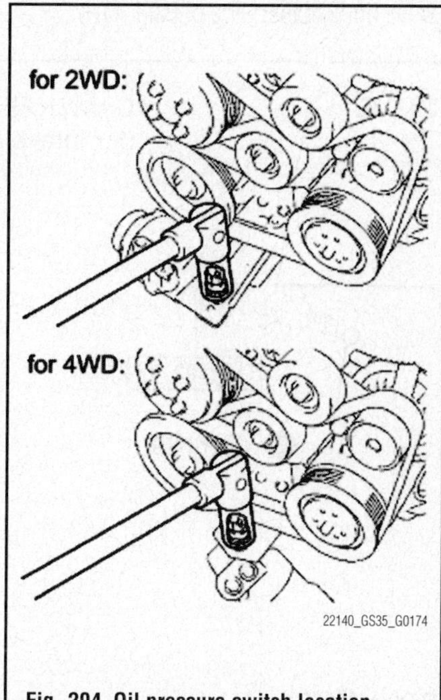

Fig. 204 Oil pressure switch location—3.5L (2GR-FSE)

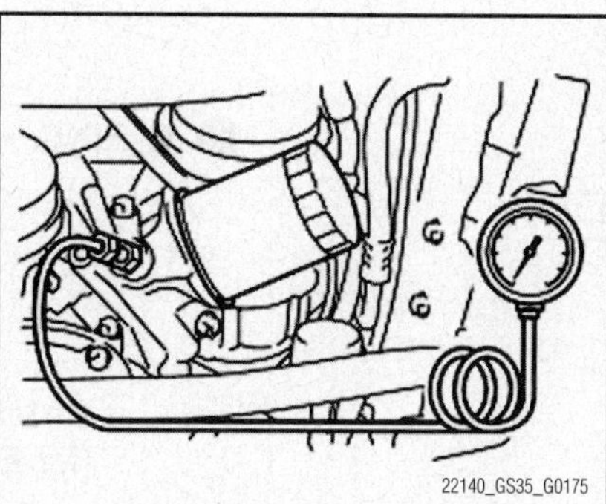

Fig. 205 Oil pressure switch location—4.3L (3UZ-FE)

Fig. 206 Oil pressure switch location—4.6L (1UR-FSE)

REMOVAL & INSTALLATION

See Figure 207.

1. Before servicing the vehicle, refer to the precautions in the beginning of this section.
2. Disconnect the negative battery cable. Wait at least 90 seconds before performing any other work.
3. Remove the engine under cover(s), if necessary.
4. Disconnect the oil pressure switch connector, located on the side of the engine block.
5. Remove the oil pressure switch.

To install:

6. Apply adhesive (Part No. 08833-00080, THREE BOND 1344, LOCTITE 242 or equivalent) to 2 or 3 threads of the oil pressure switch.
7. Install the oil pressure switch.
8. Connect the oil pressure switch connector.
9. Install the engine under cover(s), if necessary.
10. Connect the negative battery cable.
11. Start the engine and check for engine oil leaks.

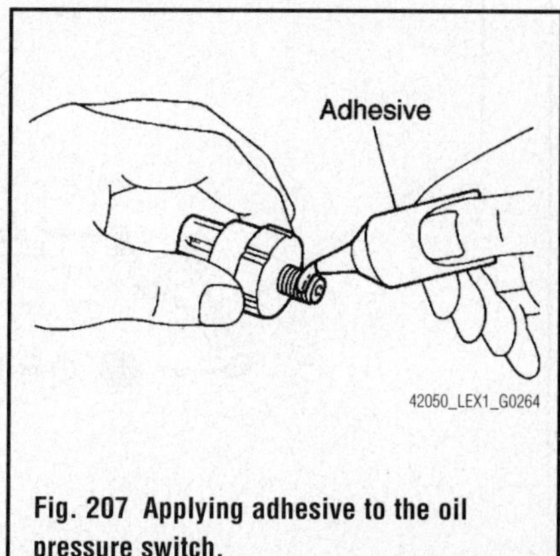

Fig. 207 Applying adhesive to the oil pressure switch.

THROTTLE POSITION SENSOR (TPS)

LOCATION

The throttle position sensor is located on the throttle body

REMOVAL & INSTALLATION

1. Remove throttle body.
2. Remove TPS connector.
3. Remove screws and throttle position sensor.

To install:

4. Reverse removal procedure.

VARIABLE CAMSHAFT TIMING OIL CONTROL SOLENOID

LOCATION

See Figures 208 through 210.

REMOVAL & INSTALLATION

3.5L (2GR-FSE)

1. Remove intake manifold.
2. Remove camshaft timing oil control valve (for Exhaust Side of Bank 1).
    a. Disconnect the oil control valve connector.
    b. Remove the bolt and oil control valve.
    c. Remove the O-ring from the oil control valve.

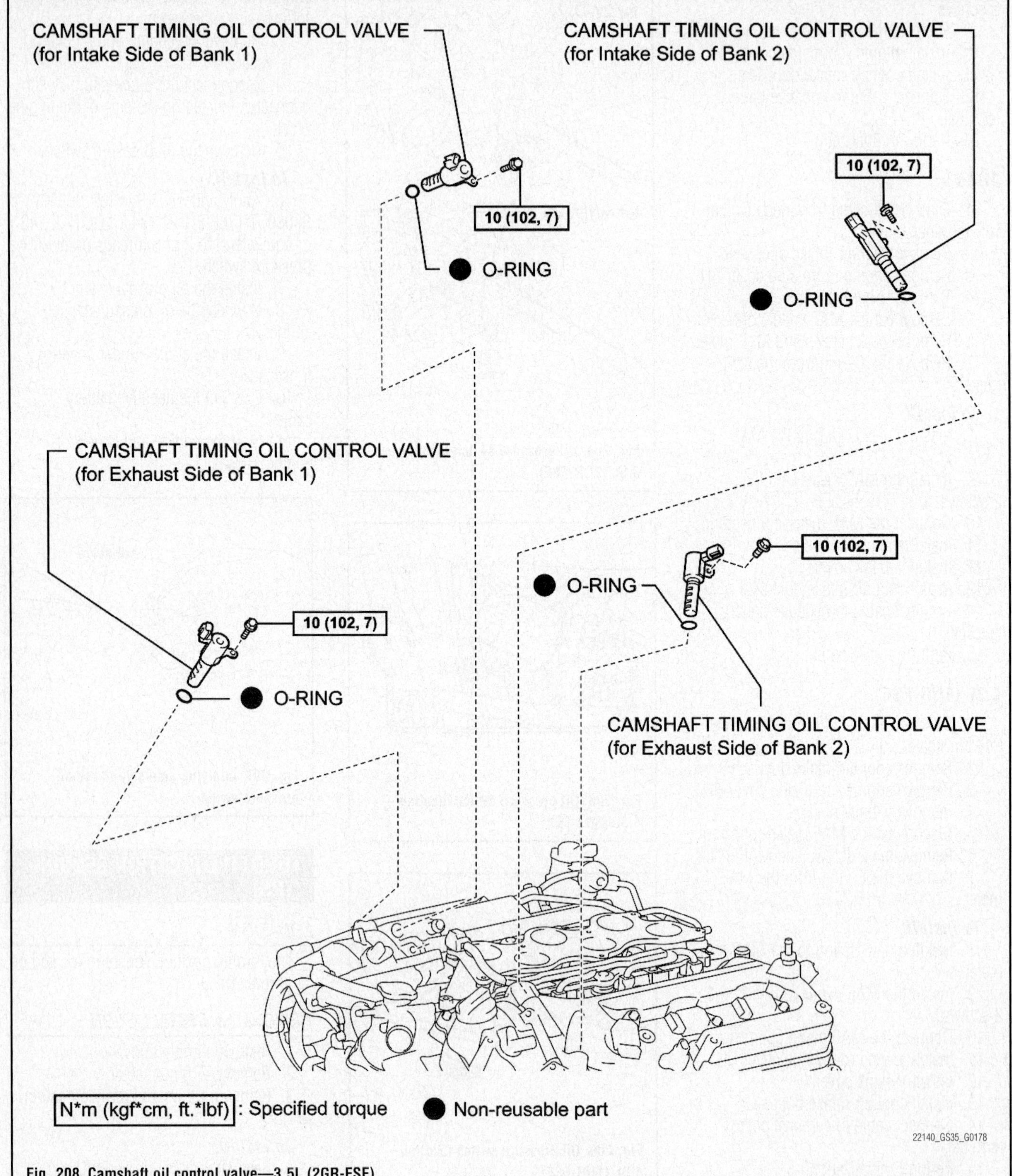

Fig. 208 Camshaft oil control valve—3.5L (2GR-FSE)

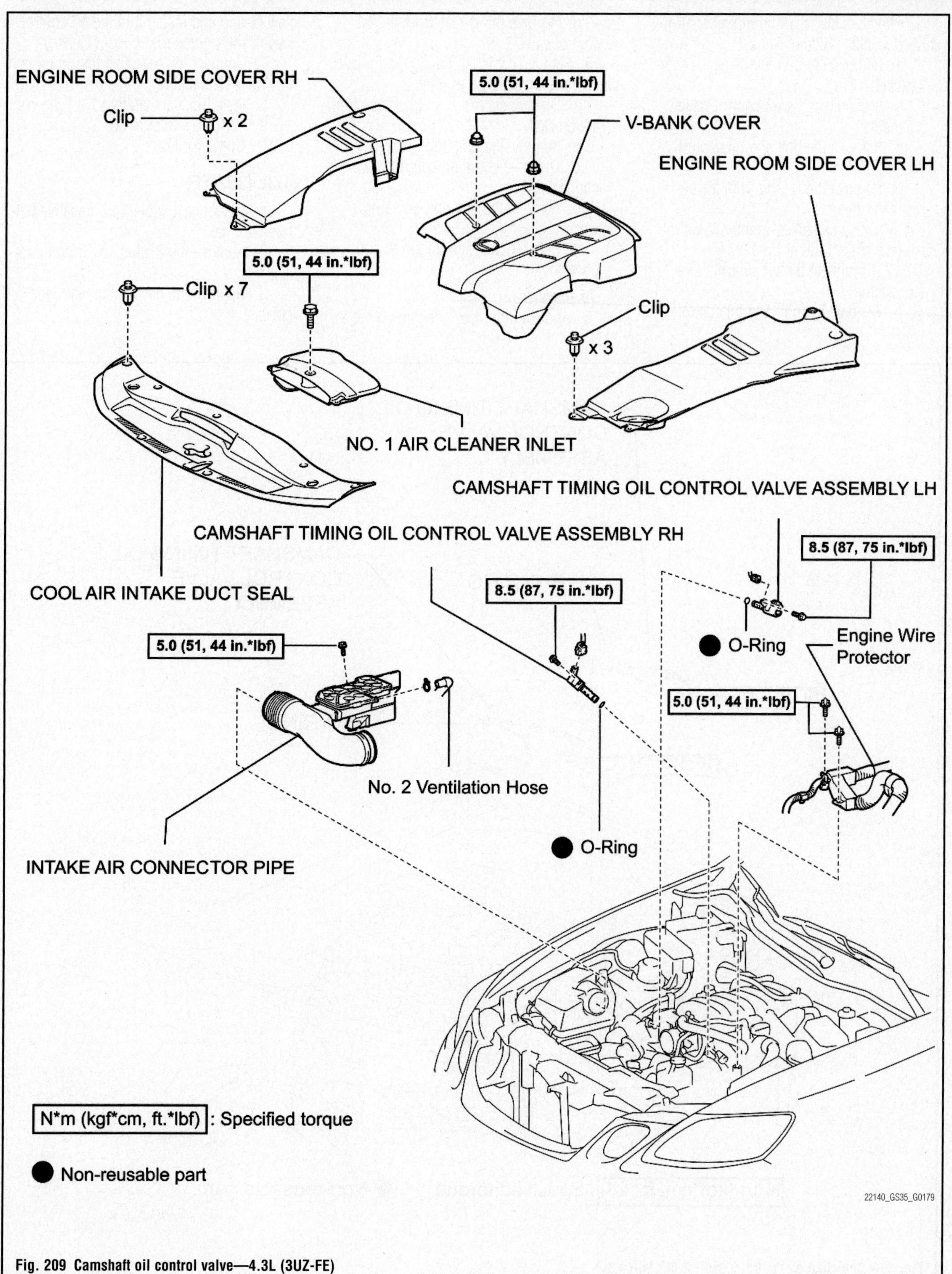

Fig. 209 Camshaft oil control valve—4.3L (3UZ-FE)

3. Remove camshaft timing oil control valve (for Intake Side of Bank 1).
   a. Remove the bolt and V-bank cover bracket.
   b. Disconnect the oil control valve connector.
   c. Remove the bolt and oil control valve.
   d. Remove the O-ring from the oil control valve.
4. Remove camshaft timing oil control valve (for Intake Side of Bank 2).
   a. Disconnect the oil control valve connector.
   b. Remove the bolt and oil control valve.
   c. Remove the O-ring from the oil control valve.
5. Remove camshaft timing oil control valve (for Exhaust Side of Bank 2).
   a. Disconnect the wiring harness from the bracket.
   b. Remove the bolt and bracket.
   c. Disconnect the oil control valve connector.
   d. Remove the bolt and oil control valve.
   e. Remove the O-ring from the oil control valve.

To install:

6. Apply a light coat of engine oil to a new O-ring and install it to the oil control valves and tighten to 7ft. lbs. (10 Nm).
7. Connect the camshaft timing oil control valve connectors.
8. Install intake manifold.
9. Perform initialization.
10. Check for leaks.

4.3L (3UZ-FE)

1. Disconnect cable from negative battery terminal.
2. Remove the 7 clips and intake duct seal.
3. Remove the nut, 2 clips and side cover.

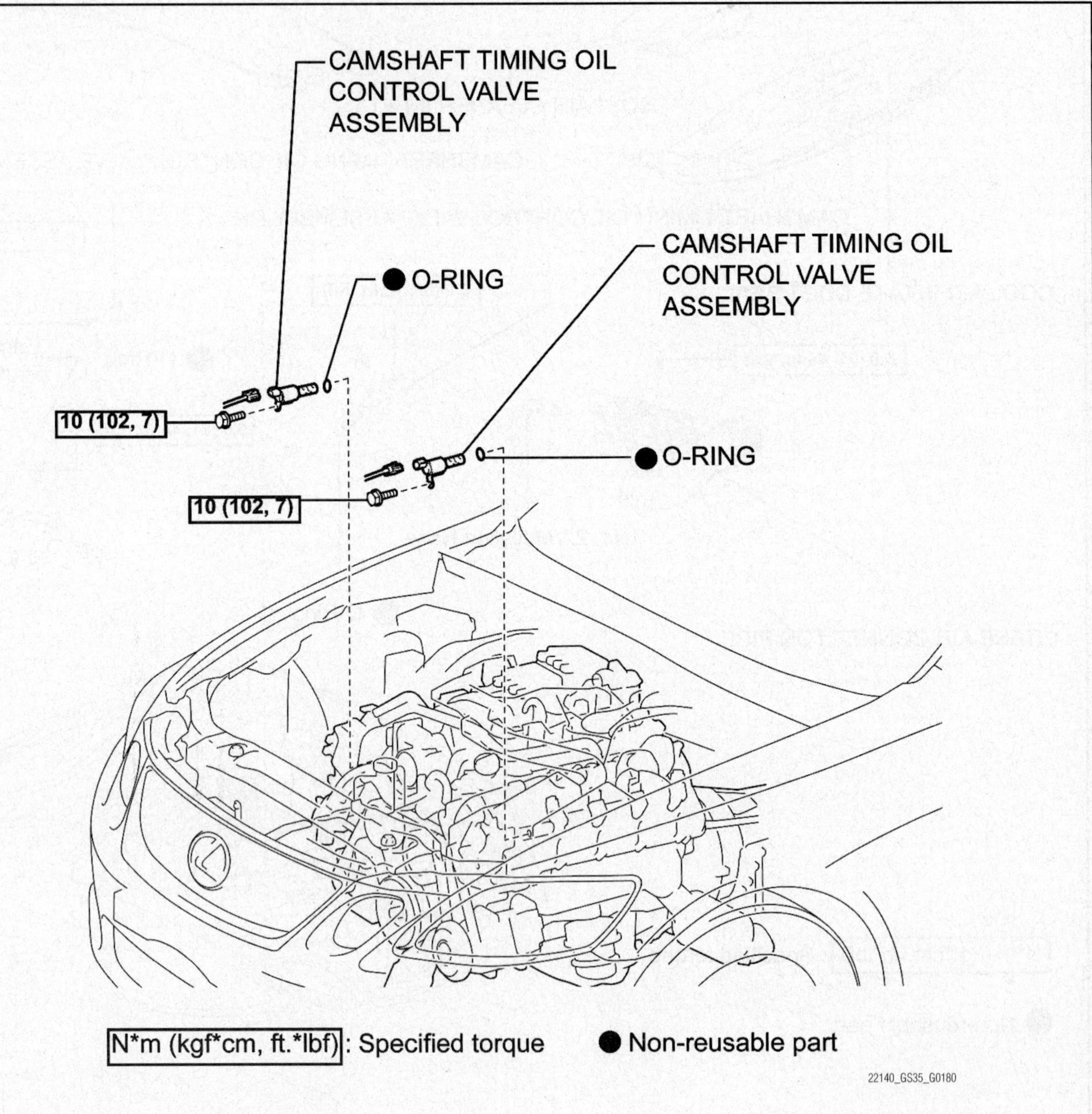

Fig. 210 Camshaft oil control valve—4.6L (1UR-FSE)

4. Remove the 3 clips and side cover.
5. Remove the 2 nuts and v-bank cover.
6. Remove the bolt and no. 1 air cleaner inlet.
7. Remove intake air connector pipe.
8. Disconnect the oil control valve connectors.
9. Remove the bolt and oil control valves.
10. Remove the O-ring from the oil control valves.

To install:

11. Install a new O-ring to the oil control valves.
12. Install the oil control valves with the bolts an tighten to 75 inch lbs. (8.5 Nm).
13. Connect the oil control valve connectors.
14. Install the intake air connector pipe with the bolt and 2 hose clamps.
15. Connect the air hose and No. 1 ventilation hose.
16. Install the air cleaner inlet and bolt.
17. Install the V-bank cover with the 2 nuts.
18. Install the side cover with the 3 clips.
19. Install the side cover with the 2 clips and nut.
20. Install the intake duct seal with the 7 clips.
21. Connect cable to negative battery terminal.
22. Perform initialization.

4.6L (1UR-FSE)

1. Disconnect cable from negative battery terminal.
2. Remove cool air intake duct seal.
3. Remove engine room side covers.
4. Remove V-bank cover sub-assembly.
5. Remove No. 1 air cleaner inlet.
6. Remove air cleaner cap sub-assembly.
7. Remove camshaft timing oil control valve assemblies.
8. Disconnect the oil control valve connectors.
9. Remove the bolt and oil control valves.
10. Remove the O-ring from the oil control valves.

To install:

11. Apply a light coat of engine oil to a new O-ring, and install it to the oil control valves.
12. Install the oil control valve with the bolts and tighten to 7 ft. lbs. (10 Nm).
13. Connect the camshaft timing oil control valve connectors.
14. Install air cleaner cap sub-assembly.
15. Install No. 1 air cleaner inlet.
16. Install V-bank cover sub-assembly.
17. Install engine room side covers.
18. Install cool air intake duct seal.
19. Connect cable to negative battery terminal.

FUEL — GASOLINE FUEL INJECTION SYSTEM

FUEL SYSTEM SERVICE PRECAUTIONS

Safety is the most important factor when performing not only fuel system maintenance but any type of maintenance. Failure to conduct maintenance and repairs in a safe manner may result in serious personal injury or death. Maintenance and testing of the vehicle's fuel system components can be accomplished safely and effectively by adhering to the following rules and guidelines.

- To avoid the possibility of fire and personal injury, always disconnect the negative battery cable unless the repair or test procedure requires that battery voltage be applied.
- Always relieve the fuel system pressure prior to disconnecting any fuel system component (injector, fuel rail, pressure regulator, etc.), fitting or fuel line connection. Exercise extreme caution whenever relieving fuel system pressure to avoid exposing skin, face and eyes to fuel spray. Please be advised that fuel under pressure may penetrate the skin or any part of the body that it contacts.
- Always place a shop towel or cloth around the fitting or connection prior to loosening to absorb any excess fuel due to spillage. Ensure that all fuel spillage (should it occur) is quickly removed from engine surfaces. Ensure that all fuel soaked cloths or towels are deposited into a suitable waste container.
- Always keep a dry chemical (Class B) fire extinguisher near the work area.
- Do not allow fuel spray or fuel vapors to come into contact with a spark or open flame.
- Always use a back-up wrench when loosening and tightening fuel line connection fittings. This will prevent unnecessary stress and torsion to fuel line piping.
- Always replace worn fuel fitting O-rings with new Do not substitute fuel hose or equivalent where fuel pipe is installed.

Before servicing the vehicle, make sure to also refer to the precautions in the beginning of this section as well.

RELIEVING FUEL SYSTEM PRESSURE

1. Before servicing the vehicle, refer to the precautions in the beginning of this section.
2. Remove the fuse for the electronic fuel pump.
3. Start the engine until the engine stalls.
4. Disconnect the negative battery terminal.
5. Place a catch-pan under the joint to be disconnected. A large quantity of fuel may be released when the joint is opened.
6. Wear eye or full-face protection.
7. Place a shop towel over the area and slowly release the joint using a wrench of the correct size.
8. Allow any fuel left in the line to bleed off slowly before fully disconnecting the joint.
9. Plug the opened lines immediately to prevent fuel spillage or the entry of dirt.
10. Dispose of the released fuel properly.
11. After connecting fuel lines, install the fuse for the fuel pump and start the engine.
12. Check for leaks and repair as needed.

FUEL FILTER

REMOVAL & INSTALLATION

Fuel filters are in the fuel tank along with the fuel pump.

FUEL RAIL & INJECTORS

REMOVAL & INSTALLATION

3.5L (2GR-FSE) Engine

See Figures 211 through 213.

1. Before servicing the vehicle, refer to the precautions in the beginning of this section.
2. Relieve fuel system pressure.
3. Remove or disconnect the following:
   - Negative battery cable. Wait at least 90 seconds before performing any other work.
   - Coolant
   - Cool air intake duct seal
   - Engine under cover
   - Right engine room side cover

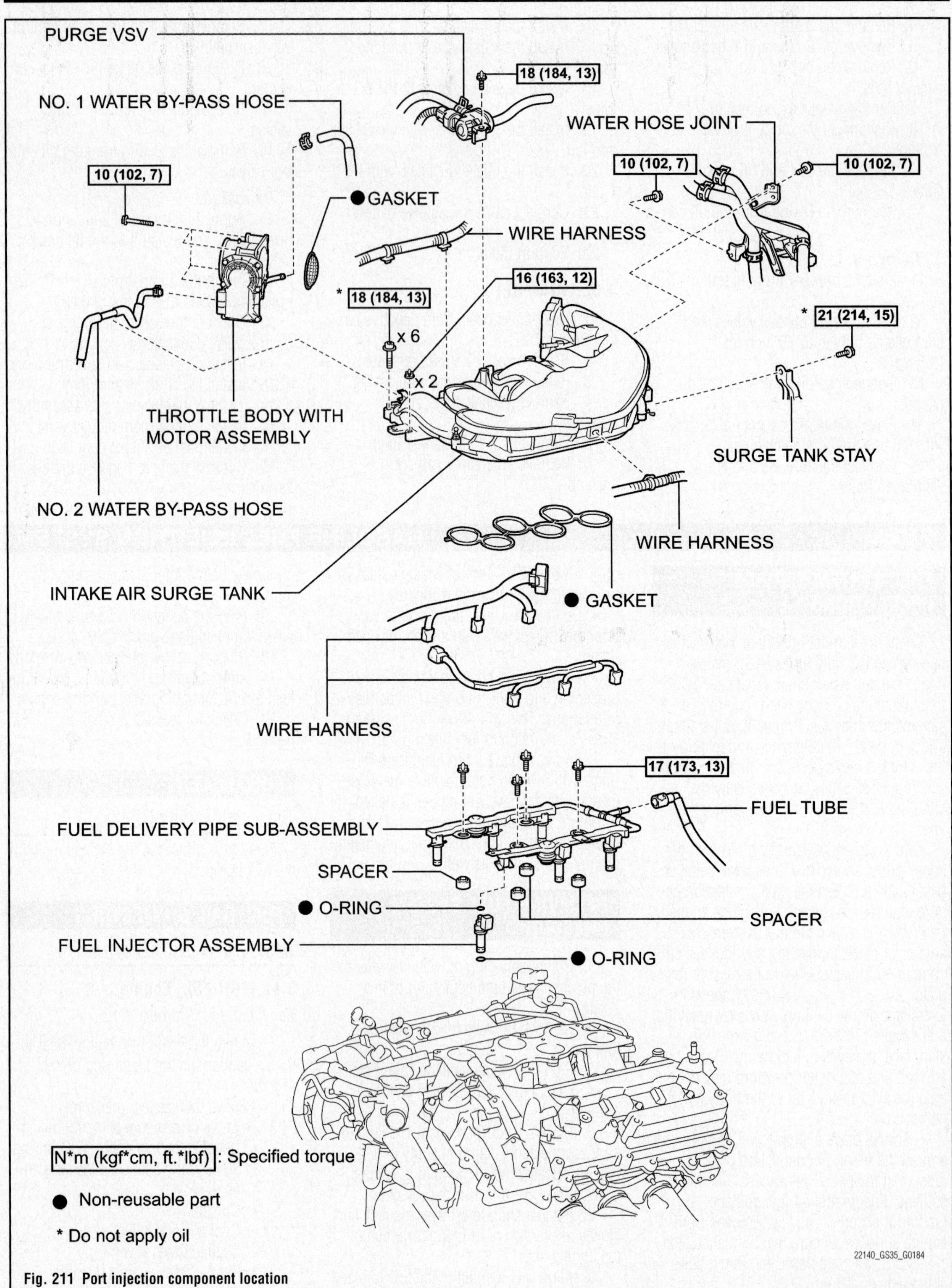

Fig. 211 Port injection component location

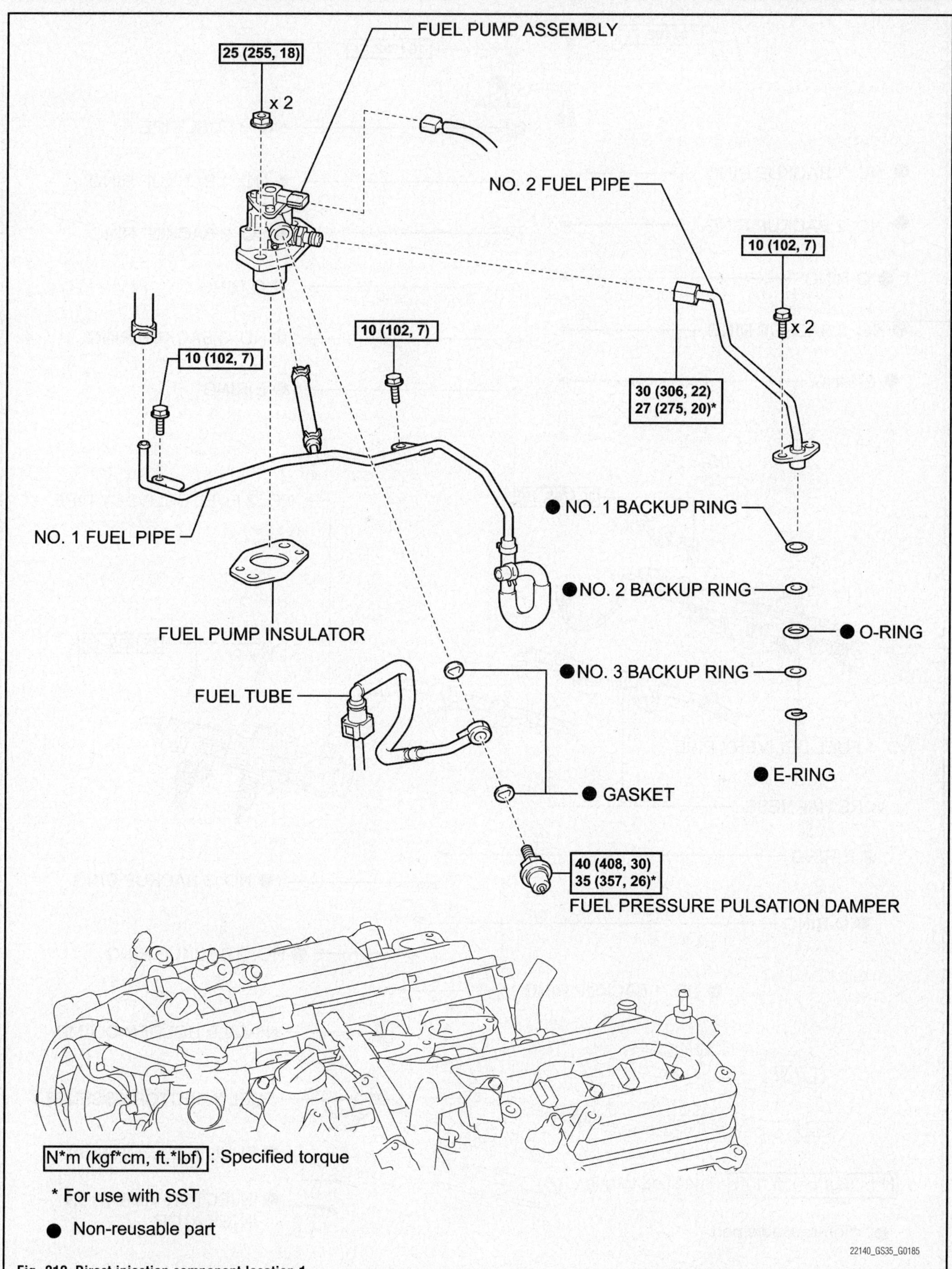

Fig. 212 Direct injection component location 1

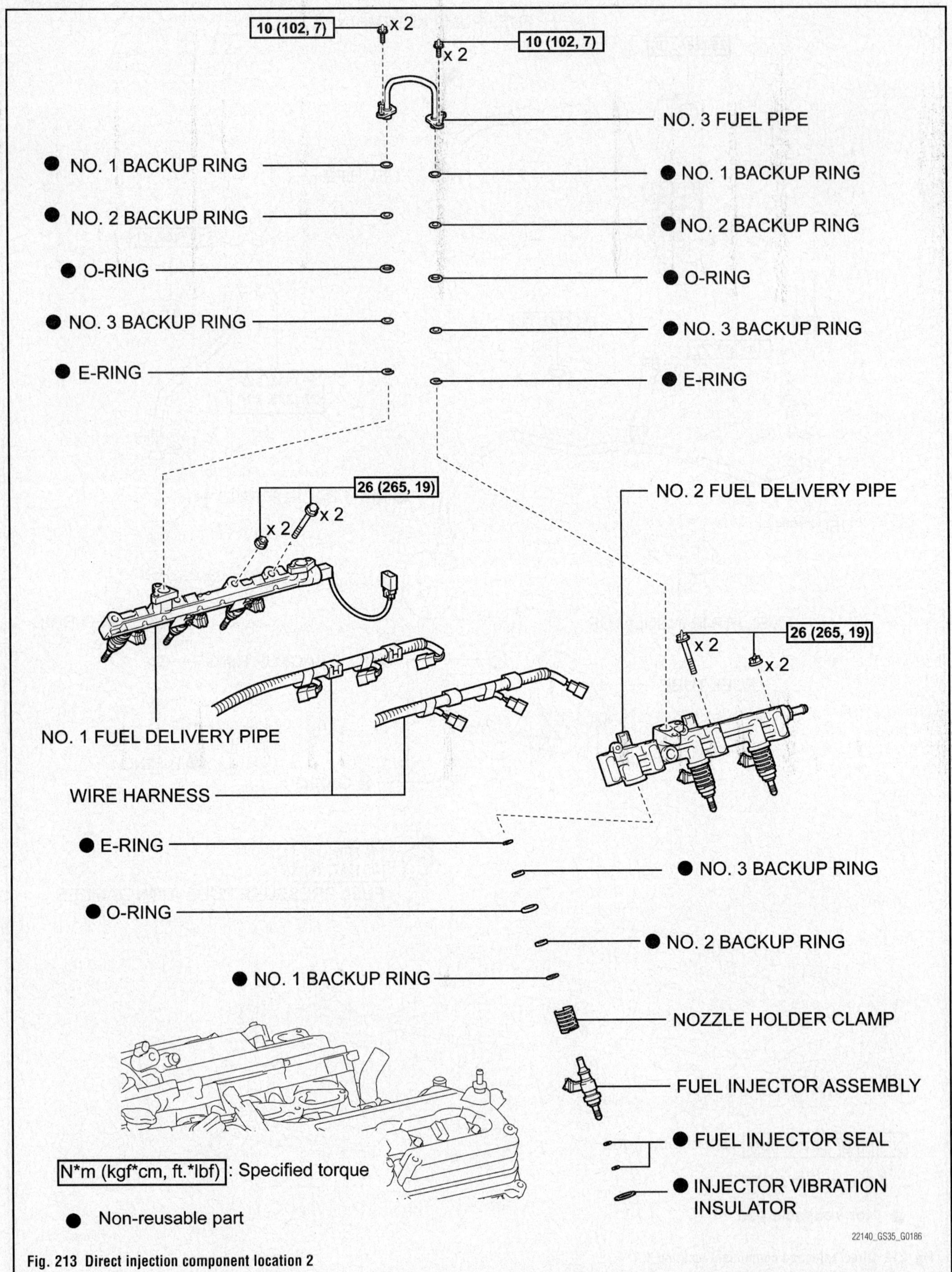

Fig. 213 Direct injection component location 2

- V-bank cover
- Intake air surge tank
- Intake manifold
- Fuel pressure pulsation damper
- No. 1 and 2 fuel pipes
- High pressure side fuel pump
- No. 2 and 3 fuel pipes
- No. 1 and 2 fuel delivery pipes
- Fuel injectors, O-rings and seals

To install:

4. Install or connect the following:
   - 2 new seals to each injector
5. Install the fuel injectors as follows:

a. Install a new O-ring, new backup rings (No. 1, No. 2, No. 3) and new E-ring to the fuel injector.

WARNING

Check that there is no foreign matter or damaged areas in the injector's O-ring groove. Check that the installation direction of the No. 1 and No. 2 backup ring are correct. Make sure the backup rings and O-ring are installed in the correct order. Check that the alignment openings of the backup rings are not overlapped or stretched. After installing the O-ring, check that it is not contaminated with foreign matter and is not damaged.

b. Install the injector nozzle holder clamp.

c. Apply gasoline to the O-ring. Install the nozzle holder clamp by aligning the protruding part of the clamp to the notch of the delivery pipe.

WARNING

Make sure there is no gap between the delivery pipe and clamp. Check that there is no foreign matter or damage in the injector insertion hole of the delivery pipe. Insert the injector straight into the delivery pipe without tilting it.

6. Install the No. 1 and 2 fuel delivery pipes as follows:

a. Install a new injector vibration insulator to the cylinder head.

b. Apply lubricant to the installation hole of the injector.

c. Insert the stud bolt into the fuel delivery pipe until the screw threads protrude enough so that a nut can be attached.

➡If an injector is dropped, replace it with a new one. Check that there is no foreign matter or damage in the injector insertion hole of the delivery pipe. Be extremely careful not to touch or strike the tips of the injectors. When inserting the fuel delivery pipe, push it in evenly without tilting it.

d. Install the fuel delivery pipe by uniformly tightening the 2 bolts and 2 nuts in several passes to 15 ft. lbs. (21 Nm).

e. Connect the 3 connectors and 2 clamps.

- No. 2 and 3 fuel pipes and all new rings and seals. Torque No. 3 fuel pipe fastener to 7 ft. lbs. (10 Nm).
- High pressure side fuel pump
- No. 1 and 2 fuel pipes. Torque fasteners to 7 ft. lbs. (10 Nm).
- Fuel pressure pulsation damper and new gasket. Torque to 28 ft. lbs. (40 Nm).
- Intake manifold
- Intake air surge tank
- Fuel main tube
- Ventilation hose
- Union to check valve hose
- Engine rear cover
- Inlet and outlet heater water hoses
- Water by-pass hose
- Cold start injector. Torque the bolts to 7 ft. lbs. (10 Nm).
- Air cleaner cap with air cleaner hose
- No. 2 ventilation hose
- V-bank cover
- Right engine room side cover
- Cool air intake duct seal
- Engine under cover
- Negative battery cable

7. Refill the cooling system. Start the engine and check for coolant and fuel leaks and proper operation.
8. Check the function of the throttle body unit
9. System initialization.

4.3L (3UZ-FE) Engines

1. Before servicing the vehicle, refer to the precautions in the beginning of this section.
2. Remove or disconnect the following:
   - V-bank cover
   - Intake air connector
   - Accelerator cable
   - Fuel pressure pulsation dampers.
   - VVT sensor connectors
   - Vacuum Switching Valve (VSV) for Evaporative Emissions (EVAP)
   - 2 nuts and accelerator cable bracket
   - 3 V-bank cover brackets
   - VSV connector for Acoustic Control Induction System (ACIS) from the No. 1 V-bank cover bracket
   - 4 bolts and 3 V-bank cover brackets
   - Engine wire from the delivery pipe
   - 2 wire clamps from the wire clamp bracket on the right-hand delivery pipe
   - 8 injector connectors
   - 4 nuts holding the delivery pipe to the intake manifold
   - 2 delivery pipes and 8 injector assemblies and 4 spacers
   - 2 O-rings, grommet and insulator from each injector

To install:

3. Install or connect the following:
   - A new insulator and grommet to each injector
   - A light coat of gasoline to new O-rings and install them to each injector
   - A light coat of gasoline on the place where a delivery pipe touches an O-ring of the injector
   - Injector, while turning the clockwise and counterclockwise, into the delivery pipe

➡Position the injector connector outward.

- The 4 spacers in position on the intake manifold
- A light coat of gasoline on the place where an intake manifold touches an O-ring
- The delivery pipes in position on the intake manifold
- Temporarily, the 3 bolts holding the delivery pipe to the intake manifold

➡Check that the injectors rotate smoothly. If the injectors do not rotate smoothly, the probable cause is incorrect installation of the O-rings. Replace the O-rings.

4. Tighten the 3 bolts holding the delivery pipe to the intake manifold. Tighten the bolts to 15 ft. lbs. (21 Nm).
5. Install or connect the following:
   - Engine wire protector with the 3 nuts
   - Injector connectors
   - Remaining components

4.6L (1UR-FSE)

See Figures 214 through 218.

1. Before servicing the vehicle, refer to the precautions in the beginning of this section.
2. Discharge fuel system pressure.
3. Disconnect cable from negative battery terminal.

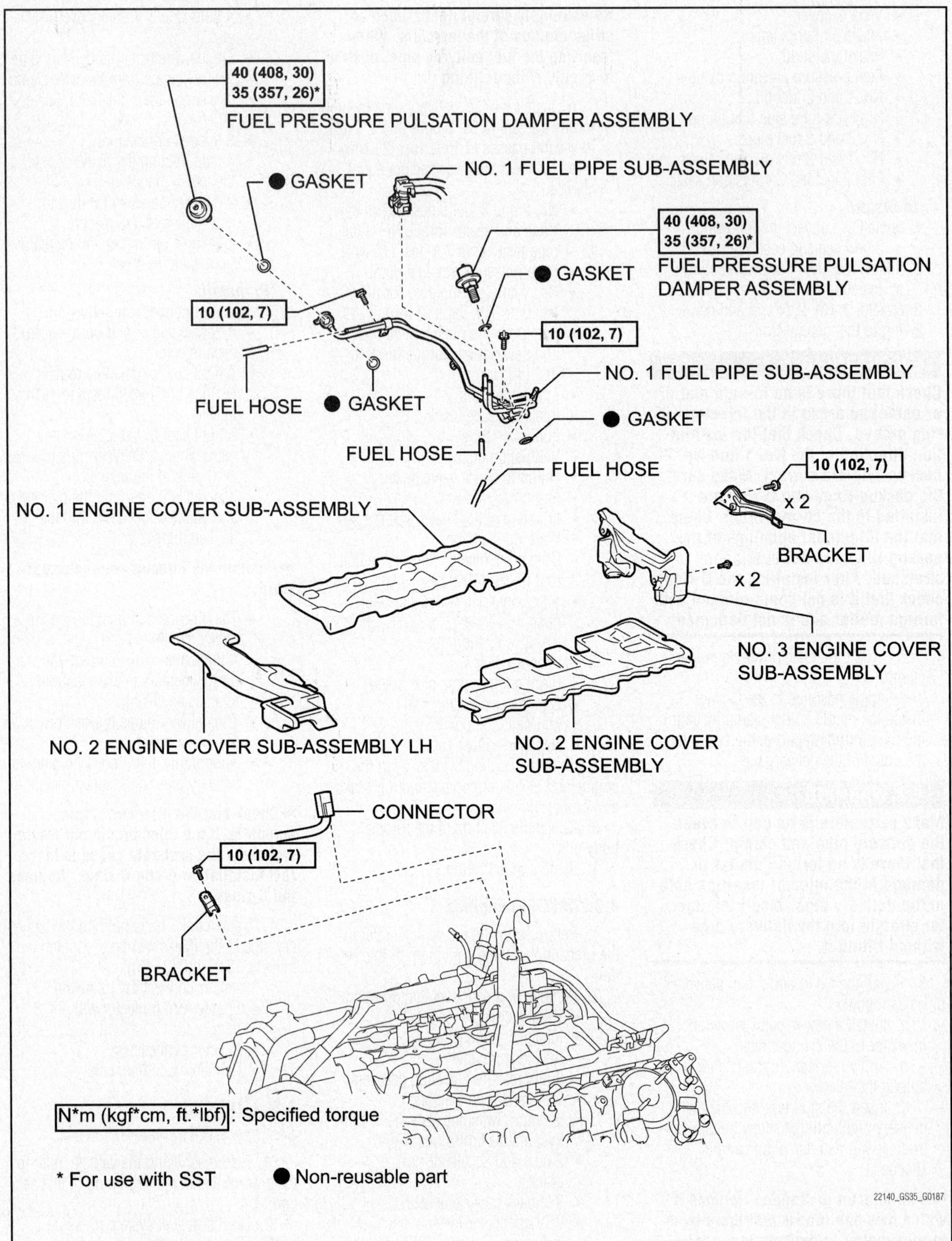

Fig. 214 Port injection component location 1

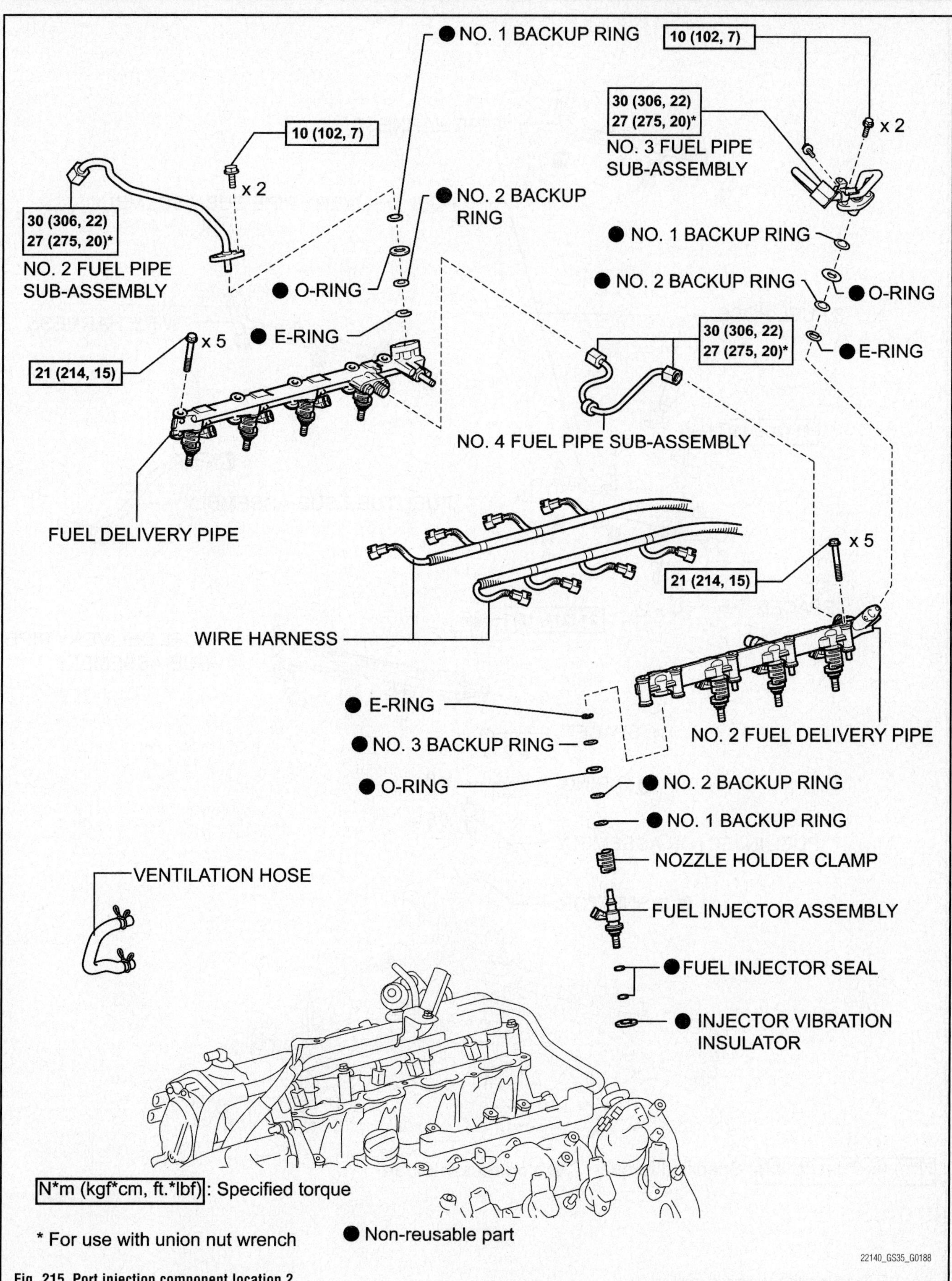

Fig. 215 Port injection component location 2

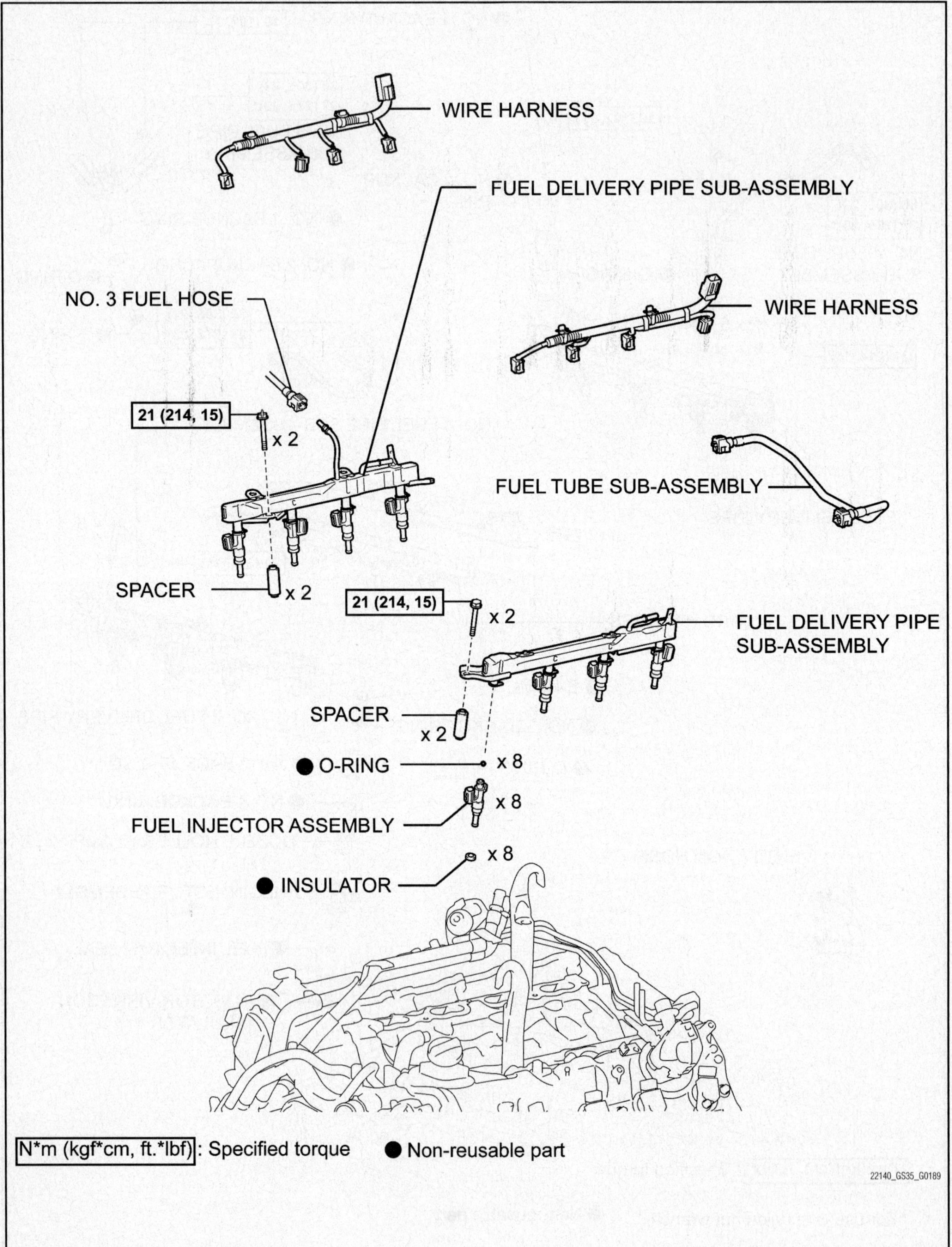

Fig. 216 Direct injection component location

4. Remove the intake manifold.
5. Remove the fuel pressure pulsation damper assembly and fuel pipe sub-assembly.
6. Remove additional fuel pipe sub-assemblies.
7. Remove the engine cover sub-assembly.
8. Remove the ventilation hose from the ventilation valve.
9. Disconnect the fuel pressure sensor connector.
10. Disconnect the injector connectors.
11. Remove the bolts and delivery pipes from the cylinder heads.

➡Make sure that the fuel delivery pipe is disconnected from the delivery pipe. Be extremely careful not to touch or strike the tips of the injectors. Pull and remove the fuel delivery pipe in a straight line without tilting it.

12. Remove the injector vibration insulators from the cylinder heads.

To install:

13. Install a new injector seals.
14. Install new injector vibration insulators to the cylinder heads.
15. Install the delivery pipe spacers for direct injection.
16. Apply lubricant to the installation injector seal and holes of the injectors.
17. Install the delivery pipe (with injector) to the intake manifold and tighten to 15 ft. lbs. (21 Nm).

➡If an injector is dropped or the tips of the injectors are struck, replace it with a new one. Check that there is no foreign matter or damage to the injector insertion hole of the cylinder head. When inserting the fuel delivery pipe, push it in evenly without tilting it.

18. Connect the connectors.
19. Connect the fuel pressure sensor connector.

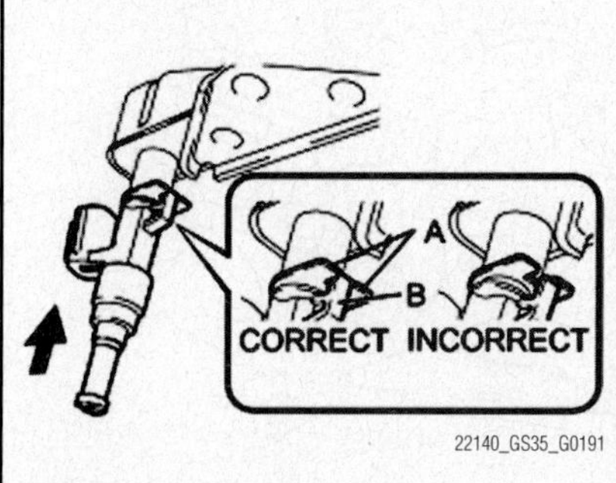

Fig. 217 Direct injection fuel injector installation

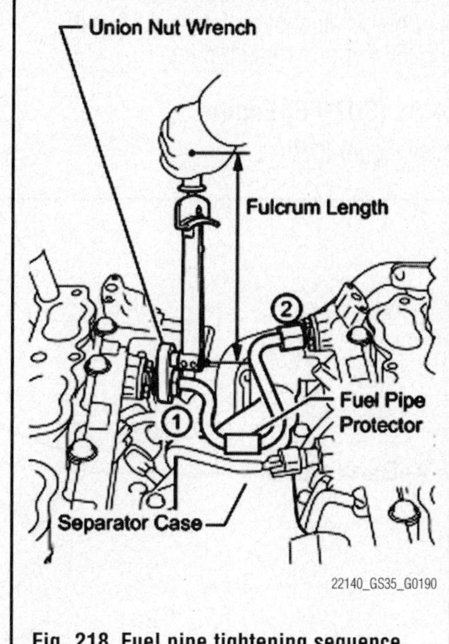

Fig. 218 Fuel pipe tightening sequence

20. Connect the fuel hoses.
21. Install the ventilation hose to the ventilation valve.
22. Temporarily install the fuel pipe.
23. Using a 19 mm union nut wrench, tighten the fuel pipe to 22 ft. lbs. (30 Nm) in the order shown.

➡After installing the fuel pipe, check that the fuel pipe protector contacts with the separator case.

24. Install the engine cover sub-assemblies.
25. Install fuel pipe sub-assemblies.
26. Install the fuel pressure pulsation damper assembly and No. 1 fuel pipe sub-assembly
27. Install the intake manifold.
28. Connect cable to negative battery terminal.
29. Add engine coolant.
30. Inspect for leaks.

FUEL PUMP

REMOVAL & INSTALLATION

1. Before servicing the vehicle, refer to the precautions in the beginning of this section.
2. Relieve the fuel system pressure.
3. Remove or disconnect the following:
   - Negative battery cable. Wait at least 90 seconds before performing any other work.
   - Rear seat bottom
   - Partition cover
   - Floor service hole cover
   - Fuel pump electrical connector
   - Fuel main tube and fuel pump tube from the top of the fuel pump
   - Mounting bolts, or retaining ring using SST 09808-14020
   - Pump, bracket and set plate as an assembly

To install:

4. Install or connect the following:
   - A new gasket on the set plate
   - Fuel pump and bracket assembly. Torque the mounting bolts to 31 inch lbs. (3.5 Nm). If equipped with a retaining ring, use SST 09808-14020 to tighten the retainer 2 full turns so that the mark on the ring lines up within the 2 marks indicated next to the ring on the fuel tank.
   - Fuel main tube and fuel pump tube to the top of the fuel pump
   - Fuel pump electrical connector
   - Floor service hole cover
   - Partition cover
   - Rear seat bottom
   - Negative battery cable
5. Start the engine; check the fuel system for leaks

THROTTLE BODY

REMOVAL & INSTALLATION

3.5L (2GR-FSE)

1. Disconnect cable from negative battery terminal.
2. Remove cool air intake duct seal.
3. Drain engine coolant.
4. Remove engine room side cover.
5. Remove V-bank cover.
6. Remove air cleaner.
7. Disconnect the ventilation hose from the cylinder head.
8. Disconnect the throttle motor connector.
9. Remove the 4 bolts and disconnect the throttle body from the intake air surge tank.
10. Disconnect the 2 water by-pass hoses from the throttle body.
11. Remove the throttle body and gasket.

To install:

12. Install a new gasket to the intake air surge tank.
13. Connect the 2 water by-pass hoses to the throttle body.
14. Install the throttle body with the 4 bolts an tighten to 7 ft. lbs. (10 Nm).
15. Connect the throttle motor connector.
16. Install the air cleaner.
17. Connect the ventilation hose to the cylinder head cover with the clamp.

18. Connect cable to negative battery terminal.
19. Add engine coolant.
20. Inspect for coolant leak.
21. Inspect function of throttle body.
22. Install V-bank cover.
23. Install engine room side cover.
24. Install cool air intake duct seal.
25. Perform initialization.

4.3L (3UZ-FE) Engine

See Figure 219.

1. Before servicing the vehicle, refer to the precautions in the beginning of this section.
2. Disconnect the negative battery cable. Wait at least 90 seconds before performing any other work.

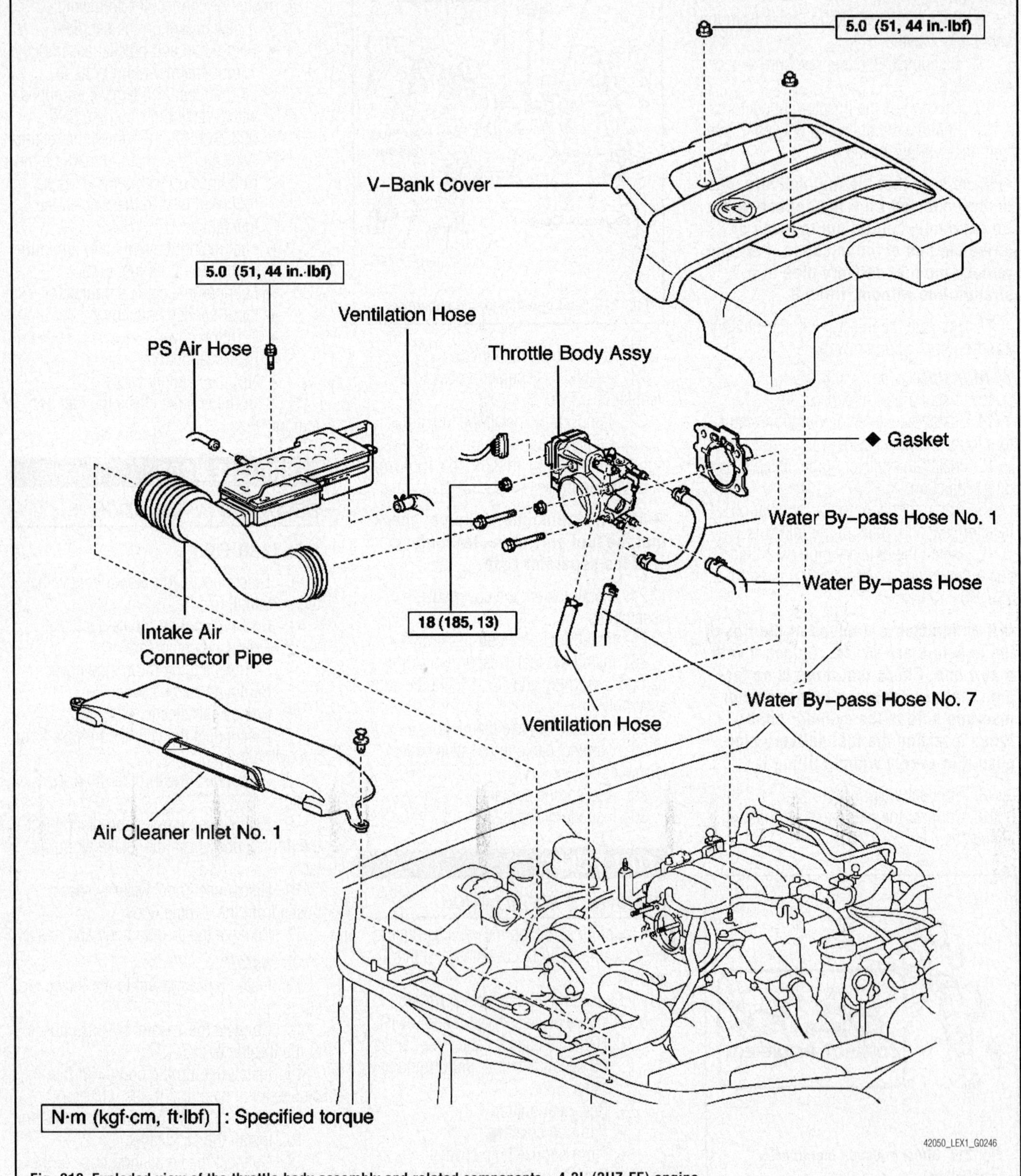

Fig. 219 Exploded view of the throttle body assembly and related components—4.3L (3UZ-FE) engine

3. Drain the engine coolant.
4. Remove the 2 nuts and V-bank cover.
5. Remove the air cleaner inlet No. 1.
6. Remove the intake air connector pipe as follows:
a. Disconnect the air hose and ventilation hose.
b. Remove the bolt.
c. Loosen the 2 hose clamps and remove the intake air connector pipe.
7. Remove the throttle body as follows:
a. Disconnect the water by-pass hose and water by-pass hose No. 7.
b. Disconnect the connector.
c. Remove the 2 bolts, 2 nuts, throttle body and gasket.

To install:

8. Install the throttle body as follows:
a. Install a new gasket and the throttle body with the 2 bolts and 2 nuts. Tighten the nuts and bolts to 13 ft. lbs. (18 Nm).
b. Connect the water by-pass hose and water by-pass hose No. 7.
c. Connect the connector.
9. Install intake air connector pipe as follows:
a. Install the intake air connector pipe with the bolt and 2 hose clamps. Tighten the hose clamps to 35 inch lbs. (4 Nm), and the bolt to 44 inch lbs. (5 Nm).
b. Connect the air hose and ventilation hose.
10. Install air cleaner inlet No. 1.
11. Install the V-bank cover and tighten the nuts to 44 inch lbs. (5 Nm).
12. Refill the engine cooling system.
13. Connect the negative battery cable.
14. Check the cooling system for leaks.
15. Inspect the throttle control motor for operating sound as follows:
a. Turn the ignition switch ON.
b. When turning the accelerator pedal position sensor lever, check the running sound of the motor. The motor should be running smoothly without friction sounds. If operation is not as specified, check the throttle control motor, wiring and ECM.
16. Inspect the throttle position sensor as follows:
a. Inspect the throttle position sensor.
b. Connect the hand-held tester (with CAN VIM) to the DLC3.
c. Turn the ignition switch ON.
d. Check that the MIL does not light up.
e. Check that, under the CURRENT DATA, THROTTLE POS (throttle valve opening percentage) is within the standard value below.
- Standard throttle valve opening percentage: 60% or more

f. If operation is not as specified, check the throttle position sensor, wiring and ECM.

4.6L (1UR-FSE) Engine

1. Before servicing the vehicle, refer to the precautions in the beginning of this section.
2. Disconnect the negative battery cable. Wait at least 90 seconds before performing any other work.
3. Remove the V-bank cover.
4. Remove engine under covers
5. Drain the engine coolant.
6. Remove the air cleaner inlet.
7. Remove the intake air connector.
8. Disconnect the throttle motor connector.
9. Remove the 4 bolts and throttle body
10. Slide the clamps, and disconnect the No. 4 and No. 5 water by-pass hoses from the throttle body

To install:

11. Connect the No. 4 water by-pass hose and No. 5 water by-pass hose to the throttle body
12. Install the throttle body with the 4 bolts and tighten to 7 ft. lbs. (10 Nm)
13. Connect the throttle motor connector.
14. Install intake air connector.
15. Install air cleaner inlet.
16. Refill the engine cooling system.
17. Connect the negative battery cable.
18. Check the cooling system for leaks.
19. Install engine under covers
20. Install V-bank cover.

HEATING & AIR CONDITIONING SYSTEM

BLOWER MOTOR

REMOVAL & INSTALLATION

See Figures 220 and 221.

1. Before servicing the vehicle, refer to the precautions in the beginning of this section.
2. Disconnect the negative battery cable. Wait 90 seconds before doing any further work while the airbag system de-energizes.
3. Remove the instrument panel assembly.
4. Remove the air conditioner unit assembly. See HEATER CORE Removal & Installation procedure in this section.
5. Remove the blower assembly as follows:
a. Disconnect the connector.
b. Remove the screw and nut.
c. Release the 2 claws and remove the blower assembly.
6. Remove the damper servo sub-assembly as follows:

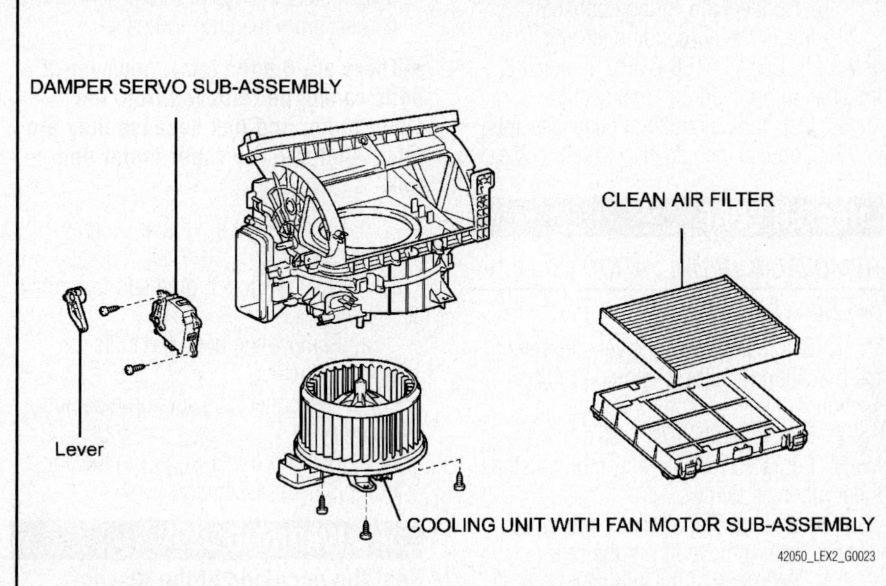

Fig. 220 Exploded view of the blower assembly components

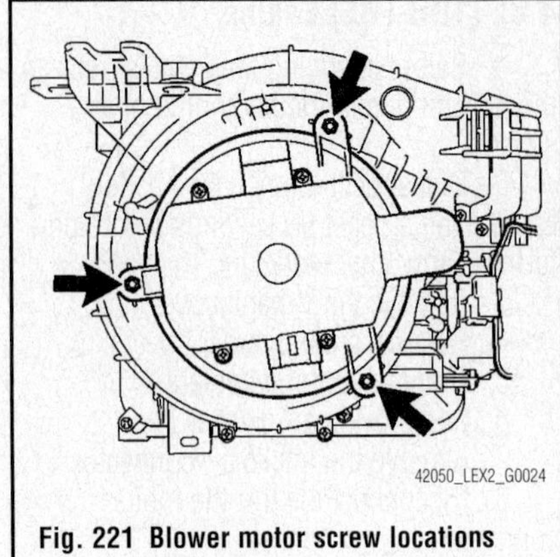

Fig. 221 Blower motor screw locations

a. Detach the claw and remove the lever.

b. Remove the 2 screws and damper servo.

7. Remove the 3 screws and the blower motor.

To install:

8. Install the blower motor to the blower unit assembly and tighten the 3 screws.

9. Install the damper servo sub-assembly as follows:

a. Install the damper servo with the 2 screws.

b. Attach the claw and install the lever.

10. Install the blower assembly as follows:

a. Install the blower assembly with the 2 claws, screw and nut. Tighten the screw to 27 inch lbs. (3 Nm); tighten the nut to 7 ft. lbs. (10 Nm).

b. Connect the wiring connector.

11. Install the air conditioner unit assembly. See HEATER CORE Removal & Installation procedure in this section.

12. Install the instrument panel assembly.

13. Connect the negative battery cable.

HEATER CORE

REMOVAL & INSTALLATION

See Figures 222 and 223.

1. Before servicing the vehicle, refer to the precautions in the beginning of this section.

2. Discharge the A/C system.

3. Set radio receiver assembly to shipment mode as follows:

a. Be sure that all discs and tapes have been removed from the unit.

b. Be sure that the engine switch off.

c. While simultaneously pressing the "SEEK UP" and "DISC" switches, turn the engine switch on (ACC).

➡The CD loading door indicator light blinks during mode setting and it remains lit after the setting is completed.

d. Turn the engine switch off.

4. Disconnect the negative battery cable. Wait 90 seconds before doing any further work while the airbag system de-energizes.

5. Drain the cooling system into a clean container for reuse.

6. Align the front wheels facing straight ahead.

7. Remove or disconnect the following:

- Cool air intake duct seal
- Left and right engine room side covers
- Left and right front pillar to front side seals, using a clip remover to detach the 3 claws
- Left and right nut, windshield wiper arms and blades
- Left and right front fender to cowl side seals by moving the component toward the center of the vehicle to detach the 2 claws

8. Remove cowl top ventilator louver assembly as follows:

a. Remove the 2 clips and detach the 5 claws.

b. Pull the ventilator louver in the direction indicated by the arrow in the illustration to detach the 10 claws and remove the ventilator louver.

9. Remove the windshield wiper motor and link assembly as follows:

a. Disconnect the connector. Then detach the 2 clamps and remove the wire harness from the cowl top panel.

➡There are 6 bolts total, however, 2 bolts cannot be removed from the wiper motor and link because they are integrated into the wiper motor and link.

b. Remove the 4 bolts and wiper motor and link.

10. Separate suction pipe sub-assembly as follows:

a. Remove the bolt, and slide the hook connector.

b. Disconnect the suction pipe sub-assembly.

c. Remove the O-ring from the suction pipe sub-assembly.

⁂ WARNING

Seal the openings of the disconnected parts using vinyl tape to prevent moisture and foreign matter from entering.

11. Separate liquid tube sub-assembly as follows:

a. Disconnect the liquid tube sub-assembly.

b. Remove the O-ring from the liquid tube subassembly.

- Inlet and outlet heater water outlet hoses
- Instrument panel assembly.
- A/C blower unit assembly
- Mounting screw and No. 2 air duct
- Upper and lower foot ducts
- Connector, mounting screw and air conditioning amplifier assembly

12. Remove the heater core from the A/C-blower assembly unit as follows:

a. Disconnect the wiring connector, remove the 3 screws and right side air outlet control servo motor.

b. Disengage 3 clamps, connector and wire harness to right side air mix control servo motor.

c. Remove the 2 screws and heater piping cover.

d. Remove the 3 screws and right side air mix control servo motor.

e. Move the A/C wiring harness out of the way of the heater radiator unit.

f. Remove the heater radiator unit sub-assembly (heater core)

To install:

13. Install or connect the following:

- Heater core to the A/C blower housing
- A/C wiring harness back into position
- Right side air mix control servo motor
- Heater piping cover
- Right side air outlet control servo motor
- Air conditioning amplifier assembly
- Upper and lower foot ducts
- No. 2 air duct
- A/C blower unit assembly. Torque the retaining nut to 7 ft. lbs. (10 Nm).
- Instrument panel assembly.
- Inlet and outlet heater water outlet hoses

14. Install the liquid tube sub-assembly as follows:

a. Remove the vinyl tape attached to the tube.

b. Sufficiently apply compressor oil to a new O-ring and the fitting surface of the liquid tube.

c. Install the O-ring on the liquid tube.

d. Install the liquid tube to the fitting hole.

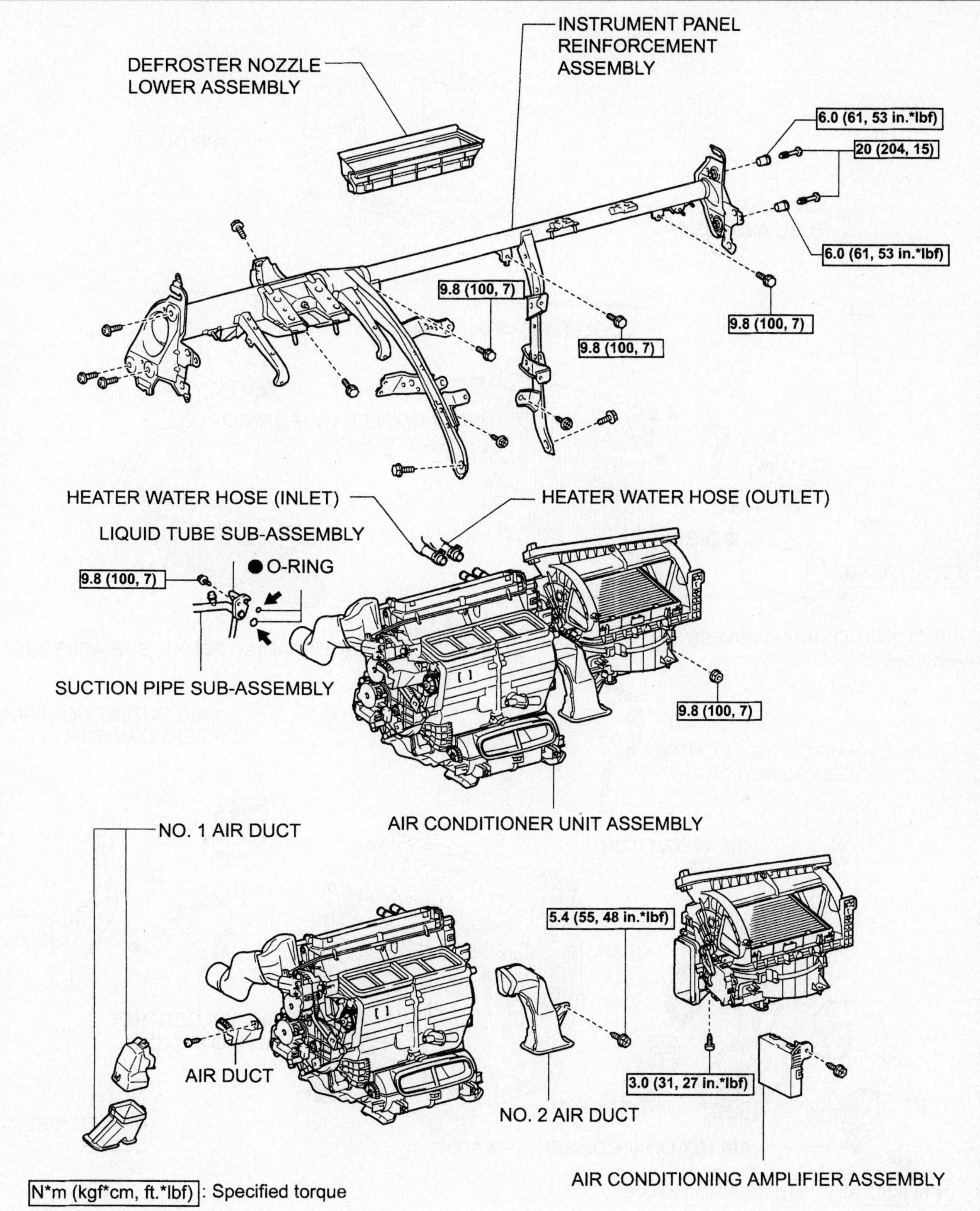

Fig. 222 Exploded view of the A/C unit assembly, instrument panel reinforcement and related components

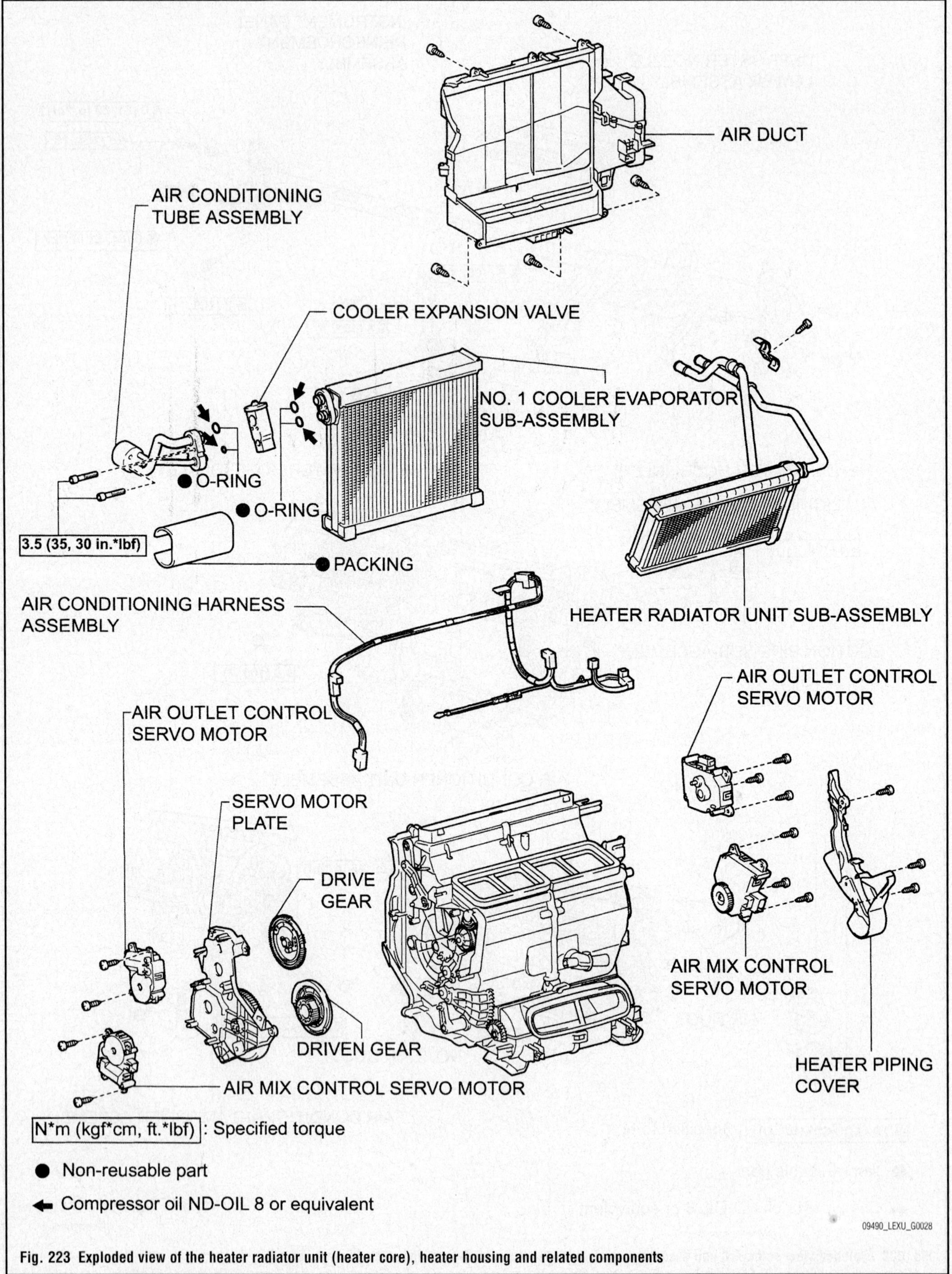

Fig. 223 Exploded view of the heater radiator unit (heater core), heater housing and related components

15. Install the suction pipe sub-assembly as follows:
  a. Remove the vinyl tape attached to the pipe.
  b. Sufficiently apply compressor oil to a new O-ring and the fitting surface of the suction pipe.
  c. Install the O-ring on the suction pipe.
  d. Move the hook connector in a counterclockwise direction.
  e. Insert the pipe joints into the fitting holes securely and tighten the bolt to 7 ft. lbs. (10 Nm).
   - Wiper motor and link assembly
   - Cowl top ventilator louver assembly
   - Left and right front fender to cowl side seals
   - Windshield wiper arms and blades
   - Left and right front pillar to front side seals
   - Left and right engine room side covers
   - Cool air intake duct seal

16. Perform system initialization procedure.
17. Refill the cooling system.
18. Connect the negative battery cable.
19. Evacuate, charge and leak test the air conditioning system refrigerant.
20. Operate the engine to normal operating temperatures; then, check the climate control operation and check for leaks.

STEERING

POWER STEERING GEAR

REMOVAL & INSTALLATION

2WD Models

See Figures 224 and 225.

1. Place front wheels facing straight ahead.
2. Disconnect cable from negative battery terminal.
3. Remove front wheels.
4. Remove engine under covers and protectors.
5. Secure the steering wheel with the seat belt in order to prevent rotation.
6. Loosen bolt (A) and remove bolt (B), then slide the steering intermediate shaft assembly No. 2.

➡Do not remove bolt (A). Do not disconnect the steering intermediate shaft assembly No. 2 from the power steering link assembly.

7. Put matchmarks on the steering intermediate shaft assembly No. 2 and the power steering link assembly.
8. Separate the intermediate shaft assembly No. 2 from the power steering link assembly.
9. Separate the tie rod ends from the steering knuckles.
10. Remove the 2 clamps to disconnect the wire harness from the bracket.
11. Disconnect 2 connectors (A) and (B) from the power steering link assembly.
12. Release the lock of connector (C) and disconnect connector (C) from the power steering link assembly.
13. Remove the 2 bolts, 2 washers, 2 nuts, and the power steering link assembly from the front suspension cross member.

To install:

14. Install the power steering link assembly with the 2 bolts, 2 washers and 2 nuts and tighten to 87 ft. lbs. (118 Nm).
15. Connect wire harness connector (C) to the power steering link assembly and securely lock the connector.
16. Connect 2 wire harness connectors (A) and (B) to the power steering link assembly.
17. Install the 2 wire harness clamps to the power steering link assembly.
18. Connect the tie rod end LH to the steering knuckle with the nut and tighten to 50 ft. lbs. (65 Nm).
19. Install a new clip.
20. Align the matchmarks on the intermediate shaft assembly No. 2 and the power steering link assembly.
21. Install bolt (A) and tighten the 2 bolts and tighten to 26 ft. lbs. (35 Nm).
22. Install engine under covers and protectors.
23. Install front wheels.
24. Connect cable from negative battery terminal.
25. Inspect and adjust front wheel alignment.
26. Initialize rotation angle sensor and calibrate torque sensor zero point.
27. Perform variable gear ratio steering system calibration (for 1UR-FSE).
28. Perform initialization.

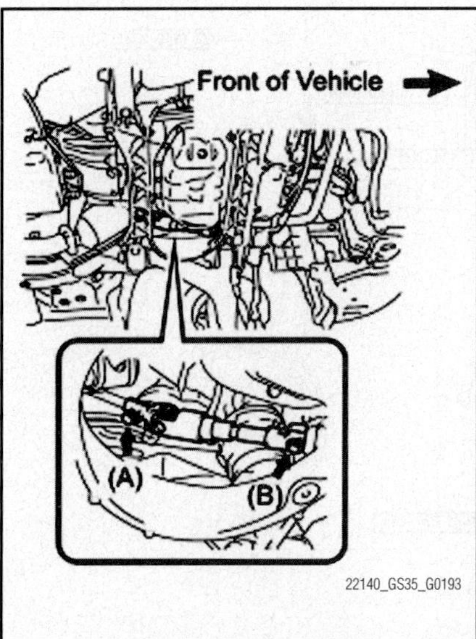

Fig. 224 Intermediate shaft removal

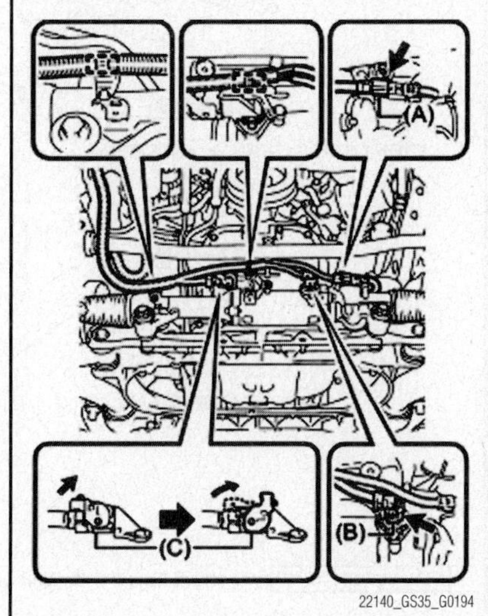

Fig. 225 Wiring harness connections

AWD Models

See Figure 226.

1. Place front wheels facing straight ahead.
2. Disconnect cable from negative battery terminal.
3. Remove front wheels.
4. Remove engine under covers and protectors.
5. Secure the steering wheel with the seat belt in order to prevent rotation.
6. Loosen bolt (A) and remove bolt (B), then slide the steering intermediate shaft assembly No. 2.

➡Do not remove bolt (A). Do not disconnect the steering intermediate shaft assembly No. 2 from the power steering link assembly.

7. Put matchmarks on the steering

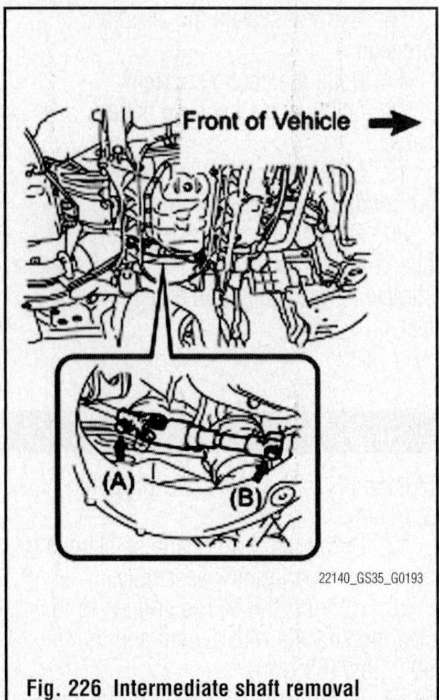

Fig. 226 Intermediate shaft removal

intermediate shaft assembly No. 2 and the power steering link assembly.

8. Separate the intermediate shaft assembly No. 2 from the power steering link assembly.
9. Separate the tie rod ends from the steering knuckles.
10. Remove engine assembly with transmission.
11. Remove the 2 bolts, 2 washers, 2 nuts, and the power steering link assembly from the front suspension cross member.

To install:

12. Install the power steering link assembly with the 2 bolts and 2 nuts and tighten to 75 ft. lbs. (102 Nm).
13. Install engine assembly with transmission.
14. Connect the tie rod end LH to the steering knuckle with the nut and tighten to 50 ft. lbs. (65 Nm).
15. Install a new clip.
16. Align the matchmarks on the intermediate shaft assembly No. 2 and the power steering link assembly.
17. Install bolt (A) and tighten the 2 bolts and tighten to 26 ft. lbs. (35 Nm).
18. Install engine under covers and protectors.
19. Install front wheels.
20. Connect cable from negative battery terminal.
21. Inspect and adjust front wheel alignment.
22. Initialize rotation angle sensor and calibrate torque sensor zero point.
23. Perform initialization.

POWER STEERING PUMP

REMOVAL & INSTALLATION

The LEXUS GS350, GS430 and GS460 models are equipped with electronic power steering. These models do not utilize a power steering (vane) pump assembly or power steering fluid.

SUSPENSION

FRONT SUSPENSION

LOWER BALL JOINT

REMOVAL & INSTALLATION

See Figures 227 and 228.

1. Before servicing the vehicle, refer to the precautions in the beginning of this section.
2. Remove the front wheel.
3. Remove engine under cover.
4. Separate the ABS speed sensor.
5. For AWD, remove the clip and castle nut, then using SST 09628-00011, separate the upper ball joint from the steering knuckle.
6. Remove the clip and castle nut, then using SST 09628-62011, separate the tie rod end.
7. Remove the strut assembly.
8. Separate the front stabilizer link assembly.
9. Remove the engine under cover.
10. Remove the lower control arm.
11. Fix the front lower control arm in a vise using aluminum plates.
12. Remove the clip and castle nut.
13. Use SST 09950-40011 (09951-04010, 09952-04010, 09953-04020, 09954-04010, 09955-04051, 09957-04010, 09958-04011) to remove the front lower ball joint from the front lower control arm.
14. Inspect the lower control arm ball joint as follows:
    a. Flip the ball joint stud back and forth 5 times.
    b. Temporarily install the nut, and use

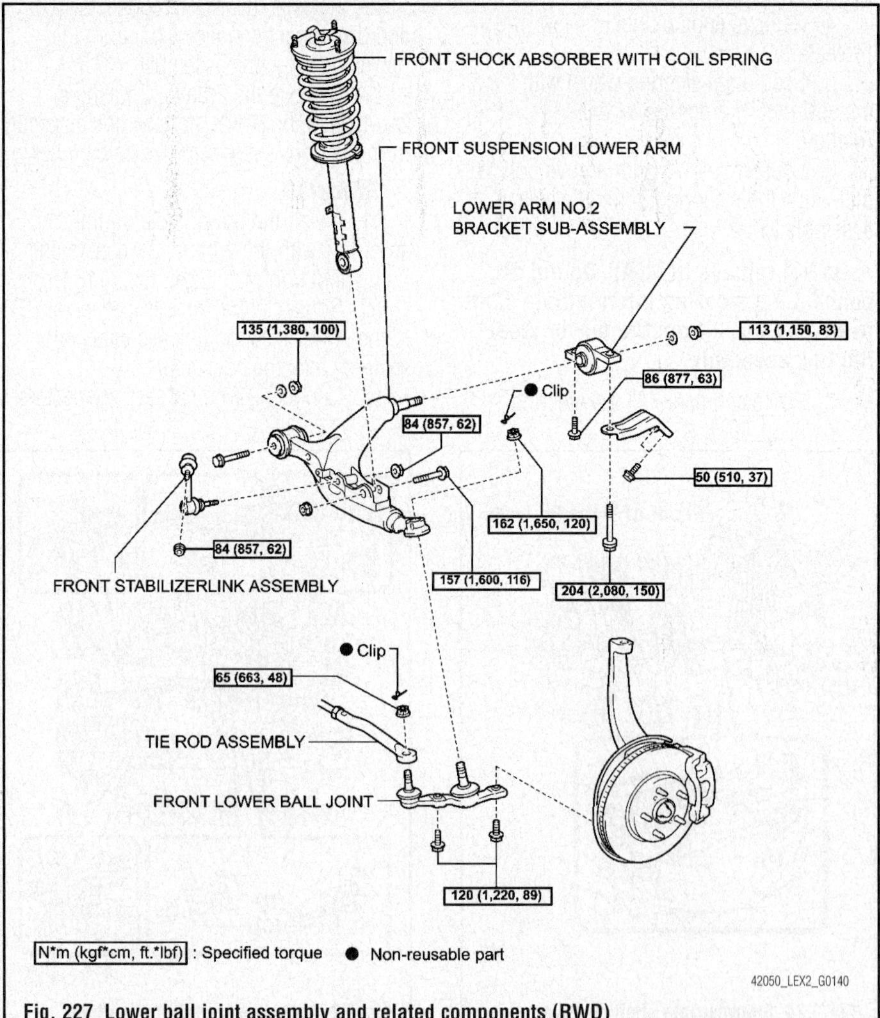

Fig. 227 Lower ball joint assembly and related components (RWD)

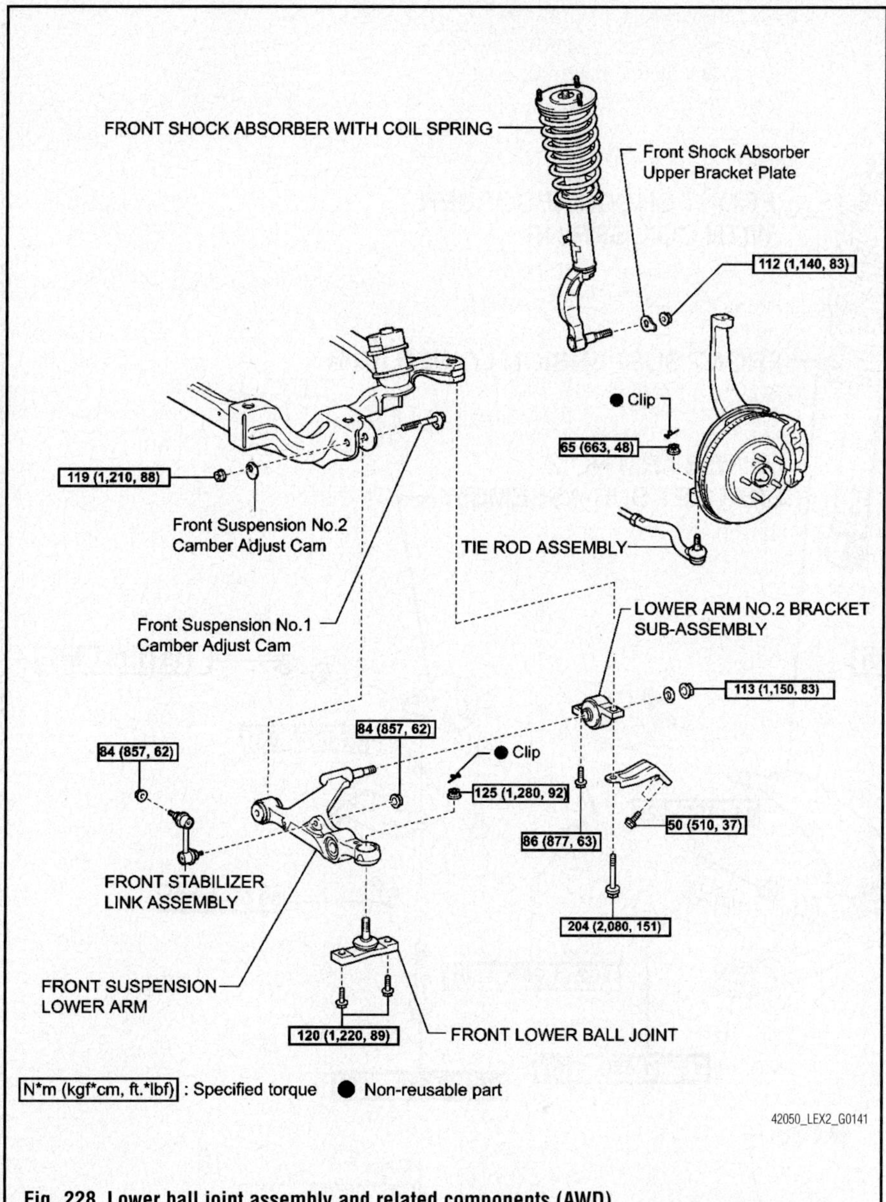

Fig. 228 Lower ball joint assembly and related components (AWD)

a torque wrench to turn the nut continuously at a rate of 3 to 5 seconds per turn. Take the torque reading on the 5th turn.

- Turning torque: 53 inch lbs (6 Nm)

c. Check the dust boots for cracks or grease leakage. If the value is not within the specified range, replace the front lower ball joint with a new one.

To install:

15. Install the front lower ball joint to the front lower control arm with the nut. Ensure that the thread and taper are free of oil or other foreign matter. Tighten the nut to 120 ft. lbs. (162 Nm) for RWD, and 92 ft. lbs. (125 Nm) for AWD.
16. Install a new clip to the front lower ball joint. Further tighten the nut up to 60° if the holes for the cotter pin are not aligned.
17. Install the lower control arm.
18. Connect the front stabilizer link assembly.
19. Install the strut assembly, but temporarily tighten the bolts at the lower control arm.
20. Connect the tie rod end to the steering knuckle Tighten the nut to 50 ft. lbs. (65 Nm). Install a new cotter pin. If the holes for the clip are not aligned, tighten the nut up to 60° further.
21. For AWD, install the steering knuckle to the front suspension upper control arm, and tighten it with the castle nut to 64 ft. lbs. (87 Nm). Install a new clip to the steering knuckle. Further tighten the nut up to 60° if the holes for the cotter pin are not aligned.
22. Connect the ABS speed sensor.
23. Stabilize the suspension as follows:
    a. Install the front wheels. Tighten the lug nuts to 76 ft. lbs. (103 Nm).
    b. Lower the vehicle and bounce it up and down several times to stabilize the front suspension.
    c. Remove the front wheels.
    d. Jack up the front suspension lower arm placing a wooden block in between. Apply a load to the front suspension so that the front suspension lower arm is placed in a horizontal position.
24. Fully tighten the strut assembly bolts at the lower control arm.
25. Fully tighten the lower control arm.
26. Install engine under cover.
27. Install the front wheel.
28. Inspect and adjust wheel alignment.

LOWER CONTROL ARM

REMOVAL & INSTALLATION

See Figure 229.

1. Before servicing the vehicle, refer to the precautions in the beginning of this section.
2. Raise the vehicle on a hoist, so that front suspension components are hanging and accessible.
3. Remove or disconnect the following:
    - Front wheels.
    - Engine under covers.
    - Brake caliper(s); Do NOT disconnect the brake hose; hang the caliper without stress on the hose
    - Tie rod end from steering knuckle
    - Stabilizer bar link from stabilizer bar
    - Height control sensor link, if equipped, from shock absorber bracket
    - Shock absorber lower mount
    - Lower control arm set bolts (loosen only)
    - Lower ball joint from No. 2 lower control arm (lower suspension arm)
    - Steering gear assembly
    - Strut bar bracket
    - No. 1 lower control arm (lower suspension arm); matchmark adjusting cam to crossmember

To install:

4. To install, reverse the removal procedure, noting the following torque settings:
    - No. 1 lower control arm (lower suspension arm) bolt to shock absorber bracket: 44 ft. lbs. (59 Nm)
    - No. 1 lower control arm (lower suspension arm) adjusting cam bolt and nut to crossmember: 127 ft. lbs. (172 Nm)

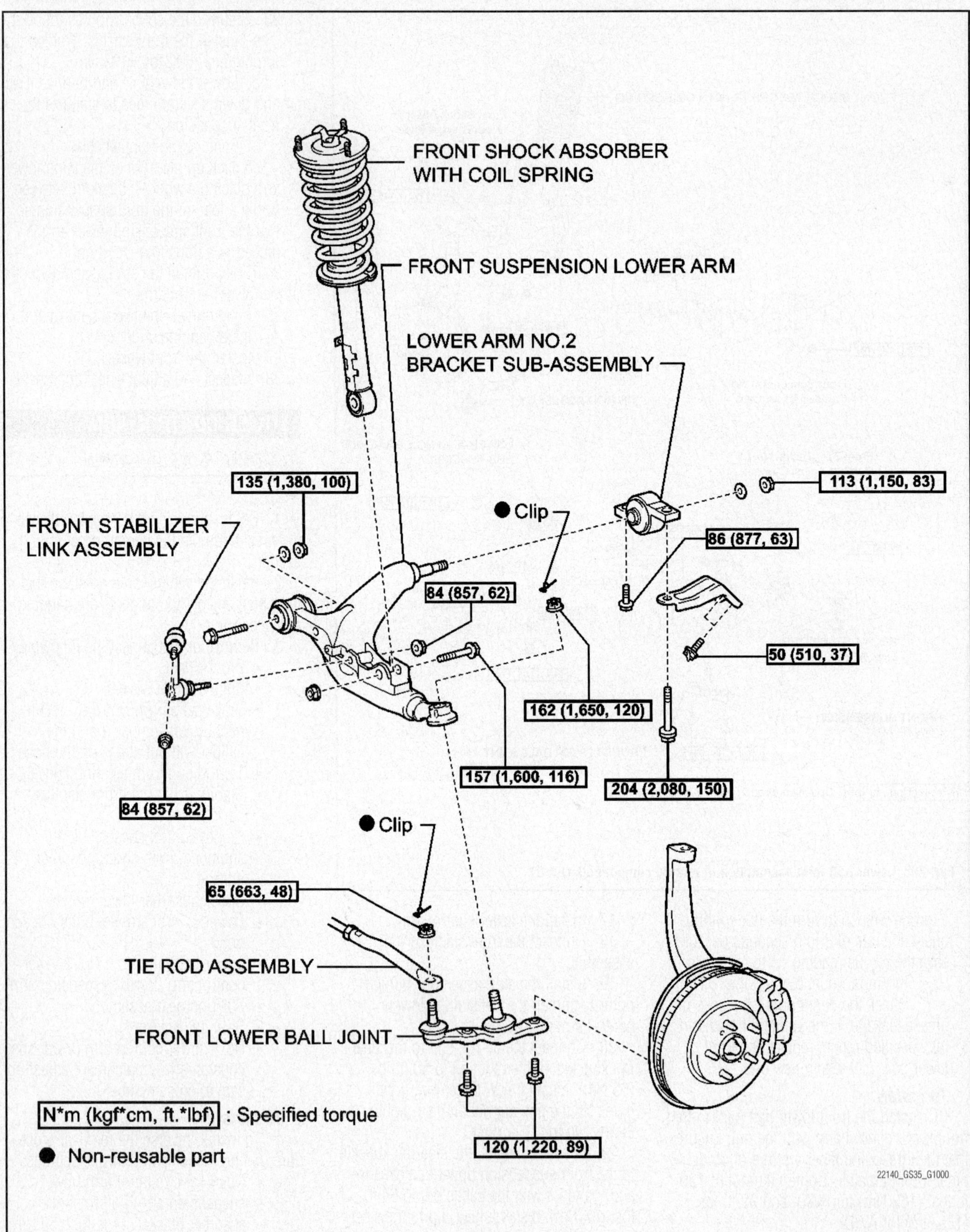

Fig. 229 Exploded view of the front lower control arm and related components

- Strut bar bracket bolts: 43 ft. lbs. (58 Nm)
- Strut bar bracket nut: 112 ft. lbs. (152 Nm)
- No. 2 lower control arm (lower suspension arm) nuts: 122 ft. lbs. (164 Nm)
- Lower shock absorber mounting bolt and nut: 116 ft. lbs. (157 Nm)
- Stabilizer bar link nut: 83 ft. lbs. (113 Nm)
- Stabilizer bar link-to-stabilizer bar bolt and nut: 43 ft. lbs. (55 Nm)
- Tie rod end nut: 64 ft. lbs. (87 Nm)
- Brake caliper bolts: 87 ft. lbs. (118 Nm)
- Front wheel nuts: 76 ft. lbs. (103 Nm)

MACPHERSON STRUT

REMOVAL & INSTALLATION

See Figure 230.

The strut removal procedure also includes the separation of the coil spring.

1. Before servicing the vehicle, refer to the precautions in the beginning of this section.
2. Remove or disconnect the following:
   - Negative battery cable.
   - Front wheel
   - Brake caliper, leaving the line attached
3. Loosen the 3 upper strut mounting nuts.
4. Loosen, but do not remove, the upper strut rod nut.

CAUTION

Do NOT remove the upper strut nut at this time.

5. Remove or disconnect the following:
   - Anti-lock Brake System (ABS) speed sensor and harness
   - Upper suspension arm from the steering knuckle
   - Stabilizer bar from the link and remove the bracket
   - Strut from the lower suspension arm.
   - 3 upper strut mounting nuts and remove the strut
6. Compress the coil spring.
7. Remove or disconnect the following:
   - Piston rod locknut
   - Suspension support, coil spring and bumper
8. If disposing the strut, perform the following procedure:
   a. Fully extend the strut rod.

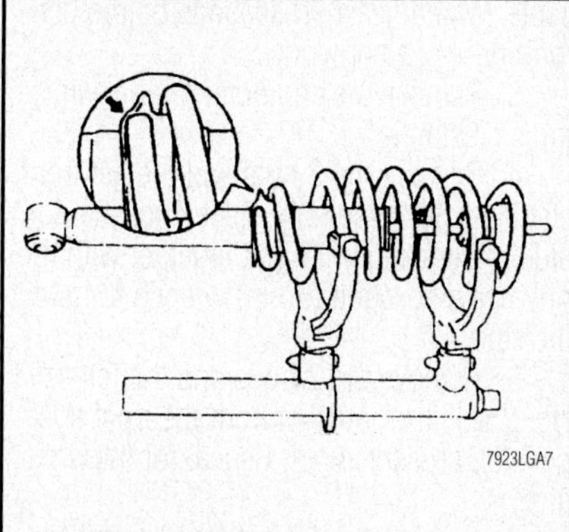

Fig. 230 Matching the spring to the seat

   b. Drill a hole near the bottom of the shock to remove the gas inside.

CAUTION

The gas is harmless, but be careful of chips that may fly up when the gas is released.

To install:

9. Install or connect the following:
   - Spring bumper
   - Coil spring
   - Suspension support to the rod and temporarily install a new nut
10. Turn the suspension support so one of the bolts on the support faces the same direction as shown in the illustration.

➡Align the bolt so a line drawn between the rod and bolt would be at 90° to the direction of the lower bushing.

11. Install or connect the following:
   - Spring compressor
   - Strut and tighten the upper retaining nuts to 41 ft. lbs. (56 Nm)
   - New upper strut rod nut to 20 ft. lbs. (27 Nm)
   - Strut to the lower arm and temporarily tighten the nut and bolt
   - Stabilizer bar bracket and tighten the bolts to 21 ft. lbs. (28 Nm)
   - The stabilizer bar to the link and tighten the bolts to 29 ft. lbs. (39 Nm)
   - Upper suspension arm to the steering knuckle. Tighten the nut to 64 ft. lbs. (87 Nm) and install a new cotter pin.
   - ABS speed sensor and tighten the bolt to 69 inch lbs. (8 Nm)
   - Caliper
   - Wheel
12. Bounce the vehicle several times to stabilize the suspension.
13. Tighten the lower strut bolt and nut to 116 ft. lbs. (157 Nm).
14. Check the front wheel alignment.

STABILIZER BAR

REMOVAL & INSTALLATION

GS350 and GS430 with RWD Non-Active Stability Control

See Figures 231 through 233.

1. Before servicing the vehicle, refer to the precautions in the beginning of this section.
2. Remove the front wheels.
3. Remove engine under cover.
4. Remove the left and right rear engine under covers.
5. Remove the 2 nuts and the left front stabilizer link assembly. Repeat for the right side. If the ball joint turns together with the nut, use a hexagon (6mm) wrench to hold the stud.
6. Remove the 2 bolts and the left front No. 2 stabilizer bracket from the front suspension crossmember. Repeat for the right side.
7. Remove the 4 bolts and the left front No. 1 stabilizer bracket from the frame. Repeat for the right side.

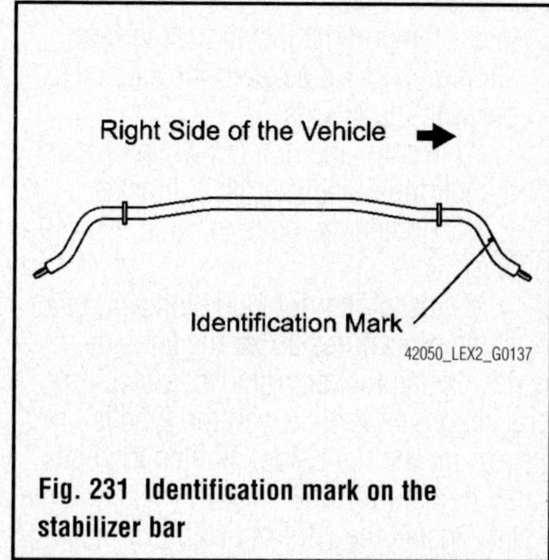

Fig. 231 Identification mark on the stabilizer bar

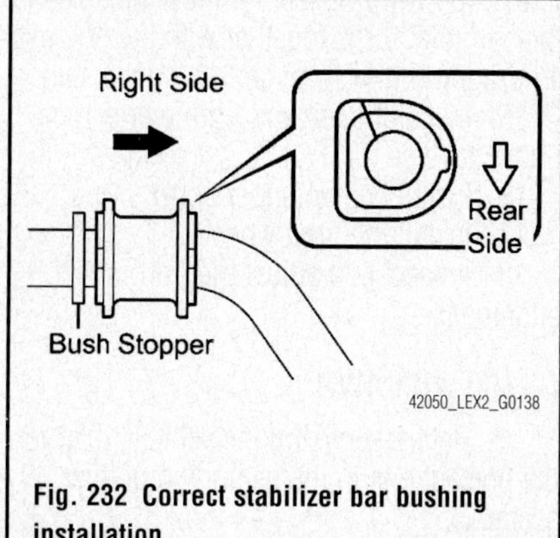

Fig. 232 Correct stabilizer bar bushing installation

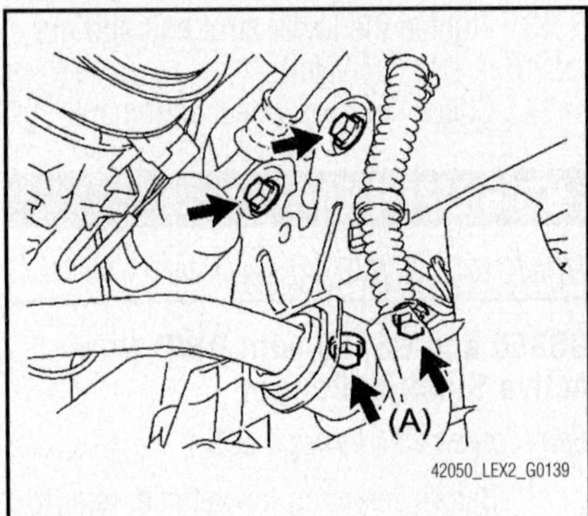

Fig. 233 Front No. 1 stabilizer bracket bolts

8. Remove the 2 front No. 1 stabilizer bar bushings from the front stabilizer bar.
9. Remove the front stabilizer bar from the vehicle.

To install:

10. Install the front stabilizer bar to the vehicle. The identification mark must be on the right side of the vehicle when installing the front stabilizer bar.
11. Install the 2 front No. 1 stabilizer bar bushings as shown in the illustration. Be sure to install the front No. 1 stabilizer bar bushings so that the cutouts face the front of the vehicle.
12. Install the left front No. 1 stabilizer bracket as follows:
    a. Temporarily install the left front stabilizer No.1 bracket to the side member with the 4 bolts.
    b. Tighten the bolt (A) while pressing the left front stabilizer No.1 bracket.
    c. Tighten the bolts to 36 ft. lbs. (49 Nm).
    d. Install the right side following the same procedures as for the left side.
13. Install the left front No. 2 stabilizer bracket on the vehicle with the 2 bolts. Repeat for the right side. Tighten the bolts to 36 ft. lbs. (49 Nm).
14. Install the left front stabilizer link assembly with the 2 nuts and tighten to 62 ft. lbs. (84 Nm). Repeat for the right side. If the ball joint turns together with the nut, use a hexagon (6mm) wrench to hold the stud.
15. Install the left and right rear engine under covers.
16. Install engine under cover.
17. Install the front wheels.
18. Inspect and adjust the front wheel alignment.

GS350 with AWD

1. Before servicing the vehicle, refer to the precautions in the beginning of this section.
2. Disconnect the negative battery cable. Wait at least 90 seconds before performing any other work.
3. Remove the engine assembly with transmission.
4. Remove the 2 nuts and the left front stabilizer link assembly. Repeat for the right side. If the ball joint turns together with the nut, use a hexagon (6mm) wrench to hold the stud.
5. Remove the 2 bolts and the left front No. 2 stabilizer bracket from the front suspension crossmember. Repeat for the right side.
6. Remove the 2 front No. 1 stabilizer bar bushings from the front stabilizer bar.
7. Remove the front stabilizer bar from the vehicle.

To install:

8. Install the front stabilizer bar to the vehicle. The identification mark must be on the right side of the vehicle when installing the front stabilizer bar.
9. Install the 2 front No. 1 stabilizer bar bushings outside the bush stoppers on the front stabilizer bar as shown in the illustration. Be sure to install the front No. 1 stabilizer bar bushings so that the cutouts face the front of the vehicle.
10. Install the left front No. 2 stabilizer bracket on the vehicle with the 2 bolts. Repeat for the right side. Tighten the bolts to 57 ft. lbs. (78 Nm).
11. Install the left front stabilizer link assembly with the 2 nuts and tighten to 62 ft. lbs. (84 Nm). Repeat for the right side.
12. Install the engine assembly with transmission
13. Connect the negative battery cable.

With Active Stability Control

See Figures 234 through 236.

1. Disconnect cable from negative battery terminal.
2. Remove front wheel.
3. Remove engine under cover.
4. Remove front fender wheel opening moulding RH.
5. Remove front fender liner RH.
6. Disconnect the 2 wire harness clamps from the bracket.
7. Using the procedures below, disconnect the ECU connector.
    a. Release the lever's lock. (*1).
    b. Press the claw and move the lever in the direction of the arrow in the illustration. (*2).
    c. Disconnect the ECU connector. (*3).

➡**When disconnecting the connector, do not apply excessive force to the wire harness.**

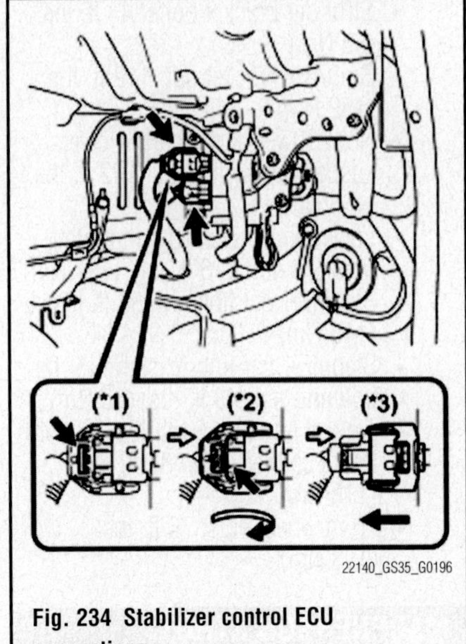

Fig. 234 Stabilizer control ECU connections

8. Remove the 2 nuts and links from the stabilizer bar.
9. Remove rear engine under covers.
10. Remove the EPS wire harness.
11. Remove the 2 bolts and bracket from the crossmember.
12. Remove front No. 2 stabilizer bracket RH.
    a. Remove the 2 bolts, bracket and control actuator RH.

➡**The front active stabilizer control actuator is very heavy. Be careful not to drop it.**

13. Remove front No. 1 stabilizer bracket LH.
    a. Disconnect the clamp from the bracket.
    b. Remove the 5 bolts and bracket from the frame.
14. Remove front No. 1 stabilizer bracket RH.
15. Remove the 2 bushings from the control actuator.
16. Install the 2 bushings.
17. Install front No. 1 stabilizer bracket LH.
    a. Push the bracket toward the outside of the vehicle against the frame, and temporarily install the bolt labeled (1).
    b. Install bolt (2). Then tighten bolt (1). Then install bolt (3) and (4). Tighten to 36 ft. lbs. (49 Nm).
    c. Install bolt (5). Tighten to 74 ft. lbs. (100 Nm).
    d. Install the wire clamp to the bracket.
18. Install front No. 1 stabilizer bracket RH.

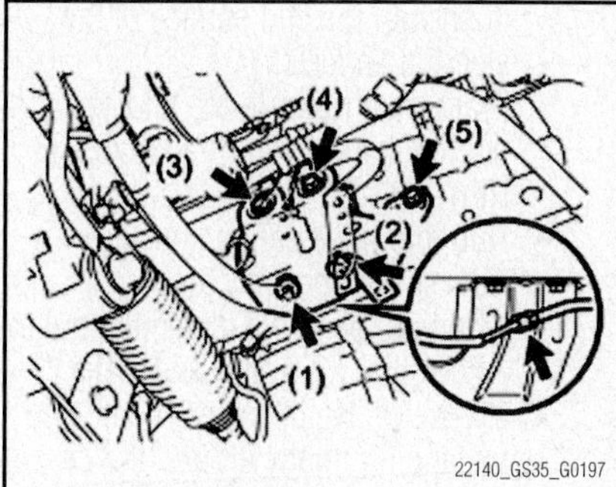

Fig. 235 Stabilizer bracket LH bolt location

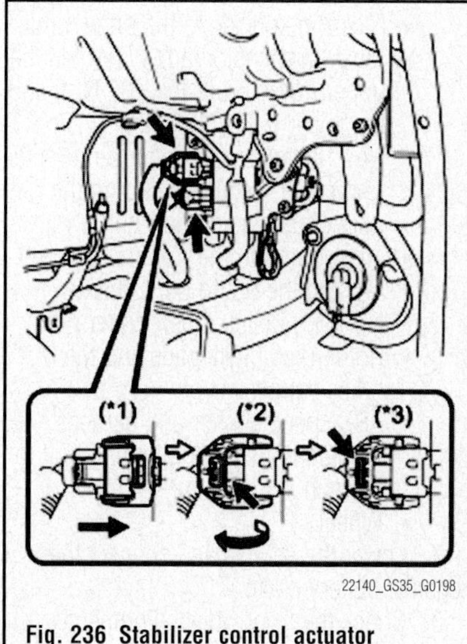

Fig. 236 Stabilizer control actuator

19. Temporarily install the control actuator, 2 bushings and 2 brackets with the 4 bolts.

➡Install the actuator so that it is on the right side of the vehicle.

20. Install front No. 2 stabilizer bracket LH and tighten to 36 ft. lbs. (49 Nm).
21. Install front No. 2 stabilizer bracket RH.
22. Install the under covers with the screws.
23. Install the link to the lower arm side and stabilizer bar side with the 2 nuts and tighten to 62 ft. lbs. (84 Nm).
24. Install front stabilizer link assembly RH.
25. Connect front active stabilizer control actuator assembly.
    a. Connect the connector to the front active stabilizer control ECU. (*1).
    b. Rotate the lever in the direction of the arrow until a "click" sound is heard. (*2).
    c. Lock the lever's lock. (*3).
26. Install the wire clamp to the bracket.
27. Install front fender liner RH.
28. Install front fender wheel opening moulding RH.
29. Install engine under cover.
30. Install front wheel.
31. Connect cable to negative battery terminal.
32. Perform initialization.

STEERING KNUCKLE

REMOVAL & INSTALLATION

1. Before servicing the vehicle, refer to the precautions in the beginning of this section.
2. Remove the front wheel.
3. Disconnect the front brake caliper assembly. Support the brake caliper securely.
4. Remove the front brake disc
5. Remove the clip and nut, then using SST 09610-20012, disconnect the tie rod end.
6. Remove the cotter pin and nut, then using SST 09628-6201 1, remove the lower ball joint from the lower control arm.
7. Remove the cotter pin and castle nut, then using SST09628-62011, separate the upper ball joint from the steering knuckle.
8. Remove the steering knuckle from the vehicle.
9. If necessary, remove the lower ball joint and hub and bearing assembly from the steering knuckle.

To install:

10. If necessary, install the hub and bearing assembly and lower ball joint to the steering knuckle.
11. Install the steering knuckle into the vehicle.
12. Connect the upper ball joint to the steering knuckle. Install castle nut and new cotter pin. Tighten to 64 ft. lbs. (87 Nm).
13. Connect the lower ball joint to lower control arm. Install castle nut and new cotter pin. Tighten to 64 ft. lbs. (87 Nm).
14. Connect the tie rod end to the steering knuckle. Install castle nut and new cotter pin. Tighten to 48 ft. lbs. (65 Nm).
15. Install the front brake disc.
16. Install the front brake caliper assembly.
17. Install the front wheel.
18. Inspect and adjust wheel alignment.

UPPER BALL JOINT

REMOVAL & INSTALLATION

The upper ball joint is an integral part of the upper arm and is not replaced separately. The upper ball joint replacement is accomplished by replacing the upper arm.

UPPER CONTROL ARM

REMOVAL & INSTALLATION

1. Before servicing the vehicle, refer to the precautions in the beginning of this section.
2. Remove or disconnect the following:
    - Negative battery cable
    - Wheel
3. Loosen the 3 upper strut mounting nuts.
4. Loosen, but do not remove, the upper strut rod nut.

⁂ CAUTION

Do NOT completely remove the upper strut nut at this time.

5. Remove or disconnect the following:
    - Brake caliper, leaving the line attached and secure it out of the way
    - Anti-lock Brake System (ABS) speed sensor and harness
    - Cotter pin and nut from the upper control arm
    - Upper control arm from the steering knuckle
    - Stabilizer bar from the link and remove the bracket
    - Cotter pin and nut from the lower control arm
    - Strut from the lower suspension arm
    - 3 upper strut mounting nuts and remove the strut
    - Mounting bolts holding the upper control arm to the frame
    - Upper control arm from the vehicle

To install:

6. Install or connect the following:
    - Upper suspension arm and tighten the mounting bolts to 39 ft. lbs. (53 Nm)
    - Strut and tighten the upper retaining nuts to 41 ft. lbs. (56 Nm). Tighten the new upper strut rod nut to 20 ft. lbs. (27 Nm).
    - Strut to the lower arm and temporarily tighten the nut and bolt
    - Stabilizer bar bracket and tighten the bolts to 21 ft. lbs. (28 Nm)
    - Stabilizer bar to the link and tighten the bolts to 29 ft. lbs. (39 Nm)
    - Upper suspension arm to the steering knuckle. Tighten the nut to 64 ft. lbs. (87 Nm) and install a new cotter pin.

- ABS speed sensor and tighten the bolt to 69 inch lbs. (8 Nm)
- Caliper
- Front wheel

7. Lower the vehicle.
8. Bounce the vehicle several times to stabilize the suspension.
9. Tighten the lower strut bolt and nut to 116 ft. lbs. (157 Nm).
10. Check the front wheel alignment.

WHEEL HUB AND BEARING

REMOVAL & INSTALLATION

1. Before servicing the vehicle, refer to the precautions in the beginning of this section.
2. Remove or disconnect the following:
   - Negative battery cable
   - Front wheel
   - Caliper, leaving the brake line connected and suspend it out of the way

⁂ WARNING

Never allow the brake caliper to hang freely from the brake hose.

- Rotor
- Anti-lock Brake System (ABS) speed sensor and harness
- Tie rod from the arm on the lower ball joint
- Upper suspension arm from the steering knuckle
- Steering knuckle from the lower control arm
- Ball joint from the steering knuckle
- Front hub grease cap

3. Clamp the hub in a soft jaw vise.
4. Using a hammer and chisel, loosen the staked part of the locknut.
5. Remove or disconnect the following:
   - Locknut
   - ABS speed sensor rotor

➡Do NOT scratch the serrations of the sensor rotor.

- Brake dust cover bolts and shift the cover toward the outside.
- Hub from the steering knuckle
- Inner bearing race from the hub shaft
- Oil seal from the knuckle
- Bearing snapring from the steering knuckle
- Bearing from the steering knuckle

To install:

6. Install or connect the following:
   - New bearing into the steering knuckle

➡If the inner race and balls come loose from the bearing outer race, be sure to install them on the same side as before.

- Snapring
- New outside inner race and tap in the new seal. Tap the seal until it is flush with the end surface of the steering knuckle.
- Brake dust cover to the knuckle and tighten the bolts to 74 inch lbs. (8 Nm)
- Hub into the steering knuckle
- ABS speed sensor rotor
- Axle hub locknut. Tighten the nut to 147 ft. lbs. (199 Nm) and stake it.
- Grease cap to the steering knuckle by tapping lightly around the circumference of the cap with a hammer
- Ball joint to the steering knuckle. Tighten the 2 bolts to 83 ft. lbs. (113 Nm).
- Steering knuckle to the upper and lower suspension arms. Tighten the upper nut to 64 ft. lbs. (87 Nm) and the lower nut to 95 ft. lbs. (127 Nm). Install a new cotter pin on the lower nut. Install the clip on the upper suspension arm nut.
- Tie rod end to the steering knuckle. Tighten the nut to 64 ft. lbs. (87 Nm) and install a new cotter pin.
- Rotor, disc brake pads and the brake caliper
- ABS speed sensor and harness. Tighten the sensor retaining bolt to 69 inch lbs. (8 Nm).
- Wheel

7. Lower the vehicle and connect the negative battery cable.
8. Check the front wheel alignment.

SUSPENSION

REAR SUSPENSION

CONTROL ARMS/LINKS

REMOVAL & INSTALLATION

Upper Control Arm No. 1

See Figure 237.

1. Remove rear wheel.
2. Remove differential No. 2 support protector.
3. Remove the bolt and nut, and separate the rear stabilizer link assembly and the load sensing valve sensor bracket from the rear No. 2 suspension arm.
4. Remove the bolt and nut, and separate the rear shock absorber with coil spring from the rear No. 2 suspension arm assembly.
5. Remove the bolt, nut washer, and separate the rear upper No. 1 control arm from the rear axle carrier.
6. Remove the bolt, nut washer, and the rear upper No. 1 control arm from the rear suspension member.

To install:

7. Temporarily install the rear upper No. 1 control arm with the bolt, nut and washer.
8. Temporarily install the rear shock absorber with coil spring with the bolt and nut.
9. Temporarily install the rear stabilizer link assembly and the load sensing valve sensor bracket to the rear No. 2 suspension arm assembly with the bolt and nut.
10. Jack up the axle carrier, with a wooden block placed between the jack and axle carrier, to apply load to the suspension so that the rear drive shaft assembly becomes level.
11. Fully tighten the nut on the rear No. 1 upper control arm assembly to 119 ft. lbs. (161 Nm).
12. Fully tighten the bolt holding the rear shock absorber with coil spring to 81 ft. lbs. (110 Nm).
13. Fully tighten the nut holding the rear stabilizer link assembly to 20 ft. lbs. (27 Nm).
14. Install differential no. 2 support protector.
15. Install rear wheel.
16. Inspect and adjust rear wheel alignment.

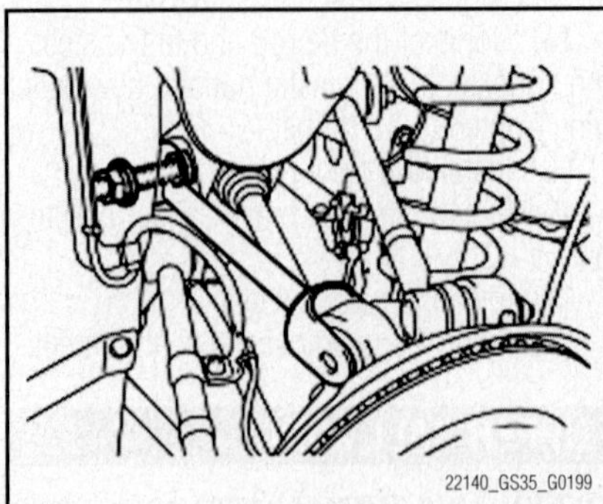

Fig. 237 Installing No. 1 upper control arm

Upper Control Arm No. 2

See Figure 238.

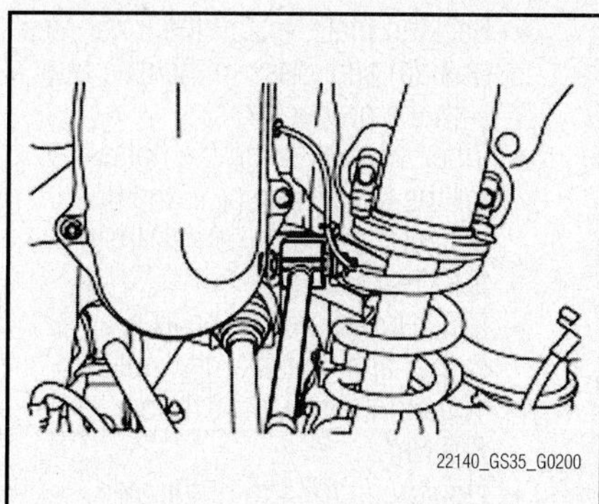

Fig. 238 Installing No. 2 upper control arm

1. Remove rear wheel.
2. Remove differential No. 2 support protector.
3. Separate rear shock absorber with coil spring.
4. Remove rear upper No. 2 control arm.
   a. Remove the nut on the axle carrier side.
   b. Using Service Tool, separate the rear upper No. 2 control arm.
5. Remove the bolt, nut, washer, and rear upper No. 2 control arm.

➡Push the axle carrier downward.

To install:

6. Install the stud of the rear upper No. 2 control arm, and temporarily tighten a new nut.
7. Install the rear upper control arm, and temporarily tighten the bolt, nut and washer.
8. Fully tighten the nut of the rear upper No. 2 control arm to 52 ft. lbs. (70 Nm).
9. Temporarily tighten rear shock absorber with coil spring.
10. Jack up the axle carrier, with a wooden block placed between the jack and axle carrier, to apply load to the suspension so that the rear drive shaft assembly becomes level.
11. Fully tighten the nut on the rear suspension member side of the rear upper No. 2 control arm to 107 ft. lbs. (145 Nm).
12. Fully tighten rear shock absorber with coil spring.
13. Install differential no. 2 support protector.
14. Install rear wheel.
15. Inspect and adjust rear wheel alignment.

SHOCK ABSORBER

REMOVAL & INSTALLATION

1. Before servicing the vehicle, refer to the precautions in the beginning of this section.
2. Remove the corresponding rear wheel(s).
3. Remove the luggage compartment trim front cover.
4. Remove the rear fender apron seal.
5. Remove the rear lower suspension arm:
   a. Remove 2 bolts, nuts and the No. 1 lower suspension arm (trailing arm).
   b. Remove the bolt and nuts and disconnect the stabilizer bar link (and height control link, if equipped) from the No. 2 lower suspension arm.
   c. Remove the bolt and nut and disconnect the strut from the No. 2 lower suspension arm.
   d. Place matchmarks on the adjusting cam and on the No. 2 lower suspension arm.
   e. Remove the nut and adjusting cams.
   f. Remove the bolt, nut and washer and the No. 2 lower suspension arm.
6. Remove the 3 upper retaining nuts. Loosen, but Do NOT remove the center stud nut.
7. Remove the lower strut mounting bolt.
8. Remove the strut assembly, with the coil spring.
9. Compress the coil spring in a suitable spring compressor.

➡Do NOT remove these items with an impact wrench; it will damage the spring compressor.

10. Remove or disconnect the following:
    - Suspension support nut
    - Washer
    - 2 cushions
    - Collar
    - Suspension support
    - Upper insulator
    - Lower cup
    - Spring bumper
11. Carefully release the spring compressor and remove the coil spring.
12. If the shock absorber is being replaced, fully extend the shock absorber rod and drill a hole to discharge the gas from the cylinder. Drill this hole about 5-8 in. (130-185mm) from lower mounting bolt hole.

To install:

13. With a new shock absorber in place, install the suspension support and compress the coil spring.
14. With a non-impact wrench, install the coil spring to the shock absorber, fitting the lower end of the spring into the recent of the spring seat on the shock absorber.
15. Install the spring bumper, lower cup, cushion, collar, upper insulator, suspension support, cushion and washer onto the shock absorber. Temporarily tighten a new nut.
16. Rotate the suspension support so the rod and one of the bolts on the suspension support are aligned with the lower shock mounting hole such that while the lower mounting bolt hole is in proper position, the 3 studs on the top of the strut assembly will align with the holes in the body.
17. Carefully remove the spring compressor.
18. Install the strut assembly into the vehicle. Torque the 2 bolts on top of the coil spring to 13 ft. lbs. (18 Nm), the 3 nuts on the top mounting studs to 47 ft. lbs. (64 Nm), and the nut in the center of the upper strut mounting to 20 ft. lbs. (27 Nm).
19. Install the bolt, nut and washer and the No. 2 lower suspension arm. Torque bolt to 81 ft. lbs. (110 Nm).
20. Position both adjusting cams, referencing the matchmarks made during removal. Install and torque the retaining nuts to 81 ft. lbs. (110 Nm).
21. Install the strut assembly lower mounting bolt to the No. 2 lower suspension arm. Torque the nut to 81 ft. lbs. (110 Nm).
22. Install the stabilizer bar link (and height control link, if equipped) to the No. 2 lower suspension arm. Torque the bolt and nut to 22 ft. lbs. (30 Nm).
23. Install the No. 1 lower suspension arm (trailing arm) and torque the bolts and nuts to 55 ft. lbs. (75 Nm).
24. Install the rear fender apron seal.
25. Install the luggage compartment trim front cover.
26. Install the rear wheel(s). Torque the wheel nuts to 76 ft. lbs. (103 Nm).

WHEEL HUB AND BEARING

REMOVAL & INSTALLATION

See Figure 239.

1. Before servicing the vehicle, refer to the precautions in the beginning of this section.
2. Remove or disconnect the following:
   - Negative battery cable.
   - Rear tire and wheel assembly
   - Brake caliper support from the rear axle carrier and support it with a piece of wire
3. Place matchmarks on the disc brake rotor and the axle hub.

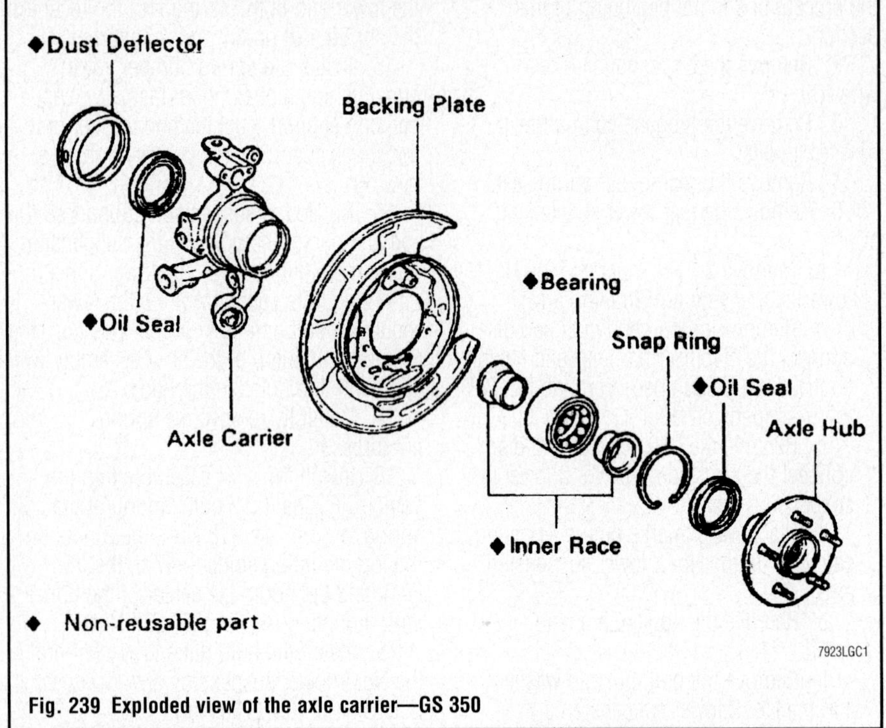

Fig. 239 Exploded view of the axle carrier—GS 350

4. Remove or disconnect the following:
   - Brake rotor
   - Speed sensor
   - Rear halfshaft
   - Parking brake shoes
   - Parking brake cable
   - Strut rod
5. Place matchmarks on the adjusting cam and rear control crossmember.
   - Nut, adjusting cam and the washer to the No. 1 control arm
   - No. 1 lower control arm from the crossmember
   - Loosen the nut holding the lower control arm to the axle carrier
   - No. 2 lower control arm from the axle carrier
   - Nut, then remove the No. 2 lower control arm from the axle carrier
   - Nut holding the upper control arm to the axle carrier
   - Axle carrier
   - Nut holding the No. 1 control arm to the axle carrier
   - No. 1 lower control arm from the axle carrier
   - Dust deflector
   - Axle hub from the carrier
   - Backing plate
   - Inner race (outside)
   - Oil seal
   - Snapring
   - Bearing

To install:

6. Install or connect the following:
   - Bearing to the axle carrier

➡If the inner races come loose from the bearing outer race, be sure to install them on the same side as before.

   - Snapring. Install the inner race (outside) and a new oil seal.
   - Backing plate. Install the inner race (inside) and press in the axle hub with the proper tools.
   - Inner oil seal. Align the holes for the speed sensor in the dust deflector and axle carrier. Install the dust deflector.
   - No. 1 lower arm to the axle carrier and install a new nut. Tighten the nut to 43 ft. lbs. (59 Nm).
   - Upper control arm to the axle carrier. Tighten the new nut and bolt to 80 ft. lbs. (109 Nm).
   - No. 2 lower control arm to the axle carrier and tighten a new nut to 110 ft. lbs. (150 Nm).
   - No. 1 lower control arm to the rear crossmember. Tighten the nut to 136 ft. lbs. (184 Nm).
   - Strut rod to the axle carrier. Tighten the nuts and bolts to 134 ft. lbs. (184 Nm).
   - Parking brake cable and slide the backing plate to the inside. Install the hex bolt and tighten it to 132 ft. lbs. (180 Nm).
   - Shoe guide plate set bolt. Tighten the bolt to 13 ft. lbs. (18 Nm).
   - 4 hub bolts and tighten them to 19 ft. lbs. (26 Nm)
   - Bolts at the speed sensor and tighten them to 69 inch lbs. (8 Nm)
   - Parking brake shoes
   - Halfshafts. Apply the brakes and tighten the locknut to 213 ft. lbs. (289 Nm).
   - Brake rotor
   - Brake caliper support to the rear axle carrier. Tighten the bolts to 77 ft. lbs. (104 Nm).
   - Rear tire and wheel assembly
   - Negative battery cable
7. Lower the vehicle and bounce it a few times to stabilize the suspension.

LEXUS

LS460

3

SPECIFICATIONS AND MAINTENANCE CHARTS

ENGINE AND VEHICLE IDENTIFICATION

Engine						
Code ①	Liters (cc)	Cu. In.	Cyl.	Fuel Sys.	Engine Type	Eng. Mfg.
1UR-FSE	4.6 (4608)	282	V8	SFI D4S	DOHC	Toyota
1UR-FSE	4.6 (4608)	282	V8	SFI D4S	DOHC	Toyota

Model Year	
Code ②	Year
7	2007
8	2008

SFI: Sequential Fuel Injection

DOHC: Double Overhead Camshaft

NA: Information not available

① Stamped on the left side of the engine block

② 10th digit of the Vehicle Identification Number (VIN)

22140_LS46_C0001

GENERAL ENGINE SPECIFICATIONS

Year	Model	Engine Displacement Liters	Engine Series ID	Net Horsepower @ rpm	Net Torque @ rpm (ft. lbs.)	Bore x Stroke (in.)	Com-pression Ratio	Oil Pressure @ rpm
2007	LS460	4.6	1UR-FSE	380@6400	367@4100	3.70x3.27	11.8:1	31 plus@2500
2008	LS460	4.6	1UR-FSE	380@6400	367@4100	3.70x3.27	11.8:1	31 plus@2500

22140_LS46_C0002

ENGINE TUNE-UP SPECIFICATIONS

Year	Engine Displacement Liters	Engine ID	Spark Plug Gap (in.)	Ignition Timing (deg.)*	Fuel Pump (psi)	Idle Speed (rpm)	Valve Clearance	
							Intake	Exhaust
2007	4.6	1UR-FSE	0.039-0.043	8-12B ①	②	700-800	HYD	HYD
2008	4.6	1UR-FSE	0.039-0.043	8-12B ①	②	700-800	HYD	HYD

NOTE: The Vehicle Emission Control Information label often reflects specification changes made during production. The label figures must be used if they differ from those in this chart.

HYD: Hydraulic Valve Lifters

① Terminals TC and CG of check connector must be connected

② Low pressure system 28-85 psi. High pressure system 508-653 psi.

22140_LS46_C0003

CAPACITIES

Year	Model	Engine Displacement Liters	Engine ID	Engine Oil with Filter (qts.)	Transmission (pts.) 5-Spd	Transmission (pts.) Auto.*	Transfer Case (pts.)	Drive Axle Front (pts.)	Drive Axle Rear (pts.)	Fuel Tank (gal.)	Cooling System (qts.)
2007	LS460	4.6	1UR-FSE	9.1	—	10.6	—	—	2.84	22.1	11.7
2008	LS460	4.6	1UR-FSE	9.1	—	10.6	—	—	2.84	22.1	11.7

22140_LS46_C0004

VALVE SPECIFICATIONS

Year	Engine Displacement Liters	Engine ID	Seat Angle (deg.)	Face Angle (deg.)	Spring Test Pressure (lbs. @ in.)	Spring Free-Length (in.)	Stem-to-Guide Clearance (in.) Intake	Stem-to-Guide Clearance (in.) Exhaust	Stem Diameter (in.) Intake	Stem Diameter (in.) Exhaust
2007	4.6	1UR-FSE	45	44.5	NA	2.035	0.0010-0.0024	0.0012-0.0026	0.2154-0.2159	0.2152-0.2157
2008	4.6	1UR-FSE	45	44.5	NA	2.035	0.0010-0.0024	0.0012-0.0026	0.2154-0.2159	0.2152-0.2157

NA: Information not available

22140_LS46_C0005

CRANKSHAFT AND CONNECTING ROD SPECIFICATIONS

All measurements are given in inches.

Year	Engine Displacement Liters	Engine ID	Crankshaft Main Brg. Journal Dia.	Crankshaft Main Brg. Oil Clearance	Crankshaft Shaft End-play	Crankshaft Thrust on No.	Connecting Rod Journal Diameter	Connecting Rod Oil Clearance	Connecting Rod Side Clearance
2007	4.6	1UR-FSE	2.6373-2.6378	①	0.0008-0.0087	3	2.0863-2.0866	0.0010-0.0020	0.0059-0.0217
2008	4.6	1UR-FSE	2.6373-2.6378	①	0.0008-0.0087	3	2.0863-2.0866	0.0010-0.0020	0.0059-0.0217

① Journal No. 1 and 5: 0.0007 - 0.0012 inch
Remaining journals: 0.0009 - 0.0015 inch

22140_LS46_C0006

PISTON AND RING SPECIFICATIONS

All measurements are given in inches.

Year	Engine Displ. Liters	Engine ID	Piston Clearance	Ring Gap Top Comp.	Ring Gap Bottom Comp.	Ring Gap Oil Control	Ring Side Clearance Top Comp.	Ring Side Clearance Bottom Comp.	Ring Side Clearance Oil Control
2007	4.6	1UR-FSE	0.0014-0.0022	0.0091-0.0130	0.0138-0.0177	0.0039-0.0157	0.0008-0.0028	0.0008-0.0024	0.0008-0.0028
2008	4.6	1UR-FSE	0.0014-0.0022	0.0091-0.0130	0.0138-0.0177	0.0039-0.0157	0.0008-0.0028	0.0008-0.0024	0.0008-0.0028

22140_LS46_C0007

TORQUE SPECIFICATIONS

All readings in ft. lbs.

Year	Engine Displacement Liters	Engine ID	Cylinder Head Bolts	Main Bearing Bolts	Rod Bearing Bolts	Crankshaft Damper Bolts	Flywheel Bolts	Manifold Intake	Manifold Exhaust	Spark Plugs	Oil Pan Drain Plug
2007	4.6	1UR-FSE	①	②	③	221	22	15	15	13	76
2008	4.6	1UR-FSE	①	②	③	221	22	15	15	13	76

① Step 1: 27 ft. lbs.
Step 2: Tighten an additional 90 degrees
Step 3: Tighten an additional 90 degrees
12 mm bolt: 15 ft. lbs.

② Inside position: 45 ft. lbs.
Outside position Step 1: 20 ft. lbs.
Outside position Step 2: Tighten an additional 90 degrees
Cylinder block Side postion: 33 ft. lbs.

③ Step 1: 30 ft. lbs.
Step 2: Plus 90 degrees

22140_LS46_C0008

WHEEL ALIGNMENT

Year	Model		Caster Range (+/-Deg.)	Caster Preferred Setting (Deg.)	Camber Range (+/-Deg.)	Camber Preferred Setting (Deg.)	Toe-in (in.)	Steering Axis Inclination (Deg.)
2007	LS460	Front	0.75	+6.62	0.75	-0.28	0.04+/-0.08	9.12+/-0.75
	w/o air susp	Rear	—	—	0.75	-1.63	0.12+/-0.08	—
	with air	Front	0.75	+7.02	0.75	-0.43	0+/-0.08	9.3+/-0.75
	suspension	Rear	—	—	0.75	-1.83	0.12+/-0.08	—
2008	LS460	Front	0.75	+6.62	0.75	-0.28	0.04+/-0.08	9.12+/-0.75
	w/o air susp	Rear	—	—	0.75	-1.63	0.12+/-0.08	—
	with air	Front	0.75	+7.02	0.75	-0.43	0+/-0.08	9.3+/-0.75
	suspension	Rear	—	—	0.75	-1.83	0.12+/-0.08	—

22140_LS46_C0009

TIRE, WHEEL AND BALL JOINT SPECIFICATIONS

Year	Model	OEM Tires		Tire Pressures (psi)		Wheel	Ball Joint	Lug Nut Torque
		Standard	Optional	Front	Rear	Size	Inspection	(ft. lbs.)
2007	LS460	P235/50R18	P235/50R18	33	33	7.5-J	①	103
2008	LS460	P235/50R18	P235/50R18	33	33	7.5-J	①	103

OEM: Original Equipment Manufacturer

PSI: Pounds Per Square Inch

STD: Standard

OPT: Optional

① Replace if any measurable movement is found.

22140_LS46_C0010

BRAKE SPECIFICATIONS

All measurements in inches unless noted

Year	Model		Brake Disc Original Thickness	Brake Disc Minimum Thickness	Maximum Runout	Minimum Lining Thickness	Brake Caliper Bracket Bolts (ft. lbs.)	Brake Caliper Mounting Bolts (ft. lbs.)
2007	LS460	F	①	②	0.0016	0.039	100	NA
		R	③	④	0.0016	0.039	63	NA
2008	LS460	F	①	②	0.0016	0.039	100	NA
		R	③	④	0.0016	0.039	63	NA

F: Front

R: Rear

① 13.14 inch disc: 1.181 inches
14.06 inch disc: 1.339 inches

② 13.14 inch disc: 1.063 inches
14.06 inch disc: 1.221 inches

③ 12.40 inch disc: 0.787 inches
13.19 inch disc: 0.866 inches

④ 12.40 inch disc: 0.709 inches
13.19 inch disc: 0.887 inches

22140_LS46_C0011

SCHEDULED MAINTENANCE INTERVALS

LEXUS—LS460

TO BE SERVICED	TYPE OF SERVICE	VEHICLE MILEAGE INTERVAL (x1000)													
		5	10	15	20	25	30	35	40	45	50	55	60	90	120
Engine oil & filter	R	✓	✓	✓	✓	✓	✓	✓	✓	✓	✓	✓	✓	✓	✓
Automatic transmission fluid	S/I			✓			✓			✓			✓	✓	✓
Ball joints & dust covers	S/I			✓			✓			✓			✓	✓	✓
Bolts & nuts on chassis & body	S/I			✓			✓			✓			✓	✓	✓
Brake linings & drums	S/I	✓	✓	✓	✓	✓	✓	✓	✓	✓	✓	✓	✓	✓	✓
Brake line pipes & hoses	S/I			✓			✓			✓			✓	✓	✓
Brake pads & discs (front & rear)	S/I	✓	✓	✓	✓	✓	✓	✓	✓	✓	✓	✓	✓	✓	✓
Brake fluid	R						✓						✓	✓	✓
Rack and pinion assembly	S/I			✓			✓			✓			✓	✓	✓
Steering linkage & boots	S/I			✓			✓			✓			✓	✓	✓
Air cleaner filter	R						✓						✓	✓	✓
Spark plugs ①	R														✓
Drive belts	S/I						✓						✓	✓	✓
Exhaust pipes & mountings	S/I			✓			✓			✓			✓	✓	✓
Fuel lines & connections	S/I						✓						✓	✓	✓
Engine coolant ②	S/I			✓			✓			✓			✓	✓	
Rear differential	S/I			✓			✓			✓			✓	✓	✓
Fuel tank cap gasket	S/I						✓						✓	✓	✓
Rotate tires	S/I			✓			✓			✓			✓		✓
Clean air conditioning filter ③	S/I			✓			✓			✓			✓		✓
Axle shaft bolts	S/I			✓			✓			✓			✓	✓	✓
Brake pad thickness and rotor runout	S/I						✓						✓	✓	✓

R: Replace S/I: Service or Inspect

① Spark plugs are replaced at 120,000 miles

② Replace engine coolant at 100,000 miles and then inspect every 15,000 miles

③ Replace air conditioning filter every 30,000 miles

FREQUENT OPERATION MAINTENANCE (SEVERE SERVICE)

If a vehicle is operated under any of the following conditions it is considered severe service:

- Extremely dusty areas.

- 50% or more of the vehicle operation is in 32°C (90°F) or higher temperatures, or constant temperatures below 0°C (32°F).

- Prolonged idling (vehicle operation in stop and go traffic).

- Frequent short running periods (engine does not warm to normal operating temperatures).

- Police, taxi, delivery usage or trailer towing usage.

Air cleaner filter: service or inspect every 5000 miles

Rear differential & transfer case oil: replace every 15,000 miles.

Ball joints & dust covers: service or inspect every 5000 miles.

Bolts & nuts on chassis & body: service or inspect every 5000 miles.

Axle shaft bolts: service or inspect every 5000 miles.

Steering linkage: service or inspect every 5000 miles.

22140_LS46_C0012

PRECAUTIONS

Before servicing any vehicle, please be sure to read all of the following precautions, which deal with personal safety, prevention of component damage, and important points to take into consideration when servicing a motor vehicle:

- Never open, service or drain the radiator or cooling system when the engine is hot; serious burns can occur from the steam and hot coolant.
- Observe all applicable safety precautions when working around fuel. Whenever servicing the fuel system, always work in a well-ventilated area. Do not allow fuel spray or vapors to come in contact with a spark, open flame, or excessive heat (a hot drop light, for example). Keep a dry chemical fire extinguisher near the work area. Always keep fuel in a container specifically designed for fuel storage; also, always properly seal fuel containers to avoid the possibility of fire or explosion. Refer to the additional fuel system precautions later in this section.
- Fuel injection systems often remain pressurized, even after the engine has been turned **OFF**. The fuel system pressure must be relieved before disconnecting any fuel lines. Failure to do so may result in fire and/or personal injury.
- Brake fluid often contains polyglycol ethers and polyglycols. Avoid contact with the eyes and wash your hands thoroughly after handling brake fluid. If you do get brake fluid in your eyes, flush your eyes with clean, running water for 15 minutes. If eye irritation persists, or if you have taken brake fluid internally, IMMEDIATELY seek medical assistance.
- The EPA warns that prolonged contact with used engine oil may cause a number of skin disorders, including cancer. You should make every effort to minimize your exposure to used engine oil. Protective gloves should be worn when changing oil. Wash your hands and any other exposed skin areas as soon as possible after exposure to used engine oil. Soap and water, or waterless hand cleaner should be used.
- All new vehicles are now equipped with an air bag system, often referred to as a Supplemental Restraint System (SRS) or Supplemental Inflatable Restraint (SIR) system. The system must be disabled before performing service on or around system components, steering column, instrument panel components, wiring and sensors. Failure to follow safety and disabling procedures could result in accidental air bag deployment, possible personal injury and unnecessary system repairs.
- Always wear safety goggles when working with, or around, the air bag system. When carrying a non-deployed air bag, be sure the bag and trim cover are pointed away from your body. When placing a non-deployed air bag on a work surface, always face the bag and trim cover upward, away from the surface. This will reduce the motion of the module if it is accidentally deployed. Refer to the additional air bag system precautions later in this section.
- Clean, high quality brake fluid from a sealed container is essential to the safe and proper operation of the brake system. You should always buy the correct type of brake fluid for your vehicle. If the brake fluid becomes contaminated, completely flush the system with new fluid. Never reuse any brake fluid. Any brake fluid that is removed from the system should be discarded. Also, do not allow any brake fluid to come in contact with a painted surface; it will damage the paint.
- Never operate the engine without the proper amount and type of engine oil; doing so WILL result in severe engine damage.
- Timing belt maintenance is extremely important. Many models utilize an interference-type, non-freewheeling engine. If the timing belt breaks, the valves in the cylinder head may strike the pistons, causing potentially serious (also time-consuming and expensive) engine damage. Refer to the maintenance interval charts for the recommended replacement interval for the timing belt, and to the timing belt section for belt replacement and inspection.
- Disconnecting the negative battery cable on some vehicles may interfere with the functions of the on-board computer system(s) and may require the computer to undergo a relearning process once the negative battery cable is reconnected.
- When servicing drum brakes, only disassemble and assemble one side at a time, leaving the remaining side intact for reference.
- Only an MVAC-trained, EPA-certified automotive technician should service the air conditioning system or its components.

BRAKES

ANTI-LOCK BRAKE SYSTEM (ABS)

GENERAL INFORMATION

This vehicle is equipped with VDIM (Vehicle Dynamics Integrated Management). It is a concept of vehicle motion control that integrates brake control, drive force control and steering control.

Conventional vehicles use a combination of independent functions such as ABS, TR(A)C, VSC, EPS and VGRS. However VDIM integrates these systems to improve "drive, turn and stop", the motion performance of the vehicle in other words.

In conventional vehicles, control starts at the limit of the vehicle. However VDIM exercises control before the limit is reached, creating a smooth vehicle response. This expands the limits of the vehicle, and increases driving pleasure. Also, with the cooperative control function with the A/T, the vehicle has smooth, natural acceleration.

The VDIM manages all functions, such as the ABS with EBD, brake assist, TR(A)C, and VSC. It is operated by the electronically controlled brake system, which regulates brake fluid pressure. The steering cooperative control function is also available. Thus the VDIM is able to perform comprehensive management.

Conventional brake control systems begin to control either the braking or motive force in order to stabilize the vehicle motion when it becomes unstable due to loss of tire traction. In contrast, in order to maintain stable vehicle control, the VDIM commences controlling the brake and steering systems in accordance with changes in balance before the vehicle becomes unstable. As a result, smooth vehicle control is achieved.

Conventional brake control systems manage all related functions, such as the ABS with EBD, brake assist, TR(A)C and VSC independently, according to the vehicle dynamics. In contrast, the VDIM provides smooth control by seamlessly integrating all those functions.

Electronically controlled brake system description: The system detects the degree of brake pedal operation with a brake pedal stroke sensor and two master cylinder pressure sensors, and calculates the optimum hydraulic brake force. The hydraulic pressure source is adjusted based on this so that the optimum hydraulic control is independently performed on all four wheels.

Meanwhile, the electronically controlled brake system performs control of the normal brakes, RBS (Regenerative Brake Cooperative Control System), ABS, TR(A)C, VSC and brake assist in accordance with the operations of the driver. There is a hydraulic backup mechanism that applies master cylinder pressure generated by human power to the wheel cylinder when the brake control stops. In addition, as a fail-safe mechanism, when the brake control is malfunctioning, the system excludes the malfunctioning sections and continues to perform brake control on the normal sections. A power backup unit (brake control power supply) is also used to ensure a stable supply of power to the system.

Function description: The yaw rate and acceleration sensor communicates with the skid control ECU via CAN communication.

The Anti-lock Brake System (ABS) helps prevent wheels from locking when the brake is applied firmly or when braking on a slippery surface. The ABS warning light comes on when the ABS system is malfunctioning.

The Electronic Brake force Distribution (EBD) control utilizes ABS, realizing proper brake force distribution between front and rear wheels in accordance with driving conditions. In addition, during cornering braking, it also controls the brake force of the right and left wheels, helping to maintain vehicle behavior. Both of the ABS and BRAKE warning lights come on to indicate a malfunction in the EBD system.

BA (Brake Assist): The primary purpose of the brake assist system is to provide auxiliary brake force to assist the driver who cannot generate a brake force large enough during emergency braking, thus helping to maximize the vehicle's brake performance. The ABS warning light comes on to indicate a malfunction in the BA system.

TR(A)C (Traction Control): The TR(A)C system helps prevent the drive wheels from slipping when the driver depresses the accelerator pedal excessively while starting off or accelerating on a slippery surface.

VSC (Vehicle Stability Control): The VSC system helps prevent the vehicle from slipping sideways as a result of strong front or rear wheel skid during cornering.

PRECAUTIONS

⁂ CAUTION

While the battery is connected, even if the engine switch is off, the brake control system activates when the brake pedal is depressed or the door courtesy switch turns on. Therefore during servicing of the brake system components, do not operate the brake pedal and open/close the doors while the battery is connected.

⁂ CAUTION

The Techstream must be used when bleeding air. If not used, the air bleeding will be incomplete, which is hazardous and may lead to an accident.

- Certain components within the ABS system are not intended to be serviced or repaired individually.
- When adjusting the brake fluid level, turn the engine switch on (IG) and then add fluid until the fluid level is at the MAX line.
- Care must be taken to replace each part properly as it could affect the performance of the brake system and result in a driving hazard. Replace the parts with those having the same part number or equivalent.
- If the vehicle is equipped with a mobile communication system, refer to the precautions.
- Care must be taken when using magnets as they could affect the performance of the speed sensors.
- Since the brake line is classified as one of the critical safety related parts, be sure to disassemble the components if a brake fluid leak is found. If any abnormality is found, replace the component with a new one.
- Do not damage or deform the brake tubes when removing or installing them. When installing a brake tube or flexible hose, ensure that they are free from twisting or bending.
- Flexible hoses must be free from absorber oil, grease, etc. If the cap of the flexible hose does not match the groove on the bracket, twist the hose slightly to insert it.
- After installing a brake tube and flexible hose, ensure that they do not interfere with any other components.
- When installing a brake tube to a plastic clamp, ensure that the brake tube is not loose or being pinched.
- Do not reuse a clip or plastic clamp removed from a flexible hose.
- Do not allow brake fluid to adhere to any painted surface such as the vehicle body. If brake fluid leaks onto any painted surface, immediately clean it off.
- Do not use rubber hoses or other parts not specifically specified for and ABS system. When using repair kits, replace all parts included in the kit. Partial or incorrect repair may lead to functional problems and require the replacement of components.
- Lubricate rubber parts with clean, fresh brake fluid to ease assembly. Do not use shop air to clean parts; damage to rubber components may result.
- Use only DOT 3 brake fluid from an unopened container.
- If any hydraulic component or line is removed or replaced, it may be necessary to bleed the entire system.
- A clean repair area is essential. Always clean the reservoir and cap thoroughly before removing the cap. The slightest amount of dirt in the fluid may plug an orifice and impair the system function. Perform repairs after components have been thoroughly cleaned; use only denatured alcohol to clean components. Do not allow ABS components to come into contact with any substance containing mineral oil; this includes used shop rags. When removing brake components, cover the brake tube connections to prevent foreign matter such as dust or dirt from entering the tubes.
- The Anti-Lock control unit is a microprocessor similar to other computer units in the vehicle. Ensure that the ignition switch is **OFF** before removing or installing controller harnesses. Avoid static electricity discharge at or near the controller.
- If any arc welding is to be done on the vehicle, the control unit should be unplugged before welding operations begin.

BRAKES — BLEEDING THE BRAKE SYSTEM

BLEEDING PROCEDURE

Proprietary testing equipment is required. The Techstream must be used during fluid replacement. If not used, the fluid replacement will be incomplete, which is hazardous and may lead to an accident.

BRAKES — FRONT DISC BRAKES

BRAKE CALIPER

REMOVAL & INSTALLATION

See Figure 1.

**** WARNING**

While the battery is connected, even if the engine switch is off, the brake control system activates when the brake pedal is depressed or the door courtesy switch turns on. Therefore during servicing of the brake system components, do not operate the brake pedal and open/close the doors while the battery is connected.

**** WARNING**

After the engine switch is turned off, the HDD navigation system requires approximately 6 minutes to record various types of memory and settings. As a result, after turning the engine switch off, wait 6 minutes or more before disconnecting the cable from the negative battery terminal.

1. Remove the right cowl top ventilator louver.
2. Disconnect cable from negative battery terminal.

**** CAUTION**

Wait at least 90 seconds after disconnecting the cable from the negative battery terminal to prevent airbag and seat belt pretensioner activation.

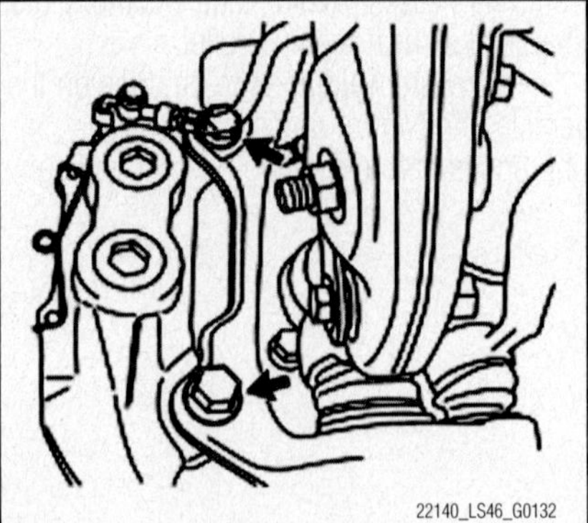

Fig. 1 Front disc brake caliper

3. Remove the front wheel.
4. Remove front disc brake pad kit.
5. Drain brake fluid.
6. Remove the union bolt and gasket from the disc brake caliper, then disconnect the flexible hose from the disc brake caliper.
7. Remove the 2 bolts and disc brake caliper from the knuckle.

To install:

8. Installation is the reverse of removal, noting the following:
    a. Tighten the brake caliper bolts to 100 ft. lbs. (135 Nm).
    b. Tighten the flexible hose union bolt to 29 ft. lbs. (39 Nm).
9. Fill reservoir and bleed brake system.

DISC BRAKE PADS

REMOVAL & INSTALLATION

See Figures 2 through 6.

**** WARNING**

While the battery is connected, even if the engine switch is off, the brake control system activates when the brake pedal is depressed or the door courtesy switch turns on. Therefore during servicing of the brake system components, do not operate the brake pedal and open/close the doors while the battery is connected.

**** WARNING**

After the engine switch is turned off, the HDD navigation system requires approximately 6 minutes to record various types of memory and settings. As a result, after turning the engine switch off, wait 6 minutes or more before disconnecting the cable from the negative battery terminal.

1. Disconnect cable from negative battery terminal.

**** CAUTION**

Wait at least 90 seconds after disconnecting the cable from the negative battery terminal to prevent airbag and seat belt pretensioner activation.

2. Remove the front wheel.
3. Remove front disc brake pad kit, as follows:
    a. While pressing the area labeled A, push the hole pin (labeled B) toward the

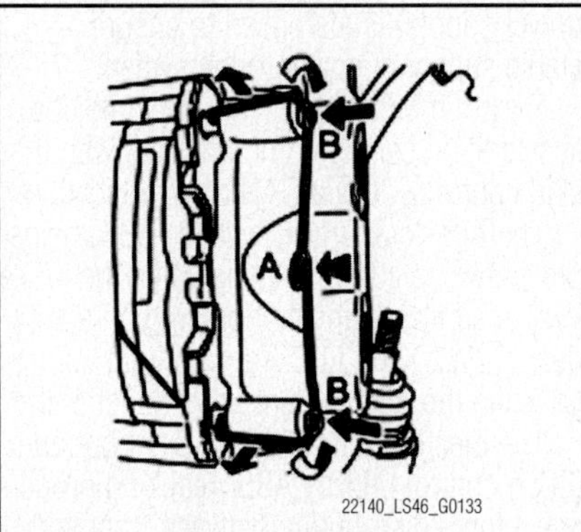

Fig. 2 Press on area "A", push the hole pin "B", and remove pin hole clip

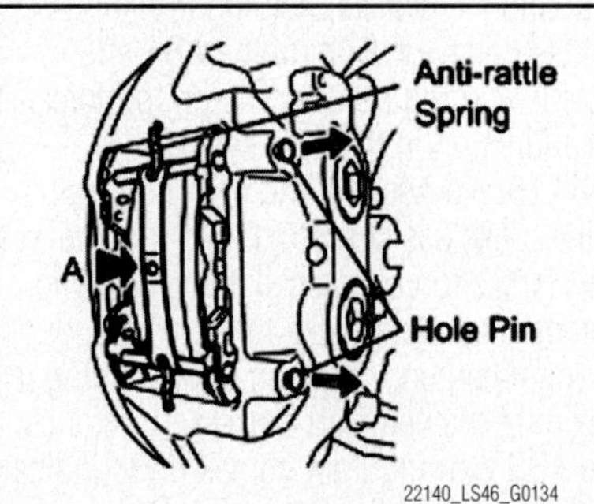

Fig. 3 Press on area "A" and remove the 2 front disc brake anti-rattles with hole pins and anti-rattle spring

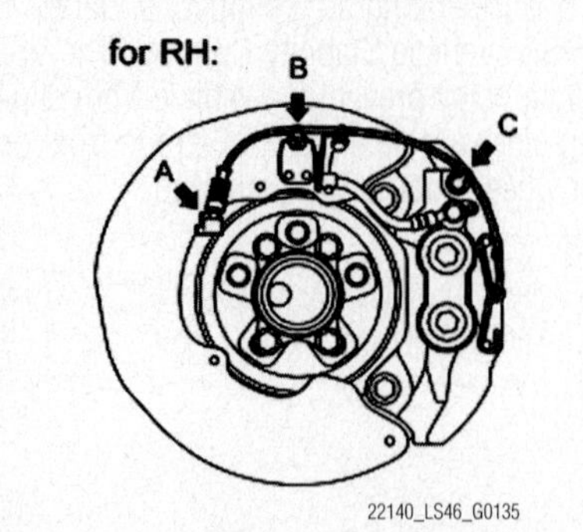

Fig. 4 Front pad wear indicator, clamp, and bleeder plug cap

brake caliper, and remove the pin hold clip.

➡**The pin hold clip can be used again if it has sufficient rebound; no deformation or wear; and has had all rust, dirt and foreign particles cleaned off.**

b. While pressing the area labeled A, remove the 2 front disc brake anti-rattles with hole pins and front disc brake anti-rattle spring.

➡**The anti-rattle spring can be used again if it has sufficient rebound; no deformation, cracks or wear; and has had all rust, dirt and foreign particles cleaned off.**

c. Remove the front disc brake anti-rattle spring.

d. Remove the 2 pads from the disc brake caliper.

e. Remove the No. 1 and No. 2 anti-squeal shims and anti-rattle spring from each pad.

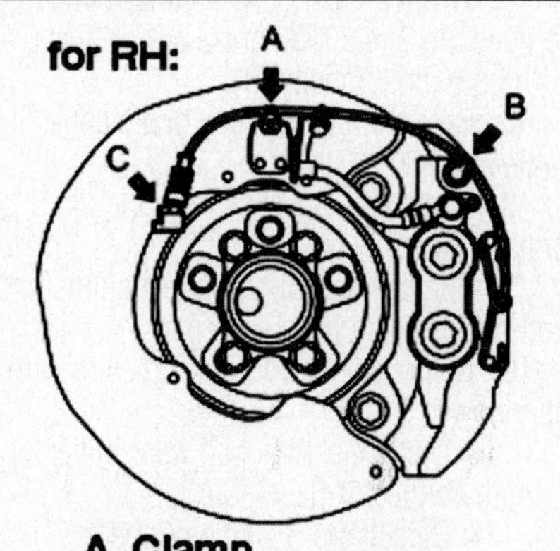

Fig. 5 Front brake clamp, bleeder plug cap, and pad wear indicator wire connector

4. Remove the pad wear indicator wire as follows:

a. Disconnect the pad wear indicator wire connector (labeled A).

b. Detach the clamp (labeled B) and bleeder plug cap (labeled C).

c. Remove the retainer and pad wear indicator wire from the inner pad.

To install:

➡**When replacing worn pads, the anti-squeal shims must be replaced together with the pad.**

➡**When installing the shim, make sure its arrow is pointing in the direction of disc rotation for forward movement.**

➡**Install each shim in the correct position and direction.**

➡**Install each pad as shown in the illustration.**

➡**There should be no oil or grease on the friction surface of the pads and the disc.**

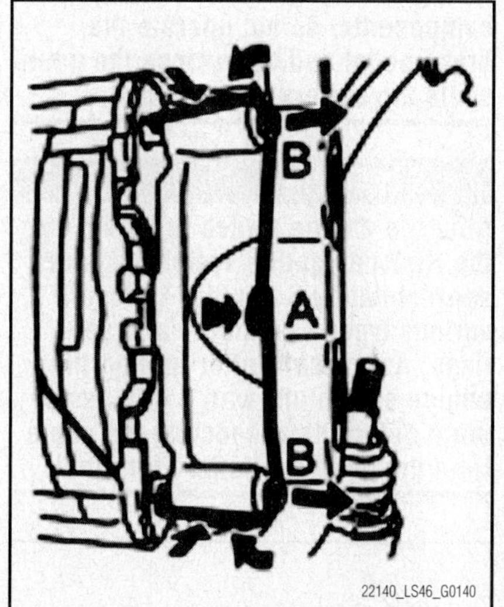

Fig. 6 Front anti-rattle hole pin installation

➡**When the brake pedal is first depressed after replacing the brake pad, DTC C1341, C1342, C1343 and/or C1344 may be output. As there is no malfunction, delete the DTC(s).**

5. Install the pad wear indicator wire as follows:

a. Install the pad wear indicator wire and a new retainer to the inner pad.

6. Attach the clamp (labeled A) and bleeder plug cap (labeled B).

7. Connect the pad wear indicator wire connector (labeled C).

8. Install the anti-rattle spring to each pad. Install the spring lock securely in the groove of the pad.

9. Apply disc brake grease to the sides of the 2 No. 1 anti-squeal shims that contact the disc brake pad.

➡**Do not apply grease to the sides of the 2 No. 1 anti-squeal shims that contact the No. 2 anti-squeal shims.**

10. Install the No. 1 and No. 2 anti-squeal shims to each pad.

11. Install the 2 pads to the disc brake caliper.

12. Install the front disc brake anti-rattle spring between the 2 pads.

➡**The anti-rattle spring and pin hold clip can be used again if they have sufficient rebound; no deformation, cracks or wear; and have had all rust, dirt and foreign particles cleaned off.**

13. Install the front disc brake anti-rattle spring to the disc brake caliper.

14. While pressing the area labeled A, install the 2 front disc brake anti-rattles with hole pins.

15. While pressing the area labeled A, slightly pull out the hole pin (labeled B) from the brake caliper, and install the pin hold clip.

16. Install the front wheel.

BRAKE CALIPER

REMOVAL & INSTALLATION

See Figure 7.

**** WARNING**

While the battery is connected, even if the engine switch is off, the brake control system activates when the brake pedal is depressed or the door courtesy switch turns on. Therefore during servicing of the brake system components, do not operate the brake pedal and open/close the doors while the battery is connected.

**** WARNING**

After the engine switch is turned off, the HDD navigation system requires approximately 6 minutes to record various types of memory and settings. As a result, after turning the engine switch off, wait 6 minutes or more before disconnecting the cable from the negative battery terminal.

1. Remove the right cowl top ventilator louver.
2. Disconnect cable from negative battery terminal.

**** CAUTION**

Wait at least 90 seconds after disconnecting the cable from the negative battery terminal to prevent airbag and seat belt pretensioner activation.

3. Remove the rear wheel.
4. Remove disc brake pad kit.
5. Drain brake fluid.
6. Remove the union bolt and gasket from the disc brake caliper, then disconnect the flexible hose.
7. Remove the 2 bolts and disc brake caliper from the knuckle.

To install:

8. The remainder of installation is the reverse of removal, noting the following:
   a. Tighten the brake caliper bolts to 63 ft. lbs. (86 Nm).
   b. Tighten the flexible hose union bolt to 29 ft. lbs. (39 Nm).
9. Fill reservoir and bleed brake system.

DISC BRAKE PADS

REMOVAL & INSTALLATION

See Figures 8 through 12.

**** WARNING**

While the battery is connected, even if the engine switch is off, the brake control system activates when the brake pedal is depressed or the door courtesy switch turns on. Therefore during servicing of the brake system components, do not operate the brake pedal and open/close the doors while the battery is connected.

**** WARNING**

After the engine switch is turned off, the HDD navigation system requires approximately 6 minutes to record various types of memory and settings. As a result, after turning the engine switch off, wait 6 minutes or more before disconnecting the cable from the negative battery terminal.

1. Remove the right cowl top ventilator louver.
2. Disconnect cable from negative battery terminal.

**** CAUTION**

Wait at least 90 seconds after disconnecting the cable from the negative battery terminal to prevent airbag and seat belt pretensioner activation.

3. Remove the rear wheel.
4. Remove disc brake pad kit, as follows:
5. While pressing the area labeled A, push the hole pin (labeled B) toward the brake caliper, and remove the pin hold clip.

➡The pin hold clip and anti-rattle spring can be used again if they have sufficient rebound; no deformation or wear; and have had all rust, dirt and foreign particles cleaned off.

6. While pressing the area labeled A, remove the 2 rear disc brake anti-rattles with hole pins (labeled B).
7. Remove the rear disc brake anti-rattle spring.
8. Remove the 2 pads from the disc brake caliper.
9. Remove the 2 anti-squeal shims and anti-rattle spring from each pad.
10. Remove the pad wear indicator wire as follows:
    a. Disconnect the pad wear indicator wire connector (labeled A).
    b. Detach the 3 clamps (labeled B) and bleeder plug cap (labeled C).
    c. Remove the retainer and pad wear indicator wire from the inner pad.

To install:

➡When replacing worn pads, the anti-squeal shims must be replaced together with the pad.

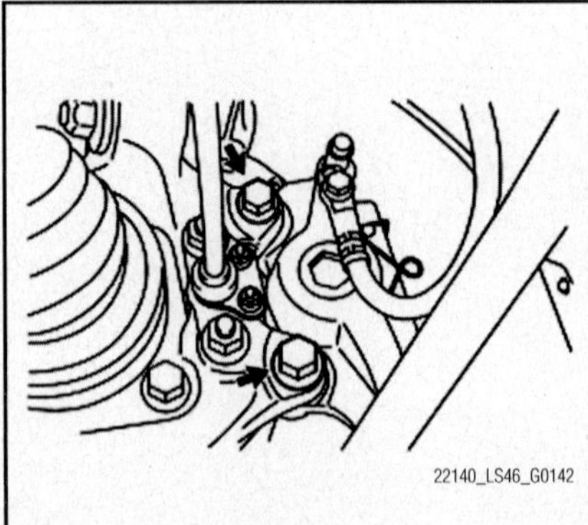

Fig. 7 Rear disc brake caliper and bolts

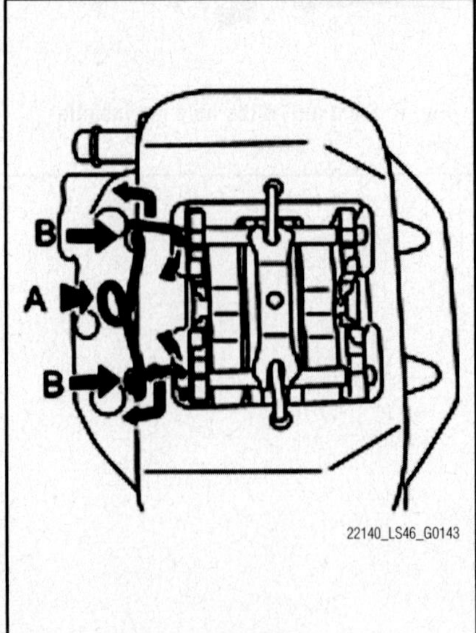

Fig. 8 Pin hole clip removal

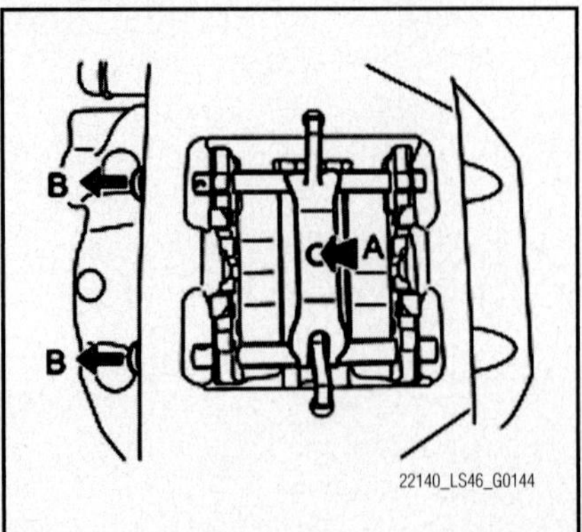

Fig. 9 Rear disc brake anti-rattle removal

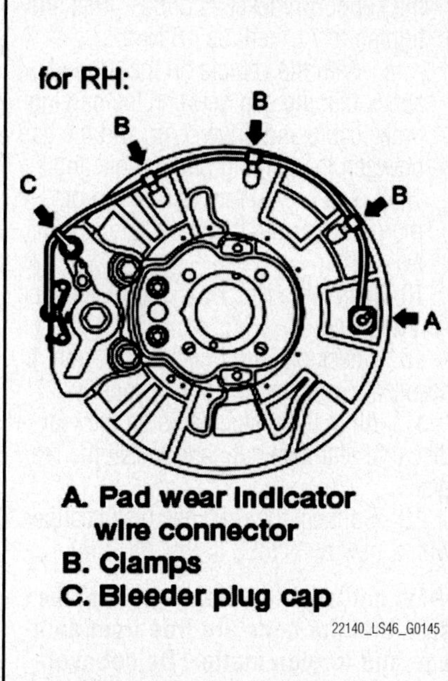

Fig. 10 Pad wear indicator wire connector, clamps, and bleeder plug cap

➡**When installing the shim, make sure its arrow is pointing in the direction of disc rotation for forward movement.**

➡**Install each shim in the correct position and direction.**

➡**Install each pad as shown in the illustration.**

➡**There should be no oil or grease on the friction surface of the pads and the disc.**

➡**When the brake pedal is first depressed after replacing the brake pad, DTC C1341, C1342, C1343 and/or C1344 may be output. As there is no malfunction, delete the DTC(s).**

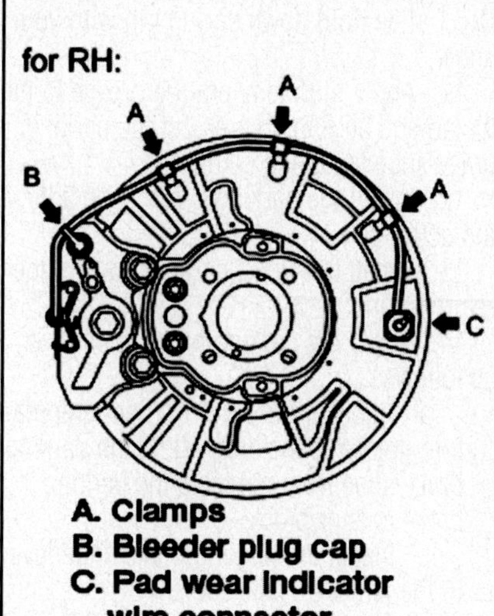

Fig. 11 Rear brake clamps, bleeder plug cap, and pad wear indicator connector

11. Install the pad wear indicator wire as follows:
    a. Install the pad wear indicator wire and a new retainer to the inner pad.
12. Attach the 3 clamps (labeled A) and bleeder plug cap (labeled B).
13. Connect the pad wear indicator wire connector (labeled C).
14. Install the anti-rattle spring to each pad. Install the spring lock securely in the groove of the pad.
15. Apply disc brake grease to the sides of the 2 No. 1 anti-squeal shims that contact the disc brake pad.

➡**Do not apply grease to the sides of the 2 No. 1 anti-squeal shims that contact the No. 2 anti-squeal shims.**

16. Install the No. 1 and No. 2 anti-squeal shims to each pad.
17. Install the 2 pads to the disc brake caliper.

➡**The anti-rattle spring and pin hold clip can be used again if they have sufficient rebound; no deformation, cracks or wear; and have had all rust, dirt and foreign particles cleaned off.**

18. Install the rear disc brake anti-rattle spring to the disc brake caliper.
19. While pressing the area labeled A, install the 2 rear disc brake anti-rattles with hole pins (labeled B).
20. While pressing the area labeled A, slightly pull out the hole pin (labeled B) from the brake caliper, and install the pin hold clip.
21. Install the front wheel.

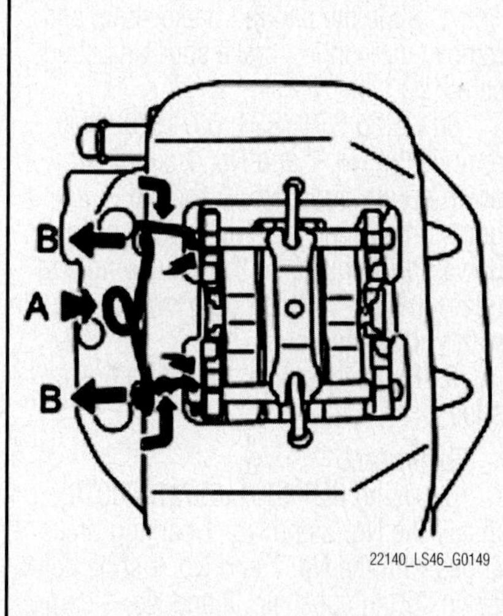

Fig. 12 Rear anti-rattle hole pin installation

BRAKES

PARKING BRAKE

PARKING BRAKE SHOES

REMOVAL & INSTALLATION

See Figures 13 through 16.

WARNING

While the battery is connected, even if the engine switch is off, the brake control system activates when the brake pedal is depressed or the door courtesy switch turns on. Therefore during servicing of the brake system components, do not operate the brake pedal and open/close the doors while the battery is connected.

WARNING

After the engine switch is turned off, the HDD navigation system requires approximately 6 minutes to record various types of memory and settings. As a result, after turning the engine switch off, wait 6 minutes or more before disconnecting the cable from the negative battery terminal.

1. Remove the right cowl top ventilator louver.
2. Disconnect cable from negative battery terminal.

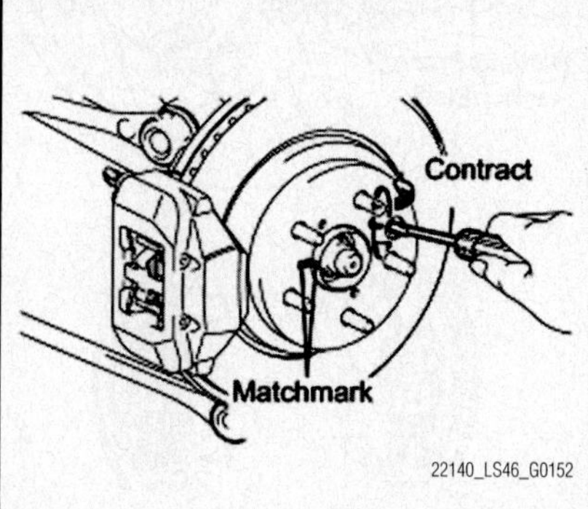

Fig. 13 Rear disc and rear axle hub matchmarks

CAUTION

Wait at least 90 seconds after disconnecting the cable from the negative battery terminal to prevent airbag and seat belt pretensioner activation.

3. Remove the 2 bolts and disconnect the rear disc brake caliper. Hang the caliper with wire or equivalent.
4. Put matchmarks on the rear disc and rear axle hub.
5. Remove the rear disc.

➡If the rear disc is difficult to remove, turn the adjustment screw in the contraction direction as shown in the illustration to make the disc easier to remove.

6. Using SST (SST: 09703-30011), remove the No. 2 and No. 1 parking brake shoe return tension springs.
7. Slide the parking brake shoe, and remove the parking brake shoe adjusting screw set.
8. Using SST (SST: 09718-00010), remove the No. 2 and No. 1 shoe hold down spring cups, No. 2 and No. 1 compression springs and No. 1 shoe hold down spring pin. Use the service hole to retain the No. 1 shoe hold down spring pin with your finger.
9. Remove the No. 2 parking brake shoe

To install:

10. Using SST (SST: 09718-00010), install the No. 2 and No. 1 parking brake shoes with the No. 2 and No. 1 shoe hold down spring cups, No. 2 and No. 1 compression springs and No. 1 shoe hold down spring pin. Use the service hole to retain the No. 1 shoe hold down spring pin with your finger.
11. Apply high temperature grease to the thread and all joining areas of the parking brake shoe adjusting screw set.
12. Install the parking brake shoe adjusting screw set.
13. Install the No. 2 parking brake shoe return spring.
14. Install the parking brake shoe lever, as follows:
    a. Apply a thin layer of high temperature grease to the area where the parking brake shoe lever contacts the parking brake anchor block.
    b. Install the parking brake shoe lever to the parking brake cable.

➡Take care to install the correct parking brake shoe lever because the direction of the pin is different between the left and right.

    c. Check that the parking brake inner cable and parking brake shoe lever are securely connected by trying to pull them apart.
    d. Install the parking brake cable to the support bracket with the 2 nuts and tighten to 71 inch lbs. (8 Nm).
    e. With the vehicle on the ground, check that the gap between the parking brake cable and drive shaft, and the gap between the parking brake cable and shaft boot are 10 mm (0.394 in.) or more. Make sure that the cable is not twisted.
15. Install the No. 1 parking brake shoe return spring.
16. Check parking brake installation. If necessary, reinstall the parts properly.
17. Align the matchmarks on the rear disc and rear axle hub, and install the rear disc.
18. Connect the rear disc brake caliper with 2 new bolts to63 ft. lbs. (86 Nm).

➡Do not twist the brake hose, Make sure that the bolts are free from damage and foreign matter. Do not overtighten the bolts.

19. Adjust parking brake shoe clearance.
20. Install the rear wheel
21. Perform parking brake shoe bedding.

ADJUSTMENT

See Figure 17.

1. Remove the rear wheel.
2. Temporarily install the hub nuts.
3. With the engine switch on, operate the electric parking brake switch to release the parking brake, and then turn the engine switch off.
4. Remove the shoe adjusting hole plug, and rotate the rear disc so that the service hole is aligned with the adjusting screw.
5. Using a screwdriver, turn the adjusting screw of the parking brake shoe in the expansion direction until the disc locks.
6. With the engine switch on (IG), operate the electric parking brake switch to lock and release the parking brake. Repeat again. Then turn the engine switch off.

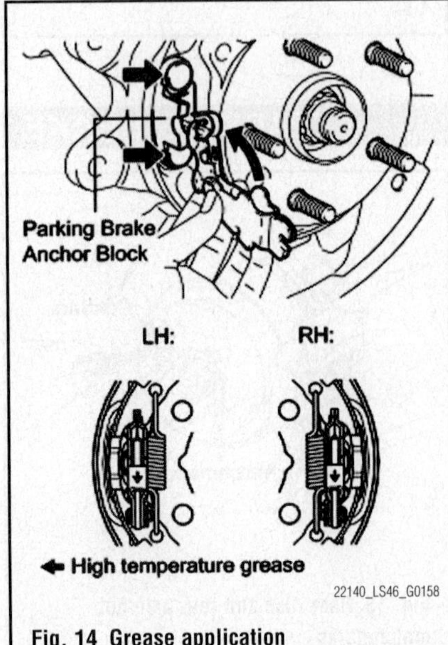

Fig. 14 Grease application

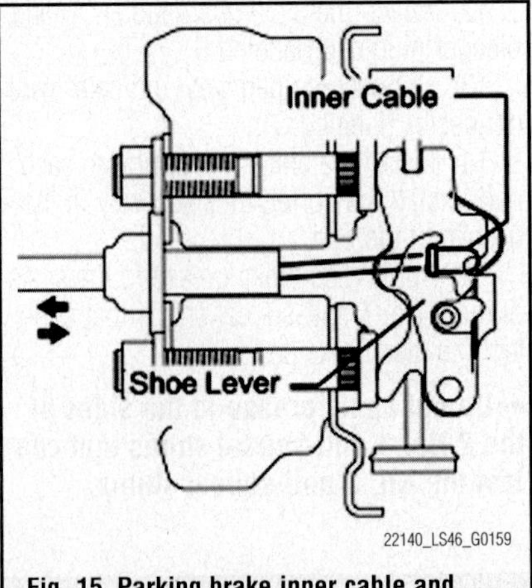

Fig. 15 Parking brake inner cable and shoe lever connection

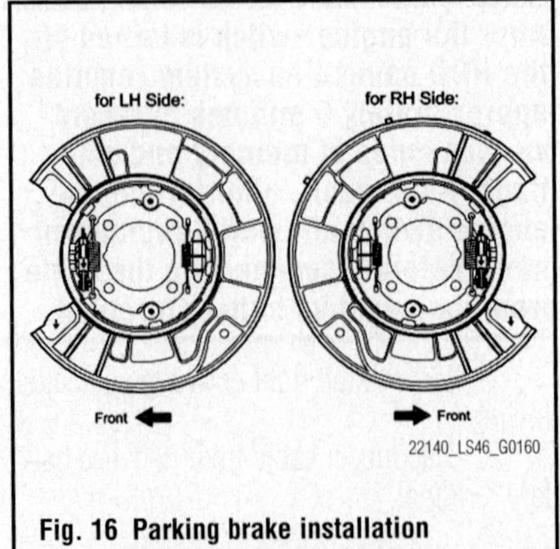

Fig. 16 Parking brake installation

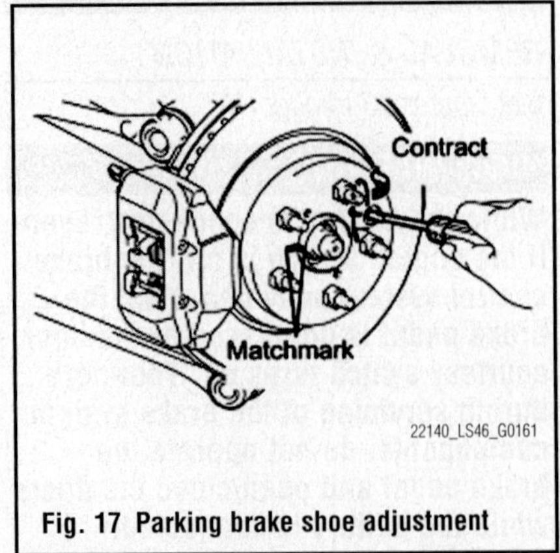

Fig. 17 Parking brake shoe adjustment

➡**Make sure that the parking brake is released.**

7. Turn the adjusting screw again in the expansion direction to lock the disc.

8. Loosen the adjusting screw so that the rear disc can rotate slightly.

➡**The standard number of return notches is 7.**

9. Check that there is no brake drag.
10. Install the shoe adjusting hole plug and remove the hub nuts.
11. Install the rear wheel.

CHASSIS ELECTRICAL

AIR BAG (SUPPLEMENTAL RESTRAINT SYSTEM)

GENERAL INFORMATION

** CAUTION

These vehicles are equipped with an air bag system. The system must be disarmed before performing service on, or around, system components, the steering column, instrument panel components, wiring and sensors. Failure to follow the safety precautions and the disarming procedure could result in accidental air bag deployment, possible injury and unnecessary system repairs. Furthermore, if a mistake is made when servicing SRS parts, they may fail to operate when required.

SERVICE PRECAUTIONS

The vehicle is equipped with a Supplemental Restraint System (SRS), which consists of a steering pad, front passenger airbag, curtain shield airbag, front seat side airbag, driver side knee airbag, front passenger side knee airbag, rear seat side airbag, seat cushion airbag, seat belt pretensioner, center airbag sensor, front airbag sensor, side airbag sensor, rear airbag sensor, occupant classification ECU and seat position airbag sensor.

Before starting work, wait at least 90 seconds after the engine switch is turned off and after the cable of the negative battery terminal is disconnected. (SRS parts are equipped with a backup power source. If work is started within 90 seconds of turning the engine switch off and disconnecting the cable from the negative battery terminal, SRS parts may deploy.)

When the cable is disconnected from the negative battery terminal, the memory settings of each system will be cleared. Because of this, be sure to write down the settings of each system before starting work. When work is finished, reset the settings of each system as before. Never use a backup power supply from outside the vehicle to avoid erasing the memory in a system.

After the engine switch is turned off, the HDD navigation system requires approximately 6 minutes to record various types of memory and settings. As a result, after turning the engine switch off, wait 6 minutes or more before disconnecting the cable from the negative (-) battery terminal.

If the vehicle is equipped with a mobile communication system, refer to the precaution in the INTRODUCTION section.

Do not expose SRS parts directly to hot air or flames.

Be sure to perform the initialization of the occupant classification ECU if any of the following conditions occur: The occupant classification ECU is replaced, accessories (seatback tray, seat cover, etc.) are installed onto the vehicle, the passenger seat is removed from the vehicle, and then reinstalled or replaced, the passenger airbag ON/OFF indicator light (OFF) comes on when the passenger seat is not occupied, the vehicle is brought to the workshop for repair due to an accident or collision. If the initialization is not performed, the SRS may not operate properly.

Malfunction symptoms of SRS parts are difficult to confirm. DTCs are the most important source of information when troubleshooting. During troubleshooting, always confirm DTCs before disconnecting the cable from the negative battery terminal.

For minor collisions where SRS parts do not deploy, always inspect the SRS parts.

If an SRS part has been dropped, or if there are any cracks, dents or other defects in the case, bracket or connector, replace the SRS part with a new one.

Use an ohmmeter/voltmeter with high impedance (10 kohms/V minimum) for troubleshooting the electrical circuits.

Do not place an intact undeployed airbag face down on a solid surface. The airbag will propel into the air if accidentally deployed and may result in personal injury or death.

When carrying or handling an undeployed airbag, the trim side (face) of the airbag should be pointing towards the body to minimize possibility of injury if accidental deployment occurs. Failure to do this may result in personal injury or death.

Replace airbag system components with OEM replacement parts. Substitute parts may appear interchangeable, but internal differences may result in inferior occupant protection. Failure to do so may result in occupant personal injury or death.

Wear safety glasses, rubber gloves, and long sleeved clothing when cleaning powder residue from vehicle after an airbag deployment. Powder residue emitted from a deployed airbag can cause skin irritation. Flush affected area with cool water if irritation is experienced. If nasal or throat irritation is experienced, exit the vehicle for fresh air until the irritation ceases. If irritation continues, see a physician.

Do not use a replacement airbag that is not in the original packaging. This may result in improper deployment, personal injury, or death.

The factory installed fasteners, screws and bolts used to fasten airbag components have a special coating and are specifically designed for the airbag system. Do not use substitute fasteners. Use only original equipment fasteners listed in the parts catalog when fastener replacement is required.

During, and following, any child restraint anchor service, due to impact event or vehicle repair, carefully inspect all mounting hardware, tether straps, and anchors for proper installation, operation, or damage. If a child restraint anchor is found damaged in any way, the anchor must be replaced. Failure to do this may result in personal injury or death.

Deployed and non-deployed airbags may or may not have live pyrotechnic material within the airbag inflator.

Do not dispose of driver/passenger/curtain airbags or seat belt tensioners unless you are sure of complete deployment. Refer to the Hazardous Substance Control System for proper disposal.

Dispose of deployed airbags and tensioners consistent with state, provincial, local, and federal regulations.

After any airbag component testing or service, do not connect the battery negative cable. Personal injury or death may result if the system test is not performed first.

If the vehicle is equipped with the Occupant Classification System (OCS), do not connect the battery negative cable before performing the OCS Verification Test using the scan tool and the appropriate diagnostic information. Personal injury or death may result if the system test is not performed properly.

Never replace both the Occupant Restraint Controller (ORC) and the Occupant Classification Module (OCM) at the same time. If both require replacement, replace one, then perform the Airbag System test before replacing the other.

Both the ORC and the OCM store Occupant Classification System (OCS) calibration data, which they transfer to one another when one of them is replaced. If both are replaced at the same time, an irreversible fault will be set in both modules and the OCS may malfunction and cause personal injury or death.

If equipped with OCS, the Seat Weight Sensor is a sensitive, calibrated unit and must be handled carefully. Do not drop or handle roughly. If dropped or damaged, replace with another sensor. Failure to do so may result in occupant injury or death.

If equipped with OCS, the front passenger seat must be handled carefully as well. When removing the seat, be careful when setting on floor not to drop. If dropped, the sensor may be inoperative, could result in occupant injury, or possibly death.

If equipped with OCS, when the passenger front seat is on the floor, no one should sit in the front passenger seat. This uneven force may damage the sensing ability of the seat weight sensors. If sat on and damaged, the sensor may be inoperative, could result in occupant injury, or possibly death.

After work on the SRS is completed, perform the SRS warning light check.

DISARMING THE SYSTEM

To avoid personal injury when working on vehicles equipped with an air bag, the negative battery cable must be disconnected and at least 90 seconds must elapse before working on the system. Failure to do so may result in deployment of the air bag.

After the engine switch is turned off, the HDD navigation system requires approximately 6 minutes to record various types of memory and settings. As a result, after turning the engine switch off, wait 6 minutes or more before disconnecting the cable from the negative (-) battery terminal.

When the cable is disconnected from the negative battery terminal, the memory settings of each system will be cleared. Because of this, be sure to write down the settings of each system before starting work. When work is finished, reset the settings of each system as before. Never use a backup power supply from outside the vehicle to avoid erasing the memory in a system.

ARMING THE SYSTEM

To arm the system after service is finished, connect the negative battery cable.

DRIVETRAIN

AUTOMATIC TRANSMISSION ASSEMBLY

REMOVAL & INSTALLATION

See Figures 18 and 19.

**** WARNING**

After the engine switch is turned off, the HDD navigation system requires approximately 6 minutes to record various types of memory and settings. As a result, after turning the engine switch off, wait 6 minutes or more before disconnecting the cable from the negative battery terminal.

1. Remove the right cowl top ventilator louver.
2. Disconnect cable from negative battery terminal.

**** CAUTION**

Wait at least 90 seconds after disconnecting the cable from the negative battery terminal to prevent airbag and seat belt pretensioner activation.

3. Remove the propeller with center bearing shaft.
4. Remove the 2 bolts and the No. 1 exhaust pipe support bracket.
5. Drain automatic transmission fluid.
6. Remove the nut and disconnect the floor shift gear shifting rod.
7. Remove the nut and the right transmission control shaft lever.
8. Remove the right and left rear engine under cover.
9. Remove the 4 bolts and disconnect the front stabilizer bar from the rear brace.
10. Remove the 4 bolts and the left front suspension member reinforcement from the vehicle.
11. Remove the 4 bolts and the right front suspension member reinforcement from the vehicle.
12. Remove the steering sliding yoke with shaft sub-assembly.
13. Remove the No. 2 steering intermediate shaft assembly.
14. Remove the left exhaust manifold sub-assembly.
15. Remove the generator.
16. Remove the engine oil level dipstick guide.
17. Remove the No. 1 exhaust manifold heat insulator.
18. Remove the right exhaust manifold sub-assembly.
19. Remove the No. 3 exhaust manifold heat insulator.
20. Remove the starter assembly.
21. Disconnect the oil cooler tube inlet and outlet, as follows:
    a. Remove the 2 bolts and 2 clamps.
    b. Loosen the union nuts of the oil cooler tube inlet and outlet and disconnect them.
22. Support the automatic transmission with a transmission jack.
23. Remove the 4 bolts and disconnect the rear engine mounting member.
24. Remove the 4 nuts and the rear engine mounting member.
25. Disconnect the wire harness and connector, as follows:

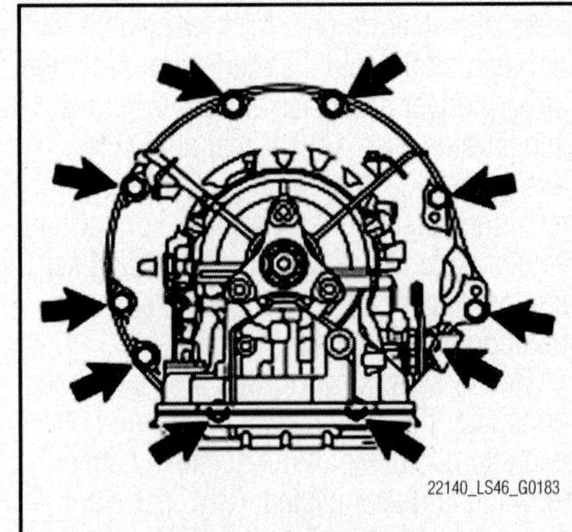

Fig. 18 Transmission bolt positions

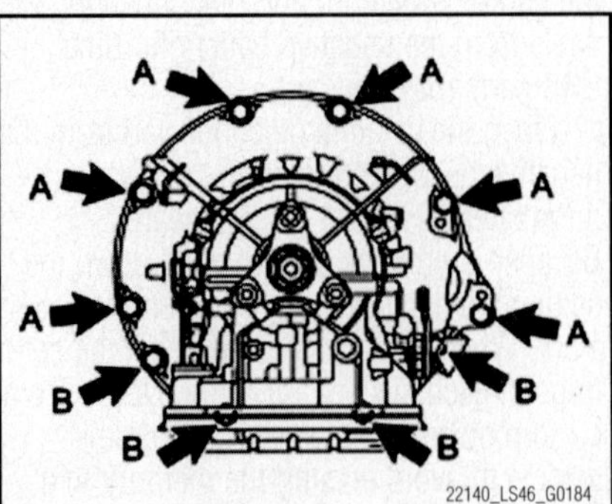

Fig. 19 Transmission bolt tightening sequence

a. Remove the bolt and disconnect the ground wire.

b. Tilt the automatic transmission downward.

➡Take care so that the rear side of the engine assembly does not come into contact with the body.

c. Disconnect the park/neutral position switch connector.

d. Disconnect the transmission wire connector: detach the claw, press down the lever, and then disconnect the transmission wire connector.

e. Detach the 5 harness clamps from the automatic transmission.

26. Remove the automatic transmission assembly, as follows:

a. Remove the flywheel housing side cover.

b. Turn the crankshaft to gain access to each bolt.

c. Hold the crankshaft pulley nut with a wrench and remove the 6 bolts.

d. Remove the 10 bolts.

e. Separate and remove the automatic transmission.

To install:

27. Install the automatic transmission to the engine with the 10 bolts tighten the A bolts to 52 ft. lbs. (71 Nm) and B bolts to 27 ft. lbs. (37 Nm).

28. Install the 6 torque converter clutch mounting bolts and tighten to 35 ft. lbs. (48 Nm). First install the black colored bolt and then the remaining 5 bolts.

29. Install the flywheel housing side cover.

30. Connect the park/neutral position switch connector.

31. Connect the transmission wire connector. Push up the lever until the claw of the transmission wire connector makes a connection sound.

32. Attach the 5 harness clamps to the automatic transmission.

33. Install the engine rear mounting member to the automatic transmission assembly with the 4 nuts and tighten to 10 ft. lbs. (14 Nm).

34. Install the engine rear mounting member to the body with the 4 bolts and tighten to 19 ft. lbs. (26 Nm).

35. Tilt up the automatic transmission assembly.

36. Install the bolt and ground wire to the automatic transmission assembly and tighten to 48 inch lbs. (5.4 Nm).

37. Connect the oil cooler tube inlet and outlet, as follows:

a. Temporarily install the oil cooler tube inlet and outlet.

b. Install the 2 clamps with the 2 bolts and tighten to 44 inch lbs. (5 Nm).

c. Tighten union nuts of the oil cooler tube inlet and outlet to32 ft. lbs (44 Nm), without SST, or 29 ft. lbs. (39 Nm) with SST: 09023-12701. Use a torque wrench with a fulcrum length of 460 mm (18.11 in.).

38. Install the starter assembly.

39. Install the No. 3 exhaust manifold heat insulator.

40. Install the right exhaust manifold sub-assembly.

41. Install the No. 1 exhaust manifold heat insulator.

42. Install the engine oil level dipstick guide.

43. Install the generator.

44. Install the left exhaust manifold sub-assembly.

45. Install the No. 2 steering intermediate shaft assembly.

46. Install the steering sliding yoke with shaft sub-assembly.

47. Install the right and left front suspension member reinforcement to the vehicle with the 4 bolts each and tighten to 37 ft. lbs. (50 Nm).

48. Connect the stabilizer bar to the rear brace with the 4 bolts and tighten to 36 ft. lbs. (49 Nm).

49. Install the right and left rear engine under cover.

50. Install the right transmission control shaft lever with the nut and tighten to 12 ft. lbs. (16 Nm).

51. Temporarily connect the floor shift gear shifting rod to the connecting rod swivel with the nut. The nut will be tightened to a torque specification during the shift lever position adjustment procedure.

52. Install the No. 1 exhaust pipe support bracket with the 2 bolts and tighten to 32 ft. lbs. (43 Nm).

53. Install the propeller with center bearing shaft.

54. Connect the cable to negative battery terminal.

55. Install the cowl top ventilator louver.

56. Adjust shift lever position.

57. Add the automatic transmission fluid.

58. Check automatic transmission system precaution.

59. Perform initialization.

REAR AXLE SHAFT, BEARING & SEAL

REMOVAL & INSTALLATION

See Figures 20 through 26.

1. Remove the rear wheel.

2. Remove the bolt and load sensing valve sensor bracket from the toe control link sub-assembly.

3. Remove the bolt and nut, and then disconnect the rear stabilizer link assembly from the rear No. 2 suspension arm assembly.

4. Using SST (SST: 09930-00010) and a hammer, release the staked part of the rear axle shaft nut. Release the staked part of the nut completely, otherwise the screw of the drive shaft may be damaged.

5. While applying the brakes, remove the rear axle shaft nut.

6. Remove the 2 bolts and separate the rear disc brake caliper assembly.

➡Hang the caliper with wire or equivalent. Do not damage the brake hose.

7. Disconnect the rear speed sensor.

8. Remove the rear disc.

9. Remove the No. 2 and No. 1 parking brake shoe return tension springs.

10. Remove the parking brake shoe adjusting screw set.

11. Remove the No. 1 and No. 2 parking brake shoe assemblies.

12. Remove the parking brake shoe lever.

13. Disconnect the rear shock absorber assembly, for coil suspension.

14. Disconnect the pneumatic cylinder with rear shock absorber assembly, for air suspension.

15. Remove the rear No. 2 suspension arm assembly, as follows:

a. Using a plastic-faced hammer or equivalent, strike the part labeled A from the rear of the vehicle to maintain the clearance at the slide pin area.

b. Remove the nut, washer, bolt and No. 2 suspension arm from the axle carrier.

c. Remove the nut, rear suspension attachment sub-assembly, and rear No. 2 suspension toe adjust plate, and then remove the rear No. 2 suspension arm assembly.

16. Remove the rear No. 1 suspension arm assembly, as follows:

a. Remove the nut from the rear No. 1 suspension arm.

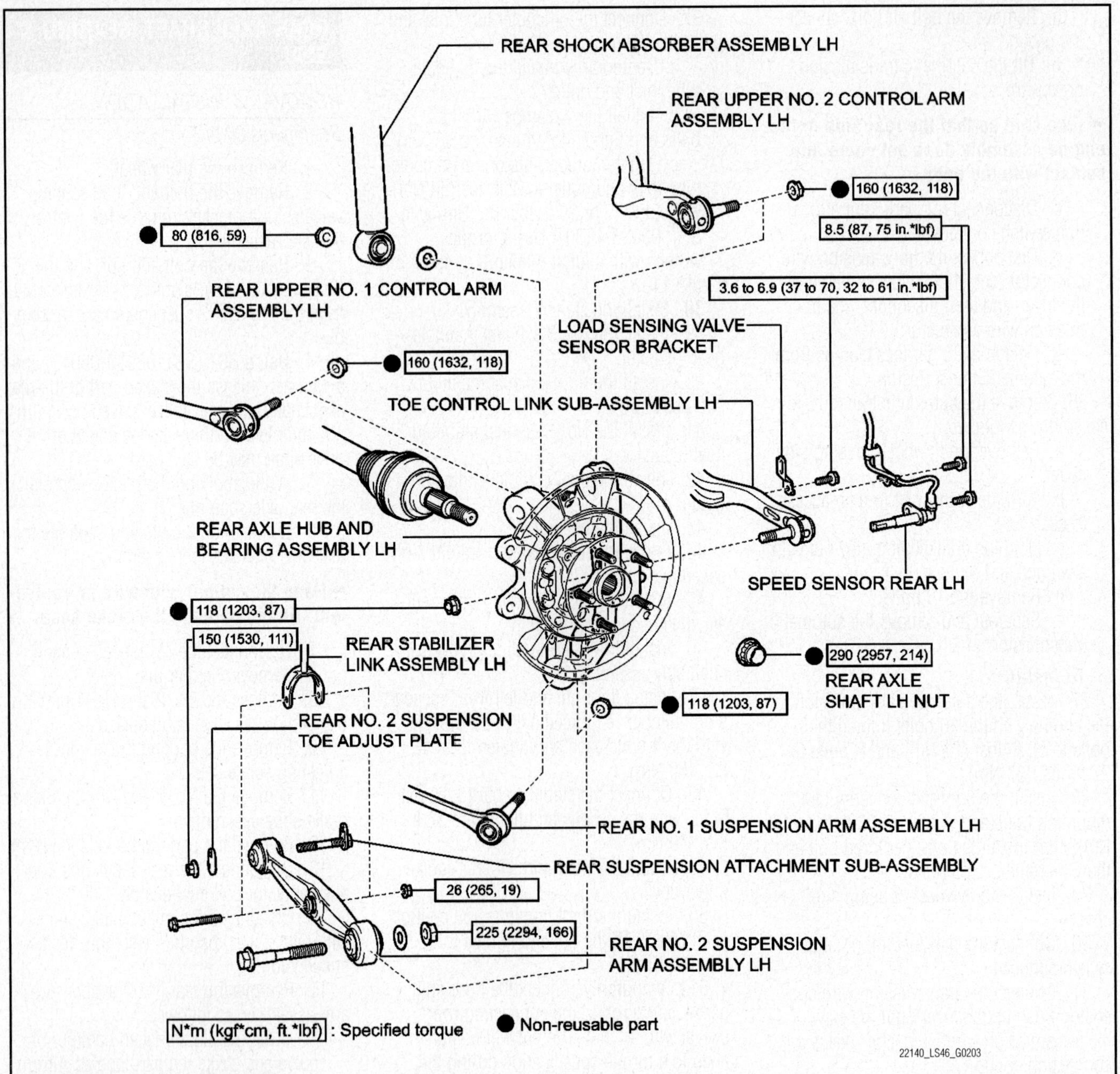

REAR SHOCK ABSORBER ASSEMBLY LH
REAR UPPER NO. 2 CONTROL ARM ASSEMBLY LH
160 (1632, 118)
8.5 (87, 75 in.*lbf)
80 (816, 59)
REAR UPPER NO. 1 CONTROL ARM ASSEMBLY LH
3.6 to 6.9 (37 to 70, 32 to 61 in.*lbf)
LOAD SENSING VALVE SENSOR BRACKET
160 (1632, 118)
TOE CONTROL LINK SUB-ASSEMBLY LH
REAR AXLE HUB AND BEARING ASSEMBLY LH
SPEED SENSOR REAR LH
118 (1203, 87)
150 (1530, 111)
REAR STABILIZER LINK ASSEMBLY LH
290 (2957, 214)
REAR AXLE SHAFT LH NUT
118 (1203, 87)
REAR NO. 2 SUSPENSION TOE ADJUST PLATE
REAR NO. 1 SUSPENSION ARM ASSEMBLY LH
REAR SUSPENSION ATTACHMENT SUB-ASSEMBLY
26 (265, 19)
225 (2294, 166)
REAR NO. 2 SUSPENSION ARM ASSEMBLY LH
N*m (kgf*cm, ft.*lbf) : Specified torque
Non-reusable part
22140_LS46_G0203

Fig. 20 Rear axle hub components

b. Install 2 spacers (SST spacer B) onto the rear No. 1 suspension arm so that there is a space of approximately 1 mm between the arm and spacers, SST: 09960-20010.

➡Make sure to install the spacers (SST spacer B) as the steering knuckle spacer may shift.

➡As SST may become damaged, make sure the space between the arm and spacers is not 1 mm or less.

c. Using SST (SST: 09960-20010), disconnect the rear No. 1 suspension arm from the axle carrier.

➡Do not damage the ball joint dust cover. As the dust cover may be damaged, adjust SST with the center nut so that the body and crow are parallel. Make sure to tie the string of SST to the vehicle to prevent SST from dropping.

17. Disconnect the toe control link sub-assembly, as follows:

a. Remove the nut on the rear axle carrier side.

b. Install 2 spacers (SST spacer B) onto the toe control link so that there is a space of approximately 1 mm between the arm and spacers. SST: 09960-20010.

A
Arm
Slide Pin
Axle Carrier
22140_LS46_G0185

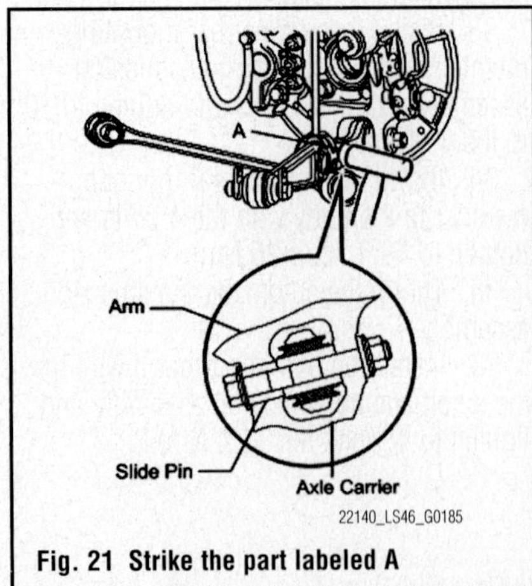

Fig. 21 Strike the part labeled A

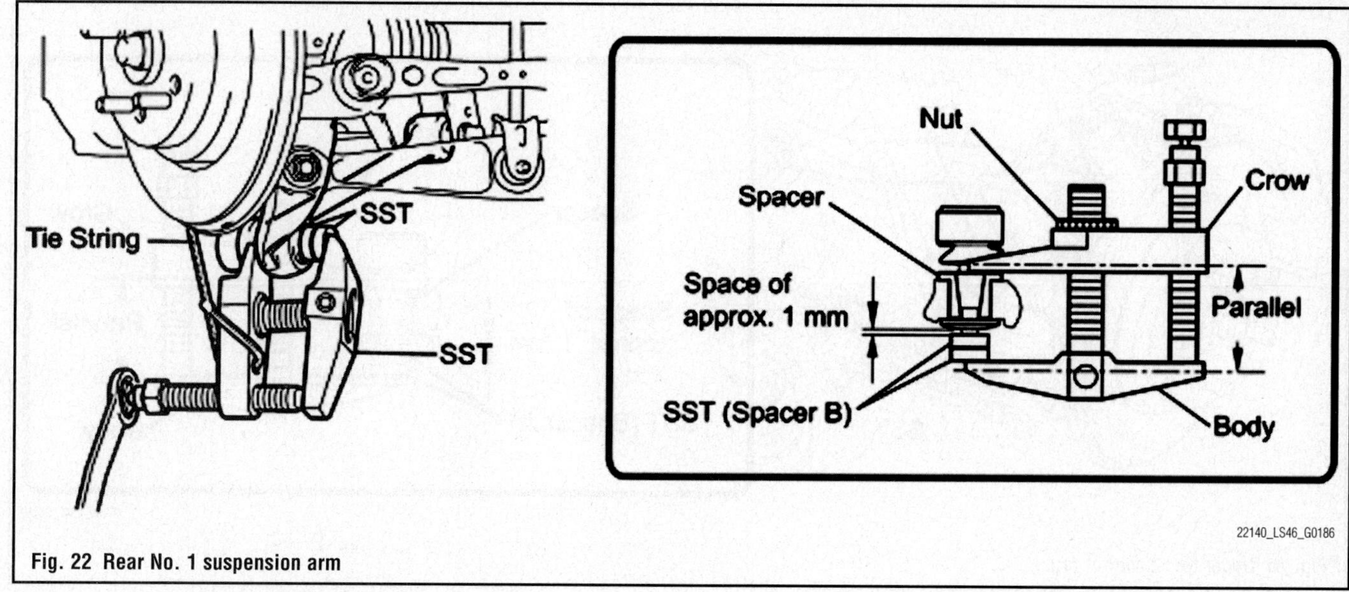

Fig. 22 Rear No. 1 suspension arm

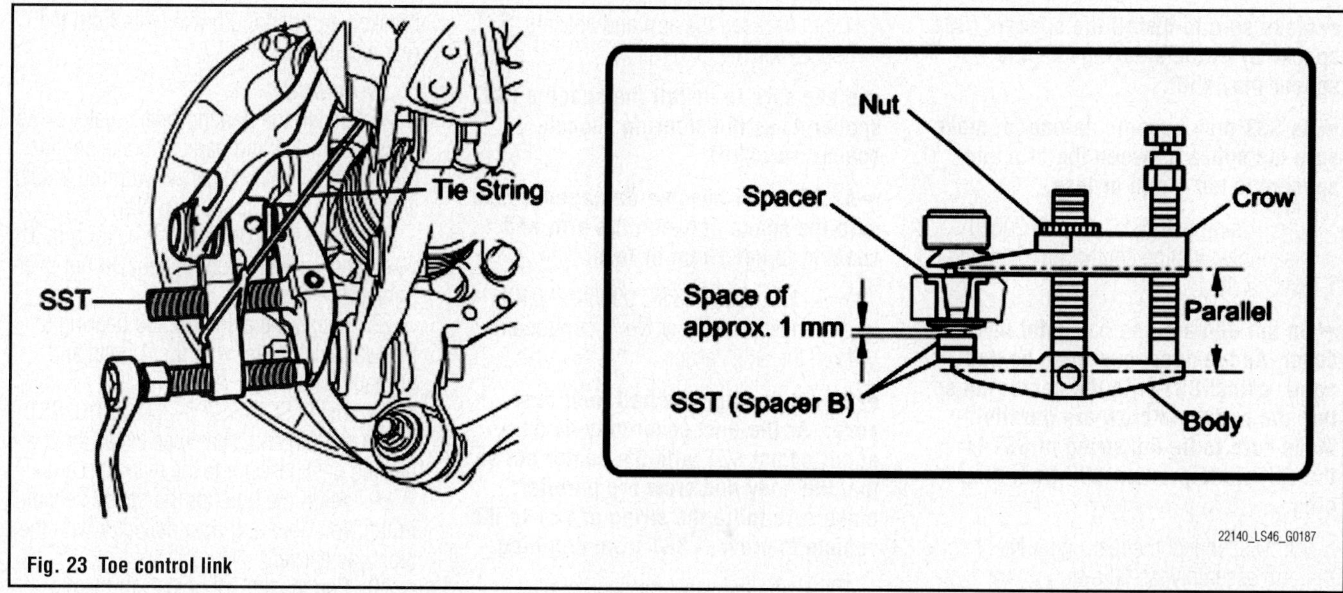

Fig. 23 Toe control link

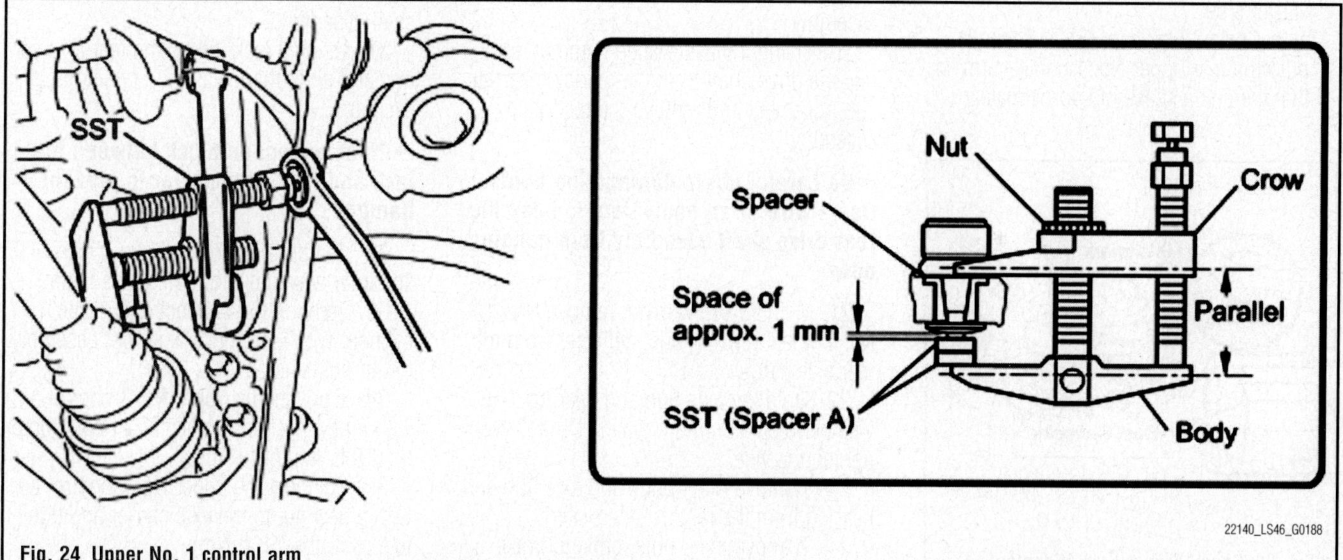

Fig. 24 Upper No. 1 control arm

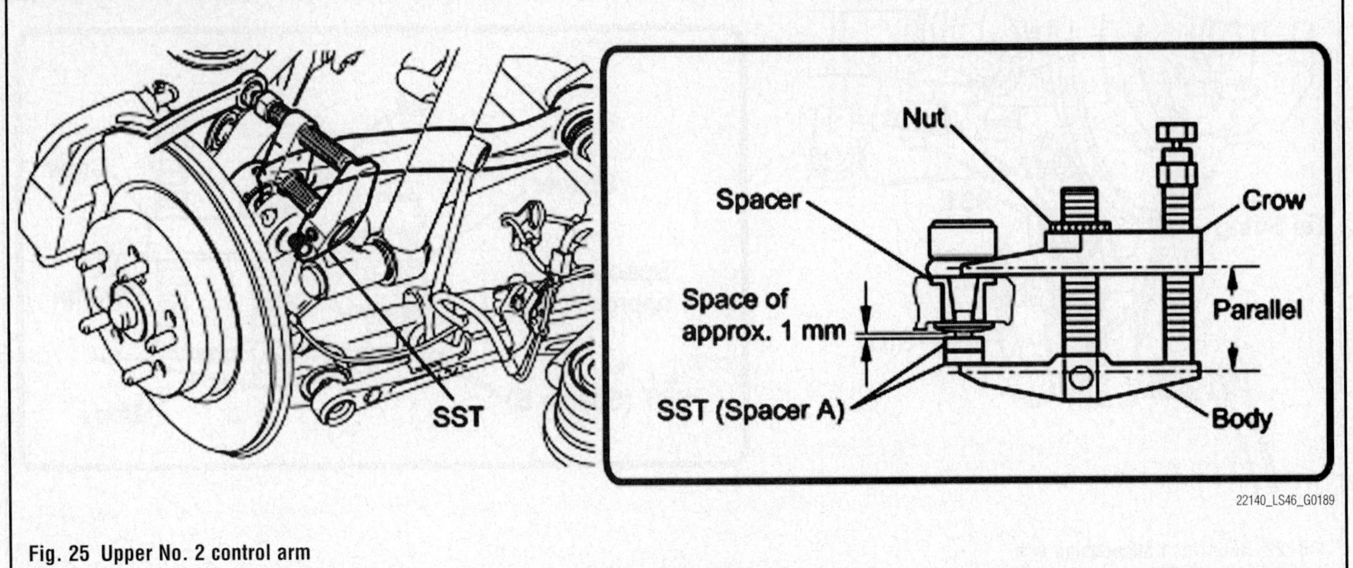

Fig. 25 Upper No. 2 control arm

➡Make sure to install the spacers (SST spacer B) as the steering knuckle spacer may shift.

➡As SST may become damaged, make sure the space between the arm and spacers is not 1 mm or less.

c. Using SST (SST: 09960-20010), disconnect the toe control link from the axle carrier.

➡Do not damage the ball joint dust cover. As the dust cover may be damaged, adjust SST with the center nut so that the body and crow are parallel. Make sure to tie the string of SST to the vehicle to prevent SST from dropping.

18. Disconnect the rear upper No. 1 control arm assembly, as follows:

a. Remove the nut on the rear axle carrier side.

b. Install 2 spacers (SST spacer A) onto the rear upper No. 1 control arm so that there is a space of approximately 1 mm between the arm and spacers. SST: 09960-20010.

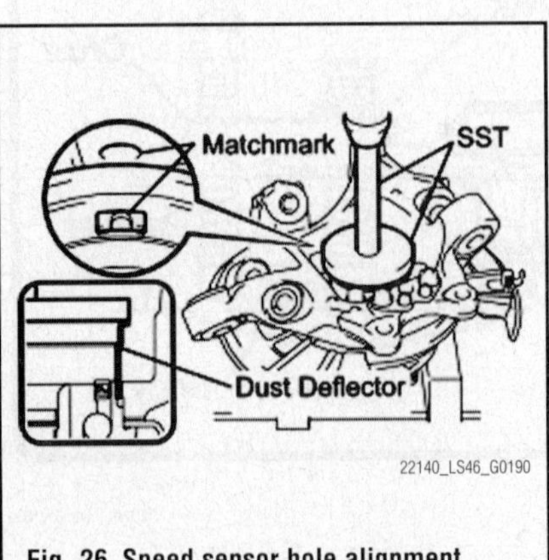

Fig. 26 Speed sensor hole alignment

➡Make sure to install the spacers (SST spacer A) as the steering knuckle spacer may shift.

➡As SST may become damaged, make sure the space between the arm and spacers is not 1 mm or less.

c. Using SST (SST: 09960-20010), disconnect the upper No. 1 control arm from the axle carrier.

➡Do not damage the ball joint dust cover. As the dust cover may be damaged, adjust SST with the center nut so that the body and crow are parallel. Make sure to tie the string of SST to the vehicle to prevent SST from dropping.

19. Using the same procedure, disconnect the rear upper No. 2 control arm assembly.

20. Using a plastic-faced hammer, separate the drive shaft from the rear axle carrier sub-assembly and remove the rear axle assembly.

➡Be careful not to damage the boots. Use a wire or an equivalent to keep the rear drive shaft assembly from hanging down.

21. Using a screwdriver, remove the No. 1 wheel bearing dust deflector from the rear axle carrier.

22. Hold the axle hub and bearing in a vise between aluminum plates. Do not overtighten the vise.

23. Remove the 4 bolts and axle hub and bearing from the rear axle carrier.

24. Remove the 2 nuts, parking brake anchor block, parking brake cable support bracket, and parking brake plate from the rear axle carrier.

To install:

25. Install the parking brake plate, cable support bracket and parking brake anchor block to the rear axle carrier with the 2 nuts and tighten to 56 ft. lbs. (76 Nm).

26. Hold the axle hub and bearing in a vise between aluminum plates. Do not overtighten the vise.

27. Install the axle hub and bearing to the rear axle carrier with the 4 bolts and tighten to 72 ft. lbs. (97 Nm).

28. Using SST (SST: 09950-70010. SST: 09951-01000) and a hammer, install the No. 1 bearing dust deflector to the rear axle carrier.

29. Align the hole for the speed sensor in the No. 1 bearing dust deflector with the rear axle carrier.

30. Engage the spline part of the rear axle carrier to the spline part of the driver shaft assembly.

31. Using a jack, lift up the rear axle carrier, and align the installation positions of each arm.

➡Place a wooden block between the jack and rear axle carrier to prevent damage.

32. Temporarily connect rear shock absorber assembly, for coil suspension.

33. Temporarily connect pneumatic cylinder with rear shock absorber assembly, for air suspension.

34. Connect the upper No. 1 control arm with a new nut to the axle carrier and tighten to 118 ft. lbs. (160 Nm).

35. Connect the upper No. 2 control arm with a new nut to the axle carrier and tighten to 118 ft. lbs. (160 Nm).

36. Connect the control link with a new nut to the axle carrier and tighten to 87 ft. lbs. (118 Nm).

37. Temporarily tighten the No. 2 suspension arm, as follows:

a. Install the stud of the suspension arm, and temporarily install a washer, nut and the bolt to the axle carrier.

b. Temporarily install the No. 2 suspension arm with the No. 2 suspension toe adjust plate, rear suspension attachment and nut to the suspension member.

38. Connect the rear No. 1 suspension arm with a new nut to the axle carrier and tighten to 87 ft. lbs. (118 Nm).

39. Apply high temperature grease.

40. Install the parking brake shoe lever.

41. Install the No. 1 and No. 2 parking brake shoe assembly.

42. Check parking brake installation.

43. Install a new rear axle shaft nut and tighten to 214 ft. lbs. (290 Nm).

44. Inspect rear axle hub bearing looseness.

45. Inspect rear axle hub runout.

46. Install rear disc.

47. Adjust parking brake shoe clearance.

48. Connect rear speed sensor.

49. Connect the rear disc brake caliper assembly with new 2 bolts and tighten to 63 ft. lbs. (86 Nm).

➡Do not twist the flexible hose. Make sure the screw parts are free from foreign matter and are not damaged. Be careful not to overtighten the bolts, as the rear axle carrier is made of aluminum.

50. Using a chisel and a hammer, stake the axle shaft nut.

51. Install the load sensing valve sensor bracket to the toe control link sub-assembly with the bolt. Tighten to 32 to 61 inch lbs. (3.6 to 6.9 Nm).

52. Stabilize the suspension.

53. Connect the rear shock absorber assembly, for coil suspension.

54. Connect the pneumatic cylinder with rear shock absorber assembly, for air suspension.

55. Insert the bolt from the front of the vehicle, and then install the rear stabilizer link assembly with the nut and tighten to 19 ft. lbs. (26 Nm).

56. Tighten the rear No. 2 suspension arm, as follows:

a. Tighten the nut on the rear suspension member side to 111 ft. lbs. (150 Nm).

b. Tighten the nut on the rear axle carrier side to 166 ft. lbs. (225 Nm).

57. Install the rear wheel.

58. Perform parking brake shoe bedding.

59. Inspect and adjust rear wheel alignment.

60. Check the ABS speed sensor signal.

61. Adjust headlight aiming.

62. Adjust the object recognition camera.

REAR HALFSHAFT

REMOVAL & INSTALLATION

See Figure 27.

⁂ WARNING

After the engine switch is turned off, the HDD navigation system requires approximately 6 minutes to record various types of memory and settings. As a result, after turning the engine switch off, wait 6 minutes or more before disconnecting the cable from the negative battery terminal.

1. Remove the right cowl top ventilator louver.

2. Disconnect cable from negative battery terminal.

⁂ CAUTION

Wait at least 90 seconds after disconnecting the cable from the negative battery terminal to prevent airbag and seat belt pretensioner activation.

3. Remove the rear wheel.

4. Inspect the rear upper No. 1 and No. 2 control arm ball joint rattle.

5. Inspect the rear No. 1 suspension arm ball joint rattle.

6. Inspect the toe control link ball joint rattle.

7. Remove the bolt and disconnect the left and right load sensing valve sensor brackets, with air suspension.

8. Remove the nuts and No. 2 and No. 1 differential support protectors.

9. Remove the bolts and the left and right rear suspension member brace.

10. Remove the right and left rear axle shaft nuts.

11. Remove the bolts and rear speed sensors.

12. Remove the right pad wear indicator wire.

13. Remove parking brake shoe adjusting hole plug.

14. Disconnect the right and left rear disc brake caliper assemblies.

15. Remove the rear disc.

16. Remove the No. 2 and No. 1 parking brake shoe return springs.

17. Remove the parking brake shoe adjusting screw set.

18. Remove the No. 2 and No. 1 parking brake shoe assemblies.

19. Remove the parking brake shoe lever.

20. Remove the 2 nuts and parking cable support bracket.

21. Disconnect the left and right rear stabilizer link assemblies.

22. Disconnect the left and right rear shock absorber assemblies, with coil suspension.

23. Disconnect the left and right pneumatic cylinder with rear shock absorber assemblies, with air suspension.

24. Disconnect the right and left rear No. 2 suspension arm assemblies.

25. Disconnect the right and left rear No. 1 suspension arm assemblies.

26. Disconnect the left and right toe control link sub-assemblies.

27. Disconnect the right and left upper rear No. 1 control arm assemblies.

28. Disconnect the right and left upper rear No. 2 control arm assemblies.

29. Using a plastic-faced hammer, separate the drive shaft from the left and right rear axle assembly, and then remove the rear axle carrier.

➡Be careful not to damage the boot. Use a wire or an equivalent to keep the rear drive shaft assembly from hanging down.

30. Using SST (SST: 09520-01010, SST: 09520-24010), remove the left and right rear drive shafts.

➡Be careful not to damage the oil seal, inboard joint boot and drive shaft dust cover, or drop the drive shaft.

To install:

31. Install the left and right rear drive shaft assemblies, as follows:

a. Coat the spline of the inboard joint shaft with gear oil.

b. Set the shaft snap ring with the opening side facing downward.

c. Align the shaft splines and install the drive shaft with a brass bar and hammer.

➡Be careful not to damage the drive shaft dust cover, boot and oil seal. Move the drive shaft while keeping it level.

32. Install the rear drive shaft to the left and right rear axle assemblies.

➡Be careful not to damage the drive shaft boot.

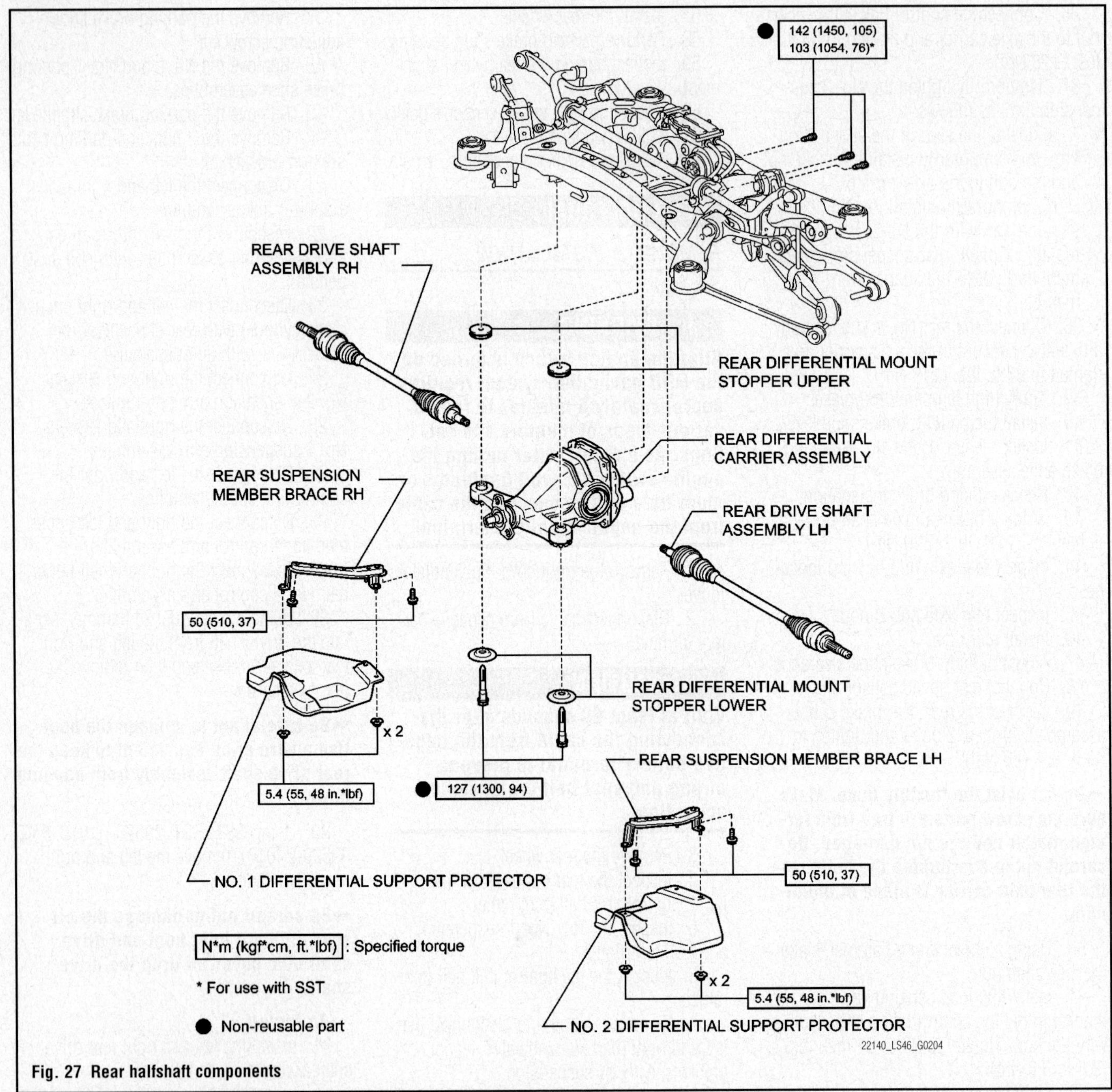

Fig. 27 Rear halfshaft components

33. Connect the right and left upper rear No. 1 control arm assemblies.
34. Connect the right and left upper rear No. 2 control arm assemblies.
35. Temporarily connect the right and left rear No. 2 suspension arm assemblies.
36. Connect the left and right toe control link sub-assemblies.
37. Connect the left and right rear No. 1 suspension arm assemblies.
38. Temporarily connect the right and left rear shock absorber assemblies, with coil suspension.
39. Temporarily connect the right and left rear pneumatic cylinder with shock absorber assemblies, with air suspension.
40. Connect the left and right rear stabilizer link assemblies.
41. Install the parking brake anchor block and parking cable support bracket with the 2 nuts and tighten to 56 ft. lbs. (76 Nm).
42. Apply high temperature grease.
43. Install the parking brake shoe lever.
44. Install the No. 1 and No. 2 parking brake shoe assemblies.
45. Install the parking brake shoe adjusting screw set.
46. Install the No. 1 and No. 2 parking brake shoe return springs.
47. Check parking brake installation.
48. Install the right and left rear axle shaft nuts.
49. Inspect rear axle hub bearing looseness.
50. Inspect rear axle hub runout.
51. Install rear disc.
52. Adjust parking brake shoe clearance.
53. Install parking brake shoe adjusting hole plug.
54. Connect the right and left rear disc brake caliper assemblies.
55. Install the rear speed sensors with the bolts and tighten to 75 inch lbs. (8.5 Nm).

56. Install the right pad wear indicator wire.

57. Stake the right and left rear axle shaft nuts.

58. Stabilize the suspension.

59. Tighten the left and right rear No. 2 suspension arm assemblies.

60. Connect the right and left rear shock absorber assemblies.

61. Connect the right and left rear pneumatic cylinder with shock absorber assemblies, with air suspension.

62. Install the right and left rear suspension member brace with the bolts and tighten to 37 ft. lbs. (50 Nm).

63. Install the No. 2 and No. 1 differential support protector with the nuts and tighten to 48 inch lbs. (5.4 Nm).

64. Connect the left and right load sensing valve sensor brackets with the bolts and tighten to 32 to 61 inch lbs. (3.6 to 6.9 Nm).

65. Install the rear wheel.

66. Connect cable to negative battery terminal.

67. Install the cowl top ventilator louver protector.

68. Perform parking brake shoe bedding.

69. Perform initialization.

70. Check the suspension control system for air suspension.

71. Inspect and adjust the rear wheel alignment.

72. Check ABS speed sensor signal.

73. Adjust the headlight.

74. Adjust the object recognition camera.

ENGINE COOLING

THERMOSTAT

REMOVAL & INSTALLATION

See Figures 28 through 30.

1. Remove the V-bank cover sub-assembly.
2. Remove the air cleaner inlet cover.
3. Remove the No. 1 air cleaner inlet.
4. Remove the right engine room side cover.
5. Remove the No. 1 engine under cover.

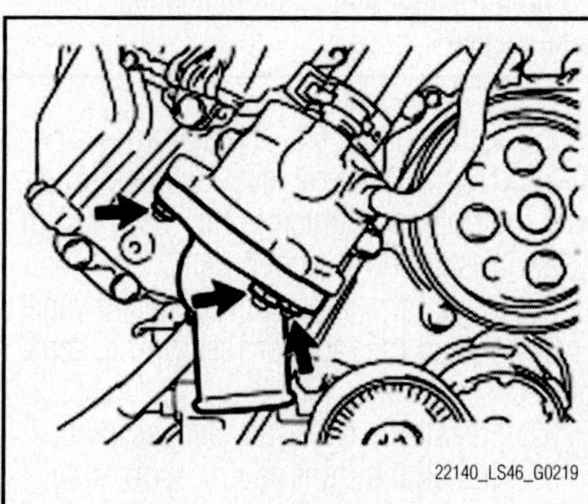

Fig. 28 Thermostat and water inlet

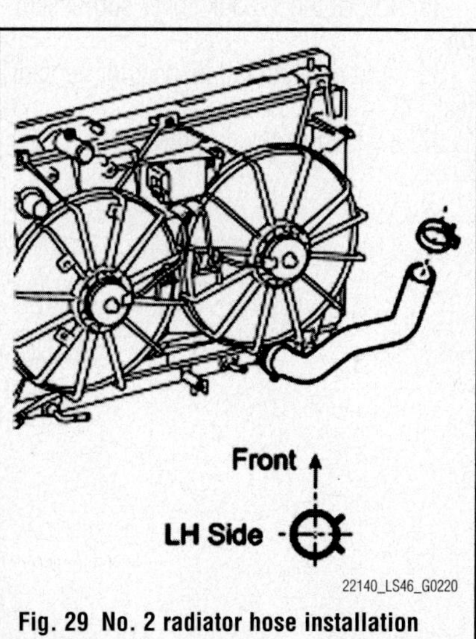

Fig. 29 No. 2 radiator hose installation

6. Drain engine coolant.
7. Remove the radiator reservoir assembly, as follows:
   a. Disconnect the 2 reservoir hoses.
   b. Remove the 2 bolts and reservoir.
8. Disconnect the No. 2 radiator hose.
9. Remove the 3 nuts, water inlet with thermostat and gasket.

To install:

10. Install a new gasket and the water inlet with thermostat with the 3 nuts and tighten to 7 ft. lbs. (10 Nm).
11. Connect the No. 2 radiator hose. The direction of the hose clamp is indicated in the illustration.
12. Install the radiator reservoir with the 2 bolts and tighten to 44 inch lbs. (5 Nm).
13. Connect the 2 reservoir hoses. The direction of the hose clamp is indicated in the illustration.
14. Add coolant and inspect for coolant leak.
15. Install the No. 1 air cleaner inlet.
16. Install the right engine room side cover.
17. Install the air cleaner inlet cover.
18. Install the V-bank cover sub-assembly.
19. Install the No. 1 engine under cover.

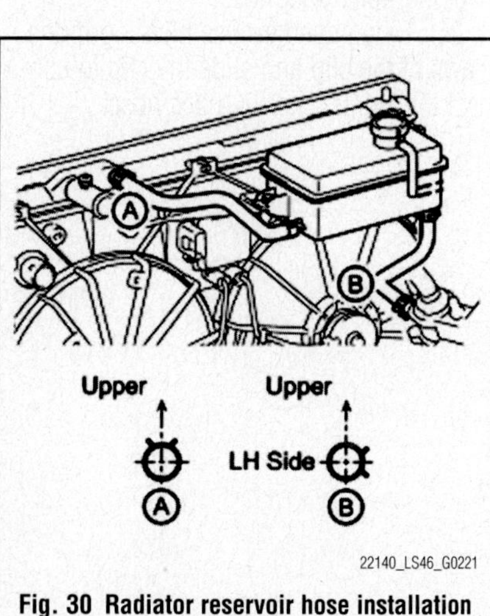

Fig. 30 Radiator reservoir hose installation

WATER PUMP

REMOVAL & INSTALLATION

See Figures 31 through 33.

1. Remove the No. 1 engine under cover.
2. Drain engine coolant.
3. Remove the V-bank cover sub-assembly.
4. Remove the air cleaner inlet cover.
5. Remove the right engine room side cover.
6. Remove the No. 1 air cleaner inlet.
7. Remove the radiator reservoir assembly.
8. Remove the intake air connector pipe.
9. Remove the V-ribbed belt.
10. Disconnect the bolt and the oil level dipstick guide.
11. Disconnect the No. 2 radiator hose.
12. Using needle-nose pliers, grip the claws of the clip and slide the clip to disconnect the No. 5 water by-pass hose.
13. Using needle-nose pliers, grip the claws of the clip and slide the clip to disconnect the water inlet hose.
14. Using needle-nose pliers, grip the claws of the clip and slide the clip to disconnect the No. 3 water by-pass hose.
15. Remove the 3 bolts, water inlet housing and gasket.
16. Using SST (SST: 09960-10010), hold the water pump pulley.
17. Remove the 4 bolts and water pump pulley.
18. Remove the 9 bolts, water pump and gasket.

To install:

19. Install the water pump and gasket with the 9 bolts as shown. Tighten as follows:
   - A: 15 ft. lbs. (20 Nm)
   - B: 17 ft. lbs. (23 Nm)
   - C: 35 ft. lbs. (47 Nm)

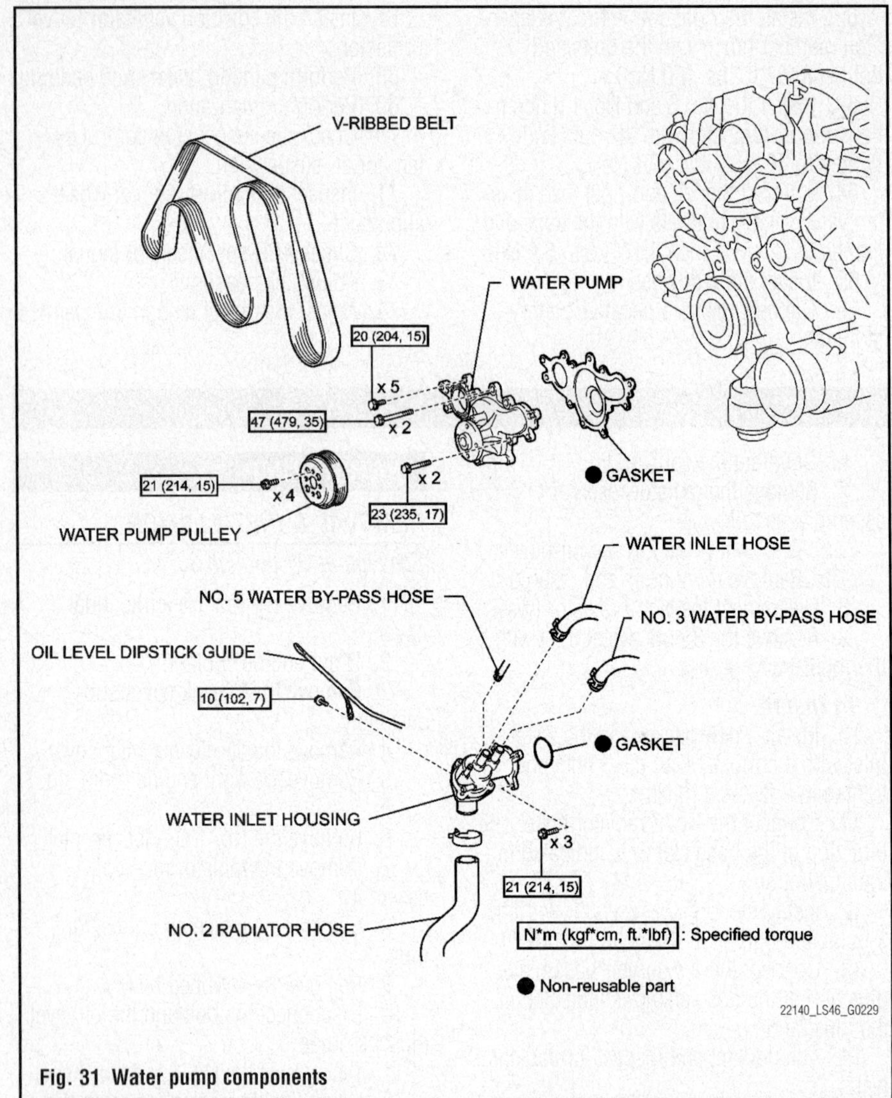

Fig. 31 Water pump components

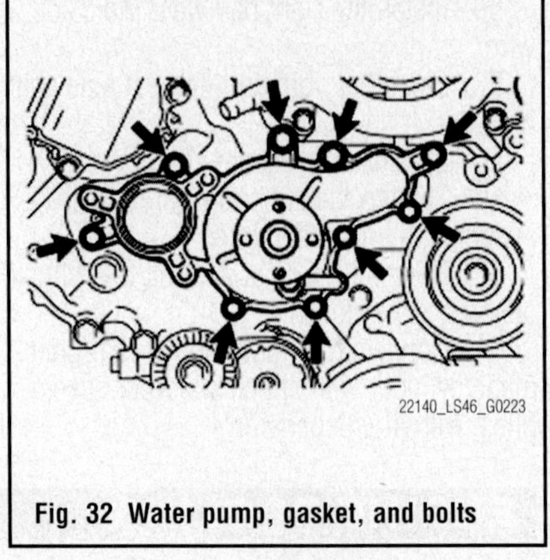

Fig. 32 Water pump, gasket, and bolts

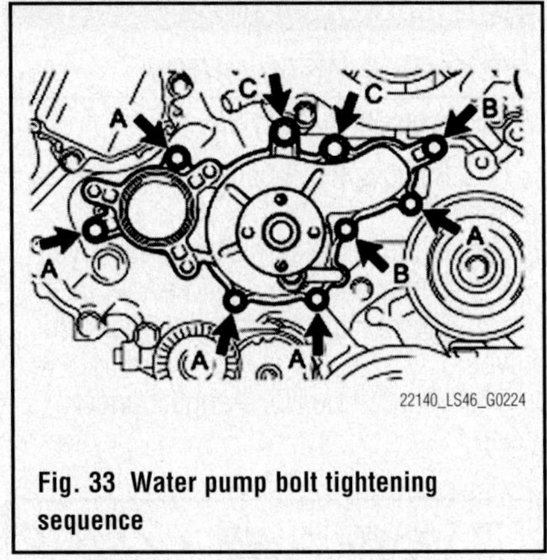

Fig. 33 Water pump bolt tightening sequence

20. Temporarily install the water pump pulley with the 4 bolts.
21. Using SST (SST: 09960-10010), hold the pulley and tighten the 4 bolts to15 ft. lbs. (21 Nm).
22. Install a new gasket to the water pump.
23. Install the water inlet housing with the 3 bolts and tighten to15 ft. lbs. (21 Nm).
24. Using needle-nose pliers, grip the claws of the clip and slide the clip to connect the No. 3 water by-pass hose.
25. Using needle-nose pliers, grip the claws of the clip and slide the clip to connect the water inlet hose.
26. Using needle-nose pliers, grip the claws of the clip and slide the clip to connect the No. 5 water by-pass hose.
27. Connect the No. 2 radiator hose.
28. Connect the oil level dipstick guide with the bolt and tighten to7 ft. lbs. (10 Nm).
29. Install the V-ribbed belt.
30. Install the intake air connector pipe.
31. Install the radiator reservoir assembly.
32. Install the No. 1 air cleaner inlet.
33. Install the right engine room side cover.
34. Install the air cleaner inlet cover.
35. Install the V-bank cover sub-assembly.
36. Add coolant and inspect for coolant leak.
37. Install the No. 1 engine under cover.

ENGINE ELECTRICAL

CHARGING SYSTEM

ALTERNATOR

REMOVAL & INSTALLATION

WARNING

After the engine switch is turned off, the HDD navigation system requires approximately 6 minutes to record various types of memory and settings. As a result, after turning the engine switch off, wait 6 minutes or more before disconnecting the cable from the negative battery terminal.

1. Remove the right cowl top ventilator louver.
2. Disconnect cable from negative battery terminal.

CAUTION

Wait at least 90 seconds after disconnecting the cable from the negative battery terminal to prevent airbag and seat belt pretensioner activation.

3. Remove the V-bank cover sub-assembly.
4. Remove the air cleaner inlet cover.
5. Remove the No. 1 air cleaner inlet.
6. Remove the No. 1 engine under cover.
7. Remove the front lower suspension member protector.
8. Remove the No. 2 engine under cover.
9. Remove the intake air connector pipe.
10. Remove the V-ribbed belt.
11. Remove the generator, as follows:
    a. Remove the nut, and disconnect the harness from the +B terminal.
    b. Disconnect the generator connector.
    c. Remove the 2 bolts and 2 nuts.
    d. Using an E8 Torx®socket wrench, remove the 2 stud bolts and generator.

To install:

12. Using an E8 Torx®socket wrench, set the generator with the 2 stud bolts and tighten to 7 ft. lbs. (10 Nm).
13. Install the generator with the 2 bolts and 2 nuts and tighten to 32 ft. lbs. (43 Nm).
14. Connect the generator connector.
15. Connect the harness to the +B terminal with the nut and tighten to 9 ft. lbs. (12 Nm).
16. The remainder of installation is the reverse of removal.
17. Perform initialization.

ENGINE ELECTRICAL

DISTRIBUTORLESS IGNITION SYSTEM

FIRING ORDER

Firing order for 4.6L (1UR-FSE) engine: 1–8–4–3–6–5–7–2

IGNITION COIL

REMOVAL & INSTALLATION

See Figure 34.

WARNING

After the engine switch is turned off, the HDD navigation system requires approximately 6 minutes to record various types of memory and settings. As a result, after turning the engine switch off, wait 6 minutes or more before disconnecting the cable from the negative battery terminal.

1. Remove the right cowl top ventilator louver.
2. Disconnect cable from negative battery terminal.

CAUTION

Wait at least 90 seconds after disconnecting the cable from the negative battery terminal to prevent airbag and seat belt pretensioner activation.

3. Remove the V-bank cover sub-assembly.
4. Remove the air cleaner inlet cover.
5. Remove the right and left engine room side cover.
6. Remove the No. 1 air cleaner inlet.

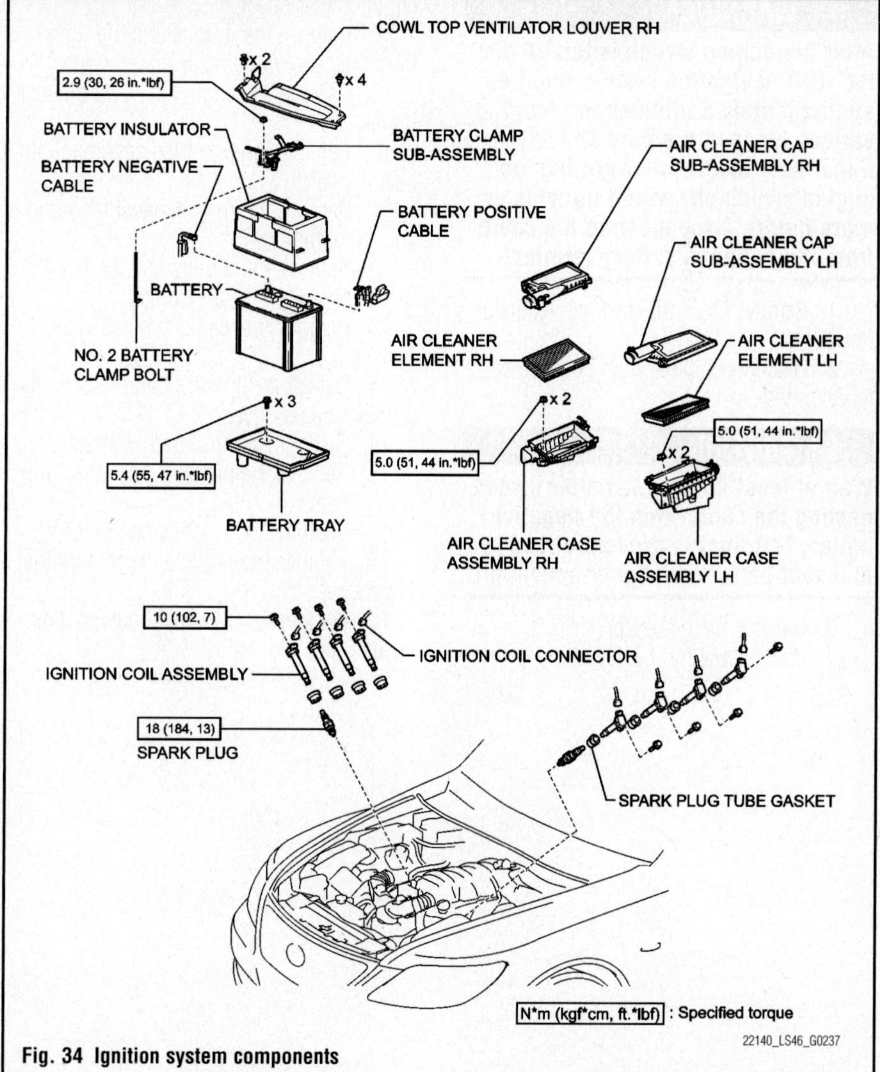

Fig. 34 Ignition system components

7. Remove the skid control ECU.
8. Remove the skid control ECU bracket.
9. Remove the battery and tray.
10. Remove the right and left air cleaner assembly.
11. Remove the ignition coil assembly, as follows:
   a. Disconnect the 8 ignition coil connectors.
   b. Remove the 8 bolts.
   c. Using a screwdriver with the tip taped, pry at the cutouts to remove the 8 ignition coils together with the 8 spark plug tube gaskets. Do not damage the cylinder head cover when removing the spark plug tube gasket.
   d. Using a 16 mm plug wrench, remove the 8 spark plugs.

To install:

12. Using a 16 mm plug wrench, install the 8 spark plugs and tighten to 13 ft. lbs. (18 Nm).
13. Perform a visual inspection on the spark plug tube gasket. There should be not scratches or deformation on the upper surface or inner or outer lip.
14. Slide a new spark plug tube gasket onto the ignition coil.
15. After installing the spark plug tube gasket, firmly insert the ignition coil.
16. Install the 8 bolts and tighten to 7 ft. lbs. (10 Nm).
17. Connect the 8 ignition coil connectors.
18. The remainder of installation is the reverse of removal.
19. Perform initialization.

IGNITION TIMING

ADJUSTMENT

The engines covered in this section are equipped with a Distributorless Ignition System (DIS). No timing adjustments are possible.

SPARK PLUGS

REMOVAL & INSTALLATION

See Ignition Coil.

ENGINE ELECTRICAL — STARTING SYSTEM

STARTER

REMOVAL & INSTALLATION

⁂ WARNING

After the engine switch is turned off, the HDD navigation system requires approximately 6 minutes to record various types of memory and settings. As a result, after turning the engine switch off, wait 6 minutes or more before disconnecting the cable from the negative battery terminal.

1. Remove the right cowl top ventilator louver.
2. Disconnect cable from negative battery terminal.

⁑ CAUTION

Wait at least 90 seconds after disconnecting the cable from the negative battery terminal to prevent airbag and seat belt pretensioner activation.

3. Remove the generator.
4. Remove the front exhaust pipe.
5. Remove the right and left rear engine under cover.
6. Disconnect the front stabilizer bar.
7. Remove the screw, clip and right engine under cover.
8. Remove the 4 bolts and the right front suspension member reinforcement.
9. Remove the engine oil level dipstick guide.
10. Remove the No. 1 exhaust manifold heat insulator.
11. Remove the right exhaust manifold sub-assembly.
12. Remove the 3 bolts and the No. 3 exhaust manifold heat insulator.
13. Remove the starter assembly, as follows:
14. Detach the 11 claws and remove the terminal upper cover.
15. Disconnect the starter connector.
16. Remove the nut and disconnect the wire harness.
17. Remove the 2 bolts and starter.
18. Remove the nut and starter terminal lower cover.
19. Remove the flywheel housing side cover.

To install:

20. Installation is the reverse of removal, noting the following:
   a. Tighten the starter terminal lower cover nut to 7 ft. lbs. (10 Nm).
   b. Tighten the starter bolts to 27 ft. lbs. (37 Nm).
   c. Tighten the wire harness nut to 88 inch lbs. (9.8 Nm).
   d. Perform initialization.

SOLENOID OR RELAY REPLACEMENT

1. Remove the starter.
2. Remove the nut, and disconnect the lead wire from the starter magnetic switch.
3. Remove the 2 nuts and starter magnetic switch assembly.

To install:

4. Temporarily install the magnetic switch by hooking its tip to the upper side of the pinion drive lever.
5. Install the 2 nuts and tighten to 66 inch lbs. (7.5 Nm).
6. Connect the lead wire to the terminal with the nut and tighten to 57 inch lbs. (5.5 Nm).
7. Install the starter.

ENGINE MECHANICAL

ACCESSORY DRIVE BELTS

ACCESSORY BELT ROUTING

See Figure 35.

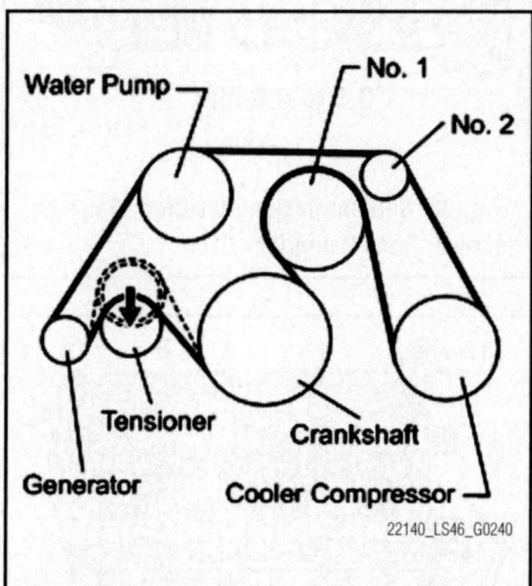

Fig. 35 Accessory belt routing

INSPECTION

Inspect the drive belt for signs of glazing or cracking. A glazed belt will be perfectly smooth from slippage, while a good belt will have a slight texture of fabric visible. Cracks will usually start at the inner edge of the belt and run outward. All worn or damaged drive belts should be replaced immediately.

ADJUSTMENT

These engines are equipped with automatic belt tensioners. Adjusting the belt tension is not possible or necessary.

REMOVAL & INSTALLATION

See Figure 35.

1. Remove the V-bank cover sub-assembly.
2. Remove the air cleaner inlet cover.
3. Remove the No. 1 air cleaner inlet.
4. Remove the intake air connector pipe.
5. Remove the V-ribbed belt, as follows:
6. Rotate the tensioner pulley counterclockwise to loosen the belt tension. The pulley bolt for the belt tensioner has a left-handed thread.
7. While turning the belt tensioner counterclockwise, align the holes. Insert a bar of φ5 mm (0.20 in.) into the holes to fix the belt tensioner in place.
8. Remove the V belt.

To install:

9. Install the V belt as shown. Check that the drive belt is properly set to each pulley.
10. Rotate the tensioner pulley counterclockwise, and then remove the fix bar.
11. Install the intake air connector pipe.
12. Install the No. 1 air cleaner inlet.
13. Install the air cleaner inlet cover.
14. Install the V-bank cover sub-assembly.

CAMSHAFT AND VALVE LIFTERS

REMOVAL & INSTALLATION

See Figures 37 through 50.

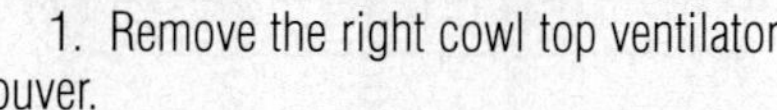

WARNING

After the engine switch is turned off, the HDD navigation system requires approximately 6 minutes to record various types of memory and settings. As a result, after turning the engine switch off, wait 6 minutes or more before disconnecting the cable from the negative battery terminal.

1. Remove the right cowl top ventilator louver.
2. Disconnect cable from negative battery terminal.

CAUTION

Wait at least 90 seconds after disconnecting the cable from the negative battery terminal to prevent airbag and seat belt pretensioner activation.

3. Discharge fuel system pressure.
4. Remove timing chain cover sub-assembly.
5. Set No. 1 cylinder to TDC/compression.

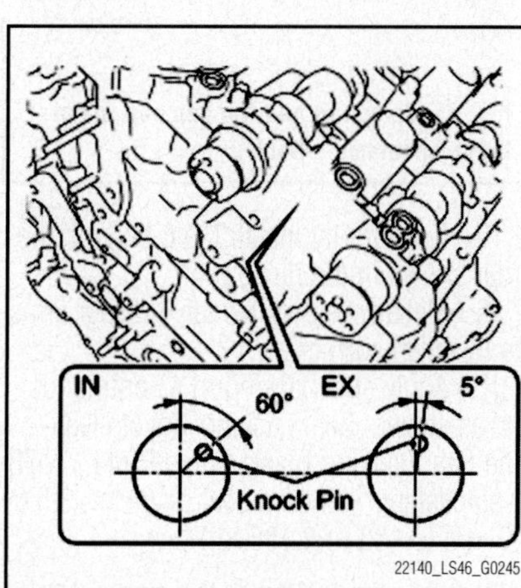

Fig. 37 Camshaft positioning—Bank 1

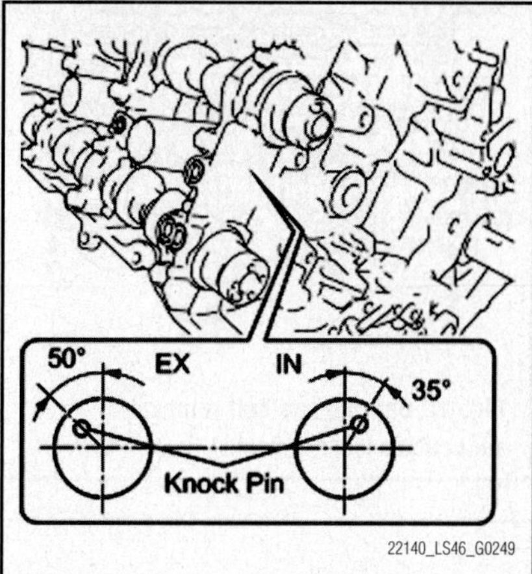

Fig. 38 Camshaft positioning—Bank 2

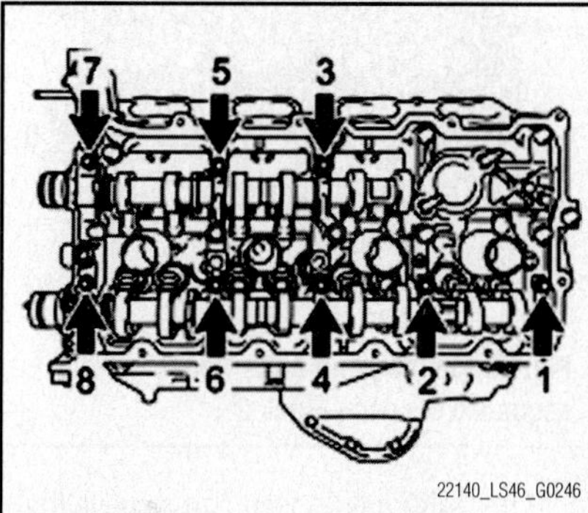

Fig. 39 Bearing cap bolt removal sequence first—Bank 1

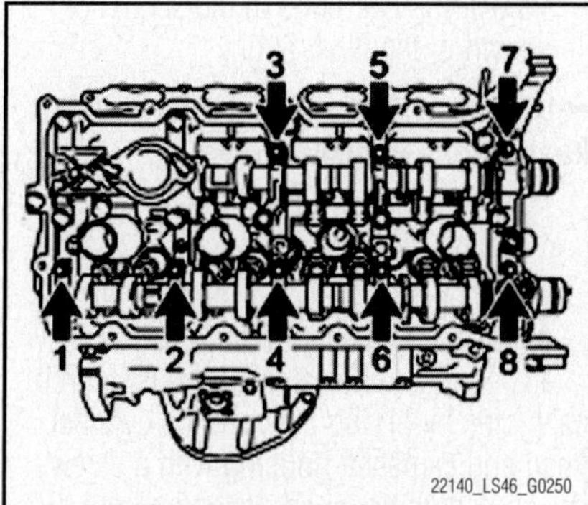

Fig. 40 Bearing cap bolt removal sequence first—Bank 2

6. Remove No. 1 chain tensioner assemblies.
7. Remove No. 1 chain vibration damper.
8. Remove chain sub-assembly.
9. Remove No. 3 chain tensioner assembly.
10. Remove camshaft bearing caps:
    a. Make sure that the knock pin of the camshaft is positioned as shown in the illustration.

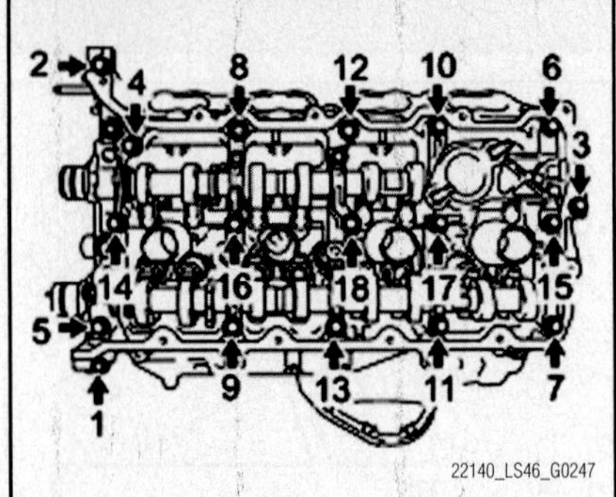

Fig. 41 Bearing cap bolt removal sequence second—Bank 1

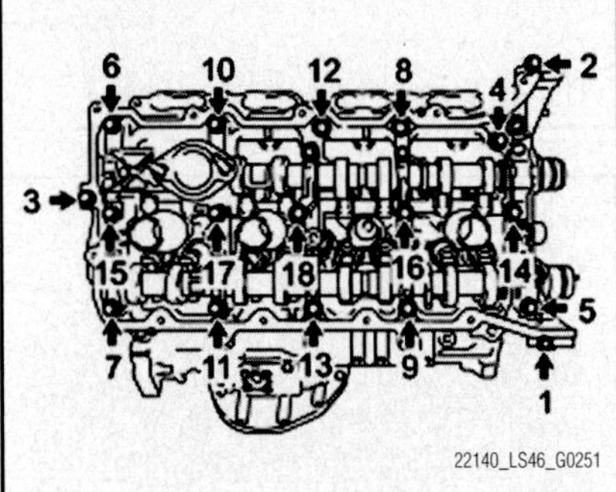

Fig. 42 Bearing cap bolt removal sequence second—Bank 2

b. Uniformly loosen and remove the 8 bearing cap bolts in the sequence shown in the illustration

c. Uniformly loosen and remove the 18 bearing cap bolts in the sequence shown in the illustration

➡Uniformly loosen the bolts while keeping the camshaft level

d. Remove the 7 bearing caps.

e. Remove the No. 3 and No. 4 camshafts from Bank 1, and the No. 1 and No. 2 camshafts from Bank 2.

11. Remove the camshaft housing from each side by prying between the cylinder head and camshaft housing with a screwdriver with its tip taped. Be careful not to damage the contact surfaces of the cylinder head and camshaft housing.

To install:

12. Apply a light coat of engine oil to the camshaft journals, camshaft housing and bearing caps.

13. Install the No. 3 and No. 4 camshafts to the Bank 1 camshaft housing, and the No. 1 and No. 2 camshafts to the Bank 2 camshaft housing.

14. Confirm the marks and numbers on the camshaft bearing caps and place them in their proper positions and directions.

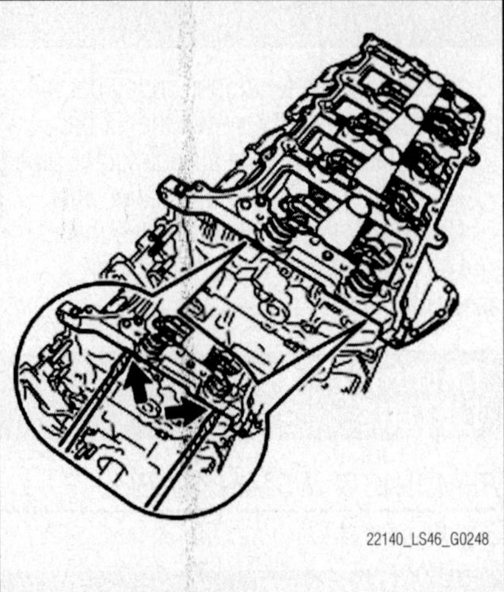

Fig. 43 Camshaft housing removal

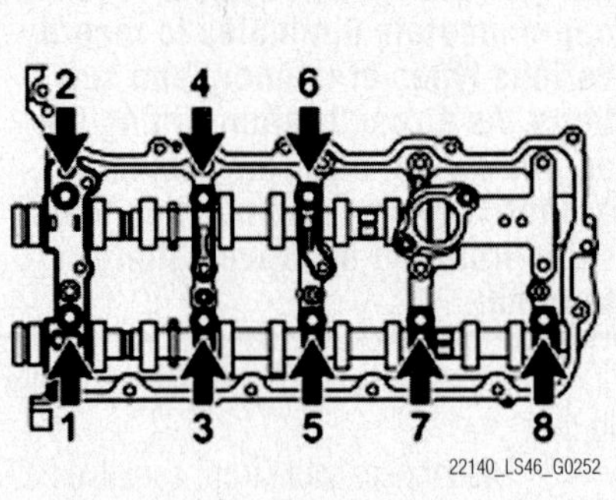

Fig. 44 Camshaft bearing cap bolt installation sequence—Bank 1

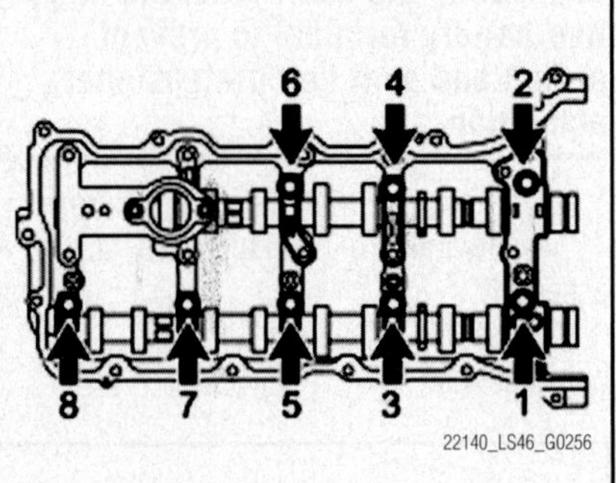

Fig. 45 Camshaft bearing cap bolt installation sequence—Bank 2

15. Temporarily install the 8 bolts in the order shown in the illustration.

16. Make sure that the valve rocker arms are installed correctly.

17. Apply seal packing in a continuous line as shown. Seal packing: Toyota Genuine Seal Packing Black, Three Bond 1207B or equivalent. Standard seal diameter: 3.5 to 4.0 mm (0.138 to 0.158 in.)

➡Remove any oil from the contact surface. Install the camshaft housing within 3 minutes and tighten the bolts within 15 minutes after applying seal packing. Do not start the engine for at least 2 hours after the installation.

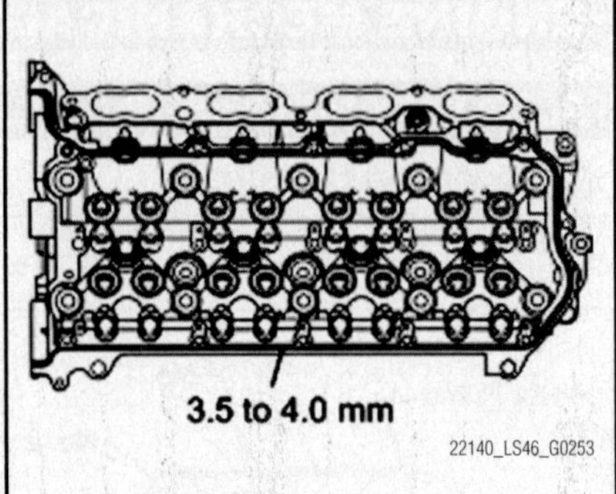

Fig. 46 Sealant packing location, Bank 1 shown, Bank 2 similar

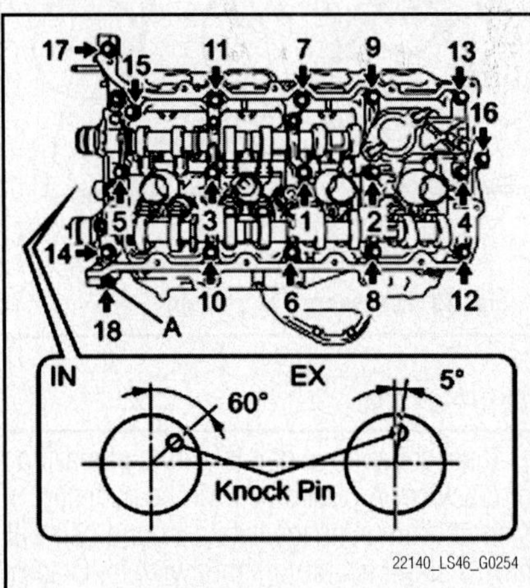

Fig. 47 Camshaft housing bolt tightening sequence—Bank 1

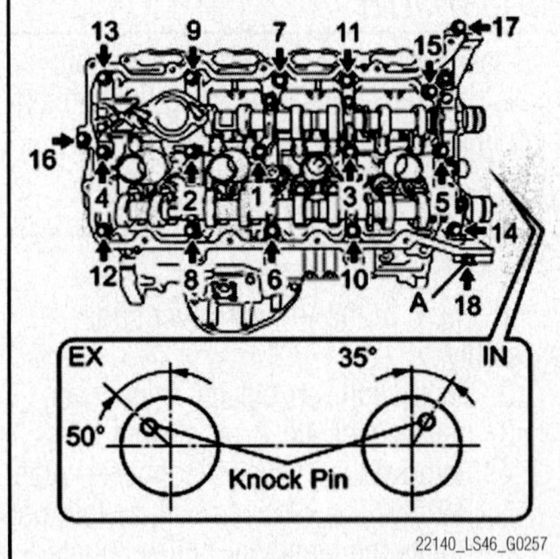

Fig. 48 Camshaft housing bolt tightening sequence—Bank 2

18. Install the camshaft housing, and install the 12 bolts and tighten to 7 ft. lbs. (10 nm) and 22 ft. lbs. (30 Nm), in the order shown.

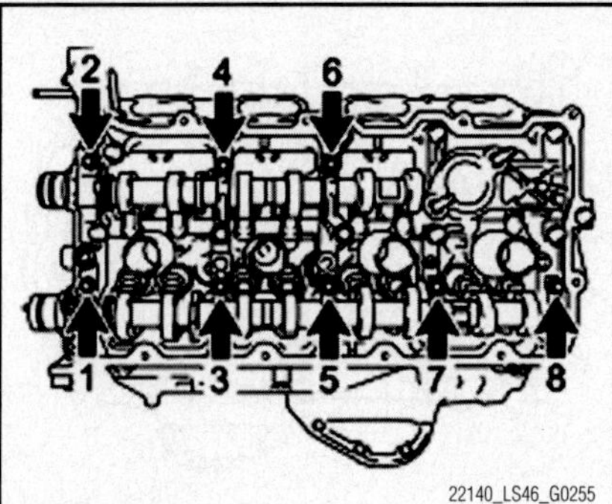

Fig. 49 Camshaft bearing cap bolt tightening sequence—Bank 1

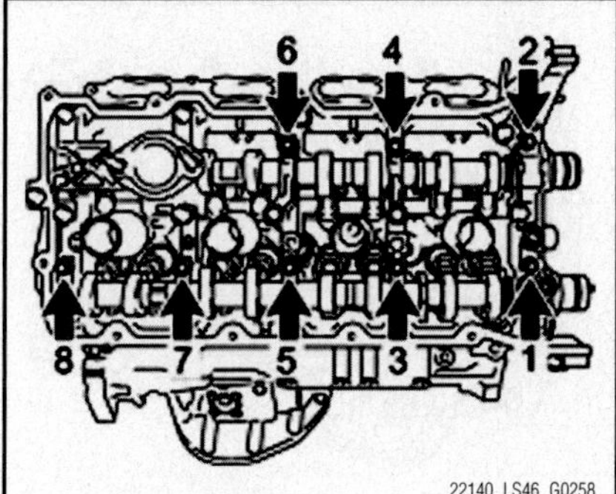

Fig. 50 Camshaft bearing cap bolt tightening sequence—Bank 2

➡Make sure that each knock pin of the camshafts is positioned as shown in the illustration before installing the camshaft housing.

19. Tighten the 8 bolts to 12 ft. lbs. (16 Nm) in the order shown in the illustration
20. Install No. 3 chain tensioner assembly.
21. Install timing chain.
22. Install chain tensioner slipper.
23. Install No. 1 chain tensioner assembly.
24. Install No. 1 chain vibration damper.
25. Tighten camshaft timing gear assembly.
26. Check No. 1 cylinder to TDC/compression.
27. Install timing chain cover.
28. Connect cable from negative battery terminal.

CRANKSHAFT DAMPER

REMOVAL & INSTALLATION

1. Before servicing the vehicle, refer to the Precautions Section.
2. Remove the V-bank cover sub-assembly.
3. Remove the air cleaner inlet cover sub-assembly.
4. Remove the No. 1 air cleaner inlet.
5. Remove the intake air connector pipe.
6. Remove the V-ribbed belt.
7. Remove the resonator bracket sub-assembly.
8. Remove the crankshaft pulley, as follows:
   a. Using SST (SST: 09213-54015, SST: 09330-00021), loosen the crankshaft pulley set bolt.
   b. Using the pulley set bolt and SST (09950-50013), remove the crankshaft pulley.

To install:

9. Align the pulley set key with the key groove of the pulley, and slide on the pulley.
10. Using SST (SST: 09213-54015, SST: 09330-00021), install the pulley bolt and tighten to 221 ft. lbs. (300 Nm).
11. The remainder of installation is the reverse of removal.

CRANKSHAFT FRONT SEAL

REMOVAL & INSTALLATION

1. Before servicing the vehicle, refer to the Precautions Section.
2. Remove the V-bank cover sub-assembly.
3. Remove the air cleaner inlet cover sub-assembly.
4. Remove the No. 1 air cleaner inlet.
5. Remove the intake air connector pipe.
6. Remove the V-ribbed belt.
7. Remove the resonator bracket sub-assembly.
8. Remove the crankshaft pulley.
9. Remove the crankshaft timing gear key from the crankshaft.
10. Remove the oil seal with a screwdriver with its tip taped. Do not damage the surface of the oil seal press fit hole and crankshaft.

To install:

11. Using SST (SST: 09223-22010, SST: 09506-35010) tap in a new oil seal until its surface is flush with the timing chain case edge.

➡Keep the lip free from foreign matter. Do not tap oil seal at an angle.

12. Install the crankshaft timing gear key.
13. Install the crankshaft pulley.
14. Install the resonator bracket with the bolt and tighten to 15 ft. lbs. (20 Nm).
15. Install the V-ribbed belt.
16. Install the intake air connector pipe.
17. Install the No. 1 air cleaner inlet.
18. Install the air cleaner inlet cover sub-assembly.
19. Install the V-bank cover sub-assembly.

CYLINDER HEAD

REMOVAL & INSTALLATION

See Figures 51 through 62.

✱✱ WARNING

After the engine switch is turned off, the HDD navigation system requires approximately 6 minutes to record various types of memory and settings. As a result, after turning the engine switch off, wait 6 minutes or more before disconnecting the cable from the negative battery terminal.

1. Remove the right cowl top ventilator louver.
2. Disconnect cable from negative battery terminal.

✱✱ CAUTION

Wait at least 90 seconds after disconnecting the cable from the negative battery terminal to prevent airbag and seat belt pretensioner activation.

3. Discharge fuel system pressure.
4. Remove the camshafts.
5. Remove the No. 1 valve rocker arm sub-assembly.

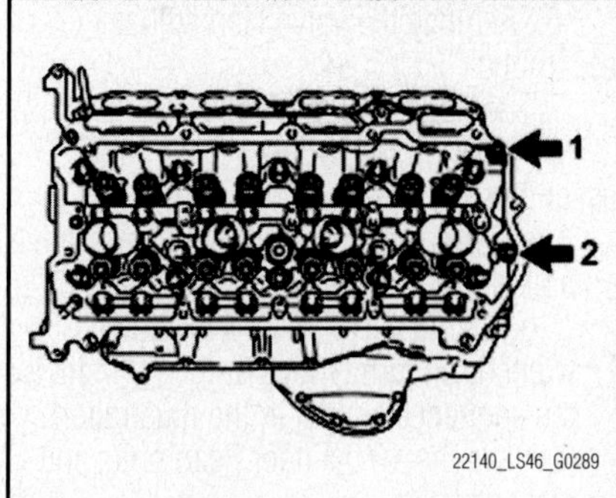

Fig. 51 Cylinder head 2 bolt removal sequence—Bank 1

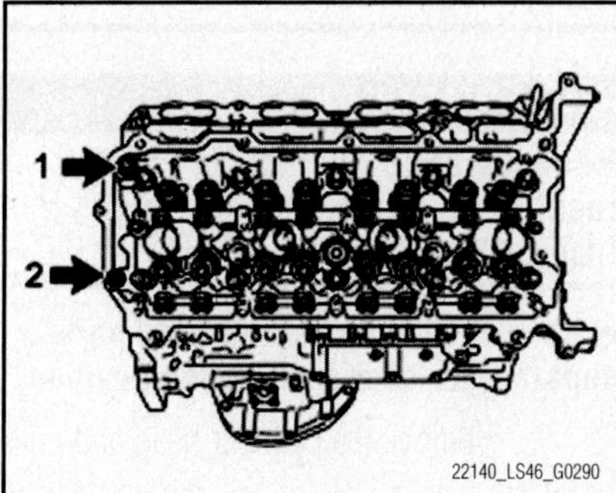

Fig. 52 Cylinder head 2 bolt removal sequence—Bank 2

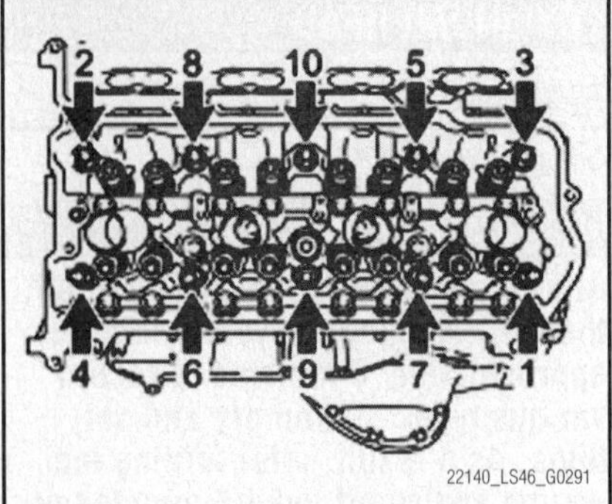

Fig. 53 Cylinder head 10 bolt removal sequence—Bank 1

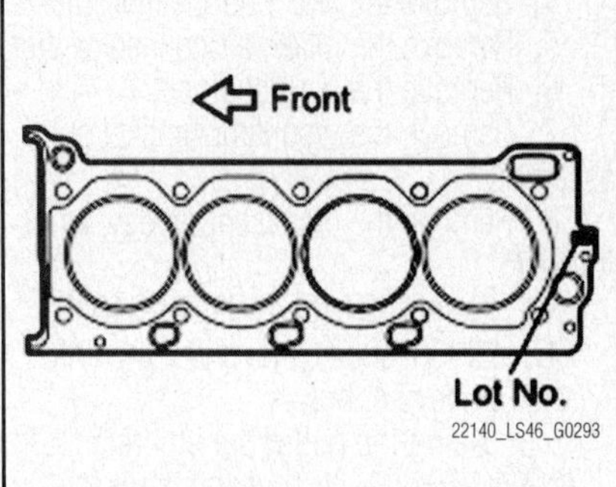

Fig. 55 Position the cylinder head gasket with Lot No. stamp upward—Bank 1

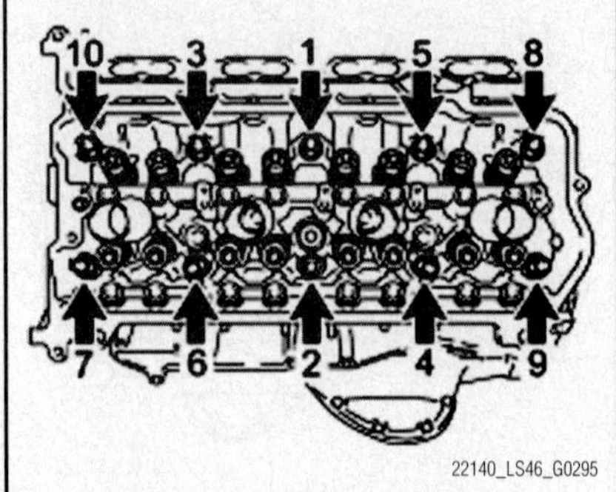

Fig. 57 Cylinder head 10 bolt tightening sequence—Bank 1

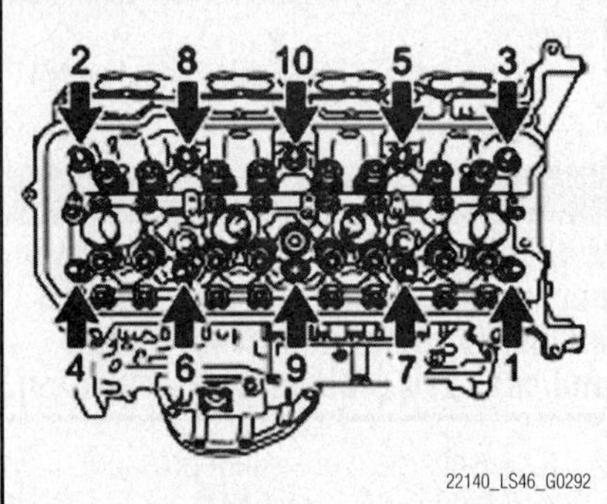

Fig. 54 Cylinder head 10 bolt removal sequence—Bank 2

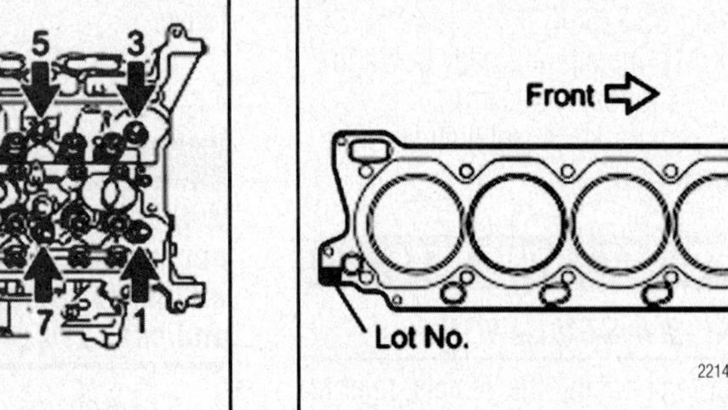

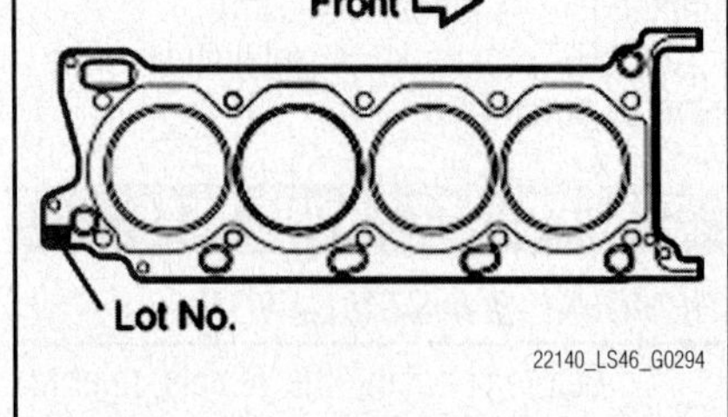

Fig. 56 Position the cylinder head gasket with Lot No. stamp upward—Bank 2

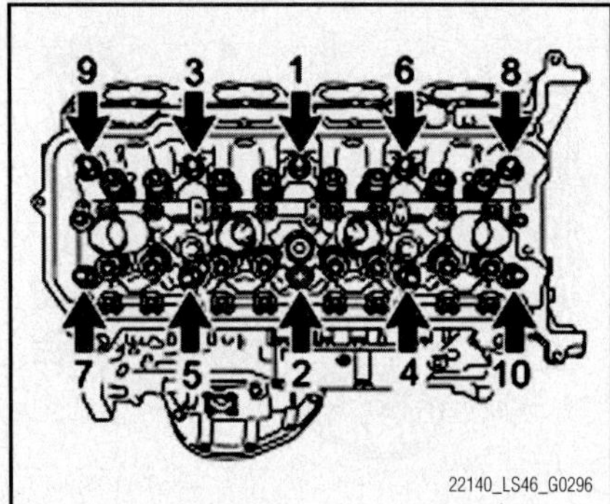

Fig. 58 Cylinder head 10 bolt tightening sequence—Bank 2

6. Remove the valve lash adjuster assembly.
7. Remove the valve stem cap.
8. Remove the cylinder head sub-assemblies, as follows:
  a. Uniformly loosen and remove the 2 bolts in the sequence shown.
  b. Using a 10 mm bi-hexagon wrench, uniformly loosen the 10 bolts in the sequence shown in the illustration. Remove the 10 cylinder head bolts and plate washers.

⁂ WARNING

Be careful not to drop washers into the cylinder head.

⁂ WARNING

Head warpage or cracking could result from removing bolts in an incorrect order.

➡Be sure to keep the removed parts separate for each installation position.

  c. Remove the cylinder head and gasket.

To install:

9. Inspect cylinder head set bolt.
10. Inspect cylinder head sub-assembly.
11. Check the piston protrusions for each cylinder:
  a. Clean the cylinder block with solvent.
  b. Set the piston of the cylinder to be measured to slightly ATDC.
12. Place the cylinder head gasket on the cylinder block surface with the front face of the Lot No. stamp upward.

⁂ WARNING

Be careful of the installation direction.

⁂ WARNING

Gently place the cylinder head in order not to damage the gasket with the bottom part of the head. Place the cylinder head on the cylinder block.

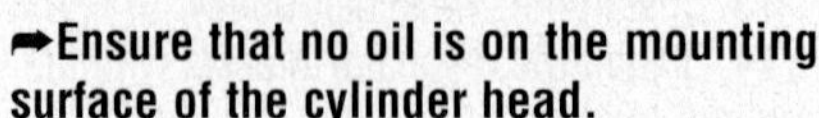

➡Ensure that no oil is on the mounting surface of the cylinder head.

13. Apply a light coat of engine oil to the threads and under the heads of the cylinder head bolts
14. The cylinder head bolts are tightened in 3 progressive steps:
  a. Using a 10 mm bi-hexagon wrench, install and uniformly tighten the 10 cylinder head bolts with the plate washers in several steps, in the sequence shown in the illustration. Tighten to 27 ft. lbs. (38 Nm).

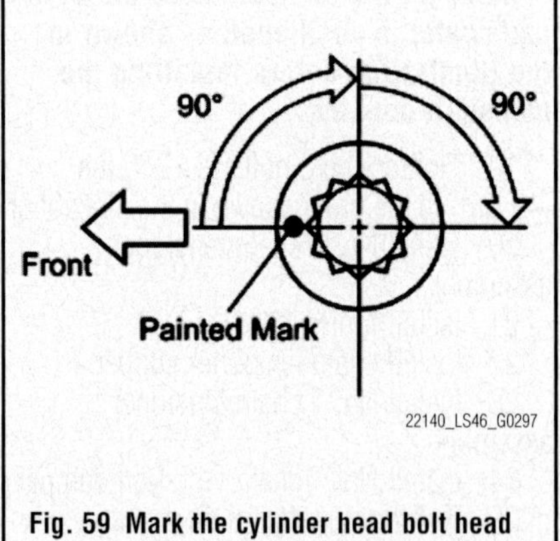

Fig. 59 Mark the cylinder head bolt head

  b. Mark the cylinder head bolt head with paint as shown in the illustration.
  c. Tighten the cylinder head bolts another 90° in the sequence shown above.
  d. Again, tighten the cylinder head bolts by an additional 90° in the sequence shown above
  e. Check that the painted marks are now facing rearward.

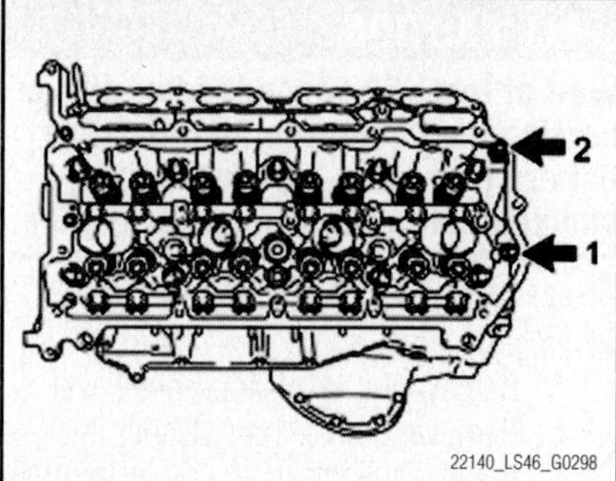

Fig. 60 Cylinder head 2 bolt tightening sequence—Bank 1

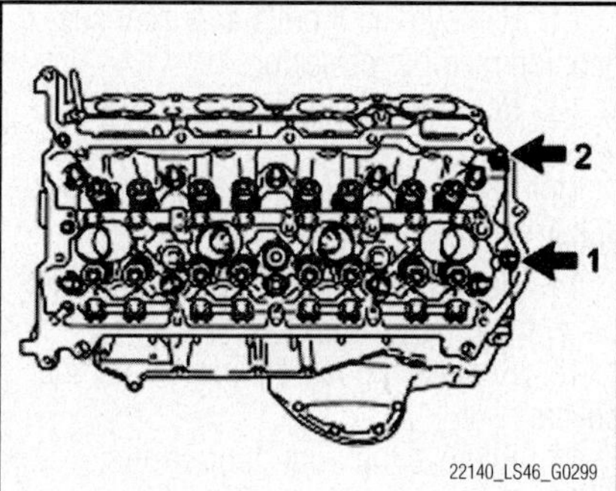

Fig. 61 Cylinder head 2 bolt tightening sequence—Bank 2

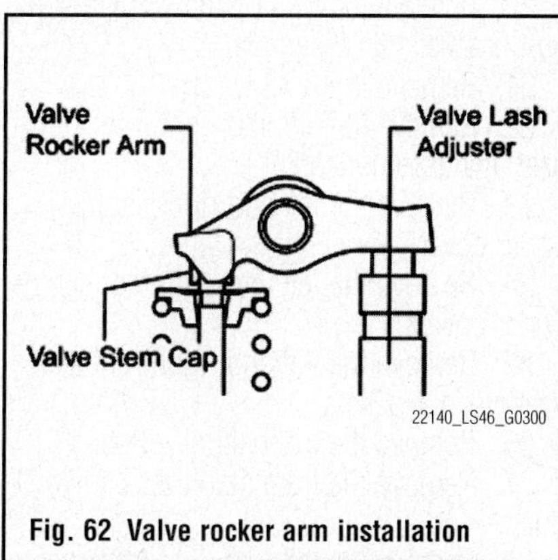

Fig. 62 Valve rocker arm installation

15. Uniformly install the 2 bolts in the sequence shown in the illustration. Tighten to 15 ft. lbs. (21 Nm).
16. Apply a light coat of engine oil to the valve stem caps and install the 16 valve stem caps to each cylinder head.

➡Be sure to inspect the valve lash adjuster before installing it.

17. Install the 16 lash adjusters to each cylinder head.

➡Install the lash adjuster at the same place it was removed from.

18. Apply engine oil to the lash adjuster tips and valve stem cap ends.
19. Make sure that the valve rocker arms are installed as shown in the illustration.
20. Install the camshaft.

ENGINE ASSEMBLY

REMOVAL & INSTALLATION

See Figures 63 and 64.

1. Before servicing the vehicle, refer to the Precautions Section.
2. Place front wheels facing straight ahead.
3. Discharge fuel system pressure.

⁂ WARNING

After the engine switch is turned off, the HDD navigation system requires approximately 6 minutes to record various types of memory and settings. As a result, after turning the engine switch off, wait 6 minutes or more before disconnecting the cable from the negative battery terminal.

4. Remove the right cowl top ventilator louver.
5. Disconnect cable from negative battery terminal.

⁂ CAUTION

Wait at least 90 seconds after disconnecting the cable from the negative battery terminal to prevent airbag and seat belt pretensioner activation.

6. Remove the V-bank cover sub-assembly.
7. Remove the air cleaner inlet cover sub-assembly.
8. Remove the clips and engine room side covers.
9. Remove engine under covers.
10. Remove the 8 bolts and front suspension member protector.
11. Drain engine oil.
12. Drain engine coolant.
13. Using needle-nose pliers, grip the claws of the clip and slide the clip to disconnect the oil cooler hoses.

➡Place a container under the connection before disconnecting the oil cooler hose because oil in the hose may spill out.

14. Remove the front wheel.
15. Remove the front bumper cover.
16. Remove the 6 bolts, 4 nuts and front suspension lower crossmember.
17. Remove the No. 1 air cleaner inlet.
18. Remove the intake air connector pipe.

Fig. 63 Ground wire

19. Remove the V-ribbed belt.
20. Remove the right and left air cleaner assembly.
21. Remove radiator hoses.
22. Remove engine room ECU outlet duct.
23. Disconnect the No. 3 fuel hose.
24. Disconnect the No. 1 fuel pipe sub-assembly.
25. Disconnect the hoses and connectors, as follows:
    a. Remove the ECM box cover (upper).
    b. Disconnect the 3 ECT connectors and 4 ECM connectors from the ECM box.
    c. Detach the clamp and disconnect the cooler compressor connector.
    d. Detach the clamp and disconnect the 3 connectors with front controller.
    e. Remove the bolt and disconnect the ground wire.
    f. Remove the 2 nuts and disconnect the wires with No. 1 engine room junction block.
    g. Disconnect the 2 heater hoses.
    h. Disconnect the purge line hose.
26. Remove the front exhaust pipe.
27. Remove the No. 1 exhaust pipe support bracket sub-assembly.
28. Remove the propeller shaft.
29. Remove the front stabilizer bar.
30. Remove the right and left front suspension member reinforcement.
31. Remove the steering sliding yoke with shaft sub-assembly.
32. Remove the No. 2 steering intermediate shaft assembly.
33. Remove the power steering link assembly.
34. Remove the No. 2 exhaust manifold heat insulator.
35. Remove the exhaust manifold sub-assembly (for Bank 1).
36. Remove the 2 bolts, 2 nuts and 2 stud bolts and disconnect the cooler

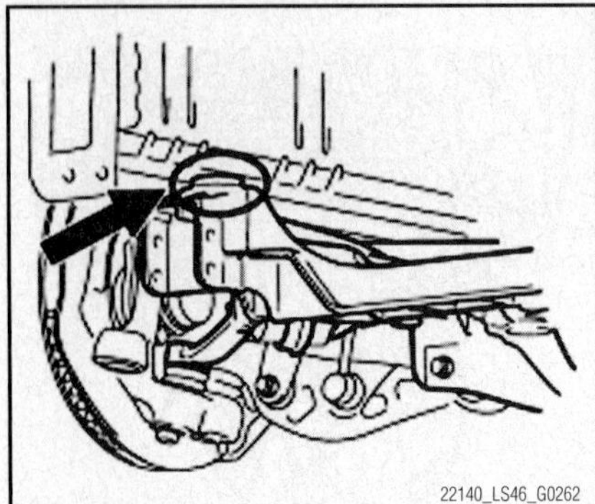

Fig. 64 Engine and transmission matchmark

compressor. It is not necessary to completely remove the compressor. With the hoses connected to the compressor, hang the compressor on the vehicle body with a rope.

37. Disconnect the front No. 2 lower suspension arm.
38. Disconnect floor shift gear shifting rod sub-assembly.
39. Remove engine and transmission, as follows:
    a. Place a mark (with spray, etc.) over the front right vehicle side attachment area of the crossmember, which is indicated in the illustration.
    b. Place a mark (with spray, etc.) over the front left vehicle side attachment area of the crossmember.
    c. Set an engine lifter underneath the engine. Place the engine on wooden blocks or an equivalent so that the engine is level.
    d. Remove the 4 engine rear mounting member's bolts.
    e. Remove the 4 bolts shown in the illustration.
    f. Operate the engine lifter, then slowly remove the engine from the vehicle.

➡Make sure that the engine is clear of all wiring and hose.

➡While lowering the engine from the vehicle, do not allow it to contact the vehicle.

    g. Install the 2 engine hangers with the 2 bolts. No. 1 engine hanger part number: 12281-38030 or 12281-38050, bolt part number: 90129-14120. Tighten to 32 ft. lbs. (43 Nm).
    h. Attach an engine sling device and hang the engine with a chain block. When hanging the engine, make sure to hang the engine with the sling device's hanging angle at 50° or less. If not, the engine or engine hangers may be damaged.

To install:

40. Place the engine on an engine lifter.
41. Place the engine on wooden blocks or an equivalent so that the engine is level.
42. Remove the 2 bolts and 2 engine hangers.
43. Operate the engine lifter, then install the engine to the vehicle. Make sure that the engine is clear of all wiring and hoses.
44. Align the crossmember to the marks on the vehicle, and temporarily install the engine and transmission with crossmember with the 4 bolts.

✲✲ WARNING

Make sure the crossmember is aligned to the vehicle marks as accurately as possible. If not performed accurately, the suspension alignment may become extremely misaligned.

45. Tighten the 4 bolts to 150 ft. lbs. (204 Nm).
46. Install the 4 engine rear mounting member's bolts and tighten to 19 ft. lbs. (26 Nm).
47. The remainder of installation is the reverse of removal, including the following:
48. Perform initialization.
49. Add engine oil, coolant, and automatic transmission fluid.
50. Inspect for oil, coolant, fuel, and exhaust gas leaks.
51. Check and adjust front wheel alignment.
52. Check shift lever position.
53. Check ignition timing and idle speed.
54. Check CO/HC.
55. Check engine oil level.

EXHAUST MANIFOLD

REMOVAL & INSTALLATION

See Figures 65 through 70.

1. Before servicing the vehicle, refer to the Precautions Section.

✲✲ WARNING

After the engine switch is turned off, the HDD navigation system requires approximately 6 minutes to record various types of memory and settings. As a result, after turning the engine switch off, wait 6 minutes or more before disconnecting the cable from the negative battery terminal.

2. Remove the right cowl top ventilator louver.
3. Disconnect cable from negative battery terminal.

✲✲ CAUTION

Wait at least 90 seconds after disconnecting the cable from the negative battery terminal to prevent airbag and seat belt pretensioner activation.

4. Remove the V-bank cover sub-assembly.
5. Remove the air cleaner inlet cover.
6. Remove the No. 1 air cleaner inlet.
7. Remove the intake air connector pipe.
8. Remove the V-ribbed belt.
9. Remove the clips and engine room side covers.
10. Remove the No 1 engine under cover.
11. Remove the 8 bolts and front suspension member protector.
12. Remove the No 2 engine under cover.
13. Remove the No. 1 and No. 2 differential support protectors.
14. Remove the No. 6 and No. 5 rocker panel molding protector.
15. Remove the rear floor side member covers.
16. Remove the front floor covers.
17. Remove the front floor center brace.
18. Disconnect the heated oxygen sensor.
19. Remove the front exhaust pipe assembly.
20. Remove the No. 1 exhaust pipe support bracket sub-assembly.
21. Remove the front wheel.
22. Remove the left and right front stabilizer link assemblies.
23. Remove the left and right engine under covers.
24. Remove the left and right rear engine under covers.
25. Remove the left and right front stabilizer No. 1 brackets.
26. Remove the front stabilizer bar.
27. Remove the front stabilizer bar No. 1 bush.
28. Remove the left and right front suspension member reinforcement.
29. Remove the generator.
30. Remove the steering sliding yoke with shaft sub-assembly.
31. Remove the No. 2 steering intermediate shaft assembly.
32. Remove the engine oil level dipstick guide.
33. Remove the No. 1 exhaust manifold heat insulator, as follows:
    a. Disconnect the air fuel ratio sensor connector.
    b. Remove the 3 bolts and No. 1 exhaust manifold heat insulator.
34. Remove the 8 nuts, right exhaust manifold, and gasket.

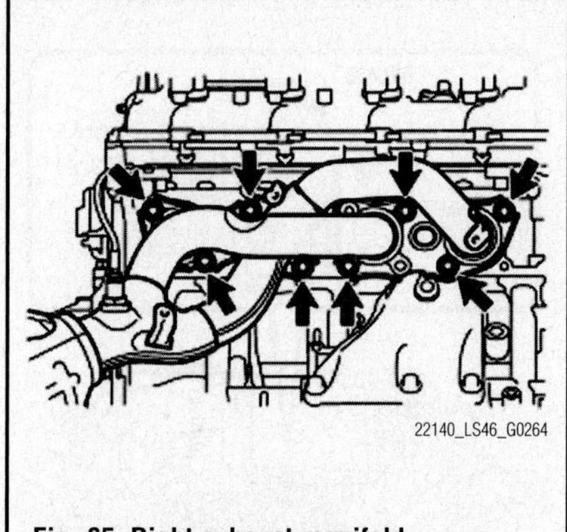

Fig. 65 Right exhaust manifold

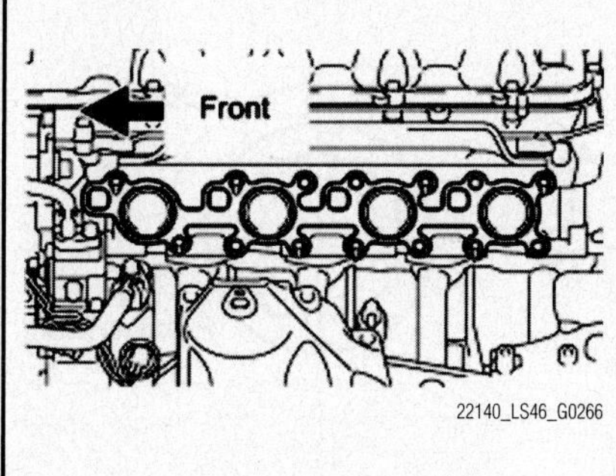

Fig. 67 Left exhaust manifold gasket installation

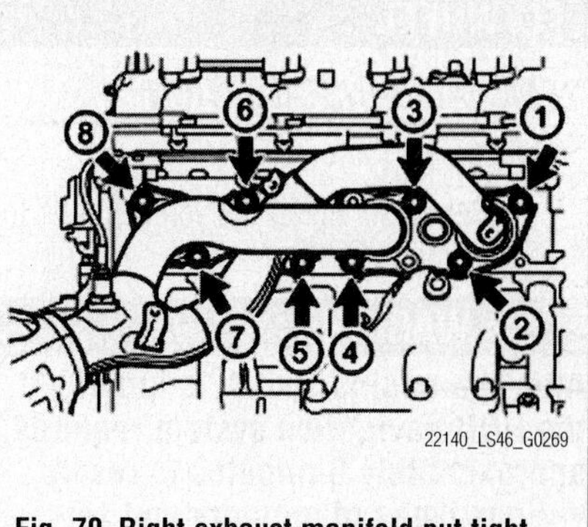

Fig. 70 Right exhaust manifold nut tightening sequence

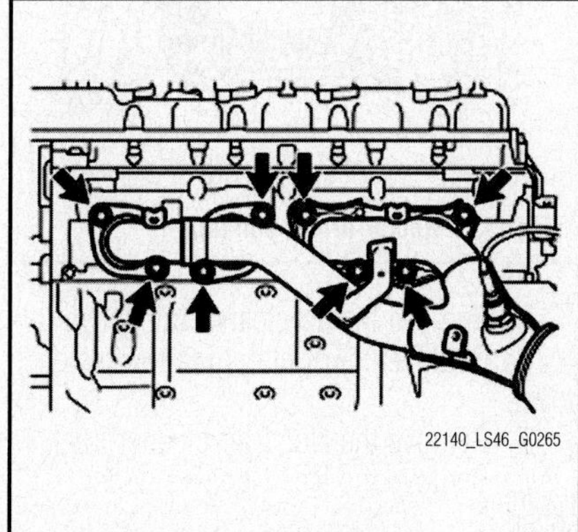

Fig. 66 Left exhaust manifold

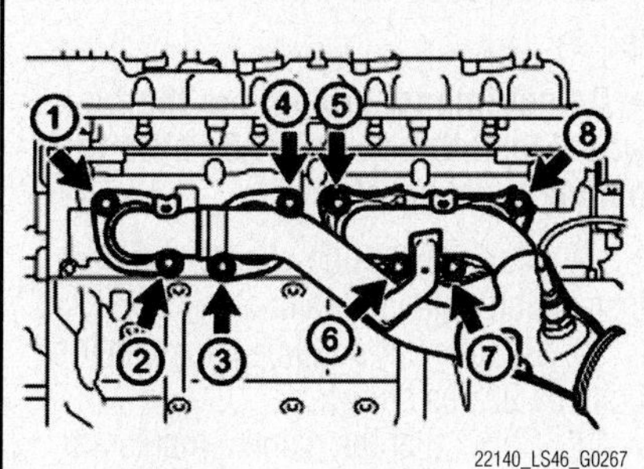

Fig. 68 Left exhaust manifold nut tightening sequence

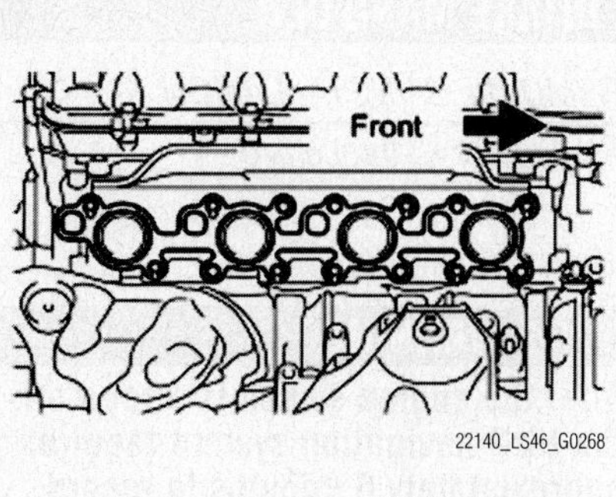

Fig. 69 Right exhaust manifold gasket installation

35. Remove the No. 2 exhaust manifold heat insulator, as follows:
  a. Disconnect the air fuel ratio sensor connector.
  b. Remove the 3 bolts and No. 2 exhaust manifold heat insulator.
36. Remove the 8 nuts, left exhaust manifold, and gasket.
37. Remove the Bank 1 and Bank 2 air fuel ratio sensors.

To install:

38. Install the Bank 1 and Bank 2 air fuel ratio sensors.
39. Install the left exhaust manifold sub-assembly, as follows:
  a. Install a new gasket.
  b. Install the left exhaust manifold, install 8 new nuts in the order shown in the illustration, and tighten to 15 ft. lbs. (21 Nm).
40. Install the No. 2 exhaust manifold heat insulator, as follows:
  a. Install the exhaust manifold heat insulator with the 3 bolts and tighten to 7 ft. lbs. (10 Nm).
  b. Connect the air fuel ratio sensor connector.
41. Install the right exhaust manifold sub-assembly, as follows:
  a. Install a new gasket.
  b. Install the right exhaust manifold, install 8 new nuts in the order shown in the illustration, and tighten to 15 ft. lbs. (21 Nm).
42. Install the No. 1 exhaust manifold heat insulator, as follows:
  a. Install the No. 1 exhaust manifold heat insulator with the 3 bolts and tighten to 7 ft. lbs. (10 Nm).
  b. Connect the air fuel ratio sensor connector.
43. Install the engine oil level dipstick guide.
44. Install the No. 2 steering intermediate shaft assembly.
45. Install the steering sliding yoke with shaft sub-assembly.
46. Install the generator.
47. Install the V-ribbed belt.
48. Install the left and right front suspension member reinforcement.
49. Install the front stabilizer bar No. 1 bush.
50. Temporarily tighten the front stabilizer bar.
51. Tighten the left and right front stabilizer No. 1 brackets.
52. Install the left and right rear engine under covers.
53. Install the left and right engine under covers.
54. Install the front exhaust pipe assembly.
55. Install the front floor center brace.
56. Connect the heated oxygen sensor.
57. Connect the cable to the negative battery terminal.
58. Install the right cowl top ventilator louver.
59. Inspect for exhaust gas leak.
60. Install the front floor covers.
61. Install the rear floor side member covers.
62. Install the No. 6 and No. 5 rocker panel molding protector.
63. Install the No. 1 and No. 2 differential support protectors.
64. Install the No 2 engine under cover.
65. Install the 8 bolts and front suspension member protector.
66. Install the No 1 engine under cover.
67. Install the front wheel.
68. Install the clips and engine room side covers.
69. Install the intake air connector pipe.
70. Install the No 1 air cleaner inlet.
71. Install the air cleaner inlet cover.
72. Install the V-bank cover sub-assembly.
73. Perform initialization.

FLYWHEEL

REMOVAL & INSTALLATION

See Figures 71 and 72.

1. Before servicing the vehicle, refer to the Precautions Section.

WARNING

After the engine switch is turned off, the HDD navigation system requires approximately 6 minutes to record various types of memory and settings. As a result, after turning the engine switch off, wait 6 minutes or more before disconnecting the cable from the negative battery terminal.

2. Remove the right cowl top ventilator louver.
3. Disconnect cable from negative battery terminal.

CAUTION

Wait at least 90 seconds after disconnecting the cable from the negative battery terminal to prevent airbag and seat belt pretensioner activation.

4. Remove the transmission assembly.
5. Remove the drive plate and ring gear sub-assembly, as follows:
   a. Using SST (SST: 09213-54015, SST: 09330-00021) hold the crankshaft.
   b. Remove the 10 bolts, spacer plate, ring gear and sensor rotor.

To install:

6. Using SST (SST: 09213-54015, SST: 09330-00021) hold the crankshaft.
7. Install the sensor rotor, ring gear and spacer plate on the crankshaft.
8. Uniformly install and tighten 10 new bolts in the sequence shown in the illustration to 22 ft. lbs. (30 Nm).

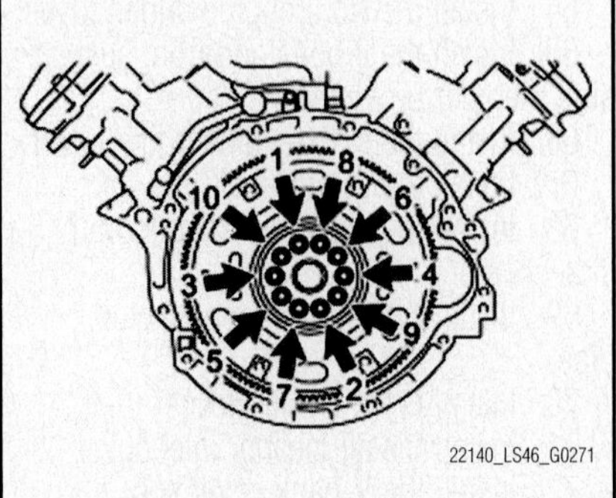

Fig. 71 Flywheel bolt tightening sequence

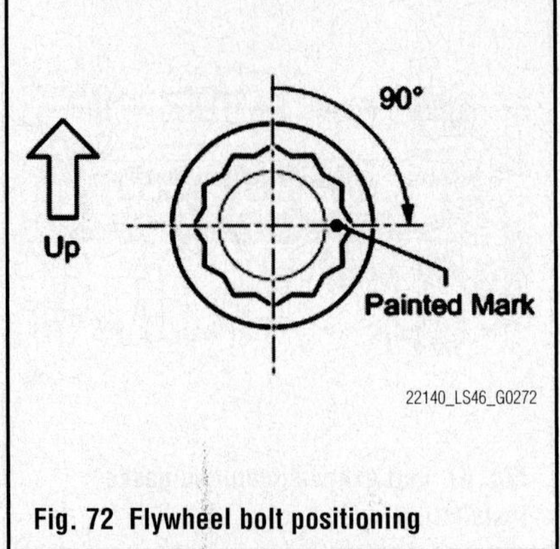

Fig. 72 Flywheel bolt positioning

➡Do not reuse the flywheel installation bolts.

➡Do not impact or damage the flywheel installation bolts. Be sure to handle them carefully.

9. Mark the upside of each flywheel installation bolt with paint.
10. Retighten the flywheel installation bolts by 90° as shown.
11. Check that the painted marks are now at a 90° angle to the upside.
12. Install the transmission.
13. Connect the cable to the negative battery terminal.

INTAKE MANIFOLD

REMOVAL & INSTALLATION

See Figures 73 through 76.

1. Before servicing the vehicle, refer to the Precautions Section.

WARNING

After the engine switch is turned off, the HDD navigation system requires approximately 6 minutes to record various types of memory and settings. As a result, after turning the engine switch off, wait 6 minutes or more before disconnecting the cable from the negative battery terminal.

2. Remove the right cowl top ventilator louver.
3. Disconnect cable from negative battery terminal.

CAUTION

Wait at least 90 seconds after disconnecting the cable from the negative battery terminal to prevent airbag and seat belt pretensioner activation.

4. Remove the V-bank cover sub-assembly.

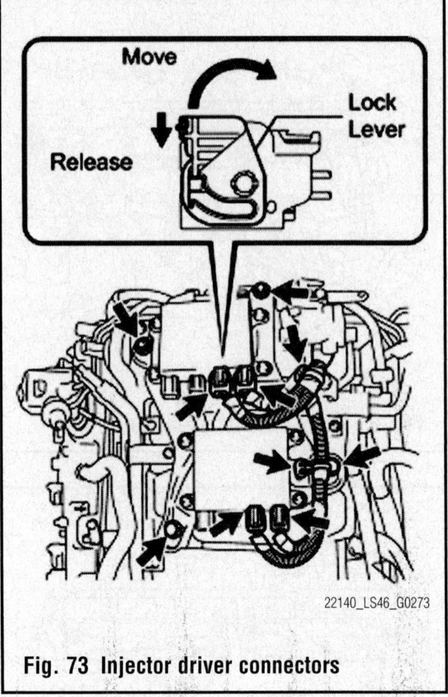

Fig. 73 Injector driver connectors

5. Remove the air cleaner inlet cover.
6. Remove the right engine room side cover.
7. Remove the No. 1 air cleaner inlet.
8. Remove the No. 1 engine under cover.
9. Remove the front lower suspension member protector.
10. Remove the No. 2 engine under cover.
11. Drain engine coolant.
12. Remove the intake air connector pipe.
13. Remove the throttle body.
14. Remove the air cleaner assemblies.
15. Disconnect the engine wire.
16. Remove the injector driver, as follows:
    a. Disconnect the 4 injector driver connectors. To disconnect the injector driver connectors, push the claw downward and move the lock lever to release the lock.
    b. Disconnect the 2 clamps from the injector driver.
    c. Remove the 2 bolts, 2 nuts and injector driver from the intake manifold.
17. Slide the 4 clamps, and disconnect the heater water inlet hose, heater water outlet hose, water inlet hose, and No. 3 water by-pass hose from the water by-pass pipe and remove the 2 bolts and water by-pass pipe.
18. Disconnect the fuel vapor feed hose and No. 2 fuel vapor feed hose and remove the bolt and purge VSV.
19. Remove the intake manifold, as follows:
    a. Disconnect the PCV hose from intake manifold.

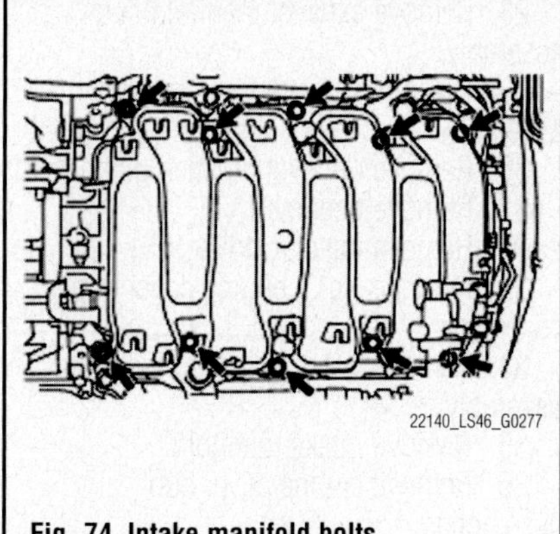

Fig. 74 Intake manifold bolts

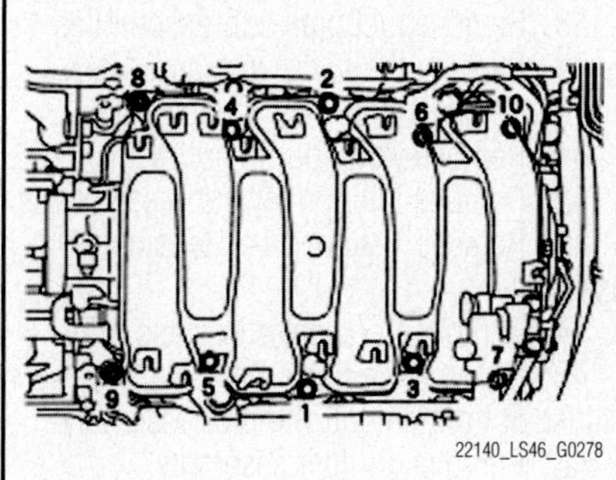

Fig. 75 Intake manifold bolt tightening sequence

b. Remove the 8 bolts, 2 nuts and intake manifold.

c. Remove the 2 gaskets from the intake manifold.

To install:

20. Install 2 new gaskets to the intake manifold.
21. Temporarily install the intake manifold with the 2 nuts and 8 bolts. Then tighten the 2 nuts and 8 bolts uniformly in the order shown in the illustration to 15 ft. lbs. (21 Nm).
22. Connect the PCV hose to the intake manifold.
23. Install the purge VSV with the bolt and tighten to 15 ft. lbs. (21 Nm).
24. Connect the fuel vapor feed hose to the intake manifold.
25. Connect the No. 2 fuel vapor feed hose to the purge VSV.
26. Install the water by-pass pipe to the intake manifold with the 2 bolts.
27. Connect the heater water inlet hose, heater water outlet hose, water inlet hose, and No. 3 water by-pass hose to the water by-pass pipe with the 4 clamps.
28. Install the injector driver to the intake manifold by installing the 2 bolts and 2 nuts in the order shown in the illustration and tighten to 7 ft. lbs. (10 Nm).

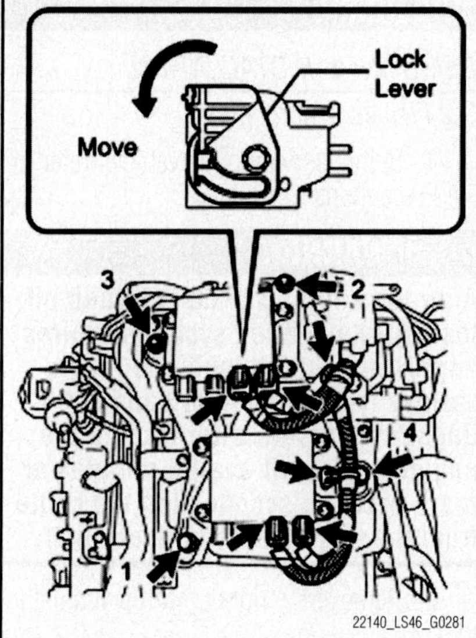

Fig. 76 Injector driver connector installation

29. Connect the 4 wire harness connectors to the injector driver. Then move the lock lever as shown in the illustration to lock the connectors.
30. Connect the engine wire.
31. Install the air cleaner assemblies.
32. Install the throttle body.
33. Install the intake air connector pipe.
34. Install the No. 1 air cleaner inlet.
35. Connect the cable to the negative battery terminal.
36. Add engine coolant and inspect for coolant leak.
37. Install the right engine room side cover.
38. Install the air cleaner inlet cover.
39. Install the V-bank cover sub-assembly.
40. Install the No. 2 engine under cover.
41. Install the front lower suspension member protector.
42. Install the No. 1 engine under cover.
43. Install the right cowl top ventilator louver.
44. Perform initialization.

OIL PAN

REMOVAL & INSTALLATION

See Figures 77 through 80.

➡It may be possible to remove the oil pan from the vehicle while the engine is still in the vehicle. If not. the engine and transmission must be removed as a unit, then separated.

1. Before servicing the vehicle, refer to the precautions in the beginning of this section.

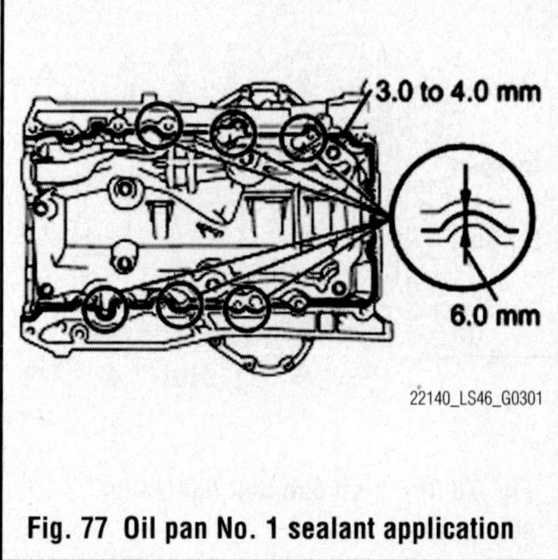

Fig. 77 Oil pan No. 1 sealant application

2. Remove engine/transmission assembly, as detailed in the Engine Assembly Removal section, above.
3. Remove the oil pan protector with the 2 bolts.
4. Remove the 15 bolts and 2 nuts and remove the No. 2 oil pan.
5. Insert the blade of an oil pan seal cutter between the oil pans. Cut through the applied sealer and remove the No. 2 oil pan.
6. Remove the 14 bolts and 2 nuts and remove the No. 1 oil pan.
7. Remove the oil pan by prying between the oil pan and cylinder block with a screwdriver with its tip taped.

➡Be sure to clean the bolts and stud bolts, and check the threads for cracks or other damage.

➡Be careful not to damage the contact surfaces of the oil pans.

To install:

8. Apply the seal packing to oil pan No. 1, as shown. Seal packing: Toyota Genuine Seal Packing Black, Three Bond 1207B or equivalent. Standard seal diameter: 3.0 to 4.0 mm (0.118 to 0.156 in.). Application position from inside edge of oil pan. 6.0 mm (0.236 in.).

➡Remove any oil from the contact surface.

➡Install the oil pan within 3 minutes and tighten the bolts and nuts within 15 minutes after applying seal packing.

➡Do not start the engine for at least 2 hours after installing.

9. Install the oil pan sub-assembly with the 14 bolts and 2 nuts.
10. Uniformly tighten the bolts and nuts in several passes to the following:
    - Bolts A to 7 ft. lbs. (10 Nm).
    - Bolts B 26 ft. lbs. (35 Nm).
    - Nut to 26 ft. lbs. (35 Nm).

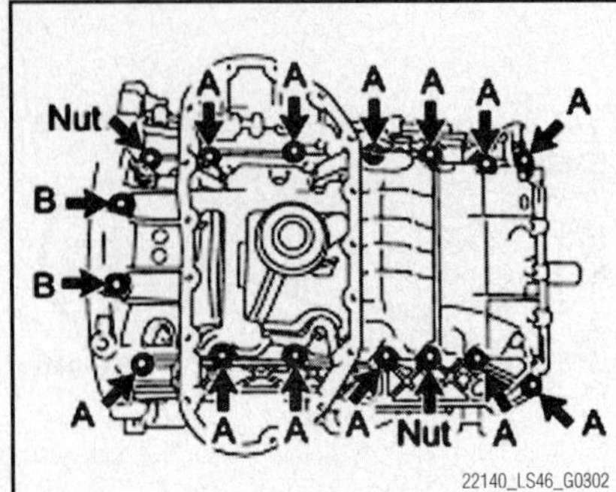

Fig. 78 No. 1 oil pan bolt tightening sequence

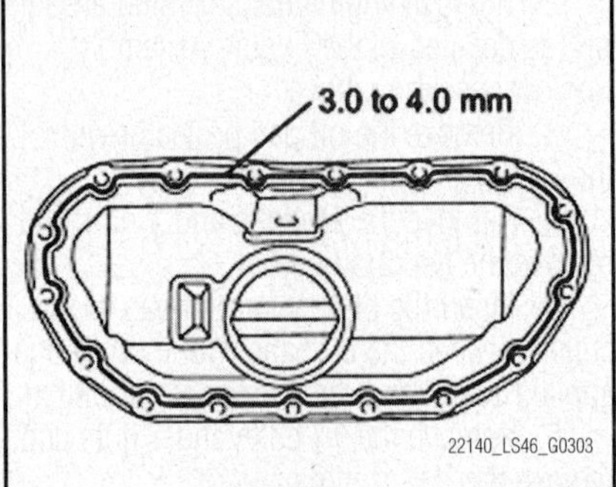

Fig. 79 Oil pan No. 2 sealant application

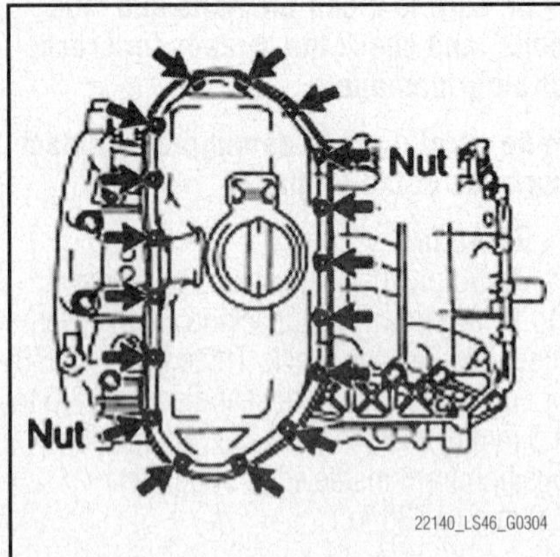

Fig. 80 No. 2 oil pan bolt tightening sequence

11. Apply seal packing to oil pan No. 2, as shown.
12. Install the oil pan with the 15 bolts and 2 nuts. Uniformly tighten the bolts and nuts in several passes to 7 ft. lbs. (10 Nm).
13. Install the oil pan protector with the 2 nuts and tighten to 44 inch lbs. (5 Nm). Install the oil pan protector with its protrusion facing the bottom of the engine.
14. Install engine/transaxle assembly, including adding fluids as detailed in the Engine Assembly Installation section, above.

OIL PUMP

REMOVAL & INSTALLATION

See Figures 81 through 87.

1. Before servicing the vehicle, refer to the Precautions Section.

**** WARNING**

After the engine switch is turned off, the HDD navigation system requires approximately 6 minutes to record various types of memory and settings. As a result, after turning the engine switch off, wait 6 minutes or more before disconnecting the cable from the negative battery terminal.

2. Remove the right cowl top ventilator louver.
3. Disconnect cable from negative battery terminal.

**** CAUTION**

Wait at least 90 seconds after disconnecting the cable from the negative battery terminal to prevent airbag and seat belt pretensioner activation.

4. Remove the V-bank cover sub-assembly.
5. Remove the air cleaner inlet cover.
6. Remove the No. 1 air cleaner inlet.
7. Remove the engine room side covers.
8. Remove the battery, nut, clamp, bolts, insulator, and tray.
9. Remove the No. 1 and No. 2 engine under covers.
10. Remove the front lower suspension member protector.
11. Remove the rear engine under covers.
12. Drain engine oil and coolant.
13. Remove the intake air connector pipe.
14. Remove the air cleaner assemblies.
15. Remove the radiator reservoir.
16. Remove the V-ribbed belt.
17. Disconnect the radiator hoses.
18. Remove engine room ECU outlet duct.
19. Remove skid control ECU.
20. Remove skid control ECU bracket.
21. Disconnect engine wire.
22. Remove front exhaust pipe assembly.
23. Disconnect front stabilizer bar.
24. Remove the left front suspension member reinforcement.
25. Remove steering sliding yoke with shaft sub-assembly.
26. Remove No. 2 steering intermediate shaft assembly.
27. Remove No. 2 exhaust manifold heat insulator.
28. Remove exhaust manifold sub-assembly.
29. Disconnect cooler compressor assembly.
30. Remove oil level dipstick guide.
31. Remove generator.
32. Remove injector driver.
33. Remove No. 1 engine cover sub-assembly.
34. Remove water by-pass pipe sub-assembly.
35. Remove intake manifold.
36. Remove engine cover sub-assemblies.
37. Remove fuel pressure pulsation damper assembly.
38. Remove fuel pipe sub-assemblies.
39. Remove the fuel pump assembly.
40. Remove inlet water housing.
41. Remove water pump pulley.
42. Remove idler pulley sub-assemblies.
43. Remove V-ribbed belt tensioner assembly.
44. Remove front water by-pass joint.
45. Remove the Bank 1 and Bank 2 camshaft timing control motor assembly.
46. Remove oil filter assembly.
47. Remove the resonator bracket sub-assembly.
48. Remove crankshaft pulley.
49. Remove ignition coil assembly.
50. Remove cylinder head cover sub-assemblies.
51. Remove timing chain cover sub-assembly.
52. Remove the oil pump gasket from the cylinder block.
53. Remove the O-ring from the cylinder block.
54. Remove the inlet water pipe.
55. Remove the 2 O-rings from the inlet water pipe.

To install:

56. Apply soapy water to 2 new O-rings and install them to the inlet water pipe.
57. Install the inlet pipe to the No. 1 heat exchanger cover.
58. Install a new oil pump gasket.
59. Install a new O-ring.
60. Apply seal packing in a continuous line to the timing chain cover as shown in the illustration. Seal packing: Toyota Genuine Seal Packing Black, Three Bond 1207B or equivalent.

➡When the contact surfaces are wet, wipe them with an oil-free cloth before applying seal packing.

➡Install the chain cover within 3 minutes and tighten the bolts within 10 minutes after applying seal packing.

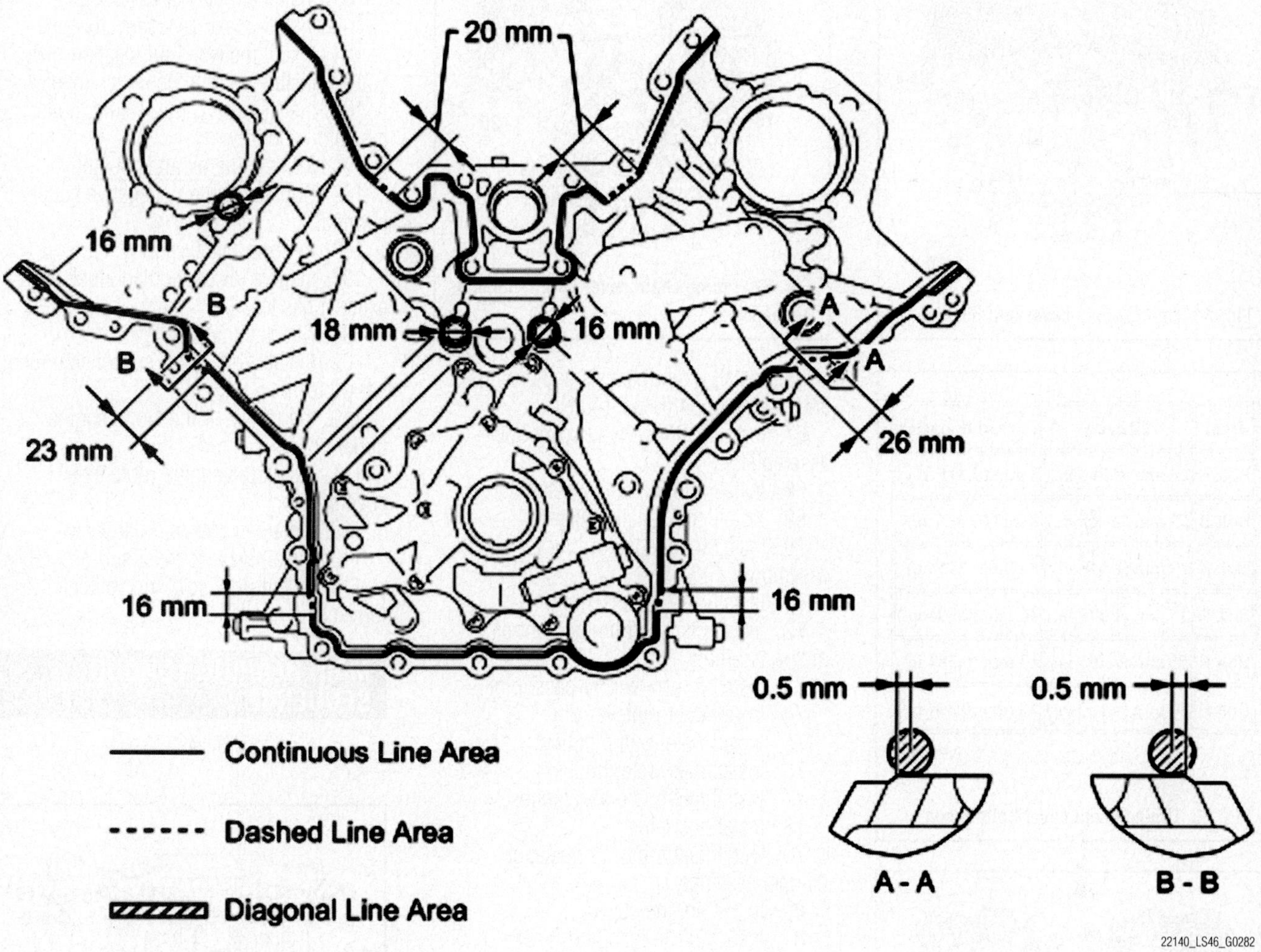

Fig. 81 Timing chain cover sealant application

• Area	Seal packing diameter	Application position from inside edge of cover
Continuous Line Area	3.0 to 4.0 mm (0.1181 to 0.1575 in.)	2.5 mm (0.098 in.)
Dashed Line Area	6.4 mm (0.2520 in.) or more, or within OK area shown in illustration	0.5 mm (0.020 in.)
Diagonal Line Area	3.0 to 4.0 mm (0.1181 to 0.1575 in.)	5.5 mm (0.217 in.)

22140_LS46_G0283

Fig. 82 Sealant application specifications

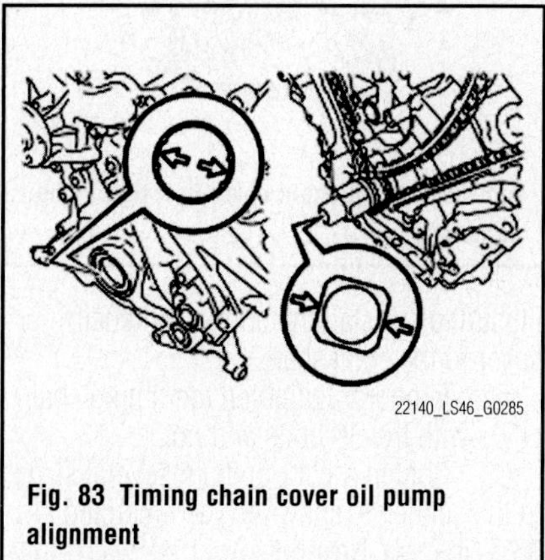

Fig. 83 Timing chain cover oil pump alignment

➡Do not start the engine for at least 2 hours after installing.

61. Align the oil pump's drive rotor spline and the crankshaft as shown in the

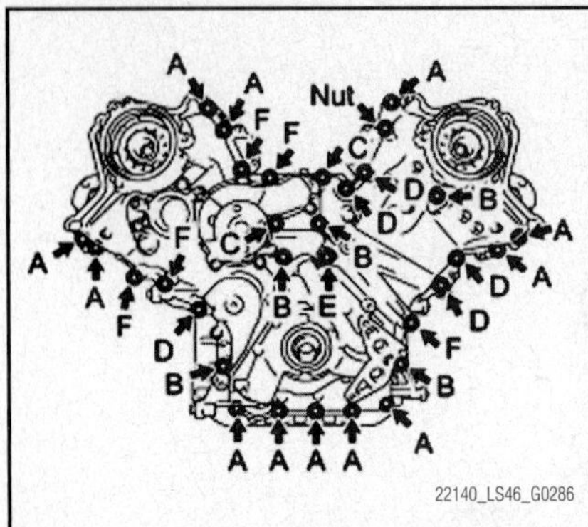

Fig. 84 Timing chain cover bolt position

Item	Length	Thread diameter
Bolt A	25 mm (0.984 in.)	8 mm (0.315 in.)
Bolt B	55 mm (2.165 in.)	8 mm (0.315 in.)
Bolt C	70 mm (2.756 in.)	8 mm (0.315 in.)
Bolt D	35 mm (1.378 in.)	10 mm (0.394 in.)
Bolt E	55 mm (2.165 in.)	10 mm (0.394 in.)
Bolt F	80 mm (3.150 in.)	10 mm (0.394 in.)

22140_LS46_G0284

Fig. 85 Timing chain cover bolt legend

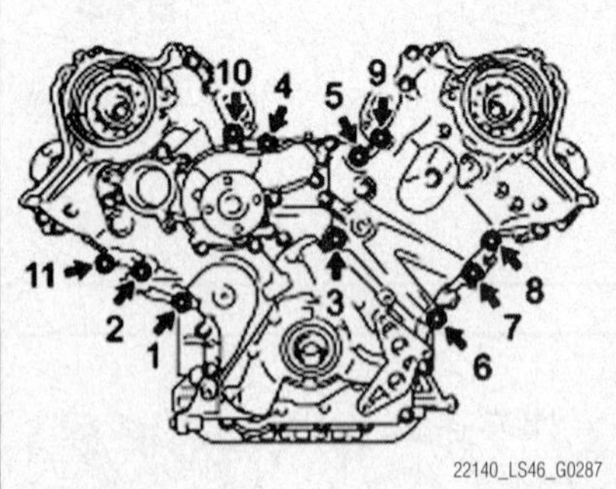

Fig. 86 Timing chain cover bolt tightening sequence 1

illustration. Install the spline and chain cover to the crankshaft.

62. Temporarily tighten the timing chain cover with the 30 bolts and nut.
63. Tighten the 11 bolts in several steps, in the sequence shown in the illustration to 35 ft. lbs. (47 Nm).
64. Temporarily tighten the belt tensioner with the standard bolt and 6 mm hexagon wrench bolt.
65. Tighten the 21 bolts and nut in several steps, in the sequence shown in the illustration to 17 ft. lbs. (23 Nm).

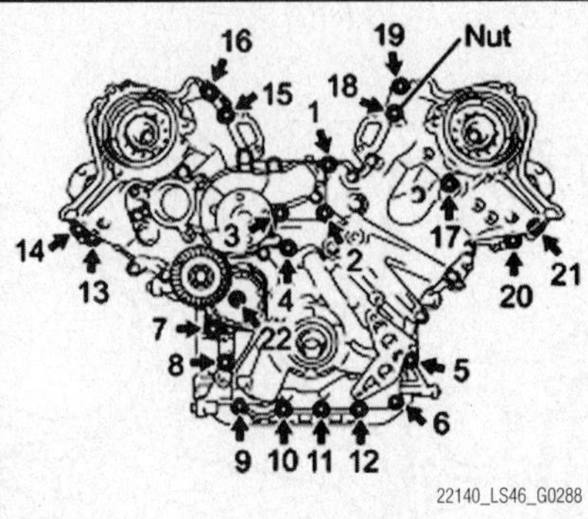

Fig. 87 Timing chain cover bolt tightening sequence 2

66. Install 2 new gaskets and 2 plugs and tighten to 34 ft. lbs. (46 Nm).
67. Install cylinder head cover sub-assemblies.
68. Install ignition coil assembly.
69. Install crankshaft pulley.
70. Install the resonator bracket sub-assembly.
71. Install oil filter assembly.
72. Install camshaft timing control motor assembly.
73. Install front water by-pass joint.
74. Install idler pulleys.
75. Install water pump pulley.
76. Install inlet water housing.
77. Install fuel pump assemblies.
78. Install fuel pipes.
79. Install fuel pressure pulsation damper assembly.
80. Install engine covers.
81. Install intake manifold.
82. Install water by-pass pipe sub-assembly.
83. Install injector driver.
84. Install generator.
85. Install oil level dipstick guide.
86. Connect cooler compressor assembly.
87. Install exhaust manifold sub-assembly.
88. Install No. 2 exhaust manifold heat insulator.
89. Install No. 2 steering intermediate shaft assembly.
90. Install steering sliding yoke with shaft sub-assembly.
91. Install left front suspension member reinforcement.
92. Connect front stabilizer bar.
93. Install front exhaust pipe assembly.
94. Connect engine wire.
95. Install engine room ECU outlet duct.
96. Install skid control ECU.
97. Install V-ribbed belt.
98. Install radiator assembly and hoses.
99. Install air cleaner assemblies.
100. Install intake air connector pipe.
101. Install the No. 1 air cleaner inlet.
102. Install rear engine under covers.
103. Install battery, nut, clamp, bolts, insulator, and tray.
104. Add engine oil and coolant.
105. Connect cable to negative battery terminal.
106. Perform initialization.
107. Inspect for oil, coolant, fuel, and exhaust gas leaks.
108. Check engine oil level.
109. Install front lower suspension member protector.
110. Install No. 1 and No. 2 engine under covers.
111. Install the engine room side covers.
112. Install air cleaner inlet cover.
113. Install V-bank cover sub-assembly.
114. Install right cowl top ventilator louver.

MAIN BEARING TORQUE SEQUENCE

See Figures 88 and 89.

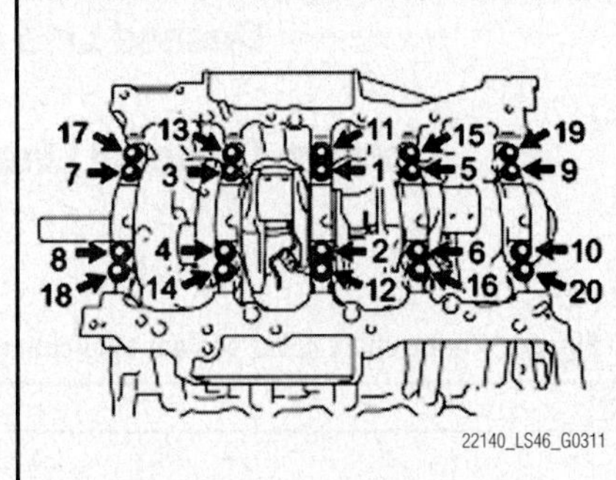

Fig. 88 Main bearing torque sequence—First

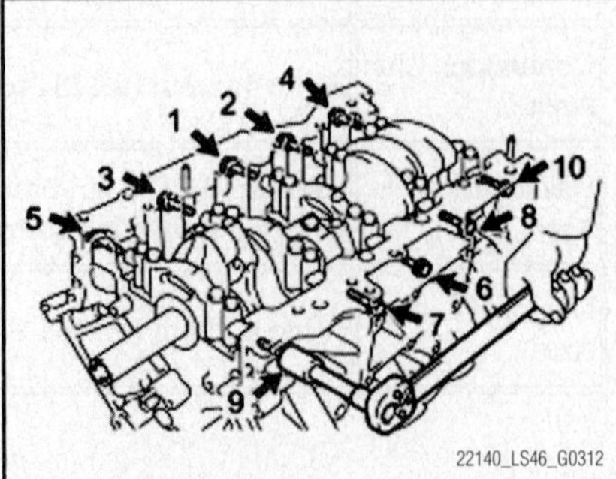

Fig. 89 Main bearing torque sequence—Second

PISTON AND RING

POSITIONING

See Figure 90.

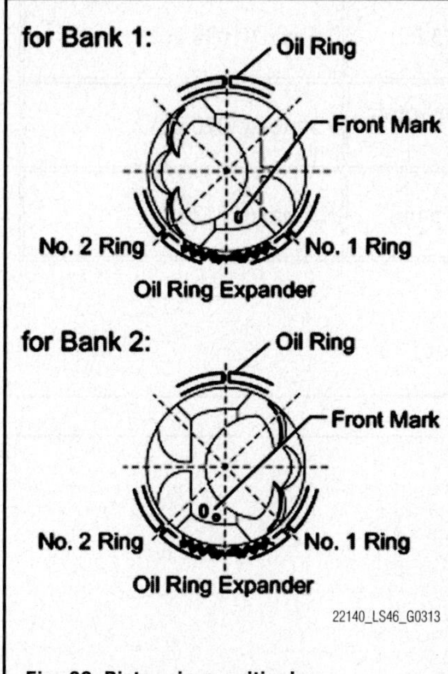

Fig. 90 Piston ring positioning

REAR MAIN SEAL

REMOVAL & INSTALLATION

1. Before servicing the vehicle, refer to the Precautions Section.

WARNING

After the engine switch is turned off, the HDD navigation system requires approximately 6 minutes to record various types of memory and settings. As a result, after turning the engine switch off, wait 6 minutes or more before disconnecting the cable from the negative battery terminal.

2. Remove the right cowl top ventilator louver.
3. Disconnect cable from negative battery terminal.

CAUTION

Wait at least 90 seconds after disconnecting the cable from the negative battery terminal to prevent airbag and seat belt pretensioner activation.

4. Remove the automatic transmission.
5. Remove the drive plate and ring gear sub-assembly.
6. Remove the oil seal with a screwdriver with its tip taped.

➡Do not damage the surface of the oil seal press fit hole and crankshaft.

To install:

Using SST (SST 09223-15030, 09950-70010), tap in a new oil seal until its surface is flush with the oil seal retainer edge.

➡Keep the lip free from foreign matter.

➡ Do not tap on the oil seal at an angle.

7. Install the drive plate and ring gear sub-assembly.
8. Install the automatic transmission.
9. Connect cable to negative battery terminal.

TIMING CHAIN COVER AND SEAL

REMOVAL & INSTALLATION

See Figures 91 through 99.

1. Before servicing the vehicle, refer to the Precautions Section.

WARNING

After the engine switch is turned off, the HDD navigation system requires approximately 6 minutes to record various types of memory and settings. As a result, after turning the engine switch off, wait 6 minutes or more before disconnecting the cable from the negative battery terminal.

2. Remove the right cowl top ventilator louver.
3. Disconnect cable from negative battery terminal.

CAUTION

Wait at least 90 seconds after disconnecting the cable from the negative battery terminal to prevent airbag and seat belt pretensioner activation.

4. Remove the V-bank cover sub-assembly.
5. Remove the air cleaner inlet cover.
6. Remove the No. 1 air cleaner inlet.
7. Remove the engine room side covers.
8. Remove the battery, nut, clamp, bolts, insulator, and tray.
9. Remove the No. 1 and No. 2 engine under covers.
10. Remove the front lower suspension member protector.
11. Remove the rear engine under covers.
12. Drain engine oil and coolant.
13. Remove the intake air connector pipe.
14. Remove the air cleaner assemblies.
15. Remove the radiator reservoir.
16. Remove the V-ribbed belt.
17. Disconnect the radiator hoses.
18. Remove engine room ECU outlet duct.
19. Remove skid control ECU.
20. Remove skid control ECU bracket.
21. Disconnect engine wire.
22. Remove front exhaust pipe assembly.
23. Disconnect front stabilizer bar.
24. Remove the left front suspension member reinforcement.
25. Remove steering sliding yoke with shaft sub-assembly.
26. Remove No. 2 steering intermediate shaft assembly.
27. Remove No. 2 exhaust manifold heat insulator.
28. Remove exhaust manifold sub-assembly.
29. Disconnect cooler compressor assembly.
30. Remove oil level dipstick guide.
31. Remove generator.
32. Remove injector driver.
33. Remove No. 1 engine cover sub-assembly.
34. Remove water by-pass pipe sub-assembly.
35. Remove intake manifold.
36. Remove engine cover sub-assemblies.
37. Remove fuel pressure pulsation damper assembly.
38. Remove fuel pipe sub-assemblies.
39. Remove the fuel pump assembly.
40. Remove inlet water housing.
41. Remove water pump pulley.
42. Remove idler pulley sub-assemblies.
43. Remove V-ribbed belt tensioner assembly.
44. Remove front water by-pass joint.
45. Remove the Bank 1 and Bank 2 camshaft timing control motor assembly.
46. Remove oil filter assembly.
47. Remove the resonator bracket sub-assembly.
48. Remove crankshaft pulley.

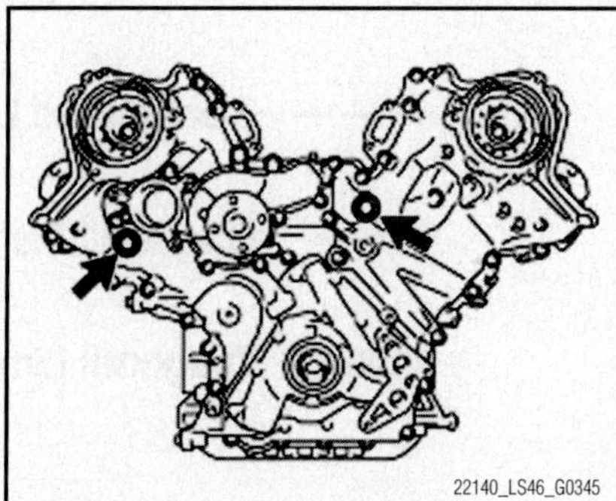

Fig. 91 Timing chain cover plugs and gaskets

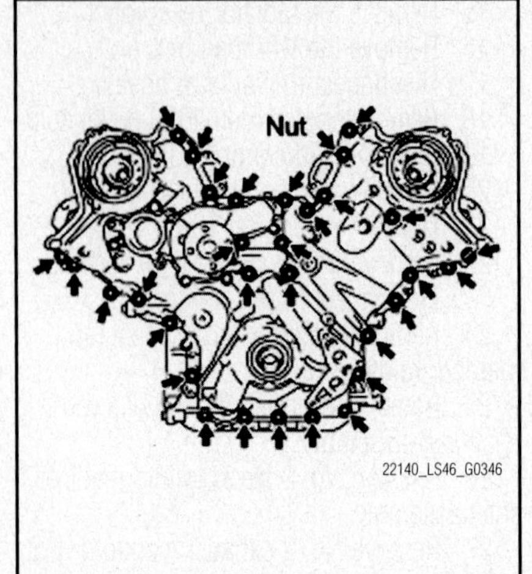

Fig. 92 Timing chain cover bolt removal

• Area	Seal packing diameter	Application position from inside edge of cover
Continuous Line Area	3.0 to 4.0 mm (0.1181 to 0.1575 in.)	2.5 mm (0.098 in.)
Dashed Line Area	6.4 mm (0.2520 in.) or more, or within OK area shown in illustration	0.5 mm (0.020 in.)
Diagonal Line Area	3.0 to 4.0 mm (0.1181 to 0.1575 in.)	5.5 mm (0.217 in.)

22140_LS46_G0283

Fig. 94 Sealant application specifications

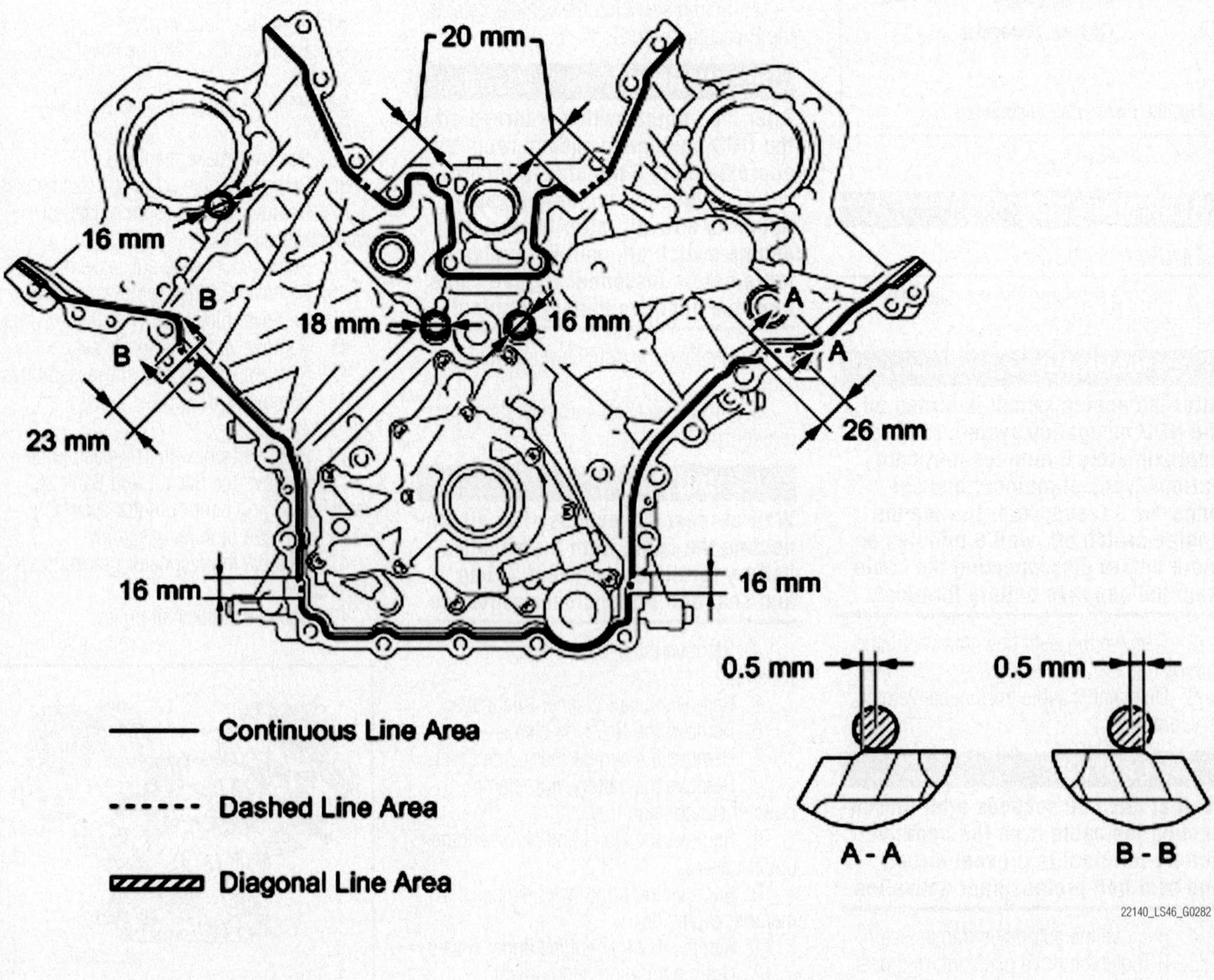

20 mm
16 mm
B
B
18 mm
16 mm
A
A
23 mm
26 mm
16 mm
16 mm
0.5 mm
0.5 mm
Continuous Line Area
Dashed Line Area
Diagonal Line Area
A - A
B - B
22140_LS46_G0282

Fig. 93 Timing chain cover sealant application

Fig. 95 Timing chain cover oil pump alignment

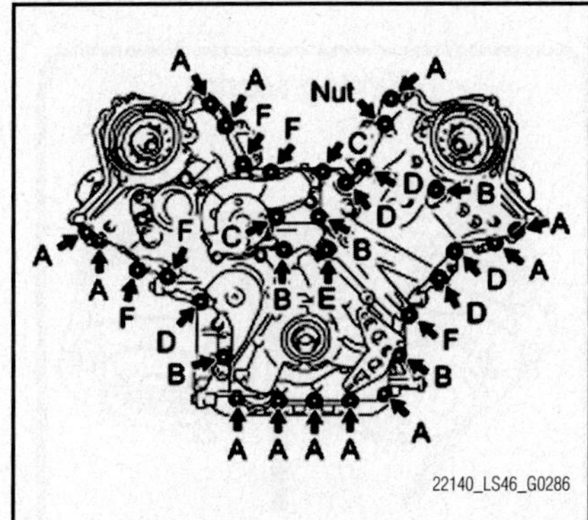

Fig. 96 Timing chain cover bolt position

49. Remove ignition coil assembly.

50. Remove cylinder head cover sub-assemblies.

51. Remove timing chain cover sub-assembly, as follows:

a. Remove the 2 plugs and 2 gaskets.

b. Remove the 30 bolts and nut shown in the illustration.

c. Remove the timing chain cover by prying between the timing chain cover and cylinder head and cylinder block with a screwdriver with its tip taped as shown in the illustration.

To install:

52. Apply seal packing in a continuous line to the timing chain cover as shown in the illustration. Seal packing: Toyota Genuine Seal Packing Black, Three Bond 1207B or equivalent.

➡When the contact surfaces are wet, wipe them with an oil-free cloth before applying seal packing.

➡Install the chain cover within 3 minutes and tighten the bolts within 10 minutes after applying seal packing.

➡Do not start the engine for at least 2 hours after installing.

Item	Length	Thread diameter
Bolt A	25 mm (0.984 in.)	8 mm (0.315 in.)
Bolt B	55 mm (2.165 in.)	8 mm (0.315 in.)
Bolt C	70 mm (2.756 in.)	8 mm (0.315 in.)
Bolt D	35 mm (1.378 in.)	10 mm (0.394 in.)
Bolt E	55 mm (2.165 in.)	10 mm (0.394 in.)
Bolt F	80 mm (3.150 in.)	10 mm (0.394 in.)

22140_LS46_G0284

Fig. 97 Timing chain cover bolt legend

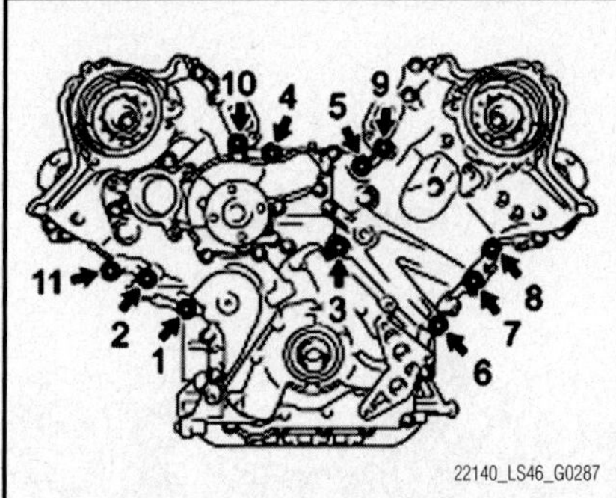

Fig. 98 Timing chain cover bolt tightening sequence 1

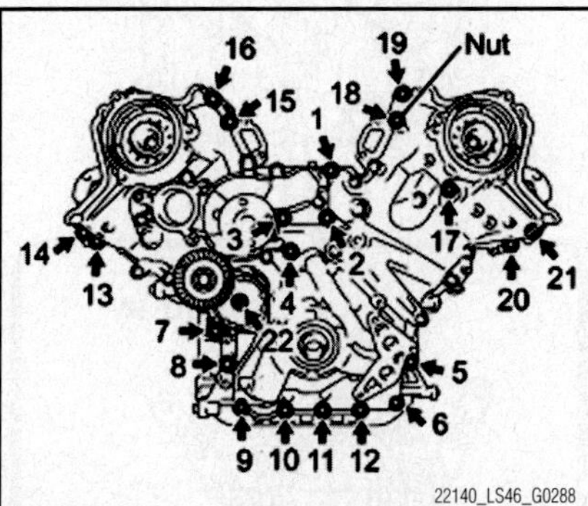

Fig. 99 Timing chain cover bolt tightening sequence 2

53. Align the oil pump's drive rotor spline and the crankshaft as shown in the illustration. Install the spline and chain cover to the crankshaft.

54. Temporarily tighten the timing chain cover with the 30 bolts and nut.

55. Tighten the 11 bolts in several steps, in the sequence shown in the illustration to 35 ft. lbs. (47 Nm).

56. Temporarily tighten the belt tensioner with the standard bolt and 6 mm hexagon wrench bolt.

57. Tighten the 21 bolts and nut in several steps, in the sequence shown in the illustration to 17 ft. lbs. (23 Nm).

58. Install 2 new gaskets and 2 plugs and tighten to 34 ft. lbs. (46 Nm).

59. Install cylinder head cover sub-assemblies.

60. Install ignition coil assembly.

61. Install crankshaft pulley.

62. Install the resonator bracket sub-assembly.

63. Install oil filter assembly.

64. Install camshaft timing control motor assembly.

65. Install front water by-pass joint.

66. Install idler pulleys.

67. Install water pump pulley.

68. Install inlet water housing.

69. Install fuel pump assemblies.

70. Install fuel pipes.

71. Install fuel pressure pulsation damper assembly.

72. Install engine covers.

73. Install intake manifold.

74. Install water by-pass pipe sub-assembly.

75. Install injector driver.

76. Install generator.

77. Install oil level dipstick guide.

78. Connect cooler compressor assembly.

79. Install exhaust manifold sub-assembly.

80. Install No. 2 exhaust manifold heat insulator.

81. Install No. 2 steering intermediate shaft assembly.

82. Install steering sliding yoke with shaft sub-assembly.

83. Install left front suspension member reinforcement.

84. Connect front stabilizer bar.

85. Install front exhaust pipe assembly.

86. Connect engine wire.

87. Install engine room ECU outlet duct.

88. Install skid control ECU.

89. Install V-ribbed belt.

90. Install radiator assembly and hoses.

91. Install air cleaner assemblies.

92. Install intake air connector pipe.

93. Install the No. 1 air cleaner inlet.

94. Install rear engine under covers.

95. Install battery, nut, clamp, bolts, insulator, and tray.

96. Add engine oil and coolant.

97. Connect cable to negative battery terminal.

98. Perform initialization.

99. Inspect for oil, coolant, fuel, and exhaust gas leaks.

100. Check engine oil level.

101. Install front lower suspension member protector.

102. Install No. 1 and No. 2 engine under covers.

103. Install the engine room side covers.

104. Install air cleaner inlet cover.

105. Install V-bank cover sub-assembly.

106. Install right cowl top ventilator louver.

TIMING CHAIN AND SPROCKETS

REMOVAL & INSTALLATION

See Figures 100 through 127.

1. Before servicing the vehicle, refer to the Precautions Section.
2. Discharge fuel system pressure.

WARNING

After the engine switch is turned off, the HDD navigation system requires approximately 6 minutes to record various types of memory and settings. As a result, after turning the engine switch off, wait 6 minutes or more before disconnecting the cable from the negative battery terminal.

3. Remove the right cowl top ventilator louver.
4. Disconnect cable from negative battery terminal.

CAUTION

Wait at least 90 seconds after disconnecting the cable from the negative battery terminal to prevent airbag and seat belt pretensioner activation.

5. Remove timing chain cover sub-assembly.

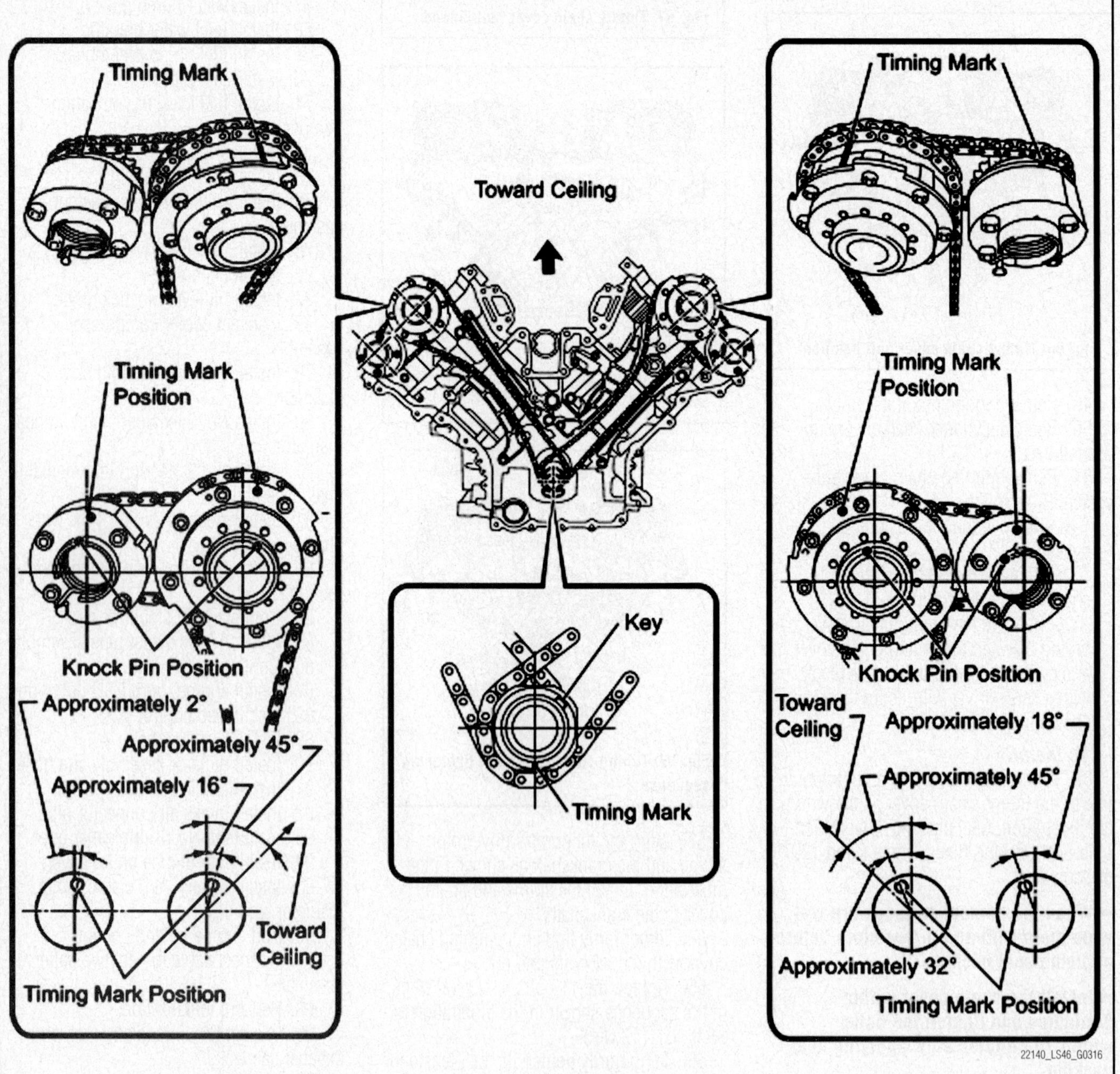

Fig. 100 Timing chain alignment

Fig. 101 No. 1 chain tensioner stopper plate and plunger—Bank 1

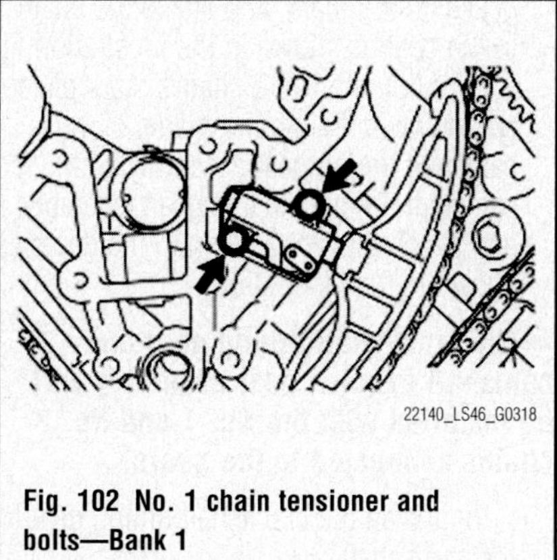

Fig. 102 No. 1 chain tensioner and bolts—Bank 1

6. Set the No. 1 cylinder to TDC/ Compression, as follows:
   a. Temporarily tighten the pulley set bolt.
   b. Rotate the crankshaft clockwise so that the timing marks on the crankshaft timing gear and camshaft timing gears are as shown.

➡If the timing marks do not align, rotate the crankshaft clockwise again and align the timing marks.

7. Remove the No. 1 chain tensioner assembly (Bank 1), as follows:
   a. Move the stopper plate upward to release the lock, and push the plunger deep into the tensioner.
   b. Move the stopper plate downward to set the lock, and insert a hexagon wrench into the stopper plate hole.
   c. Remove the 2 bolts and chain tensioner.
8. Remove the Bank 1 chain tensioner slipper.
9. Remove the 2 bolts and the No. 1 chain vibration damper (Bank 1).
10. Remove the chain assembly, as follows (Bank 1):
   a. While pushing down the No. 3 chain tensioner, insert a pin of 1.0 mm

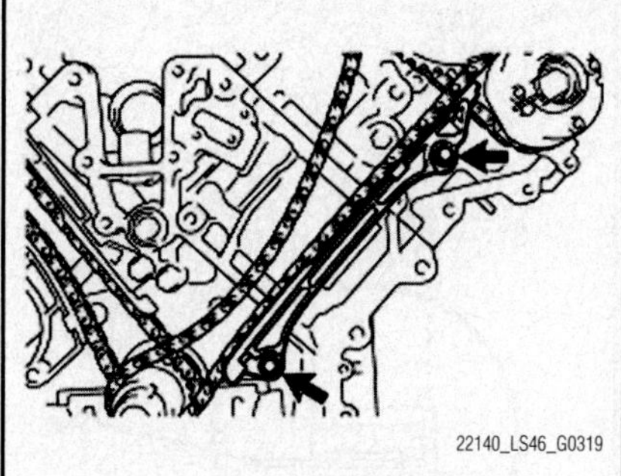

Fig. 103 No. 1 chain vibration damper and bolts—Bank 1

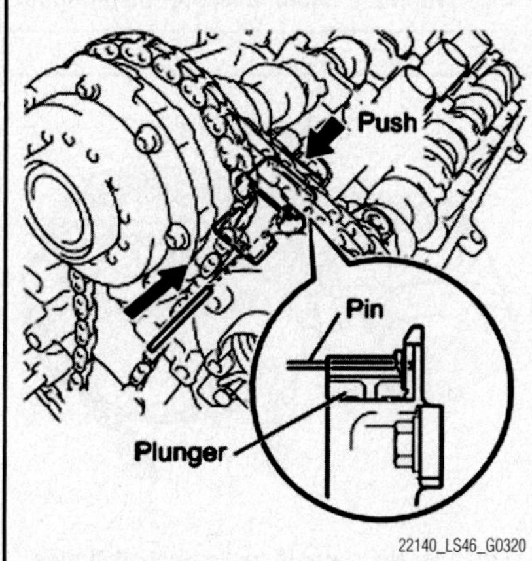

Fig. 104 Secure the No. 3 chain tensioner—Bank 1

(0.039 in.) into the hole to hold it in place.
   b. Hold the hexagonal portion of the camshaft with a wrench and loosen the bolt with a 12mm hexagon wrench.

➡Do not disassemble the camshaft timing gear. Be careful not to damage the cylinder head with the wrench.

   c. Hold the hexagonal portion of the camshaft with a wrench and loosen the bolt.
   d. Remove the 2 bolts. Then with the No. 1 and No. 2 chains still connected to the gears, remove the camshaft timing gear assembly, camshaft timing exhaust gear assembly and crankshaft timing sprocket.
   e. Remove the No. 1 and No. 2 chains from the gears.
11. Remove the 2 bolts and No. 3 chain tensioner.
12. Remove the No. 1 chain tensioner assembly (Bank 2), as follows:
   a. Move the stopper plate upward to release the lock, and push the plunger deep into the tensioner.

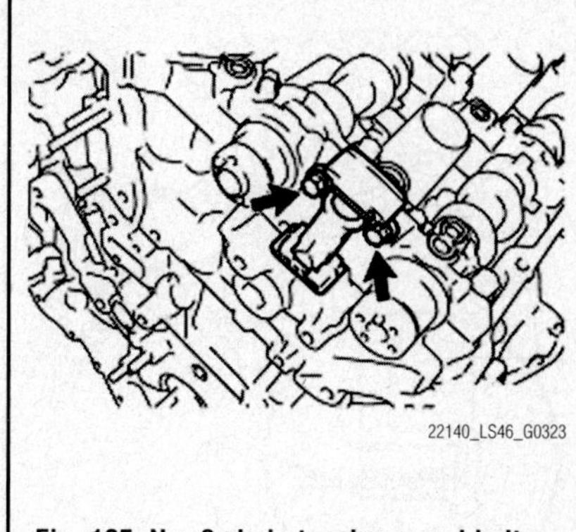

Fig. 105 No. 3 chain tensioner and bolts

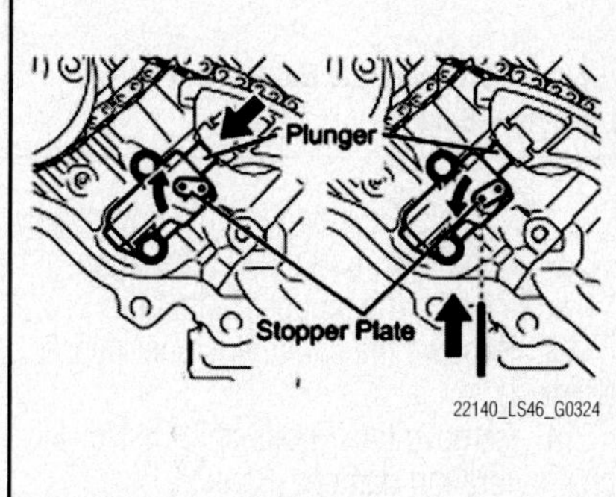

Fig. 106 No. 1 chain tensioner stopper plate and plunger—Bank 2

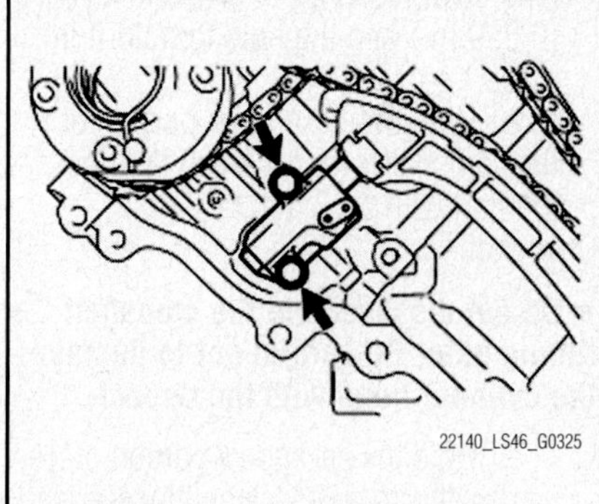

Fig. 107 No. 1 chain tensioner and bolts—Bank 2

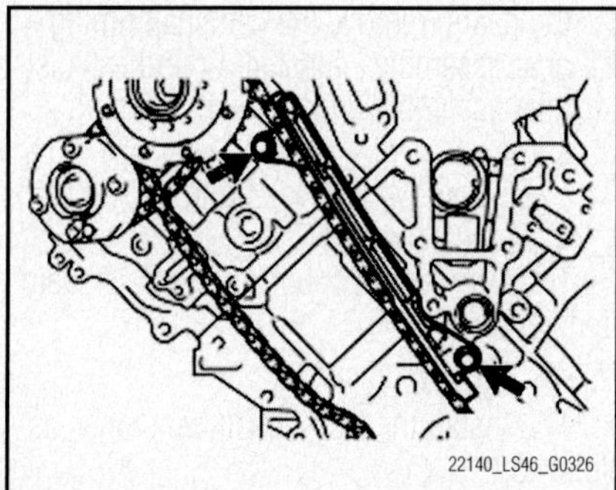

Fig. 108 No. 1 chain vibration damper and bolts—Bank 2

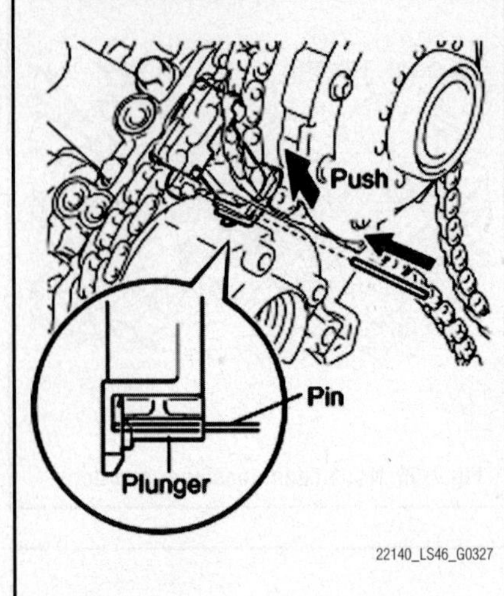

Fig. 109 Secure the No. 2 chain tensioner—Bank 2

b. Move the stopper plate downward to set the lock, and insert a hexagon wrench into the stopper plate hole.

13. Remove the chain tensioner slipper (Bank 2).

14. Remove the 2 bolts and the No. 1 chain vibration damper (Bank 2).

15. Remove the chain assembly, as follows (Bank 2):

a. While raising up the No. 2 chain tensioner, insert a pin of 1.0 mm (0.039 in.) into the hole to hold it in place.

b. Hold the hexagonal portion of the camshaft with a wrench and loosen the bolt with a 12 mm hexagon wrench.

➡Do not disassemble the camshaft timing gear. Be careful not to damage the cylinder head with the wrench.

c. Hold the hexagonal portion of the camshaft with a wrench and loosen the bolt.

d. Remove the 2 bolts. Then with the No. 1 and No. 2 chains still connected to the gears, remove the camshaft timing gear assembly, camshaft timing exhaust gear assembly and crankshaft timing sprocket.

e. Remove the No. 1 and No. 2 chains from the gears.

16. Remove the 2 bolts and No. 2 chain tensioner.

To install:

17. Install the No. 2 chain tensioner, as follows:

a. Install the No. 2 chain tensioner with the 2 bolts and tighten to 7 ft. lbs. (10 Nm).

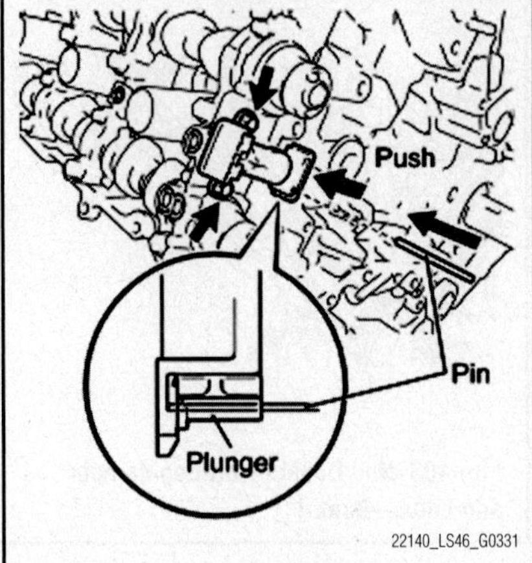

Fig. 110 No. 2 chain tensioner installation

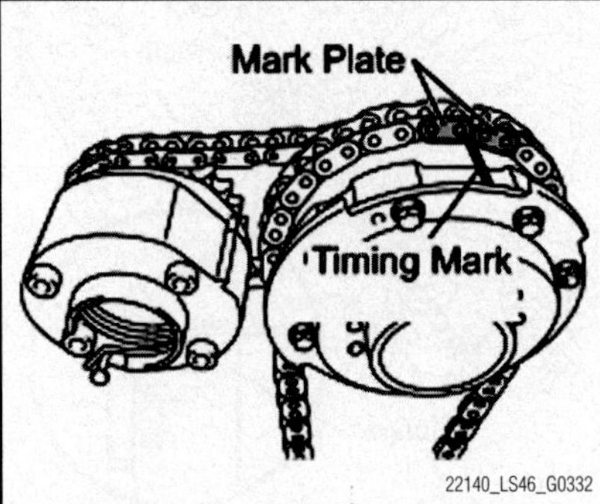

Fig. 111 No. 1 chain and camshaft timing gear alignment—Bank 2

Fig. 112 Chain and camshaft timing gear alignment—Bank 2

b. While raising up the No. 2 chain tensioner, insert a pin of 1.0 mm (0.039 in.) into the hole to hold it in place.

18. Install the No. 1 chain sub-assembly (Bank 2), as follows:

a. Align the No. 1 chain's orange mark plates with the camshaft timing gear's timing mark, and attach the chain to the gear as shown in the illustration.

b. Align the No. 1 chain's orange mark plate with the crankshaft timing gear's timing mark, and attach the chain to the gear as shown in the illustration.

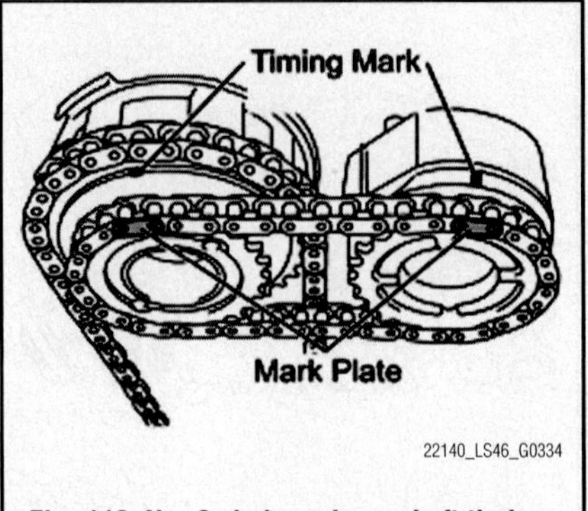

Fig. 113 No. 2 chain and camshaft timing gear alignment—Bank 2

c. Align the No. 2 chain's mark plates (yellow) with the timing marks of the camshaft timing gear assembly and camshaft timing exhaust gear assembly, and attach the No. 2 chain to the gears as shown in the illustration.

➡The crankshaft timing gear and camshaft exhaust gear assembly will be installed with the No. 1 and No. 2 chains connected to the gears.

d. Install the crankshaft timing gear to the crankshaft.

e. Align and attach the knock pin of the No. 1 camshaft with the pin hole of the camshaft timing gear assembly.

f. Using the hexagonal portion of the No. 2 camshaft, align and attach the knock pin of the No. 2 camshaft with the pin hole of the camshaft timing exhaust gear assembly.

g. Remove the pin from the No. 2 chain tensioner.

19. Install the No. 1 chain vibration damper (Bank 2) with the 2 bolts and tighten to 15 ft. lbs. (21 Nm).

20. Install the chain tensioner slipper (Bank 2). If you cannot install the chain tensioner slipper due to the tension of the chain, use the hexagonal portion of the camshaft to loosen the chain, and then install the chain tensioner slipper.

21. Install the No. 1 chain tensioner assembly (Bank 2), as follows:

a. Move the stopper plate upward to release the lock, and push the plunger deep into the tensioner.

b. Move the stopper plate downward to set the lock, and insert a hexagon wrench into the hole of the stopper plate.

c. Install the chain tensioner with the 2 bolts and tighten to 7 ft. lbs. (10 Nm).

d. Remove the hexagon wrench from the chain tensioner.

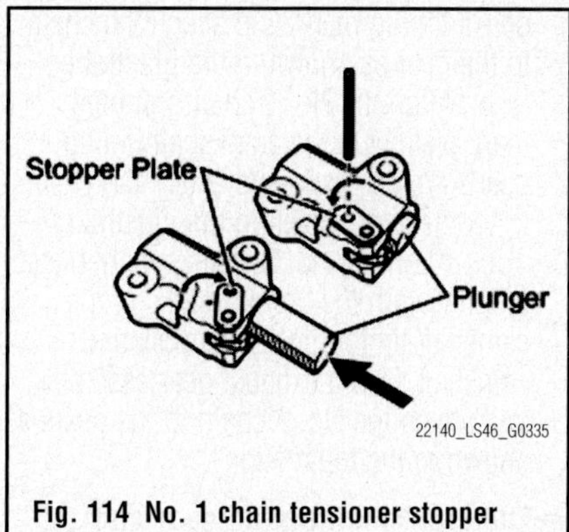

Fig. 114 No. 1 chain tensioner stopper plate

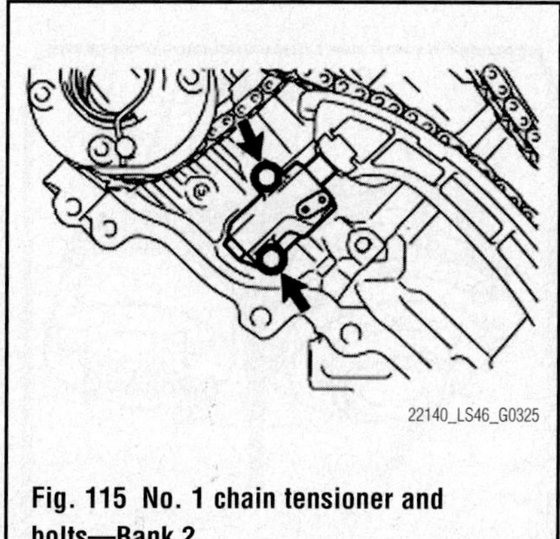

Fig. 115 No. 1 chain tensioner and bolts—Bank 2

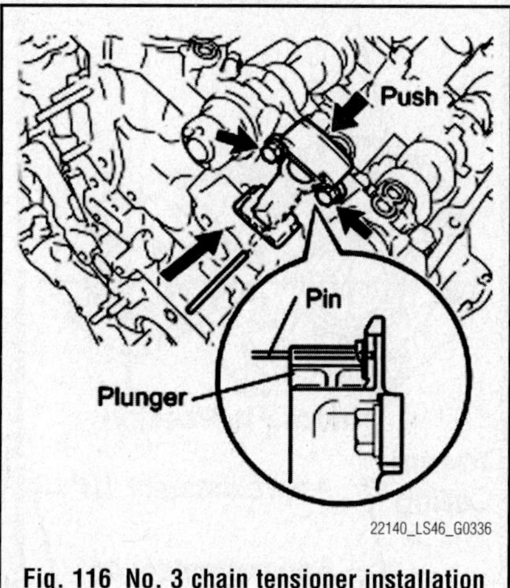

Fig. 116 No. 3 chain tensioner installation

22. Install the No. 3 chain tensioner, as follows:

a. Install the No. 3 chain tensioner with the 2 bolts and tighten to 7 ft. lbs. (10 Nm).

b. While pushing down the No. 3 chain tensioner, insert a pin of 1.0 mm (0.039 in.) into the hole to hold it in place.

23. Install the No. 1 chain sub-assembly (Bank 1), as follows:

a. Align the No. 1 chain's orange mark plates with the camshaft timing

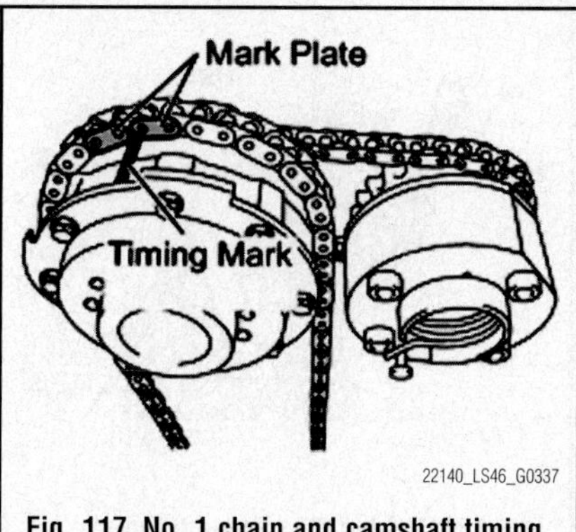

Fig. 117 No. 1 chain and camshaft timing gear alignment—Bank 1

Fig. 118 Chain and camshaft timing gear alignment—Bank 1

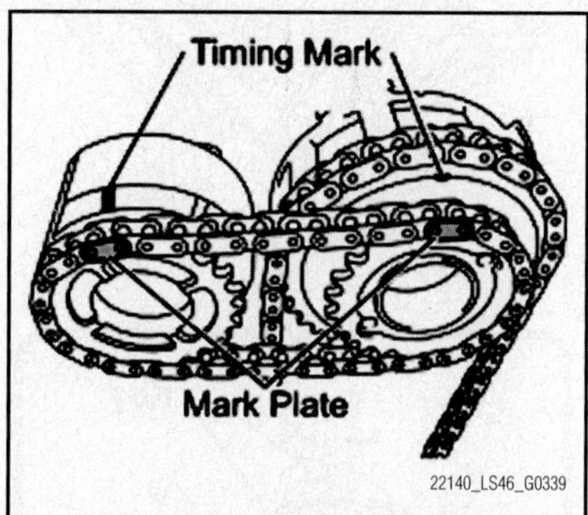

Fig. 119 No. 2 chain and camshaft timing gear alignment—Bank 1

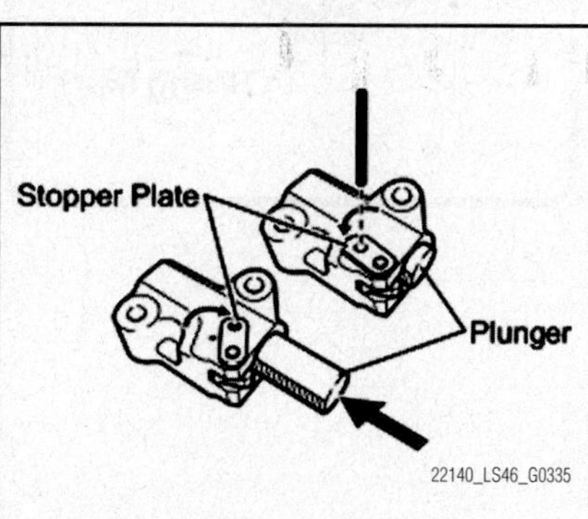

Fig. 120 No. 1 chain tensioner stopper plate

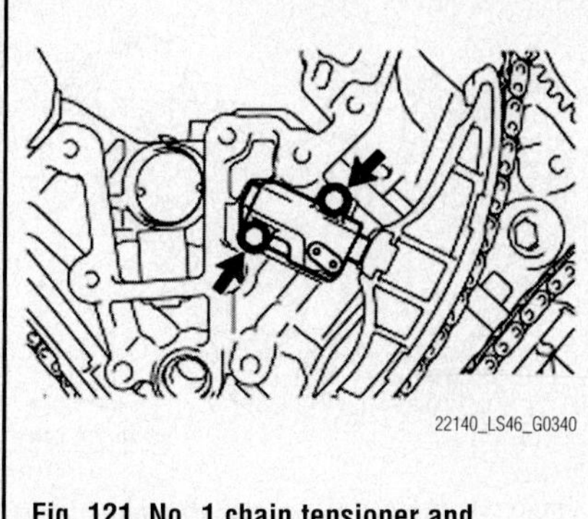

Fig. 121 No. 1 chain tensioner and bolts—Bank 1

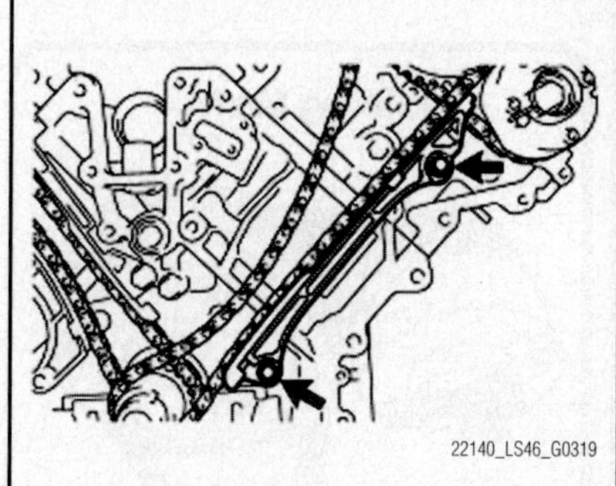

Fig. 122 No. 1 chain vibration damper and bolts—Bank 1

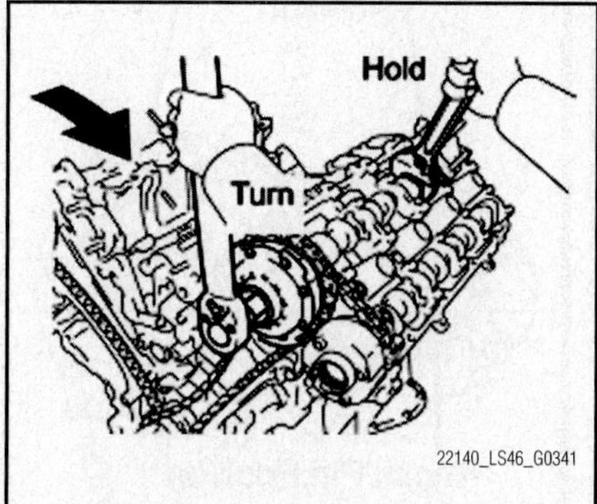

Fig. 123 Hold the hexagonal portion of the No. 3 camshaft—Bank 1

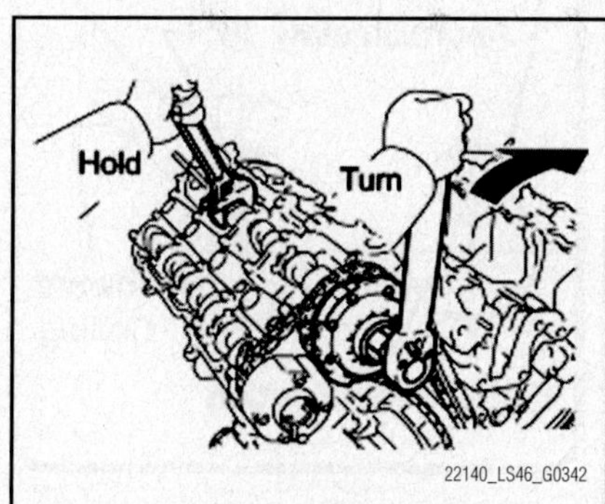

Fig. 124 Hold the hexagonal portion of the No. 1 camshaft—Bank 2

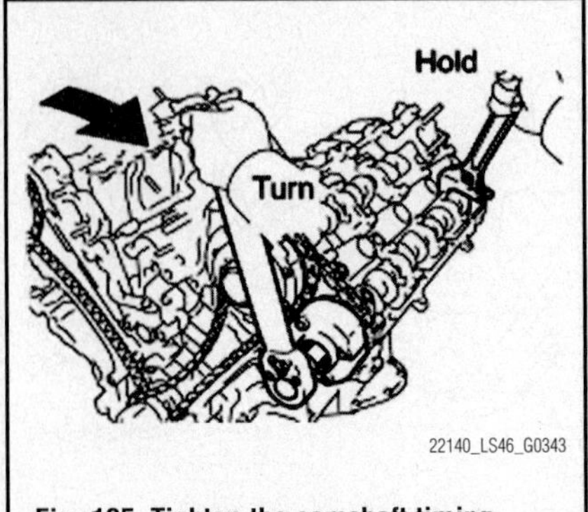

Fig. 125 Tighten the camshaft timing exhaust gear assembly—Bank 1

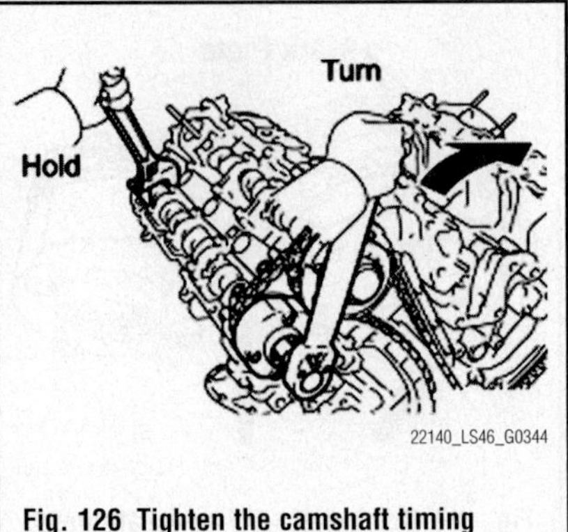

Fig. 126 Tighten the camshaft timing exhaust gear assembly—Bank 2

gear's timing mark, and attach the chain to the gear as shown in the illustration.

b. Align the No. 1 chain's orange mark plate with the crankshaft timing gear's timing mark, and attach the chain to the gear as shown in the illustration.

c. Align the No. 2 chain's mark plates (yellow) with the timing marks of the camshaft timing gear assembly and camshaft timing exhaust gear assembly, and attach the No. 2 chain to the gears as shown in the illustration.

➡**The crankshaft timing gear and camshaft exhaust gear assembly will**

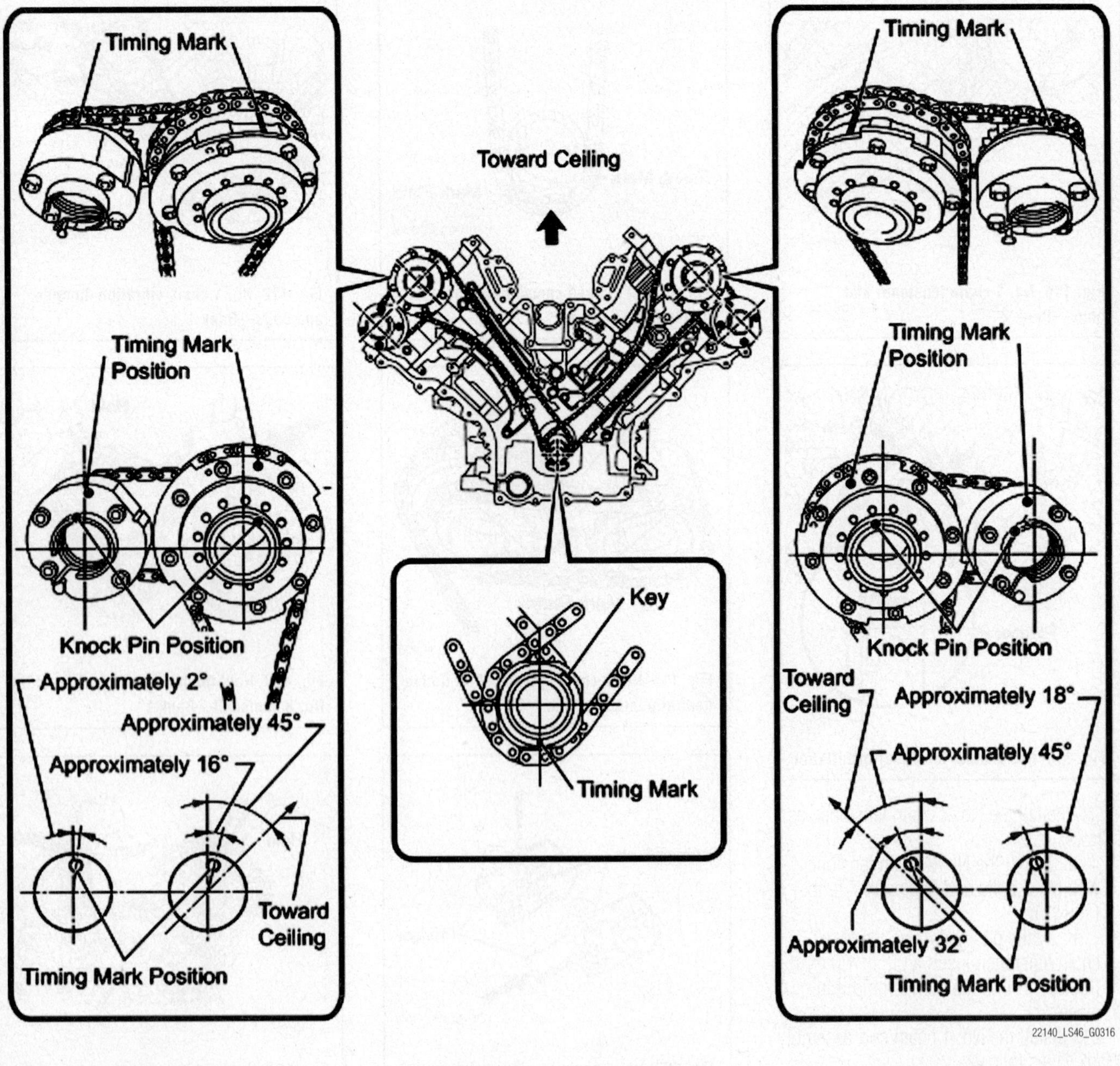

Fig. 127 Timing chain alignment

be installed with the No. 1 and No. 2 chains connected to the gears.

d. Install the crankshaft timing gear to the crankshaft.

e. Align and attach the knock pin of the No. 3 camshaft with the pin hole of the camshaft timing gear assembly.

f. Using the hexagonal portion of the No. 4 camshaft, align and attach the knock pin of the No. 4 camshaft with the pin hole of the camshaft timing exhaust gear assembly. Because the gears' timing mark positions may shift due to looseness of the No. 1 chain, use the hexagonal portion of the camshaft to hold the No. 3 camshaft in place until the No. 1 chain tensioner is installed.

g. Remove the pin from the No. 2 chain tensioner.

24. Install the chain tensioner slipper (Bank 1). If you cannot install the chain tensioner slipper due to the tension of the chain, use the hexagonal portion of the camshaft to loosen the chain and install the chain tensioner.

25. Install the No. 1 chain tensioner assembly (Bank 1), as follows:

a. Move the stopper plate upward to release the lock, and push the plunger deep into the tensioner.

b. Move the stopper plate downward to set the lock, and insert a hexagon wrench into the hole of the stopper plate.

c. Install the chain tensioner and gasket with the 2 bolts and tighten 7 ft. lbs. (10 Nm).

26. Install the No. 1 chain vibration damper (Bank 1), as follows:

a. Install the vibration damper with the 2 bolts and tighten to 15 ft. lbs. (21 Nm).

b. Remove the hexagon wrench from the No. 1 chain tensioner.

27. Tighten the camshaft timing gear assembly, as follows:

a. Using a wrench, hold the hexagonal portion of the No. 3 camshaft (for Bank 1).

b. Using a wrench, hold the hexagonal portion of the No. 1 camshaft (for Bank 2).

c. Using a 12 mm socket hexagon wrench, tighten the camshaft timing gear assembly with a new bolt to 58 ft. lbs. (79 Nm).

d. Using a wrench to hold the hexagonal portion of the No. 4 camshaft (for Bank 1), tighten the camshaft timing exhaust gear assembly with the bolt to 74 ft. lbs. (100 Nm).

e. Using a wrench to hold the hexagonal portion of the No. 2 camshaft (for Bank 2), tighten the camshaft timing exhaust gear assembly with the bolt to 74 ft. lbs. (100 Nm).

28. Set the No. 1 cylinder to TDC/Compression, as follows:

a. Temporarily tighten the pulley bolt.

b. Rotate the crankshaft clockwise and check that the timing marks on the crankshaft timing gear and camshaft timing gears are as shown in the illustration.

c. Remove the crankshaft pulley bolt.

29. Install the timing chain cover.

30. Connect cable to negative battery terminal.

VALVE COVERS

REMOVAL & INSTALLATION

See Figures 128 through 135.

1. Before servicing the vehicle, refer to the Precautions Section.
2. Remove engine, if necessary. It may be possible to remove the valve covers while the engine is still in the vehicle.
3. Remove the engine assembly.
4. Install on engine stand.
5. Remove the oil filler cap and gasket.
6. Remove the spark plugs and ignition coil assembly.
7. Remove the VVT sensor.
8. Remove the camshaft position sensor.
9. Remove the crankshaft position sensor.
10. Remove the camshaft timing oil control valve.
11. Remove the cylinder block water drain cock sub-assembly.
12. Remove the camshaft timing control motor from Bank 1 and Bank 2.
13. Remove the oil filter element.
14. Remove the oil filter bracket.
15. Remove the 4 nuts, front water bypass joint and 2 gaskets.
16. Using a screwdriver, pry out the 8 spark plug tube gaskets.
17. Remove the Bank 1 and Bank 2 cylinder head covers, as follows:

a. Remove the fuel pump spacer gasket.

b. Remove the 15 bolts, 2 seal washers, cylinder head cover and gasket.

c. Remove the 4 gaskets and 2 O-rings from the Bank 1 and Bank 2 camshaft bearing caps (For Bank 1: No. 2, No. 3, No. 7), (For Bank 2: No. 1, No. 3, No. 6).

To install:

18. Install the Bank 1 and Bank 2 cylinder head covers, as follows:

a. Install 4 new gaskets and 2 new O-rings to the Bank 1 and Bank 2

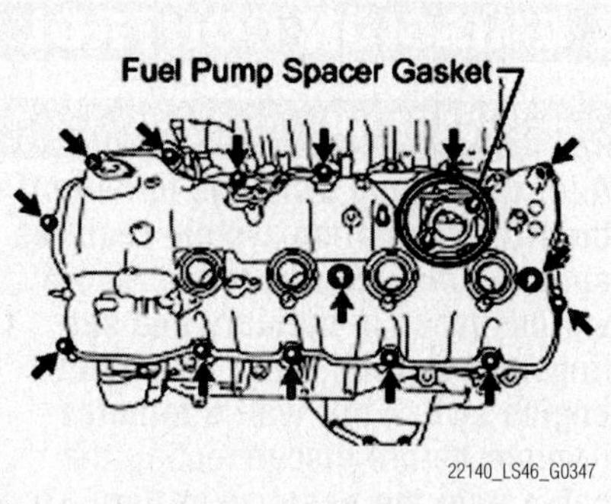

Fig. 128 Cylinder head cover bolt removal—Bank 1

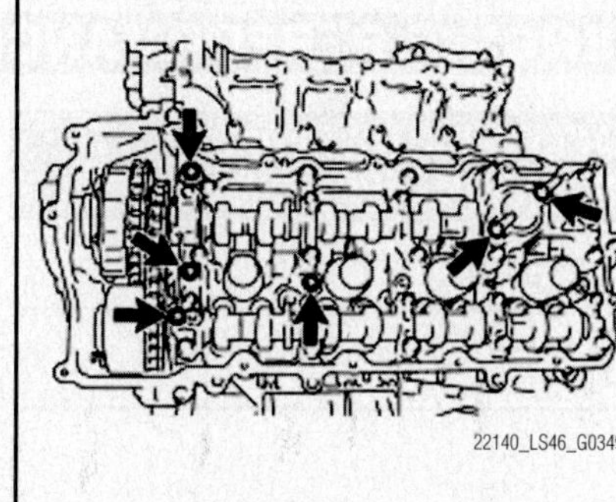

Fig. 130 Camshaft bearing cap gaskets and O-rings—Bank 1

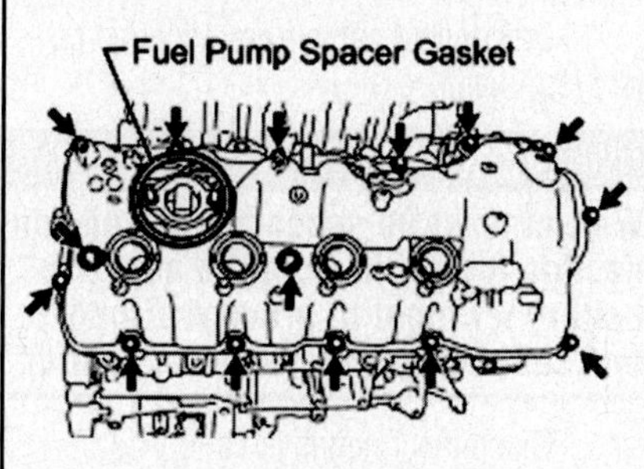

Fig. 129 Cylinder head cover bolt removal—Bank 2

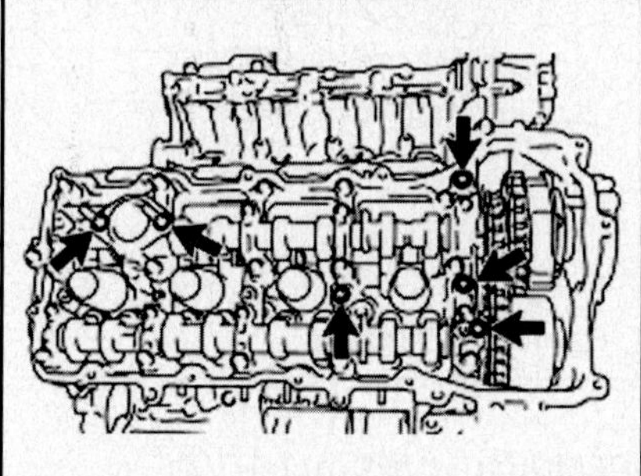

Fig. 131 Camshaft bearing cap gaskets and O-rings—Bank 2

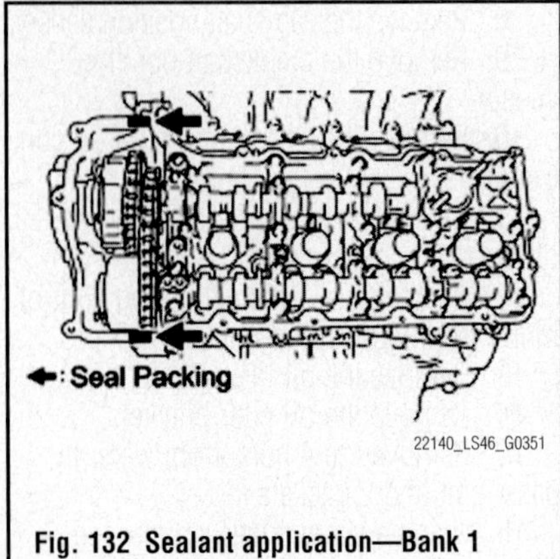

Fig. 132 Sealant application—Bank 1

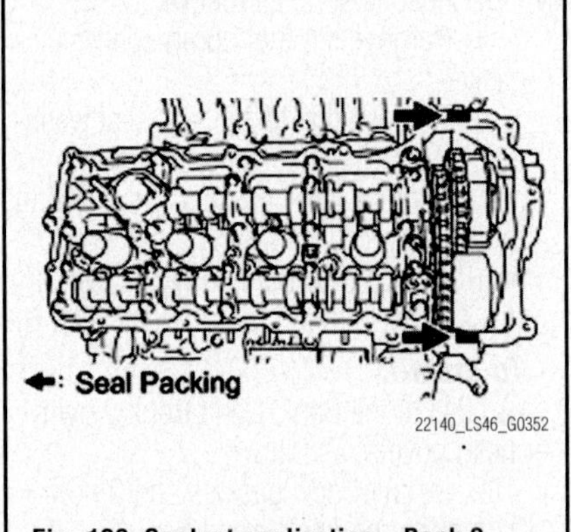

Fig. 133 Sealant application—Bank 2

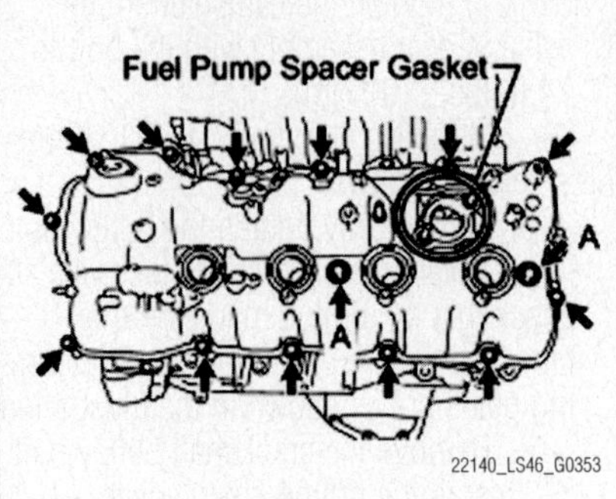

Fig. 134 Cylinder head cover bolt tightening sequence—Bank 1

Fig. 135 Cylinder head cover bolt tightening sequence—Bank 2

camshaft bearing caps (For Bank 1: No. 2, No. 3, No. 7), (For Bank 2: No. 1, No. 3, No. 6).

b. Install a new gasket to each cylinder head cover. Remove any oil from the contact surface.

c. Apply seal packing as shown. Seal packing: Toyota Genuine Seal Packing Black, Three Bond 1207B or equivalent.

➡Remove any oil from the contact surface.

➡Install the cylinder head cover within 3 minutes and tighten the bolts within 15 minutes after applying seal packing.

➡Do not start the engine for at least 2 hours after the installation.

d. Install each cylinder head cover with 2 new seal washers and the 15 bolts and tighten to 15 ft. lbs. (21 Nm), and 9 ft. lbs. (12 Nm).

e. Install the fuel pump spacer gasket to each cylinder head cover.

19. The remainder of installation is the reverse of removal.

VALVE LASH

ADJUSTMENT

No adjustment is necessary.

ENGINE PERFORMANCE & EMISSION CONTROL

ACCELERATOR PEDAL POSITION (APP) SENSOR

LOCATION

See Figure 136.

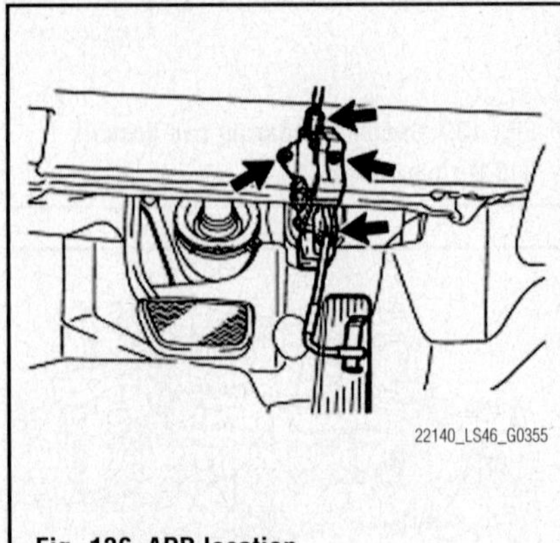

Fig. 136 APP location

REMOVAL & INSTALLATION

1. Before servicing the vehicle, refer to the Precautions Section.
2. Discharge fuel system pressure.

⁂ WARNING

After the engine switch is turned off, the HDD navigation system requires approximately 6 minutes to record various types of memory and settings. As a result, after turning the engine switch off, wait 6 minutes or more before disconnecting the cable from the negative battery terminal.

3. Remove the right cowl top ventilator louver.
4. Disconnect cable from negative battery terminal.

⁂ CAUTION

Wait at least 90 seconds after disconnecting the cable from the negative battery terminal to prevent airbag and seat belt pretensioner activation.

5. Disconnect negative battery cable.
6. Disconnect the accelerator pedal position sensor connector.
7. Remove the 3 nuts and accelerator pedal.

➡Do not disassemble the accelerator pedal assembly.

To install:

8. Installation is the reverse of removal. Tighten the accelerator pedal position sensor nuts to 48 inch lbs. (5.4 Nm).
9. Perform initialization.

CAMSHAFT POSITION (CMP) SENSOR

LOCATION

See Figure 137.

REMOVAL & INSTALLATION

1. Remove V-bank cover sub-assembly.
2. Disconnect the camshaft position sensor connector.
3. Remove the bolt and camshaft position sensor.

To install:

4. Installation is the reverse of removal. Tighten the camshaft position sensor bolt to 7 ft. lbs. (10 Nm).

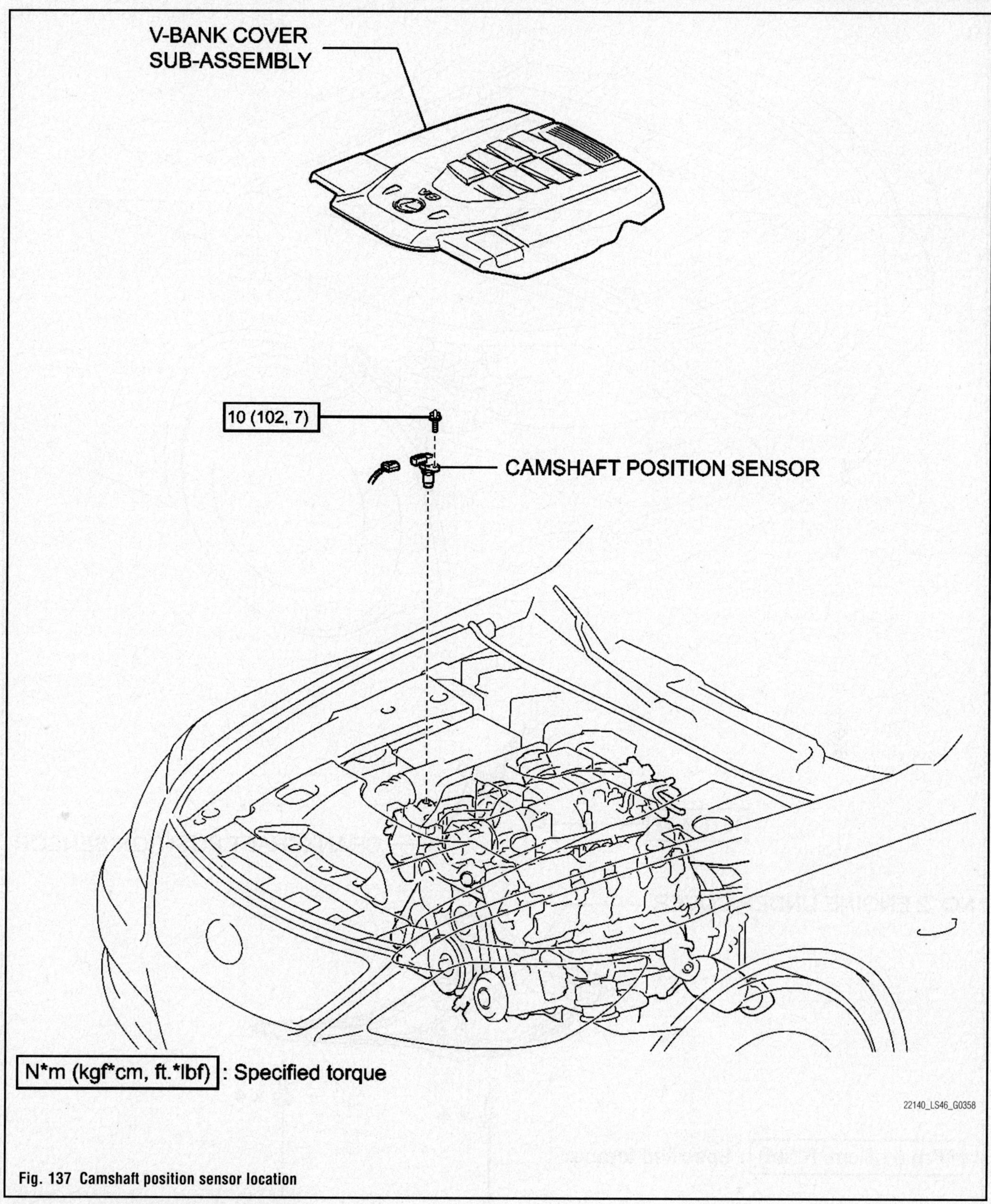

Fig. 137 Camshaft position sensor location

CRANKSHAFT POSITION (CKP) SENSOR

LOCATION

See Figure 138.

REMOVAL & INSTALLATION

See Figure 139.

1. Remove No. 2 engine under cover.
2. Disconnect the crankshaft position sensor connector.
3. Remove the bolt and crankshaft position sensor.

To install:

4. Install the crankshaft position sensor with the bolt and tighten to 7 ft. lbs. (10 Nm).

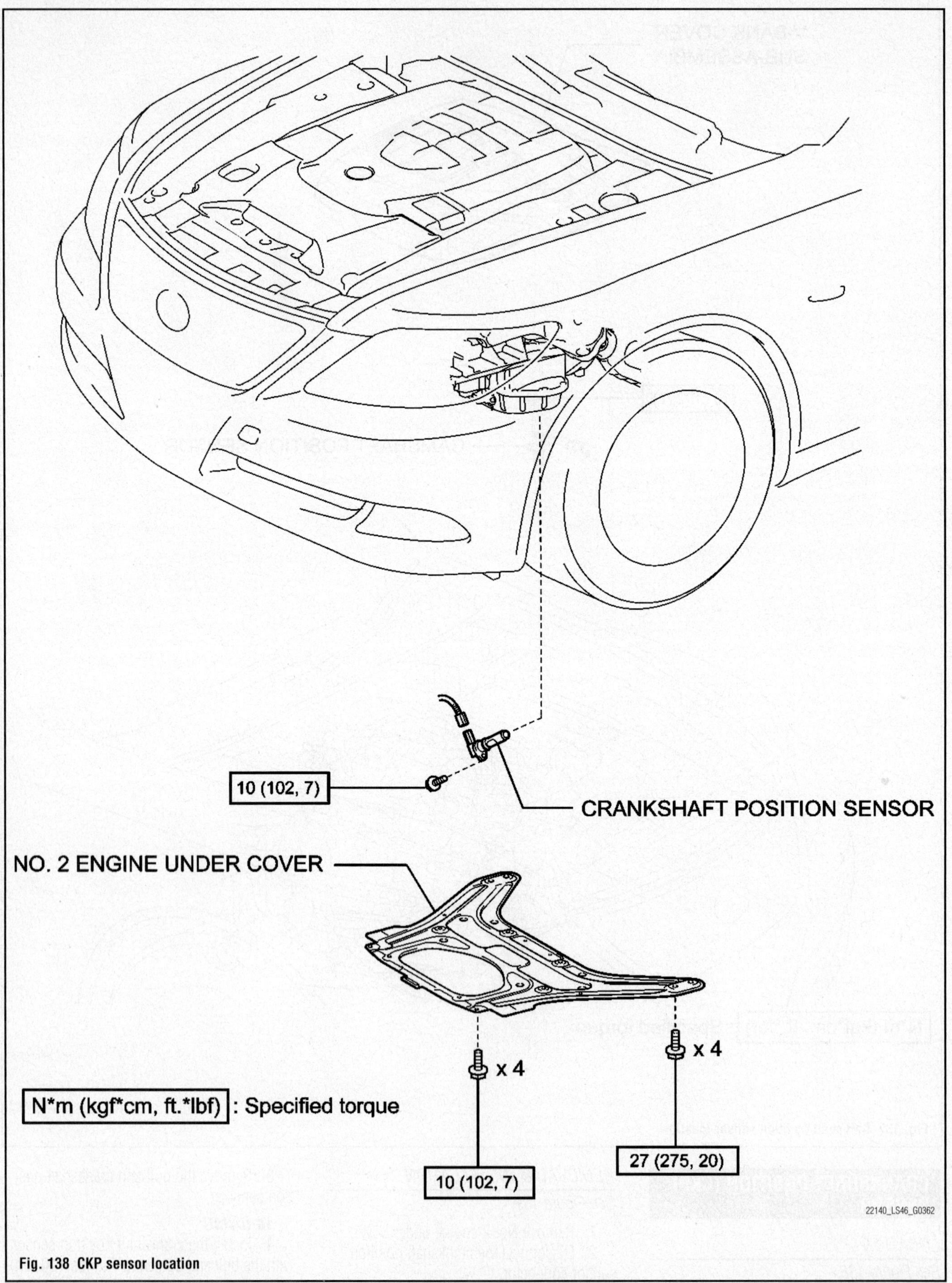

Fig. 138 CKP sensor location

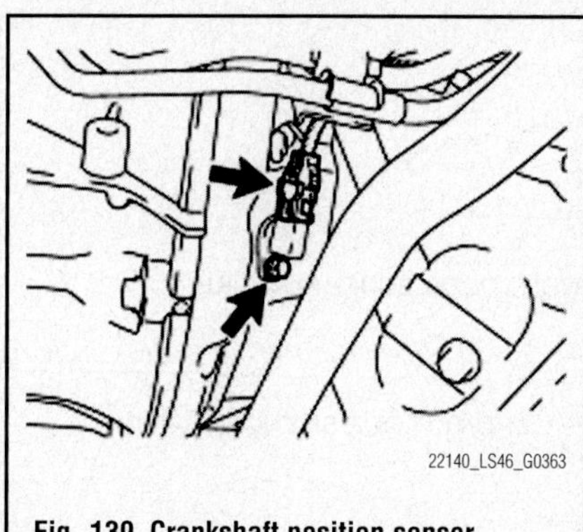

Fig. 139 Crankshaft position sensor

5. Connect the crankshaft position sensor connector.
6. Inspect for oil leak.
7. Install No. 2 engine under cover.

ELECTRONIC CONTROL MODULE (ECM)

LOCATION

See Figure 140.

REMOVAL & INSTALLATION

See Figure 140.

1. Before servicing the vehicle, refer to the Precautions Section.

⁂ WARNING

After the engine switch is turned off, the HDD navigation system requires approximately 6 minutes to record various types of memory and settings. As a result, after turning the engine switch off, wait 6 minutes or more before disconnecting the cable from the negative battery terminal.

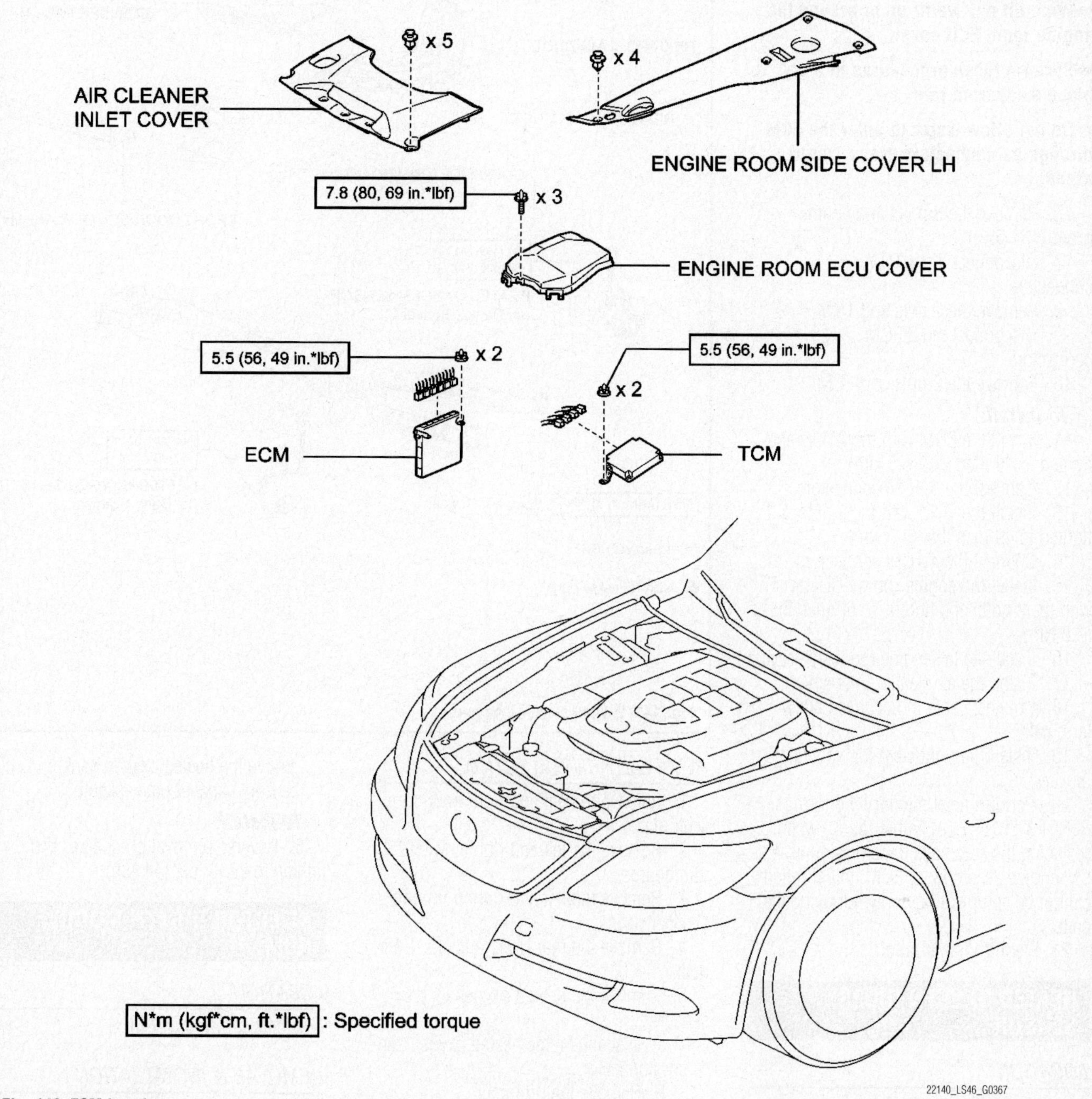

x 5
x 4
AIR CLEANER INLET COVER
ENGINE ROOM SIDE COVER LH
7.8 (80, 69 in.*lbf) x 3
ENGINE ROOM ECU COVER
5.5 (56, 49 in.*lbf) x 2
5.5 (56, 49 in.*lbf) x 2
ECM
TCM
N*m (kgf*cm, ft.*lbf) : Specified torque
22140_LS46_G0367

Fig. 140 ECM location

2. Remove the right cowl top ventilator louver.

3. Disconnect cable from negative battery terminal.

⁂ CAUTION

Wait at least 90 seconds after disconnecting the cable from the negative battery terminal to prevent airbag and seat belt pretensioner activation.

4. Remove the air cleaner inlet cover.

5. Remove the left engine room side cover.

➡Wipe off any water on or around the engine room ECU cover.

➡Perform these procedures in a dry place away from rain, etc.

➡Do not allow water to enter the ECM through its connectors areas, screw areas, etc.

6. Remove the 3 bolts and engine room ECU cover.

7. Disconnect the 4 TCM connectors.

8. Remove the 2 nuts and TCM.

9. Disconnect the 6 ECM connectors.

10. Remove the 2 nuts and ECM.

To install:

11. Install the ECM with the 2 nuts and tighten to 49 inch lbs. (5.5 Nm).

12. Connect the 6 ECM connectors.

13. Install the TCM with the 2 nuts and tighten to 49 inch lbs. (5.5 Nm).

14. Connect the 4 TCM connectors.

15. Install the engine room ECU cover with the 3 bolts and tighten to 69 inch lbs. (7.8 Nm).

16. Install the left engine room side cover.

17. Install the air cleaner inlet cover.

18. Connect cable to negative battery terminal.

19. Install the right cowl top ventilator louver.

20. Perform reset memory. Perform the RESET MEMORY (AT initialization) when replacing the automatic transmission assembly, engine assembly or ECM. Initialization cannot be completed by only removing the battery.

21. Perform initialization.

HEATED OXYGEN (HO2S) SENSOR

LOCATION

See Figure 141.

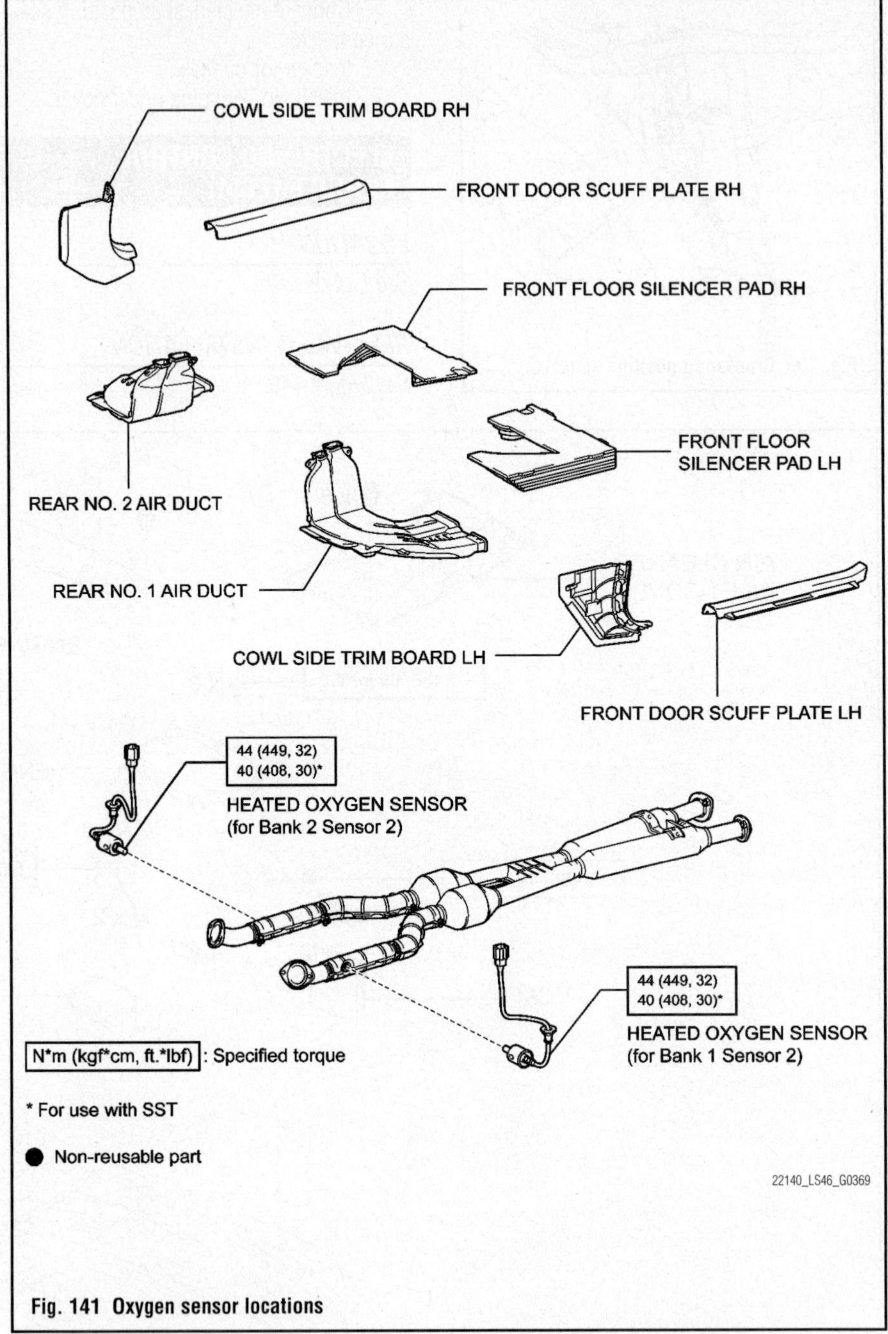

Fig. 141 Oxygen sensor locations

REMOVAL & INSTALLATION

1. Remove the left and right front door scuff plates.

2. Remove the left and right cowl side trim boards.

3. Remove the left and right front floor silencer pads.

4. Remove the rear No. 2 and No. 1 air duct.

5. Remove the heated oxygen sensors:

   a. Disconnect the heated oxygen sensor connector.

   b. Remove the grommet.

   c. Using SST (SST: 09224-00010), remove the heated oxygen sensor from the front exhaust pipe assembly.

To install:

6. Reverse removal procedure, and tighten to 32 ft. lbs. (44 Nm).

INJECTOR DRIVER MODULE (IDM)

LOCATION

See Figures 142 and 143.

REMOVAL & INSTALLATION

See Figures 144 and 145.

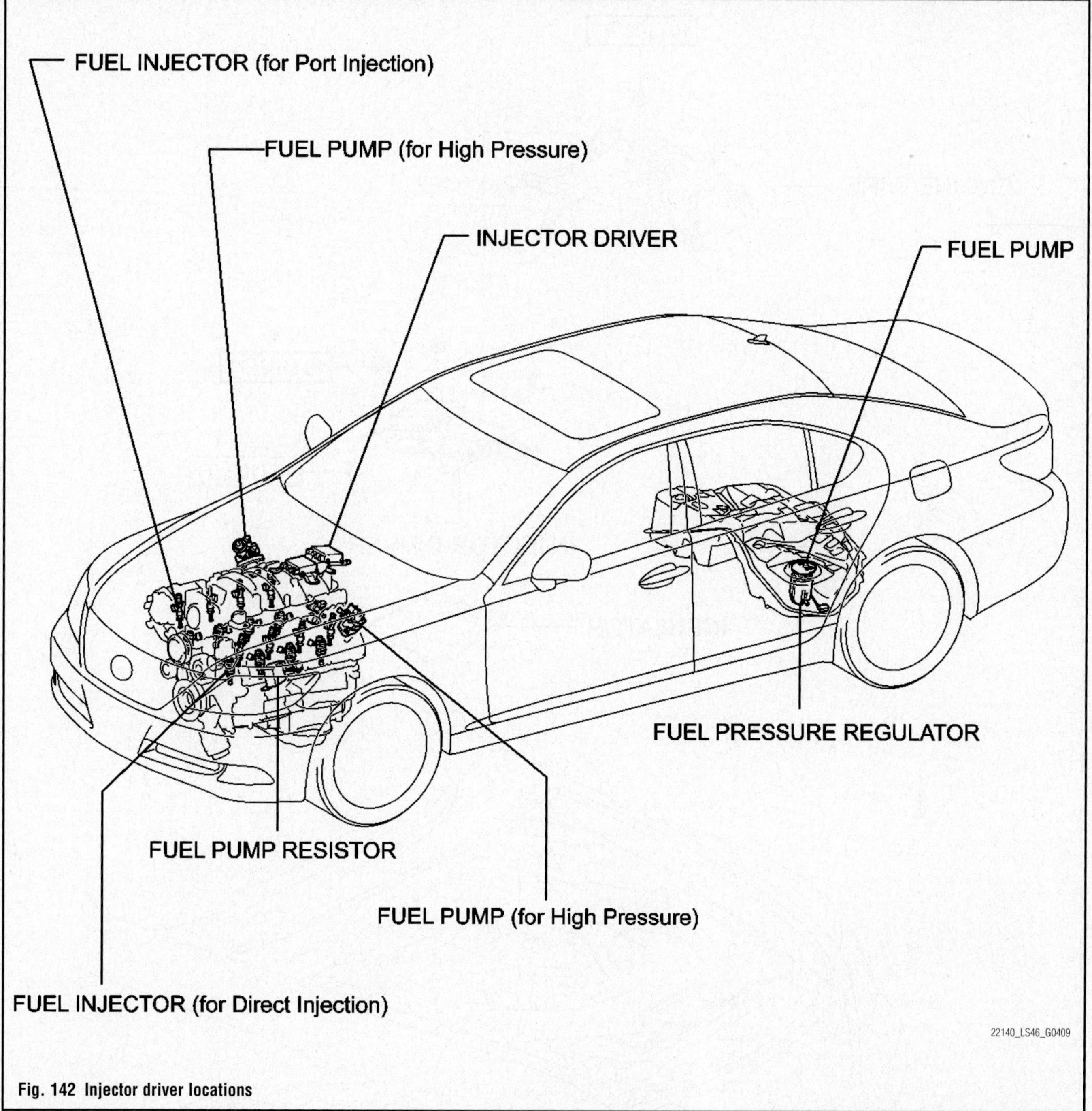

Fig. 142 Injector driver locations

1. Before servicing the vehicle, refer to the Precautions Section.

WARNING

After the engine switch is turned off, the HDD navigation system requires approximately 6 minutes to record various types of memory and settings. As a result, after turning the engine switch off, wait 6 minutes or more before disconnecting the cable from the negative battery terminal.

2. Remove the right cowl top ventilator louver.
3. Disconnect cable from negative battery terminal.

CAUTION

Wait at least 90 seconds after disconnecting the cable from the negative battery terminal to prevent airbag and seat belt pretensioner activation.

4. Remove the V-bank cover sub-assembly.
5. Remove the air cleaner inlet cover.
6. Remove the left and right engine room side covers.
7. Remove the No. 1 air cleaner inlet.
8. Remove the intake air connector pipe.
9. Remove left and right air cleaner caps.
10. Disconnect the engine wire.
11. Remove the injector driver, as follows:
    a. Disconnect the 8 injector driver connectors (6 connectors with wire harness locks and the 2 connectors without wire harness locks) from the 2 injector drivers. To release each connector lock, push the claw downward ("Release" in

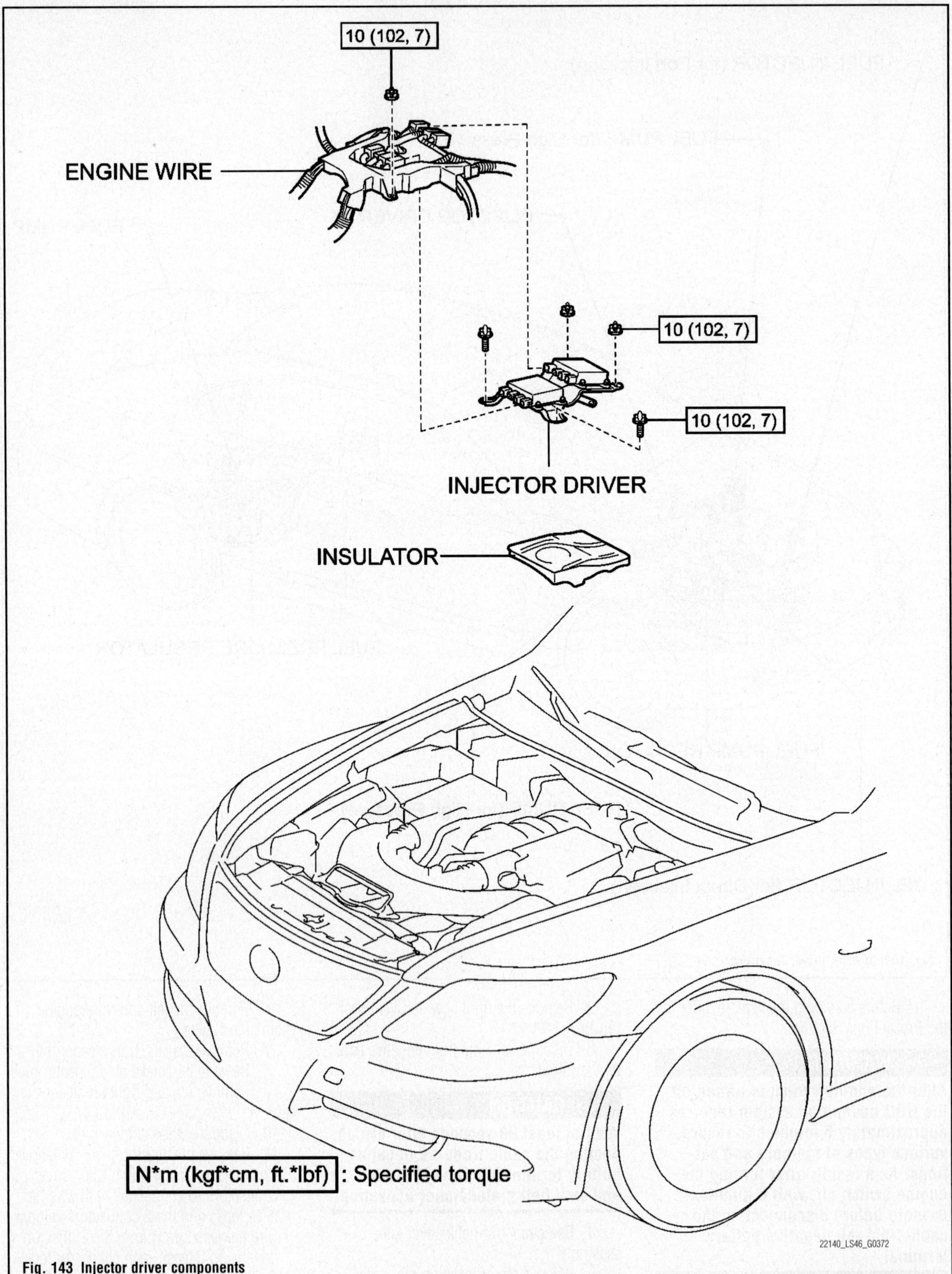

Fig. 143 Injector driver components

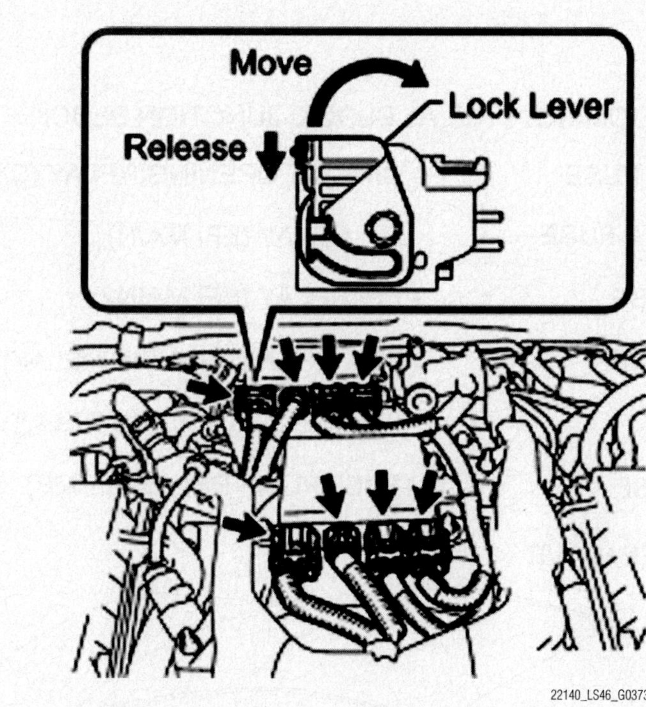

Fig. 144 Disconnect the injector driver connectors

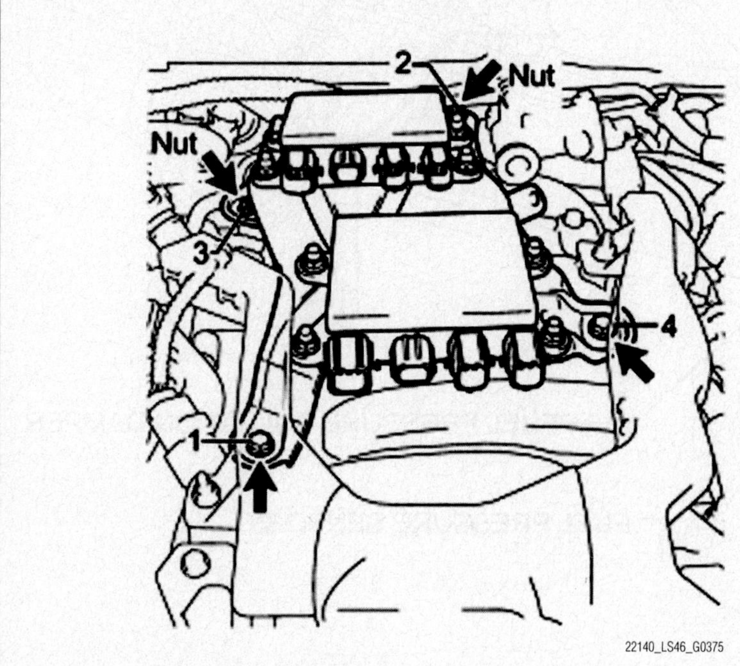

Fig. 145 Injector driver bolt and nut tightening sequence

the illustration) and move the lock lever ("Move" in the illustration).

b. Remove the 2 bolts, 2 nuts and 2 injector drivers with bracket. Be careful not to drop or strike the injector driver.

c. Remove the 8 nuts and 2 injector drivers from the bracket.

To install:

➡Be careful not to drop or strike the injector driver.

➡The injector driver is grounded at the bolt and nut. To ensure that it is grounded, clean all oil and foreign matter from the installation areas of the injector driver and engine before installing the injector driver.

12. Install the 2 injector drivers to the bracket with the 8 nuts and tighten to 66 inch lbs. (7.5 Nm).
13. Temporarily install the 2 injector drivers (with bracket) with the 2 bolts and 2 nuts.
14. Tighten the 2 bolts and 2 nuts in the order shown to 7 ft. lbs. (10 Nm).
15. Connect the 8 injector driver connectors securely. Move the lock lever in the opposite direction from removal.
16. The remainder of installation is the reverse of removal.
17. Perform initialization.

INJECTION CONTROL PRESSURE (ICP) SENSOR

LOCATION

See Figures 146 and 147.

REMOVAL & INSTALLATION

See Figure 148.

1. Before servicing the vehicle, refer to the Precautions Section.

⁂ WARNING

After the engine switch is turned off, the HDD navigation system requires approximately 6 minutes to record various types of memory and settings. As a result, after turning the engine switch off, wait 6 minutes or more before disconnecting the cable from the negative battery terminal.

2. Remove the right cowl top ventilator louver.
3. Disconnect cable from negative battery terminal.

⁂ CAUTION

Wait at least 90 seconds after disconnecting the cable from the negative battery terminal to prevent airbag and seat belt pretensioner activation.

4. Remove the intake manifold.
5. Remove the No. 2 engine cover.
6. Remove the fuel pressure sensor, as follows:
   a. Disconnect the connector.
   b. Remove the fuel pressure sensor and gasket.

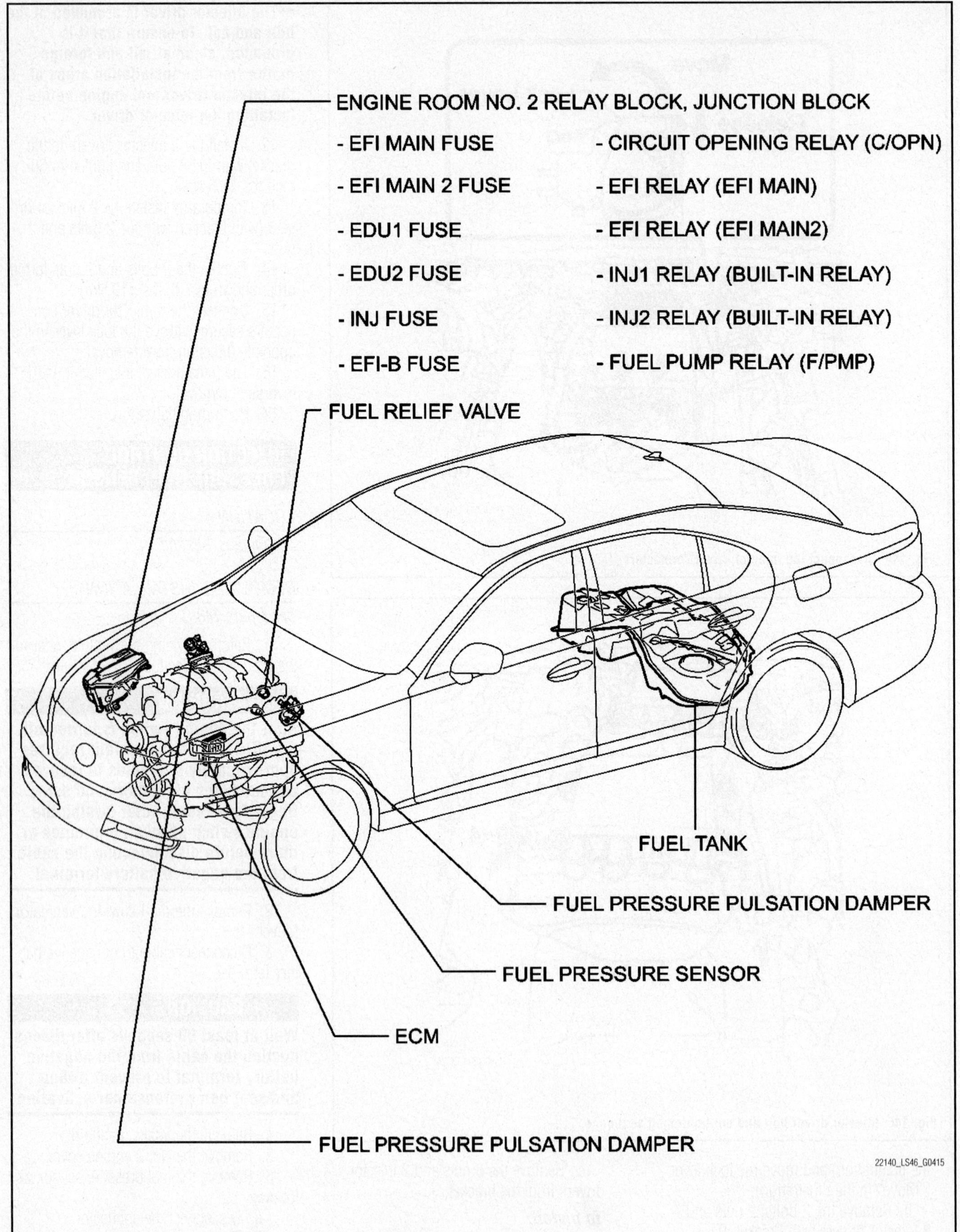

Fig. 146 Injector control pressure sensor (Fuel pressure sensor) location

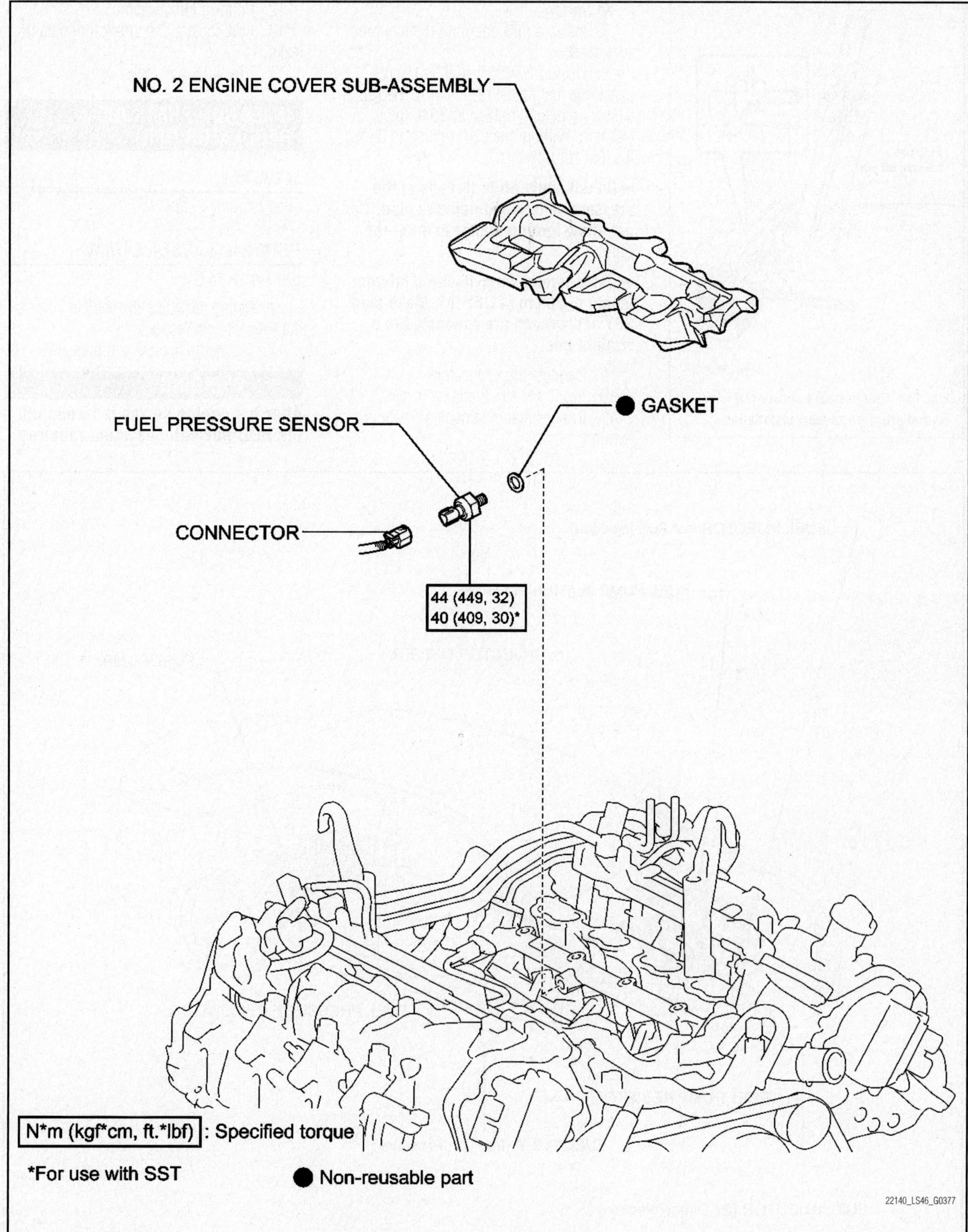

Fig. 147 Injector control pressure sensor (Fuel pressure sensor) components

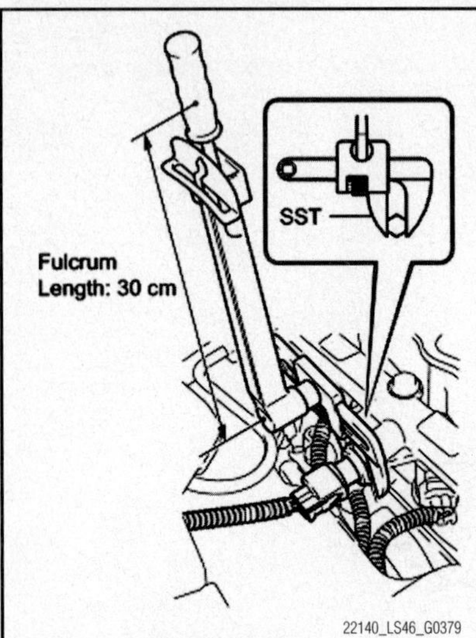

Fig. 148 Fuel pressure sensor (Injector control pressure sensor) installation

To install:

7. Install a new gasket to the fuel pressure sensor.
8. Using SST (SST: 09922-10010), install the fuel pressure sensor to the No. 1 delivery pipe and tighten to 30 ft. lbs. (40 Nm). Without the SST, tighten to 32 ft. lbs. (44 Nm).

➡Do not apply oil to threads of the pressure sensor and delivery pipe. Do not allow foreign matter to enter the pipe.

➡Use a torque wrench with a fulcrum length of 30 cm (11.81 in). Make sure SST and wrench are connected in a straight line.

9. Connect the connector.
10. Install the No. 2 engine cover.
11. Connect cable to negative battery terminal.
12. Perform initialization.
13. Add coolant and check for coolant leaks.
14. Check for fuel leaks.

INJECTION PRESSURE REGULATOR (IPR)

LOCATION

See Figures 149 and 150.

REMOVAL & INSTALLATION

See Figure 151.

1. Before servicing the vehicle, refer to the Precautions Section.
2. Discharge fuel system pressure.

⁂ WARNING

After the engine switch is turned off, the HDD navigation system requires

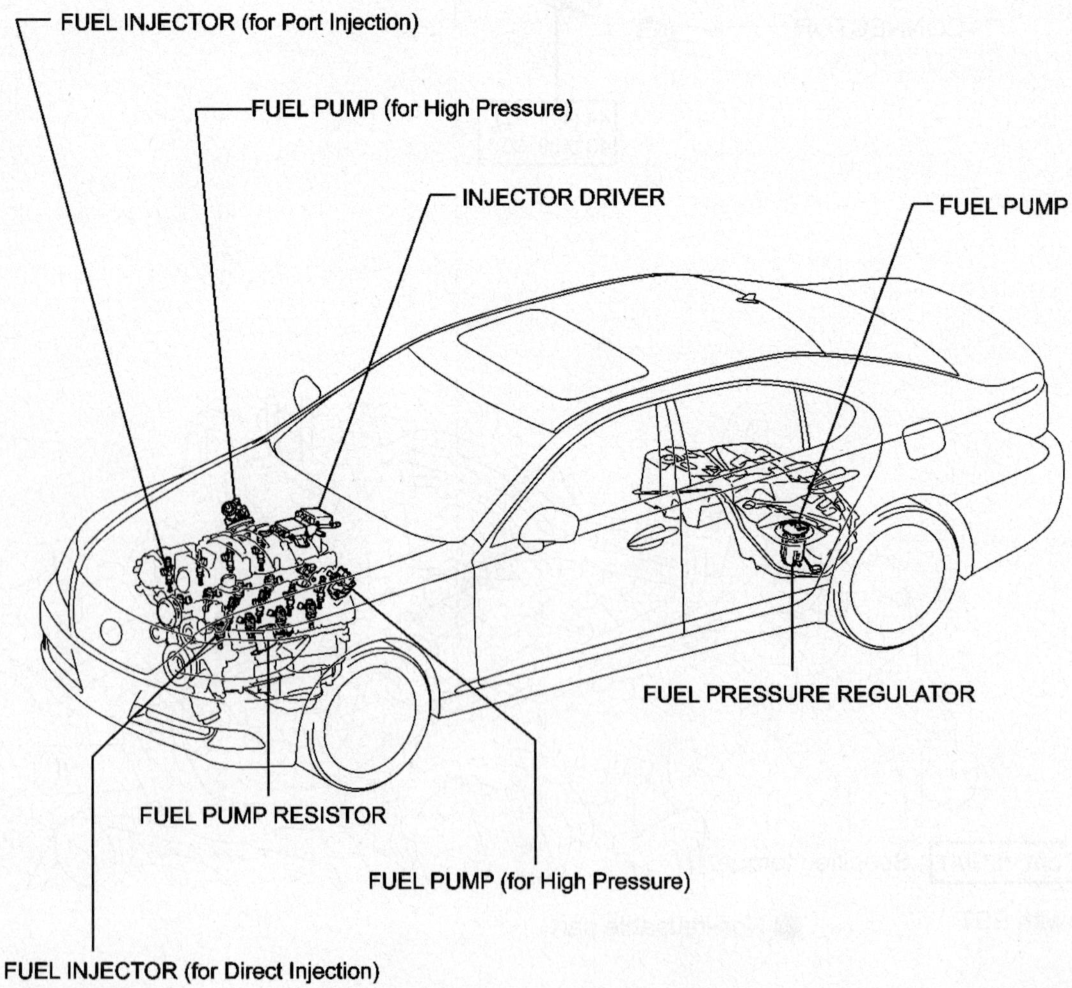

FUEL INJECTOR (for Port Injection)
FUEL PUMP (for High Pressure)
INJECTOR DRIVER
FUEL PUMP
FUEL PRESSURE REGULATOR
FUEL PUMP RESISTOR
FUEL PUMP (for High Pressure)
FUEL INJECTOR (for Direct Injection)
22140_LS46_G0409

Fig. 149 Injector pressure regulator (Fuel pressure regulator) location

approximately 6 minutes to record various types of memory and settings. As a result, after turning the engine switch off, wait 6 minutes or more before disconnecting the cable from the negative battery terminal.

3. Remove the right cowl top ventilator louver.

4. Disconnect cable from negative battery terminal.

CAUTION

Wait at least 90 seconds after disconnecting the cable from the negative battery terminal to prevent airbag and seat belt pretensioner activation.

5. Remove the left rear seat assembly.

6. Remove the rear seat cushion assembly.

7. Remove the rear floor No. 2 service hole cover and disconnect the connector from the fuel suction with pump and gauge tube assembly.

8. Remove the fuel suction with pump and gauge tube from the fuel tank.

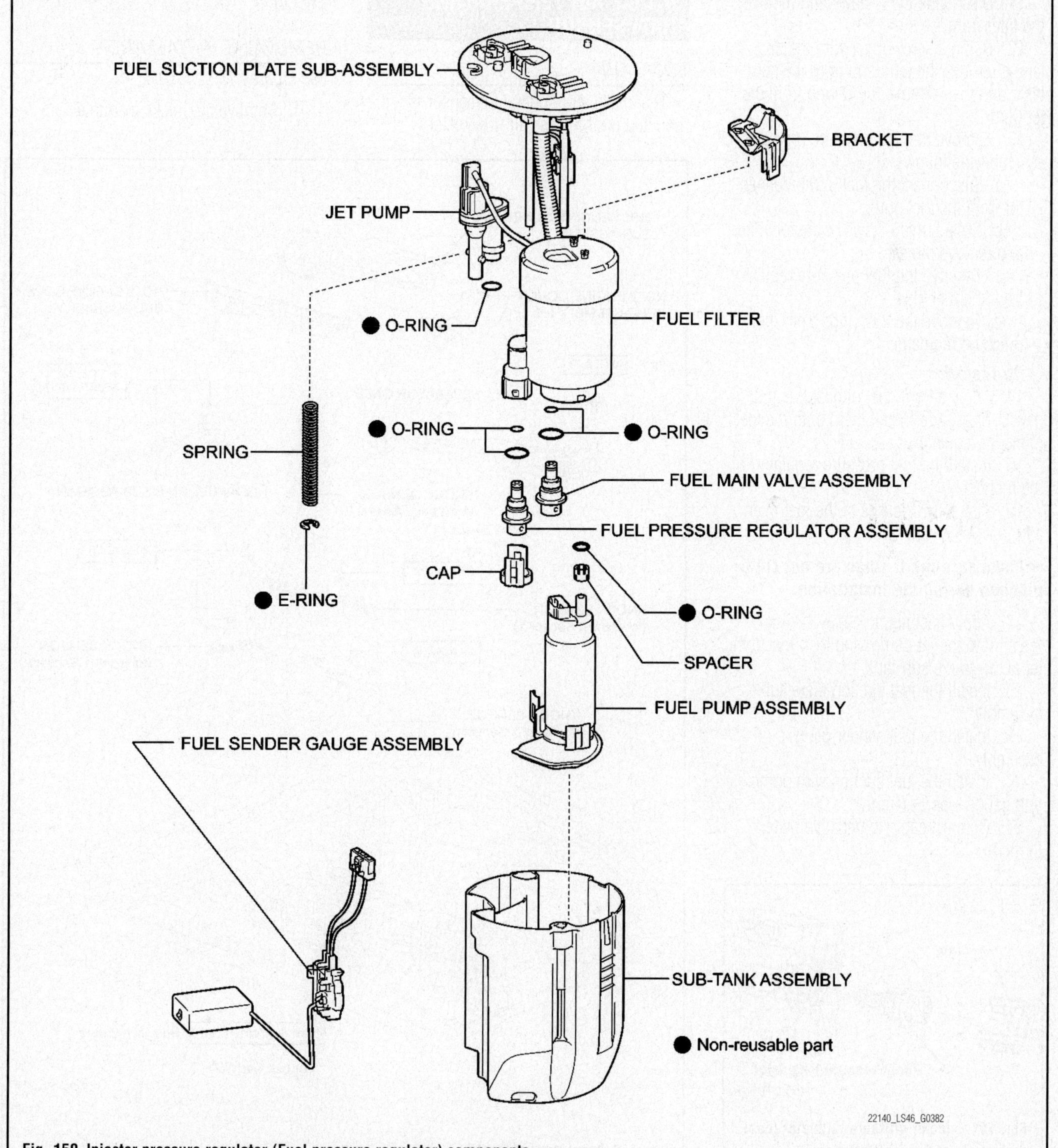

Fig. 150 Injector pressure regulator (Fuel pressure regulator) components

9. Remove the fuel sender gauge assembly.
10. Remove the fuel suction plate sub-assembly:
    a. Disconnect the fuel pump connector from the fuel suction plate.
    b. Using needle nose pliers, remove the E-ring.
    c. Remove the fuel suction plate and spring from the sub-tank.
11. Disengage the 2 claws and remove the filter from the sub-tank.
12. Using a screwdriver with the tip taped, remove the jet pump from the sub-tank, and then remove the O-ring from the jet pump.
13. Remove the fuel pressure regulator assembly, as follows:
    a. Disconnect the fuel pump connector from the fuel pump.
    b. Detach the 2 claws of the fuel filter and remove the cap.
    c. Remove the fuel pressure regulator from the fuel filter.
    d. Remove the 2 O-rings from the pressure regulator.

To install:

14. Apply a light coat of gasoline to 2 new O-rings, and install them onto the fuel pressure regulator.
15. Install the fuel pressure regulator to the fuel filter.
16. Attach the 2 claws of the fuel filter and install the cap.

➡Make sure the O-rings are not cut or pinched during the installation.

17. Apply gasoline to a new O-ring, install it to the jet pump, and then install the jet pump to the sub-tank.
18. Install the fuel suction plate sub-assembly.
19. Install the fuel sender gauge assembly.
20. Install the fuel suction with pump and gauge tube assembly.
21. Connect cable to negative battery terminal.

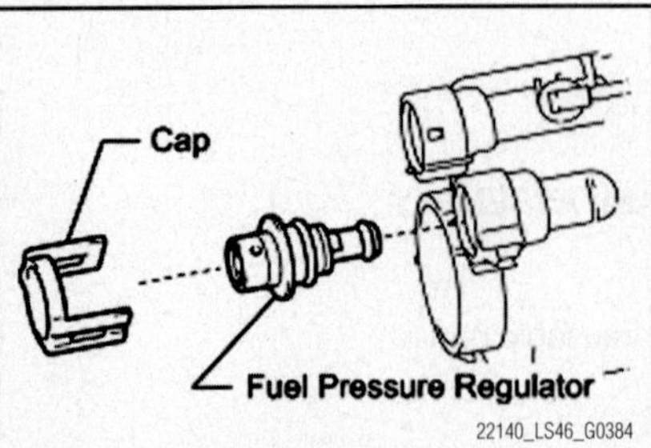

Fig. 151 Injector pressure regulator (Fuel pressure regulator) installation

22. Install the right cowl top ventilator louver.
23. Perform initialization.
24. Check for fuel leaks.
25. Install the rear floor No. 2 service hole cover with new butyl tape.
26. Install the left rear seat assembly.
27. Install the rear seat cushion assembly.

INTAKE AIR TEMPERATURE (IAT) SENSOR

LOCATION

The intake air temperature sensor is mounted on the mass air flow meter.

REMOVAL & INSTALLATION

1. Remove air flow meter.
2. Remove intake air temperature sensor.

To install:

3. Reverse removal procedure.

KNOCK SENSOR (KS)

LOCATION

See Figure 152.

REMOVAL & INSTALLATION

See Figures 153 and 154.

1. Remove the intake manifold.

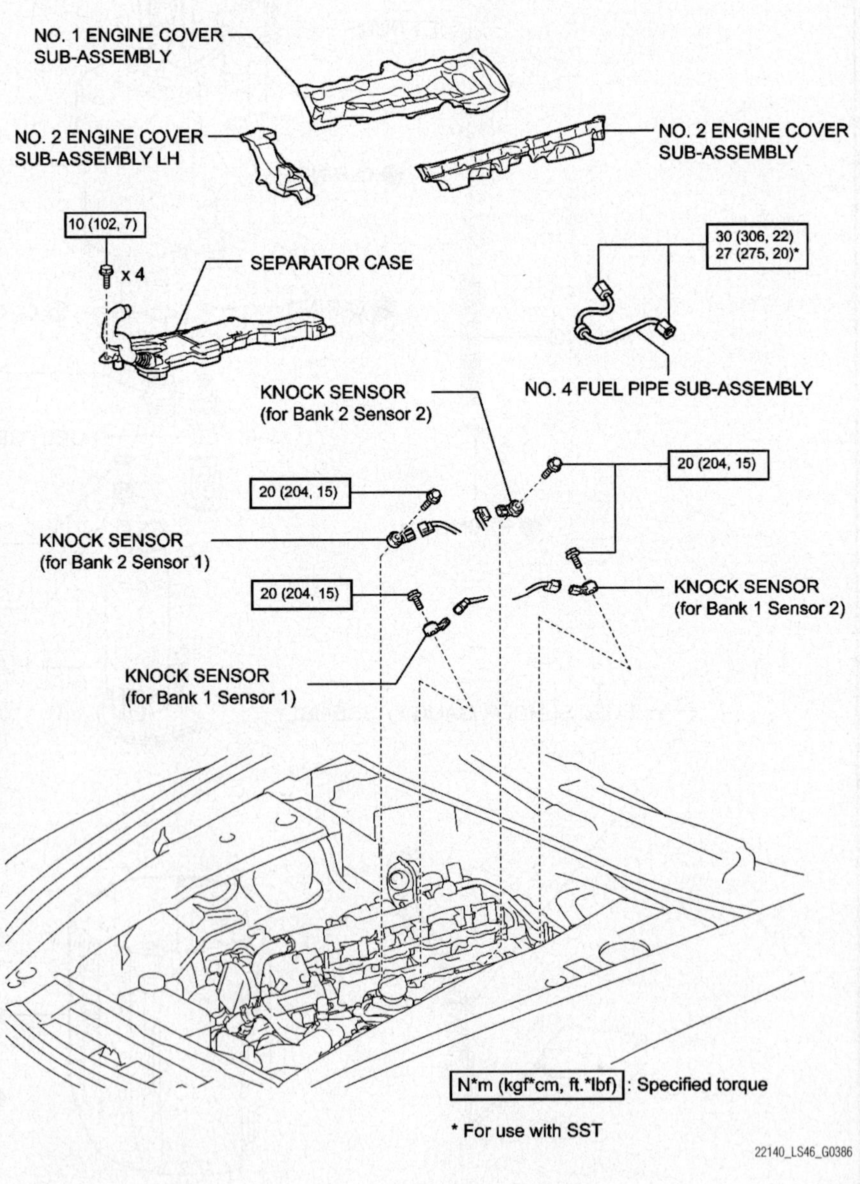

Fig. 152 Knock sensor location

2. Remove the engine cover sub-assembly.

3. Remove the No. 4 fuel pipe sub-assembly.

4. Remove the separator case, as follows:

a. Disconnect the fuel pressure sensor connector.

b. Remove the 4 bolts and separator case.

5. Disconnect the 4 knock sensor connectors.

6. Remove the 4 bolts and 4 knock sensors.

To install:

7. Install the 4 knock sensors with the 4 bolts so that the sensors are angled as

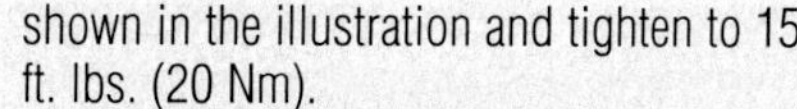

shown in the illustration and tighten to 15 ft. lbs. (20 Nm).

➡The acceptable installation angle of the knock control sensors is between 10° upwards and downwards from the horizontal position.

8. Connect the 4 knock sensor connectors.

9. Install the separator case with the 4 bolts and tighten to 7 ft. lbs. (10 Nm).

10. Connect the fuel pressure sensor connector.

11. Install the No. 4 fuel pipe sub-assembly.

12. Install the engine cover sub-assembly.

13. Install the intake manifold.

MALFUNCTION INDICATOR LIGHT (MIL)

RESET PROCEDURES

Clear the DTC codes.

MASS AIR FLOW (MAF) SENSOR

LOCATION

See Figure 155.

REMOVAL & INSTALLATION

1. Before servicing the vehicle, refer to the Precautions Section.

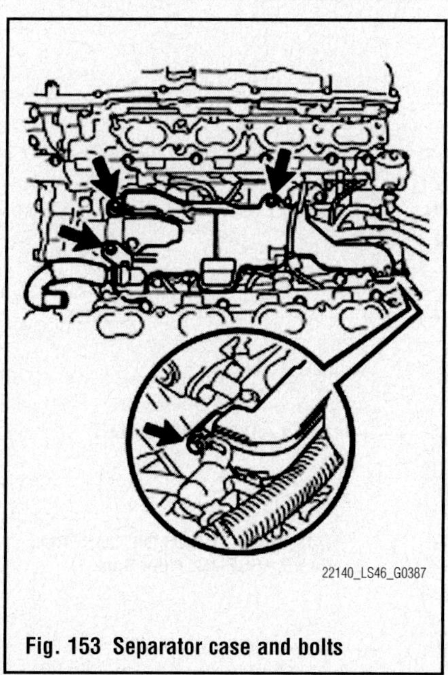

Fig. 153 Separator case and bolts

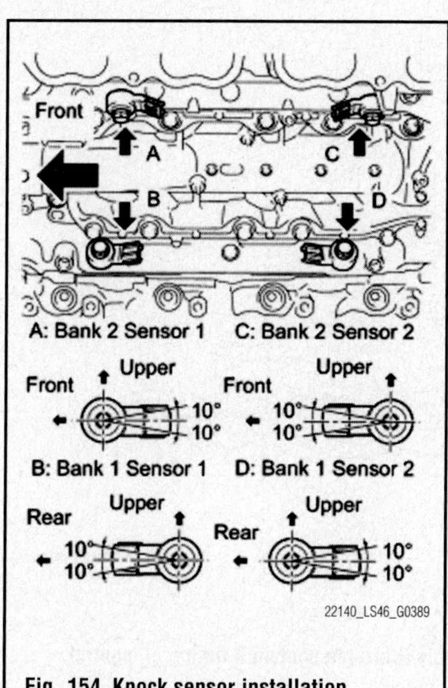

Fig. 154 Knock sensor installation

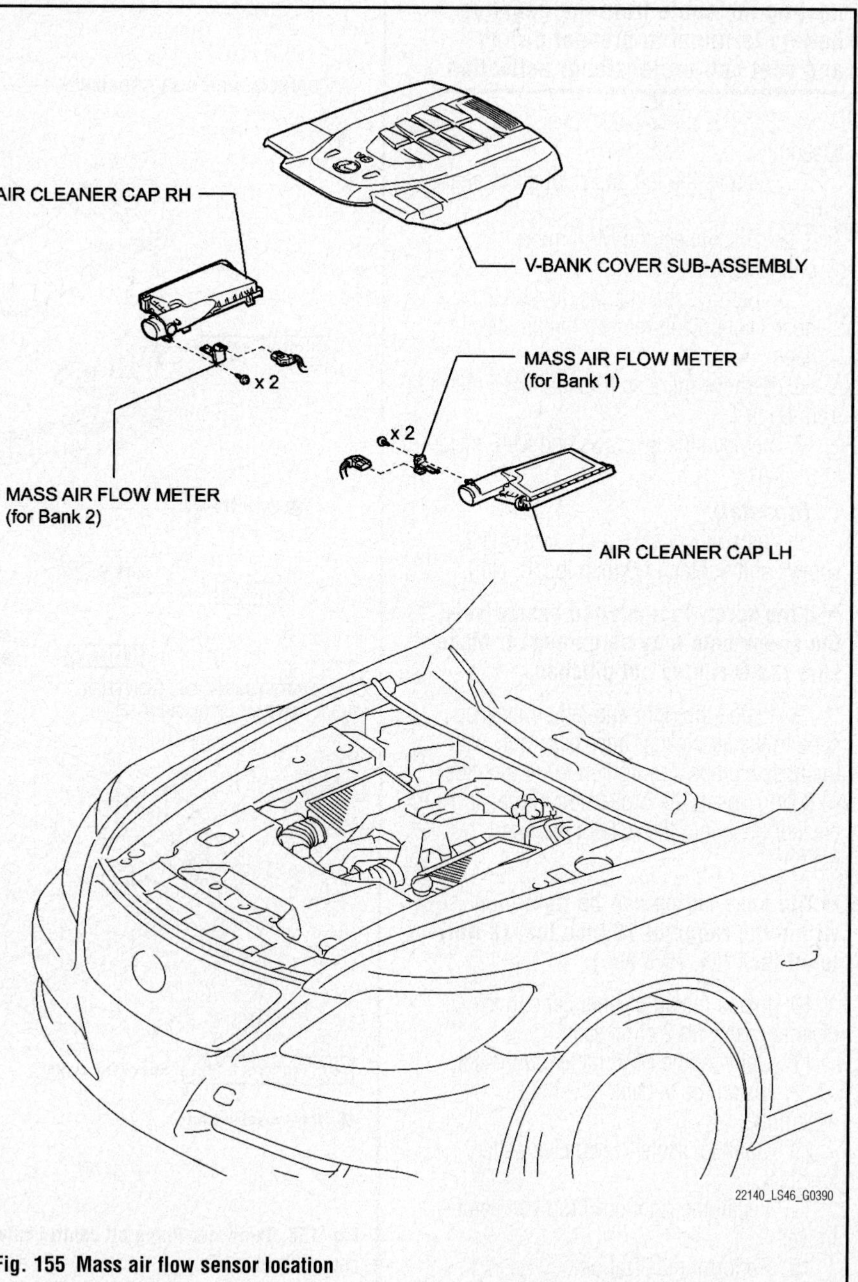

Fig. 155 Mass air flow sensor location

✲✲ WARNING

After the engine switch is turned off, the HDD navigation system requires approximately 6 minutes to record various types of memory and settings. As a result, after turning the engine switch off, wait 6 minutes or more before disconnecting the cable from the negative battery terminal.

2. Remove the right cowl top ventilator louver.
3. Disconnect cable from negative battery terminal.

✲✲ CAUTION

Wait at least 90 seconds after disconnecting the cable from the negative battery terminal to prevent airbag and seat belt pretensioner activation.

4. Remove the V-bank cover sub-assembly.
5. Remove the left and right air cleaner caps:
   a. Disconnect the MAF meter connector.
   b. Remove the clamps, loosen the hose clamps and remove the air cleaner caps.
6. Remove the 2 screws and MAF meter from Bank 1.
7. Remove the 2 screws and MAF meter from Bank 2.

To install:

8. Install each MAF meter with the 2 screws and tighten to 9 inch lbs. (1 Nm).

➡If the screw is tightened excessively, the screw hole may be damaged. Make sure the O-ring is not pinched.

9. Install the right and left air cleaner caps to the air cleaner hose assembly with the hose clamps and tighten to 34 inch lbs. (3.8 Nm). Insert the protrusion of the air cleaner hose into the hole of the hose clamp.

➡The hose clamp can be tightened within the range of 18 inch lbs. (2 Nm) to 49 inch lbs. (5.5 Nm).

10. Install the air cleaner caps to the air cleaner case with 2 clips each.
11. Connect the MAF meter connector.
12. Install the V-bank cover sub-assembly.
13. Connect cable to negative battery terminal.
14. Install the right cowl top ventilator louver.
15. Perform initialization.

THROTTLE POSITION SENSOR (TPS)

LOCATION

The throttle position sensor is located on the throttle body.

REMOVAL & INSTALLATION

1. Remove throttle body.
2. Remove TPS connector.
3. Remove screws and throttle position sensor.

To install:

4. Reverse removal procedure.

VARIABLE CAMSHAFT TIMING OIL CONTROL SOLENOID

LOCATION

See Figure 156.

REMOVAL & INSTALLATION

See Figures 157 and 158.

1. Before servicing the vehicle, refer to the Precautions Section.

✲✲ WARNING

After the engine switch is turned off, the HDD navigation system requires

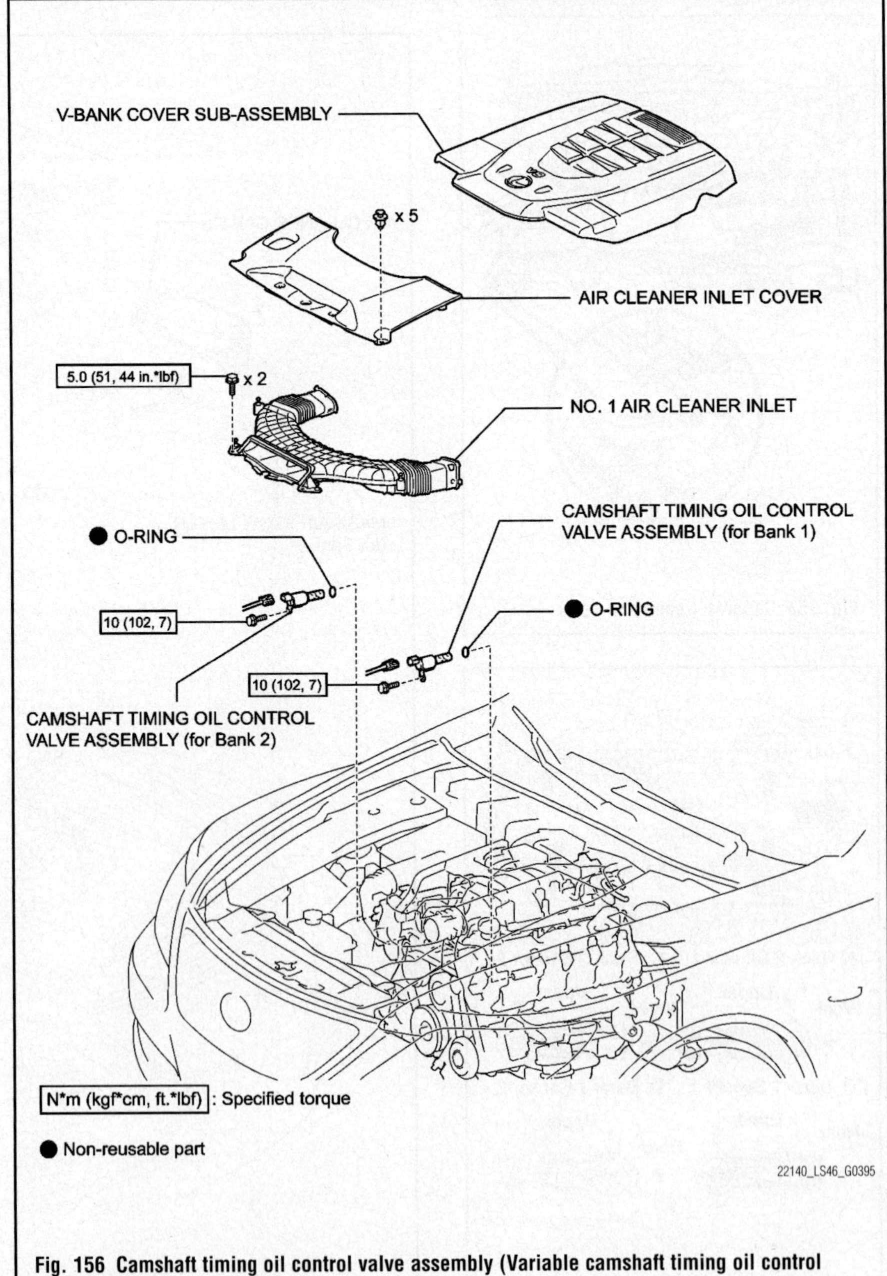

Fig. 156 Camshaft timing oil control valve assembly (Variable camshaft timing oil control solenoid) location

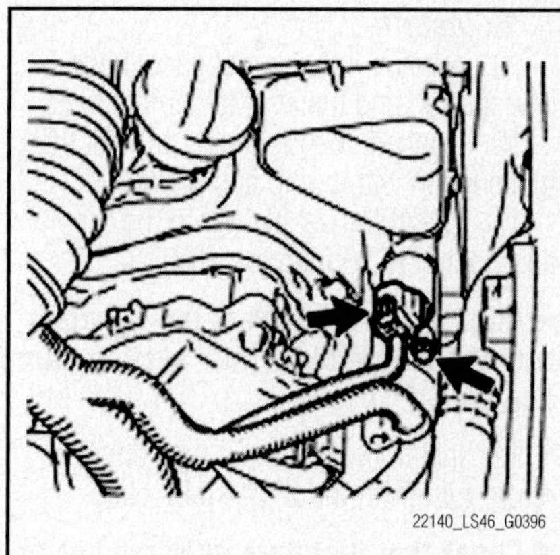

Fig. 157 Camshaft timing oil control valve—Bank 1

approximately 6 minutes to record various types of memory and settings. As a result, after turning the engine switch off, wait 6 minutes or more before disconnecting the cable from the negative battery terminal.

2. Remove the right cowl top ventilator louver.
3. Disconnect cable from negative battery terminal.

⁂ CAUTION

Wait at least 90 seconds after disconnecting the cable from the negative battery terminal to prevent airbag and seat belt pretensioner activation.

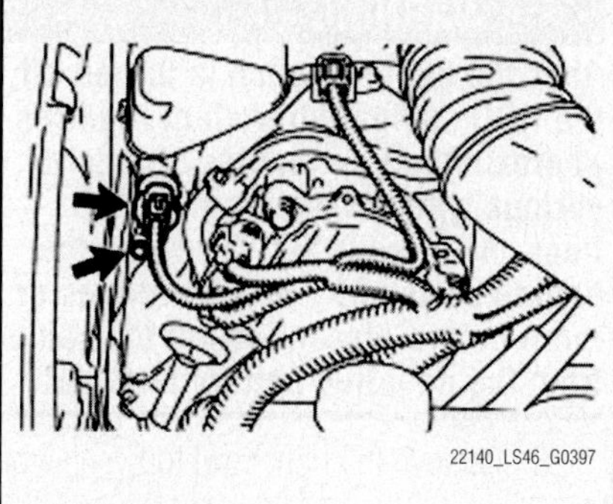

Fig. 158 Camshaft timing oil control valve—Bank 2

4. Remove the V-bank cover sub-assembly.
5. Remove the air cleaner inlet cover.
6. Remove the No. 1 air cleaner inlet.
7. Remove the camshaft timing oil control valve assembly from Bank 1 and Bank 2, as follows:
   a. Disconnect the oil control valve connector.
   b. Remove the bolt and oil control valve.
   c. Remove the O-ring from the oil control valve.

To install:

8. Install the camshaft timing oil control valve assembly to Bank 1 and Bank 2, as follows:
   a. Apply a light coat of engine oil to a new O-ring, and install it to the oil control valve.
   b. Install the oil control valve with the bolt and tighten to 7 ft. lbs. (10 Nm).

➡Do not allow foreign matter to contact the oil seal face of the oil control valve. Be careful that the O-ring is in good condition.

   c. Connect the camshaft timing oil control valve connector.
9. Install the No. 1 air cleaner inlet.
10. Install the air cleaner inlet cover.
11. Install the V-bank cover sub-assembly.
12. Connect cable to negative battery terminal.
13. Install the right cowl top ventilator louver.
14. Perform initialization.
15. After the engine is warmed up, perform the driving test. Then check that the oil control valve operates normally.

FUEL SYSTEM

GASOLINE FUEL INJECTION SYSTEM

FUEL SYSTEM SERVICE PRECAUTIONS

Before servicing any vehicle, please be sure to read all of the following precautions, which deal with personal safety, prevention of component damage, and important points to take into consideration when servicing a motor vehicle:

- Observe all applicable safety precautions when working around fuel. Whenever servicing the fuel system, always work in a well-ventilated area. Do not allow fuel spray or vapors to come in contact with a spark, open flame, or excessive heat (a hot drop light, for example). Keep a dry chemical fire extinguisher near the work area. Always keep fuel in a container specifically designed for fuel storage; also, always properly seal fuel containers to avoid the possibility of fire or explosion. Refer to the additional fuel system precautions later in this section.
- Fuel injection systems often remain pressurized, even after the engine has been turned **OFF**. The fuel system pressure must be relieved before disconnecting any fuel lines. Failure to do so may result in fire and/or personal injury.
- All new vehicles are now equipped with an air bag system, often referred to as a Supplemental Restraint System (SRS) or Supplemental Inflatable Restraint (SIR) system. The system must be disabled before performing service on or around system components, steering column, instrument panel components, wiring and sensors. Failure to follow safety and disabling procedures could result in accidental air bag deployment, possible personal injury and unnecessary system repairs.
- Disconnecting the negative battery cable on some vehicles may interfere with the functions of the on-board computer system(s) and may require the computer to undergo a relearning process once the negative battery cable is reconnected.

After the engine switch is turned off, the HDD navigation system requires approximately 6 minutes to record various types of memory and settings. As a result, after turning the engine switch off, wait 6 minutes or more before disconnecting the cable from the negative (-) battery terminal.

Before inspecting and repairing the fuel system, disconnect the cable from the negative (-) battery terminal. Wait at least 90 seconds after disconnecting the cable from the negative (-) battery terminal to prevent airbag and seat belt pretensioner activation.

Keep gasoline away from rubber or leather parts.

1. Check that there are no fuel leaks from the fuel system after doing any maintenance or repairs.

RELIEVING FUEL SYSTEM PRESSURE

⁂ CAUTION

Perform the following procedures to prevent fuel from spilling out before removing any fuel system parts.

⁂ CAUTION

Do not disconnect any part of the fuel system until you have discharged the fuel system pressure.

⁂ CAUTION

Pressure will still remain in the fuel line even after performing the following procedures. Even after discharging the fuel pressure, place a piece of cloth or equivalent over fittings as you separate them to reduce the risk of fuel spray on yourself or in the engine compartment.

⁂ CAUTION

Allow the engine to cool down before performing the following procedures, as the high pressure side's fuel pressure does not decrease until the engine cools down.

➡Perform these procedures with the engine coolant temperature at 140°F (60°C) or less.

1. Disconnect the fuel pump connector.
2. Remove the left rear seat assembly.
3. Remove the rear floor service hole cover.
4. Disconnect the fuel pump connector.
5. Start the engine.
6. After the engine stops, turn the ignition switch off.

➡DTC P3190, P3191, P0171, P0172 and/or P0A0F may be set.

7. Loosen the fuel tank cap, then discharge the pressure in the fuel tank completely.
8. Reconnect the fuel pump connector.
9. Install the rear seat.
10. Install the rear floor service hole cover.
11. Check that there are no fuel leaks from the fuel system after doing any maintenance or repairs.

FUEL FILTER

REMOVAL & INSTALLATION

Fuel filters are in the fuel tank along with the fuel pump.

FUEL PUMP

REMOVAL & INSTALLATION

1. Before servicing the vehicle, refer to the Precautions Section.
2. Discharge fuel system pressure.

⁂ WARNING

After the engine switch is turned off, the HDD navigation system requires approximately 6 minutes to record various types of memory and settings. As a result, after turning the engine switch off, wait 6 minutes or more before disconnecting the cable from the negative battery terminal.

3. Remove the right cowl top ventilator louver.
4. Disconnect cable from negative battery terminal.

⁂ CAUTION

Wait at least 90 seconds after disconnecting the cable from the negative battery terminal to prevent airbag and seat belt pretensioner activation.

5. Remove the left rear seat assembly, for power seat or ottoman.
6. Remove the rear seat cushion assembly, for fixed seat.
7. Remove the rear floor No. 2 service hole cover and disconnect the connector from the fuel suction with pump and gauge tube assembly,
8. Remove the fuel suction with pump and gauge tube assembly, as follows:
   a. Disconnect the fuel main tube and fuel return tube. Remove the 2 tube joint clips and 2 fuel tubes.

➡Remove any dirt and foreign matter on the fuel tube joint.

➡Do not allow any scratches or foreign matter on the parts when disconnecting them, as the fuel tube joint contains the O-rings that seal the plug.

➡Perform this work by hand. Do not use any tools.

➡Do not forcibly bend, twist or turn the nylon tube.

➡Protect the disconnected part by covering it with a plastic bag and tape after disconnecting the fuel tubes.

   b. Remove the 8 bolts and set plate.
   c. Disconnect the fuel hose.
   d. Remove the fuel suction with pump and gauge tube from the fuel tank.

➡Make sure that the sender gauge arm does not bend. Do not damage the fuel suction with pump and gauge tube.

   e. Remove the gasket.

To install:

9. Apply a light coat of gasoline to a new gasket, and install it to the fuel tank.
10. Connect the fuel hose, and set the fuel suction with pump and gauge.
11. Install the set plate with the 8 bolts and tighten to 53 inch lbs. (6 Nm).

➡Align the protrusion of the set plate with the cutout of the fuel suction pump and gauge.

12. Install the fuel main tube and fuel return tube with the 2 tube joint clips.

➡Check that there are no scratches or foreign objects on the connecting parts, the fuel tube joint is inserted securely, and the tube joint clips are on the collars of the fuel tube joints.

13. Connect the connector to the fuel suction with pump and gauge tube assembly.
14. The remainder of installation is the reverse of removal.
15. Perform initialization.
16. Check for fuel leaks.

FUEL RAIL & INJECTORS

REMOVAL & INSTALLATION

See Figures 159 through 164.

1. Before servicing the vehicle, refer to the Precautions Section.
2. Discharge fuel system pressure.

⁂ WARNING

After the engine switch is turned off, the HDD navigation system requires approximately 6 minutes to record various types of memory and settings. As a result, after turning the engine switch off, wait 6 minutes or more before disconnecting the cable from the negative battery terminal.

3. Remove the right cowl top ventilator louver.
4. Disconnect cable from negative battery terminal.

⁂ CAUTION

Wait at least 90 seconds after disconnecting the cable from the negative battery terminal to prevent airbag and seat belt pretensioner activation.

5. Remove the intake manifold.
6. Remove the fuel pressure pulsation damper assembly, for direct injection.
7. Remove the fuel pipe sub-assembly.
8. Remove additional fuel pipes, hoses, and tubes, as necessary.

9. Remove the No. 1, No. 2, and left-hand No. 2 engine covers, as necessary.

10. Remove the ventilation hose from the ventilation valve, for direct injection.

11. Disconnect the fuel pressure sensor connector, for direct injection.

12. Disconnect the injector connectors.

13. Disconnect the wire harness clamps, for port injection.

14. Remove the bolts and delivery pipes from the cylinder heads.

➡Make sure that the fuel delivery pipe is disconnected from the delivery pipe. Be extremely careful not to touch or strike the tips of the injectors. Pull and remove the fuel delivery pipe in a straight line without tilting it.

15. Remove the delivery pipe spacers, for port injection.

16. Remove the injector vibration insulators from the cylinder heads.

To install:

17. Install a new injector seals.

18. Connect the wire harness clamps and pipe spacers, for port injection.

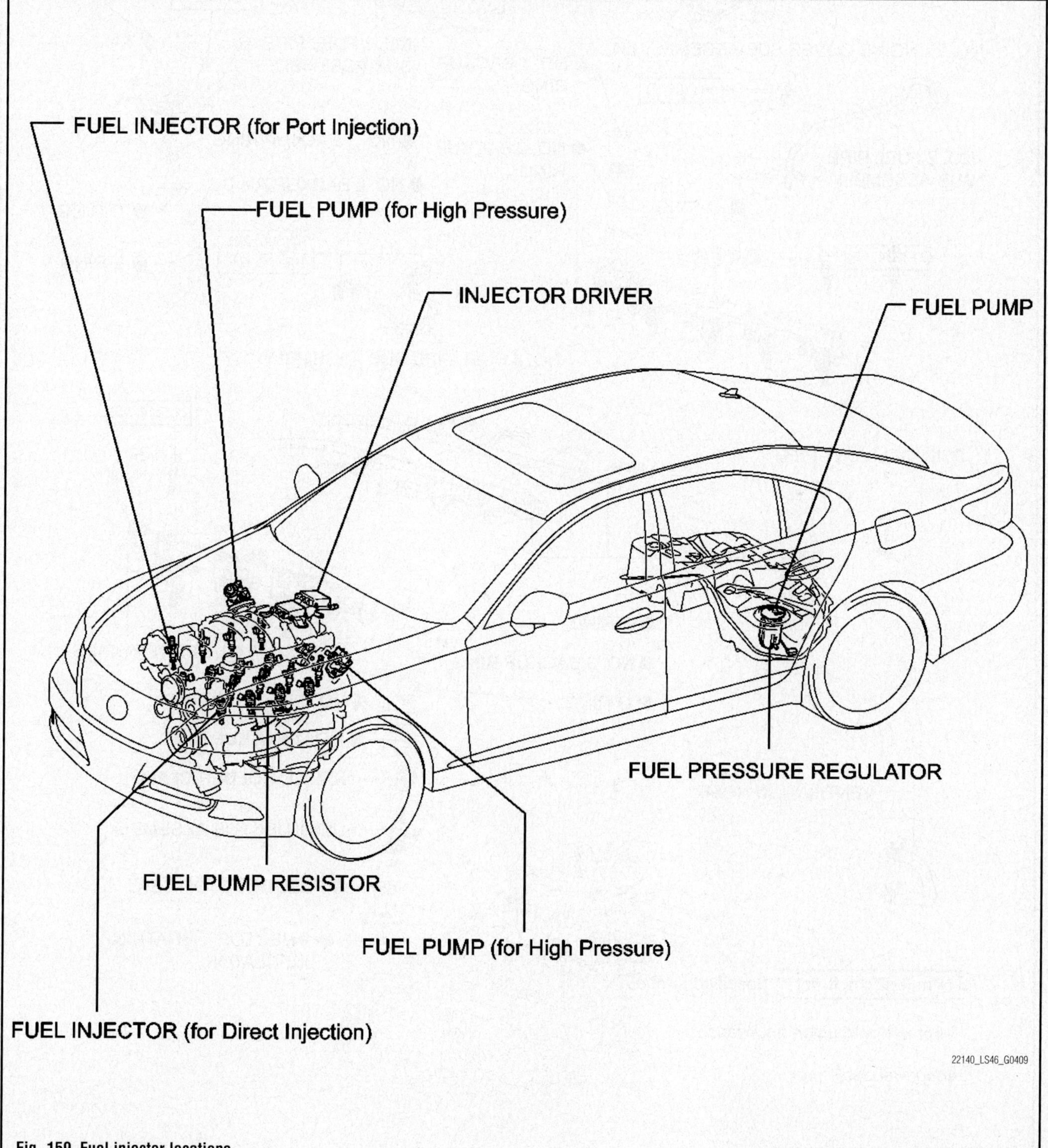

Fig. 159 Fuel injector locations

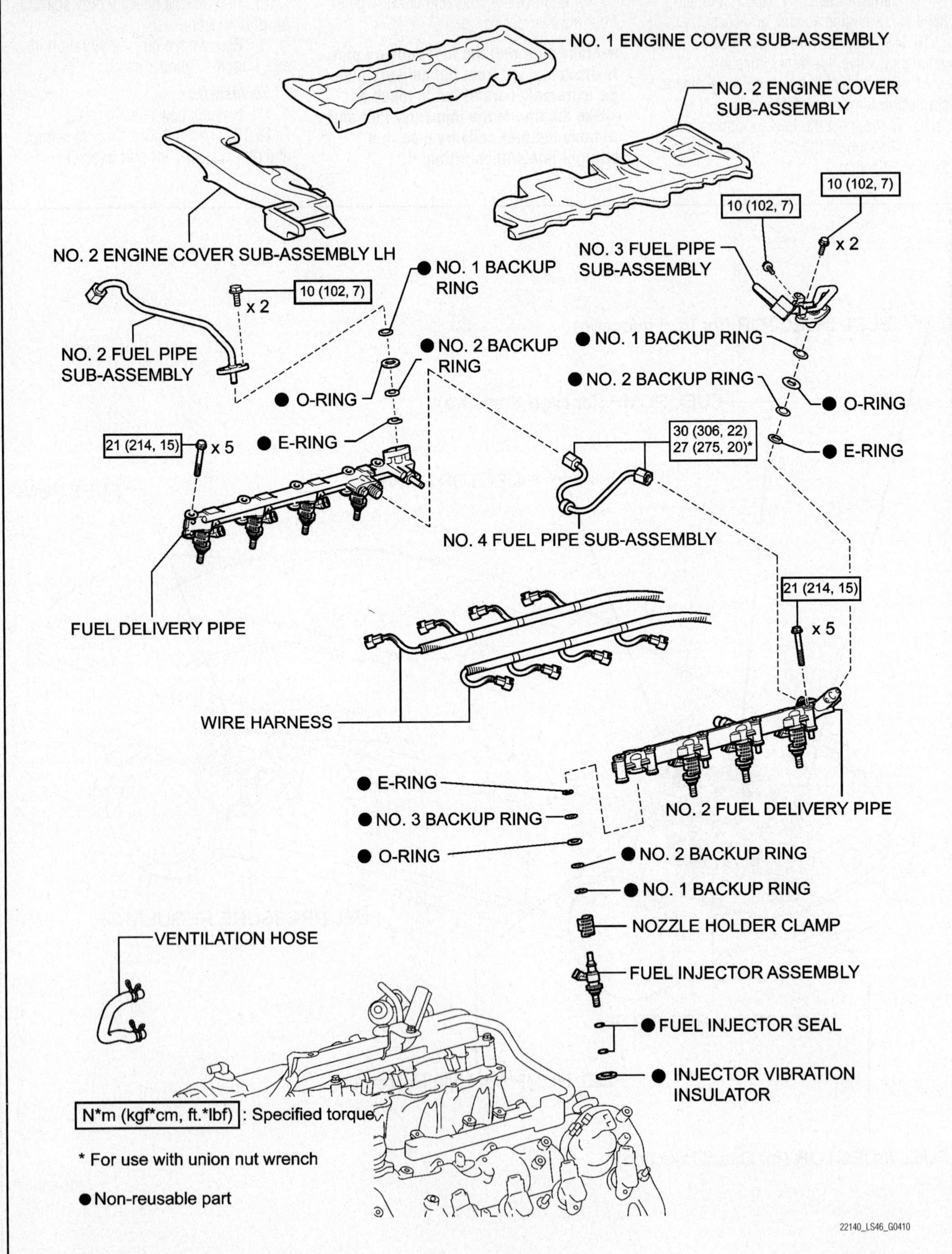

Fig. 160 Direct injection components

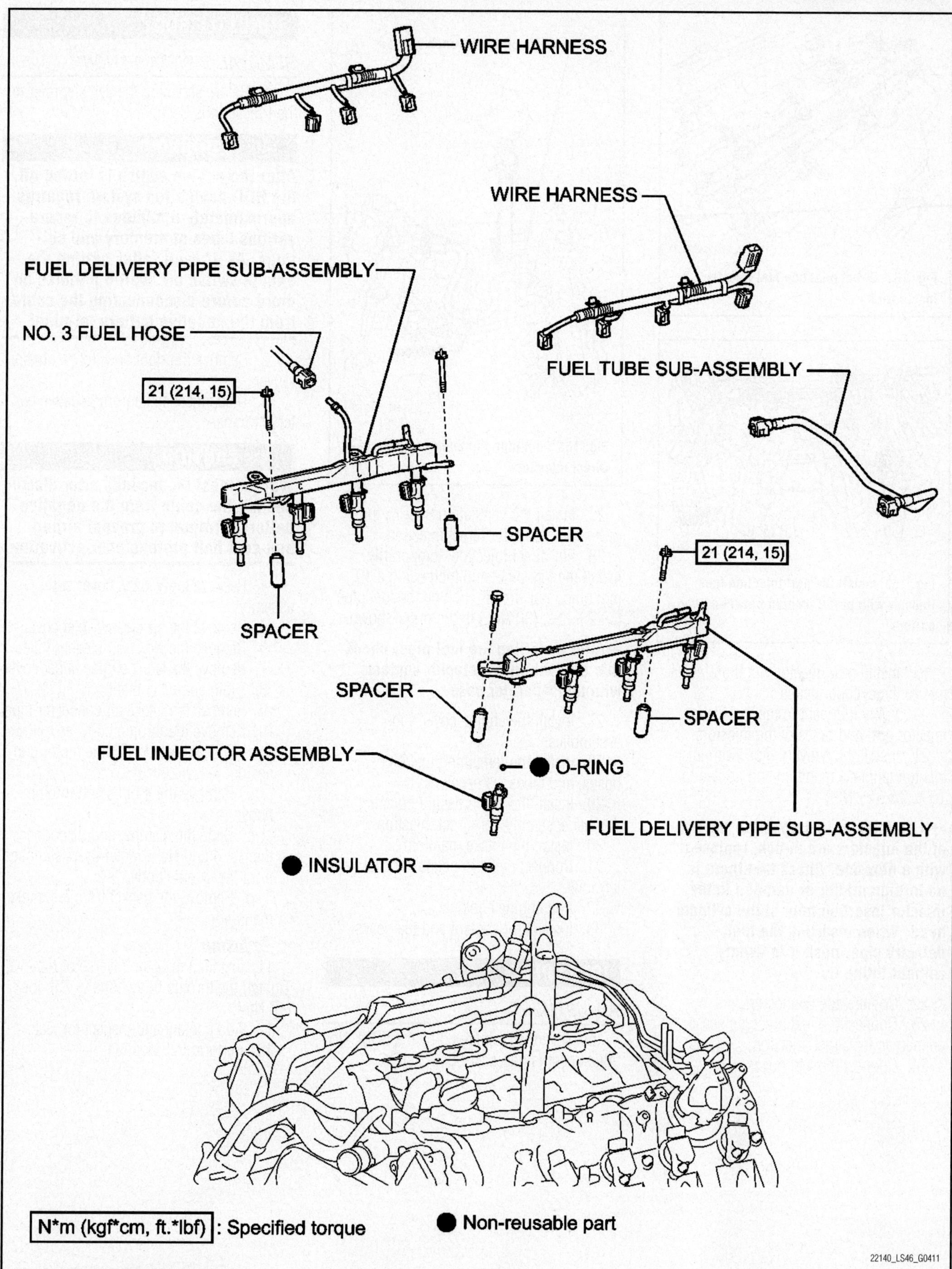

Fig. 161 Port injection components

Fig. 162 Direct injection fuel injector installation

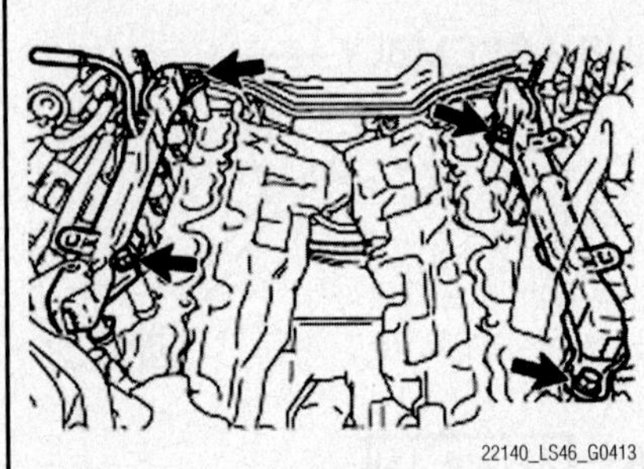

Fig. 163 Install the port injection fuel injector with part B located between parts A

19. Install new injector vibration insulators to the cylinder heads.
20. Apply lubricant to the installation injector seal and holes of the injectors.
21. Install the delivery pipe (with injector) to the intake manifold and tighten to 15 ft. lbs. (21 Nm).

➡If an injector is dropped or the tips of the injectors are struck, replace it with a new one. Check that there is no foreign matter or damage to the injector insertion hole of the cylinder head. When inserting the fuel delivery pipe, push it in evenly without tilting it.

22. Connect the connectors.
23. Connect the fuel pressure sensor connector, for direct injection.
24. Connect the fuel hoses.

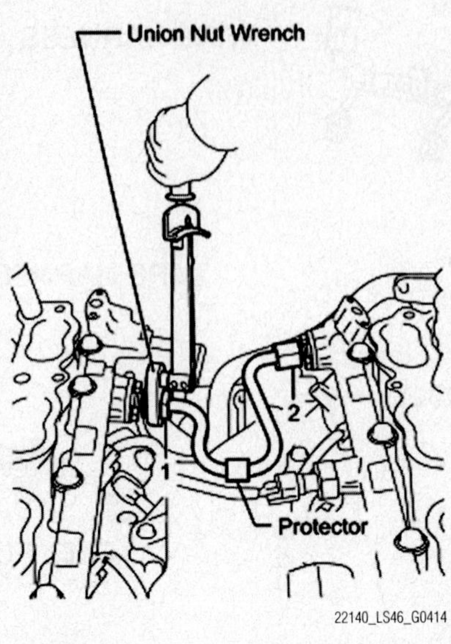

Fig. 164 Fuel pipe tightening sequence—Direct injection

25. Install the ventilation hose to the ventilation valve, for direct injection.
26. For direct injection, temporarily install the fuel pipe, and then using a 19 mm union nut wrench, tighten the fuel pipe to 22 ft. lbs. (30 Nm,) in the order shown.

➡After installing the fuel pipe, check that the fuel pipe protector contacts with the separator case.

27. Install the engine cover sub-assemblies.
28. Install the remaining fuel pipes, hoses, and tubes, as necessary.
29. Install the fuel pressure pulsation damper assembly, for direct injection.
30. Install the intake manifold.
31. Connect cable to negative battery terminal.
32. Add engine coolant.
33. Inspect for coolant and fuel leaks.

IDLE SPEED

ADJUSTMENT

Idle speed is maintained by the ECM. No adjustment is necessary or possible.

THROTTLE BODY

REMOVAL & INSTALLATION

1. Before servicing the vehicle, refer to the Precautions Section.

⁂ WARNING

After the engine switch is turned off, the HDD navigation system requires approximately 6 minutes to record various types of memory and settings. As a result, after turning the engine switch off, wait 6 minutes or more before disconnecting the cable from the negative battery terminal.

2. Remove the right cowl top ventilator louver.
3. Disconnect cable from negative battery terminal.

⁂ CAUTION

Wait at least 90 seconds after disconnecting the cable from the negative battery terminal to prevent airbag and seat belt pretensioner activation.

4. Remove the V-bank cover sub-assembly.
5. Remove the air cleaner inlet cover.
6. Remove the No. 1 air cleaner inlet.
7. Remove the No. 1 engine under cover.
8. Drain engine coolant.
9. Remove the intake air connector pipe.
10. Remove the throttle body, as follows:
    a. Disconnect the throttle motor connector.
    b. Remove the 4 bolts and throttle body.
    c. Slide the clamps, and disconnect the No. 4 and No. 5 water by-pass hoses from the throttle body.
    d. Remove the gasket from the intake manifold.

To install:

11. Installation is the reverse of removal. Tighten the throttle body bolts to 7 ft. lbs. (10 Nm).
12. Add coolant and inspect for leak.
13. Perform initialization.

HEATING & AIR CONDITIONING SYSTEM

BLOWER MOTOR

REMOVAL & INSTALLATION

See Figure 165.

1. Before servicing the vehicle, refer to the Precautions Section.
2. Remove the air conditioning unit.
3. Remove the blower assembly, as follows:
   a. Remove the screw.
   b. Disconnect the connector.
   c. Detach the claw and remove the blower unit.
4. Remove the 3 screws and damper servo.
5. Remove the 3 screws and blower with fan motor.

To install:

6. Installation is the reverse of removal.

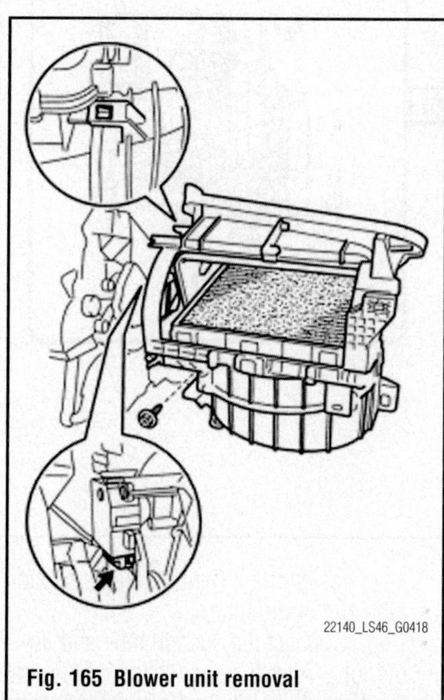

Fig. 165 Blower unit removal

HEATER CORE

REMOVAL & INSTALLATION

See Figures 166 and 167.

1. Before servicing the vehicle, refer to the Precautions Section.
2. Recover refrigerant from system.

WARNING

After the engine switch is turned off, the HDD navigation system requires approximately 6 minutes to record various types of memory and settings. As a result, after turning the engine switch off, wait 6 minutes or more before disconnecting the cable from the negative battery terminal.

3. Remove the right cowl top ventilator louver.
4. Disconnect cable from negative battery terminal.

CAUTION

Wait at least 90 seconds after disconnecting the cable from the negative battery terminal to prevent airbag and seat belt pretensioner activation.

5. Remove tube sub-assembly, as follows:
   a. Remove the bolt.
   b. Remove plate.
   c. Disconnect the suction tube and liquid tube.

➡Do not use a screwdriver or similar tool to disconnect the tube. Seal the opening of the disconnected parts to prevent moisture and foreign matter from entering them.

6. Using pliers, grip the claws of the clip and slide the clip, and disconnect the heater water inlet and outlet hoses.
7. Remove the instrument panel safety pad.
8. Remove the bolt, screw detach the claws and remove the air ducts.
9. Remove the steering column.
10. Remove the windshield wiper motor.
11. Remove the instrument panel reinforcement assembly, as follows:
    a. Disconnect the clamps, connectors and wire harness.
    b. Remove the bolts and nuts, and disconnect the ground wire and junction block.
    c. Remove the 9 bolts and 2 screws.
    d. Using a T40 Torx® socket, remove the 2 Torx® bolts (passenger side) and 3 bolts (driver side).
    e. For passenger side, using a 12 mm hexagon wrench, remove the 2 collars and instrument panel reinforcement with spacer.

➡The Torx® bolts on the passenger side can be removed with the collar for adjustment. Remove the nut and air conditioning unit.

12. Remove the rear left-hand mode servo motor.
13. Remove the 2 screws, 3 clamps and PTC heater.

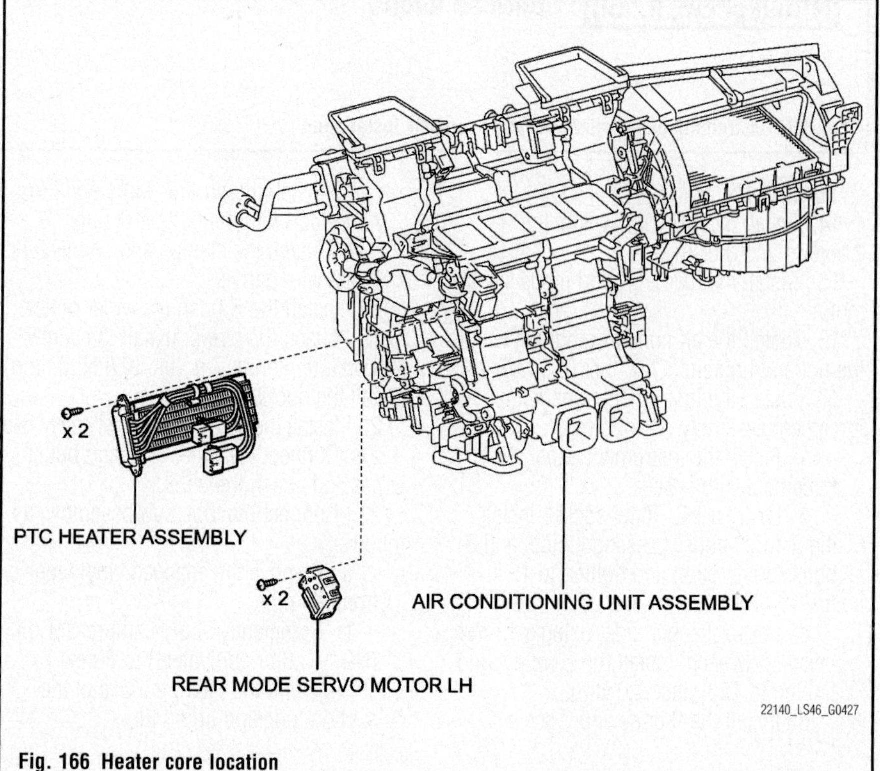

Fig. 166 Heater core location

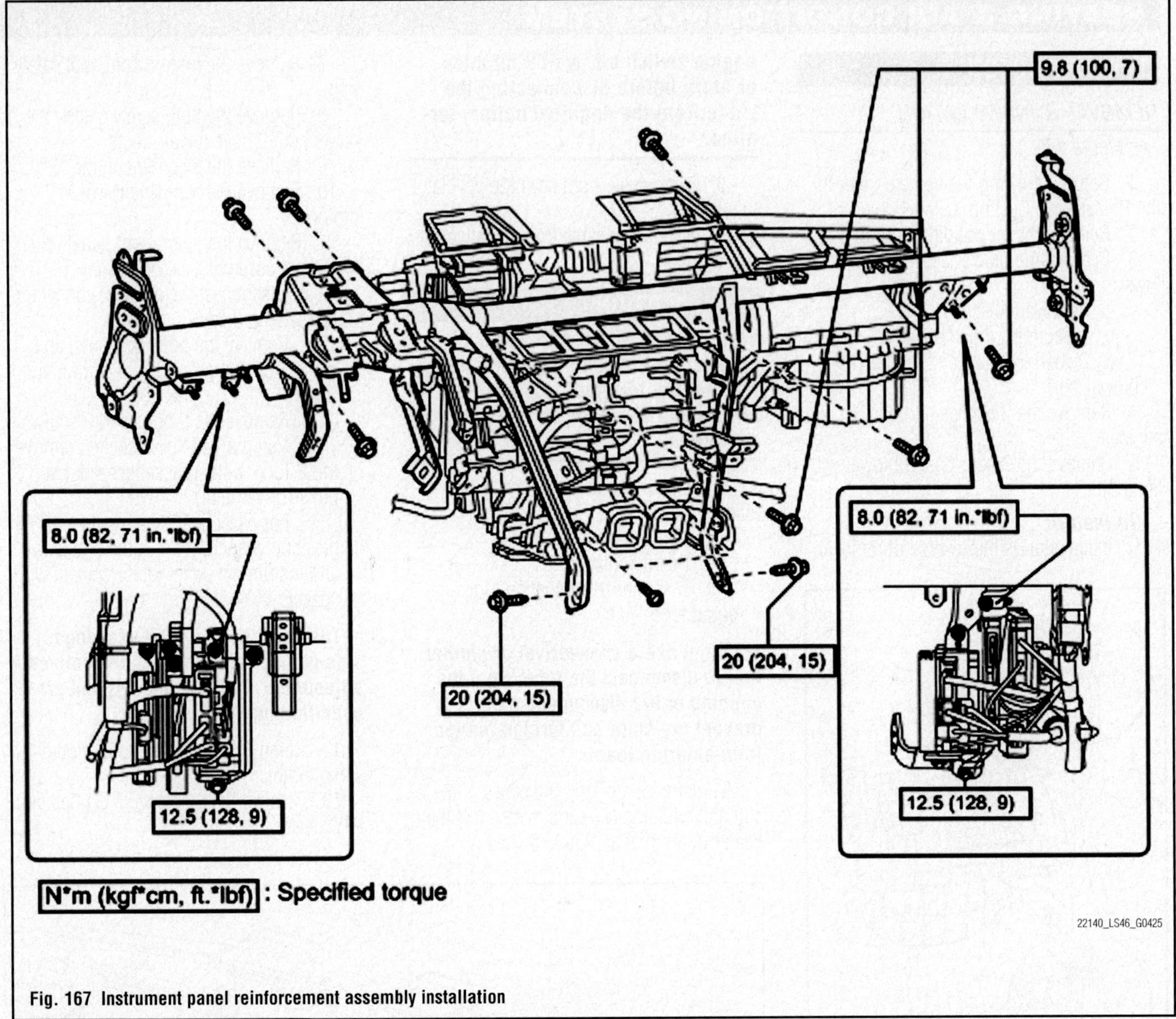

Fig. 167 Instrument panel reinforcement assembly installation

To install:

14. Install the PTC heater with the 2 screws and 3 clamps.
15. Install the rear left-hand mode servo motor.
16. Install the air conditioning unit with the nut and tighten to 7 ft. lbs. (9.8 Nm).
17. Install the instrument panel reinforcement assembly, as follows:
    a. Install the instrument panel reinforcement with spacer.
    b. Using a T40 Torx® socket, install the 2 Torx® bolts (passenger side) and 3 bolts (driver side) and tighten to 15 ft. lbs. (20 Nm).
    c. For passenger side, using a 12 mm hexagon wrench, install the 2 collars and tighten to 15 ft. lbs. (20 Nm).
    d. Install the 9 bolts and 2 screws.
    e. Install the ground wires and junction block with the bolts and nuts.
    f. Attach the clamps and connectors to the wire harness.
18. Install the windshield wiper motor.
19. Attach the claws, install the screw and bolt, tighten to 7 ft. lbs. (9.8 Nm), and install the ducts.
20. Install the instrument panel safety pad.
21. Connect the water inlet and outlet hoses and attach the clips.
22. Connect the tube sub-assembly, as follows:
    a. Remove the attached vinyl tape from the pipe.
    b. Sufficiently apply compressor oil (ND-OIL 8 or equivalent) to 2 new O-rings and the fitting surface of the suction tube and liquid tube.
    c. Install the 2 O-rings on the suction tube and liquid tube.
    d. Connect the suction tube and liquid tube. After the connection, check that the claw of the piping clamp is engaged.
    e. Install the plate with the bolt and tighten to 7 ft. lbs. (9.8 Nm).
23. Connect the cable to negative battery terminal.
24. Install the right cowl top ventilator louver.
25. Add engine coolant and compressor oil.
26. Charge refrigerant.
27. Warm up the engine.
28. Check for coolant and refrigerant leaks.
29. Check the SRS warning light.
30. Perform initialization.

STEERING

POWER RACK & PINION STEERING GEAR

REMOVAL & INSTALLATION

See Figures 168 and 169.

1. Before servicing the vehicle, refer to the Precautions Section.
2. Position the front wheels straight ahead.

⁂ WARNING

After the engine switch is turned off, the HDD navigation system requires approximately 6 minutes to record various types of memory and settings. As a result, after turning the engine switch off, wait 6 minutes or more before disconnecting the cable from the negative battery terminal.

3. Remove the right cowl top ventilator louver.
4. Disconnect cable from negative battery terminal.

⁂ CAUTION

Wait at least 90 seconds after disconnecting the cable from the negative battery terminal to prevent airbag and seat belt pretensioner activation.

5. Remove front wheels.
6. Remove the No. 1 and No. 2 engine under cover.
7. Remove the front lower suspension member protector.
8. Disconnect the steering sliding yoke with shaft sub-assembly.
9. Disconnect the left and right tie rod assemblies.
10. Using the same procedure as above, disconnect the right tie rod assembly.
11. Remove the power steering link assembly, as follows:
    a. Detach the clips and disconnect the connectors.

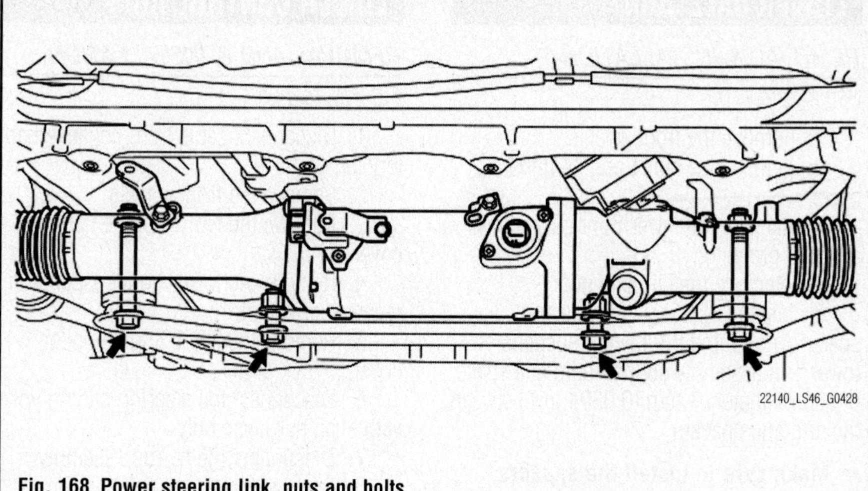

Fig. 168 Power steering link, nuts and bolts

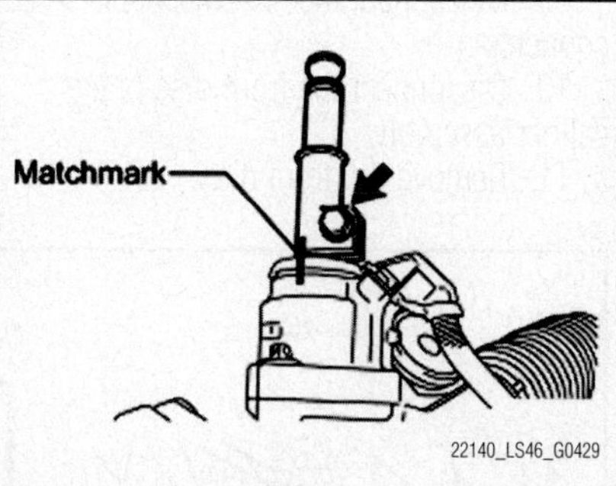

Fig. 169 Matchmarks on the intermediate shaft and power steering link—Without VGRS

    b. Remove the 4 bolts, 4 nuts and power steering link from the front frame and rack housing bracket.
12. For vehicles without VGRS, remove steering intermediate shaft, as follows:
    a. Put the matchmarks on the intermediate shaft and power steering link.
    b. Remove the bolt and intermediate shaft from the power steering link.

To install:

13. For vehicles without VGRS, align the matchmarks on the power steering link and intermediate shaft, install the bolt and tighten to 26 ft. lbs. (35 Nm).
14. Install the power steering link assembly, as follows:
    a. Install the power steering link and rack housing bracket to the front frame with the 4 bolts and 4 nuts and tighten to 52 ft. lbs. (70 Nm).
    b. Connect the connectors and attach the clips.
15. The remainder of installation is the reverse of removal, noting the following:
    a. Perform initialization.
    b. Check suspension control system.
    c. Inspect and adjust front wheel alignment.
    d. Adjust headlight aiming.
    e. Inspect the SRS warning light.
    f. Adjust object recognition camera.
    g. Initialize rotation angle sensor and calibrate torque sensor zero point.
    h. Perform variable gear ratio steering system calibration, if necessary.

POWER STEERING PUMP

REMOVAL & INSTALLATION

This vehicle is equipped with electronic power steering. These models do not utilize a power steering (vane) pump assembly or power steering fluid.

LOWER BALL JOINT

REMOVAL & INSTALLATION

See Figure 170.

1. Remove the front wheels.
2. Remove the No. 1 engine under cover.
3. Remove the lower front suspension member protector.
4. Remove the clip and nut.
5. Install 2 spacers (Service Tool spacer A) onto the front No. 1 suspension lower arm assembly so that there is a space of approximately 1 mm (0.0394 in.) between the arm and spacers.

➡ Make sure to install the spacers (Service Tool spacer A) as the steering knuckle spacer may shift, as Service Tool may become damaged, make sure the space between the arm and spacers is not 1 mm (0.0394 in.) or less.

6. Using Service Tool, disconnect the front No. 1 suspension lower arm assembly from the steering knuckle.

To install:

7. Install the front No. 1 suspension lower arm to the steering knuckle with the nut and tighten to 107 ft. lbs. (145 nm).
8. Install a new clip.

➡If it is necessary to align the holes for the clips after installing the nuts, the nuts can be tightened up to 60 ° more.

9. Install the lower front suspension member protector.
10. Install the No. 1 engine under cover.
11. Install the front wheels.
12. Inspect front end alignment, if necessary.

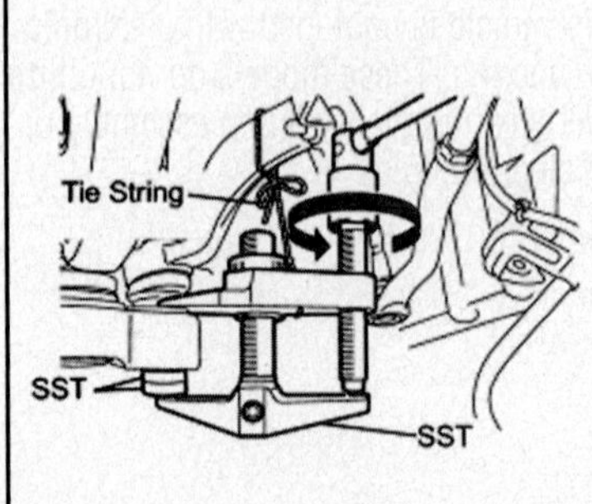

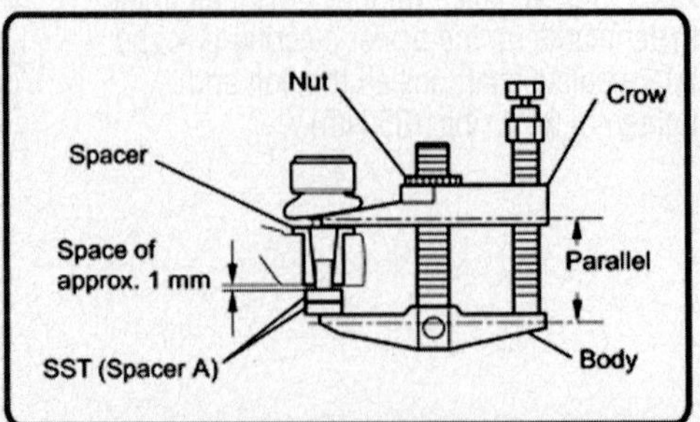

Fig. 170 Remove lower ball joint

LOWER CONTROL ARM

REMOVAL AND & INSTALLATION

See Figure 171.

1. Disconnect cable from negative battery terminal.
2. Remove the front wheels.
3. Remove the No. 1 engine under cover.
4. Remove the lower front suspension member protector.
5. Remove the No. 2 engine under cover.
6. Disconnect the steering sliding yoke with shaft sub-assembly.
7. Disconnect the tie rod assembly.
8. Remove the power steering link assembly.
9. Disconnect the speed sensor connector.
10. Disconnect the front disc brake caliper assembly.
11. Remove the front disc.
12. Remove the front disc brake dust cover.
13. Remove the front No. 1 suspension lower arm assembly.
14. Remove the clip and nut.
    a. Install 2 spacers (Service Tool spacer A) onto the front No. 1 suspension lower arm assembly so that there is a space of approximately 1 mm (0.0394 in.) between the arm and spacers.

➡ Make sure to install the spacers (Service Tool spacer A) as the steering knuckle spacer may shift, as Service Tool may become damaged, make sure the space between the arm and spacers is not 1 mm (0.0394 in.) or less.

    b. Using Service Tool, disconnect the front No. 1 suspension lower arm assembly from the steering knuckle.
    c. Remove the bolt and nut, and then remove the No. 1 suspension lower arm.
15. Remove the bolt, and then remove the bracket of the front height control sensor.

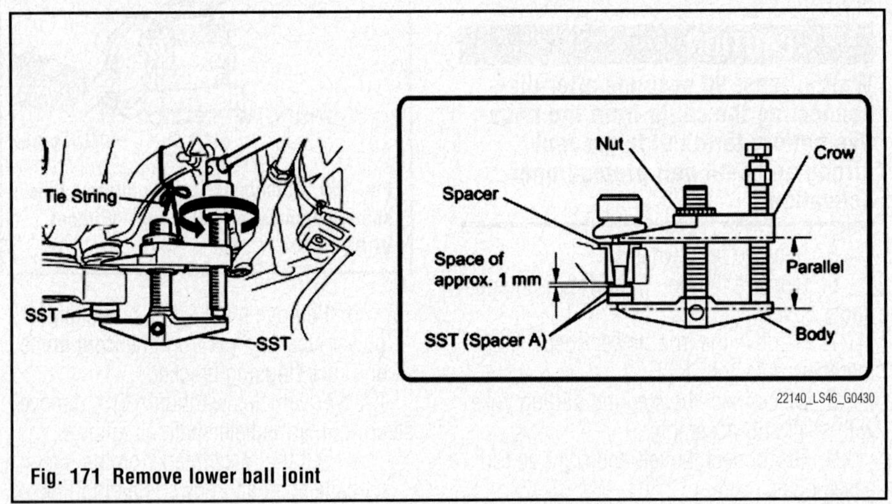

Fig. 171 Remove lower ball joint

16. Remove the 2 nuts and front stabilizer link assembly.
    a. Remove the front No. 2 suspension lower arm assembly.
    b. Support the steering knuckle with a jack and wooden block.
    c. Remove the bolt and nut, and then disconnect the bottom side of the pneumatic front with shock absorber cylinder assembly from the front No. 2 suspension lower arm.
    d. Remove the nut and clip.
    e. Install 2 spacers (Service Tool spacer A) onto the front No. 2 suspension lower arm assembly so that there is a space of approximately 1 mm (0.0394 in.) between the arm and spacers.

➡ Make sure to install the spacers (Service Tool spacer A) as the steering knuckle spacer may shift, as Service Tool may become damaged, make sure the space between the arm and spacers is not 1 mm (0.0394 in.) or less.

f. Using Service Tool, disconnect the front No. 2 suspension lower arm assembly from the steering knuckle.

17. Remove the bolt and nut, and then remove the No. 2 suspension lower arm.

To install:

18. Temporarily tighten the front No. 2 suspension lower arm assembly.

a. Insert the bolt from the front of the vehicle. Then temporarily install the front No. 2 suspension lower arm with the nut.

b. Temporarily install the front No. 2 suspension lower arm to the steering knuckle with the nut.

19. Install the bottom side of the pneumatic front with shock absorber cylinder assembly to the No. 2 suspension lower arm. Then insert the bolt from the rear of the vehicle and temporarily install it with the nut.

20. Install the front stabilizer link with the 2 nuts and tighten to 62 ft. lbs. (84 Nm).

21. Install the bracket of the front height control sensor to the front No. 2 suspension lower arm with the bolt and tighten to 48 inch lbs. (5.4 Nm).

22. Temporarily tighten front No. 1 suspension lower arm assembly.

a. Insert the bolt from the back of the vehicle. Temporarily install the front No. 1 suspension lower arm with the nut.

b. Temporarily install the front No. 1 suspension lower arm to the steering knuckle with the nut.

➡Tighten the nuts after the vehicle is stabilized.

23. Install front disc brake dust cover.

24. Install front disc.

25. Connect front disc brake caliper assembly.

26. Connect speed sensor connector.

27. Install power steering link assembly.

28. Connect tie rod assembly.

29. Connect steering sliding yoke with shaft sub-assembly.

30. Connect cable to negative battery terminal.

31. Install cowl top ventilator louver.

32. Stabilize suspension.

33. Tighten front No. 2 suspension lower arm assembly.

a. Tighten the installation nut of the front No. 2 suspension lower arm assembly to 107 ft. lbs. (145 Nm).

b. Tighten the installation nut of the steering knuckle to 107 ft. lbs. (145 Nm).

c. Tighten the installation bolt of the shock absorber cylinder to 80 ft. lbs. (108 Nm).

34. Install the front No. 1 suspension lower arm to the steering knuckle with the nut and tighten to 107 ft. lbs. (145 nm).

35. Install a new clip.

➡If it is necessary to align the holes for the clips after installing the nuts, the nuts can be tightened up to 60 ° more.

36. Install No. 2 engine under cover.

37. Install front suspension member protector lower.

38. Install No. 1 engine under cover.

39. Install front wheels.

40. Inspect front end alignment, if necessary.

MACPHERSON STRUT

REMOVAL & INSTALLATION

See Figure 172.

1. Remove the front wheel.

2. Disconnect the speed sensor connector.

3. Disconnect the front disc brake caliper assembly.

4. Remove the front disc.

5. Remove the front disc brake dust cover.

6. Disconnect the tie rod assembly.

7. Disconnect the front No. 1 suspension lower arm assembly.

8. Disconnect the front height control sensor sub-assembly.

9. Remove the front stabilizer link assembly.

10. Disconnect the front No. 2 suspension lower arm assembly.

11. Remove the air cleaner inlet cover sub-assembly

12. Remove the engine room side cover.

13. Remove the front shock absorber assembly.

a. Loosen the lock nut of the front shock absorber.

➡Do not remove the lock nut. Loosen the lock nut only when disassembling the front shock absorber with coil spring.

b. Remove the 3 nuts from the upper side of the front suspension support.

c. Slowly lower the jack. Remove the bolt from the lower side to remove the front shock absorber with coil spring.

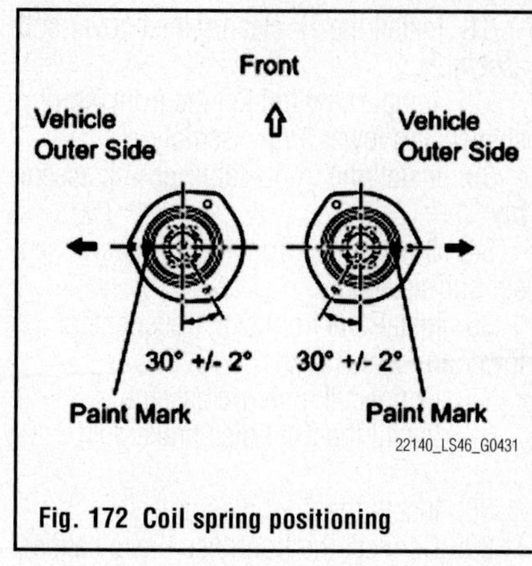

Fig. 172 Coil spring positioning

14. Install Service Tool to the front coil spring so that the distance between the upper and lower hooks is as wide as possible within the installation area.

15. Compress the front coil spring until it can be moved freely.

16. Remove the nut.

17. Remove the collar from the shock absorber.

18. Remove the suspension support assembly with the insulator upper from the shock absorber.

19. Remove the coil spring from the shock absorber.

To install:

20. Using Service Tool, compress the coil spring.

21. Install the coil spring so that the end comes to the stepped portion of the lower coil spring insulator.

22. Align the bolt of the front suspension support and the cutout of the front coil spring insulator, and install the front coil spring insulator on the front suspension support.

23. Align the width across flat of the piston rod end and of the front suspension support to install the front shock absorber.

24. Install the suspension support assembly with the upper insulator to the shock absorber.

25. Install the collar to the shock absorber.

26. Temporarily install a new lock nut to the front shock absorber.

27. Adjust the front suspension support so that the bolts come to the positions as shown in the illustration, and remove Service Tool from the front coil spring.

28. Install the front shock absorber on the vehicle by installing the 3 nuts on the suspension support side and tighten nuts to 49 ft. lbs. (67 Nm).

29. Tighten the lock nut to 20 ft. lbs. (28 Nm).

30. Install the engine room side cover.

31. Install the air cleaner inlet cover sub-assembly.
32. Temporarily tighten the front No. 2 suspension lower arm assembly.
33. Install the front stabilizer link assembly.
34. Connect the front height control sensor sub-assembly.
35. Install the front No. 1 suspension lower arm assembly.
36. Connect the tie rod assembly.
37. Install the front disc brake dust cover.
38. Install the front disc.
39. Connect the front disc brake caliper assembly.
40. Connect the speed sensor connector.
41. Stabilize the suspension:
    a. Install the front tires.
    b. Lower the vehicle and bounce it up and down several times to stabilize the front suspension.
    c. Remove the front tires.
    d. Jack up the front suspension lower arm with a wooden block between the jack and front suspension lower arm. Apply a load to the front suspension so that the front suspension lower arm is placed in a horizontal position.
42. Tighten the front No. 2 suspension lower arm assembly.
43. Install the front wheel.
44. Inspect and adjust the front wheel and headlight alignment.

STEERING KNUCKLE

REMOVAL & INSTALLATION

1. Remove the front wheel.
2. Disconnect the speed sensor connector.
3. Disconnect the front disc brake caliper assembly.
4. Remove the front disc.
5. Remove the front disc brake dust cover.
6. Disconnect the tie rod assembly.
7. Disconnect the front No. 1 suspension lower arm assembly.
8. Disconnect the front height control sensor sub-assembly.
9. Remove the front stabilizer link assembly.
10. Disconnect the front shock absorber assembly.
11. Disconnect the front No. 2 suspension lower arm assembly.
12. Remove the steering knuckle sub-assembly.
13. Remove the front axle hub sub-assembly.

To install:

14. Install the front axle hub sub-assembly.
15. Install the steering knuckle sub-assembly.
16. Connect the front No. 2 suspension lower arm assembly.
17. Connect the front shock absorber assembly.
18. Install the front stabilizer link assembly.
19. Connect the front height control sensor sub-assembly.
20. Connect the front No. 1 suspension lower arm assembly.
21. Connect the tie rod assembly.
22. Install the front disc brake dust cover.
23. Install the front disc.
24. Connect the front disc brake caliper assembly.
25. Connect the speed sensor connector.
26. Install the front wheel.

STABILIZER BAR

REMOVAL & INSTALLATION

See Figure 173.

1. Remove the front wheel.
2. Remove the 8 bolts and the No. 2 engine under cover.
3. Remove the 2 nuts and the front stabilizer link.
4. Remove the front stabilizer link assembly.
5. Remove the 2 screws and the engine under cover sub-assembly from the front suspension member reinforcement.
6. Remove the engine under cover sub-assembly.
7. Remove the 2 screws and the engine under cover rear.
8. Remove the 2 bolts and the front No. 1 stabilizer bracket from the front suspension crossmember.
9. Remove the front No. 1 stabilizer bracket.
10. Remove the front stabilizer bar from the vehicle.

To install:

11. Temporarily install the front stabilizer bar with the 4 bolts and the front No. 1 stabilizer bracket.

➡Install the front stabilizer bar so that the identification mark face vehicle right side.

12. Tighten the front No. 1 stabilizer bracket with the 2 bolts to 36 ft. lbs. (49 Nm).
13. Tighten the front No. 1 stabilizer bracket.
14. Install the engine under cover rear with the 2 screws.
15. Install the engine under cover rear.
16. Install the engine under cover sub-assembly with the 2 screws.
17. Install the engine under cover sub-assembly.
18. Install the front stabilizer link with the 2 nuts and tighten to 62 ft. lbs. (84 nm).
19. Install the No. 2 engine under cover with the 8 bolts.
20. Install front wheel.
21. Inspect and adjust front wheel and headlight alignment.

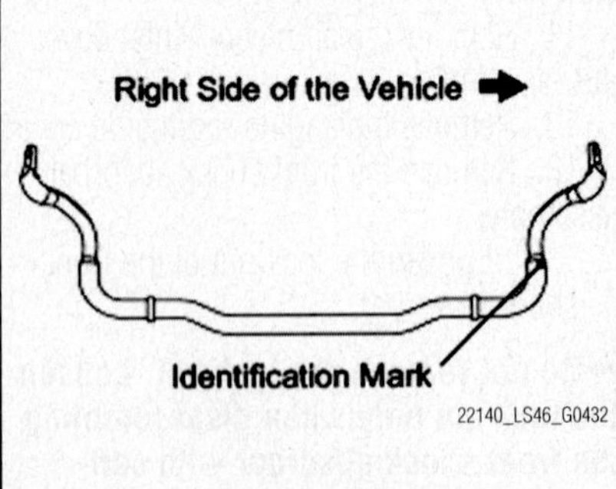

Fig. 173 Stabilizer bar identification marks

WHEEL HUB AND BEARING

REMOVAL & INSTALLATION

1. Remove the front wheel.
2. Remove the steering knuckle sub-assembly.
3. Remove the 4 bolts and front axle hub from the steering knuckle.

To install:

4. Install the front axle hub to the steering knuckle with the 4 bolts and tighten to 48 ft. lbs. (65 Nm).
5. Install the steering knuckle sub-assembly.
6. Install the front wheel.

CONTROL ARMS/LINKS

REMOVAL & INSTALLATION

Lower

1. Remove the rear wheel.
2. Remove the No. 2 differential support protector.
3. Remove the rear speed sensor.
4. Remove the load sensing valve sensor bracket.
5. Remove the rear stabilizer link assembly.
6. Disconnect the rear shock absorber assembly.
7. Remove the rear No. 2 suspension arm assembly.
   a. Using a plastic-faced hammer or equivalent, strike the part labeled from the rear of the vehicle to maintain the clearance at the slide pin area.
   b. Remove the nut, washer, bolt and no. 2 suspension arm from the axle carrier.
   c. Remove the nut, no. 2 suspension toe adjust plate, rear suspension attachment and no. 2 suspension arm from the suspension member.
8. Remove the rear No. 1 suspension arm assembly.
   a. Remove the nut on the rear axle carrier side.
   b. Install 2 spacers (Service Tool spacer B) onto the rear No. 1 suspension arm so that there is a space of approximately 1 mm (0.039 in.) between the arm and spacers.

➡Make sure to install the spacers (Service Tool spacer B) as the axle carrier spacer may shift, as Service Tool may become damaged, make sure the space between the arm and spacers is not 1 mm (0.039 in.) or less.

   c. Using Service Tool, disconnect the rear No. 1 control arm from the axle carrier.
   d. Remove the bolt, nut and No. 1 suspension arm from the suspension member.

To install:

9. Temporarily install the rear No. 2 suspension arm assembly.
   a. Temporarily install the No. 2 suspension arm with the no. 2 suspension toe adjust plate, rear suspension attachment and nut to the suspension member.
   b. Install the stud of the suspension arm, and temporarily install a washer, nut and the bolt to the axle carrier.
10. Temporarily install the rear No. 1 suspension arm assembly.
   a. Temporarily install the No. 1 suspension arm with the bolt and nut to the suspension member.
   b. Install the stud of the suspension arm, and temporarily install a new nut to the axle carrier.
11. Tighten rear No. 1 suspension arm assembly bolt to 53 ft. lbs. (72 Nm) and the nut to 87 ft. lbs. (118 Nm).
12. Temporarily connect the rear shock absorber assembly.
13. Install the rear stabilizer link assembly.
14. Stabilize the suspension.
15. Tighten the rear No. 2 suspension arm assembly to 111 ft. lbs. (150 Nm) on member side and 166 ft. lbs. (225 Nm) for carrier side.
16. Tighten connect rear shock absorber assembly to 59 ft. lbs. (80 Nm).
17. Install the rear speed sensor.
18. Install the load sensing valve sensor bracket.
19. Install the No. 2 differential support protector.
20. Install the rear wheel.
21. Inspect and adjust rear wheel alignment, headlights and object recognition camera.

Upper

See Figures 174 through 177.

1. Remove the rear wheel.
2. Remove the rear disc brake caliper assembly.
3. Remove the rear stabilizer link assembly.
4. For vehicles without air suspension, disconnect the rear shock absorber assembly.
5. For vehicles with air suspension, disconnect the pneumatic cylinder with rear shock absorber assembly.
6. Remove the rear No. 2 suspension arm assembly.
7. Disconnect the rear No. 1 suspension arm assembly.
8. Disconnect the toe control link sub-assembly.
9. Remove the rear upper control arm

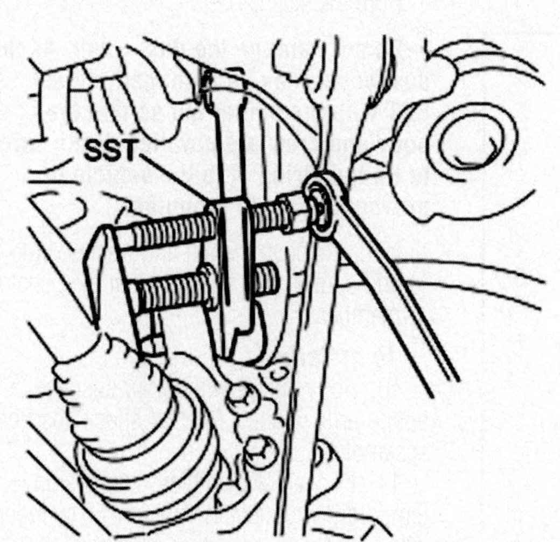

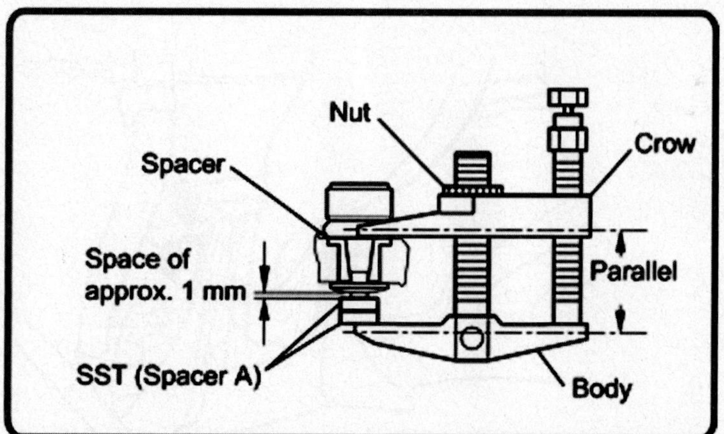

Fig. 174 Install 2 spacers onto the rear upper control arm—No. 1 control arm

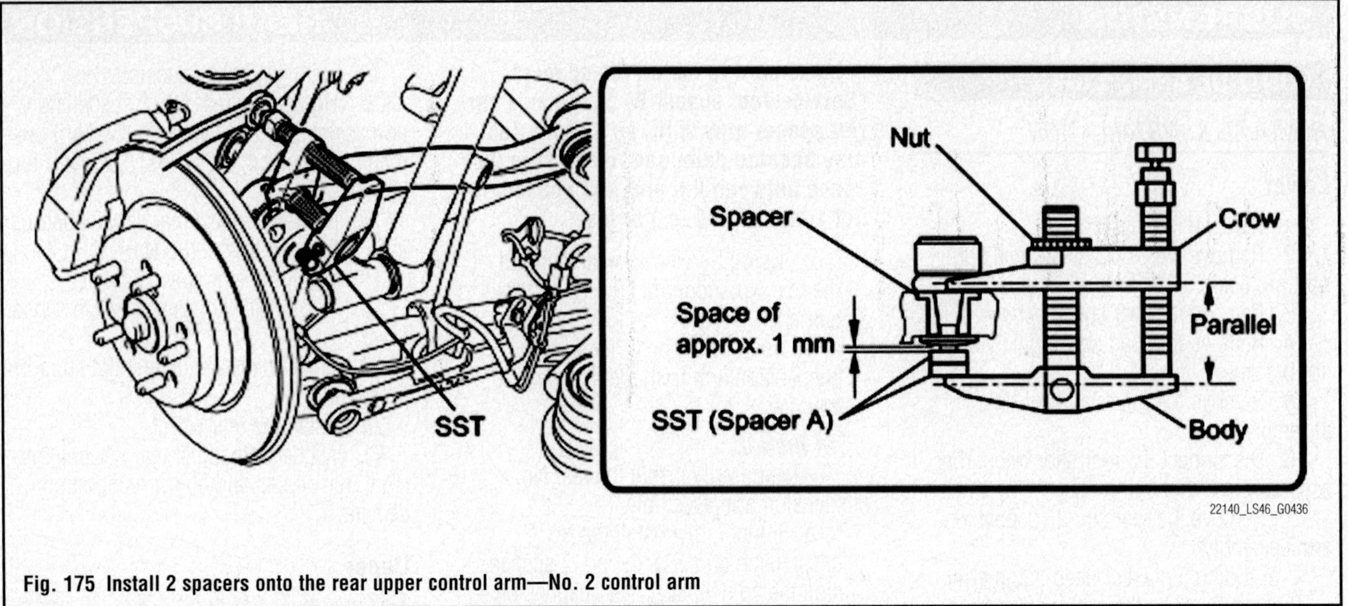

Fig. 175 Install 2 spacers onto the rear upper control arm—No. 2 control arm

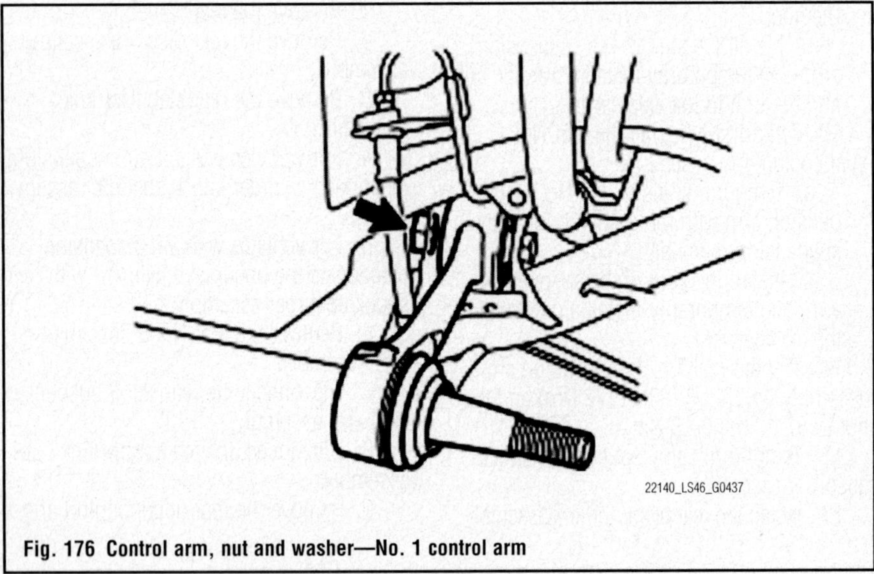

Fig. 176 Control arm, nut and washer—No. 1 control arm

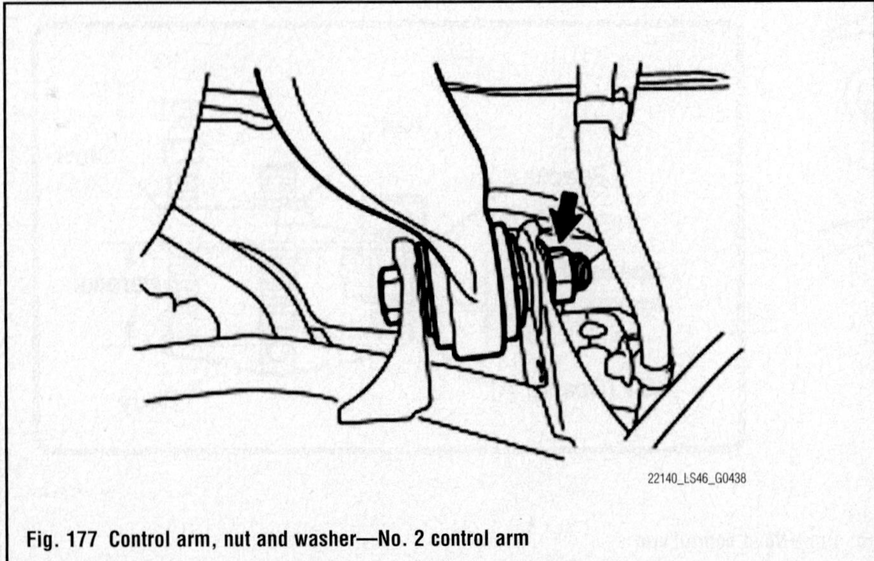

Fig. 177 Control arm, nut and washer—No. 2 control arm

assembly (use the same procedure for the No. 1 and No. 2 control arms), as follows:

a. Remove the nut on the rear axle carrier side.

b. Install 2 spacers (SST spacer A) onto the rear upper control arm so that there is a space of approximately 1 mm (0.039 in.) between the arm and spacers. SST: 09960-20010.

➡**Make sure to install the spacers (SST spacer A) as the axle carrier spacer may shift.**

➡**As SST may become damaged, make sure the space between the arm and spacers is not 1 mm (0.039 in.) or less.**

c. Using SST (SST: 09960-20010), disconnect the rear upper control arm from the axle carrier.

➡**Do not damage the dust cover. As the dust cover may be damaged, adjust SST with the center nut so that the body and crow are parallel. Make sure to tie the string of to the vehicle to prevent SST from dropping.**

d. Remove the nut and washer, bolt and control arm from the rear suspension member.

To install:

10. For vehicles without air suspension, temporarily connect the rear shock absorber assembly.

11. For vehicles with air suspension, temporarily connect the pneumatic cylinder with rear shock absorber assembly.

12. Temporarily install the rear upper control arm assembly (use the same proce-

dure for the No. 1 and No. 2 control arms), as follows:

a. Temporarily install the control arm with the bolt, nut and washer to the suspension member.

b. Install the stud of the control arm, and temporarily install a new nut to the axle carrier.

➡**Push the axle carrier downward.**

13. Tighten the nuts on the rear upper No. 1 control arm to 111 ft. lbs. (150 Nm) for suspension member side, and 118 ft. lbs. (160 Nm) for axle carrier side.
14. Tighten the nuts on the rear upper No. 2 control arm to 166 ft. lbs. (225 Nm) for suspension member side, and 118 ft. lbs. (160 Nm) for axle carrier side.
15. Temporarily install the rear No. 2 suspension arm.
16. Connect the toe control link sub-assembly.
17. Connect the rear No. 1 suspension arm.
18. Install rear disc brake caliper assembly.
19. Stabilize the suspension.
20. Tighten the rear No. 2 suspension arm assembly.
21. For vehicles without air suspension, connect the rear shock absorber assembly.
22. For vehicles with air suspension, connect the pneumatic cylinder with rear shock absorber assembly.
23. Install the rear stabilizer link assembly.
24. Install the rear wheel.
25. Check the suspension control system.
26. Inspect and adjust the rear wheel alignment.
27. Adjust the headlight.
28. Adjust the object recognition camera.

MACPHERSON STRUTS

REMOVAL & INSTALLATION

1. Remove the cowl top ventilator louver.
2. Disconnect cable from negative battery terminal.
3. Remove the rear wheel.
4. Remove the rear seat cushions, ottomans and seat backs.
5. Remove No. 6 rocker panel moulding protector.
6. Remove the 3 screws, 11 nuts and 2 clips from the liner.
7. Remove the rear door scuff plates.
8. Remove the rear seat side garnish moldings.
9. Remove the package tray trim panel assembly.
10. Remove the rear speed sensor.
11. Remove the load sensing valve sensor bracket.
12. Disconnect the rear upper No. 2 control arm assembly.
13. Disconnect the rear upper No. 1 control arm assembly.
14. Disconnect the toe control link sub-assembly.
15. Remove the rear axle shaft nut.
16. Remove the 3 nuts and cap.
17. Remove the 3 nuts on the upper side of the shock absorber.
18. Remove the nut from the shock absorber lower side.
19. Angle the axle carrier's upper tip toward the vehicle's outer side, and remove the shock absorber.

To install:

20. Install the rear shock absorber assembly and tighten to 47 ft. lbs. (64 Nm).
21. Temporarily install the rear shock absorber lower side with a new nut and washer to the axle carrier.
22. Tighten the new lock nut to 20 ft. lbs. (27 Nm).
23. Install the cap with the 3 nuts to the shock absorber and tighten to 10 ft. lbs. (14 Nm).
24. Connect the toe control link sub-assembly.
25. Connect the rear upper no. 2 control arm assembly.
26. Connect the rear upper no. 1 control arm assembly.
27. Tighten the rear axle shaft nut to 214 ft. lbs. (290 Nm) and stake.
28. Install the load sensing valve sensor bracket.
29. Install the rear speed sensor.
30. Install the package tray trim panel assembly.
31. Install the rear seat side garnish moldings.
32. Install the rear door scuff plate.
33. Install the rear seat cushions, ottomans and seat backs.
34. Install the liner with the 3 screws, 11 nuts, 2 clips to the vehicle side.
35. Install the No. 6 rocker panel moulding protector.
36. Stabilize the suspension.
37. Tighten the rear shock absorber assembly to 59 ft. lbs. (80 Nm).
38. Connect cable to negative battery terminal.
39. Install the cowl top ventilator louver.
40. Perform initialization, if necessary.
41. Install the rear wheel.
42. Inspect and adjust rear wheel alignment, and headlights.

STABILIZER BAR

REMOVAL & INSTALLATION

See Figures 178 through 180.

1. Before servicing the vehicle, refer to the Precautions Section.

⁂ WARNING

After the engine switch is turned off, the HDD navigation system requires approximately 6 minutes to record various types of memory and settings. As a result, after turning the engine switch off, wait 6 minutes or more before disconnecting the cable from the negative battery terminal.

2. Remove the right cowl top ventilator louver.
3. Disconnect cable from negative battery terminal.

⁂ CAUTION

Wait at least 90 seconds after disconnecting the cable from the negative battery terminal to prevent airbag and seat belt pretensioner activation.

4. Remove the rear wheel.
5. Remove the luggage compartment mat sub-assembly.
6. Remove the left deck trim side board.
7. Remove rear floor finish plate.
8. Remove the left side luggage compartment trim cover assembly.
9. Remove the No. 2 and No. 1 differential support protector.
10. Remove the No. 6 and No. 5 rocker panel molding protector.
11. Remove the left and right rear floor side member cover.
12. Remove the left and right rear wheel house liners.
13. Remove the No. 1 floor under cover.
14. Remove the left and right rear speed sensors.
15. Disconnect the No. 1 actuator harness clamp.
16. Remove the parking brake cable.
17. Remove the front floor center brace.
18. Remove the heated oxygen sensor (for Sensor 2).
19. Remove the left and right tailpipes.
20. Remove the No. 1 front floor heat insulator.

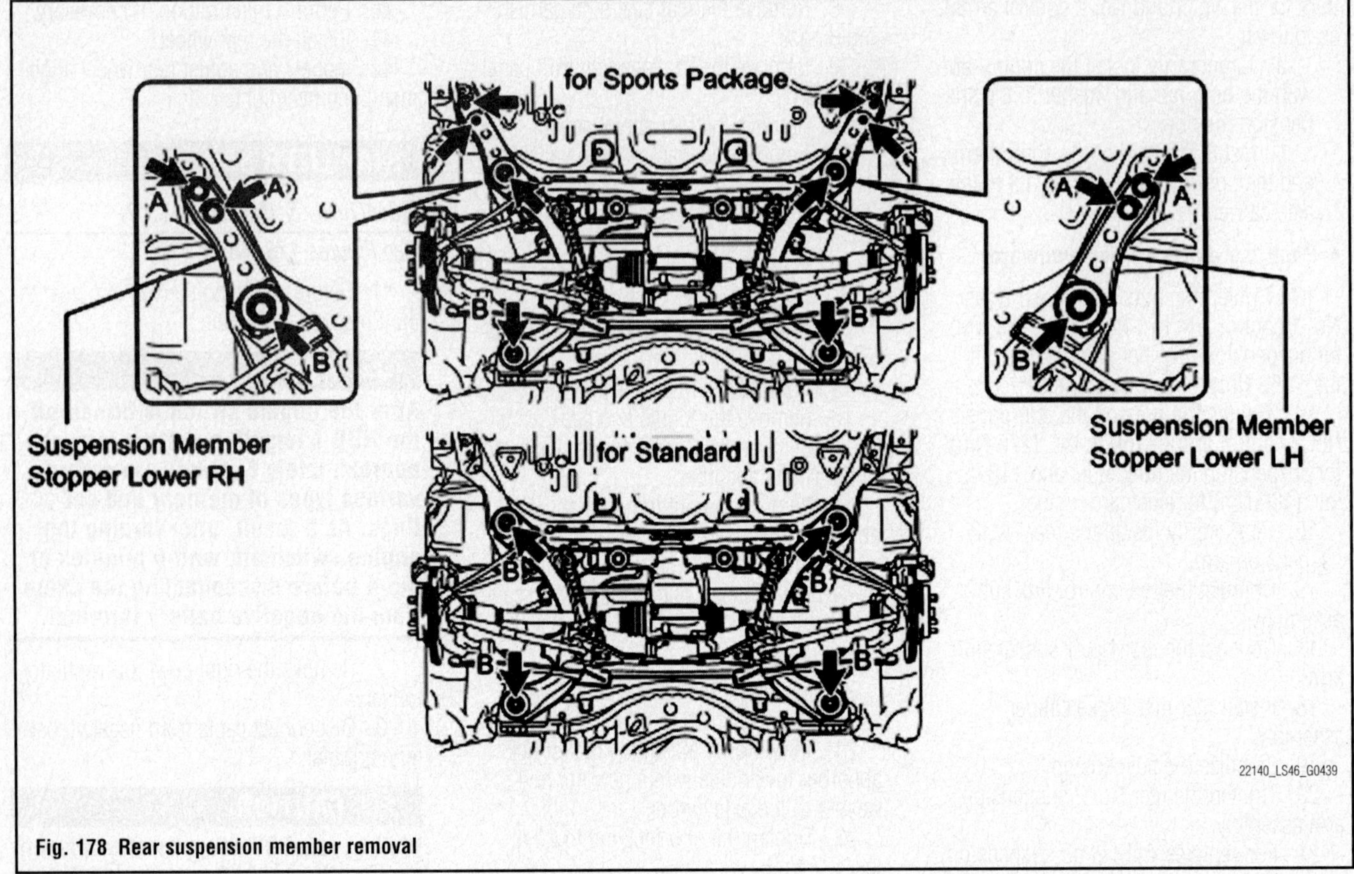

Fig. 178 Rear suspension member removal

21. Remove the propeller shaft heat insulator.

22. Remove the propeller with center bearing shaft assembly.

23. Remove the left and right rear disc brake caliper assemblies.

24. Remove the left rear stabilizer link assembly, as follows:

    a. Remove the nut and disconnect the stabilizer link from the stabilizer bar.

    b. Remove the bolt, nut and stabilizer link from the rear No. 2 suspension arm.

25. Using the same procedure as above, remove the right rear stabilizer link assembly.

26. For vehicles without air suspension, disconnect the left and right rear shock absorber assemblies.

27. For vehicles with air suspension, disconnect the left and right pneumatic cylinder with rear shock absorber assemblies.

28. Support the rear suspension cross-member with the jack.

29. Remove the rear suspension member sub-assembly for sports package:

    a. Remove the 4 bolts labeled A.

    b. Remove the 4 bolts labeled B and 2 suspension member stoppers.

    c. Slowly lower the jack, and disconnect the suspension member from the vehicle.

30. Remove the rear suspension member sub-assembly for standard package:

    a. Remove the 4 bolts.

31. Slowly lower the jack, and disconnect the suspension member from the vehicle.

32. Remove the 4 bolts and the rear stabilizer bar from the suspension member.

➡The stabilizer bracket and bush are built onto the stabilizer bar. If the bracket and/or bush detach from the bar, replace the bar.

To install:

33. Install the rear stabilizer bar, as follows:

    a. Make sure the bracket's arrows face the front of the vehicle.

    b. First temporarily install bolt 1. Then install bolts 2, 3 and 4. Then tighten bolt 5. Tighten the bolts to 35 ft. lbs. (48 Nm).

34. Install the rear stabilizer link assemblies with the nuts and bolts and tighten as follows:

    - Stabilizer bar side: 66 ft. lbs. (89 Nm).
    - No. 2 suspension arm side: 19 ft. lbs. (26 Nm).

35. Slowly raise the suspension member jack, and install the suspension member to the body.

36. For sports package:

    a. Install the 2 member stoppers with the 4 bolts labeled B, and install the 4 bolts labeled A, and tighten the A bolts to 14 ft. lbs. (19 Nm), and the B bolts to 94 ft. lbs. (127 Nm).

37. For standard package:

    a. Install the 4 bolts labeled B and tighten to 94 ft. lbs. (127 Nm).

38. For vehicles with air suspension, connect the left and right pneumatic cylinder with rear shock absorber assemblies.

39. For vehicles without air suspension, connect the left and right rear shock absorber assemblies.

40. Install the left and right rear disc brake caliper assemblies.

41. Install the propeller with center bearing shaft assembly.

42. Inspect and adjust No. 2 and No. 3 joint angle.

43. Install the propeller shaft heat insulator.

44. Install the No. 1 front floor heat insulator.

45. Install the left and right tailpipes.

46. Install the front exhaust pipe assembly.

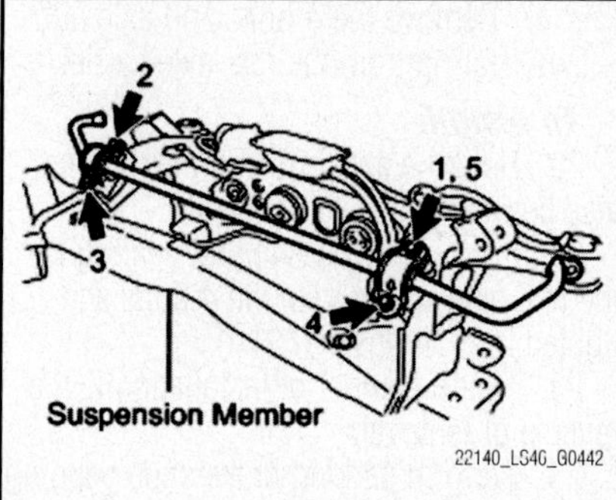

Fig. 179 Rear stabilizer bar bolt tightening sequence

47. Install the front floor center brace.
48. Install the heated oxygen sensor (for Sensor 2).
49. Inspect for gas leak.
50. Install the parking brake cable.
51. Connect the No. 1 actuator harness clamp.
52. Install the left and right rear speed sensors.
53. Install the left and right rear wheel house liners.
54. Install the left and right rear floor side member cover.
55. Install the No. 6 and No. 5 rocker panel molding protector.
56. Install the No. 2 and No. 1 differential support protector.
57. Install the No. 1 floor under cover.
58. Install the left side luggage compartment trim cover assembly.
59. Install the rear floor finish plate.
60. Install the left deck trim side board.
61. Install the luggage compartment mat sub-assembly.
62. Install the rear wheel.
63. Connect cable to negative battery terminal.
64. Install the right cowl top ventilator louver.
65. Perform initialization.
66. Check the suspension control system.
67. Inspect and adjust the rear wheel alignment.
68. Check the speed sensor signal.
69. Adjust the headlight.
70. Adjust the object recognition camera.

WHEEL HUB AND BEARING

REMOVAL & INSTALLATION

See Figure 181.

1. Remove the rear wheel.
2. Remove the bolt and load sensing valve sensor bracket from the toe control link sub-assembly.
3. Remove the bolt and nut, and then disconnect the rear stabilizer link assembly from the rear No. 2 suspension arm assembly.

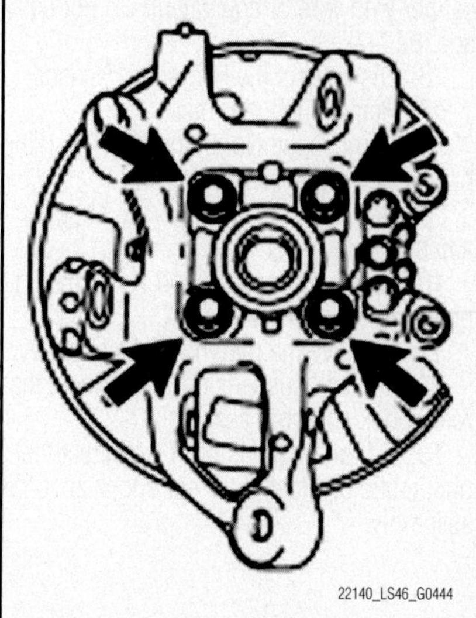

Fig. 181 Axle hub and bearing bolts

4. Using SST (SST: 09930-00010) and a hammer, release the staked part of the rear axle shaft nut, and while applying the brakes, remove the rear axle shaft nut.

**** WARNING**

Release the staked part of the nut completely, otherwise the screw of the drive shaft may be damaged.

5. Remove the 2 bolts and separate the rear disc brake caliper assembly. Hang the

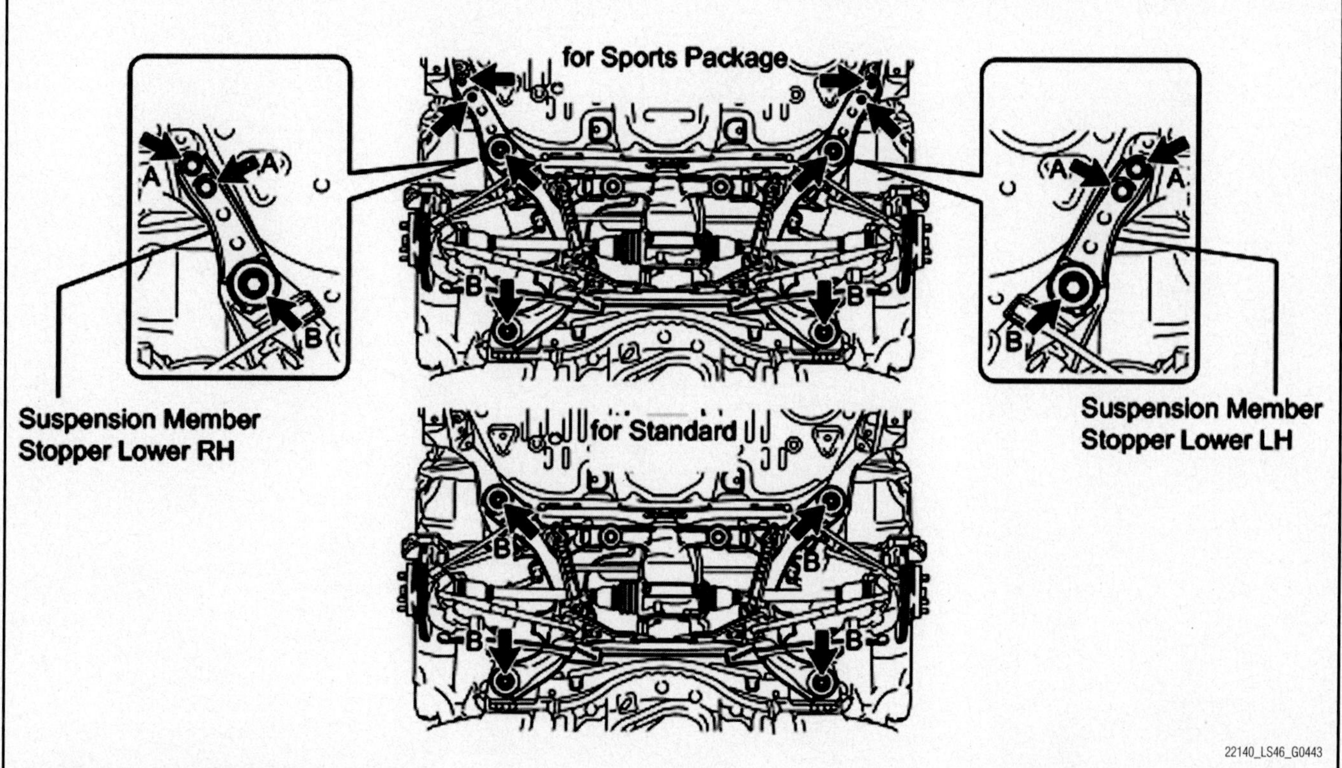

Fig. 180 Rear suspension member installation

caliper with wire or equivalent. Do not damage the brake hose.

6. Disconnect the rear speed sensor.
7. Remove the rear disc.
8. Remove the No. 2 and No. 1 Parking brake shoe return tension spring.
9. Remove the parking brake shoe adjusting screw set.
10. Remove the No. 1 and No. 2 parking brake shoe assemblies.
11. Remove the parking brake shoe lever.
12. For coil suspension, disconnect the rear shock absorber assembly.
13. For air suspension, disconnect the pneumatic cylinder with rear shock absorber assembly.
14. Remove the rear No. 2 suspension arm assembly.
15. Disconnect the rear No. 1 suspension arm assembly.
16. Disconnect the toe control link subassembly.
17. Disconnect the rear upper No. 1 and No. 2 control arm assembly.
18. Remove the rear axle assembly.
19. Remove the rear No. 1 wheel bearing dust deflector
    a. Remove the rear axle hub and bearing assembly, as follows:
    b. Hold the axle hub and bearing in a vise between aluminum plates. Do not overtighten the vise.
    c. Remove the 4 bolts and axle hub and bearing from the rear axle carrier.

To install:

20. Hold the axle hub and bearing in a vise between aluminum plates.
21. Install the axle hub and bearing to the rear axle carrier with the 4 bolts and tighten to 72 ft. lbs. (97 Nm).
22. The remainder of installation is the reverse of removal.
23. Perform parking brake shoe bedding.
24. Inspect and adjust rear wheel alignment.
25. Check the speed sensor signal.
26. Adjust headlight aiming.
27. Adjust the object recognition camera.

LEXUS

GS450h

4

SPECIFICATIONS AND MAINTENANCE CHARTS

ENGINE AND VEHICLE IDENTIFICATION

Engine							Model Year	
Code ①	Liters (cc)	Cu. In.	Cyl.	Fuel Sys.	Engine Type	Eng. Mfg.	Code ②	Year
2GR-FSE	3.5 (3456)	210.8	6	SFI	DOHC	Lexus	7	2007
2GR-FSE	3.5 (3456)	210.8	6	SFI	DOHC	Lexus	8	2008

SFI: Sequential Fuel Injection

DOHC: Double Overhead Camshaft

① Stamped on the left side of the engine block

② 10th digit of the Vehicle Identification Number (VIN)

22140_LEX2_C0001

GENERAL ENGINE SPECIFICATIONS

Year	Model	Engine Displacement Liters	Engine Series ID	Net Horsepower @ rpm	Net Torque @ rpm (ft. lbs.)	Bore x Stroke (in.)	Com-pression Ratio	Oil Pressure @ rpm
2007	GS450h	3.5	2GR-FSE	292@6400	267@4800	3.70x3.27	11.8:1	36.4@2500
2008	GS450h	3.5	2GR-FSE	292@6400	267@4800	3.70x3.27	11.8:1	36.4@2500

22140_LEX2_C0002

ENGINE TUNE-UP SPECIFICATIONS

Year	Engine Displacement Liters	Engine ID	Spark Plug Gap (in.)	Ignition Timing (deg.)	Fuel Pump (psi)	Idle Speed (rpm)	Valve Clearance Intake	Valve Clearance Exhaust
2007	3.5	2GR-FSE	0.039-0.043	8-12B①	28-85②	950-1050	0.0010-0.0024	0.0012-0.0026
2008	3.5	2GR-FSE	0.039-0.043	8-12B①	28-85②	950-1050	0.0010-0.0024	0.0012-0.0026

NOTE: The Vehicle Emission Control Information label often reflects specification changes made during production. The label figures must be used if they differ from those in this chart.

B: Before top dead center

① With terminals TC and CG of DLC3 connected

② For high pressure pump the readings should be 508-653.

22140_LEX2_C0003

CAPACITIES

Year	Model	Engine Displacement Liters	Engine ID	Engine Oil with Filter (qts.)	Automatic Transmission (qts)	Rear Drive Axle (pts.) ①	Fuel Tank (gal.)	Cooling System (qts.)
2007	GS450h	3.5	2GR-FSE	6.7	5.7	2.7-3.0	17.2	9.8 ②
2008	GS450h	3.5	2GR-FSE	6.7	5.7	2.7-3.0	17.2	9.8 ②

① Synthetic GL-5 (75W-90) or equivalent

②The use of genuine Toyota engine coolant is recommended or similar
ethylene glycol based non-silicate, non-amine, non- nitrite, and non- borate coolant

22140_LEX2_C0005

FLUID SPECIFICATIONS

Year	Model	Engine Displacement Liters (cc)	Engine ID/VIN	Engine Oil	Auto. Trans. ①	Drive Axle ②	Power Steering Fluid	Brake Master Cylinder
2007	GS450h	3.5	2GR-FSE	5W-30	ATF-WS	75W-90	N.A	DOT 3
2008	GS450h	3.5	2GR-FSE	5W-30	ATF-WS	75W-90	N.A	DOT 3

N/A: Not Applicable

DOT: Department Of Transpotation

① The use of genuine Toyota ATF-WS is recommended

② Synthetic GL-5 (75W-90) or equivalent

③The use of genuine Toyota engine coolant is recommended or similar
ethylene glycol based non-silicate, non-amine, non- nitrite, and non- borate coolant

22140_LEX2_C0004

VALVE SPECIFICATIONS

Year	Engine Displacement Liters	Engine ID	Seat Angle (deg.)	Face Angle (deg.)	Inner Spring free length (in.) ①	Spring Installed Height (in.)	Stem-to-Guide Clearance (in.)		Stem Diameter (in.)	
							Intake	Exhaust	Intake	Exhaust
2007	3.5	2GR-FSE	45	40.5	2.0354	N/A	0.0010-0.0024	0.0012-0.0026	0.2154-0.2159	0.2152 0.2157
2008	3.5	2GR-FSE	45	40.5	2.0354	N/A	0.0010-0.0024	0.0012-0.0026	0.2154-0.2159	0.2152 0.2157

N/A: Not Available

① If the free length is not as specified , replace the inner spring

22140_LEX2_C0006

CAMSHAFT AND BEARING SPECIFICATIONS CHART

All measurements are given in inches.

Year	Engine Displ. Liters	Engine ID/VIN	Journal Dia.	Brg. Oil Clearance	Shaft End-play	Runout	Journal Bore	Lobe Height	
								Intake	Exhaust
2007	3.5	2GR-FSE	①	②	0.0031-0.0051	0.0016	N/A	1.7448-1.7487	1.7457-1.7496
2008	3.5	2GR-FSE	①	②	0.0031-0.0051	0.0016	N/A	1.7448-1.7487	1.7457-1.7496

① Journal 1: 1.4152-1.4157
All Others: 1.0220-1.0226

② Oil clearance 1: 0.0016-0.0031
All Others: 0.0010-0.0024

N/A: Not Available

22140_LEX2_C0008

CRANKSHAFT AND CONNECTING ROD SPECIFICATIONS

All measurements are given in inches.

Year	Engine Displacement Liters	Engine ID	Crankshaft				Connecting Rod		
			Main Brg. Journal Dia.	Main Brg. Oil Clearance	Shaft End-play	Thrust on No.	Journal Diameter	Oil Clearance	Side Clearance
2007	3.5	2GR-FSE	2.4011-2.4016	①	0.0016-0.0094	2	2.0863-2.0866	0.0015-0.0026	0.0059-0.0118
2008	3.5	2GR-FSE	2.4011-2.4016	①	0.0016-0.0094	2	2.0863-2.0866	0.0015-0.0026	0.0059-0.0118

① Journals 1 and 4: 0.0006 - 0.0013 in.
Journals 2 and 3: 0.0010 - 0.0018 in.

22140_LEX2_C0007

PISTON AND RING SPECIFICATIONS

All measurements are given in inches.

Year	Engine Displ. Liters	Engine ID	Piston Clearance	Ring Gap			Ring Side Clearance		
				Top Comp.	Bottom Comp.	Oil Control	Top Comp.	Bottom Comp.	Oil Control
2007	3.5	2GR-FSE	0.0017-0.0020	0.0091-0.0130	0.01138-0.0177	0.0039-0.0157	0.0008-0.0028	0.0008-0.0024	0.0008-0.0028
2008	3.5	2GR-FSE	0.0017-0.0020	0.0091-0.0130	0.01138-0.0177	0.0039-0.0157	0.0008-0.0028	0.0008-0.0024	0.0008-0.0028

22140_LEX2_C0009

TORQUE SPECIFICATIONS

All readings in ft. lbs.

Year	Engine Displacement Liters	Engine ID	Cylinder Head Bolts	Main Bearing Bolts	Rod Bearing Bolts	Crankshaft Damper Bolts	Flywheel Bolts	Manifold		Spark Plugs	Oil Pan Drain Plug
								Intake	Exhaust		
2007	3.5	2GR-FSE	①	②	③	184	61	15	15	13	30
2008	3.5	2GR-FSE	①	②	③	184	61	15	15	13	30

① Step 1: 12 point bolts to 27 ft. lbs.
Step 2: 12 point bolts plus 90 degrees
Step 3: 12 point bolts plus 90 degrees
Step 4: Tighten the two remaining bolts to 22 ft. lbs.

② Step 1: 12 point cap bolts to 45 ft. lbs.
Step 2: 12 point cap bolts plus 90 degrees
Step 3: Hex head side bolts to 19 ft. lbs.

③ Step 1: 30 ft. lbs.
Step 2: Plus 90 degrees

22140_LEX2_C0010

WHEEL ALIGNMENT

Year	Model		Caster Range (+/-Deg.)	Caster Preferred Setting (Deg.)	Camber Range (+/-Deg.)	Camber Preferred Setting (Deg.)	Toe-in (in.)
2006	GS450h	F	0.75	+7.43	0.75	-0.43	0.04+/-0.08
		R	N/A	N/A	0.75	-1.32	0.12+/-0.08
2007	GS450h	F	0.75	+7.43	0.75	-0.43	0.04+/-0.08
		R	N/A	N/A	0.75	-1.32	0.12+/-0.08

N/A: Not Applicable
F: Front
R: Rear

22140_LEX2_C0011

TIRE, WHEEL AND BALL JOINT SPECIFICATIONS

Year	Model	OEM Tires		Tire Pressures (psi)		Wheel	Ball Joint	Lug Nut Torque
		Standard	Optional	Front	Rear	Size	Inspection	(ft. lbs.)
2007	GS450h	P245/40R18	245/40ZR18	35	35	8-JJ	①	76
2008	GS450h	P245/40R18	245/40ZR18	35	35	8-JJ	①	76

OEM: Original Equipment Manufacturer

PSI: Pounds Per Square Inch

STD: Standard

OPT: Optional

① Replace if any measurable movement is found.

22140_LEX2_C0012

BRAKE SPECIFICATIONS

All measurements in inches unless noted

Year	Model		Brake Disc Original Thickness	Brake Disc Minimum Thickness	Brake Disc Maximum Runout	Minimum Lining Thickness	Brake Caliper Bracket Bolts (ft. lbs.)	Brake Caliper Mounting Bolts (ft. lbs.)
2007	GS450h	F	1.181	1.063	0.0020	0.039	N/A	57
		R	0.709	0.650	0.0020	0.039	40	18
2008	GS450h	F	1.181	1.063	0.0020	0.039	N/A	57
		R	0.709	0.650	0.0020	0.039	40	18

N/A: Not Applicable:

F: Front

R: Rear

22140_LEX2_C0013

SCHEDULED MAINTENANCE INTERVALS

2007-08 Lexus Hybrid GS450h

TO BE SERVICED	TYPE OF SERVICE	VEHICLE MILEAGE INTERVAL (x1000)												
		5	10	15	20	25	30	35	40	45	50	55	60	65
Engine oil & filter	R	✓	✓	✓	✓	✓	✓	✓	✓	✓	✓	✓	✓	✓
Automatic transmission fluid	R												✓	
Ball joints & dust covers	S/I	✓	✓	✓	✓	✓	✓	✓	✓	✓	✓	✓	✓	✓
Bolts & nuts on chassis & body	S/I	✓	✓		✓	✓		✓	✓		✓	✓		✓
Brake line pipes & hoses	S/I			✓			✓			✓			✓	
Brake fluid	R						✓						✓	
Brake pads & discs (front & rear)	S/I	✓	✓	✓	✓	✓	✓	✓	✓	✓	✓	✓	✓	✓
Drive belts	R												✓	
Propeller shaft grease	S/I	✓	✓	✓	✓	✓	✓	✓	✓	✓	✓	✓	✓	✓
Steering knuckle & chassis grease	S/I	✓	✓	✓	✓	✓	✓	✓	✓	✓	✓	✓	✓	✓
Steering linkage	S/I	✓	✓	✓	✓	✓	✓	✓	✓	✓	✓	✓	✓	✓
Air cleaner filter	R						✓						✓	
Air conditioner filter	R						✓							
Spark plugs	R	Replace at 120,000 miles												
Exhaust pipes & mountings	S/I			✓			✓			✓			✓	
Fuel lines & connections	S/I												✓	
Engine coolant	R	Replace at 120,000 miles												
Rear differential fluid	R												✓	
Rotate Tires	S/I	✓	✓	✓	✓	✓	✓	✓	✓	✓	✓	✓	✓	✓

R: Replace S/I: Service or Inspect

FREQUENT OPERATION MAINTENANCE (SEVERE SERVICE)

If a vehicle is operated under any of the following conditions it is considered severe service:

- Extremely dusty areas.

- 50% or more of the constant operation is in 32°C (90°F) or higher temperatures, or in temperatures below 0°C (32°F).

- Prolonged idling (vehicle operation in stop and go traffic).

- Frequent short running periods (engine does not warm to normal operating temperatures).

- Police, taxi, delivery usage or trailer towing usage.

Air cleaner filter: service or inspect every 3750 miles

Engine oil & filter: replace every 3750 miles.

Ball joints & dust covers: service or inspect every 7500 miles.

Bolts & nuts on chassis & body: service or inspect every 7500 miles.

Brake pads & discs (front & rear): service or inspect every 7500 miles.

Steering knuckle & chassis grease: service or inspect every 7500 miles.

Steering linkage: service or inspect every 7500 miles.

Exhaust pipes & mountings: service or inspect every 15,000 miles.

22140_LEX2_C0014

PRECAUTIONS

Before servicing any vehicle, please be sure to read all of the following precautions, which deal with personal safety, prevention of component damage, and important points to take into consideration when servicing a motor vehicle:

- Never open, service or drain the radiator or cooling system when the engine is hot; serious burns can occur from the steam and hot coolant.
- Observe all applicable safety precautions when working around fuel. Whenever servicing the fuel system, always work in a well-ventilated area. Do not allow fuel spray or vapors to come in contact with a spark, open flame, or excessive heat (a hot drop light, for example). Keep a dry chemical fire extinguisher near the work area. Always keep fuel in a container specifically designed for fuel storage; also, always properly seal fuel containers to avoid the possibility of fire or explosion. Refer to the additional fuel system precautions later in this section.
- Fuel injection systems often remain pressurized, even after the engine has been turned **OFF**. The fuel system pressure must be relieved before disconnecting any fuel lines. Failure to do so may result in fire and/or personal injury.
- Brake fluid often contains polyglycol ethers and polyglycols. Avoid contact with the eyes and wash your hands thoroughly after handling brake fluid. If you do get brake fluid in your eyes, flush your eyes with clean, running water for 15 minutes. If eye irritation persists, or if you have taken brake fluid internally, IMMEDIATELY seek medical assistance.
- The EPA warns that prolonged contact with used engine oil may cause a number of skin disorders, including cancer. You should make every effort to minimize your exposure to used engine oil. Protective gloves should be worn when changing oil. Wash your hands and any other exposed skin areas as soon as possible after exposure to used engine oil. Soap and water, or waterless hand cleaner should be used.
- All new vehicles are now equipped with an air bag system, often referred to as a Supplemental Restraint System (SRS) or Supplemental Inflatable Restraint (SIR) system. The system must be disabled before performing service on or around system components, steering column, instrument panel components, wiring and sensors. Failure to follow safety and disabling procedures could result in accidental air bag deployment, possible personal injury and unnecessary system repairs.
- Always wear safety goggles when working with, or around, the air bag system. When carrying a non-deployed air bag, be sure the bag and trim cover are pointed away from your body. When placing a non-deployed air bag on a work surface, always face the bag and trim cover upward, away from the surface. This will reduce the motion of the module if it is accidentally deployed. Refer to the additional air bag system precautions later in this section.
- Clean, high quality brake fluid from a sealed container is essential to the safe and proper operation of the brake system. You should always buy the correct type of brake fluid for your vehicle. If the brake fluid becomes contaminated, completely flush the system with new fluid. Never reuse any brake fluid. Any brake fluid that is removed from the system should be discarded. Also, do not allow any brake fluid to come in contact with a painted surface; it will damage the paint.
- Never operate the engine without the proper amount and type of engine oil; doing so WILL result in severe engine damage.
- Timing belt maintenance is extremely important. Many models utilize an interference-type, non-freewheeling engine. If the timing belt breaks, the valves in the cylinder head may strike the pistons, causing potentially serious (also time-consuming and expensive) engine damage. Refer to the maintenance interval charts for the recommended replacement interval for the timing belt, and to the timing belt section for belt replacement and inspection.
- Disconnecting the negative battery cable on some vehicles may interfere with the functions of the on-board computer system(s) and may require the computer to undergo a relearning process once the negative battery cable is reconnected.
- When servicing drum brakes, only disassemble and assemble one side at a time, leaving the remaining side intact for reference.
- Only an MVAC-trained, EPA-certified automotive technician should service the air conditioning system or its components.

BRAKES

ANTI-LOCK BRAKE SYSTEM (ABS)

GENERAL INFORMATION

PRECAUTIONS

- Certain components within the ABS system are not intended to be serviced or repaired individually.
- Do not use rubber hoses or other parts not specifically specified for and ABS system. When using repair kits, replace all parts included in the kit. Partial or incorrect repair may lead to functional problems and require the replacement of components.
- Lubricate rubber parts with clean, fresh brake fluid to ease assembly. Do not use shop air to clean parts; damage to rubber components may result.
- Use only DOT 3 brake fluid from an unopened container.
- If any hydraulic component or line is removed or replaced, it may be necessary to bleed the entire system.
- A clean repair area is essential. Always clean the reservoir and cap thoroughly before removing the cap. The slightest amount of dirt in the fluid may plug an orifice and impair the system function. Perform repairs after components have been thoroughly cleaned; use only denatured alcohol to clean components. Do not allow ABS components to come into contact with any substance containing mineral oil; this includes used shop rags.
- The Anti-Lock control unit is a microprocessor similar to other computer units in the vehicle. Ensure that the ignition switch is **OFF** before removing or installing controller harnesses. Avoid static electricity discharge at or near the controller.
- If any arc welding is to be done on the vehicle, the control unit should be unplugged before welding operations begin.

BLEEDING PROCEDURE

BLEEDING PROCEDURE

See Figure 1.

1. Before servicing the vehicle, refer to the precautions in the beginning of this section.

If any work is done on the brake system or if air in the brake lines is suspected, bleed the system of air.

➡Do not let brake fluid remain on painted surfaces. Wash it off immediately.

2. Fill the reservoir with brake fluid.

➡If the master cylinder has been disassembled or if the reservoir becomes empty, bleed the air from the master cylinder.

3. Bleed the brake master cylinder as follows:
   - Disconnect the brake lines from the master cylinder.
   - Slowly depress the brake pedal and hold it.
   - Block off the outer holes with your fingers, and release the brake pedal.
   - Repeat the previous 2 steps 3 or 4 times.
4. Bleed the brake line as follows:
   - Connect the vinyl tube to the brake caliper.
   - Depress the brake pedal several times, then loosen the bleeder plug with the pedal held down.
   - At the point when fluid stops coming out, tighten the bleeder plug, then release the brake pedal.
   - Repeat the previous 2 steps until all the air in the fluid has been bled out.
   - Repeat the above procedure to bleed the air out of the brake line for each wheel.

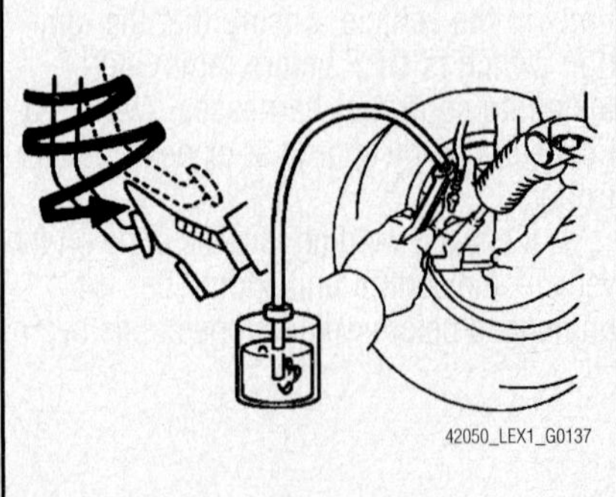

Fig. 1 Bleeding the brake line

   - Tighten the bleeder plug to 8 ft. lbs. (11 Nm).
5. Bleed the brake actuator as follows:
   - Remove the reservoir cap.
   - Install the SST 09992-00242, 09992-00350 to the reservoir.
   - Connect the vinyl tube to the bleeder plug of the brake actuator.
   - Using SST, apply the 14.2 psi (98.1kpa) of pressure to the reservoir.
   - Loosen the bleeder plug.
   - Bleed the air out of the brake actuator, tighten the bleeder plug to 74 inch lbs. (8.3 Nm).
6. Check the fluid level and add fluid if necessary.

WARNING

Clean, high quality brake fluid is essential to the safe and proper operation of the brake system. You should always buy the highest quality brake fluid that is available. If the brake fluid becomes contaminated, drain and flush the system, then refill the master cylinder with new fluid. Never reuse any brake fluid. Any brake fluid that is removed from the system should be discarded. Also, do not allow any brake fluid to come in contact with a painted surface; it will damage the paint.

BLEEDING THE ABS SYSTEM

➡After performing the usual air bleeding in the brake system, if the height or feel of the brake pedal cannot be obtained, perform air bleeding in the brake actuator assembly with a hand held tester by following procedures below. Make sure that the brake fluid in the master cylinder reservoir tank does not become empty.

1. Before servicing the vehicle, refer to the precautions in the beginning of this section.
2. Depress the brake pedal more than 20 times with the engine off.
3. Connect the hand held tester to the DLC3, then turn the ignition switch to the ON position.
4. Do not start the engine.
5. Select "AIR BLEEDING" on the hand held tester. Please refer to the Hand Held Tester Operator's Manual for further details.
6. Bleed the air out of the regular brake line in "Step1: Increase" on the hand held tester display. Perform the air bleeding by following the steps displayed on the hand held tester. Make sure that the brake fluid in the master cylinder reservoir tank does not become empty.
7. Connect the vinyl tube to either one of the bleeder plugs.
8. Depress the brake pedal several times, then loosen the bleeder plug of one of the above wheels with the pedal depressed.
9. When fluid stops coming out, tighten the bleeder plug, then release the brake pedal.
10. Repeat (2) and (3) until all air in the fluid is completely bled out.
11. Tighten the bleeder plug to 73 inch lbs. (8.3 Nm).
12. Repeat the above procedure to bleed the air out of the brake line for each wheel.
13. Bleed the air out of the suction line in "Step2: Inhalation" on the hand held tester display.
14. Connect the vinyl tube to the bleeder plug at the right front wheel or the right rear wheel and loosen the bleeder plug.
15. Operate the brake actuator assembly using the hand held tester to bleed the air.

➡The operation stops automatically in 4 seconds. At this time, be sure to release the brake pedal.

16. Check that the operation has stopped, by referring to the hand held tester display.
17. Repeat (2) and (3) until all the air in the fluid is completely bled out.
18. Tighten the bleeder plug.
19. For the rest of the wheels, bleed the air in the same way as stated in the above procedure.
20. Bleed the air out of the pressure reduction line in "Step3: Decrease" on the hand held tester display.
21. Connect a vinyl tube to either one of the bleeder plugs.
22. Loosen the bleeder plug.
23. Using the hand held tester, operate the brake actuator assembly using hand held tester, completely depress the brake pedal and keep it.

➡The operation stops automatically in 4 seconds. When performing this procedure continuously, an interval of at least 20 seconds is required. When the operation is completed, the brake pedal slightly goes down. This is a normal phenomenon caused when the solenoid opens. During this procedure, the pedal seems heavy, but completely depress it so that the brake fluid comes

out from the bleeder plug. Be sure to keep depressing the brake pedal. Never depress and release the pedal repeatedly.

24. Tighten the bleeder plug, then release the brake pedal.
25. Repeat 3 previous steps until all the air in the fluid is completely bled out.
26. Tighten the bleeder plug.
27. Repeat the above procedure to bleed the air out of the brake line for each wheel.
28. Bleed the air out of the regular brake line again in "Step4: Increase" on the hand-held tester display.
29. Connect the vinyl tube to either one of the bleeder plug.
30. Depress the brake pedal several times, then loosen the bleeder plug of one of the above wheels with the pedal depressed.
31. When fluid stops coming out, tighten the bleeder plug, then release the brake pedal.
32. Repeat the previous 2 steps until all the air in the fluid is completely bled out.
33. Tighten the bleeder plug.
34. Repeat the above procedure to bleed the air out of the brake line for each wheel.

BRAKES

FRONT DISC BRAKES

CAUTION

Dust and dirt accumulating on brake parts during normal use may contain asbestos fibers from production or aftermarket brake linings. Breathing excessive concentrations of asbestos fibers can cause serious bodily harm. Exercise care when servicing brake parts. Do not sand or grind brake lining unless equipment used is designed to contain the dust residue. Do not clean brake parts with compressed air or by dry brushing. Cleaning should be done by dampening the brake components with a fine mist of water, then wiping the brake components clean with a dampened cloth. Dispose of cloth and all residue containing asbestos fibers in an impermeable container with the appropriate label. Follow practices prescribed by the Occupational Safety and Health Administration (OSHA) and the Environmental Protection Agency (EPA) for the handling, processing, and disposing of dust or debris that may contain asbestos fibers.

BRAKE CALIPER

REMOVAL & INSTALLATION

See Figures 2 and 3.

1. Before servicing the vehicle, refer to the precautions in the beginning of this section.

➡When the brake caliper is removed from the brake disc, do not start the hybrid system. If the hybrid system is started, the brake fluid pressure may increase.

2. Remove the front wheel.
3. Drain the brake fluid.
4. Remove the pin hold clip.
5. Remove the front disc brake anti-rattle with hole pin while pushing on the anti-rattle spring.
6. Remove the front disc brake anti-rattle spring.
7. Remove the 2 pads and 2 anti-squeal shims from each pad.
8. Remove the union bolt and gasket from the disc brake caliper, then disconnect the flexible hose from the disc brake caliper.
9. Remove the 2 bolts and disc brake caliper from the knuckle.

To install:

10. Install the disc brake caliper with the 2 bolts and tighten to 57 ft. lbs. (78 Nm).
11. Connect the flexible hose with the union bolt and a new gasket. Tighten to 29 ft. lbs. (39 Nm).

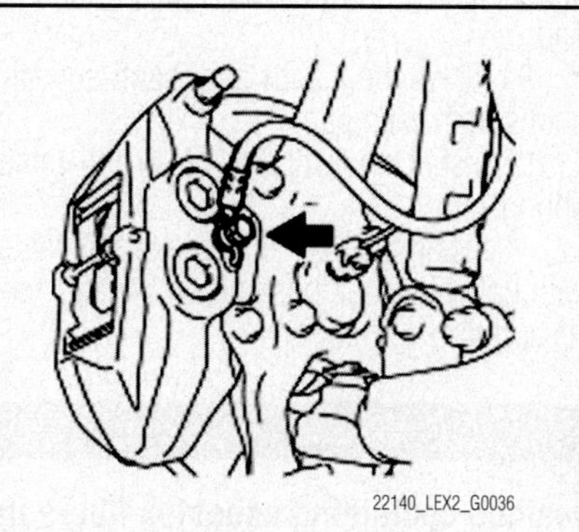

Fig. 2 Remove the flexible hose from the disc brake caliper

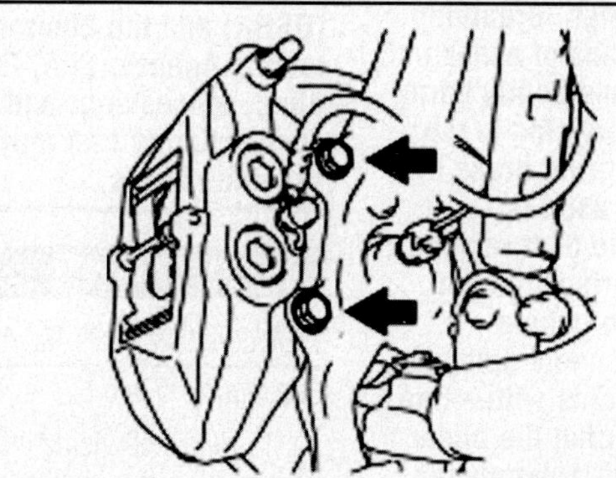

Fig. 3 Remove the 2 bolts from the disc brake caliper

12. Apply disc brake grease to the anti-squeal shims and install them to each pad.
13. Install the brake anti-rattle spring.
14. Install the hole pin while pushing on the anti-rattle spring.
15. Install the clip.
16. Disable the brake control system.
17. Bleed the air from the front brake system.
18. Add new brake fluid.
19. Check the system for leaks.
20. Install the front wheel and tighten to 76 ft. lbs. (103 Nm).
21. Check and clear any DTC.
22. Test drive the vehicle and burnish in the pads and rotor.

DISC BRAKE PADS

REMOVAL & INSTALLATION

See Figure 4.

1. Before servicing the vehicle, refer to the precautions in the beginning of this section.

➡When the brake caliper is removed from the brake disc, do not start the hybrid system. If the hybrid system is started, the brake fluid pressure may increase.

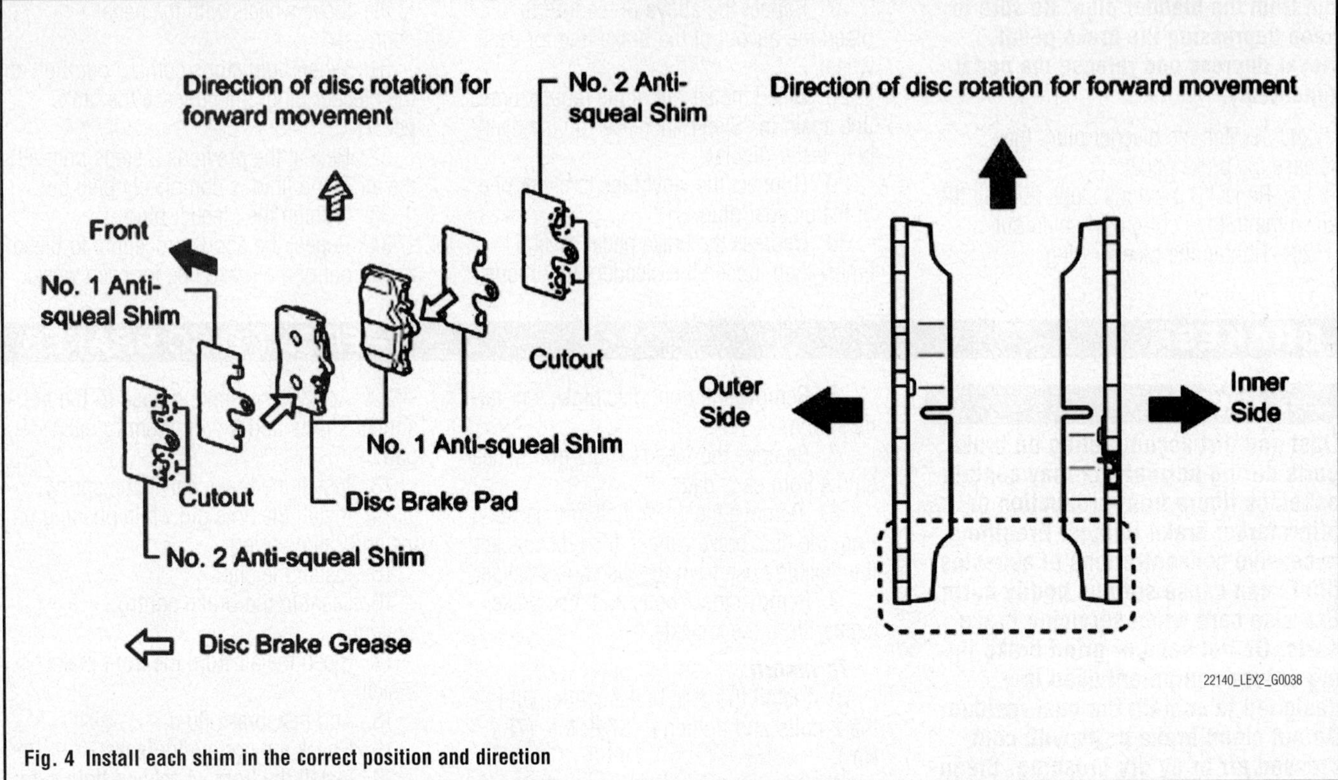

Fig. 4 Install each shim in the correct position and direction

2. Remove the front wheel.
3. Drain the brake fluid and push back brake caliper piston
4. Remove the pin hold clip.
5. Remove the front disc brake anti-rattle with hole pin while pushing on the anti-rattle spring.
6. Remove the front disc brake anti-rattle spring.
7. Remove the 2 pads and 2 anti-squeal shims from each pad.

To install:

8. Apply disc brake grease to the anti-squeal shims and install them to each pad.
9. Install the 2 pads and 2 anti-squeal shims from each pad.
10. Install the front disc brake anti-rattle spring.
11. Install the front disc brake anti-rattle with hole pin while pushing on the anti-rattle spring.
12. Install the pin hold clip.
13. Disable the brake control system.
14. Bleed the air from the front brake system.
15. Add new brake fluid.
16. Check the system for leaks.
17. Install the front wheel and tighten to 76 ft. lbs. (103 Nm).
18. Check and clear any DTC.
19. Test drive the vehicle and burnish in the pads and rotor.

⁂ CAUTION

Dust and dirt accumulating on brake parts during normal use may contain asbestos fibers from production or aftermarket brake linings. Breathing excessive concentrations of asbestos fibers can cause serious bodily harm. Exercise care when servicing brake parts. Do not sand or grind brake lining unless equipment used is designed to contain the dust residue. Do not clean brake parts with compressed air or by dry brushing. Cleaning should be done by dampening the brake components with a fine mist of water, then wiping the brake components clean with a dampened cloth. Dispose of cloth and all residue containing asbestos fibers in an impermeable container with the appropriate label. Follow practices prescribed by the Occupational Safety and Health Administration (OSHA) and the Environmental Protection Agency (EPA) for the handling, processing, and disposing of dust or debris that may contain asbestos fibers.

BRAKE CALIPER

REMOVAL & INSTALLATION

See Figures 5 and 6.

1. Before servicing the vehicle, refer to the precautions in the beginning of this section.
2. Remove or disconnect the following:
   - Wheels
   - Brake line at the caliper
   - Anti-squeal springs
   - Mounting bolts, while holding the sliding pin with a wrench
   - Caliper assembly

To install:

3. Install or connect the following:
   - Caliper. Tighten the mounting bolts to 40 ft. lbs. (54 Nm).
   - Anti-squeal springs
   - Connect the brake line with 2 new gaskets and tighten the union bolt to 22 ft. lbs. (30 Nm)

➡Install the flexible hose lock securely in the lock hole in the disc brake cylinder assembly rear.

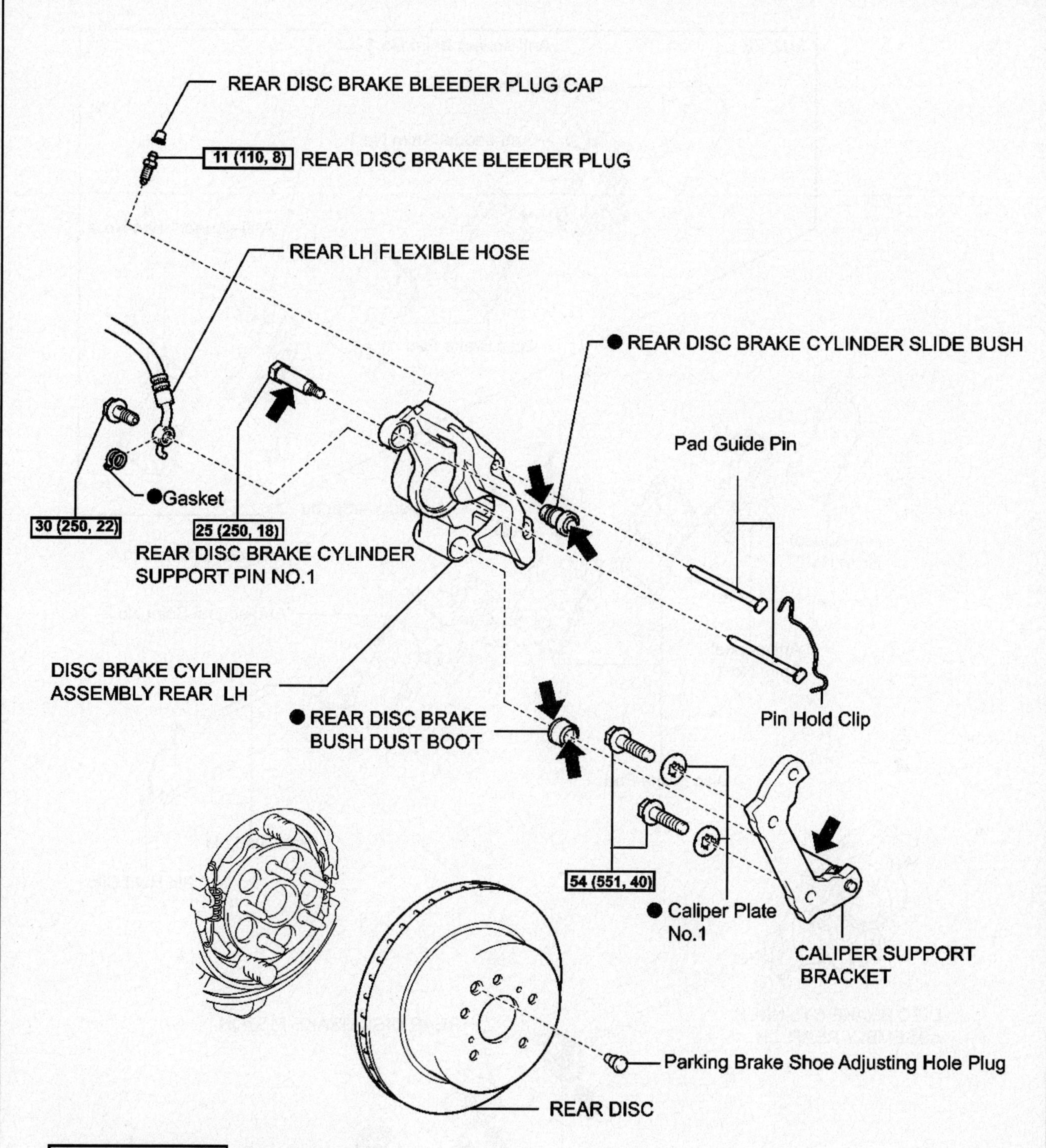

Fig. 5 Rear disc brake caliper components—GS 450h

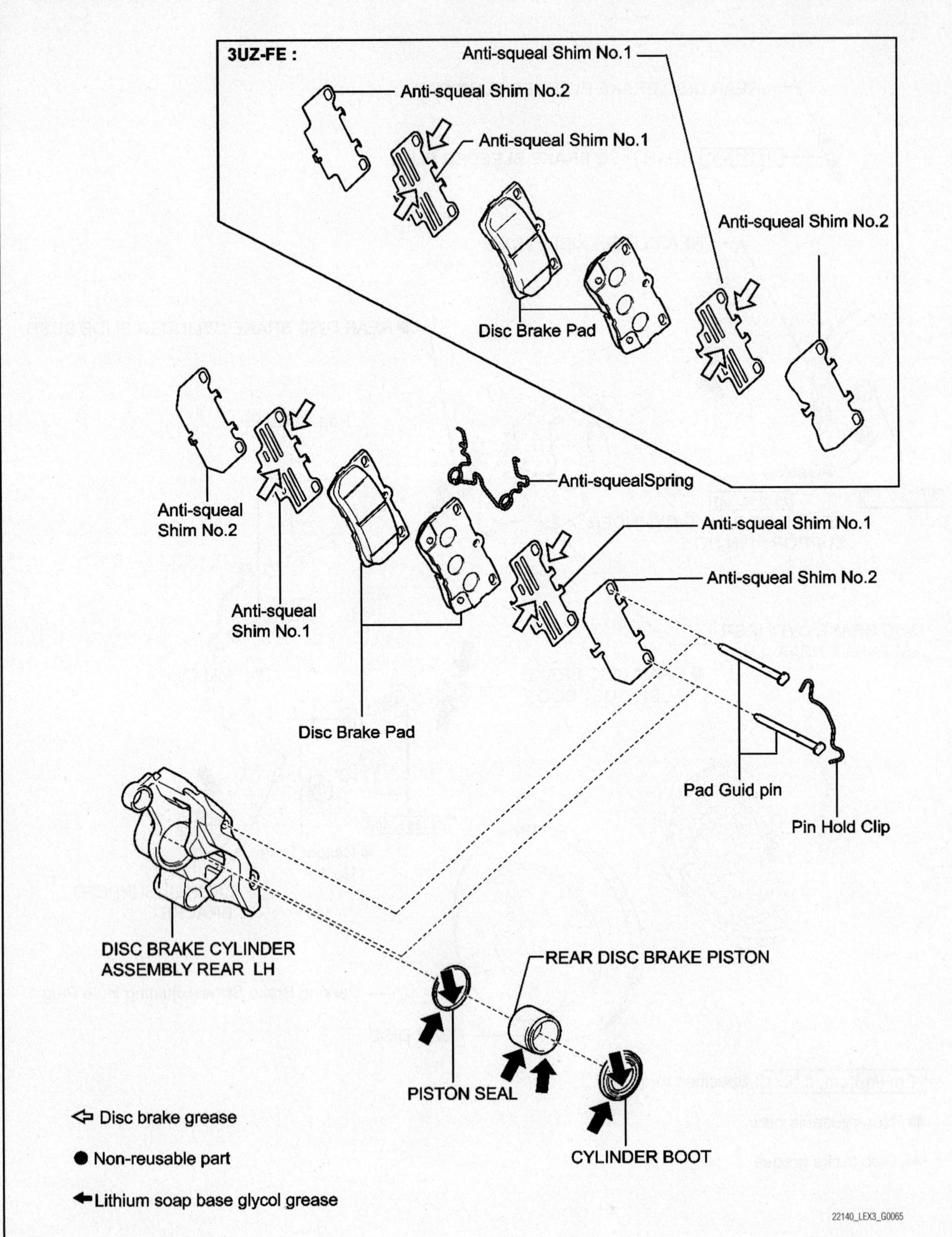

Fig. 6 Rear disc brake pads components—GS 450h

4. With ECB brake system disable the 2 ABS motor relays from No.3 block. before bleeding the system. Install scanner to verify system is off.

5. Bleed the brake system of any air present.

6. Install relays if previously removed.

7. Check and clear DTC with scanner.

8. Check for fluid leaks.

9. Install the rear wheels and tighten to 76 ft. lbs. (103 Nm).

10. Road test the vehicle and burnish in brakes.

DISC BRAKE PADS

REMOVAL & INSTALLATION

See Figure 7.

1. Before servicing the vehicle, refer to the precautions in the beginning of this section.

2. Rear wheels.

3. Remove the pin hold clip and disengage the engaged parts of anti-squeal spring.

4. Remove the 2 pad guide pins and anti-squeal spring.

5. Remove the 2 brake pads with the anti-squeal shims.

6. Remove the 2 anti-squeal shims from each pad.

7. Remove the union bolt and gasket from the rear disc brake cylinder, and then disconnect the flexible hose.

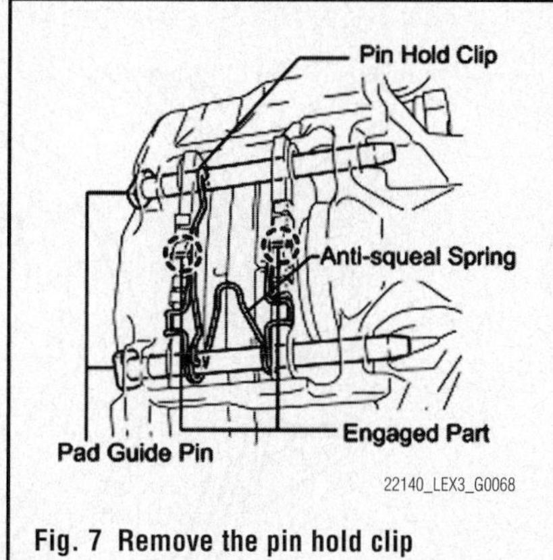

Fig. 7 Remove the pin hold clip

8. Remove the cylinder slide pin and then tilt the disc brake cylinder toward the rear of the vehicle.

9. Remove the 2 bolts and rear disc brake cylinder together with the caliper support bracket.

10. Remove the 2 No. 1 caliper plate from the caliper support bracket.

To install:

11. Install 2 new caliper plate No.1 to the caliper support bracket.

12. Install the cylinder assembly rear together with the caliper support bracket with the 2 bolts.

13. Tighten the support bolts to 40 ft. lbs. (54 Nm).

14. Install the cylinder assembly rear with the cylinder slide pin. Tighten to 18 ft. lbs. (25 Nm).

15. Apply disc brake grease to both sides of the 2 anti-squeal shims

16. Install the 4 anti-squeal shims to each of the 2 brake pads.

17. Apply disc brake grease to both sides of the 2 anti-squeal shim No.1.

18. Install the 4 anti-squeal shims to each of the 2 brake pads.

19. Install the 2 brake pads to the cylinder assembly rear.

20. Install the 2 pad guide pins, anti-squeal spring and pin hold clip as shown in the illustration.

21. Connect the flexible hose with the union bolt and a new gasket. Tighten bolt to 22 ft. lbs. (30 Nm).

➡Install the flexible hose lock securely in the lock hole in the disc brake cylinder assembly rear.

22. With ECB brake system disable the 2 ABS motor relays from No.3 block. before bleeding the system. Install scanner to verify system is off.

23. Bleed the brake system of any air present.

24. Install relays if previously removed.

25. Check and clear DTC with scanner.

26. Check for fluid leaks.

27. Install the rear wheels and tighten to 76 ft. lbs. (103 Nm).

28. Test drive the vehicle and burnish in the pads and rotor.

PARKING BRAKE CABLES

ADJUSTMENT

See Figure 8.

1. Depress the parking brake pedal. Hold the No. 1 wire adjusting nut using a wrench and loosen the lock nut.

2. Release the parking brake pedal.

3. Turn the No. 1 wire adjusting nut until the parking brake pedal travel meets the specification.

4. Standard parking brake pedal travel: 7 to 9 notches at 67.5 ft. lbs. (300 N).

5. Hold the No. 1 wire adjusting nut using a wrench or equivalent tool and tighten the lock nut to 53 inch. lbs. (6.0 Nm).

6. Count the number of clicks after depressing and releasing the parking brake pedal 3 or 4 times.

7. Check whether the parking brake drags or not.

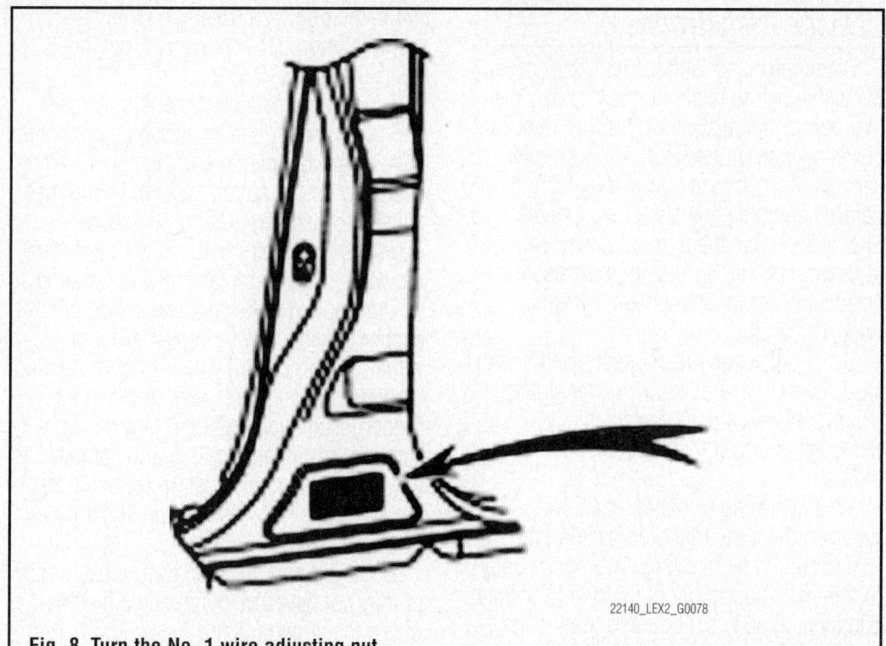

Fig. 8 Turn the No. 1 wire adjusting nut

8. When operating the parking brake pedal, check that the parking brake indicator light comes on.

PARKING BRAKE SHOES

ADJUSTMENT

See Figure 9.

1. Remove the rear wheel.
2. Temporarily install the hub nuts.
3. Remove the hole plug, and turn the adjuster and expand the shoes until the disc locks.
4. Contract the shoe adjuster until the disc can rotate smoothly. Standard: Return 7–8 notches
5. Check that shoe is no brake drag.
6. Install the hole plug.

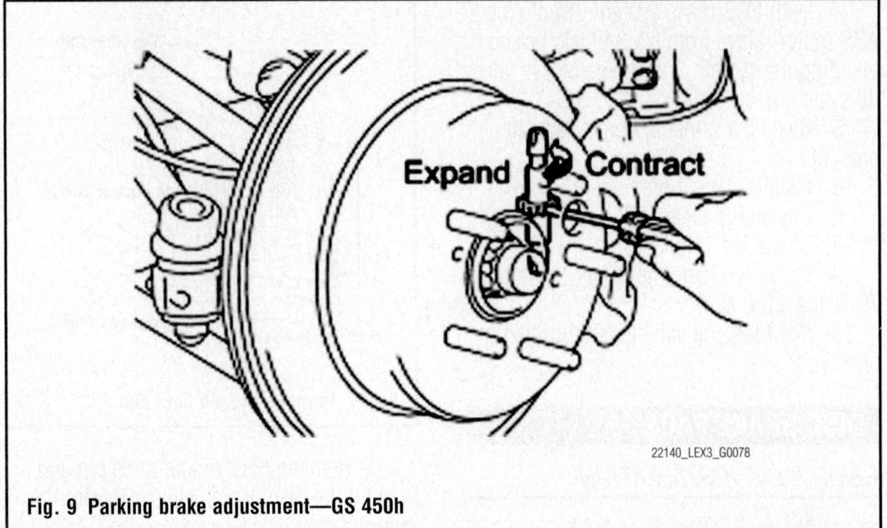

Fig. 9 Parking brake adjustment—GS 450h

CHASSIS ELECTRICAL

AIR BAG (SUPPLEMENTAL RESTRAINT SYSTEM)

GENERAL INFORMATION

CAUTION

These vehicles are equipped with an air bag system. The system must be disarmed before performing service on, or around, system components, the steering column, instrument panel components, wiring and sensors. Failure to follow the safety precautions and the disarming procedure could result in accidental air bag deployment, possible injury and unnecessary system repairs.

SERVICE PRECAUTIONS

Disconnect and isolate the battery negative cable before beginning any airbag system component diagnosis, testing, removal, or installation procedures. Allow system capacitor to discharge for two minutes before beginning any component service. This will disable the airbag system. Failure to disable the airbag system may result in accidental airbag deployment, personal injury, or death.

Do not place an intact undeployed airbag face down on a solid surface. The airbag will propel into the air if accidentally deployed and may result in personal injury or death.

When carrying or handling an undeployed airbag, the trim side (face) of the airbag should be pointing towards the body to minimize possibility of injury if accidental deployment occurs. Failure to do this may result in personal injury or death.

Replace airbag system components with OEM replacement parts. Substitute parts may appear interchangeable, but internal differences may result in inferior occupant protection. Failure to do so may result in occupant personal injury or death.

Wear safety glasses, rubber gloves, and long sleeved clothing when cleaning powder residue from vehicle after an airbag deployment. Powder residue emitted from a deployed airbag can cause skin irritation. Flush affected area with cool water if irritation is experienced. If nasal or throat irritation is experienced, exit the vehicle for fresh air until the irritation ceases. If irritation continues, see a physician.

Do not use a replacement airbag that is not in the original packaging. This may result in improper deployment, personal injury, or death.

The factory installed fasteners, screws and bolts used to fasten airbag components have a special coating and are specifically designed for the airbag system. Do not use substitute fasteners. Use only original equipment fasteners listed in the parts catalog when fastener replacement is required.

During, and following, any child restraint anchor service, due to impact event or vehicle repair, carefully inspect all mounting hardware, tether straps, and anchors for proper installation, operation, or damage. If a child restraint anchor is found damaged in any way, the anchor must be replaced. Failure to do this may result in personal injury or death.

Deployed and non-deployed airbags may or may not have live pyrotechnic material within the airbag inflator.

Do not dispose of driver/passenger/curtain airbags or seat belt tensioners unless you are sure of complete deployment. Refer to the Hazardous Substance Control System for proper disposal.

Dispose of deployed airbags and tensioners consistent with state, provincial, local, and federal regulations.

After any airbag component testing or service, do not connect the battery negative cable. Personal injury or death may result if the system test is not performed first.

If the vehicle is equipped with the Occupant Classification System (OCS), do not connect the battery negative cable before performing the OCS Verification Test using the scan tool and the appropriate diagnostic information. Personal injury or death may result if the system test is not performed properly.

Never replace both the Occupant Restraint Controller (ORC) and the Occupant Classification Module (OCM) at the same time. If both require replacement, replace one, then perform the Airbag System test before replacing the other.

Both the ORC and the OCM store Occupant Classification System (OCS) calibration data, which they transfer to one another when one of them is replaced. If both are replaced at the same time, an irreversible fault will be set in both modules and the OCS may malfunction and cause personal injury or death.

If equipped with OCS, the Seat Weight Sensor is a sensitive, calibrated unit and must be handled carefully. Do not drop or handle roughly. If dropped or damaged, replace with another sensor. Failure to do so may result in occupant injury or death.

If equipped with OCS, the front passenger seat must be handled carefully as well. When removing the seat, be careful when setting on floor not to drop. If dropped, the sensor may be inoperative, could result in occupant injury, or possibly death.

If equipped with OCS, when the passenger front seat is on the floor, no one should sit in the front passenger seat. This uneven force may damage the sensing ability of the seat weight sensors. If sat on and damaged, the sensor may be inoperative, could result in occupant injury, or possibly death.

DISARMING THE SYSTEM

To avoid personal injury when working on vehicles equipped with an air bag, the negative battery cable must be disconnected and at least 90 seconds must elapse before working on the system. Failure to do so may result in deployment of the air bag.

ARMING THE SYSTEM

To rearm the air bag system, simply reconnect the battery cable(s).

DRIVETRAIN

HYBRID PRECAUTIONS

CAUTION

Engineers must undergo special training to be able to perform high-voltage system inspection and servicing.

CAUTION

All high-voltage wire harnesses are colored orange. The HV battery and other high-voltage components have "High Voltage" caution labels. Do not carelessly touch these wires and components.

CAUTION

Before inspecting or servicing the high-voltage system, be sure to follow safety measures, such as wearing insulated gloves and removing the service plug to prevent electrocution. Carry the removed service plug in your pocket to prevent other technicians from reinstalling it while you are servicing the vehicle.

CAUTION

After removing the service plug, wait 10 minutes before touching any of the high-voltage connectors and terminals.

CAUTION

Be sure to install the service plug before starting the hybrid system. Starting the hybrid system with the service plug removed may damage the vehicle.

CAUTION

Before wearing insulated gloves, make sure that they are not cracked, ruptured, torn, or damaged in any way. Do not wear wet insulated gloves.

CAUTION

When servicing the vehicle, do not carry metal objects like mechanical pencils or scales that can be dropped accidentally and cause a short circuit.

CAUTION

Before touching a bare high-voltage terminal, wear insulated gloves and use an electrical tester to ensure that the terminal is not charged with electricity (approximately 0 V).

1. After disconnecting or exposing a high-voltage connector or terminal, insulate it immediately using insulation tape.

CAUTION

The screw of a high-voltage terminal should be tightened firmly to the specified torque. Both insufficient and excessive torque can cause failure.

CAUTION

Use a "CAUTION: HIGH VOLTAGE. DO NOT TOUCH DURING OPERATION." sign to notify other engineers that a high-voltage system is being inspected and/or repaired.

CAUTION

Do not place the battery upside down while removing and installing it.

CAUTION

After servicing the high-voltage system and before reinstalling the service plug, check again that you have not left a part or tool inside, that the high-voltage terminal screws are firmly tightened, and that the connectors are correctly connected.

2. The LEXUS GS450h automatically turns the engine ON and OFF when the READY light on the instrument panel is ON. To avoid injury, remove the key from the vehicle before inspecting or servicing the engine compartment.

HV AUTOMATIC TRANSMISSION ASSEMBLY

REMOVAL & INSTALLATION

See Figures 10 through 22.

1. Before servicing the vehicle, refer to the precautions in the beginning of this section.
2. Discharge the fuel system.
3. Discharge the A/C system refrigerant.
4. Turn the front wheels in the straight ahead position.
5. Remove the luggage compartment floor mat.

CAUTION

Remove the service plug grip to interrupt a high voltage circuit at the time of the check. Keep the removed service plug grip in your pocket to prevent other technicians from accidentally reconnecting it while you are servicing the vehicle.

CAUTION

All the high voltage wiring connectors are colored in orange.

6. Wear insulated gloves. Remove the service plug grip after sliding the lever of the service plug grip.

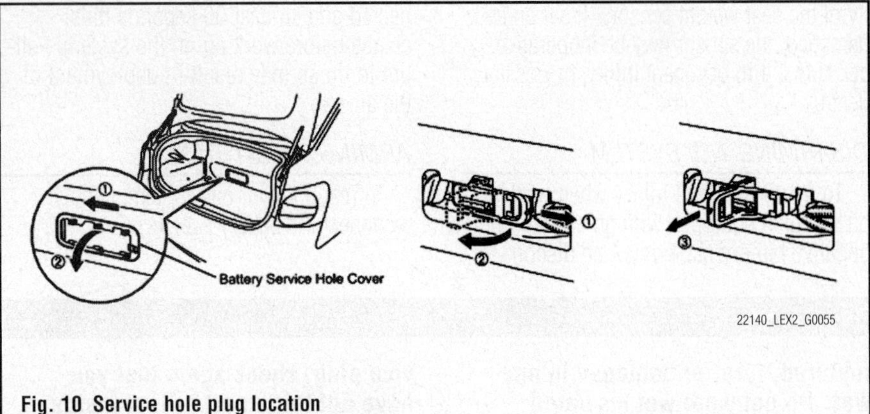

Fig. 10 Service hole plug location

⁂ WARNING

After removing the service plug grip, do not operate the power switch as it may damage the hybrid vehicle control ECU.

7. Remove the engine room covers.
8. Remove the connector cover assembly.
9. Disconnect the negative battery cable.
10. Using the voltmeter, measure the voltage between the terminals of the 2 phase connectors (N-P).

Standard voltage: 0 volts. Use measuring range of DC 750 V or more on the voltmeter.

11. Install the connector cover.
12. Drain the engine oil and engine coolant.
13. Remove the 7 clips and intake duct seal.
14. Remove the 3 clips and side cover.
15. Hold the front of the V-bank cover and raise it to disengage the 2 clips on the front of the cover. Continue to raise the cover to disengage the clip on the rear of the cover and remove the cover

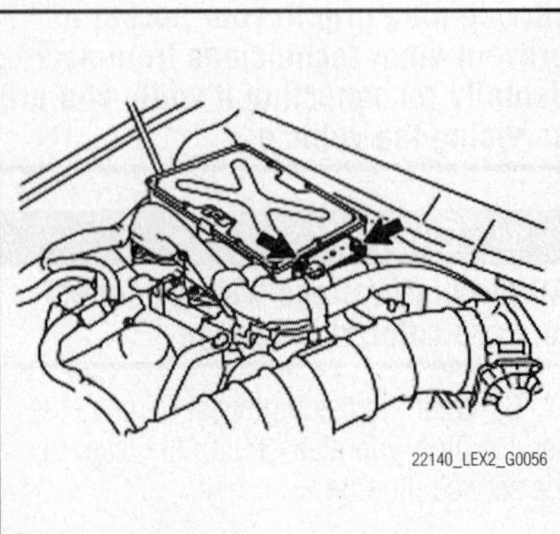

Fig. 11 Connector cover assembly

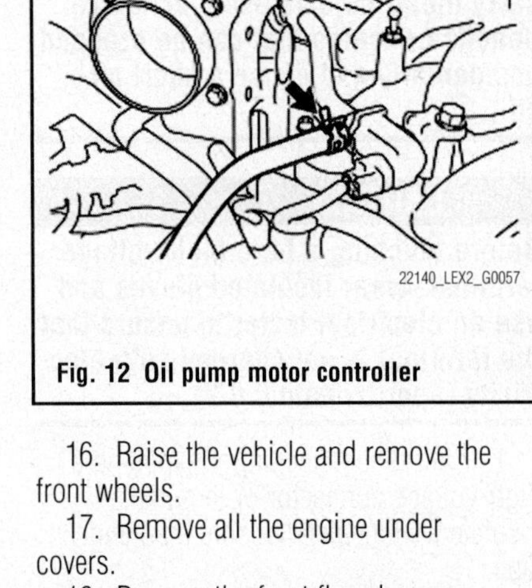

Fig. 12 Oil pump motor controller

16. Raise the vehicle and remove the front wheels.
17. Remove all the engine under covers.
18. Remove the front floor brace.

Fig. 13 Remove the hybrid vehicle control ECU

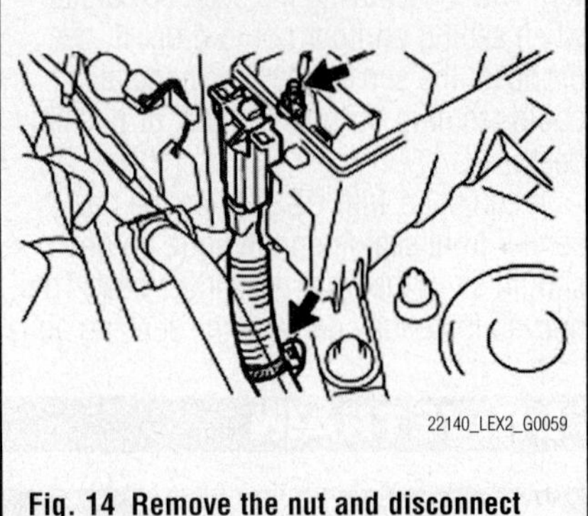

Fig. 14 Remove the nut and disconnect the No. 4 engine wire

19. Remove the front exhaust pipe assembly.
20. Make match mark and remove the propeller shaft and bearing.
21. Remove the air cleaner unit.
22. Remove the oil pump motor controller.
23. Disconnect the heater inlet and outlet hoses.
24. Disconnect the inverter cooling hose.
25. Disconnect the No. 8 inverter cooling hose.
26. Disconnect the A/C hoes.
27. Raise the lock lever and disconnect the 2 connectors. Remove the hybrid vehicle control ECU.
28. Disconnect the engine wiring as follows:
    a. Disconnect the clamp from the body.
    b. Remove the nut and disconnect the No. 4 engine wire.

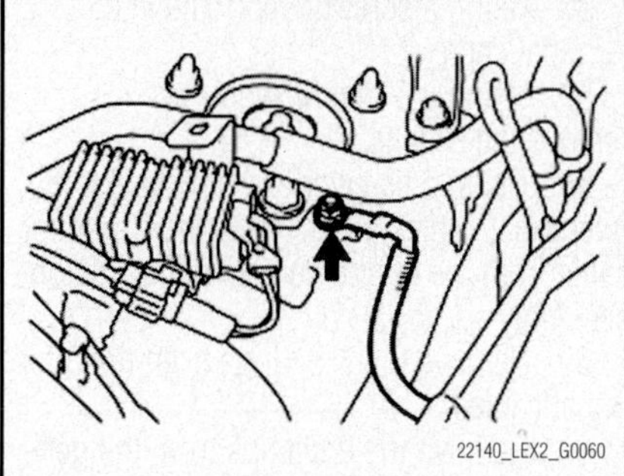

Fig. 15 Remove the nut and disconnect the No. 2 engine wire

c. Disconnect the 2 clamps.

d. Disconnect the 2 connectors from the compressor assembly.

e. Remove the nut and disconnect the No. 2 wire.

f. Disconnect the clamp from the body.

29. Remove the bolt and disconnect the No. 3 engine wire.

a. Disconnect the clamp from the body.

b. Remove the bolt and disconnect the No. 3 engine wire.

c. Remove bolt (A) to disconnect the ground wire from the bracket.

d. Remove the 2 clamps to disconnect the wire harness from the bracket.

e. Disconnect 2 connectors (C) and (D) from the power steering link assembly.

f. Release the lock of connector (E) and disconnect connector (E) from the power steering link assembly.

g. Remove bolt (B) and the power steering ground wire from the power steering link assembly.

h. Remove the nut and ground wire.

30. Disconnect the clamp and connector from the heater water pump.

(B) (C) (E) (A) (D)
Clamp
22140_LEX2_G0061

Fig. 16 A,B,C,D location graphic description

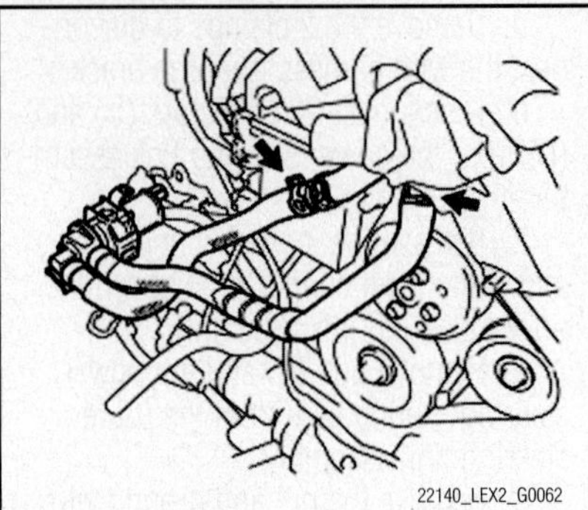

Fig. 17 Disconnect the 2 hoses and remove the water pump

31. W/ Front Active Stabilizer Control Actuator:

32. Remove the 8 bolts and fold back the front fender liner RH.

33. Remove the bolt and disconnect the ground wire.

34. Disconnect the wire harness clamp from the bracket.

35. Using the procedures below, disconnect the connector from the front active stabilizer control ECU.

36. Release the lever's locks (1), (2) and (3)

37. Disconnect the 2 hoses and remove the water pump.

38. Disconnect the fuel hose from the pipe.

39. Remove the 4 bolts from the front suspension protector.

40. Secure the steering wheel with the seat belt in order to prevent rotation.

41. Loosen bolt (A) and remove bolt (B), then slide the No. 2 steering intermediate shaft assembly.

42. Put matchmarks on the No. 2 steering intermediate shaft assembly and the power steering link assembly.

43. Disconnect the No. 2 steering intermediate shaft assembly from the power steering link assembly.

44. Remove the nut and disconnect the height control sensor link.

45. Remove the front struts.

46. Remove the front lower ball joint assembly.

47. Remove the nut and separate the floor shift gear shifting rod sub-assembly.

48. Disconnect the shift lever position sensor connector, transmission wire connector, transmission revolution sensor connector, engine oil level sensor connector, motor revolution sensor connector, and generator revolution sensor connector.

49. Remove the bolt and ground cable.

50. Separate the engine and transmission

51. Disconnect the shift lever position sensor connector, transmission wire connector, transmission revolution sensor connector, engine oil level sensor connector, motor revolution sensor connector, and generator revolution sensor connector.

52. Disconnect the transmission wiring harness as follows:

a. Remove the bolt and ground cable.

b. Remove the 3 bolts and wire harness. Separate the wire harness clamps from the hybrid vehicle transmission assembly.

53. Disengage the 6 claws and remove the motor cable protector.

54. Remove the generator cable.

55. Remove the 2 bolts and connector cover.

56. Using a screwdriver, remove the 3 terminal caps.

57. Using a 5 mm hexagon wrench, remove the 3 bolts.

58. Using a screwdriver, disconnect the 3 terminal cables and remove the generator cable from the hybrid vehicle transmission assembly.

59. Remove the 3 O-rings from the 3 terminal cables.

60. Remove the 2 clamps from the generator cable.

61. Using a screwdriver, disconnect the 3 terminal cables and remove the motor cable from the hybrid vehicle transmission assembly.

62. Remove the 4 clamps from the motor cable.

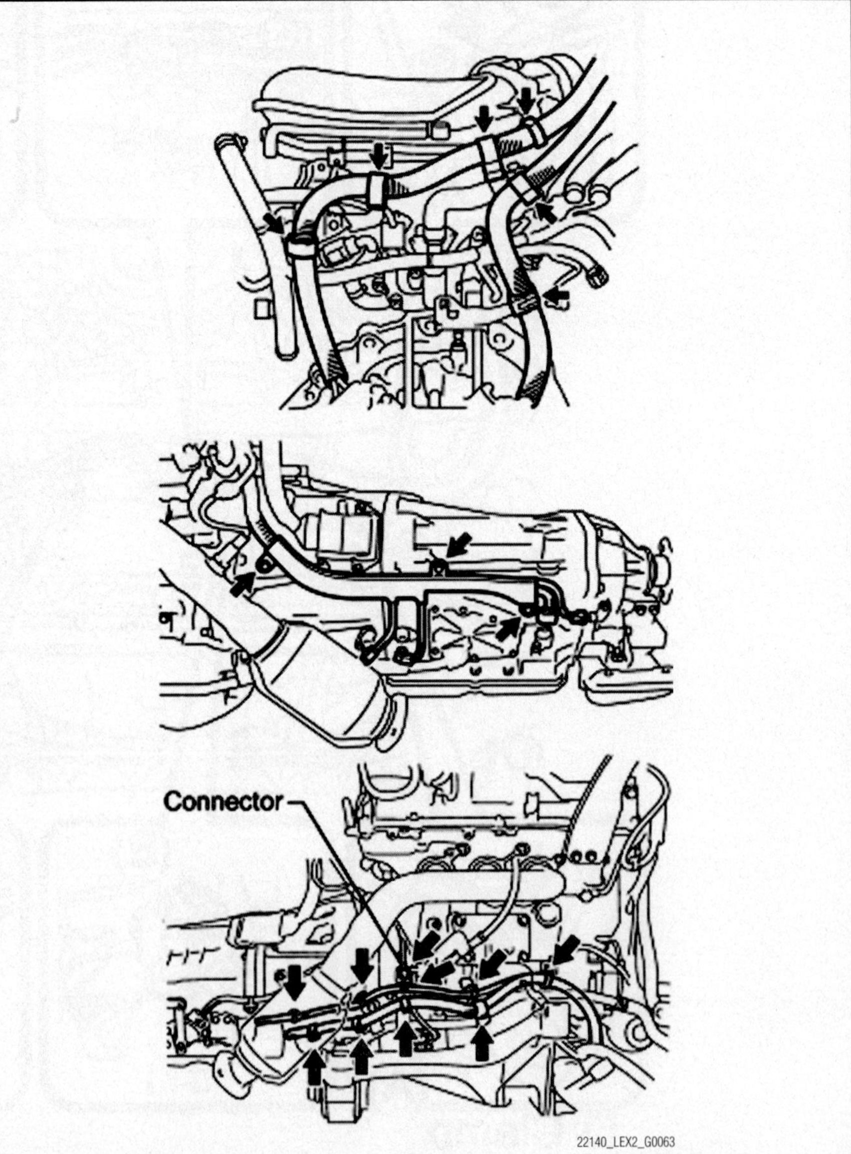

Fig. 18 Separate the wire harness clamps from the hybrid vehicle transmission assembly

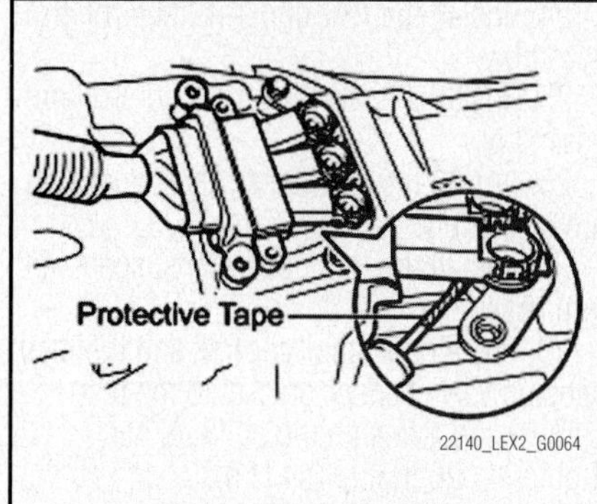

Fig. 19 Disconnect the 3 terminal cables and remove the generator cable

Fig. 20 Remove the 6 bolts and rear engine mounting insulator

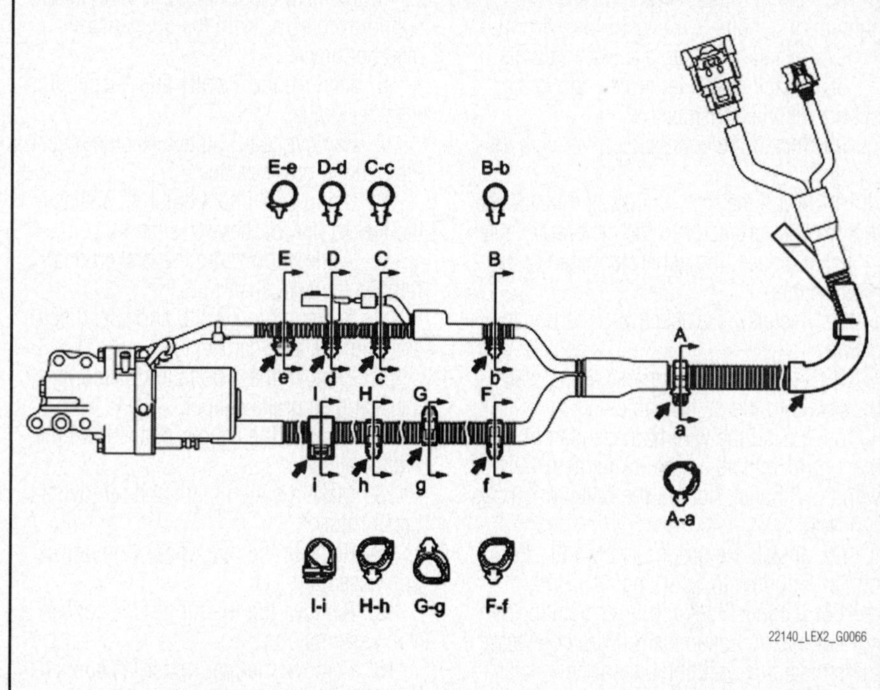

Fig. 21 Align the matchmarks and install 9 new clamps and wire harness protector to the wire

63. Remove the 6 bolts and rear engine mounting insulator from the hybrid vehicle transmission assembly.

64. Remove the transmission mass damper.

65. Remove the LF and RH the water pipe and hose sub-assembly.

66. Remove the oil cooler tubes.

67. Remove the nut and transmission control shaft lever.

68. Using a screwdriver, remove the transmission breather plug.

69. Disconnect the 4 clamps and clip.

70. Remove the 2 bolts and transmission breather hose.

71. Remove the 2 bolts and starter hole insulator.

72. Remove the flywheel housing side cover.

73. Remove the 9 bolts and hybrid vehicle transmission assembly.

74. Do not use excessive force to pry out the transmission assembly when separating it from the engine to prevent the knock pins from being damaged.

To install:

75. Align the matchmarks and install 9 new clamps and wire harness protector to the wire.

76. Install the hybrid vehicle transmission to the engine with the 9 bolts shown in the illustration.

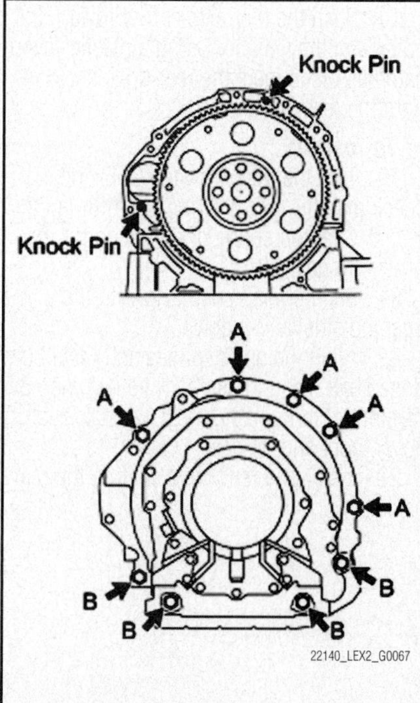

Fig. 22 Tighten the 9 bolts shown in the illustration

77. Tighten bolt (A) to 52 ft. lbs. (71 Nm).

78. Tighten bolt (A) to 27 ft. lbs. (37 Nm).

79. Install the flywheel housing side cover.

80. Install the starter hole insulator with the 2 bolts and tighten to 43 ft. lbs. (58 Nm).

81. Install the transmission breather hose with the 2 bolts. Tighten to 62 inch. lbs. (7 Nm).

82. Install the transmission breather plug with new o-ring.

83. Install the transmission control shaft lever with the nut and tighten to 12 ft. lbs. (16 Nm).

84. Install the oil cooler tube and 4 new gaskets with the 2 bolts. Tighten to 20 ft. lbs. (28 Nm).

85. Install the 3 bolts to the water pipe and hose sub-assembly in the order shown in the illustration.

86. Install the trans mission damper and tighten the mounting bolts to 44 ft. lbs. (60 Nm).

87. Install the rear engine mounting insulator assembly with the 6 bolts. Tighten to 44 ft. lbs. (60 Nm).

88. Align the matchmarks and install the clamps to the motor cable.

89. Install the 3 O-rings to the terminal cables.

90. Using a 5 mm hexagon wrench, tighten the 3 bolts to 71 inch. lbs. (8 Nm).

91. Install the 2 cable mounting bolts.

92. Install new 3 terminal caps to the hybrid vehicle transmission.

93. Install the connector cover with the 2 bolts.

94. Align the matchmarks and install the clamps to the generator cable.

95. Install the 3 O-rings to the terminal cables.

96. Using a 5 mm hexagon wrench, tighten the 3 bolts to 71 inch. lbs. (8 Nm).
97. Install the 2 cable mounting bolts.
98. Install new 3 terminal caps to the hybrid vehicle transmission.
99. Install the connector cover with the 2 bolts.
100. Align the matchmarks and install the motor cable protector to the generator cable.
101. Connect the generator cable to the motor cable.
102. Install the bracket with the nut and clamp.
103. Align the matchmarks and install the clamp to the motor cable.
104. Install the wire harness clamps to the hybrid vehicle transmission assembly with the 3 bolts. Tighten the bolts to 7 ft. lbs (10 Nm).
105. Install the ground cable with the bolt and tighten to 13 ft. lbs. (18 Nm).
106. Connect the shift lever position sensor connector, transmission wire connector, transmission revolution sensor connector, engine oil level sensor connector, generator revolution connector, and motor revolution connector.
107. If the hybrid vehicle transmission assembly is replaced, initialize the learning value of the transmission

REAR HALFSHAFT

REMOVAL & INSTALLATION

See Figure 23.

1. Before servicing the vehicle, refer to the precautions in the beginning of this section.
2. Turn the power switch OFF, and check that the "READY" indicator turns off.
3. Remove the rear wheel.
4. Remove the bolt, nut, load sensing valve sensor bracket and stabilizer link.
5. Remove the 2 nuts and differential support protector from the suspension member brace
6. Remove the 2 bolts and suspension member brace.
7. Remove the 2 bolts and disconnect the parking brake cable.
8. Using SST and a hammer, release the staked part of the axle shaft nut.
9. While depressing the brake pedal, remove the axle shaft nut.
10. Remove the 2 bolts and speed sensor from the axle carrier.
11. Remove the 2 bolts and disconnect the rear disc brake caliper.
12. Remove the caliper plates from the brake caliper.
13. Make a match mark and remove the brake rotor.
14. Remove the rear upper (2) control arm assembly.
15. Remove the rear upper (1) control arm assembly.
16. Remove the rear upper (1) suspension arm assembly.
17. Remove the rear upper (2) suspension arm assembly.
18. Push the rear axle carrier toward the outside of the vehicle. Using a plastic-faced hammer, disconnect the rear drive shaft from the rear axle carrier.

To install:

19. Turn the power switch OFF, and check that the "READY" indicator turns off.
20. Coat the spline of the inboard joint shaft with gear oil.
21. Set the shaft snap ring with the opening side facing down.
22. Align the shaft splines and install the drive shaft with a brass bar and hammer. Be careful not to damage the drive shaft dust cover, boot and oil seal.
23. Install the rear drive shaft to the rear axle carrier.
24. Install the rear upper (2) control arm assembly.
25. Install the rear upper (1) control arm assembly.
26. Install the rear upper (1) suspension arm assembly.
27. Install the rear upper (2) suspension arm assembly.
28. Install the stabilizer link and the load sensing valve sensor bracket to the rear suspension arm with the bolt and nut. Tighten to 20 ft. lbs. (54 Nm).
29. Install the brake rotor.
30. Install the rear brake caliper assembly and tighten the bolts to 40 ft. lbs. (54 Nm).
31. Install the speed sensor to the rear axle carrier. Tighten to 75 inch. lbs. (8.5 Nm).
32. Install the rear drive axles and tighten the new axle shaft nuts to 214 ft. lbs. (290 Nm). Using a chisel and a hammer, stake the axle shaft nut.
33. Check and add fluid to differential if needed.
34. Install the parking brake cable with the 2 bolts and tighten to 14 ft. lbs. (19 Nm).
35. Stabilize the rear suspension.
36. Tighten the rear upper (2) control arm assembly to 119 ft. lbs. (161 Nm).
37. Tighten the rear upper (1) control arm assembly to 119 ft. lbs. (161 Nm).
38. Tighten the rear upper (1) suspension arm assembly to 74 ft. lbs. (105 Nm).
39. Tighten the rear upper (2) suspension arm assembly to 74 ft. lbs. (105 Nm).
40. Install the member cover with the 2 screws.
41. Install the No2 support protector.
42. Install the rear wheel and tighten to 76 ft. lbs. (103 Nm).
43. Inspect and adjust the rear wheel alignment.
44. Inspect and adjust the headlights.

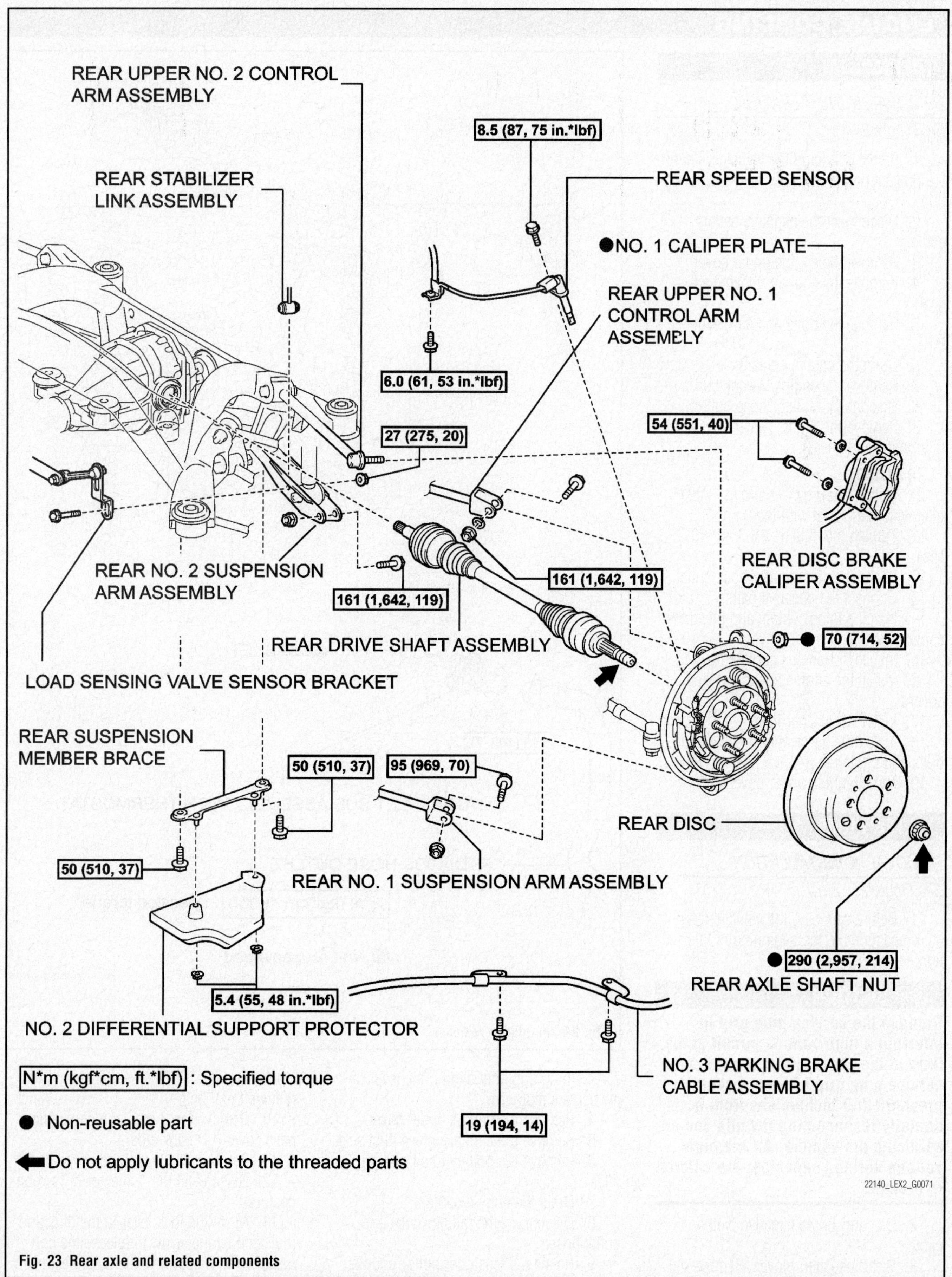

Fig. 23 Rear axle and related components

ENGINE COOLING

THERMOSTAT

REMOVAL & INSTALLATION

See Figure 24.

1. Before servicing the vehicle, refer to the precautions in the beginning of this section.
2. Disconnect the negative battery cable.
3. Remove the engine under cover.
4. Remove the cool air intake duct seal.
5. Remove the bolts and air cleaner inlet.
6. Drain the engine coolant.
7. Remove the engine V-bank cover.
8. Disconnect radiator hose outlet.
9. Remove the 3 nuts, water inlet with thermostat and gasket.

To install:

10. Install a new gasket and the water inlet with thermostat with the 3 nuts.
11. Tighten the nuts to 7 ft. lbs. (10 Nm).
12. Reconnect the radiator hose outlet.
13. Connect the negative battery cable.
14. Install engine coolant and bleed the cooling system.
15. Inspect for engine coolant leaks.
16. Install the engine cover and side covers.
17. Install air cleaner inlet.
18. Install the air cleaner inlet with the bolts and tighten to 44 inch. lbs. (4 Nm).
19. Install engine under cover.

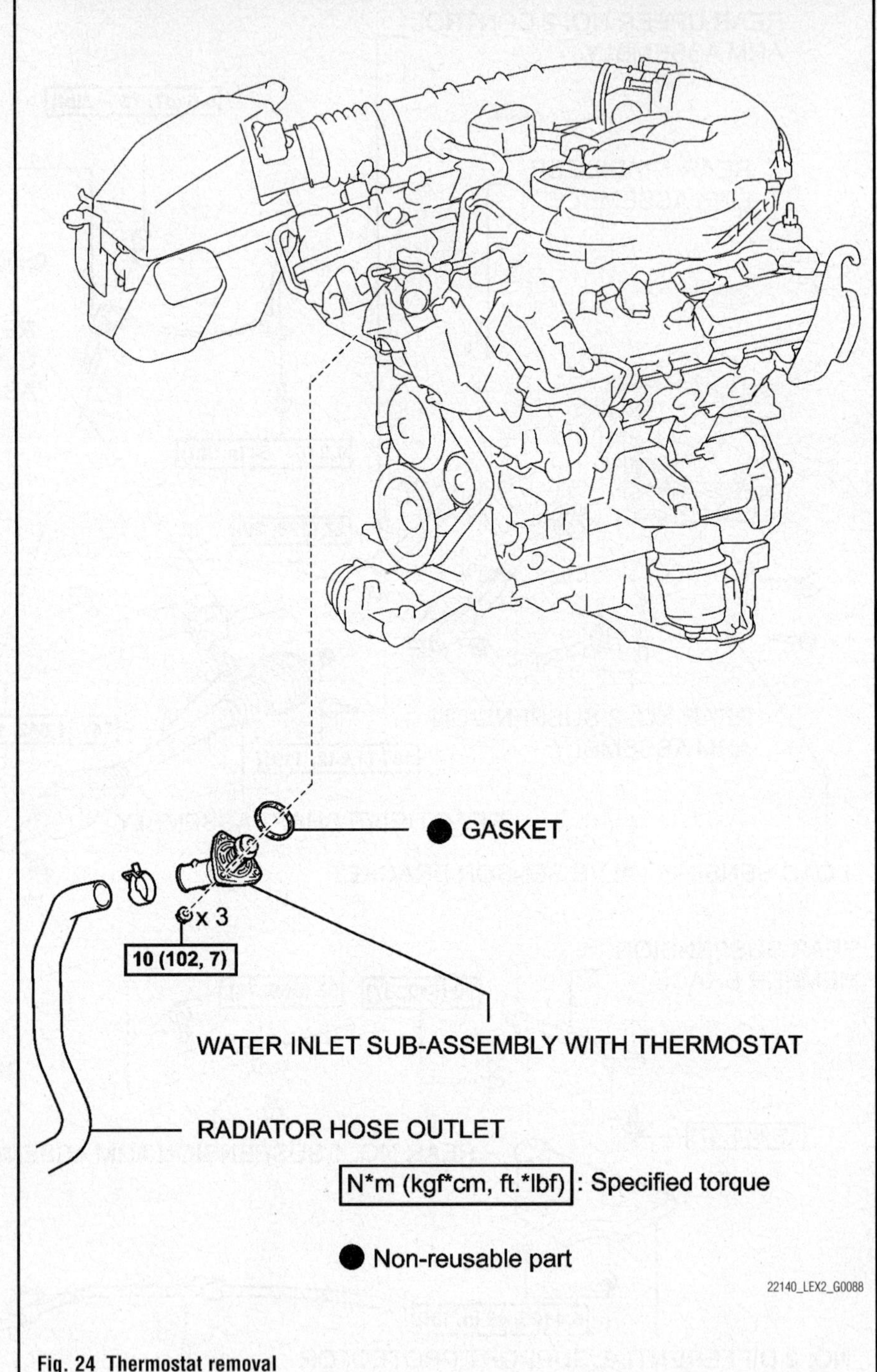

Fig. 24 Thermostat removal

WATER PUMP

REMOVAL & INSTALLATION

See Figure 25.

1. Before servicing the vehicle, refer to the precautions in the beginning of this section.

CAUTION

Remove the service plug grip to interrupt a high voltage circuit at the time of the check. Keep the removed service plug grip in your pocket to prevent other technicians from accidentally reconnecting it while you are servicing the vehicle. All the high voltage wiring connectors are colored in orange.

2. Disconnect the negative battery cable.
3. Wear insulated gloves. Remove the service plug grip after sliding the lever of the service plug grip.
4. Remove the engine under cover.
5. Remove the cool air intake duct seal.
6. Remove the bolts and air cleaner inlet.
7. Drain the engine coolant.
8. Disconnect the radiator inlet and outlet hoses.
9. Rotate the tensioner pulley counterclockwise to loosen the belt tension. Then remove the V belt.
10. Remove the 3 nuts and engine harness cover (LH bank side).
11. Remove the 3 nuts and cover.
12. Disconnect the 4 injector driver connectors.
13. Move the lock lever in the direction indicated by the arrow to release the connector lock. Disconnect the 3 connectors

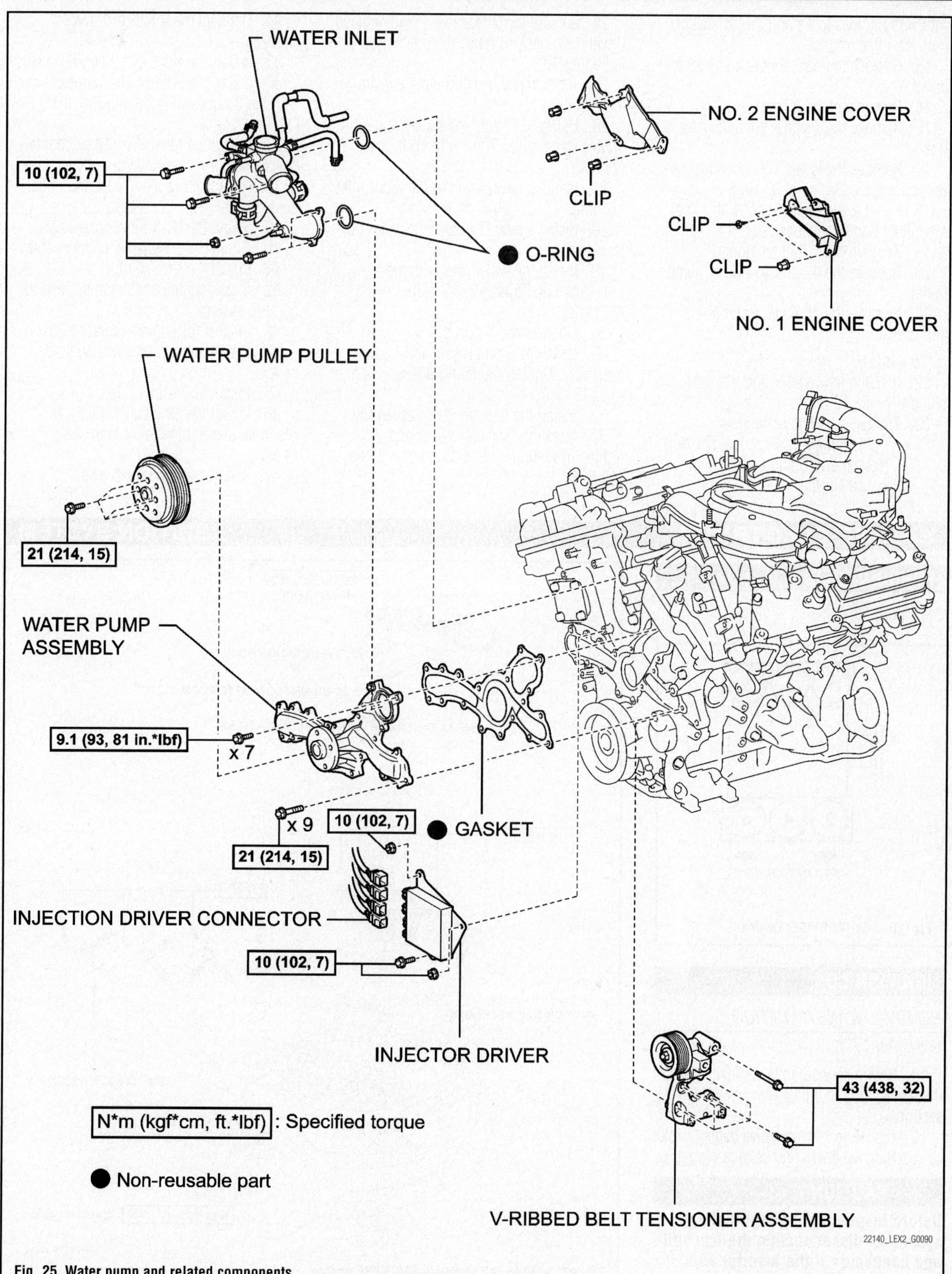

Fig. 25 Water pump and related components

with wire harness locks and the connector from the injector driver.

14. Remove the bolt, 2 nuts and injector driver.

15. Disconnect the 5 hoses.

16. Remove the 4 bolts, nut and water inlet.

17. Remove the water inlet housing gasket and water outlet pipe's O-ring.

18. Remove the compressor with motor assembly. Carefully set aside.

19. Remove the V-belt tensioner.

20. Remove the 4 bolts and water pump pulley.

21. Remove the 16 bolts, water pump and gasket.

To install:

22. Install a new gasket and the water pump with the 16 bolts.

23. Tighten the bolts as follows:

- Tighten bolt A to 15 ft. lbs. (21 Nm).
- Tighten bolt B and C to 81 inch. lbs. (9 Nm).

24. Be sure to replace the 2 bolts labeled C with new ones or reuse them after applying adhesive.

25. Temporarily install the pulley with the 4 bolts.

26. Using SST, hold the pulley and tighten the 4 bolts. Tighten to 15 ft. lbs. (21 Nm).

27. Install a new gasket to the water inlet housing.

28. Install a new O-ring to the water outlet pipe.

29. Install the water inlet with the 4 bolts and nut. Tighten to 15 ft. lbs. (21 Nm).

30. Connect the 5 hoses.

31. Install the compressor with motor assembly. Tighten mounting bolts to 32 ft. lbs. (43 Nm).

32. Install the belt tensioner assembly.

33. Install the injector driver and tighten mounting nuts and bolt to 7 ft. lbs. (10 Nm).

34. Connect the 4 injector driver connectors.

35. Install the No 1 and 2 engine covers.

36. Install the engine wire harness cover with the 3 nuts and tighten to 7 ft. lbs. (10 Nm).

37. Rotate the tensioner pulley counterclockwise, and then install the V-belt.

38. Reconnect radiator inlet and outlet hoses.

39. Install the No 1 air cleaner inlet.

40. Connect the negative battery cable.

41. Install the service plug.

42. Install engine coolant and bleed the cooling system.

43. Inspect for engine coolant leaks.

44. Install the engine cover and side covers.

45. Install air cleaner inlet.

46. Install the air cleaner inlet with the bolts and tighten to 44 inch. lbs. (4 Nm).

47. Install engine under cover.

FIRING ORDER

See Figure 26.

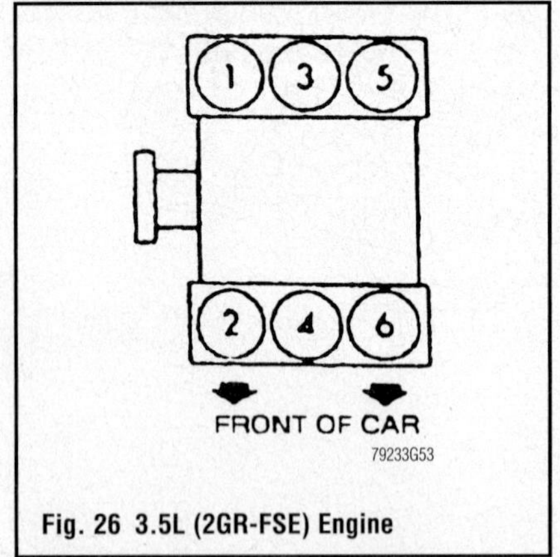

Fig. 26 3.5L (2GR-FSE) Engine

IGNITION COIL

REMOVAL & INSTALLATION

See Figure 27.

1. Before servicing the vehicle, refer to the precautions in the beginning of this section.
2. Disconnect the negative battery cable.
3. Remove the hybrid service plug grip.

⁂ CAUTION

Before inspecting the high-voltage system or disconnecting the low voltage connector of the inverter with converter assembly, take safety

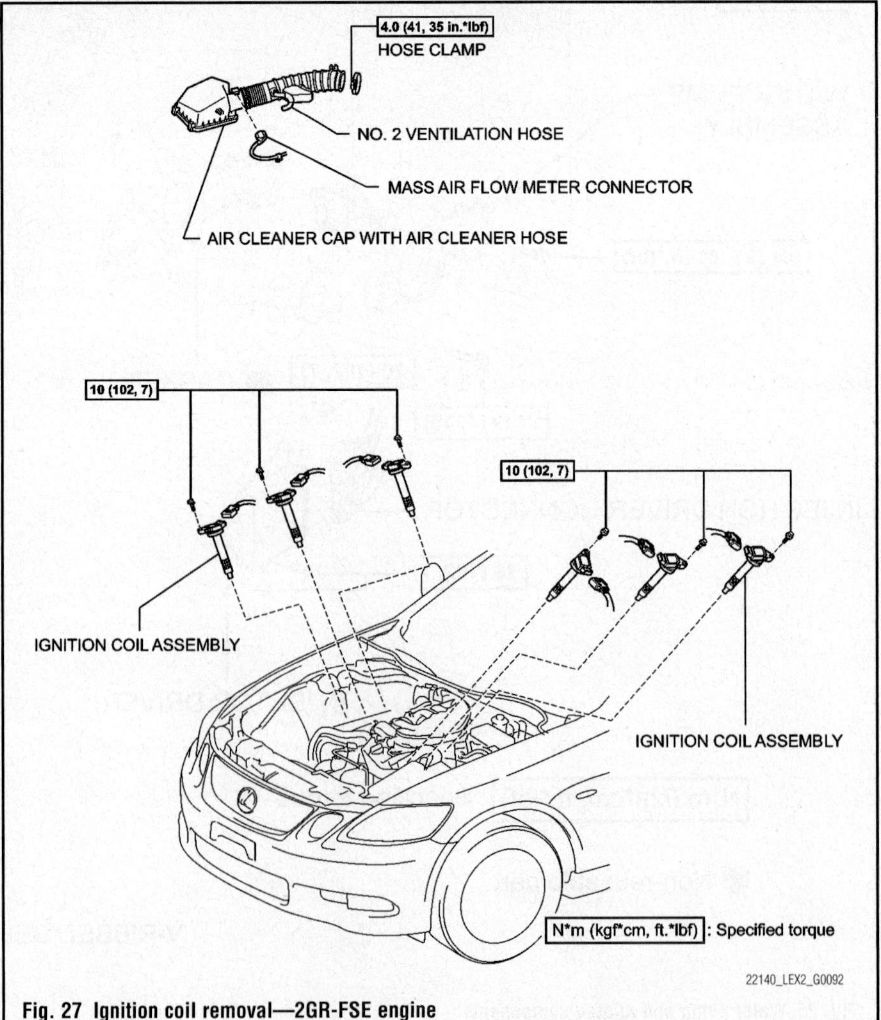

Fig. 27 Ignition coil removal—2GR-FSE engine

precautions, such as wearing insulated gloves and removing the service plug grip to prevent electrical shocks. After removing the service plug grip, put it in your pocket to prevent other technicians from accidentally reconnecting it while you are working on the high-voltage system.

⁂ CAUTION

After disconnecting the service plug grip, wait for at least 10 minutes before touching any of the high-voltage connectors or terminals. After waiting, check the voltage at the inspection point in the inverter with converter assembly. The voltage should be 0 V before beginning work

4. Remove the cool air intake duct seal.
5. Remove the engine room side cover.
6. Remove the V-bank cover.
7. Disconnect the No 2 ventilation hose.
8. Remove the air cleaner cap and hose.
9. Detach the 3 wire harness clamps and disconnect the noise filter connector.
10. Disconnect the 6 ignition coil connectors.
11. Remove the 6 bolts and the 6 ignition coils.

To install:

12. Install the 6 ignition coils with the 6 bolts and tighten to 7 ft. lbs. (10 Nm).
13. Connect the 6 ignition coil connectors.
14. Attach the 3 wire harness clamps and connect the noise filter connector.
15. Install the air cleaner cap and hose.
16. Reconnect the No 2 ventilation hose.
17. Install the V-bank cover.
18. Install the engine room side cover.
19. Install the cool air intake duct seal.
20. Install the service plug grip.
21. Connect the negative battery cable.

SPARK PLUGS

REMOVAL & INSTALLATION

See Figure 27.

1. Before servicing the vehicle, refer to the precautions in the beginning of this section.
2. Disconnect the negative battery cable.
3. Remove the hybrid service plug grip.

⁂ CAUTION

Before inspecting the high-voltage system or disconnecting the low voltage connector of the inverter with converter assembly, take safety precautions, such as wearing insulated gloves and removing the service plug grip to prevent electrical shocks. After removing the service plug grip, put it in your pocket to prevent other technicians from accidentally reconnecting it while you are working on the high-voltage system.

⁂ CAUTION

After disconnecting the service plug grip, wait for at least 10 minutes before touching any of the high-voltage connectors or terminals. After waiting, check the voltage at the inspection point in the inverter with converter assembly. The voltage should be 0 V before beginning work

4. Remove the cool air intake duct seal.
5. Remove the engine room side cover.
6. Remove the V-bank cover.
7. Disconnect the No 2 ventilation hose.
8. Remove the air cleaner cap and hose.
9. Detach the 3 wire harness clamps and disconnect the noise filter connector.
10. Disconnect the 6 ignition coil connectors.
11. Remove the 6 bolts and the 6 ignition coils.
12. Remove the 6 spark plugs.

To install:

13. Check that the spark plug gap is between: 0.039–0.043 inch.
14. Install the 6 spark plugs and tighten to 13 ft. lbs. (16 Nm).
15. Install the 6 ignition coils with the 6 bolts and tighten to 7 ft. lbs. (10 Nm).
16. Connect the 6 ignition coil connectors.
17. Attach the 3 wire harness clamps and connect the noise filter connector.
18. Install the air cleaner cap and hose.
19. Reconnect the No 2 ventilation hose.
20. Install the V-bank cover.
21. Install the engine room side cover.
22. Install the cool air intake duct seal.
23. Install the service plug grip.
24. Connect the negative battery cable.

ENGINE MECHANICAL

➡Disconnecting the negative battery cable may interfere with the functions of the on board computer systems and may require the computer to undergo a relearning process, once the negative battery cable is reconnected.

ACCESSORY DRIVE BELTS

ACCESSORY BELT ROUTING

See Figure 28.

INSPECTION

See Figure 29.

Visually check the drive belt for excessive wear, frayed cords, etc. If any defect has been found, replace the drive belt.

ADJUSTMENT

The drive belt system is equipped with an automatic belt tensioner.

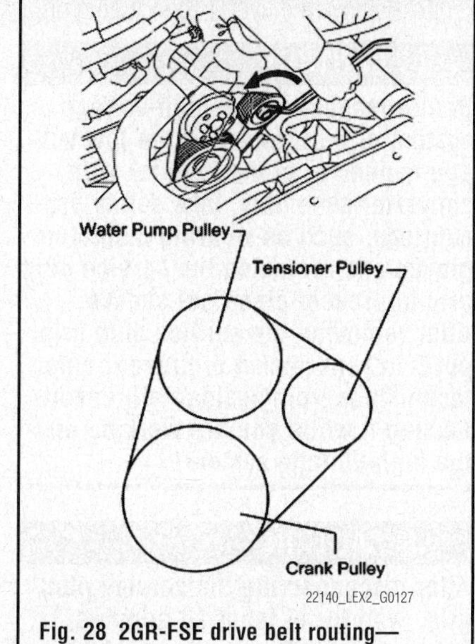

Fig. 28 2GR-FSE drive belt routing—GS450h

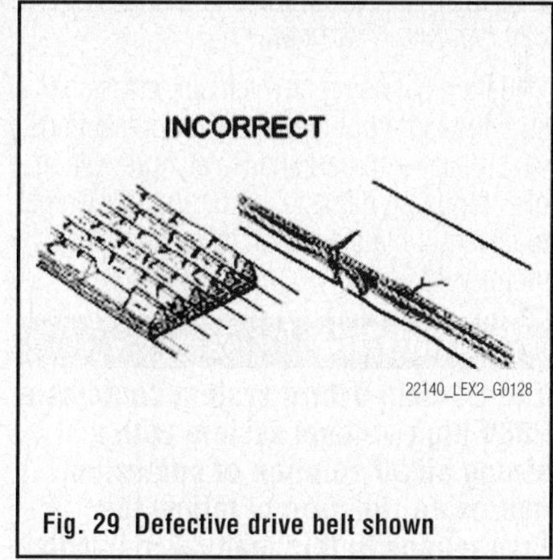

Fig. 29 Defective drive belt shown

REMOVAL & INSTALLATION

See Figure 30.

1. Before servicing the vehicle, refer to the precautions in the beginning of this section.
2. Remove the cool air duct seal.

Fig. 30 Dive belt removal shown

3. Remove the No. 1 air cleaner inlet.
4. Rotate the tensioner pulley counterclockwise to loosen the belt tension. Then remove the V belt.

➡**The pulley bolt for the belt tensioner has a left-handed thread. Do not perform this procedure from underneath the vehicle, as the tensioner may be damaged.**

To install:

5. Rotate the tensioner pulley counterclockwise, and then install the V belt.

➡**After installing a new belt, run the engine for approximately 5 minutes and then recheck the tension.**

6. Install the air cleaner inlet with the bolt. And tighten to 44 inch. lbs. (5.0 Nm).
7. Install the cool air duct seal.

CAMSHAFT AND VALVE LIFTERS

REMOVAL & INSTALLATION

See Figures 31 through 47.

Before servicing any vehicle, please be sure to read all of the following precautions, which deal with personal safety, prevention of component damage, and important points to take into consideration when servicing a motor vehicle:

✻✻ CAUTION

The GS450h hybrid system contains a 288V high-voltage system with a strong alkali solution of potassium hydroxide. Be sure to follow the instructions in this manual to handle the system correctly. Failure to do so may result in serious injury or electrocution. Engineer must undergo special training to be able to perform high-voltage system inspection and servicing.

✻✻ CAUTION

All high-voltage wire harness connectors are colored orange. The HV battery and other high-voltage components have "High Voltage" caution labels. Do not carelessly touch these wires and components.

✻✻ CAUTION

Before inspecting or servicing the high-voltage system, be sure to follow safety measures, such as wearing insulated gloves and removing the service plug to prevent electrocution. Carry the removed service plug in your pocket to prevent other technicians from reinstalling it while you are servicing the vehicle.

✻✻ CAUTION

After removing the service plug, wait 10 minutes before touching any of the high-voltage connectors and terminals.

Before servicing any vehicle, please be sure to read all of the following precautions, which deal with personal safety, prevention of component damage, and important points to take into consideration when servicing a motor vehicle:

1. Discharge the fuel system.
2. Recover the refrigerant from the A/C system using a refrigerant recovery unit.
3. Place the front wheels in the straight ahead position.
4. Disconnect the negative battery cable.
5. Remove the hybrid service plug grip.

✻✻ CAUTION

Before inspecting the high-voltage system or disconnecting the low voltage connector of the inverter with converter assembly, take safety precautions, such as wearing insulated gloves and removing the service plug grip to prevent electrical shocks. After removing the service plug grip, put it in your pocket to prevent other technicians from accidentally reconnecting it while you are working on the high-voltage system.

✻✻ CAUTION

After disconnecting the service plug grip, wait for at least 10 minutes before touching any of the high-voltage connectors or terminals. After waiting, check the voltage at the inspection point in the inverter with converter assembly. The voltage should be 0 V before beginning work.

6. Remove the luggage floor mat.
7. Remove the service plug cover.
8. Remove the service plug.
9. Disconnect the negative battery cable.
10. Drain the engine oil and coolant.
11. Remove the cool air intake duct seal.
12. Remove the engine room side cover.
13. Remove the V-bank cover.
14. Remove hoses and electrical connectors.
15. Remove the air intake surge tank sub-assembly.
16. Remove the ignition coil connectors and coils.
17. Remove the ventilation valve.
18. Remove the 4 bolts and the 4 VVT sensors.
19. Remove the 4 bolts and 4 oil control valves.
20. Remove the oil check valve bolt, oil pipe union and oil pipe.
21. Remove valve covers.
22. Remove the drive belt.
23. Remove the crankshaft pulley.
24. Remove the timing cover.
25. Remove the timing components.
26. Remove the camshaft timing gear assembly (Bank 1) as follows:
   - While raising up the No. 2 chain tensioner, insert a pin of 0.039 inch. (1.0 mm) into the hole to fix it in place.
   - Hold the hexagonal portion of the camshaft with a wrench, and remove the 2 bolts and 2 camshaft timing gears.
   - Remove the No. 2 chain.
   - Remove the No. 2 Chain tensioner

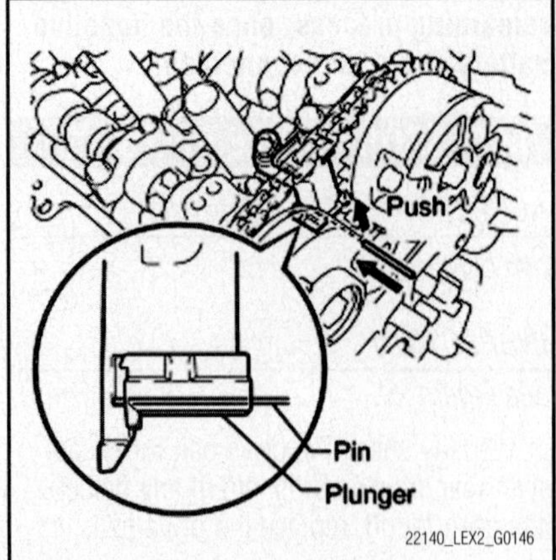

Fig. 31 Insert a pin of 0.039 inch. (1.0 mm) into the hole to fix it in place

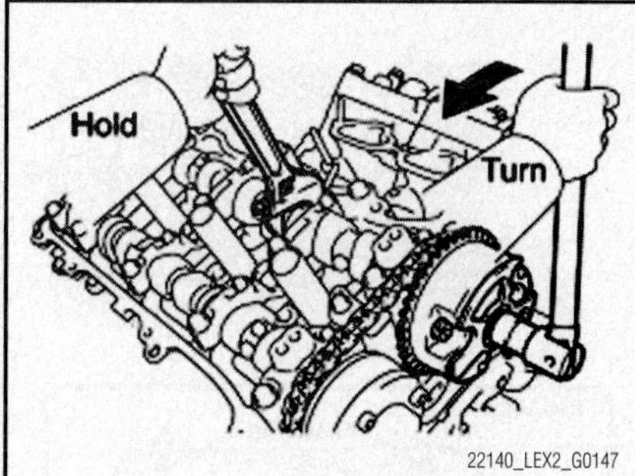

Fig. 32 Hold the hexagonal portion of the camshaft with a wrench, and remove the 2 bolts and 2 camshaft timing gears

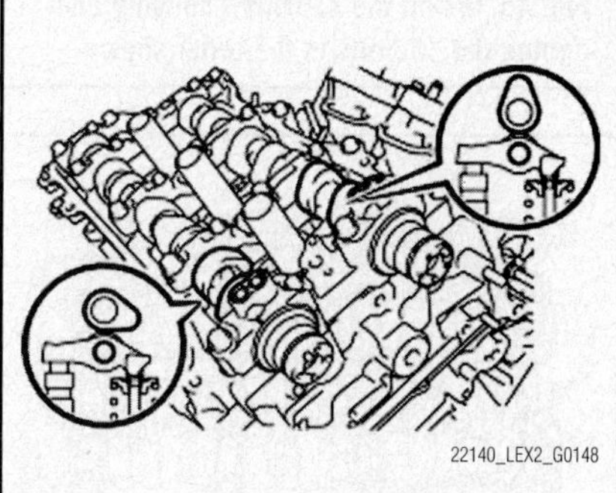

Fig. 33 Make sure that the knock pin of the camshaft is positioned as shown

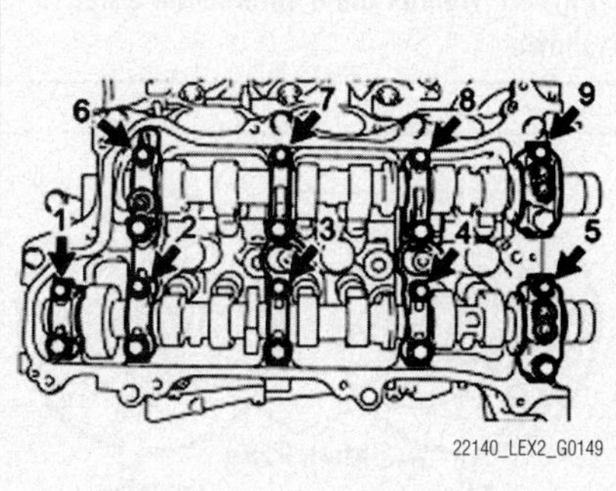
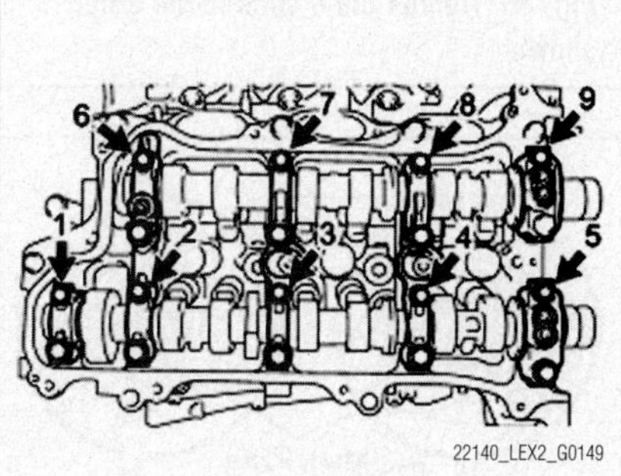

Fig. 34 Uniformly loosen and remove the 9 bearing cap bolts in the sequence shown

27. Remove the camshaft for (Bank 1) as follows:

- Remove the 3 gaskets.
- Make sure that the knock pin of the camshaft is positioned as shown in the illustration.
- Uniformly loosen and remove the 9 bearing cap bolts in the sequence shown in the illustration.
- Uniformly loosen and remove the 14 bearing cap bolts in the sequence shown in the illustration.
- Remove the 6 bearing caps.
- Remove the No. 1 and No. 2 camshafts.

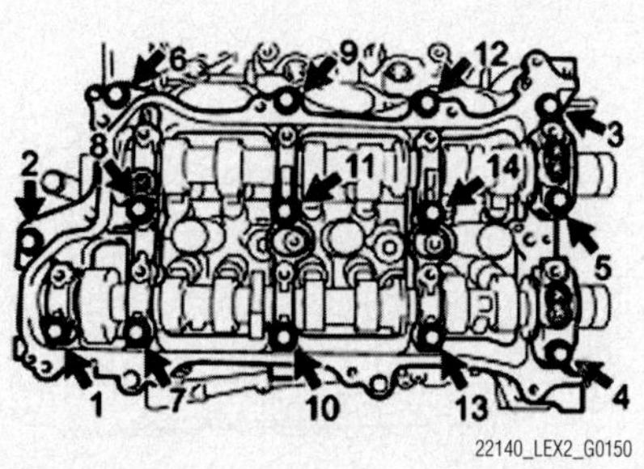

Fig. 35 Uniformly loosen and remove the 14 bearing cap bolts in the sequence shown

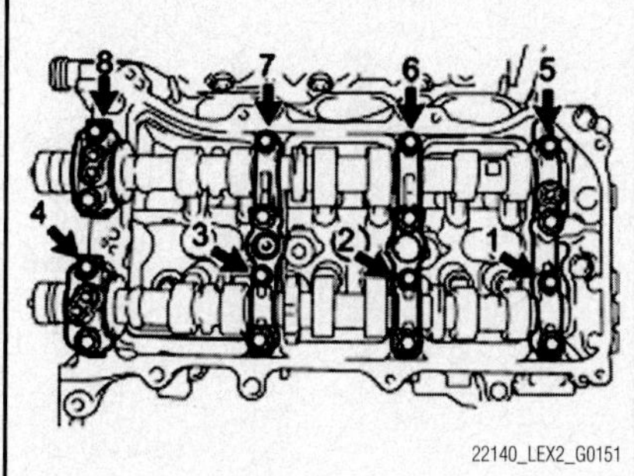

Fig. 36 Uniformly loosen and remove the 8 bearing cap bolts in the sequence shown

28. Remove the camshaft timing gear assembly (Bank 2) as follows:

- While raising up the No. 2 chain tensioner, insert a pin of 0.039 inch. (1.0 mm) into the hole to fix it in place.
- Hold the hexagonal portion of the camshaft with a wrench, and remove the 2 bolts and 2 camshaft timing gears.
- Remove the No. 3 chain.
- Remove the No. 3 chain tensioner.

29. Remove the camshaft for (Bank 2) as follows:

- Remove the 3 gaskets.
- Make sure that the knock pin of the camshaft is positioned as shown in the illustration.
- Uniformly loosen and remove the 8 bearing cap bolts in the sequence shown in the illustration.
- Uniformly loosen and remove the 13 bearing cap bolts in the sequence shown in the illustration.
- Remove the 5 bearing caps.
- Remove the No. 3 and No. 4 camshafts.

To install:

30. Apply engine oil to the camshaft journals, camshaft housings and bearing caps.

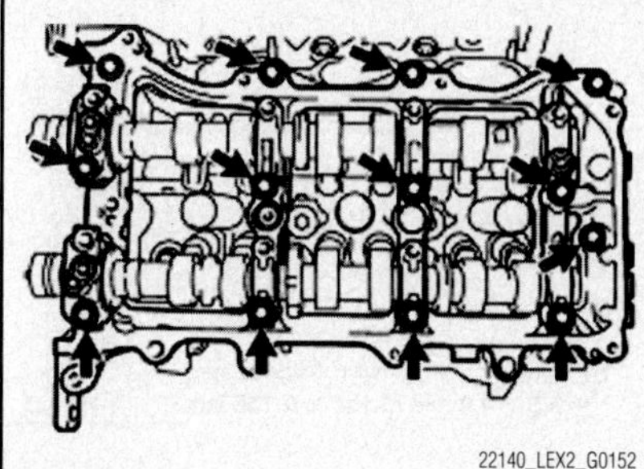

Fig. 37 Uniformly loosen and remove the 13 bearing cap bolts in the sequence shown

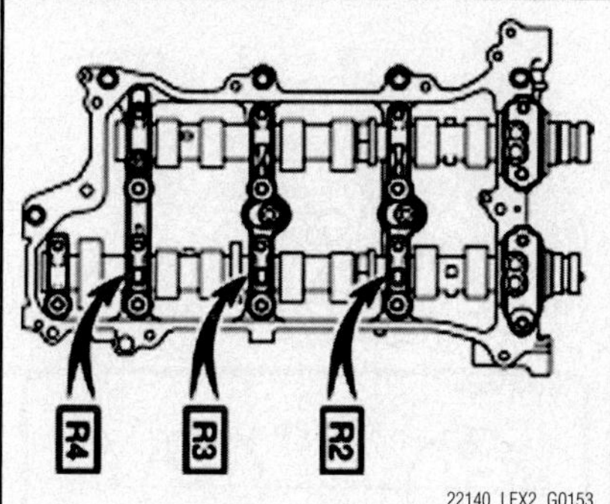

Fig. 38 Confirm the marks and numbers on the camshaft bearing caps

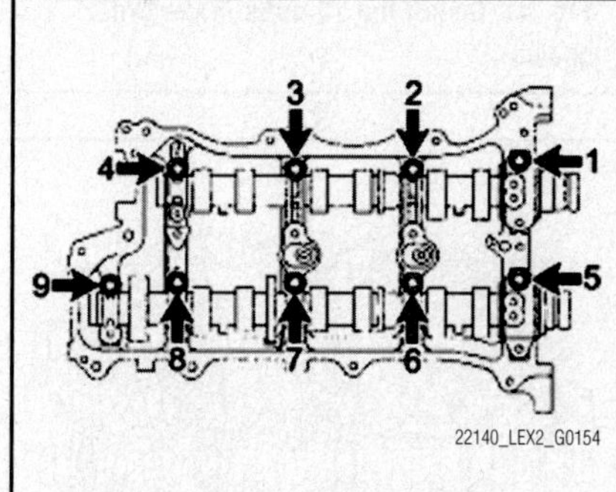

Fig. 39 Temporarily install the 8 bolts in the order shown

31. Install camshaft bearing cap for (Bank 1).

32. Install the camshaft and No. 2 camshaft to the camshaft housing.

33. Confirm the marks and numbers on the camshaft bearing caps and place them each in their proper position and direction.

34. Temporarily install the 8 bolts in the order shown in the illustration.

35. Tighten bolts to 7 ft. lbs. (10 Nm).

36. Install camshaft assembly sub-assembly.

37. Apply seal packing in a continuous line as shown in the illustration.

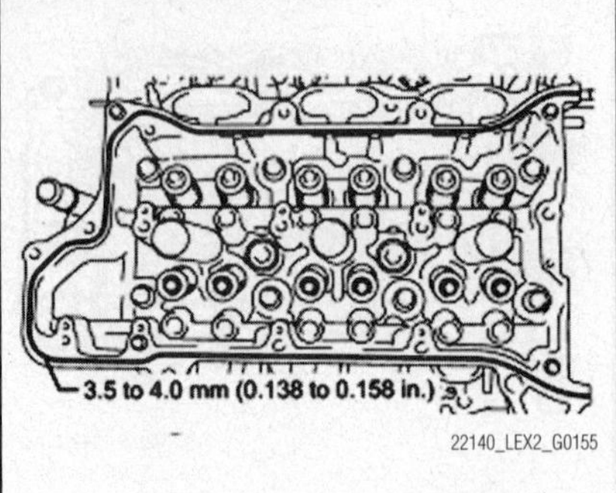

Fig. 40 Apply seal packing in a continuous line

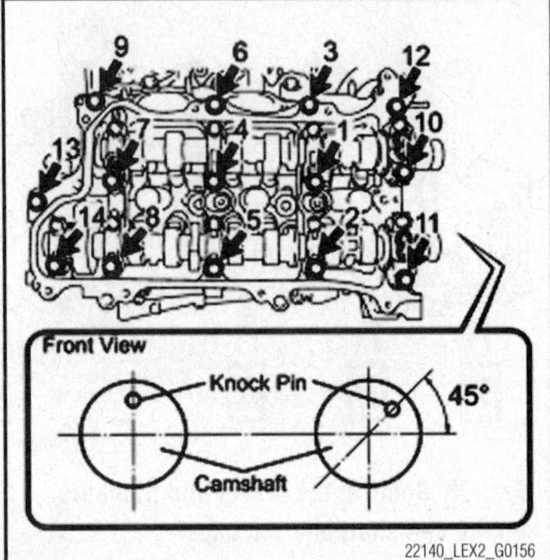

Fig. 41 Install the 12 bolts in the order shown

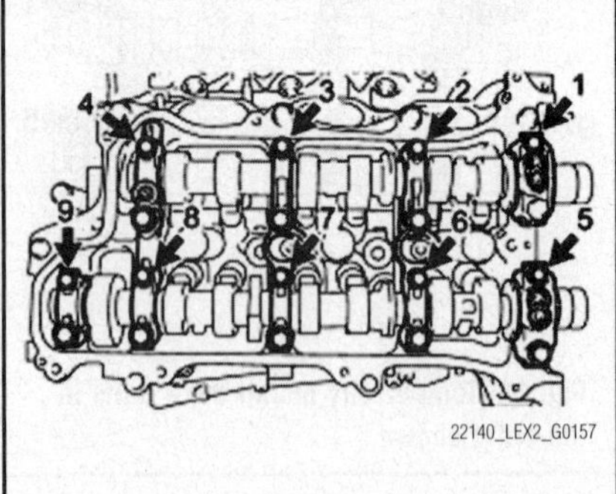

Fig. 42 Tighten the 9 bolts in the order shown

38. Install the camshaft housing, and install the 12 bolts in the order shown in the illustration. Tighten to 21 ft. lbs. (28 Nm).

➡Make sure that the knock pin of the camshaft is positioned as shown in the illustration before installing the camshaft housing.

39. Tighten the 9 bolts in the order shown in the illustration. Tighten bolts to 12 ft. lbs. (16 Nm).

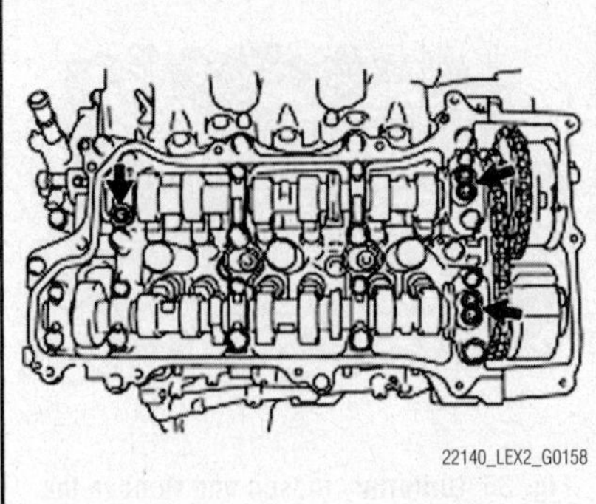

Fig. 43 Install 3 new gaskets as shown

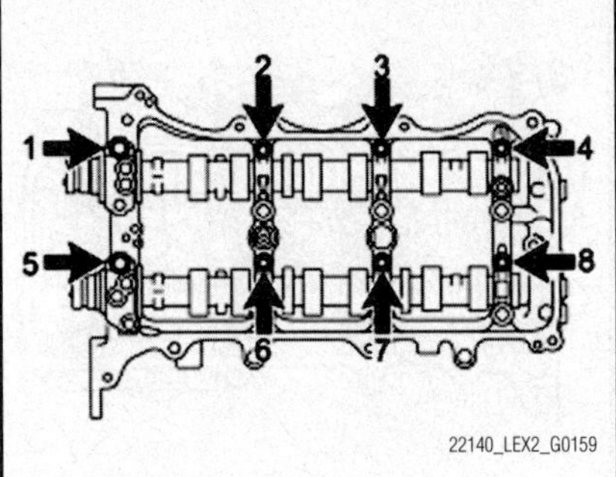

Fig. 44 Temporarily install the 8 bolts in the order shown

40. Install 3 new gaskets as shown in the illustration.

41. Apply engine oil to the camshaft journals, camshaft housings and bearing caps.

42. Install camshaft bearing cap for (Bank 1).

43. Install the camshaft and No. 2 camshaft to the camshaft housing.

44. Confirm the marks and numbers on the camshaft bearing caps and place them each in their proper position and direction.

45. Install the camshaft No. 3 and camshaft No. 4 to the camshaft housing.

46. Confirm the marks and numbers on the camshaft bearing caps and place them each in their proper position and direction.

47. Temporarily install the 8 bolts in the order shown in the illustration. Tighten bolts to 7 ft. lbs. (10 Nm).

48. Install the camshaft sub-assembly.

49. Apply seal packing in a continuous line.

50. Install the camshaft housing and tighten the 13 bolts in the order shown in the illustration. Tighten to 21 ft. lbs. (28 Nm).

➡Make sure that the knock pin of the camshaft is positioned as shown in the illustration before installing the camshaft housing.

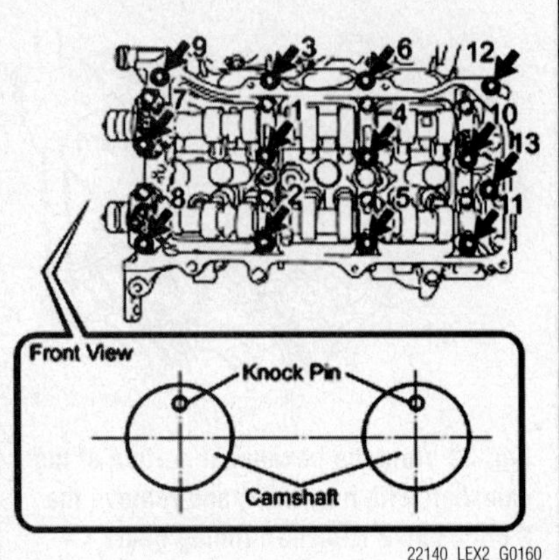

Fig. 45 Install the camshaft housing and tighten the 13 bolts in the order shown

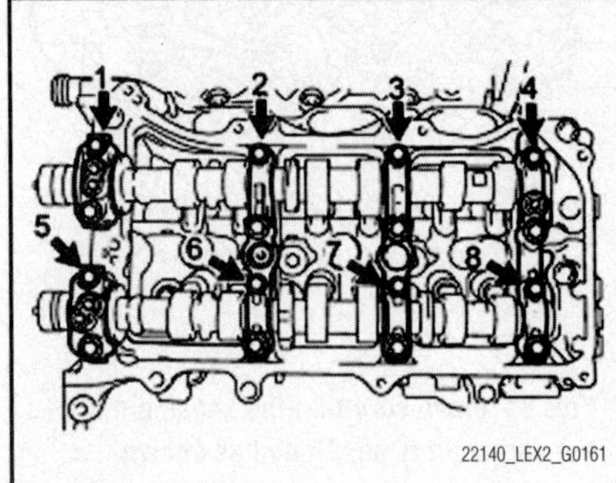

Fig. 46 Tighten the 8 bolts in the order shown

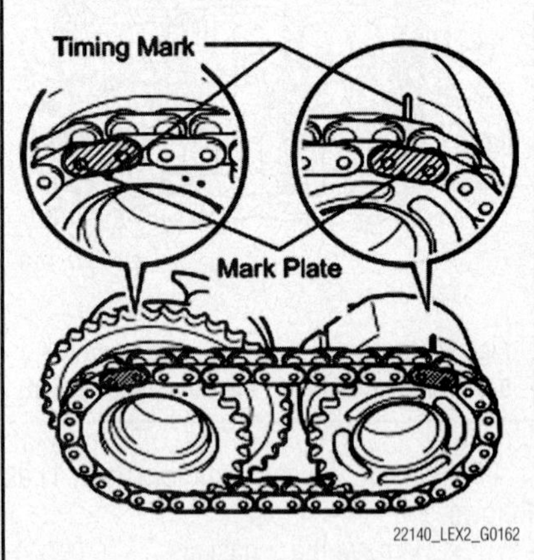

Fig. 47 Align the mark plate (yellow) with the timing marks (1 dot mark) of the camshaft timing gears

51. Tighten the 8 bolts in the order shown in the illustration. Tighten to 12 ft. lbs. (16 Nm).

52. Install the chain tensioner and tighten the mounting the bolt to 15 ft. lbs. (21 Nm).

53. While pushing in the tensioner,

insert a pin of 1.0 mm (0.039 in.) into the hole to fix it in place.

54. Install the camshaft timing gears and No.2 chain for (Bank 1) as follows:

55. Align the mark plate (yellow) with the timing marks (1 dot mark) of the camshaft timing gears as shown in the illustration.

56. Apply a small amount of engine oil to the bolt threads and bolt-seating surface.

57. Align the knock pin of the camshaft with the pin hole of the camshaft timing gear. Install the camshaft timing gear and camshaft timing exhaust gear with the No. 2 chain installed

58. Hold the hexagonal portion of the camshaft with a wrench, and tighten the 2 bolts to 74 ft. lbs. (100 Nm).

59. Remove the pin from the No. 2 chain tensioner.

60. Install the chain tensioner and tighten the mounting the bolt to 15 ft. lbs. (21 Nm).

61. While pushing in the tensioner, insert a pin of 1.0 mm (0.039 in.) into the hole to fix it in place.

62. Install the camshaft timing gears No. 2 for (Bank 2) as follows:

63. Align the mark plate (yellow) with the timing marks (1 dot mark) of the camshaft timing gears as shown in the illustration.

64. Apply a small amount of engine oil to the bolt threads and bolt-seating surface.

65. Align the knock pin of the camshaft with the pin hole of the camshaft timing gear. Install the camshaft timing gear and camshaft timing exhaust gear with the No. 2 chain installed

66. Hold the hexagonal portion of the camshaft with a wrench, and tighten the 2 bolts to 74 ft. lbs. (100 Nm).

67. Remove the pin from the No. 2 chain tensioner.

68. Install the timing components.

69. Install the timing cover.

70. Install the crankshaft pulley.

71. Install the drive belt.

72. Install the 4 bolts and 4 oil control valves.

73. Install the 4 bolts and the 4 VVT sensors.

74. Install the oil check valve bolt, oil pipe union and oil pipe.

75. Install the ventilation valve.

76. Install the head cover with the 14 bolts as follows:

- Tighten bolt A to 15 ft. lbs. (21 Nm)
- Tighten the remainder of the bolts to 7ft. lbs. (10 Nm).

➡Do not start the engine for at least 2 hours after the installation.

77. Install the oil control valve filter RH and LH to the oil pipe union. Install new gaskets and temporarily install the oil pipe (on the head cover side).

78. Install a new gasket and temporarily install the oil pipe (on the cylinder head side) with the oil pipe check valve bolt.

79. Tighten the oil pipe union (on the head cover side) tighten to 44 ft. lbs. 60 Nm).

80. Tighten the oil check valve bolt (on the cylinder head side) tighten to 44 ft. lbs. 60 Nm).

81. Install the 4 oil control valves with the 4 bolts and tighten to 7 ft. lbs. (10 Nm).

82. Install the 4 VVT sensors with the 4 bolts. Tighten 7 ft. lbs. (10 Nm).

83. Apply adhesive around the ventilation valve and install.

84. Install the 6 spark plugs.

85. Install the ignition coils.

86. Install the air intake surge tank sub-assembly.

87. Install all hoses and electrical connectors.

88. Install the service plug and cover.

89. Connect the negative battery cable.

90. Install the luggage floor mat.

CRANKSHAFT FRONT SEAL

REMOVAL & INSTALLATION

See Figures 48 and 49.

CAUTION

The GS450h hybrid system contains a 288V high-voltage system with a strong alkali solution of potassium hydroxide. Be sure to follow the instructions in this manual to handle the system correctly. Failure to do so may result in serious injury or electrocution. Engineer must undergo special training to be able to perform high-voltage system inspection and servicing.

CAUTION

All high-voltage wire harness connectors are colored orange. The HV battery and other high-voltage components have "High Voltage" caution labels. Do not carelessly touch these wires and components.

CAUTION

Before inspecting or servicing the high-voltage system, be sure to follow safety measures, such as wearing insulated gloves and removing the service plug to prevent electrocution. Carry the removed service plug in your pocket to prevent other technicians from reinstalling it while you are servicing the vehicle.

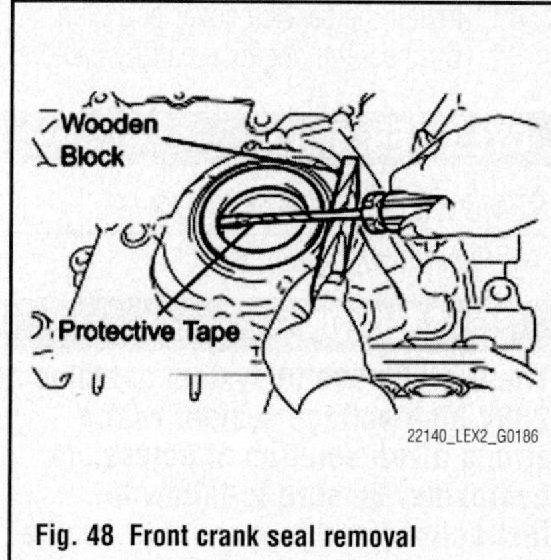

Fig. 48 Front crank seal removal

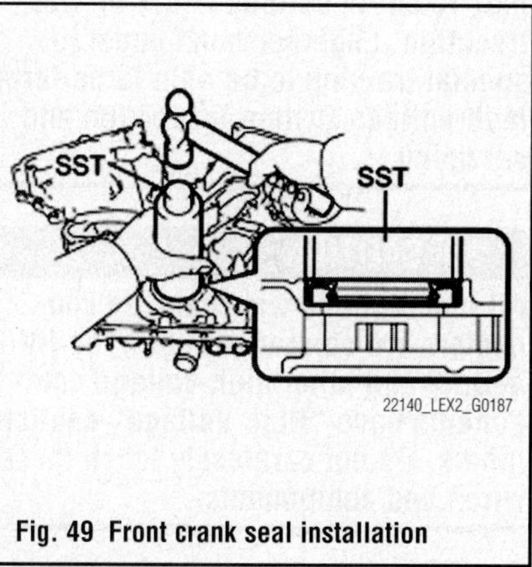

Fig. 49 Front crank seal installation

CAUTION

After removing the service plug, wait 10 minutes before touching any of the high-voltage connectors and terminals.

1. Remove the service cover and plug.
2. Disconnect the negative battery cable.
3. Remove the drive belt.
4. Using SST, loosen the crankshaft pulley set bolt.
5. Using the pulley set bolt and SST, remove the crankshaft pulley.
6. Using a screwdriver and wooden block, pry out the oil seal.

To install:

7. Install the front crankshaft seal and drive it into the cover with a seal driver.
8. Align the pulley set key with the key groove of the pulley, and slide on the pulley.
9. Install the pulley bolt and tighten bolt to 184 ft. lbs. (240 Nm).
10. Install the drive belt.

11. Install the service cover and plug.
12. Connect the negative battery cable.

CYLINDER HEAD

REMOVAL & INSTALLATION

See Figures 50 through 57.

⁂ CAUTION

The GS450h hybrid system contains a 288V high-voltage system with a strong alkali solution of potassium hydroxide. Be sure to follow the instructions in this manual to handle the system correctly. Failure to do so may result in serious injury or electrocution. Engineer must undergo special training to be able to perform high-voltage system inspection and servicing.

⁂ CAUTION

All high-voltage wire harness connectors are colored orange. The HV battery and other high-voltage components have "High Voltage" caution labels. Do not carelessly touch these wires and components.

⁂ CAUTION

Before inspecting or servicing the high-voltage system, be sure to follow safety measures, such as wearing insulated gloves and removing the service plug to prevent electrocution. Carry the removed service plug in your pocket to prevent other technicians from reinstalling it while you are servicing the vehicle.

⁂ CAUTION

After removing the service plug, wait 10 minutes before touching any of the high-voltage connectors and terminals.

1. Before servicing the vehicle, refer to the Precautions Section.
2. Remove the service grip.
3. Disconnect the negative battery cable.
4. Drain the cooling system.
5. Drain the engine oil.
6. Relieve the fuel system pressure.
7. Remove or disconnect the following:
    - Timing chain
    - Timing chain vibration damper
    - Upper intake
    - VVT sensor
    - Oil control valves
    - Oil check valve bolt, oil pipe union and oil pipe
    - Valve covers
    - Intake manifold
    - Water outlet pipe
    - Rear water by-pass joint
    - Camshafts
    - Camshaft housing
    - Rocker arms
    - Valve lash adjusters
    - Valve stem cap
8. Loosen the cylinder head mounting bolts in several steps in the sequence shown.

➡**Head warpage or cracking could result from removing bolts in an incorrect order.**

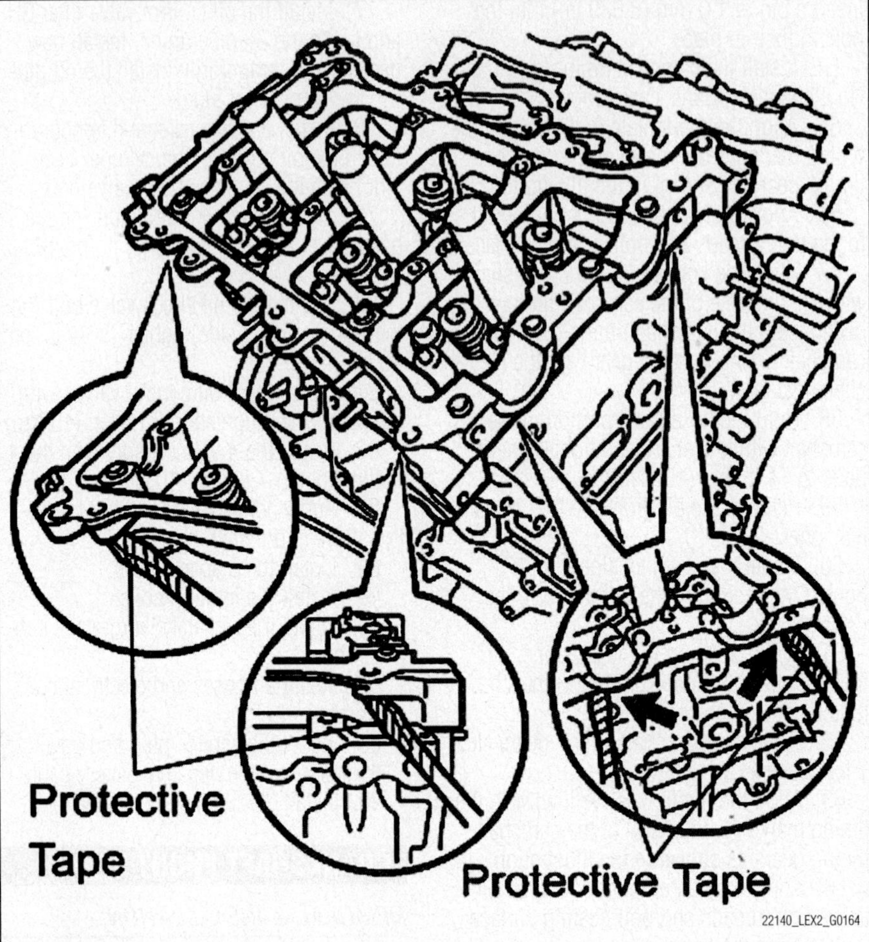

Fig. 50 Location of the prying points when removing the camshaft housing—3.5L (2GR-FSE) engine

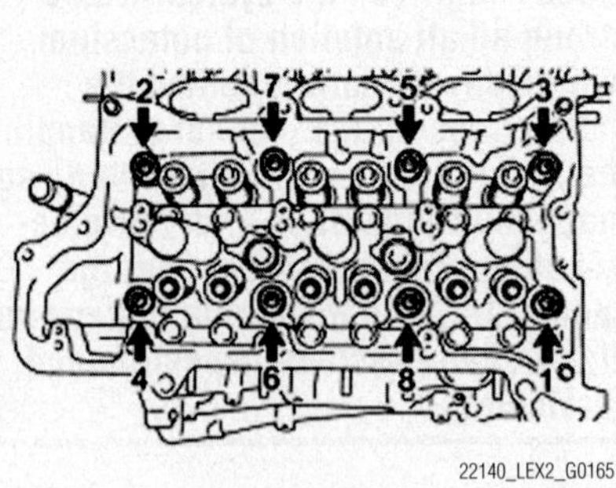

Fig. 51 Loosen the cylinder head mounting bolts in several steps (Bank 1)

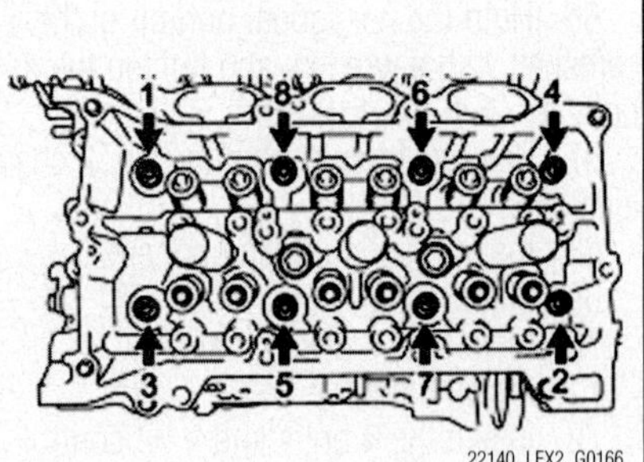

Fig. 52 Loosen the cylinder head mounting bolts in several steps (Bank 2)

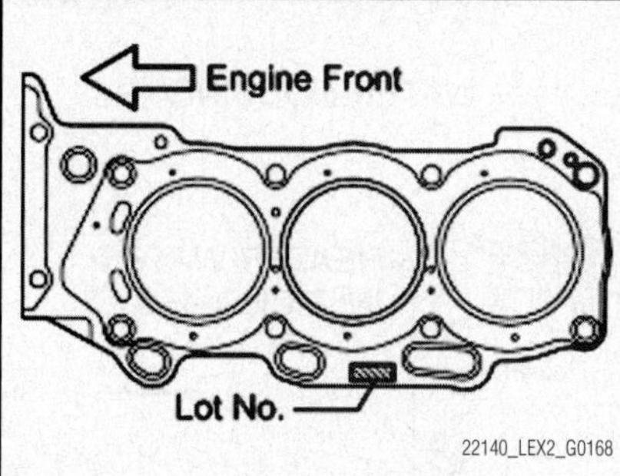

Fig. 53 Place the cylinder head gasket on the cylinder block surface with the front face of the Lot No. stamp upward.

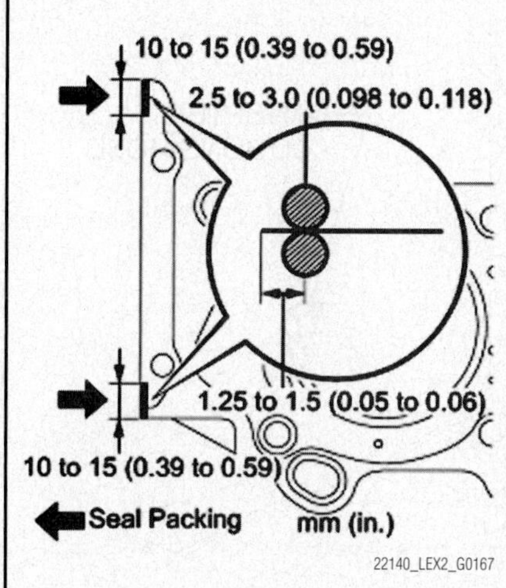

Fig. 54 Apply a continuous line of the seal packing to a new cylinder head

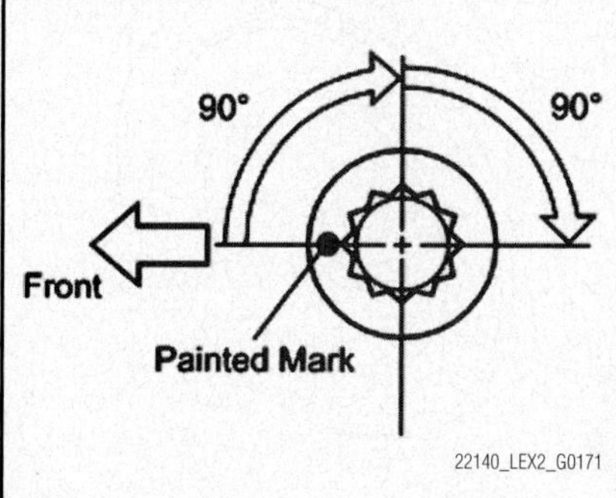

Fig. 55 Mark the cylinder head bolt heads with paint as shown in the illustration

9. Remove the head bolts and plate washers.

10. Remove the cylinder head and gasket.

To install:

11. Install a new cylinder head gasket with the Lot number stamp upper side facing upward.

12. Apply a continuous line of the seal

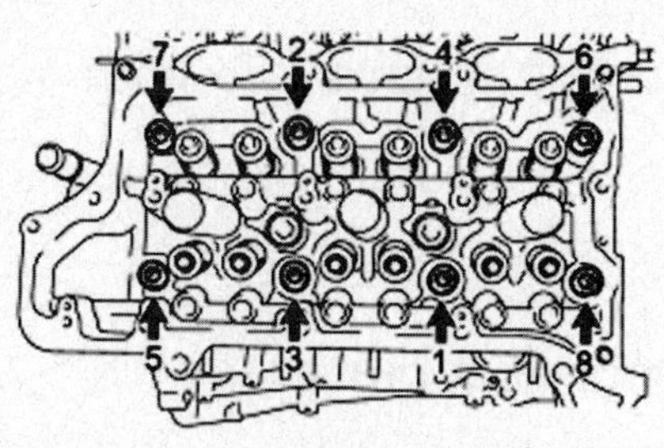

Fig. 56 Cylinder head tightening sequence—3.5L (2GR-FSE) engine (Bank 1)

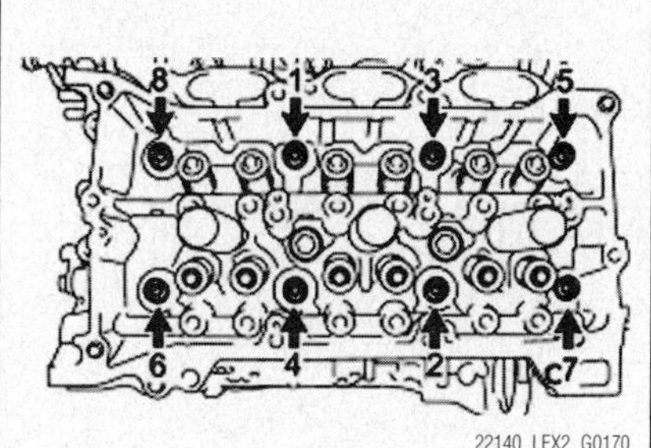

Fig. 57 Cylinder head tightening sequence—3.5L (2GR-FSE) engine (Bank 2)

packing to a new cylinder head gasket as shown in the illustration.

13. Install the cylinder head. Apply a light coat of engine oil to the threads and tighten the bolts in sequence as follows:

- Step 1: 27 ft. lbs. (36 Nm)
- Step 2: Plus 90 degrees
- Plus an additional 90 degrees

14. Install or connect the following:

- Valve stem cap
- Valve lash adjusters
- Rocker arms

15. Install the camshaft housing as follows:

- Apply 0.138—0.177 in. (3.5—4.5 mm) wide bead of sealant to the contact surface
- Install the camshaft housing and tighten the bolts in sequence to 18 ft. lbs. (25 Nm).
- Install or connect the following:
- Camshafts
- Water outlet pipe. Tighten bolts to 7 ft. lbs. (10 Nm).
- Rear water by-pass joint. Tighten the bolts and nuts to 7 ft. lbs. (10 Nm).
- Intake manifold. Tighten bolts to 15 ft. lbs. (21 Nm).
- Valve covers. Tighten bolts to 15 ft. lbs. (21 Nm).
- Upper intake
- VVT sensor
- Oil control valves
- Oil check valve bolt, oil pipe union and oil pipe
- Timing chain vibration damper
- Timing chain

16. Install the service plug grip

17. Connect the negative battery cable.

18. Refill the cooling system to the correct level.

19. Refill the engine with oil to the correct level.

20. Start then engine and check for leaks.

ENGINE ASSEMBLY

REMOVAL & INSTALLATION

See Figures 58 through 67.

Before servicing any vehicle, please be sure to read all of the following precautions, which deal with personal safety, prevention of component damage, and important points to take into consideration when servicing a motor vehicle:

✲✲ CAUTION

The GS450h hybrid system contains a 288V high-voltage system with a strong alkali solution of potassium hydroxide. Be sure to follow the instructions in this manual to handle the system correctly. Failure to do so may result in serious injury or electrocution. Engineer must undergo special training to be able to perform high-voltage system inspection and servicing.

✲✲ CAUTION

All high-voltage wire harness connectors are colored orange. The HV battery and other high-voltage components have "High Voltage" caution labels. Do not carelessly touch these wires and components.

✲✲ CAUTION

Before inspecting or servicing the high-voltage system, be sure to follow safety measures, such as wearing insulated gloves and removing the service plug to prevent electrocution. Carry the removed service plug in your pocket to prevent other technicians from reinstalling it while you are servicing the vehicle.

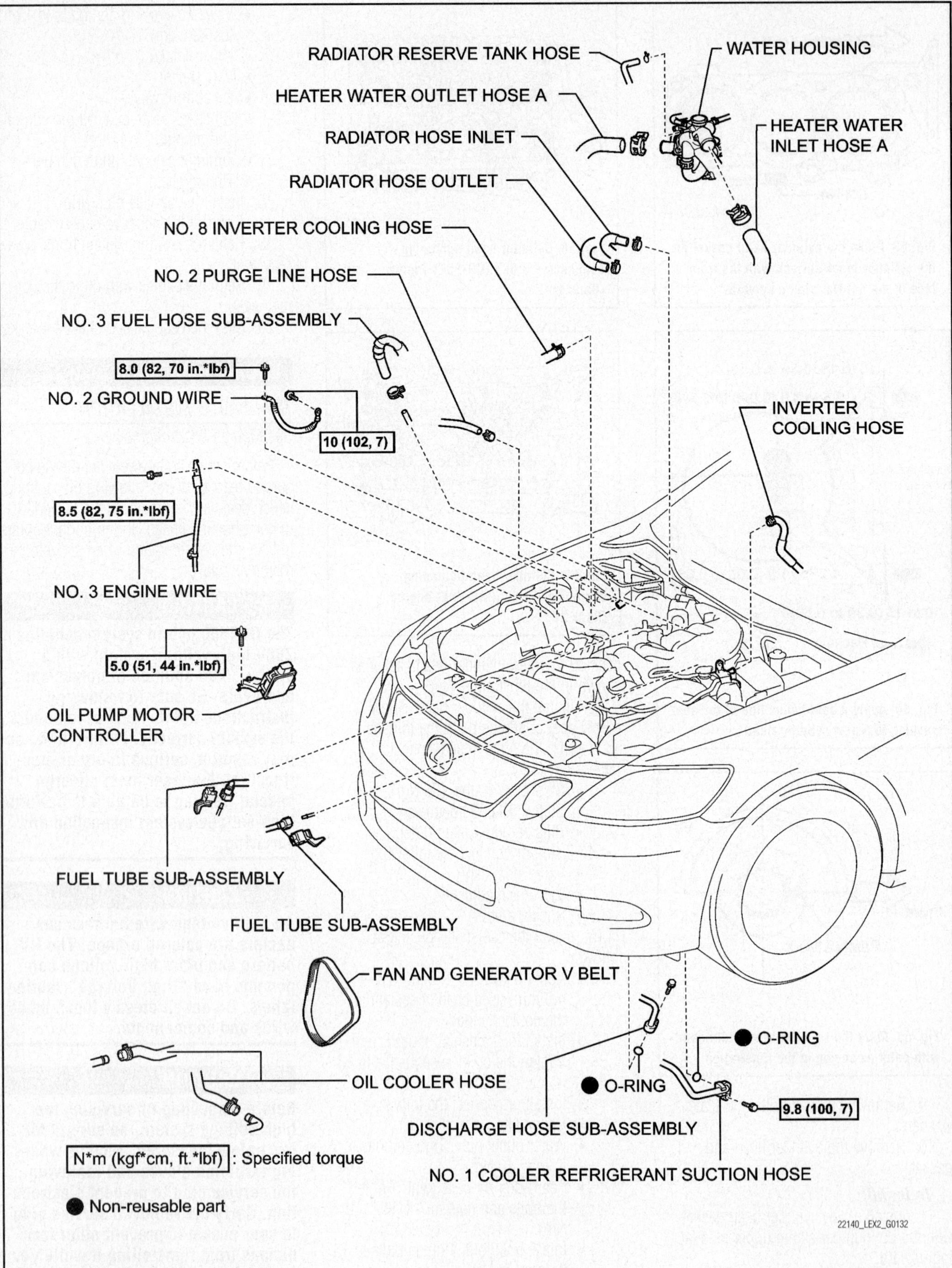

Fig. 58 Engine assembly components (1)

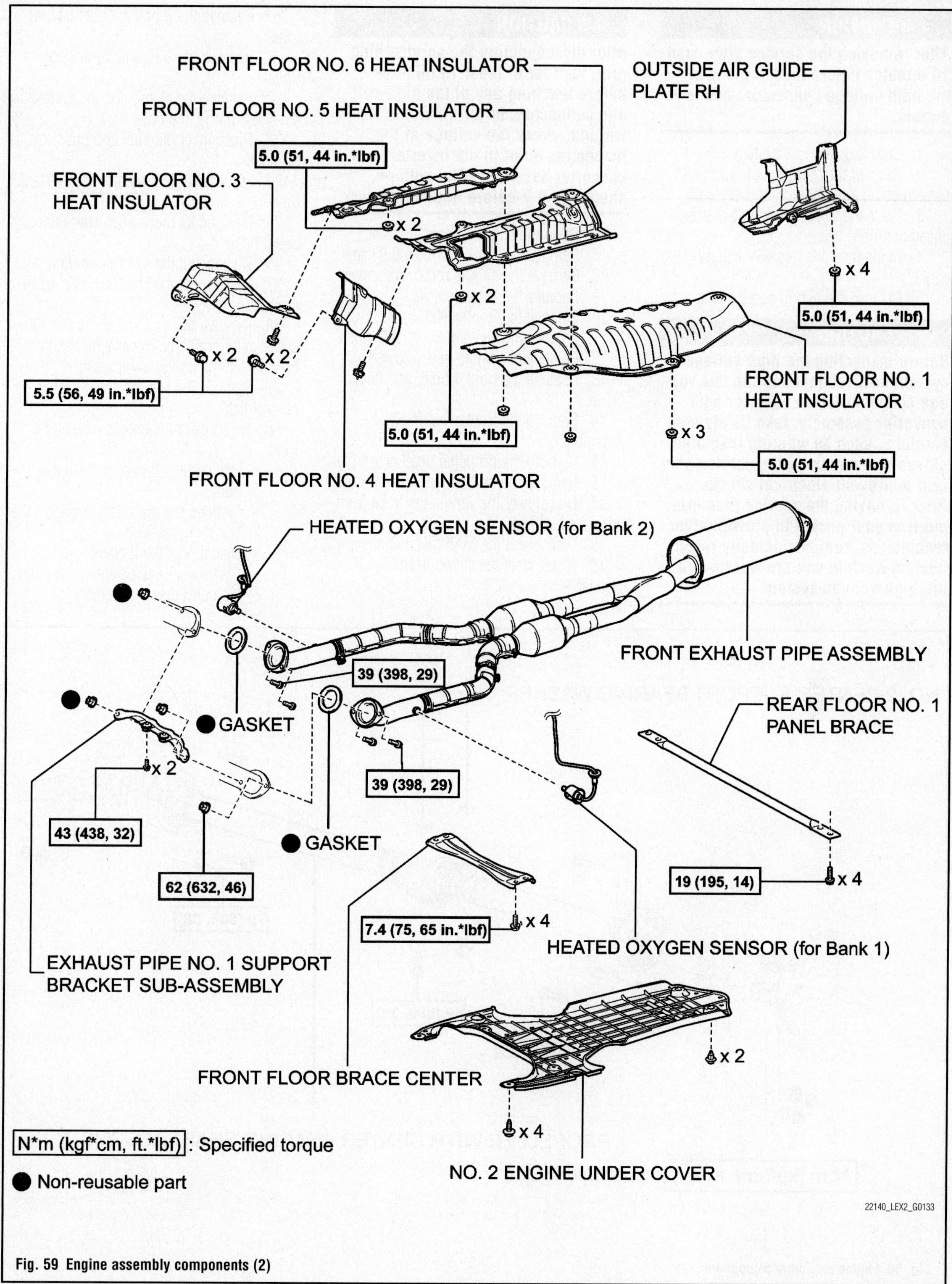

Fig. 59 Engine assembly components (2)

CAUTION

After removing the service plug, wait 10 minutes before touching any of the high-voltage connectors and terminals.

1. Discharge the fuel system.
2. Recover the refrigerant from the A/C system using a refrigerant recovery unit.
3. Place the front wheels in the straight ahead position.
4. Disconnect the negative battery cable.
5. Remove the hybrid service plug grip.

CAUTION

Before inspecting the high-voltage system or disconnecting the low voltage connector of the inverter with converter assembly, take safety precautions, such as wearing insulated gloves and removing the service plug grip to prevent electrical shocks. After removing the service plug grip, put it in your pocket to prevent other technicians from accidentally reconnecting it while you are working on the high-voltage system.

CAUTION

After disconnecting the service plug grip, wait for at least 10 minutes before touching any of the high-voltage connectors or terminals. After waiting, check the voltage at the inspection point in the inverter with converter assembly. The voltage should be 0 V before beginning work.

6. Drain the engine oil and coolant.
7. Remove the cool air intake duct seal.
8. Remove the engine room side cover.
9. Remove the V-bank cover.
10. Remove the front wheels.
11. Remove the air cleaner cap and hose.
12. Remove all engine under covers.
13. Remove the front 4 bolts and floor brace.
14. Remove the front exhaust pipe assembly.
15. Remove the propeller shaft assembly.
16. Remove the bolt and air cleaner inlet.
17. Disconnect the ventilation hose from the cylinder head.
18. Disconnect the MAF meter connector.
19. Disconnect the clamp from the air cleaner.
20. Remove the 3 bolts, 2 clamps and case.
21. Remove the oil pump motor controller
22. Remove the clamp and disconnect the No. 2 purge line hose.
23. Disconnect the inlet and outlet radiator hoses.
24. Disconnect the radiator reserve tank hose.
25. Disconnect the heater water inlet and outlet hose.
26. Disconnect the inverter cooling hose.
27. Disconnect the No. 8 inverter cooling hose.
28. Remove the bolt and disconnect the cooler refrigerant suction and discharge hoses from the cooler compressor.
29. Remove the Hybrid Vehicle (HV) control ECU.
30. Disconnect the engine wire as follows:
   - Disconnect the clamp from the body.
   - Remove the nut and disconnect the No. 4 engine wire.
   - Disconnect the 2 clamps.
   - Disconnect the 2 connectors from the compressor assembly.

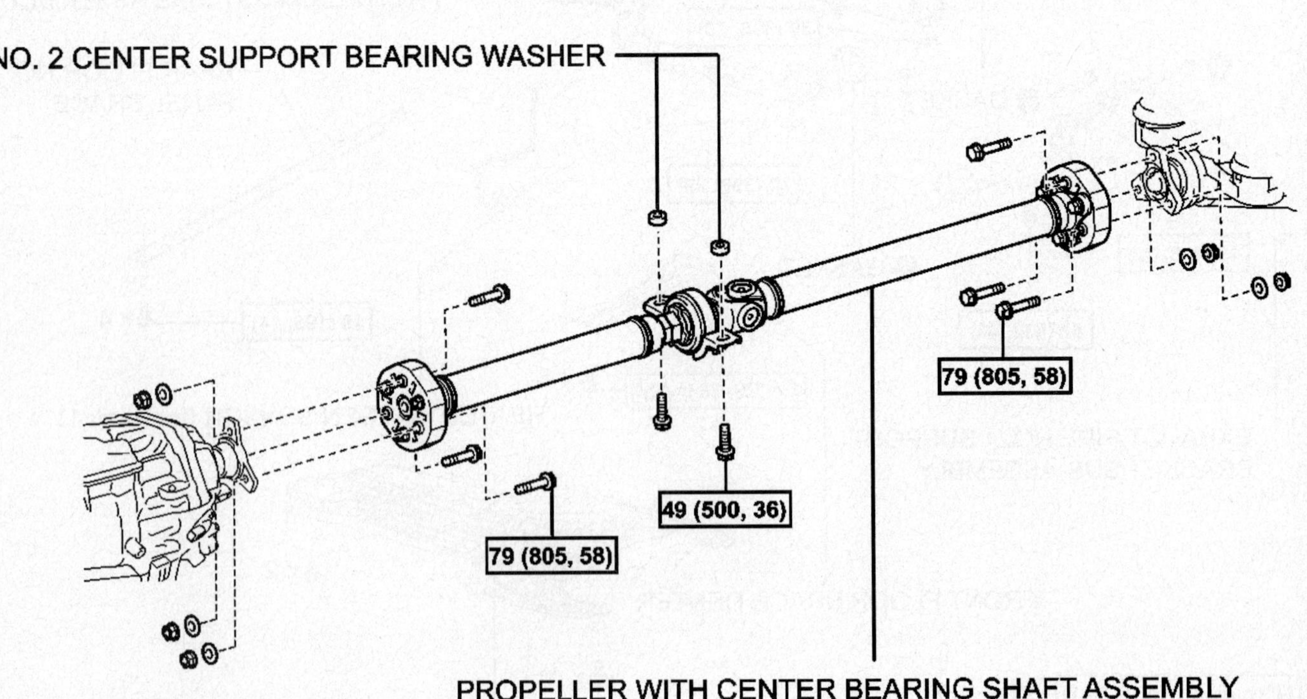

Fig. 60 Engine assembly components (3)

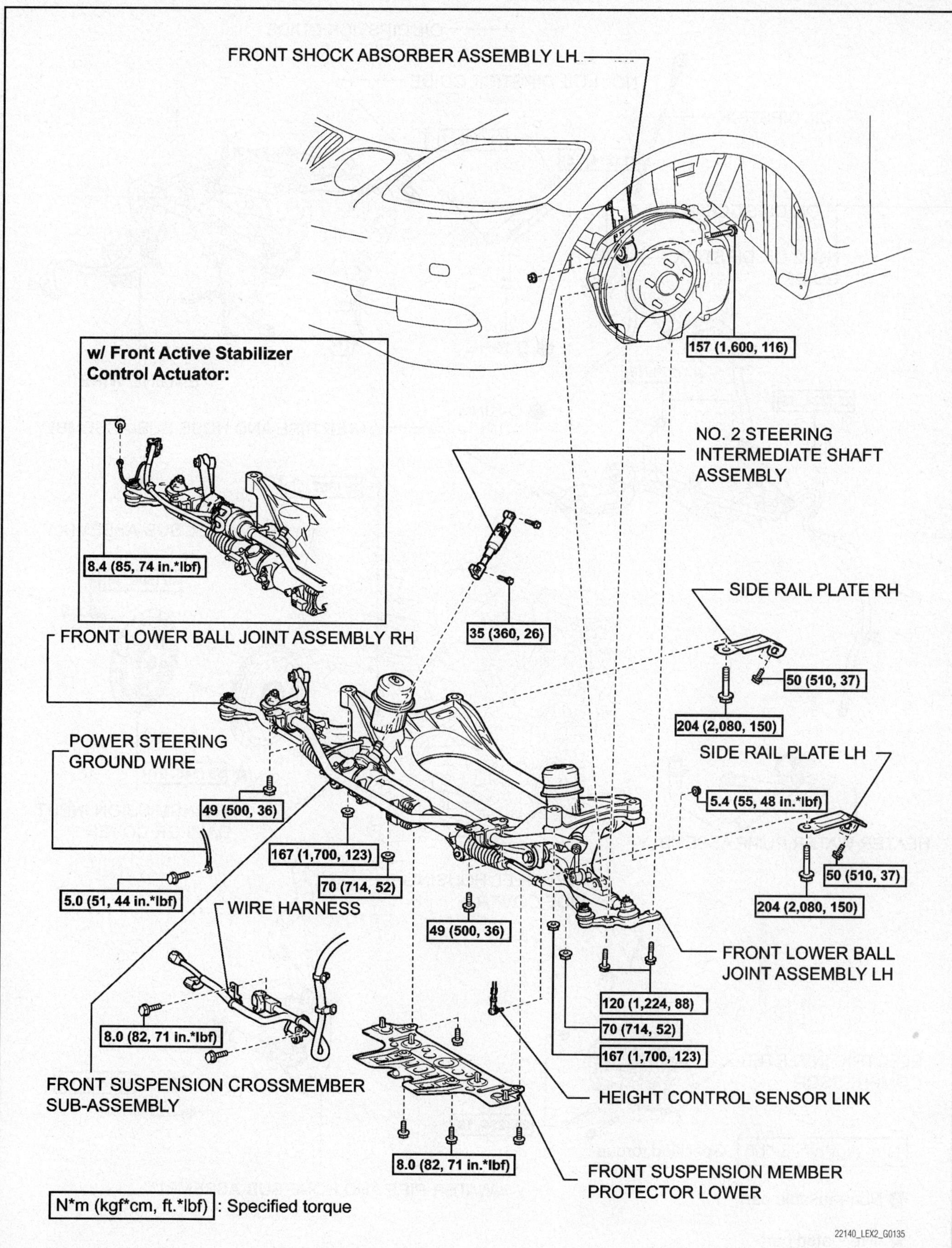

Fig. 61 Engine assembly components (4)

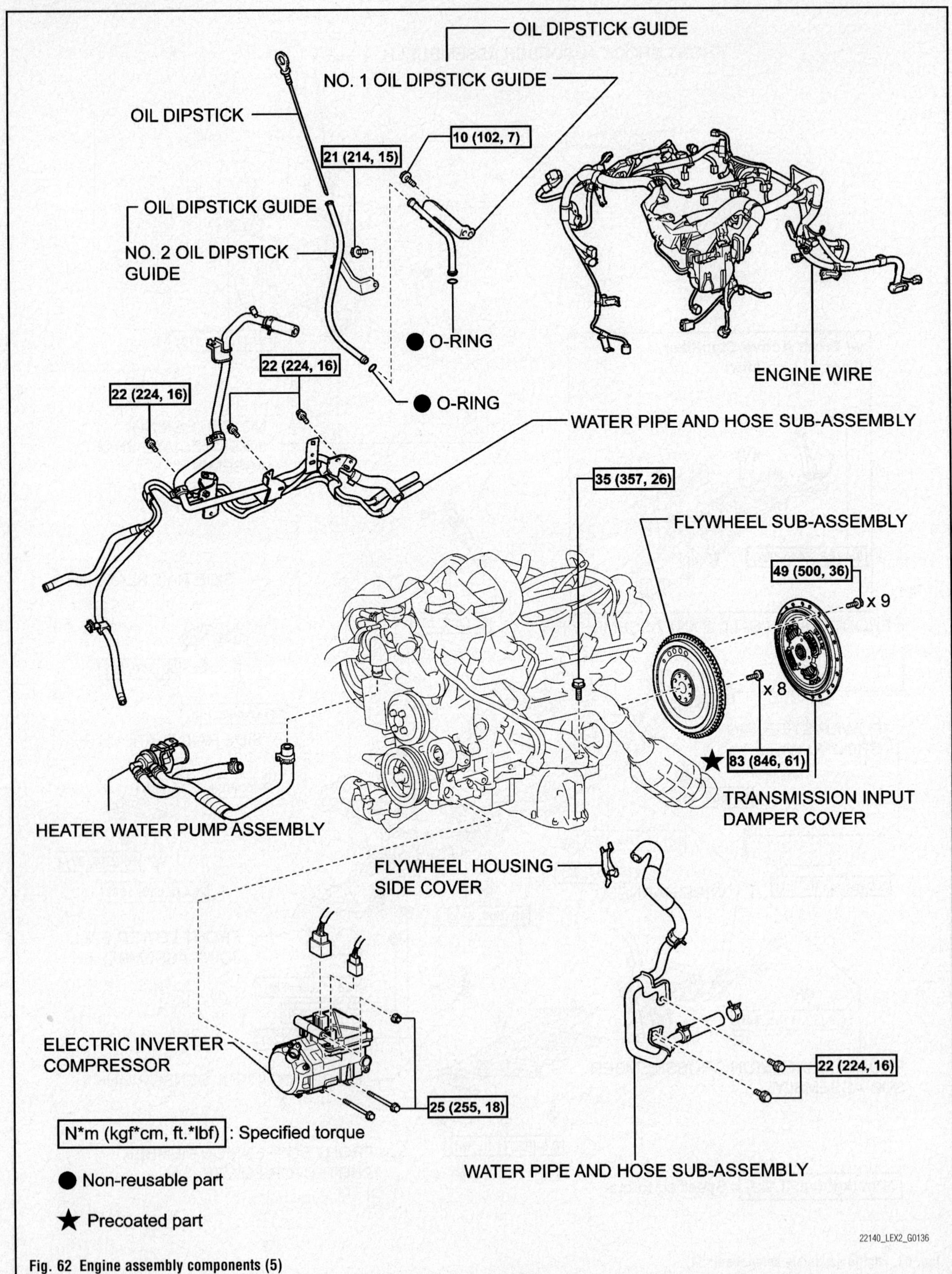

Fig. 62 Engine assembly components (5)

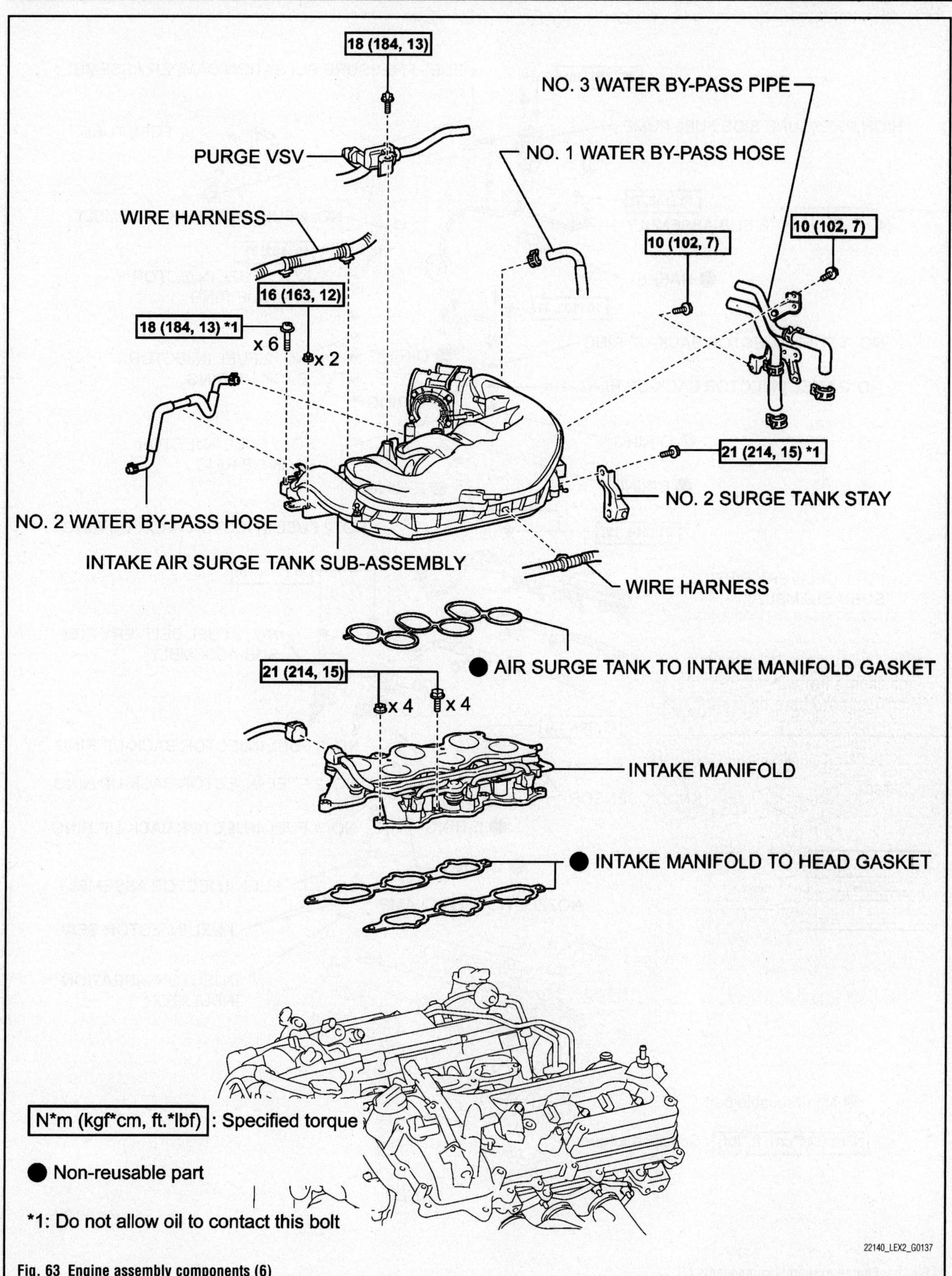

Fig. 63 Engine assembly components (6)

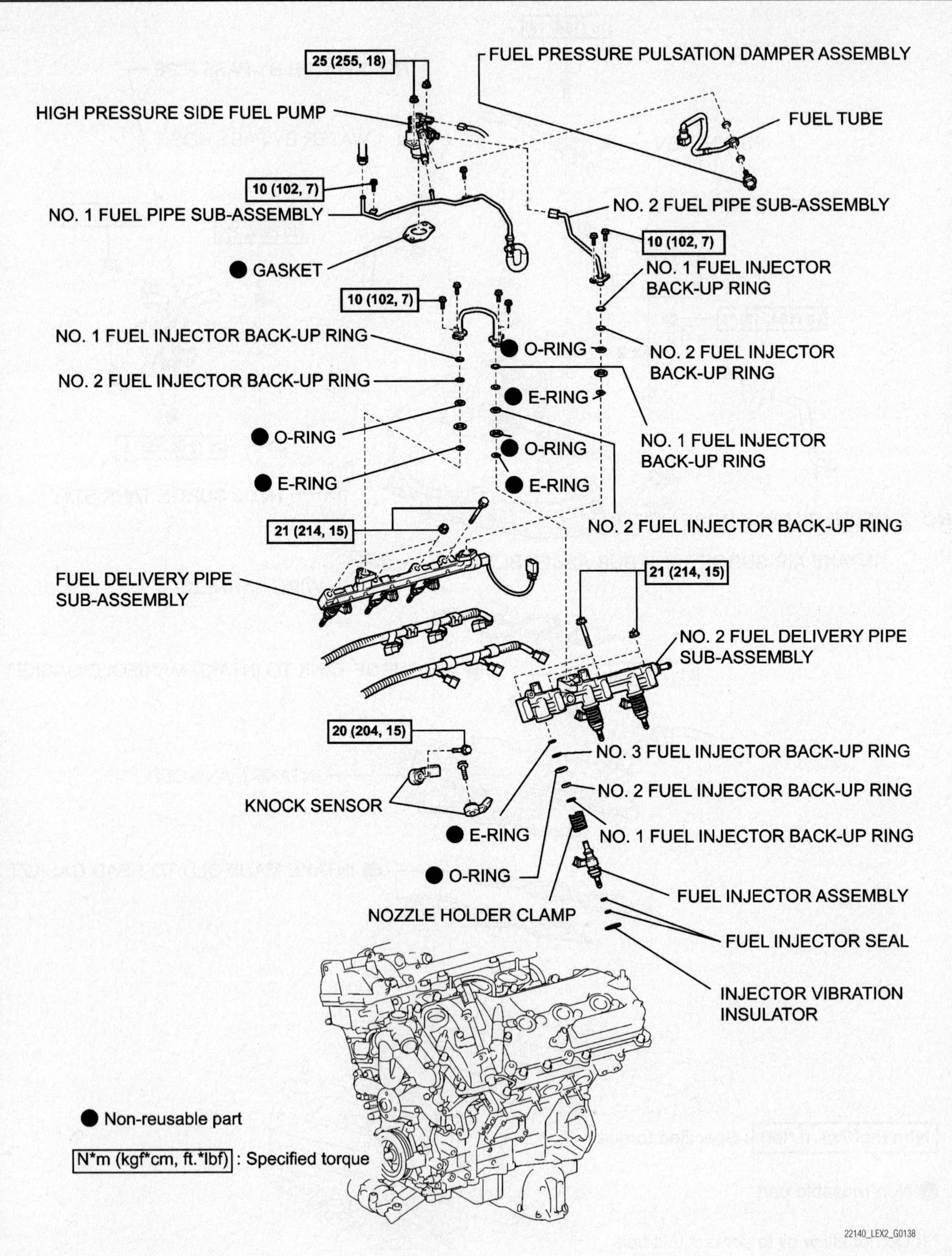

Fig. 64 Engine assembly components (7)

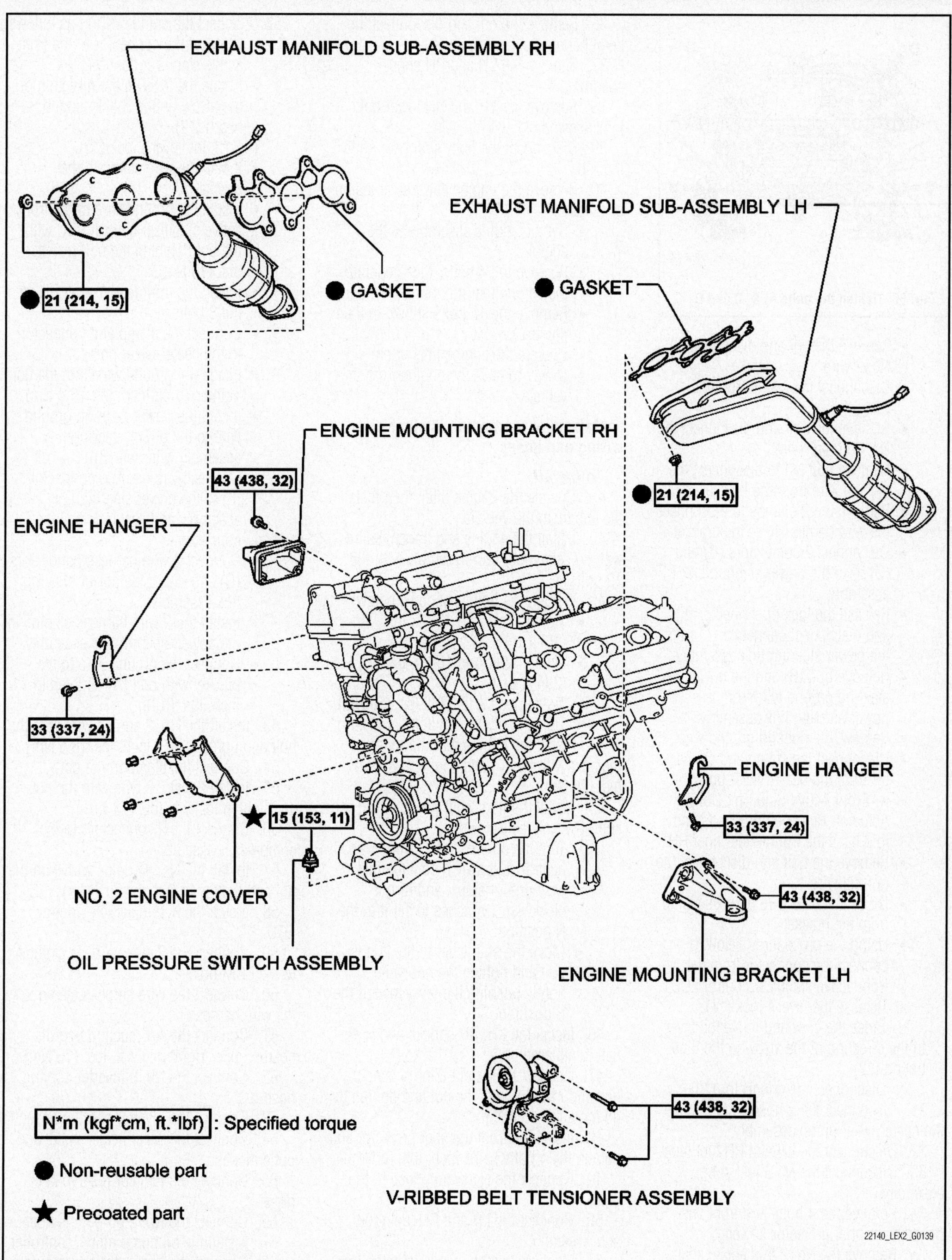

Fig. 65 Engine assembly components (8)

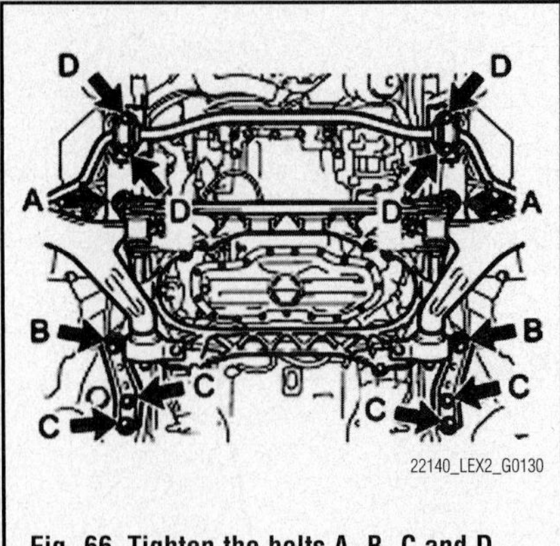

Fig. 66 Tighten the bolts A, B, C and D

- Remove the nut and disconnect the No. 2 wire.
- Disconnect the clamp from the body.
- Remove the bolt and disconnect the No. 3 engine wire.
- Remove bolt (A) to disconnect the ground wire from the bracket.
- Remove the 2 clamps to disconnect the wire harness from the bracket.
- Disconnect 2 connectors (C) and (D) from the power steering link assembly.
- Release the lock of connector (E) and disconnect connector (E) from the power steering link assembly.
- Remove bolt (B) and the power steering ground wire from the power steering link assembly.
- Remove the nut and ground wire.
- Disconnect the clamp and connector from the heater water pump.
- w/ Front Active Stabilizer Control Actuator: Remove the 8 bolts and fold back the front fender liner RH.
- Remove the bolt and disconnect the ground wire.
- Disconnect the wire harness clamp from the bracket.
- Using the procedures below, disconnect the connector from the front active stabilizer control ECU

a. Release the lever's lock (*1).

b. Press the claw and move the lever in the direction of the arrow in the illustration (*2).

c. Disconnect the connector (*3).

31. Disconnect the 2 hoses and remove the heater water pump assembly

32. Disconnect the LH and RH fuel tube.

33. Disconnect the No.3 fuel sub-assembly.

34. Remove the 4 bolts and front suspension member protector Lower.

35. Remove the No. 2 Immediate shaft assembly.

36. Remove the nut and disconnect the height control sensor link.

37. Remove the LH and RH shock absorber.

38. Remove the LH and RH lower ball joint assembly.

39. Disconnect the floor shift gear shifting rod.

40. Remove the engine and transmission as follows:

- Set the engine assembly to the engine.
- Remove the 4 bolts, then separate the engine rear mounting member.
- Remove the 12 bolts shown in the illustration.
- Operate the engine lifter, then slowly remove the engine from the vehicle.

➡Make sure the engine is clear of all wiring and hoses.

To install:

41. Operate the engine lifter, then install the engine to the vehicle.

42. Install the engine and transmission with crossmember sub-assembly with the 12 bolts.

43. Tighten the bolts as shown below:

- Tighten bolt A to 123 ft. lbs. (167 Nm).
- Tighten bolt A to 150 ft. lbs. (204 Nm).
- Tighten bolt A to 37 ft. lbs. (50 Nm).
- Tighten bolt A to 36 ft. lbs. (49 Nm).

44. Install the engine rear mounting member with the 4 bolts. Tighten to 19 ft. lbs. (26 Nm).

45. Install the floor shift gear shifting rod sub-assembly as follows:

- Turn the control shaft lever of the neutral start switch counterclockwise until it stops, and turn it clockwise 2 notches to set it to the N position.
- Move the shift lever to the N position and tighten the nut while lightly pushing the lever toward the R position.

46. Install the LH, RH shocks and lower ball joints assembly.

47. Connect the height control sensor link and install it with the nut and tighten to 48 inch. lbs. (5.4 Nm).

48. Install the front member protector and tighten the 4 bolts to 71 inch. lbs. (8 Nm).

49. Connect the No. 3 fuel hose to the pipe.

50. Connect the LH and RH fuel tube sub-assembly.

51. Connect the 2 hoses and install the water pump.

52. Connect the engine wiring as follows

- Connect the connector to the front active stabilizer control (*1).
- Rotate the lever in the direction of the arrow until a "click" sound is heard (*2).
- Lock the lever's lock (*3).
- Install the wire clamp to the bracket.
- Install the ground wire to the front active stabilizer control ECU with the bolt. Tighten the bolt to 7 ft. lbs. (10 Nm).
- Install the front fender liner RH with the 8 bolts.
- Connect the clamp and connector to the heater water pump.
- Install the ground wire with the nut. Tighten to 48 inch. lbs. (5.4 Nm).
- Install the power steering ground wire to the power steering link assembly with bolt (B).
- Connect wire harness connector (E) to the power steering link assembly and securely lock the connector.
- Connect 2 wire harness connectors (C) and (D) to the power steering link assembly.
- Install the 2 wire harness clamps to the power steering link assembly.
- Connect the ground wire to the bracket with bolt (A). Tighten to 71 inch. lbs. (8 Nm).

53. Install the No. 3 engine wire with the bolt and tighten to 75 inch. lbs. (8.5 Nm).

54. Connect the clamp to the body.

55. Install the No. 2 wire with the nut and tighten to 10 ft. lbs. (13 Nm).

56. Connect the 2 connectors to the compressor assembly.

57. Install the No. 4 engine wire with the nut and tighten to 9 ft. lbs. (12 Nm).

58. Install the hybrid vehicle control ECU.

59. Connect the 3 wire harness clamps to the ECM bracket.

60. Connect the wire harness clamp to the wire harness.

61. Connect the A/C suction and discharge hose. Tighten to 7 ft. lbs. (10 Nm).

62. Connect the No. 8 inverter cooling hose.

63. Connect the inverter cooling hose.

64. Connect the water heater outlet and outlet hoses.

65. Connect the radiator reserve tank hose.

66. Connect the No. 2 purge line hose.

67. Install the oil pump motor controller.

68. Connect the connector and securely lock the lock lever.

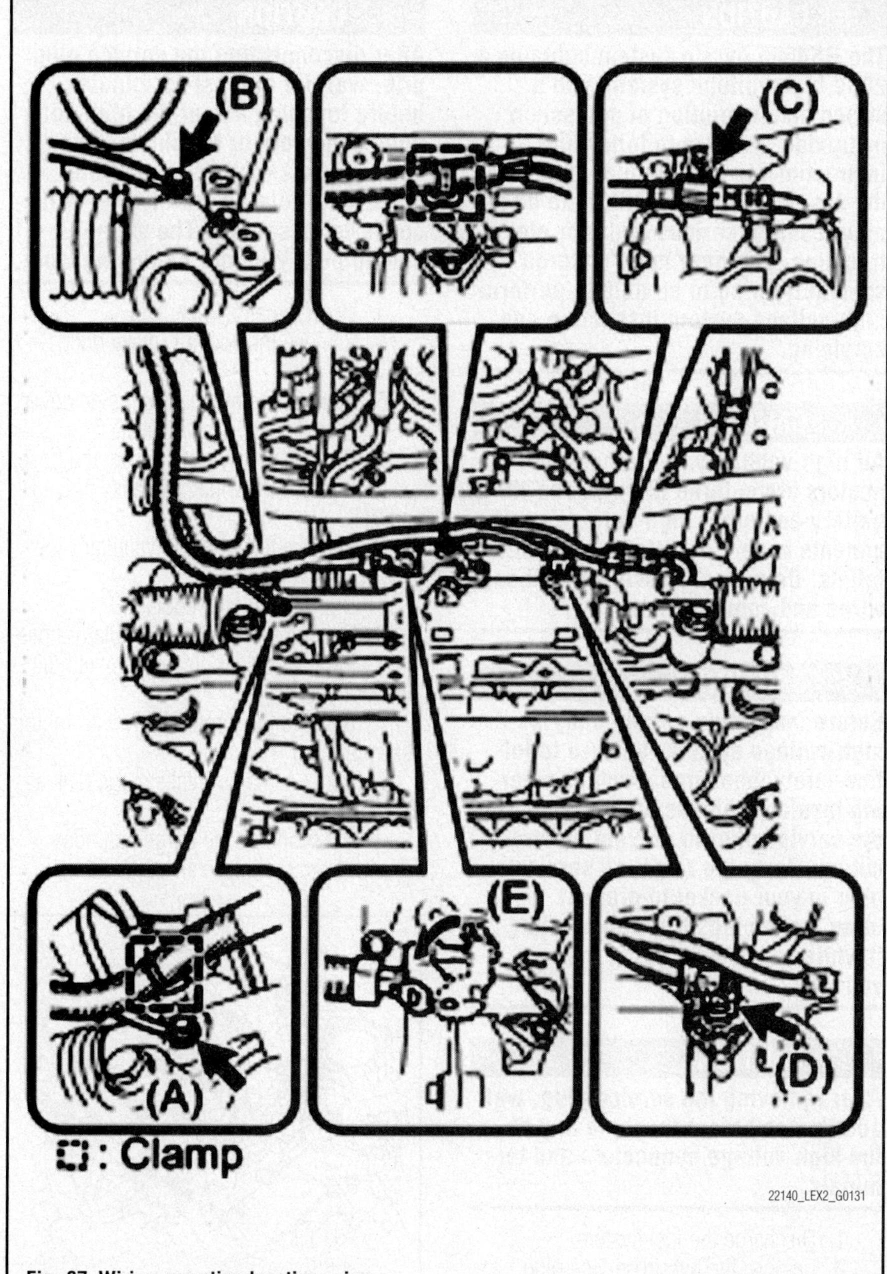

Fig. 67 Wiring mounting locations view

69. Install air cleaner component.
70. Install the propeller shaft assembly.
71. Install the exhaust pipe assembly.
72. Install the floor brace with the 4 bolts and tighten to 65 inch. lbs. (7.4 Nm).
73. Install engine oil.
74. Install the engine coolant and bleed system.
75. Vacuum and recharge A/C system, add refrigerant if any was lost.
76. Install the service plug grip.
77. Connect the negative battery cable.
78. Install the battery service hole cover.
79. Install the luggage compartment floor mat.
80. Certain systems need to be initialized after reconnecting the cable to the negative (-) battery terminal.
81. Inspect for engine oil and coolant leaks.
82. Inspect the A/C system for leaks.
83. Check the fuel system for leaks/
84. Inspect for exhaust leaks.
85. Install engine under covers.
86. Install the vehicles front wheels.
87. Check and adjust the front wheel alignment.
88. Check ignition timing and engine speed.
89. Install all engine room covers.
90. Install the intake duct seal with the 7 clips.

EXHAUST MANIFOLD

REMOVAL & INSTALLATION

See Figures 68 and 69.

Before servicing any vehicle, please be sure to read all of the following precautions, which deal with personal safety, prevention of component damage, and important points to take into consideration when servicing a motor vehicle:

**** CAUTION**

The GS450h hybrid system contains a 288V high-voltage system with a strong alkali solution of potassium hydroxide. Be sure to follow the instructions in this manual to handle the system correctly. Failure to do so may result in serious injury or electrocution. Engineer must undergo special training to be able to perform high-voltage system inspection and servicing.

**** CAUTION**

All high-voltage wire harness connectors are colored orange. The HV battery and other high-voltage components have "High Voltage" caution labels. Do not carelessly touch these wires and components.

**** CAUTION**

Before inspecting or servicing the high-voltage system, be sure to follow safety measures, such as wearing insulated gloves and removing the service plug to prevent electrocution. Carry the removed service plug in your pocket to prevent other technicians from reinstalling it while you are servicing the vehicle.

**** CAUTION**

After removing the service plug, wait 10 minutes before touching any of the high-voltage connectors and terminals.

1. Remove the service plug grip.
2. Disconnect the negative battery cable.
3. Remove any components to access exhaust manifold.
4. Remove the dipstick for RH exhaust manifold.
5. Disconnect the heated oxygen sensor connector.

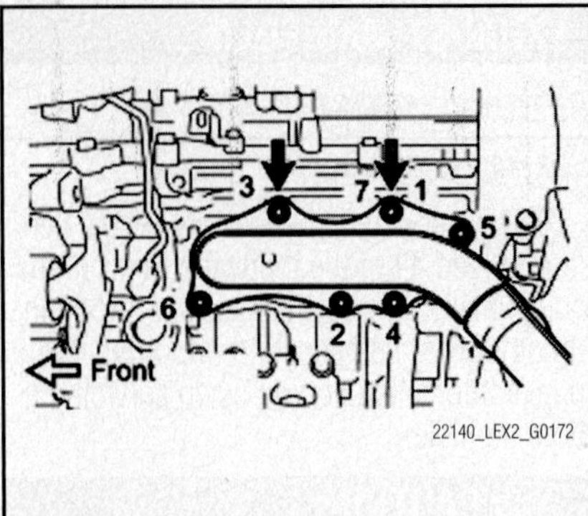

Fig. 68 Exhaust manifold tightening sequence—LH

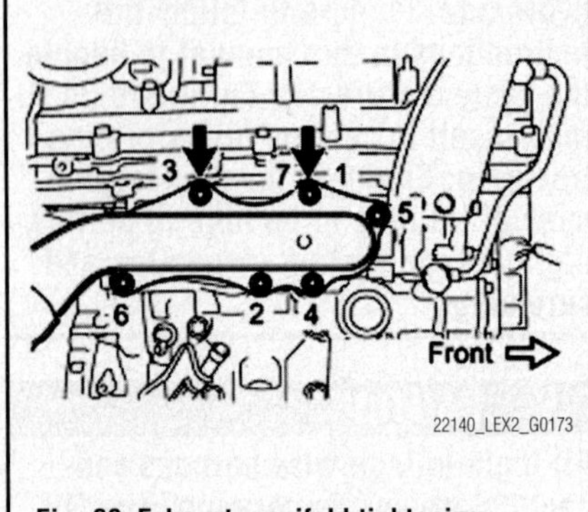

Fig. 69 Exhaust manifold tightening sequence—RH

6. Disconnect the exhaust pipe front ends from the exhaust manifold and remove the 2 gaskets.
7. Remove the 6 nuts, exhaust manifold and gasket.

To install:

8. Install a new gasket.
9. Install the exhaust manifold to the cylinder head with the 6 nuts in the order shown in the illustration.
10. Tighten the mounting nuts to 15 ft. lbs. (21 Nm).
11. Install the dipstick for RH exhaust manifold.
12. Install 2 new gaskets and the exhaust pipe front ends to the exhaust manifolds with the 4 bolts and 2 nuts. Tighten to 29 ft. lbs. (39 Nm).
13. Install any components that may have been removed to access exhaust manifold.
14. Remove the service plug grip.
15. Disconnect the negative battery cable.
16. Check for exhaust leaks.

INTAKE MANIFOLD

REMOVAL & INSTALLATION

See Figures 70 through 72.

**** CAUTION**

The GS450h hybrid system contains a 288V high-voltage system with a strong alkali solution of potassium hydroxide. Be sure to follow the instructions in this manual to handle the system correctly. Failure to do so may result in serious injury or electrocution. Engineer must undergo special training to be able to perform high-voltage system inspection and servicing.

**** CAUTION**

All high-voltage wire harness connectors are colored orange. The HV battery and other high-voltage components have "High Voltage" caution labels. Do not carelessly touch these wires and components.

**** CAUTION**

Before inspecting or servicing the high-voltage system, be sure to follow safety measures, such as wearing insulated gloves and removing the service plug to prevent electrocution. Carry the removed service plug in your pocket to prevent other technicians from reinstalling it while you are servicing the vehicle.

**** CAUTION**

After removing the service plug, wait 10 minutes before touching any of the high-voltage connectors and terminals.

1. Discharge the fuel system.
2. Remove the hybrid service plug grip.
3. Disconnect the negative battery cable.

**** CAUTION**

Before inspecting the high-voltage system or disconnecting the low voltage connector of the inverter with converter assembly, take safety precautions, such as wearing insulated gloves and removing the service plug grip to prevent electrical shocks. After removing the service plug grip, put it in your pocket to prevent other technicians from accidentally reconnecting it while you are working on the high-voltage system.

**** CAUTION**

After disconnecting the service plug grip, wait for at least 10 minutes before touching any of the high-voltage connectors or terminals. After waiting, check the voltage at the inspection point in the inverter with converter assembly. The voltage should be 0 V before beginning work.

4. Drain the engine coolant.
5. Remove the cool air intake duct seal.
6. Remove the engine room side cover.
7. Remove the V-bank cover.
8. Remove the air cleaner cap and hose.
9. Remove the wiper arm assembly LH and RH
10. Remove the cowl top ventilator louver sub-assembly.
11. Drain the engine coolant.
12. Disconnect the No. 2 ventilator hose.
13. Remove the air cleaner cap with air cleaner hose.
14. Remove the throttle body with motor assembly.
15. Remove the air intake surge tank as follows:
    - Disconnect the purge line hose from the intake air surge tank.

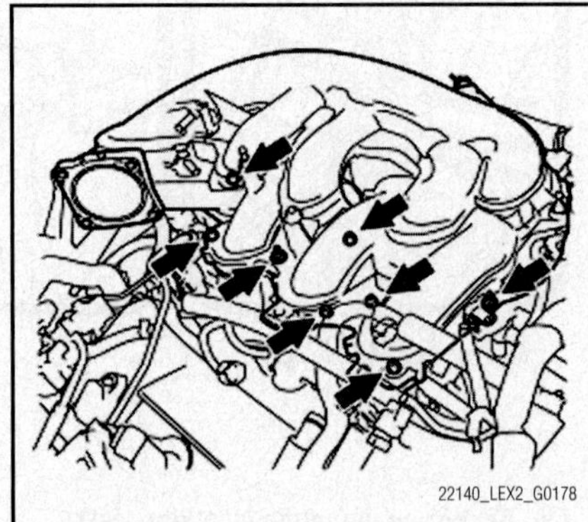

Fig. 70 Air intake surge tank removal

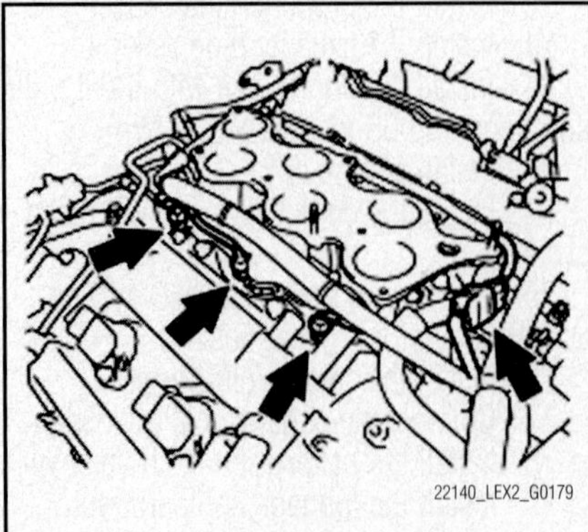

Fig. 71 Disconnect the 4 connectors

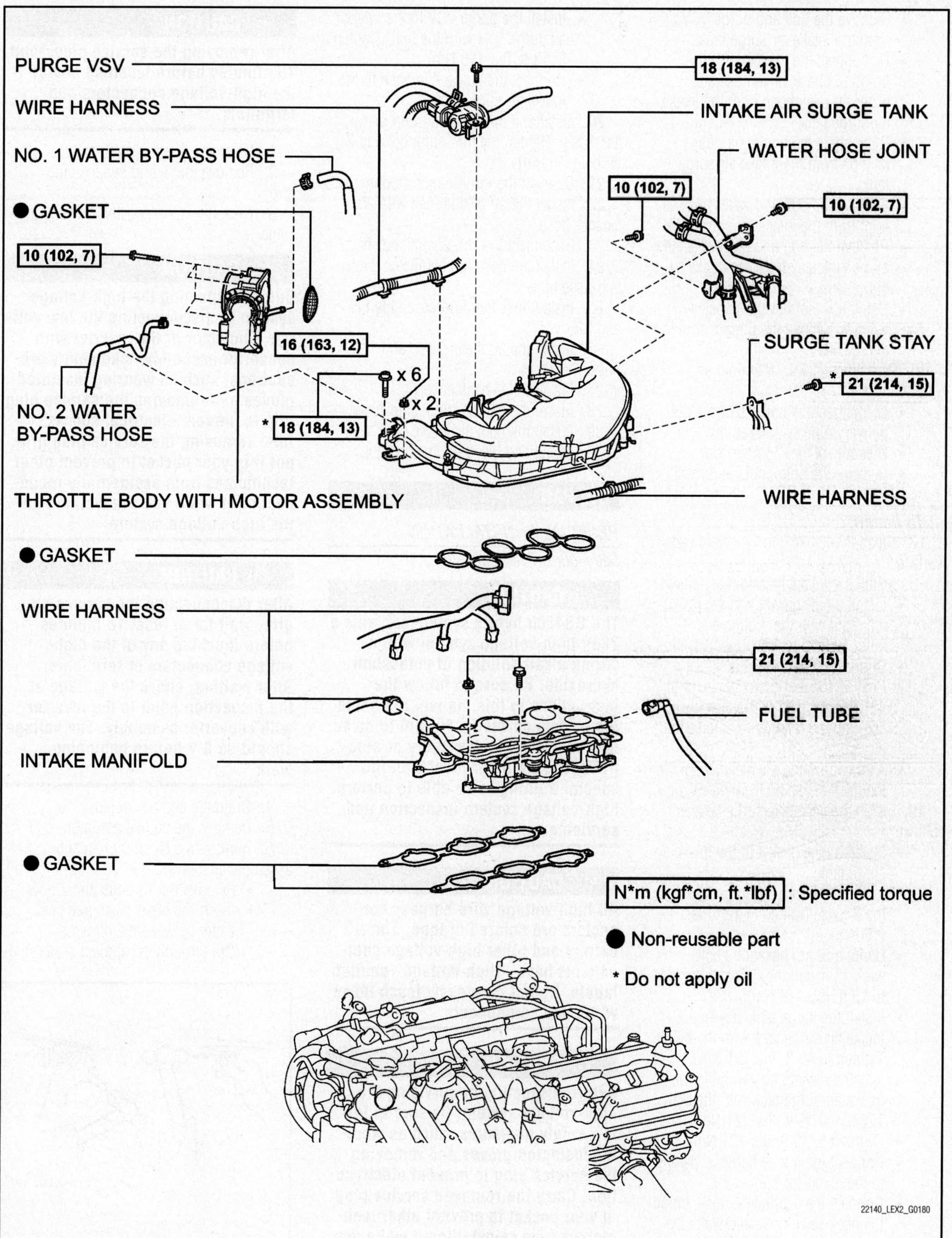

Fig. 72 Intake manifold and related parts

- Remove the bolt and purge VSV from the intake air surge tank.
- Disconnect the ventilation hose, union to check valve hose and water by-pass hose from the intake air surge tank.
- Disconnect the 4 wire harness clamps from the intake air surge tank.
- Remove the bolt and water hose joint from the intake air surge tank.
- Remove the bolt and disconnect the surge tank stay from the intake air surge tank.
- Using a 5 mm hexagon socket wrench, remove the 6 bolts, 2 nuts and gasket.

16. Remove the intake manifold as follows:
    - Disconnect the fuel tube from the delivery pipe sub-assembly.
    - Disconnect the 4 connectors.
    - Remove the bolts, nuts and manifold.

To install:

17. Clean the intake and cylinder head surface.
18. Install the intake manifold as follows:
    - Install a new gasket and the intake manifold with the 4 bolts and 4 nuts. Tighten to 15 ft. lbs. (21 Nm).
    - Connect the 4 connectors.
    - Push in the fuel tube connector to the delivery pipe sub-assembly and push up the retainer to engage the claws.
    - After connecting the fuel tube, align the paint marks on the delivery pipe.
19. Install the air intake surge tank as follows:
    - Install a new gasket to the intake air surge tank.
    - Install the intake air surge tank with the 2 nuts. Tighten to 12 ft. lbs. (16 Nm).
    - Using a 5 mm hexagon socket wrench, install the 6 bolts. Tighten to 13 ft. lbs. (18 Nm).
    - Install the surge tank stay to the intake air surge tank with the bolt. Tighten to 15 ft. lbs. (21 Nm).
    - Install the water hose joint to the intake air surge tank with the bolt. Tighten to 15 ft. lbs. (21 Nm).
    - Tighten to 15 ft. lbs. (21 Nm).
    - Connect the 4 wire harness clamps to the intake air surge tank.
    - Connect the ventilation hose, union to check valve hose and water by-pass hose to the intake air surge tank.
    - Install the purge VSV to the intake air surge tank with the bolt. Tighten to 13 ft. lbs. (18 Nm).
    - Connect the purge line hose to the intake air surge tank.
20. Install the throttle body with motor assembly. Tighten the mounting bolts to 7 ft. lbs. (10 Nm).
21. Connect the throttle motor connector.
22. Install the air cleaner cap with air cleaner hose.
23. Reconnect the No. 2 ventilator hose.
24. Install the cowl top ventilator louver sub-assembly.
25. Install the wiper arm assembly LH and RH
26. Install the air cleaner cap and hose.
27. Install the V-bank cover.
28. Install the engine room side cover.
29. Install the cool air intake duct seal.
30. Install and bleed engine coolant.

OIL PAN

REMOVAL & INSTALLATION

See Figures 73 through 76.

**** CAUTION**

The GS450h hybrid system contains a 288V high-voltage system with a strong alkali solution of potassium hydroxide. Be sure to follow the instructions in this manual to handle the system correctly. Failure to do so may result in serious injury or electrocution. Engineer must undergo special training to be able to perform high-voltage system inspection and servicing.

**** CAUTION**

All high-voltage wire harness connectors are colored orange. The HV battery and other high-voltage components have "High Voltage" caution labels. Do not carelessly touch these wires and components.

**** CAUTION**

Before inspecting or servicing the high-voltage system, be sure to follow safety measures, such as wearing insulated gloves and removing the service plug to prevent electrocution. Carry the removed service plug in your pocket to prevent other technicians from reinstalling it while you are servicing the vehicle.

**** CAUTION**

After removing the service plug, wait 10 minutes before touching any of the high-voltage connectors and terminals.

1. Discharge the fuel system.
2. Remove the hybrid service plug grip.
3. Disconnect the negative battery cable.

**** CAUTION**

Before inspecting the high-voltage system or disconnecting the low voltage connector of the inverter with converter assembly, take safety precautions, such as wearing insulated gloves and removing the service plug grip to prevent electrical shocks. After removing the service plug grip, put it in your pocket to prevent other technicians from accidentally reconnecting it while you are working on the high-voltage system.

**** CAUTION**

After disconnecting the service plug grip, wait for at least 10 minutes before touching any of the high-voltage connectors or terminals. After waiting, check the voltage at the inspection point in the inverter with converter assembly. The voltage should be 0 V before beginning work.

4. Drain the engine coolant.
5. Remove the engine assembly.
6. Remove the No. 2 oil pan sub-assembly as follows:
    - Remove the 15 bolts and 2 nuts.
    - Insert the blade of oil pan seal cutter between the oil pans.
    Cut through the applied sealer and

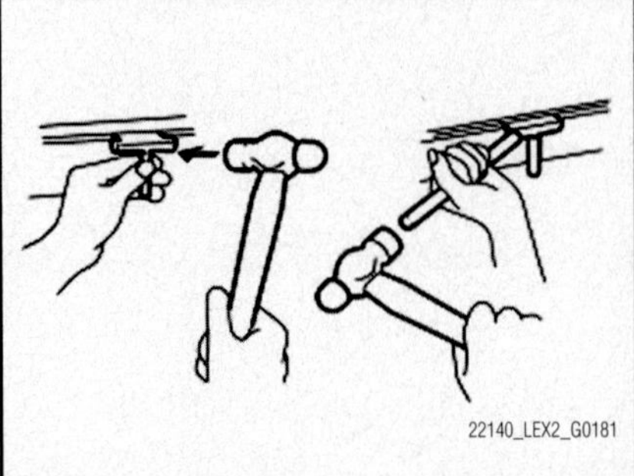

Fig. 73 Cut through the applied sealer and remove the No. 2 oil pan sub-assembly

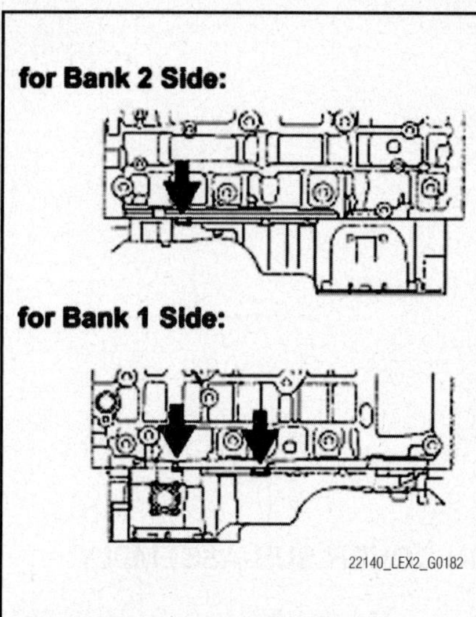

Fig. 74 Remove the oil pan by prying between the oil pan and cylinder block

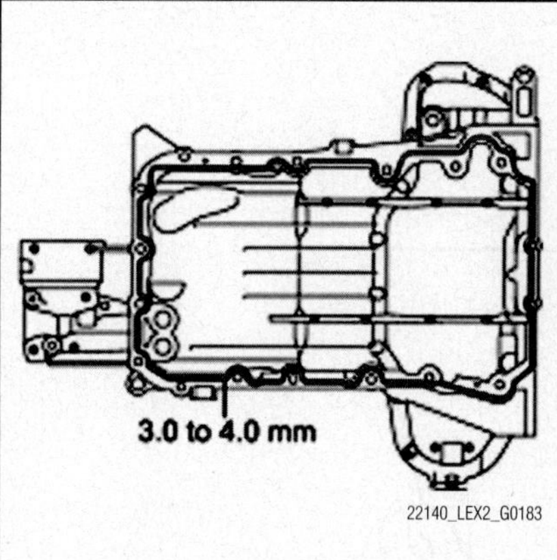

Fig. 75 Standard seal diameter: 0.118 to 0.156 inch. (3.0 to 4.0 mm)

remove the No. 2 oil pan sub-assembly.

- Remove the 16 bolts and 2 nuts.
- Remove the oil pan by prying between the oil pan and cylinder block with a screwdriver.

To install:

➡Be sure to clean the bolts and stud bolts and check the threads for cracks or other damage.

7. Install oil pan sub-assembly as follows:
   - Apply seal packing in a continuous line as shown in the illustration. Toyota Genuine Seal Packing Block, Three Bond 1207B or equivalent.
   - Install the oil pan with the 16 bolts and 2 nuts. Tighten the bolts to 7 ft. lbs. (10 Nm). Tighten the nuts to 15 ft. lbs. (21 Nm).
8. Install the No. 2 oil pan sub-assembly as follows:

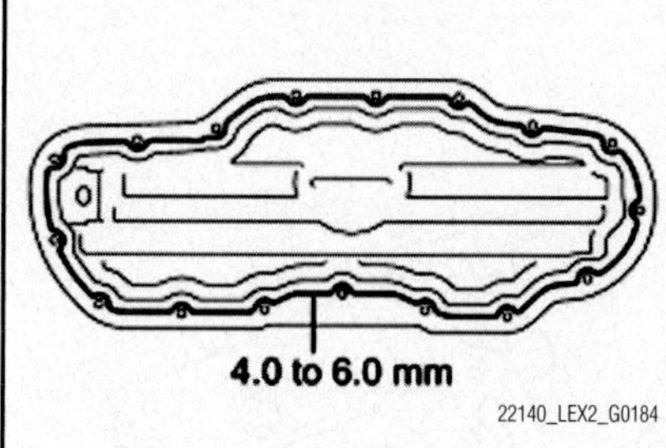

Fig. 76 Standard seal diameter: 0.156 to 0.236 inch. (4.0 to 6.0 mm)

- Apply seal packing in a continuous line as shown in the illustration.
- Install the oil pan sub-assembly with the 15 bolts and 2 nuts. Tighten the bolts and nuts to 7 ft. lbs. (10 Nm).

OIL PUMP

REMOVAL & INSTALLATION

See Figure 77.

⁂ CAUTION

The GS450h hybrid system contains a 288V high-voltage system with a strong alkali solution of potassium hydroxide. Be sure to follow the instructions in this manual to handle the system correctly. Failure to do so may result in serious injury or electrocution. Engineer must undergo special training to be able to perform high-voltage system inspection and servicing.

⁂ CAUTION

All high-voltage wire harness connectors are colored orange. The HV battery and other high-voltage components have "High Voltage" caution labels. Do not carelessly touch these wires and components.

⁂ CAUTION

Before inspecting or servicing the high-voltage system, be sure to follow safety measures, such as wearing insulated gloves and removing the service plug to prevent electrocution. Carry the removed service plug in your pocket to prevent other technicians from reinstalling it while you are servicing the vehicle.

⁂ CAUTION

After removing the service plug, wait 10 minutes before touching any of the high-voltage connectors and terminals.

1. Remove the hybrid service plug grip.
2. Disconnect the negative battery cable.

⁂ CAUTION

Before inspecting the high-voltage system or disconnecting the low voltage connector of the inverter with converter assembly, take safety precautions, such as wearing insulated gloves and removing the service plug grip to prevent electrical shocks. After removing the service plug grip, put it in your pocket to prevent other technicians from accidentally reconnecting it while you are working on the high-voltage system.

⁂ CAUTION

After disconnecting the service plug grip, wait for at least 10 minutes before touching any of the high-voltage connectors or terminals. After waiting, check the voltage at the inspection point in the inverter with converter assembly. The voltage should be 0 V before beginning work.

3. Remove the service cover and plug.
4. Disconnect the negative battery cable.
5. Remove the drive belt.
6. Using SST, loosen the crankshaft pulley set bolt.
7. Remove the oil pump assembly from the timing cover.

To install:

8. Install the oil pump assembly to the timing cover.
9. Install the front crankshaft seal and drive it into the cover with a seal driver.
10. Align the pulley set key with the key groove of the pulley, and slide on the pulley.
11. Install the pulley bolt and tighten bolt to 184 ft. lbs. (240 Nm).
12. Install the drive belt.
13. Install the service cover and plug.
14. Connect the negative battery cable.

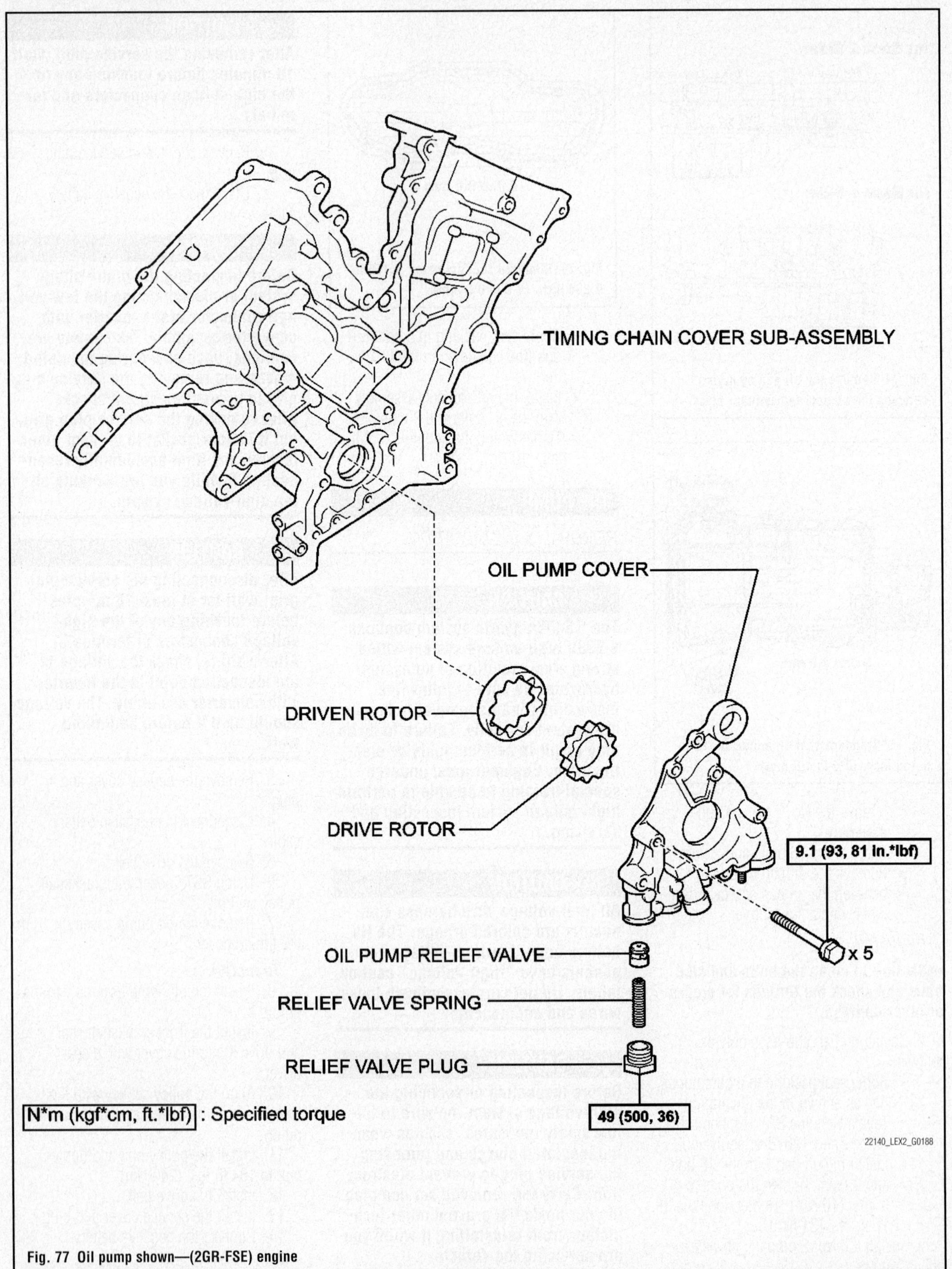

Fig. 77 Oil pump shown—(2GR-FSE) engine

PISTON AND RING

POSITIONING

See Figures 78 through 80.

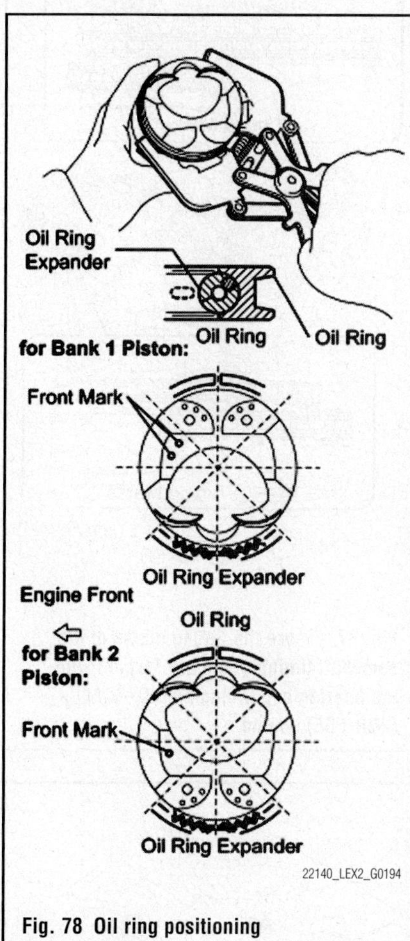

Fig. 78 Oil ring positioning

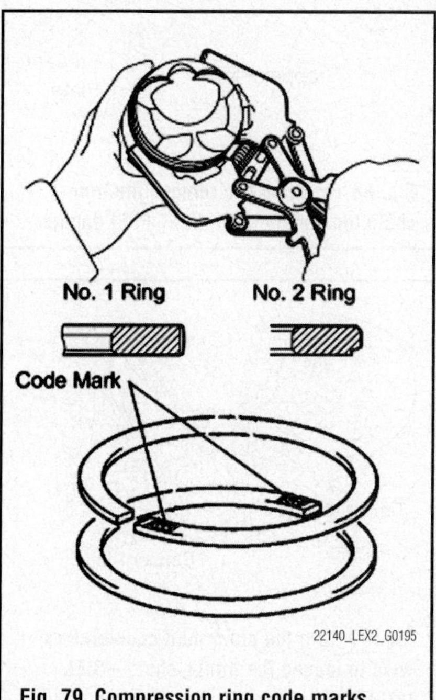

Fig. 79 Compression ring code marks

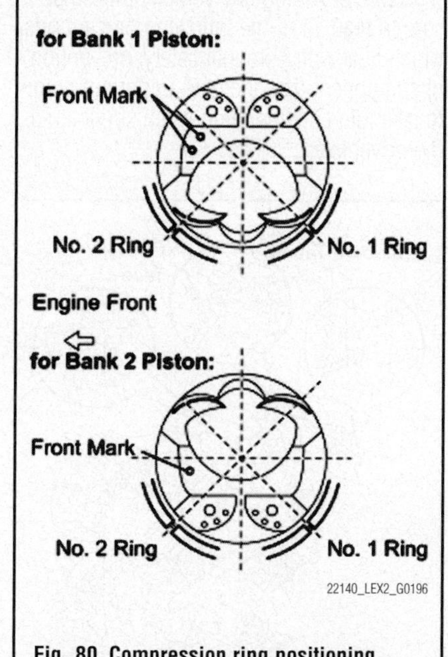

Fig. 80 Compression ring positioning

REAR MAIN SEAL

REMOVAL & INSTALLATION

See Figures 81 through 83.

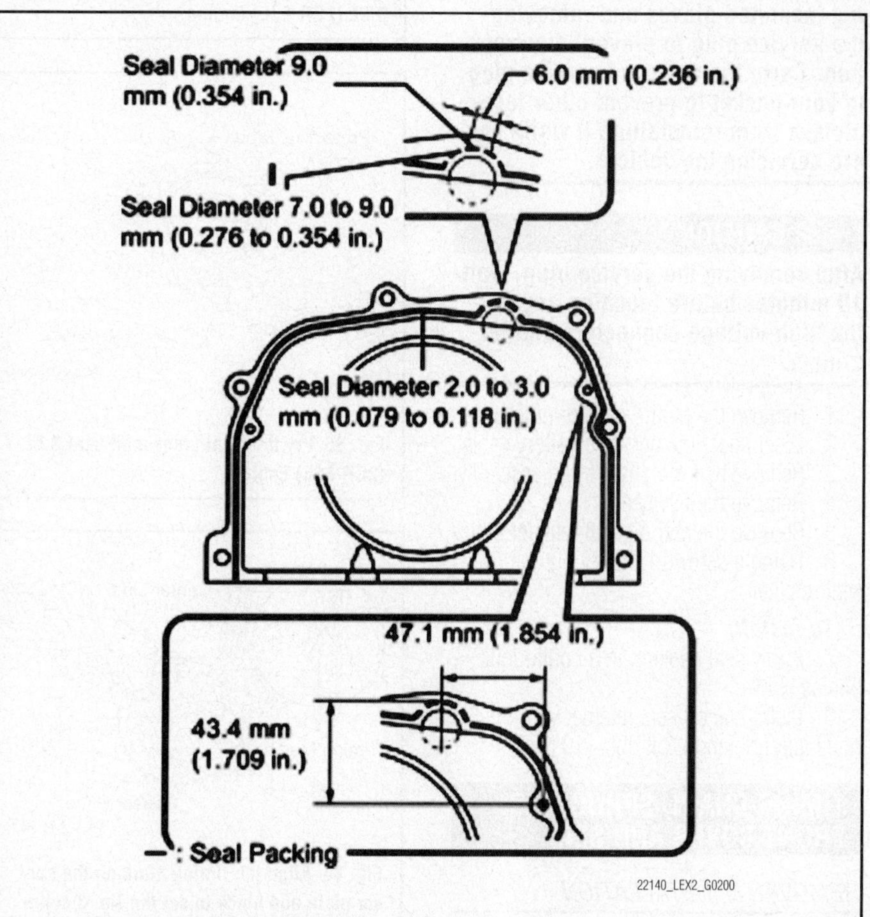

Fig. 82 Apply seal packing in a continuous line

CAUTION

The GS450h hybrid system contains a 288V high-voltage system with a strong alkali solution of potassium hydroxide. Be sure to follow the instructions in this manual to handle the system correctly. Failure to do so may result in serious injury or electrocution. Engineer must undergo special training to be able to perform high-voltage system inspection and servicing.

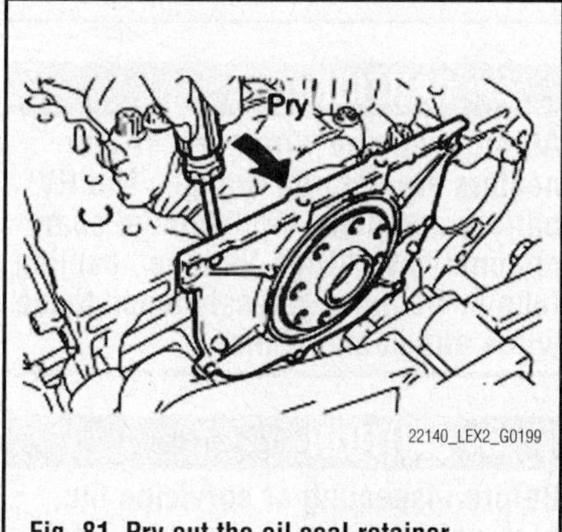

Fig. 81 Pry out the oil seal retainer

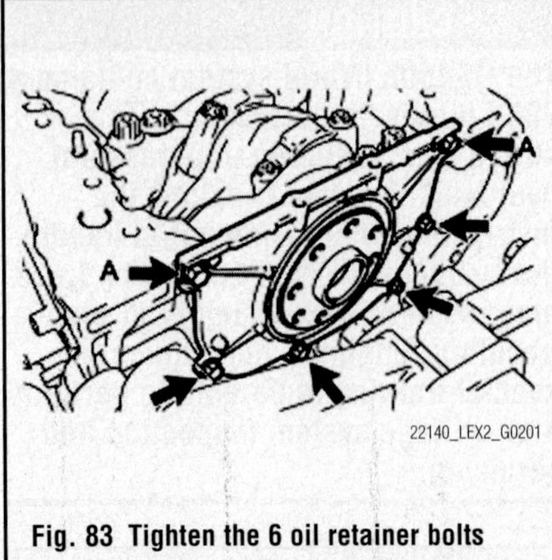

Fig. 83 Tighten the 6 oil retainer bolts

⁂ CAUTION

All high-voltage wire harness connectors are colored orange. The HV battery and other high-voltage components have "High Voltage" caution labels. Do not carelessly touch these wires and components.

⁂ CAUTION

Before inspecting or servicing the high-voltage system, be sure to follow safety measures, such as wearing insulated gloves and removing the service plug to prevent electrocution. Carry the removed service plug in your pocket to prevent other technicians from reinstalling it while you are servicing the vehicle.

⁂ CAUTION

After removing the service plug, wait 10 minutes before touching any of the high-voltage connectors and terminals.

1. Remove the hybrid service plug grip.
2. Disconnect the negative battery cable.
3. Remove hybrid transmission assembly.
4. Remove the flywheel.
5. Remove the rear oil seal retainer bolts.
6. Using a screwdriver, pry out the oil seal retainer.

To install:

7. Apply seal packing in a continuous line as shown.
8. Install the oil seal retainer with the 6 bolts and tighten to 7 ft. lbs. (10 Nm).

TIMING CHAIN, SPROCKETS, FRONT COVER AND SEAL

REMOVAL & INSTALLATION

See Figures 84 through 96.

Before servicing any vehicle, please be sure to read all of the following precautions, which deal with personal safety, prevention of component damage, and important points to take into consideration when servicing a motor vehicle:

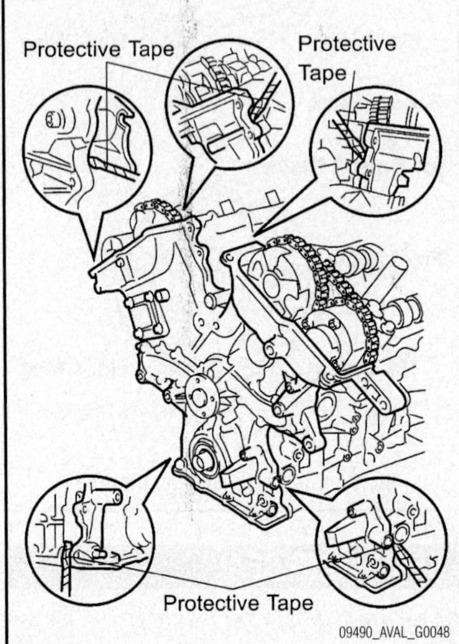

Fig. 84 Pry the front cover at the locations 3.5L (2GR-FSE) Engine

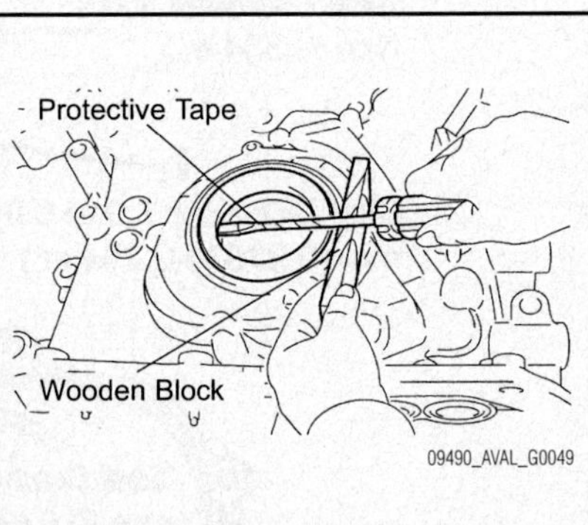

Fig. 85 Pry the front crankshaft seal 3.5L (2GR-FSE) Engine

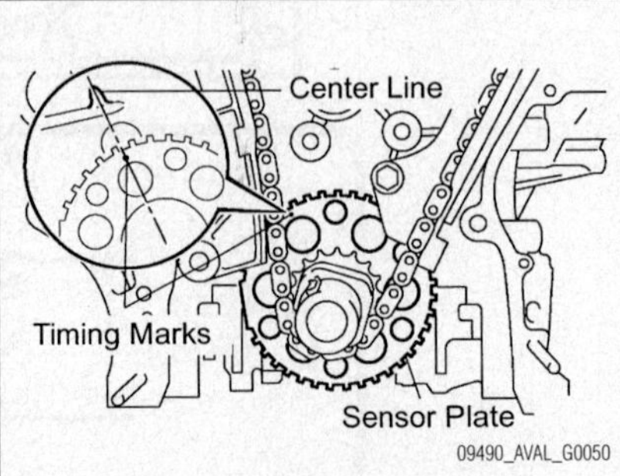

Fig. 86 Align the timing mark on the sensor plate and block to set the No. 1 cylinder at TDC—3.5L (2GR-FSE) engine

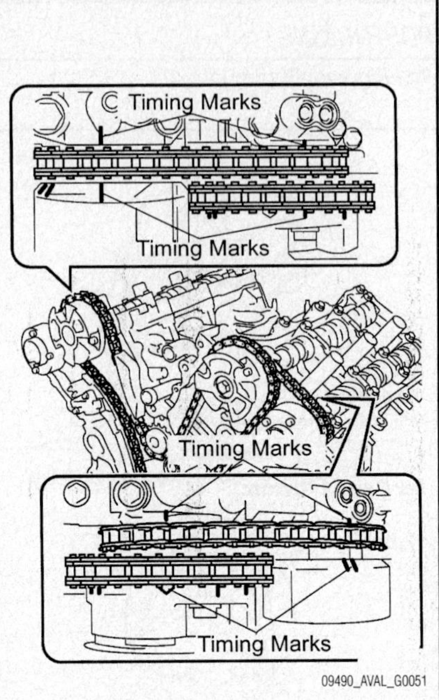

Fig. 87 Ensure the timing marks of the camshaft timing gears are aligned with the bearing cap timing marks—3.5L (2GR-FSE) engine

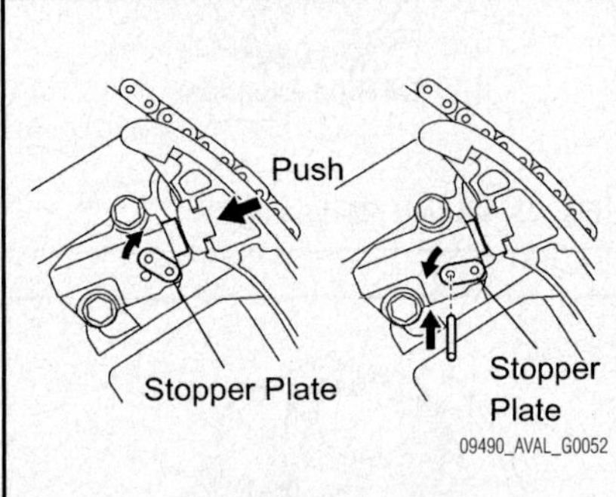

Fig. 88 Procedure to remove the timing chain tensioner—3.5L (2GR-FSE) engine

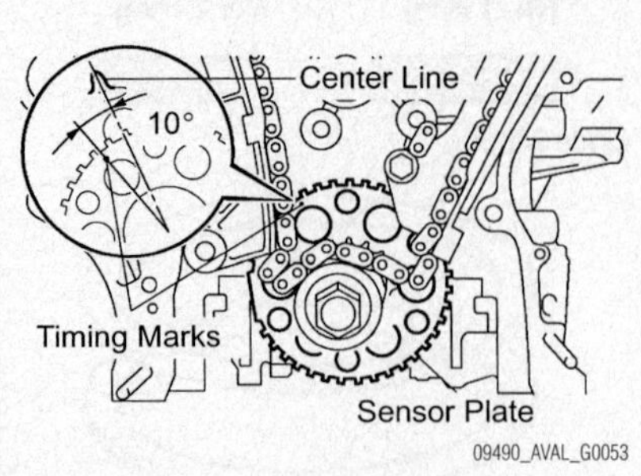

Fig. 89 Turn the crankshaft counterclockwise to loosen the timing chain—3.5L (2GR-FSE) engine

Fig. 90 Align the timing chain marked links with the camshaft timing gear marks to install—3.5L (2GR-FSE) engine

CAUTION

The GS450h hybrid system contains a 288V high-voltage system with a strong alkali solution of potassium hydroxide. Be sure to follow the instructions in this manual to handle the system correctly. Failure to do so may result in serious injury or electrocution. Engineer must undergo special training to be able to perform high-voltage system inspection and servicing.

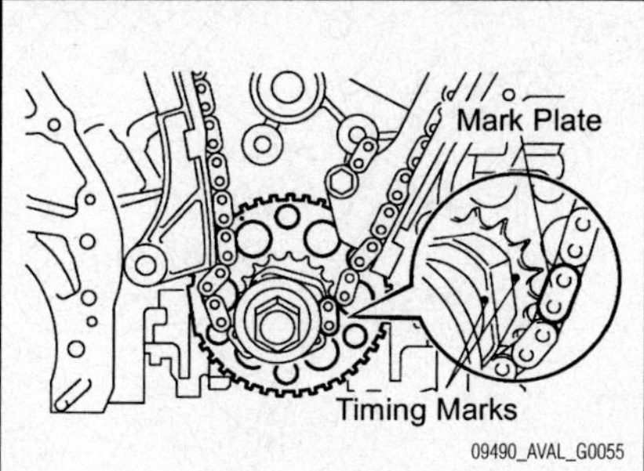

Fig. 91 Align the timing chain marked link with the crankshaft gear to install—3.5L (2GR-FSE) engine

CAUTION

All high-voltage wire harness connectors are colored orange. The HV

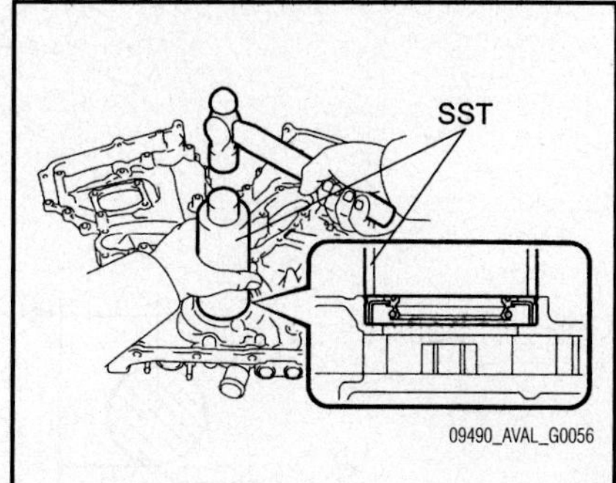

Fig. 92 Tap a new oil seal into place in the front cover

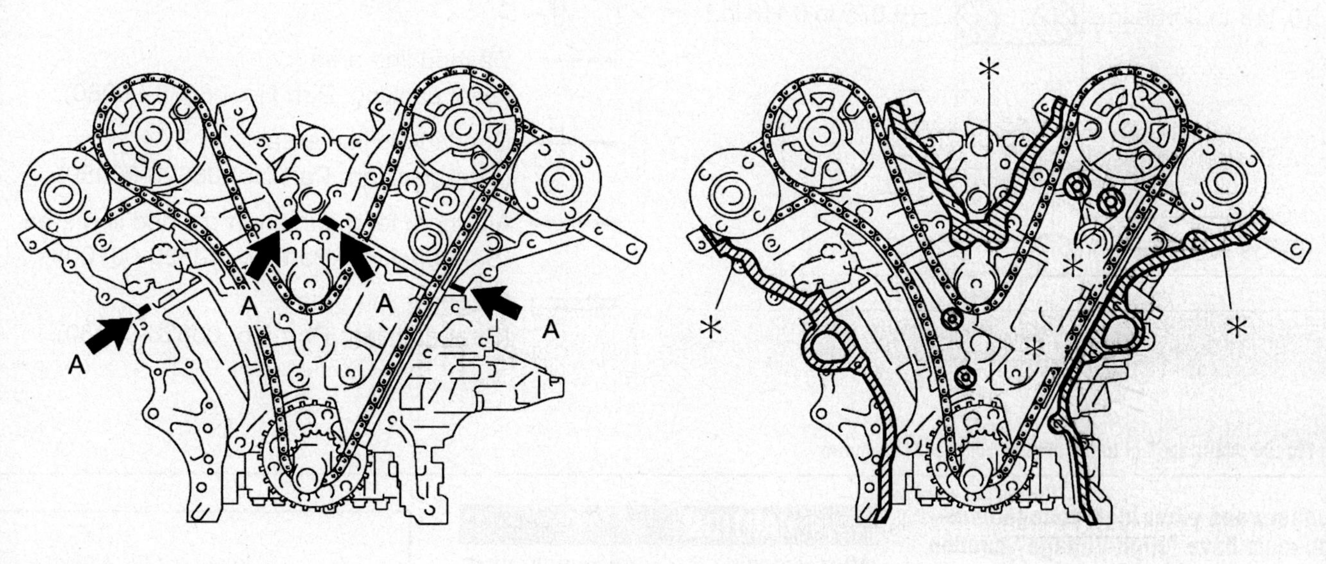

Fig. 93 Apply sealant to the engine block as shown

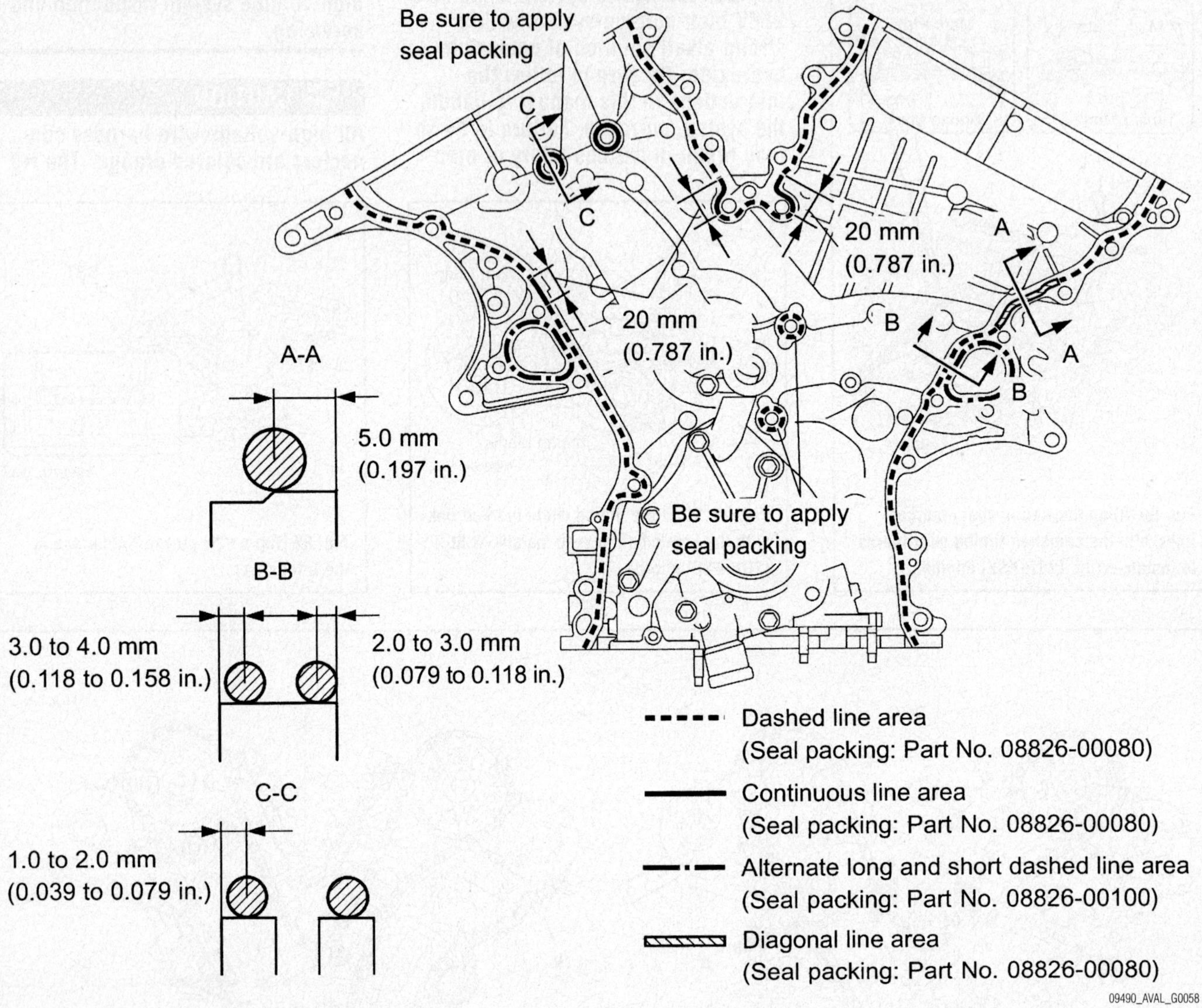

Be sure to apply seal packing
C
C
20 mm (0.787 in.)
20 mm (0.787 in.)
A
A
B
B
A-A
5.0 mm (0.197 in.)
Be sure to apply seal packing
B-B
3.0 to 4.0 mm (0.118 to 0.158 in.)
2.0 to 3.0 mm (0.079 to 0.118 in.)
C-C
1.0 to 2.0 mm (0.039 to 0.079 in.)
Dashed line area (Seal packing: Part No. 08826-00080)
Continuous line area (Seal packing: Part No. 08826-00080)
Alternate long and short dashed line area (Seal packing: Part No. 08826-00100)
Diagonal line area (Seal packing: Part No. 08826-00080)
09490_AVAL_G0058

Fig. 94 Apply sealant to the engine front cover as shown

battery and other high-voltage components have "High Voltage" caution labels. Do not carelessly touch these wires and components.

CAUTION

Before inspecting or servicing the high-voltage system, be sure to follow safety measures, such as wearing insulated gloves and removing the service plug to prevent electrocution. Carry the removed service plug in your pocket to prevent other technicians from reinstalling it while you are servicing the vehicle.

CAUTION

After removing the service plug, wait 10 minutes before touching any of the high-voltage connectors and terminals.

1. Discharge the fuel system.
2. Recover the refrigerant from the A/C system using a refrigerant recovery unit.
3. Place the front wheels in the straight ahead position.
4. Disconnect the negative battery cable.
5. Remove the hybrid service plug grip.

CAUTION

Before inspecting the high-voltage system or disconnecting the low volt-

Drive Rotor
Spline
Crankshaft
09490_AVAL_G0059

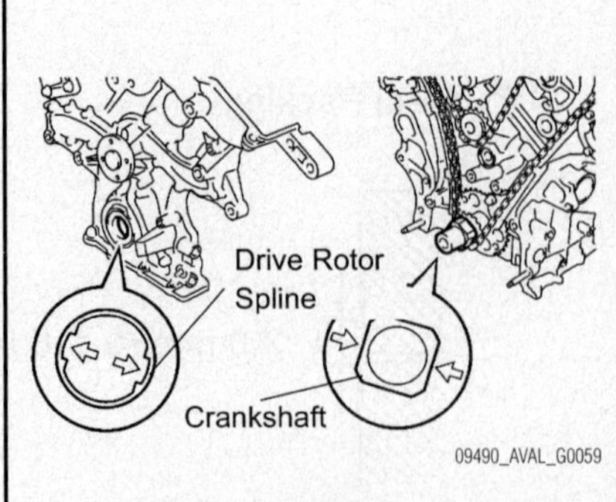

Fig. 95 Correct orientation of the oil pump rotor and crankshaft during installation of the front cover—3.5L (2GR-FSE) engine

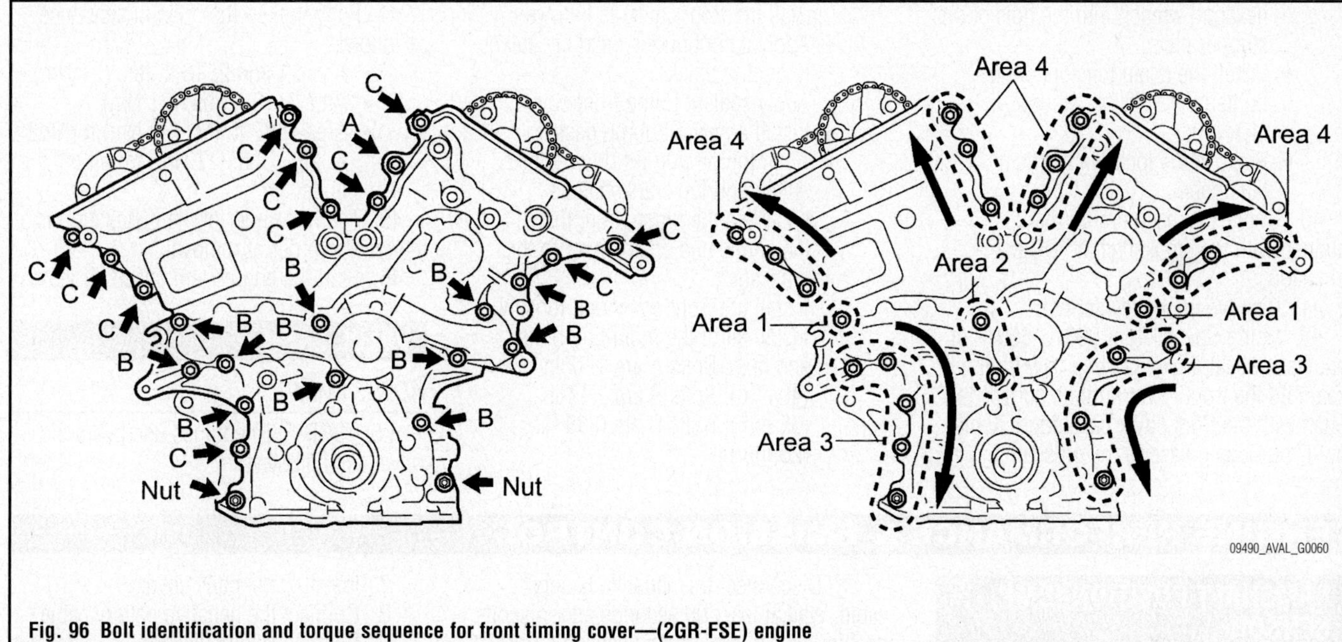

Fig. 96 Bolt identification and torque sequence for front timing cover—(2GR-FSE) engine

age connector of the inverter with converter assembly, take safety precautions, such as wearing insulated gloves and removing the service plug grip to prevent electrical shocks. After removing the service plug grip, put it in your pocket to prevent other technicians from accidentally reconnecting it while you are working on the high-voltage system.

✲✲ CAUTION

After disconnecting the service plug grip, wait for at least 10 minutes before touching any of the high-voltage connectors or terminals. After waiting, check the voltage at the inspection point in the inverter with converter assembly. The voltage should be 0 V before beginning work.

6. Remove the luggage floor mat.
7. Remove the service plug cover.
8. Remove the service plug.
9. Disconnect the negative battery cable.
10. Drain the engine oil and coolant.
11. Remove the engine/transaxle assembly from the vehicle.
12. Remove the transaxle.
13. Remove the oil dipstick tube.
14. Remove the driveplate.
15. Install the engine to a suitable engine stand.
16. Remove or disconnect the following:
    - Idler pulley
    - Right-side engine mounting bracket
    - Accessory drive belt tensioner
    - Water pump pulley
    - No. 2 timing gear cover
    - Engine mounting stay and bracket
    - Water inlet housing
17. Using Special Tool 09213-70011, hold the crankshaft pulley and loosen the pulley bolt.
18. Using the pulley bolt and Special Tool 09950-50013, remove the crankshaft pulley
19. Remove or disconnect the following:
    - Upper and lower oil pans
    - O-rings from the oil pump
    - Air intake surge tank
    - Ignition coils
    - Oil pipes
    - Cylinder head cover
20. Remove the mounting bolts from the front cover.
21. Using a suitable pry tool with the tip covered with protective tape, pry the front cover in the specified locations to remove
22. Pry the front crankshaft seal from the front cover.
23. Temporarily tighten the pulley set bolt.
24. Set the timing mark on the crank angle sensor plate to the right-hand block bore center line to put the No. 1 cylinder at TDC.
25. Check that the timing marks of the camshaft timing gears are aligned with the timing marks of the bearing caps as shown in the illustration. If not, turn the crankshaft one complete revolution (360°) and align the timing marks as shown.
26. Move the stopper plate upward to release the lock, and push the plunger deep into the tensioner.
27. Move the stopper plate downward to set the lock, and insert a hexagon wrench into the stopper plate's hole.
28. Remove the chain tensioner.
29. Remove the chain tensioner slipper
30. Turn the crankshaft 10° counterclockwise to loosen the chain off the crankshaft timing gear.
31. Remove the timing chain from the crank timing gear and place it on the crankshaft.
32. Turn the camshaft timing gear on the right-side bank clockwise (approximately 60°). Be sure to loosen the chain between the center banks.
33. Remove the timing chain

To install:

34. Align the orange marked links and timing mark as shown in the illustration and install the timing chain.
35. Turn the camshaft timing gear on the right-side bank counterclockwise to tighten the chain between banks.
36. Align the yellow marked link and timing mark as shown in the illustration and install the chain onto the crankshaft timing gear.
37. Temporarily tighten the crankshaft pulley set bolt.
38. Install the chain tensioner slipper.
39. Install the chain tensioner as follows:
    - Move the stopper plate upward to release the lock, and push the plunger deep into the tensioner.
    - Move the stopper plate downward to set the lock, and insert a

hexagon wrench into the hole of the stopper plate.

- Install the chain tensioner and tighten the bolts to 7 ft. lbs. (10 Nm).
- Remove the lock pin of chain tensioner.

40. Check that each timing mark is aligned with the crankshaft at TDC compression.

41. Remove the pulley set bolt.

42. Using Special Tool 09316-60011 or equivalent seal driver, tap in a new front oil seal into the front cover until its surface is flush with the front cover case edge. Apply multi-purpose grease to the oil seal lip.

43. Install the front cover as follows:

- Apply a continuous bead of sealant to engine block.
- Apply sealant to the front cover.
- Install a new oil pump gasket.
- Align the oil pump's drive rotor spline and the crankshaft as shown in the illustration. Install the spline and chain cover to the crankshaft.
- Install the front cover and loosely install all of the mounting bolts and nuts. Bolts A are 1.57 in. (40 mm); Bolts B are 2.17 in. (55 mm); Bolts C are 0.98 in. (25 mm).

44. Fully tighten the bolts in sequence as follows:

- Areas 1 and 2: 15 ft. lbs. (21 Nm)
- Area 3: 15 ft. lbs. (21 Nm)
- Area 4: 32 ft. lbs. (43 Nm) for Bolt A; 15 ft. lbs. (21 Nm) for all other bolts

45. The remainder of the installation is the reverse order of removal

46. Start the engine and check for leaks

VALVE LASH

ADJUSTMENT

3.5L (2GR-FSE) engines use hydraulic valve lash adjusters.

ENGINE PERFORMANCE & EMISSION CONTROL

CAMSHAFT POSITION (CMP) SENSOR

LOCATION

The camshaft position sensor also referred to as a Variable Valve Timing (VVT) sensor has four locations. The exhaust and intake sensor (bank 1 and bank 2). Refer to component views.

REMOVAL & INSTALLATION

See Figures 97 through 100.

1. Disconnect the negative battery cable. Wait at least 90 seconds after disconnecting the cable from the negative (-) battery terminal to prevent airbag and seat belt pretensioner activation.
2. Remove the V-bank engine cover.
3. Disconnect the sensor connector.
4. Remove the bolt and sensor.

To install:

5. Install the sensor with the bolt and tighten to 7 ft. lbs. (10 Nm).
6. Connect the sensor connector.
7. Install the V-bank cover.
8. Connect the negative battery cable.
9. Certain systems need to be initialized after disconnecting and reconnecting the cable from the negative (-) battery terminal.

CRANKSHAFT POSITION (CKP) SENSOR

LOCATION

The crankshaft position sensor is located behind the A/C electric inverter converter on the left of engine block.

REMOVAL & INSTALLATION

See Figure 101.

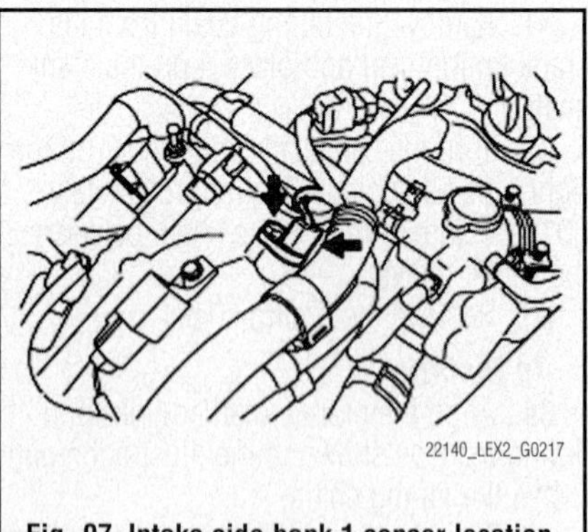

Fig. 97 Intake side bank 1 sensor location

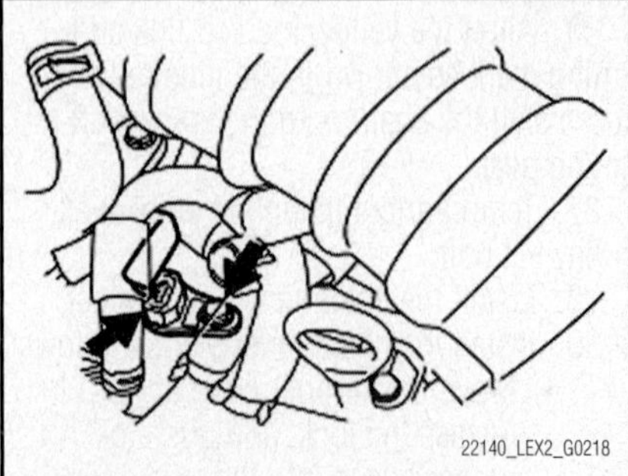

Fig. 98 Exhaust side bank 1 sensor location

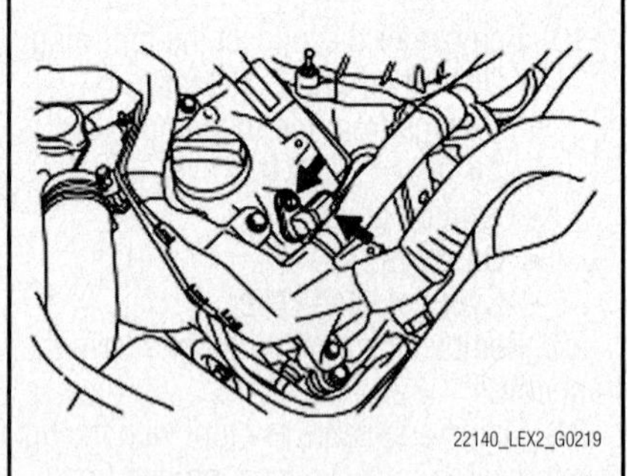

Fig. 99 Intake side bank 2 sensor location

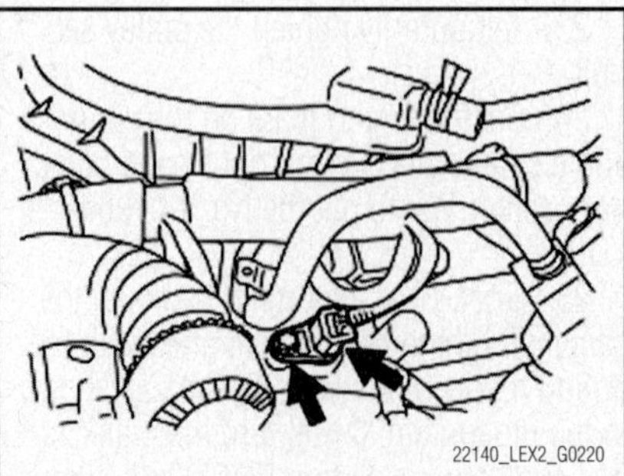

Fig. 100 Exhaust side bank 2 sensor location

Wire Harness Side:

E47 CKP Sensor

Front View

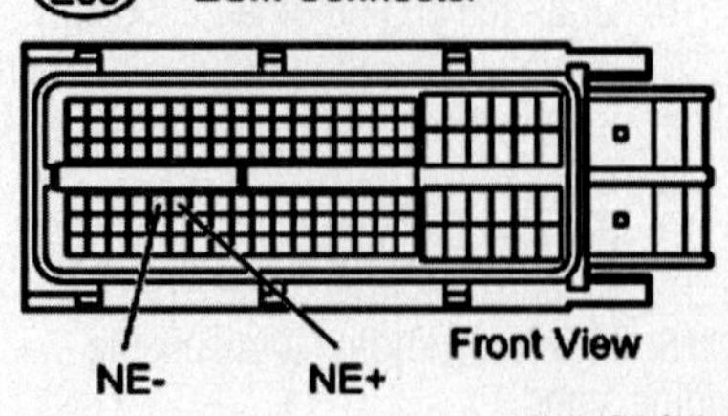

Fig. 101 Crankshaft position sensor location view

1. Remove the service plug grip.
2. Disconnect the negative battery cable.
3. Remove the bolt and air cleaner inlet.
4. Evacuate the A/C system.
5. Remove the engine under covers.
6. Disconnect the A/C lines.
7. Remove the electric A/C inverter compressor.
8. Disconnect the sensor connector.
9. Remove the bolt and sensor.

To install:

10. Apply a coat of engine oil to the O-ring of the sensor.
11. Install the sensor with the bolt and tighten to 7 ft. lbs. (10 Nm).
12. Check compressor oil and add if low.
13. Install the electric A/C compressor and tighten the mounting bolts to 18 ft. (2.4 Nm).
14. Install the A/C lines and tighten to 7 ft. lbs. (10 Nm).
15. Install the engine under covers.
16. Install the bolt and air cleaner inlet.
17. Install the service grip.
18. Connect the negative battery cable.

HYBRID VEHICLE ELECTRONIC CONTROL UNIT (ECU)

LOCATION

The ECU is located next to the No.1 Relay box under the hood of the vehicle. It is mounted above the ECM.

REMOVAL & INSTALLATION

See Figures 102 and 103.

⁂ CAUTION

When removing and installing the ECM and hybrid vehicle control ECU, be sure to distinguish the wire harnesses of the ECM and hybrid vehicle control ECU. If the wire harnesses are incorrectly connected, damage may occur to the engine.

1. Disconnect the negative battery cable.
2. Remove the cool air intake duct seal.
3. Remove the left engine room LH side cover.
4. Remove engine room side cover.
5. Remove the radiator reservoir assembly.
6. Remove the HV ECU as follows:
7. Detach the side of the wire harness clamp labeled A from the wire harness.
8. Detach the 3 wire harness clamps.
9. Remove the nut, 4 screws and bracket.

Fig. 102 Remove the hybrid vehicle control ECU

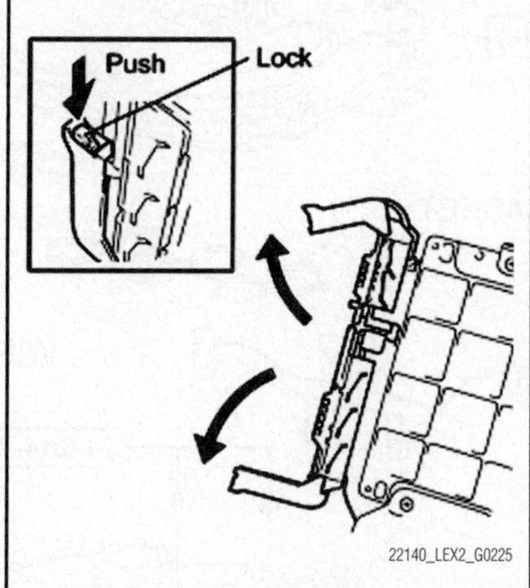

Fig. 103 Disconnect the ECM connectors as shown

10. Remove the 2 screws and ECU.
11. Raise the lock lever and disconnect the 2 connectors. Remove the hybrid vehicle control ECU.
12. Remove the ECM as follows:
13. Remove the 2 nuts and ECM with bracket.
14. Disconnect the 2 ECM connectors.
15. Raise the 2 levers while pushing the locks on the 2 levers, and remove the 2 ECM connectors.

To install:

16. Install the ECM to the lower bracket and tighten to 27 inch. lbs. (3.0 Nm).
17. Connect the 2 ECM connectors.
18. Install the ECM with bracket with the 2 nuts and tighten to 49 inch. lbs. (5.5 Nm).
19. Install the HV control unit as follows:
20. Connect the 2 connectors to the hybrid vehicle control ECU and push each lock lever down to lock the connectors.
21. Install the ECU with the 2 screws and tighten to 27 inch. lbs. (3.0 Nm).
22. Install the bracket with the 4 screws and nut.
23. Attach the 3 wire harness clamps.
24. Install the radiator reservoir assembly and tighten the mounting bolts to 44 inch. lbs. (5.0 Nm).
25. Install the engine room No.1 relay cover.
26. Install the engine room LH side cover.
27. Install the cool air duct seal.
28. Connect the negative battery cable.
29. Some vehicle systems require initialization after reconnecting the cable to the negative battery terminal.

ENGINE COOLANT TEMPERATURE (ECT) SENSOR

LOCATION

The Engine Coolant Temperature (ECT) Sensor is located at the rear of the engine in the cylinder head.

REMOVAL & INSTALLATION

See Figure 104.

1. Before servicing the vehicle, refer to the precautions in the beginning of this section.
2. Discharge the fuel system pressure.
3. Disconnect the negative battery cable.

⁂ CAUTION

Remove the service plug grip to interrupt a high voltage circuit at the time of the check. Keep the removed service plug grip in your pocket to prevent other technicians from accidentally reconnecting it while you are servicing the vehicle. All the high voltage wiring connectors are colored in orange.

4. Wear insulated gloves. Remove the service plug grip after sliding the lever of the service plug grip.
5. Remove the service plug grip.
6. Disconnect the negative battery cable.
7. Remove or disconnect the following:
   - Cool air intake seal
   - LH and RH engine room cover
   - Engine V-bank cover
   - LH and RH front upper fender protector
   - LH and RH fender cowl seal
   - LH and RH wiper arm assembly
   - Cowl top ventilator louver

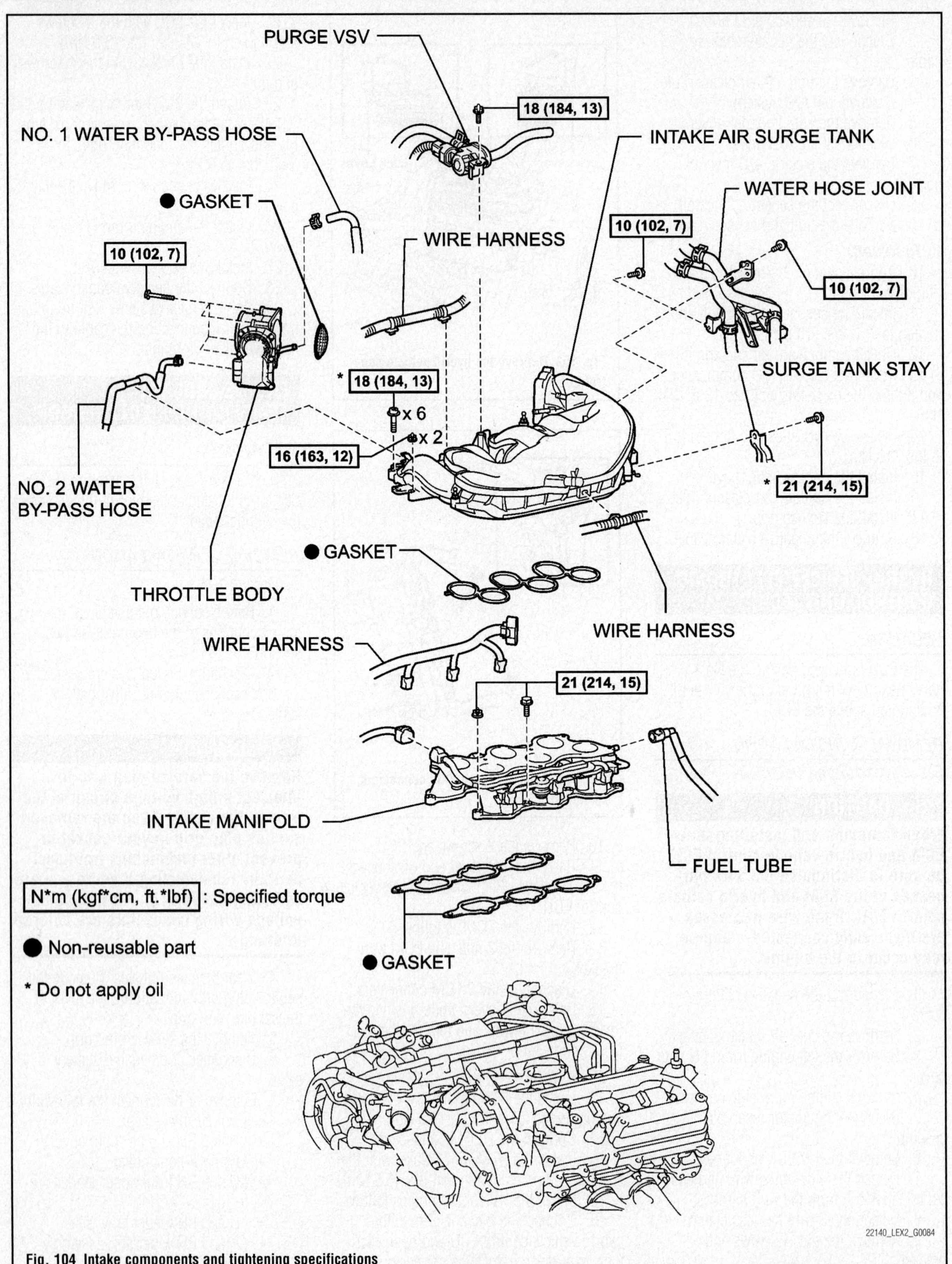

Fig. 104 Intake components and tightening specifications

- Drain engine coolant
- No 2 ventilation hose
- Air cleaner cap and hose
- Throttle body
- Intake air surge tank
- Intake manifold

8. Disconnect the coolant temperature sensor connector.
9. Remove the coolant temperature sensor.

To install:

10. Install a new gasket to the sensor.
11. Install the sensor and tighten to 15 ft. lbs. (20 Nm).
12. Connect the sensor connector.
13. Install or Reconnect the following:
    - Intake manifold
    - Intake air surge tank
    - Throttle body
    - Air cleaner cap and hose
    - Drain engine coolant
    - Cowl top ventilator louver
    - LH and RH wiper arm assembly
    - LH and RH fender cowl seal
    - LH and RH front upper fender protector
    - Engine V-bank cover
    - LH and RH engine room cover
    - Cool air intake seal
14. Connect the negative battery cable.
15. Wear insulated gloves and install the service plug.
16. Install the engine coolant, bleed system and check for leaks.

HEATED OXYGEN (HO2S) SENSOR

LOCATION

See Figure 105.

The Heated Oxygen (HO2S) Sensors are located in the front exhaust pipe assembly.

REMOVAL & INSTALLATION

1. Remove the service plug grip.
2. Disconnect the negative battery cable.
3. Remove front console upper garnish.
4. Remove the console box plate.
5. Remove the RH and LH instrument panel finish panel ends.
6. Remove the heated O2 sensor as follows:
    - Disconnect the sensor connector.
    - Remove the grommet and pass the sensor connector out of the cabin through the floor panel.
    - Using a O2 socket remove the sensor from the front exhaust pipe.

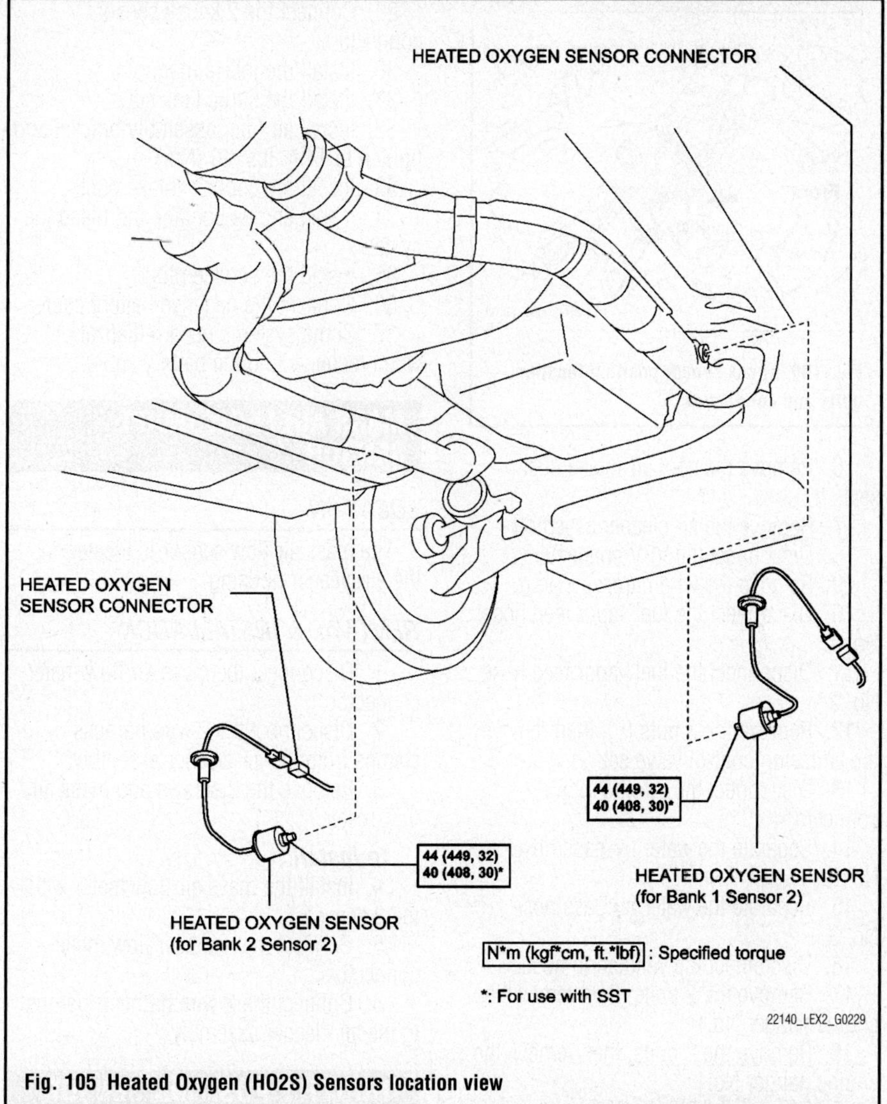

Fig. 105 Heated Oxygen (HO2S) Sensors location view

To install:

7. Install the O2 sensor and tighten to 30 ft. lbs. (40 Nm).
8. Reverse the removal procedure at this point.

INTAKE AIR TEMPERATURE (IAT) SENSOR

LOCATION

The intake air temperature sensor is integral to the mass air flow sensor. The intake air temperature sensor is located in the air cleaner housing.

REMOVAL & INSTALLATION

1. Disconnect the mass air flow meter connector.
2. Disconnect the 2 wire harness clamps from the air cleaner assembly.
3. Remove the 2 screws and mass air flow meter.

To install:

4. Install the mass air flow meter with the 2 screws.
5. Connect the mass air flow meter connector.
6. Connect the 2 wire harness clamps to the air cleaner assembly.

KNOCK SENSOR (KS)

LOCATION

The knock sensors are located under the intake manifold.

REMOVAL & INSTALLATION

See Figure 106.

1. Remove the service plug.
2. Disconnect the negative battery cable.
3. Discharge the fuel system.
4. Remove the engine room covers.
5. Drain the engine coolant.

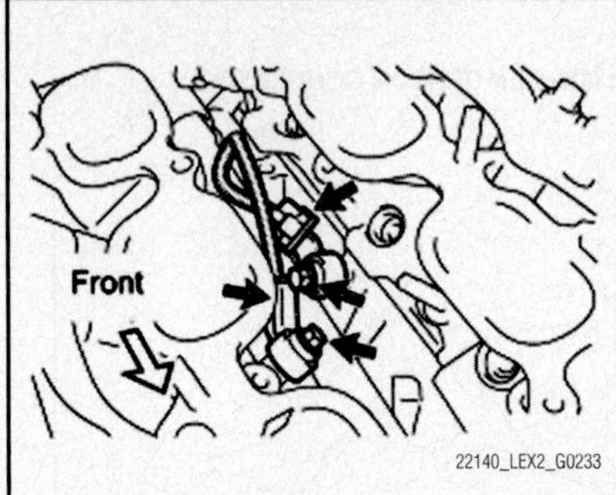

Fig. 106 Knock sensor position location, bolts and connectors

6. Remove the cool air intake duct seal.
7. Remove the air cleaner assembly.
8. Disconnect the VSV connector.
9. Remove the wire harness clamp.
10. Disconnect the fuel vapor feed hose No. 1.
11. Disconnect the fuel vapor feed hose No. 2.
12. Remove the 2 nuts (E), then remove the emission control valve set.
13. Disconnect the throttle motor connector.
14. Separate the water by-pass hose No. 2.
15. Separate the water by-pass hose No. 3.
16. Disconnect the ventilation hose.
17. Remove the 2 bolts, then remove the engine hanger No.1
18. Remove the 2 bolts, then remove the engine hanger No.1
19. Remove the 2 bolts, then remove the surge tank stay No. 1 (B).
20. Remove the 2 bolts, then remove the surge tank stay No. 2 (C).
21. Disconnect the ground cable connector.
22. Using a socket hexagon wrench 8 mm, remove the 4 bolts.
23. Remove the 2 nuts, then remove the emission control valve bracket and the intake air surge tank.
24. Remove the intake air surge tank.
25. Remove the intake manifold.
26. Disconnect the 2 knock sensor connectors.
27. Remove the 2 nuts, and then remove the 2 knock sensors.

To install:

28. Install the 2 knock sensors so that it is horizontal as shown in the illustration. Then install the 2 bolts and tighten to 15 ft. lbs. (20 Nm).
29. Connect the 2 knock sensor connectors.
30. Install the intake manifold.
31. Install the surge tank stay.
32. Install air filter assembly bracket and tighten to 14 ft. lbs. 20 (Nm).
33. Install the cool air intake duct.
34. Install engine coolant and bleed the system.
35. Install the service plug.
36. Connect the negative battery cable.
37. Some systems need initialization when reconnecting the battery cable.

MASS AIR FLOW (MAF) SENSOR

LOCATION

The mass air flow sensor is located in the air cleaner housing.

REMOVAL & INSTALLATION

1. Disconnect the mass air flow meter connector.
2. Disconnect the 2 wire harness clamps from the air cleaner assembly.
3. Remove the 2 screws and mass air flow meter.

To install:

4. Install the mass air flow meter with the 2 screws.
5. Connect the mass air flow meter connector.
6. Connect the 2 wire harness clamps to the air cleaner assembly.

THROTTLE POSITION SENSOR (TPS)

LOCATION

The throttle position sensor is integral to the throttle body assembly.

REMOVAL & INSTALLATION

See Figure 107.

1. Disconnect the negative battery cable.
2. Remove the cool air intake duct seal.
3. Drain the engine coolant.
4. Remove the V-bank cover.
5. Remove the air cleaner inlet.
6. Disconnect the No. 2 ventilation hose.
7. Disconnect the air cleaner cap and hose.
8. Disconnect the MAF meter connector.
9. Disconnect the clamp from the air cleaner.

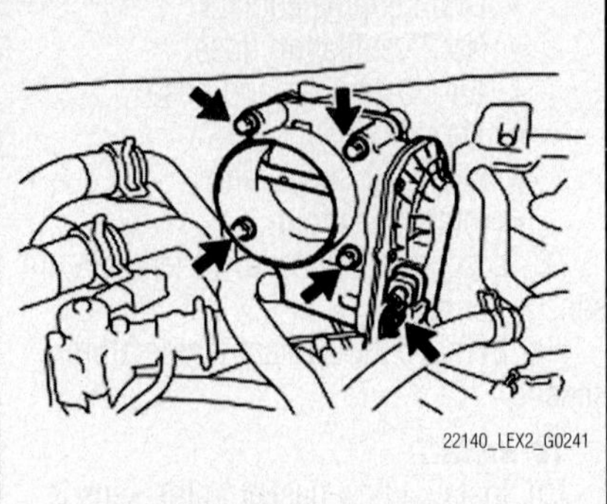

Fig. 107 Remove the 4 bolts and disconnect the throttle body

10. Disconnect the purge line hose.
11. Disconnect the 4 clamps.
12. Remove the hose clamp and air cleaner cap with air cleaner hose.
13. Remove the throttle body as follows:
    - Disconnect the throttle motor connector.
    - Remove the 4 bolts and disconnect the throttle body from the intake air surge tank.
    - Disconnect the 2 water by-pass hoses from the throttle body.
    - Remove the throttle body and gasket.

To install:

14. Install a new gasket to the intake air surge tank.
15. Connect the 2 water by-pass hoses to the throttle body.
16. Install the throttle body with the 4 bolts and tighten to 7 ft. lbs. (10 Nm).
17. Connect the throttle motor connector.
18. Install the air cleaner cap with air cleaner hose assembly with the 4 clamps and hose clamp.
19. Install the purge line hose to the air cleaner hose.
20. Connect the MAF meter connector and clamp to the air cleaner.
21. Connect the ventilation hose to the cylinder head cover with the clamp.
22. Add engine coolant.
23. Connect the negative battery cable.
24. Check the for coolant leaks.
25. Check the function of the throttle body.
26. Install the air cleaner inlet.
27. Install the engine room covers.
28. Install the cool air duct seal.
29. Certain systems need to be initialized after disconnecting and reconnecting the cable from the negative (-) battery terminal.

FUEL SYSTEM SERVICE PRECAUTIONS

Safety is the most important factor when performing not only fuel system maintenance but any type of maintenance. Failure to conduct maintenance and repairs in a safe manner may result in serious personal injury or death. Maintenance and testing of the vehicle's fuel system components can be accomplished safely and effectively by adhering to the following rules and guidelines.

- To avoid the possibility of fire and personal injury, always disconnect the negative battery cable unless the repair or test procedure requires that battery voltage be applied.
- Always relieve the fuel system pressure prior to disconnecting any fuel system component (injector, fuel rail, pressure regulator, etc.), fitting or fuel line connection. Exercise extreme caution whenever relieving fuel system pressure to avoid exposing skin, face and eyes to fuel spray. Please be advised that fuel under pressure may penetrate the skin or any part of the body that it contacts.
- Always place a shop towel or cloth around the fitting or connection prior to loosening to absorb any excess fuel due to spillage. Ensure that all fuel spillage (should it occur) is quickly removed from engine surfaces. Ensure that all fuel soaked cloths or towels are deposited into a suitable waste container.
- Always keep a dry chemical (Class B) fire extinguisher near the work area.
- Do not allow fuel spray or fuel vapors to come into contact with a spark or open flame.
- Always use a back-up wrench when loosening and tightening fuel line connection fittings. This will prevent unnecessary stress and torsion to fuel line piping.
- Always replace worn fuel fitting O-rings with new Do not substitute fuel hose or equivalent where fuel pipe is installed.

Before servicing the vehicle, make sure to also refer to the precautions in the beginning of this section as well.

RELIEVING FUEL SYSTEM PRESSURE

1. Connect the Techstream to the DLC3.
2. Set the vehicle to the "INSPECTION MODE
3. Start the engine.
4. Disconnect the fuel pump connector.
5. After the engine has stopped on its own, turn the power switch OFF.
6. Loosen the fuel tank cap, then discharge the pressure in the fuel tank completely.
7. Connect the fuel pump connector.

➡DTC P3190, P3191, P0171, P0172 and/or P0A0F may be set.

FUEL FILTER

REMOVAL & INSTALLATION

The fuel filter is integral to the fuel pump assembly inside the fuel tank.

FUEL INJECTORS

REMOVAL & INSTALLATION

Direct Fuel Injection

See Figures 108 and 109.

CAUTION

The GS450h hybrid system contains a 288V high-voltage system with a strong alkali solution of potassium hydroxide. Be sure to follow the instructions in this manual to handle the system correctly. Failure to do so may result in serious injury or electrocution. Engineer must undergo special training to be able to perform high-voltage system inspection and servicing.

CAUTION

All high-voltage wire harness connectors are colored orange. The HV battery and other high-voltage components have "High Voltage" caution labels. Do not carelessly touch these wires and components.

CAUTION

Before inspecting or servicing the high-voltage system, be sure to follow safety measures, such as wearing insulated gloves and removing the service plug to prevent electrocution. Carry the removed service plug in your pocket to prevent other technicians from reinstalling it while you are servicing the vehicle.

CAUTION

After removing the service plug, wait 10 minutes before touching any of the high-voltage connectors and terminals.

CAUTION

Before inspecting the high-voltage system or disconnecting the low voltage connector of the inverter with converter assembly, take safety precautions, such as wearing insulated gloves and removing the service plug grip to prevent electrical shocks. After removing the service plug grip, put it in your pocket to prevent other technicians from accidentally reconnecting it while you are working on the high-voltage system.

CAUTION

After disconnecting the service plug grip, wait for at least 10 minutes before touching any of the high-voltage connectors or terminals. After waiting, check the voltage at the inspection point in the inverter with converter assembly. The voltage should be 0 V before beginning work.

1. Discharge the fuel system.
2. Remove the hybrid service plug grip.
3. Disconnect the negative battery cable.
4. Drain the engine coolant.
5. Remove the cool air intake duct seal.
6. Remove the engine room side cover.
7. Remove the V-bank cover.
8. Remove the air cleaner cap and hose.
9. Remove the wiper arm assembly LH and RH
10. Remove the cowl top ventilator louver sub-assembly.
11. Drain the engine coolant.
12. Disconnect the No. 2 ventilator hose.
13. Remove the air cleaner cap with air cleaner hose.
14. Remove the throttle body with motor assembly.
15. Remove the air intake surge tank.
16. Remove the intake manifold.
17. Remove the water hose joint.
18. Remove the fuel pressure pulsation damper.
19. Remove the No. 1 fuel pipe.
20. Disconnect the No 2 fuel pipe.

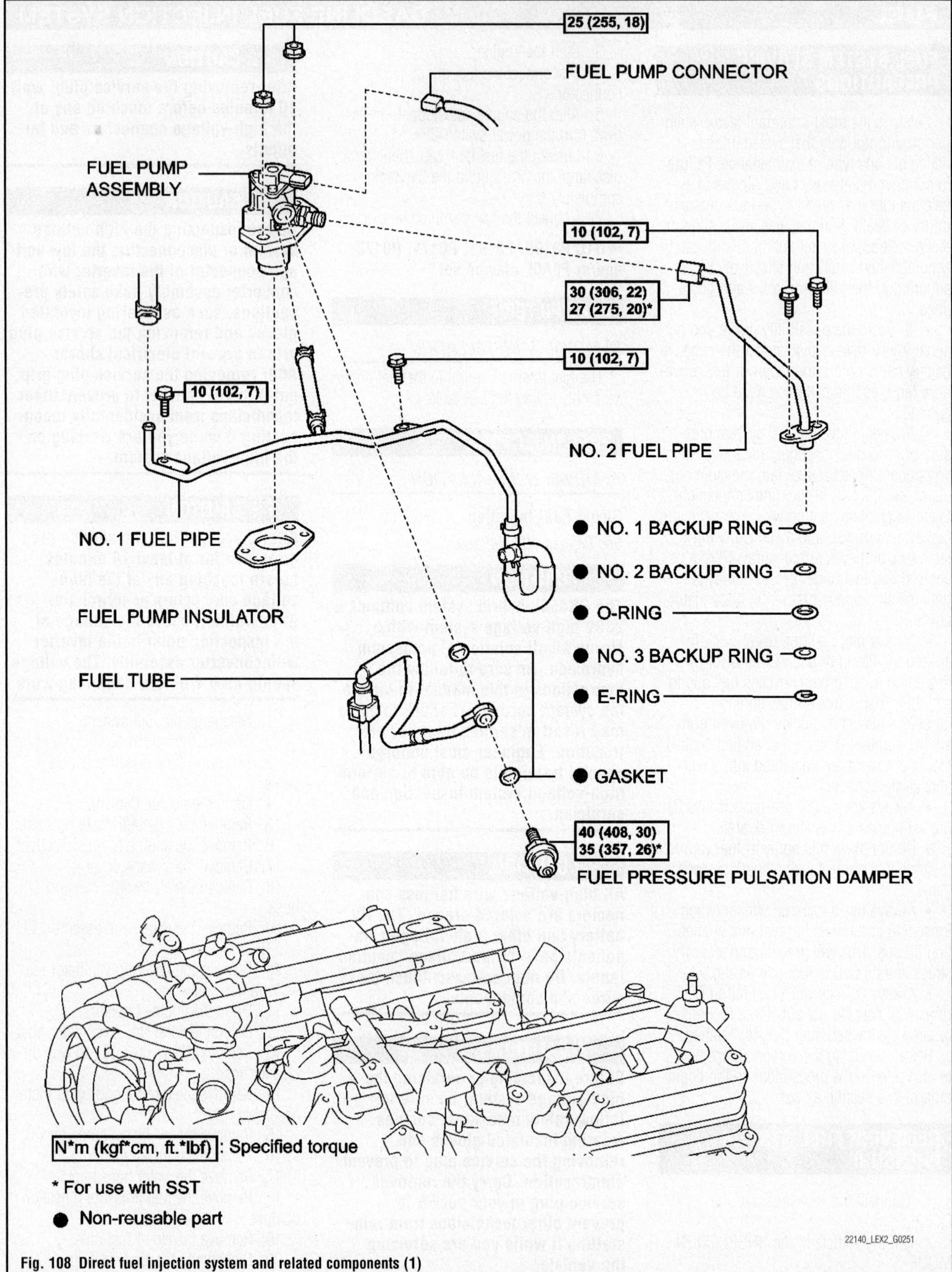

Fig. 108 Direct fuel injection system and related components (1)

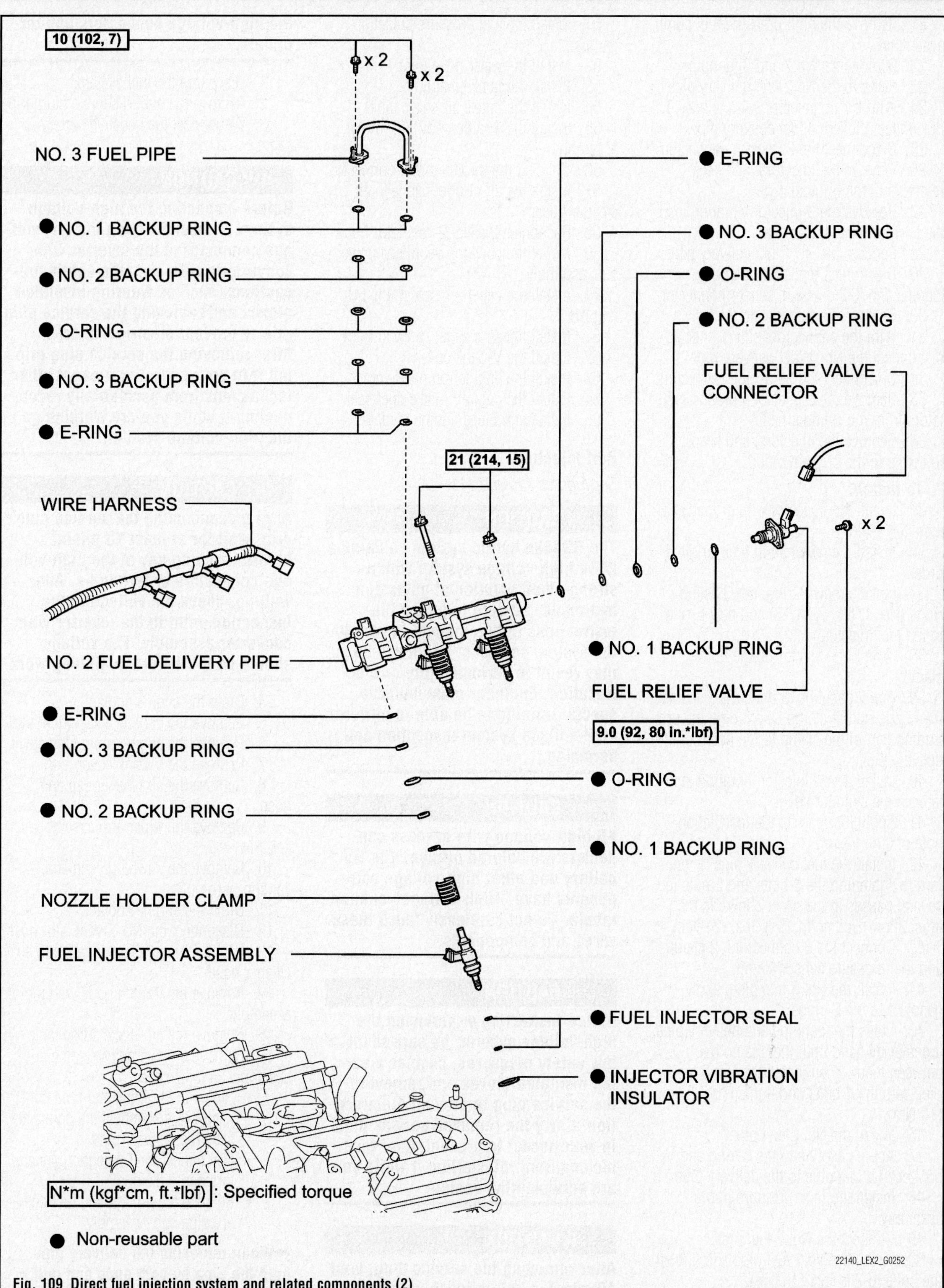

Fig. 109 Direct fuel injection system and related components (2)

21. Remove the high pressure fuel pump assembly.
22. Remove the No.2 and 3 fuel pipe.
23. Remove the No.2 fuel delivery pipe.
24. With the connectors still connected, disconnect the No. 2 fuel delivery pipe.
25. Disconnect the 3 injector connectors.
26. Remove the injectors, and keep injectors in the original order.
27. Remove the 3 injector vibration insulators from the cylinder head.
28. Remove the No.1 fuel delivery pipe.
29. Disconnect the 2 wire harness clamps and fuel pressure sensor connector.
30. Remove the 2 bolts and 2 nuts.
31. With the connectors still connected, disconnect the No. 1 fuel delivery pipe.
32. Disconnect the 3 injector connectors.
33. Remove the 3 injector vibration insulators from the cylinder head.
34. Remove the injectors, and keep injectors in the original order.

To install:

35. Install the injectors with properly installed seals.
36. Install the injectors in their original order.
37. Install a new O-ring, new backup rings (No. 1, No. 2, No. 3) and new E-ring to the fuel injector.
38. Install the injector nozzle holder clamp.
39. Apply gasoline to the O-ring. Install the nozzle holder clamp by aligning the protruding part of the clamp to the notch of the delivery pipe.
40. Install a new injector vibration insulator to the cylinder head.
41. Apply lubricant to the installation hole of the injector.
42. Install the fuel delivery pipe by uniformly tightening the 2 bolts and 2 nuts in several passes in the order shown in the illustration. Tighten to 15 ft. lbs. (21 Nm).
43. Connect the 3 connectors, 2 clamps and fuel pressure sensor clamp.
44. Install the No. 3 fuel pipe, apply gasoline to the O-rings.
45. Press the fuel pipe and delivery pipe together by hand until there is no gap between them. Then install the No. 3 fuel pipe with the 4 bolts and tighten to 7 ft. lbs. (10 Nm).
46. Install the No. 2 fuel pipe.
47. Apply gasoline to the O-ring and connect the fuel pipe to the delivery pipe.
48. Install the high pressure pump assembly.
49. Connect the No. 2 fuel pipe.
50. Install the No.1 fuel pipe and tighten the bolts to 7 ft. lbs. (10 Nm).
51. Install the fuel pressure pulsation damper.
52. Install the water hose joint.
53. Install the intake manifold.
54. Install the intake air surge tank.
55. Install the throttle body with motor assembly.
56. Connect the throttle motor connector.
57. Install the air cleaner cap with air cleaner hose.
58. Reconnect the No. 2 ventilator hose.
59. Install the cowl top ventilator louver sub-assembly.
60. Install the wiper arm assembly LH and RH
61. Install the air cleaner cap and hose.
62. Install the V-bank cover.
63. Install the engine room side cover.
64. Install the cool air intake duct seal.
65. Install and bleed engine coolant.

Port Injection

See Figures 110 and 111.

**** CAUTION**

The GS450h hybrid system contains a 288V high-voltage system with a strong alkali solution of potassium hydroxide. Be sure to follow the instructions in this manual to handle the system correctly. Failure to do so may result in serious injury or electrocution. Engineer must undergo special training to be able to perform high-voltage system inspection and servicing.

**** CAUTION**

All high-voltage wire harness connectors are colored orange. The HV battery and other high-voltage components have "High Voltage" caution labels. Do not carelessly touch these wires and components.

**** CAUTION**

Before inspecting or servicing the high-voltage system, be sure to follow safety measures, such as wearing insulated gloves and removing the service plug to prevent electrocution. Carry the removed service plug in your pocket to prevent other technicians from reinstalling it while you are servicing the vehicle.

**** CAUTION**

After removing the service plug, wait 10 minutes before touching any of the high-voltage connectors and terminals.

1. Discharge the fuel system.
2. Remove the hybrid service plug grip.
3. Disconnect the negative battery cable.

**** CAUTION**

Before inspecting the high-voltage system or disconnecting the low voltage connector of the inverter with converter assembly, take safety precautions, such as wearing insulated gloves and removing the service plug grip to prevent electrical shocks. After removing the service plug grip, put it in your pocket to prevent other technicians from accidentally reconnecting it while you are working on the high-voltage system.

**** CAUTION**

After disconnecting the service plug grip, wait for at least 10 minutes before touching any of the high-voltage connectors or terminals. After waiting, check the voltage at the inspection point in the inverter with converter assembly. The voltage should be 0 V before beginning work.

4. Drain the engine coolant.
5. Remove the cool air intake duct seal.
6. Remove the engine room side cover.
7. Remove the V-bank cover.
8. Remove the air cleaner cap and hose.
9. Remove the wiper arm assembly LH and RH
10. Remove the cowl top ventilator louver sub-assembly.
11. Drain the engine coolant.
12. Disconnect the No. 2 ventilator hose.
13. Remove the air cleaner cap with air cleaner hose.
14. Remove the throttle body with motor assembly.
15. Remove the air intake surge tank.
16. Remove the fuel delivery pipe sub-assembly as follows:
    - Pinch and pull the fuel tube connector to disconnect the connector from the delivery pipe.
    - Disconnect the 6 injector connectors and 2 clamps.
    - Remove the 4 bolts and fuel delivery pipe sub-assembly.

➡When removing the delivery pipe, hold the pipe by both ends and pull it straight upward.

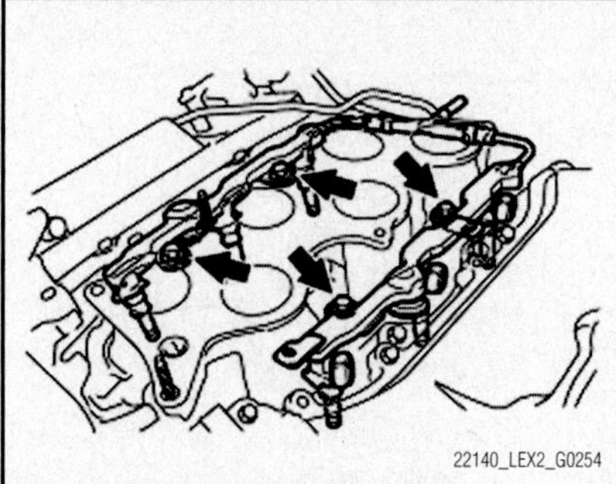

Fig. 110 Remove the 4 bolts and fuel delivery pipe sub-assembly

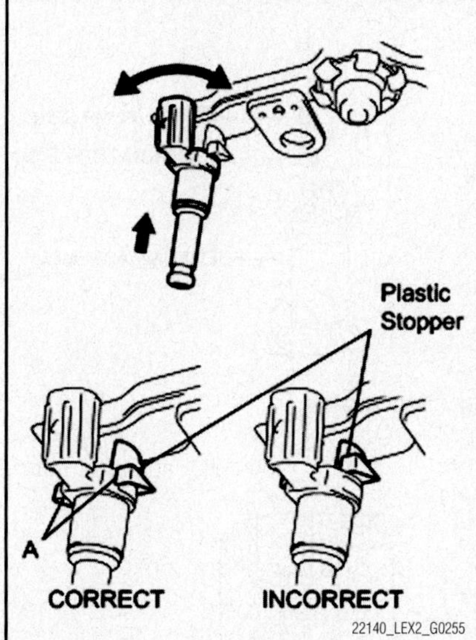

Fig. 111 Install the fuel injector assemblies to the fuel delivery pipe

- Remove the 4 delivery pipe spacers from the intake manifold.
- Remove the 6 fuel injectors from the fuel delivery pipe. For reinstallation, attach a tag or label to the injector shaft.

To install:

17. Apply gasoline to 2 new O-rings and install them to the injector.
18. Install the fuel injector assemblies to the fuel delivery pipe.
19. Install the fuel delivery pipe as follows:
    - Install the 4 delivery pipe spacers to the intake manifold.
    - Install the delivery pipe (with injector) to the intake manifold.
    - Install the 4 bolts and tighten to 13 ft. lbs. (17 Nm).
    - Connect the 6 injector connectors and 2 clamps.
    - Push in the fuel tube connector to the delivery pipe and push up the retainer to engage the claws.
    - After connecting the fuel tube, align the paint marks on the delivery pipe.
20. Install the intake air surge tank.
21. Install the throttle body with motor assembly.
22. Connect the throttle motor connector.
23. Install the air cleaner cap with air cleaner hose.
24. Reconnect the No. 2 ventilator hose.
25. Install the cowl top ventilator louver sub-assembly.
26. Install the wiper arm assembly LH and RH
27. Install the air cleaner cap and hose.
28. Install the V-bank cover.
29. Install the engine room side cover.
30. Install the cool air intake duct seal.
31. Install and bleed engine coolant.

FUEL PUMP

REMOVAL & INSTALLATION

See Figures 112 through 116.

Before servicing any vehicle, please be sure to read all of the following precautions, which deal with personal safety, prevention of component damage, and important points to take into consideration when servicing a motor vehicle:

CAUTION

The GS450h hybrid system contains a 288V high-voltage system with a strong alkali solution of potassium hydroxide. Be sure to follow the instructions in this manual to handle the system correctly. Failure to do so may result in serious injury or electrocution. Engineer must undergo special training to be able to perform high-voltage system inspection and servicing.

CAUTION

All high-voltage wire harness connectors are colored orange. The HV battery and other high-voltage components have "High Voltage" caution labels. Do not carelessly touch these wires and components.

CAUTION

Before inspecting or servicing the high-voltage system, be sure to follow safety measures, such as wearing insulated gloves and removing the service plug to prevent electrocution. Carry the removed service plug in your pocket to prevent other technicians from reinstalling it while you are servicing the vehicle.

CAUTION

After removing the service plug, wait 10 minutes before touching any of the high-voltage connectors and terminals.

1. Discharge the fuel system.
2. Remove the service plug.
3. Disconnect the negative battery cable.
4. Remove the rear seat cushion assembly.
5. Remove the clip and room partition pad.
6. Remove the service hole cover and disconnect the 2 connectors.
7. Disconnect the fuel main tubes and fuel return vent tube.
8. Remove the 2 tube joint clips and fuel tubes as shown below.
9. Remove the 8 bolts and set plate.
10. Disconnect the fuel tube.
11. Remove the fuel suction tube with pump and gauge from the fuel tank.
12. Disconnect the fuel sender gauge connector.
13. Press down on the sender gauge claw labeled A. Then slide the sender gauge upward.
14. Using needle nozzle pliers, remove the E-ring.

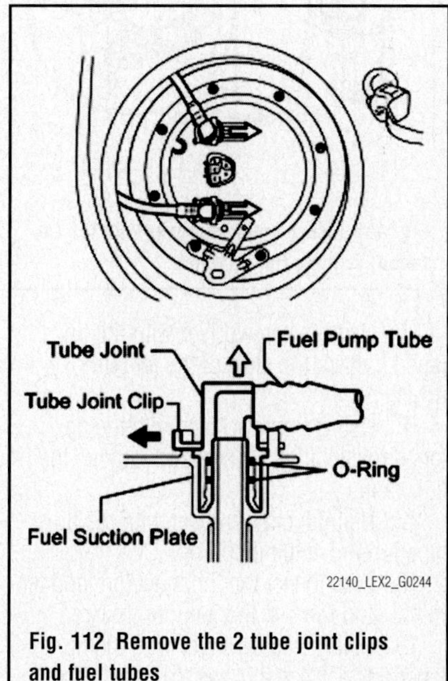

Fig. 112 Remove the 2 tube joint clips and fuel tubes

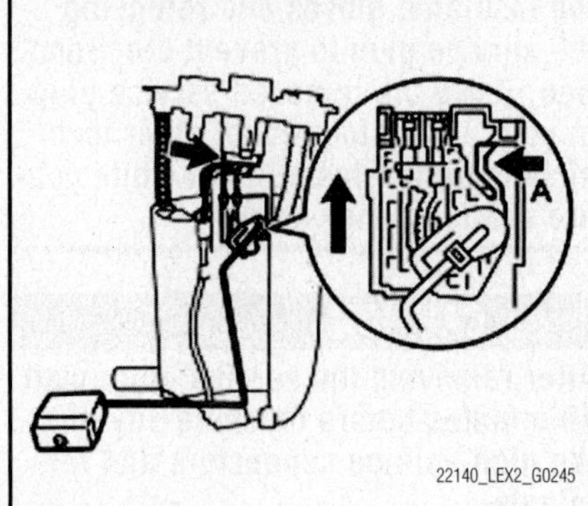

Fig. 113 Press down on the sender gauge claw labeled A. Then slide the sender gauge upward

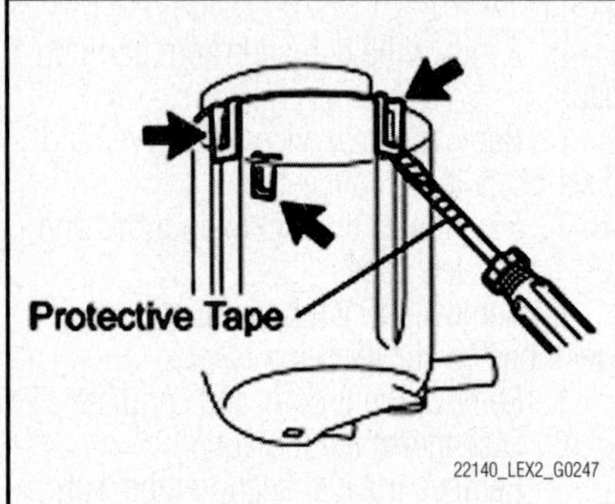

Fig. 114 Detach the 3 claws and remove the sub-tank

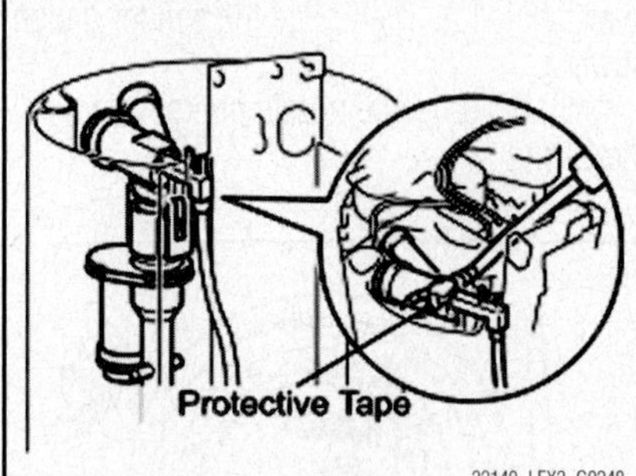

Fig. 115 Using a clip remover with the tip taped, remove the jet pump

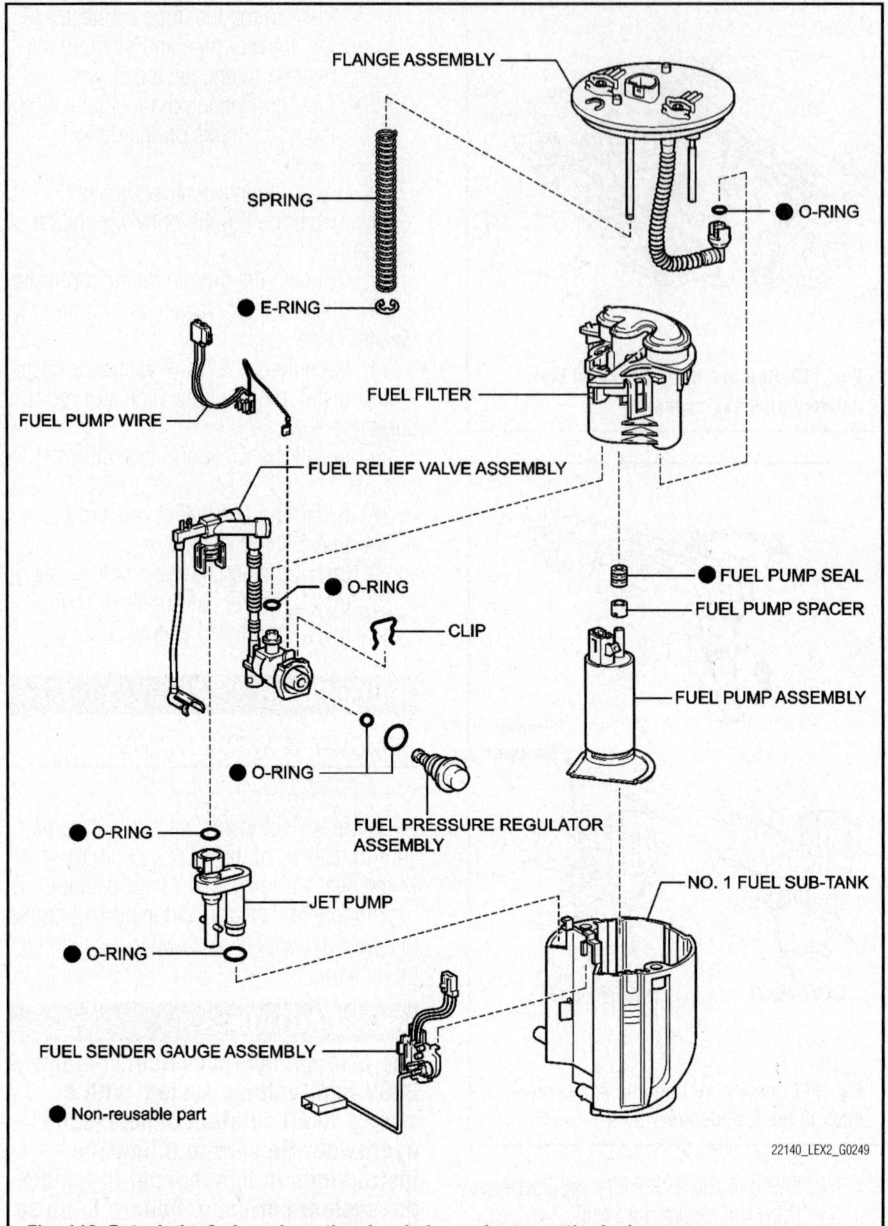

Fig. 116 Detach the 2 claws from the claw holes and remove the fuel pump

15. Using a screwdriver with the tip taped, detach the claw of the jet pump nozzle.
16. Using a screwdriver with the tip taped, detach the 3 claws and remove the sub-tank.
17. Using a clip remover with the tip taped, remove the jet pump.
18. Disconnect the 2 connectors and terminal, and remove the fuel pump wire.
19. Using a screwdriver with the tip taped, detach the 2 claws from the claw holes and remove the fuel pump.
20. Remove the fuel pump seal from the fuel pump.

To install:

21. Install the fuel pump spacer to the fuel pump.
22. Apply a light coat of gasoline to a new seal, and install it to the fuel pump.
23. Install the fuel pump to the fuel filter.
24. Connect the 2 connectors and terminal.
25. Apply gasoline to a new O-ring and install it to the jet pump.
26. Install the jet pump while aligning it to the installation position of the sub-tank.
27. Connect the jet pump nozzle to the sub-tank.
28. Attach the 3 claws to the claw holes and install the sub-tank.
29. Install a new E-ring.
30. Set the fuel sender gauge to the sub-tank. Then slide the sender gauge downward to install it.
31. Connect the fuel sender gauge connector.
32. Install the fuel suction with pump and gauge tube assembly.
33. Install the service hole cover.
34. Install the partition pad.
35. Install the rear seat cushion assembly.
36. Install the service plug grip.
37. Connect the negative battery cable.
38. Check for fuel leaks.
39. Certain systems need to be initialized after disconnecting and reconnecting the cable from the negative (-) battery terminal.

FUEL TANK

REMOVAL & INSTALLATION

See Figures 117 and 118.

CAUTION

The GS450h hybrid system contains a 288V high-voltage system with a strong alkali solution of potassium hydroxide. Be sure to follow the instructions in this manual to handle the system correctly. Failure to do so may result in serious injury or electrocution. Engineer must undergo special training to be able to perform high-voltage system inspection and servicing.

CAUTION

All high-voltage wire harness connectors are colored orange. The HV battery and other high-voltage components have "High Voltage" caution labels. Do not carelessly touch these wires and components.

CAUTION

Before inspecting or servicing the high-voltage system, be sure to follow safety measures, such as wearing insulated gloves and removing the service plug to prevent electrocution. Carry the removed service plug in your pocket to prevent other technicians from reinstalling it while you are servicing the vehicle.

CAUTION

After removing the service plug, wait 10 minutes before touching any of the high-voltage connectors and terminals.

1. Discharge the fuel system.
2. Remove the service plug.
3. Disconnect the negative battery cable.
4. Remove the rear seat cushion assembly.
5. Remove the clip and room partition pad.
6. Remove the service hole cover and disconnect all the electrical connectors.
7. Disconnect the fuel main tubes and fuel return vent tube.
8. Remove the 2 tube joint clips and fuel tubes.
9. Remove the exhaust pipe.
10. Remove the 3 nuts, 2 clips and floor LH and RH covers.

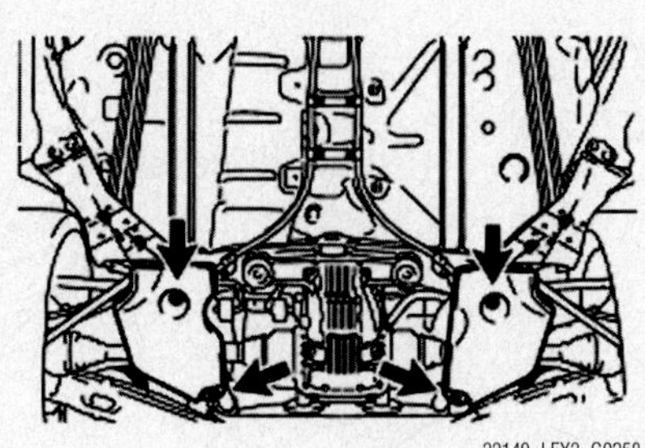

Fig. 117 Remove both differential support protectors

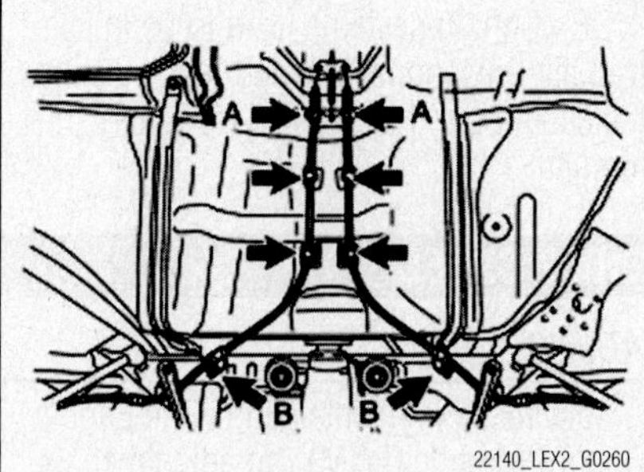

Fig. 118 Install the 2 parking brake cables with the 4 bolts

11. Remove the 5 nuts and heat insulator.
12. Remove the RH outside air guide plate.
13. Remove the propeller shaft assembly.
14. Remove both differential support protectors.
15. Remove the RH rear floor side member cover.
16. Remove the RH and LH rear suspension member brace.
17. Detach the 2 parking brake cables from the 4 clamps.
18. Remove the 4 bolts and disconnect the 2 parking brake cables from the parking brake equalizer.
19. Disconnect the fuel tank return vent tube.
20. Disconnect the fuel tank main tube RH.
21. Disconnect the fuel tank main tube LH.
22. Remove the clamp and disconnect the vent hose.
23. Remove the fuel tube clamp and disconnect the canister tube.
24. Disconnect the fuel vapor containment valve connector and 2 clamps.
25. Loosen the 2 fuel filler pipe bolts.
26. Place a mission jack under the fuel tank.
27. Remove the 4 bolts and 2 fuel tank bands.
28. Slowly lower the mission jack slightly.
29. Remove the clamp, disconnect the fuel filler pipe from the fuel tank, and lower the mission jack.
30. Remove the fuel tank from the mission jack.

To install:

31. Set the fuel tank on a mission jack and raise the fuel tank.
32. Install the clamp, and connect the fuel filler pipe to the fuel tank.
33. Install the 2 fuel tank bands with the 4 bolts and tighten to 29 ft. lbs. (39 Nm).
34. Tighten the 2 fuel filler pipe bolts to 13 ft. lbs. (18 Nm).
35. Connect the fuel vapor containment valve connector and 2 clamps.
36. Connect the canister tube and install the clamp.
37. Connect the fuel tank vent hose and install the clamp.
38. Connect the fuel tank main tube (LH).
39. Connect the fuel tank main tube (RH).
40. Connect the fuel return vent tube and install the retainer.
41. Connect the 2 parking brake cables to the parking brake equalizer.
42. Install the 2 parking brake cables with the 4 bolts. Tighten bolt (A) to 53 inch. lbs. (6 Nm). Tighten bolt (B) to 14 ft. lbs. (19 Nm).
43. Attach the 2 parking brake cables to the 4 clamps.
44. Install the RH and LH rear suspension member brace with the bolts. Tighten to 14 ft. lbs. (19 Nm).
45. Install the LH and RH member covers with the 3 bolts.
46. Install the differential supporter with the 2 nuts.
47. Install the propeller shaft assembly.
48. Install the RH outside air guide plate and tighten the nuts to 48 inch. (5.8 Nm).
49. Install the floor insulator and tighten the nuts to 48 inch. (5.8 Nm).
50. Install the LH and RH floor cover center with the 3 nuts.
51. Install 2 new gaskets and the exhaust pipe front ends to the exhaust manifolds with the 4 bolts and 2 nuts. Tighten to 29 ft. lbs. (39 Nm).
52. Install all fuel and electrical connections to top of the fuel tank.
53. Install the partition pad with the clip.
54. Install the seat cushion assembly.
55. Install the service plug grip.
56. Connect the negative battery cable.

57. Check for fuel leaks.

58. Certain systems need to be initialized after disconnecting and reconnecting the cable from the negative (-) battery terminal.

IDLE SPEED

ADJUSTMENT

Idle speed is maintained by the Engine Control Module (ECM). No adjustment is necessary or possible.

THROTTLE BODY

REMOVAL & INSTALLATION

See Figure 119.

1. Disconnect the negative battery cable.
2. Remove the cool air intake duct seal.
3. Drain the engine coolant.
4. Remove the V-bank cover.
5. Remove the air cleaner inlet.
6. Disconnect the No. 2 ventilation hose.
7. Disconnect the air cleaner cap and hose.
8. Disconnect the MAF meter connector.
9. Disconnect the clamp from the air cleaner.
10. Disconnect the purge line hose.
11. Disconnect the 4 clamps.
12. Remove the hose clamp and air cleaner cap with air cleaner hose.
13. Remove the throttle body as follows:
    - Disconnect the throttle motor connector.
    - Remove the 4 bolts and disconnect the throttle body from the intake air surge tank.
    - Disconnect the 2 water by-pass hoses from the throttle body.
    - Remove the throttle body and gasket.

To install:

14. Install a new gasket to the intake air surge tank.
15. Connect the 2 water by-pass hoses to the throttle body.
16. Install the throttle body with the 4 bolts and tighten to 7 ft. lbs. (10 Nm).
17. Connect the throttle motor connector.

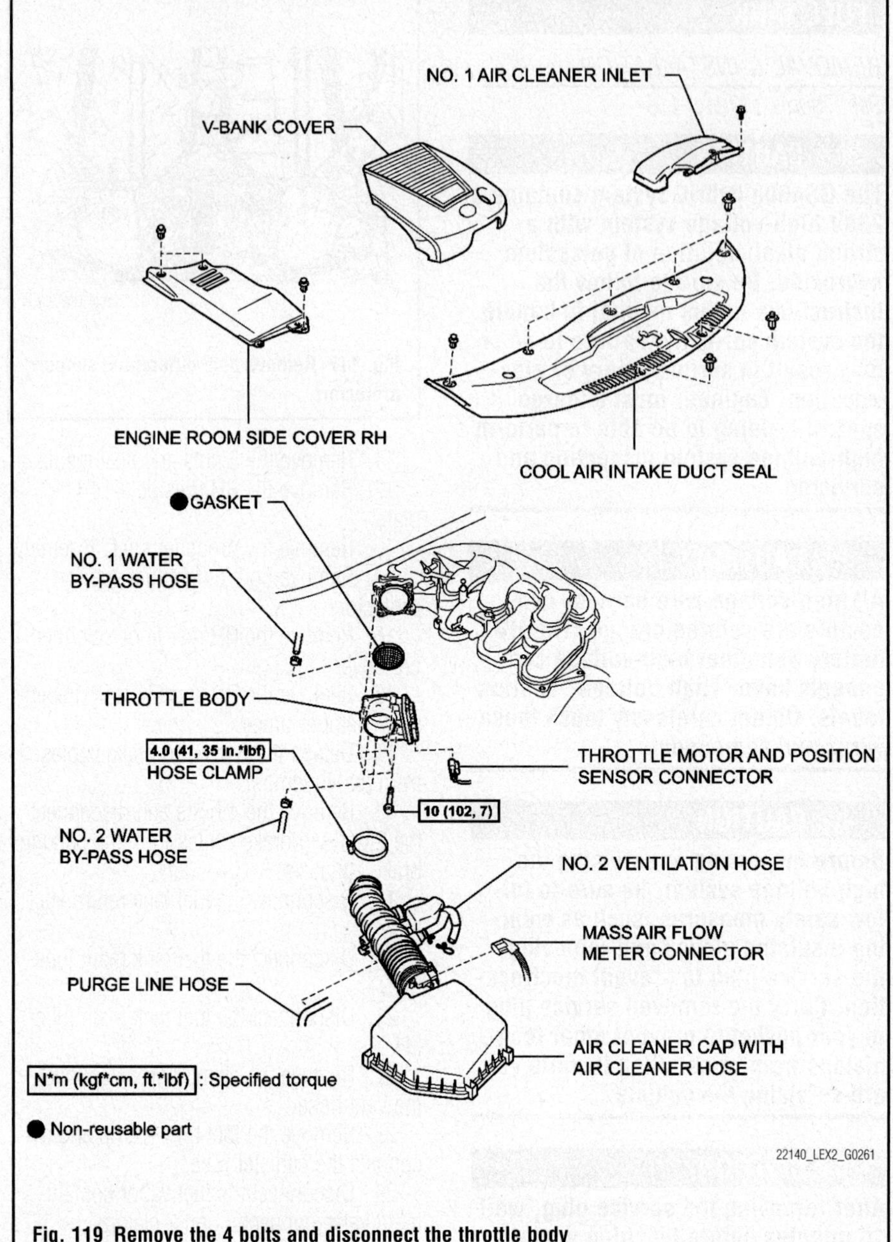

Fig. 119 Remove the 4 bolts and disconnect the throttle body

18. Install the air cleaner cap with air cleaner hose assembly with the 4 clamps and hose clamp.
19. Install the purge line hose to the air cleaner hose.
20. Connect the MAF meter connector and clamp to the air cleaner.
21. Connect the ventilation hose to the cylinder head cover with the clamp.
22. Add engine coolant.
23. Connect the negative battery cable.
24. Check the for coolant leaks.
25. Check the function of the throttle body.
26. Install the air cleaner inlet.
27. Install the engine room covers.
28. Install the cool air duct seal.
29. Certain systems need to be initialized after disconnecting and reconnecting the cable from the negative (-) battery terminal.

HEATING & AIR CONDITIONING SYSTEM

BLOWER MOTOR

REMOVAL & INSTALLATION

See Figure 120.

1. Before servicing any vehicle, please be sure to read all of the precautions.
2. Remove the service plug
3. Disconnect the negative battery cable.
4. Remove the A/C unit.
5. Remove the blower unit assembly.
6. Disconnect the connector.
7. Remove the 3 screws and fan motor.

To install:

8. Install the fan motor with the 3 screws.
9. Reconnect the connector.
10. Install the blower unit assembly.
11. Install the A/C unit.
12. Install the service plug.
13. Connect the negative battery cable.

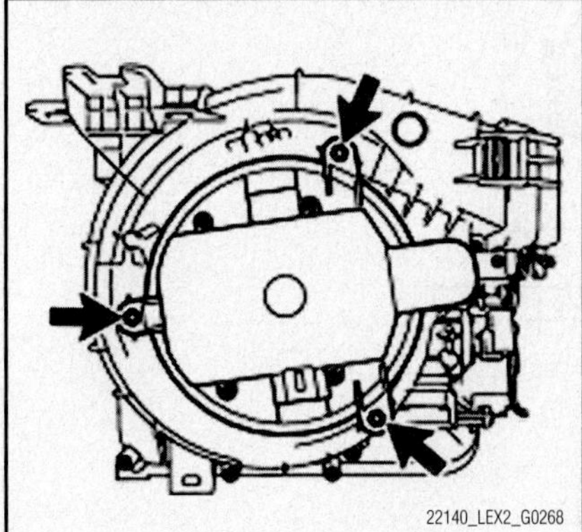

Fig. 120 Remove the 3 screws and fan motor

HEATER CORE

REMOVAL & INSTALLATION

See Figures 121 and 122.

Before servicing any vehicle, please be sure to read all of the following precautions, which deal with personal safety, prevention of component damage, and important points to take into consideration when servicing a motor vehicle.

✻✻ CAUTION

The GS450h hybrid system contains a 288V high-voltage system with a strong alkali solution of potassium hydroxide. Be sure to follow the instructions in this manual to handle the system correctly. Failure to do so may result in serious injury or electrocution. Engineer must undergo special training to be able to perform high-voltage system inspection and servicing.

✻✻ CAUTION

All high-voltage wire harness connectors are colored orange. The HV battery and other high-voltage components have "High Voltage" caution labels. Do not carelessly touch these wires and components.

✻✻ CAUTION

Before inspecting or servicing the high-voltage system, be sure to follow safety measures, such as wearing insulated gloves and removing the service plug to prevent electrocution. Carry the removed service plug in your pocket to prevent other technicians from reinstalling it while you are servicing the vehicle.

✻✻ CAUTION

After removing the service plug, wait 10 minutes before touching any of the high-voltage connectors and terminals.

1. Disable air bag system.
2. Remove the service plug.
3. Before servicing the vehicle, refer to the Precautions section.
   - Discharge the A/C system.
   - Set radio receiver assembly to shipment mode as follows:
   - Be sure that all discs and tapes have been removed from the unit.
   - Be sure that the engine switch off.
   - While simultaneously pressing the "SEEK UP" and "DISC" switches, turn the engine switch on (ACC).
   - Turn the engine switch off.
4. Disconnect the negative battery cable. Wait 90 seconds before doing any further work while the airbag system de-energizes.
5. Drain the cooling system into a clean container for reuse.
6. Align the front wheels facing straight ahead.
7. Remove or disconnect the following:
   - Cool air intake duct seal
   - Left and right engine room side covers
   - Left and right front pillar to front side seals, using a clip remover to detach the 3 claws
   - Left and right nut, windshield wiper arms and blades
   - Left and right front fender to cowl side seals by moving the component toward the center of the vehicle to detach the 2 claws
8. Remove cowl top ventilator louver assembly as follows:
   - Remove the 2 clips and detach the 5 claws.
   - Pull the ventilator louver in the direction indicated by the arrow in the illustration to detach the 10 claws and remove the ventilator louver.
9. Remove the windshield wiper motor and link assembly as follows:
   - Disconnect the connector. Then detach the 2 clamps and remove the wire harness from the cowl top panel.

➡There are 6 bolts total, however, 2 bolts cannot be removed from the wiper motor and link because they are integrated into the wiper motor and link.

   - Remove the 4 bolts and wiper motor and link.
10. Separate suction pipe sub-assembly as follows:
   - Remove the bolt, and slide the hook connector.
   - Disconnect the suction pipe sub-assembly.
   - Remove the O-ring from the suction pipe sub-assembly

✻✻ WARNING

Seal the openings of the disconnected parts using vinyl tape to prevent moisture and foreign matter from entering.

11. Separate liquid tube sub-assembly as follows:
   - Disconnect the liquid tube sub-assembly.
   - Remove the O-ring from the liquid tube subassembly.
   - Inlet and outlet heater water outlet hoses
   - Instrument panel assembly.
   - A/C blower unit assembly
   - Mounting screw and No. 2 air duct

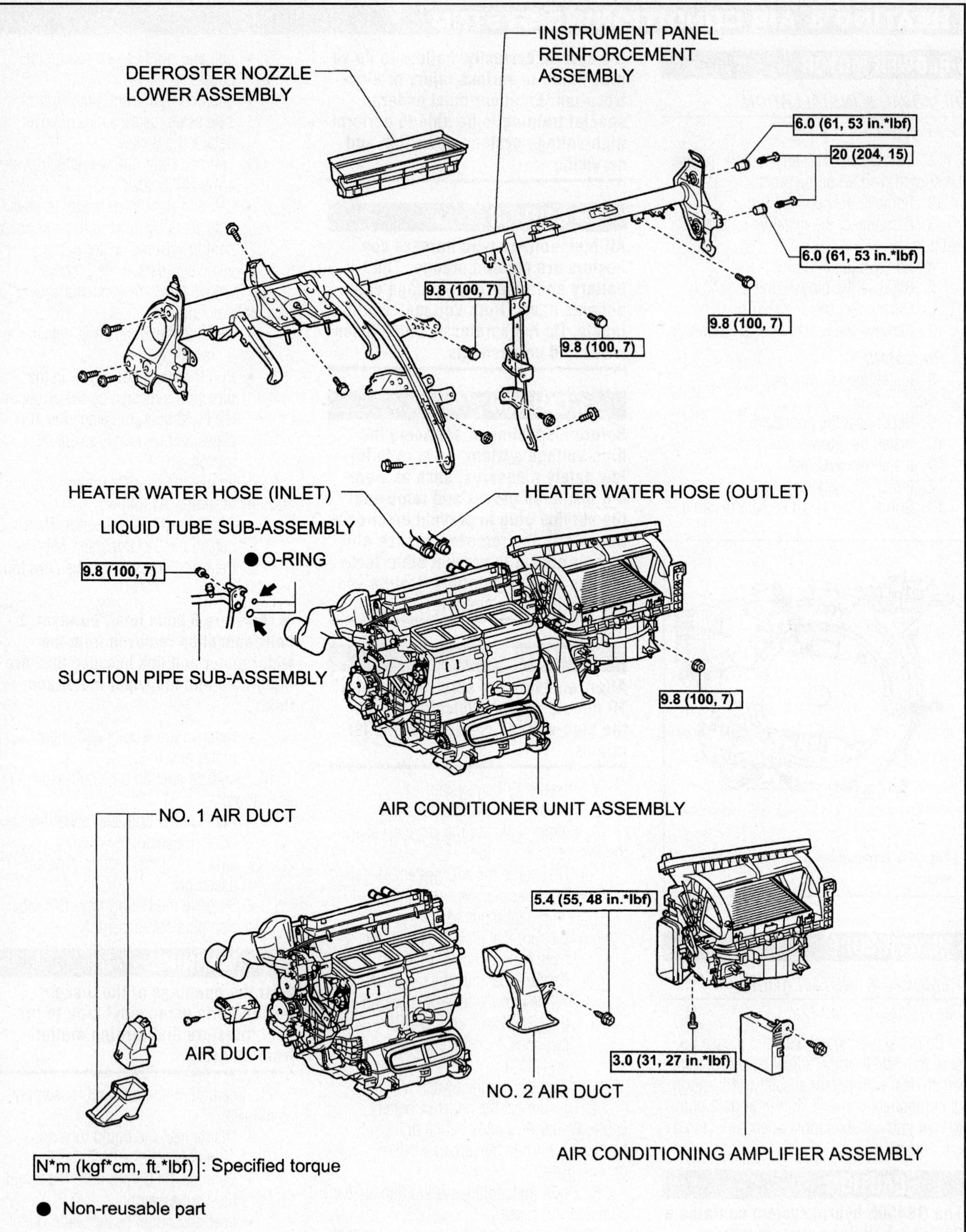

Fig. 121 Exploded view of the A/C unit assembly, instrument panel reinforcement and related components—GS450h

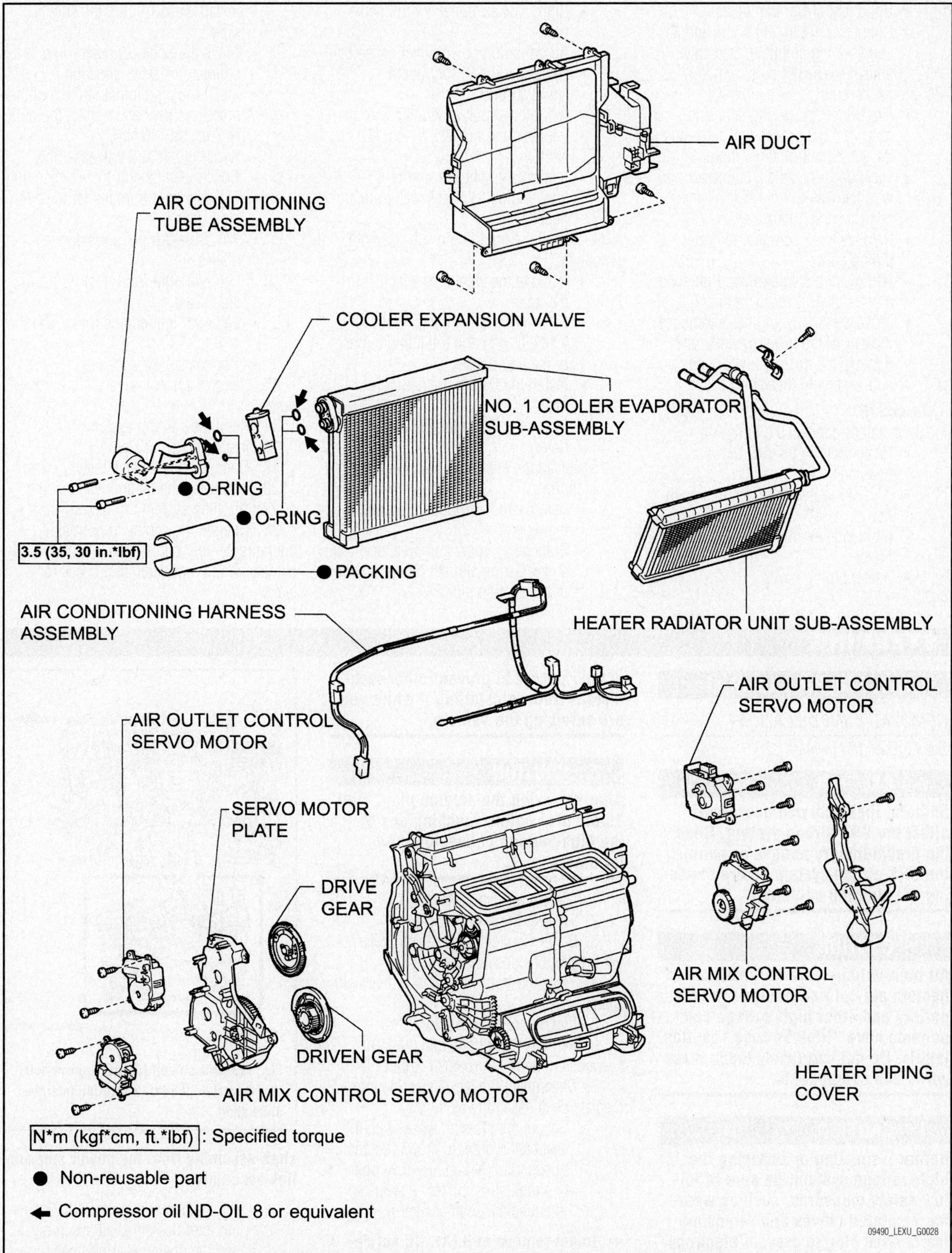

Fig. 122 Exploded view of the heater radiator unit (heater core), heater housing and related components—GS450h

- Upper and lower foot ducts
- Connector, mounting screw and air conditioning amplifier assembly

12. Remove the heater core from the A/C-blower assembly unit as follows:
   - Disconnect the wiring connector, remove the 3 screws and right side air outlet control servo motor.
   - Disengage 3 clamps, connector and wire harness to right side air mix control servo motor.
   - Remove the 2 screws and heater piping cover.
   - Remove the 3 screws and right side air mix control servo motor.
   - Move the A/C wiring harness out of the way of the heater radiator unit.
   - Remove the heater radiator unit sub-assembly (heater core)

To install:

13. Install or connect the following:
   - Heater core to the A/C blower housing
   - A/C wiring harness back into position
   - Right side air mix control servo motor
   - Heater piping cover
   - Right side air outlet control servo motor
   - Air conditioning amplifier assembly
   - Upper and lower foot ducts
   - No. 2 air duct
   - A/C blower unit assembly. Torque the retaining nut to 7 ft. lbs. (10 Nm).
   - Instrument panel assembly.
   - Inlet and outlet heater water outlet hoses
14. Install the liquid tube sub-assembly as follows:
   - Remove the vinyl tape attached to the tube.
   - Sufficiently apply compressor oil to a new O-ring and the fitting surface of the liquid tube.
   - Install the O-ring on the liquid tube.
   - Install the liquid tube to the fitting hole.
15. Install the suction pipe sub-assembly as follows:
   - Remove the vinyl tape attached to the pipe.
   - Sufficiently apply compressor oil to a new O-ring and the fitting surface of the suction pipe.
   - Install the O-ring on the suction pipe.
   - Move the hook connector in a counterclockwise direction.
   - Insert the pipe joints into the fitting holes securely and tighten the bolt to 7 ft. lbs. (10 Nm).
   - Wiper motor and link assembly
   - Cowl top ventilator louver assembly
   - Left and right front fender to cowl side seals
   - Windshield wiper arms and blades
   - Left and right front pillar to front side seals
   - Left and right engine room side covers
   - Cool air intake duct seal
16. Perform system initialization procedure.
17. Refill the cooling system.
18. Install the service plug.
19. Connect the negative battery cable.
20. Evacuate, charge and leak test the air conditioning system refrigerant.
21. Operate the engine to normal operating temperatures; then, check the climate control operation and check for leaks.

STEERING

POWER STEERING GEAR

REMOVAL & INSTALLATION

See Figures 123 through 125.

CAUTION

Some of these service operations affect the SRS airbag system. Read the precautionary notices concerning the SRS airbag system before servicing the steering column.

CAUTION

All high-voltage wire harness connectors are colored orange. The HV battery and other high-voltage components have "High Voltage" caution labels. Do not carelessly touch these wires and components.

CAUTION

Before inspecting or servicing the high-voltage system, be sure to follow safety measures, such as wearing insulated gloves and removing the service plug to prevent electrocution. Carry the removed service plug in your pocket to prevent other technicians from reinstalling it while you are servicing the vehicle.

CAUTION

After removing the service plug, wait 10 minutes before touching any of the high-voltage connectors and terminals.

1. Disable air bag system.
2. Remove the service plug.
3. Before servicing the vehicle, refer to the Precautions section.
4. Place the front wheels facing straight ahead.
5. Engine under cover No. 2.
6. Remove the 4 bolts and the front suspension member protector lower.
7. Disconnect the No. 2 steering intermediate shaft assembly as follows:
   - Secure the steering wheel with the seat belt in order to prevent rotation.
   - Loosen bolt (A) and remove bolt (B), then slide the No. 2 steering intermediate shaft assembly.

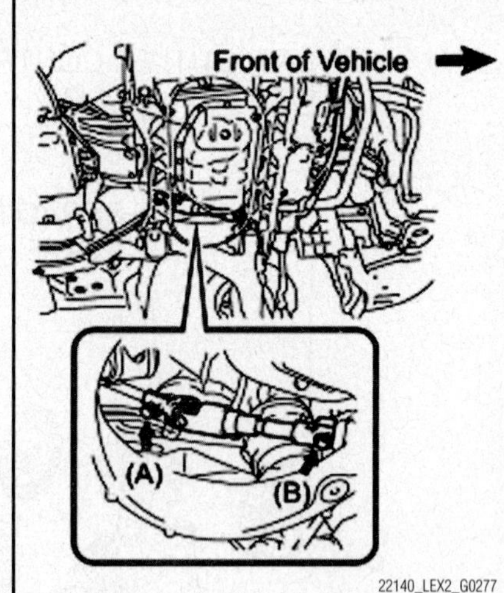

Fig. 123 Loosen bolt (A) and remove bolt (B), then slide the No. 2 steering intermediate shaft

➡ Do not remove bolt (A). Do not disconnect the No. 2 steering intermediate shaft assembly from the power steering link assembly.

   - Put matchmarks on the No. 2 steering intermediate shaft assembly and the power steering link assembly.

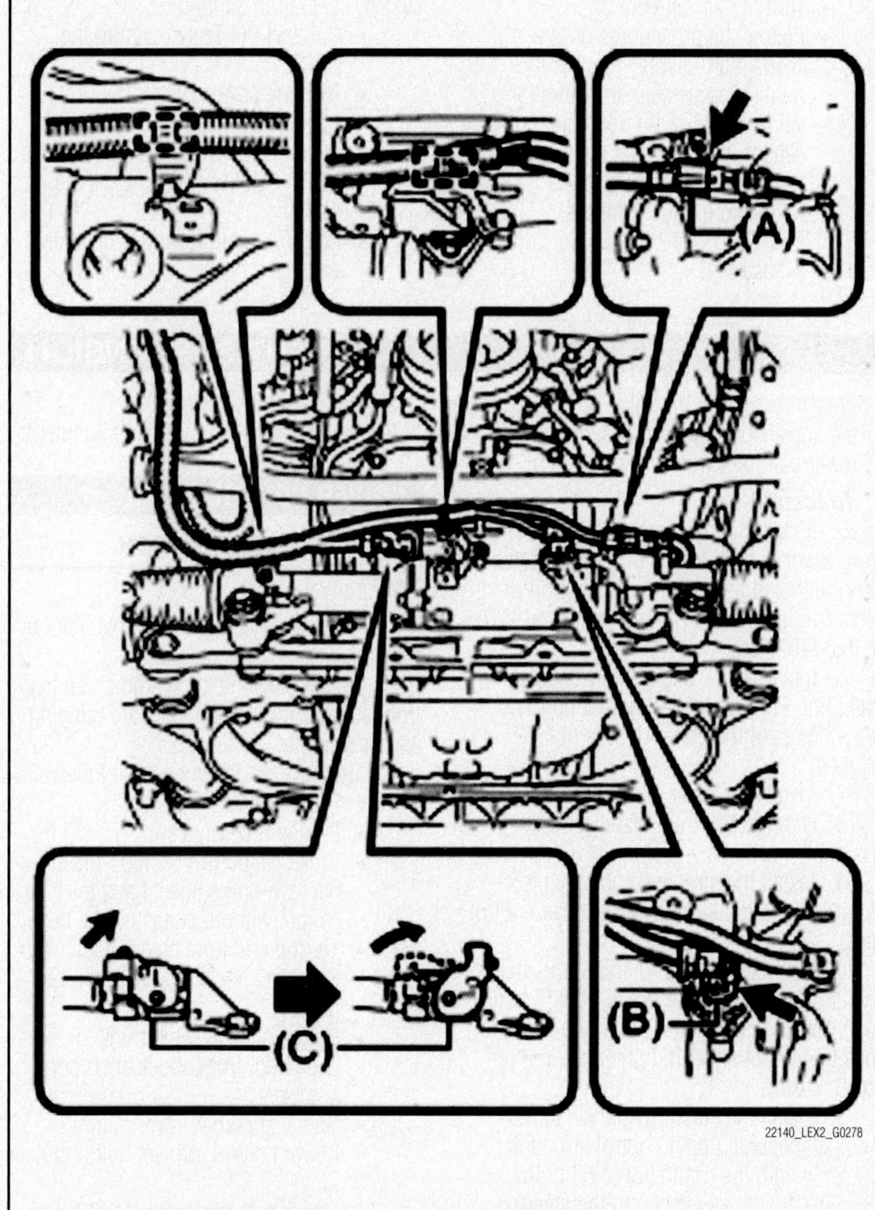

Fig. 124 Disconnect wire harness connector (A), (B) and (C) to the power steering

- Disconnect the No. 2 steering intermediate shaft assembly from the power steering link assembly.

8. Disconnect the LH and RH tie rod ends.

9. Remove the rack and pinion assembly as follows:
   - Remove the 2 clamps to disconnect the wire harness from the bracket.
   - Disconnect 2 connectors (A) and (B) from the power steering link assembly.
   - Release the lock of connector (C) and disconnect connector (C) from the power steering link assembly.
   - Remove the 2 bolts, 2 washers, 2 nuts, and power steering link assembly from the front suspension crossmember.

To install:

10. Install the power steering link assembly with the 2 bolts, 2 washers and 2 nuts. Tighten to 87 ft. lbs. (118 Nm).

11. Connect wire harness connector (C) to the power steering link assembly and securely lock the connector.

12. Connect 2 wire harness connectors (A) and (B) to the power steering link assembly.

13. Install the 2 wire harness clamps to the power steering link assembly.

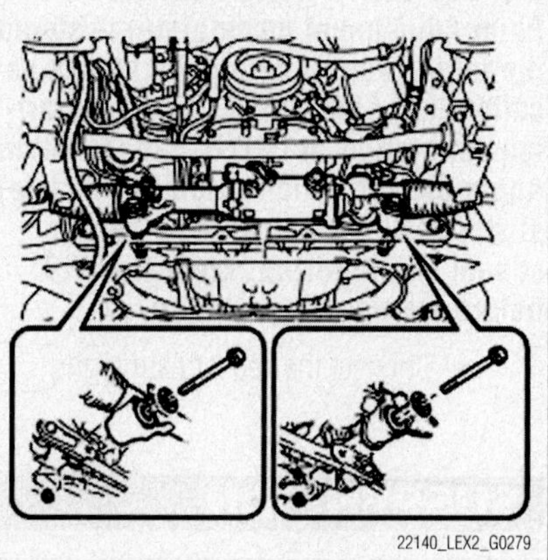

Fig. 125 Remove the 2 bolts, 2 washers, 2 nuts, and power steering link assembly from the front suspension crossmember.

14. Connect the tie rod end LH and RH to the steering knuckle with the nut. Tighten to 50 ft. lbs. (65 Nm).

15. Install a new clip. If the holes for the clip are not aligned, tighten the nut up to 60° further.

16. Connect the No. 2 steering intermediate shaft as follow:
    - Align the matchmarks on the No. 2 intermediate shaft and the power steering link assembly.
    - Install bolt (A) and tighten the 2 bolts. Tighten to 26 ft. lbs. (35 Nm).
    - Install the front suspension member protector lower to the front suspension crossmember with the 4 bolts. Tighten to 71 inch. lbs. (8 Nm).

17. Install the engine under cover.

18. Install the front wheels and tighten to 76 ft. lbs. (103 Nm).

19. Install service plug.

20. Connect the negative battery cable.

21. Inspect and adjust front wheel alignment.

22. Initialize the rotation angle sensor and calibrate torque sensor zero point as follow:
    - Inspection before calibration
    - Connect the Techstream to the DLC3.
    - Turn the power switch ON (IG).
    - Turn the Techstream on.
    - Check the IG power supply voltage on the Techstream. Enter the following menus: Chassis / EMPS / Data List.
    - Rotation angle sensor calibration value clear, rotation angle sensor initialization, and torque sensor zero point calibration

➡If DTC C1516 (Torque Sensor Zero Point Adjustment Incomplete) is stored, the torque sensor zero point cannot be calibrated. Clear the DTC before starting calibration. If DTC C1526 (Rotation Angle Sensor Initialization Incomplete) is stored, the rotation angle sensor cannot be initialized. Clear the DTC before starting initialization.

- Connect the Techstream to the DLC3.
- Turn the power switch ON (IG).
- Turn the Techstream on.
- Follow the procedures on the Techstream display to clear the rotation angle sensor calibration value, initialize the rotation angle sensor, and calibrate the torque sensor zero point. Enter the following menus: Chassis / EMPS / Utility / Torque Sensor Adjustment.

23. Perform variable gear ratio steering system calibration as follows:
    - Connect the Techstream to the DLC3.
    - Turn the power switch ON (IG).
    - Turn the Techstream on.
    - Enter the following menus: Chassis / VGRS / Utility / Steering Angle Adjust.
    - Perform the following procedures displayed on the Techstream.

SUSPENSION

FRONT SUSPENSION

COIL SPRING

REMOVAL & INSTALLATION

Refer to strut removal procedure also includes the separation of the coil spring.

LOWER BALL JOINT

REMOVAL & INSTALLATION

See Figure 126.

1. Before servicing the vehicle, refer to the precautions section.
2. Remove the front wheel.
3. Remove engine under cover.
4. Separate the ABS speed sensor.
5. For AWD, remove the clip and castle nut, then using SST 09628-00011, separate the upper ball joint from the steering knuckle.
6. Remove the clip and castle nut, then using SST 09628-62011, separate the tie rod end.
7. Remove the strut assembly.
8. Separate the front stabilizer link assembly.
9. Remove the engine under cover.
10. Remove the lower control arm.
11. Fix the front lower control arm in a vise using aluminum plates.
12. Remove the clip and castle nut.
13. Use SST 09950-40011 (09951-04010, 09952-04010, 09953-04020, 09954-04010, 09955-04051, 09957-04010, 09958-04011) to remove the front lower ball joint from the front lower control arm.
14. Inspect the lower control arm ball joint as follows:
    - Flip the ball joint stud back and forth 5 times.
    - Temporarily install the nut, and use a torque wrench to turn the nut continuously at a rate of 3 to 5 seconds per turn. Take the torque reading on the 5th turn.
    - Turning torque: 53 inch lbs (6 Nm)

    a. Check the dust boots for cracks or grease leakage. If the value is not within the specified range, replace the front lower ball joint with a new one.

To install:

15. Install the front lower ball joint to the front lower control arm with the nut. Ensure that the thread and taper are free of oil or other foreign matter. Tighten the nut to 120 ft. lbs. (162 Nm).
16. Install a new clip to the front lower ball joint. Further tighten the nut up to 60° if the holes for the cotter pin are not aligned.
17. Install the lower control arm.
18. Connect the front stabilizer link assembly.
19. Install the strut assembly, but temporarily tighten the bolts at the lower control arm.
20. Connect the tie rod end to the steering knuckle Tighten the nut to 50 ft. lbs. (65 Nm). Install a new cotter pin. If the holes for the clip are not aligned, tighten the nut up to 60° further.
21. Install the steering knuckle to the front suspension upper control arm, and tighten it with the castle nut to 64 ft. lbs. (87 Nm). Install a new clip to the steering knuckle. Further tighten the nut up to 60° if the holes for the cotter pin are not aligned.
22. Connect the ABS speed sensor.
23. Stabilize the suspension as follows:

    a. Install the front wheels. Tighten the lug nuts to 76 ft. lbs. (103 Nm).

    b. Lower the vehicle and bounce it up and down several times to stabilize the front suspension.

    c. Remove the front wheels.

    d. Jack up the front suspension lower arm placing a wooden block in between. Apply a load to the front suspension so that the front suspension lower arm is placed in a horizontal position.
24. Fully tighten the strut assembly bolts at the lower control arm.
25. Fully tighten the lower control arm.
26. Install engine under cover.
27. Install the front wheel.
28. Inspect and adjust wheel alignment.

LOWER CONTROL ARM

REMOVAL & INSTALLATION

See Figure 127.

1. Before servicing the vehicle, refer to the precautions section.
2. Raise the vehicle on a hoist, so that front suspension components are hanging and accessible.
3. Remove or disconnect the following:
    - Front wheels.
    - Engine under covers.
    - Brake caliper(s); Do NOT disconnect the brake hose; hang the caliper without stress on the hose
    - Tie rod end from steering knuckle
    - Stabilizer bar link from stabilizer bar
    - Height control sensor link, if equipped, from shock absorber bracket
    - Shock absorber lower mount
    - Lower control arm set bolts (loosen only)
    - Lower ball joint from No. 2 lower control arm (lower suspension arm)
    - Steering gear assembly
    - Strut bar bracket
    - No. 1 lower control arm (lower suspension arm); matchmark adjusting cam to crossmember

To install:

4. To install, reverse the removal procedure, noting the following torque settings:
    - No. 1 lower control arm (lower suspension arm) bolt to shock absorber bracket: 44 ft. lbs. (59 Nm)
    - No. 1 lower control arm (lower suspension arm) adjusting cam bolt and nut to crossmember: 127 ft. lbs. (172 Nm)

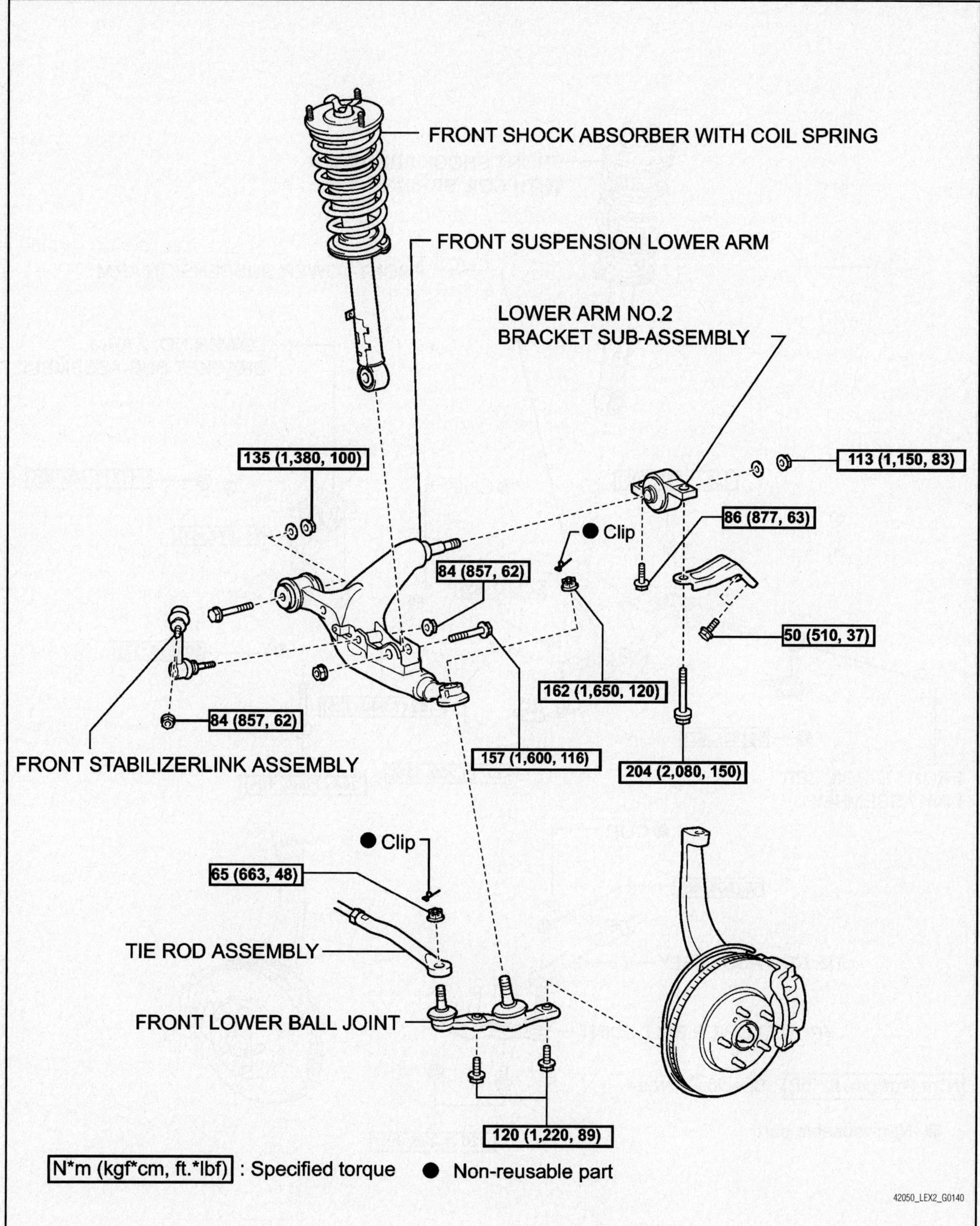

Fig. 126 Lower ball joint assembly and related components—GS450h

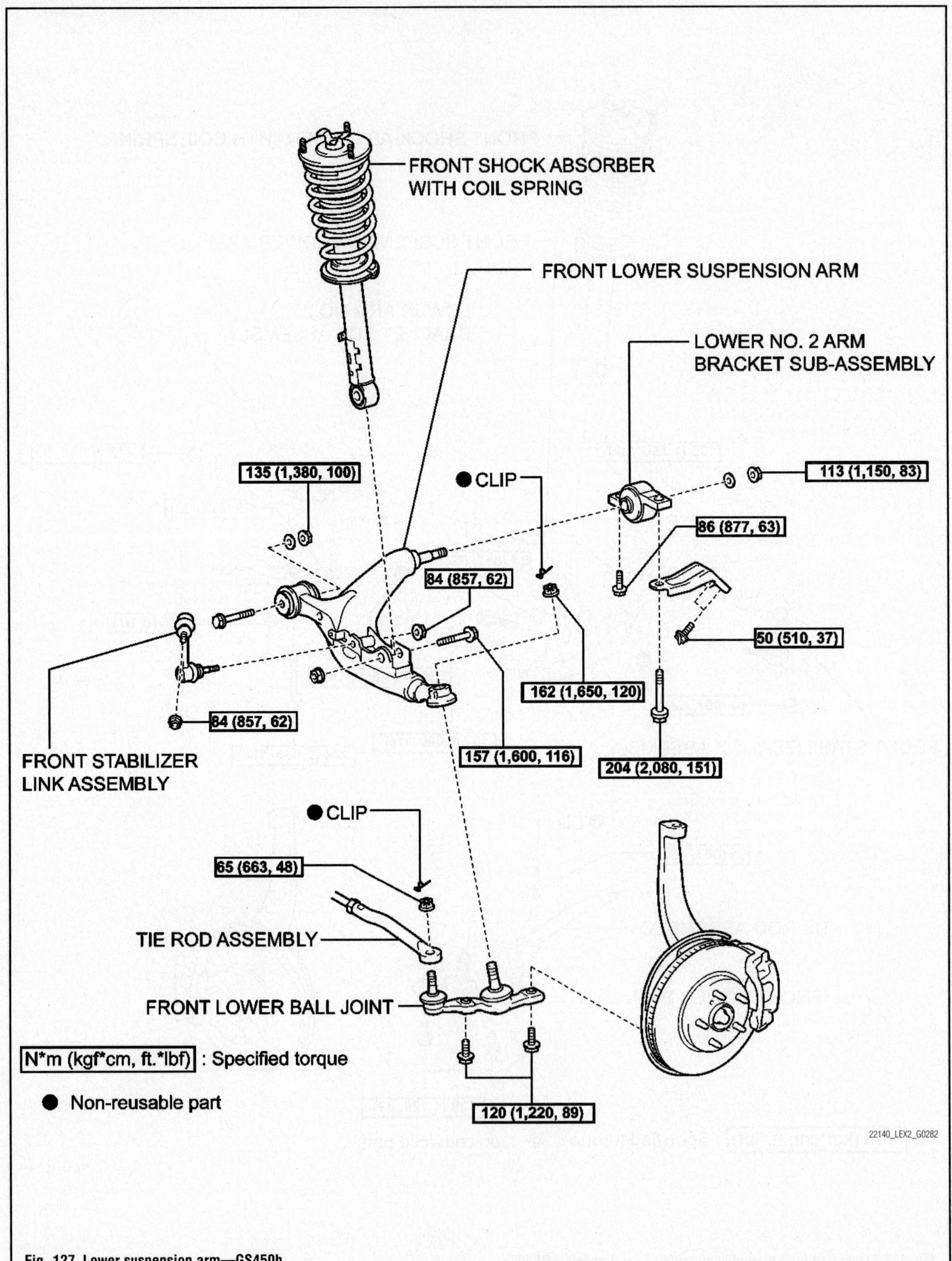

Fig. 127 Lower suspension arm—GS450h

- Strut bar bracket bolts: 43 ft. lbs. (58 Nm)
- Strut bar bracket nut: 112 ft. lbs. (152 Nm)
- No. 2 lower control arm (lower suspension arm) nuts: 122 ft. lbs. (164 Nm)
- Lower shock absorber mounting bolt and nut: 116 ft. lbs. (157 Nm)
- Stabilizer bar link nut: 83 ft. lbs. (113 Nm)
- Stabilizer bar link-to-stabilizer bar bolt and nut: 43 ft. lbs. (55 Nm)
- Tie rod end nut: 64 ft. lbs. (87 Nm)
- Brake caliper bolts: 87 ft. lbs. (118 Nm)
- Front wheel nuts: 76 ft. lbs. (103 Nm)

MACPHERSON STRUT

REMOVAL & INSTALLATION

See Figure 128.

1. Before servicing the vehicle, refer to the precautions section.
2. Remove or disconnect the following:
   - Negative battery cable.
   - Front wheel
   - Brake caliper, leaving the line attached
3. Loosen the 3 upper strut mounting nuts.
4. Loosen, but do not remove, the upper strut rod nut.

⁂ CAUTION

Do NOT remove the upper strut nut at this time.

5. Remove or disconnect the following:
   - Anti-lock Brake System (ABS) speed sensor and harness
   - Upper suspension arm from the steering knuckle
   - Stabilizer bar from the link and remove the bracket
   - Strut from the lower suspension arm.
   - 3 upper strut mounting nuts and remove the strut
6. Compress the coil spring.
7. Remove or disconnect the following:
   - Piston rod locknut
   - Suspension support, coil spring and bumper
8. If disposing the strut, perform the following procedure:
   a. Fully extend the strut rod.

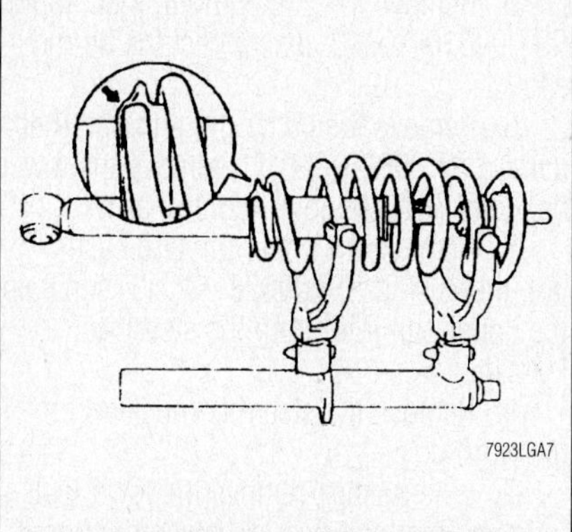

Fig. 128 Matching the spring to the seat

   b. Drill a hole near the bottom of the shock to remove the gas inside.

⁂ CAUTION

The gas is harmless, but be careful of chips that may fly up when the gas is released.

To install:

9. Install or connect the following:
   - Spring bumper
   - Coil spring
   - Suspension support to the rod and temporarily install a new nut
10. Turn the suspension support so one of the bolts on the support faces the same direction as shown in the illustration.

➡Align the bolt so a line drawn between the rod and bolt would be at 90°to the direction of the lower bushing.

11. Install or connect the following:
   - Spring compressor
   - Strut and tighten the upper retaining nuts to 41 ft. lbs. (56 Nm)
   - New upper strut rod nut to 20 ft. lbs. (27 Nm)
   - Strut to the lower arm and temporarily tighten the nut and bolt
   - Stabilizer bar bracket and tighten the bolts to 21 ft. lbs. (28 Nm)
   - The stabilizer bar to the link and tighten the bolts to 29 ft. lbs. (39 Nm)
   - Upper suspension arm to the steering knuckle. Tighten the nut to 64 ft. lbs. (87 Nm) and install a new cotter pin.
   - ABS speed sensor and tighten the bolt to 69 inch lbs. (8 Nm)
   - Caliper
   - Wheel
12. Bounce the vehicle several times to stabilize the suspension.
13. Tighten the lower strut bolt and nut to 116 ft. lbs. (157 Nm).
14. Check the front wheel alignment.

STABILIZER BAR

REMOVAL & INSTALLATION

See Figures 129 through 131.

1. Before servicing the vehicle, refer to the precautions section.
2. Remove the front wheels.
3. Remove engine under cover.
4. Remove the left and right rear engine under covers.
5. Remove the 2 nuts and the left front stabilizer link assembly. Repeat for the right side. If the ball joint turns together with the nut, use a hexagon (6mm) wrench to hold the stud.
6. Remove the 2 bolts and the left front No. 2 stabilizer bracket from the front suspension crossmember. Repeat for the right side.
7. Remove the 4 bolts and the left front No. 1 stabilizer bracket from the frame. Repeat for the right side.
8. Remove the 2 front No. 1 stabilizer bar bushings from the front stabilizer bar.
9. Remove the front stabilizer bar from the vehicle.

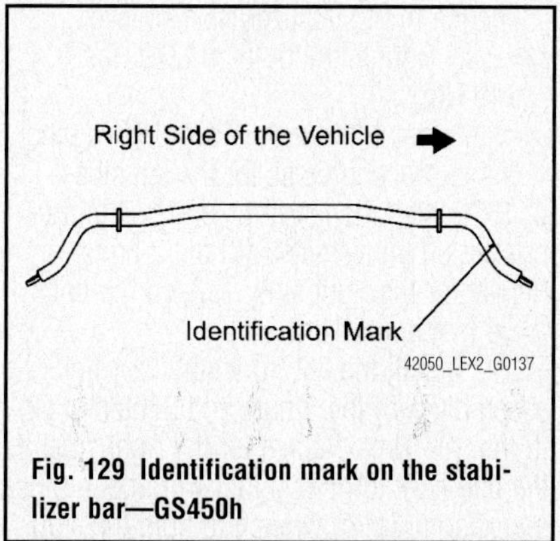

Fig. 129 Identification mark on the stabilizer bar—GS450h

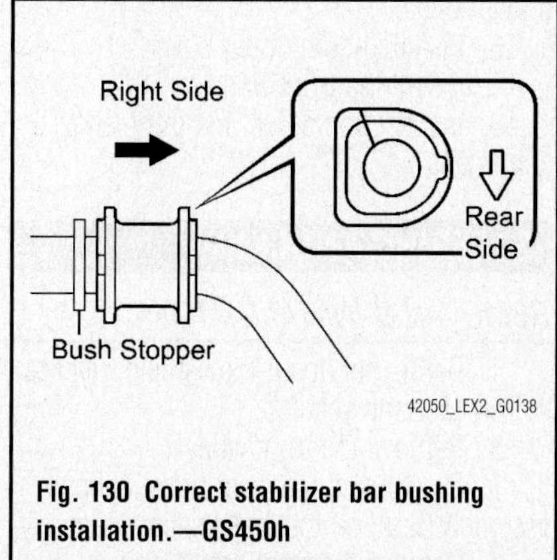

Fig. 130 Correct stabilizer bar bushing installation.—GS450h

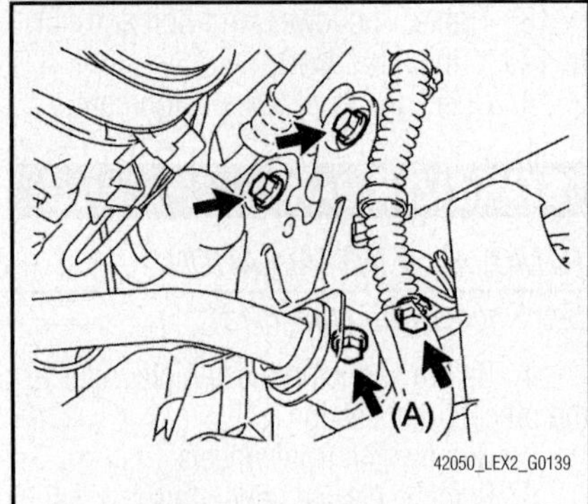

Fig. 131 Front No. 1 stabilizer bracket bolts—GS450h

To install:

10. Install the front stabilizer bar to the vehicle. The identification mark must be on the right side of the vehicle when installing the front stabilizer bar.
11. Install the 2 front No. 1 stabilizer bar bushings as shown in the illustration. Be sure to install the front No. 1 stabilizer bar bushings so that the cutouts face the front of the vehicle.
12. Install the left front No. 1 stabilizer bracket as follows:
    a. Temporarily install the left front stabilizer No.1 bracket to the side member with the 4 bolts.
    b. Tighten the bolt (A) while pressing the left front stabilizer No.1 bracket.
    c. Tighten the bolts to 36 ft. lbs. (49 Nm).
    d. Install the right side following the same procedures as for the left side.
13. Install the left front No. 2 stabilizer bracket on the vehicle with the 2 bolts. Repeat for the right side. Tighten the bolts to 36 ft. lbs. (49 Nm).
14. Install the left front stabilizer link assembly with the 2 nuts and tighten to 62 ft. lbs. (84 Nm). Repeat for the right side. If the ball joint turns together with the nut, use a hexagon (6mm) wrench to hold the stud.
15. Install the left and right rear engine under covers.
16. Install engine under cover.
17. Install the front wheels.
18. Inspect and adjust the front wheel alignment.

STEERING KNUCKLE

REMOVAL & INSTALLATION

1. Before servicing the vehicle, refer to the precautions section.
2. Remove the front wheel.
3. Disconnect the front brake caliper assembly. Support the brake caliper securely.
4. Remove the front brake disc
5. Remove the clip and nut, then using SST 09610-20012, disconnect the tie rod end.
6. Remove the cotter pin and nut, then using SST 09628-6201 1, remove the lower ball joint from the lower control arm.
7. Remove the cotter pin and castle nut, then using SST09628–62011, separate the upper ball joint from the steering knuckle.
8. Remove the steering knuckle from the vehicle.
9. If necessary, remove the lower ball joint and hub and bearing assembly from the steering knuckle.

To install:

10. If necessary, install the hub and bearing assembly and lower ball joint to the steering knuckle.
11. Install the steering knuckle into the vehicle.
12. Connect the upper ball joint to the steering knuckle. Install castle nut and new cotter pin.
13. Connect the lower ball joint to lower control arm. Install castle nut and new cotter pin.
14. Connect the tie rod end to the steering knuckle. Install castle nut and new cotter pin.
15. Install the front brake disc
16. Install the front brake caliper assembly.
17. Install the front wheel.
18. Inspect and adjust wheel alignment.

UPPER BALL JOINT

REMOVAL & INSTALLATION

The upper ball joint is an integral part of the upper arm and is not replaced separately. The upper ball joint replacement is accomplished by replacing the upper arm.

UPPER CONTROL ARM

REMOVAL & INSTALLATION

See Figure 132.

1. Before servicing the vehicle, refer to the precautions section.
2. Remove or disconnect the following:
    - Negative battery cable
    - Wheel
3. Loosen the 3 upper strut mounting nuts.
4. Loosen, but do not remove, the upper strut rod nut.

⁂ CAUTION

Do NOT completely remove the upper strut nut at this time.

5. Remove or disconnect the following:
    - Brake caliper, leaving the line attached and secure it out of the way
    - Anti-lock Brake System (ABS) speed sensor and harness
    - Cotter pin and nut from the upper control arm
    - Upper control arm from the steering knuckle
    - Stabilizer bar from the link and remove the bracket
    - Cotter pin and nut from the lower control arm
    - Strut from the lower suspension arm
    - 3 upper strut mounting nuts and remove the strut
    - Mounting bolts holding the upper control arm to the frame
    - Upper control arm from the vehicle

To install:

6. Install or connect the following:
    - Upper suspension arm and tighten the mounting bolts to 39 ft. lbs. (53 Nm)
    - Strut and tighten the upper retaining nuts to 41 ft. lbs. (56 Nm). Tighten the new upper strut rod nut to 20 ft. lbs. (27 Nm).
    - Strut to the lower arm and temporarily tighten the nut and bolt
    - Stabilizer bar bracket and tighten the bolts to 21 ft. lbs. (28 Nm)
    - Stabilizer bar to the link and tighten the bolts to 29 ft. lbs. (39 Nm)
    - Upper suspension arm to the steering knuckle. Tighten the nut to 64 ft. lbs. (87 Nm) and install a new cotter pin.
    - ABS speed sensor and tighten the bolt to 69 inch lbs. (8 Nm)
    - Caliper
    - Front wheel
7. Lower the vehicle.
8. Bounce the vehicle several times to stabilize the suspension.
9. Tighten the lower strut bolt and nut to 116 ft. lbs. (157 Nm).
10. Check the front wheel alignment.

WHEEL BEARINGS

REMOVAL & INSTALLATION

1. Before servicing the vehicle, refer to the precautions section.
2. Remove or disconnect the following:
    - Negative battery cable
    - Front wheel
    - Caliper, leaving the brake line connected and suspend it out of the way

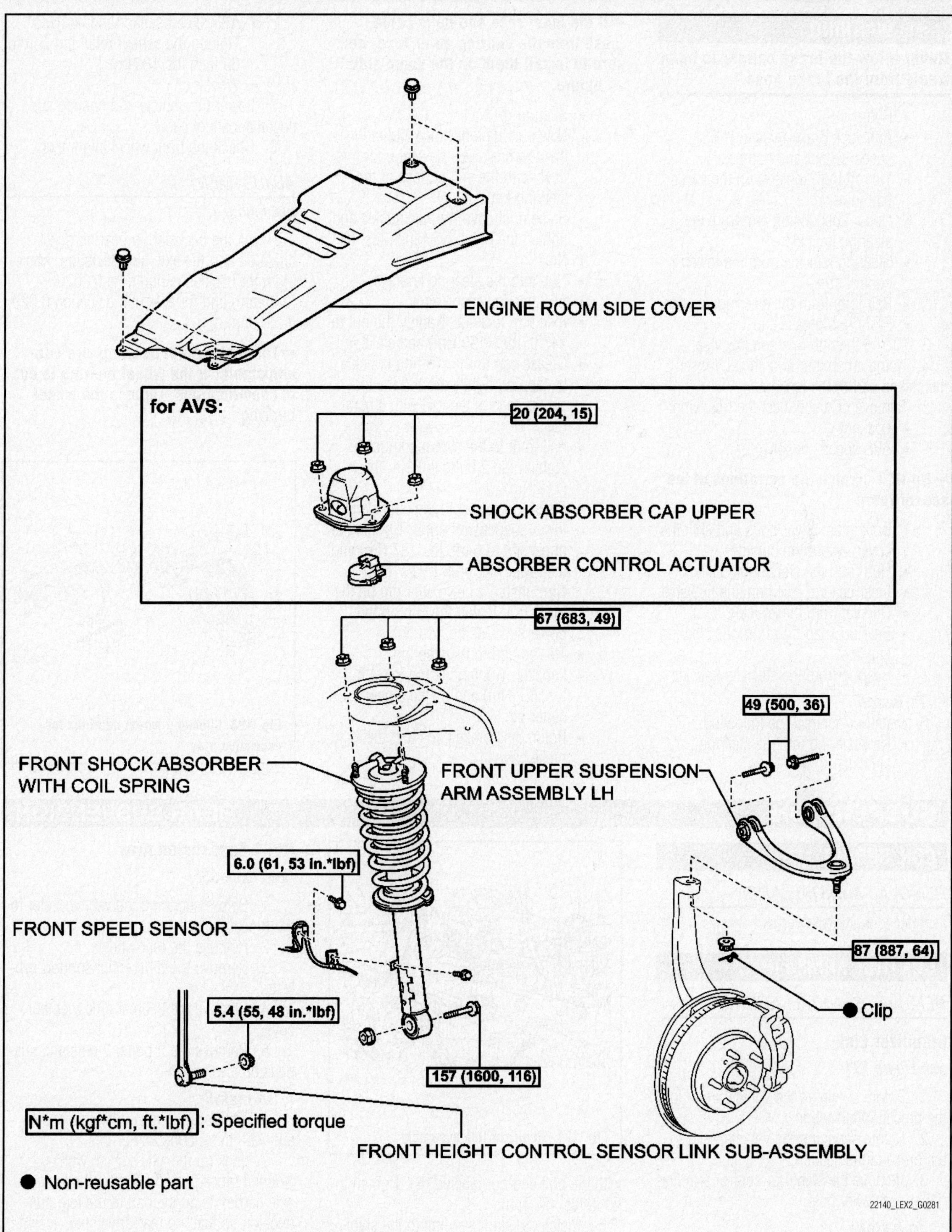

Fig. 132 Front upper suspension arm—GS450h

⁂ WARNING

Never allow the brake caliper to hang freely from the brake hose.

- Rotor
- Anti-lock Brake System (ABS) speed sensor and harness
- Tie rod from the arm on the lower ball joint
- Upper suspension arm from the steering knuckle
- Steering knuckle from the lower control arm
- Ball joint from the steering knuckle
- Front hub grease cap

3. Clamp the hub in a soft jaw vise.
4. Using a hammer and chisel, loosen the staked part of the locknut.
5. Remove or disconnect the following:
   - Locknut
   - ABS speed sensor rotor

➡Do NOT scratch the serrations of the sensor rotor.

- Brake dust cover bolts and shift the cover toward the outside.
- Hub from the steering knuckle
- Inner bearing race from the hub shaft
- Oil seal from the knuckle
- Bearing snap ring from the steering knuckle
- Bearing from the steering knuckle

To install:

6. Install or connect the following:
   - New bearing into the steering knuckle

➡If the inner race and balls come loose from the bearing outer race, be sure to install them on the same side as before.

- Snap ring
- New outside inner race and tap in the new seal. Tap the seal until it is flush with the end surface of the steering knuckle.
- Brake dust cover to the knuckle and tighten the bolts to 74 inch lbs. (8 Nm)
- Hub into the steering knuckle
- ABS speed sensor rotor
- Axle hub locknut. Tighten the nut to 147 ft. lbs. (199 Nm) and stake it.
- Grease cap to the steering knuckle by tapping lightly around the circumference of the cap with a hammer
- Ball joint to the steering knuckle. Tighten the 2 bolts to 83 ft. lbs. (113 Nm).
- Steering knuckle to the upper and lower suspension arms. Tighten the upper nut to 64 ft. lbs. (87 Nm) and the lower nut to 95 ft. lbs. (127 Nm). Install a new cotter pin on the lower nut. Install the clip on the upper suspension arm nut.
- Tie rod end to the steering knuckle. Tighten the nut to 64 ft. lbs. (87 Nm) and install a new cotter pin.
- Rotor, disc brake pads and the brake caliper
- ABS speed sensor and harness. Tighten the sensor retaining bolt to 69 inch lbs. (8 Nm).
- Wheel

7. Lower the vehicle and connect the negative battery cable.
8. Check the front wheel alignment.

ADJUSTMENT

See Figure 133.

Check the backlash in bearing shaft direction and the axle hub deviation. Maximum for backlash should be 0.0020 in. (0.05mm) and for axle hub deviation 0.020 in. (0.05mm).

➡The front wheel bearings are non-adjustable. If the wheel bearing is out of specifications, replace the wheel bearing.

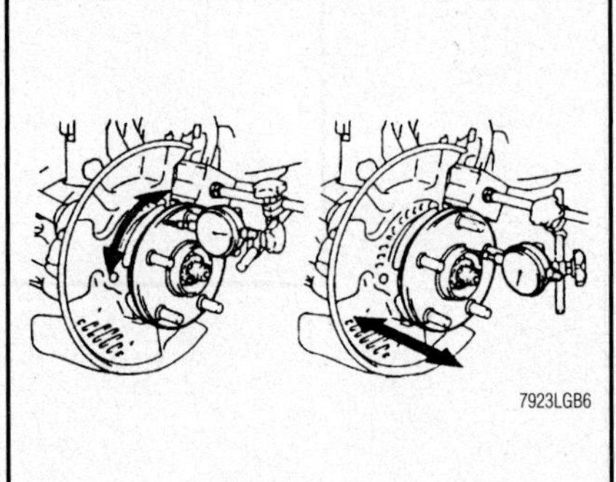

Fig. 133 Checking wheel bearings for excessive play

SUSPENSION

COIL SPRING

REMOVAL & INSTALLATION

Refer to rear strut overhaul.

CONTROL ARMS/LINKS

REMOVAL & INSTALLATION

Stabilizer Link

See Figure 134.

1. Before servicing the vehicle, refer to the precautions section.
2. Before servicing the vehicle, refer to the precautions section.
3. Remove the stabilizer control arm link upper and lower nuts.

To install:

4. Install the LH and RH stabilizer link, bracket to the rear No. 2 suspension arm with the bolt and nut labeled (B). Tighten to 20 ft. lbs. (27 Nm).
5. Install the stabilizer link to the stabilizer with the nut labeled (A). Tighten to 66 ft. lbs. (89 Nm).

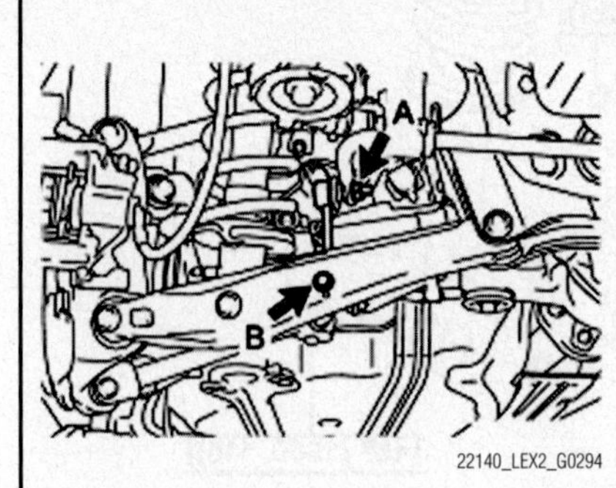

Fig. 134 Stabilizer link assembly

REAR SUSPENSION

No. 1 Suspension Arm

See Figure 135.

1. Before servicing the vehicle, refer to the Precautions Section.
2. Remove the rear wheels.
3. Remove the differential support protector.
4. Remove the LH rear side member cover.
5. Remove the 2 bolts, 2 nuts and suspension arm.

To install:

6. Temporarily install the suspension arm with the 2 bolts and 2 nuts.
7. Jack up the axle carrier, with a wooden block placed between the jack and axle carrier, to apply load to the rear suspension so that the rear drive shaft assembly becomes level.
8. Tighten the No. 1 rear suspension

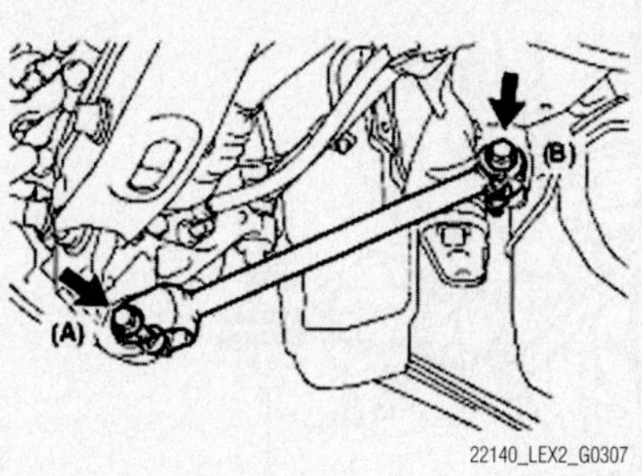

Fig. 135 Tighten the No. 1 rear suspension arm bolt (A) and (B)

arm. Tighten bolt (A) to 70 ft. lbs. (95 Nm).

9. Install the member cover with the 2 screws.

10. Install the NO. 2 differential support protector.

11. Install the rear wheel and tighten to 76 ft. lbs. (103 Nm).

12. Inspect and adjust the rear wheel alignment.

13. Check the head light adjustment.

No. 2 Suspension Arm

See Figure 136.

1. Before servicing the vehicle, refer to the Precautions Section.
2. Remove the rear wheels.
3. Remove the differential support protector.
4. Remove the LH rear member brace.
5. Remove the rear No. 2 suspension arm as follows:
   - Remove the bolt and nut, and disconnect the rear stabilizer link and the load sensing valve sensor bracket.
   - Remove the bolt and nut, and disconnect the rear shock absorber with coil spring.
   - Remove the nut and bolt on the axle carrier side, and disconnect the rear No. 2 suspension arm.

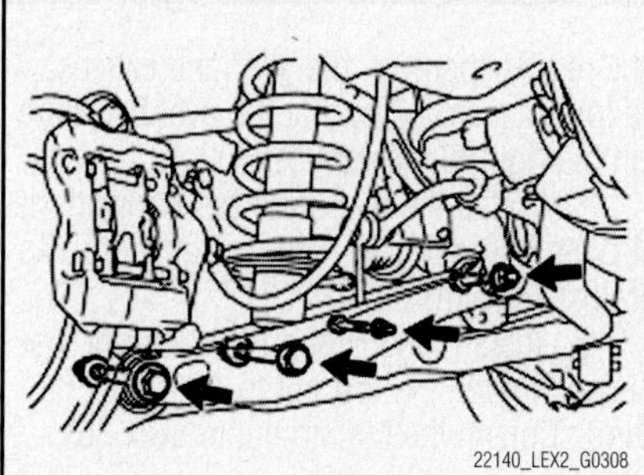

Fig. 136 Remove the rear No. 2 suspension arm

   - Remove the bolt, nut, and washer on the rear suspension member side, and remove the rear No. 2 suspension arm.

To install:

6. Temporarily install the rear suspension No. 2 arm with the bolt, nut and washer as follows:
   - Temporarily install the rear suspension No. 2 arm with the bolt and nut.
   - Temporarily install the rear shock absorber with coil spring with the bolt and nut.
   - Temporarily install the stabilizer link and the load sensing valve sensor bracket to the rear suspension No. 2 arm with the bolt and nut.
   - Jack up the axle carrier, with a wooden block placed between the jack and axle carrier, to apply load to the rear suspension so that the rear drive shaft becomes level.

7. Tighten the rear suspension arm as follows:
   - Tighten the bolt on the axle carrier side to 119 ft. lbs. (161 Nm).
   - Tighten the bolt holding the rear shock absorber with coil spring to 81 ft. lbs. (110 Nm).
   - Tighten the nut holding the rear stabilizer link to 20 ft. lbs. (27 Nm).
   - Tighten the nut on the rear suspension member side to 103 ft. lbs. (140 Nm).

8. Install the member cover with the 2 screws.

9. Install the No. 2 differential support protector.

10. Install the rear wheel and tighten to 76 ft. lbs. (103 Nm).

11. Inspect and adjust the rear wheel alignment.

12. Check the head light adjustment.

Upper No. 1 Control Arm

See Figure 137.

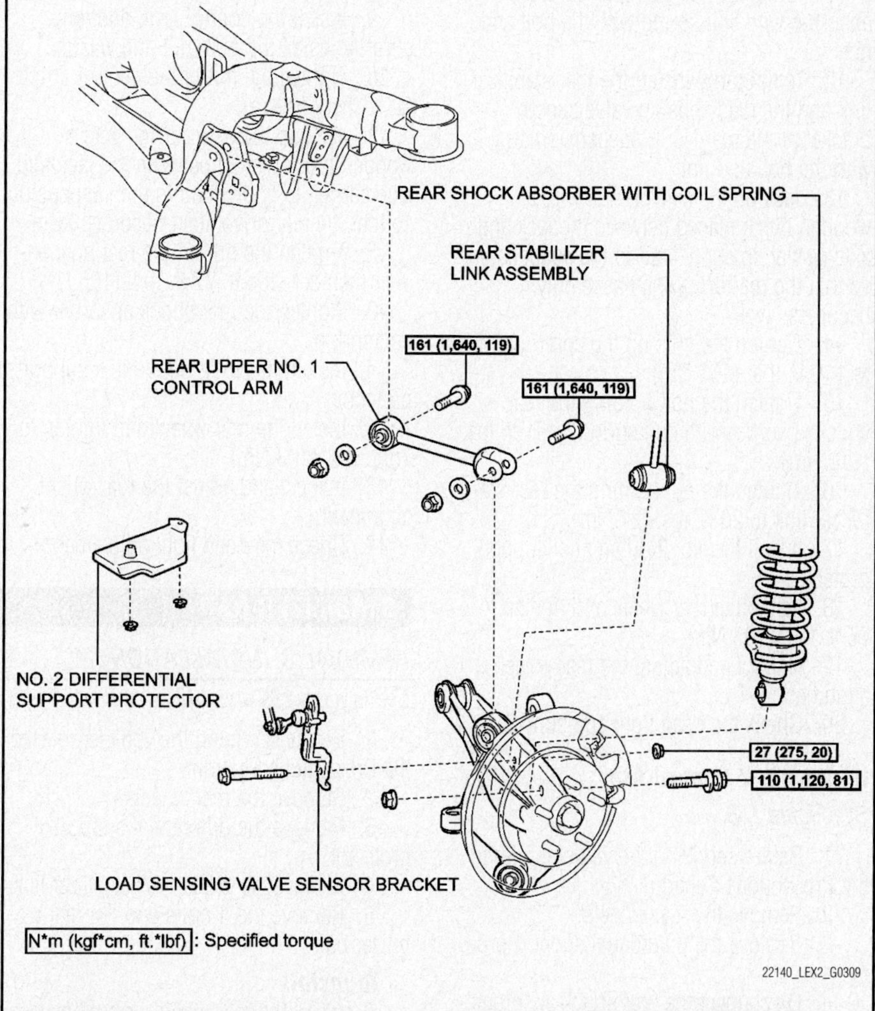

Fig. 137 Remove the bolts, nuts, washers and control arm from the rear suspension member

1. Before servicing the vehicle, refer to the Precautions Section.
2. Remove the rear wheels.
3. Remove the differential support protector.
4. Disconnect the rear shock absorber with coil spring.
5. Remove the bolt and nut, and separate the rear stabilizer link and the load sensing valve sensor bracket from the rear No. 2 suspension arm.
6. Remove the bolt and nut, and separate the rear shock absorber with coil spring from the rear No. 2 suspension arm.
7. Remove the bolts, nuts washers and the control arm from the rear axle carrier.
8. Remove the bolts, nuts, washers and control arm from the rear suspension member.

To install:

9. Temporarily install the control arm with the bolts, nuts and washers.
10. Temporarily install the rear upper No. 1 control arm with the bolts, nuts and washers.
11. Temporarily install the rear shock absorber with coil spring with the bolt and nut.
12. Temporarily install the rear stabilizer link and the load sensing valve sensor bracket to the rear No. 2 suspension arm with the bolt and nut.
13. Jack up the axle carrier, with a wooden block placed between the jack and axle carrier, to apply load to the suspension so that the rear drive shaft assembly becomes level
14. Tighten the nuts on the control arm to 119 ft. lbs. (161 Nm).
15. Tighten the bolt holding the rear shock absorber with coil spring to 81 ft. lbs. (110 Nm).
16. Tighten the nut holding the rear stabilizer link to 20 ft. lbs. (27 Nm).
17. Install the No. 2 differential support protector.
18. Install the rear wheel and tighten to 76 ft. lbs. (103 Nm).
19. Inspect and adjust the rear wheel alignment.
20. Check the head light adjustment.

Upper No. 2 Control Arm

See Figure 138.

1. Before servicing the vehicle, refer to the Precautions Section.
2. Remove the rear wheels.
3. Remove the differential support protector.
4. Disconnect the rear shock absorber with coil spring.

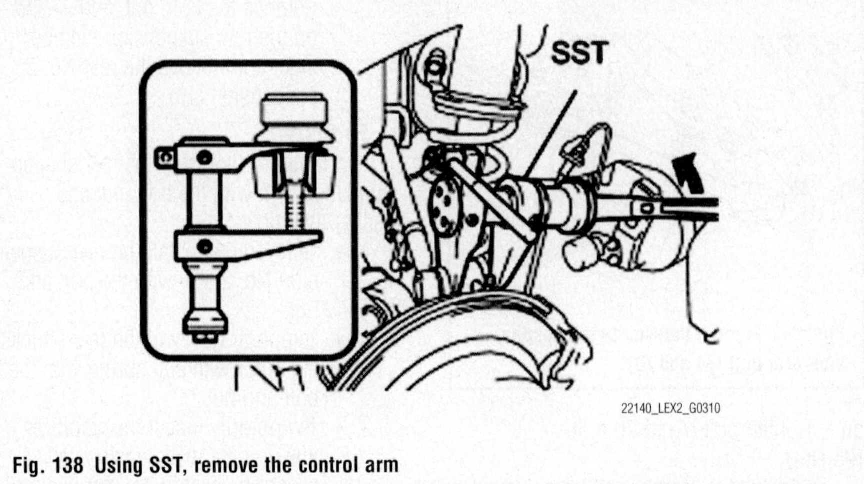

Fig. 138 Using SST, remove the control arm

5. Remove the nut on the axle carrier side.
6. Using SST, remove the control arm.

To install:

7. Remove the bolt, nut, washer, and control arm.
8. Install the stud of the control arm, and temporarily install a new nut.
9. Install the control arm, and temporarily install the bolt, nut and washer.
10. Tighten the nut of the control arm to 52 ft. lbs. (70 Nm).
11. Jack up the axle carrier, with a wooden block placed between the jack and axle carrier, to apply load to the suspension so that the rear drive shaft becomes level.
12. Tighten the nut on the rear suspension member side to 77 ft. lbs. (105 Nm).
13. Tighten the rear shock absorber with coil spring.
14. Install the No. 2 differential support protector.
15. Install the rear wheel and tighten to 76 ft. lbs. (103 Nm).
16. Inspect and adjust the rear wheel alignment.
17. Check the head light adjustment.

STABILIZER BAR

REMOVAL & INSTALLATION

See Figures 139 and 140.

1. Before servicing the vehicle, refer to the Precautions Section.
2. Remove the rear wheels.
3. Remove the differential support protector.
4. Remove the LH and RH stabilizer links.
5. Remove the 4 bolts and the rear stabilizer bar.

To install:

6. Insert the rear stabilizer bar between the rear suspension member and exhaust tailpipe assembly so that the oval holes are on the right of the vehicle.
7. Install the rear stabilizer bar with the 4 bolts in order from 1 to 4. Tighten to 24 ft. lbs. (32 Nm).
8. Install the LH and RH stabilizer links. Tighten the top nut to 20 ft. lbs. (27 Nm). Tighten the bottom nut to 66 ft. lbs. (89 Nm).
9. Install the NO. 2 differential support protector.

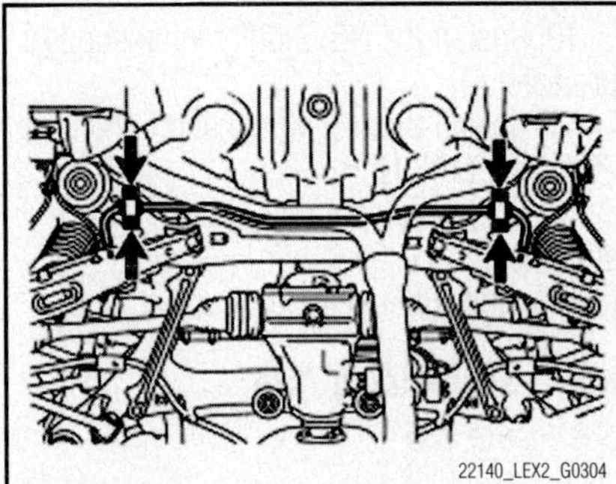

Fig. 139 Remove the 4 bolts and the rear stabilizer bar

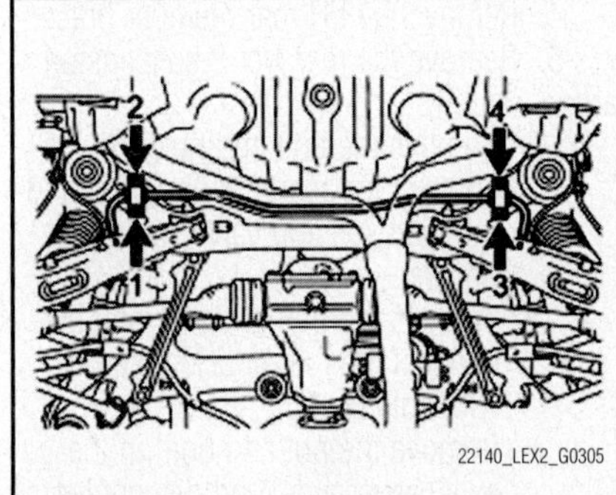

Fig. 140 Install the rear stabilizer bar with the 4 bolts in order from 1 to 4

10. Install the rear wheel and tighten to 76 ft. lbs. (103 Nm).

11. Inspect and adjust the rear wheel alignment.

12. Check the head light adjustment.

STRUT & SPRING ASSEMBLY

REMOVAL & INSTALLATION

See Figures 141 through 143.

1. Before servicing the vehicle, refer to the precautions section
2. Remove the luggage compartment floor mat.
3. Remove the LH and RH luggage compartment side trim covers.
4. Remove the tool box.
5. Remove the rear floor finish plate.
6. Remove the front luggage compartment trim cover.
7. Remove the luggage compartment trim cover inner LH.
8. Remove the rear wheel.
9. Remove the front luggage compartment trim cover.
10. Remove the 2 bolts and brace.
11. Remove the No. 2 suspension arm as follows:
    - Loosen the nut on the rear suspension member side.
    - Remove the bolt and nut and disconnect the rear stabilizer link and the load sensing valve sensor bracket.
    - Remove the bolt and nut and disconnect the rear shock absorber.
    - Remove the nut and bolt on the rear axle carrier side.
12. Remove rear shock absorber cap and disconnect the connector.
13. Turn the absorber control actuator counterclockwise approximately 40° to remove it.
14. Remove the 3 nuts on the upper side of the rear shock absorber with coil spring.
15. Remove the 2 bolts and the rear shock absorber with coil spring from the body.

To install:

16. Temporarily install the rear shock absorber with rear coil spring with the 2 bolts.
17. Install the 3 nuts on the upper side of the rear shock absorber with coil spring. Tighten to 55 ft. lbs. (74 Nm).
18. Tighten the lock nut to 15 ft. lbs. (18 Nm).
19. Tighten the 2 bolts 16 ft. lbs. (21 Nm).
20. Check that the control rod of the rear shock absorber is in the position shown in the illustration.
21. Install the absorber control actuator to the actuator support bracket. Turn the actuator clockwise approximately 40° until a click is felt.
22. Connect the connector and install the shock absorber cap.
23. Temporarily install the rear No. 2 suspension arm with the bolt and nut.
24. Temporarily install the rear shock absorber with rear coil spring with the bolt and nut.
25. Temporarily install the rear stabilizer link and load sensing valve sensor bracket to the rear No. 2 suspension arm with the bolt and nut.
26. Stabilize the suspension.
27. Tighten the No. 2 Suspension arm as follows:
    - Tighten the bolt on the axle carrier side. Tighten to 119 ft. lbs. (161 Nm).
    - Tighten the bolt holding the rear shock absorber with coil spring. Tighten to 81 ft. lbs. (110 Nm).
    - Tighten the nut holding the rear stabilizer link to 20 ft. lbs. (27 Nm).
    - Tighten the nut on the rear suspension member side. Tighten to 103 ft. lbs. (140 Nm).
28. Install the LH rear suspension brace.
29. Install the NO. 2 differential support protector.
30. Install the rear wheel and tighten to 76 ft. lbs. (103 Nm).
31. Inspect and adjust the rear wheel alignment.
32. Check the head light adjustment.

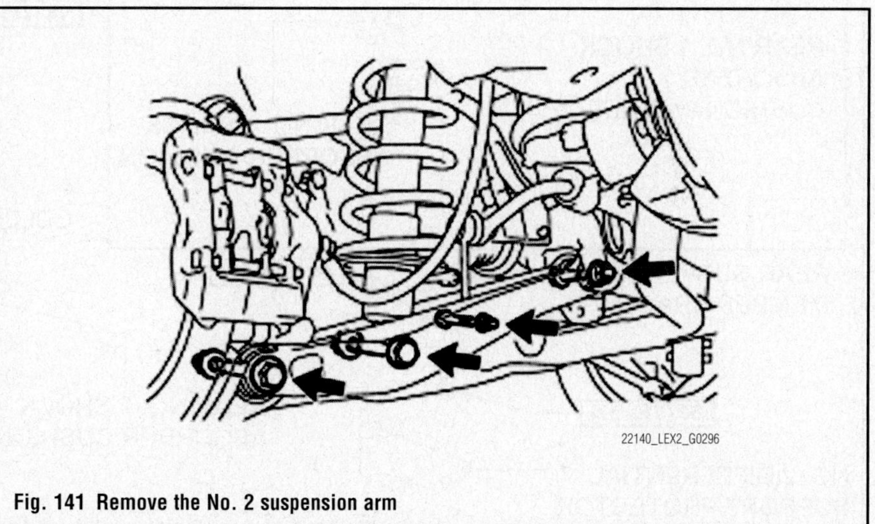

Fig. 141 Remove the No. 2 suspension arm

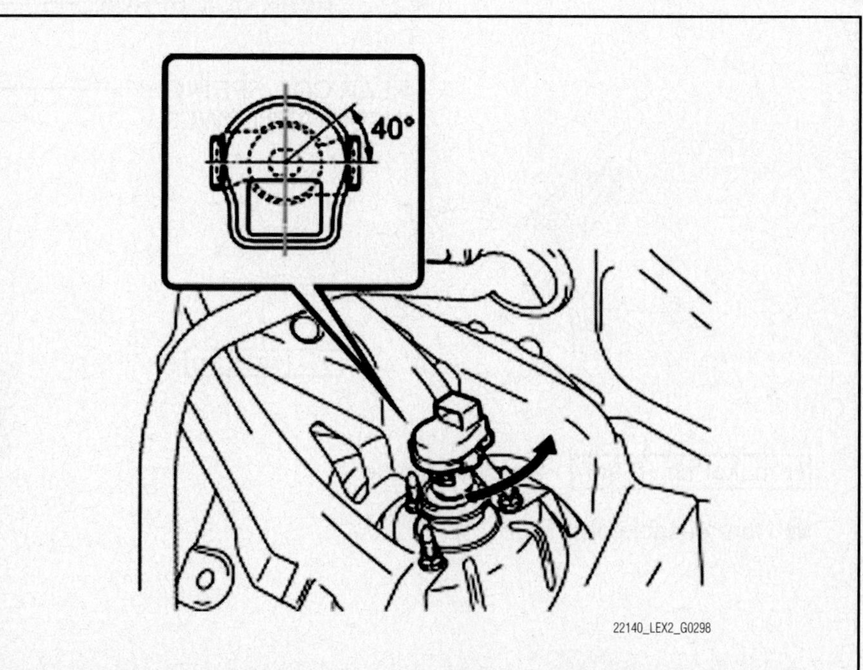

Fig. 142 Turn the absorber control actuator counterclockwise approximately 40° to remove it

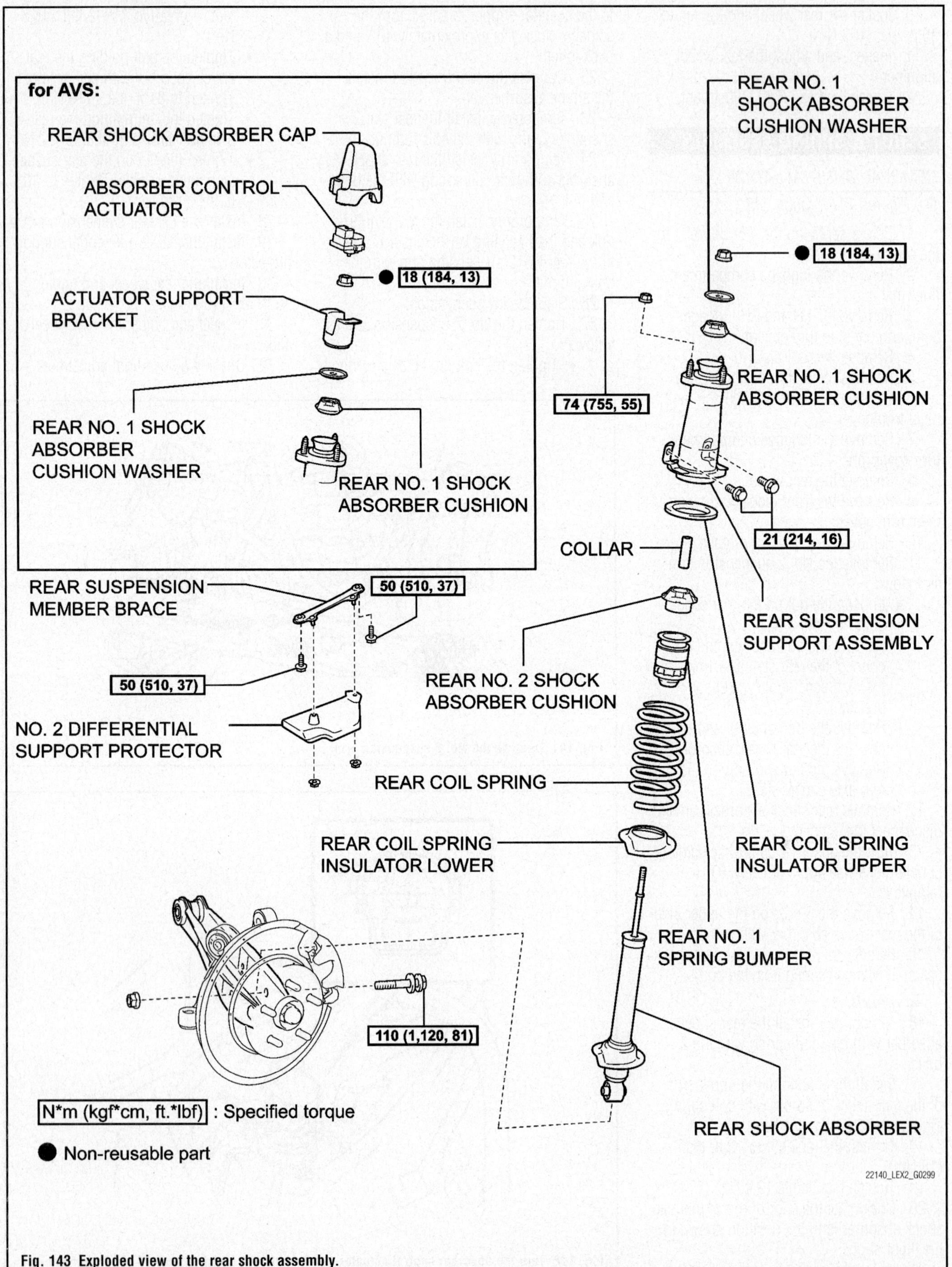

Fig. 143 Exploded view of the rear shock assembly.

OVERHAUL

See Figures 144 through 146.

1. Before servicing the vehicle, refer to the Precautions Section.
2. Install SST to the rear coil spring so that the distance between the upper and lower hooks is as wide as possible within the installation area.
3. Compress the rear coil spring until it can be moved freely.
4. Remove the nut.
5. Remove the rear No. 1 shock absorber cushion washer, rear No. 1 shock absorber cushion, rear suspension support, rear coil spring insulator upper, collar, rear No. 2 shock absorber cushion, and rear No. 1 spring bumper.

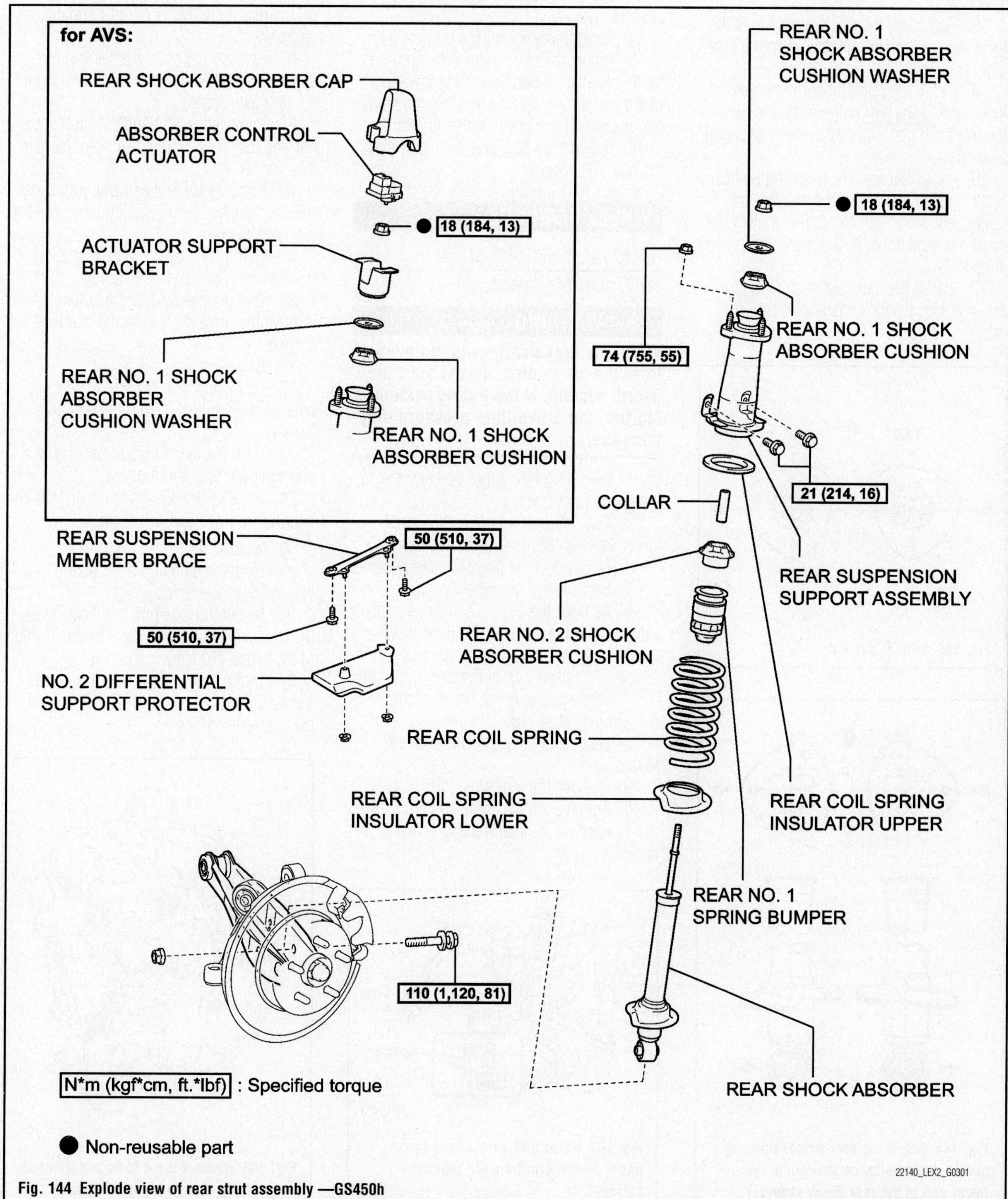

Fig. 144 Explode view of rear strut assembly —GS450h

6. Remove the rear coil spring and rear coil spring insulator lower from the rear shock absorber. Release the SST and remove the rear coil spring.

To install:

7. Using SST, compress the rear coil spring.
8. Compress the coil spring and fit the lower end of the spring into the spring seat gap.
9. Install the rear coil spring insulator lower and fit the rear coil spring end into the recessed part of the rear shock absorber lower seat.
10. Install the rear No. 1 spring bumper.
11. Install the rear coil spring insulator upper, rear No. 1 and No. 2 shock absorber cushions, and collar to the rear suspension support.
12. Install the rear suspension support and rear No. 1 shock absorber cushion washer.

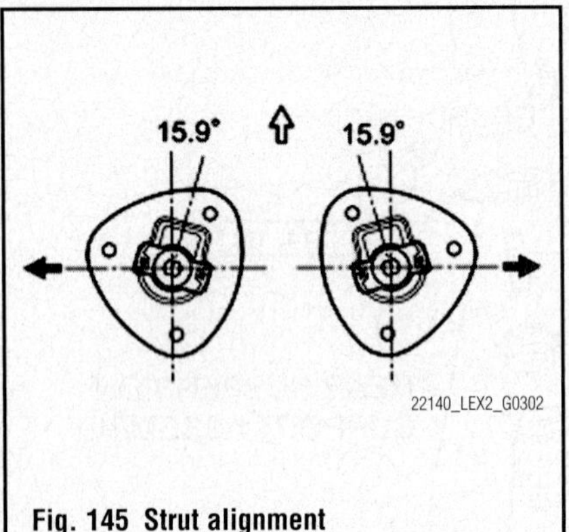

Fig. 145 Strut alignment

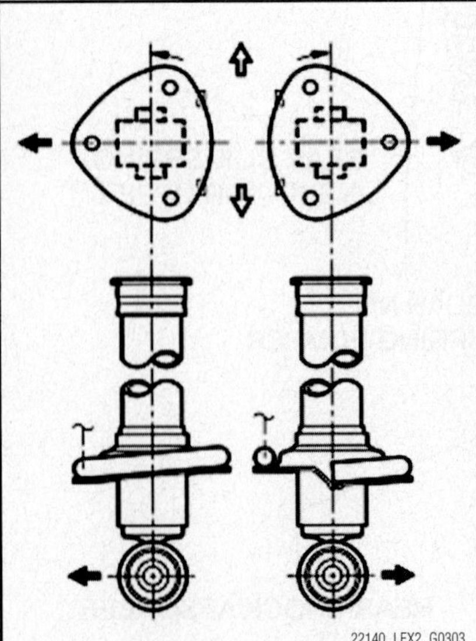

Fig. 146 Adjust the rear suspension support to the installation position of the lower part of the rear shock absorber

13. Align the width across flat on the piston rod end of the shock absorber and the width across flat on the suspension support bracket. Then align the center of the actuator support bracket and the stud bolt as shown in the illustration to install the rear suspension support to the rear shock absorber (for AVS).
14. Temporarily install the nut.
15. Adjust the rear suspension support to the installation position of the lower part of the rear shock absorber so that the studs are positioned as shown in the illustration.
16. Release the SST and remove it from the rear coil spring.

WHEEL BEARINGS

REMOVAL & INSTALLATION

See Figures 147 through 149.

⁂ WARNING

When the brake caliper is removed from the brake disc, do not start the hybrid system. If the hybrid system is started, the brake fluid pressure may increase.

1. Before servicing the vehicle, refer to the precautions section.
2. Turn the power switch OFF, and check that the "READY" indicator turns off.
3. Disconnect the negative battery cable.
4. Remove the rear wheel.
5. Remove the bolt, nut, load sensing valve sensor bracket and stabilizer link.
6. Using a punch and a hammer, release the staked part of the axle shaft nut.
7. While applying the brakes, remove the rear axle shaft nut.
8. Disconnect the rear brake caliper assembly.
9. Remove the rear brake disc.
10. Remove the rear speed sensor.
11. Remove the rear parking brake shoes.

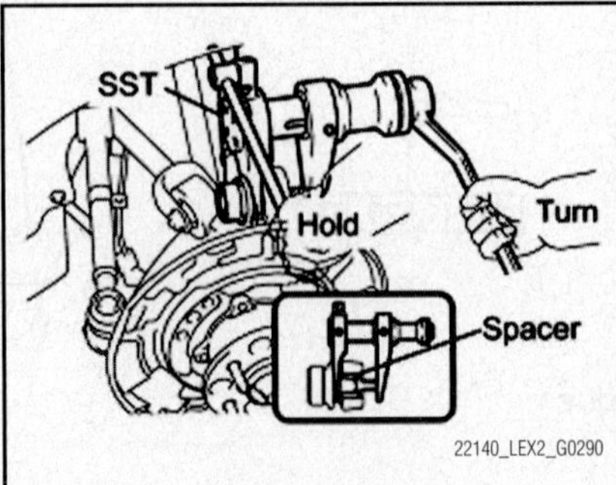

Fig. 147 Using SST, remove the rear upper control arm from the rear axle carrier.

12. Remove the 2 nuts and the parking brake anchor block from the rear axle carrier.
13. Remove the nut from the rear upper control arm.
14. Using SST, remove the rear upper control arm from the rear axle carrier.
15. Jack up the rear axle so that the bolt on the rear upper No. 1 control arm can be removed.
16. Remove the bolt, washer and nut, and the rear upper No. 1 control arm from the rear axle carrier.
17. Remove the bolt and nut, and separate the rear suspension arm from the rear axle carrier.
18. Remove the bolt and nut, and separate the rear No. 2 suspension arm from the rear axle carrier
19. Using SST, disconnect the toe control link from the rear axle carrier.
20. Using a plastic-faced hammer, remove the drive shaft from the rear axle carrier.
21. Remove the rear axle.
22. Using a screwdriver, remove the bearing dust deflector from the rear axle carrier.
23. Hold the axle hub and bearing in a vise between aluminum plates.
24. Remove the 4 bolts and axle hub and bearing from the rear axle carrier.

To install:

25. Hold the axle hub and bearing in a vise between aluminum plates.
26. Install the axle hub and bearing to the rear axle carrier with the 4 bolts. Tighten to 52 ft. lbs. (70 Nm).
27. Using a seal driver and a hammer, install the No. 2 bearing dust deflector to the rear axle carrier.

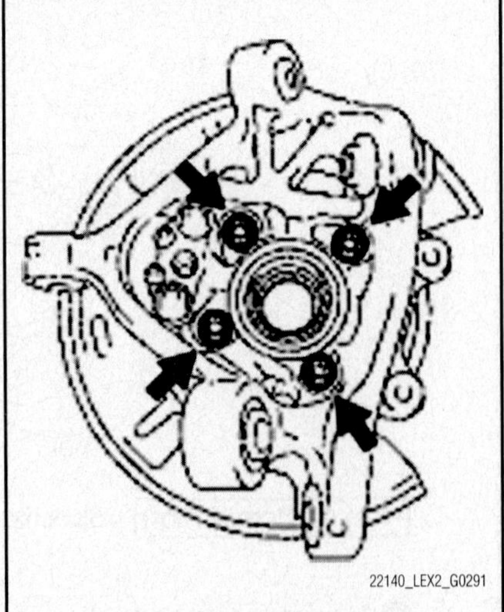

Fig. 148 Remove the 4 bolts and axle hub and bearing from the rear axle carrier

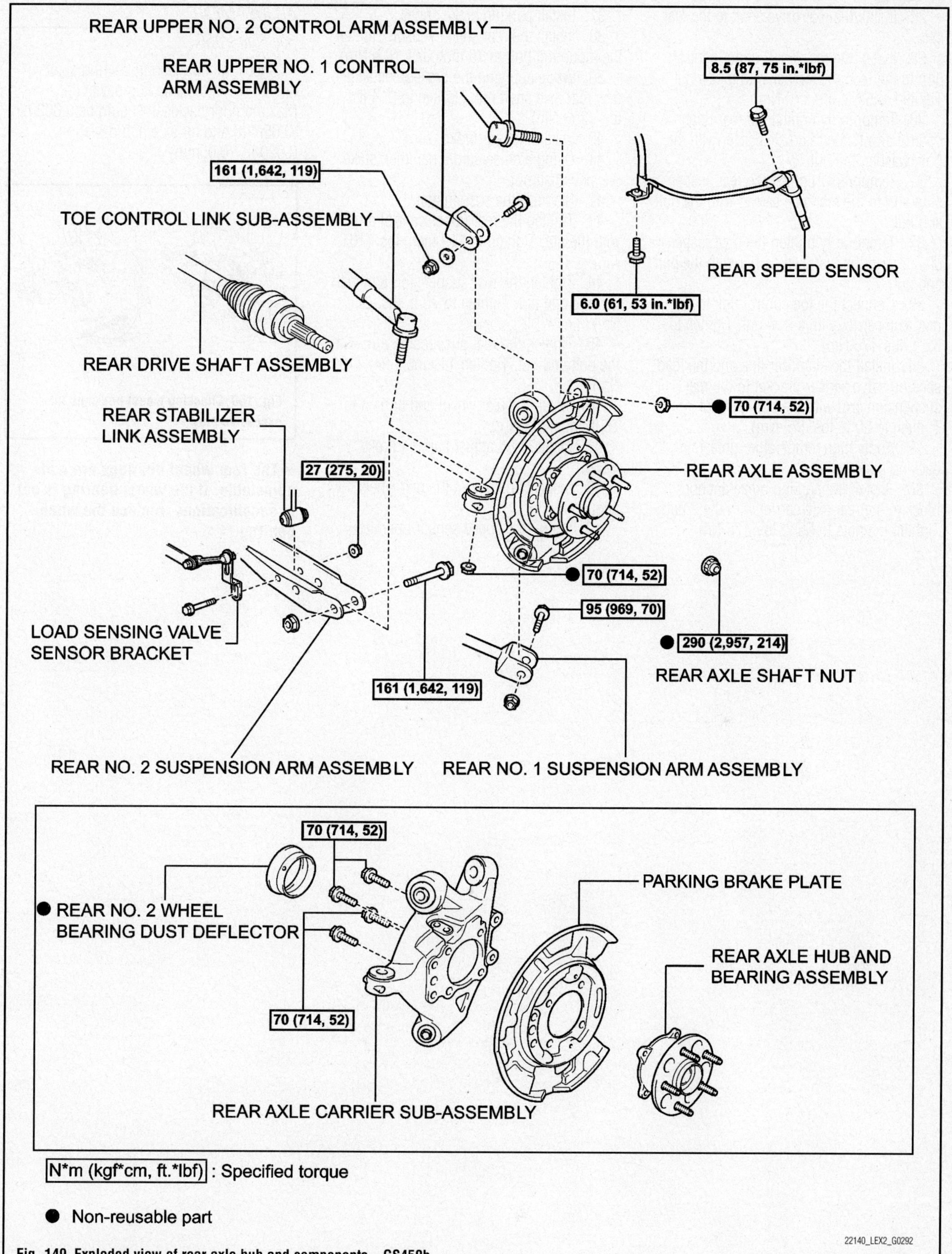

REAR UPPER NO. 2 CONTROL ARM ASSEMBLY
REAR UPPER NO. 1 CONTROL ARM ASSEMBLY
8.5 (87, 75 in.*lbf)
161 (1,642, 119)
TOE CONTROL LINK SUB-ASSEMBLY
REAR SPEED SENSOR
6.0 (61, 53 in.*lbf)
REAR DRIVE SHAFT ASSEMBLY
REAR STABILIZER LINK ASSEMBLY
70 (714, 52)
REAR AXLE ASSEMBLY
27 (275, 20)
70 (714, 52)
95 (969, 70)
LOAD SENSING VALVE SENSOR BRACKET
290 (2,957, 214)
REAR AXLE SHAFT NUT
161 (1,642, 119)
REAR NO. 2 SUSPENSION ARM ASSEMBLY
REAR NO. 1 SUSPENSION ARM ASSEMBLY

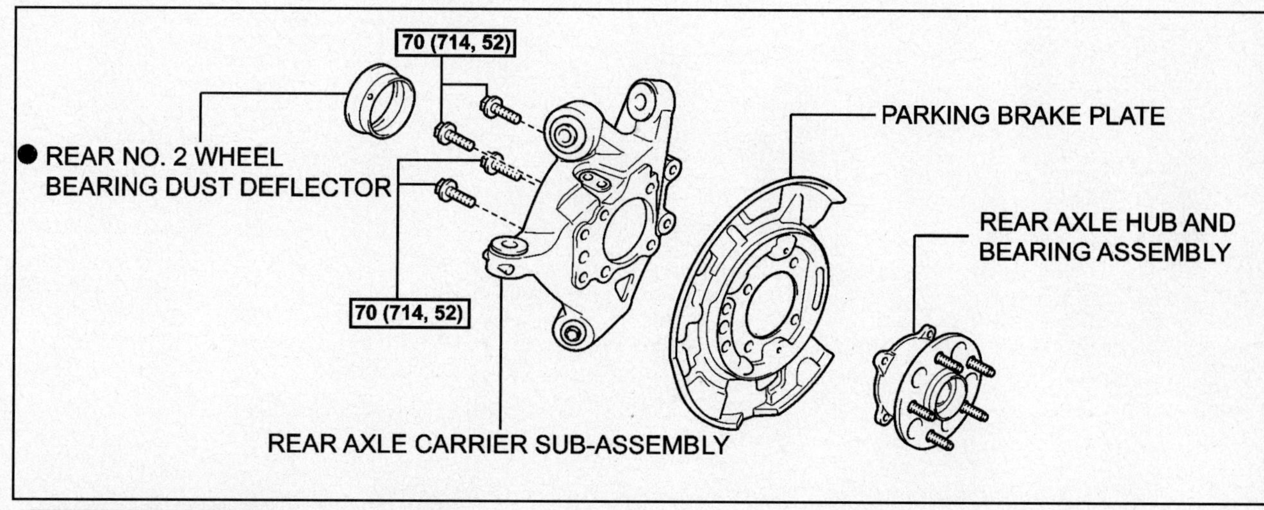

N*m (kgf*cm, ft.*lbf) : Specified torque

● Non-reusable part

22140_LEX2_G0292

Fig. 149 Exploded view of rear axle hub and components—GS450h

28. Install the rear drive shaft to the rear axle.

29. Install the rear No. 2 upper control arm to the rear axle carrier with a new nut. Tighten to 52 ft. lbs. (70 Nm).

30. Temporarily tighten the rear upper control arm to the rear axle carrier with the bolt, washer and nut.

31. Temporarily tighten the rear suspension arm to the rear axle carrier with the bolt and nut.

32. Temporarily tighten the rear suspension arm to the rear axle carrier with the bolt and nut.

33. Connect the toe control link to the rear axle carrier with a new nut. Tighten to 52 ft. lbs. (70 Nm).

34. Install the stabilizer link and the load sensing valve sensor bracket to the rear suspension arm with the bolt and nut. Tighten to 20 ft. lbs. (27 Nm).

35. Apply high temperature grease to backing plate.

36. Install the parking brake anchor block to the rear axle carrier with the 2 nuts. Tighten the nuts to 56 ft. lbs. (76 Nm).

37. Install parking brake shoes.

38. Install the rear speed sensor. Tighten the mounting bolt to 75 inch. lbs. (8.5 Nm).

39. While applying the brakes, install a new rear axle shaft nut. Tighten to 214 ft. lbs. (290 Nm).

40. Install the rear brake disc.

41. Using a chisel and a hammer, stake the axle shaft nut.

42. Stabilize the suspension.

43. Tighten the rear upper control arm with the nut. Tighten to 119 inch. lbs. (161 Nm).

44. Tighten the rear suspension arm with the bolt and nut. Tighten to 70 ft. lbs. (95 Nm).

45. Tighten the rear suspension arm with the bolt and nut. Tighten 119 inch. lbs. (161 Nm).

46. Install the rear wheel and tighten to 76 ft. lbs. (103 Nm).

47. Inspect and adjust parking brake travel.

48. Inspect and adjust the rear wheel alignment.

49. Verify rear speed sensor operation.

ADJUSTMENT

See Figure 150.

Check the backlash in bearing shaft direction and the axle hub deviation. Maximum for backlash should be 0.0020 in. (0.05mm) and for axle hub deviation 0.020 in. (0.05mm).

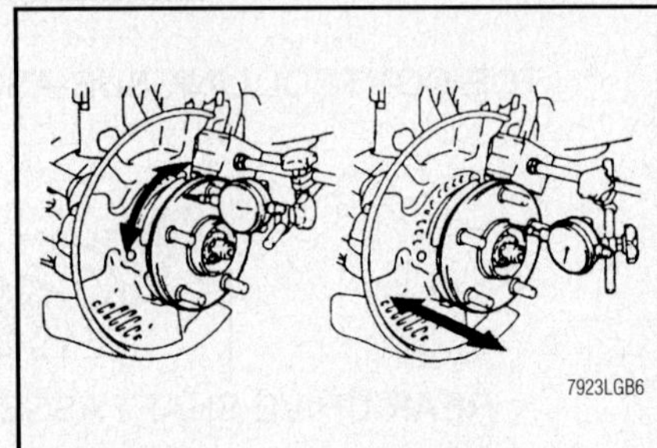

Fig. 150 Checking wheel bearings for excessive play

➡The rear wheel bearings are non-adjustable. If the wheel bearing is out of specifications, replace the wheel bearing.

LEXUS

SC 430 • GS 300

5

SPECIFICATIONS AND MAINTENANCE CHARTS

ENGINE AND VEHICLE IDENTIFICATION

Engine						
Code ①	Liters	Cu. In.	Cyl.	Fuel Sys.	Engine Type	Eng. Mfg.
3GR-FSE	3.0	183	V6	SFI	DOHC	Toyota
3UZ-FE	4.3	262	V8	SFI	DOHC	Toyota

Model Year	
Code ②	Year
6	2006
7	2007
8	2008

SFI: Sequential Multi-port Fuel Injection

DOHC: Double Overhead Camshaft

① Located on the timing belt cover

② 10th digit of the VIN

22140_LEX3_C0001

GENERAL ENGINE SPECIFICATIONS

All measurements are given in inches.

Year	Model	Engine Displacement Liters (cc)	Engine ID/VIN	Fuel System Type	Net Horsepower @ rpm	Net Torque @ rpm (ft. lbs.)	Bore x Stroke (in.)	Com-pression Ratio	Oil Pressure @ rpm
2006	GS 300	3.0 (2995)	3GR-FSE	SFI	245@6200	230@3600	3.44x3.26	11.5:1	55.5@6000
	SC 430	4.3 (4293)	3UZ-FE	SFI	288@5600	317@3400	3.58x3.25	10.5:1	43-85@3000
2007	SC 430	4.3 (4293)	3UZ-FE	SFI	288@5600	317@3400	3.58x3.25	10.5:1	43-85@3000
2008	SC 430	4.3 (4293)	3UZ-FE	SFI	288@5600	317@3400	3.58x3.25	10.5:1	43-85@3000

SFI : Sequential Multi-port Fuel Injection

22140_LEX3_C0002

ENGINE TUNE-UP SPECIFICATIONS

Year	Engine Displacement Liters	Engine ID/VIN	Spark Plug Gap (in.)	Ignition Timing (deg.)	Fuel Pump (psi)	Idle Speed (rpm)	Valve Clearance Intake	Valve Clearance Exhaust
2006	3.0	3GR-FSE	0.039-0.043	8-12B ①	28-85	600-700	HYD	HYD
	4.3	3UZ-FE	0.039-0.043	8-12B	44-50	700-800	0.006-0.010	0.010-0.014
2007	4.3	3UZ-FE	0.039-0.043	8-12B	44-50	700-800	0.006-0.010	0.010-0.014
2008	4.3	3UZ-FE	0.039-0.043	8-12B	44-50	700-800	0.006-0.010	0.010-0.014

NOTE: The Vehicle Emission Control Information label often reflects specification changes made during production. The label figures must be used if they differ from those in this chart.

B: Before top dead center

HYD: Hydraulic Valve Lifters

① Terminals TC and CG of check connector must be connected

22140_LEX3_C0003

CAPACITIES

Year	Model	Engine Displacement Liters	Engine ID/VIN	Engine Oil with Filter	Transmission (pts.) ① Auto.	Transmission (pts.) ① Manual	Drive Axle (pts.)	Fuel Tank (gal.)	Cooling System (qts.)
2006	GS 300	3.0	3GR-FSE	②	③	—	④	18.7	9.6
	SC 430	4.3	3UZ-FE	5.5	3.6	—	2.8	19.8	10.5
2007	SC 430	4.3	3UZ-FE	5.5	3.6	—	2.8	19.8	10.5
2008	SC 430	4.3	3UZ-FE	5.5	3.6	—	2.8	19.8	10.5

NOTE: All capacities are approximate. Add fluid gradually and check to be sure a proper fluid level is obtained.

① Specification is for transmission drain and refill, not overhaul.

② 2WD models: 6.6 qts.
AWD models: 6.7 qts.

③ 2WD models: 3.2 pts.
AWD models: 5.6 pts.

④ 2WD models: 2.4 pts.
AWD models front: 1.4 pts.
AWD models rear: 2.2 pts.

22140_LEX3_C0005

FLUID SPECIFICATIONS

Year	Model	Engine Displacement Liters	Engine ID/VIN	Engine Oil	Auto. Trans. ①	Drive Axle ②	Power Steering Fluid	Brake Master Cylinder	Engine Coolant ③
2006	GS 300	3.0	3GR-FSE	5W-30	ATF-WS	75W-90	Dexron II or III	DOT 3	Toyota coolant
	SC 430	4.3	3UZ-FE	5W-30	ATF-WS	75W-90	Dexron II or III	DOT 3	Toyota coolant
2007	SC 430	4.3	3UZ-FE	5W-30	ATF-WS	75W-90	Dexron II or III	DOT 3	Toyota coolant
2008	SC 430	4.3	3UZ-FE	5W-30	ATF-WS	75W-90	Dexron II or III	DOT 3	Toyota coolant

DOT: Department Of Transpotation

① The use of genuine Toyota ATF-WS is recommended

② Synthetic GL-5 (75W-90) or equivalent

③The use of genuine Toyota engine coolant is recommended or similar
ethylene glycol based non-silicate, non-amine, non- nitrite, and non- borat coolant

22140_LEX3_C0004

VALVE SPECIFICATIONS

Year	Engine Displacement Liters	Engine ID/VIN	Seat Angle (deg.)	Face Angle (deg.)	Spring Test Pressure (lbs. @ in.)	Spring Free-Length (in.)	Stem-to-Guide Clearance (in.)		Stem Diameter (in.)	
							Intake	Exhaust	Intake	Exhaust
2006	3.0	3GR-FSE	45	NA	NA	1.831	0.0010-0.0024	0.0012-0.0026	0.2154-0.2159	0.2152-0.2158
	4.3	3UZ-FE	45	44.5	45.9-50.7@ 1.3795	2.130	0.0010-0.0024	0.0012-0.0026	0.2154-0.2159	0.2152-0.2157
2007	4.3	3UZ-FE	45	44.5	45.9-50.7@ 1.3795	2.130	0.0010-0.0024	0.0012-0.0026	0.2154-0.2159	0.2152-0.2157
2008	4.3	3UZ-FE	45	44.5	45.9-50.7@ 1.3795	2.130	0.0010-0.0024	0.0012-0.0026	0.2154-0.2159	0.2152-0.2157

NA: Not Available

22140_LEX3_C0007

CAMSHAFT AND BEARING SPECIFICATIONS CHART

All measurements are given in inches.

Year	Engine Displ. Liters	Engine ID/VIN	Journal Dia.	Brg. Oil Clearance	Shaft End-play	Runout	Journal Bore	Lobe Height	
								Intake	Exhaust
2006	3.0	3GR-FSE	N.A	①	0.0031 0.0051	0.0016	NA	1.7303-1.7342	1.7366-1.74.5
	4.3	3UZ-FE	1.0612-1.0618	0.0039	0.0024 0.0039	0.0031	NA	1.7303-1.7342	1.6783-1.6823
2007	4.3	3UZ-FE	1.0612-1.0618	0.0039	0.0024 0.0039	0.0031	NA	1.7303-1.7342	1.6783-1.6823
2008	4.3	3UZ-FE	1.0612-1.0618	0.0039	0.0024 0.0039	0.0031	NA	1.7303-1.7342	1.6783-1.6823

NA: Not Available

① Intake Journal 1: 0.0016 - 0.0031 inch.
All Others: 0.00098 - 0.0024 inch.

22140_LEX3_C0006

CRANKSHAFT AND CONNECTING ROD SPECIFICATIONS

All measurements are given in inches.

Year	Engine Displacement Liters	Engine ID/VIN	Crankshaft Main Brg. Journal Dia.	Crankshaft Main Brg. Oil Clearance	Crankshaft Shaft End-play	Crankshaft Thrust on No.	Connecting Rod Journal Diameter	Connecting Rod Oil Clearance	Connecting Rod Side Clearance
2006	3.0	3GR-FSE	2.4011-2.4016	0.0005-0.0014	0.0016-0.0094	NA	2.0863-2.0866	0.0018-0.0026	0.0059-0.0157
	4.3	3UZ-FE	2.6373-2.6378	①	0.0008-0.0087	3	2.0465-2.0472	0.0008-0.0019	0.0063-0.0138
2007	4.3	3UZ-FE	2.6373-2.6378	①	0.0008-0.0087	3	2.0465-2.0472	0.0008-0.0019	0.0063-0.0138
2008	4.3	3UZ-FE	2.6373-2.6378	①	0.0008-0.0087	3	2.0465-2.0472	0.0008-0.0019	0.0063-0.0138

NA: Not Available

① Journal No. 1 and 5: 0.0007 - 0.0013 inch
Remaining journals: 0.0011 - 0.0018 inch

22140_LEX3_C0010

PISTON AND RING SPECIFICATIONS

All measurements are given in inches.

Year	Engine Displacement Liters	Engine ID/VIN	Piston Clearance	Ring Gap Top Compression	Ring Gap Bottom Compression	Ring Gap Oil Control	Ring Side Clearance Top Compression	Ring Side Clearance Bottom Compression	Ring Side Clearance Oil Control
2006	3.0	3GR-FSE	0.0006-0.0014	0.0091-0.0130	0.0197-0.0236	0.0039-0.0157	0.0008-0.0028	0.0008-0.0024	0.0008-0.0026
	4.3	3UZ-FE	0.0023-0.0040	0.0118-0.0197	0.0157-0.0236	0.0059-0.0197	0.0012-0.0031	0.0008-0.0024	SNUG
2007	4.3	3UZ-FE	0.0023-0.0040	0.0118-0.0197	0.0157-0.0236	0.0059-0.0197	0.0012-0.0031	0.0008-0.0024	SNUG
2008	4.3	3UZ-FE	0.0023-0.0040	0.0118-0.0197	0.0157-0.0236	0.0059-0.0197	0.0012-0.0031	0.0008-0.0024	SNUG

22140_LEX3_C0009

TORQUE SPECIFICATIONS

All readings in ft. lbs.

Year	Engine Displacement Liters	Engine ID/VIN	Cylinder Head Bolts	Main Bearing Bolts	Rod Bearing Bolts	Crankshaft Damper Bolts	Flywheel Bolts	Manifold Intake	Manifold Exhaust	Spark Plugs
2006	3.0	3GR-FSE	①	②	③	184	30	15	15	13
	4.3	3UZ-FE	④	⑤	⑥	181	⑦	13	32	13
2007	4.3	3UZ-FE	④	⑤	⑥	181	⑦	13	32	13
2008	4.3	3UZ-FE	④	⑤	⑥	181	⑦	13	32	13

① Step 1: 27 ft. lbs.
Step 2: Tighten an additional 90 degrees
Step 3: Tighten an additional 90 degrees
14mm head bolt: 22 ft. lbs.

② 12mm head bolts: 19 ft. lbs.
16-point bolts:
Step 1: 45 ft. lbs.
Step 2: Plus an additional 90 degrees

③ Step 1: 28 ft. lbs.
Step 2: Plus 90 degrees

④ Step 1: 44 ft. lbs.
Step 2: Plus 90 degrees

⑤ Nuts:
Step 1: 20 ft. lbs.
Step 2: Plus 90 degrees
Bolts: 36 ft. lbs.

⑥ Step 1: 18 ft. lbs.
Step 2: Plus 90 degrees

⑦ Step 1: 36 ft. lbs.
Step 2: Plus 90 degrees

22140_LEX3_C0008

WHEEL ALIGNMENT

Year	Model		Caster Range (+/-Deg.)	Caster Preferred Setting (Deg.)	Camber Range (+/-Deg.)	Camber Preferred Setting (Deg.)	Toe-in (in.)	Steering Axis Inclination (Deg.)
2006	GS 300 ①	F	0.75	②	0.75	③	0 +/- 0.04	④
		R	—	—	0.75	⑤	0.12 +/- 0.08	—
	GS 300 ⑥	F	0.75	+7.43	0.75	-0.43	0 +/- 0.04	9.43
		R	—	—	0.75	⑤	0.12 +/- 0.08	—
	GS 430	F	0.75	+7.38	0.75	-0.38	0 +/- 0.04	9.38
		R	—	—	0.75	-1.32	0.12 +/- 0.08	—
	SC 430	F	0.75	+7.92	0.75	-0.58	0.06 +/- 0.08	9.16
		R	—	—	0.50	-1.17	0.06 +/- 0.08	—
2007	GS 430	F	0.75	+7.38	0.75	-0.38	0 +/- 0.04	9.38
		R	—	—	0.75	-1.32	0.12 +/- 0.08	—
	SC 430	F	0.75	+7.92	0.75	-0.58	0.06 +/- 0.08	9.16
		R	—	—	0.50	-1.17	0.06 +/- 0.08	—
2008	GS 430	F	0.75	+7.38	0.75	-0.38	0 +/- 0.04	9.38
		R	—	—	0.75	-1.32	0.12 +/- 0.08	—
	SC 430	F	0.75	+7.92	0.75	-0.58	0.06 +/- 0.08	9.16
		R	—	—	0.50	-1.17	0.06 +/- 0.08	—

① Except air suspension

② 2WD: +7.38
AWD: +4.88

③ 2WD: -0.38
AWD: -0.42

④ 2WD: 9.38
AWD: 11.18

⑤ 2WD: -1.17
AWD: -1.05

⑥ With air suspension

22140_LEX3_C0011

TIRE, WHEEL AND BALL JOINT SPECIFICATIONS

		OEM Tires		Tire Pressures (psi)		Wheel	Ball Joint	Lug
Year	Model	Standard	Optional	Front	Rear	Size	Inspection	Nuts
2006	GS 300	225/50R17 94W	P245/40R18 93Y	33	33	8-JJ	①	76
	SC 430	245/40ZR18	—	33	33	8-JJ	①	76
2007	SC 430	245/40ZR18	—	33	33	8-JJ	①	76
2008	SC 430	245/40ZR18	—	33	33	8-JJ	①	76

OEM: Original Equipment Manufacturer

PSI: Pounds Per Square Inch

① Lower ball joint: 0.016 inch. Upper ball joint: standard turning torque 9-30 inch.

22140_LEX3_C0012

BRAKE SPECIFICATIONS

All measurements in inches unless noted

		Front Brake Disc			Rear Brake Disc			Minimum	Brake Caliper Bracket	Brake Caliper Mounting
Year	Model	Original Thickness	Minimum Thickness	Maximum Run-out	Original Thickness	Minimum Thickness	Maximum Run-out	Lining Thickness	Bolts (ft. lbs.)	Bolts (ft. lbs.)
2006	GS 300	1.260	1.181	0.0020	0.413	0.039	0.0020	①	②	—
	SC 430	1.260	1.181	0.0020	0.472	0.413	0.0020	0.0390	③	—
2007	SC 430	1.260	1.181	0.0020	0.472	0.413	0.0020	0.0390	③	—
2008	SC 430	1.260	1.181	0.0020	0.472	0.413	0.0020	0.0390	③	—

① Front: 0.039 inches
Rear: 0.690 inches

② Front: 58 ft. lbs.
Rear: 40 ft. lbs.

③ Front: 87 ft. lbs.
Rear: 77 ft. lbs.

22140_LEX3_C0013

SCHEDULED MAINTENANCE INTERVALS
2006-08 Lexus- GS300 and SC430

TO BE SERVICED	TYPE OF SERVICE	VEHICLE MILEAGE INTERVAL (x1000)												
		7.5	15	22.5	30	37.5	45	52.5	60	67.5	75	82.5	90	97.5
Engine oil & filter	R	✓	✓	✓	✓	✓	✓	✓	✓	✓	✓	✓	✓	✓
A/C filter (if equipped) ①	S/I	✓	✓	✓	✓	✓	✓	✓	✓	✓	✓	✓	✓	✓
Automatic transaxle fluid & filter	S/I		✓		✓		✓		✓		✓		✓	
Ball joints & dust covers	S/I		✓		✓		✓		✓		✓		✓	
Bolts & nuts on chassis & body	S/I		✓		✓		✓		✓		✓		✓	
Brake fluid ②	S/I		✓		✓		✓		✓		✓		✓	
Brake line pipes & hoses	S/I		✓		✓		✓		✓		✓		✓	
Brake pads & discs (front & rear)	S/I		✓		✓		✓		✓		✓		✓	
Differential oil	S/I		✓		✓		✓		✓		✓		✓	
Driveshaft boots (if equipped)	S/I		✓		✓		✓		✓		✓		✓	
Steering gear housing oil	S/I		✓		✓		✓		✓		✓		✓	
Steering linkage	S/I		✓		✓		✓		✓		✓		✓	
Air filter	R				✓				✓				✓	
Exhaust pipes & mountings	S/I				✓				✓				✓	
Fuel lines & connections	S/I				✓				✓				✓	
Engine coolant	R						✓					✓		
Fuel tank cap gasket	R								✓					
Spark plugs	R								✓					
Charcoal canister	S/I								✓					
Drive belts	S/I								✓					
Valve clearance	S/I								✓					

R: Replace S/I: Service or Inspect

① Replace at 15,000 miles.

② Replace at 30,000 miles (unless previously replaced).

FREQUENT OPERATION MAINTENANCE (SEVERE SERVICE)

If a vehicle is operated under any of the following conditions it is considered severe service

- Extremely dusty areas.

- 50% or more of the vehicle operation is in 32°C (90°F) or higher temperatures, or constant operation in temperatures below 0°C (32°F).

- Prolonged idling (vehicle operation in stop and go traffic).

- Frequent short running periods (engine does not warm to normal operating temperatures).

- Police, taxi, delivery usage or trailer towing usage.

Oil & oil filter: change every 3750 miles.

Ball joints & dust covers: service or inspect every 7500 miles.

Bolts & nuts on chassis & body: service or inspect every 7500 miles.

Brake pads & discs (front & rear): service or inspect every 7500 miles.

Driveshaft boots (if equipped): service or inspect every 7500 miles.

Steering linkage: service or inspect every 7500 miles.

Air filter: service or inspect every 15,000 miles.

Automatic transmission fluid & filter: replace every 15,000 miles.

Differential oil: replace every 15,000 miles.

Exhaust pipes & mountings: service or inspect every 15,000 miles.

Drive belts: service or inspect at 60,000 miles & every 7500 miles thereafter.

Timing belts: replace every 60,000 miles.

22140_LEX3_C0014

PRECAUTIONS

Before servicing any vehicle, please be sure to read all of the following precautions, which deal with personal safety, prevention of component damage, and important points to take into consideration when servicing a motor vehicle:

- Never open, service or drain the radiator or cooling system when the engine is hot; serious burns can occur from the steam and hot coolant.
- Observe all applicable safety precautions when working around fuel. Whenever servicing the fuel system, always work in a well-ventilated area. Do not allow fuel spray or vapors to come in contact with a spark, open flame, or excessive heat (a hot drop light, for example). Keep a dry chemical fire extinguisher near the work area. Always keep fuel in a container specifically designed for fuel storage; also, always properly seal fuel containers to avoid the possibility of fire or explosion. Refer to the additional fuel system precautions later in this section.
- Fuel injection systems often remain pressurized, even after the engine has been turned **OFF**. The fuel system pressure must be relieved before disconnecting any fuel lines. Failure to do so may result in fire and/or personal injury.
- Brake fluid often contains polyglycol ethers and polyglycols. Avoid contact with the eyes and wash your hands thoroughly after handling brake fluid. If you do get brake fluid in your eyes, flush your eyes with clean, running water for 15 minutes. If eye irritation persists, or if you have taken brake fluid internally, IMMEDIATELY seek medical assistance.
- The EPA warns that prolonged contact with used engine oil may cause a number of skin disorders, including cancer. You should make every effort to minimize your exposure to used engine oil. Protective gloves should be worn when changing oil. Wash your hands and any other exposed skin areas as soon as possible after exposure to used engine oil. Soap and water, or waterless hand cleaner should be used.
- All new vehicles are now equipped with an air bag system, often referred to as a Supplemental Restraint System (SRS) or Supplemental Inflatable Restraint (SIR) system. The system must be disabled before performing service on or around system components, steering column, instrument panel components, wiring and sensors. Failure to follow safety and disabling procedures could result in accidental air bag deployment, possible personal injury and unnecessary system repairs.
- Always wear safety goggles when working with, or around, the air bag system. When carrying a non-deployed air bag, be sure the bag and trim cover are pointed away from your body. When placing a non-deployed air bag on a work surface, always face the bag and trim cover upward, away from the surface. This will reduce the motion of the module if it is accidentally deployed. Refer to the additional air bag system precautions later in this section.
- Clean, high quality brake fluid from a sealed container is essential to the safe and proper operation of the brake system. You should always buy the correct type of brake fluid for your vehicle. If the brake fluid becomes contaminated, completely flush the system with new fluid. Never reuse any brake fluid. Any brake fluid that is removed from the system should be discarded. Also, do not allow any brake fluid to come in contact with a painted surface; it will damage the paint.
- Never operate the engine without the proper amount and type of engine oil; doing so WILL result in severe engine damage.
- Timing belt maintenance is extremely important. Many models utilize an interference-type, non-freewheeling engine. If the timing belt breaks, the valves in the cylinder head may strike the pistons, causing potentially serious (also time-consuming and expensive) engine damage. Refer to the maintenance interval charts for the recommended replacement interval for the timing belt, and to the timing belt section for belt replacement and inspection.
- Disconnecting the negative battery cable on some vehicles may interfere with the functions of the on-board computer system(s) and may require the computer to undergo a relearning process once the negative battery cable is reconnected.
- When servicing drum brakes, only disassemble and assemble one side at a time, leaving the remaining side intact for reference.
- Only an MVAC-trained, EPA-certified automotive technician should service the air conditioning system or its components.

BRAKES — ANTI-LOCK BRAKE SYSTEM (ABS)

GENERAL INFORMATION

Anti-lock Brake System (ABS) helps prevent the wheels from locking when the brakes are applied firmly or when braking on a slippery surface.

Electronic Brake Force Distribution (EBD) control utilizes the ABS, realizing the proper brake force distribution between the front and rear wheels in accordance with the driving conditions. In addition, during cornering and braking, it also controls the brake forces of the right and left wheels, helping to maintain the vehicle balance.

Brake Assist (BA). The primary purpose of the brake assist system is to provide an auxiliary brake force to assist a driver who cannot generate a large brake force during emergency braking, thus helping improve the vehicle's brake performance.

The Traction Control (TRAC) system helps prevent the drive wheels from slipping if the driver depresses the accelerator pedal excessively when starting off or accelerating on a slippery surface.

The Vehicle Stability Control (VSC) system helps prevent the vehicle from slipping sideways as a result of strong front wheel skid or strong rear wheel skid during cornering.

1. **ABS WITH EBD & BA & TRAC & VSC OPERATION:**

The skid control ECU calculates vehicle stability tendency based on the signal of the speed sensor, yaw rate sensor and steering angle sensor. The ECU judges whether the control of engine output torque by the electronic control throttle and wheel brake pressure by the brake actuator can start based on the calculation results.

The SLIP indicator light blinks and the skid control buzzer sounds to inform the driver that the VSC system is operating. The SLIP indicator light also blinks when TRAC is operating, and the operation being performed is displayed.

2. **FAIL-SAFE FUNCTION**

When a failure occurs in the ABS & BA & TRAC & VSC systems, the ABS warning light turns on and the ABS & BA & TRAC & VSC operations are prohibited. In addition to this, when there is a failure that disables the EBD operation, the brake warning light also turns on and the EBD operation is prohibited. If a control is prohibited due to a malfunction during its operation, the control will be cut off gradually so that vehicle stability will not change suddenly.

BLEEDING PROCEDURE

See Figure 1.

If any work is done on the brake system or if air in the brake lines is suspected, bleed the system of air.

➡Do not let brake fluid remain on painted surfaces. Wash it off immediately.

1. Fill the reservoir with brake fluid.

➡If the master cylinder has been disassembled or if the reservoir becomes empty, bleed the air from the master cylinder.

2. Bleed the brake master cylinder as follows:
   - Disconnect the brake lines from the master cylinder.
   - Slowly depress the brake pedal and hold it.
   - Block off the outer holes with your fingers, and release the brake pedal.
   - Repeat the previous 2 steps 3 or 4 times.
3. Bleed the brake line as follows:
   - Connect the vinyl tube to the brake caliper.
   - Depress the brake pedal several times, then loosen the bleeder plug with the pedal held down.
   - At the point when fluid stops coming out, tighten the bleeder plug, then release the brake pedal.
   - Repeat the previous 2 steps until all the air in the fluid has been bled out.
   - Repeat the above procedure to bleed the air out of the brake line for each wheel.
   - Tighten the bleeder plug to 8 ft. lbs. (11 Nm).
4. Bleed the brake actuator as follows:
   - Remove the reservoir cap.
   - Install the SST 09992-00242, 09992-00350 to the reservoir.
   - Connect the vinyl tube to the bleeder plug of the brake actuator.
   - Using SST, apply the 14.2 psi (98.1kpa) of pressure to the reservoir.
   - Loosen the bleeder plug.
   - Bleed the air out of the brake actuator, tighten the bleeder plug to 74 inch lbs. (8.3 Nm).
5. Check the fluid level and add fluid if necessary.

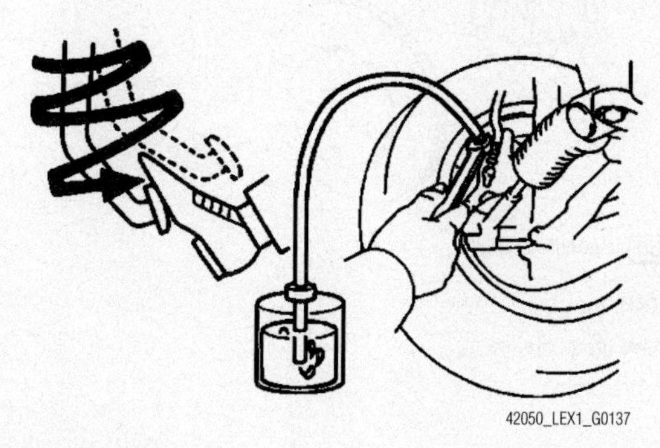

Fig. 1 Bleeding the brake line

BLEEDING THE ABS SYSTEM

➡After performing the usual air bleeding in the brake system, if the height or feel of the brake pedal cannot be obtained, perform air bleeding in the brake actuator assembly with a hand held tester by following procedures below. Make sure that the brake fluid in the master cylinder reservoir tank does not become empty.

1. Depress the brake pedal more than 20 times with the engine off.
2. Connect the hand held tester to the DLC3, then turn the ignition switch to the ON position.
3. Do not start the engine.
4. Select "AIR BLEEDING" on the hand held tester. Please refer to the Hand Held Tester Operator's Manual for further details.
5. Bleed the air out of the regular brake line in "Step1: Increase" on the hand held tester display. Perform the air bleeding by following the steps displayed on the hand held tester. Make sure that the brake fluid in the master cylinder reservoir tank does not become empty.
6. Connect the vinyl tube to either one of the bleeder plugs.
7. Depress the brake pedal several times, then loosen the bleeder plug of one of the above wheels with the pedal depressed.
8. When fluid stops coming out, tighten the bleeder plug, then release the brake pedal.
9. Repeat (2) and (3) until all air in the fluid is completely bled out.
10. Tighten the bleeder plug to 73 inch lbs. (8.3 Nm).
11. Repeat the above procedure to bleed the air out of the brake line for each wheel.
12. Bleed the air out of the suction line in "Step2: Inhalation" on the hand held tester display.
13. Connect the vinyl tube to the bleeder plug at the right front wheel or the right rear wheel and loosen the bleeder plug.
14. Operate the brake actuator assembly using the hand held tester to bleed the air.

➡The operation stops automatically in 4 seconds. At this time, be sure to release the brake pedal.

15. Check that the operation has stopped, by referring to the hand held tester display.
16. Repeat (2) and (3) until all the air in the fluid is completely bled out.
17. Tighten the bleeder plug.
18. For the rest of the wheels, bleed the air in the same way as stated in the above procedure.
19. Bleed the air out of the pressure reduction line in "Step3: Decrease" on the hand held tester display.
20. Connect a vinyl tube to either one of the bleeder plugs.
21. Loosen the bleeder plug.
22. Using the hand held tester, operate the brake actuator assembly using hand held tester, completely depress the brake pedal and keep it.

➡The operation stops automatically in 4 seconds. When performing this procedure continuously, an interval of at least 20 seconds is required. When the operation is completed, the brake pedal slightly goes down. This is a normal phenomenon caused when the solenoid opens. During this procedure, the pedal seems heavy, but completely depress it so that the brake fluid comes out from the bleeder plug. Be sure to keep depressing the brake pedal. Never depress and release the pedal repeatedly.

23. Tighten the bleeder plug, then release the brake pedal.
24. Repeat 3 previous steps until all the air in the fluid is completely bled out.
25. Tighten the bleeder plug.
26. Repeat the above procedure to bleed the air out of the brake line for each wheel.
27. Bleed the air out of the regular brake line again in "Step4: Increase" on the hand-held tester display.
28. Connect the vinyl tube to either one of the bleeder plug.
29. Depress the brake pedal several times, then loosen the bleeder plug of one of the above wheels with the pedal depressed.
30. When fluid stops coming out, tighten the bleeder plug, then release the brake pedal.
31. Repeat the previous 2 steps until all the air in the fluid is completely bled out.
32. Tighten the bleeder plug.
33. Repeat the above procedure to bleed the air out of the brake line for each wheel.

CAUTION

Dust and dirt accumulating on brake parts during normal use may contain asbestos fibers from production or aftermarket brake linings. Breathing excessive concentrations of asbestos fibers can cause serious bodily harm. Exercise care when servicing brake parts. Do not sand or grind brake lining unless equipment used is designed to contain the dust residue. Do not clean brake parts with compressed air or by dry brushing. Cleaning should be done by dampening the brake components with a fine mist of water, then wiping the brake components clean with a dampened cloth. Dispose of cloth and all residue containing asbestos fibers in an impermeable container with the appropriate label. Follow practices prescribed by the Occupational Safety and Health Administration (OSHA) and the Environmental Protection Agency (EPA) for the handling, processing, and disposing of dust or debris that may contain asbestos fibers.

BRAKE CALIPER

REMOVAL & INSTALLATION

GS 300

See Figure 2.

1. Before servicing the vehicle, refer to the precautions in the beginning of this section.
2. Remove or disconnect the following:
   - Wheels
   - Brake line at the caliper
   - Anti-squeal springs
   - Mounting bolts, while holding the sliding pin with a wrench
   - Caliper assembly

To install:

3. Install or connect the following:
   - Caliper. Hold the sliding pin and tighten the mounting bolts to 25 ft. lbs. (34 Nm).
   - Anti-squeal springs
   - Brake line with 2 new gaskets and tighten the union bolt to 22 ft. lbs. (30 Nm)
4. Bleed the brake system.
   - Wheels
5. Check and if necessary fill the master cylinder reservoir.

SC 430

See Figure 3.

1. Before servicing the vehicle, refer to the precautions in the beginning of this section.
2. Remove or disconnect the following:
   - Front wheels
   - Brake line at the caliper

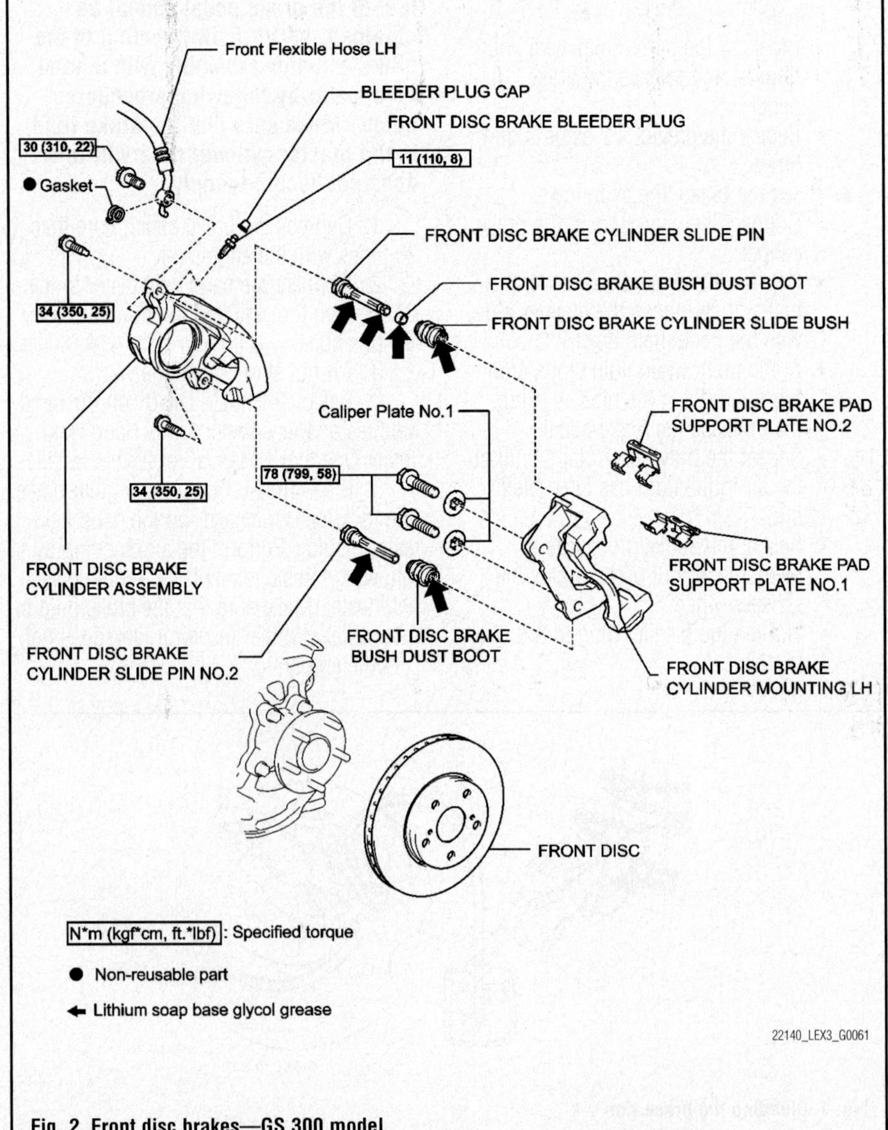

Fig. 2 Front disc brakes—GS 300 model

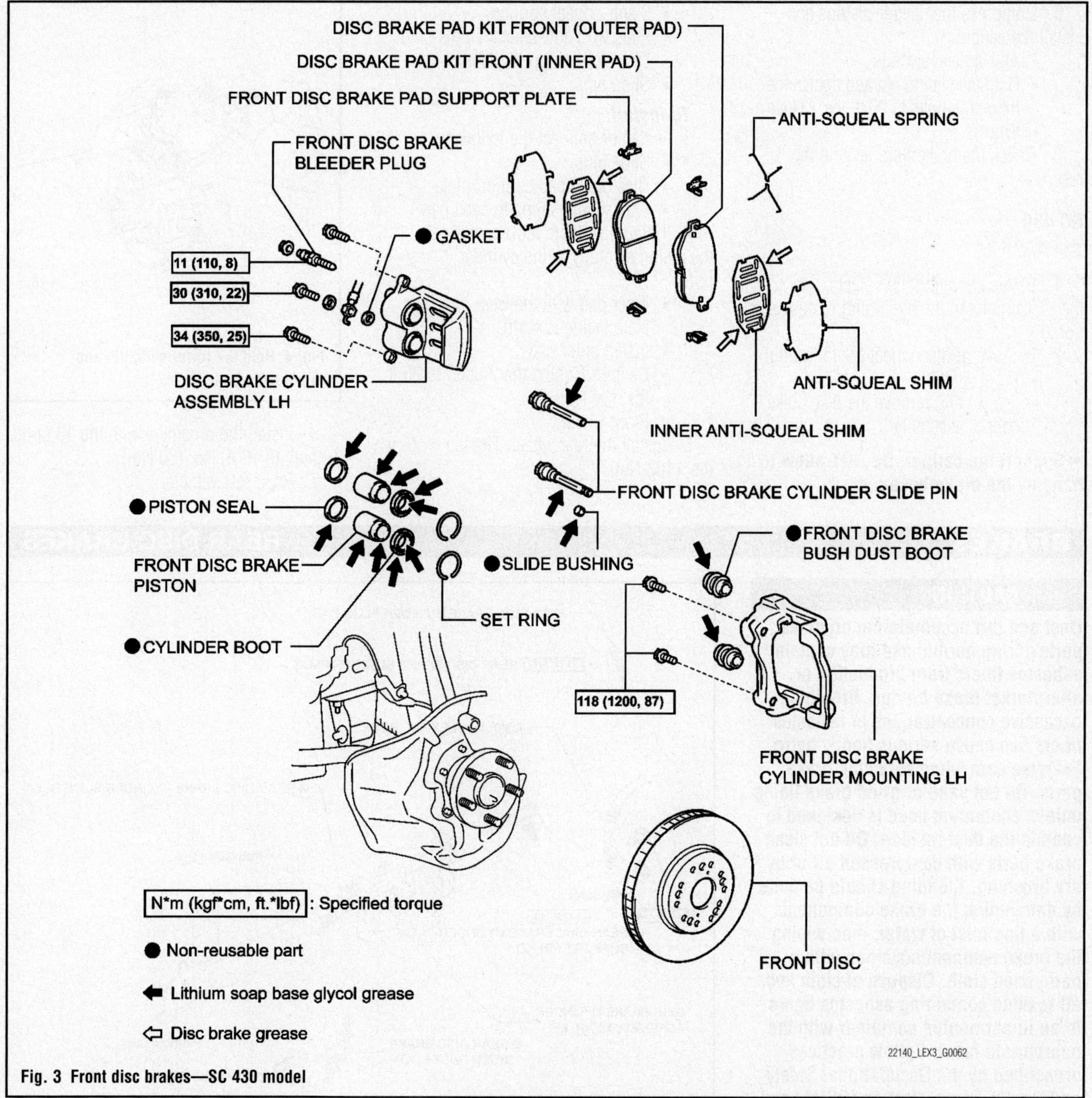

Fig. 3 Front disc brakes—SC 430 model

- 2 bolts to the holding the caliper to the steering knuckle
- Caliper assembly

To install:

3. Install or connect the following:
   - Caliper. Tighten the 2 bolts to 87 ft. lbs. (118 Nm)
   - Brake line with 2 new gaskets and tighten the union to 22 ft. lbs. (30 Nm)
4. Refill the reservoir as necessary and bleed the brake system.
5. Install the rear wheel. Tighten to 76 ft. lbs. (103 Nm).
6. Install the ornament with the 10 bolts. Tighten to 15 ft. lbs. (20 Nm).
7. Road test vehicle.

DISC BRAKE PADS

REMOVAL & INSTALLATION

GS 300

1. Before servicing the vehicle, refer to the precautions in the beginning of this section.
2. Remove or disconnect the following:
   - Wheels
3. Hold the sliding pin on the lower mounting bolt and remove the bolt. Swivel the caliper upward and out of the way.
   - Anti-squeal springs
   - Brake pads, retainers and anti-squeal shims

To install:

4. Install or connect the following:
   - Pad support plates and the pad wear indicator plate on the inside pad
   - Both pads (and anti-squeal shims) with the wear indicator plates facing downward

5. Compress the caliper pistons and install the caliper.
- Anti-squeal springs
- Hold the sliding pin and tighten the mounting bolts to 25 ft. lbs. (34 Nm)
- Wheels

6. Check the brake fluid level in the reservoir.

SC 430

See Figure 4.

1. Before servicing the vehicle, refer to the precautions in the beginning of this section.
2. Remove or disconnect the following:
- Front wheel
- 2 bolts and remove the disc brake caliper assembly

➡Support the caliper. Do NOT allow to hang by the brake hose

- 2 anti-squeal springs
- Pads with anti-squeal shims
- Disc brake support plates
- Slide pins

To install:

3. Install or connect the following:
- Slide pins
- Disc brake pad support plate
- Anti-squeal shims to each pad

4. Using a suitable tool, compress the piston carefully in the cylinder bores
- Inner pad with the wear indicator plate facing upward
- Install outer pad
- Caliper. Tighten the 2 bolts to 25 ft. lbs. (34 Nm)
- Front wheel

5. Install the rear wheel. Tighten to 76 ft. lbs. (103 Nm).

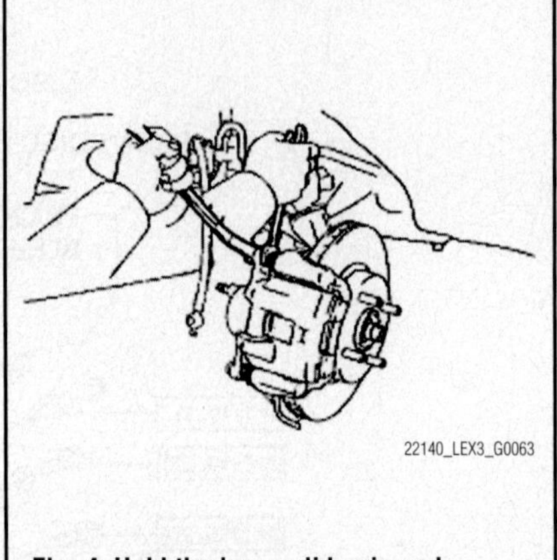

Fig. 4 Hold the lower slide pin and remove the bolt

6. Install the ornament with the 10 bolts. Tighten to 15 ft. lbs. (20 Nm).
7. Road test vehicle.

⁂ CAUTION

Dust and dirt accumulating on brake parts during normal use may contain asbestos fibers from production or aftermarket brake linings. Breathing excessive concentrations of asbestos fibers can cause serious bodily harm. Exercise care when servicing brake parts. Do not sand or grind brake lining unless equipment used is designed to contain the dust residue. Do not clean brake parts with compressed air or by dry brushing. Cleaning should be done by dampening the brake components with a fine mist of water, then wiping the brake components clean with a dampened cloth. Dispose of cloth and all residue containing asbestos fibers in an impermeable container with the appropriate label. Follow practices prescribed by the Occupational Safety and Health Administration (OSHA) and the Environmental Protection Agency (EPA) for the handling, processing, and disposing of dust or debris that may contain asbestos fibers.

BRAKE CALIPER

REMOVAL & INSTALLATION

GS 300

See Figures 5 and 6.

1. Before servicing the vehicle, refer to the precautions in the beginning of this section.

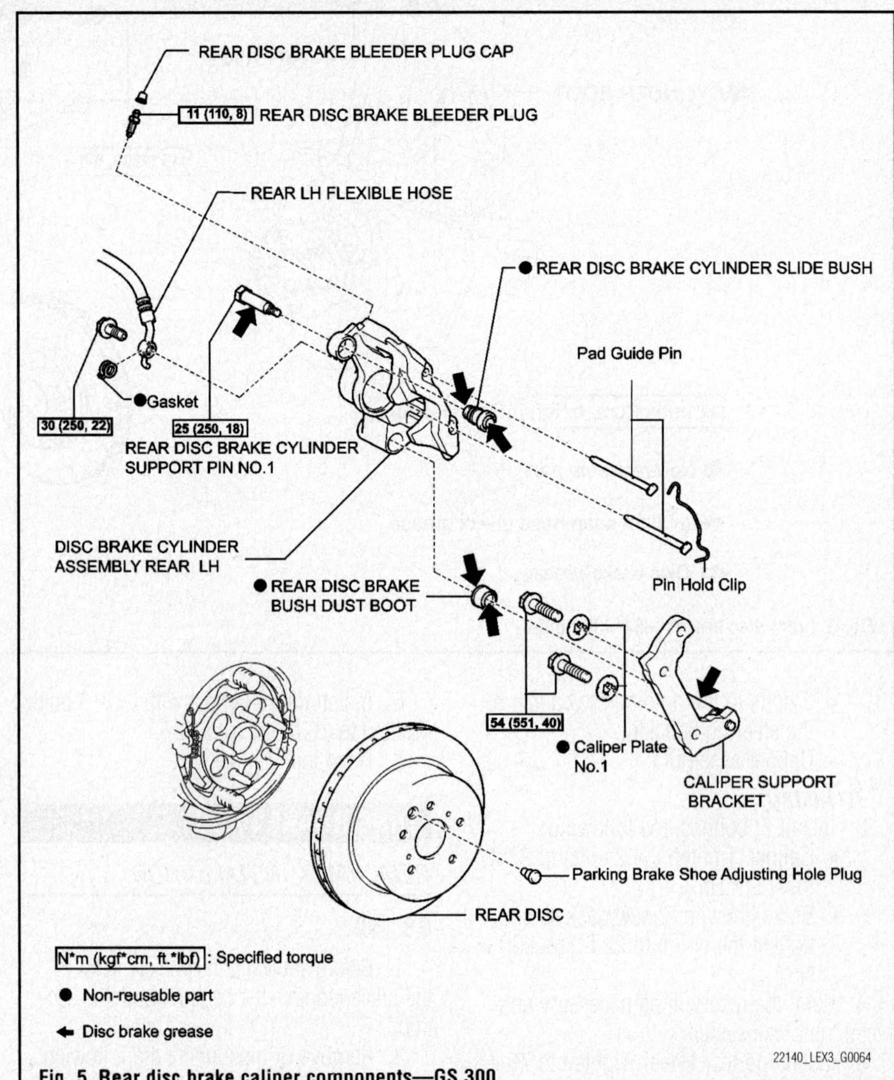

Fig. 5 Rear disc brake caliper components—GS 300

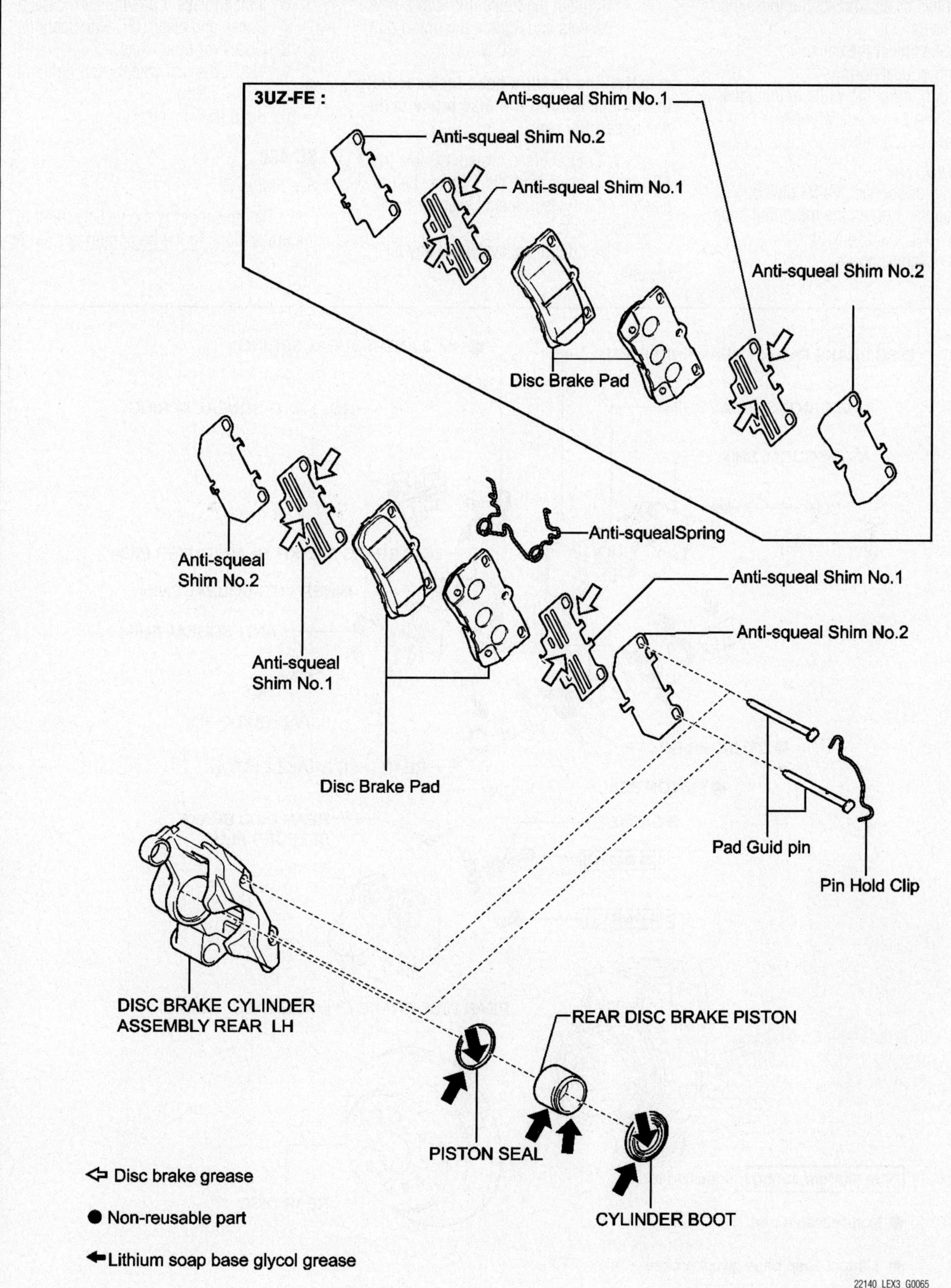

Fig. 6 Rear disc brake pads components—GS 300

2. Remove or disconnect the following:
- Wheels
- Brake line at the caliper
- Anti-squeal springs
- Mounting bolts, while holding the sliding pin with a wrench
- Caliper assembly

To install:

3. Install or connect the following:
- Caliper. Tighten the mounting bolts to 40 ft. lbs. (54 Nm).
- Anti-squeal springs
- Connect the brake line with 2 new gaskets and tighten the union bolt to 22 ft. lbs. (30 Nm)

➡Install the flexible hose lock securely in the lock hole in the disc brake cylinder assembly rear.

4. With ECB brake system disable the 2 ABS motor relays from No.3 block. before bleeding the system. Install scanner to verify system is off.
5. Bleed the brake system of any air present.
6. Install relays if previously removed.
7. Check and clear DTC with scanner.
8. Check for fluid leaks.
9. Install the rear wheels and tighten to 76 ft. lbs. (103 Nm).
10. Road test vehicle.

SC 430

See Figure 7.

1. Before servicing the vehicle, refer to the precautions in the beginning of this section.

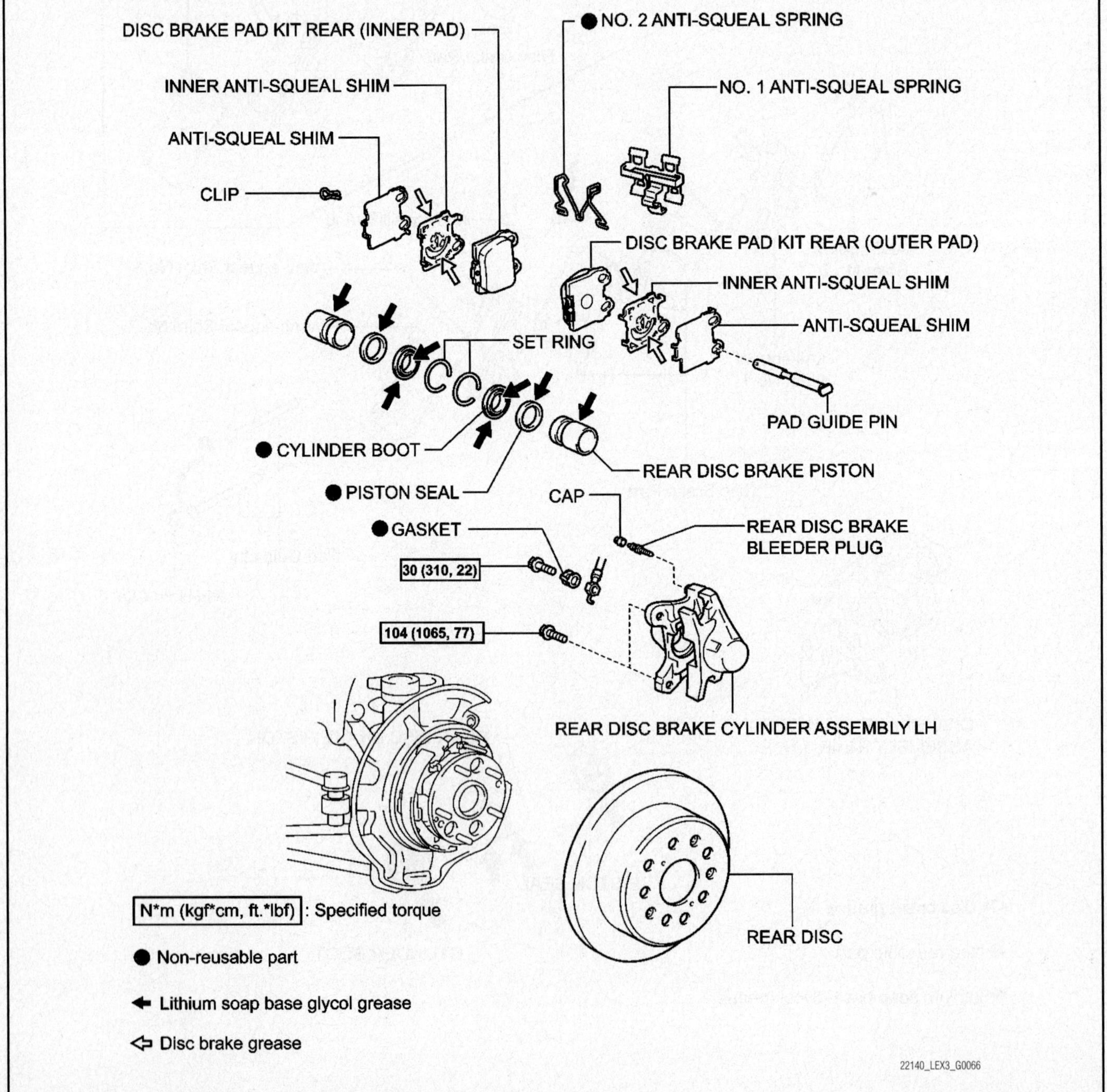

Fig. 7 Rear disc brake pads and components—SC 430

2. Remove or disconnect the following:
- Rear wheels
- Brake line at the caliper, then plug it
- Mounting bolts and the caliper assembly

To install:

3. Install the caliper on the torque plate with the 2 installation bolts. Tighten the mounting bolts to 77 ft. lbs. (104 Nm).
4. Connect the brake line with 2 new gaskets and tighten the union to 22 ft. lbs. (30 Nm).
5. Install the rear wheel. Tighten to 76 ft. lbs. (103 Nm).
6. Install the ornament with the 10 bolts. Tighten to 15 ft. lbs. (20 Nm).
7. Road test vehicle.
8. Refill the reservoir as necessary and bleed the brake system.
9. Road text vehicle and burnish brake linings.

DISC BRAKE PADS

REMOVAL & INSTALLATION

GS 300

See Figure 8.

1. Before servicing the vehicle, refer to the precautions in the beginning of this section.
2. Rear wheels.
3. Remove the pin hold clip and disengage the engaged parts of anti-squeal spring.
4. Remove the 2 pad guide pins and anti-squeal spring.

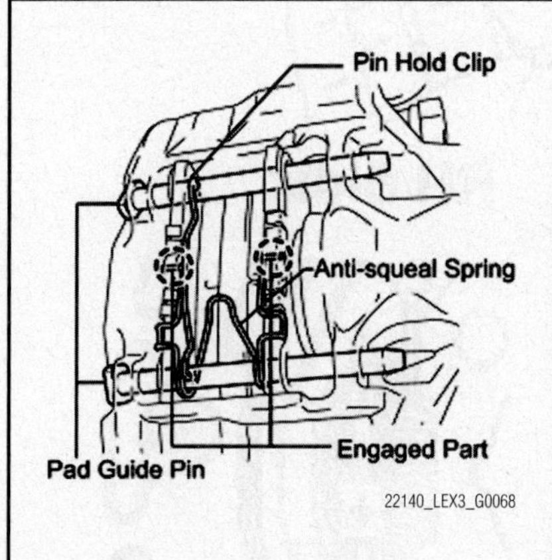

Fig. 8 Remove the pin hold clip

5. Remove the 2 brake pads with the anti-squeal shims.
6. Remove the 2 anti-squeal shims from each pad.
7. Remove the union bolt and gasket from the rear disc brake cylinder, and then disconnect the flexible hose.
8. Remove the cylinder slide pin and then tilt the disc brake cylinder toward the rear of the vehicle.
9. Remove the 2 bolts and rear disc brake cylinder together with the caliper support bracket.
10. Remove the 2 No. 1 caliper plate from the caliper support bracket.

To install:

11. Install 2 new caliper plate No.1 to the caliper support bracket.
12. Install the cylinder assembly rear together with the caliper support bracket with the 2 bolts.
13. Tighten the support bolts to 40 ft. lbs. (54 Nm).
14. Install the cylinder assembly rear with the cylinder slide pin. Tighten to 18 ft. lbs. (25 Nm).
15. Apply disc brake grease to both sides of the 2 anti-squeal shims
16. Install the 4 anti-squeal shims to each of the 2 brake pads.
17. Apply disc brake grease to both sides of the 2 anti-squeal shim No.1.
18. Install the 4 anti-squeal shims to each of the 2 brake pads.
19. Install the 2 brake pads to the cylinder assembly rear.
20. Install the 2 pad guide pins, anti-squeal spring and pin hold clip as shown in the illustration.
21. Connect the flexible hose with the union bolt and a new gasket. Tighten bolt to 22 ft. lbs. (30 Nm).

➡Install the flexible hose lock securely in the lock hole in the disc brake cylinder assembly rear.

22. With ECB brake system disable the 2 ABS motor relays from No.3 block. before bleeding the system. Install scanner to verify system is off.
23. Bleed the brake system of any air present.
24. Install relays if previously removed.
25. Check and clear DTC with scanner.
26. Check for fluid leaks.
27. Install the rear wheels and tighten to 76 ft. lbs. (103 Nm).
28. Road test vehicle.

SC 430

See Figure 9.

1. Before servicing the vehicle, refer to the precautions in the beginning of this section.
2. Remove or disconnect the following:
- Rear wheel
- Anti-squeal springs
- Clip and guide pin
- Disc pads
- 4 anti-squeal shims from each pad

To install:

3. Install or connect the following:
- Apply disc brake grease to both sides of the inner anti-squeal shims
- Install 2 shims on each pad

➡Make sure that the arrows on the anti-squeal shims face the direction of wheel rotation.

4. Using a suitable tool, compress the piston carefully in the cylinder bores
- Install pads
- Anti-squeal springs
5. Install the rear wheel. Tighten to 76 ft. lbs. (103 Nm).
6. Install the ornament with the 10 bolts. Tighten to 15 ft. lbs. (20 Nm).
7. Road test vehicle.

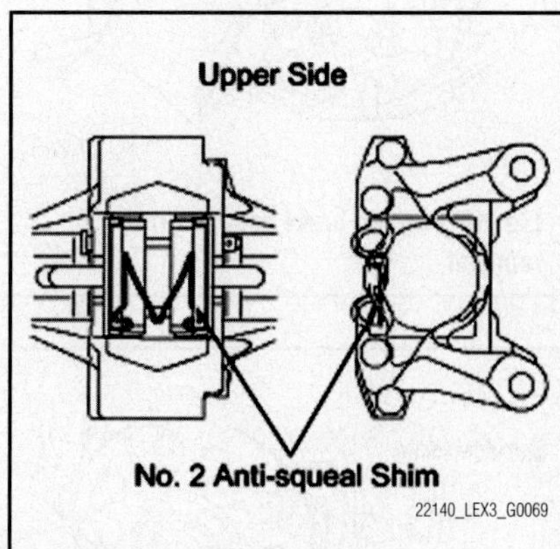

Fig. 9 Rear pads, spring and anti-squeal shims

PARKING BRAKE SHOES

REMOVAL & INSTALLATION

GS 300

See Figures 10 through 13.

1. Remove the rear wheels.
2. Remove the 2 bolts and separate the rear disc brake caliper assembly.

➡Hang the caliper with wire or equivalent.

3. Remove the caliper mounting bolts.
4. Release the parking brake.
5. Put matchmarks on the rear disc and rear axle hub.
6. Remove the rear disc.

➡If the disc cannot be removed easily, turn the shoe adjuster until the wheel turns smoothly.

7. Using SST 09703-30011, remove both parking brake shoe return springs.
8. Slide the parking brake shoe, and remove the parking brake shoe adjusting screw set.
9. Using SST 09718-00010, remove the shoe hold down spring No. 1 cup, compression No.1 spring and shoe hold down spring No.1 pin.
10. Remove the parking brake No. 1 shoe assembly.
11. Repeat for No. 2 shoe.
12. Remove the parking brake shoe lever.

To install:

13. Apply a thin layer of high temperature grease to the area where the parking brake plate contacts the parking brake shoe
14. Apply a thin layer of high temperature grease to the area where the parking brake shoe lever contacts the parking brake anchor block.
15. Install the parking brake shoe lever to the parking brake cable assembly.

➡Take care to install the correct parking brake shoe lever because the direction of the pin is different between left and right.

16. Using SST, install the parking brake No.2 shoe assembly with the shoe hold down spring No. 1 cup, compression No. 1 spring and shoe hold down spring No. 1 pin.
17. Repeat for No. 2 shoe.
18. Apply high temperature grease to the thread and all joining areas of the parking brake shoe adjusting screw set.
19. Install the parking brake shoe adjusting screw set.
20. Install the both parking brake shoe return springs.
21. Make sure that all the parts are installed properly.

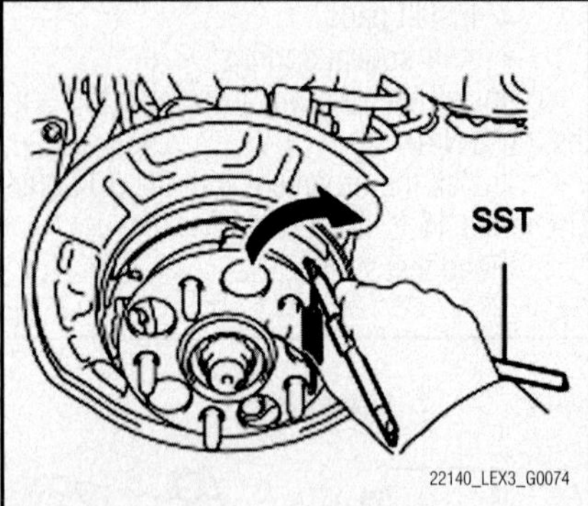

Fig. 10 Parking brake return spring removal

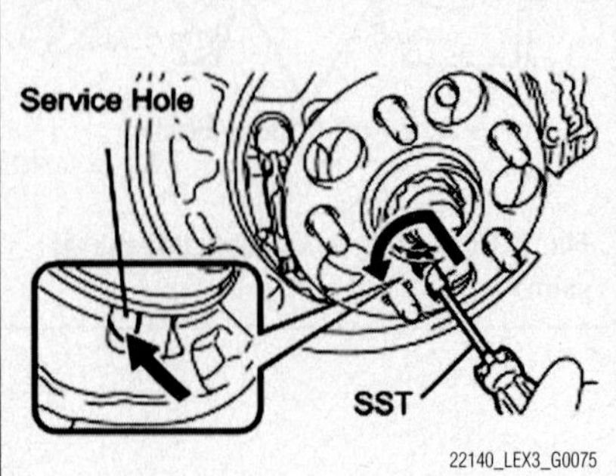

Fig. 11 Removal of parking brake hold down spring

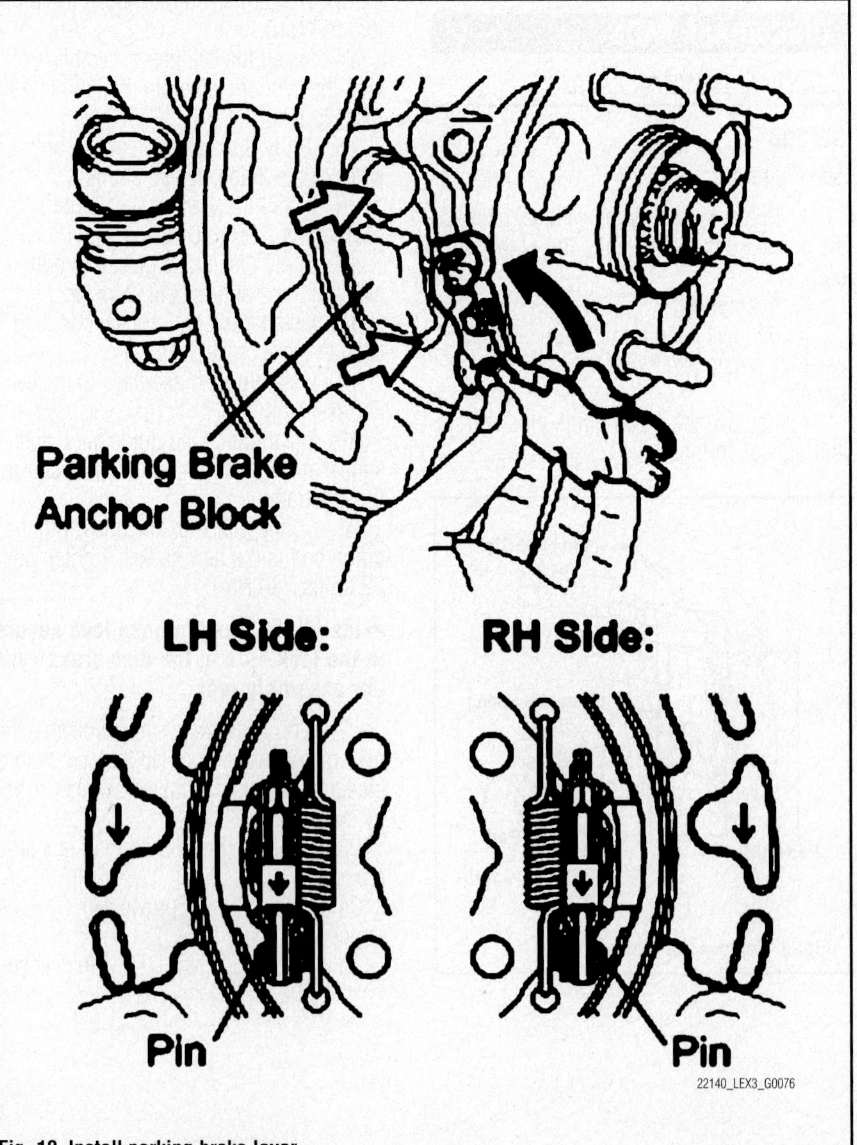

Fig. 12 Install parking brake lever

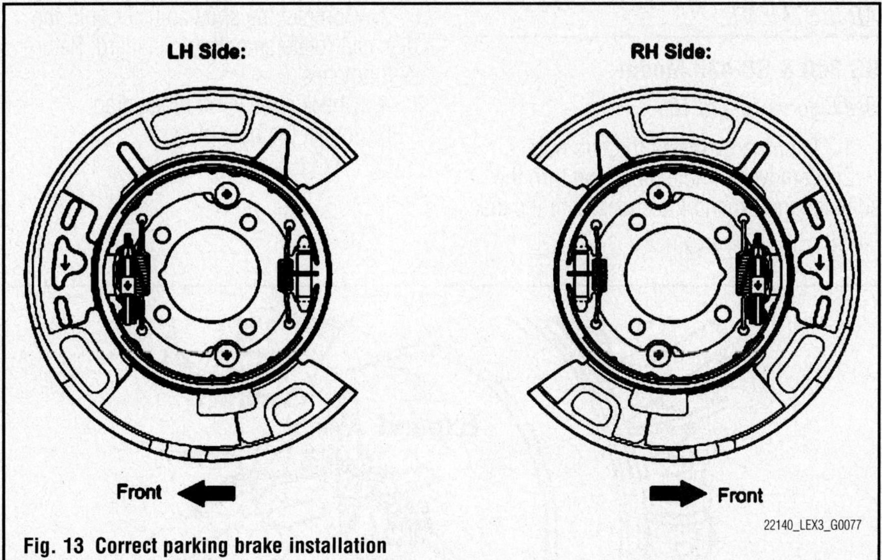

Fig. 13 Correct parking brake installation

22. Align the matchmarks on the rear disc and rear axle hub.
23. Adjust parking brake clearance.
24. Install rear brake caliper and tighten mounting bolts to 40 ft. lbs. (54 Nm).
25. Install the rear wheel. Tighten to 76 ft. lbs. (103 Nm).
26. Inspect parking brake travel and adjust as needed.

SC 430

See Figures 14 through 16.

1. Remove the rear wheels.
2. Remove the 2 bolts and separate the rear disc brake caliper assembly.

➡Hang the caliper with wire or equivalent.

3. Remove the caliper mounting bolts.
4. Release the parking brake.
5. Put matchmarks on the rear disc and rear axle hub.
6. Remove the rear disc.

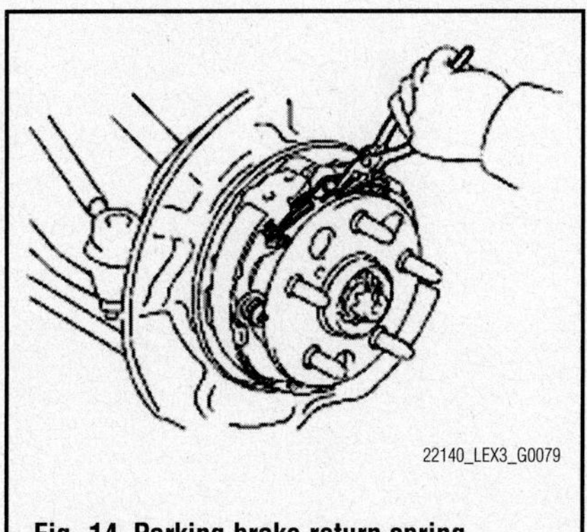

Fig. 14 Parking brake return spring removal

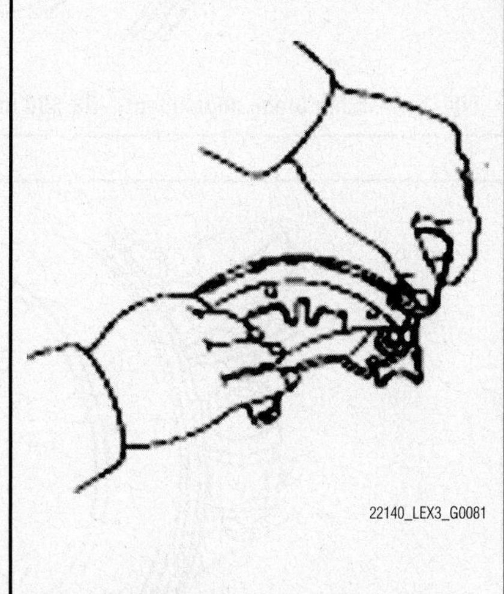

Fig. 15 Remove the parking brake lever

➡If the disc cannot be removed easily, turn the shoe adjuster until the wheel turns smoothly.

7. Using needle nose pliers remove both parking brake shoe return springs.
8. Remove the strut.
9. Remove the screw set.
10. Slide the shoe No. 1 to remove the 2 cups, shoe hold-down spring, pin and shoe.
11. Repeat for No. 2 shoe.
12. Disconnect the No. 3 parking brake cable from the parking brake shoe lever.
13. Using a screwdriver, remove the C-washer and shim.
14. Remove the lever.

To install:

15. Apply high temperature grease to the contact surface of the shoe and backing plate.
16. Install parking brake shoe lever.
17. Install the correct size shim.
18. Install a new C-washer.
19. Connect the No. 3 parking brake cable to the parking brake shoe lever.
20. Install the No. 2 shoe with the 2 cups, shoe hold-down spring and pin.
21. Repeat for No. 1 shoe.
22. Apply high temperature grease to the adjusting bolt.
23. Install the parking brake adjusting screw.
24. Using needle-nose pliers, install the tension spring.
25. Apply high temperature grease to the contact surface of the strut and shoe return spring.
26. Install the strut.

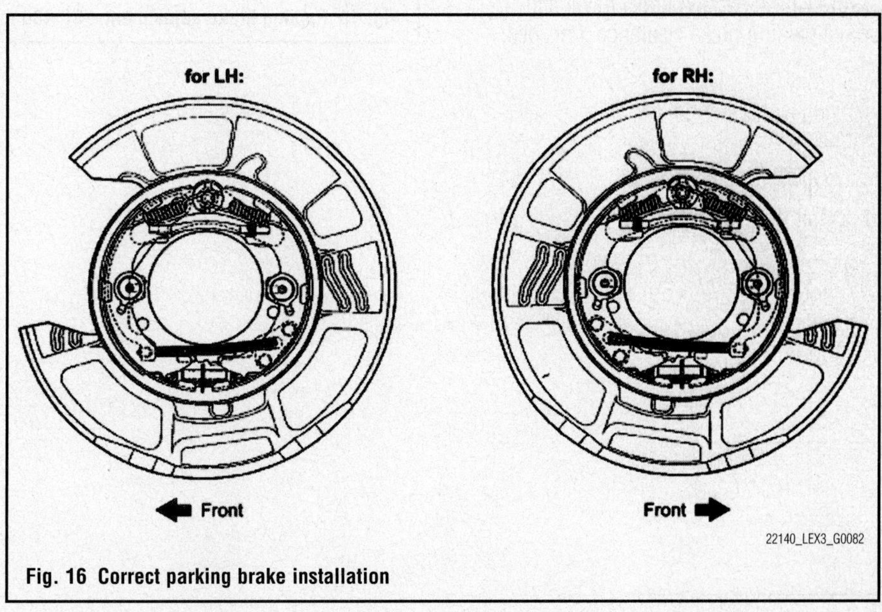

Fig. 16 Correct parking brake installation

27. Using needle-nose pliers, install the 2 shoe return springs.
28. Check that each part is installed properly

➡There should be no oil or grease adhering to the friction surface of the shoe lining and disc.

29. Align the matchmarks and install the disc.
30. Temporarily install the hub nuts.
31. Remove the hole plug, and then turn the adjuster to expand the shoe adjuster until the disc locks.
32. Contract the shoe adjuster until the disc can rotate smoothly. Standard : Return 8 notches.
33. Check that the shoe has no brake drag.
34. Install the hole plug.
35. Depress the parking brake pedal.
36. Drive the vehicle at approximately 31 mph (50 km/h) for approximately 0.25 miles.
37. Repeat the previous 2 steps 2 or 3 times.
38. Install rear brake caliper and tighten mounting bolts to 77 ft. lbs. (104 Nm).
39. Install the rear wheel. Tighten to 76 ft. lbs. (103 Nm).
40. Install the ornament with the 10 bolts. Tighten to 15 ft. lbs. (20 Nm).
41. Inspect parking brake travel and adjust as needed.
42. Depress the parking brake pedal.
43. Drive the vehicle at approximately 31 mph (50 km/h) for approximately 0.25 miles.
44. Repeat the previous 2 steps 2 or 3 times.
45. Check parking brake travel and adjust parking brake clearance if needed.

ADJUSTMENT

GS 300 & SC 430 Model

See Figures 17 and 18.

1. Temporarily install the hub nuts.
2. Remove the hole plug, and turn the adjuster and expand the shoes until the disc locks.
3. Contract the shoe adjuster until the disc can rotate smoothly. Standard: Return 7–8 notches
4. Check shoe is no brake drag.
5. Install the hole plug.

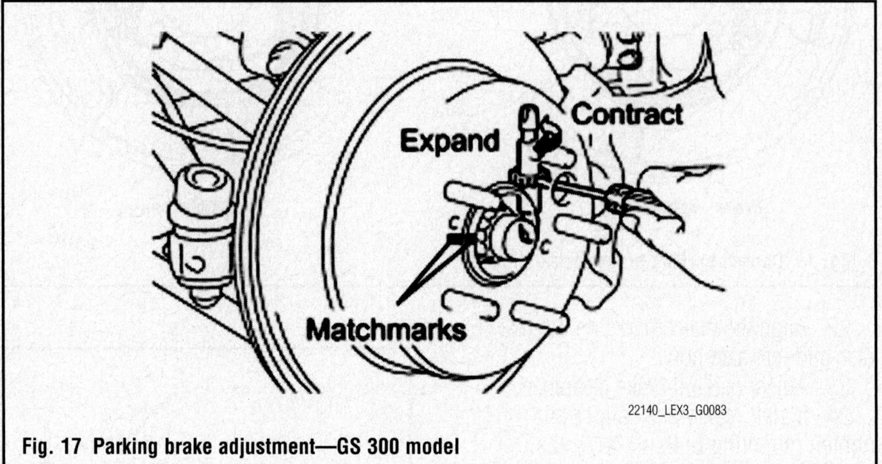

Fig. 17 Parking brake adjustment—GS 300 model

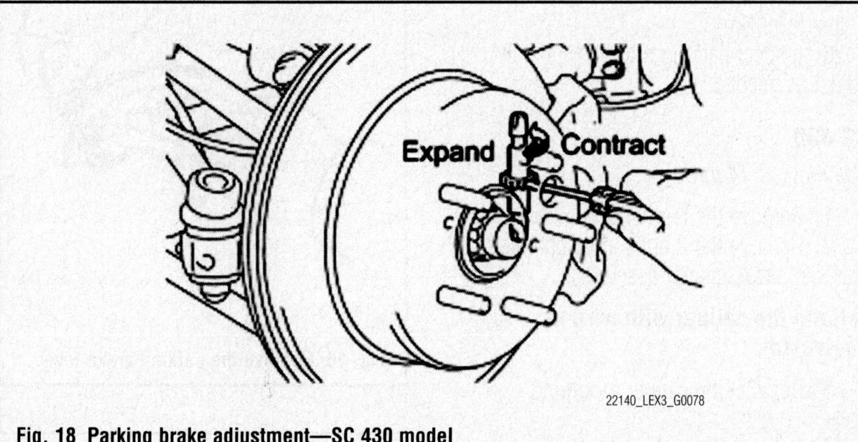

Fig. 18 Parking brake adjustment—SC 430 model

GENERAL INFORMATION

⁂ CAUTION

Some vehicles are equipped with an air bag system. The system must be disarmed before performing service on, or around, system components, the steering column, instrument panel components, wiring and sensors. Failure to follow the safety precautions and the disarming procedure could result in accidental air bag deployment, possible injury and unnecessary system repairs.

SERVICE PRECAUTIONS

Disconnect and isolate the battery negative cable before beginning any airbag system component diagnosis, testing, removal, or installation procedures. Allow system capacitor to discharge for two minutes before beginning any component service. This will disable the airbag system. Failure to disable the airbag system may result in accidental airbag deployment, personal injury, or death.

Do not place an intact undeployed airbag face down on a solid surface. The airbag will propel into the air if accidentally deployed and may result in personal injury or death.

When carrying or handling an undeployed airbag, the trim side (face) of the airbag should be pointing towards the body to minimize possibility of injury if accidental deployment occurs. Failure to do this may result in personal injury or death.

Replace airbag system components with OEM replacement parts. Substitute parts may appear interchangeable, but internal differences may result in inferior occupant protection. Failure to do so may result in occupant personal injury or death.

Wear safety glasses, rubber gloves, and long sleeved clothing when cleaning powder residue from vehicle after an airbag deployment. Powder residue emitted from a deployed airbag can cause skin irritation. Flush affected area with cool water if irritation is experienced. If nasal or throat irritation is experienced, exit the vehicle for fresh air until the irritation ceases. If irritation continues, see a physician.

Do not use a replacement airbag that is not in the original packaging. This may result in improper deployment, personal injury, or death.

The factory installed fasteners, screws and bolts used to fasten airbag components have a special coating and are specifically designed for the airbag system. Do not use substitute fasteners. Use only original equipment fasteners listed in the parts catalog when fastener replacement is required.

During, and following, any child restraint anchor service, due to impact event or vehicle repair, carefully inspect all mounting hardware, tether straps, and anchors for proper installation, operation, or damage. If a child restraint anchor is found damaged in any way, the anchor must be replaced. Failure to do this may result in personal injury or death.

Deployed and non-deployed airbags may or may not have live pyrotechnic material within the airbag inflator.

Do not dispose of driver/passenger/curtain airbags or seat belt tensioners unless you are sure of complete deployment. Refer to the Hazardous Substance Control System for proper disposal.

Dispose of deployed airbags and tensioners consistent with state, provincial, local, and federal regulations.

After any airbag component testing or service, do not connect the battery negative cable. Personal injury or death may result if the system test is not performed first.

If the vehicle is equipped with the Occupant Classification System (OCS), do not connect the battery negative cable before performing the OCS Verification Test using the scan tool and the appropriate diagnostic information. Personal injury or death may result if the system test is not performed properly.

Never replace both the Occupant Restraint Controller (ORC) and the Occupant Classification Module (OCM) at the same time. If both require replacement, replace one, then perform the Airbag System test before replacing the other.

Both the ORC and the OCM store Occupant Classification System (OCS) calibration data, which they transfer to one another when one of them is replaced. If both are replaced at the same time, an irreversible fault will be set in both modules and the OCS may malfunction and cause personal injury or death.

If equipped with OCS, the Seat Weight Sensor is a sensitive, calibrated unit and must be handled carefully. Do not drop or handle roughly. If dropped or damaged, replace with another sensor. Failure to do so may result in occupant injury or death.

If equipped with OCS, the front passenger seat must be handled carefully as well. When removing the seat, be careful when setting on floor not to drop. If dropped, the sensor may be inoperative, could result in occupant injury, or possibly death.

If equipped with OCS, when the passenger front seat is on the floor, no one should sit in the front passenger seat. This uneven force may damage the sensing ability of the seat weight sensors. If sat on and damaged, the sensor may be inoperative, could result in occupant injury, or possibly death.

DISARMING THE SYSTEM

To avoid personal injury when working on vehicles equipped with an air bag, the negative battery cable must be disconnected and at least 90 seconds must elapse before working on the system. Failure to do so may result in deployment of the air bag.

ARMING THE SYSTEM

To rearm the air bag system, simply reconnect the battery cable(s).

CLOCKSPRING CENTERING

See Figure 19.

1. Center the clockspring spiral cable as follows:
   - Check that the ignition switch is OFF.
   - Check that the battery negative terminal is disconnected. Do not start the operation for 90 seconds after removing the terminal.
   - Turn the cable counterclockwise by hand until it becomes harder to turn.
   - Then rotate the cable clockwise about 2.5 turns to align the marks. The cable will rotate about 2.5 turns to both left and right of the center.

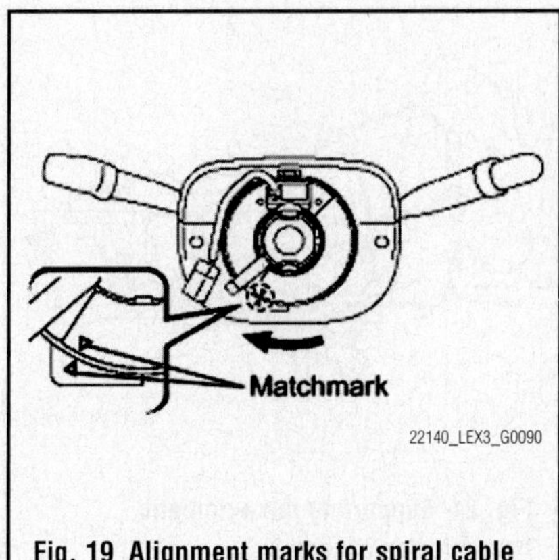

Fig. 19 Alignment marks for spiral cable

DRIVETRAIN

AUTOMATIC TRANSMISSION ASSEMBLY

REMOVAL & INSTALLATION

GS 300 (A760H) & (A960E) Model Transmissions

See Figures 20 through 23.

1. Before servicing the vehicle, refer to the precautions in the beginning of this section.
2. Disconnect the negative battery cable. Wait at least 90 seconds before performing any other work.
3. Drain the engine coolant.
4. Drain transmission fluid.
5. Remove the front exhaust pipe assembly.
6. Remove the left manifold assembly.
7. Remove the starter motor.
8. Remove the front propeller shaft and rear shaft center bearing assembly.
9. Turn the crankshaft to gain access to each bolt.
10. Hold the crankshaft pulley nut with a wrench and remove the 6 torque converter clutch setting bolts.
11. Remove the 2 bolts and exhaust pipe No.1 support bracket sub-assembly from the automatic transmission assembly.
12. Set the shift lever to the neutral position.
13. Remove the nut and separate the floor shift gear shifting rod sub-assembly.
14. Support the automatic transmission assembly with a transmission jack.
15. Remove the 4 bolts and separate the engine rear mounting member from the body.
16. Tilt down the automatic transmission assembly.
17. Disconnect the neutral start switch connector, transmission wire connector and 2 transmission revolution sensor connectors.
18. Remove the 6 bolts and separate the wire harness clamps from the automatic transmission assembly.
19. Remove the bolt and water by-pass hose clamp.
20. Separate the 2 water by-pass hoses from the transmission oil cooler.
21. Remove the 9 bolts and transmission assembly.
22. Remove the 3 bolts and engine rear mounting member from the automatic transmission assembly.
23. Remove the 3 bolts and engine mounting insulator rear No.1 from the automatic transmission assembly.
24. Remove the 2 bolts and dynamic damper from the automatic transmission assembly.
25. Remove the nut and transmission control shaft lever LH from the neutral start switch assembly.
26. Put matchmarks on the torque converter and transmission case. Remove torque converter.
27. Carefully remove the transmission assembly.

To install:

28. Aligning the matchmarks on the transmission case and torque converter, engage the splines of the input shaft and turbine runner.
29. Engage the splines of the stator shaft and stator while the torque converter clutch.

➡Turn the torque converter clutch approximately 180°.

30. Turn the torque converter clutch and align the matchmarks on the torque converter clutch and transmission case to engage the key of the oil pump drive gear into the slot on the torque converter clutch.
31. Install the transmission control shaft to the neutral start switch assembly with the nut. Tighten to 12 ft. lbs. (16 Nm).
32. Install the engine mounting insulator rear No.1 to the automatic transmission assembly with the 4 bolts. Tighten the bolts to 9 ft. lbs. (12 Nm).
33. Install the automatic transmission assembly to the engine with the 9 bolts.

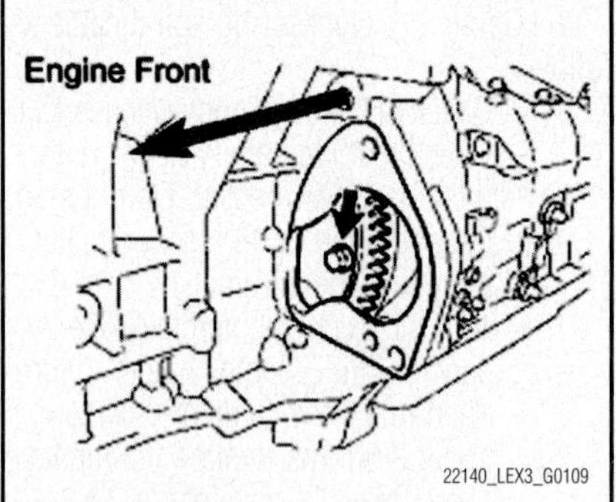

Fig. 20 Torque converter clutch setting bolt view

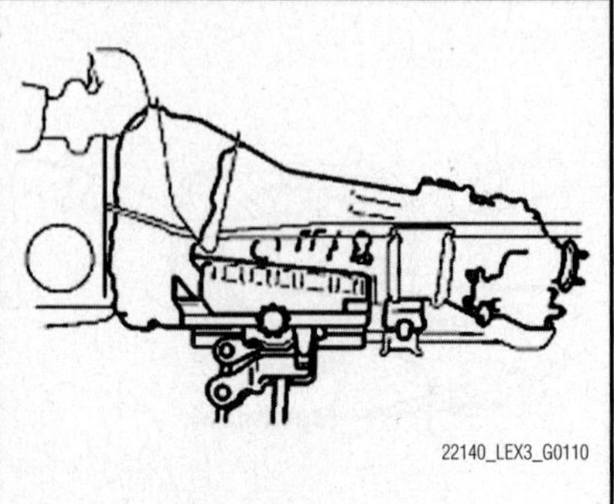

Fig. 21 Supporting the automatic transmission assembly

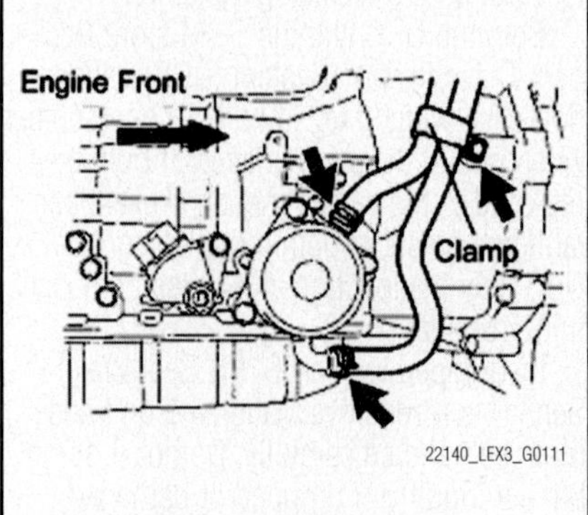

Fig. 22 Water by-pass hoses shown

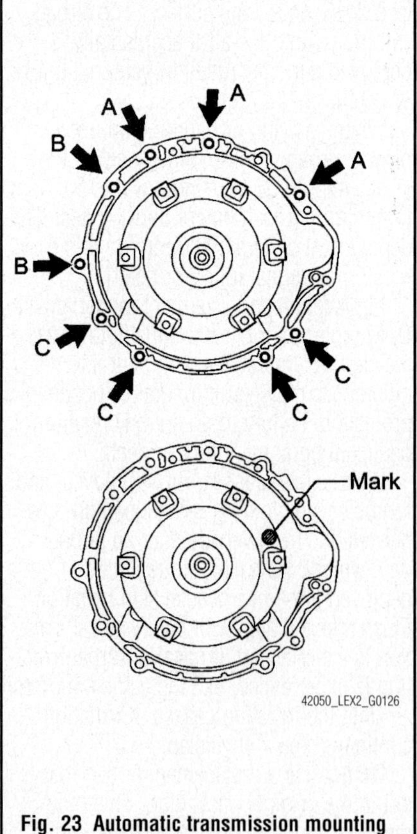

Fig. 23 Automatic transmission mounting bolts—2006 GS 300

34. Torque bolts A and B to 52 ft. lbs. (71 Nm) and bolt C to 27 ft. lbs. (37 Nm)

➡ **Make sure that the mark is as shown in the illustration.**

35. Install the 2 water by-pass hoses to the transmission assembly.
36. Install the water by-pass hose clamp with the bolt. Tighten bolt to 15 ft. lbs. (20 Nm).
37. Install the wire harness clamps to the automatic transmission assembly with the 2 bolts. Tighten to 7 ft. lbs. (10 Nm).
38. Tilt down the automatic transmission assembly.
39. Connect the neutral start switch connector, transmission wire connector and 2 transmission revolution sensor connectors.
40. Install the engine rear mounting member to the automatic transmission assembly with the 4 nuts.
41. 10 ft. lbs. (13 Nm).
42. Install the engine rear mounting member to the body with the 4 bolts. Tighten the bolts to 19 ft. lbs. (26 Nm).
43. Temporarily tighten the floor shift gear shifting rod sub-assembly with the nut.
44. Install the exhaust pipe No.1 support bracket sub-assembly to the automatic transmission assembly with the 2 bolts. Tighten to 32 ft. lbs.(43 Nm).
45. Install the 6 torque converter clutch setting bolts. Tighten the converter bolts to 30 ft. lbs. (41 Nm).

➡ **First install the black colored bolt and then the remaining 5 bolts.**

46. Install the starter motor and tighten the 2 mounting bolts to 43 ft. lbs. (58 Nm).
47. Install left exhaust manifold.
48. Install front propeller shaft and rear center bearing assembly. Tighten the flange mounting bolts to 54 ft. lbs. (74 Nm).
49. Connect the negative battery cable.
50. Adjust the shifter lever position.
51. Install the front exhaust pipe assembly. Tighten the front mounting bolts to 46 ft. lbs. (62 Nm).
52. Install the front exhaust pipe to the tail exhaust pipe with the 2 compression springs and 2 bolts. Tighten the 2 rear bolts to 32 ft. lbs. (43 Nm).
53. Install the O2 sensor.
54. Install engine under cover air guide bracket.

Add engine coolant.

55. Add transmission fluid.
56. Check for any leaks.
57. adjust fluid levels as needed.
58. Install engine under cover.
59. Reset memory.

SC 430 (A761E) Model Transmission

See Figures 24 through 28.

1. Before servicing the vehicle, refer to the precautions in the beginning of this section.
2. Disconnect the negative battery cable.
3. Wait at least 90 seconds after disconnecting the cable from the negative (-) battery terminal to prevent airbag and seat belt pre-tensioner activation.
4. Remove the engine cover.
5. Disconnect heated O2 sensor 2.
6. Remove the front floor center brace.
7. Remove the front and center exhaust assembly.
8. Remove the exhaust pipe support bracket.
9. Remove the catalyst converter.
10. Remove the 4 bolts and the front heat insulator.
11. Remove the 4 bolts and the parking brake heat insulator.
12. Remove the propeller with center bearing shaft assembly.
13. Remove the drain plug and gasket, and drain the ATF.

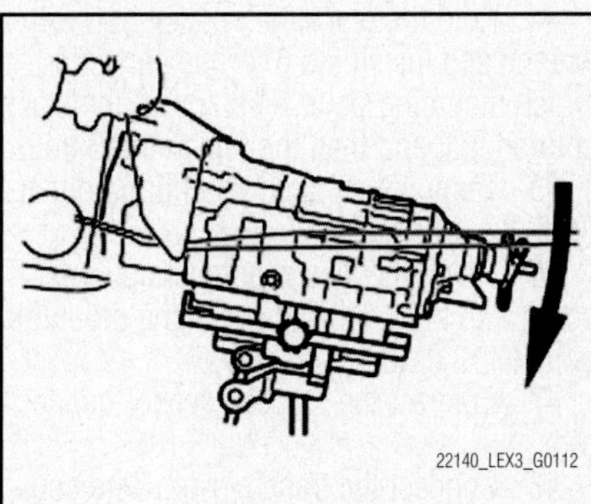

Fig. 24 Tilt down the automatic transmission

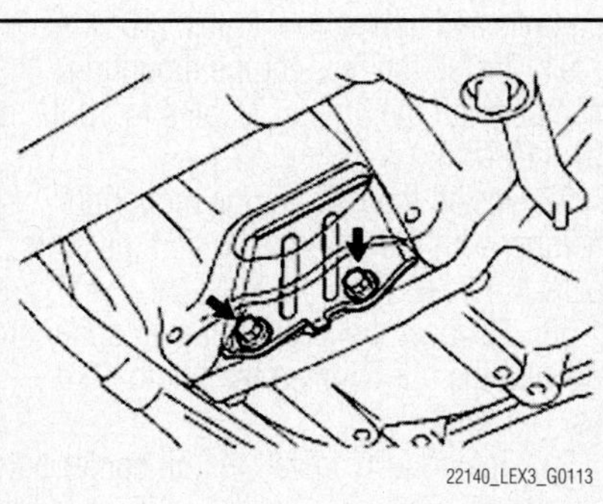

Fig. 25 Flywheel housing under cover and bolts

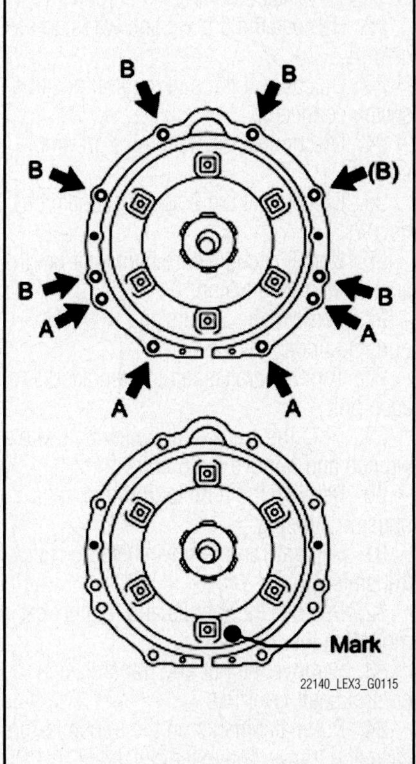

Fig. 27 Automatic transmission mounting bolts—SC 430

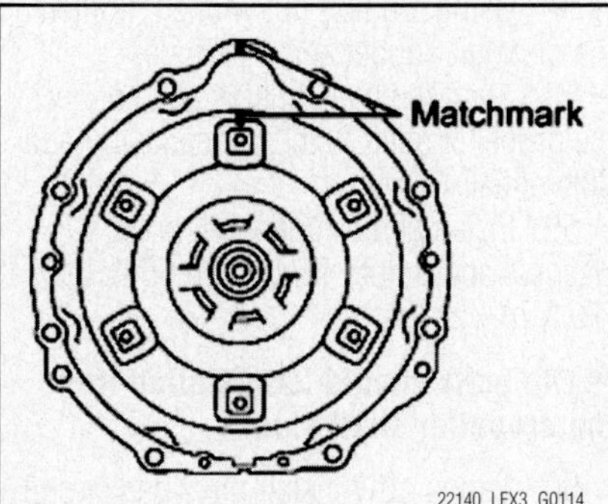

Fig. 26 Matchmarks on the transmission case and torque converter

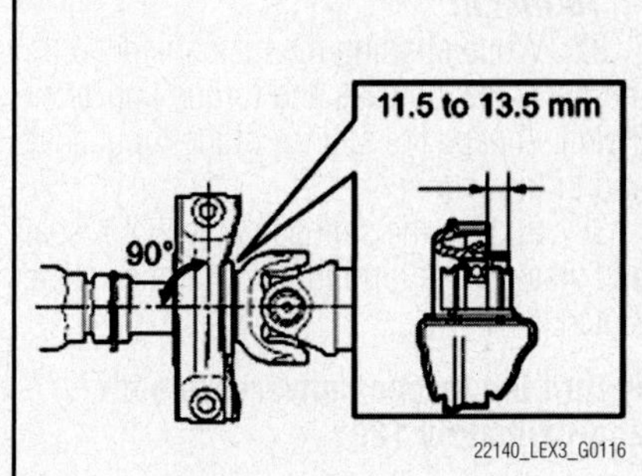

Fig. 28 Adjust the center bearing assembly as shown

14. Remove the 3 bolts and 3 clamps.

15. Using a union nut wrench, disconnect the oil cooler inlet tube.

16. Using a union nut wrench, disconnect the oil cooler outlet tube.

17. Remove the nut and disconnect the floor shift gear shifting rod.

18. Support the automatic transmission with a transmission jack.

19. Remove the 4 bolts and rear engine mounting member.

20. Remove the 4 nuts and rear engine mounting member.

21. Slowly tilt the transmission downward so that the wire harness can be disconnected. Take care so that the cooling fan does not come into contact with the fan shroud.

22. Remove the 5 bolts and wire harness clamps.

23. Disconnect the park/neutral position switch connector.

24. Disconnect the transmission wire connector.

25. Disconnect the 2 speed sensor connectors.

26. Disconnect the wire harness from the automatic transmission.

27. Remove the 2 bolts and flywheel housing under cover.

28. Turn the crankshaft to gain access to each bolt.

29. Hold the crankshaft pulley bolt with a wrench and remove the 6 bolts.

30. Remove the 9 bolts from the transmission housing.

31. Separate and remove the automatic transmission.

32. Remove the 4 bolts and engine rear mounting insulator.

33. Remove the nut and transmission control shaft lever RH.

34. Put matchmarks on the transmission case and torque converter clutch.

35. Remove the torque converter clutch from the automatic transmission.

36. Carefully remove the transmission assembly.

To install:

37. While aligning the matchmarks on the transmission case and torque converter clutch, engage the splines of the input shaft and turbine runner.

38. Engage the splines of the stator shaft and stator while turning the torque converter clutch.

➡Turn the torque converter clutch approximately 180°.

39. Turn the torque converter clutch and align the matchmarks on the torque converter clutch and transmission case to engage the key of the oil pump drive gear into the slot on the torque converter clutch.

40. Install the transmission control shaft lever RH with the nut. Tighten the nut to 12 ft. lbs (16 Nm).

41. Install the mounting insulator with the 4 bolts. Tighten the bolts to 9 ft. lbs. (12 Nm).

42. Install the automatic transmission assembly to the engine with the 9 bolts.

43. Torque bolts A and B to 52 ft. lbs. (71 Nm) and bolt C to 27 ft. lbs. (37 Nm)

➡Make sure that the mark is as shown in the illustration.

44. Hold the crankshaft pulley bolt with a wrench and install the 6 torque converter clutch mounting bolts. First install the black colored bolt and then the remaining 5 bolts.

45. Tighten the clutch mounting bolts to 35 ft. lbs. (48 Nm).

46. Install the fly wheel housing under cover with the 2 bolts. Tighten the mounting bolts to 13 ft. lbs. (18 Nm).

47. Connect the 2 speed sensor connectors.

48. Connect the transmission wire connector.

49. Connect the park/neutral position switch connector.

50. Install the wire harness clamp mounting bolts and tighten to 7 ft. lbs. (10 Nm).

51. Install the rear engine mounting member with the 4 nuts. Tighten to 10 ft. lbs. (13 Nm).

52. Install the rear engine mounting member with the 4 bolts. Tighten the bolts to 19 ft. lbs. (26 Nm).

53. Connect the floor shift gear shifting rod with the nut. Tighten the nut to 10 ft. lbs. (13 Nm).

54. Temporarily install the oil cooler outlet tube.

55. Temporarily install the oil cooler inlet tube.

56. Install the 3 clamps with the 3 bolts and tighten to 44 inch. (5.0 Nm).

57. Using a union nut wrench, tighten the oil cooler outlet and inlet tube.

58. Align the matchmarks and connect the propeller shaft to the transmission and differential.

59. Install and torque the 6 bolts, 6 washers and 6 nuts. Tighten to 58 ft. lbs. (79 Nm).

➡The bolts should be installed from the propeller shaft side.

60. Temporarily install the 2 center support bearing set bolts with the adjusting washers. Use the adjusting washers which were removed.

61. Adjust the center bearing assembly in relation to the front and rear of the vehicle. Make sure that the distance between the center bearing housing's rear end and cushion's rear end is 0.492 +-0.04 inch. (12.5 +-1 mm) as shown in the illustration.

62. Check that the center line of the bracket is at right angles to the shaft axial direction.

63. Tighten the 2 center support bearing set bolts to 27 ft. lbs. (37 Nm).

➡When performing operations which involve the removal and installation of the propeller shaft, always check the joint angle. Make adjustments if necessary.

64. Install parking brake cable insulator and tighten retaining bolts to 48 inch. (5.8 Nm).

65. Install the front floor insulator and tighten retaining bolts to 48 inch. (5.8 Nm).

66. Install a new gasket and the center exhaust pipe with the 2 nuts, 2 bolts and 2 rings. Tighten to 32 ft. lbs. (43 Nm).

67. Install 2 new gaskets and the front exhaust pipe with the 4 nuts and 4 bolts. Tighten to 32 ft. lbs. (43 Nm).

68. Reconnect the O2 sensor.

69. Install the floor center brace with the 4 bolts and tighten to 15 ft. lbs. (20 Nm).

TRANSFER CASE ASSEMBLY

REMOVAL & INSTALLATION

GS 300 With AWD

See Figure 29.

1. Before servicing the vehicle, refer to the precautions in the beginning of this section.

2. Disconnect the negative battery cable. Wait at least 90 seconds before performing any other work.

3. Remove or disconnect the following:
   - Automatic transmission assembly
   - Transfer case oil pan

4. Remove the transfer valve body assembly as follows:

   a. Remove the 4 bolts and the transfer valve body assembly

   b. Disconnect the transfer wire connector from the transfer control solenoid.

   c. Remove the bolt and the transfer wire.
   - Transfer extension housing using SST 09950-40011
   - 8 bolts (in order), clamp and transfer case. Use a plastic hammer to tap the transfer case to remove it.
   - Transfer rear output shaft assembly

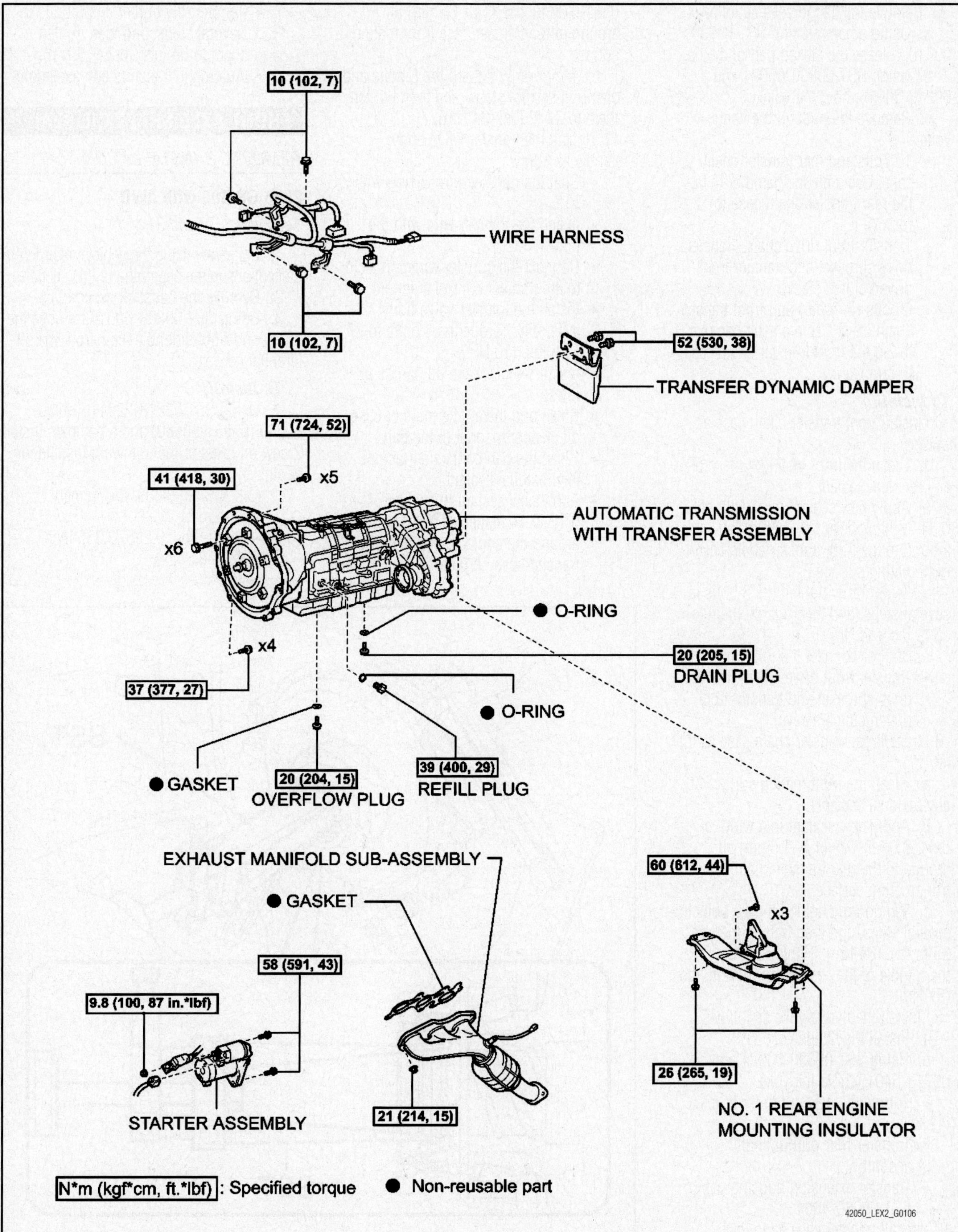

Fig. 29 Exploded view of the transfer case and related parts—2006 GS 300 with AWD

5. Remove the flange yoke as follows:
a. Using a hammer and SST 09930-00010, release the staked part of the nut.
b. Using SST 09330-00021 and 09213-54015, hold the yoke.
c. Remove the nut and the flange yoke.
- 10 bolts and rear transfer chain case. Use a plastic hammer to tap the rear transfer chain case to remove it.
- Transfer front drive chain, transfer drive sprocket and transfer front driven clutch sleeve
- 6 bolts (in order) and front transfer chain case. Use a plastic hammer to tap the front transfer chain case to remove it.

To install:

6. Install front transfer chain case as follows:
a. Clean the front mating surface of any residual sealant.
b. Apply new sealant in a width of 0.04–0.06 inches (1.0–1.5mm) all around to the front transfer chain case front mating surface.
c. Temporarily tighten the 6 bolts in several steps, and then tighten them to 25 ft. lbs. (34 Nm).
7. Install or connect the following:
- Transfer front drive chain, transfer drive sprocket and transfer front driven clutch sleeve
8. Install rear transfer chain case as follows:
a. Clean the front mating surface of any residual sealant.
b. Apply new sealant in a width of 0.04–0.06 inches (1.0–1.5mm) all around to the rear transfer chain case front mating surface.
c. Temporarily tighten the 10 bolts in several steps, and then tighten them to 25 ft. lbs. (34 Nm). Tighten the bolt in the middle of the case cover to 42 ft. lbs. (57 Nm).
9. Install the flange yoke as follows:
a. Install the flange yoke.
b. Using SST 09330-00021 and 09213-54015, hold the yoke.
c. Torque the new nut to 91 ft. lbs. (123 Nm).
- Transfer rear output shaft assembly
- Transfer extension housing using SST 09950-40011
10. Install transfer case as follows:
a. Clean the front mating surface of any residual sealant.
b. Apply new sealant in a width of 0.04–0.06 inches (1.0–1.5mm) all around to the transfer case front mating surface.
c. Temporarily tighten the 8 bolts and clamp in several steps, and then tighten them to 25 ft. lbs. (34 Nm).
11. Install the transfer valve body assembly as follows:
- Coat the transfer wire O-ring with ATF.
- Install the transfer wire with the bolt.
- Connect the transfer wire connector to the transfer control solenoid.
- Install the transfer valve body assembly. Torque the 4 bolts to 87 inch lbs. (10 Nm).
- Install transfer case oil pan as follows:
- Clean and install the transfer case oil cleaner magnet to the pan.
- Clean the pan contact surface of any residual sealant.
- Apply new sealant in a width of 0.12–0.14 inches (3.0–3.5mm) all around the contact surface of the transfer case oil pan.
- Temporarily tighten the 9 bolts in several steps, and then tighten them to 65 inch lbs. (7.5 Nm).
- Automatic transmission assembly.

FRONT AXLE SHAFT OIL SEAL

REMOVAL & INSTALLATION

2006 GS 300 with AWD

See Figures 30 and 31.

1. Before servicing the vehicle, refer to the precautions in the beginning of this section.
2. Remove the halfshaft assemblies.
3. Using SST 09308-00010, remove the 2 (left and right side) oil seals from the differential carrier.

To install:

4. Using SST 09613-22011, 09950-60010 (09951-00560) and a hammer, install 2 new oil seals to the following installation depth:
- Left Side: 0.039–0.079 inch (1–2mm)
- Right Side: 0.295–0.335 inch (7.5–8.5mm)

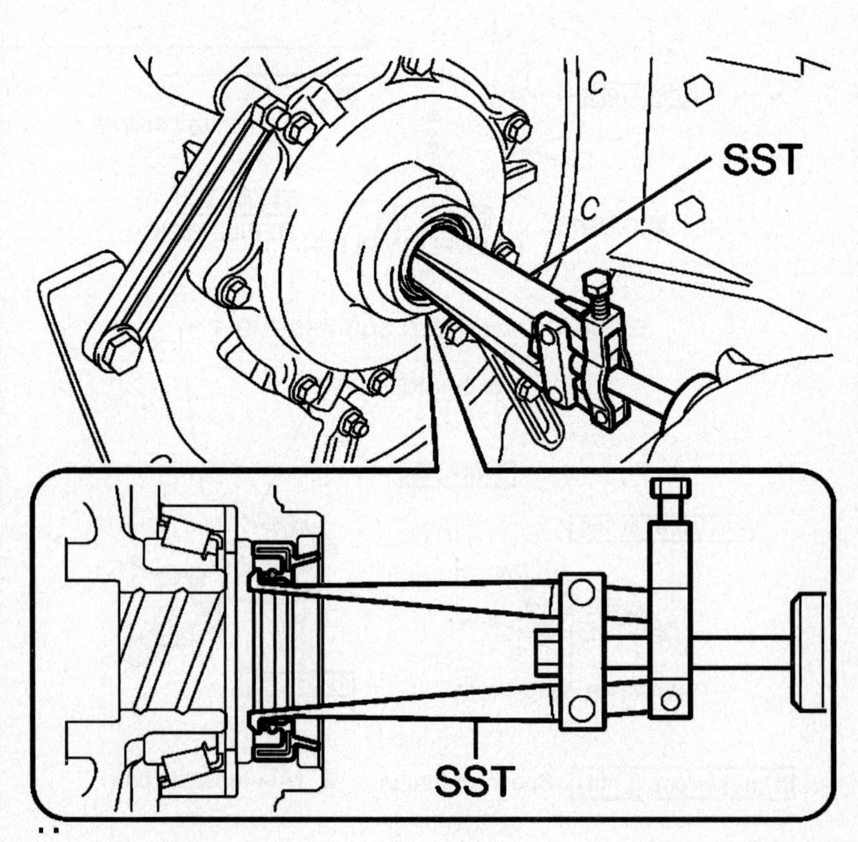

Fig. 30 Using SST 09308-00010 to remove the oil seals from the differential carrier.—AWD

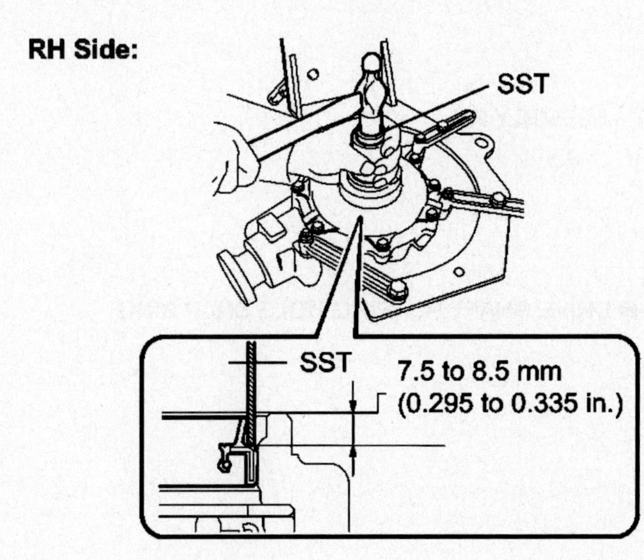

Fig. 31 Installation depth of the 2 new oil seals—AWD

5. Apply MP grease to the oil seal lips.
6. Install the halfshaft assemblies.

FRONT HALFSHAFT

REMOVAL & INSTALLATION

GS 300 with AWD

See Figure 32.

1. Before servicing the vehicle, refer to the precautions in the beginning of this section.
2. Disconnect the negative battery cable. Wait at least 90 seconds before performing any other work.
3. Remove the front wheel.
4. Separate the front speed sensor.
5. Separate the front disc brake caliper assembly.
6. Remove the front brake rotor.
7. Using SST 09930-00010 and a hammer, release the staked part of the front axle hub nut and remove. Release the staked part of the nut completely, otherwise the threads of the halfshaft may be damaged.
8. Separate the front lower ball joint assembly.
9. Separate the tie rod assembly.
10. Using a plastic hammer, separate the front axle assembly from the halfshaft assembly. Be careful not to damage the halfshaft boot, dust cover and oil seal

➡Be careful not to drop the halfshaft assembly.

11. Using SST 09520-01010, 09520-24010 (09520-32040), remove the left front halfshaft assembly.
12. Using water pump pliers, remove the bearing bracket hole snap ring.
13. Remove the bolt and right front halfshaft assembly from the halfshaft bearing bracket.
14. Inspect the halfshaft as follows:
    a. Check that there is no excessive play in the outboard joint.
    b. Check that the inboard joint slides smoothly in the thrust direction.
    c. Check that there is no excessive play in the radial directions of the inboard joint.
    d. Check the boot for damage.

To install:

15. Install the left front halfshaft assembly as follows:
16. Coat the spline of the inboard joint shaft assembly with gear oil.
17. Set the shaft snap ring with the opening side facing down.
18. Align the shaft splines and install the halfshaft assembly with a brass bar and hammer. Move the drive shaft assembly while keeping it level.

➡It is possible to determine if the inboard joint shaft is properly engaged (the shaft is in contact with the pinion shaft, and the snap ring is engaged in the pinion gear) based on the sound or feeling when the shaft is driven in.

19. Install the right front halfshaft assembly as follows:
20. Coat the spline of the inboard joint shaft assembly with gear oil.
21. Install the halfshaft assembly.
22. Using water pump pliers, install a new bearing bracket hole snap ring.
23. Install a new bolt and tighten it to 24 ft. lbs. (32 Nm).
24. Install the front halfshaft assembly to the front axle assembly. Be careful not to damage the halfshaft boot.
25. Install the tie rod assembly.
26. Install the front lower ball joint assembly.
27. Install the front brake rotor.
28. Install the front disc brake caliper assembly.
29. Using a socket wrench (30mm), install a new axle hub nut and tighten it to 217 ft. lbs. (294 Nm).
30. Using a chisel and hammer, stake the front axle hub nut.
31. Install the front speed sensor.
32. Install the front wheel.
33. Inspect and adjust the front wheel alignment.
34. Connect the negative battery cable.
35. Check ABS speed sensor signal.

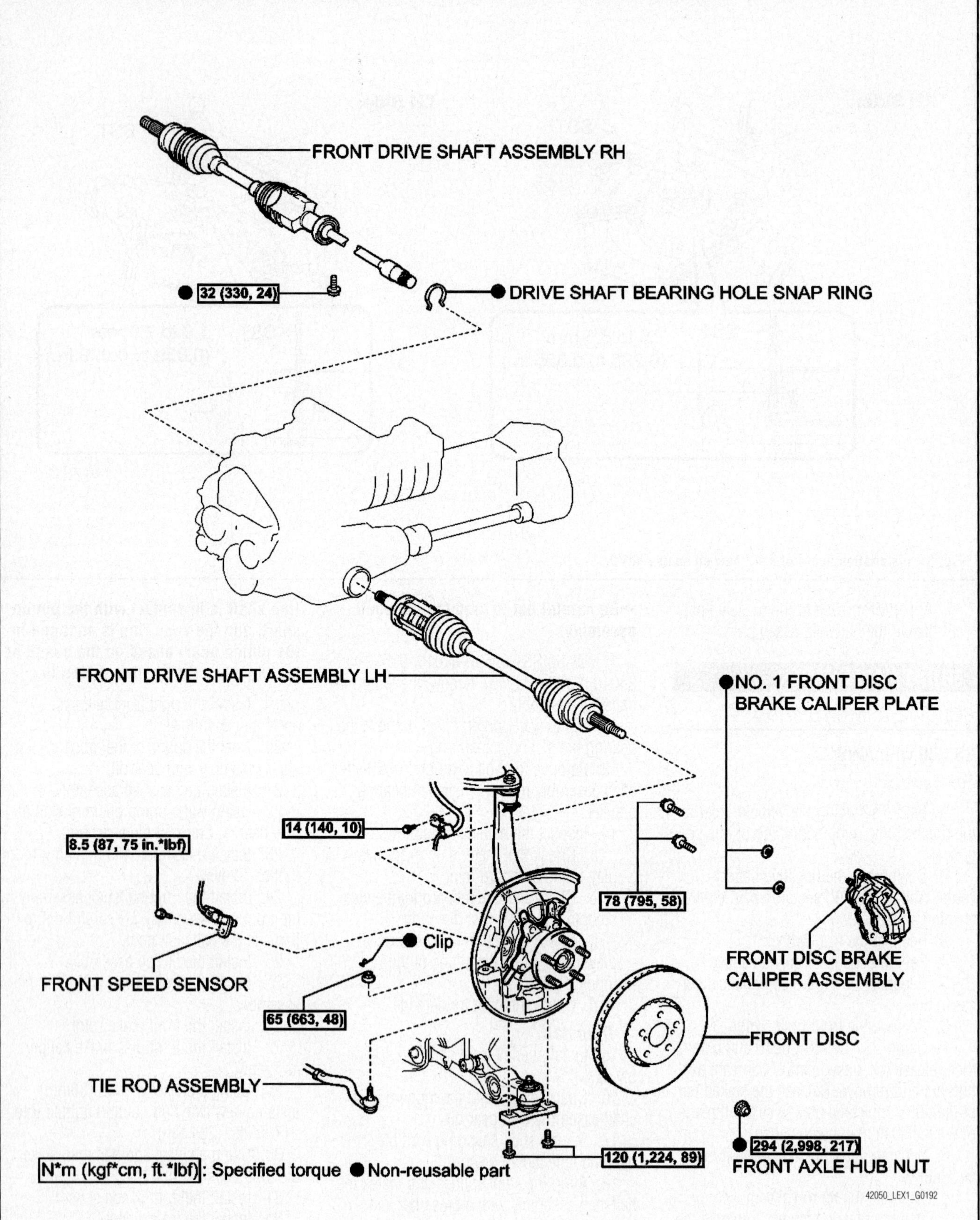

Fig. 32 Exploded view of the halfshaft (drive shaft) assembly—AWD

REAR AXLE SHAFT, BEARING & SEAL

REMOVAL & INSTALLATION

GS 300

See Figures 33 and 34.

1. Raise and safely support the vehicle.
2. Remove the rear tire.
3. Drain the differential oil, install plug and tighten to 36 ft. lbs. (49 Nm).
4. Remove the drive shaft assembly.
5. Using seal removal adapter and slide hammer. Remove rear side gear shaft oil seal.

To install:

6. Using seal driver SST 09308-00010 and a hammer, install new seal.
7. Oil seal drive in depth are as follows:
   a. (2WD)-0.020 to 0.020 inch. (-0.5 to 0.5 mm)
   b. (4WD) 0 to 0.039 inch. (0 to 1.0 mm)
8. Apply MP grease to the oil seal lip.
9. Install rear drive shaft.
10. Install new rear axle shaft nut and tighten to 214 ft. lbs. (290 Nm).
11. Add differential oil. Hypoid gear oil API GL-5 or equivalent. 80W-90.
12. Install rear tire. Tighten to 76 ft. lbs. (103 Nm).
13. Check that speed sensors are working properly.

SC 430

See Figure 35.

1. Raise and safely support the vehicle.
2. Remove the rear tire.
3. Drain the differential oil, install plug and tighten to 36 ft. lbs. (49 Nm).
4. Remove right rear suspension member brace.
5. Remove the bolt and nut, and disconnect the stabilizer link and the height control link from the rear No. 2 suspension arm.
6. Remove the 2 bolts and nuts, and disconnect the No. 1 suspension arm.
7. Disconnect the speed sensor connector and mounting bolt.
8. Push the rear axle carrier toward the outside of the vehicle. Using a plastic-faced hammer, disconnect the rear drive shaft from the rear axle carrier.
9. Using seal removal adapter and slide hammer. Remove rear side gear shaft oil seal.

To install:

10. Using seal driver and a hammer, install new seal.

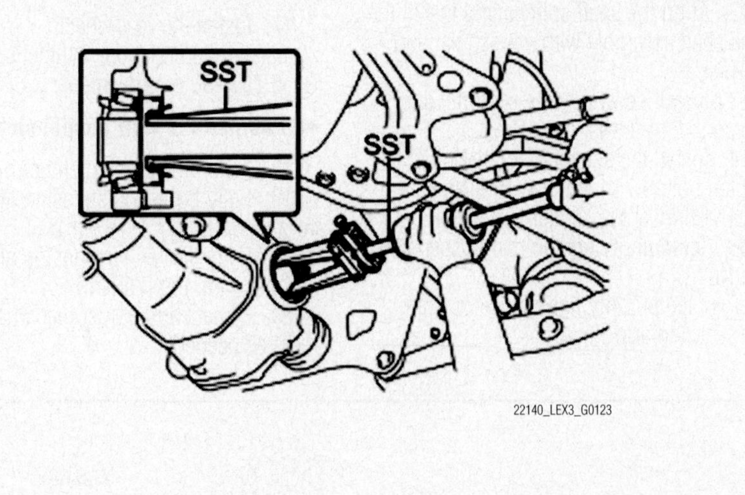

Fig. 33 Seal removal shown with slide hammer and adapter

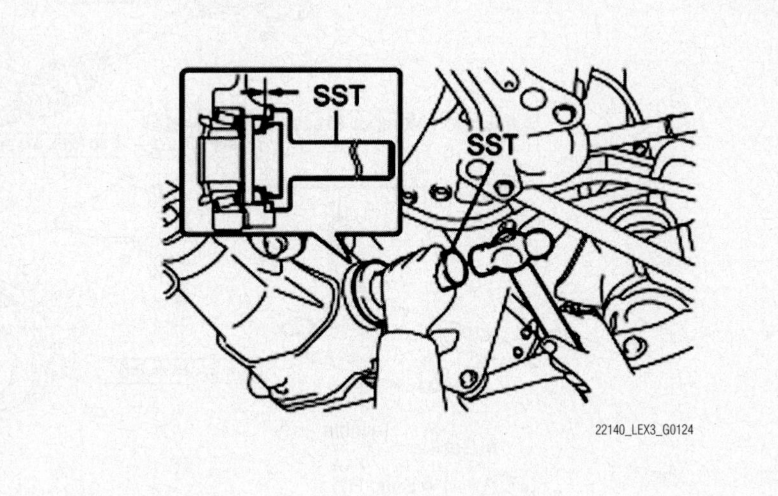

Fig. 34 Seal Installation shown

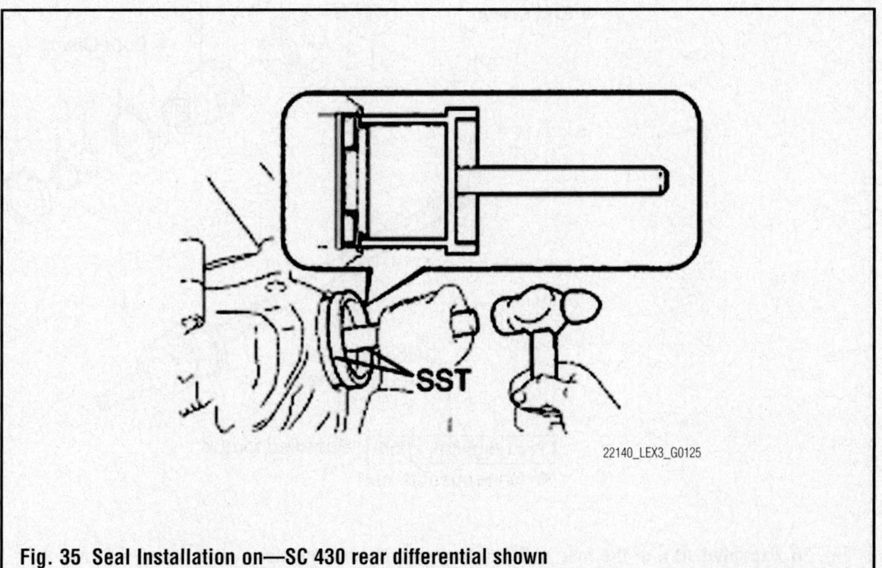

Fig. 35 Seal Installation on—SC 430 rear differential shown

11. Apply MP grease to the oil seal lip.
12. Align the shaft splines and install the drive shaft assembly with a brass bar and hammer.
13. Install new rear axle shaft nut and tighten to 214 ft. lbs. (290 Nm).
14. Install the sensor with the bolt. Tighten bolt to 71 inch. lbs. (8.0 Nm).
15. Connect the sensor connector.
16. Temporarily tighten rear suspension arm No. 2.
  • Temporarily tighten rear suspension arm No. 1.

➡**Do NOT torque nuts.**

17. Install the rear wheel.
Install differential oil. Synthetic gear oil GL-5 75W-90 or equivalent

➡**Stabilize the rear suspension**

18. Install the cross member brace
19. Fully tighten suspension arm No. 2. Torque: 81 ft. lbs. (110 Nm).
20. Fully tighten suspension arm No. 1. Torque: 55 ft. lbs (75 Nm).
21. Inspect and adjust rear wheel alignment as needed.
22. Check that speed sensors are working properly.

REAR HALFSHAFT

REMOVAL & INSTALLATION

GS 300

See Figure 36.

1. Before servicing the vehicle, refer to the precautions in the beginning of this section.

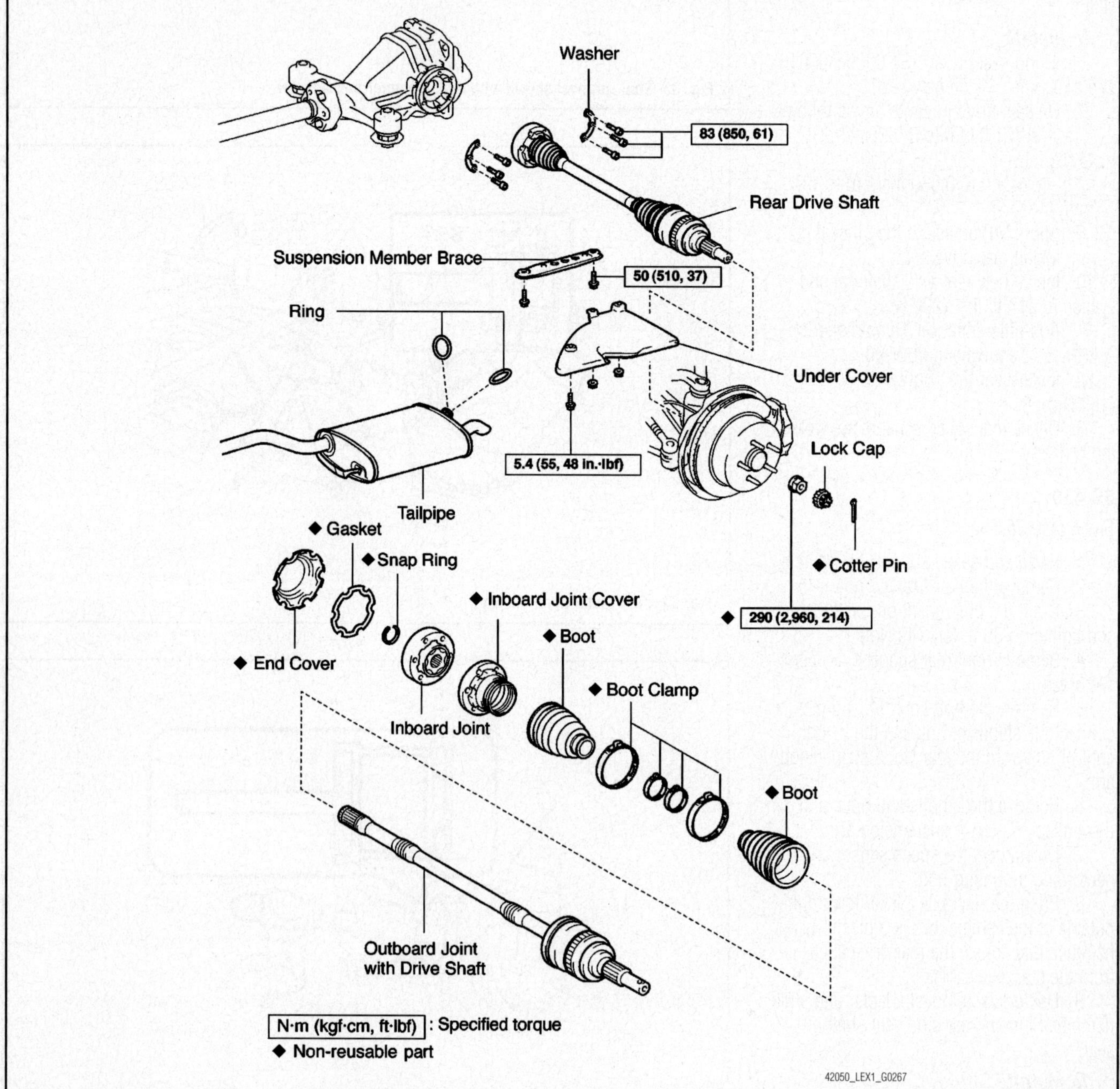

Fig. 36 Exploded view of the rear halfshaft and related components.

2. Remove or disconnect the following:
   - Negative battery cable
   - Rear tire and wheel assembly
   - Cotter pin, locknut cap, and locknut
   - Height control sensor, if equipped
   - 2 exhaust pipe support brackets, if necessary

3. Place matchmarks on the halfshaft and the side gear shaft. Remove the 6 hex bolts and 2 washers.

4. Hold the inboard joint side of the halfshaft so the outboard joint side does not bend too much. Tap the end of the halfshaft with a rubber mallet to loosen it from the axle hub and remove the halfshaft.

To install:

5. Insert the outboard joint side of the half-shaft through the axle hub. Align the match-marks on the side gear shaft and the halfshaft.

6. Coat the threads with clean oil and install the hex bolts. Tighten the bolts to 61 ft. lbs. (83 Nm).

7. Install or connect the following:
   - Exhaust pipe support brackets, if removed, and tighten to 14 ft. lbs. (19 Nm)
   - Bearing locknut, if removed, and have a helper apply the brakes. Tighten the locknut to 214 ft. lbs. (290 Nm).
   - Lock cap and a new cotter pin
   - Height control sensor, if removed
   - Rear tire and wheel assembly
   - Negative battery cable

SC 430

See Figure 37.

1. Before servicing the vehicle, refer to the precautions in the beginning of this section.

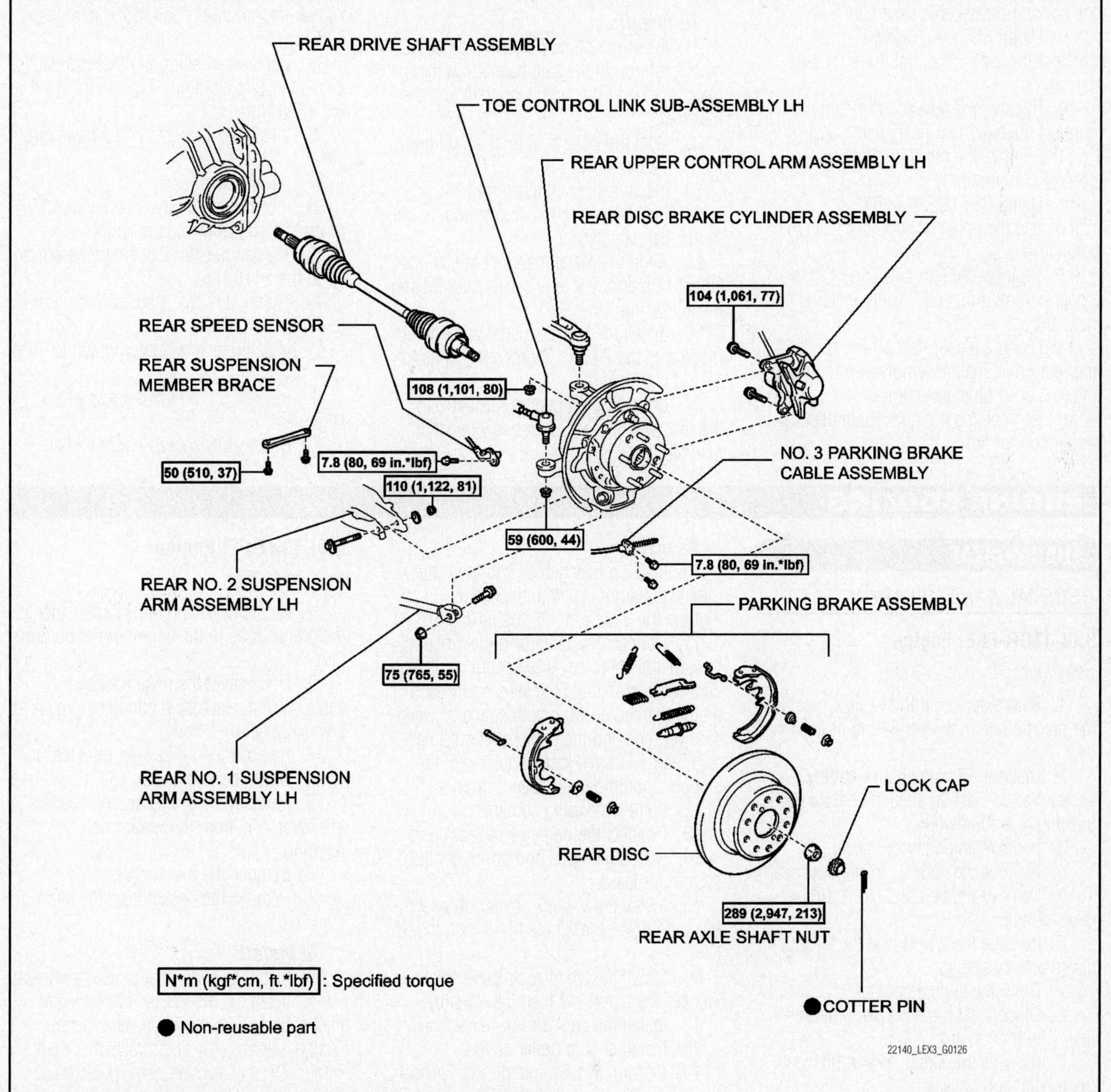

Fig. 37 Exploded view rear drive shaft assembly and related parts—SC 430

2. Raise and safely support the vehicle securely.
3. Remove the cotter pin and lock cap.
4. While depressing the brake pedal, remove the nut.
5. Remove the 2 bolts, and disconnect the rear disc brake cylinder.

➡Use a wire or equivalent to keep the brake cylinder from hanging down by the flexible hose.

6. Remove the rear brake disc.
7. Remove the bolts and disconnect the speed sensor from the axle carrier.
8. Remove the 2 bolts, and disconnect the parking brake cable assembly.
9. Using SST 09610-20012, disconnect the toe control link from the axle carrier.
10. Remove the bolt and nut, and disconnect the No.1 rear suspension arm.
11. Remove the nut to the rear upper control arm assembly.
12. Using SST 09628-00011, disconnect the upper control arm from the axle carrier.
13. Connect the rear axle carrier to the upper control arm, and loosely install the nut.
14. Using a plastic-faced hammer, tap the rear drive shaft assembly's tip to detach the rear drive shaft assembly.
15. Remove the nut and disconnect the upper control arm.
16. Place matchmarks on the adjusting cam and rear suspension arm.
17. Remove the nut, No. 2 camber adjusting cam and No. 1 camber adjusting cam. Then disconnect the rear suspension arm and remove the axle carrier assembly.
18. Remove the 2 bolts and the rear suspension member brace.
19. Push the rear axle carrier toward the outside of the vehicle. Using a plastic-faced hammer, disconnect the rear drive shaft from the rear axle carrier.
20. Using a slide hammer and adapter remove the axle assembly.

To install:

21. Insert the outboard joint side of the halfshaft through the axle hub. Align the matchmarks on the side gear shaft and the halfshaft.
22. Coat the threads with clean oil and install the hex bolts.
23. Install axle shaft assembly.
24. Install the rear No. 2 suspension arm to the rear axle carrier.
25. Align the matchmarks of the camber adjust cam and rear axle carrier, and loosely install the nut.
26. Install the rear suspension member brace with the 2 bolts. Tighten the bolts to 37 ft. lbs. (50 Nm).
27. Connect the upper control arm to the rear axle carrier, and loosely install a new nut.
28. Connect the toe control link LH to the rear axle carrier, and install a new nut. Tighten to 44 ft. lbs. (59 Nm).
29. Tighten the rear upper control arm assembly to 80 ft. lbs. (108 Nm).
30. Connect the rear suspension arm and loosely install the bolt and a new nut.
31. Install the parking brake cable assembly and tighten backing plate bolts to 69 inch. lbs. (7.8 Nm).
32. Inspect and adjust parking brake pedal travel.
33. Connect the brake cylinder with the 2 bolts and tighten to 77 ft. lbs. (104 Nm).
34. Install the sensor and connector tighten the mounting bolt to 71 inch. lbs. (8.0 Nm).
35. While depressing the brake pedal, install the axle shaft nut. Tighten to 214 ft. lbs. (290 Nm).
36. Install the lock cap and a new cotter pin.
37. Lower the vehicle.
38. Press down on the vehicle several times to stabilize the suspension.
39. Tighten the No. 2 suspension arm to 81 ft. lbs. (110 Nm).
40. Tighten the No. 1 suspension arm to 55 ft. lbs. (74 Nm).
41. Install rear wheel and tighten to 76 ft. lbs. (103 Nm).
42. Inspect and adjust the rear wheel alignment.
43. Check ABS speed sensor signal.

ENGINE COOLING

THERMOSTAT

REMOVAL & INSTALLATION

3.0L (3GR-FSE) Engine

See Figure 38.

1. Before servicing the vehicle, refer to the precautions in the beginning of this section.
2. Disconnect the negative battery cable. Wait at least 90 seconds before performing any other work.
3. Remove engine under cover.
4. Remove the cool air intake duct seal.
5. Remove the bolt and No. 1 air cleaner inlet.
6. Remove the 2 nuts and the V-bank cover sub-assembly.
7. Drain the engine coolant.
8. Disconnect the hose from the water inlet with thermostat.
9. Remove the 3 nuts, water inlet with thermostat and gasket.

To install:

10. Install a new gasket and the water inlet with thermostat with the 3 nuts. Tighten the 3 nuts to 7 ft. lbs. (10 Nm).
11. Connect the radiator outlet hose to the radiator assembly and secure it with the clip. Make sure that the claws on the clip are positioned outward with space between the hose end and the clip end should be less than 1–5mm. Hose contact with the stopper should be less than 2.5mm.
12. Refill the engine coolant.
13. Connect the negative battery cable.
14. Start the engine and check cooling system for leaks.
15. Install the V-bank cover sub-assembly and tighten the 2 nuts to 44 inch lbs. (5 Nm).
16. Install the No. 1 air cleaner inlet. Tighten the bolt to 44 inch lbs. (5 Nm).
17. Install the cool air intake duct seal.
18. Install engine under cover.
19. Perform system initialization procedure.

4.3L (3UZ-FE) Engine

See Figure 39.

1. Before servicing the vehicle, refer to the precautions in the beginning of this section.
2. Disconnect the negative battery cable. Wait at least 90 seconds before performing any other work.
3. Remove the air cleaner inlet No. 1.
4. Drain the engine coolant.
5. Remove the 3 nuts and disconnect the water inlet from the water inlet housing.
6. Remove the thermostat.
7. Remove the gasket from the thermostat.

To install:

8. Install a new gasket to the thermostat.
9. Insert the thermostat into the water inlet housing with the jiggle valve facing straight upward. The jiggle valve may be set within 30° of either side of the prescribed position.

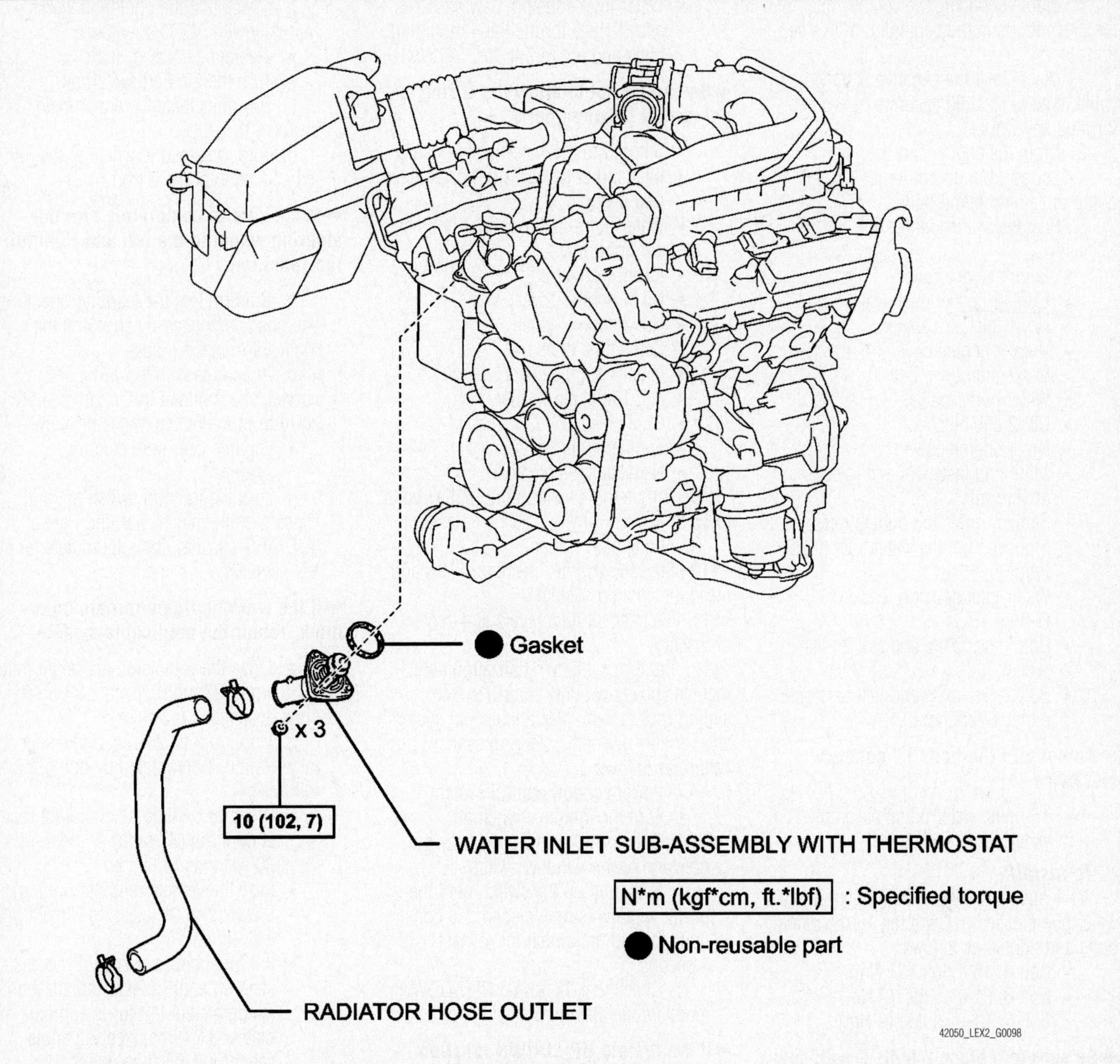

Gasket
x 3
10 (102, 7)
WATER INLET SUB-ASSEMBLY WITH THERMOSTAT
N*m (kgf*cm, ft.*lbf) : Specified torque
Non-reusable part
RADIATOR HOSE OUTLET
42050_LEX2_G0098

Fig. 38 Exploded view of thermostat assembly and related components—3GR-FSE engine

Upward
30°
30°
Jiggle
Valve
42050_LEX1_G0165

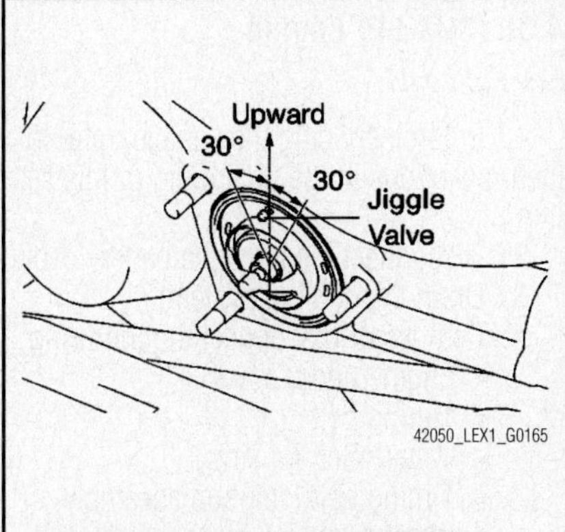

Fig. 39 Correct alignment of the thermostat jiggle valve—3UZ-FE engine

10. Install the water inlet with the 3 nuts and tighten to 14 ft. lbs. (19 Nm).
11. Refill the engine coolant.
12. Connect the negative battery cable.
13. Start the engine and check cooling system for leaks.
14. Install the air cleaner inlet No. 1.
15. Check the coolant level and add if necessary.

WATER PUMP

REMOVAL & INSTALLATION

3.0L (3GR-FSE) Engine

See Figure 40.

A
B
A
A
A
A
A
A
A
A
42050_LEX2_G0101

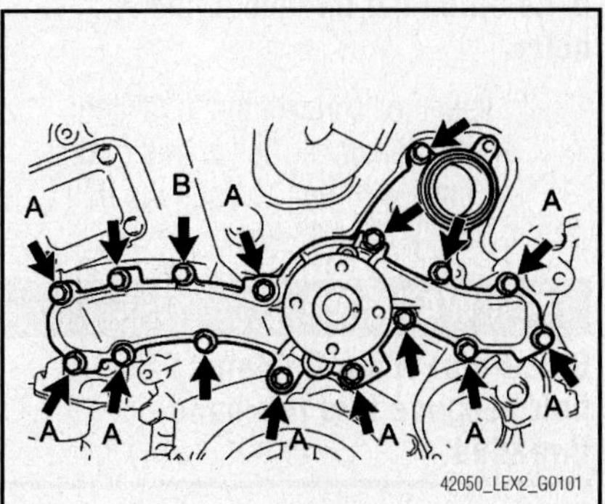

Fig. 40 Water pump mounting bolts—3.0L (3GR-FSE) engine

1. Before servicing the vehicle, refer to the precautions in the beginning of this section.

2. Disconnect the negative battery cable. Wait at least 90 seconds before performing any work.

3. Drain the engine coolant.

4. Loosen, but do not remove water pump pulley mounting bolts.

5. Remove or disconnect the following:

- Engine under cover
- Cool air intake duct seal
- No. 1 inlet air cleaner
- V-bank engine cover
- Serpentine drive belt
- Water pump pulley
- No. 2 engine cover
- No. 1 engine cover
- Mounting fasteners and injector driver unit
- Radiator inlet and outlet hoses
- 5 hoses from the water inlet housing
- Water inlet housing, gasket and O-ring
- Bolt, cover plate and No. 2 idler pulley
- Bolt, cover plate and belt tensioner pulley (marking: L)

➡Do not turn the bolt "L" counterclockwise.

- 16 bolts, water pump assembly and water pump gasket

To install:

6. Install a new water pump gasket and water pump assembly with the 16 mounting bolts and tighten as follows:

- Bolt A: 15 ft. lbs. (21 Nm)
- Bolt B: 81 inch lbs. (9 Nm)
- Bolt C: 81 inch lbs. (9 Nm)

➡Be sure to replace 2 bolts C with new ones or reuse them after applying adhesive 1344. Make sure that there is no oil on the threads of the A bolts.

7. Install or connect the following:

- Belt tensioner pulley and cover plate with bolt. Torque the bolt to 32 ft. lbs. (43 Nm).

⁂ WARNING

Be careful when tightening the bolt because it is left-hand threaded.

- No. 2 idler pulley, cover plate and bolt. Torque the bolt to 32 ft. lbs. (43 Nm).
- Water inlet housing, new gasket and O-ring. Torque the 4 mounting bolts and nut to 7 ft. lbs. (10 Nm).

➡Be careful not to allow the O-ring to get caught between parts.

- 5 hoses to the water inlet housing
- Inlet and outlet radiator hoses and new clamps
- Water pump pulley with 4 bolts finger tight only
- Injector driver unit
- No. 1 engine cover
- No. 2 engine cover
- Serpentine drive belt
- V-bank engine cover
- No. 1 inlet air cleaner
- Cool air intake duct seal
- Engine under cover
- Negative battery cable

8. Torque the 4 water pump pulley bolts to 15 ft. lbs. (21 Nm)

9. Add engine coolant.

10. Start the engine, check for leaks and bleed the cooling system.

11. Recheck all fluid levels and add if necessary.

12. Perform system initialization (which includes power window control system, sliding roof system, clearance sonar system and variable gear ratio steering system) procedure as follows:

- Power window control system

a. Turn the ignition switch on.

b. Open power window halfway by pressing power window switch.

c. Fully pull up the switch until the power window is fully closed and continue to hold the switch for at least 1 second.

d. Check that the AUTO UP / DOWN function operates normally.

➡If the remote UP / DOWN function does not operate after the conditions 1), 2), or 3) is satisfied, the power window regulator master switch may have a malfunction.

- Sliding roof system

e. Turn the ignition switch on.

f. If the sliding roof is opened, close it fully.

g. Push the open switch of the slide switch, or the up switch of the tilt switch on the personal light, making the sliding roof tilt up approximately 1 second, tilt down, slide open, slide close.

h. Sliding roof stops at the fully closed position.

i. Finish the initialization.

j. Check that the operation works normally with AUTO operation.

- Clearance sonar system

k. Turn the ignition switch on.

l. Turn the clearance sonar main switch ON.

m. Turn the steering wheel to the full left and right lock position.

➡Make sure to completely turn the steering wheel to the left and right full lock position.

n. Confirm that the learning operation has been completed by checking the multi-information display.

o. At an area with few turns and curves, and minimal traffic, drive at 20 km/h or more for 5 minutes or more.

- Variable gear ratio steering system

p. Turn the ignition switch on, and check that the master warning light and VSC/ABS warning lights illuminate for a few seconds.

➡If the warning lights remain on or blink, repair the applicable system.

q. Drive the vehicle on a straight road at 35 km/h (22 mph) or more for 5 seconds or longer.

r. Confirm that steering angle sensor initialization is completed by doing the following:

- Drive the vehicle on a straight road at 60 km/h (37 mph) or more for 30 seconds or longer.
- Stop the vehicle (engine running).
- Slowly turn the steering wheel from lock to lock.
- If it turns approximately 2.7 turns, steering angle sensor initialization is completed. If it turns approximately 3.2 turns, steering angle sensor initialization is not completed.

13. Road test the vehicle.

4.3L (3UZ-FE) Engine

See Figure 41.

1. Before servicing the vehicle, refer to the precautions in the beginning of this section.

2. Disconnect the negative battery cable.

3. Drain the cooling system.

4. Remove or disconnect the following:

- Engine under cover
- Timing belt
- Water inlet housing
- Timing belt idler sub assembly
- Water pump

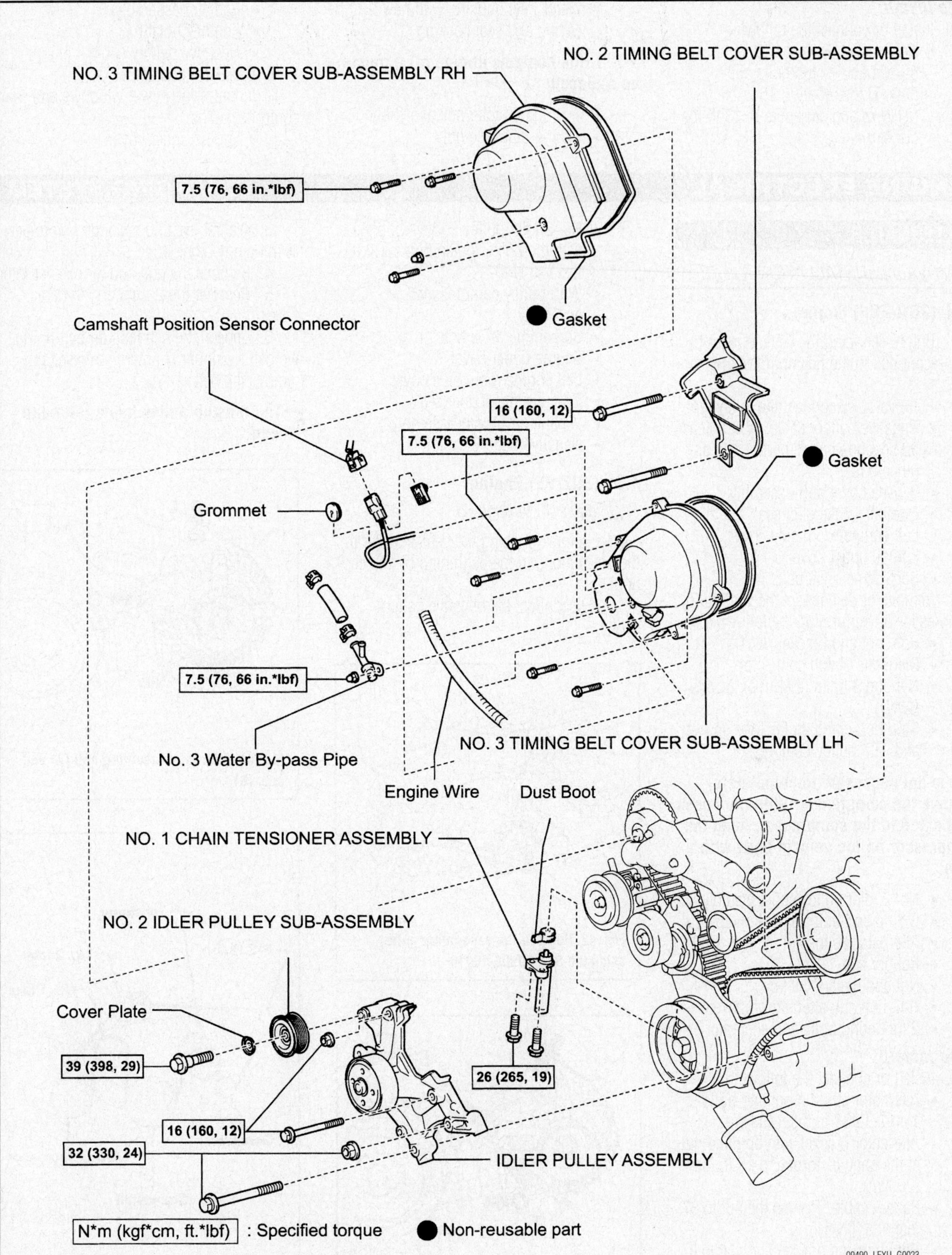

Fig. 41 Exploded view of water pump assembly—3UZ-FE Engine

To install:

5. Install or connect the following:
   - New gasket, water pump with 5 bolts and 2 stud bolts and nuts. Tighten bolt to 15 ft lbs (21 Nm) and stud nuts to 13 ft. lbs (18 Nm)
   - Water inlet housing with new O-ring and seal packing

➡If O-ring contacts engine oil it must be replaced

   - Timing belt idler sub assembly
   - Timing belt assembly
   - Radiator assembly
   - Engine coolant
   - Negative battery cable
6. Check for coolant leaks
7. Initialize the power windows and seat control systems
   - Engine under cover

ENGINE ELECTRICAL

CHARGING SYSTEM

ALTERNATOR

REMOVAL & INSTALLATION

3.0L (3GR-FSE) Engine

1. Before servicing the vehicle, refer to the precautions in the beginning of this section.
2. Remove or disconnect the following:
   - Negative battery cable. Wait at least 90 seconds before performing any other work.
   - V-bank cover sub-assembly
   - Cool air intake duct seal
   - Left engine room side cover
   - Engine under cover
   - Serpentine drive belt
3. Remove or disconnect the with pulley compressor by performing the following:
   - Bolt, nut and bracket (AWD)
   - Magnetic clutch connector
   - Nut and 3 bolts (2WD); or 2 bolts (AWD)
   - Stud bolt using an E8 Torx®socket and with pulley compressor

➡It is not necessary to completely remove the compressor. With the hoses connected to the compressor, hang the compressor on the vehicle body with a rope.

   - No. 2 idler pulley sub-assembly
   - Clamp and bolt from alternator
   - Alternator connector
   - Rubber cap
   - Nut and battery cable.
   - Nut, bolt and alternator bracket
   - 2 mounting bolts and alternator

To install:

4. Install or connect the following:
   - Alternator and 2 mounting bolts. Torque to 32 ft. lbs. (43 Nm).
   - Alternator bracket with bolt and nut at the engine. Torque to 15 ft. lbs. (20 Nm).
   - Battery cable. Tighten the nut to 87 inch lbs. (10 Nm).
   - Wire harness bracket and clamp to the alternator
   - Alternator wire to terminal.
   - Terminal cap
   - No. 2 idler pulley and cover plate with the bolt. Tighten bolt to 32 ft. lbs. (43 Nm).
   - With pulley compressor assembly
   - Serpentine drive belt
   - Engine under cover
   - Left engine room side cover
   - Cool air intake duct seal
   - V-bank cover sub-assembly
   - Negative battery cable

4.3L (3UZ-FE) Engine

See Figures 42 through 45.

1. Before servicing the vehicle, refer to the precautions in the beginning of this section.
2. Disconnect the negative battery cable.

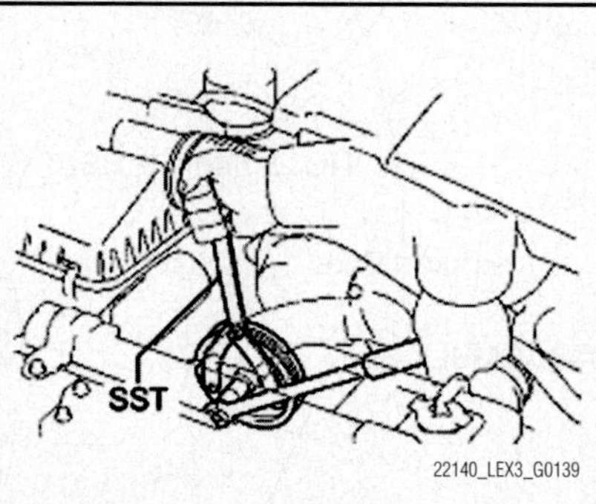

Fig. 42 Remove the vane pump pulley using the SST 09960-10010

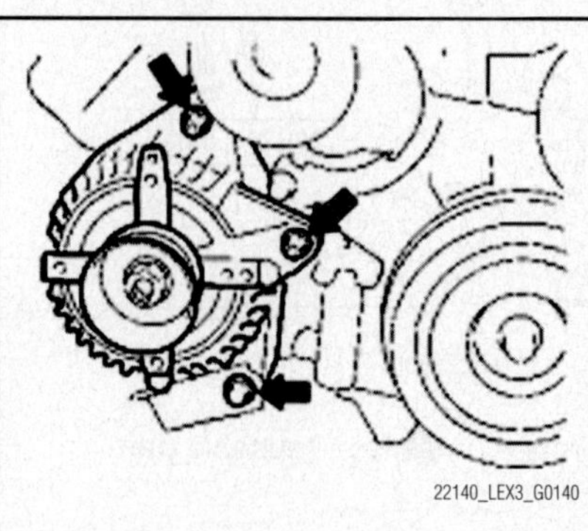

Fig. 43 Remove the 2 nuts, bolt and the alternator

3. Wait at least 90 seconds before performing any other work.
4. Remove the bolt and air cleaner inlet.
5. Remove the 2 nuts and V-bank cover.
6. Loosen the belt tension by turning the belt tensioner counterclockwise, and remove the belt.

➡The tension pulley has a left hand thread.

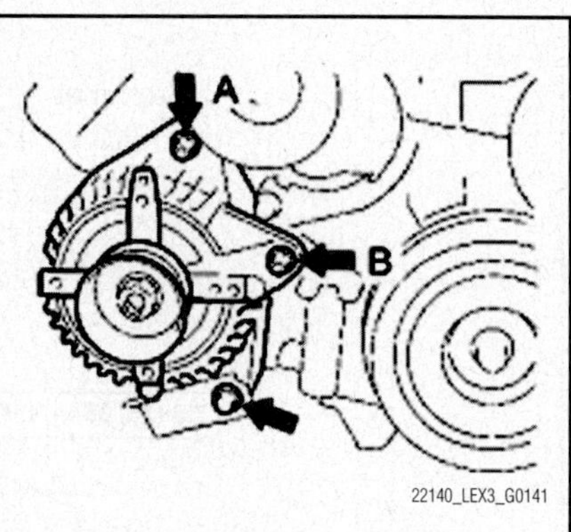

Fig. 44 Alternator mounting bolt (A) and nuts (B)

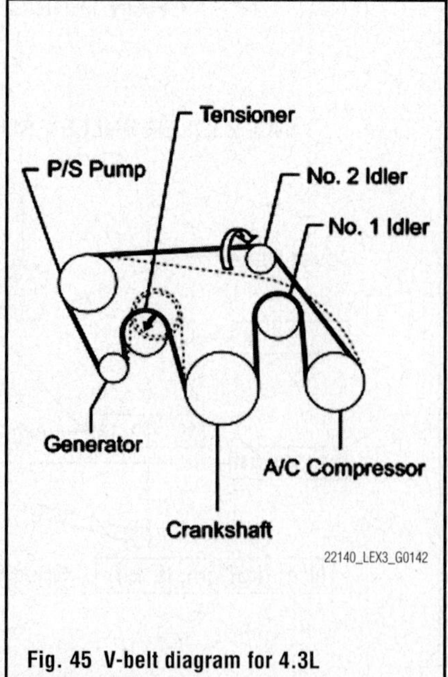

Fig. 45 V-belt diagram for 4.3L (3UZ-FE) Engine

7. Using SST 09960-10010, remove the nut and vane pump pulley.

8. Disconnect the generator connector.

9. Remove the rubber cap and nut, and disconnect the generator wire.

10. Remove the bolt and disconnect the engine wire bracket.

11. Remove the 2 nuts, bolt and the alternator.

To install:

12. Install the generator with the 2 nuts and bolt.

a. Tighten the 2 nuts (B) to 11 ft. lbs. (15 Nm).

b. Tighten the bolt (A) to 29 ft. lbs. (39 Nm).

13. Connect the engine wire bracket with the bolt.

14. Using SST, install the vane pump pulley with the nut. Tighten the nut to 32 ft. lbs. (43 Nm).

15. Set the V-belt to everything except the No. 2 idler pulley, as shown in the illustration.

16. Loosen the V-belt by turning the belt tensioner counterclockwise.

17. Then set the V-belt to the idler pulley.

18. Install engine undercover.

19. Install the V-bank cover with the 2 nuts. Tighten the nuts to 44 inch lbs. (5.0 Nm).

20. Install the air cleaner inlet with the bolt and tighten to 44 inch lbs. (5.0 Nm).

21. Connect the negative battery cable.

22. Certain systems need to be initialized after disconnecting and reconnecting the cable from the negative (-) battery terminal.

ENGINE ELECTRICAL — DISTRIBUTORLESS IGNITION SYSTEM

ADJUSTMENT

The engines covered in this section are equipped with a Distributorless Ignition System (DIS). No timing adjustments are possible.

FIRING ORDERS

See Figures 46 and 47.

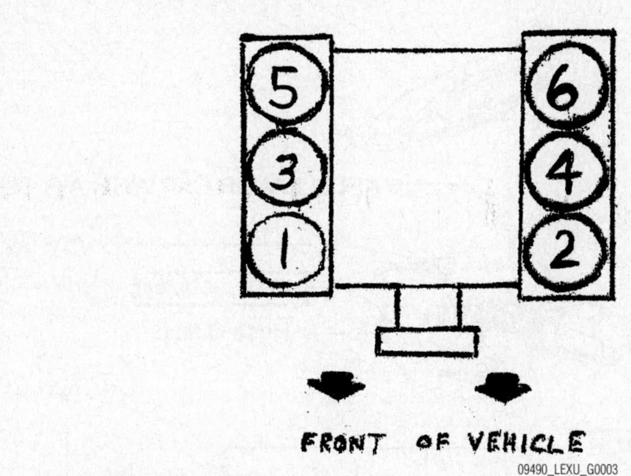

Fig. 46 3.0L (3GR-FSE) Engine
Firing order: 1-2-3-4-5-6
Distributorless ignition system (one coil per cylinder)

Fig. 47 4.3L (3UZ-FE) Engine
Firing order: 1-8-4-3-6-5-7-2
Distributorless ignition system (one coil on each cylinder)

IGNITION COIL

REMOVAL & INSTALLATION

3.0L (3GR-FSE) Engine

See Figures 48 and 49.

1. Before servicing the vehicle, refer to the precautions in the beginning of this section.

2. Disconnect the negative battery cable. Wait at least 90 seconds before performing any other work.

3. Remove cool air intake duct seal.

4. Remove the right and left engine room side covers.

5. Remove V-bank cover.

6. Disconnect the ventilation hose from the cylinder head.

7. Disconnect the Mass Air Flow (MAF) meter connector.

8. Disconnect the clamp from the air cleaner.

9. Disconnect the VSV from the EVAP.

10. Disconnect the 4 clamps.

11. Remove the hose clamp and air cleaner cap with air cleaner hose.

12. Disconnect the 2 wire harness clamps.

13. Disconnect the ignition coil connector.

14. Remove the bolt and ignition coil.

To install:

15. Install the ignition coil and bolt, tighten to 7 ft. lbs. (10 Nm).

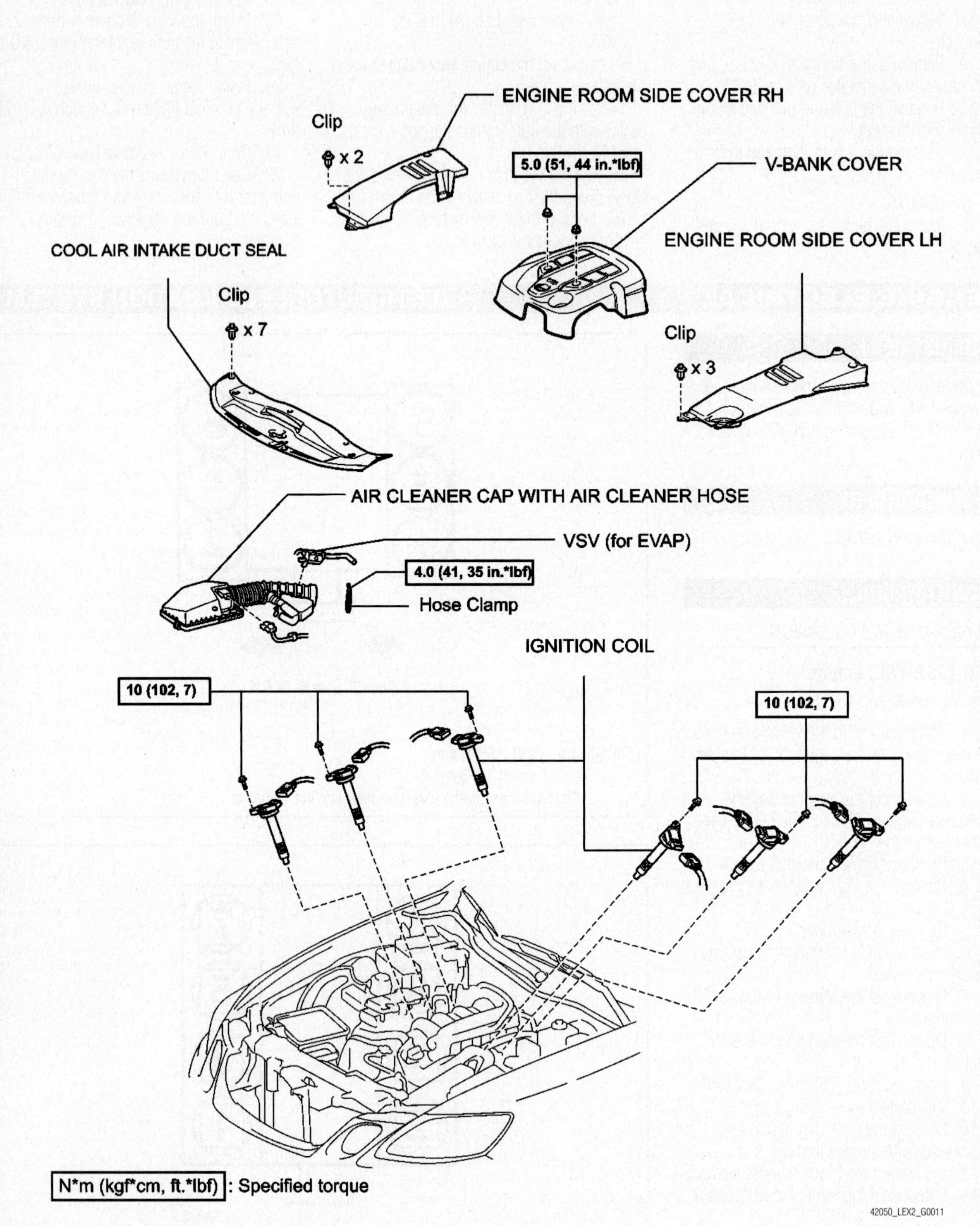

Fig. 48 V-bank cover, ignition coil and related components 3.0L (3GR-FSE) engine

Fig. 49 Disconnect the MAF meter connector and clamp from the air cleaner 3.0L (3GR-FSE) engine

16. Connect the ignition coil connector.
17. Connect the 2 wire harness clamps.
18. Install the air cleaner cap with the air cleaner hose assembly with the 4 clamps and hose clamp.

➡Be sure to install the air cleaner assembly so that the screw part of the hose clamp is as shown in the illustration.

19. Install the VSV for EVAP to the air cleaner hose
20. Connect the MAF meter connector and clamp to the air cleaner.
21. Connect the ventilation hose to the cylinder head cover with the clamp.
22. Install V-bank cover.
23. Install the right and left engine room side covers
24. Install cool air intake duct seal.
25. Connect the negative battery cable
26. Perform initialization procedure.

4.3L (3UZ-FE) Engine

See Figure 50.

1. Before servicing the vehicle, refer to the precautions in the beginning of this section.
2. Disconnect the negative battery cable. Wait at least 90 seconds before performing any other work.
3. Remove V-bank cover.
4. Remove intake air connector pipe.
5. Disconnect the ignition coil connector.
6. Remove the bolts and pull out the ignition coil.

To install:

7. Connect the ignition coil to the spark plug, attach the ignition coil to the cylinder head cover. Install the bolt and tighten to 66 inch lbs. (7.5 Nm).
8. Connect the ignition coil connector.
9. Install intake air connector pipe.
10. Install V-bank cover.
11. Connect the negative battery cable.
12. Perform system initialization procedure.

IGNITION TIMING

INSPECTION

3.0L (3GR-FSE) Engine

See Figures 51 and 52.

1. Warm up the engine and stop the engine.

➡A warmed up engine should have an engine coolant temperature of over 80°C (176°F), have an engine oil temperature of 60°C (140°F), and the engine rpm should be stabilized.

2. When using the intelligent tester, perform the following:
   a. Connect the intelligent tester to the DLC3.
   b. Start the engine and idle it.
   c. Push the intelligent tester main switch ON.
   d. Enter the following items: Powertrain / Engine and ECT / Data list / IGN Advance.
   e. Ignition timing should measure 5–15°BTDC at idle.

➡Refer to the intelligent tester operator's manual for further details.

3. When not using the intelligent tester, perform the following:
   a. Remove the V-bank cover sub-assembly.
   b. Connect the tester probe of a timing light to the wire of the ignition connector for No. 1 cylinder. Use a timing light which can detect the first signal.
   c. Using SST 09843-18040, connect terminals TC and CG of the DLC3.

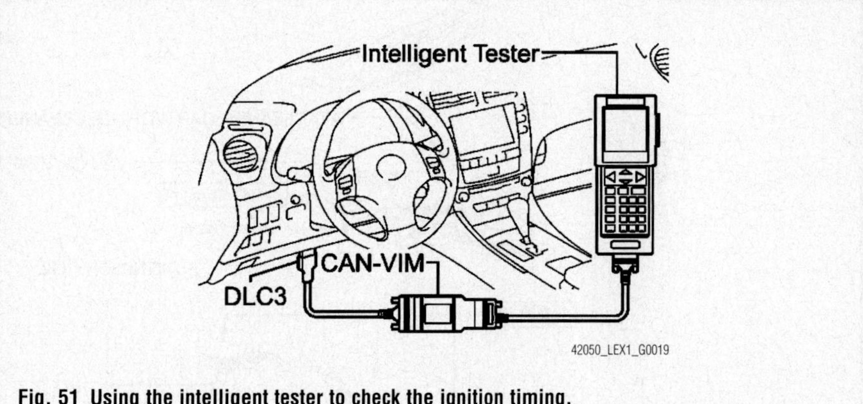

Fig. 51 Using the intelligent tester to check the ignition timing.

22140_LEX3_G0143

Fig. 50 Ignition coil and mounting bolt view

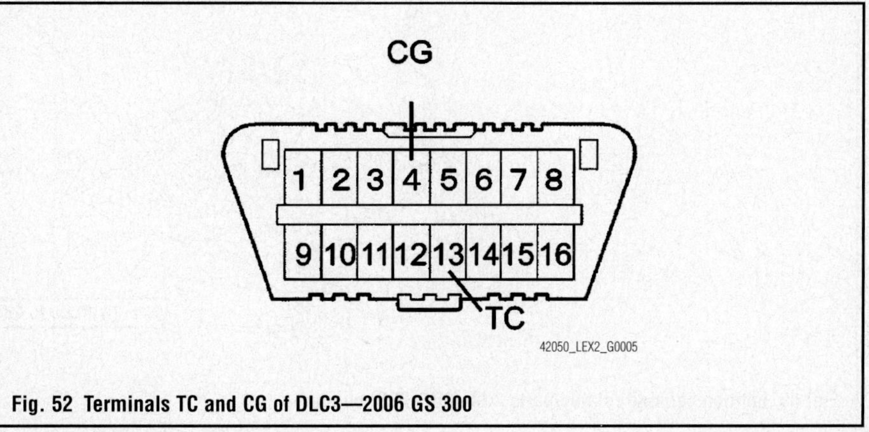

Fig. 52 Terminals TC and CG of DLC3—2006 GS 300

✻✻ CAUTION

Confirm the terminal numbers before connecting them. Connecting the wrong terminals can damage the engine. When checking the ignition timing, the transmission should be in neutral.

d. Using a timing light, check the ignition timing. Ignition timing should measure 8–12°BTDC at idle.

e. Remove the SST from the DLC3.

f. Check the ignition timing. Ignition timing should measure 5–15°BTDC at idle.

g. Check that the ignition timing advances immediately when the engine speed is increased.

h. Disconnect the timing light from the engine.

i. Install the V-bank cover sub-assembly.

4.3L (3UZ-FE) Engine

1. Warm up the engine and stop the engine.

➡A warmed up engine should have an engine coolant temperature of over 80°C (176°F), have an engine oil temperature of 60°C (140°F), and the engine rpm should be stabilized.

2. Connect a hand-held tester to the DLC3.
3. Enter DATA LIST MODE on the hand-held tester. Please refer to the hand-held tester operator's manual for further details.
4. Start and idle the engine.
5. Check the ignition timing. The ignition timing should measure 8–12°BTDC @ idle (Shift lever on N).
6. Disconnect the hand-held tester from the DLC3.

ADJUSTMENT

The ignition timing is controlled by the Powertrain Control Module (PCM). No adjustment is necessary or possible.

SPARK PLUGS

REMOVAL & INSTALLATION

3.0L (3GR-FSE) & 4.3L (3UZ-FE) engines

See Figures 53 and 54.

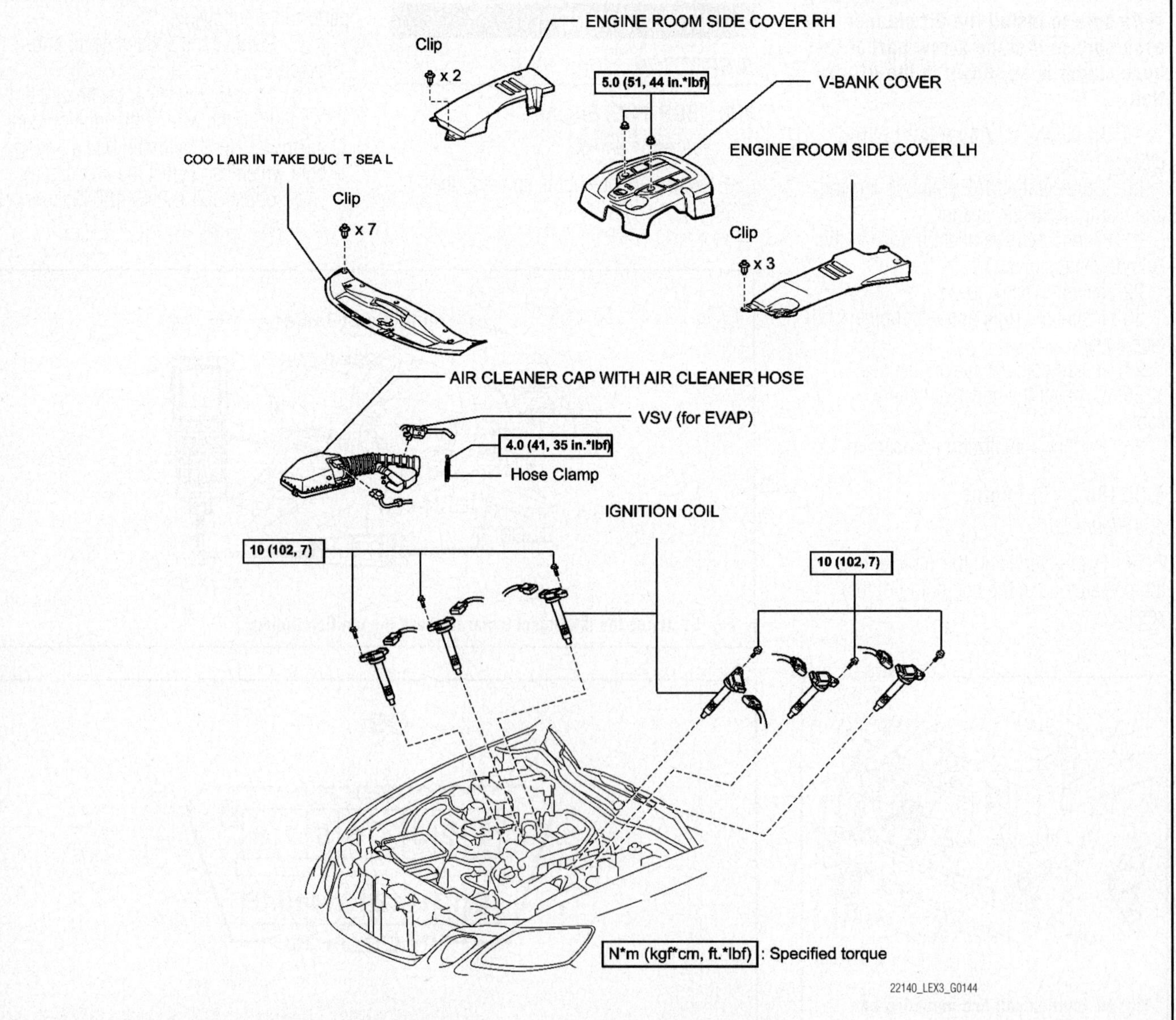

Fig. 53 Ignition coil and related parts 3.0L (3GR-FSE) engine

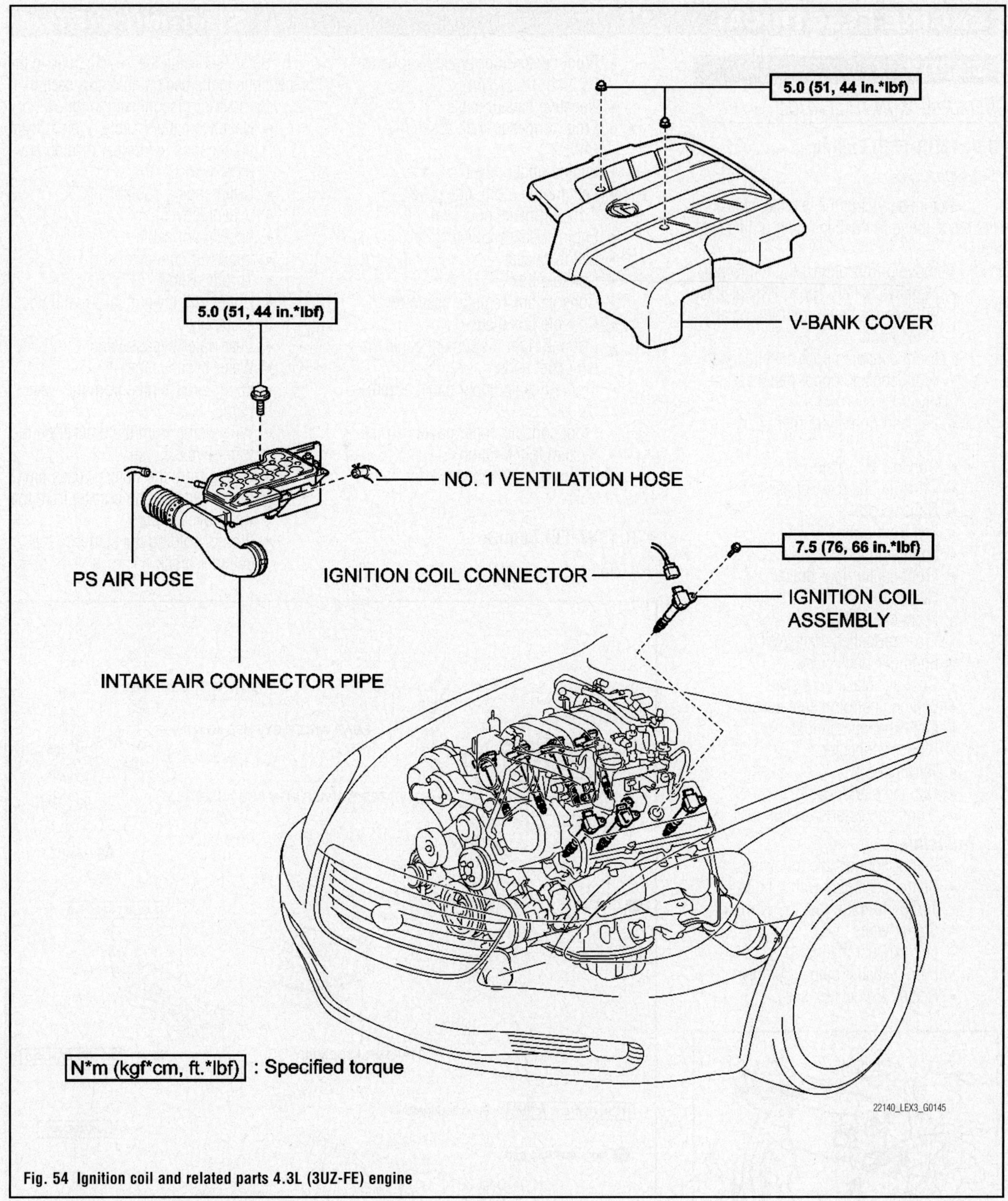

Fig. 54 Ignition coil and related parts 4.3L (3UZ-FE) engine

1. Before servicing the vehicle, refer to the precautions in the beginning of this section.
2. Disconnect the negative battery cable. Wait at least 90 seconds before performing any other work.
3. Remove the ignition coil(s).
4. Remove the spark plug(s).

To install:

5. Install the spark plug(s). Torque the spark plugs to 13 ft. lbs. (19 Nm).
6. Install the ignition coil(s). Tighten to 66 inch lbs. (7.5 Nm).
7. Connect the negative battery cable.
8. Perform system initialization procedure.

STARTER

REMOVAL & INSTALLATION

3.0L (3GR-FSE) Engine

See Figure 55.

1. Before servicing the vehicle, refer to the precautions in the beginning of this section.
2. Remove or disconnect the following:
   - Negative battery cable. Wait at least 90 seconds before performing any other work.
   - Front console upper panel garnish
   - Front console upper panel assembly
   - Left and right instrument panel finish panel ends
   - Console box plate
   - Console box register assembly
   - Console box
   - Left front seat
   - Engine under cover
   - Front center floor brace
   - Heated oxygen sensor
   - Front exhaust pipe
   - Front propeller shaft (AWD)
   - Engine V-bank cover
   - Cool air intake duct seal
   - Left engine room side cover
   - Left exhaust manifold
   - Starter connector
   - Terminal cap
   - Nut and starter cable
   - 2 bolts and starter motor

To install:

3. Install or connect the following:
   - Starter assembly with 2 bolts. Torque to 43 ft. lbs. (58 Nm).
   - Starter wires
   - Left exhaust manifold
   - Front exhaust pipe assembly
   - Heated oxygen sensor
   - Front center floor brace. Torque to 65 inch lbs. (7 Nm).
   - Negative battery cable
   - Front propeller shaft assembly (AWD)
   - Engine under cover
   - Left engine room side cover
   - Cool air intake duct seal
   - Engine V-bank cover
   - Left front seat
   - Console box
   - Console box register assembly
   - Console box plate
   - Left and right instrument panel finish panel ends
   - Front console upper panel assembly
   - Front console upper panel garnish
   - System initialization
4. Start the vehicle and check for exhaust leaks.

4.3L (3UZ-FE) Engine

See Figure 56.

1. Before servicing the vehicle, refer to the precautions in the beginning of this section.
2. Remove or disconnect the following:
   - Negative battery cable. Wait at least 90 seconds before performing any other work
   - Drain engine coolant
   - V-bank cover
   - Accelerator cable
   - Intake air connector
   - Throttle Body
   - Upper and lower intake manifold assembly
   - Rear water bypass joint
   - Water bypass pipe
   - Water bypass pipe from the water pump
   - Wire clamp from the bracket on the water bypass pipe
   - O-ring from the water bypass pipe
   - Water bypass pipe bracket from the water bypass pipe
   - 2 bolts holding the starter to the cylinder block

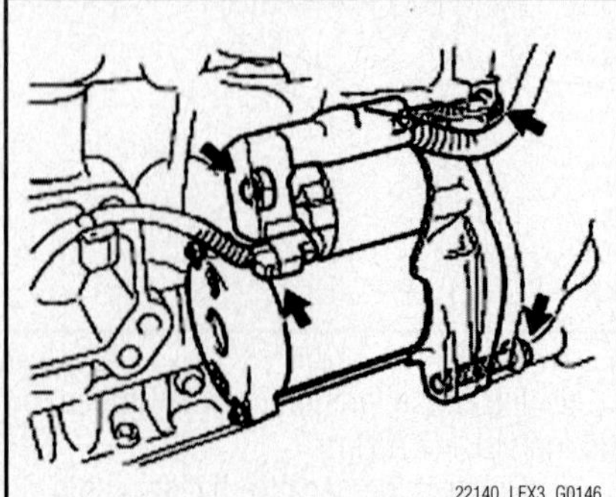

Fig. 55 Starter removal, bolts and electrical locations shown

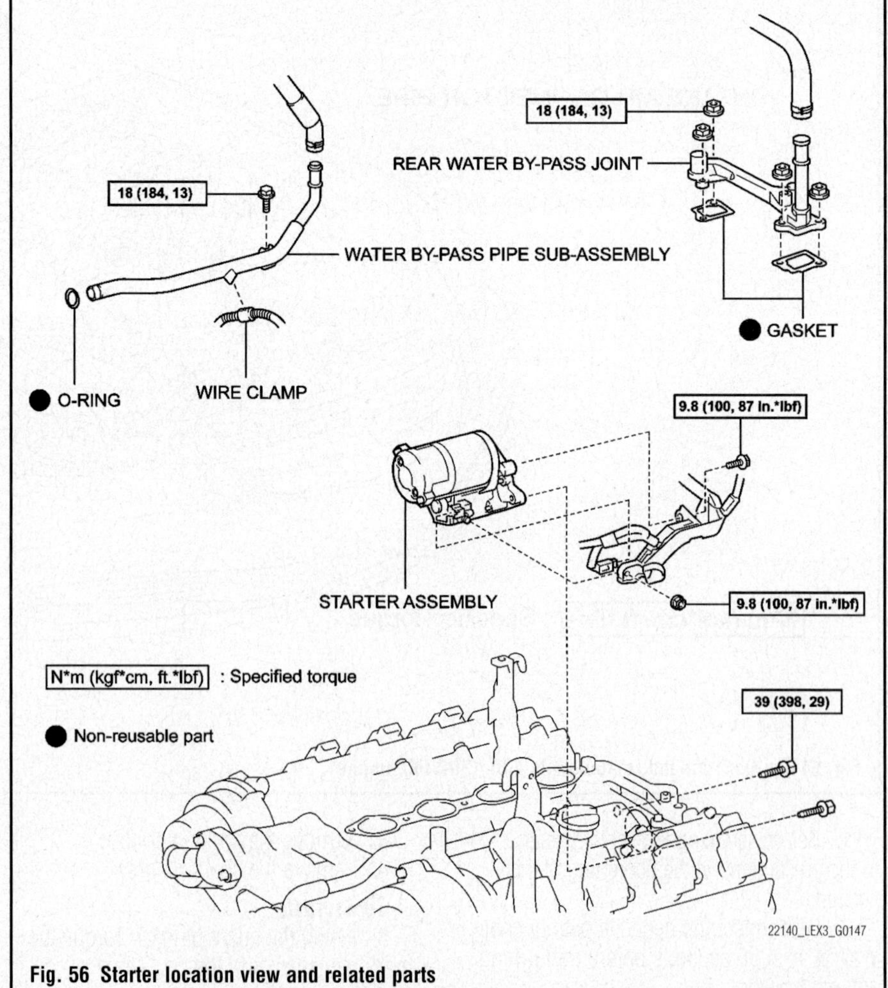

Fig. 56 Starter location view and related parts

- Starter connector
- Starter from the cylinder block
- Nut, and disconnect the starter wire
- Starter

To install:

3. Install or connect the following:
    - Wire clamp to the wire bracket with the bolt. Tighten to 87 inch lbs. (9.8 Nm).
    - Starter wire with the nut. Tighten to 87 inch lbs.
    - Starter connector
    - Starter with the 2 bolts. Torque the bolts to 29 ft. lbs. (39 Nm).
    - Water bypass pipe bracket to the water bypass pipe
    - O-ring to the water bypass pipe
    - Water bypass pipe
    - Wire clamp to the bracket on the water bypass pipe
    - Water bypass pipe bolts. Torque the bolts to 13 ft. lbs. (18 Nm).
    - Rear water bypass joint
    - Intake manifold assembly
    - Throttle body
    - Intake air connector
    - Accelerator cable
    - V-bank cover

ENGINE MECHANICAL

ACCESSORY DRIVE BELTS

ACCESSORY BELT ROUTING

See Figures 57 and 58.

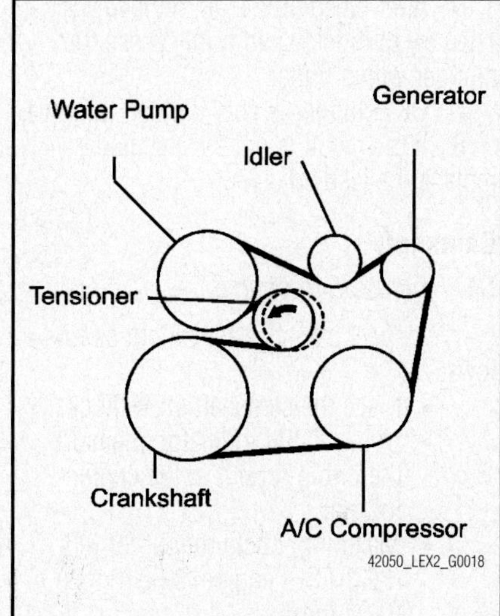

Fig. 57 Serpentine drive belt routing—3.0L (3GR-FSE) engine

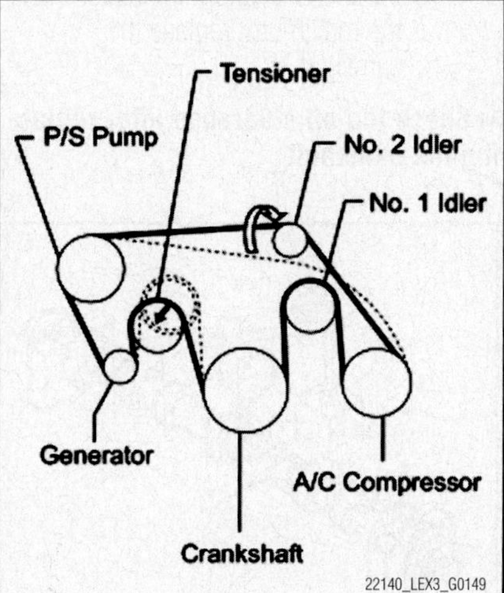

Fig. 58 Serpentine drive belt routing—4.3L (3UZ-FE) engines

INSPECTION

See Figure 59.

Inspect the drive belt for signs of glazing or cracking. A glazed belt will be perfectly smooth from slippage, while a good belt will have a slight texture of fabric visible. Cracks will usually start at the inner edge of the belt and run outward. All worn or damaged drive belts should be replaced immediately.

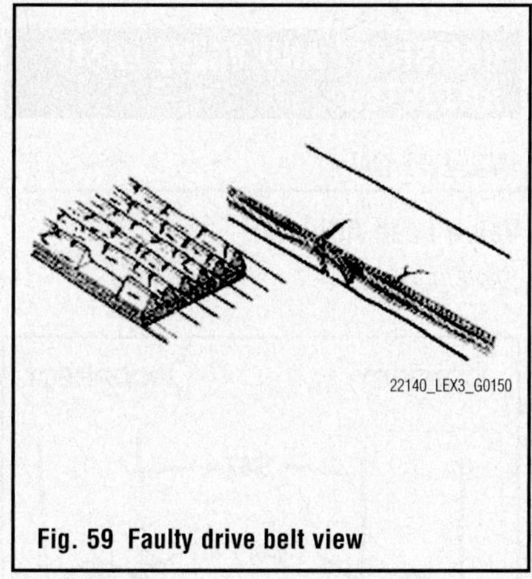

Fig. 59 Faulty drive belt view

ADJUSTMENT

These engines are equipped with automatic belt tensioners. Adjusting the belt tension is not possible or necessary.

REMOVAL & INSTALLATION

3.0L (3GR-FSE) Engine

See Figures 60 and 61.

1. Before servicing the vehicle, refer to the precautions in the beginning of this section.
2. Disconnect the negative battery cable. Wait at least 90 seconds before performing any other work.
3. Remove the cool air intake duct seal.
4. Remove the air cleaner inlet.
5. Remove the V-bank cover.
6. While releasing the belt tension by turning the belt tensioner counterclockwise, and remove the drive belt from the belt tensioner.

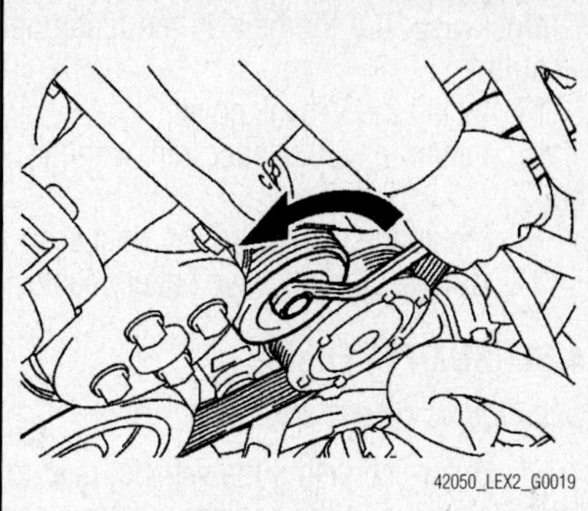

Fig. 60 Loosening the drive belt tension—3GR-FSE engine

Fig. 61 Fix the belt tensioner in place by inserting a 5mm bihexagon wrench into the holes.—3GR-FSE engine

7. While turning the belt tensioner counterclockwise, align with its holes, and then insert a 5mm bihexagon wrench into the holes to fix the belt tensioner.
8. Visually check the drive belt for excessive wear, frayed cords, chunks missing from its ribs, etc. If any defect has been found, replace the belt.
9. Check that nothing gets caught in the tensioner by turning it clockwise and counterclockwise. If a malfunction exists, replace the tensioner.

To install:

10. Install the drive belt.
11. While turning the belt tensioner counterclockwise, remove the bar.
12. Put the backside of the drive belt on the tensioner pulley and idler pulley. Check that the belt is properly set to each pulley.
13. If it is difficult to install the drive belt, perform the following procedure:
    a. Put the belt on every part except the tensioner pulley as shown in the routing illustration.
    b. While releasing the belt tension by turning the belt tensioner counterclockwise, put the belt on the tensioner pulley.
14. Install the V-bank cover.
15. Install the air cleaner inlet with the bolt.
16. Install the cool air intake duct seal.
17. Connect the negative battery cable.

4.3L (3UZ-FE) Engine

See Figures 62 and 63.

1. Before servicing the vehicle, refer to the precautions in the beginning of this section.
2. Disconnect the negative battery cable. Wait at least 90 seconds before performing any other work.

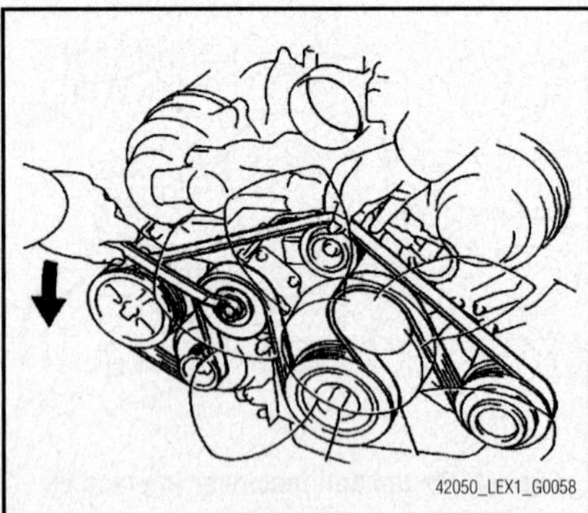

Fig. 62 Loosening the drive belt tension—3UZ-FE engine

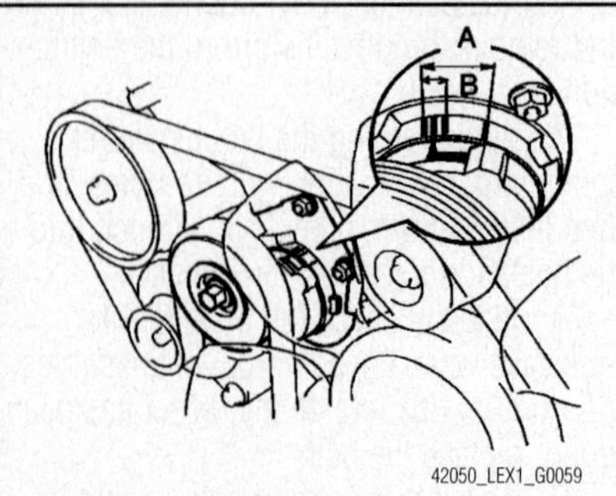

Fig. 63 Check that the mark is within range B as shown on the tensioner pulley—3UZ-FE engines

3. Remove the air cleaner inlet.
4. Loosen the belt tension by turning the belt tensioner counterclockwise.

➡The pulley bolt for the belt tensioner has a left hand thread.

5. Remove the drive belt.
6. Visually check the drive belt for excessive wear, frayed cords, chunks missing from its ribs, etc. If any defect has been found, replace the belt.

To install:

7. Set the drive belt to everything except the idler pulley No. 2, as shown in the routing illustration.
8. Loosen the belt by turning the belt tensioner counterclockwise.
9. Then set the belt to the idler pulley.
10. After a new belt has been installed, check that the mark is within range B as shown in the illustration.
11. Install the air cleaner inlet.
12. Connect the negative battery cable.

CAMSHAFT AND VALVE LASH ADJUSTER

INSPECTION

Valve Lash Adjuster

See Figure 64.

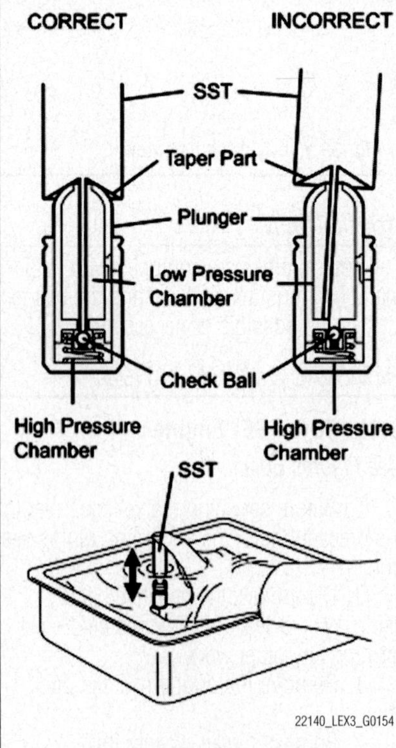

Fig. 64 Valve lash adjuster inspection shown

1. Place the lash adjuster into a container full of engine oil.
2. Insert SST 09276-75010 tip into the lash adjuster's plunger and use the tip to press down on the check ball inside the plunger.
3. Squeeze the SST and lash adjuster together to move the plunger up and down 5 to 6 times.
4. Check the movement of the plunger and bleed the air.
5. OK: Plunger moves up and down.

➡When bleeding high-pressure air from the compression chamber, make sure that the tip of the SST is actually pressing the checkball as shown in the illustration. If the checkball is not pressed, air will not bleed.

6. After bleeding the air, remove SST. Then try to quickly and firmly press the plunger with a finger.
7. OK: Plunger is very difficult to move.
8. If the result is not as specified, replace the lash adjuster.

Camshaft

See Figures 65 through 67.

1. Inspect camshaft for runout as follows:
    - Place the camshaft on V-blocks.
    - Using a dial indicator, measure the circle runout at the center journal.
    - Maximum circle runout for 3.0L (3GR-FSE) engine: 0.0016 inch. (0.04 mm)
    - Maximum circle runout for 4.3L (3UZ-FE) engine: 0.0031 inch. (0.08 mm)
    - If the circle runout is greater than the maximum, replace the camshaft.

➡Check the oil clearance after replacing the camshaft.

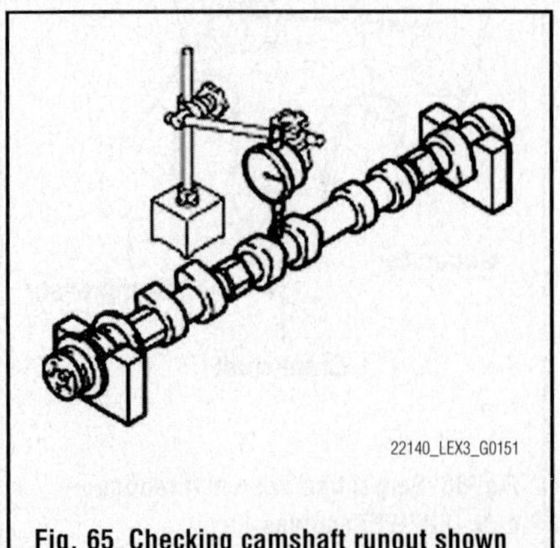

Fig. 65 Checking camshaft runout shown

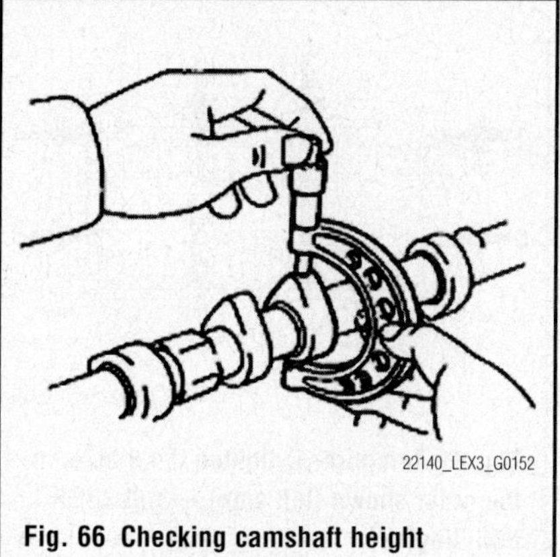

Fig. 66 Checking camshaft height

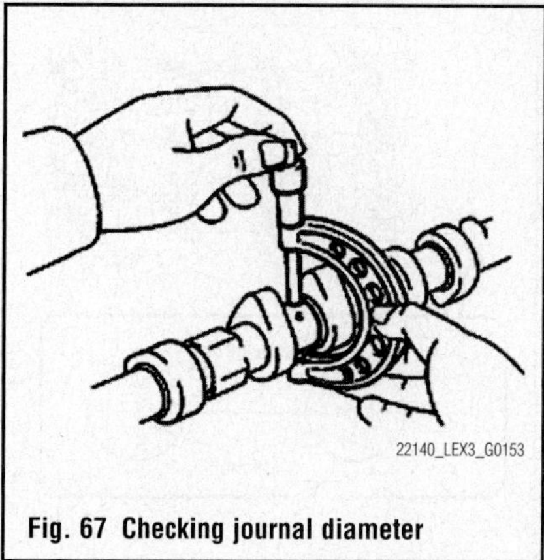

Fig. 67 Checking journal diameter

2. Using a micrometer, measure the cam lobe height and the journal diameter.
3. Standard cam lobe height for 3.0L (3GR-FSE) engine is as follows:
    a. Intake 1.7303 to 1.7342 inch. (43.950 to 44.050 mm).
    b. Exhaust 1.7366 to 1.74.5 inch. (44.110 to 44.210 mm).
4. Standard cam lobe height for 4.3L (3UZ-FE) engine is as follows:
    a. 1.6776 to 1.6815 inch. (42.610 to 42.710 mm).
5. Standard journal diameter for 3.0L (3GR-FSE) engine is as follows:
    a. No.1 Journal 1.4152 to 1.4157 inch. (35.946 to 35.960 mm).
    b. All other journals 1.0220 to 1.0226 inch. (25.959 to 25. 975 mm).
6. Standard journal diameter for 4.3L (3UZ-FE) engine is as follows:
    a. 1.0612 to 1.0618 inch. (26.954 to 26.970).
7. If the journal diameter is not as specified, check the oil clearance.

REMOVAL & INSTALLATION

3.0L (3GR-FSE) Engine

See Figures 68 through 84.

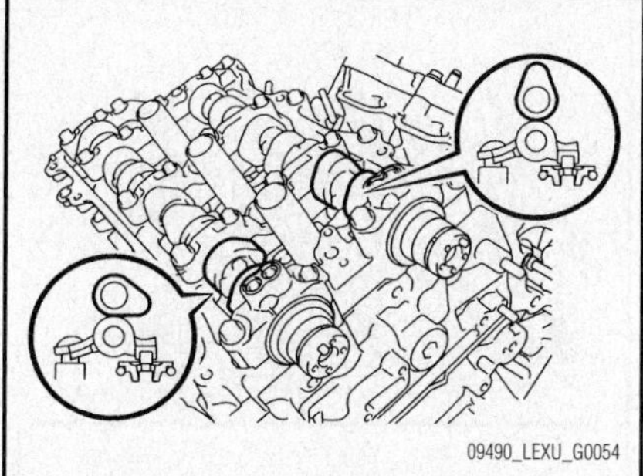

Fig. 68 Check that the camshafts are positioned as shown (right side)—3.0L (3GR-FSE) Engine

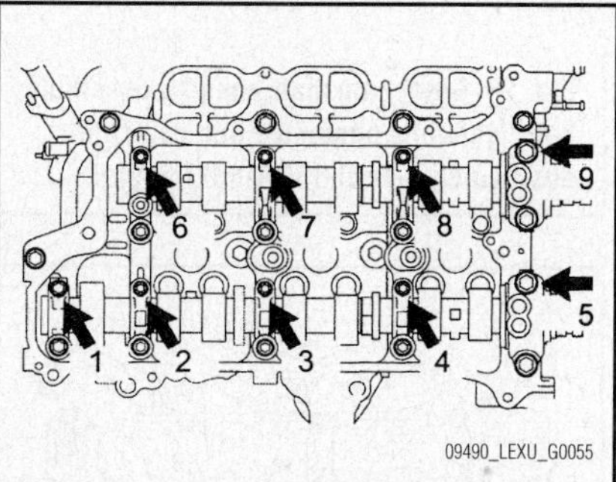

Fig. 69 Right side camshaft bearing inner bolt loosening sequence—3.0L (3GR-FSE) Engine

Fig. 70 Right side camshaft bearing outer bolt loosening sequence—3.0L (3GR-FSE) Engine

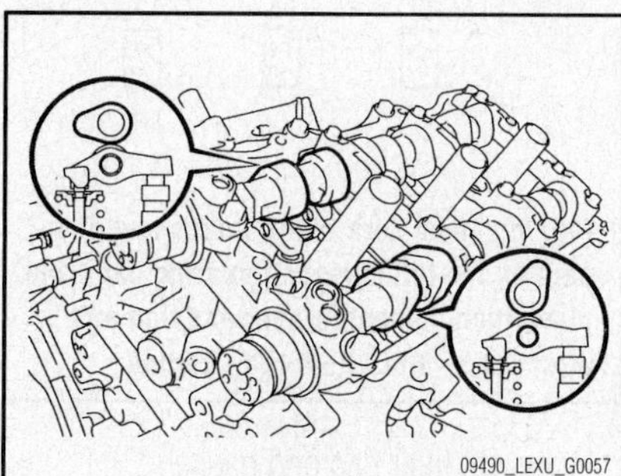

Fig. 71 Check that the camshafts are positioned as shown (left side)—3.0L (3GR-FSE) Engine

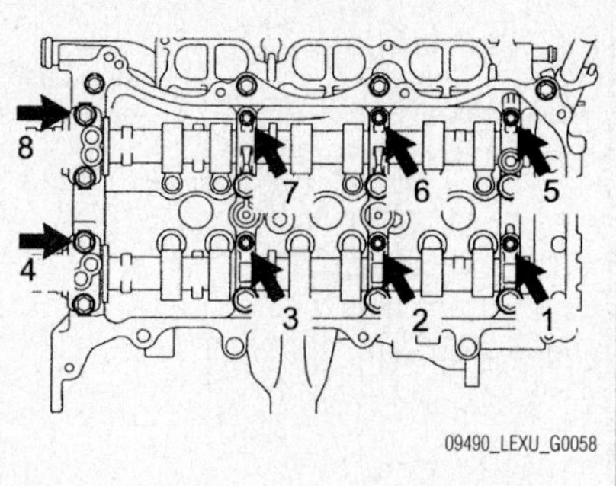

Fig. 72 Left side camshaft bearing inner bolt loosening sequence—3.0L (3GR-FSE) Engine

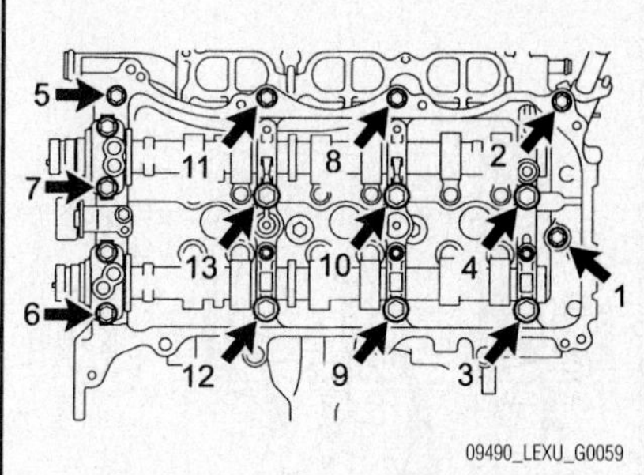

Fig. 73 Left side camshaft bearing outer bolt loosening sequence—3.0L (3GR-FSE) Engine

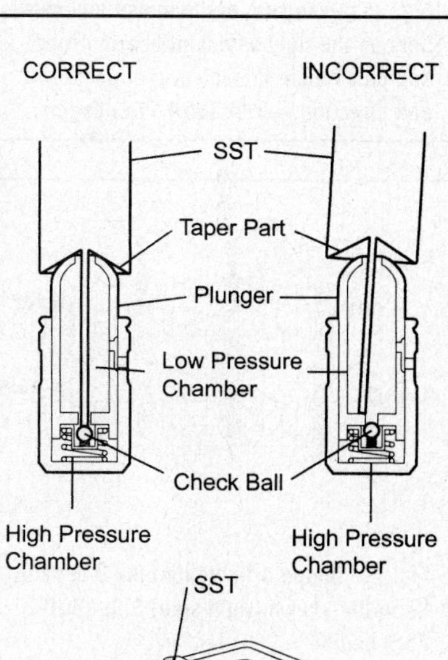

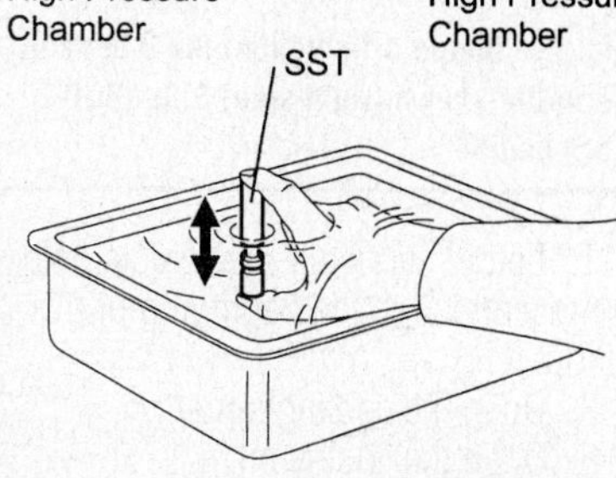

Fig. 74 Bleeding air from the valve lash adjuster assembly—3.0L (3GR-FSE) Engine

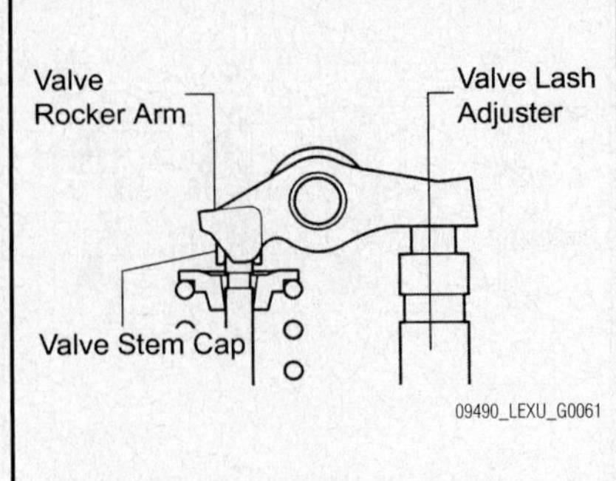

Fig. 75 Correct installation of the valve rocker arm assembly—3.0L (3GR-FSE) Engine

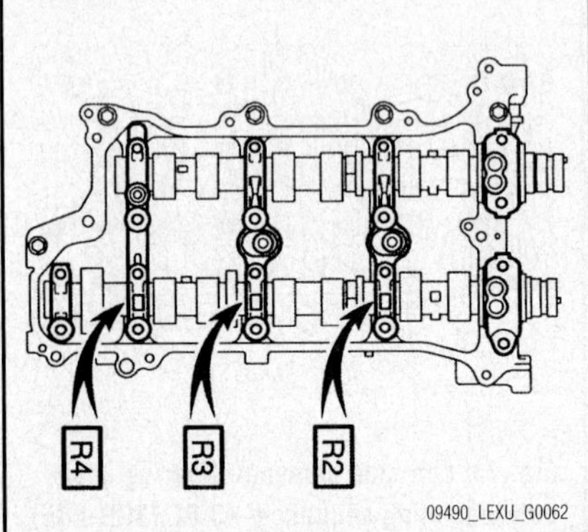

Fig. 76 Make sure of the marks and numbers on the right camshaft bearing caps and place them in each proper position and direction.—3.0L (3GR-FSE) Engine

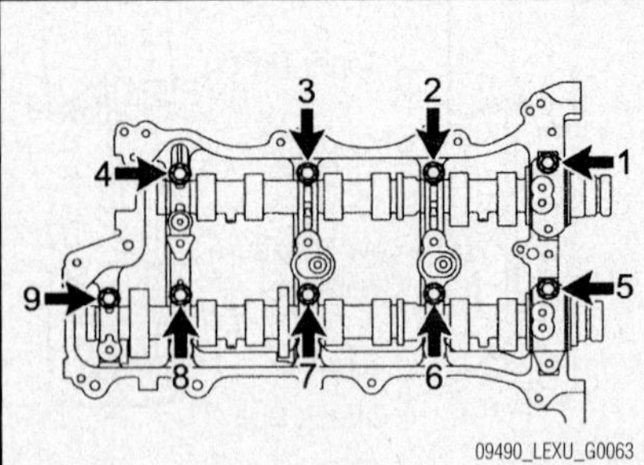

Fig. 77 Temporarily tighten the 9 bolts in the order shown (right side) 3.0L (3GR-FSE) Engine

1. Before servicing the vehicle, refer to the precautions in the beginning of this section.
2. Drain the cooling system.
3. Drain the engine oil.
4. Relieve the fuel system pressure.
5. Remove or disconnect the following:
   - Negative battery cable. Wait at least 90 seconds before performing any other work

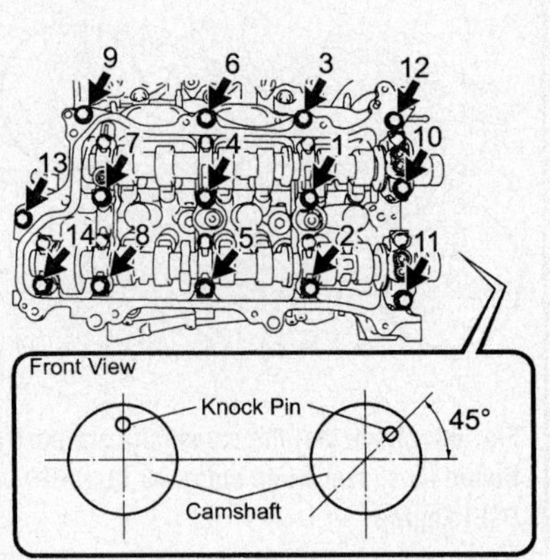

Fig. 78 Right camshaft housing assembly bolt tightening sequence and camshaft positioning.—3.0L (3GR-FSE) Engine

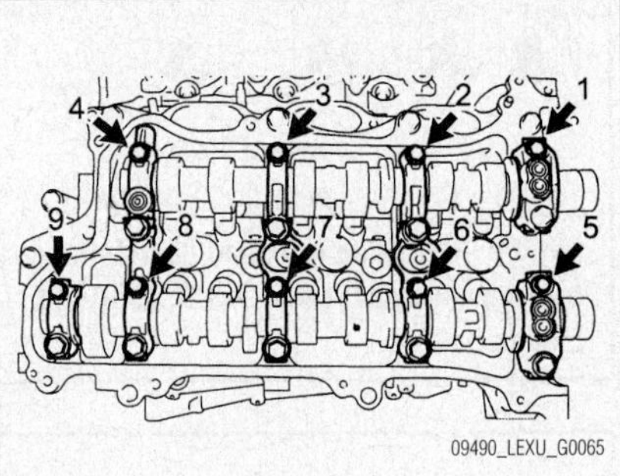

Fig. 79 Right side camshaft bearing inner bolt tightening sequence—3.0L (3GR-FSE) Engine

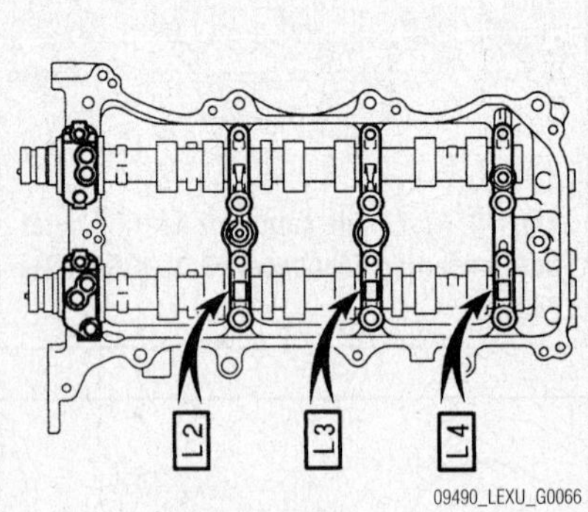

Fig. 80 Make sure of the marks and numbers on the left camshaft bearing caps and place them in each proper position and direction.—3.0L (3GR-FSE) Engine

   - Oil filler cap and gasket
   - Radiator cap
   - No. 1 and 2 engine hangers
   - Spark plugs
   - Ventilation valve

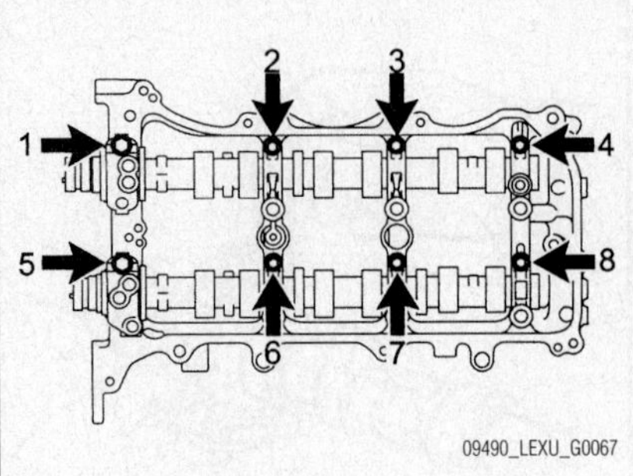

Fig. 81 Temporarily tighten the 8 bolts in the order shown (left side).—3.0L (3GR-FSE) Engine

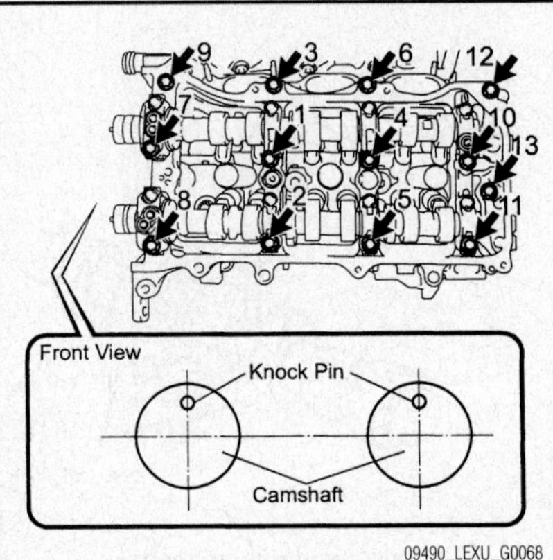

Fig. 82 Left camshaft housing assembly bolt tightening sequence and camshaft positioning.—3.0L (3GR-FSE) Engine

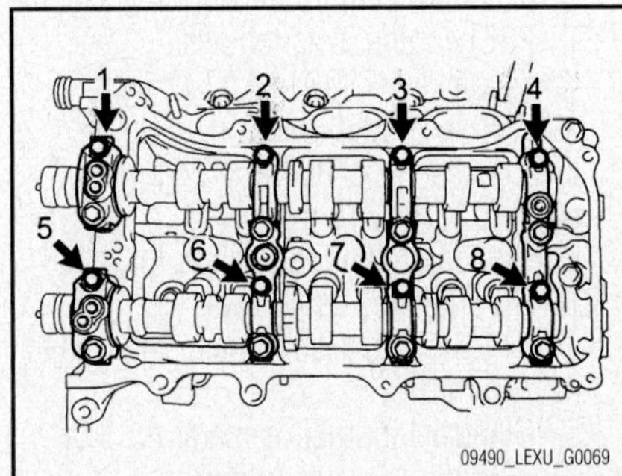

Fig. 83 Left side camshaft bearing inner bolt tightening sequence—3.0L (3GR-FSE) Engine

   - 4 camshaft position sensors
   - 4 camshaft timing oil control valves
   - Crankshaft position sensor
   - Left and right side oil check valve bolt, oil pipe union and oil pipe
   - Left and right side oil control valve filter and gaskets
   - Cylinder block water drain cocks
   - Oil filter element
   - Water inlet and thermostat assembly

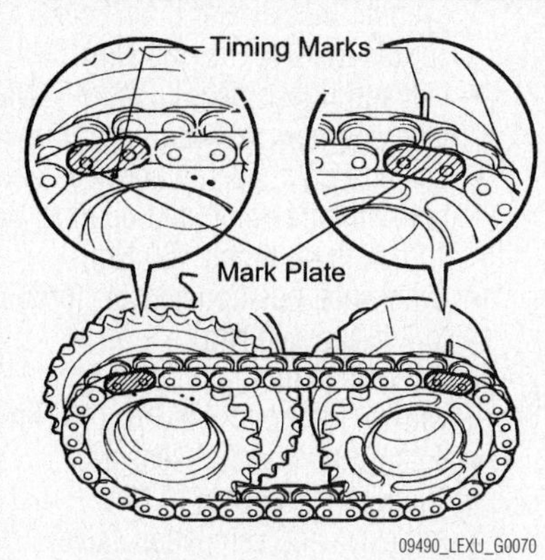

Fig. 84 Align the mark plate with the timing marks of the camshaft timing gears (right side shown, left side similar)—3.0L (3GR-FSE) Engine

- Rear water by-pass joint
- Cylinder head cover and gaskets
- Timing chain cover, timing chain and timing chain sprockets
- No. 2 chain tensioner assembly

6. Remove the camshaft timing gears and No. 2 chain (right) as follows:
   a. While raising the No. 2 chain tensioner, insert a pin of 1.0mm (0.039 in.) into the hole to fix the No. 2 chain tensioner.
   b. Hold the hexagonal portion of the camshaft with a wrench, and remove the 2 bolts and 2 camshaft timing gear assemblies.
   c. Remove the No. 2 chain.
   - No. 2 chain tensioner
7. Remove the camshaft bearing cap (right) as follows:
   a. Check that the camshafts are positioned as shown in the illustration.
   b. Uniformly loosen and remove the 9 bearing cap bolts in the sequence shown in the illustration.
   c. Uniformly loosen and remove the 14 bearing cap bolts in the sequence shown in the illustration.
   d. Remove the 6 bearing caps.
   - Camshaft
   - No. 2 camshaft
   - Camshaft housing sub-assembly (right) by prying between the cylinder head and camshaft housing sub-assembly (right) with a screwdriver.

WARNING

Be careful not to damage the contact surfaces of the cylinder head and camshaft housing.

8. Remove the camshaft timing gears and No. 2 chain (left) as follows:
   a. While pushing down the No. 3 chain tensioner, insert a pin of 1.0mm (0.039 in.) into the hole to fix the No. 3 chain tensioner.
   b. Hold the hexagonal portion of the camshaft with a wrench, and remove the 2 bolts and 2 camshaft timing gear assemblies.
   c. Remove the No. 2 chain.
   - No. 3 chain tensioner
9. Remove the camshaft bearing cap (left) as follows:
   a. Check that the camshafts are positioned as shown in the illustration.
   b. Uniformly loosen and remove the 8 bearing cap bolts in the sequence shown in the illustration.
   c. Uniformly loosen and remove the 13 bearing cap bolts in the sequence shown in the illustration.
   d. Remove the 5 bearing caps.
   - No. 3 camshaft
   - No. 4 camshaft
   - Camshaft housing sub-assembly (left) by prying between the cylinder head and camshaft housing sub-assembly (left) with a screwdriver.

WARNING

Be careful not to damage the contact surfaces of the cylinder head and camshaft housing.

- Valve rocker arms. Arrange the removed rocker arms in the correct order.
- Valve lash adjusters. Arrange the removed valve lash adjusters in the correct order.

To install:

10. Install valve lash adjusters as follows:

WARNING

Keep the lash adjuster free of dirt and foreign objects. Only use clean engine oil.

   a. Place the lash adjuster into a container filled with engine oil.
   b. Insert the SST's tip (09276-75010) into the lash adjuster's plunger and use the tip to press down on the check ball inside the plunger as is shown.
   c. Squeeze the SST and lash adjuster together to move the plunger up and down 5 to 6 times.
   d. Check the movement of the plunger and bleed the air.
   e. After bleeding the air, remove SST. Then, try to quickly and firmly press the plunger with a finger. If the result is not as specified, replace the lash adjuster.
   f. Install the lash adjusters.

➡Install the lash adjuster to the same place it was removed from.

11. Install No. 1 valve rocker arm assembly as follows:
   a. Apply engine oil to the lash adjuster tip and valve stem cap end.
   b. Make sure that the valve rocker arms are installed as shown in the illustration.
12. Install right side camshaft bearing cap as follows:
   a. Apply engine oil to the camshaft journals, camshaft housing and bearing caps.
   b. Install the camshaft and camshaft No. 2 to the right camshaft housing.
   c. Make sure of the marks and numbers on the camshaft bearing caps and place them in each proper position and direction.
   d. Temporarily tighten the 9 bolts to 7 ft. lbs. (10 Nm) in the order shown in the illustration.
13. Install the right side camshaft housing assembly as follows:
   a. Make sure that the valve rocker arm is installed as shown in the illustration.
   b. Apply seal packing in a continuous line approximately 0.138–0.158 inches (3.5–4.0mm) wide.

➡Remove any oil from the contact surface. Install the camshaft housing assembly within 3 minutes. Do not start the engine for at least 2 hours after installing.

   c. Install the right camshaft housing and tighten the 14 bolts to 21 ft. lbs. (28 Nm) in the order shown in the illustration.

WARNING

When installing the camshaft housing, it is necessary to correctly position the camshafts as shown in the illustration. Failure to correctly position these parts may result in damage due to contact between the pistons and valves. If a camshaft is rotated with a piston at TDC, valve contact will occur. If any of the bolts are loosened during installation, remove the camshaft housing, clean the installation surfaces, and reapply seal packing. If the camshaft housing is removed because any of the bolts are loosened during installation, make sure that the previously applied seal packing does not enter any oil passages.

d. Tighten the 9 bolts to 12 ft. lbs. (16 Nm) in the order shown in the illustration.

14. Install left side camshaft bearing cap as follows:

a. Apply engine oil to the camshaft journals, camshaft housing and bearing caps.

b. Install camshaft No. 3 and camshaft No. 4 to the left camshaft housing.

c. Make sure the marks and numbers on the camshaft bearing caps and place them in each proper position and direction.

d. Temporarily tighten the 8 bolts to 7 ft. lbs. (10 Nm) in the order shown in the illustration.

15. Install the left side camshaft housing assembly as follows:

a. Make sure that the valve rocker arm is installed as shown in the illustration.

b. Apply seal packing in a continuous line approximately 0.197–0.217 inches (5.0–5.5mm) wide.

➡Remove any oil from the contact surface. Install the camshaft housing assembly within 3 minutes. Do not start the engine for at least 2 hours after installing.

c. Install the left camshaft housing and tighten the 13 bolts to 21 ft. lbs. (28 Nm) in the order shown in the illustration.

⁂ WARNING

When installing the camshaft housing, it is necessary to correctly position the camshafts as shown in the illustration. Failure to correctly position these parts may result in damage due to contact between the pistons and valves. If a camshaft is rotated with a piston at TDC, valve contact will occur. If any of the bolts are loosened during installation, remove the camshaft housing, clean the installation surfaces, and reapply seal packing. If the camshaft housing is removed because any of the bolts are loosened during installation, make sure that the previously applied seal packing does not enter any oil passages.

d. Tighten the 8 bolts to 12 ft. lbs. (16 Nm) in the order shown in the illustration.

16. Install No. 2 chain tensioner assembly as follows:

a. Install the No. 2 chain tensioner with the bolt. Torque the bolt to 15 ft. lbs. (21 Nm).

b. While pushing in the tensioner, insert a pin of 0.039 in. (1.0mm) into the hole to fix it.

17. Install right side camshaft timing gears and No. 2 chain.

a. Align the mark plate (yellow) with the timing marks (1-dot mark) of the camshaft timing gears as shown in the illustration.

b. Apply a light coat of engine oil to the bolt threads and bolt-seating surface.

c. Align the knock pin of the camshaft with the pin hole of the camshaft timing gear. Install the camshaft timing gear and camshaft timing exhaust gear (right) with the No. 2 chain installed.

d. Hold the hexagonal portion of the camshaft with a wrench, and tighten the 2 bolts to 74 ft. lbs. (100 Nm)

e. Remove the pin from the chain tensioner.

18. Install No. 3 chain tensioner assembly as follows:

a. Install the chain tensioner with the bolt. Torque the bolt to 15 ft. lbs. (21 Nm).

b. While pushing in the tensioner, insert a pin of 0.039 in. (1.0mm) into the hole to hold it.

19. Install left side camshaft timing gears and No. 2 chain.

a. Align the mark plate (yellow) with the timing marks (2-dot mark) of the camshaft timing gears as shown in the illustration.

b. Apply a light coat of engine oil to the bolt threads and bolt-seating surface.

c. Align the knock pin of the camshaft with the pin hole of the camshaft timing gear. Install the camshaft timing gear and camshaft timing exhaust gear (left) with the No. 2 chain installed.

d. Hold the hexagonal portion of the camshaft with a wrench, and tighten the 2 bolts to 74 ft. lbs. (100 Nm)

e. Remove the pin from the chain tensioner.

20. Install or connect the following:

- Timing chain cover, timing chain and timing chain sprockets
- Cylinder head cover and gaskets
- New gaskets, O-ring and rear water by-pass joint. Torque to 7 ft. lbs. (10 Nm).
- New gasket, water inlet and thermostat assembly. Torque to 7 ft. lbs. (10 Nm).
- Oil filter element
- Adhesive sealer, left and right side cylinder block water drain cocks. Torque to 22 ft. lbs. (30 Nm).
- Left and right side oil control valve filter and new gaskets
- Left and right side oil check valve bolt, oil pipe union and oil pipe. Torque to 44 ft. lbs. (60 Nm).
- Crankshaft position sensor. Torque to 7 ft. lbs. (10 Nm).
- 4 camshaft timing oil control valves. Torque to 7 ft. lbs. (10 Nm).
- 4 camshaft position sensors. Torque to 7 ft. lbs. (10 Nm).
- Adhesive and ventilation valve. Torque to 20 ft. lbs. (27 Nm).
- Spark plugs. Torque to 13 ft. lbs. (18 Nm).
- Engine hangers. Torque to 24 ft. lbs. (33 Nm).
- Radiator cap
- Oil filler cap and gasket
- Negative battery cable

21. Refill the coolant and engine oil. Start the engine and check for leaks or abnormal conditions. Perform and road test. Then, recheck for leaks and recheck fluid levels.

4.3L (3UZ-FE) Engine

See Figures 85 through 91.

1. Before servicing the vehicle, refer to the precautions in the beginning of this section.

2. Relieve the fuel pressure from the fuel lines.

3. Remove or disconnect the following:

- Engine under cover
- Drain engine coolant and engine oil
- V-bank cover
- Air cleaner assembly
- Intake air pipe
- Radiator
- Throttle body
- Upper and lower intake manifold assembly
- Timing belt from camshaft pulleys
- Camshaft position sensor and LH timing belt rear plates
- RH timing belt rear plates

⁂ WARNING

Do NOT drop anything inside timing belt cover during this procedure. Keep oil, water and dust from timing belt.

- Power steering pump from engine mount (Do NOT disconnect hoses)
- Catalytic converters
- Water inlet housing assembly
- Water bypass pipe, front bypass joint, and rear bypass joint

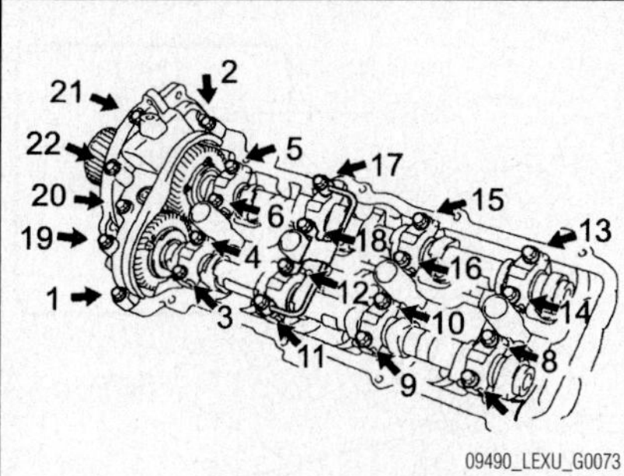

Fig. 85 LH camshaft bearing cap bolt loosening sequence—4.3L (3UZ-FE) Engine

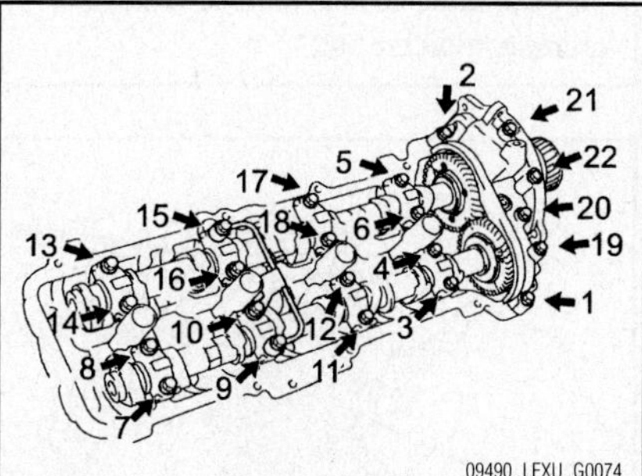

Fig. 86 RH camshaft bearing cap bolt loosening sequence—4.3L (3UZ-FE) Engine

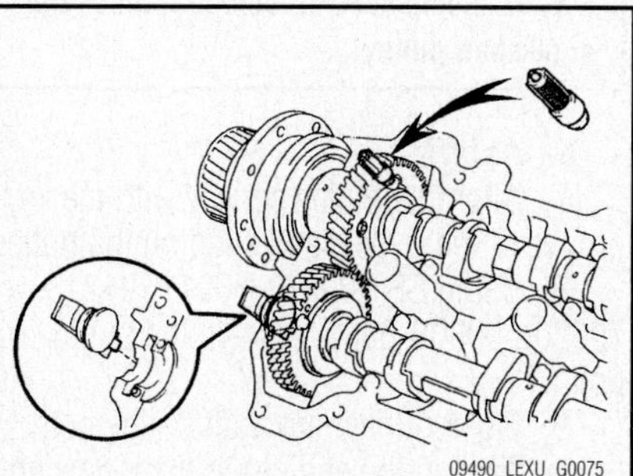

Fig. 87 Installing the camshaft housing plug and oil control valve filter for the LH camshaft—4.3L (3UZ-FE) Engine

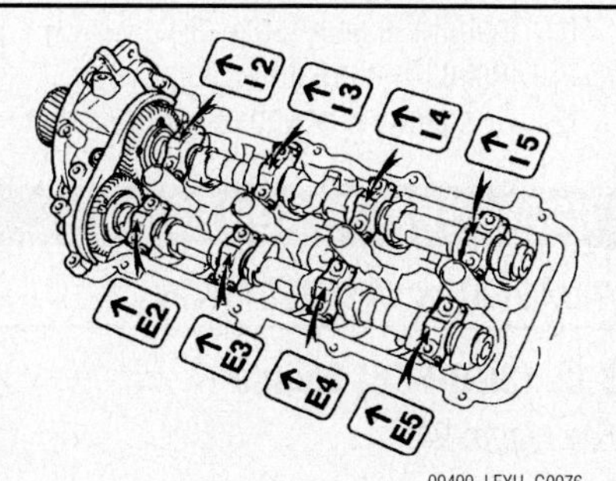

Fig. 88 Installing the camshaft bearing caps in sequence on the camshaft (left shown)—4.3L (3UZ-FE) Engine

- Ignition coils
- Variable valve timing (VVT) sensors
- Engine hangers
- Oil dipsticks and guides for engine oil and transmission fluid
- Cylinder head covers
- Spark plugs
- Semi-circular plugs and camshaft housing plugs
- Camshaft timing oil control valve

✽✽ WARNING

Since the thrust clearance of the camshaft is small, the camshaft must be kept level during removal steps. If it is not kept level, the portion of the cylinder head receiving the shaft thrust may crack or be damaged, causing the camshaft to later seize or break. Follow the procedure carefully to avoid this damage.

4. Check the crankshaft pulley position and ensure the timing mark of the pulley is aligned with the centers of the crankshaft pulley bolt and the No. 2 timing belt idler pulley bolt.

✽✽ WARNING

Having the crankshaft pulley at the wrong angle can cause the piston head and valve head to come into contact with each other during camshaft removal. Always set the crankshaft pulley at the described angle.

5. Using a special wrench, rotate the camshaft timing tube from left to right about 2-3 times, within only a 25°range of movement. Use a waste cloth to collect oil from the camshaft timing oil control valve installation hole.
6. Remove the LH camshafts first. With a hex wrench on the hex portion of the camshaft, rotate so that a 6mm service bolt can be inserted into the bolt hole in the rear face of the camshaft pulley into order to secure the camshaft in place. The bolt should be about 0.63-0.79 in. (16-20mm) long.
7. Align the timing mark (2 dots) of the camshaft drive gear by turning the camshaft with a hex wrench until the timing mark aligns.
8. Now, uniformly loosen the 22 camshaft bearing cap bolts, in several passes, following the sequence shown.
9. Remove the 22 bearing cap bolts, 4 seal washers, oil feed pipe, 9 bearing caps, the camshaft housing plug, the oil control valve filter, and both LH camshafts. Keep all parts in order for proper reinstallation.
10. Remove the RH camshafts. With a hex wrench on the hex portion of the camshaft, rotate so that a 6mm service bolt can be inserted into the bolt hole in the rear face of the camshaft pulley into order to secure the camshaft in place. The bolt should be about 0.63-0.79 in. (16-20mm) long.
11. Align the timing mark (1 dot) of the camshaft main gear about 10° angle by turning the camshaft with a hex wrench until the timing mark aligns.
12. Now, uniformly loosen the 22 camshaft bearing cap bolts, in several passes, following the sequence shown.

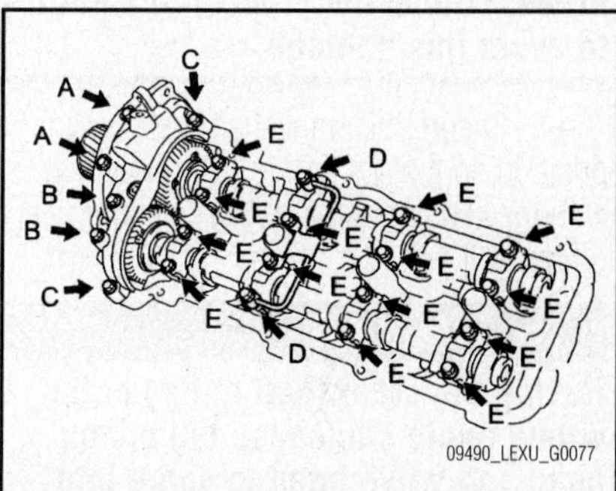

Fig. 89 Identifying the locations of camshaft bearing caps bolts on the LH camshaft—4.3L (3UZ-FE) Engine

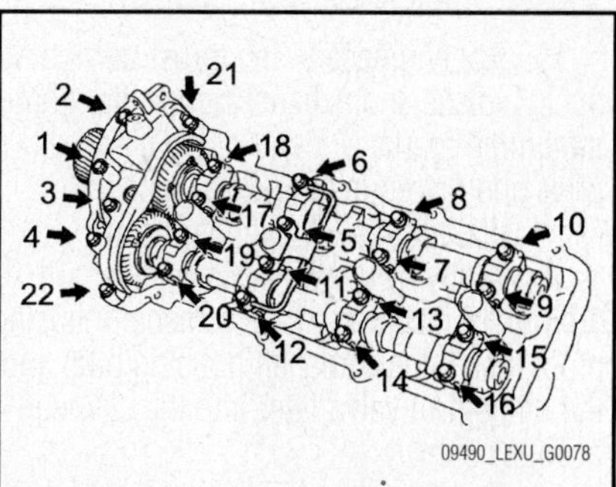

Fig. 90 LH camshaft bearing cap bolt tightening sequence—4.3L (3UZ-FE) Engine

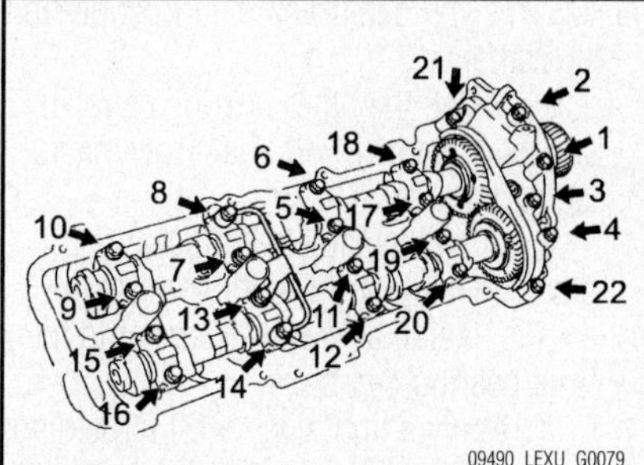

Fig. 91 RH camshaft bearing cap bolt tightening sequence—4.3L (3UZ-FE) Engine

13. Remove the 22 bearing cap bolts, 4 seal washers, oil feed pipe, 9 bearing caps, the camshaft housing plug, the oil control valve filter, and both LH camshafts. Keep all parts in order for proper reinstallation.

To install:

✱✱ WARNING

Since the thrust clearance of the camshaft is small, the camshaft must be kept level during removal steps. If it is not kept level, the portion of the cylinder head receiving the shaft thrust may crack or be damaged, causing the camshaft to later seize or break. Follow the procedure carefully to avoid this damage.

14. Ensure the crankshaft pulley is in position so that its timing mark is in line with the centers of the pulley bolt and idler pulley bolt.

✱✱ WARNING

Having the crankshaft pulley at the wrong angle can cause the piston head and valve head to come into contact with each other during camshaft removal. Always set the crankshaft pulley at the described angle.

15. Apply grease to the thrust portion of the LH intake and exhaust camshafts. Align the timing marks (2 dots) of the camshaft drive and driven main gears. Place the camshafts into the LH cylinder head.
16. Apply new seal packing material around the opening of the camshaft housing plug. Install the camshaft housing plug and the oil control valve filter into the cylinder head, as shown.
17. Remove any old packing material, then install new packing material around the mounting edge (not in the grooves) of the front bearing cap.
18. Position the front bearing cap in place. This will determine the thrust portion of the camshaft.
    a. Install the other bearing caps, in sequence shown, with the arrow marks facing forward.
    b. Push the camshaft oil seal into place by pushing from the front of the engine. Install a new seal washer to the front bearing cap bolts.
    c. Apply a light coat of oil to bearing cap bolt threads and under the heads of the bearing cap bolts "D" and "E", as shown. Do NOT apply engine oil under the heads of bearing cap bolts "A", "B" and "C".
    d. Bolt lengths vary for each bearing cap. Refer to the illustration for each of the following bolts:
    - Bolt "A" with seal washer is 3.70 in. (94mm)
    - Bolt "B" with seal washer is 2.83 in. (72mm)
    - Bolt "C" is 0.98 in. (25mm)
    - Bolt "D" is 2.05 in. (52mm)
    - Bolt "E" is 1.50 in. (38mm)
19. Install the oil feed pipe and the 22 bearing cap bolts in their respective locations. Uniformly tighten the 22 bearing cap bolts, in several passes, following the sequence as shown. Torque bolt "C" to a final torque of 66 inch lbs. (7.5 Nm). Torque all other bearing cap bolts to a final torque of 12 ft. lbs. (16 Nm).
20. Remove the service bolt installed in the rear face of the gear.
21. Repeat this entire procedure for the RH camshafts. Use the illustrations given for the LH camshafts, as the sequences are the same on the RH camshafts.
22. Check and adjust valve lash. See VALVE LASH.
23. Install or reconnect the following:
    - Camshaft timing oil control valve.
    - Semi-circular plugs in rear ends of each cylinder head
    - Cylinder head covers, with new gaskets and packing material
    - Engine hangers
    - Variable valve timing sensors
    - Oil dipsticks and tubes
    - All remaining components in reverse of removal procedure.

CRANKSHAFT DAMPER

REMOVAL & INSTALLATION

See Figures 92 and 93.

1. Before servicing the vehicle, refer to the precautions in the beginning of this section.
2. Disconnect the negative battery cable. Wait at least 90 seconds before performing any other work.
3. It may be necessary to remove the radiator assembly for access.
4. Remove the accessory drive belt(s).
5. Using SST 09213-70011 (91213-70020), 09330-00021, loosen the crankshaft pulley bolt.
6. Using SST 09950-50013 (09951-05010, 09952-05010, 09953-05010, 09954-05021), remove the crankshaft pulley.

➡Before using SST, apply lubricating oil on the threads and tip of the center bolt.

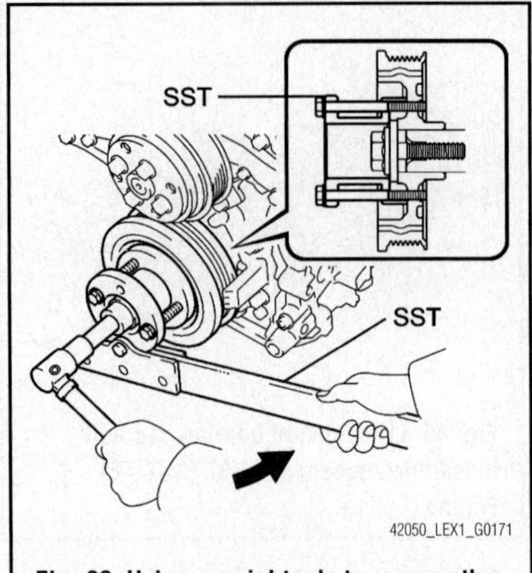

Fig. 92 Using special tools to remove the crankshaft pulley bolt

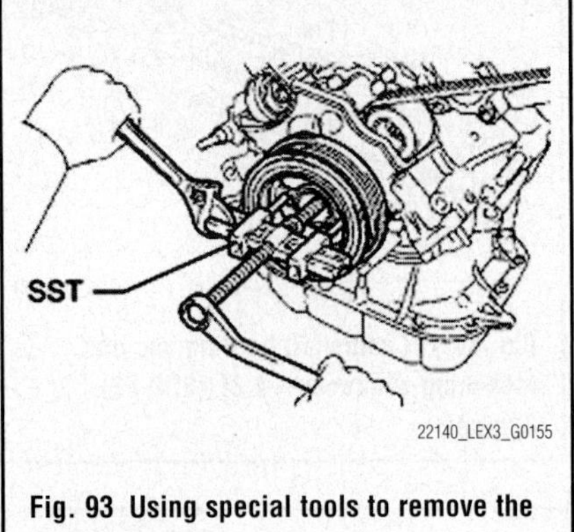

Fig. 93 Using special tools to remove the crankshaft pulley

To install:

7. Align the pulley set key with the key groove of the pulley, and slide on the pulley.
8. Using SST 09213-70011 (91213-70020), 09330-00021, install the pulley bolt.
9. Tighten the crankshaft pulley bolt (depending on the engine) to the following values:
    - 3.0L (3GR-FSE) engine: 184 ft. lbs. (250 Nm).
    - 4.3L (3UZ-FE) engines: 181 ft. lbs. (245 Nm).
10. Install the accessory drive belt(s).
11. Install the radiator, if removed.
12. Connect the negative battery cable.

CRANKSHAFT FRONT SEAL

REMOVAL & INSTALLATION

3.0L (3GR-FSE) Engine

See Figure 94.

1. Before servicing the vehicle, refer to the precautions in the beginning of this section.
2. Disconnect the negative battery

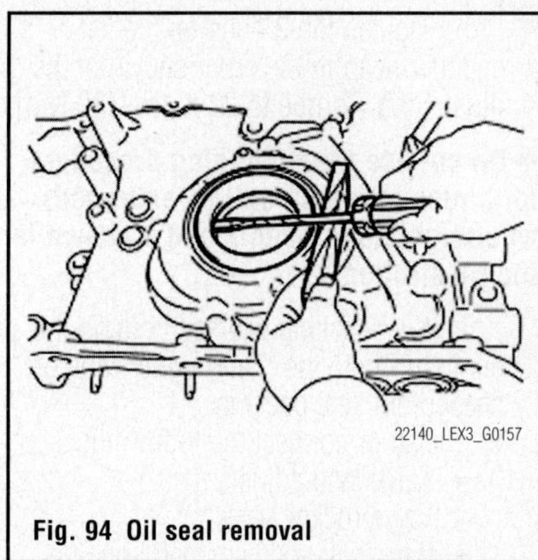
Fig. 94 Oil seal removal

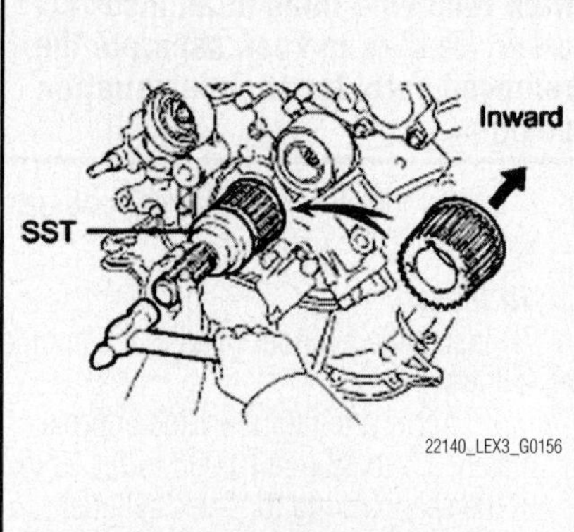

Fig. 95 Crankshaft pulley installation.

cable. Wait at least 90 seconds before performing any other work.

3. It may be necessary to remove the radiator assembly for access.
4. Remove the accessory drive belt.
5. Using SST 09213-70011 (91213-70020), 09330-00021, loosen the crankshaft pulley bolt.
6. Using SST 09950-50013 (09951-05010, 09952-05010, 09953-05010, 09954-05021), remove the crankshaft pulley.
7. Using a screwdriver with its tip taped, pry out the timing chain case oil seal.

➡**After the removal, check the crankshaft for damage. If it is damaged, smooth the surface with 400-grit sandpaper.**

To install:

8. Using SST 09223-22010, 09506-35010 and a hammer, tap in the oil seal until its surface is flush with the rear oil seal retainer edge. Keep the lip free of foreign matter. Do not tap the oil seal at an angle.
9. Install the crankshaft pulley.
10. Install the accessory drive belt.
11. Install the radiator assembly.
12. Inspect for oil leaks.
13. Connect the negative battery cable.

4.3L (3UZ-FE) Engine

See Figure 95.

1. Before servicing the vehicle, refer to the precautions in the beginning of this section.
2. Disconnect the negative battery cable. Wait at least 90 seconds before performing any other work.
3. It may be necessary to remove the radiator assembly for access.
4. Remove the accessory drive belt.
5. Using SST 09213-70011 (91213-70020), 09330-00021, loosen the crankshaft pulley bolt.
6. Remove the timing belt.
7. Using SST 09950-50013 (09951-05010, 09952-05010, 09953-05010, 09954-05021), remove the crankshaft pulley.
8. Using SST 09950–50013 (09951–05010, 09952–05010, 09953–05010, 09953–05020, 09954–05011), remove the timing pulley.

➡**Do not turn the timing pulley.**

9. Using a knife, cut the oil seal lip.
10. Using a screwdriver with its tip taped, pry out the oil seal.
11. After the removal, check if the crankshaft is not damaged. If it is damaged, smooth the surface with 400–grit sandpaper.

To install:

12. Apply MP grease to a new oil seal lip. Keep the lip free from foreign materials.
13. Using SST 09316–60011 (09316–00011) and a hammer, tap in the oil seal until its surface is flush with the oil pump edge. Wipe off any extra grease on the crankshaft. Be careful not to tap the oil seal at an angle.
14. Align the timing pulley set key with the key groove of the pulley.
15. Face the timing pulley's flange side inward. Using SST 09223–46011 and a hammer, tap in the timing pulley.
16. Install the timing belt.
17. Connect the negative battery cable.
18. Check for leaks.

CYLINDER HEAD

REMOVAL & INSTALLATION

3.0L (3GR-FSE) Engine

See Figures 96 through 100.

1. Before servicing the vehicle, refer to the precautions in the beginning of this section.
2. Drain the cooling system.
3. Drain the engine oil.
4. Relieve the fuel system pressure.
5. Remove or disconnect the following:
   - Negative battery cable. Wait at least 90 seconds before performing any other work
   - Oil filler cap and gasket
   - Radiator cap
   - No. 1 and 2 engine hangers
   - Spark plugs
   - Ventilation valve
   - 4 camshaft position sensors
   - 4 camshaft timing oil control valves
   - Crankshaft position sensor
   - Left and right side oil check valve bolt, oil pipe union and oil pipe

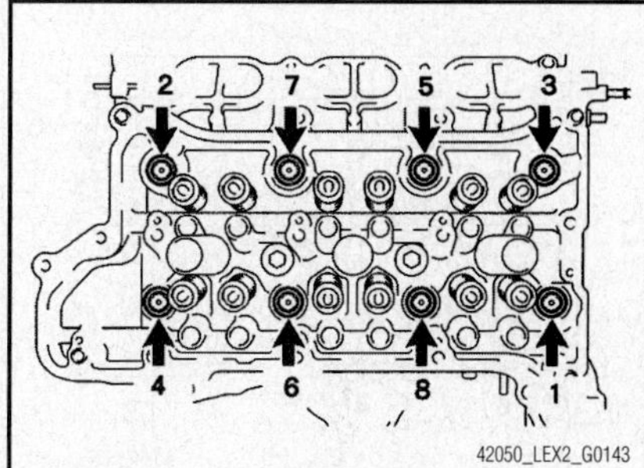

Fig. 96 Cylinder head bolt loosening sequence (right side)—3.0L (3GR-FSE) engine

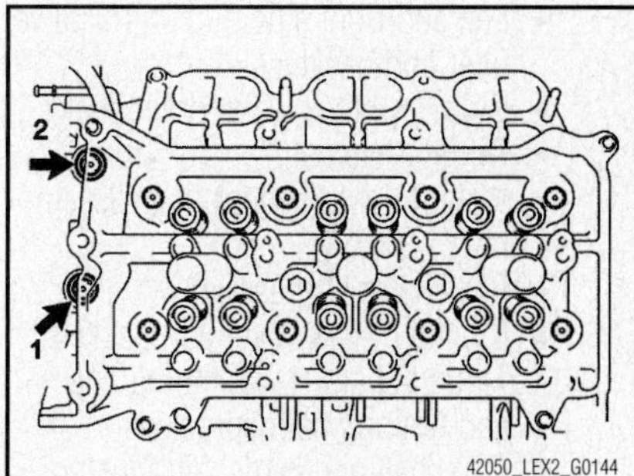

Fig. 97 Cylinder head 14mm bolt loosening sequence (left side)—3.0L (3GR-FSE) engine

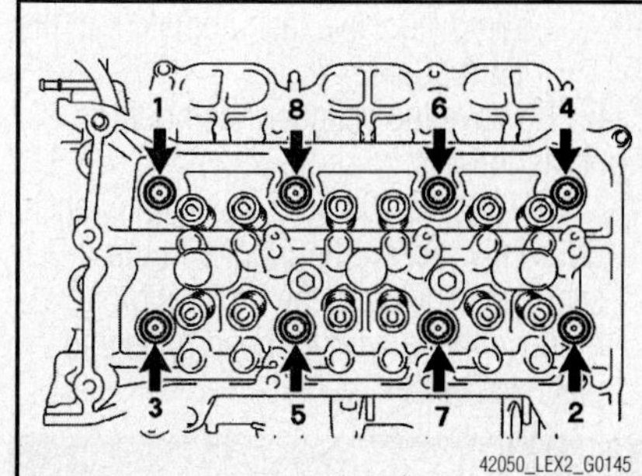

Fig. 98 Cylinder head bolt loosening sequence (left side)—3.0L (3GR-FSE) engine

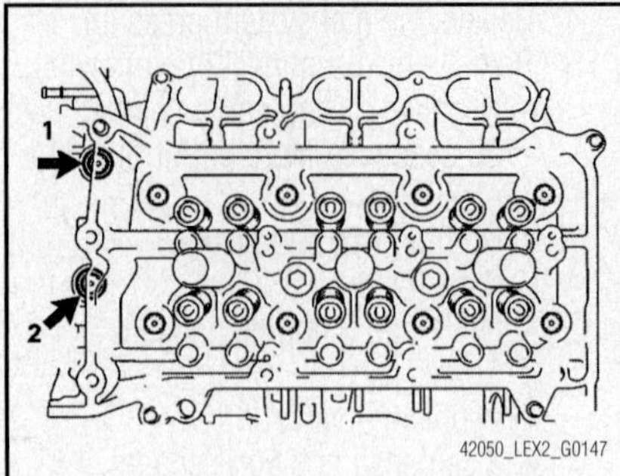

Fig. 99 Cylinder head 14mm bolt tightening sequence (left side)—3.0L (3GR-FSE) engine

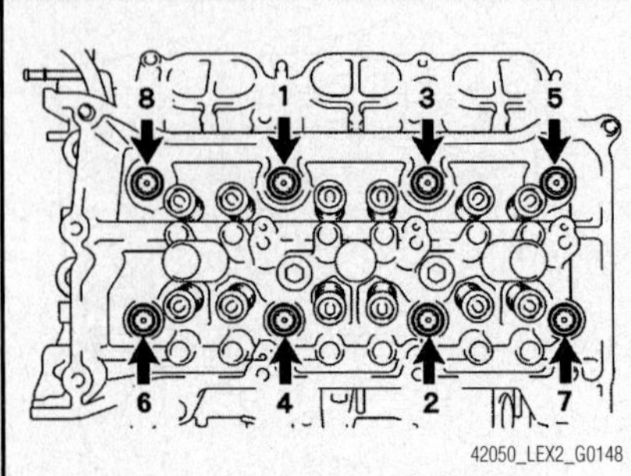

Fig. 100 Cylinder head bolt tightening sequence (left side)—3.0L (3GR-FSE) engine

- Left and right side oil control valve filter and gaskets
- Cylinder block water drain cocks
- Oil filter element
- Water inlet and thermostat assembly
- Rear water by-pass joint
- Cylinder head cover and gaskets
- Timing chain cover, timing chain and timing chain sprockets
- Camshaft and camshaft housing assembly
- Valve rocker arms. Arrange the removed rocker arms in the correct order.
- Valve lash adjusters. Arrange the removed valve lash adjusters in the correct order.

6. Remove the cylinder head (left or right) as follows:
   a. Using a 10mm bi-hexagon wrench, uniformly loosen the 8 bolts in the sequence shown in the illustration. Remove the 8 cylinder head bolts and plate washers.

**** WARNING**

Be careful not to drop washers into the cylinder head. Cylinder head warpage or cracking could result from removing bolts in an incorrect order. Be sure to keep separate the removed parts for each installation position.

   b. Remove the cylinder head and gasket.

To install:

7. Install the cylinder head to the engine as follows:
   a. Apply a continuous line approximately 2.5 to 3.0mm (0.098 to 0.118 in.) of the seal packing to a new cylinder head gasket.

➡Remove any oil from the contact surface. Install the cylinder head gasket within 3 minutes after applying the seal packing. Install the cylinder head bolt within 15 minutes after applying the seal packing. Do not apply engine oil within 2 hours of installation.

   b. Place the cylinder head gasket on the cylinder block surface with the Lot No. stamp upward.

**** WARNING**

Be careful of the installation direction. Gently place the cylinder head in order not to damage the gasket with the bottom part of the head.

   c. Place the cylinder head on the cylinder block.

**** WARNING**

Be careful not to allow oil to adhere to the bottom part of the cylinder head.

   d. Apply a light coat of engine oil to the threads and under the heads of the cylinder head bolts.
   e. Using a 10mm bi-hexagon wrench, install and uniformly tighten the 8 cylinder head bolts with the plate washers to 27 ft. lbs. (36 Nm) in the sequence shown in the illustration. If any of the bolts does not meet the torque, replace it.
   f. Mark the forward edge of each bolt with paint, then retighten each bolt, in proper sequence, an additional 90 degrees. Check that each painted mark is now at a 90 degrees angle to the front. The paint mark should have been applied to the bolt in the 9 o'clock position and should now be in the 12 o'clock position.
   g. Tighten each bolt again, in proper sequence, an additional 90 degrees. Check that each painted mark is now facing rearward.
   h. Tighten the 2 bolts on the left cylinder head in the order shown in the illustration. Torque to 22 ft. lbs. (30 Nm).

➡Do not use the tightening procedure for a plastic region bolt (if equipped) when tightening bolts 1 and 2 shown in the illustration.

   i. Seal packing will seep out on the engine's front side. Thoroughly wipe off seeped out seal packing.

8. Install or connect the following:
   - Valve lash adjusters
   - Valve rocker arms
   - Camshaft and camshaft housing assembly
   - Timing chain cover, timing chain and timing chain sprockets
   - Cylinder head cover
   - New gaskets, O-ring and rear water by-pass joint. Torque to 7 ft. lbs. (10 Nm).
   - New gasket, water inlet and thermostat assembly. Torque to 7 ft. lbs. (10 Nm).
   - Oil filter element
   - Adhesive sealer, left and right side cylinder block water drain cocks. Torque to 22 ft. lbs. (30 Nm).
   - Left and right side oil control valve filter and gaskets
   - Left and right side oil check valve bolt, oil pipe union and oil pipe. Torque to 44 ft. lbs. (60 Nm).
   - Crankshaft position sensor. Torque to 7 ft. lbs. (10 Nm).
   - 4 camshaft timing oil control valves
   - 4 camshaft position sensors
   - Adhesive and ventilation valve
   - Spark plugs. Torque to 13 ft. lbs. (18 Nm).
   - Engine hangers. Torque to 24 ft. lbs. (33 Nm).
   - Radiator cap
   - Oil filler cap and gasket
   - Negative battery cable

9. Refill the coolant and engine oil. Start the engine and check for leaks or abnormal conditions. Perform and road test. Then, recheck for leaks and recheck fluid levels.

4.3L (3UZ-FE) Engine

See Figures 101 through 103.

1. Before servicing the vehicle, refer to the precautions in the beginning of this section.
2. Relieve the fuel system pressure.
3. Remove or disconnect the following, as applicable to each engine:
   - Engine under cover
   - Drain engine coolant

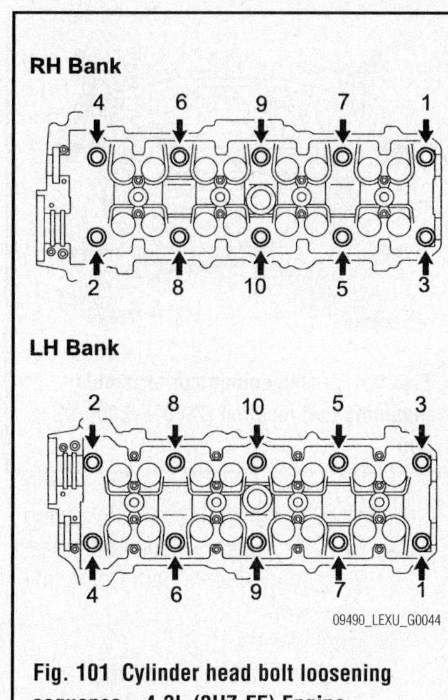

Fig. 101 Cylinder head bolt loosening sequence—4.3L (3UZ-FE) Engine

- V-bank cover
- Air cleaner assembly
- Intake air pipe
- Radiator (if necessary)
- Throttle body
- Upper and lower intake manifold assembly
- Camshaft position sensor and LH timing belt rear plates
- RH timing belt rear plates

WARNING

Do NOT drop anything inside timing belt cover during this procedure. Keep oil, water and dust from timing belt.

- Power steering pump from engine mount (Do NOT disconnect hoses)
- Catalytic converters
- Water inlet housing assembly
- Water bypass pipe, front bypass joint, and rear bypass joint
- Ignition coils
- Variable valve timing (VVT) sensors
- Engine hangers
- Oil dipsticks and guides for engine oil and transmission fluid
- Cylinder head covers
- Spark plugs
- Camshafts

WARNING

Since the thrust clearance of the camshaft is small, the camshaft must be kept level during removal. If not, the portion of the cylinder head receiving the camshaft thrust may crack or be otherwise damaged, causing the camshaft to later seize or break. Follow the camshaft removal procedure carefully as given in this section.

- Both oxygen sensor connectors
- Ground wire from LH cylinder head
- Engine wire bracket for oxygen sensor on LH cylinder head

4. Uniformly loosen the 10 cylinder head bolts on each cylinder head, in several passes, following loosening sequence as shown.

WARNING

Use care so that no bolts or washers are dropped into the recesses or enclosed portions of the cylinder head or block.

5. Carefully lift the cylinder head from the locating dowels on the engine block. Place cylinder heads on wooden blocks on the workbench.

6. If necessary, exhaust manifolds may be removed from the cylinder heads at this time.

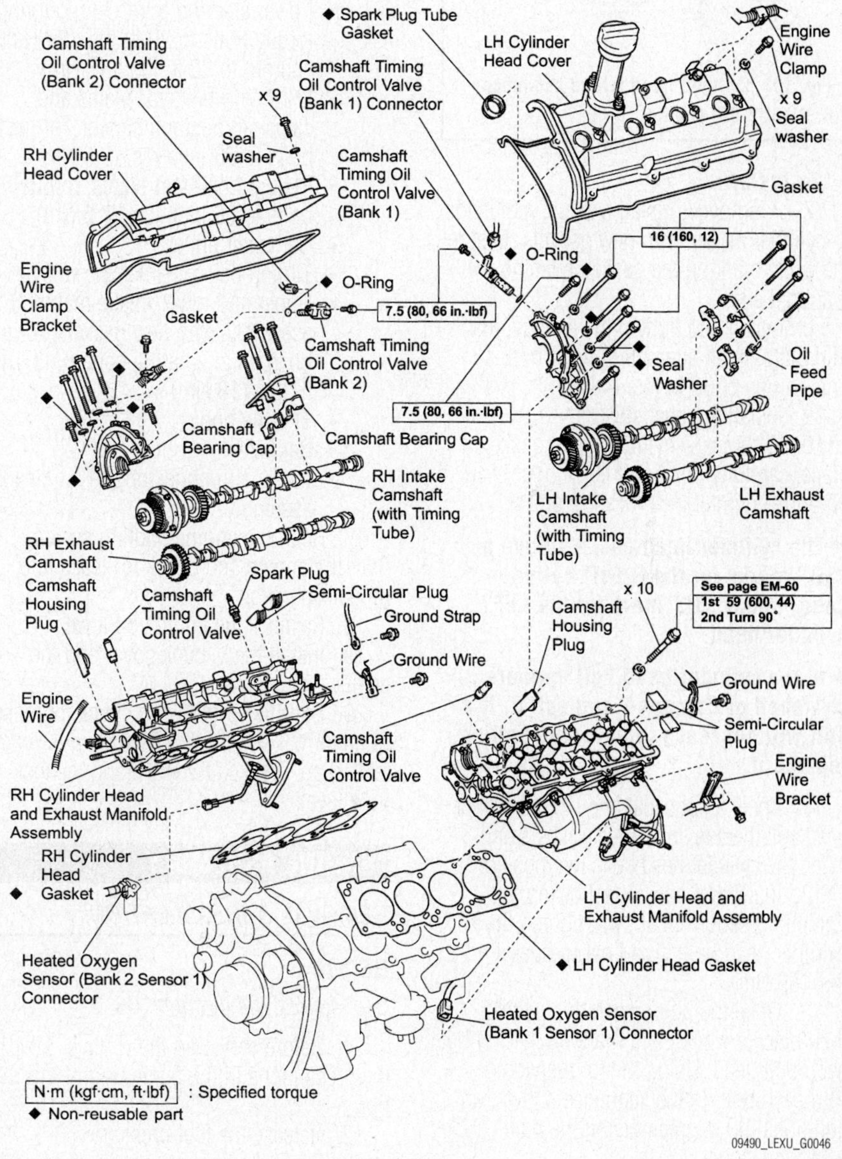

Fig. 102 Exploded view of the cylinder heads and related components—4.3L (3UZ-FE) Engine

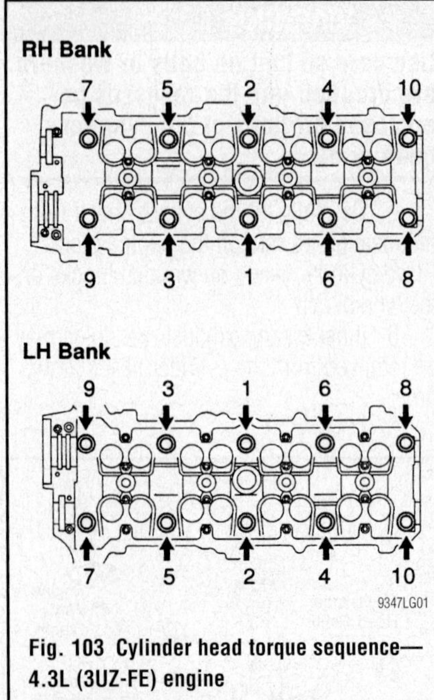

Fig. 103 Cylinder head torque sequence—4.3L (3UZ-FE) engine

To install:

7. If removed, install exhaust manifolds to cylinder heads, with new gaskets. Ensure the white mark on the gasket is facing the manifold side.
8. Install and tighten the exhaust manifold bolts, in an alternating pattern, to a final torque of 32 ft. lbs. (44 Nm).
9. Install the heat shields.
10. With a new cylinder head gasket in place, carefully position the cylinder head onto the engine block locating dowels.

➡The cylinder head gaskets have a "3R" marks for the RIGHT cylinder head, and a "3L" mark for the LEFT cylinder head.

➡If any cylinder head bolt appears stretched or damaged, replace it. If a bolt will not reach final torque setting, replace it.

11. Apply a light coat of oil to the cylinder head bolt threads. Install the washers and insert the cylinder head bolts into position.
12. In several passes, following the tightening sequence shown, tighten the cylinder head bolts to a final torque of 44 ft. lbs. (59 Nm).
13. Once the bolts reach this setting, then place a white paint mark on the front of each bolt head. Using the torque wrench, turn each bolt, in the sequence shown, an additional 90 degrees, using the paint mark as a reference.
14. Install or reconnect the following:
    - Engine wiring and ground straps
    - Oxygen sensor wire bracket on LH cylinder head
    - Spark plugs
    - Camshafts
15. Inspect and adjust the valve lash.
16. Install or reconnect the following:
    - Cylinder head covers, with new gaskets
    - Engine hangers
    - VVT sensors
    - Engine and transmission dipsticks and tubes
    - Ignition coils
    - Water bypass joints and pipe; torque nuts and bolt to 13 ft. lbs. (18 Nm)
    - Water inlet housing assembly
    - Catalytic converters, with new gaskets and new nuts; torque nuts to 46 ft. lbs. (62 Nm)
    - Power steering pump to mounting; torque bolts to 29 ft. lbs. (39 Nm) and nut to 32 ft. lbs. (43 Nm)
    - LH timing belt rear plates and camshaft position sensor; torque bolts to 66 inch lbs. (7.5 Nm)
    - RH right belt rear plates; torque bolts to 66 inch lbs. (7.5 Nm)
    - Camshaft pulleys
    - Timing belt to camshaft pulleys
    - Upper and lower intake manifold assembly, with new gaskets; torque bolts, in alternating pattern, to 13 ft. lbs. (18 Nm)
    - Throttle body
    - Radiator
    - Intake air connector and air cleaner assembly
17. Refill the engine cooling system. Start the engine and check for leaks and proper operation.
18. Recheck the engine oil level.
19. Install the V-bank cover and the engine under cover.
20. Start the engine and check for leaks or abnormal conditions. Perform and road test. Then, recheck for leaks and recheck fluid levels.

ENGINE ASSEMBLY

REMOVAL & INSTALLATION

GS 300

See Figures 104 through 108.

1. Before servicing the vehicle, refer to the precautions in the beginning of this section.
2. Release the fuel pressure.
3. Drain the engine coolant and engine oil.
4. Drain transmission fluid.
5. Drain differential fluid (AWD).

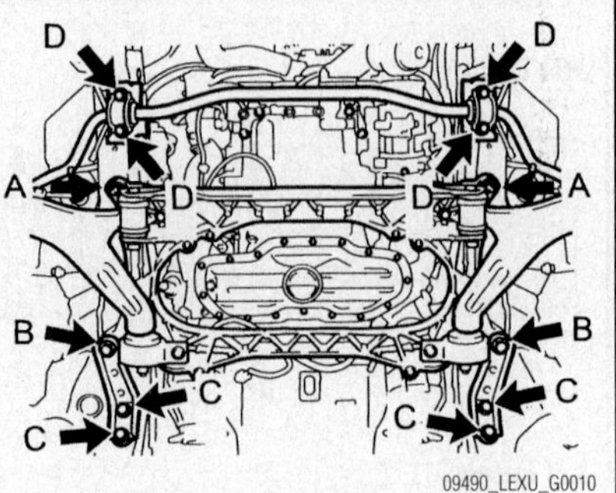

Fig. 104 Crossmember sub-assembly mounting bolt location (2WD)—2006 GS 300

6. Remove or disconnect the following:
    - Negative battery cable. Wait at least 90 seconds before performing any other work.
    - Front wheels
    - Hood insulator pad and the hood
    - 7 clips and cool air intake duct seal
    - Nut, 2 clips and right engine room side cover
    - 3 clips and left engine room side cover
    - 2 nuts and V-bank cover
    - Engine under cover
    - 4 screws, 2 grommets, 2 spacers and No. 2 engine under cover (2WD)
    - Front exhaust pipe assembly
    - Propeller with center bearing shaft assembly
    - Front propeller shaft assembly (AWD)
    - Bolt and air cleaner inlet
    - Ventilation hose from the cylinder head
    - Mass Air Flow (MAF) meter connector
    - Clamp from the air cleaner
    - EVAP VSV
    - Hose clamp
    - 3 bolts and air cleaner case

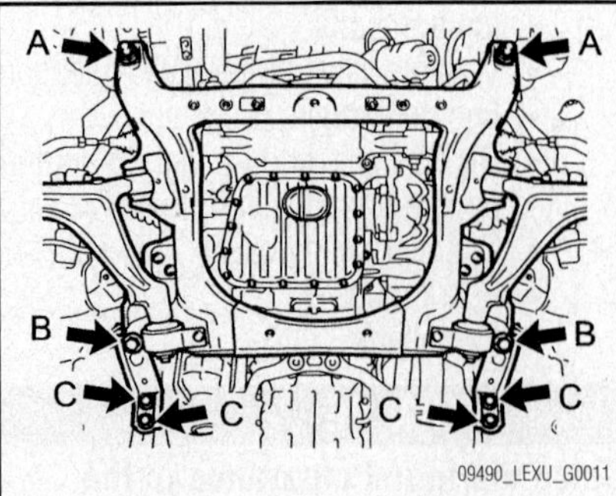

Fig. 105 Crossmember sub-assembly mounting bolt location (AWD)—2006 GS 300

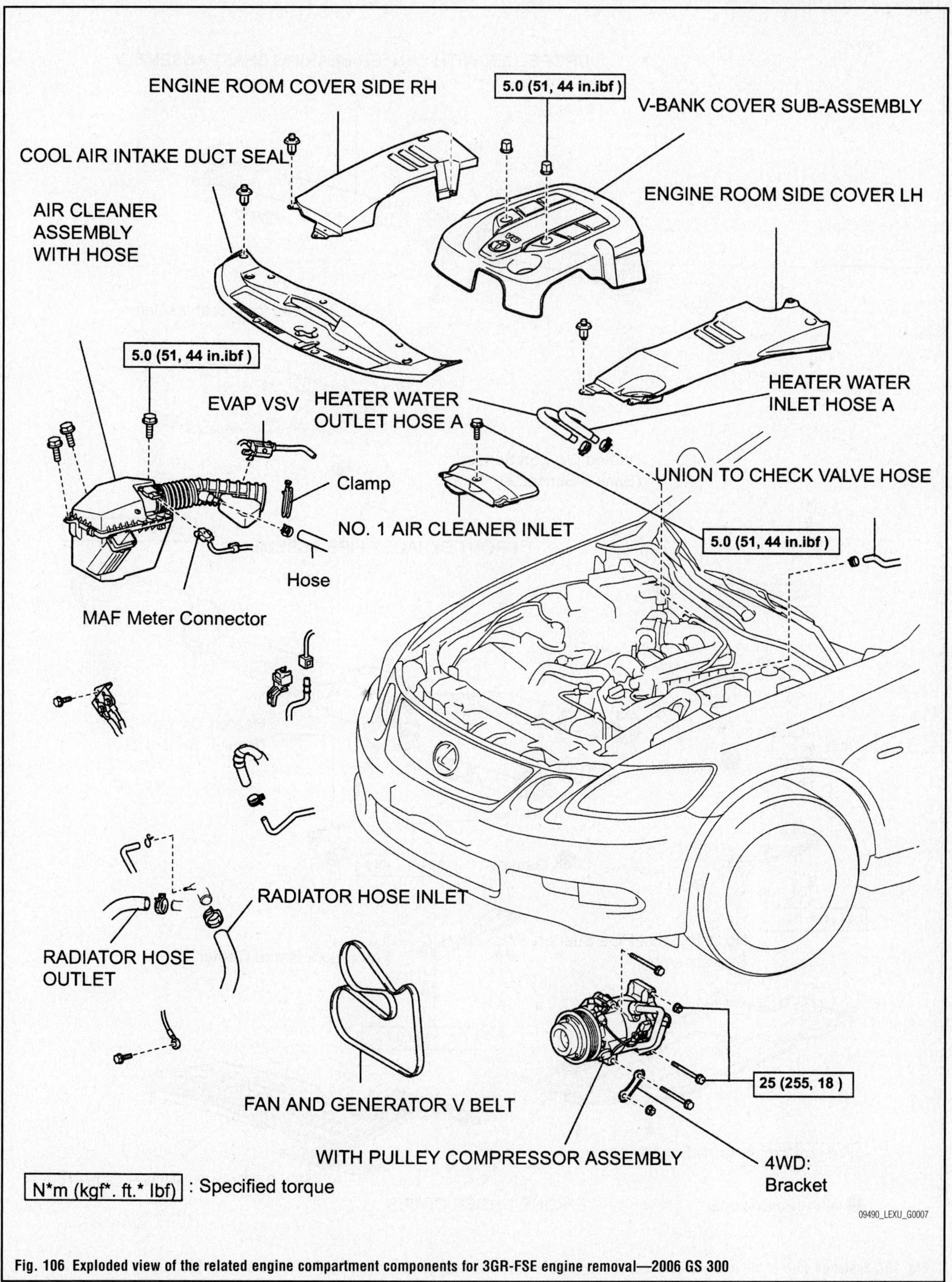

Fig. 106 Exploded view of the related engine compartment components for 3GR-FSE engine removal—2006 GS 300

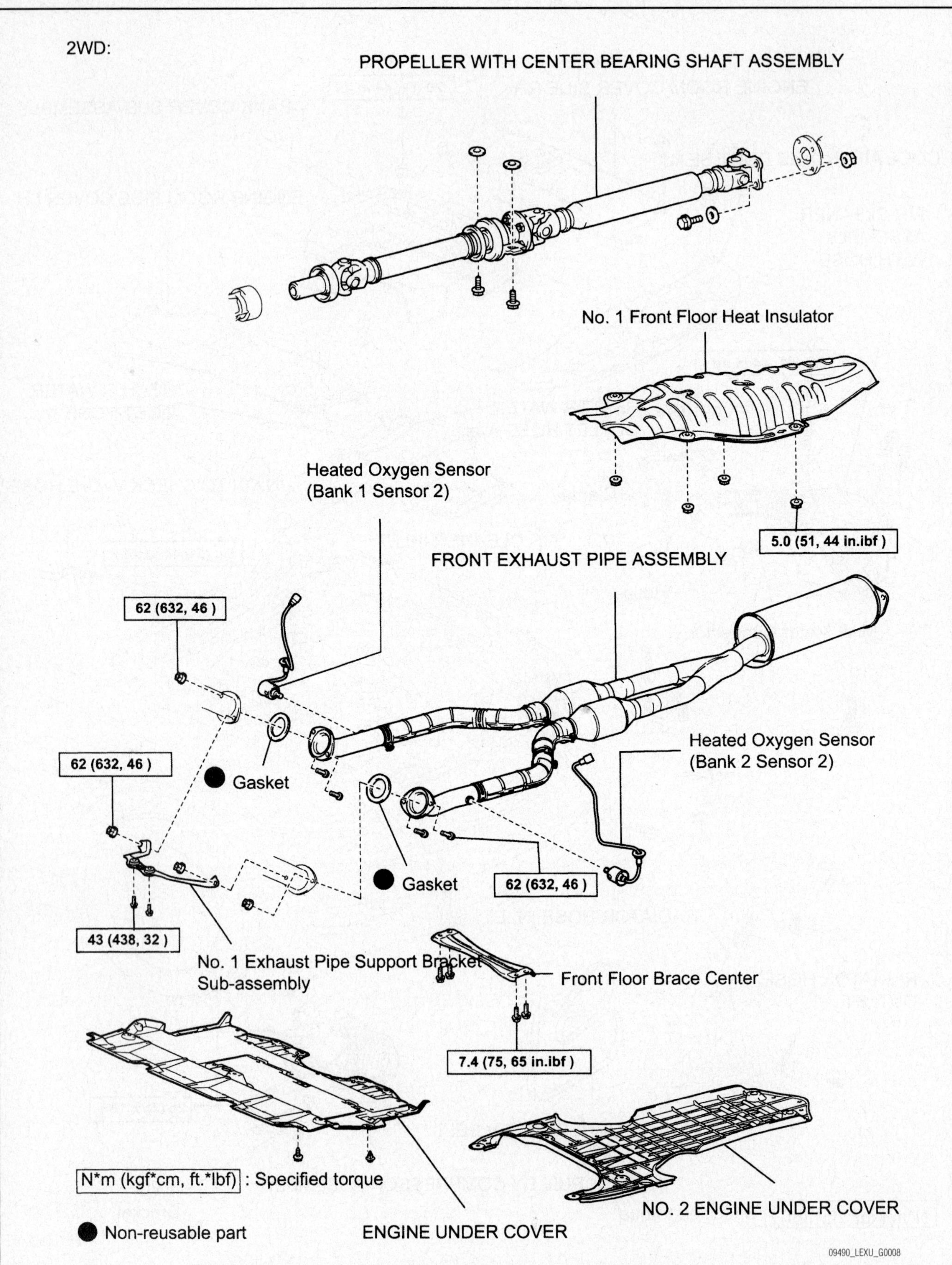

Fig. 107 Exploded view of the related underbody components for 3GR-FSE engine removal (cont.)—2006 GS 300 (2WD)

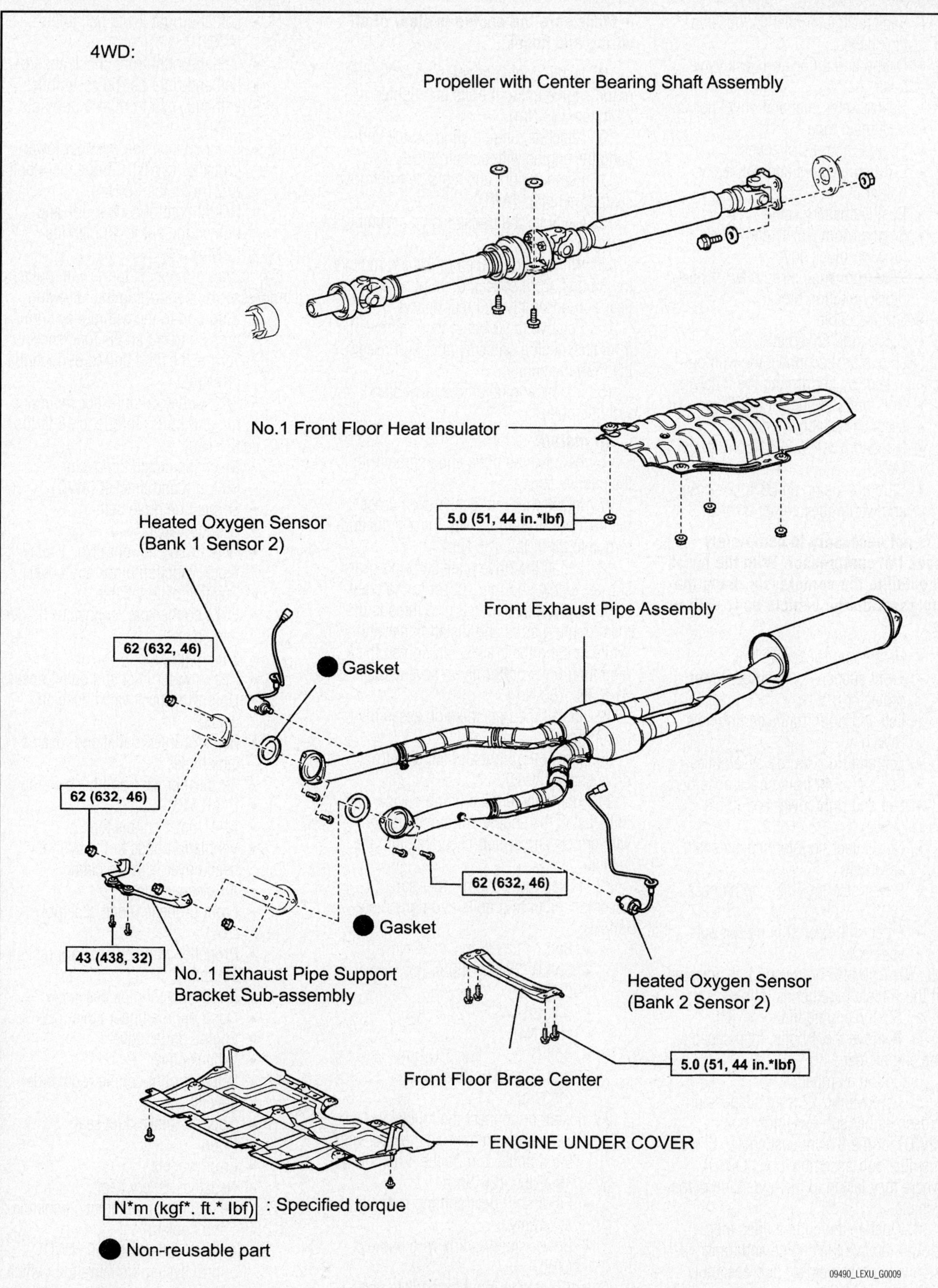

Fig. 108 Exploded view of the related underbody components for 3GR-FSE engine removal (cont.)—2006 GS 300 (AWD)

- Radiator inlet, outlet and reserve tank hoses
- Clamp and union-to-check valve hose
- Heater water inlet and outlet hoses
- Fuel main tube
- Engine room ECM cover
- ECM connectors and connector holder
- Positive battery cable
- Engine room No. 1 relay block, junction block cover
- Wire from engine room No. 1 relay block, junction block
- Ground cable
- Serpentine drive belt

7. Remove or disconnect the with pulley compressor by performing the following:
   - Bolt, nut and bracket (AWD)
   - Magnetic clutch connector
   - Nut and 3 bolts (2WD); or 2 bolts (AWD)
   - Stud bolt using an E8 Torx®socket and with pulley compressor

➡It is not necessary to completely remove the compressor. With the hoses connected to the compressor, hang the compressor on the vehicle body with a rope.

   - Height control sensor link
   - Front suspension member lower protector (2WD)
   - Left and right front axle hub nuts (AWD)
   - Left and right tie rod assemblies
   - Left and right front shock absorbers
   - Left and right lower ball joints (AWD)
   - No. 2 steering intermediate shaft assembly
   - Power steering link wire harness (2WD)
   - Floor shift gear shifting rod sub-assembly

8. Remove the engine and transmission from the vehicle together as follows:
   a. Set the engine lifting device.
   b. Remove the 4 bolts, then separate the engine rear mounting member.
   c. Remove the 12 front suspension crossmember sub-assembly bolts (2WD); or the 8 front suspension crossmember sub-assembly bolts (AWD) where they fasten to the underside of the vehicle
   d. Operate the engine lifter, then slowly remove the engine and transmission with crossmember sub-assembly from the vehicle.

➡Make sure the engine is clear of all wiring and hoses.

9. Install engine hangers on each side of the engine using 4 bolts and tighten to 24 ft. lbs. (33 Nm).
10. Attach an engine sling device and hang the engine with a chain block.
11. Remove right and left side front drive shaft assemblies (AWD).
12. Separate the transmission assembly from the engine.
13. While holding the crankshaft, remove the 8 bolts, front spacer, drive plate and ring gear sub-assembly and rear spacer.
14. Remove the 2 bolts, then remove the front suspension crossmember sub-assembly from the engine.
15. Fix the engine onto engine stand with the bolts.

To install:

16. Remove the bolts and engine from the engine stand.
17. Install the front suspension crossmember sub-assembly with the 2 bolts and torque to 26 ft. lbs. (35 Nm).
18. Install the drive plate and ring gear sub-assembly and rear spacer to the crankshaft. Apply adhesive to 2 or 3 threads of the mounting bolts and install finger tight. While holding the crankshaft, tighten the 8 mounting bolts uniformly in several steps to 61 ft. lbs. (83 Nm).
19. Install the transmission assembly to the engine.
20. Install the right and left side front drive shafts (AWD).
21. Using the engine lifting device, carefully install the engine and transmission with crossmember sub-assembly into the vehicle.
22. Install the crossmember sub-assembly mounting bolts and tighten as follows:
    - Bolt A: 123 ft. lbs. (167 Nm)
    - Bolt B (2WD): 150 ft. lbs. (204 Nm)
    - Bolt B (AWD): 121 ft. lbs. (165 Nm)
    - Bolt C: 37 ft. lbs. (50 Nm)
    - Bolt D (2WD): 36 ft. lbs. (49 Nm)
23. Install or connect the following:
    - Engine rear mounting member with the 4 bolts. Torque the 4 bolts to 19 ft. lbs. (26 Nm).
    - Floor shift gear shifting rod sub-assembly
    - Power steering link wire harness (2WD)
    - No. 2 steering intermediate shaft assembly
    - Left and right lower ball joints (AWD)
    - Left and right front shock absorbers
    - Left and right tie rod assemblies
    - Left and right front axle hub nuts (AWD)
    - Front suspension member lower protector (2WD). Torque the 4 bolts to 71 inch lbs. (8 Nm).
    - Height control sensor link and tighten the nut to 48 inch lbs. (5 Nm).
24. Install or connect to the with pulley compressor by performing the following:
    - Stud bolt to the with pulley compressor using an E8 Torx®socket. Torque the stud bolt to 89 inch lbs. (10 Nm).
    - With pulley compressor mounting nut and bolts. Torque to 18 ft. lbs. (25 Nm).
    - Magnetic clutch connector
    - Bolt, nut and bracket (AWD)
    - Serpentine drive belt
    - Ground cable
    - Wire to engine room No. 1 relay block, junction block and cover
    - Positive battery cable
    - ECM connectors, connector holder and cover
    - Fuel main tube
    - Heater water inlet and outlet hoses
    - Union-to-check valve hose and clamp
    - Radiator inlet, outlet and reserve tank hoses
    - Air cleaner case and hose clamp
    - EVAP VSV
    - MAF meter connector
    - Ventilation hose to the cylinder head cover with the clamp
    - Air cleaner inlet
    - Front propeller shaft assembly (AWD)
    - Propeller with center bearing shaft assembly
    - Front exhaust pipe assembly
    - No. 2 engine under cover (2WD)
    - Engine under cover
    - V-bank cover
    - Left and right engine room side covers
    - Cool air intake duct seal
    - Hood
    - Front wheels
    - Negative battery cable
25. Add engine oil, automatic transmission fluid and coolant
26. Add front differential oil (AWD)
27. Perform system initialization (which includes power window control system, sliding roof system, clearance sonar system

and variable gear ratio steering system) procedure as follows:

- Power window control system
- Turn the ignition switch on.
- Open power window halfway by pressing power window switch.
- Fully pull up the switch until the power window is fully closed and continue to hold the switch for at least 1 second.
- Check that the AUTO UP / DOWN function operates normally.

➡If the remote UP / DOWN function does not operate after the conditions 1, 2, or 3 is satisfied, the power window regulator master switch may have a malfunction.

28. Sliding roof system:
   - Turn the ignition switch on.
   - If the sliding roof is opened, close it fully.
   - Push the open switch of the slide switch, or the up switch of the tilt switch on the personal light, making the sliding roof tilt up approximately 1 second, tilt down, slide open, slide close.
   - Sliding roof stops at the fully closed position.
   - Finish the initialization.
   - Check that the operation works normally with AUTO operation.
29. Clearance sonar system:
   - Turn the ignition switch on.
   - Turn the clearance sonar main switch ON.
   - Turn the steering wheel to the full left and right lock position.

➡Make sure to completely turn the steering wheel to the left and right full lock position.

- Confirm that the learning operation has been completed by checking the multi-information display.
- At an area with few turns and curves, and minimal traffic, drive at 20 km/h or more for 5 minutes or more.

30. Variable gear ratio steering system:
   - Turn the ignition switch on, and check that the master warning light and VSC/ABS warning lights illuminate for a few seconds.

➡If the warning lights remain on or blink, repair the applicable system.

- Drive the vehicle on a straight road at 35 km/h (22 mph) or more for 5 seconds or longer.

31. Confirm that steering angle sensor initialization is completed by doing the following:
   - Drive the vehicle on a straight road at 60 km/h (37 mph) or more for 30 seconds or longer.
   - Stop the vehicle (engine running).
   - Slowly turn the steering wheel from lock to lock.
   - If it turns approximately 2.7 turns, steering angle sensor initialization is completed. If it turns approximately 3.2 turns, steering angle sensor initialization is not completed.
32. Check shift lever position as follows:
   - Remove the nut and disconnect the shifting rod.
   - Turn the control shaft lever of the neutral start switch counterclockwise until it stops, and turn it clockwise 2 notches to set it to the N position.
   - Move the shift lever to the N position and tighten the nut while lightly pushing the lever toward the R position.
   - After adjustment, check that the shift lever moves smoothly and the shift lever and gear operate correctly.
33. Inspect and adjust front end alignment
34. Check idle speed and ignition timing.

➡Do not start the engine for at least 1 hour after installing.

35. Recheck all fluid levels and add if necessary.
36. Start engine and check for fluid, fuel and exhaust leaks.
37. Road test the vehicle.

SC 430

See Figures 109 and 110.

1. Before servicing the vehicle, refer to the precautions in the beginning of this section.
2. Relieve the fuel pressure from the fuel lines.
3. Drain the engine coolant from the cooling system.
4. Remove or disconnect the following:
   - Battery negative cable
   - Hood sub-assembly
   - V-Bank cover
   - Air cleaner inlet
   - Air cleaner assembly and connector pipe
   - Engine undercover and drain the oil
   - Drain Automatic Transmission fluid
   - Remove radiator
   - Fuel sub-pipe assembly
   - V-belts
   - Engine wire from ECU
   - Alternator wires
   - PS hose from # 1 oil pan bolt
   - Ground strap from the body
   - PS Air hose
   - Fuel vapor feed hose
   - Heater inlet and outlet water hose
   - PS pump reservoir

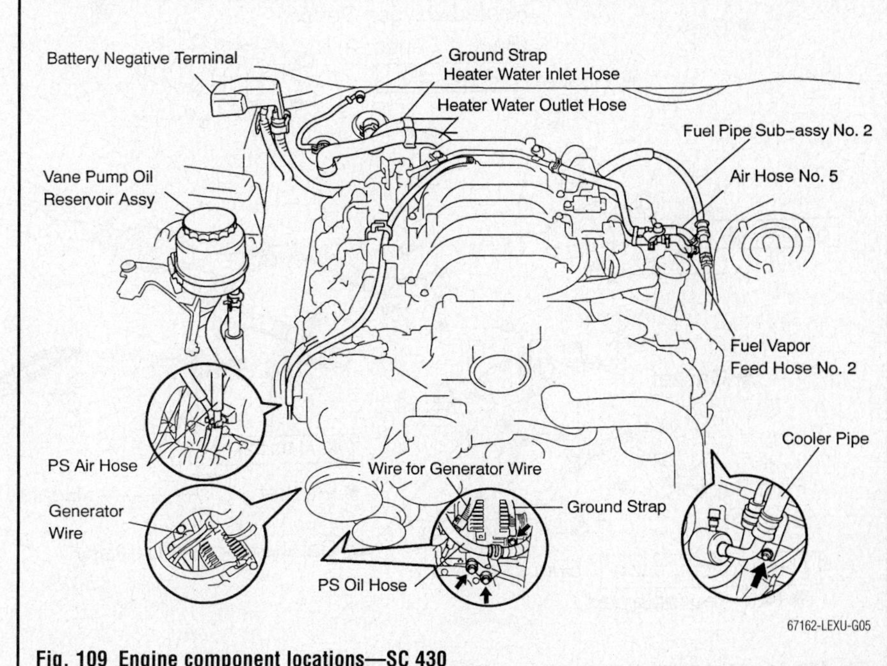

Fig. 109 Engine component locations—SC 430

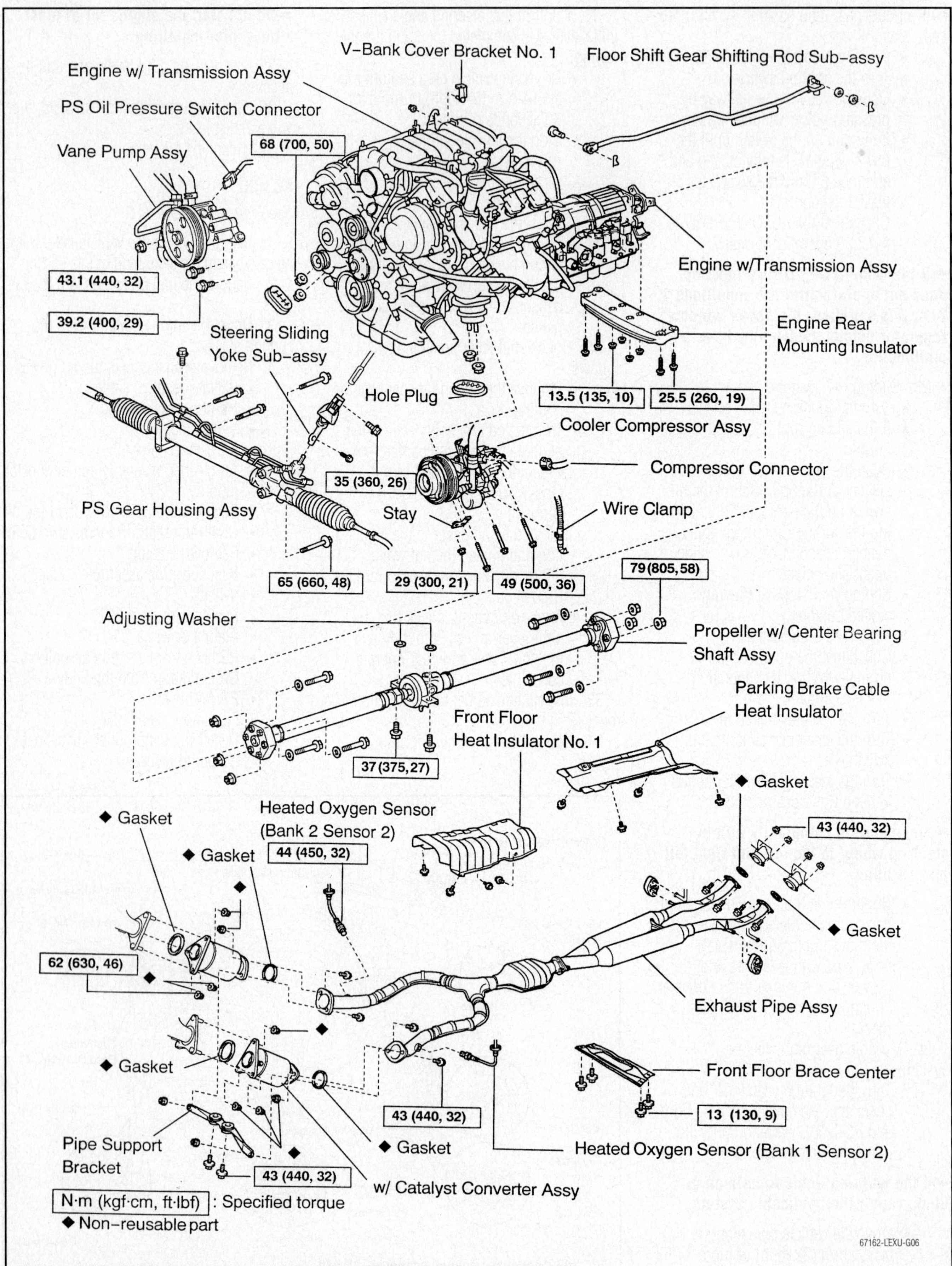

Fig. 110 Engine removal components, exploded view—SC 430

- RH engine under cover
- PS oil switch connector
- PS pump from the engine
- A/C Compressor electrical connector and clamp
- A/C compressor from the engine (Do NOT disconnect hoses)
- Support compressor assembly
- Front floor brace assembly
- Exhaust pipe assembly
- Front suspension member brace sub-assembly
- Catalytic converter assembly
- Front floor heat insulator
- Parking brake cable heat insulator
- Drive shaft W/Center bearing assembly
- Steering sliding yoke assembly
- The bolts and disconnect the 2 PS oil tubes from the

front frame.

- The 4 bolts securing the PS gear housing from the front frame
- Suspend the PS gear housing securely
- V-bank cover bracket from number engine hanger
- Install engine chain hoist to the engine hangers and support engine
- Remove 2 hole plugs
- 4 nuts holding the engine mounting insulator to the front suspension cross member
- 4 nuts and bolts from the rear engine mounting member
- Remove engine slowly and carefully. Make sure engine is clear of all wiring, hoses and cables.

To install:

5. Lower engine and transmission assembly onto the engine compartment
6. Install or connect the following:
   - front engine mounting bolts in front suspension crossmember
   - Rear engine mounting member with 4 bolts and nuts. Torque to: bolts 19 ft. lbs (25.5 Nm), Nuts 10 ft. lbs (13.5 Nm)
   - 4 nuts holding engine mounting brackets to front crossmember. Torque to: 50 ft. lbs. (68 Nm)
   - Front V-bank cover bracket to engine hanger with nut
   - Sliding PS gear housing yoke to intermediate shaft
   - PS gear housing with 4 bolts. Torque to 48 ft. lbs (65 Nm)
   - PS oil tube with bolts
   - Steering sliding yoke sub-assembly with 2 bolts. Torque to 26 ft. lbs (35 Nm)
   - Floor gear shift rod sub-assembly
   - Drive shaft W/Center bearing shaft assembly
   - Parking brake cable heat insulator
   - Front floor heat insulator
   - Catalytic converter assembly
   - Front suspension member sub-assembly. Torque to 43 ft. lbs. (58Nm)
   - Exhaust pipe assembly
   - Front floor brace center
   - A/C compressor, stay and wire bracket with 3 bolts and nuts. Torque bolts to 36 ft. lbs (49 Nm) and nuts to 21 ft. lbs. (29 Nm)
   - A/C wiring connector and clamps
   - PS pump with 2 bolts and nuts. Alternately tighten the bolts and nuts: bolts to 29 ft. lbs. (39 Nm) and nuts to 32 ft. lbs. (43 Nm)
   - PS oil pressure switch connector
   - PS air hoses to the PS pump
   - PS pump reservoir assembly
   - Heater inlet and outlet water hoses
   - Vapor feed hose
   - Upper PS air hose
   - Engine wire from ECM box
   - Alternator wires and clamp
   - Ground cable to alternator
   - PS hose to #1 oil pan bolt
   - Ground strap to body
   - V-belts
   - Air cleaner assembly and connector pipes
   - V-bank cover
   - Hood sub-assembly
   - Battery negative cable
   - Automatic transmission fluid
   - Engine coolant
   - Engine oil
   - Start engine
   - Inspect for coolant, oil and transmission fluid leaks
   - Adjust engine settings
   - Test drive vehicle

EXHAUST MANIFOLD

REMOVAL & INSTALLATION

3.0L (3GR-FSE) Engine

See Figure 111.

1. Before servicing the vehicle, refer to the precautions in the beginning of this section.
2. Remove or disconnect the following:
   - Negative battery cable. Wait at least 90 seconds before performing any other work
   - Cool air intake duct seal
   - Left and right engine room side covers
   - V-bank cover
   - Engine under cover
   - No. 2 engine under cover (2WD)
   - Left and right rear engine under cover (2WD)
   - Front exhaust pipe assembly
   - Front lower suspension member protector (2WD)
3. Remove the oil dipstick guide assembly as follows:
   a. Remove the oil level gauge.
   b. Remove the bolt, then remove the No. 2 oil dipstick guide.
   c. Remove the O-ring from the No. 2 oil dipstick guide.
   d. Remove the bolt and clamp (2WD), then remove the No. 1 oil dipstick guide.
   e. Remove the O-ring from the No. 1 oil dipstick guide.
   - Oxygen sensor wiring connectors
   - 6 nuts, exhaust manifold and gasket (left and right sides)

To install:

4. Install the exhaust manifold(s) as follows:
   a. Install a new exhaust manifold gasket.
   b. Install the exhaust manifold with 6 new nuts and torque to 15 ft. lbs. (21 Nm).

⁂ WARNING

Do not damage the stud bolt when installing the exhaust manifold. Be sure to tighten either of nuts A first as shown in the illustration.

5. Install the oil dipstick guide assembly as follows:
   a. Install a new O-ring to the oil dipstick guide.

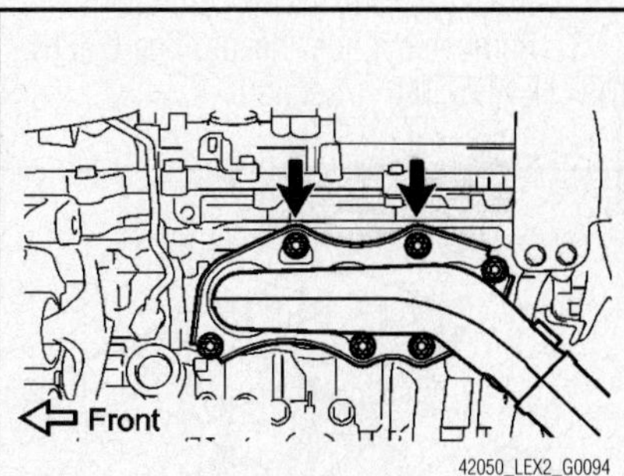

Fig. 111 When tightening the exhaust manifold nuts, start with the nuts depicted by the arrows as shown (left side illustrated, right side the same)—3.0L (3GR-FSE) engine

b. Apply a light coat of engine oil to the O-ring.

c. Push in the oil dipstick guide end into the guide hole.

d. Install the No. 1 oil dipstick guide with the bolt and torque to 7 ft. lbs. (10 Nm).

e. Connect the clamp (2WD).

f. Install the No. 2 oil dipstick guide with the bolt and torque to 15 ft. lbs. (21 Nm).

g. Install the oil dipstick.

6. Install or connect the following:
- Oxygen sensor wiring connectors
- Front lower suspension member protector (2WD) and 4 bolts. Torque the 4 bolts to 71 inch lbs. (8 Nm).
- Front exhaust pipe assembly
- Left and right rear engine under cover (2WD)
- No. 2 engine under cover (2WD)
- Engine under cover
- V-bank cover
- Left and right engine room side covers
- Cool air intake duct seal
- Negative battery cable

7. Start the engine and check for exhaust leaks and proper operation.

8. System initialization

4.3L (3UZ-FE) Engine

See Figures 112 and 113.

1. Before servicing the vehicle, refer to the precautions in the beginning of this section.

2. Remove or disconnect the following:
- Negative battery cable
- Engine under cover
- Coolant
- V-bank cover
- Air cleaner assembly (if needed for access to exhaust manifold)

3. Remove cylinder heads. See CYLINDER HEADS in this section.

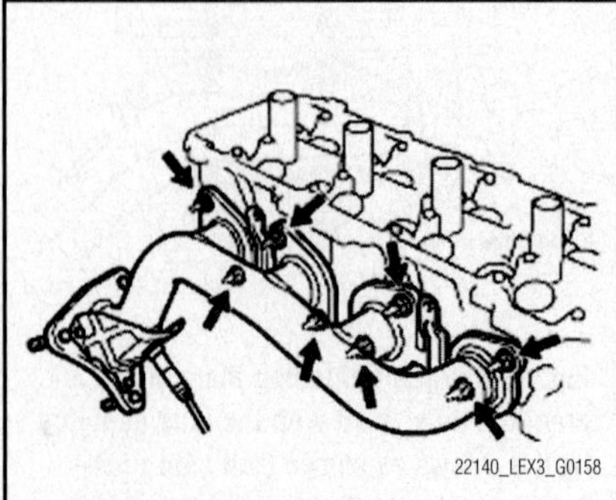

Fig. 112 Exhaust manifold view

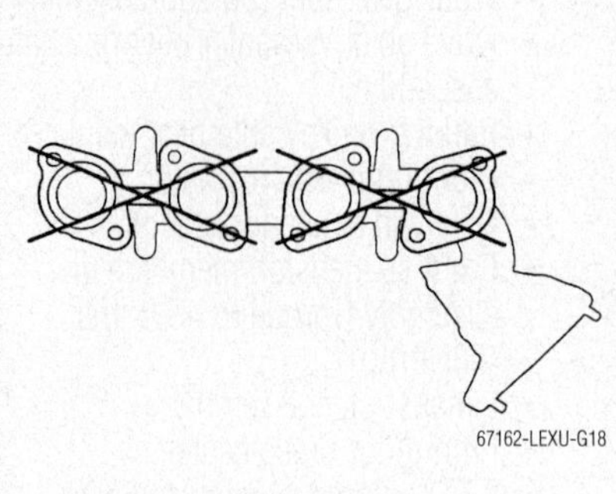

Fig. 113 Measuring exhaust manifold warpage

4. Remove 4 bolts and heat shield from exhaust manifold.

5. Remove 8 nuts and remove exhaust manifold and gasket.

To install:

6. To install, reverse the removal procedure. Install new manifold gasket and new retaining nuts. Tighten the exhaust manifold nuts to 32 ft. lbs. (44 Nm).

7. Install the manifold heat shields.

8. Refill the engine cooling system. Start the engine and check for leaks.

FLYWHEEL

REMOVAL & INSTALLATION

See Figure 114.

1. Before servicing the vehicle, refer to the precautions in the beginning of this section.

2. Disconnect the negative battery cable. Wait at least 90 seconds before performing any other work

3. Remove the engine or transmission assembly.

4. Using SST 09213-54015 (91651-60855), 09330-00021, hold the crankshaft.

5. Remove the 8 bolts and remove the rear spacer, drive plate and front spacer.

To install:

6. Using SST 09213-54015 (91651-60855), 09330-00021, hold the crankshaft.

7. Clean the bolts and bolt holes.

8. Apply adhesive (Part No. 08833-00070, THREE BOND 1324 or equivalent) to 2 or 3 threads of the bolts.

9. Install the rear spacer, drive plate and front spacer.

10. Install and uniformly tighten the 8 bolts in several passes, in the sequence shown, to the following torque values:
- 4.3L (3UZ-FE) engine—36 ft. lbs. (49 Nm) plus an additional 90°
- 3.0L (3GR-FSE) engine—30 ft. lbs. (41 Nm)

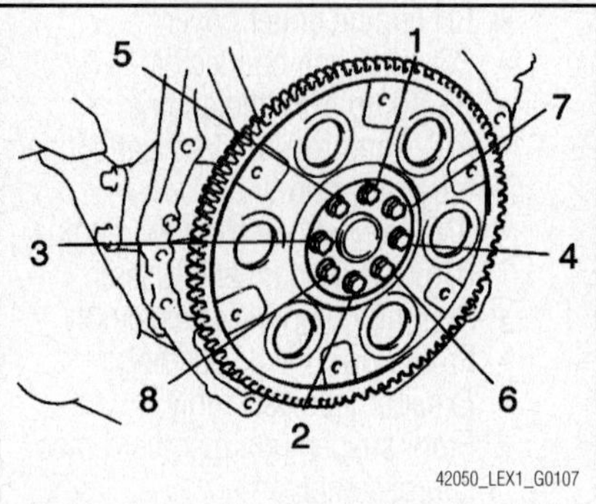

Fig. 114 Bolt tightening sequence for flywheel (drive plate).

- All other engines—61 ft. lbs. (83 Nm)

➡Do not start the engine within an hour after installing.

11. Install the engine or transmission assembly.

12. Connect the negative battery cable.

INTAKE MANIFOLD

REMOVAL & INSTALLATION

3.0L (3GR-FSE) Engine

See Figure 115.

1. Before servicing the vehicle, refer to the precautions in the beginning of this section.

2. Relieve fuel system pressure.

3. Remove or disconnect the following:
- Negative battery cable. Wait at least 90 seconds before performing any other work.
- Coolant
- Cool air intake duct seal
- Left and right engine room side covers
- V-bank cover
- No. 1 air cleaner inlet

4. Remove the air cleaner assembly with hose as follows:

a. Disconnect the ventilation hose from the cylinder head.

b. Disconnect the MAF meter connector.

c. Disconnect the clamp from the air cleaner.

d. Disconnect the EVAP VSV.

e. Loosen the hose clamp.

f. Remove the 3 bolts and air cleaner case.

g. No. 1 water by-pass hose.

h. No. 2 water by-pass hose.

i. Rear engine cover.

j. No. 3 water by-pass pipe.

k. No. 4 water by-pass pipe.

5. Remove intake air surge tank as follows:

a. Remove the 2 bolts and intake manifold stay.

b. Remove the 4 bolts and 2 surge tank stays.

c. Using a 5mm hexagon socket wrench, remove the 7 bolts, 2 nuts and gasket.

6. Remove the 4 bolts, 2 nuts, intake manifold and gasket.

WARNING

Cover the cylinder head intake port to prevent foreign matter from entering it.

To install:

7. Place new gaskets onto the intake manifold and position the intake manifold between the cylinder heads. Tighten the nuts and bolts to 15 ft. lbs. (21 Nm).

8. Install the intake air surge tank as follows:

a. Install a new gasket to the intake air surge tank.

b. Install the intake air surge tank with the 2 nuts. Torque the 2 nuts to 13 ft. lbs. (18 Nm).

c. Using a 5mm hexagon socket wrench, install the 7 bolts. Torque the bolts to 12 ft. lbs. (16 Nm).

d. Install the 2 surge tank stay to the intake air surge tank. Torque the 4 bolt to 15 ft. lbs. (21 Nm).

e. Install the intake manifold stay with 2 bolts and torque to 7 ft. lbs. (10 Nm).

9. Install or connect the following:

- No. 4 water by-pass pipe
- No. 3 water by-pass pipe
- Rear engine cover
- No. 2 water by-pass hose
- No. 1 water by-pass hose
- Air cleaner assembly
- No. 1 air cleaner inlet
- V-bank cover
- Left and right engine room side covers
- Cool air intake duct seal
- Negative battery cable

10. Refill the cooling system. Start the engine and check for leaks and proper operation.

11. Check the function of the throttle body unit

12. System initialization

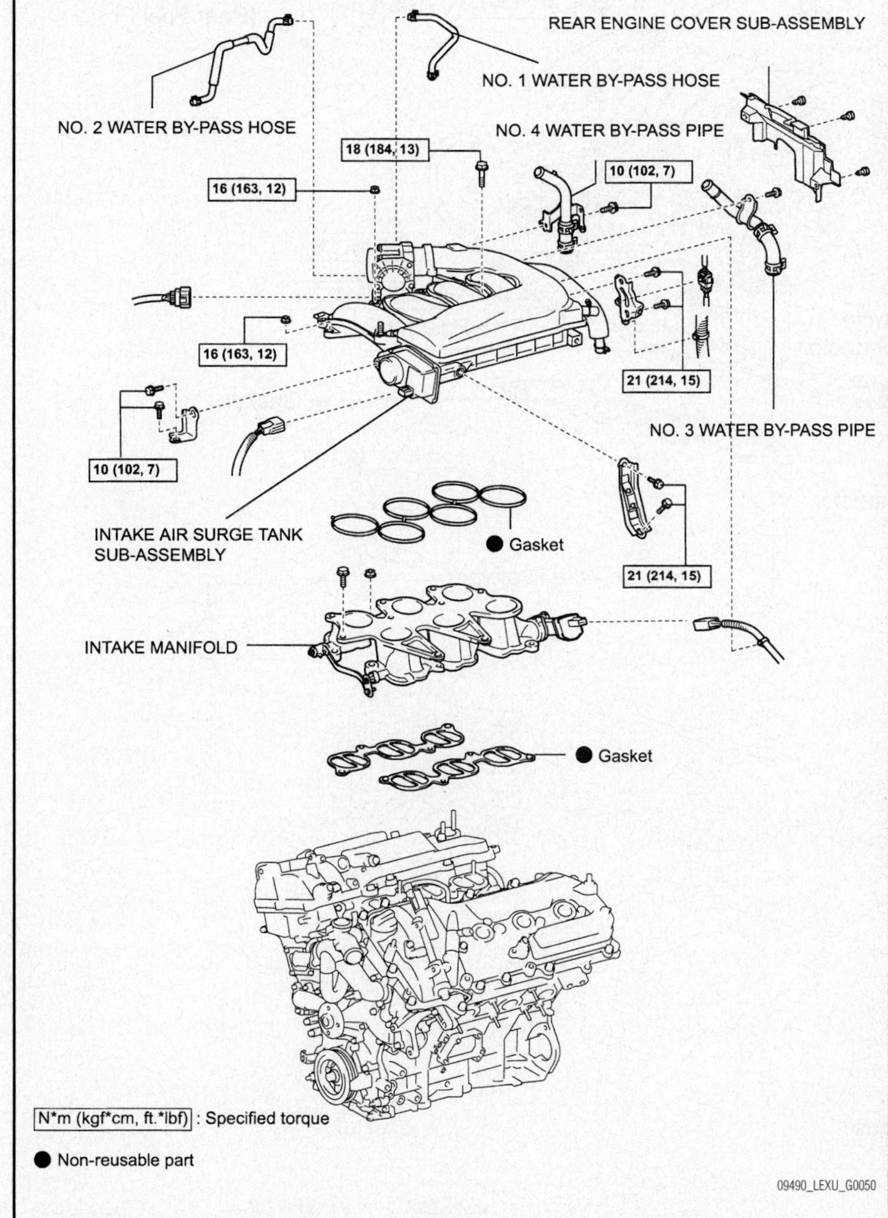

Fig. 115 Exploded view of the intake manifold mounting and related components—3.0L (3GR-FSE) engine

4.3L (3UZ-FE) Engine

See Figure 116.

1. Before servicing the vehicle, refer to the precautions in the beginning of this section.

2. Properly relieve the fuel system pressure.

3. Remove or disconnect the following:

- Negative battery cable
- Engine under cover
- Coolant
- V-bank cover
- Air cleaner assembly and connectors
- Throttle body assembly

4. Disconnect the fuel inlet hose (rear fuel pipe) from the fuel main tube.

5. Remove and disconnect the following:

- VSV connector for EVAP
- EVAP hose from VSV
- VSV from upper intake manifold
- 4 V-bank cover brackets
- Engine wiring protector (LH side) from the upper intake manifold and camshaft bearing cap
- 2 wire clamps (RH side) from the brackets on the delivery pipe
- Engine wire protector (rear side) from the rear water bypass joint and the RH cylinder head
- VSV connector for the ACIS
- 8 injector connectors

6. Remove the 6 bolts and 4 nuts and remove the upper and lower intake manifold assembly.

7. If necessary, the upper intake manifold can be disassembled from the lower

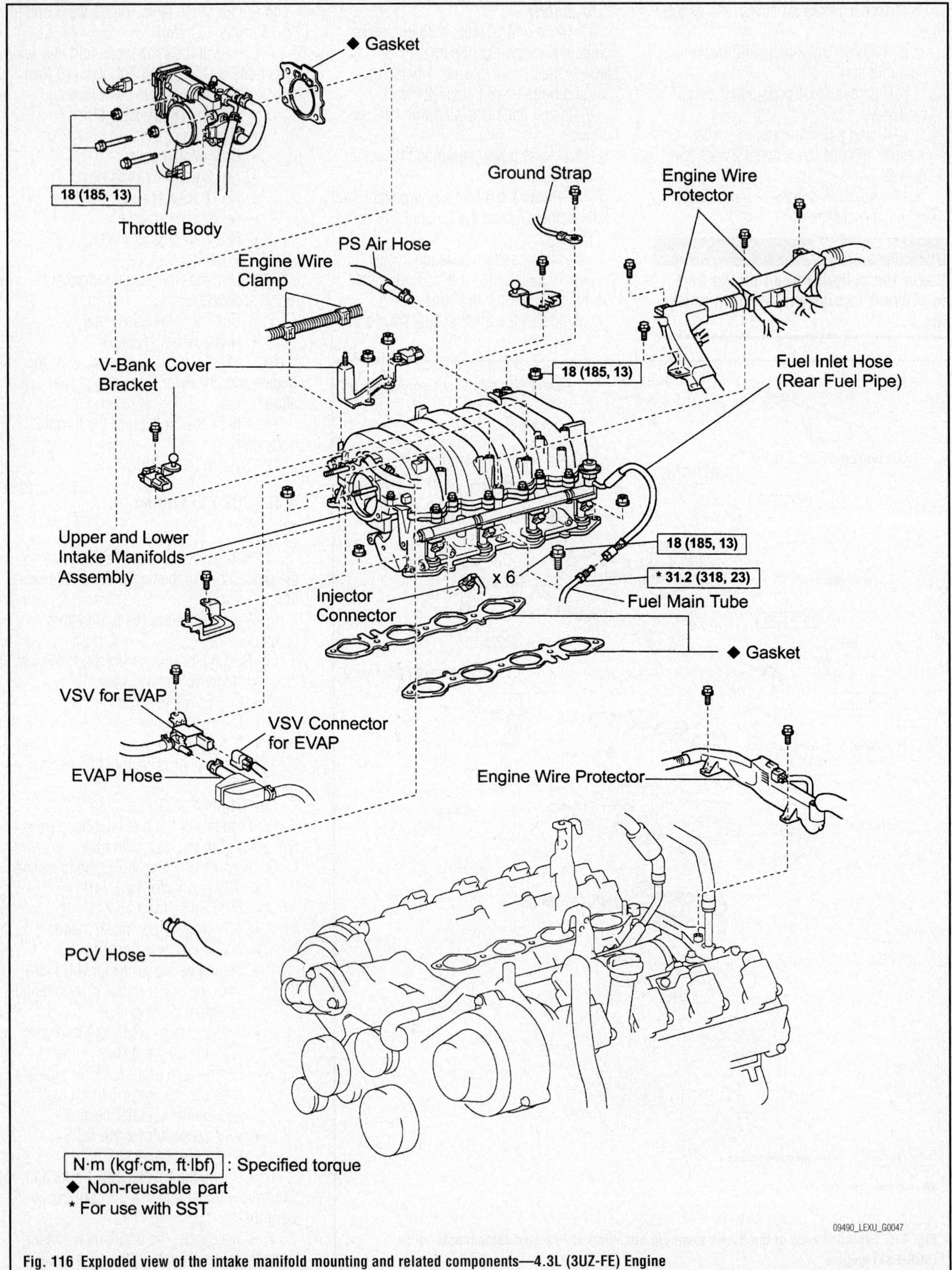

Fig. 116 Exploded view of the intake manifold mounting and related components—4.3L (3UZ-FE) Engine

intake manifold by removing or disconnecting the following:

- Vacuum hose for VSV from air control valve actuator
- Vacuum tank hose from lower intake manifold
- Vacuum hose (VSV for ACIS) from clamp
- Wire clamp from lower intake manifold
- Vacuum tank and VSV assembly from ACIS
- Air control valve actuator
- 15 bolts and 5 nuts to remove upper intake manifold from lower intake manifold

To install:

➡Always be sure to use new gaskets at each component mounting.

8. Reassemble the upper and lower intake manifold in reverse of disassembly procedure given. Torque the upper intake manifold-to-lower manifold bolts and nuts to 13 ft. lbs. (18 Nm).
9. Install the vacuum tank and VSV assembly. Torque the nuts to 13 ft. lbs. (18 Nm).
10. Reconnect all of the wiring connectors, clamps and vacuum hoses in reverse of the removal procedure.
11. With new gaskets on the cylinder heads (white marks facing outward), position the upper and lower intake manifold assembly into position. Install the 6 bolts and 4 nuts and torque them to 13 ft. lbs. (18 Nm).
12. Reinstall and reconnect all remaining components in reverse of the removal procedure.
13. Refill the cooling system. Start the engine and check for leaks and proper operation.

MAIN BEARING TORQUE SEQUENCE

See Figures 117 and 118.

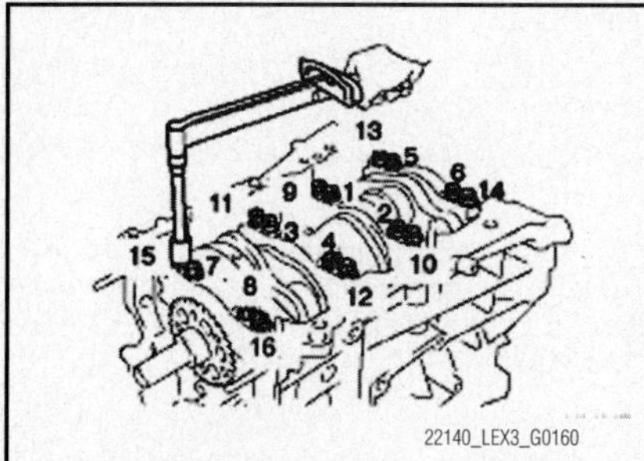

Fig. 117 Main bearing torque sequence—3.0L (3GR-FSE) Engine

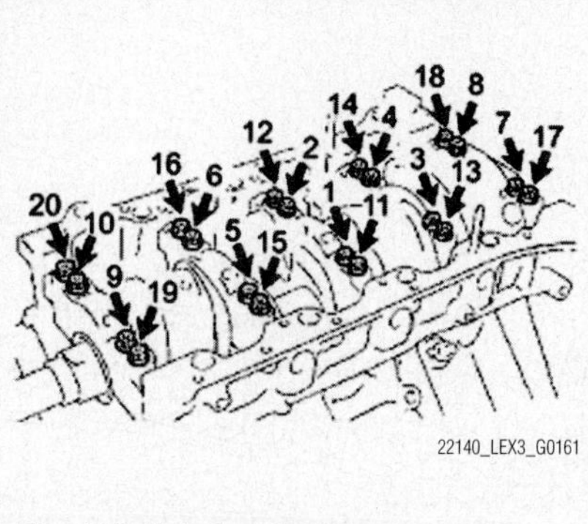

Fig. 118 Main bearing torque sequence—4.3L (3UZ-FE) Engine

OIL PAN

REMOVAL & INSTALLATION

3.0L (3GR-FSE) Engine

See Figures 119 through 121.

➡The No. 1 oil pan cannot be removed with the engine in the vehicle. The engine and transmission must be removed as a unit, then separated. See ENGINE ASSEMBLY section. It may be possible to remove the No. 2 oil pan from the vehicle while the engine is still in the vehicle.

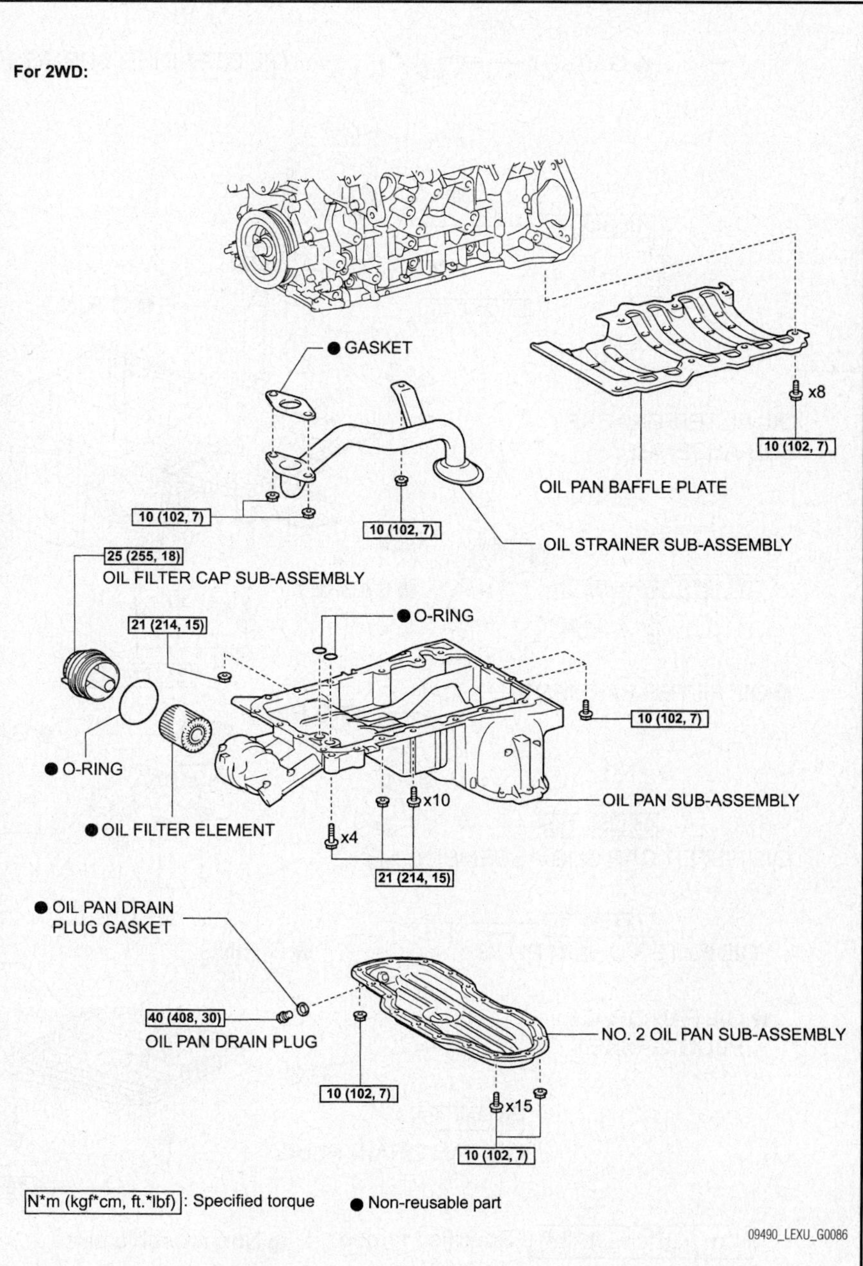

Fig. 119 Exploded view of the oil pan and related components (2WD)—3.0L (3GR-FSE) Engine

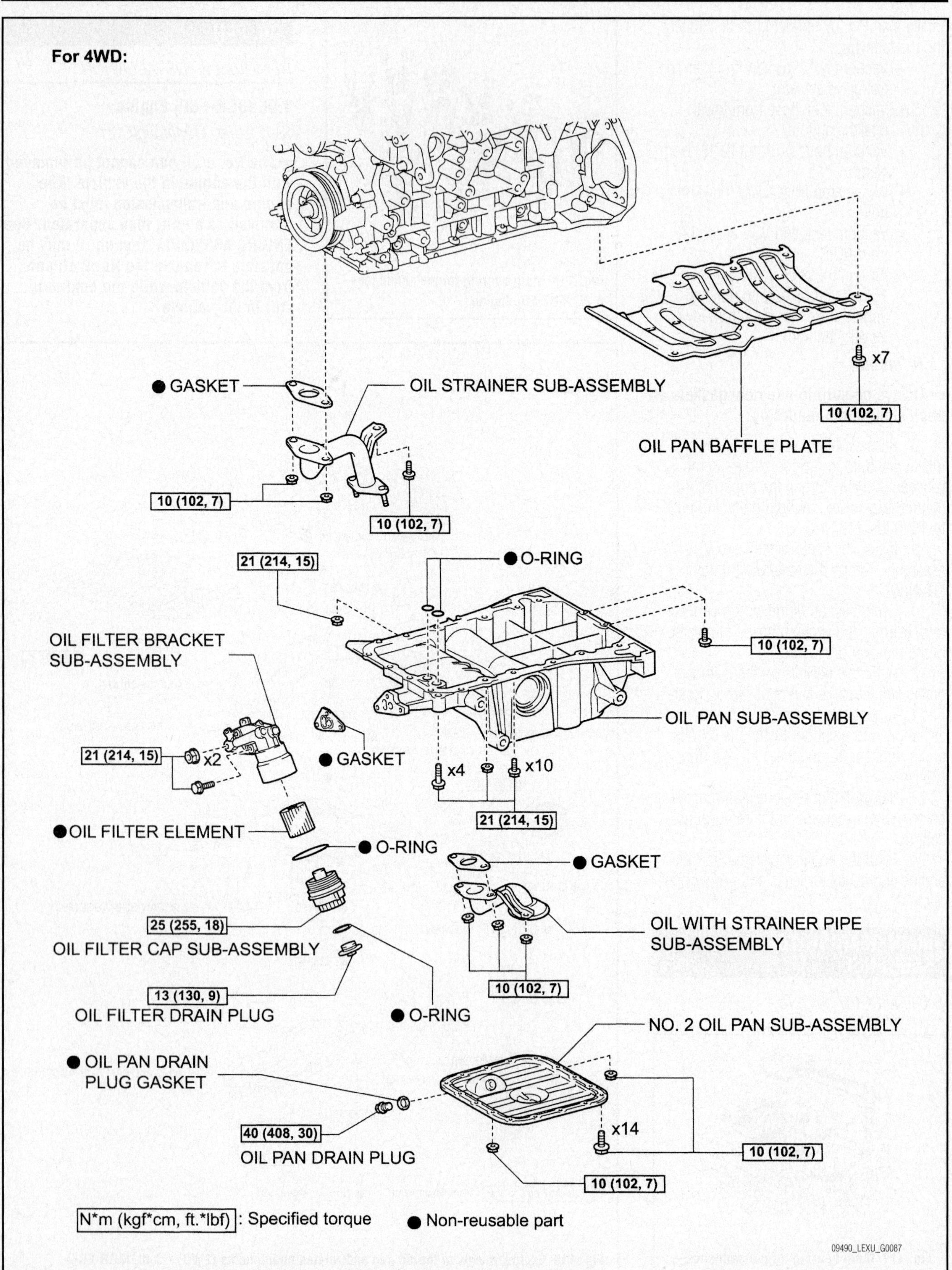

Fig. 120 Exploded view of the oil pan and related components (AWD)—3.0L (3GR-FSE) Engine

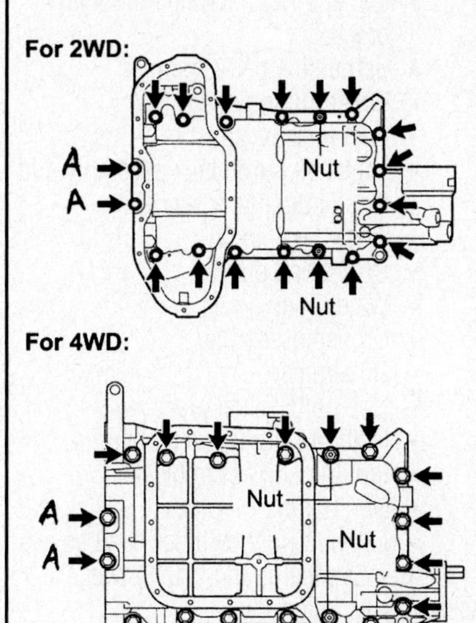

Fig. 121 Location of the A bolts on the oil pan assembly—3.0L (3GR-FSE) Engine

1. Before servicing the vehicle, refer to the precautions in the beginning of this section.
2. Drain the engine oil.
3. Remove or disconnect the following:
   - Negative battery cable from the battery
   - Engine/transmission assembly
   - Oil filter element
   - Oil filter bracket (AWD)
4. Remove the No. 2 oil pan sub-assembly as follows:
   a. Remove the 15 bolts and 2 nuts (2WD).
   b. Remove the 14 bolts and 2 nuts (AWD).
   c. Insert the blade of a prying tool between the oil pans. Cut through the applied sealer and remove the No. 2 oil pan sub-assembly.

➡Be careful not to damage the contact surfaces of the oil pans.

5. Remove the oil with strainer pipe sub-assembly (AWD) as follows:
   a. Remove the 3 mounting nuts.
   b. Remove the oil with strainer pipe and gasket.
6. Remove the oil pan sub-assembly as follows:
   a. Remove the 16 bolts and 2 nuts.

➡Be sure to clean the bolts and stud bolts and check the threads for cracks or other damage.

   b. Remove the oil pan by prying between the oil pan and cylinder block with a screwdriver.

⁂ WARNING

Be careful not to damage the contact surfaces of the cylinder block and oil pan.

   c. Remove the 2 O-rings.

To install:

7. Install the oil pan sub-assembly as follows:
   a. When replacing a stud bolt, install it by using an E6 Torx®socket wrench. Tighten the stud bolt to 35 inch lbs. (4 Nm).
   b. Apply seal packing in a continuous line of 0.118–0.156 inches (3.0–4.0mm) in diameter.

➡Remove any oil from the contact surface. Install the oil pan within 3 minutes after applying seal packing. Do not start the engine for at least 2 hours after installing.

   c. Install the oil pan with the 16 bolts and 2 nuts. Tighten the A bolts to 7 ft. lbs. (10 Nm). Tighten the remaining bolts to 15 ft. lbs. (21 Nm).
8. Install the oil with strainer pipe sub-assembly (AWD) as follows:
   a. Install a new gasket.
   b. Install the oil with strainer pipe with the 3 nuts. Tighten to 7 ft. lbs. (10 Nm).
9. Install the No. 2 oil pan sub-assembly as follows:
   a. Apply seal packing in a continuous line of 0.118–0.156 inches (3.0–4.0mm) in diameter.

➡Remove any oil from the contact surface. Install the oil pan No. 2 within 3 minutes after applying seal packing. Do not start the engine for at least 2 hours after installing.

   b. Install the oil pan with the 15 bolts and 2 nuts (2WD). Tighten the bolts to 7 ft. lbs. (10 Nm).
   c. Install the oil pan with the 14 bolts and 2 nuts (AWD). Tighten the bolts to 7 ft. lbs. (10 Nm).
10. Install or connect the following:
    - New gasket and oil filter bracket (AWD). Tighten the bolt and 2 nuts to 15 ft. lbs. (21 Nm).
    - Oil filter element
    - Engine/transmission assembly
    - Negative battery cable from the battery
    - Engine with oil

4.3L (3UZ-FE) Engine

See Figure 122.

➡The No. 1 oil pan cannot be removed with the engine in the vehicle. The engine and transmission must be removed as a unit, then separated. It may be possible to remove the No. 2 oil pan from the vehicle while the engine is still in the vehicle.

1. Before servicing the vehicle, refer to the precautions in the beginning of this section.
2. Remove or disconnect the following:
   - Engine/transmission assembly
   - Oil dipstick and guide
   - 12 bolts and 2 nuts. Use a gasket-cutting tool to separate the No. 2 (lower) oil pan. Be careful not to damage the No. 1 pan while performing this procedure.
   - 3 bolts and 2 nuts; remove the baffle plate
   - 17 bolts, then carefully pry off the No. 1 oil pan

➡There are slots for inserting the prybar.

To install:

3. Install or connect the following:
   - No. 1 pan. Apply a ⅛ inch (3–4mm) bead on RTV sealant to the pan mating surface. Bolts: 12mm: 66 inch lbs. (8mm); 14mm: 21 ft. lbs. (28 Nm)
   - Baffle plate. Tighten the bolts and nuts to 66 inch lbs. (8 Nm).
   - RTV sealant to the pan mating surface
   - No. 2 oil pan. Tighten the bolts to 66 inch lbs. (8 Nm)
   - Dipstick and guide
   - Engine/transaxle assembly
   - All fluids

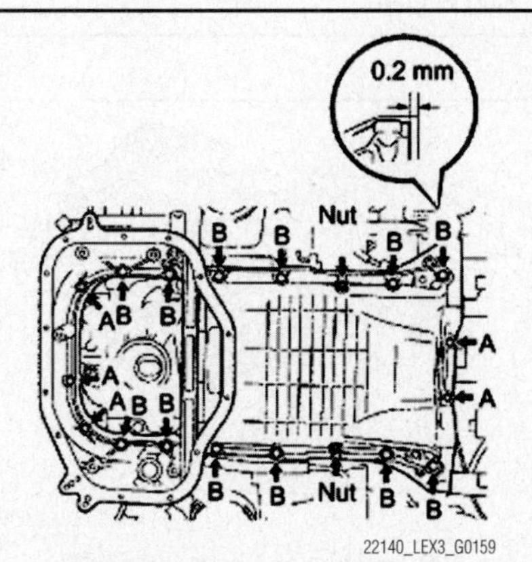

Fig. 122 Location of the 10 mm bolts (A) and 12 mm (B)

OIL PUMP

REMOVAL & INSTALLATION

3.0L (3GR-FSE) Engine

See Figures 123 through 128.

➡The oil pump cannot be removed with the engine in the vehicle. The engine and transmission must be removed as a unit, then separated.

Fig. 123 Location of mounting bolts for oil pump cover behind the timing chain cover—3.0L (3GR-FSE) Engine

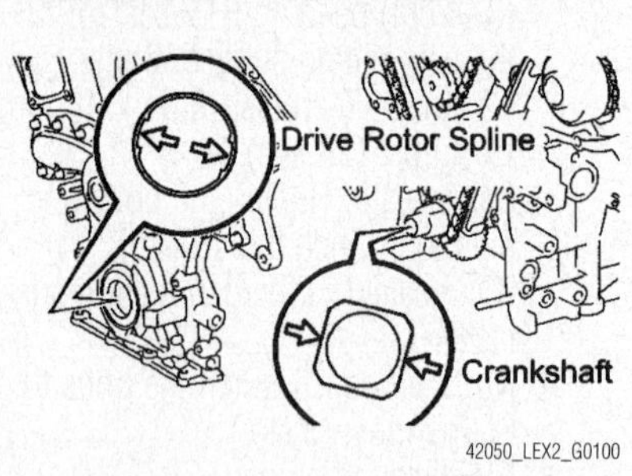

Fig. 124 Align the oil pump drive rotor spline and the crankshaft—3.0L (3GR-FSE) Engine

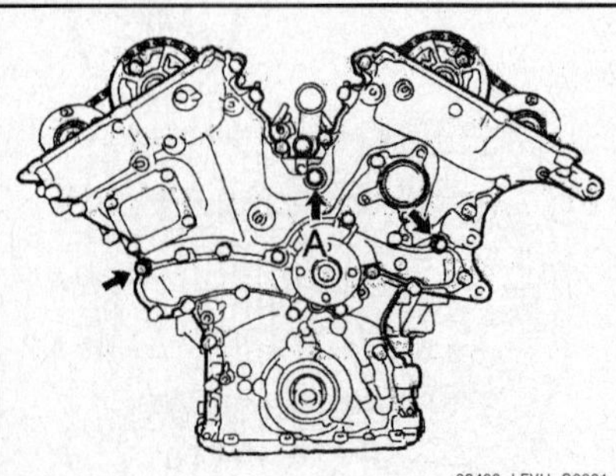

Fig. 125 Location of 3 bolts to be tightened first (location of bolt A shown)—3.0L (3GR-FSE) Engine

1. Before servicing the vehicle, refer to the precautions in the beginning of this section.
2. Remove or disconnect the following:
   - Engine/transmission assembly
   - Front differential assembly (AWD)
   - Serpentine drive belt
   - No. 2 idler pulley
   - Alternator
   - A/C compressor unit, if necessary

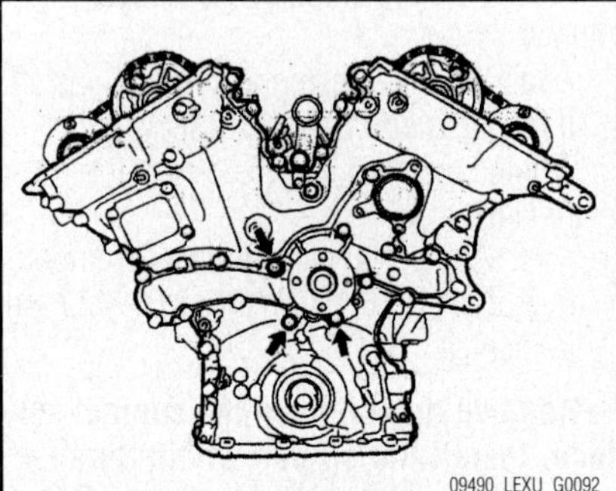

Fig. 126 Location of 3 bolts to be tightened second—2.5L (4GR-FSE), 3.0L (3GR-FSE) and 3.5L (2GR-FSE) Engines

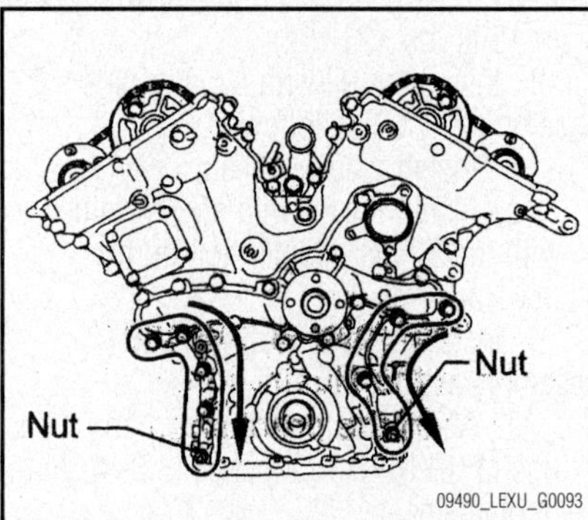

Fig. 127 Location of 7 bolts and 2 nuts to be tightened third—2.5L (4GR-FSE), 3.0L (3GR-FSE) and 3.5L (2GR-FSE) Engines

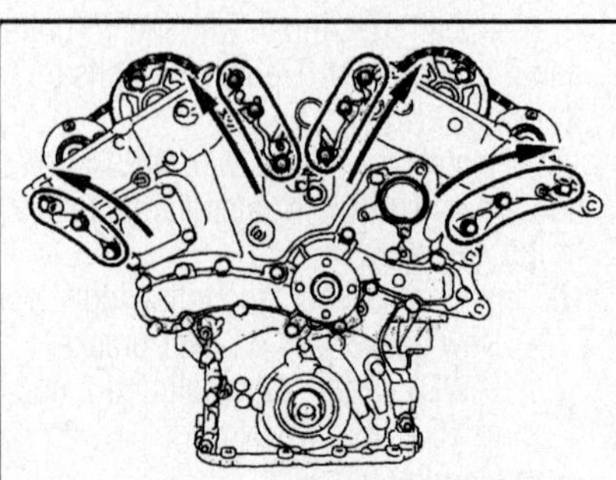

Fig. 128 Location of 12 bolts to be tightened fourth—2.5L (4GR-FSE), 3.0L (3GR-FSE) and 3.5L (2GR-FSE) Engines

   - Left and right engine mounting brackets
   - Serpentine belt tensioner
   - Water pump pulley
   - Fuel injector driver
   - Intake air surge tank assembly and No. 2 surge tank stay
   - Water hose joint
   - Crankshaft pulley
   - Water inlet
   - Oil pan assembly
   - Oil strainer
   - No. 1 and 2 fuel pipes
   - High pressure side fuel pump
   - Ignition coil assembly
   - No. 1 and 2 oil pipes
   - Left and right cylinder head covers
3. Remove the timing chain cover assembly as follows:
   a. Remove bolt and wiring harness clamp bracket.
   b. Remove 25 mounting bolts and 2 mounting nuts.
   c. Remove the timing chain cover by prying between the timing chain cover and cylinder head or cylinder block with a screwdriver.
   d. Remove the gasket.

➡The oil pump assembly is incorporated into the back of the timing chain cover. The oil pump assembly can be disassembled from the back of the timing chain cover for inspection purposes.

To install:

4. Install the timing chain cover assembly as follows:
   a. Apply seal packing in a continuous line of 0.197–0.217 inches (5.0–5.5mm) in diameter to the engine at the seam where the cylinder head meets the camshaft bearing cap assembly and the cylinder head meets the cylinder block

➡Be sure to clean, degrease and dry the contact surfaces before applying the seal packing. Install the component within 3 minutes after applying seal packing. Do not start the engine for at least 2 hours after installing.

   b. Apply seal packing in a continuous line of 0.138–0.158 inches (3.5–4.0mm) in diameter to the timing chain cover
   c. Install a new gasket
   d. Align the oil pump drive rotor spline and the crankshaft. Install the spline and chain cover to the crankshaft.
   e. Temporarily tighten the timing chain cover with the 25 bolts and nuts.
   f. Fully tighten the 3 bolts shown in the illustration. Torque bolt A to 32 ft.

lbs. (43 Nm). Torque the 2 remaining bolts to 15 ft. lbs (21 Nm).

g. Fully tighten the 3 bolts shown in the illustration. Torque the bolts to 15 ft. lbs (21 Nm).

h. Fully tighten the 7 bolts and 2 nuts shown in the illustration. Torque the bolts to 15 ft. lbs (21 Nm).

➡Be sure to tighten the bolts and nuts in order of upper to lower.

i. Fully tighten the 12 bolts shown in the illustration. Torque the bolts to 15 ft. lbs (21 Nm).

➡Be sure to tighten the bolts in order of lower to upper.

j. Install the bolt and wiring harness bracket. Torque the bolts to 7 ft. lbs (10 Nm).

5. Install or connect the following:
   - Left and right cylinder head covers
   - No. 1 and 2 oil pipes
   - Ignition coil assembly
   - High pressure side fuel pump
   - No. 1 and 2 fuel pipes
   - Oil strainer
   - Oil pan assembly
   - Water inlet
   - Crankshaft pulley. Torque the bolt to 192 ft. lbs (260 Nm).
   - Water hose joint
   - Intake air surge tank assembly and No. 2 surge tank stay
   - Fuel injector driver
   - Water pump pulley. Torque the bolt to 15 ft. lbs (21 Nm).
   - Serpentine belt tensioner
   - Left and right engine mounting brackets
   - A/C compressor unit, if necessary
   - Alternator
   - No. 2 idler pulley
   - Serpentine drive belt
   - Front differential assembly (AWD)
   - Engine/transmission assembly

4.3L (3UZ-FE) Engine

See Figures 129 and 130.

➡The oil pump cannot be removed with the engine in the vehicle. The engine and transmission must be removed as a unit, then separated.

1. Before servicing the vehicle, refer to the precautions in the beginning of this section.
2. Remove or disconnect the following:
   - Engine/transmission assembly
   - Timing belt
   - Idler pulleys
   - Crankshaft timing pulley

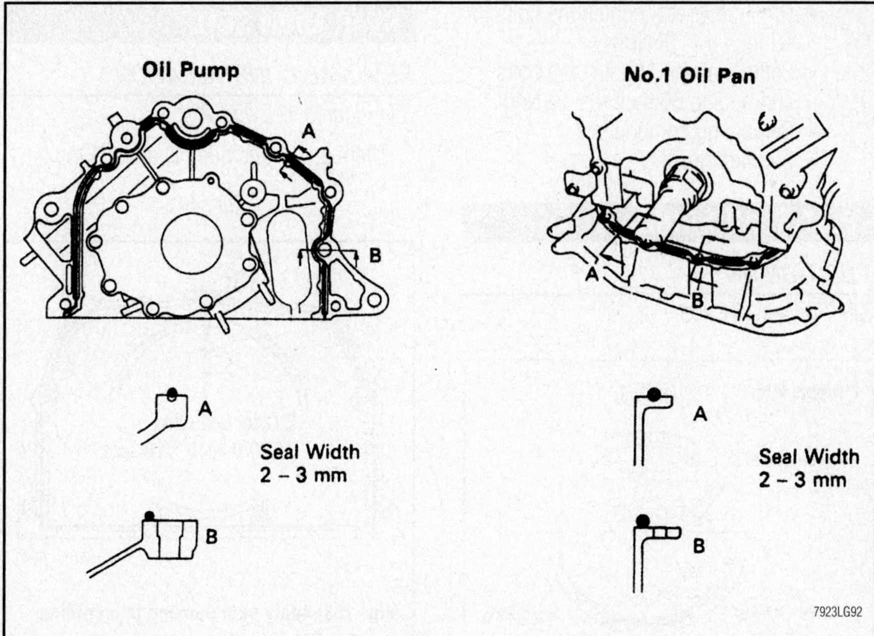

Fig. 129 Apply sealant to the oil pump and the No. 1 oil pan, as shown, before installing the oil pump—4.3L (3UZ-FE) Engines

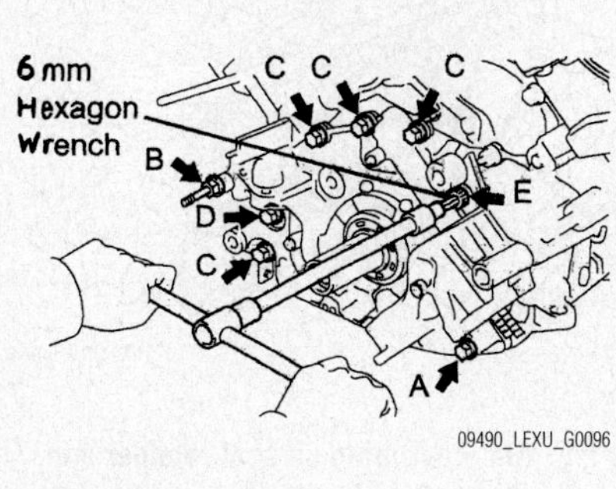

Fig. 130 4.3L engine oil pump mounting bolt locations, according to bolt lengths—(A) 1.97 in. (50mm), (B) 4.17 in. (106mm), (C) 1.18 in. (30mm), (D) 1.73 in. (44mm) and (E) 1.10 in. (28mm)

   - Oil dipstick and guide
   - Oil level sensor lead
   - 4 bolts and lift off the oil level sensor. Be careful not to drop this sensor.
   - Main Oxygen (O_2) sensor bracket, if necessary
   - Oil filter and filter bracket assembly by removing the stud bolt and 2 nuts
   - Engine Crankshaft Position (CKP) sensor. Remove the sensor by removing the bolt.
   - 12 bolts and 2 nuts from the No. 2 oil pan
   - No. 2 (lower) oil pan. Use a gasket-cutting tool
   - 2 bolts and 3 nuts and drop down the baffle plate
   - Oil strainer
   - No. 1 oil pan. There are slots for inserting the prybar.
   - 8 bolts holding the oil pump to the engine

➡Make certain to observe bolt position during removal. The bolts are different lengths and sizes. Record their position for proper reassembly.

   - Oil pump from the engine block
   - O-ring from the block

To install:

➡Prior to installing the oil pump, lubricate the gears with clean engine oil.

3. Install or connect the following:
   - A 2–3mm wide (0.08–0.12 in.) bead of RTV sealant to the oil pump
   - New O-ring in position on the block
   - Oil pump on the engine
   - The 8 bolts in their correct locations. Tighten the bolts with 12mm or 6mm heads to 12 ft. lbs. (16 Nm) and the bolts with 14mm heads to 22 ft. lbs. (30 Nm).
   - A ⅛inch (3–4mm) bead of RTV sealant to the pan mating surface.
   - No. 1 pan. Bolts–10mm: 66 inch lbs. (8 Nm); 12mm: 21 ft. lbs. (28 Nm)

- Oil strainer and tighten the bolts to 66 inch lbs. (8 Nm)
- Baffle plate and tighten the bolts and nuts to 66 inch lbs. (8 Nm)
- Remaining components
- Engine/transaxle

PISTON AND RING

POSITIONING

See Figures 131 and 132.

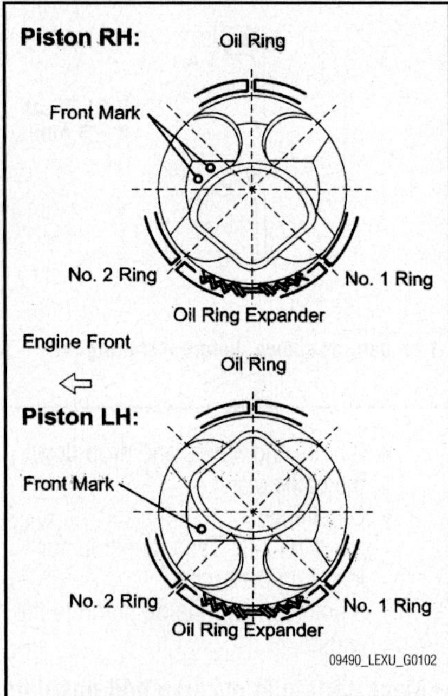

Fig. 131 Piston ring positioning—3.0L (3GR-FSE) Engine

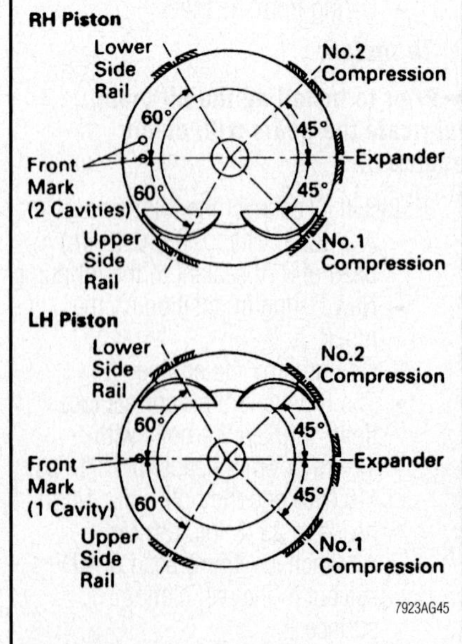

Fig. 132 Piston ring positioning—4.3L (3UZ-FE) Engine

REAR MAIN SEAL

REMOVAL & INSTALLATION

See Figures 133 and 134.

There are 2 methods (1) and (2) to replace the oil seal.

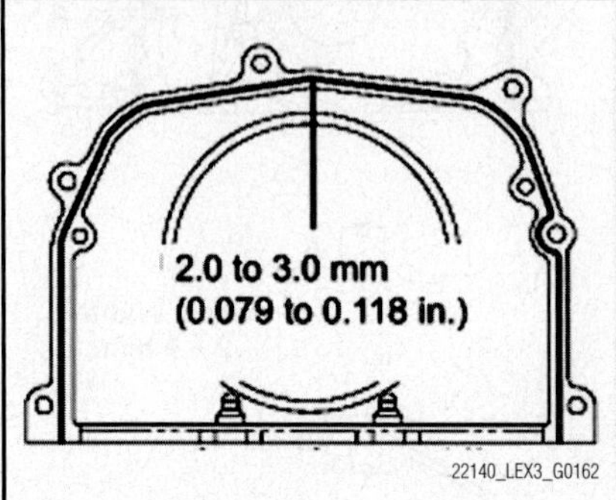

Fig. 133 Apply seal packing in a continuous bead as shown

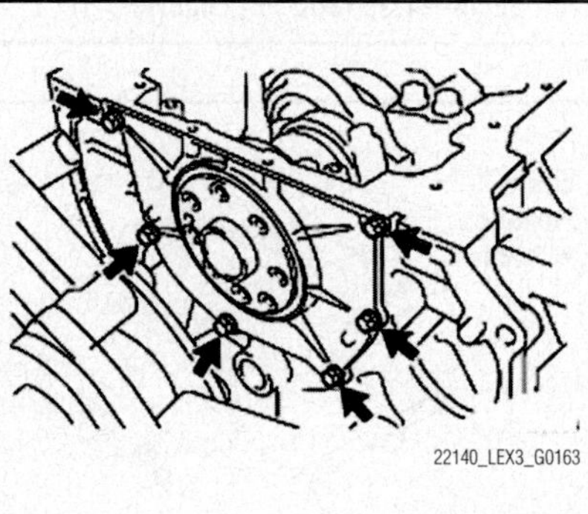

Fig. 134 Rear main oil seal retainer and mounting bolts

Method (1)

1. If the rear oil seal retainer is removed from the cylinder block, perform the following:
   - Using a screwdriver and hammer, tap out the oil seal.
   - Using SST 09223-15030, 09950-70010 (09951-07100) and a hammer, tap in a new oil seal until its surface is flush with the rear oil seal retainer edge.
   - Apply MP grease to the oil seal lip.
   - Apply seal packing in a continuous bead to the retainer.
   - Install the oil seal retainer and tighten the bolts to 7 ft. lbs. (10 Nm).

Method (2)

2. If the rear seal retainer is installed on the cylinder block, perform the following:
   - Using a knife, cut off the oil seal lip.
   - Using a screwdriver, pry out the oil seal.

**** CAUTION**

Be careful not to damage the crankshaft. Tape the screwdriver tip.

- Apply MP grease to a new oil seal lip.
- Using SST 09223-15030, 09950-70010 (09951-07100) and a hammer, tap in the oil seal until its surface is flush with the rear oil seal retainer edge.

TIMING BELT FRONT COVER

REMOVAL & INSTALLATION

4.3L (3UZ-FE) Engine

See Figures 135 through 139.

1. Before servicing the vehicle, refer to the precautions in the beginning of this section.
2. Disconnect the negative battery cable. Wait at least 90 seconds before performing any other work.
3. Remove air cleaner inlet No. 1.
4. Drain engine coolant.

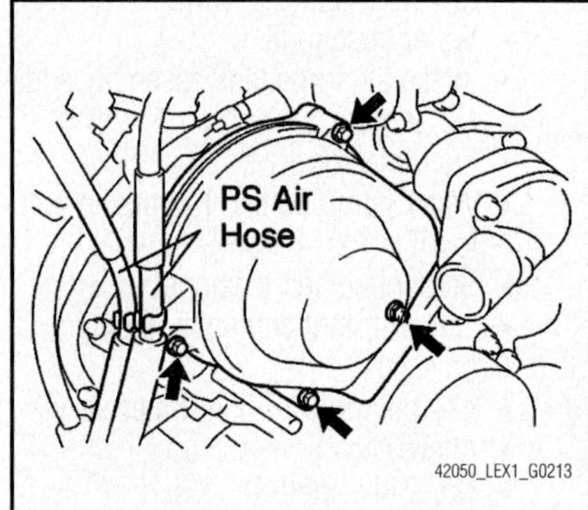

Fig. 135 Removal of timing belt cover No. 2.—4.3L (3UZ-FE) engine

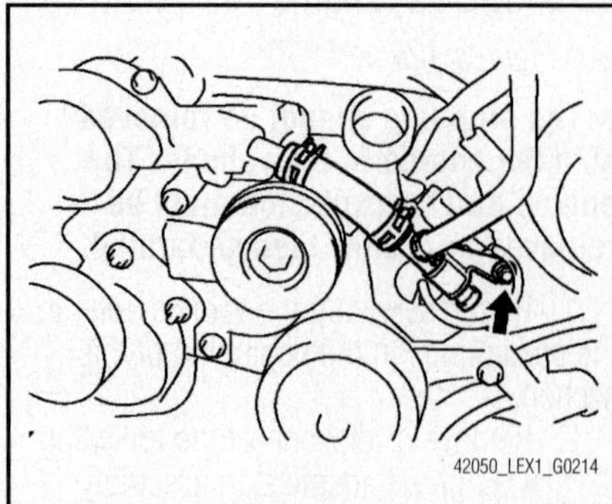

Fig. 136 Disconnect the 2 water by-pass hoses as shown—4.3L (3UZ-FE) engine

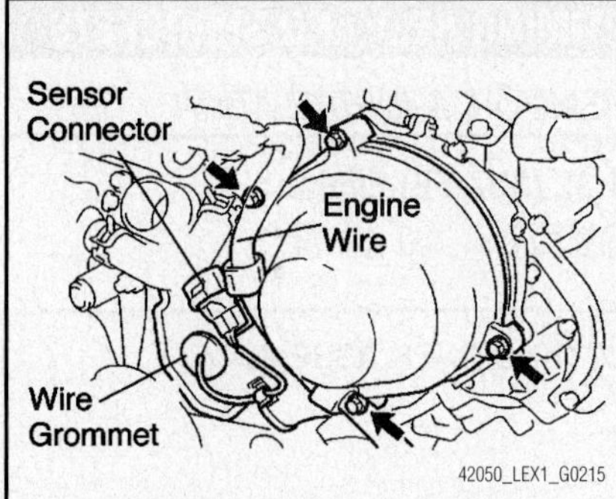

Fig. 137 Left timing belt cover sub-assembly No. 3.—4.3L (3UZ-FE) engine

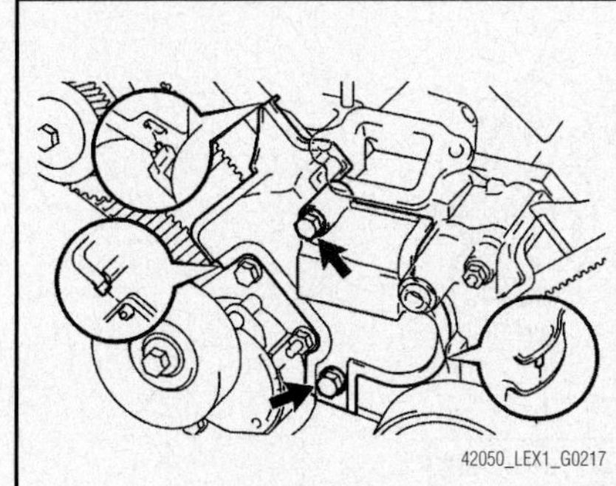

Fig. 138 Timing belt cover sub-assembly No. 2.—4.3L (3UZ-FE) engine

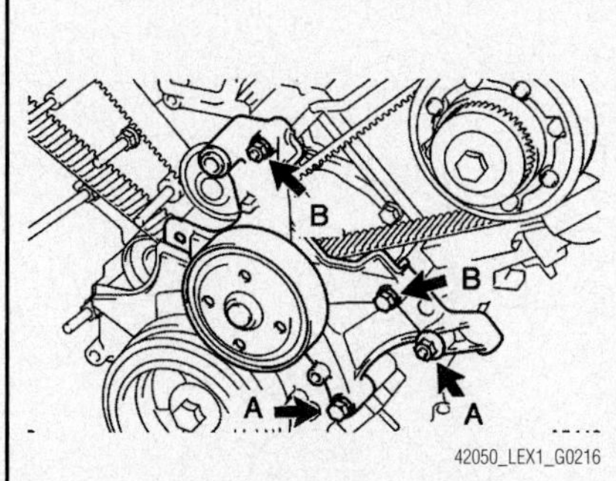

Fig. 139 Idler pulley assembly with 2 bolts and 2 nuts—4.3L (3UZ-FE) engine

5. Remove the V-bank cover.
6. Remove the intake air connector pipe.
7. Remove the engine under cover No. 1.
8. Disconnect radiator hose No. 1.
9. Disconnect radiator hose No. 2.
10. Disconnect oil cooler inlet tube No. 1.
11. Disconnect oil cooler outlet tube No. 1.
12. Remove the air cleaner assembly.
13. Remove the radiator assembly.
14. Remove the accessory drive belt
15. Remove the 2 bolts, nut and vane pump assembly. Pump should be removed with the hoses connected and then hang with a rope or wire on the body's side.
16. Remove the alternator.
17. Remove the A/C compressor assembly. The cooler compressor with the magnetic clutch should be removed with the low-pressure and high-pressure hoses connected and then hang with a rope or wire on the body's side.
18. Disconnect the 2 PS air hoses from the clamp on the timing belt cover No. 2.
19. Remove the cap nut, 3 bolts, timing belt cover No. 2 and gasket.
20. Remove the left timing belt cover sub-assembly No. 3 as follows:
    a. Remove the cap nut, and disconnect the No. 3 water bypass pipe from the cover.
    b. Disconnect the 2 water by-pass hoses as shown in the illustration.
    c. Disconnect the engine wire from the 2 wire clamps.
    d. Disconnect the camshaft position sensor connector.
    e. Disconnect the camshaft position sensor wire from the wire clamp on the cover.
    f. Remove the wire grommet from the cover.
    g. Remove the 4 bolts.
    h. Disconnect the cover from the timing plate and camshaft bearing cap.
    i. Disconnect the wire clamp for the sensor from the cover.
    j. Remove the connector holder from the sensor connector.
    k. Remove the left timing belt cover No. 3 and gasket.
21. Remove the 2 bolts and timing belt cover sub-assembly No. 2.
22. Remove the bolt, 2 nuts and drive belt tensioner.
23. Remove the 2 bolts, 2 nuts and idler pulley assembly.
24. Remove the crankshaft pulley (damper).
25. Remove the 4 bolts and timing belt No. 1 cover.

To install:

26. Install the timing belt No. 1 cover with the 4 bolts and tighten to 66 inch lbs. (7.5 Nm).
27. Install the crankshaft pulley (damper).
28. Install the idler pulley assembly with the 2 bolts and 2 nuts. Tighten the bolts and nuts as follows:
    - 14mm head bolt A: 24 ft. lbs. (32 Nm)
    - 12mm head bolt B: 12 ft. lbs. (16 Nm)

    a. Each bolt length is indicated below. Bolt Length:
    - 4.49 inches (114mm) for 14mm head (A)
    - 4.17 inches (106mm) for 12mm head (B)
29. Install the drive belt tensioner with the bolt and 2 nuts. Tighten the bolt and nuts to 12 ft. lbs. (16 Nm). Use a bolt that is 4.18 inches (106mm) in length.
30. Fit the timing belt cover sub-assembly No. 2, matching the claws and pin with each part. Install the timing belt cover with the 2 bolts and tighten to 12 ft. lbs. (16 Nm). Use bolts that are 4.17 inches (106mm) in length.
31. Install the left timing belt cover sub-assembly No. 3 as follows:
    a. Install the gasket to the cover.
    b. Run the camshaft position sensor wire through the cover hole.
    c. Install the cover with the 4 bolts and tighten to 66 inch lbs. (7.5 Nm).
    d. Install the wire grommet to the cover.
    e. Install the sensor connector to the sensor holder.
    f. Connect the sensor connector.
    g. Install the sensor wire to the wire clamp on the cover.
    h. Install the engine wire to the 2 wire clamps on the cover.
    i. Connect the 2 water by-pass hoses, as shown in the illustration.
    j. Install the No. 3 water by-pass pipe to the cover with the cap nut and tighten to 66 inch lbs. (7.5 Nm).
32. Install gasket to timing belt cover No. 2. Install the cover with the cap nut and 3 bolts and tighten to 66 inch lbs. (7.5 Nm).
33. Install the 2 PS air hoses to the clamp on the timing belt cover No. 2.
34. Install the A/C compressor, stay and wire bracket with the 3 bolts and nut. Tighten the bolts to 36 ft. lbs. (49 Nm). Tighten the nut to 21 ft. lbs. (29 Nm).
35. Remove the alternator.
36. Install the vane pump assembly with the 2 bolts and nuts. Alternately tighten the bolts and nut. Tighten the bolts to 29 ft. lbs. (39.2 Nm). Tighten the nut to 32 ft. lbs. (43 Nm).
37. Install the accessory drive belt
38. Install the radiator assembly.
39. Install the air cleaner assembly and tighten the fasteners to 44 inch lbs. (5 Nm).
40. Connect oil cooler outlet tube No. 1.
41. Connect oil cooler inlet tube No. 1.
42. Connect radiator hose No. 2.
43. Connect radiator hose No. 1.

44. Install air cleaner inlet No. 1 and tighten the fasteners to 44 inch lbs. (5 Nm).
45. Install the intake air connector pipe and tighten the fasteners to 44 inch lbs. (5 Nm).
46. Install the V-bank cover and tighten the fasteners to 44 inch lbs. (5 Nm).
47. Refill the engine cooling system and check for leaks.
48. Install the engine under cover No. 1.
49. Connect the negative battery cable.

TIMING BELT AND SPROCKETS

REMOVAL & INSTALLATION

4.3L (3UZ-FE) Engines

See Figures 140 and 141.

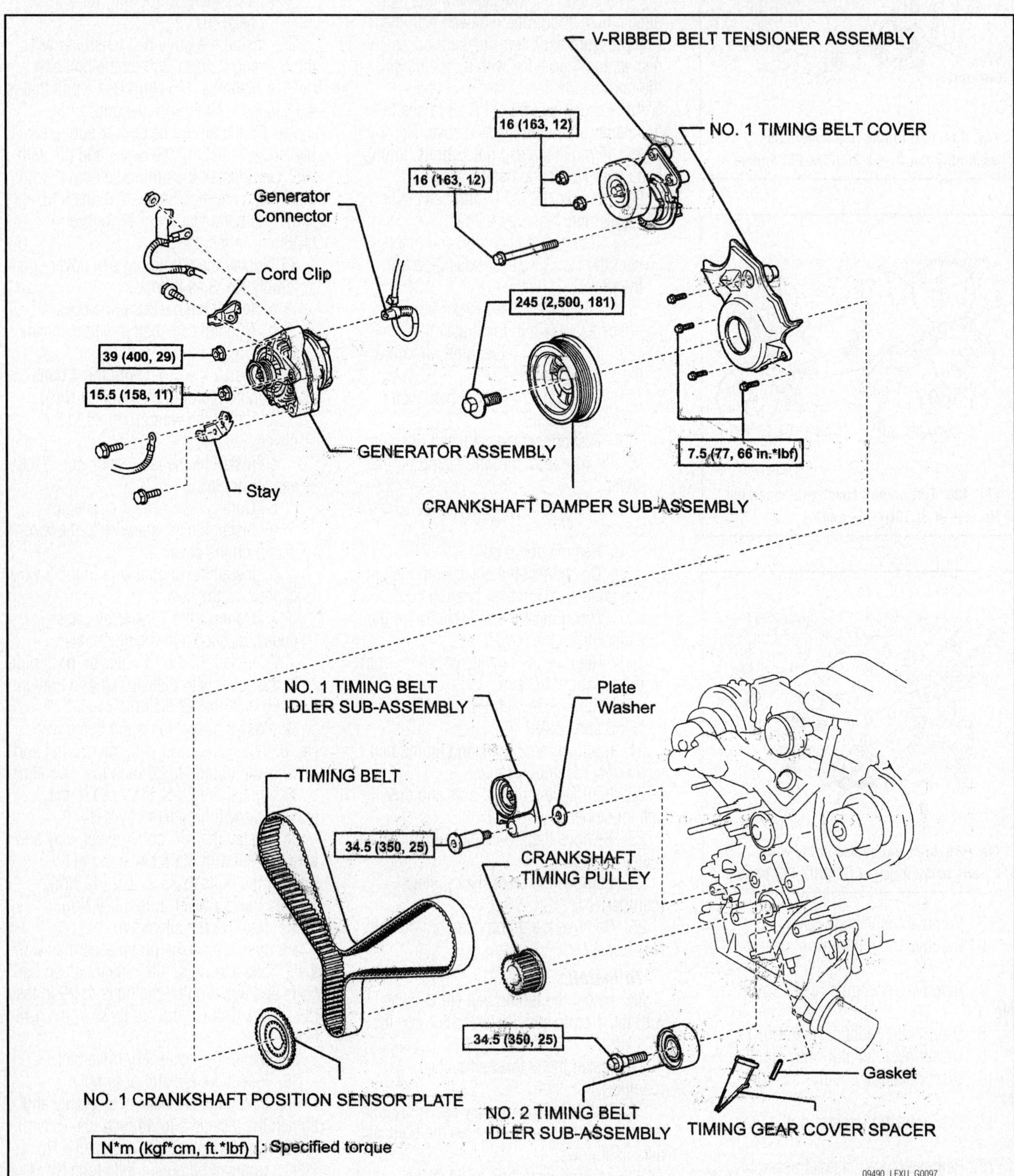

Fig. 140 Exploded view of the timing belt and cover assembly and related components—4.3L (3UZ-FE) Engine

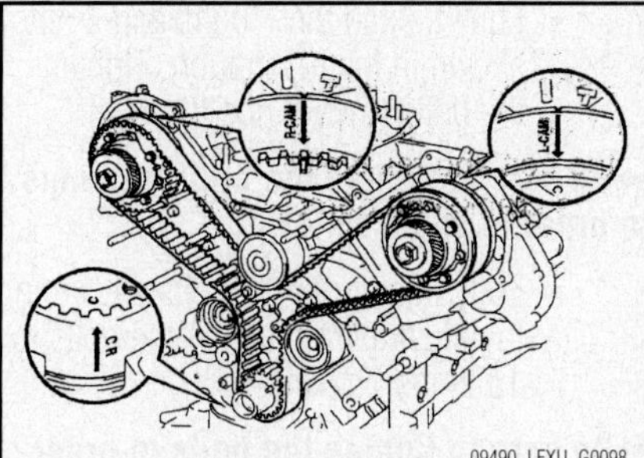

Fig. 141 Timing belt sprocket mark alignment for belt installation—4.3L (3UZ-FE) Engine

1. Remove all necessary components for access to the right-hand side No. 3 and No. 2, and left-hand side No. 2 timing belt covers, then remove the covers.
2. Turn the crankshaft pulley and align its groove with the timing mark **0** of the No. 1 timing cover. Check that the timing marks of the camshaft timing pulleys and timing belt rear plates are aligned. If not, turn the crankshaft 1 full revolution (360°).
3. Loosen crankshaft pulley bolt, then set the No. 1 cylinder to about 50° ATDC on the compression stroke. With the crankshaft pulley notch aligned with the **0** mark on the timing belt cover, turn the crankshaft pulley about 50° clockwise and put timing mark of the crank pulley in line with the centers of the pulley bolt and the No. 2 timing belt idler pulley bolt.
4. Remove or disconnect the following, as applicable for the vehicle:
   - Timing belt tensioner. Using the proper tool, loosen the tension between the left side and right side timing pulleys by slightly turning the left side camshaft clockwise.
   - Timing belt from the camshaft timing pulleys
   - Power steering pump pulley
   - Alternator
   - Drive belt tensioner
   - Bolt and timing pulleys, using the proper tool
   - Bolt and the crankshaft pulley with the proper tool.
   - Fan bracket
   - Mounting bolts and the No. 1 timing belt cover
   - 2 upper and lower timing belt covers
   - Timing belt guide (No. 1 crank position sensor plate)
   - Timing belt
   - No. 1 and No. 2 timing belt idler pulleys, if necessary.

➡If the timing belt is to be reused, draw a directional arrow on the timing belt in the direction of engine rotation (clockwise) and place matchmarks on the timing belt and crankshaft gear to match the drilled mark on the pulley.

To install:

5. Align the installation mark on the timing belt with the drilled mark of the crankshaft timing pulley. Install the timing belt on the crankshaft timing pulley, No. 1 idler pulley and the No. 2 idler pulley.

➡If the old timing belt is being reinstalled, be sure the directional arrow is facing in the original direction and that the belt and crankshaft gear matchmarks are properly aligned.

6. Install the timing belt guide (No. 1 crank angle sensor plate) with the cup side facing forward. Replace the timing belt cover spacer.
7. Install the No. 1 timing belt cover and tighten the mounting bolts.
8. Align the pulley set key on the crankshaft with the key groove of the pulley. Install the pulley, using the proper tool to tap in the pulley. Tighten the pulley bolt to 181 ft. lbs. (245 Nm).
9. Align the knock pin on the right side camshaft with the knock pin of the timing pulley. Slide on the timing pulley with the right side mark facing forward. Tighten the bolt to 80 ft. lbs. (108 Nm).
10. Align the knock pin on the left side camshaft with the knock pin of the timing pulley. Slide on the timing pulley with the left side mark facing forward. Tighten the bolt to 80 ft. lbs. (108 Nm).
11. Turn the crankshaft pulley and align its groove with the **0** timing mark on the No. 1 timing belt cover. Using the proper tool, turn the crankshaft timing pulley and align the timing marks of the camshaft timing pulley and the timing belt rear plate.
12. Install the timing belt to the left side camshaft timing pulley by:
    a. Using the proper tool, slightly turn the left side timing pulley clockwise. Align the installation mark of the timing belt with the timing mark of the camshaft timing pulley and hang the timing belt on the left side camshaft pulley.
    b. Using the proper tool, align the timing marks of the left side camshaft pulley and the timing belt rear plate.
    c. Check that the timing belt has tension between crankshaft timing pulley and the left side camshaft pulley.
13. Install the timing belt to the right side camshaft timing pulley by:
    a. Using the proper tool, slightly turn the right side timing pulley clockwise. Align the installation mark of the timing belt with the timing mark of the camshaft timing pulley and hang the timing belt on the right side camshaft pulley.
    b. Using the proper tool, align the timing marks of the right side camshaft pulley and the timing belt rear plate.
    c. Check that the timing belt has tension between the crankshaft timing pulley and the right side camshaft pulley.
14. The timing belt tensioner must be set prior to installation. The tensioner can be set as follows:
    a. Place a plate washer between the tensioner and a block. Using a suitable press, press in the pushrod using 220–2205 lbs. (100–1000kg) of pressure.
    b. Align the holes of the pushrod and housing, pass the proper tool (0.05 in. Allen wrench) through the holes to keep the setting position of the pushrod.
    c. Release the press and install the dust boot on the tensioner.
15. Install the tensioner and tighten the bolts to 20 ft. lbs. (26 Nm). Remove the tool from the tensioner.
16. Turn the crankshaft pulley two complete revolutions from TDC-to-TDC. Always turn the crankshaft clockwise. Check that each pulley aligns with the timing marks.
17. Install all remaining components in the reverse order of removal.

TIMING CHAIN COVER AND SEAL

REMOVAL & INSTALLATION

3.0L (3GR-FSE) Engine

See Figures 142 through 145.

1. Before servicing the vehicle, refer to the precautions in the beginning of this section.

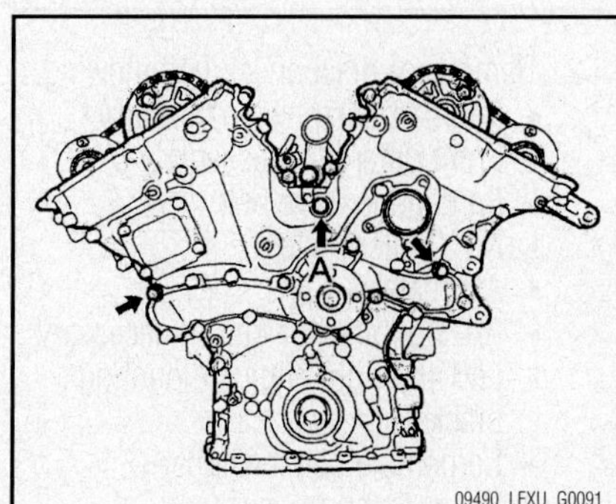

Fig. 142 Location of 3 bolts to be tightened first (location of bolt A shown)—3.0L (3GR-FSE) Engine

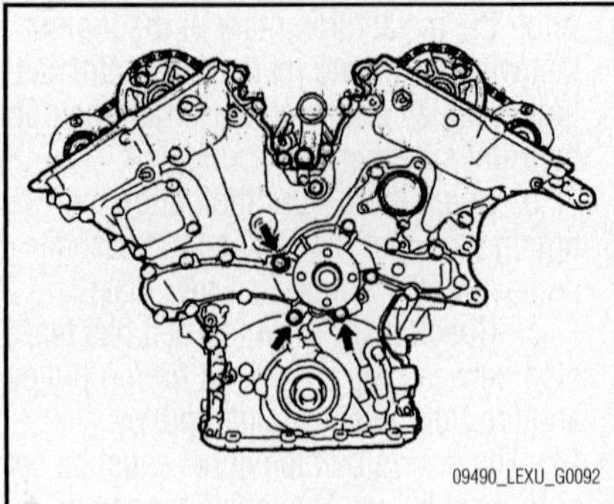

Fig. 143 Location of 3 bolts to be tightened second—3.0L (3GR-FSE) Engine

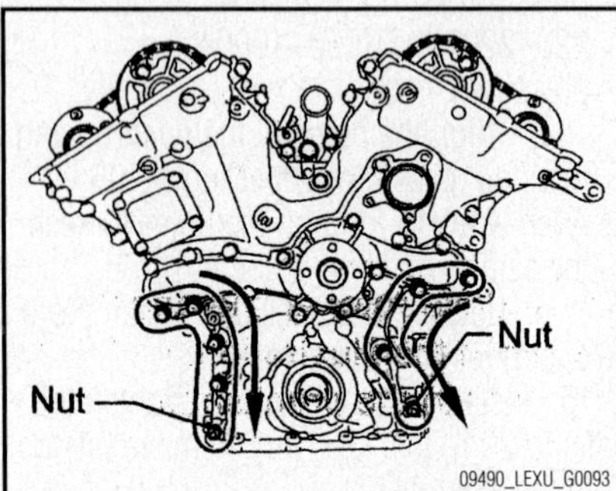

Fig. 144 Location of 7 bolts and 2 nuts to be tightened third—3.0L (3GR-FSE) Engine

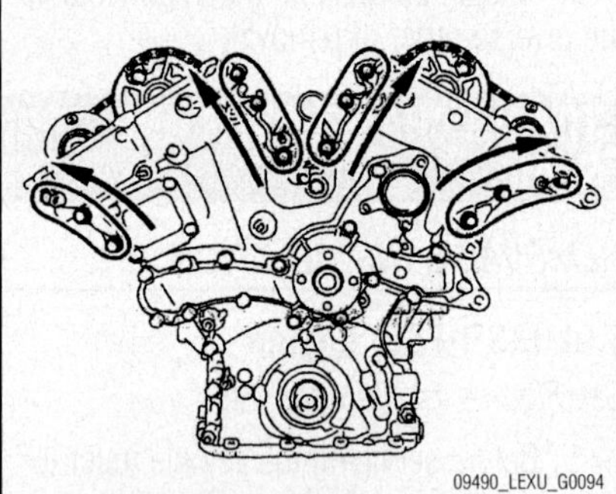

Fig. 145 Location of 12 bolts to be tightened fourth—3.0L (3GR-FSE) Engine

2. Remove or disconnect the following:
   - Engine/transmission assembly
   - Front differential assembly (AWD)
   - Serpentine drive belt
   - No. 2 idler pulley
   - Alternator
   - A/C compressor unit, if necessary
   - Left and right engine mounting brackets
   - Serpentine belt tensioner
   - Water pump pulley
   - Fuel injector driver
   - Intake air surge tank assembly and No. 2 surge tank stay
   - Water hose joint
   - Crankshaft pulley
   - Water inlet
   - Oil pan assembly
   - Oil strainer
   - No. 1 and 2 fuel pipes
   - High pressure side fuel pump
   - Ignition coil assembly
   - No. 1 and 2 oil pipes
   - Left and right cylinder head covers
3. Remove the timing chain cover assembly as follows:
   - Remove bolt and wiring harness clamp bracket.
   - Remove 25 mounting bolts and 2 mounting nuts.
   - Remove the timing chain cover by prying between the timing chain cover and cylinder head or cylinder block with a screwdriver.
   - Remove the gasket.
   - Remove the timing chain case oil seal.

To install:

4. Install the timing chain cover assembly as follows:
   - Install a new timing chain case oil seal.
   - Apply seal packing in a continuous line of 0.197–0.217 inches (5.0–5.5mm) in diameter to the engine at the seam where the cylinder head meets the camshaft bearing cap assembly and the cylinder head meets the cylinder block

➡Be sure to clean, degrease and dry the contact surfaces before applying the seal packing. Install the component within 3 minutes after applying seal packing. Do not start the engine for at least 2 hours after installing.

   - Apply seal packing in a continuous line of 0.138–0.158 inches (3.5–4.0mm) in diameter to the timing chain cover.
   - Install a new gasket.
   - Align the oil pump drive rotor spline and the crankshaft. Install the spline and chain cover to the crankshaft.
   - Temporarily tighten the timing chain cover with the 25 bolts and nuts.
   - Fully tighten the 3 bolts shown in the illustration. Tighten bolt A to 32 ft. lbs. (43 Nm). Torque the 2 remaining bolts to 15 ft. lbs (21 Nm).
   - Fully tighten the 3 bolts shown in the illustration. Tighten the bolts to 15 ft. lbs (21 Nm).
   - Fully tighten the 7 bolts and 2 nuts shown in the illustration. Tighten the bolts to 15 ft. lbs (21 Nm).

➡Be sure to tighten the bolts and nuts in order of upper to lower.

   - Fully tighten the 12 bolts shown in the illustration. Tighten the bolts to 15 ft. lbs (21 Nm).

➡Be sure to tighten the bolts in order of lower to upper.

   - Install the bolt and wiring harness bracket. Tighten the bolts to 7 ft. lbs (10 Nm).
5. Install or connect the following:
   - Left and right cylinder head covers
   - No. 1 and 2 oil pipes
   - Ignition coil assembly
   - High pressure side fuel pump
   - No. 1 and 2 fuel pipes
   - Oil strainer
   - Oil pan assembly
   - Water inlet
   - Crankshaft pulley. Torque the bolt to 192 ft. lbs (260 Nm).
   - Water hose joint
   - Intake air surge tank assembly and No. 2 surge tank stay
   - Fuel injector driver
   - Water pump pulley. Torque the bolt to 15 ft. lbs (21 Nm).
   - Serpentine belt tensioner
   - Left and right engine mounting brackets
   - A/C compressor unit, if necessary
   - Alternator
   - No. 2 idler pulley
   - Serpentine drive belt
   - Front differential assembly (AWD)
   - Engine/transmission assembly

TIMING CHAIN AND SPROCKETS

REMOVAL & INSTALLATION

3.0L (3GR-FSE) Engine

See Figures 146 through 148.

➡The timing chain cannot be removed with the engine in the vehicle. The engine and transmission must be removed as a unit, then separated.

1. Before servicing the vehicle, refer to the precautions in the beginning of this section.
2. Remove or disconnect the following:
   - Engine/transmission assembly
   - Front differential assembly (AWD)
   - Serpentine drive belt
   - No. 2 idler pulley

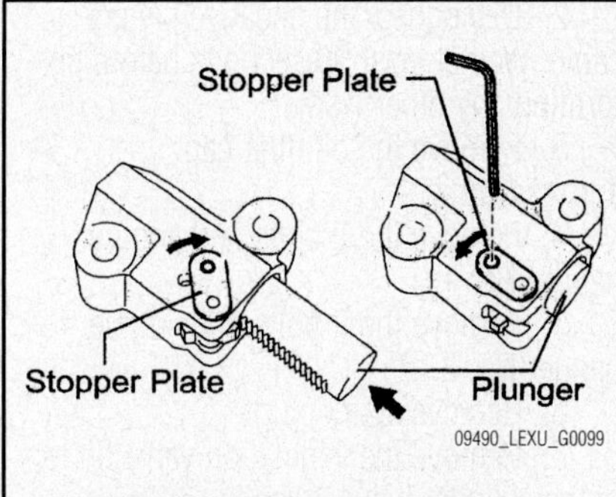

Fig. 146 Chain tensioner component showing stopper plate and plunger—3.0L (3GR-FSE) Engine

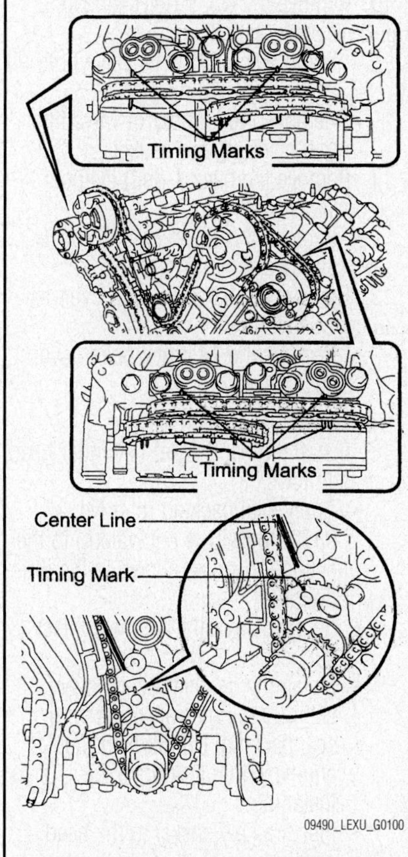

Fig. 147 Aligning the timing marks at the block bore centerline and camshaft bearing caps—3.0L (3GR-FSE) Engine

- Alternator
- A/C compressor unit, if necessary
- Left and right engine mounting brackets
- Serpentine belt tensioner
- Water pump pulley
- Fuel injector driver
- Intake air surge tank assembly and No. 2 surge tank stay
- Water hose joint
- Crankshaft pulley

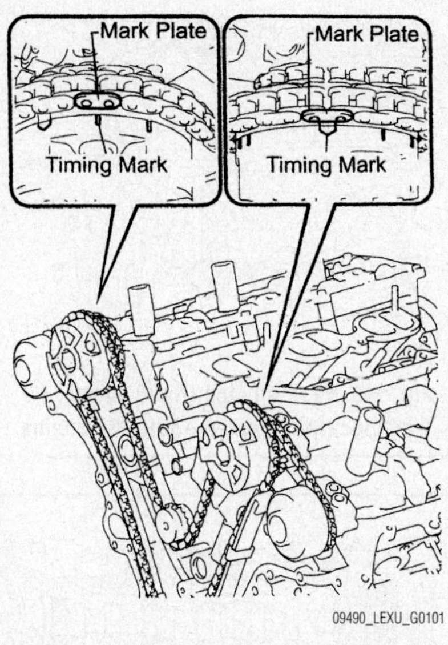

Fig. 148 Aligning the mark plates to the timing marks—3.0L (3GR-FSE) Engine

- Water inlet
- Oil pan assembly
- Oil strainer
- No. 1 and 2 fuel pipes
- High pressure side fuel pump
- Ignition coil assembly
- No. 1 and 2 oil pipes
- Left and right cylinder head covers

3. Remove the timing chain cover and seal.
4. Set the No. 1 cylinder to TDC/ compression as follows:
   - Temporarily tighten the pulley set bolt.
   - Set the timing mark on the crank angle sensor plate to the right block bore center line (TDC/ compression).
   - Check that the timing marks of the camshaft timing gears are aligned with the timing marks of the bearing cap. If not, turn the crankshaft 1 revolution (360 degrees) and align the timing marks.
5. Remove the No. 1 chain tensioner assembly as follows:
   - Move the stopper plate upward to release the lock, and push the plunger deep into the tensioner.
   - Move the stopper plate downward to set the lock, and insert a hexagon wrench into the stopper plate's hole.
   - Remove the 2 bolts and chain tensioner.
   - Chain tensioner slipper
6. Remove the timing chain as follows:
   - Turn the crankshaft counterclockwise 10 degrees to loosen the chain of the crankshaft timing sprocket.
   - Remove the pulley set bolt.
   - Remove the chain from the crankshaft timing sprocket and place it on the crankshaft.

   a. Turn the camshaft timing gear assembly on the right bank clockwise (approx. 60 degrees). Be sure to loosen the chain between the banks.

   b. Remove the timing chain.
7. Remove or disconnect the following:
   - No. 2 idle gear shaft, sprocket and No. 1 idle gear shaft
   - 2 bolts and No. 1 chain vibration damper
   - Two No. 2 vibration dampers
   - Crankshaft timing sprocket and 2 pulley set keys

To install:

8. Install or connect the following:
   - No. 1 chain vibration damper and 2 bolts. Torque to 17 ft. lbs. (23 Nm).
   - Two No. 2 vibration dampers
   - Crankshaft timing sprocket and 2 pulley set keys
9. Install the idle sprocket assembly as follows:
   - Apply a light coat of engine oil to the rotating surface of the No. 1 idle gear shaft.
   - Temporarily install the No. 1 idle gear shaft and idle sprocket with the No. 2 idle gear shaft while aligning the knock pin of the No. 1 idle gear with the knock pin groove of the cylinder block.

➡Be careful of the idle gear direction.

   - Using a 10mm hexagon wrench, tighten the No. 2 idle gear shaft to 44 ft. lbs. (60 Nm). Check that the idle sprocket turns smoothly.
10. Install the timing chain as follows:
    - Align the mark plate and timing mark and install the chain.
    - Do not pass the chain over the crankshaft, just put it on it.
    - Turn the camshaft timing gear assembly on the right bank counterclockwise to tighten the chain between the banks.

➡When the idle sprocket is reused, align the chain plate with the mark where the plate had been in order to tighten the chain between the banks.

- Align the mark plate and timing mark and install the chain onto the crankshaft timing sprocket.
- Temporarily tighten the pulley set bolt.
- Turn the crankshaft clockwise to set it to the right block bore center line (TDC/compression).
- Chain tensioner slipper

11. Install the No. 1 chain tensioner assembly as follows:
    a. Move the stopper plate upward to release the lock, and push the plunger deep into the tensioner.
    b. Move the stopper plate downward to set the lock, and insert a hexagon wrench into the hole of the stopper plate.
    c. Install the chain tensioner with the 2 bolts and torque to 7 ft. lbs. (10 Nm).
    d. Remove the lock pin of the chain tensioner. Check that each timing mark is aligned with the crankshaft at the TDC/compression.
    e. Remove the pulley set bolt.
12. Install or connect the following:
    - Timing chain cover assembly
    - Left and right cylinder head covers
    - No. 1 and 2 oil pipes
    - Ignition coil assembly
    - High pressure side fuel pump
    - No. 1 and 2 fuel pipes
    - Oil strainer
    - Oil pan assembly
    - Water inlet
    - Crankshaft pulley. Torque the bolt to 192 ft. lbs (260 Nm).
    - Water hose joint
    - Intake air surge tank assembly and No. 2 surge tank stay
    - Fuel injector driver
    - Water pump pulley. Torque the bolt to 15 ft. lbs (21 Nm).
    - Serpentine belt tensioner
    - Left and right engine mounting brackets
    - A/C compressor unit, if necessary
    - Alternator
    - No. 2 idler pulley
    - Serpentine drive belt
    - Front differential assembly (AWD)
    - Engine/transmission assembly

VALVE COVERS

REMOVAL & INSTALLATION

3.0L (3GR-FSE) Engine

See Figures 149 through 152.

1. Before servicing the vehicle, refer to the precautions in the beginning of this section.

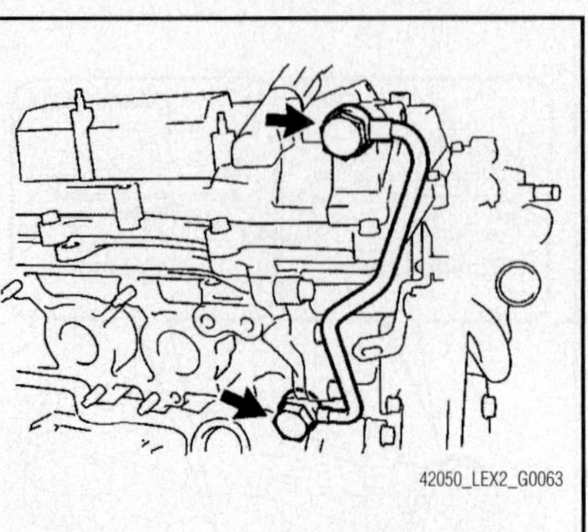

Fig. 149 No. 2 oil pipe (right) shown, No. 1 oil pipe (left) similar—3GR-FSE engine

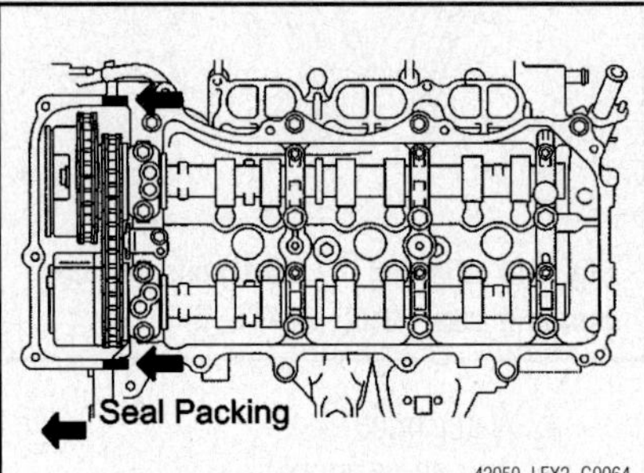

Fig. 150 Correct seal packing application (bank 1 shown, bank 2 similar)—3GR-FSE Engine

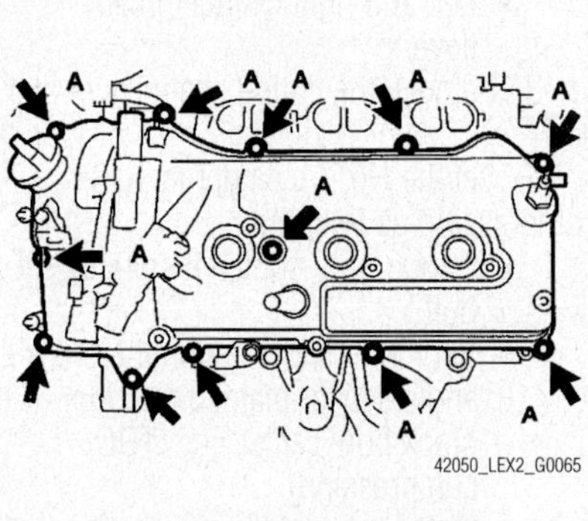

Fig. 151 Cylinder head bolt locations (bank 1)—3GR-FSE engine

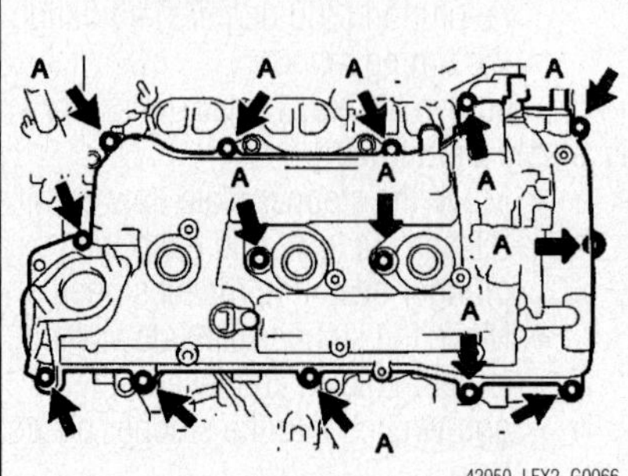

Fig. 152 Cylinder head bolt locations (bank 2)—3GR-FSE engine

2. Disconnect the negative battery cable. Wait at least 90 seconds before performing any other work.
3. Remove the oil filler cap sub-assembly.
4. Remove the 2 bolts and engine hanger No. 1.
5. Remove the 2 bolts and engine hanger No. 2.
6. Remove spark plugs.
7. Remove the ventilation valve on top of the cylinder head cover.
8. Remove the 4 bolts and 4 camshaft position sensors.
9. Remove the 4 bolts and 4 camshaft timing oil control valves.
10. Remove the No. 1 (left) and No. 2 (right) oil pipes as follows:
    - Remove the oil check valve bolt, oil pipe union and oil pipe.
    - Remove the oil control valve filter (left or right) and gaskets.
11. Remove the bank 1 (left) cylinder head cover as follows:
    - Remove the 12 bolts, head cover and gasket.
12. Remove the bank 2 (right) cylinder head cover as follows:
    - Remove the 14 bolts, head cover and gasket.

To install:

13. Install the bank 1 (left) cylinder head cover as follows:
    - Apply seal packing (part No. 08826-00080 or equivalent) to the cylinder head as shown in the illustration.
    - Remove any oil from the contact surface.
    - Install the cylinder head cover within 3 minutes after applying seal packing. Do not start the engine within 2 hours after installing.
    - Install a new gasket to the head cover.
    - Install the cylinder head cover with the 12 bolts. Tighten the bolts (A) to 7 ft. lbs. (10 Nm), and tighten the remaining bolts (except A) to 15 ft. lbs. (21 Nm).
14. Install the bank 2 (right) cylinder head cover as follows:
    - Apply seal packing (part No. 08826-00080 or equivalent) to the cylinder head as shown in the illustration.
    - Remove any oil from the contact surface.
    - Install the cylinder head cover within 3 minutes after applying seal

packing. Do not start the engine within 2 hours after installing.
- Install a new gasket to the head cover.
- Install the cylinder head cover with the 14 bolts. Tighten the bolts (A) to 7 ft. lbs. (10 Nm), and tighten the remaining bolts (except A) to 15 ft. lbs. (21 Nm).

15. Install the No. 1 (left) and No. 2 (right) oil pipes as follows:
- Make sure that there is no foreign matter on the mesh of the oil control valve filter (left or right). Be careful not to touch the mesh when installing the oil control valve filter.
- Install the oil control valve filter (left or right) to the oil pipe union. Install new gaskets and temporarily install the oil pipe (on the head cover side).
- Install a new gasket and temporarily install the oil pipe (on the cylinder head side) with the oil check valve bolt.
- Tighten the oil pipe union (on the head cover side) to 44 ft. lbs. (60 Nm).
- Tighten the oil check valve bolt (on the cylinder head side) to 44 ft. lbs. (60 Nm).

16. Install the 4 camshaft timing oil control valves with the 4 bolts and tighten to 7 ft. lbs. (10 Nm).
17. Install the camshaft position sensors.
18. Install the 4 camshaft position sensors with the 4 bolts and tighten to 7 ft. lbs. (10 Nm).
19. Apply adhesive (part No. 08833-00070, THREE BOND 1324 or equivalent) around the ventilation valve and install the valve. Tighten the ventilation valve to 20 ft. lbs. (27 Nm).
20. Install spark plugs.
21. Install engine hanger No. 2 and tighten the 2 bolts to 24 ft. lbs. (33 Nm).
22. Install engine hanger No. 1 and tighten the 2 bolts to 24 ft. lbs. (33 Nm).
23. Install the oil filler cap sub-assembly.
24. Connect the negative battery cable.

4.3L (3UZ-FE) Engine

See Figures 153 and 154.

1. Before servicing the vehicle, refer to the precautions in the beginning of this section.
2. Disconnect the negative battery cable. Wait at least 90 seconds before performing any other work.
3. Remove the V-bank cover.

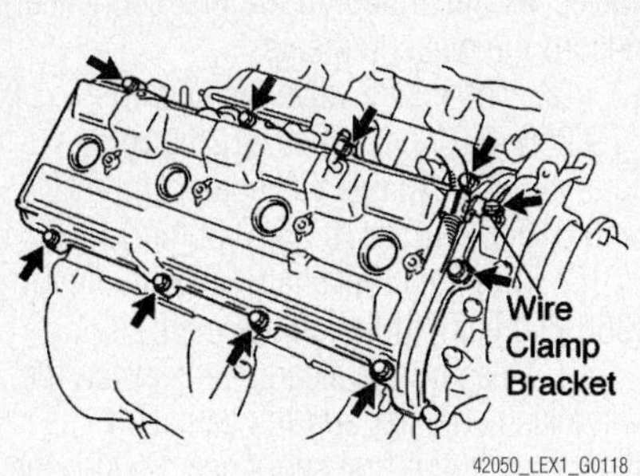

Fig. 153 Right cylinder head cover and mounting bolts—3UZ-FE engine

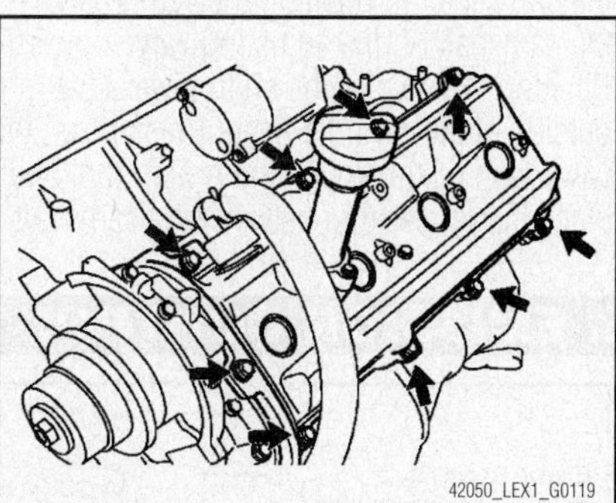

Fig. 154 Left cylinder head cover and mounting bolts—3UZ-FE engine

4. Remove the right cylinder head cover as follows:
- Remove the bolt, and disconnect the wire clamp bracket on the engine wire from the camshaft bearing cap.
- Remove the EVAP from the cylinder head cover sub-assembly.
- Remove the 9 bolts, 9 seal washers, cylinder head cover sub-assembly and gasket.

5. Remove the left cylinder head cover as follows:
- Disconnect the wire clamp from the wire bracket on the cylinder head cover.
- Remove the EVAP from the cylinder head cover.
- Remove the 9 bolts, 9 seal washers, cylinder head cover and gasket.

To install:

6. Install the left and right cylinder head cover as follows:
- Install the gasket to the cylinder head cover.
- Install the seal washer to the bolt.
- Install the cylinder head cover with the 9 bolts. Uniformly tighten the bolts in several passes to 53 inch lbs. (6 Nm).

7. Install the hose bracket of the EVAP to the cylinder head cover.
8. Install the wire clamp to the bracket on the cylinder head cover.
9. Install the V-bank cover.
10. Connect the negative battery cable.

VALVE LASH

ADJUSTMENT

3.0L (3GR-FSE) Engine

The 3.0L (3GR-FSE) engine is equipped with hydraulic valves which are not adjustable.

4.3L (3UZ-FE) Engine

See Figures 155 and 156.

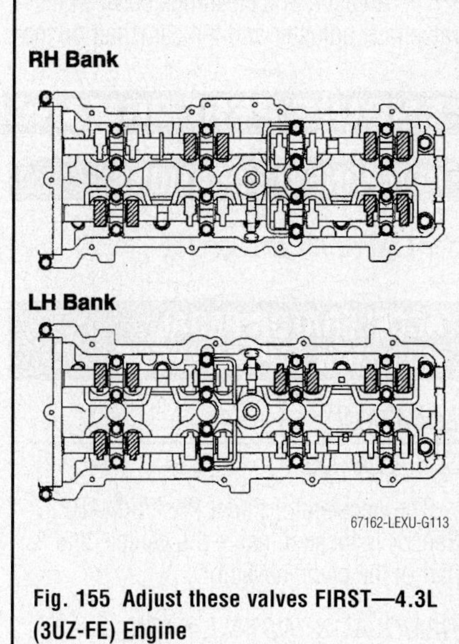

Fig. 155 Adjust these valves FIRST—4.3L (3UZ-FE) Engine

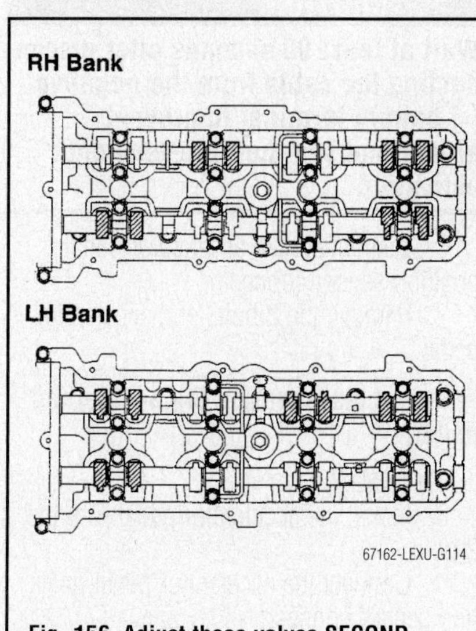

Fig. 156 Adjust these valves SECOND—4.3L (3UZ-FE) Engine

1. Before servicing the vehicle, refer to the precautions in the beginning of this section.

2. Remove or disconnect the following, as applicable:

- Negative battery cable
- V-bank cover
- Intake air connector pipe
- Ignition coils
- No. 3 timing belt covers
- Spark plug wires
- Cylinder head covers

3. Turn the crankshaft pulley and align its groove with the timing mark **0** of the No. 1 timing cover. Check that the timing marks of the camshaft timing pulleys and timing belt rear plates are aligned. If not, turn the crankshaft 1 revolution (360 degrees) and align the mark.

4. Measure the clearance between the valve lash adjuster and the camshaft on the valves, as illustrated, in the first sequence. Record the measurements.

a. The intake valve lash cold is 0.006–0.010 in. (0.15–0.25mm).

b. The exhaust valve lash cold is 0.010–0.014 in. (0.25–0.35mm).

5. Turn the crankshaft 1 full revolution (360 degrees) and align the mark.

6. Measure the clearance between the valve lash adjuster and the camshaft , as illustrated in the first sequence. Record the measurements.

7. If necessary, remove the camshafts.

8. Remove the adjusting shim and turn the crankshaft to position the cam lobe of the camshaft on the adjusting valve upward. Position the hole in the shim toward the outside of the cylinder head. Press down the valve lash adjuster with the proper tool and place the proper tool between the camshaft and the valve lash adjuster. Remove the tool.

9. Remove the adjusting shim with the proper tool.

10. Determine the thickness of the replacement shim as follows:

a. T = Thickness of the used shim

b. A = Measured valve lash

c. N = Thickness of new shim

d. Intake: N = T + (A - 0.006–0.010 in. (0.15–0.25mm))

e. Exhaust: N = T + (A - 0.010–0.014 in. (0.25–0.35mm))

11. Recheck the valve lash. Install the cylinder head covers.

12. Connect the spark plug wires and install the No. 3 timing belt covers.

13. Install or reconnect all other components in reverse of removal procedure.

14. Connect the negative battery cable.

ENGINE PERFORMANCE & EMISSION CONTROL

COMPONENT LOCATIONS

See Figures 157 through 162.

ACCELERATOR PEDAL POSITION (APP) SENSOR

LOCATION

See Figure 163.

The Accelerator Pedal Position (APP) Sensor is located inside the vehicle and is part of the pedal assembly.

REMOVAL & INSTALLATION

CAUTION

Wait at least 90 seconds after disconnecting the cable from the negative (-) battery terminal to prevent airbag and seat belt pre-tensioner activation.

1. Disconnect the accelerator pedal position sensor connector.

2. Remove the 2 bolts and accelerator pedal.

➡Do not disassemble the accelerator pedal.

To install:

3. Install the accelerator pedal with the 2 bolts.

4. Connect the accelerator pedal position sensor connector.

5. Perform initialization procedure. Certain systems need to be initialized after

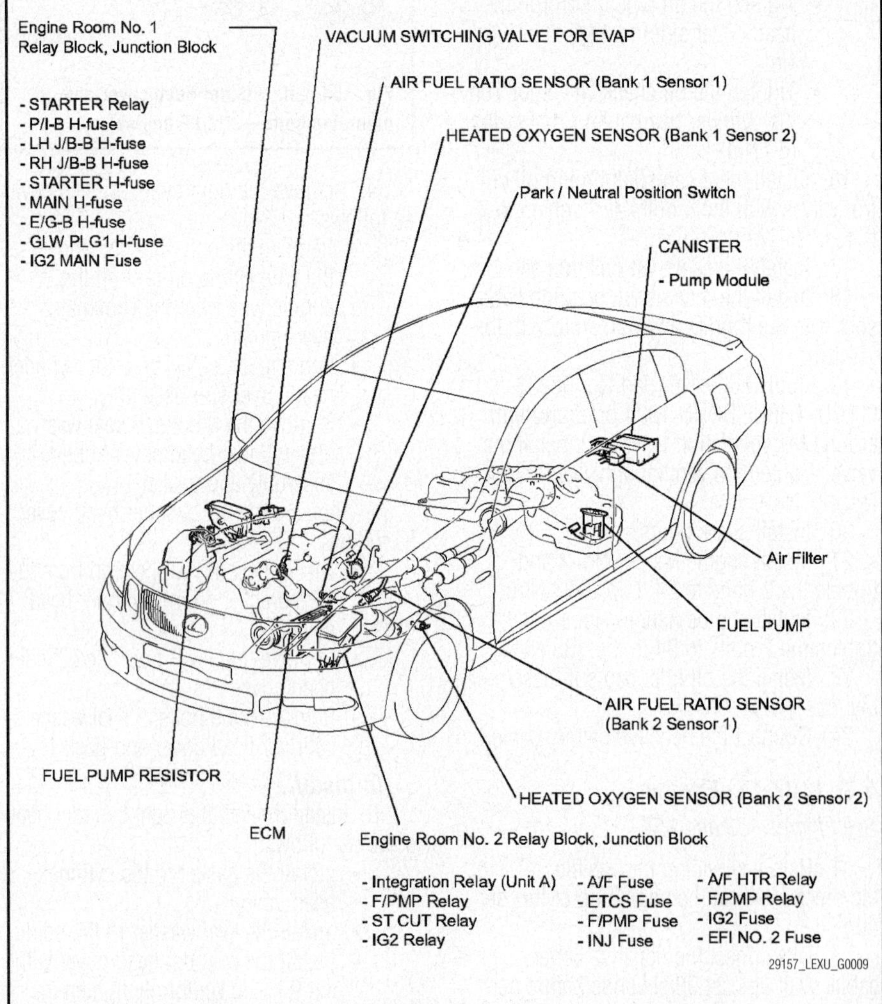

Fig. 157 Vehicle component locations—2006 GS300

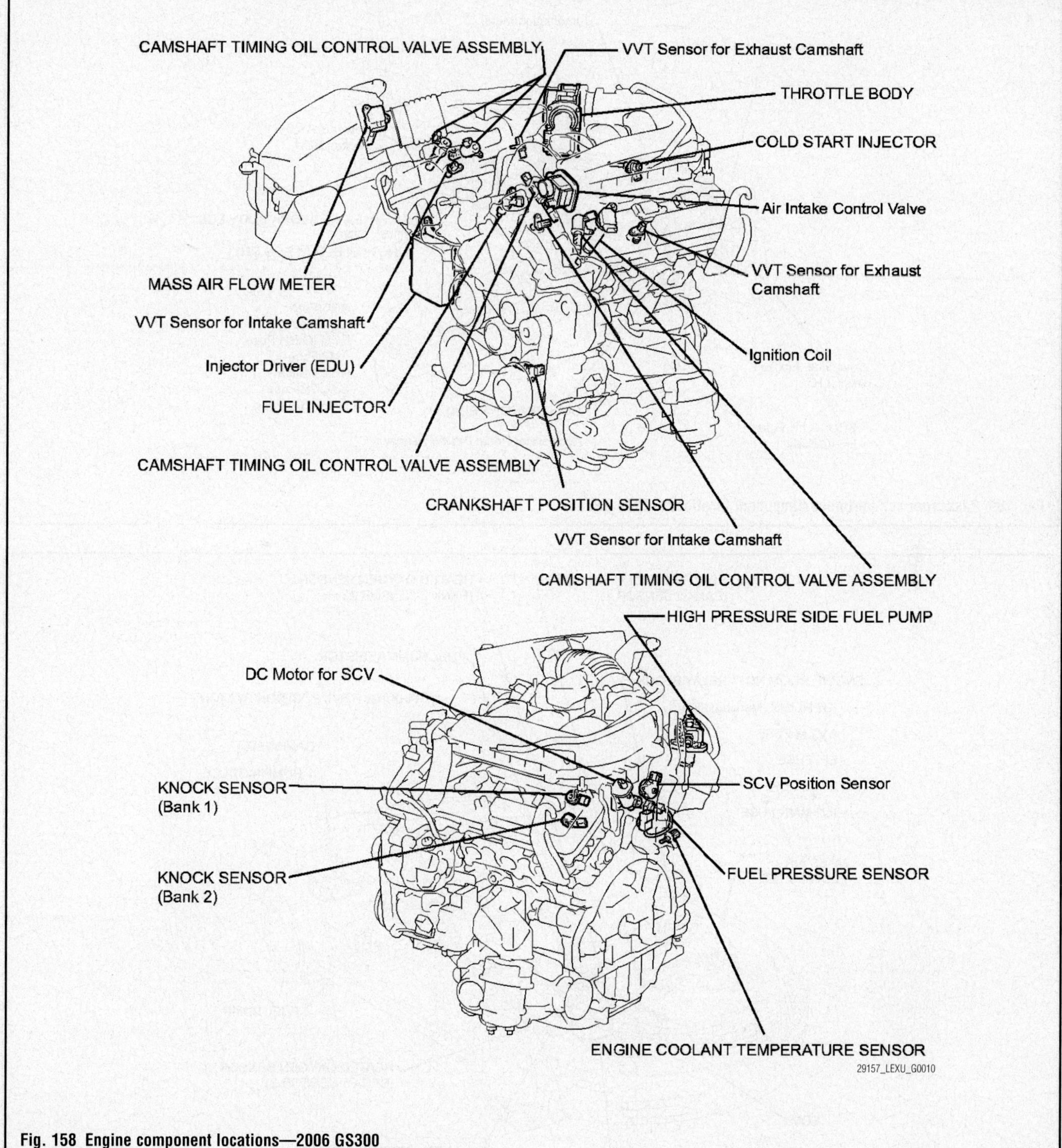

Fig. 158 Engine component locations—2006 GS300

disconnecting and reconnecting the cable from the negative (-) battery terminal.

AIR FUEL RATIO SENSOR

LOCATION

3.0L (3GR-FSE) ENGINE

See Figure 164.

Air fuel ratio sensor location view 3.0L (3GR-FSE) engine.

REMOVAL & INSTALLATION

3.0L (3GR-FSE) Engine

See Figure 165.

1. Remove the battery.
2. Remove the 3 bolts and disconnect the ground cable terminal.
3. Release the locks of the 2 power steering ECU assembly connectors and disconnect the connectors.
4. Remove the power steering ECU assembly.
5. Remove the 2 bolts and the battery tray.
6. Disconnect the sensor connector.
7. Using an 02 sensor socket remove the air fuel ratio sensor

Combination Meter
Stop Light Switch
Power Source
Control ECU
MULTIPLEX NETWORK BODY ECU
(COWL SIDE
JUNCTION BLOCK ECU RH)
- ACC Fuse
- IGN Fuse
- ECU-IG RH Fuse
- AM2 Fuse
- RH-IG Fuse
- GAUGE Fuse
- STOP SW Fuse
Cowl Side Junction
Block LH
DLC3
ACCELERATOR PEDAL
- ECU-IG LH Fuse
- LH-IG Fuse
- Accelerator Pedal Position Sensor
29157_LEXU_G0011

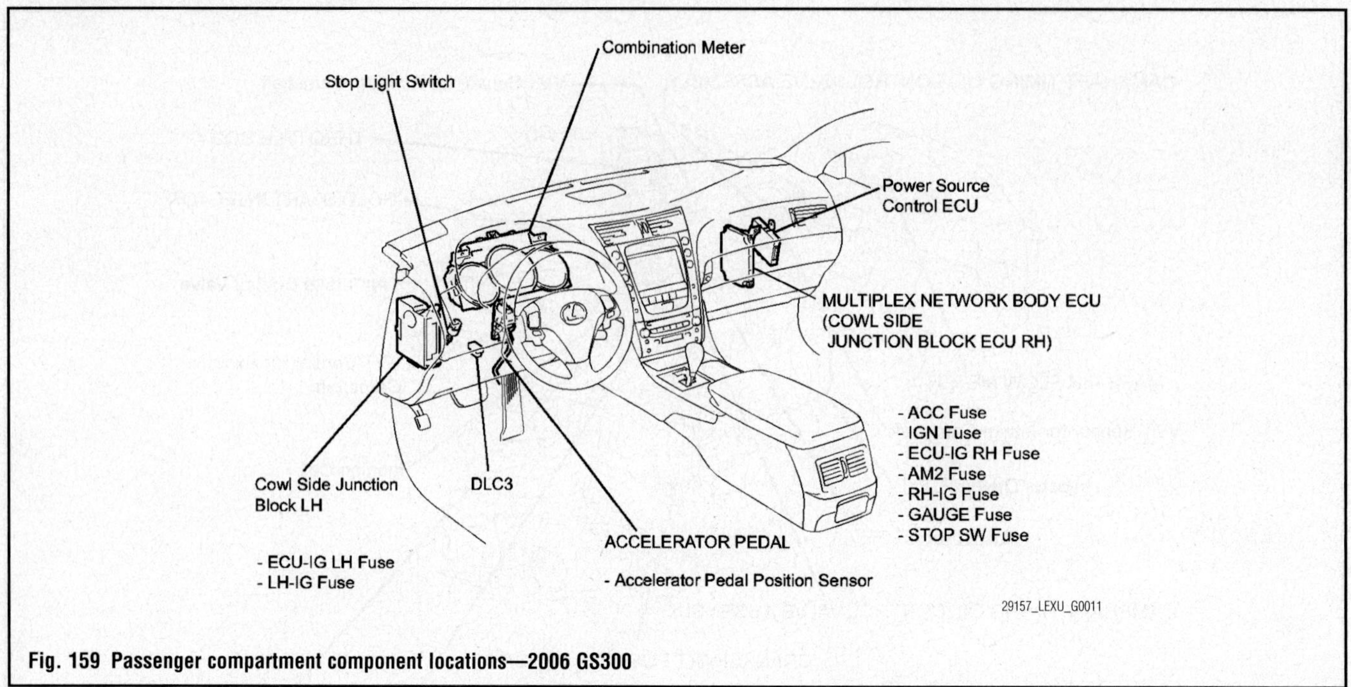

Fig. 159 Passenger compartment component locations—2006 GS300

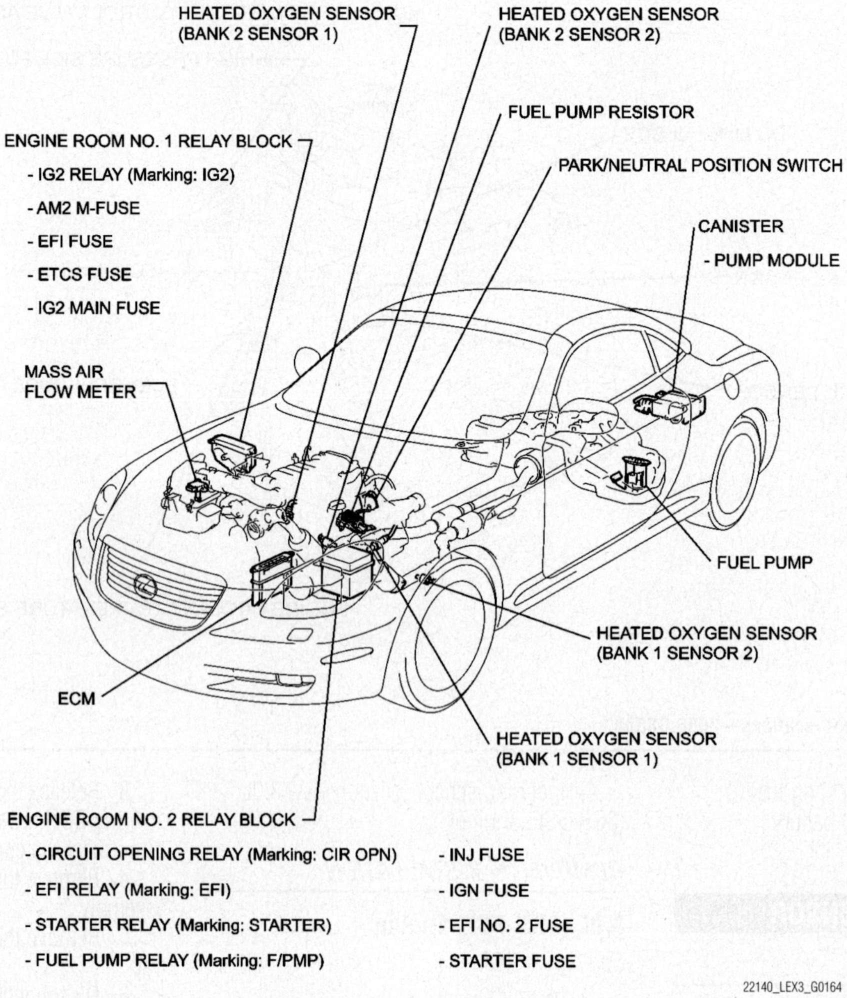

HEATED OXYGEN SENSOR
(BANK 2 SENSOR 1)
HEATED OXYGEN SENSOR
(BANK 2 SENSOR 2)
FUEL PUMP RESISTOR
ENGINE ROOM NO. 1 RELAY BLOCK
- IG2 RELAY (Marking: IG2)
- AM2 M-FUSE
- EFI FUSE
- ETCS FUSE
- IG2 MAIN FUSE
PARK/NEUTRAL POSITION SWITCH
CANISTER
- PUMP MODULE
MASS AIR
FLOW METER
FUEL PUMP
HEATED OXYGEN SENSOR
(BANK 1 SENSOR 2)
ECM
HEATED OXYGEN SENSOR
(BANK 1 SENSOR 1)
ENGINE ROOM NO. 2 RELAY BLOCK
- CIRCUIT OPENING RELAY (Marking: CIR OPN)
- EFI RELAY (Marking: EFI)
- STARTER RELAY (Marking: STARTER)
- FUEL PUMP RELAY (Marking: F/PMP)
- INJ FUSE
- IGN FUSE
- EFI NO. 2 FUSE
- STARTER FUSE
22140_LEX3_G0164

Fig. 160 Vehicle component locations—SE 430 model

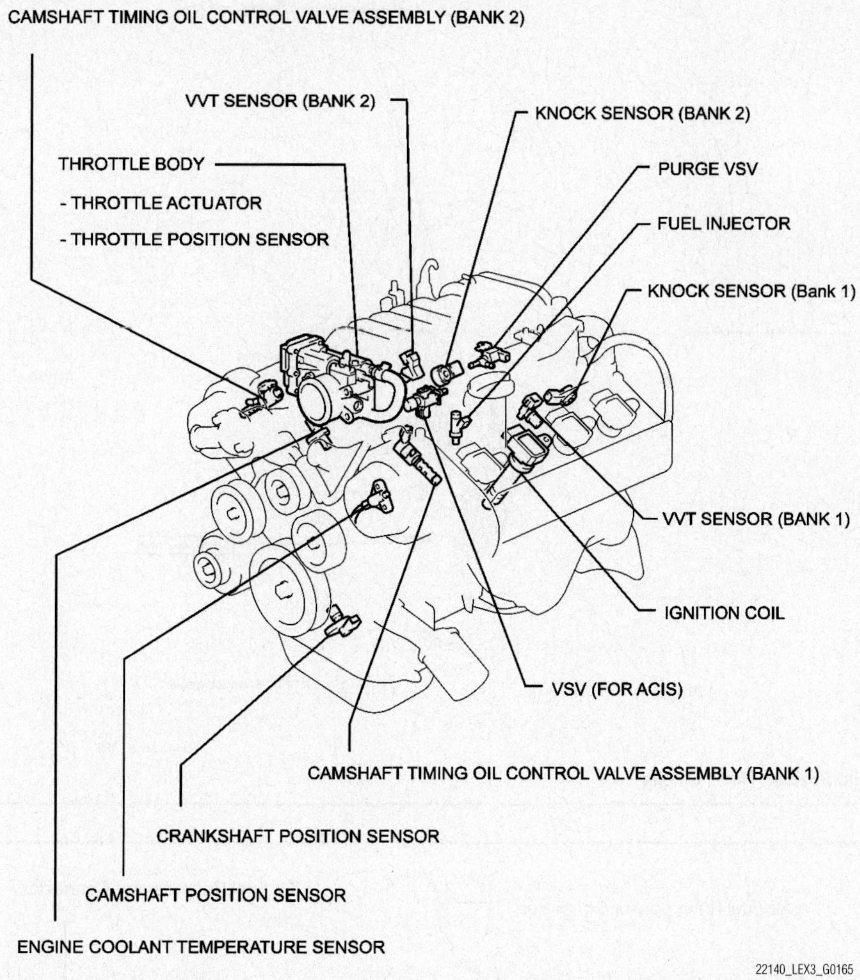

CAMSHAFT TIMING OIL CONTROL VALVE ASSEMBLY (BANK 2)
VVT SENSOR (BANK 2)
KNOCK SENSOR (BANK 2)
THROTTLE BODY
- THROTTLE ACTUATOR
- THROTTLE POSITION SENSOR
PURGE VSV
FUEL INJECTOR
KNOCK SENSOR (Bank 1)
VVT SENSOR (BANK 1)
IGNITION COIL
VSV (FOR ACIS)
CAMSHAFT TIMING OIL CONTROL VALVE ASSEMBLY (BANK 1)
CRANKSHAFT POSITION SENSOR
CAMSHAFT POSITION SENSOR
ENGINE COOLANT TEMPERATURE SENSOR
22140_LEX3_G0165

Fig. 161 Engine component locations—SE 430 model

COMBINATION METER
DLC3
ACCELERATOR PEDAL
- ACCELERATOR PEDAL POSITION SENSOR
DRIVER SIDE JUNCTION BLOCK
- IG1 RELAY (Marking: IG1)
- D-IG FUSE
- STOP FUSE
STOP LIGHT SWITCH
22140_LEX3_G0166

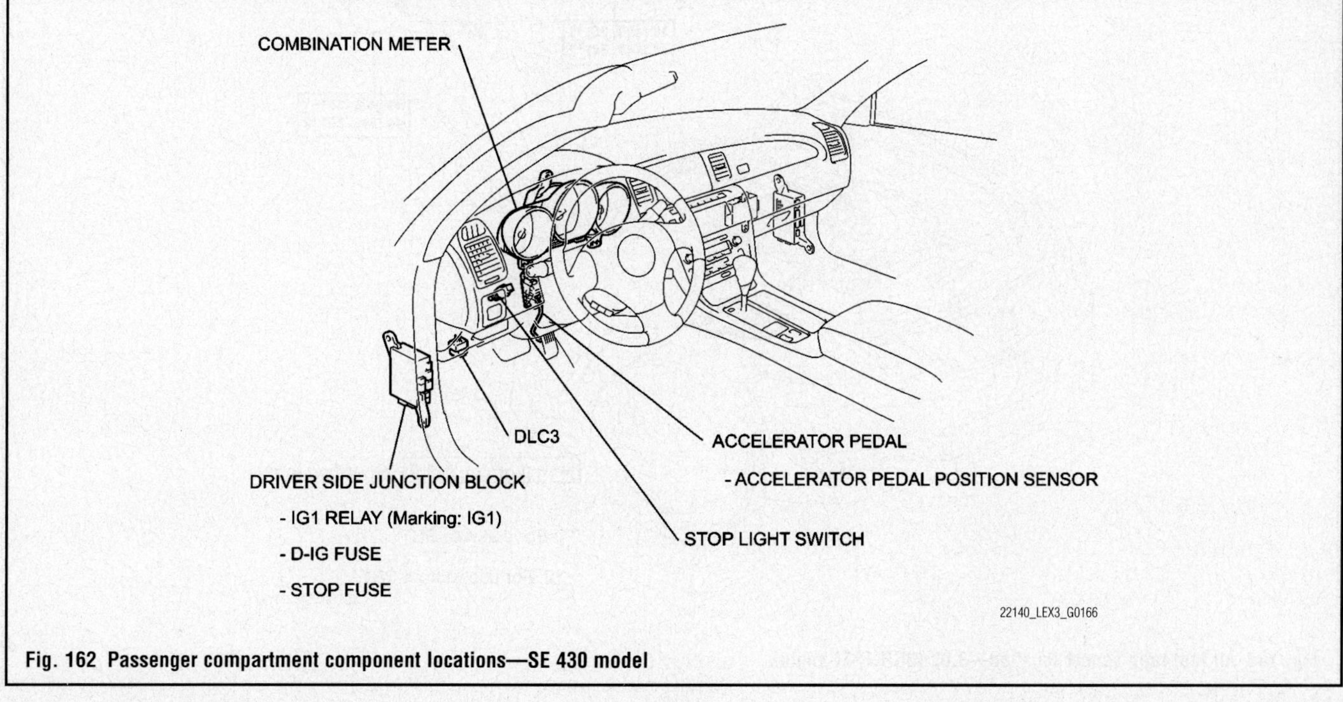

Fig. 162 Passenger compartment component locations—SE 430 model

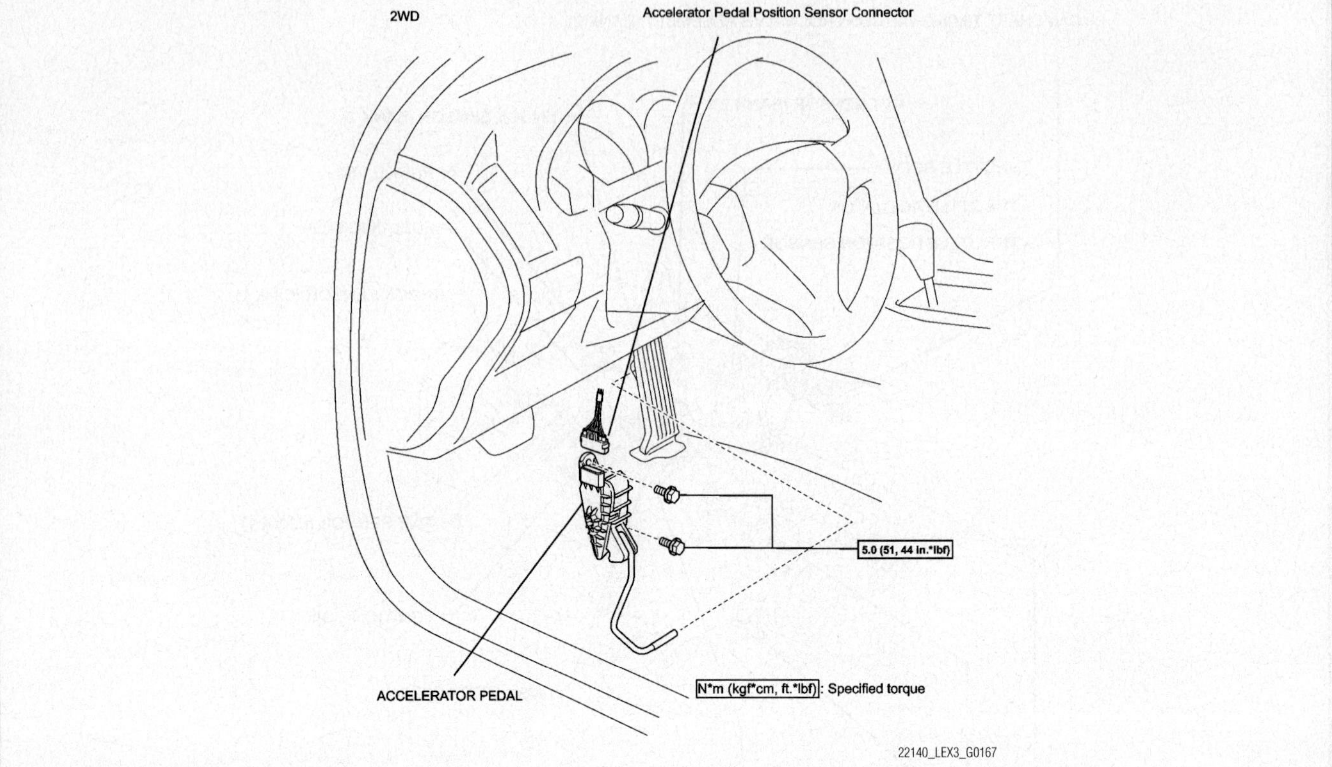

Fig. 163 Accelerator Pedal Position (APP) Sensor

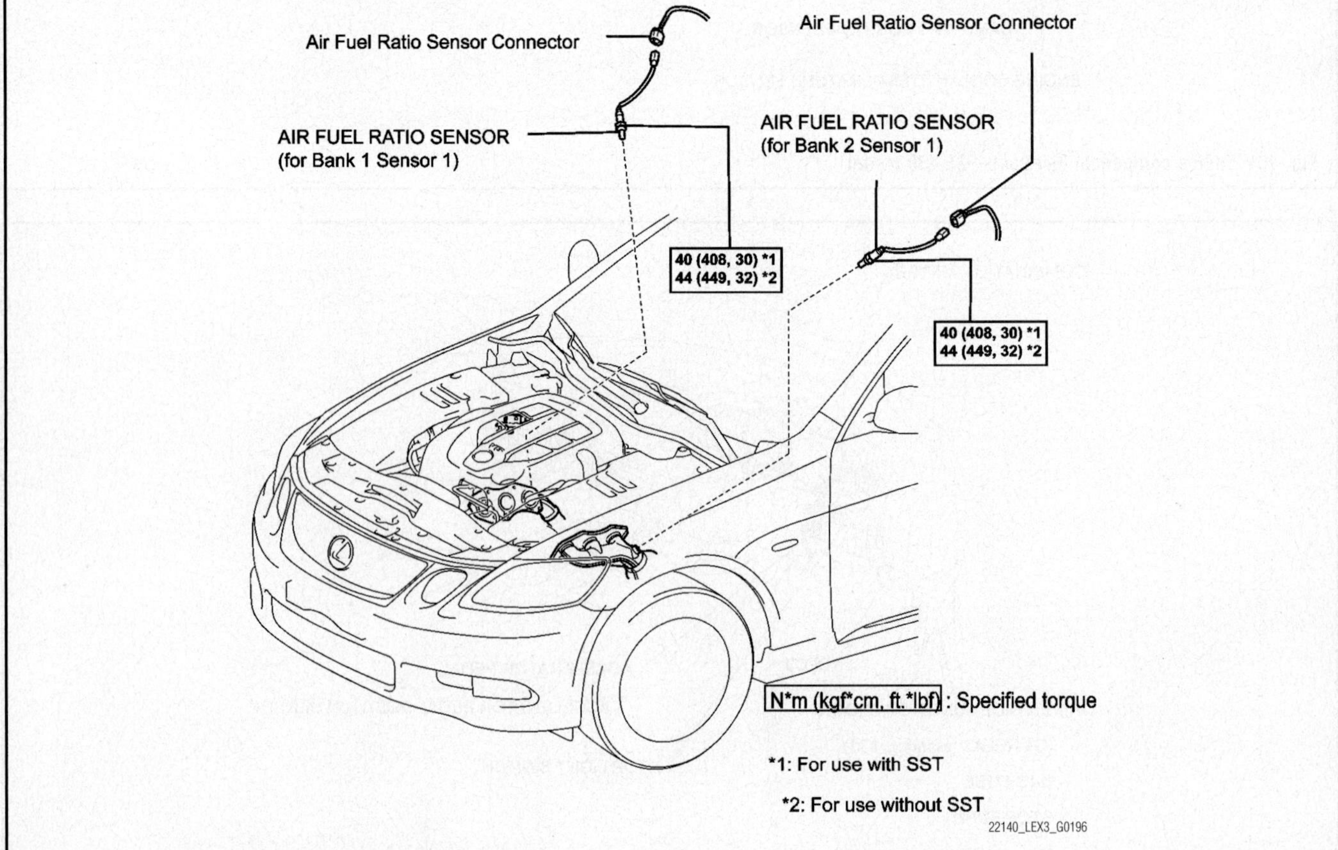

Fig. 164 Air fuel ratio sensor location—3.0L (3GR-FSE) engine

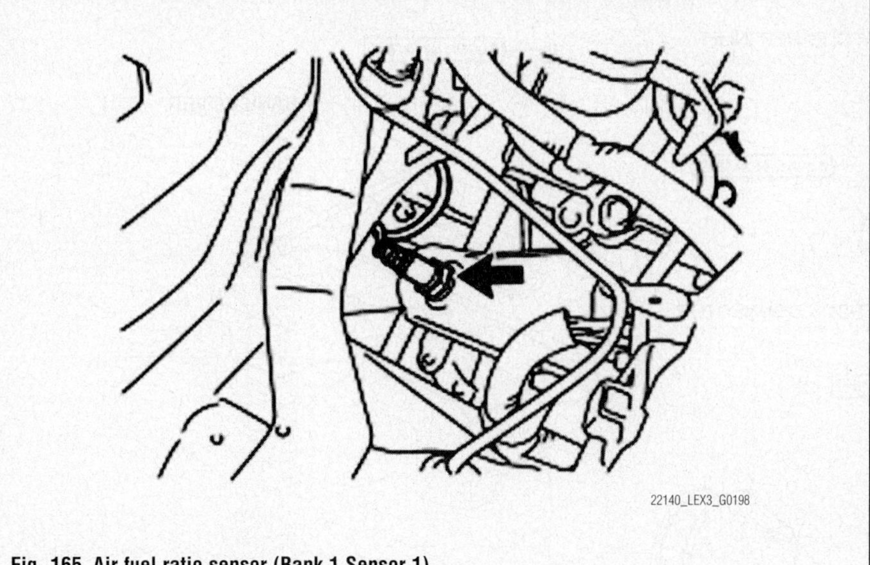

Fig. 165 Air fuel ratio sensor (Bank 1 Sensor 1)

To install:

8. Using a O2 socket install the sensor to the exhaust manifold. Tighten to 32 ft. lbs. (44 Nm).
9. Connect the sensor connector.
10. Install the battery tray and tighten the 2 mounting bolts to 44 ft. lbs. (51 Nm).
11. Connect the 2 power steering ECU assembly connectors and securely lock the connectors.
12. Install the power steering ECU assembly to the battery tray with the 2 bolts. Tighten the bolts to 44 ft. lbs. (51 Nm).
13. Install the ground cable terminal and tighten the bolt to 44 ft. lbs. (51 Nm).
14. Install the battery tray support No.1.
15. Install the battery.
16. Install the battery clamp with the 2 nuts and tighten to 26 inch. lbs. (2.9 Nm).
17. Connect the negative and positive battery cables to the battery. Tighten the cables to 50 inch. lbs. (5.6 Nm).

➡Some systems need initialization when disconnecting the cable from the negative battery terminal.

4.3L (3UZ-FE) Engine

1. Disconnect the negative battery cable.
2. Wait at least 90 seconds after disconnecting the cable from the negative (-) battery terminal to prevent airbag and seat belt pre-tensioner activation.
3. Remove both under covers.
4. Disconnect the sensor connector.
5. With a O2 socket remove the heated O2 sensor.
6. For (Bank 2) disconnect the sensor clamp.

To install:

7. Using O2 socket install the sensor and tighten to 32 ft. lbs. 44 Nm).
8. Connect the sensor connector.
9. For (Bank 2) connect the sensor clamp.
10. Install the undercover with the 18 bolts and 5 clips.
11. Install the No. 2 under cover with the 5 bolts.
12. Connect the negative battery cable.
13. Certain systems need to be initialized after disconnecting and reconnecting the cable from the negative (-) battery terminal.

CAMSHAFT POSITION (CMP) SENSOR

LOCATION

See Figure 166.

Camshaft Position (CMP) Sensor location SC 430 (3UZ-FE) engine.

REMOVAL & INSTALLATION

See Figure 167.

✲✲ CAUTION

Wait at least 90 seconds after disconnecting the cable from the negative (-) battery terminal to prevent airbag and seat belt pre-tensioner activation.

1. Disconnect the negative battery cable.
2. Drain the engine coolant.
3. Remove the 7 clips and intake duct seal.
4. Remove the 3 clips and side cover.
5. Remove the 2 nuts and V-bank cover.
6. Remove the radiator hose inlet.
7. Remove the cap nut and disconnect the No. 3 water by-pass pipe from the timing belt cover.
8. Disconnect the 2 water by-pass hoses from the No. 3 water by-pass pipe.
9. Disconnect the engine wire from the 2 wire clamps.
10. Disconnect the camshaft position sensor connector.
11. Disconnect the camshaft position sensor wire from the wire clamp on the timing belt cover.
12. Remove the wire grommet from the timing belt cover.
13. Remove the 4 bolts.
14. Disconnect the timing belt cover from the timing plate and camshaft bearing cap.
15. Disconnect the wire clamp for the sensor from the timing belt cover.
16. Remove the connector holder from the sensor connector.
17. Remove the timing belt cover.
18. Remove the bolt, stud bolt and camshaft position sensor.

To install:

19. Install the sensor with the bolt and stud bolt. Tighten to 66 inch. lbs. (7.5 Nm).
20. Install the gasket to the cover.
21. Run the camshaft position sensor wire through the cover hole.
22. Install the cover with the 4 bolts and tighten to 66 inch. lbs. (7.5 Nm).
23. Install the wire grommet to the cover.
24. Install the sensor connector to the sensor holder.
25. Connect the sensor connector.
26. Install the sensor wire to the wire clamp on the cover.
27. Install the engine wire to the 2 wire clamps on the cover.
28. Connect the 2 water by-pass hoses as shown in the illustration.
29. Install the No. 3 water by-pass pipe to the cover with the cap nut. Tighten to 66 inch. lbs. (7.5 Nm).
30. Install the radiator inlet hose.
31. Install the V-bank cover with the 2 nuts. Tighten the nuts to 44 inch. lbs. (5.0 Nm).
32. Install the side cover with the 2 clips and nut.
33. Install the intake duct seal with the 7 clips.
34. Connect the negative battery cable.
35. Perform initialization procedure. Certain systems need to be initialized after disconnecting and reconnecting the cable from the negative (-) battery terminal.

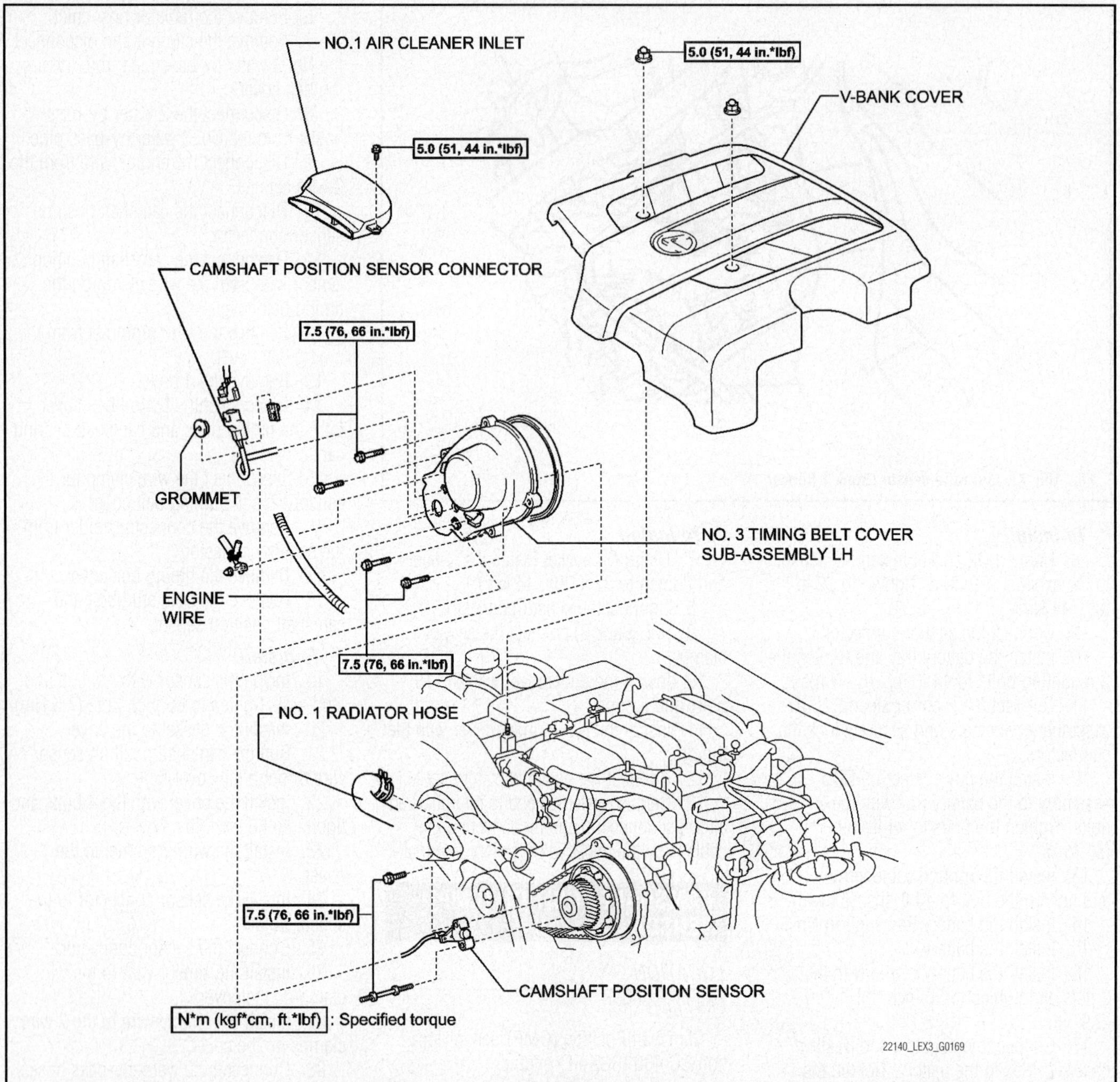

Fig. 166 Camshaft position sensor location and related parts SC 430 (3UZ-FE) engine

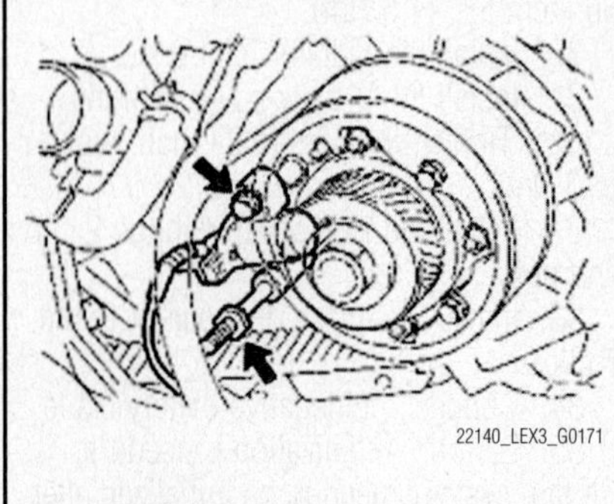

Fig. 167 Camshaft position sensor, and mounting bolts

CAMSHAFT TIMING OIL CONTROL VALVE

LOCATION

See Figures 168 and 169.

Camshaft timing oil control valve and related parts.

Camshaft timing oil control valve and related parts.

REMOVAL & INSTALLATION

3.0L (3GR-FSE) Engine

See Figure 170.

✱✱ CAUTION

Wait at least 90 seconds after disconnecting the cable from the negative (-) battery terminal to prevent airbag and seat belt pre-tensioner activation.

1. Disconnect the negative battery cable.

Removal procedure for (Bank 1) oil control valve.

2. Remove the engine cover.
3. Disconnect the oil control valve connector.

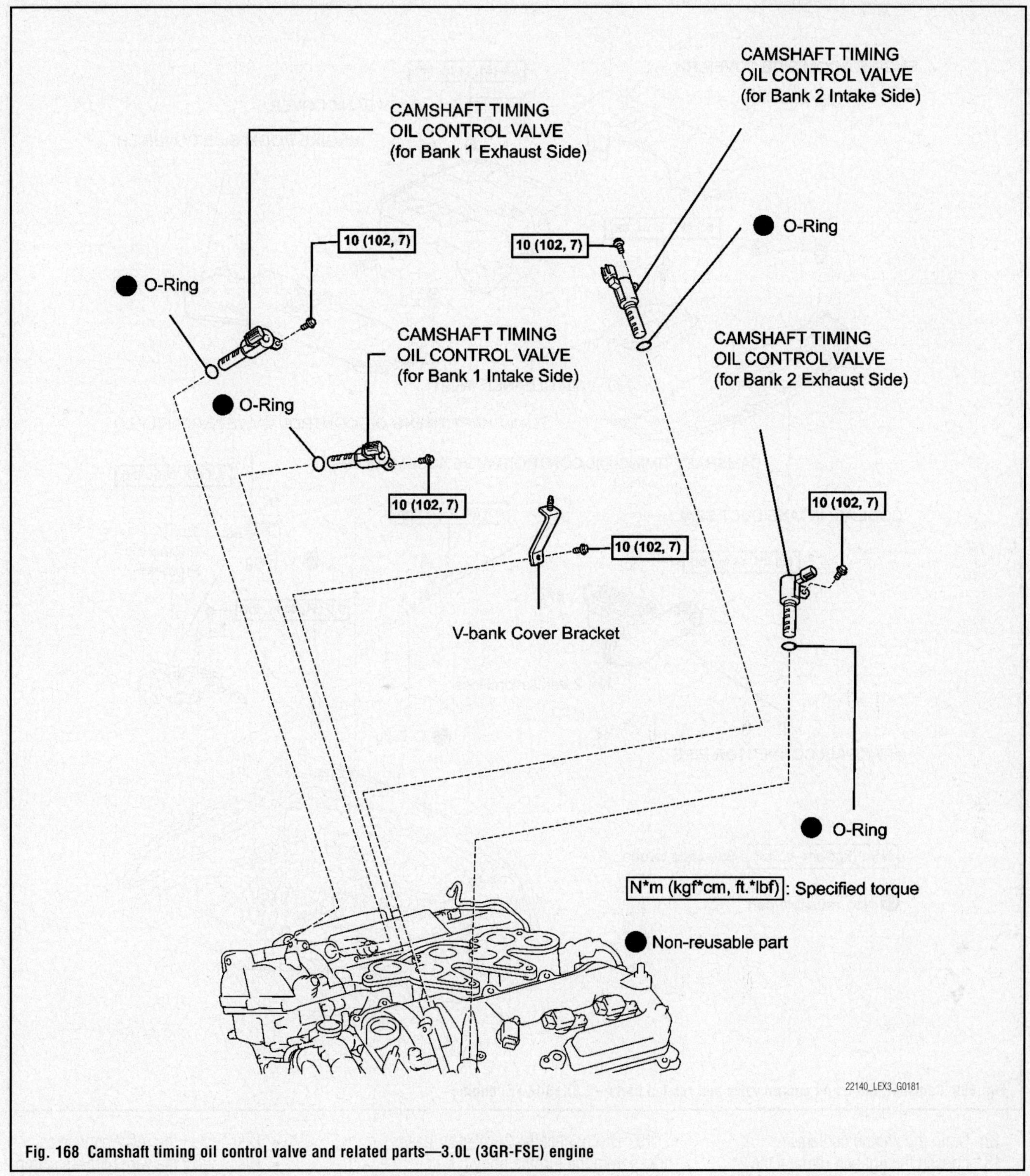

Fig. 168 Camshaft timing oil control valve and related parts—3.0L (3GR-FSE) engine

4. Remove the bolt and oil control valve. remove the bolt and V-bank cover bracket on (Bank 1) intake side.
5. Remove the O-ring from the oil control valve.

To install:

6. Apply a light coat of engine oil to a new O-ring of the oil control valve.
7. Install the oil control valve with the bolt and tighten to 7 ft. lbs. (10 Nm).

➡Be careful that the O-ring is not cracked or jumped when installing the oil control valve.

8. Connect the camshaft timing oil control valve connector.

Removal procedure for (Bank 2) oil control valve.

9. Using a clip remover, remove the 7 clips and duct seal.
10. Using a clip remover, remove the 3 clips.
11. Remove the 10 screws and under cover.

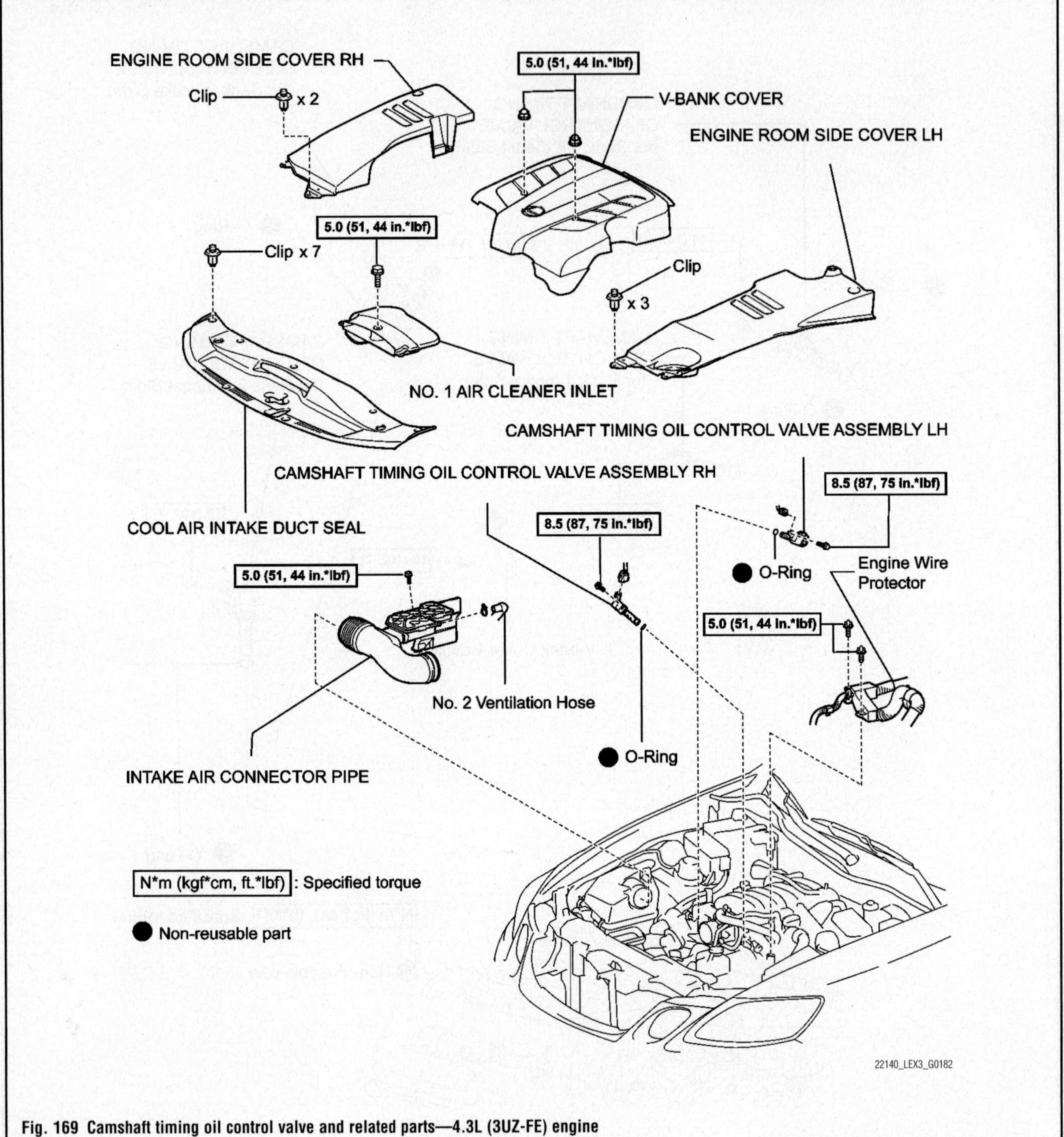

Fig. 169 Camshaft timing oil control valve and related parts—4.3L (3UZ-FE) engine

12. Drain the engine coolant.
13. Loosen the nut and remove the 2 clips and side cover.
14. Disconnect the ventilation hose from the cylinder head.
15. Disconnect the MAF meter connector
16. Disconnect the clamp from the air cleaner.
17. Disconnect the VSV (for EVAP).
18. Disconnect the 4 clamps, remove the air cleaner cap and hose.
19. Disconnect the 2 water by-pass hoses from the throttle body.
20. Disconnect the heater hoses.
21. Remove the engine rear cover.
22. Disconnect the union to the check valve hose.
23. remove the ventilation hose.
24. Disconnect the cold start injector.
25. Remove the intake plenum as follows:
    - Disconnect the throttle motor connector.
    - Disconnect the IACV connector.
    - Disconnect the wire harness clamp from the surge tank.
    - Remove the 2 bolts and intake manifold stay.
    - Disconnect the 2 wire harness clamps from the surge tank stays.
    - Remove the 4 bolts and 2 surge tank stays.
    - Remove the bolt and disconnect the water hose joint.

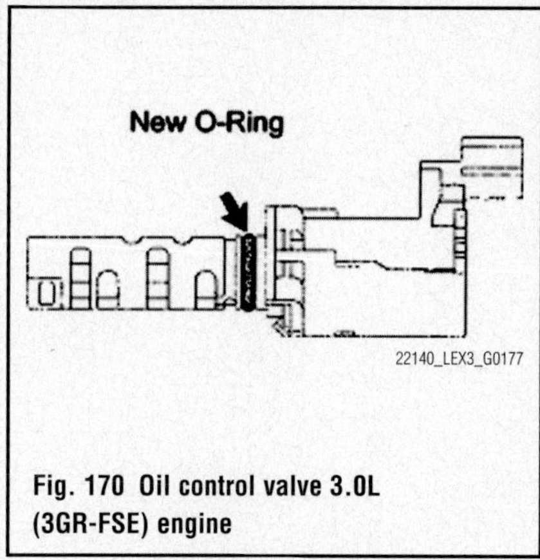

Fig. 170 Oil control valve 3.0L (3GR-FSE) engine

- Using a 5 mm hexagon socket wrench, remove the 7 bolts, 2 nuts and gasket.

26. Remove intake plenum.

➡Cover the intake manifold port to prevent foreign matter from entering it.

27. Disconnect the oil control valve connector.
28. Remove the bolt and oil control valve. (Applies to intake and exhaust oil control valve).

To install:

29. Apply a light coat of engine oil to a new O-ring of the oil control valve.
30. Install the oil control valve with the bolt and tighten to 7 ft. lbs. (10 Nm).

➡Be careful that the O-ring is not cracked or jumped when installing the oil control valve.

31. Connect the camshaft timing oil control valve connector.
32. Install a new gasket to the intake plenum.
33. Install the intake air surge tank with the 2 nuts and tighten to 13 ft. lbs. (18 Nm).
34. Using a 5 mm hexagon socket wrench, install the 7 bolts and tighten to 12 ft. lbs. (16 Nm).
35. Install the 2 surge tank stays with the 4 bolts and tighten to 15 ft. lbs. (20 Nm).
36. Connect the connector to the fuel main pipe. Push the two parts together firmly until a "click" sound is heard. Then attach the lock claws to the connector by pushing down on the connector cover.

➡Check that the connector and fuel main pipe are securely connected by trying to pull them apart.

37. Connect the ventilation hose to the ventilation valve.
38. Install the engine rear cover with the 3 clips.
39. Connect the hose to the surge tank.
40. Connect the water heater hoses.
41. Connect the 2 water by-pass hoses to the throttle body.
42. Install a new gasket and the cold start injector with the 2 bolts. Tighten the bolts to (7 ft. lbs. (10 Nm).
43. Install the air cleaner cap with air cleaner hose assembly with the 4 clamps and hose clamp.
44. Connect the ventilation hose to the cylinder head cover with the clamp.
45. Install engine coolant and bleed the system.
46. Connect the negative battery cable.
47. Check the operation of throttle body.
48. Check for coolant leaks.
49. Install the V-bank cover.
50. Install the side cover with the 2 clips and nut.
51. Install the duct seal with the 7 clips.
52. Connect the negative battery cable.
53. After the engine is warmed up, check that the maintained parts operate normally.
54. Perform initialization procedure. Certain systems need to be initialized after disconnecting and reconnecting the cable from the negative (-) battery terminal.

4.3L (3UZ-FE) Engine

See Figure 171.

⁂ CAUTION

Wait at least 90 seconds after disconnecting the cable from the negative (-) battery terminal to prevent airbag and seat belt pre-tensioner activation.

1. Disconnect the negative battery cable.
2. Remove the 7 clips and intake duct seal.
3. Remove the nuts, clips and remove both side covers.
4. Remove the 3 clips and side cover.
5. Remove the engine cover.

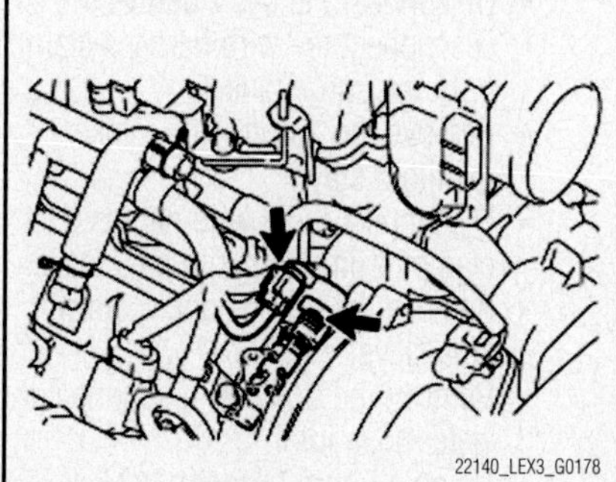

Fig. 171 Oil control valve 4.3L (3UZ-FE) engine

6. Remove the bolt and No. 1 air cleaner inlet.
7. Disconnect the air hose and No. 1 ventilation hose.
8. Loosen the 2 hose clamps and remove the intake air connector.
9. Disconnect the oil control valve connector. (Procedure for left and right valve).
10. Remove the bolt and oil control valve.

To install:

11. Install a new O-ring to the oil control valve.
12. Install the oil control valve with the bolt and tighten to 75 inch. lbs. (8.5 Nm).
13. Connect the oil control valve connector.
14. Install the intake air connector pipe with the bolt and 2 hose clamps.
15. Connect the air hose and No. 1 ventilation hose.
16. Install the air cleaner inlet and bolt. Tighten to 44 inch. lbs. (5.0 Nm).
17. Install the V-bank engine cover with the 2 nuts. Tighten to 44 inch. lbs. (5.0 Nm).
18. Install the side covers with.
19. Install the intake duct seal with the 7 clips.
20. Connect the negative battery cable.
21. After the engine is warmed up, check that the maintained parts operate normally.
22. Perform initialization procedure. Certain systems need to be initialized after disconnecting and reconnecting the cable from the negative (-) battery terminal.

COOLANT TEMPERATURE SENSOR

LOCATION

See Figures 172 and 173.

Coolant Temperature Sensor location—3.0L (3GR-FSE) engine

Coolant Temperature Sensor location—4.3L (3UZ-FE) engine

REMOVAL & INSTALLATION

3.0L (3GR-FSE) Engine

See Figure 174.

⁂ CAUTION

Wait at least 90 seconds after disconnecting the cable from the negative (-) battery terminal to prevent airbag and seat belt pre-tensioner activation.

1. Disconnect the negative battery cable.

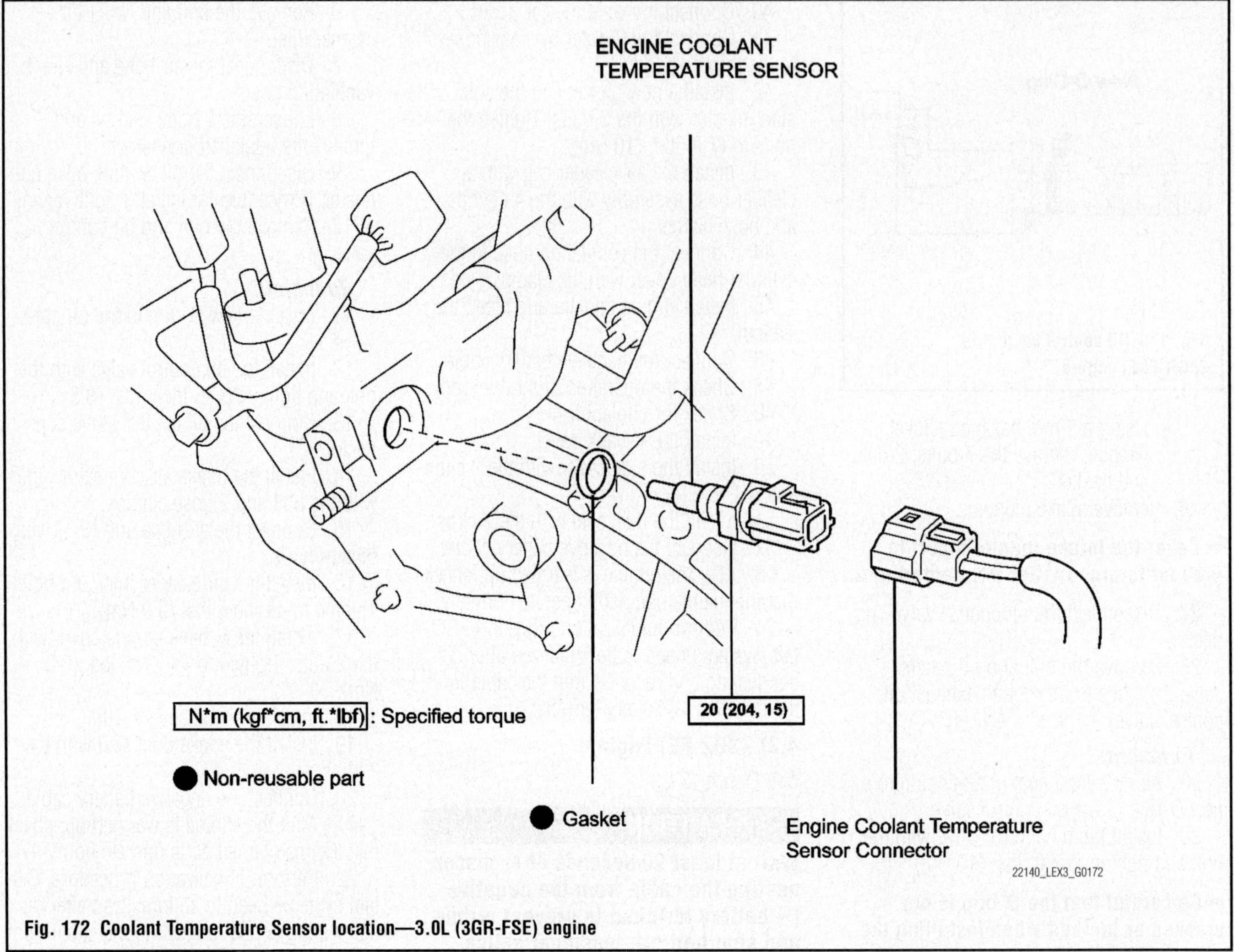

Fig. 172 Coolant Temperature Sensor location—3.0L (3GR-FSE) engine

2. Using a clip remover, remove the 7 clips and duct seal.
3. Using a clip remover, remove the 3 clips.
4. Remove the 10 screws and under cover.
5. Drain the engine coolant.
6. Remove the right engine side cover.
7. Remove the 2 nuts and V-bank cover.
8. Disconnect the ventilation hose from the cylinder head.
9. Disconnect the MAF meter connector.
10. Disconnect the clamp from the air cleaner.
11. Disconnect the cold start injector connector.
12. Remove the 2 bolts, gasket, cold start injector from the surge tank.
13. Disconnect the 2 water by-pass hoses from the throttle body.
14. Remove the 3 clips and engine rear cover.
15. Disconnect the hose from the surge tank.
16. Disconnect the ventilation hose from the ventilation valve
17. Remove the fuel pipe clamp.
18. Pinch and pull the fuel tube's connector to disconnect it from the fuel main pipe.
19. Remove the intake plenum as follows:
   - Disconnect the throttle motor connector.
   - Disconnect the IACV connector.
   - Disconnect the wire harness clamp from the surge tank.
   - Remove the 2 bolts and intake manifold stay.
   - Disconnect the 2 wire harness clamps from the surge tank stays.
   - Remove the 4 bolts and 2 surge tank stays.
   - Remove the bolt and disconnect the water hose joint.
   - Using a 5 mm hexagon socket wrench, remove the 7 bolts, 2 nuts and gasket.
20. Remove intake plenum.

➡Cover the intake manifold port to prevent foreign matter from entering it.

21. Remove the intake manifold.
22. Remove the engine coolant temperature sensor.

To install:

23. Install a new gasket to the sensor.
24. Install and tighten the sensor to 15 ft. lbs. (20 Nm).
25. Install a new gasket and the intake manifold with the 4 bolts and 4 nuts. Tighten mounting nuts and bolts to 15 ft. lbs. (20 Nm).
26. Install a new gasket to the intake plenum.
27. Install the intake air surge tank with the 2 nuts and tighten to 13 ft. lbs. (18 Nm).
28. Using a 5 mm hexagon socket wrench, install the 7 bolts and tighten to 12 ft. lbs. (16 Nm).
29. Install the 2 surge tank stays with the 4 bolts and tighten to 15 ft. lbs. (20 Nm).
30. Connect the connector to the fuel main pipe. Push the two parts together

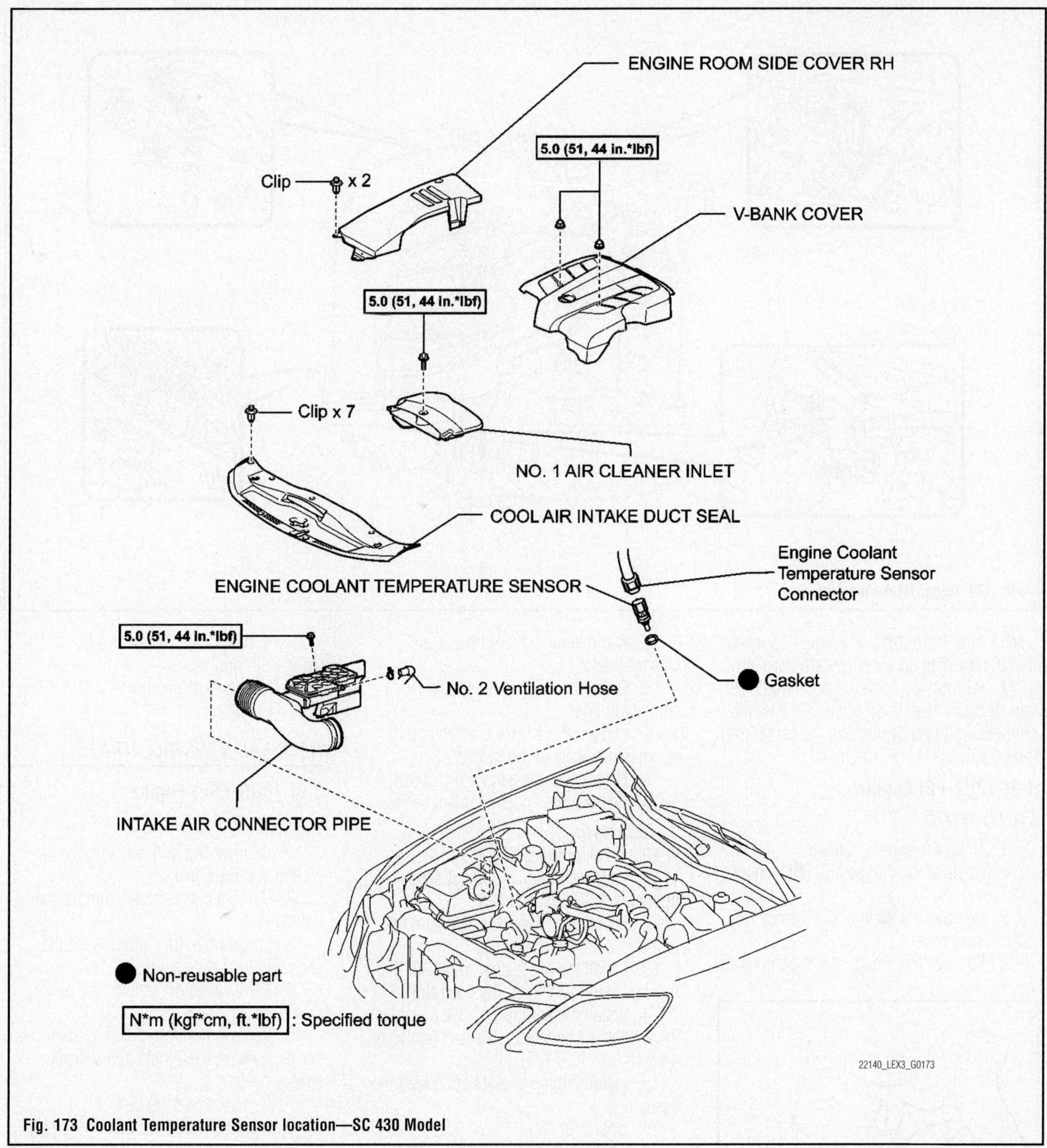

Fig. 173 Coolant Temperature Sensor location—SC 430 Model

firmly until a "click" sound is heard. Then attach the lock claws to the connector by pushing down on the connector cover.

➡**Check that the connector and fuel main pipe are securely connected by trying to pull them apart.**

31. Connect the ventilation hose to the ventilation valve.
32. Install the engine rear cover with the 3 clips.
33. Connect the hose to the surge tank.
34. Connect the water heater hoses.
35. Connect the 2 water by-pass hoses to the throttle body.
36. Install a new gasket and the cold start injector with the 2 bolts. Tighten the bolts to (7 ft. lbs. (10 Nm).
37. Install the air cleaner cap with air cleaner hose assembly with the 4 clamps and hose clamp.
38. Connect the ventilation hose to the cylinder head cover with the clamp.
39. Install engine coolant and bleed the system.
40. Connect the negative battery cable.
41. Check the operation of throttle body.
42. Check for coolant leaks.
43. Install the V-bank cover.
44. Install the side cover with the 2 clips and nut.
45. Install the duct seal with the 7 clips.

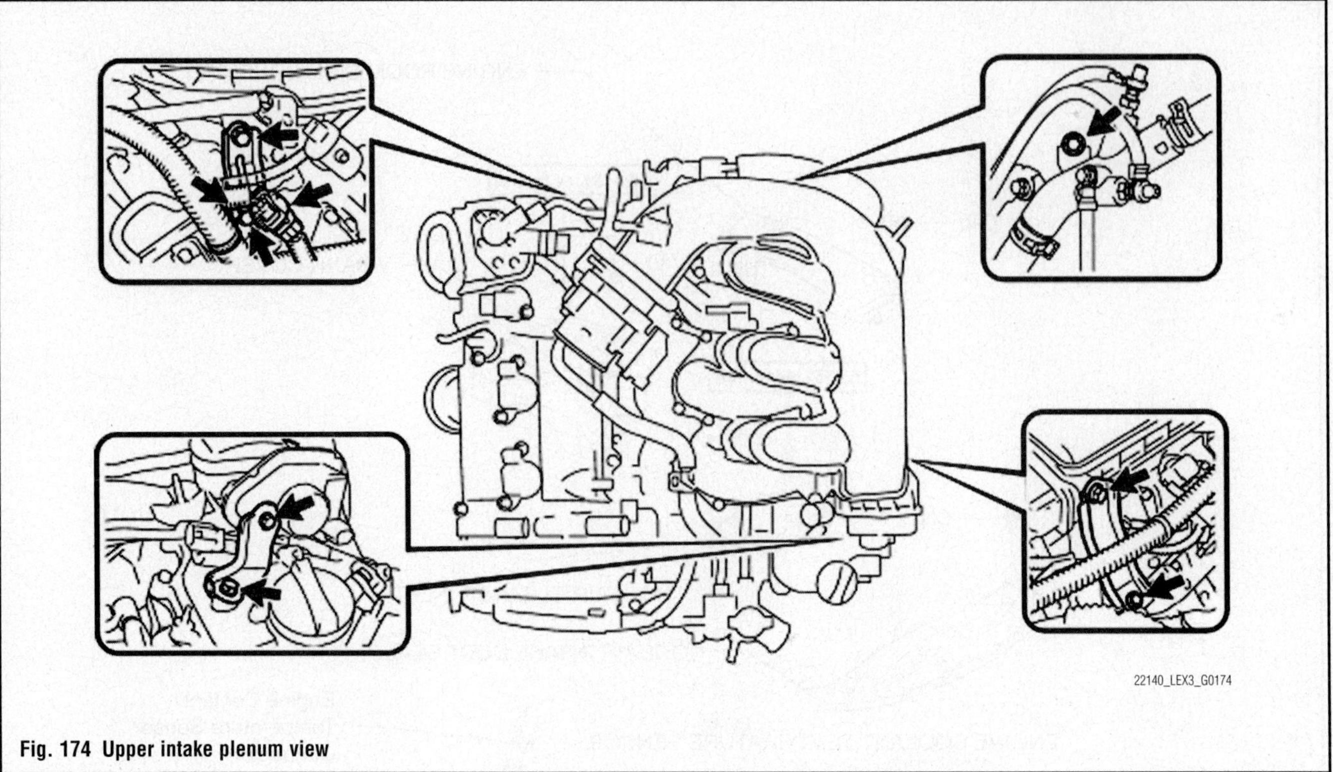

Fig. 174 Upper intake plenum view

46. After the engine is warmed up, check that the maintained parts operate normally.

47. Perform initialization procedure. Certain systems need to be initialized after disconnecting and reconnecting the cable from the negative (-) battery terminal.

4.3L (3UZ-FE) Engine

See Figure 175.

1. Drain the engine coolant.
2. Remove the 7 clips and intake duct seal.
3. Remove the 3 clips and the right engine side cover.
4. Remove the 2 nuts and V-bank cover.

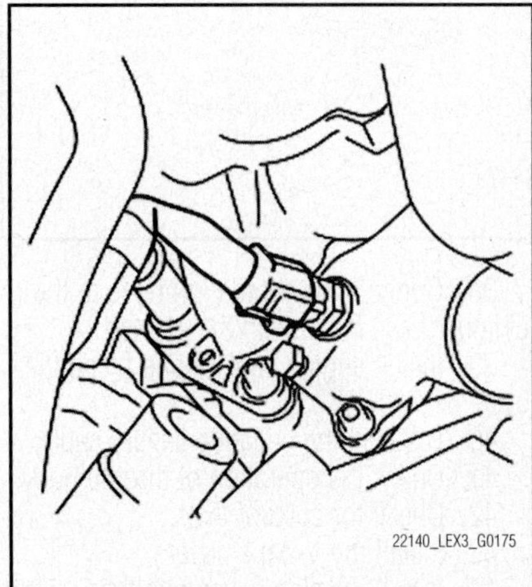

Fig. 175 Engine coolant temperature sensor

5. Remove the bolt and No. 1 air cleaner inlet.
6. Disconnect the air hose and No. 1 ventilation hose.
7. Loosen the 2 hose clamps and remove the intake air connector.
8. Disconnect the sensor connector.
9. Remove the engine coolant sensor.

To install:

10. Install a new gasket to the sensor.
11. Install the engine coolant sensor and tighten to 15 ft. lbs. (20 Nm).
12. Install the connector pipe with the bolt and 2 hose clamps.
13. Install the air cleaner inlet with the bolt. Tighten to 44 inch. lbs. (5.0 Nm).
14. Install the engine V- bank cover with the 2 clips. Install the 2 nuts and tighten to 44 inch. lbs. (5.0 Nm).
15. Install engine coolant and bleed the system.
16. Check for leaks.
17. After the engine is warmed up, check that the maintained parts operate normally.
18. Perform initialization procedure. Certain systems need to be initialized after disconnecting and reconnecting the cable from the negative (-) battery terminal.

CRANKSHAFT POSITION (CKP) SENSOR

LOCATION

See Figures 176 and 177.

Crankshaft location view—3.0L (3GR-FSE) engine.

Crankshaft location view—4.3L (3UZ-FE) engine.

REMOVAL & INSTALLATION

3.0L (3GR-FSE) Engine

See Figure 178.

1. Using a clip remover, remove the 7 clips and duct seal.
2. Using a clip remover, remove the 3 clips.
3. Loosen the nuts and remove the clips and both side covers.
4. Remove the 10 screws and under cover.
5. Remove the bolt and air cleaner inlet.
6. Remove the 2 nuts and V-bank engine cover.
7. Remove the drive belt.
8. Remove the compressor assembly to access crankshaft sensor. (support A/C assembly).
9. Disconnect the crankshaft position sensor connector.
10. Remove the bolt and sensor.

To install:

11. Apply a coat of engine oil to an O-ring of the sensor.
12. Install the sensor with the bolt and tighten to 7 ft. lbs. (10 Nm).
13. Connect the sensor connector.
14. Install the compressor with the

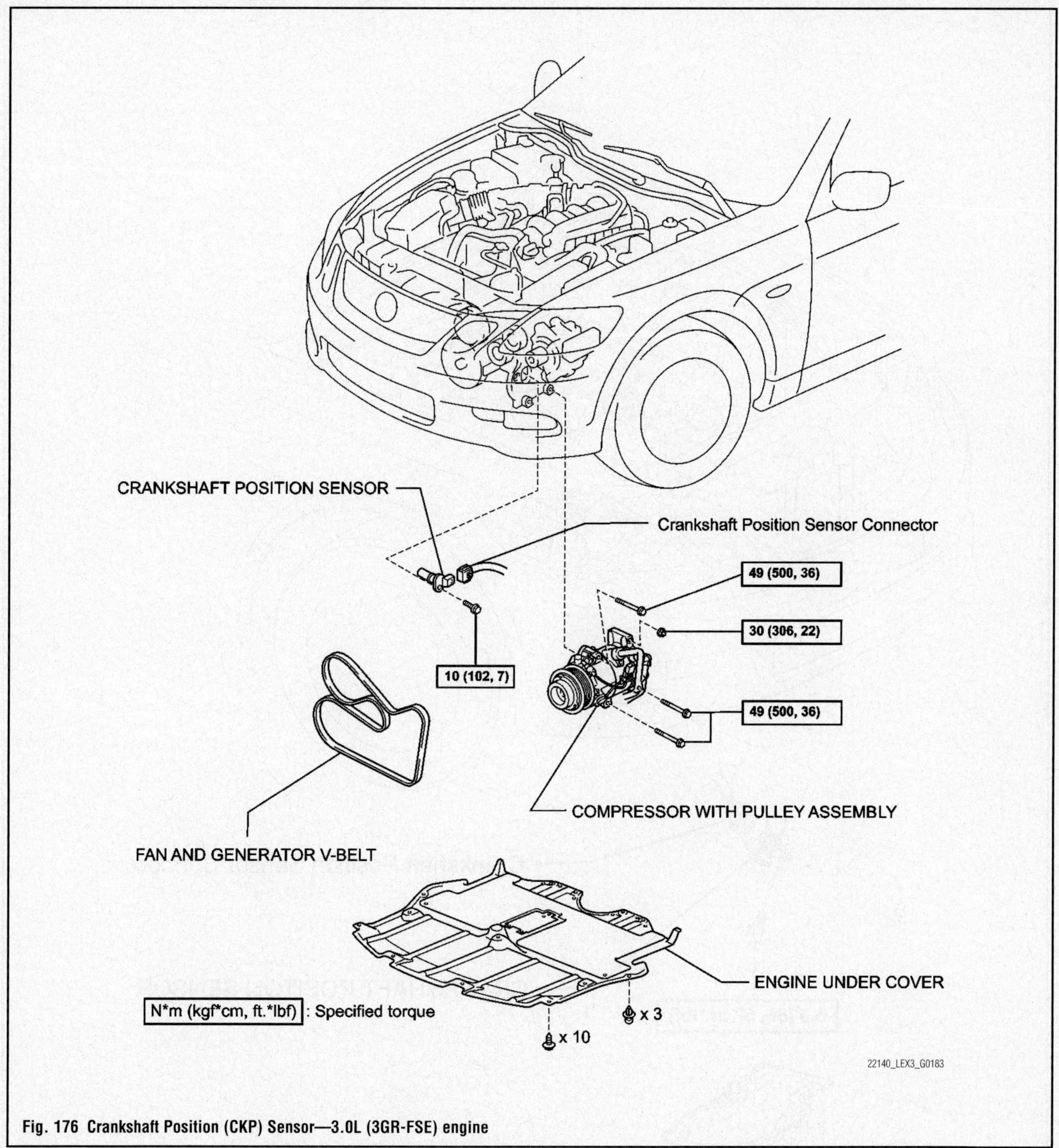

Fig. 176 Crankshaft Position (CKP) Sensor—3.0L (3GR-FSE) engine

3 bolts, nut and bracket. Tighten bolts to 36 ft. lbs. (49 Nm).

15. Install the undercover with the 10 bolts and 3 clips.
16. Install drive belt.
17. Install the V-bank engine cover with the 2 nuts. Tighten the nuts to 44 inch. lbs. (5.0 Nm).
18. Install the air cleaner inlet with the bolt. Tighten the nuts to 44 inch. lbs. (5.0 Nm).
19. Install left and right engine side covers.
20. Install the cool air intake duct seal with the 7 clips.
21. After the engine is warmed up, check that the maintained parts operate normally.
22. Perform initialization procedure. Certain systems need to be initialized after disconnecting and reconnecting the cable from the negative (-) battery terminal.

4.3L (3UZ-FE) Engine

See Figure 179.

1. Remove the engine under cover.
2. Remove the bolt and crankshaft sensor.

To install:

3. Install the crankshaft position sensor.
4. Tighten the mounting bolt to 57 inch. lbs. (6.5 Nm).

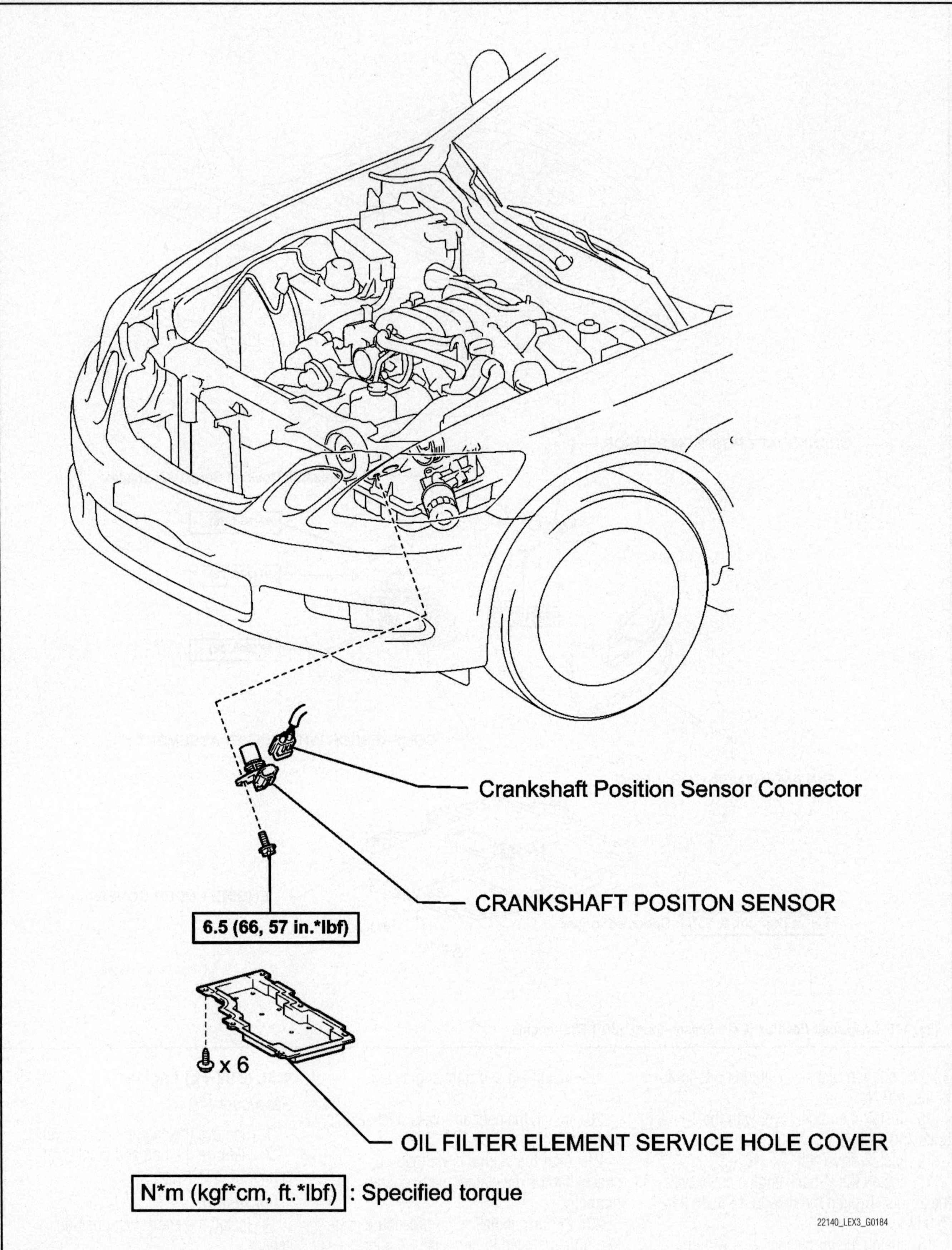

Fig. 177 Crankshaft Position (CKP) Sensor—4.3L (3UZ-FE) engine

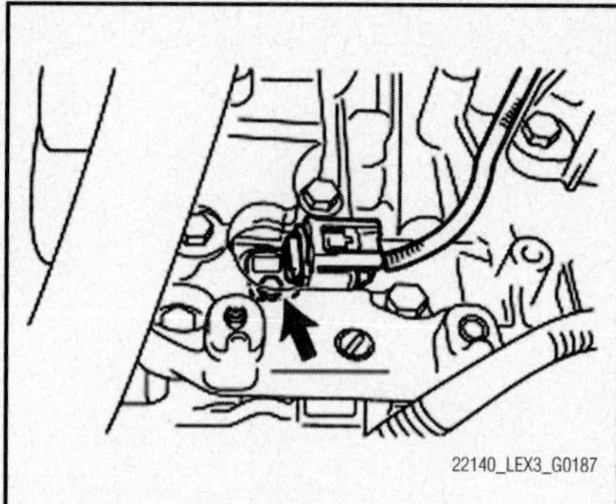

Fig. 178 Crankshaft sensor and mounting bolt—3.0L (3GR-FSE) engine

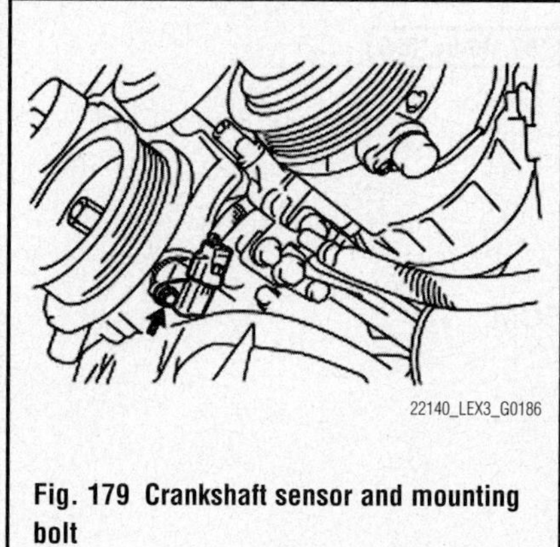

Fig. 179 Crankshaft sensor and mounting bolt

5. Install the undercover with the 18 bolts and 5 clips.
6. Perform initialization procedure. Certain systems need to be initialized after disconnecting and reconnecting the cable from the negative (-) battery terminal.

ELECTRONIC CONTROL MODULE (ECM)

LOCATION

See Figures 180 and 181.

REMOVAL & INSTALLATION

3.0L (3GR-FSE) Engine

1. Disconnect the negative battery cable.

⁂ CAUTION

Wait at least 90 seconds after disconnecting the cable from the negative (-) battery terminal to prevent airbag and seat belt pre-tensioner activation.

2. Using a clip remover, remove the 7 clips and cool air intake duct seal.
3. Remove the 3 clips and the left engine side cover.
4. Remove the 3 bolts and ECM cover.

➡Be sure to prevent water intrusion to the ECM (connectors and screw parts).

5. Disconnect the ECM connectors.
6. Using a screwdriver, disconnect the connector holder.
7. For 3.0L (3GR-FSE) engine disconnect the 6 ECM connectors.
8. Remove the 2 nuts and ECM from the ECM box.

To install:

9. Install the ECM to the ECM box. Tighten the retaining nuts to nuts 49 inch. lbs. (5.5 Nm).

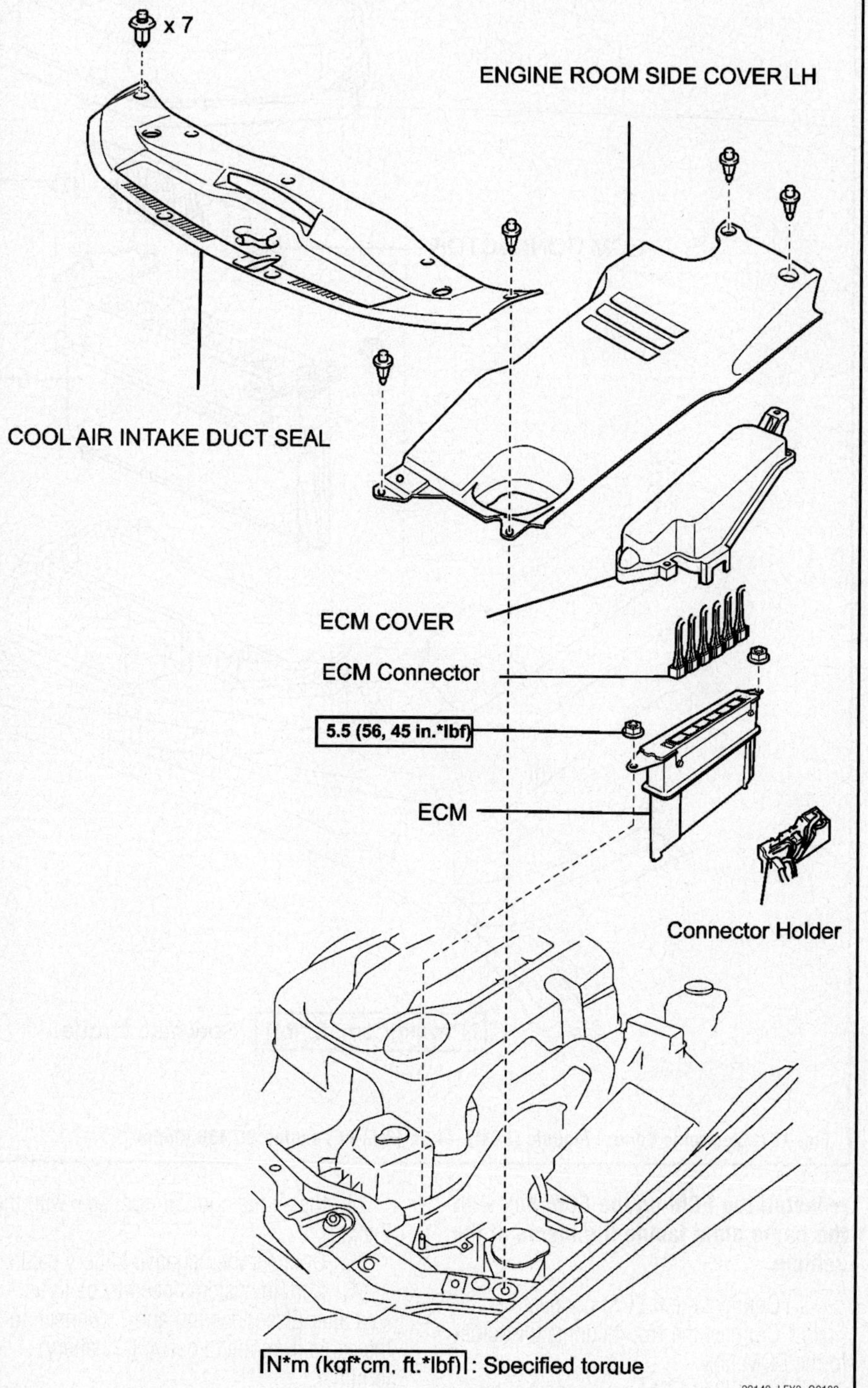

x 7
ENGINE ROOM SIDE COVER LH
COOL AIR INTAKE DUCT SEAL
ECM COVER
ECM Connector
5.5 (56, 45 in.*lbf)
ECM
Connector Holder
N*m (kgf*cm, ft.*lbf) : Specified torque
22140_LEX3_G0188

Fig. 180 Electronic Control Module (ECM)—3.0L (3GR-FSE) engine GS 300 Model

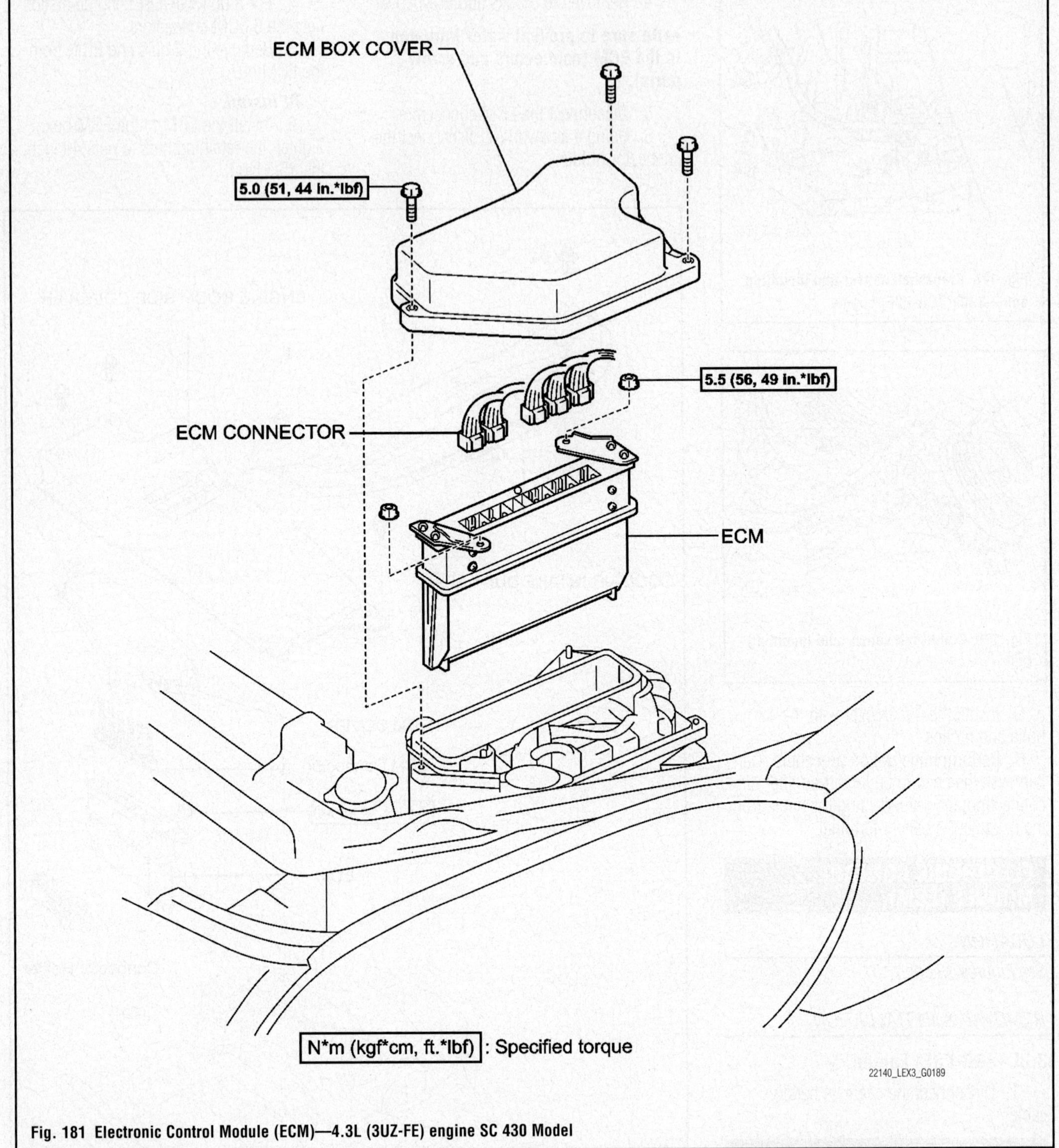

Fig. 181 Electronic Control Module (ECM)—4.3L (3UZ-FE) engine SC 430 Model

➡Install the ECM on the ECM box with the name plate facing the inside of the vehicle.

10. Connect the 6 ECM connectors.
11. Connect the No. 4 connector holder to the ECM box.
12. Install the ECM cover and tighten the 3 bolts to 49 inch. lbs. (5.5 Nm).
13. Install the left engine side cover.
14. Install the cool air duct seal with the 7 clips.
15. Connect the negative battery cable.
16. Certain systems need to be initialized after disconnecting and reconnecting the cable from the negative (-) battery terminal.
17. After the engine is warmed up, check that the maintained parts operate normally.

4.3L (3UZ-FE) Engine

1. Disconnect the negative battery cable.

CAUTION

Wait at least 90 seconds after disconnecting the cable from the negative (-) battery terminal to prevent airbag and seat belt pre-tensioner activation.

2. Using a clip remover, remove the 7 clips and cool air intake duct seal.

3. Remove the 3 clips and the left engine side cover.

4. Remove the 3 bolts and ECM cover.

➡Be sure to prevent water intrusion to the ECM (connectors and screw parts).

5. Disconnect the 5 ECM connectors.

6. Remove the 2 nuts and ECM from the ECM box.

To install:

7. Insert the ECM into the ECM box.

8. Install the ECM to the ECM box with the 2 nuts.

9. Tighten the retaining nuts to nuts 49 inch. lbs. (5.5 Nm).

10. Connect the 5 ECM connectors.

11. Install the ECM box cover with the 3 bolts.

12. Connect the negative battery cable.

13. Certain systems need to be initialized after disconnecting and reconnecting the cable from the negative (-) battery terminal.

14. After the engine is warmed up, check that the maintained parts operate normally.

ENGINE COOLANT TEMPERATURE (ECT) SENSOR

LOCATION

See Figures 182 and 183.

Coolant Temperature Sensor location—3.0L (3GR-FSE) engine

Coolant Temperature Sensor location—4.3L (3UZ-FE) engine

REMOVAL & INSTALLATION

The Engine Coolant Temperature (ECT) sensor is located on the top front area of the engine.

1. Before servicing the vehicle, refer to the precautions in the beginning of this section.

2. Disconnect the negative battery cable. Wait at least 90 seconds before performing any other work.

3. To access the ECT sensor, it may be necessary to remove the intake manifold, throttle body assembly, or only the intake air connector pipe.

4. Disconnect the ECT sensor connector.

5. Using a deep socket, remove the sensor from the top of the engine.

6. Remove the gasket from the sensor.

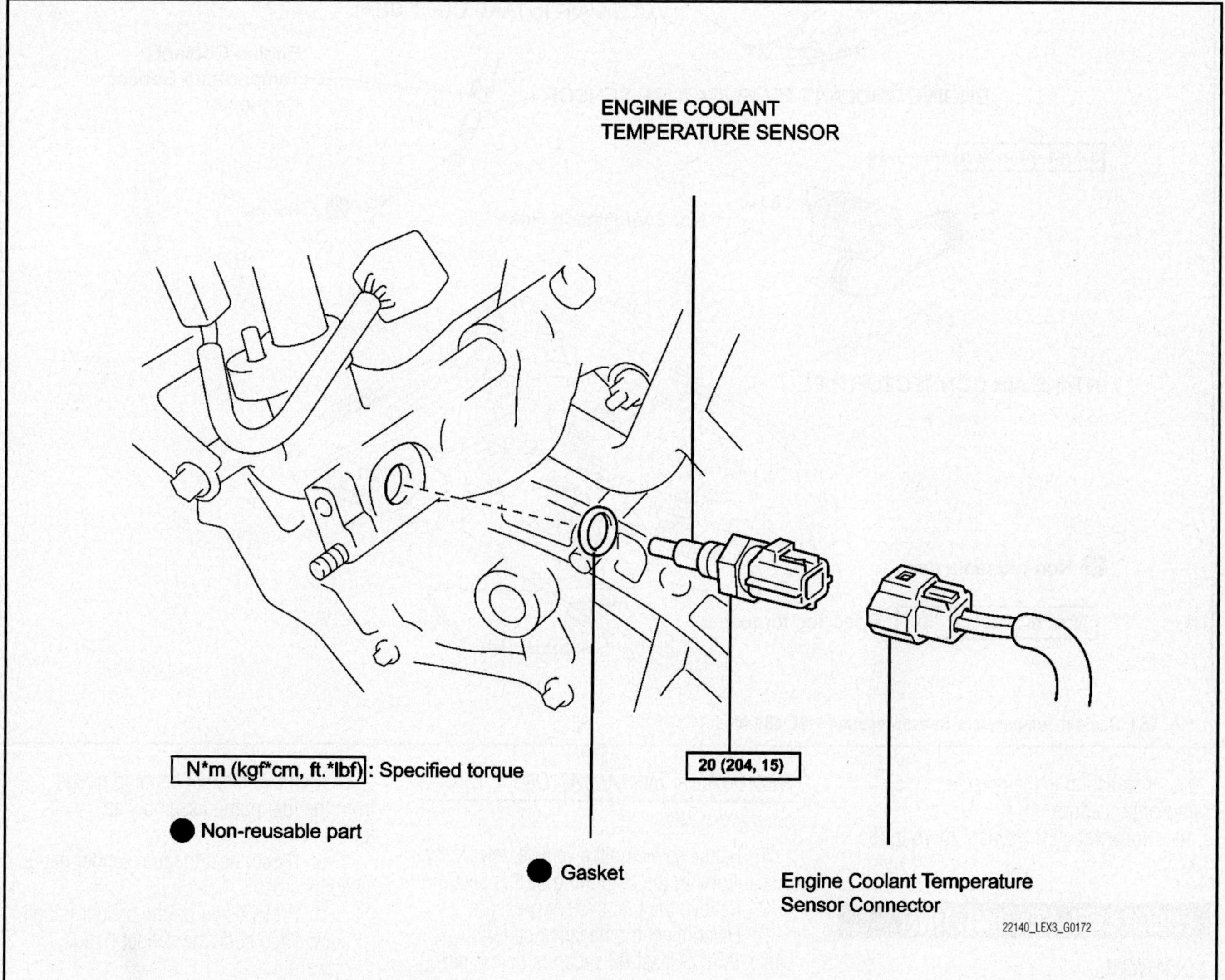

Fig. 182 Coolant Temperature Sensor location—3.0L (3GR-FSE) engine

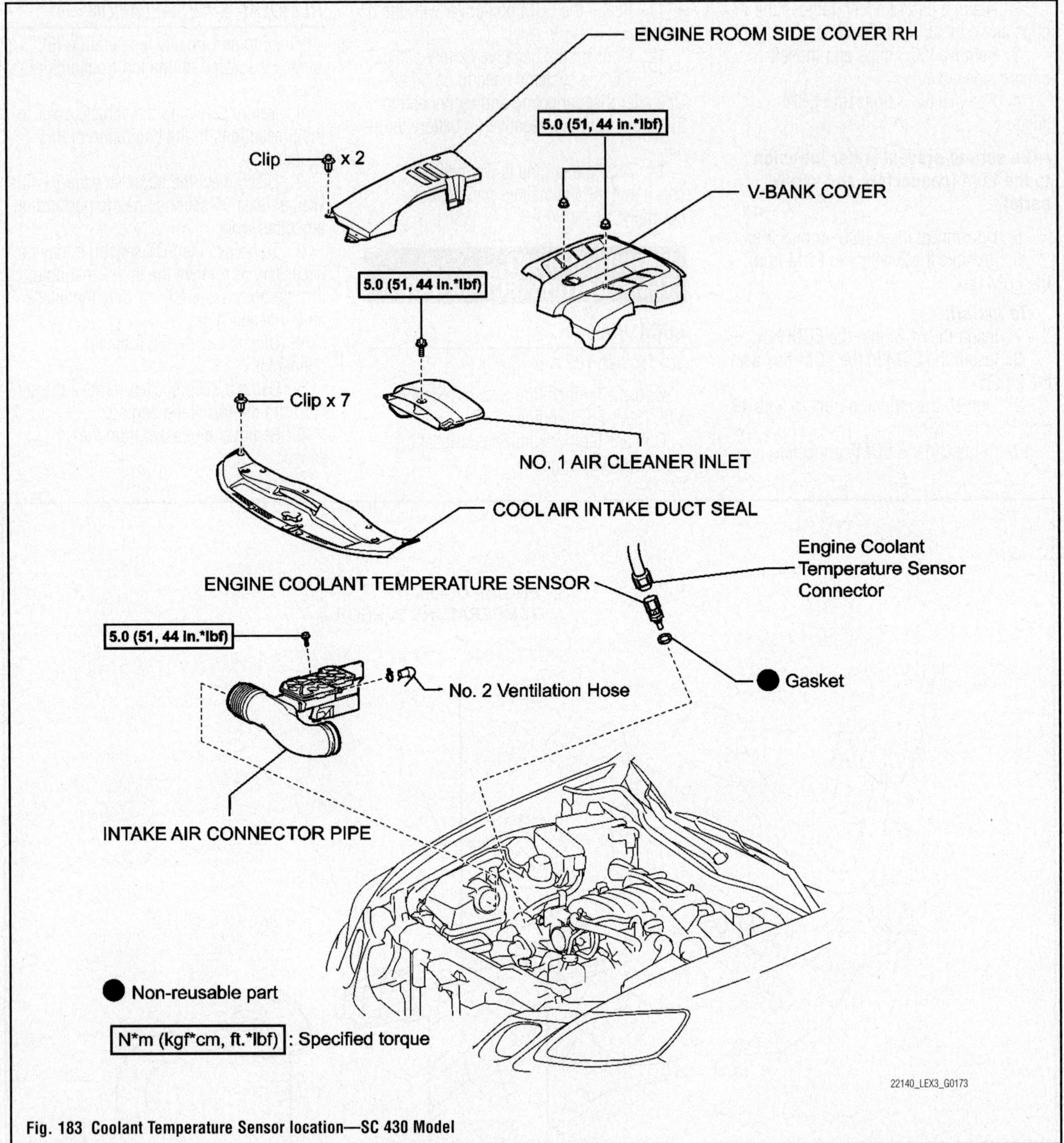

Fig. 183 Coolant Temperature Sensor location—SC 430 Model

7. Installation is the reverse of the removal procedure.
8. Tighten the ECT sensor to 15 ft. lbs. (20 Nm).

FUEL LEVEL SENDING UNIT

LOCATION

The fuel level sending unit is located inside the fuel tank, and is mounted to the fuel pump.

REMOVAL & INSTALLATION

See Figure 185.

1. Before servicing the vehicle, refer to the precautions in the beginning of this section.
2. Relieve the fuel system pressure.
3. Disconnect the negative battery cable. Wait at least 90 seconds before performing any other work
4. Remove the fuel pump and sending gauge assembly.
5. Remove the fuel sending gauge from the fuel pump assembly as follows:
   a. Disconnect the fuel sender gauge connector.
   b. Press down on the sender gauge claw. Then slide the sender gauge upward.

To install:

6. Install the fuel sending gauge to the fuel pump assembly as follows:

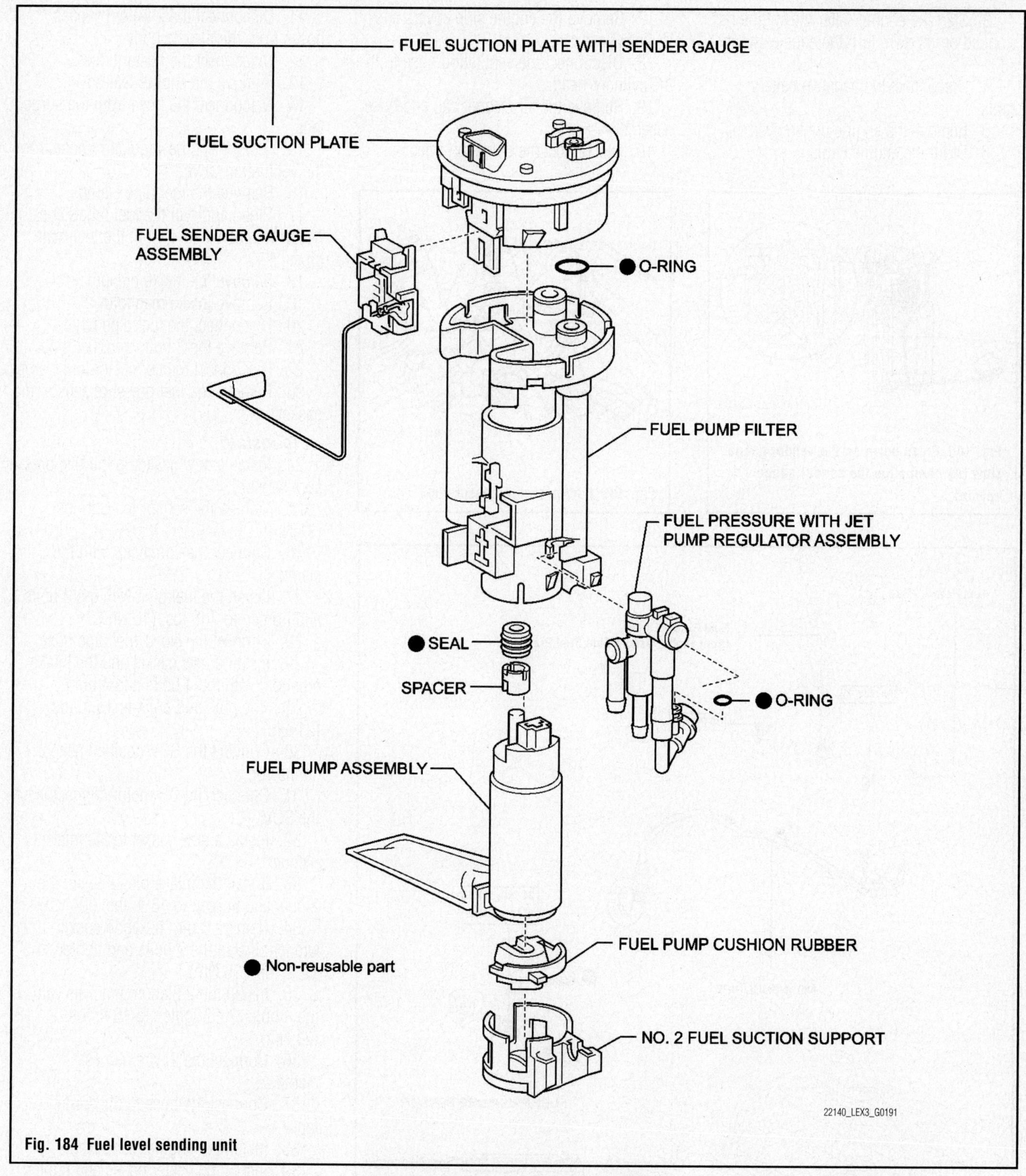

Fig. 184 Fuel level sending unit

a. Set the fuel sender gauge to the No. 1 fuel sub-tank. Then slide the sender gauge downward to install it.

b. Connect the fuel sender gauge connector.

7. Install the fuel pump and sending gauge assembly.

8. Connect the negative battery cable.

FUEL PRESSURE SENSOR

LOCATION

3.0L (3GR-FSE) Engine

See Figure 186.

Fuel pressure sensor location—3.0L (3GR-FSE) engine.

REMOVAL & INSTALLATION

3.0L (3GR-FSE) Engine

See Figure 187.

1. Using a clip remover, remove the 7 clips and cool air duct seal.
2. Disconnect the fuel pump connector.

3. Start the engine. After the engine has stopped on its own, turn the engine switch off.
4. Disconnect the negative battery cable.
5. Remove the engine undercover.
6. Drain the engine coolant.

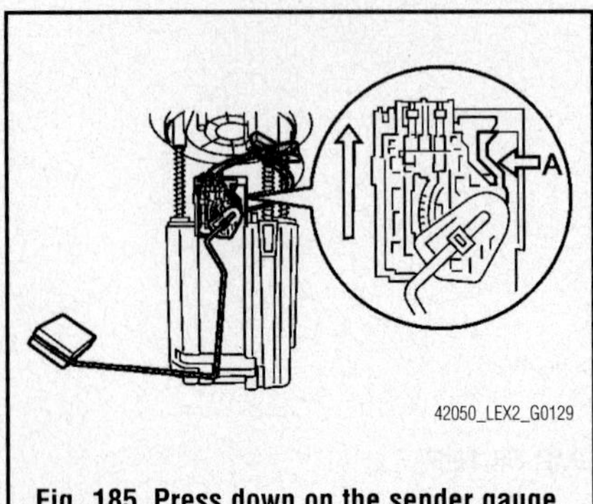

Fig. 185 Press down on the sender gauge claw (A), then slide the sender gauge upward

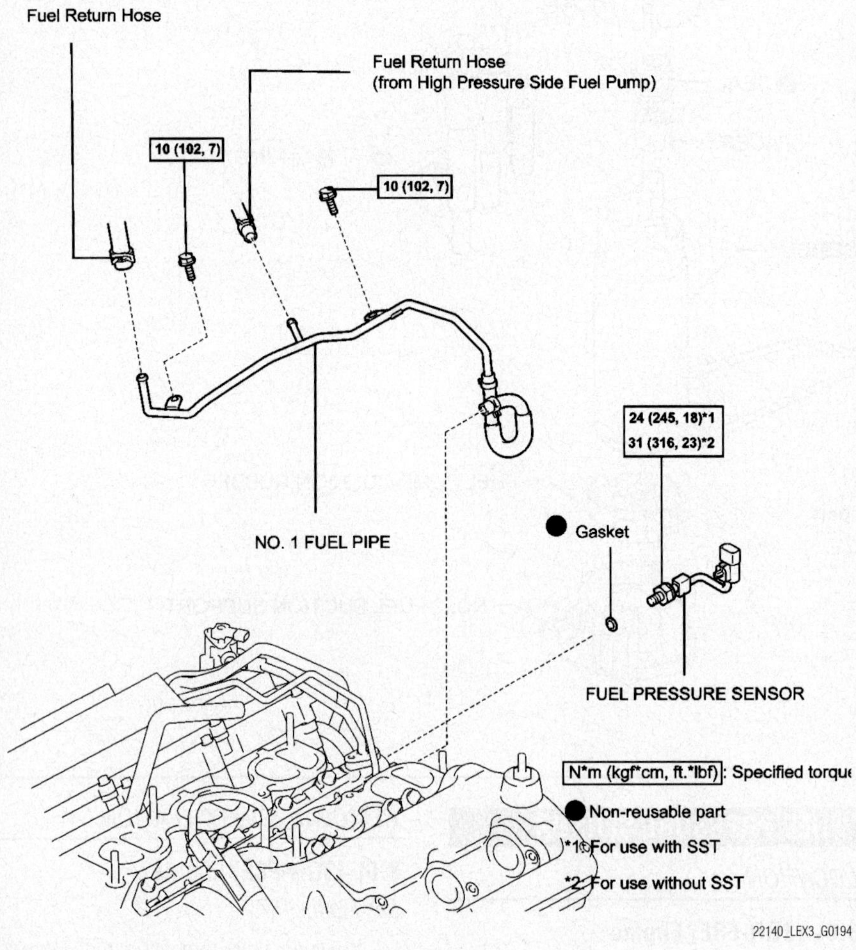

Fig. 186 Fuel pressure sensor —3.0L (3GR-FSE) engine

7. Remove the engine side covers and V- bank cover.
8. Disconnect the ventilation hose from the cylinder head.
9. Remove the air cleaner cap and hose assembly.
10. Disconnect the cold start injector.

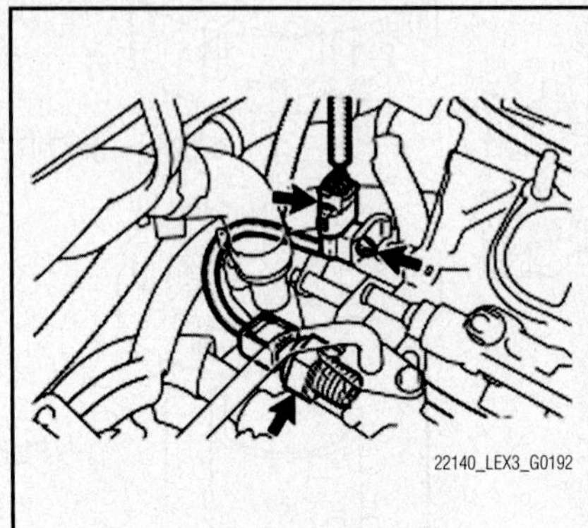

Fig. 187 Fuel pressure sensor view

11. Disconnect the 2 water by-pass hoses from the throttle body.
12. Disconnect the heater hoses.
13. Remove the engine rear cover.
14. Disconnect the hose from the surge tank.
15. Disconnect the ventilation hose from the ventilation valve.
16. Remove the fuel pipe clamp.
17. Pinch and pull the fuel tube's connector to disconnect it from the fuel main pipe.
18. Remove the intake plenum.
19. Remove intake manifold.
20. Disconnect the fuel pipe hose.
21. Remove the 2 bolts and fuel pipe.
22. Disconnect the connector and clamp.
23. Remove the fuel pressure sensor and gasket.

To install:

24. Install a new gasket to the fuel pressure sensor.
25. Tighten the sensor to 23 ft. lbs. (31 Nm).
26. Connect the connector with the clamp.
27. Install the fuel pipe with the 2 bolts and tighten to 7 ft. lbs. (10 Nm).
28. Connect the No. 1 fuel pipe hose.
29. Install a new gasket and the intake manifold with the 4 bolts and 4 nuts. Tighten the nuts and bolts to 15 ft. lbs. (21 Nm).
30. Connect the SCV position sensor connector.
31. Connect the DC motor connector for the SCV.
32. Install a new gasket to the intake plenum.
33. Install the intake plenum with the 2 nuts and tighten to 13 ft. lbs. (18 Nm).
34. Using a 5 mm hexagon socket wrench, install the 7 bolts and tighten to 12 ft. lbs. (16 Nm).
35. Install the 2 plenum brackets with the 4 bolts. And tighten to 15 ft. lbs. (21 Nm).
36. Connect the 2 wire harness clamps.
37. Connect the hose to the surge tank.
38. Install the engine rear cover.
39. Install the heater hoses and tighten.
40. Connect the water bypass hose.
41. Install the cold start injector with a new gasket. tighten the 2 bolts to 7 ft. lbs. (10 Nm).
42. Install air cleaner cap and hose.
43. Connect the ventilation hose to the cylinder head cover with the clamp.
44. Connect the negative battery cable.

45. Add engine coolant and bleed the system.
46. Check for coolant leaks.
47. Start the engine, and check that there are no fuel leaks after performing maintenance anywhere on the system.
48. Start the engine, and check that there are no oil leaks after performing maintenance.
49. Install engine side covers and V-bank cover.
50. Install cool air intake duct seal.
51. Install engine under cover.
52. After the engine is warmed up, check that the maintained parts operate normally.
53. Certain systems need to be initialized after disconnecting and reconnecting the cable from the negative (-) battery terminal.

HEATED OXYGEN (HO2S) SENSOR

LOCATION

See Figure 188.

Refer to the accompanying illustration for sensor location for the 3.0L engine.

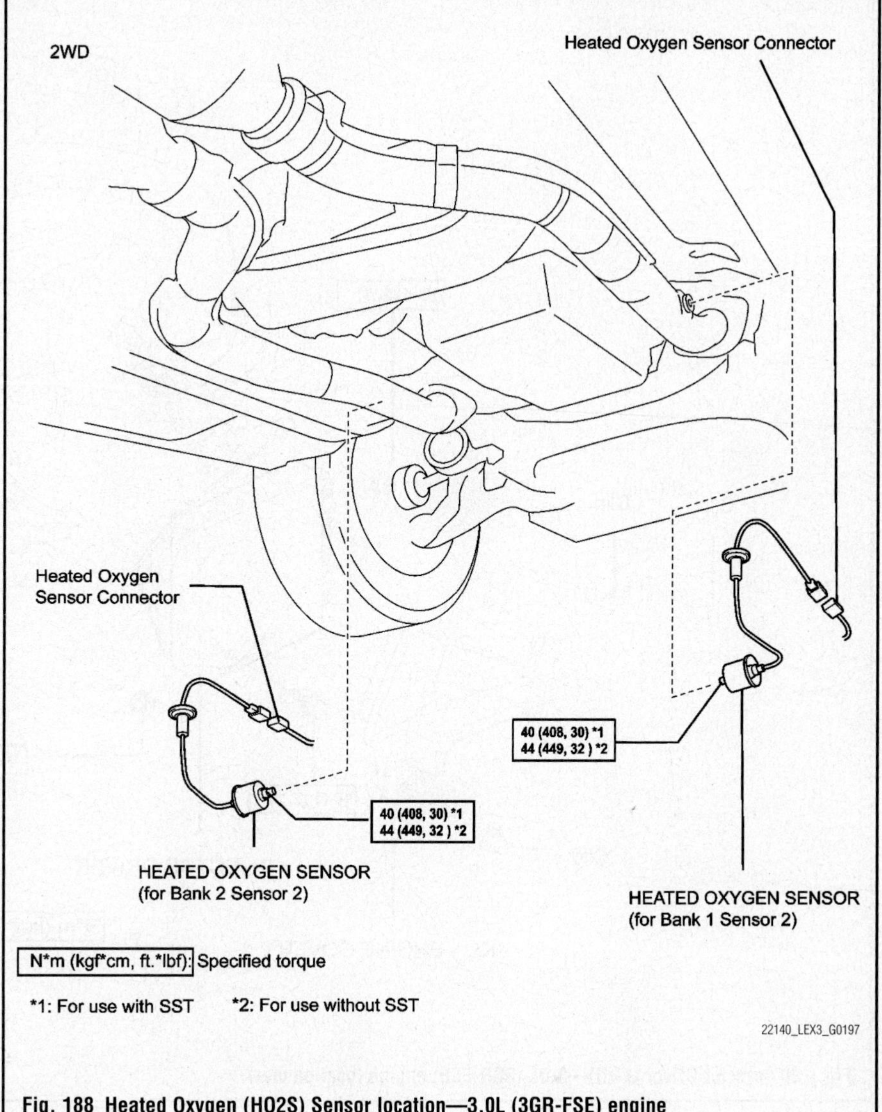

Fig. 188 Heated Oxygen (HO2S) Sensor location—3.0L (3GR-FSE) engine

REMOVAL & INSTALLATION

3.0L (3GR-FSE) Engine

Rear Sensors

1. Disconnect the negative battery cable.
2. Wait at least 90 seconds after disconnecting the cable from the negative (-) battery terminal to prevent airbag and seat belt pre-tensioner activation.
3. Remove the engine under cover.
4. For (Bank 1) remove the 2 bolts and the heat insulator.
5. For (Bank 2) disconnect the sensor wire from the 2 wire clamps.
6. Disconnect the sensor connector.
7. Using O2 socket remove the sensor.

To install:

8. Using O2 socket install the sensor and tighten to 32 ft. lbs. 44 Nm).
9. Connect the sensor connector.
10. For (Bank 1) install the heat shield and tighten mounting bolts to 13 ft. lbs. (18 Nm).
11. For (Bank 2) install the sensor wire to the 2 wire clamps.
12. Install the engine under cover.
13. Connect the negative battery cable.
14. Certain systems need to be initialized after disconnecting and reconnecting the cable from the negative (-) battery terminal.

4.3L (3UZ-FE) Engine

1. Disconnect the negative battery cable.
2. Wait at least 90 seconds after disconnecting the cable from the negative (-) battery terminal to prevent airbag and seat belt pre-tensioner activation.
3. Remove the console box.
4. Remove both engine under covers.
5. Disconnect the sensor connector.
6. Disconnect the grommet and pull the sensor connector out of the cabin through the floor panel.
7. With a O2 socket remove the heated O2 sensor.

To install:

8. Using O2 socket install the sensor and tighten to 32 ft. lbs. 44 Nm).
9. Pass the sensor connector through the floor panel and into the cabin, and install the grommet.
10. Connect the sensor connector.
11. Install the undercover with the 18 bolts and 5 clips.
12. Install the No. 2 under cover with the 5 bolts.
13. Connect the negative battery cable.
14. Certain systems need to be initialized after disconnecting and reconnecting the cable from the negative (-) battery terminal.

INJECTOR DRIVER (EDU)

LOCATION

3.0L (3GR-FSE) Engine

See Figure 189.

Injector Driver (EDU)—3.0L (3GR-FSE) engine.

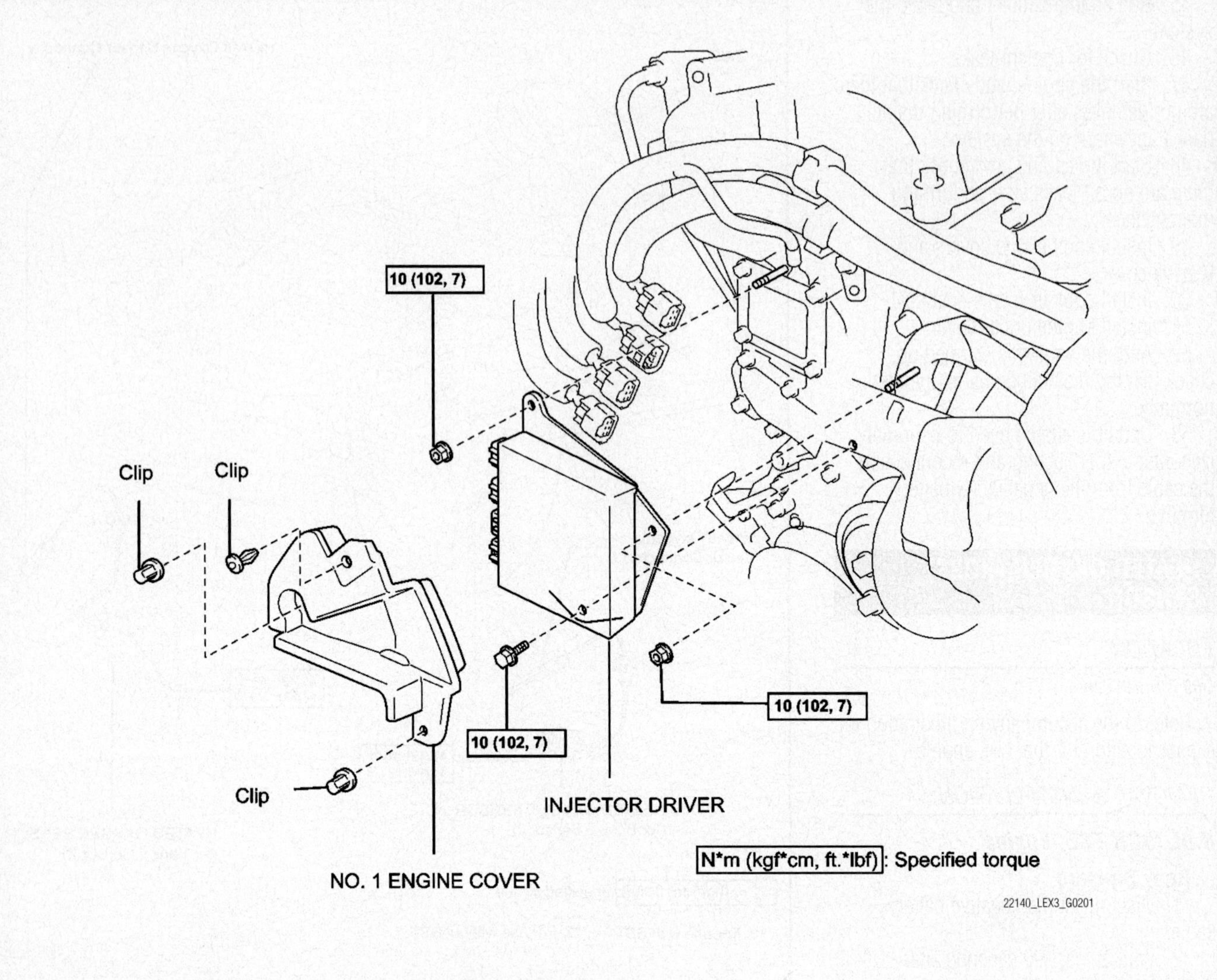

Fig. 189 Injector Driver (EDU)—3.0L (3GR-FSE) engine location view

REMOVAL & INSTALLATION

3.0L (3GR-FSE) Engine

See Figure 190.

1. Disconnect the negative battery cable.
2. Wait at least 90 seconds after disconnecting the cable from the negative (-) battery terminal to prevent airbag and seat belt pre-tensioner activation.
3. Using a clip remover, remove the 7 clips and duct seal.
4. Remove air cleaner inlet.
5. Remove the engine cover.
6. Disconnect the 4 connectors.
7. Remove the bolt, 2 nuts and injector driver.

To install:

➡The injector driver is grounded at the bolt and nut. To ensure that it is grounded, clean all oil and foreign matter from the installation areas of the injector driver and engine before installing the injector driver.

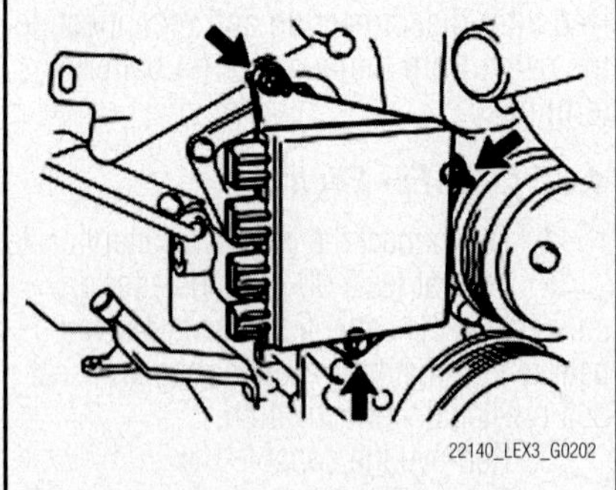

Fig. 190 Injector driver and mounting hardware

8. Install the injector driver with the bolt and 2 nuts. Tighten to 7 ft. lbs. (10 Nm).
9. Install the cover with the 3 clips.
10. Connect the 4 connectors.
11. Install the air cleaner inlet with the bolt.
12. Install the duct seal with the 7 clips.
13. Connect the negative battery cable.
14. Certain systems need to be initialized after disconnecting and reconnecting the cable from the negative (-) battery terminal

INTAKE AIR TEMPERATURE (IAT) SENSOR

LOCATION

See Figure 191.

The Intake Air Temperature (IAT) sensor is mounted on the air filter housing. The sensor is integral to the Mass Air Flow (MAF) Sensor.

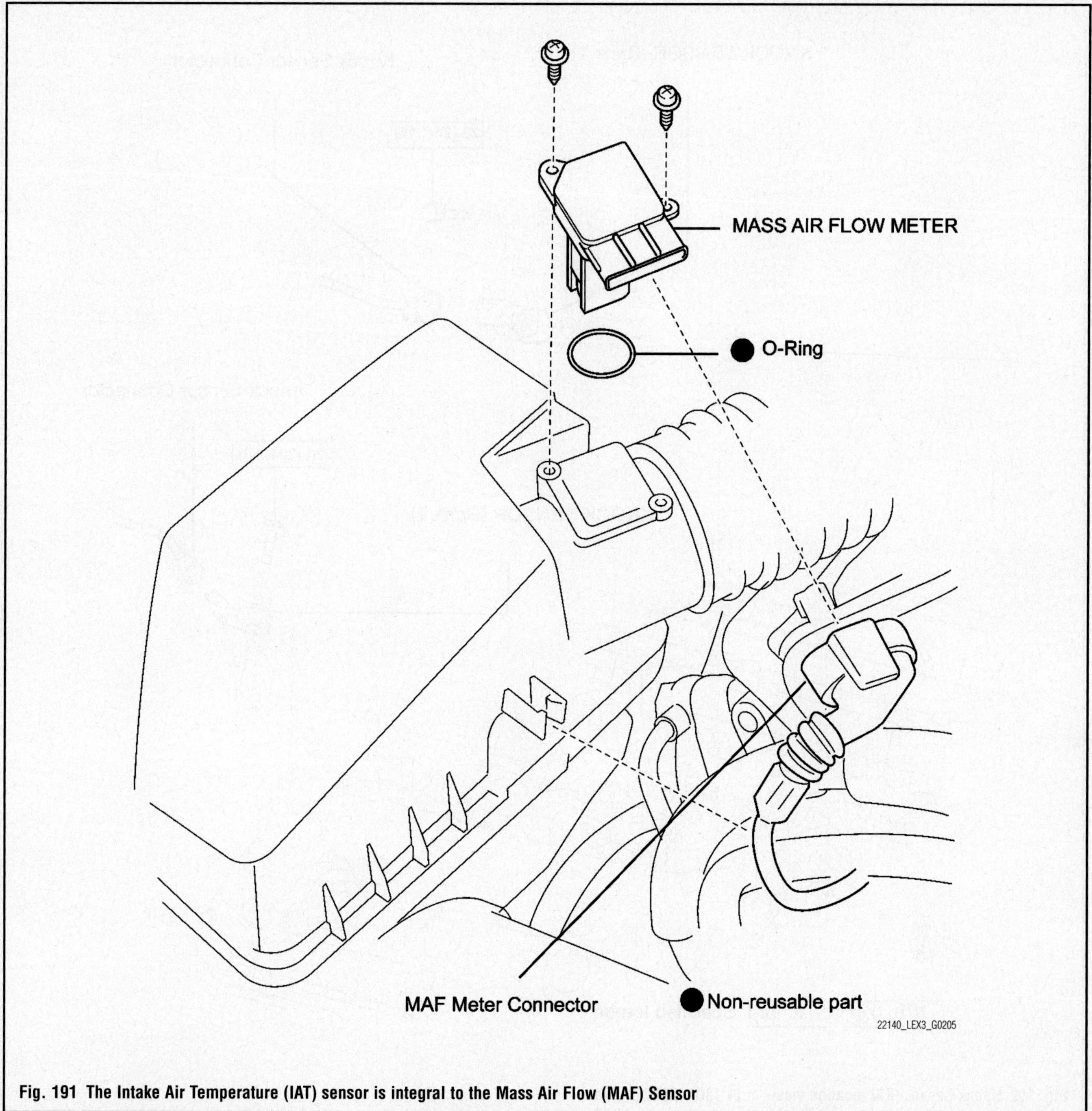

Fig. 191 The Intake Air Temperature (IAT) sensor is integral to the Mass Air Flow (MAF) Sensor

REMOVAL & INSTALLATION

1. Remove the intake cool duct seal.
2. Remove the engine covers.
3. Disconnect the MAF meter connector.
4. Remove the 2 screws and MAF meter.
5. Remove the O-ring from the MAF meter.

To install:

6. Install a new O-ring to the MAF meter.
7. Install the MAF meter with the 2 screws.
8. Connect the MAF meter connector.
9. Install engine covers
10. Install the intake duct seal with the 7 clips.

KNOCK SENSOR (KS)

LOCATION

3.0L (3GR-FSE) Engine

See Figure 192.

The knock sensors are located under the intake manifold. The sensors are mounted on the left and right cylinder heads.

4.3L (3UZ-FE) Engine

See Figure 193.

The knock sensors are located under the intake manifold. The sensors are mounted on the engine block.

REMOVAL & INSTALLATION

3.0L (3GR-FSE) Engine

See Figures 194 and 195.

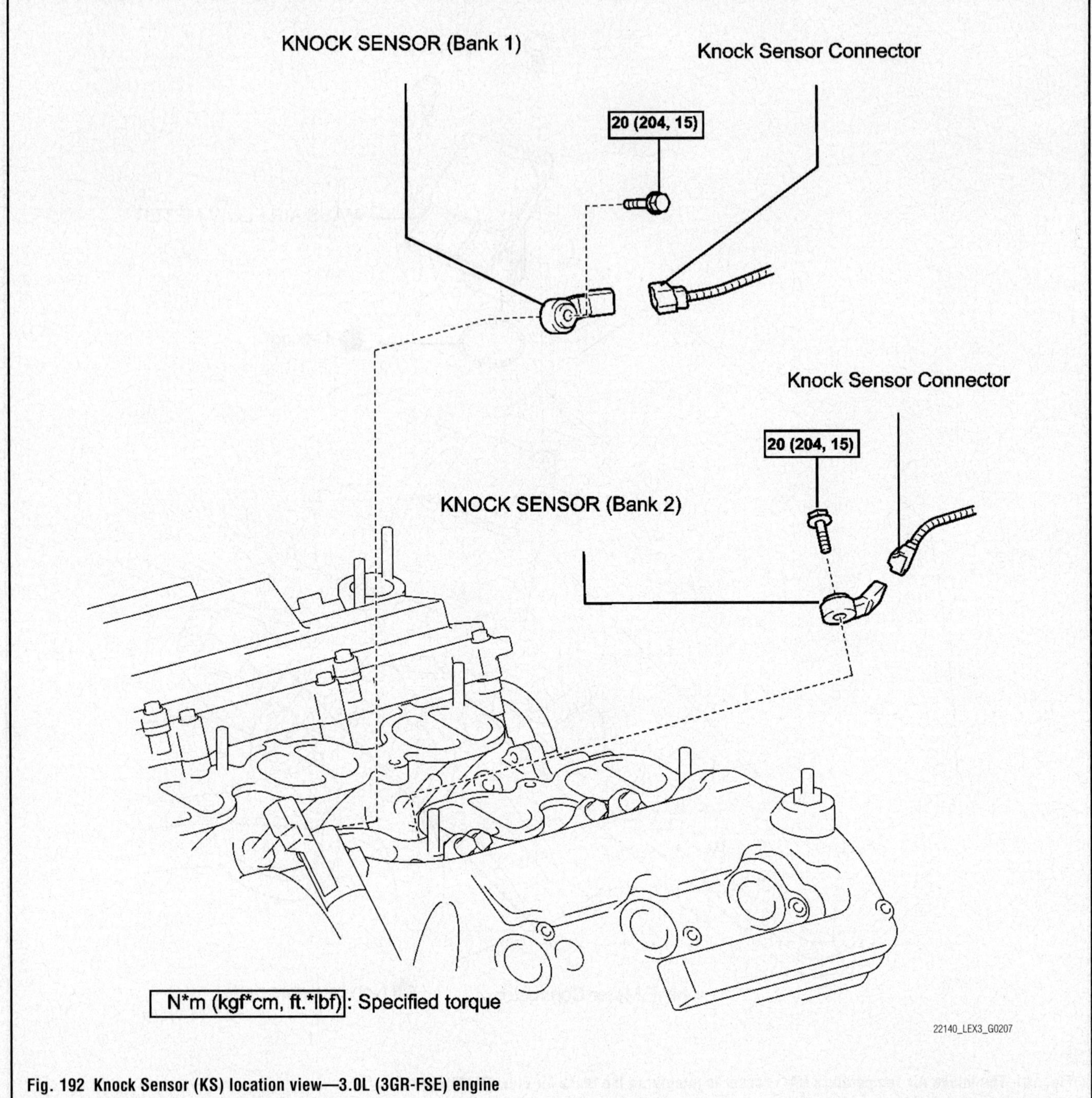

Fig. 192 Knock Sensor (KS) location view—3.0L (3GR-FSE) engine

1. Disconnect the negative battery cable.
2. Wait at least 90 seconds after disconnecting the cable from the negative (-) battery terminal to prevent airbag and seat belt pre-tensioner activation.
3. Using a clip remover, remove the 7 clips and duct seal.
4. Using a clip remover, remove the 3 clips.
5. Remove the 10 screws and under cover.
6. Drain the engine coolant.
7. Remove the engine covers.
8. Disconnect the ventilation hose from the cylinder head.
9. Remove the air cleaner housing cover and hose.
10. Disconnect cold start injector.
11. Remove the water bypass hose.
12. Disconnect the heater hoses.
13. Remove the engine rear cover.
14. Disconnect the hose from the surge tank.
15. Disconnect the fuel line main tube.
16. Remove the intake plenum.
17. Remove the intake manifold.
18. Remove the fuel pressure pulsation damper.
19. Disconnect the fuel pipe hose.
20. Remove the 2 bolts and fuel pipe.
21. Disconnect the fuel high pressure side fuel pump connector.
22. Remove the 2 bolts on the delivery pipe side.
23. Remove both fuel delivery pipes.
24. Disconnect the 2 knock sensor connectors.
25. Remove the 2 bolts and 2 knock sensors.

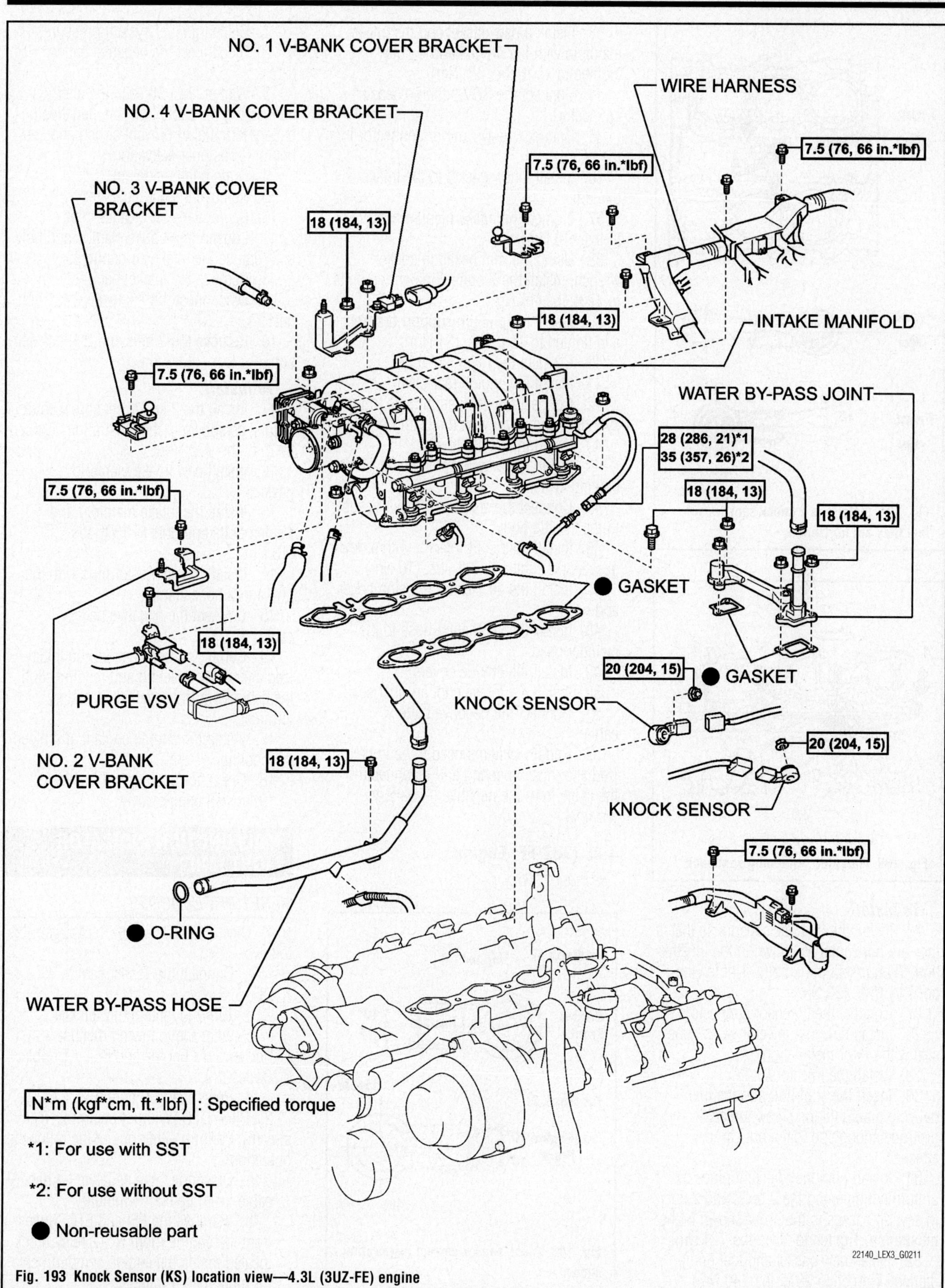

Fig. 193 Knock Sensor (KS) location view—4.3L (3UZ-FE) engine

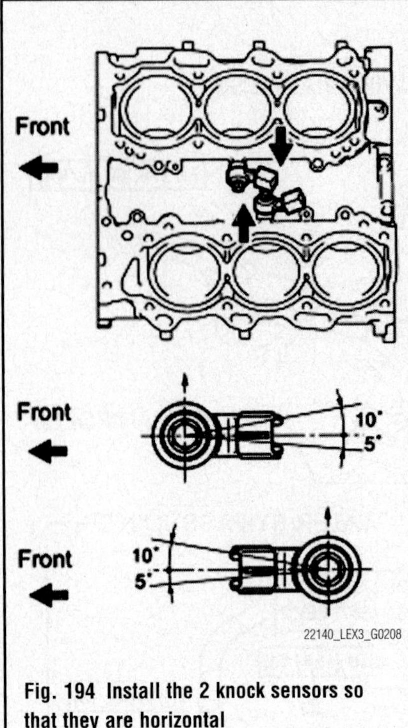

Fig. 194 Install the 2 knock sensors so that they are horizontal

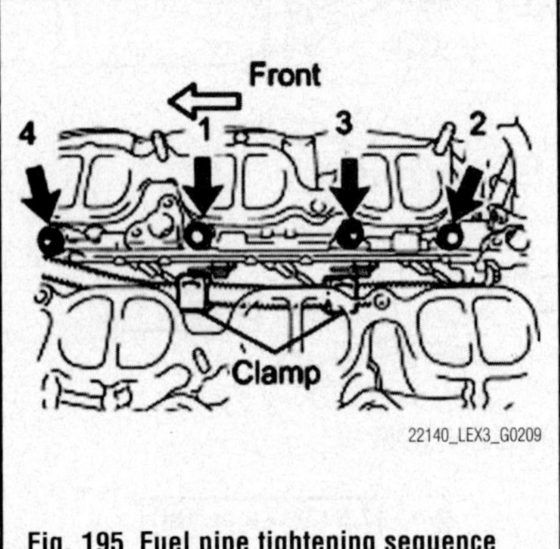

Fig. 195 Fuel pipe tightening sequence

To install:

26. Install the 2 knock sensors so that they are horizontal as shown in the illustration. Then install the 2 bolts. Tighten the bolts to 15 ft. (20 Nm).
27. Connect the 2 sensor connectors.
28. Install new fuel injector seals, lube seal with oil for installation.
29. Install the injectors.
30. Insert the stud bolt into the fuel delivery pipe until the screw threads protrude enough so that a nut can be attached.
31. Install both fuel delivery pipes by uniformly tightening the 2 bolts and 2 nuts in several passes in the order shown in the illustration. Tighten to 15 ft. lbs. (21 Nm).
32. Install new gasket and fuel pump damper. Tighten to 28 ft. lbs. (40 Nm).
33. Install a new gasket and the intake manifold with the 4 bolts and 4 nuts. Tighten to 15 ft. lbs. (21 Nm).
34. Connect the SCV position sensor connector.
35. Connect the DC motor connector for the SCV.
36. Install a new gasket to the intake plenum.
37. Tighten the intake plenum nuts to 13 ft. lbs. (18 Nm).
38. Using a 5 mm hexagon socket wrench, install the 7 bolts. Tighten the bolts to 12 ft. lbs.(16 Nm).
39. Install the plenum support brackets and tighten to 15 ft. lbs. (21 Nm).
40. Connect the main fuel tube.
41. Install the engine rear cover
42. Connect the ventilation hose to the cylinder head cover with the clamp.
43. Install the heater hoses.
44. Install the engine coolant and bleed cooling system.
45. Connect the 2 water by-pass hoses to the throttle body.
46. Install cold start injector with a new gasket and tighten to 7 ft. lbs. (10 Nm).
47. Install the air cleaner housing cover and hose.
48. Install the ventilation hose to the cylinder head.
49. Install the engine covers.
50. Install the intake cool air duct seal.
51. Connect the negative battery cable.
52. Certain systems need to be initialized after disconnecting and reconnecting the cable from the negative (-) battery terminal

4.3L (3UZ-FE) Engine

See Figure 196.

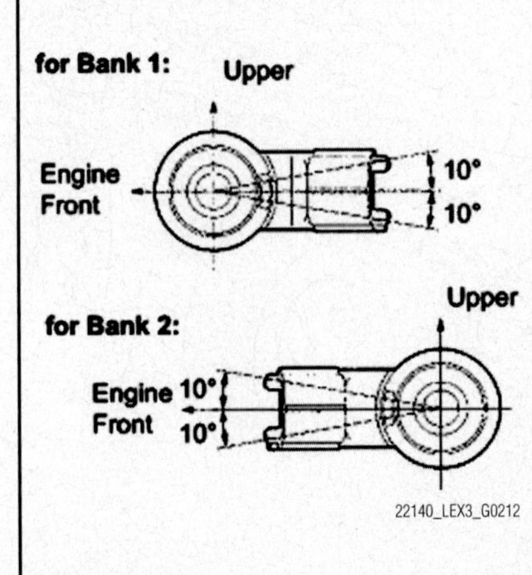

Fig. 196 Knock sensor correct installation shown

1. Discharge fuel system pressure.
2. Disconnect the negative battery cable.
3. Wait at least 90 seconds after disconnecting the cable from the negative (-) battery terminal to prevent airbag and seat belt pre-tensioner activation.
4. Drain engine coolant.
5. Remove the engine cover.
6. Remove the air cleaner inlet.
7. Loosen the 2 hose clamps and bolt, and remove the intake air connector.
8. Remove the intake manifold.
9. Disconnect the 2 knock sensor connectors.
10. Remove the 2 nuts and 2 knock sensors.

To install:

11. Install the 2 knock sensors with the 2 nuts, as shown in the illustration. Tighten to 15 ft. lbs. (20 Nm).
12. Install new intake manifold gaskets
13. Install the intake manifold and tighten bolts and nuts to 13 ft. lbs. (18 Nm).
14. Install the connector pipe with the bolt and 2 hose clamps.
15. Connect the negative battery cable.
16. Certain systems need to be initialized after disconnecting and reconnecting the cable from the negative (-) battery terminal.
17. Install the engine coolant and bleed the cooling system.
18. Check for coolant and fuel leaks.
19. Install engine cover.

MALFUNCTION INDICATOR LIGHT (MIL)

RESET PROCEDURES

1. Clear DTC (Using the Techstream) as follows:
    a. Connect the Techstream to the DLC3.
    b. Turn the ignition switch ON.
    c. Enter the following menus: Powertrain / Engine and ECT / Trouble Codes.
    d. Press the YES button.
2. Clear DTC (Without using the Techstream). Perform either one of the following operations.
    a. Disconnect the negative (-) battery cable for more than 1 minute.
    b. Remove the EFI and ETCS fuses from the engine room No. 2 relay block located inside the engine compartment for more than 1 minute.

MASS AIR FLOW (MAF) SENSOR

LOCATION

See Figure 197.

The Mass Air Flow (MAF) Sensor is located at the air filter housing assembly.

REMOVAL & INSTALLATION

1. Remove the intake cool duct seal.
2. Remove the engine covers.
3. Disconnect the MAF meter connector.
4. Remove the 2 screws and MAF meter.
5. Remove the O-ring from the MAF meter.

To install:

6. Install a new O-ring to the MAF meter.
7. Install the MAF meter with the 2 screws.
8. Connect the MAF meter connector.
9. Install engine covers
10. Install the intake duct seal with the 7 clips.

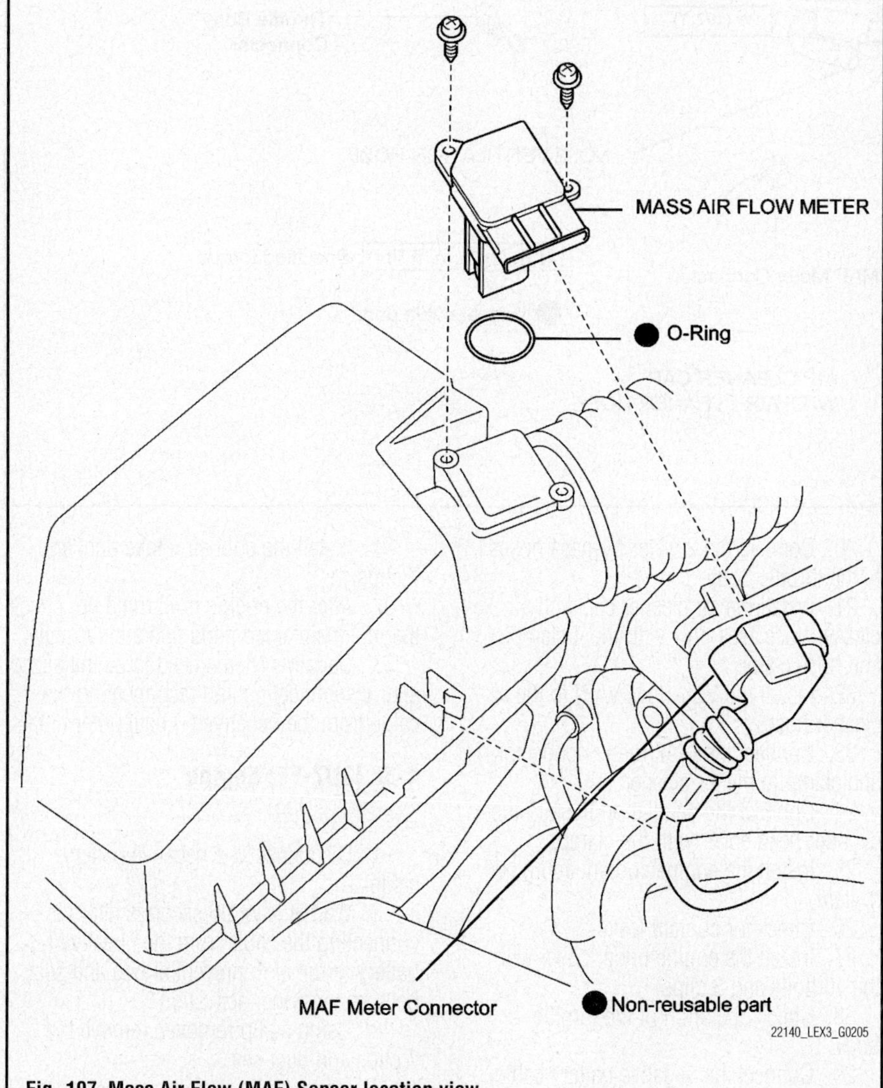

Fig. 197 Mass Air Flow (MAF) Sensor location view

OIL PRESSURE SENSOR

LOCATION

The oil pressure switch is mounted at the oil filter housing.

REMOVAL & INSTALLATION

See Figure 198.

1. Before servicing the vehicle, refer to the precautions in the beginning of this section.
2. Disconnect the negative battery cable. Wait at least 90 seconds before performing any other work
3. Remove the engine under cover(s), if necessary.

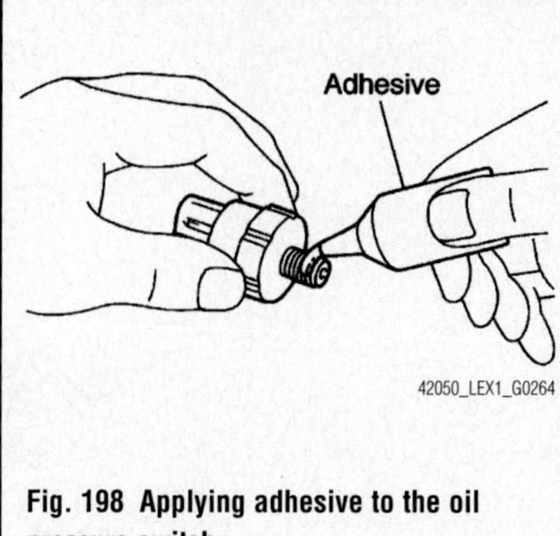

Fig. 198 Applying adhesive to the oil pressure switch.

4. Disconnect the oil pressure switch connector, located on the side of the engine block.
5. Remove the oil pressure switch.

To install:

6. Apply adhesive (Part No. 08833-00080, THREE BOND 1344, LOCTITE 242 or equivalent) to 2 or 3 threads of the oil pressure switch.
7. Install the oil pressure switch.
8. Connect the oil pressure switch connector.
9. Install the engine under cover(s), if necessary.
10. Connect the negative battery cable.
11. Start the engine and check for engine oil leaks.

THROTTLE POSITION SENSOR (TPS)

LOCATION

See Figure 199.

The Throttle Position Sensor (TPS) is mounted on the throttle body and is integral to the assembly.

REMOVAL & INSTALLATION

3.0L (3GR-FSE) Engine

1. Disconnect the negative battery cable.
2. Wait at least 90 seconds after disconnecting the cable from the negative (-) battery terminal to prevent airbag and seat belt pre-tensioner activation.
3. Using a clip remover, remove the 7 clips and duct seal.
4. Using a clip remover, remove the 3 clips.
5. Remove the 10 screws and under cover.
6. Drain the engine coolant.
7. Remove the engine covers.
8. Disconnect the ventilation hose from the cylinder head.

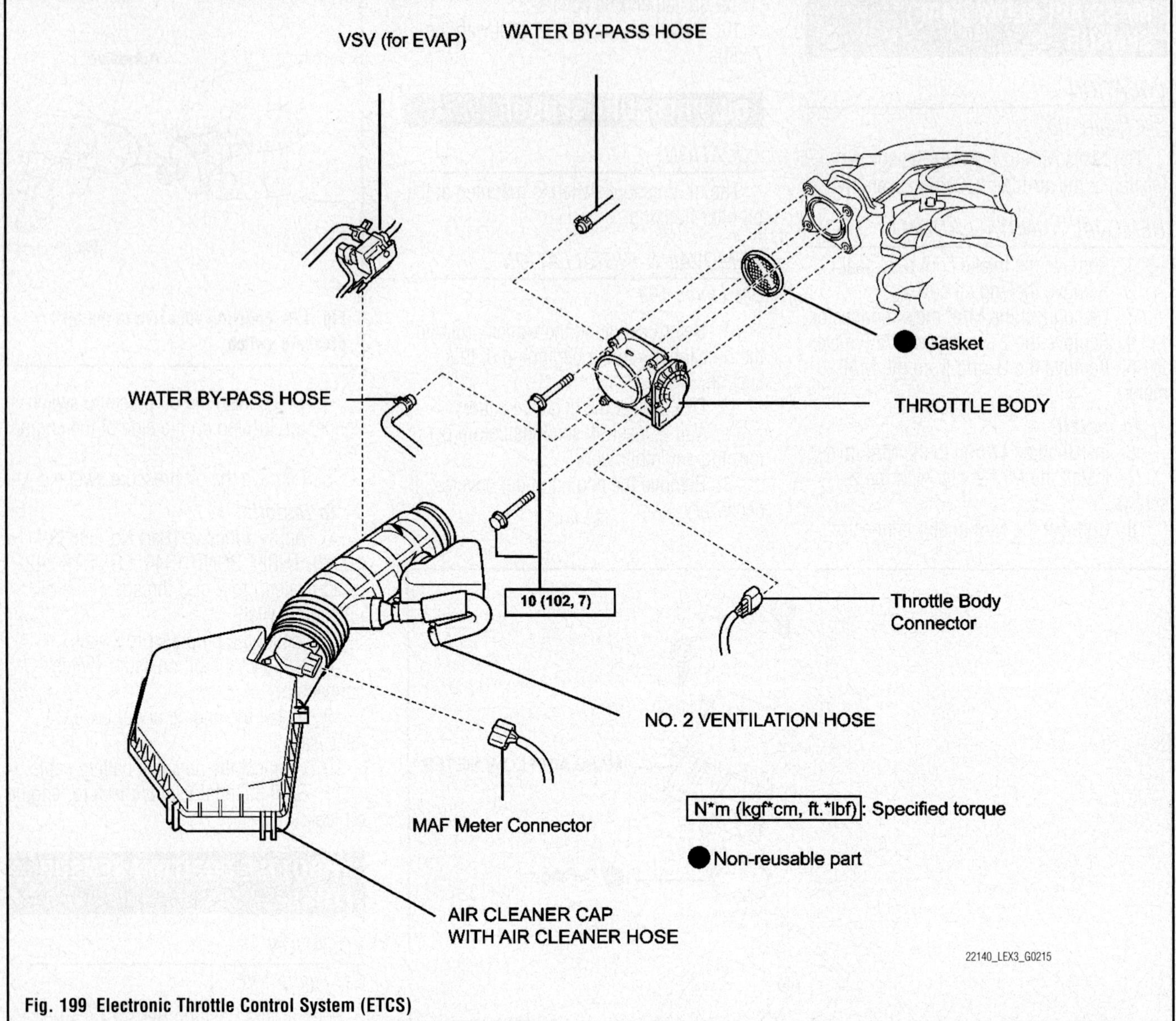

Fig. 199 Electronic Throttle Control System (ETCS)

9. Disconnect the MAF meter connector.
10. Disconnect the clamp from the air cleaner.
11. Disconnect the VSV (for EVAP).
12. Disconnect the 4 clamps.
13. Remove the hose clamp and air cleaner cap with air cleaner hose.
14. Disconnect the 2 water by-pass hoses from the throttle body.
15. Disconnect the throttle motor connector.
16. Remove the 4 bolts, throttle body and gasket.

To install:

17. Install a new gasket to the intake air surge tank.
18. Install the throttle body with the 4 bolts. Tighten to 7 ft. lbs.(10 Nm).
19. Connect the throttle motor connector.
20. Connect the 2 water by-pass hoses to the throttle body.
21. Install the air cleaner cap with air cleaner hose assembly with the 4 clamps and hose clamp.
22. Install the VSV (for EVAP) to the air cleaner hose.
23. Connect the MAF meter connector and clamp to the air cleaner.
24. Connect the ventilation hose to the cylinder head cover with the clamp.
25. Install the engine coolant and bleed system.
26. Check for coolant leaks.
27. Install the engine under cover with the 10 bolts and 3 clips.
28. Check operation of the throttle body.
29. Connect the negative battery cable.
30. Install the engine covers.
31. Install the cool air intake duct and 7 clips.
32. After the engine is warmed up, check that the maintained parts operate normally.
33. Certain systems need to be initialized after disconnecting and reconnecting the cable from the negative (-) battery terminal.

4.3L (3UZ-FE) Engine

See Figure 200.

1. Disconnect the negative battery cable.
2. Wait at least 90 seconds after disconnecting the cable from the negative (-) battery terminal to prevent airbag and seat belt pre-tensioner activation.
3. Using a clip remover, remove the 7 clips and duct seal.
4. Using a clip remover, remove the 3 clips.

5. Remove the 10 screws and under cover.
6. Drain the engine coolant.
7. Remove the engine covers.

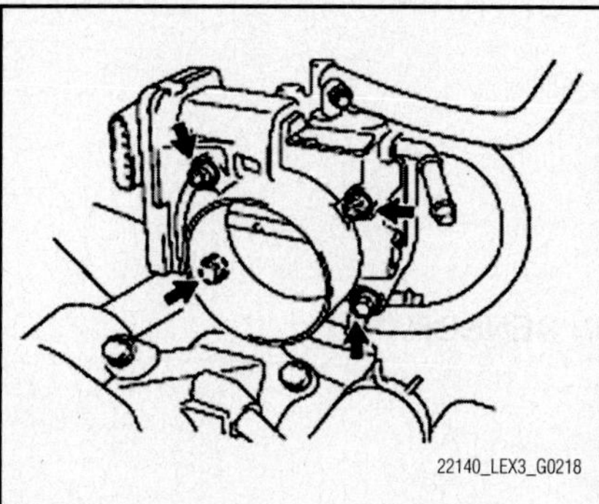

Fig. 200 Throttle body and mounting bolts.

8. Disconnect the air hose and No. 1 ventilation hose.
9. Loosen the 2 hose clamps and bolt, and remove the intake air connector.
10. Disconnect the throttle body connector.
11. Disconnect the ventilation hose.
12. Disconnect the 2 water by-pass hoses.
13. Remove the 2 bolts, 2 nuts, throttle body and gasket.

To install:

14. Install a new gasket and the throttle body with the 2 bolts and 2 nuts. Tighten to 13 ft. lbs. (18 Nm).
15. Connect the 2 water by-pass hoses.
16. Connect the ventilation hose.
17. Connect the connector.
18. Install the connector pipe with the bolt and 2 hose clamps.
19. Connect the air hose and No. 1 ventilation hose.
20. Install the air cleaner inlet with the bolt.
21. Install the engine covers.
22. Connect the negative battery cable.
23. Install the engine coolant and bleed system.
24. After the engine is warmed up, check that the maintained parts operate normally.
25. Certain systems need to be initialized after disconnecting and reconnecting the cable from the negative (-) battery terminal

VEHICLE SPEED SENSOR (VSS)

LOCATION

See Figures 201 through 203.

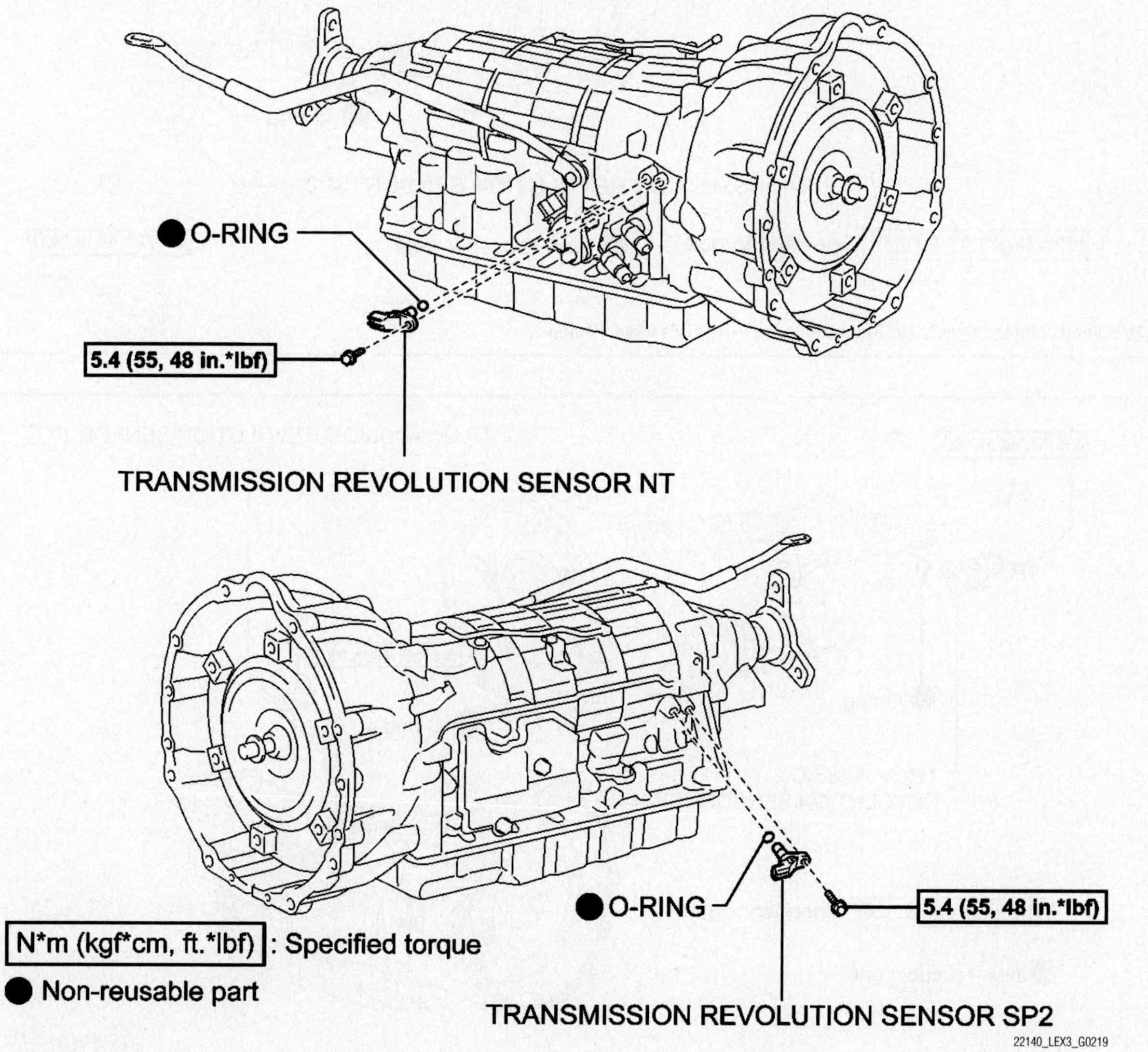

Fig. 201 Vehicle Speed Sensor (VSS) location view—(A761E) transmission

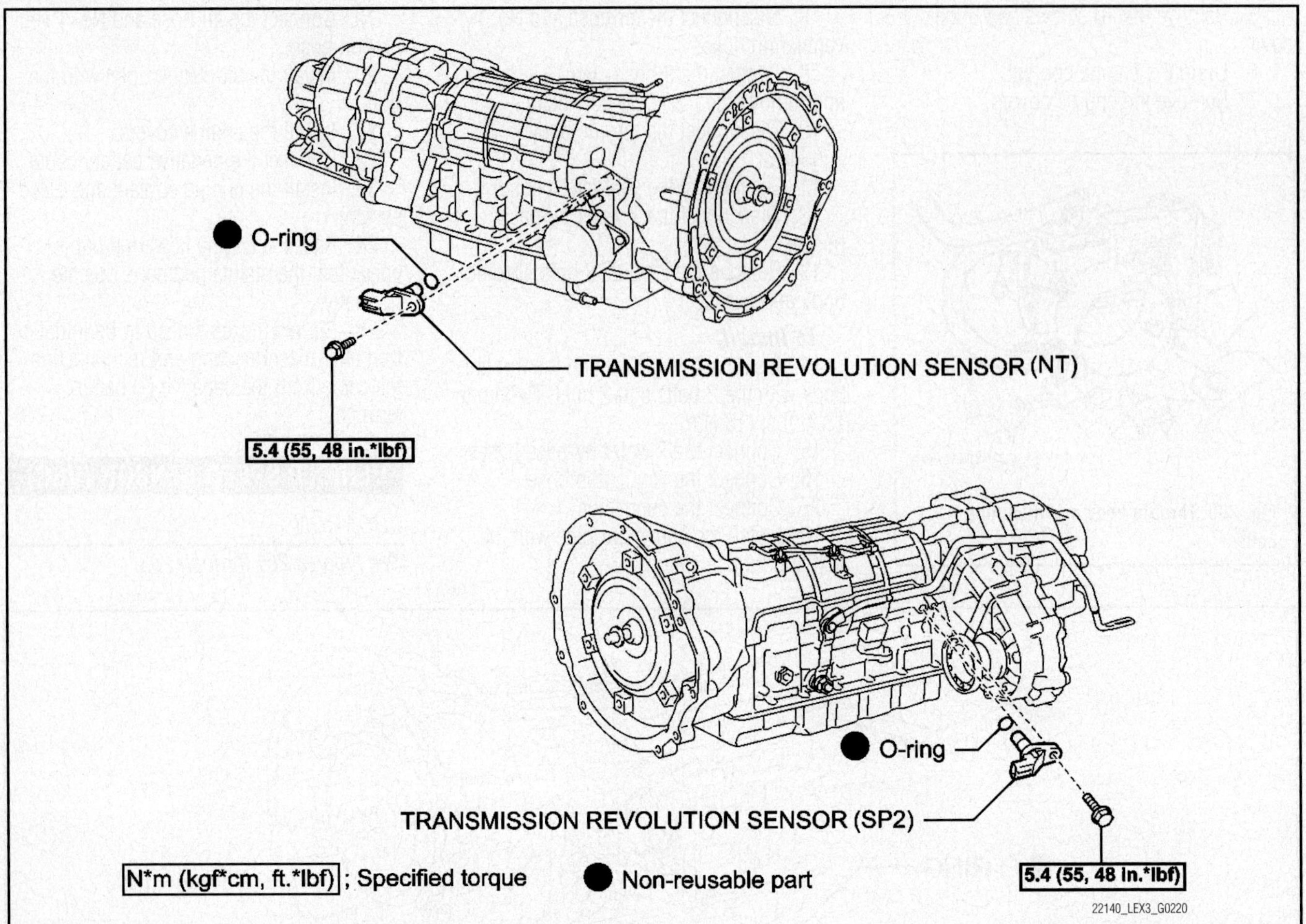

O-ring
TRANSMISSION REVOLUTION SENSOR (NT)
5.4 (55, 48 in.*lbf)
O-ring
TRANSMISSION REVOLUTION SENSOR (SP2)
5.4 (55, 48 in.*lbf)
N*m (kgf*cm, ft.*lbf) ; Specified torque
Non-reusable part
22140_LEX3_G0220

Fig. 202 Vehicle Speed Sensor (VSS) location view—(A760H) transmission

5.4 (55, 48 in.*lbf)
TRANSMISSION REVOLUTION SENSOR (NT)
5.4 (55, 48 in.*lbf)
O-ring
O-ring
TRANSMISSION REVOLUTION SENSOR (SP2)
N*m (kgf*cm, ft.*lbf) : Specified torque
Non-reusable part
22140_LEX3_G0221

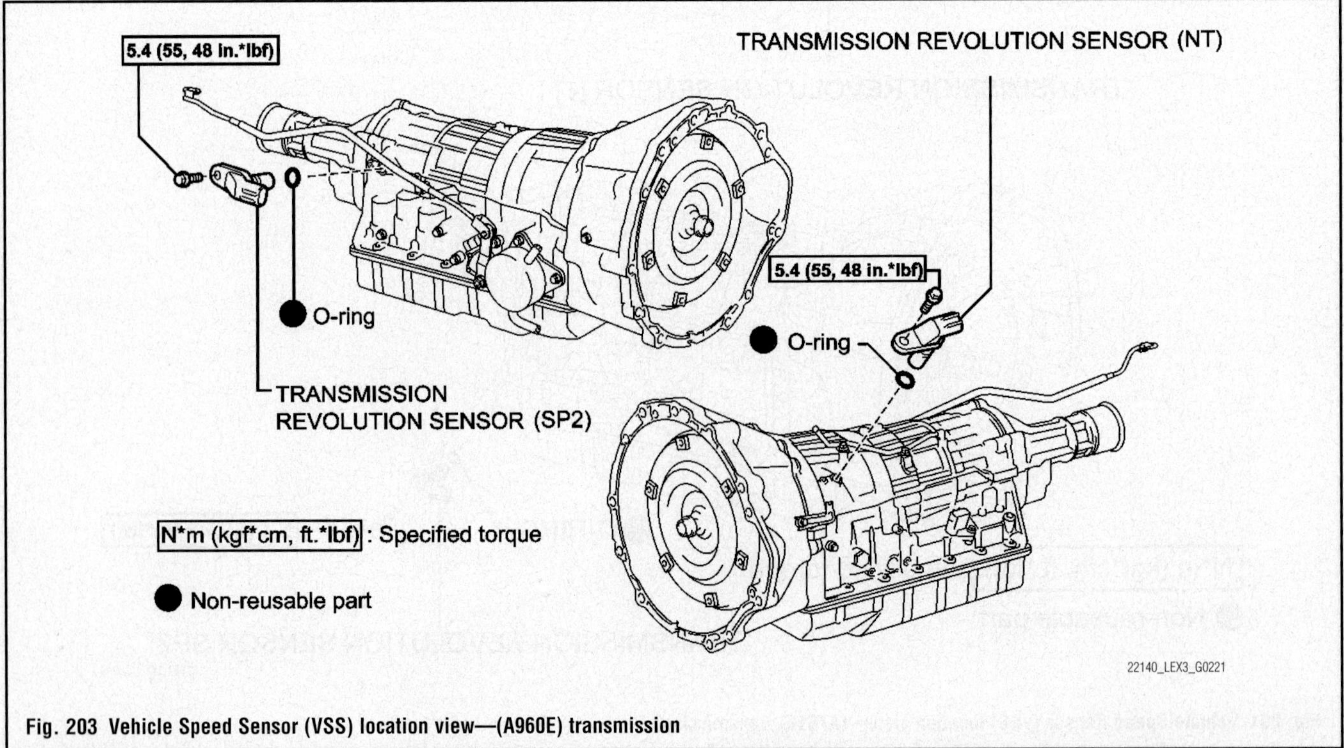

Fig. 203 Vehicle Speed Sensor (VSS) location view—(A960E) transmission

The vehicle speed sensors are located on the left front and right rear of the transmission assembly.

REMOVAL & INSTALLATION

1. Disconnect the transmission speed sensor connector.
2. Remove the bolt and transmission speed sensor.

To install:

3. Coat a new O-ring with ATF and install it to the transmission speed sensor.
4. Install the transmission speed sensor with the bolt. Tighten the mounting bolt to 48 inch. lbs. (5.4 Nm).
5. Connect the transmission speed sensor connector.

FUEL SYSTEM — GASOLINE FUEL INJECTION SYSTEM

FUEL SYSTEM SERVICE PRECAUTIONS

Safety is the most important factor when performing not only fuel system maintenance but any type of maintenance. Failure to conduct maintenance and repairs in a safe manner may result in serious personal injury or death. Maintenance and testing of the vehicle's fuel system components can be accomplished safely and effectively by adhering to the following rules and guidelines.

- To avoid the possibility of fire and personal injury, always disconnect the negative battery cable unless the repair or test procedure requires that battery voltage be applied.
- Always relieve the fuel system pressure prior to disconnecting any fuel system component (injector, fuel rail, pressure regulator, etc.), fitting or fuel line connection. Exercise extreme caution whenever relieving fuel system pressure to avoid exposing skin, face and eyes to fuel spray. Please be advised that fuel under pressure may penetrate the skin or any part of the body that it contacts.
- Always place a shop towel or cloth around the fitting or connection prior to loosening to absorb any excess fuel due to spillage. Ensure that all fuel spillage (should it occur) is quickly removed from engine surfaces. Ensure that all fuel soaked cloths or towels are deposited into a suitable waste container.
- Always keep a dry chemical (Class B) fire extinguisher near the work area.
- Do not allow fuel spray or fuel vapors to come into contact with a spark or open flame.
- Always use a back-up wrench when loosening and tightening fuel line connection fittings. This will prevent unnecessary stress and torsion to fuel line piping.
- Always replace worn fuel fitting O-rings with new. Do not substitute fuel hose or equivalent where fuel pipe is installed.

Before servicing the vehicle, make sure to also refer to the precautions in the beginning of this section as well.

RELIEVING FUEL SYSTEM PRESSURE

1. Before servicing the vehicle, refer to the precautions in the beginning of this section.
2. Remove the fuse for the electronic fuel pump.
3. Start the engine until the engine stalls.
4. Disconnect the negative battery terminal.
5. Place a catch-pan under the joint to be disconnected. A large quantity of fuel may be released when the joint is opened.
6. Wear eye or full-face protection.
7. Place a shop towel over the area and slowly release the joint using a wrench of the correct size.
8. Allow any fuel left in the line to bleed off slowly before fully disconnecting the joint.
9. Plug the opened lines immediately to prevent fuel spillage or the entry of dirt.
10. Dispose of the released fuel properly.
11. After connecting fuel lines, install the fuse for the fuel pump and start the engine.
12. Check for leaks and repair as needed.

FUEL FILTER

REMOVAL & INSTALLATION

The fuel filter is mounted in the fuel tank and is part of fuel pump unit housing.

FUEL PUMP

REMOVAL & INSTALLATION

GS 300

1. Before servicing the vehicle, refer to the precautions in the beginning of this section.
2. Relieve the fuel system pressure.
3. Remove or disconnect the following:
   - Negative battery cable. Wait at least 90 seconds before performing any other work.
   - Rear seat bottom
   - Partition cover
   - Floor service hole cover
   - Fuel pump electrical connector
   - Fuel main tube and fuel pump tube from the top of the fuel pump
   - Mounting bolts, or retaining ring using SST 09808-14020
   - Pump, bracket and set plate as an assembly

To install:

4. Install or connect the following:
   - A new gasket on the set plate
   - Fuel pump and bracket assembly. Torque the mounting bolts to 31 inch lbs. (3.5 Nm). If equipped with a retaining ring, use SST 09808-14020 to tighten the retainer 2 full turns so that the mark on the ring lines up within the 2 marks indicated next to the ring on the fuel tank.
   - Fuel main tube and fuel pump tube to the top of the fuel pump
   - Fuel pump electrical connector
   - Floor service hole cover
   - Partition cover
   - Rear seat bottom
   - Negative battery cable
5. Start the engine; check the fuel system for leaks

SC 430

1. Before servicing the vehicle, refer to the precautions in the beginning of this section.
2. Remove or disconnect the following:
   - Negative battery cable. Wait at least 90 seconds before performing any other work.
   - Rear seat bottom and seat back
   - Fuel pump tube
   - 8 bolts and fuel tank vent tube set plate
   - Fuel pump and sensor gauge assembly
   - Fuel suction hose and support
   - Fuel pump cushion rubber
   - Fuel pressure w/jet pump regulator assembly
   - Fuel suction plate w/sender gauge
   - Fuel pump and filter

To install:

3. To install, reverse the removal procedure. Install a new O-ring on the fuel jet pump regulator assembly and a new gasket on the fuel pump/sender gauge assembly. Torque the 8 bolts securing the fuel tank set plate to 52 in. lbs (6.0 Nm).
4. Inspect fuel pump operation and check for fuel leaks.

FUEL PUMP (HIGH PRESSURE SIDE)

REMOVAL & INSTALLATION

3.0L (3GR-FSE) Engine

See Figures 204 and 205.

CAUTION

Do not allow fuel to spray when removing the pipe between the high pressure side fuel pump and the fuel injector. The fuel in the pipe is highly pressurized.

1. Discharge the fuel system pressure.
2. Disconnect the negative battery cable.
3. remove the cool air intake duct seal.
4. Remove the engine under cover.
5. Drain the engine coolant.
6. Remove the right engine room side cover.
7. Remove the engine cover.

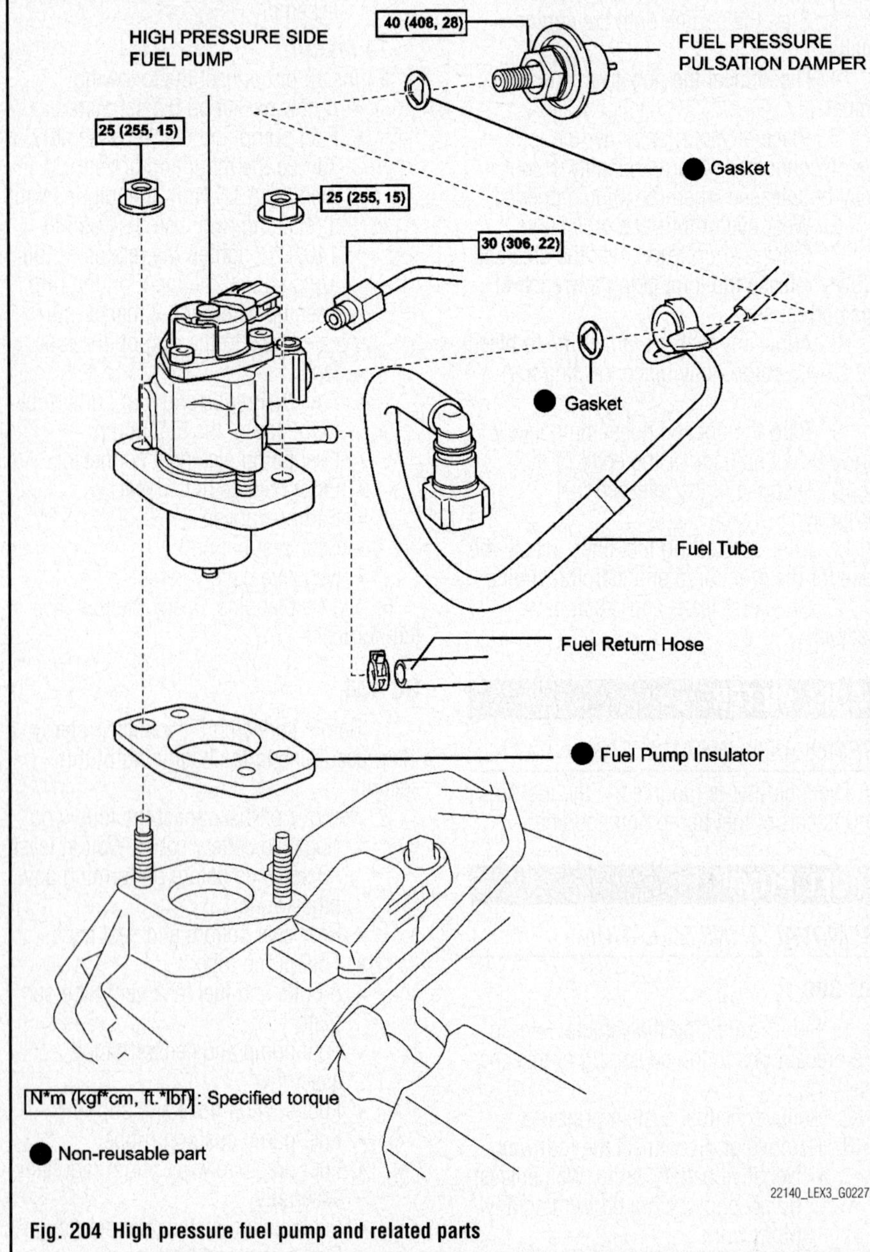

Fig. 204 High pressure fuel pump and related parts

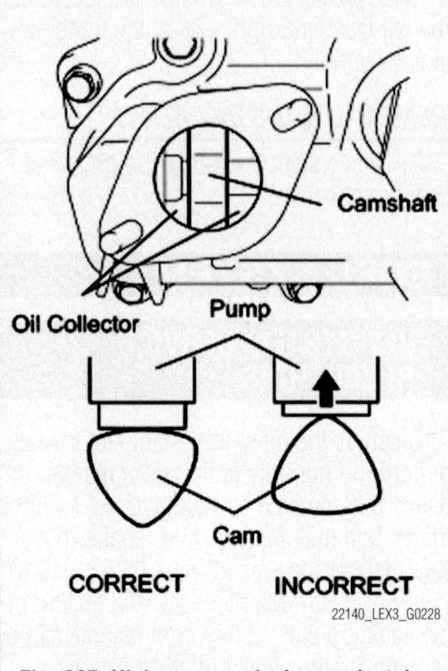

Fig. 205 High pressure fuel pump installation

8. Disconnect the cold start injector.
9. Disconnect the water bypass hose.
10. Disconnect the heater hoses.
11. Remove the rear engine cover.
12. Disconnect the main fuel tube.
13. Disconnect the hose from the surge tank.
14. Disconnect the ventilation hose.
15. Remove the intake plenum.
16. Remove the intake manifold.
17. Disconnect the fuel pipe hose.
18. Remove the 2 bolts and fuel pipe.
19. Remove the fuel pulsation damper and 2 gaskets, and disconnect the fuel main tube from the fuel pump.
20. Disconnect the fuel high pressure side fuel pump connector.
21. Fix the union bolt on the fuel pump side in place with a 21 mm wrench. Using a 19 mm union nut wrench, loosen the union and remove the fuel pipe.

➡If the union on the fuel pump side has free play, replace the fuel pump.

22. Remove the 2 bolts on the delivery pipe side.
23. Disconnect the fuel hose.
24. Remove the 2 nuts, fuel pump and fuel pump insulator.

To install:

25. Turn the crankshaft until the flat of the cam is facing the cylinder head cover's fuel pump attachment hole, as shown in the illustration.
26. Pour 30 cc of engine oil through the cylinder head cover's fuel pump attachment hole into the cylinder head oil collector.

27. Apply a coat of engine oil to the pump activation cam and pump lifter part.

➡When installing the fuel pump using the procedure described above: By not using the crankshaft pointed side to push up the pump activation surface, it is easier to install the fuel pump and No. 2 fuel pipe later.

28. Install a new fuel pump insulator to the cylinder head cover. Then pass the 2 stud bolts through the holes of the fuel pump and set it on the insulator.
29. Install the union nut of the No. 1 fuel pipe without damaging its seal surface. Tighten the nut as much as possible by hand
30. Install the 2 nuts and tighten them in several passes to 18 ft. lbs. (25 Nm).
31. Connect the fuel hose.
32. Using a 19 mm union nut wrench, connect the No. 2 fuel pipe. Tighten to 22 ft. lbs. (30 Nm).
33. Install the No. 2 fuel pipe to the delivery pipe with the 2 bolts and tighten to 7 ft. lbs. (10 Nm).
34. Connect the connector to the fuel pump.
35. Install 2 new gaskets, the fuel main tube and pulsation damper to the fuel pump.
36. Install the fuel pipe with the 2 bolts and tighten to 7 ft. lbs. (10 Nm).
37. Connect the No. 1 fuel pipe hose.
38. Install a new gasket and the intake manifold with the 4 bolts and 4 nuts. Tighten to 15 ft. lbs. (21 Nm).
39. Connect the SCV position sensor connector.
40. Connect the DC motor connector for the SCV.
41. Install a new gasket to the intake plenum.
42. Install the intake air surge tank with the 2 nuts and tighten to 13 ft. lbs. (18 Nm).
43. Using a 5 mm hexagon socket wrench, install the 7 bolts. Tighten the bolts to 12 ft. lbs. (16 Nm).
44. Install the plenum brackets and tighten to 15 ft. lbs. (21 Nm).
45. Connect the connector to the fuel main pipe. Push the two parts together firmly until a "click" sound is heard. Then attach the lock claws to the connector by pushing down on the connector cover.
46. Connect the ventilation hose to the ventilation valve.
47. Connect the hose to the surge tank.
48. Install the engine rear cover with the 3 clips.
49. Connect the heater hoses.
50. Connect the 2 water by-pass hoses to the throttle body.
51. Install the cold start injector with a new gasket and tighten the 2 bolts to 7 ft. lbs. (10 Nm).
52. Install the air cleaner cap with air cleaner hose assembly with the 4 clamps and hose clamp.
53. Install the VSV (for EVAP) to the air cleaner hose.
54. Connect the MAF meter connector and clamp to the air cleaner.
55. Connect the ventilation hose to the cylinder head cover with the clamp.
56. Connect the negative battery cable.
57. Add the engine coolant and bleed the cooling system.
58. Check for coolant and fuel leaks.
59. Install the engine cover.
60. Install the engine room side cover.
61. Install the cool air intake duct seal.
62. Install the engine undercover.
63. Certain systems need to be initialized after disconnecting and reconnecting the cable from the negative (-) battery terminal.
64. After the engine is warmed up, check that the maintained parts operate normally.

FUEL TANK

REMOVAL & INSTALLATION

3.0L (3GR-FSE) Engine

See Figure 206.

1. Before servicing the vehicle, refer to the precautions in the beginning of this section.
2. Discharge the fuel system pressure.
3. Disconnect the negative battery cable.
4. Wait at least 90 seconds after disconnecting the cable from the negative (-) battery terminal to prevent airbag and seat belt pre-tensioner activation.
5. Detach the seat cushions 2 front hooks from the vehicle body.
6. Remove the seat cushion.
7. Remove the clip and room partition pad.
8. Remove the service hole cover and disconnect the fuel pump connector.
9. Disconnect the fuel main tubes and pump tube.
10. Remove the 2 tube joint clips and fuel tubes.
11. Remove the rear exhaust assembly.
12. Remove the 3 nuts, 2 clips and the right floor cover.
13. Remove the 4 nuts and heat insulator.
14. Remove propeller shaft assembly.
15. Remove the 4 nuts and fuel tank protector.
16. Remove the both differential support protectors.
17. Remove the left rear side member cover.
18. Remove the 2 parking brake cables from the 4 clamps.
19. Remove the 4 bolts and disconnect the 2 parking brake cables.
20. Disconnect the fuel lines.
21. Disconnect the filler pipe hose from the fuel tank filler pipe.
22. Disconnect the fuel tank bracket hose.
23. Disconnect the 2 fuel emission hoses.
24. Remove the fuel tank as follows:
    - Remove the 2 nuts from the fuel tank.
    - Place a mission jack under the fuel tank.
    - Remove the 4 nuts and 2 fuel tank bands.
    - Remove the 2 nuts.
25. With a helper and the use of transmission jack carefully lower the fuel tank.
26. With a helper and the use of transmission jack carefully raise the fuel tank
27. Install the 2 fuel tank bands with the 4 bolts and tighten to 29 ft. lbs. (39 Nm).
28. Install the 2 nuts and tighten to 15 ft. lbs. (20 Nm).
29. Connect the 2 emission hoses.
30. Connect the fuel tank to filler pipe hose with the clamp.
31. Connect the fuel lines.
32. Connect the 4 clamps and tighten to 53 inch. lbs. (6.0 Nm).
33. Connect the 2 parking brake cables with the 4 bolts and tighten to 14 ft. lbs. (19 Nm).
34. Install the right and left floor side member covers. Tighten the bolts.
35. Install both differential support protectors and tighten the bolts.
36. Install the tank protector with the 4 nuts and tighten to 48 inch. lbs. (5.0 Nm).
37. Install rear propeller shaft assembly. Tighten the front mounting bolts to 58 ft. lbs. (79 Nm).
38. Temporarily install the 2 center support bearing set bolts with the adjusting washers.
39. Check that the center line of the bracket is at right angles to the shaft axial direction.
40. Tighten the center bearing mounting bolts to 36 ft. lbs. (49 Nm).
41. Tighten the front mounting bolts to 58 ft. lbs. (79 Nm).
42. Install the heat insulator with the 4 nuts.

43. Install the floor cover center with the 3 nuts.

44. Install the rear exhaust assembly.

45. Install fuel lines and connectors to fuel tank.

46. Install the both service hole covers with new butyl tape.

47. Install the partition pad with the clip.

48. Install rear seat cushion assembly.

49. Connect the negative battery cable.

50. Check for fuel leaks.

51. Certain systems need to be initialized after disconnecting and reconnecting the cable from the negative (-) battery terminal.

4.3L (3UZ-FE) Engine

See Figure 207.

1. Before servicing the vehicle, refer to the precautions in the beginning of this section.

2. Detach the seat cushions 2 front

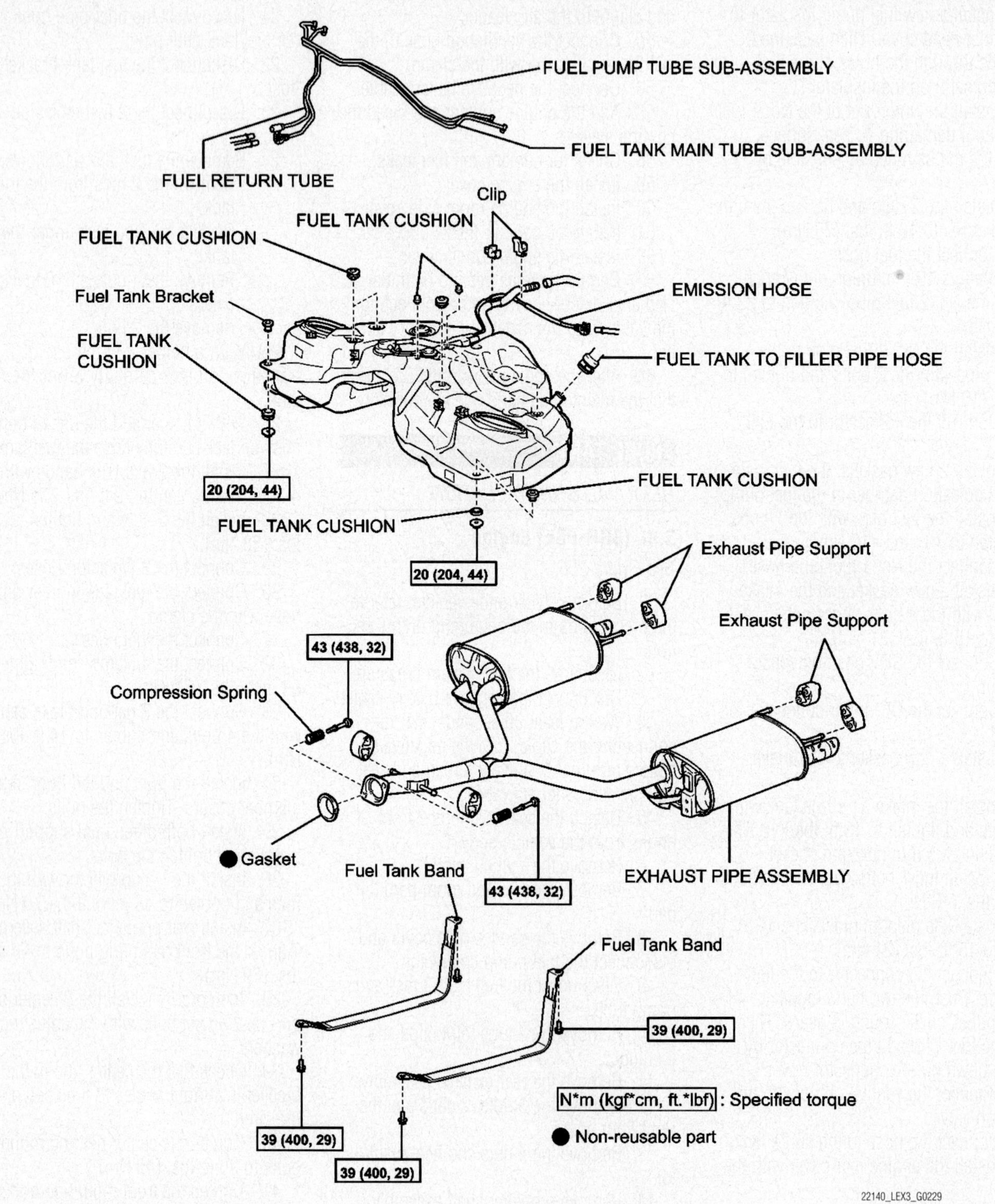

Fig. 206 Fuel tank assembly and related parts.

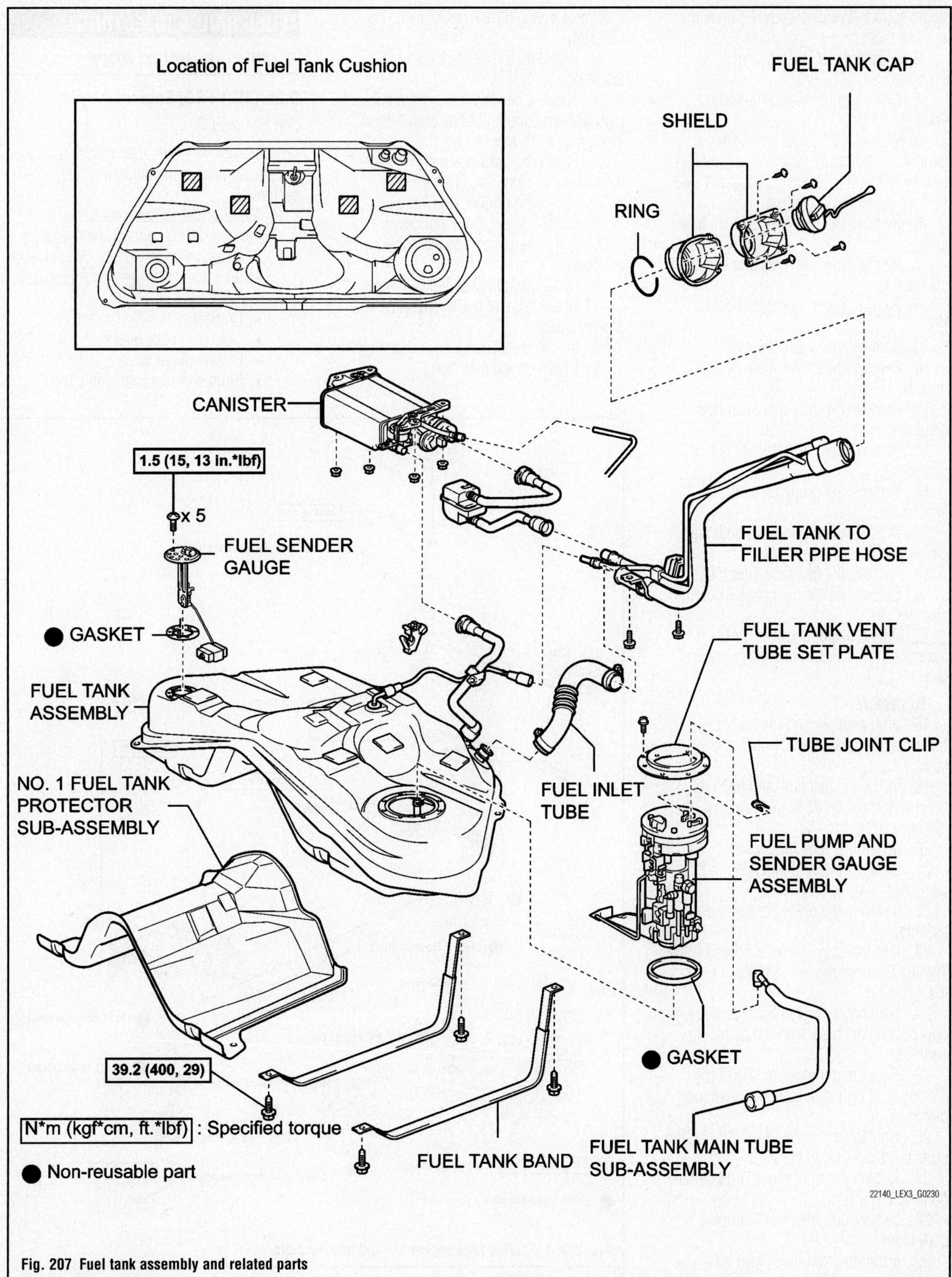

Fig. 207 Fuel tank assembly and related parts

hooks from the vehicle body. Remove the seat cushion.

3. Discharge the fuel system pressure.
4. Disconnect the negative battery cable.
5. Wait at least 90 seconds after disconnecting the cable from the negative (-) battery terminal to prevent airbag and seat belt pre-tensioner activation.
6. Remove the rear floor service hole covers.
7. Remove the fuel lines and electrical connectors.
8. Remove the front center floor brace.
9. Remove the undercover.
10. Remove the center exhaust and muffler.
11. Remove the propeller shaft heat insulator.
12. Remove both differential support protectors.
13. Remove the propeller shaft assembly.
14. Disconnect and remove both rear parking brake cable assembly.
15. Remove the fuel tank filler tube.
16. Disconnect the vent hose from the canister.
17. With a helper and the use of transmission jack carefully lower the fuel tank.

To install:

18. With a helper and the use of transmission jack carefully raise the fuel tank
19. Install the fuel tank and fuel tank band with the 4 bolts. Tighten to 29 ft. lbs. (39 Nm).
20. Install fuel tank filler tube.
21. Connect the vent line hose to the canister.
22. Install both parking brake cable assembly.
23. Install rear propeller shaft assembly. Tighten the front mounting bolts to 58 ft. lbs. (79 Nm).
24. Temporarily install the 2 center support bearing set bolts with the adjusting washers.
25. Check that the center line of the bracket is at right angles to the shaft axial direction.
26. Tighten the center bearing mounting bolts to 36 ft. lbs. (49 Nm).
27. Tighten the front mounting bolts to 58 ft. lbs. (79 Nm).
28. Install both differential support protectors.
29. Install the front floor heat insulator with the 4 bolts. Tighten to 47 inch. lbs. (5.4 Nm).
30. Install the propeller shaft heat insulator.
31. Install a new gasket and the exhaust pipe with the 2 nuts, 2 bolts and 2 rings. Tighten to 42 ft. lbs. (43 Nm).
32. Install the floor brace and tighten the 4 bolts to 10 ft. lbs. (13 Nm).
33. Install the muffler assembly.
34. Install the engine under cover.
35. Install fuel lines and electrical connectors.
36. Install both service hole covers.
37. Check for fuel pump operation and fuel leaks.
38. Install rear seat cushion assembly.
39. Check for exhaust leaks.

FUEL RAIL & INJECTORS

REMOVAL & INSTALLATION

3.0L (3GR-FSE) Engine

See Figure 208.

1. Before servicing the vehicle, refer to the precautions in the beginning of this section.
2. Relieve fuel system pressure.
3. Remove or disconnect the following:
   - Negative battery cable. Wait at least 90 seconds before performing any other work.
   - Coolant
   - Cool air intake duct seal
   - Engine under cover
   - Right engine room side cover

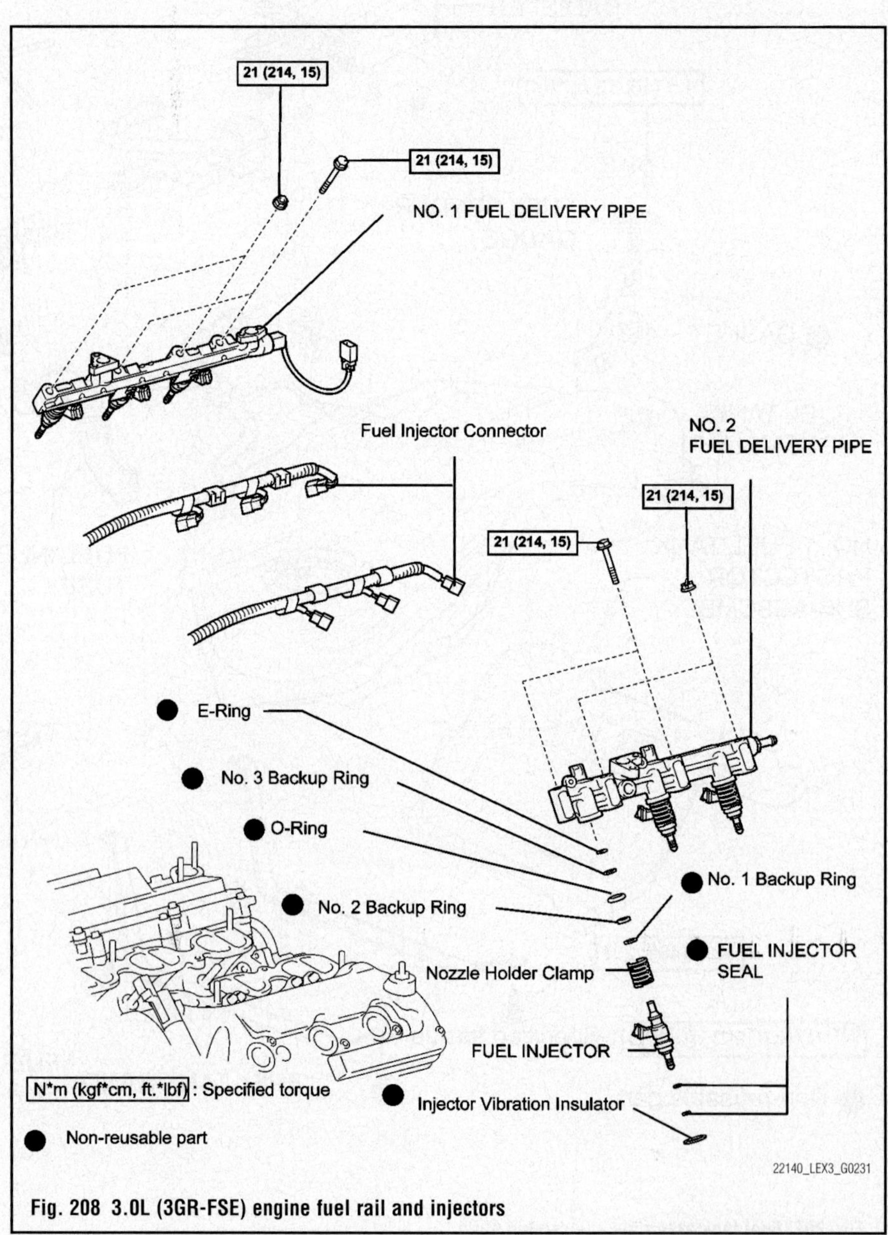

Fig. 208 3.0L (3GR-FSE) engine fuel rail and injectors

- V-bank cover
- No. 2 ventilation hose
- Air cleaner cap with air cleaner hose
- Cold start injector
- Water by-pass hose
- Inlet and outlet heater water hoses
- Engine rear cover
- Union to check valve hose
- Ventilation hose
- Fuel main tube
- Intake air surge tank
- Intake manifold
- Fuel pressure pulsation damper
- No. 1 and 2 fuel pipes
- High pressure side fuel pump
- No. 2 and 3 fuel pipes
- No. 1 and 2 fuel delivery pipes
- Fuel injectors, O-rings and seals

To install:

4. Install or connect the following:
   - 2 new seals to each injector
5. Install the fuel injectors as follows:

a. Install a new O-ring, new backup rings (No. 1, No. 2, No. 3) and new E-ring to the fuel injector.

WARNING

Check that there is no foreign matter or damaged areas in the injector's O-ring groove. Check that the installation direction of the No. 1 and No. 2 backup ring are correct. Make sure the backup rings and O-ring are installed in the correct order. Check that the alignment openings of the backup rings are not overlapped or stretched. After installing the O-ring, check that it is not contaminated with foreign matter and is not damaged.

b. Install the injector nozzle holder clamp.

c. Apply gasoline to the O-ring. Install the nozzle holder clamp by aligning the protruding part of the clamp to the notch of the delivery pipe.

WARNING

Make sure there is no gap between the delivery pipe and clamp. Check that there is no foreign matter or damage in the injector insertion hole of the delivery pipe. Insert the injector straight into the delivery pipe without tilting it.

6. Install the No. 1 and 2 fuel delivery pipes as follows:

a. Install a new injector vibration insulator to the cylinder head.

b. Apply lubricant to the installation hole of the injector.

c. Insert the stud bolt into the fuel delivery pipe until the screw threads protrude enough so that a nut can be attached.

➡If an injector is dropped, replace it with a new one. Check that there is no foreign matter or damage in the injector insertion hole of the delivery pipe. Be extremely careful not to touch or strike the tips of the injectors. When inserting the fuel delivery pipe, push it in evenly without tilting it.

d. Install the fuel delivery pipe by uniformly tightening the 2 bolts and 2 nuts in several passes to 15 ft. lbs. (21 Nm).

e. Connect the 3 connectors and 2 clamps.

- No. 2 and 3 fuel pipes and all new rings and seals. Torque No. 3 fuel pipe fastener to 7 ft. lbs. (10 Nm).
- High pressure side fuel pump
- No. 1 and 2 fuel pipes. Torque fasteners to 7 ft. lbs. (10 Nm).
- Fuel pressure pulsation damper and new gasket. Torque to 28 ft. lbs. (40 Nm).
- Intake manifold
- Intake air surge tank
- Fuel main tube
- Ventilation hose
- Union to check valve hose
- Engine rear cover
- Inlet and outlet heater water hoses
- Water by-pass hose
- Cold start injector. Torque the bolts to 7 ft. lbs. (10 Nm).
- Air cleaner cap with air cleaner hose
- No. 2 ventilation hose
- V-bank cover
- Right engine room side cover
- Cool air intake duct seal
- Engine under cover
- Negative battery cable

7. Refill the cooling system. Start the engine and check for coolant and fuel leaks and proper operation.
8. Check the function of the throttle body unit
9. System initialization

4.3L (3UZ-FE) Engine

See Figure 209.

1. Before servicing the vehicle, refer to the precautions in the beginning of this section.
2. Remove or disconnect the following:
   - V-bank cover
   - Intake air connector
   - Accelerator cable
   - Fuel pressure pulsation dampers.
   - VVT sensor connectors
   - Vacuum Switching Valve (VSV) for Evaporative Emissions (EVAP)
   - 2 nuts and accelerator cable bracket
   - 3 V-bank cover brackets
   - VSV connector for Acoustic Control Induction System (ACIS) from the No. 1 V-bank cover bracket
   - 4 bolts and 3 V-bank cover brackets
   - Engine wire from the delivery pipe
   - 2 wire clamps from the wire clamp bracket on the right-hand delivery pipe
   - 8 injector connectors
   - 4 nuts holding the delivery pipe to the intake manifold
   - 2 delivery pipes and 8 injector assemblies and 4 spacers
   - 2 O-rings, grommet and insulator from each injector

To install:

3. Install or connect the following:
   - A new insulator and grommet to each injector
   - A light coat of gasoline to new O-rings and install them to each injector
   - A light coat of gasoline on the place where a delivery pipe touches an O-ring of the injector
   - Injector, while turning the clockwise and counterclockwise, into the delivery pipe

➡Position the injector connector outward.

- The 4 spacers in position on the intake manifold
- A light coat of gasoline on the place where an intake manifold touches an O-ring
- The delivery pipes in position on the intake manifold
- Temporarily, the 3 bolts holding the delivery pipe to the intake manifold

➡Check that the injectors rotate smoothly. If the injectors do not rotate smoothly, the probable cause is incorrect installation of the O-rings. Replace the O-rings.

4. Tighten the 3 bolts holding the delivery pipe to the intake manifold. Tighten the bolts to 15 ft. lbs. (21 Nm).
5. Install or connect the following:
   - Engine wire protector with the 3 nuts
   - Injector connectors
   - Remaining components

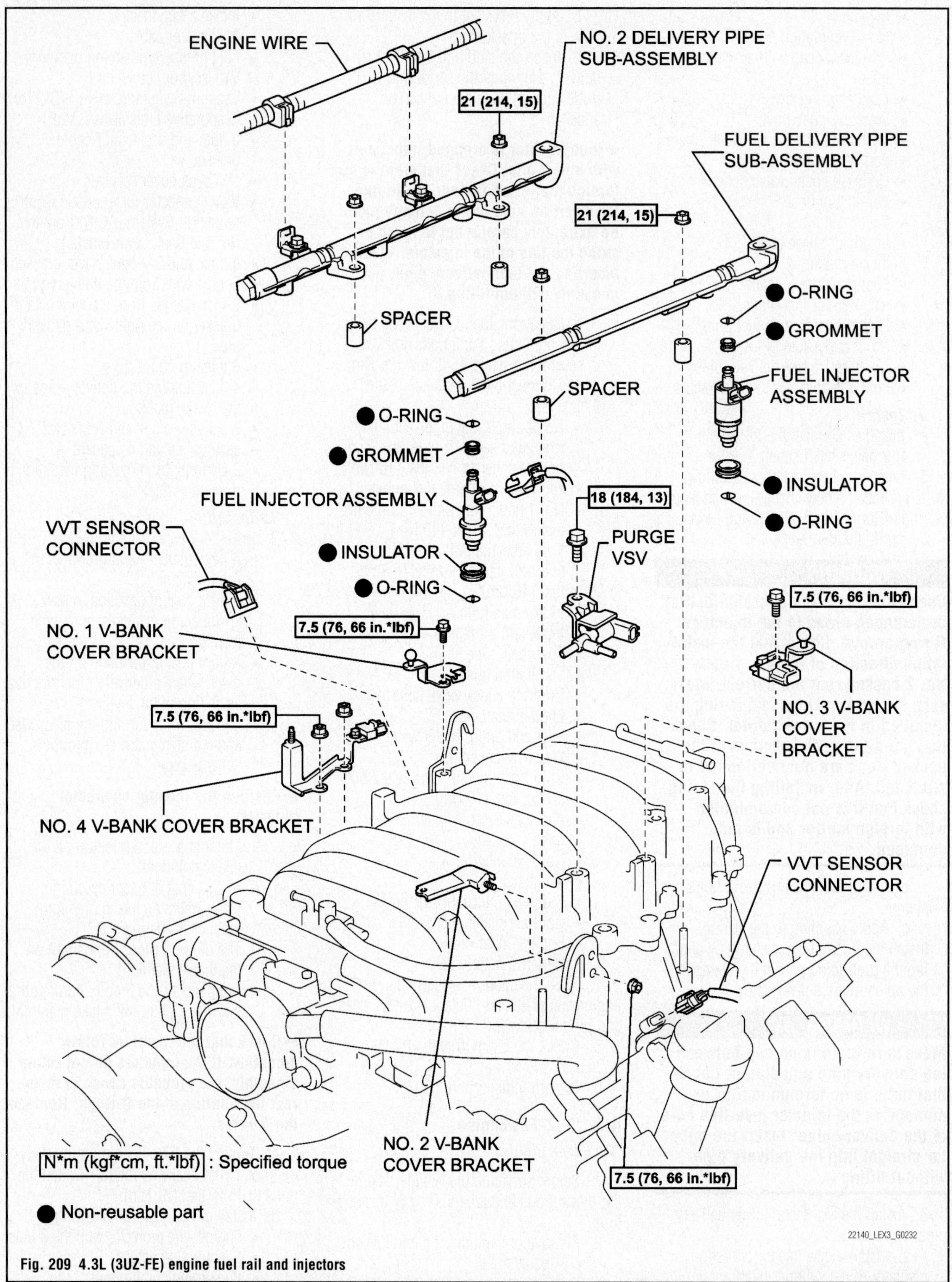

Fig. 209 4.3L (3UZ-FE) engine fuel rail and injectors

IDLE SPEED

ADJUSTMENT

Idle speed is maintained by the Engine Control Module (ECM). No adjustment is necessary or possible.

THROTTLE BODY

REMOVAL & INSTALLATION

3.0L (3GR-FSE) Engine

See Figure 210.

1. Before servicing the vehicle, refer to the precautions in the beginning of this section.
2. Disconnect the negative battery cable. Wait at least 90 seconds before performing any other work.
3. Using a clip remover, remove the 7 clips and the cool air intake duct seal.
4. Using a clip remover, remove the 3 clips, then remove the 10 screws and engine under cover.
5. Drain the engine coolant.
6. Loosen the nut and remove the 2 clips and right engine room side cover.
7. Remove the 2 nuts and V-bank cover.
8. Disconnect the No. 2 ventilation hose from the cylinder head.
9. Remove the air cleaner cap with air cleaner hose as follows:
    a. Disconnect the MAF meter connector.
    b. Disconnect the clamp from the air cleaner.
    c. Disconnect the VSV hose.
    d. Disconnect the 4 clamps.
    e. Remove the hose clamp and air cleaner cap with air cleaner hose.
10. Disconnect the 2 water by-pass hoses from the throttle body.

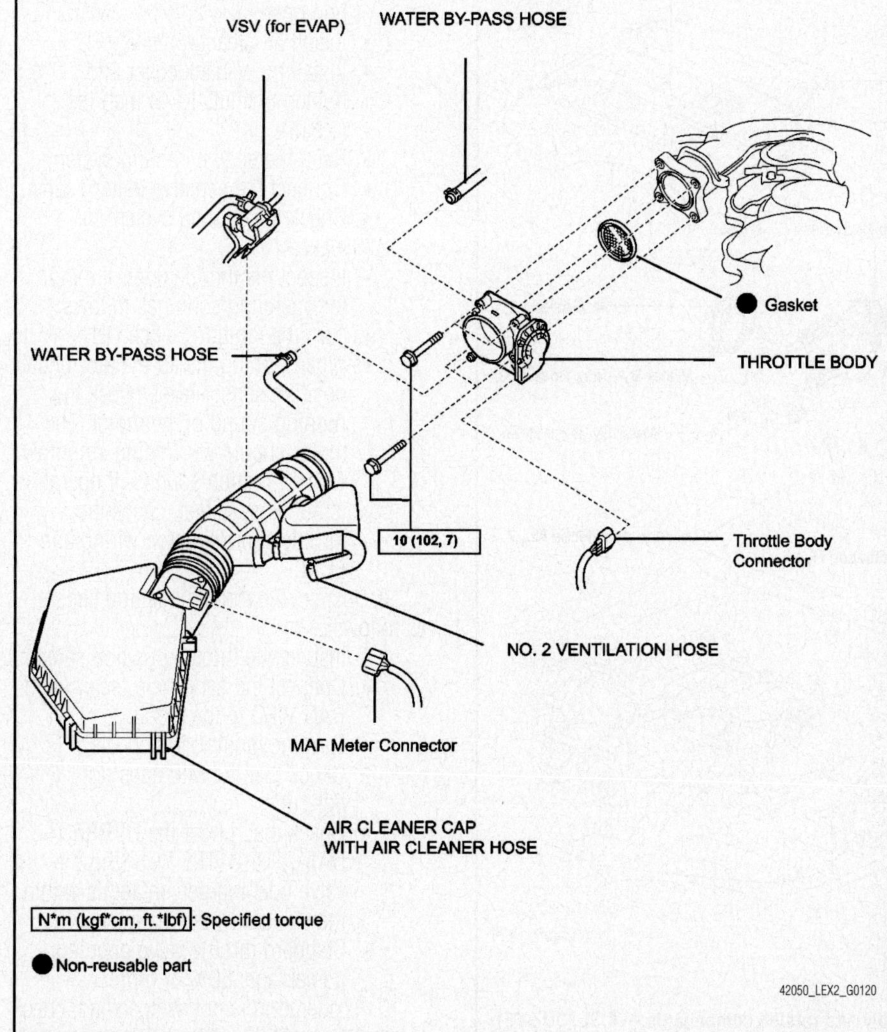

Fig. 210 Exploded view of the throttle body assembly and related components.—3.0L (3GR-FSE) engine

11. Remove the throttle body assembly as follows:
    a. Disconnect the throttle motor connector.
    b. Remove the 4 bolts, throttle body and gasket.

To install:

12. Install the throttle body assembly as follows:
    a. Install a new gasket to the intake air surge tank. Align the protrusion of the gasket on the intake air surge tank.
    b. Install the throttle body with the 4 bolts. Tighten the 4 bolts to 7 ft. lbs. (10 Nm).
    c. Connect the throttle motor connector.
13. Connect the 2 water by-pass hoses to the throttle body.
14. Install air cleaner cap with air cleaner hose as follows:
    a. Install the air cleaner cap with air cleaner hose assembly with the 4 clamps and hose clamp.
    b. Install the VSV hose to the air cleaner hose.
    c. Connect the MAF meter connector and clamp to the air cleaner.
15. Connect the No. 2 ventilation hose to the cylinder head cover with the clamp.
16. Refill the engine cooling system.
17. Check the cooling system for leaks.
18. Install the engine under cover with the 10 bolts and 3 clips.
19. Connect the negative battery cable.
20. Inspect the throttle control motor for operating sound as follows:
    a. Turn the ignition switch ON.
    b. When pressing the accelerator pedal, check the operating sound of the running motor. Make sure that no friction noises emit from the motor. If friction noise exists, replace the throttle body.
21. Inspect the throttle position sensor as follows:
    a. Connect the intelligent tester to the DLC3.
    b. Turn the engine switch on (IG).
    c. Push the intelligent tester main switch ON.
    d. Enter the following menus: DIAGNOSIS / ENHANCED OBD II / DATA LIST / THROTTLE POS AND THROTTLE POS #2.
    e. Depress the accelerator pedal. When the throttle valve is fully opened, check that the value of the "Throttle Sensor Position" is within the specification.
    - Standard throttle valve opening percentage: 60 percent or more

➡When checking the standard throttle valve opening percentage, the shift lever should be in the N position.

f. If the percentage is less than 60%, replace the throttle body.

22. Install the V-bank cover with the 2 nuts and tighten to 44 inch lbs. (5 Nm).
23. Install the right engine room side cover with the 2 clips and nut.
24. Install the cool air intake duct seal with the 7 clips.
25. Perform system initialization procedure.
26. After the engine is warmed up, check that the maintained parts operate normally.

4.3L (3UZ-FE) Engine

See Figure 211.

1. Before servicing the vehicle, refer to the precautions in the beginning of this section.
2. Disconnect the negative battery cable. Wait at least 90 seconds before performing any other work.
3. Drain the engine coolant.
4. Remove the 2 nuts and V-bank cover.
5. Remove the air cleaner inlet No. 1.
6. Remove the intake air connector pipe as follows:
    - Disconnect the air hose and ventilation hose.
    - Remove the bolt.
    - Loosen the 2 hose clamps and remove the intake air connector pipe.

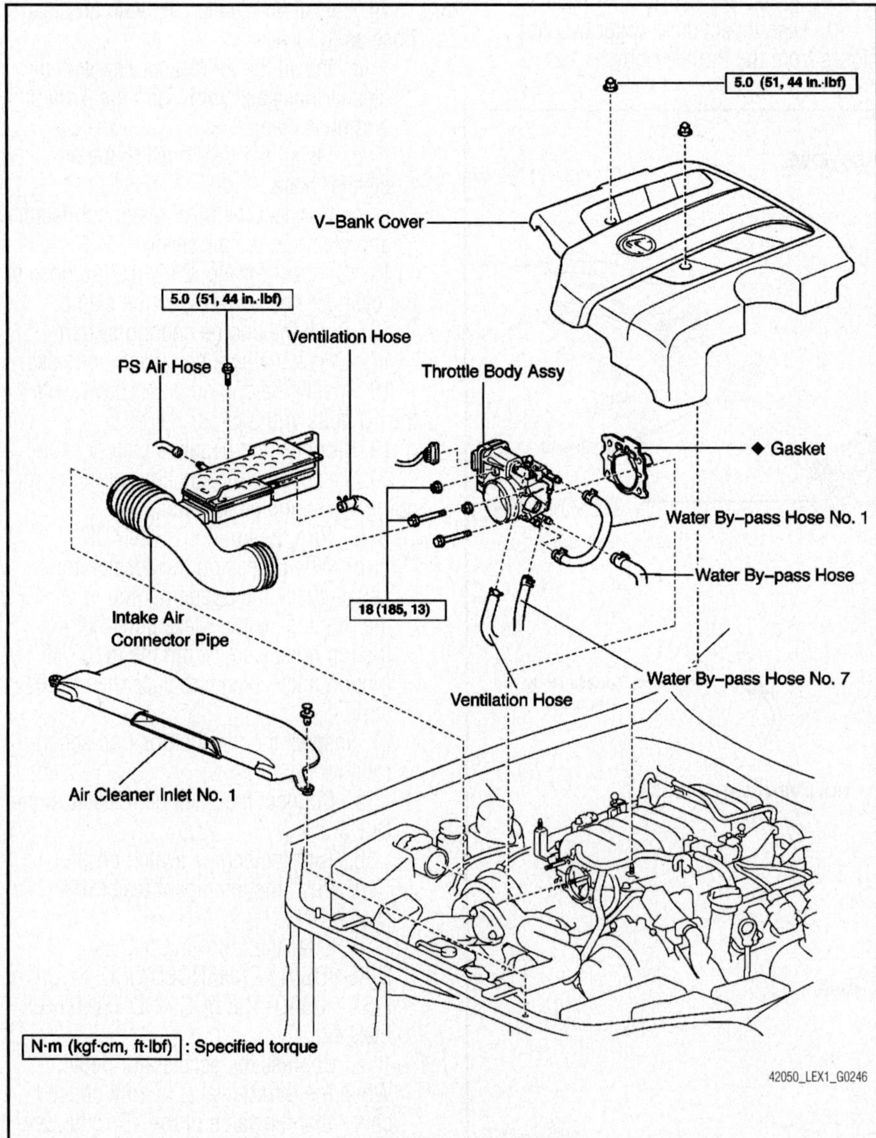

Fig. 211 Exploded view of the throttle body assembly and related components.—4.3L (3UZ-FE) engine

7. Remove the throttle body as follows:
    - Disconnect the water by-pass hose and water by-pass hose No. 7.
    - Disconnect the connector.
    - Remove the 2 bolts, 2 nuts, throttle body and gasket.

To install:

8. Install the throttle body as follows:
    - Install a new gasket and the throttle body with the 2 bolts and 2 nuts. Tighten the nuts and bolts to 13 ft. lbs. (18 Nm).
    - Connect the water by-pass hose and water by-pass hose No. 7.
    - Connect the connector.
    - Install intake air connector pipe as follows:
    - Install the intake air connector pipe with the bolt and 2 hose clamps. Tighten the hose clamps to 35 inch lbs. (4 Nm), and the bolt to 44 inch lbs. (5 Nm).
    - Connect the air hose and ventilation hose.
    - Install air cleaner inlet No. 1.
    - Install the V-bank cover and tighten the nuts to 44 inch lbs. (5 Nm).
    - Refill the engine cooling system.
    - Connect the negative battery cable.
    - Check the cooling system for leaks.
    - Inspect the throttle control motor for operating sound as follows:
    - Turn the ignition switch ON.
    - When turning the accelerator pedal position sensor lever, check the running sound of the motor. The motor should be running smoothly without friction sounds. If operation is not as specified, check the throttle control motor, wiring and ECM.
9. Inspect the throttle position sensor as follows:
    - Inspect the throttle position sensor.
    - Connect the hand-held tester (with CAN VIM) to the DLC3.
    - Turn the ignition switch ON.
    - Check that the MIL does not light up.
    - Check that, under the CURRENT DATA, THROTTLE POS (throttle valve opening percentage)is within the standard value below.
    - Standard throttle valve opening percentage: 60% or more
10. If operation is not as specified, check the throttle position sensor, wiring and ECM.

HEATING & AIR CONDITIONING SYSTEM

BLOWER MOTOR

REMOVAL & INSTALLATION

GS 300

See Figures 212 and 213.

1. Before servicing the vehicle, refer to the precautions in the beginning of this section.
2. Disconnect the negative battery cable. Wait 90 seconds before doing any further work while the airbag system de-energizes.
3. Remove the instrument panel assembly.
4. Remove the air conditioner unit assembly. (Refer to heater core R&I)
5. Remove the blower assembly as follows:
   - Disconnect the connector.
   - Remove the screw and nut.
   - Release the 2 claws and remove the blower assembly.
6. Remove the damper servo sub-assembly as follows:
   - Detach the claw and remove the lever.
   - Remove the 2 screws and damper servo.
   - Remove the 3 screws and the blower motor.

To install:

7. Install the blower motor to the blower unit assembly and tighten the 3 screws.
8. Install the damper servo sub-assembly as follows:
   - Install the damper servo with the 2 screws.
   - Attach the claw and install the lever.
   - Install the blower assembly as follows:
   - Install the blower assembly with the 2 claws, screw and nut. Tighten the screw to 27 inch lbs. (3 Nm); tighten the nut to 7 ft. lbs. (10 Nm).
   - Connect the wiring connector.
   - Install the air conditioner unit assembly. See HEATER CORE Removal & Installation procedure in this section.

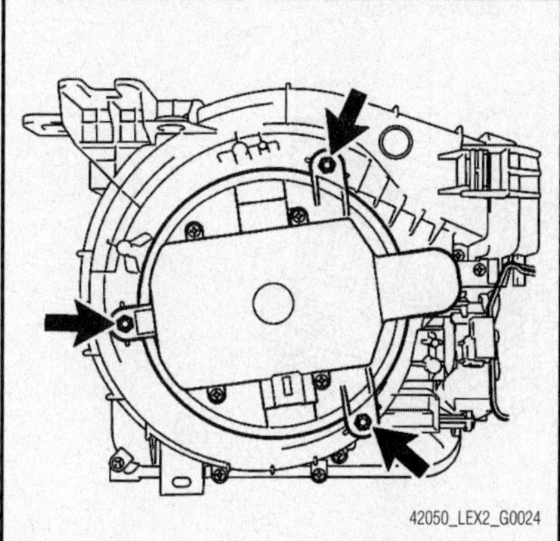

Fig. 213 Blower motor screw locations—2006 GS 300 and 2006 GS 430

   - Install the instrument panel assembly.
   - Connect the negative battery cable.

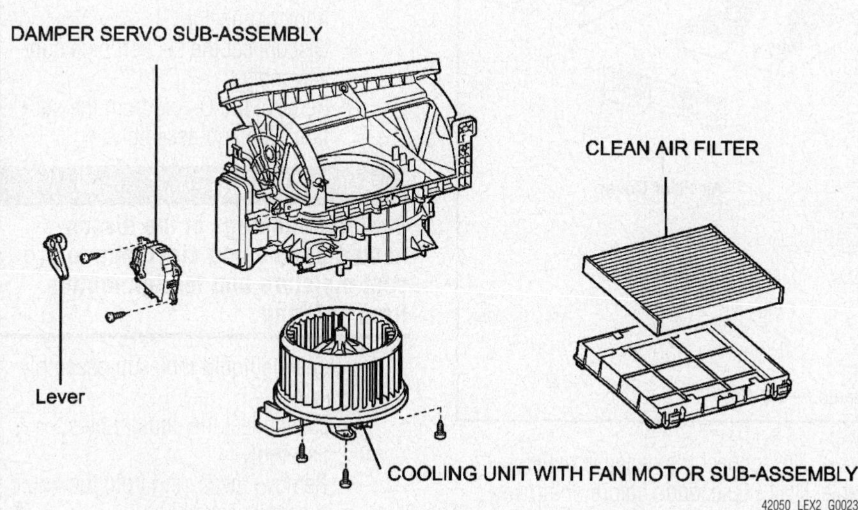

Fig. 212 Exploded view of the blower assembly components—2006 GS 300

SC 430

See Figures 214 and 215.

1. Before servicing the vehicle, refer to the precautions in the beginning of this section.
2. Disconnect the negative battery cable. Wait 90 seconds before doing any further work while the airbag system de-energizes.
3. Remove the instrument panel assembly.
4. Remove the 2 screws and air duct No. 2.
5. Remove the blower assembly as follows:
   - Disconnect the connector and clamp from the blower.
   - Using a screwdriver, detach the claw and disconnect the connector.
   - Remove the 3 screws, bolt and nut.
   - Detach the claw and remove the blower.
   - Remove the 3 screws and recirculation damper servo.
   - Remove the 3 screws and blower w/ fan motor.

To install:

6. Install blower w/ fan motor.
7. Install the recirculation damper servo.
8. Install the blower assembly.
9. Install air duct No. 2.
10. Install the instrument panel assembly.
11. Connect the negative battery cable.

HEATER CORE

REMOVAL & INSTALLATION

GS 300

See Figures 216 and 217.

1. Before servicing the vehicle, refer to the precautions in the beginning of this section.
2. Discharge the A/C system.
3. Set radio receiver assembly to shipment mode as follows:
   a. Be sure that all discs and tapes have been removed from the unit.
   b. Be sure that the engine switch off.
   c. While simultaneously pressing the "SEEK UP" and "DISC" switches, turn the engine switch on (ACC).

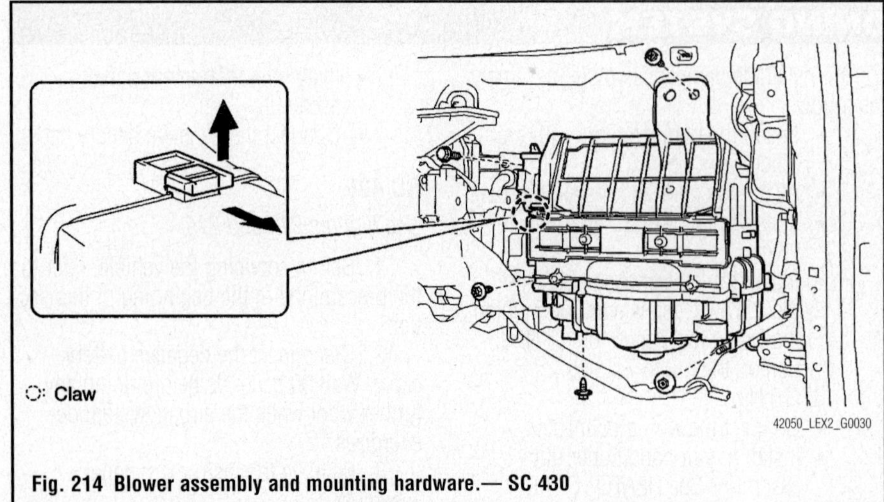

Fig. 214 Blower assembly and mounting hardware.— SC 430

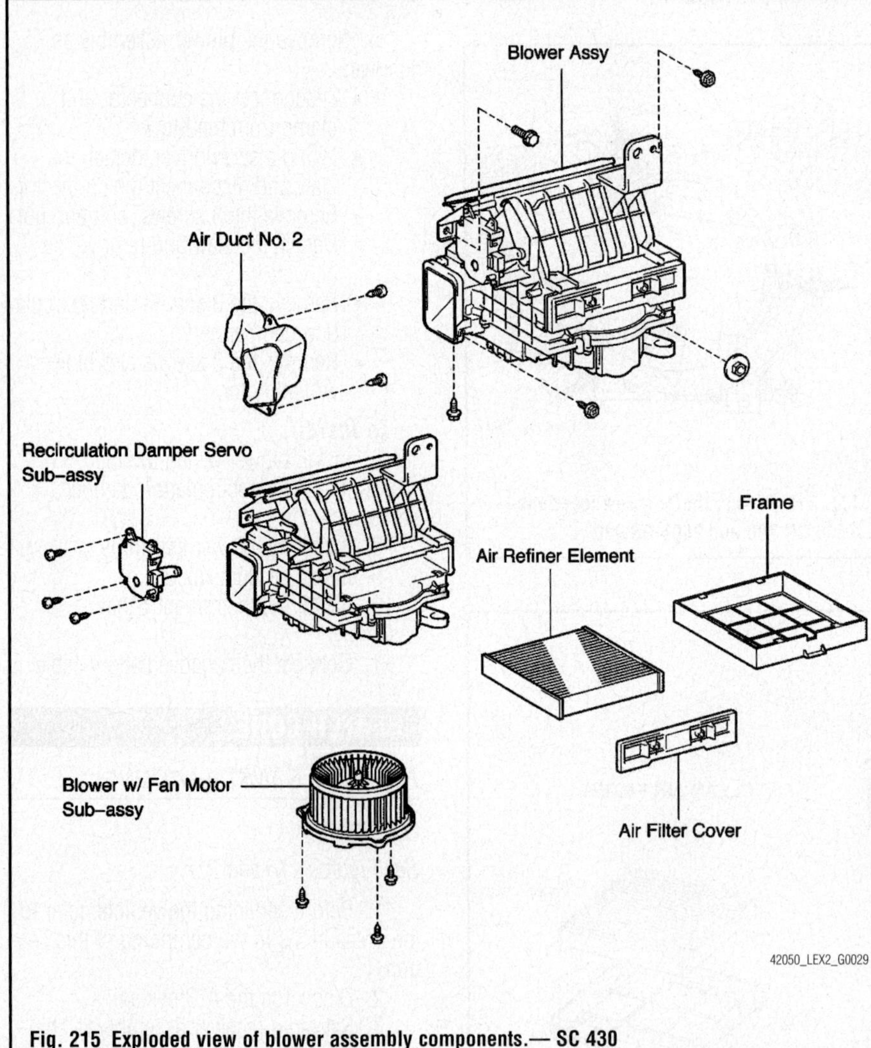

Fig. 215 Exploded view of blower assembly components.— SC 430

➡The CD loading door indicator light blinks during mode setting and it remains lit after the setting is completed.

d. Turn the engine switch off.

4. Disconnect the negative battery cable. Wait 90 seconds before doing any further work while the airbag system de-energizes.

5. Drain the cooling system into a clean container for reuse.

6. Align the front wheels facing straight ahead.

7. Remove or disconnect the following:

- Cool air intake duct seal
- Left and right engine room side covers
- Left and right front pillar to front side seals, using a clip remover to detach the 3 claws
- Left and right nut, windshield wiper arms and blades
- Left and right front fender to cowl side seals by moving the component toward the center of the vehicle to detach the 2 claws

8. Remove cowl top ventilator louver assembly as follows:

a. Remove the 2 clips and detach the 5 claws.

b. Pull the ventilator louver in the direction indicated by the arrow in the illustration to detach the 10 claws and remove the ventilator louver.

9. Remove the windshield wiper motor and link assembly as follows:

- Disconnect the connector. Then detach the 2 clamps and remove the wire harness from the cowl top panel.

➡There are 6 bolts total, however, 2 bolts cannot be removed from the wiper motor and link because they are integrated into the wiper motor and link.

- Remove the 4 bolts and wiper motor and link.

10. Separate suction pipe sub-assembly as follows:

- Remove the bolt, and slide the hook connector.
- Disconnect the suction pipe sub-assembly.
- Remove the O-ring from the suction pipe sub-assembly.

⁂ WARNING

Seal the openings of the disconnected parts using vinyl tape to prevent moisture and foreign matter from entering.

11. Separate liquid tube sub-assembly as follows:

- Disconnect the liquid tube sub-assembly.
- Remove the O-ring from the liquid tube subassembly.
- Inlet and outlet heater water outlet hoses
- Instrument panel assembly.

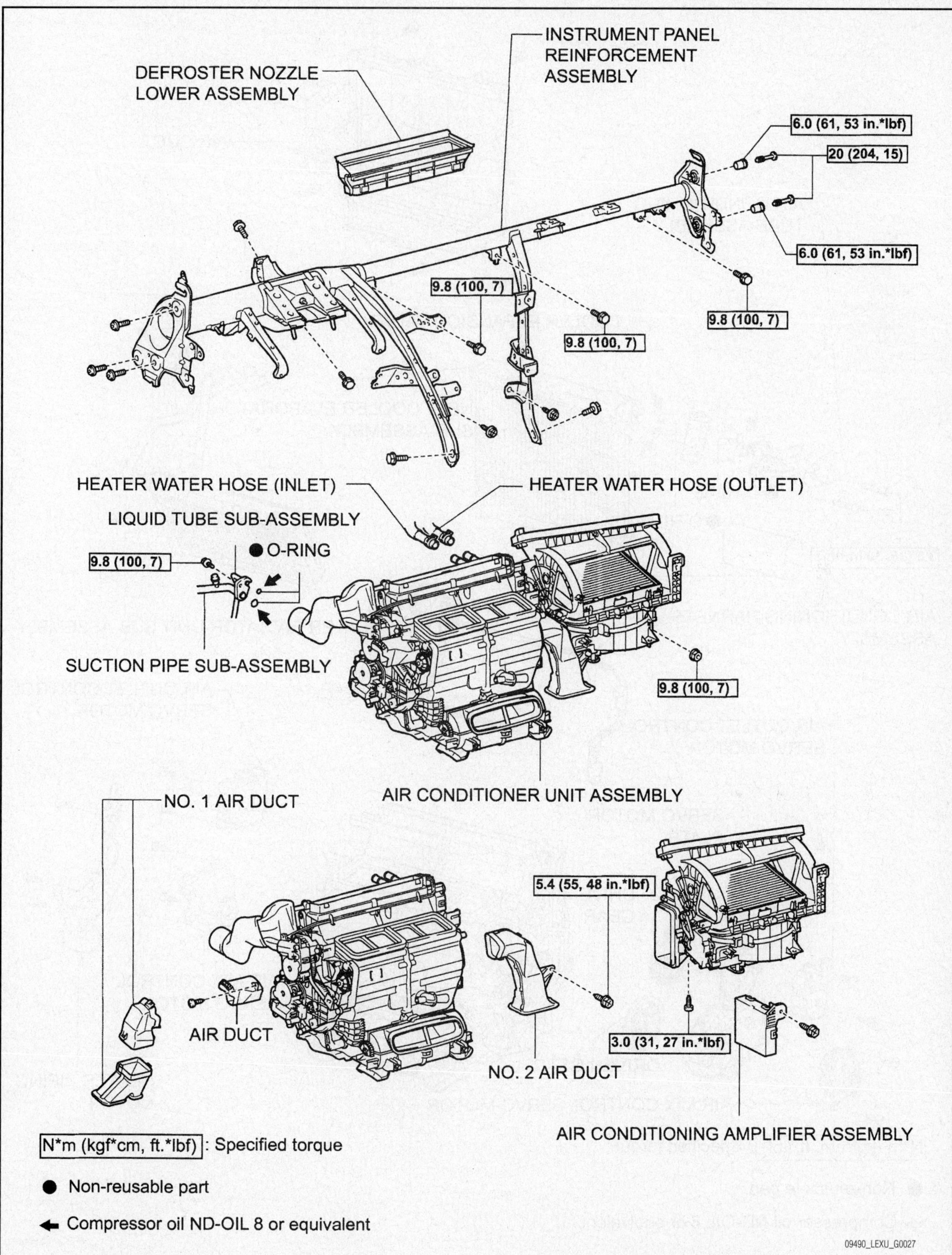

Fig. 216 Exploded view of the A/C unit assembly, instrument panel reinforcement and related components—GS 300 Model

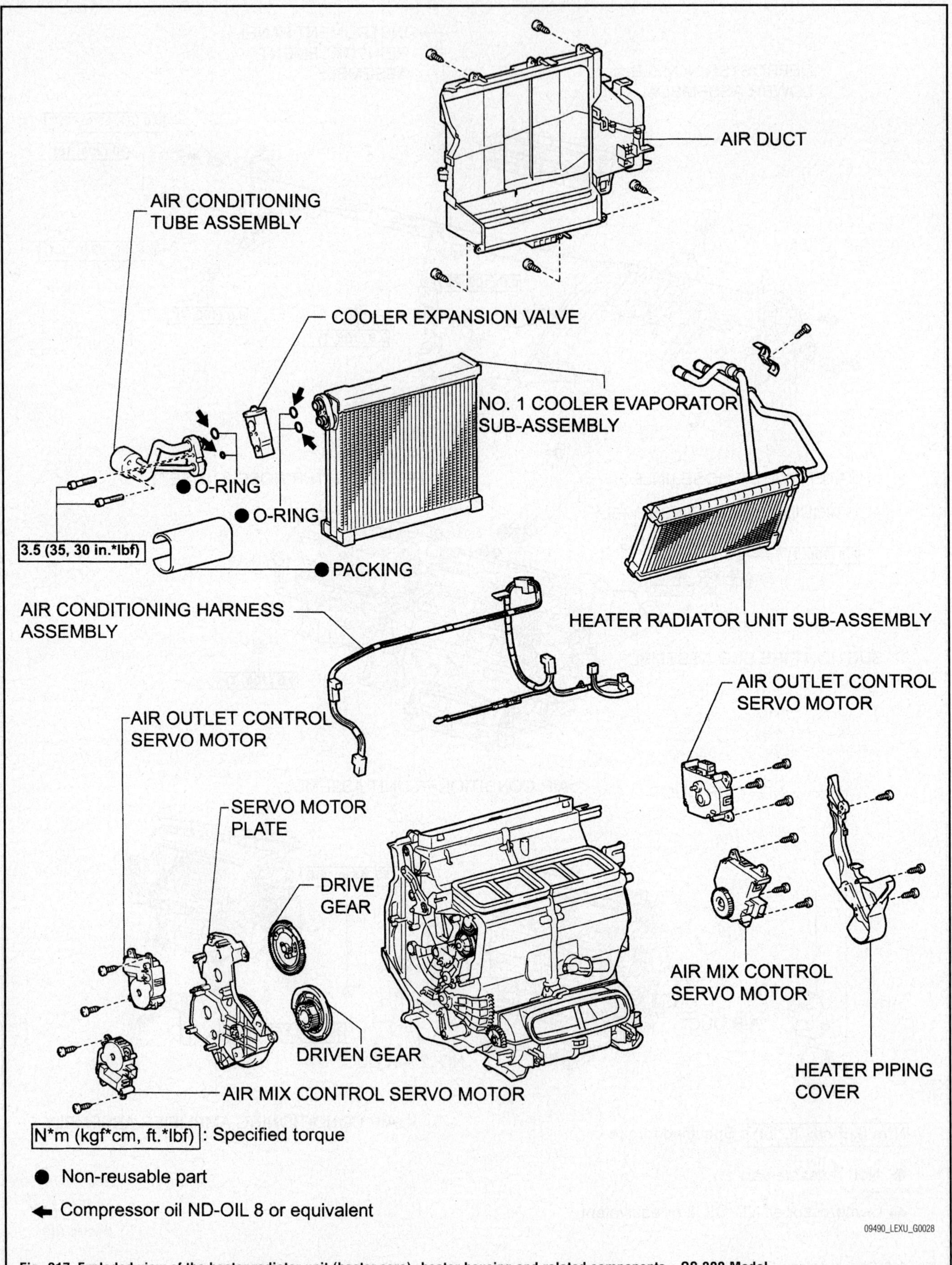

Fig. 217 Exploded view of the heater radiator unit (heater core), heater housing and related components—GS 300 Model

- A/C blower unit assembly
- Mounting screw and No. 2 air duct
- Upper and lower foot ducts
- Connector, mounting screw and air conditioning amplifier assembly

12. Remove the heater core from the A/C-blower assembly unit as follows:

a. Disconnect the wiring connector, remove the 3 screws and right side air outlet control servo motor.

b. Disengage 3 clamps, connector and wire harness to right side air mix control servo motor.

c. Remove the 2 screws and heater piping cover.

d. Remove the 3 screws and right side air mix control servo motor.

e. Move the A/C wiring harness out of the way of the heater radiator unit.

f. Remove the heater radiator unit sub-assembly (heater core)

To install:

13. Install or connect the following:
    - Heater core to the A/C blower housing
    - A/C wiring harness back into position
    - Right side air mix control servo motor
    - Heater piping cover
    - Right side air outlet control servo motor
    - Air conditioning amplifier assembly
    - Upper and lower foot ducts
    - No. 2 air duct
    - A/C blower unit assembly. Torque the retaining nut to 7 ft. lbs. (10 Nm).
    - Instrument panel assembly.
    - Inlet and outlet heater water outlet hoses
14. Install the liquid tube sub-assembly as follows:
    - Remove the vinyl tape attached to the tube.
    - Sufficiently apply compressor oil to a new O-ring and the fitting surface of the liquid tube.
    - Install the O-ring on the liquid tube.
    - Install the liquid tube to the fitting hole.
15. Install the suction pipe sub-assembly as follows:
    - Remove the vinyl tape attached to the pipe.
    - Sufficiently apply compressor oil to a new O-ring and the fitting surface of the suction pipe.
    - Install the O-ring on the suction pipe.
    - Move the hook connector in a counterclockwise direction.
    - Insert the pipe joints into the fitting holes securely and tighten the bolt to 7 ft. lbs. (10 Nm).
16. Install the following:
    - Wiper motor and link assembly
    - Cowl top ventilator louver assembly
    - Left and right front fender to cowl side seals
    - Windshield wiper arms and blades
    - Left and right front pillar to front side seals
    - Left and right engine room side covers
    - Cool air intake duct seal
17. Perform system initialization procedure.
18. Refill the cooling system.
19. Connect the negative battery cable.
20. Evacuate, charge and leak test the air conditioning system refrigerant.
21. Operate the engine to normal operating temperatures; then, check the climate control operation and check for leaks.

SC 430

See Figures 218 and 219.

➡Removal of the heater core requires removal of the entire heater air conditioning assembly.

1. Before servicing the vehicle, refer to the precautions in the beginning of this section.
2. Drain the cooling system into a clean container for reuse.
3. Discharge and recover the air conditioning system refrigerant.
4. Remove or disconnect the following:
    - A/C Suction and pressure hose bolt and plate at firewall

➡Do Not use pry tools to separate. Cap the end to prevent system contamination

- Heater hoses from the heater core

5. Set CD changer to ship mode setting using following procedure
    - Remove all CDs
    - Simultaneously press "Seek Up" and "Disc" while turning the ignition switch to "Acc"

➡When mode setting is complete "Ship" appears on the display

6. Disconnect the negative battery cable. Wait 90 seconds before doing any further work while the airbag system de-energizes.
7. Center the front wheels and steering wheel.
8. Remove the instrument panel assembly.
9. Remove the air conditioning unit assembly as follows:
    - Wire harness clamps and disconnect wire harness from air conditioning unit assembly
    - 2 screws, 3 nuts, bolt and air conditioning unit
10. Remove the heater bracket.
11. Remove the air conditioning evaporator assembly as follows:
    - Connector and clamp for the blower assembly
    - Release claw and disconnect the connector from the bracket
    - 2 screws and release the claw or the evaporator assembly
    - evaporator assembly
12. Disconnect the cooler thermistor hose.
13. Remove air duct No.1, 2, 3 and 4.
14. Remove heater core unit sub-assembly as follows:
    - Screw and clamp
    - Heater core assembly from A/C assembly

To install:

15. Install the heater core unit sub-assembly as follows:
    - Heater core into A/C assembly
    - Screw and clamp
16. Install air duct No. 1,2, 3 and 4.
17. Connect the cooler thermistor hose.
18. Install the air conditioning unit assembly as follows:
    - 2 screws and 3 nuts and air conditioning unit
    - Bolt to air conditioning unit assembly. Torque: 87 inch lbs. (9.8 Nm)
    - Clamps to the air conditioner unit assembly
19. Install the instrument panel assembly.
20. Connect the heater core hoses.
21. Connect the A/C suction and pressure hoses, attach with bolt and plate.

➡Lubricate O-rings with compressor oil

22. Connect the negative battery cable.
23. Fill cooling system with coolant.
24. Evacuate and recharge A/C system.
25. Warm up engine and inspect for coolant leaks.

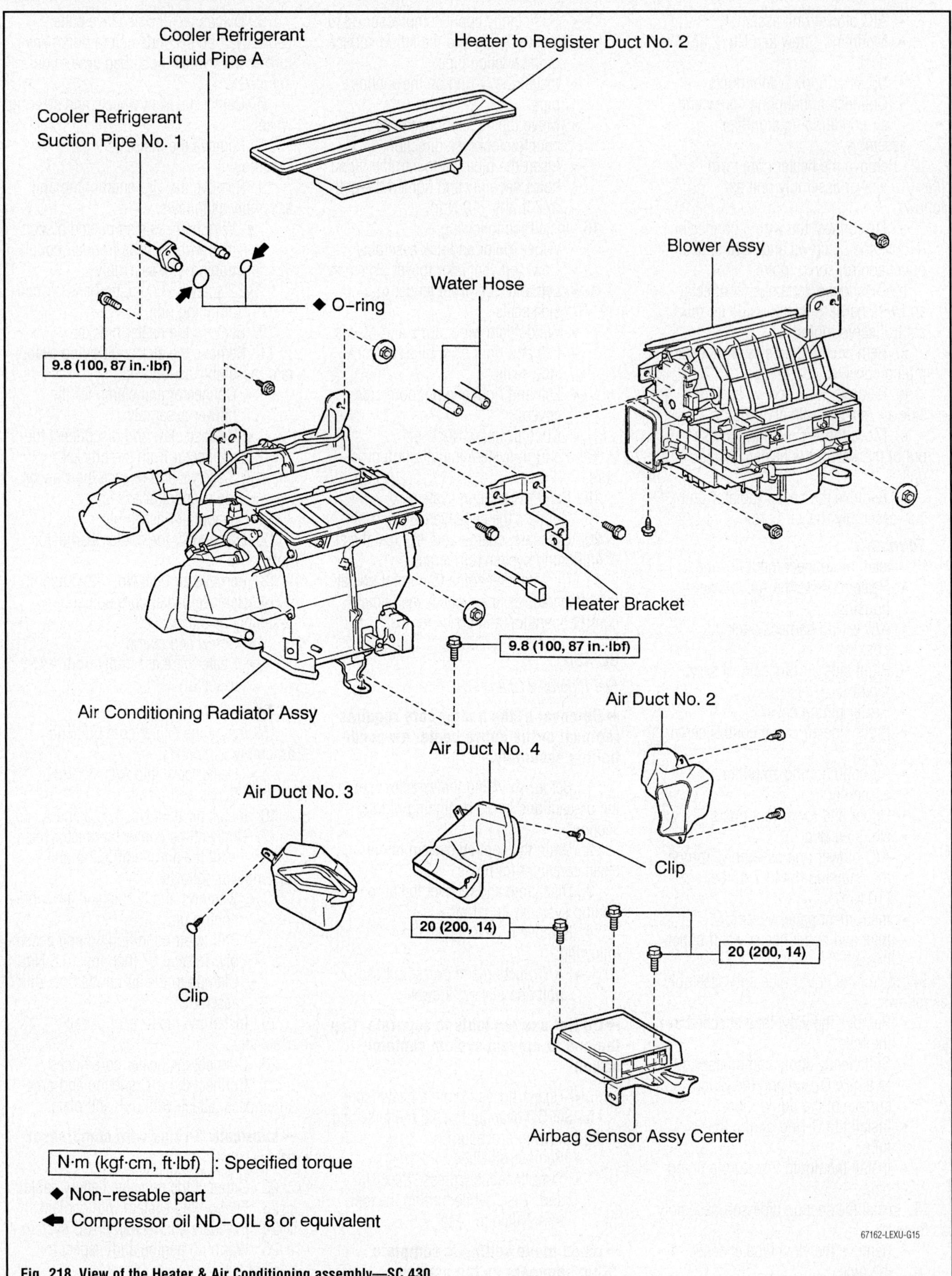

Fig. 218 View of the Heater & Air Conditioning assembly—SC 430

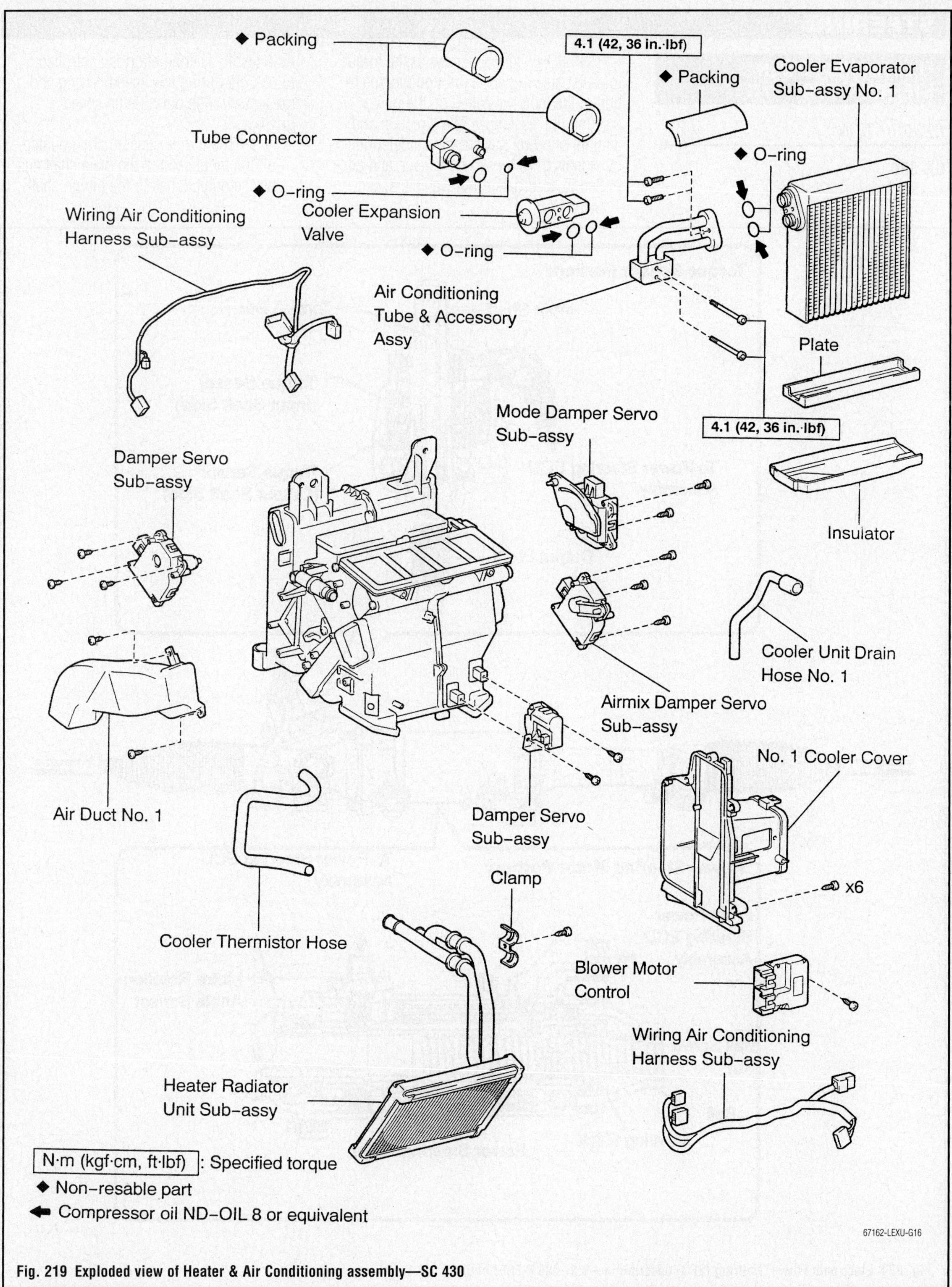

Fig. 219 Exploded view of Heater & Air Conditioning assembly—SC 430

STEERING

ELECTRONIC POWER STEERING (EPS)

DESCRIPTION

GS 300

See Figure 220.

The EPS system generates assist torque to assist steering effort through the operation of the motor installed on the power steering link assembly. The direction and amount of power assistance are determined by signals from the torque sensor, and controlled in accordance with vehicle speed. As a result, steering effort is controlled to be light during low speed driving and moderately high during high speed driving.

a. The torque sensors are mounted on the input shaft to the main shaft and on the output shaft to the pinion shaft.

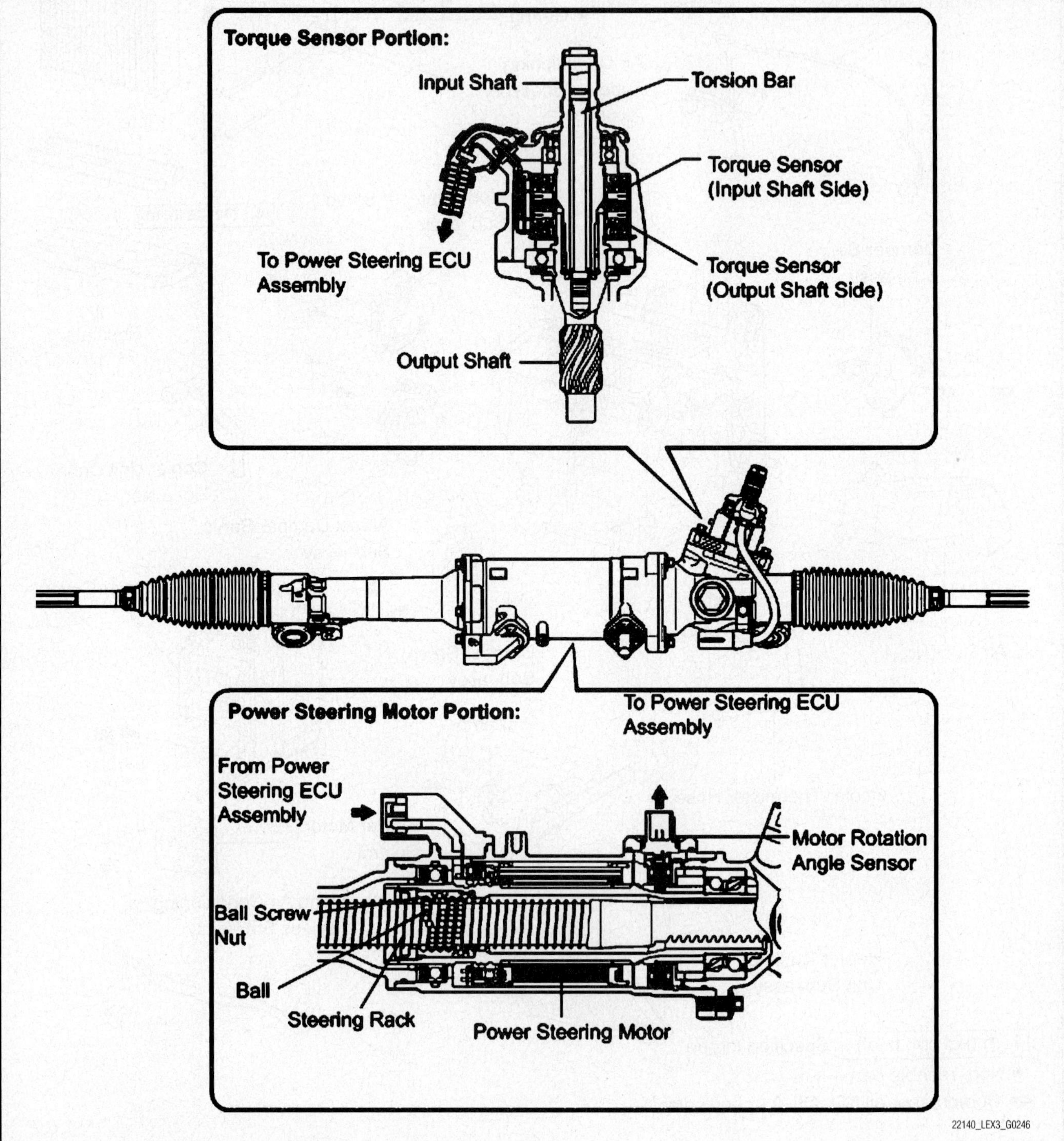

Fig. 220 Electronic Power Steering (EPS) description—3.0L (3GR-FSE) engine

The input and output shafts are joined by a torsion bar.

b. If the steering wheel is turned, the torsion bar twists, resulting in a difference in the rotation angle detected by each torque sensor. The power steering ECU assembly calculates torque based on this difference.

c. The power steering ECU assembly calculates proper assisting torque, according to vehicle speed, based on the torque obtained in step (b). Then the ECU controls the motor drive circuit in order to cause it to generate assisting torque.

d. A ball screw nut is attached to the motor shaft. It converts the rotational motion of the motor into the rectilinear motion of the steering rack via the ball.

e. The assisting force generated in the above process will reduce the steering effort required by the driver

POWER RACK & PINION STEERING GEAR

REMOVAL & INSTALLATION

GS 300

See Figures 221 through 224.

➡For the (4WD) model the engine and transmission assembly must be removed.

1. Before servicing the vehicle, refer to the precautions in the beginning of this section.
2. Set front wheels in straight-ahead position, then remove the front wheels.
3. Disconnect the negative battery cable and wait for at least 90 seconds before proceeding.
4. Remove the front wheels.
5. Remove the engine under covers.
6. Remove the 4 bolts and the front suspension member protector lower.
7. Fix the steering wheel with the seat belt or wheel holder in order to prevent rotation.

➡This operation is useful to prevent damage to the spiral cable.

8. Loosen bolt (A) and remove bolt (B), then slide the steering intermediate shaft assembly No.2.

➡ Do not remove bolt (A). Do not disconnect the steering intermediate shaft assembly No.2 from the power steering link assembly.

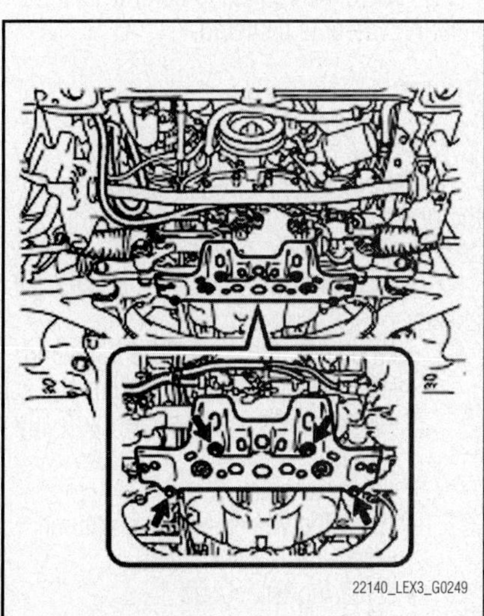

Fig. 221 Remove the front suspension member protector

22140_LEX3_G0250

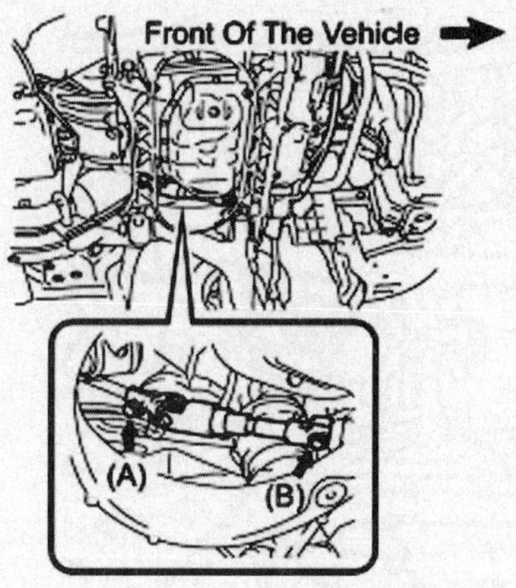

Fig. 222 Loosen bolt (A) and remove bolt (B)

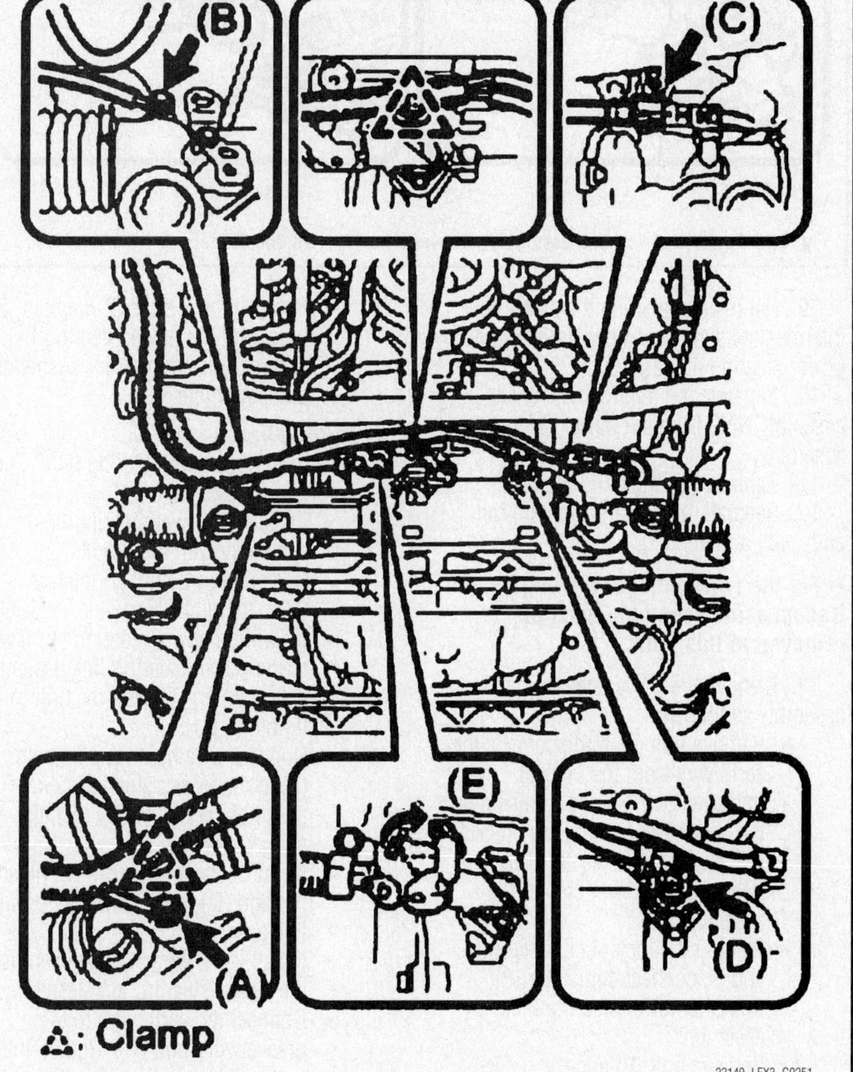

Fig. 223 Power steering link assembly removal view (1)—GS 300 Model

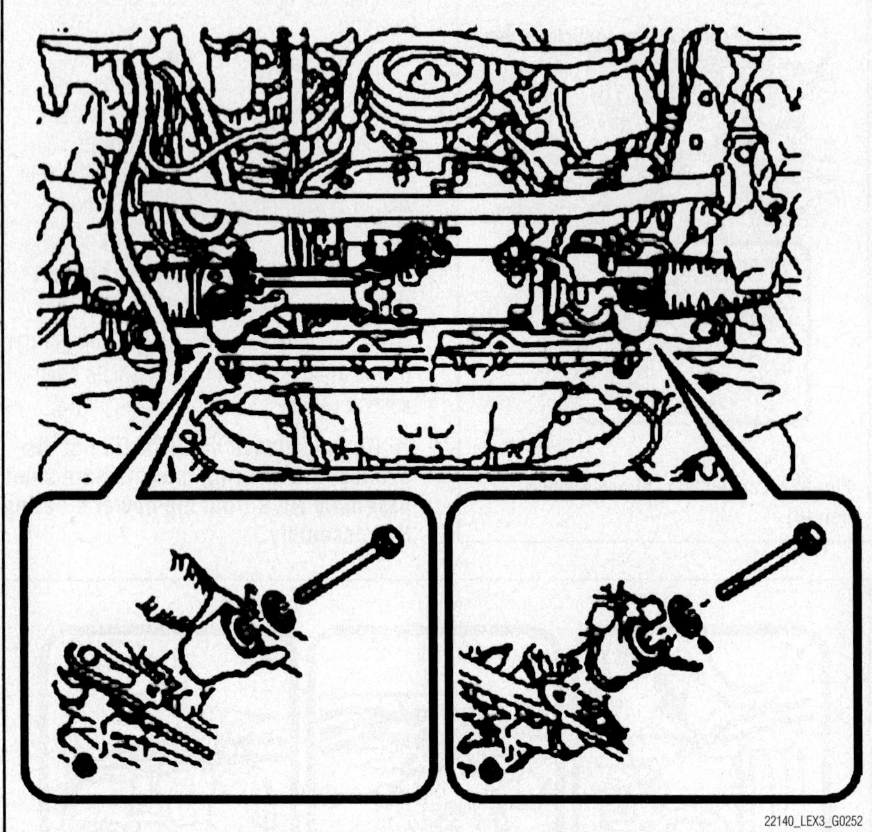

Fig. 224 Power steering link assembly removal view (2)—GS 300 Model

9. Put matchmarks on the steering intermediate shaft assembly No.2 and the power steering link assembly.

10. Separate the intermediate shaft assembly No.2 from the power steering link assembly.

11. Remove the clips and the castle nuts.

12. Separate the left and right tie rod ends with a puller.

➡**For the (4WD) model the engine and transmission assembly must be removed at this point.**

13. Remove the power steering link assembly as follows:

- Remove bolt (A) to disconnect the earth wire from the bracket.
- Remove the 2 clamps to disconnect the wire harness from the bracket.
- Disconnect 2 connectors (C) and (D) from the power steering link assembly.
- Release the lock of connector (E) and disconnect connector (E) from the power steering link assembly
- Remove bolt (B) and the power steering earth wire from the power steering link assembly.
- Remove the 2 bolts, 2 washers, 2 nuts, and the power steering link assembly from the front suspension cross member.

To install:

14. Install the power steering link assembly as follows:

- Install the power steering link assembly with the 2 bolts, 2 wasters and 2 nuts. Tighten to 87 ft. lbs. (118 Nm).
- Install the power steering earth wire to the power steering link assembly with bolt (B). Tighten the bolt to 44 ft. lbs. (5.0 Nm).
- Connect wire harness connector (E) to the power steering link assembly and securely lock the connector.
- Connect 2 wire harness connectors (C) and (D) to the power steering link assembly.
- Install the 2 wire harness clamps to the power steering link assembly.
- Connect the earth wire to the bracket with bolt (A). Tighten the bolt to 71 inch. lbs. (8.0 Nm).

15. Connect both the tie rod ends to the steering knuckle with the nut. Tighten the castle nut to 50 ft. lbs. (65 Nm). Install a new clip.

16. Align the matchmarks on the intermediate shaft assembly No.2 and the power steering link assembly.

17. Install bolt (A) and tighten the 2 bolts to 26 ft. lbs. (35 Nm).

18. Install the front suspension protector and tighten to 71 inch. lbs. (8.0 Nm).

19. Install the engine under covers.

20. Install the front wheels and tighten to 76 ft. lbs. (103 Nm).

21. Connect the negative battery cable.

22. Place the front wheels straight ahead.

23. Adjust the front wheel alignment.

24. Initialize rotation angle and calibrate torque sensor zero point as follows:

25. Clear the rotation angle sensor calibration value, initialize the rotation angle sensor, and calibrate the torque sensor zero point if any of the following has occurred:

- The power steering ECU assembly has been replaced.
- The power steering link assembly has been replaced.
- Steering effort differs between left and right.

26. Inspect the following before calibration as follows:

- Connect the intelligent tester to the DLC3.
- Turn the engine switch on (IG).
- Turn the intelligent tester on.
- Select "IG SUPPLY" from the data list.
- Check the IG power supply voltage on the tester screen.

27. Voltage reading should be between 10 and 14 volts.

28. Drive the vehicle to confirm that the steering wheel is centered.

SC 430

See Figure 225.

1. Before servicing the vehicle, refer to the precautions in the beginning of this section.

2. Place the front wheels facing straight ahead.

3. Remove or disconnect the following:

- Negative battery cable. Wait at least 90 seconds before performing any work.
- Front wheels
- Steering wheel switch & volume case
- Steering wheel pad
- Steering wheel column lower cover
- Horn button
- Steering wheel

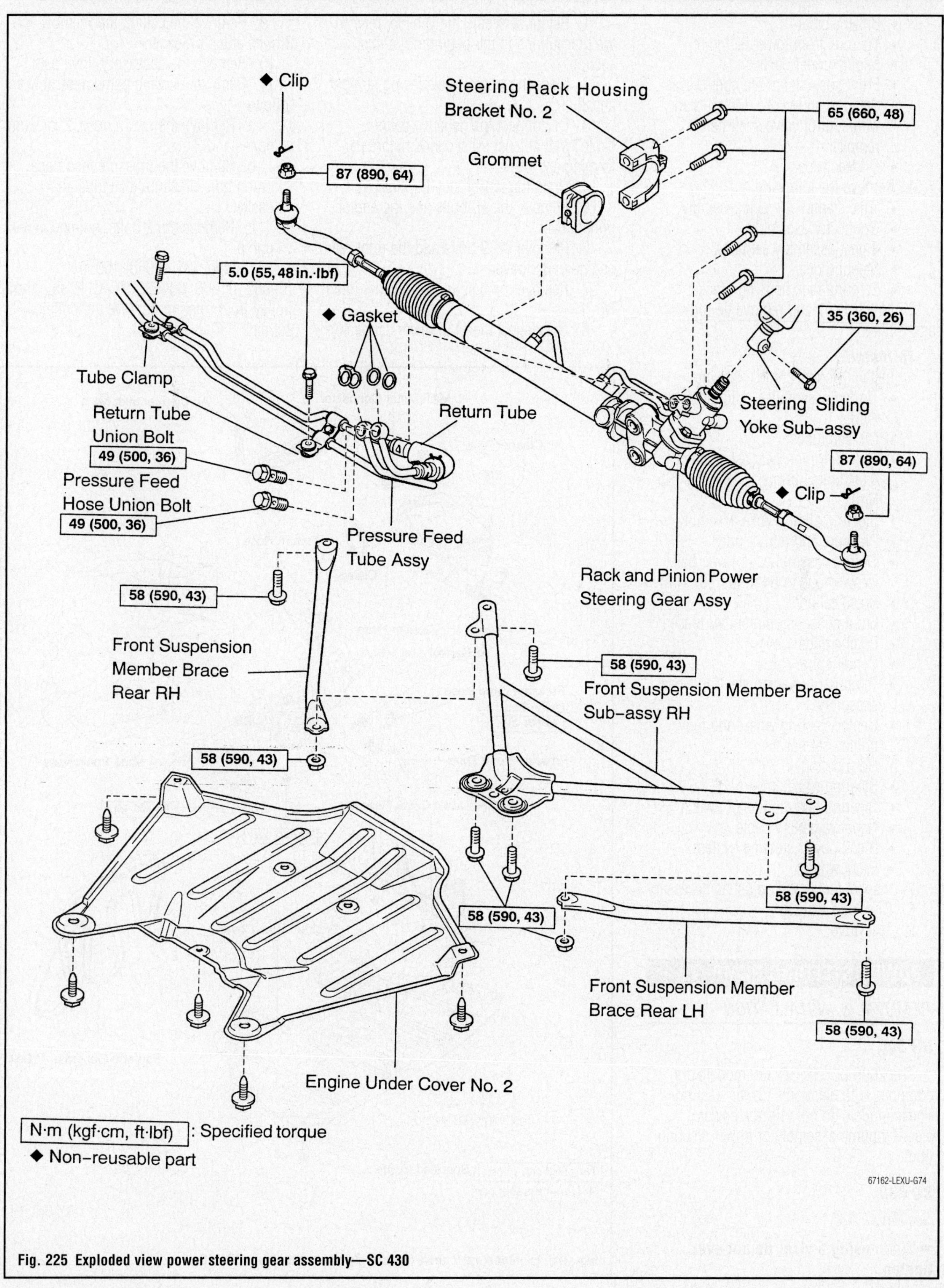

Fig. 225 Exploded view power steering gear assembly—SC 430

- Brake caliper
- Tie rods from lower ball joint
- Engine under cover
- Front suspension member braces
- Steering slide yoke. Match-mark with control valve shaft before removal
- Oil feed tubes

4. Remove the following:
   - Tube clamps on rack assembly
   - Connector assembly
   - 4 gear assembly set bolts
   - Steering gear
   - Steering rack housing bracket
5. Match-mark and remove tie rod assemblies

To install:

6. Install or connect the following:
   - Tie rod assemblies. Align match marks
   - Oil feed tubes
   - Rack & Pinion gear assembly with 4 set bolts; torque to 48 ft. lbs (65 Nm)
   - Steering sliding yoke assembly
   - Pressure and return tubes
   - Tie rod assemblies to lower ball joint; torque to 64 ft. lbs. (87 Nm)
   - Brake calipers
   - Front suspension member braces
   - Engine under cover
   - Center spiral cable
   - Temporarily tighten steering wheel assembly
   - Center steering wheel and fully tighten set nut
   - Horn button
   - Steering wheel covers
   - Steering pad modulator switch
   - Negative battery cable
   - Bleed power steering system
   - Front wheels
7. Inspect toe in and adjust as necessary
8. Check for leaks.
9. Test drive

POWER STEERING PUMP

REMOVAL & INSTALLATION

GS 300

The 2006 LEXUS GS 300 models are equipped with electronic power steering. These models do not utilize a power steering pump assembly or power steering fluid.

SC 430

See Figure 226.

➡**When using a vise, do not over tighten.**

1. Before servicing the vehicle, refer to the precautions in the beginning of this section.
2. Place the front wheels facing straight ahead.
3. Disconnect the negative battery cable. Wait at least 90 seconds before performing any work.
4. Drain the power steering fluid.
5. Remove the 16 bolts and the engine under cover.
6. Remove the 2 bolts and the right rear engine under cover.
7. Remove the bolt and air cleaner inlet No. 1.
8. Disconnect the MAF meter connector.
9. Remove the clamp, then remove the 3 bolts and air cleaner.
10. Remove the accessory drive belt.
11. Remove the vane pump assembly as follows:
    a. Remove the return hose, 2 vacuum hoses and 3 clips.
    b. Remove the pressure feed hose union bolt, pressure feed tube and gasket.
    c. Remove the 2 bolts, nut and vane pump.
12. Using SST 09960_10010 (09962_01000, 09963_01000) to stop the pulley from rotating, remove the pulley set nut.

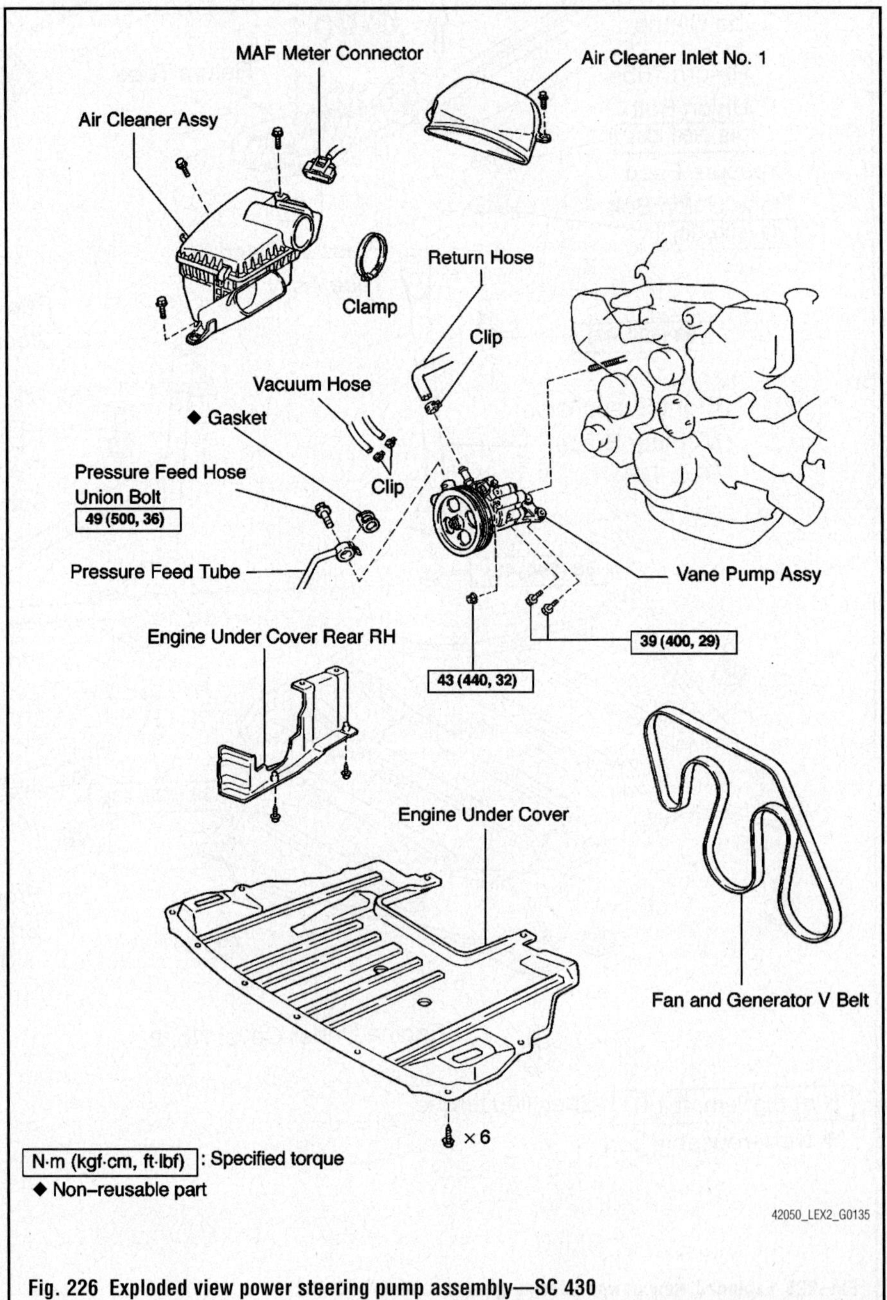

Fig. 226 Exploded view power steering pump assembly—SC 430

To install:

➡When installing the parts indicated by arrows, coat them with power steering fluid.

13. Install the vane pump pulley as follows:
    a. Install the vane pump pulley to the vane pump shaft.
    b. Using SST 09960-10010 (09962-01000, 09963-01000), stop the vane pump pulley rotation and install the nut.
    c. Tighten the nut to 32 ft. lbs. (43 Nm).
14. Install the vane pump assembly as follows:
    a. Install the power steering vane pump with the nut. Tighten the nut to 32 ft. lbs. (43 Nm).
    b. Install the 2 bolts and tighten them to 29 ft. lbs. (39 Nm).
    c. Install the pressure feed tube and a new gasket to the pressure feed hose union bolt. Tighten the union bolt to 36 ft. lbs. (49 Nm).
    d. Connect the return hose with the clip.
    e. Connect the 2 vacuum hoses with the 2 clips.
15. Install the accessory drive belt.
16. Install the air cleaner assembly with the 3 bolts and install the clamp.
17. Connect the MAF meter connector.
18. Install the air cleaner inlet with the bolt.
19. Install the undercover with the 2 bolts.
20. Install the undercover with the 16 bolts.
21. Connect the negative battery cable.
22. Bleed the power steering system and inspect for leaks.

BLEEDING

1. Check the fluid level.
2. Jack up the front of the vehicle and support it with the stands.
3. Turn the steering wheel. With the engine stopped, turn the wheel slowly from lock to lock several times.
4. Lower the vehicle.
5. Start the engine. Run the engine at idle for a few minutes.
6. Turn the steering wheel. With the engine idling, turn the wheel to left or right full lock position and keep it there for 2–3 seconds, then turn the wheel to the opposite full lock position and keep it there for 2–3 seconds.
7. Repeat last step several times.
8. Stop the engine.
9. Check for foaming or emulsification. If the system has to be bled twice specifically because of foaming or emulsification, check for fluid leaks in the system.
10. Check the fluid level.

SUSPENSION

FRONT SUSPENSION

See Figures 227 through 229.

COIL SPRING

REMOVAL & INSTALLATION

Refer to MacPherson Strut

CONTROL LINKS

REMOVAL & INSTALLATION

GS 300

See Figure 230.

1. Remove the front wheels.
2. Remove the 2 nuts and the front stabilizer link assembly LH.
3. If the ball joint turns together with the nut, use a hexagon (6 mm) wrench to hold the stud.
4. Use the same procedures for the RH.

To install:

5. Install the front stabilizer link assembly LH with the 2 nuts.
6. Tighten the nuts to 62 ft. lbs. (84 Nm).
7. Use the same procedures for the RH.
8. Install the front wheels. Tighten to 76 ft. lbs. (103 Nm).

SC 430

See Figure 231.

1. Remove the front wheels.
2. Remove the bolt and nut, and disconnect the stabilizer LH bar from the stabilizer link.

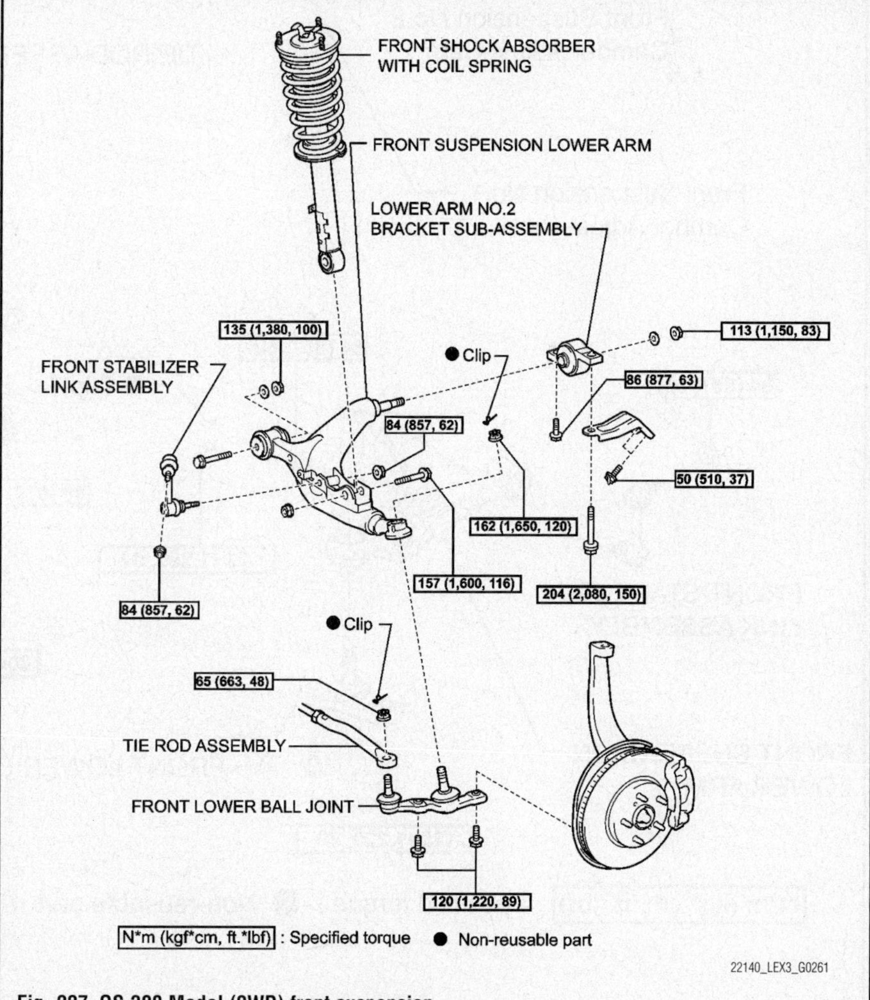

Fig. 227 GS 300 Model (2WD) front suspension

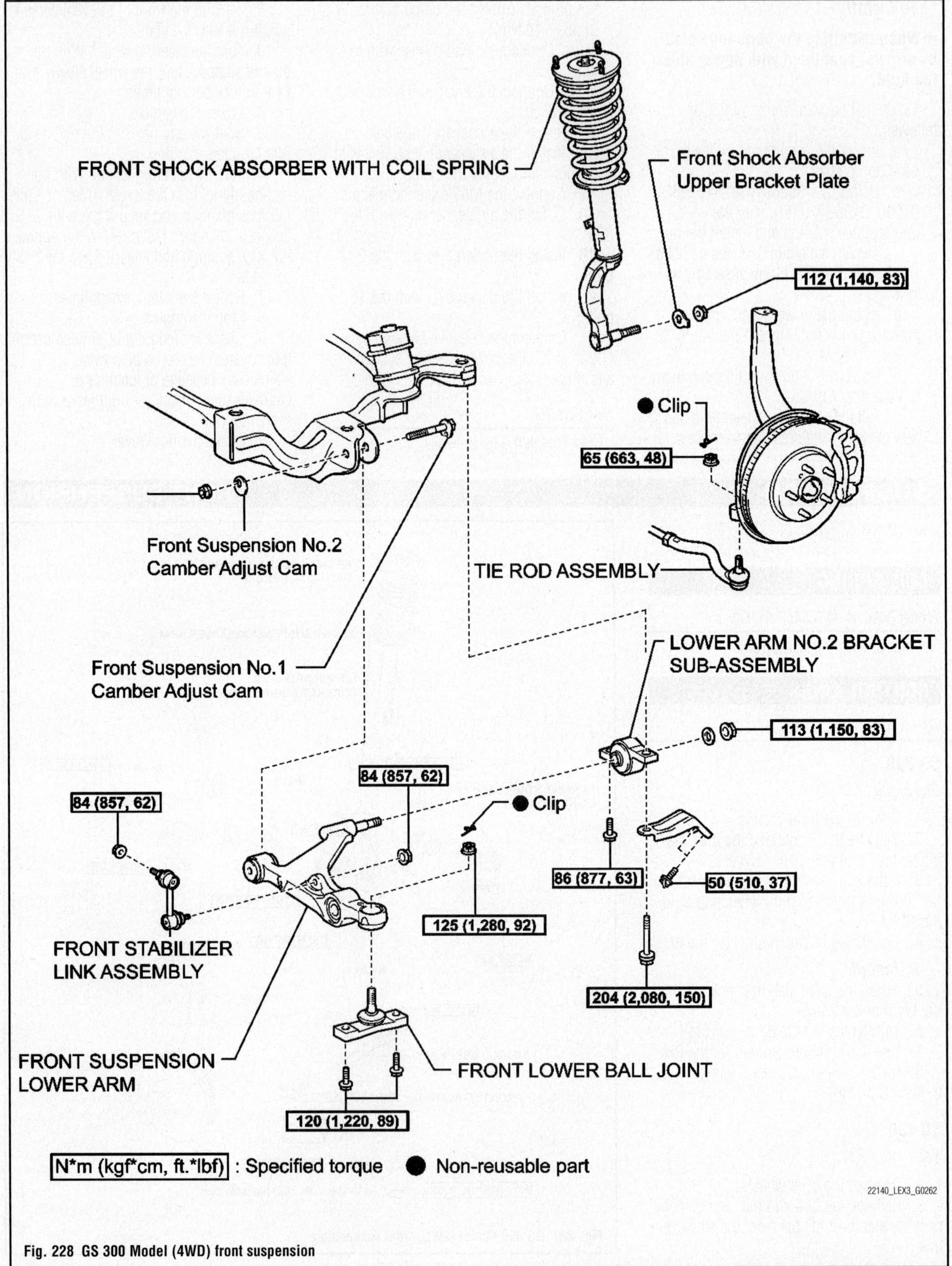

Fig. 228 GS 300 Model (4WD) front suspension

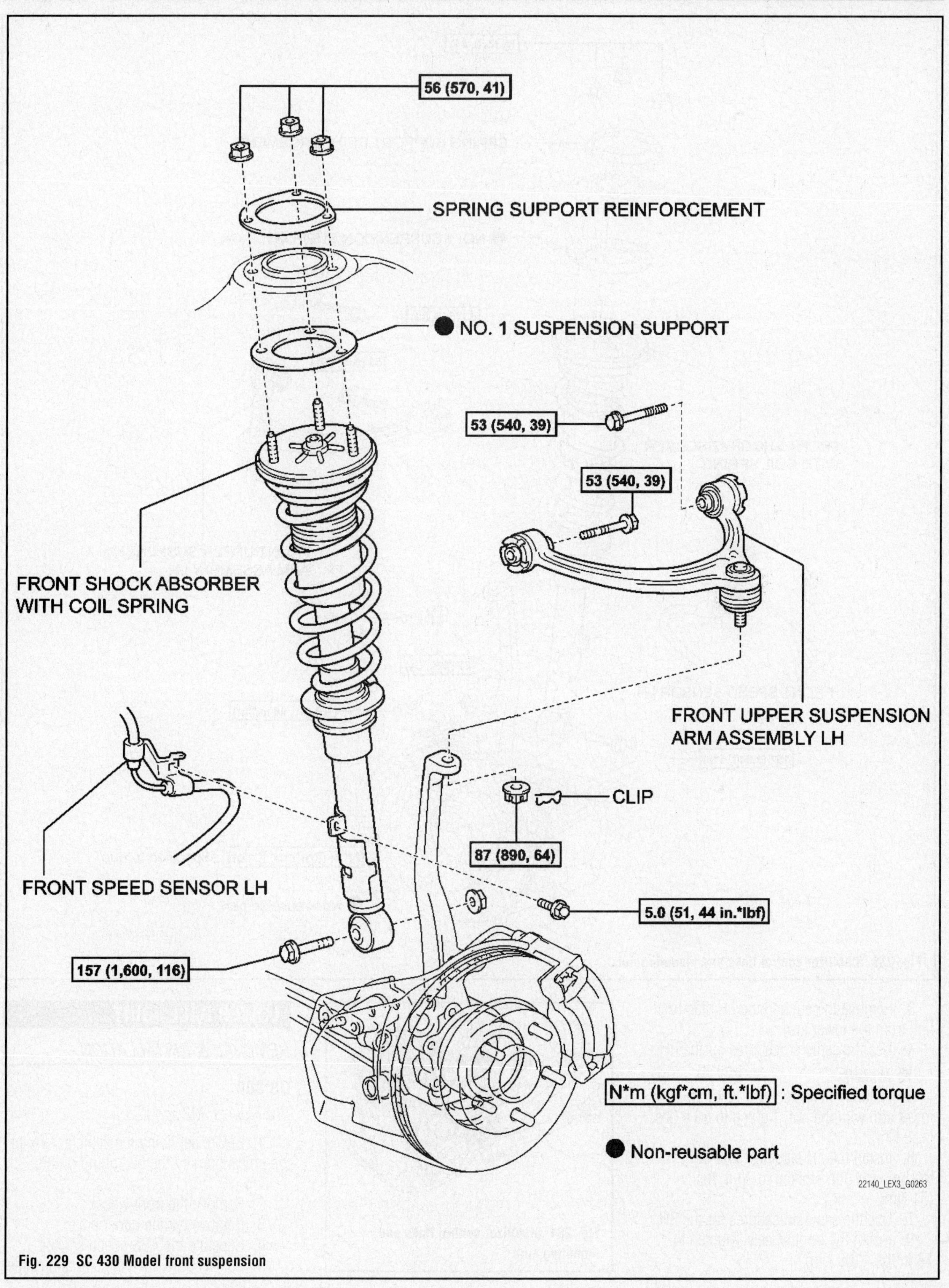

Fig. 229 SC 430 Model front suspension

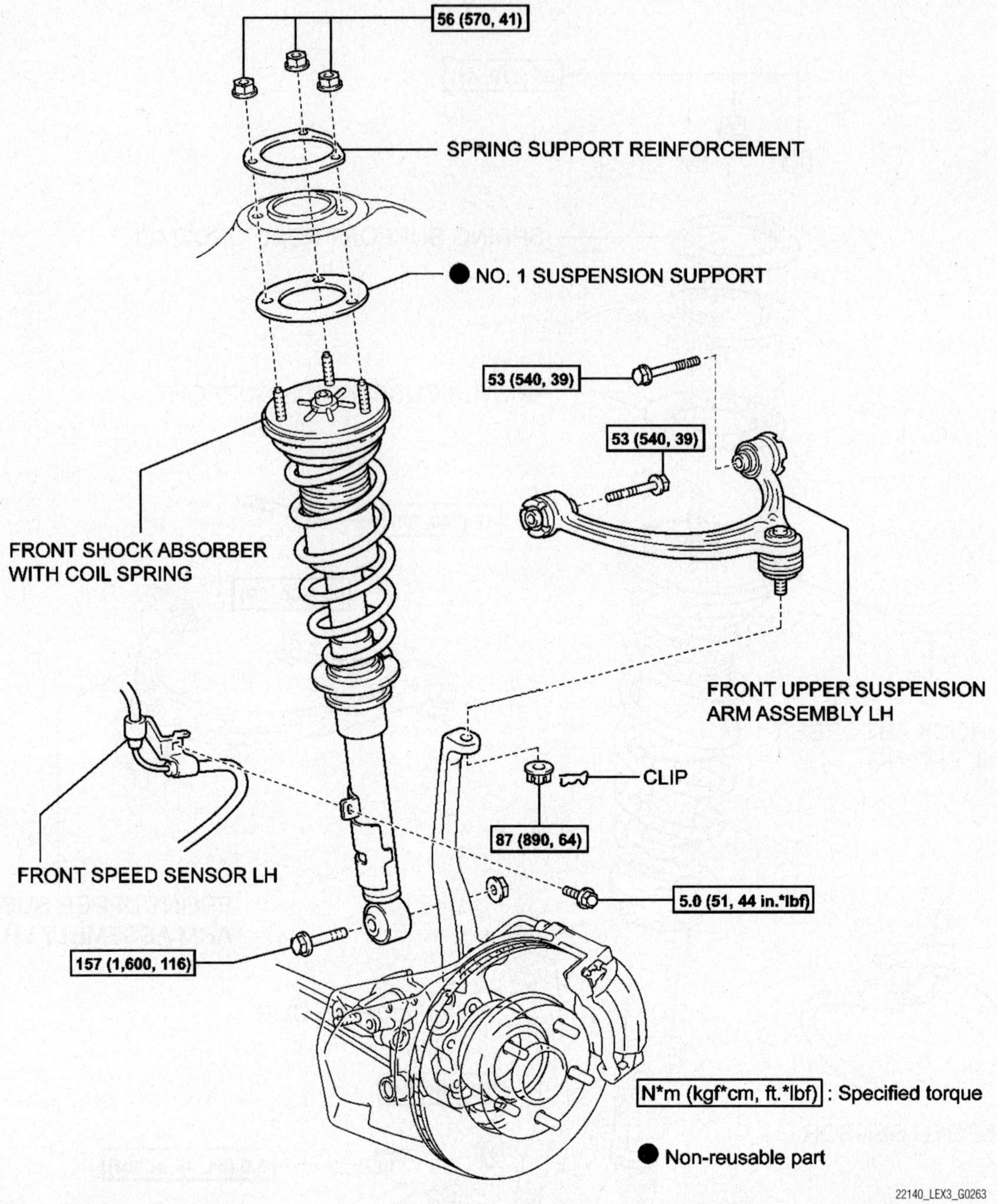

Fig. 230 Stabilizer control links and mounting nuts

3. Remove the nut and the LH stabilizer link from the lower arm.
4. Use the same procedures for the RH.

To install:

5. Install the LH stabilizer link to the lower arm with the nut. Tighten to 83 ft. lbs. (113 Nm).
6. Install the LH stabilizer link with the bolt and nut. Tighten to 38 ft. lbs (51 Nm).
7. Use the same procedures for the RH.
8. Install the front wheels. Tighten to 76 ft. lbs. (103 Nm).

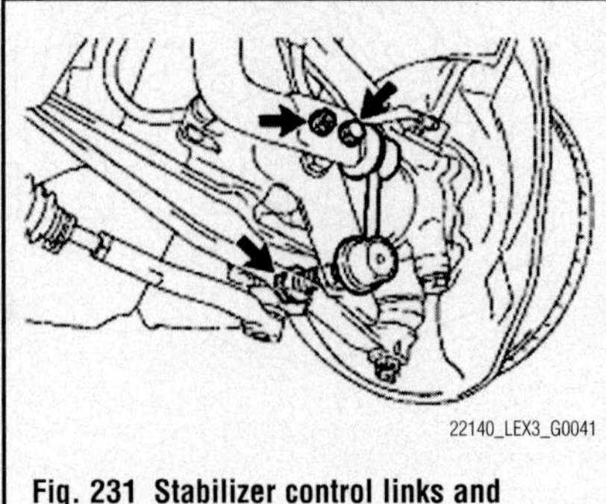

Fig. 231 Stabilizer control links and mounting nuts

LOWER BALL JOINT

REMOVAL & INSTALLATION

GS 300

See Figures 232 and 233.

1. Before servicing the vehicle, refer to the precautions in the beginning of this section.
2. Remove the front wheel.
3. Remove engine under cover.
4. Separate the ABS speed sensor.
5. For 4WD, remove the clip and castle

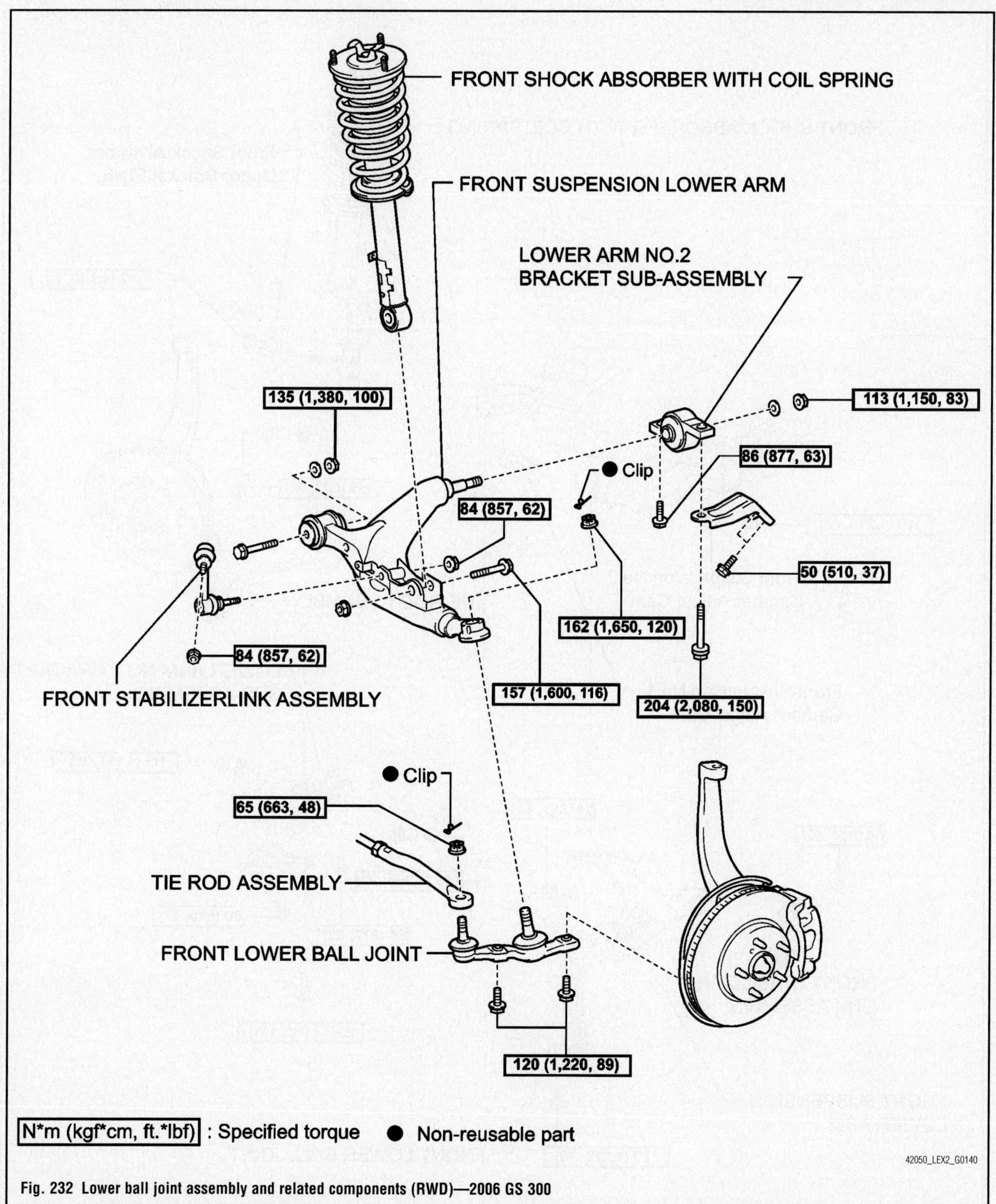

Fig. 232 Lower ball joint assembly and related components (RWD)—2006 GS 300

nut, then using SST 09628-00011, separate the upper ball joint from the steering knuckle.

6. Remove the clip and castle nut, then using SST 09628_62011, separate the tie rod end.

7. Remove the strut assembly.

8. Separate the front stabilizer link assembly.

9. Remove the engine under cover.

10. Remove the lower control arm.

11. Fix the front lower control arm in a vise using aluminum plates.

12. Remove the clip and castle nut.

13. Use SST 09950-40011 (09951-04010, 09952-04010, 09953-04020,

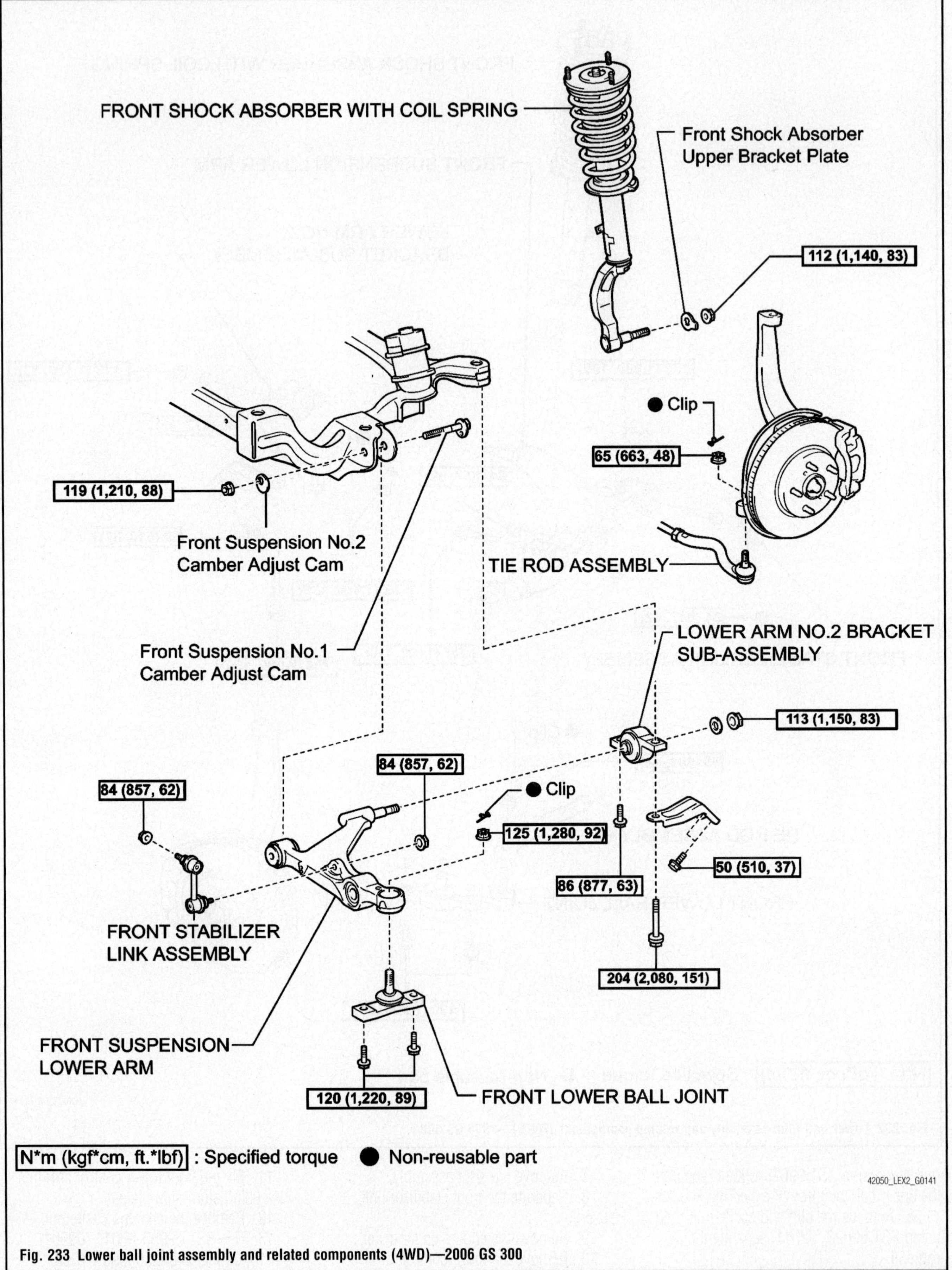

Fig. 233 Lower ball joint assembly and related components (4WD)—2006 GS 300

09954-04010, 09955-04051, 09957-04010, 09958-04011) to remove the front lower ball joint from the front lower control arm.

14. Inspect the lower control arm ball joint as follows:

- Flip the ball joint stud back and forth 5 times.
- Temporarily install the nut, and use a torque wrench to turn the nut continuously at a rate of 3 to 5 seconds per turn. Take the torque reading on the 5th turn.
- Turning torque: 53 inch lbs. (6 Nm)
- Check the dust boots for cracks or grease leakage. If the value is not within the specified range, replace the front lower ball joint with a new one.

To install:

15. Install the front lower ball joint to the front lower control arm with the nut. Ensure that the thread and taper are free of oil or other foreign matter. Tighten the nut to 120 ft. lbs. (162 Nm) for RWD, and 92 ft. lbs. (125 Nm) for AWD.

16. Install a new clip to the front lower ball joint. Further tighten the nut up to 60° if the holes for the cotter pin are not aligned.

17. Install the lower control arm.

18. Connect the front stabilizer link assembly.

19. Install the strut assembly, but temporarily tighten the bolts at the lower control arm.

20. Connect the tie rod end to the steering knuckle Tighten the nut to 50 ft. lbs. (65 Nm). Install a new cotter pin. If the holes for the clip are not aligned, tighten the nut up to 60° further.

21. For AWD, install the steering knuckle to the front suspension upper control arm, and tighten it with the castle nut to 64 ft. lbs. (87 Nm). Install a new clip to the steering knuckle. Further tighten the nut up to 60° if the holes for the cotter pin are not aligned.

22. Connect the ABS speed sensor.

23. Stabilize the suspension as follows:

- Install the front wheels. Tighten the lug nuts to 76 ft. lbs. (103 Nm).
- Lower the vehicle and bounce it up and down several times to stabilize the front suspension.
- Remove the front wheels.
- Jack up the front suspension lower arm placing a wooden block in between. Apply a load to the front suspension so that the front suspension lower arm is placed in a horizontal position.
- Fully tighten the strut assembly bolts at the lower control arm.
- Fully tighten the lower control arm.

24. Install engine under cover.

25. Install the front wheels. Tighten to 76 ft. lbs. (103 Nm).

26. Inspect and adjust wheel alignment.

SC 430

See Figure 234.

1. Before servicing the vehicle, refer to the precautions in the beginning of this section.

2. Remove the front wheel.

3. Remove the cotter pin and nut, remove the 2 bolts, then using SST 09628–62011, remove the lower ball joint from the lower control arm.

4. Remove the clip and nut, then using SST 09628–62011, disconnect the tie rod end from the lower ball joint.

5. Inspect the lower control arm ball joint as follows:

- Flip the ball joint stud back and forth 5 times.
- Temporarily install the nut, and use a torque wrench to turn the nut continuously at a rate of 2 to 4 seconds per turn. Take the torque reading on the 5th turn.
- Turning torque: 0.9–26 inch lbs (0.1–3 Nm)
- Check the dust boots for cracks or grease leakage. If the value is not within the specified range, replace the front lower ball joint with a new one.

To install:

6. Install the lower ball joint to the lower suspension arm.

7. Install the nut and a new cotter pin. Tighten the nut to 119 ft. lbs. (162 Nm).

8. Install the lower ball joint with the 2 bolts and tighten to 83 ft. lbs. (113 Nm).

9. Install the tie rod to the lower ball joint. Install the nut and clip and tighten to 64 ft. lbs. (87 Nm).

10. Install the front wheels. Tighten to 76 ft. lbs. (103 Nm).

11. Stabilize the suspension as follows:

- Lower the vehicle.
- Press down on the vehicle several times to stabilize the suspension.

12. Inspect and adjust wheel alignment.

LOWER CONTROL ARM

REMOVAL & INSTALLATION

GS 300

See Figure 235.

1. Before servicing the vehicle, refer to the precautions in the beginning of this section.

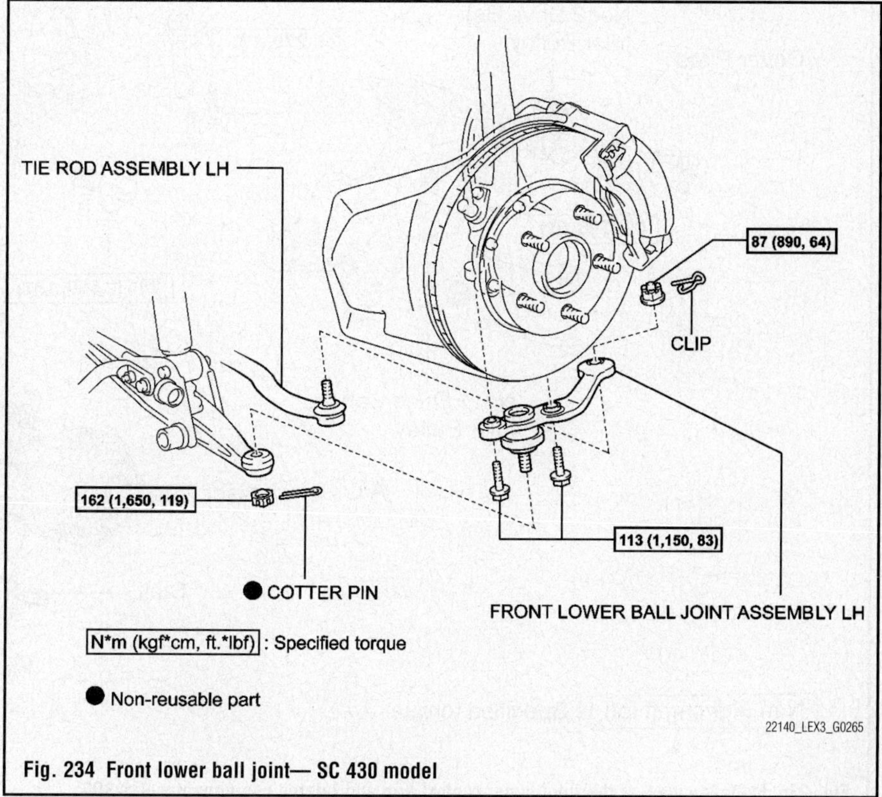

Fig. 234 Front lower ball joint— SC 430 model

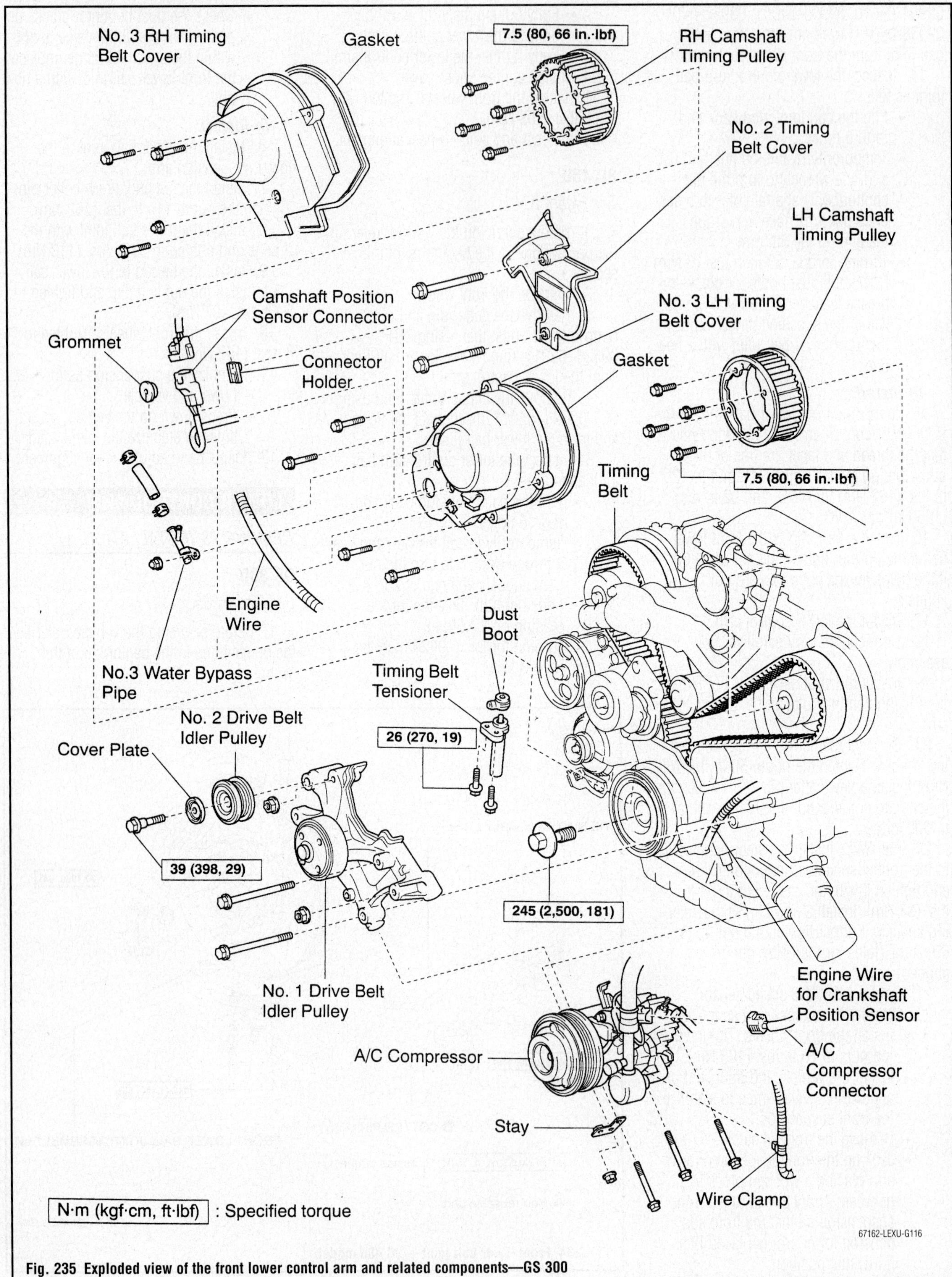

Fig. 235 Exploded view of the front lower control arm and related components—GS 300

2. Raise the vehicle on a hoist, so that front suspension components are hanging and accessible.

3. Remove or disconnect the following:

- Front wheels.
- Engine under covers.
- Brake caliper(s); Do NOT disconnect the brake hose; hang the caliper without stress on the hose
- Tie rod end from steering knuckle
- Stabilizer bar link from stabilizer bar
- Height control sensor link, if equipped, from shock absorber bracket
- Shock absorber lower mount
- Lower control arm set bolts (loosen only)
- Lower ball joint from No. 2 lower control arm (lower suspension arm)
- Steering gear assembly
- Strut bar bracket
- No. 1 lower control arm (lower suspension arm); matchmark adjusting cam to crossmember

To install:

4. To install, reverse the removal procedure, noting the following torque settings:

- No. 1 lower control arm (lower suspension arm) bolt to shock absorber bracket: 44 ft. lbs. (59 Nm)
- No. 1 lower control arm (lower suspension arm) adjusting cam bolt and nut to crossmember: 127 ft. lbs. (172 Nm)
- Strut bar bracket bolts: 43 ft. lbs. (58 Nm)
- Strut bar bracket nut: 112 ft. lbs. (152 Nm)
- No. 2 lower control arm (lower suspension arm) nuts: 122 ft. lbs. (164 Nm)
- Lower shock absorber mounting bolt and nut: 116 ft. lbs. (157 Nm)
- Stabilizer bar link nut: 83 ft. lbs. (113 Nm)
- Stabilizer bar link-to-stabilizer bar bolt and nut: 43 ft. lbs. (55 Nm)
- Tie rod end nut: 64 ft. lbs. (87 Nm)
- Brake caliper bolts: 87 ft. lbs. (118 Nm)
- Front wheel nuts: 76 ft. lbs. (103 Nm)

SC 430

See Figure 236.

1. Before servicing the vehicle, refer to the precautions in the beginning of this section.

2. Remove or disconnect the following:

- Front wheel
- Engine under covers
- Loosen 2 bolts on the suspension lower arm
- Shock absorber from mounting bracket
- Height control sensor
- Front stabilizer link
- Separate rack and pinion gear assembly
- Remove front suspension lower arm

3. Remove or disconnect the following:

- Ball joint cotter pin and bolt
- Lower ball joint from the lower arm assembly
- Shock absorber bracket
- Lower arm assembly

4. Matchmark the front and rear adjustment cams to the body and then remove the nuts and adjusting cams.

5. Lift out the lower control arm.

To install:

6. Install or connect the following:

- Shock absorber bracket. Torque: 44 ft. lbs. (59 Nm)
- Lower control arm to the body and temporarily install the adjusting cams and nuts. Do NOT tighten the nuts at this time.
- Lower control arm to the knuckle and tighten the ball joint nut to 119 ft. lbs. (162 Nm). Install a new cotter pin.
- 2 bolts on lower suspension arm to 121 ft. lbs. (164 Nm)
- Rack and pinion steering gear assembly nuts to 48 ft. lbs. (65 Nm)
- Front stabilizer link bolts and nuts (to stabilizer bar) to 38 ft. lbs. (51 Nm) and (to stabilizer link) to 116 ft. lbs. (157 Nm)
- Shock absorber assembly bolt to 116 ft. lbs. (157 Nm)
- Height control sensor
- Front wheel

7. Inspect and adjust front wheel alignment

8. Adjust height control sensor

MACPHERSON STRUT

REMOVAL & INSTALLATION

GS 300

See Figures 237 through 240.

1. Remove the front wheels.

2. Disconnect the negative battery cable.

3. Remove the 4 clips and engine room cover.

4. Remove the bolt and separate the speed sensor front from the front shock absorber with coil spring.

5. Remove the nut and separate the front suspension lower arm from the height control sensor link sub-assembly front.

6. Remove the clip and castle nut from the upper control arm.

7. Using SST 09628-00011, remove the upper ball joint from the steering knuckle.

8. Remove the 3 nuts and shock absorber cap upper. (for AVS)

9. Separate the electrical connector.

10. Turn the absorber control actuator counterclockwise 40° to remove it from the front shock absorber with coil spring.

11. Support the front suspension lower arm with a jack. Be sure to place a wooden block between the jack and the front suspension lower arm to avoid damage.

12. Loosen the bolt while holding the nut. Separate the lower part of the front shock absorber from the front suspension lower arm.

13. Loosen the lock nut of the front shock absorber.

14. Remove the 3 nuts from the upper side of the front suspension support.

15. Slowly lower the jack. Remove the bolt from the lower side to remove the front shock absorber with coil spring.

16. Remove the coil spring from strut as follows:

- Install the coil spring compressor to the front coil spring so that the 2 hooks on the compressor are fully extended.
- Fully compress the coil spring.
- Remove the lock nut.
- Remove the actuator support bracket FR. (for AVS)
- Remove the front suspension support sub-assembly with front coil spring insulator upper.
- Remove the front coil spring insulator upper from the front suspension support sub-assembly.
- Remove the front spring bumper from the front suspension support sub-assembly.
- Remove the front coil spring and compressor.

17. Remove the front coil spring insulator lower.

To install:

18. Install the coil spring to the strut as follows:

- Install the front coil spring insulator lower to the front shock

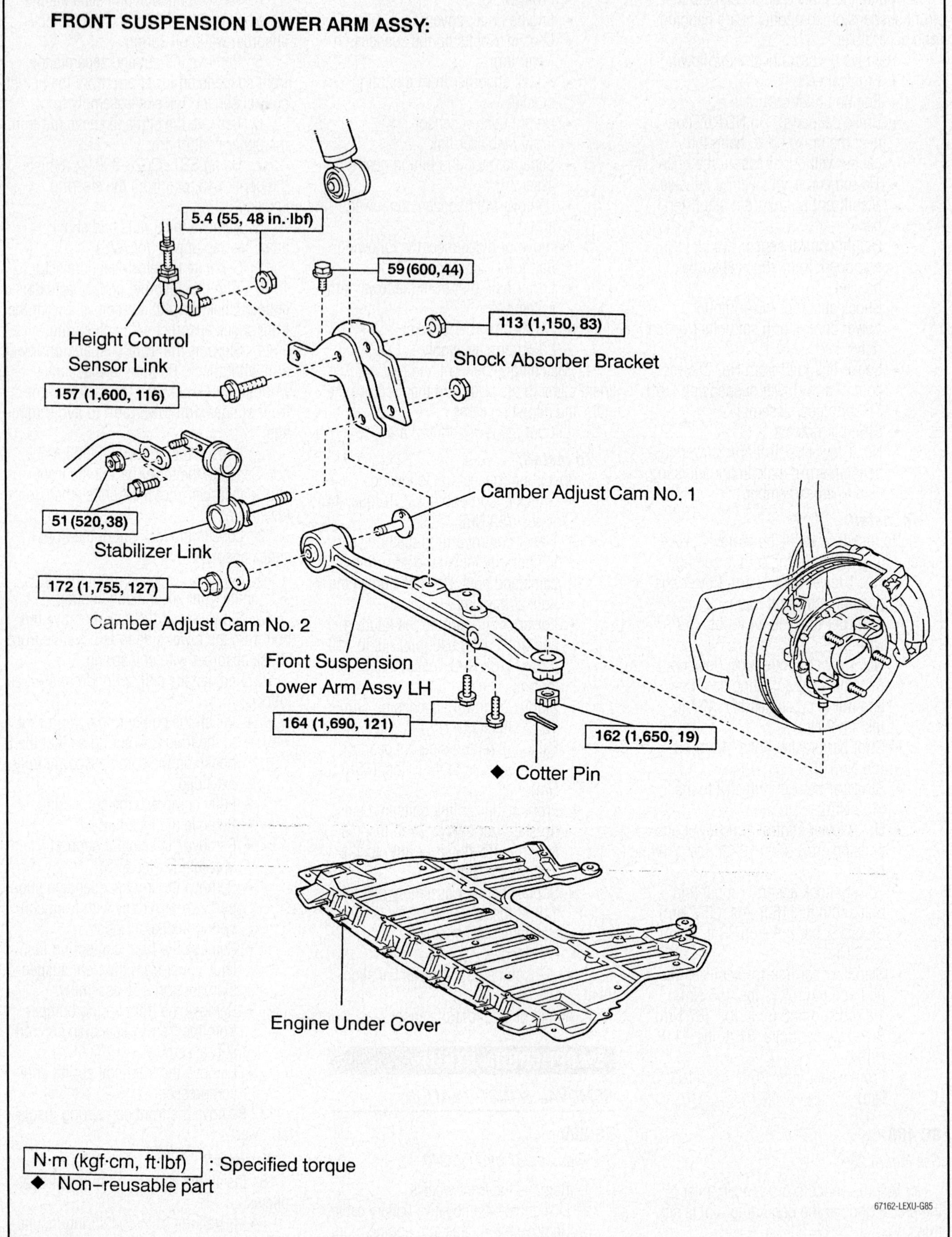

FRONT SUSPENSION LOWER ARM ASSY:

Fig. 236 Expanded view, lower front suspension arm—SC 430

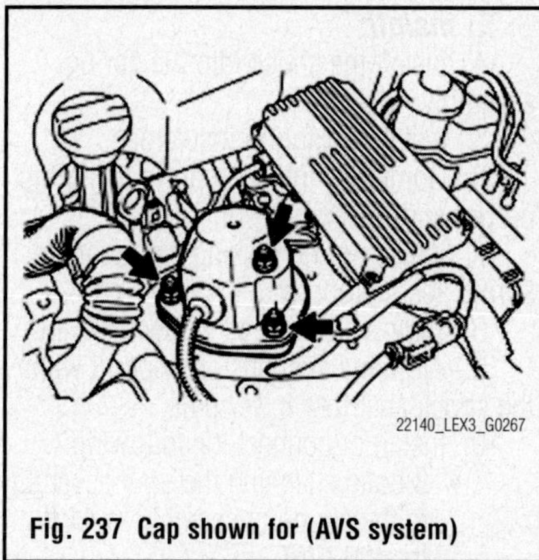

Fig. 237 Cap shown for (AVS system)

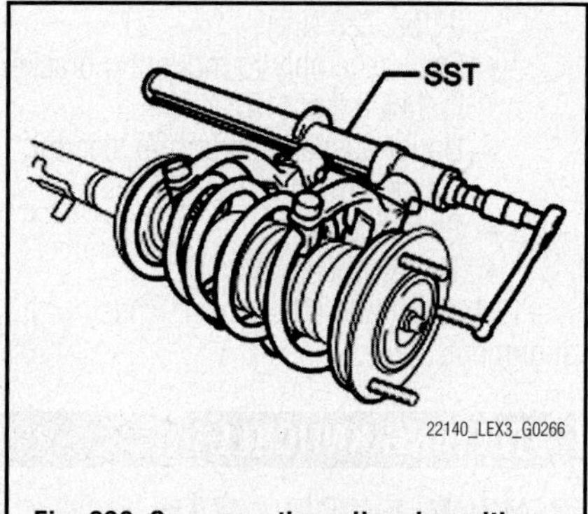

Fig. 238 Compress the coil spring with spring compressor tool

absorber so that both of the recessed parts are aligned.

- Install the front coil spring so that the end comes to the stepped portion of the spring lower seat.
- Install the front spring bumper on the front suspension support sub-assembly.
- Align the bolt of the front suspension support sub-assembly and the cutout of the front coil spring insulator upper, and install the front coil spring insulator upper on the front suspension support sub-assembly.
- Align the width across flat of the piston rod end and of the front suspension support sub-assembly to install the front shock absorber.
- Temporarily tighten a new lock nut to the front shock absorber.
- Adjust the front suspension support sub-assembly so that the bolts come to the positions as shown in the illustration, and remove the compressor from the front coil spring.

19. Install the front shock absorber with front coil spring on the vehicle by tightening the 3 nuts on the suspension support side. Tighten to 49 ft. lbs. (67 Nm).

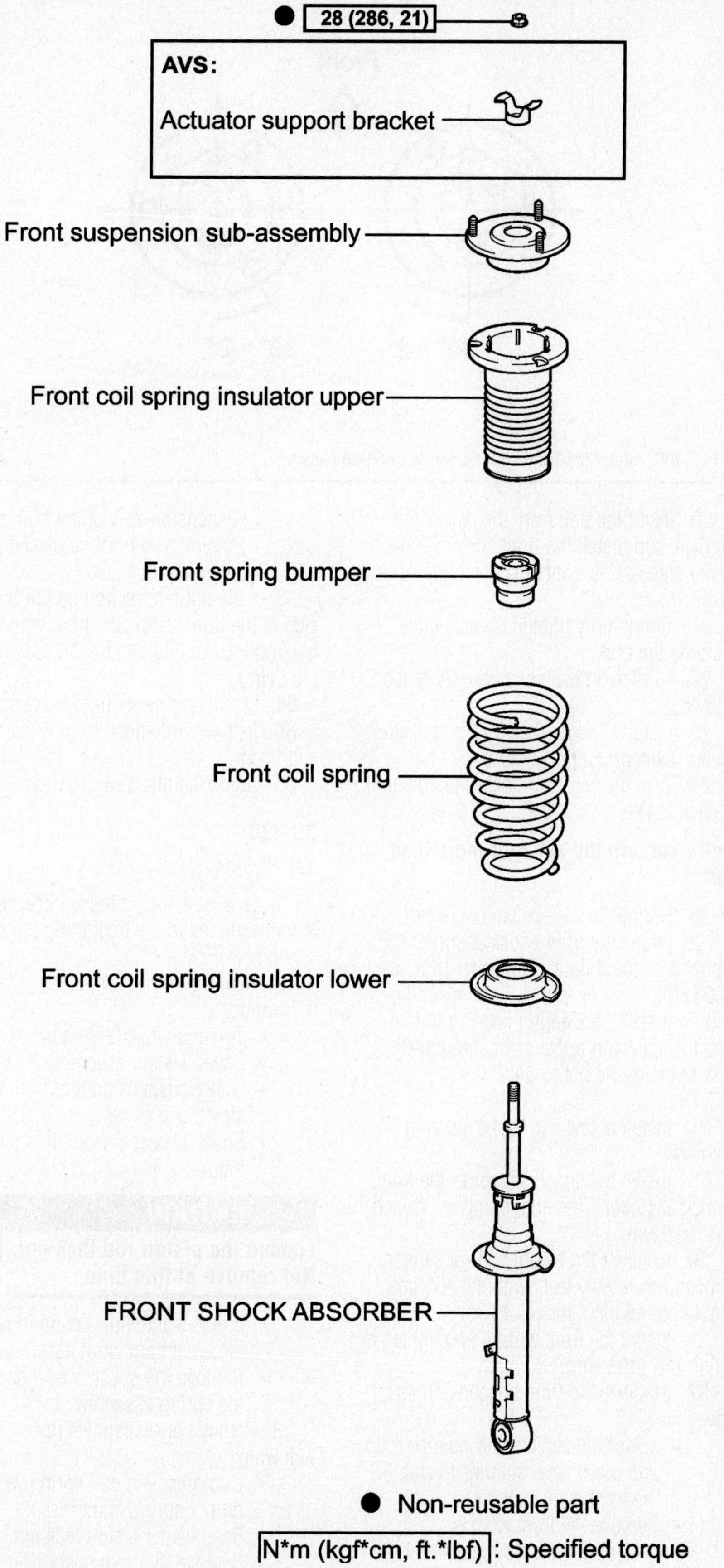

● 28 (286, 21)

AVS:
Actuator support bracket

Front suspension sub-assembly

Front coil spring insulator upper

Front spring bumper

Front coil spring

Front coil spring insulator lower

FRONT SHOCK ABSORBER

● Non-reusable part

N*m (kgf*cm, ft.*lbf) : Specified torque

22140_LEX3_G0268

Fig. 239 Exploded view of coil spring and strut

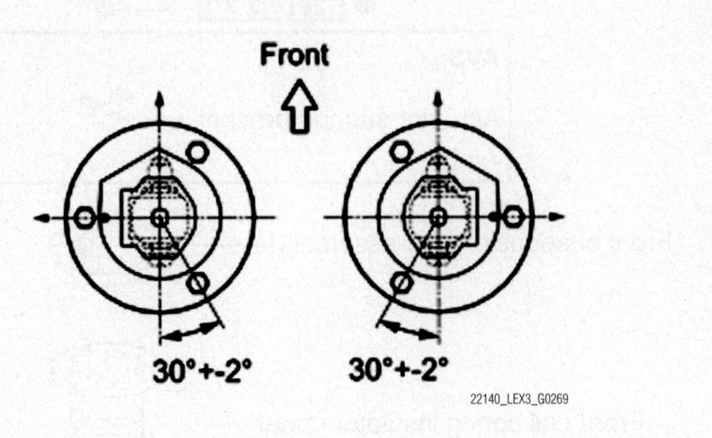

Fig. 240 Adjust the front suspension to position shown

20. Insert the bolt from the rear of the vehicle, and install the front shock absorber lower side on the front suspension lower arm.
21. Temporarily tighten the nut while holding the bolt.
22. Tighten a new lock nut to 21 ft. lbs. (28 Nm).
23. Install the absorber control actuator to the actuator support bracket.
24. Turn the actuator clockwise 40° until a click is felt.

➡Do not turn the actuator more than 40°

25. Connect the electrical connector.
26. Install the front shock absorber cap upper with the 3 nuts. Tighten to 15 ft. lbs. (20 Nm).
27. Install the steering knuckle to the front suspension upper arm, and tighten it with the castle nut to 64 ft. lbs. (87 Nm).
28. Install a new clip to the steering knuckle.
29. Install the speed sensor to the front shock absorber. Tighten the bolt to 53 inch. lbs. (6.0 Nm).
30. Connect the height control sensor link sub-assembly front with the nut, and tighten to 48 inch. lbs. (5.4 Nm).
31. Install the front wheels and tighten to 76 ft. lbs. (103 Nm).
32. Stabilize the front suspension as follows:
   - Lower the vehicle and bounce it up and down several times to stabilize the front suspension.
   - Remove the front wheels.
   - Jack up the front suspension lower arm placing a wooden block in between. Apply a load to the front suspension so that the front suspension lower arm is placed in a horizontal position.
33. Fully tighten the bolt on the lower side of the front shock absorber while holding the nut. Tighten to 116 ft. lbs. (157 Nm).
34. Install engine room side cover
35. Inspect and adjust front wheel alignment
36. Adjust height control sensor

SC 430

See Figures 241 and 242.

1. Before servicing the vehicle, refer to the precautions in the beginning of this section.
2. Remove or disconnect the following:
   - Tire and wheel assembly
   - Speed sensor assembly
   - Upper suspension arm from the steering knuckle
   - Shock absorber from the mounting bracket

⁂ CAUTION

Loosen the piston rod lock nut. Do Not remove at this time

   - 3 nuts front spring support reinforcement and strut assembly
   - Remove the shock absorber from the spring assembly
3. Remove or disconnect the following:
   - Compress the coil spring with proper spring compressor
   - Remove the piston lock nut
   - Remove the front suspension support assembly, front coil spring insulator and coil spring.

To install:

4. Install the shock into the spring assembly.
5. Install the spring insulator.
6. Compress the coil spring using the proper spring compressor.
7. Install spring making sure that the spring seats properly.
8. Temporarily install a new lock nut.
9. Align the suspension support with the shock absorber lower bolt.
10. Install or connect the following:
   - 3 bolts attaching the strut assembly to the support assembly to 41 ft. lbs (56 Nm)
   - Fully tighten piston lock nut to 21 ft. lbs. (28 Nm)
   - Strut assembly to mounting bracket to 116 ft lbs. (157 Nm)
   - Upper suspension arm to steering knuckle to 64 ft. lbs. (87 Nm)
   - Speed sensor
   - Tire & wheel
11. Inspect and adjust front wheel alignment.

STEERING KNUCKLE

REMOVAL & INSTALLATION

See Figures 243 through 245.

1. Before servicing the vehicle, refer to the precautions in the beginning of this section.
2. Remove the front wheel.
3. Disconnect the front brake caliper assembly. Support the brake caliper securely.
4. Remove the front brake disc.
5. Remove the clip and nut, then using SST 09610-20012, disconnect the tie rod end.
6. Remove the cotter pin and nut, then using SST 09628-6201 1, remove the lower ball joint from the lower control arm.
7. Remove the cotter pin and castle nut, then using SST09628–62011, separate the upper ball joint from the steering knuckle.
8. For (4WD) models remove the axle.
9. Remove the steering knuckle from the vehicle.
10. If necessary, remove the lower ball joint and hub and bearing assembly from the steering knuckle.

To install:

11. If necessary, install the hub and bearing assembly and lower ball joint to the steering knuckle.
12. Install the steering knuckle into the vehicle.

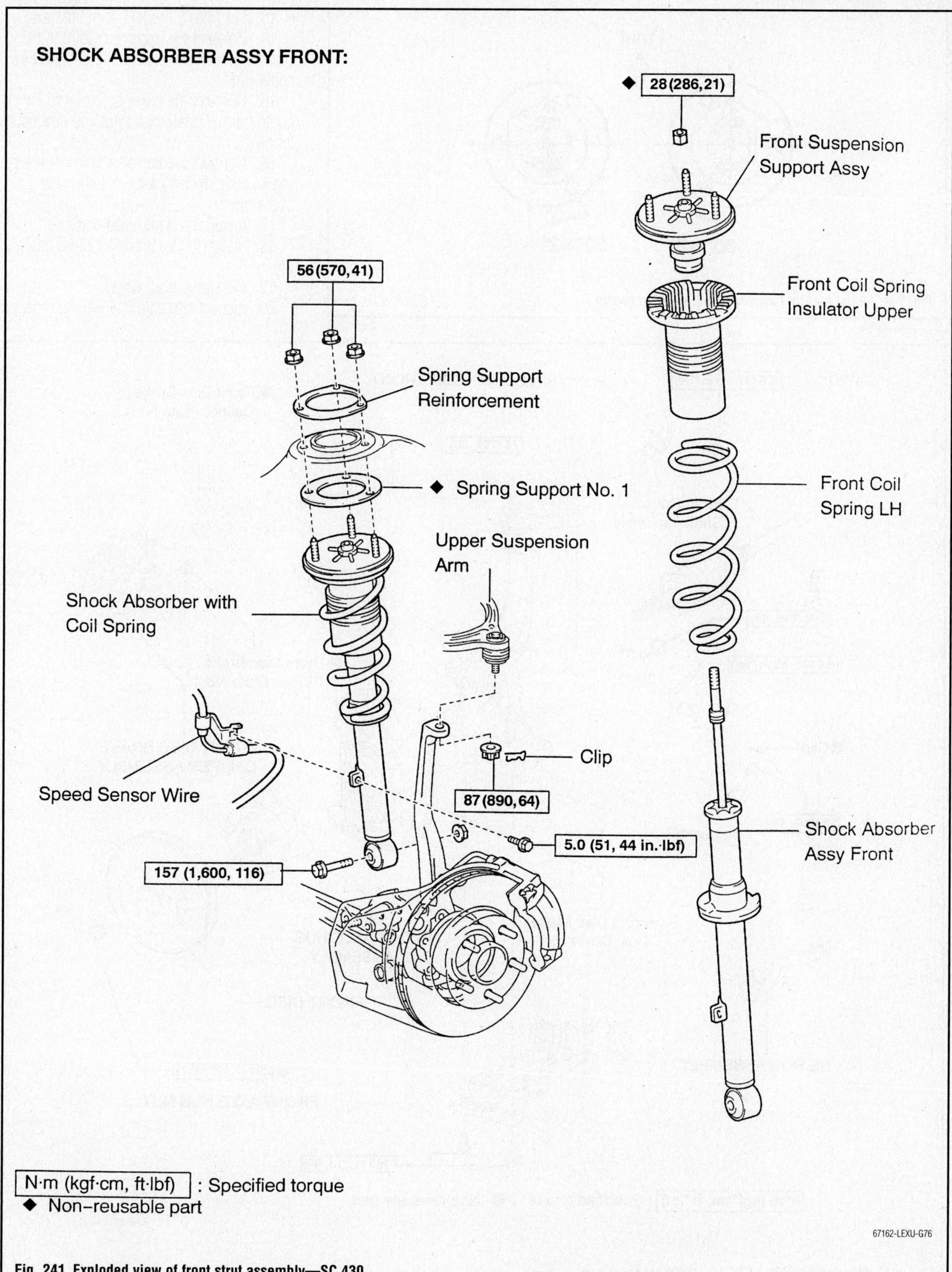

Fig. 241 Exploded view of front strut assembly—SC 430

Front
30°+-2°
30°+-2°
22140_LEX3_G0269

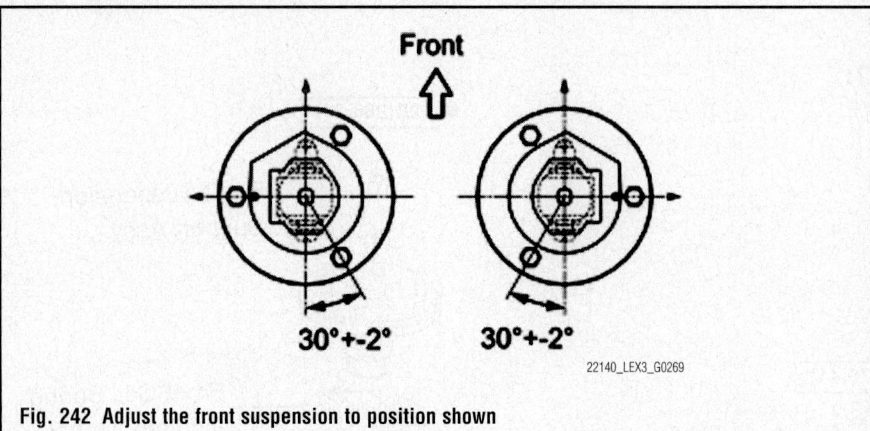

Fig. 242 Adjust the front suspension to position shown

13. For (4WD) models install the axle.
14. Connect the upper ball joint to the steering knuckle. Install castle nut and new cotter pin.
15. Connect the lower ball joint to lower control arm. Install castle nut and new cotter pin.
16. Connect the tie rod end to the steering knuckle. Install castle nut and new cotter pin.
17. Install the front brake disc.
18. Install the front brake caliper assembly.
19. Install the front wheel.
20. Inspect and adjust wheel alignment.

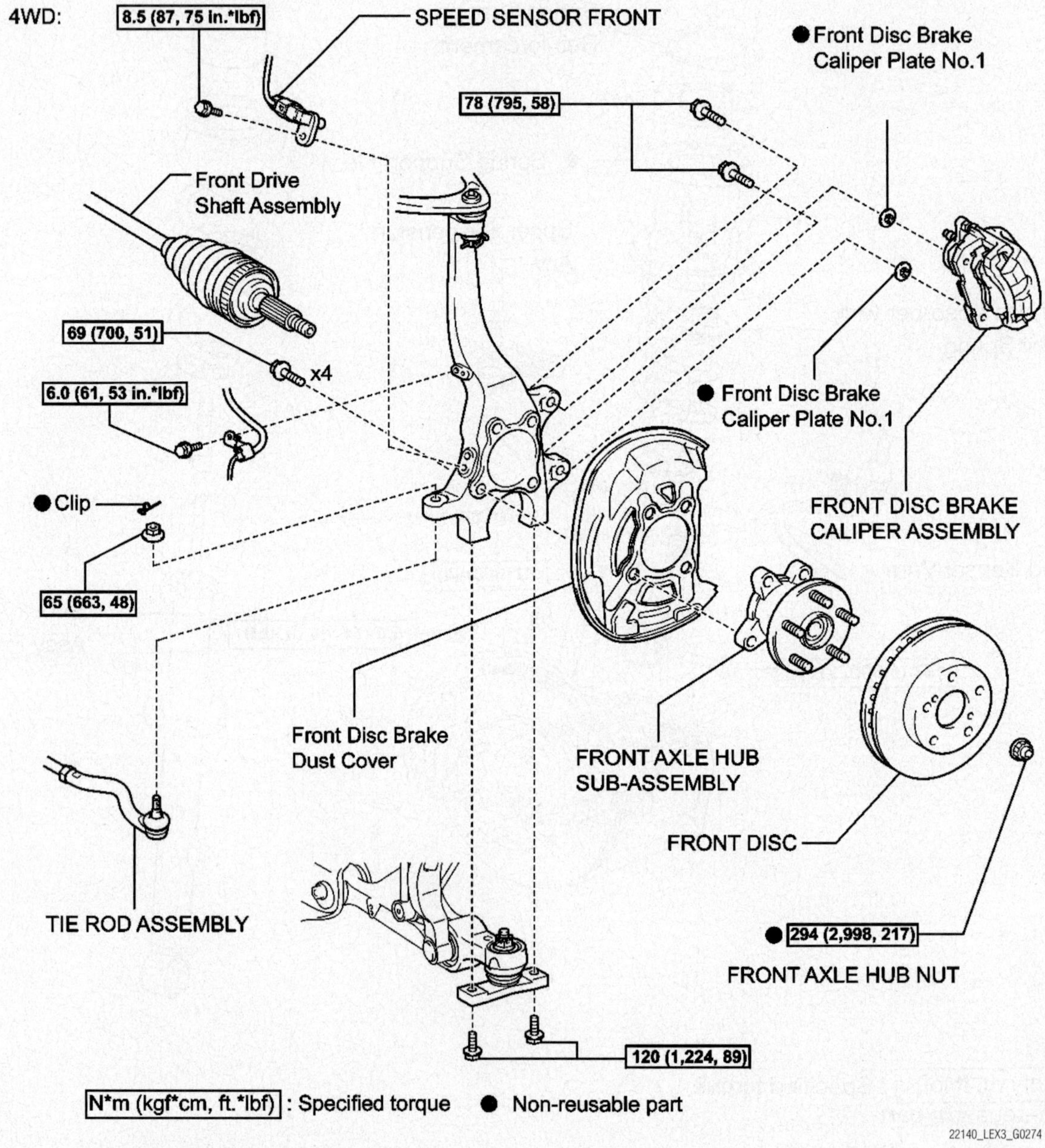

4WD:
8.5 (87, 75 in.*lbf)
SPEED SENSOR FRONT
● Front Disc Brake Caliper Plate No.1
78 (795, 58)
Front Drive Shaft Assembly
69 (700, 51)
x4
6.0 (61, 53 in.*lbf)
● Front Disc Brake Caliper Plate No.1
● Clip
FRONT DISC BRAKE CALIPER ASSEMBLY
65 (663, 48)
Front Disc Brake Dust Cover
FRONT AXLE HUB SUB-ASSEMBLY
FRONT DISC
TIE ROD ASSEMBLY
● 294 (2,998, 217)
FRONT AXLE HUB NUT
120 (1,224, 89)
N*m (kgf*cm, ft.*lbf) : Specified torque ● Non-reusable part
22140_LEX3_G0274

Fig. 243 Steering knuckle—GS 300 (4WD) model shown

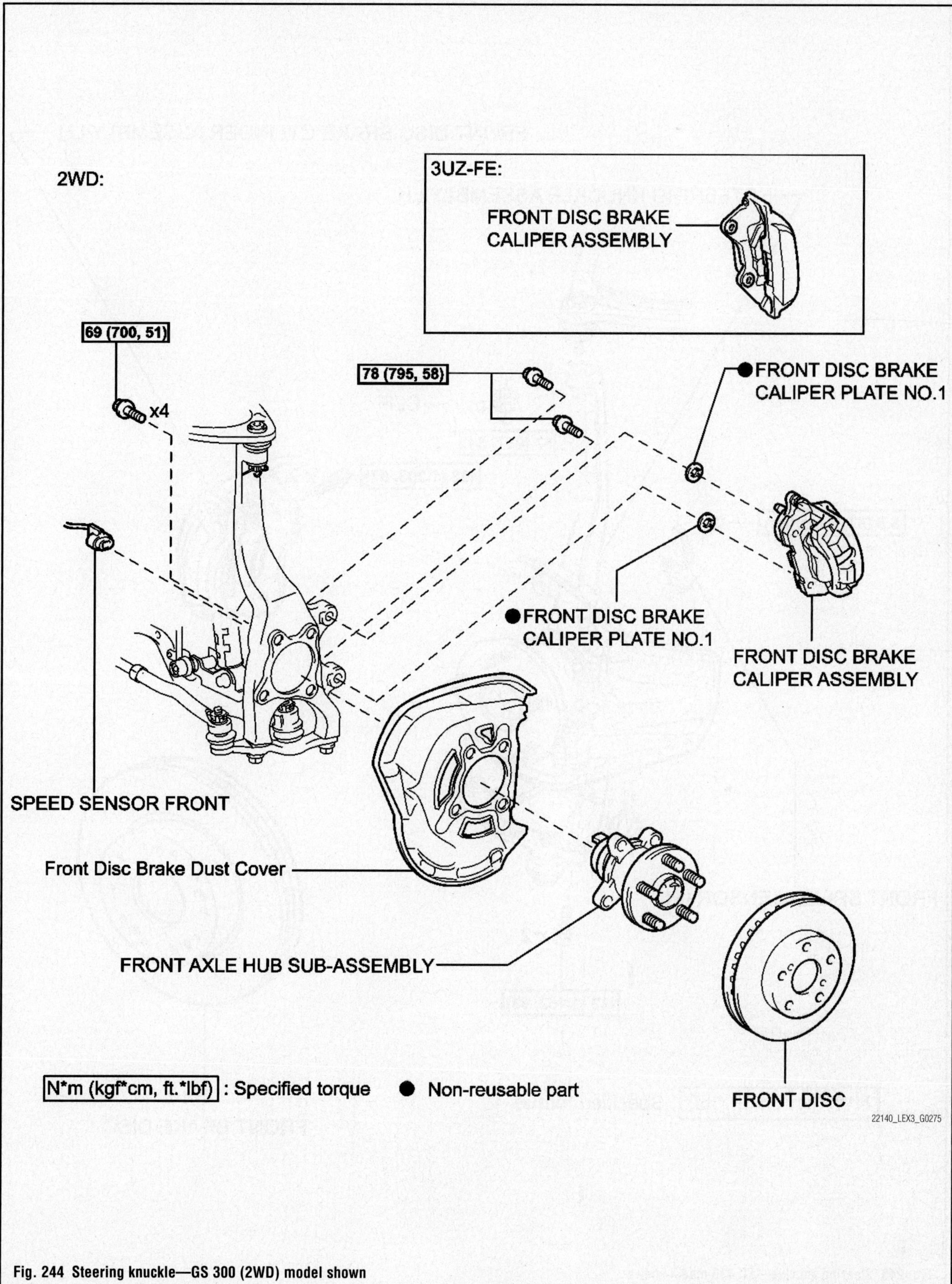

Fig. 244 Steering knuckle—GS 300 (2WD) model shown

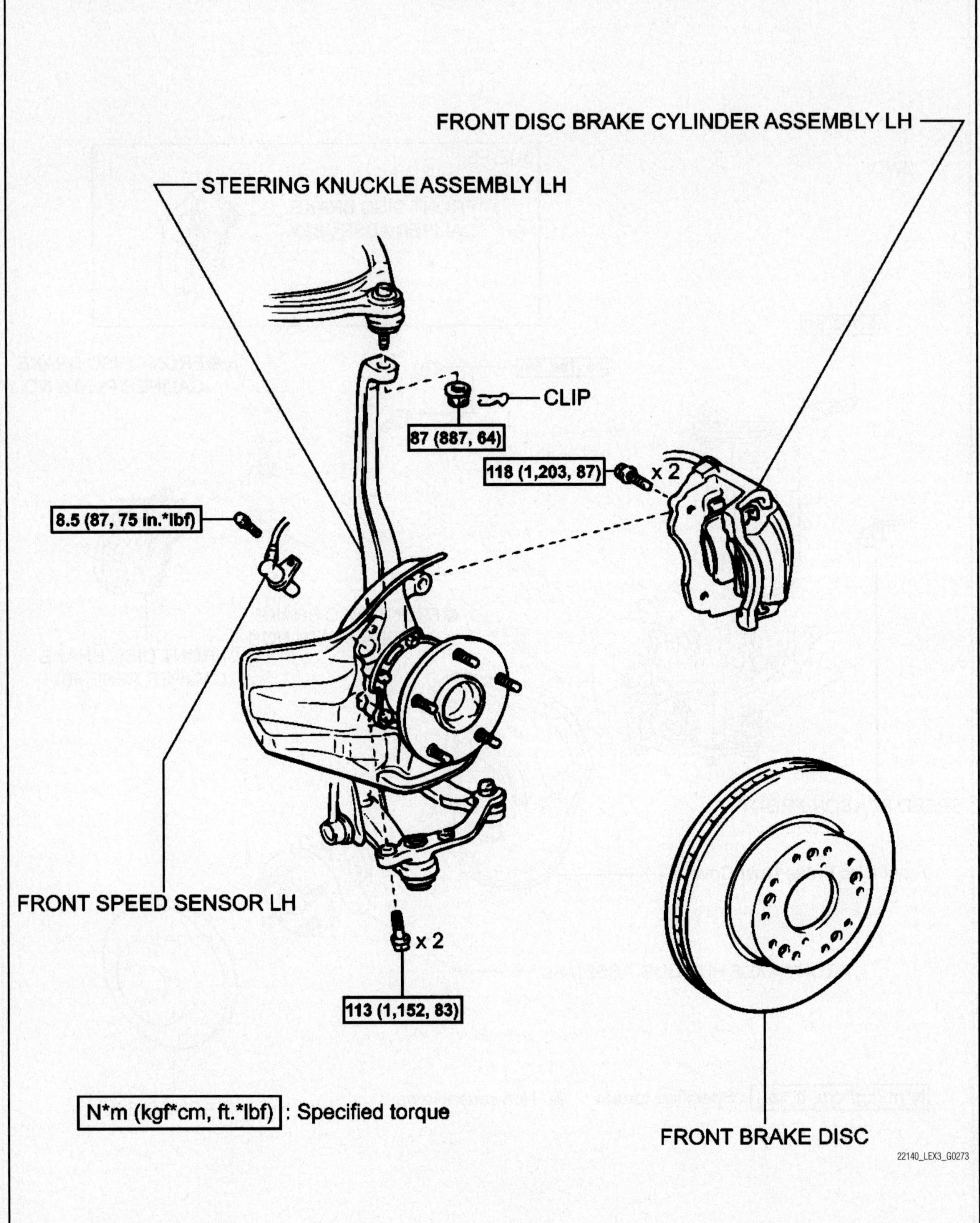

Fig. 245 Steering knuckle—SC 430 model shown

STABILIZER BAR

REMOVAL & INSTALLATION

GS 300 (2WD)

See Figures 246 through 248.

1. Before servicing the vehicle, refer to the precautions in the beginning of this section.
2. Remove the front wheels.
3. Remove engine under covers.
4. Remove the left and right rear engine under covers.
5. Remove the 2 nuts and the left front stabilizer link assembly. Repeat for the right side. If the ball joint turns together with the nut, use a hexagon (6mm) wrench to hold the stud.
6. Remove the 2 bolts and the left front No. 2 stabilizer bracket from the front suspension crossmember. Repeat for the right side.
7. Remove the 4 bolts and the left front No. 1 stabilizer bracket from the frame. Repeat for the right side.
8. Remove the 2 front No. 1 stabilizer bar bushings from the front stabilizer bar.
9. Remove the front stabilizer bar from the vehicle.

To install:

10. Install the front stabilizer bar to the vehicle. The identification mark must be on the right side of the vehicle when installing the front stabilizer bar.
11. Install the 2 front No. 1 stabilizer bar bushings as shown in the illustration. Be sure to install the front No. 1 stabilizer bar bushings so that the cutouts face the front of the vehicle.
12. Install the left front No. 1 stabilizer bracket as follows:
    a. Temporarily install the left front stabilizer No.1 bracket to the side member with the 4 bolts.
    b. Tighten the bolt (A) while pressing the left front stabilizer No. 1 bracket.
    c. Tighten the bolts to 36 ft. lbs. (49 Nm).
    d. Install the right side following the same procedures as for the left side.
13. Install the left front No. 2 stabilizer bracket on the vehicle with the 2 bolts. Repeat for the right side. Tighten the bolts to 36 ft. lbs. (49 Nm).
14. Install the left front stabilizer link assembly with the 2 nuts and tighten to 62 ft. lbs. (84 Nm). Repeat for the right side. If the ball joint turns together with the nut, use a hexagon (6mm) wrench to hold the stud.
15. Install the left and right rear engine under covers.
16. Install engine under cover.
17. Install the front wheels.
18. Inspect and adjust the front wheel alignment.

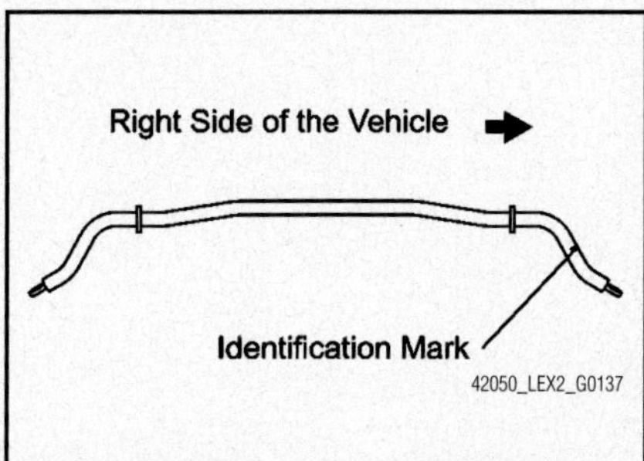

Fig. 246 Identification mark on the stabilizer bar—2006 GS 300 model (2WD)

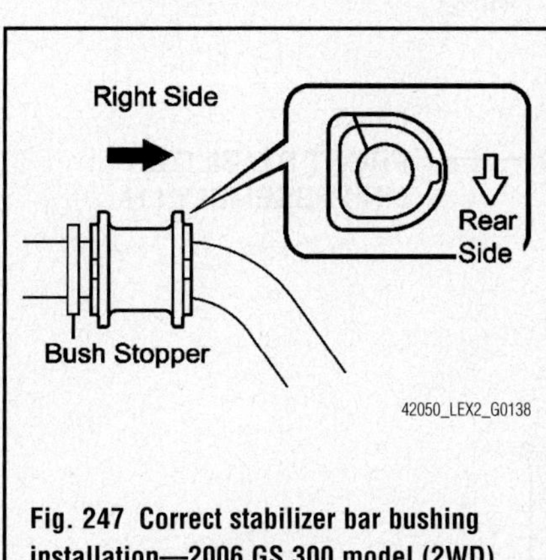

Fig. 247 Correct stabilizer bar bushing installation—2006 GS 300 model (2WD)

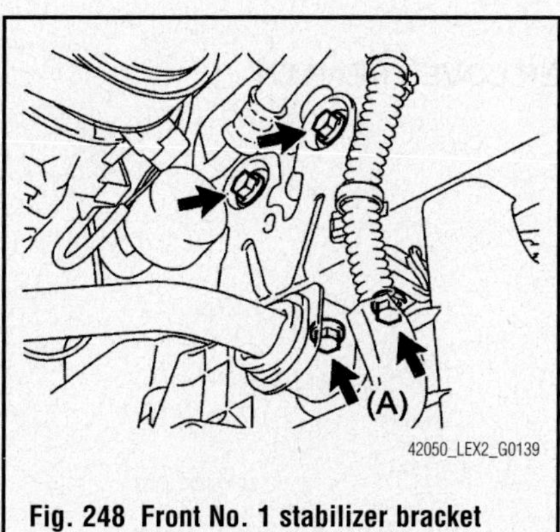

Fig. 248 Front No. 1 stabilizer bracket bolts—2006 GS 300 model (2WD)

GS 300 (4WD)

See Figure 249.

1. Before servicing the vehicle, refer to the precautions in the beginning of this section.
2. Disconnect the negative battery cable. Wait at least 90 seconds before performing any other work
3. Remove the engine assembly with transmission

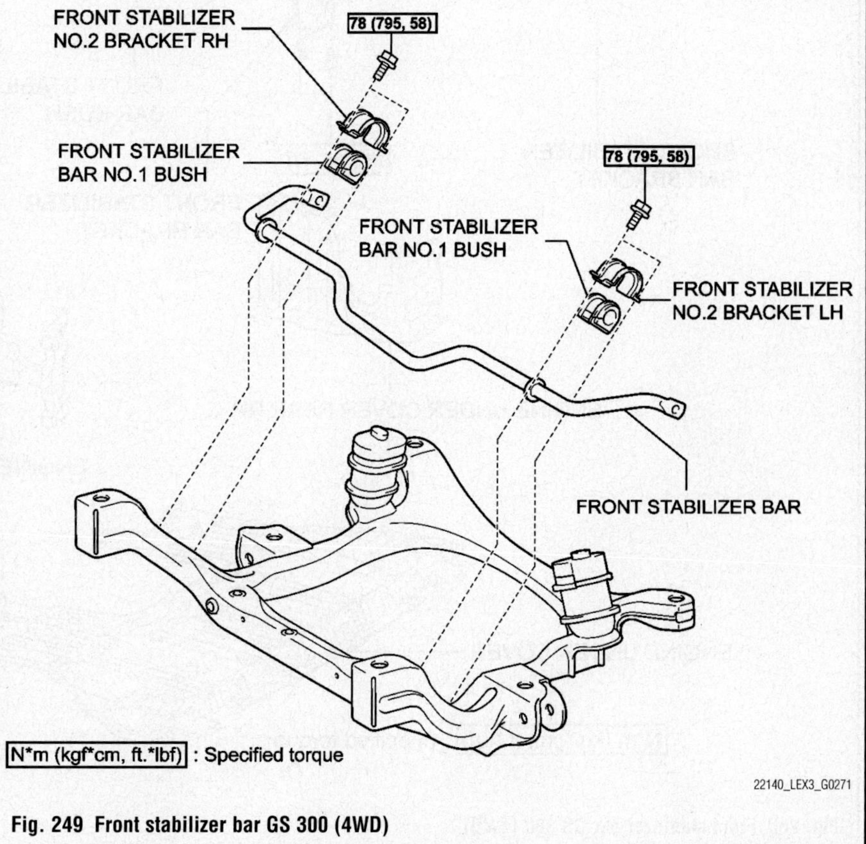

Fig. 249 Front stabilizer bar GS 300 (4WD)

4. Remove the 2 nuts and the left front stabilizer link assembly. Repeat for the right side. If the ball joint turns together with the nut, use a hexagon (6mm) wrench to hold the stud.

5. Remove the 2 bolts and the left front No. 2 stabilizer bracket from the front suspension crossmember. Repeat for the right side.

6. Remove the 2 front No. 1 stabilizer bar bushings from the front stabilizer bar.

7. Remove the front stabilizer bar from the vehicle.

To install:

8. Install the front stabilizer bar to the vehicle. The identification mark must be on the right side of the vehicle when installing the front stabilizer bar.

9. Install the 2 front No. 1 stabilizer bar bushings outside the bush stoppers on the front stabilizer bar as shown in the illustration. Be sure to install the front No. 1 stabilizer bar bushings so that the cutouts face the front of the vehicle.

10. Install the left front No. 2 stabilizer bracket on the vehicle with the 2 bolts. Repeat for the right side. Tighten the bolts to 57 ft. lbs. (78 Nm).

11. Install the left front stabilizer link assembly with the 2 nuts and tighten to 62 ft. lbs. (84 Nm). Repeat for the right side.

12. Install the engine assembly with transmission

13. Connect the negative battery cable.

SC 430

See Figure 250.

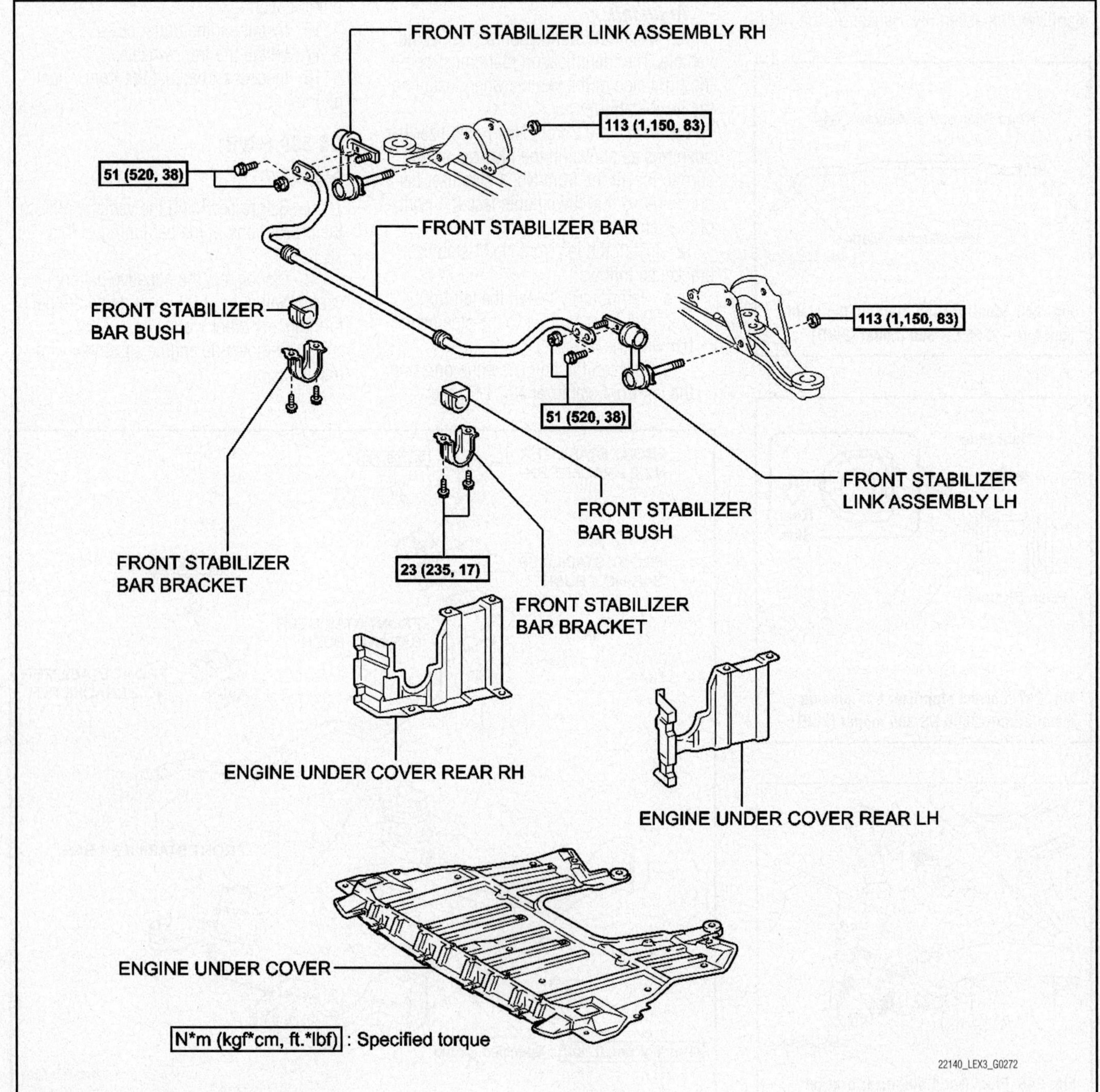

Fig. 250 Front stabilizer bar GS 300 (4WD)

1. Before servicing the vehicle, refer to the precautions in the beginning of this section.
2. Remove the front wheels.
3. Remove engine under covers.
4. Remove the bolt and nut, and disconnect the stabilizer bar from the stabilizer link.
5. Remove the nut and stabilizer link from the lower arm.
6. Remove the 4 bolts, 2 stabilizer bar brackets and stabilizer bar.
7. Remove the 2 bushings from the stabilizer bar.

To install:

8. Install the 2 bushings to the stabilizer bar.
9. Install the stabilizer bar and 2 stabilizer bar brackets with the 4 bolts.
10. Install the stabilizer link to the lower arm with the nut and tighten to 83 ft. lbs. (113 Nm).
11. Install the stabilizer link with the bolt and nut and tighten to 38 ft. lbs. (51 Nm).
12. Install engine under cover.
13. Install the front wheels.
14. Lower the vehicle.
15. Press down on the vehicle several times to stabilize the suspension.

UPPER BALL JOINT

REMOVAL & INSTALLATION

The upper ball joint is an integral part of the upper arm and is not replaced separately. The upper ball joint replacement is accomplished by replacing the upper arm.

UPPER CONTROL ARM

REMOVAL & INSTALLATION

GS 300

See Figure 251.

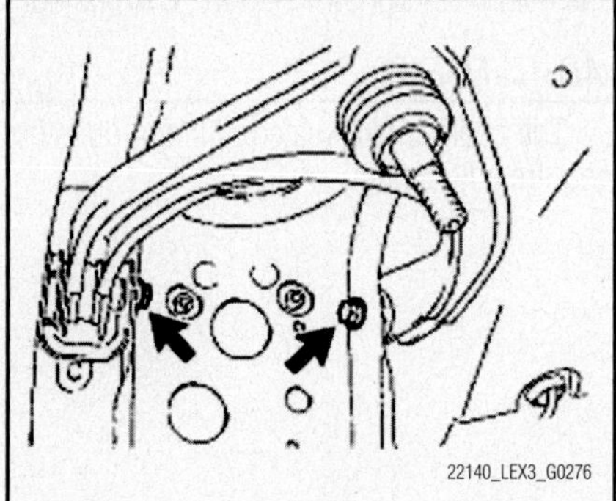

Fig. 251 Remove upper arm bolts as shown

1. Before servicing the vehicle, refer to the precautions in the beginning of this section.
2. Disconnect the negative battery cable. (For AVS system)
3. Remove engine room side cover.
4. Remove front wheel.
5. Separate the speed sensor.
6. Remove the nut and separate the front suspension lower arm from the height control sensor link sub-assembly front.
7. Remove the clip and castle nut.
8. Using SST 09628-00011, remove the upper ball joint from the steering knuckle.

➡Use wire or something similar to suspend the steering knuckle and prevent it from hanging down by the flexible hose.

9. For (4WD) Models separate the tie rod end.
10. Remove the 3 nuts and the upper strut cap. For (AVS system)
11. Turn the absorber control actuator counterclockwise 40° to remove it from the front shock absorber with coil spring.
12. Remove the front strut assembly
13. Remove the 2 bolts and front suspension upper arm.

To install:

14. Install the front suspension upper arm with the 2 bolts. Tighten the bolts to 36 ft. lbs. (49 Nm).
15. Install front strut assembly.
16. Install the front shock absorber with coil spring on the vehicle by tightening the 3 nuts on the suspension support side. Tighten nuts to 49 ft. lbs. (67 (Nm).
17. For (AVS) models install the actuator. Turn the actuator clockwise 40° until a click is felt.
18. Install the front shock absorber with coil spring to the front suspension lower arm.
19. Align the protrusion of the front shock absorber upper bracket plate and the cutout of the front shock absorber bracket lower. Then install the front shock absorber to the front door arm and loosely tighten the nut.
20. For (4WD) models install outer tie rod end, tighten the castle nut to 50 ft. lbs. (65 Nm). Install a new clip.
21. Install the front shock absorber cap upper with the 3 nuts. Tighten to 15 ft. lbs. (20 Nm).
22. Install the steering knuckle to the front suspension upper arm, and tighten it with the castle nut to 64 ft. lbs. (87 Nm).
23. Install a new clip to the steering knuckle.
24. Install the speed sensor to the front strut assembly and tighten the bolt to 53 inch. lbs. (6.0 Nm).
25. Connect the height control sensor link sub-assembly front with the nut. Tighten to 48 inch. lbs. (5.0 Nm).
26. Install engine room side cover.
27. Install the front wheel and tighten to 76 ft. lbs. (103 Nm).
28. Inspect and adjust front wheel alignment.
29. Reconnect the negative battery cable. (For AVS system)

SC 430

1. Before servicing the vehicle, refer to the precautions in the beginning of this section.
2. Raise and safely support the vehicle.
3. Remove or disconnect the following:
   - Wheel
   - Strut
   - Anti-lock Brake System (ABS) speed sensor wire harness from the upper control arm by removing the bolt.
   - Mounting bolts holding the upper control arm to the vehicle
   - Upper control arm

To install:

4. Install or connect the following:
   - Upper control arm and tighten the 2 mounting bolts to 39 ft. lbs. (53 Nm)
   - ABS speed sensor wire harness to the upper control arm with the attaching bolt. Tighten to 48 inch. lbs. (5.0 Nm).
   - Strut
   - Wheel and tighten to 76 ft. lbs. (103 Nm).
5. Lower the vehicle.
6. Stabilize the suspension.
7. Check and adjust the wheel alignment as necessary.

WHEEL HUB AND BEARING

REMOVAL & INSTALLATION

GS 300

1. Before servicing the vehicle, refer to the precautions in the beginning of this section.
2. Remove or disconnect the following:
   - Negative battery cable
   - Front wheel

- Caliper, leaving the brake line connected and suspend it out of the way

⁂ WARNING

Never allow the brake caliper to hang freely from the brake hose.

- Rotor
- Anti-lock Brake System (ABS) speed sensor and harness
- Tie rod from the arm on the lower ball joint
- Upper suspension arm from the steering knuckle
- Steering knuckle from the lower control arm
- Ball joint from the steering knuckle
- Front hub grease cap

3. Clamp the hub in a soft jaw vise.
4. Using a hammer and chisel, loosen the staked part of the locknut.
5. Remove or disconnect the following:
   - Locknut
   - ABS speed sensor rotor

➡Do NOT scratch the serrations of the sensor rotor.

- Brake dust cover bolts and shift the cover toward the outside.
- Hub from the steering knuckle
- Inner bearing race from the hub shaft
- Oil seal from the knuckle
- Bearing snapring from the steering knuckle
- Bearing from the steering knuckle

To install:

6. Install or connect the following:
   - New bearing into the steering knuckle

➡If the inner race and balls come loose from the bearing outer race, be sure to install them on the same side as before.

- Snapring
- New outside inner race and tap in the new seal. Tap the seal until it is flush with the end surface of the steering knuckle.
- Brake dust cover to the knuckle and tighten the bolts to 74 inch lbs. (8 Nm)
- Hub into the steering knuckle
- ABS speed sensor rotor
- Axle hub locknut. Tighten the nut to 147 ft. lbs. (199 Nm) and stake it.
- Grease cap to the steering knuckle by tapping lightly around the circumference of the cap with a hammer
- Ball joint to the steering knuckle. Tighten the 2 bolts to 83 ft. lbs. (113 Nm).
- Steering knuckle to the upper and lower suspension arms. Tighten the upper nut to 64 ft. lbs. (87 Nm) and the lower nut to 95 ft. lbs. (127 Nm). Install a new cotter pin on the lower nut. Install the clip on the upper suspension arm nut.
- Tie rod end to the steering knuckle. Tighten the nut to 64 ft. lbs. (87 Nm) and install a new cotter pin.
- Rotor, disc brake pads and the brake caliper
- ABS speed sensor and harness. Tighten the sensor retaining bolt to 69 inch lbs. (8 Nm).
- Wheel

7. Lower the vehicle and connect the negative battery cable.
8. Check the front wheel alignment.

SC 430

1. Before servicing the vehicle, refer to the precautions in the beginning of this section.
2. If equipped with air suspension, move the height control switch in the trunk area to the **OFF** position.
3. Remove or disconnect the following:
   - Front tire and wheel assembly
   - Brake caliper bracket from the steering knuckle, leaving the brake line connected. Support the caliper with a piece of wire.
   - Brake rotor
   - Anti-lock Brake System (ABS) speed sensor from the steering knuckle
   - Steering knuckle from the lower ball joint by removing the 2 bolts
   - Steering knuckle from the upper ball joint
   - Steering knuckle with the axle hub from the vehicle
   - Grease cap from the hub
   - Nut and the speed sensor rotor
   - 4 bolts and shift the brake dust cover towards the hub side
   - Axle hub from the steering knuckle
   - Outside inner race from the axle
   - Oil seal from the steering knuckle
   - Snapring and bearing from the steering knuckle

To install:

4. Install or connect the following:
   - Bearing in the steering knuckle
   - Snapring
   - Inner race (outside)
   - New oil seal until it is flush with the end surface of the steering knuckle
   - Brake dust cover to the steering knuckle and tighten the bolts to 74 inch lbs. (8.4 Nm)
   - Axle hub to the steering knuckle
   - ABS speed sensor
   - New nut on the axle shaft. Tighten the nut to 147 ft. lbs. (199 Nm). Stake the nut and install the grease cap.
   - Steering knuckle to the lower ball joint and tighten the bolts to 83 ft. lbs. (113 Nm)
   - Steering knuckle to the upper ball joint and tighten the nut to 48 ft. lbs. (65 Nm)
   - Brake rotor
   - Brake caliper and tighten the 2 bolts to 87 ft. lbs. (118 Nm)
   - Speed sensor to the steering knuckle
   - Front tire and wheel assembly
5. If equipped with air suspension, turn the height control switch to the **ON** position.

ADJUSTMENT

The bearing is a sealed unit and does not require adjustment.

See Figures 252 through 260.

COIL SPRING

REMOVAL & INSTALLATION

GS 300

See Figures 261 and 262.

1. Remove the rear strut assembly.
2. Install the coil spring compressor to the rear coil spring so that the distance between the upper and lower hooks is as wide as possible within the installation area.
3. Compress the rear coil spring until it can be moved freely.
4. Remove the nut.
5. Remove the rear shock absorber cushion washer No.1, rear shock absorber cushion No.1, rear suspension support assembly, rear coil spring insulator upper, collar, rear shock absorber cushion No. 2, and rear spring bumper No. 1.
6. Remove the rear coil spring and rear coil spring insulator lower from the rear shock absorber. Release the spring compressor and remove the rear coil spring.

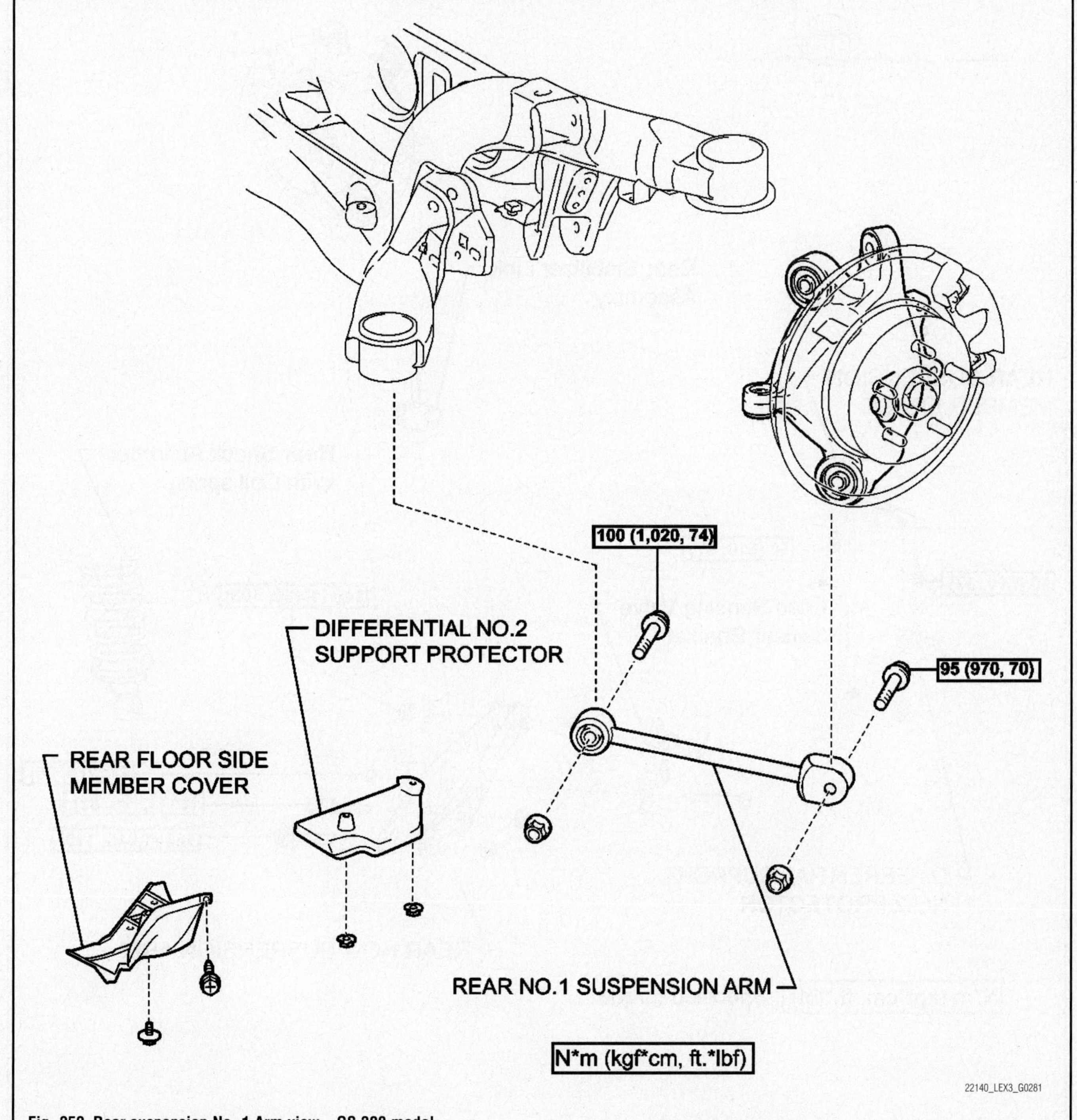

Fig. 252 Rear suspension No. 1 Arm view—GS 300 model

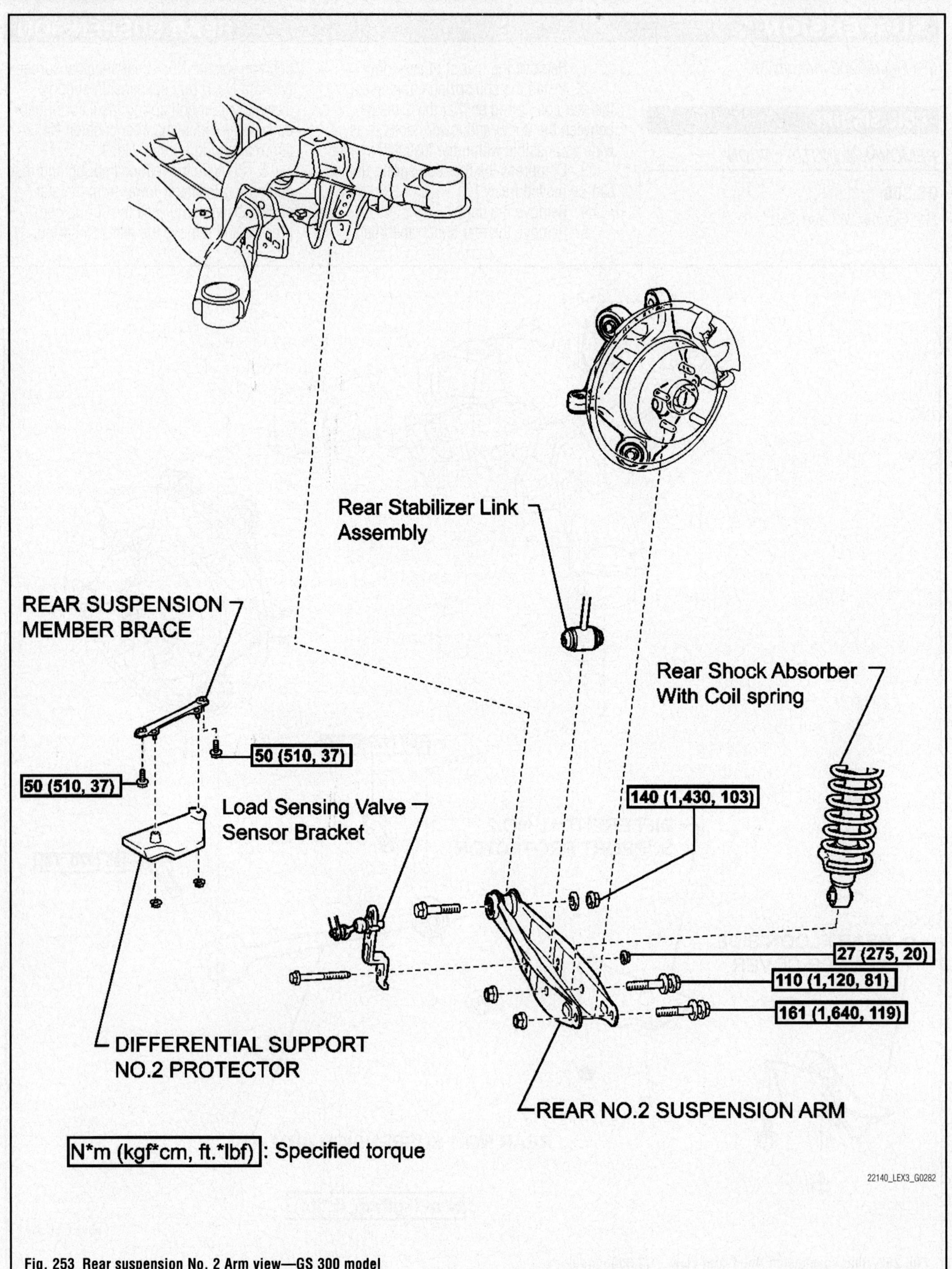

Fig. 253 Rear suspension No. 2 Arm view—GS 300 model

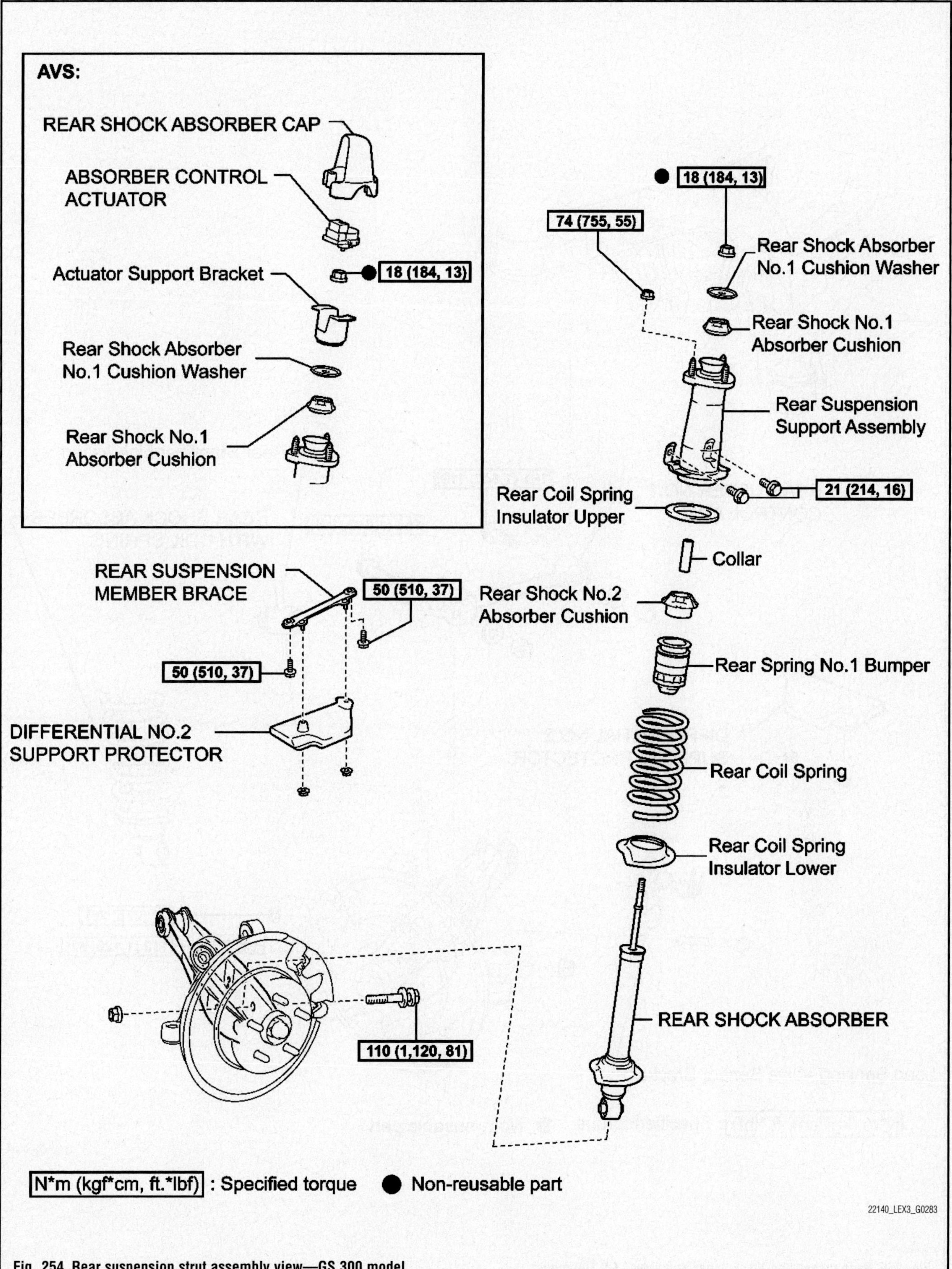

Fig. 254 Rear suspension strut assembly view—GS 300 model

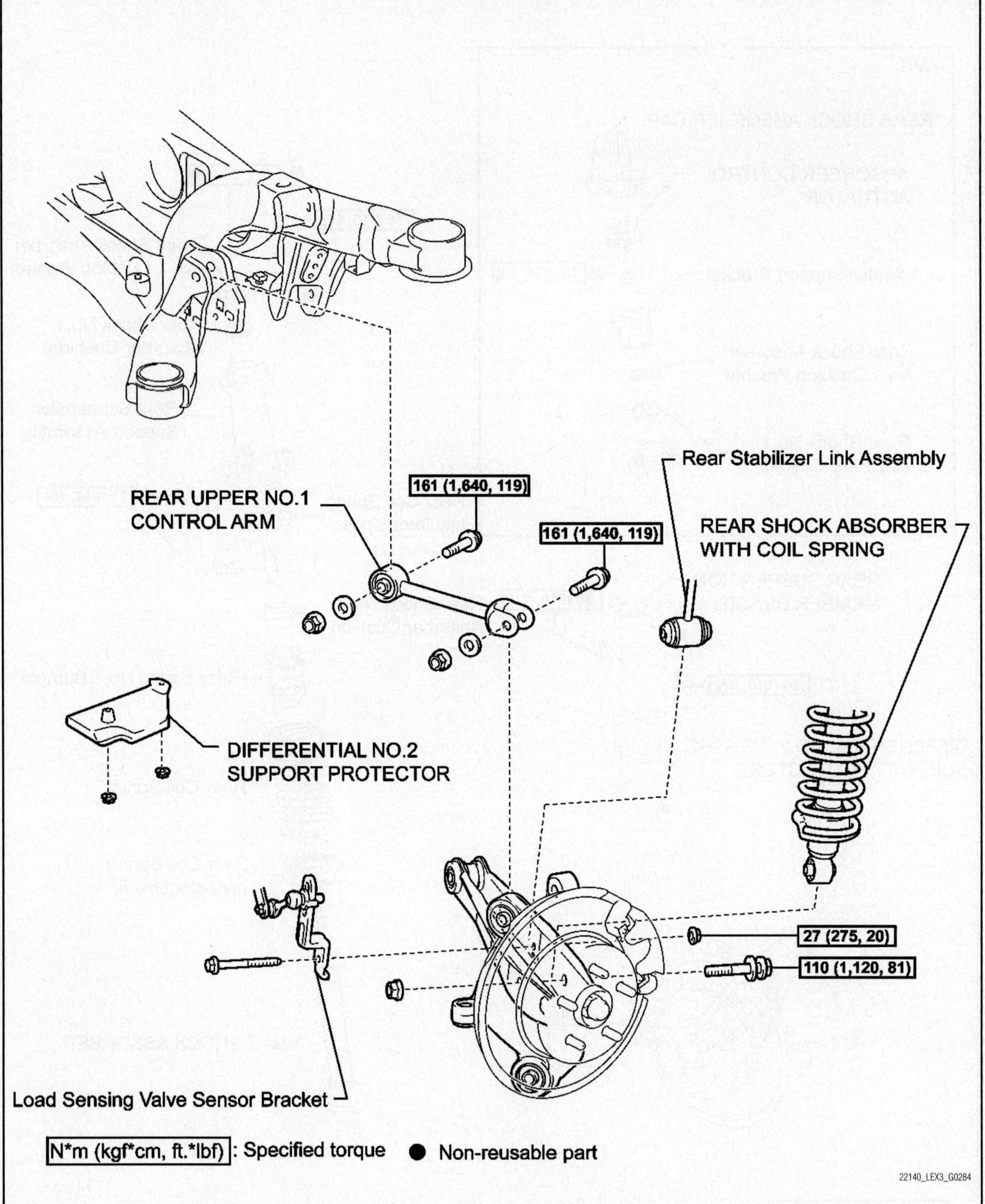

Fig. 255 Rear suspension No. 1 upper arm view—GS 300 model

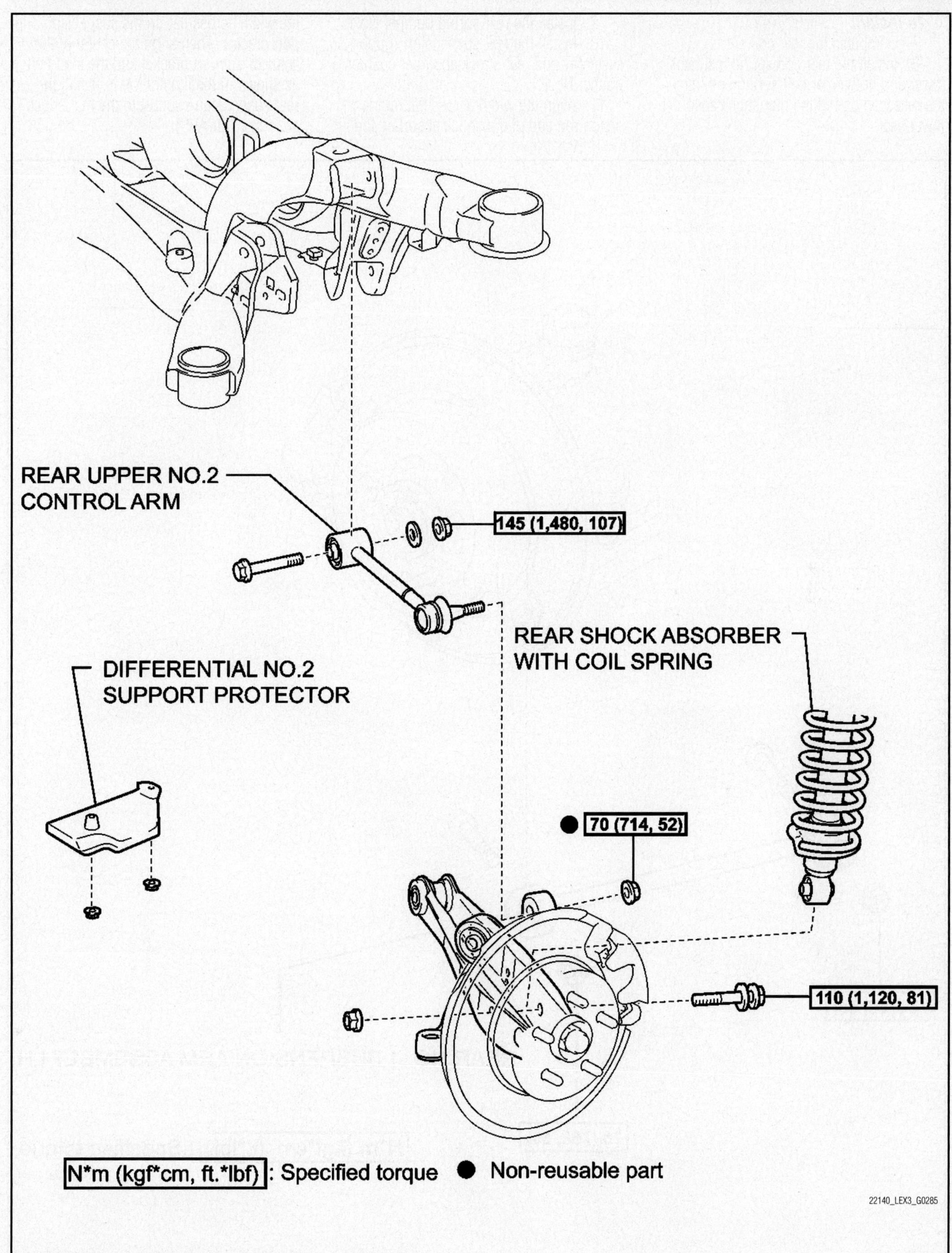

Fig. 256 Rear suspension No. 2 upper arm view—GS 300 model

To install:

7. Compress the rear coil spring.

8. Install the rear coil spring insulator lower and fit the rear coil spring end into the recessed part of the rear shock absorber lower seat.

9. Install the rear spring bumper No.1.

10. Install the rear suspension support assembly and rear shock absorber cushion washer No.1.

11. Align the width across flat on the piston rod end of the shock absorber and the width across flat on the suspension support bracket. Then align the center of the actuator support bracket and the stud bolt as shown in the illustration to install the rear suspension support to the rear shock absorber. (for AVS)

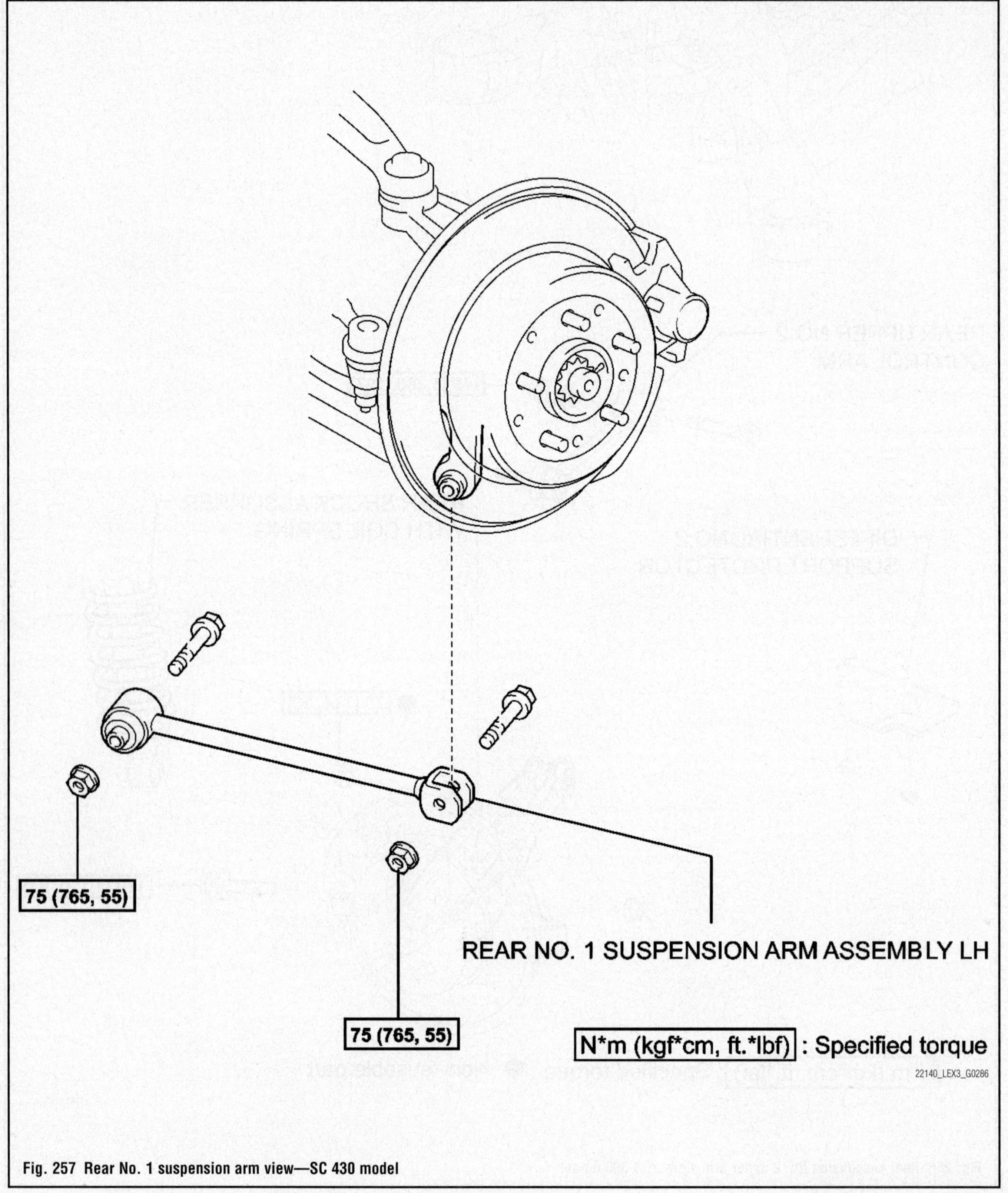

Fig. 257 Rear No. 1 suspension arm view—SC 430 model

12. Adjust the rear suspension support assembly to the installation position of the lower part of the rear shock absorber so that the studs are positioned as shown in the illustration

13. Remove the coil spring compressor from the rear coil spring.

14. Install the strut assembly.

SC 430

1. Before servicing the vehicle, refer to the precautions in the beginning of this section.

2. Remove the strut assembly.

3. Remove the coil spring from strut as follows:

- Install the coil spring compressor to the coil spring so that the 2 hooks on the compressor are fully extended.
- Fully compress the coil spring.

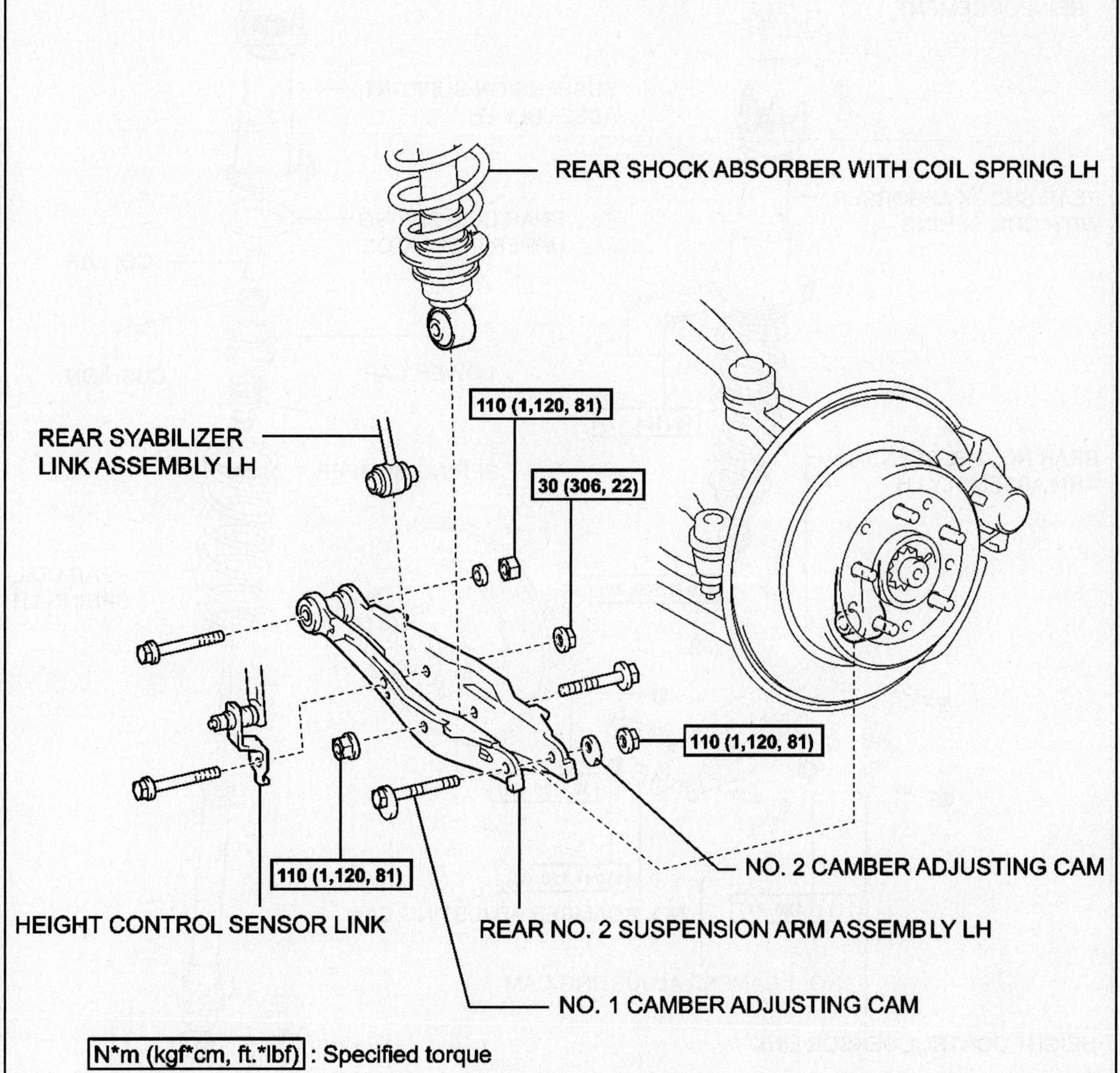

Fig. 258 Rear No. 2 suspension arm view—SC 430 model

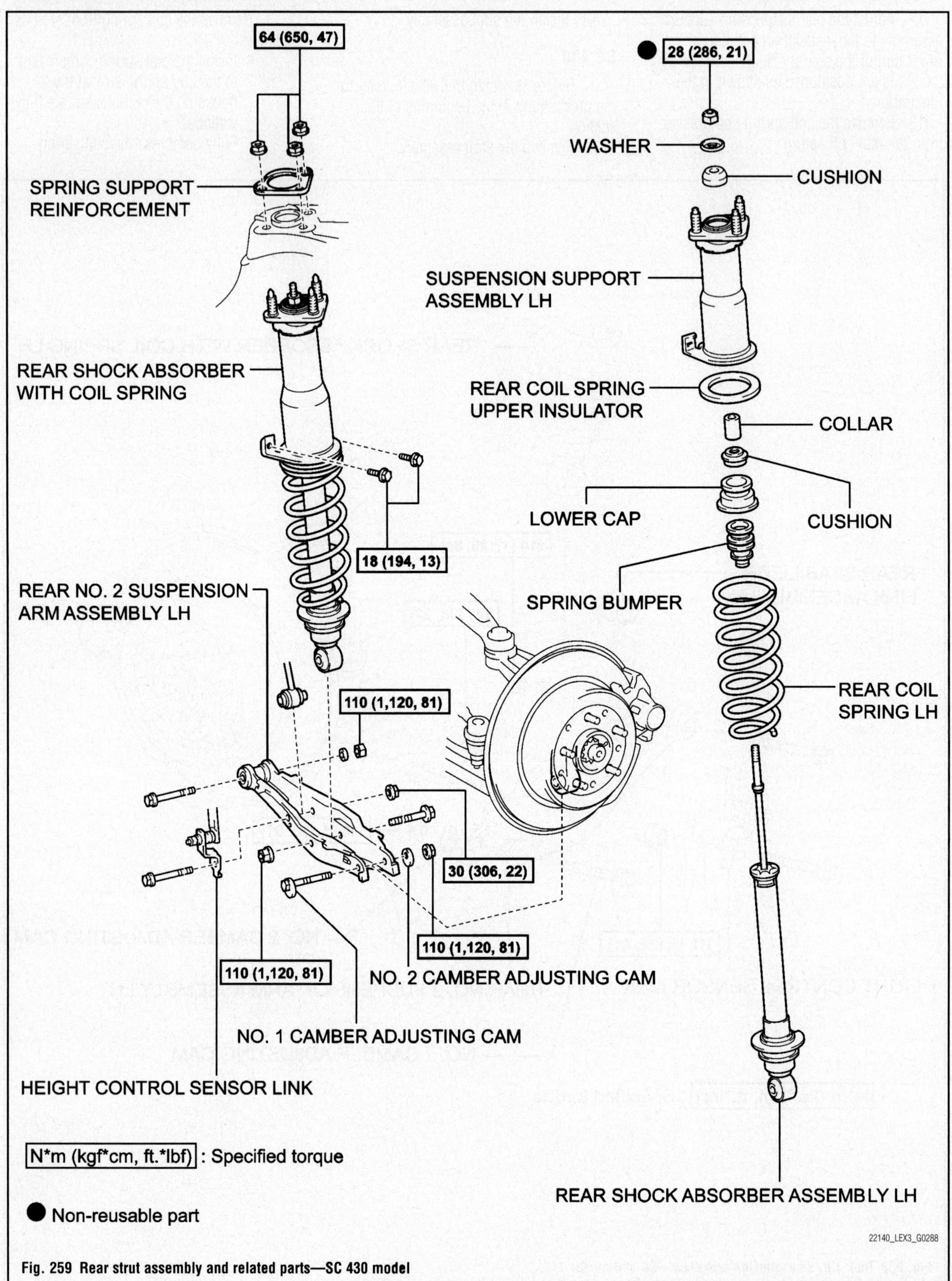

Fig. 259 Rear strut assembly and related parts—SC 430 model

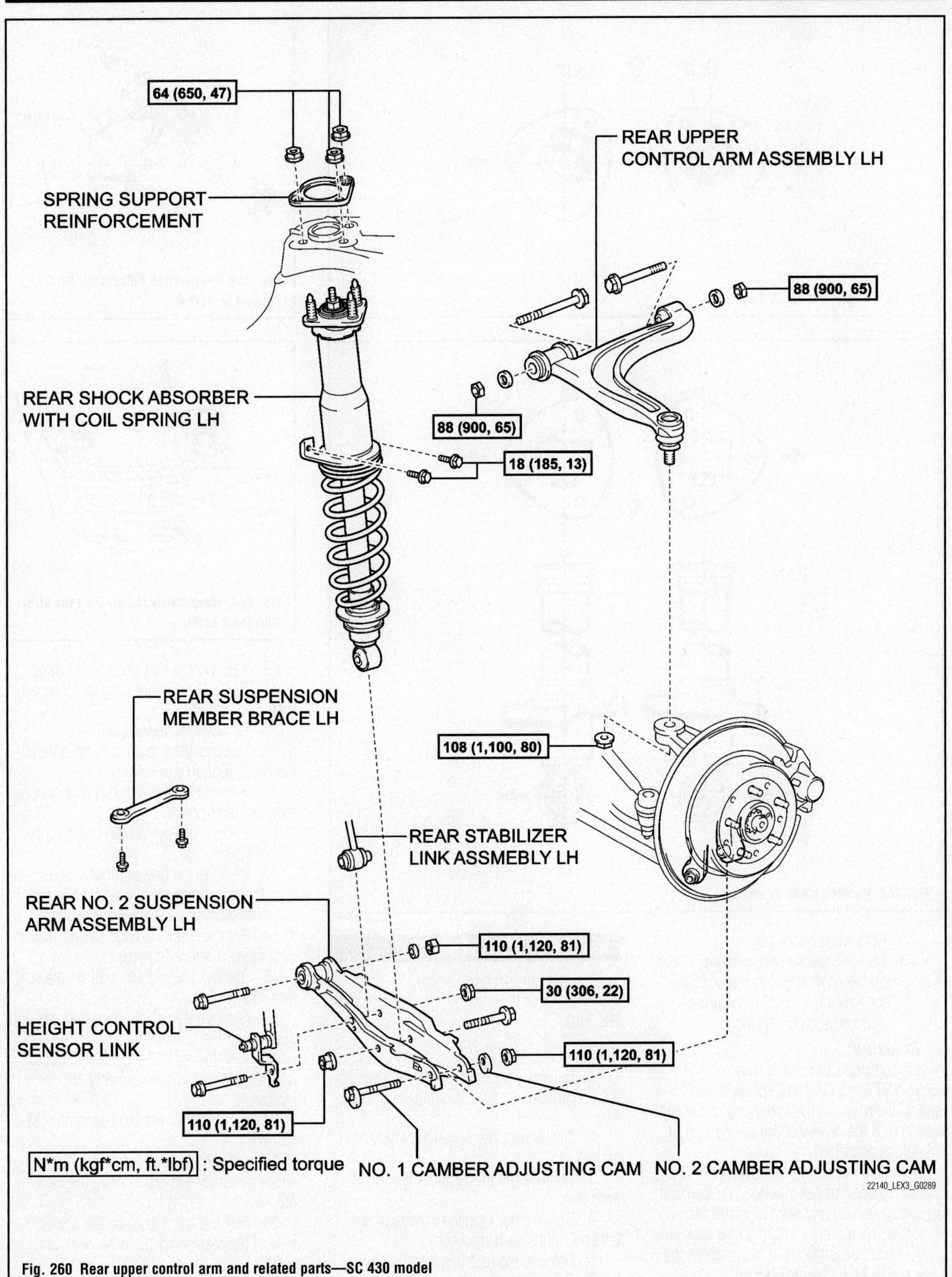

Fig. 260 Rear upper control arm and related parts—SC 430 model

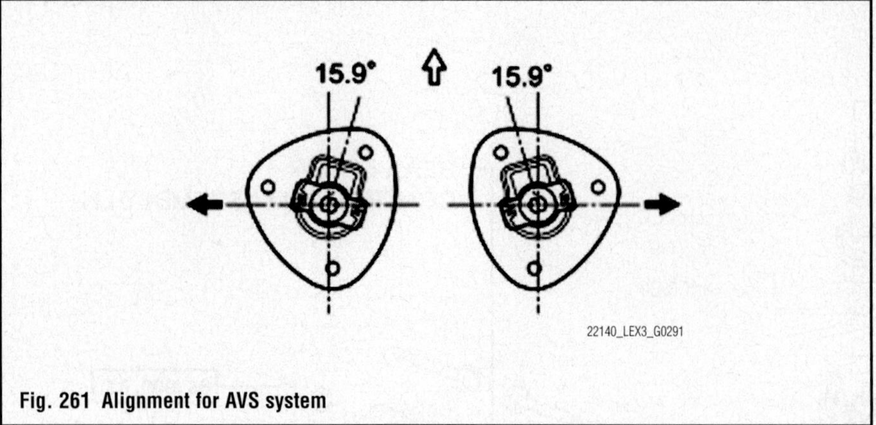

Fig. 261 Alignment for AVS system

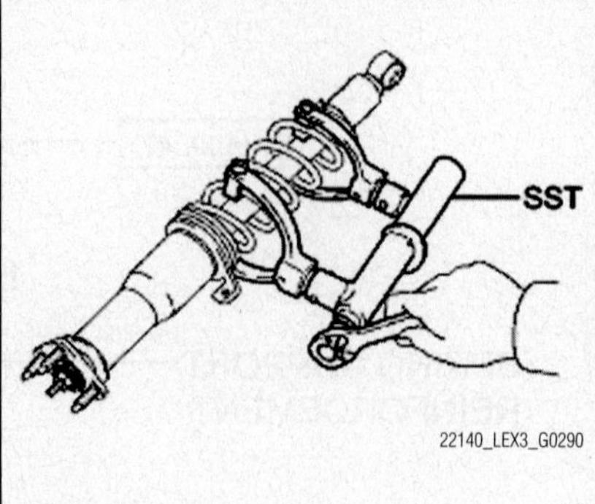

Fig. 263 Remove the differential No.2 support protector

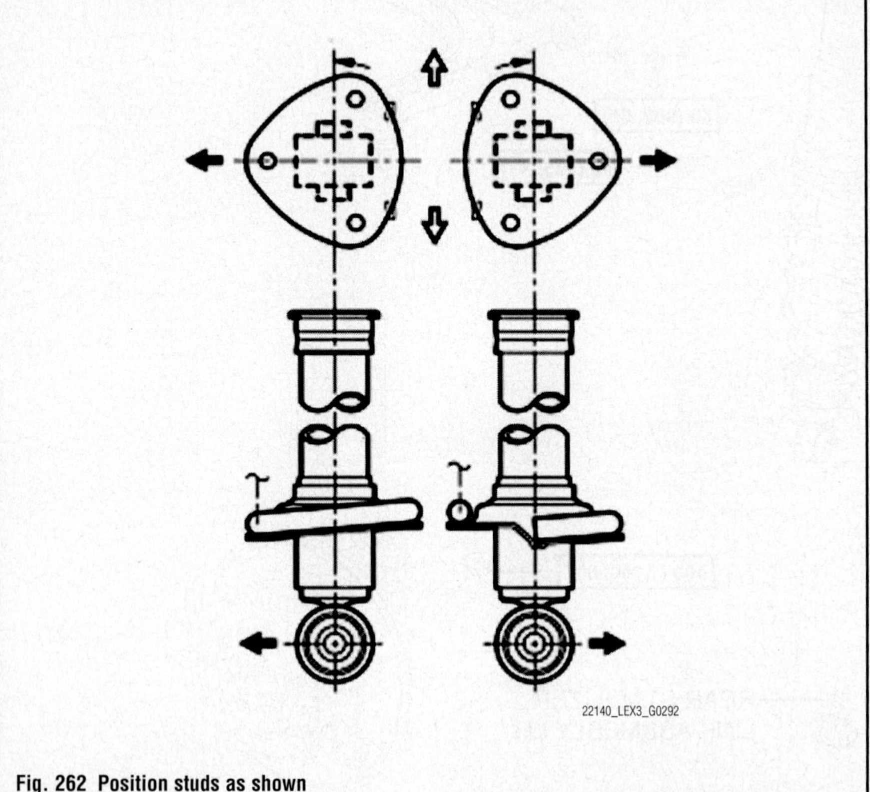

Fig. 262 Position studs as shown

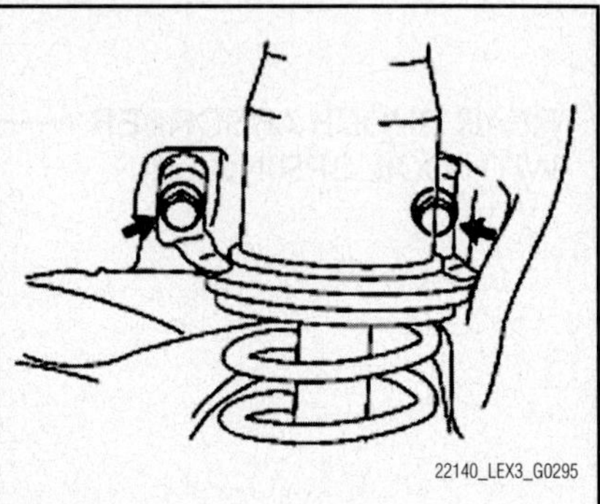

Fig. 264 Temporarily install the rear Strut with the 2 bolts

- Remove the lock nut.
- Remove the washer, cushion, collar, suspension support, upper insulator, lower cup, cushion, spring bumper and coil spring.

To install:

4. Compress the coil spring.
5. SST Install the coil spring to the shock absorber. Fit the lower end of the coil spring into the recess of the spring seat of the shock absorber.
6. Install the spring bumper, lower cup, cushion, collar, upper insulator, suspension support, cushion and washer to the shock absorber. Temporarily install a new lock nut.
7. Install assembly and then tighten the lock nut to 21 ft. lbs. (28 Nm).

MACPHERSON STRUTS

REMOVAL & INSTALLATION

GS 300

See Figures 263 and 264.

1. Before servicing the vehicle, refer to the precautions in the beginning of this section.
2. Disconnect the negative battery cable.
3. Remove the luggage compartment floor mat.
4. Remove the luggage compartment left and right side trim covers.
5. Remove the tool box board.
6. Remove the tool box.
7. Remove the rear floor finish plate.
8. Remove the inner luggage compartment trim cover.
9. Remove the rear wheel.
10. Remove the 2 nuts and the differential No.2 support protector.
11. Remove the 2 bolts and rear suspension member brace.
12. Loosen the nut on the rear suspension member side.
13. Remove the bolt and nut and disconnect the rear stabilizer link assembly and the load sensing valve sensor bracket.
14. Remove the bolt and nut and disconnect the rear shock absorber.
15. Remove the nut and bolt on the rear axle carrier side.
16. Remove rear shock absorber cap and disconnect the connector.
17. Turn the absorber control actuator counterclockwise approximately 40° to remove it.
18. Remove the rear strut assembly as follows:
19. Loosen the lock nut of the rear shock absorber. Do not remove the lock nut.
20. Remove the 3 nuts on the upper side of the rear shock absorber with coil spring.

21. Remove the 2 bolts and the rear shock absorber with coil spring from the body.

To install:

22. Temporarily install the rear shock absorber with rear coil spring with the 2 bolts.
23. Install the 3 nuts on the upper side of the rear shock absorber with coil spring. Tighten to 55 ft. lbs. (74 Nm).
24. Tighten the lock nut to 13 ft. lbs. (18 Nm).
25. Tighten the 2 bolts to 16 ft. lbs. (21 Nm).
26. Install the absorber control actuator to the actuator support bracket. Turn the actuator clockwise approximately 40° until a click is felt.
27. Connect the connector and install the shock absorber cap.
28. Temporarily install the rear suspension No.2 arm with the bolt and nut.
29. Temporarily install the rear shock absorber with rear coil spring with the bolt and nut.
30. Temporarily install the rear stabilizer link assembly and the load sensing valve sensor bracket to the rear suspension No. 2 arm with the bolt and nut.
31. Install the rear wheel and stabilize the suspension.
32. Remove the rear wheel.
33. Jack up the axle carrier, with a wooden block placed between the jack and axle carrier, to apply a load to the suspension so that the rear drive shaft assembly becomes level.
34. Fully tighten the bolt on the axle carrier side to 119 ft. lbs. (161 Nm).
35. Fully tighten the bolt holding the rear shock absorber with coil spring to 81 ft. lbs. (110 Nm).
36. Fully tighten the nut holding the rear stabilizer link assembly to 20 ft. lbs. (27 Nm).
37. Fully tighten the nut on the rear suspension member side to 103 ft. lbs. (140 Nm).
38. Install the rear suspension member brace with the 2 bolts. Tighten to 37 ft. lbs. (50 Nm).
39. Install the differential No. 2 support protector with the 2 nuts.
40. Install the rear wheel and tighten to 76 ft. lbs. (103 Nm).
41. Install the inner luggage compartment trim cover.
42. Install the rear floor finish plate.
43. Install the tool box.
44. Install the tool box board.
45. Install both side trim covers.
46. Install the compartment floor mat.
47. Check the rear wheel alignment.
48. Connect the negative battery cable.
49. Check head light alignment.

SC 430

See Figures 265 through 268.

1. Before servicing the vehicle, refer to the precautions in the beginning of this section.
2. Remove the right and left tonneau cover sub-assembly.

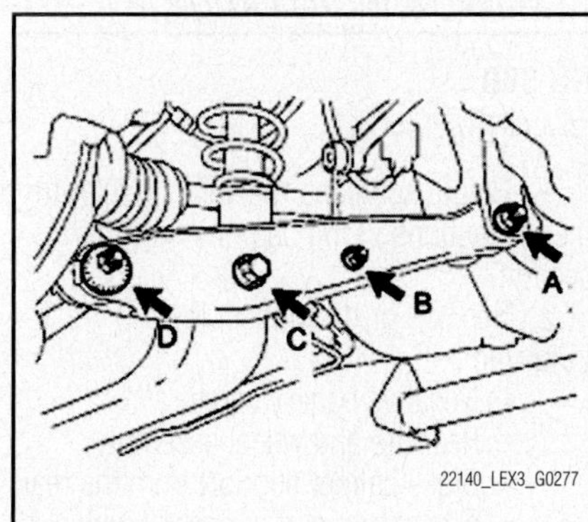

Fig. 265 No. 2 Suspension arm, nut (A) and (B) view

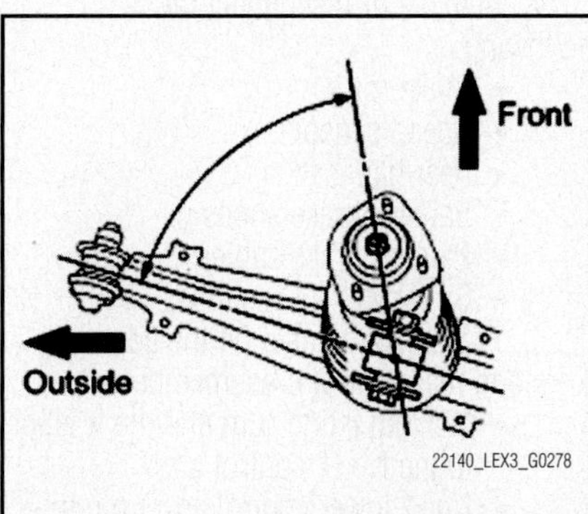

Fig. 266 Suspension support alignment view

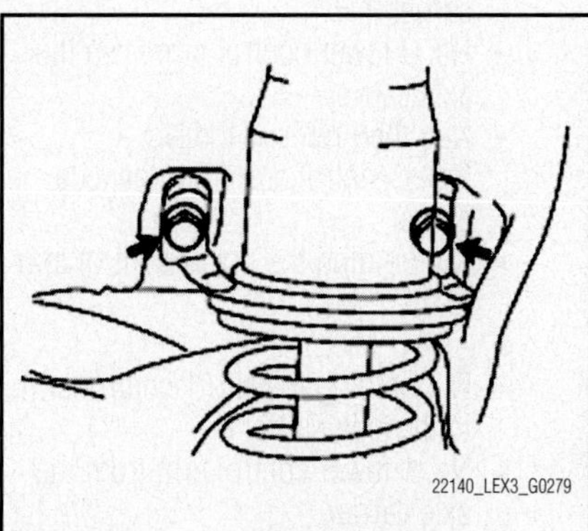

Fig. 267 Install the 2 bolts and tighten

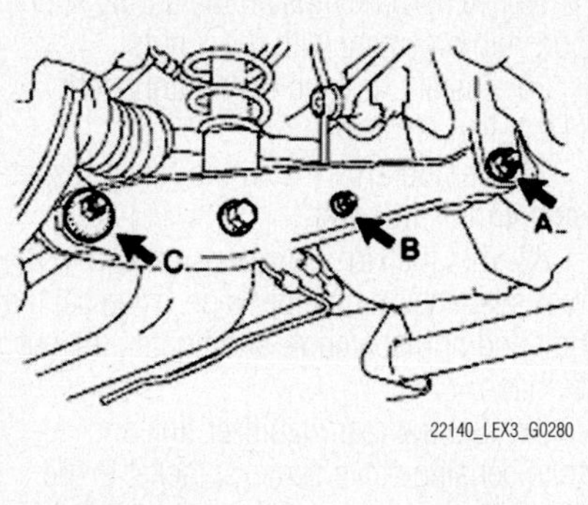

Fig. 268 Lower control arm No. 2 nuts (A), (B) and (C)

3. Remove the rear floor finish plate.
4. Remove the left and right trim covers.
5. Remove the spare tire assembly.
6. Remove the tool box cover.
7. Remove the luggage compartment trim cover rear sub-assembly.
8. Remove the luggage compartment front trim cover.
9. Remove the rear tonneau cover assembly.
10. Remove the inner lower trim cover.
11. Remove the front and both side trim covers.
12. Remove the rear wheel.
13. Disconnect the rear No. 2 Suspension arm as follows:
    - Place matchmarks on the camber adjusting cam and No. 2 suspension arm.
    - Fix the bolt on the rear suspension member side, and loosen the nut labeled (A). Do not remove the nut.
    - Fix the bolt and remove the nut labeled (B). Then disconnect the rear stabilizer link and road sensing valve sensor bracket from the No. 2 suspension arm.
    - Fix the nut and remove the bolt labeled C. Then disconnect the strut lower end from the No. 2 suspension arm.
14. Remove the 3 upper strut nuts.
15. Remove the 2 bolts and shock absorber with coil spring.

To install:

16. Temporarily install the rear strut assembly as follows:
17. Install the shock absorber to the lower suspension arm.
18. Rotate the suspension support so that the rod and a bolt on the suspension support are aligned with the lower shock absorber as shown in the illustration.

19. Temporarily install the spring support reinforcement with the 3 nuts.
20. Install the 2 bolts and tighten to (13 ft. lbs. (18 Nm).
21. Temporarily install the No. 2 suspension arm as follows:
22. Set the No. 2 suspension arm to the rear suspension member side. Then set the bolt and nut labeled A, and tighten the nut by hand.
23. Set the rear stabilizer link and road sensing valve sensor bracket to the No. 2 suspension arm. Then set the bolt and nut labeled B, and tighten the nut by hand.
24. Set the struts lower end to the No. 2 suspension arm. Then set the nut and bolt labeled C, and tighten the bolt by hand.
25. Using the bolt and nut labeled D, install the No. 1 camber adjusting cam, No. 2 camber adjusting cam and No. 2 suspension arm to the axle carrier.
26. Install the rear wheel and tighten to 76 ft. lbs. (103 Nm).
27. Stabilize the suspension by lowering vehicle and press down on the vehicle several times to stabilize the suspension.
28. Tighten the rear strut assembly as follows:
- Tighten the 3 upper nuts to 47 ft. lbs. (64 Nm).
- Tighten the strut lock nut to 21 ft. lbs. (28 Nm).
- Fix the lower strut assembly bolt and tighten the nut to 81 ft. lbs. (110 Nm).

29. Tighten the No. 2 suspension arm assembly. Fix the bolt and tighten the nuts labeled A, B and C.
30. Install the rear wheel and tighten to 76 ft. lbs. (103 Nm).
31. Inspect and adjust the rear wheel alignment

STABILIZER BAR

REMOVAL & INSTALLATION

1. Remove the rear wheels.
2. Remove the 2 nuts and left stabilizer bar link. If the ball joint turns together with the nut, use a hexagon wrench (5 mm) to hold the stud. Repeat for the right side.
3. Remove the 8 bolts, 2 No. 1 brackets and 2 bushings. Two types of bolts are used, so make sure the correct bolts are installed.
4. Remove the stabilizer bar.

To install:

5. Install the stabilizer bar.
6. Install the bushing and bracket with the 2 bolts and tighten to 14 ft. lbs. (19 Nm). Install the bushing to the inner side of the bushing stopper on the stabilizer bar.
7. Connect the stabilizer bar links and tighten the nuts to 29 ft. lbs. (39 Nm). Repeat for the right side.
8. Install the rear wheels.
9. Inspect and adjust rear wheel alignment.

WHEEL HUB AND BEARING

REMOVAL & INSTALLATION

GS 300

See Figure 269.

1. Before servicing the vehicle, refer to the precautions in the beginning of this section.
2. Remove or disconnect the following:
   - Negative battery cable.
   - Rear tire and wheel assembly
   - Brake caliper support from the rear axle carrier and support it with a piece of wire
3. Place matchmarks on the disc brake rotor and the axle hub.
4. Remove or disconnect the following:
   - Brake rotor
   - Speed sensor
   - Rear halfshaft
   - Parking brake shoes
   - Parking brake cable
   - Strut rod
5. Place matchmarks on the adjusting cam and rear control crossmember.
   - Nut, adjusting cam and the washer to the No. 1 control arm
   - No. 1 lower control arm from the crossmember
   - Loosen the nut holding the lower control arm to the axle carrier
   - No. 2 lower control arm from the axle carrier
   - Nut, then remove the No. 2 lower control arm from the axle carrier
   - Nut holding the upper control arm to the axle carrier
   - Axle carrier
   - Nut holding the No. 1 control arm to the axle carrier
   - No. 1 lower control arm from the axle carrier
   - Dust deflector
   - Axle hub from the carrier
   - Backing plate
   - Inner race (outside)
   - Oil seal
   - Snapring
   - Bearing

To install:

6. Install or connect the following:
   - Bearing to the axle carrier

➡If the inner races come loose from the bearing outer race, be sure to install them on the same side as before.

   - Snapring. Install the inner race (outside) and a new oil seal.
   - Backing plate. Install the inner race (inside) and press in the axle hub with the proper tools.
   - Inner oil seal. Align the holes for the speed sensor in the dust deflector and axle carrier. Install the dust deflector.
   - No. 1 lower arm to the axle carrier and install a new nut. Tighten the nut to 43 ft. lbs. (59 Nm).
   - Upper control arm to the axle carrier. Tighten the new nut and bolt to 80 ft. lbs. (109 Nm).
   - No. 2 lower control arm to the axle carrier and tighten a new nut to 110 ft. lbs. (150 Nm)
   - No. 1 lower control arm to the rear crossmember. Tighten the nut to 136 ft. lbs. (184 Nm).
   - Strut rod to the axle carrier. Tighten the nuts and bolts to 134 ft. lbs. (184 Nm).
   - Parking brake cable and slide the backing plate to the inside. Install the hex bolt and tighten it to 132 ft. lbs. (180 Nm).
   - Shoe guide plate set bolt. Tighten the bolt to 13 ft. lbs. (18 Nm).
   - 4 hub bolts and tighten them to 19 ft. lbs. (26 Nm)
   - Bolts at the speed sensor and tighten them to 69 inch lbs. (8 Nm)
   - Parking brake shoes
   - Halfshafts. Apply the brakes and tighten the locknut to 213 ft. lbs. (289 Nm).
   - Brake rotor
   - Brake caliper support to the rear axle carrier. Tighten the bolts to 77 ft. lbs. (104 Nm).
   - Rear tire and wheel assembly
   - Negative battery cable
7. Lower the vehicle and bounce it a few times to stabilize the suspension.

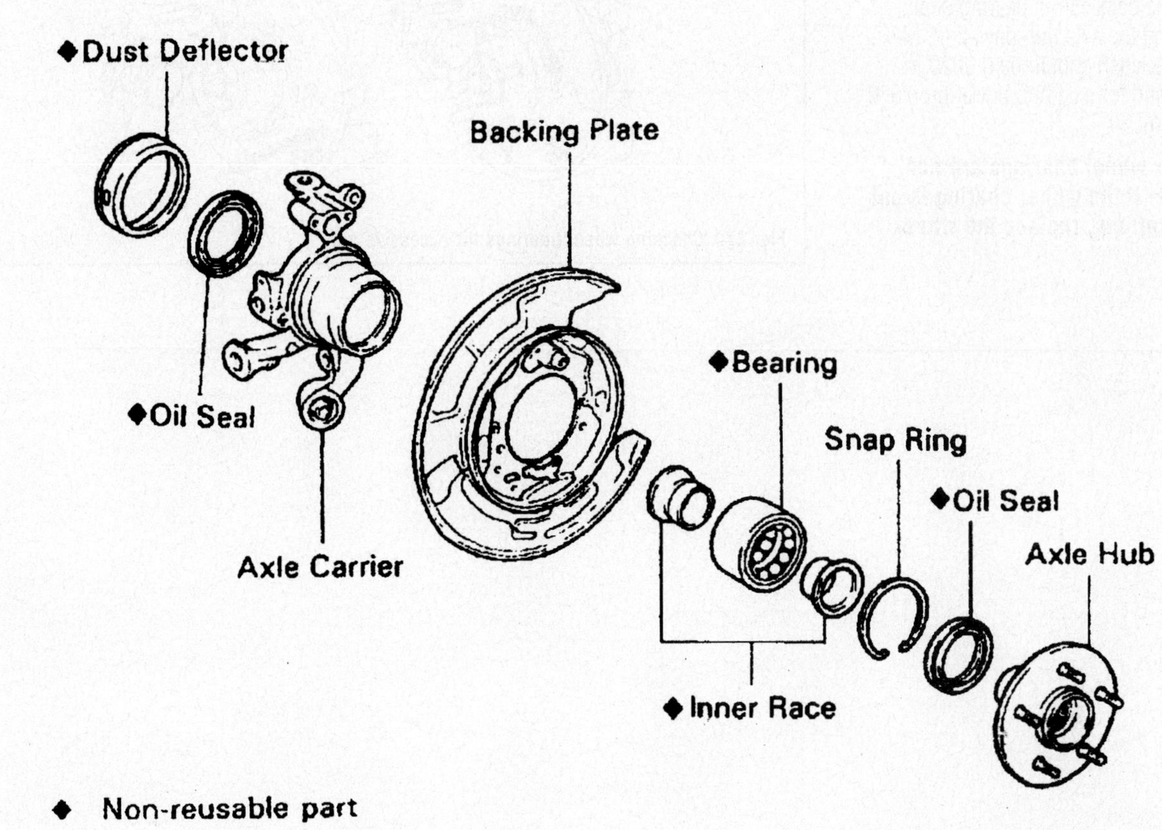

Fig. 269 Exploded view of the axle carrier—GS 300

SC 430

1. Before servicing the vehicle, refer to the precautions in the beginning of this section.
2. Remove or disconnect the following:
   - Rear tire and wheel assembly
   - Brake caliper support bracket
   - Brake rotor
   - Speed sensor
   - Rear halfshaft
   - Parking brake shoes
   - 2 bolts at the parking brake cable. Remove the 2 hub bolts and the hex bolt. Slide the backing plate to the outside and disconnect the parking brake cable.
   - Strut rod at the axle carrier
   - Nut, then press out the upper suspension arm.
   - Nut, then press out the No. 2 lower suspension arm
   - Axle carrier
   - Dust deflector and pull out the oil seal
   - Axle hub from the carrier
   - Backing plate
   - Inner race (outside) from the hub
   - Oil seal
   - Snapring
   - Bearing and inner race (inside)

To install:

3. Install or connect the following:
   - Bearing to the axle carrier

➡If the inner races come loose from the bearing outer race, be sure to install them on the same side as before.

   - Snapring, the inner race (outside) and a new oil seal
   - Backing plate. Install the inner race (inside) and press in the axle hub with the proper tools.
   - New dust deflector
   - Upper arm to the axle carrier. Tighten the nut and bolt to 65 ft. lbs (88 Nm), rear 55 ft. lbs (74 Nm)
   - No. 2 lower arm to the carrier and tighten a new nut to 81 ft. lbs. (110 Nm).
   - Toe control link with camber adjusting cams. Tighten to 36 ft. lbs. (49 Nm). Stabilize and retighten to 44 ft. lbs. (59 Nm).
   - Rear drive shaft assembly
   - Parking brake cable and brake assembly
   - Install the parking brake shoes and the ABS sensor.
   - Brake rotor
   - Brake caliper to the rear axle carrier by installing the 2 bolts.
   - Tighten rear suspension arm assembly to 81 ft. lbs. (110 Nm)
4. Inspect and adjust to rear alignment
5. Perform speed sensor signal check.

ADJUSTMENT

See Figure 270.

Check the backlash in bearing shaft direction and the axle hub deviation. Maximum for backlash should be 0.0020 in. (0.05mm) and for axle hub deviation 0.020 in. (0.05mm).

➡The rear wheel bearings are non-adjustable. If the wheel bearing is out of specifications, replace the wheel bearing.

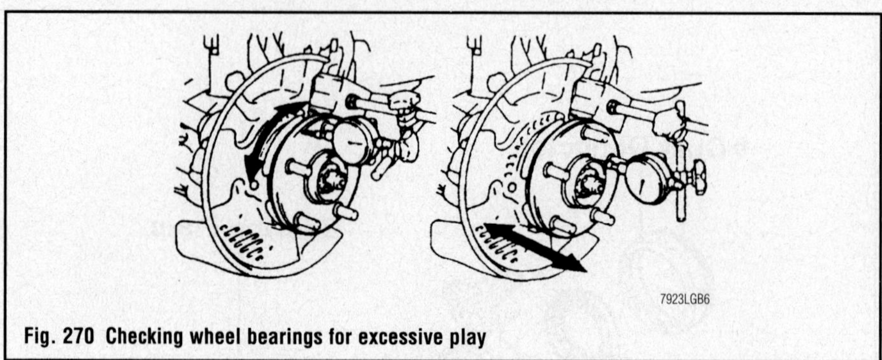

Fig. 270 Checking wheel bearings for excessive play

LEXUS

RX 350

6

SPECIFICATIONS AND MAINTENANCE CHARTS

ENGINE AND VEHICLE IDENTIFICATION

Engine							Model Year	
Code ①	Liters (cc)	Cu. In.	Cyl.	Fuel Sys.	Engine Type	Eng. Mfg.	Code ②	Year
2GR-FE	3.5 (3456)	210	6	SFI	DOHC	Toyota	7	2007
2GR-FE	3.5 (3456)	210	6	SFI	DOHC	Toyota	8	2008

SFI: Sequential Fuel Injection

DOHC: Double Overhead Camshaft

NA: Information not available

① Stamped on the left side of the engine block

② 10th digit of the Vehicle Identification Number (VIN)

22140_RX35_C0001

GENERAL ENGINE SPECIFICATIONS

Year	Model	Engine Displacement Liters	Engine Series ID	Net Horsepower @ rpm	Net Torque @ rpm (ft. lbs.)	Bore x Stroke (in.)	Com-pression Ratio	Oil Pressure @ rpm
2007	RX350	3.5	2GR-FE	270@6200	251@4700	3.70x3.27	10.8:1	36-78@3000
2008	RX350	3.5	2GR-FE	270@6200	251@4700	3.70x3.27	10.8:1	36-78@3000

22140_RX35_C0002

ENGINE TUNE-UP SPECIFICATIONS

Year	Engine Displacement Liters	Engine ID	Spark Plug Gap (in.)	Ignition Timing (deg.)*	Fuel Pump (psi)	Idle Speed (rpm)	Valve Clearance	
							Intake	Exhaust
2007	3.5	2GR-FE	0.039-0.043	N/A	44-50	650-750	NA	NA
2008	3.5	2GR-FE	0.039-0.043	N/A	44-50	650-750	NA	NA

NOTE: The Vehicle Emission Control Information label often reflects specification changes made during production.

The label figures must be used if they differ from those in this chart.

NA: Not available

22140_RX35_C0003

CAPACITIES

Year	Model	Engine Displacement Liters	Engine ID	Engine Oil with Filter (qts.)	Transmission (pts.) 5-Spd	Transmission (pts.) Auto.*	Transfer Case (pts.)	Drive Axle Front (pts.)	Drive Axle Rear (pts.)	Fuel Tank (gal.)	Cooling System (qts.)
2007	RX350	3.5	2GR-FE	6.4	—	3.7	2.0	—	2.0	19.2	8.9
2008	RX350	3.5	2GR-FE	6.4	—	3.7	2.0	—	2.0	19.2	8.9

22140_RX35_C0004

VALVE SPECIFICATIONS

Year	Engine Displacement Liters	Engine ID	Seat Angle (deg.)	Face Angle (deg.)	Spring Test Pressure (lbs. @ in.)	Spring Installed Height (in.)	Stem-to-Guide Clearance (in.) Intake	Stem-to-Guide Clearance (in.) Exhaust	Stem Diameter (in.) Intake	Stem Diameter (in.) Exhaust
2007	3.5	2GR-FE	45	44.5	NA	NA	0.0010-0.0024	0.0012-0.0026	0.2154-0.2159	0.2151-0.2157
2008	3.5	2GR-FE	45	44.5	NA	NA	0.0010-0.0024	0.0012-0.0026	0.2154-0.2159	0.2151-0.2157

NA: Information not available

22140_RX35_C0005

CRANKSHAFT AND CONNECTING ROD SPECIFICATIONS

All measurements are given in inches.

Year	Engine Displacement Liters	Engine ID	Crankshaft Main Brg. Journal Dia.	Crankshaft Main Brg. Oil Clearance	Crankshaft Shaft End-play	Crankshaft Thrust on No.	Connecting Rod Journal Diameter	Connecting Rod Oil Clearance	Connecting Rod Side Clearance
2007	3.5	2GR-FE	2.4011-2.4016	0.0010-0.0019	0.0016-0.0095	2	2.0863-2.0866	0.0018-0.0026	0.0059-0.0157
2008	3.5	2GR-FE	2.4011-2.4016	0.0010-0.0019	0.0016-0.0095	2	2.0863-2.0866	0.0018-0.0026	0.0059-0.0157

22140_RX35_C0006

PISTON AND RING SPECIFICATIONS

All measurements are given in inches.

	Engine			Ring Gap			Ring Side Clearance		
	Displ.	Engine	Piston	Top	Bottom	Oil	Top	Bottom	Oil
Year	Liters	ID	Clearance	Comp.	Comp.	Control	Comp.	Comp.	Control
2007	3.5	2GR-FE	0.0018-0.0020	0.0098-0.0138	0.0197-0.0236	0.0039-0.0157	0.0008-0.0028	0.0008-0.0024	0.0028-0.0059
2008	3.5	2GR-FE	0.0018-0.0020	0.0098-0.0138	0.0197-0.0236	0.0039-0.0157	0.0008-0.0028	0.0008-0.0024	0.0028-0.0059

22140_RX35_C0007

TORQUE SPECIFICATIONS

All readings in ft. lbs.

	Engine		Cylinder	Main	Rod	Crankshaft					Oil Pan
	Displacement	Engine	Head	Bearing	Bearing	Damper	Flywheel	Manifold		Spark	Drain
Year	Liters	ID	Bolts	Bolts	Bolts	Bolts	Bolts	Intake	Exhaust	Plugs	Plug
2007	3.5	2GR-FE	①	②	③	184	61	15	15	18	33
2008	3.5	2GR-FE	①	②	③	184	61	15	15	18	33

① Step 1: 10 mm bolts to 27 ft. lbs.
Step 2: 10mm point cap bolts plus 90 degrees
Step 3: 10mm point cap bolts plus 90 degrees
Step 4: Front bolts to 22 ft. lbs.

② Step 1: 16 cap cap bolts 45 ft. lbs.
Step 2: 16 cap bolts plus 90 degrees
Step 3: 8 side bolts to 38 ft. lbs.

② Step 1: 18 ft. lbs.
Step 2: Plus 90 degrees

22140_RX35_C0008

WHEEL ALIGNMENT

Year	Model		Caster Range (+/-Deg.)	Caster Preferred Setting (Deg.)	Camber Range (+/-Deg.)	Camber Preferred Setting (Deg.)	Toe-in (in.)	Steering Axis Inclination (Deg.)
2007	RX350 without air suspension	2WD F	0.75	+2.75	0.75	-0.67	0+/-0.2	10.75+/-0.75
		4WD F	0.75	+2.75	0.75	-0.58	0+/-0.2	10.75+/-0.75
		2WD R	—	—	0.75	-1.33	0.24+/-0.16	—
		4WD R	—	—	0.75	-0.83	0.24+/-0.16	—
	with air suspension	2WD F	0.75	+2.85	0.75	-0.67	0+/-0.2	10.7+/-0.75
		4WD F	0.75	+2.83	0.75	-0.62	0+/-0.2	10.58+/-0.75
		2WD R	—	—	0.75	-1.35	0.3+/-0.2	—
		4WD R	—	—	0.75	-0.92	0.3+/-0.2	—
2008	RX350 without air suspension	2WD F	0.75	+2.75	0.75	-0.67	0+/-0.2	10.75+/-0.75
		4WD F	0.75	+2.75	0.75	-0.58	0+/-0.2	10.75+/-0.75
		2WD R	—	—	0.75	-1.33	0.24+/-0.16	—
		4WD R	—	—	0.75	-0.83	0.24+/-0.16	—
	with air suspension	2WD F	0.75	+2.85	0.75	-0.67	0+/-0.2	10.7+/-0.75
		4WD F	0.75	+2.83	0.75	-0.62	0+/-0.2	10.58+/-0.75
		2WD R	—	—	0.75	-1.35	0.3+/-0.2	—
		4WD R	—	—	0.75	-0.92	0.3+/-0.2	—

22140_RX35_C0009

TIRE, WHEEL AND BALL JOINT SPECIFICATIONS

Year	Model	OEM Tires Standard	OEM Tires Optional	Tire Pressures (psi) Front	Tire Pressures (psi) Rear	Wheel Size	Ball Joint Inspection	Lug Nut Torque (ft. lbs.)
2007	RX350	P225/65R17	P235/55R18	30	30	6.5-J	①	76
2008	RX350	P225/65R17	R235/55R18	30	30	6.5-J	①	76

OEM: Original Equipment Manufacturer

PSI: Pounds Per Square Inch

STD: Standard

OPT: Optional

① Replace if any measurable movement is found.

22140_RX35_C0010

BRAKE SPECIFICATIONS

All measurements in inches unless noted

Year	Model		Brake Disc Original Thickness	Brake Disc Minimum Thickness	Brake Disc Maximum Runout	Minimum Lining Thickness	Brake Caliper Bracket Bolts (ft. lbs.)	Brake Caliper Mounting Bolts (ft. lbs.)
2007	RX350	F	1.102	1.024	0.0020	0.039	77	25
		R	0.394	0.335	0.0059	0.039	58	32
2008	RX350	F	1.102	1.024	0.0020	0.039	77	25
		R	0.394	0.335	0.0059	0.039	58	32

F: Front

R: Rear

22140_RX35_C0011

SCHEDULED MAINTENANCE INTERVALS

LEXUS—RX350

TO BE SERVICED	TYPE OF SERVICE	VEHICLE MILEAGE INTERVAL (x1000)													
		5	10	15	20	25	30	35	40	45	50	55	60	90	120
Engine oil & filter	R	✓	✓	✓	✓	✓	✓	✓	✓	✓	✓	✓	✓	✓	✓
Automatic transmission fluid	S/I			✓			✓			✓			✓	✓	✓
Ball joints & dust covers	S/I			✓			✓			✓			✓	✓	✓
Bolts & nuts on chassis & body	S/I			✓			✓			✓			✓	✓	✓
Brake linings & drums	S/I	✓	✓	✓	✓	✓	✓	✓	✓	✓	✓	✓	✓	✓	✓
Brake line pipes & hoses	S/I			✓			✓			✓			✓	✓	✓
Brake pads & discs (front & rear)	S/I	✓	✓	✓	✓	✓	✓	✓	✓	✓	✓	✓	✓	✓	✓
Brake fluid	R						✓						✓	✓	✓
Rack and pinion assembly	S/I			✓			✓			✓			✓	✓	✓
Steering linkage & boots	S/I			✓			✓			✓			✓	✓	✓
Air cleaner filter	R						✓						✓	✓	✓
Spark plugs ①	R														✓
Drive belts	S/I												✓	✓	✓
Exhaust pipes & mountings	S/I			✓			✓			✓			✓	✓	✓
Fuel lines & connections	S/I						✓						✓	✓	✓
Engine coolant ②	S/I			✓			✓			✓			✓	✓	
Rear differential & transfer case oil	S/I			✓			✓			✓			✓	✓	✓
Fuel tank cap gasket	S/I						✓						✓	✓	✓
Rotate tires	S/I			✓			✓			✓			✓		✓
Clean air conditioning filter ③	S/I			✓			✓			✓			✓		✓
Axle shaft bolts	S/I			✓			✓			✓			✓	✓	✓
Brake pad thickness and rotor runout	S/I						✓						✓	✓	✓

R: Replace S/I: Service or Inspect

① Spark plugs are replaced at 120,000 miles

② Replace engine coolant at 100,000 miles and then inspect every 15,000 miles

③ Replace air conditioning filter every 30,000 miles

FREQUENT OPERATION MAINTENANCE (SEVERE SERVICE)

If a vehicle is operated under any of the following conditions it is considered severe service:

- **Extremely dusty areas.**
- **50% or more of the vehicle operation is in 32°C (90°F) or higher temperatures, or constant temperatures below 0°C (32°F).**
- **Prolonged idling (vehicle operation in stop and go traffic).**
- **Frequent short running periods (engine does not warm to normal operating temperatures).**
- **Police, taxi, delivery usage or trailer towing usage.**

Air cleaner filter: service or inspect every 5000 miles

Rear differential & transfer case oil: replace every 15,000 miles.

Ball joints & dust covers: service or inspect every 5000 miles.

Bolts & nuts on chassis & body: service or inspect every 5000 miles.

Axle shaft bolts: service or inspect every 5000 miles.

Steering linkage: service or inspect every 5000 miles.

22140_RX35_C0012

PRECAUTIONS

Before servicing any vehicle, please be sure to read all of the following precautions, which deal with personal safety, prevention of component damage, and important points to take into consideration when servicing a motor vehicle:

- Never open, service or drain the radiator or cooling system when the engine is hot; serious burns can occur from the steam and hot coolant.
- Observe all applicable safety precautions when working around fuel. Whenever servicing the fuel system, always work in a well-ventilated area. Do not allow fuel spray or vapors to come in contact with a spark, open flame, or excessive heat (a hot drop light, for example). Keep a dry chemical fire extinguisher near the work area. Always keep fuel in a container specifically designed for fuel storage; also, always properly seal fuel containers to avoid the possibility of fire or explosion. Refer to the additional fuel system precautions later in this section.
- Fuel injection systems often remain pressurized, even after the engine has been turned **OFF**. The fuel system pressure must be relieved before disconnecting any fuel lines. Failure to do so may result in fire and/or personal injury.
- Brake fluid often contains polyglycol ethers and polyglycols. Avoid contact with the eyes and wash your hands thoroughly after handling brake fluid. If you do get brake fluid in your eyes, flush your eyes with clean, running water for 15 minutes. If eye irritation persists, or if you have taken brake fluid internally, IMMEDIATELY seek medical assistance.
- The EPA warns that prolonged contact with used engine oil may cause a number of skin disorders, including cancer. You should make every effort to minimize your exposure to used engine oil. Protective gloves should be worn when changing oil. Wash your hands and any other exposed skin areas as soon as possible after exposure to used engine oil. Soap and water, or waterless hand cleaner should be used.
- All new vehicles are now equipped with an air bag system, often referred to as a Supplemental Restraint System (SRS) or Supplemental Inflatable Restraint (SIR) system. The system must be disabled before performing service on or around system components, steering column, instrument panel components, wiring and sensors. Failure to follow safety and disabling procedures could result in accidental air bag deployment, possible personal injury and unnecessary system repairs.
- Always wear safety goggles when working with, or around, the air bag system. When carrying a non-deployed air bag, be sure the bag and trim cover are pointed away from your body. When placing a non-deployed air bag on a work surface, always face the bag and trim cover upward, away from the surface. This will reduce the motion of the module if it is accidentally deployed. Refer to the additional air bag system precautions later in this section.
- Clean, high quality brake fluid from a sealed container is essential to the safe and proper operation of the brake system. You should always buy the correct type of brake fluid for your vehicle. If the brake fluid becomes contaminated, completely flush the system with new fluid. Never reuse any brake fluid. Any brake fluid that is removed from the system should be discarded. Also, do not allow any brake fluid to come in contact with a painted surface; it will damage the paint.
- Never operate the engine without the proper amount and type of engine oil; doing so WILL result in severe engine damage.
- Timing belt maintenance is extremely important. Many models utilize an interference-type, non-freewheeling engine. If the timing belt breaks, the valves in the cylinder head may strike the pistons, causing potentially serious (also time-consuming and expensive) engine damage. Refer to the maintenance interval charts for the recommended replacement interval for the timing belt, and to the timing belt section for belt replacement and inspection.
- Disconnecting the negative battery cable on some vehicles may interfere with the functions of the on-board computer system(s) and may require the computer to undergo a relearning process once the negative battery cable is reconnected.
- When servicing drum brakes, only disassemble and assemble one side at a time, leaving the remaining side intact for reference.
- Only an MVAC-trained, EPA-certified automotive technician should service the air conditioning system or its components.

BRAKES

ANTI-LOCK BRAKE SYSTEM (ABS)

GENERAL INFORMATION

The Anti-lock Brake System (ABS) helps prevent wheels from locking when the brake is applied firmly or when braking on a slippery surface.

The Electronic Brake force Distribution (EBD) control utilizes ABS, realizing proper brake force distribution between front and rear wheels in accordance with driving conditions. In addition, during cornering braking, it also controls the brake force of the right and left wheels, helping to maintain vehicle behavior.

The primary purpose of the Brake Assist (BA) system is to provide an auxiliary brake force to assist the driver who cannot generate a large enough brake force during emergency braking. This system helps to maximize the brake performance of the vehicle.

The Traction Control (TRAC) system helps prevent the drive wheels from slipping if drive depresses the accelerator pedal excessively when starting off or accelerating on a slippery surface.

The Vehicle Stability Control (VSC) system helps prevent the vehicle from slipping sideways as a result of strong front wheel skid or strong rear wheel skid during cornering.

BRAKES — BLEEDING THE BRAKE SYSTEM

BLEEDING PROCEDURE

Master Cylinder

1. Before servicing the vehicle, refer to the precautions section.
   If the master cylinder has been disassembled or if the reservoir becomes empty, bleed the air from the master cylinder.
2. Remove the air cleaner assembly with hose.
3. Disconnect the brake lines from the master cylinder, using a suitable brake line wrench.
4. Have an assistant slowly depress the brake pedal and hold it.
5. Cover the outer holes with your fingers, and have your assistant release the brake pedal.
6. Repeat steps 3. and 4. several times.
7. Connect the brake lines. Tighten to 11 ft. lbs. (15 Nm).
8. Install the air cleaner assembly with hose.

Brake Lines

1. Before servicing the vehicle, refer to the precautions section.
2. Raise and safely support the vehicle.
3. Connect a piece of vinyl tubing to the brake caliper.
4. Have an assistant depress the brake pedal several times, then loosen the bleeder plug while the pedal is depressed.
5. When fluid stops coming out, tighten the bleeder plug, then release the brake pedal.
6. Repeat steps 2. and 3. until all the air in the fluid has been bled out.
7. Tighten the brake bleeder plug to 73 inch lbs. (8.3 Nm).
8. Repeat the above steps to bleed the air out of the brake line for each wheel.

BRAKES — FRONT DISC BRAKES

BRAKE CALIPER

REMOVAL & INSTALLATION

See Figure 1.

1. Before servicing the vehicle, refer to the precautions section.
2. Remove front wheel.
3. Drain brake fluid.

➡Wash the brake fluid off immediately if it adheres to any painted surfaces.

4. Remove the union bolt and the gasket from the disc brake cylinder sub-assembly, then disconnect the flexible hose.
5. Hold the front disc brake cylinder slide pin, slide pin No. 2 and remove the 2 bolts and brake cylinder assembly.

To install:

6. Install the disc brake cylinder assembly with the 2 bolts and tighten to 25 ft. lbs. (34 Nm).
7. Connect the flexible hose with the union bolt and a new gasket and tighten to 21 ft. lbs. (29 Nm).

➡Install the flexible hose lock securely in the lock hole in the disc brake cylinder sub-assembly.

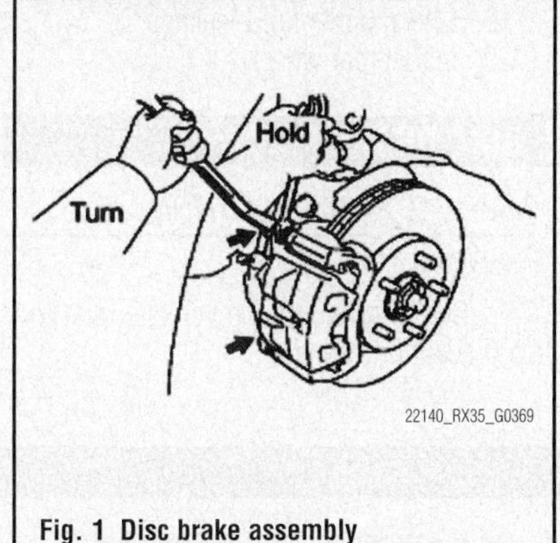

Fig. 1 Disc brake assembly

8. Fill reservoir with brake fluid.
9. Bleed master cylinder.
10. Bleed brake line.
11. Check fluid level in reservoir.
12. Check brake fluid leakage.
13. Install front wheel.

DISC BRAKE PADS

REMOVAL & INSTALLATION

See Figure 2.

1. Before servicing the vehicle, refer to the Precautions Section.
2. Remove the disc brake cylinder slide pins (upper and lower) from the disc brake cylinder mounting.
3. Remove brake cylinder.
4. Remove the 2 brake pads with the anti squeal shims from the cylinder mounting.

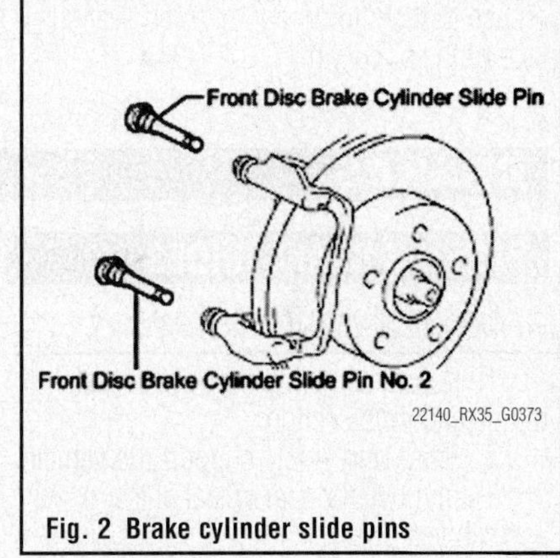

Fig. 2 Brake cylinder slide pins

To install:

5. Install the 2 brake pads with the anti squeal shims to the disc brake cylinder mounting.

BRAKE CALIPER

REMOVAL & INSTALLATION

1. Before servicing the vehicle, refer to the precautions section.
2. Remove rear wheel.
3. Drain brake fluid.

➡Wash the brake fluid off immediately if it adheres to any painted surfaces.

4. Remove the union bolt and the gasket from the disc brake cylinder subassembly, then disconnect the flexible hose.
5. Remove the No. 2 rear disc brake cylinder slide pin.
6. Remove the rear disc brake cylinder slide pin.
7. Remove the disc brake cylinder assembly.

To install:

8. Install the rear disc brake cylinder assembly.
9. Apply the lithium soap base glycol grease to the sliding part and the seal surface of the slide pin.
10. Install the rear disc brake cylinder slide pin.
11. Apply the lithium soap base glycol grease to the sliding part and the seal surface of the No. 2 slide pin.
12. Install the No. 2 rear disc brake cylinder slide pin to the bottom side of the cylinder mounting.
13. Fully tighten the rear disc brake cylinder slide pin and the No. 2 rear disc brake cylinder slide pin to 32 ft. lbs. (43 Nm).
14. Connect the flexible hose with the union bolt and a new gasket and tighten to 21 ft. lbs. (29 Nm).
15. Fill reservoir with brake fluid.
16. Bleed master cylinder.
17. Bleed brake line.
18. Check fluid level in reservoir.
19. Check brake fluid leakage.
20. Install rear wheel.

DISC BRAKE PADS

REMOVAL & INSTALLATION

See Figure 3.

1. Before servicing the vehicle, refer to the precautions section.
2. Remove the disc brake cylinder slide pins (upper and lower) from the disc brake cylinder mounting.

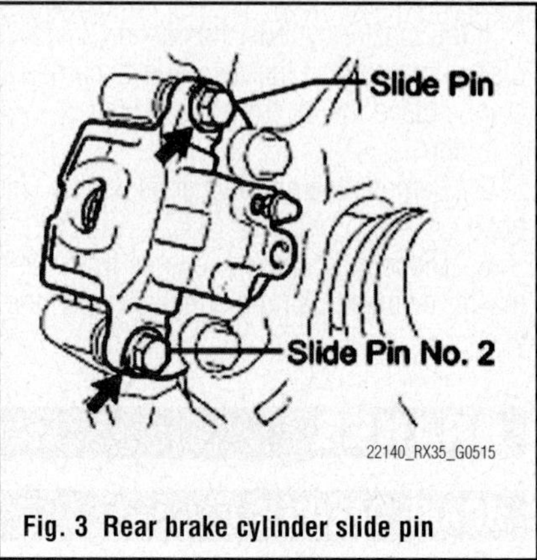

Fig. 3 Rear brake cylinder slide pin

3. Remove brake cylinder.
4. Remove the 2 brake pads with the anti squeal shims from the cylinder mounting.

To install:

5. Install the 2 brake pads with the anti squeal shims to the disc brake cylinder mounting.

PARKING BRAKE SHOES

REMOVAL & INSTALLATION

1. Before servicing the vehicle, refer to the precautions section.
2. Raise and safely support the vehicle.
3. Remove the rear wheel and tire assemblies.
4. Unbolt and remove the rear caliper, but do not disconnect the fluid line. Suspend the caliper out of the way with a piece of wire.
5. Make sure the parking brake is fully released, then remove the rear brake disc.
6. Place matchmarks on the brake disc and the axle hub.

➡If the disc cannot be easily removed, turn the shoe adjuster until the wheel turns freely.

7. Use needle-nose pliers to remove the 2 No. 1 parking brake shoe return tension springs.
8. Remove the parking brake shoe strut set, as follows:
   a. Remove the parking brake shoe strut and the parking brake shoe strut compression spring.
9. Remove the No. 1 parking brake shoe assembly, as follows:
   a. Remove the parking brake shoe hold down spring cup No. 1, parking brake shoe hold down spring and parking brake shoe hold down spring cup No. 2.
   b. FWD vehicles, remove the parking brake shoe hold down spring pin No. 1.
   c. Disconnect the parking brake shoe return spring No. 2 and remove the parking brake shoe assembly No. 1.
10. Remove the parking brake shoe adjusting screw set, as follows:
   a. Remove the parking brake shoe adjusting screw set.
   b. Remove the parking brake shoe return tension spring No. 2.
11. Remove the parking brake shoe assembly No. 2:
   a. Remove the parking brake shoe hold down spring cup No. 1, parking brake shoe hold down spring, parking brake shoe hold down spring cup No. 2 and parking brake shoe hold down spring pin No. 2.
   b. Remove the parking brake shoe assembly No. 2.
   c. Using needle-nose pliers, disconnect the parking brake cable No. 3 from the parking brake cable shoe lever.

⁂ WARNING

Be careful not to damage parking brake cable No. 3.

12. On 4WD models, separate the rear speed sensor.
13. On 4WD models, remove the rear axle shaft nut.
14. On 4WD models, remove rear axle hub and bearing assembly.
15. On 4WD models, remove parking brake shoe hold down spring pin No. 1.
16. Remove the parking brake shoe, as follows:
   a. Using a screwdriver, remove the c-washer.
   b. Remove the shim and parking brake shoe lever from the parking brake shoe assembly No. 2.
17. Inspect parking brake shoe lining thickness:
   a. Using a ruler, measure the thickness of the shoe lining. Standard thickness is 2.5 mm (0.098 in.) and minimum

thickness is 1.0 mm (0.039 in.). If the lining thickness is less than or equal to the minimum, or

To install:

18. Install the parking brake shoe, as follows:
    a. Using pliers, install the parking brake shoe lever and the shim with a new C-washer.
    b. Using a feeler gauge, measure the clearance. Standard clearance: less than 0.35 mm (0.014 in.).
    If the clearance is not within the specifications, replace the shim with one of the correct size. The shim sizes: 0.3 mm (0.012 in.), 0.9 mm (0.035 in.) or 0.6 mm (0.024 in.).
19. Apply high temperature grease to the shaded parts shown in the illustration of the backing plate which make contact with the shoe.
20. On 4WD models, perform the following:
    a. Install the parking brake shoe hold down spring pin No. 1.
    b. Install the rear axle hub and bearing.
    c. Install the rear axle shaft nut.
    d. Install the rear speed sensor.
21. Install parking brake shoe assembly No. 2, as follows:
    a. Using needle-nose pliers, connect the parking brake cable No. 3 to the parking brake cable shoe lever.

➡Be careful not to damage the parking brake cable No. 3.

    b. Install the parking brake shoe assembly No. 2 with the parking brake shoe hold down spring, parking brake shoe hold down spring cup No. 1, parking brake shoe hold down spring cup No. 2 and parking brake shoe hold down spring pin No. 2.
22. Install the parking brake shoe adjusting screw set, as follows:
    a. Apply high temperature grease to the parking brake shoe adjusting bolt and piece.
    b. Attach the parking brake shoe return tension spring No. 2 to the parking brake shoe assembly No. 1 and parking brake shoe assembly No. 2.
    c. Attach the parking brake shoe adjusting screw set to the parking brake shoe assembly No. 1 and parking brake shoe assembly No. 2.
23. Install the parking brake shoe assembly No. 1:
    a. Install the parking brake shoe assembly No. 1 with the parking brake shoe hold down spring, parking brake shoe hold down spring cup No. 2, parking brake shoe hold down spring pin No. 1.
24. Attach the parking brake shoe strut and the parking brake shoe strut compression spring to the parking brake shoe assembly No. 2 and parking brake shoe assembly No. 1.
25. Install the parking brake shoe return tension spring using needle-nose pliers.

➡First install the front side spring then the rear side spring.

26. Check that the parking brake components are properly installed.

⁂ WARNING

There should be no oil or grease on the friction surface of the shoe lining and disc.

27. For 4WD models, inspect the bearing backlash and axle hub deviation.
28. Install the rear disc, aligning the matchmarks made during removal.
29. Install the rear disc brake caliper assembly and tighten the 2 bolts to 58 ft. lbs. (78 Nm).
30. Install the rear wheel and tire assembly and tighten the lug nuts to 76 ft. lbs. (103 Nm).
31. Inspect and adjust the parking brake pedal travel.
32. For 4WD models, check the ABS speed sensor signal.

ADJUSTMENT

1. Before servicing the vehicle, refer to the Precautions Section.
2. Raise and safely support the vehicle.
3. Remove the rear wheel and tire assemblies.
4. Adjust parking brake shoe clearance, as follows:
    a. Temporarily install the hub nuts.
    b. Remove the hole plug, turn the adjuster and expand the shoes until the disc locks.
    c. Contract the shoe adjuster until the disc rotates smoothly. Standard : return 8 notches
    d. Check that the shoe has no brake drag.
    e. Install the hole plug.
5. Install the rear wheel and tire assembly and tighten the lug nuts to 76 ft. lbs. (103 Nm).

CHASSIS ELECTRICAL

AIR BAG (SUPPLEMENTAL RESTRAINT SYSTEM)

GENERAL INFORMATION

SERVICE PRECAUTIONS

The vehicle is equipped with a Supplemental Restraint System (SRS). It consists of a driver airbag, front passenger airbag, driver side knee airbag, front seat side airbag, curtain shield airbag, and front seat belt pretensioner. Failure to carry out service operations in the correct sequence could cause the SRS to unexpectedly deploy during servicing, possibly leading to a serious accident. Further, if a mistake is made in servicing the SRS, it is possible that the SRS may fail to operate when required. Before performing servicing (including removal or installation of parts, inspection or replacement), be sure to read the following items carefully, then follow the correct procedures indicated in the repair manual.

Disconnect and isolate the battery negative cable before beginning any airbag system component diagnosis, testing, removal, or installation procedures. Wait at least 90 seconds after the ignition switch is turned to the "LOCK" position and the negative (-) terminal cable is disconnected from the battery before starting the operation. This will disable the airbag system. The SRS is equipped with a backup power source, so if work is started within 90 seconds after disconnecting the negative (-) terminal cable from the battery, the SRS may be deployed. Failure to disable the airbag system may result in accidental airbag deployment, personal injury, or death.

Do not expose the steering pad, front passenger airbag assembly, driver side knee airbag assembly, center airbag sensor assembly, front airbag sensor, front seat inner belt assembly, seat position sensor, occupant classification ECU, front seat side airbag assembly, side airbag sensor, curtain shield airbag assembly, rear airbag sensor, or front seat outer belt assembly directly to hot air or flames.

Malfunction symptoms of the SRS are difficult to confirm, so DTCs are the most important source of information when troubleshooting. When troubleshooting the SRS, always inspect DTCs before disconnecting the battery.

Even in the case of a minor collision when the SRS does not deploy, the steering pad, front passenger airbag assembly, driver side knee airbag assembly, center airbag sensor assembly, front airbag sensor, front seat inner belt assembly, seat position sensor, occupant classification ECU, front seat side airbag assembly, side airbag sensor, curtain shield airbag assembly, rear airbag sensor, and front seat outer belt assembly should be inspected.

Before repair work, remove the airbag sensor if any kind of shock is likely to occur to the airbag sensor during the operation.

Never disassemble or repair any of the following parts in order to reuse them. If any of these parts have been dropped, or a defect is found (e.g. cracks, dents or any other defects) in any of the housings, brackets or connectors, then replace the part with a new one: Steering Pad, Front Passenger Airbag Assembly, Driver Side Knee Airbag Assembly, Front Seat Side Airbag Assembly, Curtain Shield Airbag Assembly, Center Airbag Sensor Assembly, Front Airbag Sensor, Front Seat Inner Belt Assembly, Seat Position Sensor, Occupant Classification ECU, Side Airbag Sensor, Rear Airbag Sensor, Front Seat Outer Belt Assembly.

Use an volt/ohmmeter with high impedance (10 kΩ/V minimum) for troubleshooting the electrical circuits.

Information labels are attached near the SRS components. Follow the instructions in the caution.

Do not place an intact undeployed airbag face down on a solid surface. The airbag will propel into the air if accidentally deployed and may result in personal injury or death.

When carrying or handling an undeployed airbag, the trim side (face) of the airbag should be pointing towards the body to minimize possibility of injury if accidental deployment occurs. Failure to do this may result in personal injury or death.

Replace airbag system components with OEM replacement parts. Substitute parts may appear interchangeable, but internal differences may result in inferior occupant protection. Failure to do so may result in occupant personal injury or death.

Wear safety glasses, rubber gloves, and long sleeved clothing when cleaning powder residue from vehicle after an airbag deployment. Powder residue emitted from a deployed airbag can cause skin irritation. Flush affected area with cool water if irritation is experienced. If nasal or throat irritation is experienced, exit the vehicle for fresh air until the irritation ceases. If irritation continues, see a physician.

Do not use a replacement airbag that is not in the original packaging. This may result in improper deployment, personal injury, or death.

The factory installed fasteners, screws and bolts used to fasten airbag components have a special coating and are specifically designed for the airbag system. Do not use substitute fasteners. Use only original equipment fasteners listed in the parts catalog when fastener replacement is required.

During, and following, any child restraint anchor service, due to impact event or vehicle repair, carefully inspect all mounting hardware, tether straps, and anchors for proper installation, operation, or damage. If a child restraint anchor is found damaged in any way, the anchor must be replaced. Failure to do this may result in personal injury or death.

Deployed and non-deployed airbags may or may not have live pyrotechnic material within the airbag inflator.

Do not dispose of driver/passenger/curtain airbags or seat belt tensioners unless you are sure of complete deployment. Refer to the Hazardous Substance Control System for proper disposal.

Dispose of deployed airbags and tensioners consistent with state, provincial, local, and federal regulations.

After any airbag component testing or service, do not connect the battery negative cable. Personal injury or death may result if the system test is not performed first.

If the vehicle is equipped with the Occupant Classification System (OCS), do not connect the battery negative cable before performing the OCS Verification Test using the scan tool and the appropriate diagnostic information. Personal injury or death may result if the system test is not performed properly.

Never replace both the Occupant Restraint Controller (ORC) and the Occupant Classification Module (OCM) at the same time. If both require replacement, replace one, then perform the Airbag System test before replacing the other.

Both the ORC and the OCM store Occupant Classification System (OCS) calibration data, which they transfer to one another when one of them is replaced. If both are replaced at the same time, an irreversible fault will be set in both modules and the OCS may malfunction and cause personal injury or death.

If equipped with OCS, the Seat Weight Sensor is a sensitive, calibrated unit and must be handled carefully. Do not drop or handle roughly. If dropped or damaged, replace with another sensor. Failure to do so may result in occupant injury or death.

If equipped with OCS, the front passenger seat must be handled carefully as well. When removing the seat, be careful when setting on floor not to drop. If dropped, the sensor may be inoperative, could result in occupant injury, or possibly death.

If equipped with OCS, when the passenger front seat is on the floor, no one should sit in the front passenger seat. This uneven force may damage the sensing ability of the seat weight sensors. If sat on and damaged, the sensor may be inoperative, could result in occupant injury, or possibly death.

DISARMING THE SYSTEM

To avoid personal injury when working on vehicles equipped with an air bag, the negative battery cable must be disconnected and at least 90 seconds must elapse before working on the system. Failure to do so may result in deployment of the air bag.

ARMING THE SYSTEM

To arm the system after service is finished, connect the negative battery cable.

CLOCKSPRING CENTERING

See Figures 4 and 5.

1. Check that the ignition switch is **OFF**.
2. Check that the battery negative (-) terminal is disconnected.

✲✲ CAUTION

After removing the terminal, wait for at least 90 seconds before starting the operation.

3. Rotate the spiral cable counterclockwise slowly by hand until it feels firm.
4. Rotate the spiral cable clockwise approximately 2.5 turns to align the marks.

➡Do not turn the spiral cable by the airbag wire harness.

➡The spiral cable will rotate approximately 2.5 turns to both the left and right from the center.

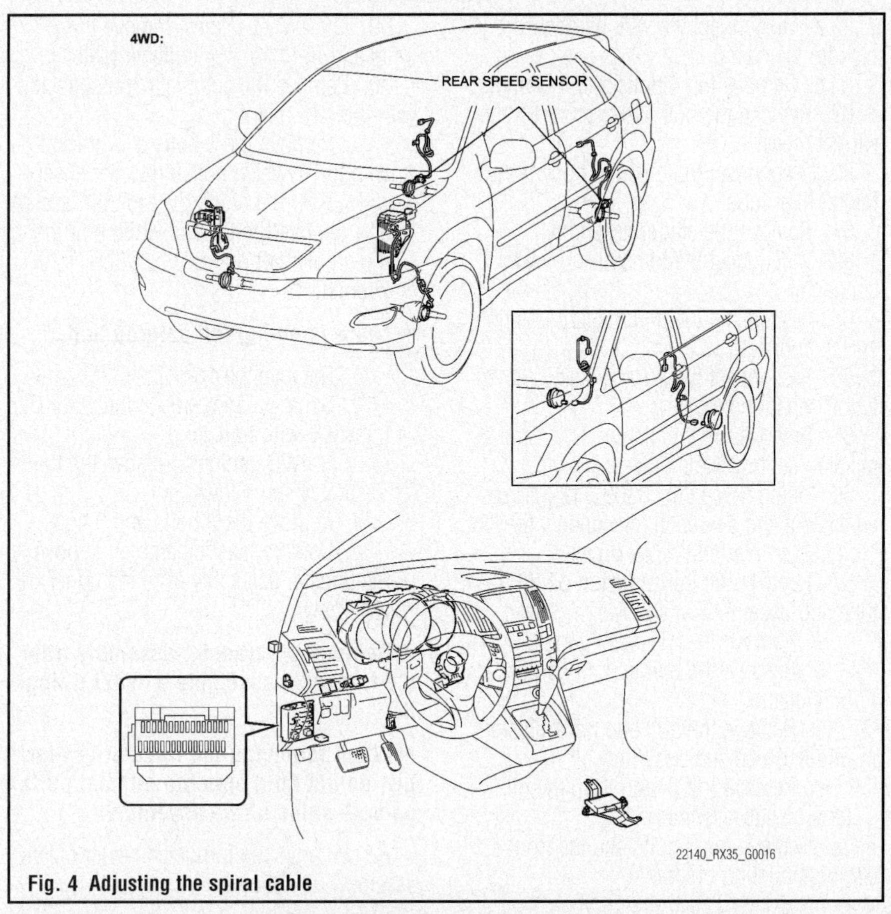

Fig. 4 Adjusting the spiral cable

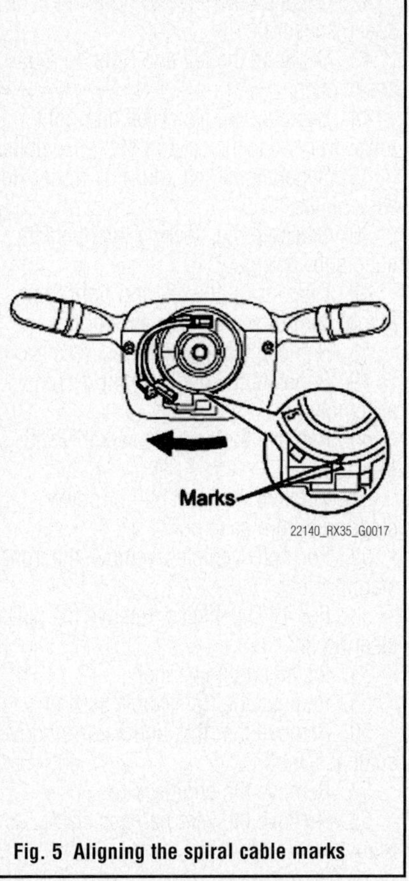

Fig. 5 Aligning the spiral cable marks

DRIVETRAIN

AUTOMATIC TRANSAXLE ASSEMBLY

REMOVAL & INSTALLATION

See Figures 6 through 15.

1. Discharge fuel system pressure.
2. Disconnect the cable from negative battery terminal.
3. Place the front wheels facing straight ahead.
4. Remove the front wheels.
5. Remove the engine under cover assembly.
6. Remove the No. 1 and 2 engine under covers .
7. Separate the right front fender splash shield sub-assembly.
8. Remove the right front fender apron seal.
9. Drain the engine oil, coolant, and automatic transaxle fluid.
10. Remove the windshield wiper link assembly.
11. Remove the cowl top panel outer sub-assembly.
12. Remove the cool air intake duct seal.
13. Remove the side engine room cover.
14. Remove the V-bank cover sub-assembly.
15. Remove the V-ribbed belt.
16. Remove the No. 2 air cleaner inlet.
17. Remove the air cleaner cap and case sub-assemblies.
18. Remove the No. 1 air cleaner inlet.
19. Remove the battery.
20. Remove the intake air resonator sub-assembly.
21. Remove the air cleaner bracket.
22. Remove the right No. 2 engine mounting stay.
23. Remove the engine moving control rod sub-assembly.
24. Disconnect the No. 1 fuel vapor feed hose.
25. Disconnect the union to check valve hose.
26. Remove the radiator inlet and outlet hoses.
27. Disconnect the oil cooler inlet and outlet hoses.
28. Disconnect the heater inlet and outlet water hoses.
29. Remove the junction block cover.
30. Remove the glove compartment door assembly.
31. Disconnect the engine wire.
32. Disconnect the transmission control cable assembly, as follows:
    a. Remove the nut from the control shaft lever.
    b. Disconnect the transmission control cable assembly from the control shaft lever.
    c. Remove the clip and disconnect the transmission control cable assembly from the control cable bracket.
    d. Disconnect the control cable from the control cable clamp.
33. Separate the fuel tube sub-assembly.
34. Disconnect the No. 1 oil reservoir to pump hose.
35. Disconnect the return tube sub-assembly.
36. For 4WD vehicles, remove the propeller w/center bearing shaft assembly.
37. Remove the oxygen sensor.
38. Remove the front No. 3 exhaust pipe sub-assembly.
39. Remove the front exhaust pipe assembly.
40. Remove the left and right front axle hub nuts.
41. Separate the left and right front stabilizer link assemblies.

42. Separate the left and right front speed sensors.

43. Separate the left and right tie rod assemblies.

44. Separate the No. 1 left and right lower front suspension arm sub-assemblies.

45. Separate the left and right front axle assemblies.

46. Separate the steering intermediate shaft sub-assembly.

47. Disconnect the left and right front height control sensor sub-assemblies.

48. Remove the alternator assembly.

49. Separate the cooler compressor assembly.

50. Remove the engine assembly with transaxle.

51. Remove the vane pump (power steering pump) assembly.

52. For 4WD vehicles, remove the front stabilizer bar.

53. For 4WD vehicles, remove the power steering link.

54. Install engine hangers.

55. Remove the front frame assembly.

56. Remove the right and left front drive shaft assemblies.

57. Remove the engine wire.

58. Remove the wire harness clamp, as follows:

a. Disconnect the wire harnesses from the clamp.

b. Remove the 4 bolts and 4 clamps.

59. Remove the bolt and disconnect the wire harness.

60. Disconnect the breather hose from the oil filler tube.

61. Remove the starter assembly.

62. Disconnect the transmission wire connector.

63. Disconnect the park/neutral position switch connector.

64. Disconnect the 2 transmission revolution sensor connectors.

65. Remove the 2 bolts and No. 1 transmission control cable bracket.

66. Using SST (SST: 09023-12701) or equivalent and a wrench, disconnect the No. 1 oil cooler inlet and outlet tubes.

67. Remove the transmission oil filler tube sub-assembly, as follows:

a. Remove the ATF level gauge.

b. Remove the bolt and oil cooler tube clamp.

c. Remove the bolt and transmission oil filter tube sub-assembly.

d. Remove the O-ring from the oil filter tube sub-assembly.

68. Remove the 3 bolts and the front engine mounting bracket.

69. For 4WD vehicles, remove the 5 bolts and right transfer stiffener plate.

70. Remove the automatic transmission with transfer, as follows:

a. Remove the 2 bolts and bracket with the flywheel housing under cover.

b. Turn the crankshaft to gain access and remove the 6 bolts while holding the crankshaft pulley bolt with a wrench.

➡There is one green colored bolt.

c. Remove the 10 bolts.

d. Separate and remove the automatic transaxle with transfer.

71. For 4WD vehicles, remove the transfer assembly, as follows:

a. Remove the 2 bolts and 6 bolts.

b. Using a plastic hammer, remove the transfer assembly from the transaxle assembly.

➡Remove the transfer assembly from the transaxle assembly without tilting it.

➡When removing the transfer assembly, do not hold onto the oil seal parts on both sides of the assembly.

72. Remove the torque converter clutch assembly, as follows:

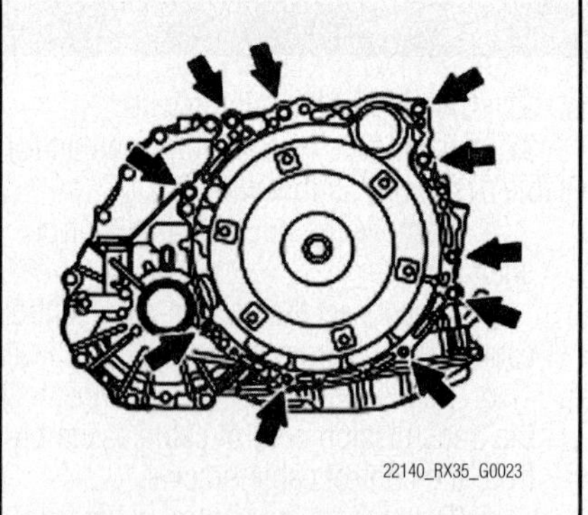
22140_RX35_G0023

Fig. 6 Separate transaxle from engine

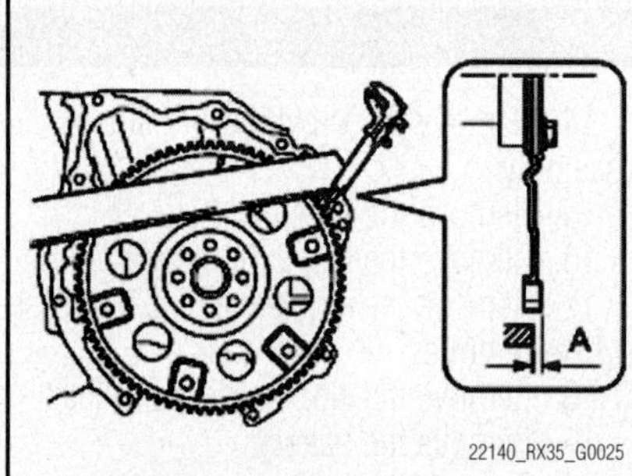

22140_RX35_G0025

Fig. 8 Measure dimension "A" between the transaxle fitting part of the engine and the converter fitting part of the drive plate

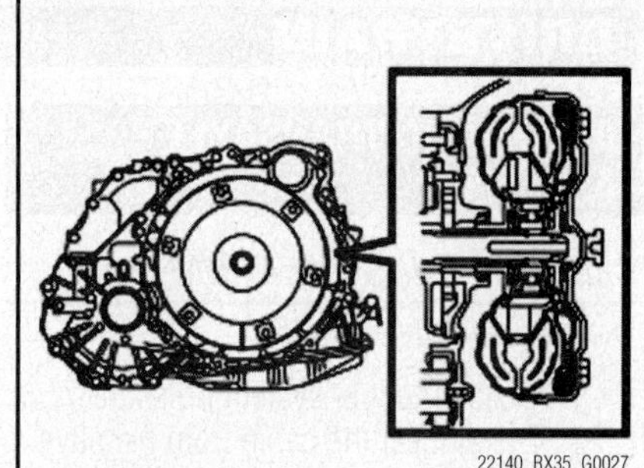
22140_RX35_G0027

Fig. 10 Engage the splines of the stator shaft and the stator

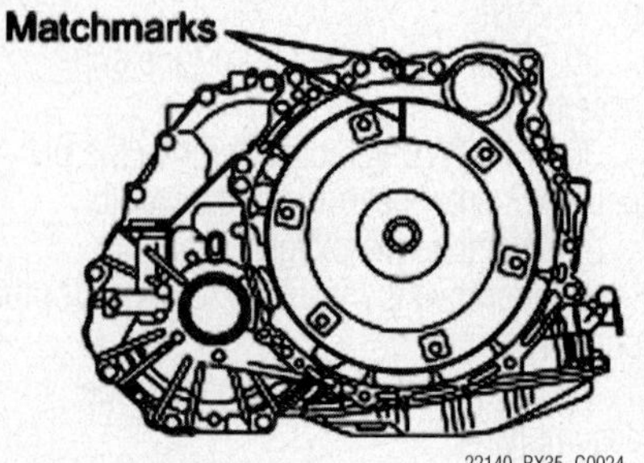

22140_RX35_G0024

Fig. 7 Put matchmarks on the transmission case and torque converter clutch assembly

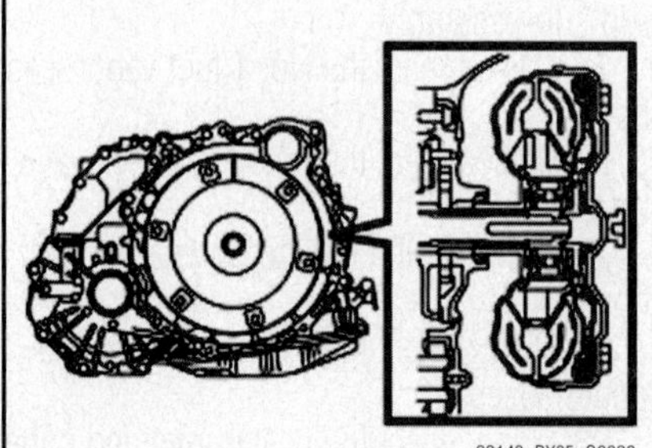
22140_RX35_G0026

Fig. 9 Align the matchmarks on the transmission case and torque converter clutch assembly

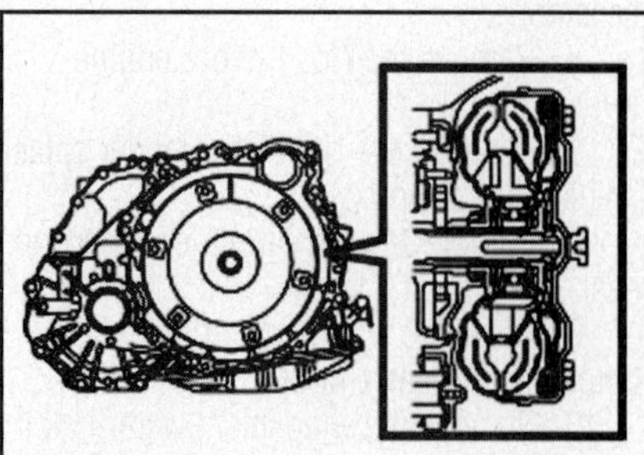
22140_RX35_G0028

Fig. 11 Align the matchmarks on the torque converter clutch assembly and transmission case

a. Put matchmarks on the transmission case and torque converter clutch assembly.

b. Remove the torque converter clutch assembly from the automatic transmission assembly.

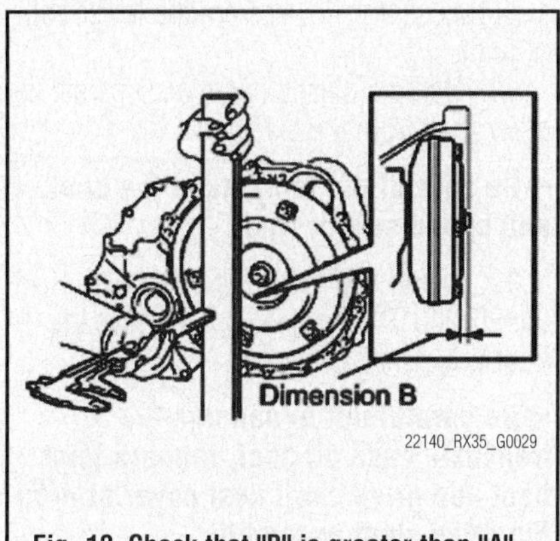

Fig. 12 Check that "B" is greater than "A"

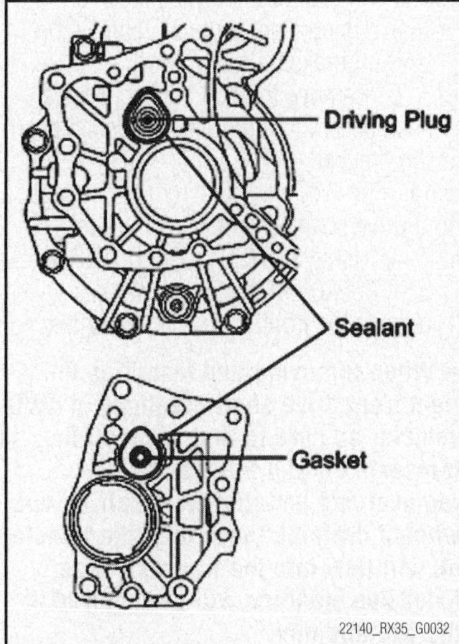

Fig. 13 Transaxle and transfer assembly sealant application

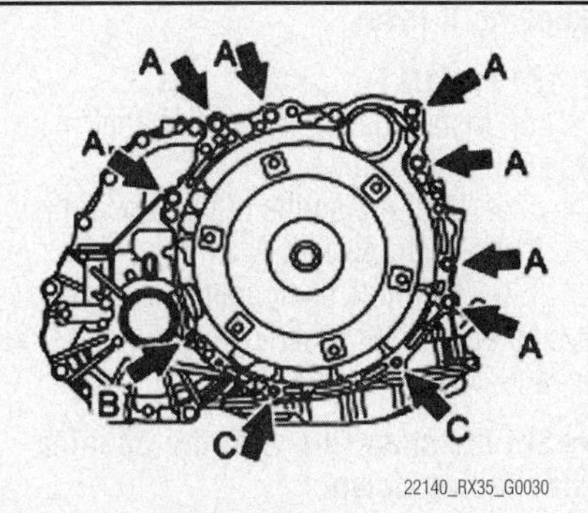

Fig. 14 Automatic transaxle bolt tightening sequence

73. Inspect torque converter clutch assembly.

To install:

74. Install the torque converter clutch assembly, as follows:

a. Install the torque converter clutch to the automatic transaxle.

b. Using vernier calipers and a straight edge, measure dimension "A" between the transaxle fitting part of the engine and the converter fitting part of the drive plate.

c. Aligning the matchmarks on the transmission case and torque converter clutch assembly, engage the splines of the input shaft and turbine runner.

➡Do not push on the torque converter when aligning the matchmarks.

d. Engage the splines of the stator shaft and the stator while turning the torque converter clutch assembly.

➡Turn the torque converter clutch assembly approximately 180°.

e. Turn the torque converter clutch assembly and align the matchmarks on the torque converter clutch assembly and transmission case to engage the key of the oil pump drive gear into the slot on the torque converter clutch assembly.

f. Using vernier calipers and a straight edge, measure dimension "B" shown in the illustration and check that "B" is greater than "A" (measured earlier). Standard: A + 1 mm (0.04 in.) or more.

➡Make sure to deduct the thickness of the straight edge.

75. For 4WD vehicles, install the transfer assembly, as follows:

a. Apply sealant (FIPG: Toyota Genuine Seal Packing 1281, Three Bond 1281 or equivalent) to the transaxle assembly and transfer assembly in continuous beaded from of 1.2 mm diameter as shown in the illustration.

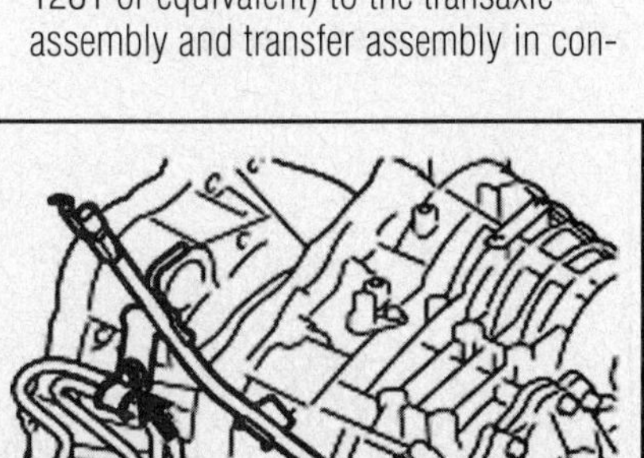

Fig. 15 Oil cooler tube cushion

➡Wipe any grease off from the attaching surfaces.

➡Install it within 10 minutes after applying the sealant.

➡Sealant stuck on the gasket, case oil seal and driving plug may cause oil leakage and seizure due to oil shortage. Care must be taken.

b. Install the transfer assembly to the transaxle assembly with the 2 bolts and 6 nuts , and tighten to 51 ft. lbs. (69 Nm).

➡Check that the gasket is installed to the transfer assembly before installing them to the transaxle assembly.

➡Install the transfer assembly to the transaxle assembly without tilting.

➡When moving the transfer assembly, do not hold the oil seal on the both sides.

76. Install the automatic transmission with transfer, as follows:

a. Install the automatic transaxle with transfer to the engine with the 10 bolts, and tighten to 47 ft. lbs. (64 Nm), 34 ft. lbs. (46 Nm), and 32 ft. lbs. (43 Nm).

b. Apply a few drops of adhesive (Toyota Genuine Adhesive 1324, Three Bond 1324 or equivalent) to each of 2 threads on the tip of the 6 torque converter clutch mounting bolts.

c. Install the 6 torque converter clutch mounting bolts and tighten to 30 ft. lbs. (41 Nm).

➡First install the green colored bolt, and then the remaining 5 bolts.

d. Install the flywheel housing under cover to the automatic transaxle with the 2 bolts and tighten to 69 inch lbs. (7.8 Nm).

77. For 4WD vehicles, install the right transfer stiffener plate with 5 bolts to the right rear transfer and engine mounting bracket and tighten to 25 ft. lbs. (34 Nm).

78. Install the front engine mounting bracket to the automatic transaxle with the 3 bolts and tighten to 47 ft. lbs. (64 Nm).

79. Install the transmission oil filler tube sub-assembly, as follows:

a. Coat a new O-ring with ATF, and install it to the transmission oil filler tube sub-assembly.

b. Install the transmission oil filler tube sub-assembly to the automatic transaxle with the bolt and tighten to 49 inch lbs. (5.5 Nm).

80. Temporarily install the No. 1 oil cooler inlet and outlet tubes.

81. Install the oil cooler tube clamp with the bolt and tighten to 48 inch lbs. (5.4 Nm).

➡Install them so that the oil cooler tube cushion is positioned as shown.

82. Install the No. 1 oil cooler inlet and outlet tubes and tighten to 25 ft. lbs. (34 Nm).

83. Install the No. 1 transmission control cable bracket with the 2 bolts and tighten to 10 ft. lbs. (12 Nm).

84. Connect the breather hose to the oil filler tube.

85. Connect the transmission wire connector.

86. Connect the park/neutral position switch connector.

87. Connect the 2 transmission revolution sensor connectors.

88. Install the starter assembly.

89. Connect the wire harness with the bolt and tighten to 14 ft. lbs. (19 Nm).

90. Install the wire harness clamp, as follows:

a. Install the 4 clamps and 4 bolts and tighten to 74 inch lbs. (8.4 Nm).

91. Connect the wire harnesses to the clamps.

92. Install the No. 2 transmission control cable bracket with the bolt and tighten to 9 ft. lbs. (12 Nm).

93. The remainder of installation is the reverse of the removal procedure, noting the following:

a. After adding engine oil, connect the negative battery cable, then add engine coolant and automatic transaxle fluid.

b. Add and then bleed power steering fluid.

c. Check for fuel, engine oil, coolant, and exhaust gas leaks.

d. Inspect the shift lever position.

e. Inspect and adjust front wheel alignment.

f. Inspect engine timing, engine idle speed, CO/HC, and throttle body.

g. Complete the rest of the installation process, then check the ABS speed sensor signal, reset memory, and perform initialization.

TRANSFER CASE ASSEMBLY

REMOVAL & INSTALLATION

The transfer case is part of the transmission/transaxle assembly and is serviced with those units.

FRONT HALFSHAFT

REMOVAL & INSTALLATION

See Figure 16.

1. Remove the engine under cover assembly.

2. Remove the No. 2 engine under cover.

3. Drain automatic transaxle fluid.

a. Remove the drain plug, gasket and drain ATF

b. Install a new gasket and the drain plug and tighten to 36 ft. lbs. (49 Nm).

4. Drain transfer oil

5. Remove the front wheel.

6. Remove the left front axle hub nut.

a. Using SST(SST: 09930-00010) or equivalent and a hammer, unstake the staked part of the left axle hub nut.

➡Loosen the staked part of the nut completely, otherwise the screw of the drive shaft may be damaged.

b. While applying the brakes, remove the left lock axle hub nut.

7. Remove the nut and separate the left stabilizer link assembly.

➡If the ball joint turns together with the nut, use a hexagon wrench (6 mm) to hold the stud.

8. Separate the left front speed sensor, as follows:

a. Remove the bolt and clip, and separate the sensor wire and hose from the shock absorber.

➡Be careful not to damage the speed sensor.

b. Remove the bolt, and separate the left speed sensor from the steering knuckle.

➡Do not allow foreign matter to adhere to the speed sensor.

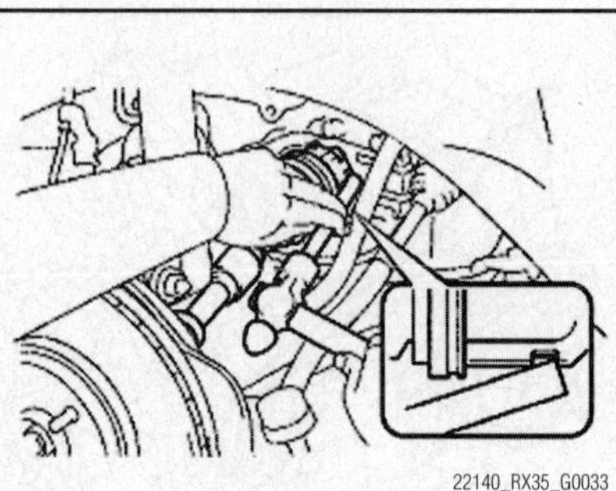
22140_RX35_G0033

Fig. 16 Align the shaft splines and install the drive shaft assembly

9. Separate the left tie rod end sub-assembly, as follows:

a. Remove the cotter pin and nut.

10. Separate the No. 1 lower front suspension arm sub-assembly, as follows:

a. Remove the bolt and 2 nuts, and separate the No. 1 front suspension arm sub-assembly lower from the lower ball joint.

11. Using a plastic hammer, separate the drive shaft from the axle hub.

➡Be careful not to damage the boot and speed sensor rotor.

12. Using SST(SST: 09520-01010, SST: 09520-24010) or equivalent, remove the front drive shaft assembly.

➡Be careful not to damage the transaxle case oil seal, inboard joint boot and drive shaft dust cover or drop the drive shaft assembly.

13. For 2WD vehicles, remove the right front drive shaft assembly, as follows:

a. Using a screwdriver, remove the bearing brake hole snap ring.

b. Remove the bolt and front drive shaft assembly from the drive shaft bearing bracket.

14. For 4WD vehicles, remove the right front drive shaft assembly, as follows:

a. Using SST(SST: 09520-01010, SST: 09520-24010) or equivalent, remove the front drive shaft assembly.

➡When removing and installing the right front drive shaft assembly in 4WD vehicle, be sure to first drain all the transaxle oil and transfer oil. If removal and installation is carried out without draining these oils, the transfer oil will flow into the transaxle side. Extensive cleaning will be required if the two oils mix.

➡Do not damage the oil seal and dust cover.

➡Move the drive shaft assembly while keeping it level.

To install:

15. Install the left front drive shaft assembly, as follows:

a. Coat the spline of the inboard joint shaft assembly with ATF.

b. Align the shaft splines and install the drive shaft assembly with a brass bar and hammer.

➡Set the snap ring with the opening side facing down.

➡Be careful not to damage the drive shaft dust cover, boot and oil seal.

➡**Move the drive shaft assembly while keeping it level.**

16. For 2WD vehicles, install the right front drive shaft assembly, as follows:
    a. Using a screwdriver, install a new bearing bracket hole snap ring.

➡**Do not damage the oil seal and boot.**

    b. Install the bolt and tighten to 24 ft. lbs. (32 Nm).
17. For 4WD vehicles, install the right front drive shaft assembly by using the same procedures as for the left side, as described above.

➡**Set the snap ring with the opening side facing downward.**

➡**Be careful not to damage the transaxle case oil seal, inboard joint boot and drive shaft dust cover.**

18. Install the left drive shaft assembly to the left front axle assembly.

➡**Be careful not to damage the outboard joint boot or speed sensor rotor.**

19. Install the lower ball joint to the lower front suspension arm sub-assembly with the bolt and nuts and tighten to 94 ft. lbs. (127 Nm).
20. Install the left tie rod end sub-assembly, as follows:
    a. Install the tie rod end to the steering knuckle with the nut and tighten to 36 ft. lbs. (49 Nm).
    b. Install a new cotter pin.

➡**If the holes for the cotter pin are not are not aligned, tighten the nut up to 60° further.**

21. Install the left front speed sensor to the steering knuckle with the bolt and tighten to 71 inch lbs. (8 Nm).

➡**Prevent foreign matter from adhering to the speed sensor.**

22. Install the flexible hose and the speed sensor to the shock absorber with the bolt and set the clip of sensor on knuckle and tighten to 14 ft. lbs. (19 Nm).

➡**Be careful not to damage the speed sensor, allow foreign matter to adhere to the speed sensor, or twist the sensor wire when installing the speed sensor.**

23. Install the left stabilizer link assembly with the nut and tighten to 55 ft. lbs. (74 Nm).

➡**If the ball joint turns together with nut. use a hexagon (6 mm) wrench to hold the stud.**

24. Using s socket wrench (30 mm), install a new left axle hub nut and tighten to 217 ft. lbs. (294 Nm).
25. Using a chisel and hammer, stake the left axle hub nut.
26. Install the front wheel and tighten the lug nuts to 76 ft. lbs. (103 Nm).
27. Add automatic transaxle fluid.
28. Inspect automatic transaxle fluid.
29. For 4wd vehicles, add transfer oil.
30. Inspect transfer oil.
31. Adjust front wheel alignment.
32. Install the No. 2 engine under cover.
33. Install the engine under cover assembly.
34. Check the ABS speed sensor signal.

REAR HALFSHAFT

REMOVAL & INSTALLATION

See Figures 17 and 18.

1. Remove the rear wheel.
2. Separate the rear speed sensor.
3. Remove the rear axle shaft nut, as follows:
    a. Using SST (SST: 09930-00010) or equivalent and a hammer, unstake part of the left axle shaft nut.

➡**Loosen the staked part of the nut completely, otherwise the screw of the drive shaft may be damaged.**

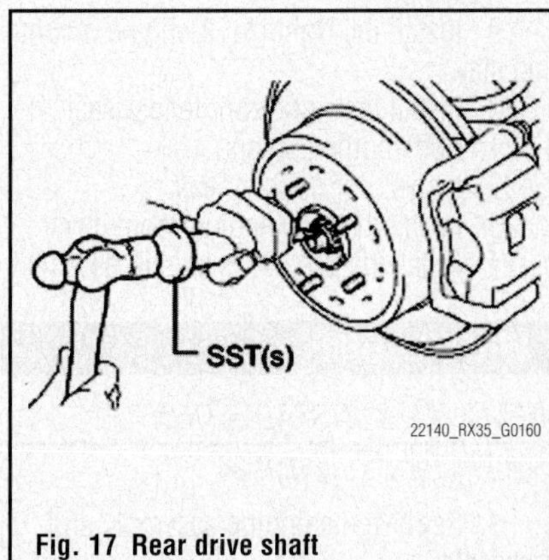

Fig. 17 Rear drive shaft

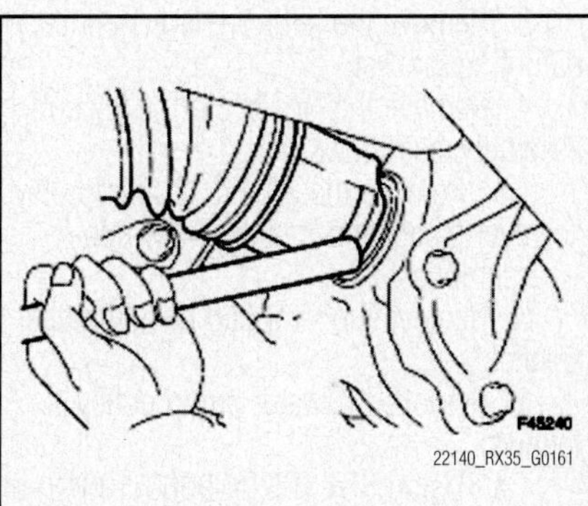

Fig. 18 Install the left rear drive shaft assembly

    b. While applying the brakes, remove the left lock axle shaft nut.
4. Separate the rear height control sensor sub-assembly, as applicable.
5. Separate the No. 1 rear suspension arm assembly.
6. Separate the rear strut rod assembly.
7. Separate the No. 2 rear suspension arm assembly.
8. Separate the left rear axle carrier sub-assembly, as follows:
    a. Push the left rear axle carrier sub-assembly toward the outside of the vehicle. Using plastic hammer, separate the left rear drive shaft assembly from the rear axle carrier sub-assembly.

➡**Be careful not to damage the joint boot and speed sensor rotor.**

➡**Do not push out the left axle carrier sub-assembly.**

9. Remove the left rear drive shaft assembly as shown in the illustration.

➡**Move the drive shaft assembly while keeping it level.**

To install:

10. Align the shaft splines and install the left rear drive shaft assembly using a brass bar and hammer.

➡**Set the snap ring with the opening side facing downward.**

➡**Be careful not to damage the oil seal boot and dust cover.**

➡**Move the drive shaft assembly while keeping it level.**

11. Push the left rear axle carrier sub-assembly toward the inside of the vehicle fitting it to the spline part of the left rear drive shaft assembly and insert.

➡**Be careful not to damage the joint boot and speed sensor rotor.**

➡**Do not push out the rear axle carrier sub-assembly excessively.**

12. Temporarily tighten the No. 2 rear suspension arm assembly.
13. Temporarily tighten the rear strut rod assembly.
14. Temporarily tighten the No. 1 rear suspension arm assembly.
15. Install the rear height control sensor sub-assembly, as applicable.
16. Install a new left axle shaft nut and tighten to 217 ft. lbs. (294 Nm).
17. Using a chisel and hammer, stake the left axle shaft nut.

➡Be careful not to damage the speed sensor.

➡Prevent foreign matter from adhering to the speed sensor.

➡Do not twist the sensor wire when installing the sensor.

18. Install the rear wheel.
19. Stabilize suspension.
20. Fully tighten the No. 2 rear suspension arm assembly.
21. Fully tighten the rear strut rod assembly.
22. Fully tighten the No. 1 rear suspension arm assembly.
23. Adjust rear wheel alignment.
24. Adjust headlight aim, as applicable.
25. Check the ABS speed sensor signal.

ENGINE COOLING

THERMOSTAT

REMOVAL & INSTALLATION

See Figures 19 and 20.

1. Drain engine coolant.
2. Remove the V-bank cover sub-assembly.
3. Remove the side engine room cover.
4. Remove the right front fender seal.
5. Remove the right No. 2 engine mounting stay.
6. Remove the engine moving control rod sub-assembly.
7. Remove the V-ribbed belt.
8. Remove the bolt, No. 2 idler pulley cover plate and No. 2 idler pulley.
9. Slide the clamp and disconnect the radiator hose outlet.
10. Remove the 2 nuts and water inlet.

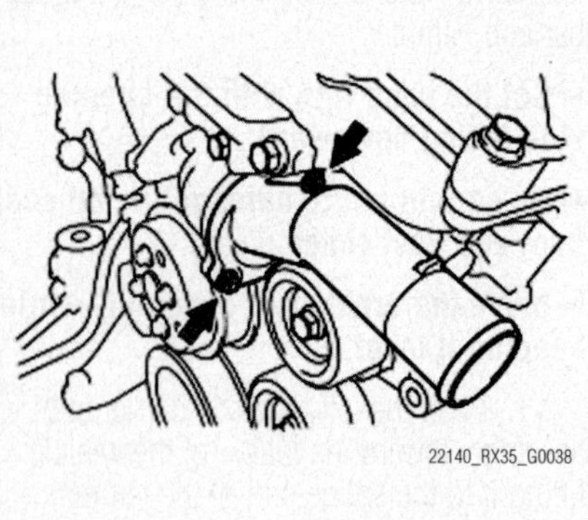

Fig. 19 Remove the 2 nuts and water inlet

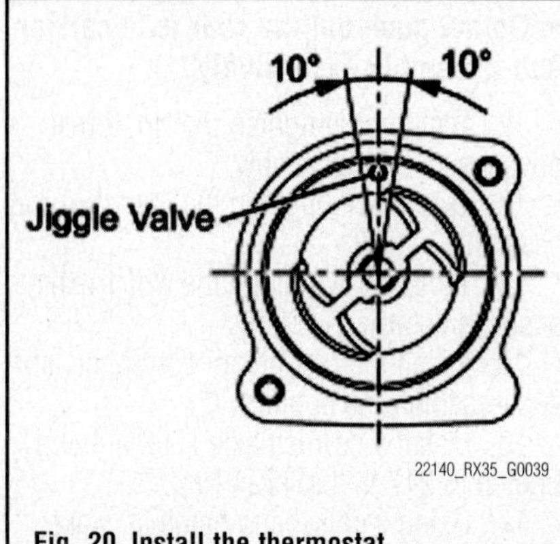

Fig. 20 Install the thermostat

11. Remove the gasket from the thermostat and remove thermostat.

To install:

12. Install a new gasket to the thermostat.
13. Install the thermostat with the jiggle valve facing up.

➡The jiggle valve may be set within 10° on either side of the prescribed position.

14. Install the water inlet with the 2 nuts and tighten to 7 ft. lbs. (10 Nm).
15. Connect the radiator hose outlet with the clamp.
16. Install the No. 2 idler pulley cover plate and No. 2 idler pulley with the bolt and tighten to 32 ft. lbs. (43 Nm).
17. Install the V-ribbed belt.
18. Install the engine moving control rod sub-assembly.
19. Install the right No. 2 engine mounting stay.
20. Install the right front fender seal.
21. Add engine coolant.
22. Inspect for coolant leak.
23. Install the side engine room cover.
24. Install the V-bank cover sub-assembly.

WATER PUMP

REMOVAL & INSTALLATION

See Figures 21 through 24.

1. Remove the engine assembly and transaxle.
2. Secure the engine.
3. Remove the left front No. 1 engine mounting bracket.
4. Remove the No. 2 idler pulley sub-assembly, as follows:
   a. Remove the 2 bolts, 2 idler pulley cover plates and 2 idler pulley sub-assemblies.
5. Remove the V-ribbed belt tensioner assembly.
6. Remove the water pump pulley, as follows:
   a. Using SST (SST: 09960-10010) or equivalent, hold the water pump pulley.
   b. Remove the 4 bolts and water pump pulley.
7. Remove the water inlet housing, as follows:
   a. Disconnect the water hose.
   b. Remove the 2 bolts, nut and water inlet housing.
   c. Remove the water inlet housing gasket and water outlet pipe O-ring.
8. Remove the 16 bolts, water pump assembly and water pump gasket.

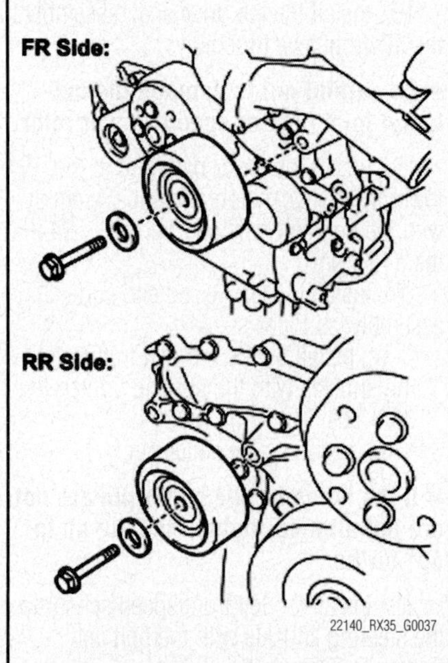

Fig. 21 Remove the 2 bolts, 2 idler pulley cover plates and idler pulley sub-assemblies

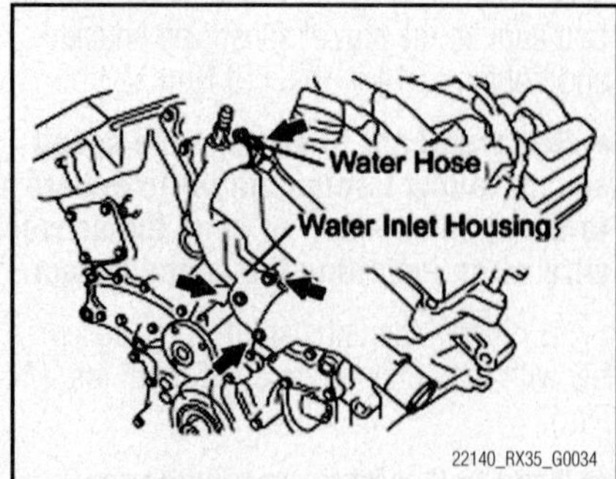

Fig. 22 Remove the 2 bolts, nut and water inlet housing

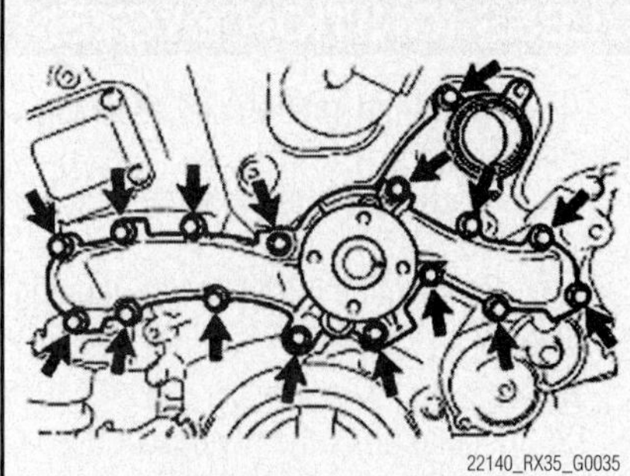

Fig. 23 Remove the 16 bolts, water pump assembly and water pump gasket

Fig. 24 Install a new water pump gasket and the water pump assembly with the 16 bolts

To install:

9. Install a new water pump gasket and the water pump assembly with the 16 bolts and tighten to 15 ft. lbs. (21 Nm), 81 inch lbs. (9.1 Nm).

➡Make sure that there is no oil on the threads of bolts A.

➡Be sure to replace 2 bolts C with new ones or reuse them after applying adhesive 1344. Adhesive: Toyota Genuine Seal Packing Black, Three Bond 1207B or equivalent.

10. Install a new water inlet housing gasket and water outlet pipe O-ring.
11. Install the water inlet housing with the 2 bolts and nut and tighten to 7 ft. lbs. (10 Nm).

➡Be careful not to allow the O-ring to get caught between the parts.

12. Connect the water hose.
13. Install the water pump pulley, as follows:
    a. Temporarily install the water pump pulley with the 4 bolts.
    b. Using SST (SST: 09960-10010) or equivalent, hold the water pump pulley.
    c. Tighten the 4 bolts to 15 ft. lbs. (21 Nm).
14. Install the V-ribbed belt tensioner assembly.
15. Install the 2 idler pulley cover plates and idler pulley sub-assemblies with the 2 bolts and tighten to 32 ft. lbs. (43 Nm).
16. Install the left front No. 1 engine mounting bracket.
17. Remove engine stand.
18. Install the engine assembly and transaxle.

ENGINE ELECTRICAL

CHARGING SYSTEM

ALTERNATOR

REMOVAL & INSTALLATION

See Figures 25 and 26.

1. Disconnect the negative battery cable.
2. Remove the right front wheel.
3. Remove the No. 1 engine under cover.
4. Remove the right front fender splash shield sub-assembly.
5. Remove the right front fender apron seal.
6. Remove the side engine room cover.
7. Remove the V-bank cover sub-assembly.
8. Remove the cool air intake duct seal.
9. Remove the No. 2 air cleaner inlet.
10. Remove the air cleaner cap sub-assembly.
11. Remove the No. 1 air cleaner inlet.
12. Disconnect the radiator reservoir tank hose.
13. Remove the hood lock control cable cover.
14. Remove the hood lock assembly.
15. Remove the upper radiator support sub-assembly.
16. Remove the fan shroud.
17. Remove the V-ribbed belt.
18. Remove the alternator assembly, as follows:
    a. Remove the terminal cap.
    b. Remove the nut and disconnect the wire harness from terminal B.
    c. Disconnect the generator connector from the generator assembly.
    d. Disconnect the connector from the compressor and magnetic clutch.
    e. Disconnect the 2 wire harness clamps.
    f. Remove the 2 bolts.
    g. Remove the bolt from the cylinder block.
    h. Disconnect the wire harness clamp and remove the generator assembly.
    i. Remove the bolt and wire harness clamp stay.
    j. Remove the bolt and bracket.

To install:

19. Install the alternator assembly, as follows:
    a. Install the bracket with the bolt and tighten to 15 ft. lbs. (20 Nm).
    b. Install the wire harness clamp stay and tighten to 74 inch lbs. (8.4 Nm).
    c. Connect the wire harness clamp.
    d. Install the alternator assembly to the cylinder block with the bolt and tighten to 15 ft. lbs. (20 Nm).
    e. Install the 2 bolts and tighten to 32 ft. lbs. (43 Nm).
    f. Connect the alternator connector to the alternator assembly.
    g. Install the alternator wire with the nut and tighten to 87 inch lbs. (9.8 Nm).
    h. Install the terminal cap.
    i. Connect the 2 wire harness clamps.
    j. Connect the magnetic clutch connector to the compressor and magnetic clutch.
20. The remainder of installation is the reverse of the removal procedure.
21. After connecting cable to negative battery terminal, perform initialization.

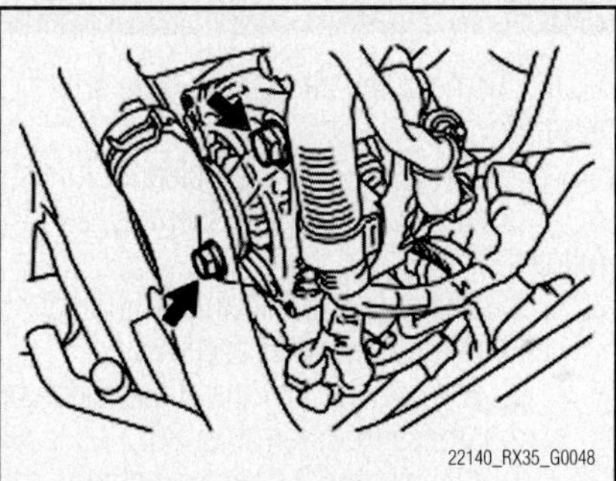

Fig. 25 Remove the 2 bolts

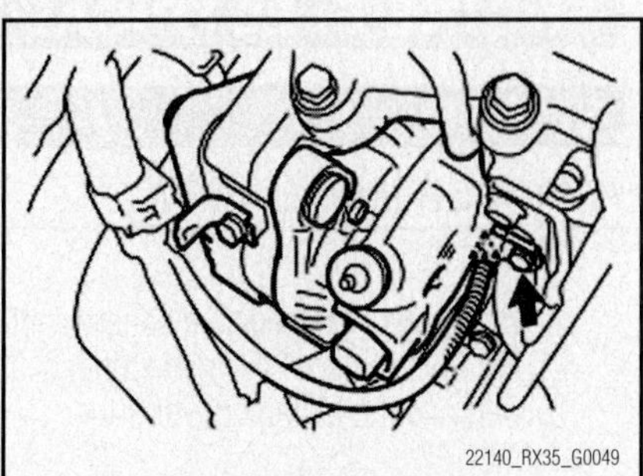

Fig. 26 Remove the bolt from the cylinder block

FIRING ORDERS

Firing order for 3.5L engine: 1–2–3–4–5–6

IGNITION COIL

REMOVAL & INSTALLATION

See Figures 27 through 29.

1. Disconnect the negative battery cable.
2. Remove the No. 1 engine under cover.

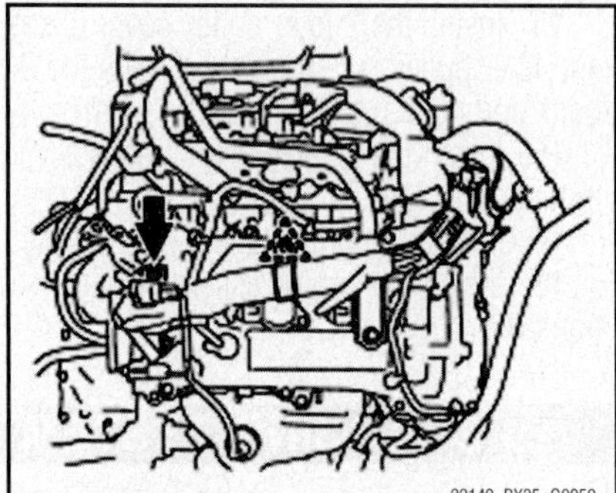

Fig. 27 Remove the bolt and disconnect the harness clamp

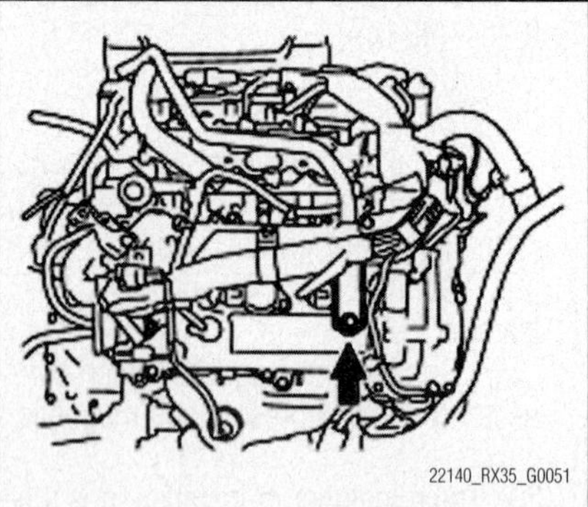

Fig. 28 Remove the bolt and No. 1 surge tank stay

3. Remove the engine room side cover.
4. Remove the V-bank cover sub-assembly.
5. Drain engine coolant.
6. Remove the windshield wiper motor and link assembly.
7. Remove the cowl top panel outer sub-assembly.
8. Remove the air cleaner cap sub-assembly.
9. Remove the intake air surge tank assembly.
10. Remove the No. 1 surge tank stay, as follows:
    a. Remove the bolt and disconnect the harness clamp.
    b. Remove the bolt and No. 1 surge tank stay.

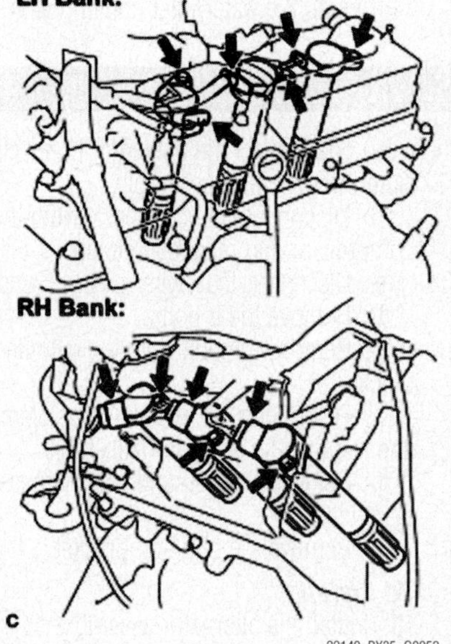

Fig. 29 Remove the 6 bolts and 6 ignition coils

11. Remove the ignition coil assembly, as follows:
    a. Disconnect the 6 ignition coil connectors.
    b. Remove the 6 bolts and 6 ignition coils.

To install:

12. Installation is the reverse of the removal procedure, noting the following:
    a. Tighten the ignition coil bolts to 7 ft. lbs. (10 Nm).
    b. Tighten the No. 1 surge tank stay bolt to 15 ft. lbs. (21 Nm).
    c. Tighten the surge tank bolt and clamp to 62 inch lbs. (7 Nm).
    d. After adding engine coolant, check for leaks.
    e. After completing installation, perform initialization procedure.

IGNITION TIMING

ADJUSTMENT

All engines are equipped with a Distributorless Ignition System (DIS). No timing adjustment is possible.

SPARK PLUGS

REMOVAL & INSTALLATION

1. Referring to the ignition coil pack removal procedure, disconnect the 6 ignition coil connectors, and remove the 6 bolts and ignition coils.
2. Using a 16 mm (0.63 in.) plug wrench, remove the spark plugs.

To install:

3. Installation is the reverse of the removal procedure. Tighten the spark plugs to 13 ft. lbs. (18 Nm).

STARTER

REMOVAL & INSTALLATION

See Figure 30.

1. Disconnect the negative battery cable.
2. Remove the cool air intake duct seal.
3. Remove the V-bank cover sub-assembly.
4. Remove the No. 2 air cleaner inlet.
5. Remove the air cleaner cap sub-assembly.
6. Remove the air cleaner case sub-assembly.
7. Remove the No. 1 air cleaner inlet.
8. Remove the starter assembly, as follows:
    a. Disconnect the terminal 50 connector from the starter assembly.
    b. Remove the nut and disconnect the wire harness from terminal 30.
    c. Remove the 2 bolts and starter assembly.

To install:

9. Install the starter assembly with the 2 bolts and tighten to 28 ft. lbs. (37 Nm).
10. Connect the wire harness to terminal 30 and install the nut, and then attach the terminal cap and tighten to 87 inch lbs. (9.8 Nm).
11. Connect the terminal 50 connector to the starter assembly.
12. The remainder of installation is the reverse of the removal procedure.

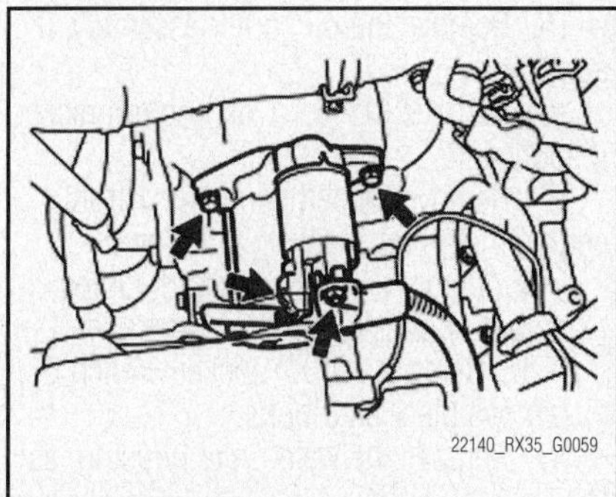
Fig. 30 Remove starter assembly

SOLENOID OR RELAY REPLACEMENT

1. Remove the nut and disconnect the lead wire from terminal C.
2. Remove the 2 screws that hold the magnetic switch assembly to the motor terminal starter kit.
3. Remove the magnetic switch assembly.
4. Remove the return spring and the plunger from the magnetic switch assembly.

To install:

5. Apply high-temperature grease to the plunger and the hook.
6. Hang the plunger hook of the magnetic switch assembly to the drive lever hook.
7. Install the plunger and the return spring.
8. Install the repair magnetic switch assembly with the 2 screws and tighten to 66 inch lbs. (7.5 Nm).
9. Connect the lead wire to terminal C with the nut and tighten to 7 ft. lbs. (10 Nm).

ENGINE MECHANICAL

ACCESSORY DRIVE BELTS

ACCESSORY BELT ROUTING

See Figure 31.

Refer to the accompanying illustration for Accessory Drive Belt routing.

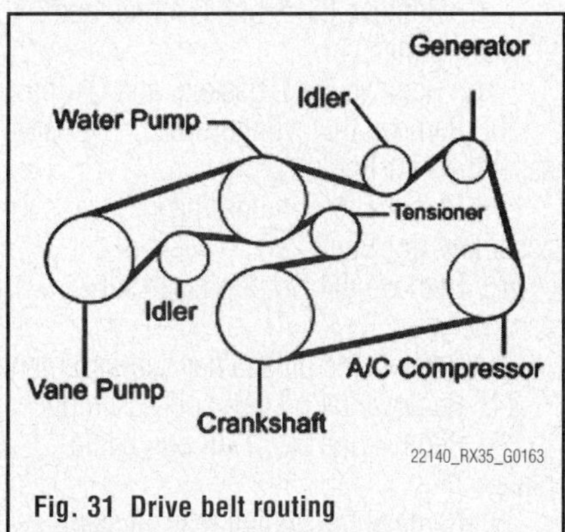

Fig. 31 Drive belt routing

INSPECTION

Visually check the V-ribbed belt for excessive wear, frayed cords, etc. All worn or damaged drive belts should be replaced immediately. Cracks on the rib side of a V-ribbed belt are considered acceptable, If the drive belt has chunks missing from its ribs, it should be replaced. After installing the V-ribbed belt, check that it fits properly in the ribbed grooves. Check to confirm that the belt has not slipped out of the grooves on the bottom of the crank pulley by hand.

ADJUSTMENT

Belt tension is maintained by an automatic tensioner. No adjustment is necessary or possible.

REMOVAL & INSTALLATION

See Figures 32 through 34.

1. Remove the right front wheel.
2. Separate the right front fender splash shield sub-assembly.
3. Remove the right front fender apron seal.
4. Remove the engine room side cover.
5. Remove the V-bank cover sub-assembly.
6. Remove the V-ribbed belt, as follows:
   a. Using SST (SST: 09249-63010) or equivalent, release the belt tension by turning the belt tensioner counterclockwise, and remove the V-ribbed belt from the belt tensioner.

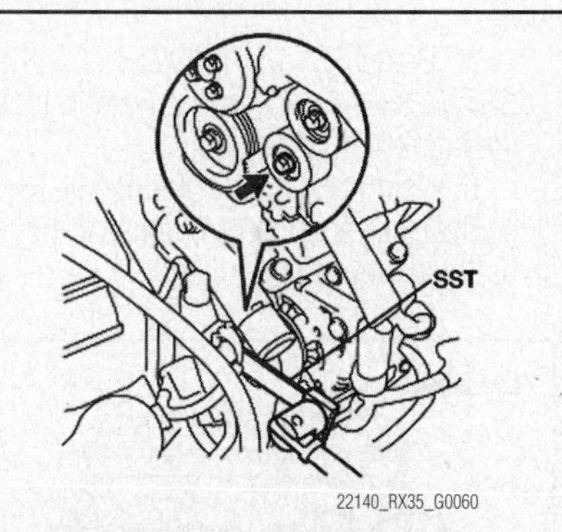

Fig. 32 Release the belt tension by turning the belt tensioner counterclockwise

   b. While turning the belt tensioner counterclockwise, align with its holes, and then insert the 5 mm bi-hexagon wrench into the holes to fix the V-ribbed belt tensioner.

To install:

7. Install the V-ribbed belt.
8. Using SST (SST: 09249-63010) or equivalent, turn the belt tensioner counterclockwise and remove the bar.
9. If it is difficult to install the V-ribbed belt, perform the following procedure:
   a. Put the V-ribbed belt on every pulley except the tensioner pulley as shown in the illustration.
   b. While releasing the belt tension by turning the belt tensioner counterclockwise, put the V-ribbed belt on the tensioner pulley.

➡Put the backside of the V-ribbed belt on the tensioner pulley and idler pulley.

➡Check that the V-ribbed belt is properly set to each pulley.

   c. After installing the V-ribbed belt, check that it fits properly in the ribbed grooves. Confirm that the belt has not slipped out of the grooves on the bottom of the crank pulley by hand.

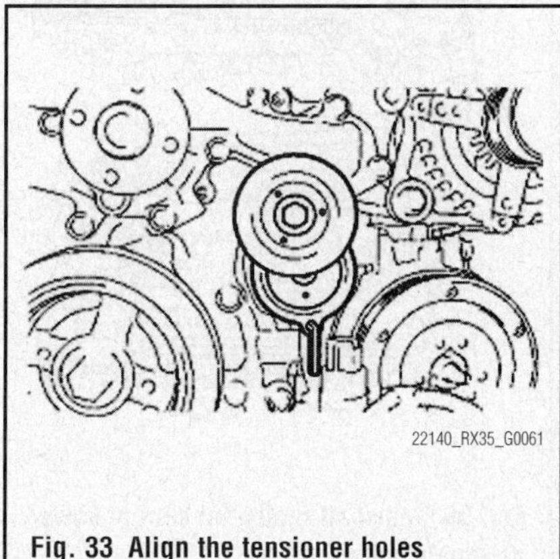
Fig. 33 Align the tensioner holes

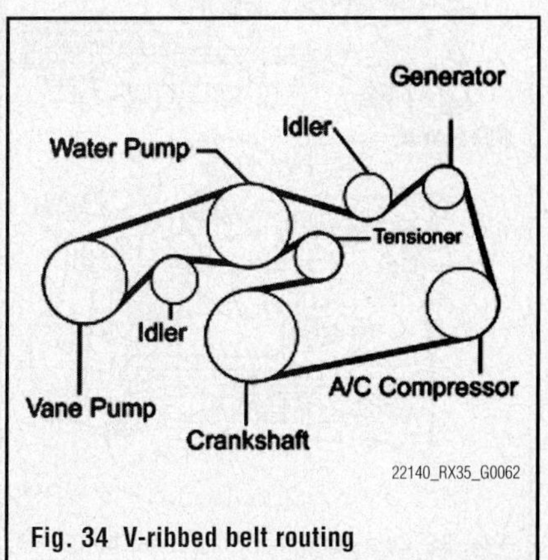

Fig. 34 V-ribbed belt routing

10. Install the V-bank cover sub-assembly.
11. Install the engine room side cover.
12. Install the right front fender apron seal.
13. Install the right front fender splash shield sub-assembly.
14. Install the right front wheel and tighten the lug nuts to 76 ft. lbs. (103 Nm).

CAMSHAFT AND VALVE LIFTERS

REMOVAL & INSTALLATION

See Figures 35 through 64.

1. Remove the engine assembly.
2. Install on engine stand.
3. Remove the oil filler cap and gasket.
4. Remove the spark plugs and ignition coil assembly.
5. Remove the drain plug and gasket.
6. Remove the ventilation valve.
7. Remove the camshaft position sensor, as follows:
    a. Remove the 4 bolts and 4 sensors.
8. Remove the camshaft timing oil control valve assembly, as follows:
    a. Remove the 4 bolts and 4 oil control valves.
9. Remove the crankshaft position sensor.
10. Remove the No. 1 oil pipe, as follows:
    a. Remove the 2 oil pipe unions and oil pipe.
    b. Remove the left hand oil control valve filter and gaskets.
11. Remove the oil pipe, as follows:
    a. Remove the bolt.
    b. Remove the 2 oil pipe unions and oil pipe.
    c. Remove the right oil control valve filter and gaskets.
12. Remove the cylinder block water drain cock sub-assembly, as follows:
    a. Remove the water drain cocks from the cylinder block.
    b. Remove the water drain cock plugs from the water drain cocks.
13. Remove the oil filter element, as follows:
    a. Remove the drain plug. Do not remove the O-ring from the oil filter cap.
    b. Connect the hose to the pipe.
    c. Insert the pipe with the hose into the oil filter cap.
    d. Make sure that the oil is completely drained and remove the pipe and O-ring.
    e. Using SST (SST: 09228-06501) or equivalent, remove the oil filter cap.
    f. Remove the oil filter element and O-ring from the oil filter cap. Do not use any tools when removing the O-ring to prevent the O-ring groove from being damaged.
14. Remove the crankshaft pulley, as follows:
    a. Using SST (SST: 09213-70011, SST: 09330-00021) or equivalent, loosen the crankshaft pulley bolt.
    b. Using SST (SST: 09950-50013) or equivalent, remove the crankshaft pulley bolt and crankshaft pulley.
15. Remove the oil cooler assembly, if applicable.
16. Remove the No. 1 oil cooler bracket, if applicable.
17. Remove the left hand No. 1 front engine mounting bracket, as follows:
    a. Remove the 6 bolts and engine mounting bracket.
    b. Using a TORX® socket wrench E8, remove the 2 stud bolts.
18. Remove the water inlet housing, as follows:
    a. Remove the 2 nuts, water inlet and thermostat.
    b. Remove the gasket.
    c. Remove the drain cock plug.
    d. Remove the drain cock.
    e. Remove the 2 stud bolts.
    f. Separate the No. 1 water by-pass hose.
    g. Remove the 2 bolts, nut, and water inlet housing.
    h. Remove the 2 O-rings.
19. Remove the water outlet, as follows:
    a. Remove the 2 bolts, 4 nuts and water outlet.
    b. Remove the 2 gaskets and O-ring.
20. Remove the cylinder head cover sub-assembly (for Bank 1).
21. Remove the cylinder head cover sub-assembly (for Bank 2).
22. Remove the No. 2 oil pan sub-assembly.
23. Remove the oil strainer sub-assembly.
24. Remove the oil pan sub-assembly.
25. Remove the No. 1 oil pan baffle plate.
26. Remove the engine rear oil seal retainer, as follows:
    a. Remove the 6 bolts.
    b. Using a screwdriver with the tip taped, pry out the oil seal retainer. Be careful not to damage the engine rear oil seal retainer.
27. Remove the engine rear oil seal, as follows:

LH Bank:
RH Bank:
22140_RX35_G0063

Fig. 35 Camshaft position sensor bolts

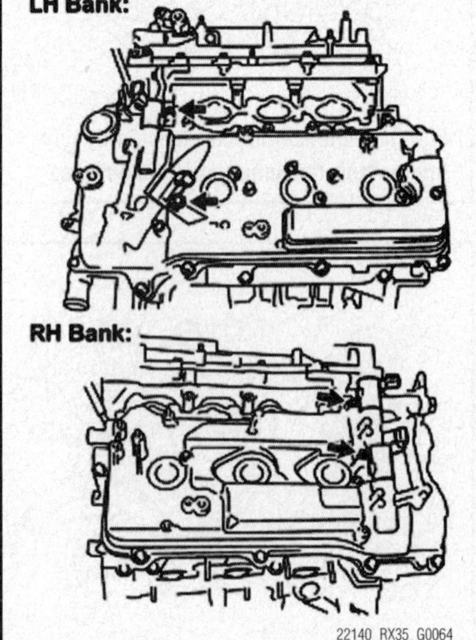

Fig. 36 Camshaft timing oil control valve assembly bolts

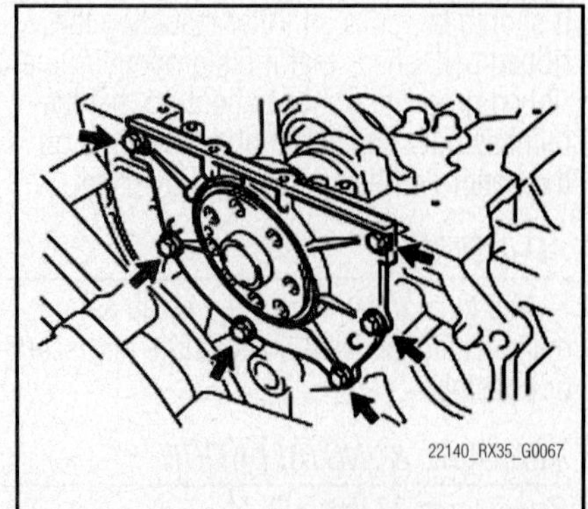

Fig. 37 Engine rear oil seal retainer bolts

a. Place the oil seal retainer on wooden blocks. Be careful not to damage the engine rear oil seal retainer.

b. Using a screwdriver with the tip taped and a hammer, tap out the oil seal.

28. Remove the water pump assembly.
29. Remove the timing chain cover sub-assembly.
30. Remove the timing chain case oil seal.
31. Set the No. 1 cylinder to TDC/compression.
32. Remove the No. 1 chain tensioner assembly.
33. Remove the chain tensioner slipper.
34. Remove the chain sub-assembly.
35. Remove the idle sprocket assembly.
36. Remove the No. 1 chain vibration damper.
37. Remove the No. 2 chain vibration damper.
38. Remove the crankshaft timing sprocket, as follows:

a. Remove the crankshaft timing sprocket from the crankshaft.

b. Remove the 2 pulley set keys from the crankshaft.

39. Remove the camshaft timing gears and No. 2 chain (for Bank 1), as follows:

a. While raising the No. 2 chain tensioner, insert a pin of 1.0 mm (0.039 in.) into the hole to hold the No. 2 chain tensioner.

b. Hold the hexagonal portion of the camshaft with a wrench, and remove the 2 bolts and 2 camshaft timing gear assemblies.

➡Be careful not to damage the cylinder head with the wrench.

➡Do not disassemble the camshaft timing gear assemblies.

c. Remove the No. 2 chain.

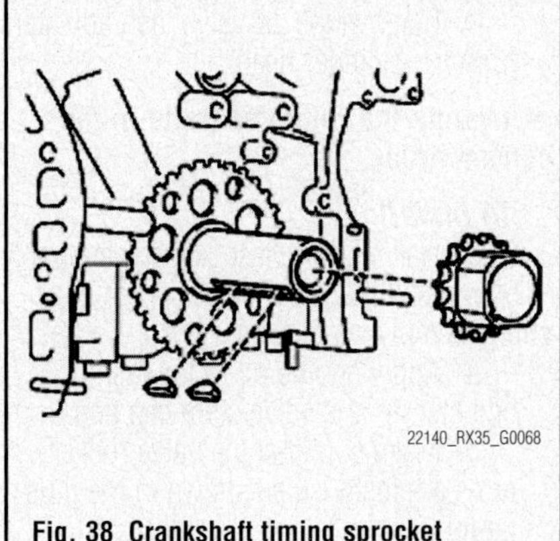

Fig. 38 Crankshaft timing sprocket

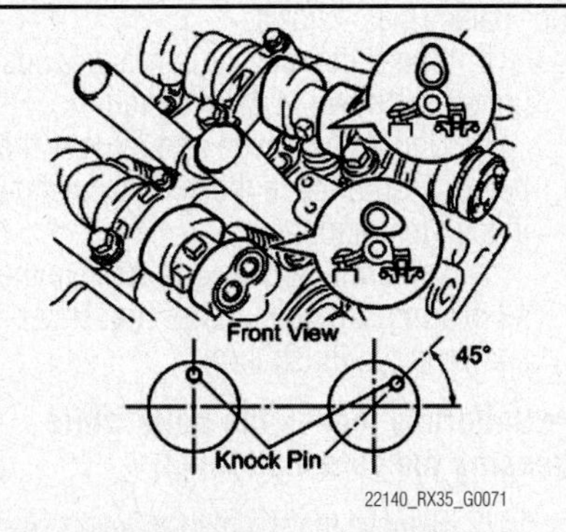

Fig. 41 Positioning bank 1 camshafts for bearing cap removal

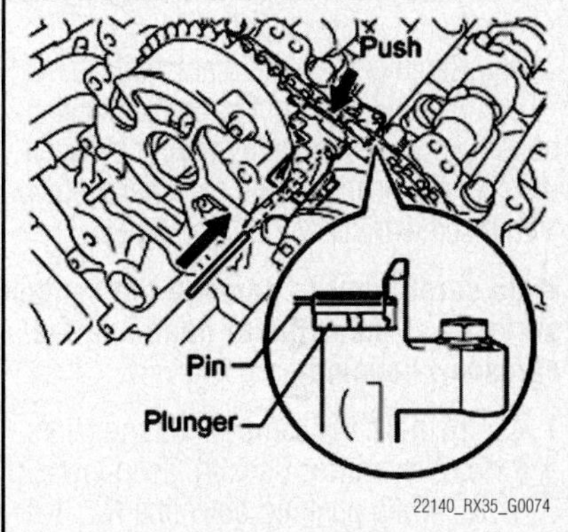

Fig. 44 Pinning No. 3 tensioner

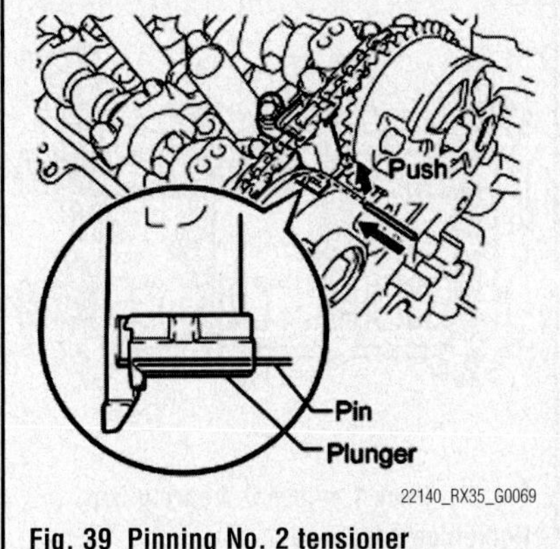
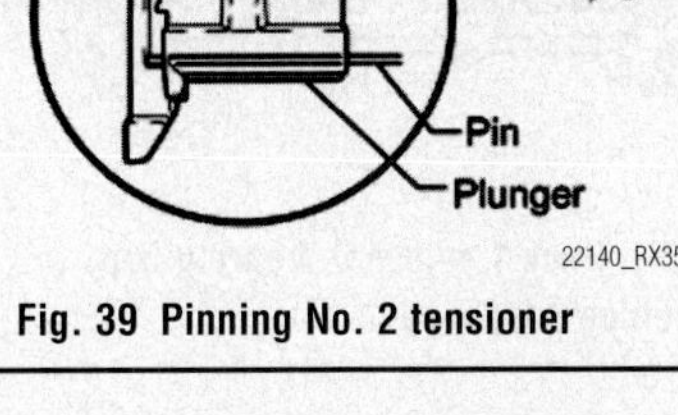

Fig. 39 Pinning No. 2 tensioner

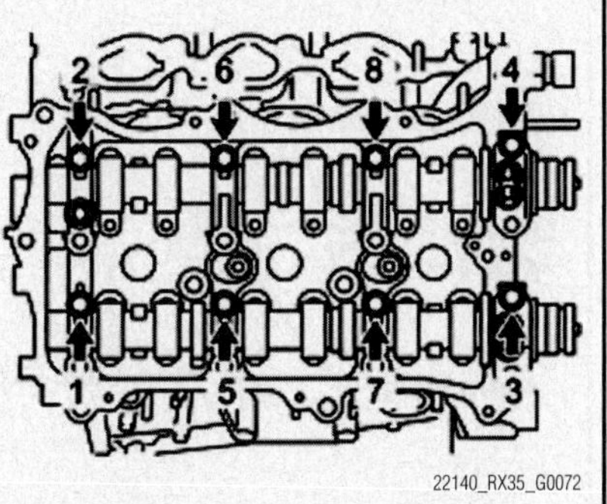

Fig. 42 Bank 1 camshaft bearing cap 8 bolt removal sequence

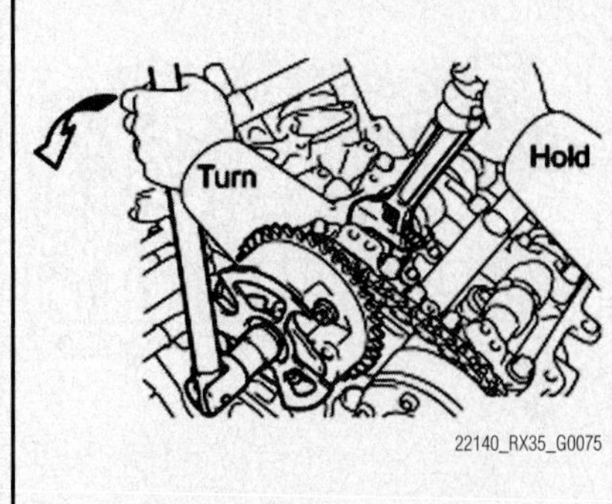

Fig. 45 Removing bank 2, No. 2 chain gear assemblies

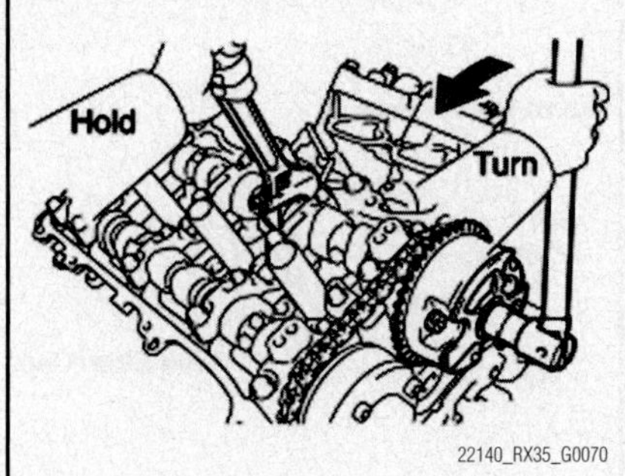

Fig. 40 Removing bank 1, No.2 chain gear assemblies

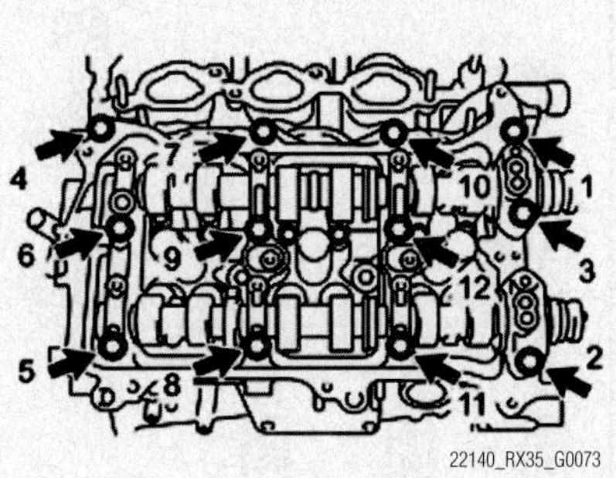

Fig. 43 Bank 1 camshaft bearing cap 12 bolt removal sequence

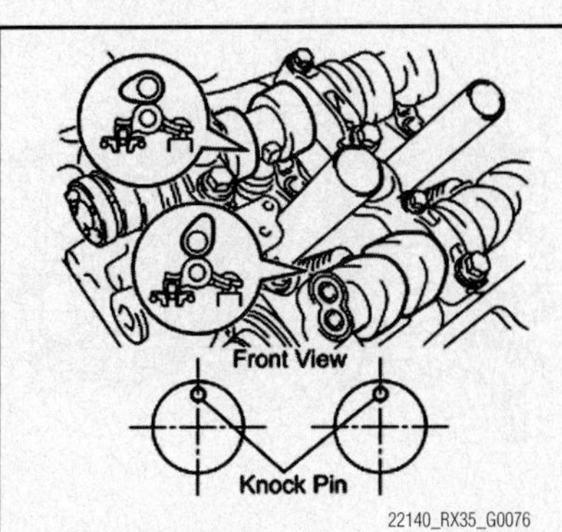

Fig. 46 Positioning bank 2 camshafts for bearing cap removal

40. Remove the bolt and No. 2 chain tensioner assembly.

41. Remove the camshaft bearing cap (for Bank 1), as follows:

a. Check that the camshafts are positioned as shown in the illustration.

b. Uniformly loosen and remove the 8 bearing cap bolts in several steps and in the sequence shown in the illustration.

c. Uniformly loosen and remove the 12 bearing cap bolts in several steps and in the sequence shown in the illustration.

➡**Uniformly loosen the bolts while keeping the camshaft level.**

d. Remove the 5 camshaft bearing caps.

42. Remove the camshaft.

43. Remove the No. 2 camshaft.

44. Remove the right hand camshaft housing sub-assembly by prying between the cylinder head and the camshaft housing with a screwdriver with the tip taped.

➡**Be careful not to damage the contact surfaces of the cylinder head and the camshaft housing.**

45. Remove the camshaft timing gears and No. 2 chain (for Bank 2), as follows:

a. While pushing down the No. 3 chain tensioner, insert a pin of 1.0 mm (0.039 in.) into the hole to fix the No. 3 chain tensioner.

b. Hold the hexagonal portion of the camshaft with a wrench, and remove the 2 bolts and 2 camshaft timing gear assemblies.

➡**Be careful not to damage the cylinder head with the wrench.**

➡**Do not disassemble the camshaft timing gear assemblies.**

c. Remove the No. 2 chain.

46. Remove the bolt and No. 3 chain tensioner.

47. Remove the camshaft bearing cap (for Bank 2), as follows:

a. Check that the camshafts are positioned as shown in the illustration.

b. Uniformly loosen and remove the 8 bearing cap bolts in the sequence shown in the illustration.

c. Uniformly loosen and remove the 13 bearing cap bolts in the sequence shown in the illustration.

➡**Uniformly loosen the bolts while keeping the camshaft level.**

d. Remove the 5 camshaft bearing caps.

48. Remove the No. 3 camshaft.

49. Remove the No. 4 camshaft.

50. Remove the left hand camshaft housing sub-assembly by prying between the cylinder head and the camshaft housing with a screwdriver with the tip taped.

➡**Be careful not to damage the contact surfaces of the cylinder head and the camshaft housing.**

51. Remove the No. 1 valve rocker arm sub-assembly, as follows:

a. Remove the 24 valve rocker arms.

➡**Arrange the removed parts in the correct order.**

52. Remove the valve lash adjuster assembly, as follows:

a. Remove the 24 valve lash adjusters from the cylinder head.

➡**Arrange the removed parts in the correct order.**

To install:

53. Install the valve lash adjuster assembly.

54. Install the No. 1 valve rocker arm sub-assembly, as follows:

a. Apply engine oil to the lash adjuster tip and valve stem cap end.

b. Make sure that the valve rocker arms are installed as shown in the illustration.

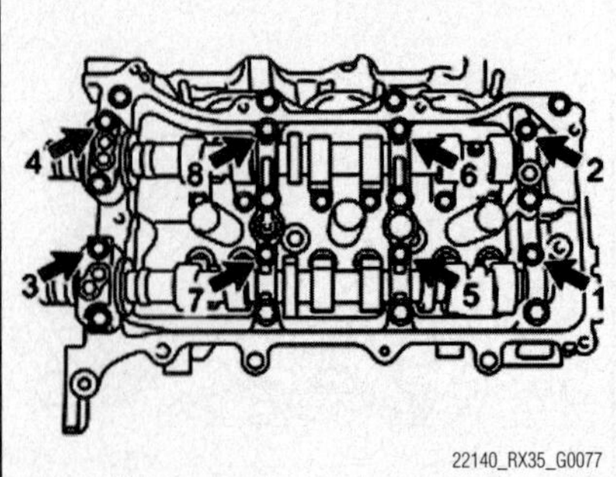

Fig. 47 Bank 2 camshaft bearing cap 8 bolt removal sequence

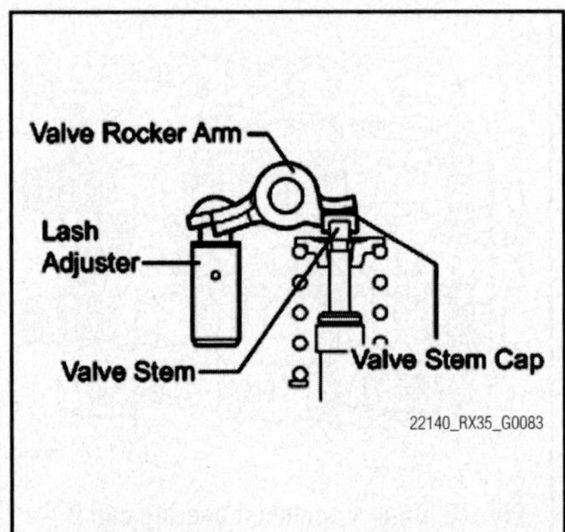

Fig. 49 Valve rocker arm installation

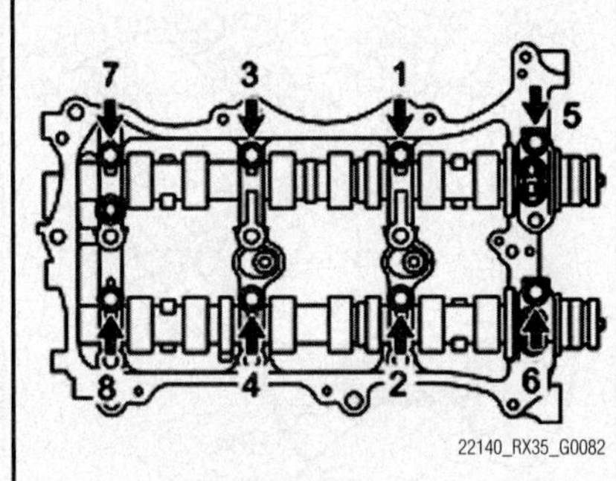

Fig. 51 Bank 1 camshaft bearing cap tightening sequence

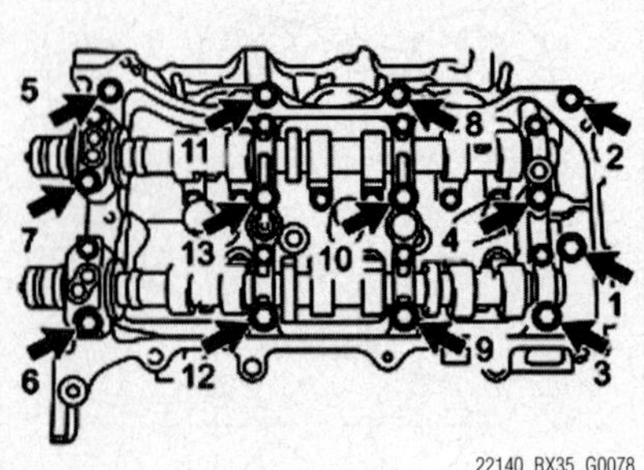

Fig. 48 Bank 2 camshaft bearing cap 13 bolt removal sequence

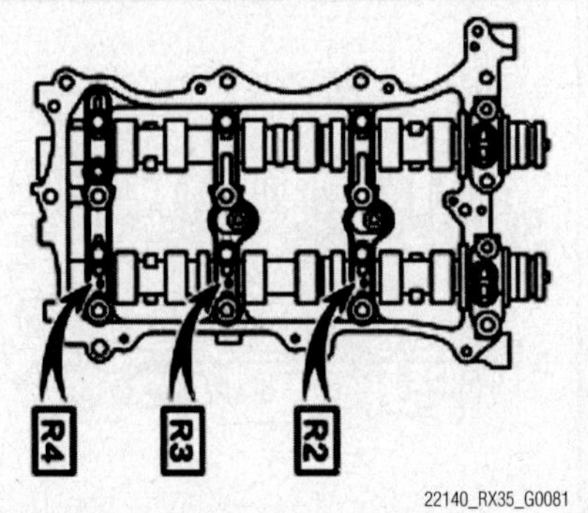

Fig. 50 Bank 1 camshaft bearing cap location

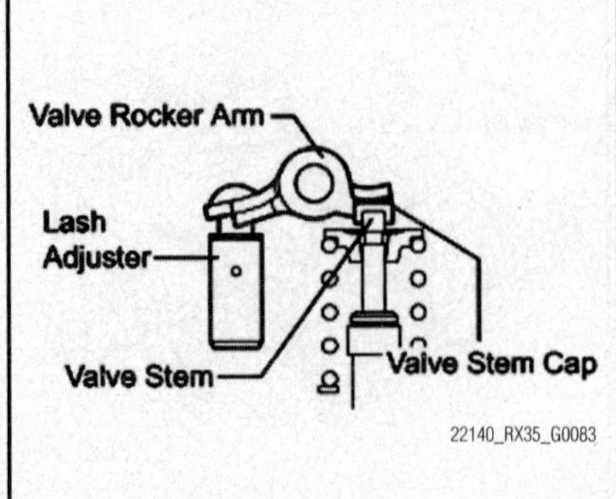

Fig. 52 Valve rocker arm installation

55. Install the camshaft bearing cap (for Bank 1), as follows:

a. Apply engine oil to the camshaft journals, camshaft housing and bearing caps.

b. Install the camshaft and No. 2 camshaft to the right camshaft housing.

c. Make sure of the marks and numbers on the camshaft bearing caps and place them in each proper position and direction.

d. Temporarily tighten the 8 bearing cap bolts to 7 ft. lbs. (10 Nm) in the order shown in the illustration.

56. Install the right camshaft housing sub-assembly, as follows:

a. Make sure that the valve rocker arm is installed as shown in the illustration.

b. Apply seal packing in a continuous line as shown in the illustration. Seal packing: Toyota Genuine Seal Packing Black, Three Bond 1207B or equivalent. Seal diameter: 3.5 to 4.5 mm (0.138 to 0.177 in.).

➡Remove any oil from the contact surface. Install the camshaft housing sub-assembly within 3 minutes. Do not start the engine for at least 2 hours after installing.

c. Install the camshaft housing and tighten the 12 bolts in the order shown in the illustration to 21 ft. lbs. (28 Nm).

➡When installing the camshaft housing, it is necessary to correctly position the camshafts as shown in the removal illustration. Failure to correctly position these parts may result in damage due to contact between the pistons and valves. If a camshaft is rotated with a piston at TDC, valve contact will occur. If any of the bolts are loosened during installation, remove the camshaft housing, clean the installation surfaces, and reapply seal packing. If the camshaft housing is removed because any of the bolts are loosened during installation, make sure that the previously applied seal packing does not enter any oil passages.

d. Complete the tightening of the 8 bolts to 12 ft. lbs. (16 Nm) in the order shown in the illustration.

57. Install the camshaft bearing cap (for Bank 2), as follows:

a. Apply engine oil to the camshaft journals, camshaft housing and bearing caps.

b. Install the No. 3 camshaft and No. 4 camshaft to the left camshaft housing.

c. Make sure of the marks and numbers on the camshaft bearing caps and place them in each proper position and direction.

d. Temporarily tighten the 8 bolts in the order shown in the illustration to 7 ft. lbs. (10 Nm).

58. Install the left camshaft housing sub-assembly, as follows:

a. Make sure that the valve rocker arm is installed as shown in the illustration.

b. Apply seal packing in a continuous line as shown in the illustration. Seal packing: Toyota Genuine Seal Packing Black, Three Bond 1207B or equivalent. Seal diameter: 3.5 to 4.5 mm (0.138 to 0.177 in.).

➡Remove any oil from the contact surface. Install the camshaft housing sub-assembly within 3 minutes. Do not start the engine for at least 2 hours after installing.

c. Install the camshaft housing and tighten the 13 bolts in the order shown in the illustration to 21 ft. lbs. (28 Nm).

➡When installing the camshaft housing, it is necessary to correctly position

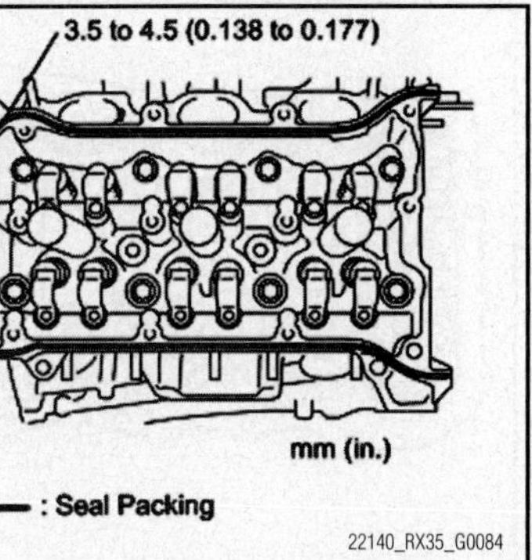

Fig. 53 Right-hand camshaft sealant application

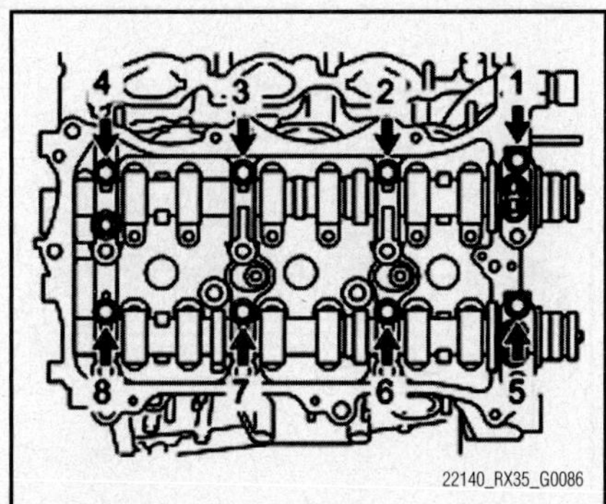

Fig. 55 Right-hand camshaft housing 8 bolt tightening sequence

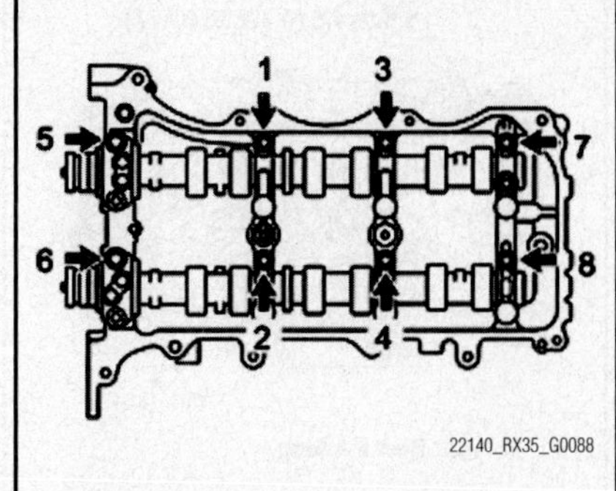

Fig. 57 Bank 2 camshaft bearing cap bolt tightening sequence

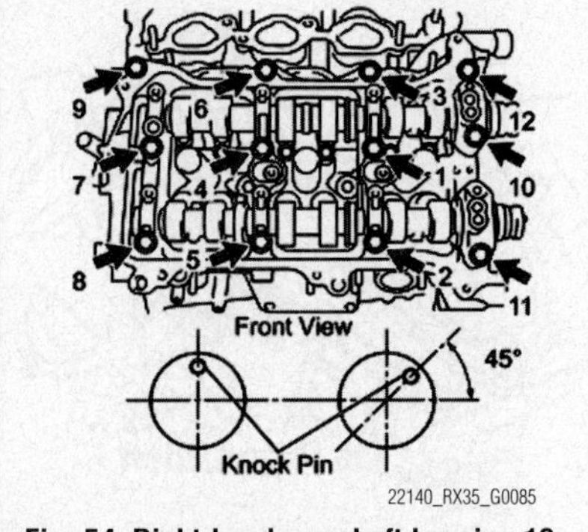

Fig. 54 Right-hand camshaft housing 12 bolt tightening sequence

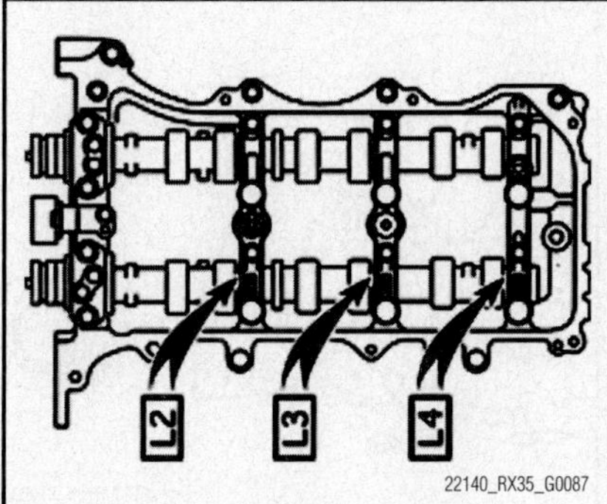

Fig. 56 Bank 2 camshaft bearing cap location

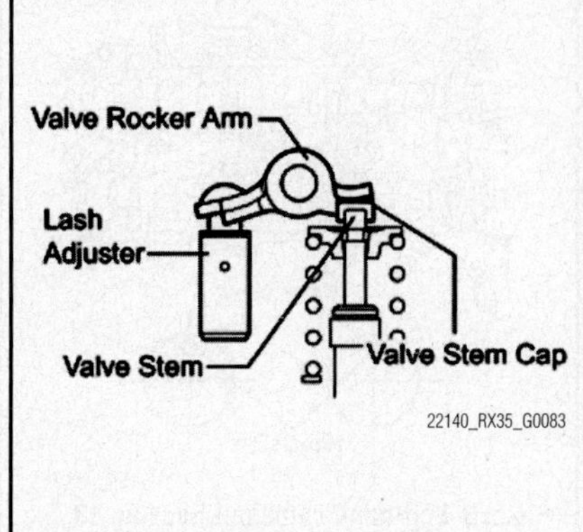

Fig. 58 Valve rocker arm installation

the camshafts as shown in the removal illustration. Failure to correctly position these parts may result in damage due to contact between the pistons and valves. If a camshaft is rotated with a piston at TDC, valve contact will occur. If any of the bolts are loosened during installation, remove the camshaft housing, clean the installation surfaces, and reapply seal packing. If the camshaft housing is removed because any of the bolts are loosened during installation, make sure that the previously applied seal packing does not enter any oil passages.

d. Complete the tightening of the 8 bolts to 12 ft. lbs. (16 Nm) in the order shown in the illustration.

59. Install the No. 2 chain tensioner assembly with the bolt and tighten to 15 ft. lbs. (21 Nm).

60. While pushing in the tensioner, insert a pin of 1.0 mm (0.039 in.) diameter into the hole to fix it.

61. Install the camshaft timing gears and No. 2 chain (for Bank 1).

62. Install the No. 3 chain tensioner assembly with the bolt and tighten to 15 ft. lbs. (21 Nm).

63. While pushing in the tensioner, insert a pin of 1.0 mm (0.039 in.) diameter into the hole to hold it.

64. Install the camshaft timing gears and No. 2 chain (for Bank 2).

65. Install the chain vibration damper with the 2 bolts and tighten to 17 ft. lbs. (23 Nm).

66. Install the 2 chain vibration dampers.

67. Install the 2 timing gear set keys and crankshaft timing sprocket as shown in the illustration.

68. Install the idle sprocket assembly.

69. Install chain sub-assembly.

70. Install the chain tensioner slipper.

71. Install the No. 1 chain tensioner assembly.

72. Install the timing chain case oil seal.

73. Install the water pump assembly.

74. Install the timing chain cover sub-assembly.

75. Install the water inlet housing.

76. Install the No. 1 left front engine mounting bracket, as follows:

a. Install the No. 1 left front engine mounting bracket with the 6 bolts and tighten to 40 ft. lbs. (54 Nm).

➡Install the water inlet and mounting bracket within 15 minutes after installing the chain cover. Do not start the engine for at least 2 hours after installation.

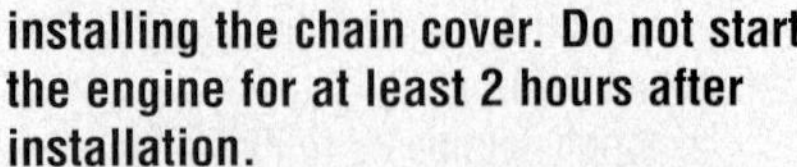

b. When replacing a stud bolt, install it by using a E8 TORX® socket wrench and tighten to 7 ft. lbs. (10 Nm).

77. Install the oil pan baffle plate with the 7 bolts and tighten to 7 ft. lbs. (10 Nm).

➡Temporarily tighten the 7 bolts. Fully tighten 2 bolts A as shown before tightening the other bolts.

78. Install the oil pan sub-assembly.

79. Install the oil strainer sub-assembly.

80. Install the No. 2 oil pan sub-assembly.

81. Install a new gasket and oil pan drain plug and tighten to 30 ft. lbs. (40 Nm).

82. Install the cylinder head cover sub-assembly (for Bank 1).

83. Install the cylinder head cover sub-assembly (for Bank 2).

84. Install water outlet.

85. Install the No. 1 oil cooler bracket, if applicable.

86. Install the oil cooler assembly, if applicable.

87. Install the crankshaft pulley.

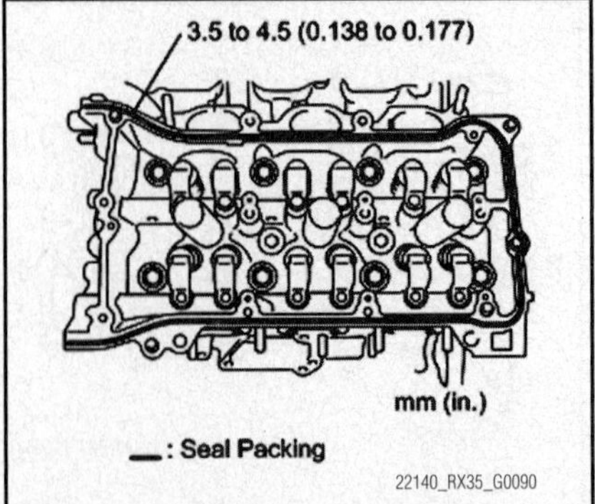

Fig. 59 Left-hand camshaft housing sealant application

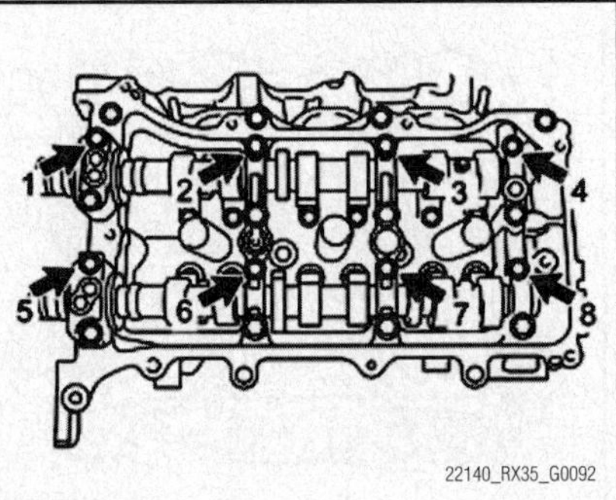

Fig. 61 Left-hand camshaft housing 8 bolt tightening sequence

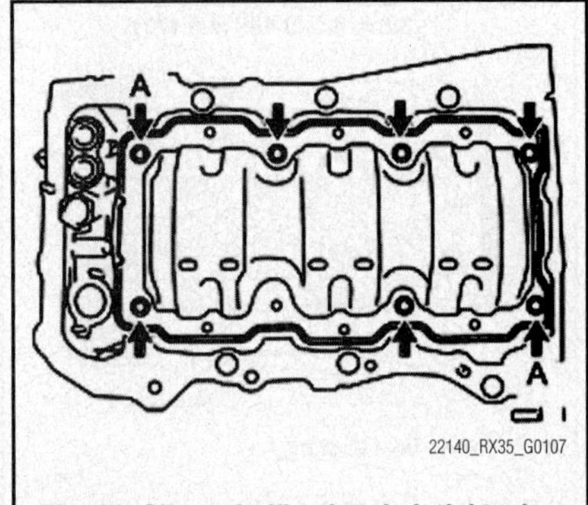

Fig. 63 Oil pan baffle plate bolt tightening sequence

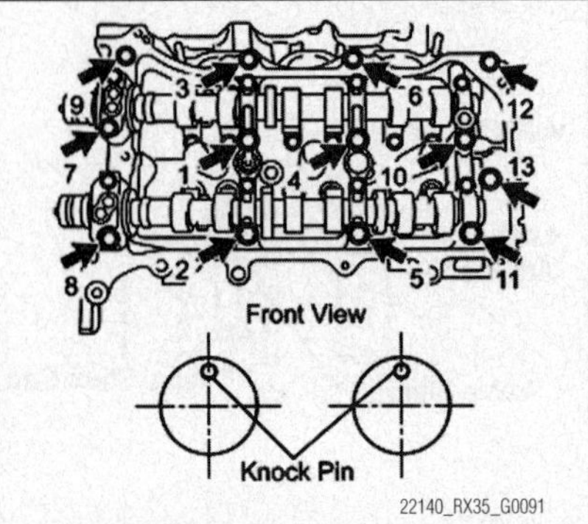

Fig. 60 Left-hand camshaft housing 13 bolt tightening sequence

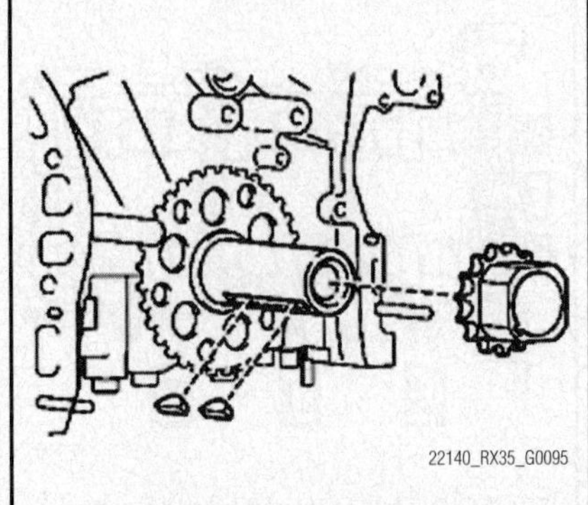

Fig. 62 Crankshaft timing sprocket

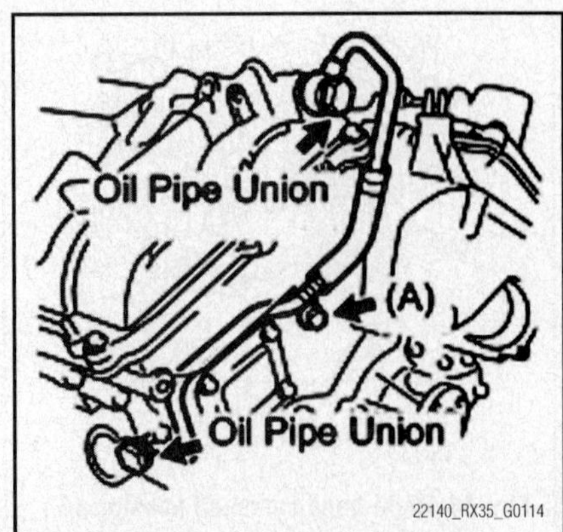

Fig. 64 Tighten bolt to cylinder head

88. Install the oil filter element.

89. Install the cylinder block water drain cock sub-assembly, as follows:

a. Apply adhesive around the drain cocks. Adhesive: Toyota Genuine Adhesive 1324, Three Bond 1324 or equivalent.

b. Install the water drain cocks and tighten to 18 ft. lbs. (25 Nm). Do not rotate the drain cocks more than 1 revolution (360°) after tightening the drain cocks with the specified torque.

c. Install the water drain cock plugs to the water drain cocks and tighten to 9 ft. lbs. (13 Nm).

90. Install the No. 1 oil pipe, as follows:

a. Make sure that there is no foreign matter on the mesh of the left hand oil control valve filter. Do not touch the mesh when installing the oil control valve filter.

b. Install the left hand oil control valve filter to the oil pipe union. Install new gaskets and temporarily install the oil pipe (on the head cover side).

c. Install a new gasket and temporarily install the oil pipe (on the cylinder head side) with the oil check valve bolt.

d. Tighten the oil pipe union (on the head cover side) to 48 ft. lbs. (65 Nm).

e. Tighten the oil pipe union (on the cylinder head side) to 48 ft. lbs. (65 Nm).

f. If the link that connects the gaskets is broken, remove the connecting link by using nippers or similar tools.

91. Install the oil pipe, as follows:

a. Make sure that there is no foreign matter on the mesh of the right hand oil control valve filter. Do not touch the mesh when installing the oil control valve filter.

b. Install the right hand oil control valve filter to the oil pipe union. Install new gaskets and temporarily install the oil pipe (on the head cover side).

c. Install a new gasket and temporarily install the oil pipe (on the cylinder head side) with the oil check valve bolt.

d. Install the bolt (A) to the cylinder head and tighten to 7 ft. lbs. (10 Nm).

e. Tighten the oil pipe union (on the head cover side) to 48 ft. lbs. (65 Nm).

f. Tighten the oil check valve bolt (on the cylinder head side) to 48 ft. lbs. (65 Nm).

g. If the link that connects the gaskets is broken, remove the connecting link by using nippers or similar tools.

92. Install the crankshaft position sensor with the bolt and tighten to 7 ft. lbs. (10 Nm).

93. Install the camshaft timing oil control valve assembly, as follows:

a. Install the 4 oil control valves with the 4 bolts and tighten to 7 ft. lbs. (10 Nm).

94. Install the camshaft position sensor, as follows:

a. Install the 4 sensors with the 4 bolts and tighten to 7 ft. lbs. (10 Nm).

95. Install the ventilation valve sub-assembly, as follows:

a. Apply adhesive (Toyota Genuine Adhesive 1324, Three Bond 1324 or equivalent) around the ventilation valve.

b. Install the ventilation valve and tighten to 20 ft. lbs. (27 Nm).

96. Install the 6 spark plugs and the ignition coil assembly.

97. Install the oil filler cap sub-assembly.

98. Remove the engine stand.

99. Install the engine assembly.

CRANKSHAFT DAMPER

REMOVAL & INSTALLATION

1. Remove the right front wheel.
2. Remove the right front fender apron seal.
3. Remove the engine room side cover.
4. Remove the V-bank cover sub-assembly.
5. Remove the V-ribbed belt.
6. Using SST (SST: 09213-70011. SST: 09330-00021), loosen the crankshaft pulley bolt.
7. Using SST (SST: 09950-50013), remove the crankshaft pulley bolt and crankshaft pulley.

To install:

8. Install crankshaft pulley, as follows:

a. Align the pulley set key with the key groove of the pulley, and slide on the pulley.

b. Using SST (SST: 09213-70011, SST: 09330-00021) , install the pulley bolt and tighten to 184 ft. lbs (250 Nm).

9. Install the V-ribbed belt.
10. Install the V-bank cover sub-assembly.
11. Install the engine room side cover.
12. Install the right front fender apron seal.
13. Install the right front wheel.

CRANKSHAFT FRONT SEAL

REMOVAL & INSTALLATION

See Figure 65.

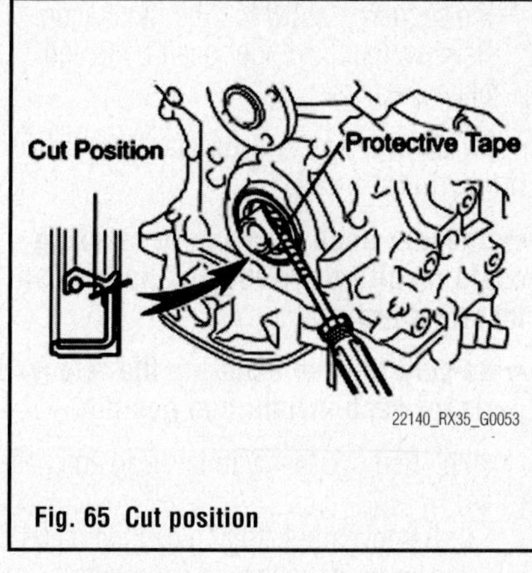

Fig. 65 Cut position

1. Remove the right front wheel.
2. Remove the right front fender apron seal.
3. Remove the engine room side cover.
4. Remove the V-bank cover sub-assembly.
5. Remove the V-ribbed belt.
6. Using SST (SST: 09213-70011. SST: 09330-00021), loosen the crankshaft pulley bolt.
7. Using SST (SST: 09950-50013), remove the crankshaft pulley bolt and crankshaft pulley.
8. Using a screwdriver with the tip taped, pry out the timing chain case oil seal.

➡After the removal, check the crankshaft for damage. If it is damaged, smooth the surface with 400-grit sandpaper.

To install:

9. Apply MP grease to a new oil seal lip.
10. Using SST (SST: 09223-22010, SST: 09506-35010) and a hammer, tap in the oil seal until its surface is flush with the timing chain cover edge.

➡Keep the lip free of foreign matter. Do not tap the oil seal at an angle.

11. Install crankshaft pulley, as follows:

a. Align the pulley set key with the key groove of the pulley, and slide on the pulley.

b. Using SST (SST: 09213-70011, SST: 09330-00021) , install the pulley bolt and tighten to 184 ft. lbs (250 Nm).

12. Install the V-ribbed belt.
13. Install the V-bank cover sub-assembly.
14. Install the engine room side cover.
15. Install the right front fender apron seal.
16. Install the right front wheel.

CYLINDER HEAD

REMOVAL & INSTALLATION

See Figures 66 through 75.

1. Remove and disassemble engine. Refer to Camshaft and Valve Lifters removal procedure.
2. Remove the right hand cylinder head sub-assembly, as follows:

a. Using a 10 mm bi-hexagon wrench, uniformly loosen the 8 bolts in the sequence shown in the illustration. Remove the 8 cylinder head bolts and plate washers.

➡**Be careful not to drop washers into the cylinder head.**

➡**Cylinder head warpage or cracking could result from removing bolts in an incorrect order.**

➡**Be sure to keep separate the removed parts for each installation position.**

b. Remove the cylinder head and gasket.

3. Remove the left hand cylinder head sub-assembly, as follows:

a. Uniformly loosen and remove the bolts in the sequence shown in the illustration.

b. Using a 10 mm bi-hexagon wrench, uniformly loosen the 8 bolts in the sequence shown in the illustration. Remove the 8 cylinder head bolts and plate washers.

➡**Be careful not to drop washers into the cylinder head.**

➡**Cylinder head warpage or cracking could result from removing bolts in an incorrect order.**

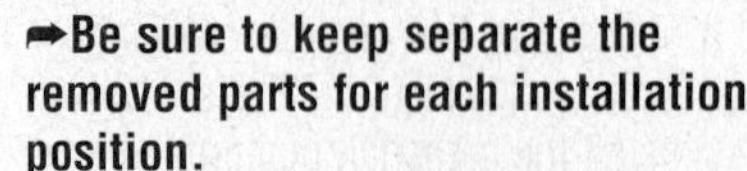

➡**Be sure to keep separate the removed parts for each installation position.**

c. Remove the cylinder head and gasket.

To install:

4. Place the right hand cylinder head gasket on the cylinder block surface with the Lot No. stamp upward.

➡**Be careful of the installation direction.**

➡**Gently place the cylinder head in order not to damage the gasket with the bottom part of the head.**

5. Place the cylinder head on the cylinder block.

➡**Be careful not to allow oil to adhere to the bottom part of the cylinder head.**

6. Apply a light coat of engine oil to the threads and under the heads of the cylinder head bolts.

7. The cylinder head bolts are tightened in 3 progressive steps:

a. Step 1: Using a 10 mm bi-hexagon wrench, install and uniformly tighten the 8 cylinder head bolts with the plate washers in several steps and in the sequence shown in the illustration. Tighten to 27 ft. lbs. (36 Nm).

b. Step 2: Mark the cylinder head bolt head with paint as shown in the illustration. Tighten the cylinder head bolts another 90°.

c. Step 3: Tighten the cylinder head bolts an additional 90°. Check that the painted mark is now facing rearward.

8. Place the left hand cylinder head gasket on the cylinder block surface with the Lot No. stamp upward.

➡**Be careful of the installation direction.**

➡**Gently place the cylinder head in order not to damage the gasket with the bottom part of the head.**

9. Place the cylinder head on the cylinder block.

➡**Be careful not to allow oil to adhere to the bottom part of the cylinder head.**

10. Apply a light coat of engine oil to the threads and under the heads of the cylinder head bolts.

11. The cylinder head bolts are tightened in 3 progressive steps:

a. Step 1: Using a 10 mm bi-hexagon wrench, install and uniformly tighten the

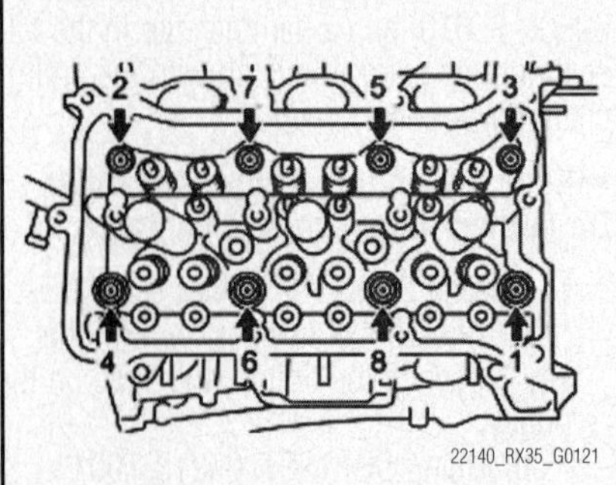

Fig. 66 Right hand cylinder head bolt removal sequence

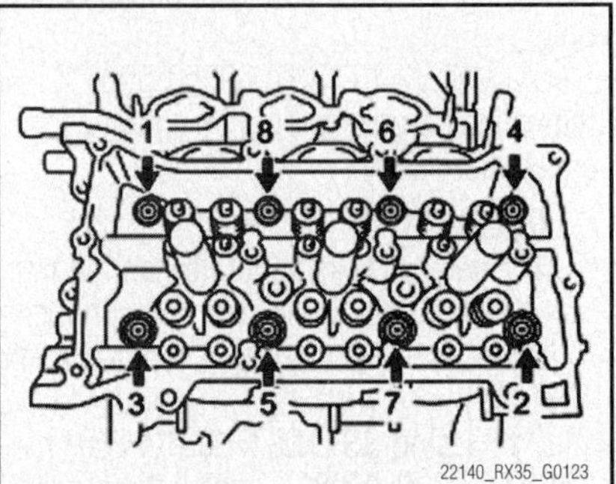

Fig. 68 Left hand cylinder head 8 bolt removal sequence

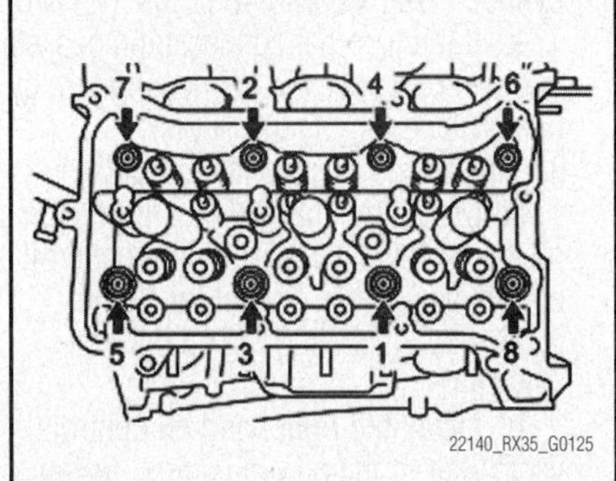

Fig. 70 Right-hand cylinder head bolt tightening sequence

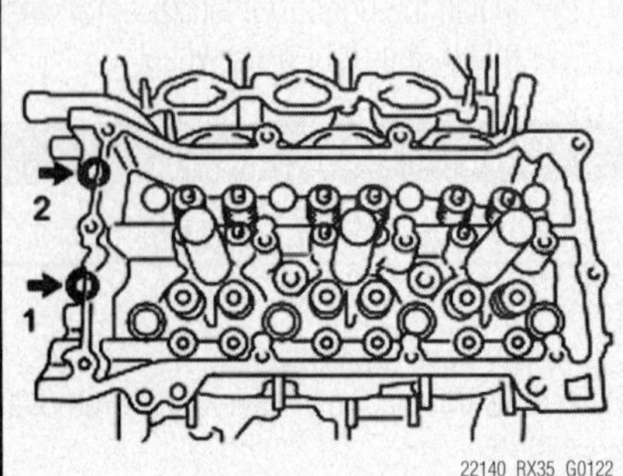

Fig. 67 Left hand cylinder head 2 bolt removal sequence

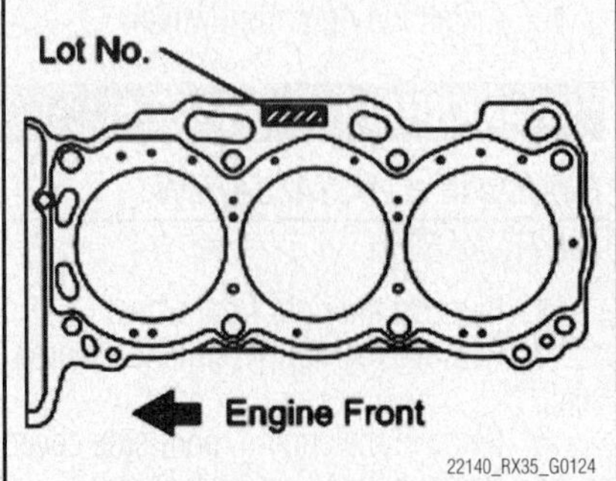

Fig. 69 Place the right-hand cylinder head gasket with Lot No. stamp upward

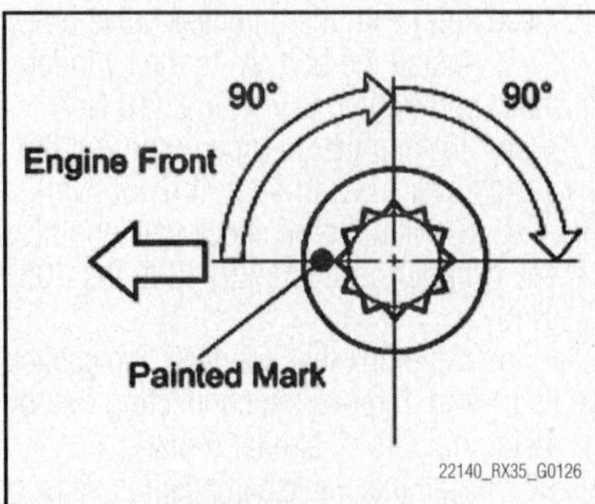

Fig. 71 Mark the cylinder head bolt and tighten another 90°

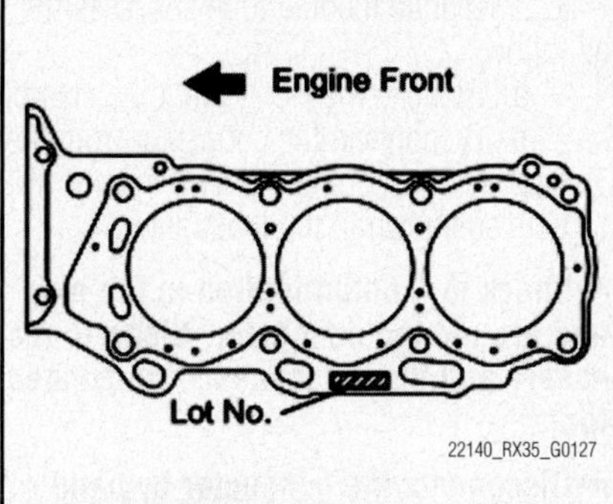

Fig. 72 Place the left-hand cylinder head with Lot No. stamp upward

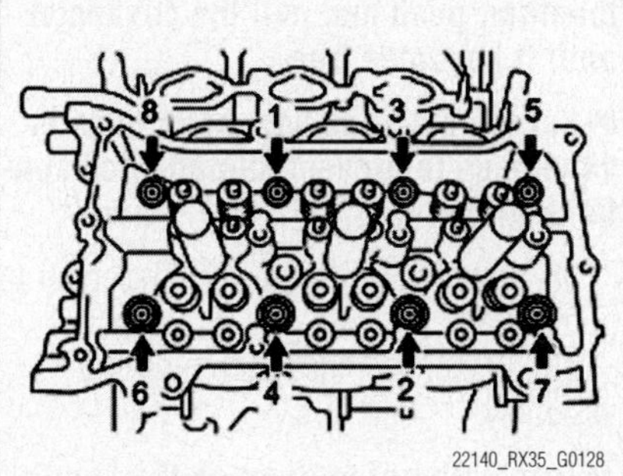

Fig. 73 Left hand cylinder head 8 bolt tightening sequence

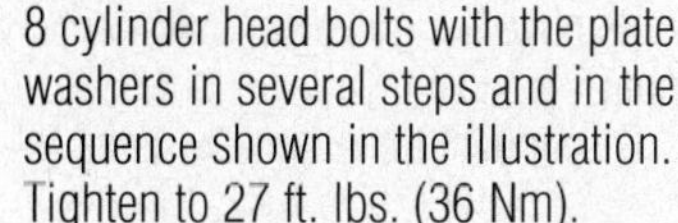

8 cylinder head bolts with the plate washers in several steps and in the sequence shown in the illustration. Tighten to 27 ft. lbs. (36 Nm).

b. Step 2: Mark the cylinder head bolt head with paint as shown in the illustration. Tighten the cylinder head bolts another 90°.

c. Step 3: Tighten the cylinder head bolts an additional 90°. Check that the painted mark is now facing rearward.

d. Tighten the 2 bolts in the order shown in the illustration to 22 ft. lbs. (30 Nm).

12. Assemble and install engine. Refer to Camshaft and Valve Lifters installation procedure.

ENGINE ASSEMBLY

REMOVAL & INSTALLATION

See Figures 76 through 107.

1. Discharge fuel system pressure.
2. Disconnect cable from negative battery terminal.
3. Place the front wheels facing straight ahead.
4. Remove the front wheels.
5. Remove the engine under cover assembly, as applicable.
6. Remove the No. 1 engine under cover.
7. Remove the No. 2 engine under cover, as applicable.
8. Separate the right front fender splash shield sub-assembly.
9. Remove the right front fender apron seal.
10. Drain engine oil.
11. Drain engine coolant.
12. Drain automatic transaxle fluid.
13. Remove the windshield wiper link assembly.
14. Remove the cowl top panel outer sub-assembly.
15. Remove the 4 clips and intake duct seal.
16. Remove the 4 clips and engine room cover side.
17. Hold the front of the V-bank cover and raise it to disengage the 2 clips on the front of the cover. Continue to raise the cover to disengage the clip on the rear of the cover and remove the cover.

➡Attempting to disengage both front and rear clips at the same time may cause the cover to break.

18. Remove the V-ribbed belt.
19. Remove the 2 bolts, 2 clamps and No. 2 air cleaner inlet.

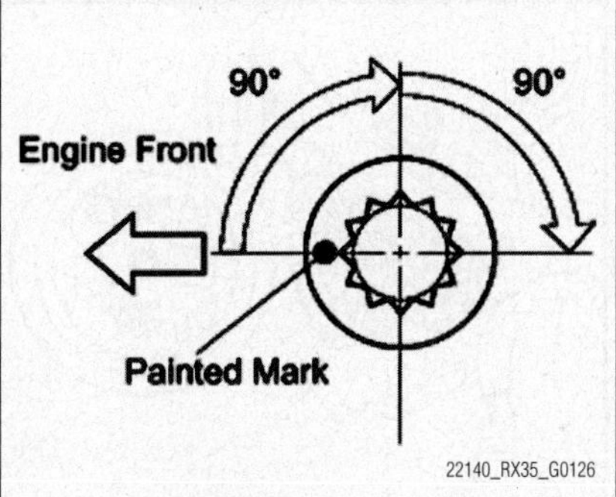

Fig. 74 Mark the cylinder head bolt and tighten another 90°

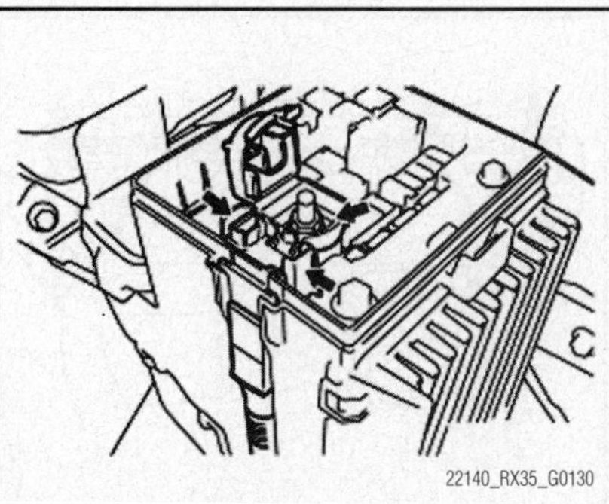

Fig. 76 Disconnect the engine wire from the engine room junction block

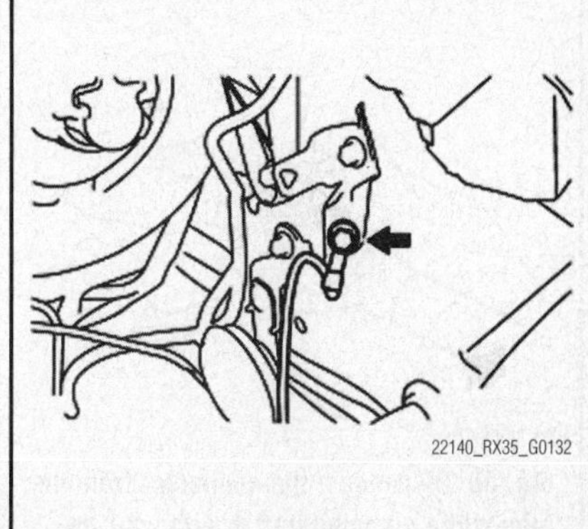

Fig. 78 Remove the bolt and ground cable

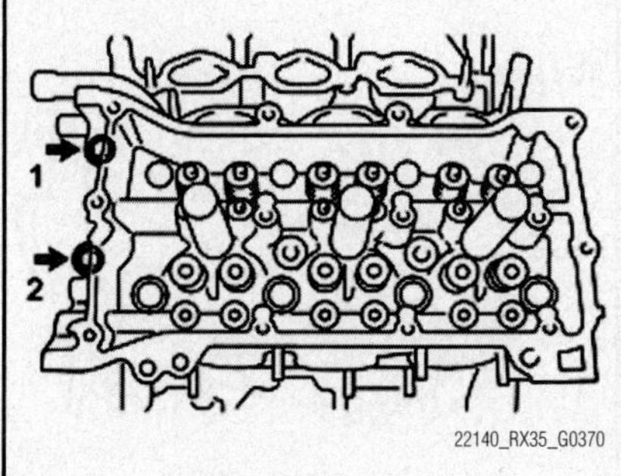

Fig. 75 Left hand cylinder head 2 bolt tightening sequence

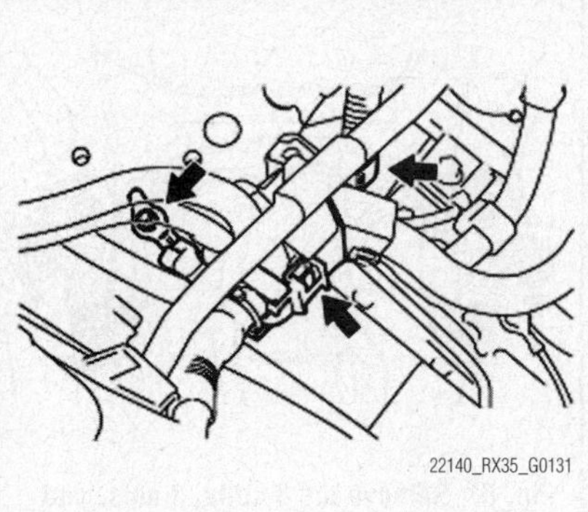

Fig. 77 Remove the bolt and 2 clamps

Fig. 79 Remove the No. 1 fuel pipe clamp

20. Remove the air cleaner cap sub-assembly.

21. Remove the air cleaner case sub-assembly, as follows:

a. Disconnect the vacuum hose.

b. Remove the 3 bolts and air cleaner case.

22. Remove the bolt and No. 1 air cleaner inlet.

23. Remove the battery and battery tray.

24. Remove the clip, bolt and intake air resonator.

25. Remove the 2 bolts and air cleaner bracket.

26. Remove the bolt, 2 nuts, and right No. 2 mounting stay.

27. Remove the 3 bolts and engine moving control rod.

28. Remove the clamp and disconnect the No. 1 fuel vapor feed hose.

29. Remove the clamp and disconnect the union to check valve hose.

30. Remove the clamps and disconnect the radiator inlet and outlet hoses.

31. Remove the clamps and disconnect the oil cooler inlet and outlet hoses.

32. Disconnect the heater inlet and outlet water hoses.

33. Remove the junction block cover.

34. Remove the glove compartment door assembly.

35. Disconnect the engine wire, as follows:

a. Disconnect the engine wire from the engine room junction block: Remove the nut and separate the wire harness. Using a screwdriver, unlock the engine room junction block. Pull the engine room junction block upward.

b. Remove the bolt and 2 clamps from the body.

c. Remove the bolt and ground cable.

d. Remove the 2 nuts and engine wire from the body.

e. Disconnect the engine wire from the ECM and passenger side junction block.

36. Disconnect the transmission control cable assembly, as follows:

a. Remove the nut from the control shaft lever.

b. Disconnect the transmission control cable assembly from the control shaft lever.

c. Remove the clip and disconnect the transmission control cable assembly from the control cable bracket.

d. Disconnect the control cable from the control cable clamp.

37. Separate the fuel tube sub-assembly, as follows:

a. Remove the No. 1 fuel pipe clamp.

b. Disconnect the connector from the tube while pinching part A with your fingers as shown in the illustration.

➡Check for contamination in the pipe and around the connector. Clean if necessary and then disconnect the connector.

➡Disconnect the connector by hand.

➡Do not bend, fold or rotate the nylon tube.

➡If the pipe and connector are stuck together, push and pull the connector until it becomes free.

➡Put the pipe and connector ends in vinyl bags to prevent damage and contamination.

38. Disconnect the No. 1 oil reservoir to pump hose.

39. Disconnect the return tube sub-assembly.

➡Take care not to damage the hose protector.

40. For 4WD vehicles, remove the propeller with center bearing shaft assembly.

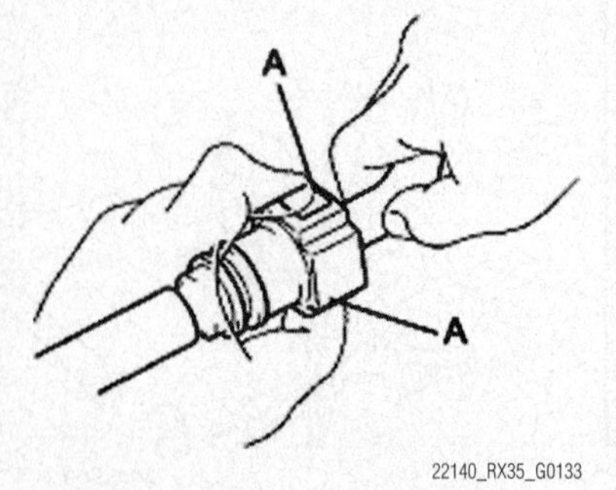

22140_RX35_G0133

Fig. 80 Disconnect the connector from the tube while pinching part A with your fingers

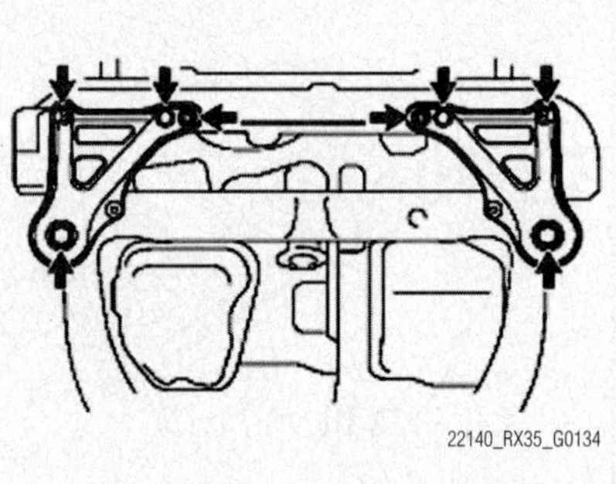

Fig. 82 Remove the 6 bolts, 2 nuts, and right and left frame side rail plates

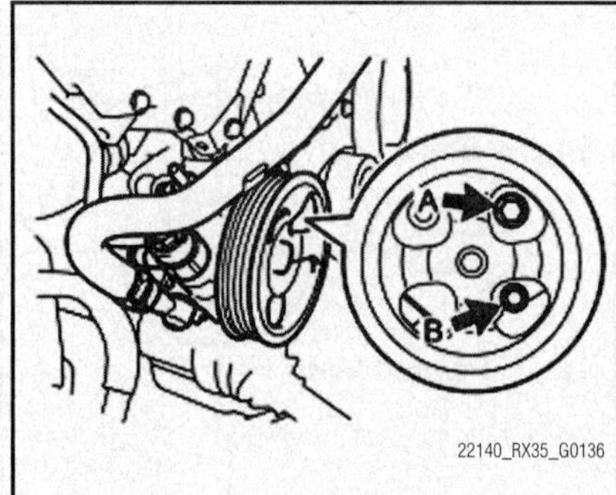

Fig. 84 Vane pump (power steering pump) removal

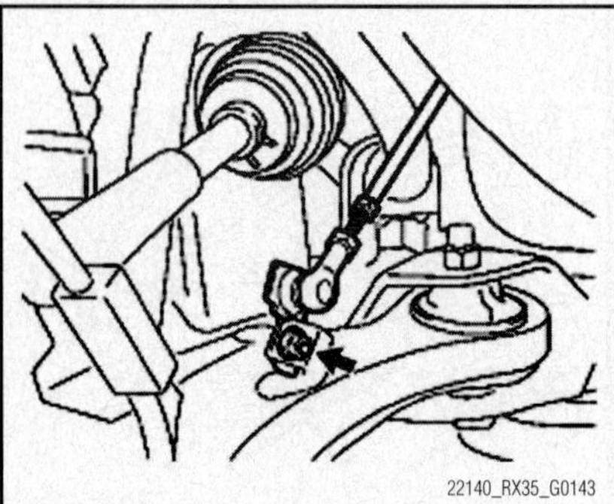

Fig. 81 Disconnect front height control sensor sub-assemblies (w/air suspension)

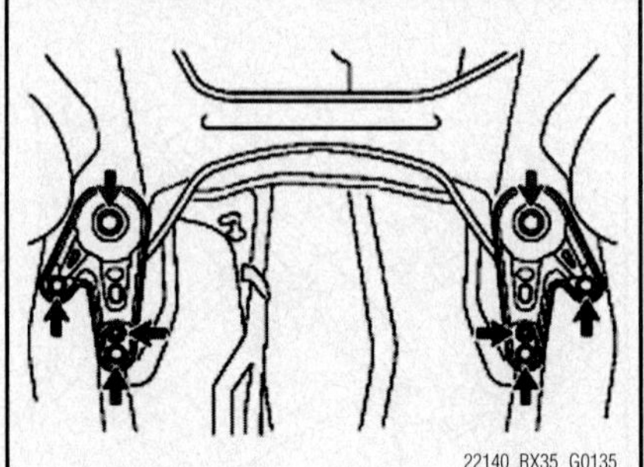

Fig. 83 Remove the 6 bolts, 2 nuts, and right and left front suspension member rear braces

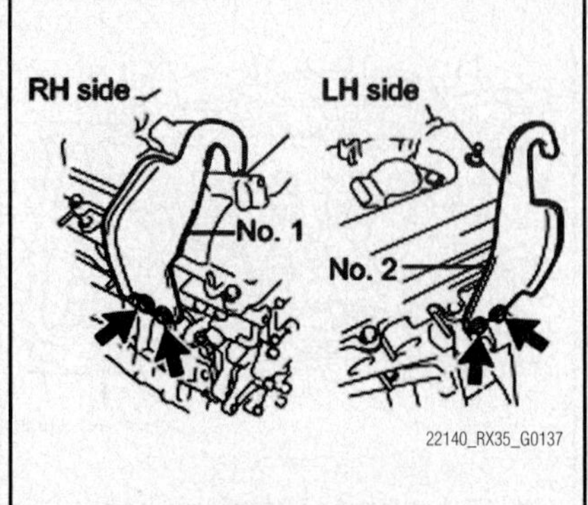

Fig. 85 Install engine hangers

41. Remove the oxygen sensor.
42. Remove the front No. 3 exhaust pipe sub-assembly.
43. Remove the front exhaust pipe assembly.
44. Disconnect the left and right front stabilizer link assemblies.
45. Remove the left and right front axle hub nuts.
46. Disconnect the left and right front speed sensors.
47. Disconnect the left and right tie rod assemblies.
48. Separate the left and right lower No. 1 front suspension arms.
49. Separate the left and right front axle assemblies.
50. Disconnect steering intermediate shaft sub-assembly.
51. Disconnect left and right front height control sensor sub-assemblies (with air suspension).
52. Remove the alternator assembly.
53. Separate the cooler compressor assembly, as follows:
   a. Remove the 2 connector clamps.
   b. Remove the 4 bolts and separate the compressor.

➡Hang up the hoses instead of detaching them.

54. Remove the engine assembly with transaxle, as follows:
   a. Set the engine lifter.
   b. Remove the 6 bolts, 2 nuts, and right and left frame side rail plates.
   c. Remove the 6 bolts, 2 nuts, and right and left front suspension member rear braces.
   d. Operate the engine lifter, then remove the engine assembly from the vehicle.

➡Make sure that the engine is clear of all wiring and hoses.

55. Remove the vane pump (power steering pump) assembly, as follows:
   a. Disconnect the power steering oil pressure switch connector.
   b. Remove the 3 pressure feed tube clamp bolts.
   c. Loosen the bolt A.
   d. Remove the bolt B and vane pump.
56. For 4WD vehicles, remove the front stabilizer bar.
57. For 4WD vehicles, remove the power steering link.
58. Install the 2 engine hangers with the 4 bolts as shown in the illustration. Part No.: No. 1 Engine hanger 12281-31120, No. 2 Engine hanger 12282-31100, Bolts 91671-10825. Tighten to 24 ft. lbs. (33 Nm).
59. Attach the engine sling device and hang the engine with the chain block.
60. Remove the front frame assembly, as follows:
   a. Disconnect the connector and clamp.
   b. Disconnect the 2 clamps.
   c. Remove the 2 nuts and disconnect the right and left engine mounting insulators.
   d. Remove the bolt and disconnect the right front engine mounting insulator.
   e. Remove the 2 bolts and separate the right rear engine mounting insulator.
61. Remove the left and right front drive shaft assemblies.
62. Remove the engine wire.
63. Remove the starter assembly.
64. Remove the front engine mounting bracket.
65. Remove the manifold stay.
66. Remove the automatic transaxle assembly.
67. For 4WD vehicles, remove the right transfer stiffener plate.
68. Remove the drive plate and ring gear sub-assembly.
69. Secure the engine onto an engine stand with the bolts.

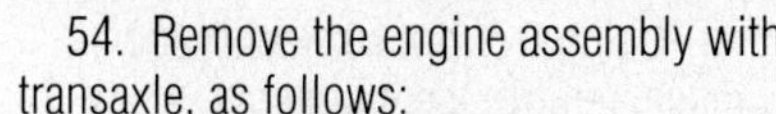

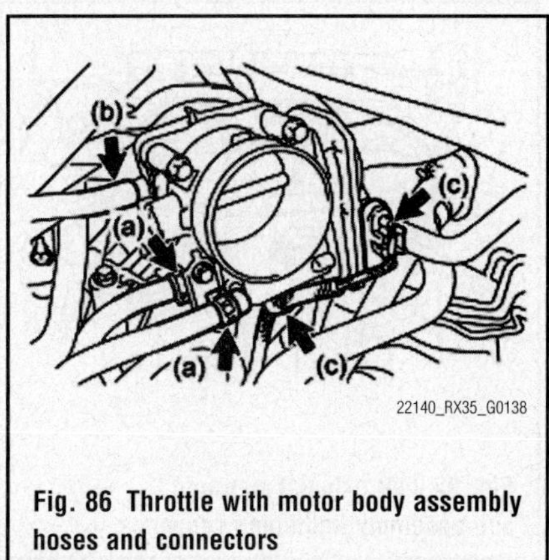

Fig. 86 Throttle with motor body assembly hoses and connectors

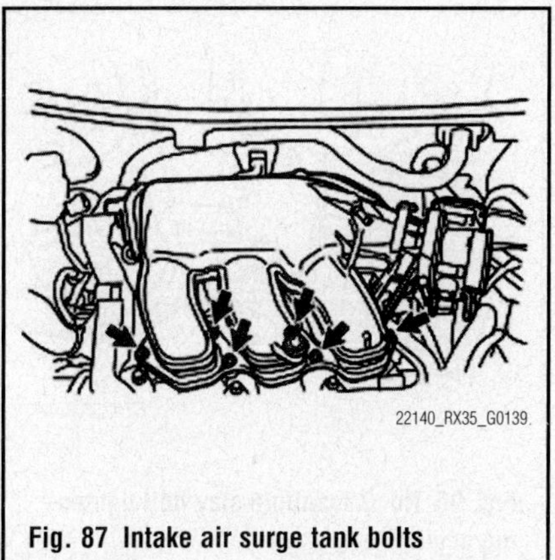

Fig. 87 Intake air surge tank bolts

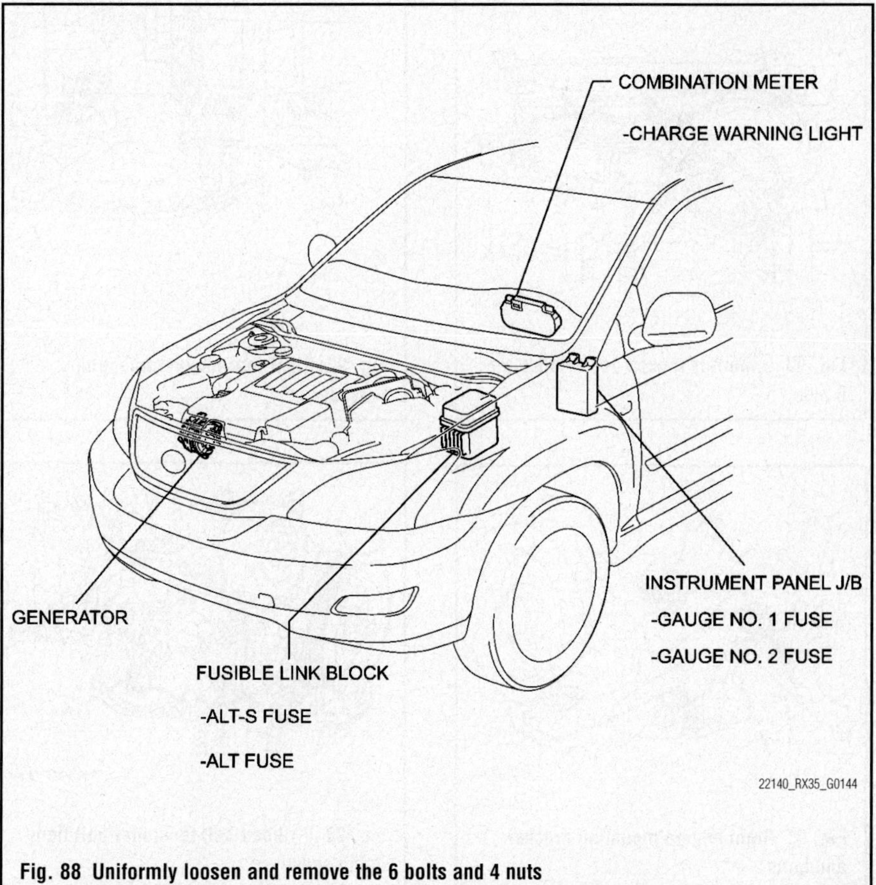

Fig. 88 Uniformly loosen and remove the 6 bolts and 4 nuts

70. Remove the 2 bolts and exhaust pipe support bracket.

71. Remove the intake air surge tank assembly, as follows:

a. Disconnect the 2 water by-pass hoses from the throttle with motor body assembly.

b. Disconnect the vapor feed hose.

c. Disconnect the throttle with motor body assembly connector and clamp.

d. Disconnect the No. 1 ventilation hose.

e. Disconnect the connector.

f. Remove the 4 bolts, No. 1 surge tank stay and throttle body bracket.

g. Using a 5 mm socket hexagon wrench, remove the 4 bolts.

h. Remove the 2 nuts and intake air surge tank.

i. Remove the gasket from the intake air surge tank.

72. Remove the 6 bolts and 6 ignition coils.

73. Remove the bolt and the right No. 2 engine mounting stay.

74. Remove the intake manifold, as follows:

a. Uniformly loosen and remove the 6 bolts and 4 nuts.

b. Remove the intake manifold and 2 gaskets.

75. Remove the right exhaust manifold sub-assembly, as follows:

a. Disconnect the A/F sensor connector clamp.

b. Uniformly loosen and remove the 6 nuts.

c. Remove the manifold and gasket.

76. Remove the oil level gauge guide sub-assembly, as follows:

a. Remove the oil level gauge.

b. Remove the 2 bolts, oil level gauge guides No. 1 and No. 2.

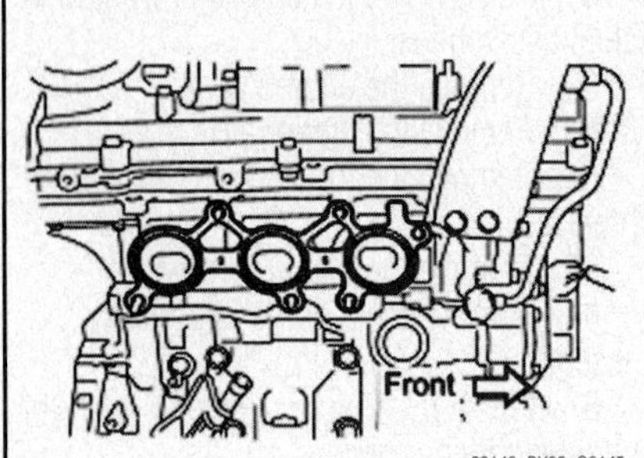

Fig. 89 Uniformly loosen and remove the 6 nuts

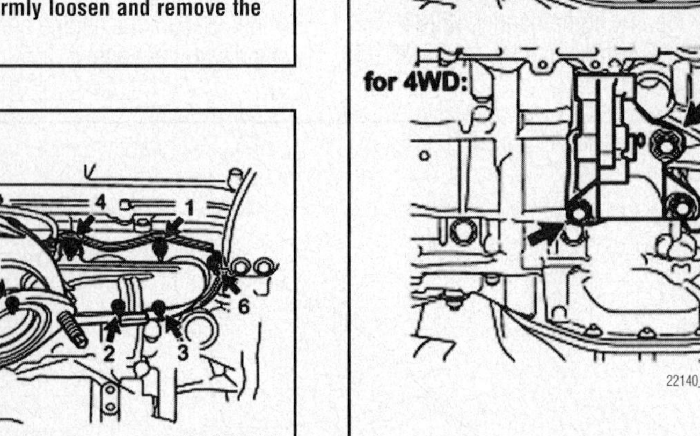

Fig. 90 Uniformly loosen and remove the 6 nuts

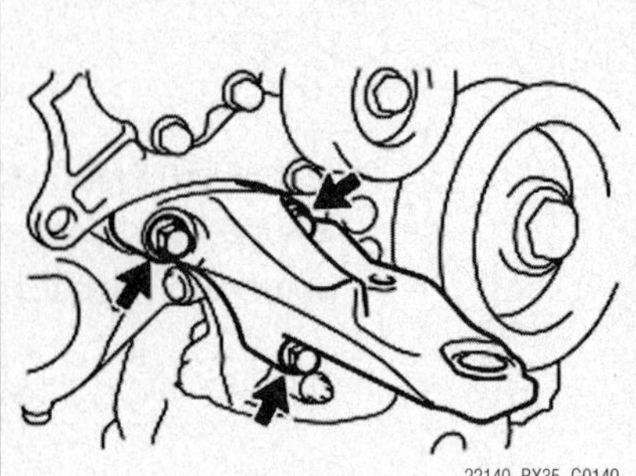

Fig. 91 Right engine mounting bracket and bolts

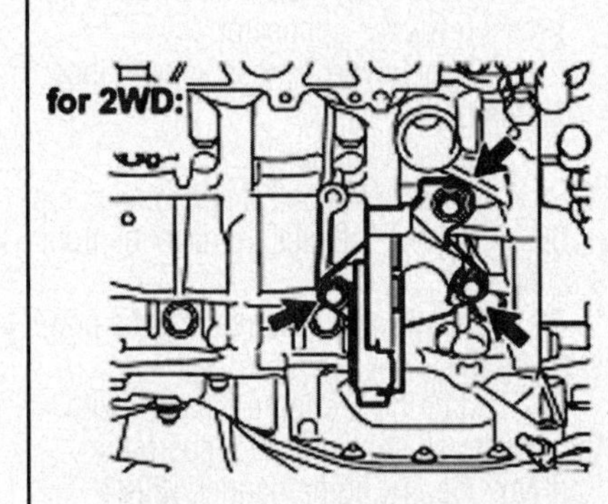

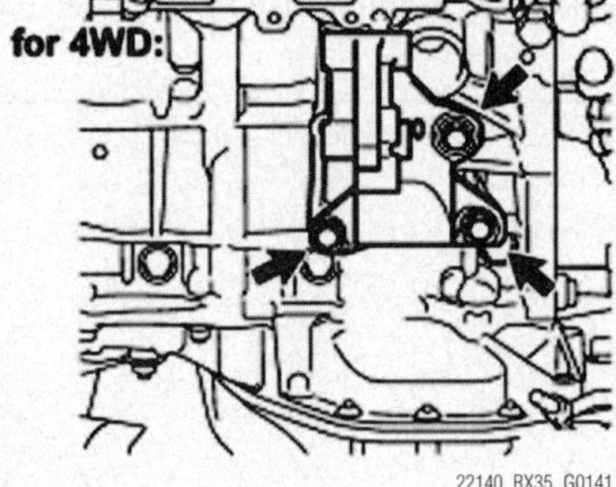

Fig. 92 Right rear engine mounting bracket

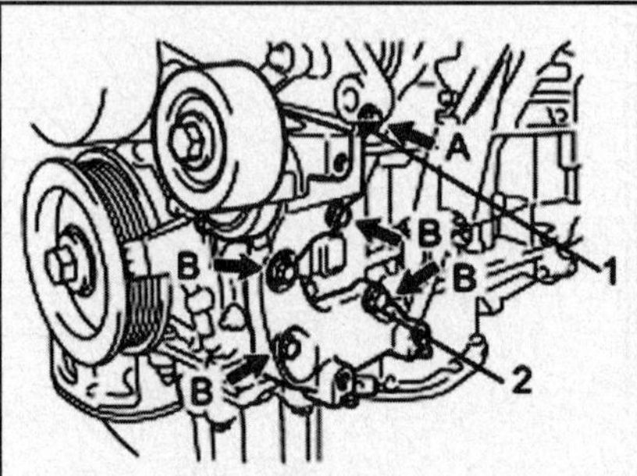

Fig. 93 V-ribbed belt tensioner bolt tightening sequence

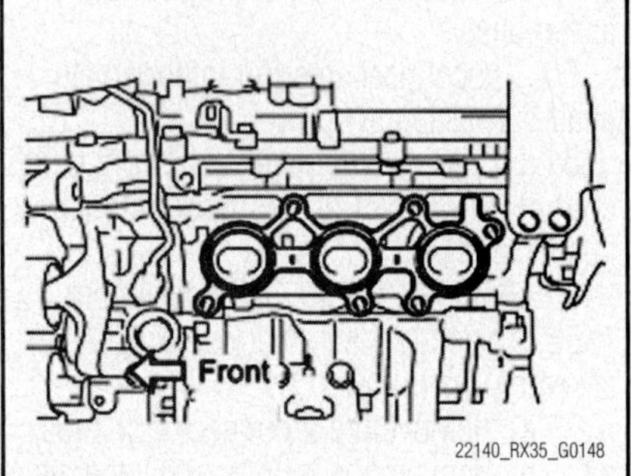

Fig. 94 Install a new gasket

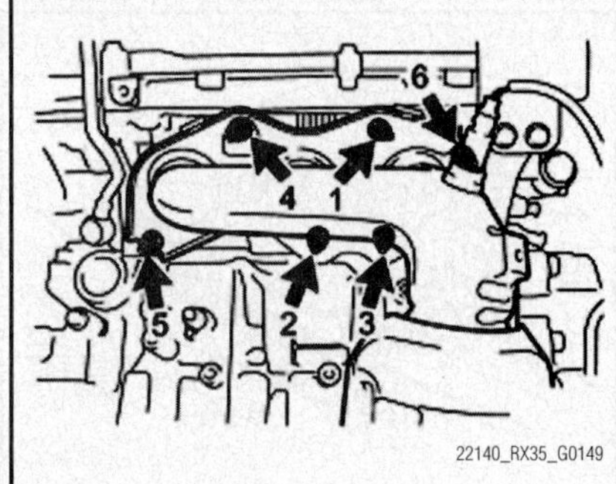

Fig. 95 Left exhaust manifold sub-assembly tightening sequence

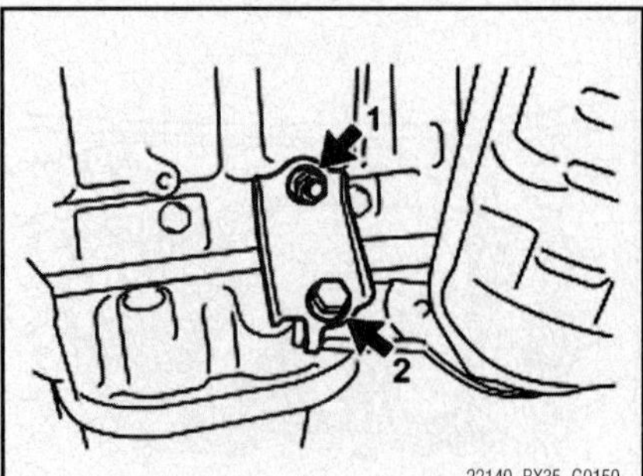

Fig. 96 No. 2 manifold stay bolt tightening sequence

c. Remove the O-rings from the oil level gauge guide.

77. Remove the bolt, nut and No. 2 manifold stay.

78. Remove the 3 bolts and No. 2 exhaust manifold heat insulator.

79. Remove the left exhaust manifold sub-assembly, as follows:

a. Uniformly loosen and remove the 6 nuts.

b. Remove the manifold and gasket.

80. Remove the 3 bolts and right engine mounting bracket.

81. Remove the 3 bolts and the right rear engine mounting bracket.

82. Remove the 5 bolts and V-ribbed belt tensioner assembly.

83. Remove the 2 bolts and No. 2 timing gear cover.

84. Remove the bolt, plate and No. 2 idler pulley sub-assembly.

85. Remove the water pump pulley.

86. Remove the 6 bolts and front left No. 1 engine mounting bracket.

87. Remove the 2 bolts and 2 radio setting condensers.

88. Remove the bolt and No. 1 vacuum switching valve.

89. Using a 24 mm deep socket wrench, remove the engine oil pressure switch assembly.

90. Remove the knock control sensor.

91. Using a 19 mm deep socket wrench, remove the EFI engine coolant temperature sensor and gasket.

To install:

92. Using a 19 mm deep socket wrench, install the engine coolant temperature sensor and a new gasket and tighten to 14 ft. lbs. (20 Nm).

93. Install the knock control sensor.

94. Install the engine oil pressure switch assembly, as follows:

a. Clean the threads of the oil pressure switch. Apply adhesive to 2 or 3 threads of the oil pressure switch. Adhesive: Toyota Genuine Adhesive 1344, Three Bond 1344 or equivalent.

b. Using a 24 mm deep socket wrench, install the oil pressure switch and tighten to 15 ft. lbs. (21 Nm).

➡Do not start the engine within 1 hour after installation to prevent oil leaks.

95. Install the bolt and No. 1 vacuum switching valve and tighten to 7 ft. lbs. (10 Nm).

96. Install the 2 bolts and 2 radio setting condensers and tighten to 7 ft. lbs. (10 Nm).

97. Install the left front No. 1 engine mounting bracket with the 6 bolts and tighten to 40 ft. lbs. (54 Nm).

98. Install the water pump pulley.

99. Install the No. 2 idler pulley sub-assembly and cover plate with the bolt and tighten to 32 ft. lbs. (43 Nm).

100. Install the No. 2 timing gear cover with the 2 bolts and tighten to 53 inch lbs. (6 Nm).

101. Install the V-ribbed belt tensioner assembly, as follows:

a. Temporarily install the V-ribbed belt tensioner with the 5 bolts. Bolt length A: 70 mm (2.76 in.), Bolt length B: 33 mm (1.30 in.).

b. Install the V-ribbed belt tensioner by tightening the bolt 1 and bolt 2 in the order shown in the illustration to 32 ft. lbs. (43 Nm).

c. Tighten the other bolts to 32 ft. lbs. (43 Nm).

102. Install the right rear engine mounting bracket with the 3 bolts and tighten to 47 ft. lbs. (64 Nm).

103. Install the right engine mounting bracket with the 3 bolts and tighten to 40 ft. lbs. (54 Nm).

104. Install the left exhaust manifold sub-assembly, as follows:

Fig. 97 Install a new gasket

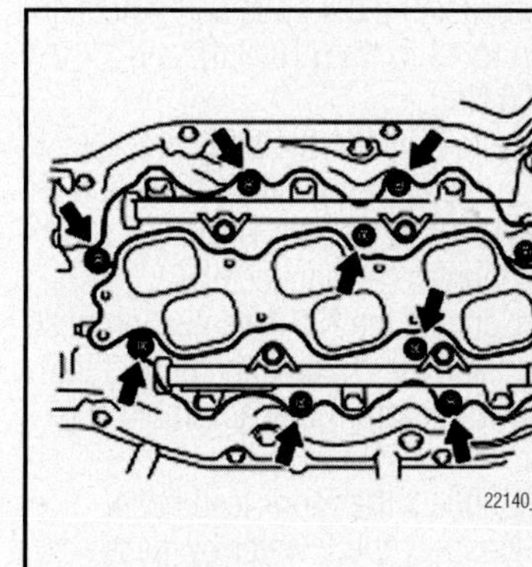

Fig. 99 Intake manifold

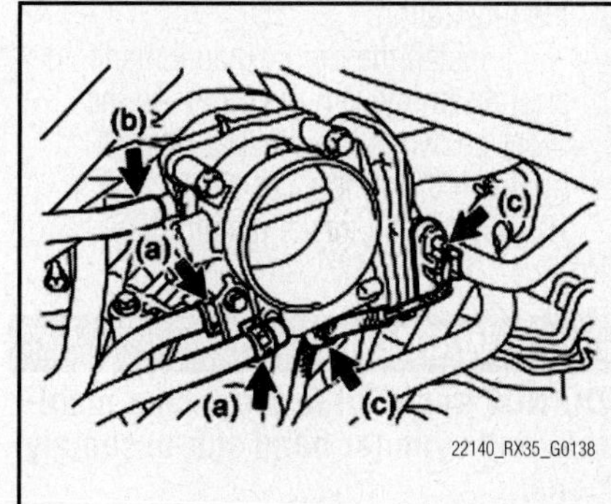

Fig. 101 Throttle with motor body assembly hoses and connectors

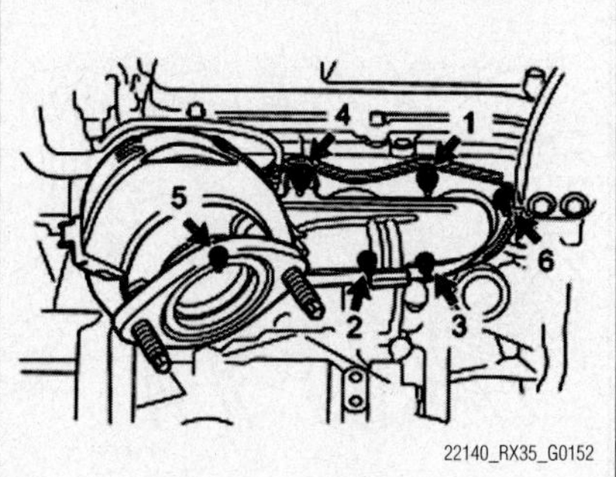

Fig. 98 Right exhaust manifold sub-assembly tightening sequence

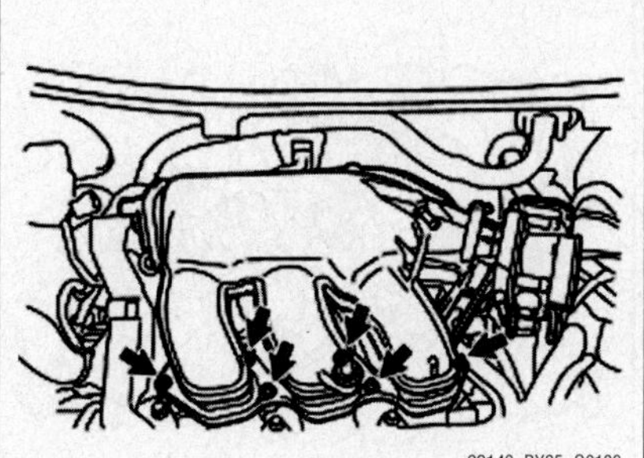

Fig. 100 Intake air surge tank bolts and nuts

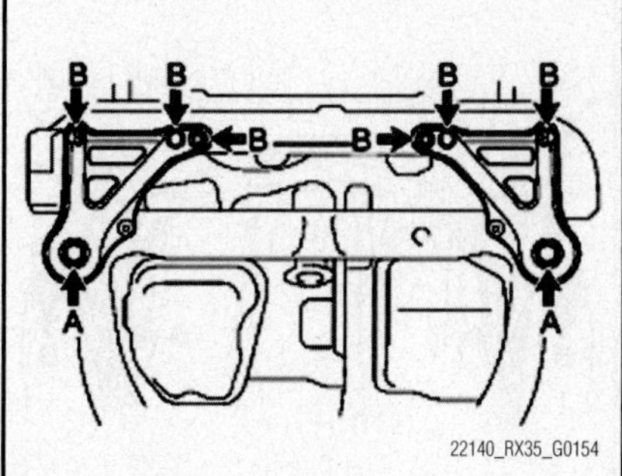

Fig. 102 Frame side rail plates, bolts and nuts

a. Install a new gasket as shown in the illustration.

b. Install the left exhaust manifold sub-assembly with the 6 nuts in the order shown in the illustration and tighten to 15 ft. lbs. (21 Nm).

105. Install the No. 2 exhaust manifold heat insulator with the 3 bolts and tighten to 75 inch lbs. (8.5 Nm).

106. Install the No. 2 manifold stay with the bolt and nut in the order shown in the illustration and tighten to 25 ft. lbs. (34 Nm).

107. Install the oil level gauge guide sub-assembly, as follows:

a. Install 2 new O-rings to the oil level gauge guide.

b. Apply a light coat of engine oil to the O-rings.

c. Push in the oil level gauge guide end into the guide hole.

d. Install the oil level gauge guide No. 1 with the bolt and tighten to 15 ft. lbs. (21 Nm).

e. Install the oil level gauge guide No. 2 with the bolt and tighten to 15 ft. lbs. (21 Nm).

f. Install the oil level gauge.

108. Install the right exhaust manifold sub-assembly, as follows:

a. Install a new gasket as shown in the illustration.

b. Install the right exhaust manifold sub-assembly with the 6 nuts in the order shown in the illustration and tighten to 15 ft. lbs. (21 Nm).

109. Install the intake manifold, as follows:

✱✱ WARNING

DO NOT apply oil to the intake manifold and cylinder head sub-assembly bolts.

a. Set a new gasket on each cylinder head.

➡Align the port holes of the gasket and cylinder head.

➡Make sure that the gasket is installed in the correct direction.

b. Set the intake manifold on the cylinder heads.

c. Install and tighten the 6 bolts and 4 nuts uniformly in several steps to 15 ft. lbs. (21 Nm).

110. Install the right No. 2 mounting stay with the bolt and tighten to 15 ft. lbs. (21 Nm).

111. Install the 6 ignition coil assemblies with the 6 bolts and tighten to 7 ft. lbs. (10 Nm).

112. Install the intake air surge tank assembly, as follows:

✱✱ WARNING

DO NOT apply oil to the surge tank and intake manifold bolts, the No. 1 surge tank stay and cylinder head cover bolts, the No. 1 surge tank stay and surge tank bolts, the throttle body bracket and cylinder head cover bolts, or the throttle body bracket and surge tank bolts.

a. Install a new gasket to the intake air surge tank.

b. Using a 5 mm hexagon socket wrench, install the 4 bolts and 2 nuts and tighten to 13 ft. lbs. (18 Nm), and 12 ft. lbs. (16 Nm).

c. Install the throttle body bracket, No. 1 surge tank stay and 4 bolts and tighten to 15 ft. lbs. (21 Nm).

d. Connect the connector.

e. Connect the No. 1 ventilation hose.

f. Install the clamp and connect the throttle with motor body assembly connector.

g. Connect the vapor feed hose.

h. Connect the 2 water by-pass hoses to the throttle with motor body assembly.

113. Install the 2 bolts and exhaust pipe support bracket and tighten to 15 ft. lbs. (21 Nm).

114. Remove engine stand.

115. Install the drive plate and ring gear sub-assembly.

116. Install the automatic transaxle assembly.

117. For 4WD vehicles, install the right transfer stiffener plate.

118. Install the bolt, nut, and manifold stay and tighten to 25 ft. lbs. (34 Nm).

119. Install the front engine mounting bracket.

120. Install the starter assembly.

121. Install the engine wire.

122. Install the left and right front drive shaft assemblies.

123. Install the front frame assembly, as follows:

a. Install the right and left engine mounting insulators with the 2 nuts and tighten to 70 ft. lbs. (95 Nm).

b. Install the right front engine mounting insulator with the bolt and tighten to 64 ft. lbs. (87 Nm).

c. Install the right rear engine mounting insulator with the 2 bolts and tighten to 58 ft. lbs. (78 Nm).

Fig. 105 Engine moving control rod bolt installation sequence

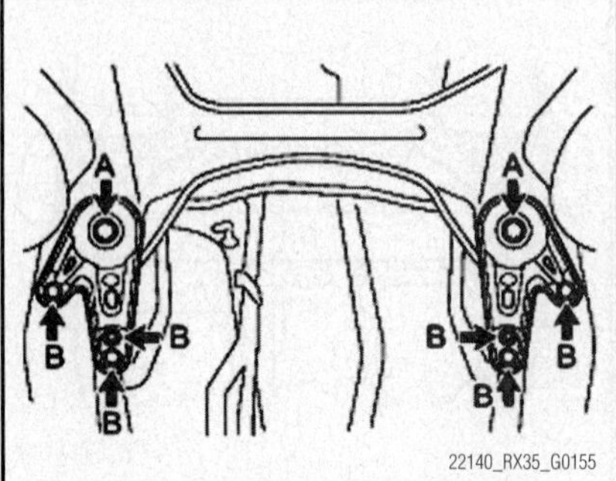

Fig. 103 Front suspension member rear braces, bolts and nuts

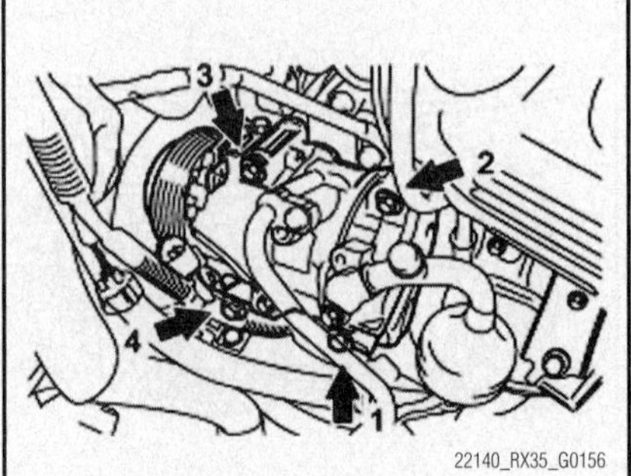

Fig. 104 Cooler compressor bolt tightening sequence

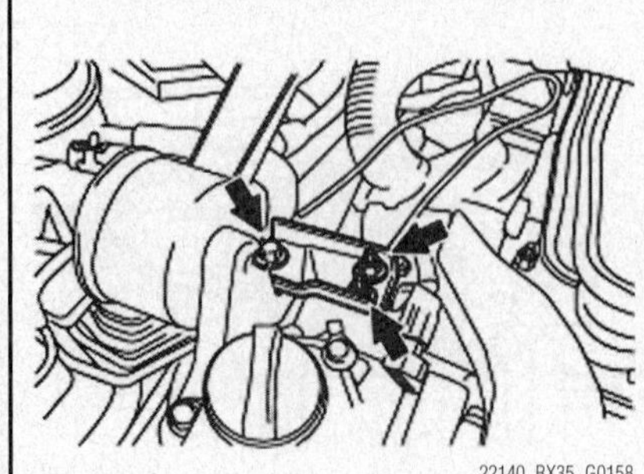

Fig. 106 Right No. 2 engine mounting stay bolt and nuts

d. Connect the connector and clamp.
e. Connect the 2 clamps.

124. For 4WD vehicles, install the power steering link.

125. For 4WD vehicles, install the front stabilizer bar.

126. Install vane pump (power steering pump) assembly, as follows:

a. Install the vane pump with the 2 bolts and nut and tighten to 32 ft. lbs. (43 Nm).
b. Install the 3 pressure feed tube clamp bolts and tighten to 69 inch lbs. (7.8 Nm).
c. Connect the power steering oil pressure switch connector.

127. Install the engine assembly with transaxle, as follows:

a. Set the engine assembly with transaxle on the engine lifter.
b. Install the engine assembly to the vehicle.
c. Install the right and left frame side rail plates with the 6 bolts and 2 nuts and tighten to 63 ft. lbs. (85 Nm), and 24 ft. lbs. (32 Nm).
d. Install the right and left front suspension member rear braces with the 6 bolts and 2 nuts and tighten to 63 ft. lbs. (85 Nm), and 24 ft. lbs. (32 Nm).

128. Install the cooler compressor assembly, as follows:

a. Temporarily install the cooler compressor with the 4 bolts.
b. Install the compressor with the 4 bolts by tightening the bolts in the order shown in the illustration to 18 ft. lbs. (25 Nm).
c. Install the 2 connector clamps.

129. Install the alternator assembly.

130. Install the nut and connect the right and left height control sensors (with

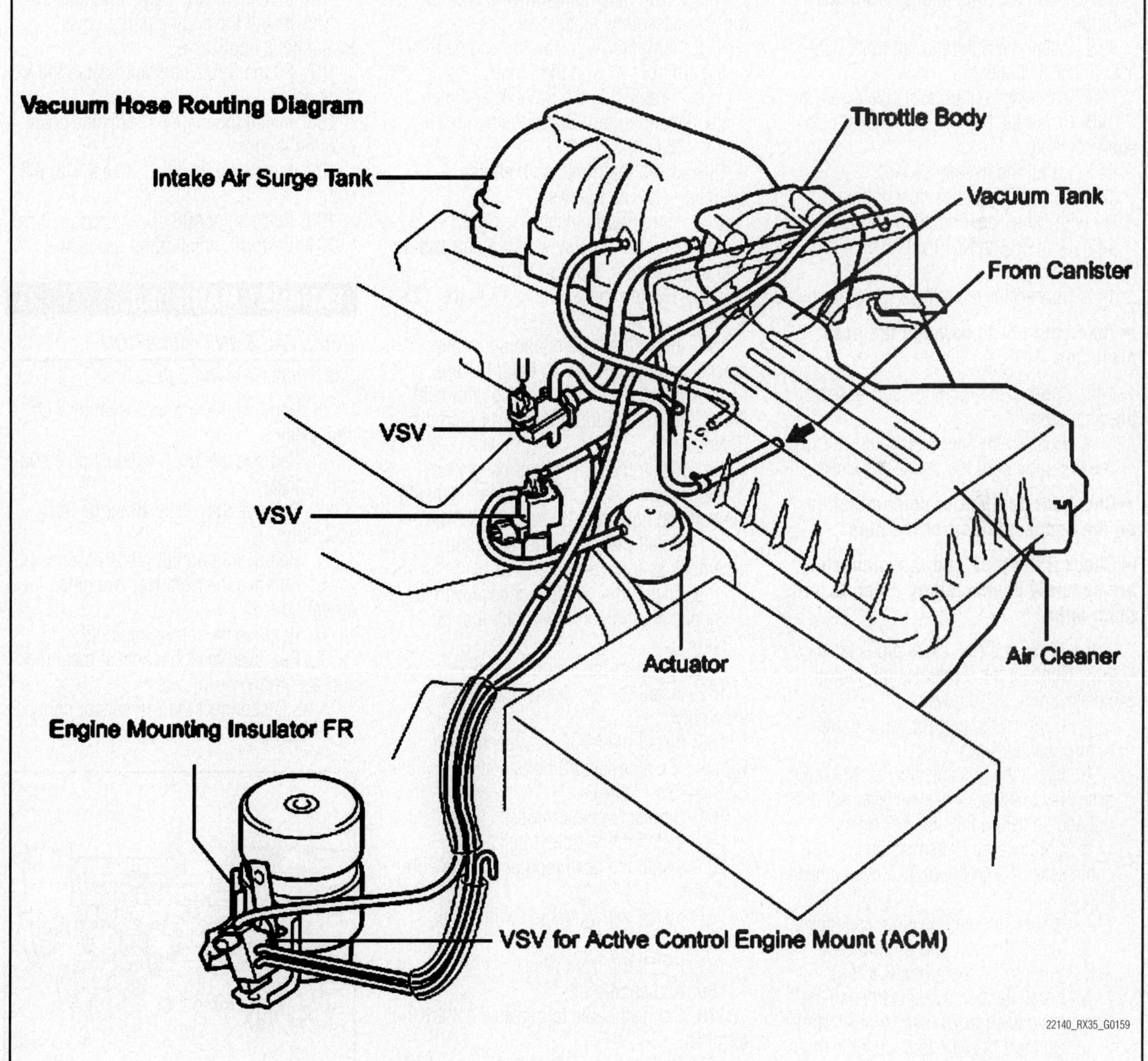

Fig. 107 Vacuum hoses

air suspension). Tighten to 48 inch lbs. (5.4 Nm).

131. Install the steering intermediate shaft sub-assembly.

132. Install the No. 1 exhaust pipe support bracket with the bolt and tighten to 15 ft. lbs. (21 Nm).

133. Install the right and left front axle assemblies.

134. Install the right and left lower No. 1 front suspension arms

135. Install the right and left tie rod assemblies.

136. Install the right and left front speed sensors.

137. Install the right and left front axle hub nuts

138. Install the right and left front stabilizer link assemblies.

139. Install the front exhaust pipe assembly.

140. Install the front No. 3 exhaust pipe sub-assembly.

141. Install the oxygen sensor.

142. For 4WD vehicles, install the propeller with center bearing shaft assembly.

143. Connect the No. 1 oil reservoir to pump hose.

144. Connect the return tube sub-assembly.

➡Take care not to damage the hose protector.

145. Connect the fuel tube sub-assembly, as follows:

a. Push in the fuel tube connector to the fuel pipe until the connector clicks.

➡Check for damage or contamination on the connected part of the pipe.

➡Check if the pipe and the connector are securely connected by trying to pull them apart.

b. Install the No. 1 fuel pipe clamp.

146. Connect the transmission control cable assembly, as follows:

a. Connect the control cable to the control cable clamp.

b. Install the transmission control cable assembly to the control shaft lever with the nut and tighten to 9 ft. lbs. (12 Nm).

c. Connect the transmission control cable assembly to the bracket with a new clip.

147. Connect the engine wire, as follows:

a. Install the bolt and ground cable and tighten to 74 inch lbs. (8.4 Nm).

b. Install the bolt and 2 clamps to the body and tighten to 74 inch lbs. (8.4 Nm).

c. Connect the wire to the engine room junction block. Then, install it with the nut and tighten to 74 inch lbs. (8.4 Nm).

d. Install the engine wire to the body with the 2 nuts and tighten to 74 inch lbs. (8.4 Nm).

148. Install the glove compartment door assembly.

149. Install the junction block cover.

150. Connect the heater inlet and outlet water hoses.

151. Install the clamps and connect the oil cooler inlet and outlet hoses.

152. Install the clamps and connect the radiator inlet and outlet hoses.

153. Install the clamp and connect the union to check valve hose.

154. Install the clamp and connect the No. 1 fuel vapor feed hose.

155. Install the engine moving control rod sub-assembly, as follows:

a. Temporarily install the engine moving control rod with the 3 bolts.

b. First install the bolts A, and then the remaining bolt B and tighten to 28 ft. lbs. (38 Nm).

156. Install the right No. 2 engine mounting stay, as follows:

a. Temporarily install the right No. 2 engine mounting stay with the bolt and tighten to 28 ft. lbs. (38 Nm).

b. Tighten the 2 nuts to 17 ft. lbs. (23 Nm).

157. Install the 2 bolts and air cleaner bracket and tighten to 9 ft. lbs. (12 Nm).

158. Install the intake air resonator with the bolt and clip and tighten to 44 inch lbs. (5 Nm).

159. Install the battery and battery tray.

160. Install the No. 1 air cleaner inlet with the bolt and tighten to 62 inch lbs. (7 Nm).

161. Install the air cleaner case sub-assembly, as follows:

a. Install the air cleaner case with the 3 bolts and tighten to 44 inch lbs. (5 Nm).

b. Connect the vacuum hose.

162. Install the air cleaner cap sub-assembly.

163. Install the No. 2 air cleaner inlet with the 2 clamps and 2 bolts and tighten to 62 inch lbs. (7 Nm).

164. Connect vacuum hoses.

165. Install the V-ribbed belt.

166. Install the cowl top panel outer sub-assembly.

167. Install the windshield wiper link assembly.

168. Install the front wheels.

169. Add engine oil.

170. Connect cable to negative battery terminal.

171. Add engine coolant and automatic transaxle fluid.

172. Check automatic transaxle fluid.

173. Add power steering fluid.

174. Bleed power steering fluid.

175. Inspect for fuel, oil, coolant and exhaust gas leaks.

176. Inspect shift lever position.

177. Inspect and adjust front wheel alignment.

178. Inspect ignition timing.

179. Inspect engine idle speed.

180. Inspect CO/HC.

181. Inspect function of throttle body.

182. Install the right front fender apron seal.

183. Install the right front fender splash shield sub-assembly.

184. Install the No. 2 engine under cover, if applicable.

185. Install the No. 1 engine under cover.

186. Install the engine under cover assembly, if applicable.

187. Fit the 3 retainers and install the V-bank cover.

188. Install the engine room side cover with the 4 clips.

189. Install the intake duct seal with the 4 clips.

190. Inspect the ABS speed sensor signal.

191. Perform initialization procedure.

EXHAUST MANIFOLD

REMOVAL & INSTALLATION

See Figures 108 through 112.

1. Remove the engine assembly with transaxle.
2. Remove the No. 1 exhaust pipe support bracket.
3. Remove the intake air surge tank assembly.
4. Remove the ignition coil assembly.
5. Remove the right No. 2 engine mounting stay.
6. Remove the intake manifold.
7. Remove the right exhaust manifold sub-assembly, as follows:

a. Disconnect the A/F sensor connector clamp.

Fig. 108 Install a new gasket

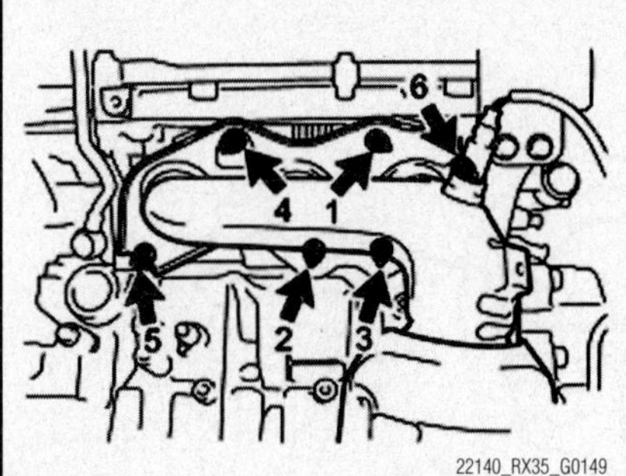

Fig. 109 Left exhaust manifold sub-assembly tightening sequence

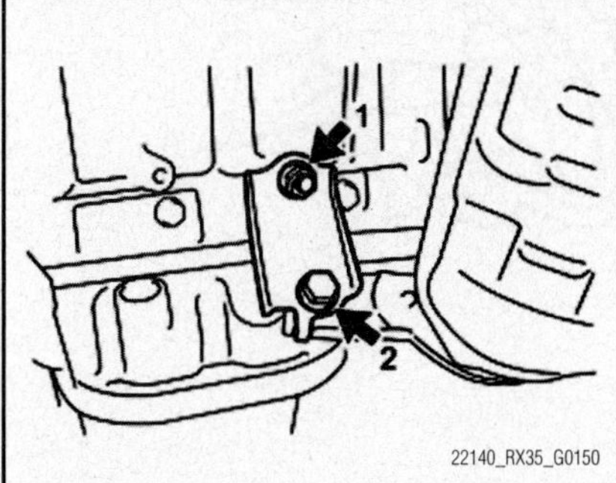

Fig. 110 No. 2 manifold stay bolt tightening sequence

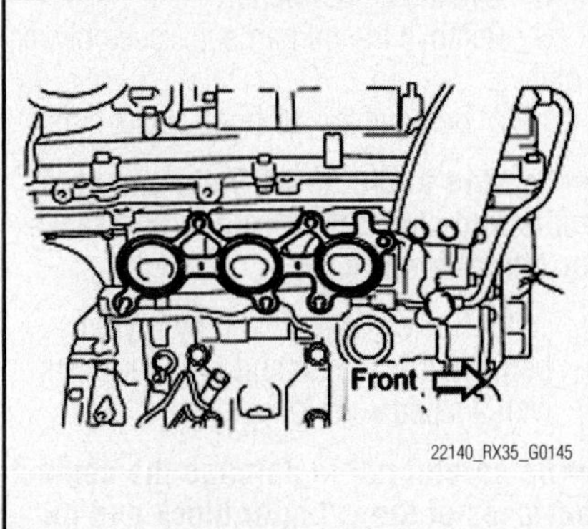

Fig. 111 Install a new gasket

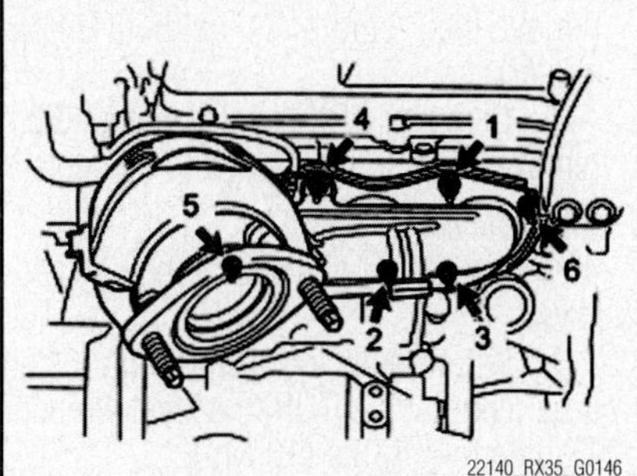

Fig. 112 Right exhaust manifold sub-assembly bolt tightening sequence

b. Uniformly loosen and remove the 6 nuts.

c. Remove the manifold and gasket.

8. Remove the oil level gauge guide sub-assembly.

9. Remove the bolt, nut and No. 2 manifold stay.

10. Remove the 3 bolts and No. 2 exhaust manifold heat insulator.

11. Remove the left exhaust manifold sub-assembly, as follows:

a. Uniformly loosen and remove the 6 nuts.

b. Remove the manifold and gasket.

To install:

12. Install the left exhaust manifold sub-assembly, as follows:

a. Install a new gasket as shown in the illustration.

b. Install the left exhaust manifold sub-assembly with the 6 nuts in the order shown in the illustration and tighten to 15 ft. lbs. (21 Nm).

13. Install the No. 2 exhaust manifold heat insulator with the 3 bolts and tighten to 75 inch lbs. (8.5 Nm).

14. Install the No. 2 manifold stay with the bolt and nut in the order shown in the illustration and tighten to 25 ft. lbs. (34 Nm).

15. Install the oil level gauge guide sub-assembly.

16. Install the right exhaust manifold sub-assembly, as follows:

a. Install a new gasket as shown in the illustration.

b. Install the right exhaust manifold sub-assembly with the 6 nuts in the order shown in the illustration and tighten to 15 ft. lbs. (21 Nm).

17. Install the intake manifold.

18. Install the right No. 2 engine mounting stay.

19. Install the ignition coil assembly.

20. Install the intake air surge tank assembly.

21. Install the No. 1 exhaust pipe support bracket.

22. Install the engine assembly with transaxle.

FLEXPLATE

REMOVAL & INSTALLATION

See Figure 113.

1. Remove the automatic transaxle assembly.

2. Remove the drive plate and ring gear sub-assembly, as follows:

a. Using SST (SST: 09213-70011, SST: 09330-00021) or equivalent tool, hold the crankshaft.

Fig. 113 Remove the 8 bolts, front spacer, drive plate and rear spacer

b. Remove the 8 bolts, front spacer, drive plate and rear spacer.

To install:

3. Install the drive plate and ring gear sub-assembly, as follows:

a. Using SST (SST: 09213-70011, SST: 09330-0002) or equivalent, hold the crankshaft.

b. Apply adhesive (Toyota Genuine Adhesive 1324, Three Bond 1324 or equivalent) to 2 or 3 threads of the mounting bolt end.

c. Install the front spacer, drive plate and rear spacer on the crankshaft.

d. Install and tighten the 8 mounting bolts uniformly in several steps. Tighten to 61 ft. lbs (83 Nm).

4. Install automatic transaxle assembly.

INTAKE MANIFOLD

REMOVAL & INSTALLATION

See Figure 114.

1. Remove the engine assembly with transaxle.

2. Remove the No. 1 exhaust pipe support bracket.

3. Remove the intake air surge tank assembly.

4. Remove the ignition coil assembly.

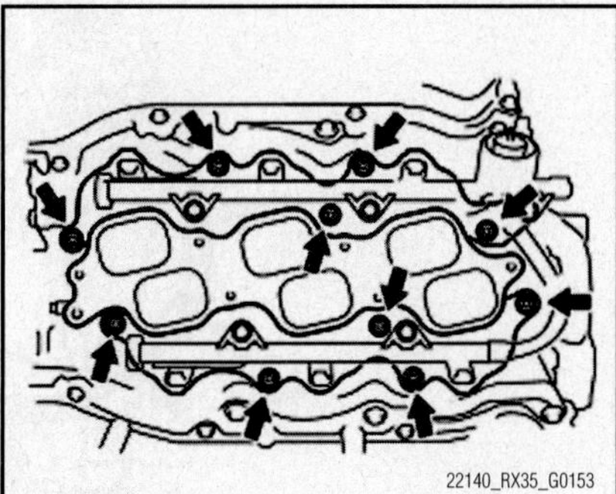

Fig. 114 Intake manifold

5. Remove the right No. 2 engine mounting stay.

6. Remove the intake manifold, as follows:

a. Uniformly loosen and remove the 6 bolts and 4 nuts.

b. Remove the intake manifold and 2 gaskets.

To install:

7. Install the intake manifold, as follows:

**** WARNING**

DO NOT apply oil to the intake manifold and cylinder head sub-assembly bolts.

a. Set a new gasket on each cylinder head.

➡Align the port holes of the gasket and cylinder head.

➡Make sure that the gasket is installed in the correct direction.

b. Set the intake manifold on the cylinder heads.

c. Install and tighten the 6 bolts and 4 nuts uniformly in several steps to 15 ft. lbs. (21 Nm).

8. Install the right No. 2 engine mounting stay.

9. Install the ignition coil assembly.

10. Install the intake air surge tank assembly.

11. Install the No. 1 exhaust pipe support bracket.

12. Install the engine assembly with transaxle.

OIL PAN

REMOVAL & INSTALLATION

See Figures 115 through 124.

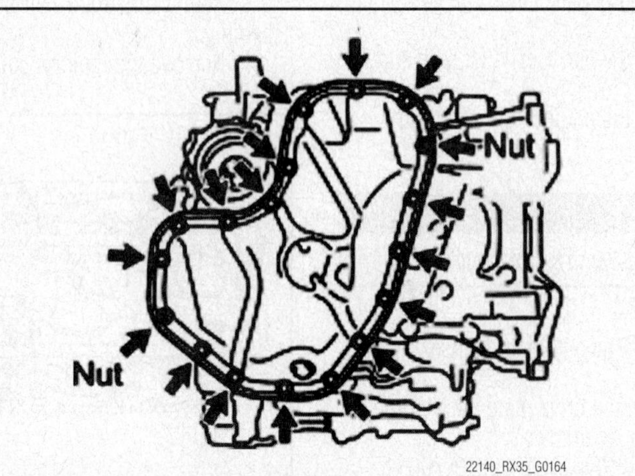

Fig. 115 No. 2 oil pan sub-assembly bolts and nuts

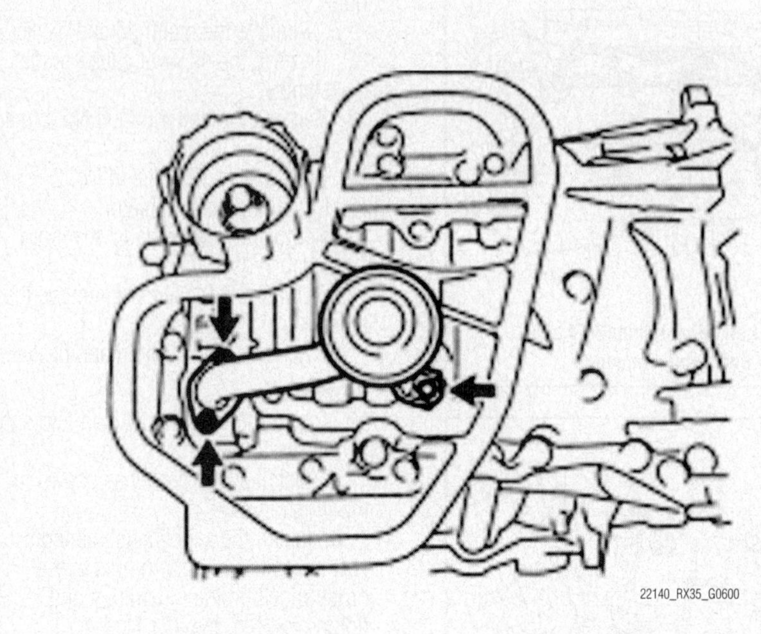

Fig. 116 Oil strainer sub-assembly

1. Remove the No. 2 oil pan sub-assembly, as follows:

a. Remove the 16 bolts and 2 nuts.

b. Insert the blade of SST (SST: 09032-00100) or equivalent tool between the oil pans. Cut through the applied sealer and remove the No. 2 oil pan sub-assembly.

➡Be careful not to damage the contact surfaces of the oil pans.

c. Using a TORX® socket wrench E6, remove the 2 stud bolts.

2. Remove the oil strainer sub-assembly, as follows:

a. Remove the bolt, 2 nuts, oil strainer and gasket.

b. Using a TORX® socket wrench E6, remove the 2 stud bolts.

3. Remove the oil pan sub-assembly, as follows:

a. Remove the 16 bolts and 2 nuts.

➡Be sure to clean the bolts and stud bolts and check the threads for cracks or other damage.

b. Remove the oil pan by prying between the oil pan and cylinder block with a taped screwdriver.

➡Be careful not to damage the contact surfaces of the cylinder block and oil pan.

c. Remove the 2 O-rings.

d. Using a TORX® socket wrench E8, remove the 2 stud bolts. (without oil cooler).

e. Using a TORX® socket wrench E8, remove the 4 stud bolts. (with oil cooler).

To install:

4. Install the oil pan sub-assembly, as follows:

a. When replacing a stud bolt, install it by using an E8 TORX® socket wrench. Tighten to 7 ft. lbs (10 Nm).

b. Install 2 new O-rings.

c. Apply seal packing in a continuous line as shown in the illustration. Seal packing: Toyota Genuine Seal Packing

Black, Three Bond 1207B or equivalent. Seal diameter: 3.0 to 4.0 mm (0.118 to 0.156 in.).

➡Remove any oil from the contact surface.

➡Install the oil pan within 3 minutes after applying seal packing.

➡Do not start the engine for at least 2 hours after installing.

d. Install the oil pan with the 16 bolts and 2 nuts and tighten to 7 ft. lbs (10 Nm), and 15 ft. lbs (21 Nm).

5. Install the oil strainer sub-assembly, as follows:

a. Using an E6 TORX® socket, install the stud bolts as shown in the illustration and tighten to 35 inch lbs (4 Nm).

b. Install a new gasket and the oil strainer with the bolt and 2 nuts and tighten to 7 ft. lbs (10 Nm).

6. Install the No. 2 oil pan sub-assembly, as follows:

a. Using an E6 TORX® socket, install the stud bolts as shown in the illustration and tighten to 35 inch lbs (4 Nm).

b. Apply seal packing in a continuous line as shown in the illustration. Seal packing: Toyota Genuine Seal Packing Black, Three Bond 1207B or equivalent. Seal diameter: 3.0 to 4.0 mm (0.118 to 0.156 in.).

➡Remove any oil from the contact surface.

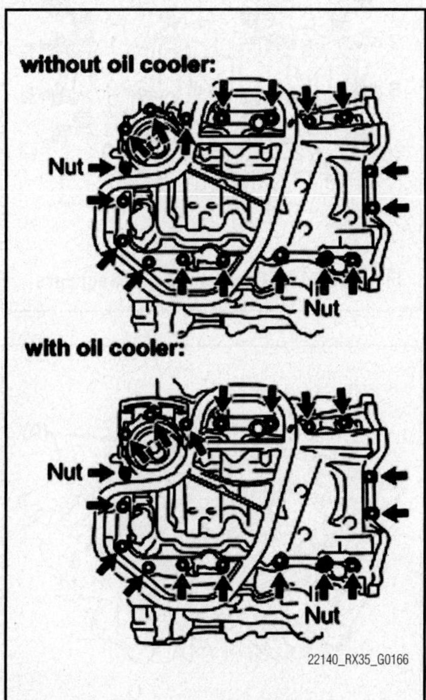

Fig. 117 Oil pan sub-assembly bolts and nuts

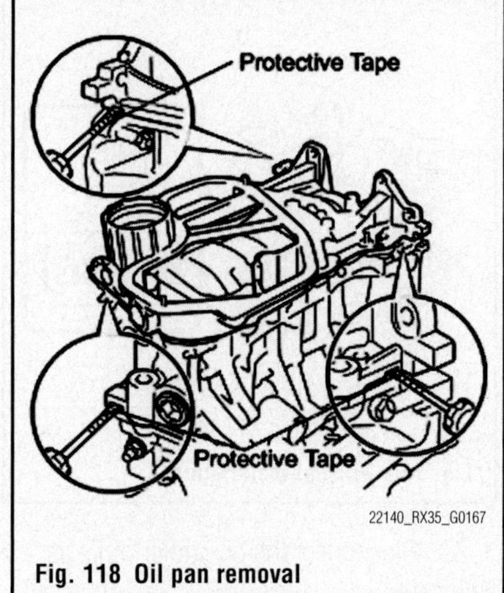

Fig. 118 Oil pan removal

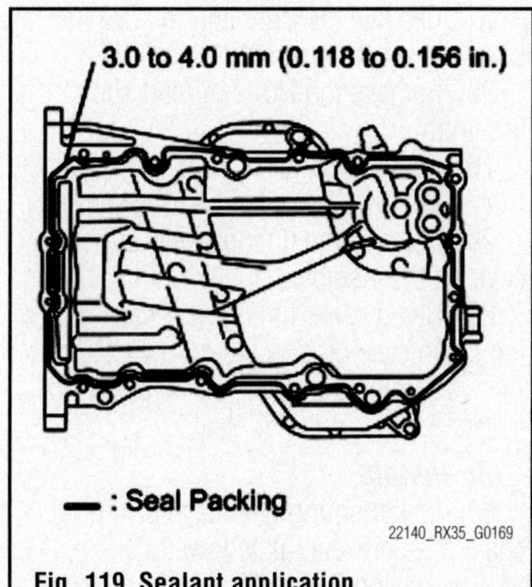

Fig. 119 Sealant application

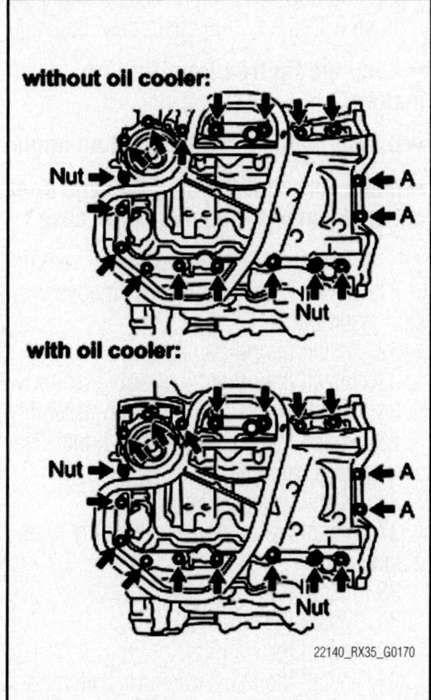

Fig. 120 Oil pan bolts and nuts

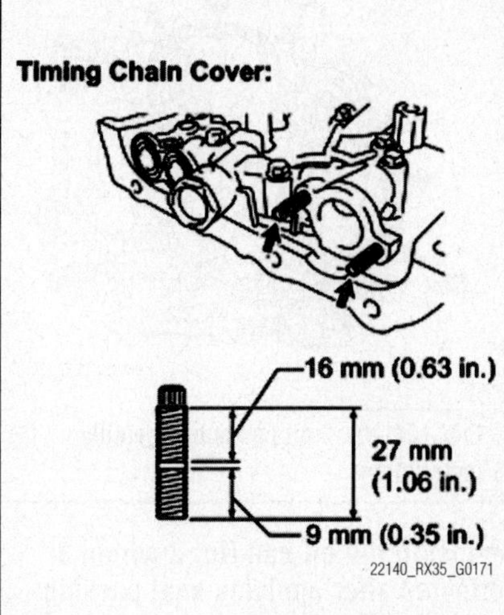

Fig. 121 Oil strainer sub-assembly installation

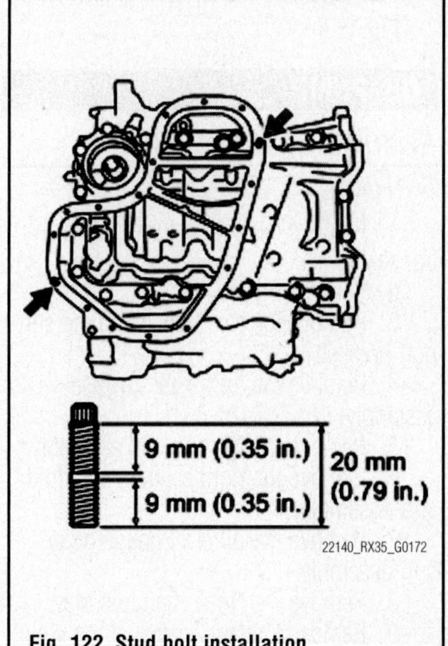

Fig. 122 Stud bolt installation

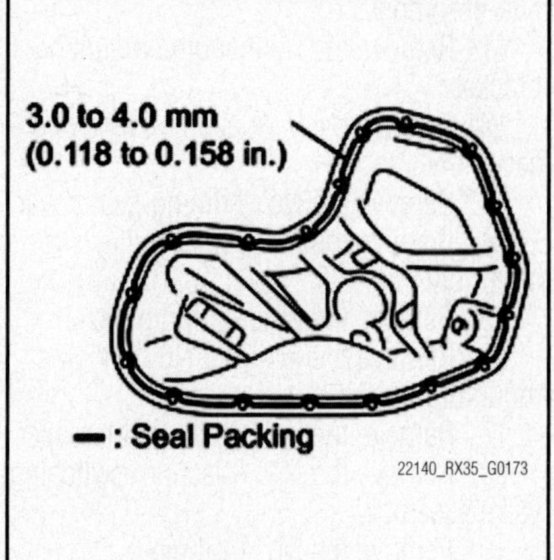

Fig. 123 Sealant application

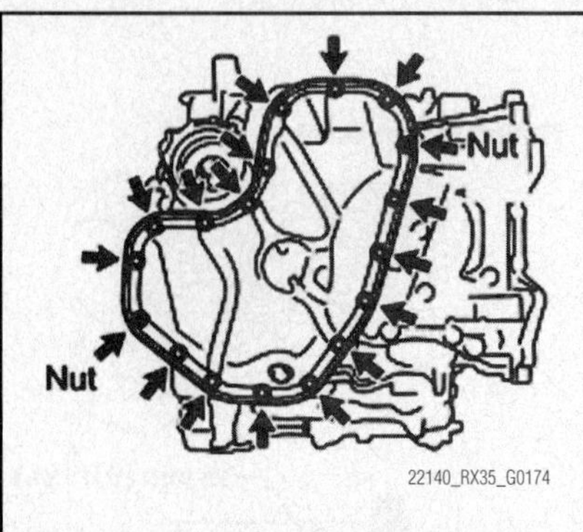

Fig. 124 No. 2 oil pan sub-assembly installation

➡**Install the oil pan No. 2 within 3 minutes after applying seal packing.**

➡**Do not start the engine for at least 2 hours after installing.**

c. Install the oil pan with the 16 bolts and 2 nuts and tighten to 7 ft. lbs (10 Nm).

OIL PUMP

REMOVAL & INSTALLATION

See Figure 125.

1. Remove the engine assembly with transaxle.
2. Secure engine.
3. Remove the No. 1 exhaust pipe support bracket.
4. Remove the intake air surge tank assembly.
5. Remove the ignition coil assembly.
6. Remove the right exhaust manifold sub-assembly.
7. Remove the oil level gauge guide sub-assembly.
8. Remove the No. 2 manifold stay.
9. Remove the No. 2 exhaust manifold heat insulator.
10. Remove the left exhaust manifold sub-assembly.
11. Remove the right engine mounting bracket.
12. Remove the V-ribbed belt tensioner assembly.
13. Remove the No. 2 timing gear cover.
14. Remove the No. 2 idler pulley sub-assembly.
15. Remove the water pump pulley.
16. Remove the left front No. 1 engine mounting bracket.
17. Remove the radio setting condenser.
18. Remove the No. 1 vacuum switching valve assembly.
19. Remove the No. 1 oil pipe.
20. Remove the oil pipe.
21. Remove the crankshaft pulley.

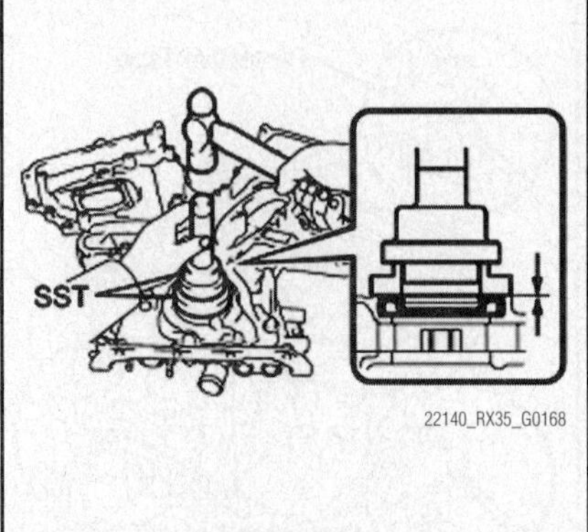

Fig. 125 Oil seal installation

22. Disconnect the oil cooler pipe, as applicable.
23. Remove the water inlet housing.
24. Remove the right and left cylinder head cover sub-assemblies.
25. Remove the No. 2 oil pan sub-assembly.
26. Remove the oil strainer sub-assembly.
27. Remove the oil pan sub-assembly.
28. Remove the timing chain or belt cover sub-assembly.
29. Remove the timing gear case or timing chain case oil seal, as follows:
    a. Using a screwdriver with the tip taped, pry out the oil seal.

To install:

30. Install timing gear case or timing chain case oil seal, as follows:
    a. Using SST (SST: 09223-22010, SST: 09506-35010) or equivalent tool, tap in a new oil seal until its surface is flush with the timing chain case edge.

➡**Keep the lip free from foreign matter.**

➡**Do not tap on the oil seal at an angle.**

➡**Make sure that the oil seal edge does not stick out of the timing chain case.**

    b. Apply MP grease to the oil seal lip.
31. Install timing chain or belt cover sub-assembly.
32. Install the oil pan sub-assembly.
33. Install the oil strainer sub-assembly.
34. Install the No. 2 oil pan sub-assembly.
35. Install the right and left cylinder head cover sub-assemblies.
36. Install water outlet.
37. Install the oil cooler assembly pipe, if applicable.
38. Install the crankshaft pulley.
39. Install the No. 1 oil pipe.
40. Install the oil pipe.
41. Install the No. 1 vacuum switching valve assembly.
42. Install the radio setting condenser.
43. Install the No 1 left front engine mounting bracket.
44. Install the water pump pulley.
45. Install the No. 2 idler pulley sub-assembly.
46. Install the No. 2 timing gear cover.
47. Install the V-ribbed belt tensioner assembly.
48. Install the right engine mounting bracket.
49. Install the left exhaust manifold sub-assembly.
50. Install the No. 2 exhaust manifold heat insulator.
51. Install the No. 2 manifold stay.
52. Install the oil level gauge guide sub-assembly.
53. Install the right exhaust manifold sub-assembly.
54. Install the ignition coil assembly.
55. Install the intake air surge tank assembly.
56. Install the No. 1 exhaust pipe support bracket.
57. Install the engine assembly with transaxle.

MAIN BEARING TORQUE SEQUENCE

See Figures 126 and 127.

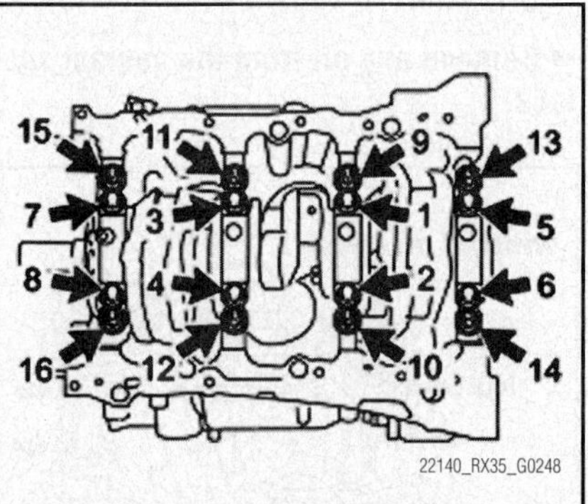

Fig. 126 Main Cap Bolt Torque Sequence

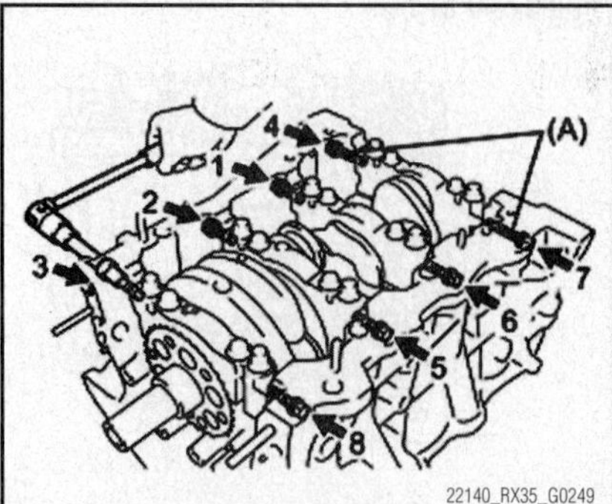

Fig. 127 Side Bolt Torque Sequence

PISTON AND RING

POSITIONING

See Figure 128.

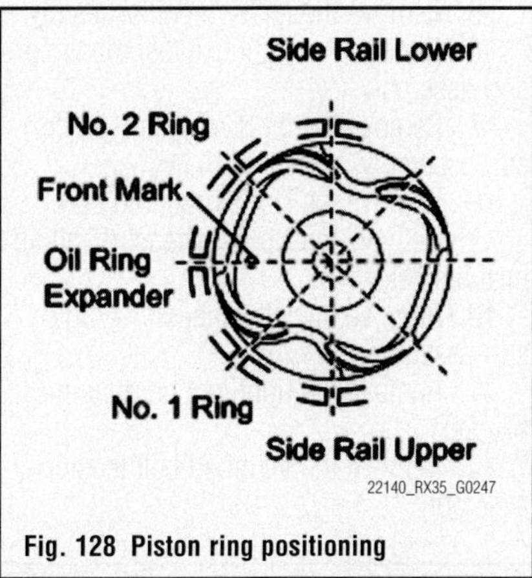

Fig. 128 Piston ring positioning

REAR MAIN SEAL

REMOVAL & INSTALLATION

See Figures 129 through 131.

1. Remove the automatic transaxle assembly.

Fig. 129 Remove the 8 bolts, front spacer, drive plate and rear spacer

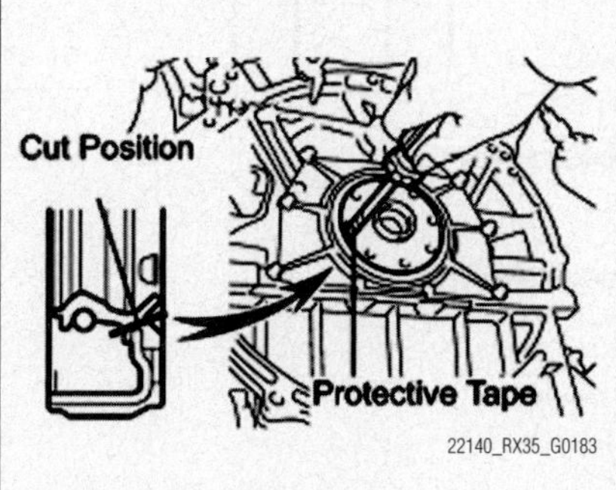

Fig. 130 Cut and pry the oil seal

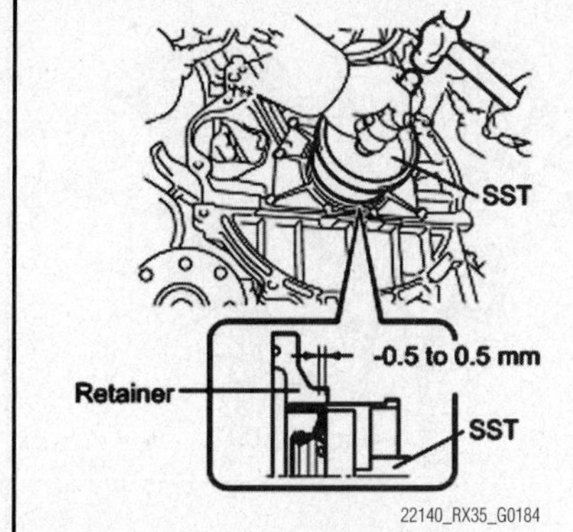

Fig. 131 Rear main seal installation

2. Remove the drive plate and ring gear sub-assembly, as follows:
   a. Using SST (SST: 09213-70011, SST: 09330-00021) or equivalent tool, hold the crankshaft.
   b. Remove the 8 bolts, front spacer, drive plate and rear spacer.
3. Remove the rear main seal, as follows:
   a. Using a knife, cut off the oil seal lip.
   b. Using a screwdriver with the tip taped, pry out the oil seal.

➡**Be careful not to damage the crankshaft.**

To install:

4. Apply MP grease to a new oil seal lip.
5. Using SST (SST: 09223-15030, SST: 09950-70010) or equivalent and a hammer, tap in the oil seal. Oil seal tap in depth: -0.5 to 0.5 mm (-0.020 to 0.020 in.)
6. Install the drive plate and ring gear sub-assembly, as follows:
   a. Using SST (SST: 09213-70011, SST: 09330-0002) or equivalent, hold the crankshaft.
   b. Apply adhesive to 2 or 3 threads of the mounting bolt end. Adhesive: Toyota Genuine Adhesive 1324, Three Bond 1324 or equivalent.
   c. Install the front spacer, drive plate and rear spacer on the crankshaft.
   d. Install and tighten the 8 mounting bolts uniformly in several steps. Tighten to 61 ft. lbs (83 Nm).
7. Install automatic transaxle assembly.

ROCKER ARMS/SHAFTS

REMOVAL & INSTALLATION

Refer to the Camshaft and Valve Lifter removal and installation procedures.

TIMING CHAIN COVER AND SEAL

REMOVAL & INSTALLATION

See Figures 132 through 139.

1. Remove the engine assembly with transaxle.

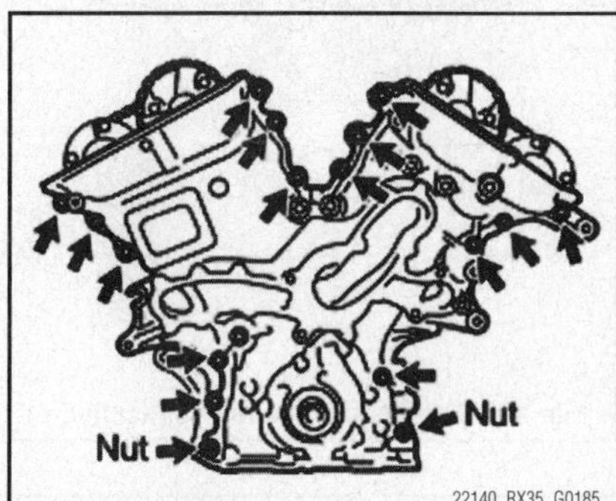

Fig. 132 Timing chain cover bolts and nuts

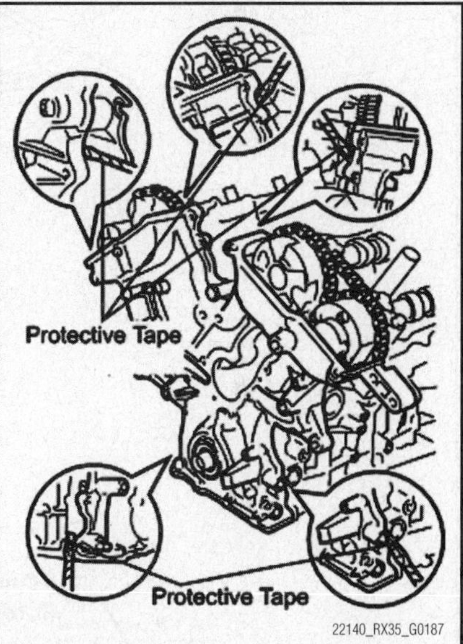

Fig. 133 Timing chain cover removal

Fig. 134 Oil seal installation

C
A
A
A
A
C
C
C
: Seal Packing
3.0 mm or more
(0.118 in.)
22140_RX35_G0186

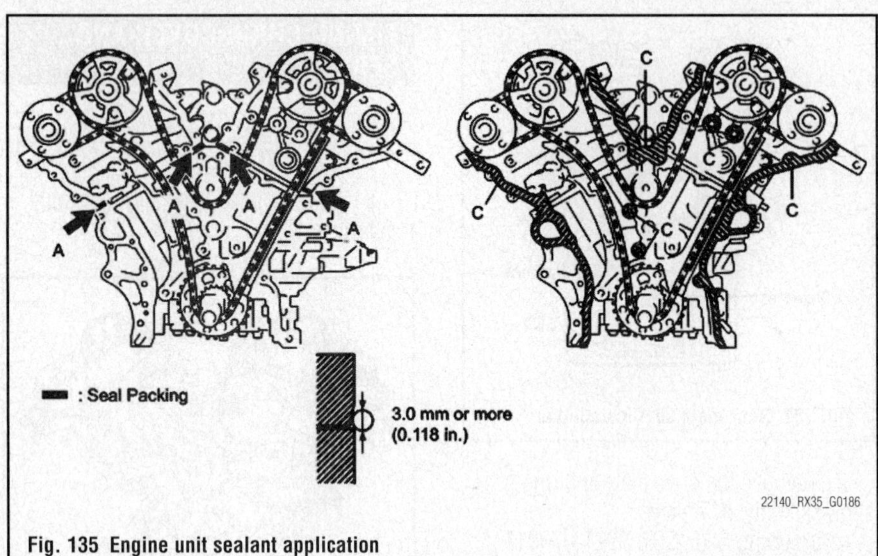

Fig. 135 Engine unit sealant application

2. Secure engine.
3. Remove the No. 1 exhaust pipe support bracket.
4. Remove the intake air surge tank assembly.
5. Remove the ignition coil assembly.
6. Remove the right exhaust manifold sub-assembly.
7. Remove the oil level gauge guide sub-assembly.
8. Remove the No. 2 manifold stay.
9. Remove the No. 2 exhaust manifold heat insulator.
10. Remove the left exhaust manifold sub-assembly.
11. Remove the right engine mounting bracket.
12. Remove the V-ribbed belt tensioner assembly.

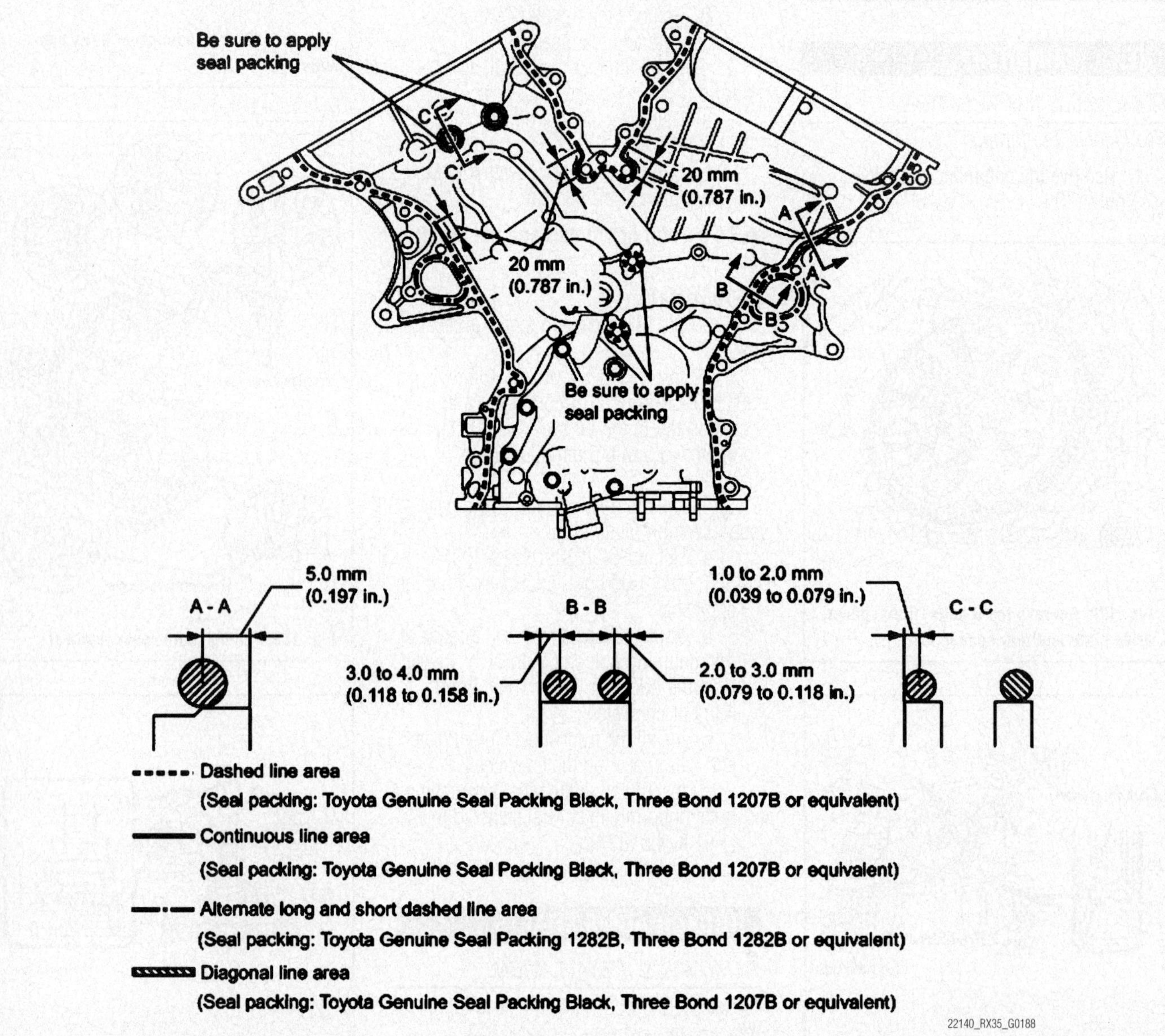

Be sure to apply
seal packing
20 mm
(0.787 in.)
20 mm
(0.787 in.)
Be sure to apply
seal packing
5.0 mm
(0.197 in.)
A - A
B - B
1.0 to 2.0 mm
(0.039 to 0.079 in.)
C - C
3.0 to 4.0 mm
(0.118 to 0.158 in.)
2.0 to 3.0 mm
(0.079 to 0.118 in.)
Dashed line area
(Seal packing: Toyota Genuine Seal Packing Black, Three Bond 1207B or equivalent)
Continuous line area
(Seal packing: Toyota Genuine Seal Packing Black, Three Bond 1207B or equivalent)
Alternate long and short dashed line area
(Seal packing: Toyota Genuine Seal Packing 1282B, Three Bond 1282B or equivalent)
Diagonal line area
(Seal packing: Toyota Genuine Seal Packing Black, Three Bond 1207B or equivalent)
22140_RX35_G0188

Fig. 136 Timing chain cover sealant application

Area	Seal Packing Diameter	Application Position from Inside Seal Line
Continuous Line Area	4.5 mm or more (0.177 in.)	3.0 to 4.0 mm (0.118 to 0.158 in.)
Alternate Long and Dashed Line Area	3.5 mm or more (0.138 in.)	2.0 to 3.0 mm (0.079 to 0.118 in.)
Dashed Line Area	3.5 mm or more (0.138 in.)	3.0 to 4.0 mm (0.118 to 0.158 in.)
Diagonal Line Area	6.0 mm or more (0.236 in.)	5.0 mm (0.197 in.)

22140_RX35_G0189

Fig. 137 Sealant application diameter and position

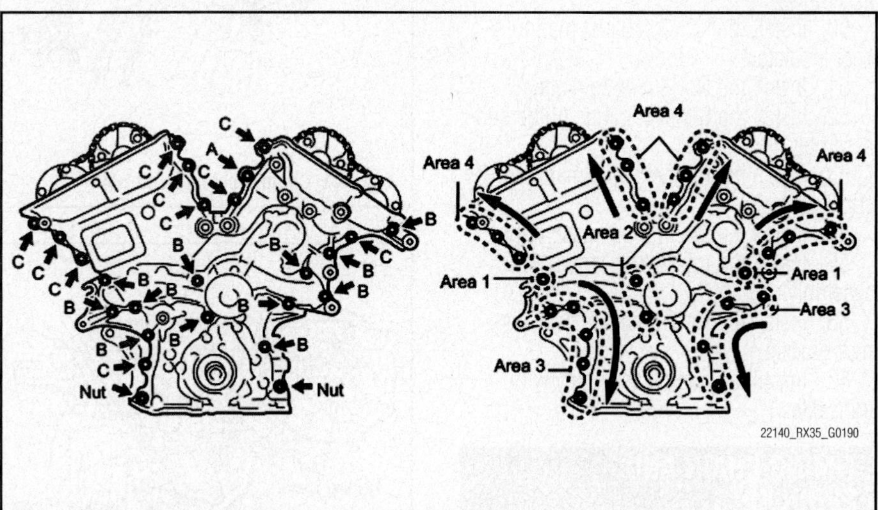

Fig. 138 Timing chain cover bolts and nuts

Item	Length
Bolt A	40 mm (1.57 in.)
Bolt B	55 mm (2.17 in.)
Bolt C	25 mm (0.98 in.)

22140_RX35_G0191

Fig. 139 Bolt length

13. Remove the No. 2 timing gear cover.
14. Remove the No. 2 idler pulley sub-assembly.
15. Remove the water pump pulley.
16. Remove the left front No. 1 engine mounting bracket.
17. Remove the radio setting condenser.
18. Remove the No. 1 vacuum switching valve assembly.
19. Remove the No. 1 oil pipe.
20. Remove the oil pipe.
21. Remove the crankshaft pulley.
22. Disconnect the oil cooler pipe, as applicable.
23. Remove the water inlet housing.
24. Remove the right and left cylinder head cover sub-assemblies.
25. Remove the No. 2 oil pan sub-assembly.
26. Remove the oil strainer sub-assembly.
27. Remove the oil pan sub-assembly.
28. Remove the timing chain cover sub-assembly, as follows:
    a. Remove the 15 bolts and 2 nuts as shown in the illustration.
    b. Remove the timing chain cover by prying between the timing chain cover and cylinder head or cylinder block with a screwdriver with the tip taped.

➡Be careful not to damage the contact surfaces of the cylinder head, cylinder block and chain cover.

    c. Remove the 4 bolts, chain cover plate and gasket.
    d. Remove the gasket.
29. Remove the timing chain case oil seal, as follows:
    a. Using a screwdriver with the tip taped, pry out the oil seal.

To install:

30. Install timing gear case or timing chain case oil seal, as follows:
    a. Using SST (SST: 09223-22010, SST: 09506-35010) or equivalent tool, tap in a new oil seal until its surface is flush with the timing chain case edge.

➡Keep the lip free from foreign matter.

➡Do not tap on the oil seal at an angle.

➡Make sure that the oil seal edge does not stick out of the timing chain case.

    b. Apply MP grease to the oil seal lip.
31. Install timing chain or belt cover sub-assembly, as follows:
    a. Apply seal packing in a continuous line to the engine unit as shown in the following illustration. Seal packing:

Toyota Genuine Seal Packing Black, Three Bond 1207B or equivalent. Seal diameter: 3.0 mm (0.118 in.).

➡Be sure to clean and degrease the contact surfaces, especially the surfaces indicated by C in the illustration.

➡When the contact surfaces are wet, wipe them with an oil-free cloth before applying seal packing.

➡Install the chain cover within 3 minutes after applying seal packing.

➡Do not start the engine for at least 2 hours after installing.

b. Apply seal packing in a continuous line to the timing chain cover as shown in the following illustration. Seal packing: Toyota Genuine Seal Packing Black, Three Bond 1207B or equivalent, Toyota Genuine Seal Packing Black, Three Bond 1282B, Three Bond 1282B or equivalent.

➡When the contact surfaces are wet, wipe them with an oil-free cloth before applying seal packing.

➡Install the chain cover within 3 minutes and tighten the bolts within 15 minutes after applying seal packing.

➡Do not start the engine for at least 2 hours after installing.

c. Install a new gasket.

d. Align the oil pump's drive rotor spline and the crankshaft as shown in the illustration. Install the spline and chain cover to the crankshaft.

e. Temporarily tighten the timing chain cover with the 23 bolts and 2 nuts.

✲✲ WARNING

Make sure that there is no oil on the bolt and nut threads.

f. Fully tighten the bolts in this order: Area 1 and Area 2. Tighten to 15 ft. lbs. (21 Nm).

g. Fully tighten the bolts in Area 3 to 15 ft. lbs. (21 Nm). Tighten the bolts and nuts in the order of upper to lower as shown in the illustration.

h. Fully tighten the bolts in Area 4 to 32 ft. lbs. (43 Nm), and to 15 ft. lbs. (21 Nm). Tighten the bolts and nuts in the order of lower to upper as shown in the illustration.

i. Install a new gasket and the chain cover plate with the 4 bolts and tighten to 81 inch lbs. (9.1 Nm).

32. Install the oil pan sub-assembly.
33. Install the oil strainer sub-assembly.
34. Install the No. 2 oil pan sub-assembly.
35. Install the right and left cylinder head cover sub-assemblies.
36. Install water outlet.
37. Install the oil cooler assembly pipe, if applicable.
38. Install the crankshaft pulley.
39. Install the No. 1 oil pipe.
40. Install the oil pipe.
41. Install the No. 1 vacuum switching valve assembly.
42. Install the radio setting condenser.
43. Install the No 1 left front engine mounting bracket.
44. Install the water pump pulley.
45. Install the No. 2 idler pulley sub-assembly.
46. Install the No. 2 timing gear cover.
47. Install the V-ribbed belt tensioner assembly.
48. Install the right engine mounting bracket.
49. Install the left exhaust manifold sub-assembly.
50. Install the No. 2 exhaust manifold heat insulator.
51. Install the No. 2 manifold stay.
52. Install the oil level gauge guide sub-assembly.
53. Install the right exhaust manifold sub-assembly.
54. Install the ignition coil assembly.
55. Install the intake air surge tank assembly.
56. Install the No. 1 exhaust pipe support bracket.
57. Install the engine assembly with transaxle.

TIMING CHAIN AND SPROCKETS

REMOVAL & INSTALLATION

See Figures 140 through 153.

1. Remove the timing chain cover and seal.

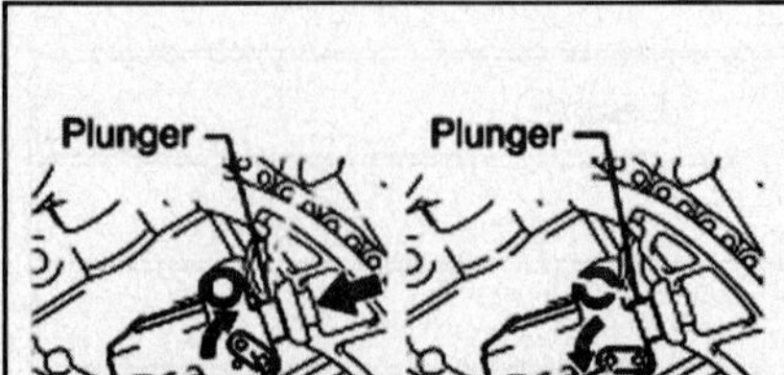

Fig. 140 Remove the No. 1 chain tensioner assembly

2. Set the No. 1 cylinder to TDC/Compression.
3. Remove the No. 1 chain tensioner assembly, as follows:

a. Move the stopper plate upward to release the lock, and push the plunger deep into the tensioner.

b. Move the stopper plate downward to set the lock, and insert a pin of 1.27 mm (0.05 in.) into the stopper plate's hole.

c. Remove the 2 bolts and chain tensioner.

4. Remove the chain tensioner slipper.

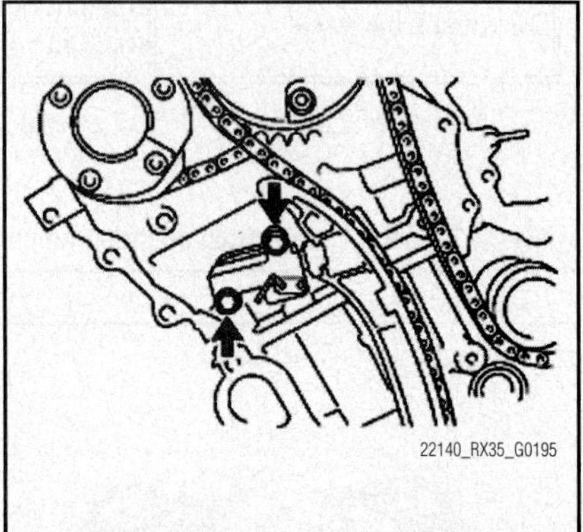

Fig. 141 Chain tensioner and bolts

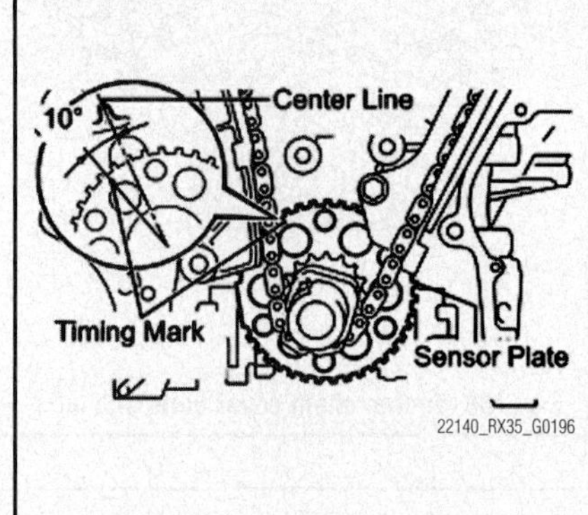

Fig. 142 Crankshaft timing sprocket

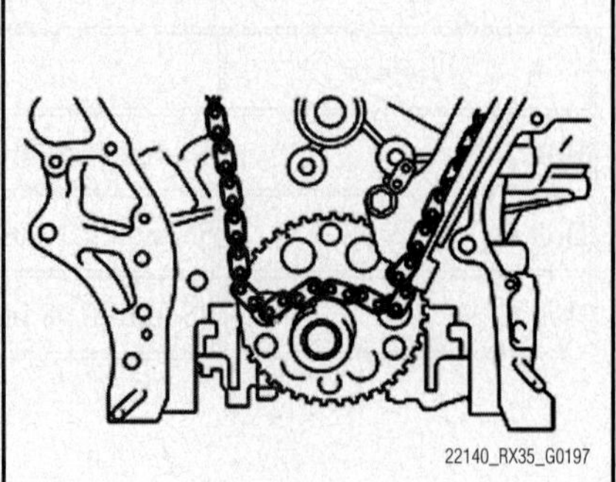

Fig. 143 Place chain on crankshaft

5. Remove the chain sub-assembly, as follows:

a. Turn the crankshaft counterclockwise 10° to loosen the chain of the crankshaft timing sprocket.

b. Remove the pulley set bolt.

c. Remove the chain from the crankshaft timing sprocket and place it on the crankshaft.

d. Turn the camshaft timing gear assembly on the right hand bank clockwise (approximately 60°) and set it as shown in the illustration. Be sure to loosen the chain between the banks.

e. Remove the chain.

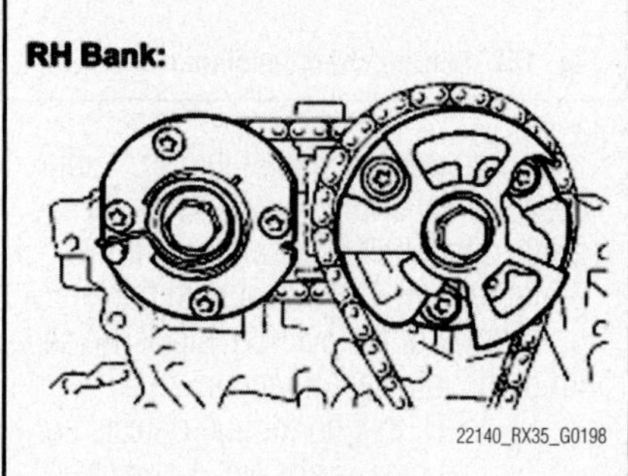

Fig. 144 Set the camshaft timing gear assembly

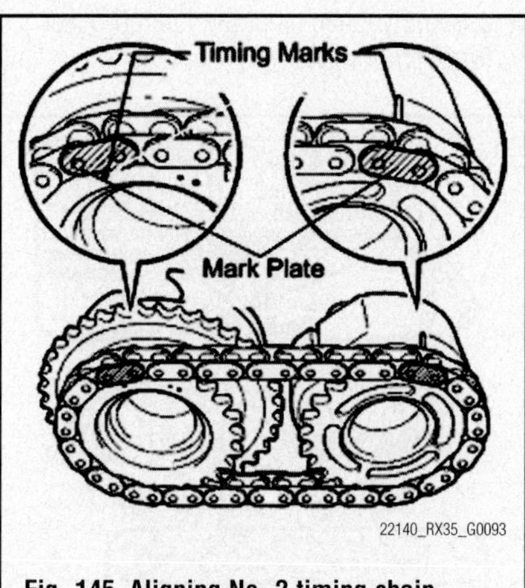

Fig. 145 Aligning No. 2 timing chain

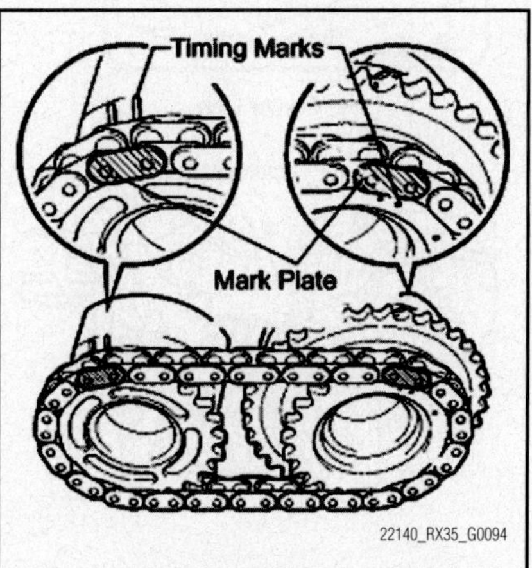

Fig. 146 Aligning No. 2 timing chain

6. Remove the idle sprocket assembly, as follows:

a. Using a 10 mm hexagon wrench, remove the No. 2 idle gear shaft, sprocket and No. 1 idle gear shaft.

7. Remove the 2 bolts and the No. 1 chain vibration damper.

8. Remove the No. 2 chain vibration damper.

9. Remove the crankshaft timing sprocket from the crankshaft.

10. Remove the 2 pulley set keys from the crankshaft.

11. Remove the camshaft timing gears and No. 2 chain (for Bank 1).

12. Remove the bolt and No. 2 chain tensioner.

13. Perform the same operations for Bank 2.

To install:

14. Install the No. 2 chain tensioner assembly with the bolt and tighten to 15 ft. lbs. (21 Nm).

15. While pushing in the tensioner, insert a pin of 1.0 mm (0.039 in.) diameter into the hole to fix it.

16. Install the camshaft timing gears and No. 2 chain (for Bank 1), as follows:

a. Align the mark plate with the timing marks of the camshaft timing gear as shown.

b. Apply a light coat of engine oil to the bolt threads and bolts seating surface.

c. Align the knock pin of the camshaft with pinhole of the camshaft timing gear. Install the camshaft timing gear and the right camshaft timing exhaust gear with the No. 2 chain installed.

d. Hold the hexagonal portion of the camshaft with the wrench and tighten the two bolts to 74 ft. lbs. (100 Nm).

Fig. 147 Crankshaft timing sprocket

e. Remove the pin from the chain tensioner.

17. Install the No. 3 chain tensioner assembly with the bolt and tighten to 15 ft. lbs. (21 Nm).

18. While pushing in the tensioner, insert a pin of 1.0 mm (0.039 in.) diameter into the hole to hold it.

19. Install the camshaft timing gears and No. 2 chain (for Bank 2), as follows:

a. Align the mark plate with the timing marks of the camshaft timing gear as shown.

b. Apply a light coat of engine oil to the bolt threads and bolts seating surface.

c. Align the knock pin of the camshaft with pinhole of the camshaft timing gear. Install the camshaft timing gear and the left camshaft timing exhaust gear with the No. 2 chain installed.

d. Hold the hexagonal portion of the camshaft with the wrench and tighten the two bolts to 74 ft. lbs. (100 Nm).

e. Remove the pin from the chain tensioner.

20. Install the chain vibration damper with the 2 bolts and tighten to 17 ft. lbs. (23 Nm).

21. Install the 2 chain vibration dampers.

22. Install the 2 timing gear set keys and crankshaft timing sprocket as shown in the illustration.

23. Install the idle sprocket assembly, as follows:

a. Apply a light coat of engine oil to the rotating surface of the No. 1 idle gear shaft.

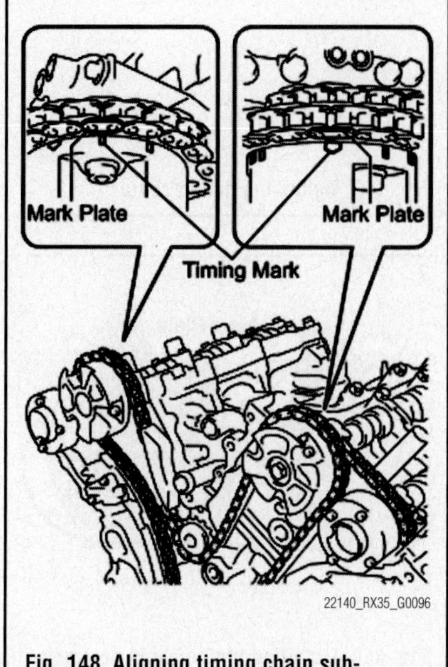

Fig. 148 Aligning timing chain sub-assembly

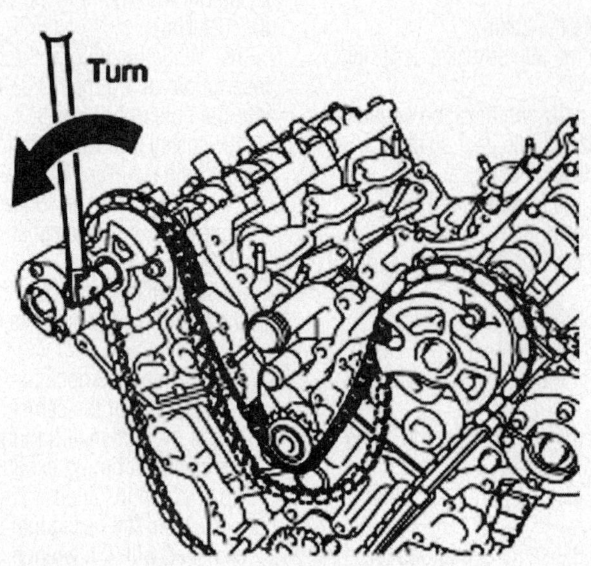

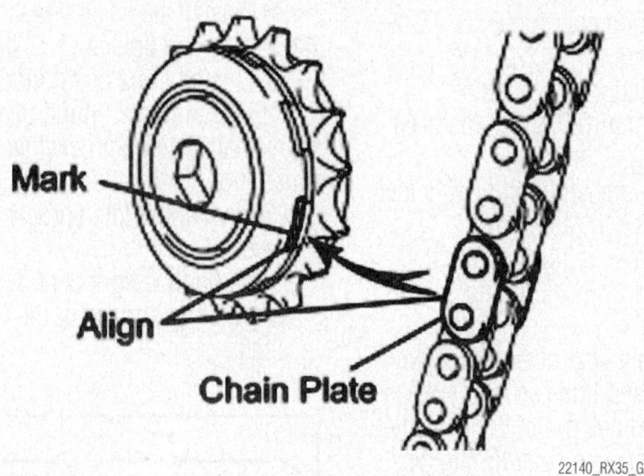

Fig. 149 Tightening timing chain

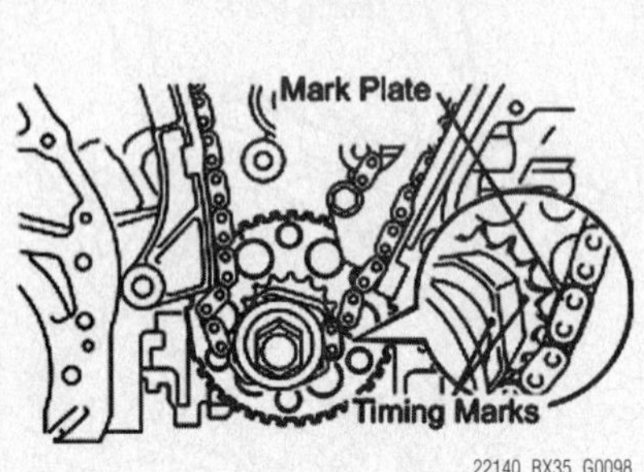

Fig. 150 Installing timing chain on crankshaft

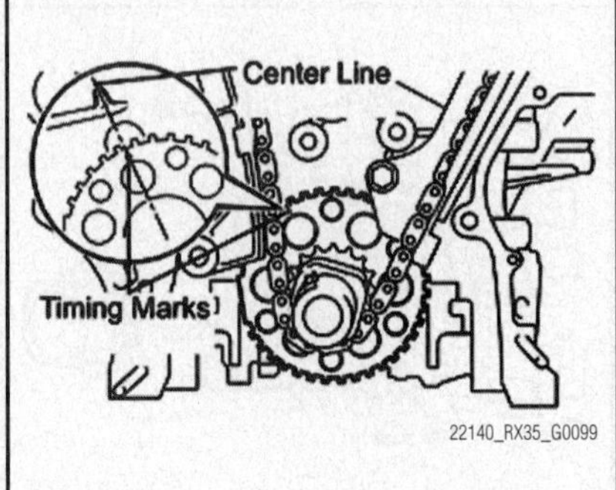

Fig. 151 Aligning timing chain on crankshaft

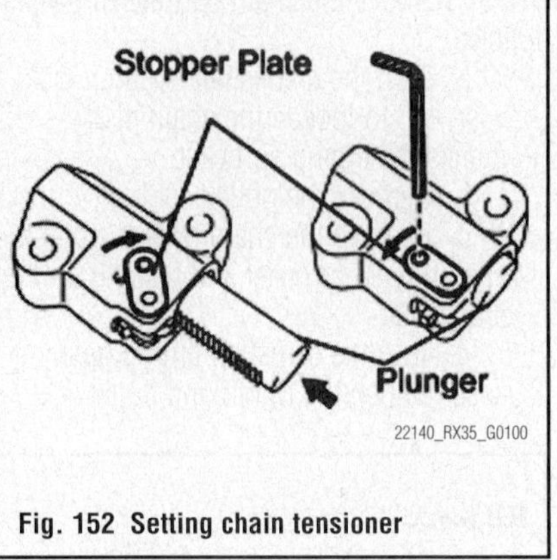

Fig. 152 Setting chain tensioner

b. Temporarily install the No. 1 idle gear shaft and idle sprocket with the No. 2 idle gear shaft while aligning the knock pin of the No. 1 idle gear with the knock pin groove of the cylinder block. Be careful of the idle gear direction.

c. Check that no foreign objects are on the idle gear shafts No. 1 and No. 2.

d. Using a 10 mm hexagon wrench, tighten the No. 2 idle gear shaft and tighten to 44 ft. lbs. (60 Nm).

e. After installing the idle sprocket assembly, check that the idle sprocket turns smoothly.

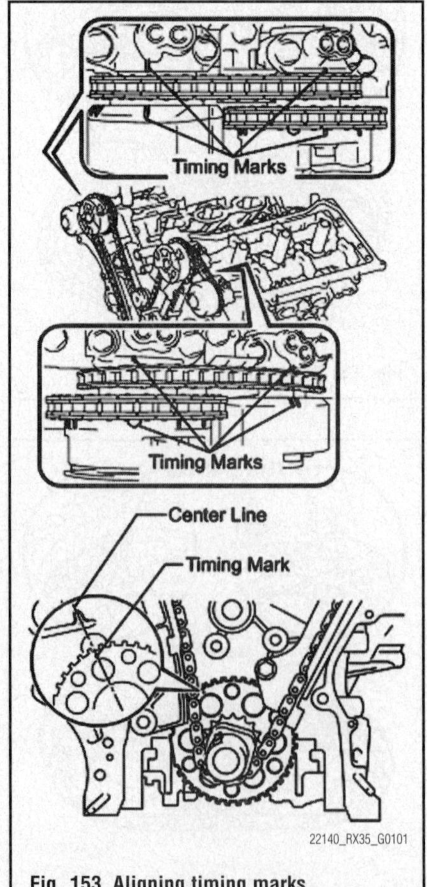

Fig. 153 Aligning timing marks

24. Install chain sub-assembly, as follows:

a. Align the mark plate and timing marks as shown in the illustration and install the chain. The camshaft mark plates are orange.

b. Do not pass the chain over the crankshaft, just temporarily place it on the crankshaft.

c. Turn the camshaft timing gear assembly on the right bank counterclockwise to tighten the chain between the banks.

➡When the idle sprocket assembly is reused, align the timing chain plate with the mark where the plate has been in order to tighten the chain between the banks.

d. Align the mark plate and timing marks as shown in the illustration and install the chain onto the crankshaft timing sprocket. The crankshaft to mark plate is yellow.

e. Temporarily tighten the pulley set bolt.

f. Turn the crankshaft clockwise to set it to the right-hand block bore more centerline. (TDC/compression).

25. Install the chain tensioner slipper.

26. Install the No. 1 chain tensioner assembly, as follows:

a. Move the stopper plate upward to release the lock, and push the plunger deep into the tensioner.

b. Move the stopper plate downward to set the lock, and insert a hexagon wrench into the hole of the stopper plate.

c. Install the chain tensioner with the 2 bolts and tighten to 7 ft. lbs. (10 Nm).

d. Remove the hexagon wrench from the chain tensioner. Check that each timing mark is aligned with the crankshaft at TDC/compression.

e. Remove the pulley set bolt.

27. Install timing chain cover and seal.

VALVE COVERS

REMOVAL & INSTALLATION

See Figures 154 through 161.

1. Remove the engine assembly.
2. Install on engine stand.
3. Remove the oil filler cap and gasket.
4. Remove the spark plugs and ignition coil assembly.
5. Remove the drain plug and gasket.
6. Remove the ventilation valve.
7. Remove the camshaft position sensor.
8. Remove the camshaft timing oil control valve assembly.
9. Remove the crankshaft position sensor.
10. Remove the No. 1 oil pipe.
11. Remove the oil pipe.
12. Remove the cylinder block water drain cock sub-assembly.
13. Remove the oil filter element.
14. Remove the crankshaft pulley.
15. Remove the oil cooler assembly, if applicable.
16. Remove the No. 1 oil cooler bracket, if applicable.

Fig. 154 Cylinder head cover sub-assembly (for Bank 1)

Fig. 155 Cylinder head cover sub-assembly (for Bank 2)

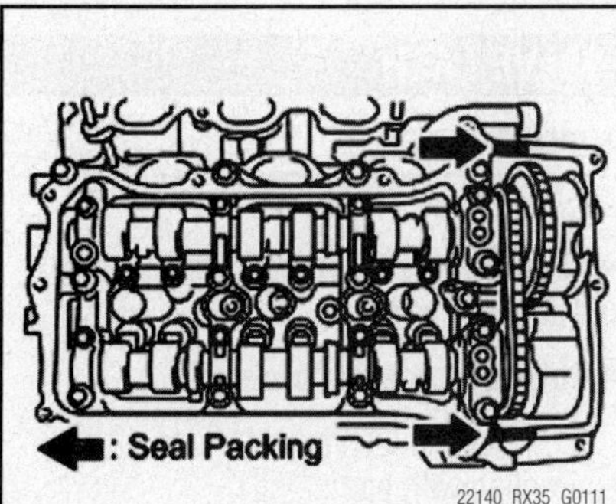

Fig. 156 Sealant application

17. Remove the left hand No. 1 front engine mounting bracket.

18. Remove the water inlet housing.

19. Remove the water outlet.

20. Remove the cylinder head cover sub-assembly (for Bank 1), as follows:

a. Remove the 12 bolts, seal washer, head cover and gasket.

b. Remove the 3 gaskets.

21. Remove the cylinder head cover sub-assembly (for Bank 2), as follows:

a. Remove the 12 bolts, seal washer, head cover and gasket. The baffle plate is located on the back of the portion shown in the illustration. Do not damage the baffle plate when removing the head cover.

b. Remove the 3 gaskets.

To install:

22. Install the cylinder head cover sub-assembly (for Bank 1), as follows:

a. Apply seal packing as shown in the illustration. Seal packing: Toyota Genuine Seal Packing Black, Three Bond 1207B or equivalent.

➡Remove any oil from the contact surface. Install the head cover within 3 minutes after applying seal packing. Do not start the engine for at least 2 hours after installing.

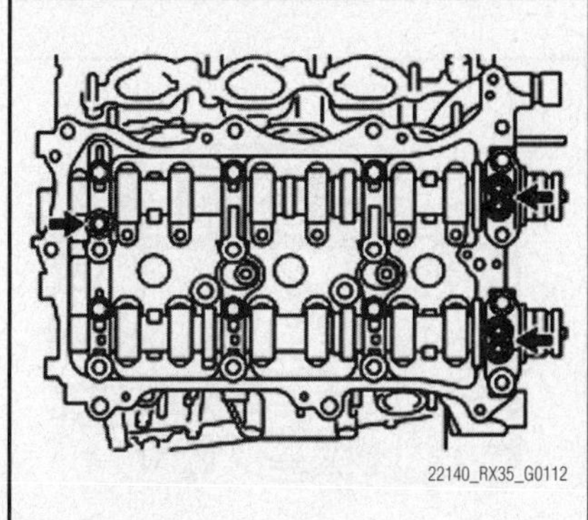

Fig. 157 Install 3 new gaskets

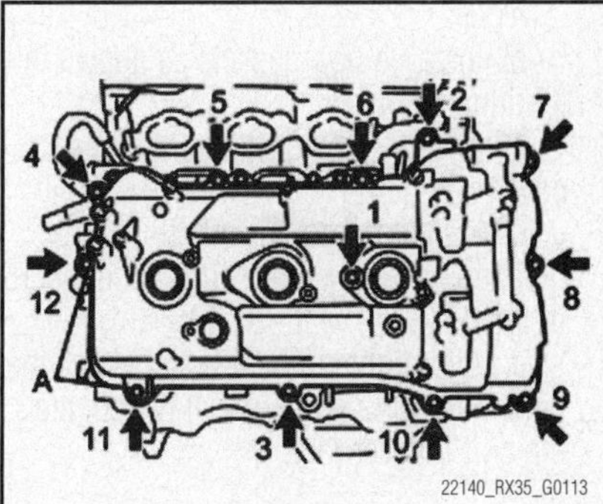

Fig. 158 Tightening torque

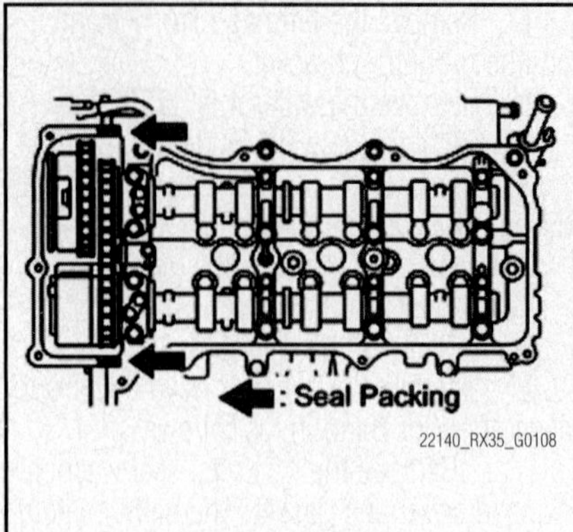

Fig. 159 Sealant application

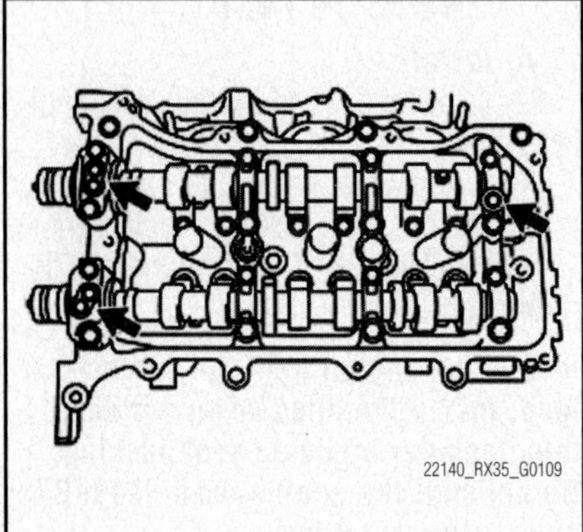

Fig. 160 Install 3 new gaskets

Fig. 161 Tightening torque

b. Install 3 new gaskets as shown in the illustration.

c. Install a new gasket to the head cover.

d. Install a head cover with the 12 bolts and a new washer and tighten to 15 ft. lbs. (21 Nm), 7 ft. lbs. (10 Nm).

e. After tightening all bolts, check the tightening torque of 1 and 11. Retighten the bolt if necessary.

23. Install the cylinder head cover sub-assembly (for Bank 2), as follows:

a. Apply seal packing as shown in the illustration. Seal packing: Toyota Genuine Seal Packing Black, Three Bond 1207B or equivalent.

➡Remove any oil from the contact surface. Install the head cover within 3 minutes after applying seal packing. Do not start the engine for at least 2 hours after installing.

b. Install 3 new gaskets as shown in the illustration.

c. Install a new gasket to the head cover.

d. Install a head cover with the 12 bolts and a new washer and tighten to 15 ft. lbs. (21 Nm), 7 ft. lbs. (10 Nm).

e. After tightening all bolts, check the tightening torque of 1 and 10. Retighten the bolt if necessary.

24. Install water outlet.

25. Install the No. 1 oil cooler bracket, if applicable.

26. Install the oil cooler assembly, if applicable.

27. Install the crankshaft pulley.

28. Install the oil filter element.

29. Install the cylinder block water drain cock sub-assembly.

30. Install the No. 1 oil pipe.

31. Install the oil pipe.

32. Install the crankshaft position sensor with the bolt and tighten to 7 ft. lbs. (10 Nm).

33. Install the camshaft timing oil control valve assembly.

34. Install the camshaft position sensor.

35. Install the ventilation valve sub-assembly.

36. Install the spark plugs and the ignition coil assembly.

37. Install the oil filler cap sub-assembly.

38. Remove the engine stand.

39. Install the engine assembly.

VALVE LASH

ADJUSTMENT

See Figure 162.

➡Keep the lash adjuster free of dirt and foreign objects.

➡Only use clean engine oil.

1. Place the lash adjuster into a container filled with engine oil.

2. Insert the SST's (SST: 09276-75010) tip into the lash adjuster's plunger and use the tip to press down on the check ball inside the plunger.

3. Squeeze the SST and lash adjuster together to move the plunger up and down 5 to 6 times.

4. Check the movement of the plunger and bleed the air. OK: Plunger moves up and down.

➡When bleeding air from the high-pressure chamber, make sure that the tip of the SST is actually pressing the check ball as shown in the illustration. If the check ball is not pressed, air will not bleed.

5. After bleeding the air, remove the SST. Then, try to press the plunger quickly and firmly with a finger. OK: Plunger is very difficult to move. If the result is not as specified, replace the lash adjuster.

6. Install the lash adjusters.

➡Install the lash adjuster to the same place where it was removed from.

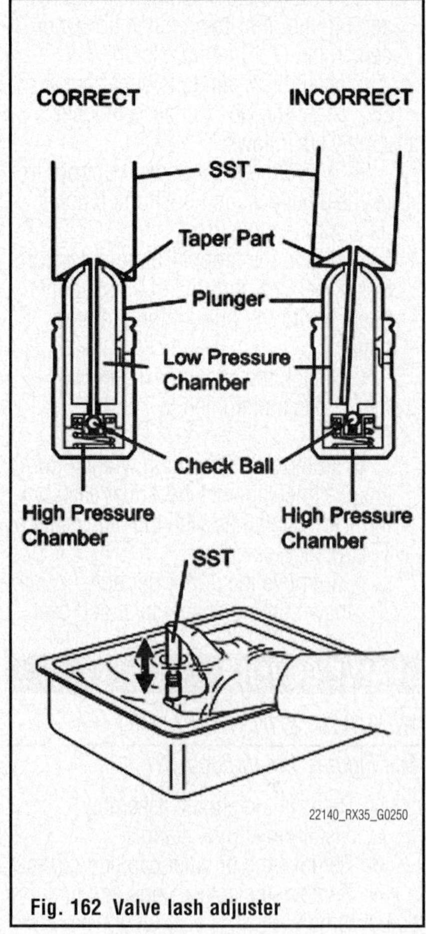

Fig. 162 Valve lash adjuster

ENGINE PERFORMANCE & EMISSION CONTROL

ACCELERATOR PEDAL POSITION (APP) SENSOR

LOCATION

See Figure 163.

REMOVAL & INSTALLATION

See Figure 164.

1. Remove left center floor carpet cover.
2. Disconnect the accelerator pedal position sensor connector.
3. Remove the 2 nuts and accelerator pedal rod.

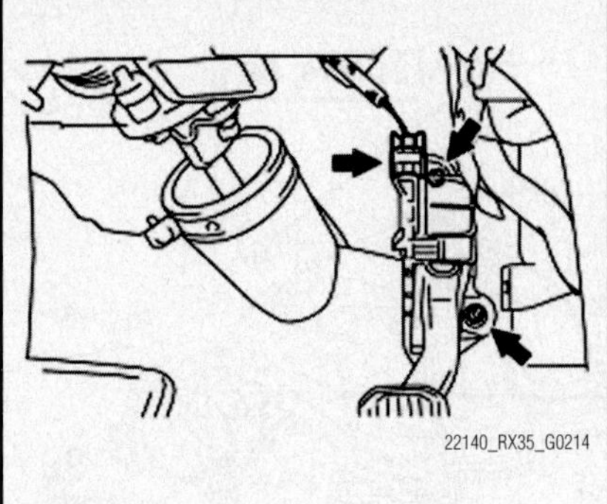

Fig. 164 Accelerator pedal position sensor

➡Avoid physical shock to the accelerator pedal assembly.

➡Do not disassemble the accelerator pedal assembly.

To install:

4. Installation is the reverse of the removal procedure. Tighten the accelerator pedal rod nuts to 43 inch lbs. (4.9 Nm).

CAMSHAFT POSITION (CMP) SENSOR

LOCATION

See Figure 165.

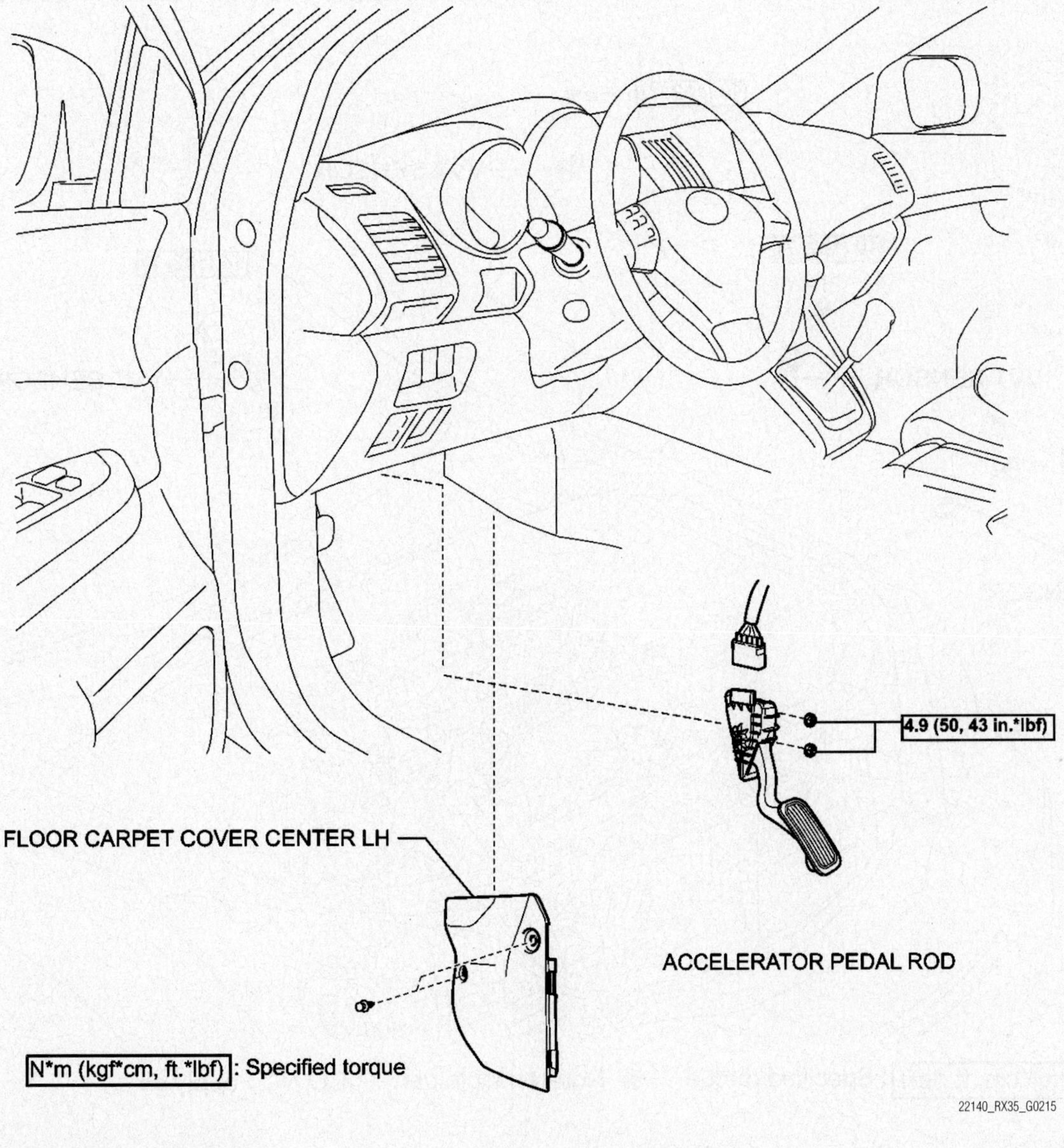

4.9 (50, 43 in.*lbf)

FLOOR CARPET COVER CENTER LH

ACCELERATOR PEDAL ROD

N*m (kgf*cm, ft.*lbf) : Specified torque

22140_RX35_G0215

Fig. 163 Accelerator pedal position sensor

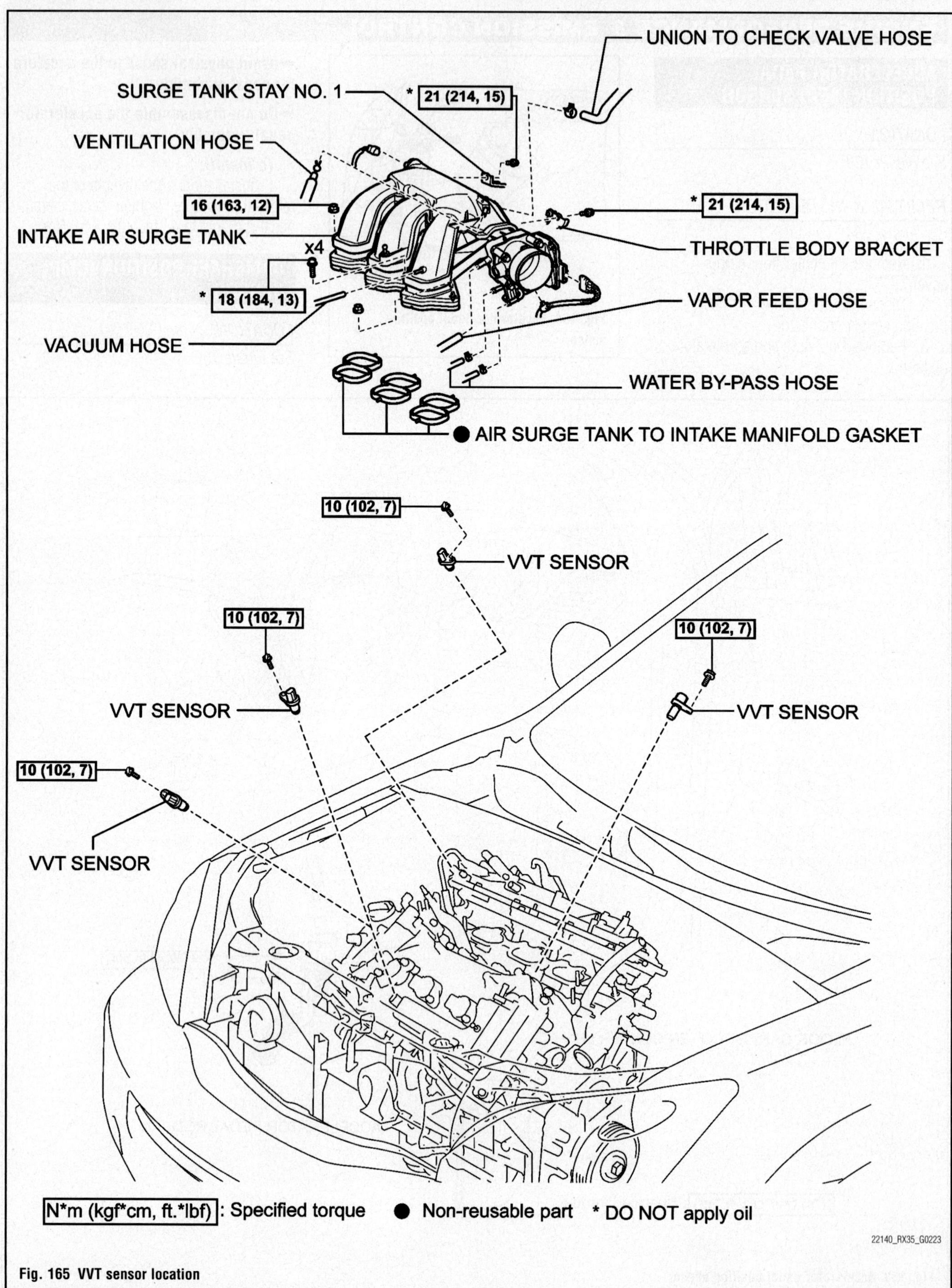

Fig. 165 VVT sensor location

REMOVAL & INSTALLATION

See Figures 166 through 169.

1. Remove the windshield wiper link assembly.
2. Remove the cowl top panel outer sub-assembly.
3. Drain engine coolant.
4. Remove the No. 1 V-bank cover bracket.
5. Remove No. 2 air cleaner inlet.
6. Remove No. 1 air cleaner inlet.
7. Remove the air cleaner cap sub-assembly.
8. Remove the air cleaner case sub-assembly.
9. Remove the intake air surge tank assembly.
10. Remove the VVT (Camshaft Position) sensor (for Bank 1 Intake Side), as follows:
    a. Disconnect the VVT sensor connector.
    b. Remove the bolt and VVT sensor.
11. Remove the VVT sensor (for Bank 1 Exhaust Side), as follows:
    a. Disconnect the VVT sensor connector.
    b. Remove the bolt and VVT sensor.
12. Remove the VVT sensor (for Bank 2 Intake Side), as follows:
    a. Disconnect the VVT sensor connector.
    b. Remove the bolt and VVT sensor.
13. Remove the VVT sensor (for Bank 2 Exhaust Side), as follows:
    a. Disconnect the VVT sensor connector.
    b. Remove the bolt and VVT sensor.

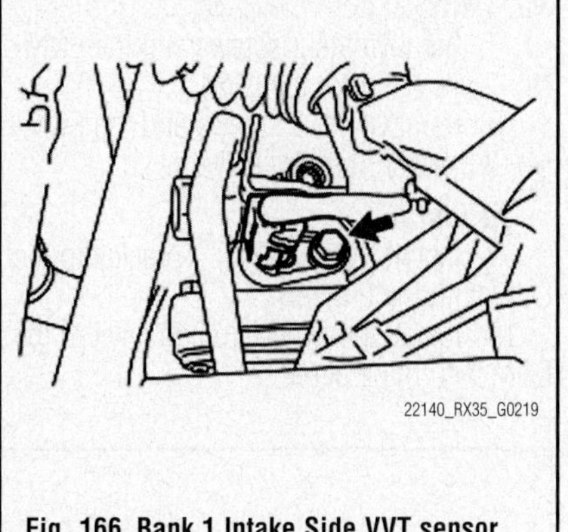

Fig. 166 Bank 1 Intake Side VVT sensor

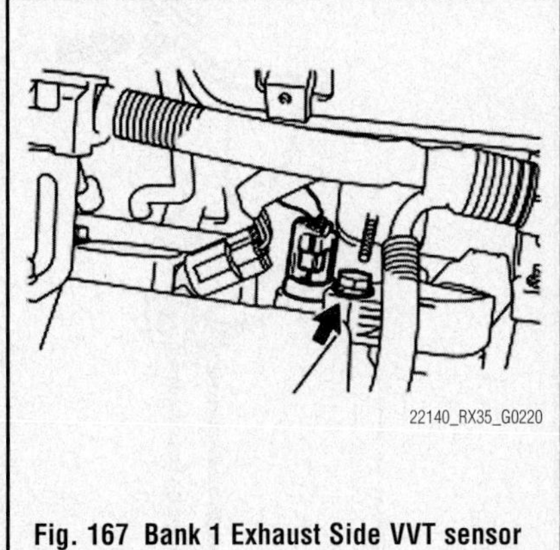

Fig. 167 Bank 1 Exhaust Side VVT sensor

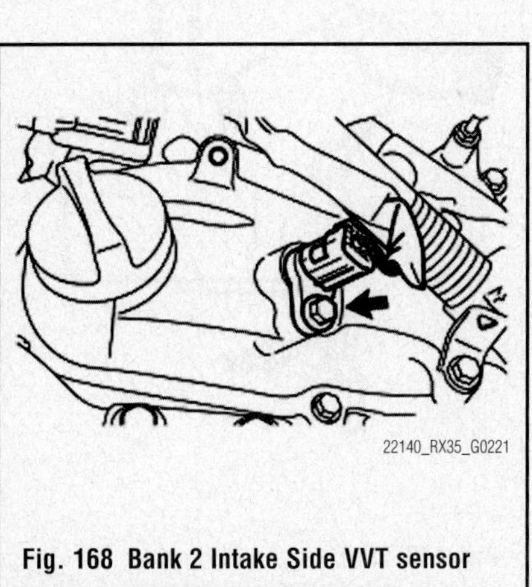

Fig. 168 Bank 2 Intake Side VVT sensor

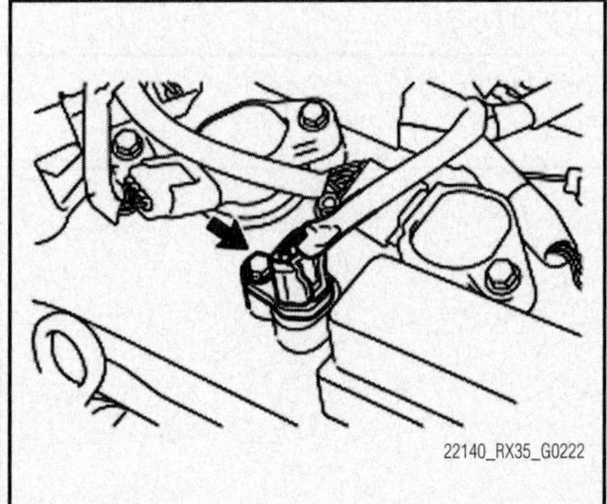

Fig. 169 Bank 2 Exhaust Side VVT sensor

CRANKSHAFT POSITION (CKP) SENSOR

LOCATION

See Figure 170.

REMOVAL & INSTALLATION

See Figure 171.

1. Remove alternator assembly.
2. Remove compressor and magnetic clutch.
3. Disconnect the crankshaft position sensor connector.
4. Remove the bolt, and then remove the crankshaft position sensor.

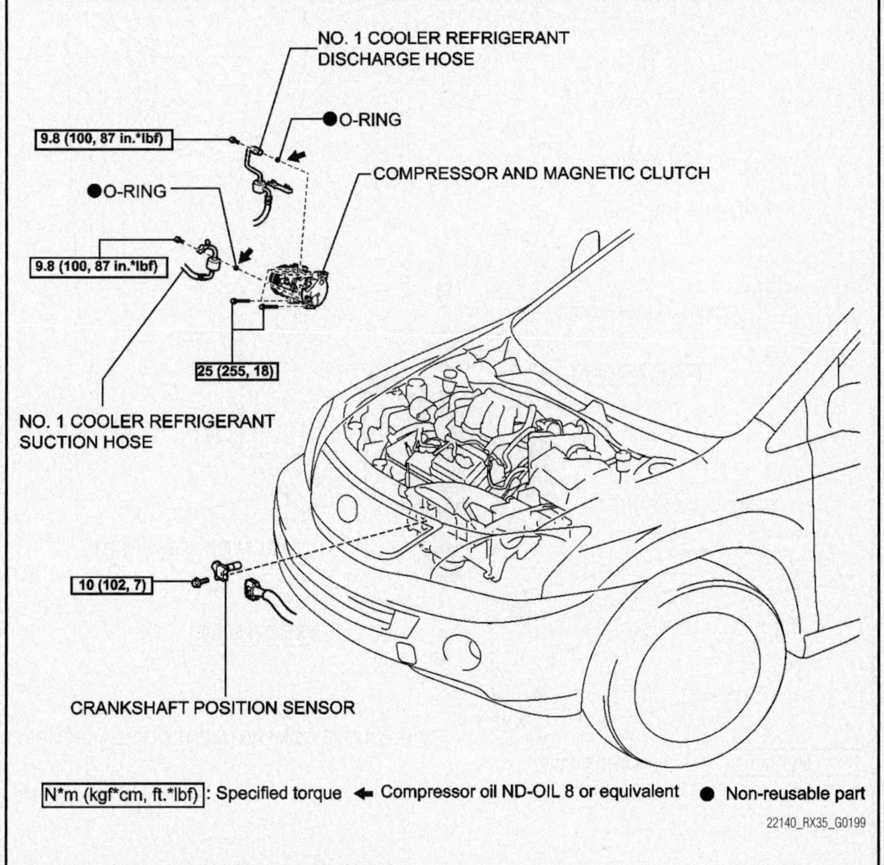

Fig. 170 Crankshaft position sensor location

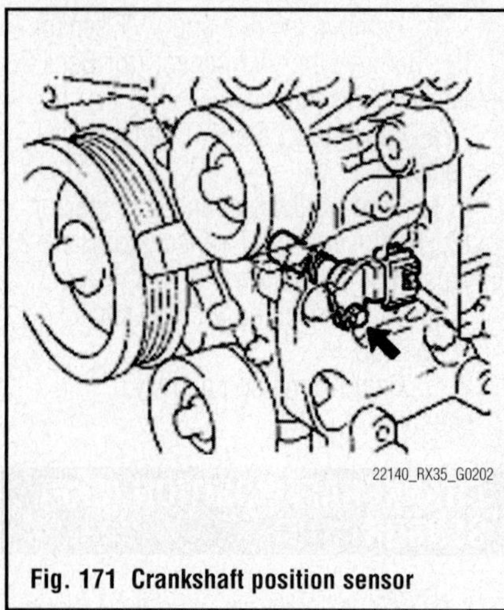

Fig. 171 Crankshaft position sensor

To install:

5. Apply a light coat of engine oil to the O-ring on the crankshaft position sensor.
6. Install the crankshaft position sensor with the bolt and tighten to 7 ft. lbs. (10 Nm).
7. Connect the crankshaft position sensor connector.
8. Install compressor and magnetic clutch.
9. Install alternator assembly.

ELECTRONIC CONTROL MODULE (ECM)

LOCATION

See Figure 172.

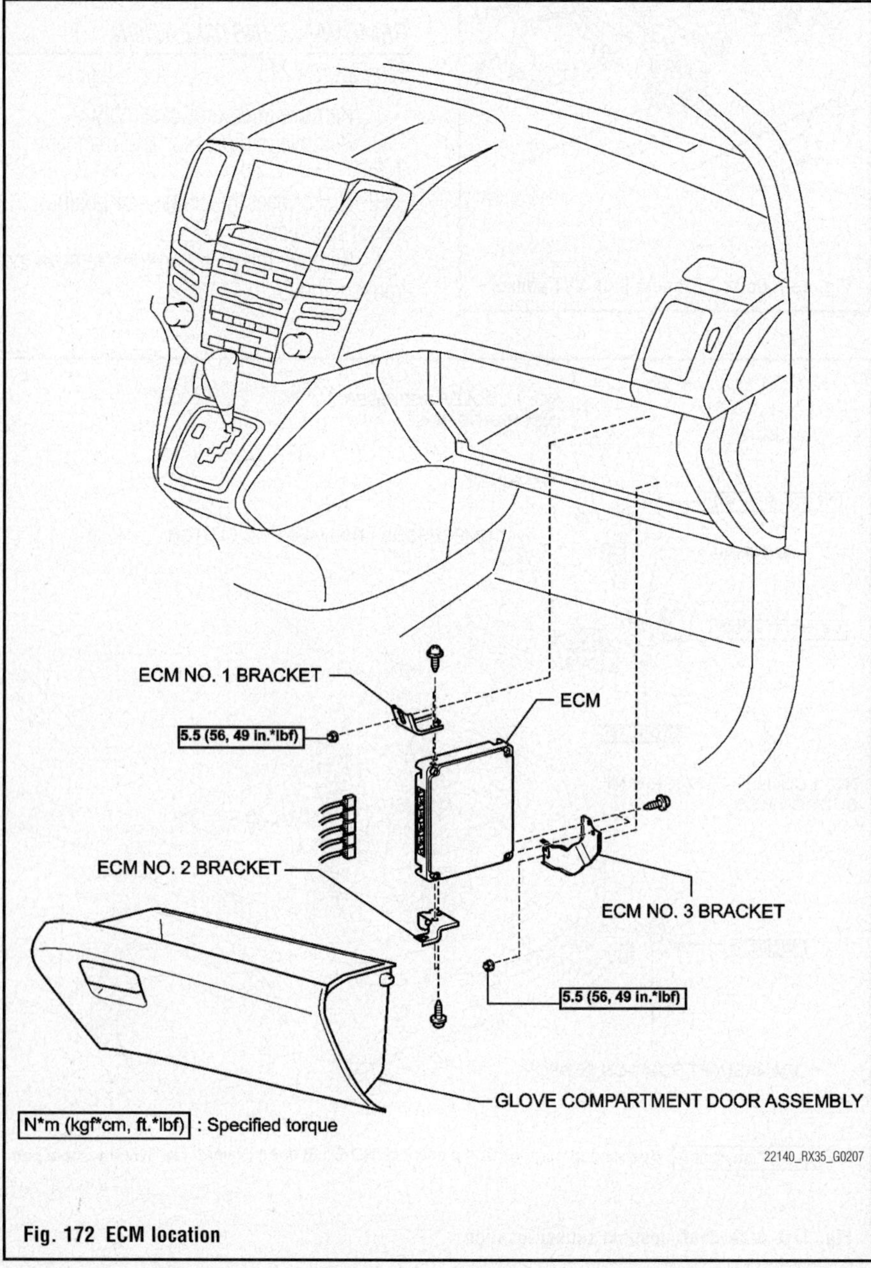

Fig. 172 ECM location

OPERATION

The ECM controls all engine computer related functions.

REMOVAL & INSTALLATION

See Figure 173.

1. Disconnect the negative battery cable.
2. Remove glove compartment door assembly.
3. Disconnect the 5 ECM connectors.
4. Disconnect the wire harness clamp.
5. Remove the 2 nuts and ECM.
6. Remove the 2 screws and the ECM No. 1 bracket from the ECM.
7. Remove the 2 screws and the ECM No. 2 bracket from the ECM.
8. Remove the 2 screws and the ECM No. 3 bracket from the ECM.

To install:

9. Install the ECM No. 1 bracket to the ECM with the 2 screws.
10. Install the ECM No. 2 bracket to the ECM with the 2 screws.

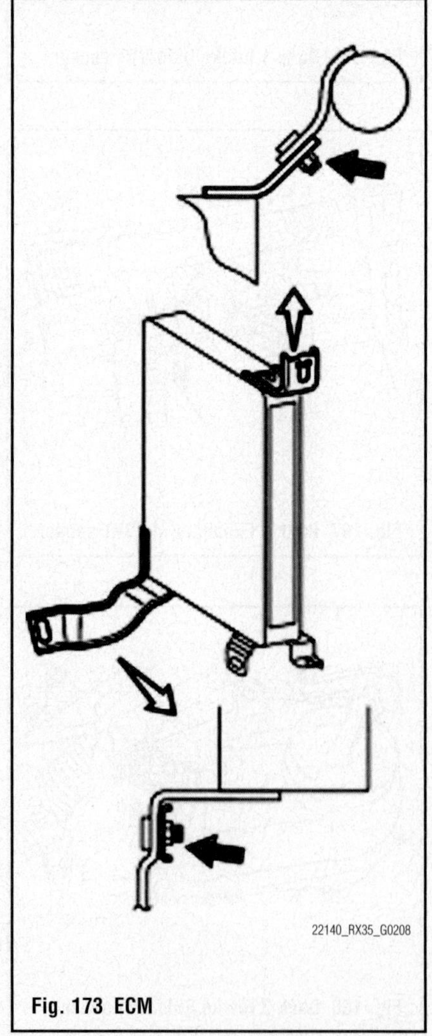

Fig. 173 ECM

11. Install the ECM No. 3 bracket to the ECM with the 2 screws.

12. Install the ECM with the 2 nuts and tighten to 49 inch lbs. (5.5 Nm).

13. Install the wire harness clamp.

14. Connect the 5 ECM connectors.

15. Install glove compartment door assembly.

16. Connect the negative battery cable.

➡After replacing the ECM on vehicles with a dynamic laser cruise control system, it is necessary to initialize the ECM so that the ECM can recognize the dynamic laser cruise control system.

17. Initialize the ECM. Be sure to perform the following procedures after replacing the ECM:

a. Turn the ignition switch to the ON position.

b. Turn the cruise main switch on.

c. With the brake pedal depressed, push the cruise control main switch to RES/ACC 3 times within 3 seconds. Check that the buzzer sounds at this time.

➡Do not turn the headlight dimmer switch on at this time because the optical axis automatic adjustment mode has already started, which may lead to an incorrect optical axis setting. If the headlight dimmer switch is turned on by mistake, readjust the optical axis.

d. Turn the ignition switch off.

ENGINE COOLANT TEMPERATURE (ECT) SENSOR

LOCATION

See Figure 174.

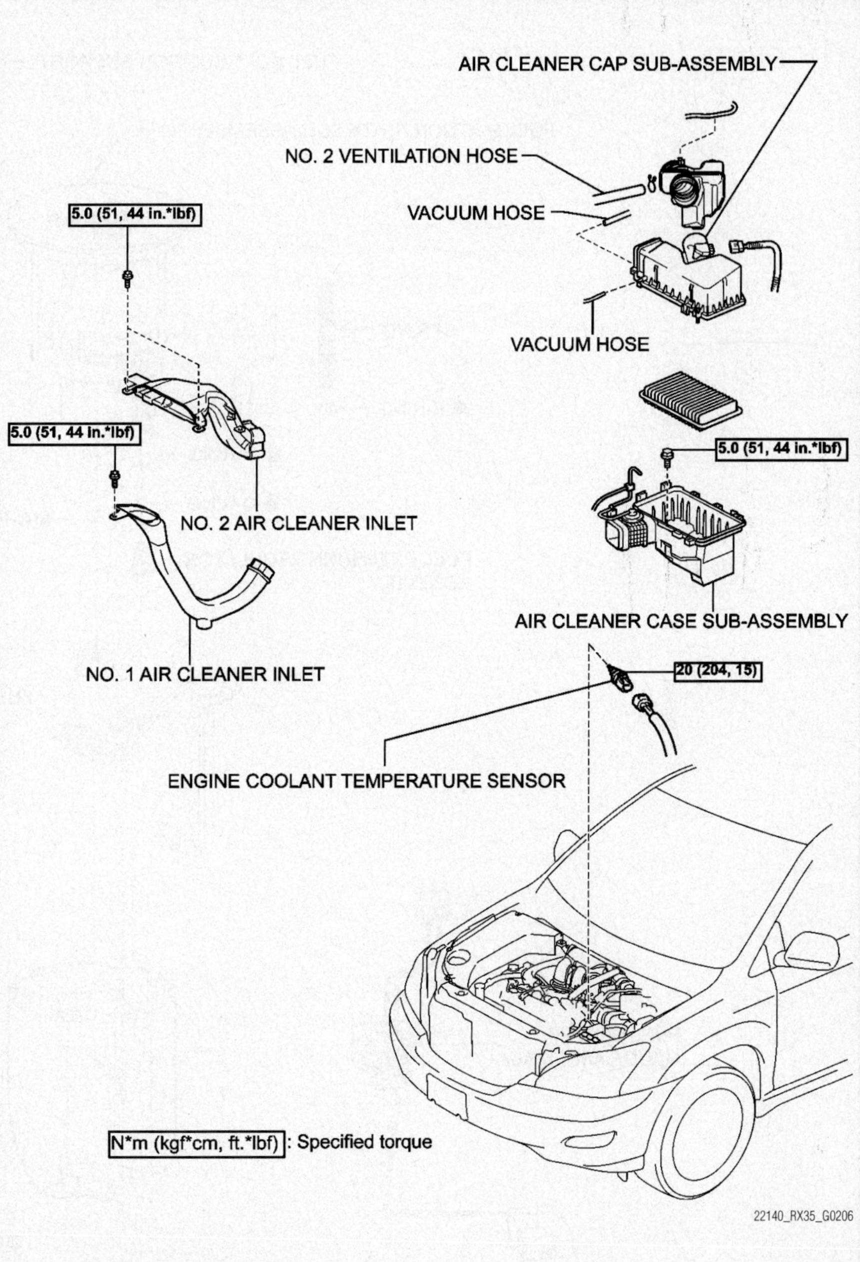

REMOVAL & INSTALLATION

See Figure 175.

1. Drain engine coolant.
2. Remove V-bank cover sub-assembly.
3. Remove No. 2 air cleaner inlet.
4. Remove No. 1 air cleaner inlet.
5. Remove air cleaner cap sub-assembly.
6. Remove air cleaner case sub-assembly.
7. Remove the engine coolant temperature sensor connector.
8. Remove the engine coolant temperature sensor.

To install:

9. Installation is the reverse of the removal procedure. Torque the engine coolant temperature sensor to 15 ft. lbs. (20Nm).

Fig. 174 Coolant temperature sensor location

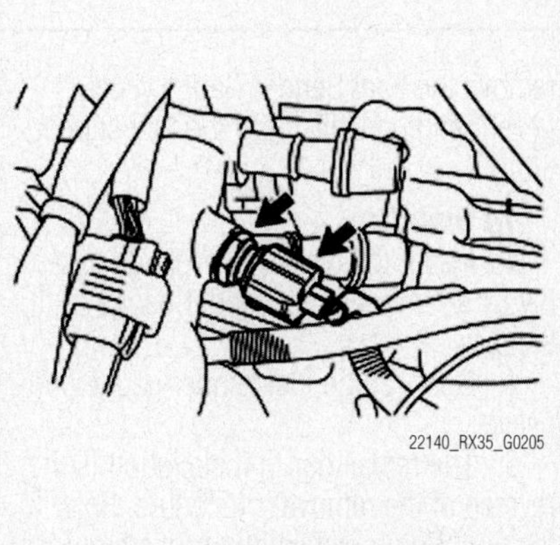

Fig. 175 Coolant temperature sensor

FUEL LEVEL SENDING UNIT

LOCATION

See Figure 176.

REMOVAL & INSTALLATION

See Figure 177.

1. Before servicing the vehicle, refer to the precautions section.

2. Remove fuel suction tube assembly with pump and gauge. Refer to the Fuel Pump removal procedure for instructions.

3. Disconnect the connector and

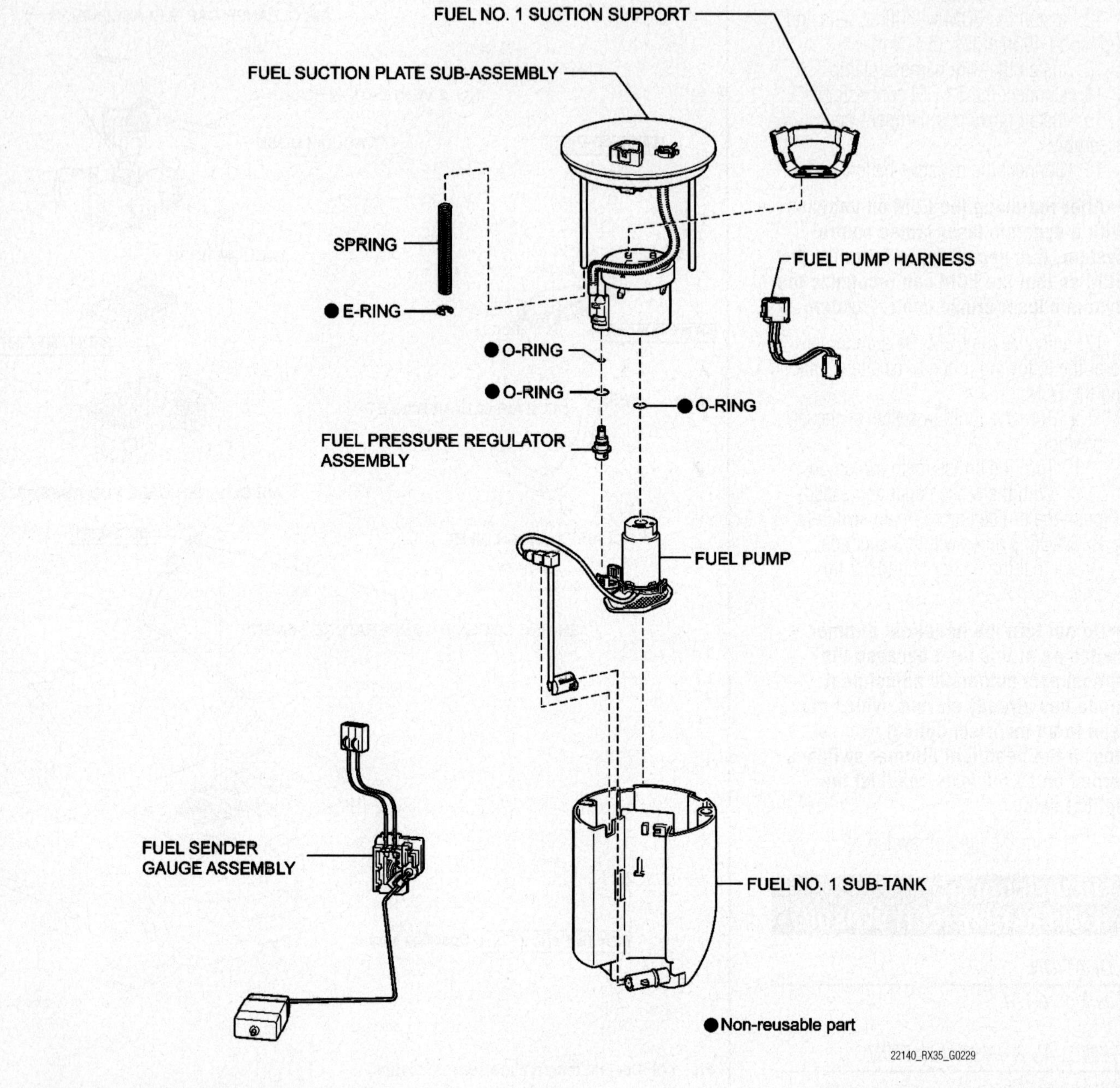

FUEL NO. 1 SUCTION SUPPORT
FUEL SUCTION PLATE SUB-ASSEMBLY
SPRING
● E-RING
● O-RING
● O-RING
● O-RING
FUEL PUMP HARNESS
FUEL PRESSURE REGULATOR ASSEMBLY
FUEL PUMP
FUEL SENDER GAUGE ASSEMBLY
FUEL NO. 1 SUB-TANK
● Non-reusable part
22140_RX35_G0229

Fig. 176 Fuel sender gauge location

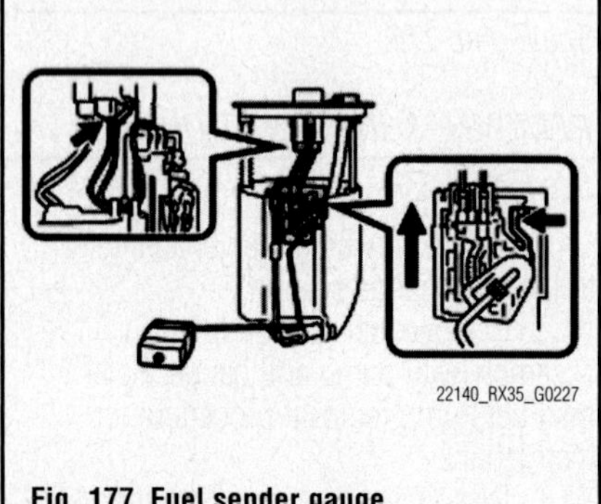

Fig. 177 Fuel sender gauge

remove the Fuel Sender Gauge (Fuel Level Sending Unit) from the fuel suction tube.

To install:

4. Install the fuel sender gauge assembly by sliding the fuel sender gauge to fit the claw.
5. Connect the fuel sender gauge connector.
6. The remainder of installation is the reverse of the removal procedure. Refer to the Fuel Pump installation procedure for instructions.

HEATED OXYGEN (HO2S) SENSOR

LOCATION

See Figure 178.

REMOVAL & INSTALLATION

See Figure 179.

1. Remove the oxygen sensor (for Bank 1):
   a. Disconnect the oxygen sensor connector and 2 clamps.

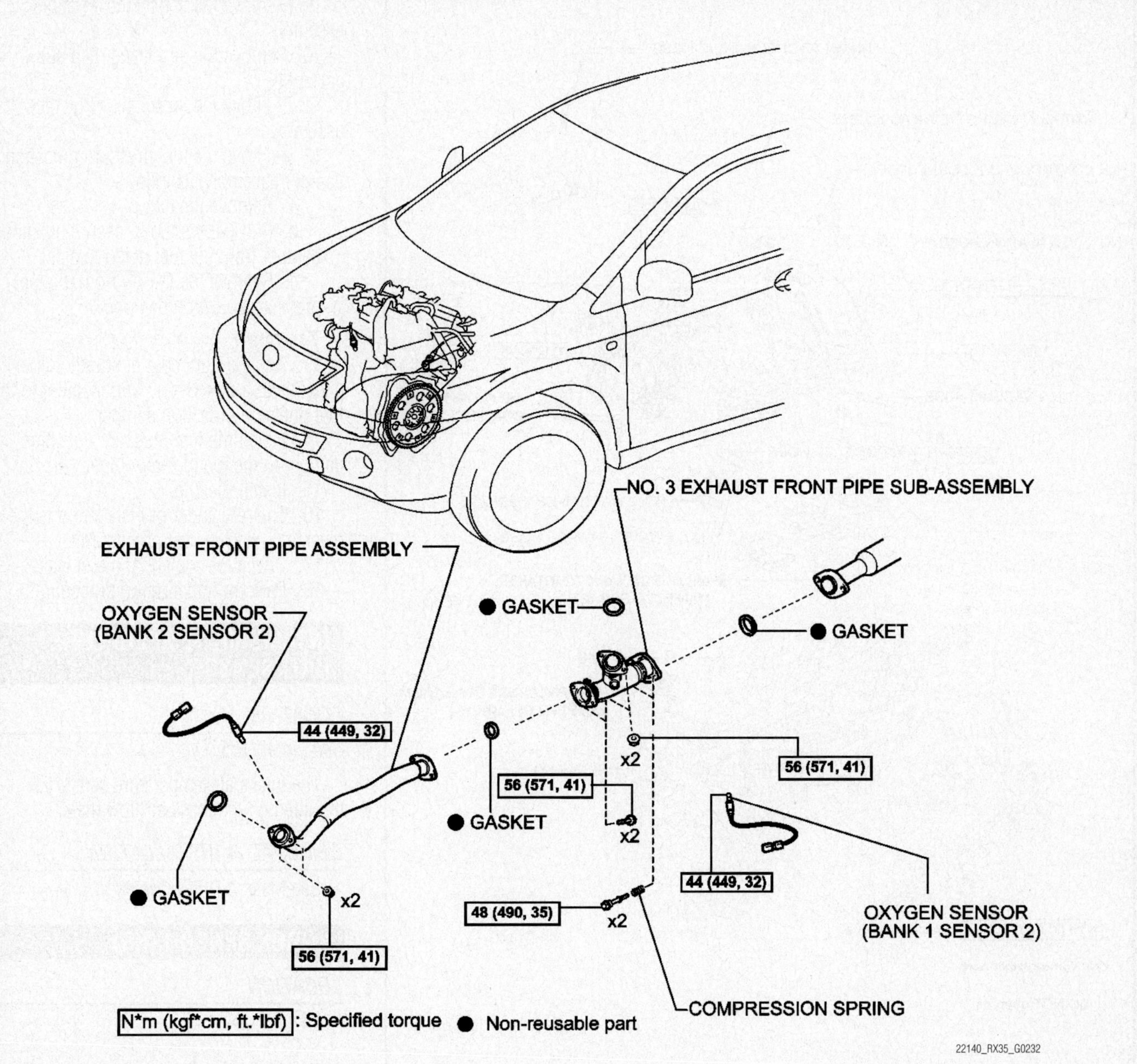

Fig. 178 Heated Oxygen Sensor location

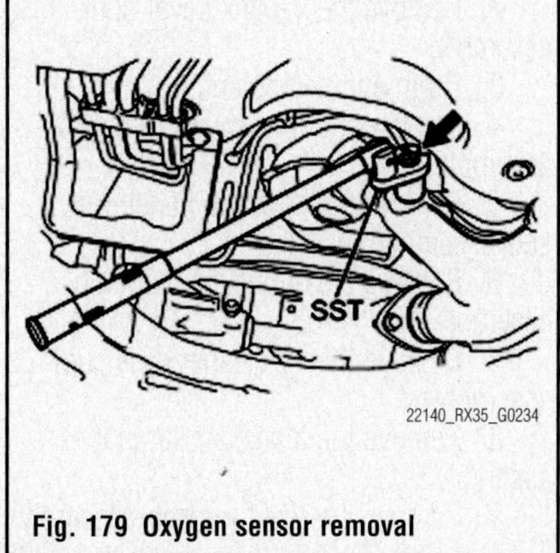

Fig. 179 Oxygen sensor removal

b. Using SST (SST: 09224-00010) or equivalent, remove the oxygen sensor from the No. 3 exhaust front pipe sub-assembly.

➡**Do not damage the oxygen sensor.**

2. Remove the No. 3 exhaust front pipe sub-assembly (for Bank 2):

a. Remove the 4 bolts, 2 nuts, 2 compression springs and No. 3 exhaust front pipe sub-assembly.

3. Remove the exhaust front pipe assembly (for Bank 2).

4. Remove the oxygen sensor (for Bank 2):

a. Using SST (SST: 09224-00010)) or equivalent, remove the oxygen sensor from the exhaust front pipe assembly.

➡**Do not damage the oxygen sensor.**

To install:

5. Installation is the reverse of the removal procedure. Tighten the oxygen sensors to 32 ft. lbs. (44 Nm).

INJECTION PRESSURE REGULATOR (IPR)

LOCATION

See Figure 180.

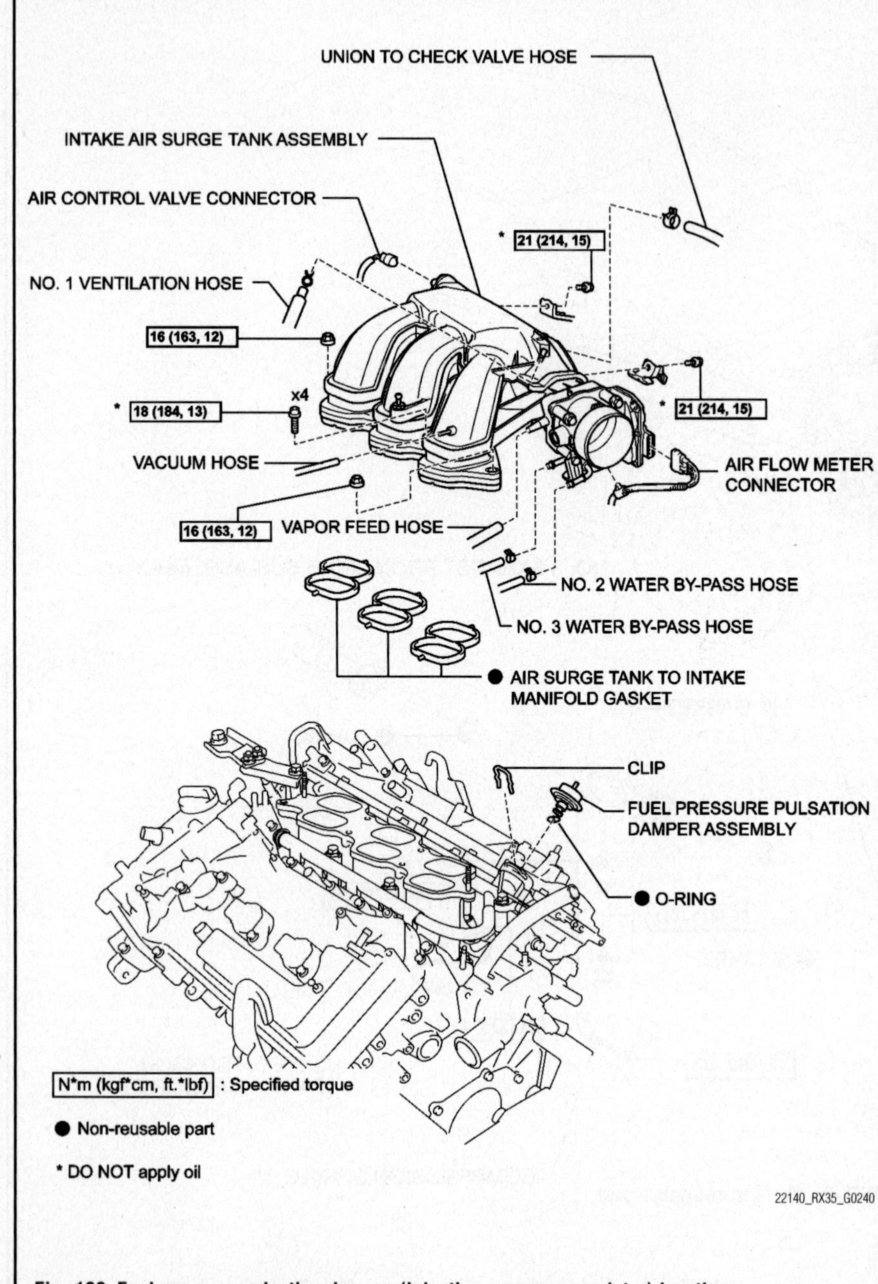

Fig. 180 Fuel pressure pulsation damper (injection pressure regulator) location

REMOVAL & INSTALLATION

See Figure 181.

1. Disconnect the negative battery cable.
2. Drain engine coolant.
3. Remove the engine room side cover,
4. Remove both front wiper arm and blade assemblies.
5. Remove the right and left front fender to cowl side seal.
6. Remove the cowl top ventilator louver sub-assembly.
7. Remove the windshield wiper link assembly.
8. Remove the cowl top panel outer sub-assembly.

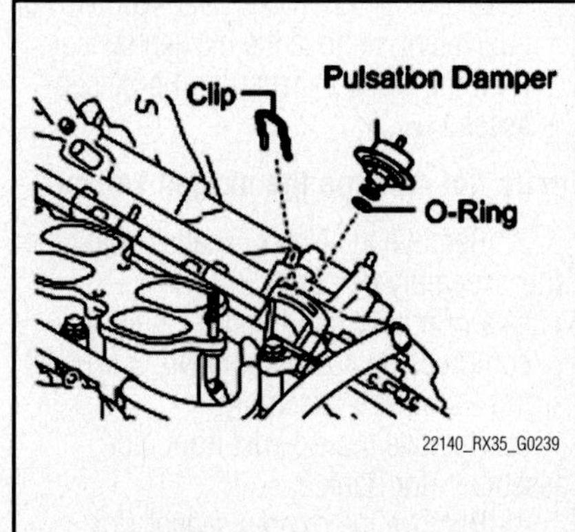

Fig. 181 Fuel pressure pulsation damper

9. Remove the V-bank cover sub-assembly.
10. Remove the air cleaner cap sub-assembly.
11. Remove the intake air surge tank assembly.
12. Remove the fuel pressure pulsation damper assembly, as follows:
    a. Remove the clip.
    b. Pull out the fuel pressure pulsation damper from the fuel delivery pipe.
    c. Remove the O-ring from the fuel pressure pulsation damper.

To install:

13. Apply a light coat of spindle oil or gasoline to a new O-ring and install it to the fuel pressure pulsation damper.
14. Install the fuel pressure pulsation damper to the fuel delivery pipe.
15. Install the clip.
16. The remainder of installation is the reverse of the removal procedure.
17. Check for coolant and fuel leaks.
18. Perform initialization procedure.

INTAKE AIR TEMPERATURE (IAT) SENSOR

LOCATION

See Figure 185.

The intake air temperature sensor is mounted on the mass air flow meter.

REMOVAL & INSTALLATION

See Mass Air Flow Meter.

KNOCK SENSOR (KS)

LOCATION

See Figure 182.

REMOVAL & INSTALLATION

See Figures 183 and 184.

1. Properly discharge the fuel system pressure.
2. Remove the V-bank cover sub-assembly.
3. Drain engine coolant.
4. Remove the windshield wiper link assembly.
5. Remove the cowl top panel outer sub-assembly.
6. Remove the air cleaner cap sub-assembly.
7. Remove the air cleaner case sub-assembly.
8. Remove the intake air surge tank assembly.
9. Remove the fuel tube sub-assembly.

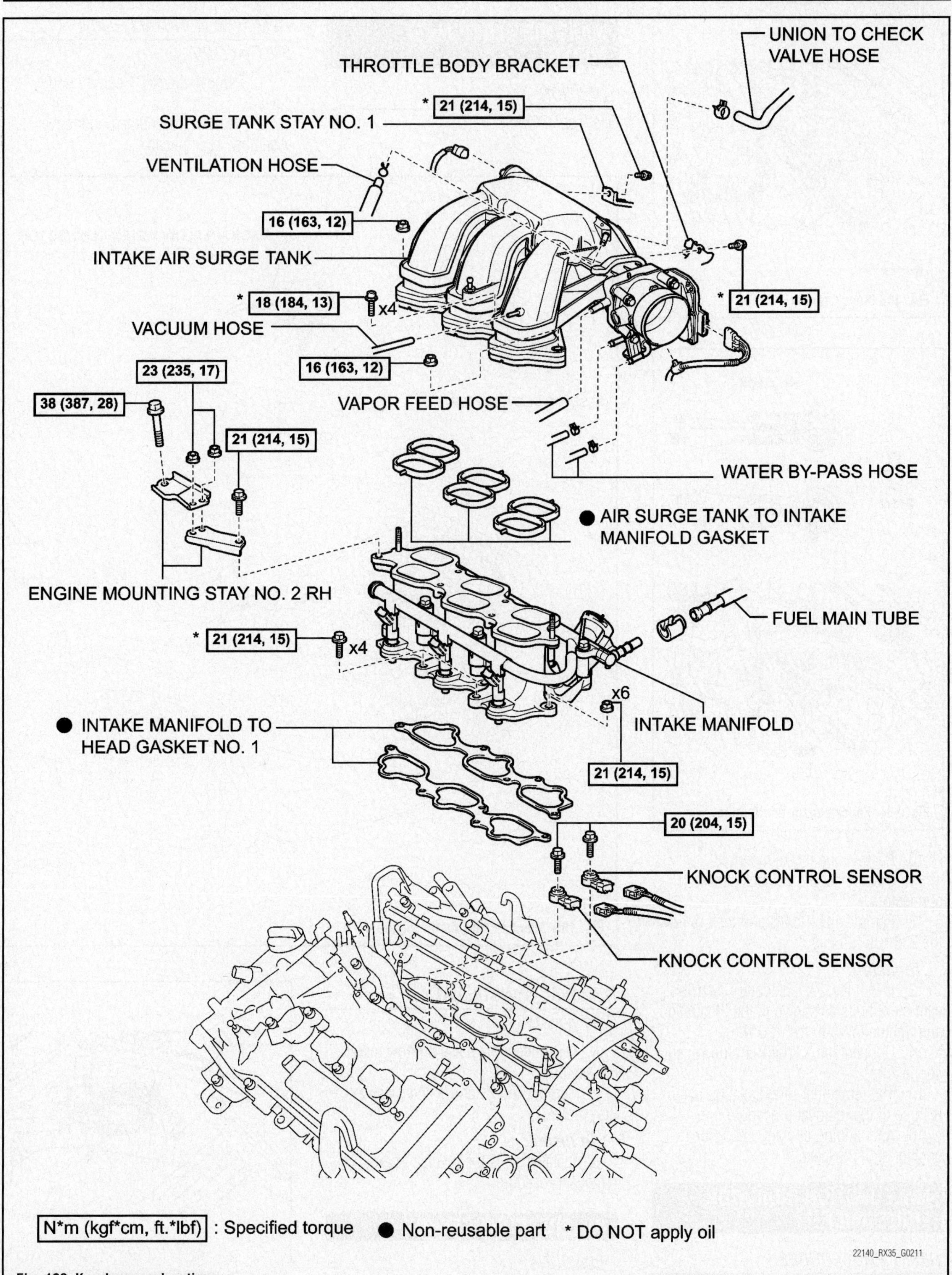

Fig. 182 Knock sensor location

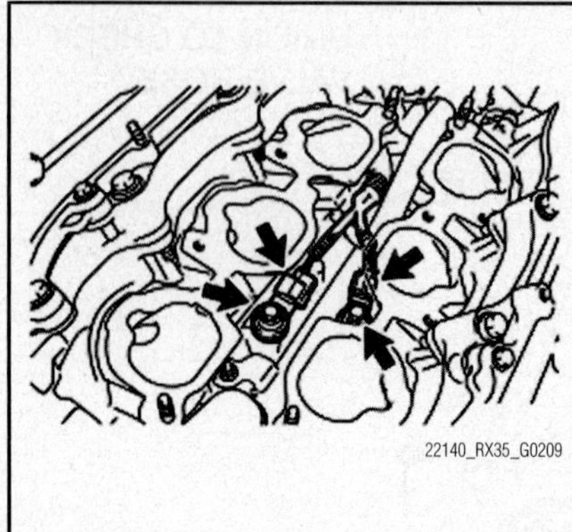

Fig. 183 Knock sensor

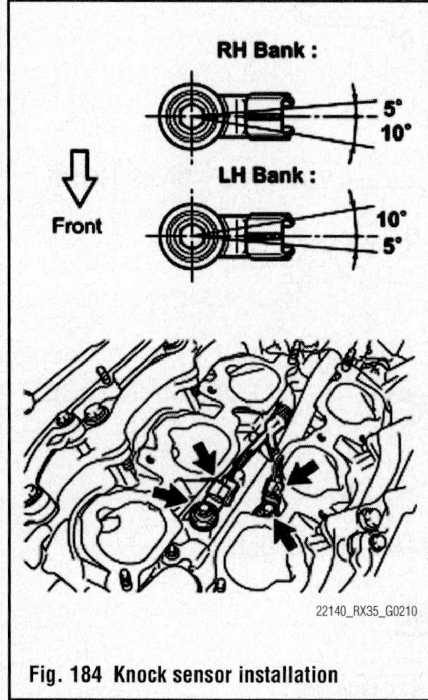

Fig. 184 Knock sensor installation

10. Remove the intake manifold.
11. Disconnect the 2 knock sensor connectors.
12. Remove the 2 bolts and then remove the 2 knock sensors.

To install:

13. Install the 2 knock control sensors with the 2 bolts as shown in the illustration and tighten to 15 ft. lbs. (20 Nm).
14. Connect the 2 knock control sensor connectors.
15. The remainder of installation is the reverse of the removal procedure.
16. After adding coolant, check for coolant and fuel leaks.

MALFUNCTION INDICATOR LIGHT (MIL)

RESET PROCEDURES

Clear the DTC codes.

MASS AIR FLOW (MAF) METER

LOCATION

See Figure 185.

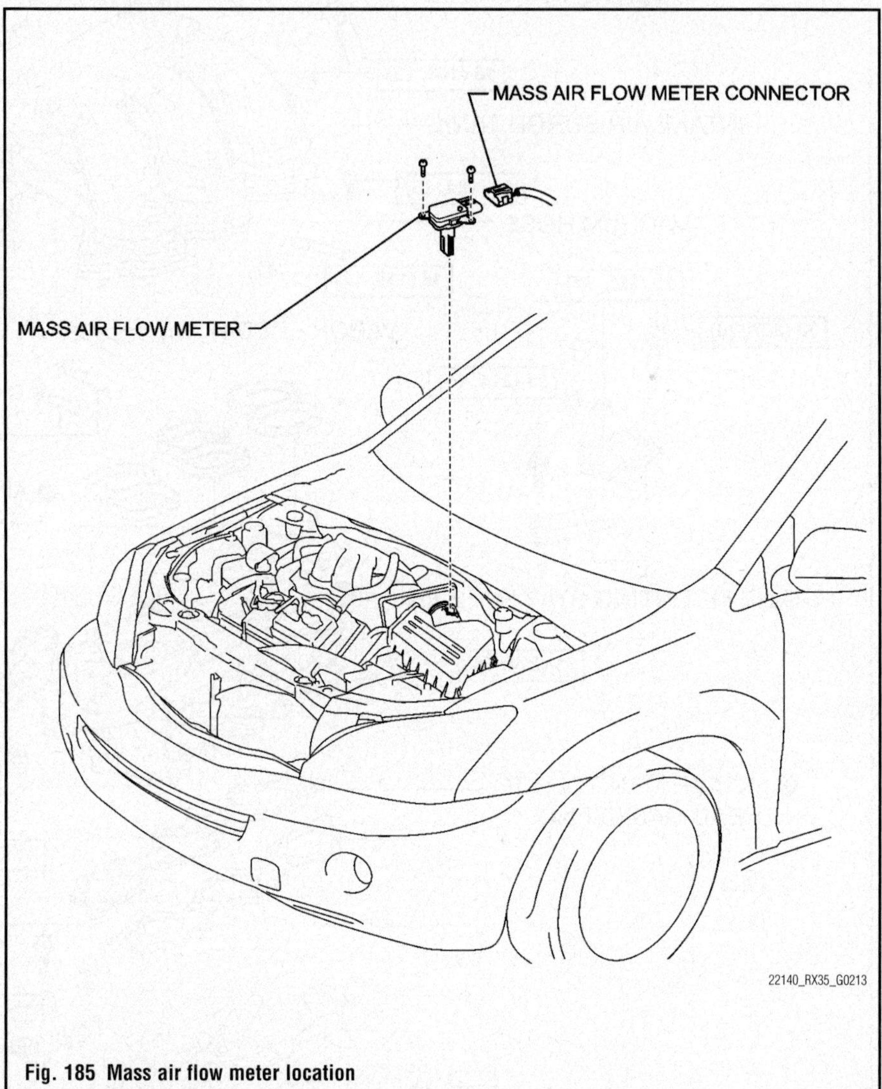

Fig. 185 Mass air flow meter location

REMOVAL & INSTALLATION

See Figure 186.

1. Disconnect the mass air flow meter connector.
2. Remove the 2 screws and mass air flow meter.

To install:

3. Installation is the reverse of the removal procedure.

OIL PRESSURE SENSOR

LOCATION

See Figure 187.

REMOVAL & INSTALLATION

See Figure 187.

1. Remove the No. 1 engine under cover.
2. Remove the engine under cover assembly, as applicable.

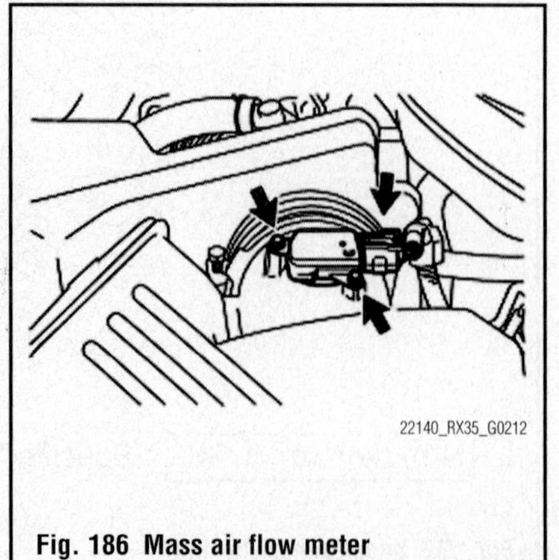

Fig. 186 Mass air flow meter

Fig. 187 Oil pressure switch

3. Remove the engine oil pressure switch assembly, as follows:
    a. Disconnect the oil pressure switch connector.
    b. Using a 24 mm deep socket wrench, remove the oil pressure switch.

To install:

4. Remove the oil pressure gauge.
5. Apply adhesive to 2 or 3 threads of the oil pressure switch. Adhesive: Toyota Genuine Adhesive 1344, Three Bond 1344 or equivalent.
6. Using a 24 mm deep socket wrench, install the oil pressure switch and tighten to 11 ft. lbs. (15 Nm).

➡Do not start the engine for at least 1 hour after installation.

7. Connect the oil pressure switch assembly connector.
8. Check for oil leaks.
9. Install the engine under cover assembly, as applicable.
10. Install the No. 1 engine under cover.

THROTTLE POSITION SENSOR (TPS)

LOCATION

See Figure 188.

The throttle position sensor is located on the throttle body.

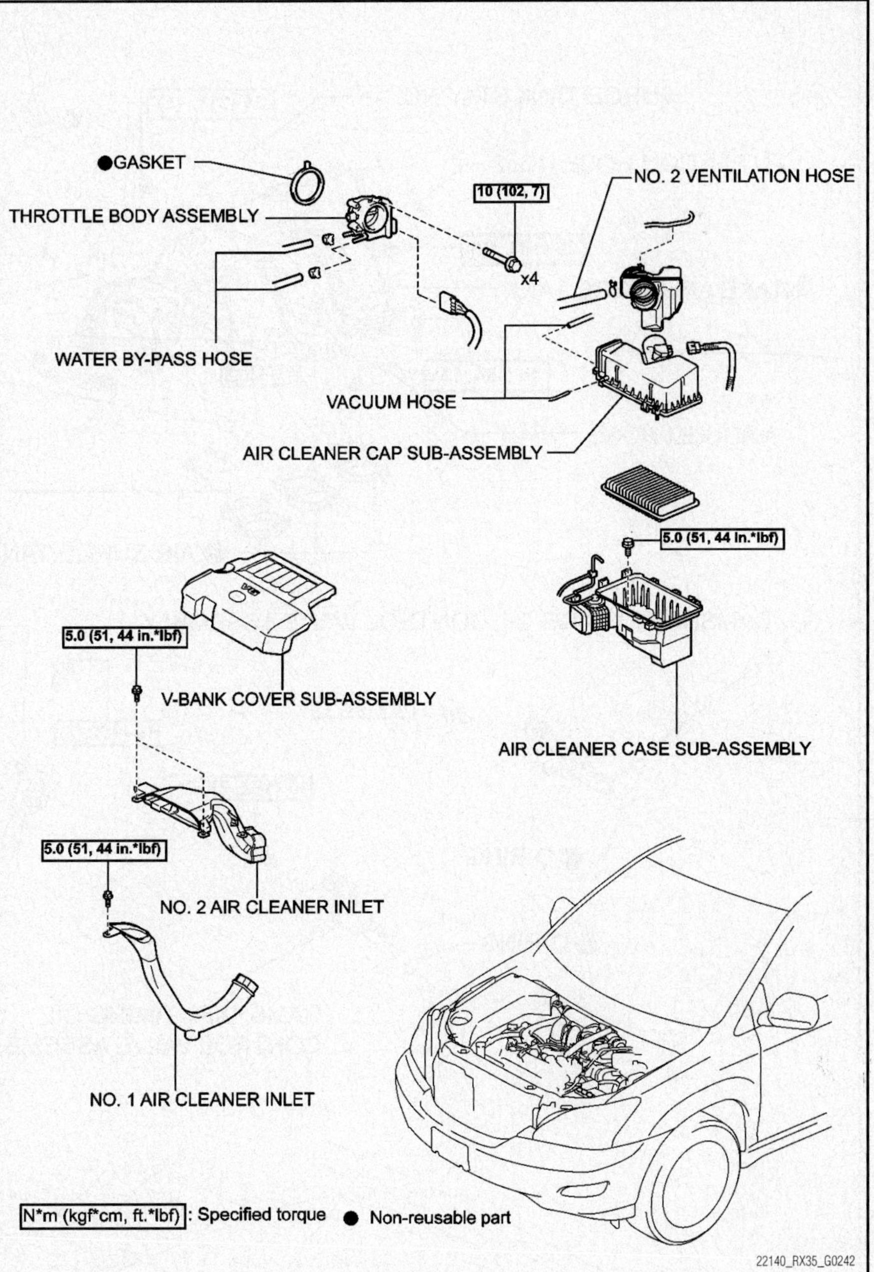

Fig. 188 Throttle body component locations

REMOVAL & INSTALLATION

Refer to the Throttle Body removal and installation procedures.

VARIABLE CAMSHAFT TIMING OIL CONTROL SOLENOID

LOCATION

See Figure 189.

REMOVAL & INSTALLATION

See Figure 190.

1. Remove the engine assembly.
2. Install on engine stand.
3. Remove the oil filler cap and gasket.
4. Remove the spark plugs and ignition coil assembly.
5. Remove the drain plug and gasket.

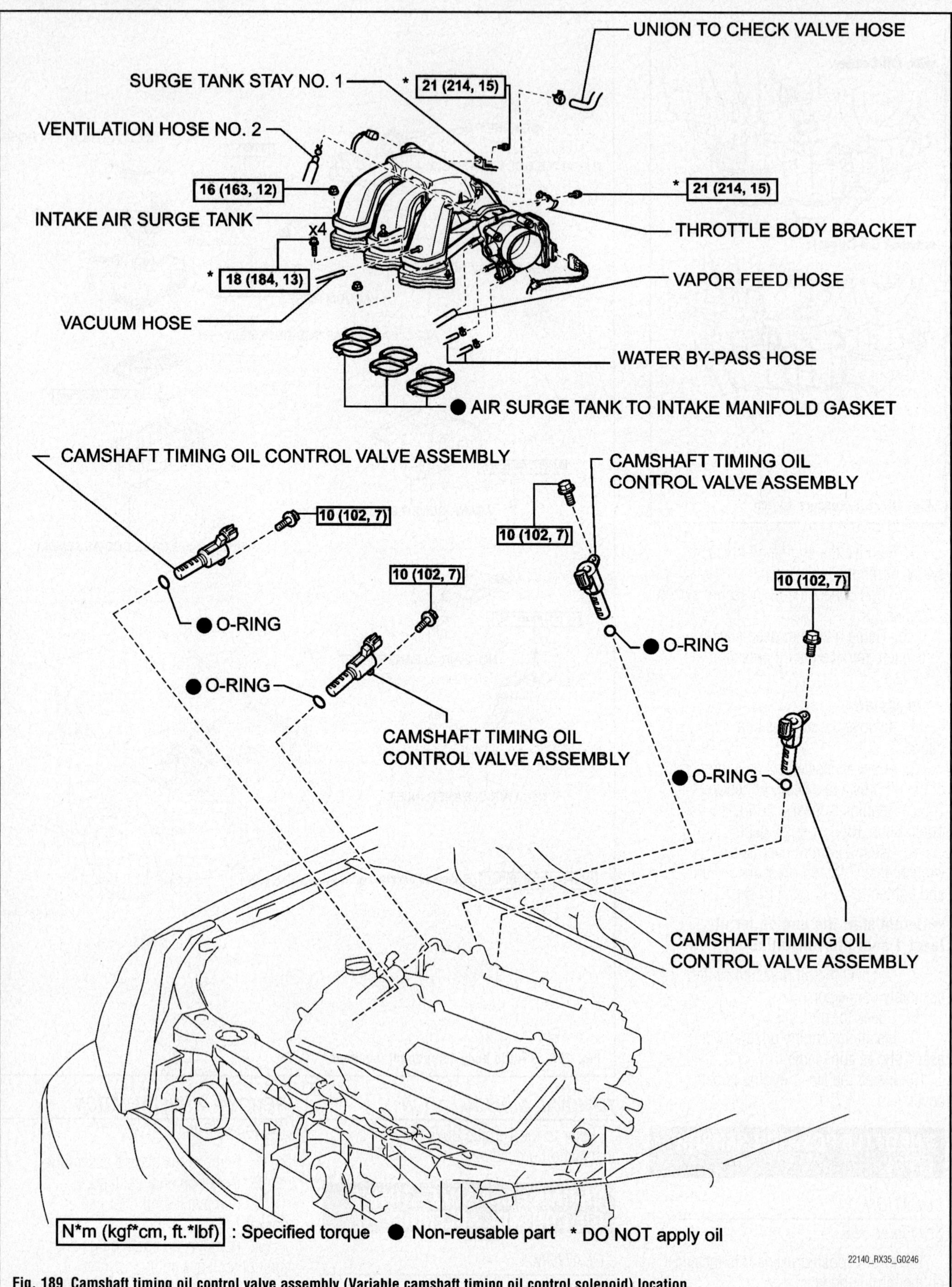

Fig. 189 Camshaft timing oil control valve assembly (Variable camshaft timing oil control solenoid) location

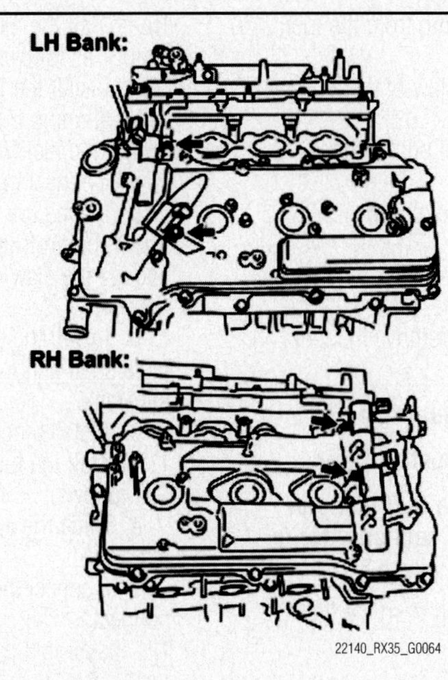

Fig. 190 Camshaft timing oil control valve assembly bolts

6. Remove the ventilation valve.
7. Remove the camshaft position sensor.
8. Remove the camshaft timing oil control valve assembly, as follows:
    a. Remove the 4 bolts and 4 oil control valves.

To install:

9. Install the camshaft timing oil control valve assembly, as follows:
    a. Install the 4 oil control valves with the 4 bolts and tighten to 7 ft. lbs. (10 Nm).
10. Install the camshaft position sensor.
11. Install the ventilation valve sub-assembly.
12. Install the spark plugs and the ignition coil assembly.
13. Install the oil filler cap sub-assembly.
14. Remove the engine stand.
15. Install the engine assembly.

FUEL SYSTEM — GASOLINE FUEL INJECTION SYSTEM

FUEL SYSTEM SERVICE PRECAUTIONS

Before servicing any vehicle, please be sure to read all of the following precautions, which deal with personal safety, prevention of component damage, and important points to take into consideration when servicing a motor vehicle:

- Observe all applicable safety precautions when working around fuel. Whenever servicing the fuel system, always work in a well-ventilated area. Do not allow fuel spray or vapors to come in contact with a spark, open flame, or excessive heat (a hot drop light, for example). Keep a dry chemical fire extinguisher near the work area. Always keep fuel in a container specifically designed for fuel storage; also, always properly seal fuel containers to avoid the possibility of fire or explosion. Refer to the additional fuel system precautions later in this section.
- Fuel injection systems often remain pressurized, even after the engine has been turned **OFF**. The fuel system pressure must be relieved before disconnecting any fuel lines. Failure to do so may result in fire and/or personal injury.
- All new vehicles are now equipped with an air bag system, often referred to as a Supplemental Restraint System (SRS) or Supplemental Inflatable Restraint (SIR) system. The system must be disabled before performing service on or around system components, steering column, instrument panel components, wiring and sensors. Failure to follow safety and disabling procedures could result in accidental air bag deployment, possible personal injury and unnecessary system repairs.
- Disconnecting the negative battery cable on some vehicles may interfere with the functions of the on-board computer system(s) and may require the computer to undergo a relearning process once the negative battery cable is reconnected.

Before inspecting and repairing the fuel system, disconnect the cable from the negative (-) battery terminal.

Keep gasoline away from rubber or leather parts.

1. Check that there are no fuel leaks from the fuel system after doing any maintenance or repairs.

RELIEVING FUEL SYSTEM PRESSURE

CAUTION

Perform the following procedures to prevent fuel from spilling out before removing any fuel system parts.

CAUTION

Pressure will still remain in the fuel line even after performing the following procedures. When disconnecting the fuel line, cover it with a shop rag or a piece of cloth to prevent fuel from spraying or coming out.

1. Remove the circuit opening relay/;
    a. Disconnect the cable from the negative (-) battery terminal.
    b. Remove the circuit opening relay from the engine room relay block.
    c. Connect the cable to the negative (-) battery terminal.
    d. Start the engine.
    e. After the engine stops, turn the ignition switch off.

➡DTC P0171/25 (fuel problem) and/or P0191/49 (fuel pressure sensor signal error) may be detected.

    f. Crank the engine again. Check that the engine does not start.
    g. Remove the fuel tank cap to discharge pressure from the fuel tank.
    h. Disconnect the cable from the negative (-) battery terminal.
    i. Install the circuit opening relay.
2. Disconnect the fuel pump connector:
    a. Fold back the floor carpet.
    b. Remove the rear floor service hole cover.
    c. Disconnect the fuel pump connector.
    d. Start the engine.
    e. After the engine stops, turn the ignition switch off.

➡DTC P0171/25 (fuel problem) and/or P0191/49 (fuel pressure sensor signal error) may be detected.

f. Crank the engine again. Check that the engine does not start.

g. Remove the fuel tank cap to discharge pressure from the fuel tank.

h. Disconnect the cable from the negative (-) battery terminal.

i. Reconnect the fuel pump connector.

j. Install the rear floor service hole cover.

k. Install the floor carpet.

3. Check that there are no fuel leaks from the fuel system after doing any maintenance or repairs.

FUEL FILTER

REMOVAL & INSTALLATION

See Figures 191 through 193.

1. Remove the fuel pump from the vehicle.
2. Remove the fuel sender gauge assembly, as follows:

a. Disconnect the connector and remove the fuel sender gauge from the fuel suction tube.

3. Remove the fuel suction plate sub-assembly, as follows:

a. Using needle nozzle pliers, remove the E-ring.

b. Disengage the 2 claws of the fuel No. 1 suction support and remove the fuel suction plate with the fuel filter from the fuel No. 1 sub-tank.

Fig. 191 Disengage the 5 claws on the filter and remove the fuel pump from the fuel filter

c. Remove the spring from the fuel suction plate.

d. Disengage the claw of the jet pump nozzle.

e. Separate the fuel pump filter hose.

4. Disconnect the fuel pump harness.
5. Using a screwdriver with its tip wrapped in protective tape, disengage the 5 claws on the filter and remove the fuel pump from the fuel filter.

➡Do not damage the fuel filter.

➡Do not remove the suction filter.

➡Do not use either the fuel pump or the suction filter if the suction filter is removed from the fuel pump.

6. Remove the O-ring from the fuel filter.

To install:

7. Apply gasoline to a new O-ring and install it to the fuel filter.
8. Engage the 5 claws on the fuel filter and install the fuel pump with the pump filter.
9. Connect the fuel pump harness connector.

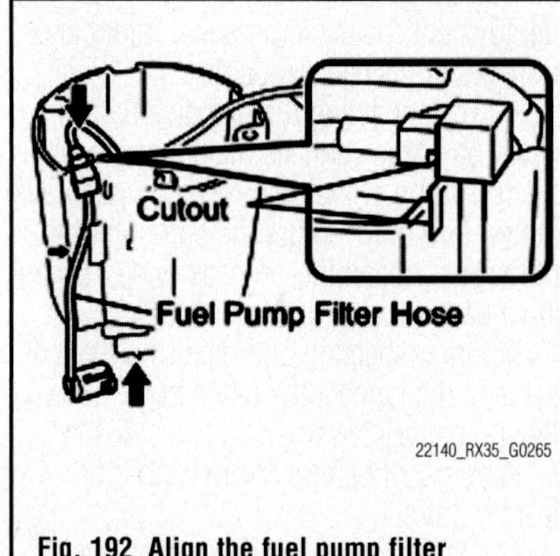

Fig. 192 Align the fuel pump filter

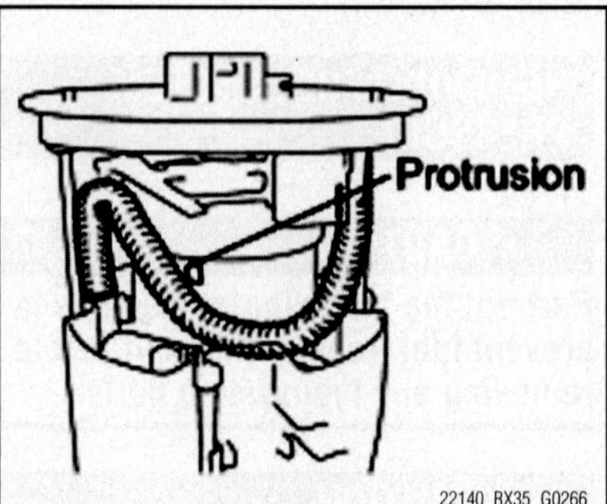

Fig. 193 Make sure that the fuel tube passes under the protrusion of the fuel filter

10. Install the fuel suction plate sub-assembly, as follows:

a. Install the fuel pump filter tube while aligning it to the installation position of the fuel No. 1 sub-tank.

b. Connect the jet pump nozzle.

c. Make sure that the fuel tube passes under the protrusion of the fuel filter, and engage the claw of the fuel No. 1 suction support.

d. Install the spring to the fuel suction plate shaft and install it to the fuel No. 1 sub-tank.

e. Install a new E-ring.

11. Install the fuel sender gauge assembly, as follows;

a. Slide the fuel sender gauge to fit the claw.

b. Connect the fuel sender gauge connector.

12. Install fuel pump to vehicle.

FUEL PUMP

REMOVAL & INSTALLATION

See Figures 194 through 196.

1. Discharge fuel system pressure.
2. Remove the deck board sub-assembly, as follows:

a. Disengage the 5 clips and turn up the front side of the deck board.

3. Remove the right and left rear seat assemblies.
4. Remove the left rear door scuff plate.
5. Remove the left deck side trim cover.
6. Remove the left rear seat side cover.
7. Remove the rear floor service hole cover, as follows:

a. Using a clip remover, remove the 2 clips and tear off the front floor carpet.

b. Remove the rear floor service hole cover.

c. Disconnect the fuel pump connector.

8. Remove the fuel suction tube assembly with pump and gauge, as follows:

a. Disconnect the fuel pump tube by removing the tube joint clip, and pull out the fuel pump tube.

➡Check if there is any dirt or mud around the connector before this operation and remove the dirt as necessary.

➡Be careful of mud because the quick connector has an O-ring which seals the pipe and connector that can be contaminated.

➡Do not use any tool in this operation.

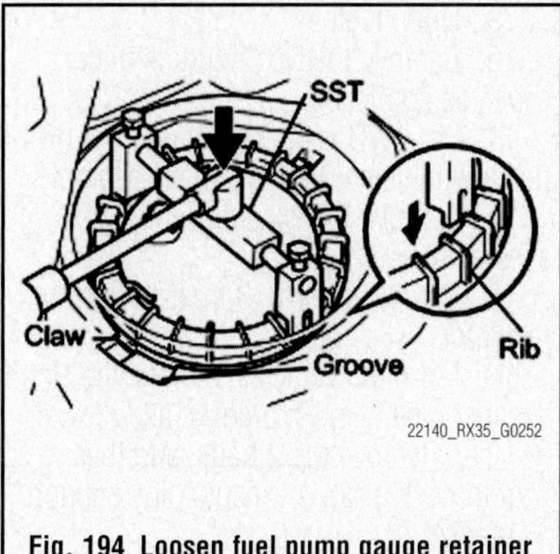

Fig. 194 Loosen fuel pump gauge retainer

➡Do not bend or twist the nylon tube. Protect the connector by covering it with a vinyl or plastic bag.

➡When the pipe and connector are stuck, push and pull the connector to release and pull the connector out carefully.

b. Using SST (SST: 09808-14020) or equivalent, loosen the fuel pump gauge retainer.

➡Loosen the retainer by turning it counterclockwise while holding the SST down. Do not allow the claw of the tank suction tube support to slip out of its groove on the fuel tank.

➡A rib on the fuel pump gauge retainer can be fitted into a tip of the SST.

c. Remove the fuel pump gauge retainer.

d. Remove the fuel suction tube with pump and gauge.

➡Be careful not to bend the arm of the fuel sender gauge.

e. Remove the gasket from the fuel tank.

To install:

9. Install the fuel suction tube assembly with pump and gauge, as follows:

a. Install a new gasket to the fuel tank.

b. Attach the fuel suction tube with pump and gauge to the fuel tank.

➡Be careful not bend the arm of the fuel sender gauge.

c. Align the keyway of the fuel suction tube support with the key of the fuel suction tube with pump and gauge.

d. Align the triangle mark on a new fuel pump gauge retainer with the "S" mark on the fuel tank while pushing down the fuel suction tube with pump and gauge, and attach the fuel pump gauge retainer.

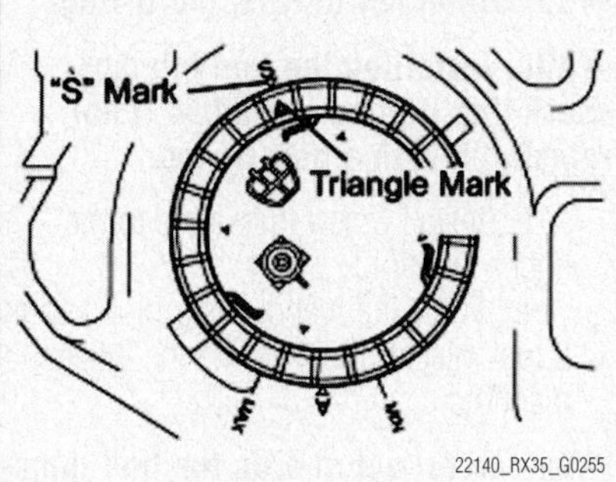

Fig. 195 Align the triangle mark on a new fuel pump

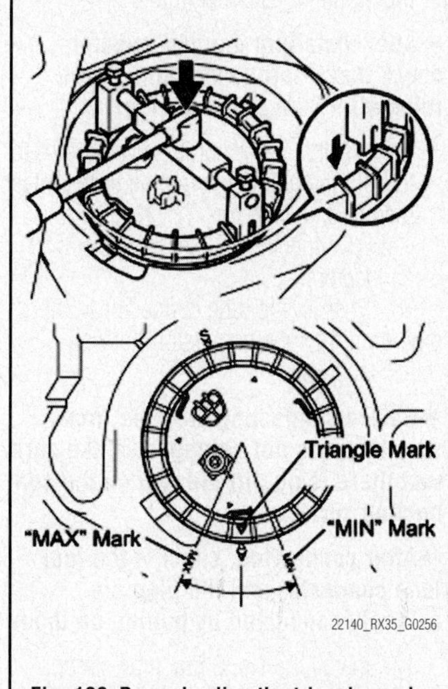

Fig. 196 Properly align the triangle mark

e. Rotate the fuel pump gauge retainer by hand, then tighten it one complete turn and another half turn using SST (SST: 09808-14020). The triangle mark on the fuel pump gauge retainer must be positioned between the "MIN" and "MAX" marks on the fuel tank.

➡Do not use other tools in this operation. Damage to the fuel pump gauge retainer and the fuel tank may result.

➡A rib on the fuel pump gauge retainer can be fitted into a tip of the SST.

f. Connect the fuel pump tube by installing the fuel pump tube and the tube joint clip.

➡Check that there are no scratch or foreign objects on the connecting part.

➡Check that the fuel tube joint is inserted securely.

➡Check that the tube joint clip is on the collar of the fuel tube joint.

➡After installing the tube joint clip, check that the fuel tube joint has not been pulled off.

10. Inspect for fuel leak.

11. Install the rear floor service hole cover, as follows:

a. Install a new butyl tape to the rear floor service hole cover.

b. Connect the fuel pump connector.

c. Install the rear floor service hole cover.

d. Install the front floor carpet with the 3 clips.

12. Install the left rear seat side cover.

13. Install the left deck side trim cover.

14. Install the left rear door scuff plate.

15. Install the right and left rear seat assemblies.

16. Install the deck board sub-assembly.

FUEL RAIL & INJECTORS

REMOVAL & INSTALLATION

See Figure 197.

1. Discharge fuel system pressure.
2. Disconnect cable from negative battery terminal.
3. Drain engine coolant.
4. Remove the engine room side cover.
5. Remove both front wiper arm and blade assemblies.
6. Remove the right and left front fender to cowl side seals.
7. Remove the cowl top ventilator louver sub-assembly.
8. Remove the windshield wiper motor and link assembly.
9. Remove the cowl top panel outer sub-assembly.
10. Remove the V-bank cover sub-assembly.
11. Remove the air cleaner cap sub-assembly.
12. Remove the intake air surge tank assembly.
13. Disconnect the fuel tube sub-assembly, as follows:

a. Remove the No. 2 fuel pipe clamp.

b. Pinch the tube connector and then pull out the fuel pipe.

➡Check that there is no dirt or other foreign objects around the connector before this operation and clean the connector as necessary.

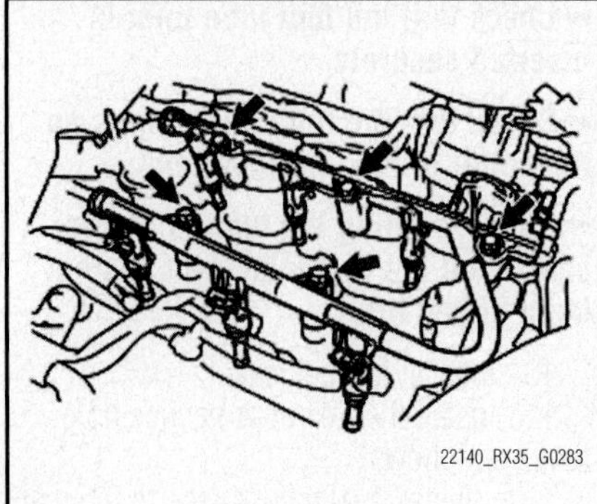

Fig. 197 Fuel delivery pipe and injectors

➡**It is necessary to prevent mud or dirt from entering the quick connector. If mud enters the connector, the O-rings may not seal properly.**

➡**Remove the quick connector by hand.**

➡**Do not bend or twist the nylon tube. Protect the connector by covering it with a plastic bag.**

➡**If the pipe and the connector are stuck, try wiggling or pushing and pulling the connector to release it and pull the connector off of the pipe carefully.**

14. Remove the fuel injector assembly, as follows:
    a. Disconnect the 6 fuel injector connectors.
    b. Remove the 5 bolts and fuel delivery pipe together with the 6 fuel injectors.

➡**Be careful not to drop the fuel injectors when removing the fuel delivery pipe.**

    c. Remove the 6 insulators from the intake manifold.
    d. Pull out the fuel injectors from the fuel delivery pipe.
    e. Remove the 6 O-rings from the injectors.

To install:

15. Install the fuel injector assembly, as follows:
    a. Apply a light coat of spindle oil or gasoline to new O-rings, and install them to each injector.
    b. Apply a light coat of spindle oil or gasoline where the fuel delivery pipe contacts the O-ring.
    c. Push the fuel injector while turning it to install the injector in the fuel delivery pipe.
    d. Position the fuel injector connector outward.

➡**Be careful not to twist the O-ring.**

➡**After installing the fuel injector, check that it turns smoothly. If not, reinstall it with a new O-ring.**

    e. Install 6 new insulators to the intake manifold.
    f. Place the fuel delivery pipe and the 6 fuel injectors together to the intake manifold.

➡**Be careful not to drop the fuel injectors when installing the fuel delivery pipe.**

    g. Temporarily install the 5 bolts which are used to hold the fuel delivery pipe to the intake manifold.

➡**After installing the fuel injector, check that it turns smoothly. If not, reinstall it with a new O-ring.**

    h. Tighten the 5 bolts which are used to hold the fuel delivery pipe to the intake manifold to 15 ft. lbs. (21 Nm).
    i. Connect the 6 fuel injector connectors.

16. Push in the tube connector to the pipe until the tube connector makes a "click" sound.

➡**Before connecting the tube, make sure that it is not damaged. Make sure that there is no dirt present on the connecting surfaces.**

➡**After connecting, check if the fuel tube connector and the pipe are securely connected by pulling on them.**

17. Install the No. 2 fuel pipe clamp.
18. The remainder of installation is the reverse of the removal procedure.
19. Check for coolant leak and fuel leak.
20. Perform initialization procedure.

FUEL TANK

REMOVAL & INSTALLATION

See Figures 198 through 206.

1. Discharge fuel system pressure.
2. Remove the deck board sub-assembly.
3. Remove the right and left rear seat assemblies.
4. Remove the left rear door scuff plate.
5. Remove the left deck side trim cover.
6. Remove the left rear seat side cover.
7. Remove the rear floor service hole cover.
8. Remove the fuel suction tube assembly with pump and gauge, as follows:
9. Drain fuel.
10. Remove the No. 2 engine under cover, as applicable.
11. For 4WD vehicles, remove the propeller with center bearing shaft assembly.
12. Remove the exhaust center pipe assembly.
13. Remove the No. 3 front floor heat insulator.
14. For 4WD vehicles, remove the No. 4 exhaust pipe support bracket, as follows:
    a. Remove the 2 bolts, and then remove the No. 4 exhaust pipe support bracket.
    b. Remove the 2 clips and unfasten the claw, and then remove the No. 1 fuel tube protector.
    c. Remove the 4 nuts, and then remove the No. 1 fuel tank protector.
    d. Remove the 3 clips (A) and 6 nuts, and then remove the No. 1 fuel tank protector.
15. Remove the fuel tank assembly, as follows:
    a. Disconnect the fuel pump tube:
    - Pinch the projection of the retainer to remove the lock claws and pull it down as shown in the illustration
    - Pull out the fuel pump tube

➡**Check if there is any dirt or mud around the connector before this operation and remove the dirt if necessary.**

➡**Be careful of mud because the quick connector can be contaminated.**

➡**Do not use any tool in this operation.**

➡**Do not bend or twist the nylon tube. Protect the connector by covering it with a vinyl or plastic bag.**

Fig. 198 No. 1 fuel tank protector

➡When the pipe and connector are stuck, push and pull the connector to release and pull the connector out carefully.

b. Disconnect the fuel tank vent hose:

- Deeply push the connector to release the locking tab
- Pinch portion A
- Pull out the connector

c. Disconnect the No. 3 fuel tank breather:

- Pinch the tube connector and then pull out the No. 3 fuel tank breather tube.

➡Check if there is any dirt or mud around the connector before this operation and remove the dirt as necessary.

➡Be careful of mud because the quick connector has an O-ring which seals the pipe and connector that can be contaminated.

➡Do not use any tool in this operation.

➡Do not bend or twist the nylon tube. Protect the connector by covering it with a vinyl or plastic bag.

➡When the pipe and connector are stuck, push and pull the connector to release and pull the connector out carefully.

d. Loosen the hose clamp bolt and disconnect the fuel tank to filler pipe hose.

e. Set up a transmission jack under the fuel tank.

f. Remove the 4 bolts, and then remove the 2 fuel tank bands.

g. Remove the 2 nuts.

h. Operate the transmission jack to remove the fuel tank.

16. Unfasten the 2 claws, and then remove the fuel pump tube.

17. Loosen the hose clamp bolt, and then remove the fuel tank to filler pipe hose.

18. Remove the fuel tank vent hose, as follows:

a. Completely push the connector to release the locking tab (*1).

b. Pinch portion A (*2).

c. Pull out the connector.

d. Unfasten the claw, and then remove the fuel tank vent hose.

e. Disconnect the No. 1 evaporation vent tube.

➡Check if there is any dirt or mud around the connector before this operation and remove the dirt as necessary.

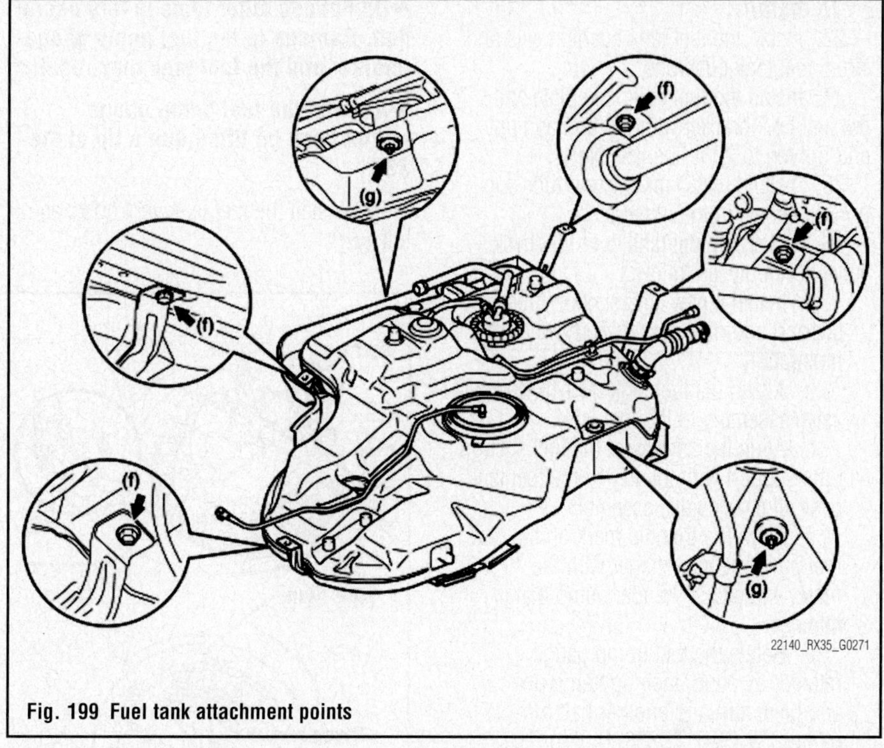

Fig. 199 Fuel tank attachment points

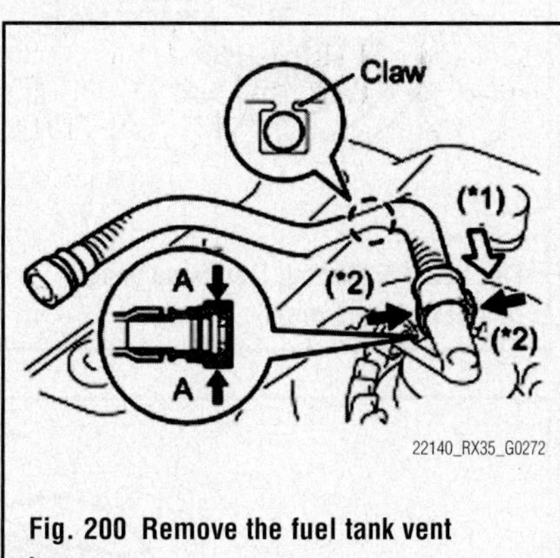

Fig. 200 Remove the fuel tank vent hose

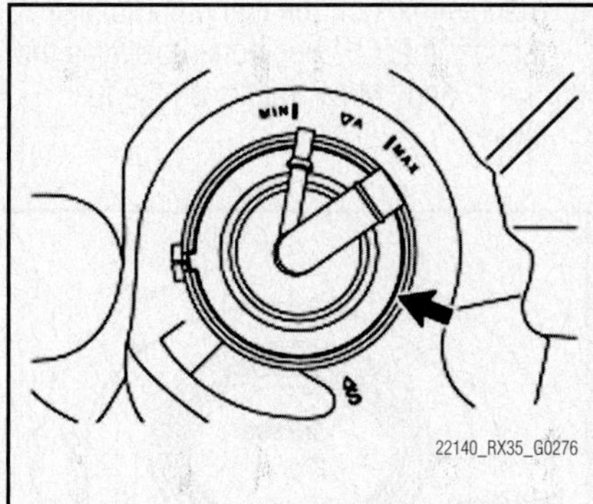

Fig. 201 Fuel tank over fill check valve assembly

➡Be careful of mud because the quick connector has an O-ring which seals the pipe and connector that can be contaminated.

➡Do not use any tool in this operation.

➡Do not bend or twist the nylon tube. Protect the connector by covering it with a vinyl or plastic bag.

➡When the pipe and connector are stuck, push and pull the connector to release and pull the connector out carefully.

19. Remove fuel tank over fill check valve assembly, as follows:

a. Using SST (SST: 09808-14020), loosen the fuel pump gauge retainer.

➡Loosen the retainer by turning it counterclockwise while holding the SST down. Do not allow the claw of the tank suction tube support to slip out of its groove on the fuel tank.

➡A rib on the fuel pump gauge retainer can be fitted into a tip of the SST.

b. Remove the fuel pump gauge retainer.

c. Remove the fuel tank over fill check valve assembly.

d. Remove the tank suction tube support and gasket from the fuel tank.

20. Remove the tank suction tube support from the fuel tank.

21. Remove the bolt and nut, and then remove the fuel tank side plate and the fuel tank bracket.

22. Remove the fuel tank cushion sets and No. 5 fuel tank cushions.

To install:

23. Install the fuel tank cushion sets and No. 5 fuel tank cushions.
24. Install the fuel tank side plate and the fuel tank bracket with the bolt and nut and tighten to 22 ft. lbs. (30 Nm).
25. Install a new tank suction tube support as shown in the illustration.
26. Install the fuel tank over fill check valve assembly, as follows:
    a. Install a new tank suction tube support and new gasket as shown in the illustration.
    b. Attach the fuel tank over fill check valve assembly to the fuel tank.
    c. Align the keyway of the fuel suction tube support with the key of the fuel tank over fill check valve assembly.
    d. Align the triangle mark on the new fuel pump gauge retainer with the "S" mark, and attach the fuel pump gauge retainer.
    e. Rotate the fuel pump gauge retainer by hand, then tighten it one complete turn and another half turn using SST (SST: 09808-14020). The triangle mark on the fuel pump gauge retainer must be positioned between the "MIN" and "MAX" marks on the fuel tank.

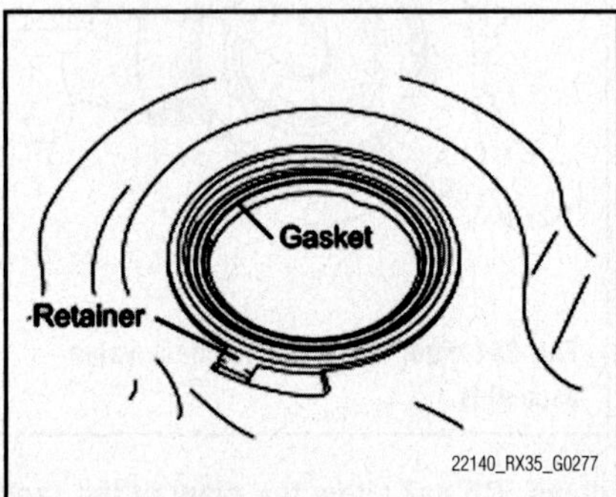

Fig. 202 Install a new tank suction tube support and new gasket

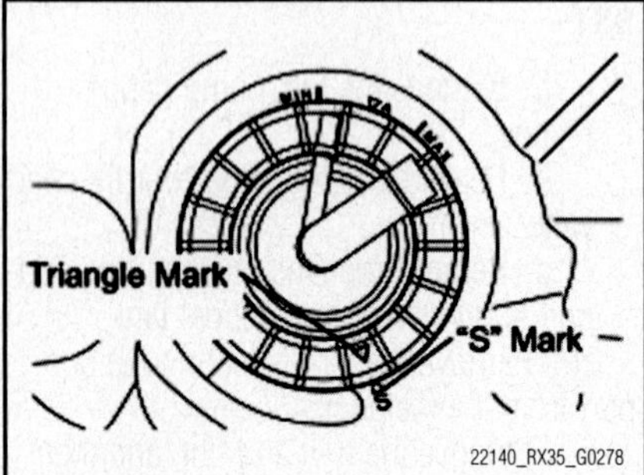

Fig. 203 Align the triangle mark on the new fuel pump gauge retainer with the "S" mark

➡Do not use other tools in this operation. Damage to the fuel pump gauge retainer and the fuel tank may result.

➡A rib on the fuel pump gauge retainer can be fitted into a tip of the SST.

27. Install the fuel tank vent hose, as follows:

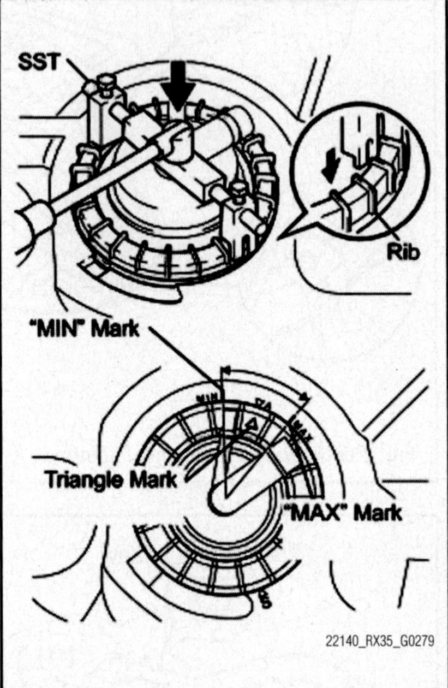

Fig. 204 Positioning fuel pump gauge retainer triangle mark

a. Connect the No. 1 evaporation vent tube:
- Align the axis of the connector with the axis of the pipe. Push the pipe into the connector until the connector makes a click sound. If the connection is too tight, apply a small amount of fresh engine oil to the tip of the pipe.

➡Check for damage or foreign objects on the parts that are to be connected.

➡After connecting, check that the fuel tube connector and the pipe are securely connected by pulling on them.

b. Connect the fuel tank vent hose:
- Align the axis of the connector with the axis of the pipe. Push the pipe into the connector until the connector makes a click sound. If the connection is too tight, apply a small amount of fresh engine oil to the tip of the pipe.

➡Check for damage or foreign objects on the parts that are to be connected.

➡After connecting, check that the fuel tube connector and the pipe are securely connected by pulling on them.

- Install the fuel tank vent hose, and fasten the claw.

28. Install the fuel tank to filler pipe hose with the hose clamp.

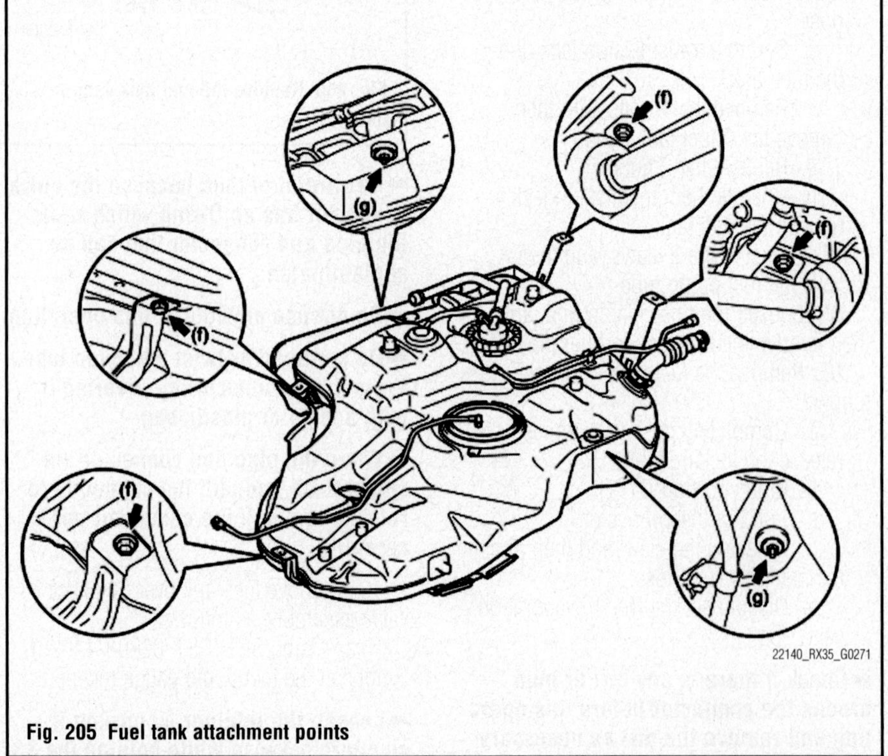

Fig. 205 Fuel tank attachment points

29. Install the fuel pump tube, and fasten the 2 claws.

30. Install the fuel tank assembly, as follows:

a. Set up the fuel tank to the transmission jack.

b. Operating the transmission jack, install the fuel tank.

c. Tighten the 2 nuts to 14 ft. lbs. (20 Nm).

d. Install the 2 fuel tank bands with the 4 bolts and tighten to 29 ft. lbs. (39 Nm).

e. Connect the fuel tank to filler pipe hose.

f. Connect the No. 3 fuel tank breather tube:

- Push the quick connector to the pipe until it makes a "click" sound.

➡**Check if there is any damage or foreign objects on the connected part.**

Fig. 206 No. 1 fuel tank protector

➡**After connecting, check if the quick connector and the pipe are securely connected by pulling on them.**

g. Connect the fuel tank vent hose.

h. Connect the fuel pump tube:

- Push the quick connector and push up the retainer to lock the claws.

➡**Check if there is any damage or foreign objects on the connected part.**

➡**After connecting, check if the quick connector and the pipe are securely connected by pulling on them.**

31. Install the No. 1 fuel tank protector sub-assembly, as follows:

a. Install the No. 1 fuel tank protector with the 6 nuts and tighten to 49 inch lbs. (5.5 Nm).

b. Install 3 new clips. (A).

32. Install the No. 1 fuel tank protector with the 4 nuts and tighten to 49 inch lbs. (5.5 Nm).

33. For 4WD vehicles, install the No. 4 exhaust pipe support bracket with the 2 bolts and tighten to 16 ft. lbs. (22 Nm).

34. The remainder of installation is the reverse of the removal procedure.

35. Add fuel, inspect for fuel leak and exhaust gas leak.

IDLE SPEED

ADJUSTMENT

Idle speed is maintained by the Powertrain Control Module (PCM). No adjustment is necessary or possible.

THROTTLE BODY

REMOVAL & INSTALLATION

See Figure 207.

1. Remove the windshield wiper link assembly.

2. Remove the cowl top panel outer sub-assembly.

3. Drain engine coolant.

4. Remove the V-bank cover sub-assembly.

5. Remove the No. 2 air cleaner inlet.

6. Remove the No. 1 air cleaner inlet.

7. Remove the air cleaner cap sub-assembly.

8. Remove the air cleaner case sub-assembly.

9. Remove the throttle body, as follows:

a. Disconnect the throttle body connector and clamp.

b. Disconnect the 2 water by-pass hoses from the throttle Body.

c. Remove the 4 bolts and throttle body.

d. Remove the throttle body gasket from the intake air surge tank.

To install:

10. Installation is the reverse of the removal procedure. Torque the throttle body bolts to 7 ft. lbs. (10 Nm).

11. Inspect for coolant leak.

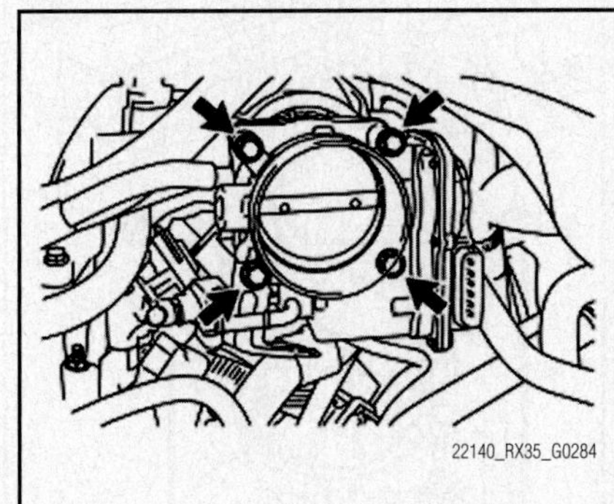

Fig. 207 Throttle body and bolts

HEATING & AIR CONDITIONING SYSTEM

BLOWER MOTOR

REMOVAL & INSTALLATION

See Figures 208 through 210.

1. Disconnect cable from negative battery terminal.
2. Remove the right hand front door scuff plate.
3. Remove the right hand cowl side trim sub-assembly.
4. Remove the No. 2 instrument panel under cover sub-assembly.
5. Remove the glove compartment door assembly.
6. Remove the screw, release the 2 claws and remove the No. 2 air duct.
7. Disconnect the No. 2 air indicator wiring harness sub-assembly, as follows:
    a. Remove the clamp and disconnect the 2 connectors.
    b. Remove blower assembly, as follows:

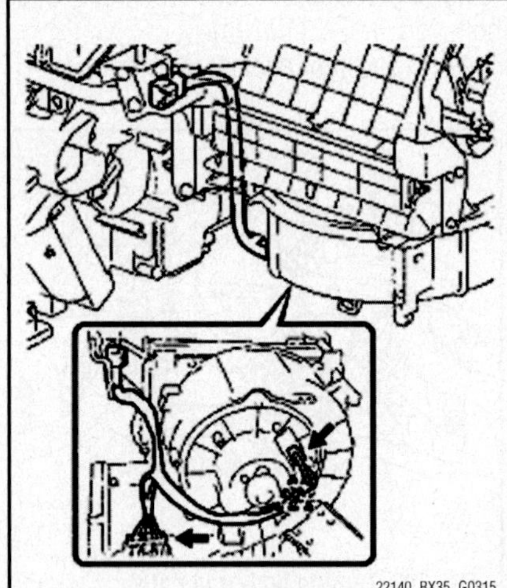

Fig. 208 No. 2 wiring air indicator harness

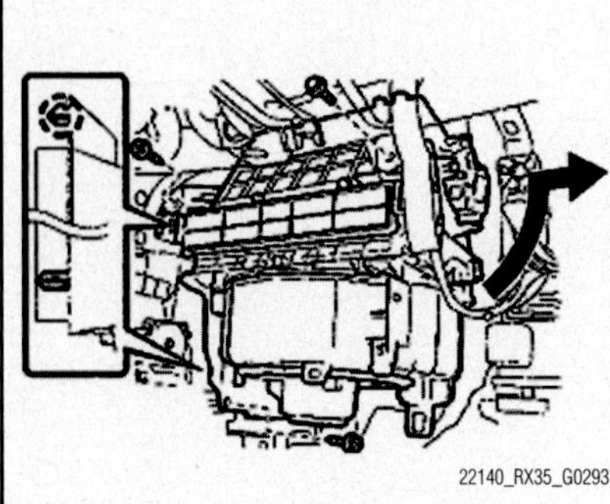

Fig. 209 Blower assembly removal

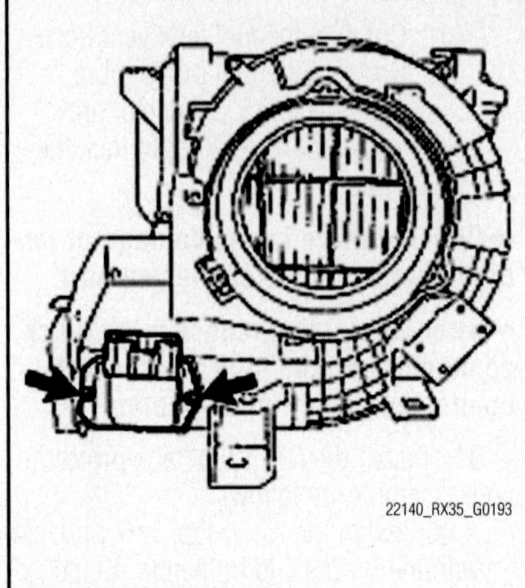

Fig. 210 Blower motor

    c. Remove the clamp and disconnect the connectors.
    d. Remove the 4 screws and the nut, and then the bracket.
    e. Remove the bolt and 2 screws.
    f. Release the claw and remove the blower assembly.
8. Remove the 2 screws and blower motor control.

To install:

9. Installation is the reverse of removal, noting the following torque specifications:
    a. Blower assembly: 87 inch lbs. (9.8 Nm).
    b. Blower assembly bracket: 87 inch lbs. (9.8 Nm).
    c. No. 2 air duct: 22 inch lbs. (2.5 Nm).
10. Perform initialization.

HEATER CORE

REMOVAL & INSTALLATION

See Figures 211 through 214.

1. Discharge refrigerant from refrigeration system.
2. Disconnect battery negative terminal.
3. Remove both front wiper arm and blade assemblies.
4. Remove the cowl top ventilator louver sub-assembly.
5. Remove the windshield wiper link assembly.
6. Remove the cowl top panel outer sub-assembly.
7. Disconnect the air conditioning tube and accessory assembly, as follows:
    a. Install SST (SST: 09870-00025) on the piping clamp.

➡Make sure the direction of the piping clamp and SST by seeing the illustration shown on the caution label.

    b. Push down SST and release the clamp lock. Do not deform the tube when pushing the SST.
    c. Pull the SST slightly and push the release lever, and then remove the piping clamp with SST.
    d. Disconnect the air conditioning tube and accessory assembly by hand.

➡Do not use tools like a screwdriver to remove the tube.

➡Cap the open fittings immediately to keep moisture or dirt out of the system.

    e. Remove the 2 O-rings from the air conditioning tube and accessory assembly.
8. Disconnect the No. 1cooler refrigerant suction hose, using SST: 09870-00015, and the same procedure as for removing the air conditioning tube and accessory assembly.
9. Using pliers, grip the claws of the clip and slide the clip, and then disconnect the heater water outlet hose. Do not apply any excessive force to the heater water outlet hose.

➡Prepare a drain pan or cloth for cooling water leaks.

10. Using the same procedure as for the heater water outlet hose, disconnect the heater water inlet hose.
11. Remove the instrument panel assembly.
12. Remove the air conditioner amplifier assembly.
13. Turn back the floor carpet, release the claw and remove the rear No. 1 and 2air ducts.
14. Remove the bolt, release the 3 claws and remove the No. 1 air duct.
15. Disconnect the transmission control cable assembly.
16. Remove the shift lever.
17. Remove the 4 nuts and instrument panel bracket No. 4.
18. Remove instrument panel brace No. 1 sub-assembly, as follows:
    a. Remove the clamp and disconnect the connector.
    b. Remove the 3 bolts and 2 nuts and instrument panel brace No. 1 sub-assembly.

19. Remove instrument panel brace No. 2 sub-assembly, as follows:
   a. Remove the clamp and disconnect the connector.
   b. Remove the 3 bolts and 2 nuts and instrument panel brace No. 2 sub-assembly.
20. Separate steering intermediate shaft sub-assembly.
21. Remove steering column assembly.
22. Remove the 3 nuts and air duct No. 1 sub-assembly.
23. Remove instrument panel reinforcement assembly, as follows:
   a. Remove the 4 bolts and 4 nuts.
   b. Remove the clamp.
   c. Disconnect the connectors.
   d. Remove the 9 bolts and the 2 nuts and then remove the instrument panel reinforcement while holding the air conditioner unit assembly.

➡**Make sure to hold the air conditioner unit assembly securely as its bracket installation parts may be damaged.**

24. Remove the 2 nuts and air conditioner unit assembly.

WARNING

Make sure to hold the air conditioner unit assembly securely as its bracket installation parts may be damaged.

25. Remove the air duct No. 2, as follows:
   a. Remove the screw.
   b. Release the 2 claws and remove the air duct No. 2.
26. Remove the wiring air indicator harness No .2 sub-assembly.
27. Remove the blower assembly.
28. Release the 4 claws and remove the center heater to register duct.
29. Remove the 3 screws and air outlet control servomotor.

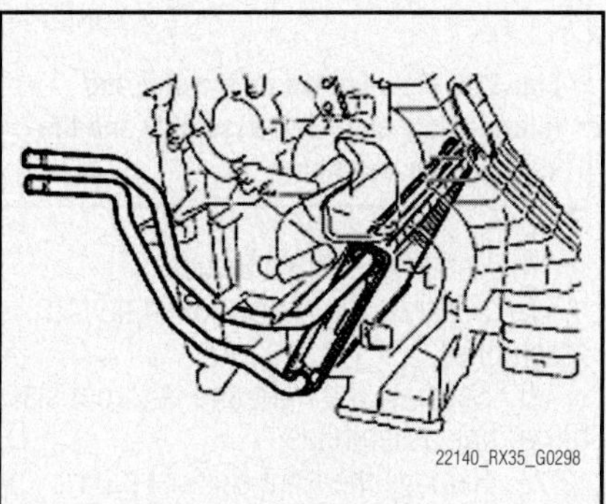

Fig. 211 Remove the heater radiator unit sub-assembly

30. Remove the air mix control servomotor.
31. Remove the evaporator temperature sensor.
32. Remove the heater radiator unit sub-assembly, as follows:
   a. Remove the 2 screws and piping cover.
   b. Remove the screw and bracket.
   c. Remove the heater radiator unit sub-assembly.

➡**Prepare a drain pan or cloth for cooling water leaks.**

To install:

33. Install the heater radiator unit sub-assembly.
34. Install the bracket with the screw.
35. Install the piping cover with the 2 screws.
36. Engage the 2 claws to install the evaporator temperature sensor.
37. Engage the clamp to install the connector.
38. Install the air mix control servomotor with the 3 screws.

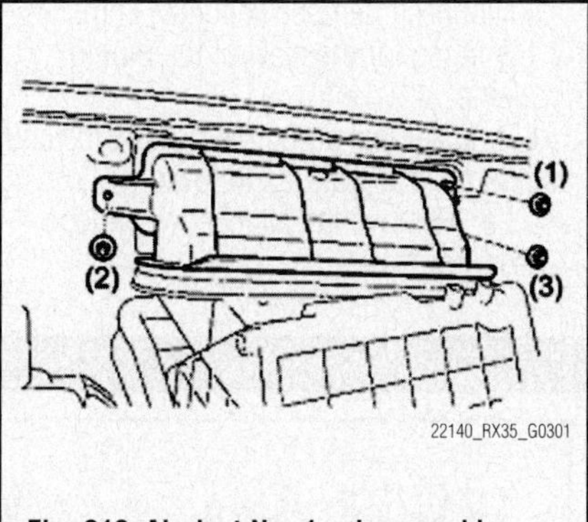

Fig. 212 Air duct No. 1 sub-assembly tightening sequence

39. Install the air outlet control servomotor with the 3 screws.
40. Engage the 4 clamps to install the heater to register duct center.
41. Engage the claw to install the blower assembly.
42. Install the wiring air indicator harness No. 2 sub-assembly with the 3 screws.
43. Connect the connectors and clamp.
44. Install the air duct No. 2 with the screw and tighten to 22 inch lbs. (2.5 Nm).
45. Temporarily tighten the air conditioner unit assembly with the 2 nuts. Hold the A/C unit securely to prevent damaging its bracket installation parts.
46. Install the instrument panel reinforcement assembly, as follows:
   a. Install the instrument panel reinforcement with the 8 bolts and tighten to 15 ft. lbs. (20 Nm).
   b. Temporarily tighten the air conditioner unit assembly with the 2 nuts and bolt.
   c. Install the 4 bolts and 4 nuts.
   d. Install the clamps.
47. Connect the connectors.
48. Install the instrument panel brace No. 2 sub-assembly with the 3 bolts and 2 nuts.
49. Install the clamps and connect the connectors.
50. Install the instrument panel brace No. 1 sub-assembly with the 3 bolts and 2 nuts.
51. Install the clamps and connect the connectors.
52. Install the No. 4 instrument panel bracket.

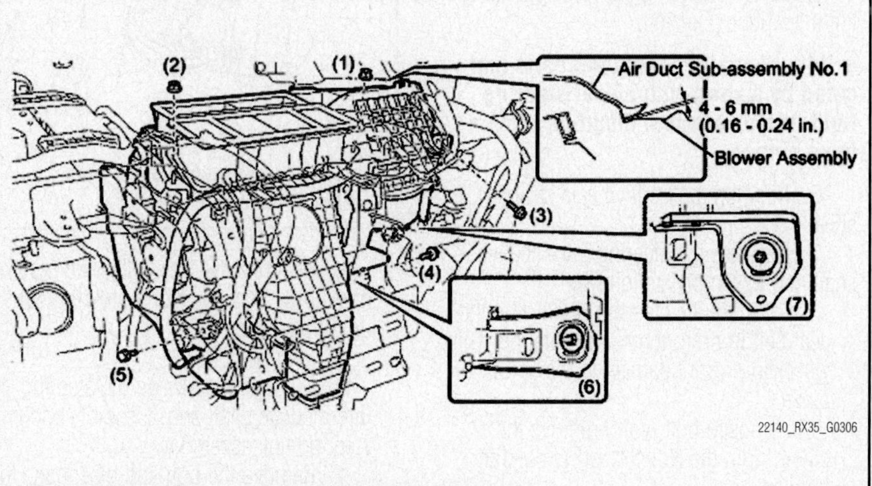

Fig. 213 Air conditioner unit assembly tightening sequence

53. Install the air duct No. 1 sub-assembly with the 3 nuts and tighten to 87 inch lbs. (9.8 Nm).

54. Fully tighten the air conditioner unit assembly with the 3 bolts and 4 nuts to 87 inch lbs. (9.8 Nm) in the order shown. Install the air conditioner unit assembly so that there is no clearance between the duct and blower assemblies.

55. Install the steering column assembly.

56. Install the steering intermediate shaft sub-assembly

57. Install the air duct No. 1 with the bolt and tighten to 87 inch lbs. (9.8 Nm).

58. Install the air conditioner amplifier assembly.

59. Install the shift lever.

60. Connect the transmission control cable assembly.

61. Inspect shift lever position.

62. Install the rear No. 1 air duct.

63. Install the rear No. 2 air duct.

64. Install the instrument panel assembly.

65. Install the heater water inlet and outlet hoses.

66. Install the No. 1 cooler refrigerant suction pipe, as follows:

a. Remove the attached vinyl tape from the hose.

b. Coat a new O-ring with compressor oil and install it to the hose. Compressor oil: ND-OIL 11 or equivalent.

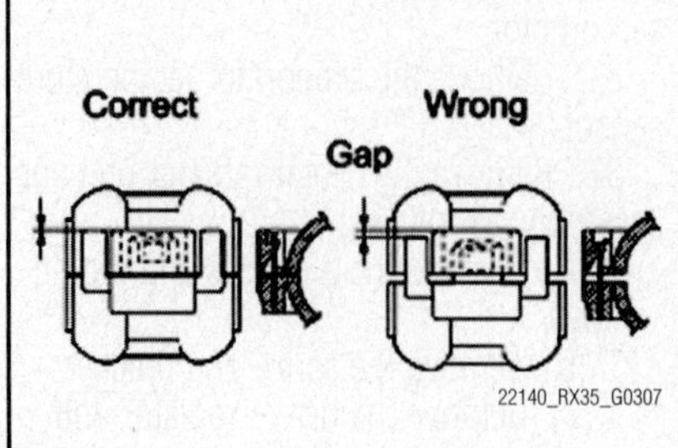

Fig. 214 No. 1 cooler refrigerant suction pipe

WARNING

Do not use any compressor oil other than ND-OIL 11 or equivalent. If any compressor oil other than ND-OIL 11 or equivalent is used, compressor motor insulation performance may decrease, resulting in leakage of electric power.

c. Install the No. 1 cooler refrigerant suction pipe and piping clamp. Be sure to connect the hose securely, and check the fitting for the claw of the piping clamp.

67. Install the air conditioning tube and accessory assembly, as follows:

a. Remove the attached vinyl tape from the hose.

b. Coat a new O-ring with compressor oil and install it to the pipe. Compressor oil: ND-OIL 11 or equivalent.

WARNING

Do not use any compressor oil other than ND-OIL 11 or equivalent. If any compressor oil other than ND-OIL 11 or equivalent is used, compressor motor insulation performance may decrease, resulting in leakage of electric power.

c. Install the air conditioning tube and accessory assembly and piping clamp. Be sure to connect the hose securely, and check the fitting for the claw of the piping clamp.

68. Install cowl top panel outer sub-assembly.

69. Install windshield wiper link assembly.

70. Install cowl top ventilator louver sub-assembly.

71. Install both front wiper arm and blade assemblies.

72. Add engine coolant.

73. Connect cable to negative battery terminal.

74. Charge refrigerant.

75. Warm up compressor.

76. Check for engine coolant leaks.

77. Check for refrigerant leaks.

78. Perform initialization.

STEERING

POWER RACK & PINION STEERING GEAR

REMOVAL & INSTALLATION

See Figures 215 through 219.

1. Before servicing the vehicle, refer to the precautions section.

➡When installing, coat the parts indicated by arrows with power steering fluid or molybdenum disulfide lithium base grease.

2. Position the front wheels straight ahead.

3. Separate the steering intermediate shaft sub-assembly, as follows:

a. Secure the steering wheel with the seat belt in order to prevent rotation. This will help prevent damaging the spiral cable.

b. Loosen bolt A and remove the clamp from the No. 1 steering column hole cover.

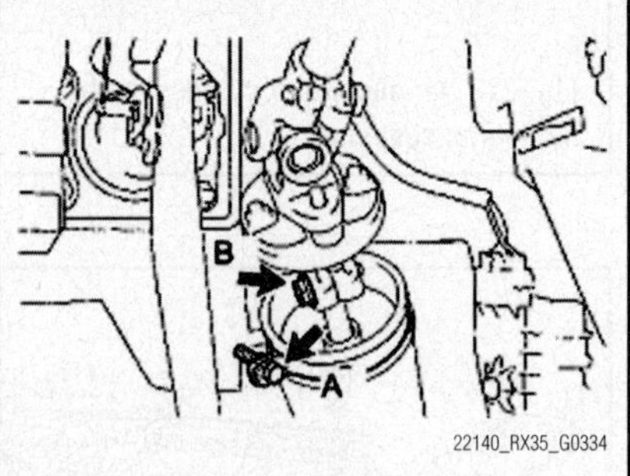

Fig. 215 Loosen bolts A and B from steering column hole covers

c. Separate the No. 2 steering column hole cover from the No. 1 steering column hole cover.

d. Loosen bolt B.

e. Put matchmarks on the steering intermediate shaft sub-assembly and the steering link assembly.

f. Remove the bolt and disengage the steering intermediate shaft sub-assembly.

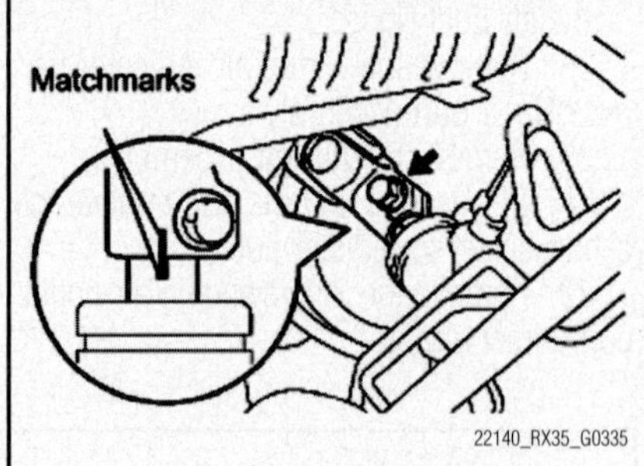

Fig. 216 Matchmarks on the steering intermediate shaft sub-assembly and the steering link assembly

4. Remove the front wheel.

5. Separate the right and left tie rod assemblies.

6. Separate the right and left front stabilizer link assemblies.

7. Remove the front exhaust pipe assembly.

8. Remove the No. 1 left front stabilizer bracket:

a. For 2 WD vehicles, remove the 2 bolts and the No. 1 stabilizer bracket.

b. For 4WD vehicles, remove the 2 bolts, the No. 1 left stabilizer bracket and the No. 2 stabilizer bracket.

9. Using the same procedure as for the left side, remove the No. 1 right front stabilizer bracket.

10. Remove the left front height control sensor sub-assembly.

11. Disconnect the return tube assembly, as follows:

a. Remove the tube clamp from the pressure feed tube assembly.

b. Using SST (SST: 09023-12701) or equivalent, disconnect the return tube assembly from the steering link assembly.

c. Remove the nut and the return tube clamp.

12. Disconnect the pressure feed tube assembly, as follows:

a. Using SST (SST: 09023-12701) or equivalent, disconnect the pressure feed tube assembly from the steering link assembly.

b. Remove the bolt and the pressure feed tube clamp.

13. Remove the 2 bolts, the nuts and the steering link assembly.

To install:

14. Install the power steering link assembly with the 2 bolts and the nuts and tighten to 52 ft. lbs. (70 Nm).

15. Using SST (SST: 09023-12701) or equivalent, connect the pressure feed tube assembly to the steering link assembly and tighten to 16 ft. lbs. (22 Nm). Use a torque wrench with a fulcrum length of 300 mm (11.81 in.). This torque value is effective when SST is parallel to a torque wrench.

16. Install the pressure feed tube assembly clamp with the bolt and tighten to 87 inch lbs. (9.8 Nm).

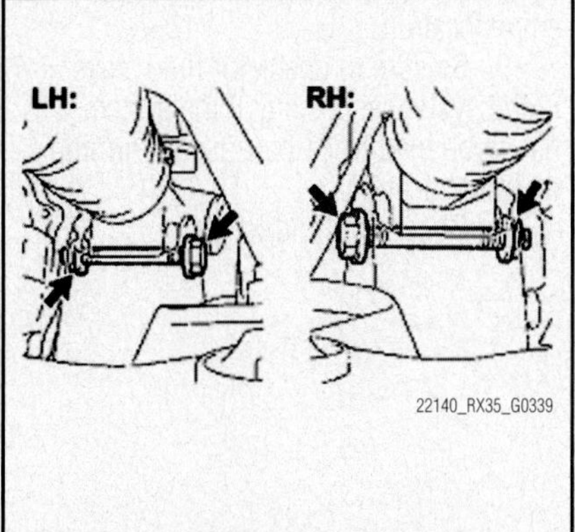

Fig. 217 Power steering link assembly

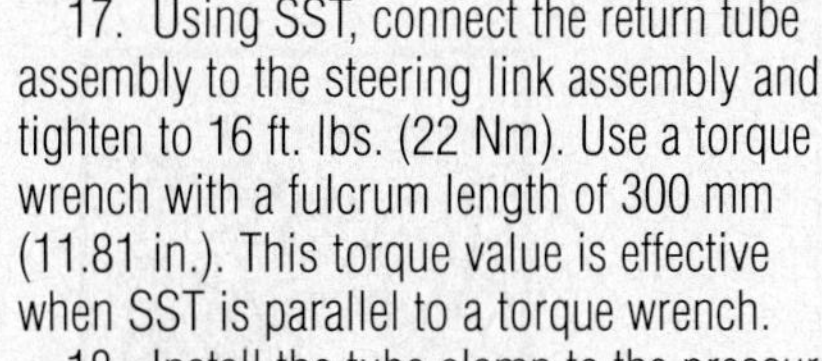

17. Using SST, connect the return tube assembly to the steering link assembly and tighten to 16 ft. lbs. (22 Nm). Use a torque wrench with a fulcrum length of 300 mm (11.81 in.). This torque value is effective when SST is parallel to a torque wrench.

18. Install the tube clamp to the pressure feed tube assembly.

19. Install the pressure feed tube clamp with the nut and tighten to 87 inch lbs. (9.8 Nm).

20. Install the left front height control sensor sub-assembly.

21. Install the No. 1 left front stabilizer bracket:

a. For 2WD vehicles, install the No. 1 left stabilizer bracket with the 2 bolts and tighten to 12 ft. lbs. (16 Nm).

b. For 4WD vehicles, install the No. 1 left stabilizer bracket and the No. 2 stabilizer bracket with the 2 bolts and tighten to 12 ft. lbs. (16 Nm).

22. Using the same procedure as for the left side, install the No. 1 right front stabilizer bracket.

23. Install the front exhaust pipe assembly.

24. Install the right and left front stabilizer link assemblies.

25. Install the right and left tie rod assemblies.

26. Install the front wheel.

27. Connect the steering intermediate shaft sub-assembly, as follows:

a. Align the matchmarks on the intermediate shaft sub-assembly and the steering link assembly.

b. Install the bolt and tighten to 26 ft. lbs. (35 Nm).

c. Tighten the bolt A to 26 ft. lbs. (35 Nm).

d. Install the steering column hole cover No. 2 to the No. 1 steering hole cover.

e. Install the clamp to the No. 1 steering column hole cover and tighten the bolt B.

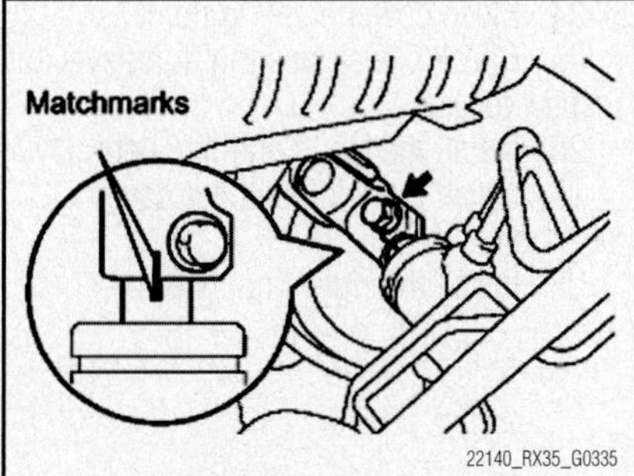

Fig. 218 Matchmarks on the steering intermediate shaft sub-assembly and the steering link assembly

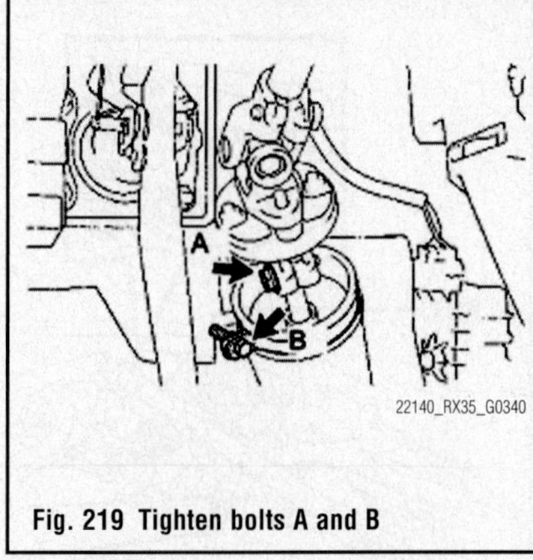

Fig. 219 Tighten bolts A and B

28. Bleed power steering fluid.

29. Inspect for power steering fluid leak.

30. Inspect for gas leak.

31. Inspect the steering wheel center point.

POWER STEERING PUMP

REMOVAL & INSTALLATION

See Figures 220 through 223.

1. Remove the right front wheel.

2. Remove the engine under cover assembly.

3. Remove the No. 2 engine under cover.

4. Drain power steering fluid.

5. Remove fan and alternator V-belt.

6. Slide the clip and disconnect the No. 1 fluid reservoir to pump hose from the vane pump (power steering pump) assembly.

7. Disconnect the pressure feed tube assembly, as follows:

a. Remove the union bolt and disconnect the pressure feed tube assembly from the vane pump assembly.

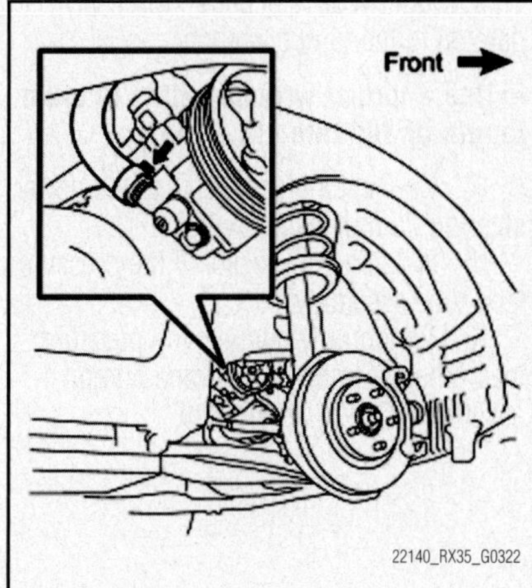

Fig. 220 Disconnect the No. 1 fluid reservoir to pump hose

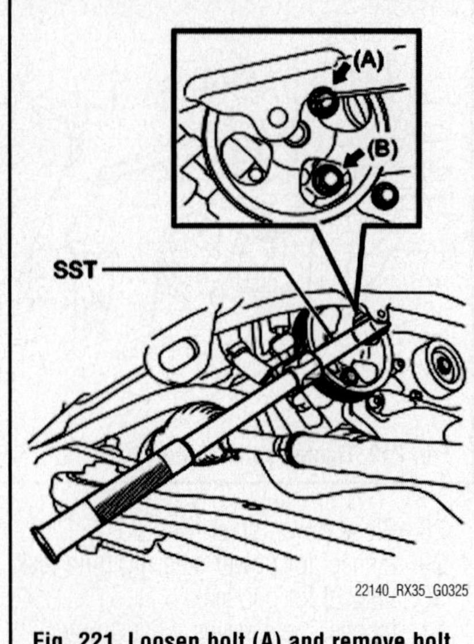

Fig. 221 Loosen bolt (A) and remove bolt (B), and then remove the vane (power steering) pump assembly

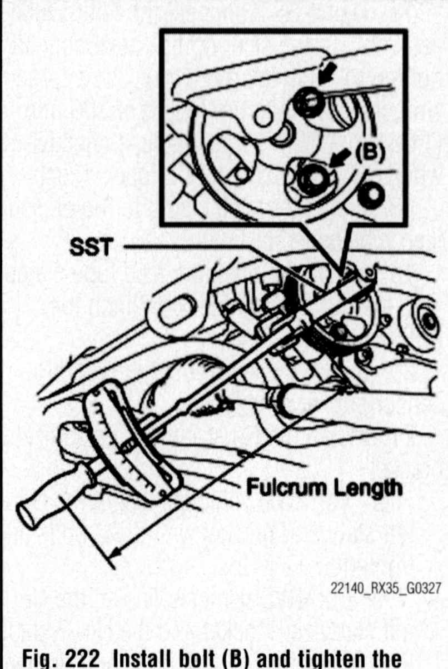

Fig. 222 Install bolt (B) and tighten the 2 bolts

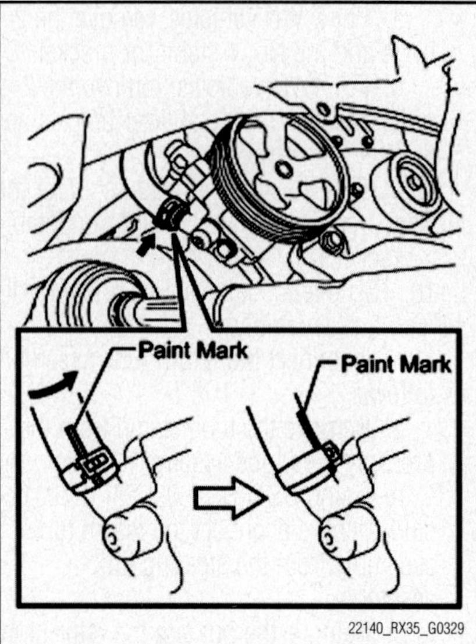

Fig. 223 Install the No. 1 fluid reservoir to pump hose

b. Remove the bolt and separate the pressure feed tube clamp.

c. Remove the gasket from the pressure feed tube assembly.

8. Disconnect the fluid pressure switch connector.

9. Using SST (SST: 09249-63010), loosen bolt (A) and remove bolt (B), and then remove the vane pump (power steering pump) assembly.

10. Remove the bolt from the vane pump assembly.

To install:

11. Temporarily install the bolt to the vane pump assembly.

12. Install the vane pump assembly.

13. Using SST (SST: 09249-63010), install bolt (B) and tighten the 2 bolts to 32 ft. lbs. (43 Nm), and 21 ft. lbs. (29 Nm). This torque value is accurate when SST is parallel to the torque wrench.

➡Use a torque wrench with a fulcrum length of 300 mm (11.81 in.).

14. Connect the connector to the power steering fluid pressure switch.

15. Install a new gasket to the pressure feed tube assembly.

16. Temporarily connect the pressure feed tube assembly to the vane pump assembly with the union bolt.

17. Install the pressure feed tube assembly clamp with the bolt and tighten to 69 inch lbs. (7.8 Nm). Install the pressure feed tube assembly clamp in the correct position.

18. Fully tighten the union bolt to 37 ft. lbs. (50 Nm). Make sure that the stopper of the pressure feed tube assembly contacts the vane pump assembly securely as shown in the illustration.

19. Connect the No. 1 fluid reservoir to pump hose to the vane pump assembly with the clip.

20. After installing the No. 1 fluid reservoir to pump hose with its paint mark and the claw of the clip facing the right of the vehicle as shown in the illustration, turn the hose counterclockwise 90 degrees so that the paint mark and claw face the front of the vehicle.

21. Install the No. 1 fluid reservoir to pump hose pushed all the way on as shown in the illustration.

22. Install fan and alternator V-belt.

23. Add power steering fluid.

24. Bleed power steering fluid.

25. Check power steering fluid level and inspect for leak.

26. Install the No. 2 engine under cover.

27. Install the engine under cover assembly.

28. Install the right front wheel.

BLEEDING

1. Check the fluid level.

2. Jack up the front of the vehicle and support it with stands.

3. Turn the steering wheel:

a. With the engine stopped, turn the steering wheel slowly from lock to lock several times.

4. Lower the vehicle.

5. Start the engine:

a. Run the engine at idle for a few minutes.

6. Turn the steering wheel:

a. With the engine idling, turn the steering wheel left or right to the full lock position and keep it in that position for 2 to 3 seconds, then turn the steering wheel to the opposite full lock position and keep it there for 2 to 3 seconds.

b. Repeat this procedure several times.

7. Stop the engine.

8. Check for foaming or emulsification.

9. Be sure to check for fluid leaks in the system especially if the system has to be bled twice because of forming or emulsification.

10. Check the fluid level.

CONTROL LINKS

REMOVAL & INSTALLATION

Refer to Stabilizer Bar removal and installation procedure.

LOWER BALL JOINT

REMOVAL & INSTALLATION

See Figure 224.

1. Remove the front wheel.
2. Remove the front axle hub nut.
3. Separate the front speed sensor.
4. Remove the 2 bolts and separate the front disc brake caliper assembly from the steering knuckle.
5. Remove the front disc.
6. Separate the tie rod end sub-assembly.
7. Separate the lower No. 1 front suspension arm sub-assembly.
8. Remove the front axle assembly, as follows:
   a. Using a plastic hammer, separate the front drive shaft assembly from the front axle hub sub-assembly.

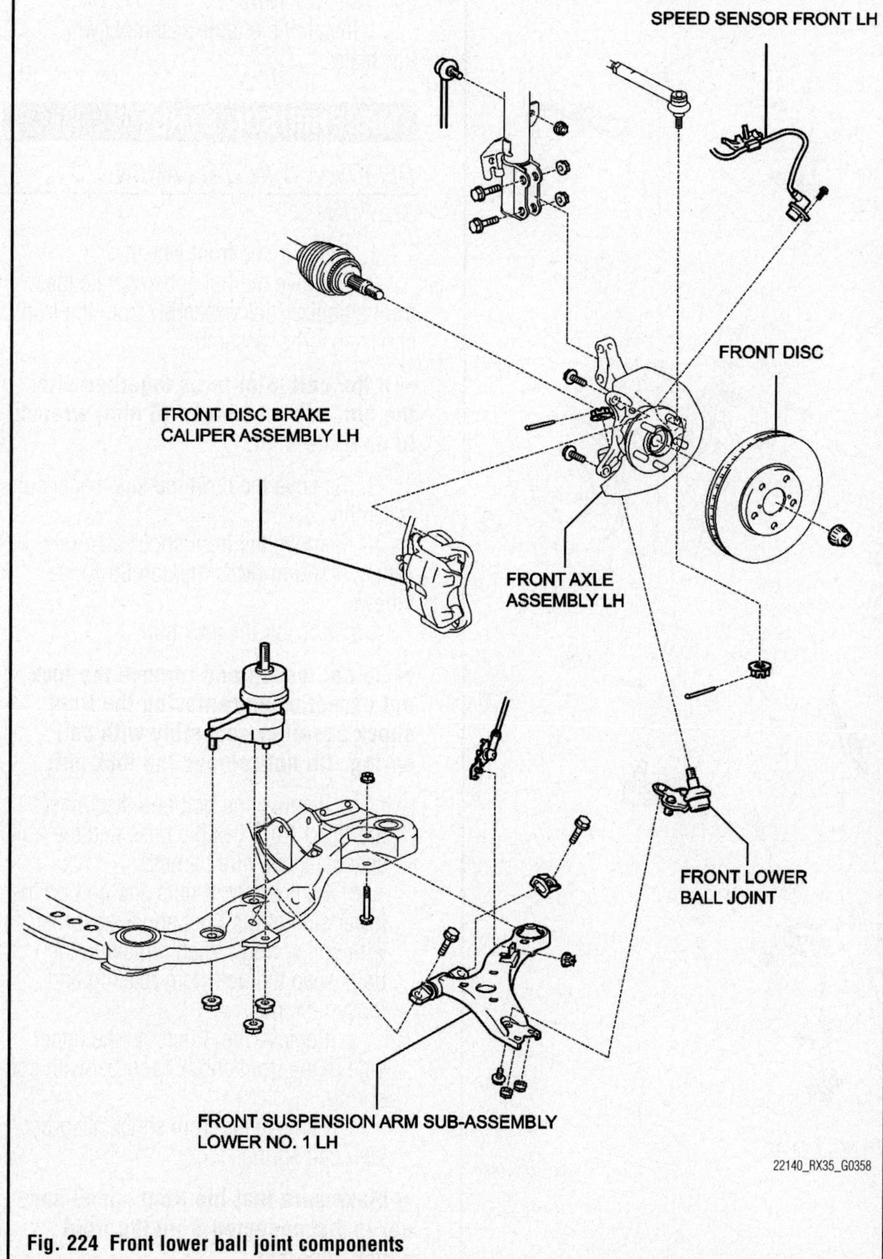

Fig. 224 Front lower ball joint components

**** WARNING**

Be careful not to damage the boot and ABS speed sensor rotor.

   b. Remove the 2 bolts, nuts and steering knuckle with the front axle hub sub-assembly.
9. Remove the front lower ball joint, as follows:
   a. Remove the cotter pin and castle nut.
   b. Using SST (SST: 09628-62011) or equivalent, remove the front lower ball joint assembly.

To install:

10. Install the front lower ball joint assembly to the steering knuckle with the nut and tighten to 91 ft. lbs. (123 Nm).

➡Do not allow oil to adhere to the screw and tapered parts.

11. Install a new cotter pin to the steering knuckle. If the holes for the cotter pin are not aligned, tighten the nut further up to 60°.
12. Install the front axle assembly, as follows:
    a. Install the 2 bolts, nuts and front axle assembly with the 2 bolts and nuts to the front shock absorber assembly and tighten to 170 ft. lbs. (230 Nm).

➡Only when reusing the bolts and nuts, apply the small amount of engine oil to the screw part of the nuts.

➡Do not excessively push out the front axle assembly.

**** WARNING**

Be careful not to damage the outboard joint boot or speed sensor rotor.

13. Install the lower No. 1 front suspension arm sub-assembly.
14. Install the tie rod end sub-assembly.
15. Install the front disc.
16. Install the front disc brake caliper assembly with the 2 bolts to the steering knuckle and tighten to 77 ft. lbs. (104 Nm).
17. Install the front speed sensor.
18. Install the front axle hub nut, as follows:
    a. Using a socket wrench (30 mm), install a new axle hub nut and tighten to 217 ft. lbs. (294 Nm).
    b. Using a chisel and hammer, stake the axle hub nut.

19. Inspect and adjust front wheel alignment.
20. Check ABS speed sensor signal.
21. Install the front wheel.

LOWER CONTROL ARM

REMOVAL & INSTALLATION

See Figures 225 and 226.

1. Remove the engine assembly with transaxle.
2. Remove the 3 nuts and the transverse engine mounting insulator.
3. Remove the 3 bolts and the nut on the lower No. 1 suspension arm and remove it from the front engine support member.
4. Remove the front lower arm bushing stopper.

To install:

5. Install the lower front No. 1 suspension arm, as follows:
   a. Install the lower front arm bushing stopper.
   b. Install the lower No. 1 suspension arm to the front engine support member with the 3 bolts and the nut but do not tighten.
   c. Tighten the 3 bolts in numerical order as shown in the illustration to 148 ft. lbs. (200 Nm), and 152 ft. lbs. (206 Nm). Begin installing the bolts from the front of the vehicle.

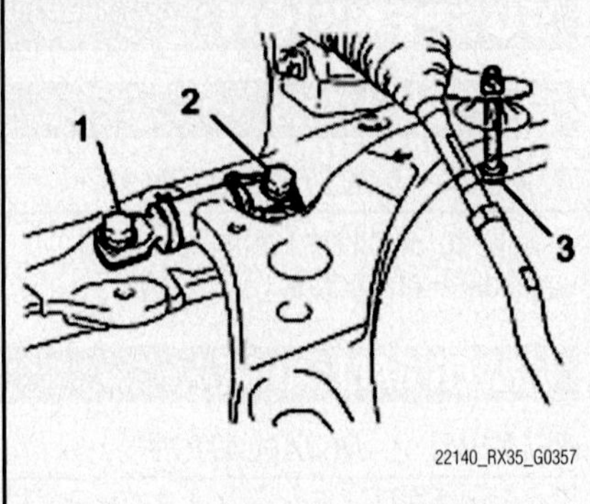

Fig. 226 Lower front No. 1 suspension arm bolt tightening sequence

6. Install the transverse engine mounting insulator with the 3 nuts and tighten to 64 ft. lbs. (87 Nm).
7. Install the engine assembly with transaxle.

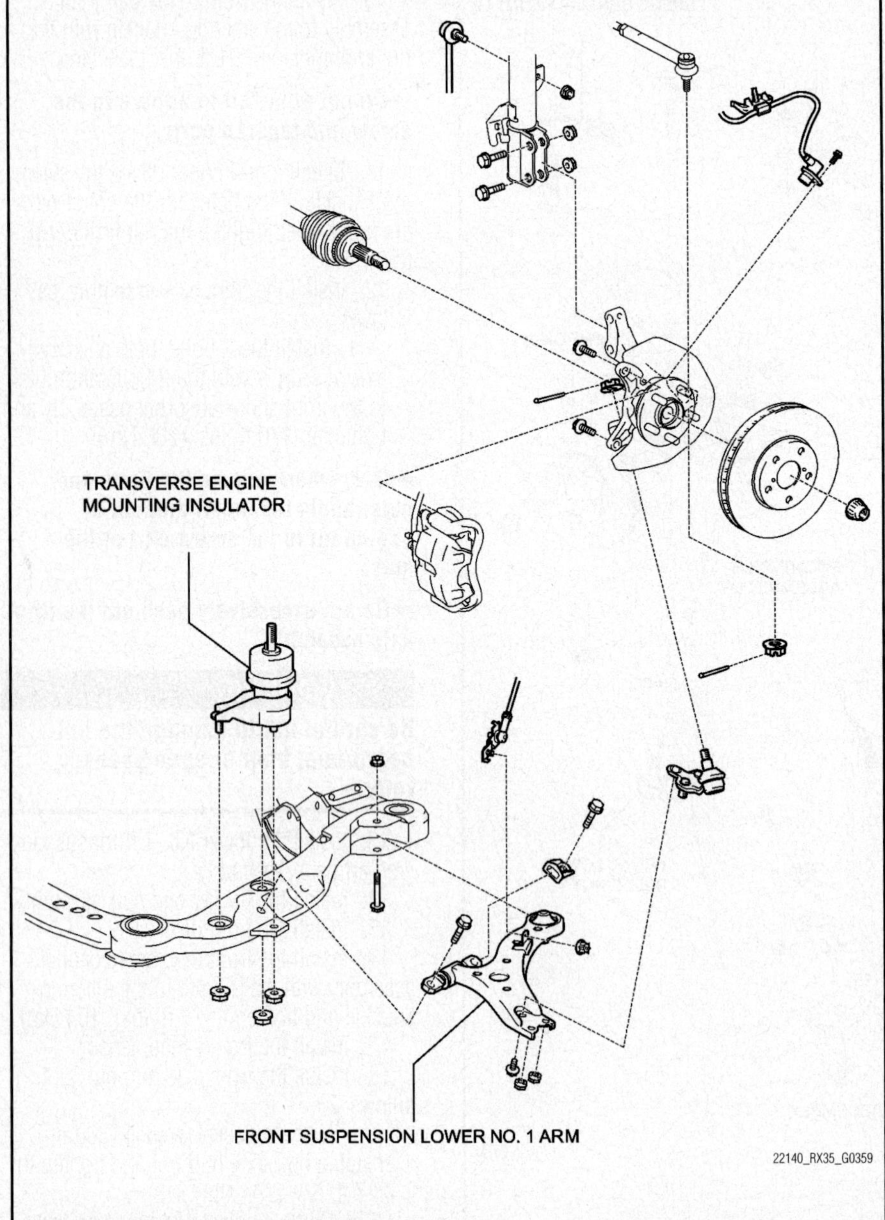

Fig. 225 Front lower control arm components

MACPHERSON STRUT

REMOVAL & INSTALLATION

See Figure 227.

1. Remove the front wheel.
2. Remove the nut and separate the front stabilizer link assembly from the front shock absorber assembly.

➡If the ball joint turns together with the nut, use a hexagon (6 mm) wrench to hold the stud.

3. Remove the cowl top side panel sub-assembly.
4. Remove the front shock absorber with coil spring (MacPherson Strut), as follows:
   a. Loosen the lock nut.

➡Do not loosen and remove the lock nut except when removing the front shock absorber assembly with coil spring. Do not remove the lock nut.

   b. Remove the bolt and disconnect the No. 1 front flexible hose and the front speed sensor wire harness.
   c. Remove the 2 nuts and bolt on the lower side of the front shock absorber with coil spring. When removing the bolts keep the bolt from rotating and loosen the nut.
   d. Remove the 3 nuts on the upper side of the front shock absorber with coil spring.
   e. Remove the front shock absorber with coil spring.

➡Make sure that the front speed sensor is disconnected from the front shock absorber.

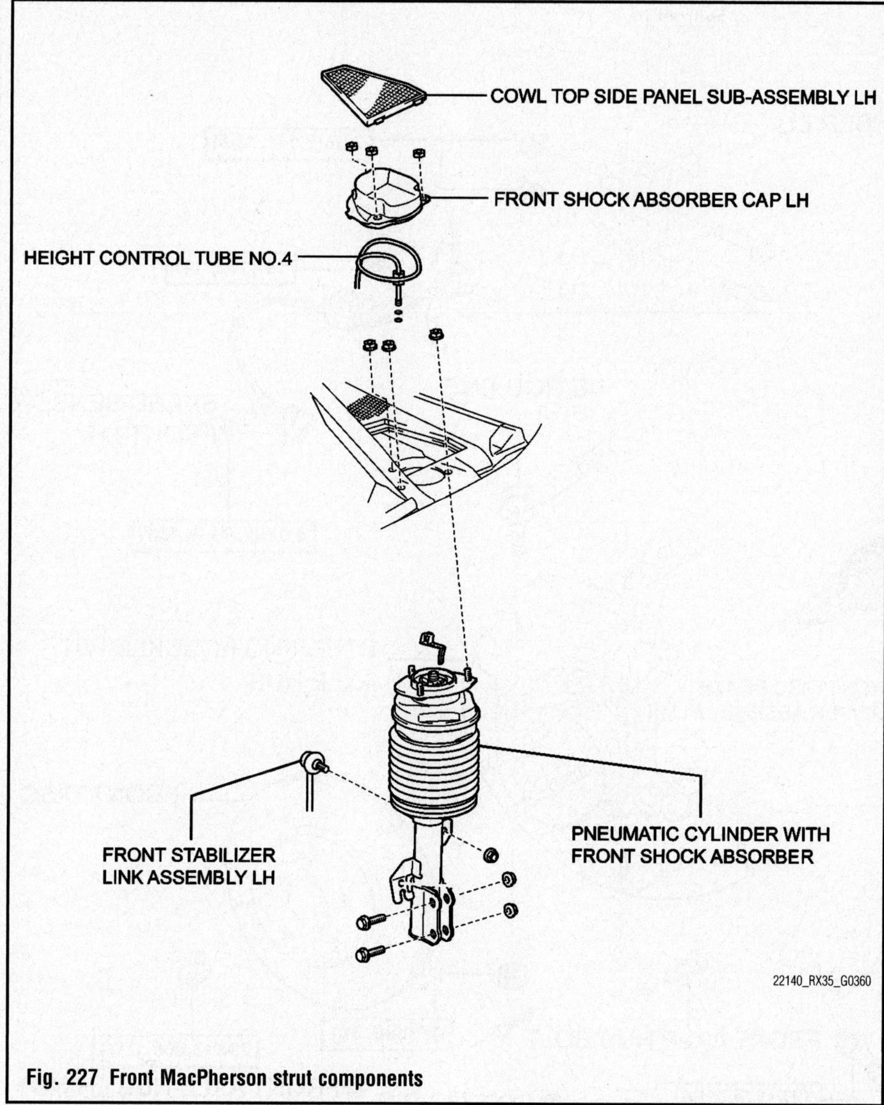

Fig. 227 Front MacPherson strut components

To install:

5. Install the front shock absorber with coil spring.

6. Install the 3 nuts to the upper side of front shock absorber with coil spring and tighten to 59 ft. lbs. (80 Nm). Do not torque the nuts until installing the bolts and nuts (lower side).

7. Install the 2 bolts and 2 nuts to the lower side of front shock absorber with coil spring and tighten to 170 ft. lbs. (230 Nm). When the installing the nuts, keep the bolts from rotating and torque the nuts.

8. Install the No. 1 front flexible hose and front speed sensor wire harness with the bolt and tighten to 14 ft. lbs. (19 Nm).

9. Fully tighten the lock nut to 36 ft. lbs. (49 Nm).

10. Install the cowl top side panel sub-assembly.

11. Install the front stabilizer link assembly with the nut and tighten to 55 ft. lbs. (74 Nm).

➡If the ball joint turns together with the nut, use a hexagon (6 mm) wrench to hold the stud.

12. Install the front wheel.

13. Inspect and adjust front wheel alignment.

SHOCK ABSORBERS

REMOVAL & INSTALLATION

Refer to MacPherson Strut removal and installation procedure.

STEERING KNUCKLE

REMOVAL & INSTALLATION

See Wheel Hub and Bearing.

STABILIZER BAR

REMOVAL & INSTALLATION

1. Remove the front wheel.

2. Remove the nuts and the left and right front stabilizer link assembly.

➡If the ball joint turns together with the nut, use a hexagon (6 mm) wrench to hold the stud.

3. For 4WD vehicles, remove the No. 2 engine under cover and the front No. 3 exhaust pipe sub-assembly.

4. Remove the bolts and the left and right front stabilizer brackets.

5. Disconnect the left and right tie rod end sub-assemblies.

6. Separate the steering intermediate shaft sub-assembly.

7. Disconnect the steering gear outlet return tube (SST: 09023-12701).

8. Disconnect the pressure feed tube assembly (SST: 09023-12701).

9. Remove the power steering link assembly.

10. Remove the No. 1 front stabilizer bar bushing, as follows:

a. Remove the 2 bushings from the stabilizer.

11. For 4WD vehicles, remove the manifold stay and the No. 2 exhaust manifold heat insulator.

12. Remove the front stabilizer bar from the vehicle.

To install:

13. Installation is the reverse of the removal procedure, noting the following:

a. Tighten the No. 1 front stabilizer bracket bolts to 12 ft. lbs. (16 Nm).

b. Tighten the front stabilizer link assembly nuts to 55 ft. lbs. (74 Nm).

14. Bleed power steering fluid and check power steering fluid leakage.

15. Inspect and adjust steering wheel center point.

16. Inspect and adjust front wheel alignment.

WHEEL HUB AND BEARING

REMOVAL & INSTALLATION

See Figure 228.

1. Remove the front wheel.

2. Remove the front axle hub nut.

3. Separate the front speed sensor.

4. Remove the 2 bolts and separate the front disc brake caliper assembly from the steering knuckle.

5. Remove the front disc.

6. Separate the tie rod end sub-assembly.

7. Separate the lower No. 1 front suspension arm sub-assembly.

8. Remove the front axle assembly, as follows:

a. Using a plastic hammer, separate

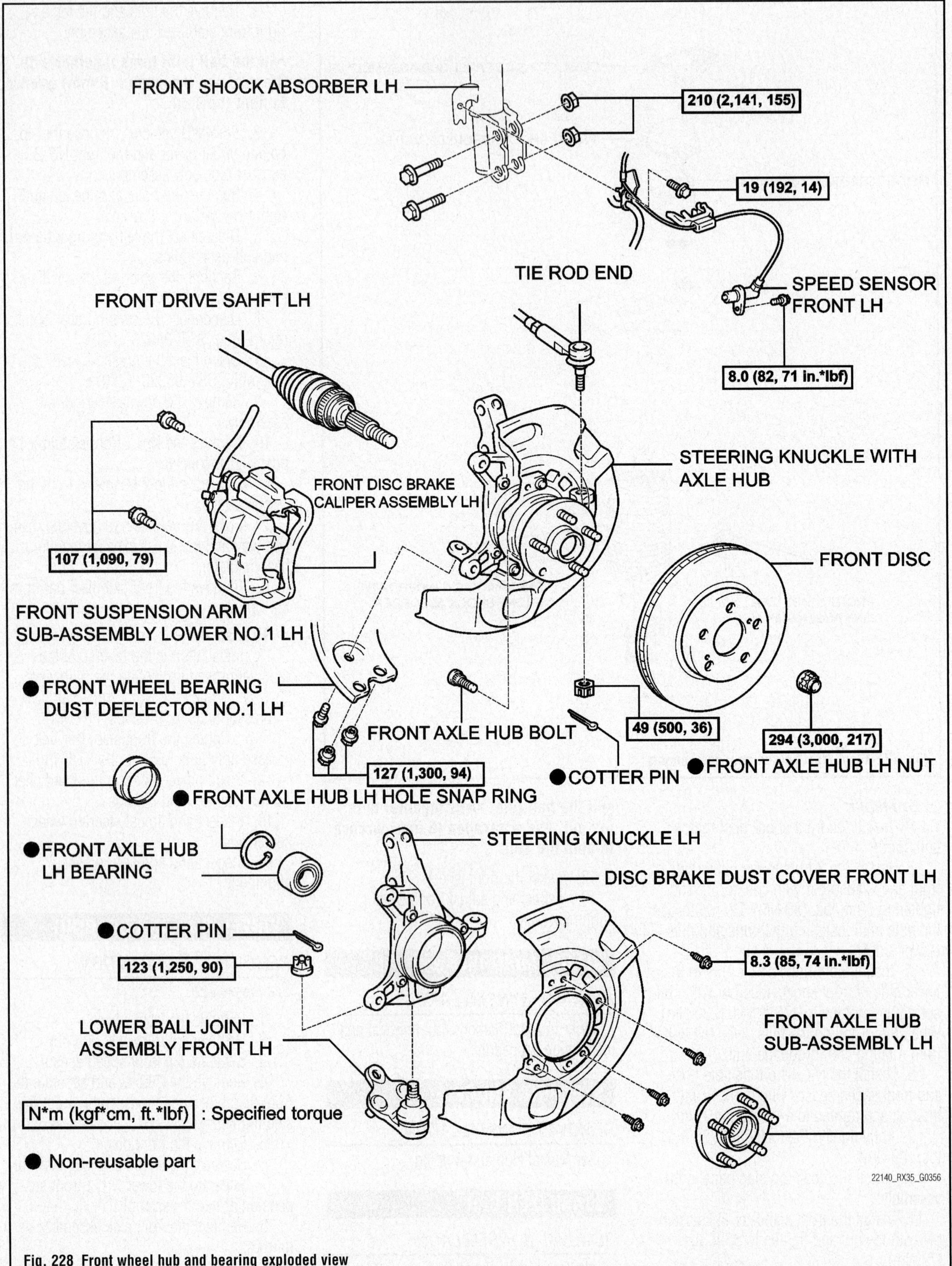

Fig. 228 Front wheel hub and bearing exploded view

the front drive shaft assembly from the front axle hub sub-assembly.

WARNING

Be careful not to damage the boot and ABS speed sensor rotor.

b. Remove the 2 bolts, nuts and steering knuckle with the front axle hub sub-assembly.

To install:

9. Install the front axle assembly, as follows:

a. Install the 2 bolts, nuts and front axle assembly with the 2 bolts and nuts to the front shock absorber assembly and tighten to 170 ft. lbs. (230 Nm).

➡Only when reusing the bolts and nuts, apply the small amount of engine oil to the screw part of the nuts.

➡Do not excessively push out the front axle assembly.

WARNING

Be careful not to damage the outboard joint boot or speed sensor rotor.

10. Install the lower No. 1 front suspension arm sub-assembly.
11. Install the tie rod end sub-assembly.
12. Install the front disc.
13. Install the front disc brake caliper assembly with the 2 bolts to the steering knuckle and tighten to 77 ft. lbs. (104 Nm).
14. Using a 30 mm socket wrench, install the front axle hub nut and tighten to 217 ft. lbs. (294 Nm).
15. Remove the 2 bolts and separate the front disc brake caliper assembly from the steering knuckle.
16. Remove the front disc.
17. Inspect bearing backlash.
18. Inspect axle hub deviation.
19. Install the front disc.
20. Install the front disc brake caliper assembly with the 2 bolts to the steering knuckle and tighten to 77 ft. lbs. (104 Nm).
21. Install the front speed sensor.
22. Install the front axle hub nut, as follows:

a. Using a socket wrench (30 mm), install a new axle hub nut and tighten to 217 ft. lbs. (294 Nm).

b. Using a chisel and hammer, stake the axle hub nut.

23. Install the front wheel.
24. Inspect and adjust front wheel alignment.
25. Check ABS speed sensor signal.

SUSPENSION — REAR SUSPENSION

CONTROL ARMS/LINKS

REMOVAL & INSTALLATION

2WD

No. 1

➡Press the height control switch to stop the vehicle height control operation before jacking up or lifting up the vehicle with air suspension.

➡Press the height control switch to operate the vehicle height control after jacking down or lifting down the vehicle with air suspension.

➡Support the rear axle carrier with a jack.

1. Remove the rear wheel.
2. Remove the rear stabilizer bar.
3. Remove the left rear No. 1 suspension arm assembly, as follows:
4. Remove the bolt and disconnect the rear suspension arm assembly No. 1 (inner side).
5. Remove the bolt, nut and the rear suspension arm assembly No. 1 (outer side) from the rear axle carrier. When removing the bolt, keep the nut from rotating.

To install:

6. Install the left rear No. 1 suspension arm assembly (inner side) with the bolt, and temporarily tighten the bolt. Ensure that the paint mark faces to the rear.
7. Connect the left rear No. 1 suspension arm assembly (outer side) to the rear axle carrier with the bolt and nut, and temporarily tighten the bolt and nut. When installing the bolt, tighten the bolt temporarily with the nut fixed.
8. Stabilize the suspension.
9. Fully tighten the left rear No. 1 suspension arm assembly, as follows:

a. Fully tighten the bolt to 89 ft. lbs. (120 Nm).

b. Fully tighten the bolt to 83 ft. lbs. (112 Nm).

10. Install the rear stabilizer bar.
11. Install the rear wheel.
12. Inspect the rear wheel alignment.

No. 2

➡Press the height control switch to stop the vehicle height control operation before jacking up or lifting up the vehicle with air suspension.

➡Press the height control switch to operate the vehicle height control after jacking down or lifting down the vehicle with air suspension.

➡Support the rear axle carrier with a jack.

1. Remove the rear wheel.
2. Remove the left rear No. 2 suspension arm assembly, as follows:

a. Remove the nut, and separate the height control sensor sub-assembly from the No. 2 suspension arm assembly (w/ height control sensor sub-assembly).

b. Remove the bolt, nut and disconnect the No. 2 rear suspension arm assembly (inner side).

c. Remove the bolt, nut and the No. 2 rear suspension arm assembly (outer side) from the rear axle carrier. When removing the bolt, keep the nut from rotating.

To install:

3. Install the rear No. 2 suspension arm assembly (inner side) with the bolt, and temporarily tighten the bolt. Ensure that the paint mark faces to the rear.
4. Connect the rear No. 2 suspension arm assembly (outer side) to the rear axle carrier with the bolt and nut, and temporarily tighten the bolt. When installing the bolt, tighten the bolt temporarily with the nut fixed.
5. Stabilize the suspension.
6. Fully tighten the left rear No. 2 suspension arm assembly, as follows:

a. Fully tighten the bolt to 89 ft. lbs. (120 Nm).

b. Fully tighten the bolt to 83 ft. lbs. (112 Nm).

7. Install the height control sensor sub-assembly and nut to the suspension arm assembly No. 2 (w/ height control sensor sub-assembly) and tighten to 48 inch lbs. (5.4 Nm).
8. Install the rear wheel.
9. Inspect the rear wheel alignment.
10. Adjust vehicle height.
11. Adjust headlight aim only.

4WD

No. 1

See Figure 229.

1. Remove rear wheel.
2. Remove exhaust pipe assembly.
3. Remove propeller w/center bearing shaft assembly.
4. Remove rear strut rod assembly.
5. Remove the height control sensor sub-assembly, and separate the height control sensor wire (w/ height control sensor sub-assembly).
    a. Remove the bolt, nut and the left and right rear No. 2 suspension arm assemblies (outer side) from the rear axle carrier. When removing the bolt, keep the nut from rotating.
6. Remove the bolt, nut and the left and right No. 1 rear suspension arm assemblies (outer side) from the rear axle carrier. When removing the bolt, keep the nut from rotating.
7. Remove the left and right speed sensors.
8. Remove the right and left rear drive shaft assemblies.
9. Support the rear suspension member with a jack.
10. Remove the 4 nuts, 2 bolts and 2 retainers from the rear suspension member.
11. Lower the rear suspension member.
12. Remove the bolt, nuts and left rear No. 1 suspension arm assembly. When removing the bolt, keep the nut from rotating.

To install:

13. Install the left rear No. 1 suspension arm assembly with the bolt and nut, and temporarily tighten the bolt. Ensure that the paint mark faces to the rear.
14. Set the No. 1 rear suspension arm assembly as shown in the illustration, and fully tighten the bolt and tighten to 59 ft. lbs. (80 Nm).
15. Raise the rear suspension member with a jack.
16. Install the rear suspension member with the 4 nuts, 2 bolts and 2 retainers. Front side: Torque: 85 ft. lbs. (115 Nm). Rear side: 134 ft. lbs. (181 Nm).

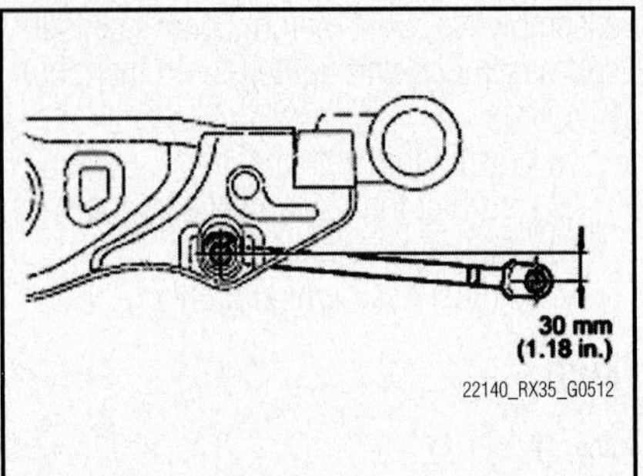

Fig. 229 Suspension arm placement

17. Install the right and left rear drive shaft assemblies.
18. Install the left and right speed sensors.
19. Install the left and right No. 1 rear suspension arm assemblies (outside) to the rear axle carrier, and temporarily tighten the bolt. When installing the bolt, tighten the bolt temporarily with the nut fixed.
20. Install the left and right No. 2 rear suspension arms (outside) to the rear axle carrier, and temporarily tighten the bolt. When installing the bolt, tighten the bolt temporarily with the nut fixed.
21. Temporarily tighten the rear strut rod assembly.
22. Stabilize the suspension.
23. Fully tighten the left and right rear No. 1 suspension arm assemblies, as follows:
    a. Fully tighten the bolt to 83 ft. lbs. (112 Nm).
24. Fully tighten the left and right rear No. 2 suspension arm assemblies, as follows:
    a. Fully tighten the bolt to 48 inch lbs. (5.4 Nm).
25. Fully tighten the rear strut rod assembly.
26. Install the rear wheel.
27. Temporarily tighten the propeller w/center bearing shaft assembly.
28. Install the exhaust pipe assembly.
29. Check for exhaust gas leak.
30. Inspect the rear wheel alignment.
31. Adjust vehicle height.
32. Adjust headlight aim only.
33. Check the ABS speed sensor signal.

No. 2

See Figure 230.

➡**Press the height control switch to stop the vehicle height control operation before jacking up or lifting up the vehicle with air suspension.**

➡**Press the height control switch to operate the vehicle height control after jacking down or lifting down the vehicle with air suspension.**

➡**Support the rear axle carrier with a jack.**

1. Remove the rear wheel.
2. Remove the left rear No. 2 suspension arm assembly, as follows:
    a. Remove the nut, and separate the height control sensor sub-assembly from the No. 2 suspension arm assembly (w/ height control sensor sub-assembly).

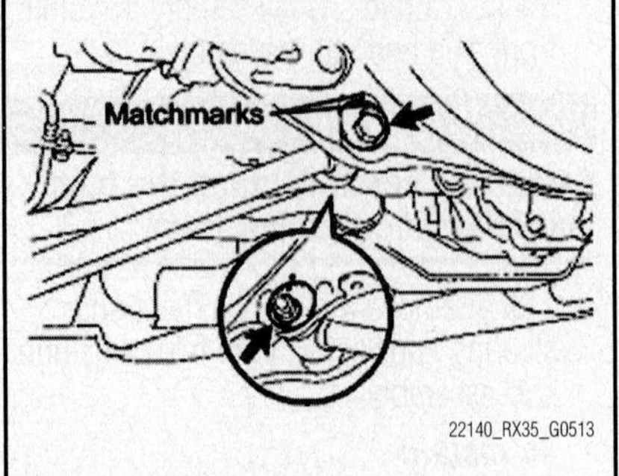

Fig. 230 Matchmarks on the adjust cams and rear suspension member sub-assembly

    b. Place matchmarks on the adjust cams and rear suspension member sub-assembly.
    c. Remove the nut, camber adjust cam and toe adjust cam, and disconnect the No. 2 rear suspension arm assembly (inner side). When removing the nut, keep the bolt from rotating.
    d. Remove the bolt, nut and the No. 2 rear suspension arm assembly (outer side) from the rear axle carrier. When removing the bolt, keep the nut from rotating.

To install:

3. Install the left No. 2 rear suspension arm (inner side) to the rear suspension member sub-assembly with the camber adjust cam and toe adjust cam, and temporarily tighten the nut. Ensure that the paint mark faces to the rear.
4. Connect the left No. 2 rear suspension arm assembly (outer side) to the rear axle carrier with the bolt and nut. When installing the bolt, tighten the bolt temporarily with the nut fixed.
5. Stabilize the suspension.
6. Fully tighten the left rear No. 2 suspension arm assembly, as follows:
    a. Align the matchmarks on the adjust cams and rear suspension member sub-assembly.
    b. Fully tighten the nut to 74 ft. lbs. (100 Nm).
    c. Fully tighten the nut to 83 ft. lbs. (112 Nm).
    d. Install the height control sensor sub-assembly and nut to the No. 2 suspension arm assembly (w/ height control sensor sub-assembly) to 48 inch lbs. (5.4 Nm).
7. Install the rear wheel.
8. Inspect the rear wheel alignment.
9. Adjust vehicle height.
10. Adjust headlight aim only.

MACPHERSON STRUTS

REMOVAL & INSTALLATION

See Figures 231 and 232.

➡**Support the rear axle carrier with a jack.**

1. Remove the tonneau cover assembly.
2. Remove the deck side trim cover.
3. Remove the rear wheel.
4. Separate the rear stabilizer link assembly, as follows:
   a. Support the rear axle carrier with a jack.
   b. Remove the nut and disconnect the stabilizer link from the shock absorber.

➡**If the ball joint turns together with the nut, use a hexagon (5 mm) wrench to hold the stud.**

5. Remove the rear shock absorber with coil spring (MacPherson strut), as follows:
   a. For 2WD vehicles, disconnect the skid control sensor connector. Remove the 2 bolts, and disconnect the flexible hose and skid control sensor wire from the shock absorber and rear axle carrier.
   b. For 4WD vehicles, remove the 3 bolts, and disconnect the flexible hose and speed sensor from the shock absorber and rear axle carrier.
   c. Loosen the 2 nuts on the lower side of the shock absorber. Do not remove the 2 bolts and 2 nuts.
   d. Support the rear axle carrier with a jack.
   e. Remove the 3 nuts.
   f. Lower the rear axle carrier, and remove the 2 nuts and 2 bolts on the lower side of the rear shock absorber.
   g. Remove the shock absorber with coil spring.

To install:

6. Install the rear shock absorber with coil spring and 3 nuts and tighten to 43 ft. lbs. (58 Nm).
7. Install the shock absorber with coil spring, 2 bolts and nuts and tighten to 133 ft. lbs. (180 Nm). Keep the bolts from rotating.
   a. For 2WD vehicles, install the flexible hose and skid control sensor wire with the 2 bolts and tighten bolt A to 14 ft. lbs. (19 Nm), and bolt B 44 inch lbs. (5 Nm). Connect the skid control sensor connector.
   b. For 4WD vehicles, install the flexible hose and speed sensor with the 3 bolts and tighten bolt A to 44 inch lbs. (5 Nm), bolt B to 14 ft. lbs. (19 Nm), and bolt C to 71 inch lbs. (8 Nm).
8. Install the stabilizer link to the shock absorber with the nut and tighten to 29 ft. lbs. (39 Nm).

➡**If the ball joint turns together with the nut, use a hexagon (5 mm) wrench to hold the stud.**

9. Install the rear wheel. Tighten the lug nuts to 76 ft. lbs. (103 Nm).
10. Inspect rear wheel alignment.

WHEEL HUB AND BEARING

REMOVAL & INSTALLATION

2WD Vehicles

See Figure 233.

1. Remove the rear wheel.
2. Separate the rear disc brake caliper assembly, as follows:
   a. Remove the bolt and separate the flexible hose from the shock absorber.
   b. Remove the 2 bolts and separate the rear disc brake caliper assembly.
3. Remove the rear disc.
4. Disconnect the skid control sensor connector.
5. Remove the 4 bolts and the rear axle hub and bearing assembly.

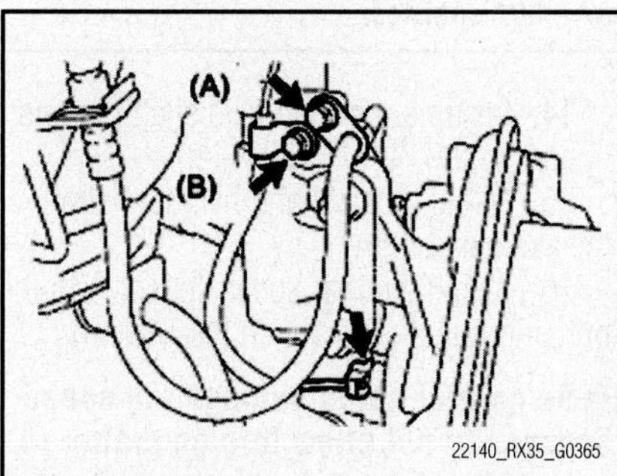

Fig. 231 Flexible hose and skid control sensor wire bolt tightening sequence—2WD vehicles

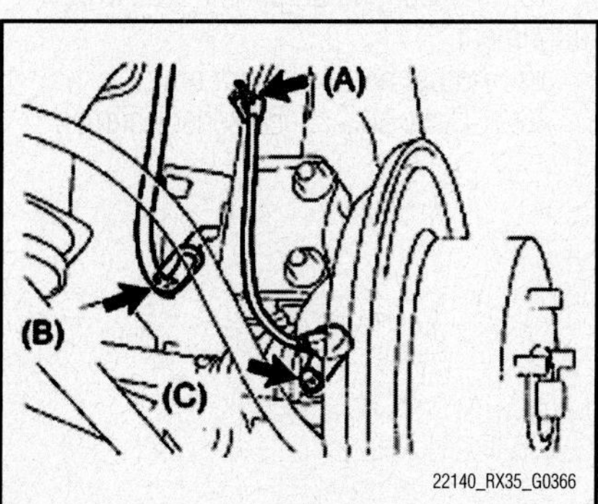

Fig. 232 Flexible hose and speed sensor bolt tightening sequence—4WD vehicles

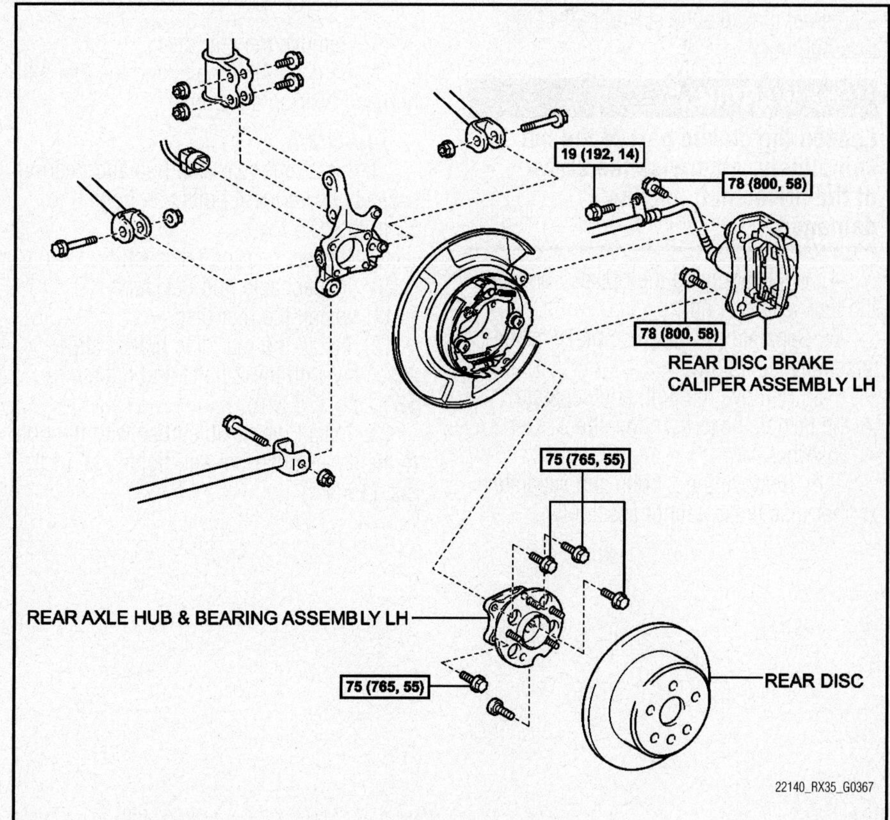

Fig. 233 Rear wheel hub and bearing exploded view—2WD vehicles

To install:

6. Install the hub and bearing assembly with the 4 bolts and tighten to 55 ft. lbs. (75 Nm).
7. Inspect bearing backlash.
8. Inspect axle hub deviation.
9. Connect the skid control sensor connector. Do not twist the sensor wire.
10. Install the rear disc.
11. Install the rear disc brake caliper assembly, as follows:
   a. Install the rear disc brake caliper with the 2 bolts and tighten to 58 ft. lbs. (78 Nm).
   b. Install the bolt with flexible hose to the shock absorber and tighten to 14 ft. lbs. (19 Nm).
12. Install the rear wheel.
13. Inspect and adjust the rear wheel alignment.
14. Adjust headlight aim only.

4WD Vehicles

See Figure 234.

1. Remove the rear wheel.
2. Remove the bolt and separate the speed sensor from the axle carrier.

➡Be careful not to damage the speed sensor. Do not allow foreign matter to adhere to the speed sensor.

3. Using SST (SST: 09930-00010) and a hammer, unstake the staked part of the axle hub nut.

⁂ WARNING

Loosen the staked part of the nut completely, otherwise the screw of the drive shaft may be damaged.

4. While applying the brakes, remove the lock axle hub nut.
5. Separate the rear disc brake caliper assembly, as follows:
   a. Remove the bolt, and separate the flexible hose from the shock absorber.
   b. Remove the 2 bolts and separate rear disc brake caliper assembly.

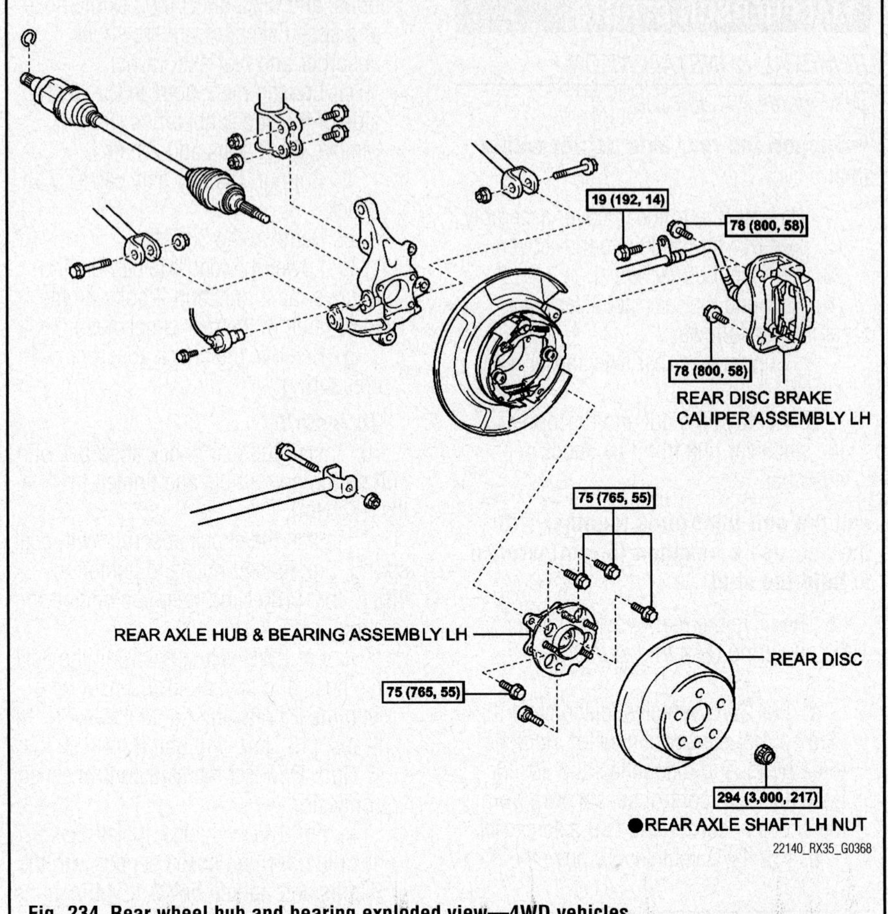

Fig. 234 Rear wheel hub and bearing exploded view—4WD vehicles

6. Remove the rear disc.
7. Remove the 4 bolts and the rear axle hub and bearing assembly.

To install:

8. Install the rear axle hub and bearing assembly with the 4 bolts and tighten to 55 ft. lbs. (75 Nm).
9. Inspect bearing backlash.
10. Inspect axle hub deviation.
11. Install the rear disc.
12. Install the rear disc brake caliper assembly with the 2 bolts and tighten to 58 ft. lbs. (78 Nm).
13. Install the flexible hose with the bolt to the shock absorber and tighten to 14 ft. lbs. (19 Nm).
14. Install a new rear axle shaft nut and tighten to 217 ft. lbs. (294 Nm).
15. Using a chisel and hammer, stake the axle shaft nut.
16. Install the rear speed sensor with the bolt and tighten to 71 inch lbs. (8 Nm).

➡Be careful not to damage the speed sensor. Do not allow foreign matter to adhere to the speed sensor.

➡Do not twist the sensor wire.

17. Install the rear wheel.
18. Inspect and adjust the rear wheel alignment.
19. Adjust headlight aim only.
20. Check ABS speed sensor signal.

LEXUS

RX 400h

7

SPECIFICATIONS AND MAINTENANCE CHARTS

ENGINE AND VEHICLE IDENTIFICATION

Engine						
Code ①	Liters (cc)	Cu. In.	Cyl.	Fuel Sys.	Engine Type	Eng. Mfg.
3MZ-FE	3.3 (3311)	202.1	6	SFI	DOHC	Toyota

Model Year	
Code ②	Year
6	2006
7	2007
8	2008

SFI: Sequential Fuel Injection

DOHC: Double Overhead Camshaft

① Stamped on the left side of the engine block

② 10th digit of the Vehicle Identification Number (VIN)

22140_RX40_C0001

GENERAL ENGINE SPECIFICATIONS

Year	Model	Engine Displacement Liters	Engine Series ID	Net Horsepower @ rpm	Net Torque @ rpm (ft. lbs.)	Bore x Stroke (in.)	Com-pression Ratio	Oil Pressure @ rpm
2006	RX 400h	3.3	3MZ-FE	268@5600	212@4400	3.62x3.27	10.8:1	36-78@3000
2007	RX 400h	3.3	3MZ-FE	268@5600	212@4400	3.62x3.27	10.8:1	36-78@3000
2008	RX 400h	3.3	3MZ-FE	268@5600	212@4400	3.62x3.27	10.8:1	36-78@3000

22140_RX40_C0002

ENGINE TUNE-UP SPECIFICATIONS

Year	Engine Displacement Liters	Engine ID	Spark Plug Gap (in.)	Ignition Timing (deg.)	Fuel Pump (psi)	Idle Speed (rpm)	Valve Clearance	
							Intake	Exhaust
2006	3.3	3MZ-FE	0.039-0.043	8-12B①	44-50	850-950	0.006-0.010	0.010-0.014
2007	3.3	3MZ-FE	0.039-0.043	8-12B①	44-50	850-950	0.006-0.010	0.010-0.014
2008	3.3	3MZ-FE	0.039-0.043	8-12B①	44-50	850-950	0.006-0.010	0.010-0.014

NOTE: The Vehicle Emission Control Information label often reflects specification changes made during production. The label figures must be used if they differ from those in this chart.

B: Before top dead center

① With terminals TC and CG of DLC3 connected

22140_RX40_C0003

CAPACITIES

Year	Model	Engine Displacement Liters	Engine ID	Engine Oil with Filter (qts.)	Transaxle (pts)	Rear Transaxle (pts.)	Rear Drive Axle (pts.)	Fuel Tank (gal.)	Cooling System (qts.)
2006	RX 400h	3.3	3MZ-FE	5.0	①	4.2	N/A	17.2	10.3
2007	RX 400h	3.3	3MZ-FE	5.0	①	4.2	N/A	17.2	10.3
2008	RX 400h	3.3	3MZ-FE	5.0	①	4.2	N/A	17.2	10.3

① With towing package: 8.8 pts.
Without towing package: 8.2 pts.

22140_RX40_C0005

FLUID SPECIFICATIONS

Year	Model	Engine Displacement Liters	Engine ID/VIN	Engine Oil	Auto. Trans. ①	Drive Axle ②	Power Steering Fluid	Brake Master Cylinder	Engine Coolant ③
2006	RX 400h	3.3	3MZ-FE	5W-30	ATF-WS	75W-90	Dexron II or III	DOT 3	Toyota coolant
2007	RX 400h	3.3	3MZ-FE	5W-30	ATF-WS	75W-90	Dexron II or III	DOT 3	Toyota coolant
2008	RX 400h	3.3	3MZ-FE	5W-30	ATF-WS	75W-90	Dexron II or III	DOT 3	Toyota coolant

DOT: Department Of Transpotation

① The use of genuine Toyota ATF-WS is recommended

② Synthetic GL-5 (75W-90) or equivalent

③The use of genuine Toyota engine coolant is recommended or similar

ethylene glycol based non-silicate, non-amine, non- nitrite, and non- borat coolant

22140_RX40_C0004

VALVE SPECIFICATIONS

All measurements in inches unless noted

Year	Engine Displacement Liters	Engine ID	Seat Angle (deg.)	Face Angle (deg.)	Spring Test Pressure (lbs. @ in.)	Spring Installed Height (in.)	Stem-to-Guide Clearance (in.) Intake	Stem-to-Guide Clearance (in.) Exhaust	Stem Diameter (in.) Intake	Stem Diameter (in.) Exhaust
2006	3.3	3MZ-FE	45	40.5	41.9-46.3@ 1.331	1.331	0.0010-0.0024	0.0012-0.0026	0.2154-0.2159	0.2152 0.2157
2007	3.3	3MZ-FE	45	40.5	41.9-46.3@ 1.331	1.331	0.0010-0.0024	0.0012-0.0026	0.2154-0.2159	0.2152 0.2157
2008	3.3	3MZ-FE	45	40.5	41.9-46.3@ 1.331	1.331	0.0010-0.0024	0.0012-0.0026	0.2154-0.2159	0.2152 0.2157

22140_RX40_C0006

CAMSHAFT AND BEARING SPECIFICATIONS CHART

All measurements are given in inches.

Year	Engine Displ. Liters	Engine ID/VIN	Journal Dia.	Brg. Oil Clearance	Shaft End-play	Runout	Journal Bore	Lobe Height Intake	Lobe Height Exhaust
2006	3.3	3MZ-FE	1.0614-1.0620	①	0.0016-0.0035	0.0024	NA	1.6981-1.7020	1.6933-1.6972
2007	3.3	3MZ-FE	1.0614-1.0620	①	0.0016-0.0035	0.0024	NA	1.6981-1.7020	1.6933-1.6972
2008	3.3	3MZ-FE	1.0614-1.0620	①	0.0016-0.0035	0.0024	NA	1.6981-1.7020	1.6933-1.6972

NA: Not Available

① Intake Journals 4 and 5: 0.0010 - 0.0022 in.

All Others: 0.0010 - 0.0024 in.

22140_RX40_C0008

CRANKSHAFT AND CONNECTING ROD SPECIFICATIONS

All measurements are given in inches.

Year	Engine Displacement Liters	Engine ID	Crankshaft Main Brg. Journal Dia.	Crankshaft Main Brg. Oil Clearance	Crankshaft Shaft End-play	Crankshaft Thrust on No.	Connecting Rod Journal Diameter	Connecting Rod Oil Clearance	Connecting Rod Side Clearance
2006	3.3	3MZ-FE	2.4011-2.4016	①	0.0016-0.0094	2	2.0863-2.0866	0.0015-0.0026	0.0059-0.0118
2007	3.3	3MZ-FE	2.4011-2.4016	①	0.0016-0.0094	2	2.0863-2.0866	0.0015-0.0026	0.0059-0.0118
2008	3.3	3MZ-FE	2.4011-2.4016	①	0.0016-0.0094	2	2.0863-2.0866	0.0015-0.0026	0.0059-0.0118

① Journals 1 and 4: 0.0006 - 0.0013 in.

Journals 2 and 3: 0.0010 - 0.0018 in.

22140_RX40_C0007

PISTON AND RING SPECIFICATIONS

All measurements are given in inches.

Year	Engine Displ. Liters	Engine ID	Piston Clearance	Ring Gap Top Comp.	Ring Gap Bottom Comp.	Ring Gap Oil Control	Ring Side Clearance Top Comp.	Ring Side Clearance Bottom Comp.	Ring Side Clearance Oil Control
2006	3.3	3MZ-FE	0.0013-0.0023	0.0118-0.0157	0.0197-0.0236	0.0059-0.0157	0.0012-0.0031	0.0008-0.0024	0.0012-0.0043
2007	3.3	3MZ-FE	0.0013-0.0023	0.0118-0.0157	0.0197-0.0236	0.0059-0.0157	0.0012-0.0031	0.0008-0.0024	0.0012-0.0043
2008	3.3	3MZ-FE	0.0013-0.0023	0.0118-0.0157	0.0197-0.0236	0.0059-0.0157	0.0012-0.0031	0.0008-0.0024	0.0012-0.0043

22140_RX40_C0009

TORQUE SPECIFICATIONS

All readings in ft. lbs.

Year	Engine Displacement Liters	Engine ID	Cylinder Head Bolts	Main Bearing Bolts	Rod Bearing Bolts	Crankshaft Damper Bolts	Flywheel Bolts	Manifold Intake	Manifold Exhaust	Spark Plugs	Oil Pan Drain Plug
2006	3.3	3MZ-FE	①	②	③	162	61	11	36	18	33
2007	3.3	3MZ-FE	①	②	③	162	61	11	36	18	33
2008	3.3	3MZ-FE	①	②	③	162	61	11	36	18	33

① Step 1: 12 point bolts to 40 ft. lbs.
Step 2: 12 point bolts plus 90 degrees
Step 3: Hex head recessed bolt to 14 ft. lbs.

② Step 1: 12 point cap bolts to 16 ft. lbs.
Step 2: 12 point cap bolts plus 90 degrees
Step 3: Hex head side bolts to 20 ft. lbs.

③ Step 1: 18 ft. lbs.
Step 2: Plus 90 degrees

22140_RX40_C0010

WHEEL ALIGNMENT

Year	Model		Caster Range (+/-Deg.)	Caster Preferred Setting (Deg.)	Camber Range (+/-Deg.)	Camber Preferred Setting (Deg.)	Toe-in (in.)
2006	RX 400h	2WD F	0.75	+2.75	0.75	-0.58	0+/-0.08
		4WD F	0.75	+2.50	0.75	-0.58	0+/-0.08
		2WD R	—	—	0.75	-1.17	0.12+/-0.08
		4WD R	—	—	0.75	-0.67	0.12+/-0.08
2007	RX 400h	2WD F	0.75	+2.75	0.75	-0.58	0+/-0.08
		4WD F	0.75	+2.50	0.75	-0.58	0+/-0.08
		2WD R	—	—	0.75	-1.17	0.12+/-0.08
		4WD R	—	—	0.75	-0.67	0.12+/-0.08
2008	RX 400h	2WD F	0.75	+2.75	0.75	-0.58	0+/-0.08
		4WD F	0.75	+2.50	0.75	-0.58	0+/-0.08
		2WD R	—	—	0.75	-1.17	0.12+/-0.08
		4WD R	—	—	0.75	-0.67	0.12+/-0.08

F: Front
R: Rear

22140_RX40_C0011

TIRE, WHEEL AND BALL JOINT SPECIFICATIONS

Year	Model	OEM Tires Standard	OEM Tires Optional	Tire Pressures (psi) Front	Tire Pressures (psi) Rear	Wheel Size	Ball Joint Inspection	Lug Nut Torque (ft. lbs.)
2006	RX 400h	P225/65R17	P235/55VR18	32	32	6.5-JJ	①	76
2007	RX 400h	P225/65R17	P235/55VR18	32	32	6.5-JJ	①	76
2008	RX 400h	P225/65R17	P235/55VR18	32	32	6.5-JJ	①	76

OEM: Original Equipment Manufacturer

PSI: Pounds Per Square Inch

STD: Standard

OPT: Optional

① Replace if any measurable movement is found.

22140_RX40_C0012

BRAKE SPECIFICATIONS

All measurements in inches unless noted

Year	Model		Brake Disc Original Thickness	Brake Disc Minimum Thickness	Brake Disc Maximum Runout	Minimum Lining Thickness	Brake Caliper Bracket Bolts (ft. lbs.)	Brake Caliper Mounting Bolts (ft. lbs.)
2006	RX 400h	F	1.102	1.024	0.0020	0.039	78	25
		R	0.394	0.335	0.0059	0.039	56	32
2007	RX 400h	F	1.102	1.024	0.0020	0.039	78	25
		R	0.394	0.335	0.0059	0.039	56	32
2008	RX 400h	F	1.102	1.024	0.0020	0.039	78	25
		R	0.394	0.335	0.0059	0.039	56	32

F: Front

R: Rear

22140_RX40_C0013

SCHEDULED MAINTENANCE INTERVALS

LEXUS—RX 400h

TO BE SERVICED	TYPE OF SERVICE	VEHICLE MILEAGE INTERVAL (x1000)												
		5	10	15	20	25	30	35	40	45	50	55	60	65
Engine oil & filter	R	✓	✓	✓	✓	✓	✓	✓	✓	✓	✓	✓	✓	✓
Automatic transmission fluid	R												✓	
Ball joints & dust covers	S/I	✓	✓	✓	✓	✓	✓	✓	✓	✓	✓	✓	✓	✓
Bolts & nuts on chassis & body	S/I	✓	✓		✓	✓		✓	✓		✓	✓		✓
Brake line pipes & hoses	S/I			✓			✓			✓			✓	
Brake fluid	R						✓						✓	
Brake pads & discs (front & rear)	S/I	✓	✓	✓	✓	✓	✓	✓	✓	✓	✓	✓	✓	✓
Propeller shaft grease	S/I	✓	✓	✓	✓	✓	✓	✓	✓	✓	✓	✓	✓	✓
Steering knuckle & chassis grease	S/I	✓	✓	✓	✓	✓	✓	✓	✓	✓	✓	✓	✓	✓
Steering linkage	S/I	✓	✓	✓	✓	✓	✓	✓	✓	✓	✓	✓	✓	✓
Air cleaner filter	R						✓						✓	
Air conditioner filter	R						✓							
Spark plugs	R	Replace at 120,000 miles												
Exhaust pipes & mountings	S/I			✓			✓			✓			✓	
Fuel lines & connections	S/I												✓	
Engine coolant	R	Replace at 120,000 miles												
Timing belt	R	Replace at 90,000 miles												
Rear differential fluid	R												✓	
Rotate Tires	S/I	✓	✓	✓	✓	✓	✓	✓	✓	✓	✓	✓	✓	✓

R: Replace   S/I: Service or Inspect

FREQUENT OPERATION MAINTENANCE (SEVERE SERVICE)

If a vehicle is operated under any of the following conditions it is considered severe service:

- Extremely dusty areas.

- 50% or more of the constant operation is in 32°C (90°F) or higher temperatures, or in temperatures below 0°C (32°F).

- Prolonged idling (vehicle operation in stop and go traffic).

- Frequent short running periods (engine does not warm to normal operating temperatures).

- Police, taxi, delivery usage or trailer towing usage.

Air cleaner filter: service or inspect every 3750 miles

Engine oil & filter: replace every 3750 miles.

Ball joints & dust covers: service or inspect every 7500 miles.

Bolts & nuts on chassis & body: service or inspect every 7500 miles.

Brake pads & discs (front & rear): service or inspect every 7500 miles.

Steering knuckle & chassis grease: service or inspect every 7500 miles.

Steering linkage: service or inspect every 7500 miles.

Exhaust pipes & mountings: service or inspect every 15,000 miles.

22140_RX40_C0014

PRECAUTIONS

⁂ CAUTION

The RX400h hybrid system contains a 288V high-voltage system with a strong alkali solution of potassium hydroxide. Be sure to follow the instructions in this manual to handle the system correctly. Failure to do so may result in serious injury or electrocution. Engineer must undergo special training to be able to perform high-voltage system inspection and servicing.

⁂ CAUTION

All high-voltage wire harness connectors are colored orange. The HV battery and other high-voltage components have "High Voltage" caution labels. Do not carelessly touch these wires and components.

⁂ CAUTION

Before inspecting or servicing the high-voltage system, be sure to follow safety measures, such as wearing insulated gloves and removing the service plug to prevent electrocution. Carry the removed service plug in your pocket to prevent anyone from reinstalling it while you are servicing the vehicle.

⁂ CAUTION

After removing the service plug, wait 5 minutes before touching any of the high-voltage connectors and terminals.

Before servicing any vehicle, please be sure to read all of the following precautions, which deal with personal safety, prevention of component damage, and important points to take into consideration when servicing a motor vehicle:

- Never open, service or drain the radiator or cooling system when the engine is hot; serious burns can occur from the steam and hot coolant.
- Observe all applicable safety precautions when working around fuel. Whenever servicing the fuel system, always work in a well-ventilated area. Do not allow fuel spray or vapors to come in contact with a spark, open flame, or excessive heat (a hot drop light, for example). Keep a dry chemical fire extinguisher near the work area. Always keep fuel in a container specifically designed for fuel storage; also, always properly seal fuel containers to avoid the possibility of fire or explosion. Refer to the additional fuel system precautions later in this section.
- Fuel injection systems often remain pressurized, even after the engine has been turned **OFF**. The fuel system pressure must be relieved before disconnecting any fuel lines. Failure to do so may result in fire and/or personal injury.
- Brake fluid often contains polyglycol ethers and polyglycols. Avoid contact with the eyes and wash your hands thoroughly after handling brake fluid. If you do get brake fluid in your eyes, flush your eyes with clean, running water for 15 minutes. If eye irritation persists, or if you have taken brake fluid internally, IMMEDIATELY seek medical assistance.
- The EPA warns that prolonged contact with used engine oil may cause a number of skin disorders, including cancer. You should make every effort to minimize your exposure to used engine oil. Protective gloves should be worn when changing oil. Wash your hands and any other exposed skin areas as soon as possible after exposure to used engine oil. Soap and water, or waterless hand cleaner should be used.
- All new vehicles are now equipped with an air bag system, often referred to as a Supplemental Restraint System (SRS) or Supplemental Inflatable Restraint (SIR) system. The system must be disabled before performing service on or around system components, steering column, instrument panel components, wiring and sensors. Failure to follow safety and disabling procedures could result in accidental air bag deployment, possible personal injury and unnecessary system repairs.
- Always wear safety goggles when working with, or around, the air bag system. When carrying a non-deployed air bag, be sure the bag and trim cover are pointed away from your body. When placing a non-deployed air bag on a work surface, always face the bag and trim cover upward, away from the surface. This will reduce the motion of the module if it is accidentally deployed. Refer to the additional air bag system precautions later in this section.
- Clean, high quality brake fluid from a sealed container is essential to the safe and proper operation of the brake system. You should always buy the correct type of brake fluid for your vehicle. If the brake fluid becomes contaminated, completely flush the system with new fluid. Never reuse any brake fluid. Any brake fluid that is removed from the system should be discarded. Also, do not allow any brake fluid to come in contact with a painted surface; it will damage the paint.
- Never operate the engine without the proper amount and type of engine oil; doing so WILL result in severe engine damage.
- Timing belt maintenance is extremely important. Many models utilize an interference-type, non-freewheeling engine. If the timing belt breaks, the valves in the cylinder head may strike the pistons, causing potentially serious (also time-consuming and expensive) engine damage. Refer to the maintenance interval charts for the recommended replacement interval for the timing belt, and to the timing belt section for belt replacement and inspection.
- Disconnecting the negative battery cable on some vehicles may interfere with the functions of the on-board computer system(s) and may require the computer to undergo a relearning process once the negative battery cable is reconnected.
- When servicing drum brakes, only disassemble and assemble one side at a time, leaving the remaining side intact for reference.
- Only an MVAC-trained, EPA-certified automotive technician should service the air conditioning system or its components.

BRAKES — ANTI-LOCK BRAKE SYSTEM (ABS)

GENERAL INFORMATION

See Figure 1.

The system detects the degree of brake pedal operation with a brake pedal stroke sensor and two master cylinder pressure sensors, and calculates the optimum hydraulic brake force. The hydraulic pressure source is adjusted based on this so that the optimum hydraulic control is independently performed on all four wheels.

Meanwhile, the Electronically Controlled Brake system performs control of the normal brakes, ABS, TRAC, VSC and brake assist in accordance with the operations of the driver.

There is a hydraulic backup mechanism that applies master cylinder pressure generated by human power to the wheel cylinder when the brake control stops. In addition, as a failsafe mechanism, when the brake control is malfunctioning, the system excludes the malfunctioning sections and continues to perform brake control on the normal sections. A power backup unit (brake control power supply) is also used to ensure a stable supply of power to the system.

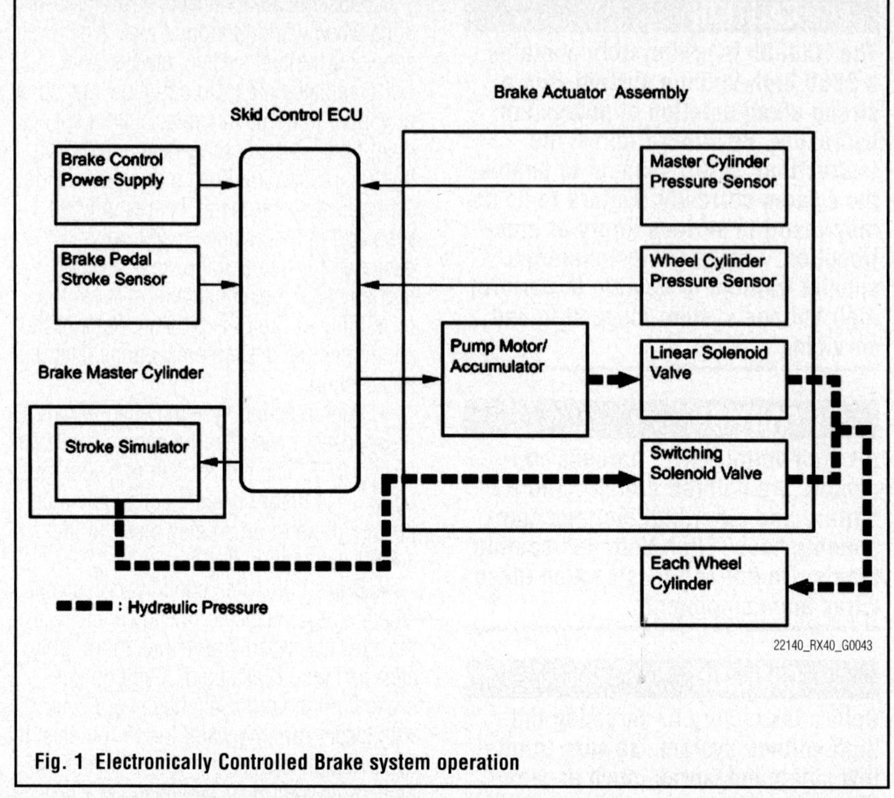

Fig. 1 Electronically Controlled Brake system operation

BRAKES — BLEEDING THE BRAKE SYSTEM

BLEEDING PROCEDURE

See Figures 2 through 4.

➡This procedure requires specialized tools. Please read through the procedure and make sure you have access to the proper equipment before beginning the bleeding procedure.

CAUTION

Never bleed air from the brake hydraulic system without using the intelligent tester. failure to use the intelligent tester could cause serious injury or an accident.

Note the following before bleeding the brake system:

- Move the shift lever to the P position and apply the parking brake before bleeding.
- Add brake fluid carefully and check that the reservoir level remains between the min and max lines while bleeding the brakes.
- Do not stand the fluid can on the reservoir inlet when bleeding the brake actuator. doing so will cause brake fluid to overflow.
- The actuator pump motor and solenoid can be operated by the driver even if the ignition switch is off.
- If the pump motor operates while air still remains inside the brake actuator hose, air will enter the actuator, making it more difficult to bleed the brakes. If there is concern about air remaining in the actuator hose, remove the two motor relays (skid control relay No. 2) until instructed to reinstall them.
- Although a buzzer may sound due to a decline in the accumulator pressure while bleeding, it is not necessary to stop bleeding.
- DTCs indicating a malfunction in the motor relays (skid control relay No. 2) or the pressure sensor are stored after bleeding. clear the DTCs when instructed during or after bleeding.

1. Add SAE J1703 or FMVSS No. 116 DOT 3 brake fluid to the max line in the reservoir.

CAUTION

Add brake fluid carefully and check that the reservoir level remains between the min and max lines while bleeding the brakes. Do not stand the fluid can on the reservoir inlet when bleeding the brake actuator. Doing so will cause brake fluid to overflow.

2. Disable the brake control (ECB). When using the intelligent tester:

➡When using the intelligent tester , refer to the intelligent tester operator's manual for further details. Bleed the air by following the steps displayed on the intelligent tester.

a. Move the shift lever to the P position and apply the parking brake.

b. Connect the intelligent tester to the DLC3 with the ignition switch **OFF** as shown in the illustration.

c. Turn the ignition switch to the ON position and turn on the intelligent tester.

➡Do not start the engine.

d. Enter the following menus: DIAGNOSIS / OBD/ MOBD / ABS/ TRAC/VSC / ECB UTILITY / ECB INVALID.

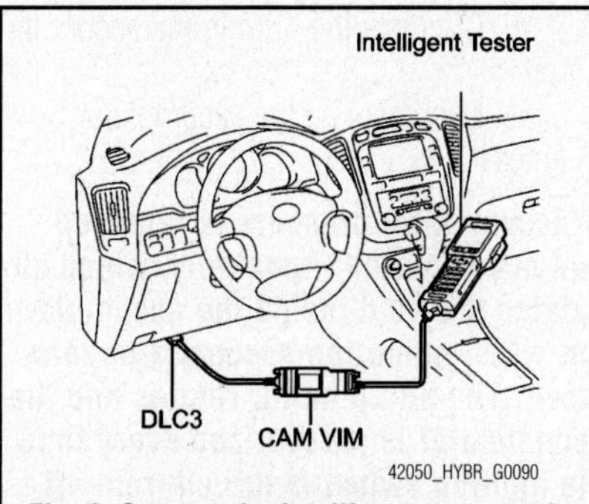

Fig. 2 Connect the intelligent tester to the DLC3 with the ignition switch OFF as shown in the illustration

WARNING

If the pump motor operates while air remains inside the brake actuator hose, air will enter the actuator, and this will make bleeding the brakes more difficult.

e. When removing the ABS motor relay: Remove the 2 ABS motor relays with the ignition switch off in order to disable brake control.

WARNING

If the pump motor operates while air remains inside the brake actuator hose, air will enter the actuator, and this will make bleeding the brakes more difficult.

➡After the brake actuator assembly has been replaced, remove the ABS motor relay before bleeding the brakes.

3. **Bleed the brake actuator hose, as follows:**

a. Connect Special Tool 09992-00242, 0992-00350 or equivalent, to the reservoir with the brake reservoir pressure adapter.

b. Using Special Tool 09023-00101, loosen the bleeder plug of the actuator.

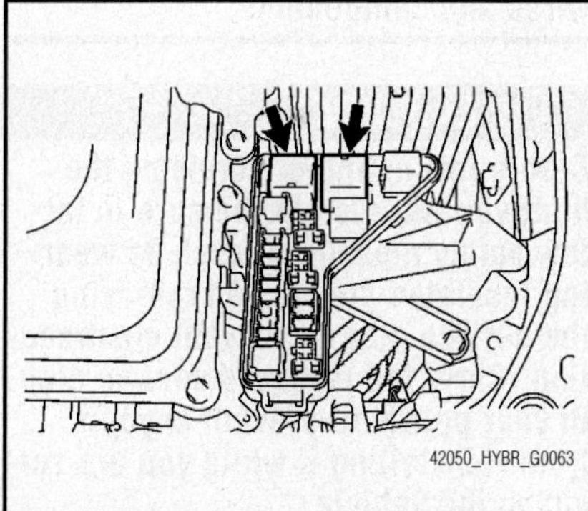

Fig. 3 Location of the 2 ABS motor relays

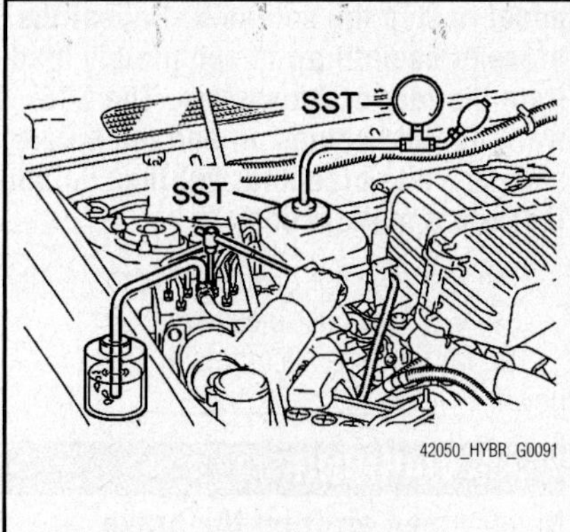

Fig. 4 View of the special tools

c. Connect a vinyl tube to the bleeder plug of the actuator.

d. Use the SST to boost pressure in the reservoir. Standard pressure is 50 to 80 kPa (0.5 to 0.8 kgf/cm2, 7.3 to 11.6 psi)

e. Drain approximately 100 cc of fluid.

f. Tighten the bleeder plug and boost the pressure in the reservoir again (50 to 80 kPa (0.5 to 0.8 kgf/cm2)). Then, loosen the bleeder plug and bleed the brake actuator hose.

➡Repeat this procedure at least 5 times.

g. When air is completely bled out from the hose between the reservoir and the actuator, tighten the bleeder plug to 74 inch lbs. (8.3 Nm).

4. **Bleed the master cylinder, as follows:**

➡If the master cylinder has been disassembled or if the reservoir becomes empty, bleed the air from the master cylinder.

a. Enter the following menus: DIAGNOSIS / OBD/MOBD / ABS/TRAC/VSC / AIR BLEEDING.

b. Select "USUAL" if the front/rear brakes are removed, installed or disassembled.

c. Select "ACTUATOR" if the actuator is removed, installed or replaced.

d. Select "MASTER CYLINDER" if the brake master cylinder or the brake stroke simulator is removed, installed or replaced.

e. Disconnect the brake lines from the master cylinder.

f. Slowly depress and hold the brake pedal (Procedure A).

g. Cover the outer holes with fingers, and release the brake pedal (Procedure B).

h. Repeat procedure A and B 3 or 4 times.

i. Connect the brake lines to the master cylinder and tighten to 11 ft. lbs. (15 Nm).

5. **Bleed the front brake system, as follows:**

➡Air can be easily bled from the front brake system if air has been bled from the master cylinder when replacing the brake master cylinder assembly.

WARNING

If brake fluid leaks onto any painted surface of the vehicle, wash or otherwise remove it completely.

➡Bleed the air by following the steps displayed on the intelligent tester.

(a) Depress the brake pedal several times and bleed the front brake system from the bleeder plugs on the front brake cylinder RH and LH.

➡Repeat the procedure until air is completely bled from the front brake system.

6. Tighten the bleeder plugs to 74 inch lbs. (8.3 Nm) after bleeding.

7. Cancel brake control (ECB) disable

a. Install the 2 motor relays (skid control relay No. 2) if they have been removed.

b. Complete brake control prevention following the prompts on the tester screen. (If brake control has been prevented using the intelligent tester.)

8. Clear the DTC(s).

9. Bleed the rear brake system, as follows:

WARNING

Never bleed air from the brake hydraulic system without using the intelligent tester. Failure to use the intelligent tester could cause serious injury or an accident.

➡Bleed the air by following the steps displayed on the intelligent tester.

a. Connect the intelligent tester to the DLC3 with the ignition switch off.

b. Check that the parking brake is applied and turn the ignition switch to the **ON** position.

c. Enter the following menus: DIAGNOSIS / OBD/ MOBD / ABS/ TRAC/VSC / ECB UTILITY / ECB INVALID.

d. With the brake pedal depressed, bleed the rear brake system from the bleeder plug on the rear disc brake cylinder LH while the pump motor and solenoid are operating.

⁂ WARNING

Keep the fluid inside the reservoir above the LOW level by replenishing.

➡Depress and hold the brake pedal. After the solenoid operates for approximately 30 seconds, release the brake pedal to stop the solenoid. Repeat the procedures until air is completely bled from the rear brake system. The ECB warning light comes on and the buzzer sounds while bleeding, but they do not indicate a malfunction.

e. Tighten the bleeder plug to 74 inch lbs. (8.3 Nm) after bleeding.

f. Enter the following menus: DIAGNOSIS / OBD/ MOBD / ABS/TRAC/VSC / ECB UTILITY / ECB INVALID.

g. With the brake pedal depressed, bleed the rear brake system from the bleeder plug on the rear disc brake cylinder RH while the pump motor and solenoid are operating.

⁂ WARNING

Keep the fluid inside the reservoir above the LOW level by replenishing.

➡Depress and hold the brake pedal. After the solenoid operates for approximately 30 seconds, release the brake pedal to stop the solenoid. Repeat the procedures until air is completely bled from the rear brake system. The ECB warning light comes on and the buzzer sounds while bleeding, but they do not indicate a malfunction.

h. Tighten the bleeder plug to 74 inch lbs. (8.3 Nm) after bleeding.

10. Perform the accumulator zero down:

⁂ WARNING

Never bleed air from the brake hydraulic system without using the intelligent tester. Failure to use the intelligent tester could cause serious injury or an accident. Be sure to perform this procedure before replacement, removal, or installation of the actuator.

➡Perform accumulator zero down by following the steps displayed on the intelligent tester.

a. Connect the intelligent tester to the DLC3 with the ignition switch **OFF**.

b. Depressurize the accumulator:

- Check that the parking brake is applied and turn the ignition switch to the **ON** position.
- Enter the following menus: DIAGNOSIS / OBD/ MOBD / ABS/TRAC/VSC / ECB UTILITY / ZERO DOWN.
- When the buzzer sounds, turn the ignition switch **OFF**.

c. Circulate the fluid in the accumulator.

d. Depressurize the accumulator 5 times.

➡Accumulator pressure is released and accumulated repeatedly, which circulates the fluid inside the accumulator, when repeating accumulator zero down. The pump motor rotates and the accumulator is pressurized every time the ignition switch is turned from off to on.

11. Check the brake fluid level:

a. After performing accumulator zero down (accumulator depressurizing), return the fluid in the accumulator back to the reservoir and then adjust the fluid level in the master cylinder reservoir to the MAX level.

➡After performing accumulator zero down (accumulator depressurizing), fluid is built up in the accumulator by turning the ignition switch to the ON position and the fluid level of the reservoir lowers. If the fluid level is adjusted without performing accumulator zero down (accumulator depressurizing), fluid is sent from the accumulator to the reservoir. The fluid level may exceed the MAX level, but it is normal.

12. Clear the DTC(s).

13. When the brake actuator assembly is replaced, perform linear valve offset learning after bleeding is completed

BRAKES — FRONT DISC BRAKES

⁂ CAUTION

Dust and dirt accumulating on brake parts during normal use may contain asbestos fibers from production or aftermarket brake linings. Breathing excessive concentrations of asbestos fibers can cause serious bodily harm. Exercise care when servicing brake parts. Do not sand or grind brake lining unless equipment used is designed to contain the dust residue. Do not clean brake parts with compressed air or by dry brushing. Cleaning should be done by dampening the brake components with a fine mist of water, then wiping the brake components clean with a dampened cloth. Dispose of cloth and all residue containing asbestos fibers in an impermeable container with the appropriate label. Follow practices prescribed by the Occupational Safety and Health Administration (OSHA) and the Environmental Protection Agency (EPA) for the handling, processing, and disposing of dust or debris that may contain asbestos fibers.

⁂ CAUTION

The RX400h hybrid system contains a 288V high-voltage system with a strong alkali solution of potassium hydroxide. Be sure to follow the instructions in this manual to handle the system correctly. Failure to do so may result in serious injury or electrocution. Engineer must undergo special training to be able to perform high-voltage system inspection and servicing.

⁂ CAUTION

All high-voltage wire harness connectors are colored orange. The HV battery and other high-voltage components have "High Voltage" caution labels. Do not carelessly touch these wires and components.

⁂ CAUTION

Before inspecting or servicing the high-voltage system, be sure to follow safety measures, such as wearing insulated gloves and removing the service plug to prevent electrocution. Carry the removed service plug in your pocket to prevent anyone from reinstalling it while you are servicing the vehicle.

⁂ CAUTION

After removing the service plug, wait 5 minutes before touching any of the high-voltage connectors and terminals.

BRAKE CALIPER

REMOVAL & INSTALLATION

See Figure 5.

1. Raise the vehicle.
2. Remove the tire and wheel assembly.
3. Disconnect the brake line from the caliper and plug it.
4. Hold the caliper slide pins and remove the mounting bolts.
5. Lift off the caliper.
6. Remove the pads and anti-squeal shims.

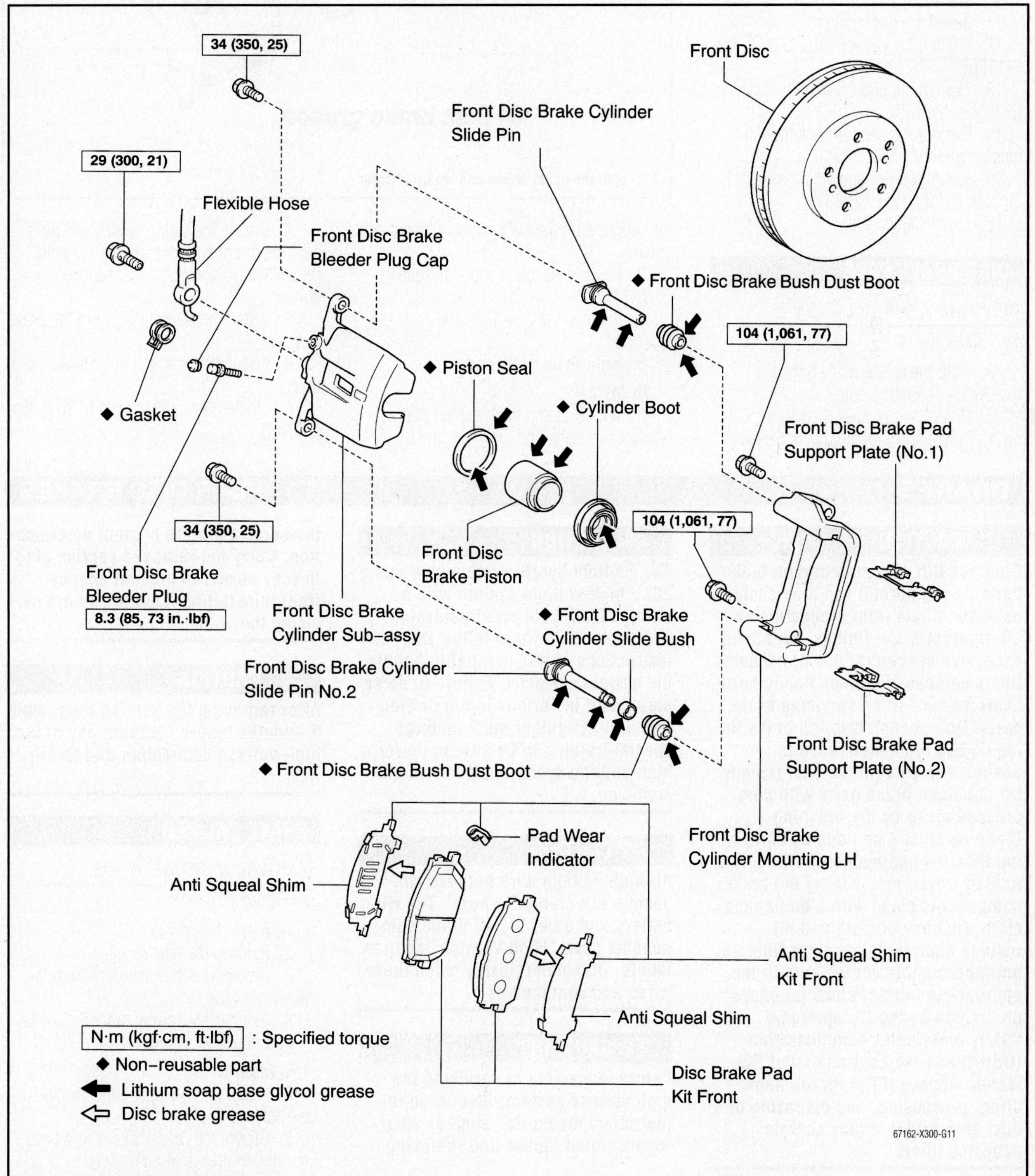

Fig. 5 Front disc brake components

7. Remove the wear indicator from the inner pad.
8. Remove the brake caliper.

To install:

9. Grease the caliper slides and bolts with lithium grease or equivalent.
10. Apply disc brake grease to the anti-squeal shims.
11. Install the brake caliper.
12. Tighten the caliper bolts to 25 ft. lbs. (34 Nm).
13. Tighten the brake line union bolt to 21 ft. lbs. (29 Nm).
14. Bleed the brake caliper and refill master cylinder.
15. Install the tire and wheel assembly. Tighten to 76 ft. lbs. (103 Nm).
16. Lower the vehicle.

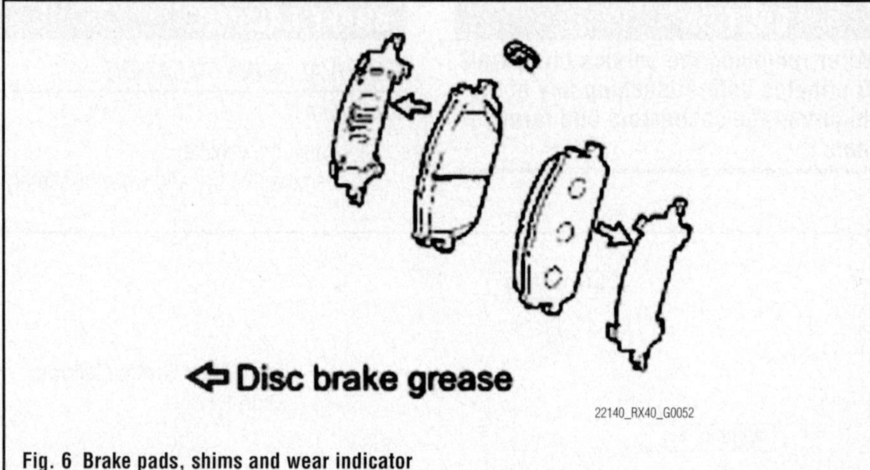

Fig. 6 Brake pads, shims and wear indicator

DISC BRAKE PADS

REMOVAL & INSTALLATION

See Figure 6.

1. Hold the caliper slide pins and remove the mounting bolts.
2. Unbolt and remove the caliper, but do NOT disconnect the fluid line. Suspend the caliper from the suspension with a piece of wire.
3. Remove the pads and anti-squeal shims.
4. Remove the wear indicator from the inner pad.
5. Remove the brake pads.

To install:

6. With C clamp push caliper piston back into its bore.
7. Grease the caliper slides and bolts with lithium grease or equivalent. Apply disc brake grease to the anti-squeal shims.
8. Install the wear indicator to the brake pad.
9. Install the inner and outer brake pads.
10. Tighten the caliper bolts to 25 ft. lbs. (34 Nm).

BRAKES — REAR DISC BRAKES

**** CAUTION**

Dust and dirt accumulating on brake parts during normal use may contain asbestos fibers from production or aftermarket brake linings. Breathing excessive concentrations of asbestos fibers can cause serious bodily harm. Exercise care when servicing brake parts. Do not sand or grind brake lining unless equipment used is designed to contain the dust residue. Do not clean brake parts with compressed air or by dry brushing. Cleaning should be done by dampening the brake components with a fine mist of water, then wiping the brake components clean with a dampened cloth. Dispose of cloth and all residue containing asbestos fibers in an impermeable container with the appropriate label. Follow practices prescribed by the Occupational Safety and Health Administration (OSHA) and the Environmental Protection Agency (EPA) for the handling, processing, and disposing of dust or debris that may contain asbestos fibers.

**** CAUTION**

The RX400h hybrid system contains a 288V high-voltage system with a strong alkali solution of potassium hydroxide. Be sure to follow the instructions in this manual to handle the system correctly. Failure to do so may result in serious injury or electrocution. Engineer must undergo special training to be able to perform high-voltage system inspection and servicing.

**** CAUTION**

All high-voltage wire harness connectors are colored orange. The HV battery and other high-voltage components have "High Voltage" caution labels. Do not carelessly touch these wires and components.

**** CAUTION**

Before inspecting or servicing the high-voltage system, be sure to follow safety measures, such as wearing insulated gloves and removing the service plug to prevent electrocution. Carry the removed service plug in your pocket to prevent anyone from reinstalling it while you are servicing the vehicle.

**** CAUTION**

After removing the service plug, wait 5 minutes before touching any of the high-voltage connectors and terminals.

BRAKE CALIPER

REMOVAL & INSTALLATION

See Figure 7.

1. Raise the vehicle.
2. Remove the rear wheel.
3. Disconnect the brake line from the caliper and plug it.
4. Remove the caliper mounting bolts.
5. Lift off the caliper.

To install:

6. Install the rear disc brake caliper assembly.
7. Grease the caliper slides and bolts with lithium grease or equivalent.

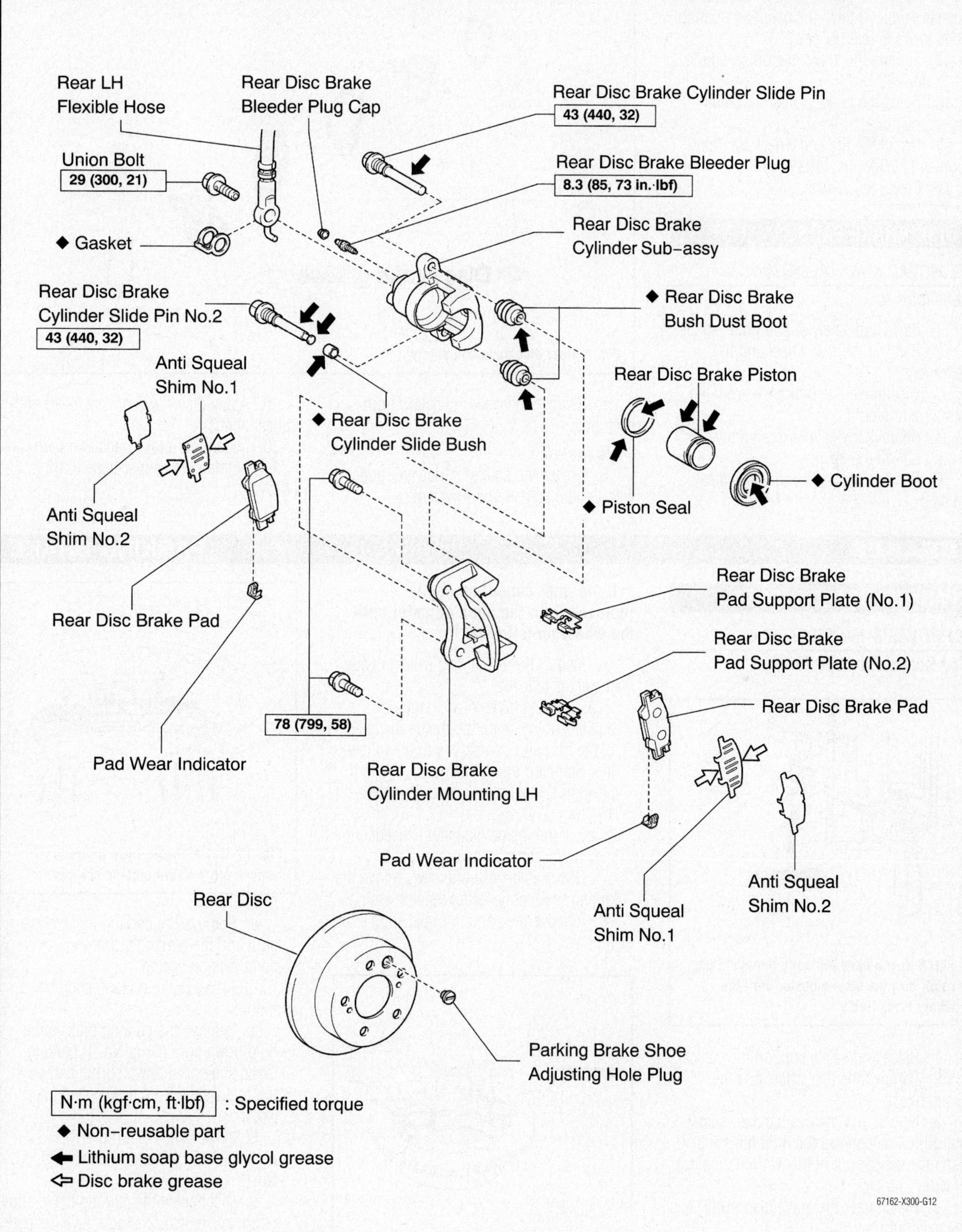

Fig. 7 Rear disc brake components

8. Apply disc brake grease to the anti-squeal shims. Tighten the caliper mounting bolts to 32 ft. lbs. (43 Nm).

9. Tighten the brake line union bolt to 21 ft. lbs. (29 Nm).

10. Bleed the brake caliper and refill master cylinder.

11. Install the tire and wheel assembly. Tighten to 76 ft. lbs. (103 Nm).

12. Lower the vehicle.

DISC BRAKE PADS

REMOVAL & INSTALLATION

See Figure 8.

1. Raise and safely support the vehicle.
2. Remove the rear wheel and tire assembly.
3. Disconnect the brake line from the caliper and plug it.
4. Remove the caliper mounting bolts.
5. Lift off the caliper.
6. Remove the pads and anti-squeal shims.

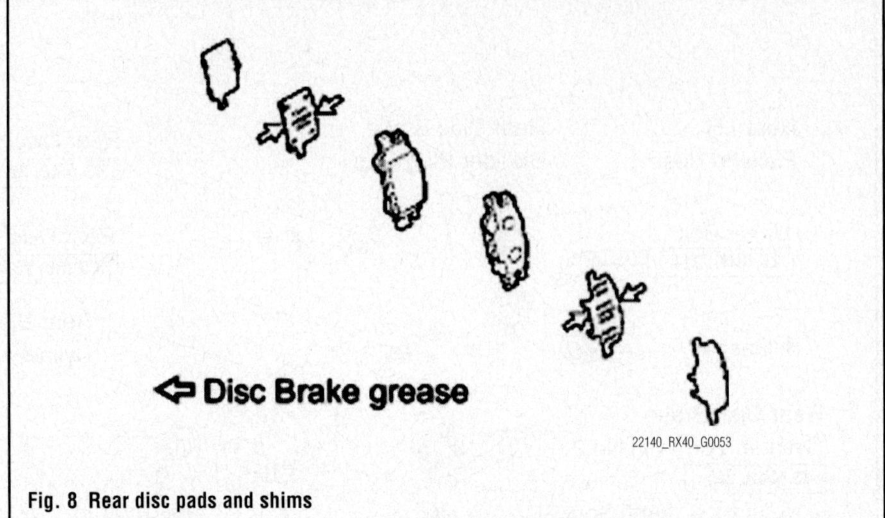

Fig. 8 Rear disc pads and shims

7. Remove the wear indicators from each pad.

To install:

8. Grease the caliper slides and bolts with lithium grease or equivalent.

9. Apply disc brake grease to the anti-squeal shims.

10. Install rear brake pads and hardware.

11. Tighten the caliper bolts to 32 ft. lbs. (43 Nm).

BRAKES — PARKING BRAKE

PARKING BRAKE SHOES

REMOVAL & INSTALLATION

See Figures 9 through 20.

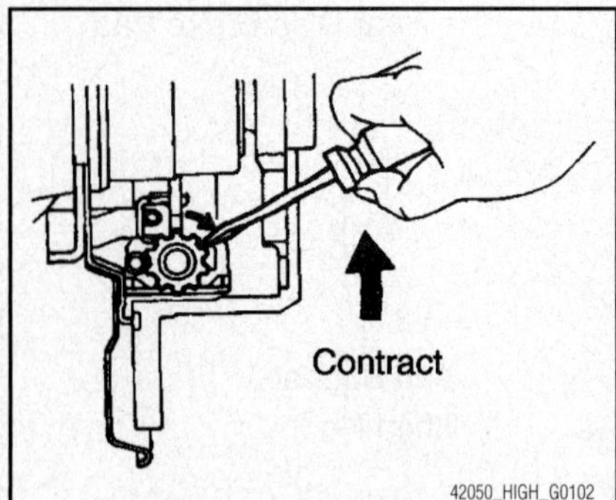

Fig. 9 If you have difficulty removing the rotor, turn the shoe adjuster until the wheel turns freely

1. Raise and safely support the vehicle.
2. Remove the rear wheel and tire assemblies.
3. Unbolt and remove the rear caliper, but do not disconnect the fluid line. Suspend the caliper out of the way with a piece of wire.
4. Matchmark the brake disc (rotor) to the axle hub.
5. Make sure the parking brake is fully released, then remove the rear brake disc (rotor).

➡If the rotor cannot be easily removed, turn the shoe adjuster until the wheel turns freely.

6. Inspect the brake disc (rotor) inside diameter, as follows:
   a. Using a brake drum gauge or equivalent, measure the inside diameter of the disc and compare with the following: Standard inside diameter: 190 mm (7.48 in.). Maximum inside diameter: 191 mm (7.52 in.)
   b. If the inside diameter exceeds the maximum, replace the brake disc.
7. Use needle-nose pliers to remove the 2 parking brake shoe return tension springs.
8. Remove the parking brake shoe strut, as follows:

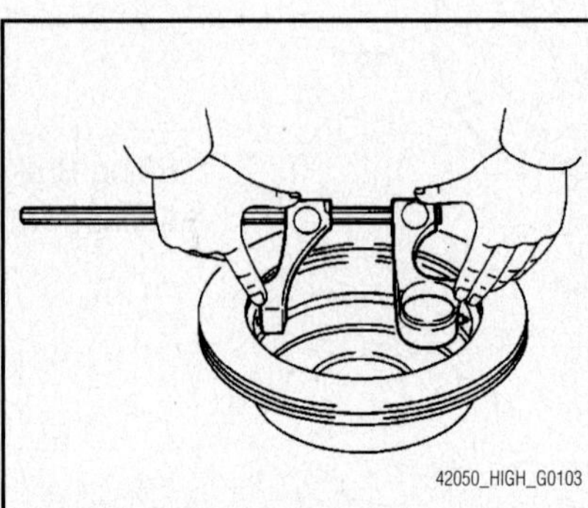

Fig. 10 Use a brake drum gauge to measure the inside of the disc

Fig. 11 Use a needle-nose pliers to remove the 2 return tension springs

   a. Remove the parking brake shoe strut and the parking brake shoe strut compression spring.
9. Remove parking brake shoe No. 1, as follows:
   a. Remove the parking brake shoe hold down spring cup No. 1, parking brake shoe hold down spring and parking brake shoe hold down spring cup No. 2.
   b. 2WD vehicles, remove the parking brake shoe hold down spring pin No. 1.
   c. Disconnect the parking brake shoe return spring No. 2 and remove the parking brake shoe assembly No. 1.
10. Remove parking brake shoe adjusting screw set:

Fig. 12 Remove the parking brake shoe hold-down spring cups

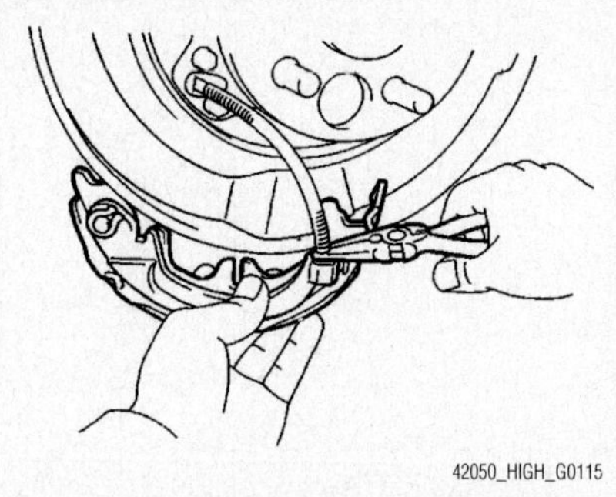

Fig. 14 Using needle-nose pliers, disconnect the parking brake cable No. 3 from the parking brake cable shoe lever

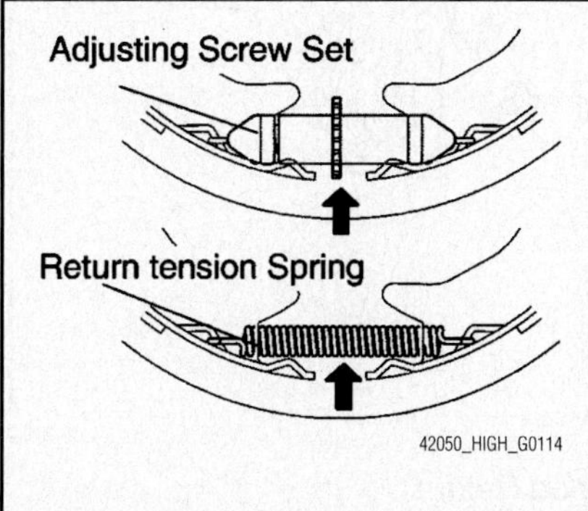

Fig. 13 Remove the parking brake shoe adjusting screw set and shoe return spring

a. Remove the parking brake shoe adjusting screw set.

b. Remove the parking brake shoe return tension spring No. 2.

11. Remove parking brake shoe assembly No. 2:

a. Remove the parking brake shoe hold down spring cup No. 1, parking brake shoe hold down spring, parking brake shoe hold down spring cup No. 2 and parking brake shoe hold down spring pin No. 2.

b. Remove the parking brake shoe assembly lh No. 2.

c. Using needle-nose pliers, disconnect the parking brake cable No. 3 from the parking brake cable shoe lever.

WARNING

Be careful not to damage parking brake cable No. 3.

12. On 4WD models, separate the rear speed sensor.

13. On 4WD models, remove the rear axle shaft nut.

14. On 4WD models, remove rear axle hub & bearing assembly

15. On 4WD models, remove parking brake shoe hold down spring pin.

16. Remove parking brake shoe type C-washer, as follows:

a. Using a screwdriver, remove the c-washer.

b. Remove the shim and parking brake shoe lever from the parking brake shoe No. 2.

17. Inspect parking brake shoe lining thickness:

a. Using a ruler, measure the thickness of the shoe lining. Standard thickness is 2.5 mm (0.098 in.) and minimum thickness is 1.0 mm (0.039 in.). If the lining thickness is less than or equal to the minimum, or If there is severe or uneven wear, replace the brake shoe.

18. Inspect brake disc and parking brake shoe lining for proper contact

a. Apply chalk to the inside surface of the disc, then grind down the brake shoe lining to fit disc.

b. If the contact between the brake disc and the shoe lining is improper, repair it using a brake shoe grinder or replace the brake shoe assembly.

To install:

19. Install the parking brake shoe type C-washer, as follows

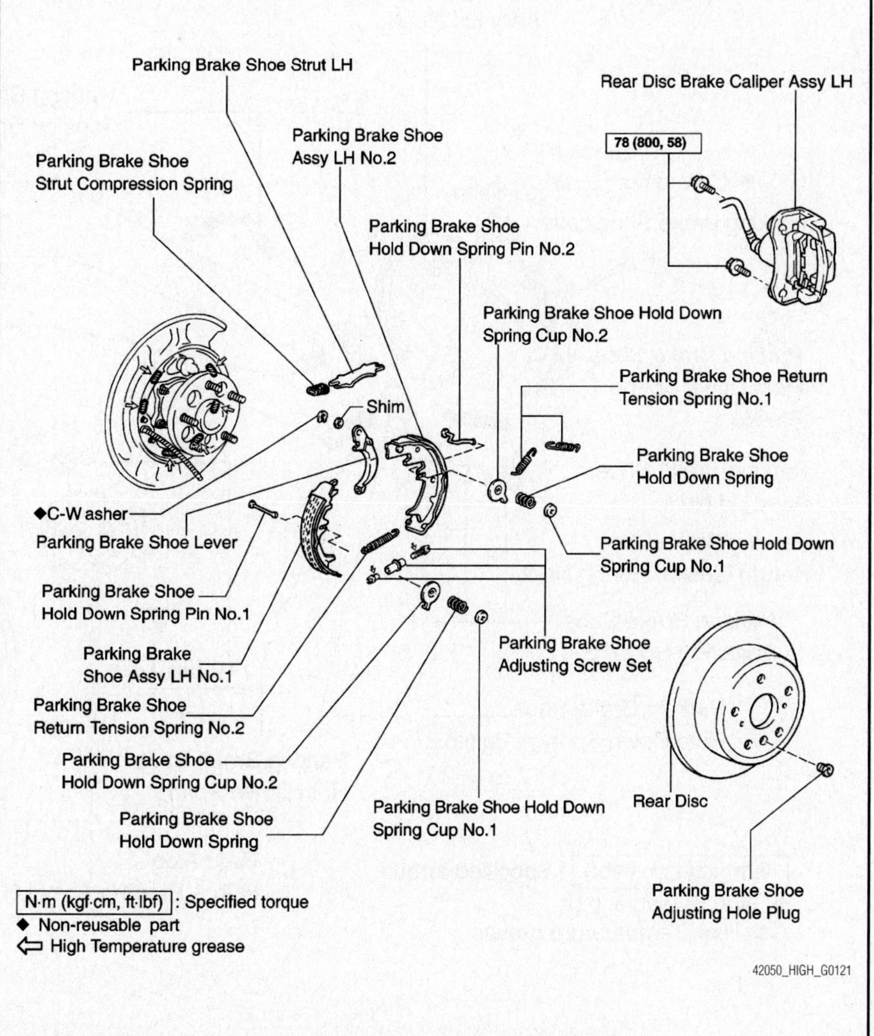

Fig. 15 Exploded view of the parking brake components—2WD vehicles

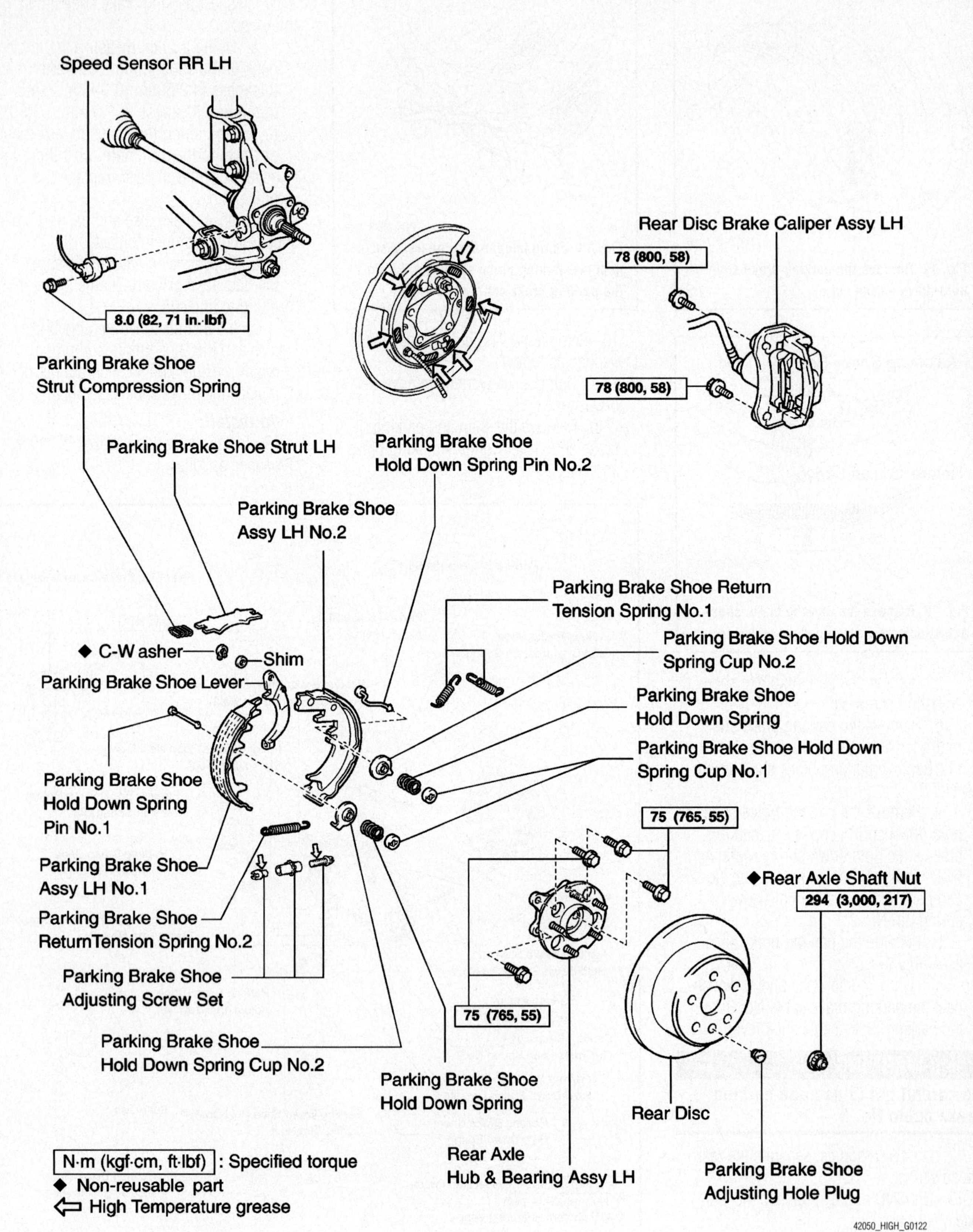

Fig. 16 Exploded view of the parking brake components—4WD vehicles

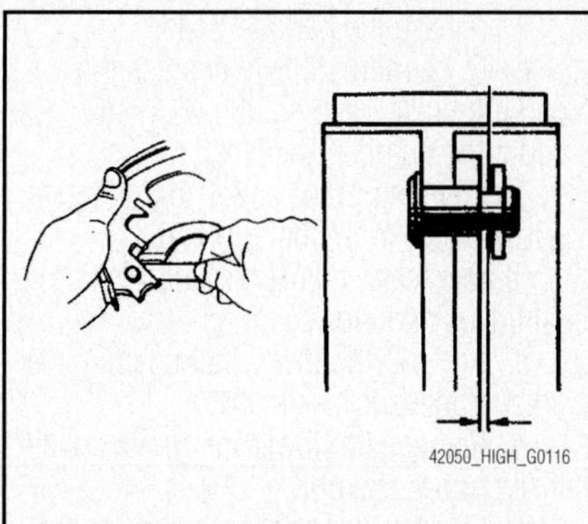

Fig. 17 Installing the parking brake shoe type C-washer

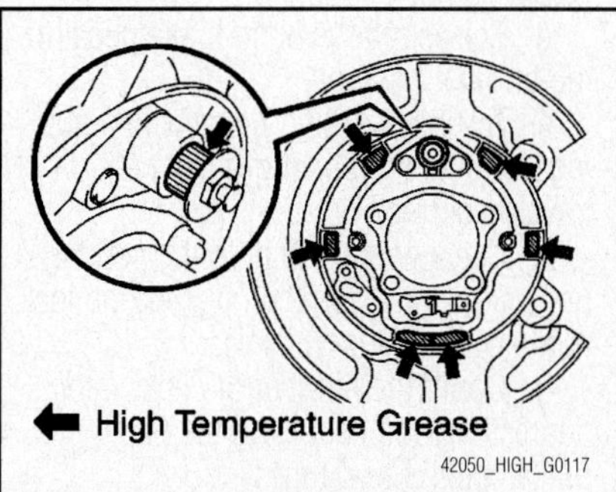

Fig. 18 Apply high temperature grease to the shaded parts of the backing plate which make contact with the shoe

a. Using a feeler gauge, measure the clearance. Standard clearance: less than 0.35 mm (0.014 in.).

If the clearance is not within the specifications, replace the shim with one of the correct size. The shim sizes: 0.3 mm (0.012 in.), 0.9 mm (0.035 in.) or 0.6 mm (0.024 in.).

b. Using pliers, install the parking brake shoe lever and the shim with a new C-washer.

20. Apply high temperature grease to the shaded parts shown in the illustration of the backing plate which make contact with the shoe.

21. On 4WD models, perform the following:

a. Install the parking brake shoe hold down spring pin.

b. Install the rear axle hub & bearing.

c. Install the rear axle shaft nut.

d. Install the rear speed sensor.

22. Install parking brake shoe No. 2, as follows:

a. Using needle-nose pliers, connect the parking brake cable No. 3 to the parking brake cable shoe lever.

➡Be careful not to damage the parking brake cable No. 3.

b. Install the parking brake shoe No. 2 with the parking brake shoe hold down spring, parking brake shoe hold down spring cup No. 1, parking brake shoe hold down spring cup No. 2 and parking brake shoe hold down spring pin No. 2.

23. Install the parking brake shoe adjusting screw set, as follows:

a. Apply high temperature grease to the parking brake shoe adjusting bolt and piece.

b. Attach the parking brake shoe return tension spring No. 2 to the parking brake shoe No. 1 and parking brake shoe assembly No. 2.

c. Attach the parking brake shoe adjusting screw set to the parking brake shoe No. 1 and parking brake shoe No. 2.

24. Install parking brake shoe No. 1:

a. For 2WD models, install the parking brake shoe hold down spring pin No. 1.

b. Install the parking brake shoe No. 1 with the parking brake shoe hold down spring, parking brake shoe hold down

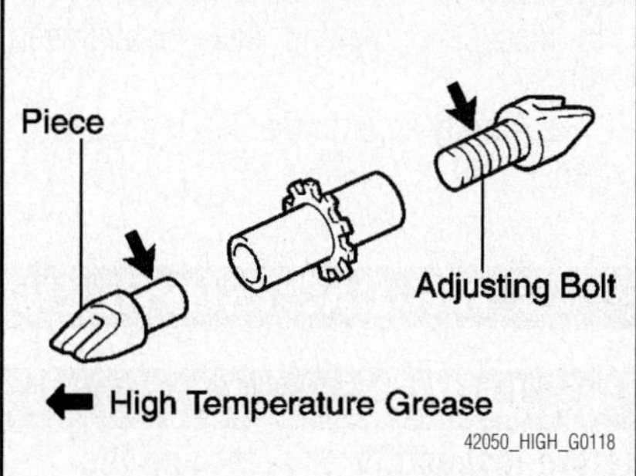

Fig. 19 Apply high temperature grease to the parking brake shoe adjusting bolt and piece

spring cup No. 2, parking brake shoe hold down spring cup No. 2.

25. attach the parking brake shoe strut and the parking brake shoe strut compression spring to parking brake shoe No. 2 and parking brake shoe No. 1.

26. Install parking brake shoe return tension spring using needle-nose pliers as shown in the illustration.

➡First install the front side spring then the rear side spring.

27. Check that the parking brake components are properly installed.

⁂ WARNING

There should be no oil or grease on the friction surface of the shoe lining and disc.

28. For 4WD models, inspect the bearing backlash and axle hub deviation.

29. Install the rear disc (rotor), aligning the matchmarks made during removal.

30. Adjust parking brake shoe clearance, as follows:

a. Temporarily install the hub nuts.

b. Remove the hole plug, turn the adjuster and expand the shoes until the disc locks.

c. Contract the shoe adjuster until the disc rotates smoothly. Standard : return 8 notches

d. Check that the shoe has no brake drag.

e. Install the hole plug.

31. Install the caliper, as outlined earlier in this section.

32. Install the rear wheel and tire assembly and tighten the lug nuts to 76 ft. lbs. (103 Nm).

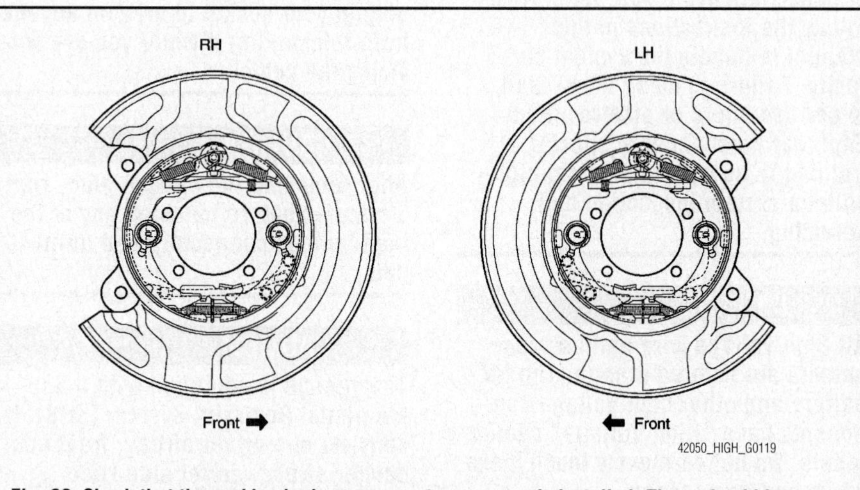

Fig. 20 Check that the parking brake components are properly installed. There should be no oil or grease on the friction surface of the shoe lining and disc

33. Inspect and adjust the parking brake pedal travel, as outlined in this section.

34. For 4WD models, check the ABS speed sensor signal.

ADJUSTMENT

See Figures 21 and 22.

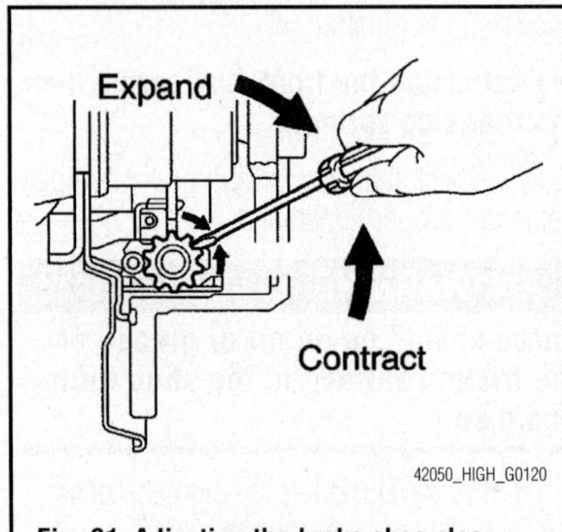

Fig. 21 Adjusting the brake shoe clearance

1. Raise and safely support the vehicle.
2. Remove the rear wheel and tire assemblies.
3. Adjust parking brake shoe clearance, as follows:
   a. Temporarily install the hub nuts.
   b. Remove the hole plug, turn the

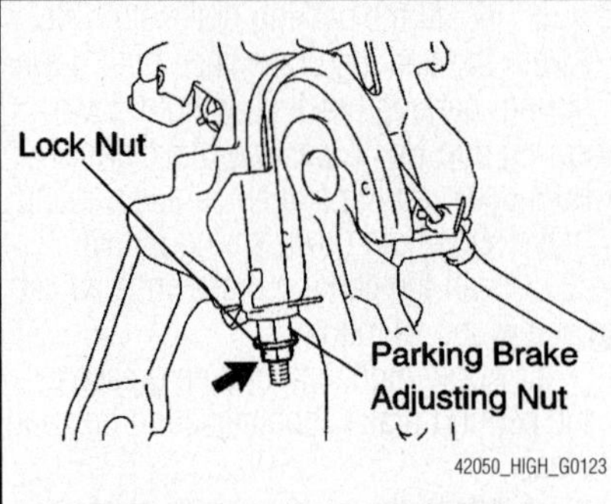

Fig. 22 Use a wrench to hold the parking brake adjusting nut secure while tightening the lock nut

adjuster and expand the shoes until the disc locks.
   c. Contract the shoe adjuster until the disc rotates smoothly. Standard : return 8 notches
   d. Check that the shoe has no brake drag.
   e. Install the hole plug.
4. Install the rotor and caliper.
5. Install the rear wheel and tire assembly and tighten the lug nuts to 76 ft. lbs. (103 Nm).
6. Inspect the parking brake pedal travel, as follows:
   a. Firmly step on the parking brake pedal.
   b. release the parking brake.
   c. Once more, slowly depress the parking brake pedal all the way, and count the number of clicks.

   The parking brake pedal should travel 5 to 7 clicks at 67 lbs. (300 N).
7. If necessary, adjust parking brake pedal travel, as follows:
   a. Remove the lower instrument panel finish panel sub-assembly.
   b. Remove the lower instrument panel insert sub- assembly.
   c. Depress the parking brake pedal 5 clicks to make room for the procedure, and loosen the lock nut with fixing adjusting nut by wrench.
   d. Release the parking brake pedal to the original position.
   e. Turn the parking brake wire adjusting nut until the parking brake pedal travel is correct.
   f. Use a wrench to hold the parking brake adjusting nut, then tighten the lock nut to 53 inch lbs. (6 Nm).
   g. Count the number of clicks after depressing and canceling the parking brake pedal 3 to 4 times.
   h. Check whether the parking brake drags or not.
   i. When operating the parking brake pedal, check that the parking brake pedal indicator light is lit.

CHASSIS ELECTRICAL

AIR BAG (SUPPLEMENTAL RESTRAINT SYSTEM)

GENERAL INFORMATION

**** CAUTION**

The RX400h hybrid system contains a 288V high-voltage system with a strong alkali solution of potassium hydroxide. Be sure to follow the instructions in this manual to handle the system correctly. Failure to do so may result in serious injury or electrocution. Engineer must undergo special training to be able to perform high-voltage system inspection and servicing.

**** CAUTION**

All high-voltage wire harness connectors are colored orange. The HV battery and other high-voltage components have "High Voltage" caution labels. Do not carelessly touch these wires and components.

**** CAUTION**

Before inspecting or servicing the high-voltage system, be sure to follow safety measures, such as wearing insulated gloves and removing the service plug to prevent electrocution. Carry the removed service plug in your pocket to prevent anyone from reinstalling it while you are servicing the vehicle.

**** CAUTION**

After removing the service plug, wait 5 minutes before touching any of the high-voltage connectors and terminals.

**** CAUTION**

The vehicle is equipped with a Supplemental Restraint System (SRS). It consists of a driver airbag, front passenger airbag, driver side knee airbag, side airbag, curtain shield airbag and front seat belt pretensioner. Failure to carry out service operations in the correct sequence could cause the SRS to unexpectedly deploy during servicing, possibly leading to a serious accident. Further, if a mistake is made in servicing the SRS, it is possible that the SRS may fail to operate when required. Before performing servicing (including removal or installation of parts, inspection or replacement), be sure to read the following items carefully, then follow the correct procedures indicated in the repair manual.

**** CAUTION**

Wait at least 90 seconds after the ignition switch is turned to the "LOCK" position and the negative (-) terminal cable is disconnected from the battery before starting the operation. (The SRS is equipped with a backup power source, so that if work

is started within 90 seconds after disconnecting the negative (-) terminal cable of the battery, the SRS may be deployed).

⁂ CAUTION

Do not expose the steering pad, front passenger airbag assembly, driver side knee airbag assembly, center airbag sensor assembly, front airbag sensor, front seat inner belt assembly, seat position sensor, occupant classification ECU, front seat side airbag assembly, side airbag sensor, curtain shield airbag assembly, rear airbag sensor or front seat outer belt assembly directly to hot air or flames.

SERVICE PRECAUTIONS

Disconnect and isolate the battery negative cable before beginning any airbag system component diagnosis, testing, removal, or installation procedures. Allow system capacitor to discharge for two minutes before beginning any component service. This will disable the airbag system. Failure to disable the airbag system may result in accidental airbag deployment, personal injury, or death.

Do not place an intact undeployed airbag face down on a solid surface. The airbag will propel into the air if accidentally deployed and may result in personal injury or death.

When carrying or handling an undeployed airbag, the trim side (face) of the airbag should be pointing towards the body to minimize possibility of injury if accidental deployment occurs. Failure to do this may result in personal injury or death.

Replace airbag system components with OEM replacement parts. Substitute parts may appear interchangeable, but internal differences may result in inferior occupant protection. Failure to do so may result in occupant personal injury or death.

Wear safety glasses, rubber gloves, and long sleeved clothing when cleaning powder residue from vehicle after an airbag deployment. Powder residue emitted from a deployed airbag can cause skin irritation. Flush affected area with cool water if irritation is experienced. If nasal or throat irritation is experienced, exit the vehicle for fresh air until the irritation ceases. If irritation continues, see a physician.

Do not use a replacement airbag that is not in the original packaging. This may result in improper deployment, personal injury, or death.

The factory installed fasteners, screws and bolts used to fasten airbag components have a special coating and are specifically designed for the airbag system. Do not use substitute fasteners. Use only original equipment fasteners listed in the parts catalog when fastener replacement is required.

During, and following, any child restraint anchor service, due to impact event or vehicle repair, carefully inspect all mounting hardware, tether straps, and anchors for proper installation, operation, or damage. If a child restraint anchor is found damaged in any way, the anchor must be replaced. Failure to do this may result in personal injury or death.

Deployed and non-deployed airbags may or may not have live pyrotechnic material within the airbag inflator.

Do not dispose of driver/passenger/curtain airbags or seat belt tensioners unless you are sure of complete deployment. Refer to the Hazardous Substance Control System for proper disposal.

Dispose of deployed airbags and tensioners consistent with state, provincial, local, and federal regulations.

After any airbag component testing or service, do not connect the battery negative cable. Personal injury or death may result if the system test is not performed first.

If the vehicle is equipped with the Occupant Classification System (OCS), do not connect the battery negative cable before performing the OCS Verification Test using the scan tool and the appropriate diagnostic information. Personal injury or death may result if the system test is not performed properly.

Never replace both the Occupant Restraint Controller (ORC) and the Occupant Classification Module (OCM) at the same time. If both require replacement, replace one, then perform the Airbag System test before replacing the other.

Both the ORC and the OCM store Occupant Classification System (OCS) calibration data, which they transfer to one another when one of them is replaced. If both are replaced at the same time, an irreversible fault will be set in both modules and the OCS may malfunction and cause personal injury or death.

If equipped with OCS, the Seat Weight Sensor is a sensitive, calibrated unit and must be handled carefully. Do not drop or handle roughly. If dropped or damaged, replace with another sensor. Failure to do so may result in occupant injury or death.

If equipped with OCS, the front passenger seat must be handled carefully as well. When removing the seat, be careful when setting on floor not to drop. If dropped, the sensor may be inoperative, could result in occupant injury, or possibly death.

If equipped with OCS, when the passenger front seat is on the floor, no one should sit in the front passenger seat. This uneven force may damage the sensing ability of the seat weight sensors. If sat on and damaged, the sensor may be inoperative, could result in occupant injury, or possibly death.

DISARMING THE SYSTEM

To avoid personal injury when working on vehicles equipped with an air bag, the negative battery cable must be disconnected and at least 90 seconds must elapse before working on the system. Failure to do so may result in deployment of the air bag.

ARMING THE SYSTEM

To arm the system after service is completed, connect the negative battery cable. If necessary, perform the initialization procedure, as outlined in the Chassis Electrical System.

CLOCKSPRING CENTERING

See Figure 23.

1. Check that the ignition switch is off.
2. Check that the battery negative (-) terminal is disconnected.
3. After removing the terminal, wait for at least 90 seconds before starting the operation.
4. Rotate the spiral cable counterclockwise slowly by hand until it feels firm.

➡Do not turn the spiral cable by the airbag wire harness.

5. Rotate the spiral cable clockwise approximately 2.5 turns to align the marks.

➡The spiral cable will rotate approximately 2.5 turns to both the left and right from the center.

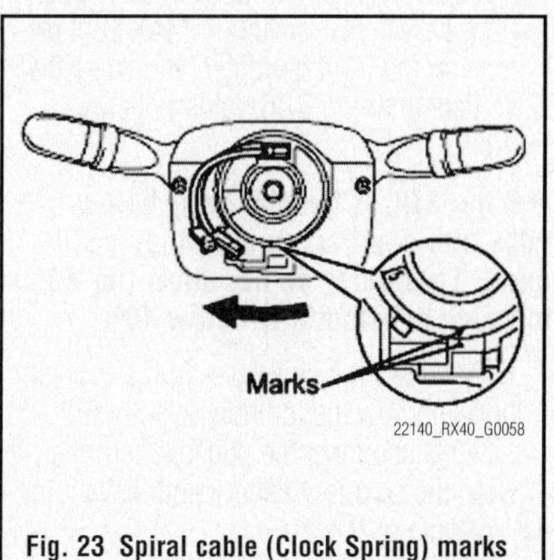

Fig. 23 Spiral cable (Clock Spring) marks

INITIALIZATION PROCEDURE

See Figure 24.

➡When disconnecting the cable from negative (-) battery terminal, initialize the following system(s) after the terminal is reconnected:

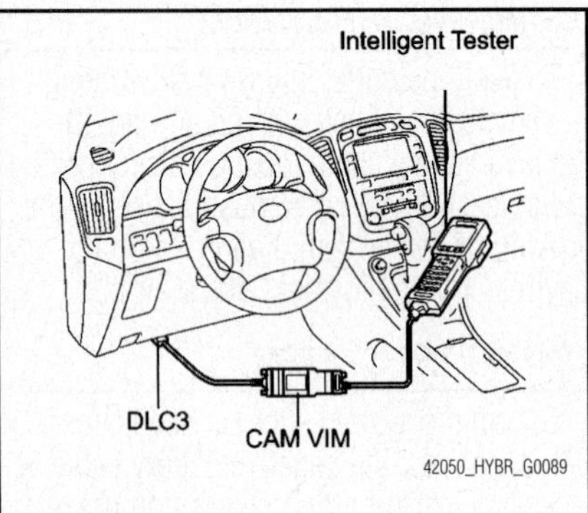

Fig. 24 Connect the intelligent tester to the DLC3

- SFI (fuel injection) system
- Power window control system
- Sunroof system

1. PERFORM INITIALIZATION

a. Apply the parking brake and chock the vehicle.

b. Connect the intelligent tester to the DLC3.

c. Put the engine in the inspection mode.

d. Select the following menu items: DIAGNOSIS / ENHANCED OBD II / DATA LIST / PRIMARY / COOLANT TEMP.

e. Warm up the engine with the A/C switch off until the coolant temperature is 83°C or more.

f. Select the following menu items: DIAGNOSIS /HV CONTROL ECU / DATALIST / SOC.

g. Move the shift lever to the N position, set the A/C switch to MAX cool or turn on the front defroster, and leave the vehicle until the SOC value is below 40%.

➡If the SOC is substantially below 40%, the auxiliary battery may go dead. Therefore, do not allow the SOC to reach substantially below 40%.

h. Move the shift lever to the P position and turn the ignition switch **OFF**.

i. Disconnect the negative terminal from the auxiliary battery and leave it for 1 minute or more.

j. Reconnect the negative terminal to the auxiliary battery and turn the ignition switch **ON**.

k. Select the following menu items: DIAGNOSIS / ENHANCED OBD II / DATALIST / PRIMARY / COOLANT TEMP.

l. Check that the engine coolant temperature is 83°C or more.

➡If the engine coolant temperature is below 83°C, start the engine by depressing the accelerator pedal with the shift lever in the P position to increase the coolant temperature.

m. Select the following menu items: DIAGNOSIS / ENHANCED OBDII / DATA LIST / CALCLOAD.

n. Move the shift lever to the D position while depressing the brake pedal with the left foot.

o. Depress the accelerator pedal with the right foot while firmly depressing the brake pedal with the left foot, and maintain an engine load value of 45% or more (60 to 70% of full acceleration) for approximately 30 seconds.

⁂ WARNING

Do not perform this step for 40 seconds or more.

p. Shift the shift lever to the P position. Turn the ignition switch off and then wait for 5 seconds before putting the vehicle into the READY-on state again.

q. Select the following menu items: DIAGNOSIS / ENHANCED OBDII / DATA LIST / ISC LEARNING.

r. Check that the air conditioner is off. Lightly depress the accelerator pedal and release it when the engine starts.

s. Check that "COMPLETE" is displayed on the intelligent tester screen.

➡If "ISC LEARNING" is not completed within 1 minute, repeat steps (o) through (r). The engine will usually stop when "ISC LEARNING" is completed. However, the engine will not stop even when "ISC LEARNING" is completed in such cases when the charge level of the auxiliary battery is decreased.

2. RESET (INITIALIZE) POWER WINDOW REGULATOR MOTOR

➡Resetting the power window motor (initializing the pulse sensor) is necessary when the battery terminal is disconnected; when the power window regulator master switch, power window regulator switch, wire harness, power window regulator and power window motor are replaced or removed/installed; or when the P FR DOOR fuse, ECU-B fuse and IG1 fuse are replaced. AUTO operation function, jam protection function and remote operation function via the power window regulator master switch do not operate without resetting.

➡Be sure to reset the power window regulator motors using the power window switch on each door because the motors cannot be reset by remote operation using the power window master switch.

⁂ WARNING

In order to prevent a large amount of current from flowing in the wires, do not simultaneously reset 2 or more power window regulator motors.

➡If the power window switch has been continuously operated for a long time, the power window regulator motor will stop so that no additional load will be applied to the motor. Continuing to operate the switch even after the motor has stopped will cause the switch to blink while restricting the AUTO UP/DOWN function and the remote UP/DOWN function. If this happens, reset the motor again after several minutes have passed.

➡Whenever disconnecting the battery terminal, reset all the other systems besides the power window control system.

a. Turn the ignition switch **ON**.

b. Halfway open the power window by pressing the power window switch.

c. Fully pull up the switch until the power window is fully closed and continue to hold the switch for approx. 1 second after the power window is fully closed.

d. Check that the AUTO UP/DOWN function operates normally. If the AUTO UP/DOWN function operates normally, reset operations have been completed at this time. If not normal, proceed with the steps.

e. Disconnect the negative battery terminal for 10 seconds.

f. Connect the battery terminal.

g. Turn the ignition switch **ON**.

h. Halfway open the power window by pressing the power window switch.

i. If the AUTO UP/DOWN function operates normally, reset operations have been completed at this time. If not normal, follow steps below.

j. Turn the ignition switch **ON**.

k. Halfway open the power window by pressing the power window switch.

l. Fully pull up the switch until the power window is fully closed and continue to hold the switch for approx. 12 seconds after the power window is fully closed.

m. Check that the AUTO UP/DOWN function operates normally.

3. RESET SLIDING ROOF DRIVE GEAR SUBASSEMBLY

➡When disconnecting the cable from the negative (-) battery terminal, or disconnecting the sliding roof drive gear or connector, sliding roof drive gear requires initialization of the pulse sensor. If a reset is not executed, the following functions do not operate: auto operation, jam protection function, key off operation, key-linked open and close, transmitter-linked open and close.

a. Turn the ignition switch **ON**.

b. If the sliding roof is opened, close it fully.

c. Push the open switch of the slide switch or the up switch of the tilt switch on the personal light, making the sliding roof operation as follows: Tilt up → Approximately 1 second → tilt down → slide open → slide close

d. Sliding roof stops at the fully closed position.

e. Finish the initialization.

f. Check that the operation works normally with AUTO operation.

➡If the following conditions occur while operating, initialization will fail.

- Ignition switch is turned off.
- Pushed switch is released while sliding roof

is operating.

- Vehicle speed is 5 km/h (3 mph) or higher.
- Battery of sliding roof ECU is 1 V or higher.
- Communication is cut off.

➡If the sliding roof cannot fully close or its position has become misaligned, perform the initialization again. If the sliding roof TILT UP switch or SLIDE CLOSE switch is pressed and held until the roof glass has either stopped moving or started moving in the opposite direction, and then the switch is held for another 10 seconds or more, perform the initialization again. If the AUTO operation function and jam protection function do not operate after the drive gear has been reset, replace the sliding roof drive gear (sliding roof ECU).

DRIVETRAIN

HYBRID TRANSAXLE ASSEMBLY

REMOVAL & INSTALLATION

See Figure 25.

⁂ CAUTION

The RX400h hybrid system contains a 288V high-voltage system with a strong alkali solution of potassium hydroxide. Be sure to follow the instructions in this manual to handle the system correctly. Failure to do so may result in serious injury or electrocution. Engineer must undergo special training to be able to perform high-voltage system inspection and servicing.

⁂ CAUTION

All high-voltage wire harness connectors are colored orange. The HV battery and other high-voltage components have "High Voltage" caution labels. Do not carelessly touch these wires and components.

⁂ CAUTION

Before inspecting or servicing the high-voltage system, be sure to follow safety measures, such as wearing insulated gloves and removing the service plug to prevent electrocution. Carry the removed service plug in your pocket to prevent anyone from reinstalling it while you are servicing the vehicle.

⁂ CAUTION

After removing the service plug, wait 5 minutes before touching any of the high-voltage connectors and terminals.

1. Before servicing the vehicle, refer to the Precautions Section.
2. Remove or disconnect the following:
   - Engine/transaxle assembly. (See Engine Removal and Installation).
   - Manifold stay
   - Transaxle damper
   - Front frame assembly
   - Halfshafts
   - Flywheel housing undercover
   - Engine wiring harnesses
   - Transaxle case cover
   - Coolant hose
   - Front engine mounting bracket
   - Transaxle oil cooler assembly
   - Transmission control cable bracket
3. Remove the 8 mounting bolts and separate the transaxle assembly from the vehicle.

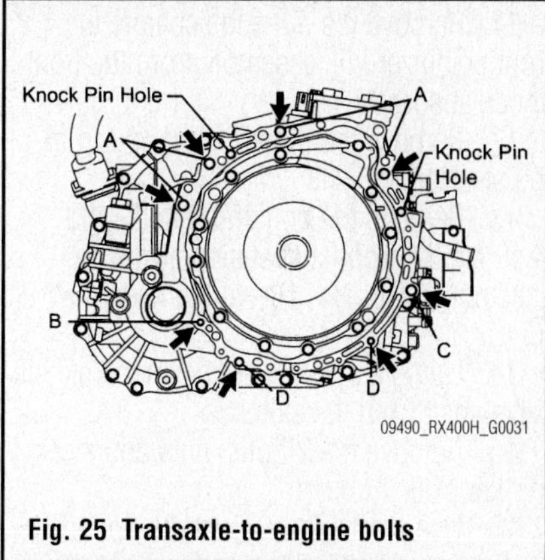

Fig. 25 Transaxle-to-engine bolts

4. Installation is the reverse of removal. Observe the following torques:
   - Transaxle-to-engine: Bolts A to 47 ft. lbs (64 Nm); Bolt B to 34 ft. lbs. (46 Nm); Bolts C to 47 ft. lbs. (64 Nm); Bolts D to 27 ft. lbs. (37 Nm)

➡Do not reuse Bolt B.

   - Front engine mounting bracket: 47 ft. lbs. (64 Nm)
   - Transaxle case cover: 74 inch lbs. (8.4 Nm)
   - Undercover: 69 inch lbs. (8 Nm)
5. If necessary, perform the initialization procedure, as outlined in the Chassis Electrical System.

HYBRID TRANSAXLE OUTER OIL SEAL

REMOVAL & INSTALLATION

See Figures 26 through 29.

1. Before servicing the vehicle, refer to the Precautions Section.
2. Disconnect the negative battery cable.
3. Remove No. 1 and No.2 engine under covers.
4. Drain the transaxle fluid.
5. Remove the front wheel.
6. Remove the axle hub nut.
7. Remove the 2 bolts and clip, and separate the sensor wire and hose from the shock absorber.
8. Remove the 2 bolts, and separate the brake caliper assembly.

➡Use a wire or an equivalent to keep the brake caliper from hanging down by the flexible hose.

9. Remove the brake rotor.
10. Support the front suspension lower arm No. 1 with a jack using a wooden block to avoid damage.
11. Remove the nut and separate the front stabilizer link assembly from the front shock absorber.
12. Remove the outer tie rod end from the steering knuckle.
13. Remove the bolt and 2 nuts, and separate the front suspension arm sub-assembly lower No. 1 from the lower ball joint.
14. Using a plastic hammer, separate the drive shaft from the axle hub.
15. Remove the 2 bolts, nuts and steering knuckle.
16. Use a screwdriver with its tip taped to disengage the 5 claws and remove the cowl top silencer pad.
17. Remove the bolt and disconnect the front flexible hose No. 2 and front speed sensor wire harness from the front strut assembly with coil spring.
18. Remove the 2 nuts on the lower side of the front shock absorber with coil spring.
19. Remove the 3 nuts on the upper side of the front shock absorber with coil spring.
20. Lower the front axle assembly, and remove the 2 bolts on the lower side of the front shock absorber.
21. Remove the front strut assembly with coil spring.
22. Remove the LH axle with a slide hammer and proper adapters.
23. Remove the RH axle as follows:
    - Using pliers, remove the bearing bracket hole snap ring.
    - Remove the bolt from the drive shaft bearing bracket.
24. Remove the RH axle with a slide hammer and proper adapters.

➡Be careful not to damage the transaxle case oil seal and inboard joint boot.

25. Disconnect the heated oxygen sensor connector.
26. Remove the 2 bolts, 2 nuts and exhaust pipe assembly front.

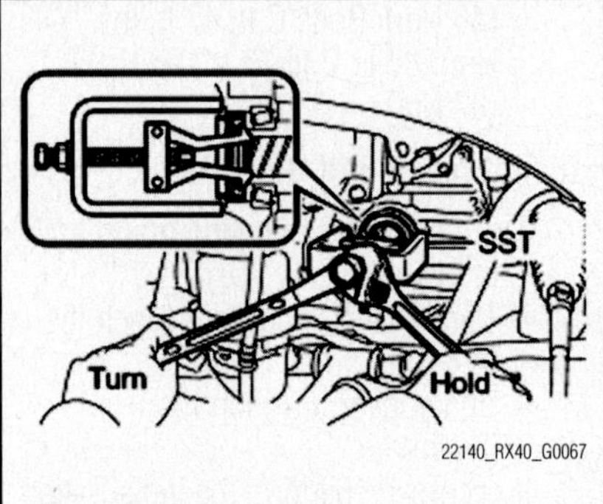

Fig. 26 LH outer transaxle oil seal removal

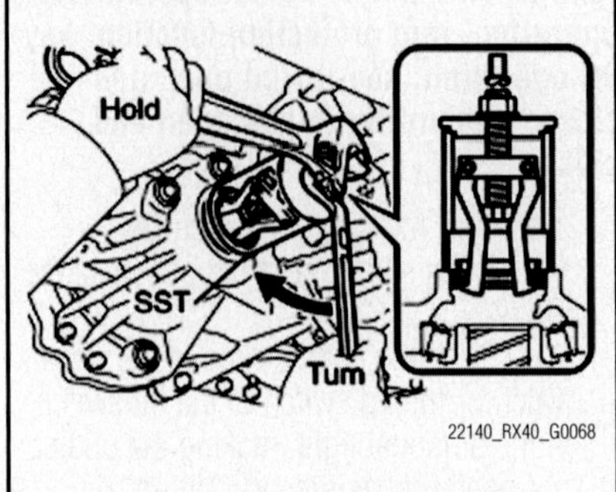

Fig. 27 RH outer transaxle oil seal removal

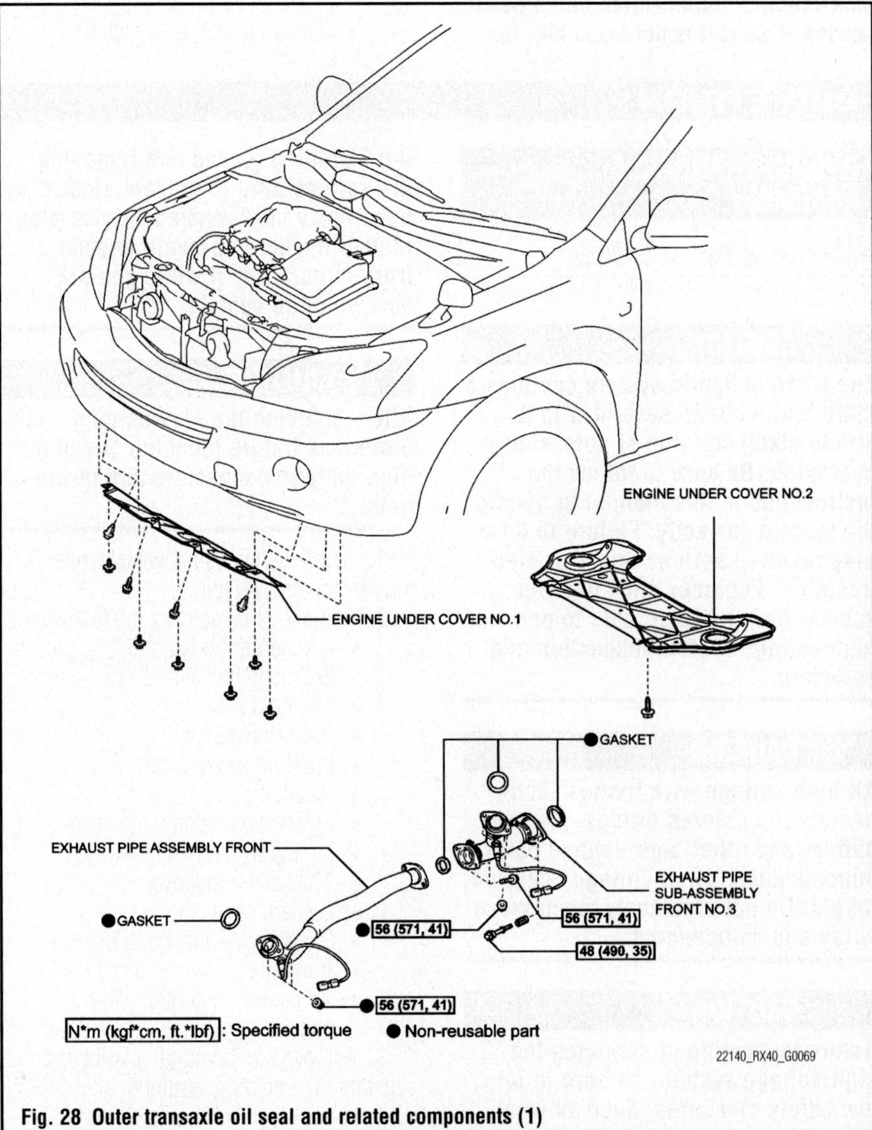

Fig. 28 Outer transaxle oil seal and related components (1)

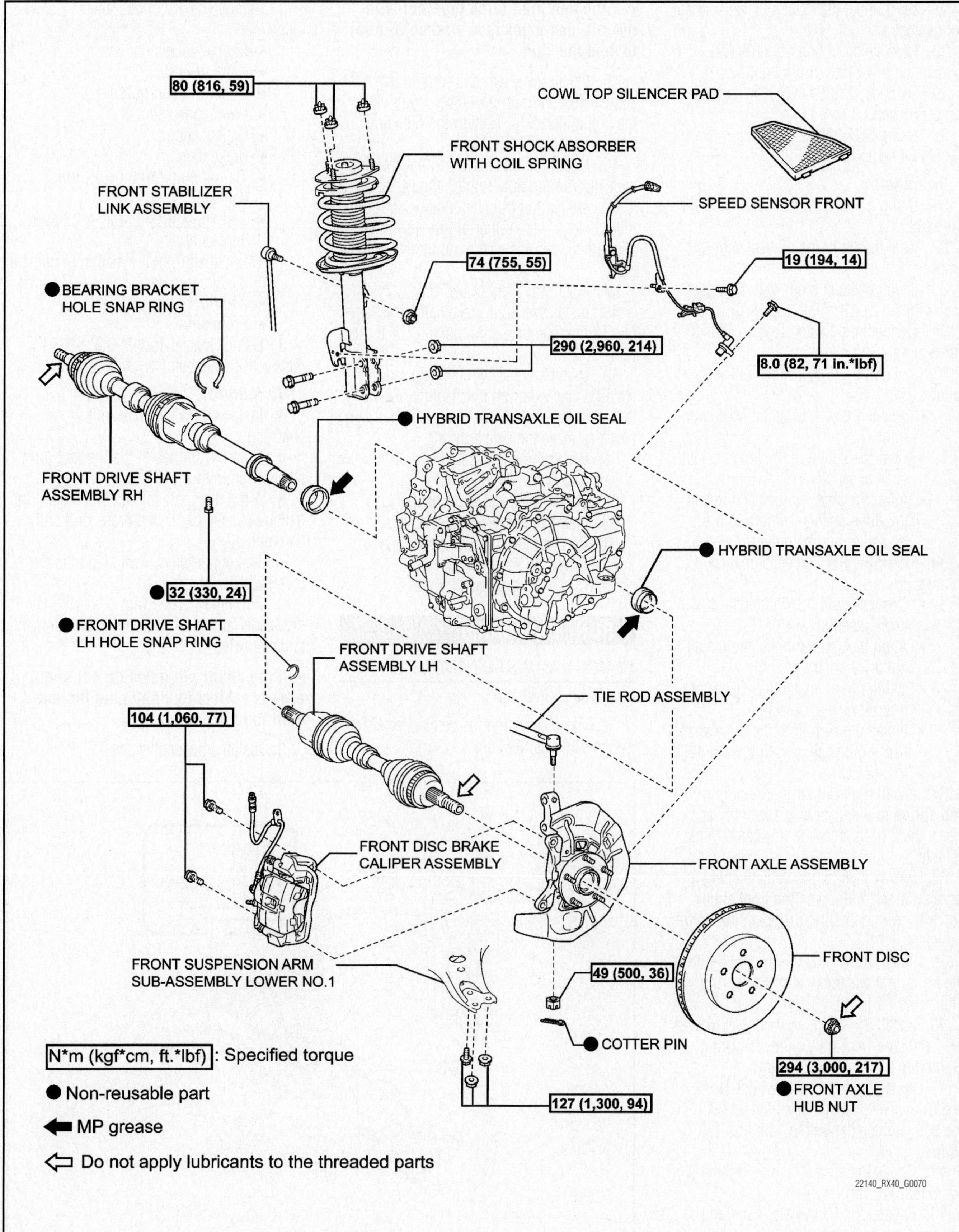

Fig. 29 Outer transaxle oil seal and related components (2)

27. Disconnect the heated oxygen sensor connector.

28. Remove the 2 bolts, 2 nuts and exhaust pipe sub-assembly front No. 3.

29. Using SST 09514-35011, remove the LH oil seal

30. Using SST 09612-30012, remove the RH oil seal.

To install:

31. Using a seal driver install LH or RH axle seal.

32. Coat the lip of the oil seal with MP grease.

33. Install exhaust pipes with new gaskets. Tighten to 41 ft. lbs. (56 Nm).

34. Connect the heated oxygen sensor connectors.

35. Install the LH axle assembly as follows:

- Install a new drive shaft hole snap ring.
- Coat the spline of the inboard joint shaft assembly with ATF.
- Align the shaft splines and install the drive shaft assembly with a brass bar and hammer.

36. Install the RH axle assembly as follows:

- Coat the spline of the inboard joint shaft assembly with ATF.
- Align the shaft splines and install the drive shaft assembly.
- Using pliers, install a new bearing bracket hole snap ring.
- Install a new bolt to the drive shaft bearing bracket and tighten to 42 ft. lbs. (32 Nm).

37. Install the front strut assembly with coil spring to the front axle assembly and insert the 2 bolts from the front side of the vehicle.

38. Slowly jack up the vehicle using a wooden block and install the front shock absorber with coil spring (upper side) to the vehicle.

39. Install the 3 nuts to the upper side of the front strut assembly with coil spring. Tighten to 59 ft. lbs. (80 Nm).

40. Install the 2 nuts to the lower side of the front strut assembly with coil spring. Tighten to 214 ft. lbs. (290 Nm).

41. Install the front flexible hose No. 2 and front speed sensor wire harness with the bolt. Tighten the bolt to 14 ft. lbs. (19 Nm).

42. Check the strut lock nut. Tighten to 36 ft. lbs. (49 Nm).

43. Engage the 5 claws to install the cowl top silencer pad.

44. Install the front stabilizer link assembly and tighten the nut to 55 ft. lbs. (74 Nm).

➡If the ball joint turns together with the nut, use a hexagon wrench (6 mm) to hold the stud.

45. Install the lower ball joint to the front suspension arm sub-assembly lower with the bolt and 2 nuts. Tighten the nuts and bolt to 94 ft. lbs. (127 Nm).

46. Install the tie rod end to the steering knuckle with the nut. Tighten the nut to 36 ft. lbs. (49 Nm) and install a new cotter pin. If the holes for the cotter pin are not aligned, tighten the castle nut up to 60° further.

47. Install the front brake rotor.

48. Install the front brake caliper assembly. Tighten the mounting bolts to 77 ft. lbs. (104 Nm).

49. Using a socket wrench (30 mm), install a new axle hub nut. Tighten the nut to 217 ft. lbs. (294 Nm). Using a chisel and hammer, stake the axle hub nut.

50. Install the speed sensor.

51. Add transaxle fluid.

52. Check front end alignment.

53. Install the engine under covers.

54. Connect the negative battery cable.

55. Check ABS sensor signal.

56. Check the axle seal for leaks.

FRONT HALFSHAFT

REMOVAL & INSTALLATION

See Figures 30 and 31.

1. Before servicing the vehicle, refer to the Precautions Section.

2. Remove or disconnect the following:

- Engine splash shields
- Front wheels
- Cotter pin and hub nut
- Front speed sensor
- Brake caliper
- Brake disc
- Tie rod end, from the steering knuckle
- Steering knuckle, from the lower control arm
- Halfshaft from the axle hub, using a plastic hammer
- Stabilizer link
- Front strut

3. Using Special Tool 095020-01010, remove the halfshaft from the transaxle.

To install:

4. Install a new halfshaft hole snapring.

5. Coat the splines of the inboard joint shaft assembly with ATF.

6. Align the shaft splines and install the halfshaft assembly with a brass drift and hammer.

7. The remainder of installation is the reverse order of removal.

8. Fill the transaxle with gear oil, install the fender apron, check front end alignment and test drive.

➡If the cotter pin holes do not align, always correct by tightening the nut until the next hole aligns.

9. Install a new cotter pin.

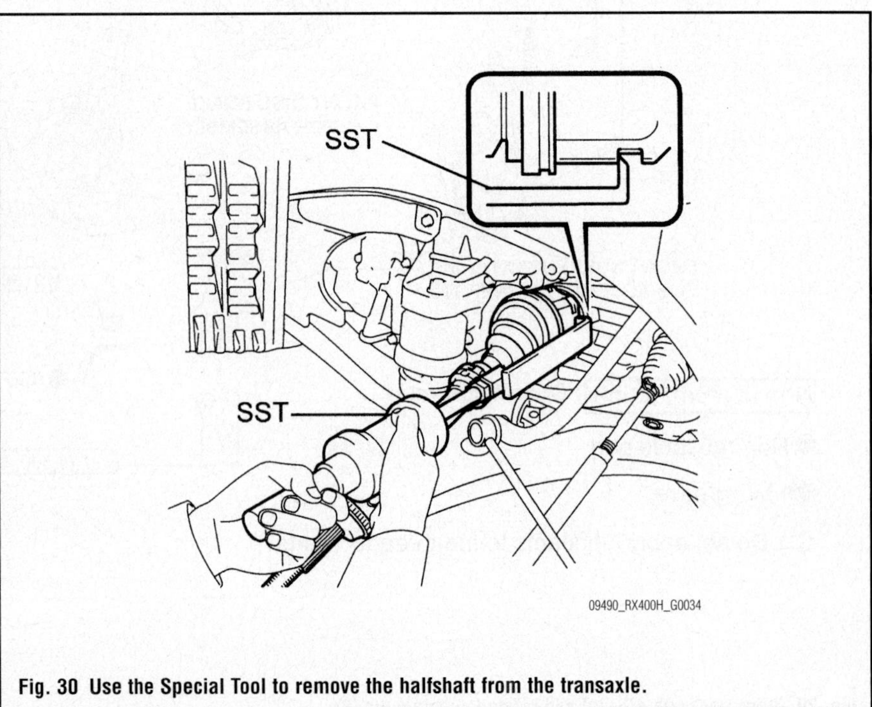

Fig. 30 Use the Special Tool to remove the halfshaft from the transaxle.

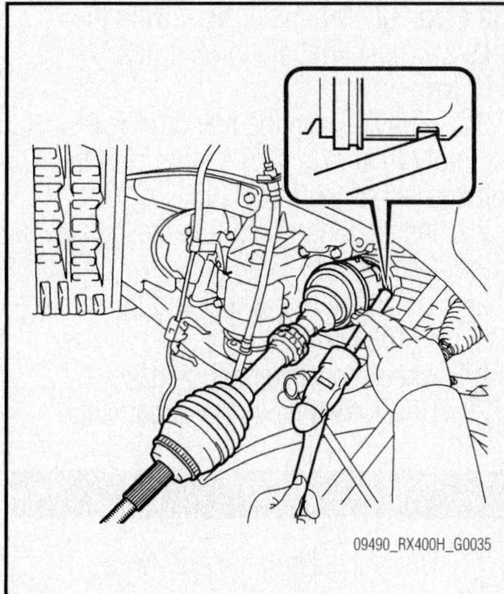

Fig. 31 Insert a brass drift into the groove to install the halfshaft.

REAR HALFSHAFT (4WD)

REMOVAL & INSTALLATION

See Figure 32.

1. Before servicing the vehicle, refer to the Precautions Section.
2. Remove the rear wheel.
3. Disconnect and remove the speed sensor.
4. Remove the brake disc and brake caliper assembly.
5. Remove the axle shaft nut.
6. Remove the nut, and separate the parking brake cable No. 3.
7. Remove the 4 bolts and the rear axle hub and bearing assembly.
8. Support the rear axle carrier with a wooden block placed between the jack and axle carrier.
9. Remove the 2 bolts, 2 nuts and strut rod assembly rear.
10. Remove the bolts and nuts, and separate the rear suspension arm assembly No. 1. and No. 2.
11. Remove the 2 bolts and 2 nuts, and the rear axle carrier sub-assembly from the shock absorber.
12. Remove the rear drive shaft assembly as follows:
    - Put matchmarks on the rear drive shaft assembly and differential side gear shaft.
    - Remove the 4 nuts, washers and rear drive shaft assembly.

To install:

13. Align the match marks and install the rear drive shaft assembly with the 4 nuts and washers. Tighten to 41 ft. lbs. (56 Nm).

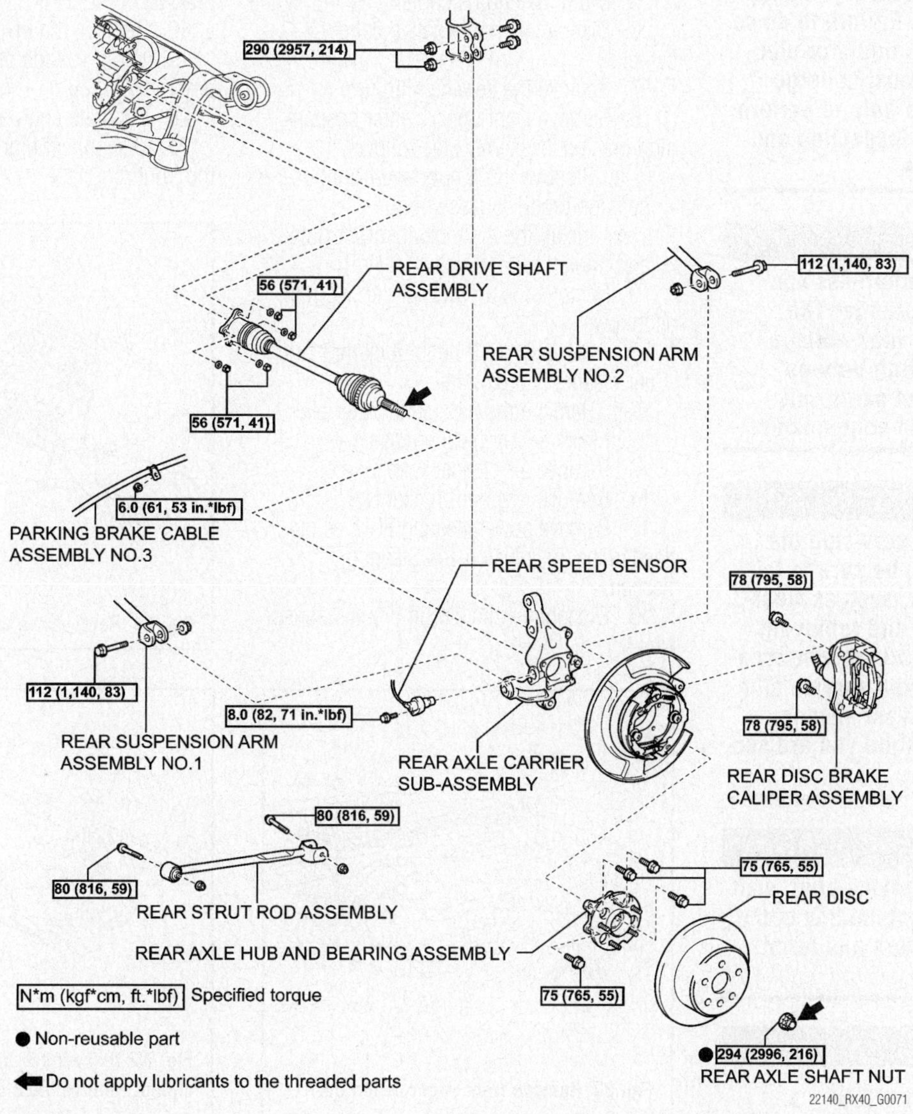

290 (2957, 214)
56 (571, 41)
REAR DRIVE SHAFT ASSEMBLY
56 (571, 41)
112 (1,140, 83)
REAR SUSPENSION ARM ASSEMBLY NO.2
6.0 (61, 53 in.*lbf)
PARKING BRAKE CABLE ASSEMBLY NO.3
REAR SPEED SENSOR
78 (795, 58)
112 (1,140, 83)
8.0 (82, 71 in.*lbf)
78 (795, 58)
REAR SUSPENSION ARM ASSEMBLY NO.1
REAR AXLE CARRIER SUB-ASSEMBLY
REAR DISC BRAKE CALIPER ASSEMBLY
80 (816, 59)
80 (816, 59)
75 (765, 55)
REAR STRUT ROD ASSEMBLY
REAR DISC
REAR AXLE HUB AND BEARING ASSEMB LY
N*m (kgf*cm, ft.*lbf) Specified torque
75 (765, 55)
● Non-reusable part
● 294 (2996, 216)
Do not apply lubricants to the threaded parts
REAR AXLE SHAFT NUT
22140_RX40_G0071

Fig. 32 Rear drive shaft and related parts

14. Install the rear axle carrier sub-assembly with the 2 bolts and nuts to the strut assembly. Tighten the 2 bolts to 214 ft. lbs. (290 Nm).

15. Temporarily tighten the rear suspension arm assembly No. 2 and No. 1 with the bolt and nut.

16. Install the hub and bearing assembly with the 4 bolts. Tighten the bolts to 55 ft. lbs. (75 Nm).

17. Install the rear brake rotor.

18. Install the rear brake caliper assembly. Tighten the mounting bolts to 58 ft. lbs. (78 Nm).

19. Install the rear speed sensor.

20. Clean the threaded parts on the drive shaft and axle shaft nut using a non-residue solvent.

21. Using a socket wrench (30mm), install a new rear axle shaft nut. Tighten to 217 ft. lbs. (294 Nm). Using a chisel and hammer, stake the rear axle shaft nut.

22. Stabilize the suspension.

23. Fully tighten the rear suspension arm assembly No. 1 and No. 2 with the bolts and nuts. Tighten to 83 ft. lbs. (112 Nm).

24. Fully tighten the rear strut rod assembly (rear axle carrier side) with the bolt and nut to 59 ft. lbs. (80 Nm).

25. Install parking brake cable assembly.

26. install the rear wheel and tighten to 76ft. lbs (103 Nm).

27. Check speed sensor signal.

28. Check the rear wheel alignment.

ENGINE COOLING

CAUTION

The RX400h hybrid system contains a 288V high-voltage system with a strong alkali solution of potassium hydroxide. Be sure to follow the instructions in this manual to handle the system correctly. Failure to do so may result in serious injury or electrocution. Engineer must undergo special training to be able to perform high-voltage system inspection and servicing.

CAUTION

All high-voltage wire harness connectors are colored orange. The HV battery and other high-voltage components have "High Voltage" caution labels. Do not carelessly touch these wires and components.

CAUTION

Before inspecting or servicing the high-voltage system, be sure to follow safety measures, such as wearing insulated gloves and removing the service plug to prevent electrocution. Carry the removed service plug in your pocket to prevent anyone from reinstalling it while you are servicing the vehicle.

CAUTION

After removing the service plug, wait 5 minutes before touching any of the high-voltage connectors and terminals.

THERMOSTAT

REMOVAL & INSTALLATION

See Figures 33 through 40.

1. Before servicing the vehicle, refer to the Precautions Section.
2. Disconnect the negative battery cable.
3. Using a clip remover, remove the engine side cover.
4. Drain the hybrid vehicle coolant.
5. Drain the engine coolant.
6. Disconnect the negative battery cable.
7. Remove the service plug grip.
8. Remove front upper center suspension brace subassembly, as follows:
   a. Remove the 4 nuts and front suspension brace subassembly.
   b. Install the 4 shock absorber nuts and tighten to 59 ft. lbs. (80 Nm).
9. Remove the air cleaner cap w/ inlet, as follows:
   a. Remove the 2 bolts, 4 clamps and air cleaner cap w/ inlet.
   b. Remove the air cleaner filter element from the air cleaner case.
10. Remove air cleaner w/resonator.
11. Remove inverter bracket No. 5.
12. Remove power steering ECU, as outlined in the Suspension and Steering Section.
13. Disconnect the circuit breaker sensor No. 1.
14. Disconnect engine room wire No. 2.
15. Remove the inverter reserve tank subassembly.
16. Disconnect water hose.
17. Disconnect the power steering ECU bracket.
18. Remove the inverter cover.
19. Verify voltage of w/converter inverter assembly is 0 v.
20. Separate engine wire No. 4.
21. Disconnect high voltage cable from the front motor.

Fig. 34 Remove the water inlet pipe

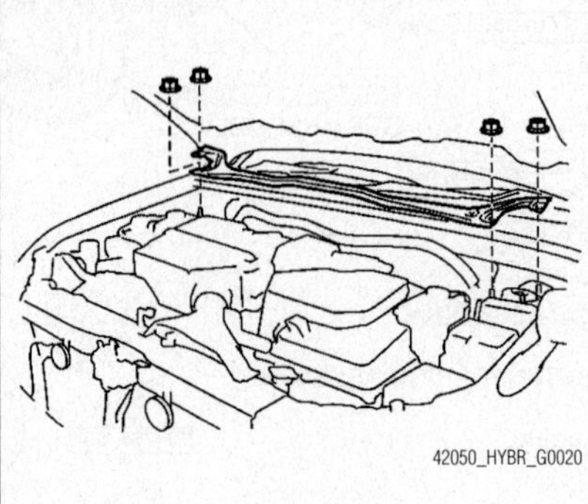

Fig. 33 Remove front upper center suspension brace sub-assembly

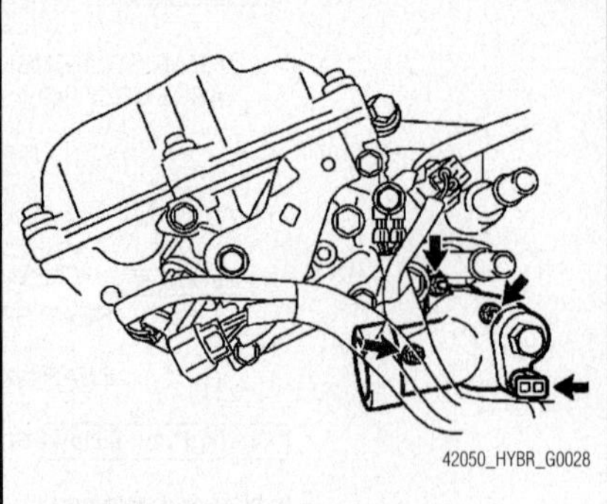

Fig. 35 Disconnect the wire harness clamp, remove the 3 nuts, and then remove the water inlet

22. Disconnect the MG ECU connector.
23. Disconnect the No. 3 wire frame.
24. Install the inverter cover.
25. Separate engine room relay block assembly
26. Remove the inverter bracket No. 4.
27. Remove w/converter inverter assembly.
28. Remove the water inlet pipe, as follows:
    a. Remove the bolt and the water inlet pipe.
    b. Remove the o-ring from the water inlet pipe.
29. Remove the water inlet, as follows:
    a. Disconnect the wire harness clamp.
    b. Remove the 3 nuts and the water inlet.
30. Remove the thermostat, as follows:
    a. Remove the thermostat.
    b. Remove the gasket from the thermostat and discard. Thoroughly clean the gasket mating surfaces.

To install:

31. Install thermostat:
    a. Install a new gasket to the thermostat.
    b. Align the jiggle valve of the thermostat and water inlet, and insert the thermostat in the water inlet housing.

➡The jiggle valve should be set within +-15°from the prescribed position.

32. Install water inlet. Tighten the 3 retaining nuts to 71 inch lbs. (8 Nm).

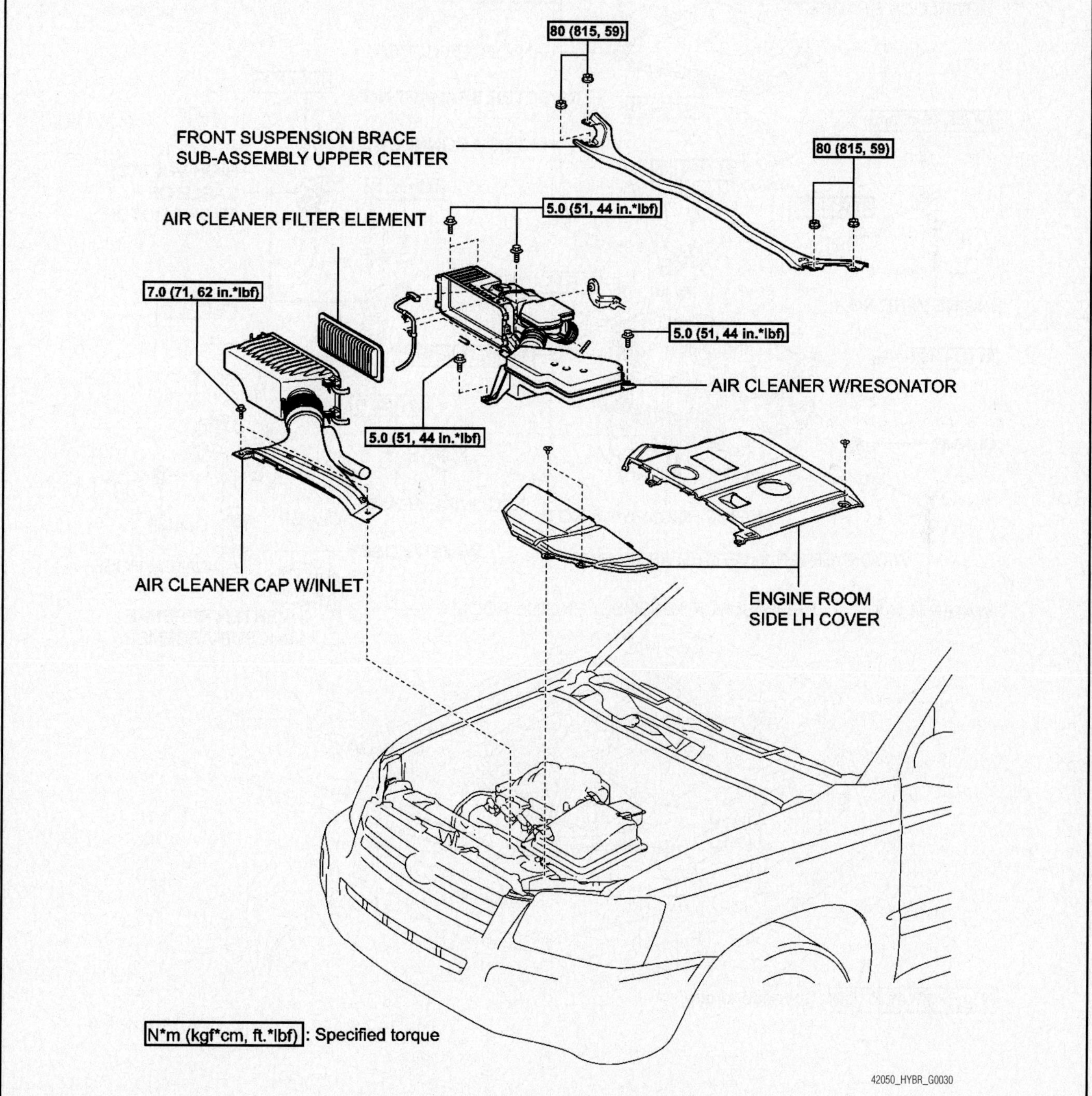

Fig. 36 Exploded view of the components you need to remove to access the thermostat (1 of 3)—All models

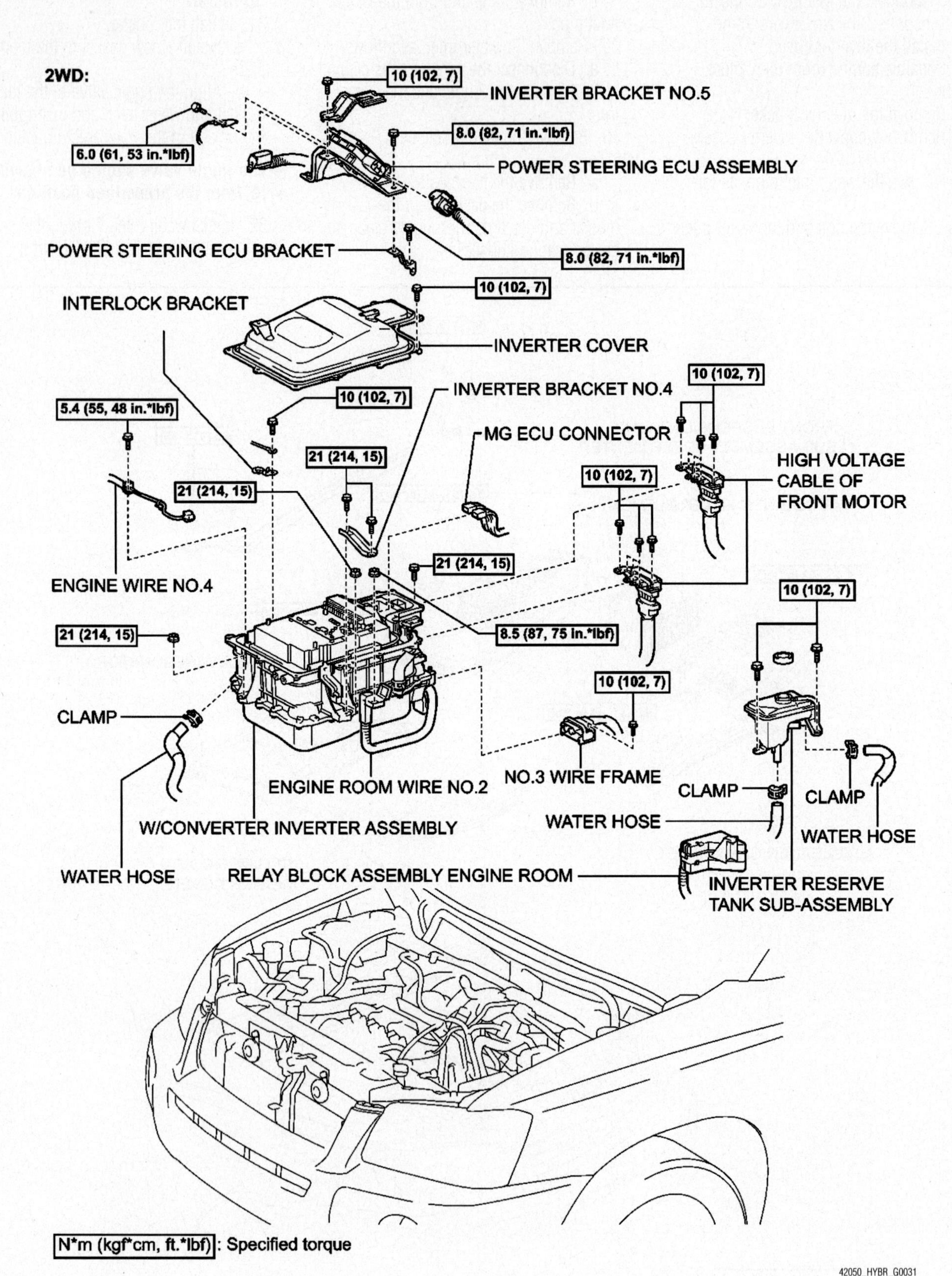

Fig. 37 Exploded view of the components you need to remove to access the thermostat (2 of 3)—2WD models

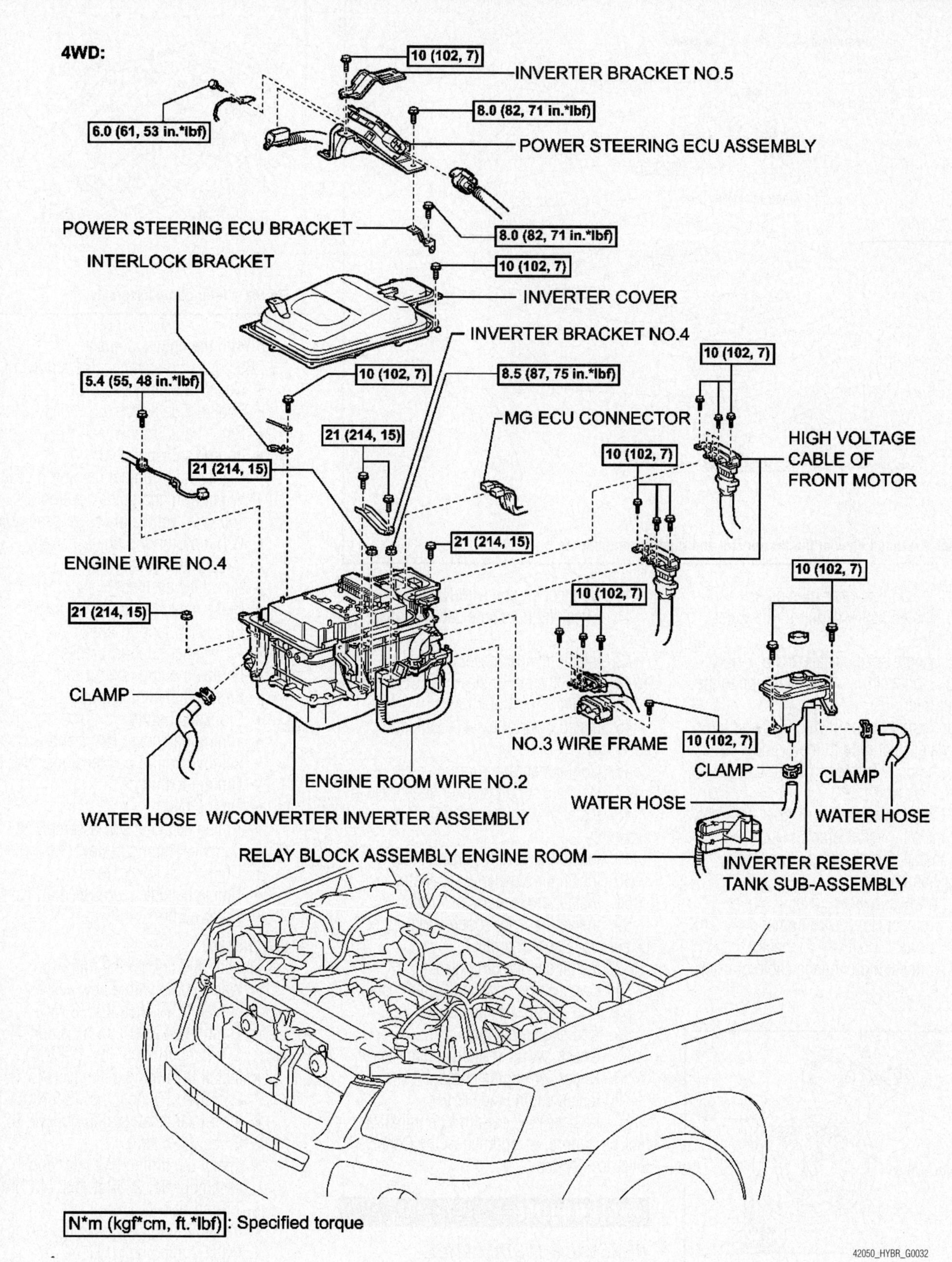

Fig. 38 Exploded view of the components you need to remove to access the thermostat (3 of 3)—4WD models

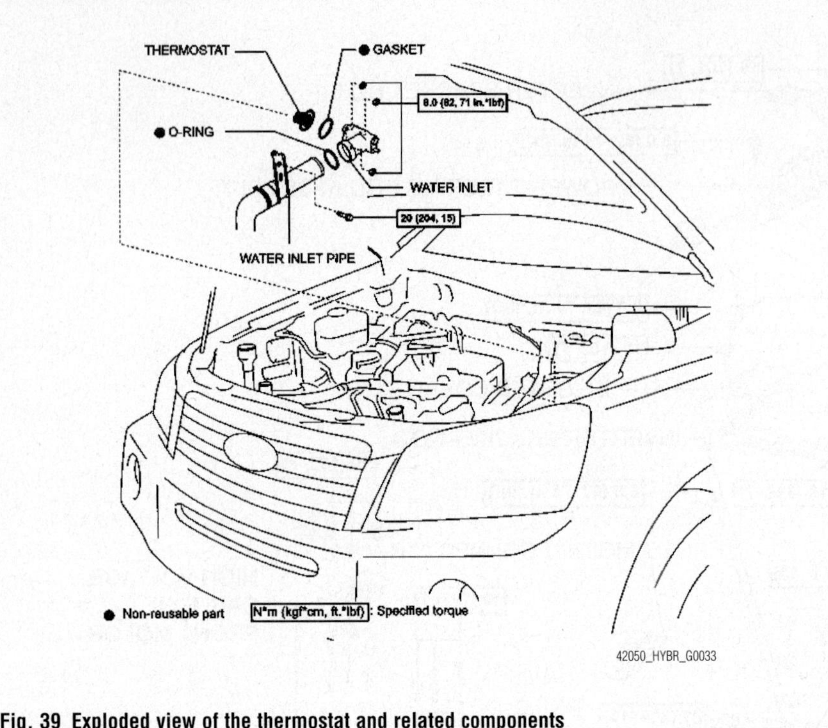

Fig. 39 Exploded view of the thermostat and related components

a. Install the water inlet pipe:

b. Install a new o-ring to the water inlet pipe.

c. Apply soapy water to the o-ring.

d. Connect the water inlet pipe to the water inlet.

e. Install the bolt which is used to fix the water inlet pipe to the cylinder head with the bolt. Tighten to 15 ft. lbs. (20 Nm).

33. Install w/converter inverter assembly.
34. Install inverter bracket No. 4.
35. Install engine compartment relay block assembly.
36. Remove inverter cover.
37. Connect No. 3 wire frame.
38. Connect MG ECU connector.
39. Connect high voltage cable of front motor

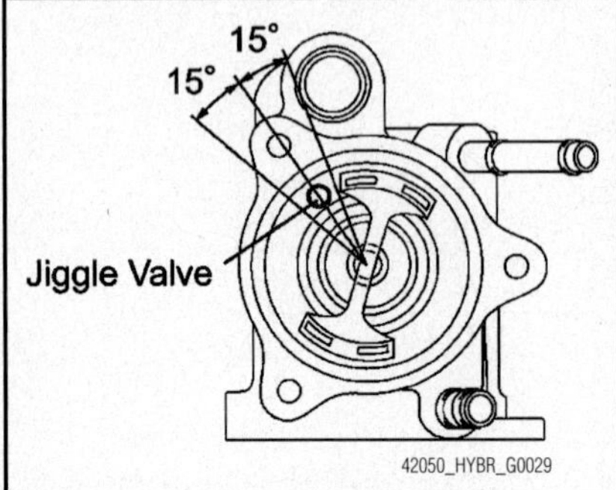

Fig. 40 Proper installed position of the jiggle valve

40. Connect engine wire No. 4.
41. Check high voltage cable connection.
42. Install inverter cover.
43. Install the power steering ECU bracket.
44. Connect the water hose.
45. Install inverter reserve tank sub-assembly.
46. Connect engine room wire No. 2.
47. Connect circuit breaker sensor No. 1.
48. Install the power steering ECU assembly.
49. Install the inverter bracket No. 5.
50. Install air cleaner w/resonator.
51. Install air cleaner cap w/inlet.
52. Install the front suspension upper center brace subassembly.
53. Install service plug grip.
54. Connect the negative battery cable.
55. Add hybrid vehicle coolant.
56. Add engine coolant.
57. Check hybrid vehicle coolant leaks.
58. Check for engine coolant leaks.
59. Install engine side cover.
60. If necessary, perform the initialization procedure, as outlined in the Chassis Electrical System.

WATER PUMP

REMOVAL & INSTALLATION

See Figure 41.

1. Before servicing the vehicle, refer to the Precautions Section.

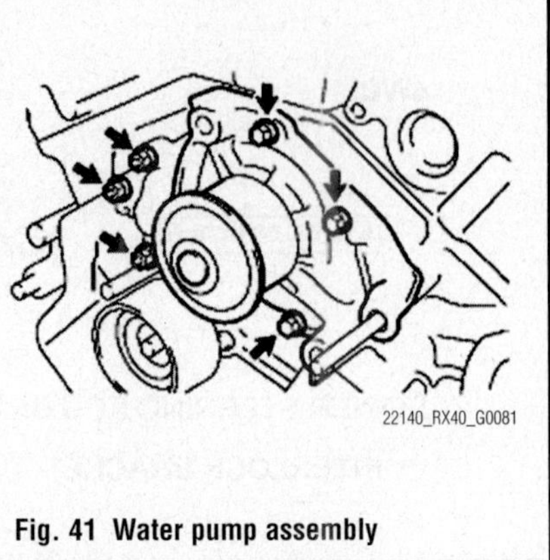

Fig. 41 Water pump assembly

2. Drain the engine coolant.
3. Remove or disconnect the following:
   - Negative battery cable
   - Engine side covers
   - Right-hand front wheel
   - Engine splash shield
   - Right-hand front fender apron seal
   - Wiper arm and blade assembly
   - Top cowl ventilator louver assembly
   - Wiper motor and link assembly
   - Battery and battery tray
   - Air intake assembly
   - Brake master cylinder reservoir
   - Reservoir support bracket
   - Air cleaner support bracket
   - Engine moving control rod
   - Right-hand engine mount
   - Crankshaft pulley
   - Timing belt cover No. 1 and 2
   - Remove timing belt tensioner No. 1
   - Timing belt guide
   - Timing belt
   - Timing belt idler sub-assembly No. 1
   - Camshaft timing pulley
   - Timing belt cover No. 3
   - Timing belt idler sub-assembly No. 2
   - Water pump

To install:

4. Install and tighten the following:
   - Water pump with a new gasket, tighten to 71 inch lbs. (8 Nm)
   - Timing belt idler sub-assembly No. 1, tighten to 25 ft. lbs. (34 Nm)
   - Install the timing belt cover No. 3, tighten to 76 inch. lbs. (8.5 Nm)
   - Install camshaft timing pulleys, 92 ft. lbs. (125 Nm)
   - Install the timing belt idler sub-assembly No. 2, 32 ft. lbs. (43 Nm)
   - Timing belt
   - Install timing belt tensioner No. 1
   - Install timing belt guild No. 2
   - Install the engine mounting bracket and tighten to 75 inch. lbs. (8.5 Nm)

- Install the engine moving control rod.
- Install the air cleaner bracket
- Timing belt No. 2 cover
- Timing belt No. 1 cover
- Crank shaft pulley bolt, tighten to 162 ft. lbs. (220 Nm).
- Install the reservoir bracket
- Install master cylinder reservoir
- Install air cleaner and resonator
- Install the air cleaner cap with inlet
- Install the cool air intake duct seal
- Install the battery carrier and battery
- Cowl assembly
- Install wiper motor and link assembly
- Install cowl top louver assembly
- Install wiper arm assembly RH and LH
- RH fender splash shield
- RH wheel

5. Refill the coolant to the correct level.
6. Start the engine and check for leaks.
7. Inspect the ignition timing.
8. Install the engine room covers.
9. Install the engine under cover.
10. If necessary, perform the initialization procedure, as outlined in the Chassis Electrical System.

ENGINE ELECTRICAL

PRECAUTIONS

**** CAUTION**

The RX400h hybrid system contains a 288V high-voltage system with a strong alkali solution of potassium hydroxide. Be sure to follow the instructions in this manual to handle the system correctly. Failure to do so may result in serious injury or electrocution. Engineer must undergo special training to be able to perform high-voltage system inspection and servicing.

**** CAUTION**

All high-voltage wire harness connectors are colored orange. The HV battery and other high-voltage components have "High Voltage" caution labels. Do not carelessly touch these wires and components.

**** CAUTION**

Before inspecting or servicing the high-voltage system, be sure to follow safety measures, such as wearing insulated gloves and removing the service plug to prevent electrocution. Carry the removed service plug in your pocket to prevent anyone from reinstalling it while you are servicing the vehicle.

**** CAUTION**

After removing the service plug, wait 5 minutes before touching any of the high-voltage connectors and terminals.

ENGINE ELECTRICAL

CHARGING SYSTEM

The HV transaxle assembly consists of the planetary gear unit, Motor, and Generator. The gear unit uses the planetary gear to split engine output in accordance with a driving request while the vehicle is driven or the HV battery is charged. The Motor assists engine output while increasing vehicle driving force. The Motor also converts the energy, which is consumed in the form of heat during normal braking, into electrical energy and recover it into the HV battery to effect regenerative braking. The Generator supplies power, which is used for charging the HV battery or driving the Motor. It also controls the stepless transmission function of the transaxle by regulating the amount of electricity generated to change Generator speed. In addition, the Generator is used as a starter Motor to start the engine. The transmission input damper absorbs the shock generated when the driving force from the engine is transmitted.

**** CAUTION**

The RX400h hybrid system contains a 288V high-voltage system with a strong alkali solution of potassium hydroxide. Be sure to follow the instructions in this manual to handle the system correctly. Failure to do so may result in serious injury or electrocution. Engineer must undergo special training to be able to perform high-voltage system inspection and servicing.

**** CAUTION**

All high-voltage wire harness connectors are colored orange. The HV battery and other high-voltage components have "High Voltage" caution labels. Do not carelessly touch these wires and components.

**** CAUTION**

Before inspecting or servicing the high-voltage system, be sure to follow safety measures, such as wearing insulated gloves and removing the service plug to prevent electrocution. Carry the removed service plug in your pocket to prevent anyone from reinstalling it while you are servicing the vehicle.

**** CAUTION**

After removing the service plug, wait 5 minutes before touching any of the high-voltage connectors and terminals.

FIRING ORDERS

See Figure 42.

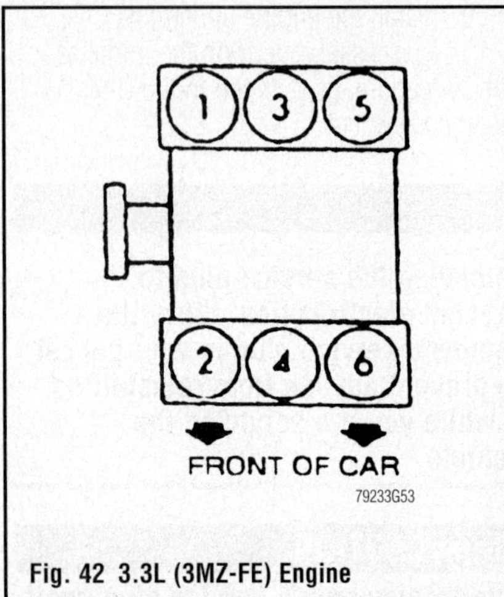

Fig. 42 3.3L (3MZ-FE) Engine

IGNITION COIL

REMOVAL & INSTALLATION

See Figure 43.

1. Disconnect the negative battery cable.
2. Remove any engine cover(s) necessary for access to the ignition coils.
3. Unfasten the retainers, detach the connectors, and then remove the ignition coil(s)
4. Installation is the reverse of the removal procedure.
5. If necessary, perform the initialization procedure, as outlined in the Chassis Electrical System.

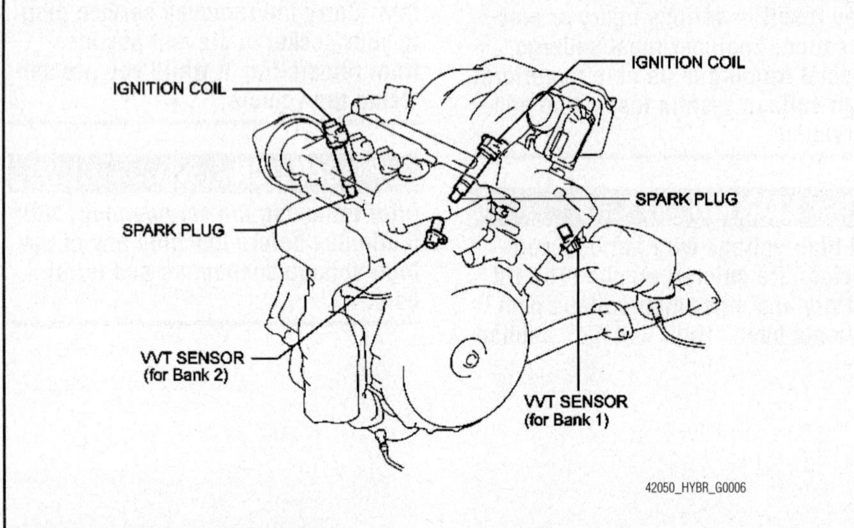

Fig. 43 Ignition system components

IGNITION TIMING

See Figure 44.

INSPECTION

1. Set the vehicle to inspection mode.
2. Warm up the engine.
3. When using the Techstream:
   a. Connect the Techstream to the DLC3.
   b. Select the following menu items: Power Train / Engine and ECT / Data List / IGN Advance
   c. Ignition timing: 8 to 12°BTDC

➡Turn all the electrical systems off. Inspect the engine idle speed with the cooling fan off.

   d. Check that the ignition timing advances immediately when the engine speed is increased.

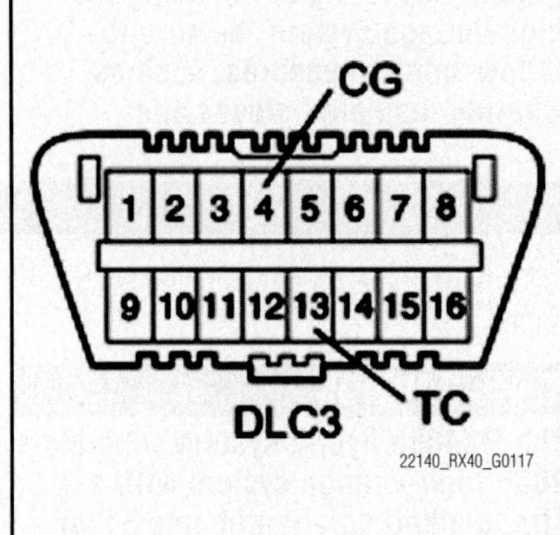

Fig. 44 DL3C connector view

4. When using a timing light:
   a. Using SST 09843-18040, connect terminals 13 (TC) and 4 (CG) of the DLC3.
   b. Pull out the red-colored wire harness as shown in the illustration.

⁂ WARNING

Make sure of the terminal numbers before connecting them. Connection with a wrong terminal can damage the engine.

   c. Connect the tester terminal of the timing light to the engine.
   d. Inspect ignition timing at idle.
   e. Ignition timing: 8 to 12° BTDC
   f. Disconnect terminals 13 (TC) and 4 (CG) of the DLC3.
   g. Inspect ignition timing at idle.
   h. Ignition timing: 7 to 24° BTDC
   i. Confirm that the ignition timing advances when the engine rpm is increased.
   j. Remove the timing light.

➡Run the engine at 1,000 to 1,300 rpm for 5 seconds, check that the engine rpm returns to idle speed.

ADJUSTMENT

All engines are equipped with a Distributorless Ignition System (DIS). No timing adjustment is possible.

SPARK PLUGS

REMOVAL & INSTALLATION

1. Disconnect the negative battery cable.
2. Remove the ignition coils, as outlined in this section.
3. Use a spark plug socket and wrench and remove the spark plugs.
4. Inspect the condition of the spark plugs, as outlined in this section.

To install:

5. Install the spark plugs and tighten to 18 ft. lbs. (25 Nm).
6. Install the ignition coils, as outlined in this section.
7. Connect the negative battery cable.
8. If necessary, perform the initialization procedure, as outlined in the Chassis Electrical System.

VVT SENSOR

REMOVAL & INSTALLATION

See Figures 45 through 49.

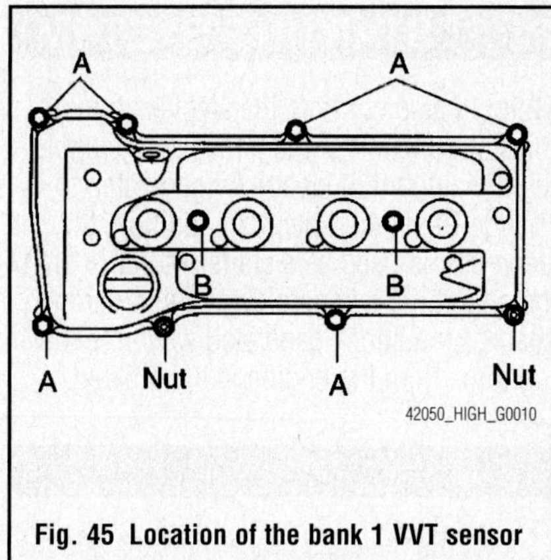

Fig. 45 Location of the bank 1 VVT sensor

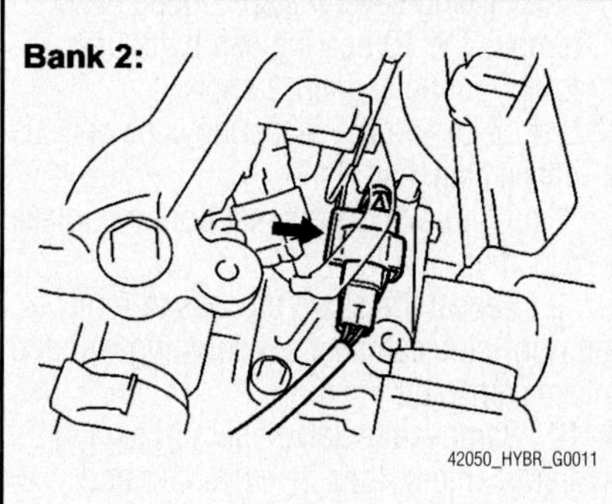

Fig. 46 Detach the heated oxygen sensor connector to access the bank 2 VVT sensor

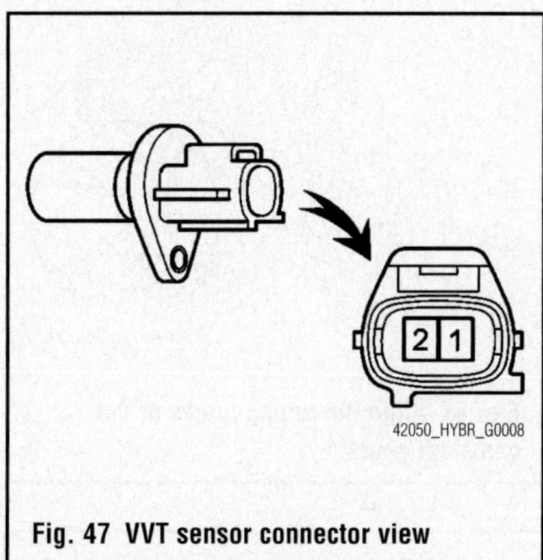

Fig. 47 VVT sensor connector view

1. Disconnect the negative battery cable.
2. Remove air cleaner cap w/ inlet.
3. Remove the bank 1 VVT sensor, as follows:
   a. Disconnect the VVT sensor (bank 1) connector.
   b. Remove the bolt, and then remove the bank 1 VVT sensor.

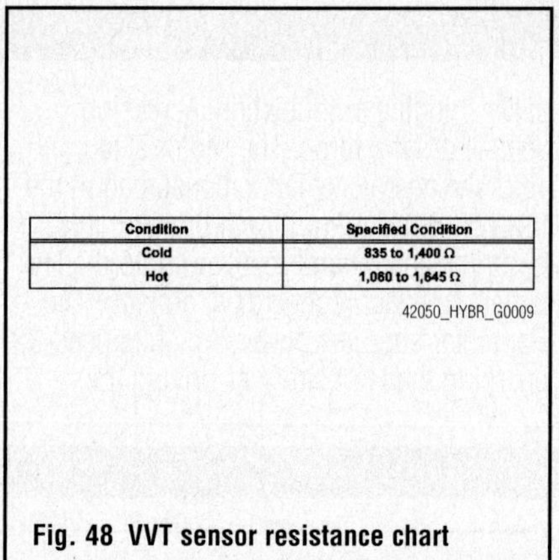

Condition	Specified Condition
Cold	835 to 1,400 Ω
Hot	1,060 to 1,645 Ω

Fig. 48 VVT sensor resistance chart

4. Remove the bank 2 VVT sensor, as follows:
   a. Disconnect the heated oxygen sensor connector (bank 2) from the wire harness bracket.
   b. Disconnect the bank 2 VVT sensor connector.
   c. Remove the bolt, and then remove the bank 2 VVT sensor.
5. Inspect the VVT sensor:
   a. Using an ohmmeter, measure the resistance according to the value(s) in the accompanying chart. Standard resistance cold is 14–122°f (-10–50°c) and hot is 122–212°f (50–100°c. If the resistance is not within specifications, replace the sensor.

➡Cold and hot refer to the temperature of the coils themselves.

To install:

6. Install the bank 1 VVT sensor:
   a. Apply a light coat of engine oil to the o-ring on the VVT sensor (bank 1).
   b. Install the VVT sensor (bank 1) with the bolt an tighten to 71 inch lbs. (8 Nm).
   c. Connect the VVT sensor (bank 1) connector.
7. Install the bank 2 VVT sensor:
   a. Apply a light coat of engine oil to the o-ring on the VVT sensor (bank 2).
   b. Install the VVT sensor (bank 2) with the bolt and tighten to 71 inch lbs. (8 Nm).
   c. Connect the VVT sensor (bank 2) connector.
   d. Install the heated oxygen sensor connector (bank 2) to the wire harness bracket.
8. Install air cleaner cap w/ inlet.
9. If necessary, perform the initialization procedure, as outlined in the Chassis Electrical System.

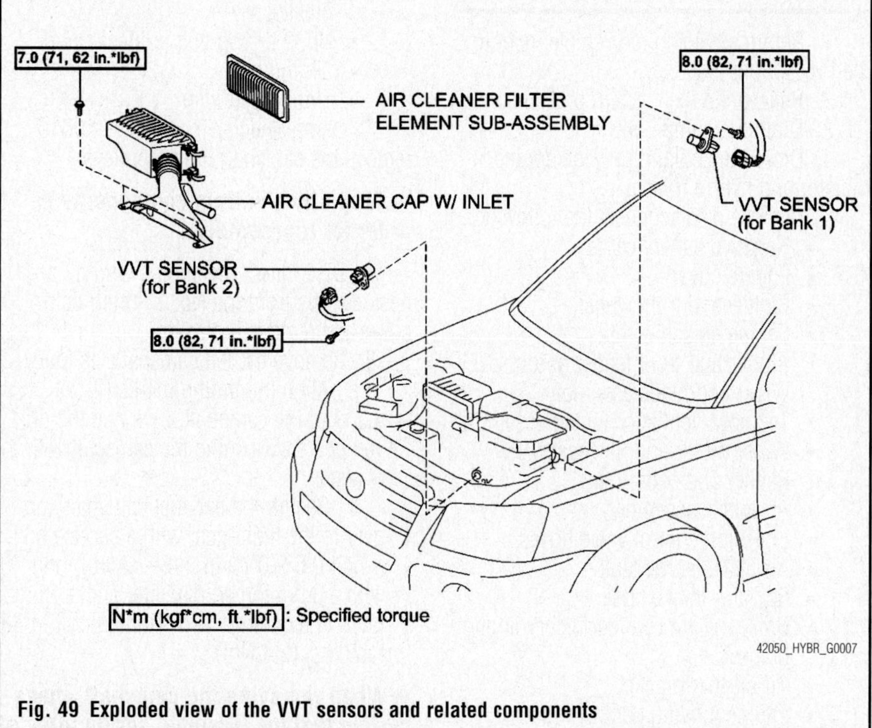

Fig. 49 Exploded view of the VVT sensors and related components

STARTER

The HV transaxle assembly consists of the planetary gear unit, Motor, and Generator. The gear unit uses the planetary gear to split engine output in accordance with a driving request while the vehicle is driven or the HV battery is charged. The Motor assists engine output while increasing vehicle driving force. The Motor also converts the energy, which is consumed in the form of heat during normal braking, into electrical energy and recover it into the HV battery to effect regenerative braking. The Generator supplies power, which is used for charging the HV battery or driving the Motor. It also controls the stepless transmission function of the transaxle by regulating the amount of electricity generated to change Generator speed. In addition, the Generator is used as a starter Motor to start the engine. The transmission input damper absorbs the shock generated when the driving force from the engine is transmitted.

ENGINE MECHANICAL

CAMSHAFT AND VALVE LIFTERS

REMOVAL & INSTALLATION

See Figures 50 through 65.

WARNING

The thrust clearance on both the intake and exhaust camshafts is very small; the camshafts must be kept level during removal. If the camshafts are removed without being kept level, the camshaft may be caught in the cylinder head, causing the head to break or the camshaft to seize.

1. Before servicing the vehicle, refer to the Precautions Section.
2. Relieve the fuel system pressure.
3. Drain the engine oil.
4. Drain the coolant from the engine radiator and hybrid transaxle.
5. Remove or disconnect the following:
    - Negative battery cable
    - Engine cover
    - Right-hand front wheel
    - Engine splash shields
    - Right-hand front fender apron seal
    - Wiper and blade assembly
    - Top cowl ventilator louver assembly
    - Wiper motor and link assembly
    - Battery and battery tray
    - Air intake assembly
    - Emission control valve hoses
    - Air intake surge tank
    - Radiator intake hose
    - Brake master cylinder reservoir and bracket
    - Air cleaner bracket
    - Engine moving control rod
    - Right-hand engine No. 2 mounting stay
    - Ignition coil
    - Valve covers
    - Crankshaft pulley
    - Timing belt No. 1 and No. 2 covers

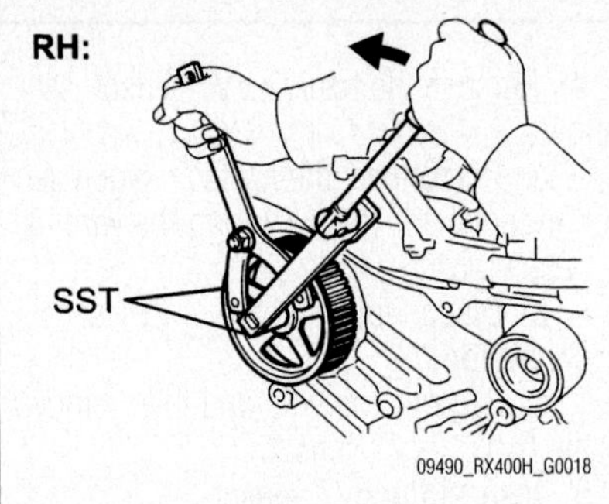

Fig. 50 Removing the right-hand camshaft timing pulley, left-hand similar.

    - Right-hand engine mounting bracket
    - No. 2 timing belt guide
    - Timing belt
    - Timing belt idler
6. Using Special Tool 09960-10010, remove the camshaft timing pulleys.

➡Keep all valvetrain components in order for reassembly.

7. Disconnect the engine wiring harness clamps from the No. 3 timing belt cover and remove the cover.
8. Remove the left camshafts as follows:
    a. Align the timing marks (2-dot mark) of the camshaft drive and the driven gears by turning the camshaft with a wrench.
    b. Secure the exhaust camshaft sub-gear to the main gear with a service bolt. A bolt 0.63–0.79 in. (16–20mm) long with a 6mm thread diameter and a 1mm pitch is recommended. Tighten bolt to 48 inch lbs. (5.4 Nm).

➡When removing the camshaft, make certain that the torsional spring force of the sub-gear has been eliminated by installing the service bolt.

    c. Using several steps, loosen and remove the 10 bearing cap bolts uniformly in the sequence shown.
    d. Remove the 5 bearing caps and the exhaust camshaft.
    e. Using several steps, loosen and remove the 10 bearing cap bolts uniformly in the sequence shown.
    f. Remove the 5 bearing caps and the intake camshaft.
    g. Remove the oil seal from the intake camshaft.
9. Repeat the same process to remove the right-side camshafts, beginning with the intake camshaft.
10. Remove the valve lifter shims and hydraulic lifters. Identify each lifter and shim as it is removed so it can be rein-

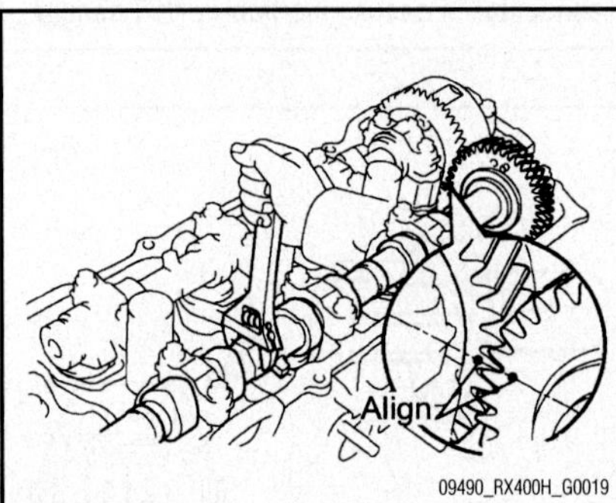

Fig. 51 Align the timing marks of the camshaft gears.

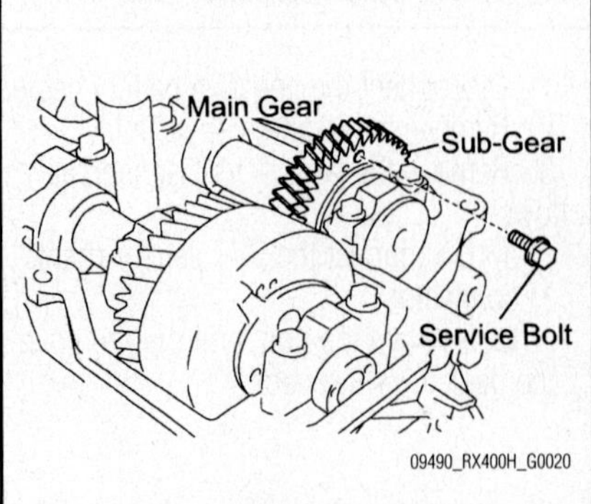

Fig. 52 Install a service bolt to secure the camshaft gears.

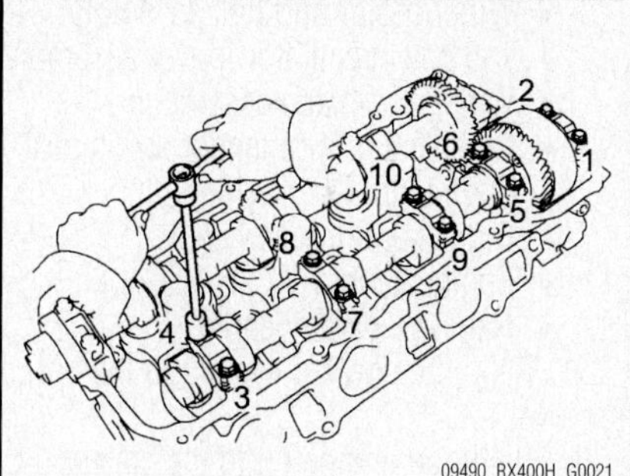

Fig. 53 Intake camshaft bearing cap loosening sequence

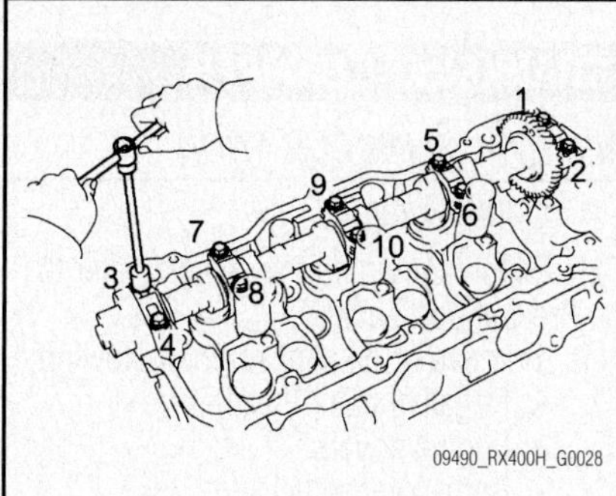

Fig. 54 Exhaust camshaft bearing cap loosening sequence

stalled in the same position. If the lifters are to be reused, store them upside down in a sealed container.

To install:

11. Install the valve lifters into their original positions and install the shims. Check valve clearance and replace the shims as necessary.

12. When reinstalling, remember that the camshafts must be handled carefully and kept straight and level to avoid damage.

13. Install the right camshafts, as follows:

a. Apply new engine oil to the thrust portion and journal of the camshaft.

b. Position the exhaust camshaft on the head so that the alignment marks are at a 90 degrees angle from vertical.

c. Apply multi-purpose grease to the lip of a new oil seal.

d. Install the oil seal to the camshaft.

e. Apply sealant to the No. 1 bearing cap.

f. Apply a light coat of clean engine oil to the bolt threads and under the bolt head. Install the bearing caps to their proper position. Tighten the bolts evenly and in several passes to 12 ft. lbs. (16 Nm) in the proper sequence.

g. Position the intake camshaft on the head so that the alignment marks are at a 90 degrees angle from vertical. The mark should be at the 9 o'clock position and must align with the marks on the other gear.

h. Apply a light coat of clean engine oil to the bolt threads and under the bolt head. Install the bearing caps to their proper position. Tighten the bolts evenly and in several passes to 12 ft. lbs. (16 Nm) in the proper sequence.

i. Remove the service bolt.

14. Install the left camshafts, as follows:

a. Apply new engine oil to the thrust portion and journal of the camshaft

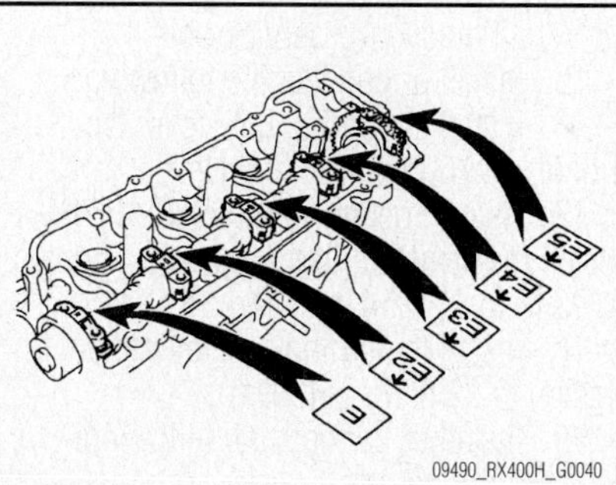

Fig. 56 Right exhaust camshaft bearing caps must be placed in their proper locations.

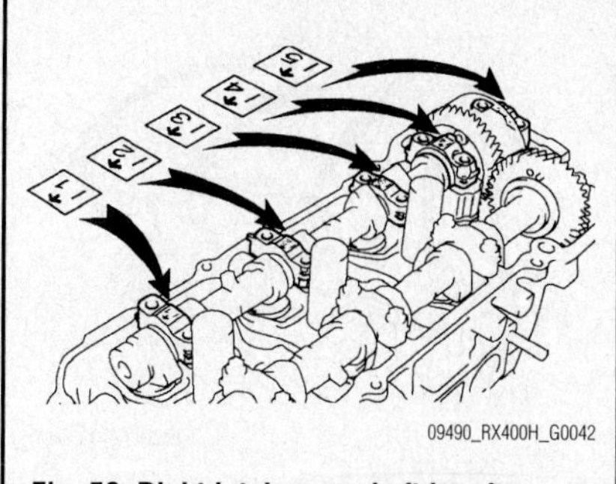

Fig. 58 Right intake camshaft bearing caps must be placed in their proper locations

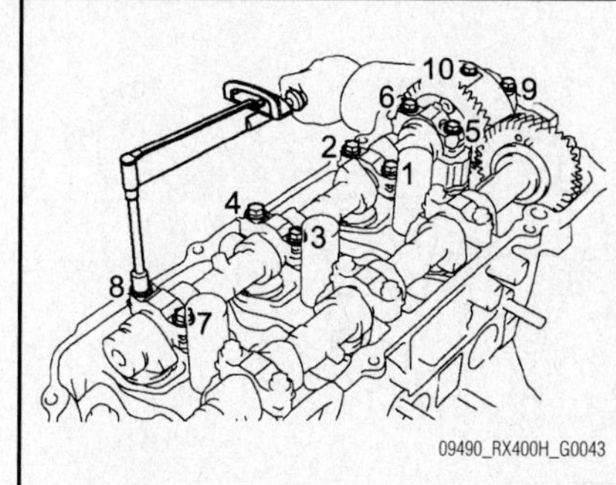

Fig. 59 Right intake camshaft bearing cap bolt tightening sequence

b. Position the exhaust camshaft on the head so that the alignment mark is at a 90 degrees angle from vertical. The mark should be at the 9 o'clock' position.

c. Apply multi-purpose grease to the oil seal lip and install the new oil seal to the camshaft.

d. Apply sealant to the No. 1 bearing cap.

e. Apply a light coat of clean engine oil to the bolt threads and under the bolt head. Install the bearing caps to their proper position. Tighten the bolts evenly

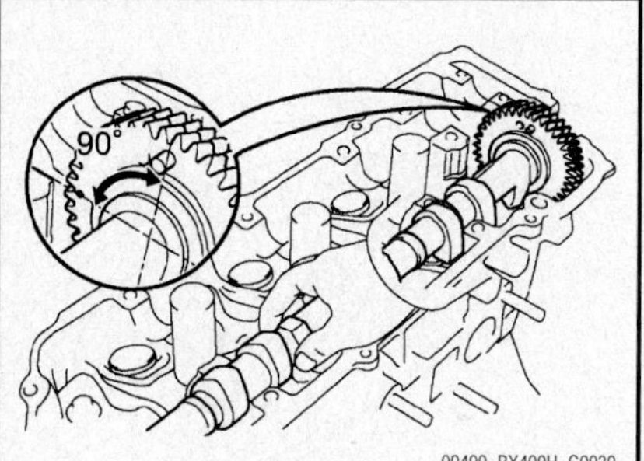

Fig. 55 Install the right exhaust camshaft with the alignment marks in the correct position.

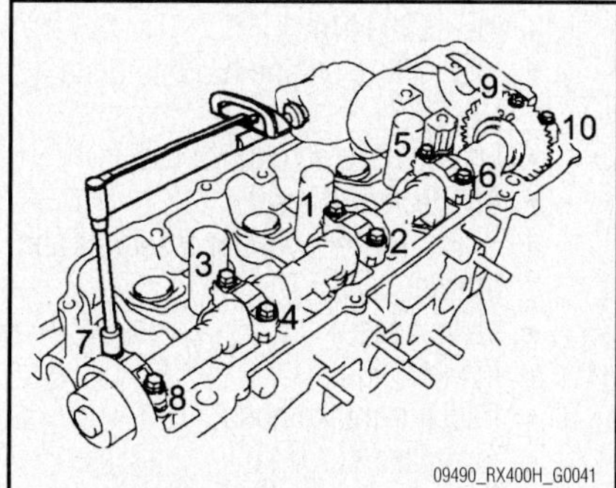

Fig. 57 Right exhaust camshaft bearing cap torque sequence

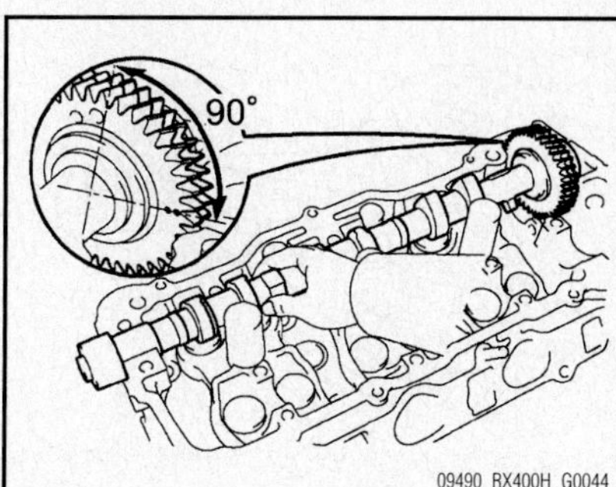

Fig. 60 Install the left exhaust camshaft with the alignment mark in the correct position.

Fig. 61 Left exhaust camshaft bearing caps must be placed in their proper locations

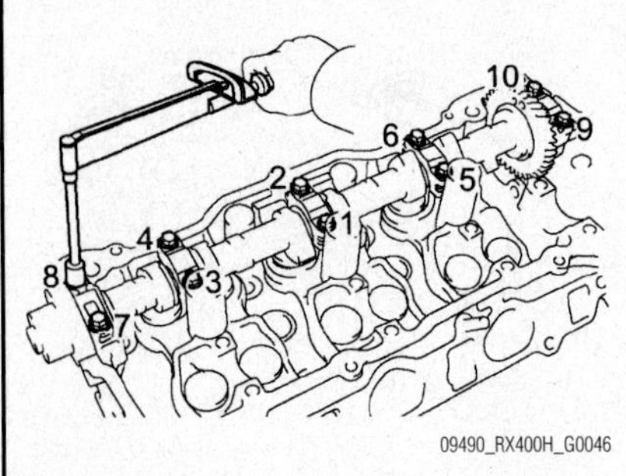

Fig. 62 Left exhaust camshaft bearing cap torque sequence

and in several passes to 12 ft. lbs. (16 Nm) in the proper sequence.

f. Position the intake camshaft on the head so that the alignment marks are at a 90 degrees angle from vertical. The mark should be at the 3 o'clock' position and must align with the marks on the exhaust camshaft gear.

g. Apply a light coat of clean engine oil to the bolt threads and under the bolt head. Install the bearing caps to their proper position. Tighten the bolts evenly and in several passes to 12 ft. lbs. (16 Nm) in the proper sequence.

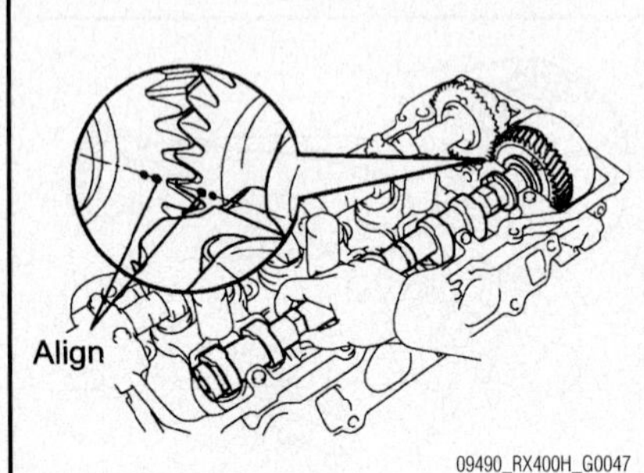

Fig. 63 Install the left intake camshaft with the alignment mark in the correct position.

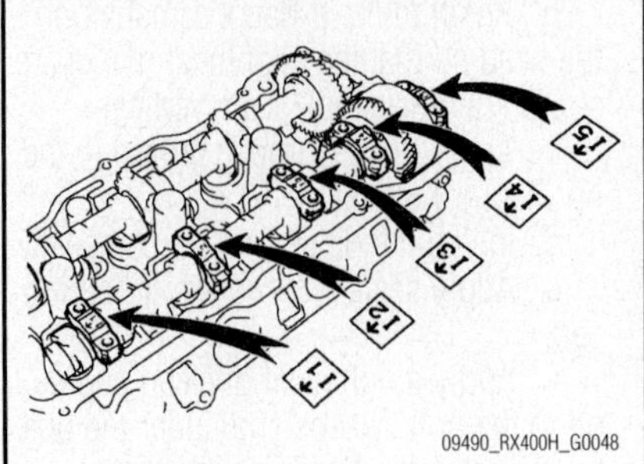

Fig. 64 Left exhaust camshaft bearing caps must be placed in their proper locations.

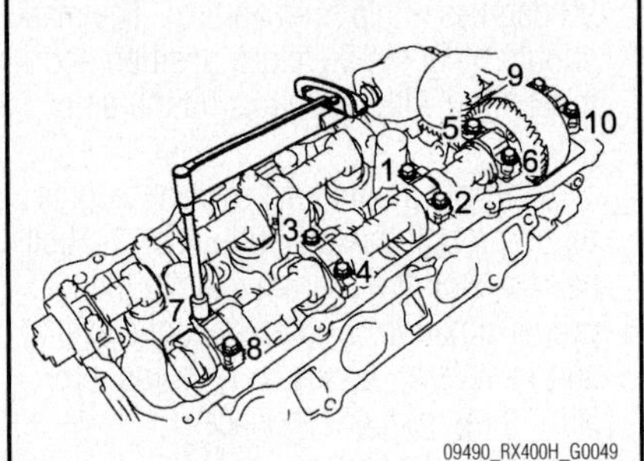

Fig. 65 Left exhaust bearing cap torque sequence

h. Remove the service bolt.

15. Install or connect the following:

16. Install the timing belt cover No. 3. Tighten to 76 inch lbs. (8.5 Nm)

17. Using Special Tool 09960-10010, install the camshaft timing pulleys. Tighten to 92 ft. lbs. (125 Nm).

18. Install the timing belt idler and tighten to 32 ft. lbs. (43 Nm).

19. Install or connect the following:

- Timing belt
- No. 2 Timing belt guide
- Right-hand engine mounting bracket
- Timing belt covers Nos. 1 and 2
- Crankshaft pulley. Tighten to 162 ft. lbs. (220 Nm).
- Right-hand engine mounting stay No. 2
- Engine moving control rod
- Air cleaner bracket
- Brake master cylinder reservoir and bracket
- Valve covers
- Ignition coil
- Radiator intake hose
- Air intake surge tank
- Emission control valve hoses
- Air intake assembly
- Battery and battery tray
- Wiper motor and link assembly
- Top cowl ventilator louver assembly
- Wiper and blade assembly
- Right-hand front fender apron seal
- Engine splash shields
- Right-hand front wheel
- Engine covers
- Negative battery cable

20. Refill the cooling system to the correct level.

21. Refill the engine oil to the correct level.

22. Start the engine and check for leaks.

23. If necessary, perform the initialization procedure, as outlined in the Chassis Electrical System.

CRANKSHAFT DAMPER

REMOVAL & INSTALLATION

See Figures 66 through 70.

1. Before servicing the vehicle, refer to the Precautions Section.
2. Remove or disconnect the following:
   - Negative battery cable
   - Engine covers
   - Right front wheel
   - Fender splash shields
   - Wiper arms
   - Top cowl ventilator louver
   - Wiper motor and linkage assembly
   - Battery and battery tray
   - Air intake assembly
   - Brake master cylinder reservoir and bracket
   - Air cleaner bracket
   - Engine moving control rod
   - Right-hand engine mounting stay No. 2
3. Use Special Tool 09213-54015 to hold the crankshaft pulley in order to loosen the pulley bolt.
4. Use Special Tool 09950-50013 to remove the crankshaft pulley.

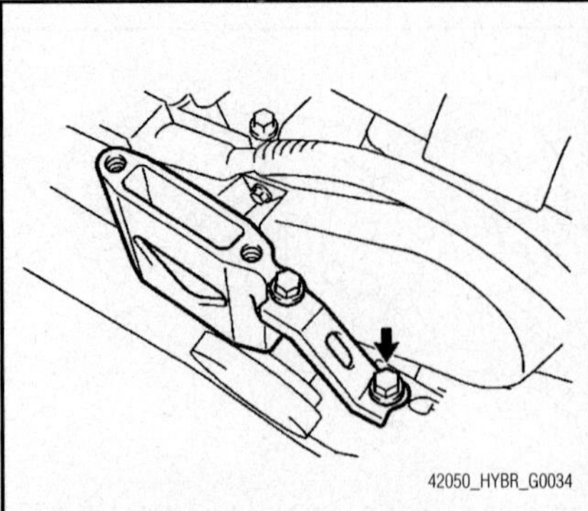

Fig. 66 Remove the right-hand engine mounting stay No. 2

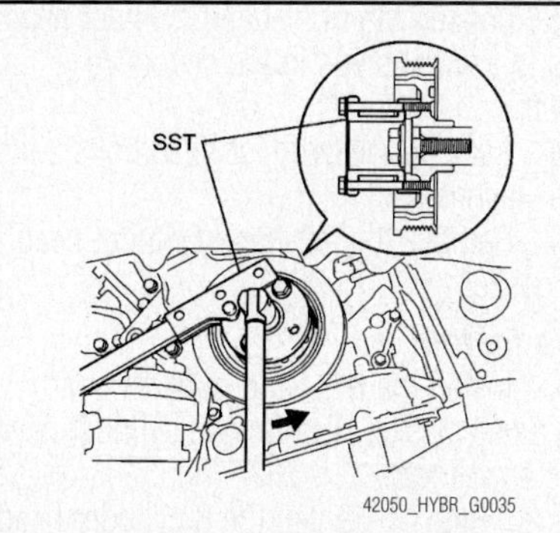

Fig. 67 Use the special tool to hold the pulley in order to loosen the pulley bolt

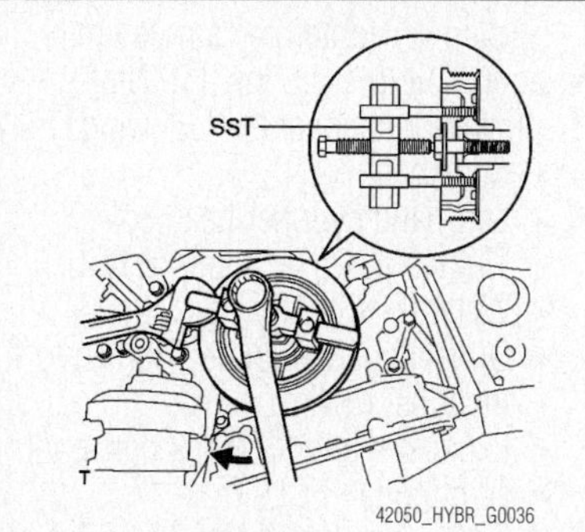

Fig. 68 Use the special tool to remove the crankshaft pulley

To install:

5. Install the crankshaft pulley, as follows:
   a. Align the keyway of the pulley with the key located on the crankshaft and slide the pulley into place.
   b. Using Special Tool 09213-54015 (91651-60855), 09330-00021, or equivalent, install the pulley bolt and tighten to 162 ft. lbs. (220 Nm).
6. The remainder of installation is the reverse of the removal procedure. Observe the following torques:
   - Right engine mount stay: 47 ft. lbs. (64 Nm)

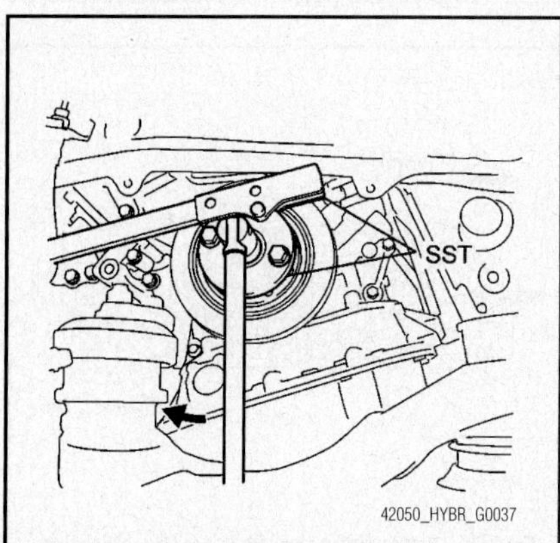

Fig. 69 Crankshaft pulley installation

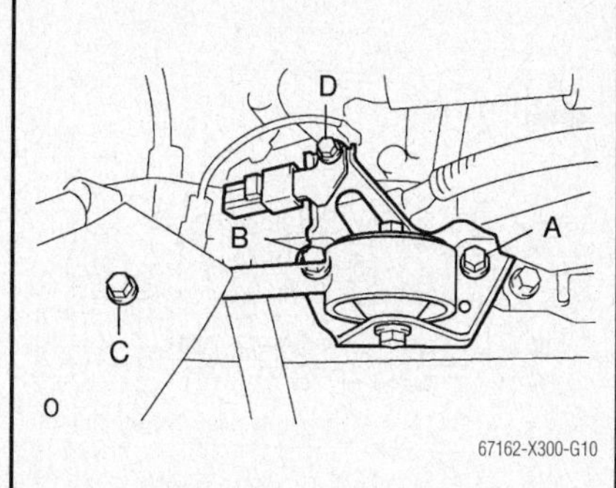

Fig. 70 Tighten the engine roll control rod bolts in this order

- Engine roll control rod: tighten first A, then B, and then C to 47 ft. lbs. (64 Nm). Torque D to 17 ft. lbs. (23 Nm)

7. If necessary, perform the initialization procedure, as outlined in the Chassis Electrical System.

CRANKSHAFT FRONT SEAL

REMOVAL & INSTALLATION

See Figures 71 and 72.

1. Before servicing the vehicle, refer to the Precautions Section.
2. Disconnect the negative battery cable.
3. Remove the timing belt.
4. Remove the timing belt plate.
5. Install the crankshaft pulley bolts to the crankshaft.
6. Using Special Tool 09950-50013 or equivalent, remove the crankshaft timing pulley.
7. Using a knife, cut off the oil seal lip.
8. Using a suitable pry tool, pry out the oil seal.

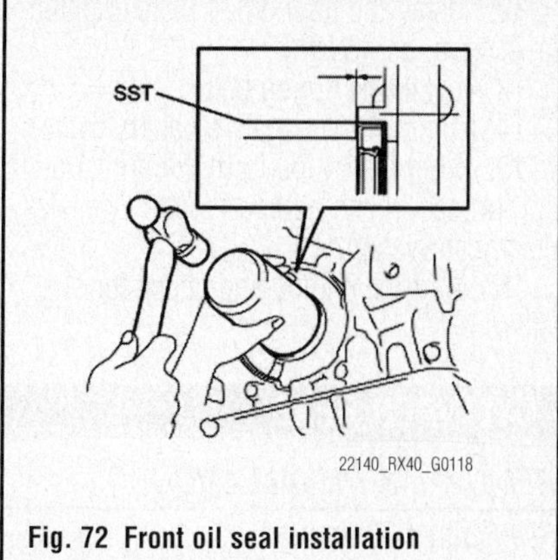

Fig. 72 Front oil seal installation

CAUTION

Use tape on the tip of the pry tool to prevent damage to the crankshaft.

To install:

9. Apply MP grease to a new oil seal lip.
10. Using Special Tool 09223-00010 or suitable seal installer and hammer, tap in a new oil seal until its surface is flush with the oil pump edge.
11. Align the keyway of the crankshaft timing pulley with the key located on the crankshaft and slide the pulley into place.

➡Do not scratch the sensor area of the crankshaft timing pulley.

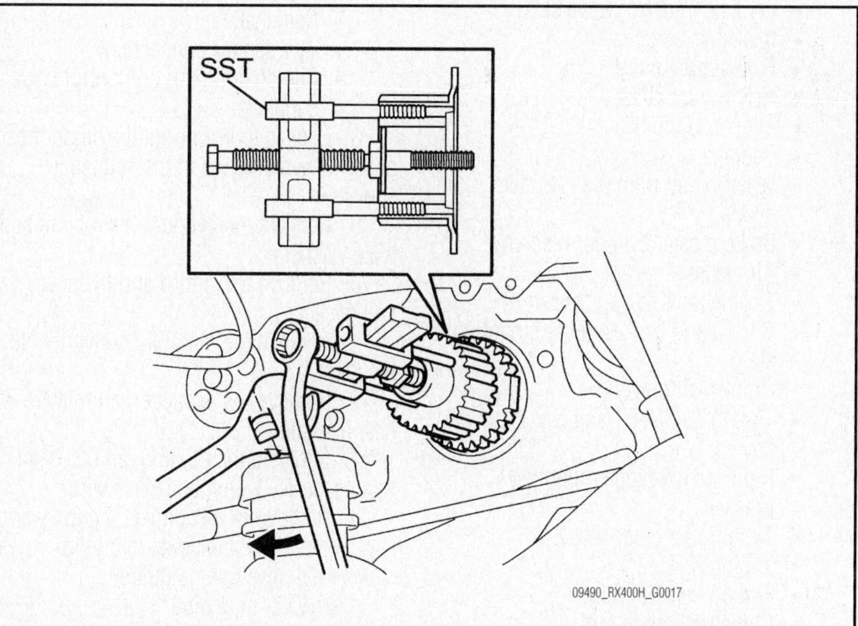

Fig. 71 Using the Special Tool or equivalent puller, remove the crankshaft timing pulley.

12. Install the timing belt plate. Tighten to 71 inch lbs. (8 Nm).
13. Install the timing belt.
14. Connect the negative battery cable.
15. If necessary, perform the initialization procedure, as outlined in the Chassis Electrical System.
16. Start the engine and check for leaks.

CYLINDER HEAD

REMOVAL & INSTALLATION

See Figures 73 through 77.

1. Before servicing the vehicle, refer to the Precautions Section.
2. Disconnect the negative battery cable.
3. Relieve the fuel system pressure.
4. Drain the engine oil.
5. Drain the coolant from the engine radiator and hybrid transaxle.
6. Remove the service plug grip, found underneath the Battery Service cover on the rear seat. Wait 5 minutes to discharge the high voltage capacitor.
7. Remove or disconnect the following:
    - Negative battery cable
    - Engine cover
    - Right-hand front wheel
    - Engine splash shields
    - Right-hand front fender apron seal
    - Wiper and blade assembly
    - Top cowl vêntilator louver assembly
    - Wiper motor and link assembly
    - Battery and battery tray
    - Air intake assembly
    - Converter with Inverter assembly
    - Emission control valve set
    - Intake air surge tank
    - Fuel supply hose
    - Heater inlet hose
    - Intake manifold
    - Radiator hoses
    - Water outlet from the cylinder heads
    - Brake master cylinder reservoir
    - Air cleaner bracket
    - Engine moving control rod
    - Right-hand No. 2 engine mounting stay
    - Crankshaft pulley
    - Timing belt No. 1 and No. 2 covers
    - Right-hand engine mounting bracket
    - Timing belt guide No. 2
    - Timing belt
    - Timing belt No. 2 idler
    - Camshaft timing pulley
    - Timing belt No. 3 cover

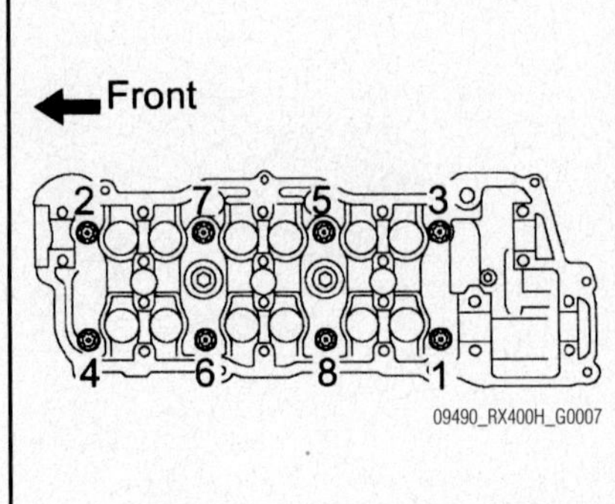

Fig. 73 Right-hand cylinder head loosening sequence

Fig. 74 Left-hand cylinder head loosening sequence

    - Front exhaust pipe assembly
    - Exhaust manifold heat insulator
    - Exhaust manifold stay
    - Right-hand exhaust manifold and gasket
    - Ignition coil
    - Right-hand cylinder head cover
    - Camshaft
    - VVT sensor connector
    - Camshaft timing oil control valve connecter
8. Loosen the right-hand cylinder head bolts in several steps in the sequence shown.
9. Remove the cylinder head bolts and plate washers.
10. Remove the right-hand cylinder head and gasket.
11. Remove the manifold converter No. 3 insulator.
12. Remove the exhaust manifold No. 2 heat insulator.
13. Separate the cooling fan ECU and hang securely with mechanic's wire.
14. Remove or disconnect the following:
    - Left-hand exhaust manifold
    - Oil level gauge guide
    - Water inlet pipe
    - Left-hand cylinder head cover
    - Camshaft
15. Loosen the right-hand cylinder head bolts in several steps in the sequence shown.
16. Remove the cylinder head bolts and plate washers.
17. Remove the left-hand cylinder head and gasket.

To install:

18. Install the left-hand cylinder head with a new gasket. Tighten the cylinder head bolts as follows:
    a. Step 1: Tighten the 8 cylinder head bolts to 40 ft. lbs. (54 Nm)
    b. Step 2: Tighten each bolt 90°
    c. Step 3: Tighten each bolt an additional 90°
    d. Step 4: Tighten the single 8mm hexagon bolt to 14 ft. lbs. (19 Nm)
19. Install or connect the following:
    - Camshaft
    - Left-hand cylinder head cover. Tighten to 71 inch lbs. (8 Nm).
    - Water inlet pipe
    - Oil level gauge guide. Tighten to 71 inch lbs. (8 Nm).
    - Exhaust manifold No. 2 converter
    - Cooling fan ECU
    - Exhaust manifold No. 2 heat insulator
    - Manifold converter No. 2 insulator

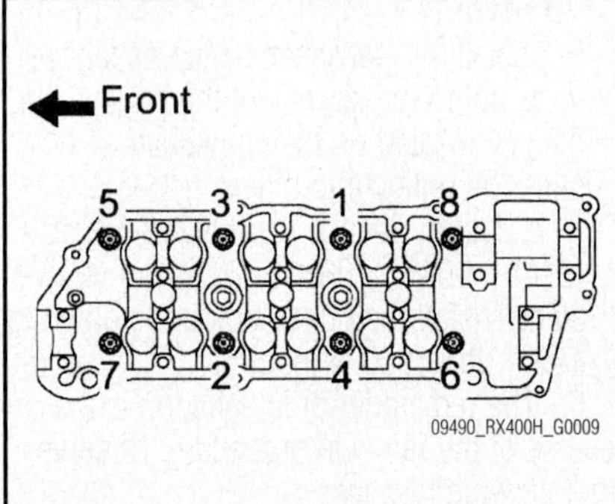

Fig. 75 Left-hand cylinder head tightening sequence

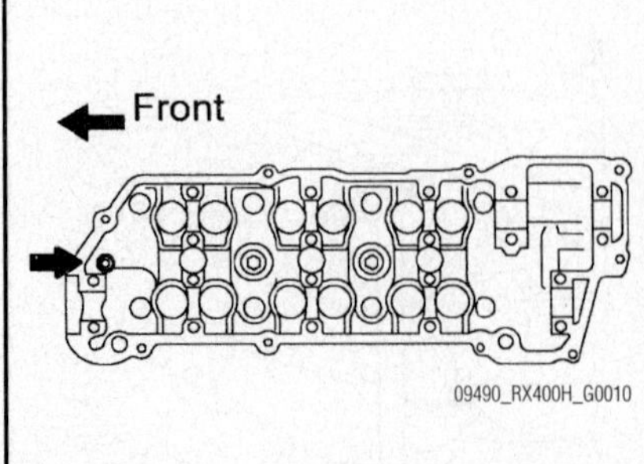

Fig. 76 8mm hexagon bolt on the cylinder head

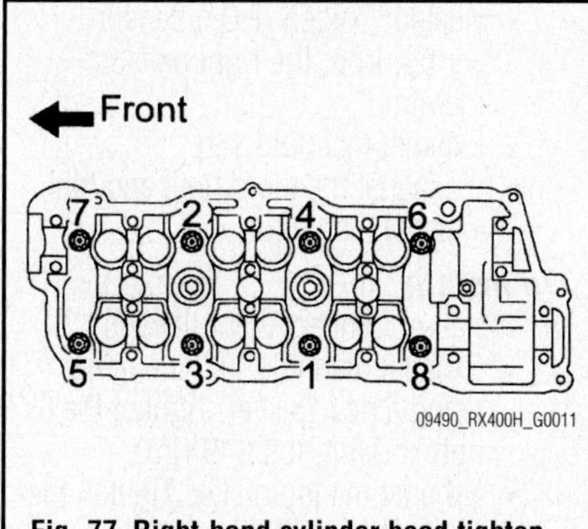

Fig. 77 Right-hand cylinder head tightening sequence

20. Install the right cylinder head with a new gasket. Tighten the cylinder head bolts as follows:
   a. Step 1: Tighten the 8 cylinder head bolts to 40 ft. lbs. (54 Nm)
   b. Step 2: Tighten each bolt 90°
   c. Step 3: Tighten each bolt an additional 90°
   d. Step 4: Tighten the single 8mm hexagon bolt to 14 ft. lbs. (19 Nm)
21. The remainder of installation is the reverse order of removal.
22. Refill the engine oil to the correct level.
23. Refill the coolant to the engine radiator and hybrid transaxle to the correct level.
24. Replace the service plug grip.
25. Connect the negative battery cable.
26. Start the engine and check for leaks.
27. If necessary, perform the initialization procedure, as outlined in the Chassis Electrical System.

ENGINE ASSEMBLY

REMOVAL & INSTALLATION

See Figures 78 and 79.

1. Before servicing the vehicle, refer to the Precautions Section.
2. Drain the coolant from the engine and hybrid assemblies.
3. Drain the engine oil.
4. Drain the transfer case fluid, if equipped.
5. Drain the hybrid transaxle fluid.
6. Discharge the fuel system pressure.
7. Remove the service plug grip, found underneath the Battery Service cover on the rear seat. Wait 5 minutes to discharge the high voltage capacitor.
8. Remove or disconnect the following:
   - Engine room side covers
   - Front wheels
   - Engine under cover assembly
   - Left and right fender splash shields
   - Left and right fender apron seals
   - Wiper arms
   - Cowl top ventilator louver
   - Wiper linkage
   - Cowl top panel outer sub-assembly
   - Battery and battery tray
   - Air intake assembly
   - Inverter support bracket No. 5
   - Power steering ECU assembly
9. Remove the inverter with converter assembly as follows:
   a. Disconnect circuit breaker sensor No. 1.
   b. Disconnect engine room wire No. 2.
   c. Remove the inverter reserve tank sub-assembly.
   d. Disconnect the coolant hose.
   e. Disconnect the power steering ECU bracket.
   f. Remove the inverter cover.
   g. Verify the voltage of w/ converter inverter assembly is 0 volts.
   h. Disconnect engine wire No. 4.
   i. Disconnect the high voltage cable of the front motor.
   j. Disconnect the Motor Generator ECU connector.
   k. Disconnect No. 3 wire frame.
   l. Install the inverter cover.
   m. Separate the engine room relay block assembly.
   n. Remove the inverter bracket No. 4.
   o. Remove the w/ converter inverter assembly.
10. Remove or disconnect the following:
   - Brake master cylinder reservoir sub-assembly
   - Master cylinder reservoir bracket
   - Air intake assembly bracket
   - Engine moving control rod
   - Right-hand engine mount
   - Compressor with the motor assembly
   - Inverter bracket No. 1
   - Transmission control cable assembly
   - Fuel vapor feed hose
   - Fuel pipe sub-assembly No. 1
   - Heater hoses
   - Radiator hoses
   - Oil cooler hoses
   - Water pump assembly
   - Glove compartment door assembly
   - Engine wire harnesses from the ECU
   - Front exhaust pipe assembly
   - Front stabilizer link assembly
   - Front axle hub nut, both sides
   - Wheel speed sensors
   - Tie rod end sub-assemblies
   - Lower control arms
11. Separate the halfshafts from the axle hub.
12. Separate the steering intermediate shaft sub-assembly.
13. Attach a lifting crane.
14. Remove the left and right hand frame side rail plates.
15. Remove the front suspension member rear braces.
16. Lift the engine/transaxle from the vehicle.
17. Installation is the reverse of removal. Observe the following torques:
   - Frame side rail plates—Bolt A: 63 ft. lbs. (85 Nm). Bolt B: 24 ft. lbs. (32 Nm). Nut C: 24 ft. lbs. (32 Nm).
   - Front suspension member brace—Bolt A: 63 ft. lbs. (85 Nm). Bolt B: 24 ft. lbs. (32 Nm). Nut C: 24 ft. lbs. (32 Nm).
   - Front exhaust pipe assembly: 41 ft. lbs. (56 Nm).
   - Front stabilizer link assembly: 55 ft. lbs. (74 Nm).
   - Steering link: 26 ft. lbs. (35 Nm)
18. If necessary, perform the initialization procedure, as outlined in the Chassis Electrical System.

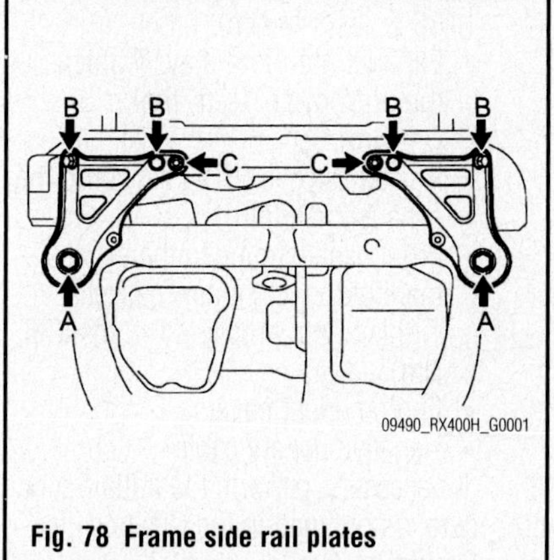

Fig. 78 Frame side rail plates

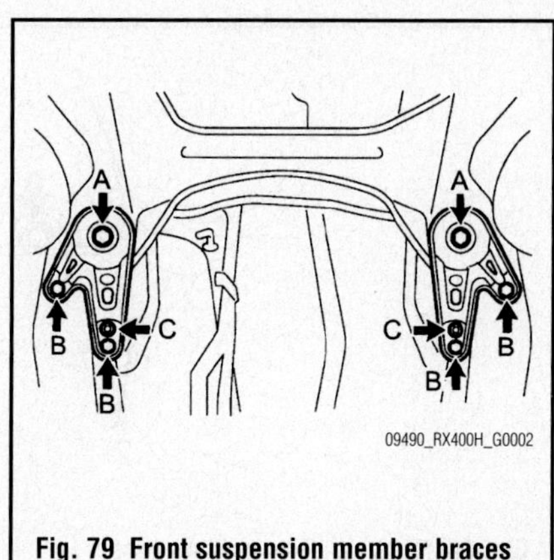

Fig. 79 Front suspension member braces

EXHAUST MANIFOLD

REMOVAL & INSTALLATION

Front

See Figure 80.

➡**Removing the oil filter helps gain access to a lower bolt in the front exhaust manifold.**

1. Before servicing the vehicle, refer to the Precautions Section.
2. Remove or disconnect the following:
   - Negative battery cable
   - Engine under covers
   - Front exhaust pipe from the exhaust manifolds, by removing the nuts

➡**Check for access to some of the manifold lower bolts, if so remove any possible.**

   - Heated Oxygen (HO_2) sensor
   - Exhaust manifold stay, by removing the bolt and nut
   - Remaining exhaust manifold nuts; then, separate the exhaust manifold from the engine

To install:

3. Install or connect the following:
   - Exhaust manifold, using a new gasket. Uniformly, tighten the bolts to 36 ft. lbs. (49 Nm).
   - Exhaust manifold stay. Tighten the nut/bolt to 15 ft. lbs. (20 Nm).
   - Heated Oxygen (HO_2) sensor to the exhaust manifold
   - Front exhaust pipe to the exhaust manifold, using a new gasket. Tighten both nuts to 41 ft. lbs. (56 Nm).
   - Engine under covers
   - Negative battery cable
4. If necessary, perform the initialization procedure, as outlined in the Chassis Electrical System.

Rear

See Figures 81 and 82.

1. Before servicing the vehicle, refer to the Precautions Section.
2. Remove or disconnect the following:
   - Negative battery cable
   - Engine under covers
   - Front exhaust pipe from both exhaust manifolds, from below the engine
   - Exhaust Gas Recirculation (EGR) pipe from the rear exhaust manifold, by removing the 4 nuts
   - Heated Oxygen (HO_2) sensor wiring, from the right exhaust manifold
   - Exhaust manifold stay
   - 6 exhaust manifold nuts and the exhaust manifold

To install:

3. Install or connect the following:
   - Exhaust manifold to the engine, using a new gasket. Tighten the 6 nuts to 36 ft. lbs. (49 Nm).
   - Exhaust manifold stay. Tighten the nut/bolt to 25 ft. lbs. (34 Nm).

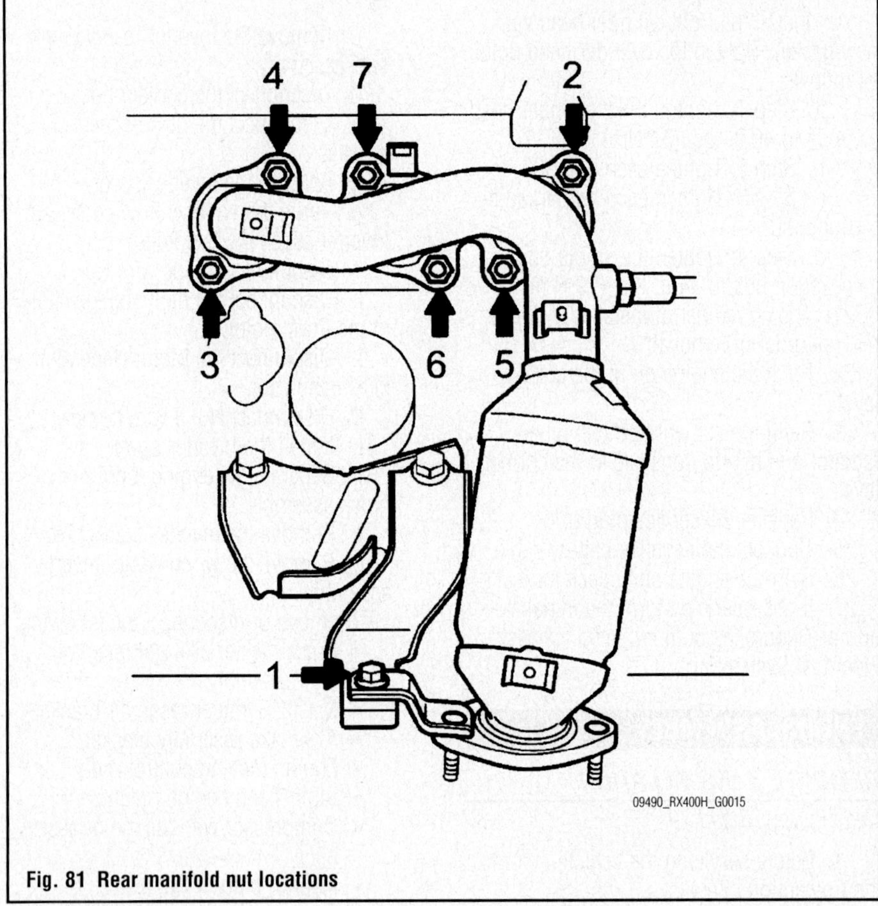

Fig. 81 Rear manifold nut locations

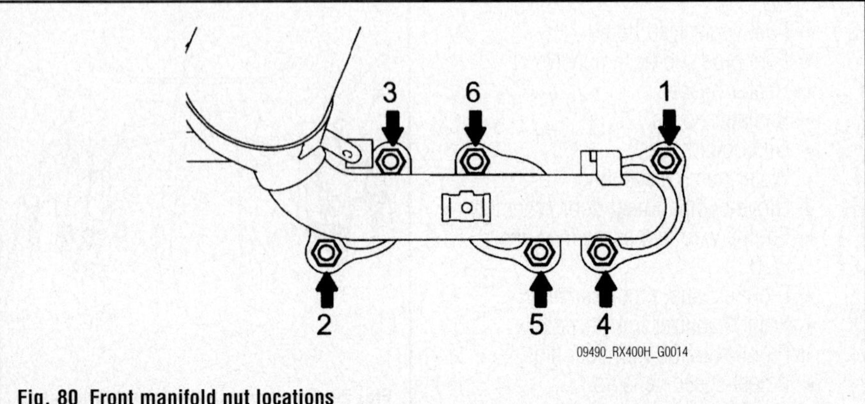

Fig. 80 Front manifold nut locations

   - HO_2 sensor wiring to the exhaust manifold
   - EGR pipe to the exhaust manifold and the engine, using new gaskets. Tighten the 4 nuts to 108 inch lbs. (12 Nm).
   - Front exhaust pipe to the exhaust manifold, use a new gasket. Tighten both nuts to 41 ft. lbs. (56 Nm).
   - Engine under covers
   - Negative battery cable
4. If necessary, perform the initialization procedure, as outlined in the Chassis Electrical System.

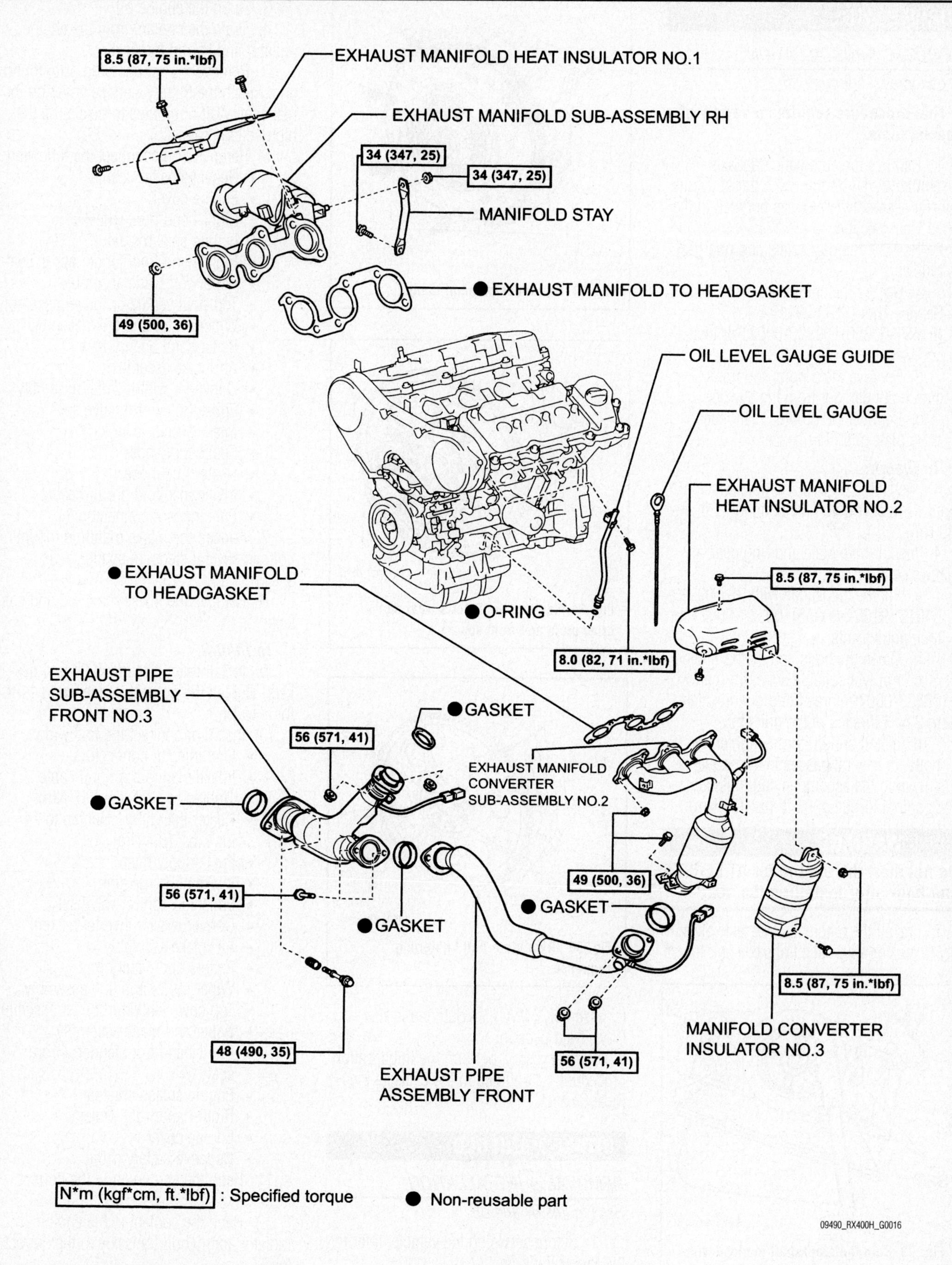

Fig. 82 Exploded view of the exhaust system

DRIVEPLATE

REMOVAL & INSTALLATION

See Figures 83 through 86.

➡This procedure requires a variety of special tools.

1. Remove the automatic transaxle assembly (2WD) or automatic transaxle and transfer assembly (4WD), as outlined in the Drive Train Section.
2. Remove the drive plate and ring gear, as follows:
   a. Secure the crankshaft with Special Service Tool (SST) 09213-54015 (91651-60855), 09330-00021 or their equivalents.
   b. Remove the 6 bolts and transmission input damper from the flywheel.
   c. Remove the 8 bolts, rear spacer, drive plate and front spacer.

To install:

3. Install the 6 bolts and transmission input damper. Tighten the bolts to 22 ft. lbs. (30 Nm).
4. Install drive plate and ring gear, as follows:
   a. Fix the crankshaft with SST 09213-54015 (91651-60855), 09330-00021, or their equivalents.
   b. Clean the bolts and the bolt holes.
   c. Apply a suitable adhesive (part No. 08833-00070, three bond or equivalent) to 2 or 3 threads of the bolt end.
   d. Install and uniformly tighten the 8 bolts, in several passes in the sequence shown in the accompanying illustration to a final torque of 61 ft. lbs. (83 Nm).

WARNING

Do not start the engine for AT LEAST one hour after installing the seal!

5. Install the automatic transaxle assembly (2WD) or automatic transaxle and transfer assembly (4WD), as outlined in the Drive Train Section.
6. If necessary, perform the initialization procedure, as outlined in the Chassis Electrical System.

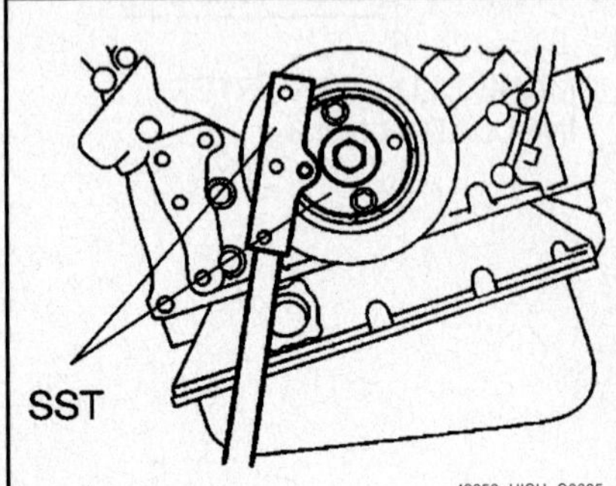

Fig. 83 Hold the crankshaft in place with the special tools as shown

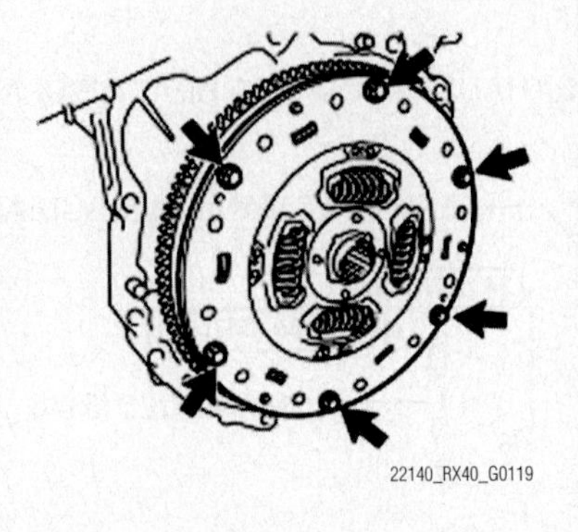

Fig. 84 Transmission input damper

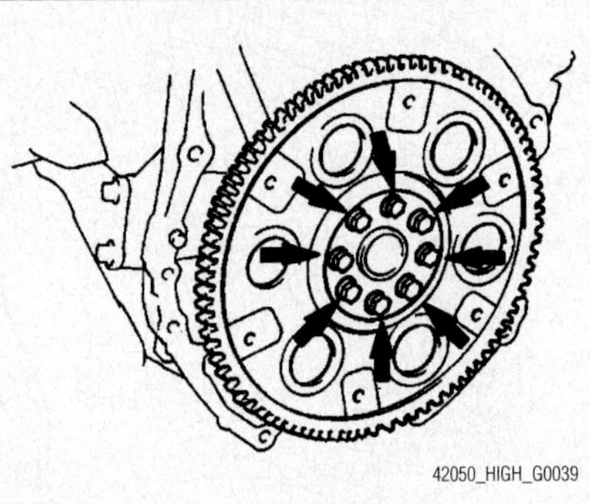

Fig. 85 Remove the 8 bolts, rear spacer, drive plate and front spacer

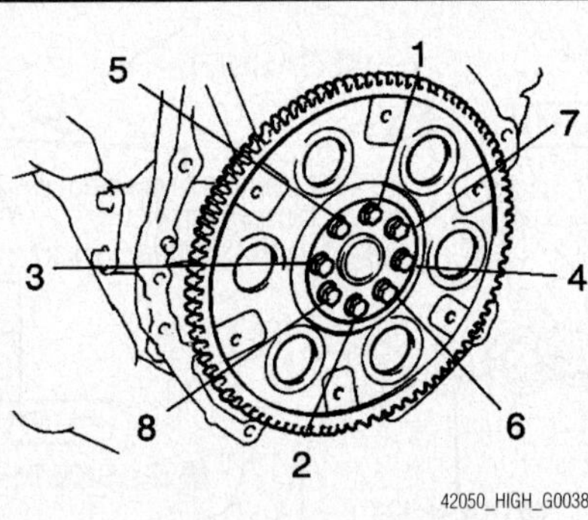

Fig. 86 Drive plate bolt tightening sequence

INTAKE MANIFOLD

REMOVAL & INSTALLATION

See Figures 87 and 88.

1. Before servicing the vehicle, refer to the Precautions Section.
2. Relieve the fuel system pressure.
3. Drain the engine oil.
4. Drain the coolant from the engine radiator and hybrid transaxle.
5. Remove the service plug grip, found underneath the Battery Service cover on the rear seat. Wait 5 minutes to discharge the high voltage capacitor.
6. Remove or disconnect the following:
   - Negative battery cable
   - Engine cover
   - Right-hand front wheel
   - Engine splash shields
   - Right-hand front fender apron seal
   - Wiper and blade assembly
   - Top cowl ventilator louver assembly
   - Wiper motor and link assembly
   - Battery and battery tray
   - Air intake assembly
   - Converter with Inverter assembly
   - Emission control valve set
   - Intake air surge tank
   - Fuel supply hose
   - Heater inlet hose
   - Intake manifold ground cable
   - Fuel injector connectors
7. Loosen the intake manifold mounting bolts in several steps, in sequence as shown.
8. Remove the intake manifold and gaskets.

To install:

9. Install the intake manifold and gaskets. Tighten the bolts in sequence to 11 ft. lbs. (15 Nm).
10. Install or connect the following:
    - Fuel injector connectors
    - Intake manifold ground cable. Tighten to 11 ft. lbs. (15 Nm).
    - Heater inlet hose. Tighten to 74 inch lbs. (8.4 Nm).
    - Fuel supply hose
    - Intake air surge tank
    - Emission control valve set
    - Converter with Inverter assembly
    - Air intake assembly
    - Battery and battery tray
    - Wiper motor and link assembly
    - Top cowl ventilator louver assembly
    - Wiper and blade assembly
    - Right-hand front fender apron seal
    - Engine splash shields
    - Right-hand front wheel
    - Engine cover
    - Negative battery cable
11. Refill the engine oil to the correct level.
12. Refill the coolant to the engine radiator and hybrid transaxle to the correct level.
13. Replace the service plug grip.

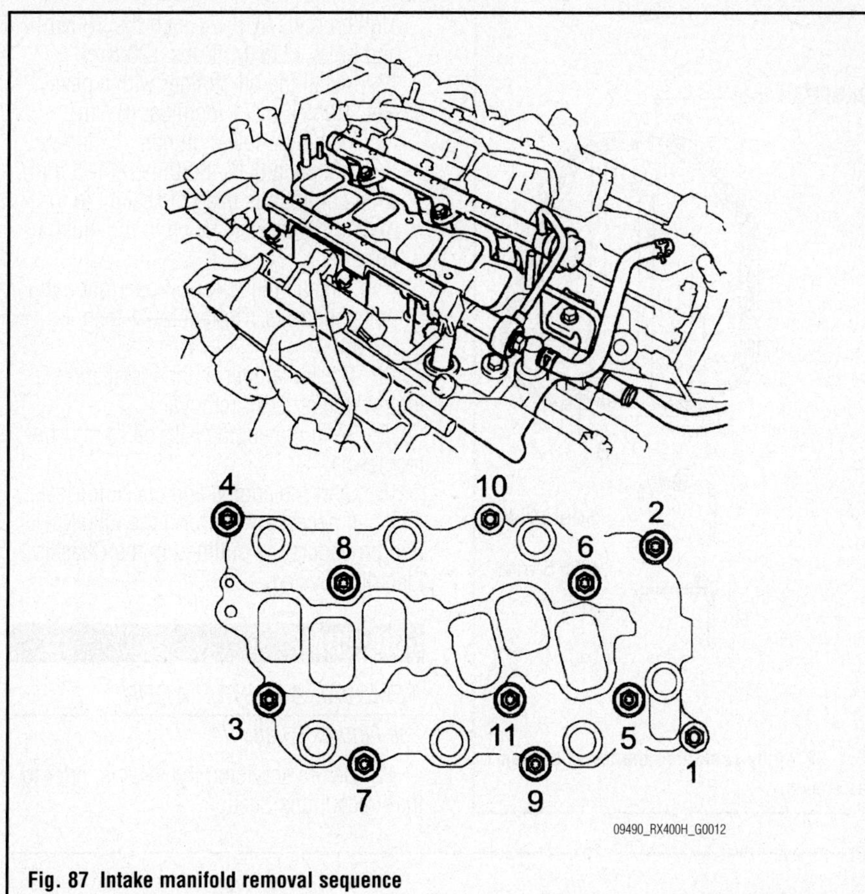

Fig. 87 Intake manifold removal sequence

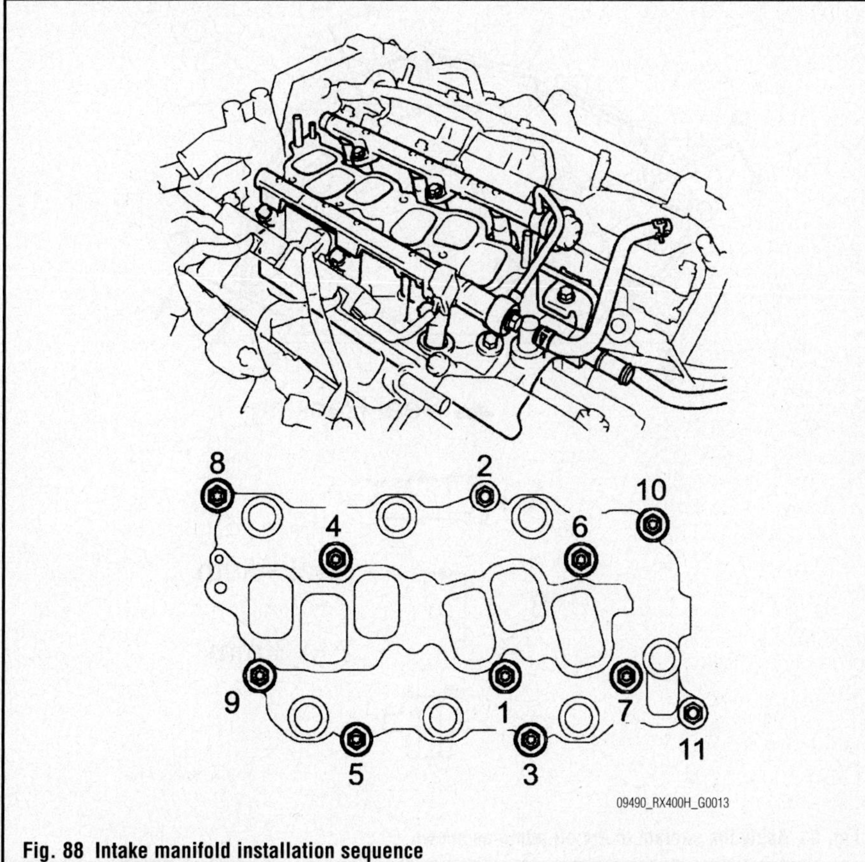

Fig. 88 Intake manifold installation sequence

14. If necessary, perform the initialization procedure, as outlined in the Chassis Electrical System.

15. Start the engine and check for leaks.

OIL PAN

REMOVAL & INSTALLATION

See Figures 89 through 92.

1. Before servicing the vehicle, refer to the Precautions Section.
2. Remove the service plug grip, found underneath the Battery Service cover on the rear seat. Wait 5 minutes to discharge the high voltage capacitor.
3. Disconnect the negative battery cable.
4. Remove or disconnect the following:
   - Engine/transaxle assembly from the vehicle
   - Right-hand exhaust manifold
   - Transaxle mass damper
   - Front frame assembly
   - Halfshafts
   - Flywheel housing undercover
   - Front engine mounting bracket
   - Transaxle assembly from the engine
   - Transmission input damper assembly
   - Flywheel
5. Install the engine to a suitable engine stand.
6. Remove or disconnect the following:
   - Remaining exhaust manifold heat shields
   - Right-hand engine mounting bracket
   - Compressor mounting bracket
   - Crankshaft pulley
   - Timing belt
   - Timing belt idler
   - Crankshaft timing pulley
   - Oil level gauge assembly

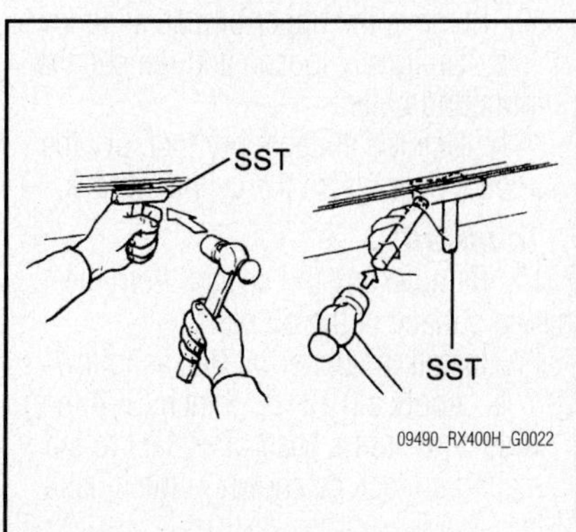

Fig. 89 Use a suitable tool to cut the sealant between the oil pans.

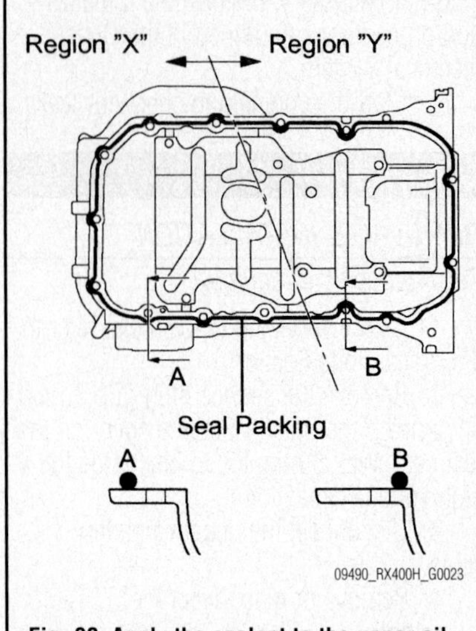

Fig. 90 Apply the sealant to the upper oil pan as shown.

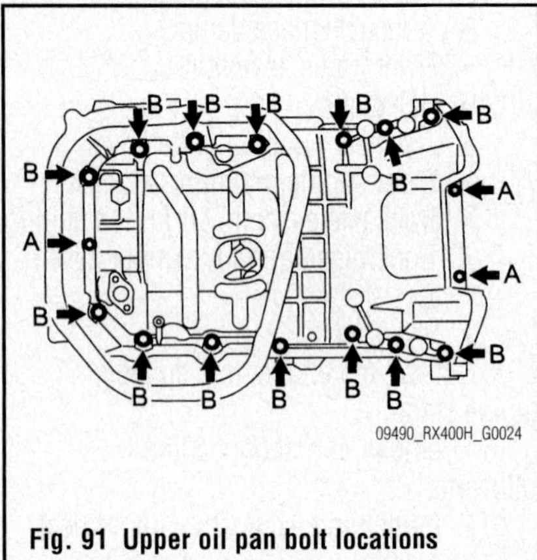

Fig. 91 Upper oil pan bolt locations

7. Remove the lower oil pan as follows:
   a. Remove the mounting bolts and nuts
   b. Using Special Tool 09032-00100 or suitable seal cutter, cut the sealant between the upper and lower oil pans.
   c. Remove the lower oil pan.
8. Remove the oil strainer and gasket.
9. Remove the upper oil pan as follows:
   a. Uniformly loosen and remove the mounting bolts.
   b. Using a suitable pry tool, pry the upper oil pan from the cylinder block.

To install:

10. Remove any old sealant from the mating surface of the oil pans.
11. Install the upper oil pan as follows:
   a. Apply a 0.12–0.16 inch (3–4 mm) wide continuous bead of sealant to the mating surface as shown in the illustration.
   b. Install the upper oil pan mounting bolts and tighten in several steps.

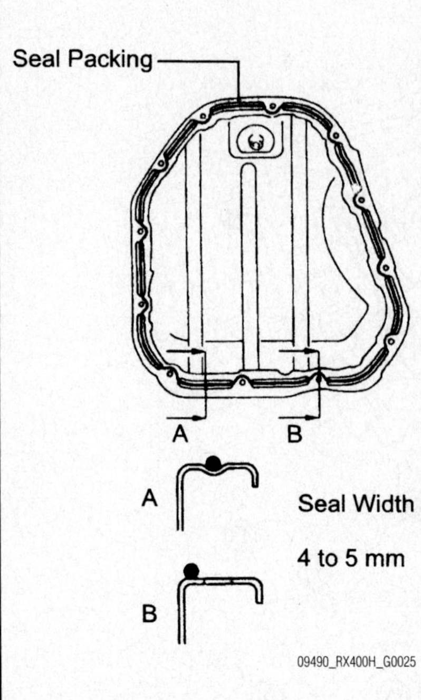

Fig. 92 Apply sealant to the lower oil pan as shown.

Tighten bolts 'A' to 71 inch lbs. (8 Nm) and bolts 'B' to 14 ft. lbs. (20 Nm).

12. Install the oil strainer with a new gasket. Tighten to 71 inch lbs. (8 Nm).
13. Install the lower oil pan as follows:
   a. Apply a 0.16–0.20 inch (4–5 mm) wide continuous bead of sealant to the mating surface as shown in the illustration.
   b. Install the lower oil pan mounting bolts and nuts. Tighten to 71 inch lbs. (8 Nm).
14. The remainder of the installation is the reverse order of removal.
15. Refill the engine with oil to the correct level.
16. Start the engine and check for leaks.
17. If necessary, perform the initialization procedure, as outlined in the Chassis Electrical System

OIL PUMP

REMOVAL & INSTALLATION

See Figures 93 and 94.

1. Before servicing the vehicle, refer to the Precautions Section.

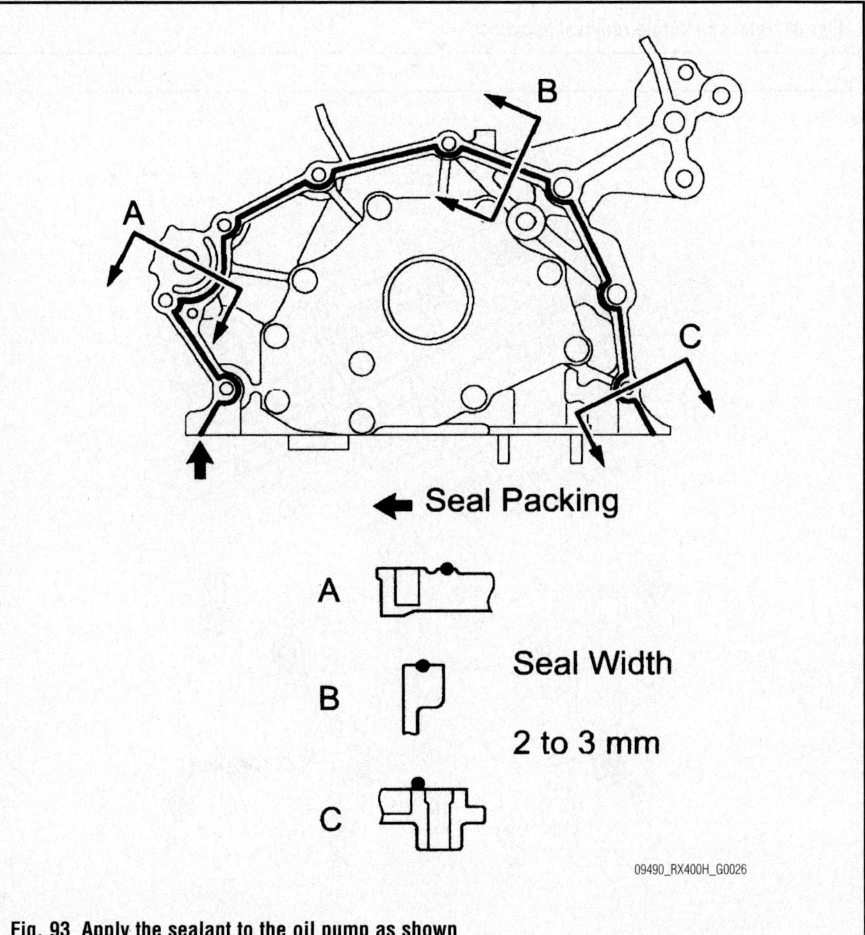

Fig. 93 Apply the sealant to the oil pump as shown

2. Remove or disconnect the following:
   - Upper and lower oil pans
   - Crankshaft Position (CKP) sensor
   - 9 oil pump bolts

➡Make a note of the position of the each bolt. When replacing the bolts into the oil pump body, place each bolt in the position from which it was removed.

   - Oil pump body, by prying between the oil pump and main bearing cap
   - O-ring from the cylinder block
   - Plug, gasket, spring and relief valve from the oil pump body
   - 9 screws, pump body cover, drive and driven rotors

To install:

3. Install or connect the following:
   - Driven rotors, drive, pump body cover, using the 9 screws
   - Oil pump relief valve, spring, gasket and the plug to the oil pump body
   - New O-ring on the cylinder block
4. Using a non residue solvent, clean both sealing surfaces to the oil pump.
5. Apply liquid sealant to the oil pump and engine block.
6. Install or connect the following:
   - Oil pump

➡Be sure to engage the splined teeth of the oil pump drive gear with the large teeth of the crankshaft.

   - 9 oil pump bolts. Tighten the bolts in several passes to 71 inch lbs. (8 Nm) for bolt 'A'; 14 ft. lbs. (20 Nm), for bolts 'B'; 32 ft. lbs. (43 Nm) for bolt 'C'
   - CKP sensor. Tighten the bolt to 71 inch lbs. (8 Nm).
   - Upper and lower oil pans
7. Refill the engine with oil to the correct level.
8. Start the engine and inspect for leaks.
9. Recheck the engine oil level.
10. If necessary, perform the initialization procedure, as outlined in the Chassis Electrical System.

INSPECTION

See Figures 95 through 97.

1. Inspect the oil pump rotor set.
2. Apply a light coat of engine oil to the oil pump rotor set and place it into the oil pump body. Check that the rotors revolve smoothly.
3. Inspect rotor tip clearance as follows:
   - Using a feeler gauge, measure the clearance between the drive and the driven rotor tips.
   - Standard tip clearance: 0.0024 to 0.0071 inch. (0.060 to 0.180 mm). (Maximum tip clearance: 0.0138 inch. (0.35 mm).
4. Inspect rotor body clearance as follows:
   - Using a feeler gauge, measure the clearance between the driven rotor and body.
   - Standard body clearance: 0.0098 to 0.0128 inch. (0.250 to 0.325 mm).

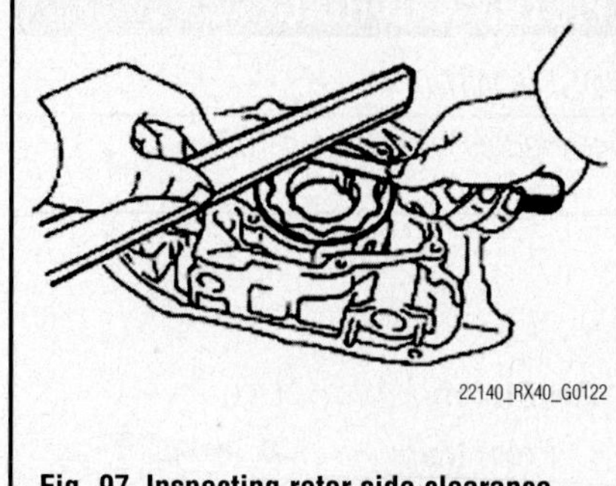

Fig. 97 Inspecting rotor side clearance

   Maximum body clearance: 0.0118 inch. (0.30 mm).

5. Inspect rotor side clearance as follows:
   - Using a feeler gauge and precision straight edge, measure the clearance between the rotors and precision straight edge.
   - Standard side clearance: 0.0012 to 0.0035 inch. (0.03 to 0.09 mm). Maximum side clearance: 0.15 mm (0.0059 in.)

MAIN BEARING TORQUE SEQUENCE

See Figures 98 and 99.

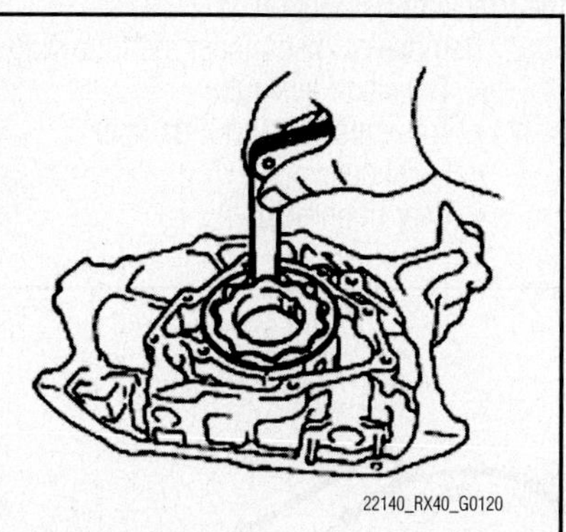

Fig. 95 Inspecting rotor tip clearance

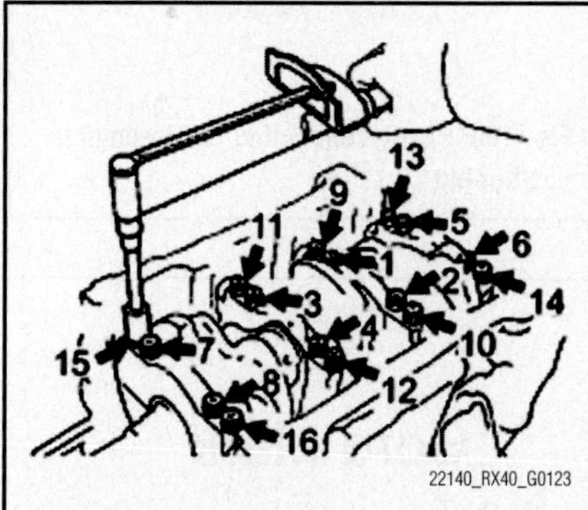

Fig. 98 Main bearing torque sequence—3.3L (3MZ-FE) engine (1)

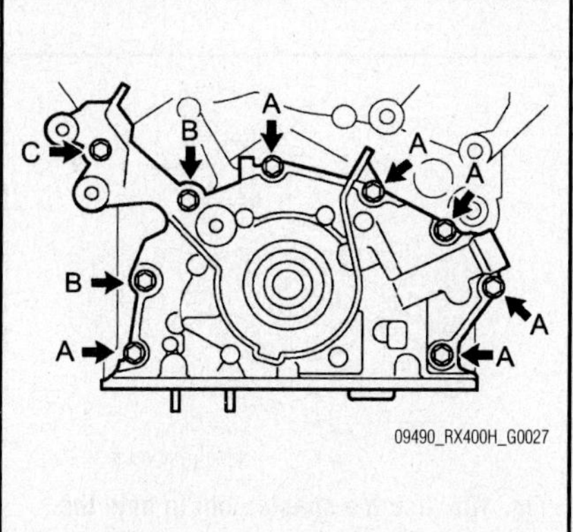

Fig. 94 Oil pump bolt locations

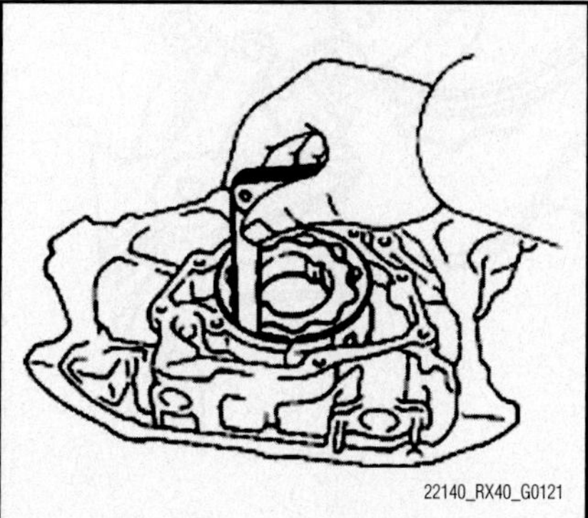

Fig. 96 Inspecting rotor body clearance

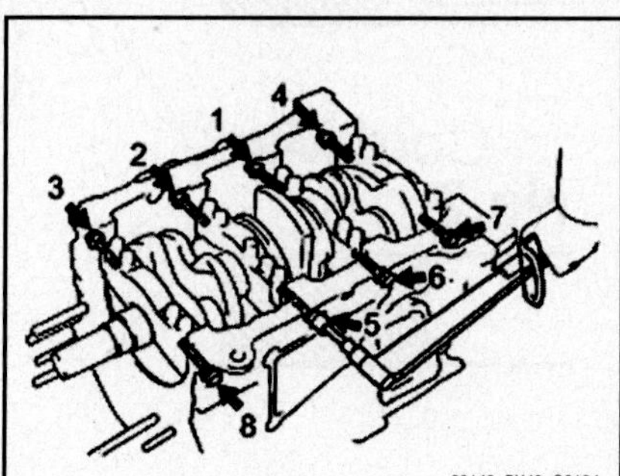

Fig. 99 Main bearing torque sequence—3.3L (3MZ-FE) engine (2)

PISTON AND RING

POSITIONING

See Figures 100 through 102.

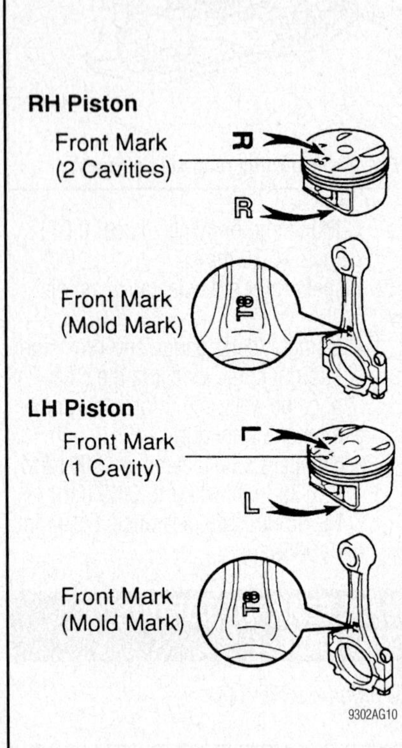

Fig. 100 Piston/connecting rod-to-engine positioning

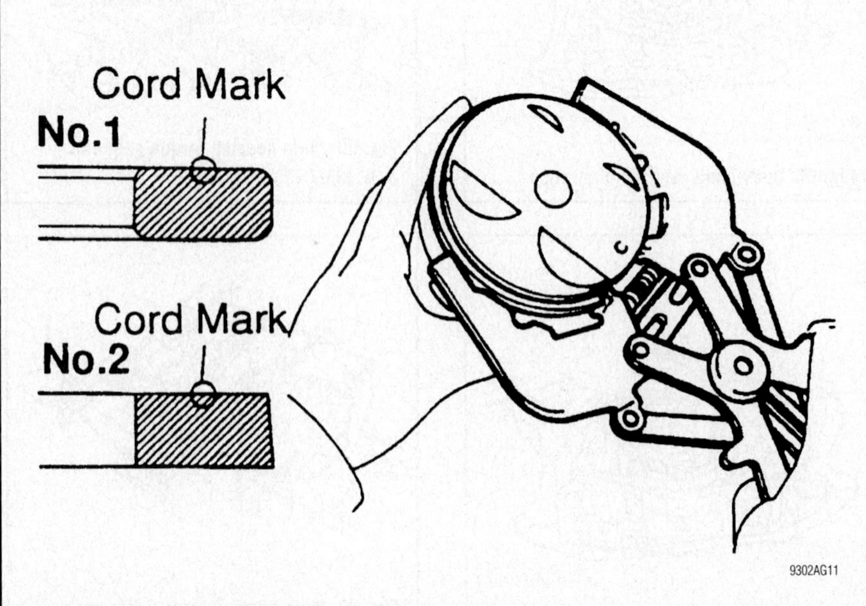

Fig. 101 Piston ring positioning

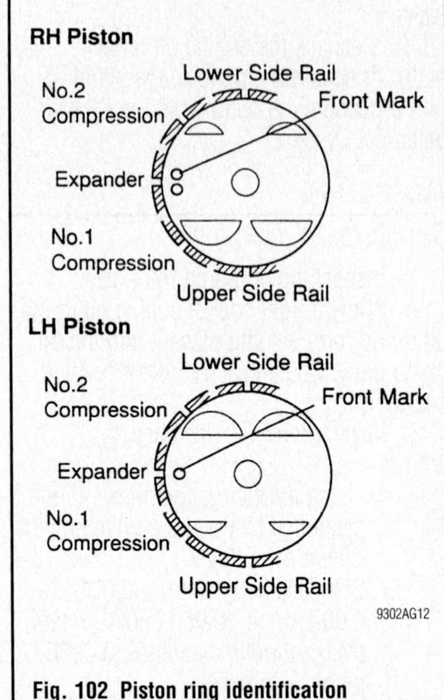

Fig. 102 Piston ring identification

REAR MAIN SEAL

REMOVAL & INSTALLATION

See Figures 103 and 104.

1. Before servicing the vehicle, refer to the Precautions Section.
2. Remove or disconnect the following:
   - Transaxle assembly
   - Transmission input damper
   - Flywheel
   - Rear main seal

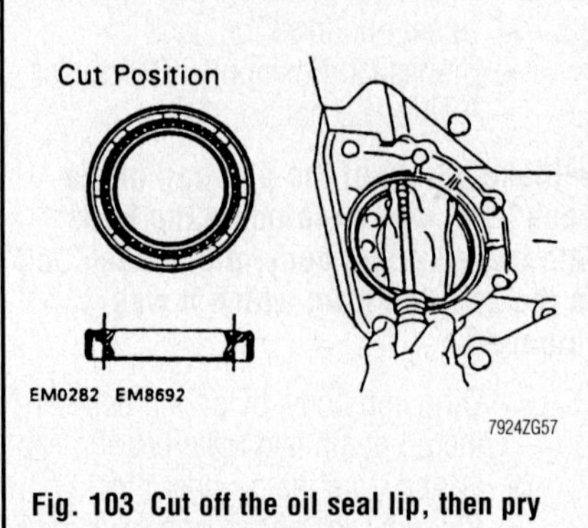

Fig. 103 Cut off the oil seal lip, then pry the seal out of the retaining plate

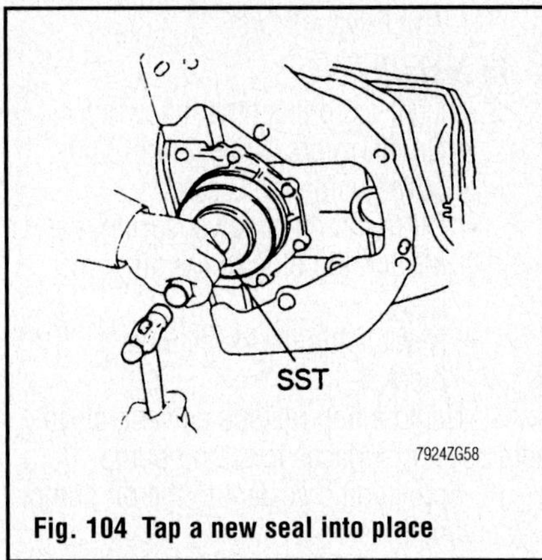

Fig. 104 Tap a new seal into place

To install:

3. Using Special Tool 09223-15030 or equivalent, tap the new seal into place until the surface is flush with the retainer edge.
4. Install or connect the following:
   - Flywheel
   - Transmission input damper
   - Transaxle assembly
5. If necessary, perform the initialization procedure, as outlined in the Chassis Electrical System.

TIMING BELT FRONT COVER

REMOVAL & INSTALLATION

See Figures 105 through 107.

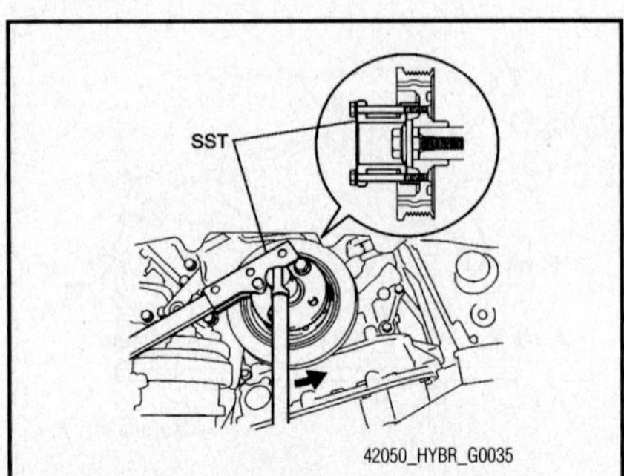

Fig. 105 Use the special tool to hold the pulley in order to loosen the pulley bolt

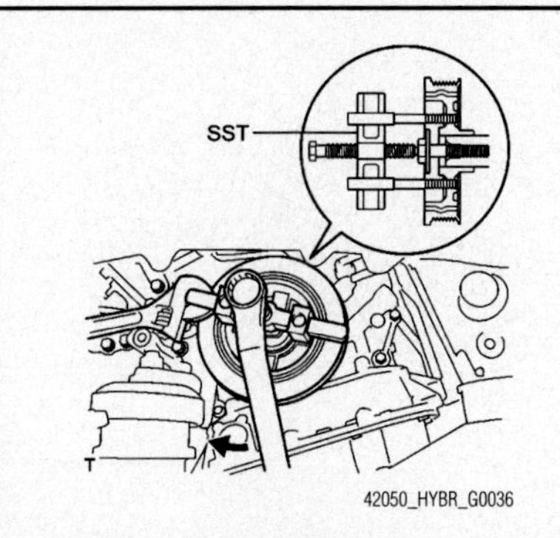

Fig. 106 Use the special tool to remove the crankshaft pulley

1. Before servicing the vehicle, refer to the Precautions Section.
2. Remove or disconnect the following:
   - Negative battery cable
   - Engine covers
   - Right front wheel
   - Fender splash shields
   - Wiper arms
   - Top cowl ventilator louver
   - Wiper motor and linkage assembly
   - Battery and battery tray
   - Air intake assembly
   - Brake master cylinder reservoir and bracket
   - Air cleaner bracket
   - Engine moving control rod
   - Right-hand engine mounting stay No. 2
3. Use Special Tool 09213-54015 to hold the crankshaft pulley in order to loosen the pulley bolt.
4. Use Special Tool 09950-50013 to remove the crankshaft pulley.
5. Remove the timing belt cover No, 1 and No. 2.

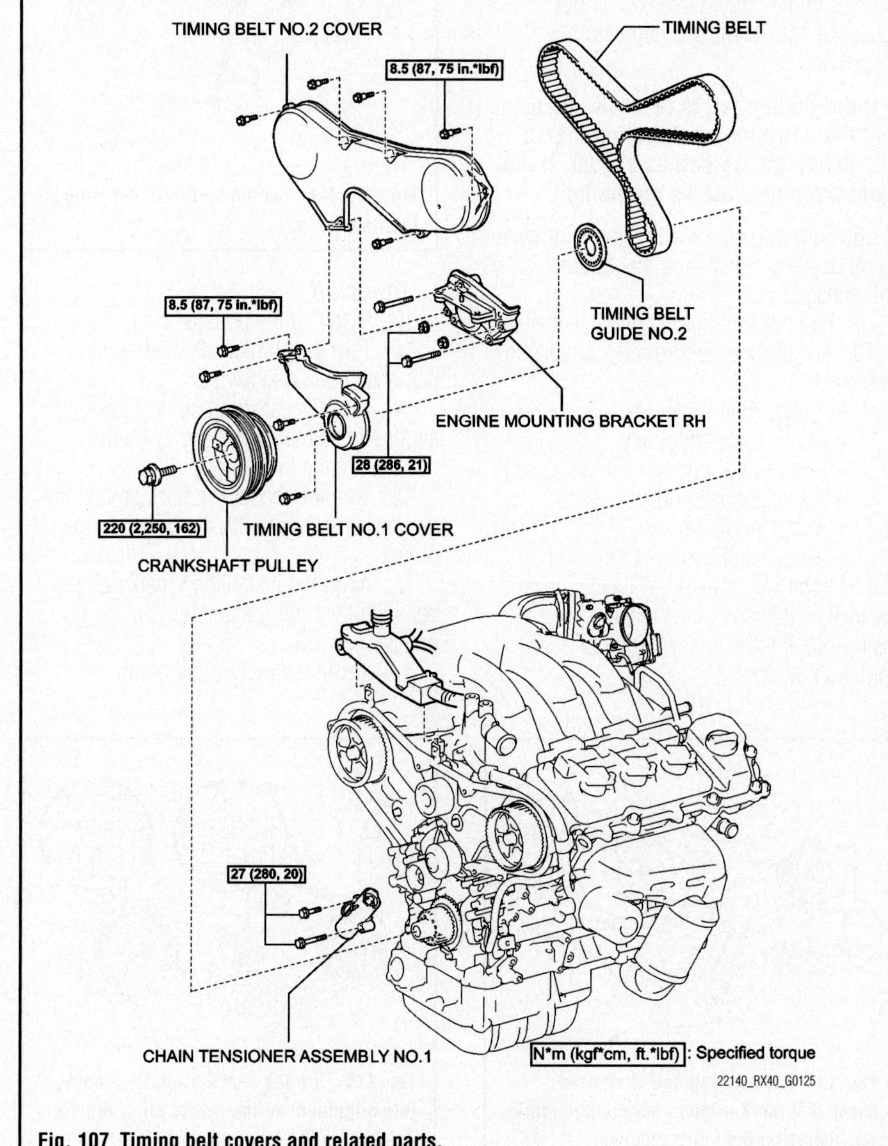

Fig. 107 Timing belt covers and related parts.

To install:

6. Install the timing belt cover No, 1 and No. 2. Tighten the mounting bolts to 75 inch. lbs. (8.5 Nm).
7. Install crankshaft pulley and tighten the bolt to 162 ft. lbs. (220 Nm).
8. Install RH engine mounting stay No. 2
9. Install engine moving control rod and tighten the mounting bolts to 47 ft. lbs. (64 Nm).
10. Install and tighten the following:
    - Air cleaner bracket
    - Brake master cylinder reservoir and bracket
    - Air intake assembly
    - Battery and battery tray
    - Wiper motor and linkage assembly
    - Top cowl ventilator louver
    - Wiper arms
    - Fender splash shields
    - Right front wheel
    - Engine covers
11. Connect the negative battery cable.

TIMING BELT AND SPROCKETS

REMOVAL & INSTALLATION

See Figures 108 through 120.

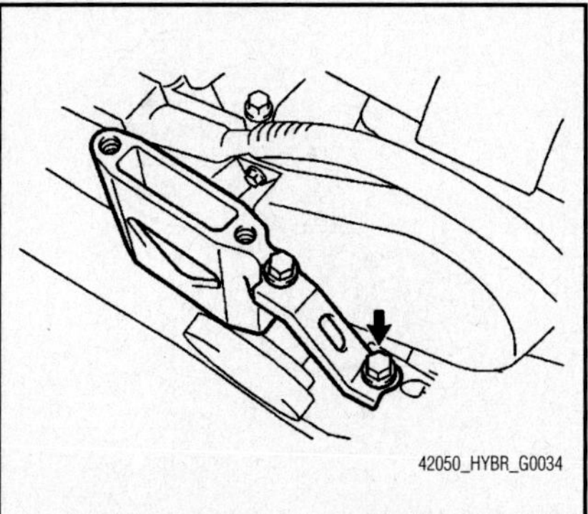

Fig. 108 Remove the right-hand engine mounting stay No. 2

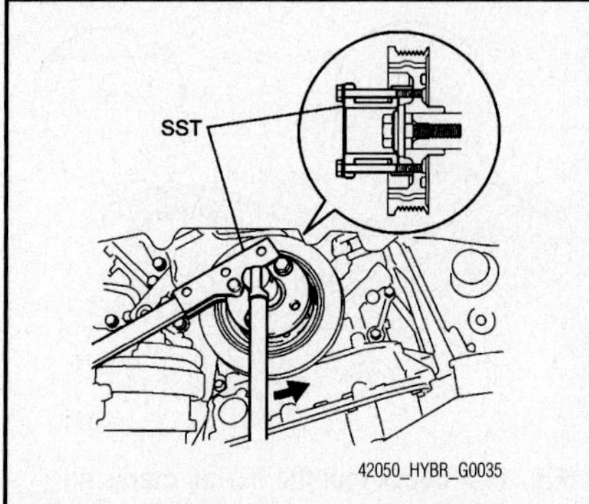

Fig. 109 Use the special tool to hold the pulley in order to loosen the pulley bolt

1. Before servicing the vehicle, refer to the Precautions Section.
2. Remove or disconnect the following:
   - Negative battery cable
   - Engine covers
   - Right front wheel
   - Fender splash shields
   - Wiper arms
   - Top cowl ventilator louver
   - Wiper motor and linkage assembly
   - Battery and battery tray
   - Air intake assembly
   - Brake master cylinder reservoir and bracket
   - Air cleaner bracket
   - Engine moving control rod
   - Right-hand engine mounting stay No. 2
3. Use Special Tool 09213-54015 to hold the crankshaft pulley in order to loosen the pulley bolt.
4. Use Special Tool 09950-50013 to remove the crankshaft pulley.
5. Remove or disconnect the following:
   - Timing belt cover No. 1
   - Timing belt cover No. 2
   - Right-hand engine mounting bracket
   - Timing belt guide No. 2
6. Temporarily install the crank pulley

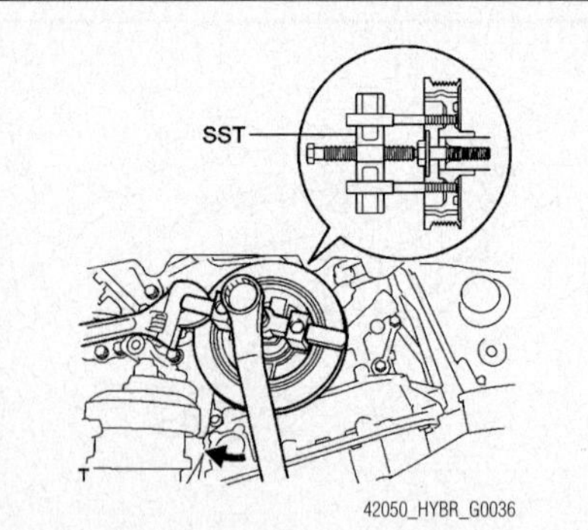

Fig. 110 Use the special tool to remove the crankshaft pulley

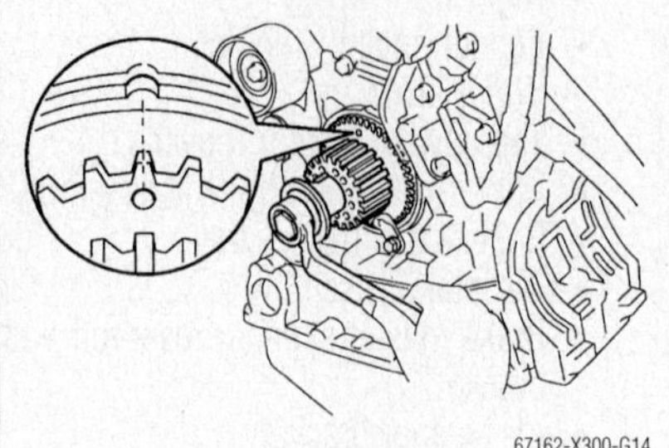

Fig. 112 Turn the crankshaft clockwise to align the timing mark on the crankshaft timing pulley with the notch in the oil pump body

bolt. Turn the crankshaft clockwise to align the timing mark on the crankshaft timing pulley with the notch in the oil pump body.

7. Check that the timing marks on the camshaft pulleys are aligned with the notches on the inner belt cover. If not, rotate the crankshaft 360 degrees clockwise.

➡If the timing belt is re-used, check that the 3 original installation marks are visible on the belt as shown. If not, paint three new marks on the belt.

8. Turn the crankshaft counterclockwise by 60 degrees. Make sure that the belt is still engaged.
9. Remove the timing belt tensioner.
10. Remove the belt from the pulleys in this order:
    - Lower idler pulley
    - Right camshaft pulley
    - Upper idler pulley
    - Left camshaft pulley
    - Water pump pulley
    - Crankshaft timing pulley
11. If the belt is being re-used, check it for wear or damage; don't twist it or turn it inside-out. If there is any doubt as to its condition, replace it.

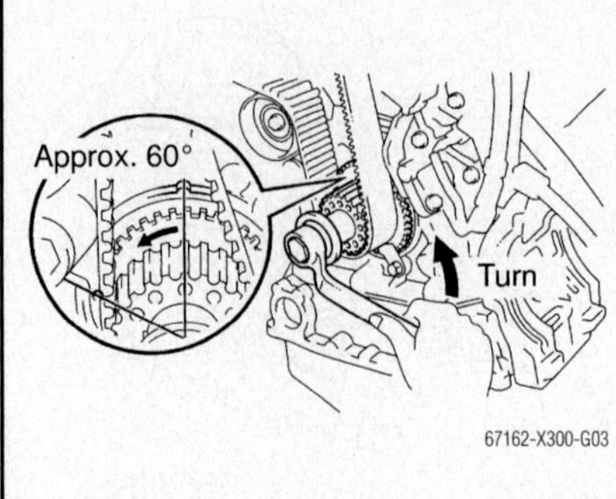

Fig. 114 Turn the crankshaft counterclockwise by 60 degrees

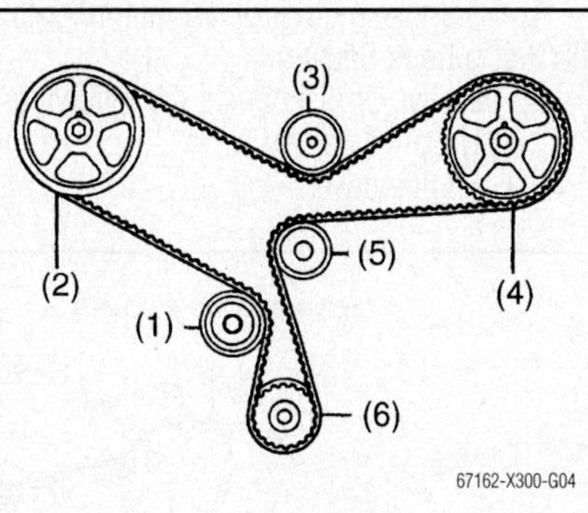

Fig. 115 Remove the belt from the pulleys in this order

To install:

12. Clean all the pulleys.
13. Turn the crankshaft another 60 degrees counterclockwise.
14. Turn the camshaft pulleys back into alignment so the marks align with the notches on the inner cover.
15. Turn the crankshaft back so that the timing mark aligns with the notch on the oil pump.
16. Align the installation marks on the belt with the timing marks on the pulleys.
17. Install the belt in this order:

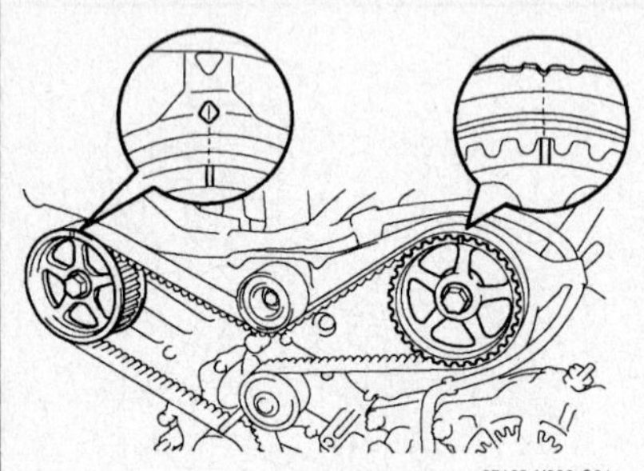

Fig. 111 Check that the timing marks on the camshaft pulleys are aligned with the notches on the inner belt cover

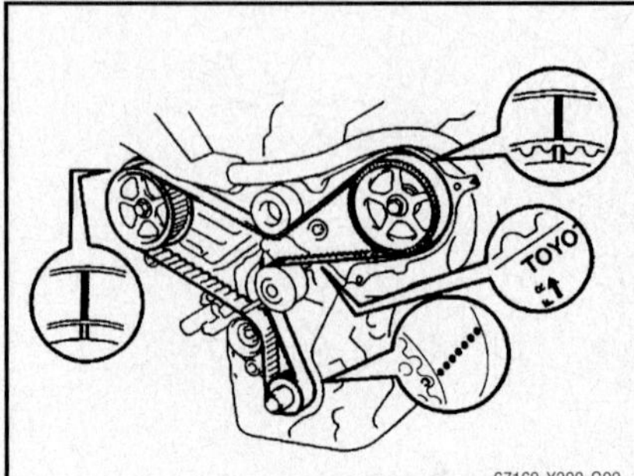

Fig. 113 If the timing belt is re-used, check that the 3 original installation marks are visible on the belt as shown

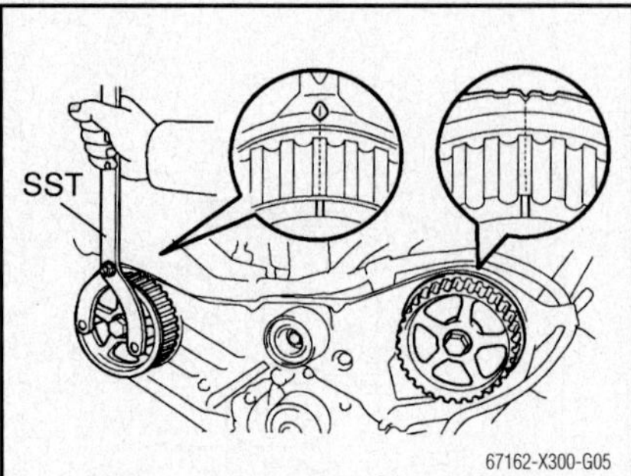

Fig. 116 Turn the camshaft pulleys back into alignment so the marks align with the notches on the inner cover

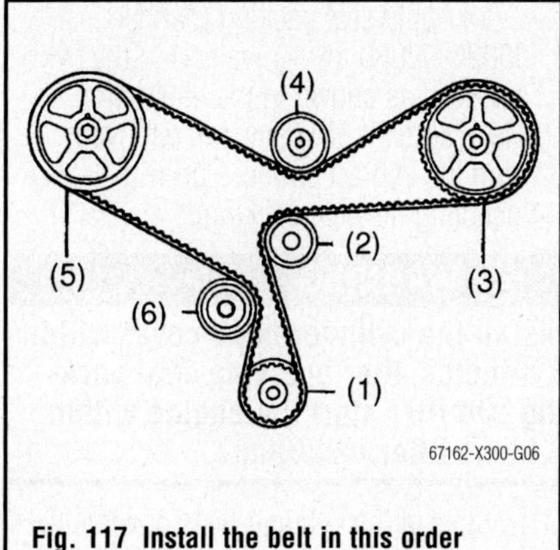

Fig. 117 Install the belt in this order

- Crankshaft
- Water pump
- Left camshaft
- Upper idler
- Right camshaft
- Lower idler

18. Set the tensioner in a press and collapse the plunger. Do not apply more that 2,205 lbs (9.8 kN) of force. Insert a suitable metal rod through the holes to hold the plunger in position.

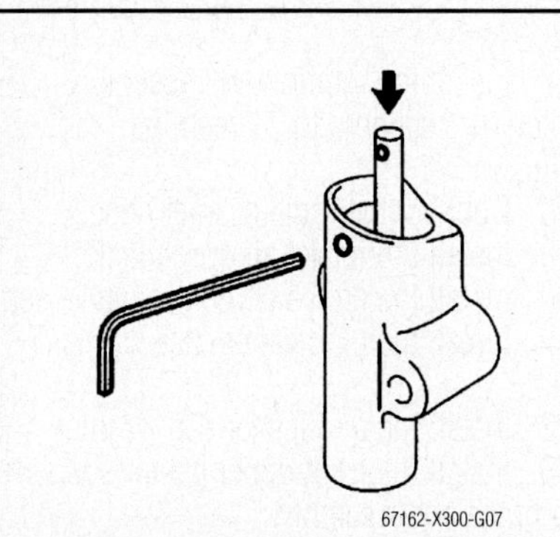

Fig. 118 Set the tensioner in a press and collapse the plunger. Do not apply more that 2,205 lbs (9.8 kN) of force. Insert a suitable metal rod through the holes to hold the plunger in position

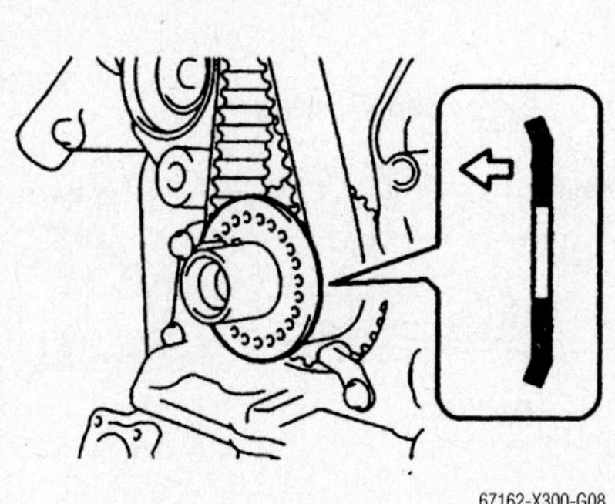

Fig. 119 Install the timing belt guide with the cupped side facing front

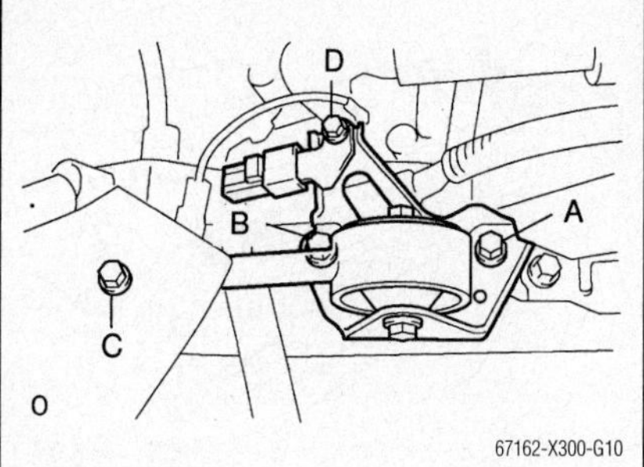

Fig. 120 Tighten the engine roll control rod bolts in this order

19. Install the tensioner and torque the 2 bolts alternately to 20 ft. lbs. (27 Nm).

**** WARNING**

Be sure to tighten to bolts alternately and evenly so the tensioner seats flat.

20. Remove the metal rod from the tensioner.
21. Turn the crankshaft 2 full revolutions clockwise (720 degrees), and align the timing mark on the crank pulley with the notch on the oil pump.
22. Check the timing marks on the camshaft pulleys for alignment with the notches on the inner cover. If they do not align, remove the belt and align the mismatched mark(s).
23. The remainder of installation is the reverse of removal. Observe the following torques:
    - Right engine mount bracket: 21 ft. lbs. (28 Nm)
    - Right engine mount insulator: 70 ft. lbs. (95 Nm)
    - Timing belt covers: 75 inch lbs. (8.5 Nm)
    - Crankshaft pulley: 162 ft. lbs. (220 Nm)
    - Alternator bracket: 21 ft. lbs. (28 Nm)
    - Right engine mount stay: 47 ft. lbs. (64 Nm)
    - Engine roll control rod: tighten first A, then B, and then C to 47 ft. lbs. (64 Nm). Torque D to 17 ft. lbs. (23 Nm)
24. If necessary, perform the initialization procedure, as outlined in the Chassis Electrical System.

TIMING BELT REAR COVER

REMOVAL & INSTALLATION

Refer to Camshaft. (Removal & Installation)

VALVE COVERS

REMOVAL & INSTALLATION

See Figures 121 through 129.

1. Disconnect the negative battery cable.
2. Using a clip remover, remove the engine side cover.
3. Drain the coolant and engine oil.
4. Remove front upper center suspension brace subassembly, as follows:
    a. Remove the 4 nuts and front suspension brace subassembly.
    b. Install the 4 shock absorber nuts and tighten to 59 ft .lbs. (80 Nm).
5. Remove the air cleaner cap w/ inlet, as follows:
    a. Remove the 2 bolts, 4 clamps and air cleaner cap w/ inlet.
    b. Remove the air cleaner filter element from the air cleaner case.
6. Remove the air cleaner case w/ resonator.
7. Remove the emission control valve set.
8. Remove the intake air surge tank.
9. Disconnect the inlet radiator hose.
10. Remove the ignition coil assembly
11. Remove the cylinder head cover subassembly, as follows:

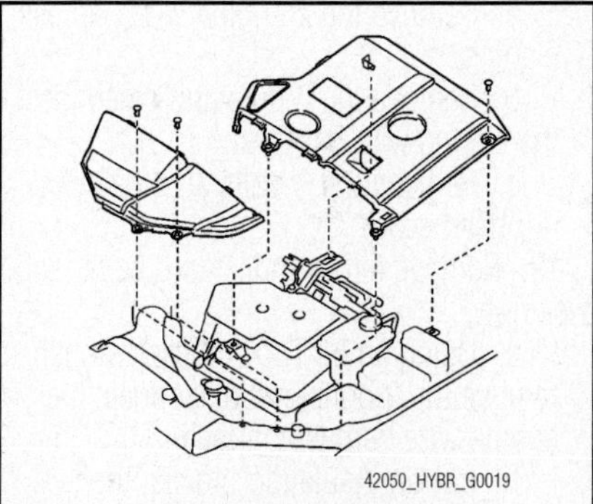

Fig. 121 Use a clip removal tool to remove the engine side cover

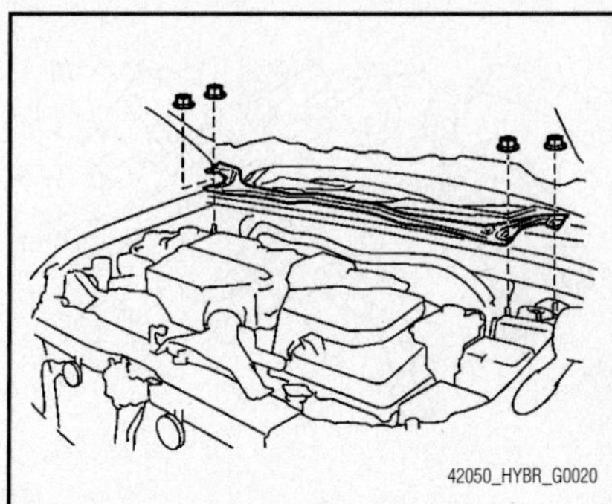

Fig. 122 Remove front upper center suspension brace subassembly

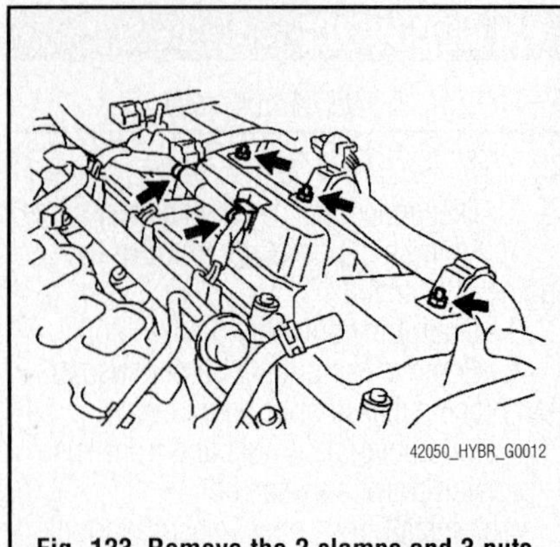

Fig. 123 Remove the 2 clamps and 3 nuts

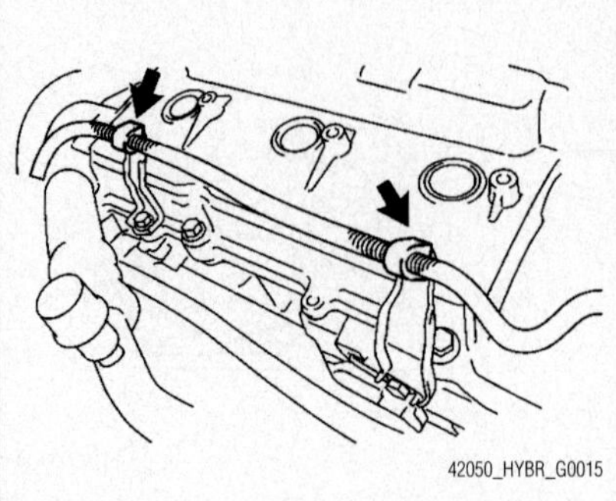

Fig. 126 Put on insulating gloves, then remove the 2 engine wire harness clamps

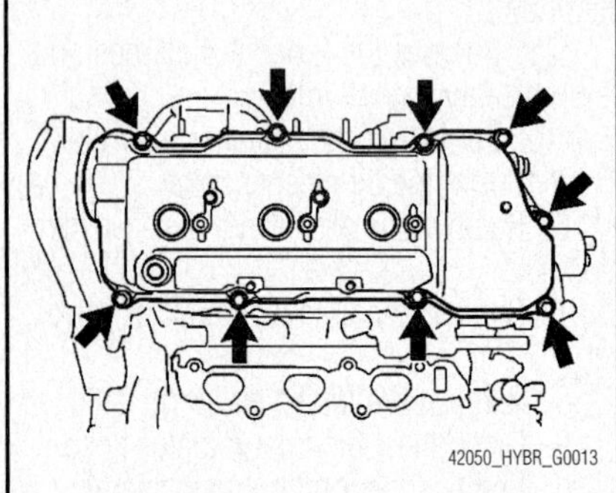

Fig. 124 Remove the 9 bolts and the cylinder head cover

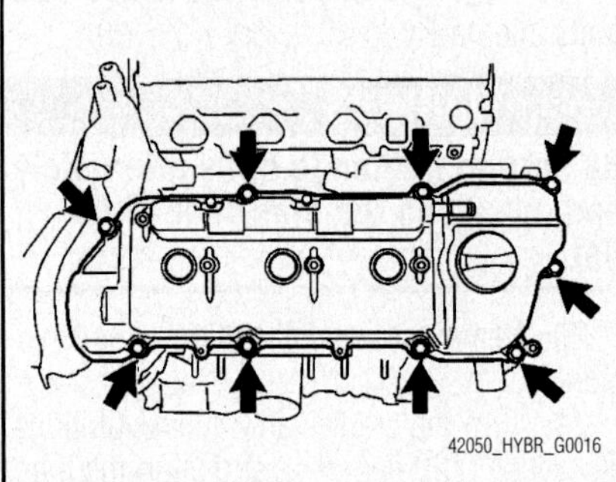

Fig. 127 Remove the 9 bolts and the cylinder head cover

a. Remove the 2 engine wire harness clamps.

b. Remove the 3 nuts and disconnect the engine wire harness.

c. Remove the 9 bolts and the cylinder head cover.

12. Remove left cylinder head cover sub-assembly:

a. Using an E6 Torx® socket wrench, remove the 2 bolts and disconnect the engine wire harness protector.

b. Put on insulating gloves, then remove the 2 engine wire harness clamps.

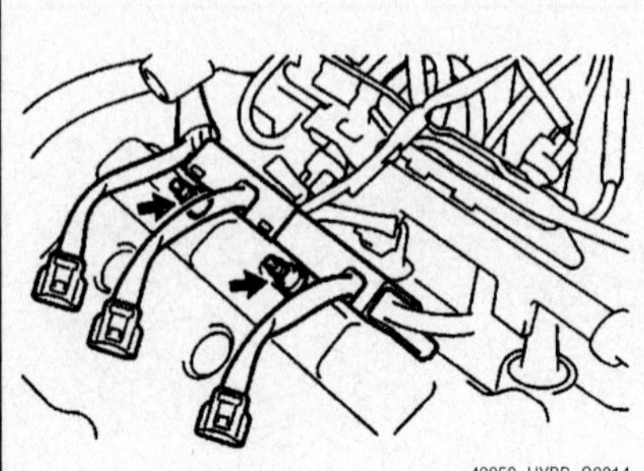

Fig. 125 Use a Torx® wrench to remove the 2 bolts, then disconnect the engine wire harness protector

c. Remove the 2 bolts and 2 brackets.

d. Remove the 9 bolts and the cylinder head cover.

13. Thoroughly clean the gasket mating surfaces.

To install:

14. Install the cylinder head cover sub-assembly, as follows:

a. Apply seal packing (part No. 08826-00080, or equivalent) to the cylinder head as shown in the illustration. Make sure all oil residue is removed from the contact surfaces prior to installing the seal packing.

WARNING

Install the cylinder head cover within 3 minutes after applying seal packing. Do NOT start the engine within 2 hours after installing.

b. Install the cylinder head cover with the 9 bolts. Tighten the bolts uniformly in several steps to a final torque of 71 inch lbs. (8 Nm).

c. Install the engine wire harness with the 3 nuts and 2 clamps. Tighten to 74 inch lbs. (8.4 Nm).

15. Install the left cylinder head cover sub-assembly:

a. Apply seal packing (part No. 08826-00080, or equivalent) to the cylinder head as shown in the illustration. Make sure all oil residue is removed from the contact surfaces prior to installing the seal packing.

WARNING

Install the cylinder head cover within 3 minutes after applying seal packing. Do NOT start the engine within 2 hours after installing.

16. Install the cylinder head cover with the 9 bolts. Tighten the bolts uniformly in several steps to a final torque of 71 inch lbs. (8 Nm).

a. Install the 2 brackets with the 2 bolts.

b. Install the 2 engine wire harness clamps.

WARNING

Wear insulating gloves.

c. Using an E6 Torx® socket wrench, install the engine wire harness protector with the 2 bolts. Tighten to 74 inch lbs. (8.4 Nm).

17. Install the ignition coil assembly. Tighten the retainers to 71 inch lbs. (8 Nm).

18. Connect the radiator inlet hose.

19. Install the intake air surge tank.

20. Install the emission control valve set.

21. Install the air cleaner case w/ resonator.

22. Install the air cleaner cap w/ inlet.

23. Install the front upper center suspension brace subassembly.

24. Add the proper and amount of engine oil and coolant to the engine.

25. Connect the negative battery cable.

26. Check for engine coolant and/or oil leaks.

27. Install the engine side cover:

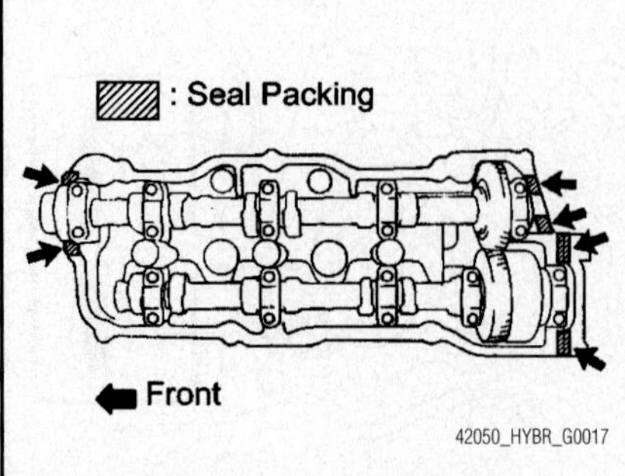

Fig. 128 Apply seal packing (part No. 08826-00080, or equivalent) to the cylinder head as shown

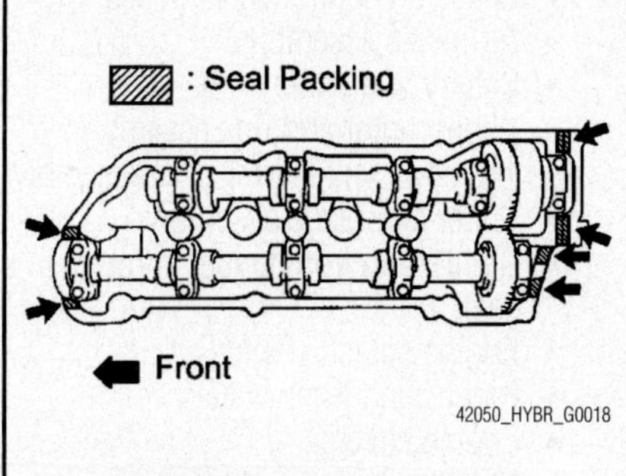

Fig. 129 Apply seal packing (part No. 08826-00080, or equivalent) to the cylinder head as shown

28. If necessary, perform the initialization procedure, as outlined in the Chassis Electrical System.

VALVE LASH

ADJUSTMENT

See Figures 130 through 132.

➡Adjust the valve clearance when the engine is cold.

1. Before servicing the vehicle, refer to the Precautions Section.
2. Relieve the fuel system pressure.
3. Drain the engine oil.
4. Drain the coolant from the engine radiator and hybrid transaxle.
5. Remove or disconnect the following:
   - Negative battery cable
   - Engine cover
   - Right-hand front wheel
   - Engine splash shields
   - Right-hand front fender apron seal
   - Wiper and blade assembly
   - Top cowl ventilator louver assembly
   - Wiper motor and link assembly
   - Battery and battery tray
   - Air intake assembly
   - Emission control valve hoses
   - Air intake surge tank
   - Radiator intake hose
   - Brake master cylinder reservoir and bracket
   - Air cleaner bracket
   - Engine moving control rod
   - Right-hand engine No. 2 mounting stay
   - Ignition coil
   - Valve covers
6. Turn the crankshaft pulley and align its groove with the timing mark **0** of the No. 1 timing cover.
7. Check that the valve lifters on the No. 1 cylinder (intake and exhaust) are loose. If not, turn the crankshaft 1 complete revolution (360 degrees).

➡All measurements should be written down. These recorded measurements will need to be used in conjunction with a mathematical formula to determine the thickness of the replacement shims.

8. Measure the clearance between the valve lifters and the camshaft. Record the measurements on valves No. 1 and 6 intake; No. 2 and 3 exhaust.
   a. The intake valve clearance cold is 0.006–0.010 in. (0.15–0.25mm).
   b. The exhaust valve clearance cold is 0.010–0.014 in. (0.25–0.35mm).
9. Turn the crankshaft ⅔ of a revolution (240 degrees). Record the measurements on valves No. 2 and 3 intake; No. 4 and 5 exhaust.

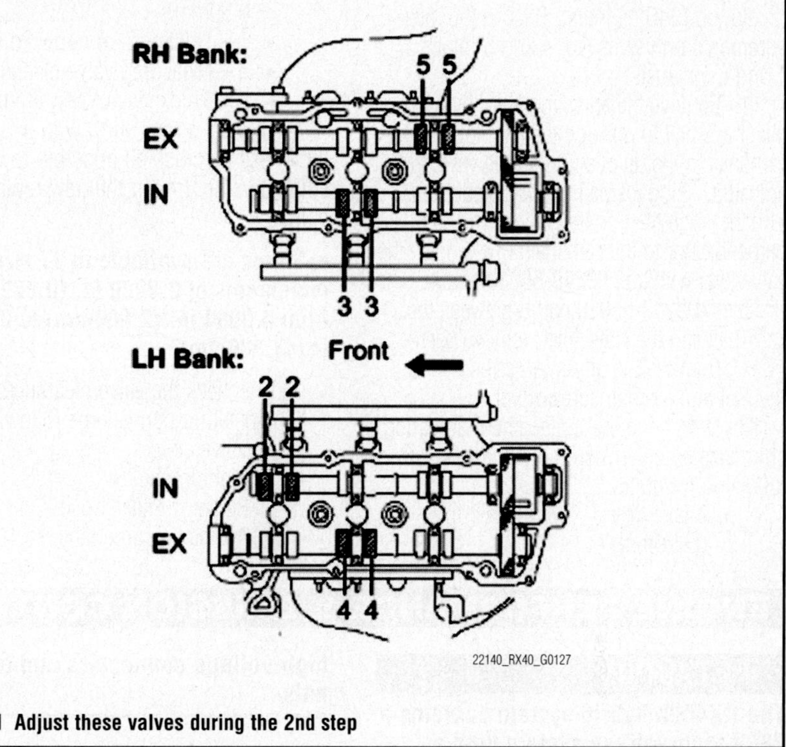

Fig. 131 Adjust these valves during the 2nd step

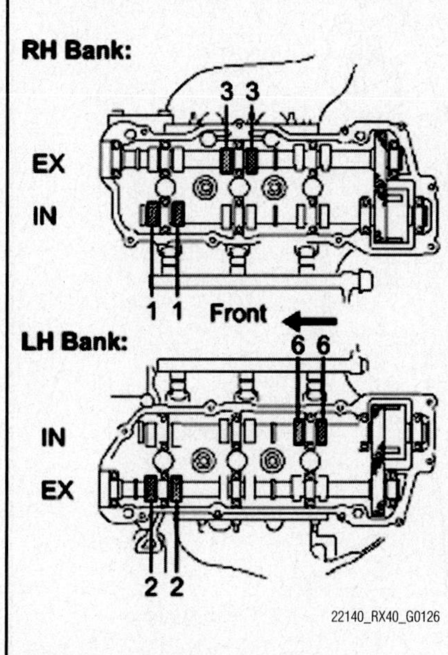

Fig. 130 Adjust these valves during the 1st step

Fig. 132 Adjust these valves during the 3rd step

10. Turn the crankshaft another ⅔ of a revolution (240 degrees). Record the measurements on valves No. 4 and 5 intake; No. 1 and 6 exhaust.

11. Remove the adjusting shim by turning the crankshaft to position the cam lobe of the camshaft in the up position on the valve to be adjusted. Using a small thin flat bladed tool, turn the valve lifter so that the notches are perpendicular to the camshaft. Press down the valve lifter with tool 09248-55010 part A. Place too 09248-55010 part B between the camshaft and the valve lifter; remove part A.

12. Remove the adjusting shim with a magnet and a small screwdriver.

13. Determine the replacement adjusting shim size by either using the charts or the following formulas:

- Intake: N = T + (A - 0.008 in./ 0.020mm)
- Exhaust: N = T + (A - 0.012 in./ 0.30mm)
- T = Thickness of removed shim
- A = Measured valve clearance
- N = Thickness of new shim

14. Select a new shim with a thickness as close as possible to the calculated value. Install the new replacement shim.

➡Shims are available in 17 sizes in increments of 0.0020 in. (0.050mm), from 0.0984 in. (2.500mm) to 0.1299 in. (3.300mm).

15. Recheck the valve clearance.

16. Install or connect the following:

- Valve covers
- Ignition coil
- Radiator intake hose
- Air intake surge tank
- Emission control valve hoses
- Air intake assembly
- Battery and battery tray
- Wiper motor and link assembly
- Top cowl ventilator louver assembly
- Wiper and blade assembly
- Right-hand front fender apron seal
- Engine splash shields
- Right-hand front wheel
- Engine covers
- Negative battery cable

17. Refill the cooling system to the correct level.

18. Refill the engine oil to the correct level.

19. Start the engine and check for leaks.

20. If necessary, perform the initialization procedure, as outlined in the Chassis Electrical System.

ENGINE PERFORMANCE & EMISSION CONTROLS

**** CAUTION**

The RX400h hybrid system contains a 288V high-voltage system with a strong alkali solution of potassium hydroxide. Be sure to follow the instructions in this manual to handle the system correctly. Failure to do so may result in serious injury or electrocution. Engineer must undergo special training to be able to perform high-voltage system inspection and servicing.

**** CAUTION**

All high-voltage wire harness connectors are colored orange. The HV battery and other high-voltage components have "High Voltage" caution labels. Do not carelessly touch these wires and components.

**** CAUTION**

Before inspecting or servicing the high-voltage system, be sure to follow safety measures, such as wearing insulated gloves and removing the service plug to prevent electrocution. Carry the removed service plug in your pocket to prevent anyone from reinstalling it while you are servicing the vehicle.

**** CAUTION**

After removing the service plug, wait 5 minutes before touching any of the high-voltage connectors and terminals.

COMPONENT LOCATIONS

See Figures 133 through 137.

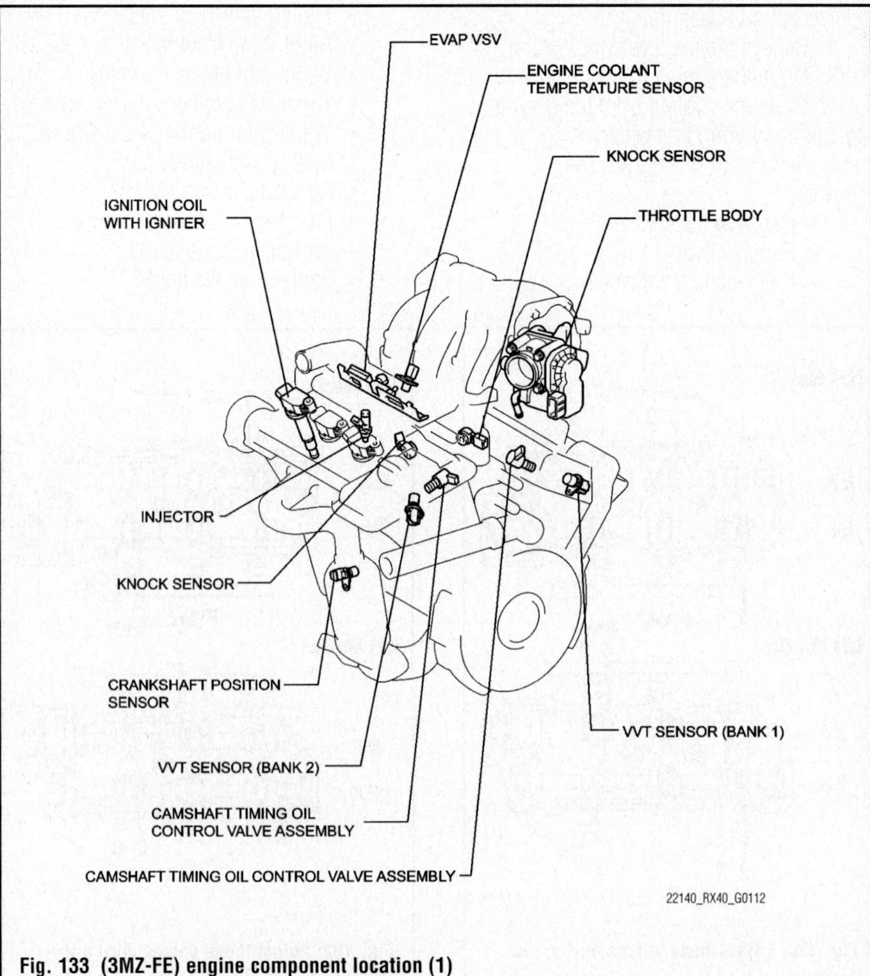

Fig. 133 (3MZ-FE) engine component location (1)

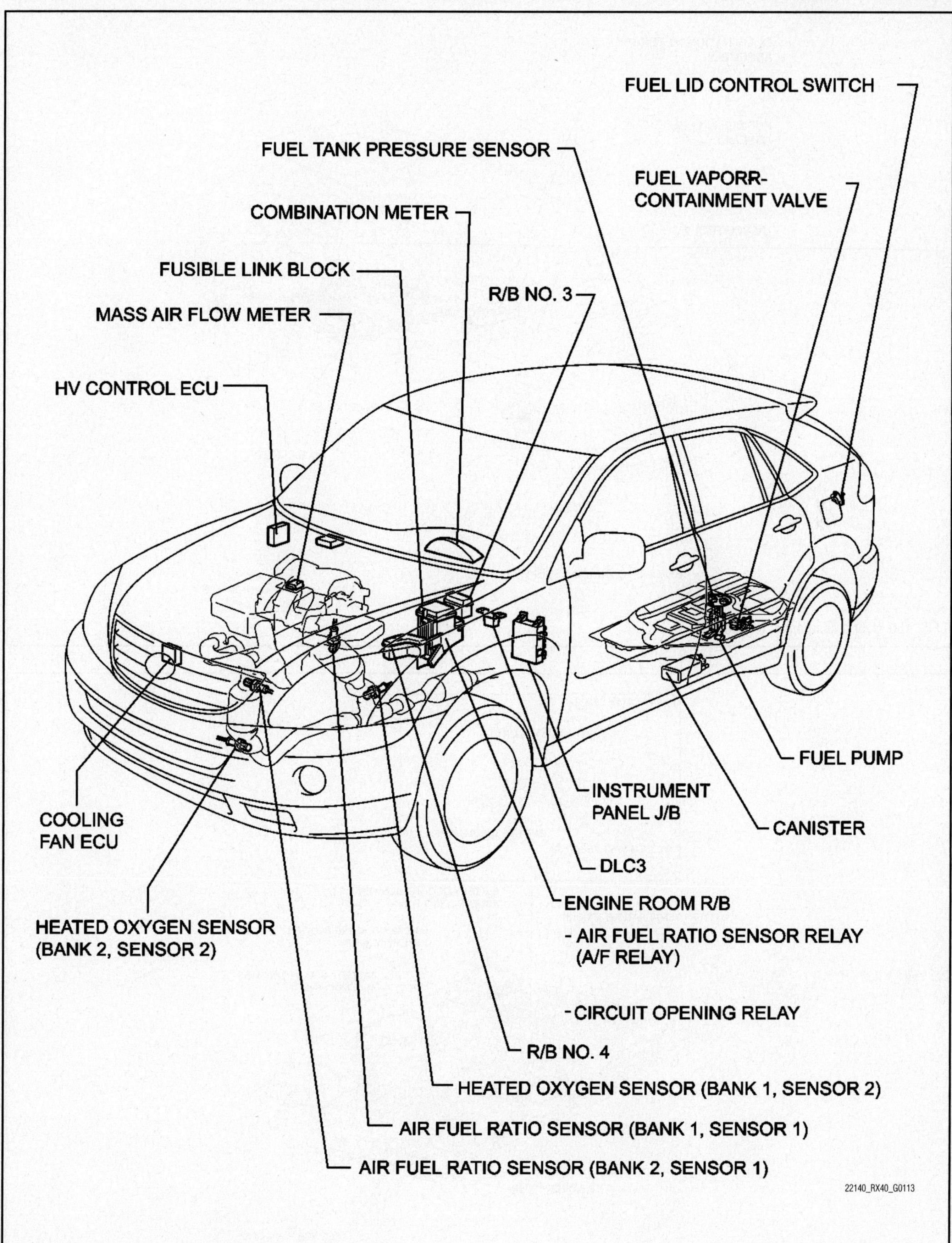

Fig. 134 (3MZ-FE) engine component location (2)

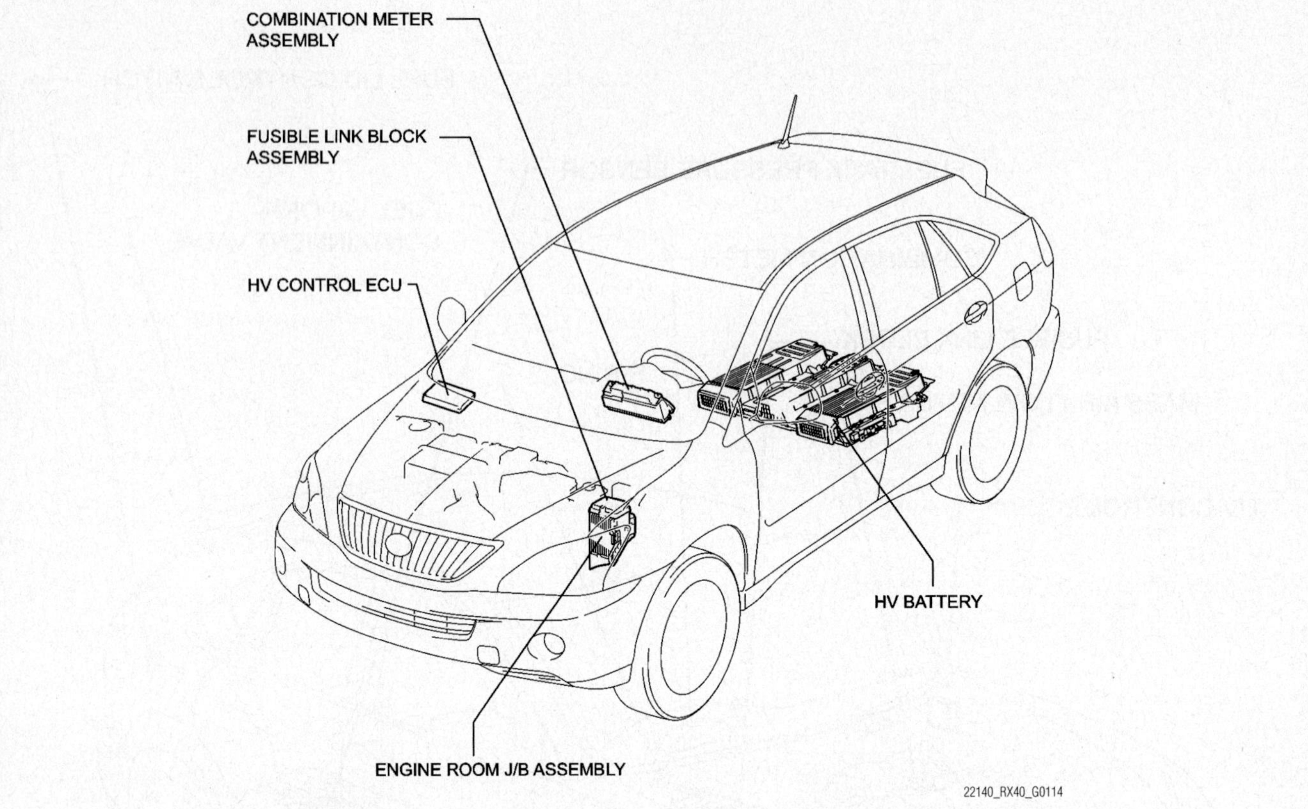

Fig. 135 Hybrid component location (1)

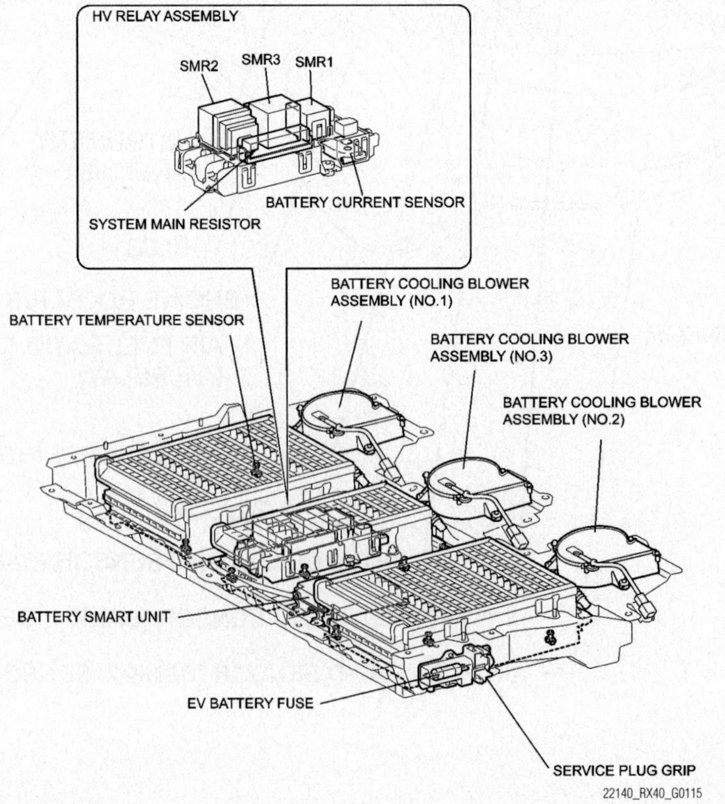

Fig. 136 Hybrid component location (2)

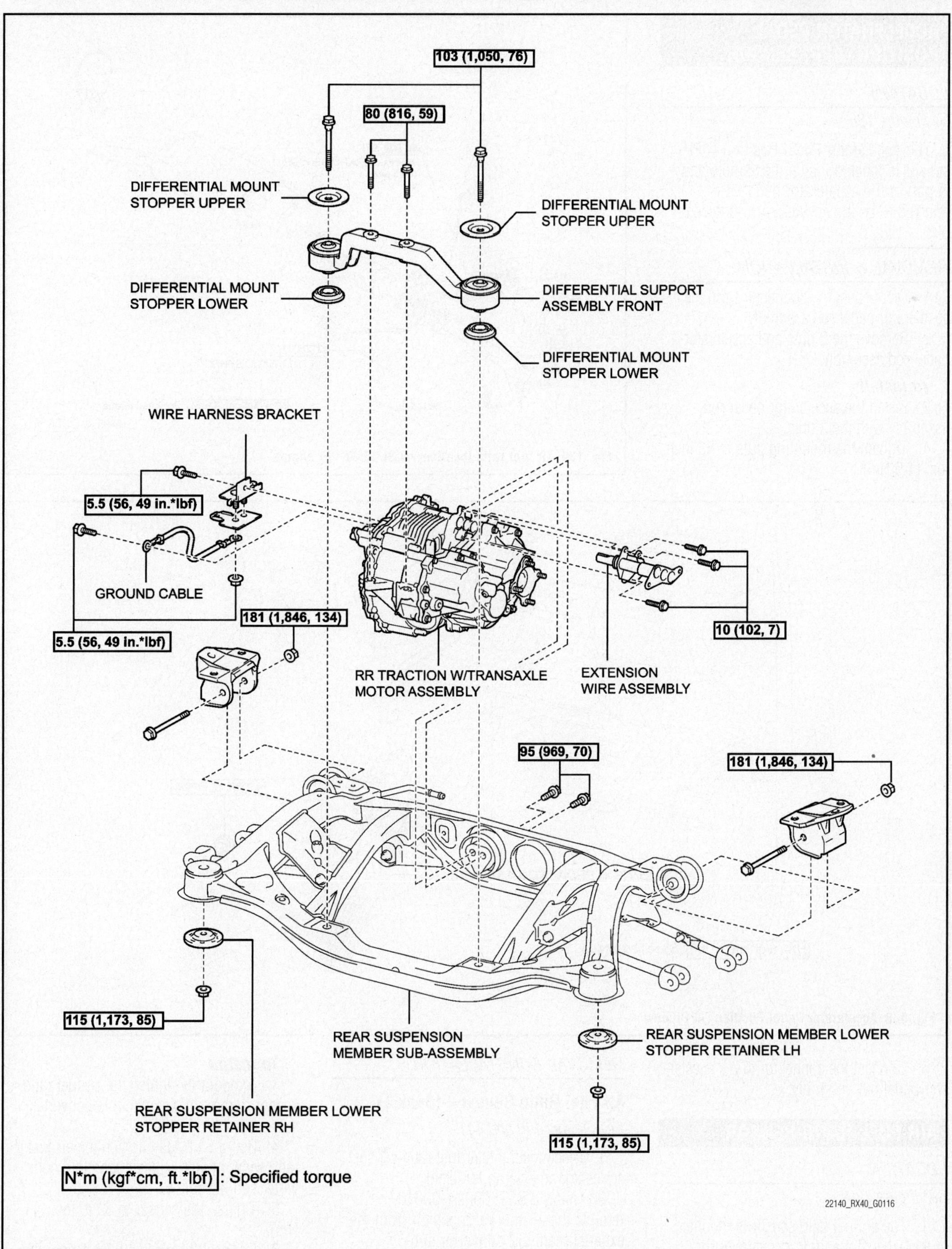

Fig. 137 Hybrid rear traction motor location

ACCELERATOR PEDAL POSITION (APP) SENSOR

LOCATION

See Figure 138.

The Accelerator Pedal Position (APP) sensor is located inside the vehicle and is part of the accelerator pedal. It is referred to as the accelerator rod assembly.

REMOVAL & INSTALLATION

1. Disconnect the connector from the accelerator pedal rod assembly.
2. Remove the 2 nuts and accelerator pedal rod assembly.

To install:

3. Install the accelerator pedal rod assembly with the 2 nuts.
4. Tighten the mounting nuts to 43 inch. lbs. (4.9 Nm).

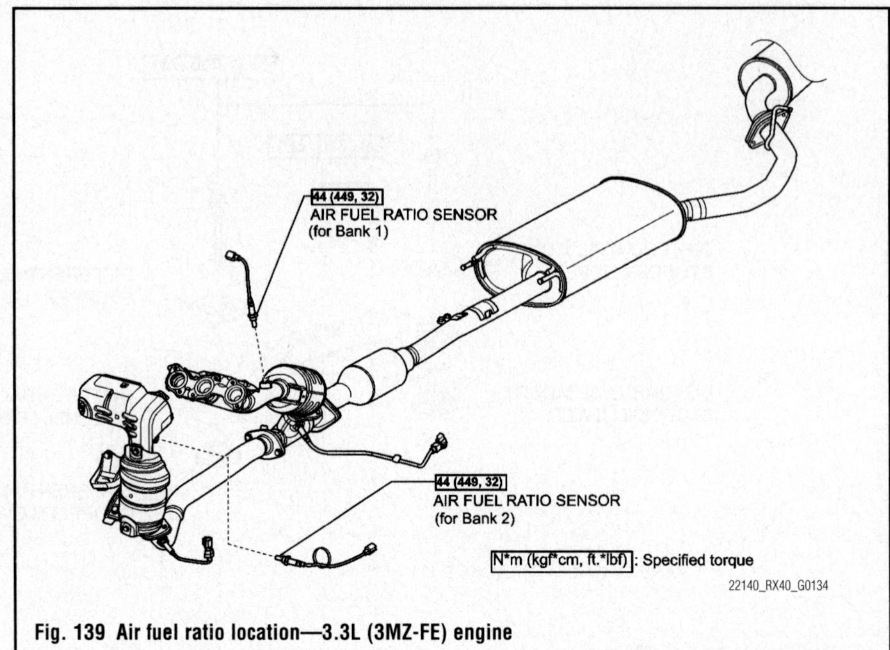

Fig. 139 Air fuel ratio location—3.3L (3MZ-FE) engine

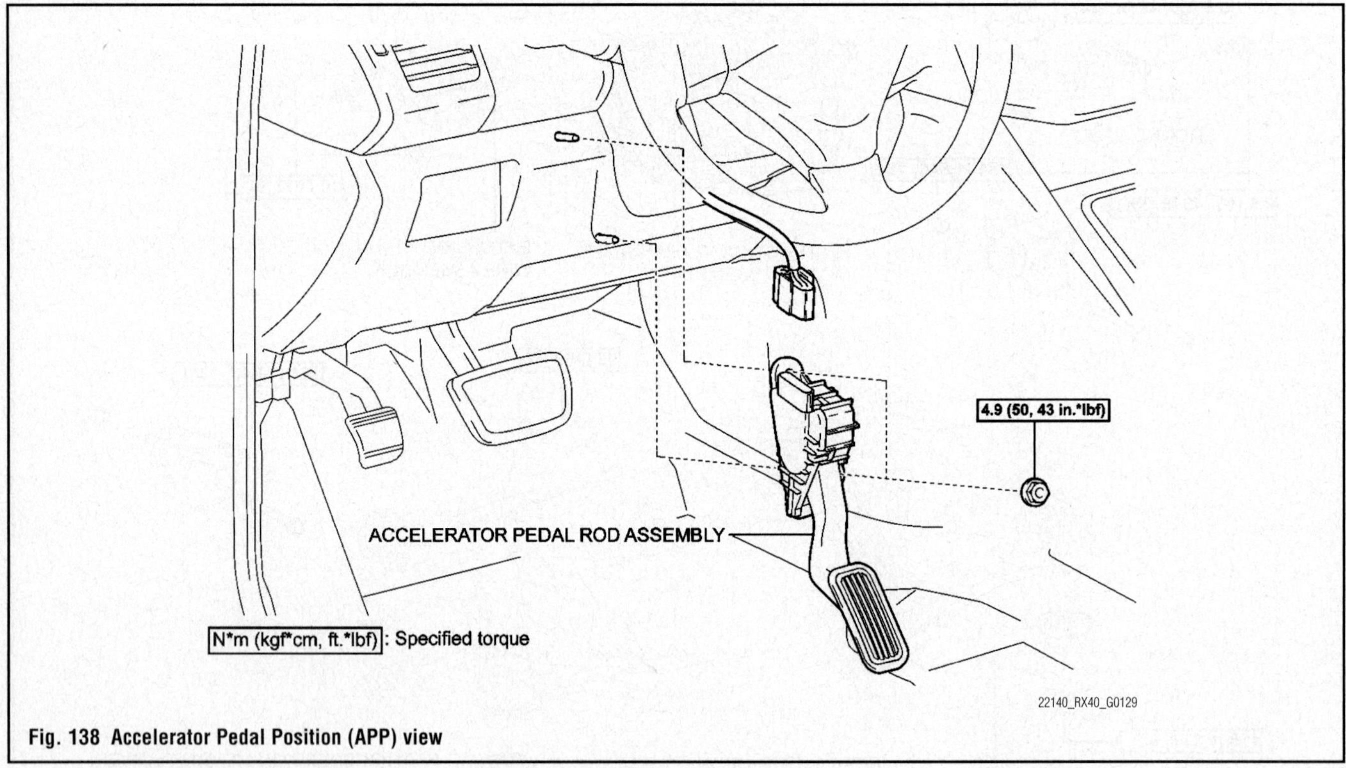

Fig. 138 Accelerator Pedal Position (APP) view

5. Connect the connector to the accelerator pedal rod assembly.

AIR FUEL RATIO SENSOR

LOCATION

See Figure 139.

1. The air fuel ratio sensor is mounted in the exhaust manifold converter sub-assembly.

REMOVAL & INSTALLATION

Air Fuel Ratio Sensor—(Bank 1)

See Figures 140 and 141.

1. Disconnect the air fuel ratio sensor connector and clamp for (Bank 1).
2. Using a SST: 09224-00010, remove the air fuel ratio sensor from the exhaust manifold converter sub-assembly.

To install:

3. Temporarily tighten the air fuel ratio sensor to the exhaust manifold converter sub-assembly.
4. Using SST, fully tighten the air fuel ratio sensor to the exhaust manifold converter sub-assembly.
5. Tighten the sensor to 32 ft. lbs. (44 Nm).
6. Connect the air fuel ratio sensor connector and clamp.

Fig. 140 Air fuel ratio sensor connector and clamp

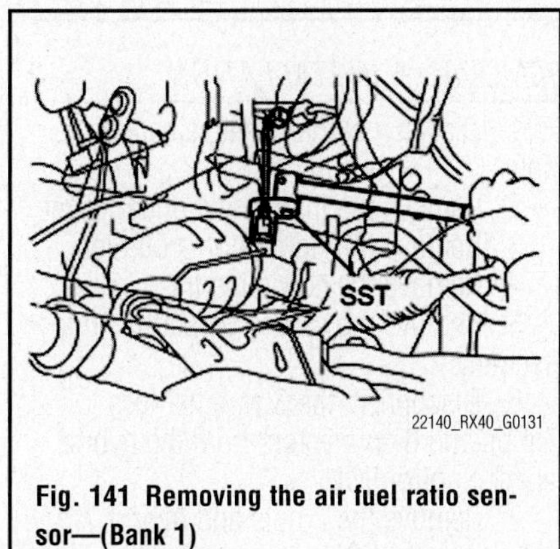

Fig. 141 Removing the air fuel ratio sensor—(Bank 1)

Air Fuel Ratio Sensor—(Bank 2)

See Figure 142.

1. Disconnect the air fuel ratio sensor connector and clamp.
2. Using a SST: 09224-00010, remove the air fuel ratio sensor from the exhaust manifold converter sub-assembly No. 2.

To install:

3. Temporarily tighten the air fuel ratio sensor to the exhaust manifold converter sub-assembly No. 2.
4. Using SST, fully tighten the air fuel ratio sensor to the exhaust manifold converter sub-assembly.
5. Tighten the sensor to 32 ft. lbs. (44 Nm).
7. Connect the air fuel ratio sensor connector and clamp.

CAMSHAFT TIMING OIL CONTROL VALVE

LOCATION

See Figure 143.

The camshaft timing oil control valves are mounted in the rear of the cylinder heads.

REMOVAL & INSTALLATION

1. Remove the engine room covers.
2. Remove the cool air intake duct seal.
3. Remove the air cleaner cap with inlet.

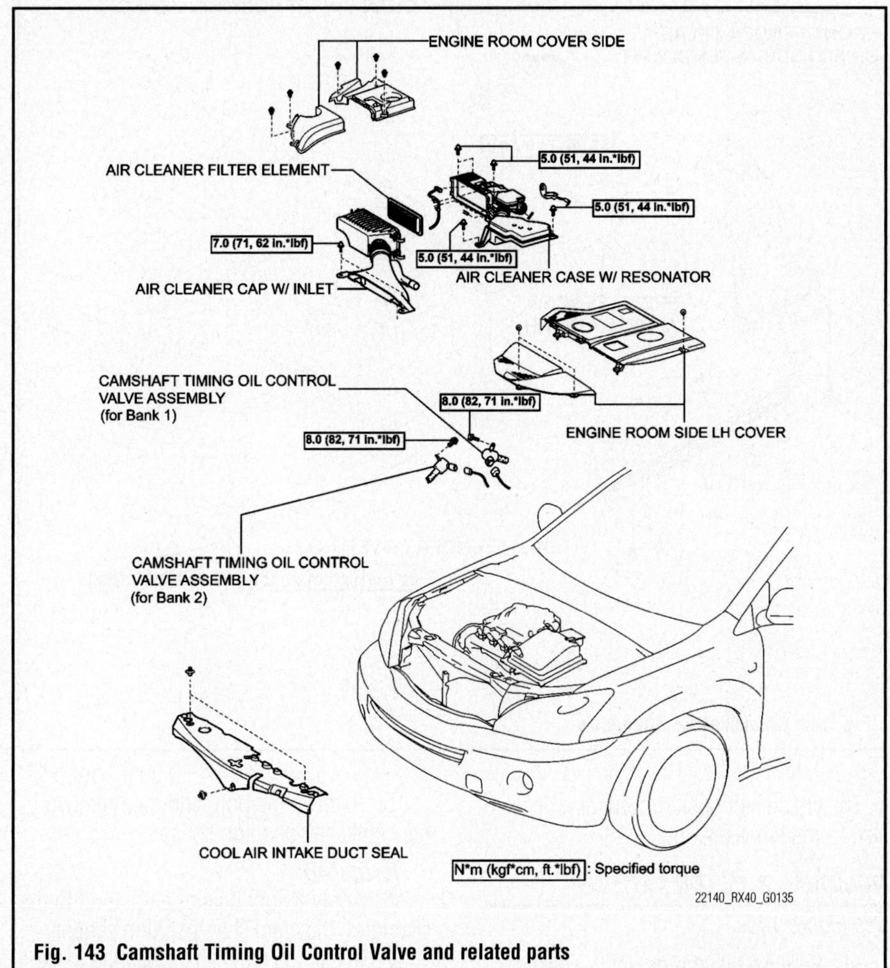

Fig. 143 Camshaft Timing Oil Control Valve and related parts

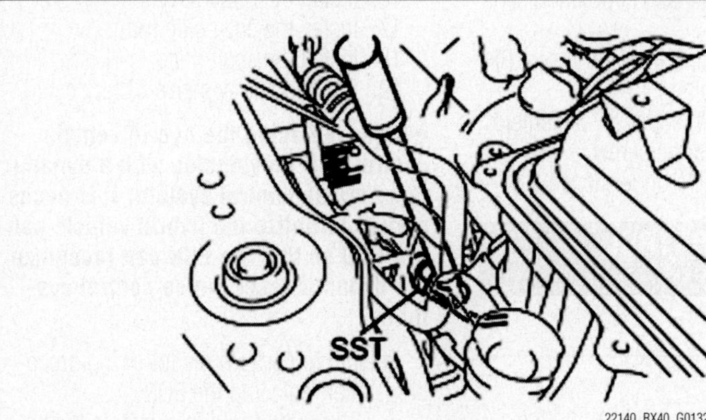

Fig. 142 Removing the air fuel ratio sensor—(Bank 2)

4. Remove the air cleaner case with resonator.
5. Disconnect the oil control valve connector.
6. Remove the bolt and oil control valve.

To install:

7. Apply a coat of engine oil to an O-ring of the sensor.
8. Install the camshaft timing oil control valve with the bolt.
9. Tighten the mounting bolt to 71 inch. (8 Nm).
10. Connect the camshaft timing oil control valve connector.

CRANKSHAFT POSITION (CKP) SENSOR

LOCATION

See Figure 144.

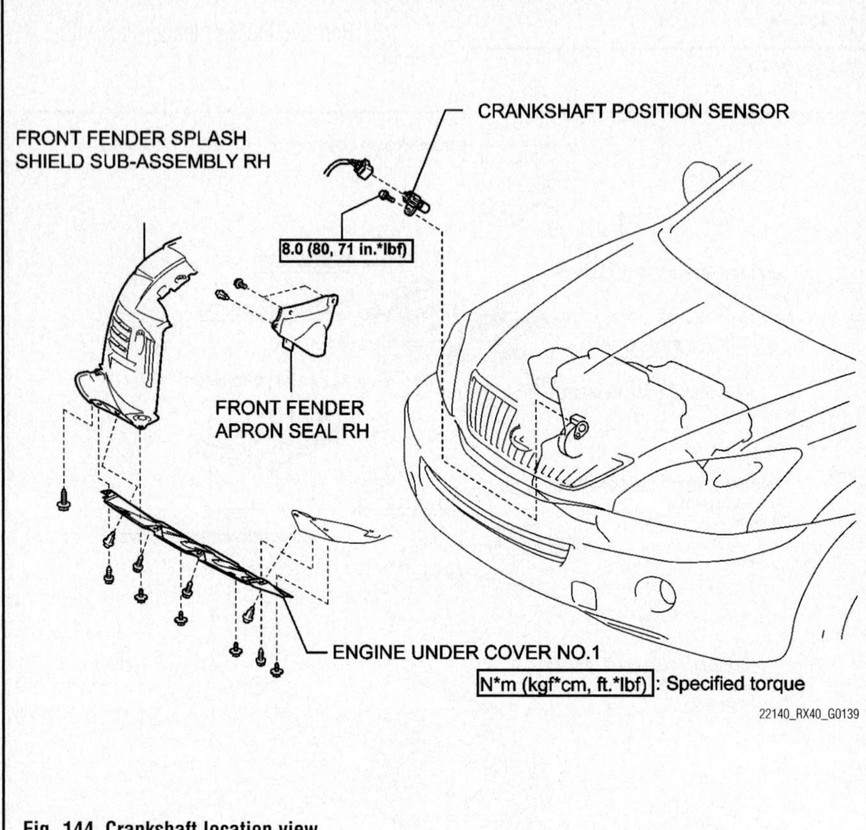

Fig. 144 Crankshaft location view

The crankshaft position sensor is located next to the crankshaft pulley.

REMOVAL & INSTALLATION

See Figure 145.

1. Remove the engine under cover.
2. Remove the screw, and then open the front fender splash shield sub-assembly RH.
3. Remove the clip and 2 bolts, and then remove the front fender apron seal RH.
4. Disconnect the crankshaft position sensor connector.

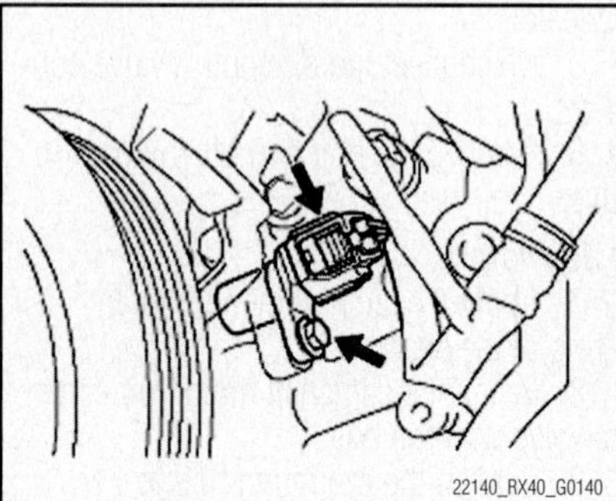

Fig. 145 Crankshaft position sensor and mounting bolt

5. Remove the bolt, and then remove the crankshaft position sensor.

To install:

6. Apply a light coat of engine oil to the O-ring on the crankshaft position sensor.
7. Install the crankshaft position sensor with the bolt.
8. Connect the crankshaft position sensor connector.
9. Install the front fender apron seal RH with the clip and 2 bolts.
10. Install the front fender splash shield sub-assembly RH with the screw.
11. Install the engine under cover.

ELECTRONIC CONTROL MODULE (ECM)

LOCATION

See Figure 146.

The Hybrid ECU is located behind the glove compartment and to the right of the blower unit.

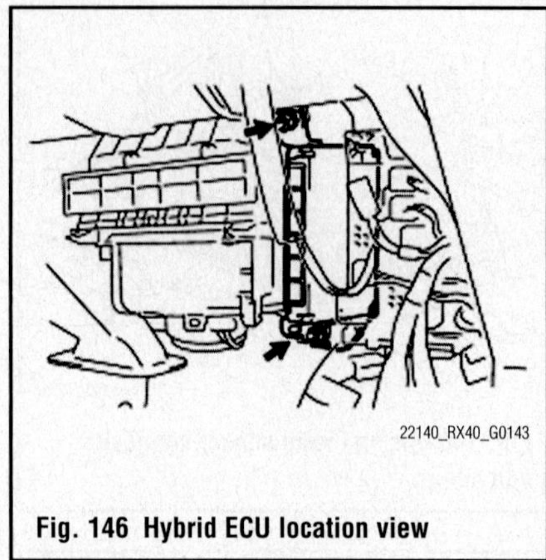

Fig. 146 Hybrid ECU location view

REMOVAL & INSTALLATION

1. Disconnect the negative battery cable.
2. Remove the instrument under cover.
3. Remove the right door scuff plate.
4. Remove the cowl side trim.
5. Remove the glove compartment door assembly.
6. Disconnect the 3 wire harness clamps and 6 connectors from the hybrid vehicle control ECU
7. Remove the 2 nuts and hybrid vehicle control ECU.
8. Remove the ECU.
9. If ECU is to be changed remove the brackets.

To install:

10. Install brackets to the ECU if previously removed.
11. Install the wire harness clamp bracket with the 2 screws.
12. Install the hybrid vehicle control ECU with the 2 nuts. Tighten the mounting nuts to 49 inch. lbs. (5.5 Nm).
13. Connect the 6 connectors to the hybrid vehicle control ECU.
14. Connect the 3 wire harness clamps.
15. Install the glove compartment door assembly.
16. Install the undercover.
17. Install the cowl side trim.
18. Install the scuff plate.
19. Connect the negative battery cable.

➡After replacing the hybrid vehicle control ECU on vehicles with a dynamic laser cruise control system, it is necessary to initialize the hybrid vehicle control ECU so that the ECU can recognize the dynamic laser cruise control system.

Be sure to perform the following procedures after replacing the ECU.

- Turn the ignition switch to the on position.
- Turn the cruise main switch on.

- With the brake pedal depressed, push the cruise control main switch to RES/ACC 3 times within 3 seconds. Check that the buzzer sounds at this time.

➡Do not turn the headlight dimmer switch on at this time because the optical axis automatic adjustment mode has already started, which may lead to an incorrect optical axis setting. If the headlight dimmer switch is turned on by mistake, readjust the optical axis.

20. Some systems need initialization when disconnecting the cable from the negative battery terminal.

ENGINE COOLANT TEMPERATURE (ECT) SENSOR

LOCATION

See Figure 147.

The Engine Coolant Temperature (ECT) sensor is located on the top front area of the engine.

REMOVAL & INSTALLATION

1. Disconnect the negative battery cable. Wait at least 90 seconds before performing any other work.
2. To access the ECT sensor, it may be necessary to remove the intake manifold, throttle body assembly, or only the intake air connector pipe.
3. Disconnect the ECT sensor connector.
4. Using a deep socket, remove the sensor from the top of the engine.
5. Remove the gasket from the sensor.
6. Installation is the reverse of the removal procedure.
7. Tighten the ECT sensor to 15 ft. lbs. (20 Nm).

FUEL LEVEL SENDING UNIT

LOCATION

The fuel level sending unit is mounted to the fuel pump assembly, and is located inside the fuel tank.

REMOVAL & INSTALLATION

See Figures 148 and 149.

1. Discharge the fuel pressure.
2. Disconnect the negative battery cable.
3. Remove fuel from fuel tank
4. Remove the center exhaust pipe.
5. Remove the front floor heat insulator.
6. Remove the fuel tank protector.
7. Remove the nut and disconnect the parking brake cable assembly
8. Separate the fuel tank wire connector from the bracket.
9. Disconnect the 2 connectors.

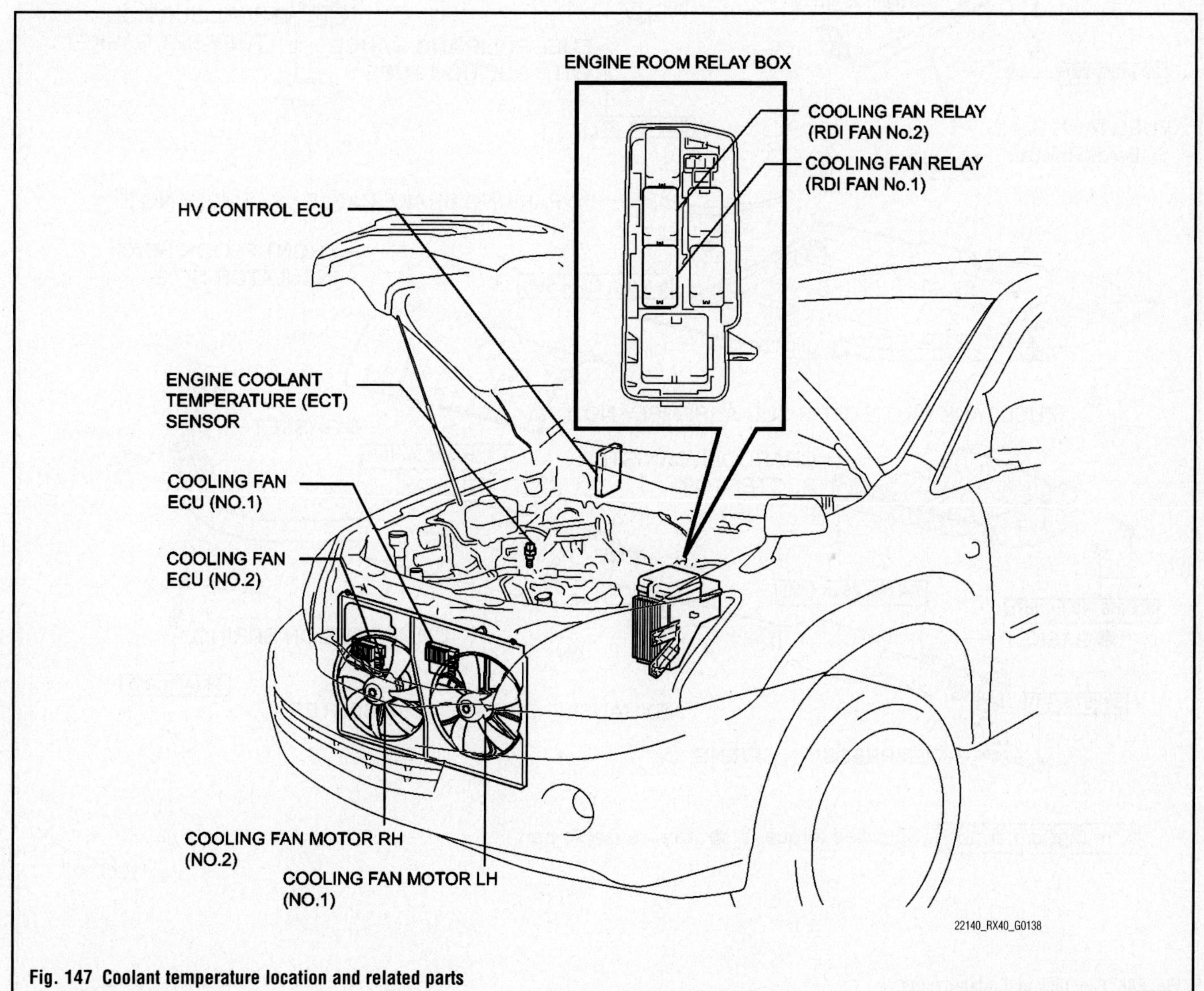

Fig. 147 Coolant temperature location and related parts

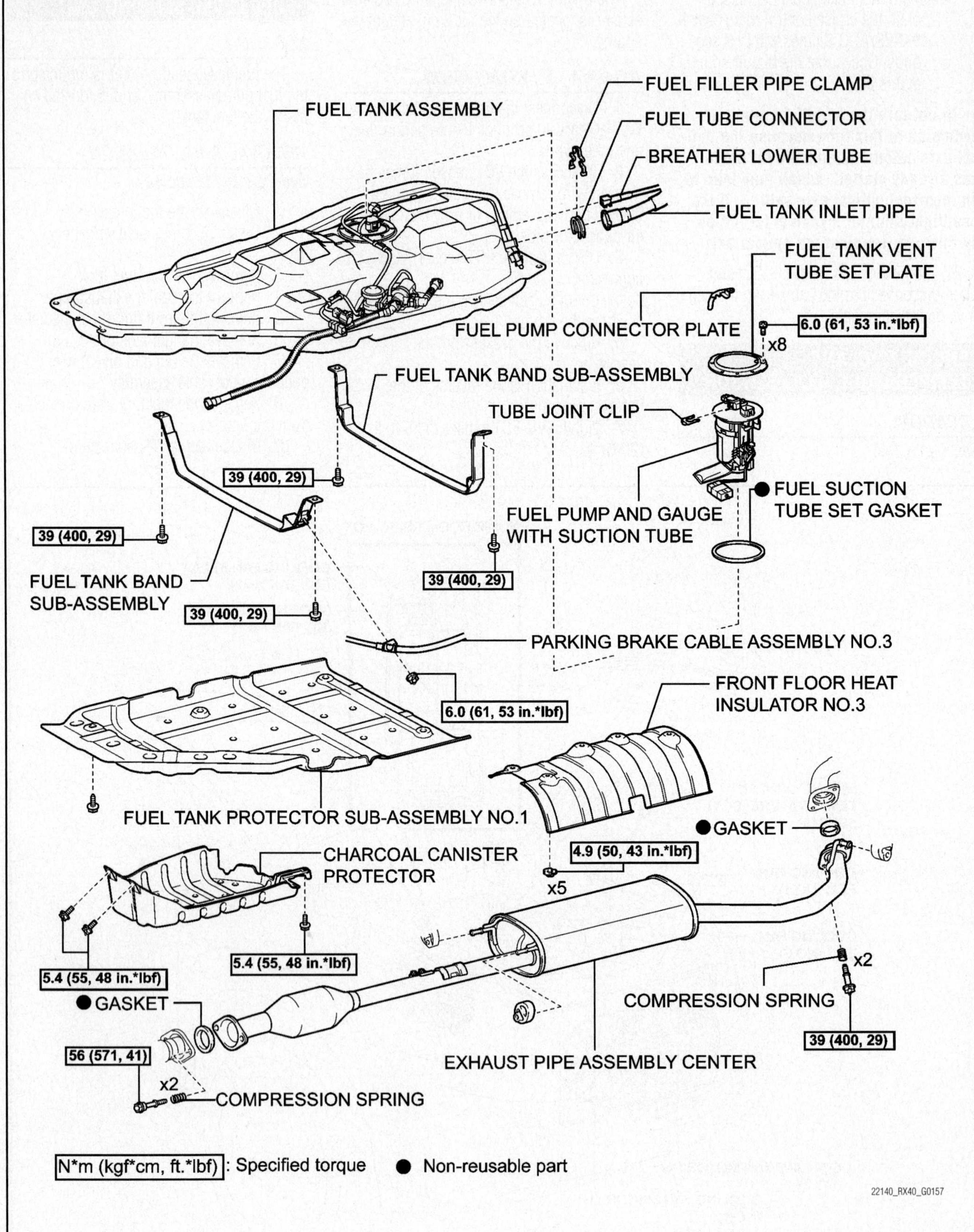

Fig. 148 Fuel tank and related parts

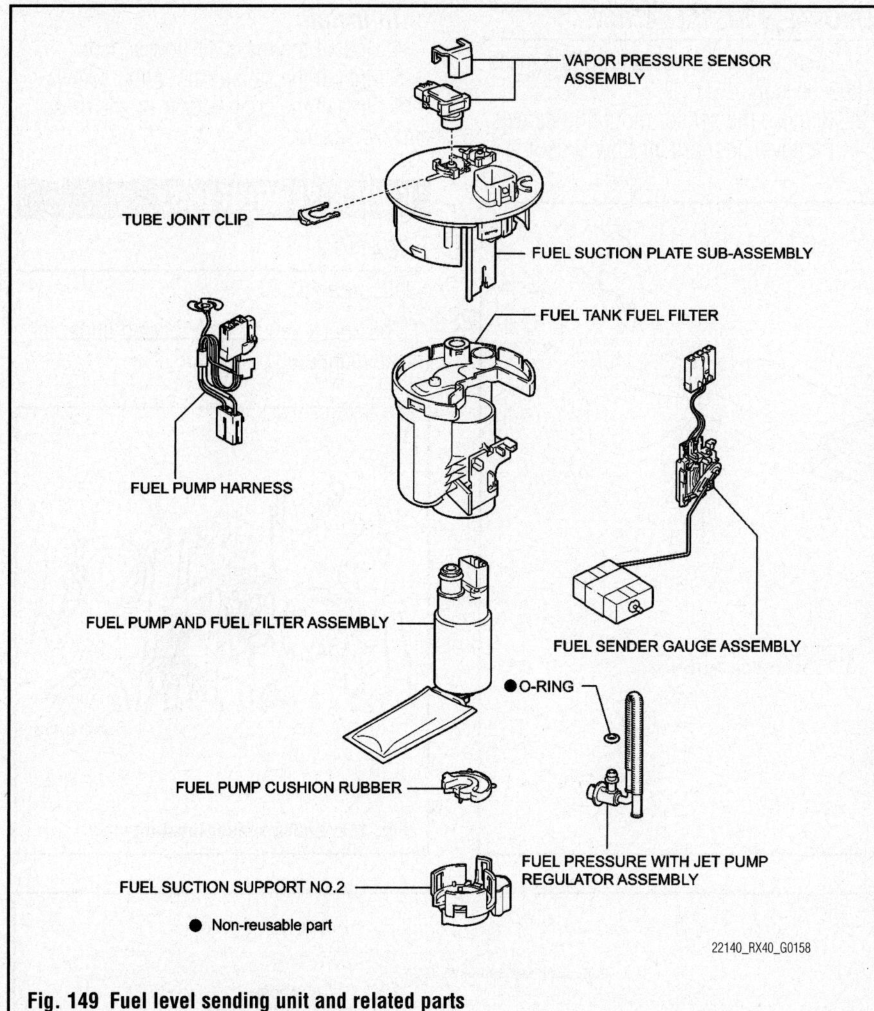

Fig. 149 Fuel level sending unit and related parts

10. Disconnect the clamp from the charcoal canister protector.
11. Remove the 3 bolts and charcoal canister protector.
12. Disconnect the charcoal canister fuel hose from the charcoal canister assembly.
13. Pinch the retainer and pull out the quick connector with the quick connector pushed to the pipe side to disconnect the charcoal canister fuel hose from the charcoal canister assembly.
14. Disconnect the main fuel supply line. Pinch the tabs of the retainer to disengage the lock claws and pull the retainer down as shown in the illustration.
15. Pinch the retainer of the breather lower tube connector, and pull out the breather lower tube connector to disconnect the breather lower tube from the fuel tank.
16. Set a transmission jack to the fuel tank.
17. Remove the fuel filler pipe clamp and fuel tube connector from the fuel tank inlet pipe.
18. Remove the 4 bolts and the fuel tank bands.
19. Operate the transmission jack, and disconnect the fuel inlet pipe.
20. Operate the transmission jack, and remove the fuel tank.
21. Disconnect the fuel pump connector.
22. Disconnect the clamp and vapor pressure sensor connector.
23. Remove the tube joint clip and clamp, and pull out the fuel pump tube.
24. Remove the 8 bolts, fuel tank vent tube set plate and fuel pump connector plate.
25. Pull out the fuel pump assembly from the fuel tank.
26. Remove the sending unit from the fuel pump assembly if applicable.

To install:

27. Install new fuel tank cushions.
28. Install the 2 fuel tube clamps of the fuel tank main tube to the fuel tank.
29. Install the fuel tank main fuel assembly.
30. Install the fuel tank wire.
31. Connect the fuel tank vent hose to the fuel tank
32. Install the 2 nuts to the fuel vapor containment valve, and tighten the nuts to 71 inch. lbs. (8 Nm).
33. Connect the connector and clamp.
34. Install a new gasket on the fuel pump assembly.
35. Install the fuel pump assembly.
36. Install the fuel tank vent tube set plate and fuel pump connector plate with the 8 bolts. Tighten the bolts to 53 inch. lbs. (6 Nm).
37. Connect the fuel tank main tube with the tube joint clip and clamp.
38. Connect the fuel tank wire connector.
39. Connect the vapor pressure sensor connector and clamp.
40. Set the fuel tank on a transmission jack.
41. Operate the transmission jack, and install the fuel tank to the vehicle.
42. Operate the transmission jack, and connect the fuel tank inlet pipe with the tube connector and clamp.
43. Install the 2 fuel tank bands with the 4 bolts and tighten to 29 ft. lbs. (39 Nm).
44. Connect the breather lower tube. Push in the tube connector to the pipe until the tube connector makes a "click" sound.
45. Install the checker to the pipe.
46. Connect the fuel tank main supply line. Push in the fuel tube connector to the pipe and push up the retainer to engage the claws.
47. Align the quick connector with the pipe, then push in the quick connector until the retainer makes a "click" sound to connect the charcoal canister fuel hose to the charcoal canister assembly.
48. Connect the clamp to the charcoal canister protector.
49. Install the charcoal canister protector with the 3 bolts. Tighten the bolts to 48 inch. lbs. (5.4 Nm).
50. Connect the fuel tank wire connectors, and install them to the bracket.
51. Install the parking brake cable assembly No. 3 with the nut.
52. Install the fuel tank protector.
53. Install the front floor heat insulator and tighten the retaining nuts to 43 inch. lbs. (4.9 Nm).
54. Install the center pipe exhaust assembly.
55. Connect the negative battery cable.
56. Add fuel and inspect for fuel leaks.
57. Inspect for exhaust leaks.
58. Some systems need initialization after reconnecting the cable to the negative battery terminal.

HEATED OXYGEN (HO2S) SENSOR

LOCATION

See Figure 150.

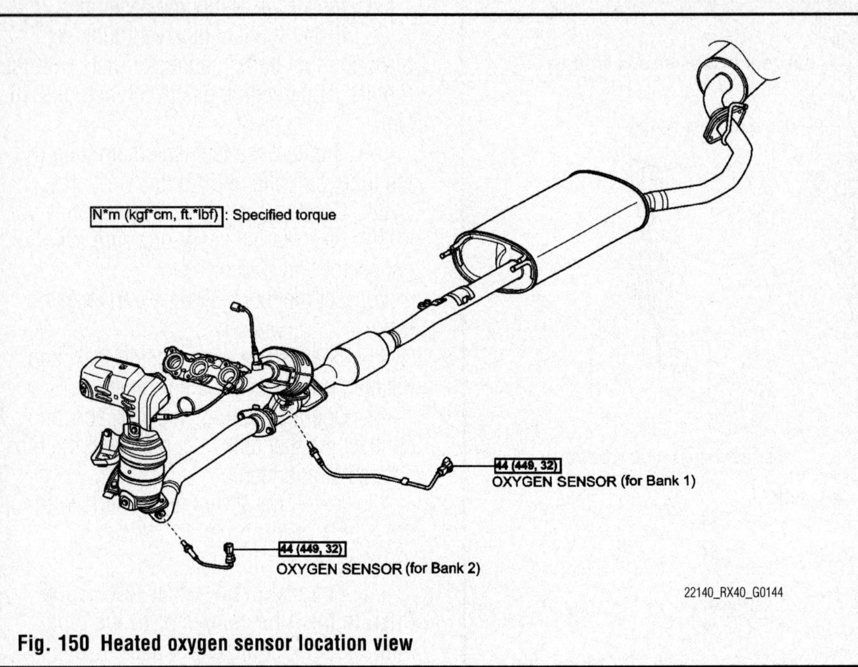

Fig. 150 Heated oxygen sensor location view

The heated oxygen sensors are located below the converter. They are mounted in the front exhaust pipes.

REMOVAL & INSTALLATION

1. Disconnect the oxygen sensor connector and clamp.
2. Using SST 09224-00010, remove the oxygen sensor from the exhaust pipe assembly front.

To install:

3. Temporarily tighten the oxygen sensor to the exhaust pipe sub-assembly front.
4. Using SST, fully tighten the oxygen sensor to the exhaust pipe sub-assembly front.
5. Tighten the sensor to 32 ft. lbs. (44 Nm),
6. Connect the oxygen sensor connector.

INTAKE AIR TEMPERATURE (IAT) SENSOR

LOCATION

See Figure 151.

The intake air temperature sensor is integral to the mass air flow sensor. The mass air flow sensor is mounted on the top of the intake plenum.

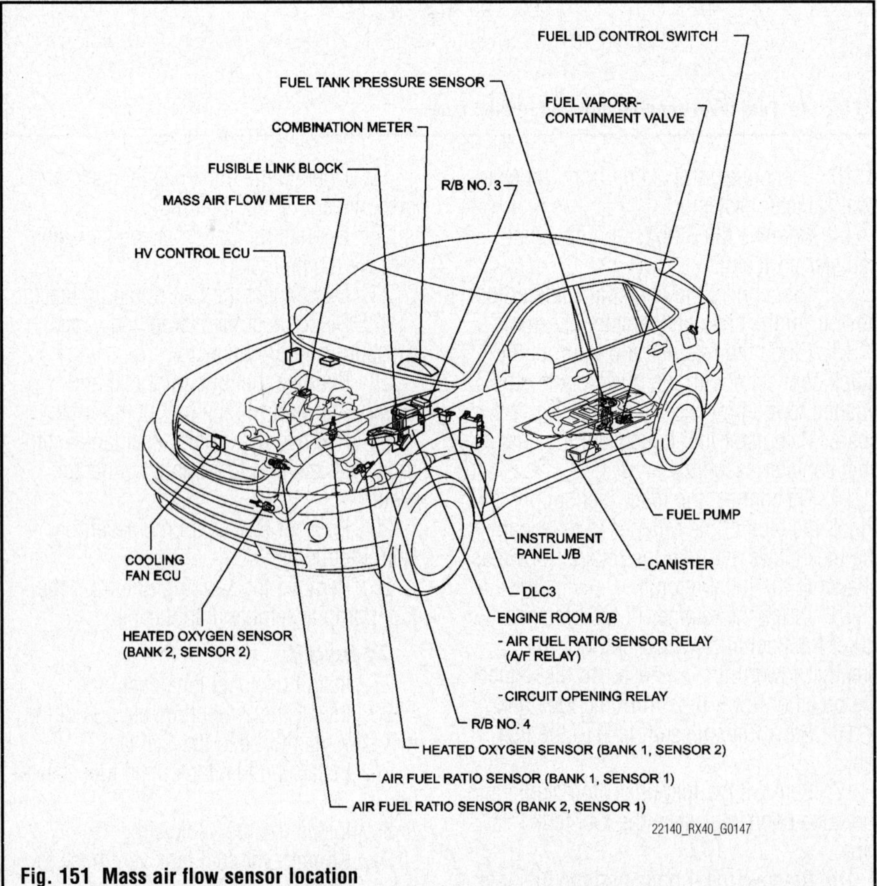

Fig. 151 Mass air flow sensor location

REMOVAL & INSTALLATION

1. Remove the connector from the mass air flow sensor.
2. Remove the sensor mounting screws.
3. Remove the mass air flow sensor.

To install:

4. Install the mass air flow sensor.
5. Install the sensor mounting screws.
6. Install the connector from the mass air flow sensor.

KNOCK SENSOR (KS)

LOCATION

See Figure 152.

The knock sensors are located under the intake manifold.

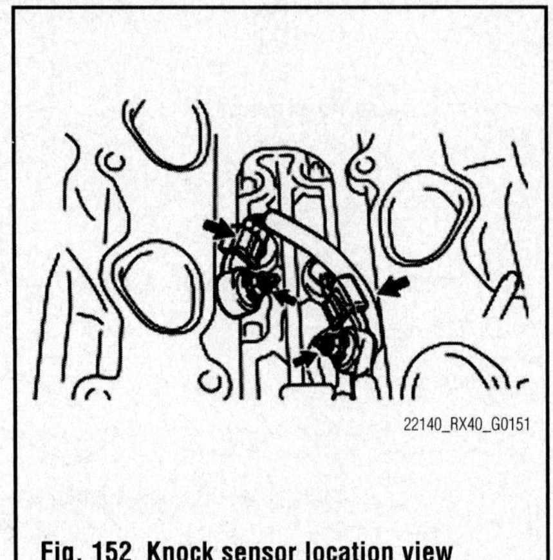

Fig. 152 Knock sensor location view

REMOVAL & INSTALLATION

See Figures 153 through 156.

1. Discharge the fuel system.
2. Remove the engine room covers.
3. Disconnect the negative battery cable.
4. Drain the engine coolant.
5. Remove LH and RH wiper arm assembly.
6. Remove the cowl top ventilator louver.
7. Remove the wiper motor and link assembly.
8. Remove the cowl top panel.
9. Remove the cool air intake duct seal.
10. Remove the air cleaner assembly.
11. Remove the brake master cylinder reservoir.
12. Remove the air cleaner bracket.
13. Remove the air filter bracket.
14. Remove engine moving control rod.
15. Disconnect the VSV connector.
16. Remove the wire harness clamp.
17. Disconnect the fuel vapor feed hose No. 1.
18. Disconnect the fuel vapor feed hose No. 2.
19. Remove the 2 nuts (E), then remove the emission control valve set.
20. Disconnect the throttle motor connector.
21. Separate the water by-pass hose No. 2.
22. Separate the water by-pass hose No. 3.
23. Disconnect the ventilation hose.

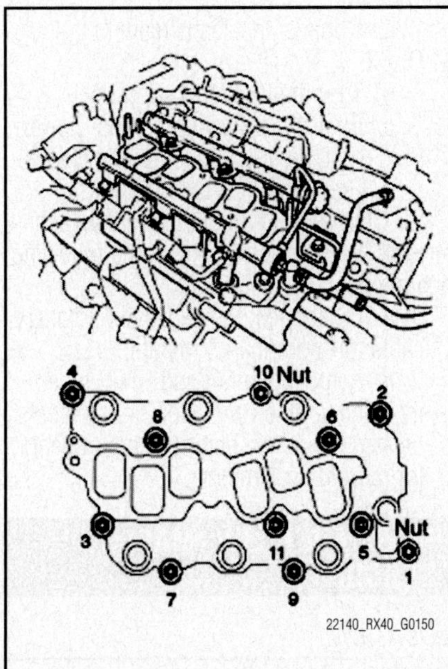

Fig. 153 Remove the 9 bolts and 2 nuts in the sequence shown

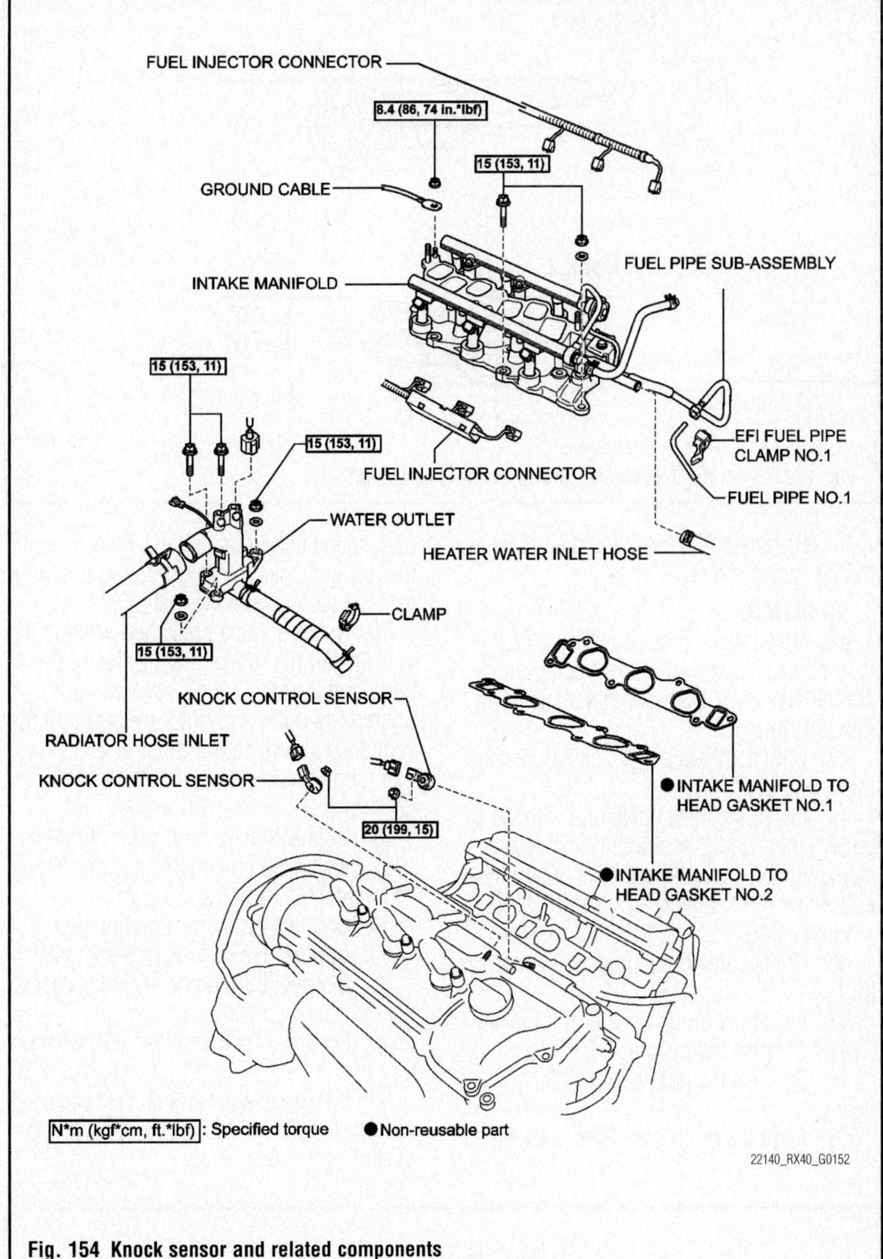

Fig. 154 Knock sensor and related components

24. Remove the 2 bolts, then remove the engine hanger No. 1
25. Remove the 2 bolts, then remove the engine hanger No. 1
26. Remove the 2 bolts, then remove the surge tank stay No. 1 (B).
27. Remove the 2 bolts, then remove the surge tank stay No. 2 (C).
28. Disconnect the ground cable connector.
29. Using a socket hexagon wrench 8 mm, remove the 4 bolts.
30. Remove the 2 nuts, then remove the emission control valve bracket and the intake air surge tank.
31. Remove the gasket from the intake air surge tank.
32. Remove the EFI fuel pipe clamp No. 1.
33. Pinch the quick connector and then pull out the fuel pipe No. 1.
34. Disconnect the heater water inlet hose.
35. Disconnect the radiator hose inlet.
36. Remove the nut and ground cable.
37. Disconnect the 6 fuel injector connectors.
38. In order to remove the intake manifold, using several steps, remove the 9 bolts and 2 nuts in the sequence shown in the illustration.
39. Remove the water outlet.
40. Disconnect the 2 knock sensor connectors.

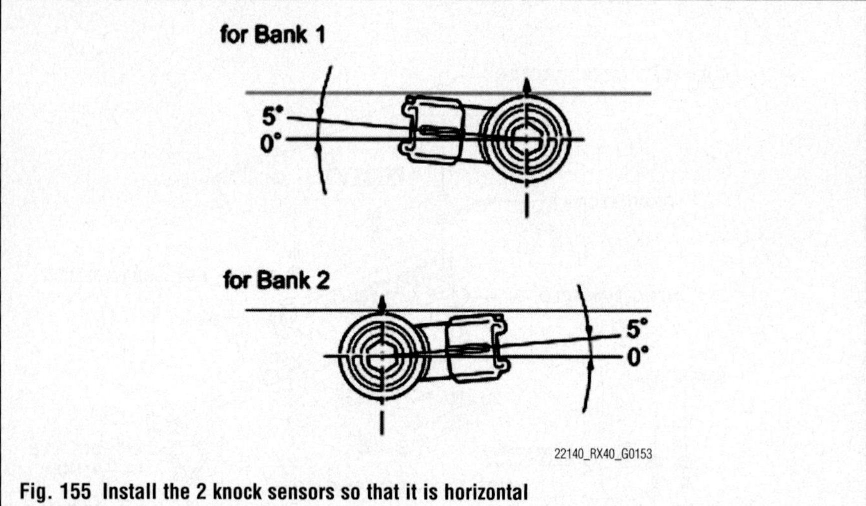

Fig. 155 Install the 2 knock sensors so that it is horizontal

41. Remove the 2 nuts, and then remove the 2 knock sensors.

To install:

42. Install the 2 knock sensors so that it is horizontal as shown in the illustration. Then install the 2 bolts and tighten to 15 ft. lbs. (20 Nm).
43. Connect the 2 knock sensor connectors.
44. Install the intake manifold with the 9 bolts, 2 nuts and 2 washers. Using several steps, tighten the bolts and nuts uniformly in the removal sequence. Tighten to 11 ft. lbs. (1 Nm).
45. Connect the 6 fuel injector connectors.
46. Install the ground cable with the nut.
47. Connect the radiator hose inlet.
48. Connect the heater water inlet hose.
49. Align the quick connector with the pipe, then push in the quick connector until the retainer makes a "click" sound to connect the fuel hose to the fuel pipe.
50. Using a socket hexagon wrench 8 mm, install the intake manifold with the 4 bolts and 2 nuts . Using several steps, tighten the bolts and nuts uniformly in the sequence shown in the illustration. Tighten the nuts and bolts to 21 ft. lbs. (28 Nm).
51. Connect the ground cable connector.
52. Install the surge tank stay brackets and tighten to 14 ft. lbs. 20 (Nm).
53. Connect the ventilation hose.
54. Connect the fuel vapor feed hose.
55. Connect both of the water by-pass hoses.
56. Reconnect the throttle motor connector.
57. Install the emission control valve set with the 2 nuts and tighten to 80 inch. lbs (8 Nm).

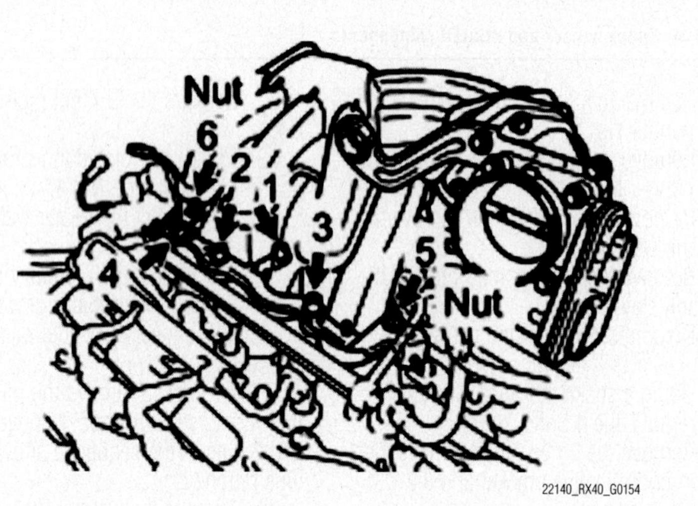

Fig. 156 Using several steps, tighten the bolts and nuts uniformly in the sequence shown

58. Reconnect both of the fuel vapor hose.
59. Connect the wire harness clamp.
60. Connect the VSV connector.
61. Install the engine moving control rod.
62. Install the reservoir bracket and tighten to 71 inch. lbs. (9 Nm).
63. Install air filter assembly bracket and tighten to 14 ft. lbs. 20 (Nm).
64. Install the 2 bolts and brake master cylinder reservoir to the bracket. Tighten to 80 inch. lbs. (9 Nm).
65. Install the air cleaner assembly.
66. Install the cool air intake duct.
67. Install engine coolant and bleed the system.
68. Connect the negative battery cable.
69. Inspect for coolant and fuel leaks.
70. Install the cowl top panel assembly.
71. Install the wiper motor and link assembly.
72. Install the top cowl ventilator louver.
73. Install the LH and RH wiper arm assembly.
74. Install the engine room covers.
75. Some system need initialization when reconnecting the battery cable

MALFUNCTION INDICATOR LIGHT (MIL)

RESET PROCEDURES

1. Clear DTC (Using the Techstream) as follows:
    a. Connect the Techstream to the DLC3.
    b. Turn the ignition switch ON.
    c. Enter the following menus: Powertrain / Engine and ECT / Trouble Codes.
    d. Press the YES button.
2. Clear DTC (Without using the Techstream). Perform either one of the following operations.
    a. Disconnect the negative (-) battery cable for more than 1 minute.
    b. Remove the EFI and ETCS fuses from the engine room No. 2 relay block located inside the engine compartment for more than 1 minute.

MASS AIR FLOW (MAF) SENSOR

LOCATION

See Figure 157.

The mass air flow sensor is mounted on the top of the intake plenum.

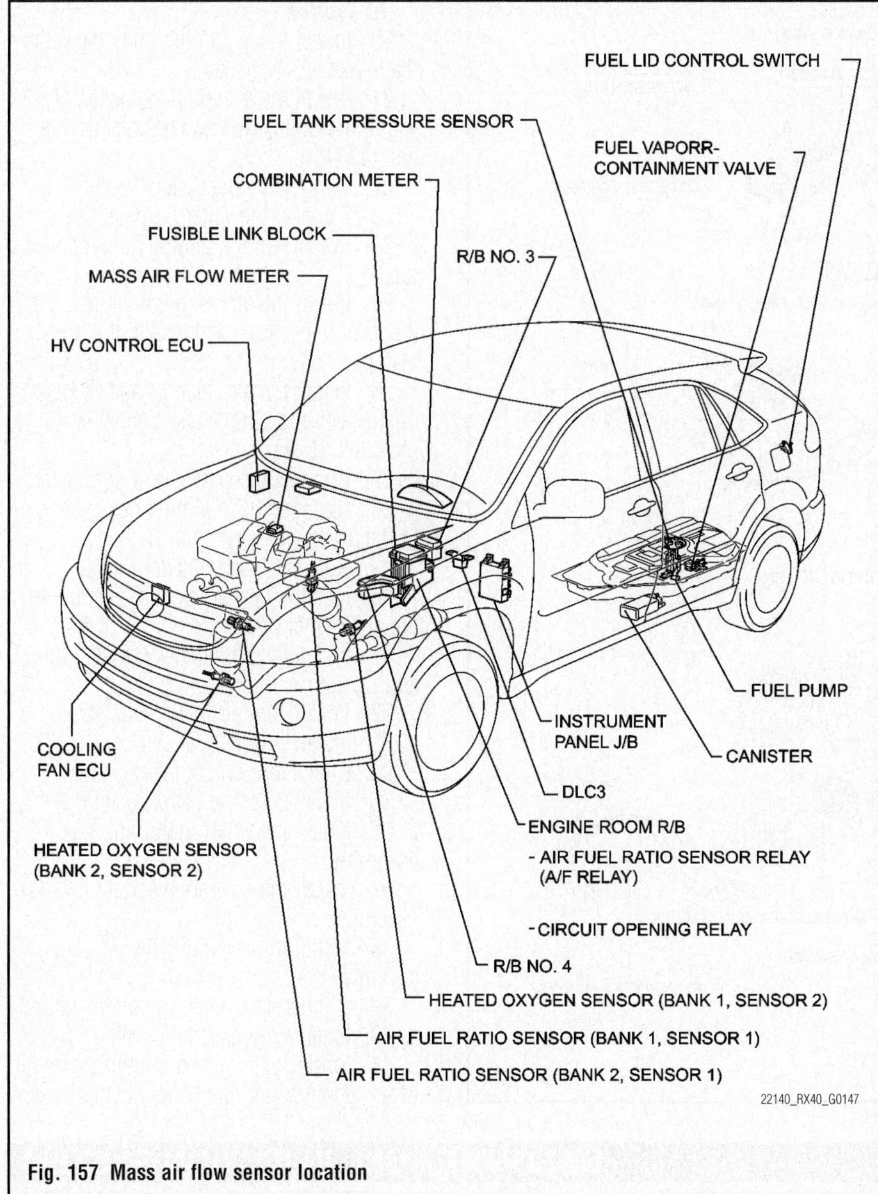

Fig. 157 Mass air flow sensor location

REMOVAL & INSTALLATION

1. Remove the connector from the mass air flow sensor.
2. Remove the sensor mounting screws.
3. Remove the mass air flow sensor.

To install:

4. Install the mass air flow sensor.
5. Install the sensor mounting screws.
6. Install the connector from the mass air flow sensor.

OIL PRESSURE SWITCH

LOCATION

See Figure 158.

The oil pressure switch is located at the front of the engine. Just to the top right of the crankshaft pulley.

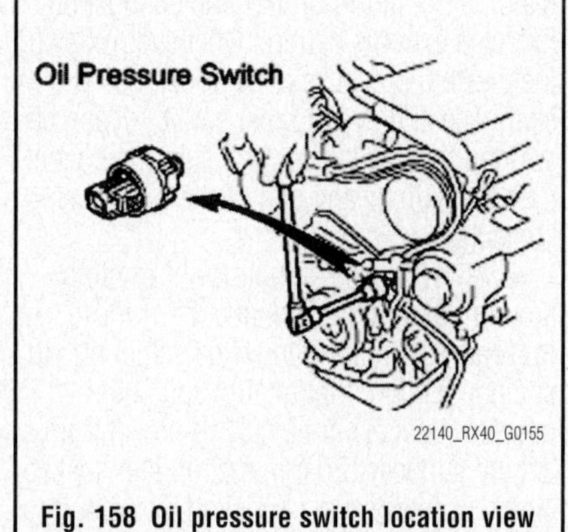

Fig. 158 Oil pressure switch location view

REMOVAL & INSTALLATION

1. Remove the engine under cover.
2. Separate front fender splash shield.
3. Remove the RH front fender apron seal.
4. Disconnect the oil pressure switch connector.
5. Using a 24 mm deep socket wrench, remove the oil pressure switch.

To install:

6. Apply adhesive to 2 or 3 threads of the oil pressure switch.
7. Using a 24 mm deep socket wrench, install the oil pressure switch.
8. Connect the oil pressure switch connector.
9. Inspect for oil leaks.
10. Install the RH front fender apron seal.
11. Install front fender splash shield.
12. Install the engine under cover.

THROTTLE POSITION SENSOR (TPS)

LOCATION

The throttle position sensor is integral to the electric throttle body assembly.

REMOVAL & INSTALLATION

See Figure 159.

1. Remove the engine covers.
2. Disconnect the negative battery cable.
3. Drain the engine coolant.
4. Remove the wiper arm assembly.
5. Remove the top cowl ventilator louver.
6. Remove the wiper motor and link assembly.
7. Remove the cowl top panel.
8. Remove the cool air intake duct seal.
9. Remove the air cleaner and cap assembly.
10. Remove the air cleaner with resonator.
11. Disconnect the MAF meter connector.
12. Disconnect the 2 wire harness clamps from the air cleaner.
13. Remove the 5 bolts from the air cleaner case w/ resonator.
14. Remove the hose clamp, and separate the air cleaner hose No. 1.
15. Disconnect the throttle motor connector.
16. Separate both of the water by-pass hoses.
17. Separate the fuel vapor feed hose.
18. Remove the 4 bolts and the throttle body assembly.
19. Remove the throttle body gasket from the intake air connector.

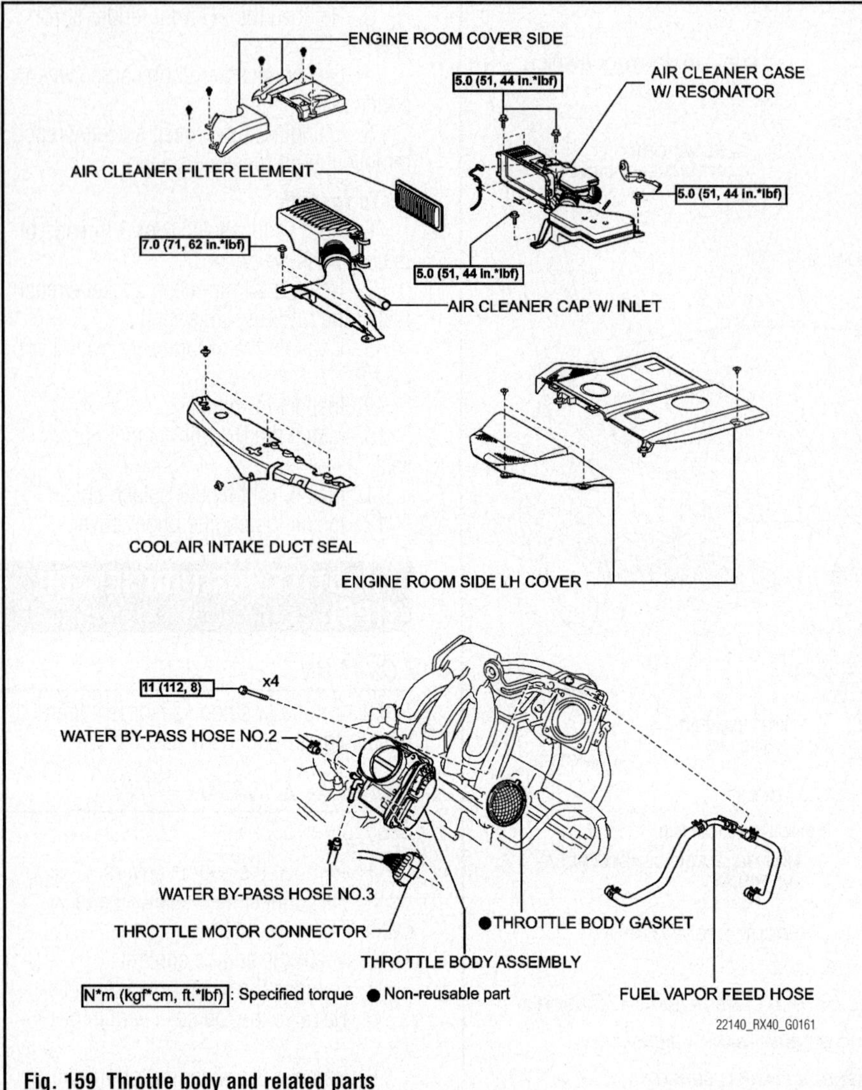

Fig. 159 Throttle body and related parts

To install:

20. Install a new throttle body gasket to the intake air connector.
21. Install the throttle body assembly with the 4 bolts. Tighten the bolts to 8 ft. lbs. (11 Nm).
22. Connect the fuel vapor feed hose.
23. Connect the water by-pass hoses.
24. Reconnect the throttle body connector.
25. Install the air cleaner hose No. 1 to the throttle body assembly with the hose clamp.
26. Install the air cleaner case w/ resonator with the 5 bolts and tighten to 44 inch. lbs. (5 Nm).
27. Connect the MAF meter connector.
28. Connect the 2 wire harness clamps to the air cleaner.
29. Connect the ventilation hose.
30. Install a new air cleaner element to the air cleaner case.
31. Install the air cleaner cap with inlet and tighten.
32. Install cool air intake duct seal.
33. Install engine coolant.
34. Check for coolant leaks.
35. Install cowl top panel assembly.
36. Install the wiper motor and link assembly.
37. Install the cowl top ventilator louver assembly.
38. Install the LH and RH wiper arm assembly.
39. Connect the negative battery cable.
40. Install engine room covers.
41. Some system need initialization when reconnecting the battery cable.

FUEL SYSTEM

GASOLINE FUEL INJECTION SYSTEM

FUEL SYSTEM SERVICE PRECAUTIONS

Safety is the most important factor when performing not only fuel system maintenance but any type of maintenance. Failure to conduct maintenance and repairs in a safe manner may result in serious personal injury or death. Maintenance and testing of the vehicle's fuel system components can be accomplished safely and effectively by adhering to the following rules and guidelines.

- To avoid the possibility of fire and personal injury, always disconnect the negative battery cable unless the repair or test procedure requires that battery voltage be applied.
- Always relieve the fuel system pressure prior to disconnecting any fuel system component (injector, fuel rail, pressure regulator, etc.), fitting or fuel line connection. Exercise extreme caution whenever relieving fuel system pressure to avoid exposing skin, face and eyes to fuel spray. Please be advised that fuel under pressure may penetrate the skin or any part of the body that it contacts.
- Always place a shop towel or cloth around the fitting or connection prior to loosening to absorb any excess fuel due to spillage. Ensure that all fuel spillage (should it occur) is quickly removed from engine surfaces. Ensure that all fuel soaked cloths or towels are deposited into a suitable waste container.
- Always keep a dry chemical (Class B) fire extinguisher near the work area.
- Do not allow fuel spray or fuel vapors to come into contact with a spark or open flame.
- Always use a back-up wrench when loosening and tightening fuel line connection fittings. This will prevent unnecessary stress and torsion to fuel line piping.
- Always replace worn fuel fitting O-rings with new. Do not substitute fuel hose or equivalent where fuel pipe is installed.

Before servicing the vehicle, make sure to also refer to the precautions in the beginning of this section as well.

⁂ CAUTION

The RX400h hybrid system contains a 288V high-voltage system with a strong alkali solution of potassium hydroxide. Be sure to follow the instructions in this manual to handle the system correctly. Failure to do so may result in serious injury or electrocution. Engineer must undergo

special training to be able to perform high-voltage system inspection and servicing.

CAUTION

All high-voltage wire harness connectors are colored orange. The HV battery and other high-voltage components have "High Voltage" caution labels. Do not carelessly touch these wires and components.

CAUTION

Before inspecting or servicing the high-voltage system, be sure to follow safety measures, such as wearing insulated gloves and removing the service plug to prevent electrocution. Carry the removed service plug in your pocket to prevent anyone from reinstalling it while you are servicing the vehicle.

CAUTION

After removing the service plug, wait 5 minutes before touching any of the high-voltage connectors and terminals.

RELIEVING FUEL SYSTEM PRESSURE

See Figure 160.

1. Before servicing the vehicle, refer to the Precautions Section.
2. Disconnect the No. 3 relay block.
3. Remove the No. 2 junction block cover.
4. Remove the C/OPN RLY.
5. Put the vehicle in Inspection Mode and start the engine.
6. Turn the ignition switch to **OFF** immediately after the engine comes to 'rough idle state'.

➡**The hybrid system has a complicated process from an 'out of gas' to 'engine stall' condition. Therefore, 'rough idle' is regarded as 'stop'.**

7. Disconnect the negative battery cable.
8. Reinstall the C/OPN RLY.

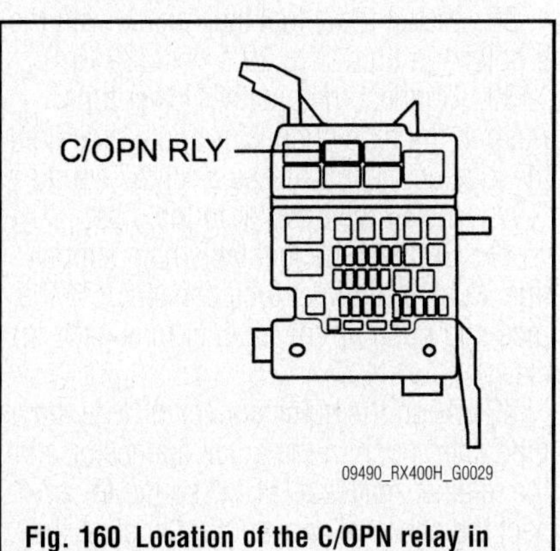

Fig. 160 Location of the C/OPN relay in the junction box.

FUEL FILTER

REMOVAL & INSTALLATION

The fuel filter is incorporated into the fuel pump assembly that is located in the fuel tank.

FUEL PUMP

REMOVAL & INSTALLATION

See Figure 161.

1. Before servicing the vehicle, refer to the Precautions Section.
2. Relieve the fuel system pressure.
3. Remove or disconnect the following:
   - Negative battery cable
   - Center exhaust pipe assembly
   - Front floor heat insulation
   - Fuel tank skid plate
   - Parking brake cable support
   - Fuel tank electrical connection
   - Charcoal canister protector and fuel hose
   - Fuel tank main supply hose
   - Fuel tank breather hose
   - Fuel tank assembly
4. Remove the fuel pump assembly from the tank as follows:
   a. Disconnect the fuel pump connector.
   b. Disconnect the clamp and vapor pressure sensor connector.
   c. Remove the hose joint clip and clamp. Pull out the fuel hose.
   d. Remove the mounting bolts and pull the fuel pump assembly from the tank.

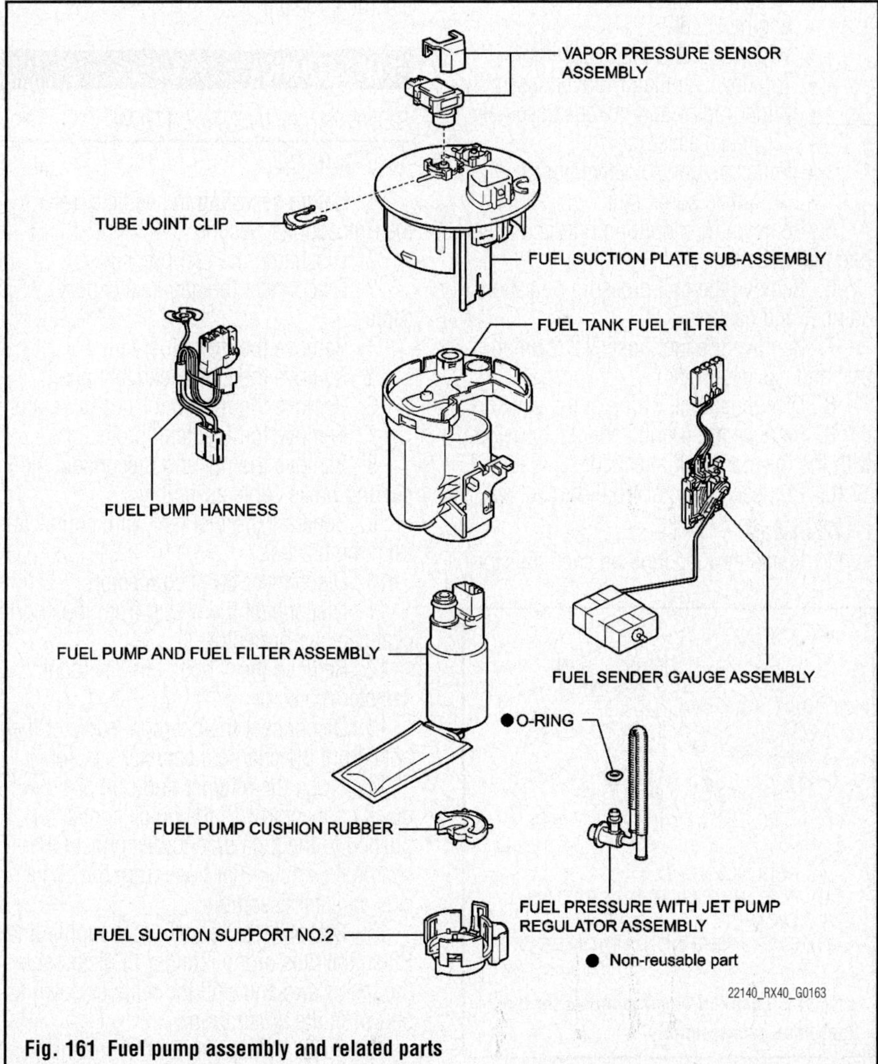

Fig. 161 Fuel pump assembly and related parts

To install:

5. Install a the fuel pump assembly into the tank with a new gasket.

➡**Align the fuel tank vent tube set plate with the cutout on the fuel pump assembly.**

6. Install the fuel connector plate and tighten the bolts to 53 inch lbs. (6 Nm).
7. The remainder of installation is the reverse order of removal.
8. If necessary, perform the initialization procedure, as outlined in the Chassis Electrical System.

FUEL RAIL & INJECTORS

REMOVAL & INSTALLATION

See Figure 162.

1. Before servicing the vehicle, refer to the Precautions Section.
2. Relieve the fuel system pressure.
3. Drain the cooling system.
4. Remove or disconnect the following:
   - Negative battery cable
   - Engine cover
   - Wiper arm and blade assembly
   - Top cowl ventilator louver assembly
   - Wiper motor and linkage assembly
   - Air intake assembly
   - Emission control valve hoses
   - Air intake surge tank
5. Remove the mounting bolt and separate fuel hose No. 1
6. Remove the fuel pressure pulsation damper and gaskets.
7. Remove the fuel hose No. 2 union bolt and gaskets.
8. Disconnect the wiring at the injectors.
9. Remove the 4 bolts and each fuel rail with the injectors still attached.
10. Pull each injector from the fuel rail..

To install:

11. Install new O-rings on each injector.

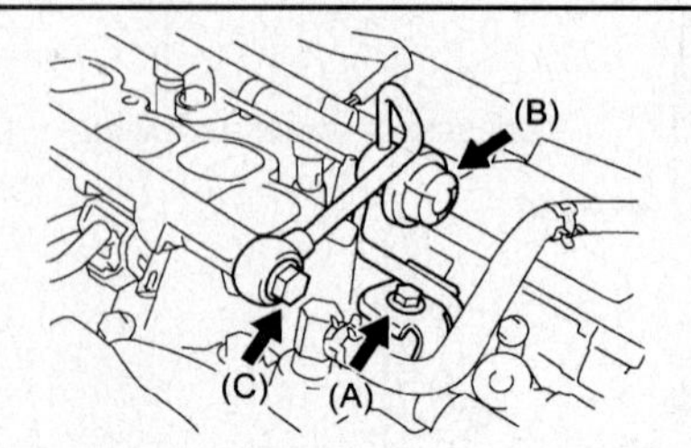

Fig. 162 Location of components for the fuel hose sub-assembly

Apply a light coating of gasoline to the O-rings and mating points on the pipes.

12. Using a twisting motion, install the injectors on the pipes.

➡**Be careful to avoid twisting the O-rings. After installation, check that the injectors turn smoothly. If not, use new O-rings.**

13. Install the pipes and injectors.
14. Loosely install the bolts and make sure that the injectors still turn freely. If not, replace the O-rings.
15. Torque the bolts to 84 inch lbs. (10 Nm).
16. The remainder of installation is the reverse of removal. Observe the following torques:
    - Fuel hose No. 2 union bolt: 24 ft. lbs. (33 Nm)
    - Pulsation damper: 24 ft. lbs. (33 Nm)
    - Fuel hose No. 1: 14 ft. lbs. (20 Nm)
17. If necessary, perform the initialization procedure, as outlined in the Chassis Electrical System.

FUEL TANK

REMOVAL & INSTALLATION

See Figure 163.

1. Before servicing the vehicle, refer to the Precautions Section.
2. Discharge the fuel pressure.
3. Disconnect the negative battery cable.
4. Remove fuel from fuel tank
5. Remove the center exhaust pipe.
6. Remove the front floor heat insulator.
7. Remove the fuel tank protector.
8. Remove the nut and disconnect the parking brake cable assembly
9. Separate the fuel tank wire connector from the bracket.
10. Disconnect the 2 connectors.
11. Disconnect the clamp from the charcoal canister protector.
12. Remove the 3 bolts and charcoal canister protector.
13. Disconnect the charcoal canister fuel hose from the charcoal canister assembly.
14. Pinch the retainer and pull out the quick connector with the quick connector pushed to the pipe side to disconnect the charcoal canister fuel hose from the charcoal canister assembly.
15. Disconnect the main fuel supply line. Pinch the tabs of the retainer to disengage the lock claws and pull the retainer down as shown in the illustration.
16. Pinch the retainer of the breather lower tube connector, and pull out the breather lower tube connector to disconnect the breather lower tube from the fuel tank.
17. Set a transmission jack to the fuel tank.
18. Remove the fuel filler pipe clamp and fuel tube connector from the fuel tank inlet pipe.
19. Remove the 4 bolts and the fuel tank bands.
20. Operate the transmission jack, and disconnect the fuel inlet pipe.
21. Operate the transmission jack, and remove the fuel tank.

To install:

22. Install new fuel tank cushions.
23. Install the 2 fuel tube clamps of the fuel tank main tube to the fuel tank.
24. Install the fuel tank main fuel assembly.
25. Install the fuel tank wire.
26. Connect the fuel tank vent hose to the fuel tank
27. Install the 2 nuts to the fuel vapor containment valve, and tighten the nuts to 71 inch. lbs. (8 Nm).
28. Connect the connector and clamp.
29. Install a new gasket on the fuel pump assembly.
30. Install the fuel pump assembly.
31. Install the fuel tank vent tube set plate and fuel pump connector plate with the 8 bolts. Tighten the bolts to 53 inch. lbs. (6 Nm).
32. Connect the fuel tank main tube with the tube joint clip and clamp.
33. Connect the fuel tank wire connector.
34. Connect the vapor pressure sensor connector and clamp.
35. Set the fuel tank on a transmission jack.
36. Operate the transmission jack, and install the fuel tank to the vehicle.
37. Operate the transmission jack, and connect the fuel tank inlet pipe with the tube connector and clamp.
38. Install the 2 fuel tank bands with the 4 bolts and tighten to 29 ft. lbs. (39 Nm).
39. Connect the breather lower tube. Push in the tube connector to the pipe until the tube connector makes a "click" sound.
40. Install the checker to the pipe.
41. Connect the fuel tank main supply line. Push in the fuel tube connector to the pipe and push up the retainer to engage the claws.
42. Align the quick connector with the pipe, then push in the quick connector until the retainer makes a "click" sound to connect the charcoal canister fuel hose to the charcoal canister assembly.

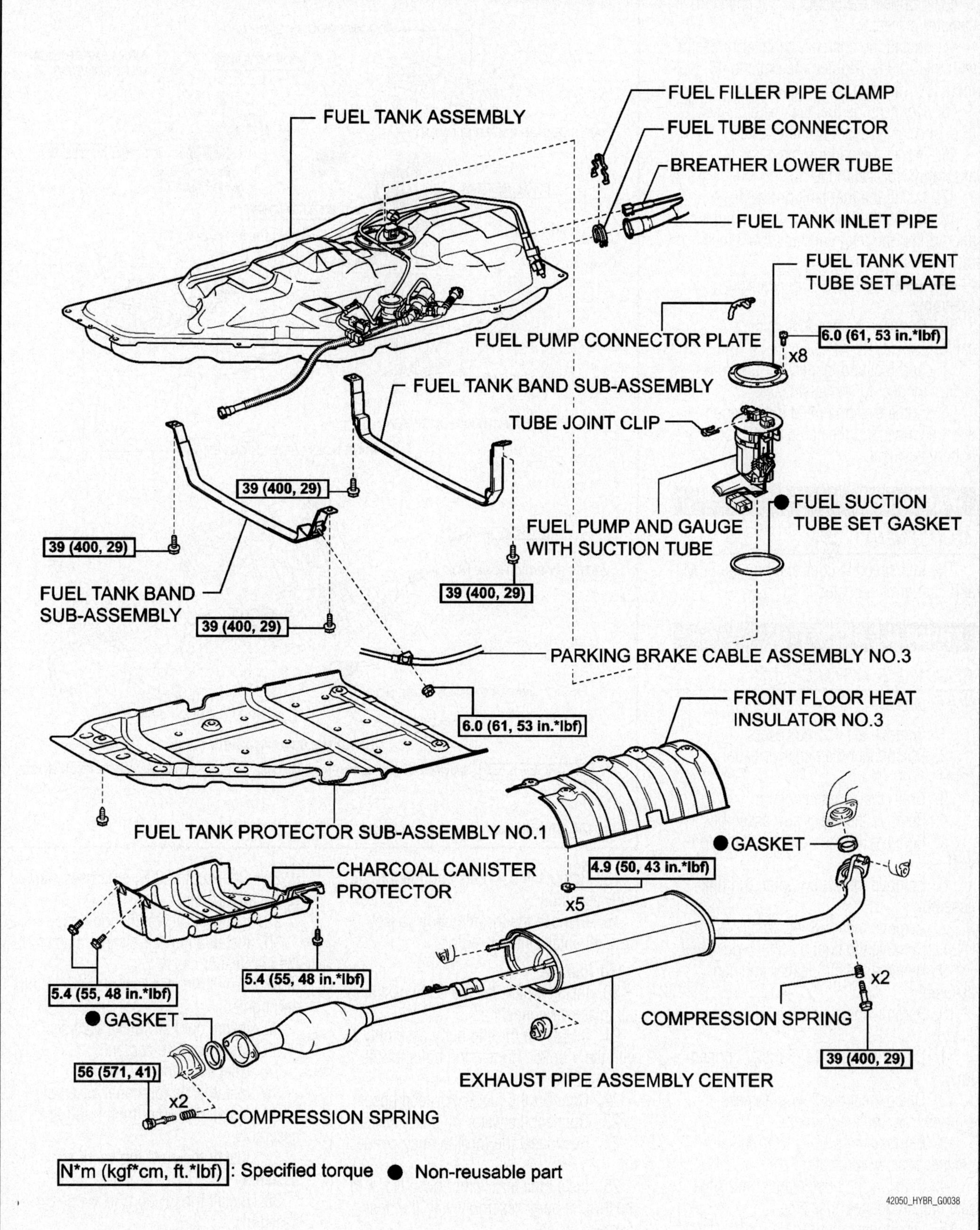

Fig. 163 Exploded view of the fuel pump and related components (1 of 2)

43. Connect the clamp to the charcoal canister protector.

44. Install the charcoal canister protector with the 3 bolts. Tighten the bolts to 48 inch. lbs. (5.4 Nm).

45. Connect the fuel tank wire connectors, and install them to the bracket.

46. Install the parking brake cable assembly No. 3 with the nut.

47. Install the fuel tank protector.

48. Install the front floor heat insulator and tighten the retaining nuts to 43 inch. lbs. (4.9 Nm).

49. Install the center pipe exhaust assembly.

50. Connect the negative battery cable.

51. Add fuel and inspect for fuel leaks.

52. Inspect for exhaust leaks.

53. Some systems need initialization after reconnecting the cable to the negative battery terminal.

IDLE SPEED

ADJUSTMENT

The idle speed is controlled by the ECM and cannot be adjusted.

THROTTLE BODY

REMOVAL & INSTALLATION

See Figure 164.

1. Remove the engine covers.
2. Disconnect the negative battery cable.
3. Drain the engine coolant.
4. Remove the wiper arm assembly.
5. Remove the top cowl ventilator louver.
6. Remove the wiper motor and link assembly.
7. Remove the cowl top panel.
8. Remove the cool air intake duct seal.
9. Remove the air cleaner and cap assembly.
10. Remove the air cleaner with resonator.
11. Disconnect the MAF meter connector.
12. Disconnect the 2 wire harness clamps from the air cleaner.
13. Remove the 5 bolts from the air cleaner case w/ resonator.
14. Remove the hose clamp, and separate the air cleaner hose No. 1.
15. Disconnect the throttle motor connector.
16. Separate both of the water by-pass hoses.
17. Separate the fuel vapor feed hose.

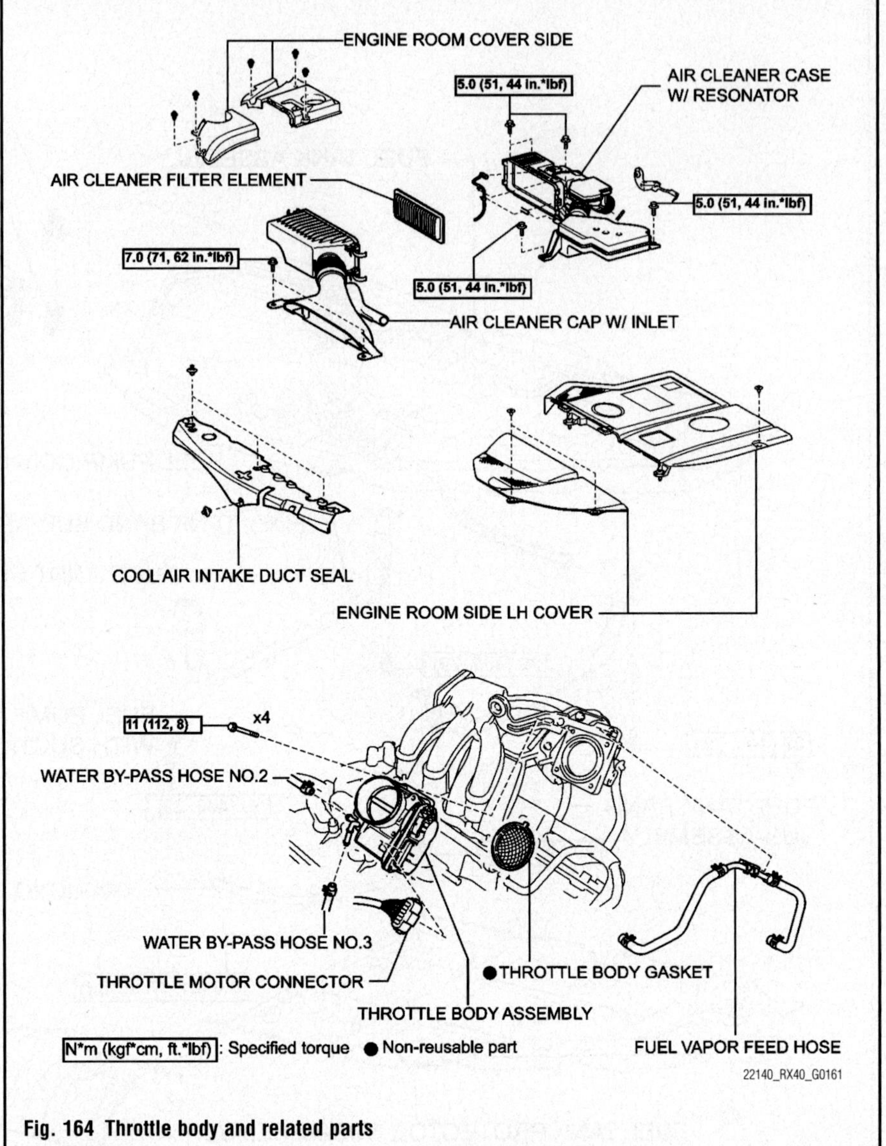

Fig. 164 Throttle body and related parts

18. Remove the 4 bolts and the throttle body assembly.
19. Remove the throttle body gasket from the intake air connector.

To install:

20. Install a new throttle body gasket to the intake air connector.
21. Install the throttle body assembly with the 4 bolts. Tighten the bolts to 8 ft. lbs. (11 Nm).
22. Connect the fuel vapor feed hose.
23. Connect the water by-pass hoses.
24. Reconnect the throttle body connector.
25. Install the air cleaner hose No. 1 to the throttle body assembly with the hose clamp.
26. Install the air cleaner case w/ resonator with the 5 bolts and tighten to 44 inch. lbs. (5 Nm).
27. Connect the MAF meter connector.
28. Connect the 2 wire harness clamps to the air cleaner.
29. Connect the ventilation hose.
30. Install a new air cleaner element to the air cleaner case.
31. Install the air cleaner cap with inlet and tighten.
32. Install cool air intake duct seal.
33. Install engine coolant.
34. Check for coolant leaks.
35. Install cowl top panel assembly.
36. Install the wiper motor and link assembly.
37. Install the cowl top ventilator louver assembly.
38. Install the LH and RH wiper arm assembly.
39. Connect the negative battery cable.
40. Install engine room covers.
41. Some system need initialization when reconnecting the battery cable.

HEATING & AIR CONDITIONING

BLOWER MOTOR

REMOVAL & INSTALLATION

See Figures 165 through 167.

1. Disconnect the negative battery cable.
2. Remove the glove compartment door, as follows:
   a. Remove the glove compartment door damper clip.
   b. Remove the 2 pins, then remove the glove compartment door.
3. Unfasten the retaining screws, detach the connector, and then remove the blower motor.
4. Installation is the reverse of the removal procedure.
5. If necessary, perform the initialization procedure, as outlined in the Chassis Electrical System.

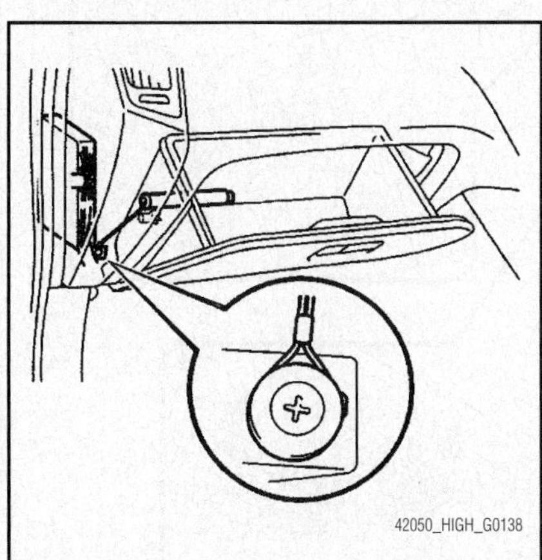

Fig. 165 Remove the glove compartment door damper clip

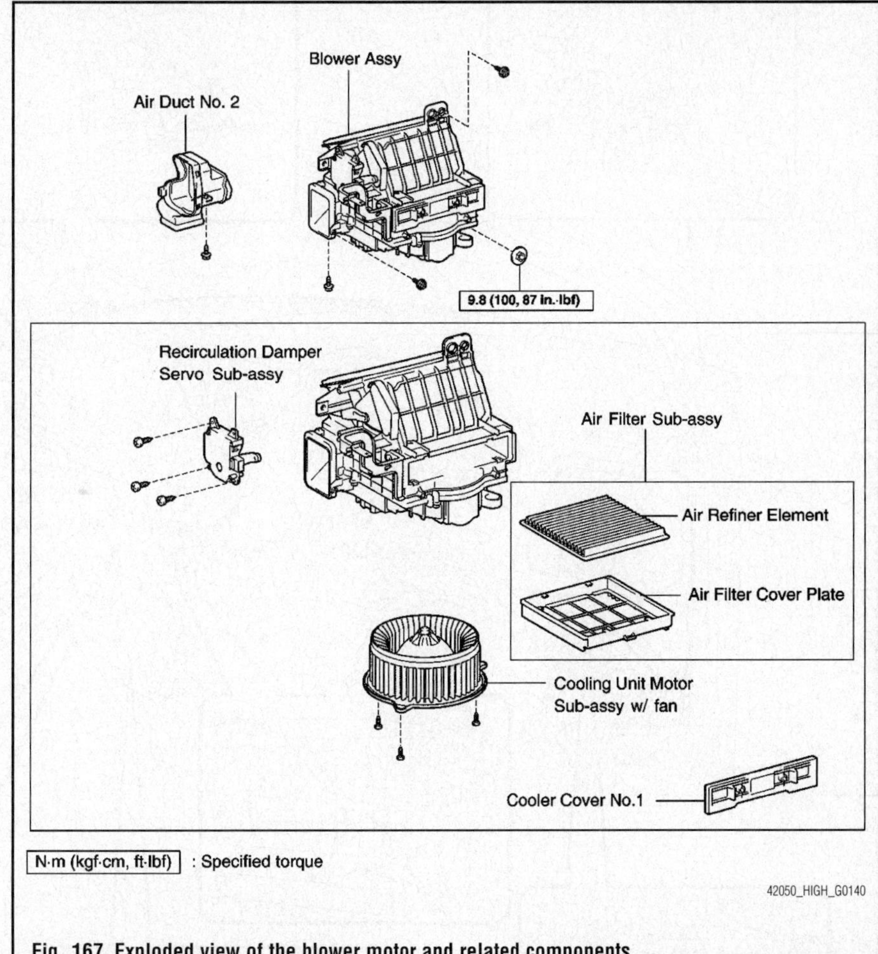

Fig. 167 Exploded view of the blower motor and related components

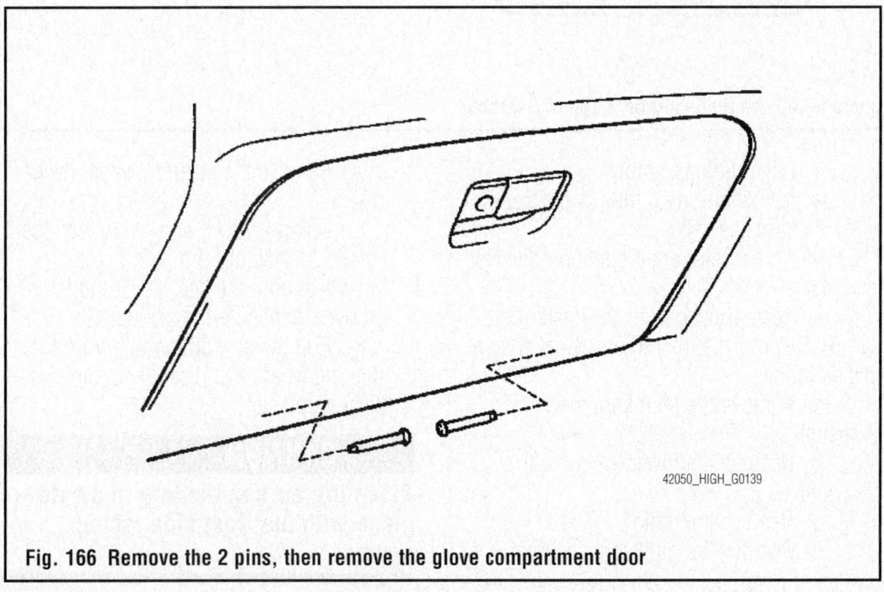

Fig. 166 Remove the 2 pins, then remove the glove compartment door

HEATER CORE

REMOVAL & INSTALLATION

See Figures 168 through 171.

CAUTION

Refer to the applicable precautions for this system before performing the following operation. Failure to follow the warnings and cautions could result in possible personal injury or death.

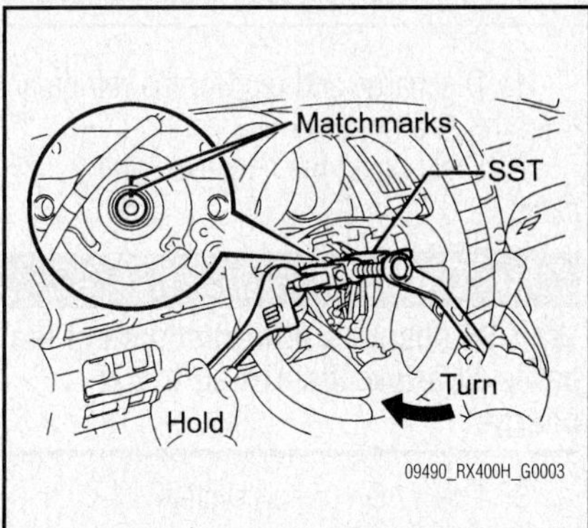

Fig. 168 Using the special tool to remove the steering wheel assembly

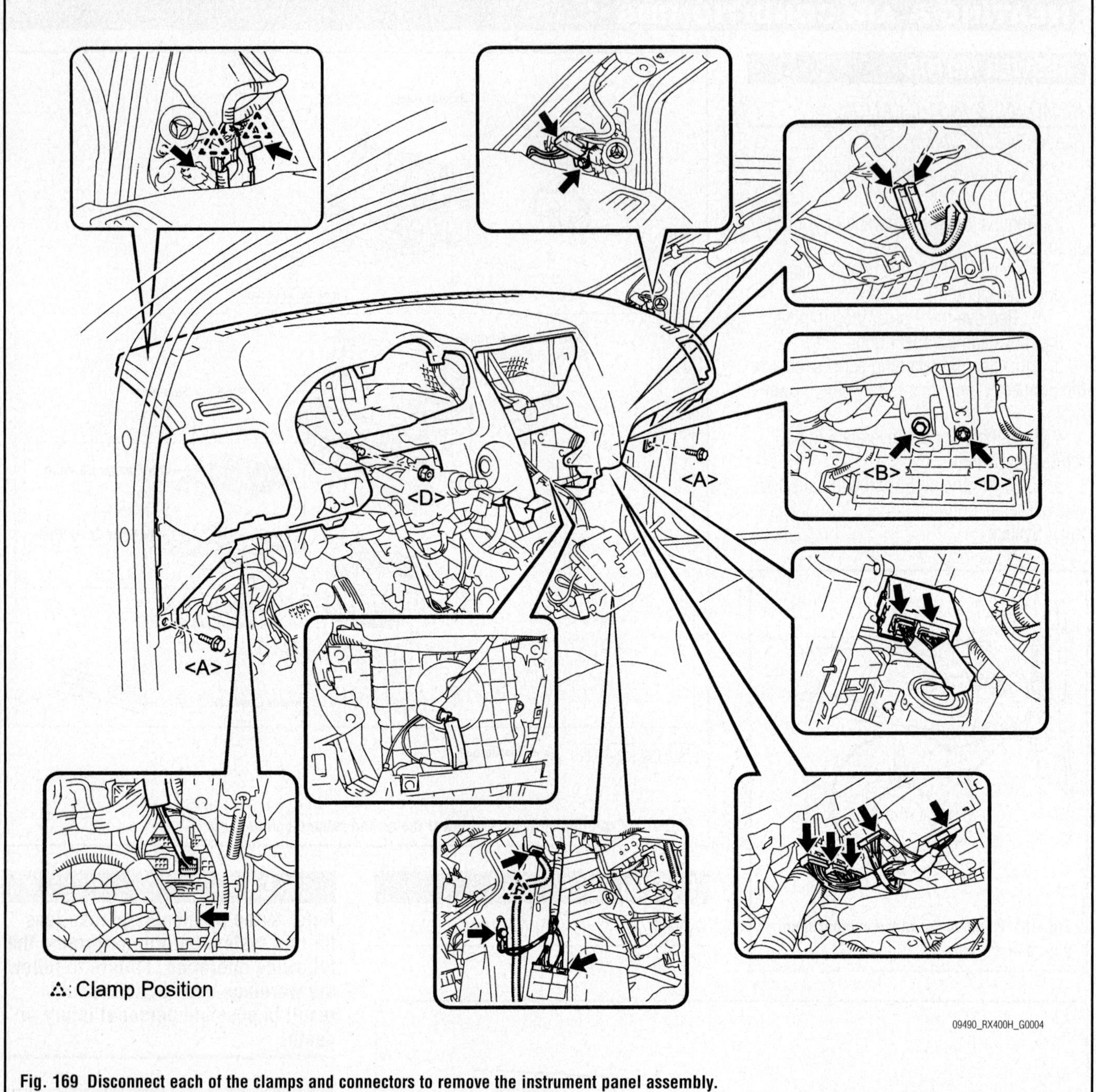

Fig. 169 Disconnect each of the clamps and connectors to remove the instrument panel assembly.

1. Discharge and recover the refrigerant from the A/C system.
2. Disconnect the negative battery cable.

CAUTION

Wait 90 seconds after disconnect the cable to allow the airbag to discharge.

3. Drain the cooling system.
4. Remove or disconnect the following:
   - Wiper arms
   - Cowl top ventilator louver assembly
   - Wiper link assembly
   - Air conditioning tube and accessory assembly
   - Cooler refrigerant suction hose No. 1
   - Heater hoses from the heater core.
5. Remove the instrument panel assembly as follows:
   a. Remove the shift knob and console panel.
   b. Remove the lower instrument panel.
   c. Remove the radio.
   d. Remove the multi-display unit.
   e. Remove the steering wheel lower covers.
   f. Using a Torx® wrench, loosen the 2 screws located at each side of the steering wheel until the screw's circumference groove catches on the screw case
   g. Pull the air bag module from the steering wheel and disconnect the electrical connector.

CAUTION

Place the air bag module in a safe place with the front side facing upward.

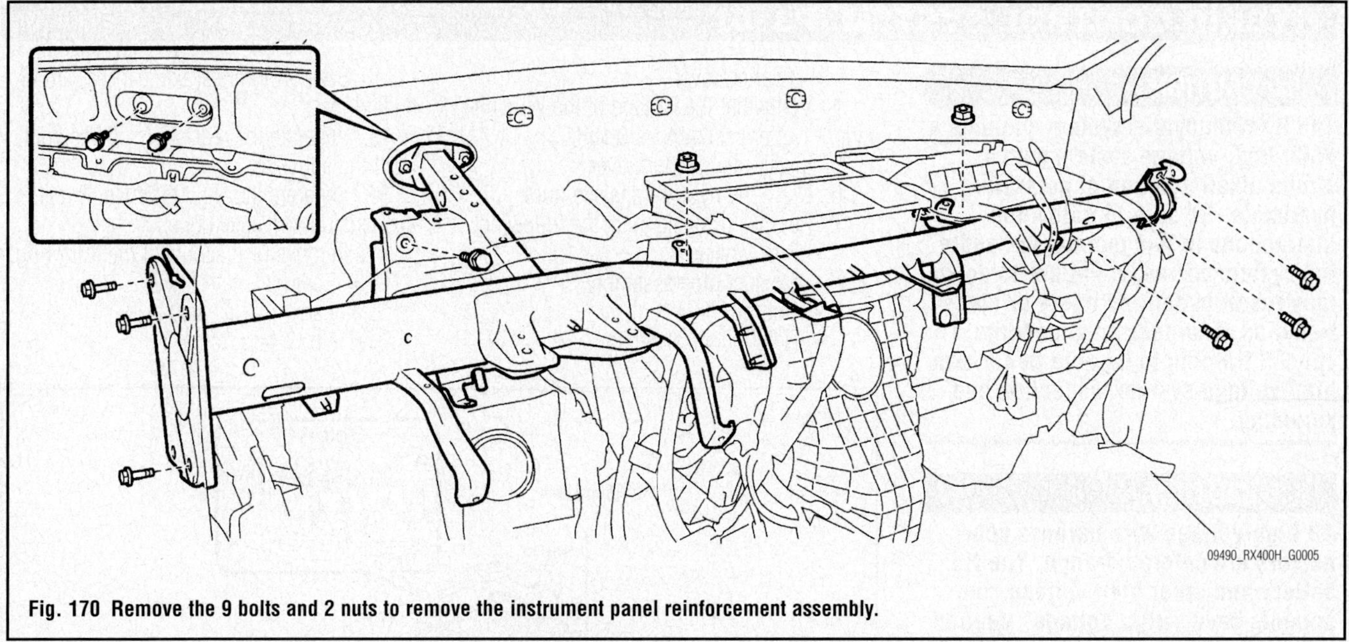

Fig. 170 Remove the 9 bolts and 2 nuts to remove the instrument panel reinforcement assembly.

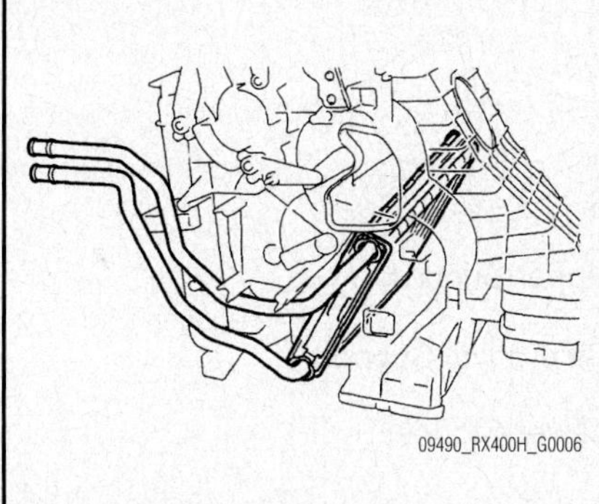

Fig. 171 Remove the heater core from the air conditioner unit.

h. Disconnect the horn connector.

i. Remove the steering pad.

j. Matchmark the steering wheel and main shaft assembly.

k. Using Special Tool 00950-50013, remove the steering wheel assembly.

l. Remove the steering column cover.

m. Remove the tilt and telescopic switch.

n. Remove the turn signal switch assembly with spiral cable assembly.

o. Remove the left-hand cowl side trim sub-assembly.

p. Remove the lower instrument panel.

q. Remove the driver side knee airbag assembly.

r. Remove the gauge cluster trim.

s. Remove the instrument gauge assembly.

t. Remove the right-hand front door scuff plate.

u. Remove the right-hand cowl side trim assembly.

v. Remove the instrument panel No. 2 under cover.

w. Remove the glove compartment door assembly.

x. Remove the center console carpet.

y. Remove the lower center instrument panel.

z. Remove the A-pillar trim.

aa. Remove the instrument panel finish plate.

bb. Remove the No. 1 register panel.

cc. Remove the instrument panel assembly.

6. Remove the hybrid vehicle control ECU.

7. Remove the air conditioner amplifier assembly.

8. Remove the rear air ducts.

9. Remove air duct No. 1.

10. Remove the No. 4 instrument panel bracket.

11. Remove the No. 1 instrument panel brace.

12. Remove the No. 2 instrument panel brace.

13. Matchmark and separate the steering intermediate shaft sub-assembly.

14. Remove the steering column assembly.

15. Remove the No. 1 air duct sub-assembly.

16. Remove the instrument panel reinforcement assembly.

17. Remove the air conditioner unit.

18. Remove the No. 2 air duct.

19. Remove the blower assembly.

20. Remove the heater to register center duct.

21. Remove the air outlet control servo motor.

22. Remove the air mix control servo motor.

23. Remove the evaporator temperature sensor.

24. Remove the heater core.

25. Installation is the reverse order of removal.

26. If necessary, perform the initialization procedure, as outlined in the Chassis Electrical System.

HYBRID CONTROL SYSTEM (P310)

✲✲ CAUTION

The RX400h hybrid system contains a 288V high-voltage system with a strong alkali solution of potassium hydroxide. Be sure to follow the instructions in this manual to handle the system correctly. Failure to do so may result in serious injury or electrocution. Engineer must undergo special training to be able to perform high-voltage system inspection and servicing.

✲✲ CAUTION

All high-voltage wire harness connectors are colored orange. The HV battery and other high-voltage components have "High Voltage" caution labels. Do not carelessly touch these wires and components.

✲✲ CAUTION

Before inspecting or servicing the high-voltage system, be sure to follow safety measures, such as wearing insulated gloves and removing the service plug to prevent electrocution. Carry the removed service plug in your pocket to prevent anyone from reinstalling it while you are servicing the vehicle.

✲✲ CAUTION

After removing the service plug, wait 5 minutes before touching any of the high-voltage connectors and terminals.

REAR TRACTION MOTOR

REMOVAL & INSTALLATION

See Figures 172 and 173.

1. Before servicing the vehicle, refer to the Precautions Section.
2. Remove the service plug grip, found underneath the Battery Service cover on the rear seat. Wait 5 minutes to discharge the high voltage capacitor.
3. Remove or disconnect the following:
   - Negative battery cable
   - Left-hand engine side cover
   - No. 5 inverter bracket
   - Power steering ECU
   - Inverter reserve tank
   - Power steering ECU bracket
   - Inverter cover
4. Verify that the voltage of the W/ Converter Inverter assembly is 0 volts.
5. Install the inverter cover.
6. Drain the hybrid transaxle fluid.
7. Remove or disconnect the following:
   - Rear wheels
   - Exhaust pipe assembly
   - Hub nut
   - Height control sensor
8. Support the rear axle carrier with a suitable jack.
9. Separate the rear strut rod assembly from the knuckle
10. Separate the No. 1 and No. 2 rear suspension arm assemblies.
11. Remove or disconnect the following:
    - Rear halfshafts
    - Parking brake cable
    - No. 3 wire frame

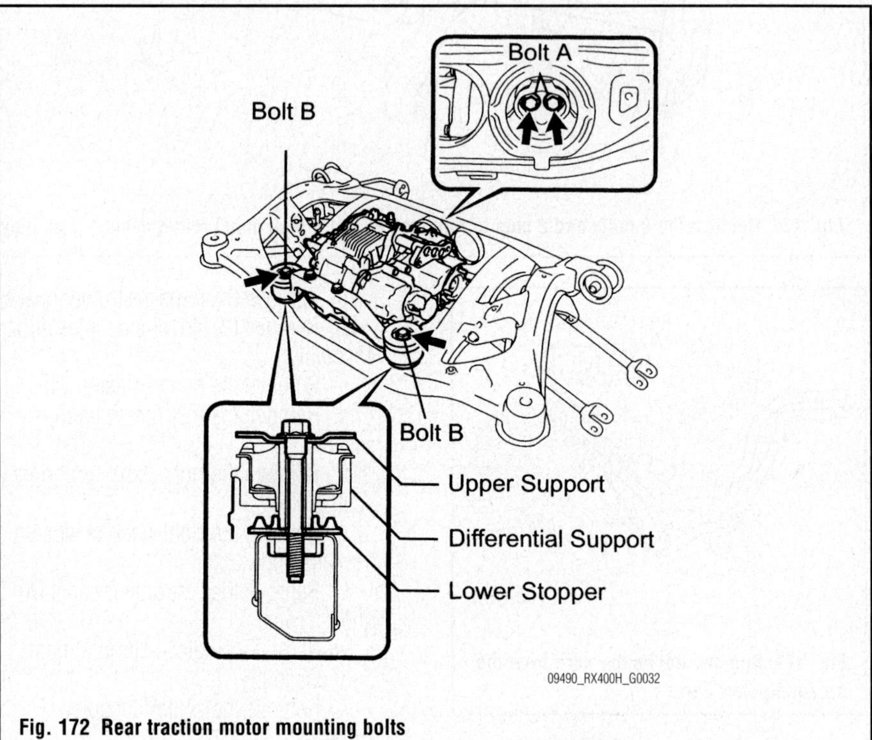

Fig. 172 Rear traction motor mounting bolts

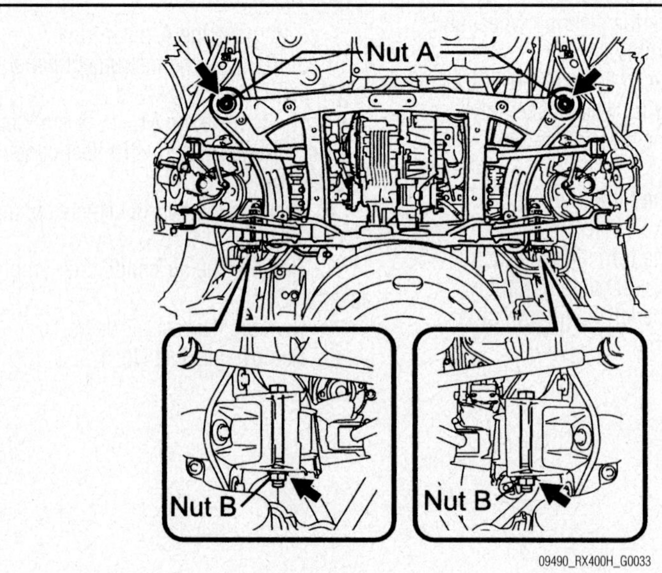

Fig. 173 Rear suspension member mounting nuts

- Rear suspension member
- Rear traction motor assembly

To install:

12. Install the rear traction motor assembly. If motor assembly was removed from the rear suspension member, tighten the mounting bolts as follows:
    a. Bolt A to 70 ft. lbs. (95 Nm).
    b. Bolt B to 76 ft. lbs. (103 Nm).
13. Slowly raise the rear suspension member into place. Tighten Nuts A to 85 ft. lbs. (115 Nm) and Nuts B to 134 ft. lbs. (181 Nm).
14. The remainder of installation is the reverse order of removal.
15. If necessary, perform the initialization procedure, as outlined in the Chassis Electrical System.

SERVICE PLUG GRIP

REMOVAL & INSTALLATION

CAUTION

The RX400h hybrid system contains a 288V high-voltage system with a strong alkali solution of potassium hydroxide. Be sure to follow the instructions in this manual to handle the system correctly. Failure to do so may result in serious injury or electrocution. Engineer must undergo special training to be able to perform high-voltage system inspection and servicing.

CAUTION

All high-voltage wire harness connectors are colored orange. The HV battery and other high-voltage components have "High Voltage" caution labels. Do not carelessly touch these wires and components.

1. Before inspecting or servicing the high-voltage system, be sure to follow safety measures, such as wearing insulated gloves and removing the service plug to prevent electrocution. Carry the removed service plug in your pocket to prevent other technicians from reinstalling it while your are servicing the vehicle.
2. After removing the service plug, wait 5 minutes before touching any of the high-voltage connectors and terminals.

➡Wear insulating gloves and protective glasses.

3. Read output DTC.
4. Disconnect the negative battery cable.
5. Remove the 2 clips, then open the battery service hole cover.
6. Wear insulated gloves, and remove the service plug grip, after sliding up the lever of the service plug grip.

To install:

7. Wear insulated gloves, then insert the service plug.
8. Push down on the grip to lock.
9. Close the battery service hole cover.
10. Connect the negative battery cable.
11. Some systems need initialization when reconnecting the negative battery cable.

WATER PUMP HYBRID ELECTRIC (P310)

REMOVAL & INSTALLATION

See Figures 174 through 176.

1. Disconnect the negative battery cable.
2. Remove the left engine room side cover.
3. Remove the transaxle side reserve tank.
4. Loosen the bleeder plug and drain the coolant from inverter cooler.
5. Loosen the bleeder plug and drain the coolant from inverter.
6. Remove the engine under cover.
7. Drain transaxle fluid if equipped with oil cooler.
8. Remove the front bumper if equipped with oil cooler.
9. Remove the frame side rail plate sub-assembly as follows:
   - Using a transmission jack, hold the front frame.
   - Remove the 3 bolts, nut and frame side rail plate sub-assembly.

➡Be sure to position the transmission jack to properly support the front frame.

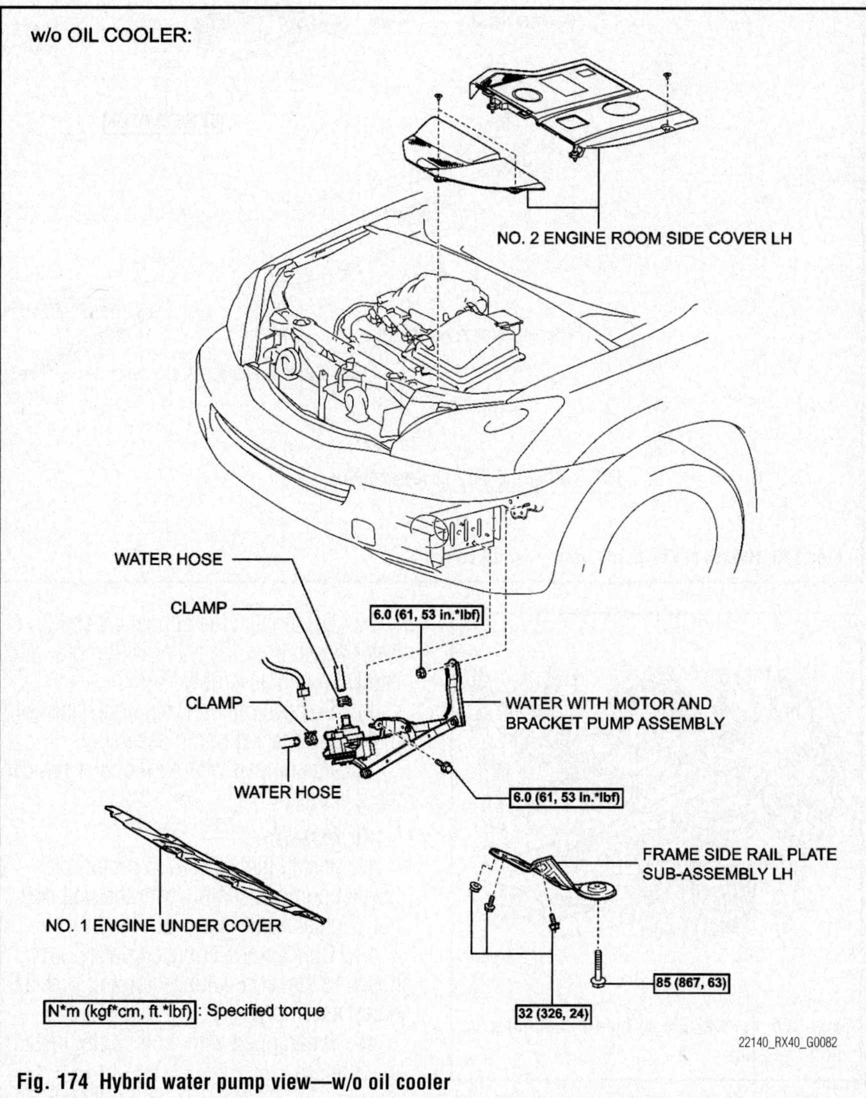

Fig. 174 Hybrid water pump view—w/o oil cooler

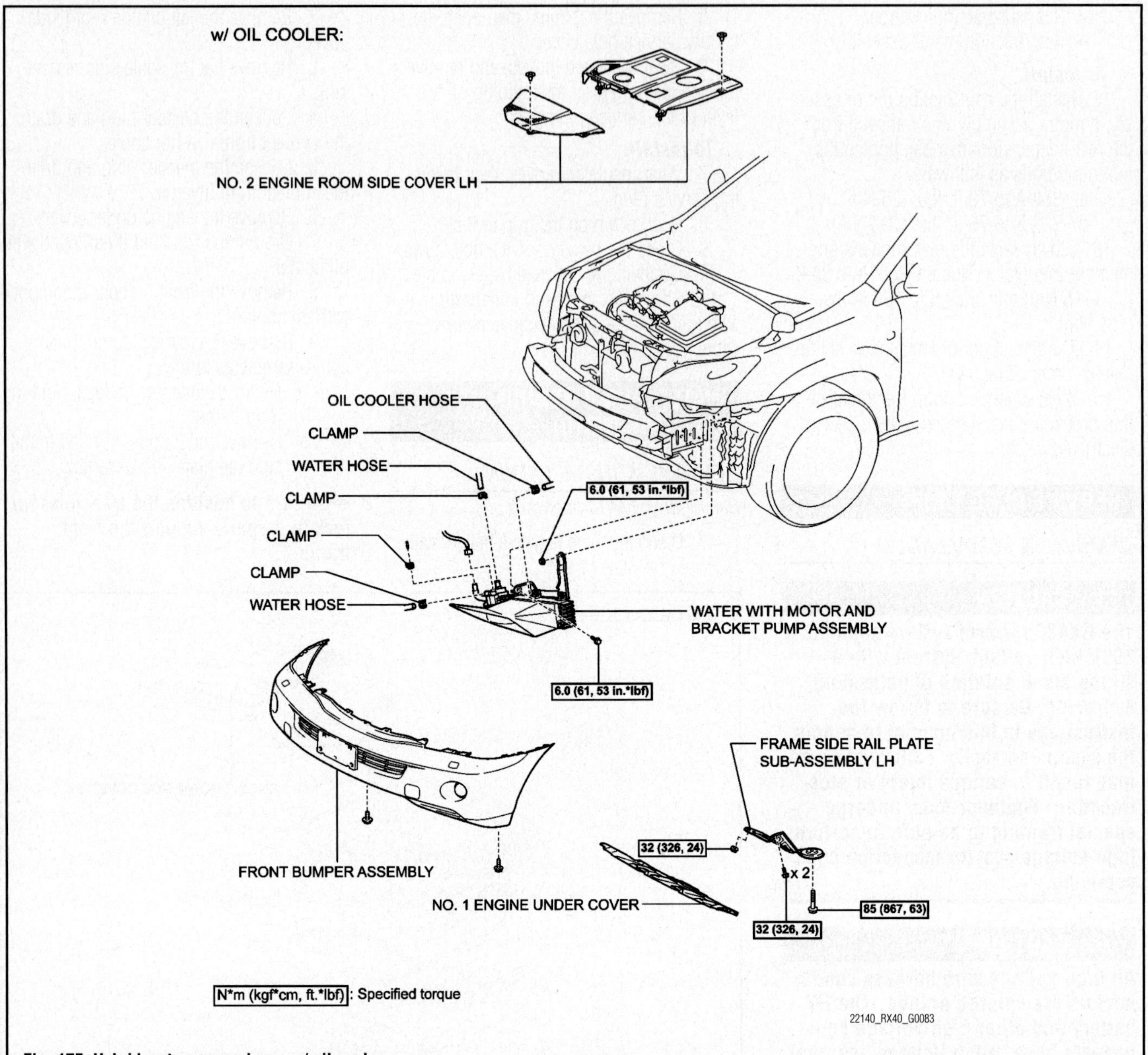

w/ OIL COOLER:

NO. 2 ENGINE ROOM SIDE COVER LH

OIL COOLER HOSE

CLAMP

WATER HOSE

CLAMP

6.0 (61, 53 in.*lbf)

CLAMP

CLAMP

WATER HOSE

WATER WITH MOTOR AND BRACKET PUMP ASSEMBLY

6.0 (61, 53 in.*lbf)

FRAME SIDE RAIL PLATE SUB-ASSEMBLY LH

32 (326, 24)

x 2

85 (867, 63)

32 (326, 24)

FRONT BUMPER ASSEMBLY

NO. 1 ENGINE UNDER COVER

N*m (kgf*cm, ft.*lbf): Specified torque

22140_RX40_G0083

Fig. 175 Hybrid water pump view—w/ oil cooler

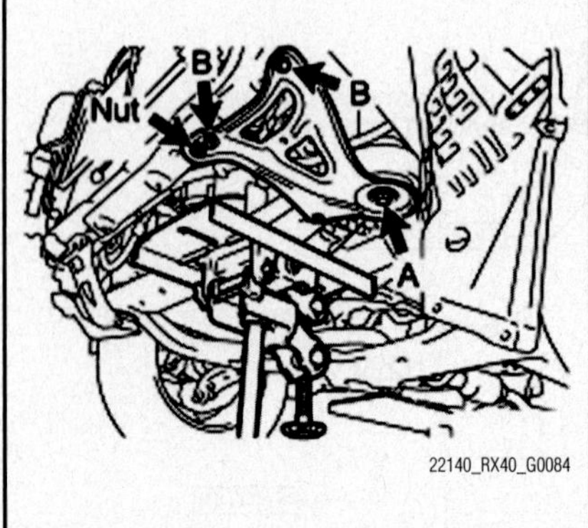

Fig. 176 Frame side rail plate mounting bolts and nut

10. Disconnect the connector and 2 water hoses from the water with motor and bracket pump assembly.
11. Remove the bolt, nut and water with motor and bracket pump assembly.
12. If equipped with a oil cooler remove the 2 hoses.

To install:

13. Install the water with motor and bracket pump assembly with the bolt and nut. Tighten to 53 inch. lbs. (6 Nm).
14. Connect the connector and 2 water hoses to the water with motor and bracket pump assembly.
15. If equipped with a oil cooler install the 2 hoses.
16. Install the frame side rail plate sub-assembly with the 3 bolts and nut.
17. Tighten the bolt (A) to 63 ft. lbs. (85 Nm).
18. Tighten the bolts (B) and nut to 24 ft. lbs. (32 Nm).
19. Install bumper if removed for oil cooler. Inspect fluid level for hybrid transaxle.
20. Add engine coolant to inverter.
21. Connect the negative battery cable.
22. Check for coolant leaks.
23. Check oil cooler lines if removed.
24. Install the engine undercover.
25. Install the engine room left side cover.

STEERING

⁂ CAUTION

The RX400h hybrid system contains a 288V high-voltage system with a strong alkali solution of potassium hydroxide. Be sure to follow the instructions in this manual to handle the system correctly. Failure to do so may result in serious injury or electrocution. Engineer must undergo special training to be able to perform high-voltage system inspection and servicing.

⁂ CAUTION

All high-voltage wire harness connectors are colored orange. The HV battery and other high-voltage components have "High Voltage" caution labels. Do not carelessly touch these wires and components.

⁂ CAUTION

Before inspecting or servicing the high-voltage system, be sure to follow safety measures, such as wearing insulated gloves and removing the service plug to prevent electrocution. Carry the removed service plug in your pocket to prevent anyone from reinstalling it while you are servicing the vehicle.

⁂ CAUTION

After removing the service plug, wait 5 minutes before touching any of the high-voltage connectors and terminals.

POWER RACK & PINION STEERING GEAR

REMOVAL & INSTALLATION

See Figures 177 and 178.

1. Before servicing the vehicle, refer to the Precautions Section.
2. Center the steering wheel to place the front wheels facing straight ahead.
3. Disconnect the negative battery cable.
4. Remove the wheels.
5. Separate the front stabilizer links LH and RH.
6. Fix the steering wheel with the seat belt in order to prevent rotation.

➡This operation is useful to prevent damage to the spiral cable.

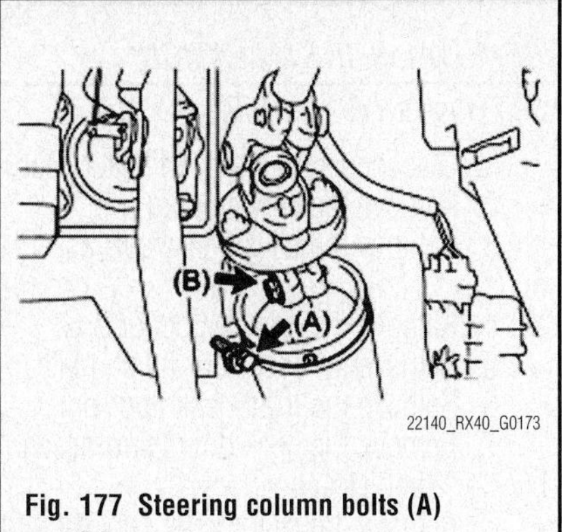

Fig. 177 Steering column bolts (A) and (B)

7. Loosen the bolt (A) and remove the clamp from the steering column hole shield.
8. Loosen the bolt (B).
9. Remove the bolt and then slide the intermediate shaft sub-assembly.

➡Do not separate the steering intermediate shaft sub-assembly.

10. Matchmark and disconnect the intermediate shaft.
11. Remove the cotter pin, nut and separate the tie rods LH and RH.
12. Use a puller and be careful not to damage any related components.
13. Remove the engine/transaxle assembly.

➡ Refer to the instructions for removal of the engine assembly.

14. Remove the 2 bolts, 2 nuts, and the power steering link assembly from the front frame assembly.

To install:

15. Install the power steering gear to the front frame assembly. Tighten the mounting bolts and nuts to 52 ft. lbs. (70 Nm).
16. Install the engine/transaxle assembly.
17. Connect the tie rod assembly LH and RH to the steering knuckle with the nut.

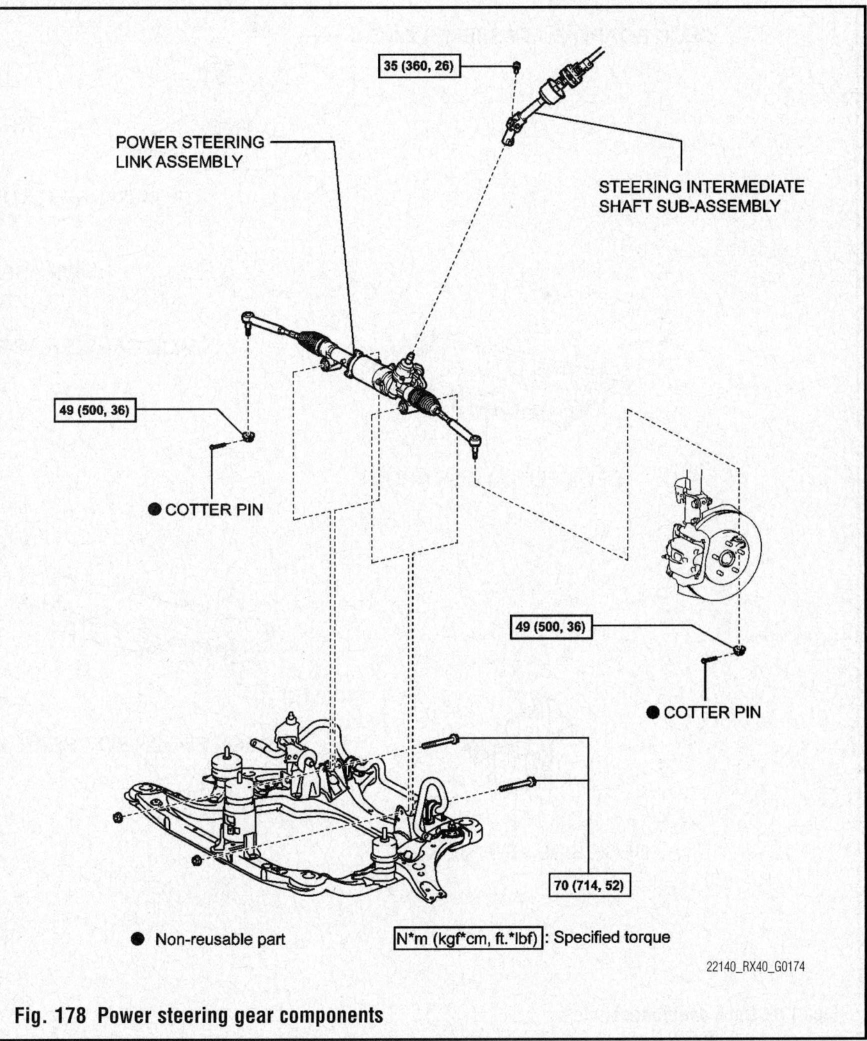

Fig. 178 Power steering gear components

18. Tighten the retaining castle nut to 36 ft. lbs. (49 Nm) and install new cotter pins.
19. Align matchmarks on the steering intermediate shaft sub-assembly and the power steering link assembly. Tighten the shaft bolt to 26 ft. lbs. (35 Nm)
20. Install the clamp to the steering column hole shield and tighten the bolt to 26 ft. lbs. (35 Nm).
21. Install the front stabilizer link assembly with the nut and tighten to 55 ft. lbs. (74 Nm).
22. Install the front wheels and tighten to 76 ft. lbs. (103 Nm).
23. Check front end alignment.
24. Perform the initialization procedure, as outlined in the Chassis Electrical System.

POWER STEERING DC-DC CONVERTER

REMOVAL & INSTALLATION

See Figures 179 and 180.

1. Disconnect the negative battery cable.
2. Remove the service plug grip.
3. Remove the deck board sub-assembly.
4. Remove the rear deck floor box.
5. Remove the rear floor finish plate.
6. Remove the front deck floor box.
7. Remove the deck board sub-assembly No. 2 and No. 3.
8. Remove the jack carrier support and jack assembly.
9. Remove the LH and RH side trim box.
10. Separate the wire harness clamp from power steering converter cover.
11. Remove the 3 nuts and the power steering converter cover.
12. Disconnect the 3 connectors from DC-DC converter assembly.
13. Remove the nut and separate wire harness clamp.
14. Remove the 4 nuts and DC-DC converter assembly.
15. Separate the earth wire clamp from the DC-DC converter assembly.
16. Disconnect the earth wire connector from the DC-DC converter assembly.
17. Remove the nut and the earth wire from the DC-DC converter assembly.

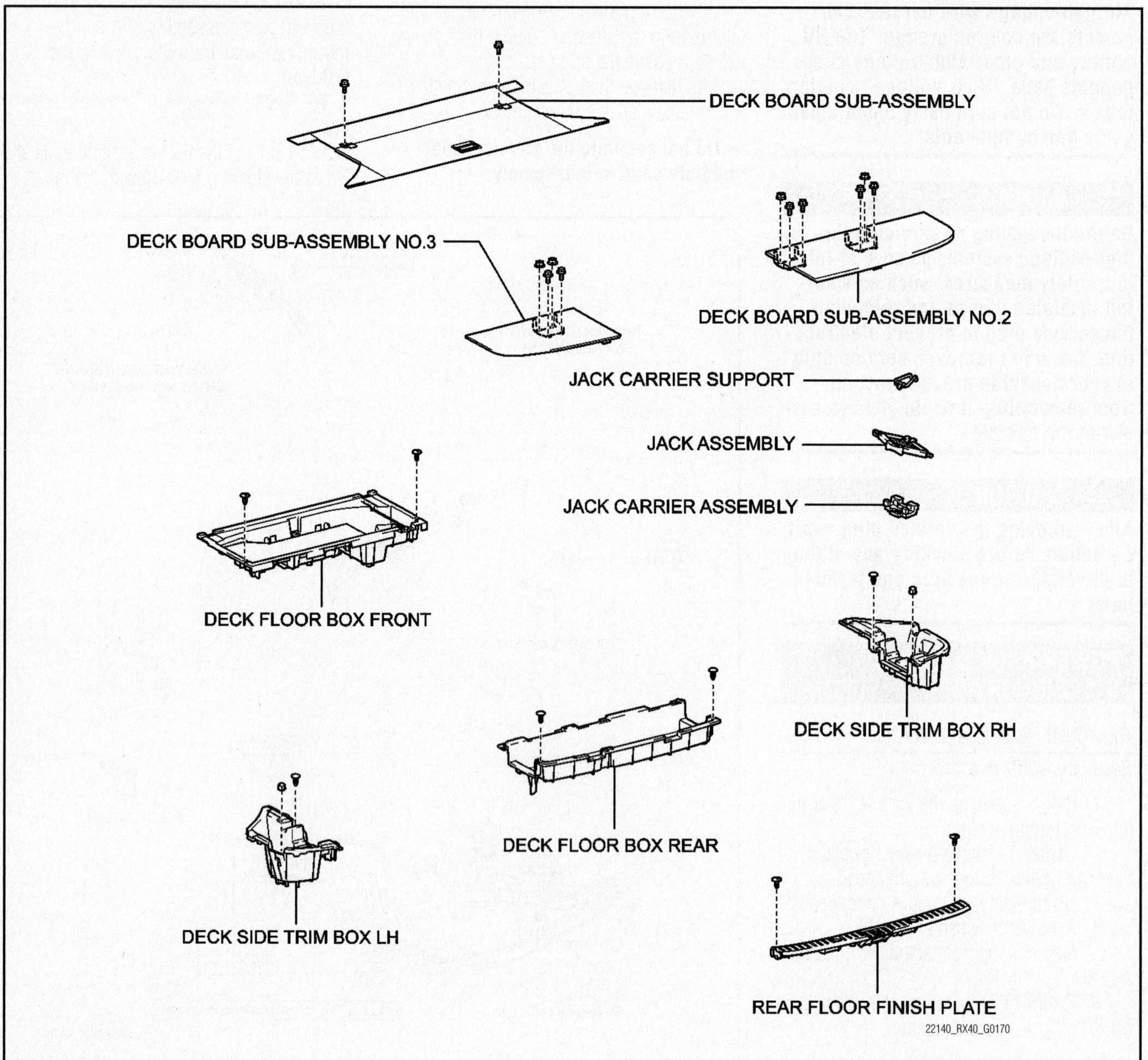

Fig. 179 Deck components view

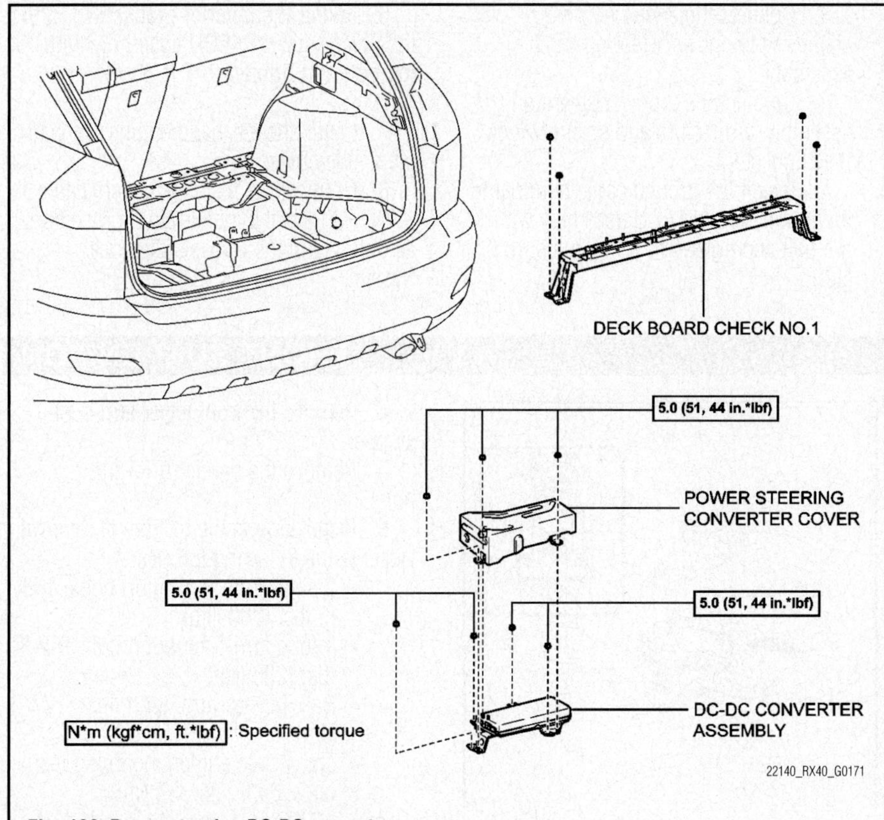

Fig. 180 Power steering DC-DC converter

To install:

18. Install the earth wire to the DC-DC converter assembly with the nut. Tighten the nut to 49 inch. lbs. (5.5 Nm).
19. Connect the earth wire clamp to the DC-DC converter assembly.
20. Connect the earth wire connector to the DC-DC converter assembly.
21. Install the DC-DC converter assembly with the 4 nuts. Tighten the nuts to 44 inch. lbs. (5.0 Nm).
22. Connect the 3 connectors to the DC-DC converter assembly.
23. Connect the wire harness clamp to the DC-DC converter assembly.
24. Install the nut and tighten to 49 inch. lbs. (5.5 Nm).
25. Check that the high voltage cable connector is securely connected to the DC-DC converter assembly.
26. Install the power steering converter cover with the 3 nuts and tighten to 44 inch. lbs. (5.0 Nm).
27. Install the wire harness clamp to the power steering converter cover.
28. Install the deck board check No. 1 with the 4 nuts.
29. Install the LH and RH side trim box.
30. Install the jack assembly and carrier support.
31. Install the deck board sub-assembly No. 2 and No. 3.
32. Install the front floor deck.
33. Install the rear floor finish plate.
34. Install the rear floor deck.
35. Install the deck board sub-assembly.
36. Install the service plug grip.
37. Connect the negative battery cable.
38. Some systems need initialization when disconnecting the cable from the negative battery terminal

POWER STEERING ECU

REMOVAL & INSTALLATION

See Figure 181.

1. Disconnect the negative battery cable.
2. Remove the left-hand engine compartment side cover.
3. Remove the bolt and inverter bracket No. 5.
4. Remove the power steering ECU assembly, as follows:
    a. Remove the bolt and ground cable terminal from the power steering ECU assembly.
    b. Release the locks of the 2 power steering ECU assembly connectors and disconnect the connectors.

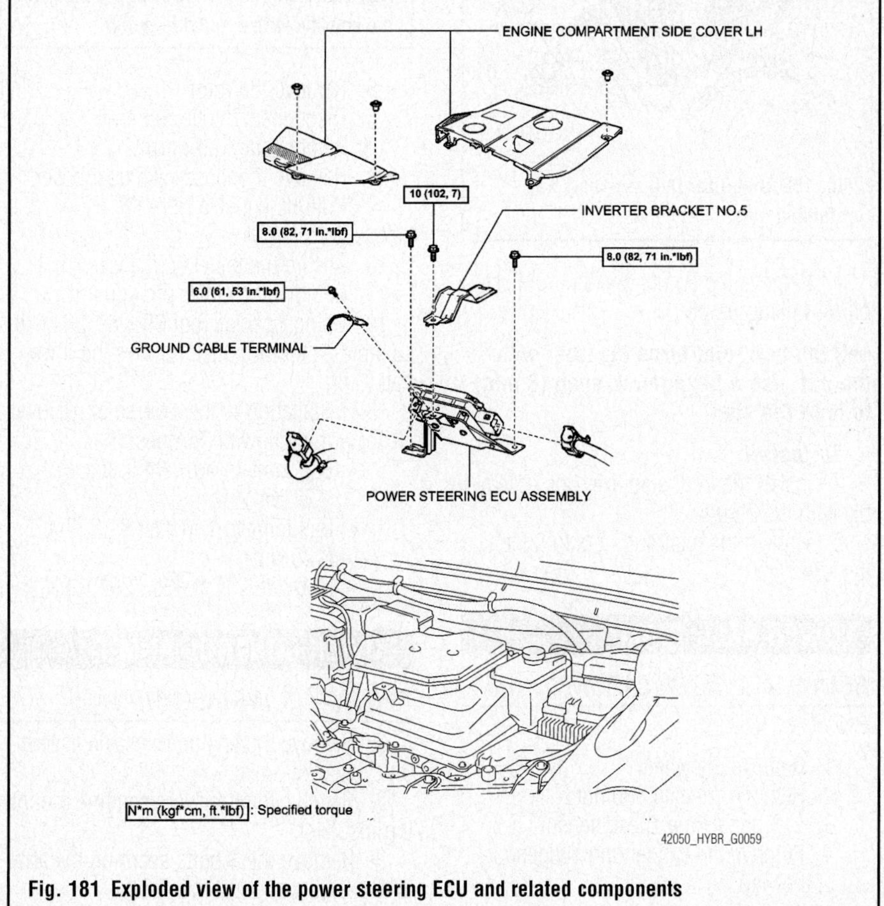

Fig. 181 Exploded view of the power steering ECU and related components

c. Separate the 2 wire harness clamps from the power steering ECU assembly.

d. Remove the 2 bolts and the power steering ECU assembly.

To install:

5. Install power steering ECU, as follows:

a. Install the power steering ECU assembly with the 2 bolts and tighten to 71 inch lbs. (8 Nm).

b. Connect the 2 wire harness clamps to the power steering ECU assembly.

c. Connect the 2 power steering ECU assembly connectors and securely lock the connectors.

d. Install the ground cable terminal to the power steering ECU assembly with the bolt and tighten to 53 inch lbs. (6.0 Nm).

6. Install the inverter bracket No. 5 to the power steering ECU assembly with the bolt and tighten to 7 ft. lbs. (10 Nm).

7. Install the left-hand engine compartment side cover.

8. Connect the negative battery cable..

9. Perform the initialization procedure, as outlined in the Chassis Electrical System.

SUSPENSION

FRONT SUSPENSION

COIL SPRING

REMOVAL & INSTALLATION

Refer to MacPherson Strut overhaul.

CONTROL LINKS

REMOVAL & INSTALLATION

See Figure 182.

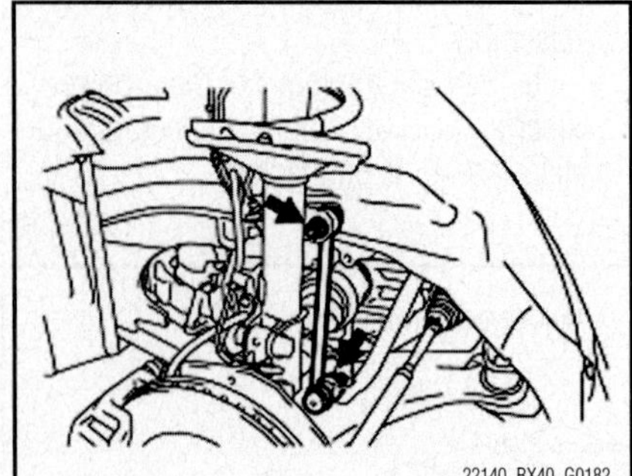

Fig. 182 Stabilizer link assembly and retaining nuts

1. Remove the 2 nuts and the front stabilizer link assembly.

➡If the ball joint turns together with the nut, use a hexagon wrench (6 mm) to hold the stud.

To install:

2. Install the front stabilizer link assembly with the 2 nuts.

3. Tighten the retaining nuts to 55 ft. lbs. (74 Nm).

LOWER BALL JOINT

REMOVAL & INSTALLATION

See Figure 183.

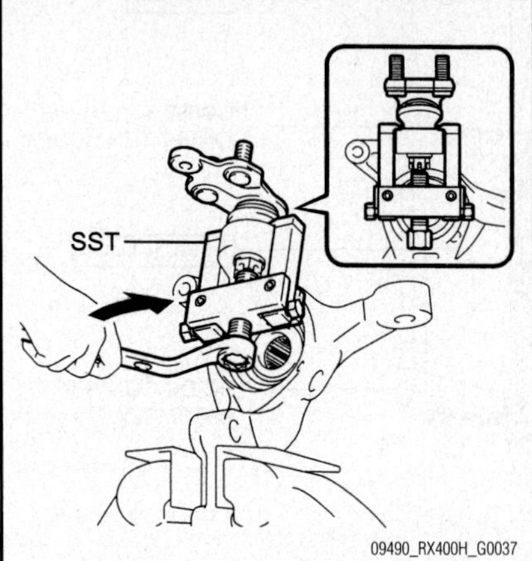

Fig. 183 Remove the lower ball joint from the knuckle with a suitable puller.

1. Remove the wheel.
2. Remove the axle hub nut.
3. Disconnect the speed sensor.
4. Remove the caliper and suspend it out of the way.
5. Remove the rotor.
6. Disconnect the tie rod end.
7. Remove the lower arm.
8. Remove the disc brake dust cover.
9. Pull the knuckle from the halfshaft.
10. Secure the knuckle in a vise.
11. Remove the cotter pin and nut.
12. Using Special Tool 09628-62011 or suitable ball joint puller, remove the lower ball joint.
13. Installation is the reverse of removal. Observe the following torques:
    - Ball joint-to-arm: 94 ft. lbs. (127 Nm)
    - Ball joint-to-knuckle: 91 ft. lbs. (123 Nm)
    - Hub nut: 217 ft. lbs. (294 Nm)

LOWER CONTROL ARM

REMOVAL & INSTALLATION

1. Remove the engine/transaxle assembly.
2. Remove the transverse engine mounting insulator.
3. Remove the 3 bolts securing the arm to the engine support member.
4. Remove the front lower arm bush stopper.
5. Remove the ball joint-to-arm bolts.
6. Installation is the reverse of removal. Observe the following torques:
    - 2 short arm-to-support bolts: 148 ft. lbs. (200 Nm)
    - 1 long arm-to-support bolt: 152 ft. lbs. (206 Nm)
    - Ball joint-to-arm: 94 ft. lbs. (127 Nm)
    - Transverse engine mounting insulator: 64 ft. lbs. (87 Nm)

MACPHERSON STRUT

REMOVAL & INSTALLATION

1. Before servicing the vehicle, refer to the Precautions Section.
2. Remove the wheel.
3. Disconnect the stabilizer bar link.
4. Loosen, don't remove, the strut locknut.
5. Disconnect the brake hose from the strut.
6. Remove the lower mounting bolts.
7. Remove the upper retaining nuts.

To install:

8. Position the strut and install the upper nuts. Torque to 59 ft. lbs. (80 Nm).
9. Install the lower bolts and torque to 214 ft. lbs. (290 Nm).
10. Connect the brake line.
11. Tighten the strut locknut to 36 ft. lbs. (49 Nm).
12. Connect the stabilizer links and torque to 55 ft. lbs. (74 Nm).
13. Install the wheel. Torque to 76 ft. lbs. (103 Nm).

STABILIZER BAR

REMOVAL & INSTALLATION

See Figure 184.

1. Before servicing the vehicle, refer to the Precautions Section.

2. Make sure the vehicle's front wheels are in the straight-ahead position.

3. Disconnect the negative battery cable.

4. Raise and safely support the vehicle.

5. Remove the remove front wheel and tire assemblies.

6. Remove the engine under cover.

7. Separate the steering intermediate shaft subassembly.

8. Separate the tie rod.

9. Remove the 2 nuts, then remove the front left stabilizer link assembly:

➡If the ball joint turns together with the nut, use a hexagon wrench (6 mm) to hold the stud.

10. Remove the 2 nuts, then remove the front right stabilizer link assembly:

➡If the ball joint turns together with the nut, use a hexagon wrench (6 mm) to hold the stud.

11. Remove the 2 bolts, then remove the left stabilizer bracket No. 1 from the front frame assembly.

12. Remove the 2 bolts, then remove the right stabilizer bracket No. 1 from the front frame assembly.

13. Remove the left and right No. 2 stabilizer brackets from the bushings.

14. Remove the 2 stabilizer bar bushings No. 1 from the stabilizer bar.

15. Remove the front No. 3 exhaust pipe subassembly

16. Remove the front exhaust pipe/

17. Remove the bolt, nut and the manifold stay.

18. Remove the power steering link.

19. Remove the front stabilizer bar from the right side of the vehicle.

20. Inspect the turning of the stabilizer link ball joint:

a. secure the front stabilizer link assembly in a vise using aluminum plates.

b. Install the nut to the front stabilizer link assembly stud.

c. If using a torque wrench, turn the nut continuously at a rate of 3 to 5 seconds per turn and take the torque reading on the 5th turn. The turning torque should be 18 inch lbs. (2.0 Nm) or less.

➡If the turning torque is not within the specified range, replace the front stabilizer link assembly with a new one.

21. Inspect the dust cover:

a. Check that the dust cover is not cracked and that there is no grease on it.

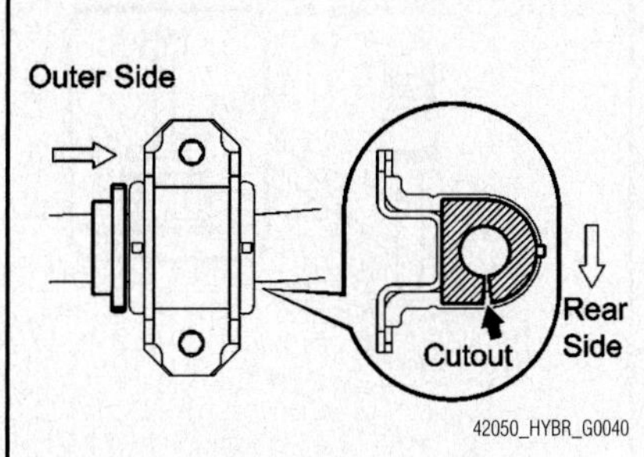

Fig. 184 Place the cutout of the front stabilizer bar bushing No. 1 as facing the rear side as shown

To install:

22. Install the front stabilizer bar by inserting it from the right side of the vehicle.

23. Install the power steering link.

24. Install the manifold stay with the bolt and nut. Tighten to 25 ft. lbs. (34 Nm).

25. Install the front exhaust pipe.

26. Install the front No. 3 exhaust pipe subassembly.

27. Install the 2 front stabilizer bar bushings No. 1 to the outer side of the bushing stopper on the front stabilizer bar.

➡Place the cutout of the front stabilizer bar bushing No. 1 as facing the rear side as shown in the illustration.

28. Install the right and left No. 2 stabilizer brackets to the stabilizer bushings.

29. Install the right and left No. 1 stabilizer bracket to the front frame assembly with the bolts.

Tighten to 12 ft. lbs. (16 Nm).

30. Install the right and left stabilizer link assemblies with the nuts and tighten to 55 ft. lbs. (75 Nm).

➡If the ball joint turns together with the nut, use a hexagon (6 mm) wrench to hold the stud.

31. Connect the tie rod.

32. Connect steering intermediate shaft subassembly.

33. Make sure the front wheels are facing straight-ahead.

34. Install the engine under cover.

35. Install the front wheels.

36. Connect the negative battery cable.

37. If necessary, perform the initialization procedure, as outlined in the Chassis Electrical System.

38. Initialize the rotation angle sensor and calibrate torque sensor zero point.

39. Inspect and adjust the front wheel alignment.

WHEEL HUB AND BEARING

REMOVAL & INSTALLATION

See Figures 185 through 191.

1. Before servicing the vehicle, refer to the Precautions Section.

2. Remove the front wheel.

3. Remove the front axle hub nut.

4. Separate the front speed sensor.

5. Separate the front disc brake caliper assembly.

6. Remove the front brake rotor.

7. Separate and remove the tie rod end.

8. Remove the bolt and 2 nuts, and separate the front suspension arm subassembly lower No. 1 from the lower ball joint.

9. Using a plastic hammer, separate the drive shaft from the axle hub.

10. Remove the 2 bolts, nuts and steering knuckle.

WARNING

Be careful not to damage the boot and speed sensor rotor.

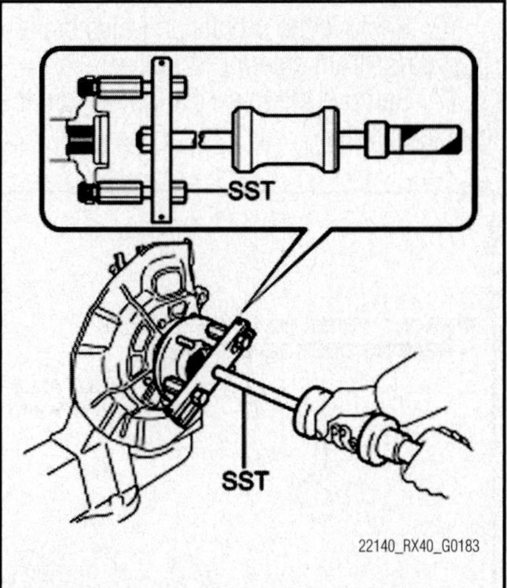

Fig. 185 Front axle hub sub-assembly removal

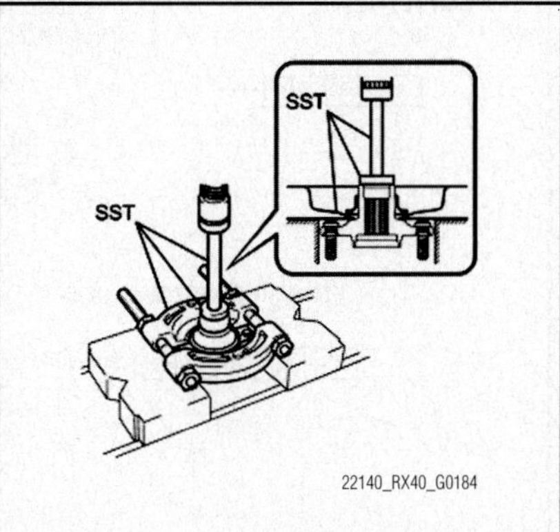

Fig. 186 Remove the bearing inner race

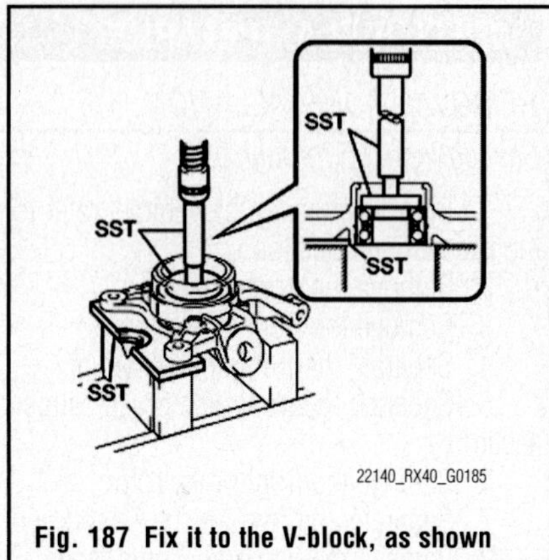

Fig. 187 Fix it to the V-block, as shown

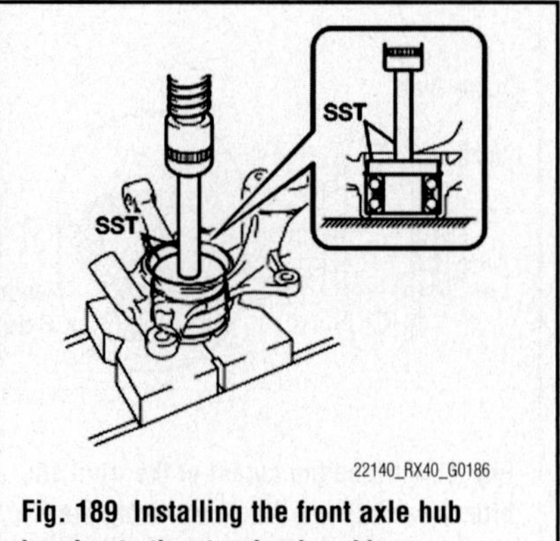

Fig. 189 Installing the front axle hub bearing to the steering knuckle

11. Using a screwdriver with its tip wrapped with vinyl tape, remove the bearing dust deflector No. 1.

12. Using snap ring pliers, remove the front axle hub hole snap ring.

13. Hold the front axle assembly between aluminum plates in a vise.

14. Using SST 09520-00031, remove the front axle hub sub-assembly.

15. Using SST and a press, remove the bearing inner race (outside) from the front axle hub sub-assembly.

16. Remove the 4 bolts and disc brake dust cover from steering knuckle.

17. Remove the lower ball joint from the hub assembly.

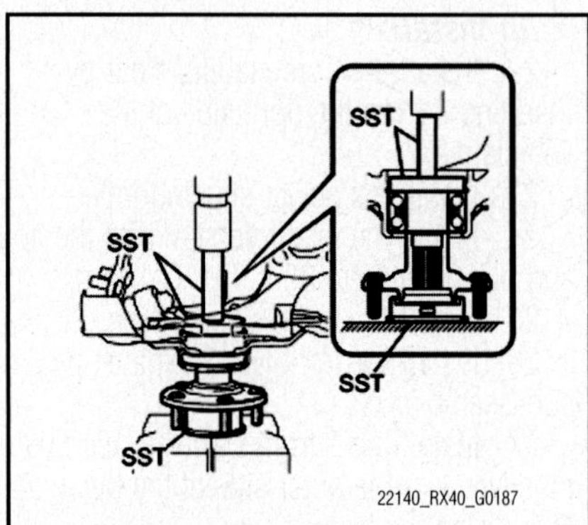

Fig. 190 Installing the front axle hub sub-assembly

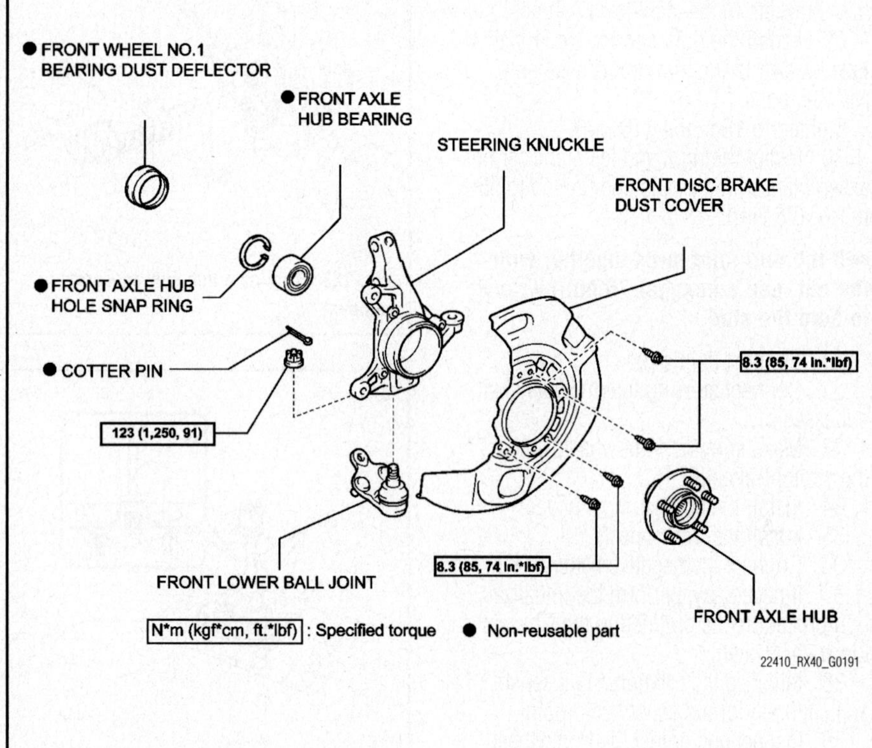

Fig. 188 Hub and bearing assembly view

18. Remove the front axle hub bearing as follows:

19. Place the bearing inner race (outside) on the front axle hub bearing.

20. Using SST and a press, press the front axle hub bearing until it contacts the SST.

21. Using SST to make the steering knuckle horizontal, fix it to the V-block, as shown in the illustration.

22. Using SST and a press, remove a front axle hub bearing to the steering knuckle.

To install:

23. Using SST and a press, install a new front axle hub bearing to the steering knuckle.

24. Install the lower ball joint into the steering knuckle and tighten castle nut to 91 ft. lbs. (93 Nm).

25. Install the disc brake dust cover to the steering knuckle with the 4 bolts. Tighten the bolts to 73 inch. lbs. (8.3 Nm).

26. Using SST and a press, install the front axle hub sub-assembly.

27. Using snap ring pliers, install a new front axle hub hole snap ring.

28. Using SST and a hammer, install the bearing dust deflector No. 1.

➡Align the hole for the speed sensor in the bearing dust deflector No. 1 with the steering knuckle.

29. Install the front axle assembly to the front drive shaft assembly.

30. Install the front axle assembly to the front shock absorber assembly with the 2 bolts and nuts. Tighten the bolts and nuts to 217 ft. lbs. (290 Nm).

31. Install the lower ball joint to the front suspension arm sub-assembly lower with the bolt and 2 nuts. Tighten the nuts and bolts to 94 ft. lbs. (127 Nm).

32. Install the tie rod end to the steering knuckle with the nut. Tighten the nut to 36 ft. lbs. (49 Nm).

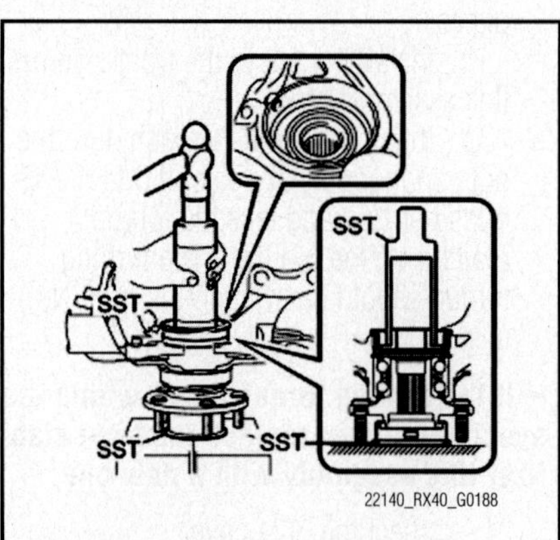

Fig. 191 Installing the bearing dust deflector No. 1.

33. Install the front brake rotor.

34. Install the front brake caliper assembly.

35. Install the front axle hub nut and tighten to 217 ft. lbs. (294 Nm). Using a chisel and hammer, stake the axle hub nut.

36. Install the speed sensor to the steering knuckle with the bolt and tighten to 71 inch. lbs. (8 Nm).

37. Install the front wheel and tighten to 76 ft. lbs. (103 Nm).

38. Adjust the front wheel alignment.

39. Check the speed sensor operation.

ADJUSTMENT

The front wheel bearing is a not adjustable.

SUSPENSION — REAR SUSPENSION

COIL SPRING

REMOVAL & INSTALLATION

Refer to MacPherson Strut overhaul.

CONTROL ARMS/LINKS

REMOVAL & INSTALLATION

Rear No. 1 Suspension Arm (2WD)

See Figures 192 and 193.

1. Remove the rear wheel.
2. Remove the Stabilizer link links RH and LH.
3. Remove the rear stabilizer bar, bracket and bushing.
4. Remove the bolt and disconnect the rear suspension arm No. 1 (inner side).
5. Remove the bolt, nut and the rear suspension arm No. 1 (outer side) from the rear axle carrier.

To install:

6. Install the rear suspension arm No. 1 (inner side) with the bolt, and temporarily tighten the bolt.
7. Connect the rear suspension arm No. 1 (outer side) to the rear axle carrier with the bolt and nut, and temporarily tighten the bolt and nut.
8. Stabilize the suspension.
9. Fully tighten the outer suspension arm bolt to 83 ft. lbs. (112 Nm).
10. Fully tighten the inner suspension arm bolt to 89 ft. lbs. (120 Nm).

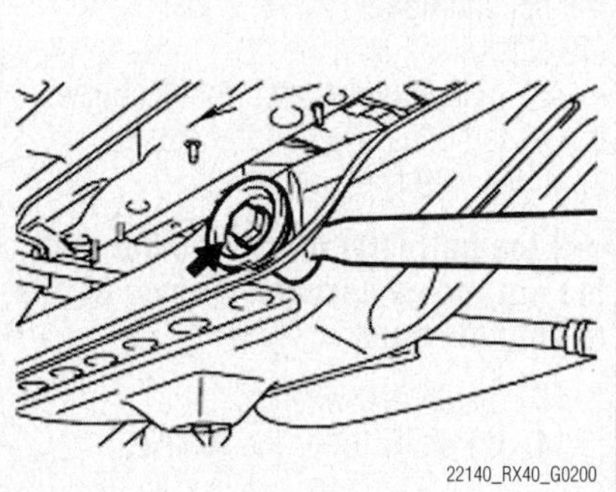

Fig. 192 Rear suspension arm No. 1 (inner side)

Rear No. 2 Suspension Arm (2WD)

See Figure 194.

1. Remove the bolt, and disconnect the rear suspension arm No. 2 (inner side).
2. Remove the bolt, nut and the rear suspension arm No. 2 (outer side) from the rear axle carrier. When removing the bolt, keep the nut from rotating.

To install:

3. Install the rear suspension arm No. 2 (inner side) with the bolt, and temporarily tighten the bolt.
4. Connect the rear suspension arm No. 2 (outer side) to the rear axle carrier with the bolt and nut, and temporarily tighten the bolt.
5. Stabilize the suspension.
6. Fully tighten the outer suspension arm bolt to 83 ft. lbs. (112 Nm).
7. Fully tighten the inner suspension arm bolt to 89 ft. lbs. (120 Nm).

Rear No. 1 Suspension Arm (4WD)

See Figure 195.

1. Remove the rear wheel.
2. Remove the nut, and separate the parking brake cable assembly No. 3.
3. Support the rear axle carrier with a jack.
4. Remove the bolt and nut, and disconnect the strut rod (front side).
5. Remove the 2 bolts, 2 nuts, and rear suspension arm No. 1.

To install:

6. Install the rear suspension arm No. 1 (inner side) with the bolt, and temporarily tighten the bolt.
7. Set the rear suspension arm No. 1 in the position shown in the illustration, and fully tighten the bolt to 59 ft. lbs. (80 Nm).
8. Install the rear suspension arm No. 1 (outer side) to the rear axle carrier, and temporarily tighten the bolt.
9. Install the strut rod (rear side), bolt and nut, and temporarily tighten the bolt.
10. Stabilize the suspension.
11. Fully tighten the outer suspension arm bolt to 83 ft. lbs. (112 Nm).
12. Fully tighten the rear strut rod to 59 ft. lbs. (80 Nm).
13. Install the parking brake cable assembly No. 3 with the nut. Tighten to 53 inch. lbs. (6 Nm).
14. Install the rear wheel and tighten to 76 ft. lbs. (103 Nm).

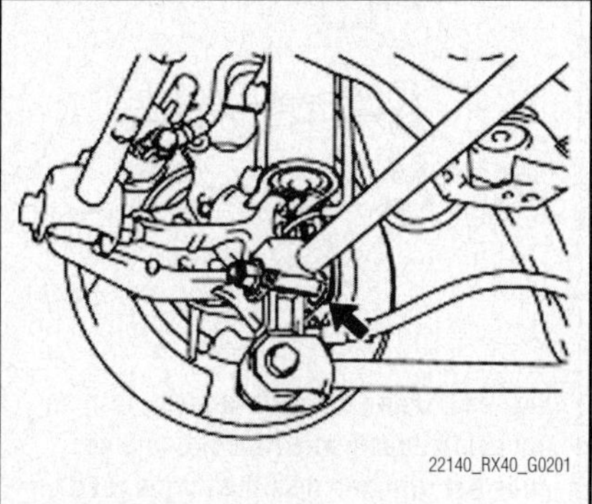

Fig. 193 Rear suspension arm No. 1 (outer side)

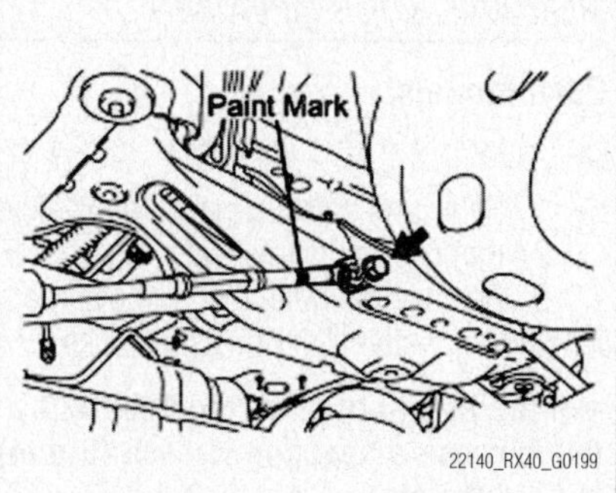

Fig. 194 Rear suspension arm No. 2 (inner side).

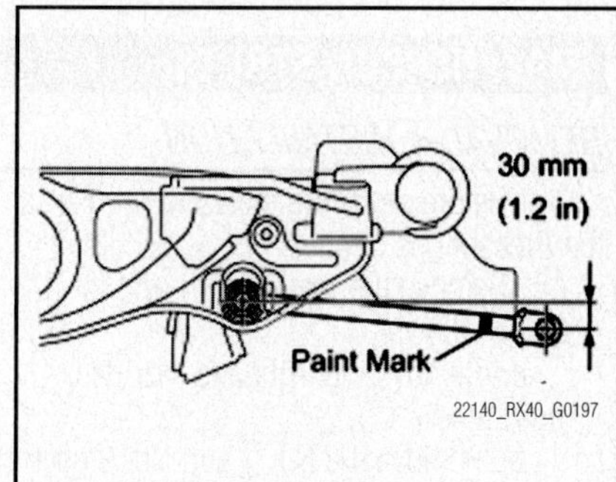

Fig. 195 Set the rear suspension arm No. 1 in the position shown

Rear No. 2 Suspension Arm (4WD)

See Figure 196.

1. Remove the rear wheel.
2. Put matchmarks on the adjust cams and rear suspension member sub-assembly.
3. Remove the nut, camber adjust cam and toe adjust cam, and disconnect the rear suspension arm No. 2 (inner side).
4. Remove the bolt, nut and the rear suspension arm No. 2 (outer side) from the rear axle carrier.

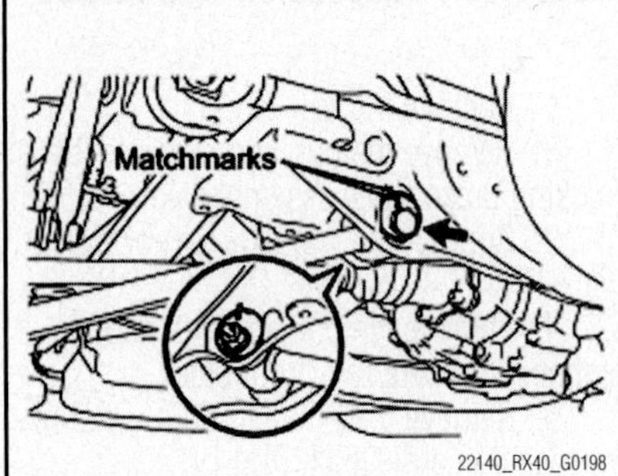

Fig. 196 Put matchmarks on the adjust cams

To install:

5. Install the rear suspension arm No. 2 (inner side) to the rear suspension member sub-assembly with the camber adjust cam and toe adjust cam, and temporarily tighten the nut.
6. Connect the rear suspension arm No. 2 (outer side) to the rear axle carrier with the bolt and nut.
7. Stabilize the suspension.
8. Connect the rear suspension arm No. 2 (outer side) to the rear axle carrier with the bolt and nut. Fully tighten the inner cam nut to 74 ft. lbs. (100 Nm).
9. Fully tighten the outer suspension arm bolt to 83 ft. lbs. (112 Nm).
10. Install the rear wheel and tighten to 76 ft. lbs. (103 Nm).

MACPHERSON STRUTS

REMOVAL & INSTALLATION

1. Before servicing the vehicle, refer to the Precautions Section.
2. Remove the rear interior trim for RX 400h as follows:
   a. Remove the front and rear deck floor box.
   b. Remove the No. 2 and No. 3 deck board assemblies.
   c. Remove rear floor finishing plate.
   d. Remove jack assembly.
   e. Remove left and right deck side trim box.
   f. Remove rear door scuff plate.
   g. Remove deck side trim cover.
   h. Remove rear floor finishing side plate.
   i. Remove deck trim side panel.
3. Remove the rear interior trim as follows:
   a. Remove the left side deck trim cover.
4. Remove the rear wheels.
5. Remove the stabilizer link from the strut.
6. On 2-wheel drive models, disconnect the skid control sensor wire and brake hose from the strut and carrier.
7. On 4-wheel drive models, disconnect the brake hose and speed sensor from the strut and carrier.
8. Loosen the 2 nuts at the lower end of the strut, but don't remove them.
9. Support the rear axle carrier with a jack.
10. Remove the 3 upper strut nuts and lower the axle.
11. Remove the lower strut bolts and nuts. Lift out the strut.

To install:

12. Position the strut and install the 3 upper nuts. Torque to 43 ft. lbs. (58 Nm).
13. Lift the axle and install the 2 lower bolts and nuts. Torque to 133 ft. lbs. (180 Nm).
14. Tighten the strut locknut to 36 ft. lbs. (49 Nm).
15. The remainder of installation is the reverse of removal. Observe the following torques:
    - Brake hose clamp: 14 ft. lbs. (19 Nm)
    - Wire-to-strut clamp: 44 inch lbs. (5 Nm)
    - Sensor clamp: 71 inch lbs. (8 Nm)
    - Stabilizer link: 29 ft. lbs. (39 Nm)

STABILIZER BAR

REMOVAL & INSTALLATION

2WD Models

See Figures 197 through 199.

1. Raise and safely support the vehicle.
2. Remove the rear wheels.
3. Remove the nuts, then remove the left and right stabilizer link assemblies.

➡If the ball joint turns together with the nut, use a hexagon wrench (5 mm) to hold the stud.

4. Remove the 4 bolts and the stabilizer bar.
5. Remove the 2 stabilizer bar brackets.
6. Remove the 2 stabilizer bushings from the rear stabilizer bar.
7. Inspect the turning of the stabilizer link ball joint:
   a. Secure the stabilizer link assembly in a vise using aluminum plates.
   b. Install the nut to the stabilizer link assembly stud.
   c. Using a torque wrench, turn the nut continuously at a rate of 3 to 5 seconds per turn and take the torque reading on the 5th turn. The turning torque should be 8.9 inch lbs. (1.0 Nm) or less.

➡If the turning torque is not within the specified range, replace the rear stabilizer link assembly with a new one.

8. Inspect the dust cover. Check that the dust cover is not cracked and that there is no grease on it.

To install:

9. Install the 2 stabilizer bushings to the outer side of the stopper ring on the stabilizer bar.
10. Install the 2 stabilizer bar brackets.
11. Install the stabilizer bar with the 4 bolts and tighten to 14 ft. lbs. (19 Nm).
12. Install the left and right stabilizer link assemblies. Tighten the nuts to 29 ft. lbs. (39 Nm).

➡If the ball joint turns together with the nut, use a hexagon wrench (5 mm) to hold the stud.

13. Install rear wheels.
14. Carefully lower the vehicle.

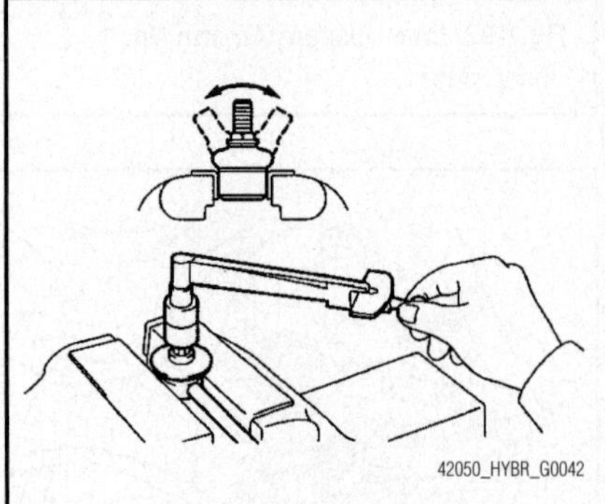

Fig. 197 Using a torque wrench, turn the nut continuously at a rate of 3 to 5 seconds per turn and take the torque reading on the 5th turn. The turning torque should be 8.9 inch lbs. (1.0 Nm) or less

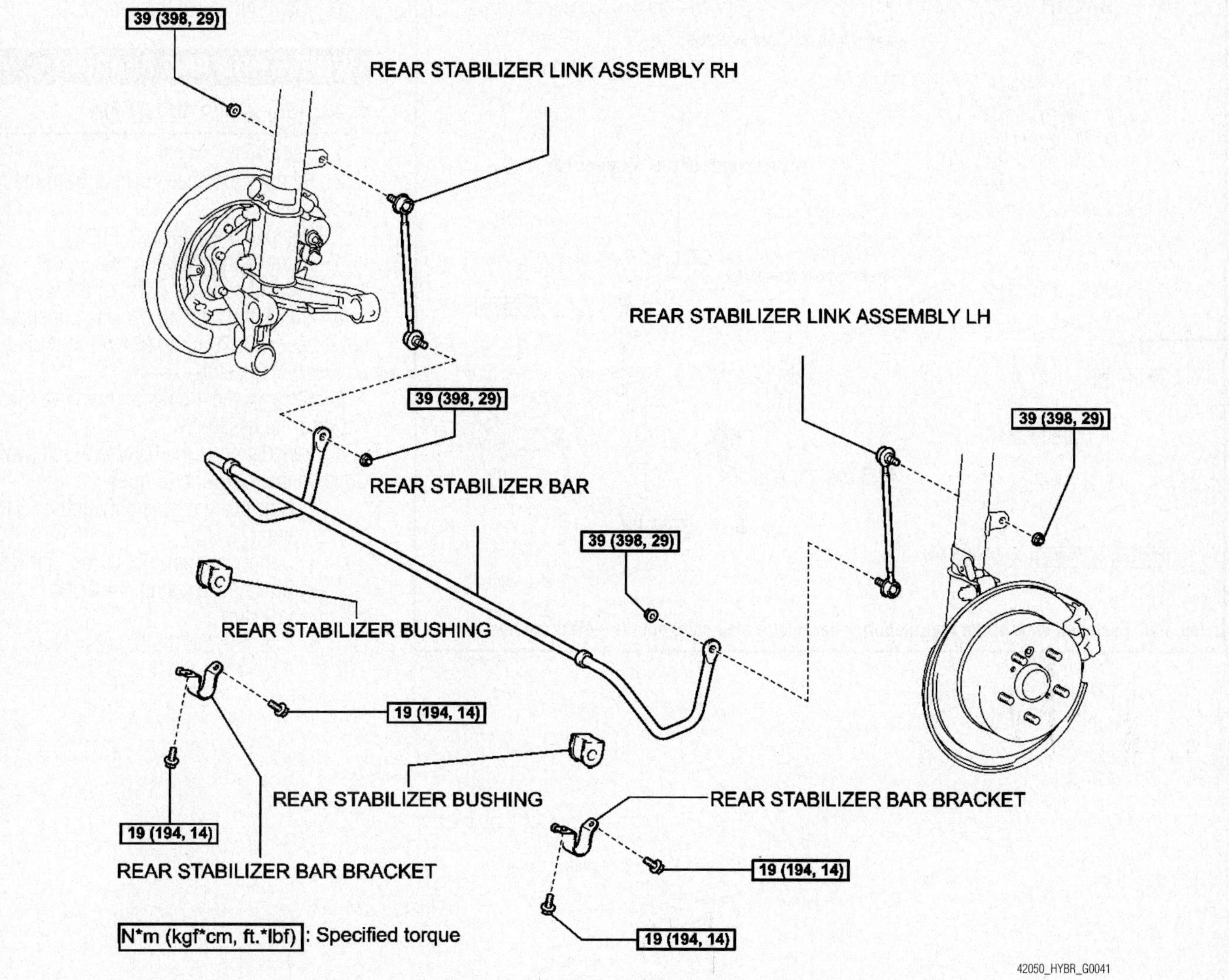

Fig. 198 Exploded view of the rear stabilizer bar and related components—2WD models

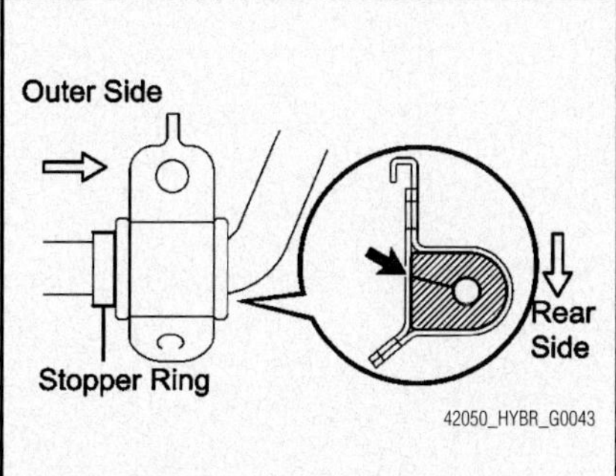

Fig. 199 Install the 2 stabilizer bushings to the outer side of the stopper ring on the stabilizer bar

4WD Models

See Figure 200.

1. Raise and safely support the vehicle.
2. Remove the rear wheels.
3. Remove the nuts, then remove the left and right stabilizer link assemblies.

➡If the ball joint turns together with the nut, use a hexagon wrench (5 mm) to hold the stud.

4. Remove the stabilizer bar:
   a. On the left side, remove the bolt.
   b. On the right side, remove the 2 bolts.
5. Remove the 2 stabilizer bar brackets.
6. Remove the 2 stabilizer bushings from the rear stabilizer bar.
7. Inspect the turning of the stabilizer link ball joint:
   a. Secure the stabilizer link assembly in a vise using aluminum plates.
   b. Install the nut to the stabilizer link assembly stud.
   c. Using a torque wrench, turn the nut continuously at a rate of 3 to 5 seconds per turn and take the torque reading on the 5th turn. The turning torque should be 8.9 inch lbs. (1.0 Nm) or less.

To install:

8. Install the stabilizer bar:
   a. Install the stabilizer bracket.
   b. On the left side, install the stabilizer bar with the bolt and tighten to 40 ft. lbs. (54 Nm).
   c. On the right side, install the stabilizer bar with the 2 bolts and tighten to 14 ft. lbs. (19 Nm).
9. Install the right and left stabilizer links and tighten the nuts to 29 ft. lbs. (39 Nm).

➡If the ball joint turns together with the nut, use a hexagon wrench (5 mm) to hold the stud.

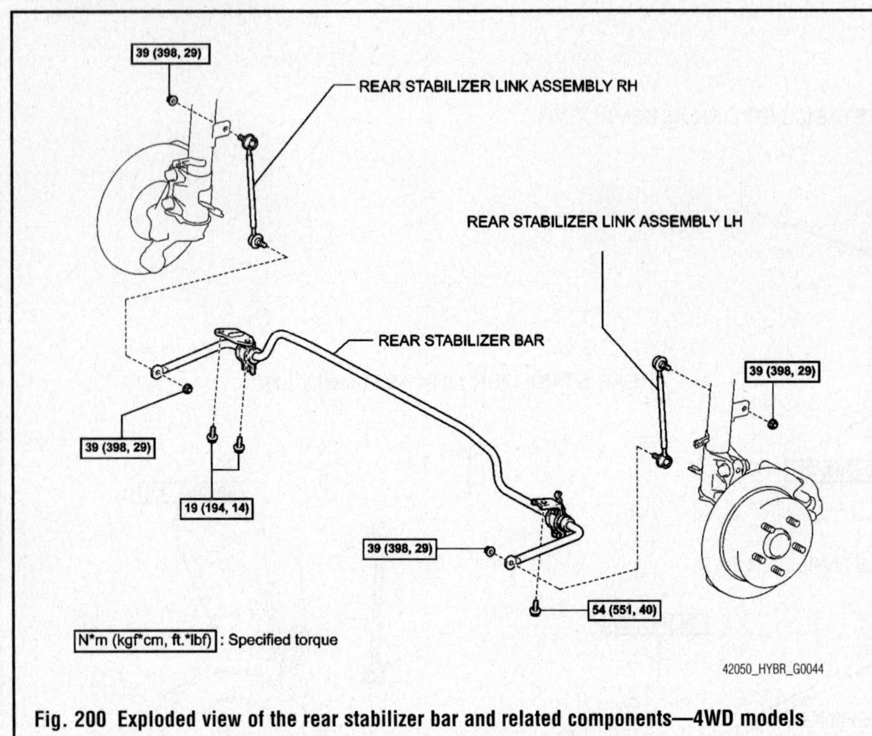

Fig. 200 Exploded view of the rear stabilizer bar and related components—4WD models

10. Install the rear wheels
11. Carefully lower the vehicle.

WHEEL HUB AND BEARING

REMOVAL & INSTALLATION

1. Remove the wheel.
2. Remove the speed sensor from the axle carrier.
3. Remove the axle shaft nut.
4. Remove the caliper and support assembly and suspend it out of the way.
5. Check the bearing backlash. It should not exceed 0.0020 in. (0.05mm). If it does, it must be replaced.
6. Remove the 4 bolts and remove the hub/bearing assembly.
7. Installation is the reverse of removal. Observe the following torques:
   - Hub/bearing assembly bolts: 55 ft. lbs. (75 Nm)
   - Caliper support: 58 ft. lbs. (78 Nm)
   - Brake hose clamp: 14 ft. lbs. (19 Nm)
   - Hub nut: 217 ft. lbs. (294 Nm)

SCION

tC • xB • xD

8

SPECIFICATIONS AND MAINTENANCE CHARTS

ENGINE AND VEHICLE IDENTIFICATION

Engine						
Code ①	Liters (cc)	Cu. In.	Cyl.	Fuel Sys.	Engine Type	Eng. Mfg.
2AZ-FE	2.4 (2362)	144.2	4	SFI	DOHC	Toyota
2ZR-FE	1.8 (1798)	109.7	4	SFI	DOHC	Toyota

Model Year	
Code ②	Year
7	2007
8	2008

SFI: Sequential Fuel Injection

DOHC: Double Overhead Camshaft

① Stamped on the left side of the engine block

② 10th digit of the Vehicle Identification Number (VIN)

22140_SCIO_C0001

GENERAL ENGINE SPECIFICATIONS

Year	Model	Engine Displacement Liters	Engine Series ID	Net Horsepower @ rpm	Net Torque @ rpm (ft. lbs.)	Bore x Stroke (in.)	Com-pression Ratio	Oil Pressure @ rpm
2007	tC	2.4	2AZ-FE	161@6000	163@4000	3.48x3.78	9.8:1	36-78@3000
2008	xB	2.4	2AZ-FE	161@6000	163@4000	3.48x3.78	9.8:1	36-78@3000
	tC	2.4	2AZ-FE	161@6000	163@4000	3.48x3.78	9.8:1	36-78@3000
	xD	1.8	2ZR-FE	128@6000	125@4400	3.17x3.48	10.0:1	36-78@3000

22140_SCIO_C0002

ENGINE TUNE-UP SPECIFICATIONS

Year	Engine Displacement Liters	Engine ID	Spark Plug Gap (in.)	Ignition Timing (deg.)	Fuel Pump (psi)	Idle Speed (rpm)	Valve Clearance	
							Intake	Exhaust
2007	2.4	2AZ-FE	0.043	5-12B	44-50	①	0.0075-0.0114	0.0150-0.0189
2008	1.8	1ZR-FE	0.043	8-12B	44-50	①	0.0060-0.0100	0.0100-0.0140
	2.4	2AZ-FE	0.043	5-12B	44-50	①	0.0075-0.0114	0.0150-0.0189

NOTE: The Vehicle Emission Control Information label often reflects specification changes made during production.

The label figures must be used if they differ from those in this chart.

B: Before top dead center

① M/T: 600-700
A/T: 650-750

22140_SCIO_C0003

CAPACITIES

Year	Model	Engine Displacement Liters	Engine ID	Engine Oil with Filter (qts.)	Transmission (pts.) Manual	Transmission (pts.) Auto	Front Drive Axle (pts.)	Fuel Tank (gal.)	Cooling System (qts.)
2007	tC	2.4	2AZ-FE	4.0	5.2	7.4	①	14.5	②
2008	xD	1.8	2ZR-FE	3.9	4.0	6.2	①	11.9	③
	xB	2.4	2AZ-FE	3.9	4.0	6.2	①	11.9	③
	tC	2.4	2AZ-FE	4.0	5.2	7.4	①	14.5	③

① Included in transmission capacity

② With Manual Transmission: 7.2 qts.
With Automatic Transmission: 7.1 qts.

③ With Manual Transmission: 4.7 qts.
With Automatic Transmission: 4.5 qts.

22140_SCIO_C0004

FLUID SPECIFICATIONS

Year	Model	Engine Displ. Liters	Engine Oil	Man. Trans.	Auto. Trans.	Drive Axle Front	Drive Axle Rear	Transfer Case	Power Steering Fluid	Brake Master Cylinder	Cooling System
2007	tC	2.4	ILSAC	①	②	NA	NA	NA	③	④	⑤
2008	All	1.8, 2.4	ILSAC	①	②	NA	NA	NA	③	④	⑤

DOT: Department Of Transpotation

NA: Not Available

① ILSAC multi-grade engine oil

② API GL-4 SAE 75W-90

③ ATF Dexron II or III

④ SAE J1703 or FMVSS No. 116 DOT3

⑤ Toyota Super Long Life Coolant

22140_SCIO_C0014

VALVE SPECIFICATIONS

Year	Engine Displacement Liters	Engine ID	Seat Angle (deg.)	Face Angle (deg.)	Spring Test Pressure (lbs. @ in.)	Spring Installed Height (in.)	Stem-to-Guide Clearance (in.) Intake	Stem-to-Guide Clearance (in.) Exhaust	Stem Diameter (in.) Intake	Stem Diameter (in.) Exhaust
2007	2.4	2AZ-FE	45	45	NA	NA	0.0010-0.0024	0.0012-0.0026	0.2154-0.2159	0.2152-0.2158
2008	1.8	2ZR-FE	NA	NA	NA	NA	0.0010-0.0024	0.0012-0.0026	0.2154-0.2159	0.2152-0.2157
	2.4	2AZ-FE	45	45	NA	NA	0.0010-0.0024	0.0012-0.0026	0.2154-0.2159	0.2152-0.2158

NA: Not Available

22140_SCIO_C0005

CAMSHAFT AND BEARING SPECIFICATIONS CHART

All measurements are given in inches.

Year	Engine Displ. Liters	Engine ID/VIN	Journal Dia.	Brg. Oil Clearance	Shaft End-play	Runout	Lobe Height Intake	Lobe Height Exhaust
2007	2.4	2AZ-FE	①	②	③	0.0012	1.8305-1.8345	1.8104-1.8143
2008	1.8	2ZR-FE	④	0.0012-0.0024	0.0016-0.0037	0.0016	1.6857-1.6896	1.7455-1.7494
	2.4	2AZ-FE	①	②	③	0.0012	1.8654-1.8664	1.8104-1.8143

① No. 1 Journal: 1.4162-1.4167 in.
All Others: 0.9039-0.9045 in.

② No. 1 Journal: 0.0003-0.0015 in.
All Others: 0.0010-0.0024 in.

③ Intake: 0.0016-0.0037 in.
Exhaust: 0.0032-0.0053 in.

④ No. 1 Journal: 1.3563-1.3569 in.
All Others: 0.9035-0.9041 in.

22140_SCIO_C0007

CRANKSHAFT AND CONNECTING ROD SPECIFICATIONS

All measurements are given in inches.

Year	Engine Displacement Liters	Engine ID	Crankshaft Main Brg. Journal Dia.	Crankshaft Main Brg. Oil Clearance	Crankshaft Shaft End-play	Crankshaft Thrust on No.	Connecting Rod Journal Diameter	Connecting Rod Oil Clearance	Connecting Rod Side Clearance
2007	2.4	2AZ-FE	2.1649-2.1654	0.0007-0.0016	0.0016-0.0095	3	1.8894-1.8898	0.0009-0.0019	0.0063-0.0143
2008	1.8	2ZR-FE	1.8893-1.8898	0.0006-0.0015	NA	3	1.8504-1.8513	0.0012-0.0024	0.0063-0.0135
	2.4	2AZ-FE	2.1649-2.1654	0.0007-0.0016	0.0016-0.0095	3	1.8894-1.8898	0.0009-0.0019	0.0063-0.0143

22140_SCIO_C0006

PISTON AND RING SPECIFICATIONS

All measurements are given in inches.

Year	Engine Displ. Liters	Engine ID	Piston Clearance	Ring Gap Top Comp.	Ring Gap Bottom Comp.	Ring Gap Oil Control	Ring Side Clearance Top Comp.	Ring Side Clearance Bottom Comp.	Ring Side Clearance Oil Control
2007	2.4	2AZ-FE	0.0008-0.0017	0.0094-0.0122	0.0130-0.0169	0.0040-0.0119	0.0008-0.0028	0.0008-0.0024	0.0008-0.0028
2008	1.8	2ZR-FE	0.0004-0.0017	0.0079-0.0118	0.0138-0.0197	0.0039-0.0157	0.0008-0.0028	0.0008-0.0024	0.0008-0.0026
	2.4	2AZ-FE	0.0008-0.0017	0.0094-0.0122	0.0130-0.0169	0.0040-0.0119	0.0008-0.0028	0.0008-0.0024	0.0008-0.0028

22140_SCIO_C0008

TORQUE SPECIFICATIONS

All readings in ft. lbs.

Year	Engine Displacement Liters	Engine ID	Cylinder Head Bolts	Main Bearing Bolts	Rod Bearing Bolts	Crankshaft Damper Bolts	Flywheel Bolts	Manifold Intake	Manifold Exhaust	Spark Plugs	Oil Pan Drain Plug
2007	2.4	2AZ-FE	①	②	③	133	④	22	27	18	18
2008	1.8	2ZR-FE	⑤	⑥	⑦	140	⑧	21	32	15	27
	2.4	2AZ-FE	①	②	③	133	④	22	27	18	18

① Step 1: 52 ft. lbs.
Step 2: Plus 90 degrees

② Step 1: 30 ft. lbs.
Step 2: Plus 90 degrees

③ Step 1: 18 ft. lbs.
Step 2: Plus 90 degrees

④ Auto Trans.: 72 ft. lbs.
Manual Trans.: 96 ft. lbs.

⑤ Step 1: 36 ft. lbs.
Step 2: Plus 90 degrees
Step 3: Plus 45 degrees

⑥ Step 1: 30 ft. lbs.
Step 2: Plus 90 degrees

⑦ Step 1: 15 ft. lbs.
Step 2: Plus 90 degrees

⑧ Manual Trans.Step 1: 38 ft. lbs.
Step 2: Plus 90 degrees
Auto Trans.: 65 ft. lbs.

22140_SCIO_C0009

WHEEL ALIGNMENT

Year	Model		Caster Range (+/-Deg.)	Caster Preferred Setting (Deg.)	Camber Range (+/-Deg.)	Camber Preferred Setting (Deg.)	Toe-in (in.)
2007	tC	F	0.75	+3.03	0.75	-0.52	0+/-0.08
		R	—	—	0.5	-0.90	0.12+/-0.08
2008	xD	F	0.75	+4.85	0.75	-0.18	0.06+/-0.08
		R	—	—	0.5	-0.95	0.13+/-0.11
	xB	F	0.75	+5.75	0.75	-0.16	0.08+/-0.08
		R	—	—	0.5	-1.42	0.08+/-0.11
	tC	F	0.75	+3.03	0.75	-0.52	0+/-0.08
		R	—	—	0.5	-0.90	0.12+/-0.08

F: Front
R: Rear

22140_SCIO_C0010

TIRE, WHEEL AND BALL JOINT SPECIFICATIONS

		OEM Tires		Tire Pressures (psi)		Wheel	Ball Joint	Lug Nut Torque
Year	Model	Standard	Optional	Front	Rear	Size	Inspection	(ft. lbs.)
2007	tC	P215/45ZR17	205/55R16	32	29	7.0-JJ	44 in. ①	76
2008	xD	P195/60R16	N/A	33	33	5.5-JJ	44 in. ①	76
	xB	P205/55R16	N/A	35	32	6.0-JJ	44 in. ①	76
	tC	P215/45ZR17	205/55R16	32	29	7.0-JJ	44 in. ①	76

OEM: Original Equipment Manufacturer

PSI: Pounds Per Square Inch

STD: Standard

OPT: Optional

① Torque required (in inch lbs.) to rotate ball joint when removed from the knuckle

22140_SCIO_C0011

BRAKE SPECIFICATIONS

All measurements in inches unless noted

			Brake Disc			Brake Drum				Brake Caliper	
			Original	Minimum	Maximum	Original Inside	Max. Wear	Maximum Machine	Minimum Lining	Bracket Bolts	Mounting Bolts
Year	Model		Thickness	Thickness	Run-out	Diameter	Limit	Diameter	Thickness	(ft. lbs.)	(ft. lbs.)
2007	tC	F	0.984	0.906	0.0020	-	-	-	0.039	79	25
		R	0.354	0.295	0.0059	-	-	-	0.039	34	29
2008	xD	F	0.866	0.748	0.0020	-	-	-	0.039	79	25
		R	-	-	-	9.000	-	9.039	0.039	-	-
	xB	F	0.984	0.866	0.0020	-	-	-	0.039	130	26
		R	0.394	0.335	0.0020	-	-	-	0.039	42	25
	tC	F	0.984	0.906	0.0020	-	-	-	0.039	79	25
		R	0.354	0.295	0.0059	-	-	-	0.039	34	29

F: Front

R: Rear

22140_SCIO_C0012

SCHEDULED MAINTENANCE INTERVALS

SCION - xD, xB and tC

TO BE SERVICED	TYPE OF SERVICE	VEHICLE MILEAGE INTERVAL (x1000)												
		5	10	15	20	25	30	35	40	45	50	55	60	65
Air cleaner filter	R						✓						✓	
Transmission fluid	S/I						✓						✓	
Ball joints & dust covers	S/I			✓			✓			✓			✓	
Bolts & nuts on chassis & body	S/I													
Brake line pipes & hoses	S/I			✓			✓			✓			✓	
Brake pads & discs/linings & drums (front & rear)	S/I	✓	✓	✓	✓	✓	✓	✓	✓	✓	✓	✓	✓	✓
Drive belts	S/I												✓	
Driveshaft boots	S/I			✓			✓			✓			✓	
Engine coolant	S/I			✓			✓			✓			✓	
Engine coolant	R	Replace at 100,000 miles												
Engine oil & filter	R	✓	✓	✓	✓	✓	✓	✓	✓	✓	✓	✓	✓	✓
Exhaust pipes & mountings	S/I			✓			✓			✓			✓	
Fuel lines & connections	S/I						✓						✓	
Propeller shaft bolt	S/I			✓			✓			✓			✓	
Radiator core & condenser	S/I			✓			✓			✓			✓	
Front differential fluid	S/I						✓						✓	
Rotate Tires	S/I	✓	✓	✓	✓	✓	✓	✓	✓	✓	✓	✓	✓	✓
Spark plugs (tC)	R	Replace at 120,000 miles												
Spark plugs (xA & xB)	R						✓						✓	
Steering linkage & gear box	S/I			✓			✓			✓			✓	

R: Replace S/I: Service or Inspect

Drivebelts: After initial inspection at 60,000 miles, inspect every 15,000 miles thereafter.

FREQUENT OPERATION MAINTENANCE (SEVERE SERVICE)

If a vehicle is operated under any of the following conditions it is considered severe service:

- Desert/Extremely dusty areas.

- Trailer towing usage.

Air cleaner filter: service or inspect every 5000 miles

Ball joints & dust covers: service or inspect every 5000 miles.

Bolts & nuts on chassis & body: service or inspect every 5000 miles.

Driveshaft boots: service or inspect every 5000 miles.

Steering linkage: service or inspect every 5000 miles.

Transmission and Front differential fluid: replace every 30,000 miles.

22140_SCIO_C0013

PRECAUTIONS

Before servicing any vehicle, please be sure to read all of the following precautions, which deal with personal safety, prevention of component damage, and important points to take into consideration when servicing a motor vehicle:

- Never open, service or drain the radiator or cooling system when the engine is hot; serious burns can occur from the steam and hot coolant.
- Observe all applicable safety precautions when working around fuel. Whenever servicing the fuel system, always work in a well-ventilated area. Do not allow fuel spray or vapors to come in contact with a spark, open flame, or excessive heat (a hot drop light, for example). Keep a dry chemical fire extinguisher near the work area. Always keep fuel in a container specifically designed for fuel storage; also, always properly seal fuel containers to avoid the possibility of fire or explosion. Refer to the additional fuel system precautions later in this section.
- Fuel injection systems often remain pressurized, even after the engine has been turned **OFF**. The fuel system pressure must be relieved before disconnecting any fuel lines. Failure to do so may result in fire and/or personal injury.
- Brake fluid often contains polyglycol ethers and polyglycols. Avoid contact with the eyes and wash your hands thoroughly after handling brake fluid. If you do get brake fluid in your eyes, flush your eyes with clean, running water for 15 minutes. If eye irritation persists, or if you have taken brake fluid internally, IMMEDIATELY seek medical assistance.
- The EPA warns that prolonged contact with used engine oil may cause a number of skin disorders, including cancer. You should make every effort to minimize your exposure to used engine oil. Protective gloves should be worn when changing oil. Wash your hands and any other exposed skin areas as soon as possible after exposure to used engine oil. Soap and water, or waterless hand cleaner should be used.
- All new vehicles are now equipped with an air bag system, often referred to as a Supplemental Restraint System (SRS) or Supplemental Inflatable Restraint (SIR) system. The system must be disabled before performing service on or around system components, steering column, instrument panel components, wiring and sensors. Failure to follow safety and disabling procedures could result in accidental air bag deployment, possible personal injury and unnecessary system repairs.
- Always wear safety goggles when working with, or around, the air bag system. When carrying a non-deployed air bag, be sure the bag and trim cover are pointed away from your body. When placing a non-deployed air bag on a work surface, always face the bag and trim cover upward, away from the surface. This will reduce the motion of the module if it is accidentally deployed. Refer to the additional air bag system precautions later in this section.
- Clean, high quality brake fluid from a sealed container is essential to the safe and proper operation of the brake system. You should always buy the correct type of brake fluid for your vehicle. If the brake fluid becomes contaminated, completely flush the system with new fluid. Never reuse any brake fluid. Any brake fluid that is removed from the system should be discarded. Also, do not allow any brake fluid to come in contact with a painted surface; it will damage the paint.
- Never operate the engine without the proper amount and type of engine oil; doing so WILL result in severe engine damage.
- Timing belt maintenance is extremely important. Many models utilize an interference-type, non-freewheeling engine. If the timing belt breaks, the valves in the cylinder head may strike the pistons, causing potentially serious (also time-consuming and expensive) engine damage. Refer to the maintenance interval charts for the recommended replacement interval for the timing belt, and to the timing belt section for belt replacement and inspection.
- Disconnecting the negative battery cable on some vehicles may interfere with the functions of the on-board computer system(s) and may require the computer to undergo a relearning process once the negative battery cable is reconnected.
- When servicing drum brakes, only disassemble and assemble one side at a time, leaving the remaining side intact for reference.
- Only an MVAC-trained, EPA-certified automotive technician should service the air conditioning system or its components.

BRAKES

ANTI-LOCK BRAKE SYSTEM (ABS)

GENERAL INFORMATION

PRECAUTIONS

- Certain components within the ABS system are not intended to be serviced or repaired individually.
- Do not use rubber hoses or other parts not specifically specified for and ABS system. When using repair kits, replace all parts included in the kit. Partial or incorrect repair may lead to functional problems and require the replacement of components.
- Lubricate rubber parts with clean, fresh brake fluid to ease assembly. Do not use shop air to clean parts; damage to rubber components may result.
- Use only DOT 3 brake fluid from an unopened container.
- If any hydraulic component or line is removed or replaced, it may be necessary to bleed the entire system.
- A clean repair area is essential. Always clean the reservoir and cap thoroughly before removing the cap. The slightest amount of dirt in the fluid may plug an orifice and impair the system function. Perform repairs after components have been thoroughly cleaned; use only denatured alcohol to clean components. Do not allow ABS components to come into contact with any substance containing mineral oil; this includes used shop rags.
- The Anti-Lock control unit is a microprocessor similar to other computer units in the vehicle. Ensure that the ignition switch is **OFF** before removing or installing controller harnesses. Avoid static electricity discharge at or near the controller.
- If any arc welding is to be done on the vehicle, the control unit should be unplugged before welding operations begin.

BRAKES

BLEEDING THE BRAKE SYSTEM

BLEEDING PROCEDURE

BLEEDING PROCEDURE

Bleeding Brake Lines

See Figure 1.

WARNING

Clean, high quality brake fluid is essential to the safe and proper operation of the brake system. You should always buy the highest quality brake fluid that is available. If the brake fluid becomes contaminated, drain and flush the system, then refill the master cylinder with new fluid. Never reuse any brake fluid. Any brake fluid that is removed from the system should be discarded. Also, do not allow any brake fluid to come in contact with a painted surface; it will damage the paint.

1. Fill the brake master cylinder reservoir with brake fluid. Use SAE J1703 or FMVSS No. 116 DOT3 brake fluid.
2. Remove the bleeder plug cap.
3. Connect a vinyl tube to either one of the bleeder plugs.
4. Depress the pedal several times, and then loosen the bleeder plug with the pedal depressed.
5. When fluid stops coming out, immediately tighten the bleeder plug. Then release the pedal.
6. Repeat steps (3) and (4) until all the air in the fluid is gone.
7. Tighten the bleeder plug to 73 inch lbs. (8.3 Nm).
8. Install the cap.
9. Repeat the procedure to bleed air from the brake line for each wheel.

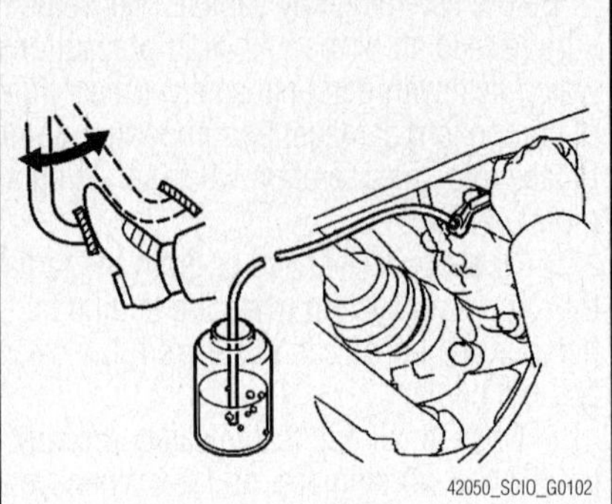

Fig. 1 Connect a vinyl tube to either one of the bleeder plugs. Depress the pedal several times, and then loosen the bleeder plug with the pedal depressed. When fluid stops coming out, immediately tighten the bleeder plug and release the pedal

BRAKES

FRONT DISC BRAKES

CAUTION

Dust and dirt accumulating on brake parts during normal use may contain asbestos fibers from production or aftermarket brake linings. Breathing excessive concentrations of asbestos fibers can cause serious bodily harm. Exercise care when servicing brake parts. Do not sand or grind brake lining unless equipment used is designed to contain the dust residue. Do not clean brake parts with compressed air or by dry brushing. Cleaning should be done by dampening the brake components with a fine mist of water, then wiping the brake components clean with a dampened cloth. Dispose of cloth and all residue containing asbestos fibers in an impermeable container with the appropriate label. Follow practices prescribed by the Occupational Safety and Health Administration (OSHA) and the Environmental Protection Agency (EPA) for the handling, processing, and disposing of dust or debris that may contain asbestos fibers.

BRAKE CALIPER

REMOVAL & INSTALLATION

See Figure 2.

1. Before servicing the vehicle, refer to the Precautions Section.
2. Remove or disconnect the following:
   - Front wheels
   - Brake line at the caliper
   - Brake pads
   - Brake pad support plate
   - Pin and sleeve boots
   - Caliper bolts
   - Caliper

To install:

3. Install or connect the following:
   - Caliper onto its mounting and install the lower mounting bolt. Torque: 79 ft. lbs. (107 Nm).
   - Pin boots, sleeve boots and brake pads. Tighten the slide pins to 25 ft. lbs. (34 Nm).
   - Brake line to the caliper with 2 new metal gaskets. Torque the brake line union bolt to 22 ft. lbs. (30 Nm).
   - Front wheels
4. Bleed the system, as outlined in this section.

DISC BRAKE PADS

REMOVAL & INSTALLATION

See Figure 3.

1. Before servicing the vehicle, refer to the Precautions Section.
2. Remove or disconnect the following:
   - Front wheels
   - Slide pins and boots
   - Brake pads from the caliper mounting

To install:

3. Install or connect the following:
   - Brake pads to the caliper mounting
   - Pin boots, sleeve boots and brake pads. Tighten the slide pins to 25 ft. lbs. (34 Nm).
   - Wheels

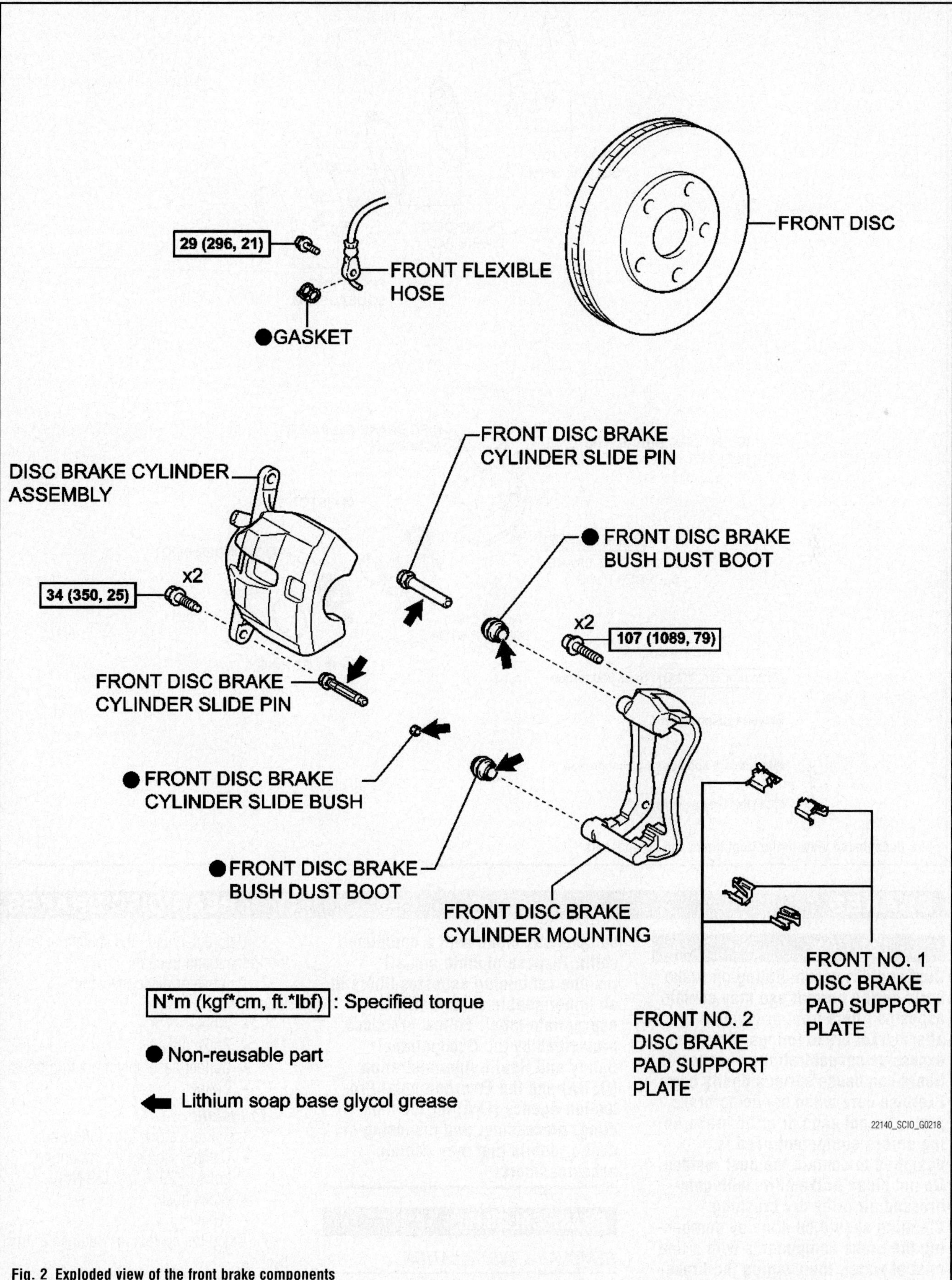

Fig. 2 Exploded view of the front brake components

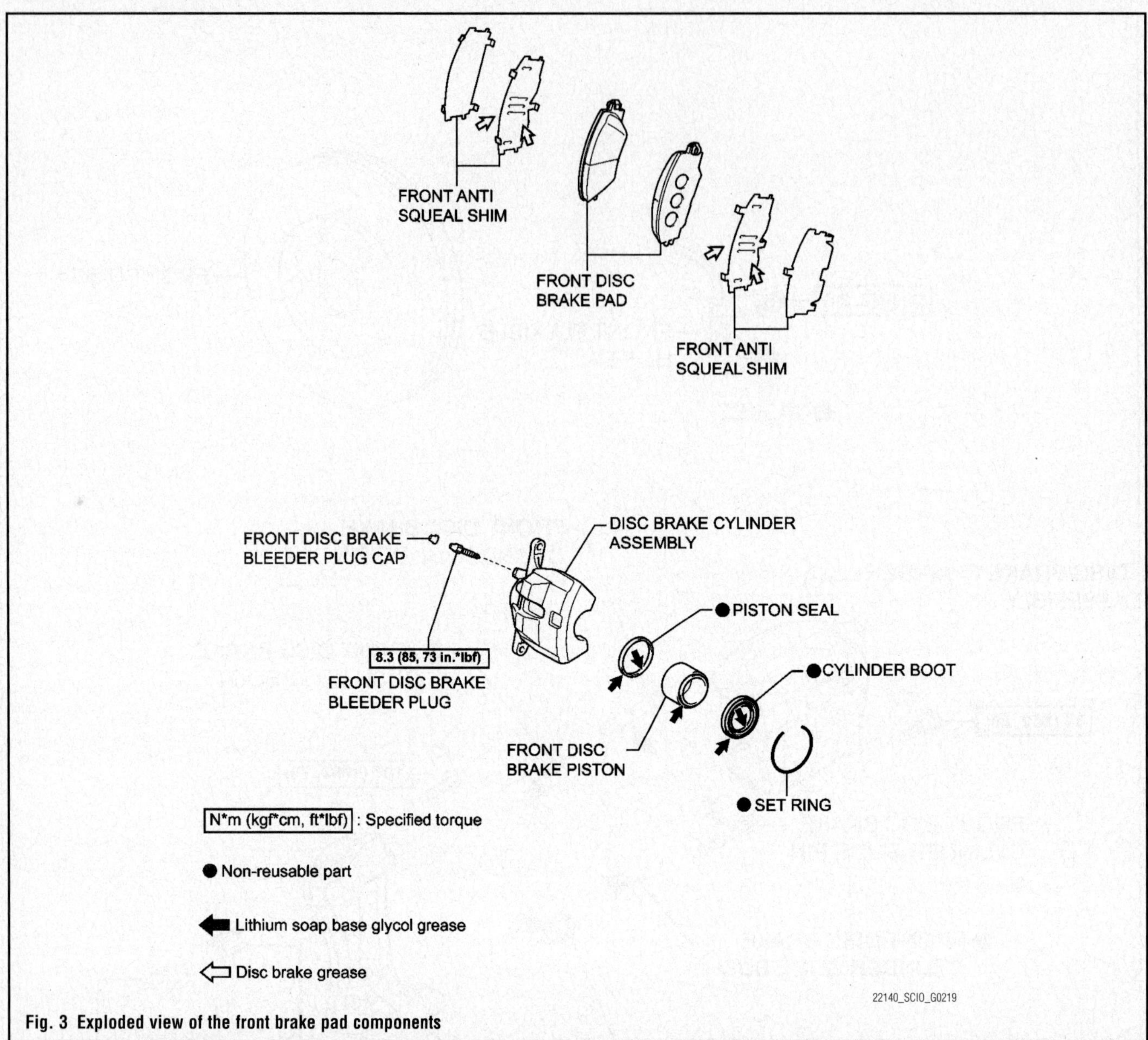

Fig. 3 Exploded view of the front brake pad components

CAUTION

Dust and dirt accumulating on brake parts during normal use may contain asbestos fibers from production or aftermarket brake linings. Breathing excessive concentrations of asbestos fibers can cause serious bodily harm. Exercise care when servicing brake parts. Do not sand or grind brake lining unless equipment used is designed to contain the dust residue. Do not clean brake parts with compressed air or by dry brushing. Cleaning should be done by dampening the brake components with a fine mist of water, then wiping the brake components clean with a dampened cloth. Dispose of cloth and all residue containing asbestos fibers in an impermeable container with the appropriate label. Follow practices prescribed by the Occupational Safety and Health Administration (OSHA) and the Environmental Protection Agency (EPA) for the handling, processing, and disposing of dust or debris that may contain asbestos fibers.

BRAKE CALIPER

REMOVAL & INSTALLATION

See Figure 4.

1. Before servicing the vehicle, refer to the Precautions Section.
2. Remove or disconnect the following:
   - Wheel
   - Brake hose
   - Caliper assembly mounting bolts
   - Caliper

To install:

3. Install or connect the following:
   - Caliper. Tighten the mounting bolts to 29 ft. lbs. (39 Nm).
   - Brake hose
   - Wheel
4. Bleed the system, as outlined in this section.

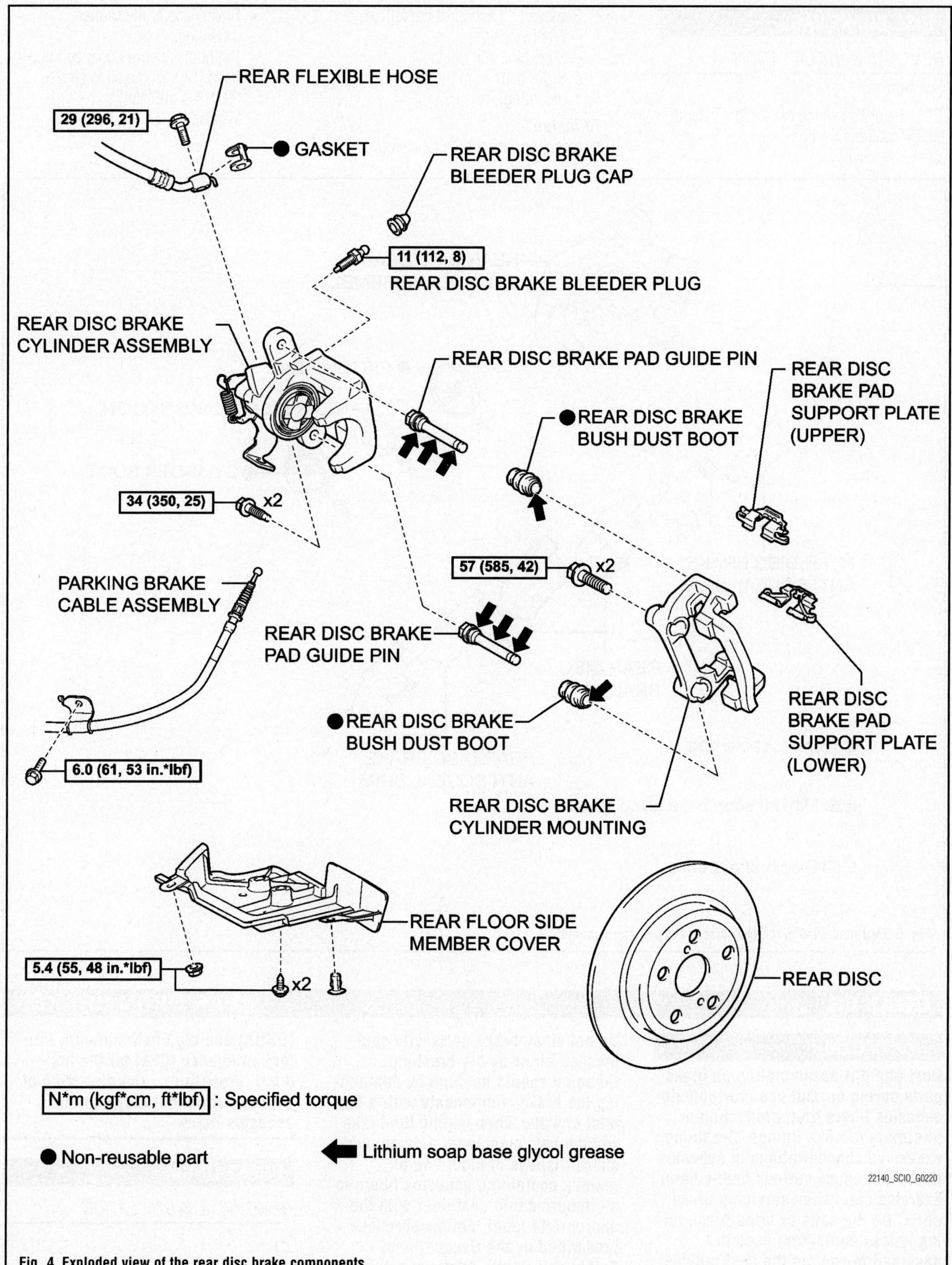

Fig. 4 Exploded view of the rear disc brake components

DISC BRAKE PADS

REMOVAL & INSTALLATION

See Figure 5.

1. Before servicing the vehicle, refer to the Precautions Section.
2. Remove or disconnect the following:
   - Front wheels
   - Slide pins and boots
   - Brake pads from the caliper mounting

To install:

3. Install or connect the following:
   - Brake pads to the caliper mounting
   - Pin boots, sleeve boots and brake pads. Tighten the slide pins to 29 ft. lbs. (39 Nm).
   - Wheels

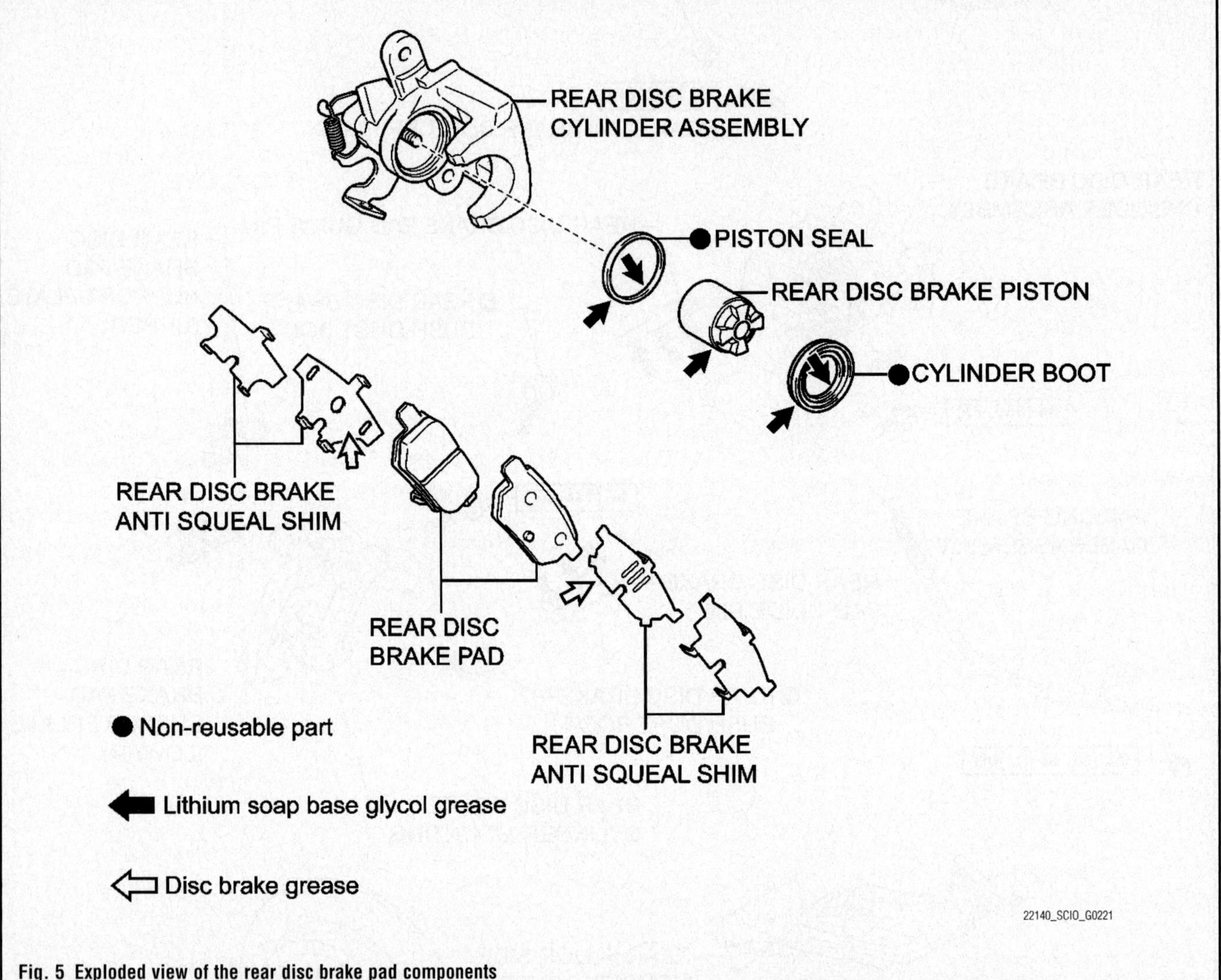

Fig. 5 Exploded view of the rear disc brake pad components

BRAKES — REAR DRUM BRAKES

CAUTION

Dust and dirt accumulating on brake parts during normal use may contain asbestos fibers from production or aftermarket brake linings. Breathing excessive concentrations of asbestos fibers can cause serious bodily harm. Exercise care when servicing brake parts. Do not sand or grind brake lining unless equipment used is designed to contain the dust residue. Do not clean brake parts with compressed air or by dry brushing. Cleaning should be done by dampening the brake components with a fine mist of water, then wiping the brake components clean with a dampened cloth. Dispose of cloth and all residue containing asbestos fibers in an impermeable container with the appropriate label. Follow practices prescribed by the Occupational Safety and Health Administration (OSHA) and the Environmental Protection Agency (EPA) for the handling, processing, and disposing of dust or debris that may contain asbestos fibers.

BRAKE DRUM

REMOVAL & INSTALLATION

xD

See Figure 6.

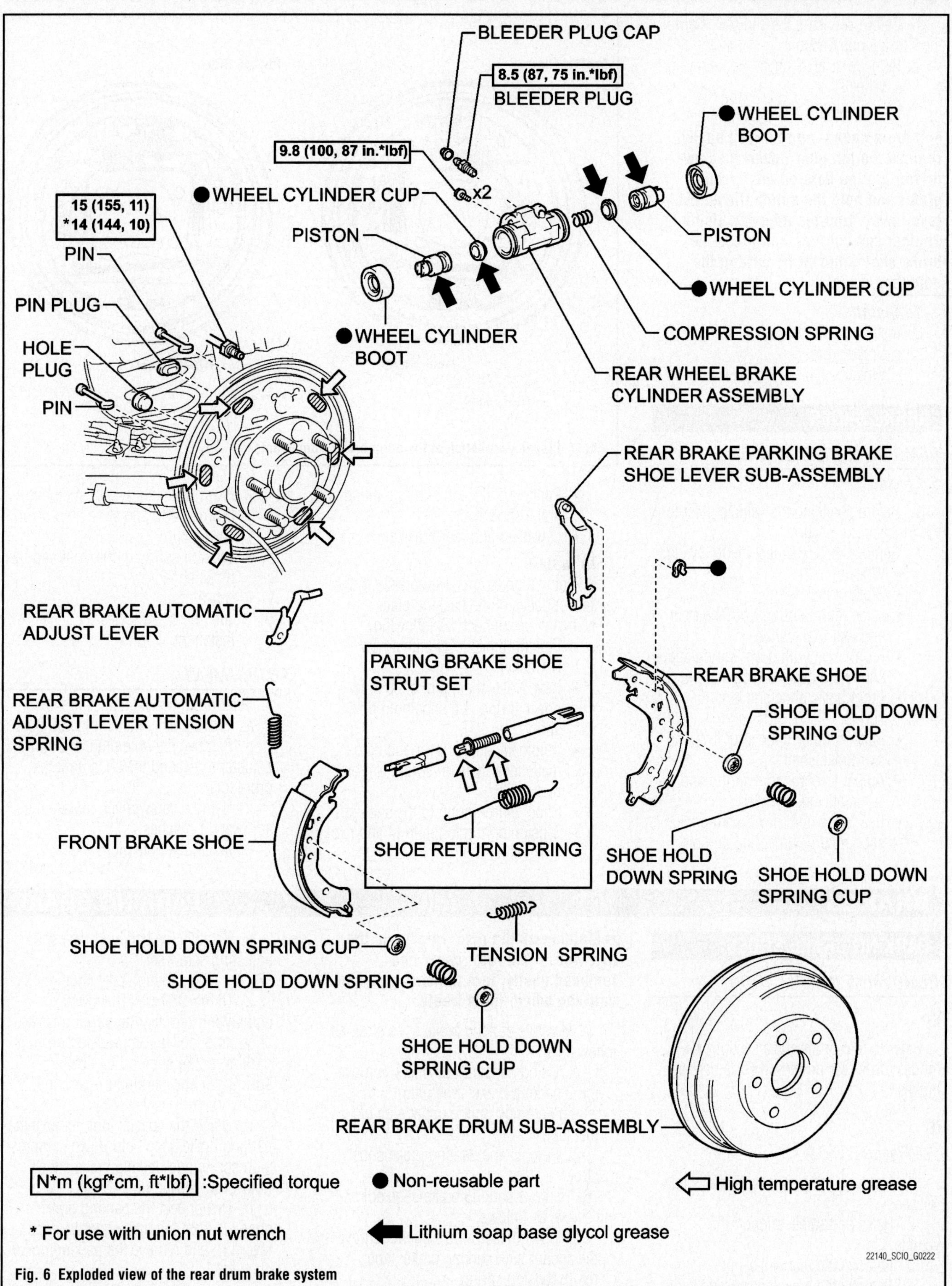

Fig. 6 Exploded view of the rear drum brake system

1. Before servicing the vehicle, refer to the Precautions Section.
2. Remove or disconnect the following:
   - Wheels
   - Brake drum

➡If drum cannot be removed easily, remove the pin plug, insert a screwdriver into the hole on the backing plate, and hold the automatic adjust lever away from the adjuster. Using another screwdriver, compress the brake shoe adjuster by turning the adjusting wheel.

To install:

3. Install or connect the following:
   - Brake drum
   - Wheel

BRAKE SHOES

REMOVAL & INSTALLATION

See Figure 7.

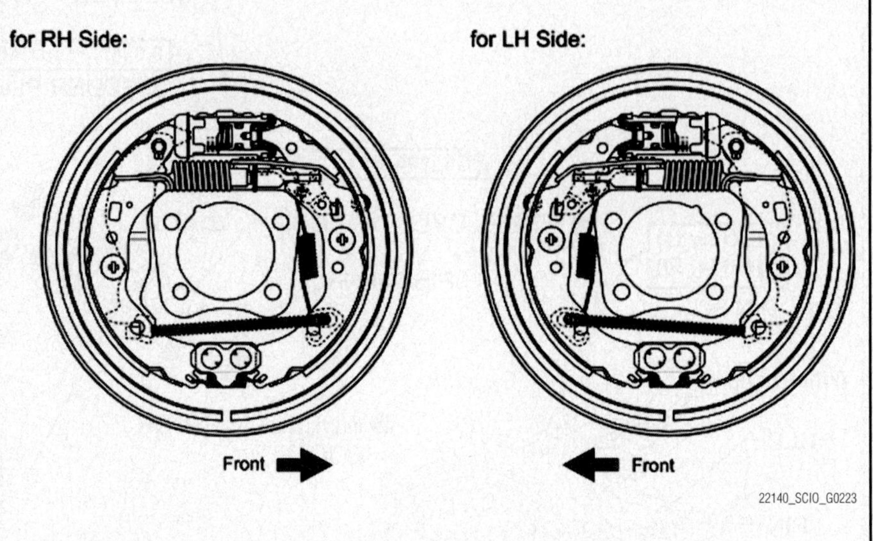

Fig. 7 Proper installation of the drum brake components

1. Before servicing the vehicle, refer to the Precautions Section.
2. Remove or disconnect the following:
   - Wheels
   - Brake drum
   - Shoe return spring from the front and rear brake shoe.
   - Front shoe hold down spring cup, shoe hold down spring and pin
   - Front brake shoe and tension spring
   - Parking brake shoe strut from the rear brake shoe
   - Adjust lever tension spring and automatic adjust lever
   - Rear shoe hold down spring cup, shoe hold down spring and pin
   - Parking brake cable from parking brake shoe lever
   - C-washer and rear brake shoe.

To install:

3. Apply high temperature grease to the contact surfaces of the backing plate.
4. Install or connect the following:
   - Parking brake lever with a new C-washer
   - Rear brake shoe, pin, shoe hold down spring and shoe hold down spring cup
   - Automatic adjust lever and automatic lever tension spring to front brake shoe
   - Shoe return spring to the shoe strut
   - Tension spring to the front and rear brake shoe
   - Front brake shoe, pin, shoe hold down spring and shoe hold down spring cup
   - Tension spring to the front and rear brake shoe
   - Brake drum
5. Adjust the brake shoes
   - Install the wheels.

ADJUSTMENT

1. Rear drum brake shoe clearance.
   a. Provisionally install the hub nuts.
   b. Remove the hole plug, and turn the adjuster to expand the shoe until the drum locks.
   c. Using a screwdriver, release the adjuster 11 notches.
   d. Install the hole plug.

PARKING BRAKE SHOES

REMOVAL & INSTALLATION

xD

Refer to rear drum brake section. xD models uses rear drum brakes as parking brakes.

tC

See Figures 8 and 9.

1. Make sure the parking brake is fully released.
2. Raise and safely support the vehicle.
3. Remove the rear wheel.
4. Remove the brake caliper and rotor.

➡Matchmark the disc (rotor) and the axle hub. If the rotor cannot be removed easily, turn the shoe adjuster until the wheel turns freely.

5. Remove parking brake shoe strut, as follows:
   a. Using needle-nose pliers, remove the 2 parking brake shoe return tension springs on the upper side of the shoe.
   b. Remove the parking brake shoe strut.
6. Remove parking brake shoe adjusting screw, as follows:
   a. Using needle-nose pliers, remove the anchor side parking brake shoe return tension spring.
   b. Remove the parking brake shoe adjusting screw set.
7. Remove parking brake shoe no. 2:
   a. Using SST 09718-00010, or equivalent, remove the parking brake shoe hold-down compression spring, parking brake shoe hold- down spring pin and parking brake shoe no. 1.
   b. Using the special tool, remove the parking brake shoe hold-down compression spring and parking brake shoe hold-down spring pin.
   c. Disconnect the parking brake shoe LH from the parking brake shoe lever LH, and remove the parking brake shoe LH.

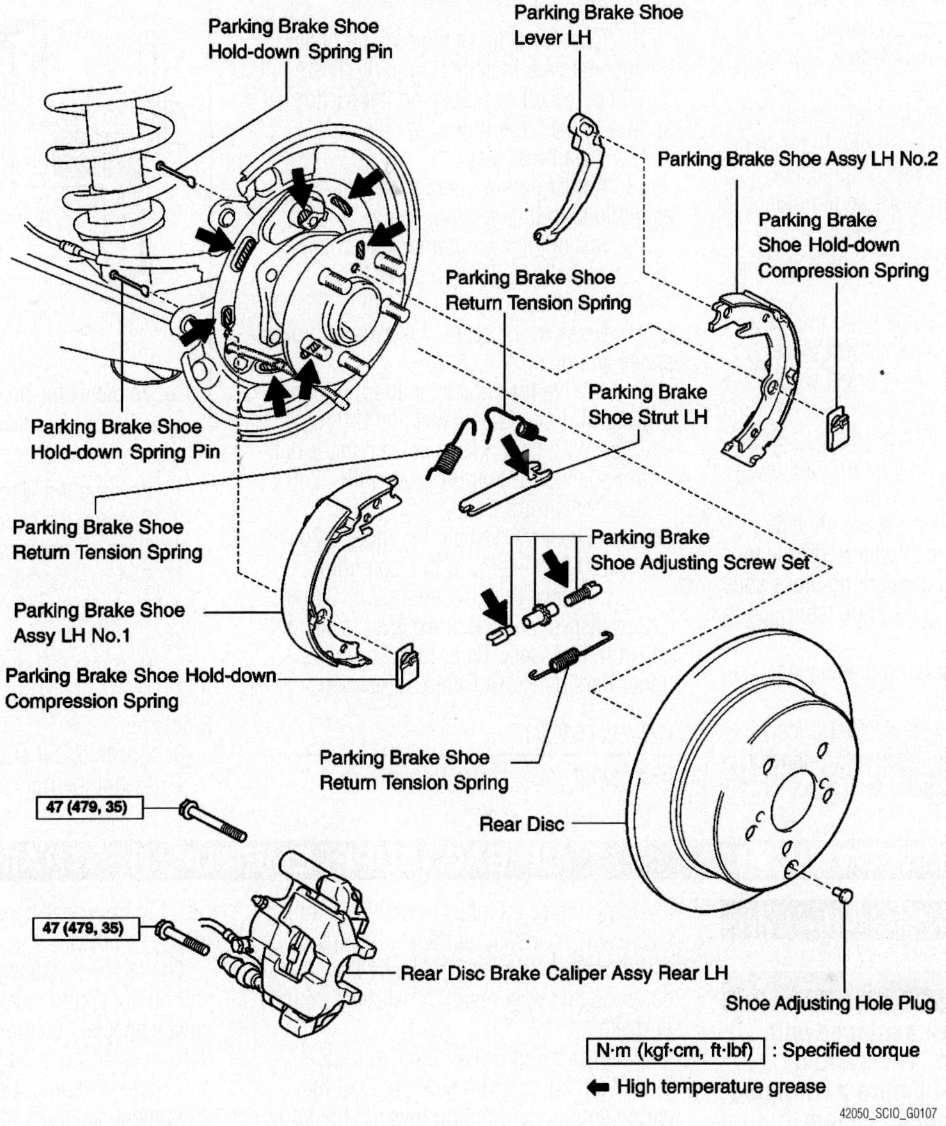

Parking Brake Shoe Hold-down Spring Pin
Parking Brake Shoe Lever LH
Parking Brake Shoe Assy LH No.2
Parking Brake Shoe Hold-down Compression Spring
Parking Brake Shoe Return Tension Spring
Parking Brake Shoe Strut LH
Parking Brake Shoe Hold-down Spring Pin
Parking Brake Shoe Return Tension Spring
Parking Brake Shoe Adjusting Screw Set
Parking Brake Shoe Assy LH No.1
Parking Brake Shoe Hold-down Compression Spring
Parking Brake Shoe Return Tension Spring
47 (479, 35)
Rear Disc
47 (479, 35)
Rear Disc Brake Caliper Assy Rear LH
Shoe Adjusting Hole Plug
N·m (kgf·cm, ft·lbf) : Specified torque
High temperature grease
42050_SCIO_G0107

Fig. 8 Exploded view of the parking brake components

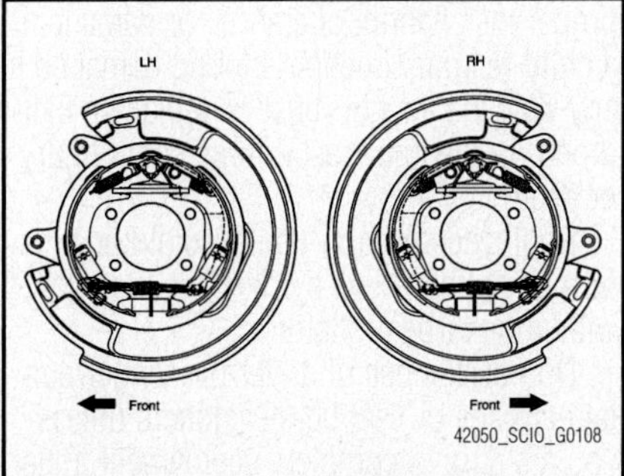

Fig. 9 Check that all of the parking brake components are installed properly. There should be no oil or grease on the friction surface of the shoe lining and disc.

8. Inspect parking brake shoe lining thickness:
   a. Using a ruler, measure the thickness of the shoe lining.
   - Standard thickness: 3.2 mm (0.126 in.)
   - Minimum thickness: 1.0 mm (0.039 in.)

9. Inspect brake disc and parking brake shoe lining for proper contact. Apply chalk to the inside surface of the disc, then grind down the brake shoe lining to fit. If the contact between the brake disc and the shoe lining is improper, repair it using a brake shoe grinder or replace the brake shoe assembly.

10. Remove the parking brake shoe lever:
    a. Using needle-nose pliers, disconnect the parking brake cable no. 3 and remove the parking brake shoe lever.

To install:

11. Apply high temperature grease to the shoe attachment surfaces of the backing plate.

12. Using needle-nose pliers, connect the parking brake cable no. 3 to the parking brake shoe lever.

13. Install parking brake shoe no. 2:
    a. Apply high temperature grease to the contact part of the parking brake shoe and parking brake shoe lever.

b. Using SST 09718-00010, or equivalent, install the parking brake shoe no.2 with the parking brake shoe hold-down compression spring and parking brake shoe hold-down spring pin.

c. Using the special tool, install the parking brake shoe no.1 with the parking brake shoe hold-down compression spring and parking brake shoe hold-down spring pin.

14. Install parking brake shoe adjusting screw set:

a. Apply high temperature grease to the adjusting bolt and support piece.

b. Install the parking brake shoe adjusting screw set.

c. Using needle-nose pliers, install the parking brake shoe return tension spring.

15. Install the parking brake shoe strut:

a. Apply high temperature grease to the contact part of the parking brake shoe strut and parking brake shoe return tension spring.

b. Install the parking brake shoe strut.

c. Using needle-nose pliers, install the 2 parking brake shoe return tension springs on the upper side of the shoe.

➡Fit the parking brake shoe return tension springs securely into the grooves of the parking brake shoe strut.

16. Check that all of the parking brake components are installed properly. There should be no oil or grease on the friction surface of the shoe lining and disc.

17. Install rear disc

18. Adjust parking brake shoe clearance, as outlined in this section.

19. Install the caliper and tighten the mounting bolts to 35 ft. lbs. (47 Nm).

20. Install rear wheel

21. Break in, or settle, the parking brake shoes and disc:

a. Drive the vehicle at about 50 km/h (31 mph) on a safe, level and dry road.

b. With the parking brake release button pushed in, pull the lever with 22 ft. lbs. (98 Nm) of force.

c. Drive the vehicle for about 0.25 miles (400 meters) in this condition.

d. Repeat this procedure 2 or 3 times.

22. Inspect parking brake lever travel, and adjust if necessary. Refer to the procedure under Parking Brake Cable Adjustment.

ADJUSTMENT

See Figure 10.

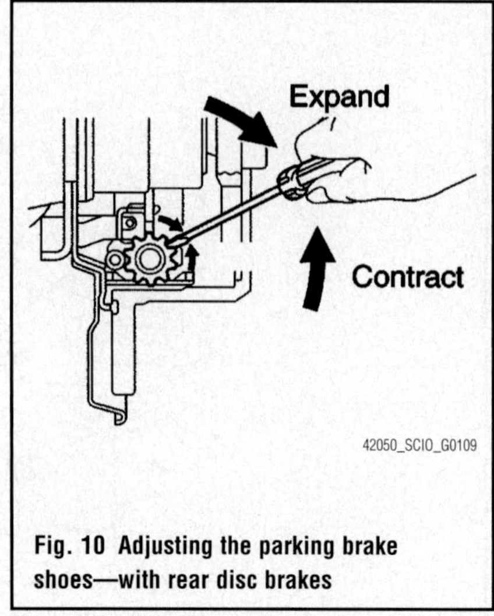

Fig. 10 Adjusting the parking brake shoes—with rear disc brakes

1. Remove the wheel and tire.
2. Remove the caliper.
3. Temporarily install the lug nuts.
4. Remove the hole plug, and turn the adjuster to expand the shoes until the disc locks.
5. Contract the shoe adjuster until the disc can rotate smoothly. Standard : return 8 notches
6. Check that there is no brake drag.
7. Install the hole plug.

CHASSIS ELECTRICAL

AIR BAG (SUPPLEMENTAL RESTRAINT SYSTEM)

GENERAL INFORMATION

⁂ CAUTION

These vehicles are equipped with an air bag system. The system must be disarmed before performing service on, or around, system components, the steering column, instrument panel components, wiring and sensors. Failure to follow the safety precautions and the disarming procedure could result in accidental air bag deployment, possible injury and unnecessary system repairs.

SERVICE PRECAUTIONS

Disconnect and isolate the battery negative cable before beginning any airbag system component diagnosis, testing, removal, or installation procedures. Allow system capacitor to discharge for two minutes before beginning any component service. This will disable the airbag system. Failure to disable the airbag system may result in accidental airbag deployment, personal injury, or death.

Do not place an intact undeployed airbag face down on a solid surface. The airbag will propel into the air if accidentally deployed and may result in personal injury or death.

When carrying or handling an undeployed airbag, the trim side (face) of the airbag should be pointing towards the body to minimize possibility of injury if accidental deployment occurs. Failure to do this may result in personal injury or death.

Replace airbag system components with OEM replacement parts. Substitute parts may appear interchangeable, but internal differences may result in inferior occupant protection. Failure to do so may result in occupant personal injury or death.

Wear safety glasses, rubber gloves, and long sleeved clothing when cleaning powder residue from vehicle after an airbag deployment. Powder residue emitted from a deployed airbag can cause skin irritation. Flush affected area with cool water if irritation is experienced. If nasal or throat irritation is experienced, exit the vehicle for fresh air until the irritation ceases. If irritation continues, see a physician.

Do not use a replacement airbag that is not in the original packaging. This may result in improper deployment, personal injury, or death.

The factory installed fasteners, screws and bolts used to fasten airbag components have a special coating and are specifically designed for the airbag system. Do not use substitute fasteners. Use only original equipment fasteners listed in the parts catalog when fastener replacement is required.

During, and following, any child restraint anchor service, due to impact event or vehicle repair, carefully inspect all mounting hardware, tether straps, and anchors for proper installation, operation, or damage. If a child restraint anchor is found damaged in any way, the anchor must be replaced. Failure to do this may result in personal injury or death.

Deployed and non-deployed airbags may or may not have live pyrotechnic material within the airbag inflator.

Do not dispose of driver/passenger/curtain airbags or seat belt tensioners unless you are sure of complete deployment. Refer to the Hazardous Substance Control System for proper disposal.

Dispose of deployed airbags and tensioners consistent with state, provincial, local, and federal regulations.

After any airbag component testing or service, do not connect the battery negative cable. Personal injury or death may result if the system test is not performed first.

If the vehicle is equipped with the Occupant Classification System (OCS), do not connect the battery negative cable before performing the OCS Verification Test using the scan tool and the appropriate diagnostic information. Personal injury or death may result if the system test is not performed properly.

Never replace both the Occupant Restraint Controller (ORC) and the Occupant Classification Module (OCM) at the same time. If both require replacement, replace one, then perform the Airbag System test before replacing the other.

Both the ORC and the OCM store Occupant Classification System (OCS) calibration data, which they transfer to one another when one of them is replaced. If both are replaced at the same time, an irreversible fault will be set in both modules and the OCS may malfunction and cause personal injury or death.

If equipped with OCS, the Seat Weight Sensor is a sensitive, calibrated unit and must be handled carefully. Do not drop or handle roughly. If dropped or damaged, replace with another sensor. Failure to do so may result in occupant injury or death.

If equipped with OCS, the front passenger seat must be handled carefully as well. When removing the seat, be careful when setting on floor not to drop. If dropped, the sensor may be inoperative, could result in occupant injury, or possibly death.

If equipped with OCS, when the passenger front seat is on the floor, no one should sit in the front passenger seat. This uneven force may damage the sensing ability of the seat weight sensors. If sat on and damaged, the sensor may be inoperative, could result in occupant injury, or possibly death.

DISARMING THE SYSTEM

Disconnect and isolate the negative battery cable. Wait 90 seconds for the system capacitor to discharge before performing any service.

ARMING THE SYSTEM

Connect the negative battery cable. Wait 2 minutes before performing and service work.

DRIVETRAIN

AUTOMATIC TRANSAXLE ASSEMBLY

REMOVAL & INSTALLATION

xB

See Figure 11.

1. Before servicing the vehicle, refer to the Precautions Section.
2. Drain the transaxle assembly.
3. Remove or disconnect the following:
   - Negative battery cable
   - Hood
   - Windshield wiper arms
   - Outer cowl top panel
   - Engine appearance cover
   - Air intake assembly
   - Transmission control cable and bracket
   - Transmission wiring connectors
   - Front wheel
   - Engine undercover
   - Front exhaust pipe assembly
   - Both halfshafts
   - Transaxle oil cooler hose
4. At this point, attach an engine crane to support the engine.
5. Remove the halfshaft heat insulator.
6. Remove the starter.
7. Support the transaxle assembly with a suitable jack.
8. Remove or disconnect the following:
   - Left-hand engine mounting bracket
   - Rear engine mount and bracket
   - Front suspension crossmember
   - Flywheel housing cover
9. Turn the crankshaft to gain access and remove the 6 torque converter mounting bolts.

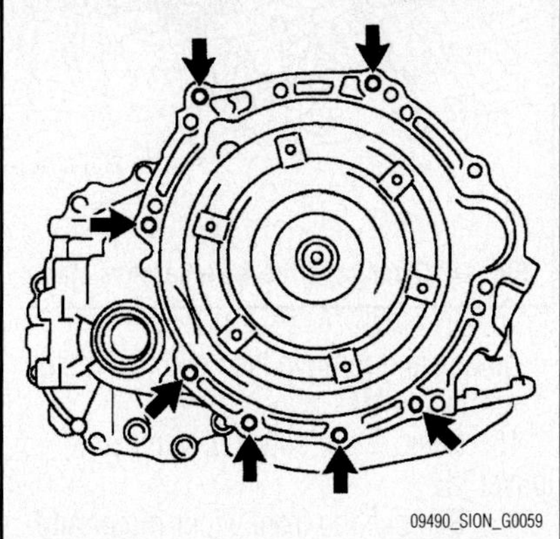

Fig. 11 Automatic transaxle mounting bolt location—xB

10. Remove the 7 transaxle mounting bolts.
11. Remove the automatic transaxle from the engine.

To install:

12. Install the transaxle assembly to the engine. Tighten the 7 mounting bolts to 22 ft. lbs. (30 Nm).
13. Install the 6 torque converter mounting bolts. Tighten to 20 ft. lbs. (27 Nm).
14. Install or connect the following:
    - Flywheel housing cover
    - Rear engine mounting bracket. Tighten to 36 ft. lbs. (49 Nm).
    - Front suspension crossmember
    - Rear engine mount. Tighten to 47 ft. lbs. (64 Nm).
    - Left-hand engine mounting bracket. Tighten to 36 ft. lbs. (49 Nm).
    - Starter
    - Halfshaft heat insulator. Tighten to 13 ft. lbs. (18 Nm).
    - Halfshafts
    - Front exhaust pipe assembly
    - Transaxle oil cooler hose
    - Engine undercover
    - Front wheel
    - Transaxle wiring connectors
    - Transaxle control cable
    - Air intake assembly
    - Engine appearance cover
    - Outer cowl top panel
    - Wiper arms
    - Hood
    - Negative battery
15. Refill the transaxle with fluid to the correct level.
16. Start the engine and check for leaks.

xD

See Figures 12 through 44.

1. Discharge the fuel system pressure.

CAUTION

The DISCHARGE FUEL SYSTEM PRESSURE procedures must be performed before disconnecting any part of the fuel system.

CAUTION

After performing the DISCHARGE FUEL SYSTEM PRESSURE procedures, pressure will remain in the fuel line. When disconnecting the fuel line, place a cloth or equivalent over fittings to reduce the risk of fuel spray.

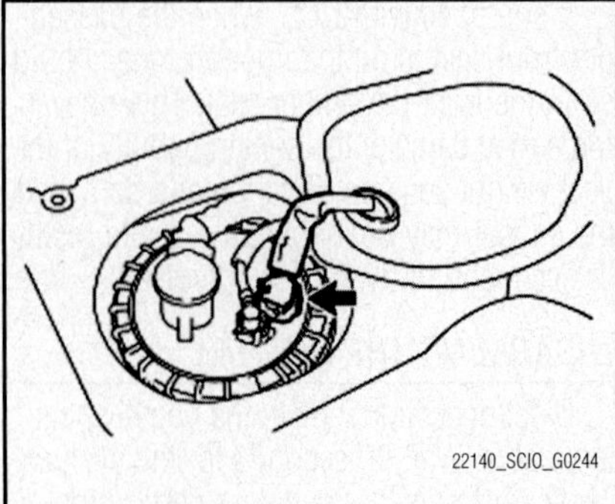

Fig. 12 Disconnect the connector from the fuel pump assembly

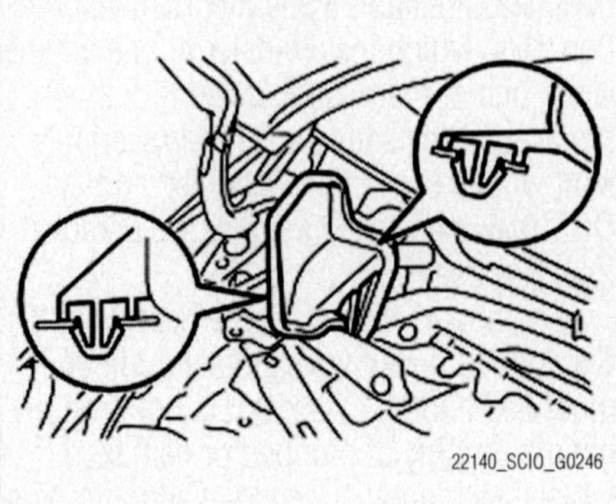

Fig. 13 Remove the front air shutter seal RH

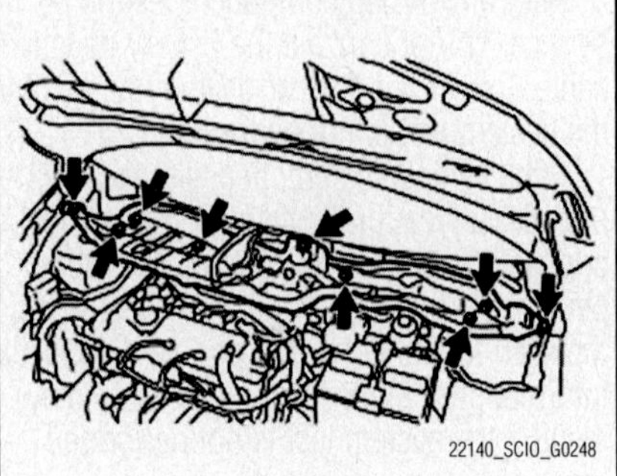

Fig. 15 Remove the 9 bolts and remove the outer cowl top panel

a. Remove the rear seat cushion assembly LH

b. Remove the rear floor service hose cover.

c. Disconnect the connector from the fuel pump assembly.

d. Start the engine. After the engine stops naturally, turn the ignition switch OFF.

e. Start the engine again and make sure that engine does not start.

f. Remove the fuel tank cap, and let the air out of the fuel tank.

g. Connect the connector.

h. Install the rear floor service hole cover with new butyl tape.

i. Install the rear seat cushion assembly LH.

2. Remove the battery.

a. Disconnect the cable from the battery terminal.

b. Loosen the nut and remove the battery clamp.

c. Remove the battery.

3. Remove the battery tray.

4. Remove the front wheels.

5. Remove the engine under cover LH.

6. Remove the engine under cover RH.

7. Drain the engine coolant.

8. Drain the manual transaxle oil (for Manual Transaxle).

9. Drain automatic transaxle oil (for Automatic Transaxle).

10. Remove the front wiper arm head cap.

11. Remove front wiper arm and blade assembly LH.

12. Remove the front wiper arm and blade assembly RH.

13. Remove the hood to cowl top seal.

a. Disengage the 7 clips and remove the hood to cowl top seal.

14. Remove the cowl top ventilator louver sub-assembly.

a. Remove the clip.

b. Disengage the 4 claws and 4

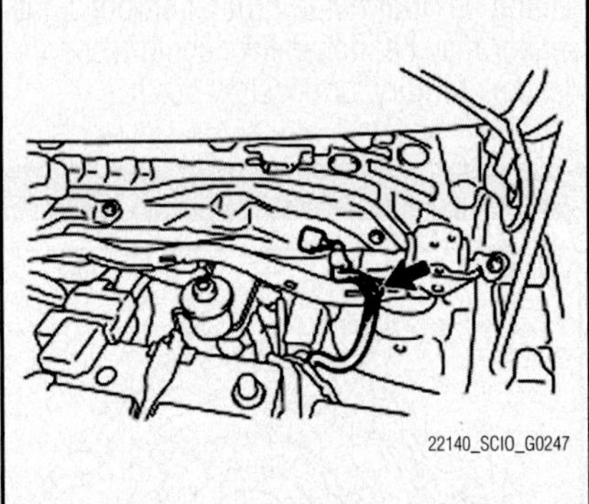

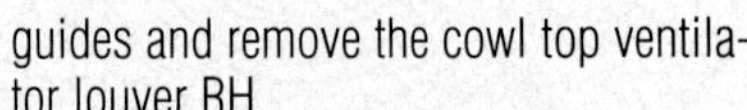
Fig. 14 Disengage the wire harness clamp

guides and remove the cowl top ventilator louver RH.

15. Remove the cowl top ventilator louver LH.

16. Remove the front wiper motor and link.

a. Disengage the clamp.

b. Remove the 2 bolts.

c. Slide the front wiper motor and link. Disengage the link from the rubber pin, then disconnect the connector and remove the front wiper motor and link.

17. Remove the front air shutter seal RH.

a. Disengage the 2 claws and remove the front air shutter seal RH.

18. Remove the outer cowl top panel.

a. Disengage the wire harness clamp.

b. Remove the 9 bolts and remove the outer cowl top panel.

19. Remove the air cleaner assembly.

a. Separate the intake air flow meter connector and 2 wire harness clamps.

b. Separate the ventilation hose from the air cleaner hose.

c. Unfasten the 2 clamps.

d. Remove the air cleaner cap sub-assembly with air cleaner hose.

e. Remove the air cleaner element.

f. Separate the wire harness clamp from the air cleaner case.

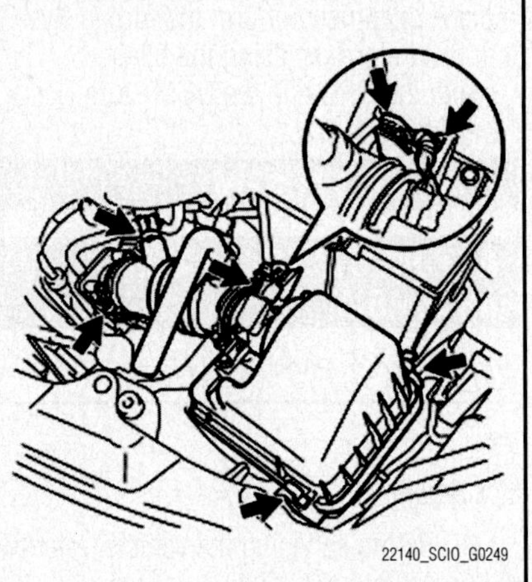

Fig. 16 Remove the air cleaner assembly

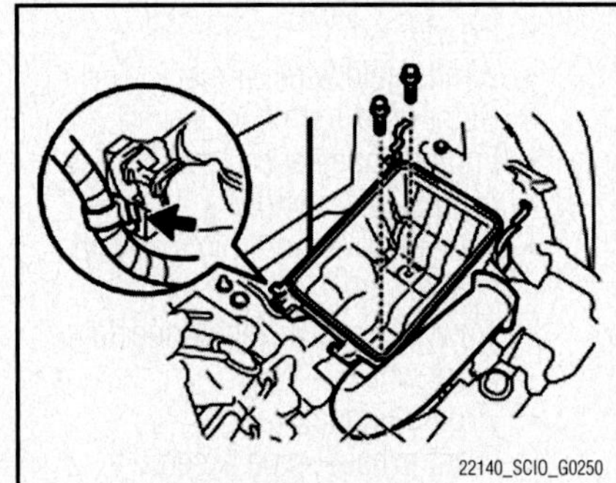

Fig. 17 Separate the wire harness clamp from the air cleaner case

g. Remove the 2 bolts and remove the air cleaner case with air cleaner inlet.

20. Remove the air cleaner bracket.

a. Separate the wire harness clamp from the air cleaner bracket.

b. Remove the 2 bolts and remove the air cleaner bracket.

21. Remove the battery carrier.

a. Separate the wire harness clamp from the battery carrier.

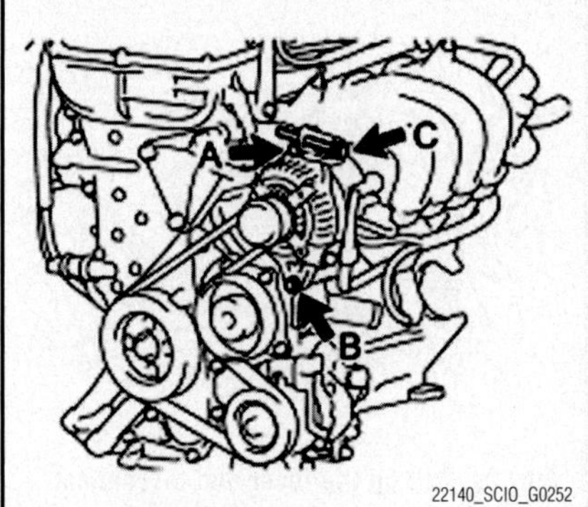

Fig. 18 Remove the fan and alternator V-belt

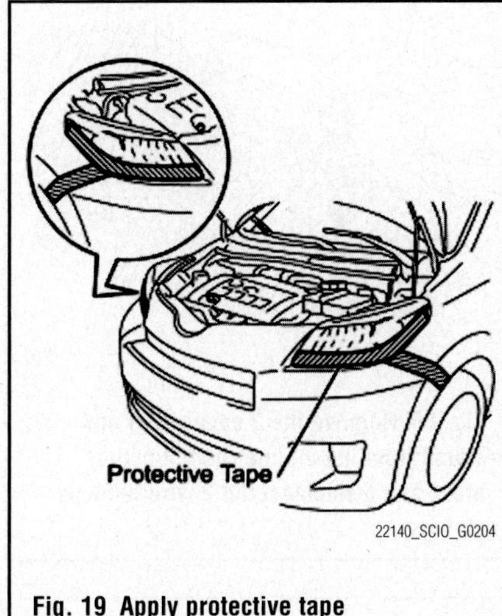

Fig. 19 Apply protective tape

b. Remove the 5 bolts and remove the battery carrier.

22. Remove the Engine cover.

a. (a) First, lift up the rear of the Engine cover to disengage the 2 fittings

b. (b) Next, lift the front of the Engine cover to disengage the 2 fittings. Then remove the Engine cover.

➡Ensure that the rubber grommets remain attached to the Engine cover. If any grommets are attached to the bolts, move them to the Engine cover.

23. Remove the fan and alternator V belt.

a. Loosen bolts A and B.

b. Turn adjusting bolt C to release the tension and remove the V belt from the pulleys.

c. Remove the fan and alternator V belt.

24. Remove the lower radiator hose.

a. Loosen the 2 clips and remove the No. 2 radiator hose.

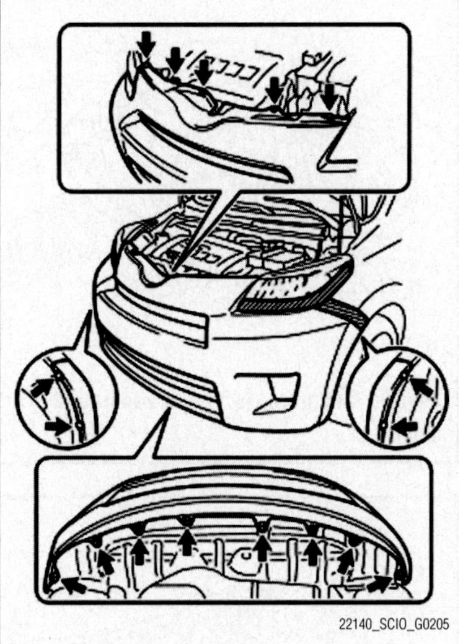

Fig. 20 Remove the 8 screws and 9 clips

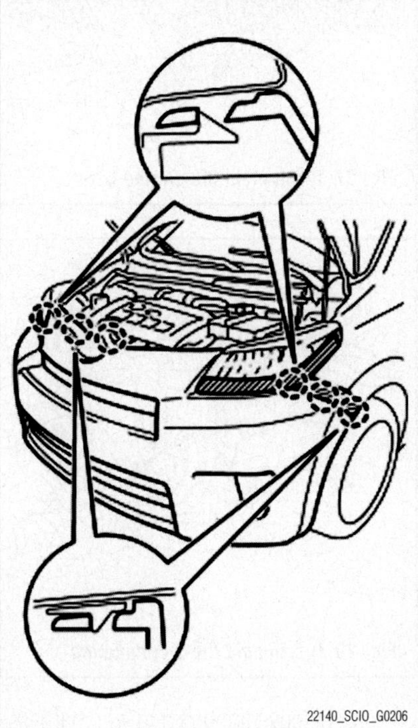

Fig. 21 Disengage the 6 claws and remove the front bumper cover

25. Remove the front bumper cover.

a. Apply protective tape, as shown in the illustration.

b. Remove the 8 screws.

c. Remove the 9 clips.

d. Disengage the 6 claws and remove the front bumper cover.

26. Disconnect the upper radiator hose.

27. Disconnect the transaxle oil cooler

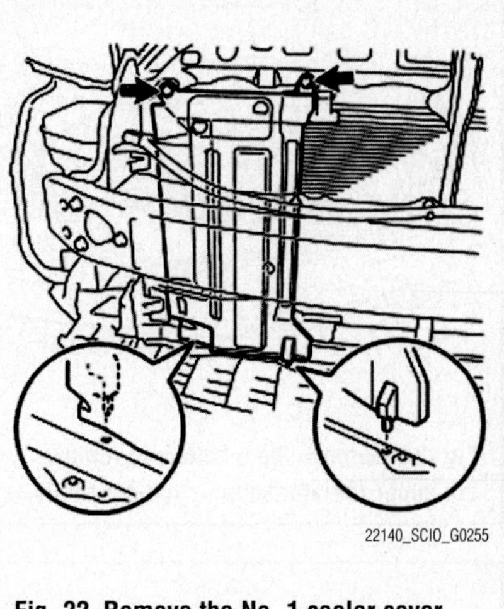

Fig. 22 Remove the No. 1 cooler cover

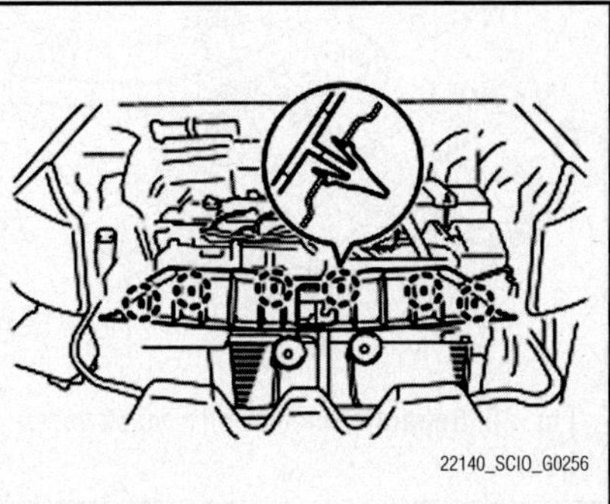

Fig. 23 Remove the upper radiator support absorber

inlet and outlet hoses from the radiator (for Automatic Transaxle).

28. Remove the hood lock assembly.

a. Separate the hood lock control cable assembly from the 2 clamps.

b. Remove the 3 bolts and remove the hood lock assembly.

29. Remove the No. 1 cooler cover.

a. Remove the 2 clips and remove the No. 1 cooler cover.

30. Remove the upper radiator support absorber.

a. Using a screwdriver with its tip wrapped in protective tape, disengage the 6 claws and remove the radiator support upper absorber.

31. Remove the upper radiator support sub-assembly.

a. Disconnect the horn connector.

b. Remove the 5 bolts and remove the upper radiator support sub-assembly.

32. Remove the radiator assembly.

a. Disconnect the cooling fan motor connector and separate the wire harness clamp.

b. Disengage the 2 claws and remove the radiator assembly from the vehicle.

Fig. 24 Remove the 5 bolts and remove the upper radiator support sub-assembly

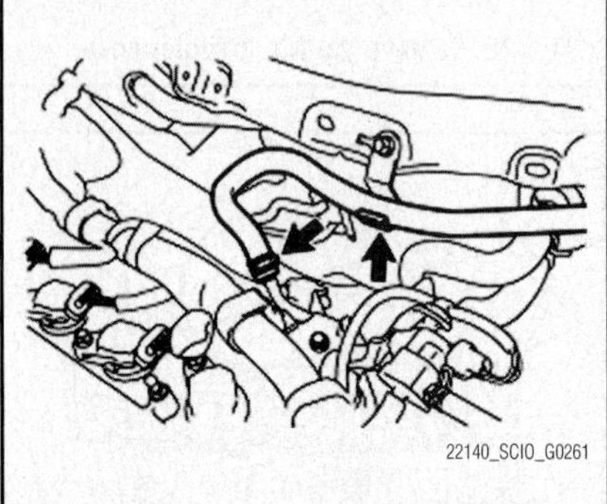
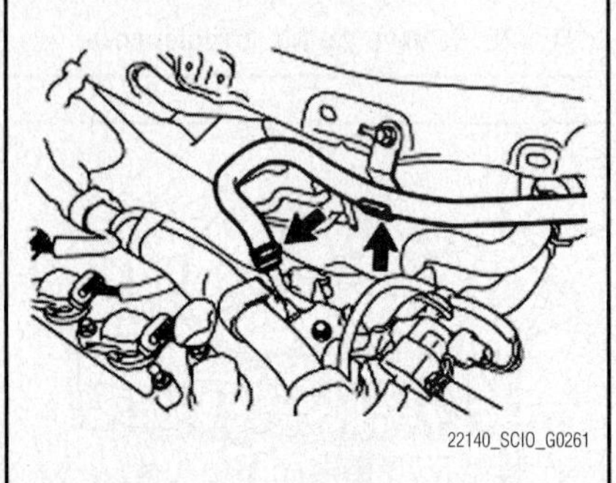

Fig. 25 Separate the union to check valve hose

⁂ CAUTION

Do not apply excessive force to the cooler condenser assembly or piping when removing the radiator assembly.

33. Separate with pulley compressor assembly.
34. Separate the transmission control cable assembly (for Automatic Transaxle).
    a. Remove the nut and disconnect the transmission control cable assembly from the control shaft lever.
35. Separate the transmission control cable assembly (for Manual Transaxle).
    a. Remove the 2 clips and the washers, and disconnect the 2 cables from the transaxle.
    b. Remove the 2 clips and disconnect the 2 cables from the control cable bracket.
36. Separate the union to check valve hose.
    a. Loosen the clip and separate the union to check valve hose.
37. Separate No. 1 fuel vapor feed hose.
    a. Separate the No. 1 fuel vapor feed hose from the vacuum switching valve assembly.
38. Disconnect the engine wire.

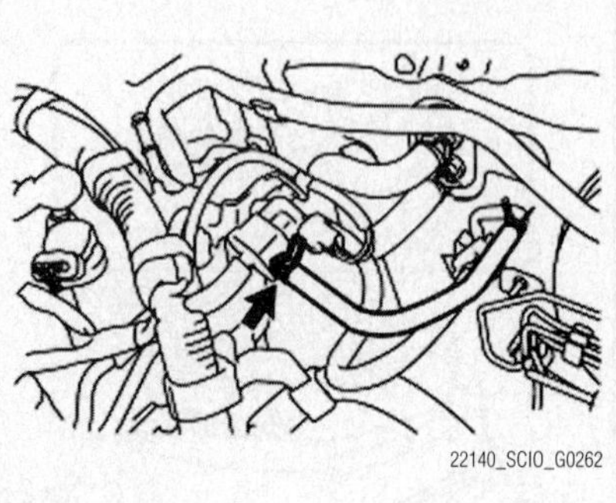

Fig. 26 Separate No. 1 fuel vapor feed hose

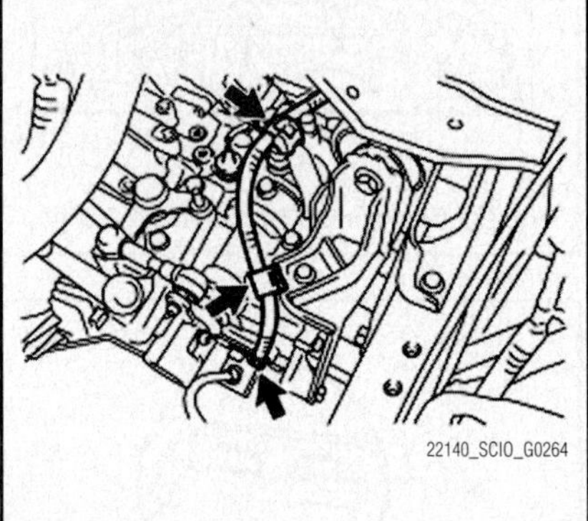

Fig. 27 Disconnect the engine wire

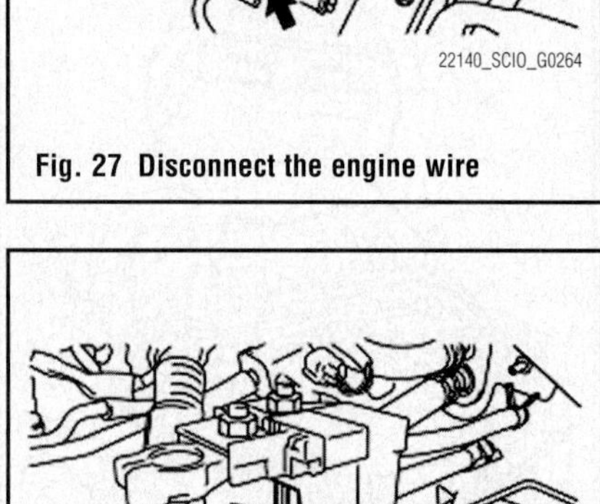

Fig. 28 Disconnect the 2 connectors

    a. Remove the bolt and wire harness clamp and separate the transaxle ground wire (for Automatic Transaxle).
    b. Remove the bolt and 2 wire harness clamp and separate the transaxle ground wire (for Manual Transaxle).
    c. Disconnect the 2 connectors.
    d. Pull up the lever and disconnect the connector from the engine control computer.
    e. Separate the 2 wire harness clamps.
    f. Remove the 2 connectors and clamp from the engine room junction block and disconnect the 2 wire harness clamps.

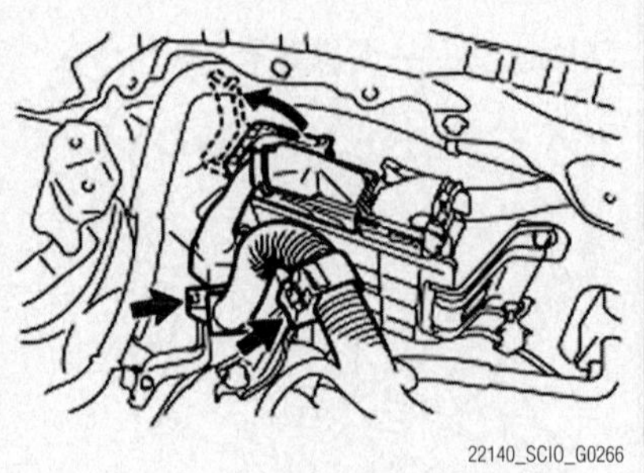

Fig. 29 Pull up the lever and disconnect the connector from the engine control computer

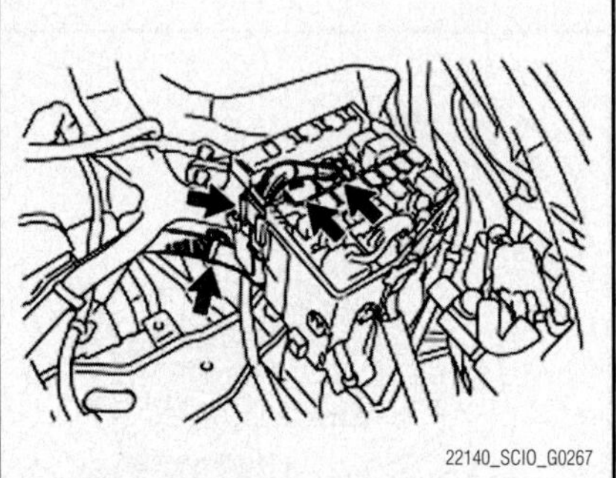

Fig. 30 Remove the 2 connectors and clamp from the engine room junction block and disconnect the 2 wire harness clamps

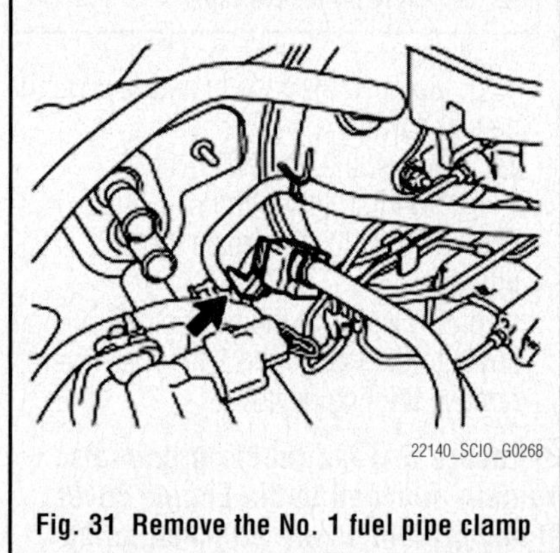

Fig. 31 Remove the No. 1 fuel pipe clamp

    g. Disconnect all wire harnesses and connectors.

⁂ CAUTION

Make sure that no wire harness is connected between the body and engine.

39. Disconnect the upper and lower heater hoses (from heater unit).
40. Disconnect the fuel tube sub-assembly.
    a. Remove the No. 1 fuel pipe clamp.

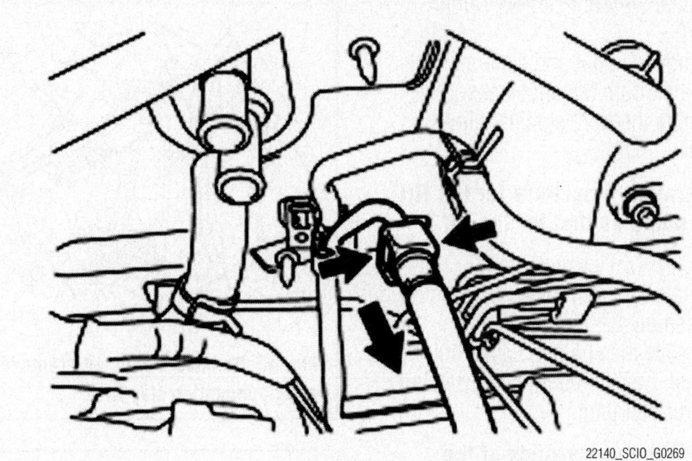

Fig. 32 Pinch the retainer, then pull the fuel tube connector out of the pipe

b. Pinch the retainer, then pull the fuel tube connector out of the pipe.

➡Remove any dirt and foreign matter from the fuel tube connector before performing this work. Do not allow any scratches or foreign matter on the parts when disconnecting, as the fuel tube connector has the O-rings that seal the pipe. Perform this work by hand. Do not use any tools. Do not forcibly bend, twist or turn the nylon tube. Protect the disconnected parts by covering them with vinyl bags after disconnecting the fuel tube. If the fuel tube connector and pipe are stuck, push and pull them to release them.

41. Remove the column hole cover silencer sheet.
   a. Remove the floor carpet and 2 clips and remove the column hole cover silencer.
42. Separate the steering sliding yoke sub-assembly.
   a. Use a seat belt to fix the steering wheel assembly, in order to avoid breakage of the spiral cable.
   b. Place matchmarks on the sliding yoke of the steering intermediate shaft assembly and the power steering.
   c. Loosen bolt A, remove bolt B and separate the steering intermediate shaft assembly.
43. Separate the No. 1 steering column hole cover sub-assembly.
   a. Remove clip A, separate clip B from the body and separate No. 1 steering column hole cover.

CAUTION

Do not damage clip B.

44. Remove the upper console panel.
   a. Disengage the 2 clips and the 6 claws and remove the upper console panel.
45. Remove the rear console box cover.
   a. Disengage the 9 claws and remove the rear console box cover.

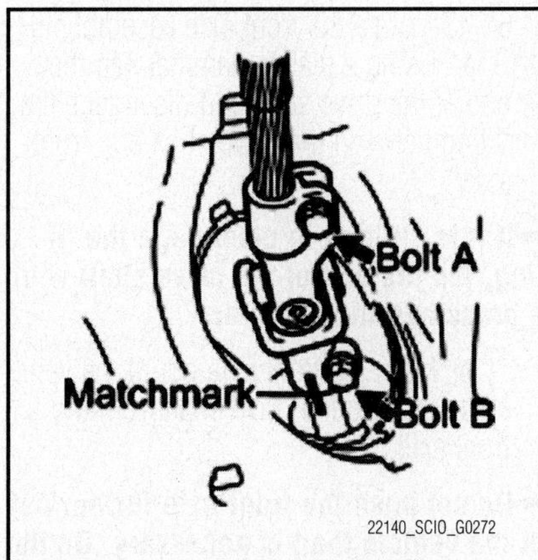

Fig. 34 Place matchmarks on the sliding yoke of the steering intermediate shaft assembly

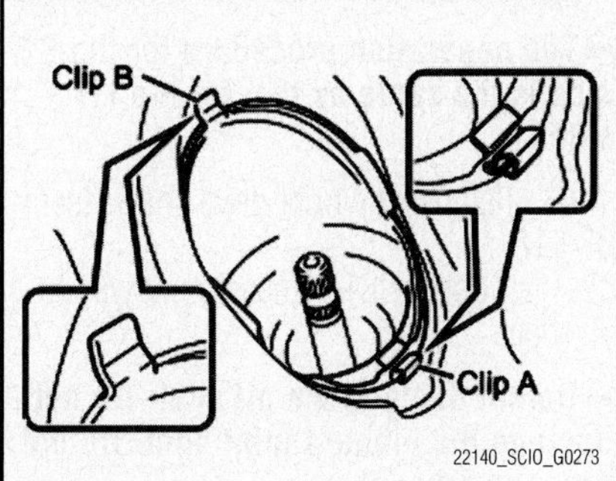

Fig. 35 Separate the No. 1 steering column hole cover sub-assembly

46. Remove the console box carpet.
47. Remove the rear console box assembly.
   a. Remove the 3 screws.
   b. Disengage the 4 claws and remove the rear console box.
48. Remove the front console box.
   a. Remove the 2 screws.
   b. Disengage the clip and 2 claws and remove the front console box.
49. Remove the front floor center brace.
   a. Remove the 2 bolts and remove the front floor center brace.
50. Remove the front exhaust pipe assembly.

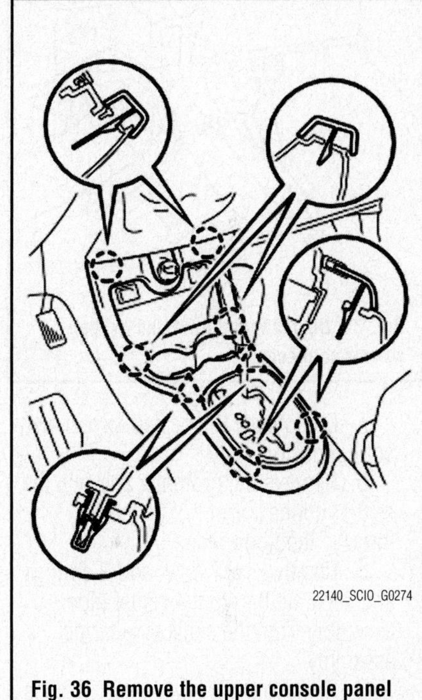

Fig. 36 Remove the upper console panel

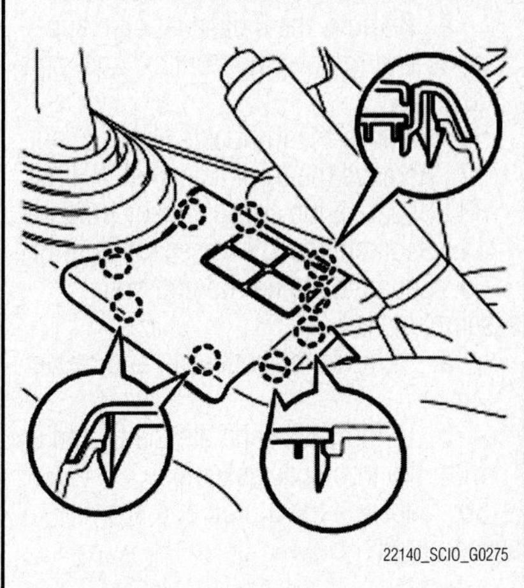

Fig. 37 Remove the rear console box cover

Fig. 38 Remove the front floor center brace

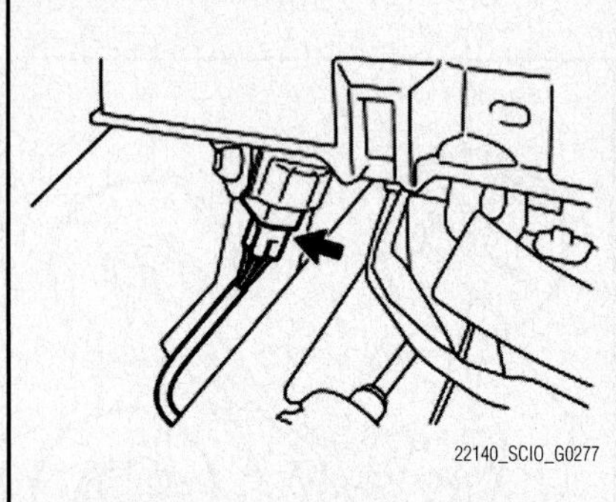

Fig. 39 Disconnect the heated oxygen sensor connector

a. Disconnect the heated oxygen sensor connector.

b. Remove the grommet and pull the sensor connector out of the cabin through the floor panel.

c. Remove the 2 bolts and 2 springs and separate the front exhaust pipe assembly from the exhaust tail pipe assembly.

d. Remove the 2 bolts and 2 springs and separate the front exhaust pipe assembly from the exhaust manifold.

e. Remove the 3 exhaust pipe supports and remove the front exhaust pipe assembly.

51. Remove the front axle shaft LH nut.
52. Remove the front axle shaft RH nut.
53. Separate the speed sensor front LH.
54. Separate the speed sensor front RH.
55. Separate the tie rod end sub-assembly LH.

a. Remove the cotter pin and castle nut.

b. Using SST, separate the tie rod end from the front axle assembly.

56. Separate the tie rod end sub-assembly RH

➡The separation procedure for the RH side is the same as that for the LH side.

57. Separate the front stabilizer link assembly LH.

a. Remove the nut and separate the stabilizer link from the shock absorber.

58. Separate the front stabilizer link assembly RH.

➡The separation procedure for the RH side is the same as that for the LH side.

59. Separate the front lower suspension arm sub-assembly LH.

a. Remove the clip and castle nut.

b. Install SST to the threaded section of the lower ball joint.

➡Make sure the upper ends of the threaded section of the lower ball joint and SST (spacer B) are aligned.

c. Using SST, separate the front lower suspension arm from the front axle assembly.

60. Separate the front lower suspension arm sub-assembly RH

➡The separation procedure for the RH side is the same as that for the LH side.

61. Separate the front axle assembly LH.

a. Using a plastic hammer, tap the end of the drive shaft and disengage the fitting between the drive shaft and front axle.

➡If it is difficult to disengage the fitting, tap the end of the drive shaft with a brass bar and hammer.

b. Push the front axle out of the vehicle to remove the drive shaft from the front axle.

➡Do not push the front axle further out of the vehicle than is necessary. Do not damage the outboard joint boot. Do not damage the speed sensor rotor. Suspend the drive shaft with a piece of rope or the equivalent.

62. Separate the front axle assembly RH.

➡The separation procedure for the RH side is the same as that for the LH side.

63. Remove the front drive shaft assembly LH.

a. Using SST, remove the drive shaft.

➡Do not damage the oil seal. Do not damage the inboard joint boot. Do not drop the drive shaft.

64. Remove the front drive shaft assembly RH.

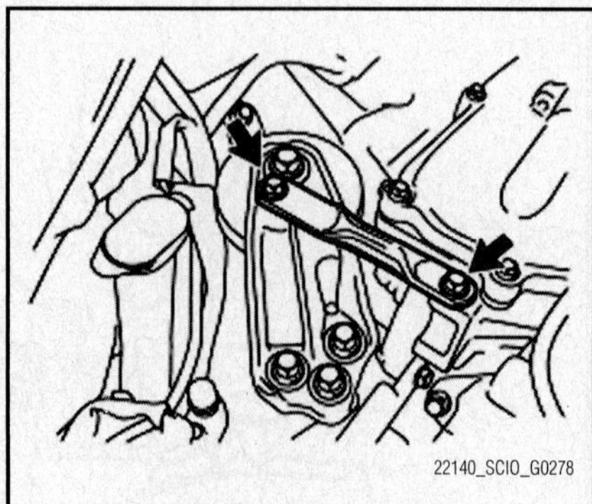

Fig. 40 Remove 2 bolts and remove the engine mounting stay RH

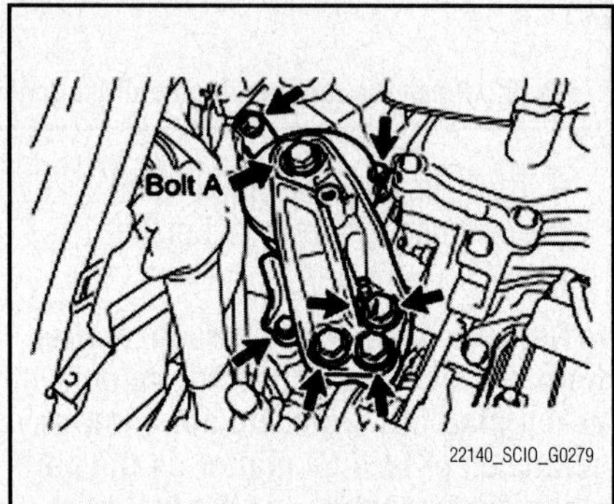

Fig. 41 Remove the 6 bolts and nut and remove the engine mounting insulator RH

a. Using a screwdriver and hammer, remove the drive shaft.

➡Do not damage the oil seal. Do not damage the inboard joint boot. Do not drop the drive shaft.

65. Remove the flywheel housing under cover (for Automatic Transaxle).

66. Remove the drive plate and torque converter clutch setting bolt (for Automatic Transaxle).

a. Remove the 6 torque converter set bolts while holding the crankshaft pulley bolt with a wrench.

67. Remove the engine assembly with transaxle.

a. Set the engine lifter.

b. Remove 2 bolts and remove the engine mounting stay RH.

c. Remove the 6 bolts and nut and remove the engine mounting insulator RH.

➡Be sure not to remove the other bolt A.

d. Remove the through bolt and nut and separate the transverse engine mounting insulator.

e. Remove the 6 bolts and remove the engine assembly with transaxle and front suspension crossmember from the vehicle.

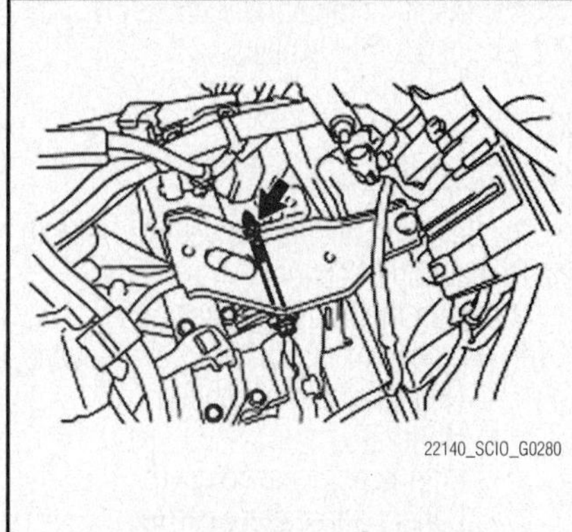
Fig. 42 Remove the through bolt and nut

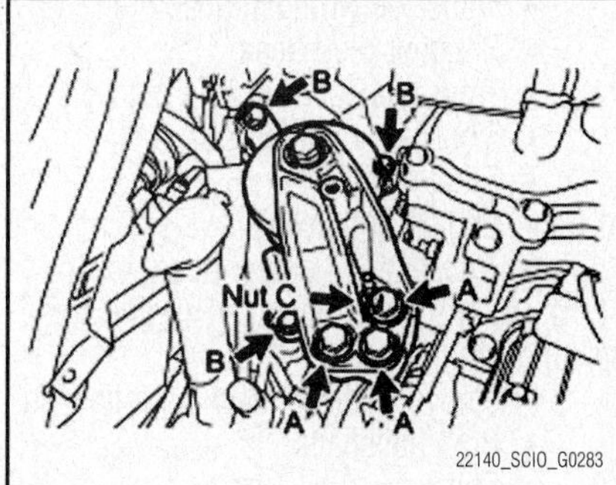

Fig. 43 Install the engine mounting insulator RH with the 6 bolts and nut

To install:

68. Install front suspension crossmember sub-assembly.
   a. Install the engine moving control rod with the through bolt. Torque: 89 ft. lbs. (120 Nm).
   b. Remove the 2 bolts and remove the 2 engine hangers.
   c. Install the air fuel ratio sensor bracket with bolt. Torque: 32 ft. lbs. (43 Nm).
69. Install transverse engine mounting insulator.
   a. Install the transverse engine mounting insulator with the 5 bolts. Torque: 38 ft. lbs. (52 Nm).
70. Install engine assembly with transaxle.
   a. Set the engine assembly with transaxle and front suspension crossmember on the engine lifter.
   b. Operate the engine lifter and lift the engine assembly with transaxle and front suspension crossmember to the position where the engine mounting insulator RH and transverse engine mounting insulator can be installed.
   c. Install the transverse engine mounting insulator with the through bolt and nut. Torque: 38 ft. lbs. (52 Nm).

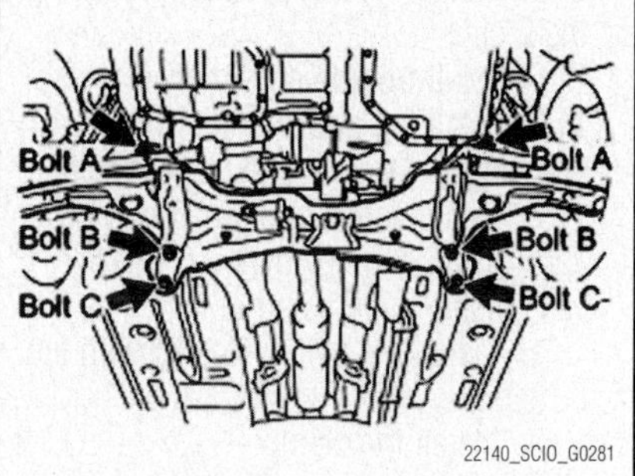

Fig. 44 Tighten bolts A, B and C on both sides in several sequences

   d. Install the engine mounting insulator RH with the 6 bolts and nut.
   e. Torque: 65 ft. lbs. (88 Nm) for Bolt A.
   f. Torque: 33 ft. lbs. (45 Nm) for Bolt B.
   g. Torque: 38 ft. lbs. (52 Nm) for Bolt C.
   h. Install the engine mounting stay RH with the 2 bolts. Torque: 19 ft. lbs. (26 Nm).
   i. Operate the engine lifter and provisionally install the engine assembly with transaxle and front suspension crossmember onto the vehicle with the 6 bolts.
   j. Insert SST into the datum holes in the front suspension crossmember RH and LH alternately and tighten bolts A, B and C on both sides in several sequences.
   k. Torque: 52 ft. lbs. (70 Nm) for Bolt A.
   l. Torque: 118 ft. lbs. (160 Nm) for Bolt B.
   m. Torque: 70 ft. lbs. (95 Nm) for Bolt C.

**** CAUTION**

Insert SST into the datum hole vertically. If impossible to insert SST vertically, loosen all bolts and then insert SST again.

71. Install drive plate and torque converter clutch setting bolt (for Automatic Transaxle).
72. Install flywheel housing under cover (for Automatic Transaxle).
73. Install front drive shaft assembly LH.
74. Install front drive shaft assembly RH.
75. Install front axle assembly LH.
76. Install front axle assembly RH.
77. Install front lower suspension arm sub-assembly LH.
78. Install front lower suspension arm sub-assembly RH.
79. Install front stabilizer link assembly LH.
80. Install front stabilizer link assembly RH.
81. Install tie rod end sub-assembly LH.
82. Install tie rod end sub-assembly RH.
83. Install speed sensor front LH.
84. Install speed sensor front RH.
85. Install front axle shaft LH nut.
86. Install front axle shaft RH nut.
87. Install front exhaust pipe assembly.
   a. Using a plastic hammer and a wooden block, tap in a new exhaust pipe gasket until its surface is flush with the exhaust manifold.
   b. Install the 3 exhaust pipe supports.
   c. Install the exhaust front pipe assembly and a new exhaust pipe gasket with the 4 compression springs and 4 bolts. Torque: 32 ft. lbs. (43 Nm).
   d. Pass the sensor connector through the floor panel and install the grommet.
   e. Connect the heated oxygen sensor connector.
88. Install front floor brace center.
89. Install front console box.
90. Install rear console box assembly.
91. Install console box rear cover.
92. Install console box carpet.
93. Install console panel upper.
94. Install shift lever knob sub-assembly (for Manual Transaxle).
95. Install No. 1 steering column hole cover sub-assembly.
96. Install steering sliding yoke sub-assembly.
97. Install column hole cover silencer sheet.
98. Install clutch release cylinder assembly (for Manual Transaxle).
   a. Install the clutch release cylinder and clutch pipe with the 4 bolts.
99. Connect fuel tube sub-assembly.
   a. Connect the fuel tube connector and fuel pipe, and install the No. 1 fuel pipe clamp.

➡Align the fuel tube connector with the pipe, then push the fuel tube connector in until the retainer makes a click sound. If the connection is tight, apply a small amount of engine oil to the tip of the pipe. After connecting, pull the pipe and connector to make sure that they are securely connected.

100. Connect heater water hoses to the heater unit.
101. Connect engine wire.
   a. Connect the 2 engine wire harness connectors and 2 wire harness clamps to the engine room junction block.

b. Connect the engine wire harness connector to the ECM.

c. Connect the 2 wire harness clamps.

d. Connect the 2 connectors.

e. Install the transaxle ground wire with the bolt and 2 wire harness clamps (for Manual Transaxle).

f. Install the transaxle ground wire with the bolt and wire harness clamp (for Automatic Transaxle).

102. Install No. 1 fuel vapor feed hose.

a. Install the No. 1 fuel vapor feed hose into the vacuum switching valve assembly.

103. Install union to check valve hose.

a. Install the union to check valve hose with the clip.

104. Install transmission control cable assembly (for Automatic Transaxle)

a. Install the transmission control cable assembly onto the control shaft lever with the nut.

b. Fix the transmission control cable assembly onto the control cable bracket with a new clip.

➡Make sure that the marking on the cable is aligned with the slit in the transmission control cable bracket before installing the cable.

105. Install transmission control cable assembly (for Manual Transaxle).

a. Install the control cable assembly onto the control cable bracket with 2 new clips.

b. Connect the control cable assembly and install the 2 washers and 2 clips.

c. Install the control cable assembly onto the shift lever retainer.

➡Install the cable with the protruding portion of the cable outer facing upward. After installing. check that the lock of the cable outer is protruding beyond portion A. as shown in the illustration.

106. Install with pulley compressor assembly.

107. Install radiator assembly.

108. Install upper radiator support sub-assembly.

109. Install upper radiator support absorber.

110. Install No. 1 cooler cover.

111. Install hood lock assembly.

112. Connect No. 2 oil cooler inlet hose (for automatic transaxle).

113. Connect No. 1 oil cooler outlet hose (for automatic transaxle).

114. Connect No. 1 radiator hose.

115. Install No. 2 radiator hose.

a. Install the No. 2 radiator hose with the 2 clips.

116. Install front bumper cover.

117. Install fan and alternator v belt.

118. Adjust fan and alternator v belt.

119. Inspect fan and alternator v belt.

120. Install engine cover.

121. Install battery carrier.

a. Install the battery carrier with the 5 bolts.

b. Install the clamp.

122. Install air cleaner bracket.

a. Install the air cleaner bracket with the 2 bolts.

b. Connect the wire harness clamp to the air cleaner bracket.

123. Install air cleaner assembly

a. Install the air cleaner case with air cleaner inlet with the 2 bolts.

b. Connect the wire harness to the air cleaner case.

c. Install the air cleaner element.

d. Install and lock the air cleaner cap with air cleaner hose and then tighten the air cleaner hose clamp.

e. Connect the ventilation hose to the air cleaner hose.

f. Connect the intake air flow meter connector and 2 wire harness clamps.

124. Install outer cowl top panel

a. Install the outer cowl top panel with the 9 bolts.

b. Connect the wire harness clamp.

125. Install front air shutter seal RH

a. Engage the 2 claws to install the front air shutter seal RH.

126. Install front wiper motor and link.

127. Install cowl top ventilator louver LH.

128. Install cowl top ventilator louver sub-assembly.

129. Install hood to cowl top seal.

130. Install front wiper arm and blade assembly LH.

131. Install front wiper arm and blade assembly RH.

132. Install front wiper arm head cap.

133. Install battery tray.

134. Install battery.

135. Add engine coolant.

136. Add automatic transaxle oil (for automatic transaxle).

137. Inspect automatic transaxle oil (for automatic transaxle).

138. Add manual transaxle oil (for manual transaxle).

139. Inspect manual transaxle oil (for manual transaxle).

140. Inspect for fuel leaks.

141. Inspect for engine oil leak..

142. Inspect for exhaust gas leak.

143. Inspect for engine coolant leak.

144. Install engine under cover RH.

145. Install engine under cover LH.

146. Install front wheels.

tC

See Figure 45.

1. Before servicing the vehicle, refer to the Precautions Section.
2. Drain the transaxle assembly.
3. Remove or disconnect the following:
   - Negative battery cable
   - Hood
   - Engine appearance cover
   - Battery and battery carrier
   - Air intake assembly
   - Transaxle control cables
   - Transaxle wiring connectors
   - Breather plug hose
   - Transaxle control cable brackets
   - Transaxle oil filler tube
   - Front wheel
   - Engine undercover
   - Transaxle oil cooler hoses
4. At this point, attach an engine crane to support the engine.
5. Remove or disconnect the following:
   - Both halfshafts
   - Starter
6. Support the transaxle with a suitable jack.
7. Remove or disconnect the following:
   - Front engine mount
   - Rear engine mount
   - Center engine crossmember
   - Left engine mount and bracket
   - Front engine mounting bracket from the transaxle
   - Rear engine mounting bracket from the transaxle
   - Flywheel housing cover
8. Turn the crankshaft to remove the 6 torque converter mounting bolts.
9. Remove the 7 transaxle mounting bolts.

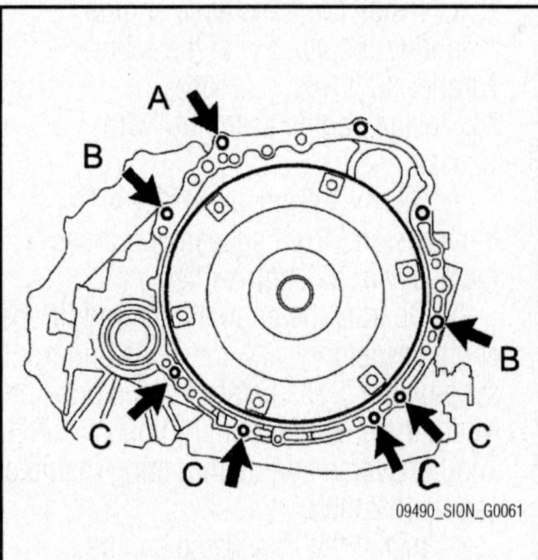

Fig. 45 Automatic transaxle mounting bolt identification—tC

10. Remove the transaxle assembly from the engine.

To install:

11. Install the transaxle. Tighten the bolts as follows:
    a. Bolt A: 47 ft. lbs. (64 Nm).
    b. Bolt B: 34 ft. lbs. (46 Nm).
    c. Bolts C: 32 ft. lbs. (44 Nm).
12. Install or connect the following:
    - Torque converter mounting bolts. Tighten to 30 ft. lbs. (41 Nm).
    - Flywheel housing cover
    - Rear engine mounting bracket. Tighten to 47 ft. lbs. (64 Nm).
    - Front engine mounting bracket. Tighten to 47 ft. lbs. (64 Nm).
    - Left engine mounting bracket. Tighten to 47 ft. lbs. (64 Nm).
    - Left engine mount
    - Center engine crossmember
    - Rear engine mount
    - Front engine mount
    - Starter
    - Halfshafts
    - Front exhaust pipe
    - Engine undercover
    - Front wheel
    - Transaxle oil cooler hoses
    - Transaxle control cable support brackets
    - Transaxle oil filler tube
    - Breather plug hose
    - Transaxle wiring connectors
    - Transaxle control cable
    - Air intake assembly
    - Battery carrier and battery
    - Engine appearance cover
    - Hood
    - Negative battery cable
13. Refill the transaxle with fluid to the correct level.
14. Start the engine and check for leaks.

MANUAL TRANSAXLE ASSEMBLY

REMOVAL & INSTALLATION

xB

See Figure 46.

1. Before servicing the vehicle, refer to the Precautions Section.
2. Drain the transaxle assembly.
3. Remove or disconnect the following:
    - Negative battery cable
    - Front wheel
    - Engine undercover
    - Hood
    - Windshield wiper arms
    - Outer cowl top panel
    - Engine appearance cover

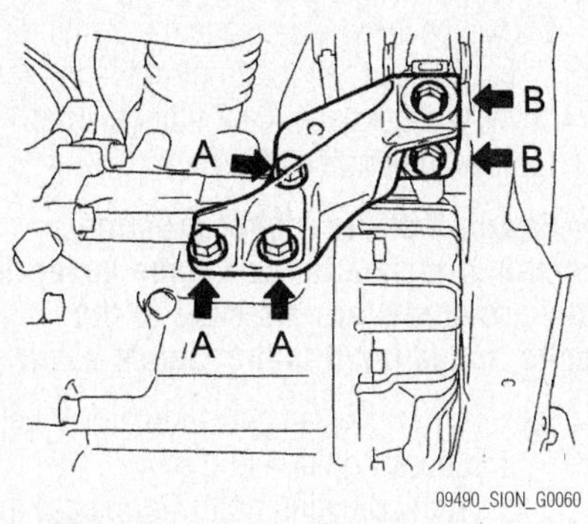

Fig. 46 Rear engine mount—Bracket-to-transaxle bolts (A) and engine mounting bolts (B)

    - Air intake assembly
    - Battery
    - Back-up light switch connector
    - Clutch release cylinder hose
    - Air cleaner mounting bracket
    - Transaxle control cables
    - Both halfshafts
    - Front floor panel brace
    - Front exhaust pipe
4. At this point, attach an engine hoist to support the engine.
5. Remove or disconnect the following:
    - Front suspension crossmember
    - Starter
6. Support the transaxle with a suitable jack.
7. Remove the left-hand engine mounting bracket.
8. Remove the rear engine mounting bracket.
9. Remove the 7 transaxle mounting bolts.
10. Remove the transaxle assembly from the engine.

To install:

11. Align the input shaft with the clutch disc and install the transaxle to the engine. Tighten the mounting bolts to 24 ft. lbs. (33 Nm).
12. Install or connect the following:
    - Engine left mounting bracket. Tighten the bracket bolts 47 ft. lbs. (64 Nm); the mount bolts to 36 ft. lbs. (49 Nm).
    - Starter
    - Front suspension crossmember
    - Halfshafts
    - Front exhaust pipe
    - Front floor panel brace
    - Transaxle control cables
    - Air cleaner mounting bracket. Tighten to 14 ft. lbs. (19 Nm).
    - Clutch release cylinder hose
    - Back-up light switch connector
    - Battery
    - Air intake assembly
    - Engine appearance cover
    - Outer cowl top panel
    - Windshield wiper arms
    - Hood
    - Engine undercover
    - Front wheel
    - Negative battery cable
13. Refill the transaxle with fluid to the correct level.
14. Start the engine and check for leaks.

xD

See Figures 12 through 26, 28 through 44 and 47.

1. Discharge the fuel system pressure.

⁂ CAUTION

The DISCHARGE FUEL SYSTEM PRESSURE procedures must be performed before disconnecting any part of the fuel system.

⁂ CAUTION

After performing the DISCHARGE FUEL SYSTEM PRESSURE procedures, pressure will remain in the fuel line. When disconnecting the fuel line, place a cloth or equivalent over fittings to reduce the risk of fuel spray.

    a. Remove the rear seat cushion assembly LH
    b. Remove the rear floor service hose cover.
    c. Disconnect the connector from the fuel pump assembly.
    d. Start the engine. After the engine stops naturally, turn the ignition switch OFF.
    e. Start the engine again and make sure that engine does not start.
    f. Remove the fuel tank cap, and let the air out of the fuel tank.
    g. Connect the connector.
    h. Install the rear floor service hole cover with new butyl tape.
    i. Install the rear seat cushion assembly LH.
2. Remove the battery.
    a. Disconnect the cable from the battery terminal.
    b. Loosen the nut and remove the battery clamp.
    c. Remove the battery.
3. Remove the battery tray.
4. Remove the front wheels.
5. Remove the engine under cover LH.

6. Remove the engine under cover RH.
7. Drain the engine coolant.
8. Drain the manual transaxle oil (for Manual Transaxle).
9. Drain automatic transaxle oil (for Automatic Transaxle).
10. Remove the front wiper arm head cap.
11. Remove front wiper arm and blade assembly LH.
12. Remove the front wiper arm and blade assembly RH.
13. Remove the hood to cowl top seal.
    a. Disengage the 7 clips and remove the hood to cowl top seal.
14. Remove the cowl top ventilator louver sub-assembly.
    a. Remove the clip.
    b. Disengage the 4 claws and 4 guides and remove the cowl top ventilator louver RH.
15. Remove the cowl top ventilator louver LH.
16. Remove the front wiper motor and link.
    a. Disengage the clamp.
    b. Remove the 2 bolts.
    c. Slide the front wiper motor and link. Disengage the link from the rubber pin, then disconnect the connector and remove the front wiper motor and link.
17. Remove the front air shutter seal RH.
    a. Disengage the 2 claws and remove the front air shutter seal RH.
18. Remove the outer cowl top panel.
    a. Disengage the wire harness clamp.
    b. Remove the 9 bolts and remove the outer cowl top panel.
19. Remove the air cleaner assembly.
    a. Separate the intake air flow meter connector and 2 wire harness clamps.
    b. Separate the ventilation hose from the air cleaner hose.
    c. Unfasten the 2 clamps.
    d. Remove the air cleaner cap sub-assembly with air cleaner hose.
    e. Remove the air cleaner element.
    f. Separate the wire harness clamp from the air cleaner case.
    g. Remove the 2 bolts and remove the air cleaner case with air cleaner inlet.
20. Remove the air cleaner bracket.
    a. Separate the wire harness clamp from the air cleaner bracket.
    b. Remove the 2 bolts and remove the air cleaner bracket.
21. Remove the battery carrier.
    a. Separate the wire harness clamp from the battery carrier.
    b. Remove the 5 bolts and remove the battery carrier.
22. Remove the Engine cover.
    a. (a) First, lift up the rear of the Engine cover to disengage the 2 fittings
    b. (b) Next, lift the front of the Engine cover to disengage the 2 fittings. Then remove the Engine cover.

➡Ensure that the rubber grommets remain attached to the Engine cover. If any grommets are attached to the bolts, move them to the Engine cover.

23. Remove the fan and alternator V belt.
    a. Loosen bolts A and B.
    b. Turn adjusting bolt C to release the tension and remove the V belt from the pulleys.
    c. Remove the fan and alternator V belt.
24. Remove the lower radiator hose.
    a. Loosen the 2 clips and remove the No. 2 radiator hose.
25. Remove the front bumper cover.
    a. Apply protective tape, as shown in the illustration.
    b. Remove the 8 screws.
    c. Remove the 9 clips.
    d. Disengage the 6 claws and remove the front bumper cover.
26. Disconnect the upper radiator hose.
27. Disconnect the oil cooler inlet and outlet hoses from the radiator (for Automatic Transaxle).
28. Remove the hood lock assembly.
    a. Separate the hood lock control cable assembly from the 2 clamps.
    b. Remove the 3 bolts and remove the hood lock assembly.
29. Remove the No. 1 cooler cover.
    a. Remove the 2 clips and remove the No. 1 cooler cover.
30. Remove the upper radiator support absorber.
    a. Using a screwdriver with its tip wrapped in protective tape, disengage the 6 claws and remove the radiator support upper absorber.
31. Remove the upper radiator support sub-assembly.
    a. Disconnect the horn connector.
    b. Remove the 5 bolts and remove the upper radiator support sub-assembly.
32. Remove the radiator assembly.
    a. Disconnect the cooling fan motor connector and separate the wire harness clamp.
    b. Disengage the 2 claws and remove the radiator assembly from the vehicle.

⁂ CAUTION

Do not apply excessive force to the cooler condenser assembly or piping when removing the radiator assembly.

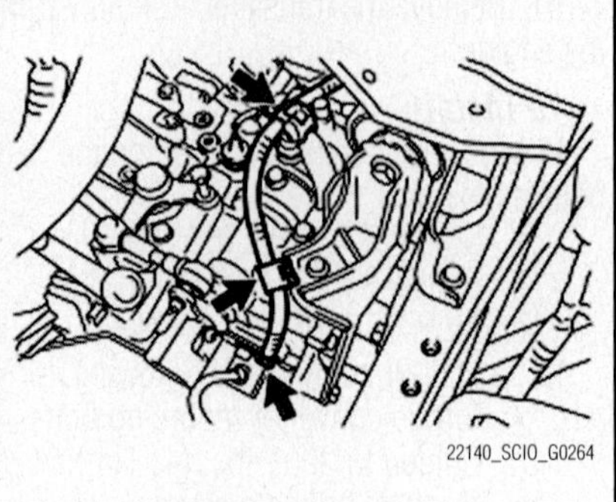

Fig. 47 Disconnect the engine wire (for Manual Transaxle)

33. Separate with pulley compressor assembly.
34. Separate the transmission control cable assembly (for Automatic Transaxle).
    a. Remove the nut and disconnect the transmission control cable assembly from the control shaft lever.
35. Separate the transmission control cable assembly (for Manual Transaxle).
    a. Remove the 2 clips and the washers, and disconnect the 2 cables from the transaxle.
    b. Remove the 2 clips and disconnect the 2 cables from the control cable bracket.
36. Separate the union to check valve hose.
    a. Loosen the clip and separate the union to check valve hose.
37. Separate No. 1 fuel vapor feed hose.
    a. Separate the No. 1 fuel vapor feed hose from the vacuum switching valve assembly.
38. Disconnect the engine wire.
    a. Remove the bolt and 2 wire harness clamp and separate the transaxle ground wire (for Manual Transaxle).
    b. Disconnect the 2 connectors.
    c. Pull up the lever and disconnect the connector from the engine control computer.
    d. Separate the 2 wire harness clamps.
    e. Remove the 2 connectors and clamp from the engine room junction block and disconnect the 2 wire harness clamps.
    f. Disconnect all wire harnesses and connectors.

⁂ CAUTION

Make sure that no wire harness is connected between the body and engine.

39. Disconnect the upper and lower heater hoses (from heater unit).

40. Disconnect the fuel tube sub-assembly.

a. Remove the No. 1 fuel pipe clamp.

b. Pinch the retainer, then pull the fuel tube connector out of the pipe.

➡Remove any dirt and foreign matter from the fuel tube connector before performing this work. Do not allow any scratches or foreign matter on the parts when disconnecting, as the fuel tube connector has the O-rings that seal the pipe. Perform this work by hand. Do not use any tools. Do not forcibly bend, twist or turn the nylon tube. Protect the disconnected parts by covering them with vinyl bags after disconnecting the fuel tube. If the fuel tube connector and pipe are stuck, push and pull them to release them.

41. Separate the clutch release cylinder assembly (for Manual Transaxle).

a. Remove the 4 bolts, then separate the clutch release cylinder.

➡Suspend the clutch release cylinder with a piece of rope so as not to overload the clutch pipe.

42. Remove the column hole cover silencer sheet.

a. Remove the floor carpet and 2 clips and remove the column hole cover silencer.

43. Separate the steering sliding yoke sub-assembly.

a. Use a seat belt to fix the steering wheel assembly, in order to avoid breakage of the spiral cable.

b. Place matchmarks on the sliding yoke of the steering intermediate shaft assembly and the power steering.

c. Loosen bolt A, remove bolt B and separate the steering intermediate shaft assembly.

44. Separate the No. 1 steering column hole cover sub-assembly.

a. Remove clip A, separate clip B from the body and separate No. 1 steering column hole cover.

⁂ CAUTION

Do not damage clip B.

45. Remove the shift lever knob sub-assembly (for Manual Transaxle).

a. Remove the shift lever knob by turning the knob counterclockwise.

46. Remove the upper console panel.

a. Disengage the 2 clips and the 6 claws and remove the upper console panel.

47. Remove the rear console box cover.

a. Disengage the 9 claws and remove the rear console box cover.

48. Remove the console box carpet.

49. Remove the rear console box assembly.

a. Remove the 3 screws.

b. Disengage the 4 claws and remove the rear console box.

50. Remove the front console box.

a. Remove the 2 screws.

b. Disengage the clip and 2 claws and remove the front console box.

51. Remove the front floor center brace.

a. Remove the 2 bolts and remove the front floor center brace.

52. Remove the front exhaust pipe assembly.

a. Disconnect the heated oxygen sensor connector.

b. Remove the grommet and pull the sensor connector out of the cabin through the floor panel.

c. Remove the 2 bolts and 2 springs and separate the front exhaust pipe assembly from the exhaust tail pipe assembly.

d. Remove the 2 bolts and 2 springs and separate the front exhaust pipe assembly from the exhaust manifold.

e. Remove the 3 exhaust pipe supports and remove the front exhaust pipe assembly.

53. Remove the front axle shaft LH nut.

54. Remove the front axle shaft RH nut.

55. Separate the speed sensor front LH.

56. Separate the speed sensor front RH.

57. Separate the tie rod end sub-assembly LH.

a. Remove the cotter pin and castle nut.

b. Using SST, separate the tie rod end from the front axle assembly.

58. Separate the tie rod end sub-assembly RH

➡The separation procedure for the RH side is the same as that for the LH side.

59. Separate the front stabilizer link assembly LH.

a. Remove the nut and separate the stabilizer link from the shock absorber.

60. Separate the front stabilizer link assembly RH.

➡The separation procedure for the RH side is the same as that for the LH side.

61. Separate the front lower suspension arm sub-assembly LH.

a. Remove the clip and castle nut.

b. Install SST to the threaded section of the lower ball joint.

➡Make sure the upper ends of the threaded section of the lower ball joint and SST (spacer B) are aligned.

c. Using SST, separate the front lower suspension arm from the front axle assembly.

62. Separate the front lower suspension arm sub-assembly RH

➡The separation procedure for the RH side is the same as that for the LH side.

63. Separate the front axle assembly LH.

a. Using a plastic hammer, tap the end of the drive shaft and disengage the fitting between the drive shaft and front axle.

➡If it is difficult to disengage the fitting, tap the end of the drive shaft with a brass bar and hammer.

b. Push the front axle out of the vehicle to remove the drive shaft from the front axle.

➡Do not push the front axle further out of the vehicle than is necessary. Do not damage the outboard joint boot. Do not damage the speed sensor rotor. Suspend the drive shaft with a piece of rope or the equivalent.

64. Separate the front axle assembly RH.

➡The separation procedure for the RH side is the same as that for the LH side.

65. Remove the front drive shaft assembly LH.

a. Using SST, remove the drive shaft.

➡Do not damage the oil seal. Do not damage the inboard joint boot. Do not drop the drive shaft.

66. Remove the front drive shaft assembly RH.

a. Using a screwdriver and hammer, remove the drive shaft.

➡Do not damage the oil seal. Do not damage the inboard joint boot. Do not drop the drive shaft.

67. Remove the flywheel housing under cover (for Automatic Transaxle).

68. Remove the drive plate and torque converter clutch setting bolt (for Automatic Transaxle).

a. (a) Remove the 6 torque converter set bolts while holding the crankshaft pulley bolt with a wrench.

69. Remove the engine assembly with transaxle.

a. Set the engine lifter.

b. Remove 2 bolts and remove the engine mounting stay RH.

c. Remove the 6 bolts and nut and remove the engine mounting insulator RH.

➡Be sure not to remove the other bolt A.

d. Remove the through bolt and nut and separate the transverse engine mounting insulator.

e. Remove the 6 bolts and remove the engine assembly with transaxle and front suspension crossmember from the vehicle.

To install:

70. Install front suspension crossmember sub-assembly.

a. Install the engine moving control rod with the through bolt. Torque: 89 ft. lbs. (120 Nm).

b. Remove the 2 bolts and remove the 2 engine hangers.

c. Install the air fuel ratio sensor bracket with bolt. Torque: 32 ft. lbs. (43 Nm).

71. Install transverse engine mounting insulator.

a. Install the transverse engine mounting insulator with the 5 bolts. Torque: 38 ft. lbs. (52 Nm).

72. Install engine assembly with transaxle.

a. Set the engine assembly with transaxle and front suspension crossmember on the engine lifter.

b. Operate the engine lifter and lift the engine assembly with transaxle and front suspension crossmember to the position where the engine mounting insulator RH and transverse engine mounting insulator can be installed.

c. Install the transverse engine mounting insulator with the through bolt and nut. Torque: 38 ft. lbs. (52 Nm).

d. Install the engine mounting insulator RH with the 6 bolts and nut.

e. Torque: 65 ft. lbs. (88 Nm) for Bolt A.

f. Torque: 33 ft. lbs. (45 Nm) for Bolt B.

g. Torque: 38 ft. lbs. (52 Nm) for Bolt C.

h. Install the engine mounting stay RH with the 2 bolts. Torque: 19 ft. lbs. (26 Nm).

i. Operate the engine lifter and provisionally install the engine assembly with transaxle and front suspension crossmember onto the vehicle with the 6 bolts.

j. Insert SST into the datum holes in the front suspension crossmember RH and LH alternately and tighten bolts A, B and C on both sides in several sequences.

k. Torque: 52 ft. lbs. (70 Nm) for Bolt A.

l. Torque: 118 ft. lbs. (160 Nm) for Bolt B.

m. Torque: 70 ft. lbs. (95 Nm) for Bolt C.

⁂ CAUTION

Insert SST into the datum hole vertically. If impossible to insert SST vertically, loosen all bolts and then insert SST again.

73. Install drive plate and torque converter clutch setting bolt (for Automatic Transaxle).

74. Install flywheel housing under cover (for Automatic Transaxle).

75. Install front drive shaft assembly LH.

76. Install front drive shaft assembly RH.

77. Install front axle assembly LH.

78. Install front axle assembly RH.

79. Install front lower suspension arm sub-assembly LH.

80. Install front lower suspension arm sub-assembly RH.

81. Install front stabilizer link assembly LH.

82. Install front stabilizer link assembly RH.

83. Install tie rod end sub-assembly LH.

84. Install tie rod end sub-assembly RH.

85. Install speed sensor front LH.

86. Install speed sensor front RH.

87. Install front axle shaft LH nut.

88. Install front axle shaft RH nut.

89. Install front exhaust pipe assembly.

a. Using a plastic hammer and a wooden block, tap in a new exhaust pipe gasket until its surface is flush with the exhaust manifold.

b. Install the 3 exhaust pipe supports.

c. Install the exhaust front pipe assembly and a new exhaust pipe gasket with the 4 compression springs and 4 bolts. Torque: 32 ft. lbs. (43 Nm).

d. Pass the sensor connector through the floor panel and install the grommet.

e. Connect the heated oxygen sensor connector.

90. Install front floor brace center.

91. Install front console box.

92. Install rear console box assembly.

93. Install console box rear cover.

94. Install console box carpet.

95. Install console panel upper.

96. Install shift lever knob sub-assembly (for Manual Transaxle).

97. Install No. 1 steering column hole cover sub-assembly.

98. Install steering sliding yoke sub-assembly.

99. Install column hole cover silencer sheet.

100. Install clutch release cylinder assembly (for Manual Transaxle).

a. Install the clutch release cylinder and clutch pipe with the 4 bolts.

101. Connect fuel tube sub-assembly.

a. Connect the fuel tube connector and fuel pipe, and install the No. 1 fuel pipe clamp.

➡Align the fuel tube connector with the pipe, then push the fuel tube connector in until the retainer makes a click sound. If the connection is tight, apply a small amount of engine oil to the tip of the pipe. After connecting, pull the pipe and connector to make sure that they are securely connected.

102. Connect heater water hoses to the heater unit.

103. Connect engine wire.

a. Connect the 2 engine wire harness connectors and 2 wire harness clamps to the engine room junction block.

b. Connect the engine wire harness connector to the ECM.

c. Connect the 2 wire harness clamps.

d. Connect the 2 connectors.

e. Install the transaxle ground wire with the bolt and 2 wire harness clamps (for Manual Transaxle).

f. Install the transaxle ground wire with the bolt and wire harness clamp (for Automatic Transaxle).

104. Install No. 1 fuel vapor feed hose.

a. Install the No. 1 fuel vapor feed hose into the vacuum switching valve assembly.

105. Install union to check valve hose.

a. Install the union to check valve hose with the clip.

106. Install transmission control cable assembly (for Automatic Transaxle)

a. Install the transmission control cable assembly onto the control shaft lever with the nut.

b. Fix the transmission control cable assembly onto the control cable bracket with a new clip.

➡Make sure that the marking on the cable is aligned with the slit in the

transmission control cable bracket before installing the cable.

107. Install transmission control cable assembly (for Manual Transaxle).
   a. Install the control cable assembly onto the control cable bracket with 2 new clips.
   b. Connect the control cable assembly and install the 2 washers and 2 clips.
   c. Install the control cable assembly onto the shift lever retainer.

➡Install the cable with the protruding portion of the cable outer facing upward. After installing. check that the lock of the cable outer is protruding beyond portion A. as shown in the illustration.

108. Install with pulley compressor assembly.
109. Install radiator assembly.
110. Install upper radiator support sub-assembly.
111. Install upper radiator support absorber.
112. Install No. 1 cooler cover.
113. Install hood lock assembly.
114. Connect No. 2 oil cooler inlet hose (for automatic transaxle).
115. Connect No. 1 oil cooler outlet hose (for automatic transaxle).
116. Connect No. 1 radiator hose.
117. Install No. 2 radiator hose.
   a. Install the No. 2 radiator hose with the 2 clips.
118. Install front bumper cover.
119. Install fan and alternator v belt.
120. Adjust fan and alternator v belt.
121. Inspect fan and alternator v belt.
122. Install engine cover.
123. Install battery carrier.
   a. Install the battery carrier with the 5 bolts.
   b. Install the clamp.
124. Install air cleaner bracket.
   a. Install the air cleaner bracket with the 2 bolts.
   b. Connect the wire harness clamp to the air cleaner bracket.
125. Install air cleaner assembly
   a. Install the air cleaner case with air cleaner inlet with the 2 bolts.
   b. Connect the wire harness to the air cleaner case.
   c. Install the air cleaner element.
   d. Install and lock the air cleaner cap with air cleaner hose and then tighten the air cleaner hose clamp.
   e. Connect the ventilation hose to the air cleaner hose.
   f. Connect the intake air flow meter connector and 2 wire harness clamps.
126. Install outer cowl top panel
   a. Install the outer cowl top panel with the 9 bolts.
   b. Connect the wire harness clamp.
127. Install front air shutter seal RH
   a. Engage the 2 claws to install the front air shutter seal RH.
128. Install front wiper motor and link.
129. Install cowl top ventilator louver LH.
130. Install cowl top ventilator louver sub-assembly.
131. Install hood to cowl top seal.
132. Install front wiper arm and blade assembly LH.
133. Install front wiper arm and blade assembly RH.
134. Install front wiper arm head cap.
135. Install battery tray.
136. Install battery.
137. Add engine coolant.
138. Add automatic transaxle oil (for automatic transaxle).
139. Inspect automatic transaxle oil (for automatic transaxle).
140. Add manual transaxle oil (for manual transaxle).
141. Inspect manual transaxle oil (for manual transaxle).
142. Inspect for fuel leaks.
143. Inspect for engine oil leak..
144. Inspect for exhaust gas leak.
145. Inspect for engine coolant leak.
146. Install engine under cover RH.
147. Install engine under cover LH.
148. Install front wheels.

tC

See Figure 48.

1. Before servicing the vehicle, refer to the Precautions Section.
2. Drain the transaxle assembly.
3. Place the front wheels facing straight ahead.
4. Remove or disconnect the following:
   - Negative battery cable
   - Steering intermediate shaft
   - Front wheel
   - Engine undercover
   - Center exhaust pipe
   - Hood
   - Engine appearance cover
   - Air intake assembly
   - Battery and battery tray
   - Starter
   - Back-up light switch
   - Clutch release cylinder assembly
   - Transaxle control cables
   - Both halfshafts

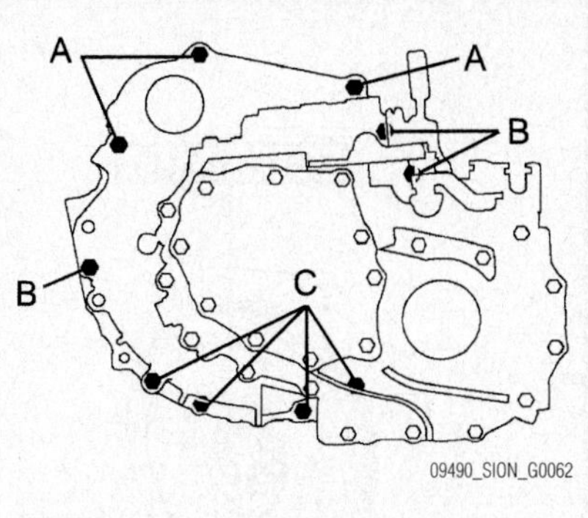

Fig. 48 Manual transaxle mounting bolt identification—tC

5. At this point, attach an engine hoist to support the engine.
6. Remove or disconnect the following:
   - Front suspension crossmember
   - Left engine mount and bracket
   - Transaxle

To install:

7. Install the transaxle to the engine. Tighten the bolts as follows:
   a. Bolts A to 47 ft. lbs. (64 Nm).
   b. Bolts B to 34 ft. lbs. (46 Nm).
   c. Bolts C to 32 ft. lbs. (44 Nm).
8. Install or connect the following:
   - Left engine mount and bracket. Tighten the bracket bolts to 38 ft. lbs. (52 Nm). Tighten the mount bolt to 64 ft. lbs. (87 Nm).
   - Front suspension crossmember
   - Halfshafts
   - Transaxle control cables
   - Clutch release cylinder assembly
   - Back-up light switch
   - Starter
   - Battery tray and battery
   - Air intake assembly
   - Engine appearance cover
   - Hood
   - Center exhaust pipe
   - Front wheel
   - Negative battery cable
9. Refill the transaxle with fluid to the correct level.
10. Start the engine and check for leaks.

CLUTCH

REMOVAL & INSTALLATION

tC and xB

See Figures 49 and 50.

1. Before servicing the vehicle, refer to the Precautions Section.
2. Remove or disconnect the following:
   - Transaxle
   - Release fork and boot

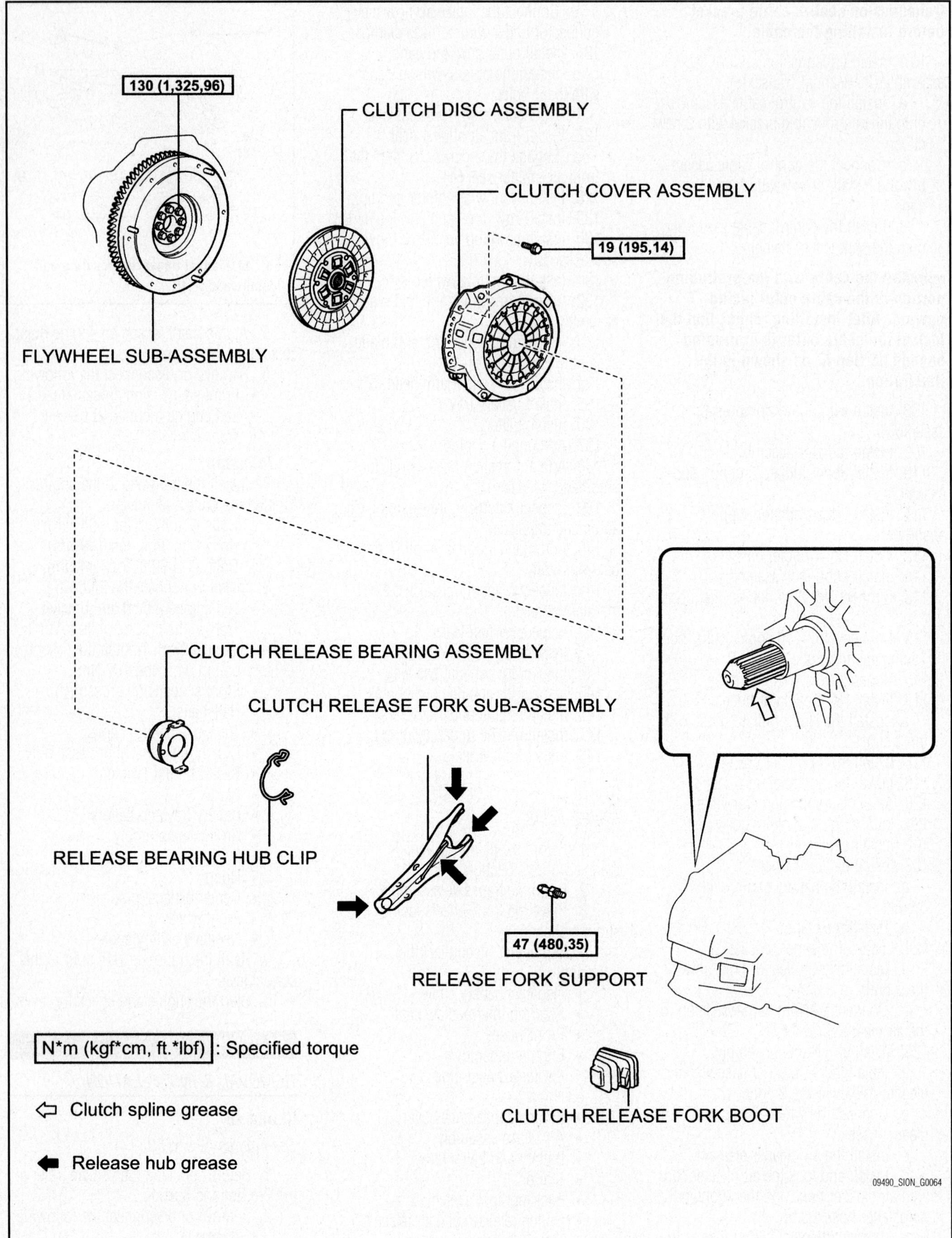

Fig. 49 Exploded view of the clutch components

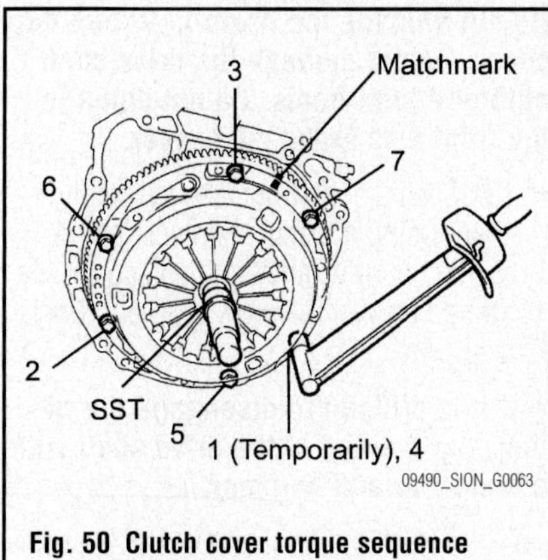

Fig. 50 Clutch cover torque sequence

- Release bearing assembly
- Release fork support

3. Matchmark the clutch cover and flywheel. Loosen the clutch cover bolts one turn at a time until the spring tension is released.
4. Remove the clutch cover.
5. Remove the clutch disc.

To install:

6. Install the clutch disc on the flywheel.
7. Align the matchmarks on the clutch cover and flywheel.
8. Evenly tighten the mounting bolts in several steps to 14 ft. lbs. (19 Nm) by following the order shown.
9. Install or connect the following:
   - Release fork
   - Release bearing hub clip
   - Release bearing assembly
   - Release fork boot
   - Transaxle

xD

See Figure 51.

1. Remove the manual transaxle assembly.
2. Remove the clutch release fork sub-assembly.
   a. Remove the clutch release fork with the clutch release bearing from the manual transaxle.
3. Remove the clutch release fork boot.
4. Remove the clutch release bearing assembly.
   a. Remove the clutch release bearing from the clutch release fork.
5. Remove the release bearing hub clip.
6. Remove the release fork support.
   a. Remove the release fork support from the manual transaxle.
7. Remove the clutch cover assembly.
   a. Put the matchmarks on the clutch cover assembly and the flywheel.
   b. Loosen each set bolt one turn at a time until the spring tension is released.

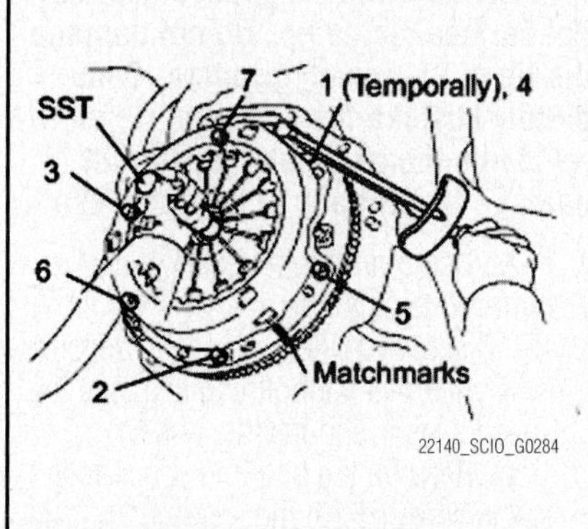

Fig. 51 Tighten the 6 bolts in order

   c. Remove the set bolts and pull off the clutch cover.

CAUTION

Do not drop the clutch disc.

8. Remove the clutch disc assembly.

To install:

9. Install the clutch disc assembly.
   a. Insert SST into the clutch disc assembly, then insert them both into the flywheel sub-assembly.

CAUTION

Insert clutch disc assembly in the correct direction.

10. Install the clutch cover assembly.
   a. Align the matchmark on the clutch cover assembly with that on the flywheel sub-assembly.
   b. Following the procedures shown in the illustration, tighten the 6 bolts in order, starting with the bolt located near the knock pin at the top. Torque: 14 ft. lbs. (19 Nm).

➡Following the order in the illustration, tighten the bolts evenly one at a time. Move SST up and down, right and left lightly after checking that the disc is in the center, and tighten the bolts.

11. Inspect and adjust the clutch cover assembly.
   a. Using a dial indicator with a roller instrument, check the diaphragm spring tip alignment.

➡Maximum non-alignment: 0.5 mm (0.020 in.)

   b. If the alignment is not as specified, using SST, adjust the diaphragm spring tip alignment.
12. Install the release fork support.
   a. Install the release fork support onto the transaxle assembly. Torque: 27 ft. lbs. (37 Nm).
13. Install the clutch release fork boot.
14. Install the release bearing hub clip.
15. Install the clutch release fork sub-assembly.
   a. Apply release hub grease to the contact surfaces of the release fork and release bearing assembly, release fork and push rod, and release fork and fork support.
   b. Install the release fork onto the release bearing assembly.
16. Install the clutch release bearing assembly.
   a. Apply clutch spline grease to the input shaft spline.
   b. Install the clutch release bearing with release fork onto the transaxle assembly.

➡After the installation, move the fork forward and backward to check that the release bearing slides smoothly.

17. Install the manual transaxle assembly.

FRONT HALFSHAFT

REMOVAL & INSTALLATION

xB and tC

See Figures 52 and 53.

1. Before servicing the vehicle, refer to the Precautions Section.
2. Drain the transaxle fluid.
3. Remove or disconnect the following:
   - Front wheel
   - Engine undercover
   - Hub nut
   - Wheel speed sensor
   - Front stabilizer bar
   - Lower control arm
   - Tie rod end
4. Using Special Tool 09520-01010, tap out the left halfshaft.

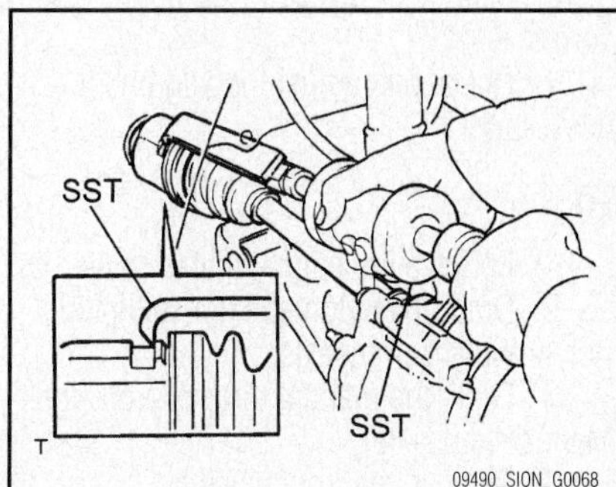

Fig. 52 Use Special Tool 09520-01010 to remove the left halfshaft—xB

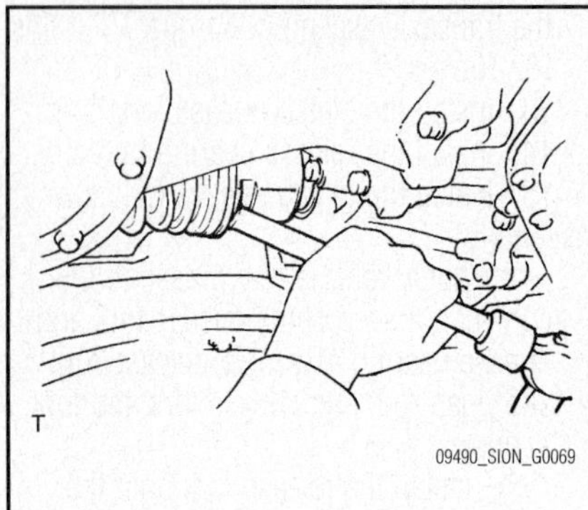

Fig. 53 Tap out the right halfshaft with a brass bar and hammer

5. Remove the right halfshaft as follows:
 a. xB: Using a brass bar and hammer, tap out the right halfshaft.
 b. tC: Remove the two mounting bolts and remove the halfshaft from the transaxle.

To install:

6. Coat the splines of the inboard joint shaft with gear oil (M/T) or ATF (A/T).
7. Align the shaft splines and tap in the left halfshaft with a brass bar and hammer.
8. Install the right halfshaft as follows:
 a. xB: Align the shaft splines and tap in the halfshaft with a brass bar and hammer.
 b. tC: Align the shaft splines and install the halfshaft to the transaxle. Tighten bolts to 47 ft. lbs. (64 Nm).
9. Install or connect the following:
 - Tie rod end. Tighten nut to 36 ft. lbs. (49 Nm).
 - Lower control arm
 - Front stabilizer arm. xB: Tighten nut to 13 ft. lbs. (18 Nm). tC: Tighten nut to 55 ft. lbs. (74 Nm).
 - Wheel speed sensor
 - New hub nut. Tighten to 159 ft. lbs. (216 Nm).
 - Engine undercover
 - Front wheel
10. Refill the transaxle with fluid to the correct level.
11. Check and adjust the alignment if necessary.

xD

1. Remove the engine under cover.
2. Drain the automatic transaxle fluid (for Automatic Transaxle).
3. Drain the manual transaxle oil (for Manual Transaxle).
4. Remove the front wheels.
5. Remove the front axle hub nut.
 a. Using SST and a hammer, release the staked part of the axle hub nut.

➡Insert SST into the groove with the flat surface facing up. Do not damage the tip of SST using grinders. Completely unstake the staked part before removing the axle hub nut. Do not damage the threads of the drive shaft.

 b. Using a 30 mm socket wrench, remove the axle hub nut.
6. Disconnect the front speed sensor.
 a. Remove the bolt and separate the speed sensor and flexible hose.
 b. Remove the bolt and separate the speed sensor from the steering knuckle.

➡Keep the speed sensor tip and installation portion free of foreign matter. Remove the speed sensor without turning it from its original installation angle.

7. Separate the front stabilizer link assembly.
 a. Remove the nut and separate the stabilizer link from the shock absorber.
8. Separate the tie rod end sub-assembly.
 a. Remove the cotter pin and castle nut.
 b. Install SST to the threaded section of the tie rod end.

➡Make sure the upper ends of the threaded section of the tie rod end and SST (spacer B) are aligned.

 c. Using SST, separate the tie rod end from the front axle assembly.

➡Make sure to tie the string of SST to the vehicle to prevent SST from dropping. Install SST so that A and B are parallel. Be sure to place the wrench on the part indicated in the illustration. Do not damage the ball joint dust cover. Do not damage the front disc brake dust cover.

9. Separate the front lower suspension arm.
 a. Remove the clip and castle nut.
 b. Install SST (spacer B) to the threaded section of the lower ball joint.

➡Make sure the upper ends of the threaded section of the lower ball joint and SST are aligned.

 c. Using SST, separate the front lower suspension arm from the front axle assembly.

➡Make sure to tie the string of SST to the vehicle to prevent SST from dropping. Install SST so that A and B are parallel. Be sure to place the wrench on the part indicated in the illustration. Do not damage the lower ball joint dust cover. Do not damage the drive shaft outboard joint boots. Do not damage the front disc brake dust cover.

10. Separate the front axle assembly.
 a. Using a plastic hammer, tap the end of the drive shaft and disengage the fitting between the drive shaft and front axle.

➡If it is difficult to disengage the fitting, tap the end of the drive shaft with a brass bar and hammer.

 b. Push the front axle out of the vehicle to remove the drive shaft from the front axle.

➡Do not push the front axle further out of the vehicle than is necessary. Do not damage the outboard joint boot. Do not damage the speed sensor rotor. Suspend the drive shaft with a piece of rope or the equivalent.

11. Remove the front drive shaft assembly LH.
 a. Using SST, remove the drive shaft.

➡Do not damage the oil seal. Do not damage the inboard joint boot. Do not drop the drive shaft.

12. Remove front drive shaft assembly RH.
 a. Using a screwdriver and hammer, remove the drive shaft.

➡Do not damage the oil seal. Do not damage the inboard joint boot. Do not drop the drive shaft.

To install:

13. Install the front drive shaft assembly LH.
 a. Coat the spline of the inboard joint with gear oil.
 b. Align the inboard joint splines and install the drive shaft with a screwdriver and hammer.

CAUTION

Face the cut area of the front drive shaft hole snap ring downward. Do not damage the oil seal. Do not damage the inboard joint boot. Confirm whether the drive shaft is securely driven in by checking the reaction force and sound.

14. Install the front drive shaft assembly RH.
 a. Use the same procedure as for the LH side.
15. Install the front axle assembly.
 a. Push the front axle out of the vehi-

cle to align the spline of the drive shaft with the front axle and insert the front axle.

CAUTION

Do not push the front axle further out of the vehicle than is necessary. Do not damage the outboard joint boot. Check for any foreign matter on the speed sensor rotor and insertion part. Do not damage the speed sensor rotor.

16. Install the front lower suspension arm.
    a. Install the lower arm onto the steering knuckle with a new castle nut. Torque: 72 ft. lbs. (98 Nm).

➡If the holes for the clip are not aligned, tighten the nut by a further turn of up to 60∞.

    b. Install a new clip.
17. Install the tie rod end sub-assembly.
    a. Install the tie rod end onto the steering knuckle with a new castle nut. Torque: 36 ft. lbs. (49 Nm).

➡If the holes for the clip are not aligned, tighten the nut by a further turn of up to 60°.

    b. Install a new cotter pin.
18. Install the front stabilizer link assembly.
    a. Install the stabilizer link with the nut. Torque: 55 ft. lbs. (74 Nm).
19. Install the front speed sensor.
    a. Install the speed sensor onto the steering knuckle with the bolt.

➡Check that the speed sensor tip and installation portion are free of foreign matter. Install the speed sensor without turning it from its original installation angle.

    b. Install the flexible hose and speed sensor with the bolt. Torque: 22 ft. lbs. (29 Nm).
    c. Install the flexible hose and speed sensor without twisting them.
20. Install the front axle hub nut.
    a. Clean the threaded parts on the drive shaft and axle hub nut using a non-residue solvent.

➡Be sure to perform this work for a new drive shaft. Keep the threaded parts free of oil and foreign objects.

    b. Using a 30 mm socket wrench, install a new axle hub nut. Torque: 160 ft. lbs. (216 Nm).
    c. Using a chisel and hammer, stake the axle hub nut.
21. Install the front wheels.
22. Add automatic transaxle fluid (for Automatic Transaxle).
23. Inspect for automatic transaxle fluid leak (for Automatic Transaxle).
24. Add manual transaxle oil (for Manual Transaxle).
25. Inspect for manual transaxle oil leak (for Manual Transaxle).
26. Inspect and adjust front wheel alignment.
27. Install the engine under cover.

ENGINE COOLING

THERMOSTAT

REMOVAL & INSTALLATION

2.4L Engine

See Figure 54.

1. Drain the engine coolant to a level below the thermostat.
2. Remove the 2 nuts then remove the water inlet from the cylinder block.
3. Remove the thermostat. Thoroughly clean the gasket mating surfaces.

To install:

4. Position a new gasket onto the thermostat, then install the thermostat with the jiggle valve facing upward.

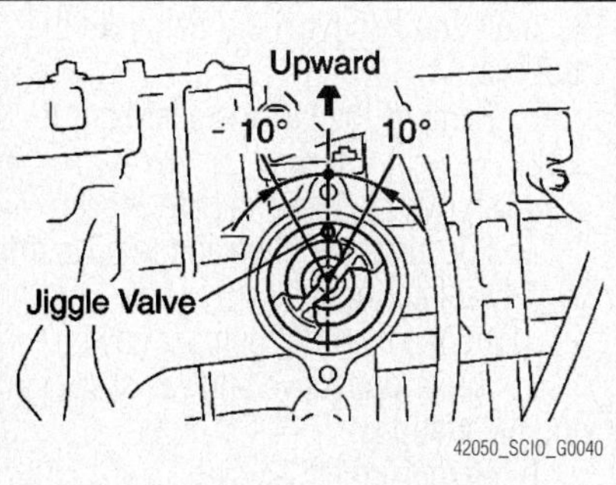

Fig. 54 Proper position of the jiggle valve when installing the thermostat—2.4L engine

➡The jiggle valve may be set within 10° of either side as shown in the accompanying illustration.

5. Install the water inlet and tighten to 80 inch lbs. (9 Nm).
6. Refill the engine cooling system to the correct level.
7. Start the engine and check for leaks.

WATER PUMP

REMOVAL & INSTALLATION

tC and xB

See Figure 55.

1. Before servicing the vehicle, refer to the Precautions Section.
2. Drain the cooling system.
3. Remove or disconnect the following:
    - Negative battery cable
    - Right-hand front fender apron
    - Engine undercover
    - Accessory drive belt
    - Alternator
4. Using Special Tool 09960-10010, remove the water pump pulley.
5. Remove the water pump assembly.

To install:

6. Install the water pump with a new gasket. Tighten to 80 inch lbs. (9 Nm).
7. Using Special Tool 09960-10010, install the water pump pulley and tighten to 19 ft. lbs. (26 Nm).

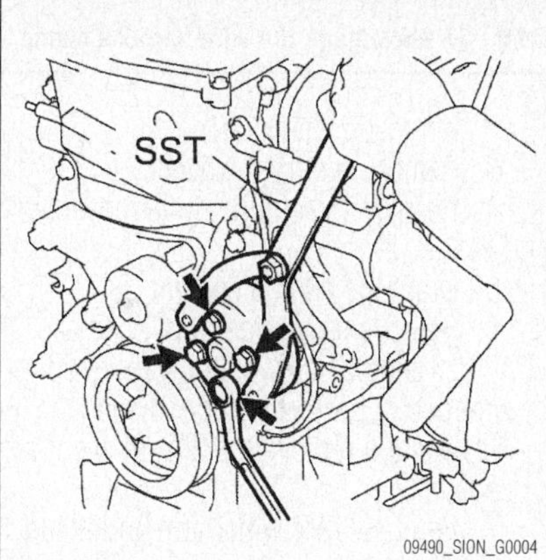

Fig. 55 Use Special Tool 09960-10010 to hold the water pump pulley while removing the mounting bolts

8. Install or connect the following:
    - Alternator
    - Accessory drive belt
    - Engine undercover
    - Right-hand front fender apron
    - Negative battery cable
9. Refill the engine cooling system to the correct level.
10. Start the engine and check for leaks.

xD

See Figures 56 through 85.

1. Discharge the fuel system pressure.

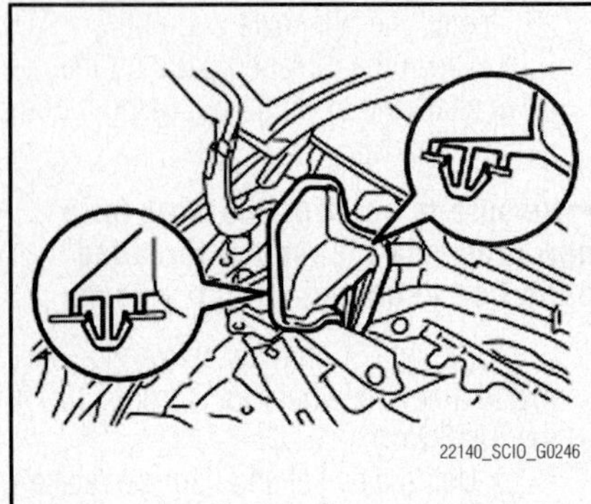

Fig. 56 Remove the front air shutter seal RH

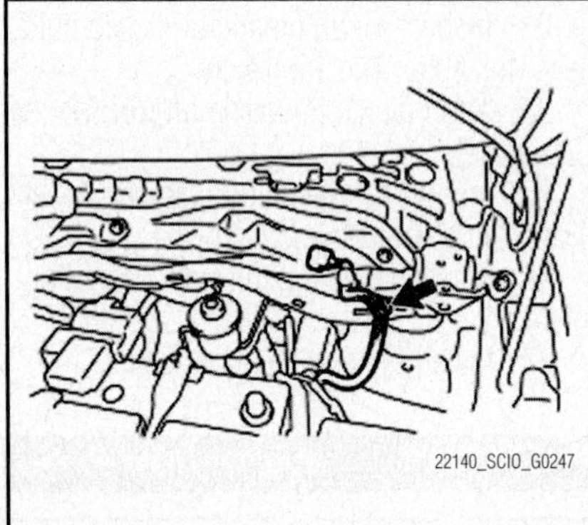

Fig. 57 Disengage the wire harness clamp

2. Remove battery
3. Remove the front wheels.
4. Remove the engine under covers LH and RH.
5. Drain the engine coolant.
6. Remove front wiper arm head cap.
   a. Using a screwdriver with its tip wrapped in protective tape, disengage the 2 claws and remove the 2 front wiper arm caps.
7. Remove front wiper arm and blade assembly LH.
   a. Operate the wiper, then stop the wiper motor in the automatic stop position.
   b. Remove the nut and front wiper arm and blade assembly LH.
8. Remove front wiper arm and blade assembly RH.

➡Use the same procedure as for the LH side.

9. Remove hood to cowl top seal.
   a. Disengage the 7 clips and remove the hood to cowl top seal.
10. Remove cowl top ventilator louver sub-assembly.
   a. Remove the clip.
   b. Disengage the 4 claws and 4 guides and remove the cowl top ventilator louver RH.

Fig. 58 Remove the 9 bolts and remove the outer cowl top panel

Fig. 59 Remove the air cleaner assembly

11. Remove cowl top ventilator louver LH.
   a. Remove the clip.
   b. Disengage the 5 claws and 5 guides and remove the cowl top ventilator louver LH.
12. Remove front wiper motor and link.
   a. Disengage the clamp.
   b. Remove the 2 bolts.
   c. Slide the front wiper motor and link. Disengage the link from the rubber pin, then disconnect the connector and remove the front wiper motor and link.
13. Remove the front air shutter seal RH.
   a. Disengage the 2 claws and remove the front air shutter seal RH.
14. Remove the outer cowl top panel.
   a. Disengage the wire harness clamp.
   b. Remove the 9 bolts and remove the outer cowl top panel.
15. Remove the air cleaner assembly.
   a. Separate the intake air flow meter connector and 2 wire harness clamps.
   b. Separate the ventilation hose from the air cleaner hose.

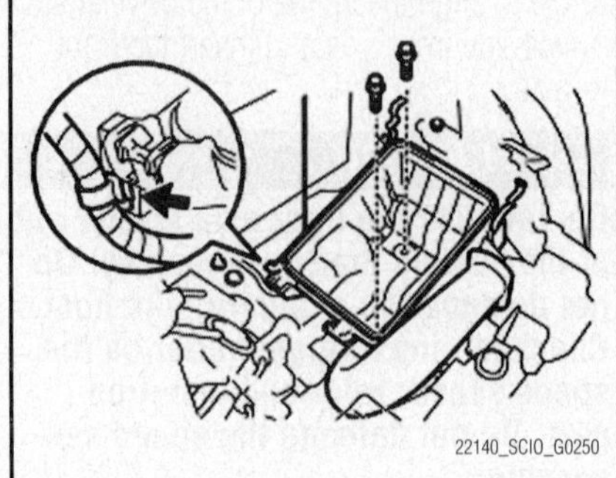

Fig. 60 Separate the wire harness clamp from the air cleaner case

   c. Unfasten the 2 clamps.
   d. Remove the air cleaner cap sub-assembly with air cleaner hose.
   e. Remove the air cleaner element.
   f. Separate the wire harness clamp from the air cleaner case.
   g. Remove the 2 bolts and remove the air cleaner case with air cleaner inlet.
16. Remove the air cleaner bracket.
   a. Separate the wire harness clamp from the air cleaner bracket.
   b. Remove the 2 bolts and remove the air cleaner bracket.
17. Remove the battery carrier.
   a. Separate the wire harness clamp from the battery carrier.
   b. Remove the 5 bolts and remove the battery carrier.
18. Remove the Engine cover.
   a. First, lift up the rear of the Engine cover to disengage the 2 fittings
   b. Next, lift the front of the Engine cover to disengage the 2 fittings. Then remove the Engine cover.

➡Ensure that the rubber grommets remain attached to the Engine cover. If any grommets are attached to the bolts, move them to the Engine cover.

19. Remove the fan and alternator V belt.
   a. Loosen bolts A and B.
   b. Turn adjusting bolt C to release the tension and remove the V belt from the pulleys.
   c. Remove the fan and alternator V belt.
20. Remove the lower radiator hose.
   a. Loosen the 2 clips and remove the No. 2 radiator hose.
21. Remove the front bumper cover.
   a. Apply protective tape, as shown in the illustration.
   b. Remove the 8 screws.
   c. Remove the 9 clips.
   d. Disengage the 6 claws and remove the front bumper cover.

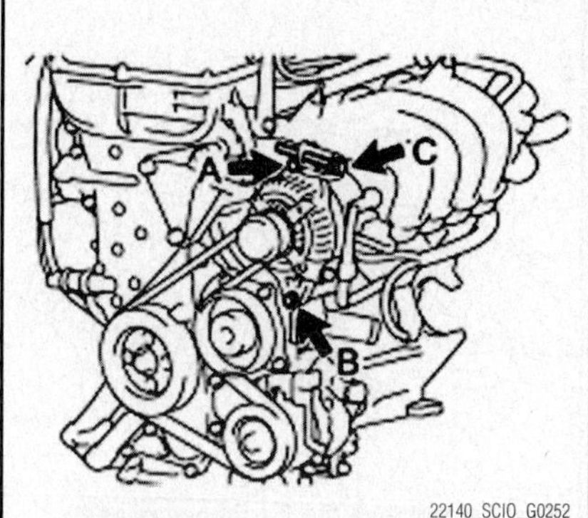

Fig. 61 Remove the fan and alternator V belt

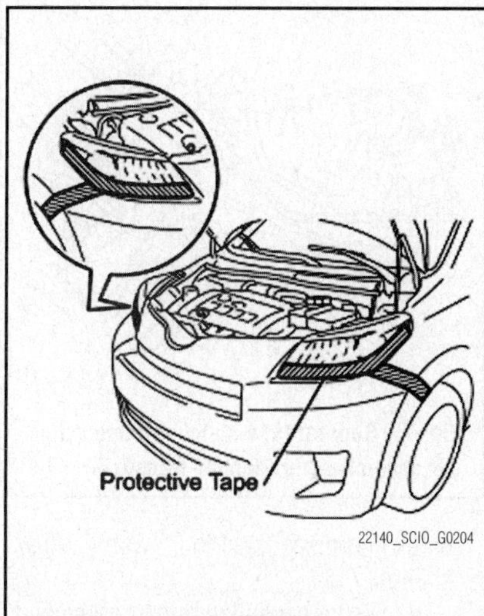

Fig. 62 Apply protective tape

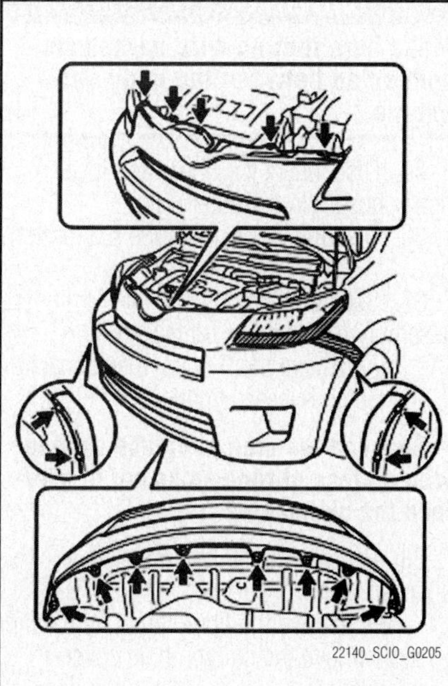

Fig. 63 Remove the 8 screws and 9 clips

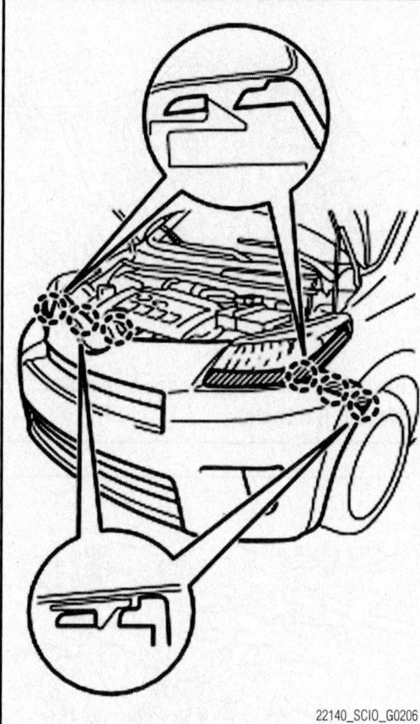

Fig. 64 Disengage the 6 claws and remove the front bumper cover

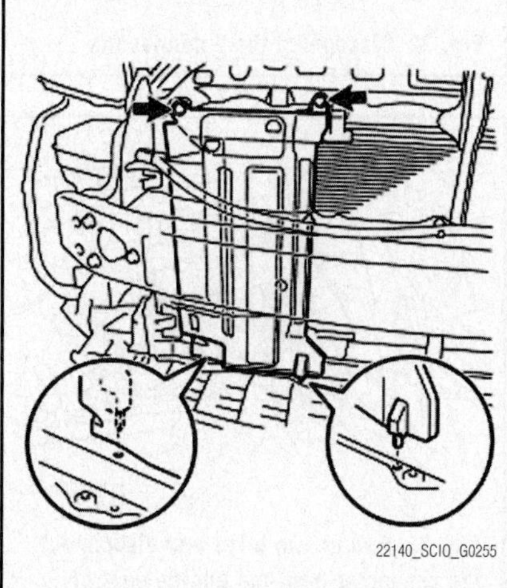

Fig. 65 Remove the No. 1 cooler cover

22. Disconnect the upper radiator hose.
23. Disconnect the oil cooler inlet and outlet hoses from the radiator (for Automatic Transaxle).
24. Remove the hood lock assembly.
a. Separate the hood lock control cable assembly from the 2 clamps.
b. Remove the 3 bolts and remove the hood lock assembly.
25. Remove the No. 1 cooler cover.
a. Remove the 2 clips and remove the No. 1 cooler cover.
26. Remove the upper radiator support absorber.

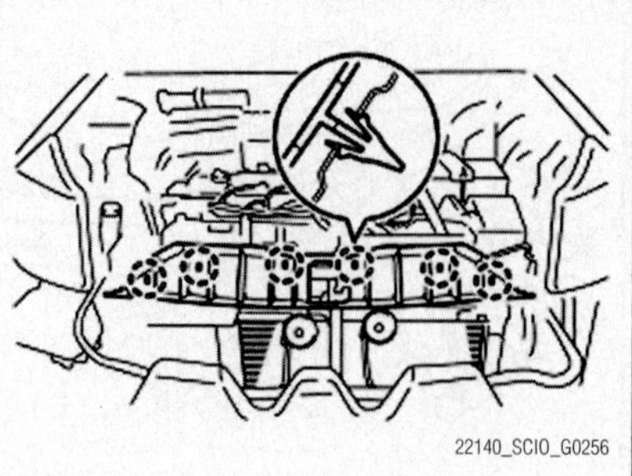

Fig. 66 Remove the upper radiator support absorber

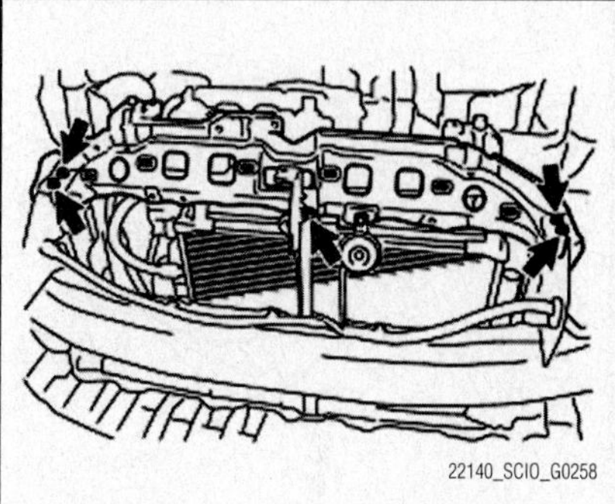

Fig. 67 Remove the 5 bolts and remove the upper radiator support sub-assembly

a. Using a screwdriver with its tip wrapped in protective tape, disengage the 6 claws and remove the radiator support upper absorber.
27. Remove the upper radiator support sub-assembly.
a. Disconnect the horn connector.
b. Remove the 5 bolts and remove the upper radiator support sub-assembly.
28. Remove the radiator assembly.
a. Disconnect the cooling fan motor connector and separate the wire harness clamp.
b. Disengage the 2 claws and remove the radiator assembly from the vehicle.

⁂ CAUTION

Do not apply excessive force to the cooler condenser assembly or piping when removing the radiator assembly.

29. Separate with pulley compressor assembly.
30. Separate the transmission control cable assembly (for Automatic Transaxle).
a. Remove the nut and disconnect the transmission control cable assembly from the control shaft lever.
31. Separate the transmission control cable assembly (for Manual Transaxle).

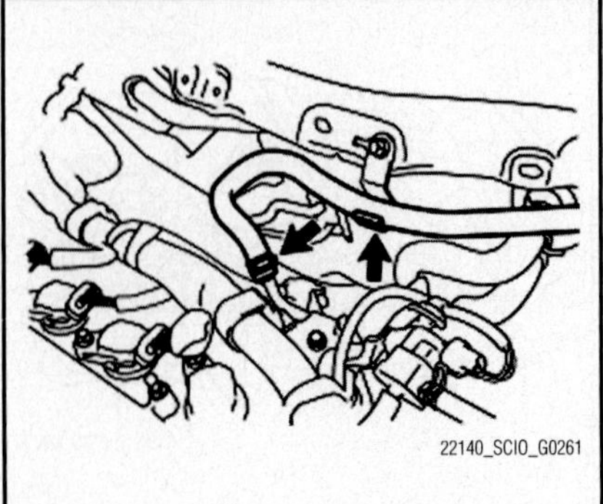

Fig. 68 Separate the union to check valve hose

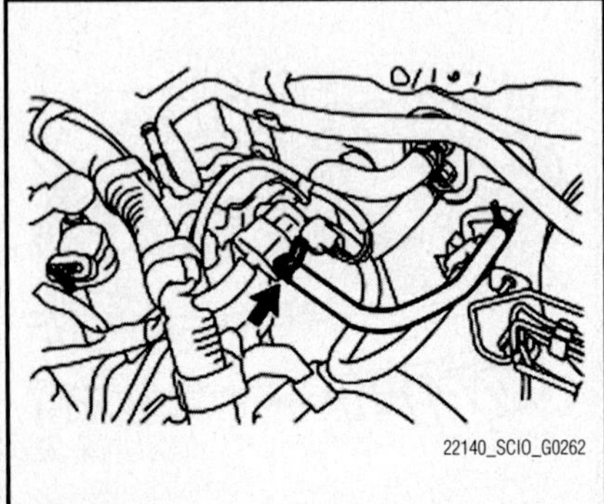

Fig. 69 Separate No. 1 fuel vapor feed hose

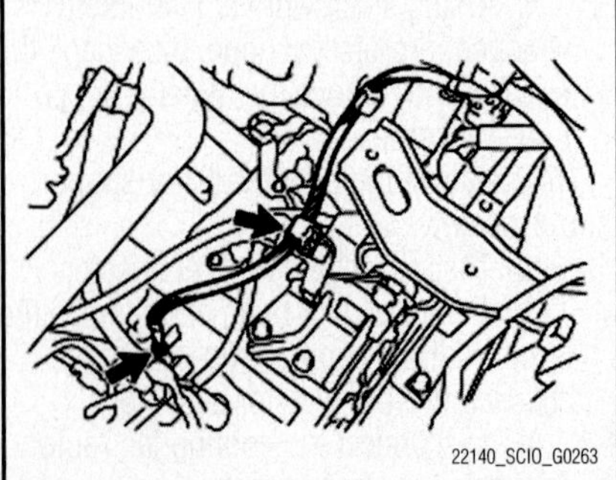

Fig. 70 Disconnect the engine wire (for Automatic Transaxle)

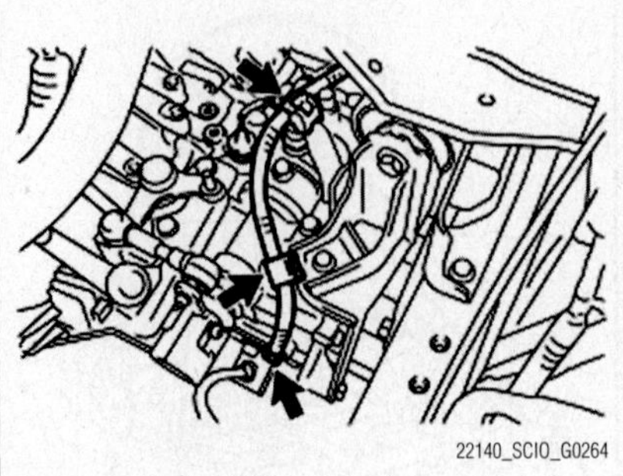

Fig. 71 Disconnect the engine wire (for Manual Transaxle)

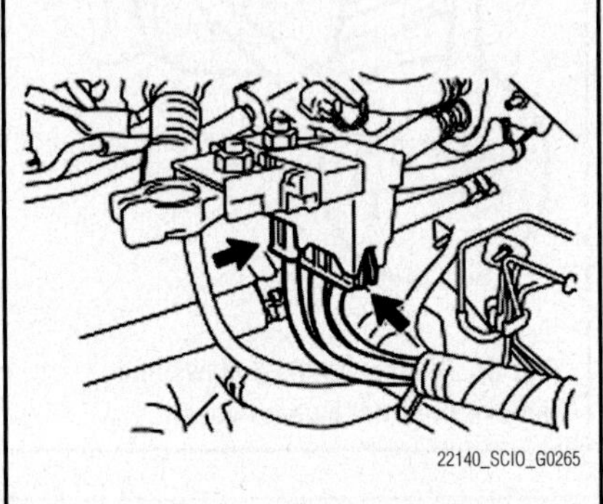

Fig. 72 Disconnect the 2 connectors

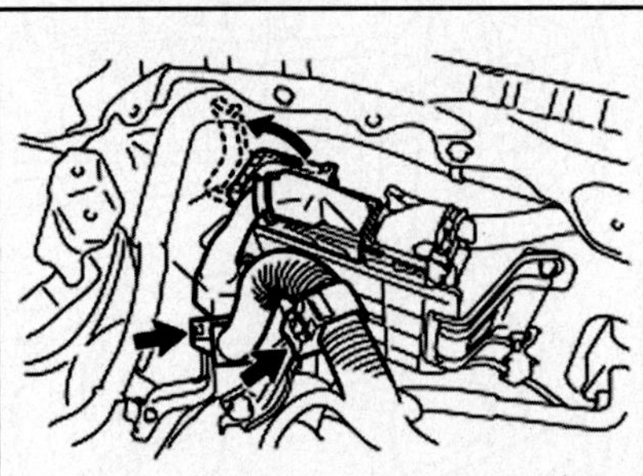

Fig. 73 Pull up the lever and disconnect the connector from the engine control computer

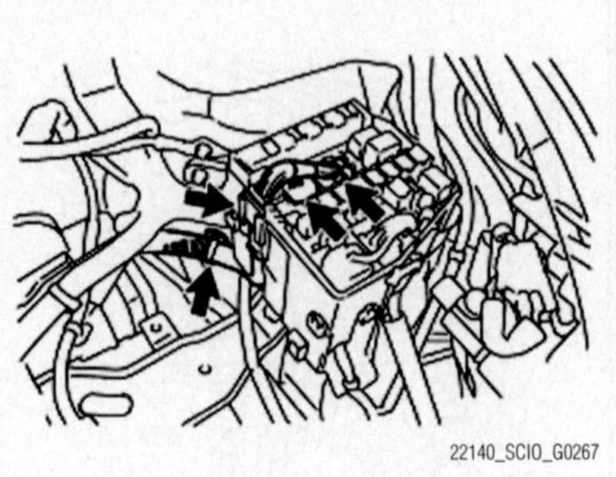

Fig. 74 Remove the 2 connectors and clamp from the engine room junction block and disconnect the 2 wire harness clamps

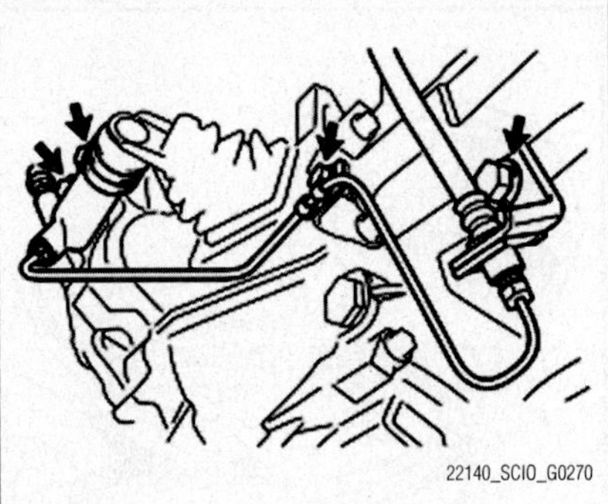

Fig. 75 Separate the clutch release cylinder assembly (for Manual Transaxle)

a. Remove the 2 clips and the washers, and disconnect the 2 cables from the transaxle.

b. Remove the 2 clips and disconnect the 2 cables from the control cable bracket.

32. Separate the union to check valve hose.

a. Loosen the clip and separate the union to check valve hose.

33. Separate No. 1 fuel vapor feed hose.

a. Separate the No. 1 fuel vapor feed hose from the vacuum switching valve assembly.

34. Disconnect the engine wire.

a. Remove the bolt and wire harness clamp and separate the transaxle ground wire (for Automatic Transaxle).

b. Remove the bolt and 2 wire harness clamp and separate the transaxle ground wire (for Manual Transaxle).

c. Disconnect the 2 connectors.

d. Pull up the lever and disconnect the connector from the engine control computer.

e. Separate the 2 wire harness clamps.

f. Remove the 2 connectors and clamp from the engine room junction block and disconnect the 2 wire harness clamps.

g. Disconnect all wire harnesses and connectors.

CAUTION

Make sure that no wire harness is connected between the body and engine.

35. Disconnect the upper and lower heater hoses (from heater unit).

36. Disconnect the fuel tube sub-assembly.

37. Separate the clutch release cylinder assembly (for Manual Transaxle).

a. Remove the 4 bolts, then separate the clutch release cylinder.

➡Suspend the clutch release cylinder with a piece of rope so as not to overload the clutch pipe.

38. Remove the column hole cover silencer sheet.

a. Remove the floor carpet and 2 clips and remove the column hole cover silencer.

39. Separate the steering sliding yoke sub-assembly.

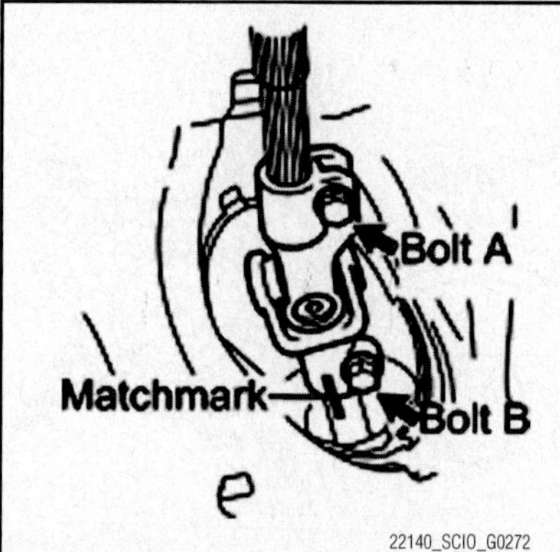

Fig. 76 Place matchmarks on the sliding yoke of the steering intermediate shaft assembly

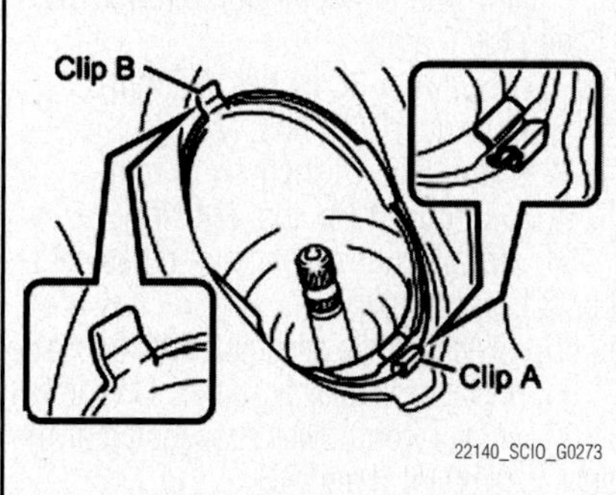

Fig. 77 Separate the No. 1 steering column hole cover sub-assembly

a. Use a seat belt to fix the steering wheel assembly, in order to avoid breakage of the spiral cable.

b. Place matchmarks on the sliding yoke of the steering intermediate shaft assembly and the power steering.

c. Loosen bolt A, remove bolt B and separate the steering intermediate shaft assembly.

40. Separate the No. 1 steering column hole cover sub-assembly.

a. Remove clip A, separate clip B from the body and separate No. 1 steering column hole cover.

CAUTION

Do not damage clip B.

41. Remove the shift lever knob sub-assembly (for Manual Transaxle).

a. Remove the shift lever knob by turning the knob counterclockwise.

42. Remove the upper console panel.

a. Disengage the 2 clips and the 6 claws and remove the upper console panel.

43. Remove the rear console box cover.

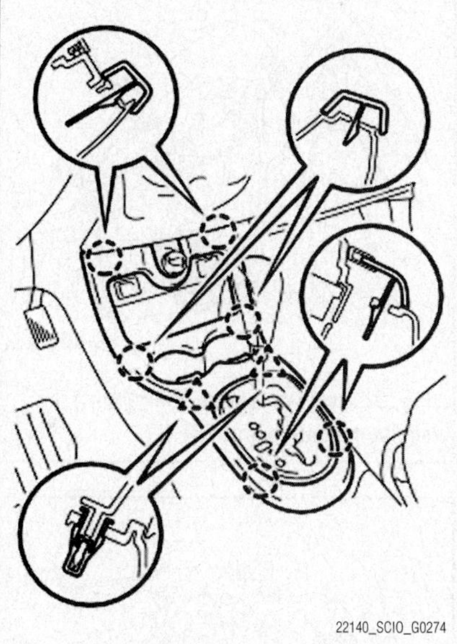

Fig. 78 Remove the upper console panel

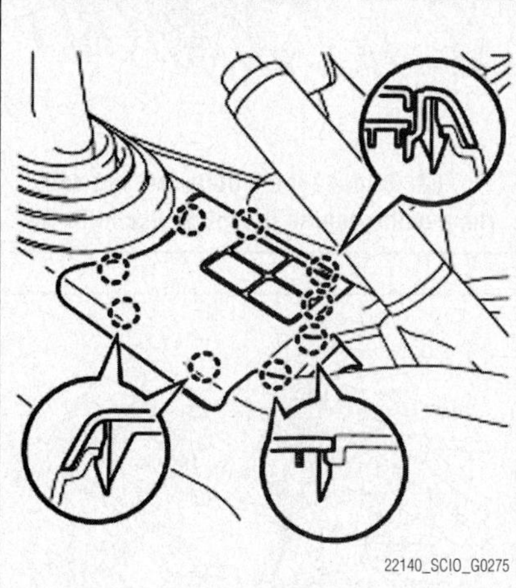

Fig. 79 Remove the rear console box cover

a. Disengage the 9 claws and remove the rear console box cover.

44. Remove the console box carpet.

45. Remove the rear console box assembly.

a. Remove the 3 screws.

b. Disengage the 4 claws and remove the rear console box.

46. Remove the front console box.

a. Remove the 2 screws.

b. Disengage the clip and 2 claws and remove the front console box.

47. Remove the front floor center brace.

a. Remove the 2 bolts and remove the front floor center brace.

48. Remove the front exhaust pipe assembly.

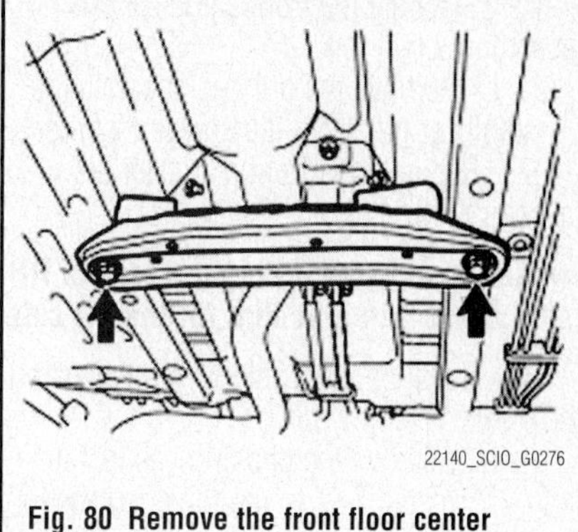

Fig. 80 Remove the front floor center brace

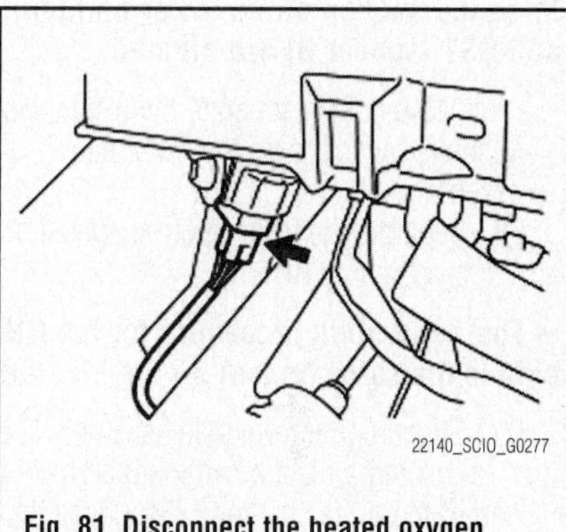

Fig. 81 Disconnect the heated oxygen sensor connector

a. Disconnect the heated oxygen sensor connector.

b. Remove the grommet and pull the sensor connector out of the cabin through the floor panel.

c. Remove the 2 bolts and 2 springs and separate the front exhaust pipe assembly from the exhaust tail pipe assembly.

d. Remove the 2 bolts and 2 springs and separate the front exhaust pipe assembly from the exhaust manifold.

e. Remove the 3 exhaust pipe supports and remove the front exhaust pipe assembly.

49. Remove the front axle shaft LH nut.

50. Remove the front axle shaft RH nut.

51. Separate the speed sensor front LH.

52. Separate the speed sensor front RH.

53. Separate the tie rod end sub-assembly LH.

a. Remove the cotter pin and castle nut.

b. Using SST, separate the tie rod end from the front axle assembly.

54. Separate the tie rod end sub-assembly RH

➡The separation procedure for the RH side is the same as that for the LH side.

55. Separate the front stabilizer link assembly LH.

a. Remove the nut and separate the stabilizer link from the shock absorber.

56. Separate the front stabilizer link assembly RH.

➡The separation procedure for the RH side is the same as that for the LH side.

57. Separate the front lower suspension arm sub-assembly LH.

a. Remove the clip and castle nut.

b. Install SST to the threaded section of the lower ball joint.

➡Make sure the upper ends of the threaded section of the lower ball joint and SST (spacer B) are aligned.

c. Using SST, separate the front lower suspension arm from the front axle assembly.

58. Separate the front lower suspension arm sub-assembly RH

➡The separation procedure for the RH side is the same as that for the LH side.

59. Separate the front axle assembly LH.

a. Using a plastic hammer, tap the end of the drive shaft and disengage the fitting between the drive shaft and front axle.

➡If it is difficult to disengage the fitting, tap the end of the drive shaft with a brass bar and hammer.

b. Push the front axle out of the vehicle to remove the drive shaft from the front axle.

➡Do not push the front axle further out of the vehicle than is necessary. Do not damage the outboard joint boot. Do not damage the speed sensor rotor. Suspend the drive shaft with a piece of rope or the equivalent.

60. Separate the front axle assembly RH.

➡The separation procedure for the RH side is the same as that for the LH side.

61. Remove the engine assembly with transaxle.

a. Set the engine lifter.

b. Remove 2 bolts and remove the engine mounting stay RH.

c. Remove the 6 bolts and nut and remove the engine mounting insulator RH.

➡Be sure not to remove the other bolt A.

d. Remove the through bolt and nut and separate the transverse engine mounting insulator.

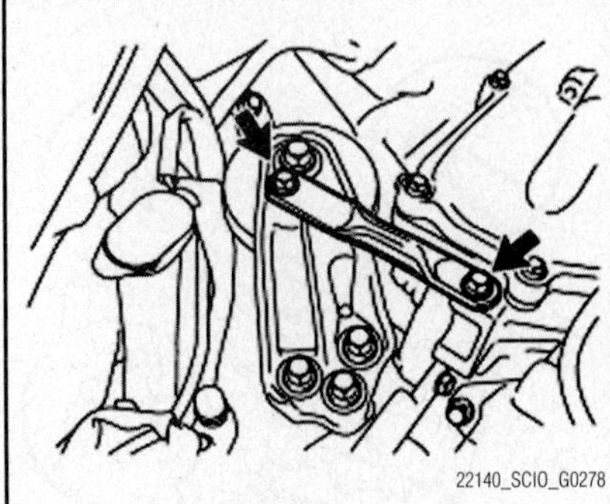

Fig. 82 Remove 2 bolts and remove the engine mounting stay RH

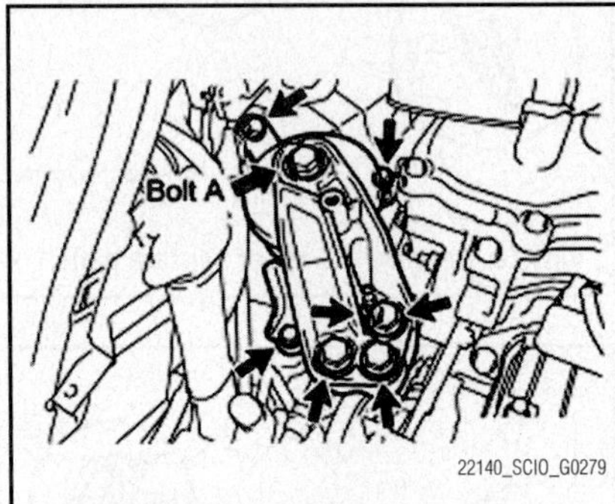

Fig. 83 Remove the 6 bolts and nut and remove the engine mounting insulator RH

Fig. 84 Remove the through bolt and nut

e. Remove the 6 bolts and remove the engine assembly with transaxle and front suspension crossmember from the vehicle.

62. Remove the water pump assembly.

a. Remove the 5 bolts and remove the water pump assembly.

b. Remove the water pump gasket from the timing chain cover.

To install:

63. Install water pump assembly.

a. Install a new water pump gasket to the timing chain cover.

➡Make sure that the installation surfaces are clean.

Fig. 85 Remove the 5 bolts and remove the water pump assembly

b. Install the water pump assembly with the 5 bolts.

- Bolt A: 1.38 inches (35 mm) Torque: 19 ft. lbs. (26 Nm).
- Bolt B: 0.71 inches (18 mm) Torque: 17 ft. lbs. (24 Nm).

64. Install front suspension crossmember sub-assembly.
65. Install engine assembly with transaxle.
66. Install front axle assembly LH and RH.
67. Install front lower suspension arm sub-assembly LH and RH.
68. Install front stabilizer link assembly LH and RH.
69. Install tie rod end sub-assembly LH and RH.
70. Install speed sensor front LH and RH.
71. Install front axle shaft LH and RH nut.
72. Install front exhaust pipe assembly.
73. Install front floor brace center.
74. Install front console box.
75. Install rear console box assembly.
76. Install console box carpet.
77. Install console box rear cover.
78. Install console panel upper.
79. Install shift lever knob sub-assembly (for Manual Transaxle).
80. Install No. 1 steering column hole cover sub-assembly.
81. Install steering sliding yoke sub-assembly.
82. Install column hole cover silencer sheet.
83. Install clutch release cylinder assembly (for Manual Transaxle).
84. Connect fuel tube sub-assembly.
85. Connect heater water inlet hose A.
86. Connect heater water outlet hose A(from heater unit).
87. Connect engine wire.
88. Install No. 1 fuel vapor feed hose.
89. Install union to check valve hose.
90. Install transmission control cable assembly (for Automatic Transaxle).

91. Install transmission control cable assembly (for Manual Transaxle).
92. Install with pulley compressor assembly.
93. Install radiator assembly.
94. Install upper radiator support sub-assembly.
95. Install upper radiator support absorber.
96. Install No. 1 cooler cover.
97. Install hood lock assembly.
98. Connect No. 2 oil cooler inlet hose (for Automatic Transaxle).
99. Connect No. 1 oil cooler outlet hose (for Automatic Transaxle).
100. Install front bumper cover.
101. Connect No. 1 radiator hose.
102. Install No. 2 radiator hose.
103. Install fan and alternator v belt.
104. Adjust fan and alternator v belt.
105. Inspect fan and alternator v belt.
106. Install engine cover.
107. Install battery carrier.
108. Install air cleaner bracket.
109. Install air cleaner assembly.
110. Install outer cowl top panel.
111. Install front air shutter seal RH.
112. Install front wiper motor and link.
113. Install cowl top ventilator louver LH.
114. Install cowl top ventilator louver sub-assembly.
115. Install hood to cowl top seal.
116. Install front wiper arm and blade assembly LH and RH.
117. Install front wiper arm head cap.
118. Install battery tray
119. Install battery.
120. Add engine coolant.
121. Inspect for fuel leak
122. Inspect for exhaust gas leak
123. Inspect for engine coolant leak.
124. Install engine under cover RH
125. Install engine under cover LH
126. Install front wheels.

ENGINE ELECTRICAL

CHARGING SYSTEM

ALTERNATOR

REMOVAL & INSTALLATION

tC and xB

1. Before servicing the vehicle, refer to the Precautions Section.
2. Remove or disconnect the following:
   - Negative battery cable
   - Right-hand front fender apron seal
   - Alternator drive belt
   - Alternator wiring harnesses
   - Alternator mounting bolts and alternator

To install:

➡Confirm the crankshaft position sensor wiring harnesses is secured in the clamp bracket on the timing chain cover.

3. Install or connect the following:
   - Alternator. Tighten upper bolt to 16 ft. lbs. (21 Nm) and lower bolt to 38 ft. lbs. (52 Nm).
   - Alternator wiring harnesses
   - Alternator drive belt
   - Right-hand front fender apron seal
   - Negative battery cable

xD

See Figures 86 and 87.

1. Disconnect the cable from negative battery terminal
2. Remove the engine under cover RH
3. Remove the Engine cover.
4. Remove the fan and alternator V belt.
5. Remove the engine mounting insulator sub-assembly RH
   a. Remove 2 bolts and remove the engine mounting stay RH .
   b. Place a wooden block on a jack underneath the engine.

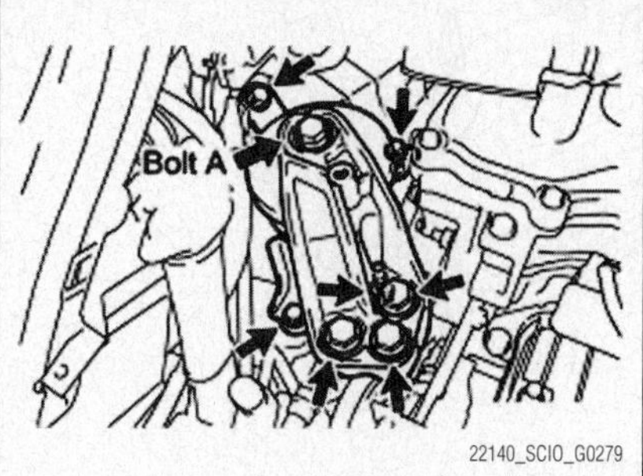

Fig. 86 Remove the 6 bolts and nut and remove the engine mounting insulator RH

Fig. 87 Tighten bolt A to the specified torque

   c. Remove the 6 bolts and nut and remove the engine mounting insulator sub-assembly RH.

➡Do not remove bolt A.

6. Remove the transverse engine mounting bracket.
   a. Remove the 3 bolts and remove the engine mounting bracket.
7. Remove the alternator assembly.
   a. Remove the 2 bolts and remove the fan belt adjusting bar.
   b. Remove the terminal cap.
   c. Remove the nut and remove terminal B.
   d. Disconnect the connector and harness clamp.
   e. Remove the bolt and remove the alternator assembly.
   f. Remove the bolt and remove the wire harness bracket.

To install:

8. Install alternator assembly.
   a. Install the wire harness bracket with the bolt.
   b. Provisionally install the alternator assembly with the bolt.
   c. Provisionally install the fan belt adjusting bar and alternator assembly with the 2 bolts.
   d. Tighten the bolt A to the specified torque. Torque: 14 ft. lbs. (19 Nm).
   e. Connect the connector and wire harness clamp.
   f. Install terminal B with the nut.
   g. Install the terminal cap.
9. Install transverse engine mounting bracket.
   a. Install the engine mounting bracket with the 3 bolts. Torque: 38 ft. lbs. (51 Nm).
10. Install engine mounting insulator sub-assembly RH.
    a. Install the engine mounting insulator sub-assembly RH with the 6 bolts and nut.
    b. Install the engine mounting stay RH with the 2 bolts. Torque: 19 ft. lbs. (26 Nm).
11. Install fan and alternator V belt.
12. Adjust fan and alternator V belt.
13. Install engine cover.
14. Install engine under cover RH
15. Connect cable to negative battery terminal.

ENGINE ELECTRICAL — IGNITION SYSTEM

FIRING ORDER

Firing order: 1–3–4–2

IGNITION COIL MODULE

REMOVAL & INSTALLATION

See Figure 89.

1. Disconnect the negative battery cable.
2. Remove the engine cover
3. Detach the 4 ignition coil connectors.
4. Remove the 4 bolts (one bolt for each coil), and remove the 4 ignition coils.

To install:

5. Install the ignition coils and secure with the bolts. Tighten to 80 inch lbs. (9 Nm).
6. Attach the 4 ignition coil connectors.
7. Install the engine cover.
8. Connect the negative battery cable.

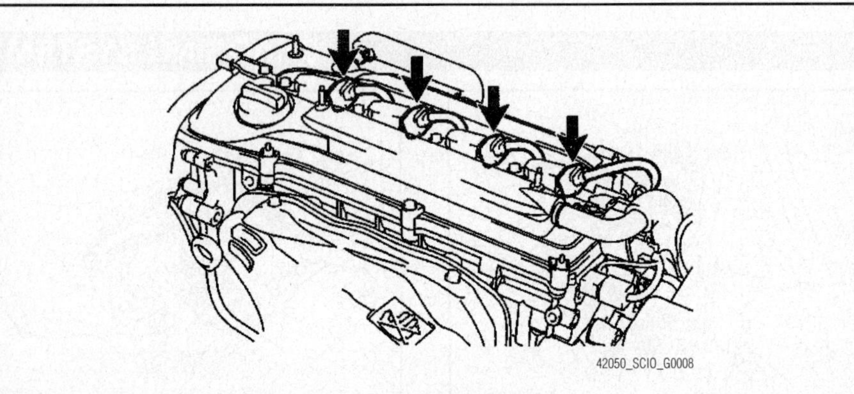

Fig. 89 Each ignition coil is secured with one mounting bolt

IGNITION TIMING

ADJUSTMENT

The ignition timing is controlled by the Powertrain Control Module (PCM). No adjustment is necessary or possible.

SPARK PLUGS

REMOVAL & INSTALLATION

1. Disconnect the negative battery cable.
2. Remove the engine cover.
3. Remove the ignition coils, as outlined in this section.
4. Use a spark plug wrench to remove the spark plugs.
5. Inspect the spark plugs, as outlined in this section.

To install:

6. Use a spark plug wrench to install and tighten the spark plugs to 14 ft. lbs. (19 Nm).
7. Install the ignition coil. Tighten the retainers to 80 inch lbs. (9 Nm). Refer to the Ignition Coil procedure in this section for more details.
8. Install the engine cover.
9. Connect the negative battery cable.

ENGINE ELECTRICAL — STARTING SYSTEM

STARTER

REMOVAL & INSTALLATION

tC and xB

See Figure 90.

1. Before servicing the vehicle, refer to the Precautions Section.
2. Remove or disconnect the following:
   - Negative battery cable
   - Starter electrical connections
   - Starter mounting bolts
   - Starter

To install:

3. Install or connect the following:
   - Starter. Tighten the mounting bolts to 27 ft. lbs. (37 Nm).
   - Starter electrical connections. Tighten the starter wire nut to 7 ft. lbs. (10 Nm).
   - Negative battery cable

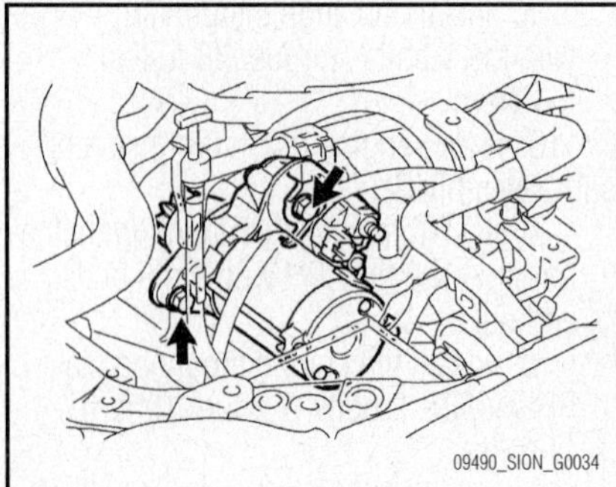

Fig. 90 Location of the starter mounting bolts—2.4L Engine

xD

1. Disconnect the cable from negative battery terminal.
2. Remove the engine under covers LH and RH.
3. Remove the flywheel housing side cover.
   a. Disengage the claw by pulling it outward and remove the flywheel housing side cover.
4. Remove the starter assembly.
   a. Remove the terminal cap.
   b. Remove the nut and disconnect terminal 30.
   c. Disconnect the connector.
   d. Remove the 2 bolts and remove the starter assembly.

To install:

5. 1. Install starter assembly.
   a. Install the starter assembly with the 2 bolts. Torque: 27 ft. lbs. (37 Nm).
   b. Connect the connector.
   c. Connect terminal 30 with the nut.
   d. Close the terminal cap.
6. Install flywheel housing side cover.
   a. Insert the protruding portion into the end of the cylinder block and while pushing it along the cylinder block, fit the claw into the cylinder block.

➡Make sure that the claw makes a click sound, indicating that it fits tightly. Replace the claw with a new one if it does not fit tightly or is deformed.

7. Install engine under cover RH.
8. Install engine under cover LH.
9. Connect cable to negative battery terminal.

ENGINE MECHANICAL

➡Disconnecting the negative battery cable may interfere with the functions of the on board computer systems and may require the computer to undergo a relearning process, once the negative battery cable is reconnected.

ACCESSORY DRIVE BELTS

ACCESSORY BELT ROUTING

See Figures 91 and 92.

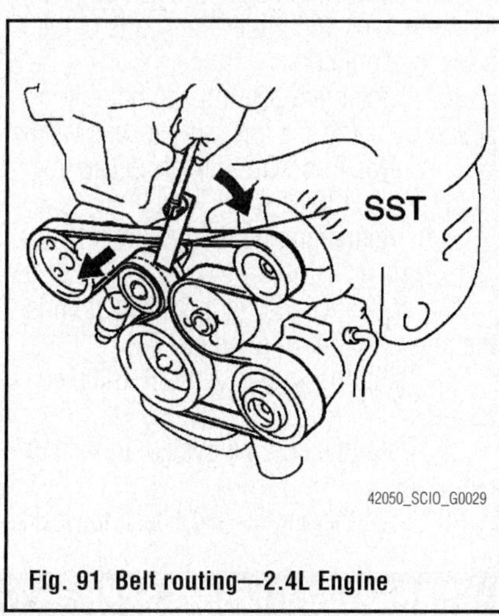

Fig. 91 Belt routing—2.4L Engine

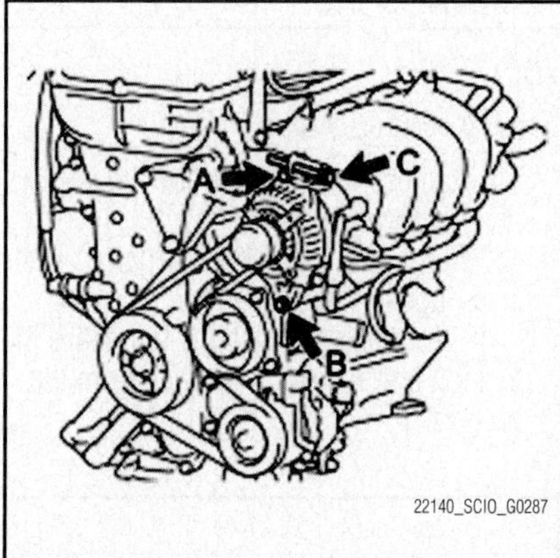

Fig. 92 Accessory belt routing—1.8L Engine

INSPECTION

Inspect the drive belt for signs of glazing or cracking. A glazed belt will be perfectly smooth from slippage, while a good belt will have a slight texture of fabric visible. Cracks will usually start at the inner edge of the belt and run outward. All worn or damaged drive belts should be replaced immediately.

REMOVAL & INSTALLATION

See Figure 93.

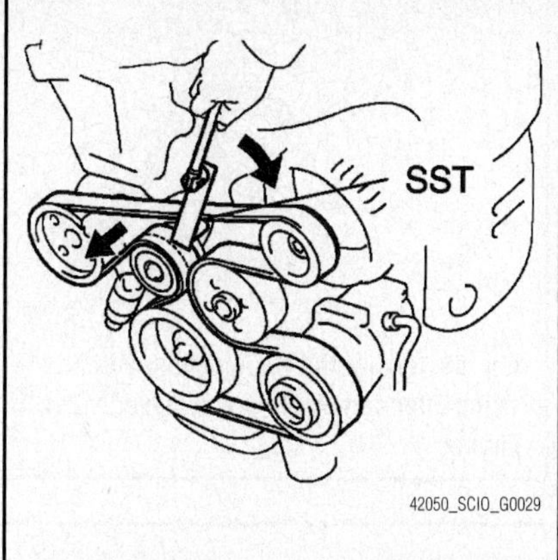

Fig. 93 Removing and installing the belt using the special tool

1. Remove the front right side fender apron seal.
2. Using SST 09249-63010, or equivalent tool, loosen the V-ribbed belt tensioner arm clockwise, then remove the fan-alternator belt.

➡Be sure to connect SST and the tools so they align using. When retracting the tensioner, turn it clockwise slowly in 3 second or more. Be sure to not apply force rapidly. After the tensioner is retracted all the way, do not apply and more force than necessary.

To install:

3. Using the special tool, loosen the V-ribbed belt tensioner arm clockwise, then install the fan and alternator V belt.
4. Install the front right side fender apron seal.

CAMSHAFT AND VALVE LIFTERS

REMOVAL & INSTALLATION

tC and xB

See Figures 94 through 101.

1. Before servicing the vehicle, refer to the Precautions Section.
2. Drain the cooling system.
3. Drain the engine oil.
4. Relieve the fuel system pressure.
5. Remove or disconnect the following:
   - Negative battery cable
   - Hood
   - Right-hand front wheel
   - Engine undercover
   - Right-hand front fender apron
   - Engine appearance cover
   - Wiper arm assembly

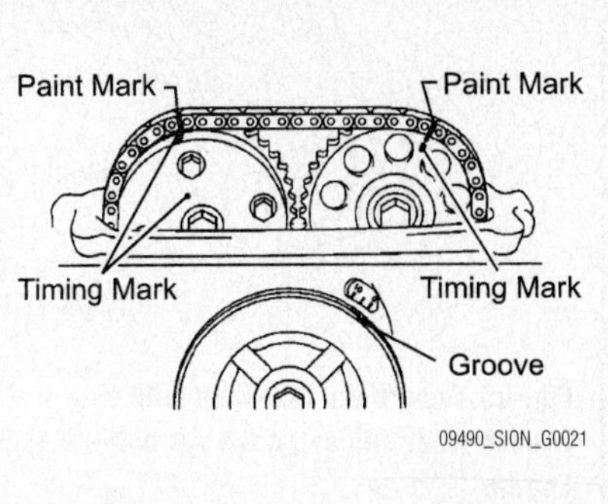

Fig. 94 Match the groove on the crankshaft pulley to the front cover timing mark. Place matchmarks on the camshaft sprocket and timing chain—2.4L Engine

   - Top cowl seal
   - Left-hand top cowl ventilator louver
   - Wiper linkage assembly
   - Top outer cowl panel
   - Left-hand cowl body mounting bracket
   - Air intake assembly
   - Throttle body
   - Fuel rail
   - Intake manifold
   - Front exhaust pipe
   - Oil dipstick guide
   - Exhaust manifold
   - Accessory drive belts
   - Alternator
   - Power steering pump
6. Support the engine with a suitable jack.
7. Remove or disconnect the following:
   - Right-hand engine mount
   - Ignition coil
   - Cylinder head cover
   - Accessory drive belt tensioner
   - Crankshaft position sensor
   - Oil pan
8. Turn the crankshaft pulley until its groove and the timing mark on the front cover are aligned to set the No. 1 cylinder to TDC.
9. Matchmark the timing chain and camshaft sprockets.
10. Holding the exhaust camshaft with a wrench, loosen the camshaft timing set bolt.
11. Using several steps, loosen the bearing cap bolts in the sequence shown. Remove the bearing caps.
12. Remove the camshaft timing set bolt while holding the exhaust camshaft in place.
13. Remove the exhaust camshaft, leaving the camshaft sprocket wrapped in the timing chain.
14. Remove the camshaft sprocket.

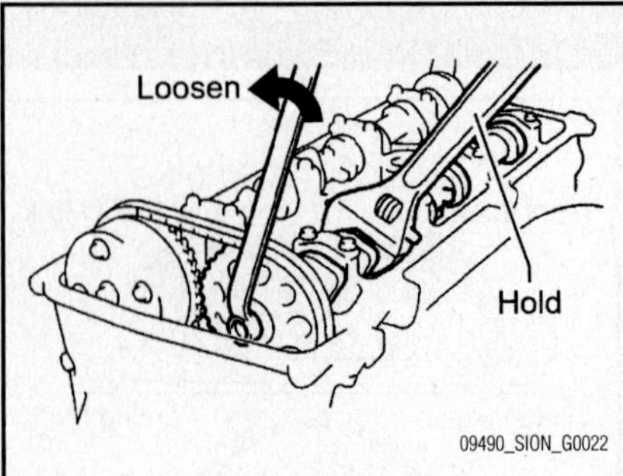

Fig. 95 Secure the camshaft with a wrench when removing the set bolt—2.4L Engine

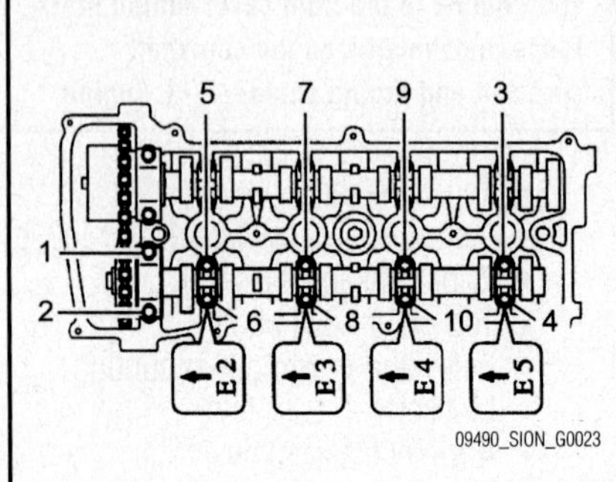

Fig. 96 Exhaust camshaft bearing cap bolt removal sequence—2.4L Engine

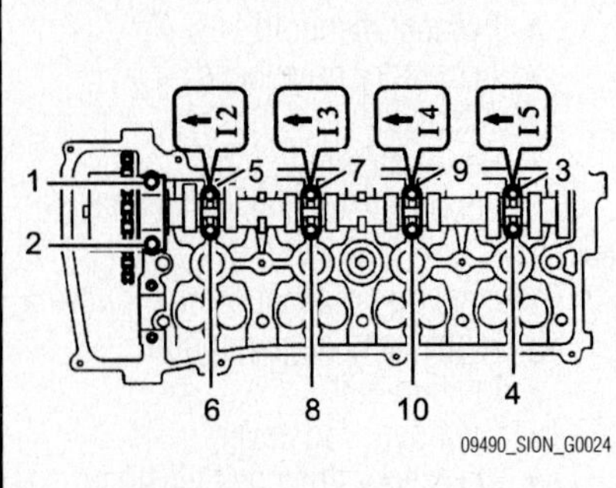

Fig. 97 Intake camshaft bearing cap bolt removal sequence—2.4L Engine

15. Holding the intake camshaft with a wrench, loosen the camshaft timing set bolt.
16. Using several steps, loosen the intake camshaft bearing cap bolts in the sequence shown. Remove the bearing caps.
17. Remove the intake camshaft, with sprocket attached, while holding the timing chain by hand.
18. Secure the timing chain with string to prevent it from falling down into the front cover.
19. Remove the valve lifters.

➡Keep all valvetrain components in order for reassembly.

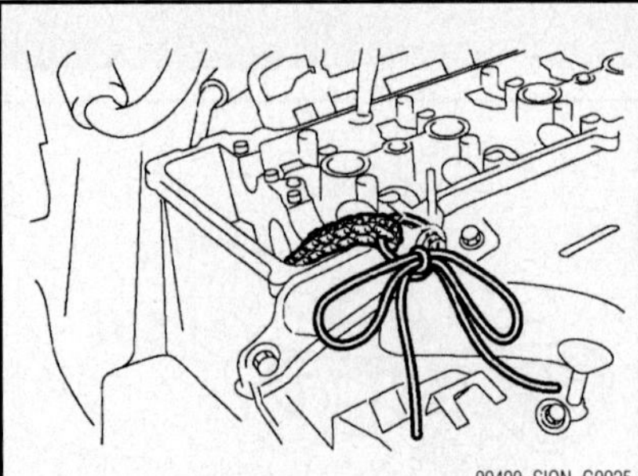

Fig. 98 Secure the timing chain with string after camshafts are removed—2.4L Engine

Fig. 99 Ensure the No. 1 cam lobes are facing the correct direction during installation—2.4L Engine

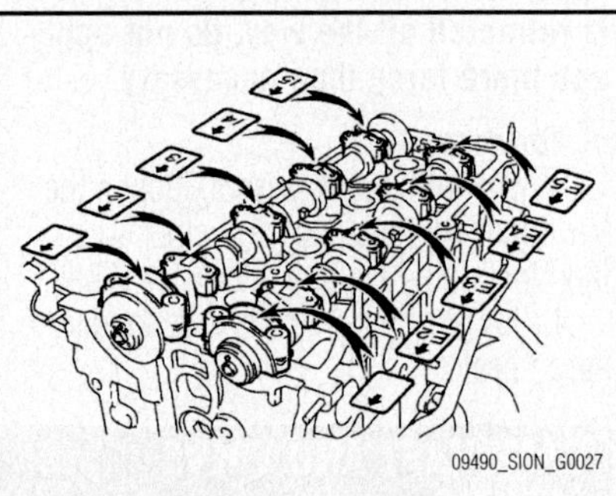

Fig. 100 Compare the markings on the camshafts to the illustration for correct installation orientation—2.4L Engine

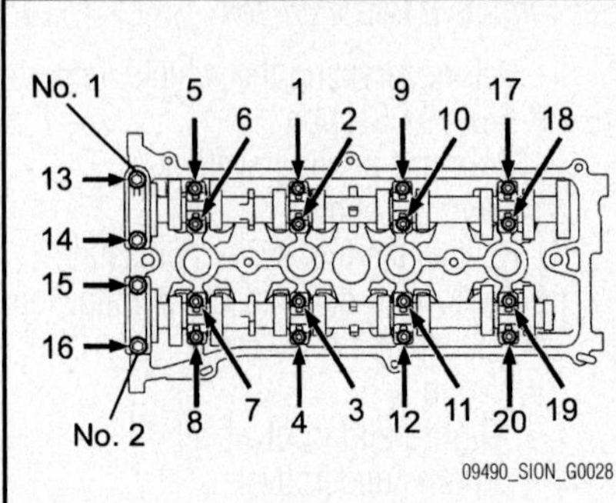

Fig. 101 Camshaft bearing cap bolt torque sequence—2.4L Engine

To install:

20. Apply a light coat of clean engine oil to each valve lifter.
21. Install the valve lifters in their original places.
22. Apply a light coat of clean engine oil to the journals of the camshafts.
23. Install the camshafts on the cylinder head with the No. 1 cam lobes facing the directions shown.
24. Examine the camshaft markings to ensure correct orientation of the camshafts for installation.
25. Apply a light coat of clean engine oil to the threads and under the heads of the bearing cap bolts.
26. Tighten the bearing cap bolts in sequence using the following torque values:
    a. No. 1 and No. 2 bearing cap to 22 ft. lbs. (30 Nm).
    b. Remaining bolts to 80 inch lbs. (9 Nm).
27. The remainder of the installation is the reverse order of removal.
28. Refill the engine with oil to the correct level.
29. Refill the cooling system to the correct level.
30. Start the engine and check for leaks.

CRANKSHAFT FRONT SEAL

REMOVAL & INSTALLATION

xD

See Figures 102 through 104.

1. Remove the engine under cover RH.
2. Remove engine cover.
3. Remove the fan and alternator v belt.
    a. Loosen bolts A and B.
    b. Turn adjusting bolt C to release the tension and remove the V belt from the pulleys.

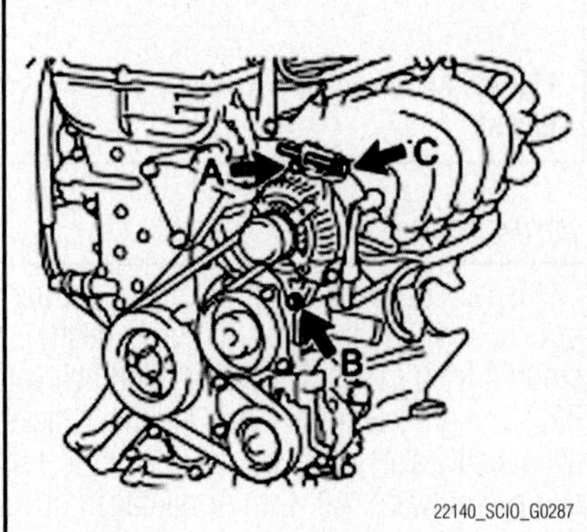

Fig. 102 Remove the fan and alternator v belt

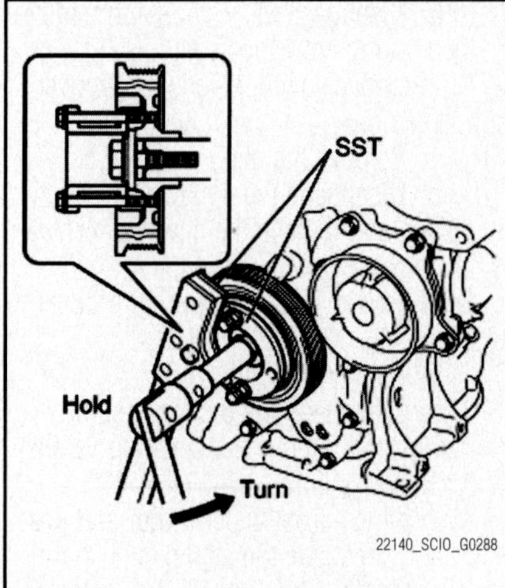

Fig. 103 Using SST, hold the pulley in place and loosen the pulley bolt

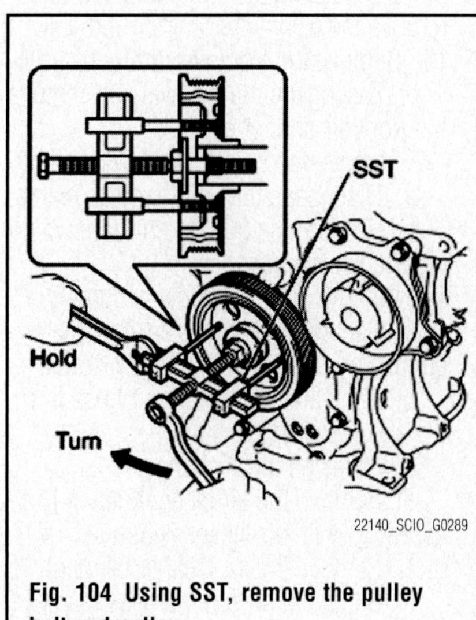

Fig. 104 Using SST, remove the pulley bolt and pulley

c. Remove the fan and alternator V belt.

4. Remove the crankshaft pulley.

a. Using SST, hold the pulley in place and loosen the pulley bolt.

b. Using SST, remove the pulley bolt and pulley.

5. Remove the timing chain or belt cover oil seal

a. Using a knife, cut off the oil seal lip.

b. Using a screwdriver with its tip taped, pry out the oil seal.

➡After the removal, check the crankshaft for damage. If it is damaged, smooth the surface with 400-grit sandpaper.

To install:

6. Install timing chain or belt cover oil seal.

a. Apply MP grease to a new oil seal lip.

b. Using SST and a hammer, tap the oil seal until its surface is flush with the timing chain cover edge.

➡Do not tap the oil seal at an angle. Keep the lip free of foreign matter.

7. Install crankshaft pulley.

a. Align the pulley set key with the key groove of the pulley.

b. Using SST, hold the pulley in place and tighten the bolt. Torque: 140 ft. lbs. (190 Nm).

8. Install fan and alternator V belt.
9. Adjust fan and alternator V belt.
10. Inspect fan and alternator V belt.
11. Install engine cover.
12. Inspect for engine oil leak.
13. Install engine under cover RH.

CYLINDER HEAD

REMOVAL & INSTALLATION

tC and xB

See Figures 105 and 106.

1. Before servicing the vehicle, refer to the Precautions Section.
2. Drain the cooling system.
3. Drain the engine oil.
4. Relieve the fuel system pressure.
5. Remove or disconnect the following:
   - Negative battery cable
   - Hood
   - Right-hand front wheel
   - Engine undercover
   - Right-hand front fender apron
   - Engine appearance cover
   - Wiper arm assembly
   - Top cowl seal
   - Left-hand top cowl ventilator louver
   - Wiper linkage assembly
   - Top outer cowl panel
   - Left-hand cowl body mounting bracket
   - Air intake assembly
   - Throttle body
   - Fuel rail
   - Intake manifold
   - Front exhaust pipe
   - Oil dipstick guide
   - Exhaust manifold
   - Accessory drive belts
   - Alternator
   - Power steering pump
6. Support the engine with a suitable jack.
7. Remove or disconnect the following:
   - Right-hand engine mount
   - Ignition coil
   - Cylinder head cover
   - Accessory drive belt tensioner

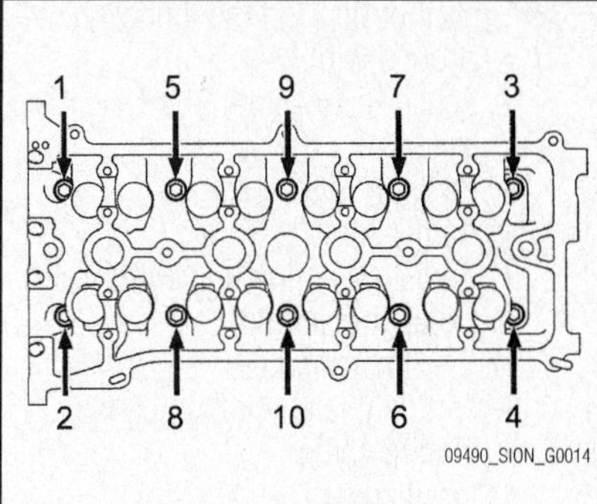

Fig. 105 Cylinder head bolt removal sequence—2.4L Engine

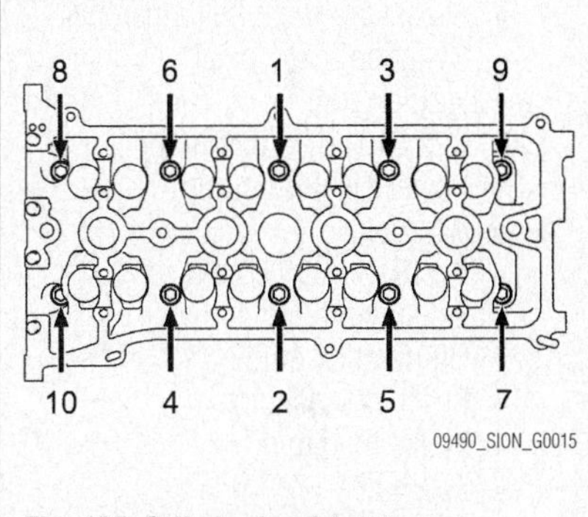

Fig. 106 Cylinder head bolt torque sequence—2.4L Engine

   - Crankshaft position sensor
   - Oil pan
8. Turn the crankshaft pulley until its groove and the timing mark on the front cover are aligned to set the No. 1 cylinder to TDC.
9. Remove or disconnect the following:
   - Camshafts
   - Crankshaft pulley
   - Front cover
   - Timing chain
   - Timing chain vibration damper
   - Camshaft timing oil control valve
   - Radiator inlet hose
   - All remaining sensors connectors
   - Ground wire
10. Loosen the cylinder head bolts in the sequence shown and remove the cylinder head and gasket.

To install:

11. Install the cylinder head with a new gasket. The Lot Number on the gasket should face upward.
12. Apply a light coat of new engine oil to the threads of the cylinder head bolts. Tighten the bolts in the sequence shown as follows:

a. Step 1: Tighten to 58 ft. lbs. (79 Nm).

b. Step 2: Plus 90 degrees.

13. Install or connect the following:
- Ground wire
- Sensor connectors to the cylinder head
- Radiator inlet hose
- Camshafts
- Camshaft timing oil control valve
- Exhaust manifold
- Oil dipstick guide
- Timing chain vibration damper
- Timing chain
- Front cover
- Crankshaft pulley. Tighten to 133 ft. lbs. (180 Nm).
- Oil pan
- Crankshaft position sensor
- Accessory drive belt tensioner
- Cylinder head cover
- Ignition coil
- Right-hand engine mount
- Power steering pump
- Alternator
- Accessory drive belts
- Exhaust manifold
- Oil dipstick guide
- Front exhaust pipe
- Intake manifold
- Fuel rail
- Throttle body

14. The remainder of the installation is the reverse order of removal.

15. Refill the engine with oil to the correct level.

16. Refill the cooling system to the correct level.

17. Start the engine and check for leaks.

xD

See Figures 13 through 30, 33 through 42, 70 and 107 through 114.

1. Discharge the fuel system pressure.

⁂ CAUTION

The DISCHARGE FUEL SYSTEM PRESSURE procedures must be performed before disconnecting any part of the fuel system.

⁂ CAUTION

After performing the DISCHARGE FUEL SYSTEM PRESSURE procedures, pressure will remain in the fuel line. When disconnecting the fuel line, place a cloth or equivalent over fittings to reduce the risk of fuel spray.

a. Remove the rear seat cushion assembly LH

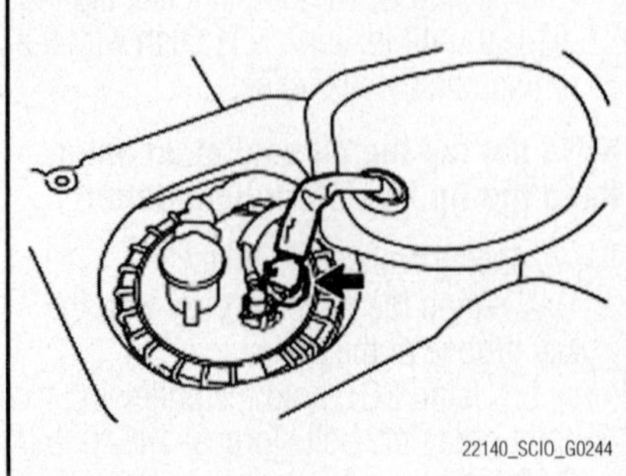

Fig. 107 Disconnect the connector from the fuel pump assembly

b. Remove the rear floor service hose cover.

c. Disconnect the connector from the fuel pump assembly.

d. Start the engine. After the engine stops naturally, turn the ignition switch OFF.

e. Start the engine again and make sure that engine does not start.

f. Remove the fuel tank cap, and let the air out of the fuel tank.

g. Connect the connector.

h. Install the rear floor service hole cover with new butyl tape.

i. Install the rear seat cushion assembly LH.

2. Remove the battery.

a. Disconnect the cable from the battery terminal.

b. Loosen the nut and remove the battery clamp.

c. Remove the battery.

3. Remove the battery tray.
4. Remove the front wheels.
5. Remove the engine under cover LH.
6. Remove the engine under cover RH.
7. Drain the engine coolant.
8. Drain the engine oil.
9. Remove front wiper arm head cap.

a. Using a screwdriver with its tip wrapped in protective tape, disengage the 2 claws and remove the 2 front wiper arm caps.

10. Remove front wiper arm and blade assembly LH.

a. Operate the wiper, then stop the wiper motor in the automatic stop position.

b. Remove the nut and front wiper arm and blade assembly LH.

11. Remove front wiper arm and blade assembly RH.

➡Use the same procedure as for the LH side.

12. Remove hood to cowl top seal.

a. Disengage the 7 clips and remove the hood to cowl top seal.

13. Remove cowl top ventilator louver sub-assembly.

a. Remove the clip.

b. Disengage the 4 claws and 4 guides and remove the cowl top ventilator louver RH.

14. Remove cowl top ventilator louver LH.

a. Remove the clip.

b. Disengage the 5 claws and 5 guides and remove the cowl top ventilator louver LH.

15. Remove front wiper motor and link.

a. Disengage the clamp.

b. Remove the 2 bolts.

c. Slide the front wiper motor and link. Disengage the link from the rubber pin, then disconnect the connector and remove the front wiper motor and link.

16. Remove the front air shutter seal RH.

a. Disengage the 2 claws and remove the front air shutter seal RH.

17. Remove the outer cowl top panel.

a. Disengage the wire harness clamp.

b. Remove the 9 bolts and remove the outer cowl top panel.

18. Remove the air cleaner assembly.

a. Separate the intake air flow meter connector and 2 wire harness clamps.

b. Separate the ventilation hose from the air cleaner hose.

c. Unfasten the 2 clamps.

d. Remove the air cleaner cap sub-assembly with air cleaner hose.

e. Remove the air cleaner element.

f. Separate the wire harness clamp from the air cleaner case.

g. Remove the 2 bolts and remove the air cleaner case with air cleaner inlet.

19. Remove the air cleaner bracket.

a. Separate the wire harness clamp from the air cleaner bracket.

b. Remove the 2 bolts and remove the air cleaner bracket.

20. Remove the battery carrier.

a. Separate the wire harness clamp from the battery carrier.

b. Remove the 5 bolts and remove the battery carrier.

21. Remove the Engine cover.

a. (a) First, lift up the rear of the Engine cover to disengage the 2 fittings

b. (b) Next, lift the front of the Engine cover to disengage the 2 fittings. Then remove the Engine cover.

➡Ensure that the rubber grommets remain attached to the Engine cover. If any grommets are attached to the bolts, move them to the Engine cover.

22. Remove the fan and alternator V belt.
a. Loosen bolts A and B.
b. Turn adjusting bolt C to release the tension and remove the V belt from the pulleys.
c. Remove the fan and alternator V belt.
23. Remove the lower radiator hose.
a. Loosen the 2 clips and remove the No. 2 radiator hose.
24. Remove the front bumper cover.
a. Apply protective tape, as shown in the illustration.
b. Remove the 8 screws.
c. Remove the 9 clips.
d. Disengage the 6 claws and remove the front bumper cover.
25. Disconnect the transaxle cooling lines.
26. Remove the hood lock assembly.
27. Remove the hood lock assembly.
a. Separate the hood lock control cable assembly from the 2 clamps.
b. Remove the 3 bolts and remove the hood lock assembly.
28. Remove the No. 1 cooler cover.
a. Remove the 2 clips and remove the No. 1 cooler cover.
29. Remove the upper radiator support absorber.
a. Using a screwdriver with its tip wrapped in protective tape, disengage the 6 claws and remove the radiator support upper absorber.
30. Remove the upper radiator support sub-assembly.
a. Disconnect the horn connector.
b. Remove the 5 bolts and remove the upper radiator support sub-assembly.
31. Remove the radiator assembly.
a. Disconnect the cooling fan motor connector and separate the wire harness clamp.
b. Disengage the 2 claws and remove the radiator assembly from the vehicle.

CAUTION

Do not apply excessive force to the cooler condenser assembly or piping when removing the radiator assembly.

32. Separate with pulley compressor assembly.
33. Separate the transmission control cable assembly (for Automatic Transaxle).
a. Remove the nut and disconnect the transmission control cable assembly from the control shaft lever.
34. Separate the transmission control cable assembly (for Manual Transaxle).
a. Remove the 2 clips and the washers, and disconnect the 2 cables from the transaxle.
b. Remove the 2 clips and disconnect the 2 cables from the control cable bracket.
35. Separate the union to check valve hose.
a. Loosen the clip and separate the union to check valve hose.
36. Separate No. 1 fuel vapor feed hose.
a. Separate the No. 1 fuel vapor feed hose from the vacuum switching valve assembly.
37. Disconnect the engine wire.
a. Remove the bolt and wire harness clamp and separate the transaxle ground wire (for Automatic Transaxle).
b. Remove the bolt and 2 wire harness clamp and separate the transaxle ground wire (for Manual Transaxle).
c. Disconnect the 2 connectors.
d. Pull up the lever and disconnect the connector from the engine control computer.
e. Separate the 2 wire harness clamps.
f. Remove the 2 connectors and clamp from the engine room junction block and disconnect the 2 wire harness clamps.
g. Disconnect all wire harnesses and connectors.

CAUTION

Make sure that no wire harness is connected between the body and engine.

38. Disconnect the upper and lower heater hoses (from heater unit).
39. Disconnect the fuel tube sub-assembly.
a. Remove the No. 1 fuel pipe clamp.
b. Pinch the retainer, then pull the fuel tube connector out of the pipe.

➡Remove any dirt and foreign matter from the fuel tube connector before performing this work. Do not allow any scratches or foreign matter on the parts when disconnecting, as the fuel tube connector has the O-rings that seal the pipe. Perform this work by hand. Do not use any tools. Do not forcibly bend, twist or turn the nylon tube. Protect the disconnected parts by covering them with vinyl bags after disconnecting the fuel tube. If the fuel tube connector and pipe are stuck, push and pull them to release them.

40. Separate the clutch release cylinder assembly (for Manual Transaxle).

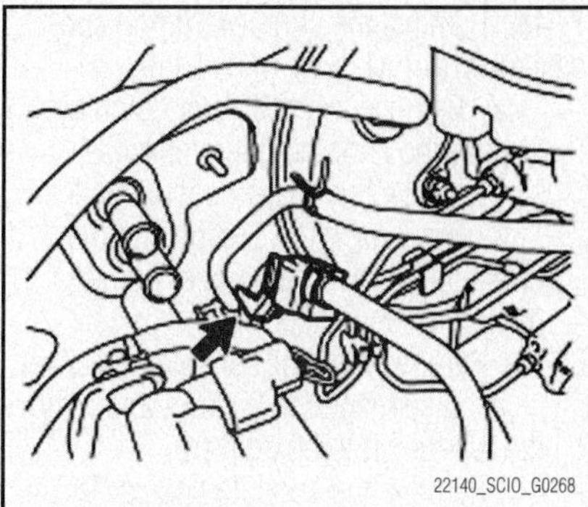

Fig. 108 Remove the No. 1 fuel pipe clamp

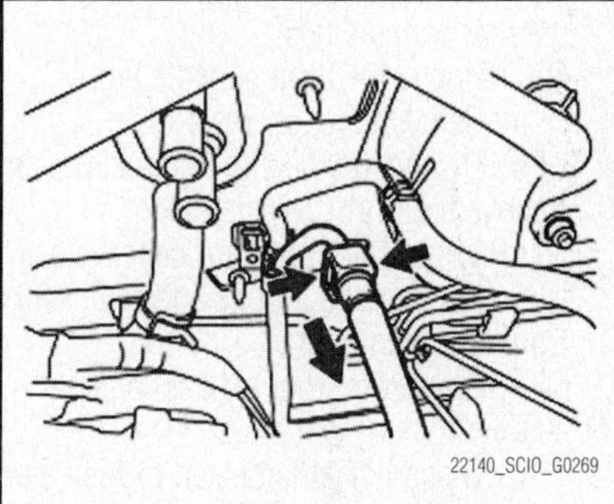

Fig. 109 Pinch the retainer, then pull the fuel tube connector out of the pipe

a. Remove the 4 bolts, then separate the clutch release cylinder.

➡Suspend the clutch release cylinder with a piece of rope so as not to overload the clutch pipe.

41. Remove the column hole cover silencer sheet.
a. Remove the floor carpet and 2 clips and remove the column hole cover silencer.
42. Separate the steering sliding yoke sub-assembly.
a. Use a seat belt to fix the steering wheel assembly, in order to avoid breakage of the spiral cable.
b. Place matchmarks on the sliding yoke of the steering intermediate shaft assembly and the power steering.
c. Loosen bolt A, remove bolt B and separate the steering intermediate shaft assembly.
43. Separate the No. 1 steering column hole cover sub-assembly.
a. Remove clip A, separate clip B from the body and separate No. 1 steering column hole cover.

CAUTION

Do not damage clip B.

44. Remove the shift lever knob sub-assembly (for Manual Transaxle).
a. Remove the shift lever knob by turning the knob counterclockwise.
45. Remove the upper console panel.
a. Disengage the 2 clips and the 6 claws and remove the upper console panel.
46. Remove the rear console box cover.
a. Disengage the 9 claws and remove the rear console box cover.
47. Remove the console box carpet.
48. Remove the rear console box assembly.
a. Remove the 3 screws.
b. Disengage the 4 claws and remove the rear console box.
49. Remove the front console box.
a. Remove the 2 screws.
b. Disengage the clip and 2 claws and remove the front console box.
50. Remove the front floor center brace.
a. Remove the 2 bolts and remove the front floor center brace.
51. Remove the front exhaust pipe assembly.
a. Disconnect the heated oxygen sensor connector.
b. Remove the grommet and pull the sensor connector out of the cabin through the floor panel.
c. Remove the 2 bolts and 2 springs and separate the front exhaust pipe assembly from the exhaust tail pipe assembly.
d. Remove the 2 bolts and 2 springs and separate the front exhaust pipe assembly from the exhaust manifold.
e. Remove the 3 exhaust pipe supports and remove the front exhaust pipe assembly.
52. Remove the front axle shaft LH nut.
53. Remove the front axle shaft RH nut.
54. Separate the speed sensor front LH.
55. Separate the speed sensor front RH.
56. Separate the tie rod end sub-assembly LH.
a. Remove the cotter pin and castle nut.
b. Using SST, separate the tie rod end from the front axle assembly.
57. Separate the tie rod end sub-assembly RH

➡The separation procedure for the RH side is the same as that for the LH side.

58. Separate the front stabilizer link assembly LH.
a. Remove the nut and separate the stabilizer link from the shock absorber.
59. Separate the front stabilizer link assembly RH.

➡The separation procedure for the RH side is the same as that for the LH side.

60. Separate the front lower suspension arm sub-assembly LH.
a. Remove the clip and castle nut.
b. Install SST to the threaded section of the lower ball joint.

➡Make sure the upper ends of the threaded section of the lower ball joint and SST (spacer B) are aligned.

c. Using SST, separate the front lower suspension arm from the front axle assembly.
61. Separate the front lower suspension arm sub-assembly RH

➡The separation procedure for the RH side is the same as that for the LH side.

62. Separate the front axle assembly LH.
a. Using a plastic hammer, tap the end of the drive shaft and disengage the fitting between the drive shaft and front axle.

➡If it is difficult to disengage the fitting, tap the end of the drive shaft with a brass bar and hammer.

b. Push the front axle out of the vehicle to remove the drive shaft from the front axle.

➡Do not push the front axle further out of the vehicle than is necessary. Do not damage the outboard joint boot. Do not damage the speed sensor rotor. Suspend the drive shaft with a piece of rope or the equivalent.

63. Separate the front axle assembly RH.

➡The separation procedure for the RH side is the same as that for the LH side.

64. Remove the engine assembly with transaxle.
a. Set the engine lifter.
b. Remove 2 bolts and remove the engine mounting stay RH.
c. Remove the 6 bolts and nut and remove the engine mounting insulator RH.

➡Be sure not to remove the other bolt A.

d. Remove the through bolt and nut and separate the transverse engine mounting insulator.

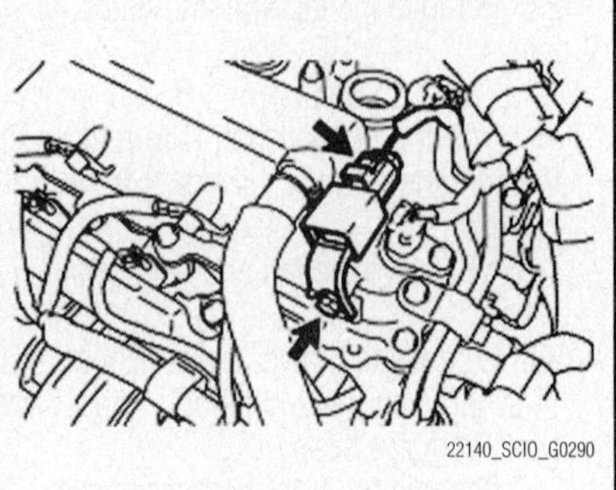

Fig. 110 Remove the radio setting condenser

e. Remove the 6 bolts and remove the engine assembly with transaxle and front suspension crossmember from the vehicle.
65. Disconnect the transaxle oil cooler inlet and outlet hoses from the radiator (for Automatic Transaxle).
66. Remove the transmission oil level gage sub-assembly (for Automatic Transaxle)
67. Separate the transmission oil filler tube sub-assembly (for Automatic Transaxle)
68. Remove the ignition coil assembly.
a. Disconnect the 4 ignition coil assembly connectors.
b. Remove the 4 bolts and remove the 4 ignition coil assemblies.
69. Remove the radio setting condenser.
a. Disconnect the radio setting condenser connector.
b. Remove the bolt and remove the radio setting condenser.
70. Remove the oil breather hose.
71. Remove the heater water hoses (from heater unit).
72. Remove the upper radiator hose.
73. Remove the intake manifold.
a. Loosen the 2 clips and separate the 2 water by-pass hoses.
b. Separate the No. 1 fuel vapor feed hose from the throttle body.
c. Loosen the clip and separate the No. 1 vacuum transmitting hose.
d. Separate the ventilation hose.
e. Separate the wire harness clamp from the intake manifold.
f. Remove the 4 bolt and 2 nuts and remove the intake manifold and manifold stay.
g. Remove the gasket from the intake manifold.
74. Remove the fuel delivery pipe sub-assembly.
75. Remove the fuel delivery pipe spacers.

76. Remove the injector vibration insulators.
77. Remove the fuel injector assembly.
a. Pull the 4 fuel injectors out of the fuel delivery pipe.
b. Place the fuel injector in a plastic bag to prevent foreign matter from entering.
78. Remove alternator assembly.
79. Remove the fan belt adjusting bar.
80. Remove the oil level dipstick
81. Remove the oil level gage guide.
82. Remove the water by-pass hose.
83. Remove the water inlet hose.
84. Remove the water by-pass pipe.
85. Remove the vacuum tube assembly.
86. Remove the exhaust manifold heat insulator.
87. Remove the manifold stay.
88. Remove the exhaust manifold.
89. Remove the wire harness clamp bracket.
90. Remove the water inlet.
91. Remove the thermostat.
92. Remove the cylinder head cover sub-assembly.
93. Set No. 1 cylinder to TDC / compression.
94. Remove the crankshaft pulley.
95. Remove the No. 1 chain tensioner assembly.
96. Remove the timing chain cover sub-assembly.
97. Remove the timing chain or belt cover oil seal.
98. Remove the No. 2 chain vibration damper.
99. Remove the chain tensioner slipper.
100. Remove the No. 1 chain vibration damper.
101. Remove the chain sub-assembly.
a. Hold the hexagonal portion of the camshaft with a wrench and turn the camshaft timing gear assembly counter-clockwise to loosen the chain on the camshaft timing gears.
b. With the chain loosened, release the chain from the camshaft timing gear assembly and rest it on the camshaft timing gear assembly.

CAUTION

Be sure to release the chain from the sprocket completely.

c. Turn the camshaft clockwise to return it to the original position and remove the chain.
102. Remove the crankshaft timing gear sprocket.
103. Remove the camshaft bearing cap.
a. Uniformly loosen and remove the 10 bearing cap bolts in the sequence shown in the illustration.
b. Uniformly loosen and remove the 15 bearing cap bolts in the sequence shown in the illustration.
c. Uniformly loosen the bolts while keeping the camshaft level.

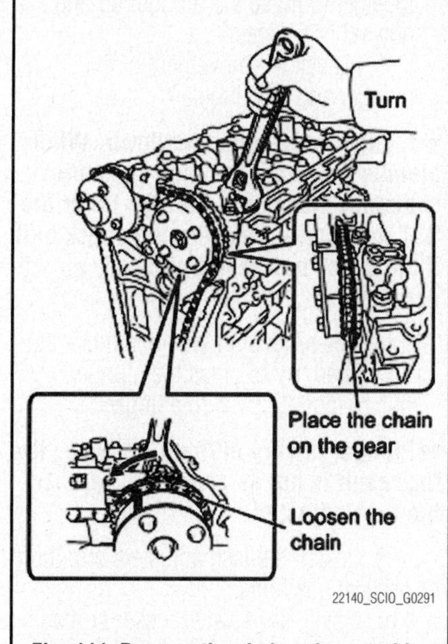

Fig. 111 Remove the chain sub-assembly

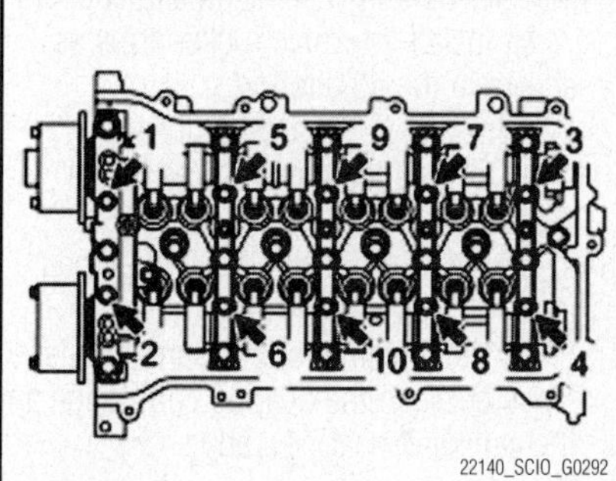

Fig. 112 Remove the 10 bearing cap bolts in the sequence

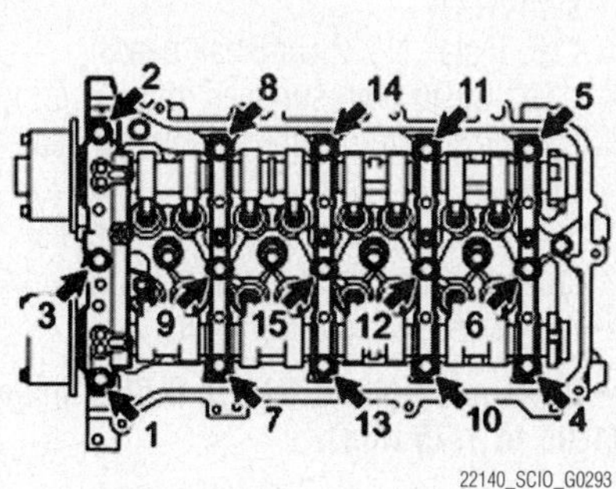

Fig. 113 Remove the 15 bearing cap bolts in the sequence

➡**If the camshaft bearing cap bolts have been loosened, reapply seal packing between the camshaft housing and cylinder head.**

d. Remove the 5 bearing caps.
104. Remove the camshaft.
a. Remove the camshaft from the camshaft housing.
105. Remove the No. 2 camshaft.
a. Remove the No. 2 camshaft from the camshaft housing.
106. Remove the No. 1 valve rocker arm sub-assembly.
a. Remove the 16 valve rocker arms.
107. Remove the valve lash adjuster assembly.
a. Remove the 16 valve lash adjusters from the cylinder head.
108. Remove the No. 1 camshaft bearing.
a. Remove the 2 No. 1 camshaft bearings.
109. Remove the No. 2 camshaft bearing.
a. Remove the 2 No. 2 camshaft bearings.
110. Remove the camshaft housing sub-assembly.
a. Remove the 2 bolts.
b. Remove the camshaft housing by prying between the cylinder head and camshaft housing with a screwdriver.

CAUTION

Be careful not to damage the contact surfaces of the cylinder head and camshaft housing. Tape the screwdriver tip before use.

111. Remove the cylinder head sub-assembly.
a. Using several steps, loosen and remove the 10 cylinder head bolts uniformly with a 10 mm bi-hexagon wrench in the sequence shown in the illustration.
b. Remove the 10 cylinder head bolts and the plate washers.

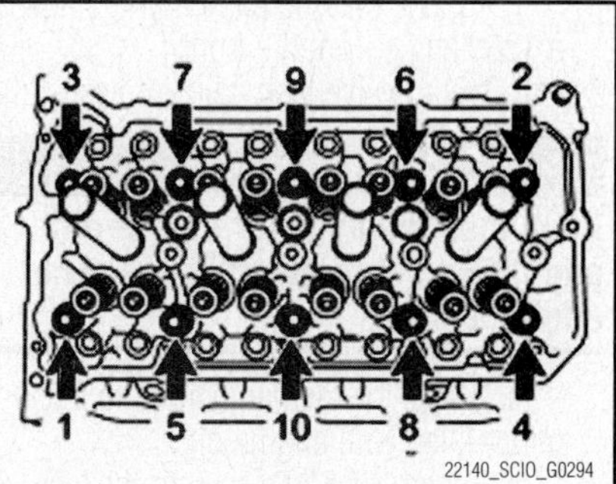

Fig. 114 Remove the 10 cylinder head bolts

**** CAUTION**

Do not drop the washers into the cylinder head.

*** CAUTION**

Head warpage or cracking could result from removing bolts in the wrong order.

c. Remove the cylinder head sub-assembly.
112. Remove the cylinder head gasket.
a. Remove the cylinder head gasket from the cylinder block.

To install:

113. Install cylinder head gasket
a. Place a new cylinder head gasket on the cylinder block with the Lot No. stamp facing upward.
b. Remove any oil from the contact surface.

➡Pay attention to the mounting orientation of the cylinder head gasket.

*** CAUTION**

Do not damage the cylinder gasket when installing the cylinder head onto the cylinder block.

114. Install cylinder head sub-assembly
a. The cylinder head bolts are tightened in 2 successive steps.
b. Apply a light coat of engine oil to the threads of the cylinder head bolts.
c. Using several steps, install and tighten the 10 cylinder head bolts and plate washers uniformly in the sequence shown in the illustration. Torque: 36 ft. lbs. (49 Nm).
d. Mark the front of the cylinder head bolt with paint.
e. Retighten the cylinder head bolts by additional 90° and one more additional 45°.
f. Check that the paint mark is now at a 135° angle from the front.
115. Install valve lash adjuster assembly

**** CAUTION**

Keep the lash adjuster free of dirt and foreign objects. Only use clean engine oil.

a. Place the lash adjuster into a container filled with engine oil.
b. Insert the SST's tip into the lash adjuster's plunger and use the tip to press down on the check ball inside the plunger.
c. Squeeze the SST and lash adjuster together to move the plunger up and down 5 or 6 times.
d. Check the movement of the plunger and bleed the air.

➡Plunger moves up and down. When bleeding air from the high-pressure chamber, make sure that the tip of the SST is actually pressing the check ball. If the check ball is not pressed, air will not bleed.

e. After bleeding the air, remove the SST. Then, try to press the plunger quickly and firmly with a finger.

➡Plunger is very difficult to move. If the result is not as specified, replace the lash adjuster.

f. Install the lash adjusters into their original positions.
116. Install No. 1 valve rocker arm sub-assembly
a. Apply engine oil to the lash adjuster tip and valve stem cap end.
b. Install the valve rocker arms as shown in the illustration.
117. Install No. 1 camshaft bearing
a. Clean the both surfaces of the bearing.
b. Install the 2 No. 1 camshaft bearings.
c. Using calipers, measure the distance between the bearing cap's edge and the camshaft bearing's edge.

➡Dimension (A—B): 0.0276 inches (0.7 mm) or less.

d. Position the bearing to the center of the bearing cap by measuring dimension A—B.
118. Install No. 2 camshaft bearing
a. Clean both surfaces of the bearing.
b. Install the 2 No. 2 camshaft bearings.
c. Using calipers, measure the distance between the bearing cap's edge and the camshaft bearing's edge.

➡Dimension (A): 0.042 to 0.068 inches (1.05 to 1.75 mm).

d. Position the bearing to the center of the bearing cap by measuring dimension A.
119. Install No. 2 camshaft
a. Clean the camshaft journals.
b. Apply a light coat of engine oil to the camshaft journals, camshaft housings and bearing caps.
c. Install the No. 2 camshaft to the camshaft housing.
120. Install camshaft
a. Clean the camshaft journals.
b. Apply a light coat of engine oil to the camshaft journals, camshaft housings and bearing caps.
c. Install the camshaft to the camshaft housing.
121. Install camshaft bearing cap
a. Apply engine oil to the camshaft journals, camshaft housing and bearing caps.
b. Check the marks and numbers on the camshaft bearing caps and place them in the proper position and direction.
c. Make sure that the knock pin of the camshaft is positioned as shown in the illustration.
d. Tighten the 10 bolts in the order shown in the illustration. Torque: 12 ft. lbs. (16 Nm).
122. Install camshaft housing sub-assembly
a. Make sure that the valve rocker arm is installed as shown in the illustration.
b. Apply seal packing in a continuous line as shown in the illustration.

➡Seal diameter: 0.138 to 0.158 inches (3.5 to 4.0 mm).

c. Remove any oil from the contact surface.
d. Install the camshaft housing sub-assembly RH within 3 minutes of applying the seal packing.
e. Do not start the engine for at least 2 hours after installing.
f. Set the camshaft and No. 2 camshaft.
g. Install the camshaft housing and tighten the 17 bolts in the order shown in the illustration. Torque: 20 ft. lbs. (27 Nm).

➡After installing the camshaft housing, make sure that the cam lobes are positioned as shown in the illustration. If any of the bolts are loosened during installation, remove the camshaft housing, clean the installation surfaces, and reapply seal packing. If the camshaft housing is removed because any of the bolts are loosened during installation, make sure that the previously applied seal packing does not enter any oil passages. After installing the camshaft housing, wipe off any seal packing that seeped out from between the housing and the cylinder head.

123. Install crankshaft timing gear or sprocket
124. Install No. 1 chain vibration damper

125. Install chain sub-assembly
126. Install chain tensioner slipper
127. Install No. 2 chain vibration damper
128. Install timing chain or belt cover oil seal
129. Install timing chain or belt cover sub-assembly
130. Install crankshaft pulley
131. Install No. 1 chain tensioner assembly
132. Install cylinder head cover sub-assembly
133. Install thermostat
134. Install water inlet
135. Install wire harness clamp bracket
136. Install exhaust manifold
137. Install manifold stay
138. Install No. 1 exhaust manifold heat insulator
139. Install air tube assembly
140. Install No. 1 water by-pass pipe
141. Install water inlet hose
142. Install water by-pass hose
143. Install oil level gage guide
144. Install oil level dipstick
145. Install fan belt adjusting bar
146. Install alternator assembly
147. Install fuel injector
148. Install injector vibration insulator
149. Install No. 1 delivery pipe spacer
150. Install fuel delivery pipe sub-assembly
151. Install intake manifold
152. Install No. 1 radiator hose
153. Install heater water inlet hose a
154. Install heater water outlet hose a (from heater unit)
155. Install No. 2 ventilation hose
156. Install radio setting condenser
157. Install ignition coil assembly
158. Install transmission oil filler tube sub-assembly (for automatic transaxle)
159. Install transmission oil level gage sub-assembly (for automatic transaxle)
160. Install No. 1 oil cooler outlet tube (for automatic transaxle)
161. Install No. 1 oil cooler inlet tube (for automatic transaxle)
162. Install No. 2 oil cooler tube clamp (for automatic transaxle)
163. Install front suspension crossmember sub-assembly
164. Install engine assembly with transaxle
165. Install front axle assembly LH
166. Install front axle assembly RH
167. Install front lower suspension arm sub-assembly LH
168. Install front lower suspension arm sub-assembly RH
169. Install front stabilizer link assembly LH
170. Install front stabilizer link assembly RH
171. Install tie rod end sub-assembly LH
172. Install tie rod end sub-assembly RH
173. Install speed sensor front LH
174. Install speed sensor front RH
175. Install front axle shaft LH nut
176. Install front axle shaft RH nut
177. Install front exhaust pipe assembly
178. Install front floor brace center
179. Install front console box
180. Install rear console box assembly
181. Install console box carpet
182. Install console box rear cover
183. Install console panel upper
184. Install shift lever knob sub-assembly (for manual transaxle)
185. Install No. 1 steering column hole cover sub-assembly
186. Install steering sliding yoke sub-assembly
187. Install column hole cover silencer sheet
188. Install clutch release cylinder assembly (for manual transaxle)
189. Connect fuel tube sub-assembly
190. Connect heater water inlet hose a
191. Connect heater water outlet hose a (from heater unit)
192. Connect engine wire
193. Install No. 1 fuel vapor feed hose
194. Install union to check valve hose
195. Install transmission control cable assembly (for automatic transaxle)
196. Install transmission control cable assembly (for manual transaxle)
197. Install with pulley compressor assembly
198. Install radiator assembly
199. Install upper radiator support sub-assembly
200. Install upper radiator support absorber
201. Install No. 1 cooler cover
202. Install hood lock assembly
203. Connect No. 1 radiator hose
204. Connect No. 2 oil cooler inlet hose (for automatic transaxle)
205. Connect No. 1 oil cooler outlet hose (for automatic transaxle)
206. Install No. 2 radiator hose
207. Install front bumper cover
208. Install fan and alternator v belt
209. Adjust fan and alternator v belt
210. Inspect fan and alternator v belt
211. Install Engine cover
212. Install battery carrier
213. Install air cleaner bracket
214. Install air cleaner assembly
215. Install outer cowl top panel
216. Install front air shutter seal RH
217. Install front wiper motor and link
218. Install cowl top ventilator louver LH
219. Install cowl top ventilator louver sub-assembly
220. Install hood to cowl top seal
221. Install front wiper arm and blade assembly LH
222. Install front wiper arm and blade assembly RH
223. Install front wiper arm head cap
224. Install battery tray
225. Install battery
226. Add engine coolant
227. Add engine oil
228. Inspect for fuel leak
229. Inspect for engine oil leak
230. Inspect for exhaust gas leak
231. Inspect for engine coolant leak
232. Install engine under cover RH
233. Install engine under cover LH
234. Install front wheels

ENGINE ASSEMBLY

REMOVAL & INSTALLATION

tC and xB

See Figure 115.

1. Before servicing the vehicle, refer to the Precautions Section.
2. Relieve the fuel system pressure.
3. Drain the engine cooling system.
4. Drain the engine oil.
5. Drain the transmission fluid.
6. Remove or disconnect the following:
   - Negative battery cable
   - Front wheels
   - Engine undercover
   - Right-hand front fender apron
   - Engine appearance cover
   - Radiator hoses
   - Transmission oil cooler hoses, if equipped
   - Radiator
   - Air intake assembly
   - Battery and battery tray
   - Battery carrier
   - Brake booster hose
   - Heater hoses
   - Transmission control cables
   - Clutch release cylinder assembly, if equipped
   - Accessory drive belt
   - Alternator
   - A/C compressor
   - Glove compartment door
   - Engine wire from the ECM and junction block
   - Engine ground wire
   - Front exhaust pipe assembly
7. Separate the intermediate steering shaft assembly.

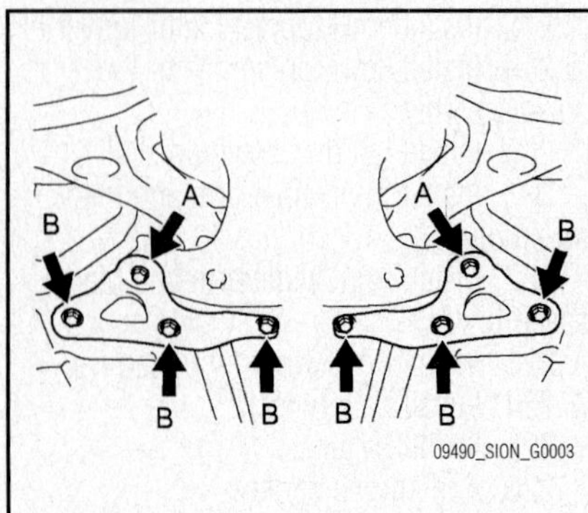

Fig. 115 Bolt identification for front suspension braces

➡Matchmark the main shaft and intermediate shaft for reassembly.

8. Remove or disconnect the following from both sides of the front suspension:
   a. Hub nut
   b. Stabilizer link
   c. Wheel speed sensor
   d. Tie rod assembly
   e. Lower control arm
9. Separate the axle assembly from the halfshaft.

➡Matchmark the halfshaft and axle hub for reassembly.

10. Disconnect the oil pump reservoir and return hose.
11. Secure the engine with a suitable engine lifting device.
12. Remove the mounting bolts from the left and right engine mounts.
13. Remove the mounting bolts from the left and right front suspension braces.
14. Remove the engine/transaxle assembly from the vehicle.
15. Installation is the reverse order of removal. Note the following torques:
   a. Left-hand engine mount bolts: 64 ft. lbs. (87 Nm).
   b. Right-hand engine mount bolts: 38 ft. lbs. (52 Nm).
   c. Front suspension braces: Bolts A to 98 ft. lbs. (133 Nm). Bolts B to 59 ft. lbs. (80 Nm).
16. Refill the transmission with fluid to the correct level.
17. Refill the engine with oil.
18. Refill the cooling system to the correct level.
19. Start the engine and check for leaks.

xD

See Figures 116 through 148.

1. Discharge the fuel system pressure.

✱✱ CAUTION

The DISCHARGE FUEL SYSTEM PRESSURE procedures must be performed before disconnecting any part of the fuel system.

✱✱ CAUTION

After performing the DISCHARGE FUEL SYSTEM PRESSURE procedures, pressure will remain in the fuel line. When disconnecting the fuel line, place a cloth or equivalent over fittings to reduce the risk of fuel spray.

   a. Remove the rear seat cushion assembly LH
   b. Remove the rear floor service hose cover.
   c. Disconnect the connector from the fuel pump assembly.
   d. Start the engine. After the engine stops naturally, turn the ignition switch OFF.
   e. Start the engine again and make sure that engine does not start.
   f. Remove the fuel tank cap, and let the air out of the fuel tank.
   g. Connect the connector.
   h. Install the rear floor service hole cover with new butyl tape.
   i. Install the rear seat cushion assembly LH.
2. Remove the battery.
   a. Disconnect the cable from the battery terminal.
   b. Loosen the nut and remove the battery clamp.
   c. Remove the battery.
3. Remove the battery tray.
4. Remove the front wheels.
5. Remove the engine under cover LH.
6. Remove the engine under cover RH.
7. Drain the engine coolant.
8. Drain the manual transaxle oil (for Manual Transaxle).
9. Drain automatic transaxle oil (for Automatic Transaxle).
10. Remove the front wiper arm head cap.
11. Remove front wiper arm and blade assembly LH.
12. Remove the front wiper arm and blade assembly RH.
13. Remove the hood to cowl top seal.
   a. Disengage the 7 clips and remove the hood to cowl top seal.
14. Remove the cowl top ventilator louver sub-assembly.
   a. Remove the clip.

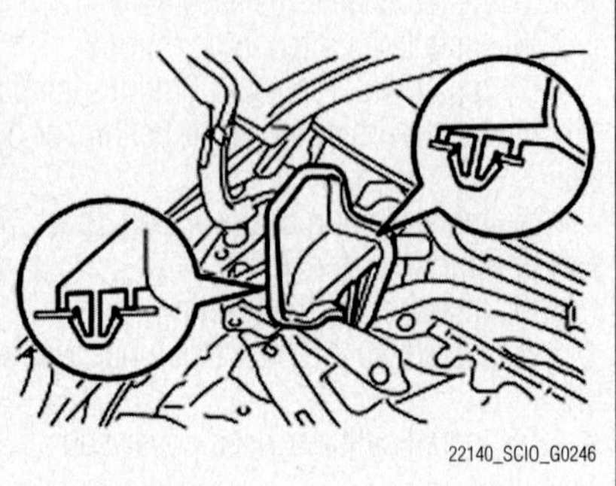

Fig. 116 Remove the front air shutter seal RH

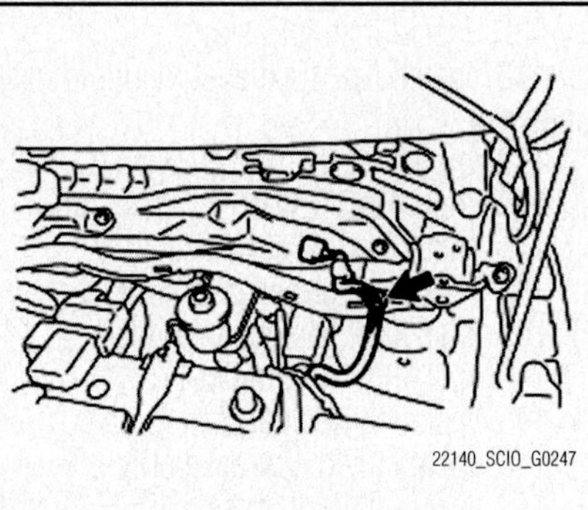

Fig. 117 Disengage the wire harness clamp

Fig. 118 Remove the 9 bolts and remove the outer cowl top panel

   b. Disengage the 4 claws and 4 guides and remove the cowl top ventilator louver RH.
15. Remove the cowl top ventilator louver LH.
16. Remove the front wiper motor and link.
   a. Disengage the clamp.
   b. Remove the 2 bolts.
   c. Slide the front wiper motor and link. Disengage the link from the rubber pin, then disconnect the connector and remove the front wiper motor and link.
17. Remove the front air shutter seal RH.

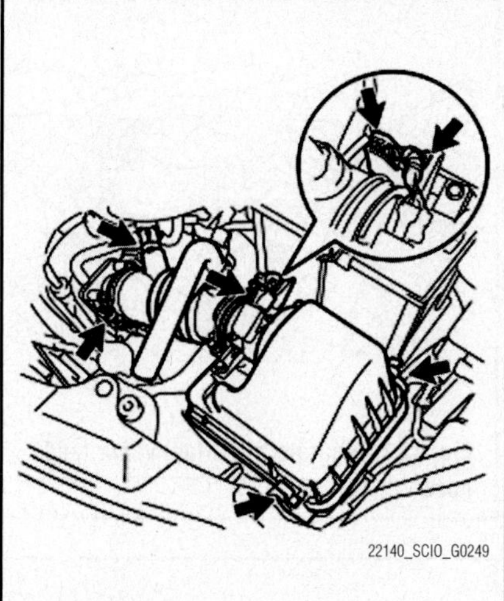

Fig. 119 Remove the air cleaner assembly

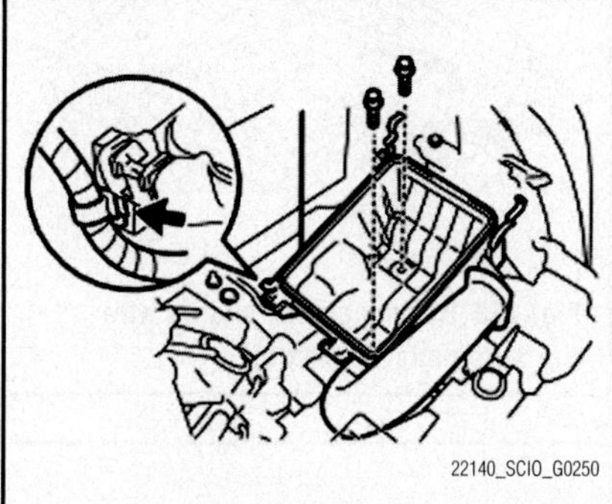

Fig. 120 Separate the wire harness clamp from the air cleaner case

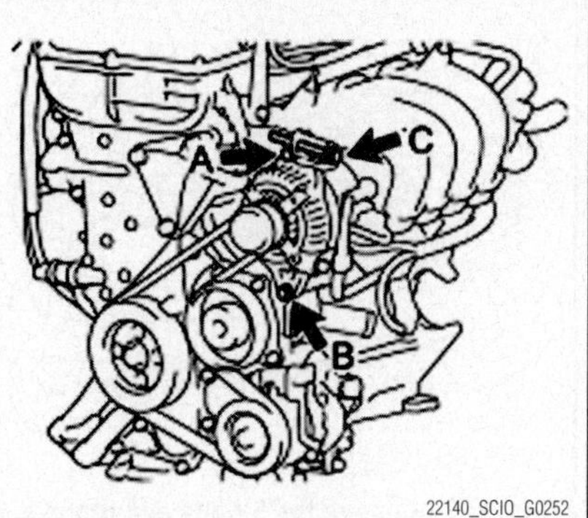

Fig. 121 Remove the fan and alternator V belt

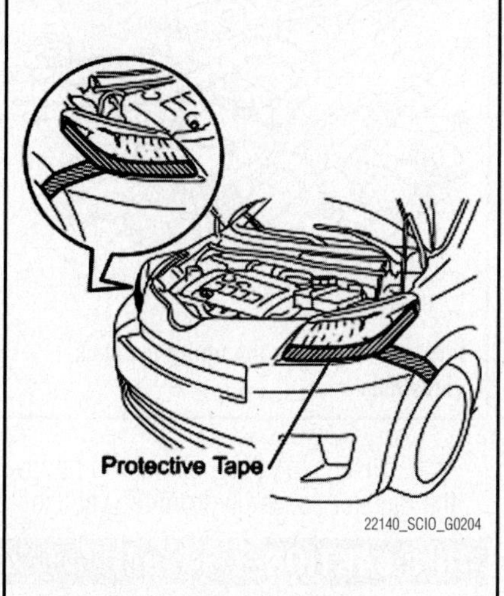

Fig. 122 Apply protective tape

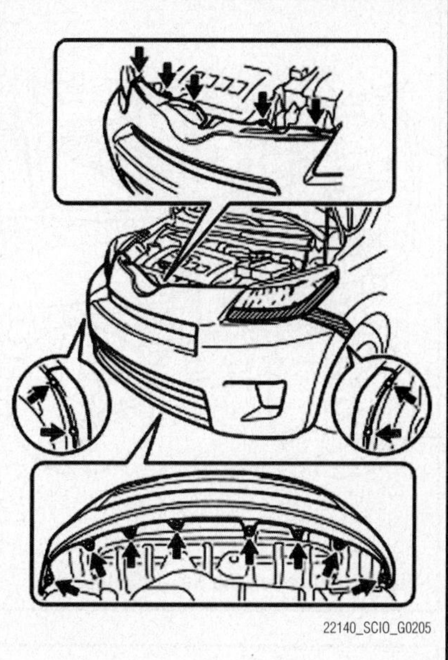

Fig. 123 Remove the 8 screws and 9 clips

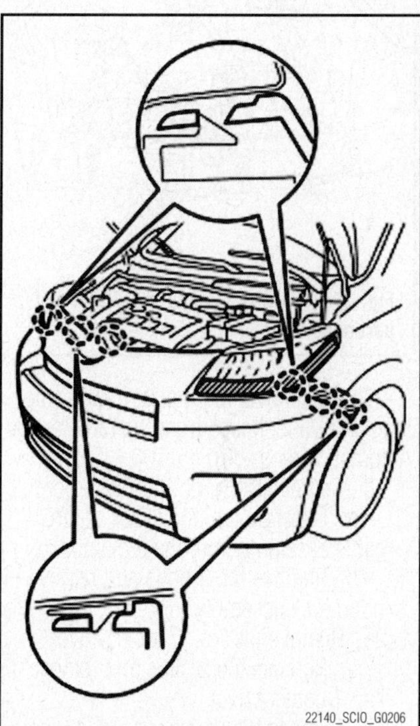

Fig. 124 Disengage the 6 claws and remove the front bumper cover

a. Disengage the 2 claws and remove the front air shutter seal RH.

18. Remove the outer cowl top panel.

a. Disengage the wire harness clamp.

b. Remove the 9 bolts and remove the outer cowl top panel.

19. Remove the air cleaner assembly.

a. Separate the intake air flow meter connector and 2 wire harness clamps.

b. Separate the ventilation hose from the air cleaner hose.

c. Unfasten the 2 clamps.

d. Remove the air cleaner cap sub-assembly with air cleaner hose.

e. Remove the air cleaner element.

f. Separate the wire harness clamp from the air cleaner case.

g. Remove the 2 bolts and remove the air cleaner case with air cleaner inlet.

20. Remove the air cleaner bracket.

a. Separate the wire harness clamp from the air cleaner bracket.

b. Remove the 2 bolts and remove the air cleaner bracket.

21. Remove the battery carrier.

a. Separate the wire harness clamp from the battery carrier.

b. Remove the 5 bolts and remove the battery carrier.

22. Remove the Engine cover.

a. (a) First, lift up the rear of the Engine cover to disengage the 2 fittings

b. (b) Next, lift the front of the Engine cover to disengage the 2 fittings. Then remove the Engine cover.

➡Ensure that the rubber grommets remain attached to the Engine cover. If any grommets are attached to the bolts, move them to the Engine cover.

23. Remove the fan and alternator V belt.

a. Loosen bolts A and B.

b. Turn adjusting bolt C to release the tension and remove the V belt from the pulleys.

c. Remove the fan and alternator V belt.

24. Remove the lower radiator hose.

a. Loosen the 2 clips and remove the No. 2 radiator hose.

25. Remove the front bumper cover.

a. Apply protective tape, as shown in the illustration.

b. Remove the 8 screws.

c. Remove the 9 clips.

d. Disengage the 6 claws and remove the front bumper cover.

26. Disconnect the upper radiator hose.

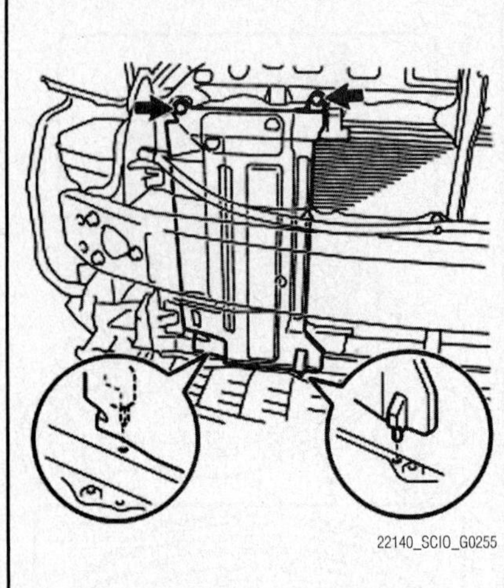

Fig. 125 Remove the No. 1 cooler cover

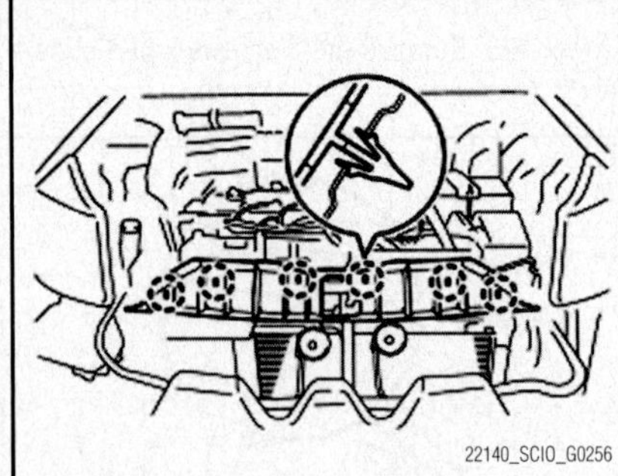

Fig. 126 Remove the upper radiator support absorber

27. Disconnect the transaxle oil cooler inlet and outlet hoses from the radiator (for Automatic Transaxle).
28. Remove the hood lock assembly.
    a. Separate the hood lock control cable assembly from the 2 clamps.
    b. Remove the 3 bolts and remove the hood lock assembly.
29. Remove the No. 1 cooler cover.
    a. Remove the 2 clips and remove the No. 1 cooler cover.
30. Remove the upper radiator support absorber.
    a. Using a screwdriver with its tip wrapped in protective tape, disengage the 6 claws and remove the radiator support upper absorber.
31. Remove the upper radiator support sub-assembly.
    a. Disconnect the horn connector.
    b. Remove the 5 bolts and remove the upper radiator support sub-assembly.
32. Remove the radiator assembly.
    a. Disconnect the cooling fan motor connector and separate the wire harness clamp.

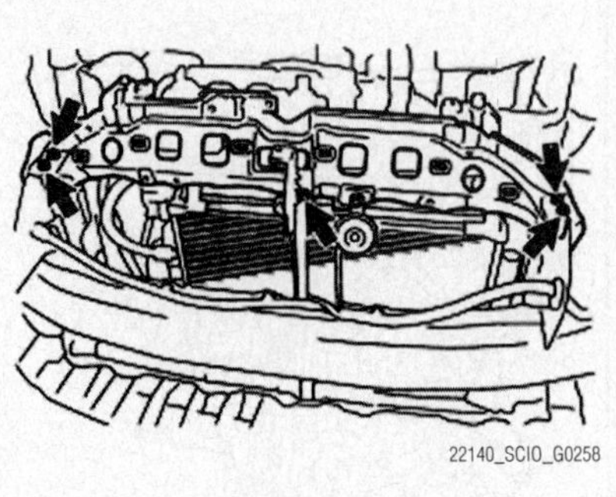

Fig. 127 Remove the 5 bolts and remove the upper radiator support sub-assembly

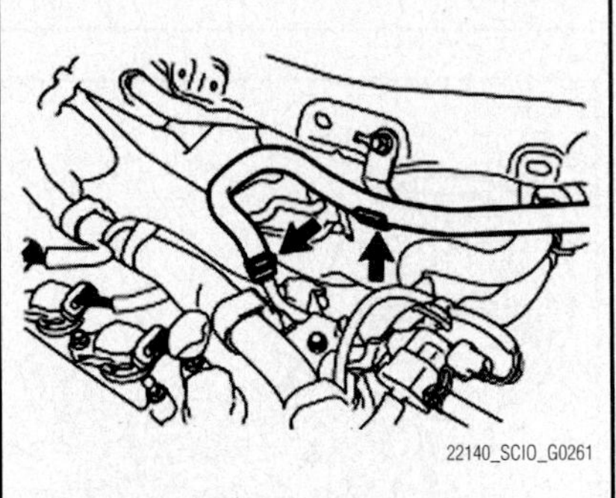

Fig. 128 Separate the union to check valve hose

    b. Disengage the 2 claws and remove the radiator assembly from the vehicle.

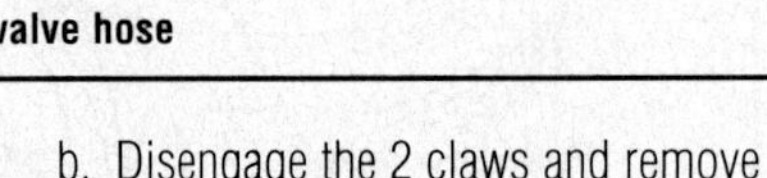

Do not apply excessive force to the cooler condenser assembly or piping when removing the radiator assembly.

33. Separate with pulley compressor assembly.
34. Separate the transmission control cable assembly (for Automatic Transaxle).
    a. Remove the nut and disconnect the transmission control cable assembly from the control shaft lever.
35. Separate the transmission control cable assembly (for Manual Transaxle).
    a. Remove the 2 clips and the washers, and disconnect the 2 cables from the transaxle.
    b. Remove the 2 clips and disconnect the 2 cables from the control cable bracket.
36. Separate the union to check valve hose.
    a. Loosen the clip and separate the union to check valve hose.
37. Separate No. 1 fuel vapor feed hose.
    a. Separate the No. 1 fuel vapor feed hose from the vacuum switching valve assembly.

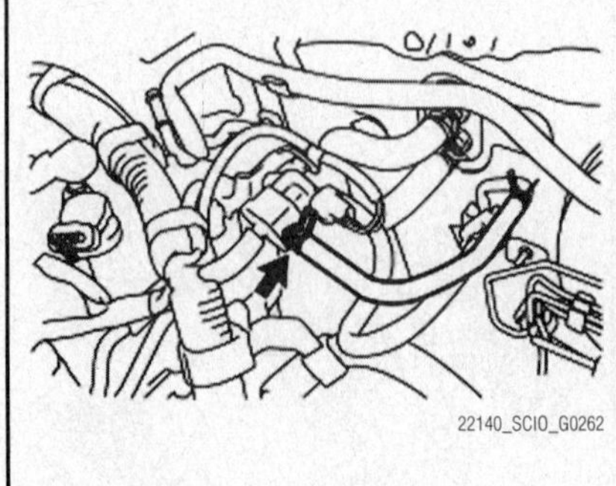

Fig. 129 Separate No. 1 fuel vapor feed hose

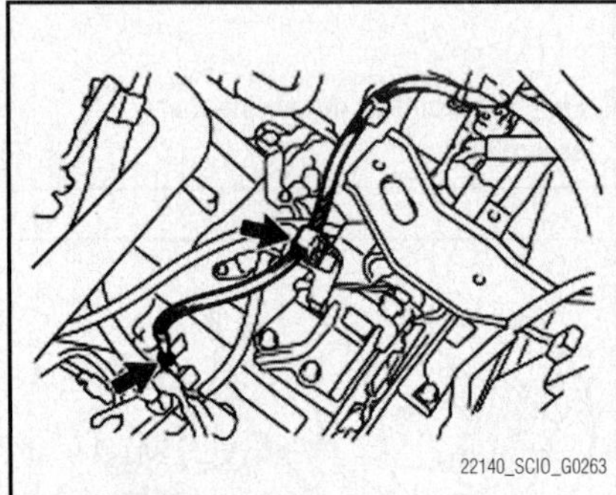

Fig. 130 Disconnect the engine wire (for Automatic Transaxle)

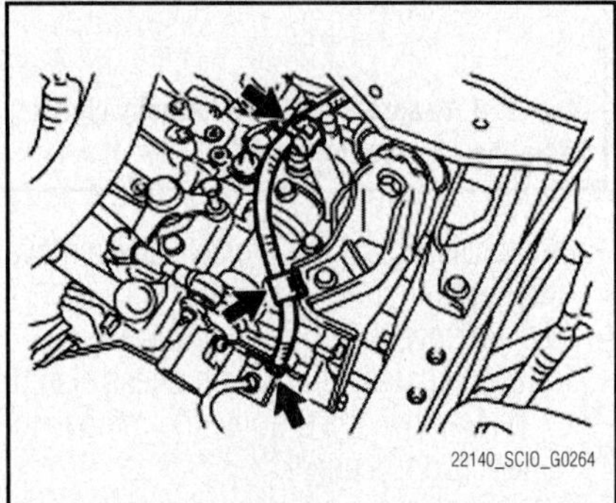

Fig. 131 Disconnect the engine wire (for Manual Transaxle)

38. Disconnect the engine wire.
    a. Remove the bolt and wire harness clamp and separate the transaxle ground wire (for Automatic Transaxle).
    b. Remove the bolt and 2 wire harness clamp and separate the transaxle ground wire (for Manual Transaxle).
    c. Disconnect the 2 connectors.
    d. Pull up the lever and disconnect the connector from the engine control computer.
    e. Separate the 2 wire harness clamps.
    f. Remove the 2 connectors and clamp from the engine room junction

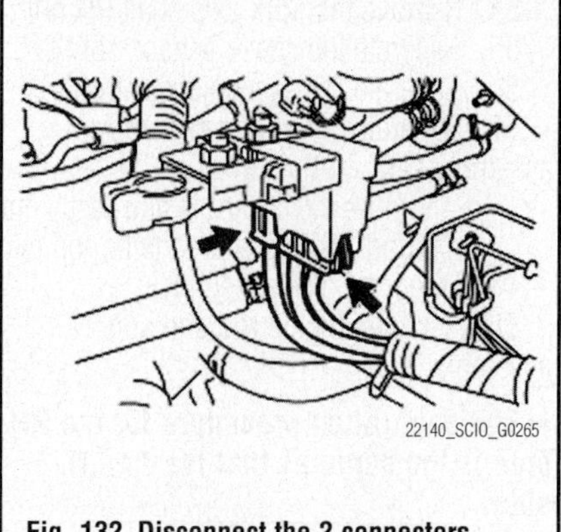

Fig. 132 Disconnect the 2 connectors

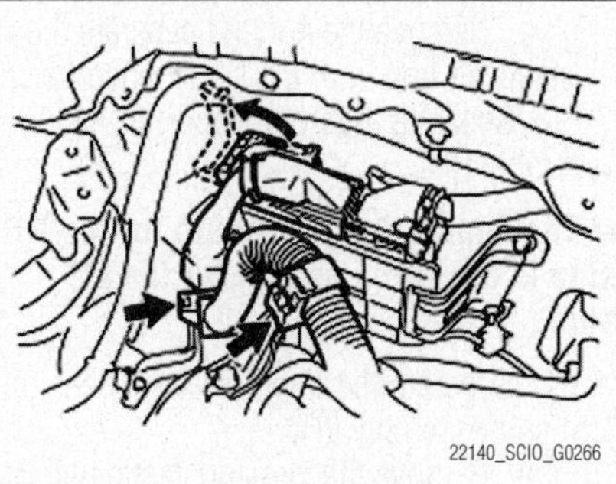

Fig. 133 Pull up the lever and disconnect the connector from the engine control computer

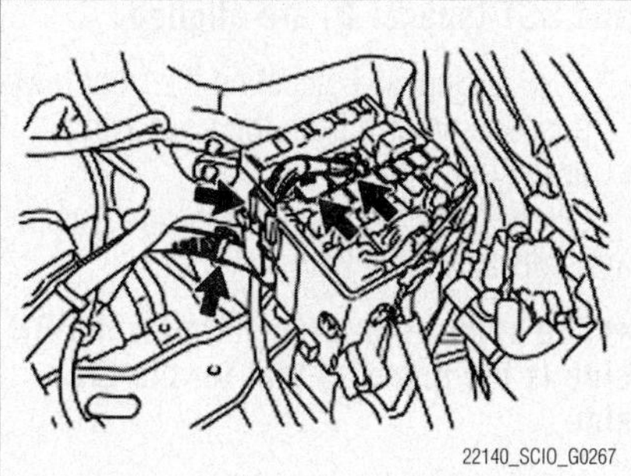

Fig. 134 Remove the 2 connectors and clamp from the engine room junction block and disconnect the 2 wire harness clamps

block and disconnect the 2 wire harness clamps.

g. Disconnect all wire harnesses and connectors.

✱✱ CAUTION

Make sure that no wire harness is connected between the body and engine.

39. Disconnect the upper and lower heater hoses (from heater unit).

40. Disconnect the fuel tube sub-assembly.

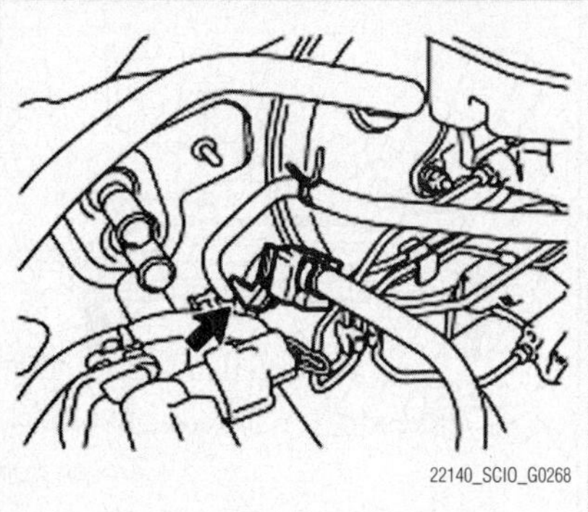

Fig. 135 Remove the No. 1 fuel pipe clamp

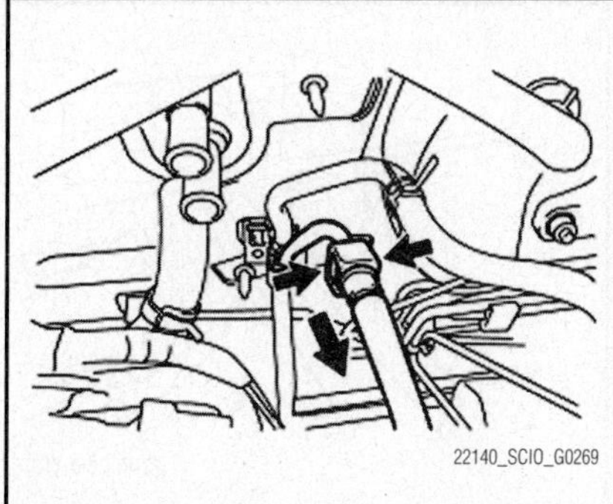

Fig. 136 Pinch the retainer, then pull the fuel tube connector out of the pipe

a. Remove the No. 1 fuel pipe clamp.

b. Pinch the retainer, then pull the fuel tube connector out of the pipe.

➡Remove any dirt and foreign matter from the fuel tube connector before performing this work. Do not allow any scratches or foreign matter on the parts when disconnecting, as the fuel tube connector has the O-rings that seal the pipe. Perform this work by hand. Do not use any tools. Do not forcibly bend, twist or turn the nylon tube. Protect the disconnected parts by covering them with vinyl bags after disconnecting the fuel tube. If the fuel tube connector and pipe are stuck, push and pull them to release them.

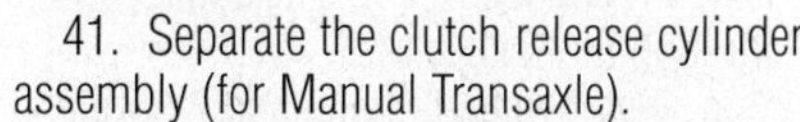

41. Separate the clutch release cylinder assembly (for Manual Transaxle).

a. Remove the 4 bolts, then separate the clutch release cylinder.

➡Suspend the clutch release cylinder with a piece of rope so as not to overload the clutch pipe.

42. Remove the column hole cover silencer sheet.

a. Remove the floor carpet and 2 clips and remove the column hole cover silencer.

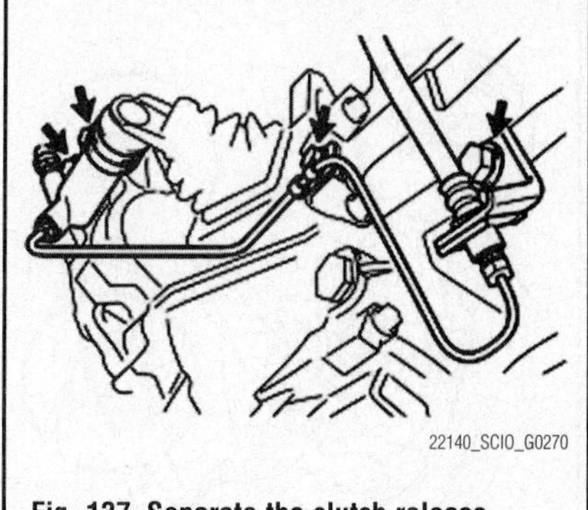

Fig. 137 Separate the clutch release cylinder assembly (for Manual Transaxle)

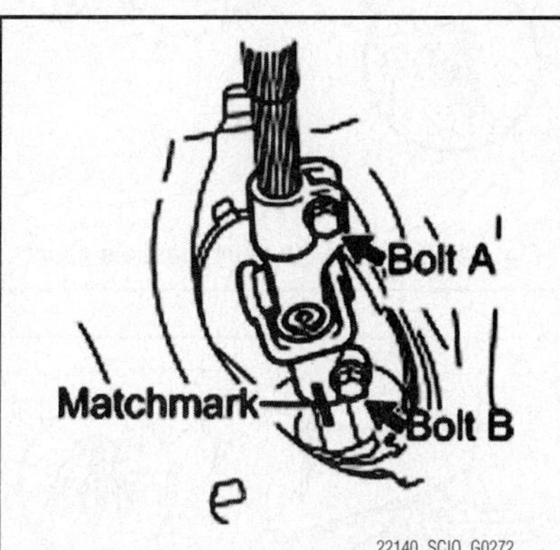

Fig. 138 Place matchmarks on the sliding yoke of the steering intermediate shaft assembly

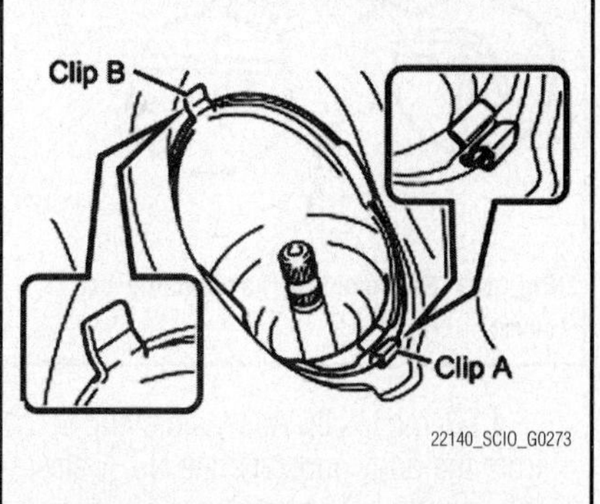

Fig. 139 Separate the No. 1 steering column hole cover sub-assembly

43. Separate the steering sliding yoke sub-assembly.

a. Use a seat belt to fix the steering wheel assembly, in order to avoid breakage of the spiral cable.

b. Place matchmarks on the sliding yoke of the steering intermediate shaft assembly and the power steering.

c. Loosen bolt A, remove bolt B and separate the steering intermediate shaft assembly.

44. Separate the No. 1 steering column hole cover sub-assembly.

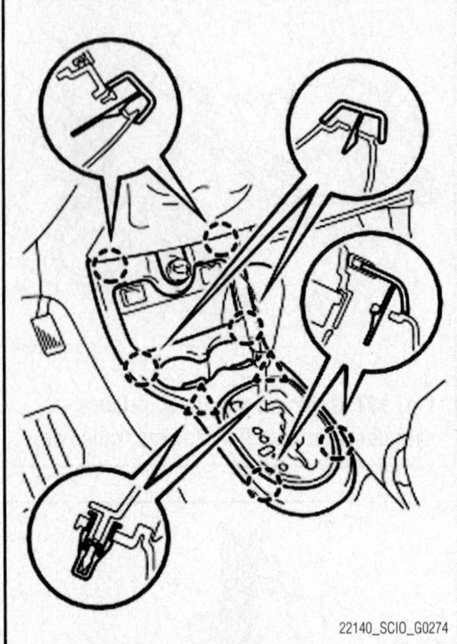
Fig. 140 Remove the upper console panel

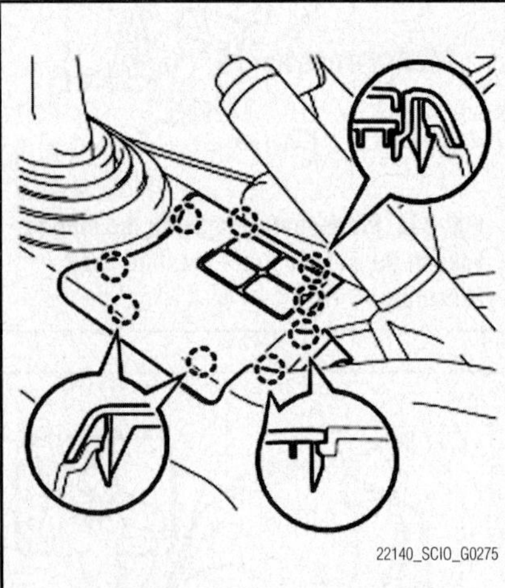
Fig. 141 Remove the rear console box cover

a. Remove clip A, separate clip B from the body and separate No. 1 steering column hole cover.

CAUTION

Do not damage clip B.

45. Remove the shift lever knob sub-assembly (for Manual Transaxle).
    a. Remove the shift lever knob by turning the knob counterclockwise.
46. Remove the upper console panel.
    a. Disengage the 2 clips and the 6 claws and remove the upper console panel.
47. Remove the rear console box cover.
    a. Disengage the 9 claws and remove the rear console box cover.
48. Remove the console box carpet.

Fig. 142 Remove the front floor center brace

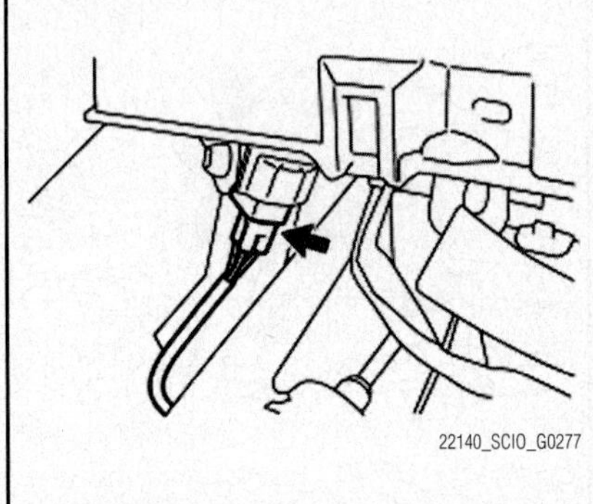
Fig. 143 Disconnect the heated oxygen sensor connector

49. Remove the rear console box assembly.
    a. Remove the 3 screws.
    b. Disengage the 4 claws and remove the rear console box.
50. Remove the front console box.
    a. Remove the 2 screws.
    b. Disengage the clip and 2 claws and remove the front console box.
51. Remove the front floor center brace.
    a. Remove the 2 bolts and remove the front floor center brace.
52. Remove the front exhaust pipe assembly.
    a. Disconnect the heated oxygen sensor connector.
    b. Remove the grommet and pull the sensor connector out of the cabin through the floor panel.
    c. Remove the 2 bolts and 2 springs and separate the front exhaust pipe assembly from the exhaust tail pipe assembly.
    d. Remove the 2 bolts and 2 springs and separate the front exhaust pipe assembly from the exhaust manifold.
    e. Remove the 3 exhaust pipe supports and remove the front exhaust pipe assembly.
53. Remove the front axle shaft LH nut.
54. Remove the front axle shaft RH nut.
55. Separate the speed sensor front LH.
56. Separate the speed sensor front RH.
57. Separate the tie rod end sub-assembly LH.
    a. Remove the cotter pin and castle nut.
    b. Using SST, separate the tie rod end from the front axle assembly.
58. Separate the tie rod end sub-assembly RH

➡The separation procedure for the RH side is the same as that for the LH side.

59. Separate the front stabilizer link assembly LH.
    a. Remove the nut and separate the stabilizer link from the shock absorber.
60. Separate the front stabilizer link assembly RH.

➡The separation procedure for the RH side is the same as that for the LH side.

61. Separate the front lower suspension arm sub-assembly LH.
    a. Remove the clip and castle nut.
    b. Install SST to the threaded section of the lower ball joint.

➡Make sure the upper ends of the threaded section of the lower ball joint and SST (spacer B) are aligned.

    c. Using SST, separate the front lower suspension arm from the front axle assembly.

62. Separate the front lower suspension arm sub-assembly RH

➡The separation procedure for the RH side is the same as that for the LH side.

63. Separate the front axle assembly LH.
    a. Using a plastic hammer, tap the end of the drive shaft and disengage the fitting between the drive shaft and front axle.

➡If it is difficult to disengage the fitting, tap the end of the drive shaft with a brass bar and hammer.

    b. Push the front axle out of the vehicle to remove the drive shaft from the front axle.

➡Do not push the front axle further out of the vehicle than is necessary. Do not damage the outboard joint boot. Do not damage the speed sensor rotor. Suspend the drive shaft with a piece of rope or the equivalent.

64. Separate the front axle assembly RH.

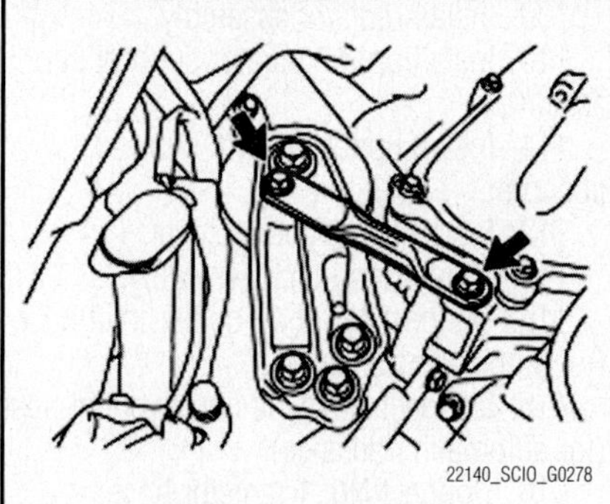
Fig. 144 Remove 2 bolts and remove the engine mounting stay RH

➡The separation procedure for the RH side is the same as that for the LH side.

65. Remove the front drive shaft assembly LH.
 a. Using SST, remove the drive shaft.

➡Do not damage the oil seal. Do not damage the inboard joint boot. Do not drop the drive shaft.

66. Remove the front drive shaft assembly RH.
 a. Using a screwdriver and hammer, remove the drive shaft.

➡Do not damage the oil seal. Do not damage the inboard joint boot. Do not drop the drive shaft.

67. Remove the flywheel housing under cover (for Automatic Transaxle).
68. Remove the drive plate and torque converter clutch setting bolt (for Automatic Transaxle).
 a. (a) Remove the 6 torque converter set bolts while holding the crankshaft pulley bolt with a wrench.
69. Remove the engine assembly with transaxle.
 a. Set the engine lifter.
 b. Remove 2 bolts and remove the engine mounting stay RH.
 c. Remove the 6 bolts and nut and remove the engine mounting insulator RH.

➡Be sure not to remove the other bolt A.

 d. Remove the through bolt and nut and separate the transverse engine mounting insulator.
 e. Remove the 6 bolts and remove the engine assembly with transaxle and front suspension crossmember from the vehicle.

To install:

70. Install front suspension crossmember sub-assembly.

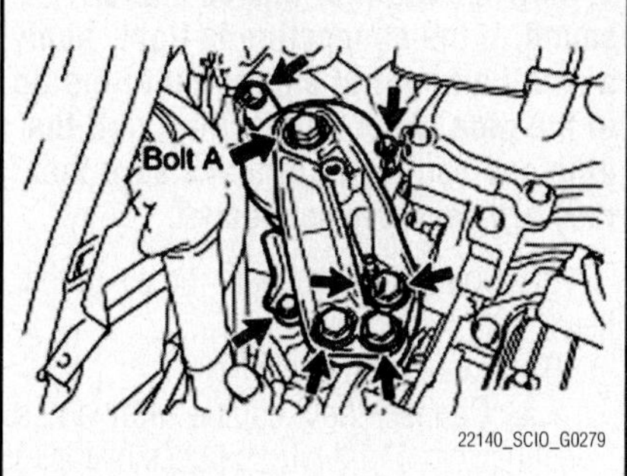

Fig. 145 Remove the 6 bolts and nut and remove the engine mounting insulator RH

Fig. 146 Remove the through bolt and nut

 a. Install the engine moving control rod with the through bolt. Torque: 89 ft. lbs. (120 Nm).
 b. Remove the 2 bolts and remove the 2 engine hangers.
 c. Install the air fuel ratio sensor bracket with bolt. Torque: 32 ft. lbs. (43 Nm).
71. Install transverse engine mounting insulator.
 a. Install the transverse engine mounting insulator with the 5 bolts. Torque: 38 ft. lbs. (52 Nm).
72. Install engine assembly with transaxle.
 a. Set the engine assembly with transaxle and front suspension crossmember on the engine lifter.
 b. Operate the engine lifter and lift the engine assembly with transaxle and front suspension crossmember to the position where the engine mounting insulator RH and transverse engine mounting insulator can be installed.
 c. Install the transverse engine mounting insulator with the through bolt and nut. Torque: 38 ft. lbs. (52 Nm).
 d. Install the engine mounting insulator RH with the 6 bolts and nut.

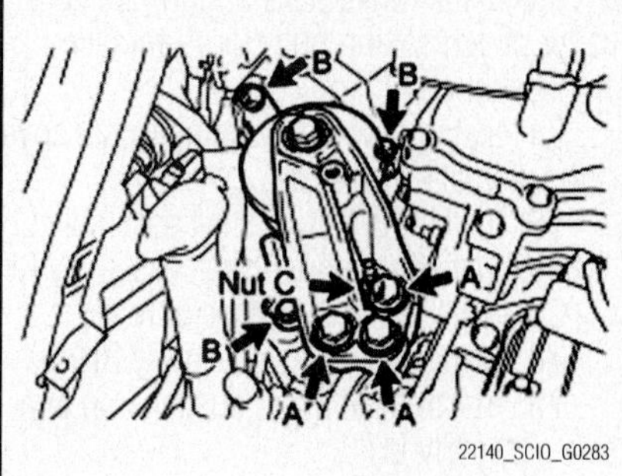

Fig. 147 Install the engine mounting insulator RH with the 6 bolts and nut

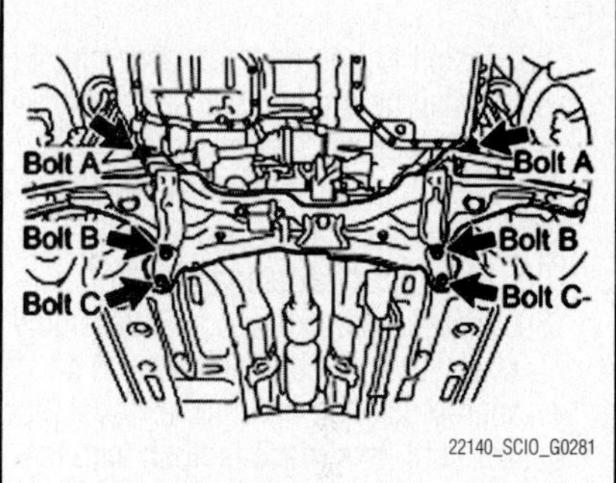

Fig. 148 Tighten bolts A, B and C on both sides in several sequences

 e. Torque: 65 ft. lbs. (88 Nm) for Bolt A.
 f. Torque: 33 ft. lbs. (45 Nm) for Bolt B.
 g. Torque: 38 ft. lbs. (52 Nm) for Bolt C.
 h. Install the engine mounting stay RH with the 2 bolts. Torque: 19 ft. lbs. (26 Nm).
 i. Operate the engine lifter and provisionally install the engine assembly with transaxle and front suspension crossmember onto the vehicle with the 6 bolts.
 j. Insert SST into the datum holes in the front suspension crossmember RH and LH alternately and tighten bolts A, B and C on both sides in several sequences.
 k. Torque: 52 ft. lbs. (70 Nm) for Bolt A.
 l. Torque: 118 ft. lbs. (160 Nm) for Bolt B.
 m. Torque: 70 ft. lbs. (95 Nm) for Bolt C.

CAUTION

Insert SST into the datum hole vertically. If impossible to insert SST vertically, loosen all bolts and then insert SST again.

73. Install drive plate and torque converter clutch setting bolt (for Automatic Transaxle).
74. Install flywheel housing under cover (for Automatic Transaxle).
75. Install front drive shaft assembly LH.
76. Install front drive shaft assembly RH.
77. Install front axle assembly LH.
78. Install front axle assembly RH.
79. Install front lower suspension arm sub-assembly LH.
80. Install front lower suspension arm sub-assembly RH.
81. Install front stabilizer link assembly LH.
82. Install front stabilizer link assembly RH.
83. Install tie rod end sub-assembly LH.
84. Install tie rod end sub-assembly RH.
85. Install speed sensor front LH.
86. Install speed sensor front RH.
87. Install front axle shaft LH nut.
88. Install front axle shaft RH nut.
89. Install front exhaust pipe assembly.
a. Using a plastic hammer and a wooden block, tap in a new exhaust pipe gasket until its surface is flush with the exhaust manifold.
b. Install the 3 exhaust pipe supports.
c. Install the exhaust front pipe assembly and a new exhaust pipe gasket with the 4 compression springs and 4 bolts. Torque: 32 ft. lbs. (43 Nm).
d. Pass the sensor connector through the floor panel and install the grommet.
e. Connect the heated oxygen sensor connector.
90. Install front floor brace center.
91. Install front console box.
92. Install rear console box assembly.
93. Install console box rear cover.
94. Install console box carpet.
95. Install console panel upper.
96. Install shift lever knob sub-assembly (for Manual Transaxle).
97. Install No. 1 steering column hole cover sub-assembly.
98. Install steering sliding yoke sub-assembly.
99. Install column hole cover silencer sheet.
100. Install clutch release cylinder assembly (for Manual Transaxle).
a. Install the clutch release cylinder and clutch pipe with the 4 bolts.
101. Connect fuel tube sub-assembly.
a. Connect the fuel tube connector and fuel pipe, and install the No. 1 fuel pipe clamp.

➡Align the fuel tube connector with the pipe, then push the fuel tube connector in until the retainer makes a click sound. If the connection is tight, apply a small amount of engine oil to the tip of the pipe. After connecting, pull the pipe and connector to make sure that they are securely connected.

102. Connect heater water hoses to the heater unit.
103. Connect engine wire.
a. Connect the 2 engine wire harness connectors and 2 wire harness clamps to the engine room junction block.
b. Connect the engine wire harness connector to the ECM.
c. Connect the 2 wire harness clamps.
d. Connect the 2 connectors.
e. Install the transaxle ground wire with the bolt and 2 wire harness clamps (for Manual Transaxle).
f. Install the transaxle ground wire with the bolt and wire harness clamp (for Automatic Transaxle).
104. Install No. 1 fuel vapor feed hose.
a. Install the No. 1 fuel vapor feed hose into the vacuum switching valve assembly.
105. Install union to check valve hose.
a. Install the union to check valve hose with the clip.
106. Install transmission control cable assembly (for Automatic Transaxle)
a. Install the transmission control cable assembly onto the control shaft lever with the nut.
b. Fix the transmission control cable assembly onto the control cable bracket with a new clip.

➡Make sure that the marking on the cable is aligned with the slit in the transmission control cable bracket before installing the cable.

107. Install transmission control cable assembly (for Manual Transaxle).
a. Install the control cable assembly onto the control cable bracket with 2 new clips.
b. Connect the control cable assembly and install the 2 washers and 2 clips.
c. Install the control cable assembly onto the shift lever retainer.

➡Install the cable with the protruding portion of the cable outer facing upward. After installing. check that the lock of the cable outer is protruding beyond portion A. as shown in the illustration.

108. Install with pulley compressor assembly.
109. Install radiator assembly.
110. Install upper radiator support sub-assembly.
111. Install upper radiator support absorber.
112. Install No. 1 cooler cover.
113. Install hood lock assembly.
114. Connect No. 2 oil cooler inlet hose (for automatic transaxle).
115. Connect No. 1 oil cooler outlet hose (for automatic transaxle).
116. Connect No. 1 radiator hose.
117. Install No. 2 radiator hose.
a.Install the No. 2 radiator hose with the 2 clips.
118. Install front bumper cover.
119. Install fan and alternator v belt.
120. Adjust fan and alternator v belt.
121. Inspect fan and alternator v belt.
122. Install engine cover.
123. Install battery carrier.
a. Install the battery carrier with the 5 bolts.
b. Install the clamp.
124. Install air cleaner bracket.
a. Install the air cleaner bracket with the 2 bolts.
b. Connect the wire harness clamp to the air cleaner bracket.
125. Install air cleaner assembly
a. Install the air cleaner case with air cleaner inlet with the 2 bolts.
b. Connect the wire harness to the air cleaner case.
c. Install the air cleaner element.
d. Install and lock the air cleaner cap with air cleaner hose and then tighten the air cleaner hose clamp.
e. Connect the ventilation hose to the air cleaner hose.
f. Connect the intake air flow meter connector and 2 wire harness clamps.
126. Install outer cowl top panel
a. Install the outer cowl top panel with the 9 bolts.
b. Connect the wire harness clamp.
127. Install front air shutter seal RH
a. Engage the 2 claws to install the front air shutter seal RH.
128. Install front wiper motor and link.
129. Install cowl top ventilator louver LH.
130. Install cowl top ventilator louver sub-assembly.
131. Install hood to cowl top seal.
132. Install front wiper arm and blade assembly LH.
133. Install front wiper arm and blade assembly RH.
134. Install front wiper arm head cap.
135. Install battery tray.
136. Install battery.
137. Add engine coolant.

138. Add automatic transaxle oil (for automatic transaxle).
139. Inspect automatic transaxle oil (for automatic transaxle).
140. Add manual transaxle oil (for manual transaxle).
141. Inspect manual transaxle oil (for manual transaxle).
142. Inspect for fuel leaks.
143. Inspect for engine oil leak..
144. Inspect for exhaust gas leak.
145. Inspect for engine coolant leak.
146. Install engine under cover RH.
147. Install engine under cover LH.
148. Install front wheels.

EXHAUST MANIFOLD

REMOVAL & INSTALLATION

tC and xB

See Figures 149 and 150.

1. Before servicing the vehicle, refer to the Precautions Section.
2. Drain the cooling system.
3. Relieve the fuel system pressure.
4. Remove or disconnect the following:
   - Negative battery cable
   - Engine appearance cover
   - Air intake assembly

Fig. 149 Exhaust manifold stays—2.4L Engine

   - Throttle body
   - Fuel rail
   - Intake manifold
   - Front exhaust pipe
   - Oil dipstick guide
   - Exhaust manifold stays
   - Exhaust manifold heat shield
   - Exhaust manifold and gasket

To install:

5. Install the exhaust manifold with a new gasket. Tighten the bolts in sequence to 27 ft. lbs. (37 Nm).
6. Install or connect the following:
   - Exhaust manifold heat shield. Tighten to 9 ft. lbs. (12 Nm).
   - Exhaust manifold stays. Tighten to 32 ft. lbs. (44 Nm).
   - Oil dipstick guide. Tighten to 80 inch lbs. (9 Nm).
   - Front exhaust pipe. Tighten to 32 ft. lbs. (44 Nm).
   - Intake manifold
   - Fuel rail
   - Throttle body
   - Air intake assembly
   - Engine appearance cover
   - Negative battery cable

xD

See Figures 151 and 152.

1. Remove the air fuel ratio sensor.
   a. Disconnect the air fuel ratio sensor connector and clamp.
   b. Using SST, remove the air fuel ratio sensor.
2. Remove the engine under cover RH.
3. Remove the front exhaust pipe assembly.
4. Drain the automatic transaxle fluid (for automatic transaxle).
5. Drain the manual transaxle oil (for manual transaxle).
6. Remove the front wheels.
7. Remove the front axle shaft RH nut
8. Remove the front speed sensor RH.
9. Remove the tie rod end sub-assembly RH.
10. Remove the front stabilizer link assembly RH.
11. Remove the No. 1 front lower suspension arm sub-assembly RH.
   a. Remove the clip and castle nut.
   b. Install SST to the threaded section of the lower ball joint.

➡Make sure the upper ends of the threaded section of the lower ball joint and SST (spacer B) are aligned.

   c. Using SST, separate the front lower suspension arm from the front axle assembly.
12. Separate the front axle assembly RH.
13. Remove the drive shaft assembly RH.
14. Remove the No. 1 exhaust manifold heat insulator.
   a. Remove the 4 bolts and remove the No. 1 exhaust manifold heat insulator.
15. Remove the manifold stay.
   a. Remove the 3 bolts and remove the manifold stay.
16. Remove the exhaust manifold.
   a. Remove the 5 nuts and remove the exhaust manifold.

To install:

17. Install exhaust manifold.
   a. Install a new exhaust manifold gasket and exhaust manifold with the 5 nuts. Torque: 16 ft. lbs. (21 Nm).
18. Install manifold support bracket.
   a. Install the manifold support bracket with the 3 bolts in the sequence. Torque: 32 ft. lbs. (43 Nm).
19. Install no. 1 exhaust manifold heat insulator.
   a. Install the No. 1 exhaust manifold heat insulator with the 4 bolts. Torque: 9 ft. lbs. (12 Nm).
20. Install drive shaft assembly RH.
21. Install front axle assembly RH.
22. Install no. 1 front suspension arm sub-assembly RH.

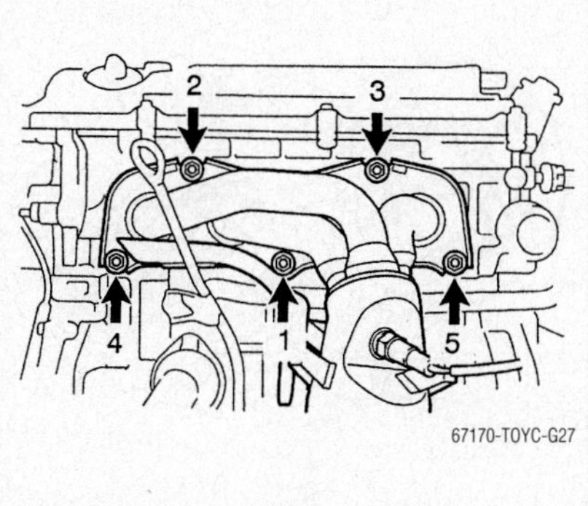

Fig. 150 Exhaust manifold torque sequence—2.4L Engine

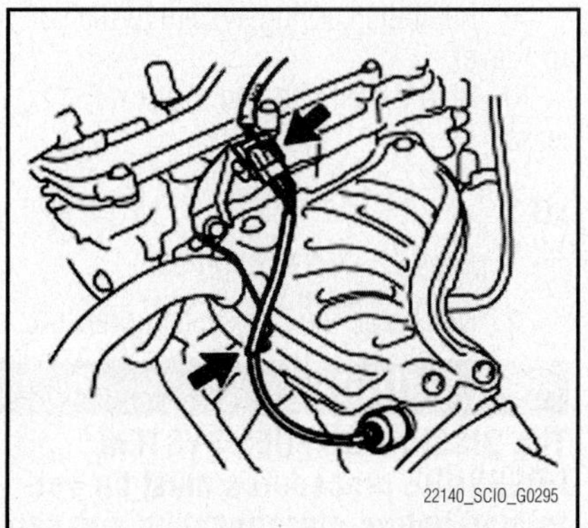

Fig. 151 Remove the air fuel ratio sensor.

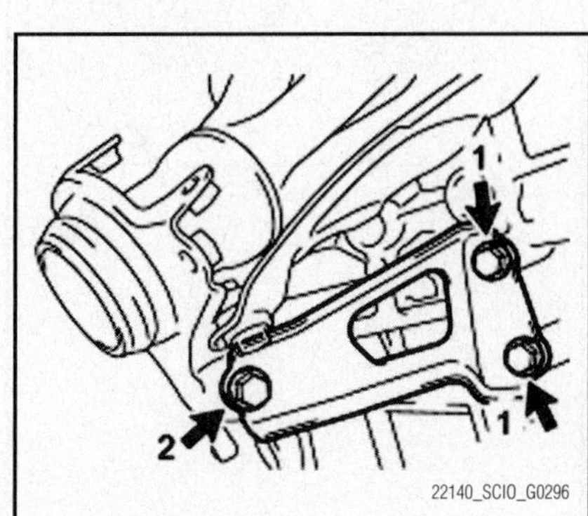

Fig. 152 Install the manifold support bracket with the 3 bolts in the sequence

23. Install front stabilizer link assembly RH.
24. Install tie rod end sub-assembly RH.
25. Install front speed sensor RH.
26. Install front axle shaft RH nut.
27. Install front wheel.
28. Install front exhaust pipe assembly.
29. Install air fuel ratio sensor.
30. Add automatic transaxle fluid (for automatic transaxle).
31. Inspect automatic transaxle fluid (for automatic transaxle).
32. Add manual transaxle oil (for manual transaxle).
33. Inspect manual transaxle oil (for manual transaxle).
34. Inspect for automatic transaxle fluid leak (for automatic transaxle).
35. Inspect for manual transaxle oil leak.
36. Inspect for exhaust gas leak.
37. Inspect and adjust front wheel alignment.
38. Install engine under cover RH.

INTAKE MANIFOLD

REMOVAL & INSTALLATION

tC and xB

See Figures 153 and 154.

1. Before servicing the vehicle, refer to the Precautions Section.
2. Drain the cooling system.
3. Relieve the fuel system pressure.
4. Remove or disconnect the following:
   - Negative battery cable

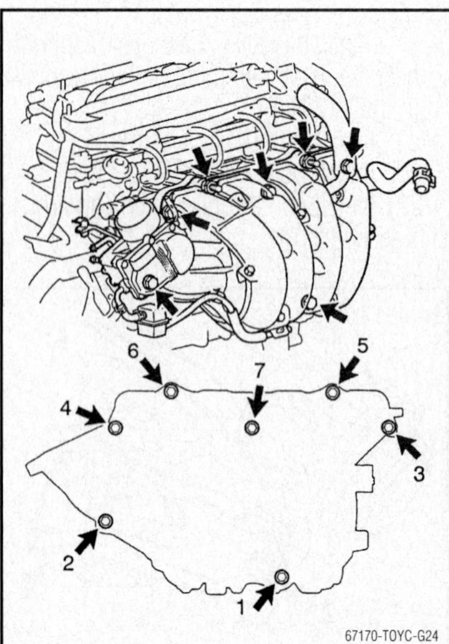

Fig. 153 Intake manifold fastener location and loosening sequence—2.4L Engine

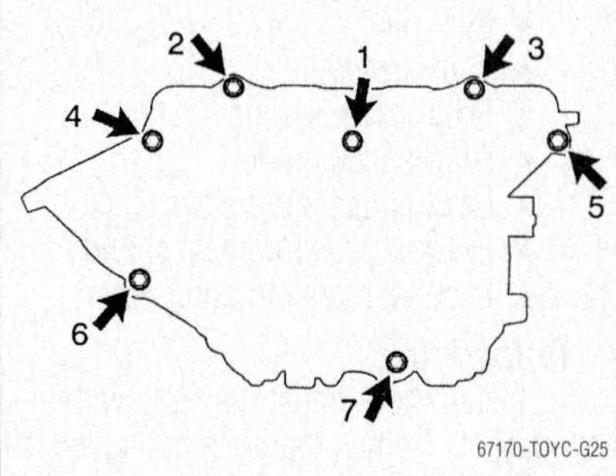

Fig. 154 Intake manifold fastener torque sequence—2.4L Engine

   - Hood
   - Engine appearance cover
   - Wiper arm assembly
   - Top cowl seal
   - Left-hand top cowl ventilator louver
   - Wiper linkage assembly
   - Top outer cowl panel
   - Left-hand cowl body mounting bracket
   - Air intake assembly
   - Throttle body
   - Fuel rail
   - Water by-pass hoses from the throttle body
   - Intake manifold and gasket

To install:

5. Install a new gasket into the intake manifold.
6. Install the intake manifold and tighten fasteners in sequence to 22 ft. lbs. (30 Nm).
7. Install or connect the following:
   - Water bypass hoses from the throttle body
   - Fuel rail
   - Throttle body. Tighten bolts to 22 ft lbs. (30 Nm).
   - Air intake assembly
8. The remainder of the installation is the reverse order of removal.
9. Refill the cooling system to the correct level.
10. Start the engine and check for leaks.

xD

See Figures 12, 155 through 158.

1. Discharge the fuel system pressure.

**** CAUTION**

The DISCHARGE FUEL SYSTEM PRESSURE procedures must be performed before disconnecting any part of the fuel system.

**** CAUTION**

After performing the DISCHARGE FUEL SYSTEM PRESSURE procedures, pressure will remain in the fuel line. When disconnecting the fuel line, place a cloth or equivalent over fittings to reduce the risk of fuel spray.

a. Remove the rear seat cushion assembly LH
b. Remove the rear floor service hose cover.
c. Disconnect the connector from the fuel pump assembly.
d. Start the engine. After the engine stops naturally, turn the ignition switch OFF.
e. Start the engine again and make sure that engine does not start.
f. Remove the fuel tank cap, and let the air out of the fuel tank.
g. Connect the connector.
h. Install the rear floor service hole cover with new butyl tape.
i. Install the rear seat cushion assembly LH.

2. Disconnect the cable from negative battery terminal.
3. Remove the engine under cover LH.
4. Remove the engine under cover RH.
5. Drain the engine coolant.
6. Remove the Engine cover.
7. Remove the air cleaner cap sub-assembly with hose.
   a. Disconnect the wire harness clamp and mass air flow meter connector.
   b. Disconnect the No. 2 ventilation hose.
   c. Loosen the hose clamp, unlock the 2 clamps and remove air cleaner cap sub-assembly with hose.
8. Separate the engine wire harness.
   a. Remove the 3 bolts, harness clamp and 4 fuel injector connectors and separate the engine wire harness.

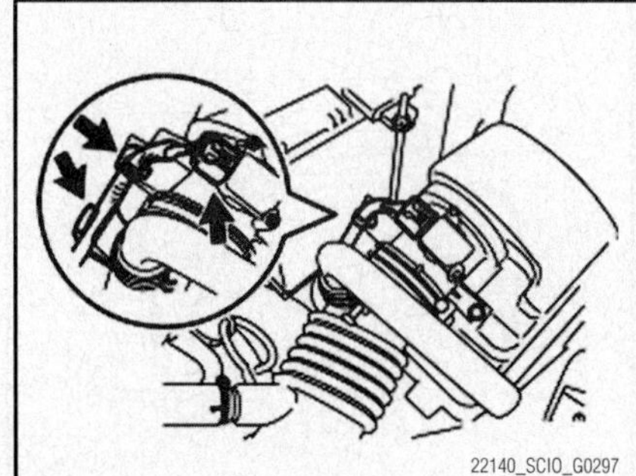

Fig. 155 Disconnect the wire harness clamp and mass air flow meter connector

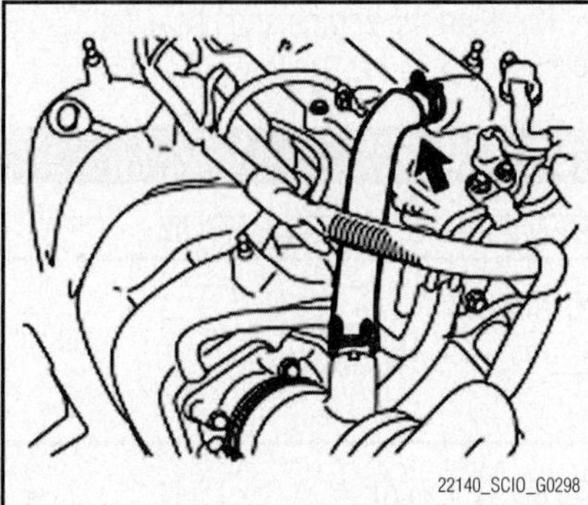

Fig. 156 Disconnect the No. 2 ventilation hose

Fig. 157 Remove air cleaner cap sub-assembly with hose

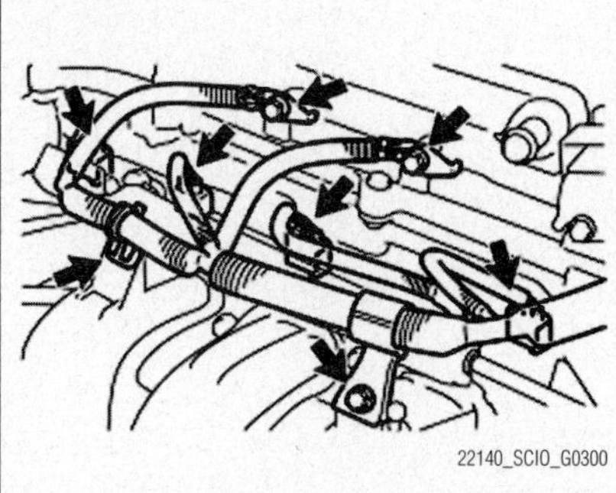

Fig. 158 Separate the engine wire harness

9. Remove the harness bracket.
10. Disconnect the No. 1 fuel vapor feed hose.
11. Disconnect the vacuum hose.
12. Remove the fuel pipe clamp.
13. Disconnect the fuel tube sub-assembly.
14. Remove fuel the delivery pipe sub-assembly.
15. Remove the No. 1 delivery pipe spacer.
16. Remove the injector vibration insulator.
17. Remove the starter assembly.
18. Remove the oil cooler tube.
   a. Using a union nut wrench, separate the inlet No. 1 oil cooler tube while holding the oil cooler tube union with a wrench.
   b. Using a union nut wrench, separate the outlet No. 1 oil cooler tube while holding the oil cooler tube union with a wrench.
   c. Remove the bolt and remove the oil cooler tube clamp.
19. Remove the transmission oil level gauge sub-assembly.
   a. Remove the transmission oil level dipstick.
   b. Disconnect the breather hose.
   c. Remove the bolt and remove the transmission oil level gauge sub-assembly.
20. Remove the engine oil level dipstick.
21. Remove the throttle body assembly.
22. Remove the intake manifold.
   a. Disconnect the 4 wire harness clamps.
   b. Disconnect the ventilation hose.
   c. Remove the 4 bolts and 2 nuts and remove the intake manifold and intake manifold stay.

To install:

23. Install intake manifold
   a. Install a new gasket onto the intake manifold.
   b. Install the intake manifold and intake manifold stay with the 4bolts and 2 nuts. Torque: 21 ft. lbs. (28 Nm).
   c. Connect the ventilation hose.
   d. Connect the 4 wire harness clamps.
24. Install throttle body assembly
25. Install transmission oil level gauge sub-assembly
   a. Install the transmission oil level gauge sub-assembly with the bolt.
   b. Connect the breather hose.
   c. Install the transmission oil level dipstick.
26. Install oil cooler tube
   a. Install the oil cooler tube clamp.
   b. Provisionally install the outlet No. 1 oil cooler tube into the oil cooler tube union.
   c. Provisionally install the inlet No. 1 oil cooler tube into the oil cooler tube union.
   d. Install the No. 2 oil cooler tube clamp onto the transmission oil filler tube sub-assembly with the bolt.
   e. Using a union nut wrench, install the inlet No. 1 oil cooler tube while holding the oil cooler tube union with a spanner.
      - Without union nut wrench - Torque: 25 ft. lbs. (34 Nm).
      - With union nut wrench - Torque: 24 ft. lbs.32 Nm).

➡This torque value can be obtained by using a torque wrench with a fulcrum length of 13.58 inches (345 mm) and a union nut wrench with a fulcrum length of 1.18 inches (30 mm). This torque value is effective when union nut wrench is parallel to a torque wrench.

27. Install engine oil level dipstick
28. Install starter assembly
29. Install injector vibration insulator
30. Install No. 1 delivery pipe spacer
31. Install fuel delivery pipe sub-assembly
32. Connect fuel tube sub-assembly
33. Install fuel pipe clamp
34. Connect vacuum hose
35. Connect No. 1 fuel vapor feed hose
36. Install harness bracket
37. Install engine wire harness
38. Install air cleaner cap sub-assembly with hose
39. Connect cable to negative battery terminal
40. Add engine coolant
41. Inspect for coolant leak
42. Inspect for fuel leak
43. Install engine cover
44. Install engine under cover RH
45. Install engine under cover LH

OIL PAN

REMOVAL & INSTALLATION

tC and xB

See Figure 159.

1. Before servicing the vehicle, refer to the Precautions Section.
2. Drain the engine oil.
3. Remove the engine assembly from the vehicle and secure to a suitable chain block and sling device.
4. Remove the oil pan mounting bolts.

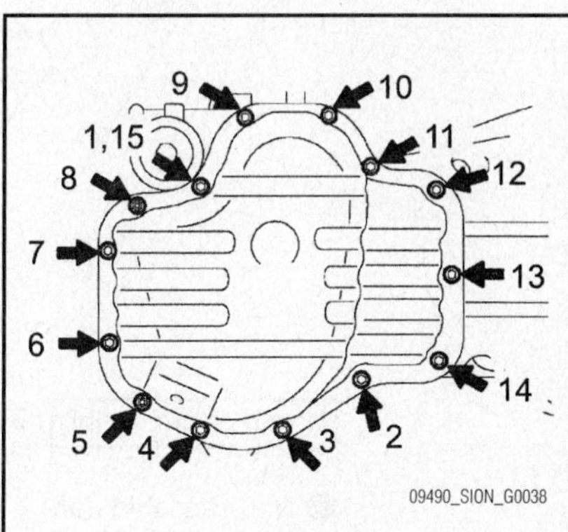

Fig. 159 Oil pan torque sequence—2.4L engine

5. Using Special Tool 09032-00100, cut off the sealant between the front cover, cylinder block and oil pan.
6. Remove the oil pan.

To install:

7. Remove any old sealant from the oil pan.
8. Apply a continuous bead of sealant to the contact surfaces of the oil pan.

➡Do not expose the sealant to engine oil for at least 2 hours after installation.

9. Install the oil pan. Tighten the bolts in sequence to 80 inch lbs. (9 Nm).
10. Install the engine assembly.
11. Fill the engine with oil to the correct level.
12. Start the engine and check for leaks.

OIL PUMP

REMOVAL & INSTALLATION

tC and xB

See Figures 160 and 161.

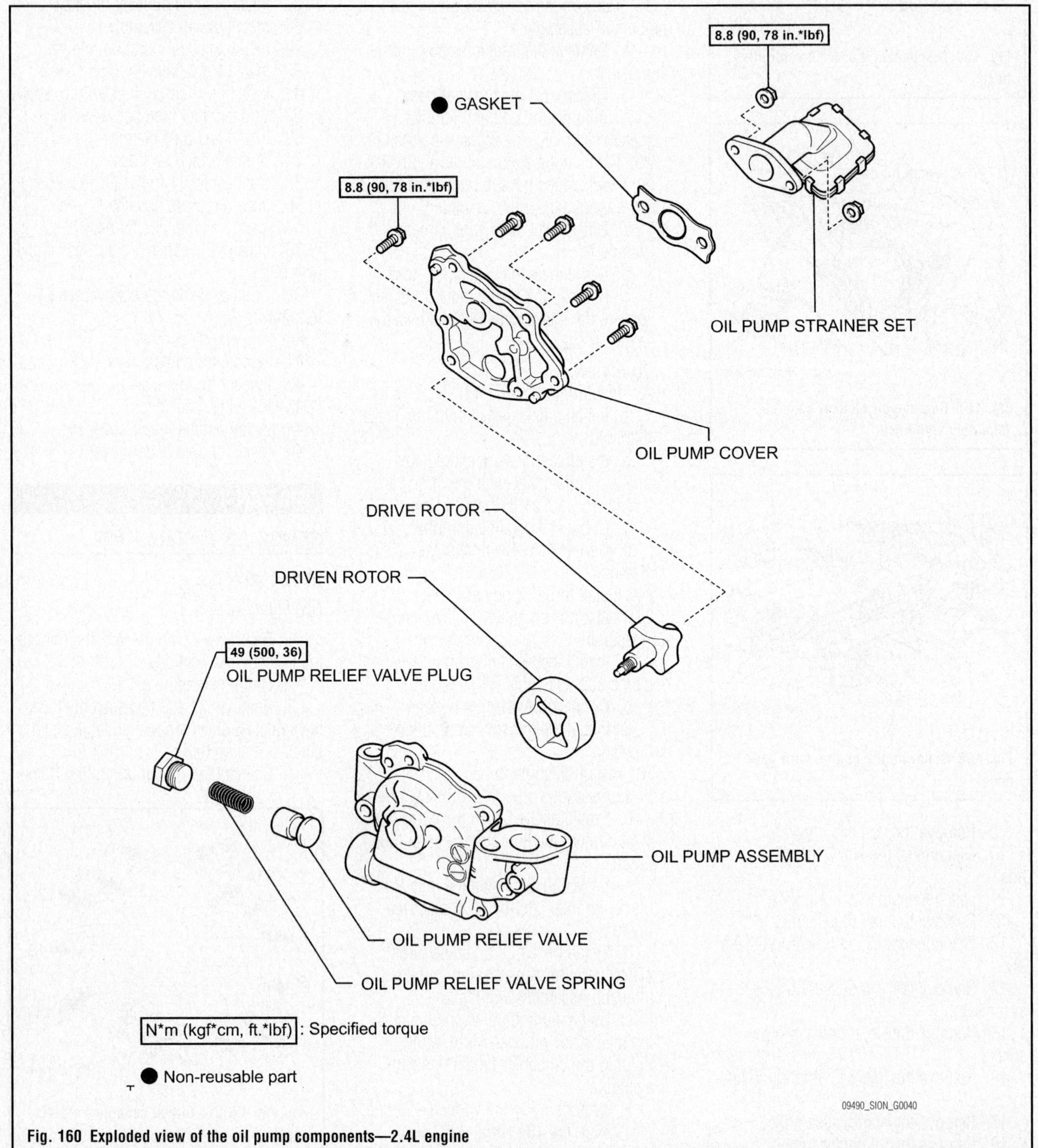

Fig. 160 Exploded view of the oil pump components—2.4L engine

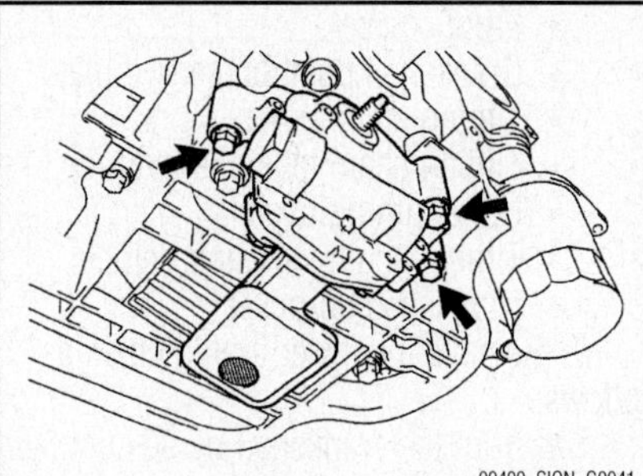

Fig. 161 Remove the three mounting bolts to remove the oil pump assembly—2.4L Engine

1. Before servicing the vehicle, refer to the Precautions Section.
2. Drain the engine oil.
3. Remove or disconnect the following:
   - Negative battery cable
   - Front cover
   - Timing chain
   - Oil pump assembly

To install:

4. Install the oil pump assembly with a new gasket. Tighten the mounting bolts to 14 ft. lbs. (19 Nm).
5. Install or connect the following:
   - Timing chain
   - Front cover
   - Negative battery cable
6. Refill the engine with oil to the correct level.
7. Start the engine and check for leaks.

xD

➡The engine must be removed from the vehicle to access the oil pump. Refer to Engine Assembly Removal and Installation.

1. Remove the engine assembly with transaxle.
2. Remove the front suspension crossmember sub-assembly.
   a. Remove the bolt and remove the air fuel ratio sensor bracket.
   b. Install the engine hangers with the bolts. Torque: 32 ft. lbs. (43 Nm).
   c. Using an engine sling device and a chain block, suspend the engine assembly with transaxle and front suspension crossmember.
   d. Remove the through bolt from the engine moving control rod and remove the front suspension crossmember.
3. Remove the No. 1 oil cooler inlet tube (for Automatic Transaxle).
4. Remove the No. 1 oil cooler outlet tube (for Automatic Transaxle).
5. Remove the No. 2 oil cooler tube clamp (for Automatic Transaxle).
6. Remove the transmission oil level gage sub-assembly (for Automatic Transaxle).
7. Separate the transmission oil filler tube sub-assembly (for Automatic Transaxle).
8. Remove the intake manifold.
9. Remove the oil level dipstick.
10. Remove the oil level gage guide.
11. Remove the alternator assembly.
12. Remove the fan belt adjusting bar.
13. Separate the water inlet hose.
14. Separate the No. 3 water by-pass hose.
15. Remove the water inlet.
16. Remove the thermostat.
17. Remove the ignition coil assembly.
18. Remove the radio setting condenser.
19. Remove No. 2 ventilation hose.
20. Remove the air tube assembly.
21. Remove the cylinder head cover sub-assembly.
22. Set the No. 1 cylinder to TDC / compression.
23. Remove the crankshaft pulley.
24. Remove the no. 1 chain tensioner assembly.
25. Remove the timing chain or belt cover sub-assembly.
26. Remove the timing chain or belt cover oil seal.
   a. Using a screwdriver and hammer, remove the oil seal

⁂ CAUTION

Be careful not to damage the timing chain or belt cover sub-assembly. Tape the screwdriver tip before use.

27. Remove the No. 2 chain vibration damper.
28. Remove the chain tensioner slipper.
29. Remove the No. 1 chain vibration damper.
30. Remove the chain sub-assembly.
31. Remove the crankshaft timing gear or sprocket.
32. Remove the No. 2 chain sub-assembly.
33. Remove the No. 2 oil pan sub-assembly.
34. Remove the oil pump assembly.
   a. Remove the 3 bolts and remove the oil pump assembly.

PISTON AND RING

POSITIONING

See Figures 162 through 165.

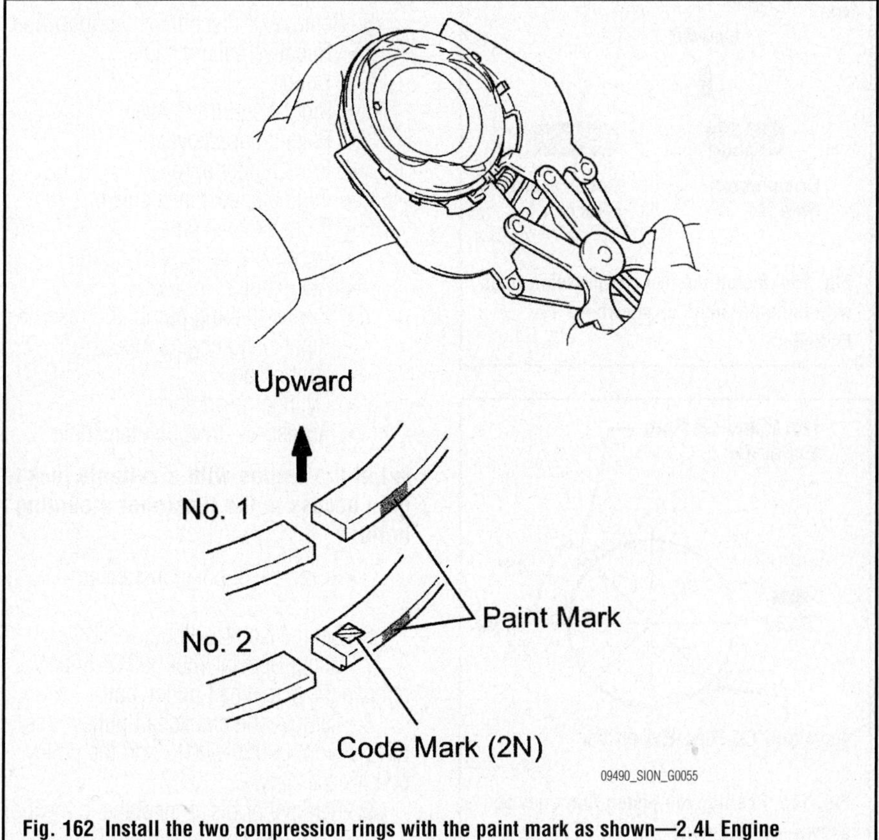

Fig. 162 Install the two compression rings with the paint mark as shown—2.4L Engine

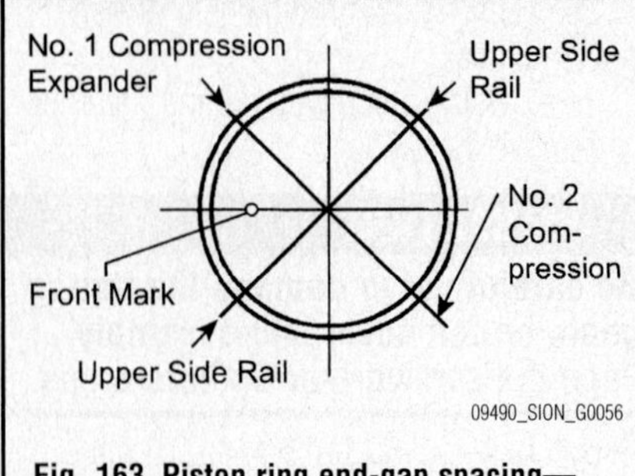

Fig. 163 Piston ring end-gap spacing—2.4L Engine

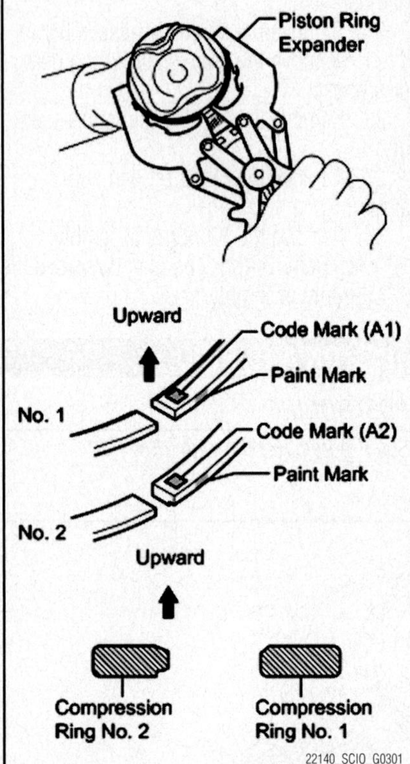

Fig. 164 Install the two compression rings with the paint mark as shown—1.8L Engine

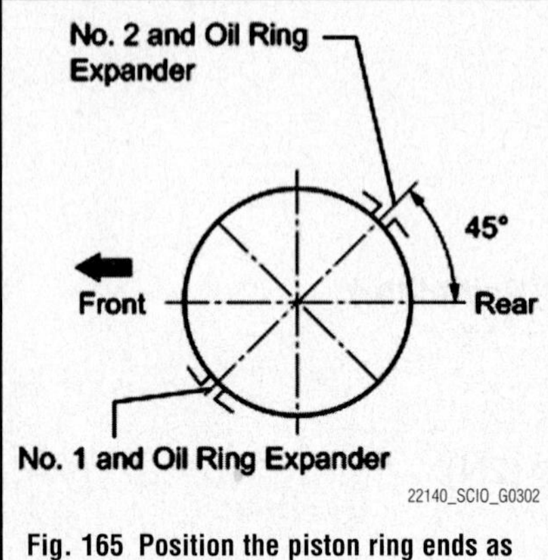

Fig. 165 Position the piston ring ends as shown—1.8L Engine

REAR MAIN SEAL

REMOVAL & INSTALLATION

1. Before servicing the vehicle, refer to the Precautions Section.
2. Remove or disconnect the following:
   - Transaxle
   - Flywheel/Driveplate
   - Oil seal

To install:

3. Using Special Tool 09223-56010 or suitable seal installer, tap in the oil until its surface is flush with the seal retainer edge.
4. Install or connect the following:
   - Flywheel/Driveplate
   - Transaxle
5. Start the engine check for leaks.

TIMING CHAIN COVER AND SEAL

REMOVAL & INSTALLATION

tC and xB

See Figures 166 through 174.

1. Before servicing the vehicle, refer to the Precautions Section.
2. Drain the engine oil.
3. Drain the cooling system.
4. Support the engine with a suitable jack.
5. Remove or disconnect the following:
   - Negative battery cable
   - Hood
   - Right-hand front wheel
   - Engine undercover
   - Front fender apron
   - Engine appearance cover
   - Front exhaust pipe
   - Accessory drive belt
   - Alternator
   - Power steering pump and reservoir
   - Right-hand engine mount
   - Ignition coil
   - Cylinder head cover
   - Accessory drive belt tensioner

➡Lift the engine with a suitable jack to gain access to the tensioner mounting bolts.

   - Crankshaft position sensor
   - Oil pan
6. Set the No. 1 cylinder to TDC.
7. Using Special Tool 09213-54015, remove the crankshaft pulley bolt.
8. Remove the crankshaft pulley. Use Special Tool 09950-50013 and the pulley bolt if necessary.
9. Remove or disconnect the following:
   - Timing chain tensioner
   - Front cover
   - Crankshaft position sensor plate
   - Timing chain guide
   - Timing chain tensioner slipper
   - Upper timing chain
   - Timing chain vibration damper
   - Crankshaft sprocket
10. Remove the lower timing chain as follows:
    a. Turn the crankshaft by 90° counterclockwise to align the adjusting hole of the oil pump drive shaft gear with the groove of the oil pump.
    b. Insert a 4 mm diameter bar into the adjusting hole of the oil pump drive shaft gear to lock in position, then remove the nut.
    c. Remove the bolt, then remove the chain tensioner plate and spring.
    d. Remove the oil pump drive gear, oil pump drive shaft gear and lower timing chain.

To install:

11. Set the crankshaft key in the left horizontal position. Turn the cutout of the drive shaft to the top.
12. Align the yellow mark links with the timing marks of the each gear as shown in the illustration.
13. Install the gears onto the crankshaft and oil pump shaft with the lower chain wrapped.
14. Temporarily tighten the oil pump drive shaft gear with the nut.
15. Insert the damper spring into the adjusting hole, then install the chain tensioner plate with the bolt. Tighten to 9 ft. lbs. (12 Nm).
16. Align the adjusting hole of the oil pump drive shaft gear with the groove of the oil pump.
17. Insert a 4 mm diameter bar into the adjusting hole of the oil pump drive shaft gear to lock in position, then tighten the nut to 22 ft. lbs. (30 Nm).
18. Turn the crankshaft clockwise by 90° to position the crankshaft key upward.
19. Install the crankshaft sprocket.
20. Turn the camshafts with a wrench on the hexagonal lobe to align the timing marks of the camshaft timing gear with each timing mark located on the No. 1 and No. 2 bearing caps as shown in the illustration.
21. Using the crankshaft pulley bolt, turn the crankshaft to position the key on the crankshaft upward.
22. Install the upper timing chain onto the crankshaft timing gear with the gold or orange mark link aligned with the timing mark on the crankshaft sprocket.

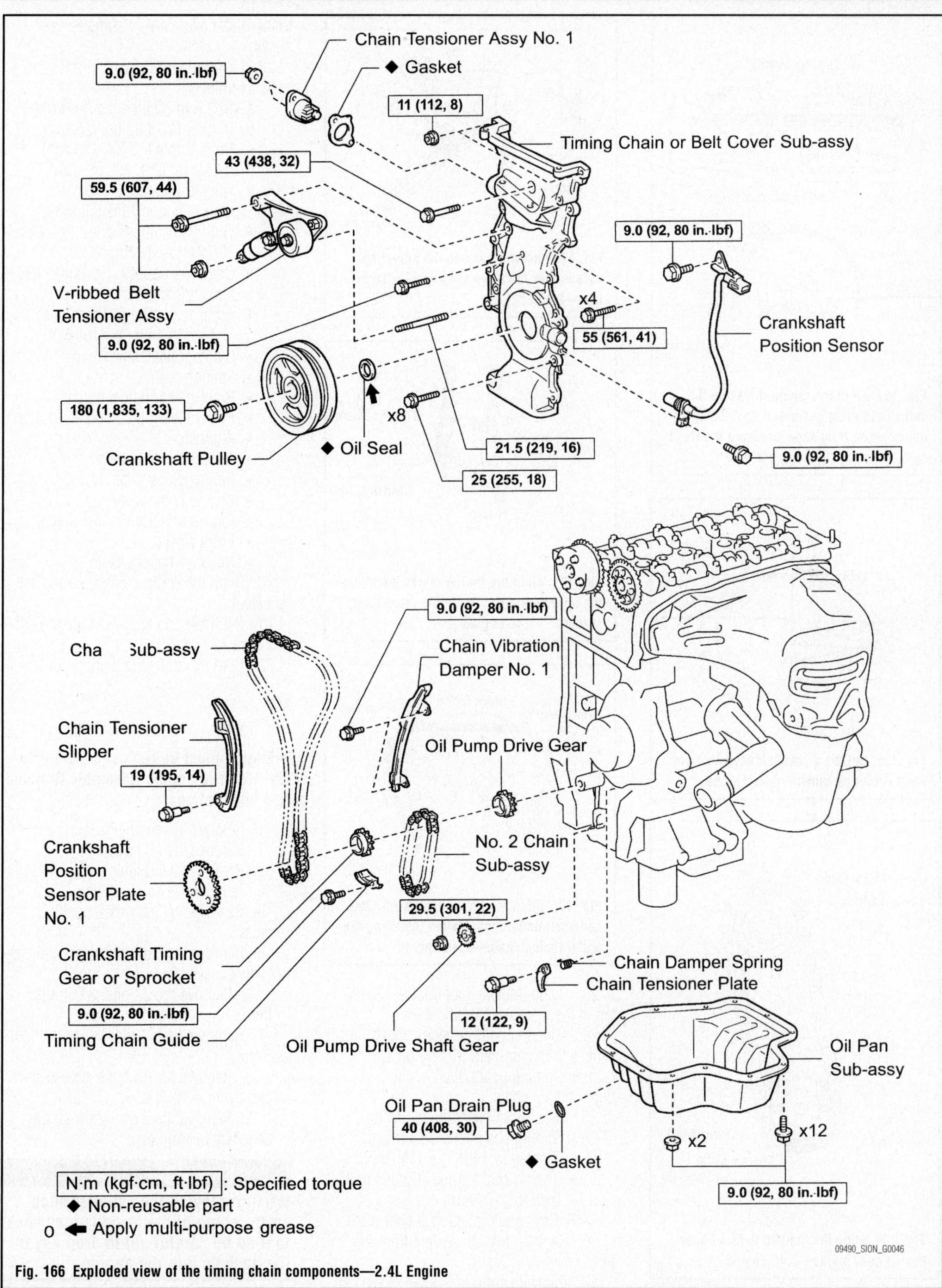

Fig. 166 Exploded view of the timing chain components—2.4L Engine

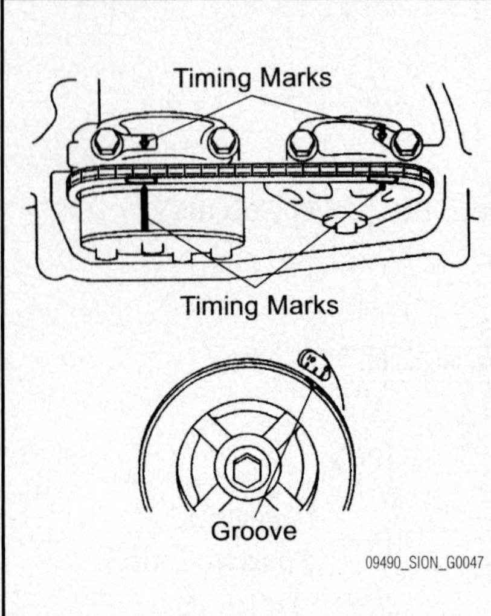

Fig. 167 Align the timing marks on the camshaft timing gears and Nos. 1 and 2 camshaft bearing caps to ensure the No. 1 cylinder is at TDC compression

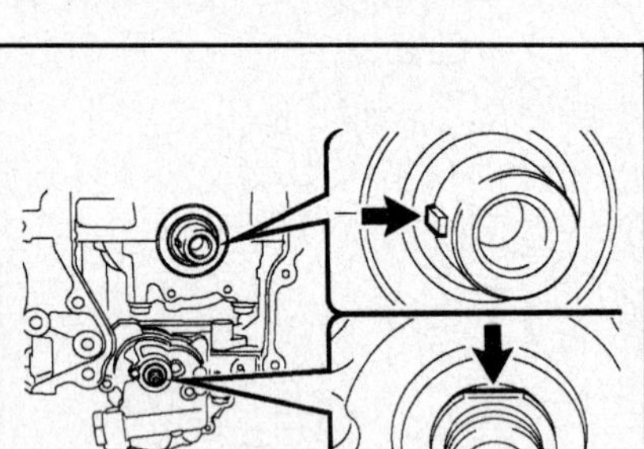

Fig. 168 Set the crankshaft key and drive shaft cutout as shown—2.4L Engine

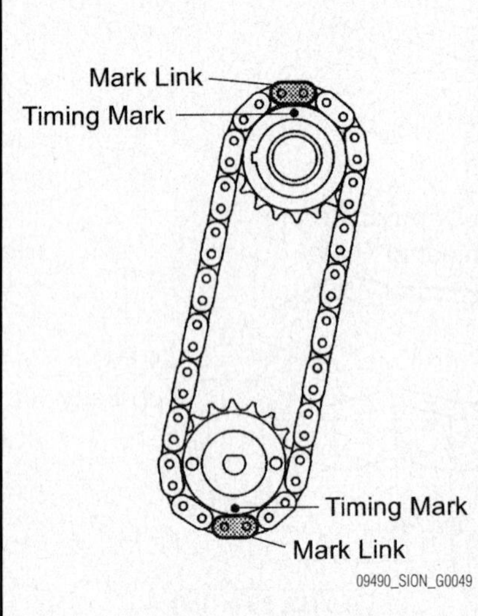

Fig. 169 Align the marked links with the timing gear marks—2.4L Engine

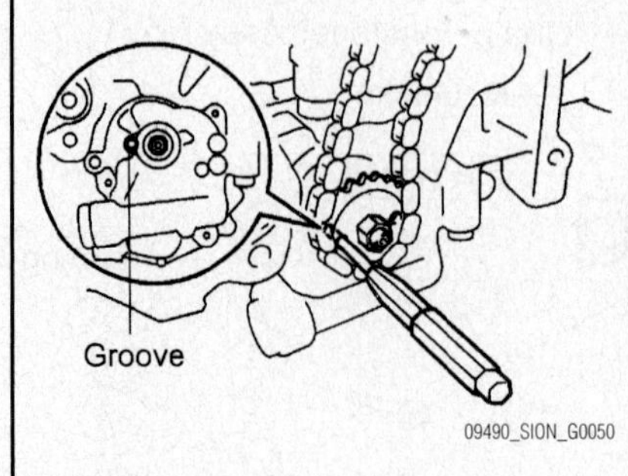

Fig. 170 Insert a bar into the adjusting hole to lock the drive shaft gear in position—2.4L Engine

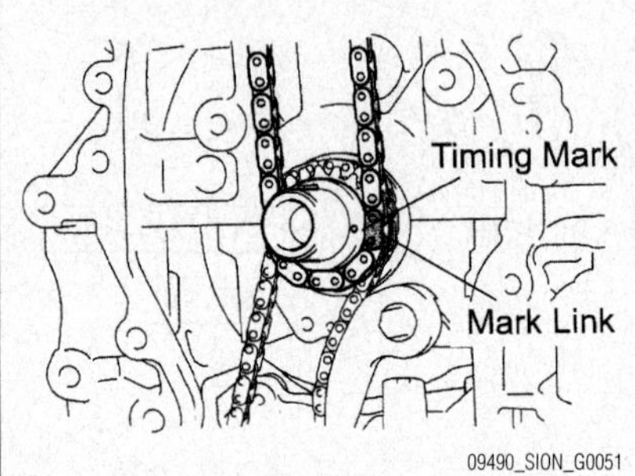

Fig. 171 Align the timing marks installing the upper timing chain on the crankshaft sprocket—2.4L Engine

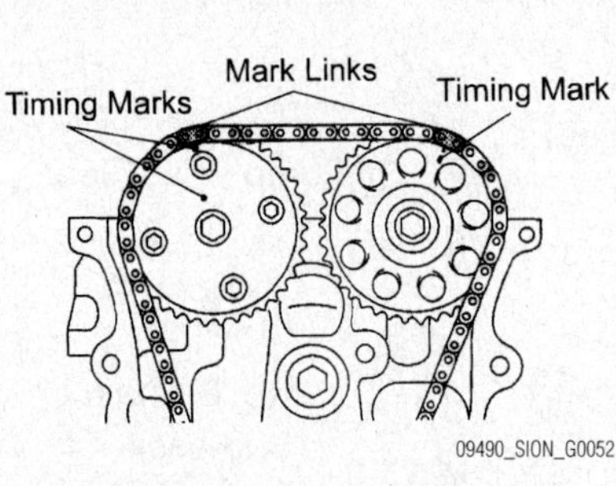

Fig. 172 Align the timing marks with the camshaft timing gears when installing the upper timing chain—2.4L Engine

23. Using Special Tool 09309-37010, tap in the crankshaft timing gear.
24. Align the gold or yellow mark links with each timing mark located on the camshaft timing gears, then install the upper timing chain.
25. Install or connect the following:
   - Timing chain tensioner slipper. Tighten to 14 ft. lbs. (19 Nm).
   - Timing chain guide. Tighten to 80 inch lbs. (9 Nm).
   - Crankshaft position sensor plate
26. Remove any old sealant from the front cover.
27. Apply a continuous bead of sealant to the front cover contact surfaces as shown.
28. Install the front cover. Tighten the bolts as follows:
   a. Bolt A to 80 inch lbs. (9 Nm).
   b. Bolts B to 18 ft. lbs. (25 Nm).
   c. Bolts C to 41 ft. lbs. (55 Nm).
   d. Bolt D to 32 ft. lbs. (43 Nm).
   e. Nuts to 8 ft. lbs. (11 Nm).
29. Install or connect the following:
   - Timing chain tensioner. Tighten to 80 inch lbs. (9 Nm).
   - Crankshaft pulley. Tighten to 133 ft. lbs. (180 Nm).
   - Accessory drive belt tensioner. Tighten to 44 ft. lbs. (60 Nm).
   - Cylinder head cover
   - Ignition coil
   - Right-hand engine mount
   - Power steering pump and reservoir
   - Alternator
   - Accessory drive belts
   - Front exhaust pipe
   - Engine undercover
   - Right-hand front fender apron
   - Front wheel
   - Negative battery cable
30. Refill the engine with oil to the correct level.
31. Refill the cooling system to the correct level.
32. Start the engine and check for leaks.

xD

See Figures 175 through 189.

➡Engine must be removed from vehicle. Refer to Engine Assembly Removal and Installation.

1. Remove the engine hanger.
2. Remove the oil filler cap.
3. Remove the oil filler cap gasket.
4. Remove the spark plugs.
5. Remove the 2camshaft position sensors.
6. Remove the camshaft timing oil control valve assembly.
   a. Remove the 2 bolts, 2 O-rings, bracket and 2 oil control valves.
7. Remove the cylinder head cover sub-assembly.
   a. Remove the 13 bolts, seal washer and cylinder head cover.
   b. Remove the 3 gaskets from the camshaft bearing cap.

CAUTION

When removing the cylinder head cover, some of the gaskets may stick to it so be careful not to drop any of the gaskets into the engine.

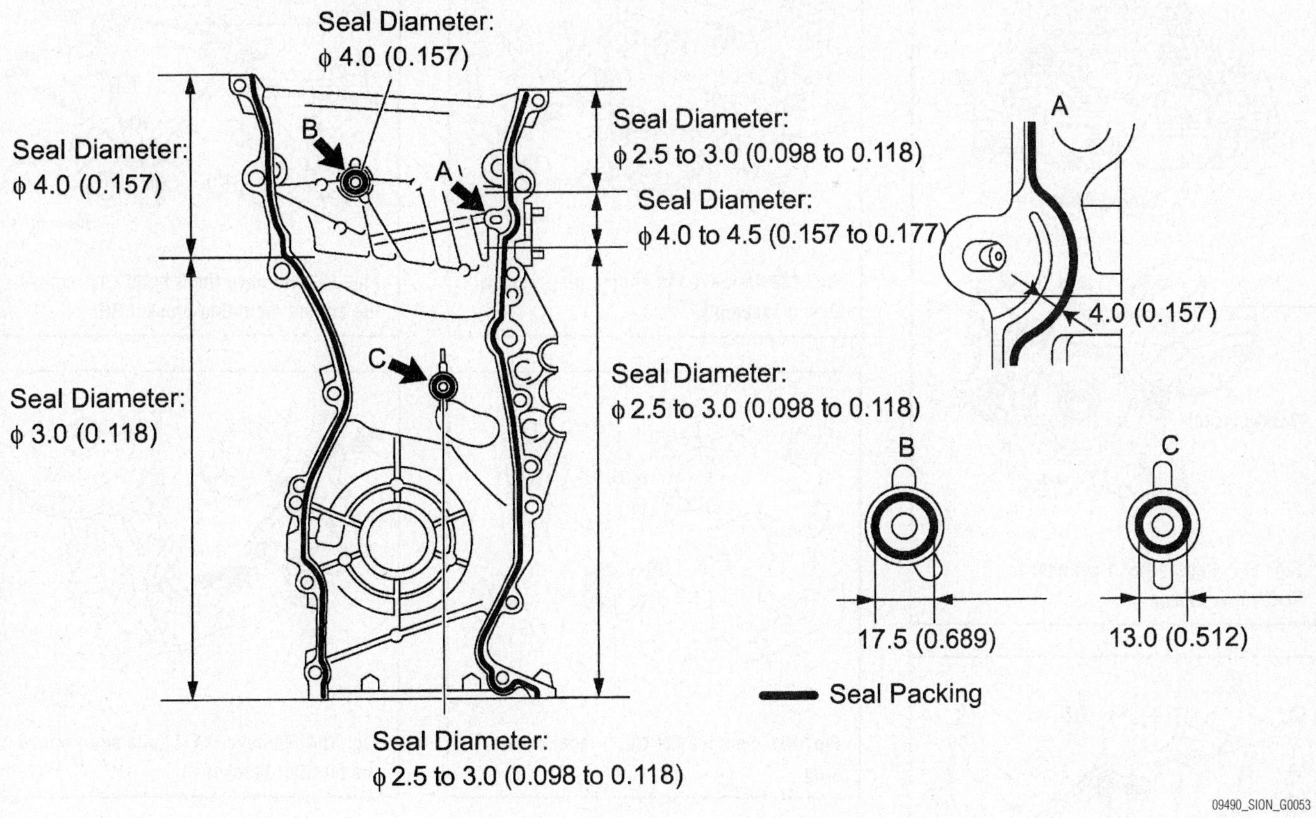

Seal Diameter:
ϕ 4.0 (0.157)
Seal Diameter:
ϕ 4.0 (0.157)
B
A
Seal Diameter:
ϕ 2.5 to 3.0 (0.098 to 0.118)
Seal Diameter:
ϕ 4.0 to 4.5 (0.157 to 0.177)
A
4.0 (0.157)
C
Seal Diameter:
ϕ 3.0 (0.118)
Seal Diameter:
ϕ 2.5 to 3.0 (0.098 to 0.118)
B
C
17.5 (0.689)
13.0 (0.512)
Seal Packing
Seal Diameter:
ϕ 2.5 to 3.0 (0.098 to 0.118)
09490_SION_G0053

Fig. 173 Apply sealant to the front cover contact surfaces shown—2.4L Engine

Nut
D
Nut
C
C
C
C
A
B
B
B
B
B
B
B
B
B
09490_SION_G0054

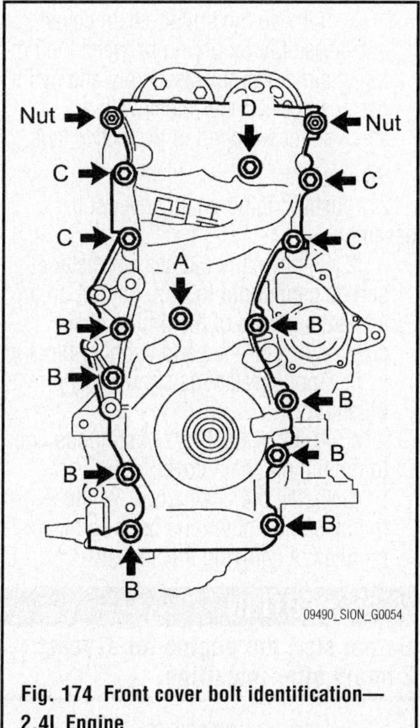

Fig. 174 Front cover bolt identification—2.4L Engine

22140_SCIO_G0303

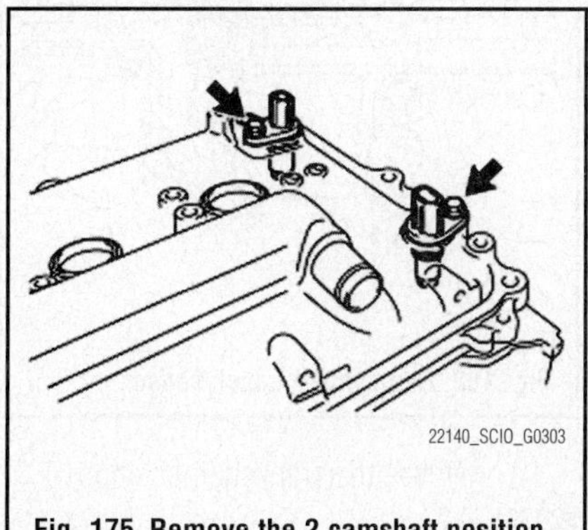

Fig. 175 Remove the 2 camshaft position sensors

22140_SCIO_G0304

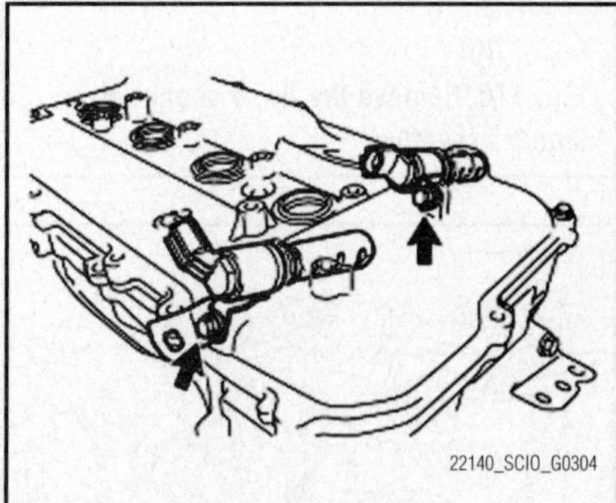

Fig. 176 Remove the camshaft timing oil control valve assembly

8. Remove the cylinder head cover gasket.
9. Set the No. 1 cylinder to TDC / compression.
   a. Turn the crankshaft pulley, and align its timing notch with the "0" timing mark on the timing chain cover sub-assembly.
   b. Check that timing marks on both the camshaft timing exhaust gear and camshaft timing gear are facing upward. If not, turn the crankshaft 1 complete revolution (360°) and align the marks as above.
10. Remove the crankshaft pulley.
   a. Using SST, hold the pulley in place and loosen the pulley bolt.
   b. Using SST, remove the pulley bolt and pulley.
11. Remove the No. 1 chain tensioner assembly.

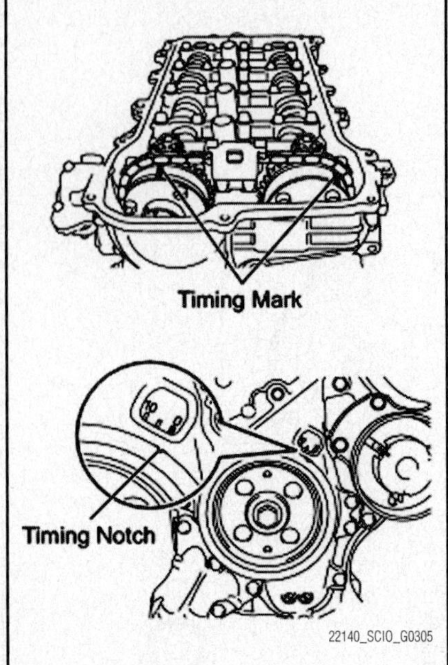

Fig. 177 Set the No. 1 cylinder to TDC/compression

Fig. 178 Remove the No. 1 chain tensioner assembly

Fig. 179 Remove the crankshaft position sensor

✱✱ CAUTION

Do not turn the crankshaft after you have removed the chain tensioner.

Fig. 180 Remove the engine oil pressure switch assembly

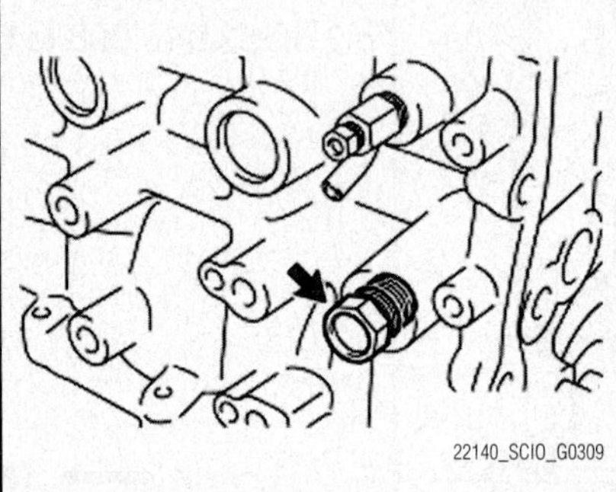

Fig. 181 Remove the No. 1 taper screw plug

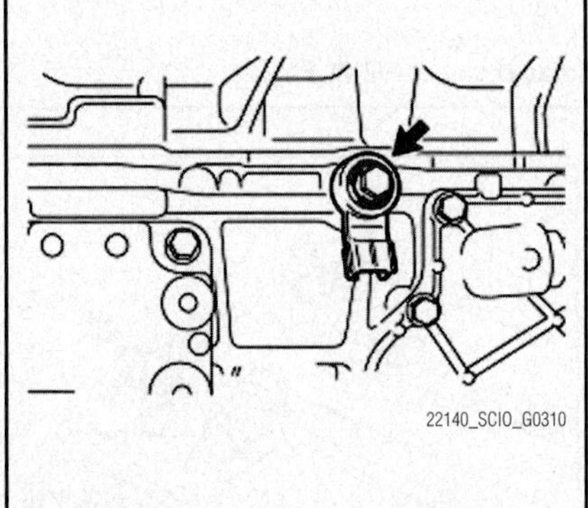

Fig. 182 Remove the knock sensor

12. Remove the crankshaft position sensor.
13. Remove the engine oil pressure switch assembly.
14. Remove the No. 1 taper screw plug.
15. Remove the knock sensor.
16. Remove the engine water temperature sensor.
17. Remove the oil filter sub-assembly.
18. Remove the water inlet.
19. Remove the thermostat.
20. Remove the timing chain cover sub-assembly.
    a. Remove the 3 bolts and remove the engine mounting bracket RH.

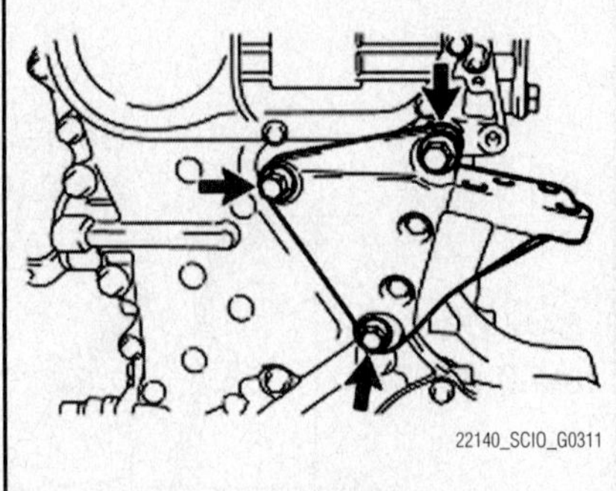

Fig. 183 Remove the 3 bolts and remove the engine mounting bracket RH

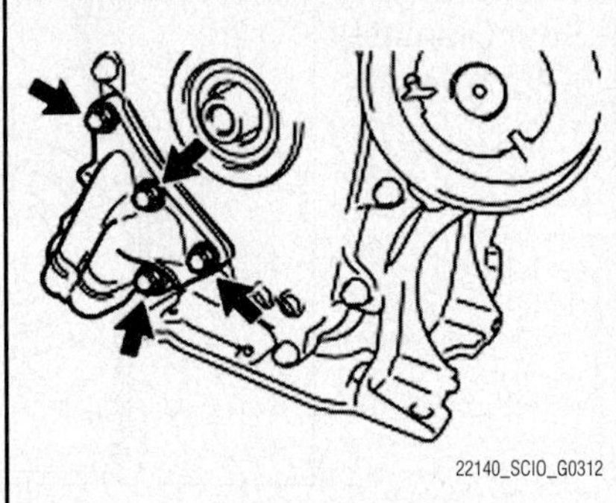

Fig. 184 Remove the 4 bolts and remove the oil filter bracket

    b. Remove the 4 bolts and remove the oil filter bracket.
    c. Remove the 2 O-rings from the timing chain cover sub-assembly.
    d. Remove the 19 bolts.
    e. Remove the timing chain cover sub-assembly by prying between the timing chain cover sub-assembly and cylinder head or cylinder block with a screwdriver wrapped in protective tape.

To install:

21. Install timing chain cover sub-assembly.
    a. Remove any old sealant material and be careful not to drop any oil on the contact surfaces of the timing chain cover, cylinder head and cylinder block.
    b. Apply sealant as shown in the illustration.
    c. Apply sealant in a continuous line to the timing chain cover.
    d. Install the chain cover within 3 minutes and tighten the bolts within 15 minutes of applying the sealant.

✱✱ CAUTION

Do not start the engine for at least 2 hours after installing.

    e. Install a new gasket.

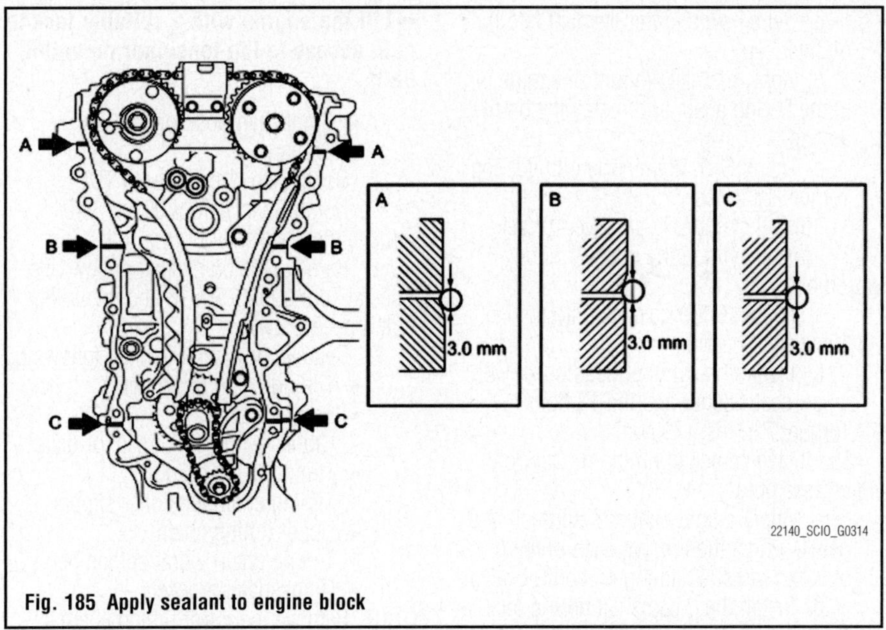

Fig. 185 Apply sealant to engine block

f. Install 2 new O-rings.

g. Install the timing chain cover with the 19 bolts as shown in the illustration.

h. Bolt A and C - Torque: 19 ft. lbs. (26 Nm)

i. Bolt B and D - Torque: 38 ft. lbs. (51 Nm)

j. Bolt E - Torque: 7 ft. lbs. (10 Nm)

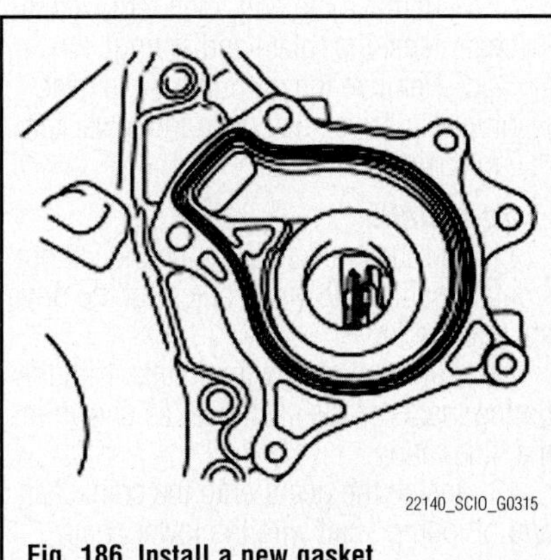

Fig. 186 Install a new gasket

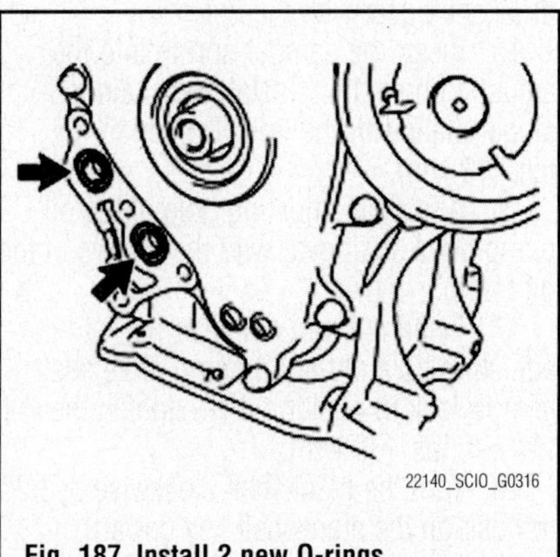

Fig. 187 Install 2 new O-rings

k. Tighten bolt D through the engine mounting bracket, pressing it against the timing chain cover sub-assembly.

l. Tighten bolt C through the oil filter bracket, pressing it against the timing chain cover sub-assembly.

➡**Bolt length:**

- Bolt A and C: 1.38 inches (35 mm)
- Bolt B: 2.16 inches (55 mm)
- Bolt D: 3.15 inches (80 mm)
- Bolt E: 1.57 inches (40 mm)

m. Install the water pump with the 3 bolts. Torque: 18 ft. lbs. (24 Nm).

22. Install thermostat.

a. Install a new gasket onto the thermostat.

b. Install the thermostat with the jiggle valve facing upward.

➡**The jiggle valve may be set within 10° on either side of the top.**

c. Install water inlet. Torque: 7 ft. lbs. (10 Nm).

23. Install crankshaft pulley.

a. Align the pin hole in the crankshaft pulley with the pin position and install the crankshaft pulley.

b. Provisionally install the bolt.

c. Using SST, tighten the bolt while holding the crankshaft pulley. Torque: 140 ft. lbs. (190 Nm).

CAUTION

Check the SST installation positions when installing them, to avoid the SST fixing bolts from coming into contact with the timing chain cover sub-assembly.

24. Install engine oil pressure switch assembly.

a. Apply adhesive to 2 or 3 threads of the oil pressure switch.

b. Install the oil pressure switch. Torque: 11 ft. lbs. (15 Nm).

25. Install engine water temperature sensor. Torque: 15 ft. lbs. (20 Nm).

26. Install knock sensor. Torque: 15 ft. lbs. (20 Nm).

27. Install No. 1 taper screw plug.

a. Apply adhesive to 2 or 3 threads of

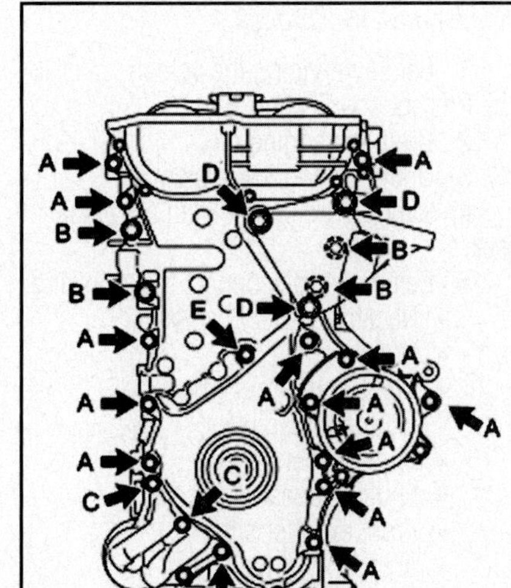

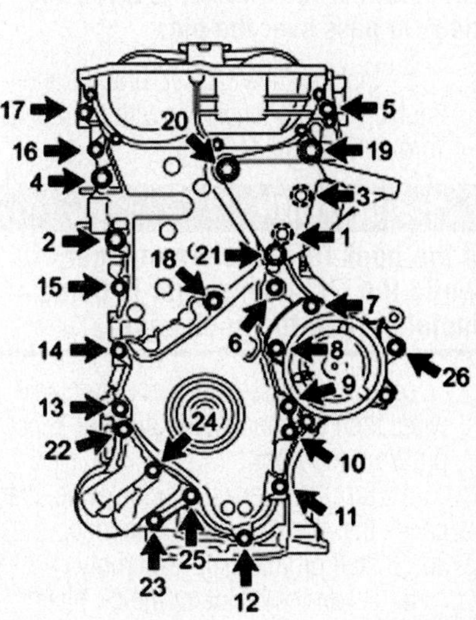

Fig. 188 Install the timing chain cover with the 19 bolts

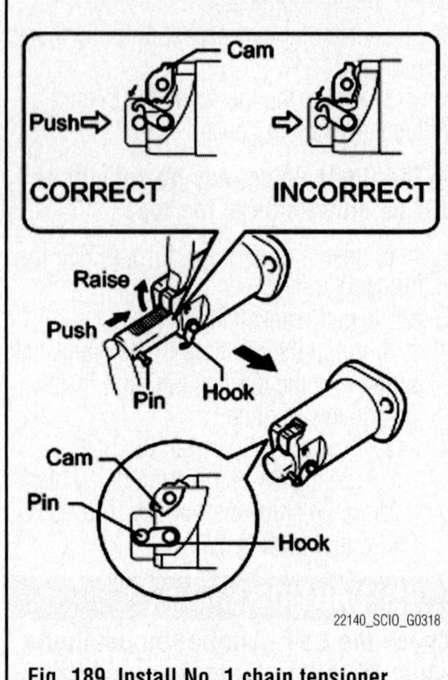

Fig. 189 Install No. 1 chain tensioner assembly

the plug, and install the plug. Torque: 32 ft. lbs. (43 Nm).

28. Install crankshaft position sensor.
    a. Apply a light coat of engine oil to the O-ring of the sensor.
    b. Install the crankshaft position sensor with the bolt. Torque: 7 ft. lbs. (10 Nm).
29. Install No. 1 chain tensioner assembly.
    a. Release the ratchet pawl, then fully push in the plunger and hook the hook on the pin so that the plunger is in the position shown in the illustration.

➡Make sure that the cam engages the first tooth of the plunger to allow the hook to pass over the pin.

    b. Install a new gasket, bracket and the No. 1 chain tensioner with the 2 nuts. Torque: 7 ft. lbs. (10 Nm).

⁂ CAUTION

If the hook releases the plunger while the chain tensioner is being installed, set the hook again.

    c. Turn the crankshaft counterclockwise, then disconnect the plunger knock pin from the hook.
    d. Turn the crankshaft clockwise, then check that the plunger is extended.
30. Install oil filter sub-assembly.
    a. Clean the inside of the oil filter cap, threads and O-ring groove.
    b. Apply a small amount of engine oil to a new O-ring and install it onto the oil filter cap.
    c. Set a new oil filter element into the oil filter cap.
    d. Apply a small amount of engine oil to the O-ring again and install the oil filter cap.
    e. Using SST, tighten the oil filter cap. Torque: 18 ft. lbs. (25 Nm).
31. Install cylinder head cover gasket.
32. Install cylinder head cover sub-assembly.
    a. Install 3 new gaskets onto the No. 1 camshaft bearing cap.
    b. Install the cylinder head cover with a new seal washer and the 13 bolts. Torque: 7 ft. lbs. (10 Nm).
33. Install camshaft timing oil control valve assembly.
    a. Apply a light coat of engine oil to 2 new O-rings, then install each of them onto the camshaft timing oil control valve.
    b. Install the 2 camshaft timing oil control valves and bracket with the 2 bolts. Torque: 7 ft. lbs. (10 Nm).
34. Install camshaft position sensor.
    a. Apply a light coat of engine oil to the O-ring of the sensor.
    b. Install the 2 sensors with the 2 bolts. Torque: 7 ft. lbs. (10 Nm).
35. Install spark plugs.
36. Install oil filler cap gasket.
37. Install oil filler cap sub-assembly.
38. Install engine hangers.

TIMING CHAIN AND SPROCKETS

REMOVAL & INSTALLATION

tC and xB

See Figures 167 through 174.

1. Before servicing the vehicle, refer to the Precautions Section.
2. Drain the engine oil.
3. Drain the cooling system.
4. Support the engine with a suitable jack.
5. Remove or disconnect the following:
    - Negative battery cable
    - Hood
    - Right-hand front wheel
    - Engine undercover
    - Front fender apron
    - Engine appearance cover
    - Front exhaust pipe
    - Accessory drive belt
    - Alternator
    - Power steering pump and reservoir
    - Right-hand engine mount
    - Ignition coil
    - Cylinder head cover
    - Accessory drive belt tensioner

➡Lift the engine with a suitable jack to gain access to the tensioner mounting bolts.

    - Crankshaft position sensor
    - Oil pan
6. Set the No. 1 cylinder to TDC.
7. Using Special Tool 09213-54015, remove the crankshaft pulley bolt.
8. Remove the crankshaft pulley. Use Special Tool 09950-50013 and the pulley bolt if necessary.
9. Remove or disconnect the following:
    - Timing chain tensioner
    - Front cover
    - Crankshaft position sensor plate
    - Timing chain guide
    - Timing chain tensioner slipper
    - Upper timing chain
    - Timing chain vibration damper
    - Crankshaft sprocket
10. Remove the lower timing chain as follows:
    a. Turn the crankshaft by 90° counterclockwise to align the adjusting hole of the oil pump drive shaft gear with the groove of the oil pump.
    b. Insert a 4 mm diameter bar into the adjusting hole of the oil pump drive shaft gear to lock in position, then remove the nut.
    c. Remove the bolt, then remove the chain tensioner plate and spring.
    d. Remove the oil pump drive gear, oil pump drive shaft gear and lower timing chain.

To install:

11. Set the crankshaft key in the left horizontal position. Turn the cutout of the drive shaft to the top.
12. Align the yellow mark links with the timing marks of the each gear as shown in the illustration.
13. Install the gears onto the crankshaft and oil pump shaft with the lower chain wrapped.
14. Temporarily tighten the oil pump drive shaft gear with the nut.
15. Insert the damper spring into the adjusting hole, then install the chain tensioner plate with the bolt. Tighten to 9 ft. lbs. (12 Nm).
16. Align the adjusting hole of the oil pump drive shaft gear with the groove of the oil pump.
17. Insert a 4 mm diameter bar into the adjusting hole of the oil pump drive shaft gear to lock in position, then tighten the nut to 22 ft. lbs. (30 Nm).
18. Turn the crankshaft clockwise by 90° to position the crankshaft key upward.
19. Install the crankshaft sprocket.

20. Turn the camshafts with a wrench on the hexagonal lobe to align the timing marks of the camshaft timing gear with each timing mark located on the No. 1 and No. 2 bearing caps as shown in the illustration.

21. Using the crankshaft pulley bolt, turn the crankshaft to position the key on the crankshaft upward.

22. Install the upper timing chain onto the crankshaft timing gear with the gold or orange mark link aligned with the timing mark on the crankshaft sprocket.

23. Using Special Tool 09309-37010, tap in the crankshaft timing gear.

24. Align the gold or yellow mark links with each timing mark located on the camshaft timing gears, then install the upper timing chain.

25. Install or connect the following:
- Timing chain tensioner slipper. Tighten to 14 ft. lbs. (19 Nm).
- Timing chain guide. Tighten to 80 inch lbs. (9 Nm).
- Crankshaft position sensor plate

26. Remove any old sealant from the front cover.

27. Apply a continuous bead of sealant to the front cover contact surfaces as shown.

28. Install the front cover. Tighten the bolts as follows:
   a. Bolt A to 80 inch lbs. (9 Nm).
   b. Bolts B to 18 ft. lbs. (25 Nm).
   c. Bolts C to 41 ft. lbs. (55 Nm).
   d. Bolt D to 32 ft. lbs. (43 Nm).
   e. Nuts to 8 ft. lbs. (11 Nm).

29. Install or connect the following:
- Timing chain tensioner. Tighten to 80 inch lbs. (9 Nm).
- Crankshaft pulley. Tighten to 133 ft. lbs. (180 Nm).
- Accessory drive belt tensioner. Tighten to 44 ft. lbs. (60 Nm).
- Cylinder head cover
- Ignition coil
- Right-hand engine mount
- Power steering pump and reservoir
- Alternator
- Accessory drive belts
- Front exhaust pipe
- Engine undercover
- Right-hand front fender apron
- Front wheel
- Negative battery cable

30. Refill the engine with oil to the correct level.

31. Refill the cooling system to the correct level.

32. Start the engine and check for leaks.

xD

See Figures 190 through 202.

➡Engine must be removed from vehicle. Refer to Engine Assembly Removal and Installation.

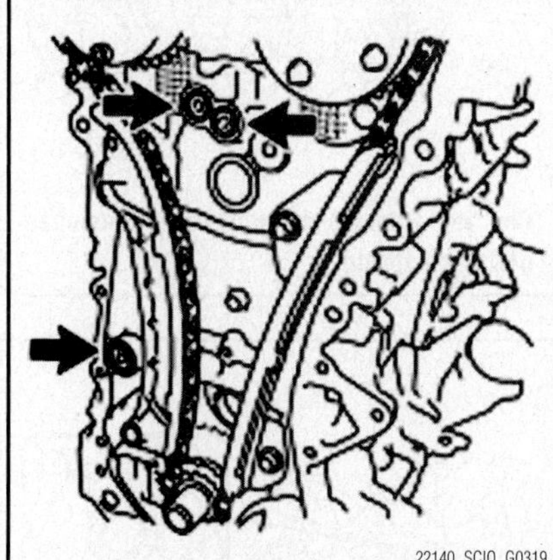

Fig. 190 Remove the 3 O-rings from the cylinder head and cylinder block.

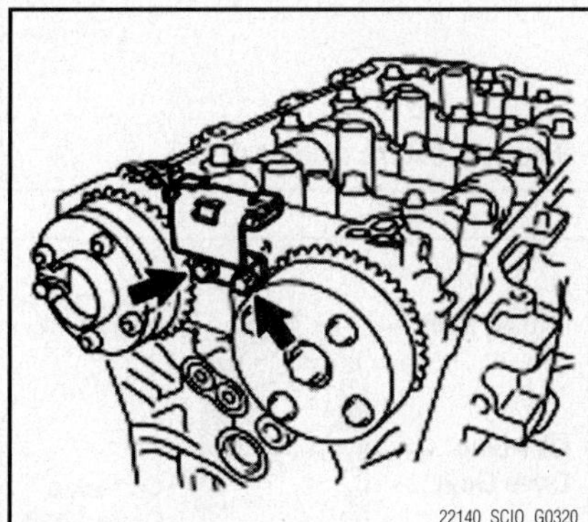

Fig. 191 Remove the No. 2 chain vibration damper

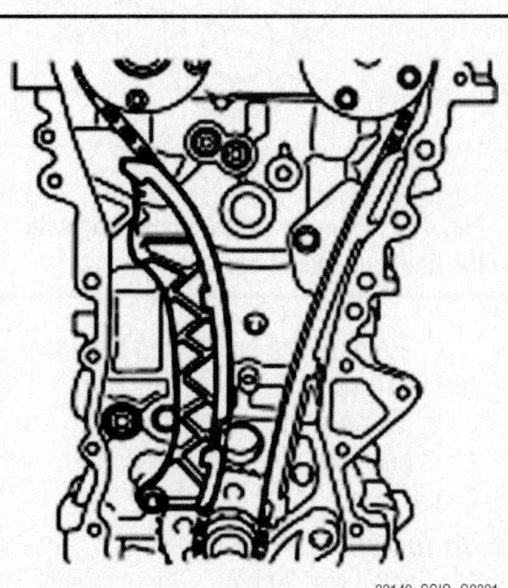

Fig. 192 Remove the chain tensioner slipper

➡Refer to Timing Chain Cover Removal and Installation.

1. Remove the 3 O-rings from the cylinder head and cylinder block.
2. Remove the 3 bolts and the water pump assembly.
3. Remove the water inlet housing.
4. Remove the alternator bracket.
5. Remove the No. 2 chain vibration damper.
6. Remove the chain tensioner slipper.
7. Remove the chain vibration damper.
8. Remove the chain sub-assembly.
   a. Hold the hexagonal portion of the camshaft with a wrench and turn the camshaft timing gear assembly counterclockwise to loosen the chain on the camshaft timing gears.

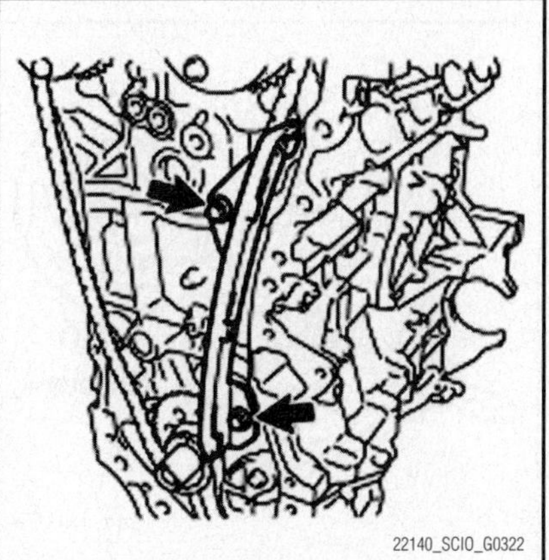

Fig. 193 Remove the chain vibration damper

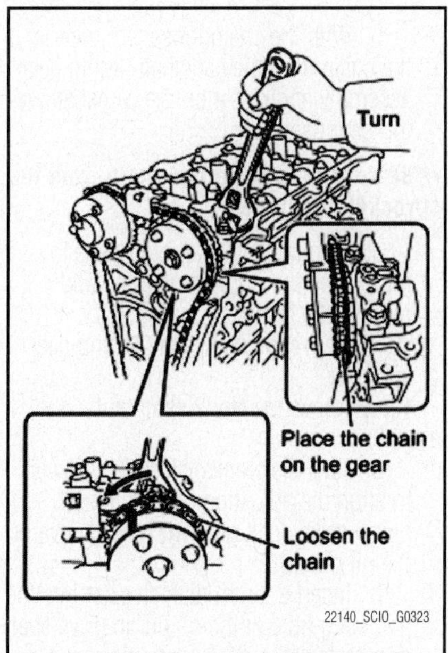

Fig. 194 Remove the chain sub-assembly

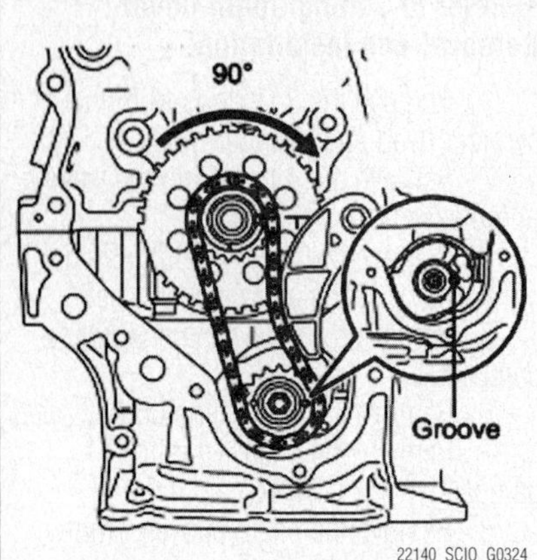

Fig. 195 Turn the crankshaft 90° clockwise to align the adjusting hole of the oil pump drive shaft gear with the groove of the oil pump

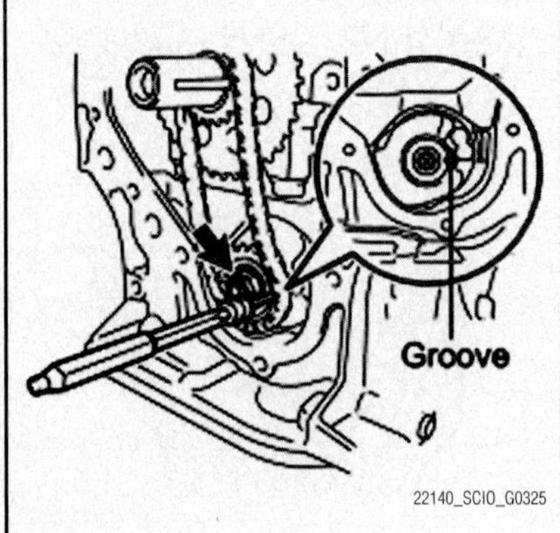

Fig. 196 Insert a 4 mm diameter bar into the adjusting hole of the oil pump drive shaft gear

b. With the chain loosened, release the chain from the camshaft timing gear assembly and rest it on the camshaft timing gear assembly.

➡Be sure to release the chain from the sprocket completely.

c. Turn the camshaft clockwise to return it to the original position and remove the chain.

9. Remove the crankshaft timing gear or sprocket.

10. Remove the No. 2 chain sub-assembly.

a. Turn the crankshaft 90° clockwise to align the adjusting hole of the oil pump drive shaft gear with the groove of the oil pump.

b. Insert a 4 mm diameter bar into the adjusting hole of the oil pump drive shaft gear to lock the gear in position, and then remove the nut.

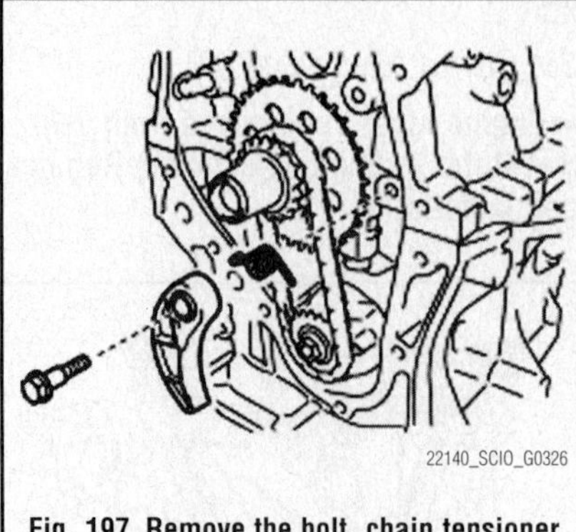

Fig. 197 Remove the bolt, chain tensioner plate and spring

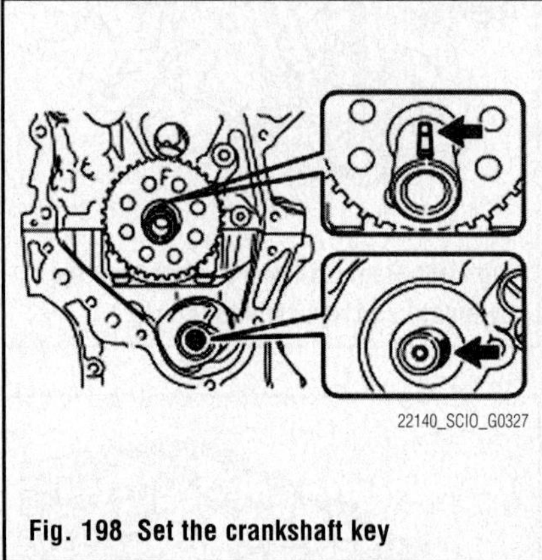

Fig. 198 Set the crankshaft key

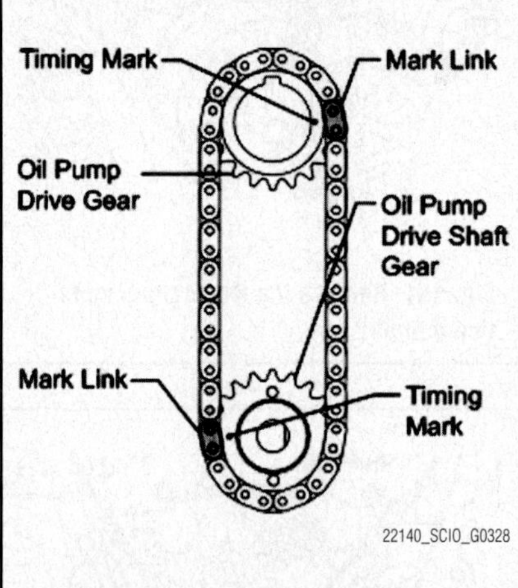

Fig. 199 Align the yellow mark links with the timing marks of each gear

c. Remove the bolt, chain tensioner plate and spring.

d. Remove the oil pump drive shaft gear, oil pump drive gear and chain.

To install:

11. Install No. 2 chain sub-assembly.

a. Set the crankshaft key as shown.

b. Turn the drive shaft so that the cutout faces right.

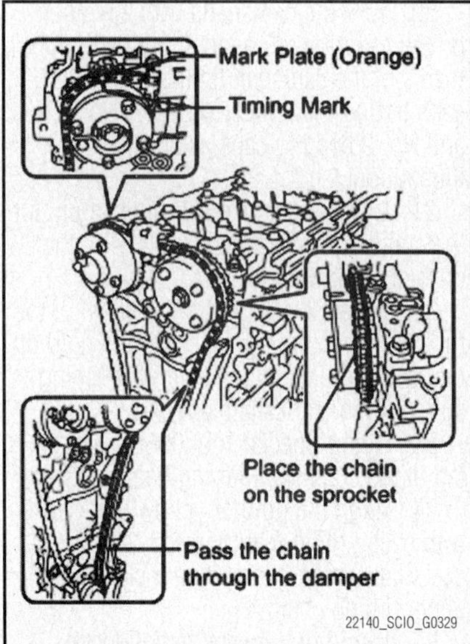

Fig. 200 Align the mark plate (orange) with the timing mark

c. Align the yellow mark links with the timing marks of each gear.

d. Install the gears onto the crankshaft and oil pump shaft with the chain on the gears.

e. Temporarily tighten the oil pump drive shaft gear with the nut.

f. Insert the damper spring into the adjusting hole, and then install the chain tensioner plate with the bolt. Torque: 7 ft. lbs. (10 Nm).

g. Align the adjusting hole of the oil pump drive shaft gear with the groove of the oil pump.

h. Insert a 4 mm diameter bar into the adjusting hole of the oil pump drive shaft gear to lock the gear in position, and then tighten the nut. Torque: 21 ft. lbs. (28 Nm).

12. Install crankshaft timing gear or sprocket.

13. Install chain vibration damper. Torque: 16 ft. lbs. (21 Nm).

14. Install chain sub-assembly.

a. Check the No. 1 cylinder TDC/compression.

b. Temporarily tighten the crankshaft pulley bolt.

c. Turn the crankshaft counterclockwise until the timing gear key is facing up.

d. Remove the crankshaft pulley bolt.

e. Check the timing mark on each camshaft timing gear.

f. Align the mark plate (orange) with the timing mark and install the chain.

➡Be sure to position the mark plate at the front of the engine. The mark plate on the camshaft side is colored orange.

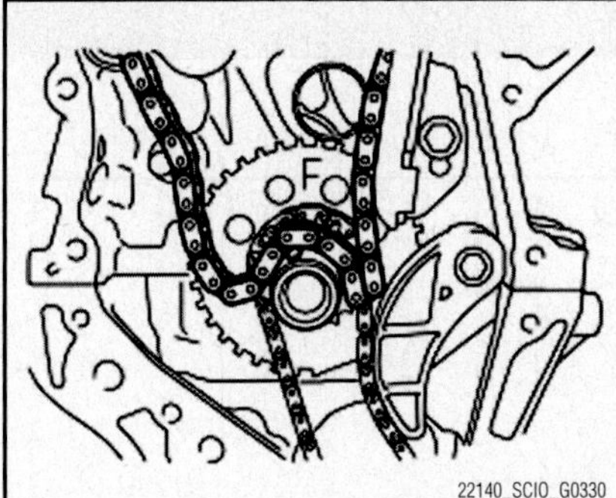

Fig. 201 Place the chain on the crankshaft without passing it around the shaft

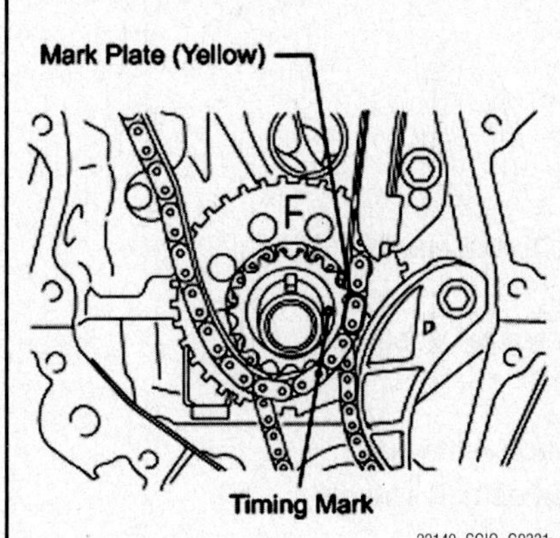

Fig. 202 Align the mark plate (yellow) and timing mark

g. Do not pass the chain around the sprocket of the camshaft timing gear assembly. Only place it on the sprocket.

h. Pass the chain through the No. 1 vibration damper.

i. Place the chain on the crankshaft without passing it around the shaft.

j. Hold the hexagonal portion of the camshaft with a wrench and turn the camshaft timing gear assembly counter-clockwise to align the mark plate (orange) and timing mark.

➡**Be sure to position the mark plate at the front of the engine. The mark plate on the camshaft side is colored orange.**

k. Hold the hexagonal portion of the camshaft with a wrench and turn the camshaft timing gear assembly clockwise.

l. When tensioning the chain, turn the camshaft timing gear assembly clockwise slowly to prevent the chain from being misaligned.

m. Align the mark plate (yellow) and timing mark and install the chain to the crankshaft timing gear. The mark plate on the crankshaft side is colored yellow.

n. Recheck each timing mark at TDC/compression.

15. Install chain tensioner slipper.
16. Install No. 2 chain vibration damper. Torque: 7 ft. lbs. (10 Nm).
17. Install alternator bracket. Torque: 16 ft. lbs. (21 Nm).
18. Install water inlet housing. Torque: 16 ft. lbs. (21 Nm).
19. Refer to Timing Chain Cover Removal and Installation.

VALVE LASH

ADJUSTMENT

tC and xB

See Figures 203 and 204.

1. Remove or disconnect the following:
   - Negative battery cable
   - Right-hand front wheel
   - Engine undercover
   - Engine appearance cover
   - Accessory drive belt
   - Power steering pump
   - Right-hand engine mount
   - Ignition coil
   - Cylinder head cover
2. Turn the crankshaft to set the No. 1 cylinder at TDC compression.

➡**Inspect the valve clearance when the engine is cold.**

3. Turn the crankshaft pulley to set the No. 1 cylinder at TDC.
4. Using a feeler gauge, measure the valve clearance shown in the illustration. Record the measurements.
5. Turn the crankshaft 1 complete revolution and set the No. 4 cylinder to TDC.
6. Using a feeler gauge, measure the valve clearance shown in the illustration. Record the measurements.
   a. Intake valve clearance is 0.0075–0.0114 inches (0.19–0.29 mm).
   b. Exhaust valve clearance is 0.0118–0.0158 inches (0.30–0.40 mm).

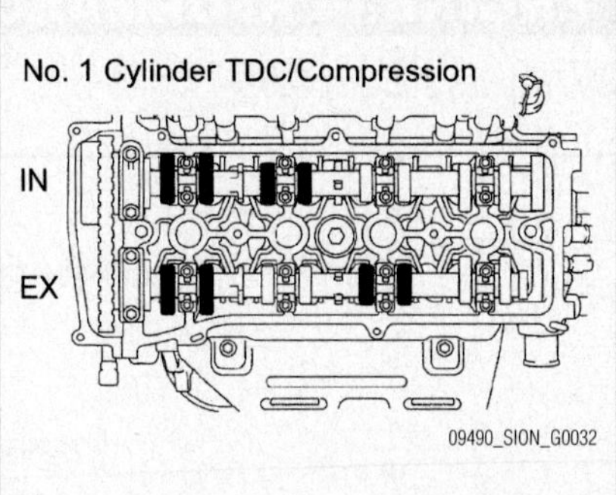

Fig. 203 Measure indicated valve lifters with No. 1 cylinder at TDC—2.4L Engine

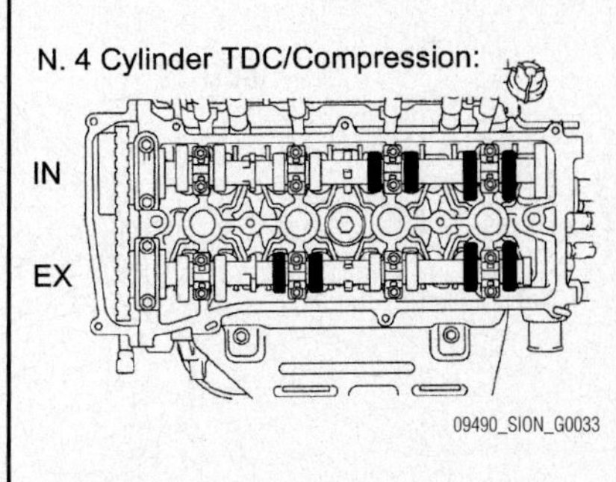

Fig. 204 Measure indicated valve lifters with No. 4 cylinder at TDC—2.4L Engine

7. To adjust the valve clearance:
8. Remove the camshafts.
9. Remove the valve lifters.
   a. Using a micrometer, measure the thickness of the removed lifter.
10. Calculate the thickness of a new lifter so that the valve clearance comes within the specified value.
    - Intake: New lifter thickness = Used lifter thickness + Recorded valve clearance measurement – 0.0095 in. (0.24 mm)
    - Exhaust: New lifter thickness = Used lifter thickness + Recorded valve clearance measurement – 0.0138 in. (0.35 mm)
11. Select a new lifter with the thickness as close to the calculated value as possible.
12. Reinstall the camshafts and check the valve clearance.
13. Install the cylinder head cover and remaining components.
14. Start the engine and check for leaks.

ENGINE PERFORMANCE & EMISSION CONTROL

See Figures 205 and 206.

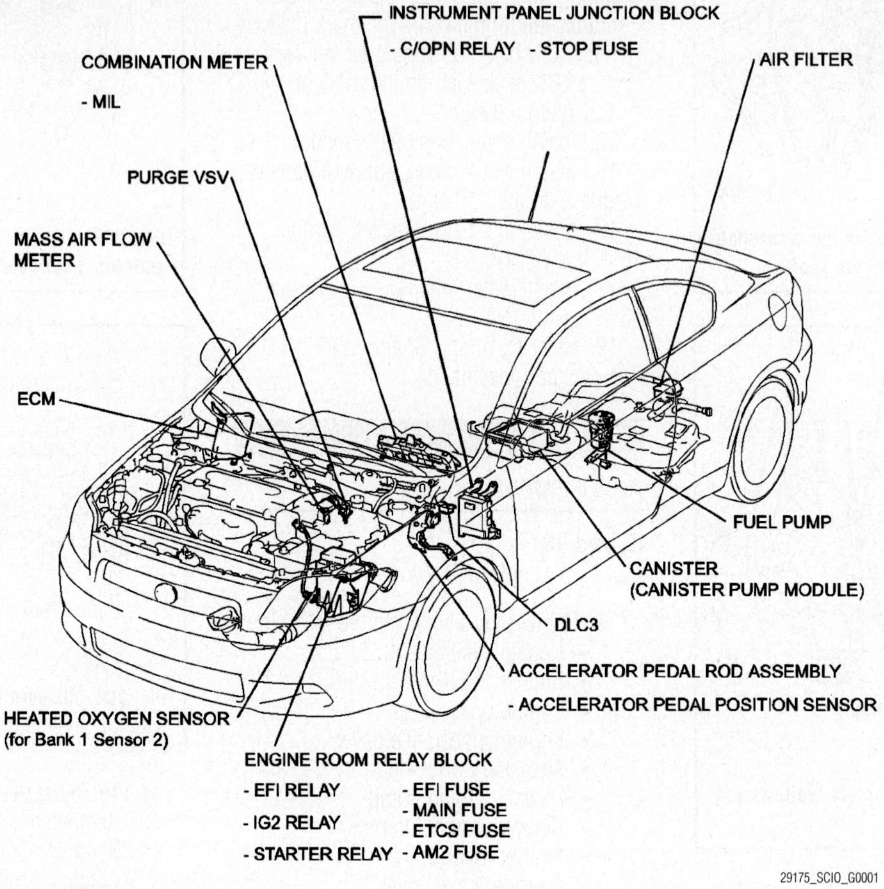

INSTRUMENT PANEL JUNCTION BLOCK
- C/OPN RELAY - STOP FUSE
COMBINATION METER
- MIL
AIR FILTER
PURGE VSV
MASS AIR FLOW METER
ECM
FUEL PUMP
CANISTER
(CANISTER PUMP MODULE)
DLC3
ACCELERATOR PEDAL ROD ASSEMBLY
- ACCELERATOR PEDAL POSITION SENSOR
HEATED OXYGEN SENSOR
(for Bank 1 Sensor 2)
ENGINE ROOM RELAY BLOCK
- EFI RELAY
- IG2 RELAY
- STARTER RELAY
- EFI FUSE
- MAIN FUSE
- ETCS FUSE
- AM2 FUSE
29175_SCIO_G0001

Fig. 205 Vehicle component locations–tC

IGNITION COIL
CAMSHAFT TIMING OIL CONTROL VALVE
FUEL INJECTOR
THROTTLE BODY
CRANKSHAFT POSITION SENSOR
CAMSHAFT POSITION SENSOR
ENGINE COOLANT TEMPERATURE SENSOR
AIR FUEL RATIO SENSOR
(for Bank 1 Sensor 1)
KNOCK SENSOR
PARK / NEUTRAL POSITION SWITCH (for A/T)
29175_SCIO_G0002

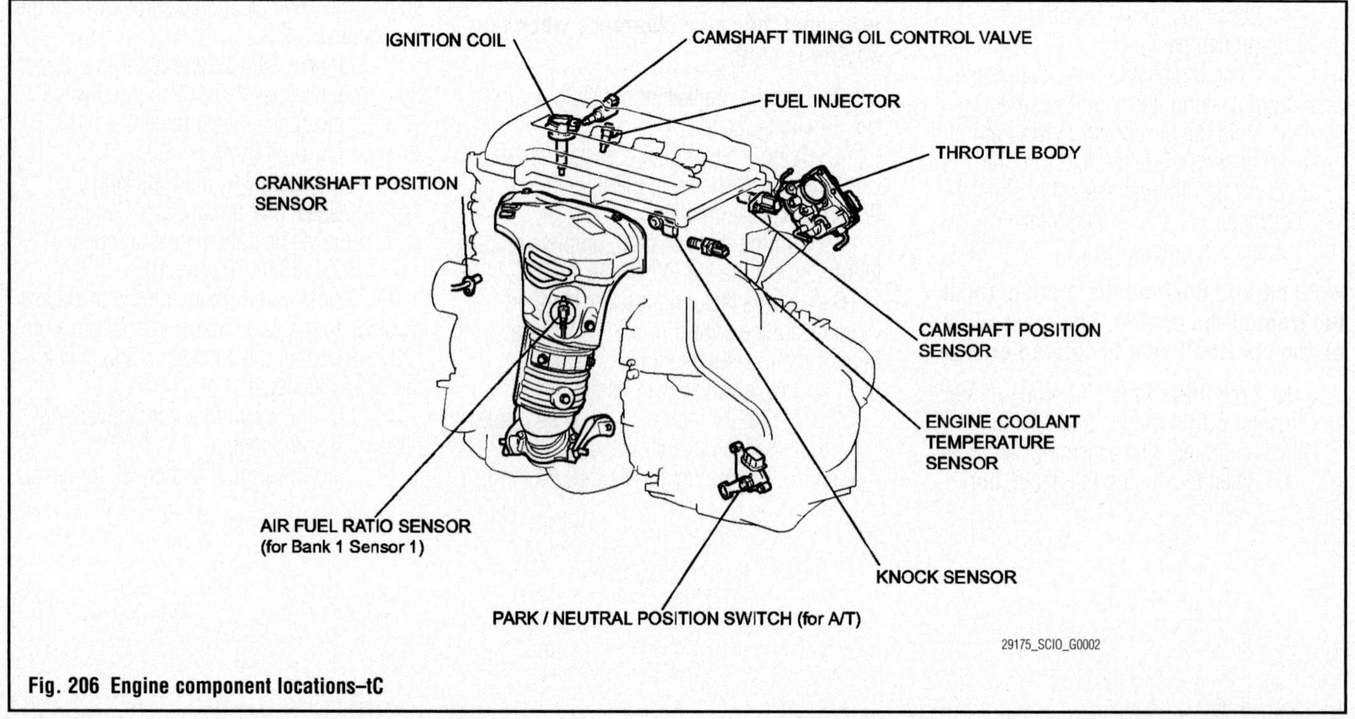

Fig. 206 Engine component locations–tC

CAMSHAFT POSITION (CMP) SENSOR

LOCATION

1.8L Engine

See Figure 207.

REMOVAL & INSTALLATION

1.8L Engine

1. Remove engine cover.
2. Remove camshaft position sensor.
    a. Disconnect the duty vacuum switching valve connector and 3 engine wire harness clamps (for exhaust camshaft side).
    b. Remove the bolt and remove the camshaft position sensor (for exhaust camshaft side).
    c. Remove the bolt and remove the camshaft position sensor (for intake camshaft side).
3. Installation is the reverse of removal.

CRANKSHAFT POSITION (CKP) SENSOR

LOCATION

1.8L Engine

See Figure 208.

REMOVAL & INSTALLATION

1.8L Engine

1. Remove engine under cover RH.

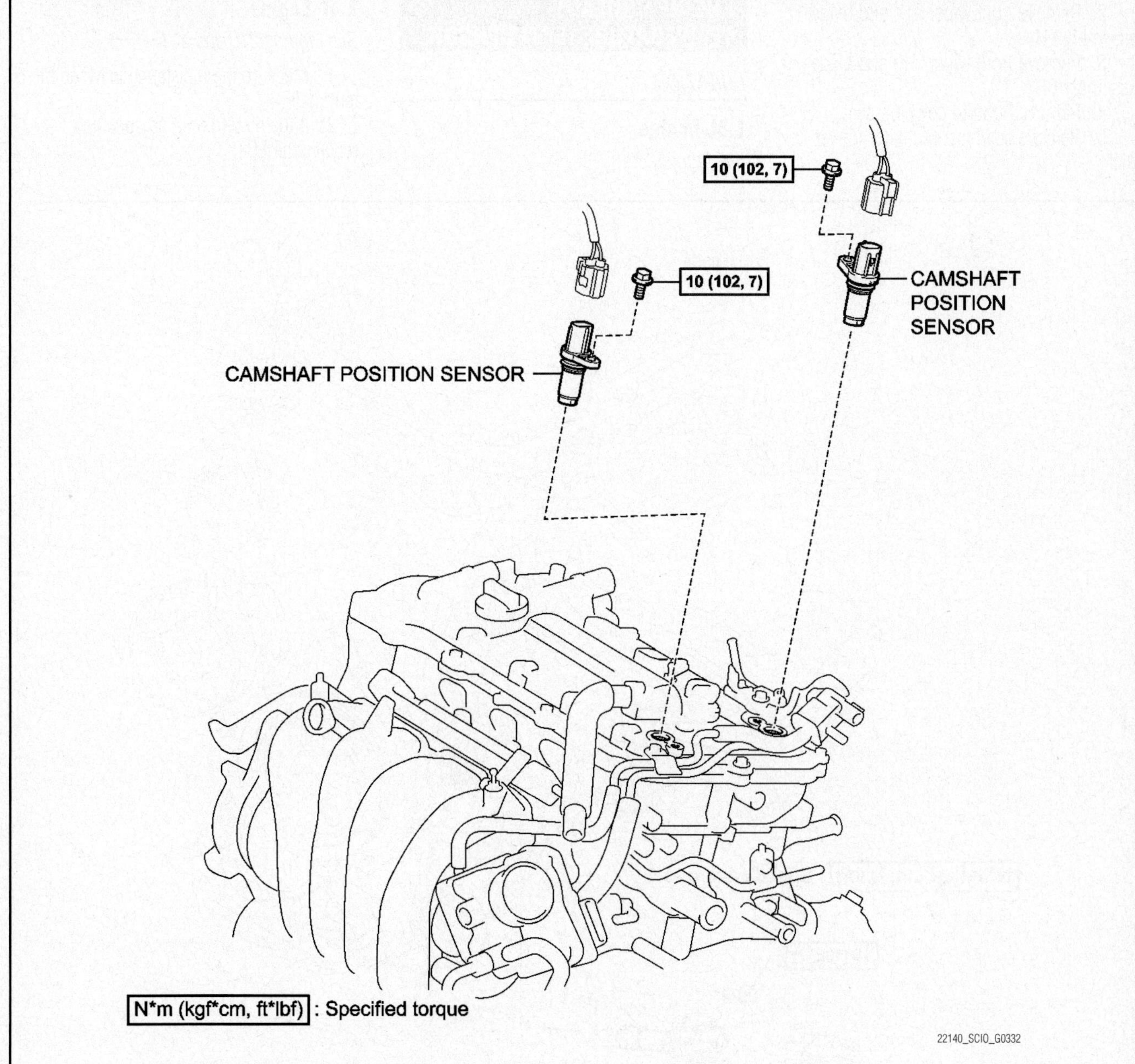

Fig. 207 Camshaft position sensor locations

2. Remove crankshaft position sensor.
 a. Disconnect the crankshaft position sensor connector.
 b. Remove the bolt and remove the crankshaft position sensor.
3. Installation is the reverse of removal.

ELECTRONIC CONTROL MODULE (ECM)

REMOVAL & INSTALLATION

1.8L Engine

1. Remove front wiper arm head cap.
2. Remove front wiper arm and blade assembly LH.
3. Remove front wiper arm and blade assembly RH.
4. Remove hood to cowl top seal.
5. Remove cowl top ventilator louver RH.
6. Remove cowl top ventilator louver LH.
7. Remove front wiper motor and link.
8. Remove front air shutter seal RH.
9. Remove outer cowl top panel.
10. Remove ECM.
 a. Remove the 2 lock knobs and harness clamp.
 b. Disconnect the 2 ECM connectors.
 c. Remove the bolt and 2 nuts and remove the ECM.
11. Installation is the reverse of removal.

ENGINE COOLANT TEMPERATURE (ECT) SENSOR

LOCATION

1.8L Engine

See Figure 209.

TESTING

1.8L Engine

1. Drain engine coolant.
2. Remove engine cover.
3. Remove engine coolant temperature sensor.
 a. Disconnect the engine coolant temperature sensor connector.
 b. Using SST, remove the engine coolant temperature sensor.
4. Installation is the reverse of removal.

2.4L Engine

See Figures 210 and 211.

1. Measure the resistance between the terminals.
2. If the result is not as specified, replace the ECT.

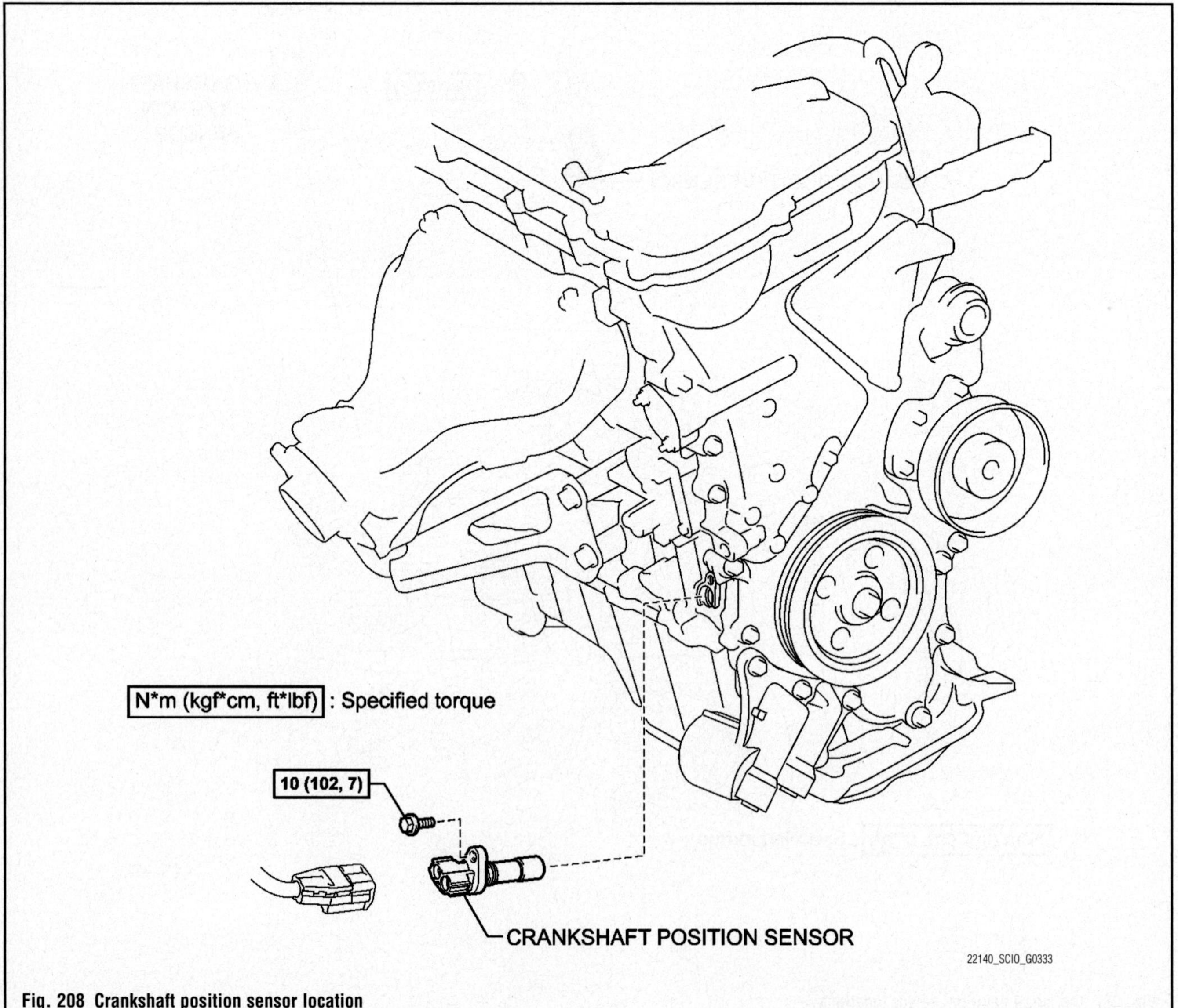

Fig. 208 Crankshaft position sensor location

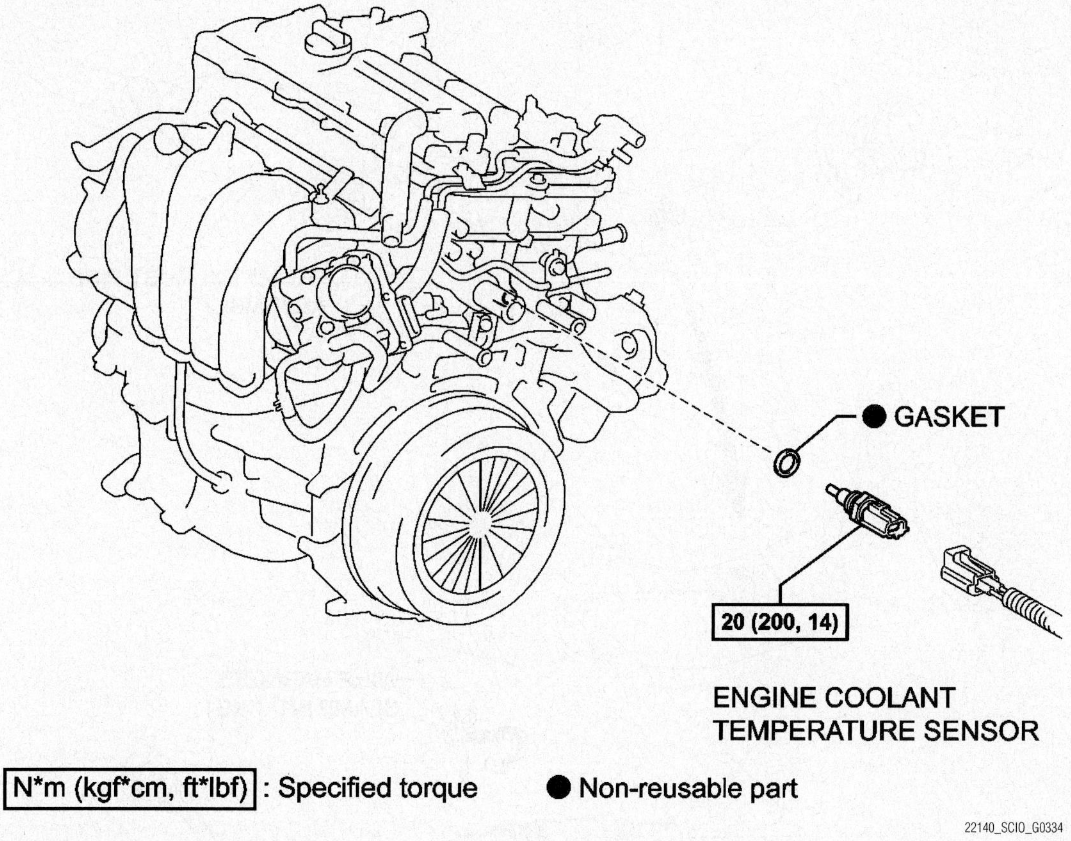

GASKET

20 (200, 14)

ENGINE COOLANT TEMPERATURE SENSOR

N*m (kgf*cm, ft*lbf) : Specified torque ● Non-reusable part

22140_SCIO_G0334

Fig. 209 Engine coolant temperature sensor location

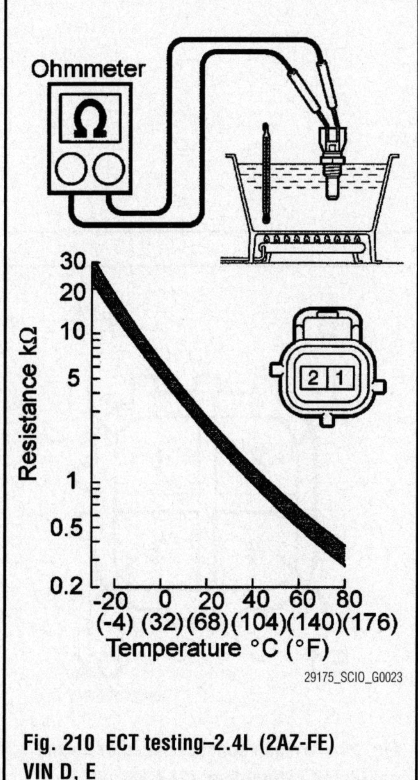

Fig. 210 ECT testing–2.4L (2AZ-FE) VIN D, E

Tester Connection	Specified Condition
1 (E2) - 2 (THW)	2.32 to 2.59 kΩ at 20°C (68°F)
1 (E2) - 2 (THW)	0.310 to 0.326 kΩ at 80°C (176°F)

29175_SCIO_G0024

Fig. 211 ECT resistance and temperature table–2.4L (2AZ-FE) VIN D, E

HEATED OXYGEN (HO2S) SENSOR

LOCATION

1.8L Engine

See Figure 212.

REMOVAL & INSTALLATION

1.8L Engine

1. Remove shift lever knob sub-assembly (for Manual Transaxle).
2. Remove upper console panel.
3. Remove rear console box cover.
4. Remove console box carpet.
5. Remove rear console box sub-assembly.
6. Remove front console box.
7. Remove heated oxygen sensor.
   a. Disconnect the heated oxygen sensor connector.
   b. Remove the grommet and pull the sensor connector out of the cabin through the floor panel.
   c. Remove the wire harness clamp bracket.
   d. Using SST, remove heated oxygen sensor.
8. Installation is the reverse of removal.

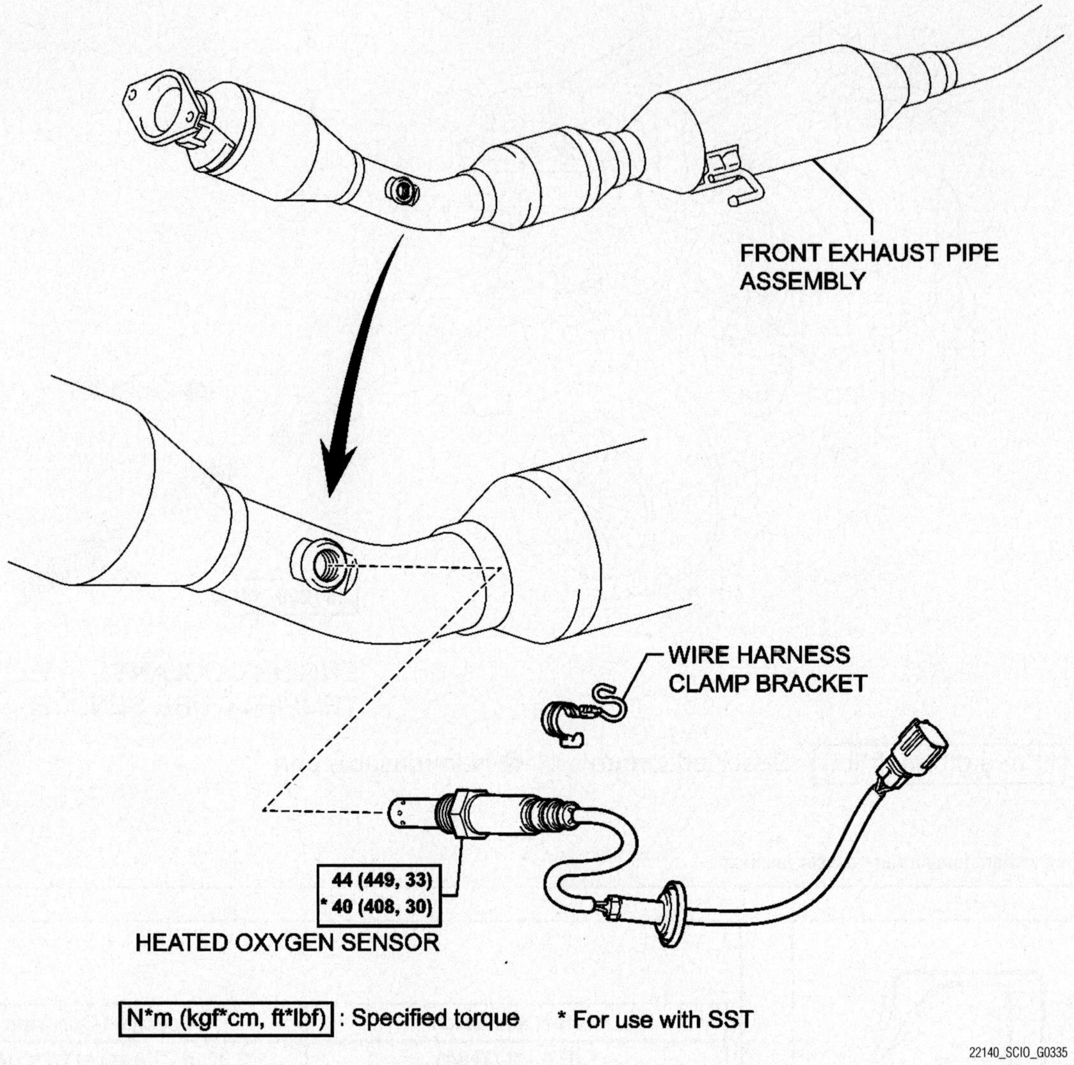

FRONT EXHAUST PIPE ASSEMBLY
WIRE HARNESS CLAMP BRACKET
44 (449, 33)
* 40 (408, 30)
HEATED OXYGEN SENSOR
N*m (kgf*cm, ft*lbf) : Specified torque
* For use with SST
22140_SCIO_G0335

Fig. 212 Heated Oxygen sensor location

TESTING

1.8L Engine

See Figure 213.

Ohmmeter
Ω
2 1
4 3
22140_SCIO_G0336

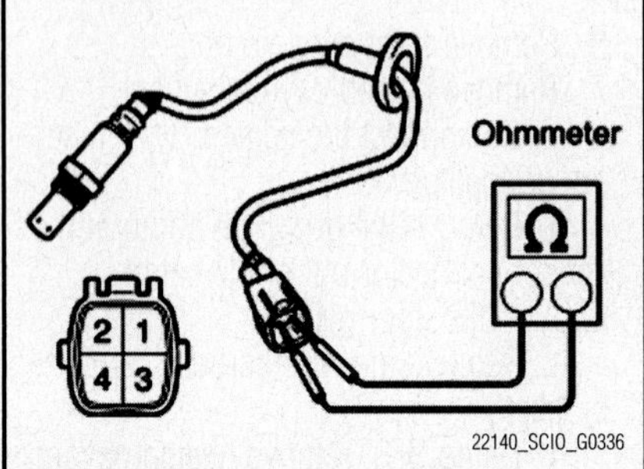

Fig. 213 Heated oxygen sensor connections for testing

1. Inspect heated oxygen sensor.
   a. Measure the resistance according to the value(s) below.
   b. Standard resistance:
   - 2 (HT) - 4 (+B) at 68°F (20°C) should read 11 to 16 Ω
   - 2 (HT) - 3 (E) always should read10 kohms or higher

   c. If the result is not as specified, replace the heated oxygen sensor.

2.4L Engine

See Figures 214 and 215.

1. Measure the resistance between terminals +B and HT. If the resistance is not as specified, replace the oxygen sensor.

+B
HT
2
1
4
3
E1
OX
29175_SCIO_G0025

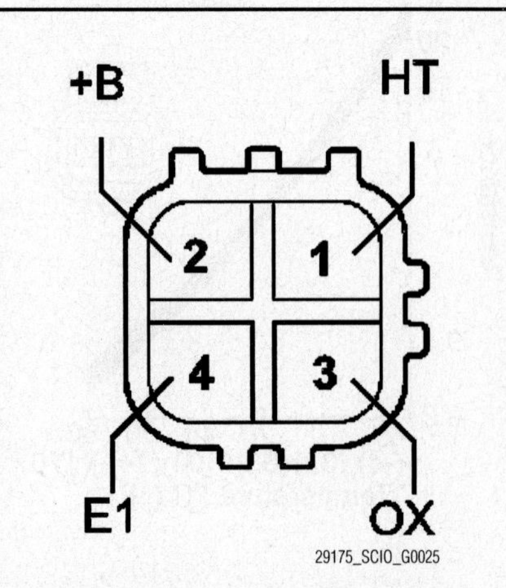

Fig. 214 Oxygen sensor connector terminal identification–2.4L (2AZ-FE) VIN D, E

20°C (68°F)	11 - 16 Ω
800°C (1,472°F)	23 - 32 Ω

29175_SCIO_G0026

Fig. 215 Oxygen sensor temperature/ resistance table–2.4L (2AZ-FE) VIN D, E

KNOCK SENSOR (KS)

LOCATION

1.8L Engine

See Figure 216.

REMOVAL & INSTALLATION

1.8L Engine

1. Remove intake manifold.
2. Remove knock sensor.
   a. Disconnect the knock sensor connector.
   b. Remove the bolt and remove the knock sensor.
3. Installation is the reverse of removal.

TESTING

1.8L and 2.4L Engine

1. Using an ohmmeter, measure the

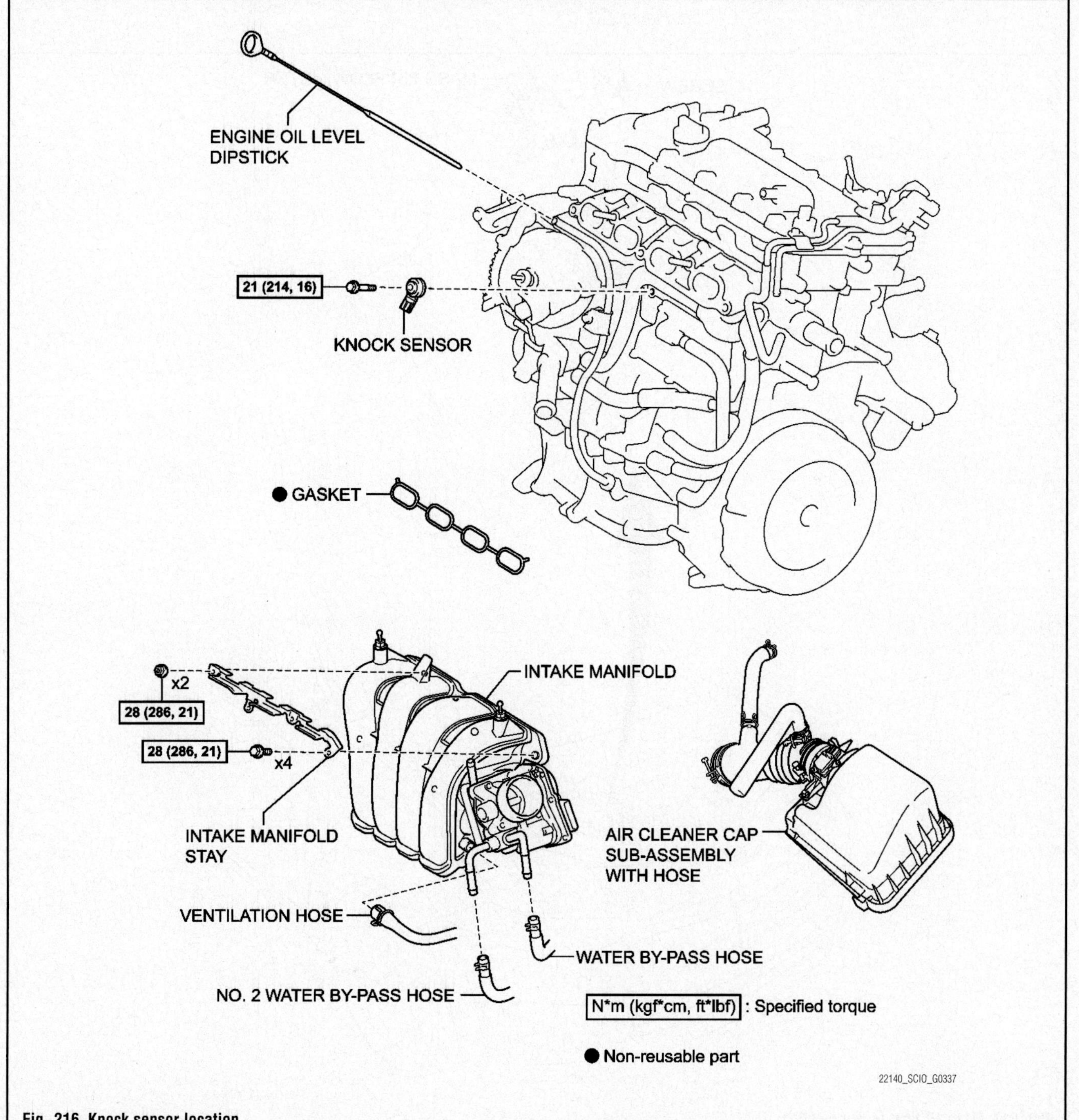

Fig. 216 Knock sensor location

resistance between terminals. Resistance should be 120–280 kohms.

2. If the resistance is not as specified, replace the sensor.

MASS AIR FLOW (MAF) SENSOR

LOCATION

1.8L Engine

See Figure 217.

REMOVAL & INSTALLATION

1.8L Engine

1. Remove mass air flow meter.
   a. Disconnect the mass air flow meter connector.
   b. Remove the 2 screws and the mass air flow meter.
2. Installation is the reverse of removal.

TESTING

2.4L Engine

See Figure 218.

1. Apply battery voltage across terminals 1 (+B) and 2 (E2G).
2. Using a voltmeter, connect the positive (+) tester probe to terminal VG, and negative (-) tester probe to terminal E2G.
3. Blow air into the MAF meter, and check that the voltage fluctuates.
4. Measure the resistance between terminals 4 (THA) and 5 (E2).
5. If the result is not as specified, replace the MAF meter.

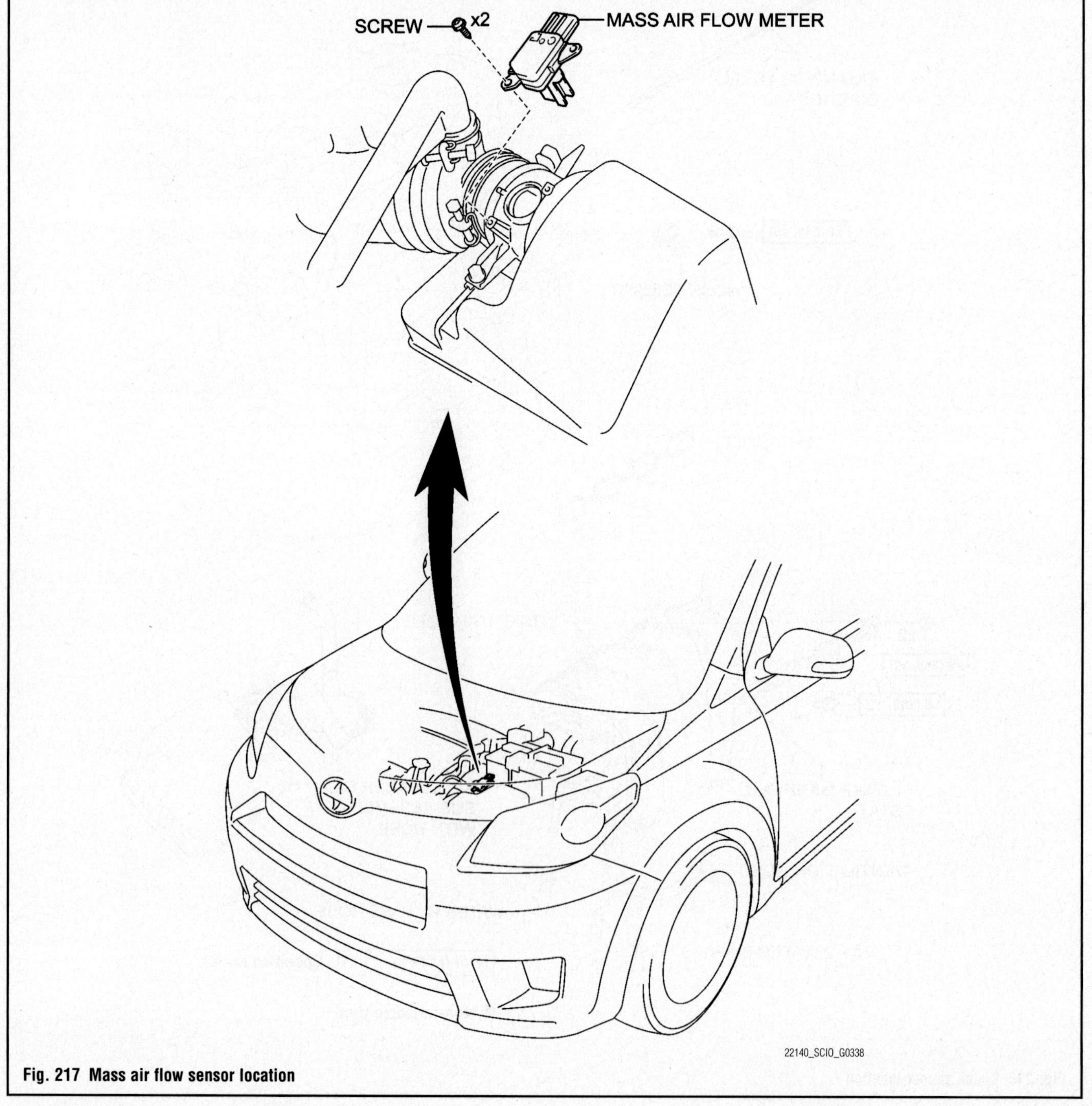

Fig. 217 Mass air flow sensor location

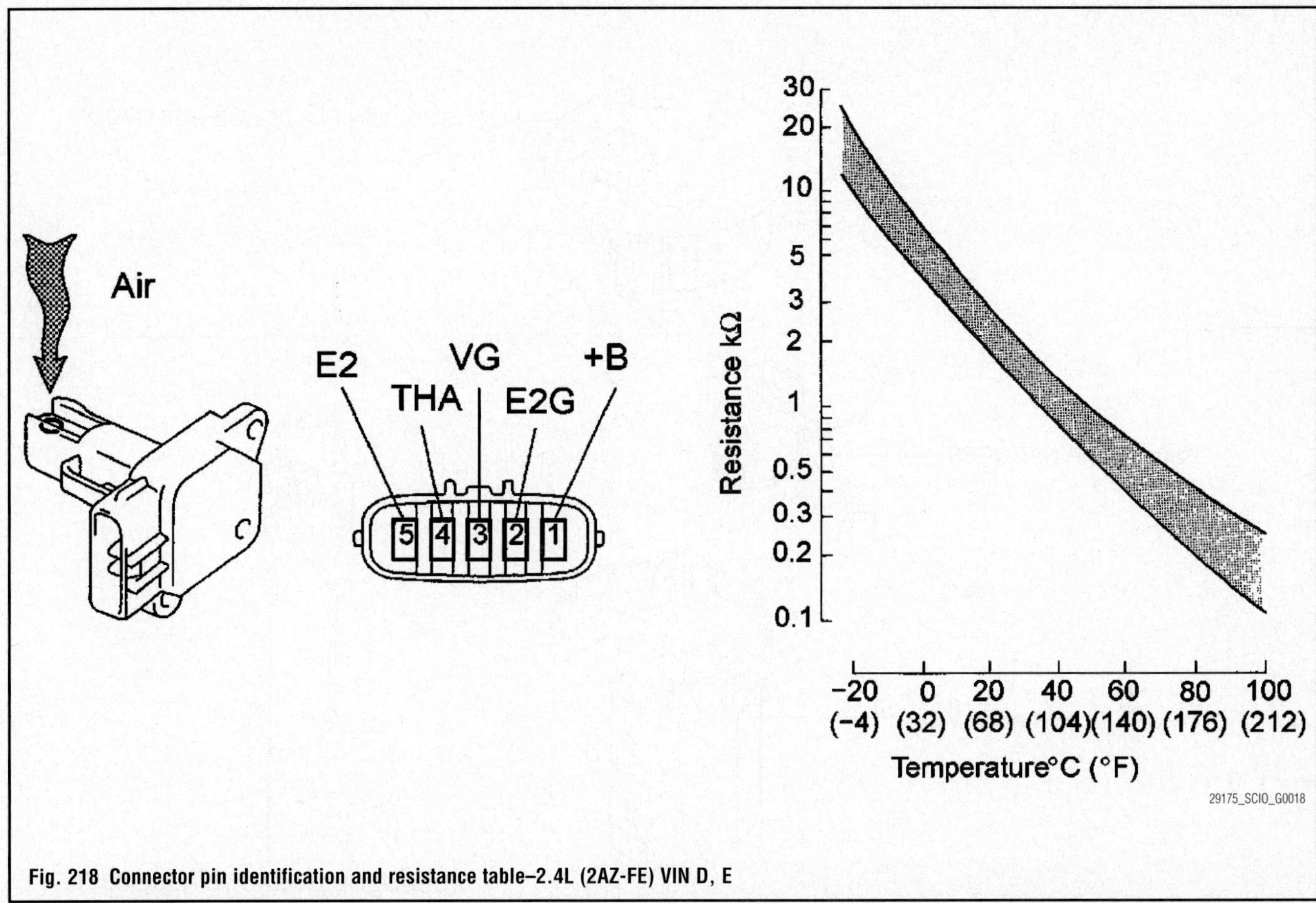

Fig. 218 Connector pin identification and resistance table–2.4L (2AZ-FE) VIN D, E

FUEL

GASOLINE FUEL INJECTION SYSTEM

FUEL SYSTEM SERVICE PRECAUTIONS

Safety is the most important factor when performing not only fuel system maintenance but any type of maintenance. Failure to conduct maintenance and repairs in a safe manner may result in serious personal injury or death. Maintenance and testing of the vehicle's fuel system components can be accomplished safely and effectively by adhering to the following rules and guidelines.

- To avoid the possibility of fire and personal injury, always disconnect the negative battery cable unless the repair or test procedure requires that battery voltage be applied.
- Always relieve the fuel system pressure prior to disconnecting any fuel system component (injector, fuel rail, pressure regulator, etc.), fitting or fuel line connection. Exercise extreme caution whenever relieving fuel system pressure to avoid exposing skin, face and eyes to fuel spray. Please be advised that fuel under pressure may penetrate the skin or any part of the body that it contacts.
- Always place a shop towel or cloth around the fitting or connection prior to loosening to absorb any excess fuel due to spillage. Ensure that all fuel spillage (should it occur) is quickly removed from engine surfaces. Ensure that all fuel soaked cloths or towels are deposited into a suitable waste container.
- Always keep a dry chemical (Class B) fire extinguisher near the work area.
- Do not allow fuel spray or fuel vapors to come into contact with a spark or open flame.
- Always use a back-up wrench when loosening and tightening fuel line connection fittings. This will prevent unnecessary stress and torsion to fuel line piping.
- Always replace worn fuel fitting O-rings with new Do not substitute fuel hose or equivalent where fuel pipe is installed.

Before servicing the vehicle, make sure to also refer to the precautions in the beginning of this section as well.

RELIEVING FUEL SYSTEM PRESSURE

1. Before servicing the vehicle, refer to the Precautions Section.
2. Remove or disconnect the following:
   - Rear seat cushion
   - Access panel
   - Fuel pump module connector
3. Start the engine and allow it to run until it stalls.
4. Turn the ignition switch to the **OFF** position.
5. Disconnect the negative battery cable.
6. Attach the fuel pump connector.

FUEL FILTER

REMOVAL & INSTALLATION

The fuel filter is in the tank as part of the fuel pump assembly.

FUEL PUMP

REMOVAL & INSTALLATION

See Figure 219.

- Negative battery cable
- Rear seat cushion assembly
- Rear seat cushion support bracket, xB only

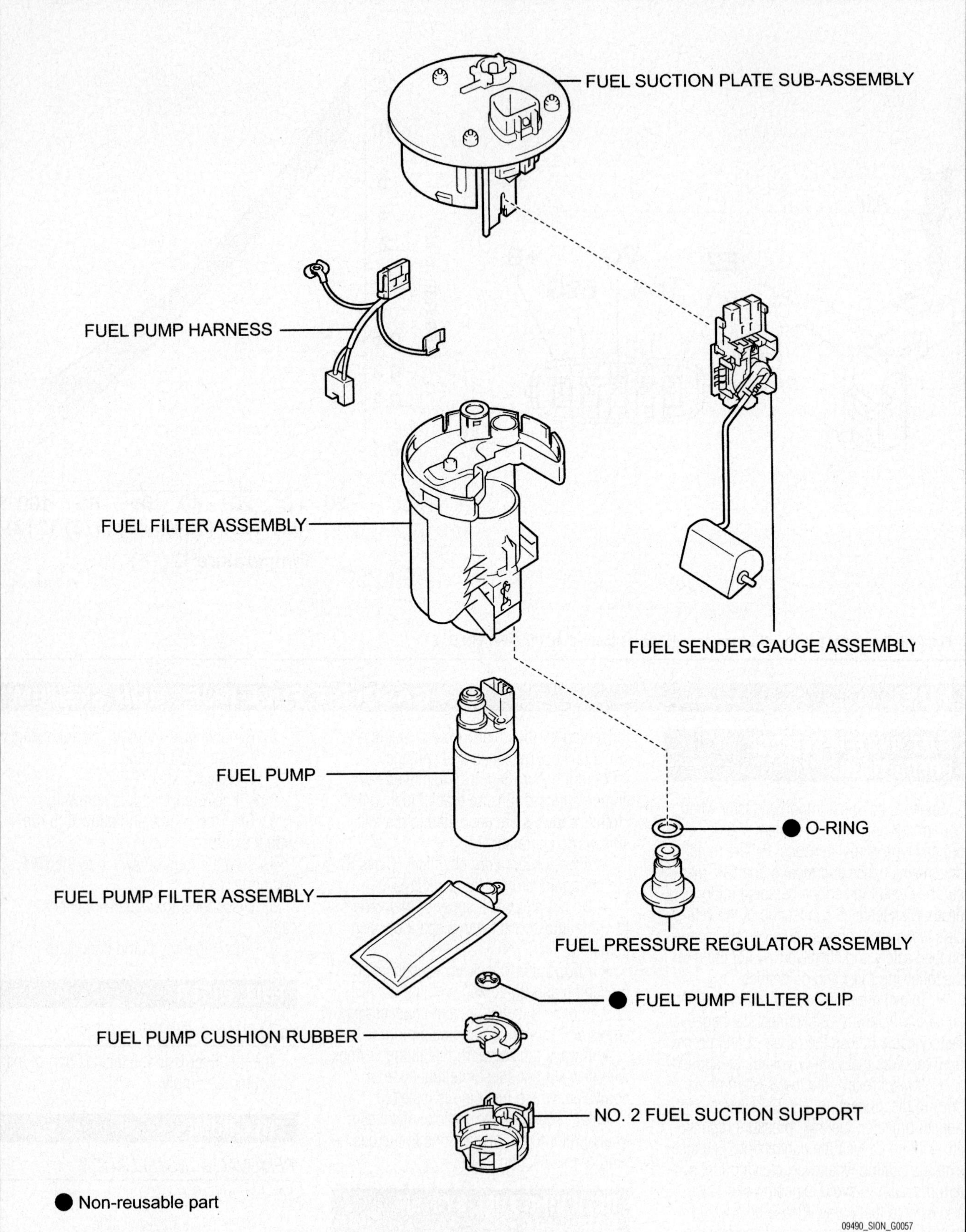

Fig. 219 Exploded view of the fuel pump assembly

- Rear floor service hole cover
- Fuel pump connector
- Fuel supply and vent hoses
- Fuel tank vent tube set plate
- Fuel pump assembly

To install:

1. Install or connect the following:
    - Fuel pump assembly with a new gasket.
    - Fuel tank vent tube set plate. Tighten bolts to 31 inch lbs. (4 Nm) for and xB models. 53 inch lbs. (6 Nm) for tC models.
    - Fuel supply and vent hoses
    - Fuel pump connector
    - Rear floor service hole cover
    - Rear seat cushion support bracket
    - Rear seat cushion
    - Negative battery cable
2. Start the engine and check for leaks.

FUEL RAIL & INJECTORS

REMOVAL & INSTALLATION

tC and xB

1. Before servicing the vehicle, refer to the Precautions Section.
2. Relieve the fuel system pressure.
3. Remove or disconnect the following:
    - Negative battery cable
    - Engine appearance cover
    - Air intake assembly
    - Fuel supply hose
    - Ventilation hose
    - Fuel injector connectors
    - Fuel rail with the injectors attached
    - Fuel injector from the fuel rail

To install:

4. Install or connect the following:
    - Injectors to the fuel rail using new O-rings
    - Fuel rail with injectors attached and torque the bolts to 15 ft. lbs. (20 Nm).
    - Fuel injector connectors
    - Ventilation hose
    - Fuel supply hose
    - Air intake assembly
    - Engine appearance cover
    - Negative battery cable
5. Start the engine and check for leaks.

xD

See Figures 220 through 229.

1. Discharge the fuel system pressure.
2. Remove the Engine cover.
3. Remove the ventilation hose.

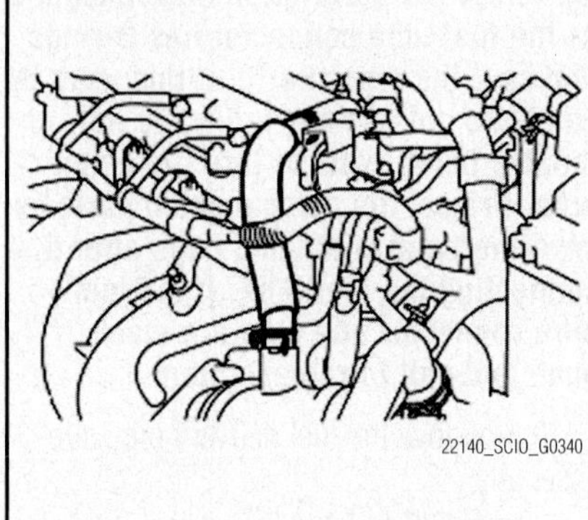

Fig. 220 Remove the ventilation hose

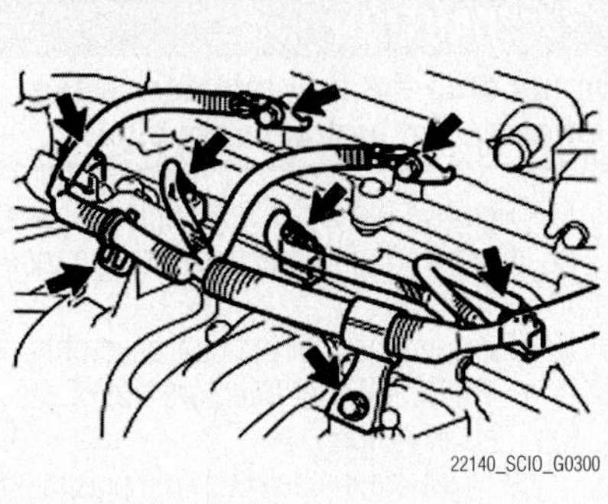

Fig. 221 Separate the engine wire harness

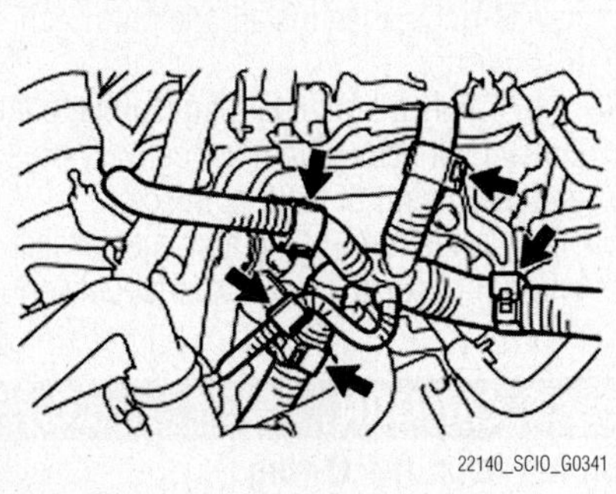

Fig. 222 Disconnect the 5 wire harness clamps

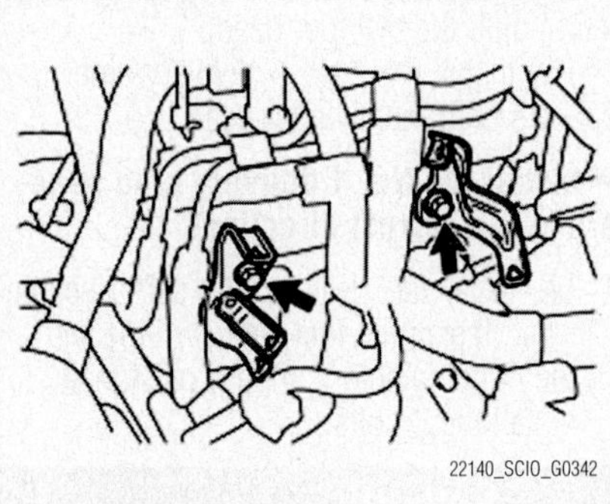

Fig. 223 Remove the 2 bolts and remove the 2 harness brackets

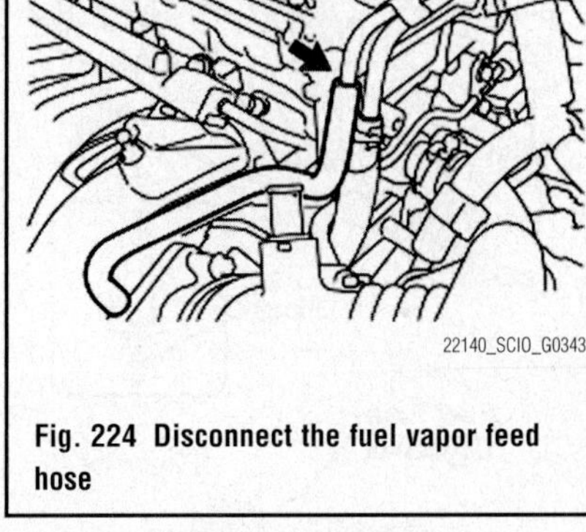

Fig. 224 Disconnect the fuel vapor feed hose

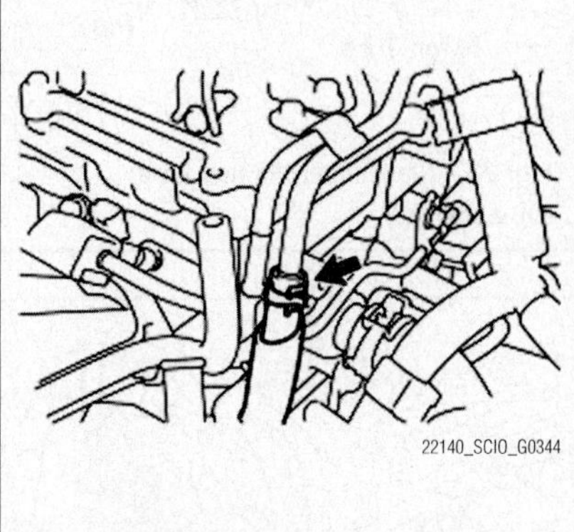

Fig. 225 Disconnect the vacuum hose

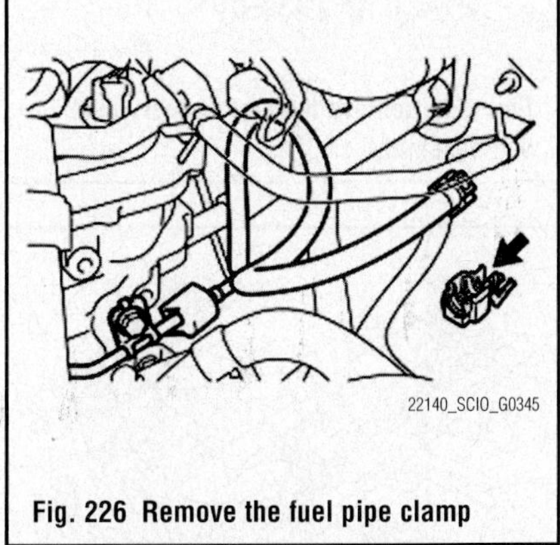

Fig. 226 Remove the fuel pipe clamp

4. Separate the engine wire harness.
    a. Remove the 3 bolts, harness clamp and 4 fuel injector connectors and separate the engine wire harness.
5. Remove the harness bracket.
    a. Disconnect the 5 wire harness clamps.
    b. Remove the 2 bolts and remove the 2 harness brackets.
6. Disconnect the fuel vapor feed hose.
7. Disconnect the vacuum hose.
8. Remove the fuel pipe clamp.
9. Disconnect the fuel tube sub-assembly.

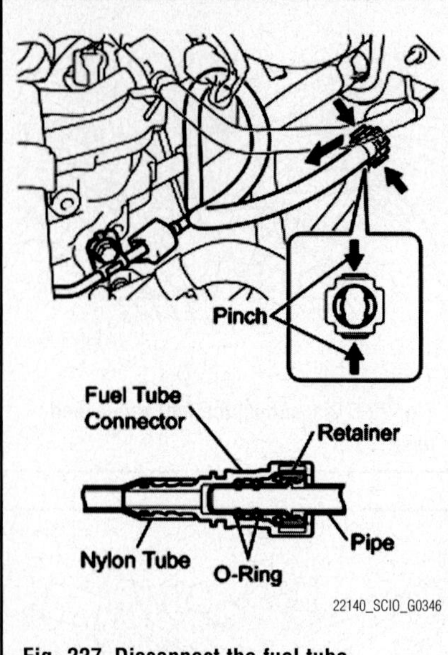

Fig. 227 Disconnect the fuel tube sub-assembly

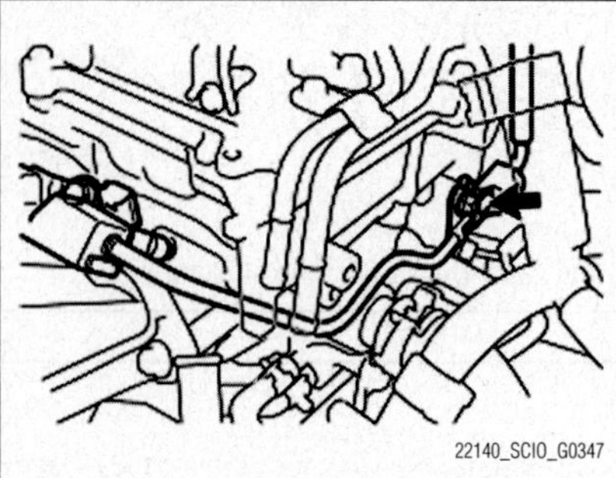

Fig. 228 Remove the fuel delivery pipe sub-assembly

Fig. 229 Remove the 2 bolts and remove the fuel delivery pipe with the 4 fuel injectors

a. Pinch the retainer of the fuel tube connector, then pull out the fuel tube connector to disconnect the fuel tube from the fuel pipe.

b. Remove any dirt and foreign matter from the fuel tube connector before performing this work.

➡Do not allow any scratches or foreign matter on the parts when disconnecting as the fuel tube connector has O-rings that seal the pipe. Perform this work by hand. Do not use any tools. Do not forcibly bend, twist or turn the nylon tube. Protect the disconnected parts by covering them with vinyl bags after disconnecting the fuel tube. If the fuel tube connector and pipe are stuck, push and pull to release them.

10. Remove the fuel delivery pipe sub-assembly.
    a. Remove the bolt.
    b. Remove the 2 bolts and remove the fuel delivery pipe with the 4 fuel injectors.

CAUTION

Do not drop the fuel injectors when removing the fuel delivery pipe.

11. Remove the 2 delivery pipe spacers.
12. Remove the 4 injector vibration insulator.
13. Remove the fuel injector assembly.
    a. Pull the 4 fuel injectors out of the fuel delivery pipe.
    b. Place the fuel injector in a plastic bag to prevent foreign matter from entering.

To install:

14. Install fuel injector assembly.
    a. Apply a light coat of gasoline to new O-rings, then install one onto each fuel injector.
    b. Apply a light coat of gasoline to the contact surfaces of the fuel delivery pipe and the O-ring of the fuel injector.
    c. While turning the fuel injector left and right, install it onto the fuel delivery pipe.

CAUTION

Do not twist the O-ring.

    d. After installing the fuel injectors, check that they turn smoothly. If they do not, replace the O-ring with a new one.
15. Install 4 new injector vibration insulators onto the cylinder head.
16. Install the 2 No. 1 delivery pipe spacers onto the cylinder head.

➡Install the No. 1 delivery pipe spacers in the correct direction.

17. Install fuel delivery pipe sub-assembly.
    a. Install the fuel delivery pipe with the 4 fuel injectors, then provisionally install the 3 bolts.

CAUTION

Do not drop the fuel injectors when installing the fuel delivery pipe.

    b. Check that the fuel injectors rotate smoothly after installing the fuel delivery pipe.
    c. Tighten the 3 bolts to the specified torque. Torque: 16 ft. lbs. (21 Nm).
18. Connect fuel tube sub-assembly.
    a. Insert the fuel tube connector into the fuel pipe until a click sound can be heard.
    b. Check that there are no scratches or foreign matter around the disconnected parts of the fuel tube connector and pipe before performing this work.
    c. After connecting the fuel tube, check that the fuel tube connector and pipe are securely connected by pulling them.
19. Install the fuel pipe clamp.
20. Connect the vacuum hose.
21. Connect the fuel vapor feed hose.
22. Install harness bracket.
    a. Install the 2 harness brackets with the 2 bolts. Torque: 9 ft. lbs. (13 Nm).
    b. Connect the 5 wire harness clamps.
23. Install engine wire harness.
    a. Install the engine wire harness with 3 bolts and harness clamp.
    b. Connect the 4 fuel injector connectors.
24. Install the ventilation hose.
25. Inspect for fuel leaks.
26. Install engine cover.

FUEL TANK

REMOVAL & INSTALLATION

xD

See Figure 230.

1. Remove the rear seat.
2. Remove the rear floor service hole cover.
3. Discharge the fuel system pressure.
4. Disconnect the fuel tank main tube sub-assembly.
5. Disconnect the fuel tank vent hose.
6. Remove the fuel pump gauge retainer.
7. Remove fuel suction with pump and gauge tube assembly.
8. Drain the fuel.
9. Remove the front floor heat insulator
10. Remove the fuel tank protector.
11. Disconnect the fuel tank main tube sub-assembly.
    a. Release the lock then pull and remove the fuel tank main tube.
    b. Remove any dirt and foreign matter from the clip before performing this work.
    c. Avoid any scratches or foreign matter on the parts when disconnecting them, as the quick connector has an O-ring that seals the plug.

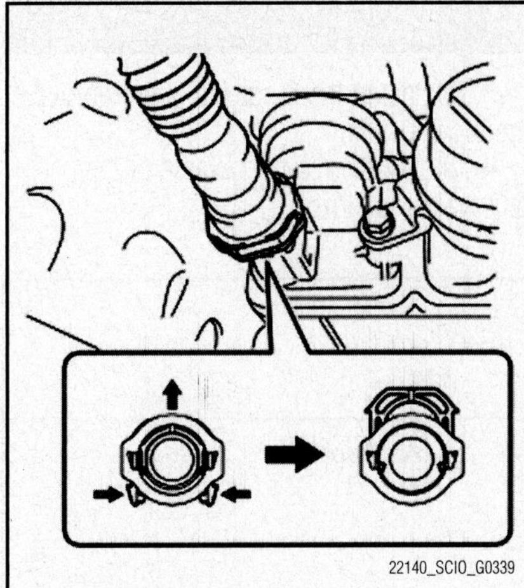

Fig. 230 Disconnect the fuel tank vent hose.

➡Perform this work by hand. Do not use any tools. Do not forcibly bend, twist or turn the nylon tube. Protect the disconnected parts by covering them with a plastic bag. If the connector and pipe are stuck, disconnect the nylon tube by turning it by hand to release them.

12. Disconnect the fuel tank vent hose.
  a. Disconnect the fuel tank vent hose from the charcoal canister assembly.
13. Disconnect the fuel tank breather hose.
  a. Release the lock then pull and remove the fuel tank breather hose.
  b. Remove any dirt and foreign matter from the clip before performing this work.
  c. Avoid any scratches or foreign matter on the parts when disconnecting them, as the quick connector has an O-ring that seals the plug.

➡Perform this work by hand. Do not use any tools. Do not forcibly bend, twist or turn the nylon tube. Protect the disconnected parts by covering them with a plastic bag. If the connector and pipe are stuck, disconnect the nylon tube by turning it by hand to release them.

14. Disconnect the lower fuel tank filler pipe sub-assembly.
15. Remove the fuel tank assembly.
  a. Remove the 4 bolts and remove the fuel tank.
  b. Remove the fuel tank main tube from the fuel tank.
  c. Remove the fuel tank vent hose from the fuel tank.

To install:

16. Install fuel tank assembly.
  a. Install the fuel vent hose onto the fuel tank.
  b. Install the fuel tank main tube onto the fuel tank.
  c. Clean the bolt hole and remove any grease.
  d. Install the fuel tank with 4 new bolts. Torque: 10 ft. lbs. (14 Nm).
17. Connect lower fuel tank filler pipe sub-assembly.
18. Connect fuel tank breather tube.
19. Connect fuel tank vent hose.
  a. Connect the fuel tank vent hose to the charcoal canister assembly.
20. Connect fuel tank main tube sub-assembly.
  a. Align the fuel tube connector with the pipe, push the fuel tube connector in until the retainer makes a click sound, then lock the cover of the connector.
21. Install fuel tank protector.
22. Install front floor heat insulator.
23. Install fuel suction with pump and gauge tube assembly.
24. Install fuel pump gauge retainer.
25. Connect fuel tank vent hose sub-assembly.
26. Connect fuel tank main tube sub-assembly.
27. Connect rear floor service hole cover.
28. Install rear seat cushion with cover pad sub-assembly LH.
29. Install rear seat.
30. Inspect for fuel leaks.

IDLE SPEED

ADJUSTMENT

Idle speed is maintained by the Powertrain Control Module (PCM). No adjustment is necessary or possible.

THROTTLE BODY

REMOVAL & INSTALLATION

xD

See Figures 231 and 232.

1. Drain the engine coolant.
2. Remove the Engine cover.
3. Remove the air cleaner cap sub-assembly with hose.
  a. Disconnect the wire harness clamp and mass air flow meter connector.
  b. Disconnect the 2 ventilation hose.
  c. Loosen the hose clamp, unlock the 2 clamps and remove air cleaner cap sub-assembly with hose.

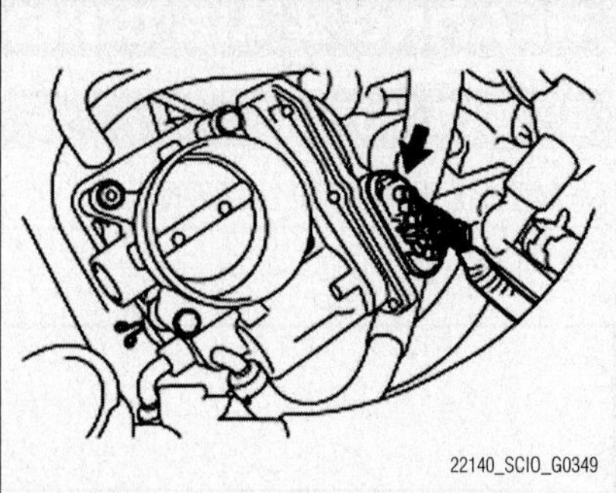

Fig. 231 Disconnect the throttle body connector

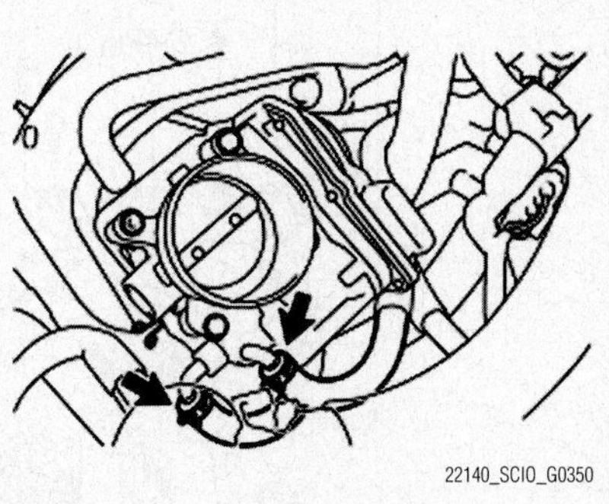

Fig. 232 Disconnect the water by-pass hoses

4. Remove the throttle body assembly.
  a. Disconnect the throttle body connector.
  b. Disconnect the water by-pass hoses.
  c. Remove the 2 bolts and 2 nuts and remove the throttle body assembly.
  d. Remove the gasket from the intake manifold.

To install:

5. Install throttle body assembly
  a. Install a new gasket onto the intake manifold.
  b. Install the throttle body assembly with the 2 bolts and 2 nuts. Torque: 7 ft. lbs. (10 Nm).
  c. Connect the water by-pass hoses.
  d. Connect the throttle body assembly connector.
6. Install air cleaner cap sub-assembly with hose
  a. Install air cleaner cap sub-assembly with hose and lock the 2 clamps.
  b. Tighten the hose clamp to the specified torque.
  c. Connect the ventilation hose.
  d. Connect the wire harness clamp and mass air flow meter connector.
7. Add engine coolant.
8. Inspect for coolant leak.
9. Install engine cover.

HEATING & AIR CONDITIONING SYSTEM

BLOWER MOTOR

REMOVAL & INSTALLATION

See Figure 233.

1. Disconnect the negative battery cable.
2. Remove any components necessary to access the blower motor.
3. Unfasten the 3 screws, then remove the blower motor.
4. Installation is the reverse of the removal procedure.

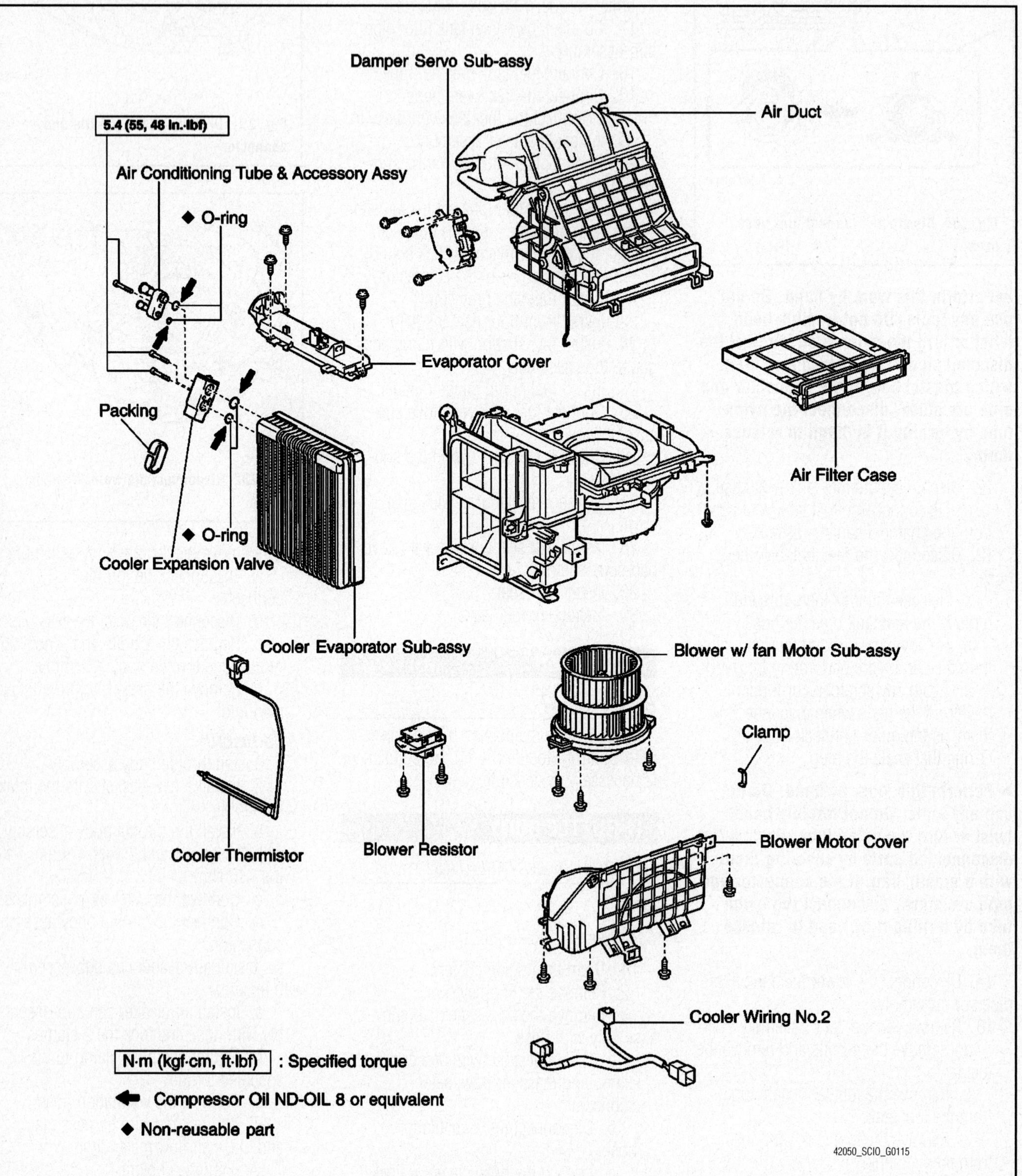

Fig. 233 Exploded view of the blower motor and related components—xB shown, others similar

HEATER CORE

REMOVAL & INSTALLATION

xB

See Figures 234 and 235.

1. Before servicing the vehicle, refer to the Precautions Section.
2. Discharge and recover the air conditioning system refrigerant.
3. Drain the cooling system.
4. Place front wheel facing straight ahead.
5. Remove or disconnect the following:
   - Negative battery cable

CAUTION

Wait 90 seconds after disconnecting the cable to allow the airbag to discharge.

- Heater hoses from the heater unit
- Heater control knobs
- Instrument cluster cover panel
- Instrument cluster
- Instrument panel speaker panel
- Front speaker
- Center instrument cluster trim panel
- Radio assembly
- A-pillar trim
- Instrument panel hole cover
- Passenger airbag connector
- Instrument panel assembly with passenger airbag attached
- Horn button assembly
- Steering wheel assembly
- Steering column cover
- Headlight dimmer switch
- Windshield wiper switch
- Front door scuff plates
- Cowl side trim boards
- Heater control assembly
- Lower center instrument cluster trim panel
- Lower center instrument panel trim panel
- Glove compartment door
- Floor console
- Floor parking brake cable assembly, if equipped with automatic transmission
- Steering column hole cover plate
- Intermediate steering shaft assembly
- Steering column assembly
- Antenna cord
- Lower instrument panel assembly with panel reinforcement attached
- Defroster nozzle assembly
- ECM
- Air conditioner blower assembly
- Defroster damper control cable
- Air-mix damper control cable
- Rear air ducts
- Heater unit assembly
- Air duct assembly from the heater unit

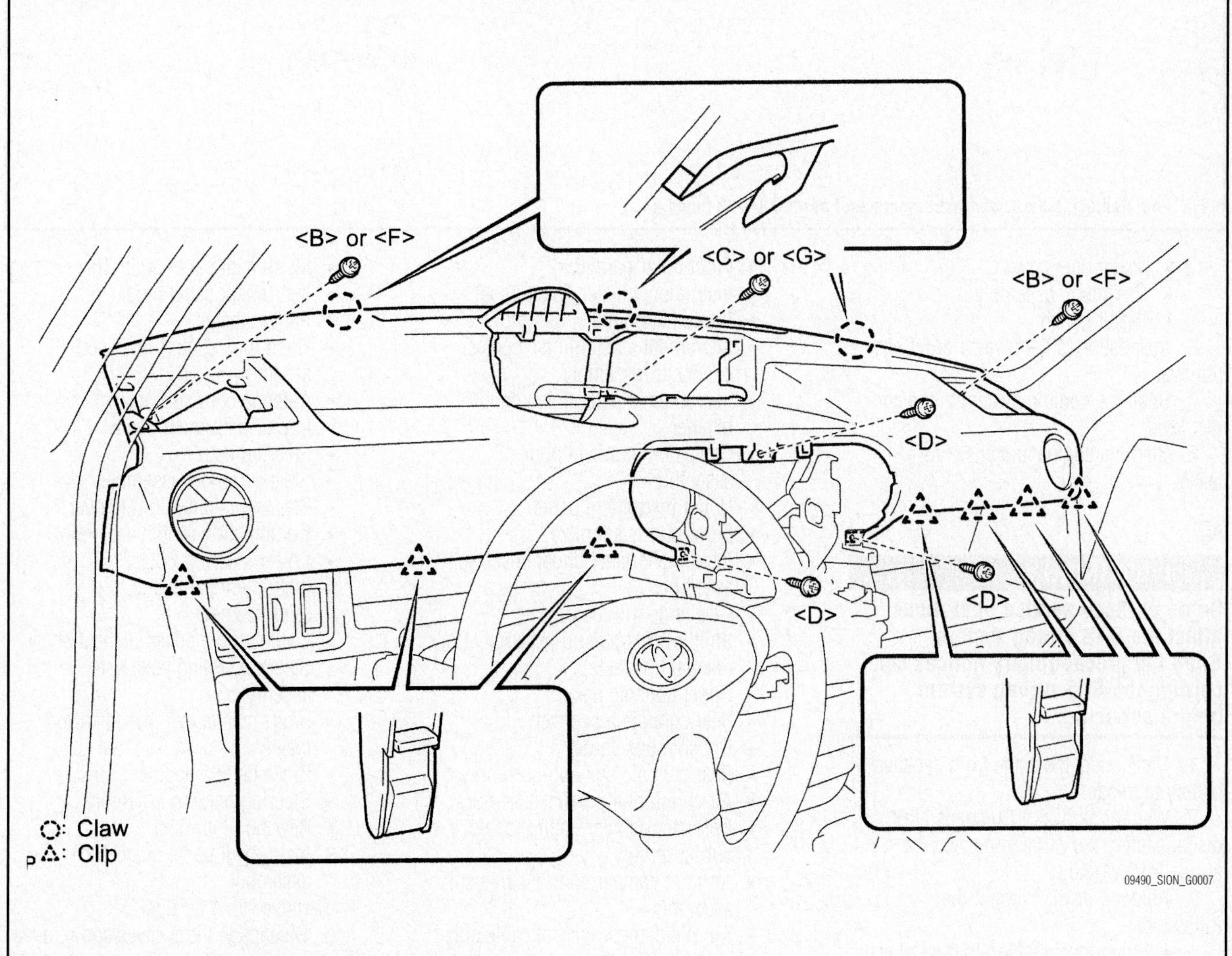

Fig. 234 Remove the instrument panel assembly—xB models

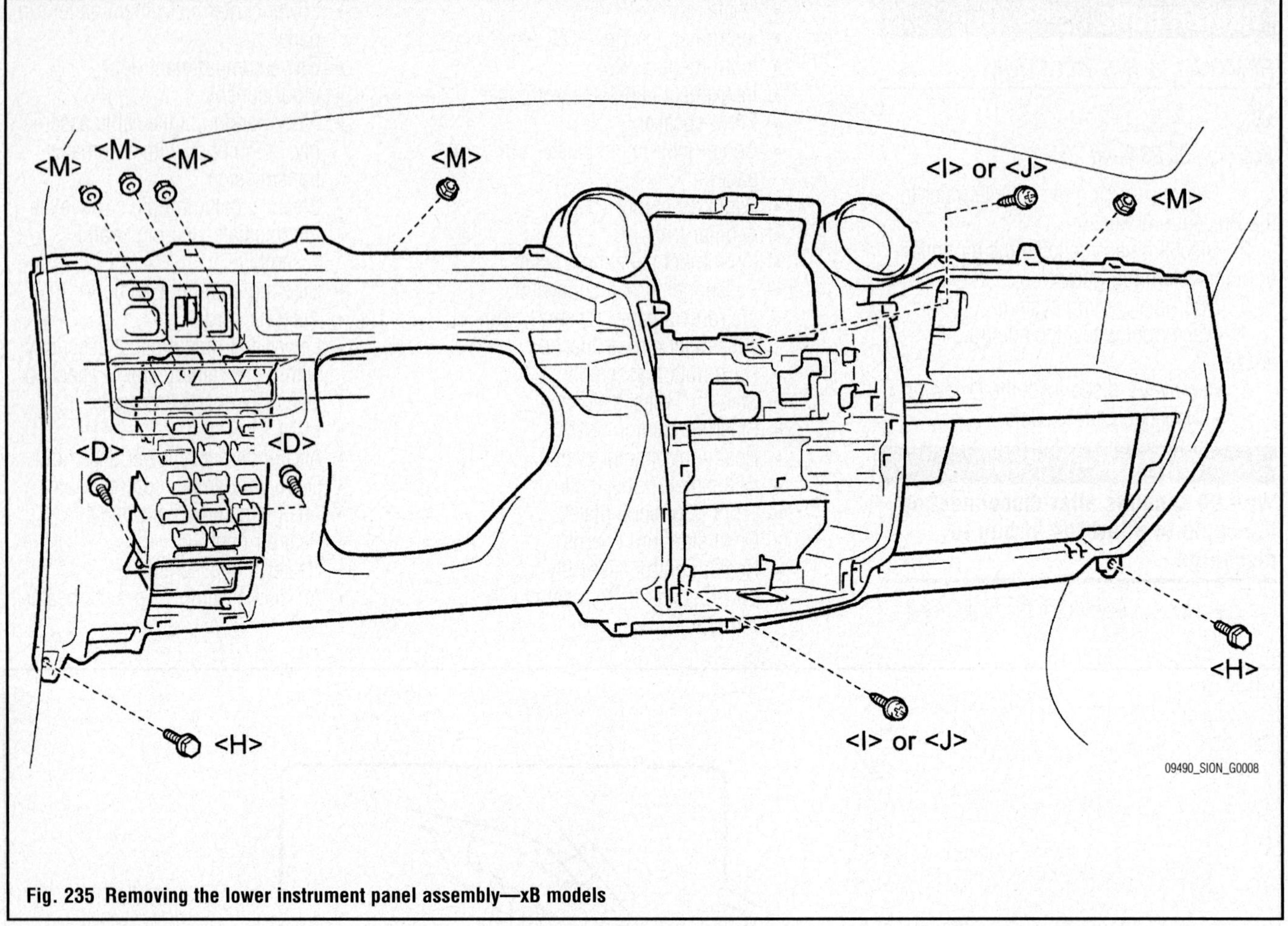

Fig. 235 Removing the lower instrument panel assembly—xB models

- Heater core cover
- Thermistor assembly
- Heater core

6. Installation is the reverse order of removal.
7. Refill the cooling system to the correct level.
8. Start the engine and check for leaks.

xD

WARNING

Some of these service operations affect the SRS airbag system. Read the precautionary notices concerning the SRS airbag system before servicing.

1. Disconnect the cable from negative battery terminal
2. Wait for at least 90 seconds after disconnecting the cable to prevent the airbag from working.
3. Remove or disconnect the following:
   - Front door opening trim weather-strip
   - Front pillar garnish
   - Instrument cluster finish panel
   - Combination meter assembly
   - Instrument cluster finish center panel sub-assembly
   - Instrument cluster finish panel retainer
   - Glove compartment door assembly
   - Upper instrument panel
   - Front door scuff plate
   - Instrument panel under cover sub-assembly
   - Cowl side trim board
   - Shift lever knob sub-assembly (for manual transaxle)
   - Upper console panel
   - Rear console box cover
   - Console box carpet
   - Rear console box sub-assembly
   - Air conditioning panel assembly
   - Defroster damper control cable sub-assembly
   - Air inlet damper control cable sub-assembly
   - Air mix damper control cable sub-assembly
   - Front console box
   - Instrument panel under tray
   - Instrument panel box
   - Radio bracket
   - Hood lock control lever sub-assembly
   - Antenna cord sub-assembly
   - Lower instrument panel
   - Steering pad
   - Steering wheel assembly
   - Steering column lower cover
   - Combination switch assembly
   - Power steering ECU
   - Instrument panel sub reinforcement
   - Column hole cover silencer sheet
   - Steering sliding yoke sub-assembly
   - Brake master cylinder push rod clevis
   - Brake pedal support
   - Steering column assembly
   - Rear no. 2 air duct
   - Instrument panel brace sub-assembly
4. Remove No. 1 air duct
   a. Disengage the 3 claws and remove the air duct.
5. Remove heater assembly

a. Disconnect the connector and 2 clamps.

b. Remove the 2 screws and heater from the air conditioning unit.

6. Installation is the reverse order of removal.

7. Refill the cooling system to the correct level.

8. Start the engine and check for leaks.

tC

See Figures 236 through 238.

1. Before servicing the vehicle, refer to the Precautions Section.
2. Discharge and recover the air conditioning system refrigerant.
3. Drain the cooling system.
4. Place front wheel facing straight ahead.
5. Remove or disconnect the following:
   - Negative battery cable

CAUTION

Wait 90 seconds after disconnecting the cable to allow the airbag to discharge.

   - Heater hoses
   - A/C hoses
   - Instrument gauge cluster hood panel
   - Instrument gauge cluster
   - Shift lever knob, if equipped with manual transmission
   - Upper front floor console panel
   - Heater control assembly
   - Heater control assembly
   - Center cluster module knob
   - Lower heater control base
   - Air conditioner amplifier assembly
   - Radio support brackets
   - Glove compartment door
   - Glove compartment door stopper
   - A-pillar trim

➡If equipped with side curtain airbags, cover the airbag with a protective cover as soon as the trim pieces are removed.

   - Passenger airbag connector
   - Instrument panel assembly with passenger airbag attached.
   - Steering wheel assembly
   - Steering column cover
   - Headlight dimmer switch
   - Windshield wiper switch assembly
   - Floor console top panel
   - Floor console
   - Front door scuff plates
   - Cowl side trim panels
   - Center heater to register duct
   - Lower instrument panel assembly
   - ECM
   - Center heater to register duct
   - Lower defroster nozzle assembly

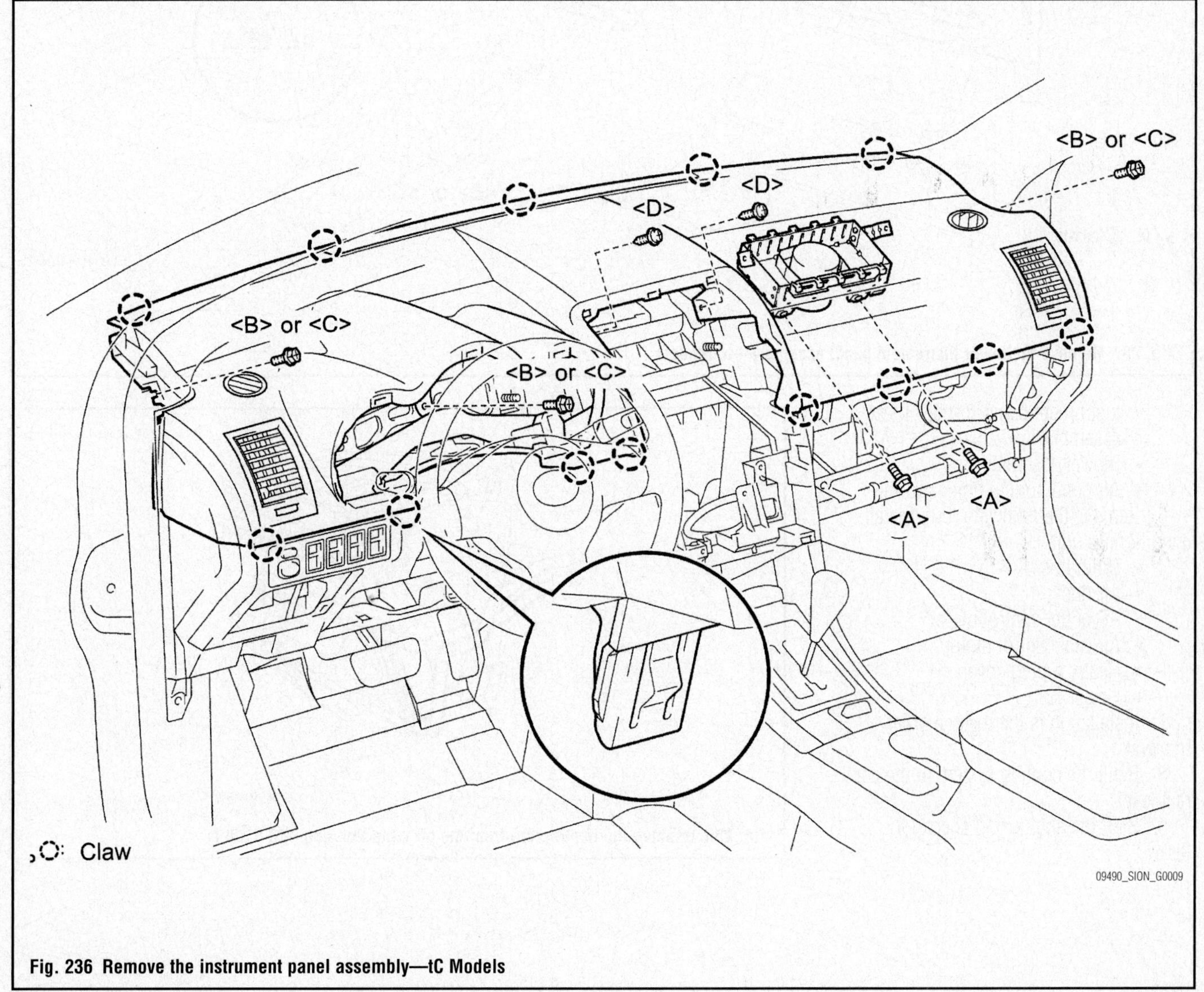

Fig. 236 Remove the instrument panel assembly—tC Models

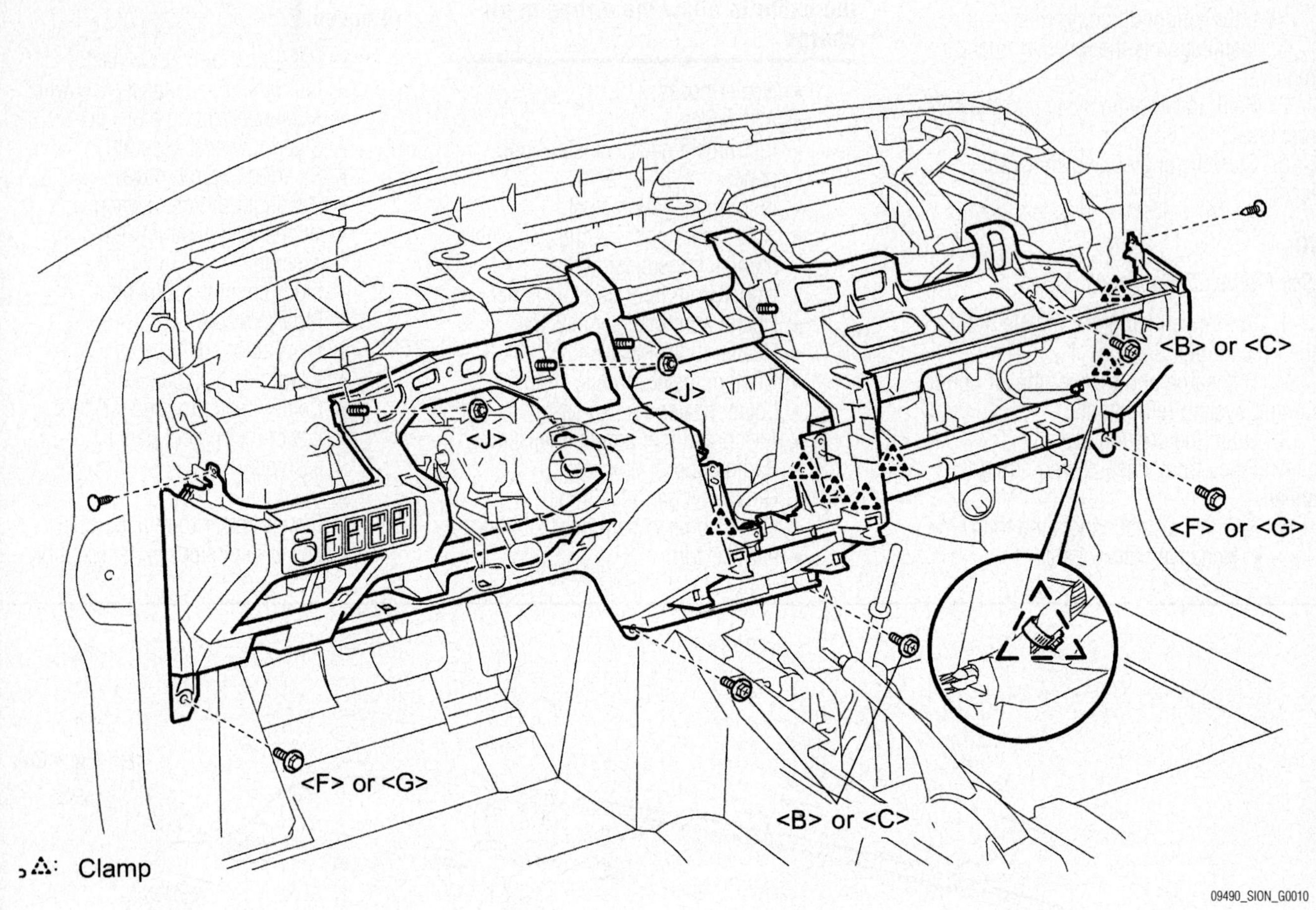

Fig. 237 Remove the lower instrument panel assembly—tC Models

- Instrument panel-to-cowl brace assembly
- Blower assembly
- Air conditioning unit assembly

6. Remove the following from the air conditioning unit:
   - Drain hose
   - Air duct
   - Air outlet control motor
   - Air-mix control motor
   - Heater piping cover
   - Heater core
7. Installation is the reverse order of removal.
8. Refill the cooling system to the correct level.
9. Start the engine and check for leaks.

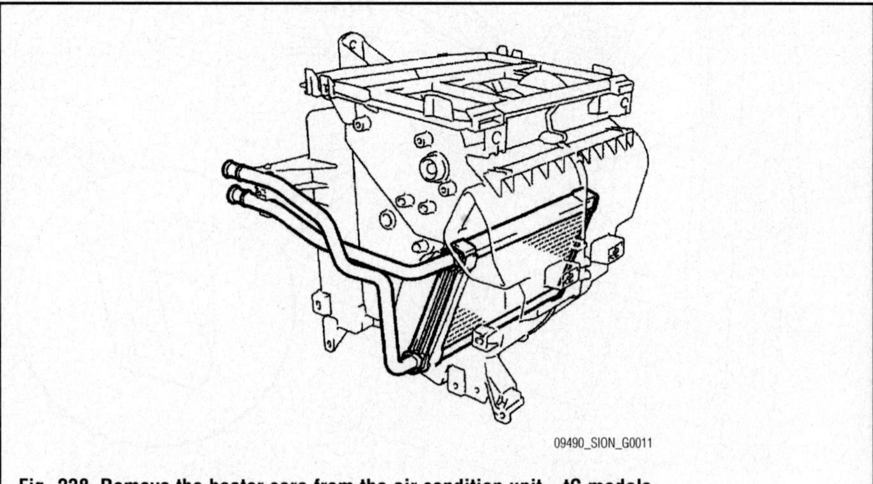

Fig. 238 Remove the heater core from the air condition unit—tC models

STEERING

POWER RACK & PINION STEERING GEAR

REMOVAL & INSTALLATION

xB and xD

See Figures 239 and 240.

1. Before servicing the vehicle, refer to the Precautions Section.
2. Face the front wheels straight ahead.
3. Drain the power steering fluid.

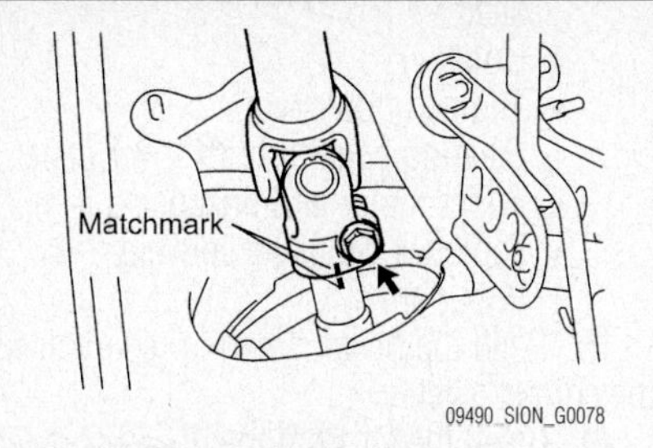

Fig. 239 Place matchmarks on the yoke and intermediate shaft

4. Remove or disconnect the following:
   - Wiper arms
   - Top cowl ventilator louvers
   - Wiper motor and linkage
   - Steering column hole cover plate
5. Matchmark the steering sliding yoke and intermediate shaft.
6. Disconnect the sliding yoke from the intermediate shaft.
7. Remove or disconnect the following:
   - Hood
   - Front wheels

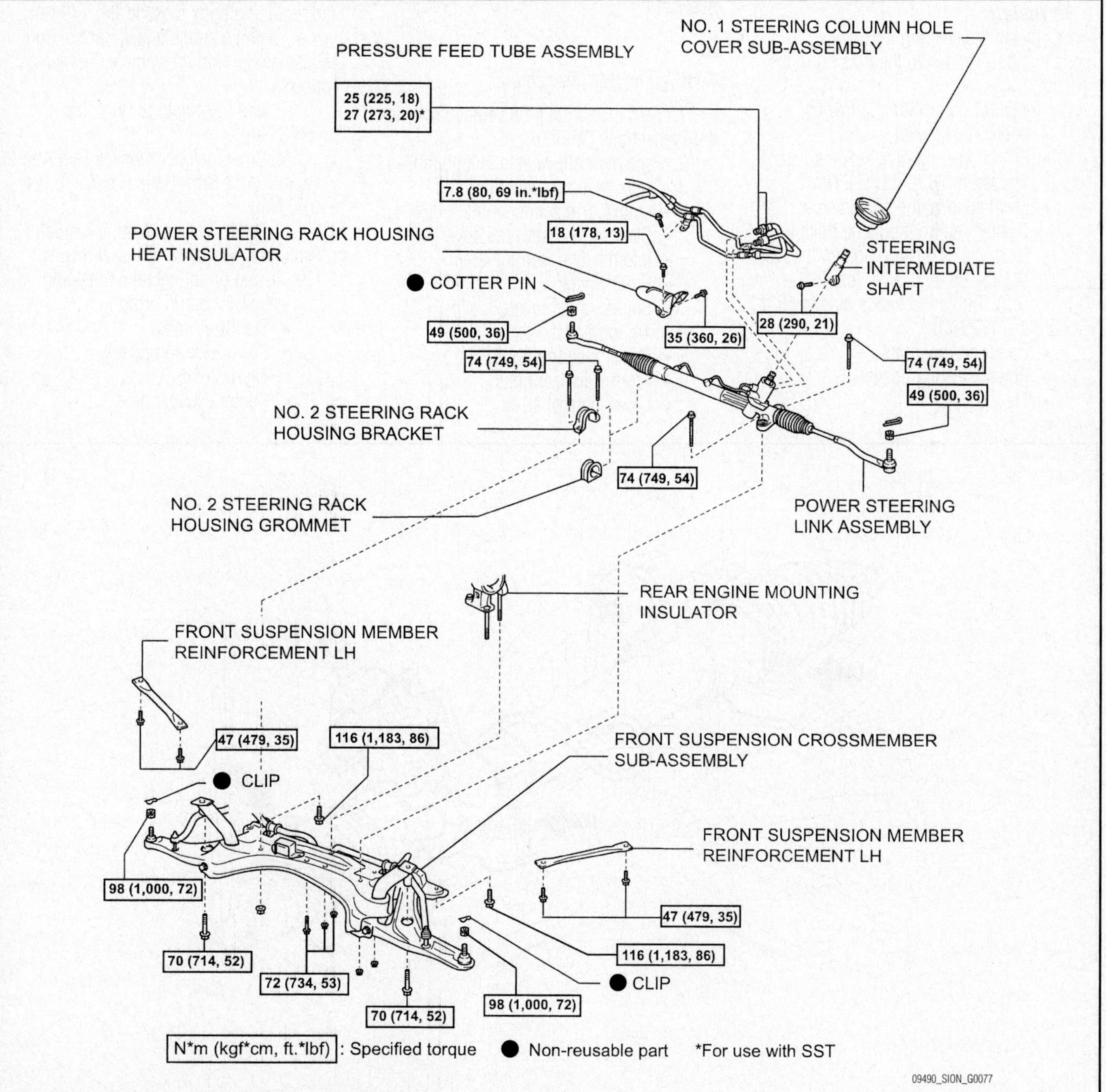

Fig. 240 Exploded view of the steering linkage components

- Tie rod ends
- Engine undercover
- Power steering hoses
- Lower control arms

8. At this point, attach an engine crane to support the engine.
9. Remove or disconnect the following:
   - Front suspension crossmember
   - Power steering rack heat insulator
   - Intermediate shaft
   - Steering gear assembly attached to the front crossmember.
10. Remove the steering rack clamps and remove the steering gear from the crossmember.

To install:

11. Install the steering gear to the crossmember. Tighten to the nuts 54 ft. lbs. (74 Nm).
12. Install or connect the following:
   - Intermediate shaft
   - Power steering rack heat insulator. Tighten to 26 ft. lbs. (35 Nm).
   - Front suspension crossmember. Tighten the front outside bolts to 52 ft. lbs. (70 Nm). Tighten the rear outside bolts to 86 ft. lbs. (116 Nm). Tighten to inside nuts to 53 ft. lbs. (72 Nm).
   - Lower control arms
   - Power steering hoses
   - Tie rod ends
   - Front wheels
   - Sliding yoke to the intermediate shaft
   - Hood
   - Engine undercover
   - Steering column hole cover plate
   - Wiper motor and linkage
   - Top cowl ventilator louvers
   - Wiper arms
13. Refill the power steering reservoir to the correct level.
14. Bleed the air from the power steering system.
15. Start the engine and check for leaks.
16. Check and adjust the alignment if necessary.

tC

See Figures 241 through 245.

1. Before servicing the vehicle, refer to the Precautions Section.
2. Face the front wheels straight ahead.
3. Drain the power steering fluid.
4. Remove or disconnect the following:
   - Steering column hole cover plate
   - Intermediate steering shaft
   - Front wheel
   - Engine undercover
   - Tie rod ends
   - Power steering hoses
   - Front stabilizer links
   - Lower control arms
   - Front floor panel brace
   - Center exhaust pipe
   - Hood
   - Engine appearance cover
5. Attach an engine crane to support the engine.
6. Remove or disconnect the following:
   - Front suspension crossmember
   - Steering gear assembly, attached to the crossmember, from the vehicle
   - Steering gear assembly from the crossmember

To install:

7. Install the power steering gear assembly to the front crossmember. Tighten the mounting bolts to 36 ft. lbs. (49 Nm).
8. Using a suitable jack, lift the front crossmember into the vehicle. Tighten as follows:
   a. Two front nuts to 98 ft. lbs (133 Nm)
   b. Crossmember bracket: Bolt A to 98 ft. lbs. (133 Nm). Bolts B to 59 ft. lbs. (80 Nm)
   c. Rear engine mounting insulator bolt and nuts to 48 ft. lbs. (65 Nm).
9. Install or connect the following:
   - Lower control arms
   - Stabilizer links
   - Power steering hoses
   - Tie rod ends
   - Center exhaust pipe

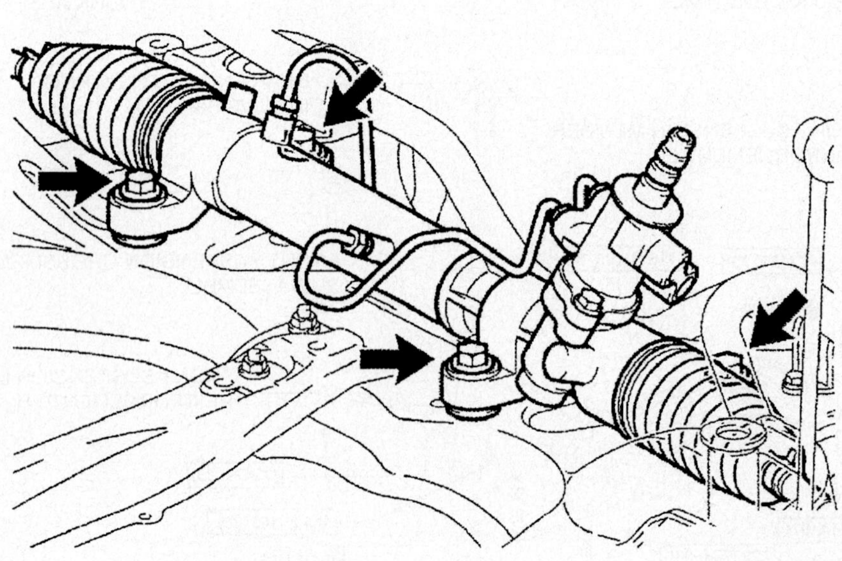

Fig. 241 Steering gear mounting bolt locations on the crossmember

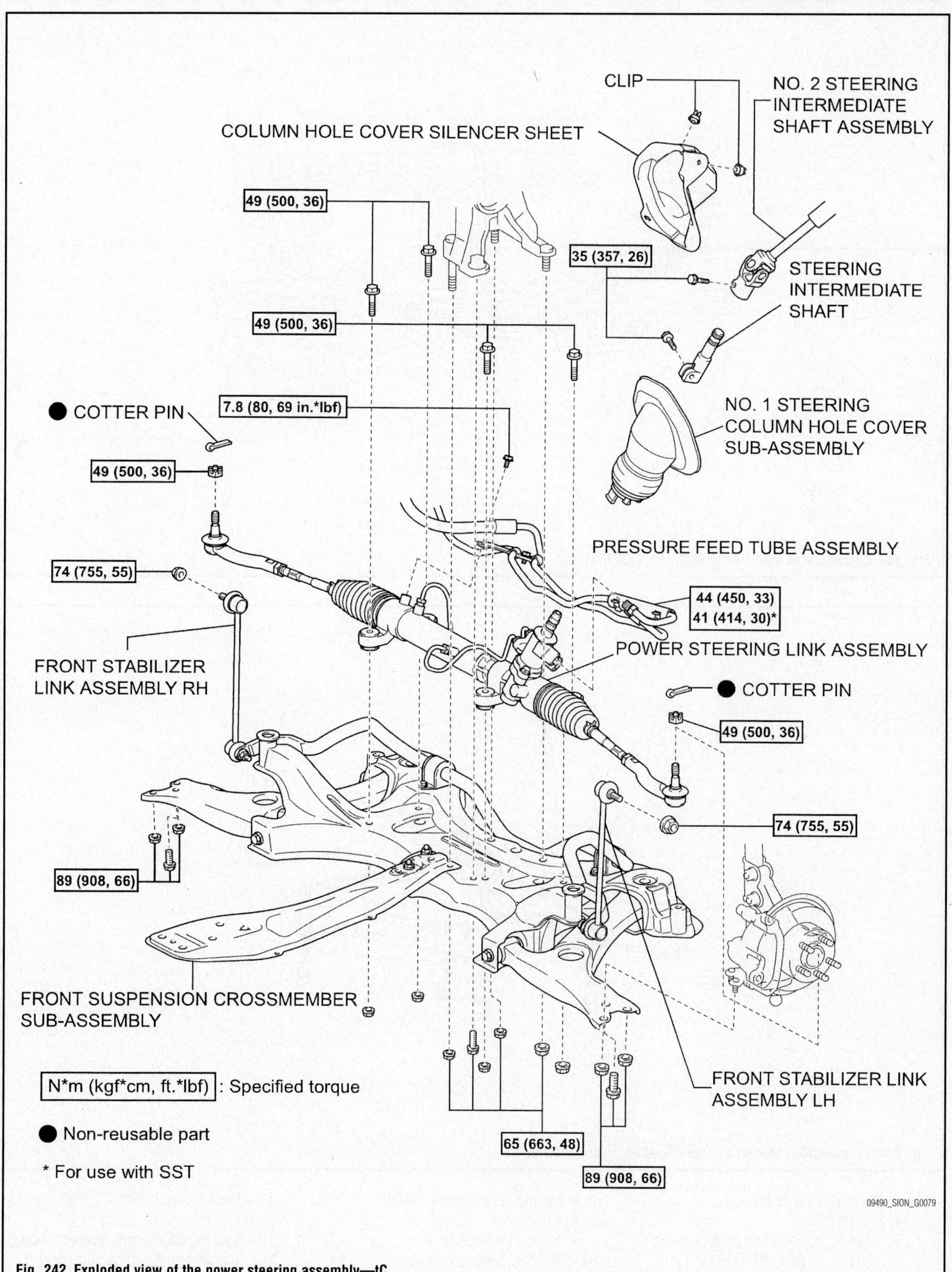

Fig. 242 Exploded view of the power steering assembly—tC

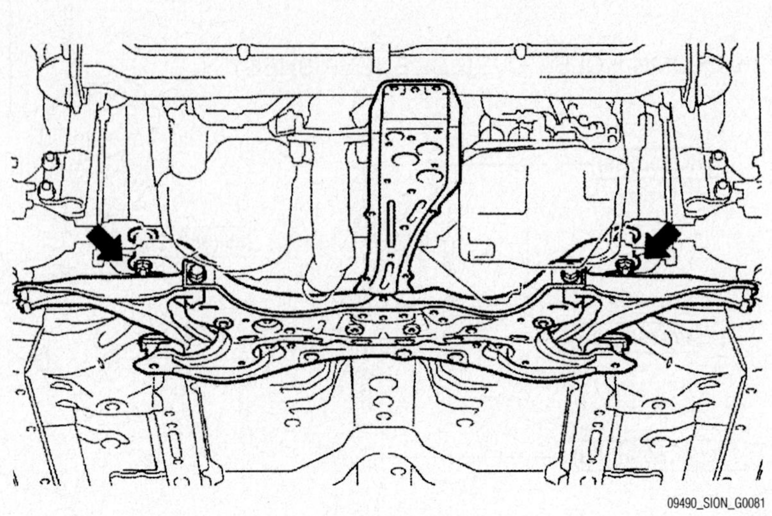

Fig. 243 Crossmember front nuts to 98 ft. lbs. (133 Nm)—tC

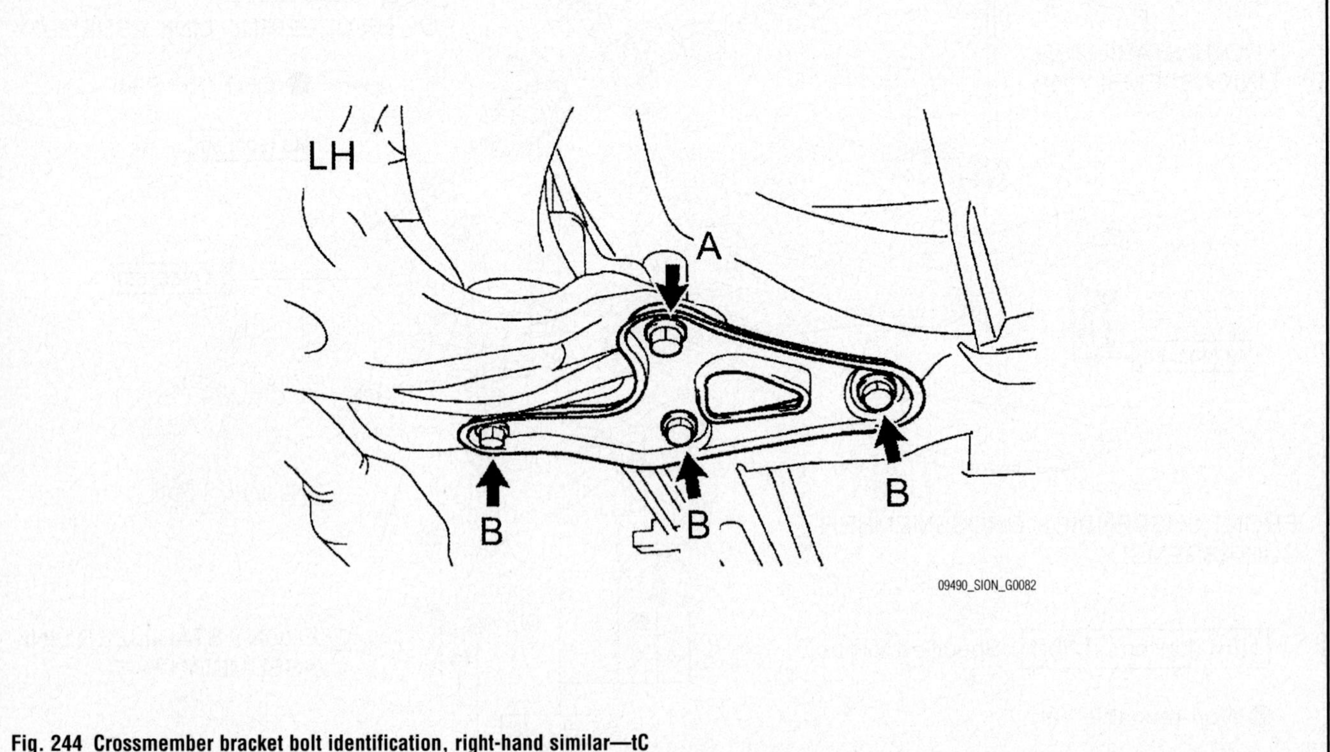

Fig. 244 Crossmember bracket bolt identification, right-hand similar—tC

- Front floor panel brace
- Front wheels
- Steering intermediate shaft. Tighten the bolts to 26 ft. lbs. (35 Nm).
- Steering column hole cover
- Engine appearance cover
- Hood
- Engine undercover

10. Refill the power steering reservoir to the correct level.
11. Bleed the air from the power steering system.
12. Start the engine and check for leaks.
13. Check and adjust the alignment if necessary.

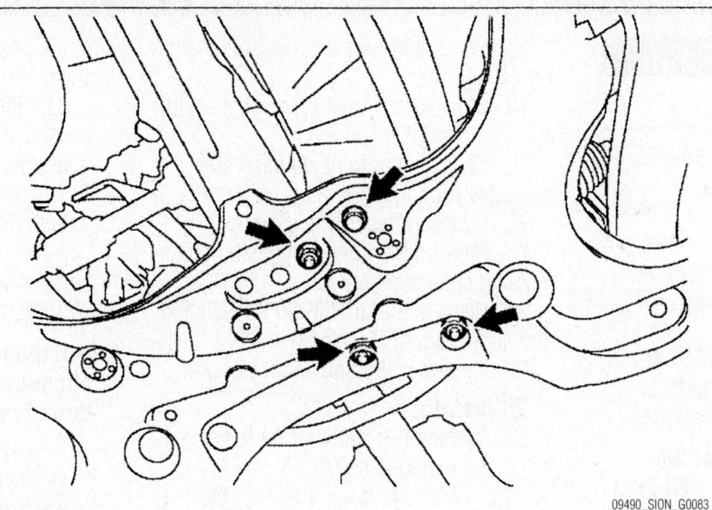

Fig. 245 Rear engine mounting insulator bolts and nuts—tC

POWER STEERING PUMP

REMOVAL & INSTALLATION

See Figures 246 and 247.

1. Raise and safely support the vehicle.
2. Remove the right front wheel and tire assembly.
3. Drain the power steering fluid into a suitable container.
4. Remove the front right fender apron seal.
5. Remove the fan-alternator belt, as outlined in the Engine Mechanical Section.
6. Remove the clip and disconnect the oil reservoir to pump hose no.1.

WARNING

Do not spill fluid on the belt.

7. Disconnect the pressure feed tube as follows:
   a. Remove the bolt and separate the pressure feed tube from the pump bracket rear.
   b. Using a wrench (27 mm) to hold the pressure port union, remove the union bolt and gasket.
8. Remove the power steering pump as follows:
   a. Disconnect the connector from the oil pressure switch.
   b. Using SST 09249-63010, or equivalent, and a deep socket (14 mm), loosen the 2 bolts.
   c. Remove the power steering pump assembly.

To install:

9. Install the power steering pump as follows:

Fig. 246 Using the special tool and a deep socket (14 mm), loosen the 2 bolts, then remove the power steering pump

   a. Temporarily install the 2 bolts to the power steering pump assembly.
   b. Install the power steering pump with the 2 bolts.
   c. Using SST 09249-63010, or equivalent, and a deep socket (14 mm), fully tighten the bolts. To 25 ft. lbs. (34 Nm).

➡Use a torque wrench with a fulcrum length of 300 mm (11.81 in.). This torque value is effective when the special tool is parallel to the torque wrench.

   d. Attach the connector to the oil pressure switch.

10. Connect the pressure feed tube as follows:
   a. Install the pressure feed tube and a new gasket to the power steering pump with the union bolt.

➡Make sure the stopper of the pressure feed tube touches the power steering pump body.

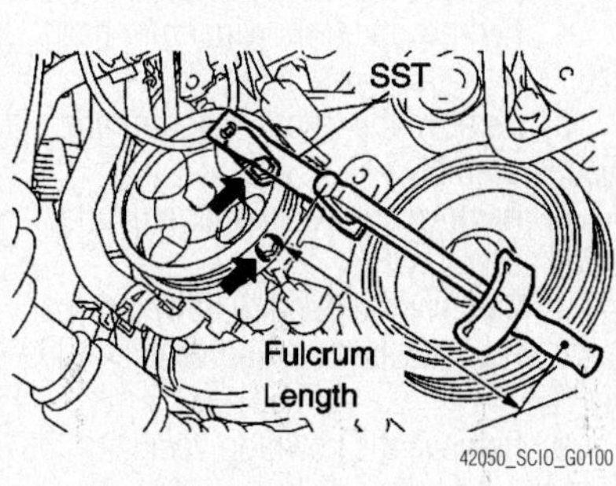

Fig. 247 Using SST 09249-63010, or equivalent, and a deep socket (14 mm), fully tighten the bolts. To 25 ft. lbs. (34 Nm). Use a torque wrench with a fulcrum length of 300 mm (11.81 in.). This torque value is effective when the special tool is parallel to the torque wrench.

   b. Using a 27mm wrench to hold the pressure port union, tighten the union bolt to 38 ft. lbs. (52 Nm).

11. Install the pressure feed tube with the bolt to the rear pump bracket and tighten to 69 inch lbs. (7.8 Nm).
12. Connect the oil reservoir to pump hose no.1 with the clip.

WARNING

Do not spill fluid on the belt.

13. Install the fan-alternator belt, as outlined in the Engine Mechanical Section.
14. Install the front fender apron seal and front wheel and tire.
15. Add power steering fluid and bleed the system, as outlined in this section.
16. Check for fluid leaks.

LOWER BALL JOINT

REMOVAL & INSTALLATION

xD

See Figures 248 through 250.

1. Remove the front wheels.
2. Remove the front lower suspension arm sub-assembly LH (for Manual Transaxle).
   a. Remove the clip and castle nut.
   b. Install SST (spacer B) to the threaded section of the lower ball joint.

➡Make sure the upper ends of the threaded section of the lower ball joint and SST (spacer B) are aligned.

   c. Using SST, separate the lower arm.
   d. Remove the 2 bolts and lower arm.
3. Remove the hood sub-assembly.
4. Remove the front wiper arm head cap.
5. Remove the front wiper arm and blade assembly LH.
6. Remove the front wiper arm and blade assembly RH.
7. Remove the hood to cowl top seal.
8. Remove the cowl top ventilator louver RH.
9. Remove the cowl top ventilator louver LH.
10. Remove the front wiper motor and link.
11. Remove the front air shutter seal RH.
12. Remove the outer cowl top panel.
13. Position the wheels facing straight ahead.
14. Remove the column hole cover silencer sheet.
15. Remove the steering sliding yoke sub-assembly.
16. Remove the steering column hole cover sub-assembly.

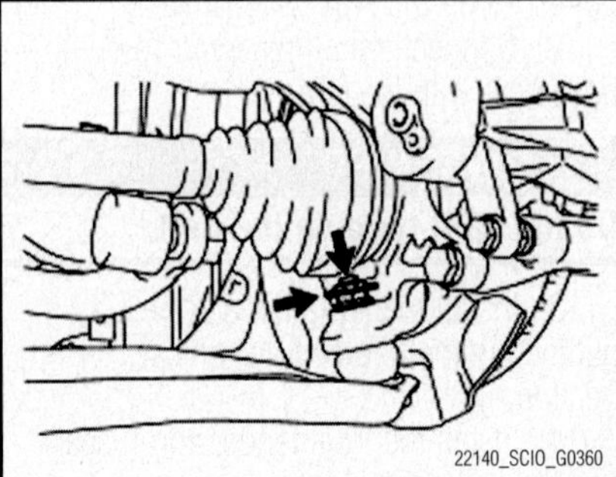

Fig. 248 Remove the front lower suspension arm sub-assembly LH

17. Separate the tie rod end sub-assembly LH and RH.
18. Separate the front lower suspension arm sub-assembly LH and RH.
19. Separate the front stabilizer link assembly LH and RH.
20. Suspend the engine assembly.
21. Remove the front suspension crossmember sub-assembly.
22. Remove the front lower suspension arm sub-assembly LH and RH.
   a. Remove the bolt and lower arm.

To install:

23. Temporarily tighten front lower suspension arm sub-assembly.
   a. Install the lower arm onto the crossmember and provisionally tighten the bolt.
24. Install front suspension crossmember sub-assembly.
25. Install front stabilizer link assembly.
26. Install front lower suspension arm sub-assembly.
27. Install tie rod end sub-assembly.
28. Install steering column hole cover sub-assembly.
29. Install steering sliding yoke sub-assembly.

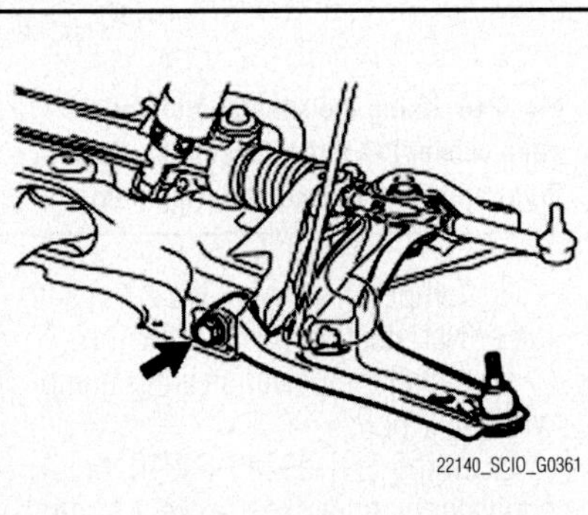

Fig. 249 Remove the front lower suspension arm sub-assembly

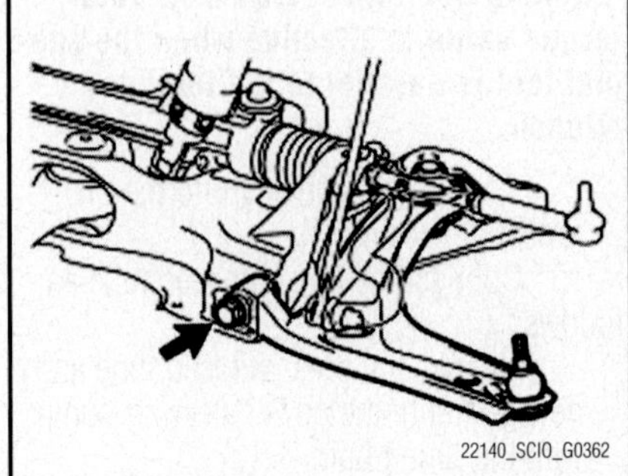

Fig. 250 Fully tighten front lower suspension arm sub-assembly

30. Install column hole cover silencer sheet.
31. Temporarily tighten front lower suspension arm sub-assembly LH (for manual transaxle).
   a. Provisionally tighten the lower arm with the 2 bolts.
   b. Install the lower arm onto the steering knuckle with a new castle nut. Torque: 72 ft. lbs. (98 Nm).

➡If the holes for the clip are not aligned, tighten the nut by a further turn of up to 60°.

   c. Install a new clip.
32. Install front wheels.
33. Position wheels facing straight ahead.
34. Stabilize suspension.
   a. Lower the vehicle from the jack.
   b. Bounce the vehicle up and down several times to stabilize the suspension.
35. Fully tighten front lower suspension arm sub-assembly.
   a. Fully tighten the 2 bolts. Bolt A: Torque: 101 ft. lbs. (137 Nm); Bolt B: Torque: 118 ft. lbs. (160 Nm).
36. Install outer cowl top panel.
37. Install front air shutter seal.
38. Install front wiper motor and link.
39. Install cowl top ventilator louver.
40. Install hood to cowl top seal.
41. Install front wiper arm and blade assembly.
42. Install front wiper arm head cap.
43. Inspect hood sub-assembly.
44. Adjust hood sub-assembly.
45. Inspect and adjust front wheel alignment.

tC

1. Before servicing the vehicle, refer to the Precautions Section.
2. Remove or disconnect the following:
   - Front wheel
   - Hub nut
   - Wheel speed sensor
   - Tie rod end
   - Lower control arm mounting bolts
3. Disconnect the halfshaft from the axle hub.
4. Remove the lower shock absorber mounting bolts.
5. Using Special Tool 09628-62011 or suitable puller, remove the lower ball joint from the steering knuckle.

To install:

➡Use a new split pin for assembly.

6. Install the ball joint and torque the mounting nut to 76 ft. lbs. (103 Nm)
7. Install the halfshaft to the front axle. Tighten the steering knuckle mounting bolts to 76 ft. lbs. (103 Nm).
8. Install or connect the following:
- Lower control arm mounting bolts. Tighten to 66 ft. lbs. (89 Nm).
- Tie rod ends. Tighten to 36 ft. lbs. (49 Nm).
- Hub nut. Tighten to 159 ft. lbs. (216 Nm).
- Front wheel

9. Check and/or adjust the wheel alignment.

LOWER CONTROL ARM

REMOVAL & INSTALLATION

xB

1. Before servicing the vehicle, refer to the Precautions Section.
2. Remove or disconnect the following:
- Front wheel
- Hood

3. Attach an engine crane to support the engine.
4. Disconnect the lower control arm from the steering knuckle.
5. Disconnect the front stabilizer bar from the lower control arm.
6. Disconnect the power steering gear assembly.
7. Disconnect the power steering hoses from the control arm.
8. Support the front crossmember with a suitable jack.
9. Disconnect the front suspension crossmember.
10. Remove the lower control arm from the front suspension crossmember.

To install:

11. Install the lower control arm to the crossmember and temporarily tighten the bolts.
12. Using a suitable jack, lift the front crossmember into the vehicle. Tighten as follows:
    a. Two front nuts to 98 ft. lbs (133 Nm)
    b. Crossmember bracket: Bolt A to 98 ft. lbs. (133 Nm). Bolts B to 59 ft. lbs. (80 Nm)
    c. Rear engine mounting insulator bolt and nuts to 48 ft. lbs. (65 Nm).
13. Install or connect the following:
- Power steering hoses
- Steering gear assembly
- Front stabilizer bar. Tighten to 13 ft. lbs. (18 Nm).
- Lower control arm to the steering knuckle. Tighten to 72 ft. lbs. (98 Nm).
- Front wheel

14. Fully tighten the lower control arm mounting on the crossmember to 97 ft. lbs. (132 Nm).

xD

See Figures 251 through 253.

1. Remove the front wheels.
2. Remove the front lower suspension arm sub-assembly LH (for Manual Transaxle).
    a. Remove the clip and castle nut.
    b. Install SST (spacer B) to the threaded section of the lower ball joint.

➡Make sure the upper ends of the threaded section of the lower ball joint and SST (spacer B) are aligned.

    c. Using SST, separate the lower arm.
    d. Remove the 2 bolts and lower arm.
3. Remove the hood sub-assembly.
4. Remove the front wiper arm head cap.
5. Remove the front wiper arm and blade assembly LH.
6. Remove the front wiper arm and blade assembly RH.
7. Remove the hood to cowl top seal.
8. Remove the cowl top ventilator louver RH.
9. Remove the cowl top ventilator louver LH.
10. Remove the front wiper motor and link.
11. Remove the front air shutter seal RH.
12. Remove the outer cowl top panel.
13. Position the wheels facing straight ahead.
14. Remove the column hole cover silencer sheet.
15. Remove the steering sliding yoke sub-assembly.
16. Remove the steering column hole cover sub-assembly.
17. Separate the tie rod end sub-assembly LH and RH.
18. Separate the front lower suspension arm sub-assembly LH and RH.
19. Separate the front stabilizer link assembly LH and RH.
20. Suspend the engine assembly.
21. Remove the front suspension crossmember sub-assembly.
22. Remove the front lower suspension arm sub-assembly LH and RH.
    a. Remove the bolt and lower arm.

To install:

23. Temporarily tighten front lower suspension arm sub-assembly.
    a. Install the lower arm onto the crossmember and provisionally tighten the bolt.
24. Install front suspension crossmember sub-assembly.
25. Install front stabilizer link assembly.
26. Install front lower suspension arm sub-assembly.
27. Install tie rod end sub-assembly.
28. Install steering column hole cover sub-assembly.
29. Install steering sliding yoke sub-assembly.

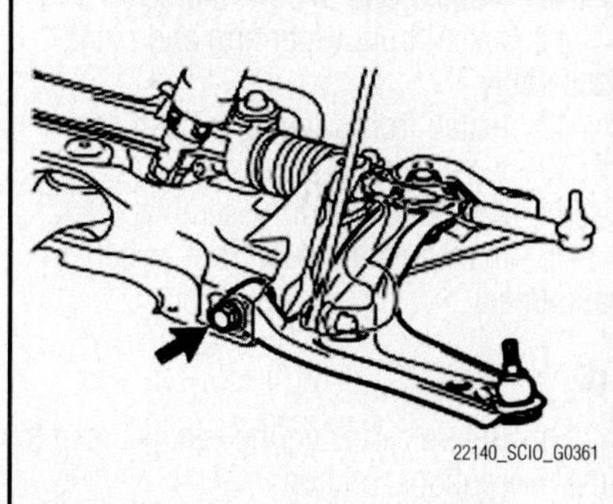

Fig. 252 Remove the front lower suspension arm sub-assembly

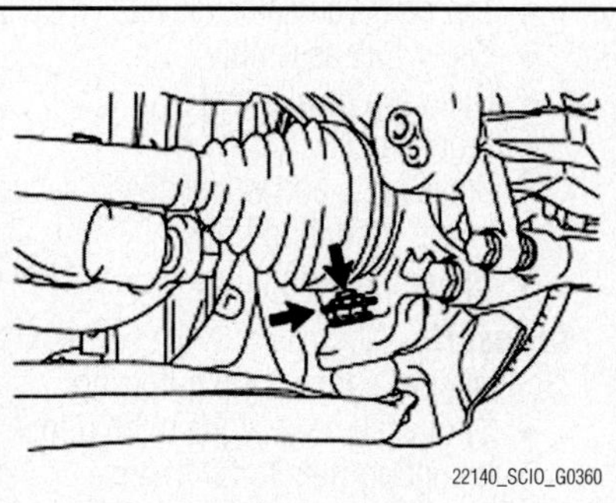

Fig. 251 Remove the front lower suspension arm sub-assembly LH

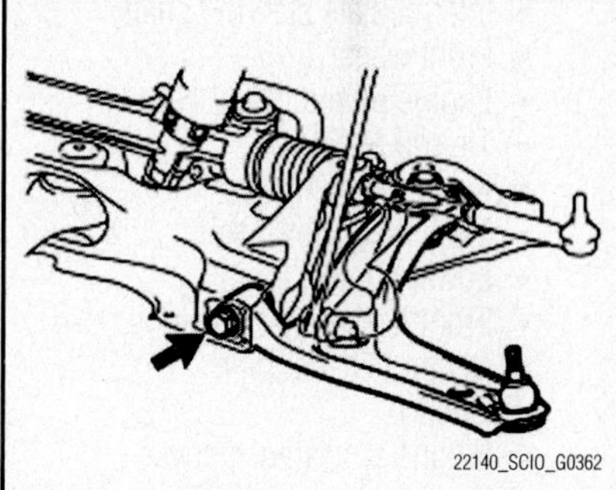

Fig. 253 Fully tighten front lower suspension arm sub-assembly

30. Install column hole cover silencer sheet.

31. Temporarily tighten front lower suspension arm sub-assembly LH (for manual transaxle).

a. Provisionally tighten the lower arm with the 2 bolts.

b. Install the lower arm onto the steering knuckle with a new castle nut. Torque: 72 ft. lbs. (98 Nm).

➡If the holes for the clip are not aligned, tighten the nut by a further turn of up to 60°.

c. Install a new clip.

32. Install front wheels.

33. Position wheels facing straight ahead.

34. Stabilize suspension.

a. Lower the vehicle from the jack.

b. Bounce the vehicle up and down several times to stabilize the suspension.

35. Fully tighten front lower suspension arm sub-assembly.

a. Fully tighten the 2 bolts. Bolt A: Torque: 101 ft. lbs. (137 Nm); Bolt B: Torque: 118 ft. lbs. (160 Nm).

36. Install outer cowl top panel.

37. Install front air shutter seal.

38. Install front wiper motor and link.

39. Install cowl top ventilator louver.

40. Install hood to cowl top seal.

41. Install front wiper arm and blade assembly.

42. Install front wiper arm head cap.

43. Inspect hood sub-assembly.

44. Adjust hood sub-assembly.

45. Inspect and adjust front wheel alignment.

tC

1. Before servicing the vehicle, refer to the Precautions Section.
2. Face the front wheels straight ahead.
3. Drain the power steering fluid.
4. Remove or disconnect the following:
   - Steering column hole cover plate
   - Intermediate steering shaft
   - Front wheel
   - Engine undercover
   - Tie rod ends
   - Power steering hoses
   - Front stabilizer links
   - Lower control arms
   - Front floor panel brace
   - Center exhaust pipe
   - Hood
   - Engine appearance cover
5. Attach an engine crane to support the engine.
6. Remove or disconnect the following:
   - Front suspension crossmember
   - Lower control arm from the crossmember

To install:

7. Install the lower control arm to the crossmember, and temporarily tighten the bolts.
8. Using a suitable jack, lift the front crossmember into the vehicle. Tighten as follows:

a. Two front nuts to 98 ft. lbs (133 Nm)

b. Crossmember bracket: Bolt A to 98 ft. lbs. (133 Nm). Bolts B to 59 ft. lbs. (80 Nm)

c. Rear engine mounting insulator bolt and nuts to 48 ft. lbs. (65 Nm).

9. Install or connect the following:
   - Lower control arms to the steering knuckle. Tighten to 66 ft. lbs. (89 Nm).
   - Stabilizer links. Tighten to 55 ft. lbs. (74 Nm).
   - Power steering hoses
   - Tie rod ends
   - Center exhaust pipe
   - Front floor panel brace
   - Front wheels
   - Steering intermediate shaft. Tighten the bolts to 26 ft. lbs. (35 Nm).
   - Steering column hole cover
   - Engine appearance cover
   - Hood
   - Engine undercover
10. Start the engine and check for leaks.
11. Check and adjust the alignment if necessary.

MACPHERSON STRUT

REMOVAL & INSTALLATION

xB

See Figure 254.

1. Before servicing the vehicle, refer to the Precautions Section.
2. Remove or disconnect the following:
   - Wiper arms
   - Top cowl ventilator louvers
   - Wiper link assembly
   - Outer top cowl panel
   - Front wheel
   - Brake hose and wheel speed sensor
   - Lower mounting bolts
   - Upper mounting nuts

To install:

3. Install or connect the following:
   - Strut assembly. Tighten the upper mounting nuts to 29 ft. lbs. (39 Nm). Lower mounting bolts to 97 ft. lbs. (132 Nm).
   - Brake hose and wheel speed sensor
   - Front wheel
4. The remainder of the installation is the reverse order of removal.
5. Check and adjust the alignment if necessary.

xD

1. Remove the front wiper arm head cap.
2. Remove the front wiper arm and blade assembly.
3. Remove the hood to cowl top seal.
4. Remove the cowl top ventilator louver.
5. Remove the front wiper motor and link.
6. Remove the front air shutter seal RH.
7. Remove the outer cowl top panel.
8. Remove the front wheels.
9. Separate the front stabilizer link assembly.

a. Remove the nut and separate the stabilizer link from the shock absorber.

10. Separate the front flexible hose.

a. Remove the bolt and separate the speed sensor and flexible hose.

11. Remove the front suspension support dust cover.
12. Remove the front shock absorber with coil spring.

a. Remove the 2 nuts and 2 bolts and separate the shock absorber with coil spring from the steering knuckle.

b. Using a socket hexagon wrench, fix the shock absorber rod and remove the nut.

c. Remove the No. 2 suspension support.

d. Remove the front shock absorber with coil spring from the vehicle.

To install:

13. Temporarily tighten front shock absorber with coil spring.

a. Provisionally tighten a new nut through No. 2 suspension support.

b. Install the front shock absorber with coil spring onto the steering knuckle.

c. Install the 2 bolts and 2 nuts. Torque: 121 ft. lbs. (164 Nm).

14. Install front flexible hose.

a. Install the flexible hose and speed sensor with the bolt. Torque: 22 ft. lbs. (29 Nm).

➡Install the flexible hose and speed sensor without twisting them.

15. Install front stabilizer link assembly

a. Install the stabilizer link with the nut. Torque: 55 ft. lbs. (74 Nm).

16. Install front wheels.

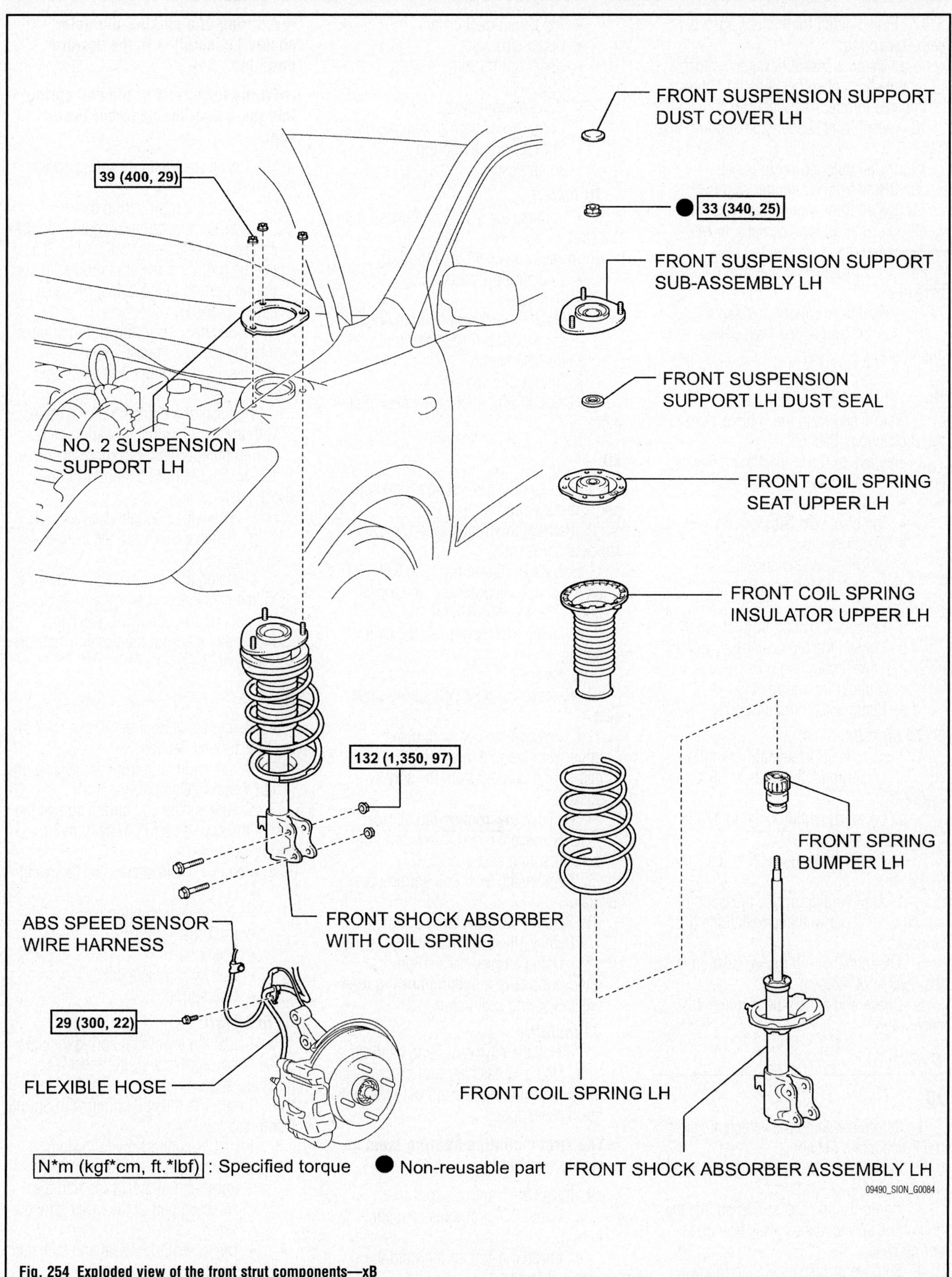

Fig. 254 Exploded view of the front strut components—xB

17. Fully tighten front shock absorber with coil spring.
    a. Using a socket hexagon wrench 6, fix the shock absorber rod and tighten the nut. Torque: 41 ft. lbs. (55 Nm).
18. Install front suspension support dust cover.
19. Install outer cowl top panel.
20. Install front air shutter seal RH.
21. Install front wiper motor and link.
22. Install cowl top ventilator louver.
23. Install hood to cowl top seal.
24. Install front wiper arm and blade assembly.
25. Install front wiper arm head cap.
26. Inspect and adjust front wheel alignment.

tC

1. Before servicing the vehicle, refer to the Precautions Section.
2. Remove or disconnect the following:
    - Front wheel
    - Wiper arms
    - Top cowl ventilator louvers
    - Stabilizer links
    - Wheel speed sensor
3. Remove the strut assembly as follows:
    a. Remove the support dust cover
    b. Loosen the top center lock nut
    c. Lower mounting nuts
    d. 3 upper mounting nuts
    e. Remove the strut assembly

To install:

4. Install the strut assembly as follows:
    a. Upper mounting nuts to 38 ft. lbs. (52 Nm)
    b. Lower mounting bolts to 177 ft. lbs. (240 Nm)
    c. Center lock nut to 35 ft. lbs. (47 Nm)
    d. Apply multipurpose grease to the center lock nut well and install the dust cover.
5. The remainder of the installation is the reverse of removal.
6. Check and adjust the alignment if necessary.

OVERHAUL

xB

1. Before servicing the vehicle, refer to the Precautions Section.
2. Remove the strut from the vehicle and install a spring compressor.
3. Compress the coil spring so that the end of the spring comes away from the spring seat.
4. Remove or disconnect the following:
    - Top center dust cover
    - Upper strut nut
    - Upper spring seat
    - Dust seal
    - Coil spring insulator
    - Compressed spring from the strut
    - Spring from the spring compressor

To install:

5. Compress the spring and install it on the strut.
6. Install or connect the following:
    - Coil spring insulator
    - Dust seal
    - Upper spring seat and the upper strut mount. Torque the nut to 25 ft. lbs. (33 Nm)
    - Strut to the vehicle
7. Check and/or adjust the wheel alignment.

xD

1. Remove the front suspension support sub-assembly.
2. Remove the front support to front shock absorber nut.
    a. Using a socket hexagon wrench 6, fix the shock absorber rod and loosen the nut.
    b. Using SST, compress the coil spring.
    c. Remove the nut.
3. Remove the front coil spring upper seat.
    a. Remove the coil spring seat upper with the strut mounting bearing and spring bumper from the shock absorber.
4. Remove the front spring bumper.
    a. Remove the spring bumper from the coil spring seat upper.
5. Remove the front coil spring upper insulator.
6. Remove the front coil spring.
7. Remove the strut mounting bearing.
    a. Using a brass bar and press, remove the strut mounting bearing from the coil spring seat upper.

To install:

8. Install the strut mounting bearing.
    a. Using a brass bar and press, install the strut mounting bearing onto the coil spring seat upper.

➡The strut mounting bearing must be securely installed.

9. Install the front coil spring.
    a. Using SST, compress the coil spring.
    b. Install the coil spring onto the shock absorber.

➡A spring of a smaller diameter should be installed in the upward direction.

➡Fit the lower end of the coil spring into the gap of the absorber lower seat.

10. Install the front coil spring upper insulator.
11. Install the front spring bumper.
    a. Install the spring bumper onto the coil spring seat upper with the air discharge groove of the spring bumper and the rib portion of the coil spring seat upper aligned.
    b. Securely insert the spring bumper into the coil spring seat upper.
12. Install the front coil spring upper seat.
    a. Install the coil spring seat upper with the strut mounting bearing and spring bumper onto the shock absorber.
13. Install the front support to front shock absorber nut.
    a. Provisionally tighten a new nut.
    b. Remove SST from the front coil spring.
    c. Using a socket hexagon wrench 6, fix the shock absorber rod and tighten the nut. Torque: 25 ft. lbs. (33 Nm).
14. Install the front suspension support sub-assembly.

tC

1. Before servicing the vehicle, refer to the Precautions Section.
2. Remove the strut from the vehicle and install a Spring Compressor Tool.
3. Compress the coil spring so that the end of the spring comes away from the spring seat.
4. Remove or disconnect the following:
    - Lock nut
    - Suspension support
    - Dust seal
    - Upper seat
    - Coil spring insulator
    - Coil spring

To install:

5. Install the lower coil spring insulator to the shock absorber so both recessed parts are aligned.
6. Compress the coil spring with Spring Compress tool.
7. Install or connect the following:
    - Coil spring to the strut. Fit the lower end of the spring into the recessed part of the lower spring seat.
    - Upper seat, dust seal and lock nut.
8. Install the strut assembly.

STABILIZER BAR

REMOVAL & INSTALLATION

xB

WARNING

This procedure requires you to suspend the engine using the proper engine lifting and holding equipment. Do not attempt this procedure unless you have the proper equipment.

1. Raise and safely support the vehicle.
2. Remove the front wheels
3. Remove the hood.
4. Suspend the engine assembly.
5. Separate the lower control arms. Refer to the procedure in this section for details.
6. Separate the stabilizer bar.
7. Separate the power steering link.
8. Separate the pressure feed tube.
9. Separate the front suspension cross-member sub-.
10. Remove the 4 bolts and the 2 bushings from the stabilizer bar bracket, then remove the front stabilizer bar.

To install:

11. Install the stabilizer bar, the 2 bushings, and the brackets and secure with the 4 bolts. Tighten to 27 ft. lbs. (327 Nm).

➡Install the bushings with the slit facing to the rear side of the vehicle. Install the bushings on the outside of each paint line.

12. Connect the front suspension cross-member.
13. Connect the pressure feed tube.
14. Connect the power steering link.
15. Connect the stabilizer bar.
16. Connect the lower control arms.
17. Install front wheel, then carefully lower the vehicle.
18. Install the hood, making sure it is aligned correctly.
19. Inspect and adjust the front wheel alignment.

xD

1. Remove the hood sub-assembly.
2. Remove the front wiper arm head cap.
3. Remove the front wiper arm and blade assembly.
4. Remove the hood to cowl top seal.
5. Remove the cowl top ventilator louver.
6. Remove the front wiper motor and link.
7. Remove the front air shutter seal RH.
8. Remove the outer cowl top panel.
9. Position the wheels facing straight ahead.
10. Remove the front wheel.
11. Remove the column hole cover silencer sheet.
12. Remove the steering sliding yoke sub-assembly.
13. Remove the steering column hole cover sub-assembly.
14. Separate the tie rod end sub-assembly.
15. Separate the front lower suspension arm sub-assembly.
16. Remove the front stabilizer link assembly.
    a. Remove the 2 nuts and stabilizer link
17. Suspend the engine assembly.
18. Remove the front suspension cross-member sub-assembly.
19. Remove the power steering gear.
20. Remove the front stabilizer bracket.
    a. Remove the 2 bolts and the stabilizer bracket.
21. Remove the front stabilizer bar.
22. Remove the front stabilizer bar bushings.

To install:

23. Install front stabilizer bar bushings.
    a. Install the stabilizer bar bush onto the stabilizer bar.

➡Install the bush onto the stabilizer so that the bush stopper of the stabilizer bar faces the outside of the vehicle.

    b. Install the bushing with its cutout facing the front of the vehicle.
24. Install front stabilizer bar
    a. Install the stabilizer onto the cross-member with the paint mark on the left side of the vehicle.
25. Install front stabilizer bracket.
    a. Provisionally tighten bolt A.
    b. Tighten the bolts to the specified torque, in the order of B then A. Torque: 35 ft. lbs. (47 Nm).
26. Install power steering gear.
27. Install front suspension crossmember sub-assembly.
28. Install front stabilizer link assembly.
    a. Install the stabilizer link with the 2 nuts. Torque: 55 ft. lbs. (74 Nm).
29. Install front lower suspension arm sub-assembly.
30. Install tie rod end sub-assembly.
31. Install steering column hole cover sub-assembly.
32. Install steering sliding yoke sub-assembly.
33. Install column hole cover silencer sheet.
34. Install front wheels.
35. Position wheels facing straight ahead.
36. Install outer cowl top panel
37. Install front air shutter seal RH.
38. Install front wiper motor and link.
39. Install cowl top ventilator louver.
40. Install hood to cowl top seal.
41. Install front wiper arm and blade assembly.
42. Install front wiper arm head cap.
43. Inspect hood sub-assembly.
44. Adjust hood sub-assembly.
45. Inspect and adjust front wheel alignment.

tC

See Figures 255 through 258.

1. Place front wheels facing straight ahead
2. Remove the column hole cover silencer sheet.
3. Separate the steering intermediate shaft.
4. Separate the steering column hole cover.
5. Remove the front wheels
6. Remove the left and right engine under covers.
7. Remove front stabilizer links:
    a. Remove the 2 nuts and front stabilizer link.

➡Use a hexagon wrench (6 mm) to hold the stud if the ball joint turns together with the nut.

8. Inspect the stabilizer link as follows:
    a. Secure the front stabilizer link in a vise.
    b. Install the nut to the stud bolt.
    c. Flip the ball joint back and forth 5 times or more.

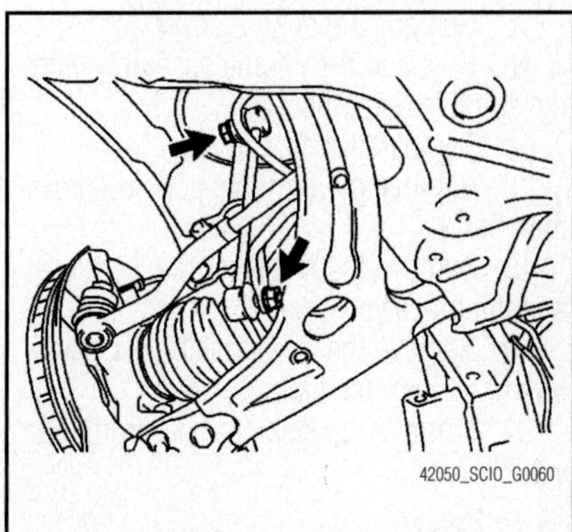

Fig. 255 Remove the stabilizer link

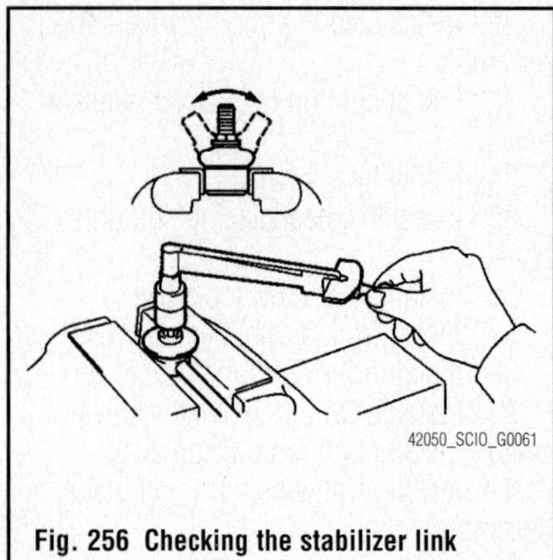

Fig. 256 Checking the stabilizer link

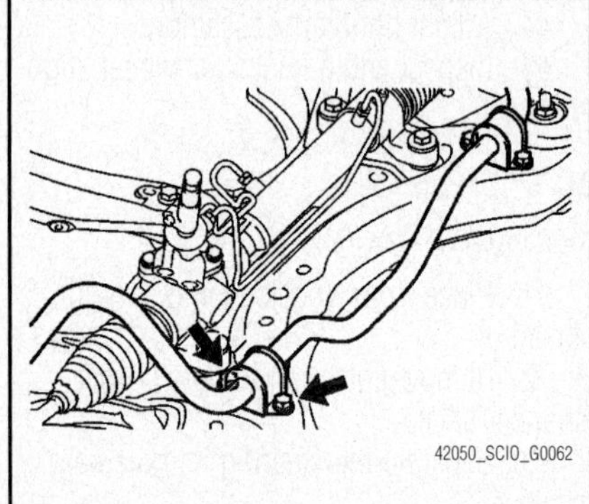

Fig. 257 Removing the stabilizer brackets

d. Use a torque wrench to turn the nut continuously at a rate of 3 to 5 seconds per turn. Take the torque reading on the 5th turn. The turning torque should be 8.9 inch lbs. (1 Nm) or less. If not, replace the link.

e. Check the dust boots for cracks or grease leakage.

9. Separate the tie rod ends.
10. Disconnect the pressure feed tube.
11. Separate the lower control arms.
12. Remove the front floor panel brace.
13. Remove the center exhaust pipe.
14. Remove the hood.
15. Remove the engine cover.
16. Suspend the engine assembly with the suitable equipment.
17. Remove the no.1 hook.
18. Remove the front suspension crossmember.
19. Remove the stabilizer bracket bolts and the brackets.
20. Remove the 2 front stabilizer bar bushings from the stabilizer bar.
21. Remove the stabilizer bar from the vehicle.

To install:

22. Install the stabilizer bar.

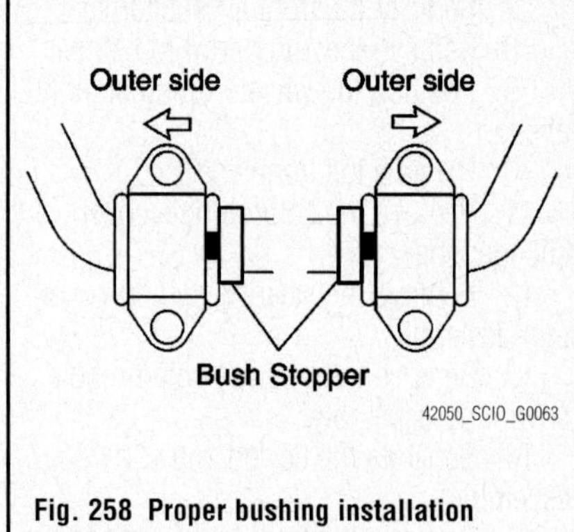

Fig. 258 Proper bushing installation position

23. Install front stabilizer bar bushings:
    a. Install the bushing to the outer side of the stopper on the stabilizer bar.

➡Place the cutout of the stabilizer bushing facing the rear side.

24. Install the stabilizer bracket to the front suspension crossmember with the 2 bolts, and tighten to 14 ft. lbs. (19 Nm).
25. Install the stabilizer brackets.
26. Install the front suspension crossmember.
27. Install the no.1 hook.
28. Connect the lower control arms.
29. Install the stabilizer links with the retaining nuts and tighten to 55 ft. lbs. (74 Nm).

➡Use a 6mm hex wrench to hold the stud if the ball joint turns together with the nut.

30. Connect pressure feed tube.
31. Connect the tie rod end.
32. Install the center exhaust pipe.
33. Install the front floor panel brace.
34. Install the front wheels
35. Connect the steering column hole cover.
36. Connect the steering intermediate shaft.
37. Install the column hole cover silencer sheet
38. Install the engine cover.
39. Add power steering fluid, then bleed the power steering system, as outlined in this section.
40. Check power steering fluid level in reservoir and add if necessary.
41. Check for fluid leaks
42. Install the hood. Check to make sure the hood is aligned properly.
43. Check and adjust front wheel alignment.
44. Install the engine under covers.

WHEEL HUB AND BEARING

REMOVAL & INSTALLATION

See Figures 259 through 261.

1. Before servicing the vehicle, refer to the Precautions Section.
2. Remove or disconnect the following:
    - Front wheel
    - Hub nut
    - Wheel speed sensor
    - Brake caliper
    - Front disc
    - Tie rod ends
    - Lower control arm
    - Axle from the hub assembly
    - Steering knuckle from strut

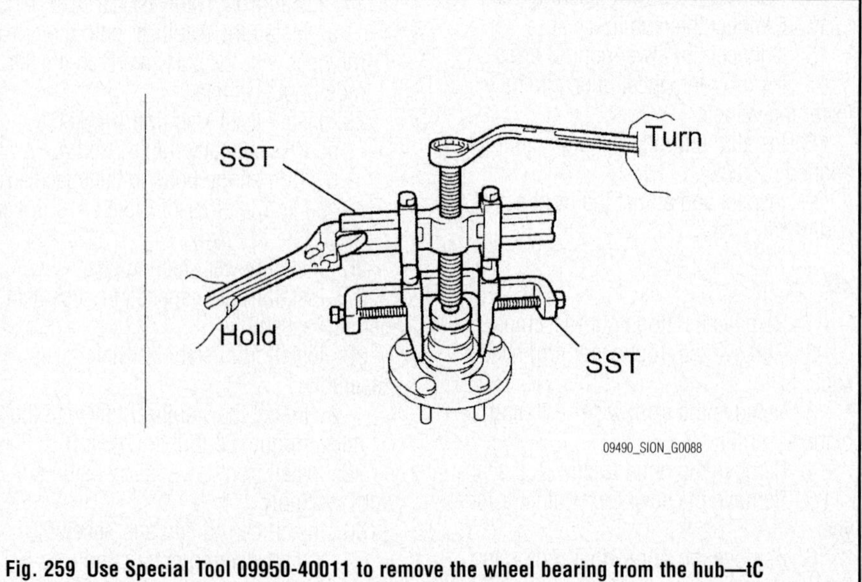

Fig. 259 Use Special Tool 09950-40011 to remove the wheel bearing from the hub—tC

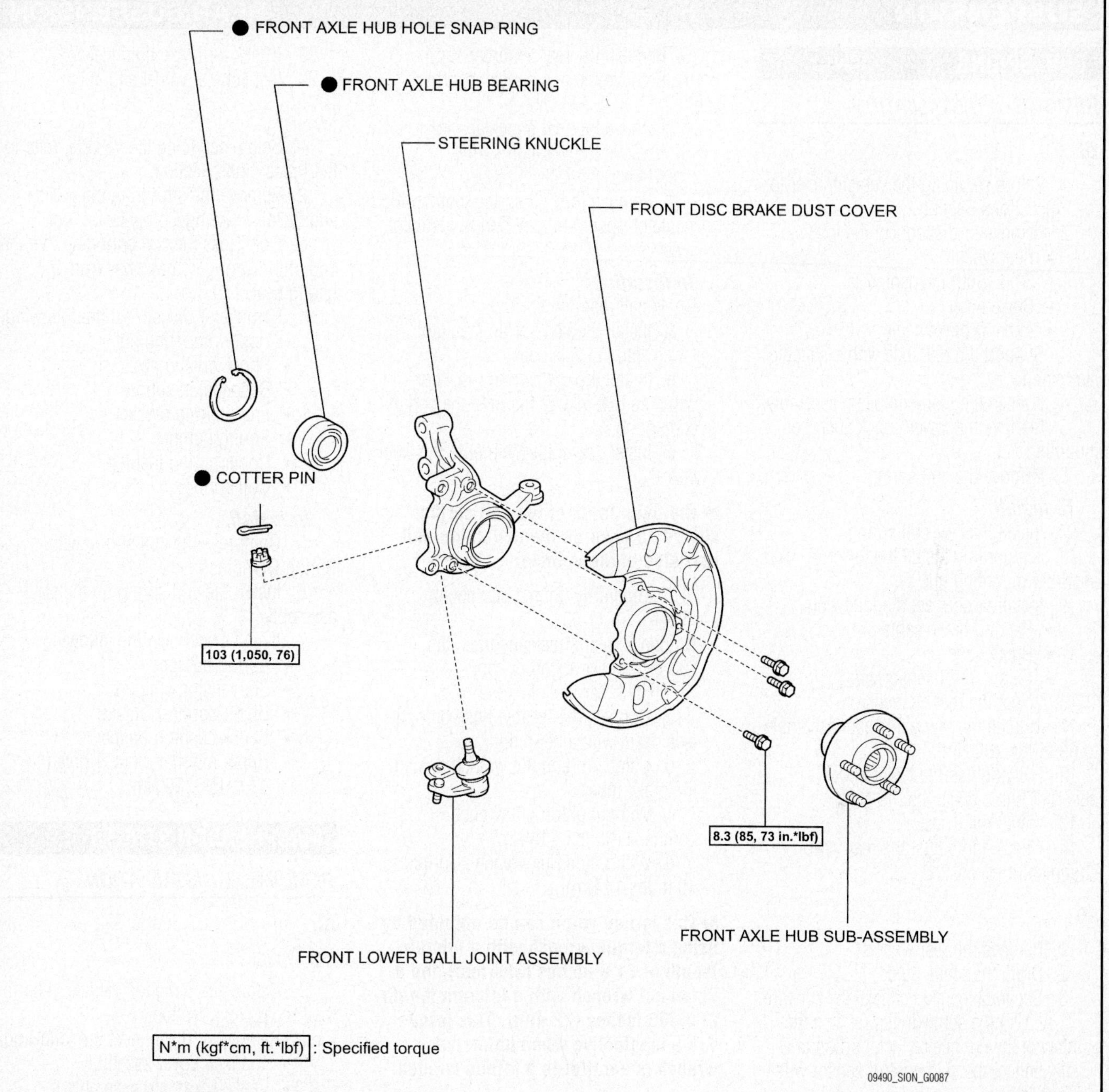

Fig. 260 Exploded view of the hub components—tC

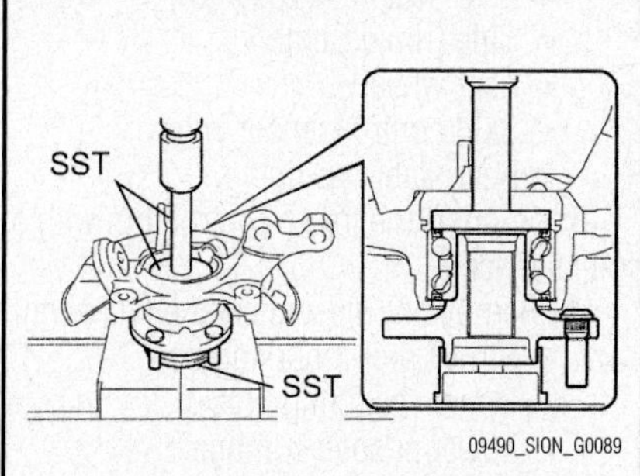

Fig. 261 Press the axle hub assembly into the steering knuckle

- Lower ball joint
- Axle hub snap ring

3. Mount the steering knuckle assembly in a vise.
4. Using Special Tool 09520-00031 to remove the axle hub from the steering knuckle.
5. Using Special Tool 09950-40011, remove the wheel bearing from the axle hub.

To install:

6. Use Special Tool 09950-60020 to press a new wheel bearing into the axle hub.
7. Using Special Tool 09608-32010, press the axle hub assembly into the steering knuckle.
8. Install or connect the following:
   - Axle hub snap ring
   - Lower ball joint
   - Steering knuckle mounting bolts
   - Axle into the hub assembly
   - Lower control arm
   - Tie rod ends
   - Front disc
   - Brake caliper
   - Wheel speed sensor
   - Hub nut
   - Front wheel

COIL SPRING

REMOVAL & INSTALLATION

xB

1. Before servicing the vehicle, refer to the Precautions Section.
2. Remove or disconnect the following:
   - Rear wheel
   - Skid control sensor wire
   - Brake hose
   - Parking brake cable
3. Support the rear axle with a suitable jack stand.
4. Loosen the rear axle beam assembly.
5. Remove the lower shock absorber mounting nut.
6. Remove the coil spring.

To install:

7. Install the rear coil spring.
8. Temporarily install the lower shock absorber mounting nut.
9. Install or connect the following:
   - Parking brake cable
   - Brake hose
   - Skid control sensor wire
10. Lower the rear suspension.
11. Tighten the rear axle beam assembly to 60 ft. lbs. (82 Nm).
12. Fully tighten the lower shock absorber nut to 36 ft. lbs (49 Nm).
13. Install the wheel.
14. Check and adjust the rear wheel alignment if necessary.

xD

1. Remove the rear wheels.
2. Drain the brake fluid.
3. Separate the skid control sensor wire.
   a. Using a screwdriver, remove the claw of the connector lock portion and disconnect the skid control sensor wire connector.

✱✱ CAUTION

Do not remove the connector cover from the connector because the skid control sensor wire may be damaged.

   b. Remove the nut and separate the skid control sensor wire.
4. Separate the rear flexible hose.
   a. Using a union nut wrench, separate the brake tube from the flexible hose.
   b. Remove the clip and disconnect the flexible hose from the axle beam.
5. Loosen the rear axle beam.
   a. Loosen the 2 bolts.

➡Do not remove the bolts.

6. Remove the rear absorber cap.
7. Remove the rear shock absorber cap.
8. Remove the rear shock absorber.
9. Remove the rear coil spring.
   a. Lower the jacks slowly.
   b. Remove the coil spring, coil spring insulator upper and coil spring insulator lower.

To install:

10. Install rear coil spring.
    a. Install the coil spring insulator lower onto the axle beam.
    b. Install the coil spring insulator upper so that its gap fits onto the end of coil spring.
    c. Install the coil spring onto the axle beam.

➡The paint mark of the coil spring should be towards the underside and rear side of the vehicle.

11. Temporarily tighten rear shock absorber.
12. Install rear shock absorber cap.
13. Install rear absorber cap.
14. Install rear flexible hose.
    a. Connect the flexible hose onto the axle beam with a new clip.
    b. Using a union nut wrench, install the brake tube.
    c. Without union nut wrench - Torque: 11 ft. lbs. (15 Nm).
    d. With union nut wrench - Torque: 10 ft. lbs.14 Nm).

➡This torque value can be obtained by using a torque wrench with a fulcrum length of 11.8 inches (300 mm) and a union nut wrench with a fulcrum length of 0.866 inches (22 mm). This torque value is effective when union nut wrench is parallel to a torque wrench.

15. Install skid control sensor wire.
    a. Install the skid control sensor wire onto the axle beam with the nut.
    b. Connect the skid control sensor wire connector.
16. Install rear wheels.
17. Stabilize suspension.
    a. Lower the vehicle from the jack.
    b. Bounce the vehicle up and down several times to stabilize the suspension.
18. Fully tighten rear axle beam.
19. Fully tighten rear shock absorber.
20. Fill reservoir with brake fluid.
21. Bleed master cylinder.
22. Bleed brake line.
23. Bleed brake actuator (w/ VSC).
24. Check fluid level in reservoir.
25. Check for brake fluid leakage.
26. Inspect rear wheel alignment.

tC

1. Before servicing the vehicle, refer to the Precautions Section.
2. Remove the strut from the vehicle and install a Spring Compressor Tool.
3. Compress the coil spring so that the end of the spring comes away from the spring seat.
4. Remove or disconnect the following:
   - Upper mounting nut
   - Shock cushion washer
   - Suspension support
   - Front spring bracket
   - Spring bumper
   - Upper spring insulator
   - Coil spring

To install:

5. Compress the coil spring with Spring Compress tool.
6. Install the coil spring on the strut assembly.
7. Install or connect the following:
   - Spring bumper
   - Suspension support
   - Upper spring insulator
   - Shock cushion washer
   - Upper mounting nut. Tighten to 41 ft. lbs. (56 Nm)

MACPHERSON STRUTS

REMOVAL & INSTALLATION

tC

See Figure 262.

1. Before servicing the vehicle, refer to the Precautions Section.
2. Remove or disconnect the following:
   - Tonneau cover assembly
   - Rear deck board assembly
   - Rear floor board
   - Rear seat cushion assembly
   - Rear seat back assembly
   - Side trim assembly
   - Rear wheel
   - Skid control sensor wire
   - Rear stabilizer link
3. Support the lower control arm with a suitable jack.
4. Remove or disconnect the following:
   - Strut lower mounting bolt
   - Upper mounting nuts
   - Lower suspension brace
   - Parking brake cable
5. Lower the jack and remove the strut assembly.

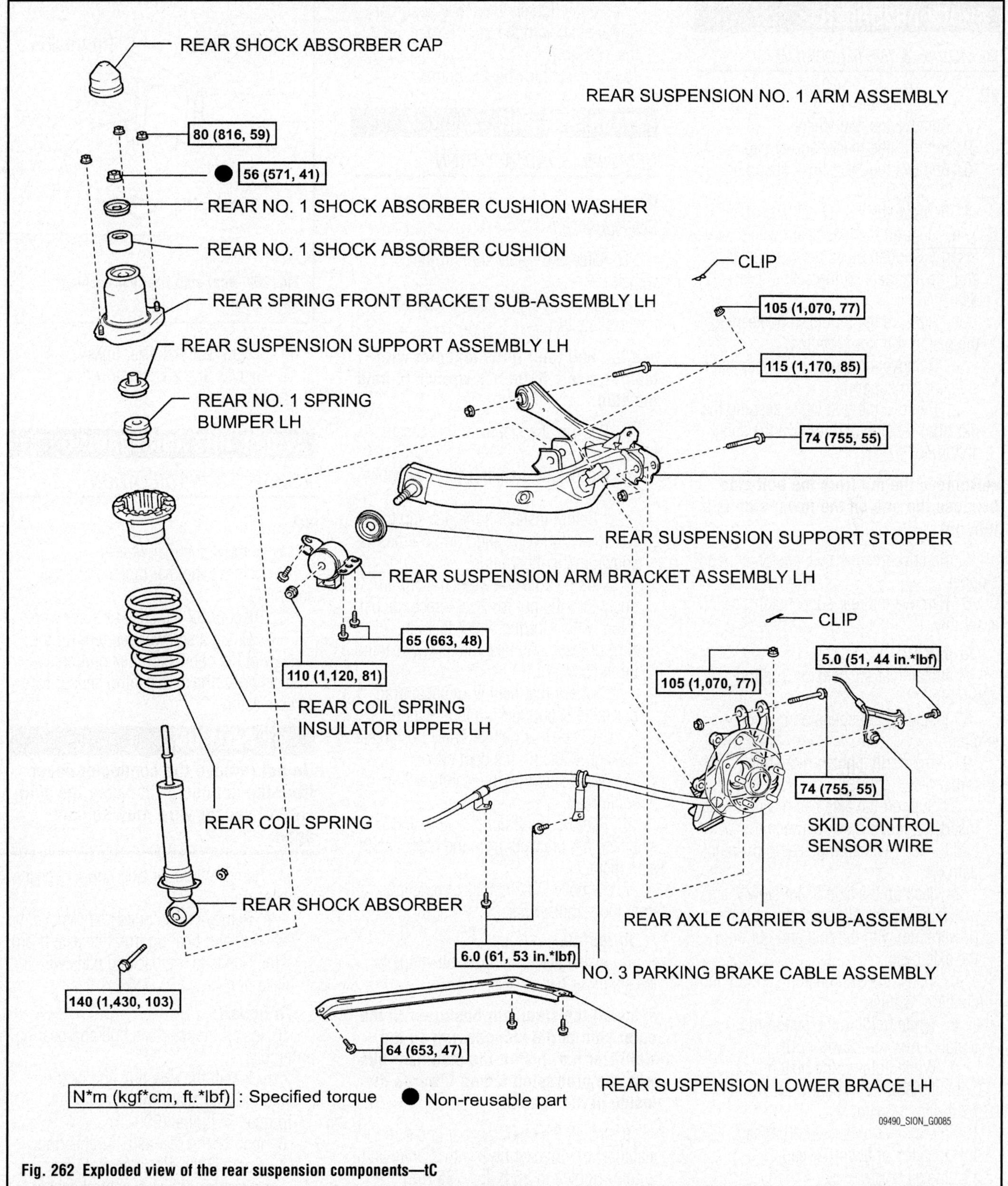

Fig. 262 Exploded view of the rear suspension components—tC

To install:

6. Install the strut assembly. Tighten the upper mounting nuts to 59 ft. lbs. (80 Nm). Temporarily tighten the lower mounting bolt.
7. Install or connect the following:
   - Parking brake cable
   - Lower suspension brace. Tighten to 47 ft. lbs. (64 Nm).
   - Rear stabilizer link
   - Skid control sensor wire
   - Rear wheel
   - Lower strut mounting bolt. Fully tighten to 103 ft. lbs. (140 Nm).
8. The remainder of the installation is the reverse order of removal.
9. Check and adjust the rear wheel alignment if necessary.

SHOCK ABSORBER

REMOVAL & INSTALLATION

xD

1. Remove the rear wheels.
2. Remove the rear absorber cap.
3. Remove the rear shock absorber cap.
4. Remove the rear shock absorber.
   a. Support the axle beam with a jack. Insert a wooden block between the jack and the rear axle spring seat to prevent damage.
   b. Remove the 2 nuts while keeping the piston rod from rotating.
   c. Remove the cushion retainer and suspension support.
   d. Remove the bolt while keeping the nut from rotating and remove the shock absorber.

➡Remove the nut from the bolt side because the one on the lower side is a jam nut.

5. Remove the rear suspension support stopper.
6. Remove the rear suspension support assembly.

To install:

7. Install rear suspension support assembly.
8. Install rear suspension support stopper.
9. Temporarily tighten rear shock absorber.
   a. Support the axle beam with a jack. Insert a wooden block between the jack and the rear axle spring seat to prevent damage.
   b. Jack up the axle beam slowly, and provisionally install the shock absorber (lower side) with the bolt and nut onto the axle beam.
   c. Install the suspension support and cushion retainer.
   d. While holding the piston rod, install a new nut (lower nut).
   e. While holding the piston rod, tighten a new nut (upper nut). Torque: 18 ft. lbs. (25 Nm).
10. Install rear shock absorber cap.
11. Install rear absorber cap.
12. Install rear wheels.
13. Stabilize suspension.
   a. Lower the vehicle from the jack.
   b. Bounce the vehicle up and down several times to stabilize the suspension.
14. Fully tighten rear shock absorber.
   a. Fully tighten the shock absorber (lower side) with the bolt. Torque: 36 ft. lbs. (49 Nm).
15. Inspect rear wheel alignment.

STABILIZER BAR

REMOVAL & INSTALLATION

tC

See Figure 263.

1. Raise and safely support the vehicle.
2. Remove the 2 nuts and the stabilizer link assemblies.

➡If the ball joint turns together with the nut, use a 5mm hex wrench to hold the stud.

3. Remove the stabilizer link assemblies.
4. Inspect the stabilizer link assembly, as follows:
   a. Before installing the nut, flip the ball joint stud back and Forth 5 times as shown in the illustration.
   b. Using a torque wrench, continuously turn the nut for 2 to 4 seconds per 1 turn with a torque of 8.7 inch lbs. (1 Nm) or less,, and take the torque reading on the 5th turn.
   c. Check that neither unusual drag nor rattle occurs during the rotation.
   d. Check that neither crack nor grease leakage exists on the dust cover.
5. Make sure that the stabilizer link is not deformed.
6. Remove the 2 bolts, 2 nuts and 2 stabilizer bar brackets from the rear suspension member.
7. Remove the 2 stabilizer bushings from the stabilizer bar.

To install:

8. Install the 2 stabilizer bushings to the stabilizer bar.

➡Install the stabilizer bush rear to the outer side of the stopper ring on the stabilizer bar. Install the stabilizer bush with the protrusion facing towards the inside of the vehicle.

9. Install the stabilizer bar and 2 rear stabilizer bar Bracket no.3 with 2 bolts and 2 nuts. Tighten to 26 ft. lbs. (35 Nm).
10. Install the rear stabilizer link assembly with the 2 nuts and tighten to 32 ft. lbs. (44 Nm).

➡If the ball joint turns together with the nut, use a 5mm hex wrench to hold the stud.

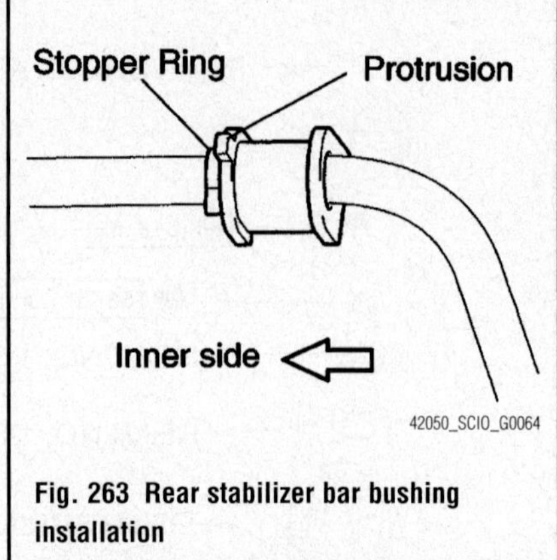

Fig. 263 Rear stabilizer bar bushing installation

11. Install rear stabilizer links.
12. Inspect and adjust rear wheel alignment.

WHEEL HUB AND BEARING

REMOVAL & INSTALLATION

xD

1. Remove the rear wheels.
2. Remove the rear brake drum sub-assembly.
3. Disconnect the speed sensor wire.
   a. Using a screwdriver, remove the claw of the connector lock portion and disconnect the skid control sensor wire connector.

CAUTION

Do not remove the connector cover from the connector because the skid control sensor wire may be damaged.

4. Remove the rear axle hub and bearing assembly.
   a. Remove the 4 bolts and remove the axle hub and bearing from the axle beam.
   b. Suspend the backing plate with a piece of rope.

To install:

5. Install the rear axle hub and bearing assembly.
   a. Install the axle hub and bearing onto the axle beam with the 4 bolts. Torque: 67 ft. lbs. (90 Nm)
6. Inspect the rear axle hub bearing.
7. Connect the speed sensor wire.
   a. Connect the skid control sensor wire connector.
8. Install the rear brake drum sub-assembly.
9. Adjust the rear drum brake shoe clearance.
10. Install the rear wheels.

LEXUS AND TOYOTA

GX470 • 4Runner

9

SPECIFICATIONS AND MAINTENANCE CHARTS

ENGINE AND VEHICLE IDENTIFICATION

Engine							Model Year	
Code ①	Liters (cc)	Cu. In.	Cyl.	Fuel Sys.	Engine Type	Eng. Mfg.	Code ②	Year
1GR-FE	4.0 (3956)	241	6	SFI	DOHC	Toyota	7	2007
2UZ-FE	4.7 (4664)	285	8	SFI	DOHC	Toyota	8	2008

SFI: Sequential Fuel Injection

DOHC: Double Overhead Camshaft

① Stamped on the left side of the engine block

② 10th digit of the Vehicle Identification Number (VIN)

22140_4RUN_C0001

GENERAL ENGINE SPECIFICATIONS

Year	Model	Engine Displacement Liters	Engine Series ID	Net Horsepower @ rpm	Net Torque @ rpm (ft. lbs.)	Bore x Stroke (in.)	Com-pression Ratio	Oil Pressure @ rpm
2007	4Runner	4.0	1GR-FE	236@5200	266@3400	3.70x3.74	10:01	45-65@3000
	4Runner	4.7	2UZ-FE	260@5400	306@3400	3.70x3.31	10:01	45-65@3000
	GX470	4.7	2UZ-FE	260@5400	306@3400	3.70x3.31	10:01	45-65@3000
2008	4Runner	4.0	1GR-FE	236@5200	266@3400	3.70x3.74	10:01	45-65@3000
	4Runner	4.7	2UZ-FE	260@5400	306@3400	3.70x3.31	10:01	45-65@3000
	GX470	4.7	2UZ-FE	260@5400	306@3400	3.70x3.31	10:01	45-65@3000

22140_4RUN_C0002

ENGINE TUNE-UP SPECIFICATIONS

Year	Engine Displacement Liters	Engine ID	Spark Plug Gap (in.)	Ignition Timing (deg.)*	Fuel Pump (psi)	Idle Speed (rpm)		Valve Clearance	
						MT	AT	Intake	Exhaust
2007	4.0	1GR-FE	0.043	N/A	38-44	—	650-750	0.006-0.010	0.011-0.015
	4.7	2UZ-FE	0.043	N/A	38-44	—	650-750	0.006-0.010	0.010-0.014
2008	4.0	1GR-FE	0.043	N/A	38-44	—	650-750	0.006-0.010	0.011-0.015
	4.7	2UZ-FE	0.043	N/A	38-44	—	650-750	0.006-0.010	0.010-0.014

NOTE: The Vehicle Emission Control Information label often reflects specification changes made during production.

The label figures must be used if they differ from those in this chart.

B: Before top dead center

* With terminals TC and E1 connected to DLC1 or for 5.7L Terminal TC and CG of DLC3 connected

22140_4RUN_C0003

CAPACITIES

Year	Model	Engine Displacement Liters	Engine ID	Engine Oil with Filter (qts.)	Transmission (qts.)		Transfer Case (pts.)	Drive Axle		Fuel Tank (gal.)	Cooling System (qts.)
					5-Spd	Auto.*		Front (pts.)	Rear (pts.)		
2007	4Runner	4.0	1GR-FE	5.5	—	11.3	3.0	3.2	6.4	23.0	10.4
	4Runner	4.7	2UZ-FE	6.5		11.3	3.0	3.0	6.4	23.0	13.0
	GX470	4.7	2UZ-FE	6.5	—	11.0	3.0	3.9	6.4	23.0	13.6
2008	4Runner	4.0	1GR-FE	5.5	—	11.3	3.0	3.2	6.4	23.0	10.4
	4Runner	4.7	2UZ-FE	6.5		11.3	3.0	3.0	6.4	23.0	13.0
	GX470	4.7	2UZ-FE	6.5	—	11.0	3.0	3.9	6.4	23.0	13.6

*After draining, add the following amounts, then fill to the cold full line

22140_4RUN_C0004

VALVE SPECIFICATIONS

Year	Engine Displacement Liters	Engine ID	Seat Angle (deg.)	Face Angle (deg.)	Spring Test Pressure (lbs. @ in.)	Spring Installed Height (in.)	Stem-to-Guide Clearance (in.)		Stem Diameter (in.)	
							Intake	Exhaust	Intake	Exhaust
2007	4.0	1GR-FE	45	44.5	41.9-46.3@ 1.311	1.311	0.0010-0.0024	0.0012-0.0026	0.2154-0.2159	0.2152-0.0158
	4.7	2UZ-FE	45	44.5	47.2-50.7@ 1.378	1.380	0.0010-0.0024	0.0012-0.0026	0.2154-0.2159	0.2152-0.2157
2008	4.0	1GR-FE	45	44.5	41.9-46.3@ 1.311	1.311	0.0010-0.0024	0.0012-0.0026	0.2154-0.2159	0.2152-0.0158
	4.7	2UZ-FE	45	44.5	47.2-50.7@ 1.378	1.380	0.0010-0.0024	0.0012-0.0026	0.2154-0.2159	0.2152-0.2157

22140_4RUN_C0005

CRANKSHAFT AND CONNECTING ROD SPECIFICATIONS

All measurements are given in inches.

Year	Engine Displacement Liters	Engine ID	Crankshaft Main Brg. Journal Dia.	Crankshaft Main Brg. Clearance	Crankshaft Shaft End-play	Crankshaft Thrust on No.	Connecting Rod Journal Diameter	Connecting Rod Oil Clearance	Connecting Rod Side Clearance
2007	4.0	1GR-FE	2.8342-2.8346	0.0007-0.0018	–	2	2.2044-2.2047	0.0010 0.0018	0.0059 0.0138
	4.7	2UZ-FE	2.6373-2.6378	①	0.0008-0.0087	3	2.0465-2.0472	0.0011-0.0021	0.0063-0.0138
2008	4.0	1GR-FE	2.8342-2.8346	0.0007-0.0018	–	2	2.2044-2.2047	0.0010 0.0018	0.0059 0.0138
	4.7	2UZ-FE	2.6373-2.6378	①	0.0008-0.0087	3	2.0465-2.0472	0.0011-0.0021	0.0063-0.0138

① Nos. 1 and 2: 0.0011-0.0018

All others: 0.0016-0.0023

22140_4RUN_C0006

PISTON AND RING SPECIFICATIONS

All measurements are given in inches.

Year	Engine Displacement Liters	Engine ID	Piston Clearance	Ring Gap Top Comp.	Ring Gap Bottom Comp.	Ring Gap Oil Control	Ring Side Clearance Top Comp.	Ring Side Clearance Bottom Comp.	Ring Side Clearance Oil Control
2007	4.0	1GR-FE	0.0031-0.0051	0.0118 0.0157	0.0157-0.0197	①	0.0008 0.0028	0.0008 0.0024	0.0028 0.0060
	4.7	2UZ-FE	0.0035-0.0044	0.0118-0.0157	0.0157-0.0217	0.0051-0.0150	0.0012-0.0031	0.0012-0.0028	SNUG
2008	4.0	1GR-FE	0.0031-0.0051	0.0118 0.0157	0.0157-0.0197	①	0.0008 0.0028	0.0008 0.0024	0.0028 0.0060
	4.7	2UZ-FE	0.0035-0.0044	0.0118-0.0157	0.0157-0.0217	0.0051-0.0150	0.0012-0.0031	0.0012-0.0028	SNUG

① No 1: 0.039, No 2: 0.043, No 3: 0.039

22140_4RUN_C0007

TORQUE SPECIFICATIONS

All readings in ft. lbs.

Year	Engine Displacement Liters	Engine ID	Cylinder Head Bolts	Main Bearing Bolts	Rod Bearing Bolts	Crankshaft Damper Bolts	Flywheel Bolts	Manifold Intake	Manifold Exhaust	Spark Plugs	Oil Pan Drain Plug
2007	4.0	1GR-FE	①	②	③	184	61	19	22	13	29
	4.7	2UZ-FE	④	⑤	⑥	181	⑦	13	33	13	29
2008	4.0	1GR-FE	①	②	③	184	61	19	22	13	29
	4.7	2UZ-FE	④	⑤	⑥	181	⑦	13	33	13	29

① Step 1: 45 ft. lbs.
Step 2: Plus 180 degrees

② Step 1: 16 main cap bolts 45 ft. lbs.
Step 2: Plus 90 degr
Step 3: 8 side main cap bolts 19 ft. lbs.

③ Step 1: 18 ft. lbs.
Step 2: Plus 90 degrees

④ Step 1: 30 ft. lbs.
Step 2: Plus 90 degrees
Step 3: Plus 90 degrees

⑤ Step 1: 20 ft. lbs.
Step 2: Plus 90 degrees

⑥ Step 1: 18 ft. lbs.
Step 2: Plus 90 degrees

⑦ Step 1: 22 ft. lbs.
Step 2: Plus 90 degrees

22140_4RUN_C0008

WHEEL ALIGNMENT

Year	Model	Caster Range (+/-Deg.)	Caster Preferred Setting (Deg.)	Camber Range (+/-Deg.)	Camber Preferred Setting (Deg.)	Toe-in (in.)	Steering Axis Inclination (Deg.)
2007	4Runner	0.75	①	0.75	②	0.08+/-0.16	③
	GX470	0.75	+3.28	0.75	-0.02	0.08+/-0.16	12.48+/-0.75
2008	4Runner	0.75	①	0.75	②	0.08+/-0.16	③
	GX470	0.75	+3.28	0.75	-0.02	0.08+/-0.16	12.48+/-0.75

Note: All algnment specifications are based on nominal ride height and standard tires

① 2WD except air suspension +3.38
2WD with air suspension +3.55
4WD except air suspension +3.22
4WD with air suspension +3.37

② 2WD except air suspension -0.47
2WD with air suspension -0.50
4WD except air suspens
4WD with air suspension -0.17

③ 2WD except air suspension 12.97+/-.075
2WD with air suspension 13.00+/-.075
4WD except air suspension 12.65+/-.075
4WD with air suspension 12.67+/-.075

22140_4RUN_C0009

TIRE, WHEEL AND BALL JOINT SPECIFICATIONS

Year	Model	OEM Tires		Tire Pressures (psi)		Wheel Size	Ball Joint Inspection	Lug Nut Torque (ft. lbs.)
		Standard	Optional	Front	Rear			
2007	4Runner	P265/70R16	P265/65R17	①	①	7-JJ/7.5JJ	②	83
	GX470	P265/65SR17	None	①	①	7.5JJ	②	83
2008	4Runner	P265/70R16	P265/65R17	①	①	7-JJ/7.5JJ	②	83
	GX470	P265/65SR17	None	①	①	7.5JJ	②	83

OEM: Original Equipment Manufacturer

① See placard on vehicle

② Upper arm ball joint turning torque: 40 inch lbs.

Lower arm ball joint turning torque: 27 inch lbs.

22140_4RUN_C0010

BRAKE SPECIFICATIONS

All measurements in inches unless noted

Year	Model		Brake Disc Original Thickness	Brake Disc Minimum Thickness	Brake Disc Maximum Runout	Minimum Lining Thickness	Brake Caliper Bracket Bolts (ft. lbs.)	Brake Caliper Mounting Bolts (ft. lbs.)
2007	4Runner	F	1.102	1.024	0.0020	0.039	—	91
		R	0.709	0.630	0.0079	0.039	77	65
	GX 470	F	1.102	1.024	0.0020	0.039	—	91
		R	0.709	0.630	0.0079	0.039	77	65
2008	4Runner	F	1.102	1.024	0.0020	0.039	—	91
		R	0.709	0.630	0.0079	0.039	77	65
	GX 470	F	1.102	1.024	0.0020	0.039	—	91
		R	0.709	0.630	0.0079	0.039	77	65

F: Front

R: Rear

22140_4RUN_C0011

SCHEDULED MAINTENANCE INTERVALS

GX470/4Runner

TO BE SERVICED	TYPE OF SERVICE	VEHICLE MILEAGE INTERVAL (x1000)													
		5	10	15	20	25	30	35	40	45	50	55	60	90	120
Engine oil & filter	R	✓	✓	✓	✓	✓	✓	✓	✓	✓	✓	✓	✓	✓	✓
Automatic transmission fluid	S/I			✓			✓			✓			✓	✓	✓
Ball joints & dust covers	S/I			✓			✓			✓			✓	✓	✓
Bolts & nuts on chassis & body	S/I			✓			✓			✓			✓	✓	✓
Brake linings & drums	S/I	✓	✓	✓	✓	✓	✓	✓	✓	✓	✓	✓	✓	✓	✓
Brake line pipes & hoses	S/I			✓			✓			✓			✓	✓	✓
Brake pads & discs (front & rear)	S/I	✓	✓	✓	✓	✓	✓	✓	✓	✓	✓	✓	✓	✓	✓
Brake fluid	R						✓						✓	✓	✓
Rack and pinion assembly	S/I			✓			✓			✓			✓	✓	✓
Steering linkage & boots	S/I			✓			✓			✓			✓	✓	✓
Air cleaner filter	R						✓						✓	✓	✓
Spark plugs ①	R														✓
Drive belts	S/I												✓	✓	✓
Exhaust pipes & mountings	S/I			✓			✓			✓			✓	✓	✓
Fuel lines & connections	S/I						✓						✓	✓	✓
Engine coolant ②	S/I			✓			✓			✓			✓	✓	
Rear differential & transfer case oil	S/I			✓			✓			✓			✓	✓	✓
Fuel tank cap gasket	S/I						✓						✓	✓	✓
Rotate tires	S/I			✓			✓			✓			✓		✓
Clean air conditioning filter ③	S/I			✓			✓			✓			✓		✓
Axle shaft bolts	S/I			✓			✓			✓			✓	✓	✓
Brake pad thickness and rotor	S/I						✓						✓	✓	✓

R: Replace S/I: Service or Inspect

① Spark plugs are replaced at 120,000 miles

② Replace engine coolant at 100,000 miles and then inspect every 15,000 miles

③ Replace air conditioning filter every 30,000 miles

FREQUENT OPERATION MAINTENANCE (SEVERE SERVICE)

If a vehicle is operated under any of the following conditions it is considered severe service:

- Extremely dusty areas.

- 50% or more of the vehicle operation is in 32°C (90°F) or higher temperatures, or constant temperatures below 0°C (32°F).

- Prolonged idling (vehicle operation in stop and go traffic).

- Frequent short running periods (engine does not warm to normal operating temperatures).

- Police, taxi, delivery usage or trailer towing usage.

Air cleaner filter: service or inspect every 5000 miles

Rear differential & transfer case oil: replace every 15,000 miles.

Ball joints & dust covers: service or inspect every 5000 miles.

Bolts & nuts on chassis & body: service or inspect every 5000 miles.

Axle shaft bolts: service or inspect every 5000 miles.

Steering linkage: service or inspect every 5000 miles.

22140_4RUN_C0012

PRECAUTIONS

Before servicing any vehicle, please be sure to read all of the following precautions, which deal with personal safety, prevention of component damage, and important points to take into consideration when servicing a motor vehicle:

- Never open, service or drain the radiator or cooling system when the engine is hot; serious burns can occur from the steam and hot coolant.
- Observe all applicable safety precautions when working around fuel. Whenever servicing the fuel system, always work in a well-ventilated area. Do not allow fuel spray or vapors to come in contact with a spark, open flame, or excessive heat (a hot drop light, for example). Keep a dry chemical fire extinguisher near the work area. Always keep fuel in a container specifically designed for fuel storage; also, always properly seal fuel containers to avoid the possibility of fire or explosion. Refer to the additional fuel system precautions later in this section.
- Fuel injection systems often remain pressurized, even after the engine has been turned **OFF**. The fuel system pressure must be relieved before disconnecting any fuel lines. Failure to do so may result in fire and/or personal injury.
- Brake fluid often contains polyglycol ethers and polyglycols. Avoid contact with the eyes and wash your hands thoroughly after handling brake fluid. If you do get brake fluid in your eyes, flush your eyes with clean, running water for 15 minutes. If eye irritation persists, or if you have taken brake fluid internally, IMMEDIATELY seek medical assistance.
- The EPA warns that prolonged contact with used engine oil may cause a number of skin disorders, including cancer. You should make every effort to minimize your exposure to used engine oil. Protective gloves should be worn when changing oil. Wash your hands and any other exposed skin areas as soon as possible after exposure to used engine oil. Soap and water, or waterless hand cleaner should be used.
- All new vehicles are now equipped with an air bag system, often referred to as a Supplemental Restraint System (SRS) or Supplemental Inflatable Restraint (SIR) system. The system must be disabled before performing service on or around system components, steering column, instrument panel components, wiring and sensors. Failure to follow safety and disabling procedures could result in accidental air bag deployment, possible personal injury and unnecessary system repairs.
- Always wear safety goggles when working with, or around, the air bag system. When carrying a non-deployed air bag, be sure the bag and trim cover are pointed away from your body. When placing a non-deployed air bag on a work surface, always face the bag and trim cover upward, away from the surface. This will reduce the motion of the module if it is accidentally deployed. Refer to the additional air bag system precautions later in this section.
- Clean, high quality brake fluid from a sealed container is essential to the safe and proper operation of the brake system. You should always buy the correct type of brake fluid for your vehicle. If the brake fluid becomes contaminated, completely flush the system with new fluid. Never reuse any brake fluid. Any brake fluid that is removed from the system should be discarded. Also, do not allow any brake fluid to come in contact with a painted surface; it will damage the paint.
- Never operate the engine without the proper amount and type of engine oil; doing so WILL result in severe engine damage.
- Timing belt maintenance is extremely important. Many models utilize an interference-type, non-freewheeling engine. If the timing belt breaks, the valves in the cylinder head may strike the pistons, causing potentially serious (also time-consuming and expensive) engine damage. Refer to the maintenance interval charts for the recommended replacement interval for the timing belt, and to the timing belt section for belt replacement and inspection.
- Disconnecting the negative battery cable on some vehicles may interfere with the functions of the on-board computer system(s) and may require the computer to undergo a relearning process once the negative battery cable is reconnected.
- When servicing drum brakes, only disassemble and assemble one side at a time, leaving the remaining side intact for reference.
- Only an MVAC-trained, EPA-certified automotive technician should service the air conditioning system or its components.

BRAKES — ANTI-LOCK BRAKE SYSTEM (ABS)

GENERAL INFORMATION

The ABS helps prevent the wheels from locking when the brakes are applied firmly or when braking on a slippery surface. The skid control ECU detects wheel lock condition by receiving vehicle speed signals from each speed sensor, and sends control signals to the pump motor and solenoid valve. The pump motor and solenoid valve avoid wheel lock by controlling the oil pressure of each wheel cylinder. The EBD control utilizes ABS, realizing proper brake force distribution between the front and rear wheels in accordance with driving conditions. In addition, when braking while cornering, it also controls the brake forces of the right and left wheels, helping to maintain vehicle behavior.

BRAKES — BLEEDING THE BRAKE SYSTEM

BLEEDING PROCEDURE

➡If any work is done on the brake system or if air is suspected in the brake lines, bleed the air from the system.

➡Do not let brake fluid remain on a painted surface. Wash it off immediately.

1. Before servicing the vehicle, refer to the precautions in the beginning of this section.
2. Check the fluid level in the reservoir after bleeding each wheel. Add DOT3 fluid, if necessary.
3. If the hydraulic brake booster was disassembled or if the reservoir becomes empty, bleed the air from the hydraulic brake booster as follows:

➡ Perform this step only if the brake booster with accumulator pump assembly is removed and/or installed.

a. Turn the ignition switch OFF, depress the brake pedal 20 times or more to release the pressure from the accumulator.
b. Fully depress the brake pedal 10 times.
c. Turn the ignition switch to the ON position and start the brake booster pump.
d. Make sure the pump operates for 8 to 14 seconds.

➡If the pump does not operate as specified, repeat the above and recheck the operating time.

4. Bleeding Front Brake Lines:
a. Turn the ignition switch to the ON position and wait until the pump motor has stopped.
b. Connect the vinyl tube to the brake caliper.
c. Depress the brake pedal several times, then loosen the bleeder plug with the pedal held down.
d. At the point when the fluid stops coming out, tighten the bleeder plug,, 8 ft. lbs. (11 Nm) then release the brake pedal.
e. Repeat procedure until all the air in the fluid has been bled out.
f. Repeat the above procedures to bleed the other brake line.
5. Bleeding Rear Brake Lines:
a. Turn the ignition switch to the ON position and depress the brake pedal.
b. Connect the vinyl tube to the brake caliper.
c. Loosen the bleeder plug and release air.

➡Brake fluid is sent through the pump, so keep the brake pedal depressed until the air is completely bled out.

d. When the air is completely bled out of the brake fluid through the bleeder plug, tighten the bleeder plug to 8 ft. lbs. (11 Nm) then release.
e. Repeat the above procedures to bleed the other brake line.
6. Bleeding Master Cylinder Solenoid is only possible with a Toyota proprietary scan system.

BRAKES — FRONT DISC BRAKES

⁂ CAUTION

Dust and dirt accumulating on brake parts during normal use may contain asbestos fibers from production or aftermarket brake linings. Breathing excessive concentrations of asbestos fibers can cause serious bodily harm. Exercise care when servicing brake parts. Do not sand or grind brake lining unless equipment used is designed to contain the dust residue. Do not clean brake parts with compressed air or by dry brushing. Cleaning should be done by dampening the brake components with a fine mist of water, then wiping the brake components clean with a dampened cloth. Dispose of cloth and all residue containing asbestos fibers in an impermeable container with the appropriate label. Follow practices prescribed by the Occupational Safety and Health Administration (OSHA) and the Environmental Protection Agency (EPA) for the handling, processing, and disposing of dust or debris that may contain asbestos fibers.

BRAKE CALIPER

REMOVAL & INSTALLATION

See Figure 1.

1. Remove the wheel.
2. Remove the anti-rattle spring from the caliper.
3. Remove the clips and anti-rattle pins.
4. Lift out the pads and shims.
5. If the caliper is being replaced, disconnect the brake line. Plug the line to prevent fluid loss.
6. Remove the caliper mounting bolts. Lift off the caliper.
7. Installation is the reverse of removal. Bleed the brakes. Observe the following torques:
   - Caliper mounting bolts: 91 ft. lbs. (123 Nm)
   - Brake line-to-caliper: 11 ft. lbs. (15 Nm)

DISC BRAKE PADS

REMOVAL & INSTALLATION

1. Raise the vehicle and support it safely.
2. Remove the wheels.
3. Remove the clip, pins and anti-rattle spring.
4. Withdraw the pads and remove the anti-squeal shims.

To install:

5. Before installing the new pads, check the disc thickness and disc runout.
6. Siphon out a small amount of brake fluid from the reservoir.
7. Press in the pistons with a hammer handle or equivalent.
8. Apply disc brake grease to both sides of the inner anti-squeal shim. Install the anti-squeal shims to the new pads.
9. Install the pads.
10. Install the anti-rattle springs and pins. Install the clip.
11. Install the wheels.
12. Check and adjust the fluid level. Apply the brake pedal several times.
13. Road-test the vehicle for proper operation.

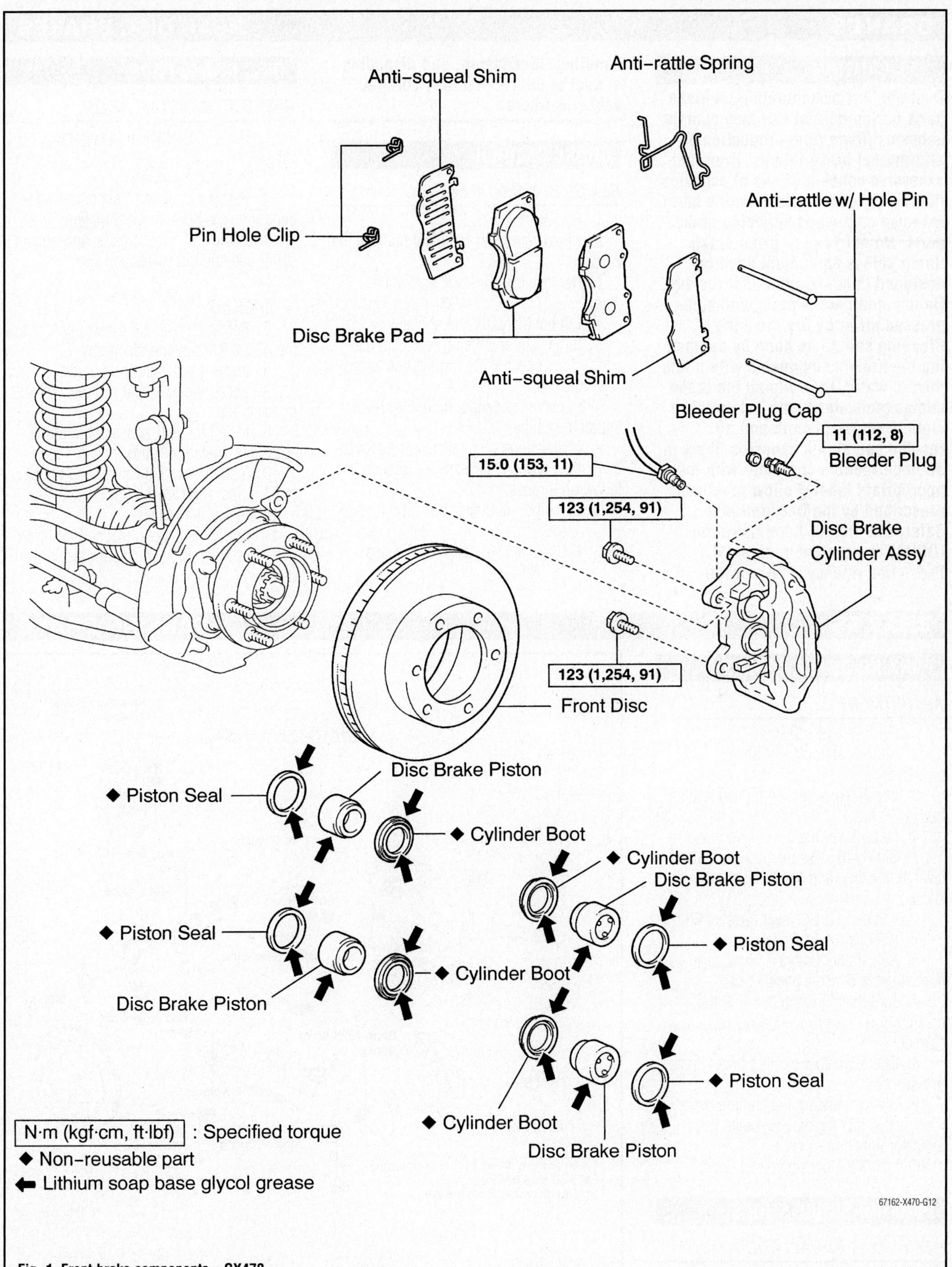

Fig. 1 Front brake components—GX470

BRAKES — REAR DISC BRAKES

** CAUTION

Dust and dirt accumulating on brake parts during normal use may contain asbestos fibers from production or aftermarket brake linings. Breathing excessive concentrations of asbestos fibers can cause serious bodily harm. Exercise care when servicing brake parts. Do not sand or grind brake lining unless equipment used is designed to contain the dust residue. Do not clean brake parts with compressed air or by dry brushing. Cleaning should be done by dampening the brake components with a fine mist of water, then wiping the brake components clean with a dampened cloth. Dispose of cloth and all residue containing asbestos fibers in an impermeable container with the appropriate label. Follow practices prescribed by the Occupational Safety and Health Administration (OSHA) and the Environmental Protection Agency (EPA) for the handling, processing, and disposing of dust or debris that may contain asbestos fibers.

BRAKE CALIPER

REMOVAL & INSTALLATION

1. Remove the wheel.
2. Remove the anti-rattle spring from the caliper.
3. Remove the clips and anti-rattle pins.
4. Lift out the pads and shims.
5. If the caliper is being replaced, disconnect the brake line. Plug the line to prevent fluid loss.
6. Remove the caliper mounting bolts. Lift off the caliper.
7. Installation is the reverse of removal. Bleed the brakes as necessary . Observe the following torques:
   - Caliper mounting bolts: 65 ft. lbs. (88 Nm)
   - Brake line-to-caliper: 23 ft. lbs. (31 Nm)

DISC BRAKE PADS

REMOVAL & INSTALLATION

1. Raise the vehicle and support it safely.
2. Remove the wheels.
3. Remove the brake caliper and suspend it so the hose is not stretched.
4. Remove the brake pads, anti-squeal shim, pad support plates and wear indicators.

To install:

5. Before installing the new pads, check the disc thickness and disc runout.
6. Install the pad support plates.
7. Install the pad wear indicator plates on each pad.
8. Install the anti-squeal shim to the outer pad. Install the pads.
9. Install the brake caliper and tighten to 65 ft. lbs. (88 Nm).
10. Install the wheels.
11. Apply the brake pedal several times.
12. Road-test the vehicle for proper operation.

BRAKES — PARKING BRAKE

PARKING BRAKE CABLES

ADJUSTMENT

1. Remove rear wheel.
2. Adjust parking brake shoe clearance.
3. Install rear wheel an tighten to 82 ft. lbs. (112 Nm)
4. Inspect parking brake lever travel.
5. Slowly depress the parking brake lever all the way, and count the number of clicks.
   a. Parking brake lever travel at 66 ft. lbs. (294 Nm) 5 to 7 clicks.
6. Adjust parking brake lever travel by removing the console panel upper.
7. Turn the adjusting nut until the parking brake lever travel becomes correct.
8. Check whether parking brake drags or not.
9. When operating the parking brake lever, check that the parking brake lever indicator light comes on.
10. Install the console panel upper.

PARKING BRAKE SHOES

REMOVAL & INSTALLATION

See Figure 2.

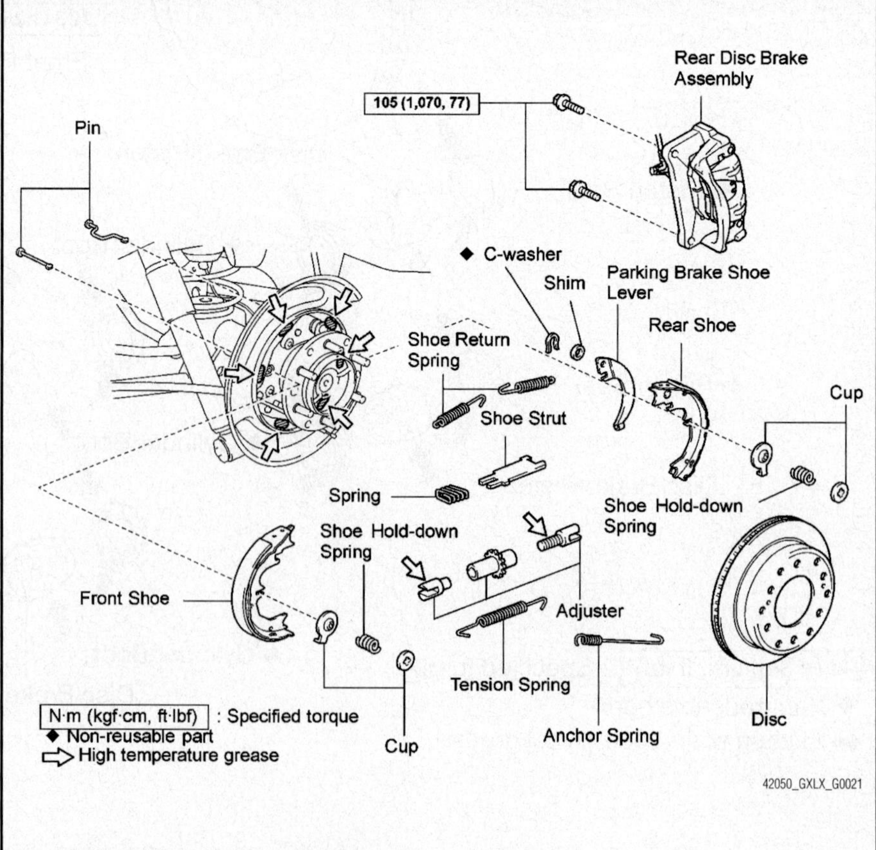

Fig. 2 Exploded view of the parking brake

1. Before servicing the vehicle, refer to the precautions in the beginning of this section.
2. Raise and safely support the vehicle.
3. Remove the rear wheel.
4. Remove the 2 mounting bolts and remove the disc brake assembly.
5. Suspend the disc brake securely and so the hose is not stretched.
6. Release the parking brake lever.
7. Place matchmarks on the disc and rear axle hub.
8. Remove the disc.

➡If the disc cannot be removed easily, turn the shoe adjuster until the wheel turns freely.

9. Using needle-nose pliers, remove the 2 shoe return springs.

➡At the time of reassembly, install the strut with the spring facing forward.

10. Slide the front shoe toward outside and remove the shoe adjuster.
11. Using a needle-nose pliers, disconnect the anchor spring and tension spring from the front shoe.
12. Using a needle-nose pliers, disconnect the anchor spring and tension spring from the rear shoe.
13. Installation is the reverse of removal.

ADJUSTMENT

1. Before servicing the vehicle, refer to the precautions in the beginning of this section.
2. Turn the adjuster and expand the shoes until the disc locks.
3. Return the adjuster 8 notches.
4. Depress the parking brake pedal with 33 ft. lbs (147 Nm).
5. Drive the vehicle at about 50 km/h (31 mph) on a safe, level and dry road for about 400 meters (0.25 mile) in this condition.
6. Repeat this procedure 2 or 3 times.

CHASSIS ELECTRICAL

AIR BAG (SUPPLEMENTAL RESTRAINT SYSTEM)

⁂ CAUTION

Some vehicles are equipped with an air bag system. The system must be disarmed before performing service on, or around, system components, the steering column, instrument panel components, wiring and sensors. Failure to follow the safety precautions and the disarming procedure could result in accidental air bag deployment, possible injury and unnecessary system repairs.

GENERAL INFORMATION

SERVICE PRECAUTIONS

Disconnect and isolate the battery negative cable before beginning any airbag system component diagnosis, testing, removal, or installation procedures. Allow system capacitor to discharge for two minutes before beginning any component service. This will disable the airbag system. Failure to disable the airbag system may result in accidental airbag deployment, personal injury, or death.

Do not place an intact undeployed airbag face down on a solid surface. The airbag will propel into the air if accidentally deployed and may result in personal injury or death.

When carrying or handling an undeployed airbag, the trim side (face) of the airbag should be pointing towards the body to minimize possibility of injury if accidental deployment occurs. Failure to do this may result in personal injury or death.

Replace airbag system components with OEM replacement parts. Substitute parts may appear interchangeable, but internal differences may result in inferior occupant protection. Failure to do so may result in occupant personal injury or death.

Wear safety glasses, rubber gloves, and long sleeved clothing when cleaning powder residue from vehicle after an airbag deployment. Powder residue emitted from a deployed airbag can cause skin irritation. Flush affected area with cool water if irritation is experienced. If nasal or throat irritation is experienced, exit the vehicle for fresh air until the irritation ceases. If irritation continues, see a physician.

Do not use a replacement airbag that is not in the original packaging. This may result in improper deployment, personal injury, or death.

The factory installed fasteners, screws and bolts used to fasten airbag components have a special coating and are specifically designed for the airbag system. Do not use substitute fasteners. Use only original equipment fasteners listed in the parts catalog when fastener replacement is required.

During, and following, any child restraint anchor service, due to impact event or vehicle repair, carefully inspect all mounting hardware, tether straps, and anchors for proper installation, operation, or damage. If a child restraint anchor is found damaged in any way, the anchor must be replaced. Failure to do this may result in personal injury or death.

Deployed and non-deployed airbags may or may not have live pyrotechnic material within the airbag inflator.

Do not dispose of driver/passenger/curtain airbags or seat belt tensioners unless you are sure of complete deployment. Refer to the Hazardous Substance Control System for proper disposal.

Dispose of deployed airbags and tensioners consistent with state, provincial, local, and federal regulations.

After any airbag component testing or service, do not connect the battery negative cable. Personal injury or death may result if the system test is not performed first.

If the vehicle is equipped with the Occupant Classification System (OCS), do not connect the battery negative cable before performing the OCS Verification Test using the scan tool and the appropriate diagnostic information. Personal injury or death may result if the system test is not performed properly.

Never replace both the Occupant Restraint Controller (ORC) and the Occupant Classification Module (OCM) at the same time. If both require replacement, replace one, then perform the Airbag System test before replacing the other.

Both the ORC and the OCM store Occupant Classification System (OCS) calibration data, which they transfer to one another when one of them is replaced. If both are replaced at the same time, an irreversible fault will be set in both modules and the OCS may malfunction and cause personal injury or death.

If equipped with OCS, the Seat Weight Sensor is a sensitive, calibrated unit and must be handled carefully. Do not drop or handle roughly. If dropped or damaged, replace with another sensor. Failure to do so may result in occupant injury or death.

If equipped with OCS, the front passenger seat must be handled carefully as well. When removing the seat, be careful when setting on floor not to drop. If dropped, the sensor may be inoperative, could result in occupant injury, or possibly death.

If equipped with OCS, when the passenger front seat is on the floor, no one should sit in the front passenger seat. This uneven force may damage the sensing ability of the seat weight sensors. If sat on and damaged, the sensor may be inoperative, could result in occupant injury, or possibly death.

DISARMING THE SYSTEM

To avoid personal injury when working on vehicles equipped with an air bag, the negative battery cable must be disconnected and at least 90 seconds must elapse before working on the system. Failure to do so may result in deployment of the air bag.

ARMING THE SYSTEM

Reconnect the negative battery cable. Wait 2 minutes for performing any service on the vehicle.

DRIVETRAIN

AUTOMATIC TRANSMISSION ASSEMBLY

REMOVAL & INSTALLATION

1. Disconnect the negative battery cable.
2. Remove the shift knob.
3. Remove the upper trim panels.
4. Remove the console.
5. Remove the snapring and remove the transfer case lever.
6. Remove the engine under-covers.
7. Remove the front suspension member brackets.
8. Disconnect the oxygen sensor.
9. Remove the exhaust pipe.
10. Remove the driveshafts.
11. Remove the drain plug.
12. Remove the transmission control cable.
13. Support the transmission with a transmission jack.
14. Remove the crossmember.
15. Disconnect all wires and lines as necessary.
16. Disconnect the breather hose.
17. Remove the bellhousing cover.
18. Turn the crankshaft as needed to access the torque converter bolts and remove them.
19. Remove the transmission-to-engine bolts.
20. Remove the transmission/transfer case assembly.
21. Separate the transfer case from the transmission as needed.
22. Remove the rear mount from the transmission.
23. Installation is the reverse of removal. Observe the following torques:
    - Rear mount: 48 ft. lbs. (65 Nm)
    - Control cable bracket: 19 ft. lbs. (25 Nm)
    - Transfer case-to-transmission: 17 ft. lbs. (24 Nm)
    - Transmission-to-engine: 17mm bolts, 53 ft. lbs. (71 Nm); 14mm bolts, 27 ft. lbs. (37 Nm)
    - Torque converter bolts: 35 ft. lbs. (48 Nm)
    - Bellhousing cover: 13 ft. lbs. (18 Nm)
    - Crossmember-to-frame: 53 ft. lbs. (72 Nm)
    - Transmission-to-crossmember: 13 ft. lbs. (18 Nm)
    - Front and rear driveshaft flanges: 65 ft. lbs. (88 Nm)
    - Suspension member brackets: 24 ft. lbs. (33 Nm)

TRANSFER CASE ASSEMBLY

REMOVAL & INSTALLATION

1. Before servicing the vehicle, refer to the precautions in the beginning of this section.
2. Drain the fluid.
3. Remove the skid plate.
4. Remove the transmission.
5. Remove the 8 bolts and 2 clamps from transfer assembly
6. Separate the transfer case from the transmission.

To install:

7. Installation is the reverse of removal. Torque the bolts to 17 ft. lbs. (24 Nm).

FRONT AXLE SHAFT, BEARING & SEAL

REMOVAL & INSTALLATION

Differential Side Seal

1. Remove front wheel.
2. Remove the 6 bolts and engine under cover assembly rear.
3. Remove the 4 bolts and no. 1 engine under cover sub-assembly.
4. Drain differential oil.
5. Remove front axle shaft LH nut.
6. Separate front speed sensor LH.
7. Separate tie rod end sub-assembly LH.
8. Separate no. 1 front suspension arm sub-assembly lower LH.
9. Remove front drive shaft assembly LH.
10. Using service tool, remove the oil seal.

To install:

11. Using SST and a hammer, install a new oil seal.
12. Coat the oil seal lip with MP grease.
13. Install front drive shaft assembly LH.
14. Connect no. 1 front suspension arm sub-assembly lower lh.
15. Connect tie rod end sub-assembly LH.
16. Connect front speed sensor LH.
17. Install front axle shaft lh nut and tighten to 173 ft. Lbs. (235 nm).
18. Fill up differential oil.
19. Install the no. 1 engine under cover sub-assembly with the 4 bolts.
20. Install the engine under cover assembly with the 6 bolts.
21. Install front wheel.
22. Inspect ABS speed sensor signal.

FRONT HALFSHAFT

REMOVAL & INSTALLATION

See Figures 3 and 4.

1. Before servicing the vehicle, refer to the precautions section.
2. Remove the wheel.
3. Drain the differential oil.
4. Remove the cotter pin and cap, then remove the hub nut.
5. Remove the speed sensor wiring harness. Remove the sensor.
6. Remove the tie rod end from the knuckle.
7. Remove the 2 bolts and separate the lower arm from the ball joint.
8. Remove the halfshaft using a slide hammer and adapter. Keep the halfshaft level when carrying it.

To install:

9. Coat the inboard end splines of the halfshaft with clean ATF.

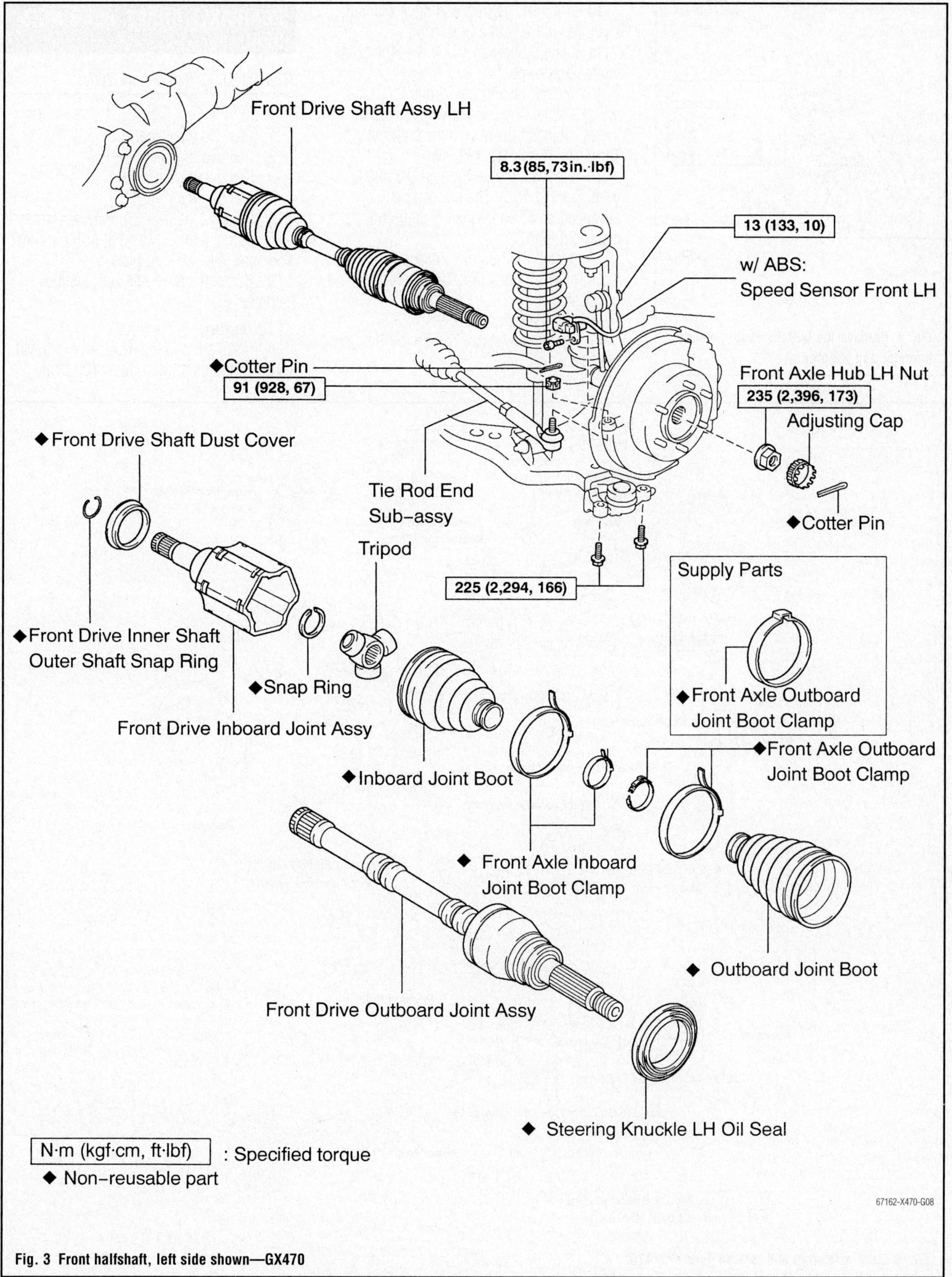

Fig. 3 Front halfshaft, left side shown—GX470

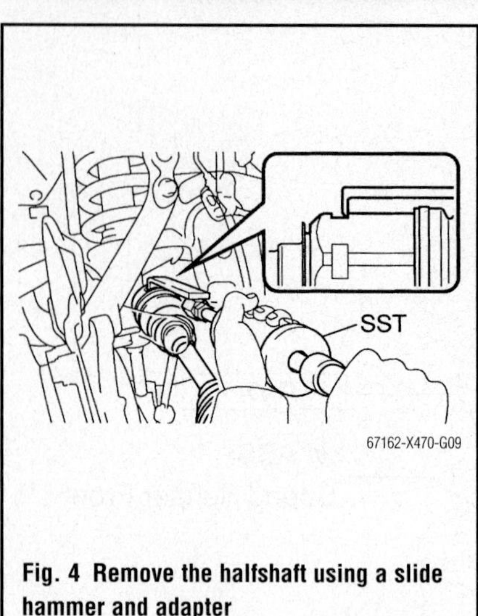

Fig. 4 Remove the halfshaft using a slide hammer and adapter

10. Align the splines and drive the half-shaft into place with a brass drift.
11. Install a new snapring with the opening facing down.
12. Install the sensor. Torque to 10 ft. lbs. (13 Nm). Connect the wire harness.
13. Connect the arm to the ball joint. Torque to 166 ft. lbs. (225 Nm).
14. Connect the tie rod end. Torque to 67 ft. lbs. (91 Nm). The nut can be advanced up to 60 degrees to align the cotter pin hole.
15. Install the hub nut. Torque to 173 ft. lbs. (235 Nm). Install the cap and a new cotter pin.
16. Fill the differential.
17. Install the wheel. Torque to 83 ft. lbs. (112 Nm).

REAR AXLE SHAFT, BEARING & SEAL

REMOVAL & INSTALLATION

See Figure 5.

1. Remove the wheel.
2. Remove the speed sensor.
3. Remove the caliper.
4. Remove the rotor.
5. Remove the parking brake assembly.
6. Remove the 4 nuts and pull out the axle shaft with backing plate.
7. Remove the oil seal with a slide hammer.

To install:

8. Installation is the reverse of removal. Torque the nuts to 89 ft. lbs. (120 Nm).

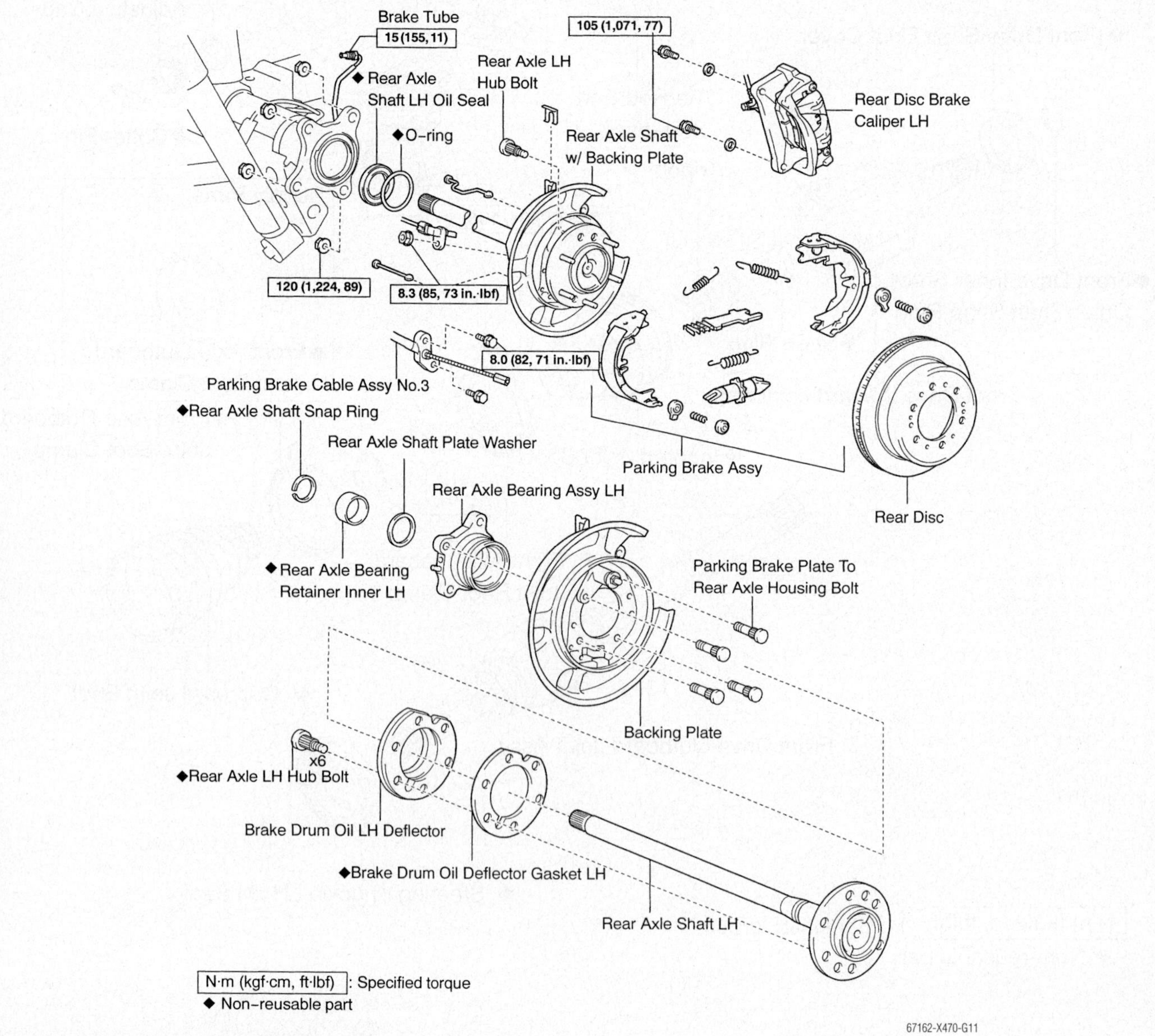

Brake Tube
15 (155, 11)
105 (1,071, 77)
Rear Axle LH Hub Bolt
◆ Rear Axle Shaft LH Oil Seal
Rear Disc Brake Caliper LH
◆ O-ring
Rear Axle Shaft w/ Backing Plate
120 (1,224, 89)
8.3 (85, 73 in.·lbf)
8.0 (82, 71 in.·lbf)
Parking Brake Cable Assy No.3
◆ Rear Axle Shaft Snap Ring
Rear Axle Shaft Plate Washer
Parking Brake Assy
Rear Axle Bearing Assy LH
Rear Disc
◆ Rear Axle Bearing Retainer Inner LH
Parking Brake Plate To Rear Axle Housing Bolt
Backing Plate
x6
◆ Rear Axle LH Hub Bolt
Brake Drum Oil LH Deflector
◆ Brake Drum Oil Deflector Gasket LH
Rear Axle Shaft LH
N·m (kgf·cm, ft·lbf) : Specified torque
◆ Non-reusable part
67162-X470-G11

Fig. 5 Rear axle shaft and related parts—GX470

ENGINE COOLING

THERMOSTAT

REMOVAL & INSTALLATION

See Figure 6.

1. Before servicing the vehicle, refer to the precautions in the beginning of this section.
2. Drain engine coolant.
3. Remove the 3 nuts and disconnect the water inlet from the water inlet housing.
4. Remove the thermostat.
5. Remove the gasket from the thermostat.

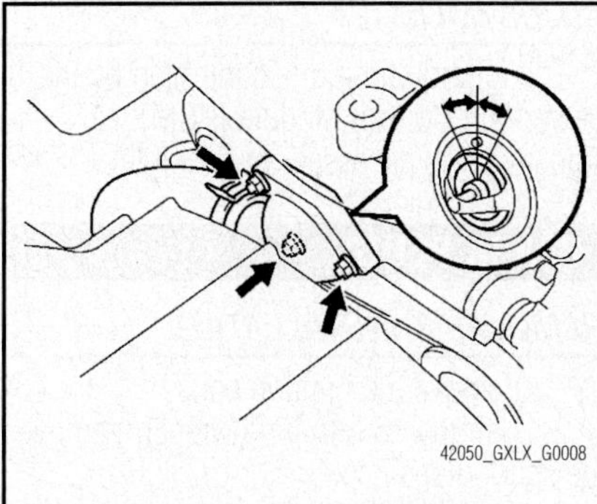

Fig. 6 Thermostat positioning and installation

To install:

6. Install a new gasket to the thermostat.
7. Insert the thermostat into the water inlet housing with the jiggle valve facing straight upward.

➡The jiggle valve may be set within 30° of either side of the prescribed position.

8. Install the water inlet with the 3 nuts and tighten to 19 Nm (14 ft. lbs.).
9. Fill with engine coolant.
10. Start engine and check for coolant leaks.
11. Recheck engine coolant level.

WATER PUMP

REMOVAL & INSTALLATION

V6 Engine

1. Remove the 4 bolts and engine under cover.
2. Drain engine coolant.
3. Remove v-bank cover.
4. Remove the 11 clips and radiator support seal upper.
5. Loosen the 4 fluid coupling bolts.
6. Remove fan and generator v belt.
7. Unfasten the hose clamp, and then separate the 2 oil cooler hoses from the fan shroud.
8. Disconnect the radiator reserve tank hose.
9. Remove the 2 fan shroud bolts and radiator reserve tank bolt.
10. Remove the 4 fluid coupling nuts, and then remove the fan with fluid coupling and fan shroud together.
11. Remove the fan pulley.
12. Disconnect ventilation hose no. 2.
13. Remove air cleaner assembly.
14. Remove water inlet.
15. Disconnect the 2 oil cooler hoses (with oil cooler).
16. Disconnect the 2 radiator hoses.
17. Disconnect the 5 water by-pass hoses.
18. Remove the 5 bolts and water inlet.
19. Remove the O-ring from the water outlet pipe.
20. Remove the gasket from the water pump.
21. Remove the 2 bolts and 2 idler pulleys.
22. Remove generator assembly.
23. Separate cooler compressor assembly.
24. Remove v-ribbed belt tensioner assembly.
25. Remove the 17 bolts, water pump and gasket.

To install:

26. Install a new gasket and the water pump with the 17 bolts and tighten 10 mm to 80 inch lbs. (9.0 Nm) 12 mm to 17 ft. lbs. (23 Nm).
27. Install v-ribbed belt tensioner assembly.
28. Install cooler compressor assembly.
29. Install generator assembly.
30. Install the 2 idler pulleys with the 2 bolts and tighten bolts to 29 ft. Lbs.
31. Install a new O-ring to the water outlet pipe.
32. Install a new gasket to the water pump.
33. Apply soapy water to the O-ring.
34. Install the water inlet with the 5 bolts and tighten bolts to 80 inch lbs. (9.0 Nm).
35. Connect the 5 water by-pass hoses.
36. Connect the 2 radiator hoses.
37. Connect the 2 oil cooler hoses (with oil cooler).
38. Put the fan pulley to the water pump.
39. Put the fan with fluid coupling and fan shroud together.
40. Temporarily install the 4 fluid coupling nuts.
41. Tighten the 2 fan shroud bolts and radiator reserve tank bolt to 44 inch lbs. (5.0 Nm).
42. Connect the reserve tank hose.
43. Fasten the 2 oil cooler hoses with the hose clamp.
44. Install fan and generator v belt.
45. Tighten the 4 fluid coupling bolts.
46. Install radiator support seal upper.
47. Add engine coolant.
48. Inspect engine coolant leak.
49. Install the V-bank cover with the 2 nuts and tighten to 66 inch lbs. (7.5Nm).
50. Install the engine under cover with the 4 bolts and tighten bolts to 21 ft. lbs. (29 Nm).

V8 Engine

1. Before servicing the vehicle, refer to the precautions section.
2. Drain the cooling system.
3. Remove or disconnect the following:
   - Negative battery cable
   - Timing belt.
   - No. 2 idler pulley
   - Radiator hose
   - Bypass hose
   - Water inlet housing assembly
   - Water pump

To install:

4. Install or connect the following:
   - Water pump. Use a new gasket and tighten the bolts to 15 ft. lbs. (21 Nm). Tighten the stud bolt and nut to 13 ft. lbs. (18 Nm).
   - Water inlet housing assembly. Use a new O-ring and apply sealant as shown. Tighten the bolts to 13 ft. lbs. (18 Nm).
   - Bypass hose
   - Radiator hose
   - No. 2 idler pulley
   - Timing belt
   - Negative battery cable
5. Fill the cooling system.
6. Start the engine and check for leaks.

ENGINE ELECTRICAL — CHARGING SYSTEM

ALTERNATOR

REMOVAL & INSTALLATION

1. Before servicing the vehicle, refer to the precautions section.
2. Disconnect the negative battery cable.
3. Remove the drive belt.
4. Disconnect the 2 oil cooler lines from the fan shroud, remove the fan shroud.
5. Remove the 4 nuts and remove the fan with the fluid coupling.
6. Disconnect the vane pump assembly.
7. Disconnect the alternator wiring.
8. Remove the nuts and bolts and remove the alternator.

To install:

9. Install the alternator. Tighten the bolt to 29 ft. lbs. (39 Nm), the upper nut to 29 ft. lbs. (39 Nm) and the side nut to 12 ft. lbs. (16 Nm).
10. Attach the alternator wiring.
11. Install the vane pump assembly.
12. Install the shroud and fluid coupling together and tighten shroud bolts. Tighten the fan coupling nuts to 21 ft. lbs. (29 Nm).

ENGINE ELECTRICAL — DISTRIBUTORLESS IGNITION SYSTEM

FIRING ORDER

4.0L Engine Firing order: 1–2–3–4–5–6
4.7L Engine Firing order: 1–8–4–3–6–5–7–2

IGNITION COIL

REMOVAL & INSTALLATION

1. Disconnect cable from negative battery terminal.

CAUTION

Wait at least 90 seconds after disconnecting the cable from the negative (-) battery terminal to prevent airbag and seat belt pretensioner activation.

2. Remove the V-bank cover.
3. Remove intake air connector pipe
   a. Disconnect the 3 hoses.
   b. Remove the 2 bolts.
   c. Loosen the 2 clamp bolts and remove the intake air connector.
4. Disconnect the 2 engine wire clamps from the cylinder head cover LH side.
5. Disconnect the connector and remove ignition coil assembly.

To install:

6. Install the ignition coil with the bolt tighten bolt to 66 inch lbs. (7.5 Nm)
7. Connect the connector.
8. Connect the 2 engine wire clamps to the cylinder head cover LH side.
9. Install intake air connector pipe.
10. Install v-bank cover sub-assembly.
11. Connect cable to negative battery terminal.
12. Perform initialization if necessary.

IGNITION TIMING

ADJUSTMENT

The ignition timing is controlled by the Powertrain Control Module (PCM). No adjustment is necessary or possible.

SPARK PLUGS

REMOVAL & INSTALLATION

1. Remove the ignition coils.
2. Using a 16 mm plug wrench, remove the spark plugs.
3. Clean the spark plugs.

To install:

4. Adjust the spark plug electrode gap. Electrode gap for new spark plug is 1.0 to 1.1 mm (0.039 to 0.043 in.).
5. Using a 16 mm plug wrench, install the spark plugs and tighten to 17.5 Nm (13 ft lbs)
6. Reinstall the ignition coils.

ENGINE ELECTRICAL — STARTING SYSTEM

STARTER

REMOVAL & INSTALLATION

See Figure 7.

V6 Engine—2WD

1. Disconnect the battery negative terminal.
2. Remove the 4 bolts and engine rear under cover assembly.
3. Remove the 3 bolts and No. 2 manifold stay.
4. Remove the bolt and disconnect the wire harness from the cylinder block.
5. Disconnect the terminal 50 connector from the starter assembly.
6. Remove the nut and disconnect the wire harness from the terminal 30.
7. Remove the 2 bolts and starter assembly.

To install:

8. Install the starter assembly with the 2 bolts and tighten to 27 ft. lbs. (37 Nm).
9. Connect the wire harness to the terminal 30 and install the nut Tighten to 7 ft. lbs. (9.8 Nm).
10. Connect the wire harness to the cylinder block and install the bolt. 10 ft. lbs. (13 Nm).
11. Install the No. 2 manifold stay with the 3 bolts and tighten to 30 ft. lbs. (40 Nm).
12. Install the engine rear under cover assembly with the 4 bolts and tighten to 21 ft. lbs. (29 Nm).
13. Connect the battery negative terminal.

V6 Engine—4WD

1. Disconnect the battery negative terminal.

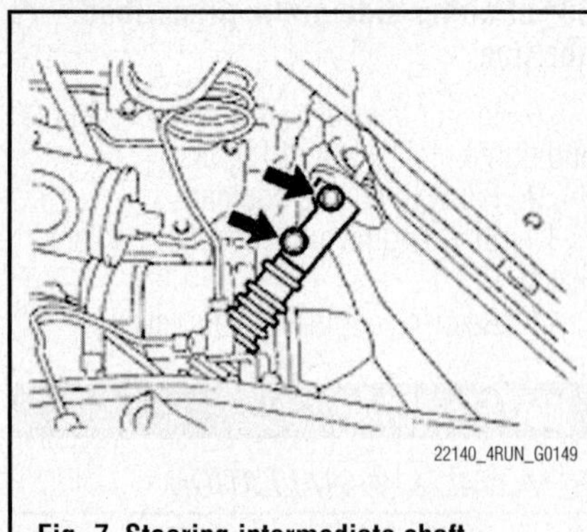

Fig. 7 Steering intermediate shaft

2. Remove the 4 bolts and engine rear under cover assembly.
3. Disconnect the heated oxygen sensor (bank 2) connector.
4. Remove the 2 bolts, 2 nuts, No. 2 front exhaust pipe assembly and 2 gaskets.

5. Remove the 5 clips and front fender rear splash shield LH.
6. Loosen the bolt A and remove the bolt B, then separate the No. 2 steering intermediate shaft sub-assembly.
7. Disconnect the terminal 50 connector from the starter assembly.
8. Remove the nut and disconnect the wire harness from the terminal 30.
9. Remove the 2 bolts and starter assembly.

To install:

10. Install the starter assembly with the 2 bolts and tighten to 27 ft. lbs. (37 Nm).
11. Connect the wire harness to the terminal 30 and install the nut and tighten to 7 ft. lbs. (9.8 Nm).
12. Install the steering sliding yoke to No. 2 steering intermediate shaft and install the bolt B and tighten the bolt B and A to 27 ft. lbs. (36 Nm).
13. Install the front fender rear splash shield LH with the 5 clips.
14. Install 2 new gaskets and No. 2 exhaust manifold with the 2 bolts and 2 nuts and tighten nut to 46 ft. lbs. (62 Nm). Tighten bolt to 35 ft. lbs. (48 Nm).
15. Connect the heated oxygen sensor (bank 2) connector.
16. Install the engine rear under cover assembly with the 4 bolts and tighten to 21 ft. lbs. (29 Nm).
17. Connect the battery negative terminal.

V8 Engine

1. Before servicing the vehicle, refer to the precautions section.
2. Drain the cooling system.
3. Relieve the fuel system pressure.
4. Remove or disconnect the following:
   - Negative battery cable
   - Engine appearance cover
   - Air intake tube
   - No. 1 & No. 2 fuel hoses
   - Ventilation hose
   - Purge VSV
   - Vacuum control valve
   - Engine wire harness
   - Throttle body water bypass hose
   - Intake manifold
   - Air pump and switching valve
   - Starter motor mounting bolts
   - Starter wiring connectors
   - Starter motor

To install:

5. Install or connect the following:
   - Starter motor
   - Starter wiring connectors. Tighten the cable nut to 86 inch lbs. (10 Nm).
   - Starter motor mounting bolts. Tighten the bolts to 29 ft. lbs. (39 Nm).
   - Air pump and switching valve
   - Intake manifold
   - Throttle body water bypass hose
   - Engine wire harness
   - Vacuum control valve
   - Purge VSV
   - Ventilation hose
   - No. 1 & No. 2 fuel hoses
   - Air intake tube
   - Engine appearance cover
   - Negative battery cable
6. Fill the cooling system.
7. Start the engine and check for leaks.
8. Perform initialization if necessary

ENGINE MECHANICAL

ACCESSORY DRIVE BELTS

ACCESSORY BELT ROUTING

See Figures 8 and 9.

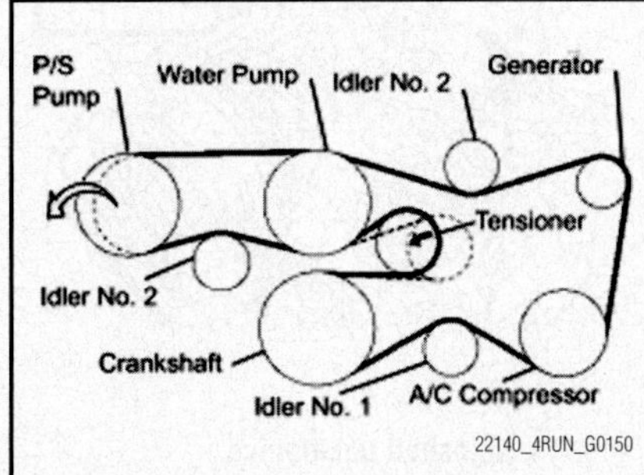

Fig. 8 Accessory drive belt routing—V6 engine

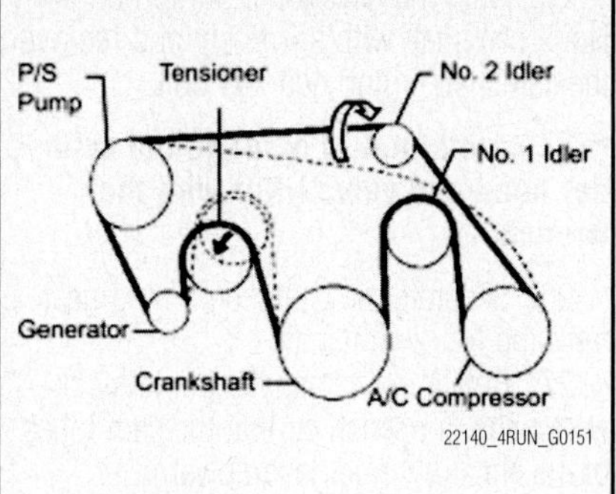

Fig. 9 Accessory drive belt routing—V8 engine

INSPECTION

Inspect the drive belt for signs of glazing or cracking. A glazed belt will be perfectly smooth from slippage, while a good belt will have a slight texture of fabric visible. Cracks will usually start at the inner edge of the belt and run outward. All worn or damaged drive belts should be replaced immediately.

ADJUSTMENT

Belt adjustment is automatic and non-adjustable.

REMOVAL & INSTALLATION

See Figure 10.

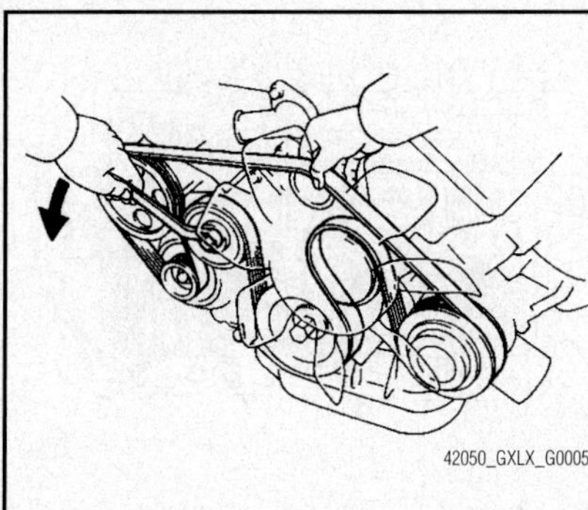

Fig. 10 Accessory drive belt replacement

1. Before servicing the vehicle, refer to the precautions in the beginning of this section.
2. Loosen the drive belt tension by turning the drive belt tensioner counterclockwise, and remove the drive belt.
3. Installation is the reverse of removal.

CAMSHAFT AND VALVE LIFTERS

REMOVAL & INSTALLATION

V6 Engine

See Figures 11 through 19.

1. Drain engine coolant.
2. Remove v-bank cover.
3. Disconnect no. 2 ventilation hose.
4. Remove air cleaner assembly.
5. Disconnect the 2 water by-pass hoses.
6. Disconnect the fuel vapor feed hose.
7. Disconnect the 2 VSV connectors.
8. Disconnect the throttle body with motor connector.
9. Separate the 3 wire harness clamps and hose clamp.
10. Remove the 2 bolts and throttle body bracket.
11. Remove the bolt and oil baffle plate.
12. Remove the 4 bolts and 2 surge tank stays.

13. Remove the 2 nuts.

14. Using an 8 mm socket hexagon wrench, remove the 4 bolts, intake air surge tank and gasket.

15. Remove ignition coil assembly.

16. Remove the 10 bolts, 3 seal washers, 2 nuts, cylinder head cover and gasket.

17. Turn the crankshaft pulley, and align the notch with the timing mark "0" of the timing chain cover.

18. Check that the timing marks of the camshaft timing gears are aligned with the timing marks of the bearing caps as shown in the illustration. If not, turn the crankshaft 1 complete revolution (360°) and align the timing marks as above.

19. Place paint marks on the No. 1 chain links that correspond with the timing marks of the camshaft timing gears.

➡Never rotate the crankshaft with the chain tensioner removed. When rotating the camshaft with the timing chain removed, rotate the crankshaft counterclockwise 40° from TDC first.

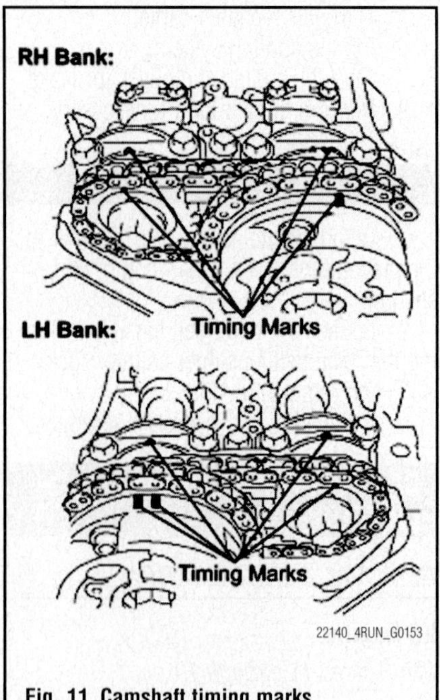

Fig. 11 Camshaft timing marks

20. Remove the 4 bolts, timing chain cover plate and gasket.

21. While turning the stopper plate of the tensioner clockwise, push in the plunger of the chain tensioner as shown in the illustration.

22. While turning the stopper plate of the tensioner counterclockwise, insert a bar of 3.5 mm (0.138 in.) into the holes on the stopper plate and tensioner to fix the stopper plate in place.

23. Remove the 2 bolts and chain tensioner.

➡As the thrust clearance of the camshaft is small, the camshaft must be kept level while it is being removed. If the camshaft is not kept level, the portion of the cylinder head which receives the shaft thrust may crack or be damaged, causing the camshaft to seize or break. To avoid this, the following steps should be carried out.

24. While raising up the No. 2 chain tensioner, insert a pin of 1.0 mm (0.039 in.) into the hole to fix it in place.

25. Hold the hexagonal portion of the No. 2 camshaft with a wrench, and remove the camshaft timing gear set bolt.

➡Be careful not to damage the cylinder head and valve lifter with the wrench.

26. Separate the camshaft timing gear from the No. 2 camshaft.

27. Rotate the camshaft counterclockwise using a wrench so that the cam lobes of the No. 1 cylinder face upward.

28. Using several steps, loosen and remove the 8 bearing cap bolts uniformly in the sequence shown in the illustration.

29. Remove the 4 bearing caps and No. 2 camshaft.

30. Remove the No. 2 chain tensioner bolt, and then remove the No. 2 chain tensioner and camshaft timing gear.

➡As the thrust clearance of the camshaft is small, the camshaft must be kept level while it is being removed. If the camshaft is not kept level, the portion of the cylinder head which receives the shaft thrust may crack or be damaged, causing the camshaft to seize or break. To avoid this, the following steps should be carried out.

31. Hold the hexagonal portion of the No. 1 camshaft with a wrench, and loosen the camshaft timing gear set bolt.

➡Be careful not to damage the cylinder head and valve lifter with the wrench. Do not disassemble the camshaft timing gear assembly.

32. Slide the camshaft timing gear and separate the No. 1 chain from the camshaft timing gear

33. Rotate the No. 1 camshaft counterclockwise using a wrench so that the cam lobes of the No. 1 cylinder face downward as shown in the illustration.

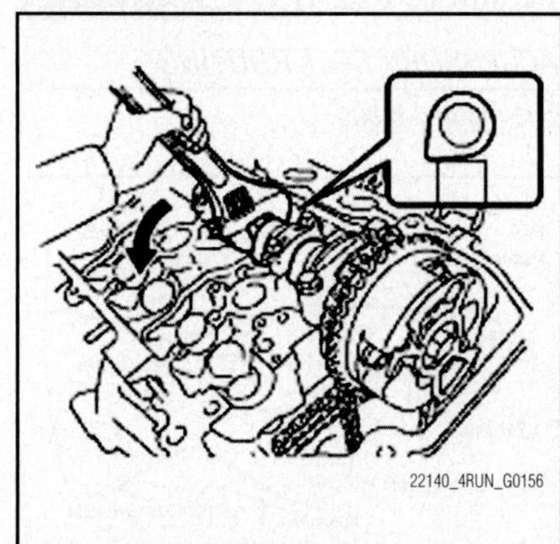

Fig. 14 Camshaft positioning

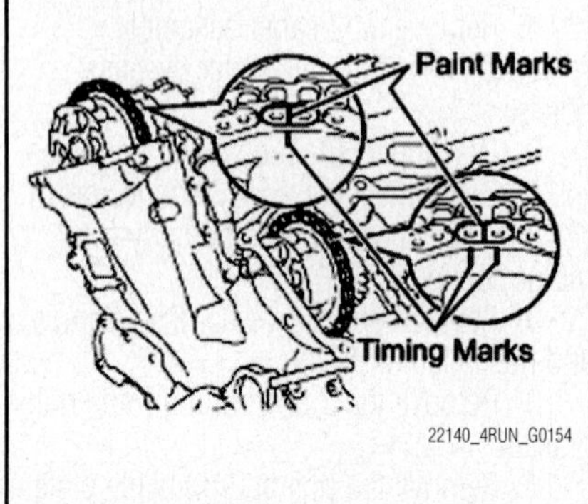

Fig. 12 Camshaft painted marks

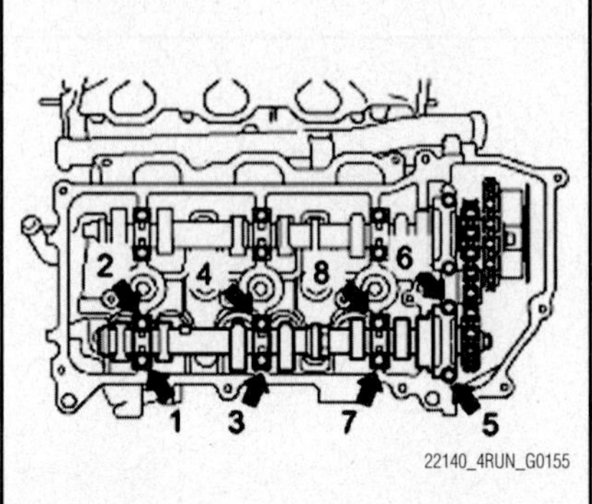

Fig. 13 Bearing cap removal sequence

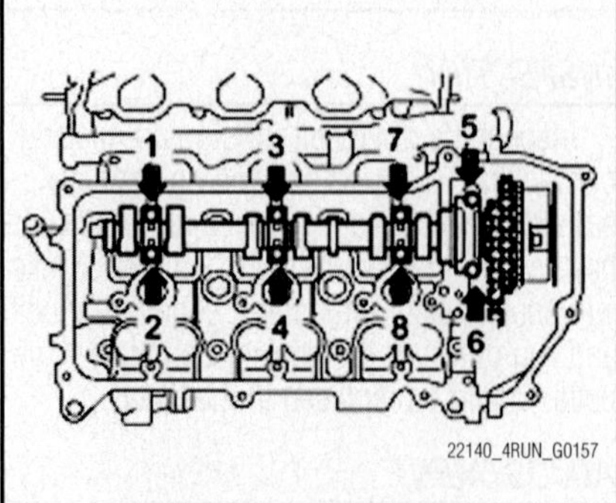

Fig. 15 Bearing cap removal sequence

34. Using several steps, loosen and remove the 8 bearing cap bolts in the sequence shown in the illustration.

35. Remove the 4 bearing caps.

36. Remove the camshaft timing gear set bolt with the No. 1 camshaft lifted up, and then remove the No. 1 camshaft and camshaft timing gear with the No. 2 chain.

37. Tie the No. 1 chain with a string.

➡Be careful not to drop anything inside the timing chain cover.

38. Remove the valve lifters.

To install:

39. Install valve lifters

➡ As the thrust clearance of the camshaft is small, the camshaft must be kept level while it is being installed. If the camshaft is not kept level, the portion of the cylinder head which receives the shaft thrust may crack or be damaged, causing the camshaft to seize or break. To avoid this, the following steps should be carried out.

40. Align the mark link (yellow) with the timing mark (1-dot mark) of the camshaft timing gear as shown in the illustration.

41. Apply new engine oil to the thrust portions and journals of the camshafts

42. Temporarily put the No. 1 chain on the No. 2 chain of the camshaft timing gear

43. Align the knock pin hole of the camshaft timing gear with the knock pin of the No. 1 camshaft, and insert the No. 1 camshaft into the camshaft timing gear.

44. Temporarily install the camshaft timing gear set bolt.

45. Set the No. 1 camshaft onto the cylinder head RH with the cam lobes of the No. 1 cylinder facing downward as shown in the illustration.

46. Install the 4 bearing caps in their proper locations

47. Apply a light coat of engine oil to the threads and under the heads of the bearing cap bolts

48. Using several steps, install the 8 bearing cap bolts uniformly in the sequence shown in the illustration, and tighten 10mm bolts to 80 inch lbs. (9.0 Nm) and 12 mm bolts to 18 ft. lbs. (24 Nm)

49. Rotate the No. 1 camshaft clockwise using a wrench so that the timing mark of the camshaft timing gear is aligned with the timing mark of the camshaft bearing cap

50. Align the paint mark of the No. 1 chain with the timing mark of the camshaft timing gear

51. Hold the hexagonal portion of the No. 1 camshaft with a wrench, and tighten the camshaft timing gear set bolt and tighten to 74 ft lbs. (100 Nm)

52. While pushing in the tensioner, insert a pin of 1.0 mm (0.039 in.) into the hole to fix it in place

53. Temporarily install the camshaft timing gear and No. 2 chain tensioner with the bolt and align the mark links (yellow) with the timing marks (1-dot mark) of the camshaft timing gears

54. Tighten the No. 2 chain tensioner bolt to 14 ft. lbs. (19 Nm).

➡As the thrust clearance of the camshaft is small, the camshaft must be kept level while it is being installed. If the camshaft is not kept level, the portion of the cylinder head which receives the shaft thrust may crack or be damaged, causing the camshaft to seize or break. To avoid this, the following steps should be carried out.

55. Set the No. 2 camshaft onto the cylinder head RH with the cam lobes of the No. 1 cylinder facing upward as shown in the illustration.

56. Install the 4 bearing caps in their proper locations.

57. Apply a light coat of engine oil to the threads and under the heads of the bearing cap bolts.

58. Using several steps, install the 8 bearing cap bolts uniformly in the sequence shown in the illustration and tighten 10mm bolts to 80 inch lbs. (9.0 Nm) and 12 mm bolts to 18 ft. lbs. (24 Nm).

59. Rotate the No. 2 camshaft clockwise using a wrench so that the knock pin of the No. 2 camshaft is aligned with the knock pin hole of the camshaft timing gear.

60. Hold the hexagonal portion of the No. 2 camshaft with a wrench, and install the camshaft timing gear set bolt and tighten to 74 ft. lbs. (100 Nm).

61. Remove the pin from the No. 2 chain tensioner.

62. While turning the stopper plate of the tensioner clockwise, push in the plunger of the tensioner.

63. While turning the stopper plate of the tensioner counterclockwise, insert a bar of φ3.5 mm (0.138 in.) into the holes on the stopper plate and tensioner to fix the stopper plate in place.

64. Install the chain tensioner with the 2 bolts and tighten to 7 ft lbs. (10 Nm).

65. Remove the bar from the chain tensioner.

66. Install a new gasket and the timing chain cover plate with the 4 bolts and tighten to 80 inch lbs.

67. Turn the crankshaft pulley 2 complete revolutions slowly, and align the notch with the timing mark "0" of the timing chain cover.

Mark Links
Timing Mark
22140_4RUN_G0160

Fig. 16 Camshaft timing gear marks

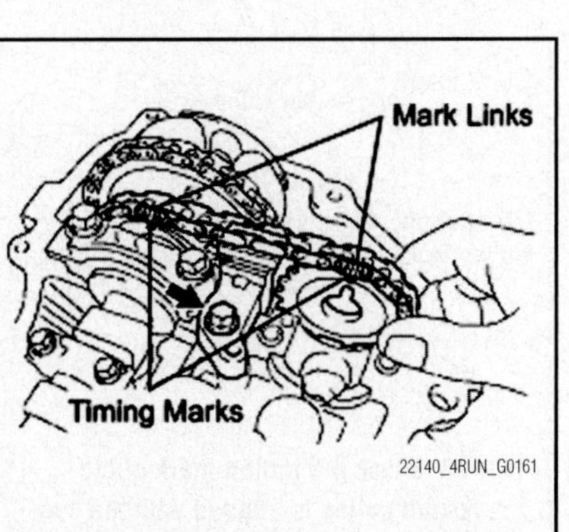

Fig. 17 Aligning camshaft timing gear marks

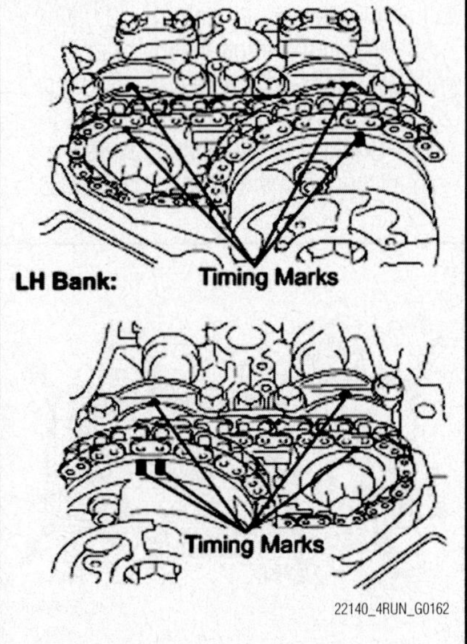

Fig. 18 Aligning camshaft timing chain marks

68. Check that the timing marks of the camshaft timing gears are aligned with the timing marks of the bearing caps as shown in the illustration.

69. Set no. 1 cylinder to TDC/compression.

70. Inspect and adjust valve clearance.

71. Remove any old packing (FIPG) material and be careful not to drop any oil on the contact surfaces of the cylinder head, timing chain cover and cylinder head cover.

72. Apply seal packing (diameter: 2 to 3 mm (0.08 to 0.12 in.)) to the cylinder head and timing chain cover.

73. Install the seal washers to the bolts.

74. Install the cylinder head cover with the 10 bolts and 2 nuts. Tighten the bolts and nuts uniformly in several steps. Tighten bolt A to 7 ft. lbs. (10 Nm) and bolt B and nut to 90 inch lbs. (9.0 Nm).

75. Install the 3 ignition coils with the 3 bolts and tighten to 7 ft. lbs. (10 Nm).

76. Connect the 3 connectors to the 3 ignition coils.

77. Install a new gasket to the intake air surge tank.

78. Using an 8 mm socket hexagon wrench, install the intake air surge tank with the 4 bolts and tighten to 21 ft. lbs. (28 Nm).

79. Install the 2 intake air surge tank nuts and tighten to 21 ft. lbs. (28 Nm)

80. Install the 3 wire harness clamps and hose clamp.

81. Connect the throttle body with motor connector.

82. Connect the 2 VSV connectors.

83. Install the 2 surge tank stays with the 4 bolts.

84. Install the oil baffle plate with the bolt and tighten to 80 inch lbs. (9.0 Nm).

85. Install the throttle body bracket with the 2 bolts and tighten to 15 ft. lbs. (21 Nm).

86. Connect the ventilation hose.

87. Connect the fuel vapor feed hose.

88. Connect the 2 water by-pass hoses.

89. Install air cleaner assembly.

90. Connect no. 2 ventilation hose.

91. Add engine coolant.

92. Install V-bank cover and tighten to 66 inch lbs. (7.5 Nm).

93. Inspect ignition timing.

V8 Engine

See Figures 20 through 33.

1. Before servicing the vehicle, refer to the precautions section.
2. Drain the cooling system.
3. Relieve the fuel system pressure.
4. Remove the V bank cover.
5. Remove the timing belt.
6. Remove the camshaft pulleys.
7. Remove the camshaft position sensor.
8. Remove the power steering pump and set it aside with the lines still attached.
9. Remove the front exhaust pipe.
10. On models with an automatic transmission, remove the oil dipstick and tube.
11. Remove the ignition coils.
12. Remove the rear timing belt plates being careful not to drop anything.
13. Disconnect the fuel inlet hose.
14. Remove the intake manifold.
15. Remove the water inlet and inlet housing. Refer to water pump removal.
16. Remove the front and rear water bypass joint.
17. Remove the engine hangers and if needed the oil dipstick and tube.
18. Remove the valve covers.

➡Since the thrust level of the camshaft is small, the camshaft must be kept level during removal. If not kept level serious damage could occur.

19. Check the timing mark of the crankshaft pulley is aligned with the center(s) of the crankshaft pulley bolt and idler pulley bolt.

➡If the crankshaft pulley is wrongly positioned, this can cause the piston to contact the head causing severe damage. Make sure the crankshaft pulley is properly positioned.

20. Release the oil from the front bearing caps using the tool illustrated. Rotate the camshaft timing tube from left to right 2 to 3 times within its VVT-I range of 25 degrees and collect the oil from the timing oil control valve installation hole using a rag.

21. Remove the left hand camshafts as follows:

a. Bring the service bolt of the sub gear up by turning the left exhaust camshaft using a wrench on the hexagon head portion of the shaft.

b. Secure the sub gear to the main gear using a 16 to 20 mm bolt with a diameter of 6mm and a thread pitch of 1mm.

c. Make sure the torsional force of the sub gear is retained by the bolt.

d. Align the 2 dot timing mark of the left side camshaft by turning the left

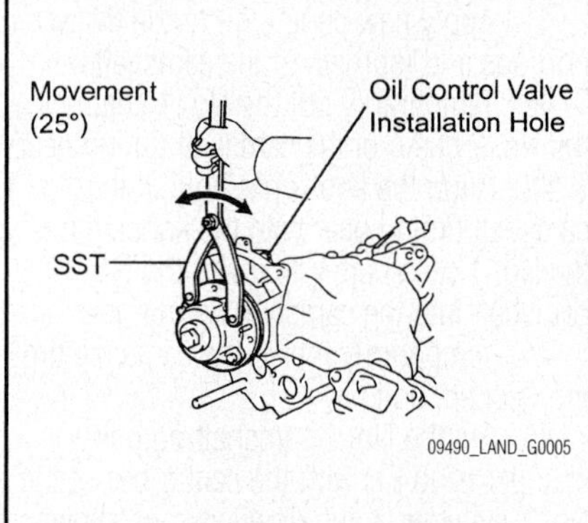

Fig. 21 Release the oil from the front bearing caps using the tool illustrated

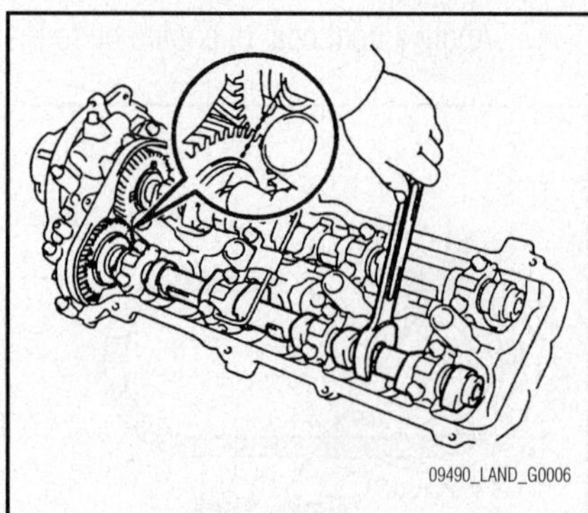

Fig. 22 Align the 2 dot timing mark of the left side camshaft by turning the left exhaust camshaft using a wrench on the hexagon head portion of the shaft

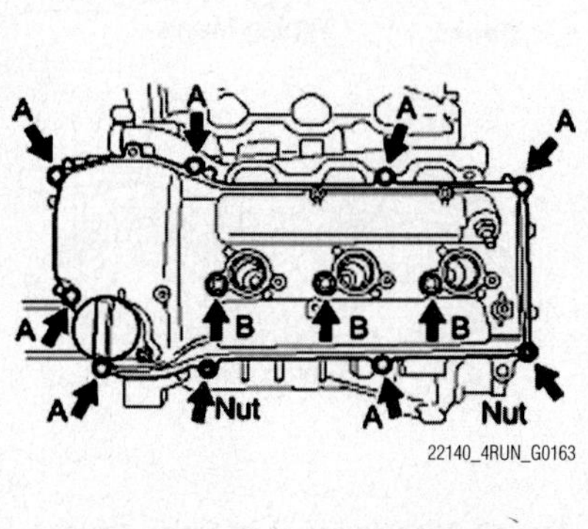

Fig. 19 Head cover tightening sequence

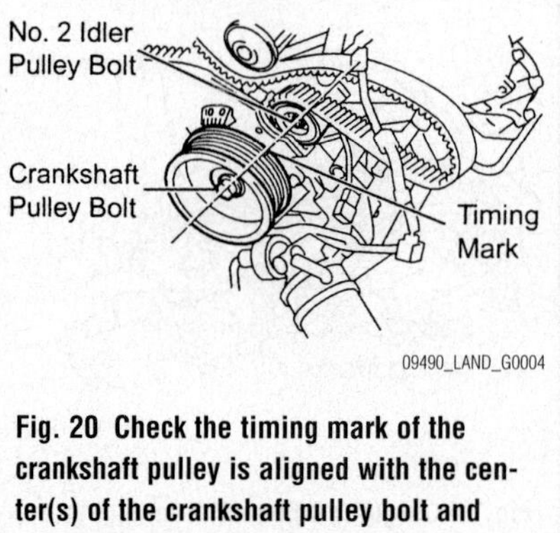

Fig. 20 Check the timing mark of the crankshaft pulley is aligned with the center(s) of the crankshaft pulley bolt and idler pulley bolt

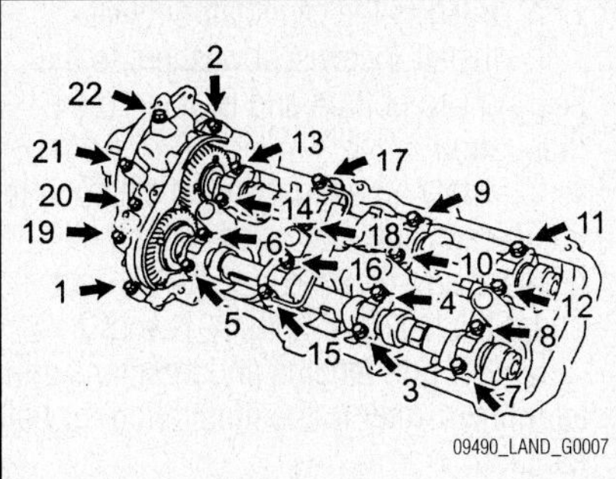

Fig. 23 Loosen the left side 22 bearing cap bolts in the sequence illustrated using several passes

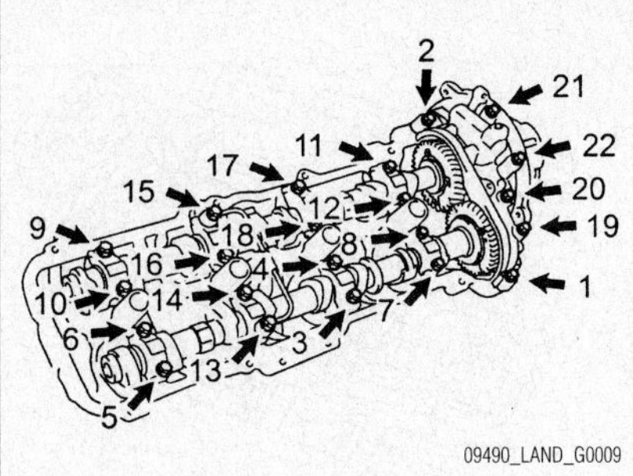

Fig. 25 Loosen the right side 22 bearing cap bolts in the sequence illustrated using several passes

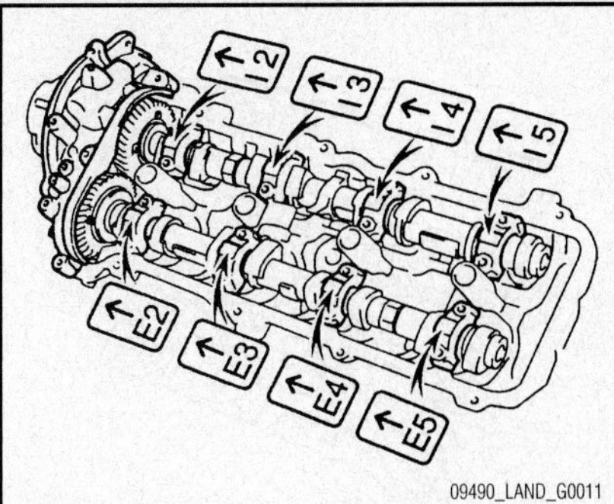

Fig. 27 Install the front bearing cap and then the other caps in the sequence illustrated on the left side camshafts

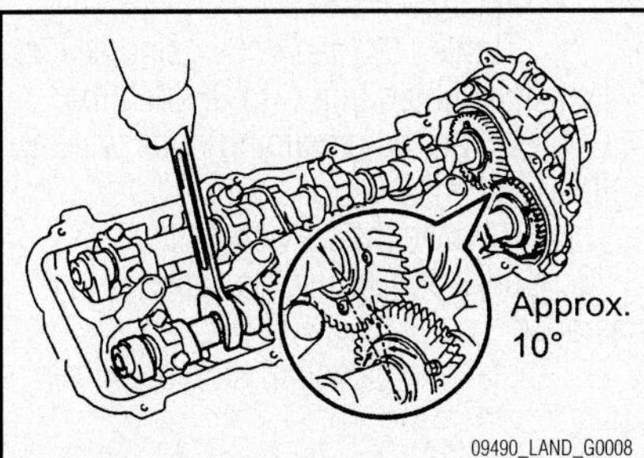

Fig. 24 Align the 1 dot timing mark of the camshaft main gear (about 10 degrees) angle by turning the right exhaust camshaft using a wrench on the hexagon head portion of the shaft

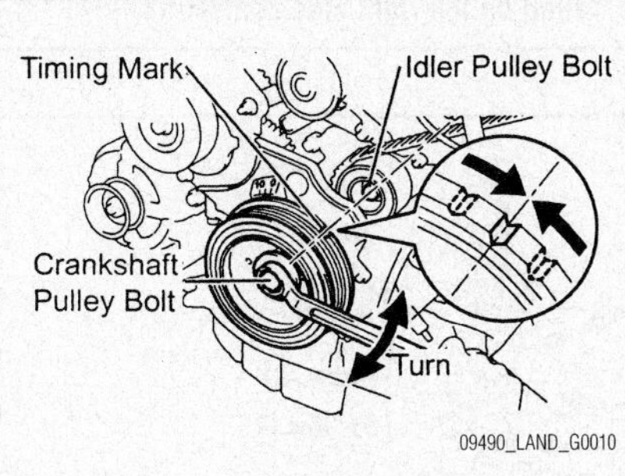

Fig. 26 Check the timing mark of the crankshaft pulley is aligned with the center(s) of the crankshaft pulley bolt and idler pulley bolt

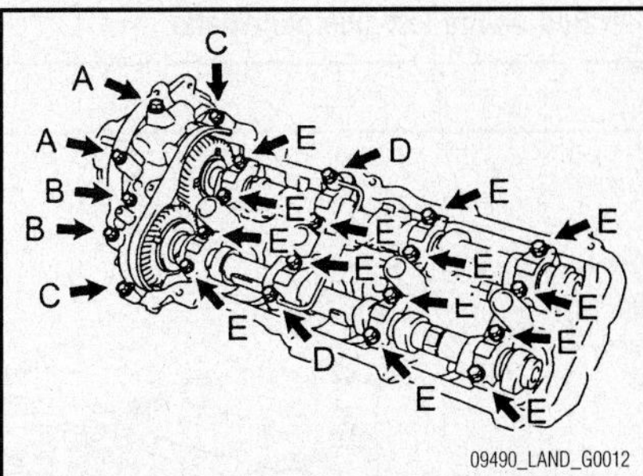

Fig. 28 Apply a light coating of clean oil to the threads and underside of the bolt heads D and E. make sure no oil gets under the heads of bolts A, B and C on the left side camshafts

exhaust camshaft using a wrench on the hexagon head portion of the shaft.

➡Mark the position of the caps so they can be reinstalled in their original positions.

e. Loosen the 22 bearing cap bolts in the sequence illustrated using several passes.

f. Remove the bolts, washers, oil feed pipe, bearing caps, camshaft housing plug, oil control valve filter and the camshafts.

22. Remove the right hand camshafts as follows:

a. Bring the service bolt of the sub gear up by turning the right exhaust camshaft using a wrench on the hexagon head portion of the shaft.

b. Secure the sub gear to the main gear using a 16 to 20 mm bolt with a diameter of 6mm and a thread pitch of 1mm.

c. Make sure the torsional force of the sub gear is retained by the bolt.

d. Align the 1 dot timing mark of the camshaft main gear (about 10 degrees) angle by turning the right exhaust camshaft using a wrench on the hexagon head portion of the shaft.

➡Mark the position of the caps so they can be reinstalled in their original positions.

e. Loosen the 22 bearing cap bolts in the sequence illustrated using several passes.

f. Remove the bolts, washers, oil feed pipe, bearing caps, camshaft housing plug, oil control valve filter and the camshafts.

To install:

23. Check the timing mark of the crankshaft pulley is aligned with the center(s) of the crankshaft pulley bolt and idler pulley bolt.

➡If the crankshaft pulley is wrongly positioned, this can cause the piston to contact the head causing severe damage. Make sure the crankshaft pulley is properly positioned.

24. Install the left side camshafts as follows:

a. Apply multipurpose grease to the thrust portion of the camshafts.

b. Align the 2 dot timing mark of the camshaft drive and driven main gears and install the camshafts.

c. Apply seal packing to the camshaft housing plug.

d. Install the camshaft housing plug on the cylinder head as illustrated. Install the strainer on the head being careful it is properly positioned.

e. Apply seal packing to the front bearing cap.

f. Install the front bearing cap and then the other caps in the sequence illustrated.

g. Push in the camshaft oil seal.

h. Install 4 new seal washers to the bearing cap bolts A and B, refer to the illustration.

i. Apply a light coating of clean oil to the threads and underside of the bolt heads D and E. make sure no oil gets under the heads of bolts A, B and C.

j. The bolt lengths and positions are as follows. refer to the illustration for bolt location:

- 94mm bolts A
- 72mm bolts B
- 25mm bolts C
- 52mm bolts D
- 38mm bolts E

k. Tighten the cap bolts using several passes. Tighten bolt C to 66 inch lbs. (7.5 Nm) an the remaining bolts to 12 ft. lbs. (16 Nm).

l. Remove the service bolt.

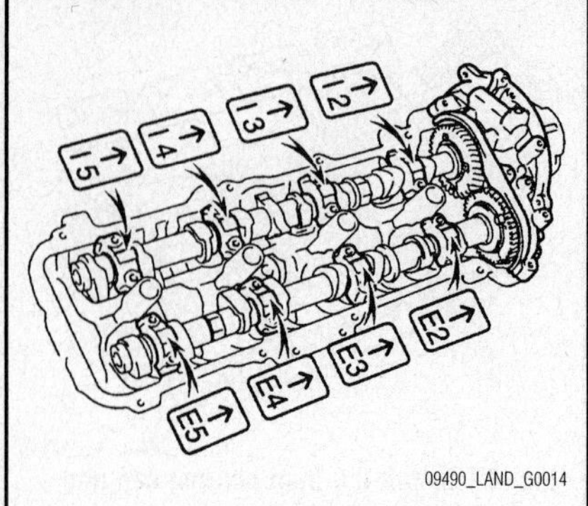

Fig. 29 Install the front bearing cap and then the other caps in the sequence illustrated on the left side camshafts

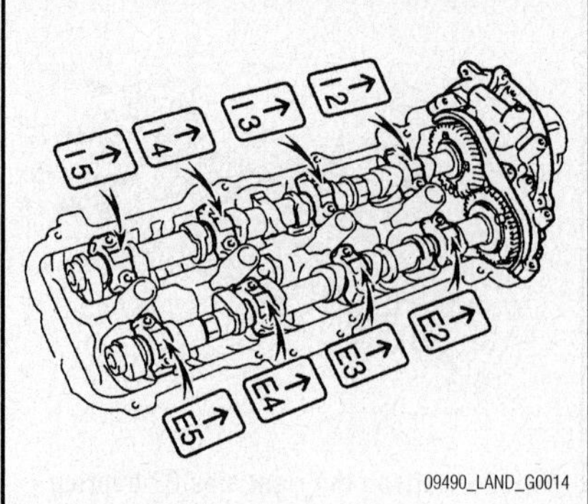

Fig. 32 Install the front bearing cap and then the other caps in the sequence illustrated on the right side camshafts

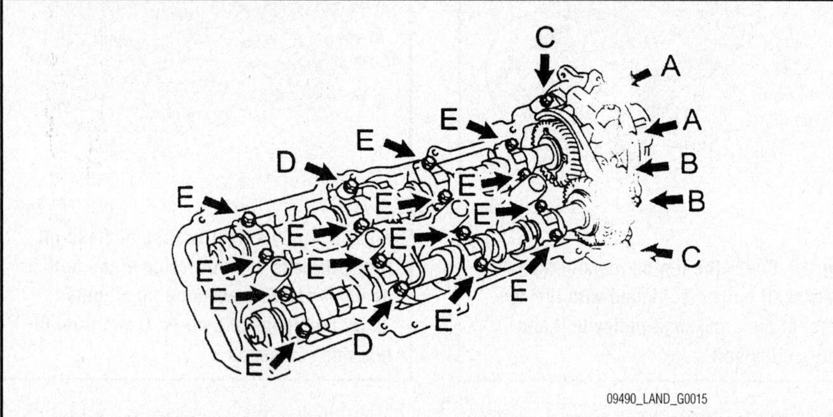

Fig. 30 Apply a light coating of clean oil to the threads and underside of the bolt heads D and E. make sure no oil gets under the heads of bolts A, B and C on the right side camshafts

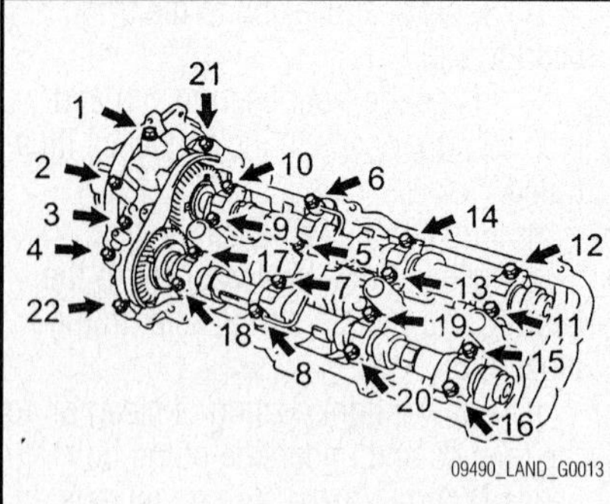

Fig. 31 Left side camshaft bolt torque sequence

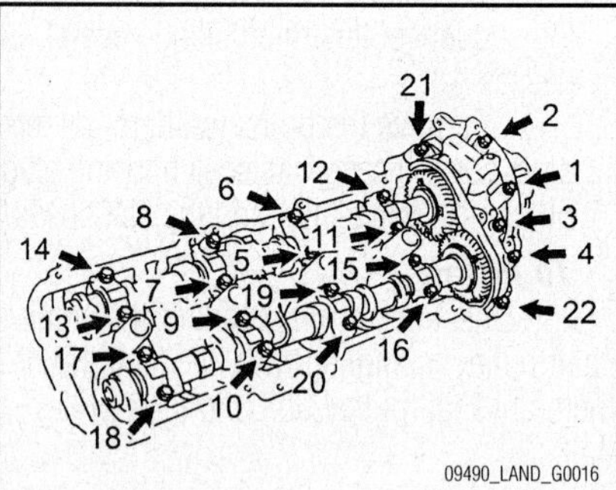

Fig. 33 Right side camshaft bolt torque sequence

25. Install the right side camshafts as follows:
   a. Apply multipurpose grease to the thrust portion of the camshafts.
   b. Align the 1 dot timing mark of the camshaft drive and driven main gears and install the camshafts.
   c. Set the 1 dot timing mark of the camshaft drive and driven gears at a 10 degree angle.
   d. Apply seal packing to the camshaft housing plug.
   e. Install the camshaft housing plug on the cylinder head as illustrated. Install the strainer on the head being careful it is properly positioned.
   f. Apply seal packing to the front bearing cap.
   g. Install the front bearing cap and then the other caps in the sequence illustrated.
   h. Push in the camshaft oil seal.
   i. Install 4 new seal washers to the bearing cap bolts A and B, refer to the illustration.
   j. Apply a light coating of clean oil to the threads and underside of the bolt heads D and E. make sure no oil gets under the heads of bolts A, B and C.
   k. The bolt lengths and positions are as follows. refer to the illustration for bolt location:
      - 94mm bolts A
      - 72mm bolts B
      - 25mm bolts C
      - 52mm bolts D
      - 38mm bolts E
   l. Tighten the cap bolts using several passes. Tighten bolt C to 66 inch lbs. (7.5 Nm) an the remaining bolts to 12 ft. lbs. (16 Nm).
   m. Remove the service bolt.
26. Check and adjust the valve clearance.
27. Install the camshaft timing control valve.
28. Install the 4 half moon plugs onto the cylinder heads.
29. Install the valve covers and tighten to 53 inch lbs. (6 Nm).
30. Install the engine hangers and tighten to 27 ft. lbs. (37 Nm).
31. Install the VVT sensors.
32. Install the oil dipstick tube and dipstick.
33. Install the ignition coils.
34. Install the water bypass joint and tighten the retainers to 13 ft. lbs. (18 Nm).
35. Install the water inlet and housing assembly.
36. Install the intake manifold.
37. Install the timing belt rear plates, right plates first, then left plates. Tighten the retainers to 66 inch lbs. (7 Nm).
38. Install the throttle body cover.
39. Install the front exhaust pipe, power steering pump.
40. Install the camshaft position sensor and camshaft timing pulleys, tighten to 25 ft. lbs. (34 Nm).
41. Install the timing belt.
42. Fill the cooling system and perform an oil change.
43. Start the vehicle and check for leaks.

CRANKSHAFT DAMPER

REMOVAL & INSTALLATION

1. Before servicing the vehicle, refer to the precautions in the beginning of this section.
2. Drain the cooling system.

3. Remove or disconnect the following:
   - Negative battery cable
   - Engine under cover
   - Engine appearance cover
   - Air intake assembly
   - Accessory drive belt
   - Cooling fan and pulley
   - Radiator
   - Drive belt idler pulley
   - Camshaft Position (CMP) sensor connector
   - Upper timing covers
   - Oil cooler pipe
   - Center timing cover
   - A/C compressor
   - Cooling fan bracket
   - Crankshaft pulley

To install:

   - Crankshaft pulley. Tighten the bolt to 181 ft. lbs. (245 Nm).
   - Cooling fan bracket. Tighten the 12mm bolts to 12 ft. lbs. (16 Nm) and the 14mm bolts to 24 ft. lbs. (32 Nm).
   - A/C compressor
   - Center timing cover
   - Oil cooler pipe
   - Upper timing covers
   - CMP sensor connector
   - Drive belt idler pulley. Tighten the bolt to 27 ft. lbs. (37 Nm).
   - Radiator
   - Cooling fan and pulley. Tighten the nuts to 16 ft. lbs. (21 Nm).
   - Accessory drive belt
   - Air intake assembly
   - Engine appearance cover
   - Engine under cover
   - Negative battery cable
4. Fill the cooling system.
5. Start the engine and check for leaks.

CRANKSHAFT FRONT SEAL

REMOVAL & INSTALLATION

1. Before servicing the vehicle, refer to the precautions in the beginning of this section.
2. Drain the cooling system.
3. Remove or disconnect the following:
   - Negative battery cable
   - Engine under cover
   - Engine appearance cover
   - Air intake assembly
   - Accessory drive belt
   - Cooling fan and pulley
   - Radiator
   - Drive belt idler pulley
   - Camshaft Position (CMP) sensor connector
   - Upper timing covers
   - Oil cooler pipe
   - Center timing cover
   - A/C compressor
   - Cooling fan bracket
   - Crankshaft pulley
   - Lower timing cover
   - Timing belt.
   - Crankshaft timing sprocket
   - Front crankshaft seal

To install:

4. Install the oil seal so that it is flush with the oil pump housing.
5. Install or connect the following:
   - Crankshaft timing sprocket
   - Timing belt
   - Lower timing cover
   - Crankshaft pulley. Tighten the bolt to 181 ft. lbs. (245 Nm).
   - Cooling fan bracket. Tighten the 12mm bolts to 12 ft. lbs. (16 Nm) and the 14mm bolts to 24 ft. lbs. (32 Nm).
   - A/C compressor
   - Center timing cover
   - Oil cooler pipe
   - Upper timing covers
   - CMP sensor connector
   - Drive belt idler pulley. Tighten the bolt to 27 ft. lbs. (37 Nm).
   - Radiator
   - Cooling fan and pulley. Tighten the nuts to 16 ft. lbs. (21 Nm).
   - Accessory drive belt
   - Air intake assembly
   - Engine appearance cover
   - Engine under cover
   - Negative battery cable
6. Fill the cooling system.
7. Start the engine and check for leaks.

CYLINDER HEAD

REMOVAL & INSTALLATION

V6 Engine

See Figures 34 through 46.

1. Discharge fuel system pressure.
2. Drain engine coolant.
3. Drain engine oil.
4. Remove timing chain.
5. Remove the 2 bolts and cool air inlet with air cleaner hose.
6. Remove the ATF level gauge.
7. Remove the bolt and pull out the oil filler tube.
8. Remove the O-ring from the oil filler tube.
9. Remove exhaust pipe assembly.
10. Remove the 3 bolts and manifold stay.
11. Disconnect the A/F sensor connector.
12. Remove the 6 nuts, exhaust manifold and gasket.
13. Remove No. 2 exhaust front pipe assembly.
14. Remove the 3 bolts and No. 2 manifold stay.
15. Disconnect the A/F sensor connector.
16. Remove the 6 nuts, exhaust manifold and gasket.
17. Disconnect No. 1 and No. 2 fuel pipe sub-assembly.
18. Disconnect the 6 fuel injector connectors.
19. Remove the 10 bolts, intake manifold and 2 gaskets.
20. Disconnect the engine coolant temperature sensor connector.
21. Disconnect the heater hose.
22. Remove the 2 bolts, 4 nuts, water by-pass joint RR and 2 gaskets.
23. Remove the O-ring from the water outlet pipe.
24. While raising the No. 2 chain tensioner, insert a pin of 1.0 mm (0.039 in.) into the hole to fix it.
25. Hold the hexagonal portion of the camshaft with a wrench, and remove the 2 bolts, camshaft timing gear, camshaft timing gear assembly and No. 2 timing chain.

➡Be careful not to damage the cylinder head and valve lifter with the wrench. Do not disassemble the camshaft timing gear assembly.

26. Remove the bolt and No. 2 chain tensioner.

➡As the thrust clearance of the camshaft is small, the camshaft must be kept level while it is being removed. If the camshaft is not kept level, the portion of the cylinder head which receives the shaft thrust may crack or be damaged, causing the camshaft to seize or break. To avoid this, the following steps should be carried out.

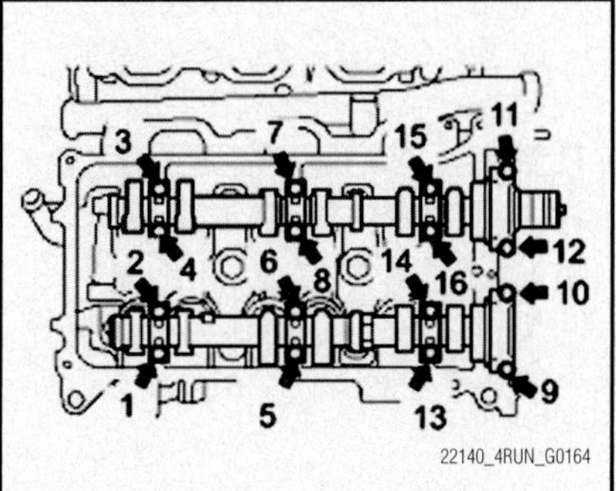

Fig. 34 Bearing cap removal sequence bank No. 1

27. Rotate the camshafts counterclockwise using a wrench so that the cam lobes of No. 1 cylinder face each direction.

28. Using several steps, loosen and remove the 16 bearing cap bolts uniformly in the sequence as shown in the illustration.

29. Remove the 8 bearing caps and 2 camshafts.

30. Remove No. 2 camshaft bearing.

31. Remove the 2 bolts and No. 1 chain vibration damper.

32. While pushing down the No. 3 chain tensioner, insert a pin of 1.0 mm (0.039 in.) into the hole to fix it.

33. Hold the hexagonal portion of the camshaft with a wrench, and remove the 2 bolts, camshaft timing gear, camshaft timing gear assembly and No. 2 timing chain.

➡Be careful not to damage the cylinder head and valve lifter with the wrench. Do not disassemble the camshaft timing gear assembly.

34. Remove the bolt and No. 3 chain tensioner.

➡As the thrust clearance of the camshaft is small, the camshaft must be kept level while it is being removed. If the camshaft is not kept level, the portion of the cylinder head which receives the shaft thrust may crack or be damaged, causing the camshaft to seize or break. To avoid this, the following steps should be carried out.

35. Using several steps, loosen and remove the 16 bearing cap bolts uniformly in the sequence as shown in the illustration.

36. Remove the 8 bearing caps and 2 camshafts.

37. Remove the 2 bolts and separate the 2 ground cables.

38. Using several steps, loosen the 8 cylinder head bolts on the cylinder head uniformly with a 10 mm bi-hexagon wrench in the sequence as shown in the illustration.

39. Remove the 8 cylinder head bolts and plate washers.

➡Be careful not to drop the plate washers into the cylinder head. Cylinder head warpage or cracking could result from removing bolts in incorrect order.

40. Lift the cylinder head from the dowels on the cylinder block, and place the cylinder head on wooden blocks on a bench.

➡Be careful not to damage the contact surfaces of the cylinder head and cylinder block.

To install:

41. Remove any old packing (FIPG) material and be careful not to drop any oil on the contact surfaces of the cylinder head and cylinder block.

42. Apply seal packing (diameter: 2.5 to 3 mm (0.098 to 0.118 in.)) to a new cylinder head gasket.

➡Install the cylinder head within 3 minutes after applying seal packing. After installing it, the cylinder head bolts must be tightened within 15 minutes. Otherwise the seal packing must be removed and reapplied.

43. Place the cylinder head gasket on the cylinder block surface with the Lot No. stamp upper side facing upward.

44. Place the cylinder head on the cylinder head gasket.

45. Install the 8 cylinder head bolts.

➡The cylinder head bolts are tightened in 2 successive steps. If any cylinder head bolt is broken or deformed, replace it.

46. Apply a light coat of engine oil to the threads of the cylinder head bolts.

47. Install the plate washer to the cylinder head bolt.

48. Using several steps, tighten each bolt uniformly with a 10 mm bi-hexagon wrench in the sequence. Tighten to 27 ft. lbs. (36 Nm). If any one of the cylinder head bolts does not meet the torque specification, replace the cylinder head bolt.

49. Retighten the cylinder head bolts 180°.

50. Install the front 2 cylinder head bolts.

51. Using several steps, install and tighten the 2 cylinder head bolts uniformly in the sequence to 22 ft. lbs. (30 Nm).

52. Install the ground cable with the bolt.

53. Perform the installation procedure for RH cylinder head.

54. Install camshafts for Bank 2.

➡As the thrust clearance of the camshaft is small, the camshaft must be kept level while it is being installed. If the camshaft is not kept level, the portion of the cylinder head which receives the shaft thrust may crack or be damaged, causing the camshaft to seize or break. To avoid this, the following steps should be carried out.

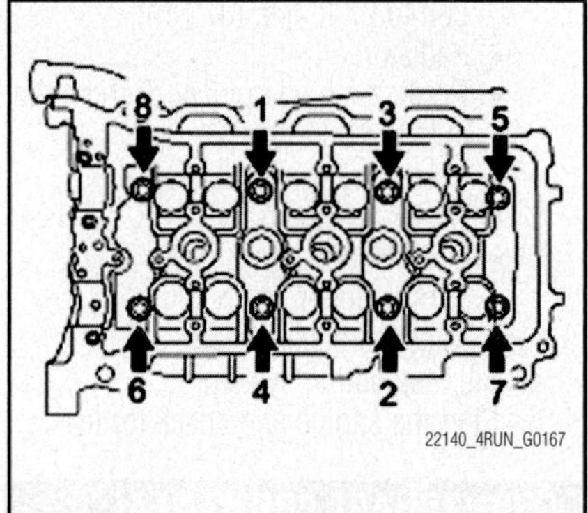

Fig. 37 Cylinder head bolt installation sequence

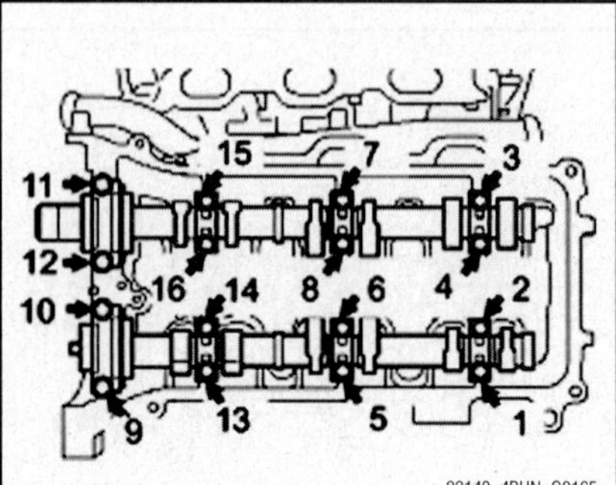

Fig. 35 Bearing cap removal sequence bank No. 2

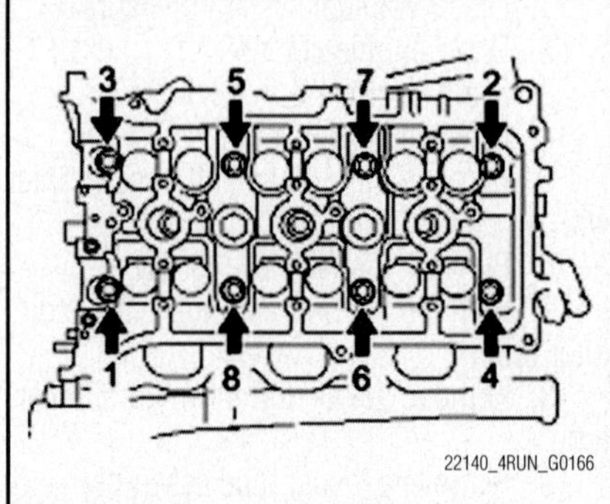

Fig. 36 Cylinder head bolt removal sequence

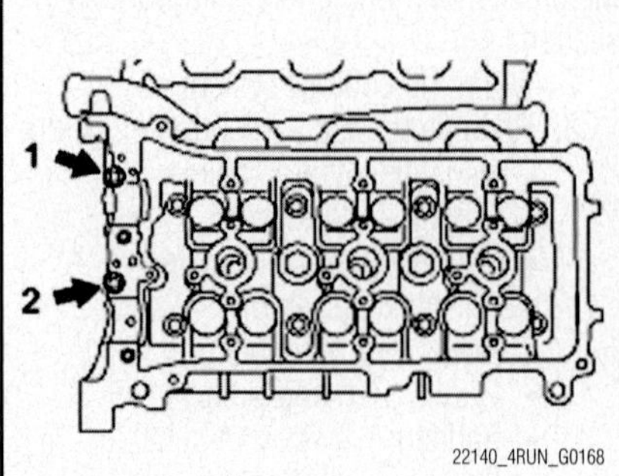

Fig. 38 Cylinder head bolt installation sequence

55. Set the crankshaft position. Install the crankshaft pulley set bolt, and turn the crankshaft, and set the crankshaft set key at the left horizontal position.

➡Setting the crankshaft at a wrong angle can cause the piston head and valve head to come into contact with each other when you install the camshaft, causing damage. So always set the crankshaft at the correct angle.

56. Apply new engine oil to the thrust portions and journals of the camshafts.

57. Place the 2 camshafts onto the cylinder head with the cam lobes of No. 2 cylinder facing each correct direction as shown in the illustration.

58. Install the 8 bearing caps in their proper locations.

59. Apply a light coat of engine oil to the threads and under the heads of the bearing cap bolts

60. Using several steps, install and tighten the 16 bearing cap bolts uniformly in the sequence and tighten 10mm bolts to 80 inch lbs. (9.0 Nm) and 12 mm to 18 ft. lbs. (24 Nm).

61. While pushing in the tensioner, insert a pin of 1.0 mm (0.039 in.) into the hole to fix it.

62. Install the No. 3 chain tensioner with the bolt and tighten to 14 ft. lbs. (19 Nm).

63. Align the mark links (yellow) with the timing marks (1-dot mark and 2-dot mark) of the camshaft timing gears.

64. Align the timing marks on the camshaft timing gears with the timing marks on the bearing caps, and install the camshaft timing gears with the chain to the LH camshafts.

65. Temporarily install the 2 camshaft timing gear bolts.

➡Do not push the camshaft timing gear assembly to the camshaft forcibly when installing it.

66. Hold the hexagonal portion of the camshaft with a wrench, and tighten the 2 bolts to 74 ft. lbs. 100 Nm).

67. Remove the pin from the No. 3 tensioner.

68. Install the No. 1 chain vibration damper with the 2 bolts and tighten to 14 ft. lbs. (19 Nm).

69. Install the No. 2 camshaft bearing to the cylinder head.

➡Clean the backside of the bearing and the contact surface of the cylinder head and prevent oil from adhering to them

70. Install camshafts for Bank 1.

➡As the thrust clearance of the camshaft is small, the camshaft must be kept level while it is being installed. If the camshaft is not kept level, the portion of the cylinder head which receives the shaft thrust may crack or be damaged, causing the camshaft to seize or break. To avoid this, the following steps should be carried out.

71. Set the crankshaft position. Install the crankshaft pulley set bolt, and turn the crankshaft, and set the crankshaft set key at the left horizontal position.

➡Setting the crankshaft at a wrong angle can cause the piston head and valve head to come into contact with each other when you install the camshaft, causing damage. So always set the crankshaft at the correct angle.

72. Apply new engine oil to the thrust portions and journals of the camshafts.

73. Place the 2 camshafts onto the cylinder head with the cam lobes of No. 2

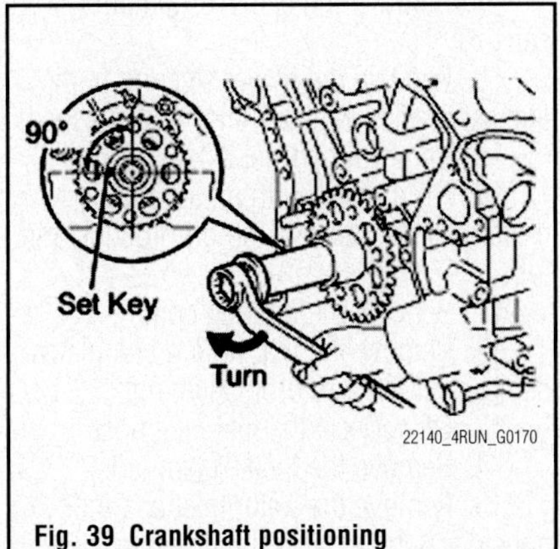

Fig. 39 Crankshaft positioning

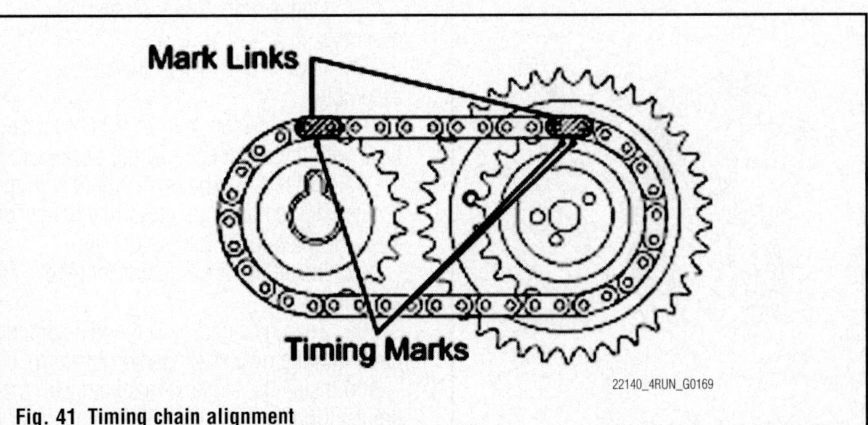

Fig. 41 Timing chain alignment

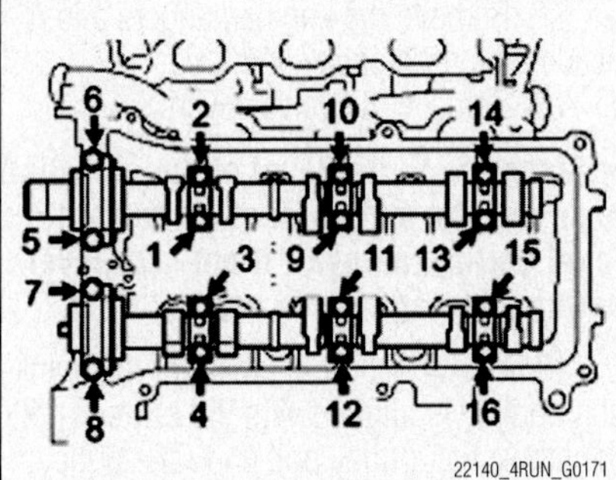

Fig. 40 Camshaft bearing cap bolt tightening sequence

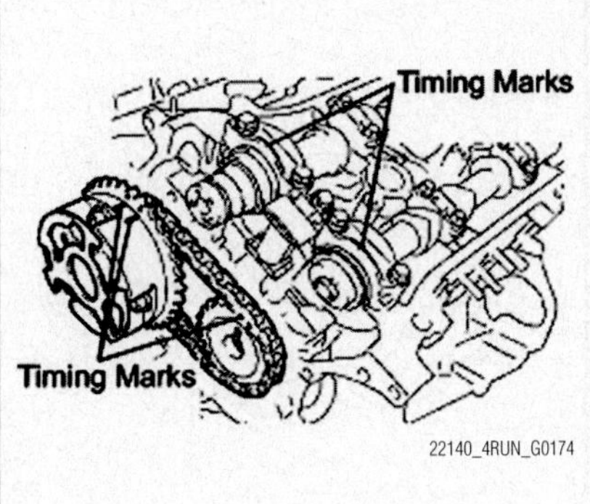

Fig. 42 Timing chain alignment marks

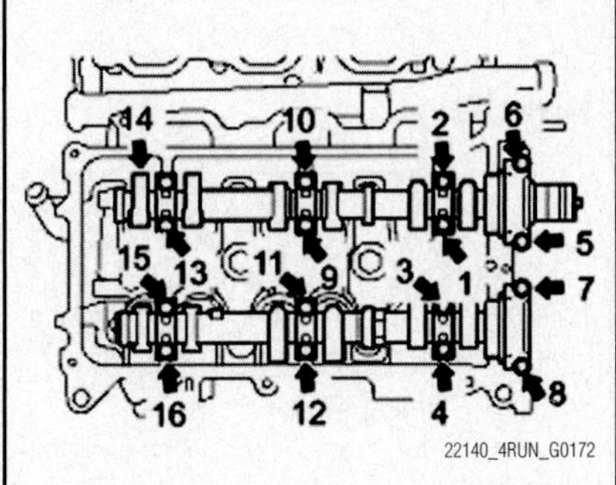

Fig. 43 Camshaft bearing cap bolt tightening sequence

cylinder facing each correct direction as shown in the illustration.

74. Install the 8 bearing caps in their proper locations.

75. Apply a light coat of engine oil to the threads and under the heads of the bearing cap bolts.

76. Using several steps, install and tighten the 16 bearing cap bolts uniformly in the sequence and tighten 10mm bolts to 80 inch lbs. (9.0 Nm) and 12 mm to 18 ft. lbs. (24 Nm).

77. Turn the camshafts clockwise until the knock pin comes to 90° position to the cylinder head.

78. While pushing in the tensioner, insert a pin of 1.0 mm (0.039 in.) into the hole to fix it.

79. Install the No. 2 chain tensioner with the bolt and tighten to 14 ft. lbs. (19 Nm).

80. Align the mark links (yellow) with the timing marks (1-dot mark and 2-dot mark) of the camshaft timing gears.

81. Align the timing marks on the camshaft timing gears with the timing marks on the bearing caps, and install the camshaft timing gears with the chain to the RH camshafts.

82. Temporarily install the 2 camshaft timing gear bolts.

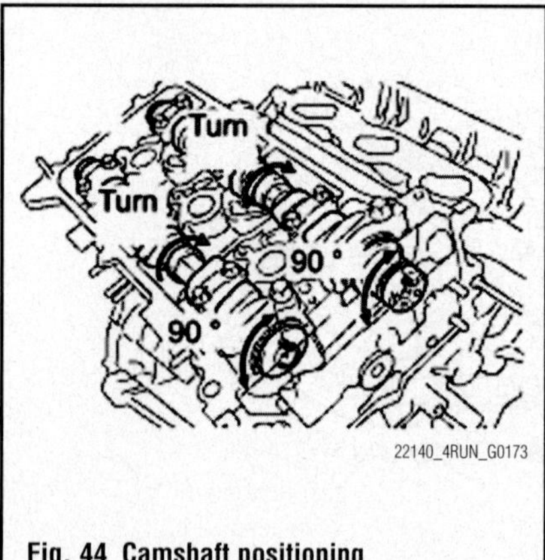

Fig. 44 Camshaft positioning

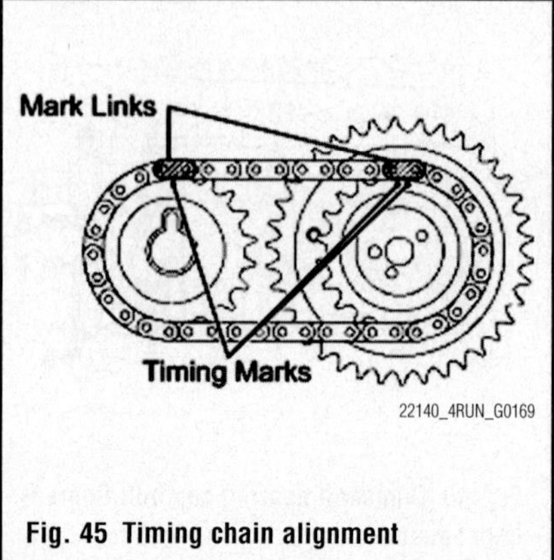

Fig. 45 Timing chain alignment

➡Do not push the camshaft timing gear assembly to the camshaft forcibly when installing it.

83. Hold the hexagonal portion of the camshaft with a wrench, and tighten the 2 bolts to 74 ft. lbs. 100 Nm).

84. Remove the pin from the No. 2 tensioner.

85. Install a new O-ring to the water outlet pipe.

86. Apply soapy water to the O-ring.

87. Install 2 new gaskets and water by-pass joint RR with the 2 bolts and 4 nuts and tighten to 80 inch lbs. (9.0 nm).

88. Connect the heater hose.

89. Connect the engine coolant temperature sensor connector.

90. Set a new intake gasket on each cylinder head.

➡Align the port holes of the gasket and cylinder head. Be careful of the installation direction.

91. Set the intake manifold on the cylinder heads.

92. Install and tighten the 10 bolts uniformly in several steps and tighten bolts to 19 ft. lbs. (26 Nm).

93. Connect the 6 fuel injector connectors.

94. Connect no. 2 fuel pipe sub-assembly.

95. Connect no. 1 fuel pipe sub-assembly.

96. Set a new gasket to the LH cylinder head with the oval shape facing backward.

97. Install the exhaust manifold with the 6 nuts. Tighten the nuts uniformly in several steps. Tighten to 22 ft .lbs. (30 Nm).

98. Install no. 2 exhaust front pipe assembly.

99. Set a new gasket to the RH cylinder head with the oval shape facing forward.

100.Install the exhaust manifold with the 6 nuts. Tighten the nuts uniformly in several steps. Tighten to 22 ft .lbs. (30 Nm).

Fig. 46 Timing chain alignment marks

101. Install the manifold stay with the 3 bolts tighten to 30 ft lbs. (40 Nm).

102. Install exhaust pipe assembly front.

103. Coat a new O-ring with ATF, and install it to the oil filler tube.

104. Push in the oil filler tube end into the oil pan lower tube.

105. Install the oil filler tube with the bolt and tighten to 9 ft. lbs. (12 Nm).

106. Install the ATF level gauge.

107. Install the cool air inlet with air cleaner hose with the 2 bolts.

108. Install timing chain assembly.

109. Add engine oil.

110. Add engine coolant.

111. Check for engine oil leakage.

112. Check for engine coolant leakage.

113. Check for fuel leakage.

114. Check for exhaust gas leakage.

115. Inspect engine idle speed.

V8 Engine

See Figure 47.

1. Before servicing the vehicle, refer to the precautions section.
2. Drain the cooling system.
3. Relieve the fuel system pressure.
4. Remove the V bank cover.
5. Remove the timing belt.
6. Remove the camshaft pulleys.
7. Remove the camshaft position sensor.
8. Remove the power steering pump and set it aside with the lines still attached.
9. Remove the front exhaust pipe.
10. On models with an automatic transmission, remove the oil dipstick and tube.
11. Remove the ignition coils.
12. Remove the rear timing belt plates being careful not to drop anything.
13. Disconnect the fuel inlet hose.
14. Remove the intake manifold.
15. Remove the water inlet and inlet housing. Refer to water pump removal.
16. Remove the front and rear water bypass joint.
17. Remove the engine hangers and if needed the oil dipstick and tube.
18. Remove the valve covers.

➡Since the thrust level of the camshaft is small, the camshaft must be kept level during removal. If not kept level serious damage could occur.

19. Check the timing mark of the crankshaft pulley is aligned with the center(s) of the crankshaft pulley bolt and idler pulley bolt.

➡If the crankshaft pulley is wrongly positioned, this can cause the piston to

contact the head causing severe damage. Make sure the crankshaft pulley is properly positioned.

20. Release the oil from the front bearing caps using the tool illustrated. Rotate the camshaft timing tube from left to right 2 to 3 times within its VVT-I range of 25 degrees and collect the oil from the timing oil control valve installation hole using a rag.

21. Remove the left hand camshafts as follows:

a. Bring the service bolt of the sub gear up by turning the left exhaust camshaft using a wrench on the hexagon head portion of the shaft.

b. Secure the sub gear to the main gear using a 16 to 20 mm bolt with a diameter of 6mm and a thread pitch of 1mm.

c. Make sure the torsional force of the sub gear is retained by the bolt.

d. Align the 2 dot timing mark of the left side camshaft by turning the left exhaust camshaft using a wrench on the hexagon head portion of the shaft.

➡Mark the position of the caps so they can be reinstalled in their original positions.

e. Loosen the 22 bearing cap bolts in the sequence illustrated using several passes.

f. Remove the bolts, washers, oil feed pipe, bearing caps, camshaft housing plug, oil control valve filter and the camshafts.

22. Remove the right hand camshafts as follows:

a. Bring the service bolt of the sub gear up by turning the right exhaust camshaft using a wrench on the hexagon head portion of the shaft.

b. Secure the sub gear to the main gear using a 16 to 20 mm bolt with a diameter of 6mm and a thread pitch of 1mm.

c. Make sure the torsional force of the sub gear is retained by the bolt.

d. Align the 1 dot timing mark of the camshaft main gear (about 10 degrees) angle by turning the right exhaust camshaft using a wrench on the hexagon head portion of the shaft.

➡Mark the position of the caps so they can be reinstalled in their original positions.

e. Loosen the 22 bearing cap bolts in the sequence illustrated using several passes.

f. Remove the bolts, washers, oil feed pipe, bearing caps, camshaft housing plug, oil control valve filter and the camshafts.

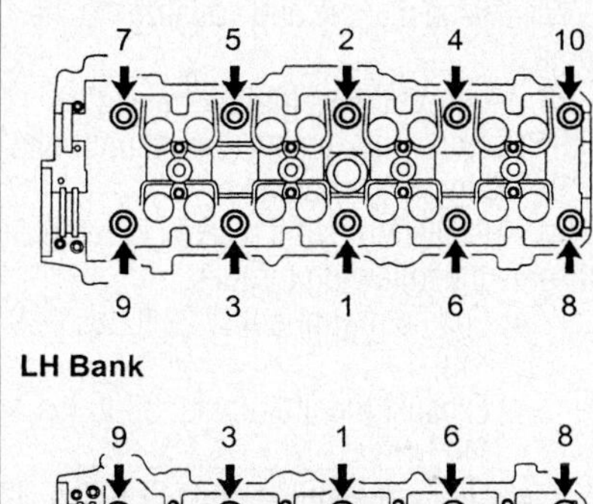

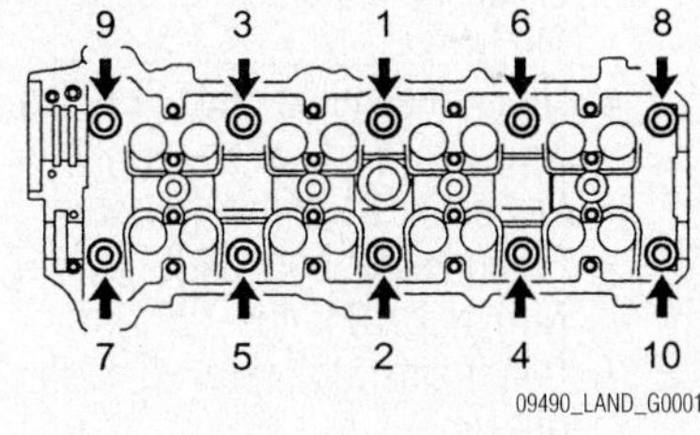

Fig. 47 Cylinder head loosening sequence—4.7L 2UZ-FE engine

23. Loosen the cylinder head bolts in the sequence shown, using several passes.

24. Remove the cylinder heads and exhaust manifolds together as an assembly.

To install:

25. Install new gaskets and the cylinder heads

26. Tighten the bolts in sequence as follows:

a. Step 1: 30 ft. lbs. (40 Nm).

b. Step 2: Plus 90 degrees.

c. Step 3: Plus 90 degrees.

27. Check the timing mark of the crankshaft pulley is aligned with the center(s) of the crankshaft pulley bolt and idler pulley bolt.

➡If the crankshaft pulley is wrongly positioned, this can cause the piston to contact the head causing severe damage. Make sure the crankshaft pulley is properly positioned.

28. Install the left side camshafts as follows:

a. Apply multipurpose grease to the thrust portion of the camshafts.

b. Align the 2 dot timing mark of the camshaft drive and driven main gears and install the camshafts.

c. Apply seal packing to the camshaft housing plug.

d. Install the camshaft housing plug on the cylinder head as illustrated. Install the strainer on the head being careful it is properly positioned.

e. Apply seal packing to the front bearing cap.

f. Install the front bearing cap and then the other caps in the sequence illustrated.

g. Push in the camshaft oil seal.

h. Install 4 new seal washers to the bearing cap bolts A and B, refer to the illustration.

i. Apply a light coating of clean oil to the threads and underside of the bolt heads D and E. make sure no oil gets under the heads of bolts A, B and C.

j. The bolt lengths and positions are as follows. refer to the illustration for bolt location:

- 94mm bolts A
- 72mm bolts B
- 25mm bolts C
- 52mm bolts D
- 38mm bolts E

k. Tighten the cap bolts using several passes. Tighten bolt C to 66 inch lbs. (7.5 Nm) an the remaining bolts to 12 ft. lbs. (16 Nm).

l. Remove the service bolt.

29. Install the right side camshafts as follows:

a. Apply multipurpose grease to the thrust portion of the camshafts.

b. Align the 1 dot timing mark of the camshaft drive and driven main gears and install the camshafts.

c. Set the 1 dot timing mark of the camshaft drive and driven gears at a 10 degree angle.

d. Apply seal packing to the camshaft housing plug.

e. Install the camshaft housing plug on the cylinder head as illustrated. Install the strainer on the head being careful it is properly positioned.

f. Apply seal packing to the front bearing cap.

g. Install the front bearing cap and then the other caps in the sequence illustrated.

h. Push in the camshaft oil seal.

i. Install 4 new seal washers to the bearing cap bolts A and B, refer to the illustration.

j. Apply a light coating of clean oil to the threads and underside of the bolt heads D and E. make sure no oil gets under the heads of bolts A, B and C.

k. The bolt lengths and positions are as follows. refer to the illustration for bolt location:

- 94mm bolts A
- 72mm bolts B
- 25mm bolts C
- 52mm bolts D
- 38mm bolts E

l. Tighten the cap bolts using several passes. Tighten bolt C to 66 inch lbs. (7.5 Nm) an the remaining bolts to 12 ft. lbs. (16 Nm).

m. Remove the service bolt.

30. Check and adjust the valve clearance.

31. Install the camshaft timing control valve.

32. Install the 4 half moon plugs onto the cylinder heads.

33. Install the valve covers and tighten to 53 inch lbs. (6 Nm).

34. Install the engine hangers and tighten to 27 ft. lbs. (37 Nm).

35. Install the VVT sensors.

36. Install the oil dipstick tube and dipstick.

37. Install the ignition coils.

38. Install the water bypass joint and tighten the retainers to 13 ft. lbs. (18 Nm).

39. Install the water inlet and housing assembly.

40. Install the intake manifold.

41. Install the timing belt rear plates, right plates first, then left plates. Tighten the retainers to 66 inch lbs. (7 Nm).

42. Install the throttle body cover.

43. Install the front exhaust pipe, power steering pump.

44. Install the camshaft position sensor and camshaft timing pulleys, tighten to 25 ft. lbs. (34 Nm).

45. Install the timing belt.

46. Fill the cooling system and perform an oil change.

47. Start the vehicle and check for leaks.

ENGINE ASSEMBLY

REMOVAL & INSTALLATION

1. Before servicing the vehicle, refer to the precautions section.
2. Remove the transmission.
3. Remove the hood.
4. Remove the V-bank cover.
5. Remove the air cleaner assembly.
6. Remove the under-covers.
7. Remove the radiator.
8. Remove the fan shroud.
9. Tag and disconnect all hoses, pipes and wires necessary for engine removal.
10. Remove the fan.
11. Remove the power steering pump and secure it out of the way.
12. Remove the alternator and secure it out of the way.
13. Remove the compressor and secure it out of the way.
14. Remove the transmission filler tube.
15. Remove the oil level sending unit.
16. Remove the exhaust manifolds.
17. Attach a crane and equalizer to the engine.
18. Support the weight of the engine with the crane and remove the mount bolts.
19. Remove the engine.
20. Installation is the reverse of removal. Observe the following torques:
    - Engine mount bolts: 28 ft. lbs. (38 Nm)
    - Exhaust manifold nuts: 33 ft. lbs. (44 Nm)
    - Oil level sending unit: 11 ft. lbs. 15 Nm)
    - Fan bolts: 21 ft. lbs. (29 Nm)
    - Compressor: bolt, 34 ft. lbs. (47 Nm); nut, 18 ft. lbs. (25 Nm)
    - Power steering pump: 32 ft. lbs. (43 Nm)
    - Hood: 10 ft. lbs. (13 Nm)

EXHAUST MANIFOLD

REMOVAL & INSTALLATION

1. Before servicing the vehicle, refer to the precautions section.
2. Attach a hoist to the engine lifting eyes.
3. Remove or disconnect the following:
    - Negative battery cable
    - Heated Oxygen (HO_2S) sensor connectors
    - Exhaust manifold heat shield
    - Exhaust front pipe
    - Motor mount
    - Motor mount bracket
    - Exhaust manifold

To install:

➡Use new exhaust manifold nuts for assembly.

4. Install or connect the following:
    - Exhaust manifold. Tighten the nuts to 32 ft. lbs. (44 Nm).
    - Motor mount bracket. Tighten the bolts to 27 ft. lbs. (36 Nm).
    - Motor mount. Tighten the fasteners to 22 ft. lbs. (30 Nm).
    - Exhaust front pipe. Tighten the nuts to 46 ft. lbs. (62 Nm).
    - Exhaust manifold heat shield
    - HO_2S sensor connectors
    - Negative battery cable
5. Start the engine and check for leaks.

FLEXPLATE

REMOVAL & INSTALLATION

See Figure 48.

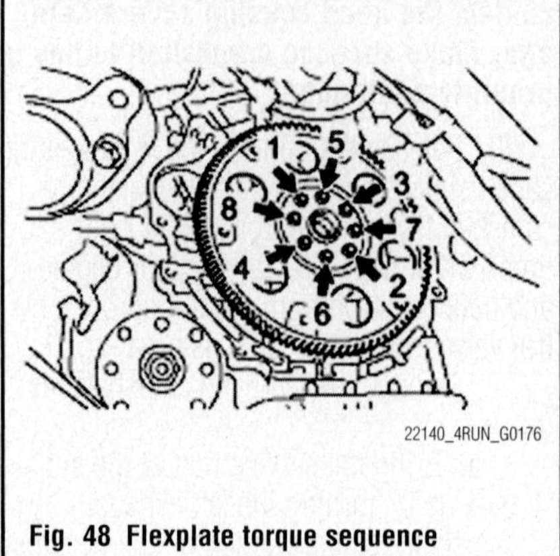

Fig. 48 Flexplate torque sequence

1. Before servicing the vehicle, refer to the precautions in the beginning of this section.
2. Remove the transmission. See transmission removal.
3. Remove the 8 bolts and the Flexplate.

To install:

4. Install the flexplate.
5. Coat two or three threads of each mounting bolt with Part No. 008833-00070 Three Bond 1324 or equivalent.
6. Tighten the bolts in sequence as follows:
    a. For V6 engine 61 ft. lbs. (83 Nm).
    b. For V8 engine 36 ft. lbs. (49 Nm) Plus 90°.
7. Install the transmission.

INTAKE MANIFOLD

REMOVAL & INSTALLATION

V6 Engine

1. Discharge fuel system pressure.
2. Drain engine coolant.
3. Drain engine oil.
4. Remove the 2 bolts and cool air inlet with air cleaner hose.
5. Remove the ATF level gauge.
6. Remove the bolt and pull out the oil filler tube.
7. Remove the O-ring from the oil filler tube.
8. Disconnect the A/F sensor connector.
9. Disconnect No. 1 and No. 2 fuel pipe sub-assembly.
10. Disconnect the 6 fuel injector connectors.
11. Remove the 10 bolts, intake manifold and 2 gaskets.

To install:

12. Set a new gasket on each cylinder head.
13. Set the intake manifold on the cylinder heads.

14. Install and tighten the 10 bolts uniformly in several steps and tighten to 19 ft. lbs. (26 Nm).
15. Connect the 6 fuel injector connectors.
16. Connect No. 2 fuel pipe sub-assembly.
17. Connect No. 1 fuel pipe sub-assembly.
18. Coat a new O-ring with ATF, and install it to the oil filler tube.
19. Push in the oil filler tube end into the oil pan lower tube.
20. Install the oil filler tube with the bolt.
21. Install the ATF level gauge.
22. Install the cool air inlet with air cleaner hose with the 2 bolts.
23. Add engine oil.
24. Add engine coolant.
25. Check for leakage.

V8 Engine

See Figure 49.

1. Discharge fuel system pressure.
2. Drain engine coolant.
3. Remove the 2 nuts and throttle body cover sub-assembly.
4. Disconnect the vacuum hoses (for the power steering idle-up and fuel pressure regulator) and ventilation hose.
5. Remove the air cleaner hose assembly.
6. Disconnect fuel hose.
7. Disconnect fuel hose No. 2.
8. Disconnect the throttle control connector.
9. Disconnect the purge VSV.
10. Disconnect the 8 injector connectors.

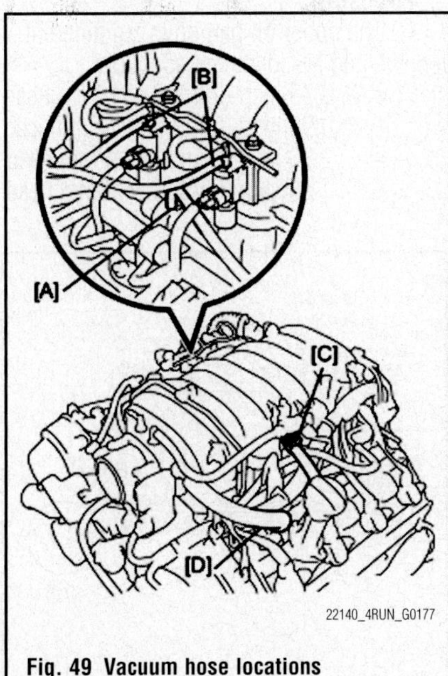

Fig. 49 Vacuum hose locations

11. Disconnect the ECT sensor connector.
12. Disconnect the 2 VSV connectors for the air injection system.
13. Disconnect the 8 ignition coil connectors.
14. Disconnect the 2 air fuel ratio sensor connectors.
15. Disconnect the vacuum hose from the fuel pressure regulator.
16. Disconnect the PCV hoses from the PCV valve on the LH cylinder head.
17. Disconnect the EVAP hose (from the charcoal canister) from the purge VSV.
18. Disconnect the 2 vacuum hoses from the VSV for the air injection system.
19. Disconnect the 2 water by-pass hoses from the throttle body.
20. Disconnect the 2 wire clamps from the wire clamp bracket on the RH delivery pipe.
21. Remove the bolt and nut holding the engine wire protector from the intake manifold and cylinder head.
22. Remove the 2 bolts and ground cables from the RH and LH cylinder heads.
23. Remove the bolt and V-bank cover bracket from the intake manifold.
24. Disconnect the engine wire from the engine hanger and wire bracket.
25. Remove the bolt and wire bracket from the intake manifold.
26. Remove the 6 bolts, 4 nuts, intake manifold assembly and 2 gaskets.
27. Remove air pump assembly w/ bracket.
28. Remove the 2 nuts and 2 knock sensors.

To install:

29. Install the 2 knock sensors with the 2 nuts and tighten to 15 ft. lbs. (20 Nm).
30. Connect the 2 knock sensor connectors.
31. Place 2 new gaskets on the intake manifold.
32. Place the intake manifold on the cylinder heads.
33. Install and uniformly tighten the 6 bolts and 4 nuts in several steps and tighten to 13 ft .lbs. (18 Nm).
34. Install the V-bank cover bracket to the intake manifold.
35. Install the wire bracket to the intake manifold with the bolt.
36. Connect the engine wire to the engine hanger and wire bracket.
37. Connect the wire protector to the intake manifold and cylinder heads with the bolt and nut.
38. Install the 2 ground cables with the 2 bolts to the RH and LH cylinder heads.
39. Connect the 2 water by-pass hoses to the throttle body.
40. Connect the 2 wire clamps to the wire clamp bracket on the RH delivery pipe.
41. Connect the vacuum hose to the fuel pressure regulator.
42. Connect the PCV hose to the PCV valve on the LH cylinder head.
43. Connect the EVAP hose (from the charcoal canister) to the purge VSV.
44. Connect the 2 vacuum hoses to the VSV for the air injection system.
45. Connect the throttle control connector.
46. Connect the 2 VSV connectors for the air injection system.
47. Connect the purge VSV connector.
48. Connect the 8 injector connectors.
49. Connect the ECT sensor connector.
50. Connect the 8 ignition coil connectors.
51. Connect the 2 air fuel ratio sensor connectors.
52. Install fuel hose No. 2.
53. Install fuel hose.
54. Install throttle body cover sub-assembly.
55. Add engine coolant.
56. Check for leaks.

OIL PAN

REMOVAL & INSTALLATION

V6 Engine

See Figures 50 through 54.

1. Before servicing the vehicle, refer to the precautions section.
2. Remove the engine from the vehicle and mount it on a stand.
3. Remove or disconnect the following:
   - Remove the drain plug and gasket
   - Remove the 10 bolts and 2 nuts

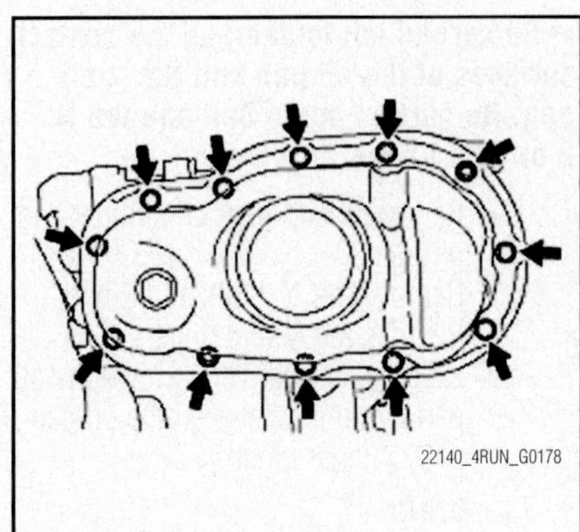

Fig. 50 No. 2 oil pan assembly

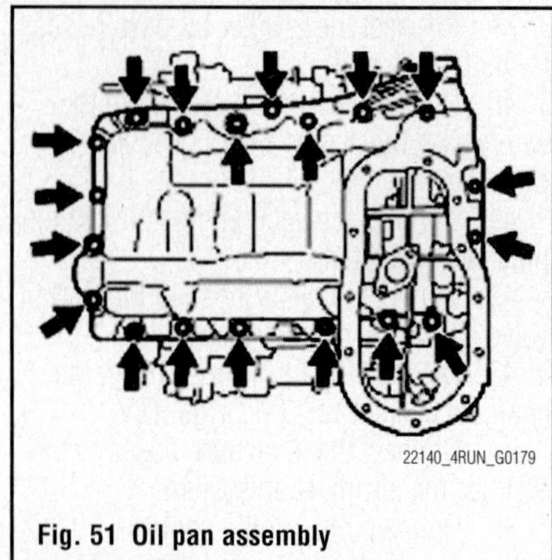

Fig. 51 Oil pan assembly

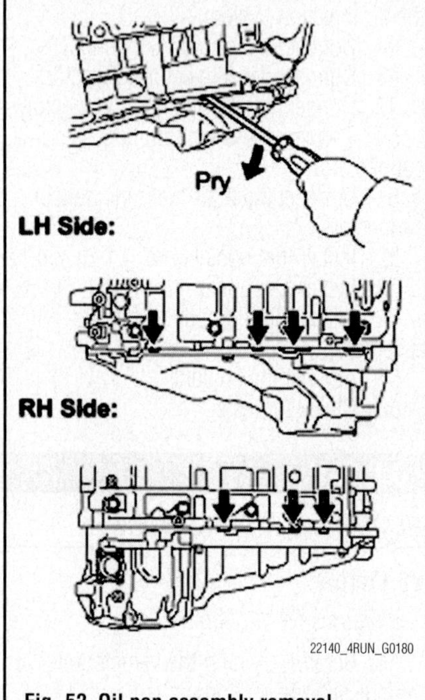

Fig. 52 Oil pan assembly removal

- Insert the blade of oil pan seal cutter between the oil pan and No. 2 oil pan, and cut off applied sealer and remove the No. 2 oil pan

➡Be careful not to damage the contact surfaces of the oil pan and No. 2 oil pan. Be careful not to damage the No. 2 oil pan flange

- Remove the 2 nuts, oil strainer and gasket
- Remove the 17 bolts and 2 nuts
- Remove the 4 stud bolts.
- Using a screwdriver, remove the oil pan by prying between the oil pan and cylinder block

To install:

4. Remove any old packing (FIPG) material and be careful not to drop any oil

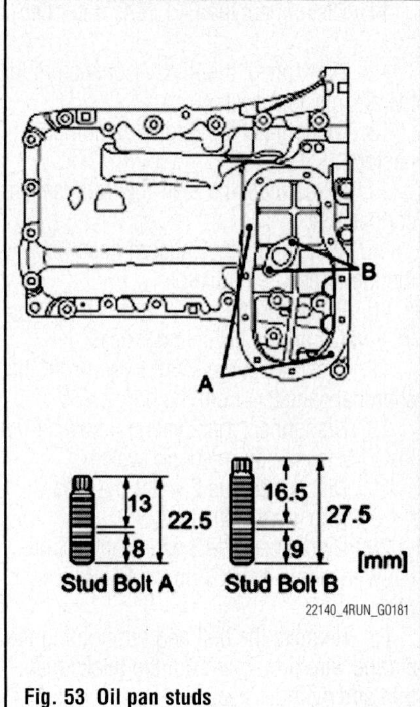

Fig. 53 Oil pan studs

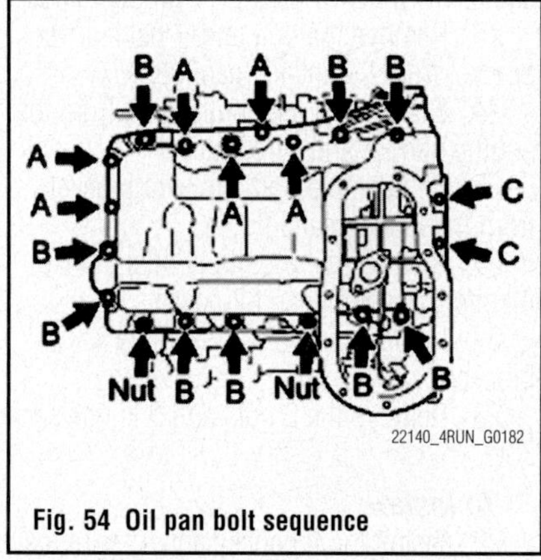

Fig. 54 Oil pan bolt sequence

on the contact surfaces of the cylinder block, rear oil seal retainer and oil pan.

5. Install the 4 stud bolts and tighten to 35 ft. lbs. (4.0 Nm).
6. Install a new O-ring to the oil pump.
7. Apply a continuous bead of seal packing (diameter: 3 to 4 mm (0.12 to 0.16 in.)) to the oil pan.

➡Install the oil pan within 3 minutes after applying seal packing. After installing it, the oil pan bolts and nuts must be tightened within 15 minutes. Otherwise the seal packing must be removed and reapplied.

8. Install the oil pan with the 17 bolts and 2 nuts. Tighten the bolts and nuts uniformly in several steps.
9. Tighten bolts as follows:
   - Bolt A 15 ft. lbs. (21 Nm)
   - Bolt B 15 ft. lbs. (21 Nm)
   - Bolt C 7 ft. lbs. (10 Nm)
   - Nut 15 ft. lbs. (21 Nm)
10. Install a new gasket and the oil strainer with the 2 nuts and tighten to 80 inch lbs. (9.0 Nm)
11. Remove any old packing (FIPG) material and be careful not to drop any oil on the contact surfaces of the oil pan and No. 2 oil pan

➡Install the No. 2 oil pan within 3 minutes after applying seal packing. After installing it, the No. 2 oil pan bolts and nuts must be tightened within 15 minutes. Otherwise the seal packing must be removed and reapplied

12. Install the No. 2 oil pan with the 10 bolts and 2 nuts. Tighten the bolts and nuts uniformly in several steps. Tighten bolts to 80 inch lbs. (9.0 nm) and nuts to 7 ft. lbs. (10 Nm)
13. Install the drain plug and a new gasket
14. Reinstall engine
15. Start engine and check for leaks

V8 Engine

See Figures 55 through 57.

1. Before servicing the vehicle, refer to the precautions section.
2. Remove the engine from the vehicle and mount it on a stand.
3. Remove or disconnect the following:
   - Oil dipstick tube
   - Lower oil pan
   - Oil pan baffle
   - Upper oil pan

To install:

4. The upper oil pan bolts are different lengths and are identified as follows:
   - A: 0.79 inch (20mm) w/10mm head
   - B: 0.98 inch (25mm) w/12mm head
   - C: 2.36 inch (60mm) w/12mm head
   - D: 1.38 inch (35mm) w/10mm head

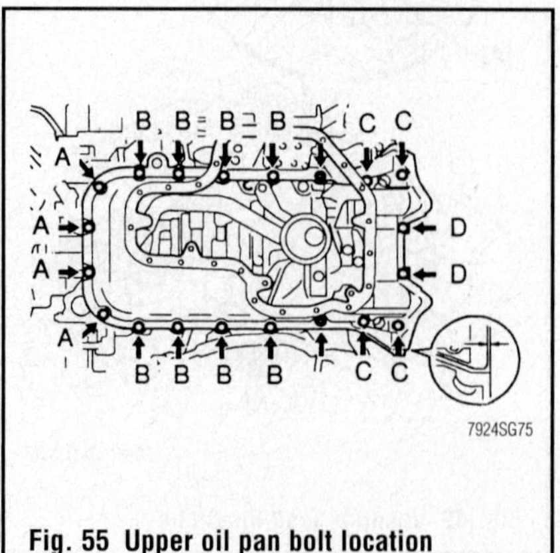

Fig. 55 Upper oil pan bolt location

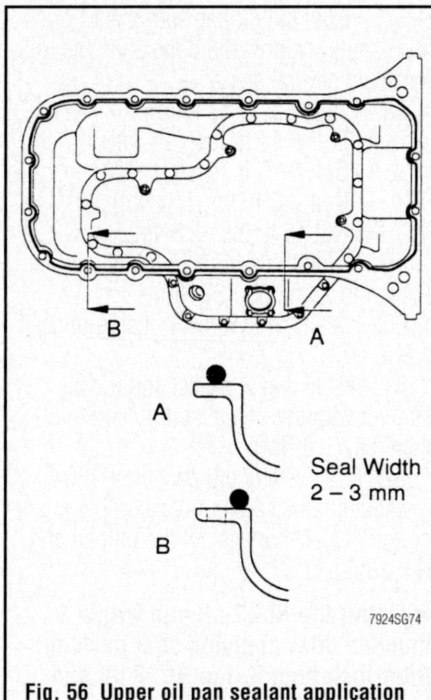

Fig. 56 Upper oil pan sealant application

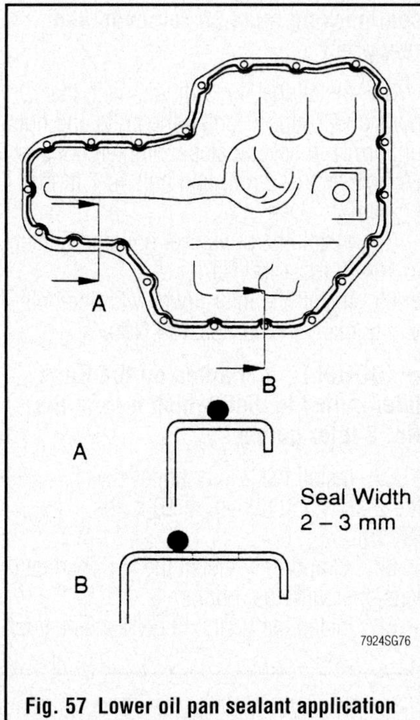

Fig. 57 Lower oil pan sealant application

5. Apply silicone sealant to the upper oil pan as shown.
6. Install the upper oil pan and tighten the fasteners in several passes to the following specifications:
   - 10mm: 66 inch lbs. (7.5 Nm)
   - 12mm: 21 ft. lbs. (28 Nm)
7. Install or connect the following:
   - Oil pan baffle. Tighten the fasteners to 66 inch lbs. (7.5 Nm).
   - Lower oil pan. Tighten the fasteners in several passes to 66 inch lbs. (7.5 Nm).
   - Oil dipstick tube
8. Install the engine.

OIL PUMP

REMOVAL & INSTALLATION

V6 Engine

See Figures 58 through 63.

1. Remove power steering link assembly.
2. Remove differential carrier assembly front.
3. Drain engine coolant.
4. Drain engine oil.
5. Remove battery.
6. Remove v-bank cover.
7. Remove the 11 clips and radiator support seal upper.
8. Loosen fan fluid coupling assembly.
9. Remove fan and generator v belt.
10. Remove fluid coupling assembly.
11. Disconnect no. 2 ventilation hose.
12. Remove air cleaner assembly.
13. Remove the oil level gauge.
14. Remove the bolt and pull out the oil level gauge guide.
15. Remove the O-ring from the oil level gauge guide.
16. Remove water inlet.
17. Disconnect the P/S oil pressure switch connector.
18. Remove the 2 bolts, and separate the vane pump.

➡Do not hit the pulley to other parts when separating the vane pump.

19. Remove generator assembly.
20. Separate cooler compressor assembly.
21. Remove the 5 bolts and V-ribbed belt tensioner.
22. Remove the 2 bolts and 2 idler pulleys.
23. Remove the bolt and idler pulley.
24. Using service tool, hold the crankshaft pulley and loosen the pulley set bolt.
25. Using the pulley set bolt and service tool, remove the crankshaft pulley.
26. Remove the 10 bolts and 2 nuts.
27. Insert the blade of oil pan seal cutter between the oil pan and No. 2 oil pan, cut off applied sealer and remove the No. 2 oil pan.

➡Be careful not to damage the contact surfaces of the oil pan and No. 2 oil pan. Be careful not to damage the No. 2 oil pan flange.

28. Remove the 2 nuts, oil strainer and gasket.
29. Remove the 4 housing bolts.
30. Remove the flywheel housing under cover.
31. Remove the 17 bolts and 2 nuts.
32. Using a screwdriver, remove the oil pan by prying between the oil pan and cylinder block.
33. Remove the O-ring from the oil pump.
34. Remove intake air surge tank.
35. Remove ignition coil assembly.
36. Remove cylinder head cover sub-assemblies.
37. Disconnect the 2 oil control valve connectors.
38. Remove the 2 bolts and 2 camshaft timing oil control valves.
39. Remove VVT sensor.
40. Remove timing chain or belt cover sub-assembly.
41. Remove timing gear case or timing chain case oil seal.
42. Remove oil pump cover and gears and relief valve.

To install:

43. Coat the relief valve with engine oil and insert the relief valve and spring into the valve hole.
44. Install the relief valve plug and tighten to 36 ft. lbs. (49 Nm).
45. Apply fresh engine oil to the drive and driven rotors.
46. Place the drive and driven rotors into the timing chain cover with the marks facing the oil pump cover side.
47. Install the oil pump cover with the 7 bolts and tighten to 80 inch lbs. (9.0 Nm).
48. Install oil pump gears and cover.
49. Install timing gear case or timing chain case oil seal.
50. Install timing chain or belt cover sub-assembly.
    a. Install a new O-ring onto the LH cylinder head.

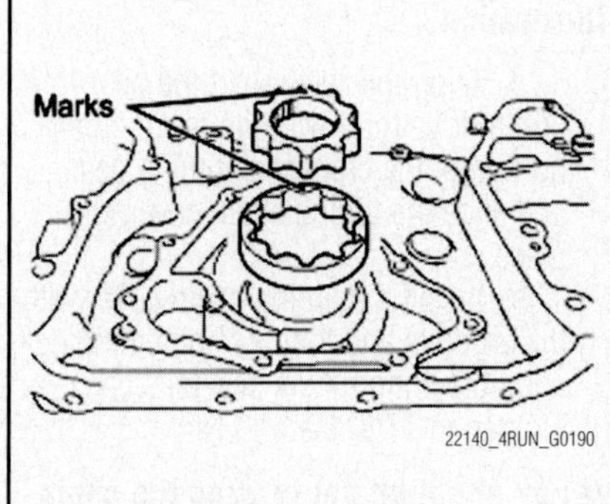

Fig. 58 Oil pump gear position

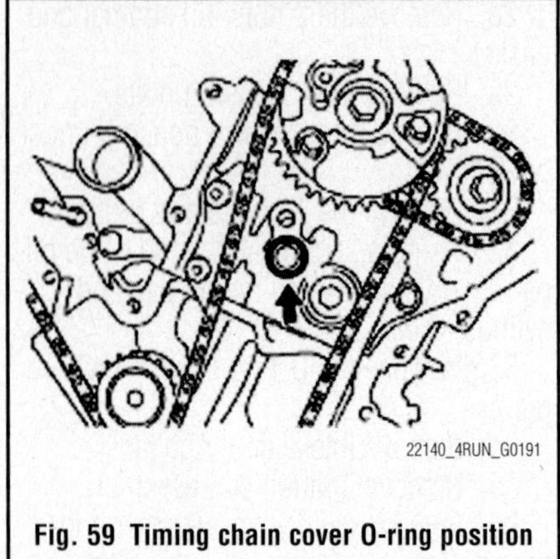

Fig. 59 Timing chain cover O-ring position

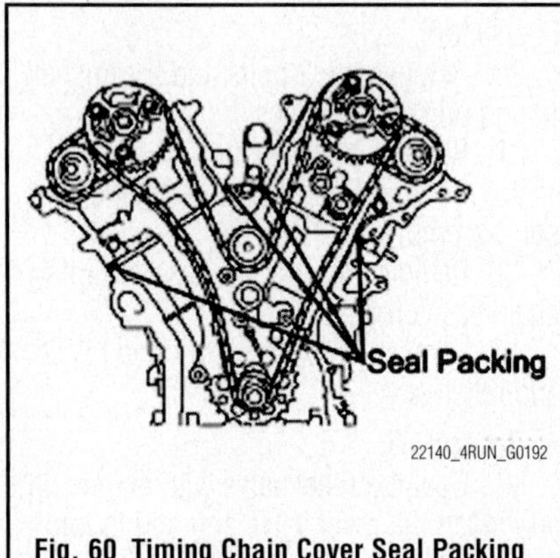

Fig. 60 Timing Chain Cover Seal Packing

b. Apply continuous beads of seal packing (diameter 3 to 4 mm (0.12 to 0.16 in.)) to the 4 locations as shown in the illustration.

c. Apply continuous beads of seal packing (diameter 3 to 4 mm (0.12 to 0.16 in.)) to the timing chain cover as shown in the illustration.

➡Install the timing chain cover within 3 minutes of applying the seal packing. Timing chain cover bolts and nuts must be tightened within 15 minutes of installation. Otherwise, the seal packing must be removed and reapplied. Do not apply seal packing to portion A shown in the illustration.

d. Align the key way of the oil pump drive rotor with the rectangular portion of the crankshaft timing gear, and slide the timing chain cover into place.

e. Install the timing chain cover with the 24 bolts and 2 nuts. Tighten the bolts and nuts uniformly in several steps. Tighten to 17 ft. lbs. (23 Nm).

➡Pay attention not to wrap the chain and slipper over the timing chain cover seal line.

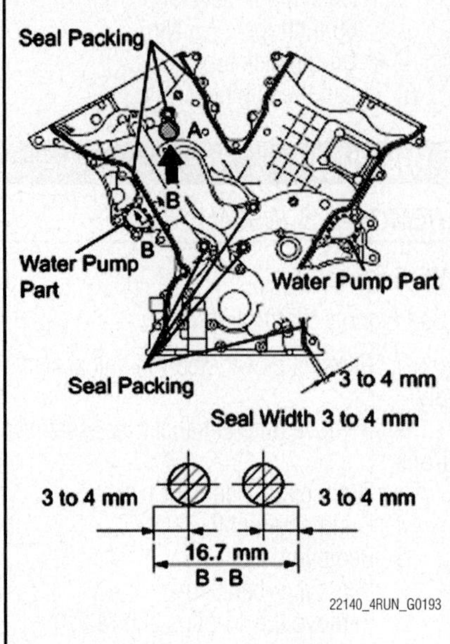

Fig. 61 Timing chain cover sealant position

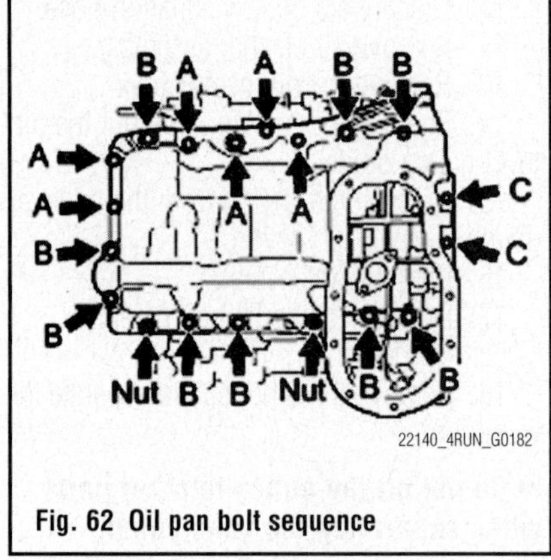

Fig. 62 Oil pan bolt sequence

51. Install VVT sensor.

52. Insert the camshaft timing oil control valves to each cylinder head, and tighten the 2 bolts to 7 ft. lbs. (10 Nm).

53. Install intake air surge tank.

54. Remove any old packing (FIPG) material and be careful not to drop any oil on the contact surfaces of the cylinder block, rear oil seal retainer and oil pan.

55. Install a new O-ring to the oil pump.

56. Apply a continuous bead of seal packing (diameter: 3 to 4 mm (0.12 to 0.16 in.)) to the oil pan.

➡Install the oil pan within 3 minutes after applying seal packing. After installing it, the oil pan bolts and nuts must be tightened within 15 minutes. Otherwise the seal packing must be removed and reapplied.

57. Install the oil pan with the 17 bolts and 2 nuts. Tighten the bolts and nuts uniformly in several steps.

58. Tighten bolts as follows:

- Bolt A 15 ft. lbs. (21 Nm)
- Bolt B 15 ft. lbs. (21 Nm)
- Bolt C 7 ft. lbs. (10 Nm)
- Nut 15 ft. lbs. (21 Nm)

59. Install the 4 housing bolts and tighten to 27 ft. lbs. (37 Nm).

60. Install the flywheel housing under cover.

61. Install a new gasket and the oil strainer with the 2 nuts and tighten to 80 inch lbs. (9.0 Nm).

62. Remove any old packing (FIPG) material and be careful not to drop any oil on the contact surfaces of the oil pan and No. 2 oil pan.

➡Install the No. 2 oil pan within 3 minutes after applying seal packing. After installing it, the No. 2 oil pan bolts and nuts must be tightened within 15 minutes. Otherwise the seal packing must be removed and reapplied.

63. Install the No. 2 oil pan with the 10 bolts and 2 nuts. Tighten the bolts and nuts uniformly in several steps. Tighten bolts to 80 inch lbs. (9.0 nm) and nuts to 7 ft. lbs. (10 Nm).

64. Install the pulley set bolt and tighten to 184 ft. lbs. (250 Nm).

65. Install the idler pulley with the bolt and tighten to 40 ft. lbs. (54 Nm).

➡"DOUBLE" is marked on the No. 1 idler pulley to distinguish it from the No. 2 idler pulley.

66. Install the 2 idler pulleys with the 2 bolts and tighten to 29 ft. lbs. (39 Nm).

67. Temporarily install the V-ribbed belt tensioner with the 5 bolts.

68. Install the V-ribbed belt tensioner by

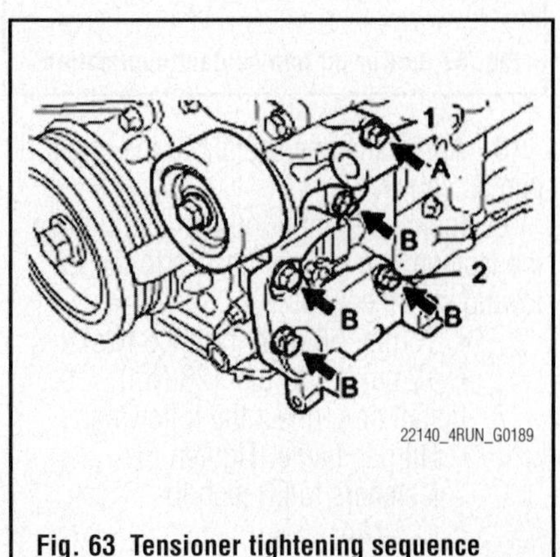

Fig. 63 Tensioner tightening sequence

tightening the bolt 1 and bolt 2 in the order shown in the illustration. Tighten to 27 ft. lbs.(36 Nm).

69. Install cooler compressor assembly.
70. Install generator assembly.
71. Install the vane pump with the 2 bolts an tighten to 32 ft. lbs. (43 Nm).
72. Connect the P/S oil pressure switch connector.
73. Install water inlet.
74. Install a new O-ring to the oil level gauge guide.
75. Apply a light coat of engine oil to the O-ring.
76. Push in the oil level gauge guide end into the guide hole of the oil pan.
77. Install the oil level gauge guide with the bolt.
78. Install the oil level gauge.
79. Install air cleaner assembly.
80. Connect No. 2 ventilation hose.
81. Install fluid coupling assembly.
82. Install fan and generator v belt.
83. Fully tighten fluid coupling assembly.
84. Install radiator support seal upper.
85. Install battery.
86. Install differential carrier assembly front as necessary.
87. Install power steering link assembly.
88. Add engine oil.
89. Add engine coolant.
90. Check for engine leaks.
91. Install the V-bank cover with the 2 nuts.

V8 Engine

See Figures 64 through 66.

1. Before servicing the vehicle, refer to the precautions section.
2. Remove the engine from the vehicle and mount it on a stand.
3. Remove or disconnect the following:

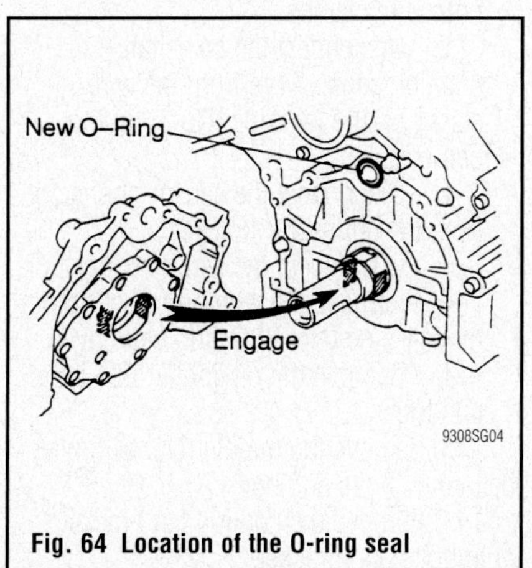

Fig. 64 Location of the O-ring seal

- Front cover
- Timing belt.
- Timing belt idler pulleys
- Crankshaft timing sprocket
- Oil dipstick tube
- Oil filter and bracket
- Crankshaft Position (CKP) sensor
- Oil pan and baffle
- Oil pump strainer
- Oil pump

To install:

4. Install a new O-ring on the engine block.
5. Apply silicone sealant to the oil pump housing as shown.
6. Install the oil pump. Tighten the bolts in several passes to the following specifications:
   - 12mm: 11 ft. lbs. (15.5 Nm)
   - 14mm: 22 ft. lbs. (30.5 Nm)
   - 6mm Hex: 11 ft. lbs. (15.5 Nm)
7. The upper oil pan bolts are different lengths and are identified as follows:
   - A: 1.38 inch (35mm) w/12mm head
   - B: 1.97 inch (50mm) w/12mm head
   - C: 4.17 inch (106mm) w/12mm head
   - D: 1.57 inch (40mm) w/14mm head
   - E: 1.18 inch (30mm) w/6mm hex head
8. Install or connect the following:
   - Oil pump pickup tube. Tighten the bolts to 66 inch lbs. (7.5 Nm).
   - Oil pan and baffle
   - CKP sensor
   - Oil filter and bracket. Tighten the bolts to 13 ft. lbs. (18 Nm).
   - Oil dipstick tube
   - Crankshaft timing sprocket
   - Timing belt idler pulleys
   - Timing belt
   - Front cover
9. Install the engine.

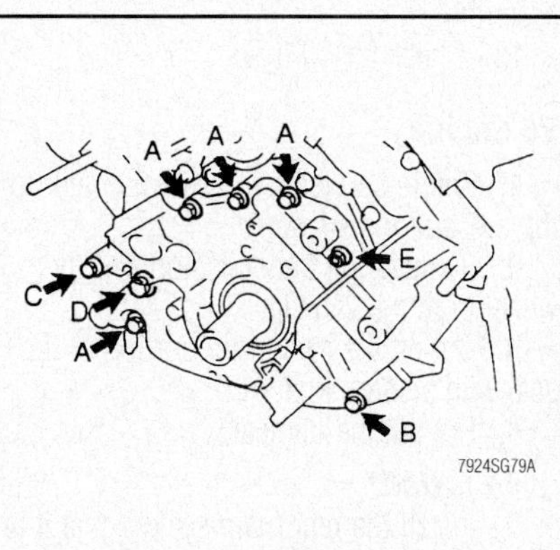

Fig. 65 Oil pump bolt location

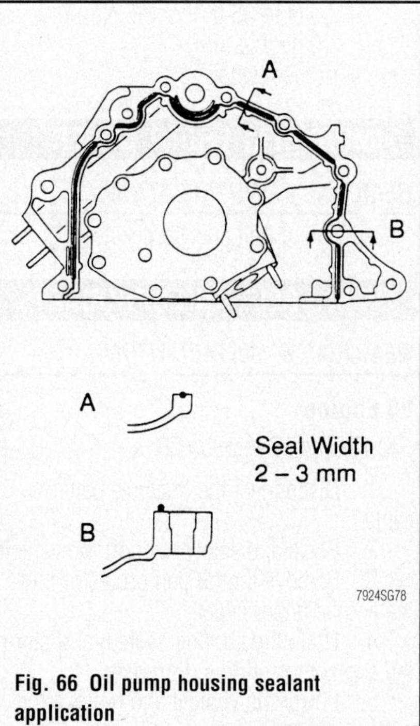

Fig. 66 Oil pump housing sealant application

PISTON AND RING

POSITIONING

See Figures 67 through 69.

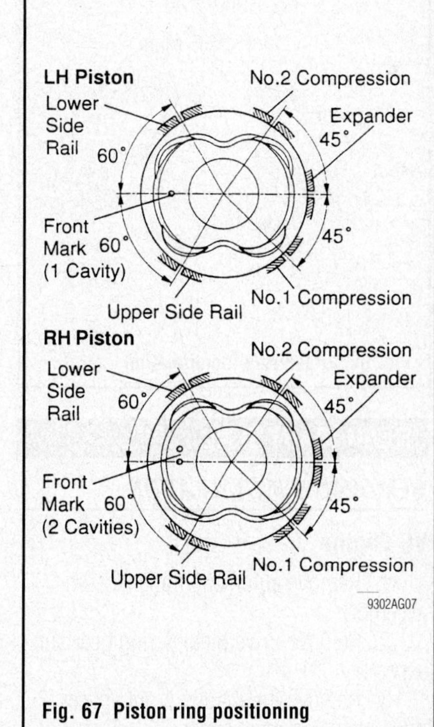

Fig. 67 Piston ring positioning

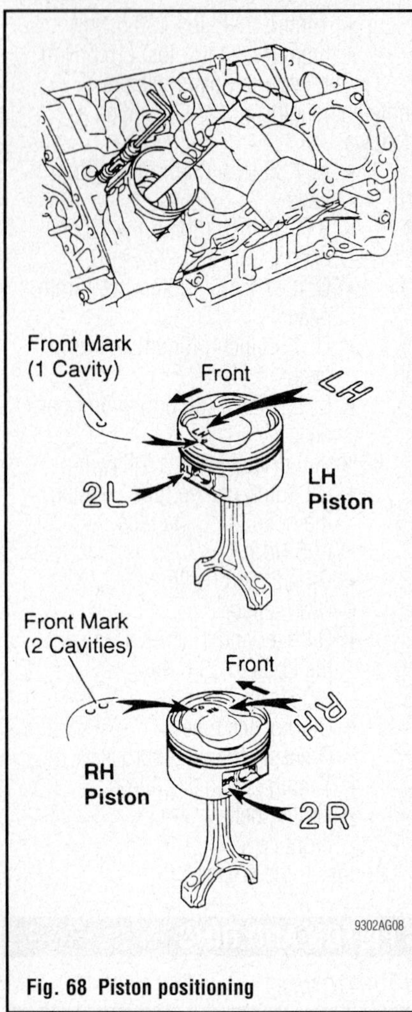

Fig. 68 Piston positioning

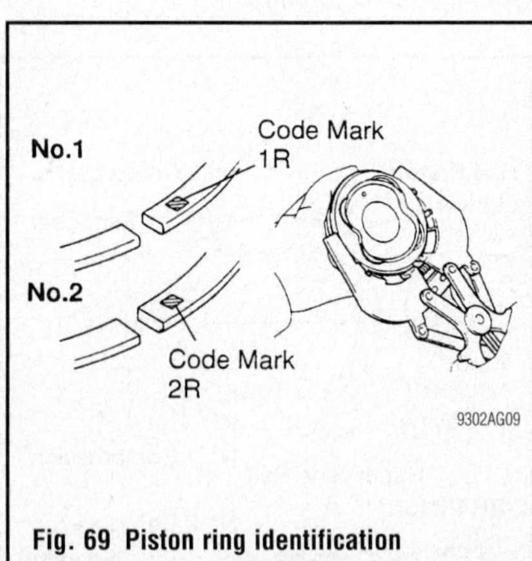

Fig. 69 Piston ring identification

REAR MAIN SEAL

REMOVAL & INSTALLATION

V6 Engine

1. Remove automatic transmission assembly.
2. Remove drive plate & ring gear sub-assembly.
3. Remove the 8 bolts, front spacer, drive plate and rear spacer.
4. Using a knife, cut off the oil seal lip.
5. Using a screwdriver, pry out the oil seal.

➡Be careful not to damage the crankshaft. Tape the screwdriver tip.

To install:

6. Apply MP grease to a new oil seal lip.
7. Using service tool and a hammer, tap in the oil seal until its surface is flush with the rear oil seal retainer edge.
8. Apply adhesive to 2 or 3 threads of the mounting bolt end.
9. Install the front spacer, drive plate and rear spacer on the crankshaft.
10. Using several steps, install and tighten the 8 mounting bolts uniformly in the sequence as shown in the illustration and tighten to 61 ft. lbs. (83 Nm).
11. Install automatic transmission assembly.
12. Check for oil leaks.

V8 Engine

1. Before servicing the vehicle, refer to the precautions section.
2. Remove the transmission and flywheel from the vehicle.
3. Cut off the rubber lip portion of the seal with a sharp knife.
4. Pry out the oil seal.

To install:

5. Install the rear main seal so that it is flush with the seal retainer housing.
6. Install or connect the following:
   - Flywheel/driveplate. Tighten the bolts to 35 ft. lbs. (48 Nm) plus a 90 degree turn.
   - Transmission

TIMING BELT FRONT COVER

REMOVAL & INSTALLATION

See Timing Belt and Sprockets

TIMING BELT AND SPROCKETS

REMOVAL & INSTALLATION

V8 Engine

See Figures 70 through 79.

1. Disconnect the negative battery cable.
2. Raise and safely support the vehicle.
3. Remove the oil pan protector and the engine under cover.
4. Drain the cooling system and store the coolant for refilling purposes.
5. Lower the vehicle and remove the battery clamp cover.
6. From the top of the engine, remove the fuel return hose, the engine cover nuts/bolts and the cover.
7. Remove the air cleaner and the intake air connector assembly.
8. Remove the cooling fan pulley by performing the following procedures:
   a. Loosen the 4 fan clutch-to-fan pulley nuts.
   b. Using a box-end wrench on the serpentine drive belt tensioner bolt, rotate the tensioner counterclockwise and remove the drive belt.

➡ The serpentine drive belt tensioner bolt is a left-hand thread.

   c. Remove the fan clutch-to-fan pulley nuts, the fan, the clutch assembly and the fan pulley.
9. Remove the radiator by performing the following procedures:
   a. Disconnect the upper, lower and reservoir hoses from the radiator.
   b. Disconnect and plug the automatic transmission oil cooler at the radiator. Disconnect the automatic transmission oil cooler hoses from the fan shroud clamp.
   c. Remove the radiator reservoir tank.
   d. Remove the fan shroud-to-radiator bolts and the shroud.
   e. Remove the 2 upper radiator-to-chassis nuts.
   f. Remove the middle radiator-to-chassis nut/bolts and brackets.
   g. Carefully, lift the radiator from the vehicle.
10. Remove the serpentine drive belt idler pulley bolt, cover plate and pulley.
11. Remove the right side (No. 3) timing belt cover.
12. Remove the left side (No. 3) timing belt cover by performing the following procedures:
   a. Disconnect the engine wire from both wire clamps.
   b. Disconnect the camshaft position sensor wire from the wire clamp on the left-side (No.3) timing belt cover.
   c. Disconnect the sensor connector from the connector bracket.
   d. Disconnect the sensor connector.
   e. Remove the wire grommet from the left-side (No. 3) timing belt cover.
   f. Remove the oil cooler tube bolts and tube.
13. Remove the middle (No. 2) timing belt cover bolts and cover.
14. Remove the cooling fan bracket nuts/bolts and bracket.

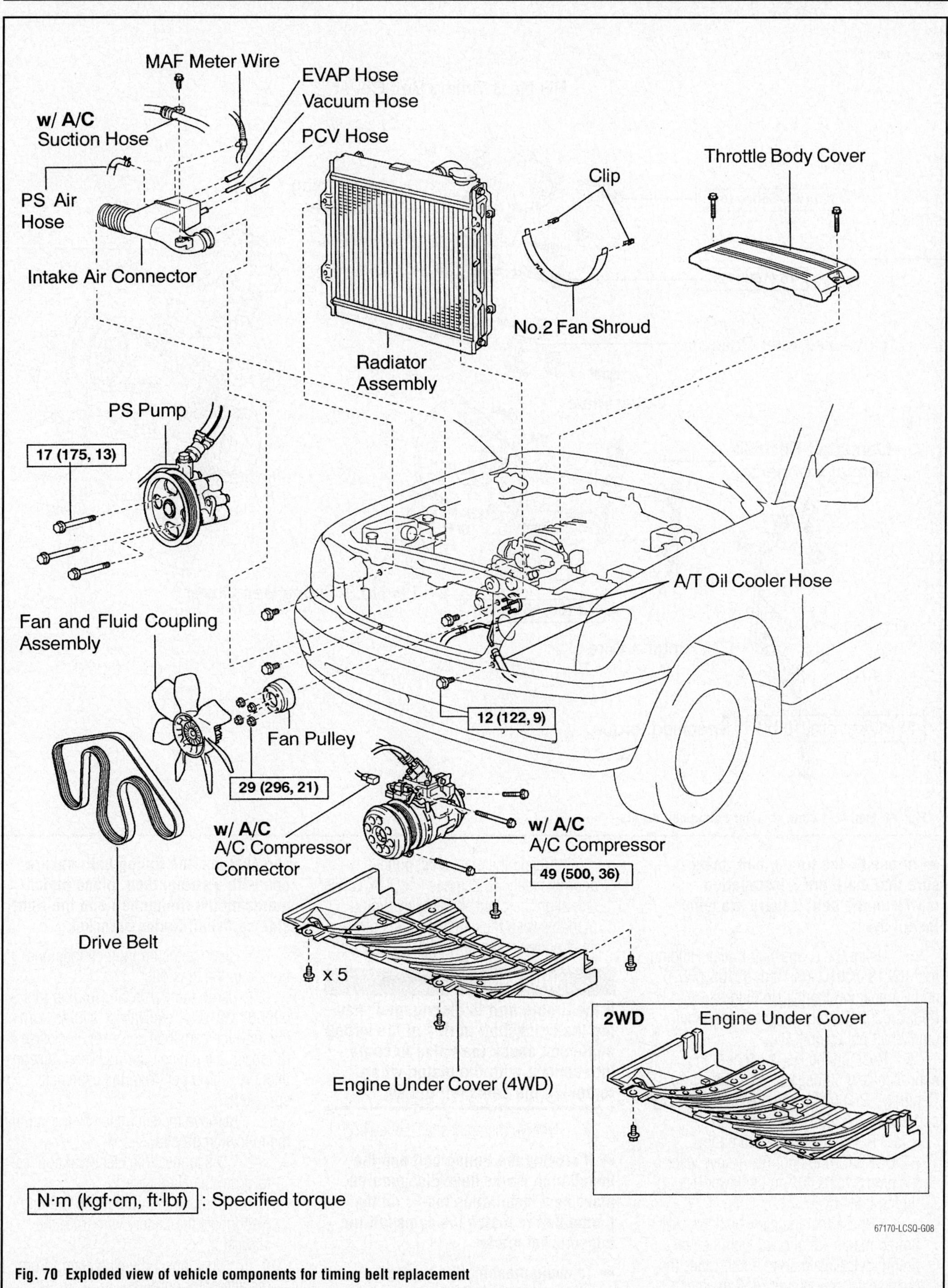

Fig. 70 Exploded view of vehicle components for timing belt replacement

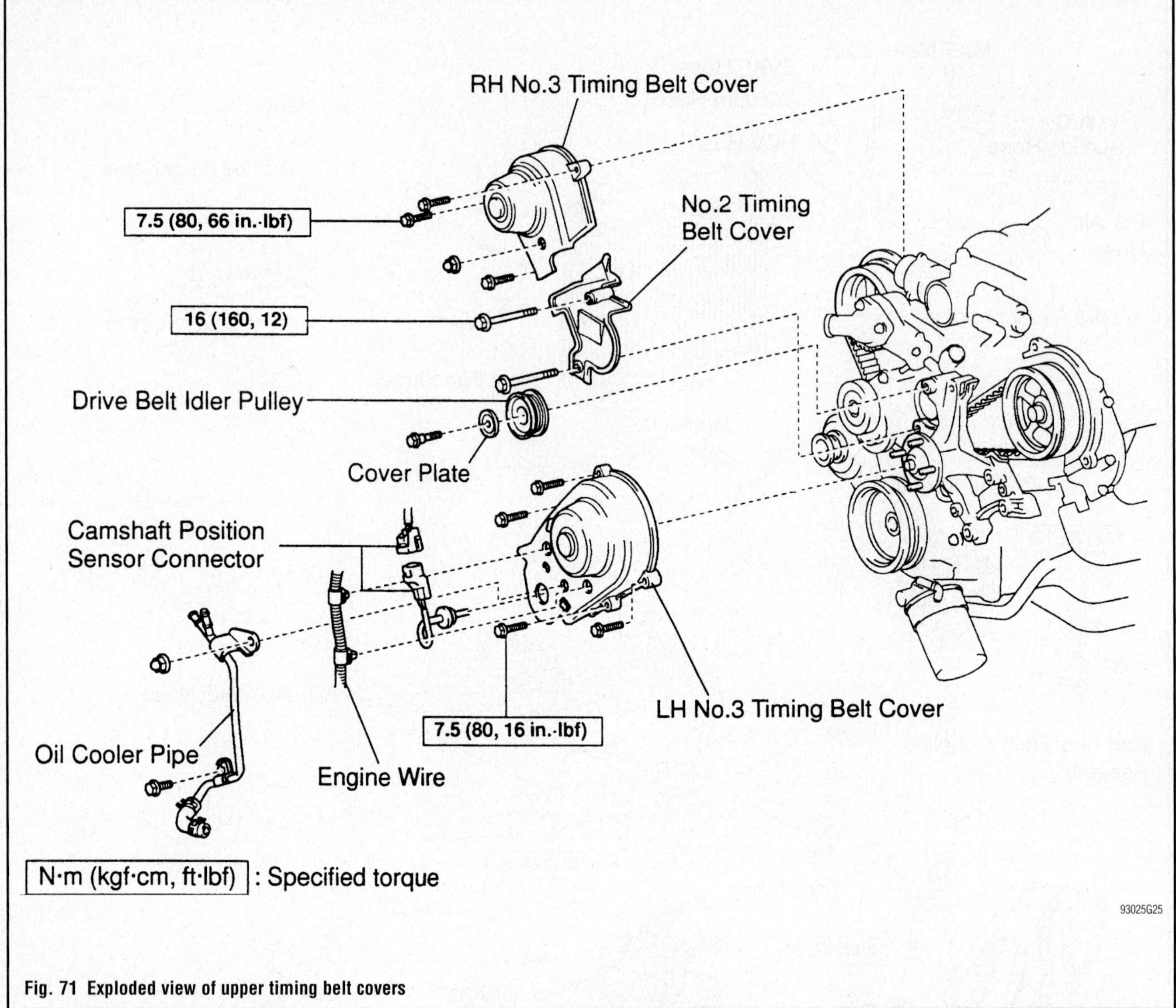

Fig. 71 Exploded view of upper timing belt covers

➡ **If reusing the timing belt, make sure that there are 3 installation marks on the belt; if there are none, install them.**

15. Using the Crankshaft Pulley Holding tool 09213-70010, Bolt tool 90105-08076 and Companion Flange Holding tool 09330-00021, or equivalent, loosen the crankshaft pulley bolt.

16. Position the No. 1 cylinder to approximately 50 degrees After Top Dead Center (ATDC) of the compression stroke by performing the following procedures:

a. Rotate the crankshaft pulley (CLOCKWISE) to align its groove with the timing mark "0" on the lower (No. 1) timing belt cover.

b. Check that the camshaft sprocket timing marks are aligned with the rear timing belt plate marks; if not, rotate the crankshaft 1 revolution (360 degrees).

c. Rotate the crankshaft pulley approximately 50 degrees (CLOCKWISE) and align the crankshaft pulley timing mark between the centers of the crankshaft pulley bolt and the idler pulley bolt.

⁂ WARNING

If the timing belt is disengaged, having the crankshaft pulley in the wrong angle can cause the valve to come into contact with the piston when removing the camshaft pulley.

17. Remove the crankshaft pulley bolt.

➡ **If reusing the timing belt and the installation marks have disappeared, place new installation marks on the timing belt to match the camshaft timing sprocket marks.**

➡ **To avoid meshing the timing sprocket and the timing belt, secure one with a string; then, place matchmarks on the timing belt and the right-side camshaft timing sprocket.**

18. Remove the timing belt tensioner bolts and the tensioner.

19. Using the Camshaft Holding tool 09960-10010, or equivalent, slightly turn the left-side camshaft sprocket clockwise to loosen the tension spring. Then, disconnect the timing belt from the camshaft sprockets.

20. Remove the alternator by performing the following procedures:

a. Disconnect the electrical connector from the alternator.

b. Remove the rubber cap/nut and disconnect the battery wire from the alternator.

c. Disconnect the wire clamp from the alternator cord clip.

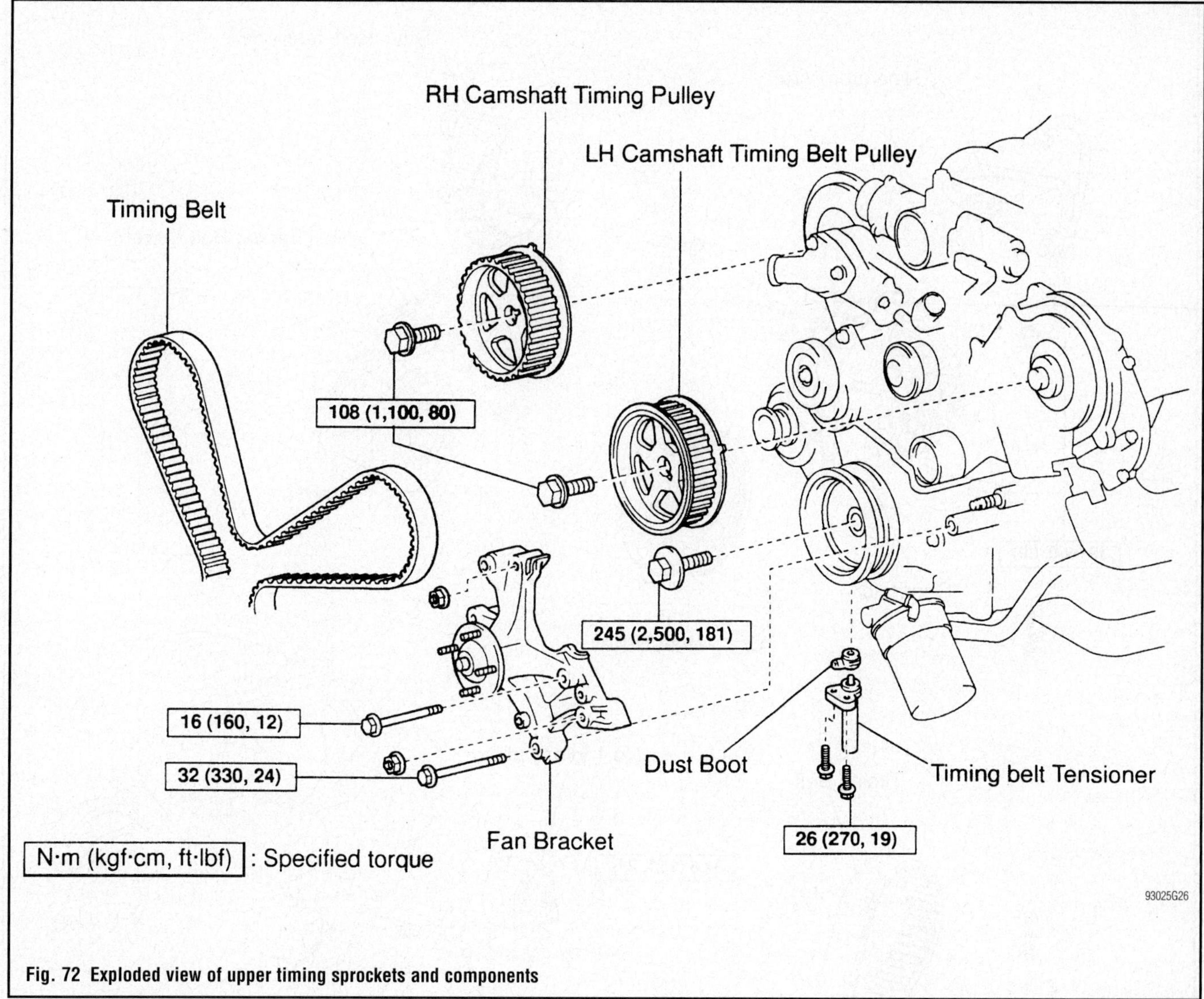

Fig. 72 Exploded view of upper timing sprockets and components

d. Remove the alternator-to-engine nuts/bolts and the alternator.

21. Remove the serpentine drive belt tensioner nuts/bolts and the tensioner.

22. Using the Crankshaft Puller Assembly tool 09950-50012, or equivalent, press the crankshaft pulley from the crankshaft.

WARNING

DO NOT rotate the crankshaft pulley.

23. Remove the lower (No. 1) timing belt cover bolts and the cover.

24. Remove the timing belt guide, spacer and the timing belt.

To install:

➡With the timing belt removed, this is a perfect opportunity to inspect and/or replace the water pump.

25. Inspect the timing belt tensioner by performing the following procedures:

a. Inspect the seal for leakage; if leakage is suspected, replace the tensioner.

b. Using both hands to hold the tensioner facing upward, strongly press the pushrod against a solid surface. If the pushrod moves, replace the tensioner.

WARNING

Never hold the tensioner with the pushrod facing downward.

c. Measure the pushrod protrusion from the housing end, it should be 0.413–0.453 in. (10.5–11.5mm). If the protrusion is not as specified, replace the tensioner.

26. Temporarily install the timing belt by performing the following procedures:

a. Align the timing belt's installation mark with the crankshaft timing sprocket.

b. Install the timing belt on the crankshaft timing sprocket, the No. 1 idler pulley and the No. 2 idler pulley.

27. Install the gasket to the timing belt cover spacer and install the cover spacer.

28. Install the timing belt guide with the cup side facing outward.

29. Install the lower (No. 1) timing belt cover.

30. Install the crankshaft pulley by performing the following procedures:

a. Align the crankshaft pulley with the crankshaft key.

b. Using the Crankshaft Installer tool 09223-46011, or equivalent, and a hammer, tap the crankshaft pulley into position.

31. Install the serpentine drive belt tensioner and torque the tensioner-to-engine bolts to 12 ft. lbs. (16 Nm).

➡To install the serpentine drive belt tensioner, use a bolt 4.18 in. (106mm) in length.

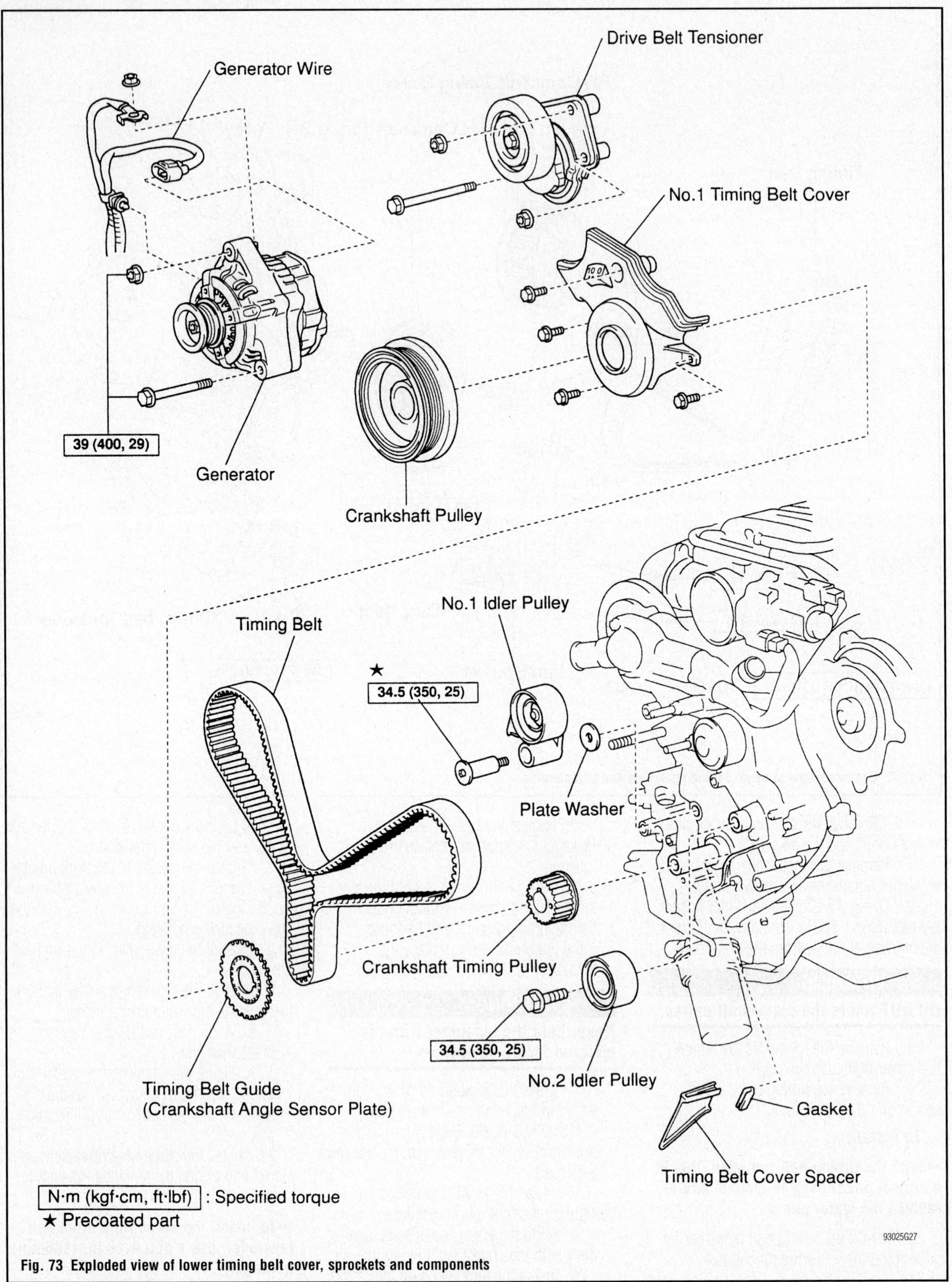

Fig. 73 Exploded view of lower timing belt cover, sprockets and components

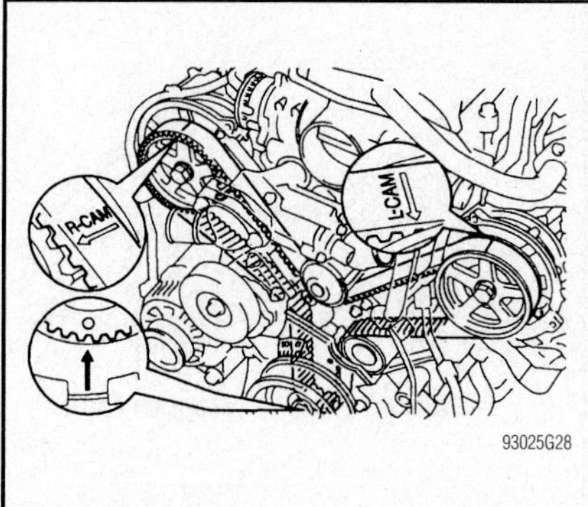

Fig. 74 Alignment of timing belt with the timing sprockets

Fig. 75 (93025G29): Approx. 50°, No.2 Idler Pulley Bolt, Timing Mark, Crankshaft Pulley Bolt, Turn

Fig. 75 Aligning of crankshaft pulley timing mark with the center line of the crankshaft pulley bolt and the idler pulley bolt

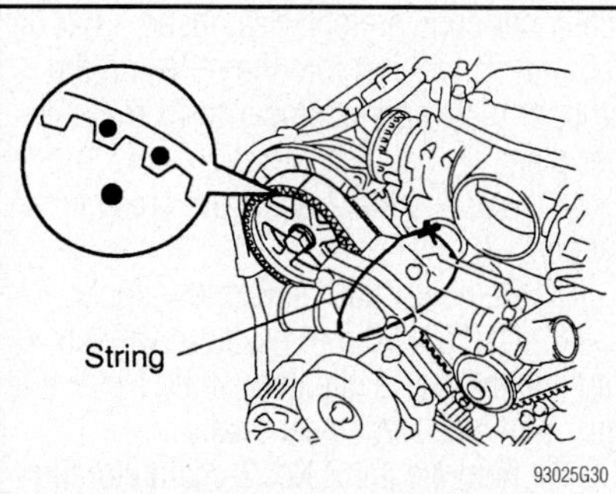

Fig. 76 Securing the timing belt with string and matchmarking the camshaft with the timing belt

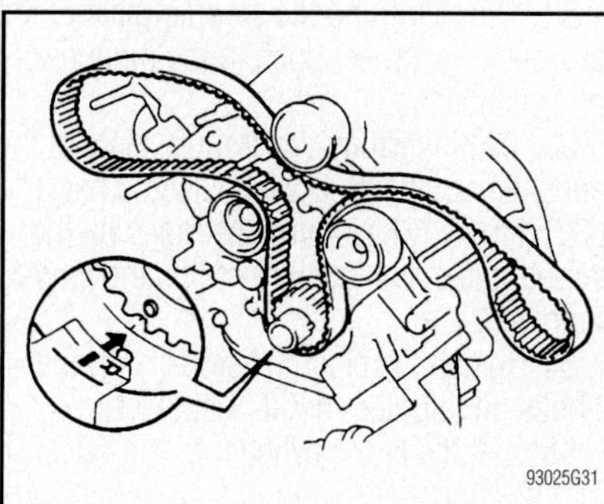

Fig. 77 Installing the timing belt on the crankshaft sprocket

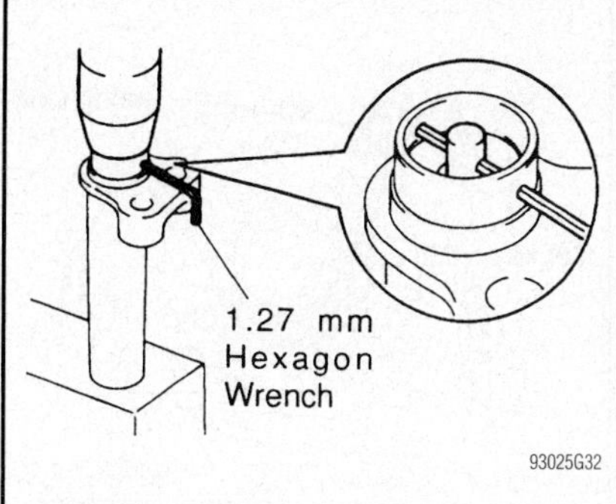

Fig. 78 Securing the timing belt tensioner pushrod

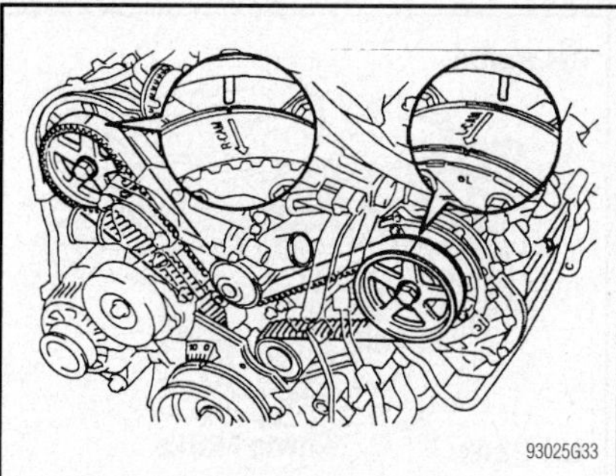

Fig. 79 Checking the TDC alignment marks after rotating the crankshaft 2 revolutions

32. Check that the crankshaft pulley's timing mark is aligned with the centers of the idler pulley and crankshaft pulley bolts.
33. Install the alternator and torque the alternator-to-engine nuts/bolts to 29 ft. lbs. (39 Nm). Connect the alternator's electrical connectors and clip.
34. Install the timing belt to the left-side camshaft by performing the following procedures:
    a. Rotate the left-side camshaft pulley to align the timing belt installation mark with the camshaft sprocket's timing mark and slide the belt onto the camshaft timing sprocket.
    b. Using the Camshaft Holding tool 09960-10010, or equivalent, slightly turn the left-side camshaft sprocket counterclockwise to place tension on the timing belt between the crankshaft sprocket and the camshaft sprocket.
35. Rotate the right-side camshaft pulley to align the timing belt installation mark with the camshaft sprocket's timing mark and slide the belt onto the camshaft timing sprocket.
36. Using a vertical press, slowly press the pushrod into the housing using 200–2205 lbs. (981–9807 N) until the holes align, then, install a 1.27mm Allen® wrench to secure the pushrod and release the press. Install the dust boot on the tensioner housing.
37. Install the timing belt tensioner and torque the bolts to 19 ft. lbs. (26 Nm).
38. Using a pair of pliers, remove the Allen® wrench from the tensioner housing.
39. Check the valve timing by performing the following procedure:
    a. Temporarily install the crankshaft pulley bolt.
    b. Slowly, rotate the crankshaft pulley 2 revolutions (CLOCKWISE) and realign the TDC marks.

➡If the pulley/sprocket timing marks do not realign, remove the timing belt and reinstall it.

40. Using the Crankshaft Pulley Holding tool 09213-70010, Bolt tool 90105-08076 and Companion Flange Holding tool 09330-00021, or equivalent, torque the crankshaft pulley bolt to 181 ft. lbs. (245 Nm).
41. Install the cooling fan bracket and torque the 12mm (head size) bolt to 12 ft. lbs. (16 Nm) and the 14mm (head size) bolt to 24 ft. lbs. (32 Nm).
42. Install the air conditioning compressor.
43. Install the middle (No. 2) timing belt cover and torque the bolts to 12 ft. lbs. (16 Nm).
44. Install the upper right-side (No. 3) timing belt cover and torque the bolts to 66 inch lbs. (7.5 Nm).
45. Install the upper left-side (No. 3) timing belt cover by performing the following procedures:
    a. Install the oil cooler tube and bolt.
    b. Feed the Camshaft Position Sensor (CPS) through the left-side (No. 3) timing belt cover hole.
    c. Install the left-side (No. 3) timing belt cover and torque the bolts to 66 inch lbs. (7.5 Nm).
    d. Install the wire grommet to the left-side (No. 3) timing belt cover.
    e. Install the sensor connector to the connector bracket and connect the sensor connector.
    f. Install the sensor wire and the engine wire to the clamps on the left-side (No. 3) timing belt cover.
46. Install the drive belt idler pulley and cover plate; then, torque the pulley bolt to 27 ft. lbs. (37 Nm).
47. To complete the installation, reverse the removal procedures.
48. Refill the cooling system and connect the negative battery cable.

TIMING CHAIN AND SPROCKETS

REMOVAL & INSTALLATION

V6 Engine

See Figures 80 through 92.

1. Remove power steering link assembly.
2. Remove differential carrier assembly front.
3. Drain engine coolant.
4. Drain engine oil.
5. Remove battery.
6. Remove v-bank cover.
7. Remove the 11 clips and radiator support seal upper.
8. Loosen fan fluid coupling assembly.
9. Remove fan and generator v belt.
10. Remove fluid coupling assembly.
11. Disconnect no. 2 ventilation hose.
12. Remove air cleaner assembly.
13. Remove the oil level gauge.
14. Remove the bolt and pull out the oil level gauge guide.
15. Remove the O-ring from the oil level gauge guide.
16. Remove water inlet.
17. Disconnect the P/S oil pressure switch connector.
18. Remove the 2 bolts, and separate the vane pump.

➡Do not hit the pulley to other parts when separating the vane pump.

19. Remove generator assembly.
20. Separate cooler compressor assembly.
21. Remove the 5 bolts and V-ribbed belt tensioner.
22. Remove the 2 bolts and 2 idler pulleys.
23. Remove the bolt and idler pulley.
24. Using service tool, hold the crankshaft pulley and loosen the pulley set bolt.
25. Using the pulley set bolt and service tool, remove the crankshaft pulley.
26. Remove the 10 bolts and 2 nuts.
27. Insert the blade of oil pan seal cutter between the oil pan and No. 2 oil pan, cut off applied sealer and remove the No. 2 oil pan.

➡Be careful not to damage the contact surfaces of the oil pan and No. 2 oil pan. Be careful not to damage the No. 2 oil pan flange.

28. Remove the 2 nuts, oil strainer and gasket.
29. Remove the 4 housing bolts.
30. Remove the flywheel housing under cover.
31. Remove the 17 bolts and 2 nuts.

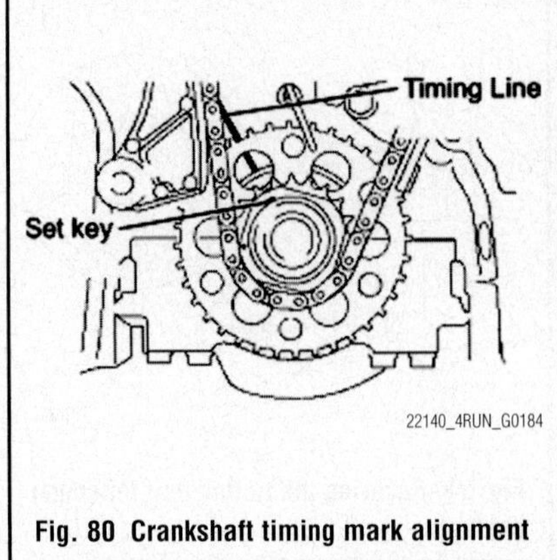

Fig. 80 Crankshaft timing mark alignment

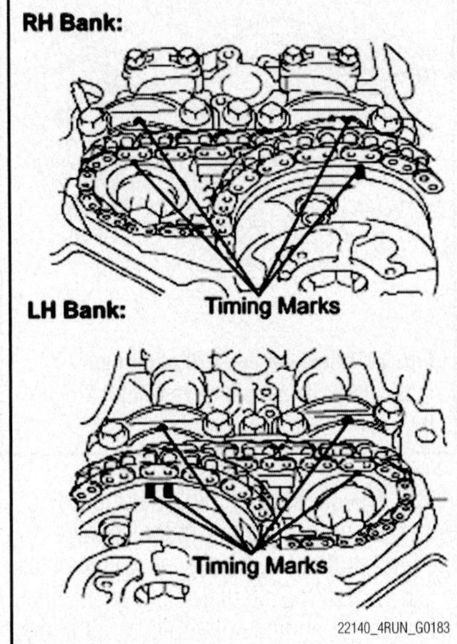

Fig. 81 Timing mark alignment

32. Using a screwdriver, remove the oil pan by prying between the oil pan and cylinder block.
33. Remove the O-ring from the oil pump.
34. Remove intake air surge tank.
35. Remove ignition coil assembly.
36. Remove cylinder head cover sub-assemblies.
37. Disconnect the 2 oil control valve connectors.
38. Remove the 2 bolts and 2 camshaft timing oil control valves.
39. Remove VVT sensor.
40. Remove timing chain or belt cover sub-assembly.
41. Remove timing gear case or timing chain case oil seal.
42. Set No. 1 cylinder to TDC by installing the crankshaft pulley set bolt, and turn the crankshaft to align the crankshaft set key with the timing line of the cylinder block.

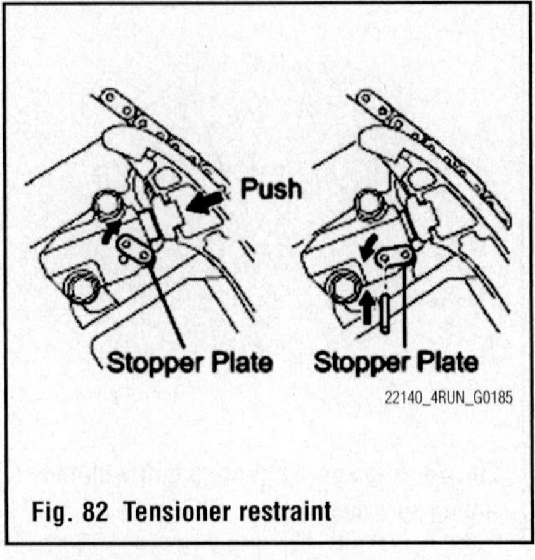

Fig. 82 Tensioner restraint

43. Check that the timing marks of the camshaft timing gears are aligned with the timing marks of the bearing caps as shown in the illustration. If not, turn the crankshaft 1 complete revolution (360°) and align the timing marks.

➡Never rotate the crankshaft with the chain tensioner removed. When rotating the camshaft with the timing chain removed, rotate the crankshaft counterclockwise 40° from the TDC first.

44. While turning the stopper plate of the tensioner clockwise, push in the plunger of the chain tensioner.
45. While turning the stopper plate of the tensioner counterclockwise, insert a bar of 3.5 mm (0.138 in.) into the holes on the stopper plate and tensioner to fix the stopper plate.
46. Remove the 2 bolts and chain tensioner.
47. Remove chain tensioner slipper.
48. Using a 10 mm hexagon wrench, remove the No. 2 idle gear shaft, No. 1 idle gear and No. 1 idle gear shaft.
49. Remove the 2 No. 2 chain vibration dampers.
50. Remove chain sub-assembly.

To install:

51. Install chain tensioner slipper.
52. While turning the stopper plate of the tensioner clockwise, push in the plunger of the tensioner.
53. While turning the stopper plate of the tensioner counterclockwise, insert a bar of φ3.5 mm (0.138 in.) into the holes on the stopper plate and tensioner to fix the stopper plate.
54. Install the chain tensioner with the 2 bolts and tighten to 7 ft. lbs, (10 Nm).
55. Set the No. 1 cylinder to the TDC / compression.
56. Align the timing marks of the camshaft timing gears and bearing caps.
57. Install the crankshaft pulley set bolt,

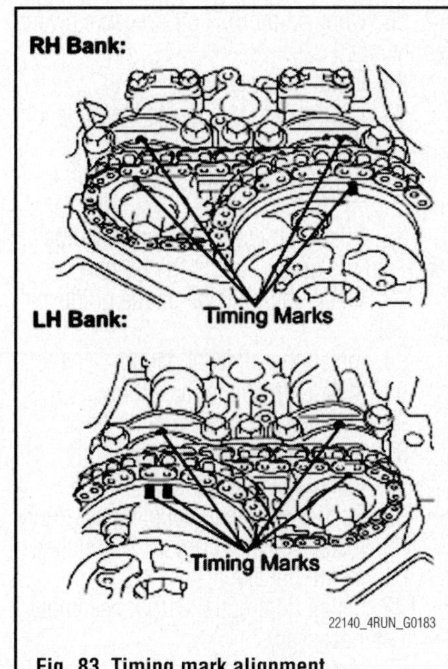

Fig. 83 Timing mark alignment

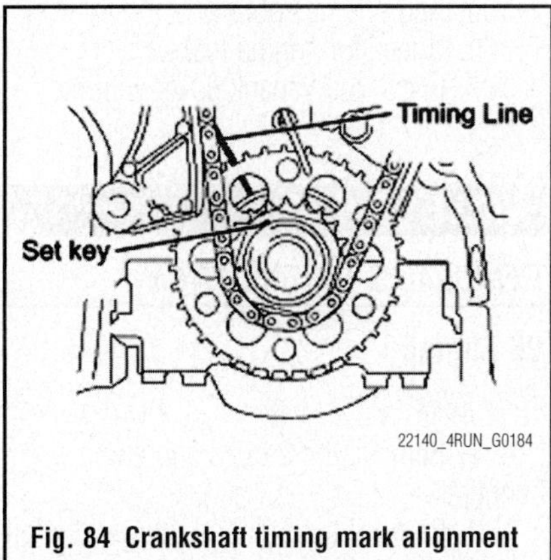

Fig. 84 Crankshaft timing mark alignment

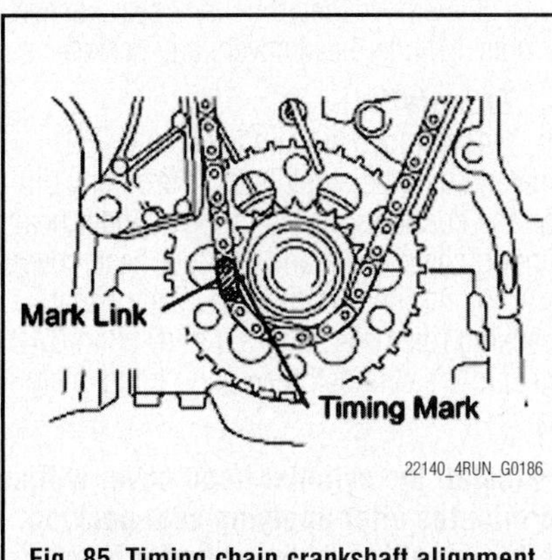

Fig. 85 Timing chain crankshaft alignment

and turn the crankshaft to align the crankshaft set key with the timing line of the cylinder block.

58. Align the mark link (yellow) with the timing mark of the crankshaft timing gear.

59. Align the mark links (orange) with

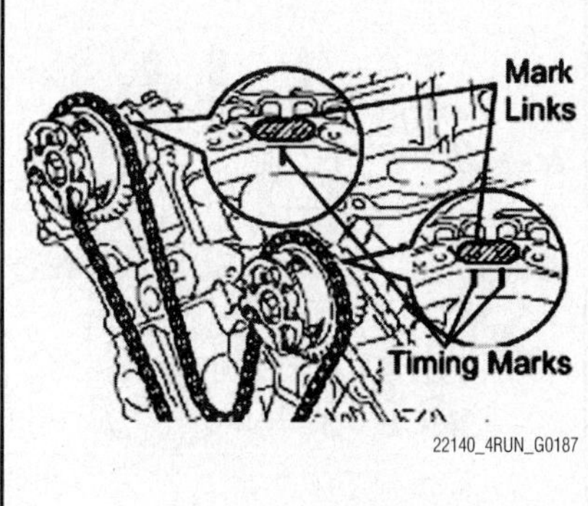

Fig. 86 Timing chain camshaft alignment

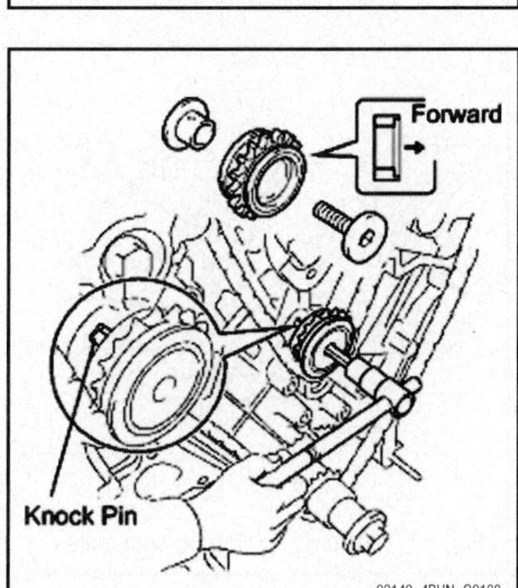

Fig. 87 Timing chain idler gear alignment

the timing marks of the camshaft timing gears, and install the chain.

60. Install the 2 No. 2 chain vibration dampers.

61. Apply a light coat of engine oil to the rotating surface of the No. 1 idle gear shaft.

62. Temporarily install the No. 1 idle gear shaft and No. 1 idle gear with the No. 2 idle gear shaft while aligning the knock pin of the No. 1 idle gear shaft with the knock pin groove of the cylinder block.

➡Be careful of the idle gear direction.

63. Using a 10 mm hexagon wrench, tighten the No. 2 idle gear shaft to 44 ft. lbs. (60 Nm).

64. Remove the bar from the chain tensioner.

65. Install timing gear case or timing chain case oil seal.

66. Install timing chain or belt cover sub-assembly.

a. Install a new O-ring onto the LH cylinder head.

b. Apply continuous beads of seal packing (diameter 3 to 4 mm (0.12 to

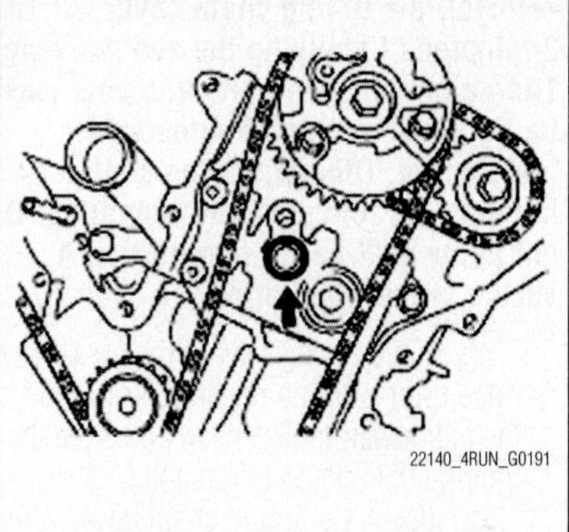

Fig. 88 Timing chain cover o-ring position

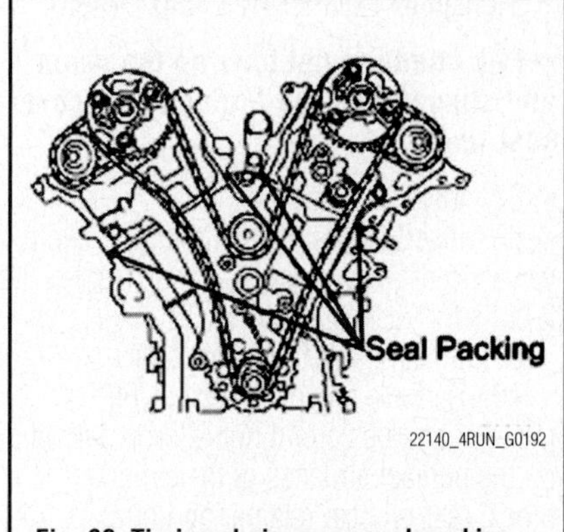

Fig. 89 Timing chain cover seal packing

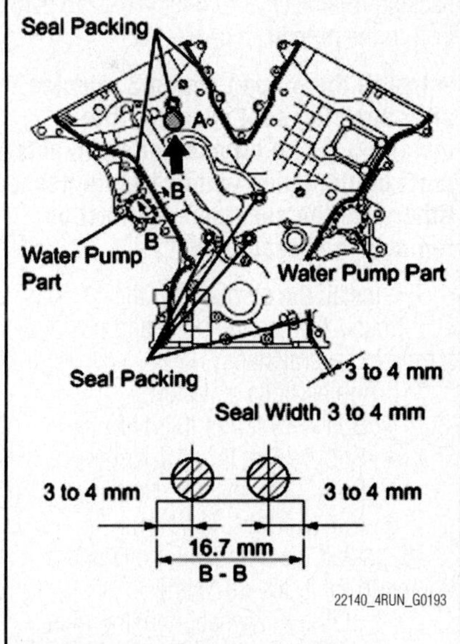

Fig. 90 Timing chain cover sealant position

0.16 in.)) to the 4 locations as shown in the illustration.

c. Apply continuous beads of seal packing (diameter 3 to 4 mm (0.12 to 0.16 in.)) to the timing chain cover as shown in the illustration.

➡Install the timing chain cover within 3 minutes of applying the seal packing. Timing chain cover bolts and nuts must be tightened within 15 minutes of installation. Otherwise, the seal packing must be removed and reapplied. Do not apply seal packing to portion A shown in the illustration.

d. Align the key way of the oil pump drive rotor with the rectangular portion of the crankshaft timing gear, and slide the timing chain cover into place.

e. Install the timing chain cover with the 24 bolts and 2 nuts. Tighten the bolts and nuts uniformly in several steps. Tighten to 17 ft. lbs. (23 Nm).

➡Pay attention not to wrap the chain and slipper over the timing chain cover seal line.

67. Install VVT sensor.
68. Insert the camshaft timing oil control valves to each cylinder head, and tighten the 2 bolts to 7 ft. lbs. (10 Nm).
69. Install intake air surge tank.
70. Remove any old packing (FIPG) material and be careful not to drop any oil on the contact surfaces of the cylinder block, rear oil seal retainer and oil pan.
71. Install a new O-ring to the oil pump.
72. Apply a continuous bead of seal packing (diameter: 3 to 4 mm (0.12 to 0.16 in.)) to the oil pan.

➡Install the oil pan within 3 minutes after applying seal packing. After installing it, the oil pan bolts and nuts must be tightened within 15 minutes. Otherwise the seal packing must be removed and reapplied.

73. Install the oil pan with the 17 bolts and 2 nuts. Tighten the bolts and nuts uniformly in several steps.
74. Tighten bolts as follows:
    - Bolt A 15 ft. lbs. (21 Nm)
    - Bolt B 15 ft. lbs. (21 Nm)
    - Bolt C 7 ft. lbs. (10 Nm)
    - Nut 15 ft. lbs. (21 Nm)
75. Install the 4 housing bolts and tighten to 27 ft. lbs. (37 Nm).
76. Install the flywheel housing under cover.
77. Install a new gasket and the oil strainer with the 2 nuts and tighten to 80 inch lbs. (9.0 Nm).
78. Remove any old packing (FIPG) material and be careful not to drop any oil on the contact surfaces of the oil pan and No. 2 oil pan.

➡Install the No. 2 oil pan within 3 minutes after applying seal packing.

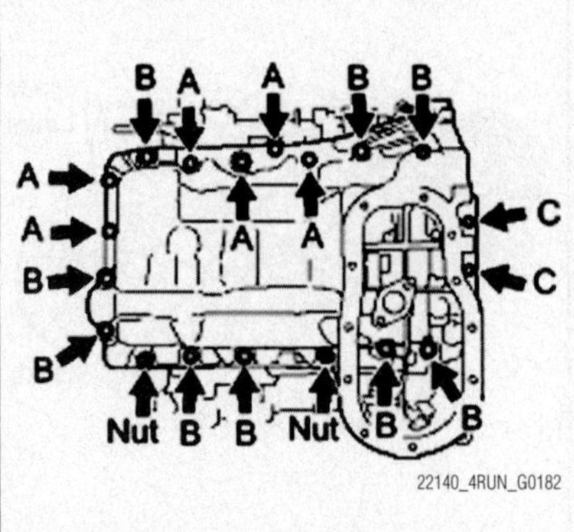

Fig. 91 Oil pan bolt sequence

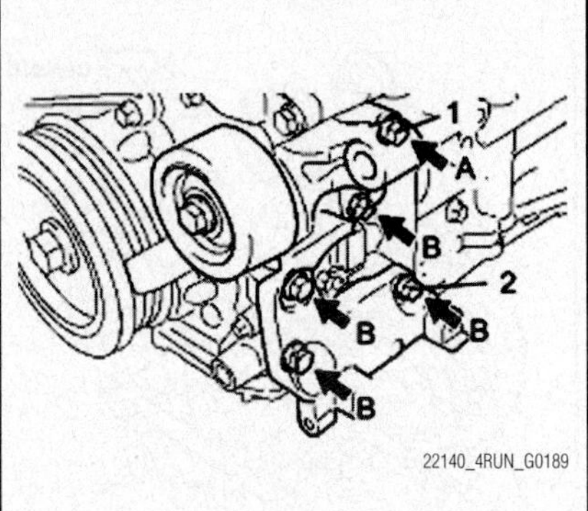

Fig. 92 Tensioner tightening sequence

After installing it, the No. 2 oil pan bolts and nuts must be tightened within 15 minutes. Otherwise the seal packing must be removed and reapplied.

79. Install the No. 2 oil pan with the 10 bolts and 2 nuts. Tighten the bolts and nuts uniformly in several steps. Tighten bolts to 80 inch lbs. (9.0 nm) and nuts to 7 ft. lbs. (10 Nm).
80. Install the pulley set bolt and tighten to 184 ft. lbs. (250 Nm).
81. Install the idler pulley with the bolt and tighten to 40 ft. lbs. (54 Nm).

➡"DOUBLE" is marked on the No. 1 idler pulley to distinguish it from the No. 2 idler pulley.

82. Install the 2 idler pulleys with the 2 bolts and tighten to 29 ft. lbs. (39 Nm).
83. Temporarily install the V-ribbed belt tensioner with the 5 bolts.
84. Install the V-ribbed belt tensioner by tightening the bolt 1 and bolt 2 in the order shown in the illustration. Tighten to 27 ft. lbs.(36 Nm).
85. Install cooler compressor assembly.
86. Install generator assembly.
87. Install the vane pump with the 2 bolts an tighten to 32 ft. lbs. (43 Nm).
88. Connect the P/S oil pressure switch connector.
89. Install water inlet.
90. Install a new O-ring to the oil level gauge guide.
91. Apply a light coat of engine oil to the O-ring.
92. Push in the oil level gauge guide end into the guide hole of the oil pan.
93. Install the oil level gauge guide with the bolt.
94. Install the oil level gauge.
95. Install air cleaner assembly.
96. Connect No. 2 ventilation hose.
97. Install fluid coupling assembly.
98. Install fan and generator v belt.
99. Fully tighten fluid coupling assembly.
100. Install radiator support seal upper.
101. Install battery.
102. Install differential carrier assembly front as necessary.
103. Install power steering link assembly.
104. Add engine oil.
105. Add engine coolant.
106. Check for engine leaks.
107. Install the V-bank cover with the 2 nuts.

VALVE COVERS

REMOVAL & INSTALLATION

V6 Engine

See Figure 93.

1. Remove the 2 bolts and engine hangers.
2. Remove the bolt and pull out the oil level gauge guide.
3. Remove the 10 bolts, 3 seal washers, 2 nuts, cylinder head covers and gaskets.

To install:

4. Remove any old packing (FIPG) material and be careful not to drop any oil on the contact surfaces of the cylinder head, timing chain cover and cylinder head cover.
5. Apply a continuous bead of seal packing (diameter: 2 to 3 mm (0.08 to 0.12 in.)) to the cylinder head and timing chain cover.

➡Install the cylinder head cover within 3 minutes after applying seal packing. After installing it, the cylinder head cover bolts and nuts must be tightened within 15 minutes. Otherwise the seal packing must be removed and reapplied.

6. Install the gasket to the cylinder head cover.
7. Install the seal washers to the bolts.
8. Install the cylinder head cover with

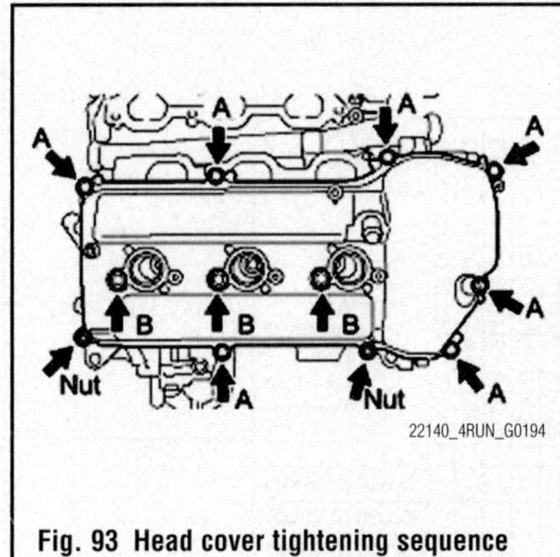

Fig. 93 Head cover tightening sequence

the 10 bolts and 2 nuts. Tighten the bolts and nuts uniformly in several steps.

9. Tighten bolt A to 7 ft. lbs (10 Nm) and bolts B and nut to 80 inch lbs, (9.0 nm).

V8 Engine

1. Before servicing the vehicle, refer to the precautions in the beginning of this section.
2. Remove battery clamp cover.
3. Remove throttle body cover.
4. Remove air cleaner and intake air connector assembly.
5. Remove the upper timing belt covers.
6. Remove ignition coils.
7. Remove the 9 bolts, seal washers and the right cylinder head cover.
8. Remove the oil dipstick for the transmission.
9. Disconnect the PCV hose.
10. Disconnect the engine wire clamp from the wire bracket on the cylinder head cover.
11. Remove the 9 bolts, 9 seal washers and the left cylinder head cover.

To install:

12. Install the left cylinder head cover. Tighten bolts to 53 inch lbs. (6.0 Nm)
13. Connect the engine wire clamp to the wire bracket on the cylinder head cover.
14. Connect the PCV hose.
15. Replace the oil dipstick for the transmission.
16. Install the right cylinder head cover. Tighten bolts to 53 inch lbs. (6.0 Nm)
17. Install the ignition coils.
18. Install the upper timing belt covers. Tighten bolts to 66 inch lbs. (7.5 Nm)
19. Install the air cleaner and intake air connector assembly.
20. Install the throttle body cover.
21. Install the battery clamp cover.

VALVE LASH

ADJUSTMENT

V6 Engine

See Figures 94 through 98.

➡Measure valve clearance with the engine cold.

1. Before servicing the vehicle, refer to the precautions section.
2. Drain the cooling system.
3. Remove or disconnect the following:

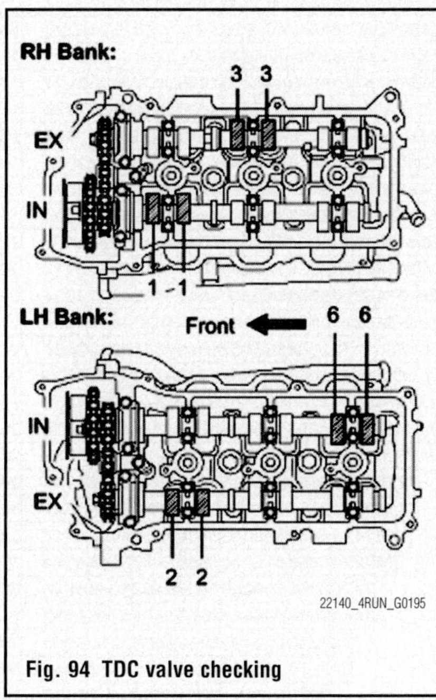

Fig. 94 TDC valve checking

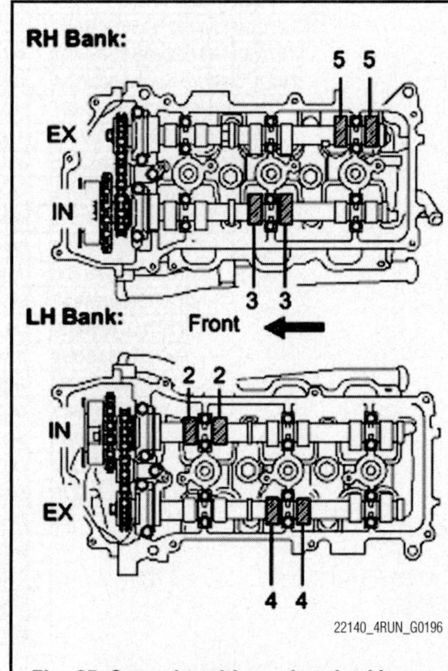

Fig. 95 Second position valve checking

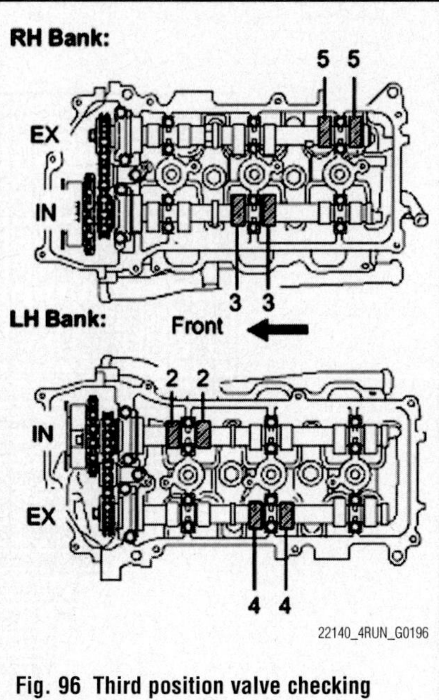

Fig. 96 Third position valve checking

- Negative battery cable
- Ignition coils
- Valve covers

4. Turn the crankshaft pulley, and align the notch with the timing mark "0" of the timing chain cover
5. Check the valves indicated in the illustration
6. Turn the crankshaft 2/3 of a revolution (240°), and check the valves indicated in the illustration
7. Turn the crankshaft 2/3 of a revolution (240°), and check the valves indicated in the illustration
8. The valve clearance specifications are as follows:
    - Intake: 0.006–0.010 in. (0.15–0.25mm)
    - Exhaust: 0.011–0.015 in. (0.29–0.39mm)
9. Record the measurements for each valve.
10. When all valve clearances have been measured, remove the camshafts if necessary.
11. Remove the valve shims and measure them. Note this measurement along with the clearance measurement recorded earlier.
12. Using the valve clearance and shim thickness measurements, find replacement shims in the Adjusting Shim Selection charts.

To install:

13. Install or connect the following:
    - Replacement valve shims
    - Camshafts
    - Valve covers

Valve Lifter Selection Chart (Intake)

Removed lifter thickness mm (in.) / Measured clearance mm (in.)	0.851 - 0.870 (0.0335 - 0.0343)	0.831 - 0.850 (0.0327 - 0.0335)	0.811 - 0.830 (0.0319 - 0.0327)	0.791 - 0.810 (0.0311 - 0.0319)	0.771 - 0.790 (0.0304 - 0.0311)	0.751 - 0.770 (0.0296 - 0.0303)	0.731 - 0.750 (0.0288 - 0.0295)	0.711 - 0.730 (0.0280 - 0.0287)	0.691 - 0.710 (0.0272 - 0.0280)	0.671 - 0.690 (0.0264 - 0.0272)	0.651 - 0.670 (0.0256 - 0.0264)	0.631 - 0.650 (0.0248 - 0.0256)	0.611 - 0.630 (0.0241 - 0.0248)	0.591 - 0.610 (0.0233 - 0.0240)	0.571 - 0.590 (0.0225 - 0.0232)	0.551 - 0.570 (0.0217 - 0.0224)	0.531 - 0.550 (0.0209 - 0.0217)	0.511 - 0.530 (0.0201 - 0.0209)	0.491 - 0.510 (0.0193 - 0.0201)	0.471 - 0.490 (0.0185 - 0.0193)	0.451 - 0.470 (0.0178 - 0.0185)	0.431 - 0.450 (0.0170 - 0.0177)	0.411 - 0.430 (0.0162 - 0.0169)	0.391 - 0.410 (0.0154 - 0.0161)	0.371 - 0.390 (0.0146 - 0.0154)	0.351 - 0.370 (0.0138 - 0.0146)	0.331 - 0.350 (0.0130 - 0.0138)	0.311 - 0.330 (0.0122 - 0.0130)	0.291 - 0.310 (0.0115 - 0.0122)	0.271 - 0.290 (0.0107 - 0.0114)	0.251 - 0.270 (0.0099 - 0.0106)	0.141 - 0.149 (0.0056 - 0.0059)	0.121 - 0.140 (0.0048 - 0.0055)	0.101 - 0.120 (0.0040 - 0.0047)	0.081 - 0.100 (0.0032 - 0.0039)	0.061 - 0.080 (0.0024 - 0.0031)	0.041 - 0.060 (0.0016 - 0.0024)	0.021 - 0.040 (0.0008 - 0.0016)	0.000 - 0.020 (0.0000 - 0.0008)
5.060 (0.1992)	72	70	68	66	64	62	60	58	56	54	52	50	48	46	44	42	40	38	36	34	32	30	28	26	24	22	20	18	16	14	12								
5.080 (0.2000)		72	70	68	66	64	62	60	58	56	54	52	50	48	46	44	42	40	38	36	34	32	30	28	26	24	22	20	18	16	14								
5.100 (0.2008)			72	70	68	66	64	62	60	58	56	54	52	50	48	46	44	42	40	38	36	34	32	30	28	26	24	22	20	18	16								
5.120 (0.2016)				72	70	68	66	64	62	60	58	56	54	52	50	48	46	44	42	40	38	36	34	32	30	28	26	24	22	20	18	06							
5.140 (0.2024)					72	70	68	66	64	62	60	58	56	54	52	50	48	46	44	42	40	38	36	34	32	30	28	26	24	22	20	08	08						
5.160 (0.2031)						72	70	68	66	64	62	60	58	56	54	52	50	48	46	44	42	40	38	36	34	32	30	28	26	24	22	10	10	08					
5.180 (0.2039)							72	70	68	66	64	62	60	58	56	54	52	50	48	46	44	42	40	38	36	34	32	30	28	26	24	12	12	10	08				
5.200 (0.2047)								72	70	68	66	64	62	60	58	56	54	52	50	48	46	44	42	40	38	36	34	32	30	28	26	14	14	12	10	08			
5.210 (0.2051)								74	72	70	68	66	64	62	60	58	56	54	52	50	48	46	44	42	40	38	36	34	32	30	28	16	14	12	10	08	06		
5.220 (0.2055)									72	70	68	66	64	62	60	58	56	54	52	50	48	46	44	42	40	38	36	34	32	30	28	16	16	14	12	10	08		
5.230 (0.2059)									74	72	70	68	66	64	62	60	58	56	54	52	50	48	46	44	42	40	38	36	34	32	30	18	16	14	12	10	08	06	
5.240 (0.2063)										72	70	68	66	64	62	60	58	56	54	52	50	48	46	44	42	40	38	36	34	32	30	18	18	16	14	12	10	08	
5.250 (0.2067)										74	72	70	68	66	64	62	60	58	56	54	52	50	48	46	44	42	40	38	36	34	32	20	18	16	14	12	10	08	06
5.260 (0.2071)											72	70	68	66	64	62	60	58	56	54	52	50	48	46	44	42	40	38	36	34	32	20	20	18	16	14	12	10	08
5.270 (0.2075)											74	72	70	68	66	64	62	60	58	56	54	52	50	48	46	44	42	40	38	36	34	22	20	18	16	14	12	10	08
5.280 (0.2079)												72	70	68	66	64	62	60	58	56	54	52	50	48	46	44	42	40	38	36	34	22	22	20	18	16	14	12	10
5.290 (0.2083)												74	72	70	68	66	64	62	60	58	56	54	52	50	48	46	44	42	40	38	36	24	22	20	18	16	14	12	10
5.300 (0.2087)													72	70	68	66	64	62	60	58	56	54	52	50	48	46	44	42	40	38	36	24	24	22	20	18	16	14	12
5.310 (0.2091)													74	72	70	68	66	64	62	60	58	56	54	52	50	48	46	44	42	40	38	26	24	22	20	18	16	14	12
5.320 (0.2094)														72	70	68	66	64	62	60	58	56	54	52	50	48	46	44	42	40	38	26	26	24	22	20	18	16	14
5.330 (0.2098)														74	72	70	68	66	64	62	60	58	56	54	52	50	48	46	44	42	40	28	26	24	22	20	18	16	14
5.340 (0.2102)															72	70	68	66	64	62	60	58	56	54	52	50	48	46	44	42	40	28	28	26	24	22	20	18	16
5.350 (0.2106)															74	72	70	68	66	64	62	60	58	56	54	52	50	48	46	44	42	30	28	26	24	22	20	18	16
5.360 (0.2110)																72	70	68	66	64	62	60	58	56	54	52	50	48	46	44	42	30	30	28	26	24	22	20	18
5.370 (0.2114)																74	72	70	68	66	64	62	60	58	56	54	52	50	48	46	44	32	30	28	26	24	22	20	18
5.380 (0.2118)																	72	70	68	66	64	62	60	58	56	54	52	50	48	46	44	32	32	30	28	26	24	22	20
5.390 (0.2122)																	74	72	70	68	66	64	62	60	58	56	54	52	50	48	46	34	32	30	28	26	24	22	20
5.400 (0.2126)																		72	70	68	66	64	62	60	58	56	54	52	50	48	46	34	34	32	30	28	26	24	22
5.410 (0.2130)																		74	72	70	68	66	64	62	60	58	56	54	52	50	48	36	34	32	30	28	26	24	22
5.420 (0.2134)																			72	70	68	66	64	62	60	58	56	54	52	50	48	36	36	34	32	30	28	26	24
5.430 (0.2138)																			74	72	70	68	66	64	62	60	58	56	54	52	50	38	36	34	32	30	28	26	24
5.440 (0.2142)																				72	70	68	66	64	62	60	58	56	54	52	50	38	38	36	34	32	30	28	26
5.450 (0.2146)																				74	72	70	68	66	64	62	60	58	56	54	52	40	38	36	34	32	30	28	26
5.460 (0.2150)																					72	70	68	66	64	62	60	58	56	54	52	40	40	38	36	34	32	30	28
5.470 (0.2154)																					74	72	70	68	66	64	62	60	58	56	54	42	40	38	36	34	32	30	28
5.480 (0.2157)																						72	70	68	66	64	62	60	58	56	54	42	42	40	38	36	34	32	30
5.490 (0.2161)																						74	72	70	68	66	64	62	60	58	56	44	42	40	38	36	34	32	30
5.500 (0.2165)																							72	70	68	66	64	62	60	58	56	44	44	42	40	38	36	34	32
5.510 (0.2169)																							74	72	70	68	66	64	62	60	58	46	44	42	40	38	36	34	32
5.520 (0.2173)																								72	70	68	66	64	62	60	58	46	46	44	42	40	38	36	34
5.530 (0.2177)																								74	72	70	68	66	64	62	60	48	46	44	42	40	38	36	34
5.540 (0.2181)																									72	70	68	66	64	62	60	48	48	46	44	42	40	38	36
5.550 (0.2185)																									74	72	70	68	66	64	62	50	48	46	44	42	40	38	36
5.560 (0.2189)																										72	70	68	66	64	62	50	50	48	46	44	42	40	38
5.570 (0.2193)																										74	72	70	68	66	64	52	50	48	46	44	42	40	38
5.580 (0.2197)																											72	70	68	66	64	52	52	50	48	46	44	42	40
5.590 (0.2201)																											74	72	70	68	66	54	52	50	48	46	44	42	40
5.600 (0.2205)																												72	70	68	66	54	54	52	50	48	46	44	42
5.620 (0.2213)																													72	70	68	56	56	54	52	50	48	46	44
5.640 (0.2220)																														72	70	58	58	56	54	52	50	48	46
5.660 (0.2228)																															72	60	60	58	56	54	52	50	48
5.680 (0.2236)																																62	62	60	58	56	54	52	50
5.700 (0.2244)																																64	64	62	60	58	56	54	52
5.720 (0.2252)																																66	66	64	62	60	58	56	54
5.740 (0.2260)																																68	68	66	64	62	60	58	56

22140_4RUN_G0197

Fig. 97 Intake valve lifter chart

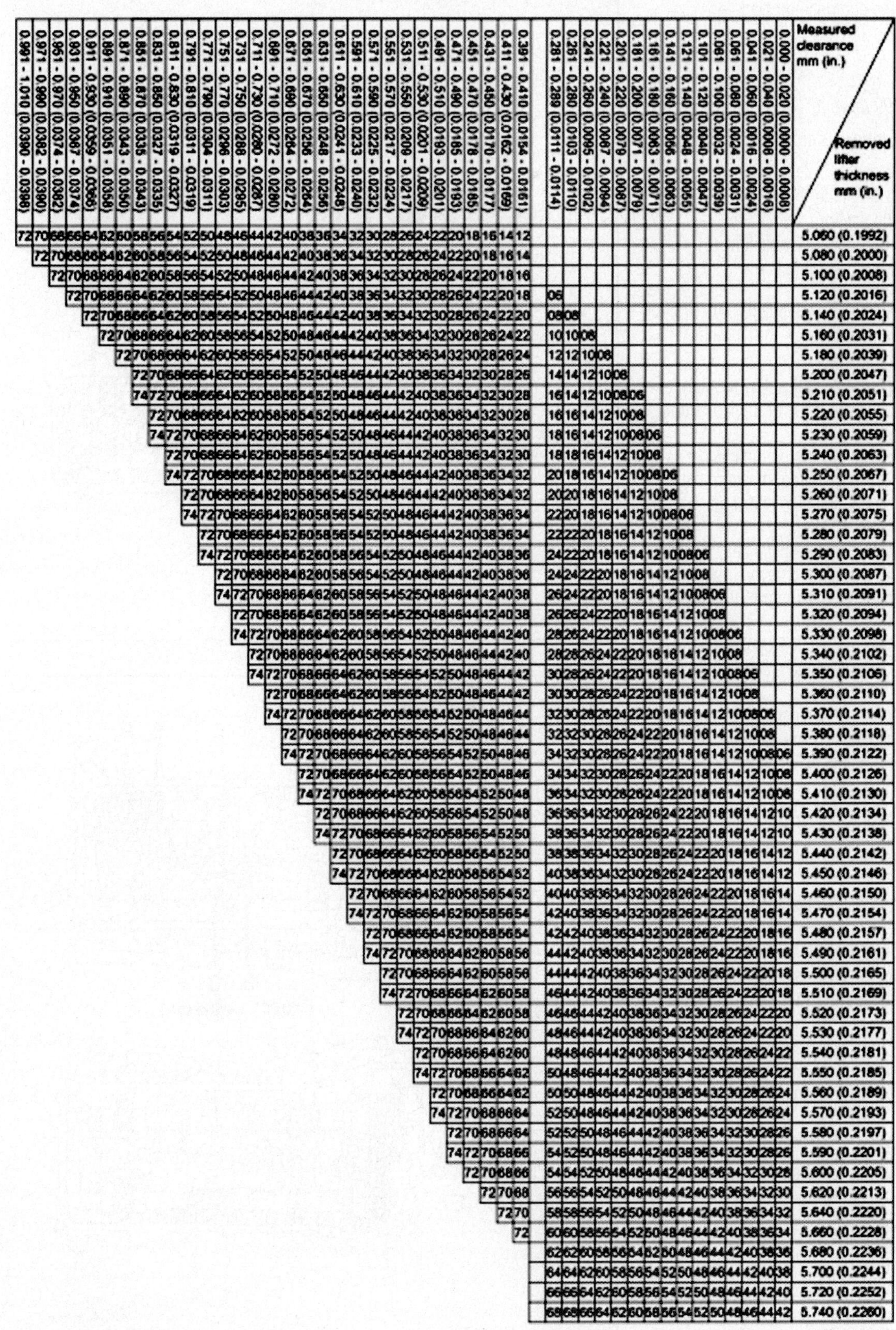

Valve Lifter Selection Chart (Exhaust)

Measured clearance mm (in.) / Removed lifter thickness mm (in.)

Removed lifter thickness mm (in.)	0.991 - 1.010 (0.0390 - 0.0398)	0.971 - 0.990 (0.0382 - 0.0390)	0.951 - 0.970 (0.0374 - 0.0382)	0.931 - 0.950 (0.0367 - 0.0374)	0.911 - 0.930 (0.0359 - 0.0366)	0.891 - 0.910 (0.0351 - 0.0358)	0.871 - 0.890 (0.0343 - 0.0350)	0.851 - 0.870 (0.0335 - 0.0343)	0.831 - 0.850 (0.0327 - 0.0335)	0.811 - 0.830 (0.0319 - 0.0327)	0.791 - 0.810 (0.0311 - 0.0319)	0.771 - 0.790 (0.0304 - 0.0311)	0.751 - 0.770 (0.0296 - 0.0303)	0.731 - 0.750 (0.0288 - 0.0295)	0.711 - 0.730 (0.0280 - 0.0287)	0.691 - 0.710 (0.0272 - 0.0280)	0.671 - 0.690 (0.0264 - 0.0272)	0.651 - 0.670 (0.0256 - 0.0264)	0.631 - 0.650 (0.0248 - 0.0256)	0.611 - 0.630 (0.0241 - 0.0248)	0.591 - 0.610 (0.0233 - 0.0240)	0.571 - 0.590 (0.0225 - 0.0232)	0.551 - 0.570 (0.0217 - 0.0224)	0.531 - 0.550 (0.0209 - 0.0217)	0.511 - 0.530 (0.0201 - 0.0209)	0.491 - 0.510 (0.0193 - 0.0201)	0.471 - 0.490 (0.0185 - 0.0193)	0.451 - 0.470 (0.0178 - 0.0185)	0.431 - 0.450 (0.0170 - 0.0177)	0.411 - 0.430 (0.0162 - 0.0169)	0.391 - 0.410 (0.0154 - 0.0161)
5.060 (0.1992)	72	70	68	66	64	62	60	58	56	54	52	50	48	46	44	42	40	38	36	34	32	30	28	26	24	22	20	18	16	14	12
5.080 (0.2000)		72	70	68	66	64	62	60	58	56	54	52	50	48	46	44	42	40	38	36	34	32	30	28	26	24	22	20	18	16	14
5.100 (0.2008)			72	70	68	66	64	62	60	58	56	54	52	50	48	46	44	42	40	38	36	34	32	30	28	26	24	22	20	18	16
5.120 (0.2016)				72	70	68	66	64	62	60	58	56	54	52	50	48	46	44	42	40	38	36	34	32	30	28	26	24	22	20	18
5.140 (0.2024)					72	70	68	66	64	62	60	58	56	54	52	50	48	46	44	42	40	38	36	34	32	30	28	26	24	22	20
5.160 (0.2031)						72	70	68	66	64	62	60	58	56	54	52	50	48	46	44	42	40	38	36	34	32	30	28	26	24	22
5.180 (0.2039)							72	70	68	66	64	62	60	58	56	54	52	50	48	46	44	42	40	38	36	34	32	30	28	26	24
5.200 (0.2047)								72	70	68	66	64	62	60	58	56	54	52	50	48	46	44	42	40	38	36	34	32	30	28	26
5.210 (0.2051)								74	72	70	68	66	64	62	60	58	56	54	52	50	48	46	44	42	40	38	36	34	32	30	28
5.220 (0.2055)									72	70	68	66	64	62	60	58	56	54	52	50	48	46	44	42	40	38	36	34	32	30	28
5.230 (0.2059)									74	72	70	68	66	64	62	60	58	56	54	52	50	48	46	44	42	40	38	36	34	32	30
5.240 (0.2063)										72	70	68	66	64	62	60	58	56	54	52	50	48	46	44	42	40	38	36	34	32	30
5.250 (0.2067)										74	72	70	68	66	64	62	60	58	56	54	52	50	48	46	44	42	40	38	36	34	32
5.260 (0.2071)											72	70	68	66	64	62	60	58	56	54	52	50	48	46	44	42	40	38	36	34	32
5.270 (0.2075)											74	72	70	68	66	64	62	60	58	56	54	52	50	48	46	44	42	40	38	36	34
5.280 (0.2079)												72	70	68	66	64	62	60	58	56	54	52	50	48	46	44	42	40	38	36	34
5.290 (0.2083)												74	72	70	68	66	64	62	60	58	56	54	52	50	48	46	44	42	40	38	36
5.300 (0.2087)													72	70	68	66	64	62	60	58	56	54	52	50	48	46	44	42	40	38	36
5.310 (0.2091)													74	72	70	68	66	64	62	60	58	56	54	52	50	48	46	44	42	40	38
5.320 (0.2094)														72	70	68	66	64	62	60	58	56	54	52	50	48	46	44	42	40	38
5.330 (0.2098)														74	72	70	68	66	64	62	60	58	56	54	52	50	48	46	44	42	40
5.340 (0.2102)															72	70	68	66	64	62	60	58	56	54	52	50	48	46	44	42	40
5.350 (0.2106)															74	72	70	68	66	64	62	60	58	56	54	52	50	48	46	44	42
5.360 (0.2110)																72	70	68	66	64	62	60	58	56	54	52	50	48	46	44	42
5.370 (0.2114)																74	72	70	68	66	64	62	60	58	56	54	52	50	48	46	44
5.380 (0.2118)																	72	70	68	66	64	62	60	58	56	54	52	50	48	46	44
5.390 (0.2122)																	74	72	70	68	66	64	62	60	58	56	54	52	50	48	46
5.400 (0.2126)																		72	70	68	66	64	62	60	58	56	54	52	50	48	46
5.410 (0.2130)																		74	72	70	68	66	64	62	60	58	56	54	52	50	48
5.420 (0.2134)																			72	70	68	66	64	62	60	58	56	54	52	50	48
5.430 (0.2138)																			74	72	70	68	66	64	62	60	58	56	54	52	50
5.440 (0.2142)																				72	70	68	66	64	62	60	58	56	54	52	50
5.450 (0.2146)																				74	72	70	68	66	64	62	60	58	56	54	52
5.460 (0.2150)																					72	70	68	66	64	62	60	58	56	54	52
5.470 (0.2154)																					74	72	70	68	66	64	62	60	58	56	54
5.480 (0.2157)																						72	70	68	66	64	62	60	58	56	54
5.490 (0.2161)																						74	72	70	68	66	64	62	60	58	56
5.500 (0.2165)																							72	70	68	66	64	62	60	58	56
5.510 (0.2169)																							74	72	70	68	66	64	62	60	58
5.520 (0.2173)																								72	70	68	66	64	62	60	58
5.530 (0.2177)																								74	72	70	68	66	64	62	60
5.540 (0.2181)																									72	70	68	66	64	62	60
5.550 (0.2185)																									74	72	70	68	66	64	62
5.560 (0.2189)																										72	70	68	66	64	62
5.570 (0.2193)																										74	72	70	68	66	64
5.580 (0.2197)																											72	70	68	66	64
5.590 (0.2201)																											74	72	70	68	66
5.600 (0.2205)																												72	70	68	66
5.620 (0.2213)																													72	70	68
5.640 (0.2220)																														72	70
5.660 (0.2228)																															72
5.680 (0.2236)																															
5.700 (0.2244)																															
5.720 (0.2252)																															
5.740 (0.2260)																															

Removed lifter thickness mm (in.)	0.281 - 0.289 (0.0111 - 0.0114)	0.261 - 0.280 (0.0103 - 0.0110)	0.241 - 0.260 (0.0095 - 0.0102)	0.221 - 0.240 (0.0087 - 0.0094)	0.201 - 0.220 (0.0079 - 0.0087)	0.181 - 0.200 (0.0071 - 0.0079)	0.161 - 0.180 (0.0063 - 0.0071)	0.141 - 0.160 (0.0056 - 0.0063)	0.121 - 0.140 (0.0048 - 0.0055)	0.101 - 0.120 (0.0040 - 0.0047)	0.081 - 0.100 (0.0032 - 0.0039)	0.061 - 0.080 (0.0024 - 0.0031)	0.041 - 0.060 (0.0016 - 0.0024)	0.021 - 0.040 (0.0008 - 0.0016)	0.000 - 0.020 (0.0000 - 0.0008)
5.060 (0.1992)															
5.080 (0.2000)															
5.100 (0.2008)															
5.120 (0.2016)	06														
5.140 (0.2024)	08	08													
5.160 (0.2031)	10	10	08												
5.180 (0.2039)	12	12	10	08											
5.200 (0.2047)	14	14	12	10	08										
5.210 (0.2051)	16	14	12	10	08	06									
5.220 (0.2055)	16	16	14	12	10	08									
5.230 (0.2059)	18	16	14	12	10	08	06								
5.240 (0.2063)	18	18	16	14	12	10	08								
5.250 (0.2067)	20	18	16	14	12	10	08	06							
5.260 (0.2071)	20	20	18	16	14	12	10	08							
5.270 (0.2075)	22	20	18	16	14	12	10	08	06						
5.280 (0.2079)	22	22	20	18	16	14	12	10	08						
5.290 (0.2083)	24	22	20	18	16	14	12	10	08	06					
5.300 (0.2087)	24	24	22	20	18	16	14	12	10	08					
5.310 (0.2091)	26	24	22	20	18	16	14	12	10	08	06				
5.320 (0.2094)	26	26	24	22	20	18	16	14	12	10	08				
5.330 (0.2098)	28	26	24	22	20	18	16	14	12	10	08	06			
5.340 (0.2102)	28	28	26	24	22	20	18	16	14	12	10	08			
5.350 (0.2106)	30	28	26	24	22	20	18	16	14	12	10	08	06		
5.360 (0.2110)	30	30	28	26	24	22	20	18	16	14	12	10	08		
5.370 (0.2114)	32	30	28	26	24	22	20	18	16	14	12	10	08	06	
5.380 (0.2118)	32	32	30	28	26	24	22	20	18	16	14	12	10	08	
5.390 (0.2122)	34	32	30	28	26	24	22	20	18	16	14	12	10	08	06
5.400 (0.2126)	34	34	32	30	28	26	24	22	20	18	16	14	12	10	08
5.410 (0.2130)	36	34	32	30	28	26	24	22	20	18	16	14	12	10	08
5.420 (0.2134)	36	36	34	32	30	28	26	24	22	20	18	16	14	12	10
5.430 (0.2138)	38	36	34	32	30	28	26	24	22	20	18	16	14	12	10
5.440 (0.2142)	38	38	36	34	32	30	28	26	24	22	20	18	16	14	12
5.450 (0.2146)	40	38	36	34	32	30	28	26	24	22	20	18	16	14	12
5.460 (0.2150)	40	40	38	36	34	32	30	28	26	24	22	20	18	16	14
5.470 (0.2154)	42	40	38	36	34	32	30	28	26	24	22	20	18	16	14
5.480 (0.2157)	42	42	40	38	36	34	32	30	28	26	24	22	20	18	16
5.490 (0.2161)	44	42	40	38	36	34	32	30	28	26	24	22	20	18	16
5.500 (0.2165)	44	44	42	40	38	36	34	32	30	28	26	24	22	20	18
5.510 (0.2169)	46	44	42	40	38	36	34	32	30	28	26	24	22	20	18
5.520 (0.2173)	46	46	44	42	40	38	36	34	32	30	28	26	24	22	20
5.530 (0.2177)	48	46	44	42	40	38	36	34	32	30	28	26	24	22	20
5.540 (0.2181)	48	48	46	44	42	40	38	36	34	32	30	28	26	24	22
5.550 (0.2185)	50	48	46	44	42	40	38	36	34	32	30	28	26	24	22
5.560 (0.2189)	50	50	48	46	44	42	40	38	36	34	32	30	28	26	24
5.570 (0.2193)	52	50	48	46	44	42	40	38	36	34	32	30	28	26	24
5.580 (0.2197)	52	52	50	48	46	44	42	40	38	36	34	32	30	28	26
5.590 (0.2201)	54	52	50	48	46	44	42	40	38	36	34	32	30	28	26
5.600 (0.2205)	54	54	52	50	48	46	44	42	40	38	36	34	32	30	28
5.620 (0.2213)	56	56	54	52	50	48	46	44	42	40	38	36	34	32	30
5.640 (0.2220)	58	58	56	54	52	50	48	46	44	42	40	38	36	34	32
5.660 (0.2228)	60	60	58	56	54	52	50	48	46	44	42	40	38	36	34
5.680 (0.2236)	62	62	60	58	56	54	52	50	48	46	44	42	40	38	36
5.700 (0.2244)	64	64	62	60	58	56	54	52	50	48	46	44	42	40	38
5.720 (0.2252)	66	66	64	62	60	58	56	54	52	50	48	46	44	42	40
5.740 (0.2260)	68	68	66	64	62	60	58	56	54	52	50	48	46	44	42

22140_4RUN_G0198

Fig. 98 Exhaust valve lifter chart

- Ignition coils
- Negative battery cable

14. Fill the cooling system.
15. Start the engine and check for leaks.

V8 Engine

See Figures 99 through 102.

➡Measure valve clearance with the engine cold.

1. Before servicing the vehicle, refer to the precautions section.
2. Drain the cooling system.
3. Remove or disconnect the following:
    - Negative battery cable
    - Ignition coils
    - Valve covers
4. Turn the crankshaft pulley to align its notch with timing mark "0" of the No. 1 timing belt cover.
5. Check only the valves indicated.
6. Using a feeler gauge, measure the clearance between the valve lifter and camshaft.
7. Record the out-of-specification valve clearance measurements. They will be used later to determine the required replacement adjusting shim.
8. Turn the crankshaft 1 complete revolution (360°) and align the camshaft timing marks
9. Check the valve clearance. The valve clearance specifications are as follows:
    - Intake: 0.006–0.010 in. (0.15–0.25mm)
    - Exhaust: 0.010–0.014 in. (0.25–0.35mm)
10. Record the measurements for each valve.
11. When all valve clearances have been measured, remove the camshafts.
12. Remove the valve shims and measure them. Note this measurement along with the clearance measurement recorded earlier.
13. Using the valve clearance and shim thickness measurements, find replacement shims in the Adjusting Shim Selection charts.

To install:

14. Install or connect the following:
    - Replacement valve shims
    - Camshafts
    - Valve covers
    - Ignition coils
    - Negative battery cable
15. Fill the cooling system.
16. Start the engine and check for leaks.

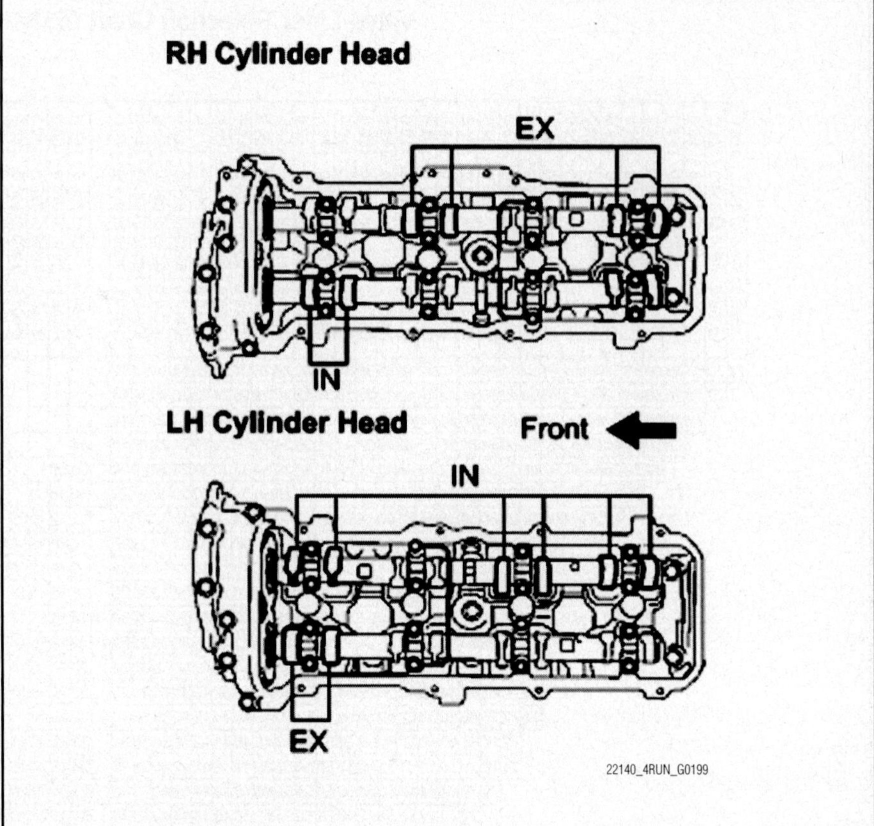

Fig. 99 TDC valve checking

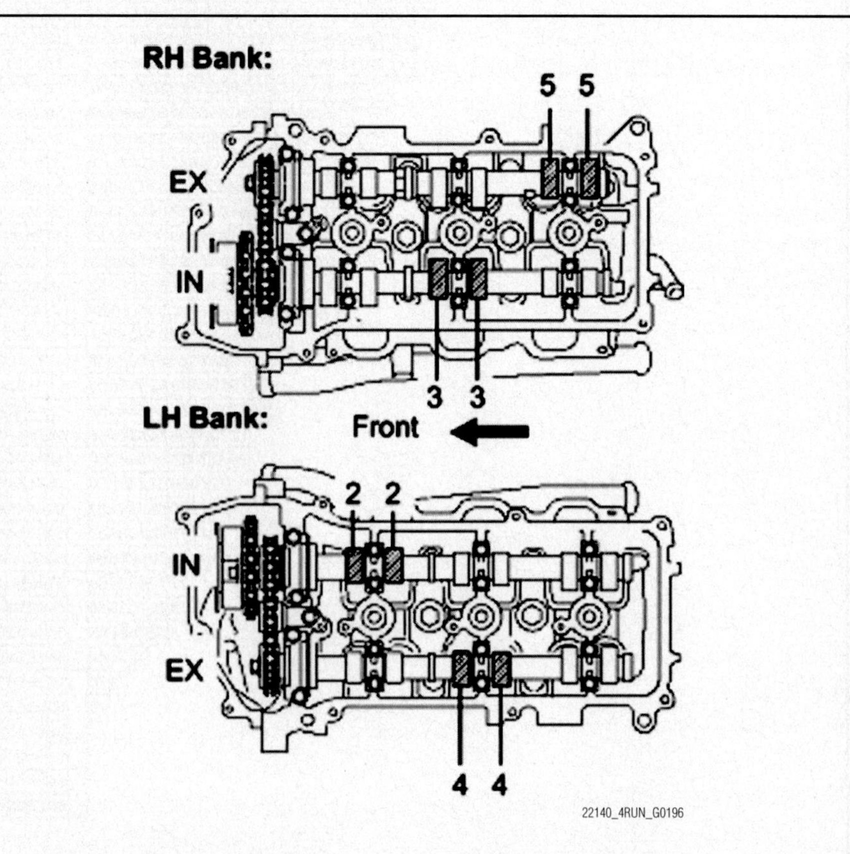

Fig. 100 Second position valve checking

Measured clearance mm (in.) \ Installed shim thickness mm (in.)	2.000 (0.0787)	2.020 (0.0795)	2.040 (0.0803)	2.060 (0.0811)	2.080 (0.0819)	2.100 (0.0827)	2.120 (0.0835)	2.140 (0.0843)	2.160 (0.0850)	2.180 (0.0858)	2.200 (0.0866)	2.210 (0.0870)	2.220 (0.0874)	2.230 (0.0878)	2.240 (0.0882)	2.250 (0.0886)	2.260 (0.0890)	2.270 (0.0894)	2.280 (0.0898)	2.290 (0.0902)	2.300 (0.0906)	2.310 (0.0909)	2.320 (0.0913)	2.330 (0.0917)	2.340 (0.0921)	2.350 (0.0925)	2.360 (0.0929)	2.370 (0.0933)	2.380 (0.0937)	2.390 (0.0941)	2.400 (0.0945)	2.410 (0.0949)	2.420 (0.0953)	2.430 (0.0957)	2.440 (0.0961)	2.450 (0.0965)	2.460 (0.0969)	2.470 (0.0972)	2.480 (0.0976)	2.490 (0.0980)	2.500 (0.0984)	2.510 (0.0988)	2.520 (0.0992)	2.530 (0.0996)	2.540 (0.1000)	2.550 (0.1004)	2.560 (0.1008)	2.570 (0.1012)	2.580 (0.1016)	2.590 (0.1020)	2.600 (0.1024)	2.620 (0.1031)	2.640 (0.1039)	2.660 (0.1047)	2.680 (0.1055)	2.700 (0.1063)	2.720 (0.1071)	2.740 (0.1079)	2.760 (0.1087)	2.780 (0.1094)	2.800 (0.1102)
0.000–0.030 (0.0000–0.0012)							00	00	00	00	02	02	04	04	06	06	08	08	10	10	12	12	14	14	16	16	18	18	20	20	22	22	24	24	26	26	28	28	30	30	32	32	34	34	36	36	38	38	40	40	42	44	46	48	50	52	54	56	58	60	62
0.031–0.050 (0.0012–0.0020)						00	00	00	00	02	04	06	06	08	08	10	10	12	12	14	14	16	16	18	18	20	20	22	22	24	24	26	26	28	28	30	30	32	32	34	34	36	36	38	38	40	40	42	42	44	44	46	48	50	52	54	56	58	60	62	64
0.051–0.070 (0.0020–0.0028)					00	00	00	00	02	04	06	08	08	10	10	12	12	14	14	16	16	18	18	20	20	22	22	24	24	26	26	28	28	30	30	32	32	34	34	36	36	38	38	40	40	42	42	44	44	46	46	48	50	52	54	56	58	60	62	64	66
0.071–0.090 (0.0028–0.0035)				00	00	00	00	02	04	06	08	10	10	12	12	14	14	16	16	18	18	20	20	22	22	24	24	26	26	28	28	30	30	32	32	34	34	36	36	38	38	40	40	42	42	44	44	46	46	48	48	50	52	54	56	58	60	62	64	66	68
0.091–0.110 (0.0036–0.0043)			00	00	00	00	02	04	06	08	10	12	12	14	14	16	16	18	18	20	20	22	22	24	24	26	26	28	28	30	30	32	32	34	34	36	36	38	38	40	40	42	42	44	44	46	46	48	48	50	50	52	54	56	58	60	62	64	66	68	70
0.111–0.130 (0.0044–0.0051)		00	00	00	00	02	04	06	08	10	12	14	14	16	16	18	18	20	20	22	22	24	24	26	26	28	28	30	30	32	32	34	34	36	36	38	38	40	40	42	42	44	44	46	46	48	48	50	50	52	52	54	56	58	60	62	64	66	68	70	72
0.131–0.149 (0.0052–0.0059)		00	00	00	02	04	06	08	10	12	14	16	16	18	18	20	20	22	22	24	24	26	26	28	28	30	30	32	32	34	34	36	36	38	38	40	40	42	42	44	44	46	46	48	48	50	50	52	52	54	54	56	58	60	62	64	66	68	70	72	74
0.150–0.250 (0.0059–0.0098)																																																													
0.251–0.270 (0.0099–0.0106)	06	08	10	12	14	16	18	20	22	24	26	28	28	30	30	32	32	34	34	36	36	38	38	40	40	42	42	44	44	46	46	48	48	50	50	52	52	54	54	56	56	58	58	60	60	62	62	64	64	66	66	68	70	72	74	76	78	80	80	80	
0.271–0.290 (0.0107–0.0114)	08	10	12	14	16	18	20	22	24	26	28	30	30	32	32	34	34	36	36	38	38	40	40	42	42	44	44	46	46	48	48	50	50	52	52	54	54	56	56	58	58	60	60	62	62	64	64	66	66	68	68	70	72	74	76	78	80	80	80		
0.291–0.310 (0.0115–0.0122)	10	12	14	16	18	20	22	24	26	28	30	32	32	34	34	36	36	38	38	40	40	42	42	44	44	46	46	48	48	50	50	52	52	54	54	56	56	58	58	60	60	62	62	64	64	66	66	68	68	70	70	72	74	76	78	80	80	80			
0.311–0.330 (0.0122–0.0130)	12	14	16	18	20	22	24	26	28	30	32	34	34	36	36	38	38	40	40	42	42	44	44	46	46	48	48	50	50	52	52	54	54	56	56	58	58	60	60	62	62	64	64	66	66	68	68	70	70	72	72	74	76	78	80	80	80				
0.331–0.350 (0.0130–0.0138)	14	16	18	20	22	24	26	28	30	32	34	36	36	38	38	40	40	42	42	44	44	46	46	48	48	50	50	52	52	54	54	56	56	58	58	60	60	62	62	64	64	66	66	68	68	70	70	72	72	74	74	76	78	80	80	80					
0.351–0.370 (0.0138–0.0146)	16	18	20	22	24	26	28	30	32	34	36	38	38	40	40	42	42	44	44	46	46	48	48	50	50	52	52	54	54	56	56	58	58	60	60	62	62	64	64	66	66	68	68	70	70	72	72	74	74	76	76	78	80	80	80						
0.371–0.390 (0.0146–0.0154)	18	20	22	24	26	28	30	32	34	36	38	40	40	42	42	44	44	46	46	48	48	50	50	52	52	54	54	56	56	58	58	60	60	62	62	64	64	66	66	68	68	70	70	72	72	74	74	76	76	78	78	80	80	80							
0.391–0.410 (0.0154–0.0161)	20	22	24	26	28	30	32	34	36	38	40	42	42	44	44	46	46	48	48	50	50	52	52	54	54	56	56	58	58	60	60	62	62	64	64	66	66	68	68	70	70	72	72	74	74	76	76	78	78	80	80	80	80								
0.411–0.430 (0.0162–0.0169)	22	24	26	28	30	32	34	36	38	40	42	44	44	46	46	48	48	50	50	52	52	54	54	56	56	58	58	60	60	62	62	64	64	66	66	68	68	70	70	72	72	74	74	76	76	78	78	80	80	80	80	80									
0.431–0.450 (0.0170–0.0177)	24	26	28	30	32	34	36	38	40	42	44	46	46	48	48	50	50	52	52	54	54	56	56	58	58	60	60	62	62	64	64	66	66	68	68	70	70	72	72	74	74	76	76	78	78	80	80	80	80	80	80										
0.451–0.470 (0.0178–0.0185)	26	28	30	32	34	36	38	40	42	44	46	48	48	50	50	52	52	54	54	56	56	58	58	60	60	62	62	64	64	66	66	68	68	70	70	72	72	74	74	76	76	78	78	80	80	80	80	80	80	80											
0.471–0.490 (0.0185–0.0193)	28	30	32	34	36	38	40	42	44	46	48	50	50	52	52	54	54	56	56	58	58	60	60	62	62	64	64	66	66	68	68	70	70	72	72	74	74	76	76	78	78	80	80	80	80	80	80	80													
0.491–0.510 (0.0193–0.0201)	30	32	34	36	38	40	42	44	46	48	50	52	52	54	54	56	56	58	58	60	60	62	62	64	64	66	66	68	68	70	70	72	72	74	74	76	76	78	78	80	80	80	80	80	80	80															
0.511–0.530 (0.0201–0.0209)	32	34	36	38	40	42	44	46	48	50	52	54	54	56	56	58	58	60	60	62	62	64	64	66	66	68	68	70	70	72	72	74	74	76	76	78	78	80	80	80	80	80	80	80																	
0.531–0.550 (0.0209–0.0217)	34	36	38	40	42	44	46	48	50	52	54	56	56	58	58	60	60	62	62	64	64	66	66	68	68	70	70	72	72	74	74	76	76	78	78	80	80	80	80	80	80	80																			
0.551–0.570 (0.0217–0.0224)	36	38	40	42	44	46	48	50	52	54	56	58	58	60	60	62	62	64	64	66	66	68	68	70	70	72	72	74	74	76	76	78	78	80	80	80	80	80	80	80																					
0.571–0.590 (0.0225–0.0232)	38	40	42	44	46	48	50	52	54	56	58	60	60	62	62	64	64	66	66	68	68	70	70	72	72	74	74	76	76	78	78	80	80	80	80	80	80	80																							
0.591–0.610 (0.0233–0.0240)	40	42	44	46	48	50	52	54	56	58	60	62	62	64	64	66	66	68	68	70	70	72	72	74	74	76	76	78	78	80	80	80	80	80	80	80																									
0.611–0.630 (0.0241–0.0248)	42	44	46	48	50	52	54	56	58	60	62	64	64	66	66	68	68	70	70	72	72	74	74	76	76	78	78	80	80	80	80	80	80	80																											
0.631–0.650 (0.0248–0.0256)	44	46	48	50	52	54	56	58	60	62	64	66	66	68	68	70	70	72	72	74	74	76	76	78	78	80	80	80	80	80	80	80																													
0.651–0.670 (0.0256–0.0264)	46	48	50	52	54	56	58	60	62	64	66	68	68	70	70	72	72	74	74	76	76	78	78	80	80	80	80	80	80	80																															
0.671–0.690 (0.0264–0.0272)	48	50	52	54	56	58	60	62	64	66	68	70	70	72	72	74	74	76	76	78	78	80	80	80	80	80	80	80																																	
0.691–0.710 (0.0272–0.0280)	50	52	54	56	58	60	62	64	66	68	70	72	72	74	74	76	76	78	78	80	80	80	80	80	80	80																																			
0.711–0.730 (0.0280–0.0287)	52	54	56	58	60	62	64	66	68	70	72	74	74	76	76	78	78	80	80	80	80	80	80	80																																					
0.731–0.750 (0.0288–0.0295)	54	56	58	60	62	64	66	68	70	72	74	76	76	78	78	80	80	80	80	80	80	80																																							
0.751–0.770 (0.0296–0.0303)	56	58	60	62	64	66	68	70	72	74	76	78	78	80	80	80	80	80	80	80																																									
0.771–0.790 (0.0304–0.0311)	58	60	62	64	66	68	70	72	74	76	78	80	80	80	80	80	80	80																																											
0.791–0.810 (0.0311–0.0319)	60	62	64	66	68	70	72	74	76	78	80	80	80	80	80	80																																													
0.811–0.830 (0.0319–0.0327)	62	64	66	68	70	72	74	76	78	80	80	80	80	80																																															
0.831–0.850 (0.0327–0.0335)	64	66	68	70	72	74	76	78	80	80	80	80																																																	
0.851–0.870 (0.0335–0.0343)	66	68	70	72	74	76	78	80	80	80																																																			
0.871–0.890 (0.0343–0.0350)	68	70	72	74	76	78	80	80	80																																																				
0.891–0.910 (0.0351–0.0358)	70	72	74	76	78	80	80	80																																																						
0.911–0.930 (0.0359–0.0366)	72	74	76	78	80	80	80																																																						
0.931–0.950 (0.0367–0.0374)	74	76	78	80	80	80																																																							
0.951–0.970 (0.0374–0.0382)	76	78	80	80	80																																																								
0.971–0.990 (0.0382–0.0390)	78	80	80	80																																																									
0.991–1.010 (0.0390–0.0398)	80	80	80																																																										
1.011–1.030 (0.0398–0.0406)	80	80																																																											
1.031–1.050 (0.0406–0.0413)	80																																																													

Intake valve clearance (Cold):
0.15 – 0.25 mm (0.006 – 0.010 in.)

EXAMPLE:
The 2.300 mm (0.0906 in.) shim is installed, and the measured clearance is 0.440 mm (0.0173 in.). Replace the 2.300 mm (0.0906 in.) shim with a No. 54 shim.

New shim thickness mm (in.)

Shim No.	Thickness	Shim No.	Thickness	Shim No.	Thickness
00	2.000 (0.0787)	28	2.280 (0.0898)	56	2.560 (0.1008)
02	2.020 (0.0795)	30	2.300 (0.0906)	58	2.580 (0.1016)
04	2.040 (0.0803)	32	2.320 (0.0913)	60	2.600 (0.1024)
06	2.060 (0.0811)	34	2.340 (0.0921)	62	2.620 (0.1031)
08	2.080 (0.0819)	36	2.360 (0.0929)	64	2.640 (0.1039)
10	2.100 (0.0827)	38	2.380 (0.0937)	66	2.660 (0.1047)
12	2.120 (0.0835)	40	2.400 (0.0945)	68	2.680 (0.1055)
14	2.140 (0.0843)	42	2.420 (0.0953)	70	2.700 (0.1063)
16	2.160 (0.0850)	44	2.440 (0.0961)	72	2.720 (0.1071)
18	2.180 (0.0858)	46	2.460 (0.0969)	74	2.740 (0.1079)
20	2.200 (0.0866)	48	2.480 (0.0976)	76	2.760 (0.1087)
22	2.220 (0.0874)	50	2.500 (0.0984)	78	2.780 (0.1094)
24	2.240 (0.0882)	52	2.520 (0.0992)	80	2.800 (0.1102)
26	2.260 (0.0890)	54	2.540 (0.1000)		

7924SG71

Fig. 101 Intake valve clearance shim selection chart

Measured clearance mm (in.) \ Installed shim thickness mm (in.)	2.000 (0.0787)	2.020 (0.0795)	2.040 (0.0803)	2.060 (0.0811)	2.080 (0.0819)	2.100 (0.0827)	2.120 (0.0835)	2.140 (0.0843)	2.160 (0.0850)	2.180 (0.0858)	2.200 (0.0866)	2.210 (0.0870)	2.220 (0.0874)	2.230 (0.0878)	2.240 (0.0882)	2.250 (0.0886)	2.260 (0.0890)	2.270 (0.0894)	2.280 (0.0898)	2.290 (0.0902)	2.300 (0.0906)
0.000−0.030 (0.0000−0.0012)												00	00	00	00	00	00	00	00	00	02
0.031−0.050 (0.0012−0.0020)											00	00	00	00	00	00	00	02	02	04	04
0.051−0.070 (0.0020−0.0028)										00	00	00	00	00	00	02	02	04	04	06	06
0.071−0.090 (0.0028−0.0035)									00	00	00	00	00	02	02	04	04	06	06	08	08
0.091−0.110 (0.0036−0.0043)								00	00	00	00	02	02	04	04	06	06	08	08	10	10
0.111−0.130 (0.0044−0.0051)							00	00	00	00	02	04	04	06	06	08	08	10	10	12	12
0.131−0.150 (0.0052−0.0059)						00	00	00	00	02	04	06	06	08	08	10	10	12	12	14	14
0.151−0.170 (0.0059−0.0067)					00	00	00	00	02	04	06	08	08	10	10	12	12	14	14	16	16
0.171−0.190 (0.0067−0.0075)				00	00	00	00	02	04	06	08	10	10	12	12	14	14	16	16	18	18
0.191−0.210 (0.0075−0.0083)			00	00	00	00	02	04	06	08	10	12	12	14	14	16	16	18	18	20	20
0.211−0.230 (0.0083−0.0091)		00	00	00	00	02	04	06	08	10	12	14	14	16	16	18	18	20	20	22	22
0.231−0.249 (0.0091−0.0098)		00	00	00	02	04	06	08	10	12	14	16	16	18	18	20	20	22	22	24	24
0.250−0.350 (0.0098−0.0138)																					
0.351−0.370 (0.0138−0.0146)	06	08	10	12	14	16	18	20	22	24	26	28	28	30	30	32	32	34	34	36	36
0.371−0.390 (0.0146−0.0154)	08	10	12	14	16	18	20	22	24	26	28	30	30	32	32	34	34	36	36	38	38
0.391−0.410 (0.0154−0.0161)	10	12	14	16	18	20	22	24	26	28	30	32	32	34	34	36	36	38	38	40	40
0.411−0.430 (0.0162−0.0169)	12	14	16	18	20	22	24	26	28	30	32	34	34	36	36	38	38	40	40	42	42
0.431−0.450 (0.0170−0.0177)	14	16	18	20	22	24	26	28	30	32	34	36	36	38	38	40	40	42	42	44	44
0.451−0.470 (0.0178−0.0185)	16	18	20	22	24	26	28	30	32	34	36	38	38	40	40	42	42	44	44	46	46
0.471−0.490 (0.0185−0.0193)	18	20	22	24	26	28	30	32	34	36	38	40	40	42	42	44	44	46	46	48	48
0.491−0.510 (0.0193−0.0201)	20	22	24	26	28	30	32	34	36	38	40	42	42	44	44	46	46	48	48	50	50
0.511−0.530 (0.0201−0.0209)	22	24	26	28	30	32	34	36	38	40	42	44	44	46	46	48	48	50	50	52	52
0.531−0.550 (0.0209−0.0217)	24	26	28	30	32	34	36	38	40	42	44	46	46	48	48	50	50	52	52	54	54
0.551−0.570 (0.0217−0.0224)	26	28	30	32	34	36	38	40	42	44	46	48	48	50	50	52	52	54	54	56	56
0.571−0.590 (0.0225−0.0232)	28	30	32	34	36	38	40	42	44	46	48	50	50	52	52	54	54	56	56	58	58
0.591−0.610 (0.0233−0.0240)	30	32	34	36	38	40	42	44	46	48	50	52	52	54	54	56	56	58	58	60	60
0.611−0.630 (0.0241−0.0248)	32	34	36	38	40	42	44	46	48	50	52	54	54	56	56	58	58	60	60	62	62
0.631−0.650 (0.0248−0.0256)	34	36	38	40	42	44	46	48	50	52	54	56	56	58	58	60	60	62	62	64	64
0.651−0.670 (0.0256−0.0264)	36	38	40	42	44	46	48	50	52	54	56	58	58	60	60	62	62	64	64	66	66
0.671−0.690 (0.0264−0.0272)	38	40	42	44	46	48	50	52	54	56	58	60	60	62	62	64	64	66	66	68	68
0.691−0.710 (0.0272−0.0280)	40	42	44	46	48	50	52	54	56	58	60	62	62	64	64	66	66	68	68	70	70
0.711−0.730 (0.0280−0.0287)	42	44	46	48	50	52	54	56	58	60	62	64	64	66	66	68	68	70	70	72	72
0.731−0.750 (0.0288−0.0295)	44	46	48	50	52	54	56	58	60	62	64	66	66	68	68	70	70	72	72	74	74
0.751−0.770 (0.0296−0.0303)	46	48	50	52	54	56	58	60	62	64	66	68	68	70	70	72	72	74	74	76	76
0.771−0.790 (0.0304−0.0311)	48	50	52	54	56	58	60	62	64	66	68	70	70	72	72	74	74	76	76	78	78
0.791−0.810 (0.0311−0.0319)	50	52	54	56	58	60	62	64	66	68	70	72	72	74	74	76	76	78	78	80	80
0.811−0.830 (0.0319−0.0327)	52	54	56	58	60	62	64	66	68	70	72	74	74	76	76	78	78	80	80	80	80
0.831−0.850 (0.0327−0.0335)	54	56	58	60	62	64	66	68	70	72	74	76	76	78	78	80	80	80	80	80	80
0.851−0.870 (0.0335−0.0343)	56	58	60	62	64	66	68	70	72	74	76	78	78	80	80	80	80	80	80	80	
0.871−0.890 (0.0343−0.0350)	58	60	62	64	66	68	70	72	74	76	78	80	80	80	80	80	80	80			
0.891−0.910 (0.0351−0.0358)	60	62	64	66	68	70	72	74	76	78	80	80	80	80	80	80					
0.911−0.930 (0.0359−0.0366)	62	64	66	68	70	72	74	76	78	80	80	80	80	80							
0.931−0.950 (0.0367−0.0374)	64	66	68	70	72	74	76	78	80	80	80	80									
0.951−0.970 (0.0374−0.0382)	66	68	70	72	74	76	78	80	80	80											
0.971−0.990 (0.0382−0.0390)	68	70	72	74	76	78	80	80	80												
0.991−1.010 (0.0390−0.0398)	70	72	74	76	78	80	80	80													
1.011−1.030 (0.0398−0.0406)	72	74	76	78	80	80	80														
1.031−1.050 (0.0406−0.0413)	74	76	78	80	80	80															
1.051−1.070 (0.0414−0.0421)	76	78	80	80	80																
1.071−1.090 (0.0422−0.0429)	78	80	80	80																	
1.091−1.110 (0.0430−0.0437)	80	80	80																		
1.111−1.130 (0.0437−0.0445)	80	80																			
1.131−1.150 (0.0445−0.0453)	80																				

Measured clearance mm (in.) \ Installed shim thickness mm (in.)	2.310 (0.0909)	2.320 (0.0913)	2.330 (0.0917)	2.340 (0.0921)	2.350 (0.0925)	2.360 (0.0929)	2.370 (0.0933)	2.380 (0.0937)	2.390 (0.0941)	2.400 (0.0945)	2.410 (0.0949)	2.420 (0.0953)	2.430 (0.0957)	2.440 (0.0961)	2.450 (0.0965)	2.460 (0.0969)	2.470 (0.0972)	2.480 (0.0976)	2.490 (0.0980)	2.500 (0.0984)
0.000−0.030 (0.0000−0.0012)	02	04	04	06	06	08	08	10	10	12	12	14	14	16	16	18	18	20	20	22
0.031−0.050 (0.0012−0.0020)	06	06	08	08	10	10	12	12	14	14	16	16	18	18	20	20	22	22	24	24
0.051−0.070 (0.0020−0.0028)	08	08	10	10	12	12	14	14	16	16	18	18	20	20	22	22	24	24	26	26
0.071−0.090 (0.0028−0.0035)	10	10	12	12	14	14	16	16	18	18	20	20	22	22	24	24	26	26	28	28
0.091−0.110 (0.0036−0.0043)	12	12	14	14	16	16	18	18	20	20	22	22	24	24	26	26	28	28	30	30
0.111−0.130 (0.0044−0.0051)	14	14	16	16	18	18	20	20	22	22	24	24	26	26	28	28	30	30	32	32
0.131−0.150 (0.0052−0.0059)	16	16	18	18	20	20	22	22	24	24	26	26	28	28	30	30	32	32	34	34
0.151−0.170 (0.0059−0.0067)	18	18	20	20	22	22	24	24	26	26	28	28	30	30	32	32	34	34	36	36
0.171−0.190 (0.0067−0.0075)	20	20	22	22	24	24	26	26	28	28	30	30	32	32	34	34	36	36	38	38
0.191−0.210 (0.0075−0.0083)	22	22	24	24	26	26	28	28	30	30	32	32	34	34	36	36	38	38	40	40
0.211−0.230 (0.0083−0.0091)	24	24	26	26	28	28	30	30	32	32	34	34	36	36	38	38	40	40	42	42
0.231−0.249 (0.0091−0.0098)	26	26	28	28	30	30	32	32	34	34	36	36	38	38	40	40	42	42	44	44
0.250−0.350 (0.0098−0.0138)																				
0.351−0.370 (0.0138−0.0146)	38	38	40	40	42	42	44	44	46	46	48	48	50	50	52	52	54	54	56	56
0.371−0.390 (0.0146−0.0154)	40	40	42	42	44	44	46	46	48	48	50	50	52	52	54	54	56	56	58	58
0.391−0.410 (0.0154−0.0161)	42	42	44	44	46	46	48	48	50	50	52	52	54	54	56	56	58	58	60	60
0.411−0.430 (0.0162−0.0169)	44	44	46	46	48	48	50	50	52	52	54	54	56	56	58	58	60	60	62	62
0.431−0.450 (0.0170−0.0177)	46	46	48	48	50	50	52	52	54	54	56	56	58	58	60	60	62	62	64	64
0.451−0.470 (0.0178−0.0185)	48	48	50	50	52	52	54	54	56	56	58	58	60	60	62	62	64	64	66	66
0.471−0.490 (0.0185−0.0193)	50	50	52	52	54	54	56	56	58	58	60	60	62	62	64	64	66	66	68	68
0.491−0.510 (0.0193−0.0201)	52	52	54	54	56	56	58	58	60	60	62	62	64	64	66	66	68	68	70	70
0.511−0.530 (0.0201−0.0209)	54	54	56	56	58	58	60	60	62	62	64	64	66	66	68	68	70	70	72	72
0.531−0.550 (0.0209−0.0217)	56	56	58	58	60	60	62	62	64	64	66	66	68	68	70	70	72	72	74	74
0.551−0.570 (0.0217−0.0224)	58	58	60	60	62	62	64	64	66	66	68	68	70	70	72	72	74	74	76	76
0.571−0.590 (0.0225−0.0232)	60	60	62	62	64	64	66	66	68	68	70	70	72	72	74	74	76	76	78	78
0.591−0.610 (0.0233−0.0240)	62	62	64	64	66	66	68	68	70	70	72	72	74	74	76	76	78	78	80	80
0.611−0.630 (0.0241−0.0248)	64	64	66	66	68	68	70	70	72	72	74	74	76	76	78	78	80	80	80	80
0.631−0.650 (0.0248−0.0256)	66	66	68	68	70	70	72	72	74	74	76	76	78	78	80	80	80	80	80	80
0.651−0.670 (0.0256−0.0264)	68	68	70	70	72	72	74	74	76	76	78	78	80	80	80	80	80	80	80	
0.671−0.690 (0.0264−0.0272)	70	70	72	72	74	74	76	76	78	78	80	80	80	80	80	80	80			
0.691−0.710 (0.0272−0.0280)	72	72	74	74	76	76	78	78	80	80	80	80	80	80	80					
0.711−0.730 (0.0280−0.0287)	74	74	76	76	78	78	80	80	80	80	80	80	80							
0.731−0.750 (0.0288−0.0295)	76	76	78	78	80	80	80	80	80	80	80									
0.751−0.770 (0.0296−0.0303)	78	78	80	80	80	80	80	80	80											
0.771−0.790 (0.0304−0.0311)	80	80	80	80	80	80	80													
0.791−0.810 (0.0311−0.0319)	80	80	80	80	80															
0.811−0.830 (0.0319−0.0327)	80	80	80																	
0.831−0.850 (0.0327−0.0335)	80																			
0.851−0.870 (0.0335−0.0343)																				
0.871−0.890 (0.0343−0.0350)																				
0.891−0.910 (0.0351−0.0358)																				
0.911−0.930 (0.0359−0.0366)																				
0.931−0.950 (0.0367−0.0374)																				
0.951−0.970 (0.0374−0.0382)																				
0.971−0.990 (0.0382−0.0390)																				
0.991−1.010 (0.0390−0.0398)																				
1.011−1.030 (0.0398−0.0406)																				
1.031−1.050 (0.0406−0.0413)																				
1.051−1.070 (0.0414−0.0421)																				
1.071−1.090 (0.0422−0.0429)																				
1.091−1.110 (0.0430−0.0437)																				
1.111−1.130 (0.0437−0.0445)																				
1.131−1.150 (0.0445−0.0453)																				

Measured clearance mm (in.) \ Installed shim thickness mm (in.)	2.510 (0.0988)	2.520 (0.0992)	2.530 (0.0996)	2.540 (0.1000)	2.550 (0.1004)	2.560 (0.1008)	2.570 (0.1012)	2.580 (0.1016)	2.590 (0.1020)	2.600 (0.1024)	2.620 (0.1031)	2.640 (0.1039)	2.660 (0.1047)	2.680 (0.1055)	2.700 (0.1063)	2.720 (0.1071)	2.740 (0.1079)	2.760 (0.1087)	2.780 (0.1094)	2.800 (0.1102)
0.000−0.030 (0.0000−0.0012)	22	24	24	26	26	28	28	30	30	32	34	36	38	40	42	44	46	48	50	52
0.031−0.050 (0.0012−0.0020)	26	26	28	28	30	30	32	32	34	34	36	38	40	42	44	46	48	50	52	54
0.051−0.070 (0.0020−0.0028)	28	28	30	30	32	32	34	34	36	36	38	40	42	44	46	48	50	52	54	56
0.071−0.090 (0.0028−0.0035)	30	30	32	32	34	34	36	36	38	38	40	42	44	46	48	50	52	54	56	58
0.091−0.110 (0.0036−0.0043)	32	32	34	34	36	36	38	38	40	40	42	44	46	48	50	52	54	56	58	60
0.111−0.130 (0.0044−0.0051)	34	34	36	36	38	38	40	40	42	42	44	46	48	50	52	54	56	58	60	62
0.131−0.150 (0.0052−0.0059)	36	36	38	38	40	40	42	42	44	44	46	48	50	52	54	56	58	60	62	64
0.151−0.170 (0.0059−0.0067)	38	38	40	40	42	42	44	44	46	46	48	50	52	54	56	58	60	62	64	66
0.171−0.190 (0.0067−0.0075)	40	40	42	42	44	44	46	46	48	48	50	52	54	56	58	60	62	64	66	68
0.191−0.210 (0.0075−0.0083)	42	42	44	44	46	46	48	48	50	50	52	54	56	58	60	62	64	66	68	70
0.211−0.230 (0.0083−0.0091)	44	44	46	46	48	48	50	50	52	52	54	56	58	60	62	64	66	68	70	72
0.231−0.249 (0.0091−0.0098)	46	46	48	48	50	50	52	52	54	54	56	58	60	62	64	66	68	70	72	74
0.250−0.350 (0.0098−0.0138)																				
0.351−0.370 (0.0138−0.0146)	58	58	60	60	62	62	64	64	66	66	68	70	72	74	76	78	80	80	80	
0.371−0.390 (0.0146−0.0154)	60	60	62	62	64	64	66	66	68	68	70	72	74	76	78	80	80	80		
0.391−0.410 (0.0154−0.0161)	62	62	64	64	66	66	68	68	70	70	72	74	76	78	80	80	80			
0.411−0.430 (0.0162−0.0169)	64	64	66	66	68	68	70	70	72	72	74	76	78	80	80	80				
0.431−0.450 (0.0170−0.0177)	66	66	68	68	70	70	72	72	74	74	76	78	80	80	80					
0.451−0.470 (0.0178−0.0185)	68	68	70	70	72	72	74	74	76	76	78	80	80	80						
0.471−0.490 (0.0185−0.0193)	70	70	72	72	74	74	76	76	78	78	80	80	80							
0.491−0.510 (0.0193−0.0201)	72	72	74	74	76	76	78	78	80	80	80	80								
0.511−0.530 (0.0201−0.0209)	74	74	76	76	78	78	80	80	80	80	80									
0.531−0.550 (0.0209−0.0217)	76	76	78	78	80	80	80	80	80	80										
0.551−0.570 (0.0217−0.0224)	78	78	80	80	80	80	80	80	80											
0.571−0.590 (0.0225−0.0232)	80	80	80	80	80	80	80													
0.591−0.610 (0.0233−0.0240)	80	80	80	80	80															
0.611−0.630 (0.0241−0.0248)	80	80	80																	
0.631−0.650 (0.0248−0.0256)	80																			
0.651−0.670 (0.0256−0.0264)																				
0.671−0.690 (0.0264−0.0272)																				
0.691−0.710 (0.0272−0.0280)																				
0.711−0.730 (0.0280−0.0287)																				
0.731−0.750 (0.0288−0.0295)																				
0.751−0.770 (0.0296−0.0303)																				
0.771−0.790 (0.0304−0.0311)																				
0.791−0.810 (0.0311−0.0319)																				
0.811−0.830 (0.0319−0.0327)																				
0.831−0.850 (0.0327−0.0335)																				
0.851−0.870 (0.0335−0.0343)																				
0.871−0.890 (0.0343−0.0350)																				
0.891−0.910 (0.0351−0.0358)																				
0.911−0.930 (0.0359−0.0366)																				
0.931−0.950 (0.0367−0.0374)																				
0.951−0.970 (0.0374−0.0382)																				
0.971−0.990 (0.0382−0.0390)																				
0.991−1.010 (0.0390−0.0398)																				
1.011−1.030 (0.0398−0.0406)																				
1.031−1.050 (0.0406−0.0413)																				
1.051−1.070 (0.0414−0.0421)																				
1.071−1.090 (0.0422−0.0429)																				
1.091−1.110 (0.0430−0.0437)																				
1.111−1.130 (0.0437−0.0445)																				
1.131−1.150 (0.0445−0.0453)																				

Exhaust valve clearance (Cold):
0.25 – 0.35 mm (0.010 – 0.014 in.)

EXAMPLE:
The 2.300 mm (0.0906 in.) shim is installed, and the measured clearance is 0.440 mm (0.0173 in.). Replace the 2.300 mm (0.0906 in.) shim with a No. 44 shim.

New shim thickness mm (in.)

Shim No.	Thickness	Shim No.	Thickness	Shim No.	Thickness
00	2.000 (0.0787)	28	2.280 (0.0898)	56	2.560 (0.1008)
02	2.020 (0.0795)	30	2.300 (0.0906)	58	2.580 (0.1016)
04	2.040 (0.0803)	32	2.320 (0.0913)	60	2.600 (0.1024)
06	2.060 (0.0811)	34	2.340 (0.0921)	62	2.620 (0.1031)
08	2.080 (0.0819)	36	2.360 (0.0929)	64	2.640 (0.1039)
10	2.100 (0.0827)	38	2.380 (0.0937)	66	2.660 (0.1047)
12	2.120 (0.0835)	40	2.400 (0.0945)	68	2.680 (0.1055)
14	2.140 (0.0843)	42	2.420 (0.0953)	70	2.700 (0.1063)
16	2.160 (0.0850)	44	2.440 (0.0961)	72	2.720 (0.1071)
18	2.180 (0.0858)	46	2.460 (0.0969)	74	2.740 (0.1079)
20	2.200 (0.0866)	48	2.480 (0.0976)	76	2.760 (0.1087)
22	2.220 (0.0874)	50	2.500 (0.0984)	78	2.780 (0.1094)
24	2.240 (0.0882)	52	2.520 (0.0992)	80	2.800 (0.1102)
26	2.260 (0.0890)	54	2.540 (0.1000)		

7924SG72

Fig. 102 Exhaust valve clearance shim selection chart

ENGINE PERFORMANCE & EMISSION CONTROL

COMPONENT LOCATIONS

See Figures 103 and 104.

ACCELERATOR PEDAL POSITION (APP) SENSOR

LOCATION

See Figure 105.

REMOVAL & INSTALLATION

1. Disconnect a accelerator pedal connector.
2. Remove the 2 nuts and accelerator pedal assembly.

To install:

➡Be care not to give a shock to the accelerator pedal assembly. Be care not to disassemble the accelerator pedal assembly.

3. Install the accelerator pedal assembly with the 2 nuts. Tighten nuts to 44 inch lbs. (5.0 Nm)
4. Connect a accelerator pedal connector.

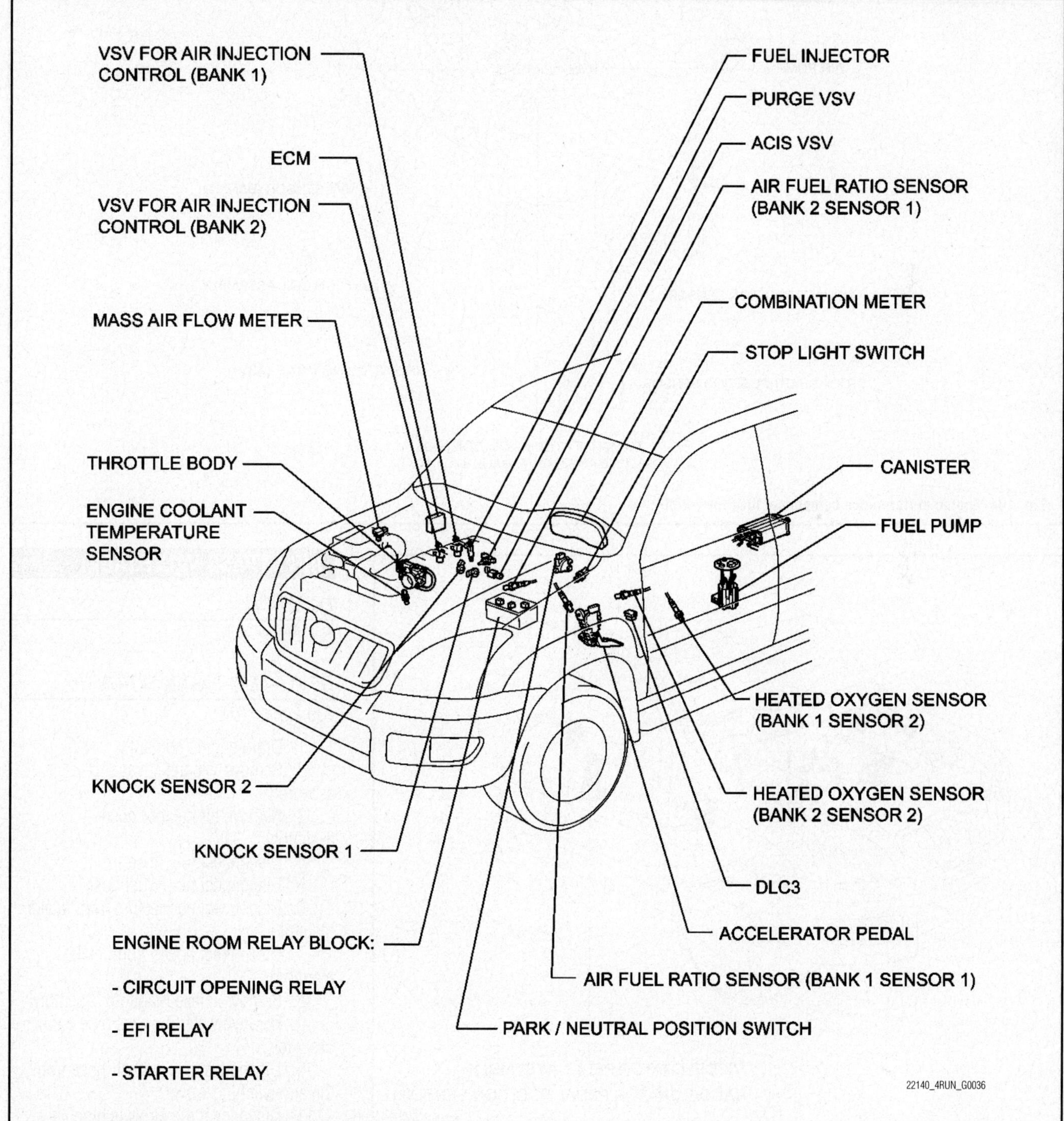

Fig. 103 Engine performance component locations–1 of 2

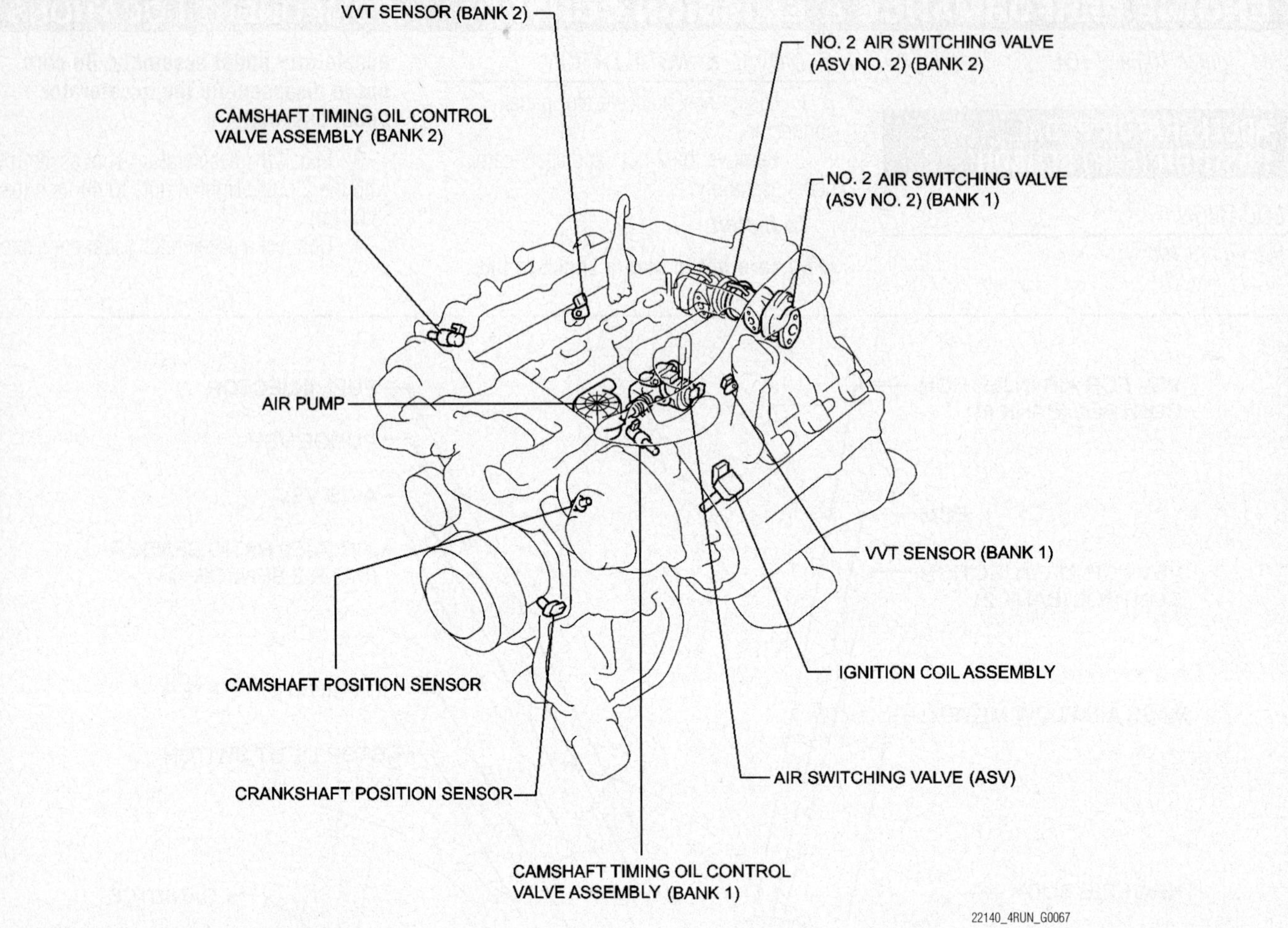

Fig. 104 Engine performance component locations–2 of 2

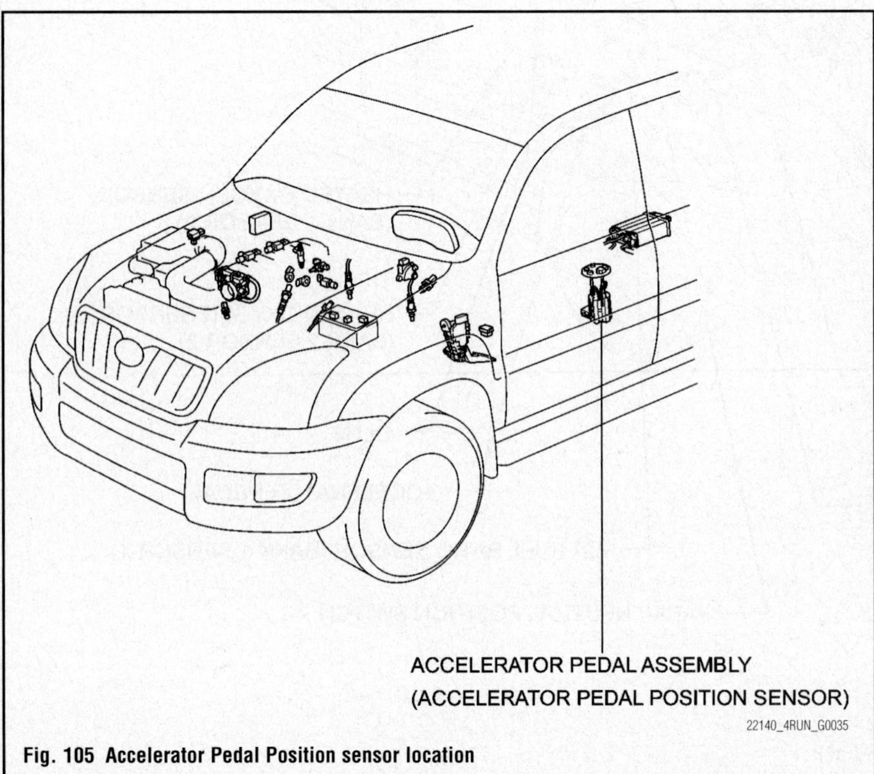

Fig. 105 Accelerator Pedal Position sensor location

AIR BYPASS VALVE (ABV)

LOCATION

See Figure 106.

REMOVAL & INSTALLATION

See Figure 107.

1. Drain engine coolant.
2. Remove V-bank cover sub-assembly.
3. Remove air cleaner hose assembly.
4. Disconnect fuel hose.
5. Disconnect no. 2 fuel hose.
6. Disconnect connectors from intake manifold.
7. Disconnect hoses from intake manifold.
8. Remove intake manifold assembly.
9. Remove water by-pass pipe sub-assembly.
10. Disconnect the no.2 air hose from the air switching valve.
11. Disconnect the air switching valve connector.

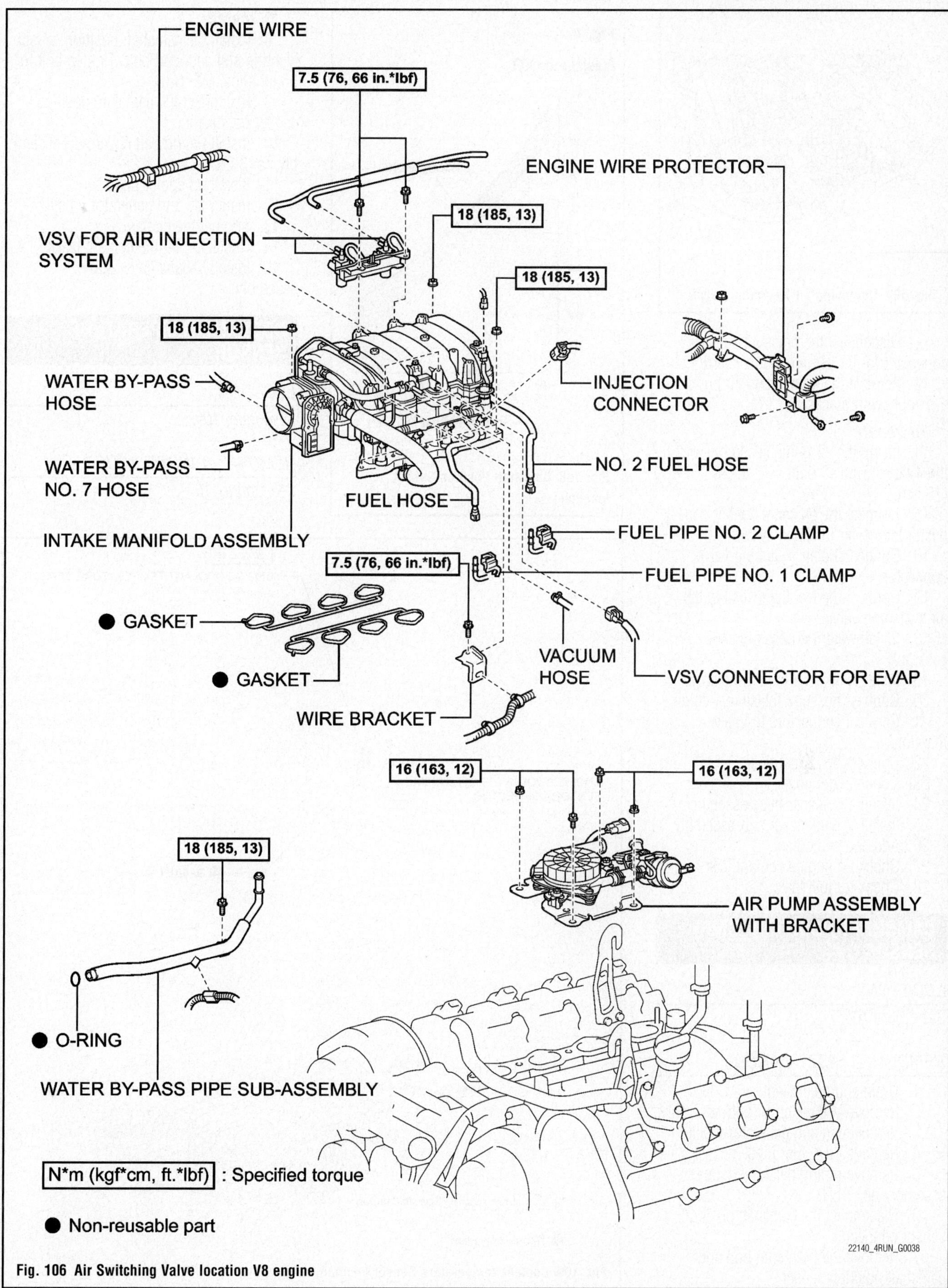

Fig. 106 Air Switching Valve location V8 engine

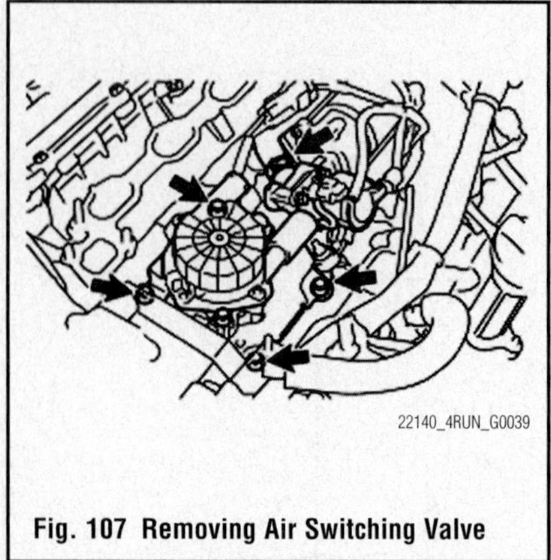

Fig. 107 Removing Air Switching Valve

12. Disconnect the pressure sensor connector for the air injection system.
13. Remove the 4 bolts and air pump assembly with bracket.

To install:

14. Install the air pump assembly with the 4 bolts. Tighten bolts to 7 ft. lbs. (10 Nm)
15. Connect the pressure sensor connector for the air injection system.
16. Connect the air switching valve connector.
17. Connect the No. 2 air hose to the air switching valve.
18. Install water by-pass pipe sub-assembly.
19. Install intake manifold assembly.
20. Connect hoses to intake manifold.
21. Connect connectors to intake manifold.
22. Connect no. 2 fuel hose.
23. Connect fuel hose.
24. Install air cleaner hose assembly.
25. Install V-bank cover sub-assembly.
26. Add engine coolant.
27. Check for engine coolant leak.
28. Check for fuel leak.

CAMSHAFT POSITION (CMP) SENSOR

LOCATION

See Figure 108.

REMOVAL & INSTALLATION

1. Drain engine coolant.
2. Remove V-bank cover sub-assembly.
3. Remove fan and generator v belt.
4. Remove oil cooler pipe.
5. Remove timing belt cover sub-assembly no. 3 LH.
6. Disconnect the camshaft position sensor connector.
7. Remove the bolt, stud bolt and camshaft position sensor.

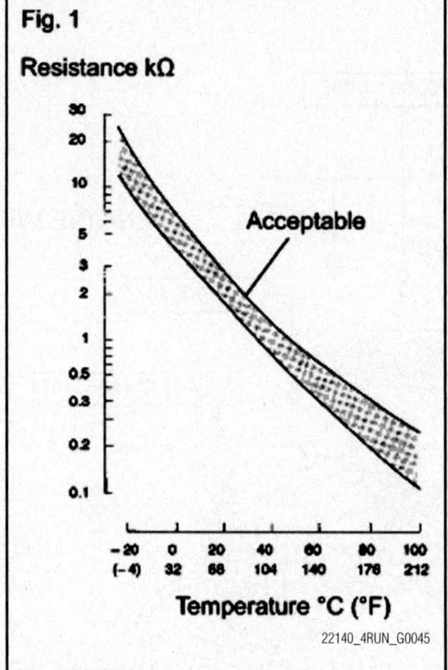

Fig. 108 Camshaft Position Sensor location

To install:

8. Install the camshaft position sensor with the bolt and stud bolt. Tighten bolt to 66 inch lbs. (7.5 Nm).
9. Reconnect the camshaft position sensor connector.
10. Install timing belt cover sub-assembly no. 3 LH.
11. Install oil cooler pipe.
12. Install fan and generator v belt.
13. Add engine coolant.
14. Check for engine coolant leaks.
15. Install V-bank cover sub-assembly.

COOLANT TEMPERATURE SENSOR

LOCATION

See Figure 109.

REMOVAL & INSTALLATION

See Figure 110.

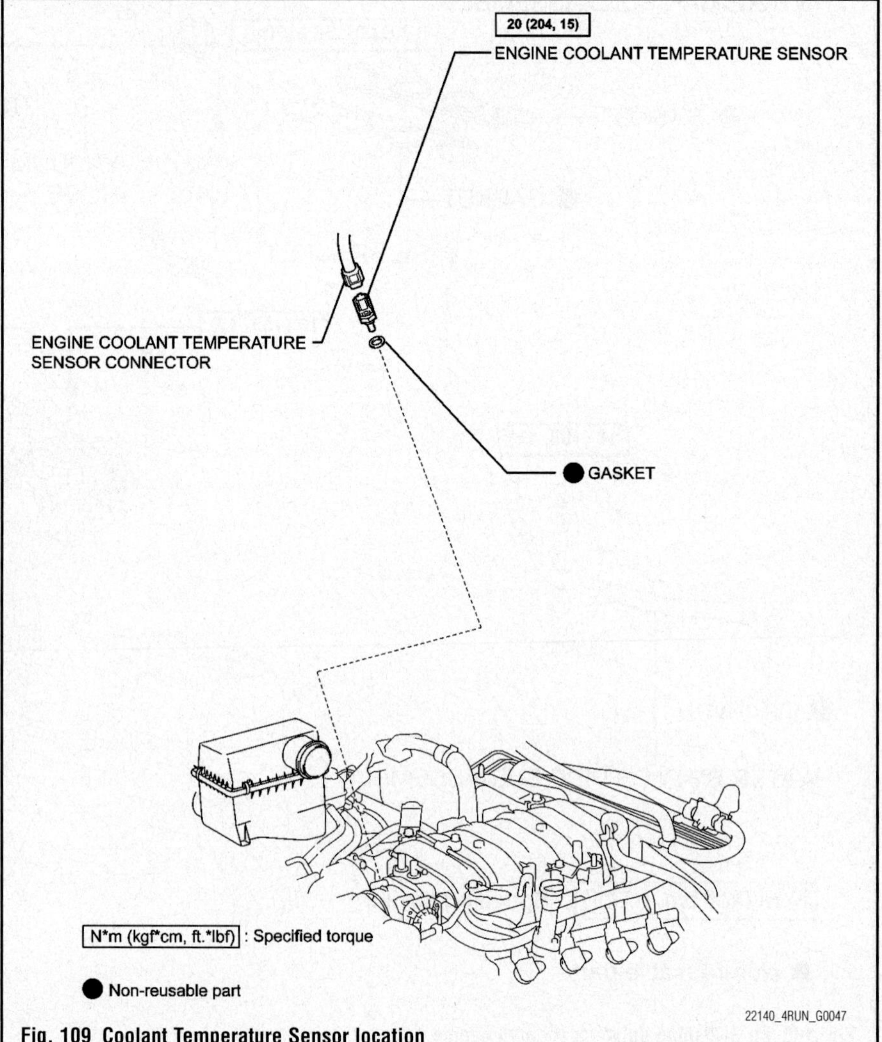

Fig. 109 Coolant Temperature Sensor location

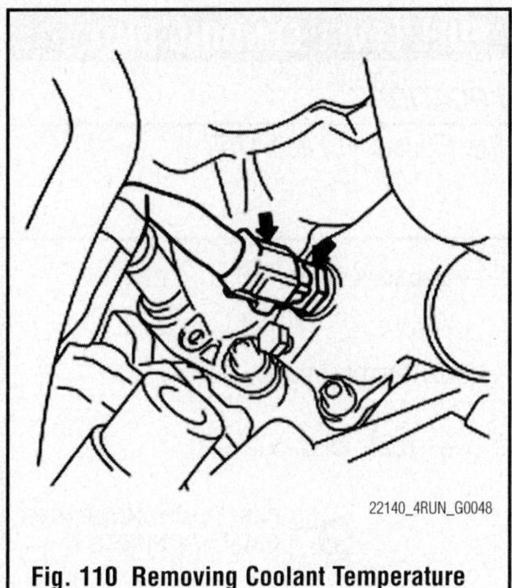

Fig. 110 Removing Coolant Temperature Sensor

1. Disconnect cable from negative battery terminal.

➡Wait at least 90 seconds after disconnecting the cable from the negative (-) battery terminal to prevent airbag and seat belt pretensioner activation.

2. Drain engine coolant.
3. Remove V-bank cover sub-assembly.
4. Remove intake air connector pipe.
5. Remove throttle body.
6. Disconnect the sensor connector.
7. Remove the sensor.
8. Remove the gasket from the sensor.
9. Install a new gasket to the sensor.
10. Install the sensor. Tighten to 15 ft. lbs. (20 Nm).
11. Install throttle body.
12. Install intake air connector pipe.
13. Install V-bank cover sub-assembly.
14. Connect cable to negative battery terminal.
15. Add engine coolant.
16. Check for engine coolant leaks.
17. Perform initialization, if necessary.

TESTING

See Figure 111.

➡If inspecting the water temperature sensor in water, prevent water from flowing into the terminals. After inspection, wipe the water off the sensor.

1. Using an ohmmeter, measure the resistance between the terminals.
   a. 2.32 to 2.59 kohms at 68°F (20°C).
   b. 0.310 to 0.326 kohms at 176°F (80°C).
2. If the result is not as specified, replace the sensor.

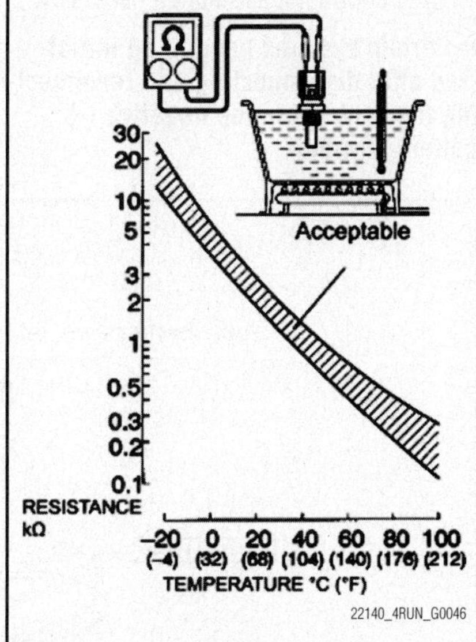

Fig. 111 Testing Coolant Temperature Sensor

CRANKSHAFT POSITION (CKP) SENSOR

LOCATION

See Figure 112.

REMOVAL & INSTALLATION

1. Disconnect cable from negative battery terminal.

➡Wait at least 90 seconds after disconnecting the cable from the negative (-) battery terminal to prevent airbag and seat belt pretensioner activation.

2. Remove No. 1 engine under cover.
3. For V6 engine, remove A/C compressor.
4. Disconnect the sensor connector.
5. Remove the bolt and sensor.

To install:

6. Install crankshaft position sensor.

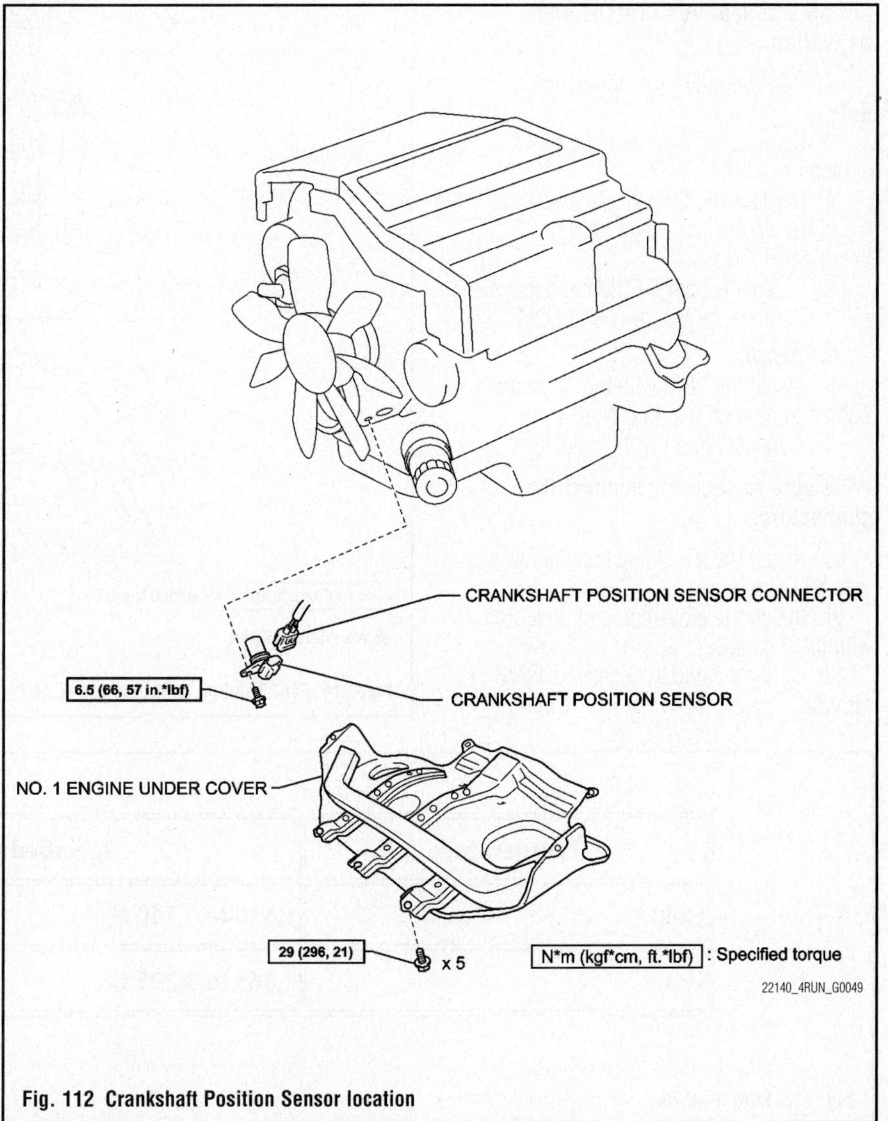

Fig. 112 Crankshaft Position Sensor location

7. Install the sensor with the bolt. Tighten bolt to 57 inch lbs. (6.5 nm).
8. Connect the sensor connector.
9. For V6 engine, install A/C compressor and charge system.
10. Install no. 1 engine under cover.
11. Connect cable to negative battery terminal.
12. Perform initialization, if necessary.

ELECTRONIC CONTROL MODULE (ECM)

LOCATION

See Figure 113.

REMOVAL & INSTALLATION

1. Disconnect cable from negative battery terminal.

➡Wait at least 90 seconds after disconnecting the cable from the negative (-) battery terminal to prevent airbag and seat belt pretensioner activation.

2. Remove glove box compartment door.
3. Remove the 2 screws and glove compartment door.
4. Remove no. 2 finish panel lower.
5. Remove the 3 screws and no. 2 finish panel lower.
6. Disconnect the 5 ECM connectors.
7. Remove the 3 screws and ECM.

To install:

8. Install the ECM with the 3 screws, tighten to 49 inch lbs. (5.5 Nm).
9. Connect the 5 ECM connectors.

➡Be sure to securely connect the connectors.

10. Install the No. 2 finish panel lower with the 3 screws.
11. Install the glove compartment door with the 2 screws.
12. Connect cable to negative battery terminal.
13. Perform initialization, if necessary.

➡Certain systems need to be initialized after disconnecting and reconnecting the cable from the negative (-) battery terminal.

FUEL LEVEL SENDING UNIT

LOCATION

See Figures 114 and 115.

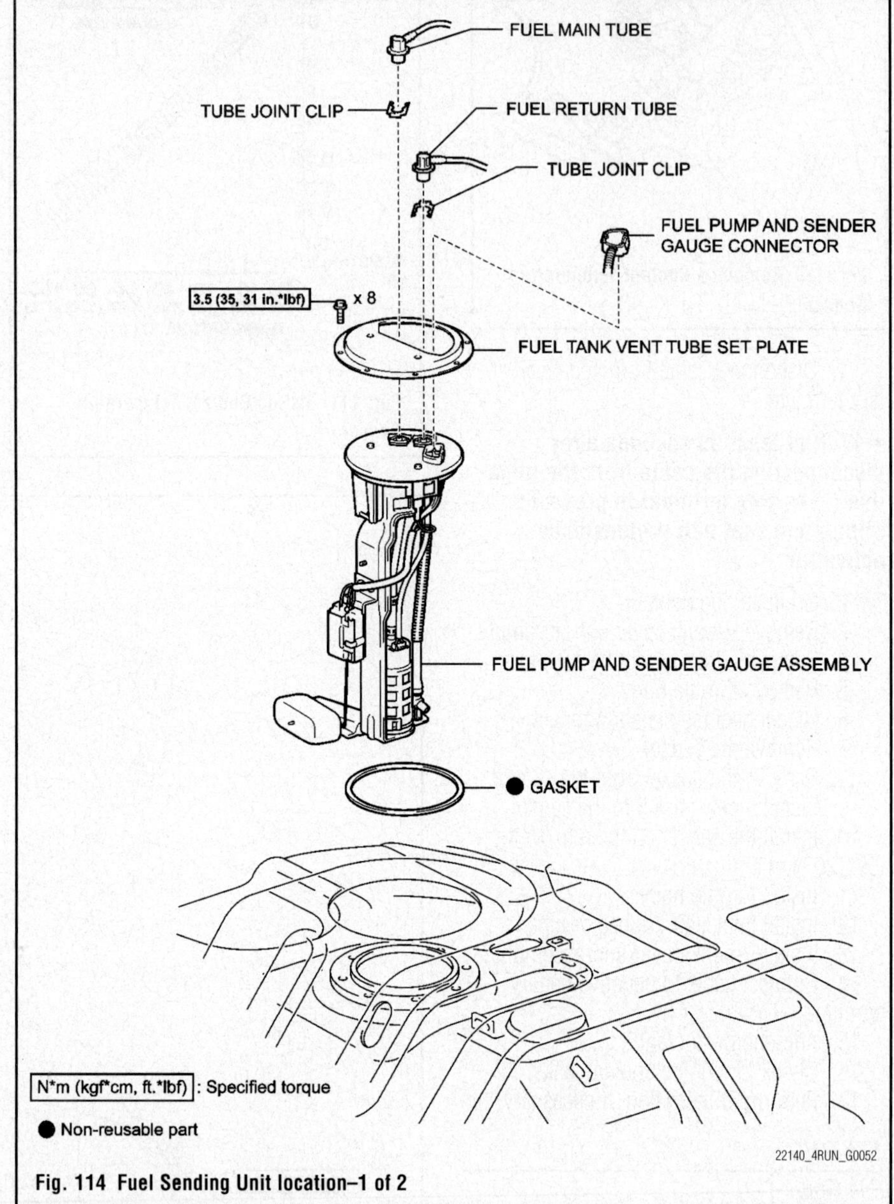

Fig. 114 Fuel Sending Unit location–1 of 2

Temperature	Specified Condition
Cold	1,630 to 2,740 Ω
Hot	2,065 to 3,225 Ω

22140_4RUN_G0050

Fig. 113 ECM location

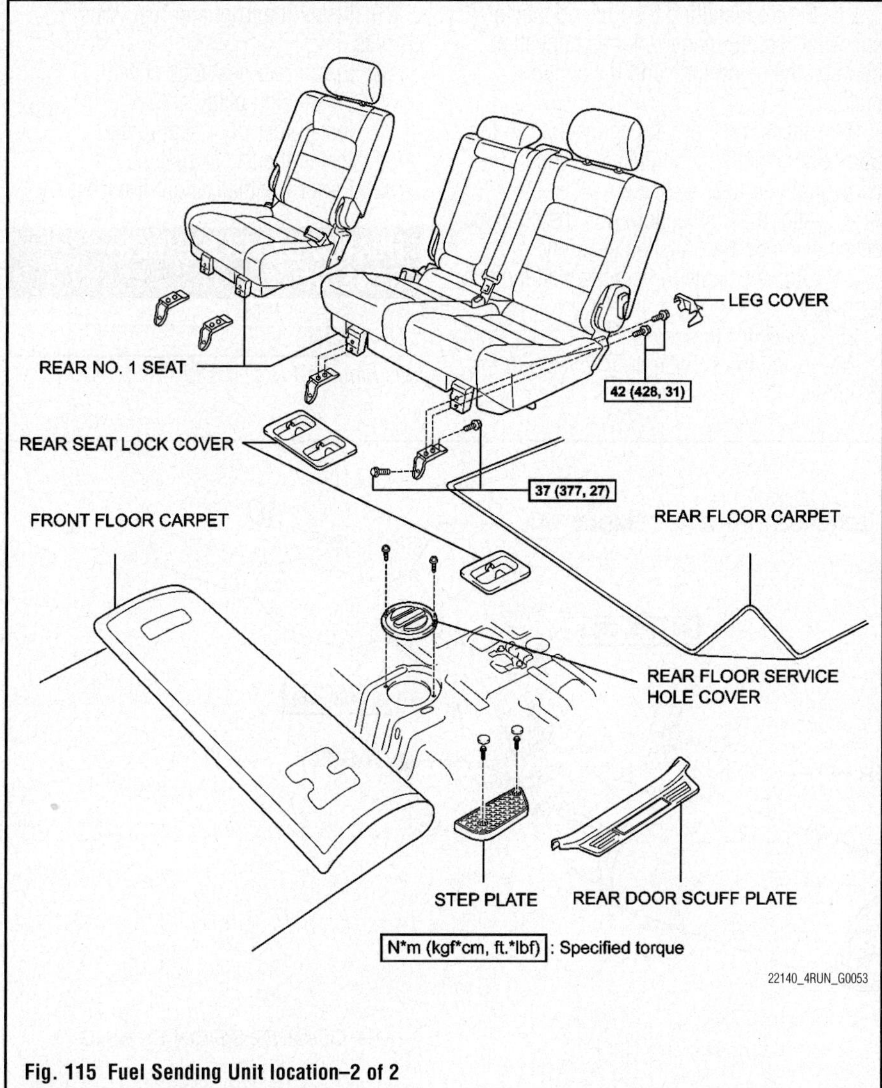

Fig. 115 Fuel Sending Unit location–2 of 2

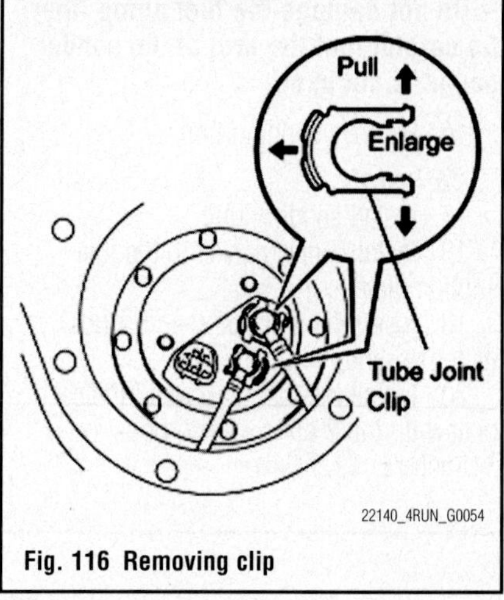

Fig. 116 Removing clip

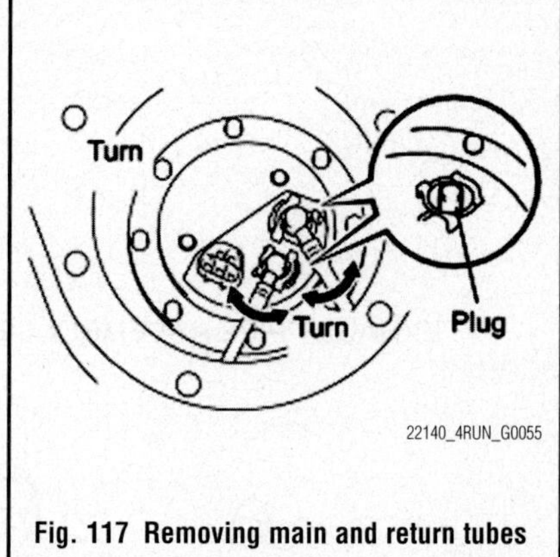

Fig. 117 Removing main and return tubes

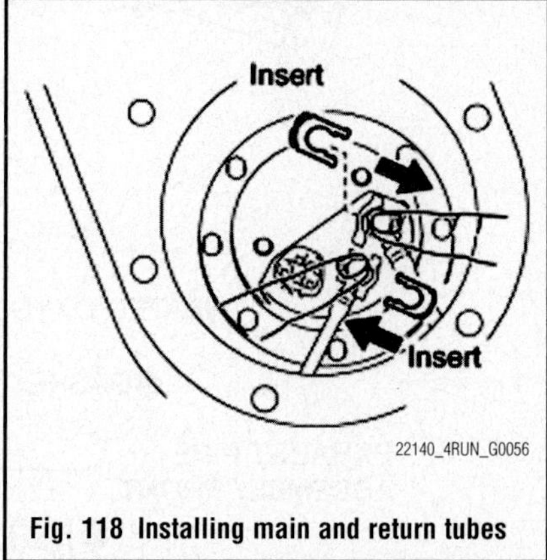

Fig. 118 Installing main and return tubes

REMOVAL & INSTALLATION

See Figures 116 through 118.

**** CAUTION**

Do not smoke or work near an open flame when working on the fuel pump.

1. Discharge fuel system pressure.
2. Disconnect cable from negative battery terminal

**** CAUTION**

Wait at least 90 seconds after disconnecting the cable from the negative (-) battery terminal to prevent airbag and seat belt pretensioner activation.

3. Remove the 2 rear seats.
4. Remove rear door scuff plate.
5. Remove step plate.
6. Remove rear seat lock cover.
7. Take off the front and rear floor carpets.
8. Remove the 2 screws and floor service hole cover.

**** CAUTION**

Prevent the retained pressure in the fuel line from splashing inside the vehicle compartment. When sealing the tube and suction plates with the O-ring of the quick connector, be careful not to damage any contact surfaces or allow foreign matter to contact any surface. Be sure to perform the disconnection by hand. Do not use tools. Do not bend or turn the nylon tube by force.

9. Disconnect fuel main tube and return tube.
10. Before the operation, remove foreign matter or dirt sticking to the tube joint clips.
11. Widen the tip of the clips with your fingers and pull them out for disconnection.
12. Pull out the fuel main tube and the return tube. If the nylon tube and the suction plate stick together, turn the nylon tube with your fingers and pull it out for disconnection.
13. After the disconnection, protect the connector with a plastic bag.
14. Remove the 8 bolts.
15. Pull out the fuel pump and sender gauge assembly.

➡Do not damage the fuel pump filter. Be careful that the arm of the sender gauge is not bent.

16. Remove sending unit.

To install:

17. Install sending unit.
18. Install a new gasket to the fuel suction plate.
19. Insert the fuel pump and sender gauge assembly into the fuel tank.
20. Install the fuel tank vent tube set plate with the 8 bolts tighten bolts to 31 inch lbs. (3.5 Nm).
21. Before installing the tube connectors, check for foreign matter on the connection between the nylon tube and the suction plate.
22. Attach the fuel tube connectors to the ports of the fuel suction plate and insert the clips until you hear a click.
23. After the connection, pull the clips to check that they are installed securely.
24. Connect cable to negative battery terminal.
25. Check for fuel leaks.
26. Install the service hole cover with the 2 screws.
27. Install the front and rear floor carpets.
28. Install rear seat lock cover.
29. Install step plate.
30. Install rear door scuff plate.
31. Install the 2 rear seats.
32. Perform initialization, if necessary.

HEATED OXYGEN (HO2S) SENSOR

LOCATION

See Figures 119 and 120.

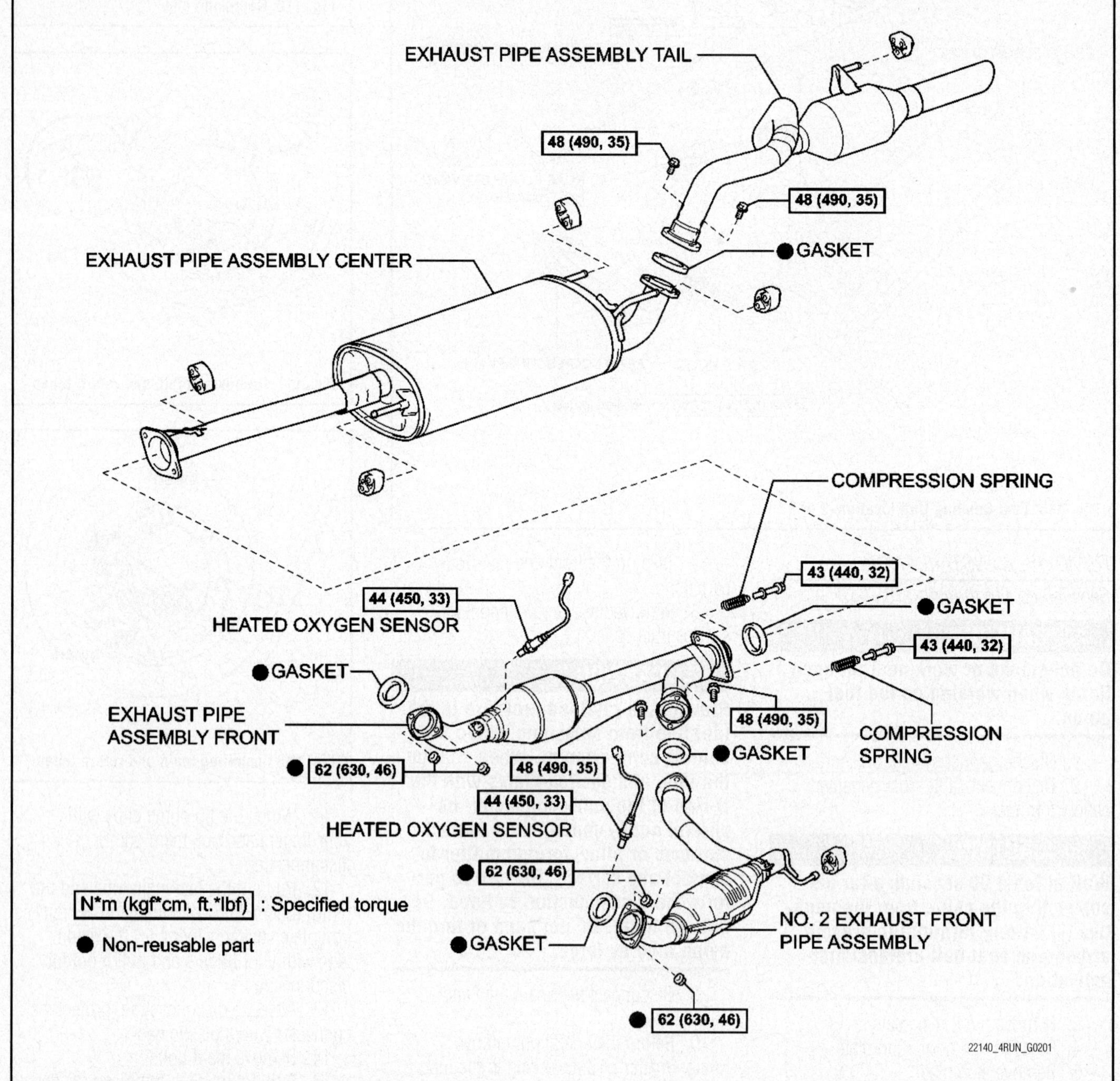

Fig. 119 Heated Oxygen Sensor location–V6 engine

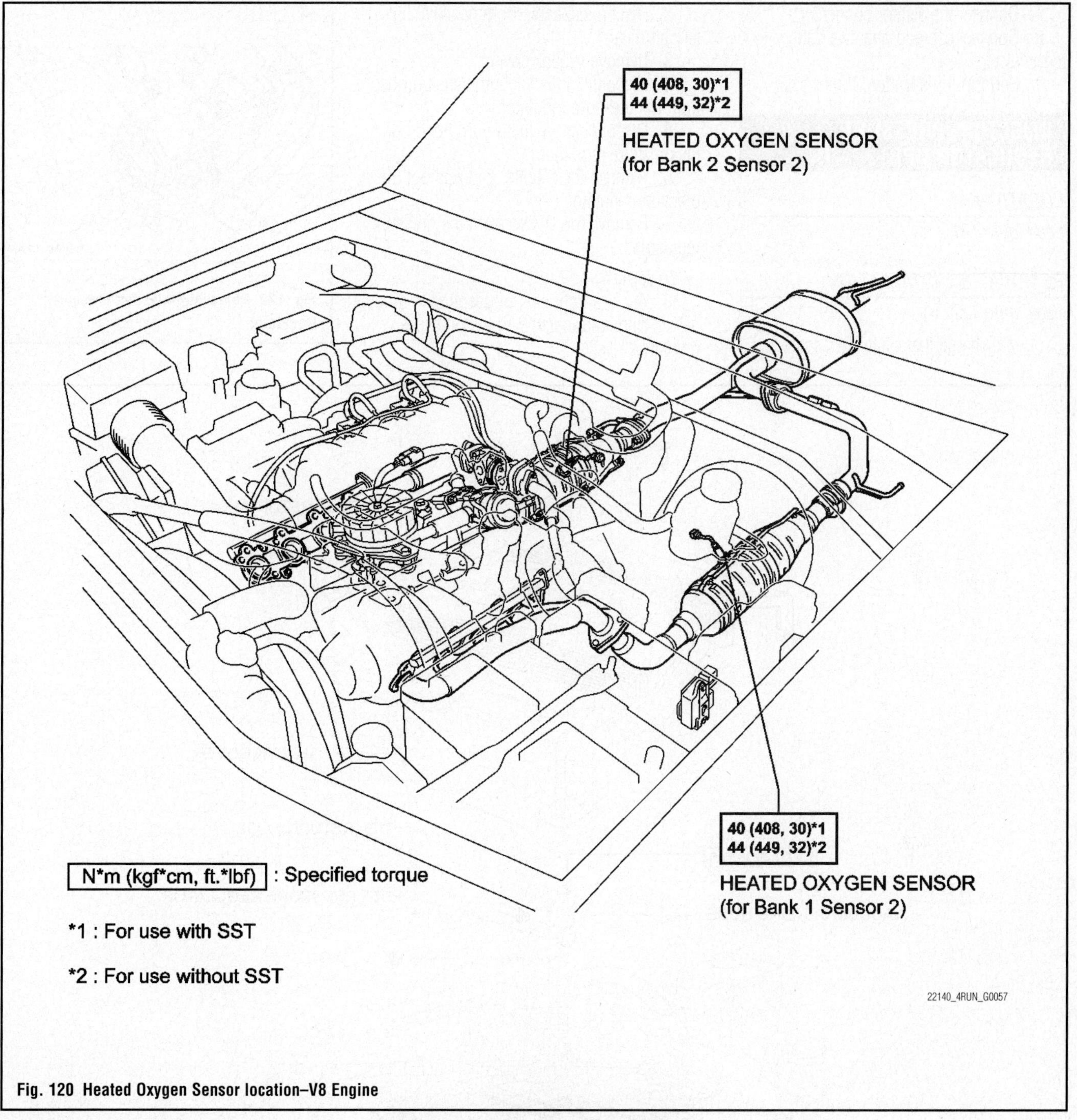

Fig. 120 Heated Oxygen Sensor location–V8 Engine

REMOVAL & INSTALLATION

**** CAUTION**

Wear protective gloves when removing the sensor. The exhaust pipe assembly is extremely hot immediately after the engine has stopped. Confirm that the exhaust pipe assembly has cooled down before removing it.

1. Disconnect cable from negative battery terminal.

**** CAUTION**

Wait at least 90 seconds after disconnecting the cable from the negative (-) battery terminal to prevent airbag and seat belt pretensioner activation.

2. Disconnect the sensor connector and remove heated oxygen sensor (for Bank 1 Sensor 2).
3. Disconnect the sensor connector and remove heated oxygen sensor (for Bank 2 Sensor 2).

To install:

4. Install heated oxygen sensor (for Bank 1 Sensor 2) and tighten to 32 ft. lbs. (44 Nm).

➡Use a torque wrench with a fulcrum length of 30 cm (11.81 in).

5. Connect the sensor connector.
6. Install heated oxygen sensor (for Bank 2 Sensor 2) and tighten to 32 ft. lbs. (44 Nm).

➡Use a torque wrench with a fulcrum length of 30 cm (11.81 in).

7. Connect the sensor connector.
8. Connect cable to negative battery terminal.
9. Perform initialization, if necessary.

INJECTION PRESSURE REGULATOR (IPR)

LOCATION

See Figure 121.

REMOVAL & INSTALLATION

See Figure 122.

1. Discharge fuel system pressure.
2. Disconnect cable from negative battery terminal.
3. Remove v-bank cover.
4. Disconnect the vacuum sensing hose from the pressure regulator.
5. Disconnect the fuel return hose from the pressure regulator
6. Remove the 2 bolts, and pull out the pressure regulator.
7. Remove the O-ring from the pressure regulator.

To install:

8. Apply a light coat of gasoline to a new O-ring, and install it to the pressure regulator.

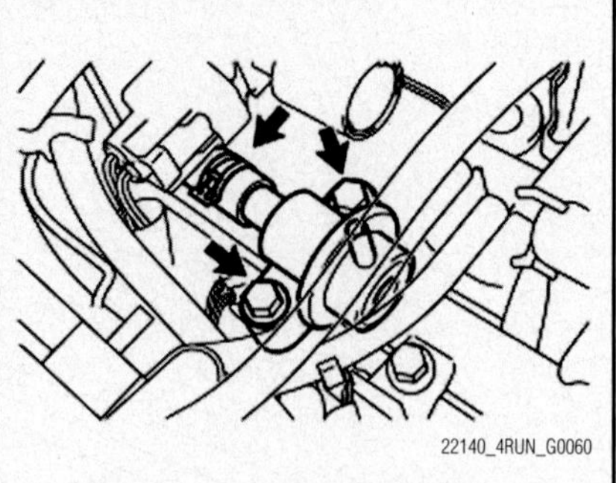

Fig. 122 Removing Injection Pressure Regulator

FUEL RETURN HOSE
V-BANK COVER
FUEL RETURN HOSE
7.5 (77, 66 in.*lbf)
FUEL PRESSURE REGULATOR
VACUUM SENSING HOSE
● O-RING
N*m (kgf*cm, ft.*lbf) : Specified torque
● Non-reusable part
22140_4RUN_G0059

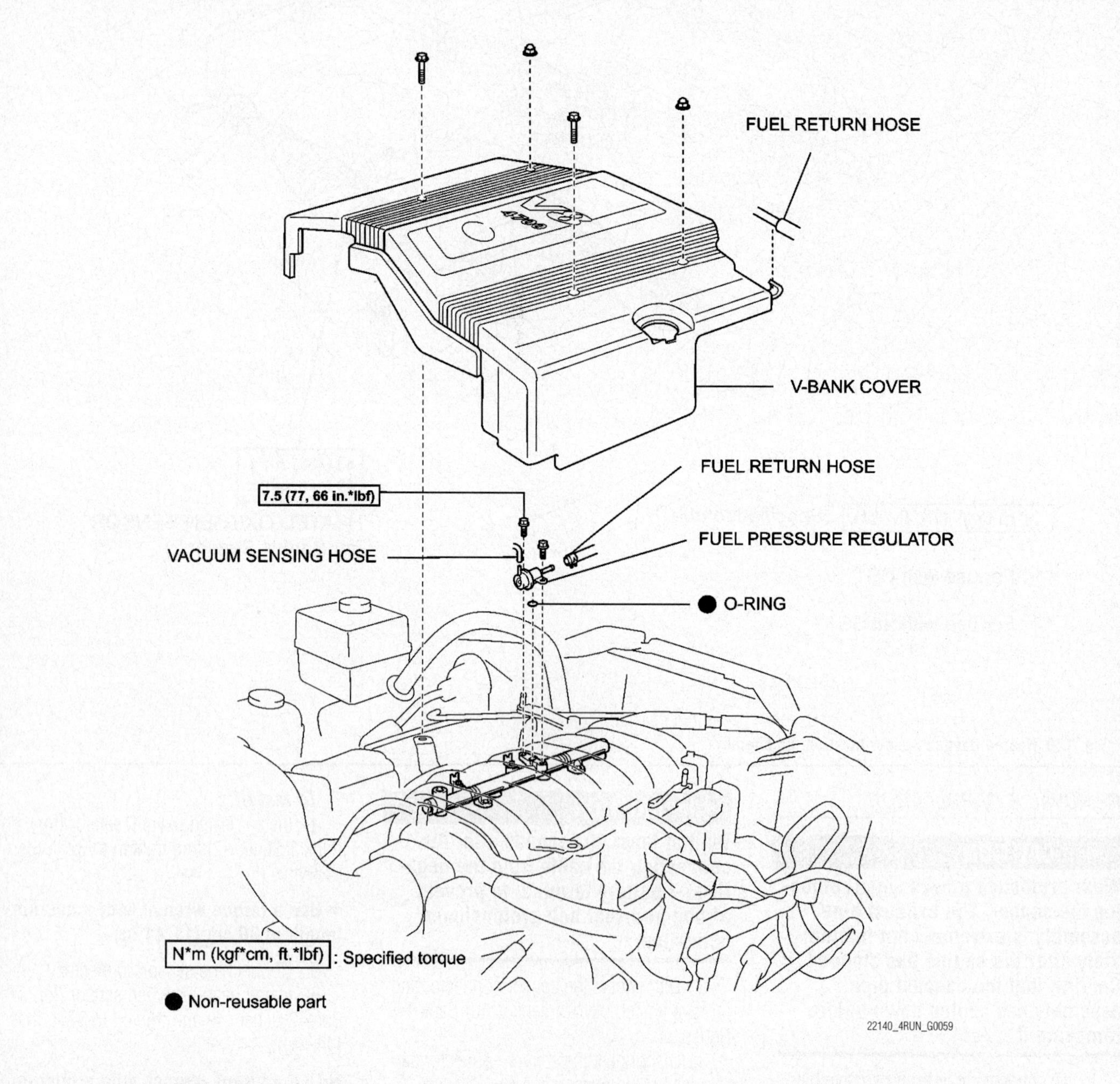

Fig. 121 Injection Pressure Regulator location

9. While turning the pressure regulator left and right, install it to the delivery pipe.

10. Install the pressure regulator with the 2 bolts and tighten bolts to 66 inch lbs. (7.5 Nm).

INTAKE AIR TEMPERATURE (IAT) SENSOR

LOCATION

The Intake Air Temperature (IAT) sensor, built into the Mass Air Flow (MAF) meter.

REMOVAL & INSTALLATION

See Mass Air Flow Meter

KNOCK SENSOR (KS)

LOCATION

See Figure 123.

REMOVAL & INSTALLATION

V6 Engine

See Figure 124.

1. Remove cylinder head sub-assembly.
2. Disconnect the heater water inlet hose.
3. Remove the 4 wire harness clamps.
4. Remove the 3 bolts and the water outlet pipe.
5. Disconnect the 2 knock sensor connectors.
6. Remove the 2 bolts and the 2 knock sensors.

To install:

7. Install the 2 knock sensors with the 2 bolts and tighten to 15 ft. lbs. (20 Nm).
8. Connect the 2 knock sensor connectors.
9. Install the 3 bolts and the water outlet pipe and tighten to 7 ft. lbs. (10 Nm).

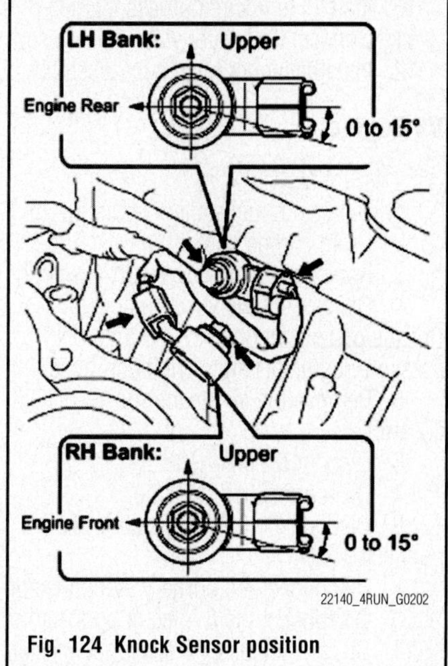

Fig. 124 Knock Sensor position

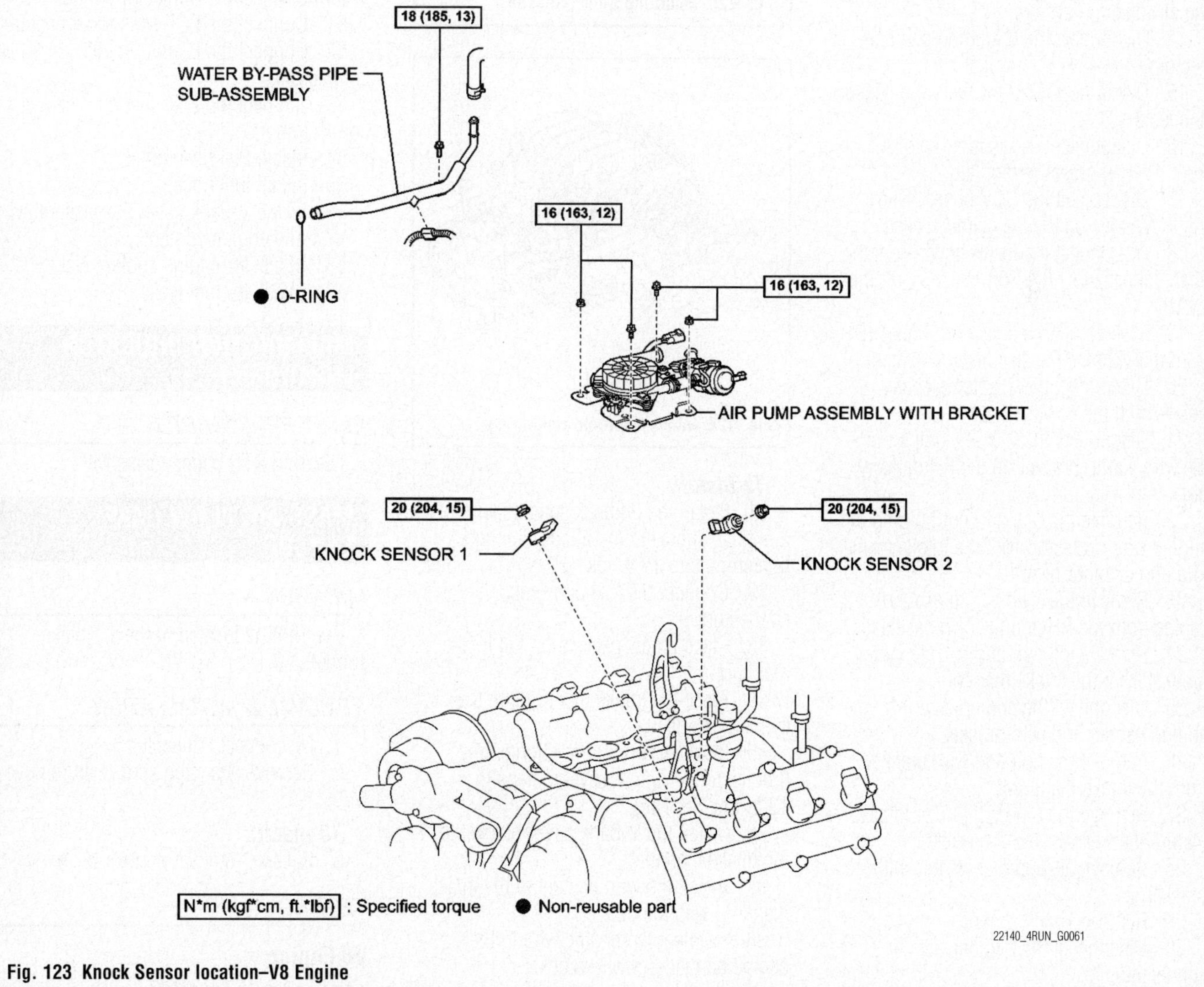

18 (185, 13)
WATER BY-PASS PIPE SUB-ASSEMBLY
● O-RING
16 (163, 12)
16 (163, 12)
AIR PUMP ASSEMBLY WITH BRACKET
20 (204, 15)
KNOCK SENSOR 1
20 (204, 15)
KNOCK SENSOR 2
N*m (kgf*cm, ft.*lbf) : Specified torque ● Non-reusable part
22140_4RUN_G0061

Fig. 123 Knock Sensor location–V8 Engine

10. Install the 4 wire harness clamps.
11. Connect the heater water inlet hose.
12. Install cylinder head sub-assembly.

V8 Engine

See Figures 125 and 126.

1. Discharge fuel system pressure.
2. Drain engine coolant.
3. Remove V-bank cover sub-assembly.
4. Disconnect the vacuum hoses (for the power steering idle-up and fuel pressure regulator) and ventilation hose.
5. Remove the air cleaner hose assembly.
6. Disconnect fuel hose.
7. Disconnect fuel hose no. 2.
8. Disconnect the throttle control connector.
9. Disconnect the purge VSV connector.
10. Disconnect the 8 injector connectors.
11. Disconnect the ECT sensor connector.
12. Disconnect the 8 ignition coil connectors.
13. Disconnect the 2 VSV connectors for the air injection system.
14. Disconnect the 8 ignition coil connectors.
15. Disconnect the 2 air fuel ratio sensor connectors.
16. Disconnect the vacuum hose [A] from the fuel pressure regulator.
17. Disconnect the PCV hoses [B] from the PCV valve on the LH cylinder head.
18. Disconnect the EVAP hose (from the charcoal canister) [C] from the VSV for the EVAP.
19. Disconnect the 2 vacuum hoses [D] from the VSV for the air injection system.
20. Disconnect the 2 water by-pass hoses from the throttle body.
21. Disconnect the 2 wire clamps from the wire clamp bracket on the RH delivery pipe.
22. Remove the bolt and nut holding the engine wire protector from the intake manifold and cylinder head.
23. Remove the 2 bolts and ground cables from the RH and LH cylinder heads.
24. Remove the bolt and V-bank cover bracket from the intake manifold.
25. Disconnect the engine wire from the engine hanger and wire bracket.
26. Remove the bolt and wire bracket from the intake manifold.
27. Remove the 6 bolts, 4 nuts, intake manifold assembly and 2 gaskets.
28. Remove air pump assembly with bracket.
29. Remove knock sensor.
30. Disconnect the 2 knock sensor connectors.

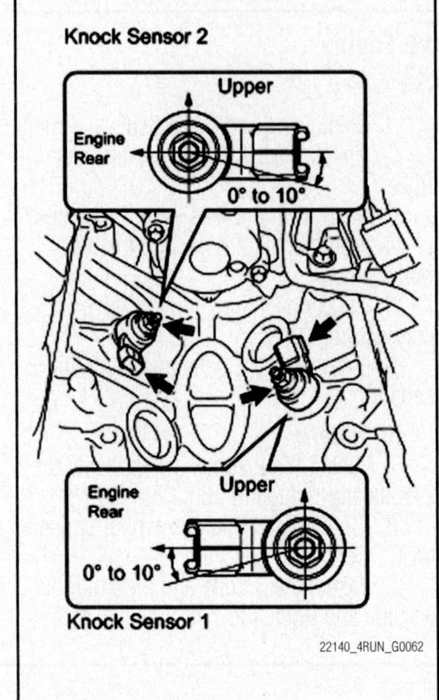

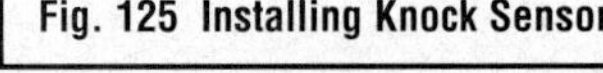
Fig. 125 Installing Knock Sensor

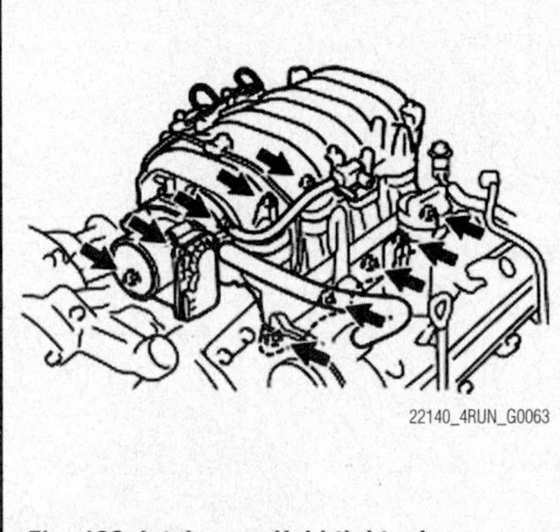

Fig. 126 Intake manifold tightening

To install:

31. Install the 2 knock sensors with the 2 nuts as shown in the illustration and tighten nuts to 15 ft. lbs. (20 Nm).
32. Connect the 2 knock sensor connectors.
33. Place 2 new gaskets on the intake manifold.
34. Place the intake manifold on the cylinder heads.
35. Install and uniformly tighten the 6 bolts and 4 nuts in several steps to 13 ft. lbs. (18 Nm).
36. Install the V-bank cover bracket to the intake manifold.
37. Install the wire bracket to the intake manifold with the bolt.
38. Connect the engine wire to the engine hanger and wire bracket.
39. Connect the wire protector to the intake manifold and cylinder heads with the bolt and nut.
40. Install the 2 ground cables with the 2 bolts to the RH and LH cylinder heads.
41. Connect the 2 water by-pass hoses to the throttle body.
42. Connect the 2 wire clamps to the wire clamp bracket on the RH delivery pipe.
43. Connect the vacuum hose to the fuel pressure regulator.
44. Connect the PCV hose to the PCV valve on the LH cylinder head.
45. Connect the EVAP hose (from the charcoal canister) to the purge VSV.
46. Connect the 2 vacuum hoses to the VSV for the air injection system.
47. Connect the throttle control connector.
48. Connect the 2 VSV connectors for the air injection system.
49. Connect the purge VSV connector.
50. Connect the 8 injector connectors.
51. Connect the ECT sensor connector.
52. Connect the 8 ignition coil connectors.
53. Connect the 2 air fuel ratio sensor connectors.
54. Install fuel hose no. 2.
55. Install fuel hose.
56. Install V-bank cover sub-assembly.
57. Add engine coolant.
58. Check for engine coolant leaks.
59. Check for fuel leaks.

MALFUNCTION INDICATOR LIGHT (MIL)

RESET PROCEDURES

Clearing DTC codes resets MIL.

MASS AIR FLOW (MAF) SENSOR

LOCATION

The MAF is located in the air intake snorkel.

REMOVAL & INSTALLATION

1. Disconnect connector.
2. Remove attaching screws an remove MAF.

To install:

3. Reverse removal procedure.

TESTING

V8 Engine

See Figures 127 and 128.

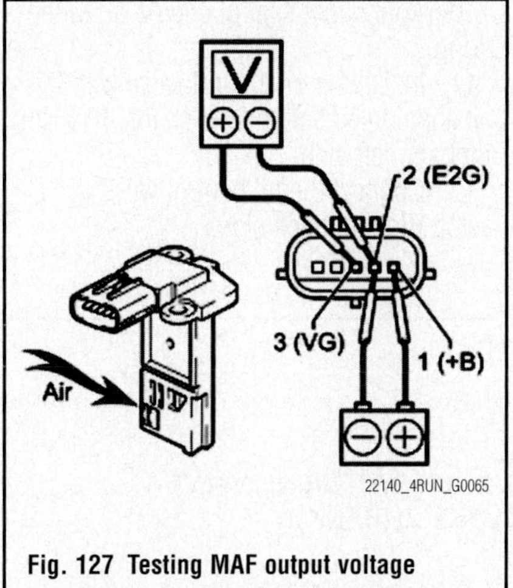

Fig. 127 Testing MAF output voltage

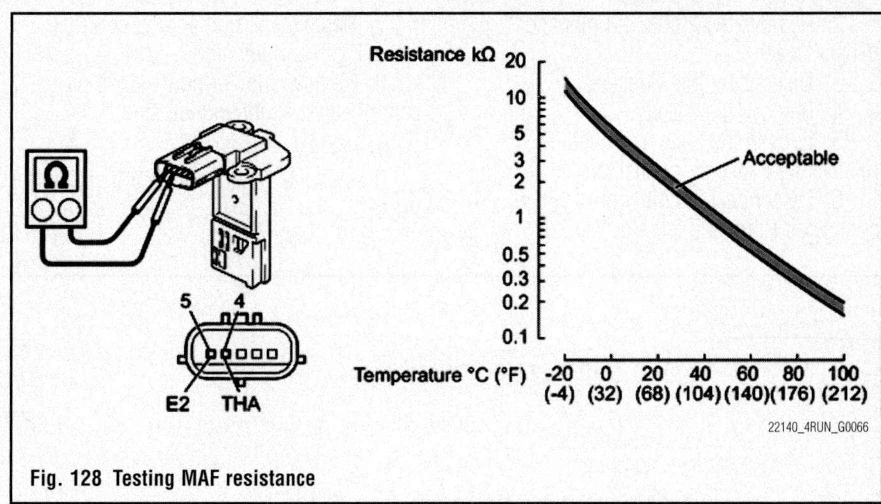

Fig. 128 Testing MAF resistance

1. Apply battery voltage across terminals 1 (+B) and 2 (E2G).
2. Connect the positive (+) tester probe to terminal 3 (VG), and the negative (-) tester probe to terminal 2 (E2G).
3. Blow air into the MAF meter, and check if the voltage fluctuates.
4. Using an ohmmeter, measure the resistance between terminals 4 (THA) and 5 (E2).
   a. Standard resistance:
      - 12.5 to 16.9 kohms at -20°C (-4°F)
      - 2.19 to 2.67 kohms at 20°C (68°F)
      - 0.50 to 0.68 kohms at 60°C (140°F)

THROTTLE POSITION SENSOR (TPS)

LOCATION

See Figure 129.

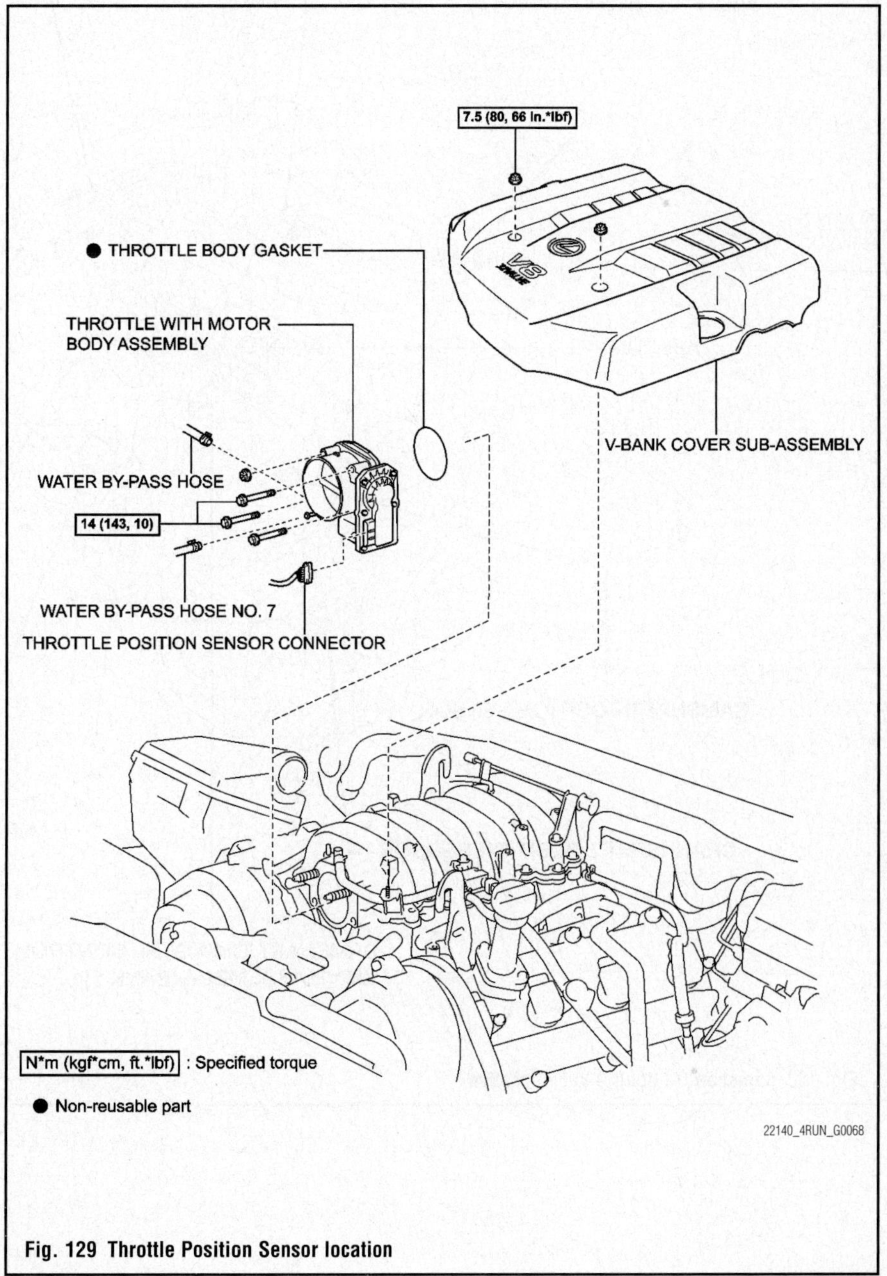

Fig. 129 Throttle Position Sensor location

REMOVAL & INSTALLATION

1. Disconnect connector.
2. Remove attaching screws and remove TPS.

To install:

3. Reverse removal procedure.

VARIABLE CAMSHAFT TIMING OIL CONTROL SOLENOID

LOCATION

See Figure 130.

REMOVAL & INSTALLATION

V8 Engine

1. Disconnect cable from negative battery terminal.

2. Remove the 2 nuts, 2 bolts and V-bank cover.
3. Disconnect the 3 hoses.
4. Remove the 2 bolts.
5. Loosen the 2 clamp bolts and remove the intake air connector.
6. Disconnect the oil control valve connector. Right and left side.
7. Remove the bolt and oil control valve. Right and left side.
8. Remove the O-ring from the oil control valve. Right and left side.

To install:

9. Install a new O-ring to the oil control valve.
10. Apply light coat of engine oil to the O-ring.
11. Install the oil control valve with the bolt and tighten bolt to 66 inch lbs. (7.5 Nm). Right and left side.
12. Connect the oil control valve connector. Right and left side.

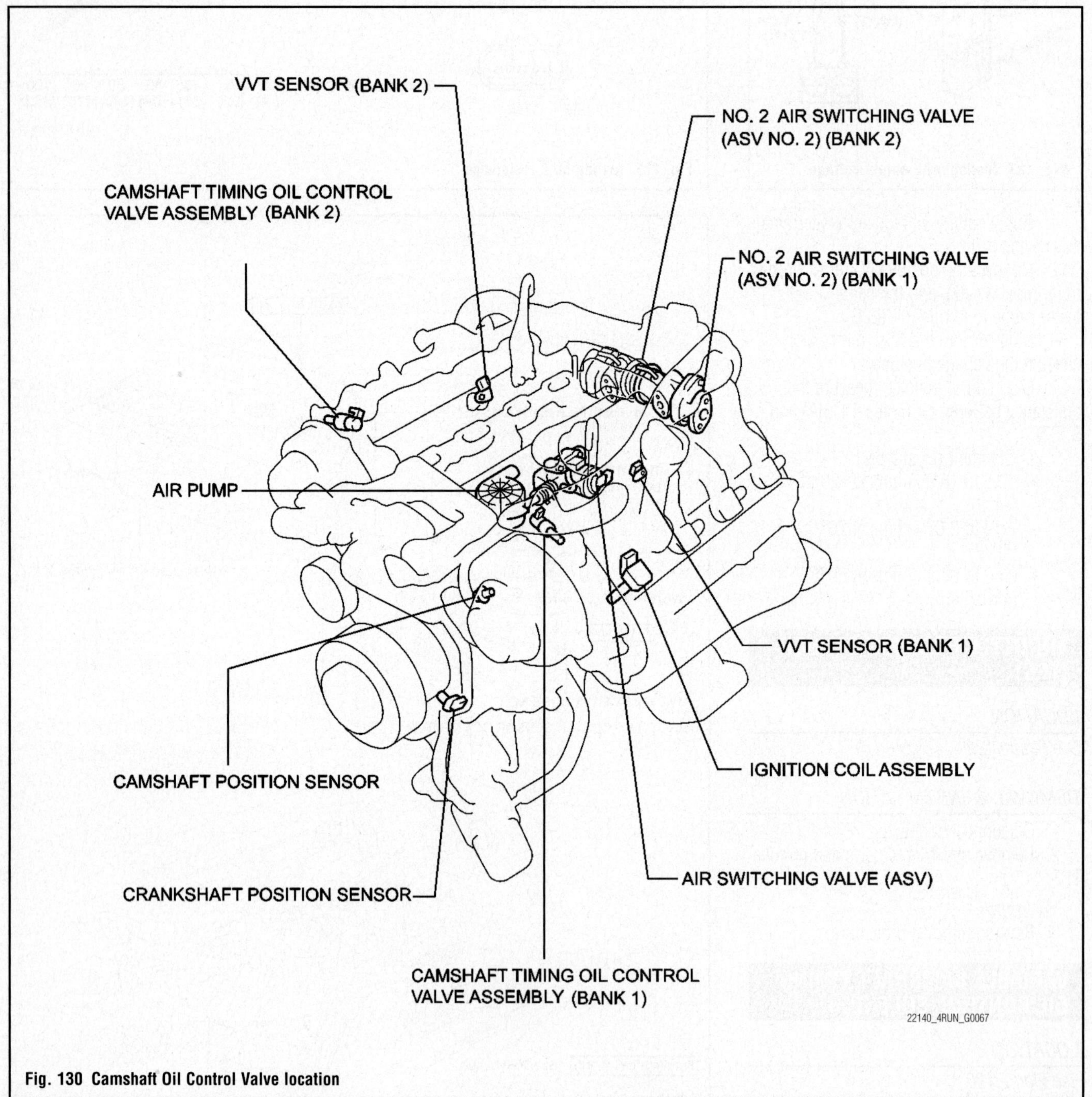

Fig. 130 Camshaft Oil Control Valve location

FUEL SYSTEM SERVICE PRECAUTIONS

Safety is the most important factor when performing not only fuel system maintenance but any type of maintenance. Failure to conduct maintenance and repairs in a safe manner may result in serious personal injury or death. Maintenance and testing of the vehicle's fuel system components can be accomplished safely and effectively by adhering to the following rules and guidelines.

- To avoid the possibility of fire and personal injury, always disconnect the negative battery cable unless the repair or test procedure requires that battery voltage be applied.
- Always relieve the fuel system pressure prior to disconnecting any fuel system component (injector, fuel rail, pressure regulator, etc.), fitting or fuel line connection. Exercise extreme caution whenever relieving fuel system pressure to avoid exposing skin, face and eyes to fuel spray. Please be advised that fuel under pressure may penetrate the skin or any part of the body that it contacts.
- Always place a shop towel or cloth around the fitting or connection prior to loosening to absorb any excess fuel due to spillage. Ensure that all fuel spillage (should it occur) is quickly removed from engine surfaces. Ensure that all fuel soaked cloths or towels are deposited into a suitable waste container.
- Always keep a dry chemical (Class B) fire extinguisher near the work area.
- Do not allow fuel spray or fuel vapors to come into contact with a spark or open flame.
- Always use a back-up wrench when loosening and tightening fuel line connection fittings. This will prevent unnecessary stress and torsion to fuel line piping.
- Always replace worn fuel fitting O-rings with new. Do not substitute fuel hose or equivalent where fuel pipe is installed.

Before servicing the vehicle, make sure to also refer to the precautions in the beginning of this section as well.

RELIEVING FUEL SYSTEM PRESSURE

1. Remove the fuel pump relay from the engine compartment relay block.
2. Start the engine and let it run until it shuts off.
3. Turn the ignition to OFF.
4. Try to start the engine and make sure it won't start.
5. Disconnect the negative battery cable.
6. Install the relay.

FUEL FILTER

REMOVAL & INSTALLATION

See Figure 131.

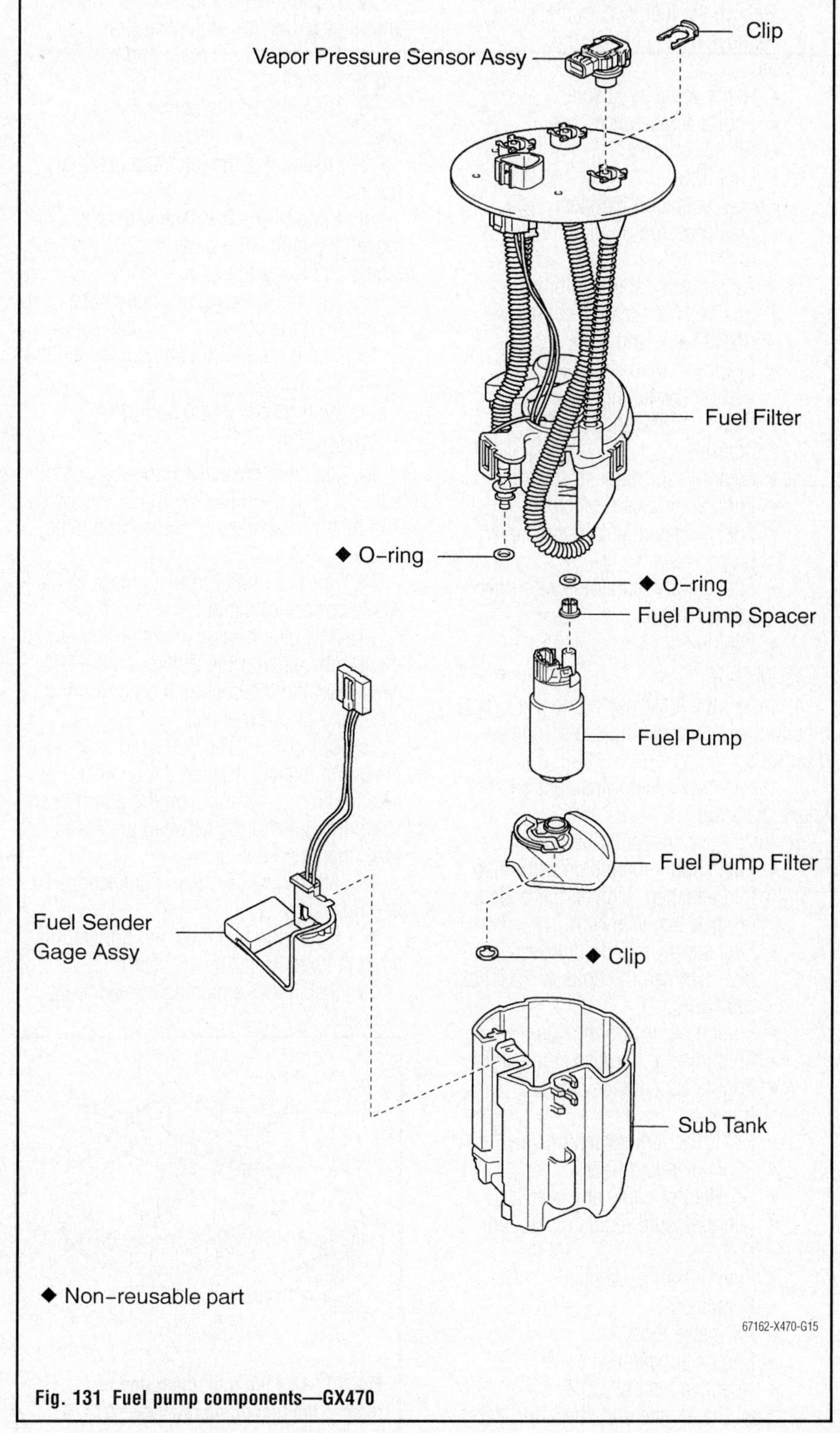

Fig. 131 Fuel pump components—GX470

The fuel filter is part of the fuel pump module unit and is not a normally replaced item.

FUEL RAIL & INJECTORS

REMOVAL & INSTALLATION

1. Before servicing the vehicle, refer to the precautions section.
2. Relieve the fuel system pressure.
3. Remove or disconnect the following:
   - Negative battery cable
   - Engine appearance cover
   - Air intake tube
   - Fuel lines
   - Fuel pulsation damper
   - Fuel pressure regulator vacuum line
   - Accelerator cable and bracket
   - Positive Crankcase Ventilation (PCV) valve and hose
   - Evaporative Emissions (EVAP) vacuum switching valve
   - Engine appearance cover brackets
   - Fuel injector harness connectors
   - Engine harness protector
   - Fuel supply manifold crossover pipe
   - Fuel supply manifolds with injectors attached
   - Fuel injectors

To install:

4. Install the fuel injectors to the supply manifold with new O-ring seals and new grommets.
5. Install new injector insulators to the intake manifold.
6. Install or connect the following:
   - Fuel supply manifolds with injectors attached. Tighten the bolts to 66 inch lbs. (7.5 Nm).
   - Fuel supply manifold crossover pipe. Tighten the bolts to 29 ft. lbs. (39 Nm).
   - Engine harness protector
   - Fuel injector harness connectors
   - Engine appearance cover brackets
   - EVAP vacuum switching valve
   - PCV valve and hose
   - Accelerator cable and bracket
   - Fuel pressure regulator vacuum line
   - Fuel pulsation damper
   - Fuel lines
   - Air intake tube
   - Engine appearance cover
   - Negative battery cable
7. Start the engine and check for leaks.

FUEL PUMP

REMOVAL & INSTALLATION

See Figures 132 through 134.

1. Before servicing the vehicle, refer to the precautions section.
2. Relieve the fuel system pressure.
3. Remove the spare tire.
4. Disconnect the fuel pump connector and remove the fuel tank protector.
5. Disconnect the main and fuel return tubes.
6. Disconnect the fuel tank vent hose.
7. Disconnect the inlet and breather hoses.
8. Support the fuel tank with a jack, loosen the tank strap bolts remove the straps and lower the tank.
9. Disconnect any necessary hoses and wiring from the pump.
10. Using the tool illustrated, loosen the pump retainer.
11. Remove the pump and gasket.

To install:

12. Install a new gasket and the pump. Make sure to align the keyway of the suction tube with the key of the suction plate No. 1.
13. Apply a multipurpose grease to the whole surface of the pump retainer.
14. Align the triangle mark on the new pump retainer with the S mark on the tank while pushing the suction tube down and attach the gauge retainer.
15. Using the same tool used to remove the pump retainer, tighten the retainer 1½ times. The triangle mark on the pump should be positioned between the A and MAX marks on the tank.
16. Attach any electrical connections and hoses.
17. Install the fuel tank and tighten the strap bolts to 45 ft. lbs. (62 Nm).
18. Install the remaining components.

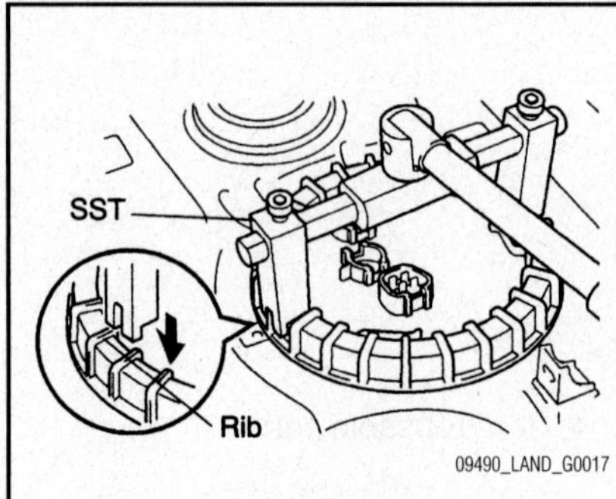

Fig. 132 Use the tool illustrated to remove the fuel pump retainer—GX470

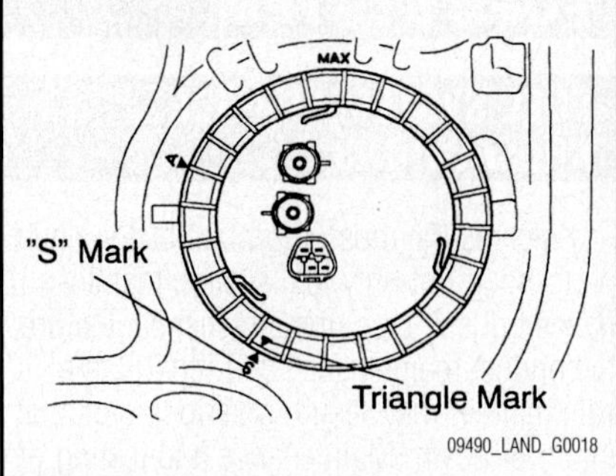

Fig. 133 Align the triangle mark on the new pump retainer with the S mark on the tank—GX470

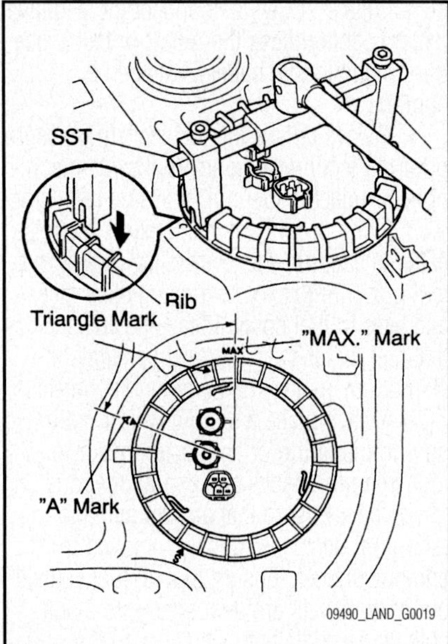

Fig. 134 The triangle mark on the pump should be positioned between the A and MAX marks on the tank when properly tightened

FUEL TANK

REMOVAL & INSTALLATION

See Figures 135 through 137.

1. Discharge fuel system pressure.
2. Disconnect cable from negative battery terminal.
3. Disconnect vent line tube.
4. Remove the bolt and bracket from the fuel tank band.
5. Remove the bolt and bracket from the body.
6. Disconnect the fuel main tube, return tube and fuel tube.
   a. With Fuel hose connector cover type disengage the lock claw by lifting up the cover, as shown in the illustration.

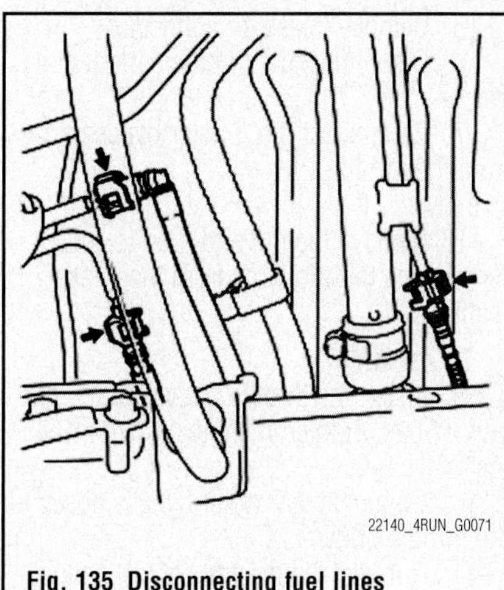

Fig. 135 Disconnecting fuel lines

Fig. 136 Removing fuel tank

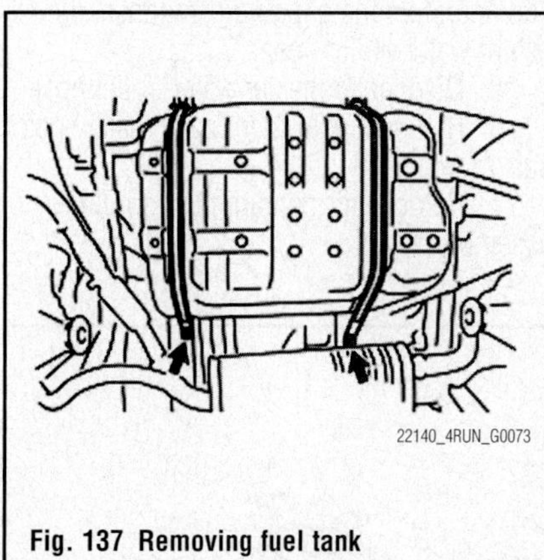

Fig. 137 Removing fuel tank

b. Check for dirt or mud on the pipe and around the connector before disconnection. Clean if necessary.

c. Disconnect the connector and pipe by hand.

d. If the connector and the pipe stuck, pinch the connector, and push and pull the pipe to disconnect it.

➡Do not use any tools.

e. Check for dirt or mud on the seal surface of the disconnected pipe. Clean if necessary.

f. To protect the disconnected pipe and connector from damage and contamination, cover it with a plastic bag.

7. Loosen the bolt of the clamp and disconnect the fuel inlet hose from the fuel inlet pipe.
8. Set up a transmission jack under the fuel tank.
9. Remove the 2 bolts and disconnect the 2 fuel tank bands from the fuel tank.
10. Slightly lower the mission jack so that the fuel pump and sender gauge connector and 2 clamps can be removed.

➡Do not lower the mission jack excessively as this may damage the connector.

11. Operate the transmission jack and remove the fuel tank.
12. Remove fuel pump and sender gauge assembly.
13. Remove fuel inlet hose.
14. Remove fuel hose.

To install:

15. Install fuel hose.
16. Install fuel inlet hose.
17. Install fuel pump and sender gauge assembly.
18. Install fuel tank assembly.
19. Operate the transmission jack so that the fuel pump and sender gauge connector and 2 clamps can be installed. Then raise the transmission jack again to install the fuel tank.
20. Install the 2 fuel tank bands with the 2 bolts and tighten to 30 ft. lbs. (40 Nm).
21. Connect the fuel main tube, return tube and fuel hose.
    a. Check that there is no damage or contamination in the connected part of the pipe.
    b. Align the axis of the connector with the axis of the pipe. Push the pipe into the connector until the connector makes a "click" sound. If the connection is tight, apply a little amount of fresh engine oil on the tip of the pipe.
    c. After having finished the connection, try to pull apart the pipe and the connector and confirm that they are securely connected.
    d. With fuel hose connector cover type attach the lock claw by lifting up the cover, as shown in the illustration.
22. Connect the fuel inlet hose to the fuel inlet pipe and tighten the bolt of the clamp to 66 inch lbs. (7.5 Nm).
23. Install the bracket to the body with the bolt and tighten bolt to 11 ft. lbs. (15 Nm).
24. Install the bracket to the fuel tank band with the bolt and tighten bolt to 11 ft. lbs. (15 Nm).
25. Connect the vent line tube to the fuel tank.
26. Check for fuel leaks.
27. Connect cable to negative battery terminal.
28. Perform initialization, if necessary.

IDLE SPEED

ADJUSTMENT

Idle speed is maintained by the Powertrain Control Module (PCM). No adjustment is necessary or possible.

THROTTLE BODY

REMOVAL & INSTALLATION

V6 Engine

1. Before servicing the vehicle, refer to the precautions in the beginning of this section.
2. Disconnect cable to negative battery terminal.
3. Remove the 2 nuts, then remove the V-bank cover.
4. Drain engine coolant.
5. Disconnect the ventilation hose No. 2.
6. Disconnect the vacuum hose.
7. Disconnect the mass air flow meter connector.
8. Remove the 2 wire harness clamps.
9. Loosen the 2 hose clamps.
10. Remove the 2 bolts, then remove the air cleaner.
11. Disconnect the water by-pass hose No. 5.
12. Disconnect the water by-pass hose No. 4.
13. Disconnect the throttle motor connector.
14. Remove the 4 bolts, then remove the throttle w/ motor body and gasket.

To install:

15. Install a new gasket and the throttle with motor body with the 4 bolts.
16. Connect the throttle motor connector.
17. Connect the water by-pass hose No. 4.
18. Connect the water by-pass hose No. 5.
19. Install the air cleaner with the 2 bolts.
20. Tighten the 2 hose clamps.
21. Install the 2 wire harness clamps.
22. Connect the mass air flow meter connector.
23. Connect the vacuum hose.
24. Connect the ventilation hose No. 2.
25. Connect cable to negative battery terminal.

26. Add engine coolant.
27. Check for engine coolant leakage.
28. Install the V-bank cover with the 2 nuts.

V8 Engine

See Figures 138 and 139.

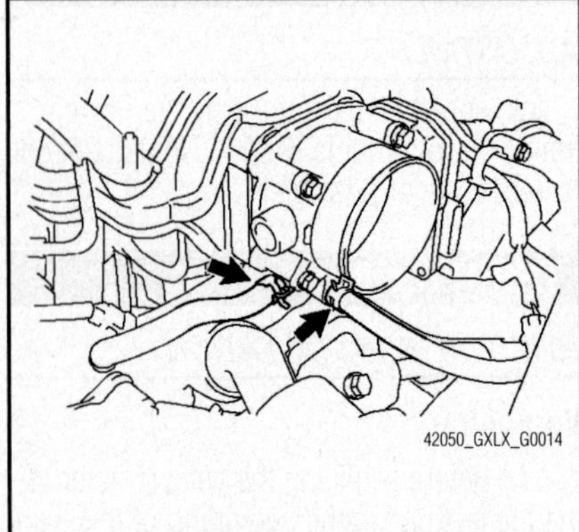

Fig. 138 Throttle body coolant hoses

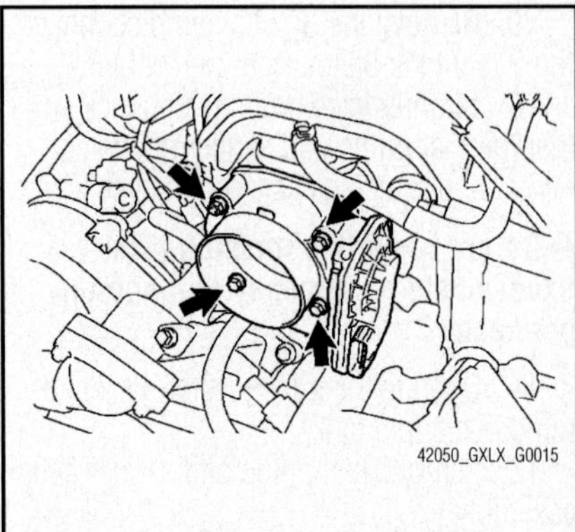

Fig. 139 Throttle body bolts and nut

1. Before servicing the vehicle, refer to the precautions in the beginning of this section.
2. Remove the 2 nuts, then remove the V-bank cover.
3. Remove throttle body cover.
4. Drain engine coolant.
5. Remove intake air connector.
6. Disconnect the throttle control connector.
7. Disconnect the 2 water bypass hoses from the throttle body.
8. Remove the nut and 3 bolts, and remove the throttle body from the intake manifold.

To install:

9. Install the throttle body with the nut and 3 bolts. Tighten them to 14 Nm (10 ft. lbs.).
10. Connect the 2 water bypass hoses to the throttle body.
11. Connect the throttle control connector.
12. Install intake air connector.
13. Fill with engine coolant.
14. Start engine and check for engine coolant leaks.
15. Install throttle body cover.

HEATING & AIR CONDITIONING SYSTEM

BLOWER MOTOR

REMOVAL & INSTALLATION

1. Before servicing the vehicle, refer to the precautions in the beginning of this section.
2. Disconnect the connector.
3. Remove the three screws and the blower motor.

To install:

4. Install the blower motor with three screws.
5. Connect the connector.

HEATER CORE

REMOVAL & INSTALLATION

See Figures 140 through 149.

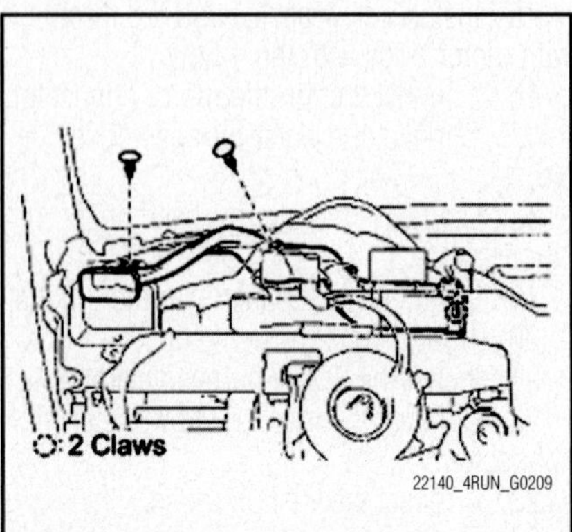

Fig. 140 Defrost nozzle duct no. 1

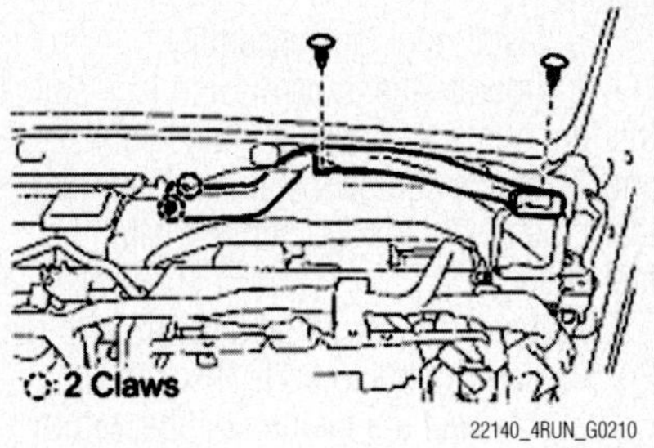

Fig. 141 Defrost nozzle duct no. 2

1. Discharge refrigerant from refrigeration system.
2. Disconnect cooler refrigerant suction pipe c.
3. Disconnect cooler refrigerant liquid pipe c.
4. Using pliers, grip the claws of the clip and slide the clip and disconnect the heater water outlet hose.
5. Disconnect heater water outlet hose.
6. Remove instrument panel safety pad sub-assembly.
7. Remove air conditioning amplifier assembly.

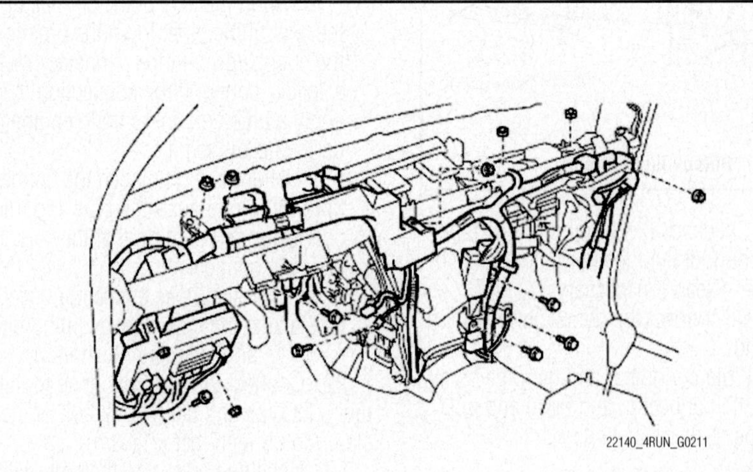

Fig. 142 Instrument panel reinforcement–1 of 4

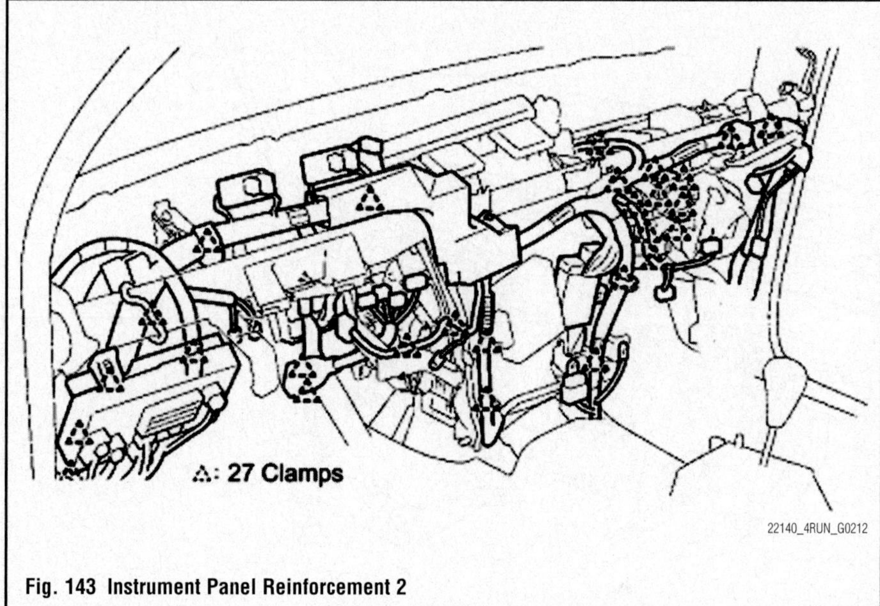

Fig. 143 Instrument Panel Reinforcement 2

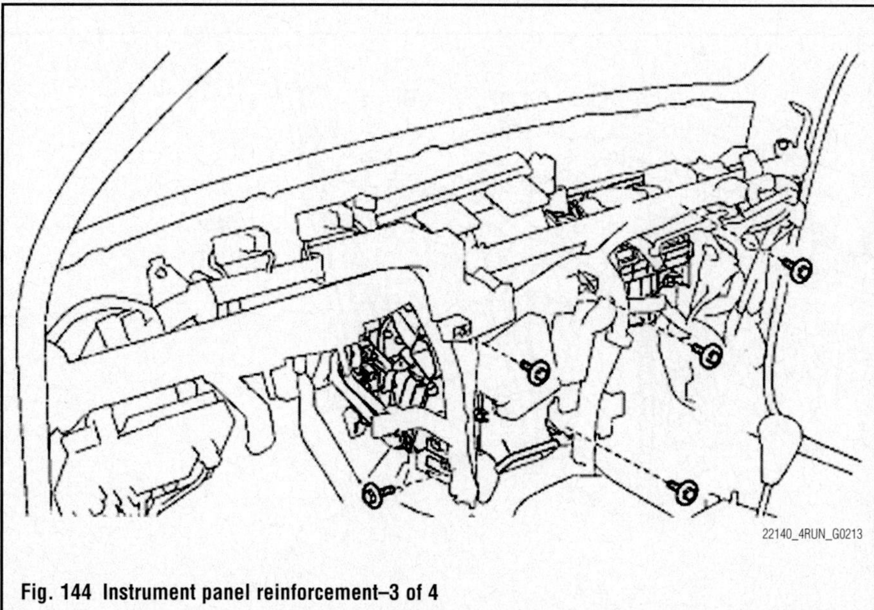

Fig. 144 Instrument panel reinforcement–3 of 4

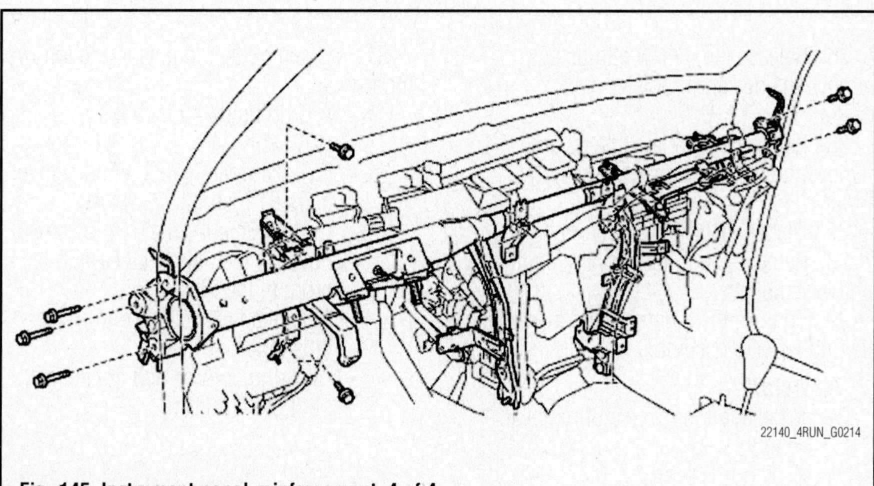

Fig. 145 Instrument panel reinforcement–4 of 4

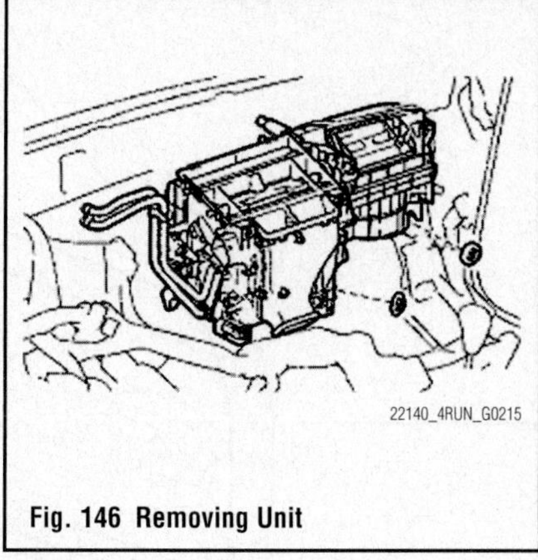

Fig. 146 Removing Unit

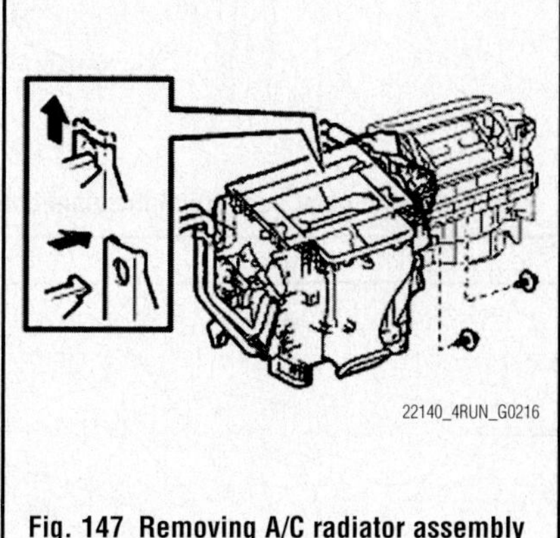

Fig. 147 Removing A/C radiator assembly

8. Remove the 2 clips.
9. Release the 2 claw fittings and remove the side defroster nozzle duct No. 1.
10. Remove the 2 clips.
11. Release the 2 claw fittings and remove the side defroster nozzle duct No. 2.
12. Remove the screw
13. Release the 2 pin fittings and remove the heater to register duct No. 1
14. Remove the screw
15. Release the 2 pin fittings and remove the heater to register duct No. 3
16. Release the 6 claw fittings and 3 clamps and remove the air duct rear No. 2.
17. Release the 6 claw fittings and 3 clamps and remove the air duct rear No. 1.
18. Remove the clip and disconnect the console box duct No. 1.
19. Release the 3 claw fittings and remove the air duct No. 1.
20. Release the 3 claw fittings and remove the air duct No. 2.
21. Remove the bolt, nut and instrument panel brace mounting brackets.

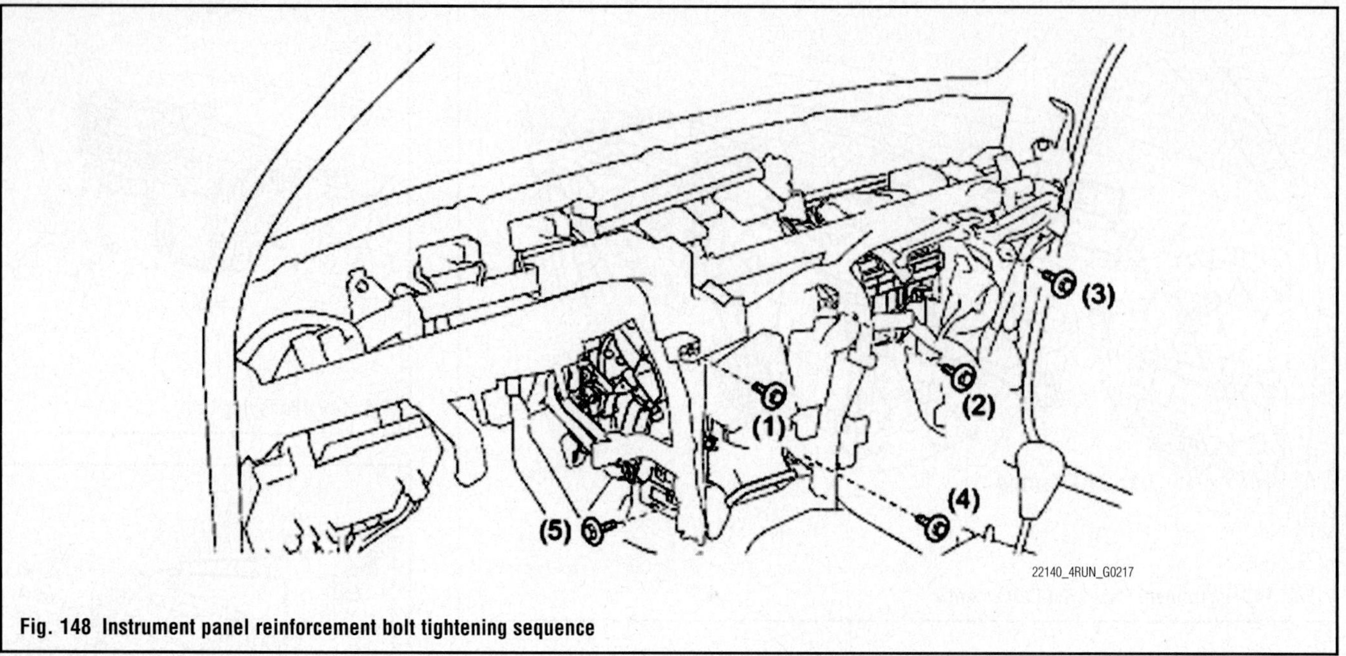

Fig. 148 Instrument panel reinforcement bolt tightening sequence

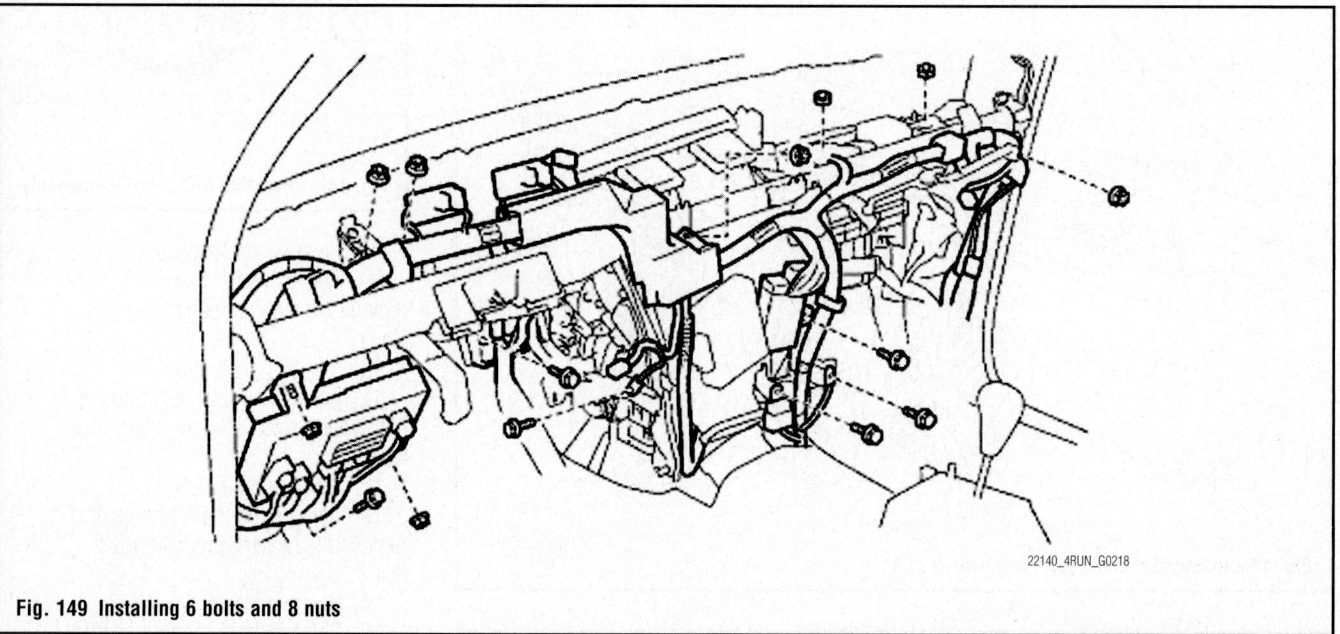

Fig. 149 Installing 6 bolts and 8 nuts

22. Release the 3 clamps and disconnect the connector.
23. Remove the 4 nuts and disconnect the steering column assembly.
24. Remove instrument panel reinforcement.
    a. Remove the 6 bolts and 8 nuts.
    b. Release the 27 clamps
    Instrument panel reinforcement–2 of 4
    c. Disconnect the connectors.
    d. Remove the 5 bolts.
    e. Remove the 7 bolts and instrument panel reinforcement.
25. Release the 4 claw fittings and remove the heater to register duct center.
26. Release the 4 claw fittings and remove the defroster nozzle assembly lower.
27. Disconnect the connectors.
28. Remove the 2 nut and air conditioner unit assembly.
29. Remove the 2 screws and air conditioning radiator assembly.

To install:

30. Installation is reverse of removal.
31. Tighten bolts and nuts to the following torque:
    - A/C unit nuts 48 inch lbs. (5.4 Nm)
    - (7) Instrument panel reinforcement bolts 87 inch lbs. (9.8 Nm)
    - (5) Instrument panel reinforcement bolts 87 inch lbs. (9.8 Nm) in the order of the illustration
    - 6 bolts and 8 nuts as shown in illustration
    - Steering column and tighten to 19 ft. lbs. (26 Nm)

AUXILIARY HEATING & AIR CONDITIONING SYSTEM

BLOWER MOTOR

REMOVAL & INSTALLATION

GX470

1. Before servicing the vehicle, refer to the precautions in the beginning of this section.
2. Remove the rear cooling unit from the vehicle.
3. Remove the 2 screws and the A/C amplifier.
4. Disconnect the connector.
5. Remove the 2 screws and the power transistor.
6. Remove the wire harness.
7. Remove the 11 screws, 3 holding springs and separate the cooling unit cases.
8. Disconnect the connector.
9. Remove the blower case.
10. Remove the 3 screws and the blower motor.

To install:

11. Installation is the reverse of the removal procedure.

HEATER CORE

REMOVAL & INSTALLATION

GX470

1. Disconnect the negative battery cable.
2. Drain the cooling system into a clean container for reuse.
3. Disconnect the heater hoses from the rear heater core.
4. Remove the front seats.
5. Remove the rear heater control assembly.
6. Remove the rear console box.
7. Remove the front console box cover.
8. Remove the lower center cluster finish panel.
9. Remove the front door scuff plates.
10. Remove the cowl side trim.
11. Remove the rear door scuff plates.
12. Remove the center pillar garnishes.
13. Slide the carpet rearward.
14. Remove the cooler bracket bolts and the bracket.
15. Remove the rear heater duct bolt/screw and the duct.
16. Disconnect the rear heater housing electrical connector.
17. Remove the 3 rear heater housing-to-chassis bolts and the heater housing.
18. Remove the heater core-to-heater housing 3 screws and 2 clamps.
19. Remove the heater core from the heater housing.

To install:

20. Install the heater core to the heater housing.
21. Install the heater core-to-heater housing 3 screws and 2 clamps.
22. Install the heater housing and the 3 rear heater housing-to-chassis bolts.
23. Connect the rear heater housing electrical connector.
24. Install the rear heater duct and the duct bolt/screw.
25. Install the cooler bracket and the bracket bolts.
26. Slide the carpet rearward.
27. Install the center pillar garnishes.
28. Install the rear door scuff plates.
29. Install the cowl side trim.
30. Install the front door scuff plates.
31. Install the lower center cluster finish panel.
32. Install the front console box cover.
33. Install the rear console box.
34. Install the rear heater control assembly.
35. Install the front seats.
36. Connect the heater hoses to the rear heater core.
37. Refill the cooling system.
38. Connect the negative battery cable.

STEERING

POWER RACK & PINION STEERING GEAR

REMOVAL & INSTALLATION

See Figure 150.

1. Before servicing the vehicle, refer to the precautions section.
2. Disconnect the battery ground cable.
3. Place the front wheels in the straight ahead position.
4. Remove the horn pad.
5. Remove the steering wheel.
6. Remove the lower steering column cover.
7. Remove the turn signal switch.
8. Remove the spiral cable assembly.
9. Remove the front wheels.
10. Remove the engine under-covers.
11. Remove the stabilizer bar.
12. Remove the tie rod ends from the knuckle.
13. Remove the steering intermediate shaft.
14. Disconnect the pressure and return lines.
15. Remove the 2 bolts and remove the steering gear assembly.

To install:

16. Position the gear and install the 2 bolts. Torque to 74 ft. lbs. (100 Nm).

➡The nuts have detents. Never turn the nuts, just the bolts.

17. Install the stabilizer bar. Torque the end links to 52 ft. lbs. (70 Nm); the clamp bolts to 30 ft. lbs. (40 Nm).
18. Connect the return line. Use a torque wrench with SST 09023-12700, or equivalent. The torque wrench should have a fulcrum length of 300mm. Torque to 31 ft. lbs. (42 Nm).
19. Connect the pressure line at the subframe. Torque to 21 ft. lbs. (28 Nm).
20. Connect the pressure line to the gear. Use a torque wrench with SST 09023-12700, or equivalent. The torque wrench should have a fulcrum length of 300mm. Torque to 31 ft. lbs. (42 Nm).
21. Connect the intermediate shaft. Torque to 26 ft. lbs. (36 Nm).
22. Connect the tie rod ends. Torque to 67 ft. lbs. (91 Nm).
23. Install the under-covers.
24. The remainder of installation is the reverse of removal.

POWER STEERING PUMP

REMOVAL & INSTALLATION

See Figure 151.

1. Before servicing the vehicle, refer to the precautions in the beginning of this section.
2. Disconnect the MAF meter connector.
3. Disconnect the hoses.
4. Remove the clamp.
5. Remove the 3 bolts and air cleaner assembly with air cleaner hose connected.

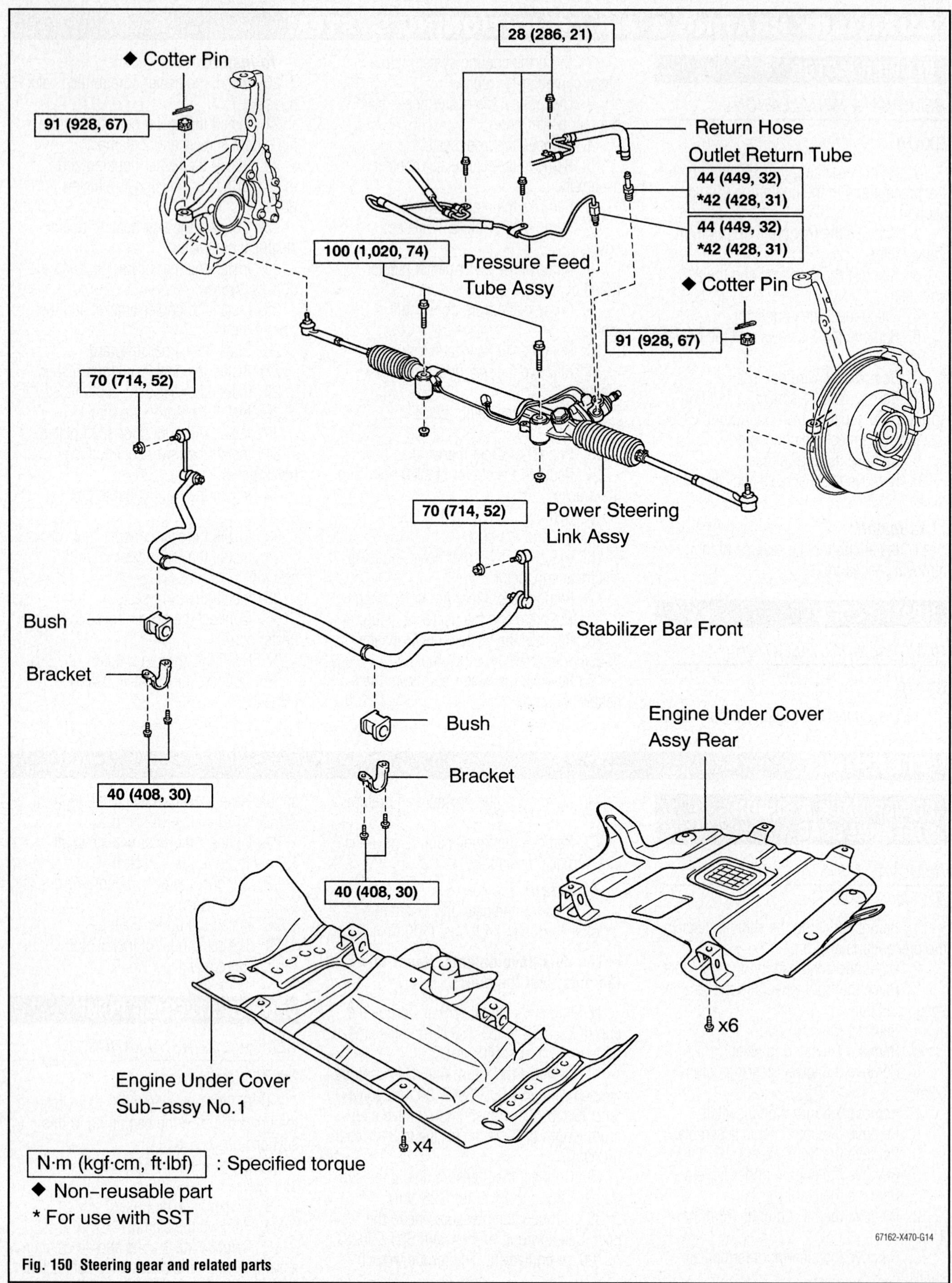

Fig. 150 Steering gear and related parts

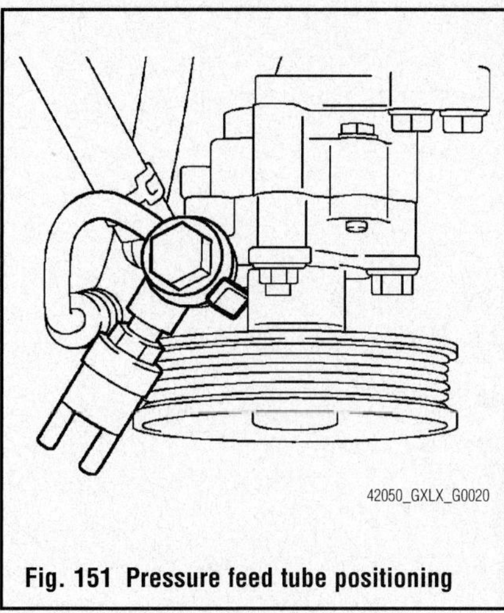

Fig. 151 Pressure feed tube positioning

6. Loosen the drive belt tension by turning the drive belt tensioner counter-clockwise, and remove the drive belt.
7. Remove the 2 clips and disconnect the 2 vacuum hoses.
8. Remove the clip and disconnect the return hose.
9. Remove the union bolt and gasket, disconnect the pressure feed tube.
10. Remove the 2 bolts, nut, stud bolt and power steering pump assembly.

To install:

11. Install the power steering pump assembly with the stud bolt.
12. Tighten the stud bolt to 22 Nm (16 ft. lbs.).
13. Install the 2 bolts and nut and tighten them to 44 Nm (33 ft. lbs.).
14. Install a new gasket and the union bolt on the pressure feed tube.

➡Make sure that the stopper of the pressure feed tube contacts the power steering pump body as shown in the illustration.

15. Tighten the union bolt to 46.5 Nm (34 ft. lbs.).
16. Connect the return hose with the clip.
17. Connect the 2 vacuum hoses and install the 2 clips.
18. Loosen the drive belt tension by turning the drive belt tensioner counter-clockwise, and install the belt.
19. Install the air cleaner assembly with air cleaner hose and the 3 bolts.
20. Install the clamp.
21. Connect the MAF meter connector.
22. Fill with power steering fluid and bleed the system.

BLEEDING

1. Before servicing the vehicle, refer to the precautions in the beginning of this section.
2. Check fluid level.
3. Jack up front of vehicle and support it with stands.
4. With the engine stopped, turn the wheel slowly from lock to lock several times.
5. Lower the vehicle.
6. Start the engine and run at idle for a few minutes.
7. With the engine idling, turn the wheel left or right to the full lock position and keep it there for 2 to 3 seconds, then turn the wheel to the opposite full lock position and keep it there for 2 to 3 seconds. Repeat several times.
8. Stop the engine.
9. Check for foaming or emulsification of the power steering fluid.
10. If the system has to be bled twice specifically because of foaming or emulsification, check for fluid leaks in the system.
11. Check fluid level.

SUSPENSION — FRONT SUSPENSION

COIL SPRING

REMOVAL & INSTALLATION

See Figure 152.

1. Remove the strut.
2. Place the strut in a compressor, such as SST 09727-30021, and compress the spring.
3. Hold the rod and remove the nut.

➡Don't use an impact wrench.

4. Remove the bushing retainer
5. Remove the upper bushing.

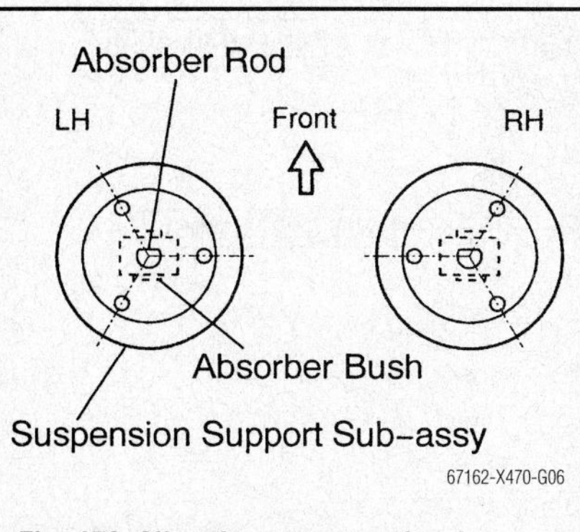

Fig. 152 Align the support, rod and bushing as shown

6. Remove the support.
7. Remove the lower bushing retainer.
8. Remove the spring.
9. Remove the lower bushing.

To install:

10. Install the new lower bushing.
11. Compress the spring and install it.
12. Install the bushing retainer.
13. Install the suspension support.
14. Install the upper bushing.
15. Install the retainer.
16. Align the support, rod and bushing as shown. Install the locknut and torque to 18 ft. lbs. (25 Nm).
17. Release the spring from the compressor and check the alignment of the parts.
18. Install the strut.

CONTROL LINKS

REMOVAL & INSTALLATION

Non-KDSS

1. Remove front disc wheel.
2. Remove the 2 nuts and the front stabilizer link assembly.

➡If the ball joint turns together with the nut, use a hexagon (6 mm) wrench to hold the stud.

3. Remove front stabilizer link assembly.

To install:

4. Installation is reverse of removal. Tighten nuts to 52 ft. lbs. (70 Nm)

LOWER BALL JOINT

REMOVAL & INSTALLATION

The lower ball joint is serviced with the lower control arm as an assembly.

LOWER CONTROL ARM

REMOVAL & INSTALLATION

See Figure 153.

1. Before servicing the vehicle, refer to the precautions section.
2. Remove the wheel.
3. Support the lower arm with a jack.
4. Remove the lower strut bolt.
5. Remove the 2 bolts and separate the lower ball joint attachment from the knuckle.
6. Place matchmarks on the camber adjusting cam and toe adjusting cam.
7. Remove the 2 nuts and remove the arm along with the cams.

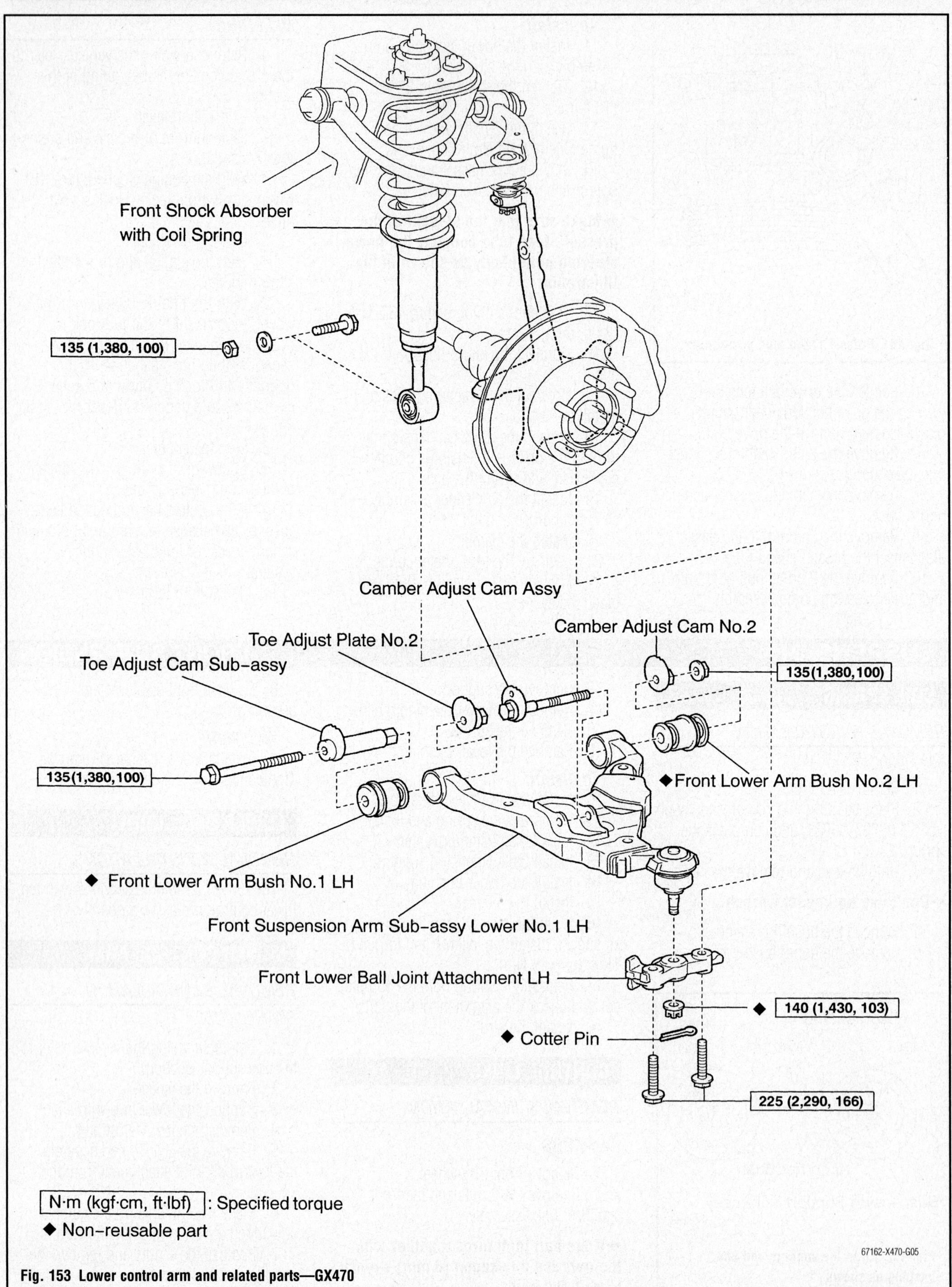

Fig. 153 Lower control arm and related parts—GX470

To install:

8. Installation is the reverse of removal. Align all matchmarks. Use new nuts and cotter pins. Don't fully tighten the control arm bolts until the vehicle is on the ground and the suspension jounced a few times. Observe the following torques:

- Lower ball joint stud: 103 ft. lbs. (140 Nm)
- Lower ball joint attachment bolts: 166 ft. lbs. (225 Nm)
- Lower arm bolts: 100 ft. lbs. (135 Nm)

MACPHERSON STRUT

REMOVAL & INSTALLATION

Non-REAS Suspension

See Figure 154.

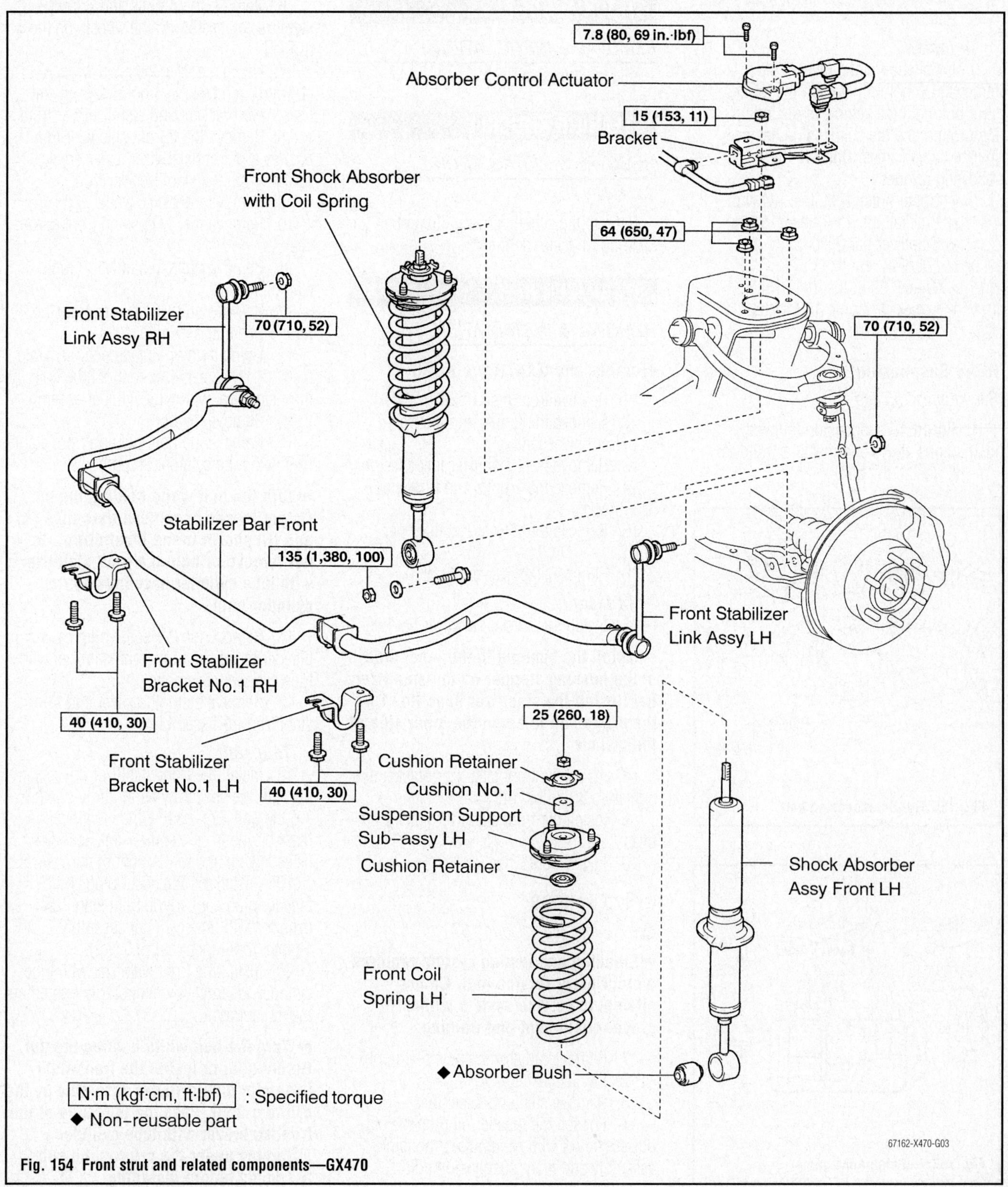

Fig. 154 Front strut and related components—GX470

1. Before servicing the vehicle, refer to the precautions section.
2. Remove the wheel.
3. Remove the stabilizer bar.
4. Remove the clamps and connector.
5. Remove the wire bracket.
6. Remove the lower strut bolt.
7. Remove the 3 upper strut nuts.
8. Remove the strut.

To install:

9. Installation is the reverse of removal. Do not fully tighten the lower strut bolt until the vehicle is resting on the ground and the suspension has been jounced a few times. Observe the following torques:
   - Upper nuts: 47 ft. lbs. (64 Nm)
   - Bracket nut: 11 ft. lbs. (15 Nm)
   - Stabilizer bar links: 52 ft. lbs. (70 Nm)
   - Wheel: 83 ft. lbs. (112 Nm)
   - Lower strut bolt: 100 ft. lbs. (135 Nm)

REAS Suspension

See Figures 155 and 156.

1. For REAS suspension follow the instructions above and install as follows:
   a. Install the bolt

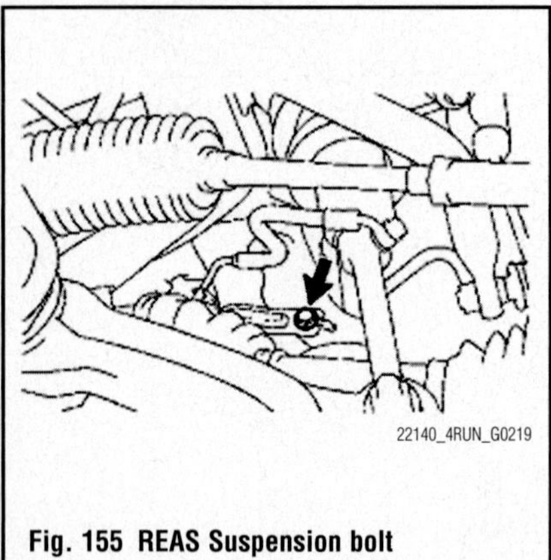

Fig. 155 REAS Suspension bolt

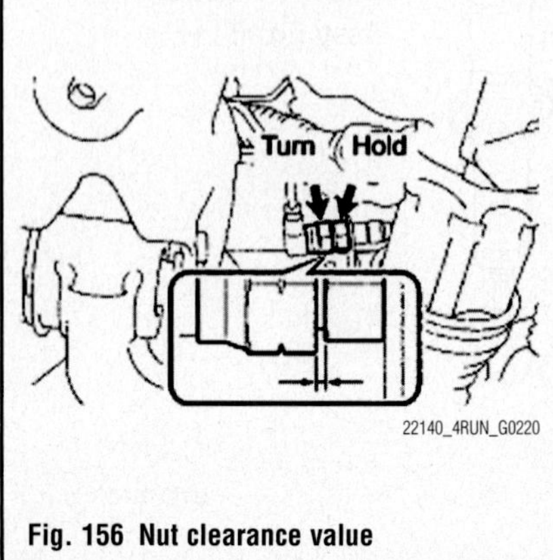

Fig. 156 Nut clearance value

➡Be sure to fit the detents attached to the bracket into a hole on the frame side

2. As shown in the illustration, tighten the nut of clearance to standard value. Tighten to 18 ft. lbs. (25 Nm)

SHOCK ABSORBERS

REMOVAL & INSTALLATION

See MacPherson Strut

STEERING KNUCKLE

REMOVAL & INSTALLATION

See Figure 157.

Refer to the accompanying illustration for Steering Knuckle mounting.

STABILIZER BAR

REMOVAL & INSTALLATION

4Runner and GX470 w/o KDSS

1. Remove front disc wheel
2. Remove the 2 nuts and the front stabilizer link assembly
3. Remove front stabilizer link assembly
4. Remove the 2 bolts and front stabilizer bracket
5. Remove the 2 front stabilizer bar bush
6. Remove stabilizer bar front

To install:

7. Install the 2 front stabilizer bar bush

➡Install the bushing to the inner side of the bushing stopper on the stabilizer bar. Install the stabilizer bush No. 1 as the protrusion to be on the inner side of the vehicle

8. Install the front stabilizer bracket and tighten to 30 ft. lbs. (40 Nm)
9. Install the front stabilizer link assemblies
10. Install front wheels

GX470 w/ KDSS

See Figures 158 through 162.

➡Bleeding and testing system requires a proprietary service tool. Do not attempt to service system without proper equipment and training.

1. Remove radiator support seal upper
2. Remove battery
3. Remove step sub-assembly
4. Loosen the bleeder plug on the stabilizer control with accumulator housing assembly and drain suspension fluid

➡Drain suspension fluid when performing the operations related to the hydraulic circuits. Draining suspension fluid decreases suspension fluid pressure.

5. Tighten the bleeder plug to 84 inch lbs. (9.5 Nm)
6. Remove the 4 bolts and 2 front stabilizer brackets with the wheels on the ground
7. Using a socket hexagon wrench (10 mm), remove the 4 bolts, 2 front stabilizer lower brackets and stabilizer bar front
8. Remove the 2 front stabilizer link bushes and front stabilizer lower bracket bushes from the stabilizer bar front
9. Remove front disc wheel
10. Remove the 7 clips and front apron seal
11. Remove bolt (A) and No. 3 tube clamp bracket.
12. Remove bolt (B) and separate the return tube sub-assembly.
13. Using a union nut wrench, separate the No. 1 front stabilizer control tube from the front stabilizer with tube cylinder assembly
14. Remove the bolt
15. Remove the bolt, nut and front stabilizer with tube cylinder assembly.

➡Turn the bolt while holding the nut. Do not loosen or remove flare nuts (A) and (B) shown in the illustration. Do not remove or hold the front stabilizer with tube cylinder assembly by the cylinder boot.

16. Remove the 2 bleeder plug caps and bleeder plugs from the front stabilizer with tube cylinder assembly.
17. Remove the bolt, spacer and front stabilizer link assembly.

To install:

18. Install the front stabilizer link assembly with the spacer by temporarily tightening the bolt. Ensure that the identification mark on the front stabilizer link assembly faces inward and to the front of the vehicle.
19. Install the 2 bleeder plugs and bleeder plug caps to the front stabilizer cylinder with tube cylinder assembly. Tighten to 84 inch lbs. (9.5 Nm).
20. Install the front stabilizer with tube cylinder assembly by temporarily tightening the nut and bolt.

➡ Turn the bolt while holding the nut. Do not hold or install the front stabilizer with tube cylinder assembly by the cylinder boot. Pass the tube side of the front stabilizer with tube cylinder assembly under the return tube sub-assembly before installing.

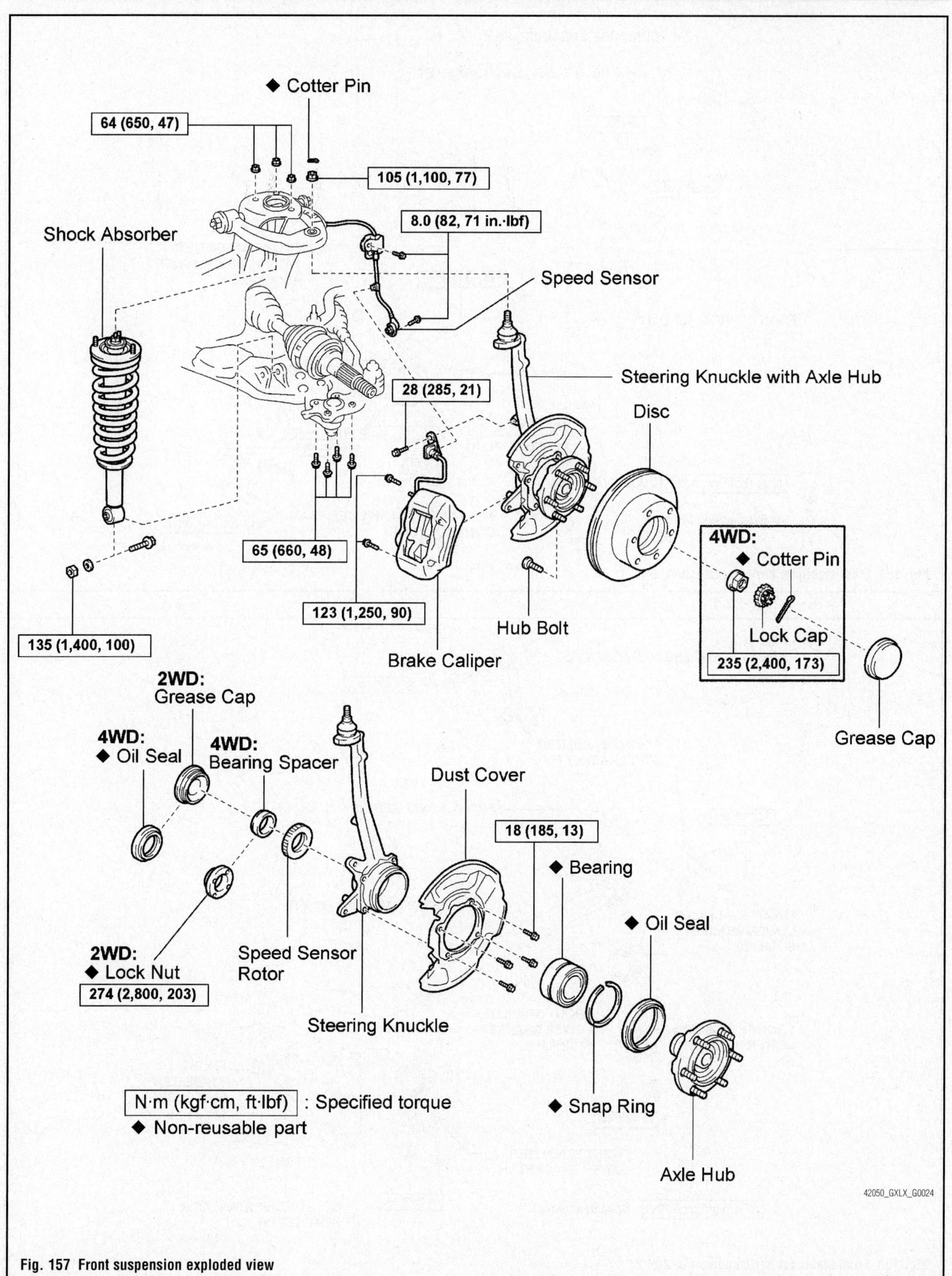

Fig. 157 Front suspension exploded view

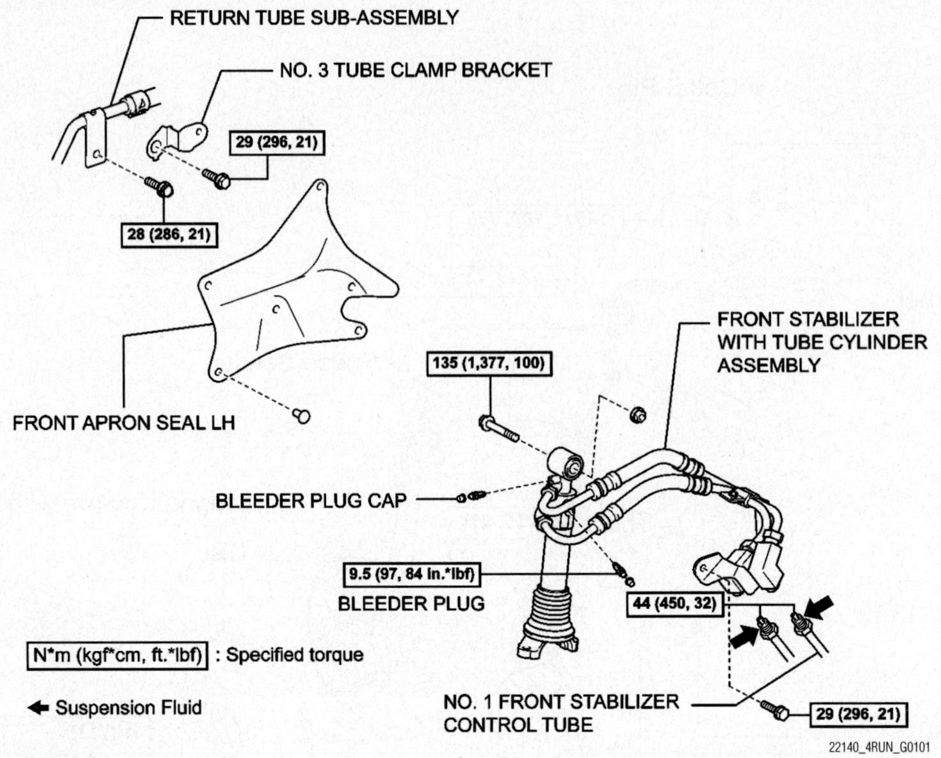

Fig. 158 Front stabilizer bar components–1 of 2

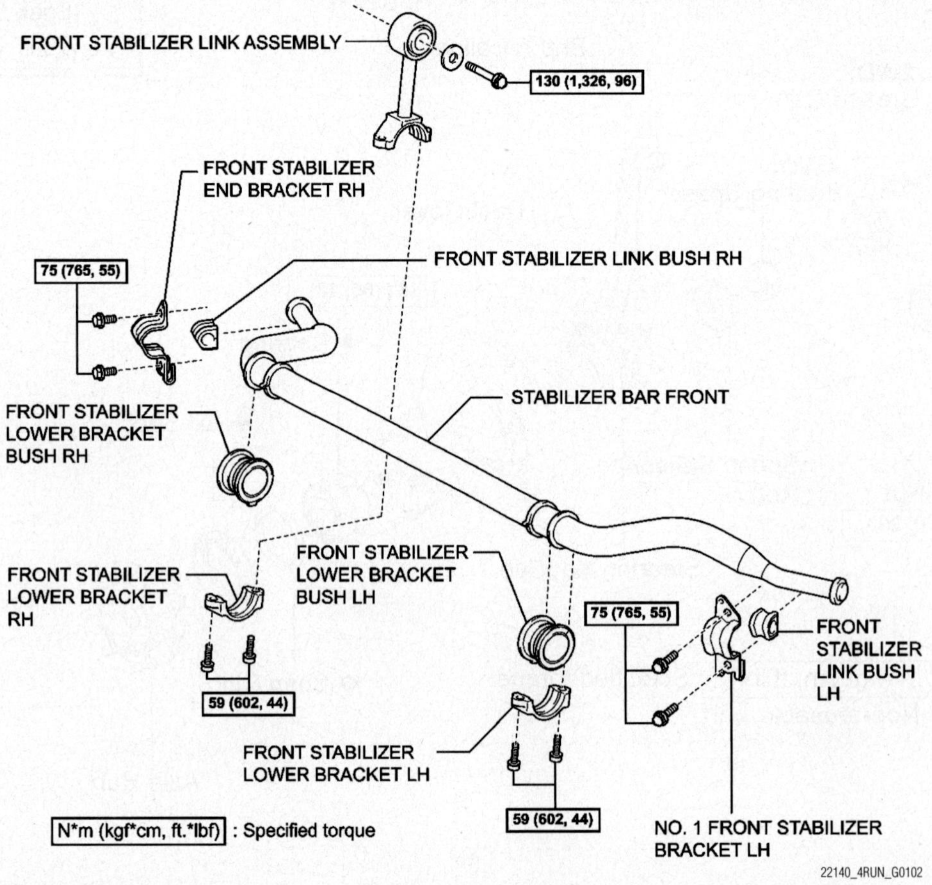

Fig. 159 Front stabilizer bar components–2 of 2

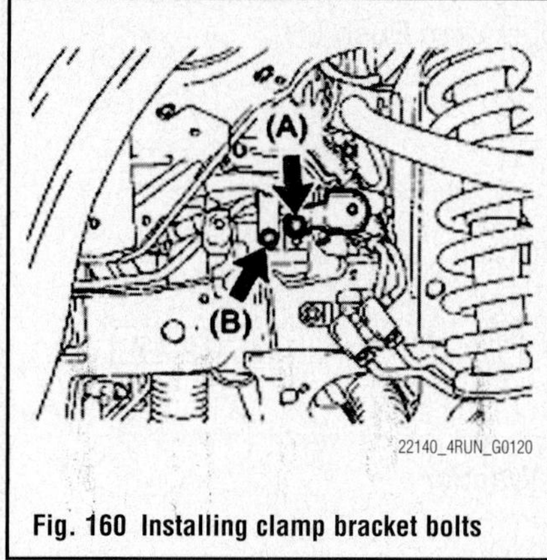

Fig. 160 Installing clamp bracket bolts

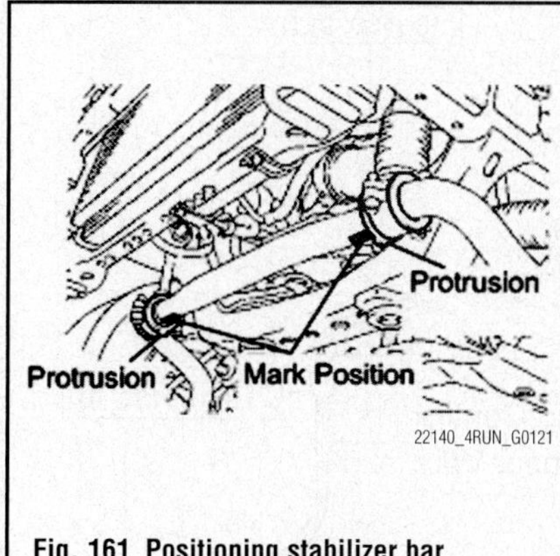

Fig. 161 Positioning stabilizer bar

21. Install the bracket by temporarily tightening the bolt.
22. Connect the No. 1 front stabilizer control tube.
23. Apply suspension fluid to the threads of the flare nuts.
24. Using a union nut wrench, connect the No. 1 front stabilizer control tube to the front stabilizer with tube cylinder assembly and tighten the flare nuts to 32 ft. lbs. (44 Nm).
25. Fully tighten the bracket bolt to 21 ft. lbs. (29 Nm)
26. Install the No. 3 tube clamp bracket with bolt (A) and tighten to 21 ft. lbs. (29 Nm).
27. Install the return tube sub-assembly with bolt (B) and tighten to 21 ft. lbs. (29 Nm).
28. Install the front apron seal LH with the 7 clips.

➡Bleeding system requires a proprietary service tool.

29. Install the 2 front stabilizer lower bracket bushes to the stabilizer bar front.
    a. Align the protrusions on the bushes with the identification marks on the front stabilizer bar front with the protrusions facing inward.

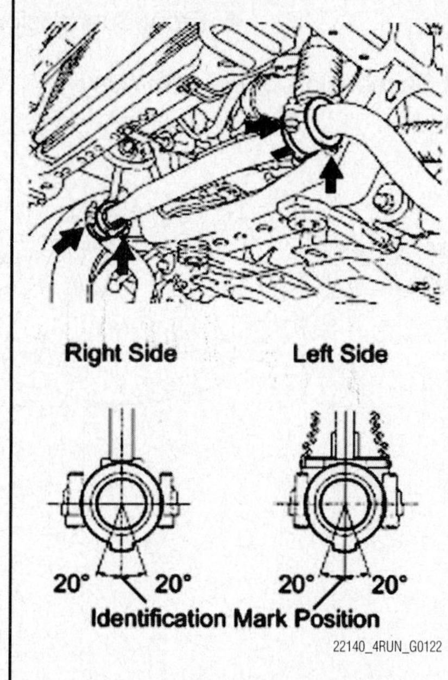

Fig. 162 Checking stabilizer bar position

➡Place the jack under the left side of the vehicle.

30. Support the stabilizer bar front with the identification marks facing down with a jack. Place a wooden block between the jack and the stabilizer bar front to prevent damage.
31. Using a socket hexagon wrench (10 mm), install the stabilizer bar front with the 4 bolts and 2 front stabilizer lower brackets and tighten to 44 ft. lbs. (59 Nm).
    a. Check that the protrusions on the front stabilizer lower bracket bushes are positioned within 20° of the identification marks
32. Install the 4 bolts, 2 front stabilizer brackets and 2 front stabilizer link bushes with the wheels on the ground and tighten to 5 ft. lbs. (75 Nm)

➡There are stamps on the front stabilizer brackets to distinguish between right and left.

33. Fully tighten the bolt on the front stabilizer with tube cylinder assembly to 100 ft. lbs. (135 Nm)

➡Tighten the bolt with the wheels on the ground.

34. Fully tighten the bolt on the front stabilizer link assembly and tighten to 96 ft. lbs. (130 Nm)

➡Tighten the bolt with the wheels on the ground.

➡System must be tested. Testing system requires a proprietary service tool.

35. Install no. 1 engine under cover sub-assembly.
36. Install spare disc wheel.
37. Install step sub-assembly lh.
38. Install battery
39. Install radiator support seal upper.

TRACK BAR

REMOVAL & INSTALLATION

1. Remove the bolt.
2. Remove the bolt, nut and lateral control rod assembly.

To install:

3. Install the lateral control rod assembly with the bolt.
4. Install the bolt and the nut.
5. Stabilize suspension.
6. Fully tighten the 2 bolts to 96 ft. lbs. (130 Nm).

UPPER BALL JOINT

REMOVAL & INSTALLATION

The upper ball joint is serviced with the upper control arm as an assembly.

UPPER CONTROL ARM

REMOVAL & INSTALLATION

See Figure 163.

1. Before servicing the vehicle, refer to the precautions section.
2. Remove the wheel.
3. Disconnect the skid control wire.
4. Support the lower arm with a jack.
5. Remove the cable bracket.
6. Disconnect the ball joint from the knuckle.
7. Remove the through-bolt, washers and nut.
8. Remove the arm.
9. Installation is the reverse of removal. Don't fully tighten the through-bolt until the vehicle is on the ground and the suspension is jounced a few times.
    - Ball joint nut: 81 ft. lbs. (110 Nm)
    - Through-bolt: 85 ft. lbs. (115 Nm)

WHEEL HUB AND BEARING

REMOVAL & INSTALLATION

See Figure 164.

1. Remove the wheel.
2. Remove the caliper.
3. Remove the hub grease cap.
4. Remove the cotter pin.
5. Remove the hub nut.
6. Remove the speed sensor.

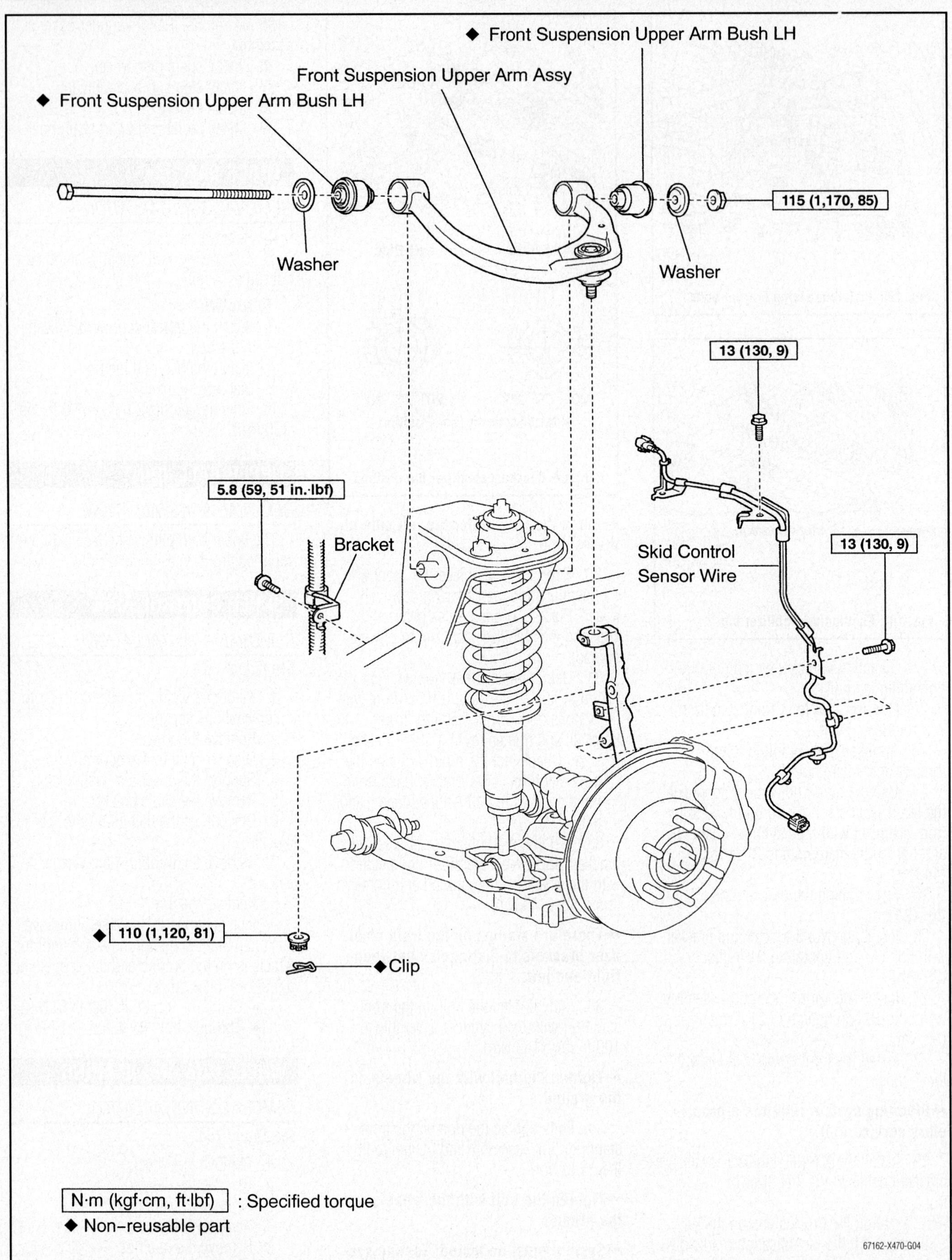

Fig. 163 Upper control arm and related parts

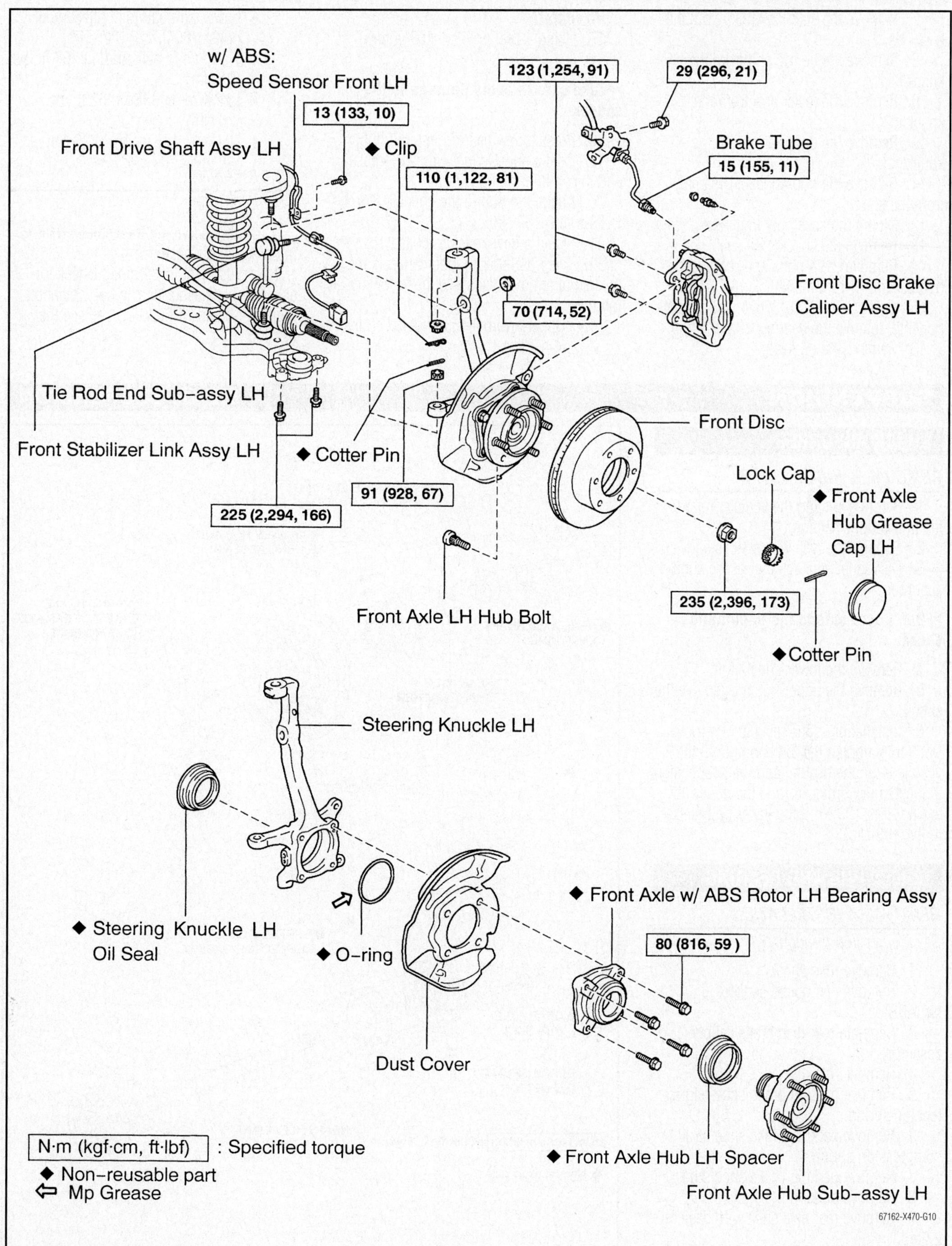

Fig. 164 Front hub and related parts

7. Remove the stabilizer links from the knuckles.
8. Remove the tie rod end from the knuckle.
9. Remove the lower arm from the knuckle.
10. Remove the upper arm from the knuckle.
11. Remove the hub/knuckle assembly from the shaft.
12. Mount the assembly in a vise.
13. Remove the knuckle oil seal.
14. Remove the 4 bolts and remove the hub assembly from the knuckle.
15. Using SST 09710-30021 and its components, remove the bearing from the hub.
16. Remove the oil seal.

To install:

17. Using a seal driver, install a new seal.

➡Take care to avoid damage to the spacer.

18. Press a new bearing into the hub.
19. Coat a new O-ring with MP grease and install it in the hub.
20. Attach the hub to the knuckle. Torque to 59 ft. lbs. (80 Nm).
21. Install a new knuckle oil seal.
22. The remainder of installation is the reverse of removal. Observe the following torques:

- Upper arm ball stud nut: 81 ft. lbs. (110 Nm)
- Lower arm ball joint attachment bolts: 166 ft. lbs. (225 Nm)
- Tie rod end ball stud nut: 67 ft. lbs. (91 Nm)
- Stabilizer end links: 52 ft. lbs. (70 Nm)
- Hub nut: 173 ft. lbs. (235 Nm)

ADJUSTMENT

See Figure 166.

1. Before servicing the vehicle, refer to the precautions section.

No adjustment is possible. Check for axle hub backlash and axle hub deviation. If either exceeds 0.0020 in., replace the bearing.

SUSPENSION

REAR SUSPENSION

SHOCK ABSORBER

REMOVAL & INSTALLATION

1. Before servicing the vehicle, refer to the precautions section.
2. Support the axle with a jackstand.
3. Disconnect the actuator at the shock absorber.

➡Don't over-extend the pneumatic shock.

4. Remove the lower shock bolt.
5. Remove the upper nut and remove the shock.
6. Installation is the reverse of removal. Don't fully tighten the lower bolt until the vehicle is on the ground and the suspension jounced a few times. Torque the upper nut to 18 ft. lbs. (25 Nm); the lower bolt to 72 ft. lbs. (98 Nm).

WHEEL HUB AND BEARING

REMOVAL & INSTALLATION

See Figures 165 through 171.

1. Remove rear wheel.
2. Remove rear speed sensor (w/ ABS).
3. Separate rear disc brake caliper assembly.
4. Remove rear disc.
5. Remove parking brake shoe return tension spring.
6. Remove parking brake shoe strut compression spring.
7. Remove parking brake shoe strut.
8. Remove parking brake shoe.
9. Remove rear axle shaft with backing plate.
10. Remove the 4 nuts and rear axle shaft with backing plate.

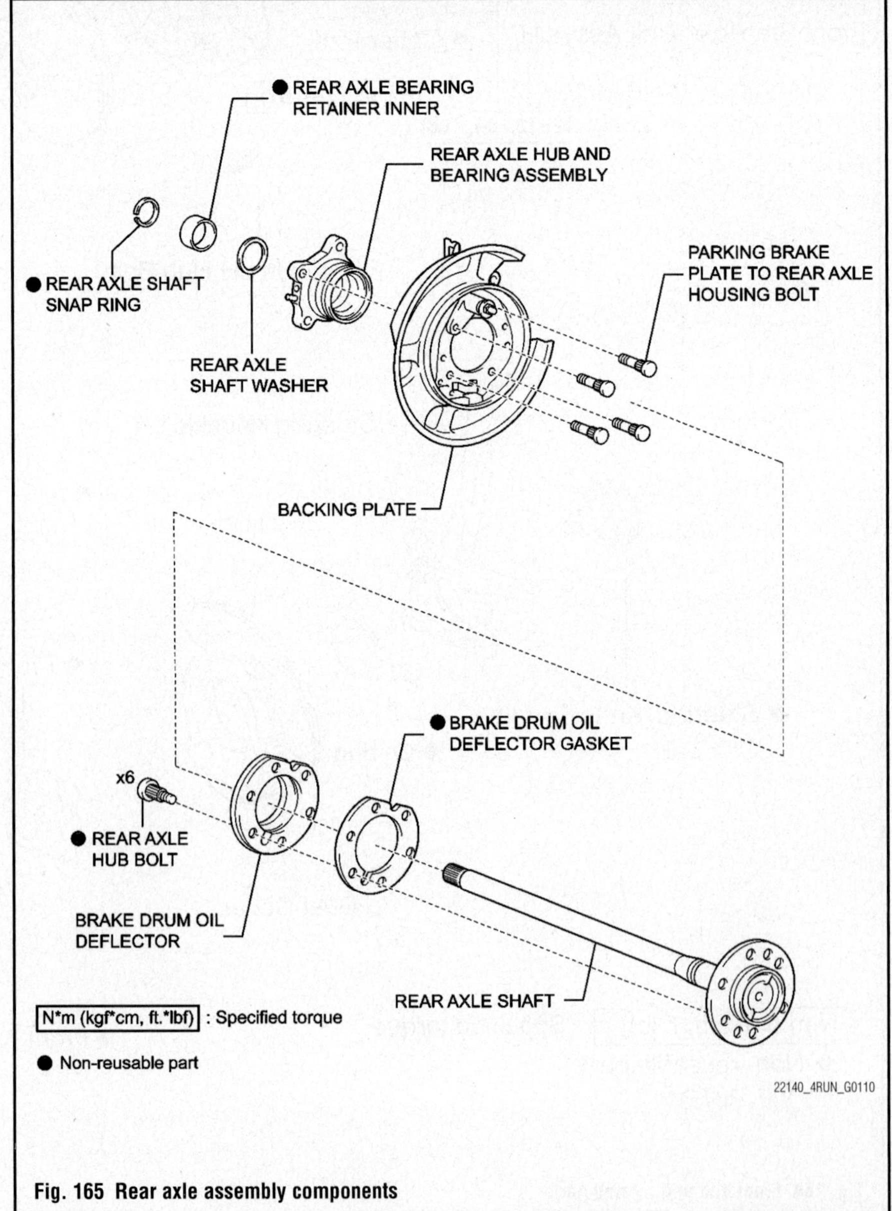

Fig. 165 Rear axle assembly components

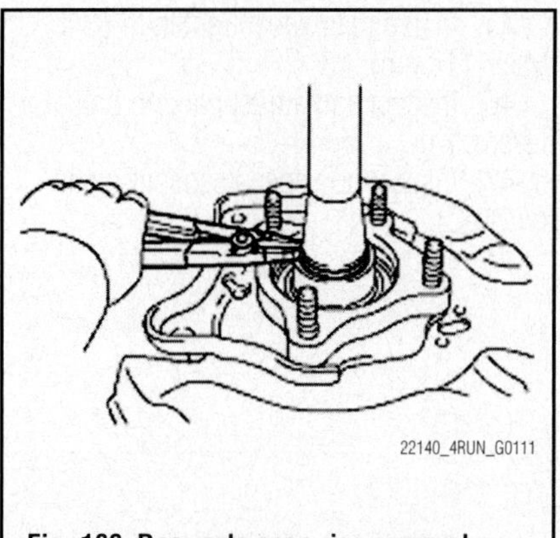
Fig. 166 Rear axle snap ring removal

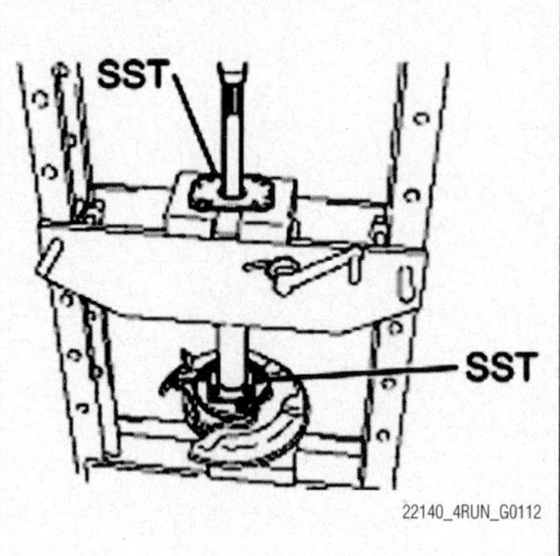

Fig. 167 Removing rear axle shaft from bearing

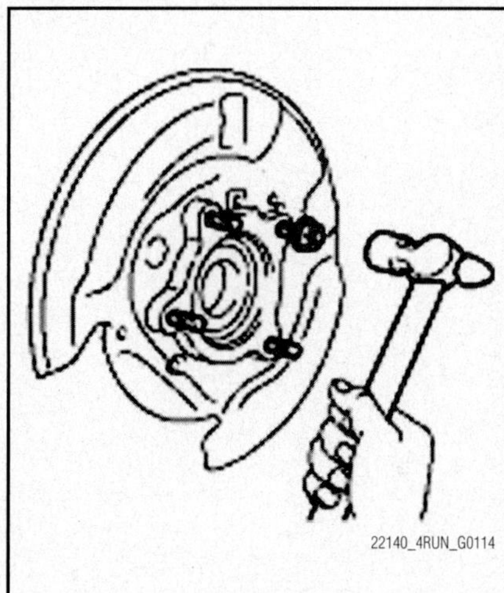
Fig. 168 Removing parking brake plate

11. Remove the O-ring.
12. Using a snap ring expander, remove the snap ring.
13. Remove the rear axle shaft from bearing.
14. Remove the rear axle bearing retainer inner from the rear axle bearing assembly.

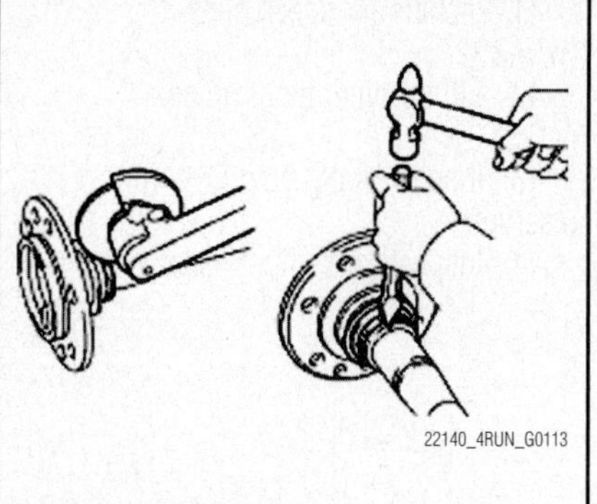
Fig. 169 Removing rear axle bearing inner race

15. Remove the rear axle shaft washer from the rear axle bearing assembly.
16. Attach the 4 nuts to the parking brake plate to rear axle housing bolts.
17. Using a hammer, remove the 4 parking brake plate to rear axle housing bolts and rear axle bearing assembly.

➡Do not reuse the nuts previously removed from the vehicle.

18. Remove the 6 hub bolts.
19. Remove brake drum oil deflector.
20. Remove brake drum oil deflector gasket.
21. Remove rear axle bearing oil seal.
22. Grind the rear axle bearing inner race surface using a grinder, then chisel them out with a chisel.
23. Remove the rear axle shaft oil seal from the rear axle shaft.

To install:

24. Install a new deflector gasket and deflector to the rear axle shaft.

➡Align the 2 notches.

25. Pass the 6 bolts through the axle hub and install.
26. Install rear axle hub and bearing assembly.
27. Install the rear axle shaft plate washer onto the rear axle shaft.
28. Position the backing plate on a rear axle bearing assembly, and install the 4 parking brake plate to rear axle housing bolts using 2 socket wrenches and a press.
29. Install the rear axle shaft plate washer onto the rear axle shaft.
30. Install a new rear axle bearing retainer inner to the rear axle shaft.
31. With and press and the appropriate tool, install the rear axle shaft to the rear axle bearing assembly.

➡Do not damage the speed sensor rotor.

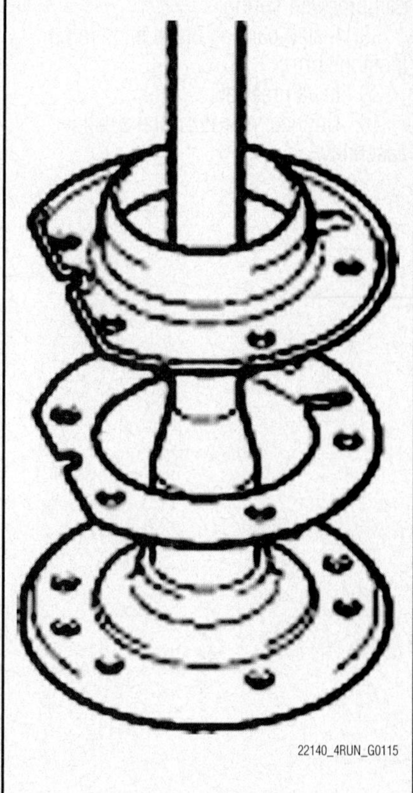
Fig. 170 Installing rear axle deflector gaskets

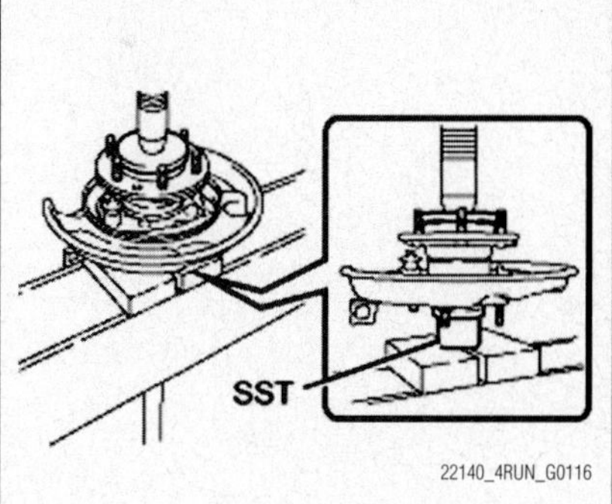

Fig. 171 Assembling rear axle shaft-to-axle bearing assembly

32. Using a snap ring expander, install a new rear axle shaft snap ring.
33. Install a new O-ring.
34. Install the rear axle shaft with backing plate with the 4 nuts, tighten to 89 ft. lbs. (120 Nm).

➡Do not damage the speed sensor rotor. Inspect no damage and no foreign matter at the speed sensor rotor.

35. Install parking brake shoe.
36. Install parking brake shoe strut.

37. Install parking brake shoe strut compression spring.
38. Install parking brake shoe return tension spring.
39. Install rear disc.
40. Connect rear disc brake caliper assembly.
41. Install rear speed sensor (w/ ABS).
42. Fill up differential oil as necessary.
43. Inspect brake fluid level in reservoir.
44. Inspect brake fluid leakage.
45. Install rear wheel tighten to 83 ft. Lbs. (112 nm).
46. Inspect and adjust parking brake lever travel.
47. Inspect abs speed sensor signal (w/ ABS).

TOYOTA

Avalon

10

SPECIFICATIONS AND MAINTENANCE CHARTS

ENGINE AND VEHICLE IDENTIFICATION

Engine							Model Year	
Code ①	Liters (cc)	Cu. In.	Cyl.	Fuel Sys.	Engine Type	Eng. Mfg.	Code ②	Year
2GR-FE	3.5 (3456)	210	6	SFI	DOHC	Toyota	7	2007
2GR-FE	3.5 (3456)	210	6	SFI	DOHC	Toyota	8	2008

SFI: Sequential Fuel Injection

DOHC: Double Overhead Camshaft

NA: Information not available

① Stamped on the left side of the engine block

② 10th digit of the Vehicle Identification Number (VIN)

22140_AVAL_C0001

GENERAL ENGINE SPECIFICATIONS

Year	Model	Engine Displacement Liters	Engine Series ID	Net Horsepower @ rpm	Net Torque @ rpm (ft. lbs.)	Bore x Stroke (in.)	Com-pression Ratio	Oil Pressure @ rpm
2007	Avalon	3.5	2GR-FE	270@6200	251@4700	3.70x3.27	10.8:1	36-78@3000
2008	Avalon	3.5	2GR-FE	270@6200	251@4700	3.70x3.27	10.8:1	36-78@3000

22140_AVAL_C0002

ENGINE TUNE-UP SPECIFICATIONS

Year	Engine Displacement Liters	Engine ID	Spark Plug Gap (in.)	Ignition Timing (deg.)*	Fuel Pump (psi)	Idle Speed (rpm)	Valve Clearance	
							Intake	Exhaust
2007	3.5	2GR-FE	0.039-0.043	N/A	44-50	650-750	NA	NA
2008	3.5	2GR-FE	0.039-0.043	N/A	44-50	650-750	NA	NA

NOTE: The Vehicle Emission Control Information label often reflects specification changes made during production.

The label figures must be used if they differ from those in this chart.

NA: Not available

22140_AVAL_C0003

CAPACITIES

Year	Model	Engine Displacement Liters	Engine ID	Engine Oil with Filter (qts.)	Transmission (qts.) 5-Spd	Transmission (qts.) Auto.*	Transfer Case (pts.)	Drive Axle Front (pts.)	Drive Axle Rear (pts.)	Fuel Tank (gal.)	Cooling System (qts.)
2007	Avalon	3.5	2GR-FE	6.4	—	6.8	—	—	—	18.5	8.8
2008	Avalon	3.5	2GR-FE	6.4	—	6.8	—	—	—	18.5	8.8

22140_AVAL_C0004

VALVE SPECIFICATIONS

Year	Engine Displacement Liters	Engine ID	Seat Angle (deg.)	Face Angle (deg.)	Spring Test Pressure (lbs. @ in.)	Spring Installed Height (in.)	Stem-to-Guide Clearance (in.) Intake	Stem-to-Guide Clearance (in.) Exhaust	Stem Diameter (in.) Intake	Stem Diameter (in.) Exhaust
2007	3.5	2GR-FE	45	44.5	NA	NA	0.0010-0.0024	0.0012-0.0026	0.2154-0.2159	0.2151-0.2157
2008	3.5	2GR-FE	45	44.5	NA	NA	0.0010-0.0024	0.0012-0.0026	0.2154-0.2159	0.2151-0.2157

NA: Information not available

22140_AVAL_C0005

CRANKSHAFT AND CONNECTING ROD SPECIFICATIONS

All measurements are given in inches.

Year	Engine Displacement Liters	Engine ID	Crankshaft Main Brg. Journal Dia.	Crankshaft Main Brg. Oil Clearance	Crankshaft Shaft End-play	Crankshaft Thrust on No.	Connecting Rod Journal Diameter	Connecting Rod Oil Clearance	Connecting Rod Side Clearance
2007	3.5	2GR-FE	2.4011-2.4016	0.0010-0.0019	0.0016-0.0095	2	2.0863-2.0866	0.0018-0.0026	0.0059-0.0157
2008	3.5	2GR-FE	2.4011-2.4016	0.0010-0.0019	0.0016-0.0095	2	2.0863-2.0866	0.0018-0.0026	0.0059-0.0157

22140_AVAL_C0006

PISTON AND RING SPECIFICATIONS

All measurements are given in inches.

Year	Engine Displ. Liters	Engine ID	Piston Clearance	Ring Gap Top Comp.	Ring Gap Bottom Comp.	Ring Gap Oil Control	Ring Side Clearance Top Comp.	Ring Side Clearance Bottom Comp.	Ring Side Clearance Oil Control
2007	3.5	2GR-FE	0.0018-0.0020	0.0098-0.0138	0.0197-0.0236	0.0039-0.0157	0.0008-0.0028	0.0008-0.0024	0.0028-0.0059
2008	3.5	2GR-FE	0.0018-0.0020	0.0098-0.0138	0.0197-0.0236	0.0039-0.0157	0.0008-0.0028	0.0008-0.0024	0.0028-0.0059

22140_AVAL_C0007

TORQUE SPECIFICATIONS

All readings in ft. lbs.

Year	Engine Displacement Liters	Engine ID	Cylinder Head Bolts	Main Bearing Bolts	Rod Bearing Bolts	Crankshaft Damper Bolts	Flywheel Bolts	Manifold Intake	Manifold Exhaust	Spark Plugs	Oil Pan Drain Plug
2007	3.5	2GR-FE	①	②	③	184	61	15	15	13	30
2008	3.5	2GR-FE	①	②	③	184	61	15	15	13	30

① Step 1: 10 mm bolts to 27 ft. lbs.

Step 2: 10mm point cap bolts plus 90 degrees

Step 3: 10mm point cap bolts plus 90 degrees

Step 4: Front bolts to 22 ft. lbs.

③ Step 1: 16 cap bolts to 45 ft. lbs.

Step 2: 16 cap bolts plus 90 degrees

Step 3: 8 side bolts to 38 ft. lbs.

③ Step 1: 18 ft. lbs.

Step 2: Plus 90 degrees

22140_AVAL_C0008

WHEEL ALIGNMENT

Year	Model		Caster Range (+/-Deg.)	Caster Preferred Setting (Deg.)	Camber Range (+/-Deg.)	Camber Preferred Setting (Deg.)	Toe-in (in.)	Steering Axis Inclination (Deg.)
2007	Avalon XL	Front	0.75	2.65	0.75	-0.67	0+/-0.04	12.25+/-0.75
		Rear	—	—	0.75	-1.15	0.16+/-0.08	—
	Avalon Touring	Front	0.75	2.72	0.75	-0.72	0+/-0.04	12.37+/-0.75
		Rear	—	—	0.75	-1.22	0.16+/-0.08	—
	Avalon XLS	Front	0.75	2.70	0.75	-0.72	0+/-0.04	12.33+/-0.75
		Rear	—	—	0.75	-1.22	0.16+/-0.08	—
	Avalon Limited	Front	0.75	2.80	0.75	-0.72	0+/-0.04	12.33+/-0.75
		Rear	—	—	0.75	-1.25	0.16+/-0.08	—
2008	Avalon XL	Front	0.75	2.65	0.75	-0.67	0+/-0.04	12.25+/-0.75
		Rear	—	—	0.75	-1.15	0.16+/-0.08	—
	Avalon Touring	Front	0.75	2.72	0.75	-0.72	0+/-0.04	12.37+/-0.75
		Rear	—	—	0.75	-1.22	0.16+/-0.08	—
	Avalon XLS	Front	0.75	2.70	0.75	-0.72	0+/-0.04	12.33+/-0.75
		Rear	—	—	0.75	-1.22	0.16+/-0.08	—
	Avalon Limited	Front	0.75	2.80	0.75	-0.72	0+/-0.04	12.33+/-0.75
		Rear	—	—	0.75	-1.25	0.16+/-0.08	—

22140_AVAL_C0009

TIRE, WHEEL AND BALL JOINT SPECIFICATIONS

Year	Model	OEM Tires Standard	OEM Tires Optional	Tire Pressures (psi) Front	Tire Pressures (psi) Rear	Wheel Size	Ball Joint Inspection	Lug Nut Torque (ft. lbs.)
2007	Avalon	P215/60R16	P215/55R17	29	29	6.5-JJ	①	76
2008	Avalon	P215/60R16	P215/55R17	29	29	6.5-JJ	①	76

OEM: Original Equipment Manufacturer

PSI: Pounds Per Square Inch

STD: Standard

OPT: Optional

① Replace if any measurable movement is found.

22140_AVAL_C0010

BRAKE SPECIFICATIONS

All measurements in inches unless noted

Year	Model		Brake Disc Original Thickness	Brake Disc Minimum Thickness	Brake Disc Maximum Runout	Minimum Lining Thickness	Brake Caliper Bracket Bolts (ft. lbs.)	Brake Caliper Mounting Bolts (ft. lbs.)
2007	Avalon	F	1.102	0.983	0.0020	0.039	79	25
		R	0.390	0.334	0.0059	0.039	46	32
2008	Avalon	F	1.102	0.983	0.0020	0.039	79	25
		R	0.390	0.334	0.0059	0.039	46	32

F: Front

R: Rear

22140_AVAL_C0011

SCHEDULED MAINTENANCE INTERVALS

TOYOTA—AVALON

TO BE SERVICED	TYPE OF SERVICE	VEHICLE MILEAGE INTERVAL (x1000)													
		5	10	15	20	25	30	35	40	45	50	55	60	90	120
Engine oil & filter	R	✓	✓	✓	✓	✓	✓	✓	✓	✓	✓	✓	✓	✓	✓
Automatic transmission fluid	S/I			✓			✓			✓			✓	✓	✓
Ball joints & dust covers	S/I			✓			✓			✓			✓	✓	✓
Bolts & nuts on chassis & body	S/I			✓			✓			✓			✓	✓	✓
Brake linings & drums	S/I	✓	✓	✓	✓	✓	✓	✓	✓	✓	✓	✓	✓	✓	✓
Brake line pipes & hoses	S/I			✓			✓			✓			✓	✓	✓
Brake pads & discs (front & rear)	S/I	✓	✓	✓	✓	✓	✓	✓	✓	✓	✓	✓	✓	✓	✓
Brake fluid	R						✓						✓	✓	✓
Rack and pinion assembly	S/I			✓			✓			✓			✓	✓	✓
Steering linkage & boots	S/I			✓			✓			✓			✓	✓	✓
Air cleaner filter	R						✓						✓	✓	✓
Spark plugs ①	R														✓
Drive belts	S/I												✓	✓	✓
Exhaust pipes & mountings	S/I			✓			✓			✓			✓	✓	✓
Fuel lines & connections	S/I						✓						✓	✓	✓
Engine coolant ②	S/I			✓			✓			✓			✓	✓	
Fuel tank cap gasket	S/I						✓						✓	✓	✓
Rotate tires	S/I			✓			✓			✓			✓		✓
Clean air conditioning filter ③	S/I			✓			✓			✓			✓		✓
Axle shaft bolts	S/I			✓			✓			✓			✓	✓	✓
Brake pad thickness and rotor runout	S/I						✓						✓	✓	✓

R: Replace S/I: Service or Inspect

① Spark plugs are replaced at 120,000 miles

② Replace engine coolant at 100,000 miles and then inspect every 15,000 miles

③ Replace air conditioning filter every 30,000 miles

FREQUENT OPERATION MAINTENANCE (SEVERE SERVICE)

If a vehicle is operated under any of the following conditions it is considered severe service:

- Extremely dusty areas.

- 50% or more of the vehicle operation is in 32°C (90°F) or higher temperatures, or constant temperatures below 0°C (32°F).

- Prolonged idling (vehicle operation in stop and go traffic).

- Frequent short running periods (engine does not warm to normal operating temperatures).

- Police, taxi, delivery usage or trailer towing usage.

Air cleaner filter: service or inspect every 5000 miles

Rear differential & transfer case oil: replace every 15,000 miles.

Ball joints & dust covers: service or inspect every 5000 miles.

Bolts & nuts on chassis & body: service or inspect every 5000 miles.

Axle shaft bolts: service or inspect every 5000 miles.

Steering linkage: service or inspect every 5000 miles.

22140_AVAL_C0012

PRECAUTIONS

Before servicing any vehicle, please be sure to read all of the following precautions, which deal with personal safety, prevention of component damage, and important points to take into consideration when servicing a motor vehicle:

- Never open, service or drain the radiator or cooling system when the engine is hot; serious burns can occur from the steam and hot coolant.
- Observe all applicable safety precautions when working around fuel. Whenever servicing the fuel system, always work in a well-ventilated area. Do not allow fuel spray or vapors to come in contact with a spark, open flame, or excessive heat (a hot drop light, for example). Keep a dry chemical fire extinguisher near the work area. Always keep fuel in a container specifically designed for fuel storage; also, always properly seal fuel containers to avoid the possibility of fire or explosion. Refer to the additional fuel system precautions later in this section.
- Fuel injection systems often remain pressurized, even after the engine has been turned **OFF**. The fuel system pressure must be relieved before disconnecting any fuel lines. Failure to do so may result in fire and/or personal injury.
- Brake fluid often contains polyglycol ethers and polyglycols. Avoid contact with the eyes and wash your hands thoroughly after handling brake fluid. If you do get brake fluid in your eyes, flush your eyes with clean, running water for 15 minutes. If eye irritation persists, or if you have taken brake fluid internally, IMMEDIATELY seek medical assistance.
- The EPA warns that prolonged contact with used engine oil may cause a number of skin disorders, including cancer. You should make every effort to minimize your exposure to used engine oil. Protective gloves should be worn when changing oil. Wash your hands and any other exposed skin areas as soon as possible after exposure to used engine oil. Soap and water, or waterless hand cleaner should be used.
- All new vehicles are now equipped with an air bag system, often referred to as a Supplemental Restraint System (SRS) or Supplemental Inflatable Restraint (SIR) system. The system must be disabled before performing service on or around system components, steering column, instrument panel components, wiring and sensors. Failure to follow safety and disabling procedures could result in accidental air bag deployment, possible personal injury and unnecessary system repairs.
- Always wear safety goggles when working with, or around, the air bag system. When carrying a non-deployed air bag, be sure the bag and trim cover are pointed away from your body. When placing a non-deployed air bag on a work surface, always face the bag and trim cover upward, away from the surface. This will reduce the motion of the module if it is accidentally deployed. Refer to the additional air bag system precautions later in this section.
- Clean, high quality brake fluid from a sealed container is essential to the safe and proper operation of the brake system. You should always buy the correct type of brake fluid for your vehicle. If the brake fluid becomes contaminated, completely flush the system with new fluid. Never reuse any brake fluid. Any brake fluid that is removed from the system should be discarded. Also, do not allow any brake fluid to come in contact with a painted surface; it will damage the paint.
- Never operate the engine without the proper amount and type of engine oil; doing so WILL result in severe engine damage.
- Timing belt maintenance is extremely important. Many models utilize an interference-type, non-freewheeling engine. If the timing belt breaks, the valves in the cylinder head may strike the pistons, causing potentially serious (also time-consuming and expensive) engine damage. Refer to the maintenance interval charts for the recommended replacement interval for the timing belt, and to the timing belt section for belt replacement and inspection.
- Disconnecting the negative battery cable on some vehicles may interfere with the functions of the on-board computer system(s) and may require the computer to undergo a relearning process once the negative battery cable is reconnected.
- When servicing drum brakes, only disassemble and assemble one side at a time, leaving the remaining side intact for reference.
- Only an MVAC-trained, EPA-certified automotive technician should service the air conditioning system or its components.

BRAKES — ANTI-LOCK BRAKE SYSTEM (ABS)

GENERAL INFORMATION

The Anti-lock Brake System (ABS) helps prevent wheels from locking when the brake is applied firmly or when braking on a slippery surface.

The ABS warning light comes on when the ABS system is malfunctioning.

The Electronic Brake force Distribution (EBD) control utilizes ABS, realizing proper brake force distribution between front and rear wheels in accordance with driving conditions. In addition, during cornering braking, it also controls the brake force of the right and left wheels, helping to maintain vehicle behavior.

Both of the ABS and BRAKE warning lights come on to indicate a malfunction in the EBD system.

Based on the signals received from the 4 wheel speed sensors, the skid control ECU calculates the speed of each wheel and deceleration, and checks wheel slipping condition. According to the slipping condition, the skid control ECU controls the pressure holding valve and pressure reduction valve in order to adjust the fluid pressure of each wheel cylinder

The primary purpose of the Brake Assist (BA) system is to provide an auxiliary brake force to assist the driver who cannot generate a large enough brake force during emergency braking. This system helps to maximize the brake performance of the vehicle. The ABS warning light comes on to indicate a malfunction in the BA (brake assist) system.

BRAKES — BLEEDING THE BRAKE SYSTEM

BLEEDING PROCEDURE

Bleeding the Master Cylinder

If the master cylinder is reinstalled or if the reservoir becomes empty, bleed the air from the master cylinder. To prevent brake fluid from adhering, cover nearly painted surfaces with a shop rag or a piece of cloth.

1. Using a union nut wrench (10 mm), disconnect the 2 brake lines from the master cylinder, using a suitable brake line wrench.
2. Have an assistant slowly depress the brake pedal and hold it.
3. Cover the 2 outer holes with your fingers, and have your assistant release the brake pedal.
4. Repeat the previous 2 steps 3 or 4 times.
5. Using a union nut wrench (10 mm), connect the 2 the brake lines to the master cylinder and tighten to 11 ft. lbs. (15 Nm).

➡Use a torque wrench with a fulcrum length of 250 mm (9.84 in.).

➡This torque value is effective when the union nut wrench is parallel to the torque wrench.

Bleeding the Brake Lines

➡Bleed air from the brake line of the wheel farthest from the master cylinder.

1. Raise and safely support the vehicle.
2. Connect a vinyl tube to the bleeder plug.
3. Have an assistant depress the brake pedal several times, then loosen the bleeder plug while the pedal is depressed.
4. When fluid stops coming out, tighten the bleeder plug, then release the brake pedal.
5. Repeat steps 3 and 4 until all the air in the fluid has been bled out.
6. Tighten the brake bleeder plug to 73 inch lbs. (8.3 Nm).
7. Repeat the above steps to bleed the air out of the brake line for each wheel.

BRAKES — FRONT DISC BRAKES

BRAKE CALIPER

REMOVAL & INSTALLATION

See Figure 1.

1. Before servicing the vehicle, refer to the Precautions Section.
2. Remove the front wheel.
3. Drain brake fluid.

✻✻ WARNING

Do not let brake fluid sit on painted surfaces, as it will eat through the paint. Wash it off immediately.

4. Remove the union bolt and gasket from the disc brake cylinder assembly, then disconnect the front brake flexible hose.
5. Hold the front disc brake cylinder slide pin and remove the 2 bolts and disc brake cylinder assembly.

➡Remove the disc brake cylinder assembly while holding both of the brake pads or the anti-squeal springs may fall off the brake pads.

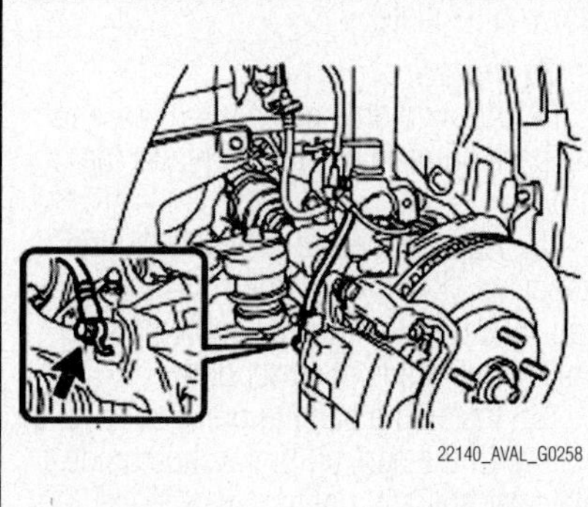

Fig. 1 Remove the union bolt and gasket

To install:

6. Install the disc brake cylinder assembly with the 2 bolts and tighten to 25 ft. lbs. (34 Nm).

➡Be sure that the anti-squeal springs are installed to the front disc brake pads.

7. Connect the flexible hose with the union bolt and a new gasket and tighten to 21 ft. lbs. (29 Nm).

➡Install the front brake flexible hose lock securely in the lock hole in the disc brake cylinder.

8. Fill reservoir with brake fluid.
9. Bleed brake line.
10. Inspect for brake fluid leak.
11. Inspect brake fluid level in reservoir.
12. Install the front wheel and tighten the lug nuts to 76 ft. lbs. (103 Nm).

DISC BRAKE PADS

REMOVAL & INSTALLATION

See Figure 2.

1. Before servicing the vehicle, refer to the Precautions Section.
2. Remove the 2 front disc brake cylinder slide pins (upper and lower) from the front disc brake cylinder mounting.
3. Remove brake cylinder.
4. Remove the 2 anti-squeal springs.
5. Remove the 2 brake pads from the front disc brake cylinder mounting.

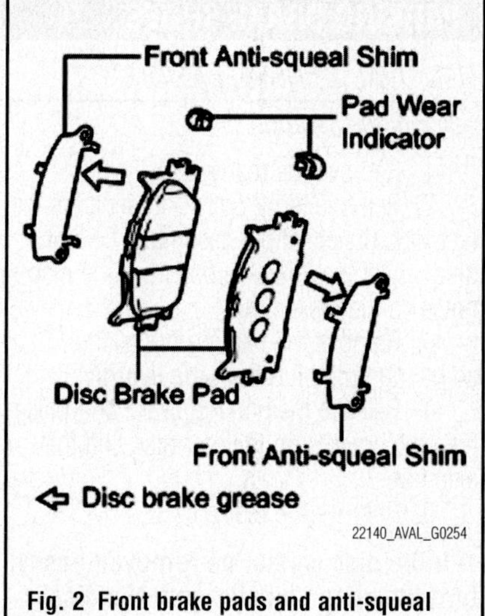

Fig. 2 Front brake pads and anti-squeal shims

To install:

6. Install the 2 brake pads with front anti-squeal shims to the front disc brake cylinder mounting.

➡When replacing worn pads, the front anti-squeal springs must be replaced at the same time.

➡Be sure to install the anti-squeal springs into the front disc brake pad installation holes as far as they will go.

7. Install the 2 front disc brake cylinder slide pins (upper and lower) from the front disc brake cylinder mounting.
8. Install the brake cylinder.

BRAKES — REAR DISC BRAKES

BRAKE CALIPER

REMOVAL & INSTALLATION

1. Before servicing the vehicle, refer to the Precautions Section.
2. Remove the rear wheel.
3. Drain brake fluid.

WARNING

Do not let brake fluid sit on painted surfaces, as it will eat through the paint. Wash it off immediately.

4. Remove the union bolt and the gasket from the rear disc brake cylinder assembly, then disconnect the rear brake flexible hose.
5. Hold the 2 rear disc brake cylinder slide pins and remove the 2 bolts and rear disc brake cylinder assembly.

To install:

6. Install the rear disc brake cylinder assembly with the 2 bolts and tighten to 20 ft. lbs. (77 Nm).
7. Connect the rear brake flexible hose with the union bolt and a new gasket and tighten to 24 ft. lbs. (33 Nm).
8. Fill reservoir with brake fluid.
9. Bleed brake line.
10. Inspect for brake fluid leak.
11. Inspect brake fluid level in reservoir.
12. Install the rear wheel and tighten the lug nuts to 76 ft. lbs. (103 Nm).

DISC BRAKE PADS

REMOVAL & INSTALLATION

1. Before servicing the vehicle, refer to the Precautions Section.
2. Remove the 2 rear disc brake cylinder slide pins (upper and lower) from the rear disc brake cylinder mounting.
3. Remove brake cylinder.
4. Remove the 2 brake pads with the rear anti-squeal shims.

To install:

5. Installation is the reverse of removal procedure.

BRAKES — PARKING BRAKE

PARKING BRAKE SHOES

REMOVAL & INSTALLATION

See Figures 3 through 7.

1. Remove the rear wheel.
2. Remove the 2 bolts and separate the rear disc brake caliper assembly. Do not disconnect the flexible hose from the disc brake caliper assembly.
3. Remove the parking brake shoe adjusting hole plug from the rear disc.
4. Release the parking brake and place the matchmarks on the rear disc and the axle hub.
5. Remove the rear disc.

➡If the disc cannot be removed easily, turn the shoe adjuster until the disc turns freely.

6. Using needle-nose pliers, remove the 2 parking brake shoe return tension No. 1 springs.

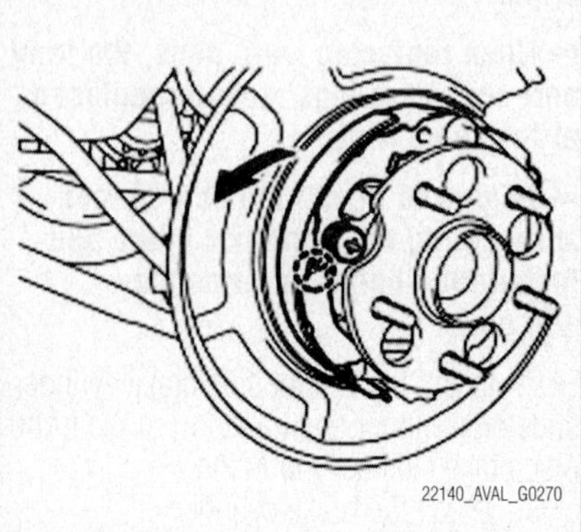

Fig. 3 Remove the No. 1 parking brake shoe assembly

7. Remove the parking brake shoe strut and the parking brake shoe strut compression spring.
8. Remove the No. 1 parking brake shoe assembly, as follows:
   a. Release the claw of the parking brake shoe hold down spring No. 2 cup.
   b. Remove the No. 1 parking brake shoe assembly as shown in the illustration.
   c. Remove the parking brake shoe hold down spring No. 1 cup, the parking brake shoe hold down spring, the parking brake shoe hold down spring No. 2 cup, and the parking brake shoe hold down spring No. 1 pin.
9. Remove the parking brake shoe adjusting screw set.
10. Remove the parking brake shoe return tension No. 2 spring.
11. Remove the No. 2 parking brake shoe assembly, as follows:

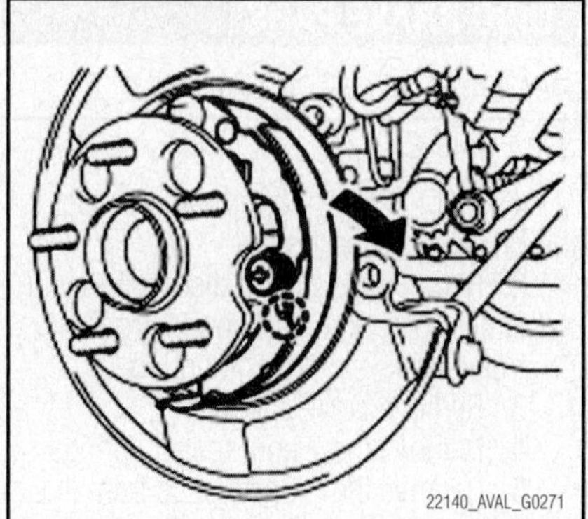

Fig. 4 Remove the No. 2 parking brake shoe assembly

   a. Release the claw of the parking brake shoe hold down spring No. 2 cup.
   b. Remove the No. 2 parking brake shoe assembly as shown in the illustration.
   c. Remove the parking brake shoe hold down spring No. 1 cup, the parking brake shoe hold down spring, the parking brake shoe hold down spring No. 2 cup, and the parking brake shoe hold down spring No. 2 pin.
   d. Using needle-nose pliers, disconnect the No. 3 parking brake cable assembly from the parking brake shoe lever.

➡Be careful not to damage the No. 3 parking brake cable assembly.

12. Using a screwdriver, remove the C-washer, shim and the parking brake shoe lever.
13. Remove the parking brake shoe guide plate set bolt and the parking brake shoe guide plate.

To install:

14. Apply high temperature grease to the backing plate where it contacts the shoe.
15. Apply adhesive (Toyota Genuine Adhesive 1344, Three Bond 1344 or equivalent) to the threads of the parking brake shoe guide plate set bolt.
16. Install the parking brake shoe guide plate with the parking brake shoe guide plate set bolt and tighten to 13 ft. lbs. (18 Nm).
17. Install the parking brake shoe lever and shim to the No. 2 parking brake shoe assembly with a new C-washer.

Shim Thickness	Shim Thickness
0.3 mm (0.012 in.)	0.9 mm (0.035 in.)
0.6 mm (0.024 in.)	-

22140_AVAL_G0272

Fig. 5 Shim thickness

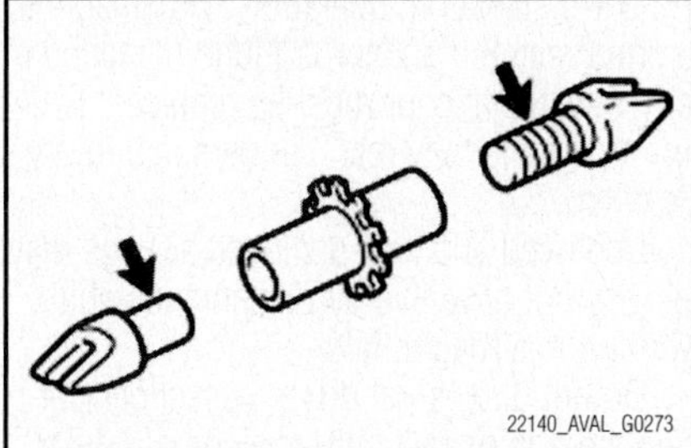

Fig. 6 Apply high temperature grease to the parking brake shoe adjusting screw set

18. Using a feeler gauge, measure the clearance between the No. 2 parking brake shoe assembly and parking brake shoe lever. Standard clearance: Less than 0.35 mm (0.014 in.). If the clearance is not as specified, replace the shim with one of the correct size.

19. Install the No. 2 parking brake shoe assembly as follows:

 a. Using needle-nose pliers, connect the No. 3 parking brake cable assembly to the parking brake shoe lever.

 b. Install the No. 2 parking brake shoe assembly with the parking brake shoe hold down spring No. 2 pin, the parking brake shoe hold down spring No. 2 cup, the parking brake shoe hold down spring and the parking brake shoe hold down spring No. 1 cup.

 c. Engage the claw of the parking brake shoe hold down spring No. 2 cup to the No. 2 parking brake shoe assembly.

20. Install the parking brake shoe adjusting screw set, as follows:

 a. Apply high temperature grease to the parking brake shoe adjusting screw set as shown in the illustration.

 b. Install the parking brake shoe return tension No. 2 spring to the No. 1 parking brake shoe assembly and the No. 2 parking brake shoe assembly.

 c. Install the parking brake shoe adjusting screw set to the No. 1 parking brake shoe assembly and the No. 2 parking brake shoe assembly.

21. Install the No. 1 parking brake shoe assembly as follows:

 a. Install the No. 1 parking brake shoe assembly with the parking brake shoe hold down spring No. 1 pin, parking brake shoe hold down spring No. 2 cup, parking brake shoe hold down spring and parking brake shoe hold down spring No. 1 cup.

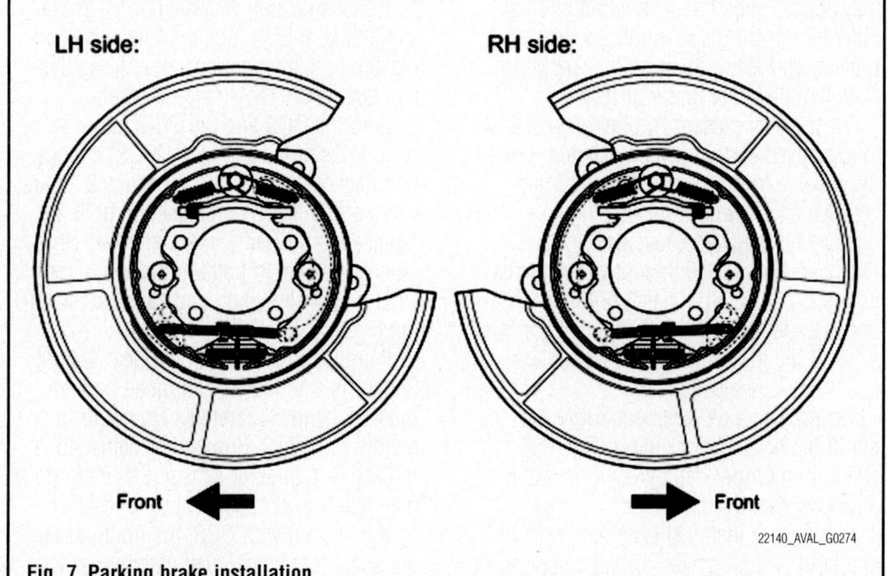

Fig. 7 Parking brake installation

 b. Engage the claw of the parking brake shoe hold down spring No. 2 cup to the No. 1 parking brake shoe assembly.

22. Attach the parking brake shoe strut and the parking brake shoe strut compression spring to the No. 1 parking brake shoe assembly and No. 2 parking brake shoe assembly.

23. Using needle-nose pliers, install the 2 parking brake shoe return tension No. 1 springs. First install the front side spring and then the rear side spring.

24. Inspect parking brake installation and check that each part is installed properly.

➡There should be no oil or grease on the friction surfaces of the shoe linings and discs.

25. Install the rear disc.

26. Install the parking brake shoe adjusting hole plug.

27. Adjust parking brake shoe clearance.

28. Install the rear disc brake caliper assembly with the 2 bolts and tighten to 46 ft. lbs. (62 Nm).

29. Install the rear wheel.

30. Adjust the parking brake pedal travel.

31. Bed in parking brake shoes to discs, as follows:

 a. Drive the vehicle at about 31 mph (50 km/h) on a safe, level and dry road.

 b. Depress the parking brake pedal with 34 lbs. (150 N) of force.

32. Drive the vehicle about 0.25 miles (400 m) in this condition.

 a. Repeat this procedure 3 times using 5-minute intervals between each procedure to prevent the parking brake assembly from overheating.

33. Remove the rear wheel.

34. Adjust parking brake shoe clearance.

35. Adjust the parking brake pedal travel.

36. Install the rear wheel and tighten the lug nuts to 76 ft. lbs. (103 Nm).

ADJUSTMENT

1. Adjust parking brake shoe clearance, as follows:

 a. Temporarily install the hub nuts.

 b. Remove the shoe adjusting hole plug, turn the adjuster and expand the shoes until the disc locks.

 c. Contract the shoe adjuster until the disc rotates smoothly. Standard: returns 8 notches

 d. Check that the disc has no brake drag.

 e. Install the shoe adjusting hole plug.

GENERAL INFORMATION

SERVICE PRECAUTIONS

The vehicle is equipped with a Supplemental Restraint System (SRS). It consists of a driver airbag, front passenger airbag, driver side knee airbag, front seat side airbag and curtain shield airbag. Failure to carry out service operations in the correct sequence could cause the SRS to unexpectedly deploy during servicing, possibly leading to a serious accident. Further, if a mistake is made in servicing the SRS, it is possible that the SRS may fail to operate when required. Before performing servicing (including removal or installation of parts, inspection or replacement), be sure to read the following items carefully, then follow the correct procedures indicated in the repair manual.

Disconnect and isolate the battery negative cable before beginning any airbag system component diagnosis, testing, removal, or installation procedures. Wait at least 90 seconds after the ignition switch is turned off and the negative (-) terminal cable is disconnected from the battery before starting the operation. The SRS is equipped with a backup power source, so if work is started within 90 seconds after disconnecting the negative (-) terminal cable from the battery, the SRS may be deployed. Failure to disable the airbag system may result in accidental airbag deployment, personal injury, or death.

Do not expose the steering pad, front passenger airbag assembly, driver side knee airbag assembly, center airbag sensor assembly, front airbag sensor, front seat side airbag assembly, side airbag sensor, curtain shield airbag assembly, rear airbag sensor, seat position airbag sensor or occupant classification ECU directly to hot air or flames.

Malfunction symptoms of the SRS are difficult to confirm, so DTCs are the most important source of information when troubleshooting. When troubleshooting the SRS, always inspect DTCs before disconnecting the battery.

Even in the case of a minor collision when the SRS does not deploy, the steering pad, front passenger airbag assembly, driver side knee airbag assembly, center airbag sensor assembly, front airbag sensor, front seat side airbag assembly, side airbag sensor, curtain shield airbag assembly, rear airbag sensor, seat position airbag sensor and occupant classification ECU should be inspected.

Before repair work, remove the airbag sensor if any kind of shock is likely to occur to the airbag sensor during the operation.

Never disassemble or repair any of the following parts in order to reuse them. If any of these parts have been dropped, or a defect is found (e.g. cracks, dents or any other defects) in any of the housings, brackets or connectors, then replace the part with a new one: Steering Pad, Front Passenger Airbag Assembly, Driver Side Knee Airbag Assembly, Front Seat Side Airbag Assembly, Curtain Shield Airbag Assembly, Center Airbag Sensor Assembly, Front Airbag Sensor, Front Seat Inner Belt Assembly, Seat Position Sensor, Occupant Classification ECU, Side Airbag Sensor, Rear Airbag Sensor, Front Seat Outer Belt Assembly.

Use an volt/ohmmeter with high impedance (10 Kohms/V minimum) for troubleshooting the electrical circuits.

Information labels are attached near the SRS components. Follow the instructions in the caution.

Do not place an intact undeployed airbag face down on a solid surface. The airbag will propel into the air if accidentally deployed and may result in personal injury or death.

When carrying or handling an undeployed airbag, the trim side (face) of the airbag should be pointing towards the body to minimize possibility of injury if accidental deployment occurs. Failure to do this may result in personal injury or death.

Replace airbag system components with OEM replacement parts. Substitute parts may appear interchangeable, but internal differences may result in inferior occupant protection. Failure to do so may result in occupant personal injury or death.

Wear safety glasses, rubber gloves, and long sleeved clothing when cleaning powder residue from vehicle after an airbag deployment. Powder residue emitted from a deployed airbag can cause skin irritation. Flush affected area with cool water if irritation is experienced. If nasal or throat irritation is experienced, exit the vehicle for fresh air until the irritation ceases. If irritation continues, see a physician.

Do not use a replacement airbag that is not in the original packaging. This may result in improper deployment, personal injury, or death.

The factory installed fasteners, screws and bolts used to fasten airbag components have a special coating and are specifically designed for the airbag system. Do not use substitute fasteners. Use only original equipment fasteners listed in the parts catalog when fastener replacement is required.

During, and following, any child restraint anchor service, due to impact event or vehicle repair, carefully inspect all mounting hardware, tether straps, and anchors for proper installation, operation, or damage. If a child restraint anchor is found damaged in any way, the anchor must be replaced. Failure to do this may result in personal injury or death.

Deployed and non-deployed airbags may or may not have live pyrotechnic material within the airbag inflator.

Do not dispose of driver/passenger/curtain airbags or seat belt tensioners unless you are sure of complete deployment. Refer to the Hazardous Substance Control System for proper disposal.

Dispose of deployed airbags and tensioners consistent with state, provincial, local, and federal regulations.

After any airbag component testing or service, do not connect the battery negative cable. Personal injury or death may result if the system test is not performed first.

If the vehicle is equipped with the Occupant Classification System (OCS), do not connect the battery negative cable before performing the OCS Verification Test using the scan tool and the appropriate diagnostic information. Personal injury or death may result if the system test is not performed properly.

Never replace both the Occupant Restraint Controller (ORC) and the Occupant Classification Module (OCM) at the same time. If both require replacement, replace one, then perform the Airbag System test before replacing the other.

Both the ORC and the OCM store Occupant Classification System (OCS) calibration data, which they transfer to one another when one of them is replaced. If both are replaced at the same time, an irreversible fault will be set in both modules and the OCS may malfunction and cause personal injury or death.

If equipped with OCS, the Seat Weight Sensor is a sensitive, calibrated unit and must be handled carefully. Do not drop or handle roughly. If dropped or damaged, replace with another sensor. Failure to do so may result in occupant injury or death.

If equipped with OCS, the front passenger seat must be handled carefully as well. When removing the seat, be careful when setting on floor not to drop. If dropped, the sensor may be inoperative, could result in occupant injury, or possibly death.

If equipped with OCS, when the passenger front seat is on the floor, no one should sit in the front passenger seat. This uneven force may damage the sensing ability of the seat weight sensors. If sat on and damaged, the sensor may be inoperative, could result in occupant injury, or possibly death.

DISARMING THE SYSTEM

To avoid personal injury when working on vehicles equipped with an air bag, the negative battery cable must be disconnected and at least 90 seconds must elapse before working on the system. Failure to do so may result in deployment of the air bag.

ARMING THE SYSTEM

To arm the system after service is finished, connect the negative battery cable.

CLOCKSPRING CENTERING

See Figures 8 and 9.

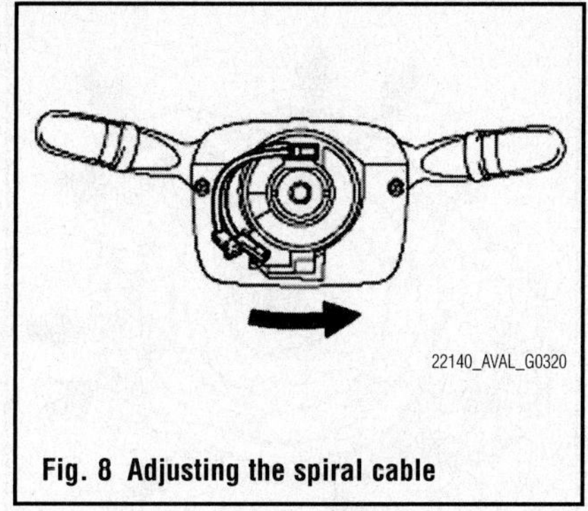

Fig. 8 Adjusting the spiral cable

1. Before servicing the vehicle, refer to the Precautions Section.
2. Check that the ignition switch is **OFF**.
3. Check that the battery negative (-) terminal is disconnected.

CAUTION

After removing the terminal, wait for at least 90 seconds before starting the operation.

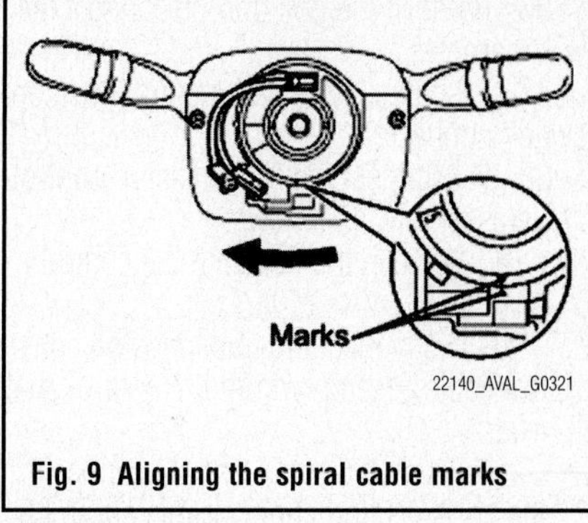

Fig. 9 Aligning the spiral cable marks

4. Rotate the spiral cable counterclockwise slowly by hand until it feels firm.
5. Rotate the spiral cable clockwise approximately 2.5 turns to align the marks.

➡Do not turn the spiral cable by the airbag wire harness.

➡The spiral cable will rotate approximately 2.5 turns to both the left and right from the center.

DRIVETRAIN

AUTOMATIC TRANSAXLE ASSEMBLY

REMOVAL & INSTALLATION

See Figures 10 through 14.

➡When the automatic transaxle is replaced, the transaxle's compensation code must be input into the

➡TCM. When the TCM is replaced, the existing transaxle compensation codes must be input into the new TCM.

➡If the valve body assembly has been replaced, initialize the TCM and perform a road test to allow the TCM to learn.

➡If the shift solenoid SL1 has been replaced, perform a road test to allow the TCM to learn.

➡If the TCM or transaxle has been replaced, register the transmission compensation code in the TCM.

➡Perform RESET MEMORY (AT initialization) when replacing the automatic transmission assembly, engine assembly or ECM.

1. Discharge the fuel system pressure.
2. Disconnect cable from negative battery terminal.
3. Place front wheels facing straight ahead.
4. Remove front wheels.
5. Remove the right and left engine under covers.
6. Remove the right front fender apron seal.
7. Drain engine coolant.
8. Drain automatic transaxle fluid, as follows:
    a. Remove the refill plug and gasket.
    b. Using a 6 mm socket hexagon wrench, remove the overflow plug and gasket.
    c. Using a 6 mm socket hexagon wrench, remove the No. 1 transmission oil filler tube.
    d. Drain the automatic transaxle fluid.
    e. Using a 6 mm socket hexagon wrench, install the No. 1 transmission oil filler tube and tighten to 15 inch lbs. (1.7 Nm).
    f. Using a 6 mm socket hexagon wrench, install the overflow plug with a new gasket and tighten to 30 ft. lbs. (40 Nm).
    g. Install the refill plug with a new gasket and tighten to 36 ft. lbs. (49 Nm).
9. Remove the right and left windshield wiper arm and blade assemblies.
10. Remove the cowl top ventilator louver sub-assembly.
11. Remove the windshield wiper motor and link assembly.
12. Remove the outer cowl top panel.
13. Remove the battery.
14. Remove the cool air intake duct seal.
15. Remove the V-bank cover sub-assembly.
16. Remove the V-ribbed belt.
17. Remove the radiator reserve tank assembly.
18. Remove the right No. 2 engine mounting stay.
19. Remove the engine moving control rod.
20. Remove the No. 2 air cleaner inlet.
21. Remove the air cleaner cap with air cleaner hose.
22. Remove the air cleaner case.
23. Remove the No. 1 air cleaner inlet.
24. Remove the intake air resonator sub-assembly.
25. Disconnect the No. 1 fuel vapor feed hose.
26. Remove the intake air surge tank assembly.
27. Separate the fuel tube sub-assembly.
28. Disconnect the radiator inlet and outlet hoses.
29. Disconnect the No. 1 oil cooler inlet and outlet hoses.
30. Disconnect the heater water inlet hose A.
31. Disconnect the heater water outlet hose A.
32. Remove the ECM.
33. Remove the starter assembly.
34. Remove the upper relay block cover.
35. Disconnect the engine wire.

36. Remove the bolt and disconnect the wire harness.
37. Disconnect the park/neutral position switch connector.
38. Disconnect the transmission control cable assembly, as follows:
   a. Remove the nut from the control shaft lever.
   b. Disconnect the transmission control cable assembly from the control shaft lever.
   c. Remove the clip and disconnect the transmission control cable assembly from the No. 1 transmission control cable bracket.
39. Disconnect the No. 1 oil reservoir to pump hose.
40. Disconnect the steering gear outlet return tube.
41. Remove the front No. 1 exhaust pipe support bracket.
42. Remove the front exhaust pipe assembly.
43. Separate the left and right front stabilizer link assemblies.
44. Remove the left and right front axle shaft nuts.
45. Separate the left and right front speed sensors.
46. Separate the left and right tie rod end sub-assemblies.
47. Separate the left and right front lower No. 1 suspension arm sub-assemblies.
48. Separate the left and right front axle assemblies.
49. Remove the No. 1 exhaust pipe support bracket.
50. Remove the 2 bolts, exhaust pipe support bracket and flywheel housing under cover.
51. Remove drive plate and torque converter clutch setting bolt, as follows:
   a. Turn the crankshaft to gain access and remove the 6 bolts while holding the crankshaft pulley bolt with a wrench.
52. Separate the steering intermediate shaft assembly.
53. Remove the alternator assembly.
54. Separate the cooler compressor assembly.
55. Remove the engine assembly with transaxle.
56. Remove the vane pump assembly.
57. Remove the front frame assembly.
58. Remove the left and right front drive shaft assemblies.
59. Remove the engine wire.
60. Remove the automatic transaxle assembly, as follows:
   a. Remove the bolt and breather bracket from the left camshaft housing sub-assembly.

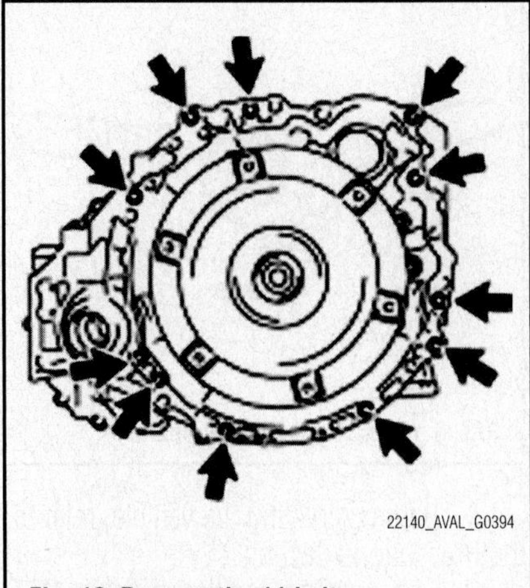

Fig. 10 Remove the 11 bolts

   b. Remove the breather hose from the breather bracket.
   c. Remove the 11 bolts.
   d. Separate and remove the automatic transaxle.
61. Remove the 3 bolts and engine mounting front bracket.
62. Remove the 2 bolts and TCM from the transaxle.
63. Remove the speedometer driven hole cover sub-assembly, as follows:
   a. Remove the bolt and hole cover from the transaxle case.
   b. Remove the O-ring from the hole cover.
64. Remove the wire harness clamp bracket, as follows:
   a. Remove the 4 bolts and 4 clamp brackets.
65. Remove the 2 bolts and No. 1 transmission control cable bracket.
66. Separate the No. 1 oil cooler inlet and outlet hoses.
67. Remove the torque converter clutch assembly from the automatic transaxle assembly.

To install:

68. Inspect the torque converter clutch assembly.
69. Install the torque converter clutch assembly, as follows:
   a. Engage the splines of the input shaft and turbine runner.
   b. Engage the splines of the stator shaft and the stator while turning the torque converter clutch assembly.

➡If the stator shaft splines are difficult to engage with the stator splines, move the torque converter back approximately 10 mm (0.40 in.) and engage the splines while rotating the torque converter.

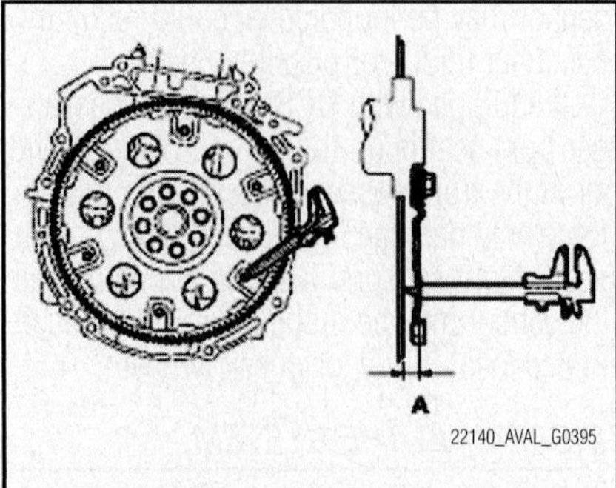

Fig. 11 Measure dimension "A" between the transaxle fitting part of the engine and the converter fitting part of the drive plate

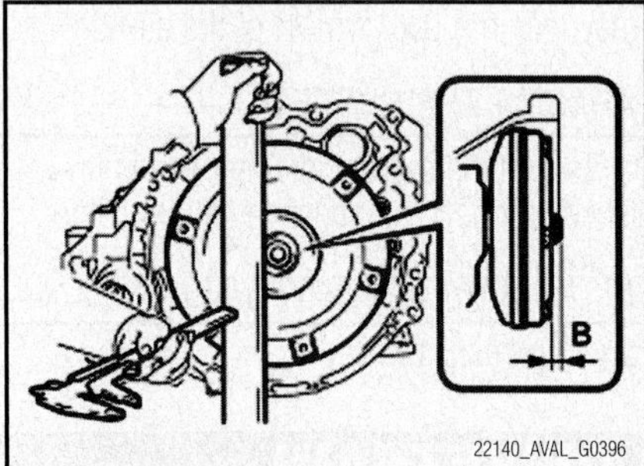

Fig. 12 Measure dimension "B" shown in the illustration and check that "B" is greater than "A"

   c. Turn the torque converter clutch assembly to engage the key of the oil pump drive gear into the slot on the torque converter clutch assembly.
   d. Using a vernier caliper and a straightedge, measure dimension "A" between the transaxle fitting part of the engine and the converter fitting part of the drive plate.
   e. Using a vernier caliper and a straightedge, measure dimension "B" shown in the illustration and check that "B" is greater than "A" (measured in previous step). Standard: A + 1 mm (0.04 in.) or more.

➡Make sure to deduct the thickness of the straightedge.

➡If the transaxle is installed to the engine with the torque converter not sufficiently inserted, the torque converter may be damaged.

➡The standard installation depth is 14 mm (0.55 in.) or more.

70. Install the No. 1 oil cooler inlet and outlet hoses.
71. Install the No. 1 transmission control cable bracket with the 2 bolts and tighten to 9 ft. lbs. (12 Nm).

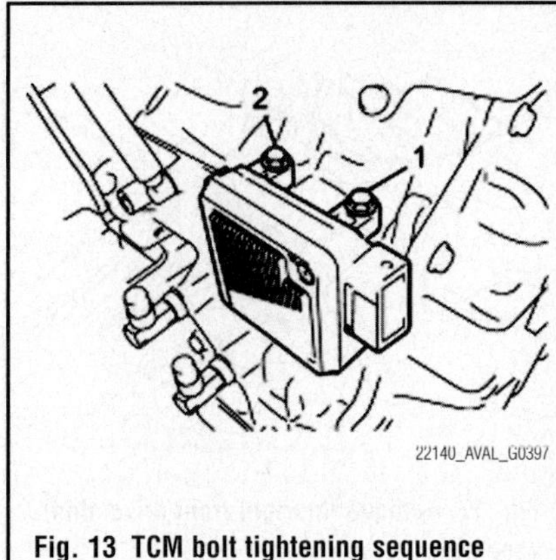

Fig. 13 TCM bolt tightening sequence

72. Install the 4 clamp brackets and 4 bolts and tighten to 74 inch lbs. (8.4 Nm).

73. Install the speedometer driven hole cover sub-assembly, as follows:

a. Coat a new O-ring with ATF, and install it to the hole cover.

b. Install the hole cover to the transaxle case with the bolt and tighten to 49 inch lbs. (5.5 Nm).

74. Install the TCM, as follows:

a. Install the TCM to the automatic transaxle assembly.

b. Install and tighten the 2 bolts in the order shown in the illustration to 8 ft. lbs. (11 Nm).

75. Install the engine mounting front bracket to the automatic transaxle with the 3 bolts and tighten to 47 ft. lbs. (64 Nm).

76. Install the automatic transaxle assembly, as follows:

a. Install the automatic transaxle to the engine with the 11 bolts and tighten A to 47 ft. lbs. (64 Nm), B to 47 ft. lbs. (64 Nm), C to 34 ft. lbs. (46 Nm), an D to 32 ft. lbs. (43 Nm).

➡Make sure that the 2 knock pins are installed on the engine before installing the transaxle.

b. Install the breather bracket to the camshaft housing sub-assembly with the bolt and tighten to 49 inch lbs. (5.5 Nm).

c. Install the breather hose to the breather bracket.

77. Connect the park/neutral position switch connector.

78. Connect the wire harness with the bolt and tighten to 9 ft. lbs. (12 Nm).

79. Install the engine wire.

80. Install the left and right front drive shaft assemblies.

81. Install the front frame assembly.

82. Install the vane pump assembly.

83. Install the engine assembly with transaxle.

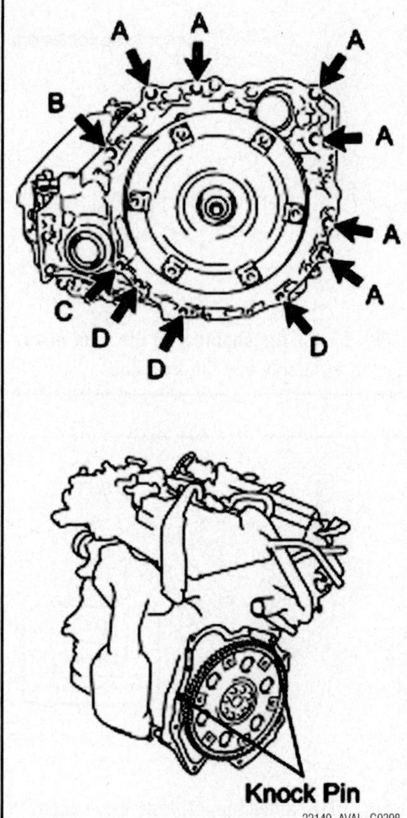

Fig. 14 Automatic transaxle bolt sequence

84. Install the cooler compressor assembly.

85. Install the alternator assembly.

86. Install the steering intermediate shaft assembly.

87. Install the drive plate and torque converter clutch setting bolt, as follows:

a. Apply a few drops of adhesive (Toyota Genuine Adhesive 1324, Three Bond 1324 or equivalent) to 2 threads on the tip of the 6 torque converter clutch mounting bolts.

b. Install the 6 torque converter clutch mounting bolts and tighten to 30 ft. lbs. (41 Nm). First install the black colored bolt, and then the remaining 5 bolts.

88. Install the flywheel housing under cover and exhaust pipe support bracket to the automatic transaxle with the 2 bolts tighten to 69 inch lbs. (7.8 Nm).

89. Install the No. 1 exhaust pipe support bracket.

90. Install the left and right front axle assemblies.

91. Install the left and right front lower No. 1 suspension arm sub-assemblies.

92. Install the left and right tie rod end sub-assemblies.

93. Install the left and right front speed sensors.

94. Install the left and right front stabilizer link assemblies.

95. Install the left and right front axle shaft nuts.

96. Install the front exhaust pipe assembly.

97. Install the No. 1 exhaust pipe support bracket.

98. Connect the steering gear outlet return tube.

99. Connect the No. 1 oil reservoir to pump hose.

100. Connect the transmission control cable assembly, as follows:

a. Install the transmission control cable assembly to the control shaft lever with the nut and tighten to 10 ft. lbs. (13 Nm).

b. Connect the transmission control cable assembly with a new clip to the No. 1 transmission control cable bracket.

101. Connect the engine wire.

102. Install the upper relay block cover.

103. Install the starter assembly.

104. Install the ECM.

105. Connect the heater water outlet hose A.

106. Connect the heater water inlet hose A.

107. Connect the No. 1 oil cooler inlet and outlet hoses.

108. Install the radiator inlet and outlet hoses.

109. Connect the fuel tube sub-assembly.

110. Install the intake air surge tank assembly.

111. Connect the No. 1 fuel vapor feed hose.

112. Install the intake air resonator sub-assembly.

113. Install the No. 1 air cleaner inlet.

114. Install the air cleaner case.

115. Install the air cleaner cap with air cleaner hose.

116. Install the No. 2 air cleaner inlet.

117. Connect the vacuum hoses.

118. Install the engine moving control rod.

119. Install the right No. 2 engine mounting stay.

120. Install the radiator reserve tank assembly.

121. Install the V-ribbed belt.

122. Install the battery.

123. Install the outer cowl top panel.

124. Install the windshield wiper motor and link assembly.

125. Install the cowl top ventilator louver sub-assembly.

126. Install the right and left windshield wiper arm and blade assemblies.

127. Install the front wheels.
128. Add coolant and power steering fluid.
129. Bleed the power steering system.
130. Add automatic transaxle fluid.
131. Inspect and adjust shift lever position.
132. Inspect for coolant leak, fuel leak, and exhaust gas leak.
133. Inspect ignition timing.
134. Inspect the engine idle speed.
135. Check the function of throttle body.
136. Install the V-bank cover sub-assembly.
137. Install the cool air intake duct seal.
138. Inspect CO/HC.
139. Inspect the ABS speed sensor signal.
140. Inspect and adjust front wheel alignment.
141. Install the right front fender apron seal.
142. Install the right and left engine under covers.

HALFSHAFT

REMOVAL & INSTALLATION

See Figures 15 through 19.

1. Remove the engine under cover.
2. Remove the drain plug and gasket, and then drain the automatic transaxle fluid.
3. Install a new gasket and drain plug and tighten to 36 ft. lbs. (49 Nm).
4. Remove front wheel.
5. Using SST (SST: 09930-00010) and hammer, release the staked part of the front axle hub nut.

➡Loosen the staked part of the nut completely, otherwise the screw of the drive shaft may be damaged.

6. While applying the brakes, remove the front axle hub nut.
7. Remove the nut and separate the front stabilizer link assembly.

➡If the ball joint turns together with the nut, use a hexagon wrench (6mm) to hold the stud.

8. Remove the bolt and clip, and separate the speed sensor wire and flexible hose from the shock absorber.
9. Remove the bolt and separate the front speed sensor from the steering knuckle.

➡Do not allow foreign matter to adhere to the speed sensor. Be careful not to damage the speed sensor.

10. Separate tie rod end sub-assembly, as follows:

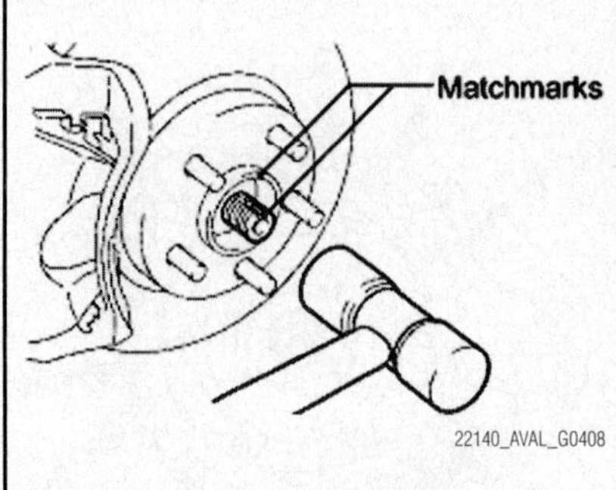

Fig. 15 Put matchmarks on the front drive shaft assembly and the axle hub

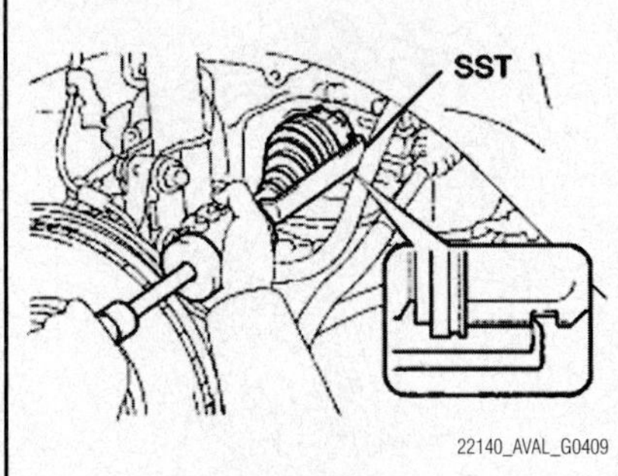

Fig. 16 Remove the left front drive shaft assembly

a. Remove the cotter pin and nut.
b. Using SST (SST: 09628-62011) or equivalent, separate the tie rod end sub-assembly from the steering knuckle.

➡Do not damage the ball joint dust cover.

11. Remove the bolt and 2 nuts, and separate the lower No. 1 front suspension arm sub-assembly from the lower ball joint.
12. Put matchmarks on the front drive shaft assembly and the axle hub.
13. Using a plastic hammer, separate the front drive shaft assembly from the front axle hub sub-assembly.

➡Be careful not to damage the drive shaft boot and speed sensor rotor.

14. Remove the front drive shaft assembly(s), as follows:
a. For left front drive shaft, use SST (SST: 09520-01010, SST: 09520-24010) or equivalent, and remove the left front drive shaft assembly.

➡Be careful not to damage the drive shaft dust cover, boot and oil seal. Be careful not to drop the drive shaft assembly.

b. For right front drive shaft, use a

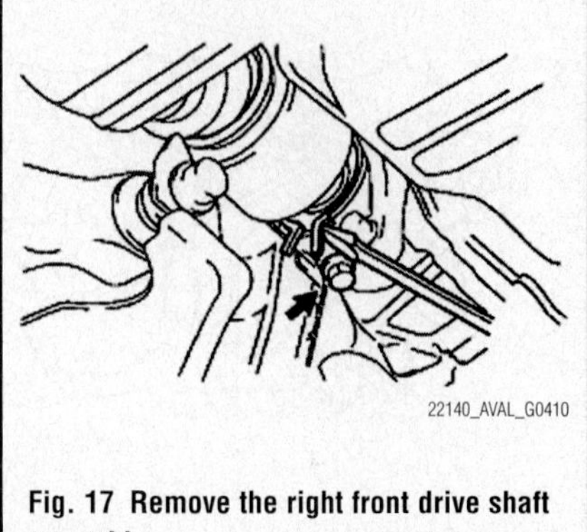

Fig. 17 Remove the right front drive shaft assembly

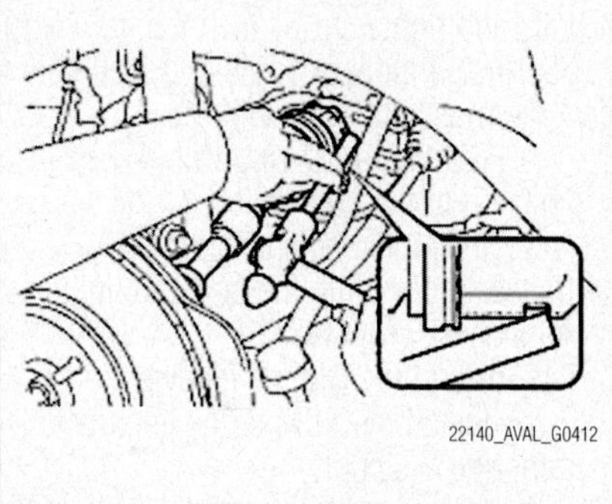

Fig. 18 Install the left front drive shaft assembly

screwdriver and remove the bearing bracket hole snap ring.
c. Remove the bolt and right front drive shaft assembly from the drive shaft bearing bracket.

➡Do not damage the boot and oil seal.

15. Secure the front axle hub bearing. The hub bearing could be damaged if it is subjected to the vehicle's full weight, such as moving the vehicle with the drive shaft removed. If it is necessary to place the vehicle's weight on the hub bearing, first support it with SST (SST: 09608-16042).

To install:

16. Install the front drive shaft assembly(s), as follows:
a. Coat the spline of the inboard joint shaft assembly with automatic transaxle fluid.
b. For the left front drive shaft, align the shaft splines and install the drive shaft assembly with a brass bar and hammer.

➡Set the shaft snap ring with the opening side facing down.

c. For the right front drive shaft, install the drive shaft and use a screwdriver to install a new bearing bracket

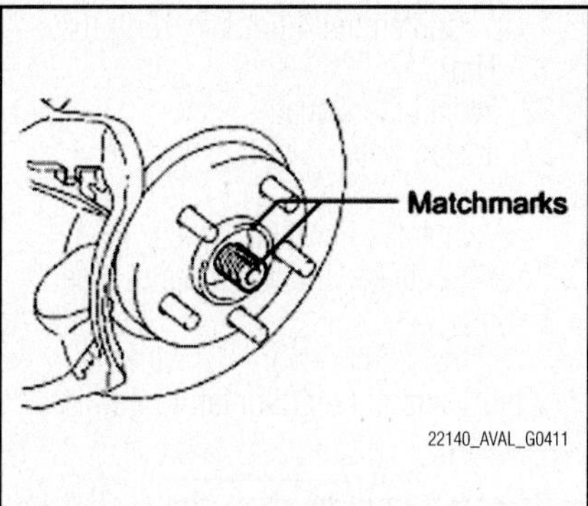

Fig. 19 Align the matchmarks and install the front drive shaft assembly

hole snap ring, and install a new bolt tightened to 24 ft. lbs. (32 Nm).

➡Be careful not to damage the drive shaft dust cover, boot and oil seal.

➡Move the drive shaft assembly while keeping it level.

17. Align the matchmarks and install the front drive shaft assembly to the front axle hub sub-assembly.

➡Be careful not to damage the drive shaft boot and speed sensor rotor.

18. Install the lower ball joint to the lower No. 1 front suspension arm sub-assembly with the bolt and 2 nuts and tighten to 55 ft. lbs. (75 Nm).
19. Install the tie rod end sub-assembly to the steering knuckle with the nut and tighten to 36 ft. lbs. (49 Nm).
20. Install a new cotter pin. If the holes for the cotter pin are not aligned, tighten the nut up to 60° further.
21. Install the front speed sensor to the steering knuckle with the bolt and tighten to 71 inch lbs. (8 Nm).
22. Install the flexible hose and the speed sensor to the shock absorber with the bolt and set the sensor clip on the knuckle and tighten to 14 ft. lbs. (19 Nm).

➡Be careful not to damage the speed sensor. Do not allow foreign matter to adhere to the speed sensor. Do not twist the sensor wire when installing the speed sensor.

23. Install the stabilizer link assembly with the nut and tighten to 55 ft. lbs. (74 Nm).

➡If the ball joint turns together with the nut, use a hexagon wrench (6 mm) to hold the stud.

24. Clean the threaded parts on the drive shaft and axle hub nut using a non-residue solvent.

➡Be sure to perform this work for a new drive shaft. Keep the threaded parts free of oil and foreign objects.

25. Using a socket wrench (30 mm), install a new axle hub nut and tighten to 217 ft. lbs. (294 Nm).
26. Using a chisel and hammer, stake the front axle hub nut.
27. Install front wheel.
28. Add automatic transaxle fluid.
29. Inspect automatic transaxle fluid.
30. Inspect and adjust front wheel alignment.
31. Install the engine under cover.
32. Check the ABS speed sensor signal.

ENGINE COOLING

THERMOSTAT

REMOVAL & INSTALLATION

See Figure 20.

1. Drain engine coolant.
2. Remove the V-bank cover sub-assembly.
3. Remove the engine moving control rod.
4. Remove the engine mounting control bracket.
5. Remove the left front No. 1 engine mounting bracket.

Fig. 20 Radiator jiggle valve

6. Remove the fan and alternator V-belt.
7. Remove the No. 2 idler pulley sub-assembly.
8. Separate the radiator hose outlet.
9. Remove the 2 bolts and water inlet.
10. Remove the thermostat.

To install:

11. Install a new gasket to the thermostat.
12. Install the thermostat with the jiggle valve facing up.

➡The jiggle valve may be set within 10° of either side of the prescribed position.

13. Install the water inlet and tighten to 7 ft. lbs. (10 Nm).
14. The remainder of installation is the reverse of removal.
15. After installation, inspect for coolant leak.

WATER PUMP

REMOVAL & INSTALLATION

See Figures 21 and 22.

1. Drain engine coolant.
2. Remove the right front wheel.
3. Remove the right engine under cover.
4. Remove the V-bank cover sub-assembly.
5. Remove the engine moving control rod.
6. Remove the engine mounting control bracket.
7. Remove the left front No. 1 engine mounting bracket.
8. Remove the fan and alternator V-belt.
9. Separate the radiator hose outlet.
10. Remove water inlet housing, as follows:
    a. Separate the water hose.
    b. Remove the 2 bolts, nut and water inlet housing.
    c. Remove the water inlet housing gasket No. 1 and water outlet pipe O-ring.
11. Remove crankshaft pulley.
12. Using SST (SST: 09960-10010), hold the water pump pulley.

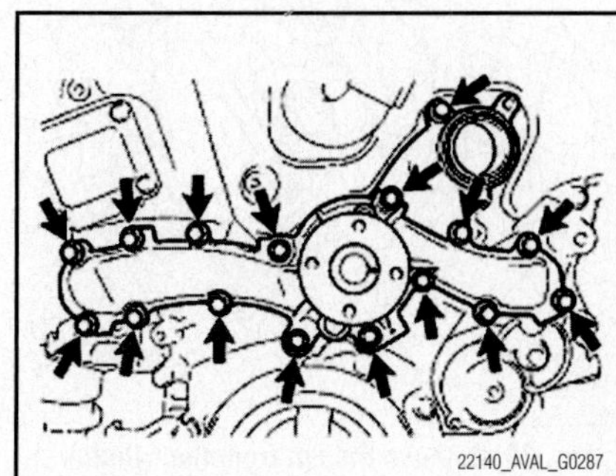

Fig. 21 Remove the 16 bolts, water pump assembly and water pump gasket

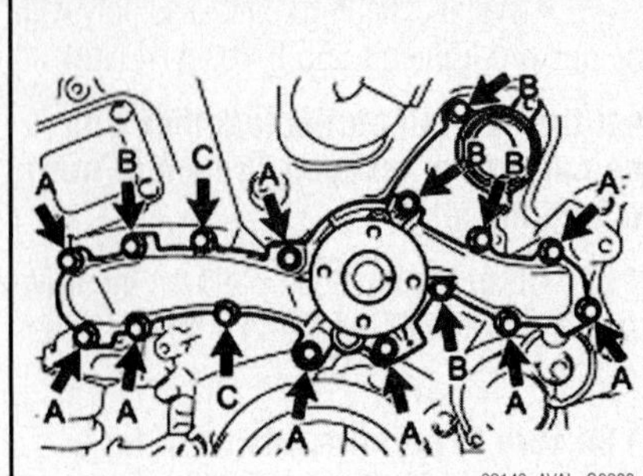

Fig. 22 Water pump assembly bolt position

13. Remove the 4 bolts and water pump pulley.
14. Remove the No. 2 idler pulley sub-assembly, as follows:
    a. Remove the bolt, No. 2 idler pulley cover plate and No. 2 idler pulley sub-assembly.
    b. Remove the bolt and idler pulley.

➡Be careful when loosening the bolt because it is left-hand threaded.

15. Separate the vane pump assembly.
16. Remove the water pump assembly, as follows:
    a. Remove the 16 bolts, water pump assembly and water pump gasket.

To install:

17. Install a new water pump gasket and the water pump assembly with the 16 bolts and tighten to 15 ft. lbs. (21 Nm), and 81 inch lbs. (9.1 Nm).

➡Make sure that there is no oil on the threads of the A bolts.

➡Be sure to replace the 2 C bolts with new ones or reuse them after applying adhesive (Part No. 08833-00080, three bond 1344 or equivalent).

18. Install vane pump assembly.
19. Install the idler pulley with the bolt and tighten to 32 ft. lbs. (43 Nm).

➡Be careful when tightening the bolt because it is left-hand threaded.

20. Install the No. 2 idler pulley cover plate and No. 2 idler pulley sub-assembly with the bolt tighten to 32 ft. lbs. (43 Nm).
21. Install the water pump pulley, as follows:
22. Temporarily install the water pump pulley with the 4 bolts.
    a. Using SST (SST: 09960-10010), hold the water pump pulley.
    b. Tighten the 4 bolts to 15 ft. lbs. (21 Nm).
23. Install crankshaft pulley.
24. Install water inlet housing, as follows:
    a. Install a new No. 1 water inlet housing gasket and water outlet pipe O-ring.
    b. Install the water inlet with the 2 bolts and nut and tighten to 7 ft. lbs. (10 Nm).

➡Be careful not to allow the O-ring to get caught between the parts.

    c. Install the water hose.
25. Install the radiator hose outlet.
26. Install the fan and alternator V-belt.
27. Install the left front No. 1 engine mounting bracket.
28. Install the engine mounting control bracket.
29. Install the engine moving control rod.
30. Install the V-bank cover sub-assembly.
31. Add engine coolant.
32. Inspect for coolant leak.
33. Install the right engine under cover.
34. Install the right front wheel.

ENGINE ELECTRICAL

CHARGING SYSTEM

ALTERNATOR

REMOVAL & INSTALLATION

See Figures 23 and 24.

1. Disconnect the negative battery cable.
2. Remove the V-bank cover sub-assembly.
3. Remove the V-ribbed belt.
4. Remove the alternator assembly, as follows:
    a. Disconnect the wire harness clamp.
    b. Remove the terminal cap.
    c. Remove the nut and disconnect the wire harness from terminal B.
    d. Disconnect the alternator connector from the alternator assembly.
    e. Remove the nut from the cylinder block.
    f. Remove the 2 bolts and alternator assembly.
    g. Remove the bolt and wire harness clamp stay.
    h. Remove the bolt and bracket.

To install:

5. Install the alternator assembly, as follows:
    a. Install the bracket with the bolt and tighten to 15 ft. lbs. (20 Nm).
    b. Install the wire harness clamp stay and tighten to 74 inch lbs. (8.4 Nm).
    c. Install the alternator assembly with the 2 bolts and tighten to 32 ft. lbs. (43 Nm).
    d. Install the nut to the cylinder block and tighten to 15 ft. lbs. (20 Nm).
    e. Connect the alternator connector to the alternator assembly.
    f. Install the alternator wire with the nut and tighten to 87 inch lbs. (9.8 Nm).
    g. Install the terminal cap.
    h. Connect the wire harness clamp.
6. Install the V-ribbed belt.
7. Install the V-bank cover sub-assembly.
8. Connect the negative battery cable.
9. Perform initialization.

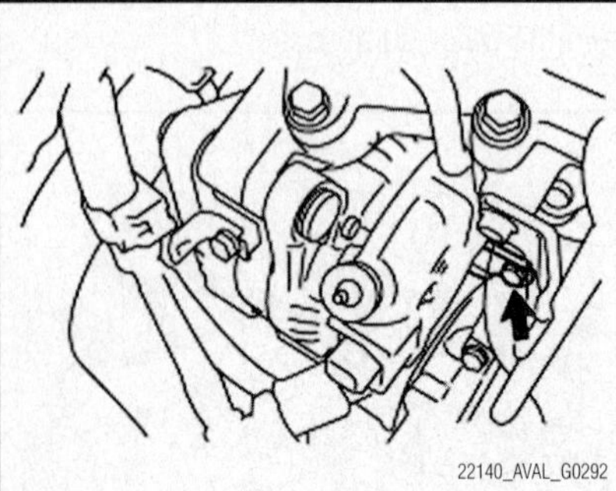

Fig. 23 Remove the nut from the cylinder block

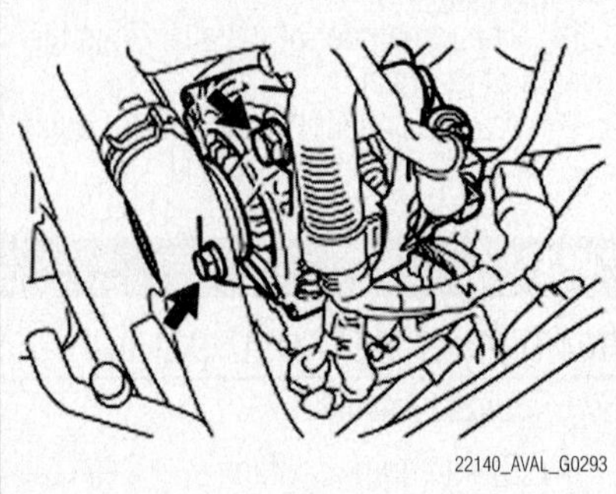

Fig. 24 Remove the 2 bolts and alternator assembly

FIRING ORDER

Firing order for 3.5L engine: 1–2–3–4–5–6

IGNITION COIL

REMOVAL & INSTALLATION

See Figure 25.

1. Disconnect the negative battery cable.
2. Remove the V-bank cover.
3. Remove the intake air surge tank.
4. Disconnect the 6 ignition coil connectors.
5. Remove the 6 bolts and 6 ignition coils.

To install:

6. Installation is the reverse of removal. Torque the ignition coils to 66 inch lbs. (7.5 Nm).

IGNITION TIMING

INSPECTION

1. Warm up the engine.
2. Using SST: 09843-18040, connect terminals 13 (TC) and 4 (CG) of the DLC3.

WARNING

Confirm the terminal numbers before connecting them. Connecting the wrong terminals can damage the engine.

➡Turn off all electrical systems before connecting the terminals.

➡Perform this inspection after the cooling fan motor is turned off.

3. Remove the v-bank cover.
4. Pull out the red lead wire harness.
5. Connect the tester terminal of the timing light to the red lead wire.

➡Use a timing light which can detect the first signal.

6. Check the ignition timing at idle. Standard ignition timing: 8 to 12° BTDC at idle.

➡When checking the ignition timing, the transmission should be in the neutral position.

➡Run the engine at 1000 to 1300 rpm for 5 seconds, and then check that the engine rpm returns to idle speed.

7. Disconnect terminals 13 (TC) and 4 (CG) of the DLC3.
8. Check the ignition timing at idle. Standard ignition timing: 7 to 24° BTDC at idle.
9. Confirm that the ignition timing advances immediately when the engine rpm is increased.
10. Remove the timing light from the engine.

ADJUSTMENT

All engines are equipped with a Distributorless Ignition System (DIS). No timing adjustment is possible.

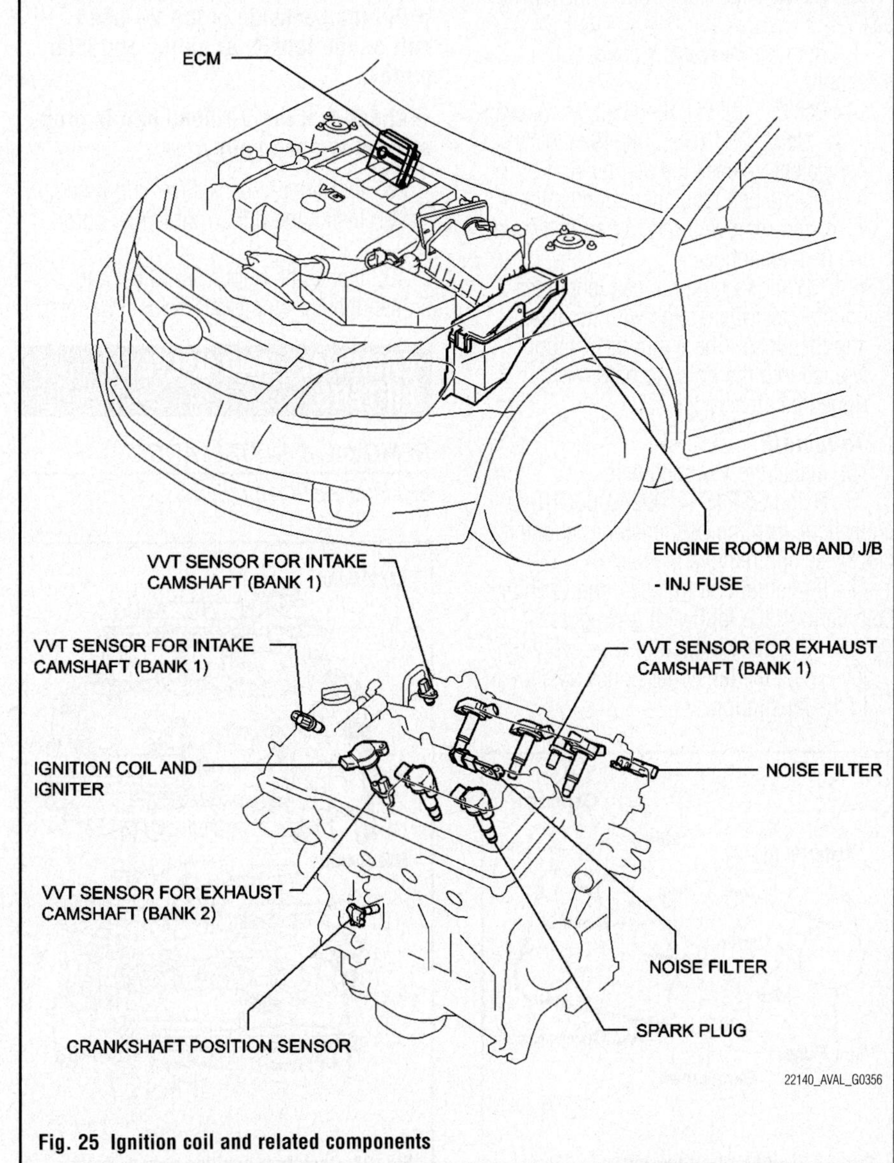

Fig. 25 Ignition coil and related components

SPARK PLUGS

REMOVAL & INSTALLATION

See Figure 25.

1. Disconnect the negative battery cable.
2. Remove the V-bank cover.
3. Remove the intake air surge tank.
4. Disconnect the 6 ignition coil connectors.
5. Remove the 6 bolts and 6 ignition coils.
6. Using a 16 mm (0.63 in.) plug wrench, remove the spark plugs.

To install:

7. Installation is the reverse of removal, noting the following:
   a. Torque the ignition coils to 66 inch lbs. (7.5 Nm) and the spark plugs to 13 ft. lbs (18 Nm).

STARTER

REMOVAL & INSTALLATION

1. Disconnect the negative battery cable.
2. Remove the air cleaner assembly with hose.
3. Remove the No. 1 air cleaner inlet.
4. Remove the starter assembly, as follows:
    a. Disconnect terminal 50 of the connector from the starter assembly.
    b. Open the terminal cap, remove the nut and disconnect the wire harness from terminal 30.
    c. Remove the 2 bolts and the starter assembly.

To install:

5. Install the starter assembly with the 2 bolts and tighten to 26 ft. lbs. (37 Nm).
6. Connect the wire harness to terminal 30 and install the nut and tighten to 87 inch lbs. (9.8 Nm).
7. Cover the nut with the cap.
8. Connect terminal 50 to the starter assembly.
9. Install the No. 1 air cleaner inlet.
10. Install the air cleaner assembly with hose.
11. Connect the negative battery cable.
12. Perform initialization.

ENGINE MECHANICAL

ACCESSORY DRIVE BELTS

ACCESSORY BELT ROUTING

See Figure 26.

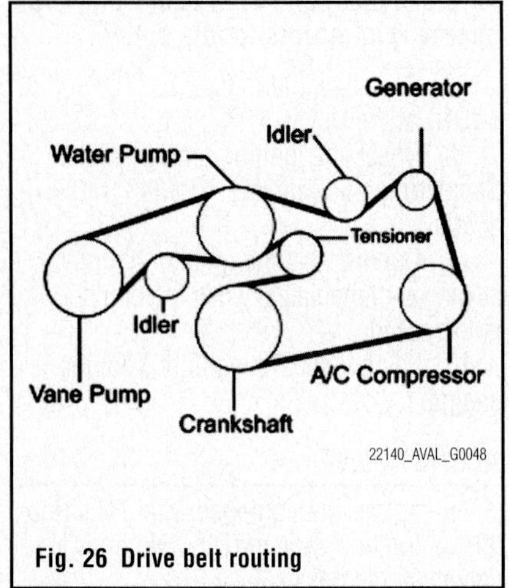

Fig. 26 Drive belt routing

Refer to the accompanying illustration for drive belt routing.

INSPECTION

Visually check the V-ribbed belt for excessive wear, frayed cords, etc. All worn or damaged drive belts should be replaced immediately. Cracks on the rib side of a V-ribbed belt are considered acceptable, If the drive belt has chunks missing from its ribs, it should be replaced. After installing the V-ribbed belt, check that it fits properly in the ribbed grooves. Check to confirm that the belt has not slipped out of the grooves on the bottom of the crank pulley by hand.

ADJUSTMENT

Belt tension is maintained by an automatic tensioner. No adjustment is necessary or possible.

REMOVAL & INSTALLATION

See Figure 27.

1. Remove the right front wheel.
2. Remove the right front fender apron seal.
3. Remove the V-bank cover sub-assembly.
4. Remove the V-ribbed belt, as follows:
    a. Using SST (SST: 09249-63010) or equivalent, release the belt tension by turning the belt tensioner counterclockwise, and remove the V-ribbed belt from the belt tensioner.
    b. While turning the belt tensioner counterclockwise, align with its holes, and then insert the 5 mm bi-hexagon wrench into the holes to hold the V-ribbed belt tensioner.

To install:

5. Install the V-ribbed belt.
6. Using SST (SST: 09249-63010) or equivalent, turn the belt tensioner counterclockwise and remove the bar.
7. If it is difficult to install the V-ribbed belt, perform the following procedure:
    a. Put the V-ribbed belt on every pulley except the tensioner pulley as shown in the illustration.

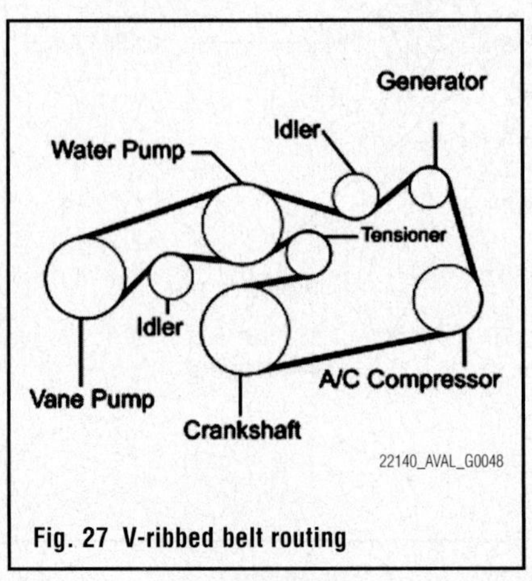

Fig. 27 V-ribbed belt routing

    b. While releasing the belt tension by turning the belt tensioner counterclockwise, put the V-ribbed belt on the tensioner pulley.

➡Put the backside of the V-ribbed belt on the tensioner pulley and idler pulley.

➡Check that the V-ribbed belt is properly set to each pulley.

8. Install the V-bank cover sub-assembly.
9. Install the right front fender apron seal.
10. Install the right front wheel and tighten the lug nuts to 76 ft. lbs. (103 Nm).

CAMSHAFT AND VALVE LIFTERS

REMOVAL & INSTALLATION

See Figures 28 through 57.

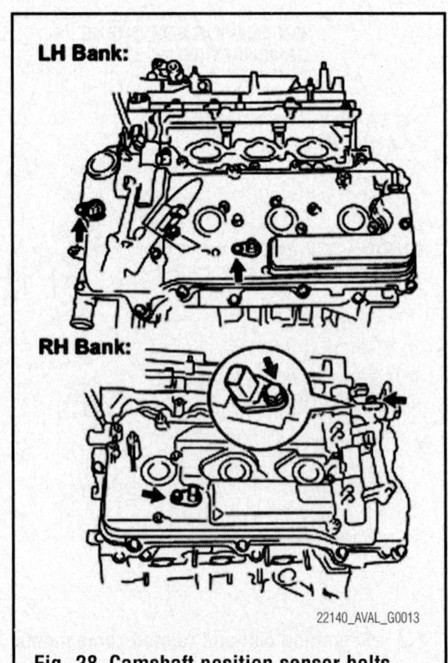

Fig. 28 Camshaft position sensor bolts

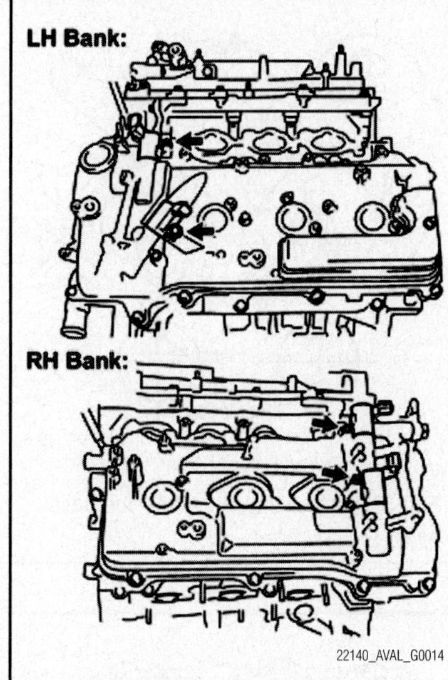

Fig. 29 Camshaft timing oil control valve assembly bolts

1. Before servicing the vehicle, refer to the Precautions Section.
2. Remove the engine assembly.
3. Install on engine stand.
4. Remove the oil filler cap and gasket.
5. Remove the spark plugs and ignition coil assembly.
6. Remove the drain plug and gasket.
7. Remove the ventilation valve.
8. Remove the 4 bolts and 4 camshaft position sensors.
9. Remove the 4 bolts and 4 camshaft timing oil control valves.
10. Remove the bolt and crankshaft position sensor.
11. Remove the No. 1 oil pipe.
12. Remove the oil pipe.
13. Remove the cylinder block water drain cock sub-assembly, as follows:
    a. Remove the water drain cocks from the cylinder block.
    b. Remove the water drain cock plugs from the water drain cocks.
14. Remove the oil filter.
15. Remove the crankshaft pulley, as follows:
    a. Using SST (SST: 09213-70011, SST: 09330-00021) or equivalent, loosen the crankshaft pulley bolt.
    b. Using SST (SST: 09950-50013) or equivalent, remove the crankshaft pulley bolt and crankshaft pulley.
16. Remove the 6 bolts and the left hand No. 1 front engine mounting bracket.
17. Remove the water inlet housing, as follows:

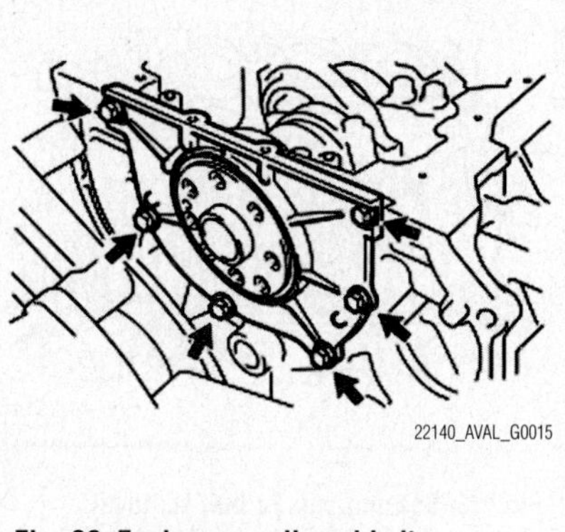

Fig. 30 Engine rear oil seal bolts

    a. Remove the 2 nuts, water inlet and thermostat.
    b. Remove the gasket.
    c. Remove the drain cock plug.
    d. Remove the drain cock.
    e. Remove the 2 stud bolts.
    f. Remove the 2 bolts, nut, and water inlet housing.
    g. Remove the 2 O-rings.
18. Remove the water outlet, as follows:
    a. Remove the 2 bolts, 4 nuts and water outlet.
    b. Remove the 2 gaskets and O-ring.
19. Remove the 12 bolts, cylinder head cover sub-assembly (for Bank 1), and gasket.
20. Remove the 12 bolts, cylinder head cover sub-assembly (for Bank 2), and gasket.
21. Remove the No. 2 oil pan sub-assembly.
22. Remove the oil strainer sub-assembly.
23. Remove the oil pan sub-assembly.
24. Remove the No. 1 oil pan baffle plate.
25. Remove the engine rear oil seal, as follows:
    a. Remove the 6 bolts.
    b. Using a screwdriver with the tip taped, pry out the oil seal retainer. Be careful not to damage the oil seal retainer.
26. Remove the water pump assembly.
27. Remove the timing chain cover sub-assembly.
28. Set the No. 1 cylinder to TDC/compression.
29. Remove the No. 1 chain tensioner assembly.
30. Remove the chain tensioner slipper.
31. Remove the chain sub-assembly.
32. Remove the idle sprocket assembly.
33. Remove the No. 1 chain vibration damper.
34. Remove the No. 2 chain vibration damper.

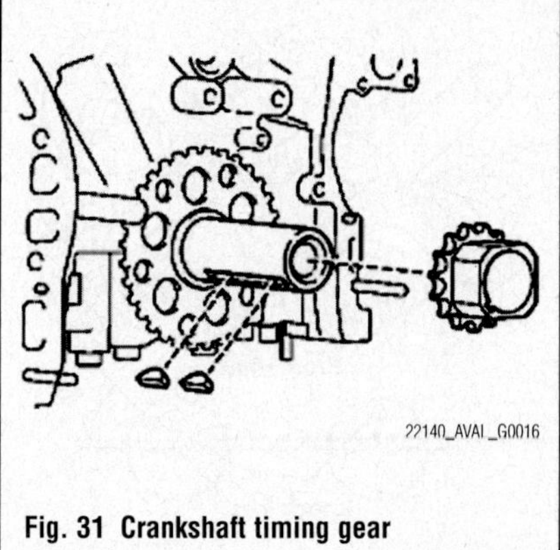

Fig. 31 Crankshaft timing gear

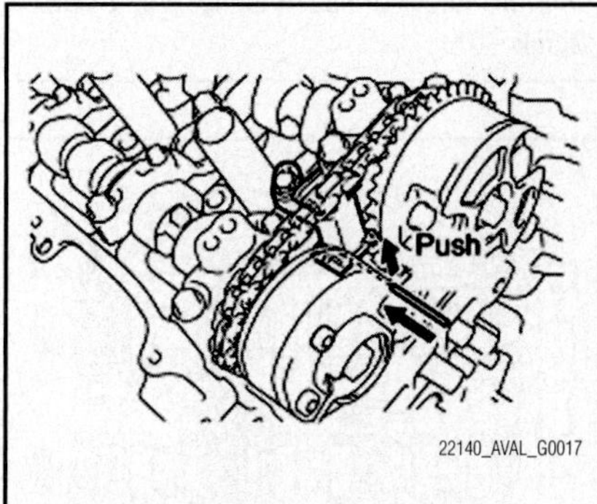

Fig. 32 Pinning No. 2 tensioner—right-hand bank

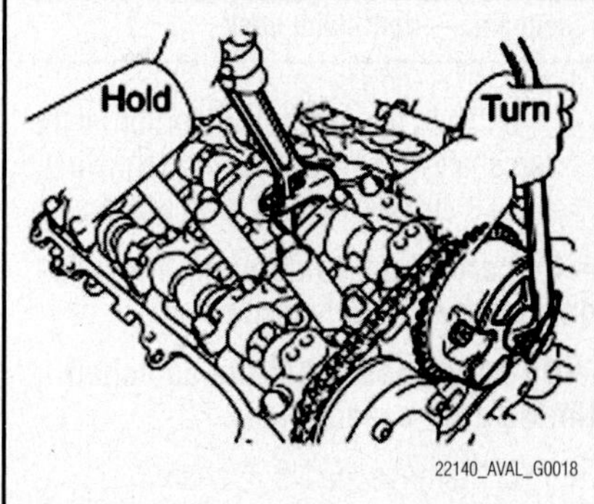

Fig. 33 Removing gear assemblies—right-hand bank

35. Remove the crankshaft timing gear, as follows:
    a. Remove the pulley set bolt.
    b. Remove the crankshaft timing gear from the crankshaft.
    c. Remove the 2 pulley set keys from the crankshaft.
36. Remove the camshaft timing gears and No. 2 chain (for right-hand bank), as follows:
    a. While raising the No. 2 chain tensioner, insert a pin of 1.0 mm (0.039 in.) into the hole to hold the No. 2 chain tensioner.

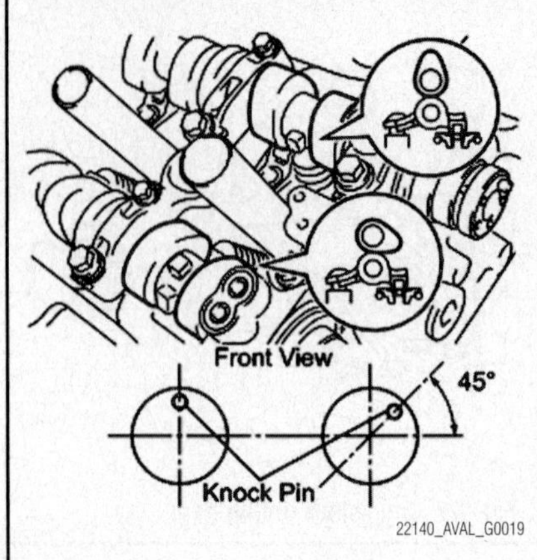

Fig. 34 Knock pin positioning—right-hand bank

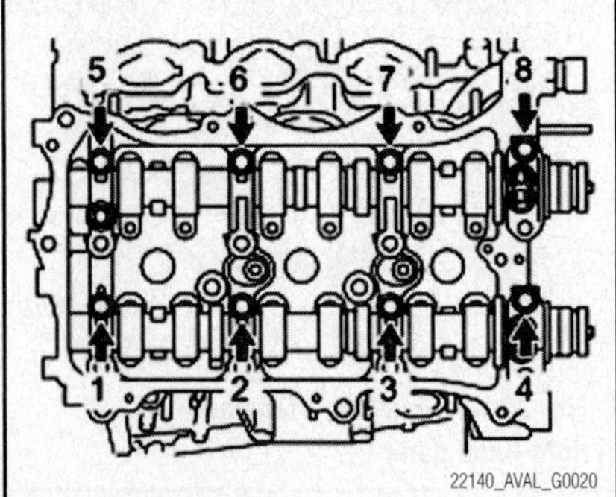

Fig. 35 Bearing cap 8 bolt removal sequence—right-hand bank

b. Hold the hexagonal portion of the camshaft with a wrench, and remove the 2 bolts and 2 camshaft timing gears.

➡Be careful not to damage the cylinder head with the wrench.

➡Do not disassemble the camshaft timing gear assemblies.

c. Remove the No. 2 chain.

37. Remove the bolt and No. 2 chain tensioner assembly.

38. Remove the camshaft bearing cap, as follows:

a. Remove the 3 gaskets.

b. Make sure that the knock pin of the camshaft is positioned as shown in the illustration.

c. Uniformly loosen and remove the 8 bearing cap bolts in the sequence shown in the illustration.

d. Uniformly loosen and remove the 12 bearing cap bolts in the sequence shown in the illustration. Loosen the bolts while keeping the camshaft level

e. Remove the 5 camshaft bearing caps.

39. Remove the No. 1 camshaft.

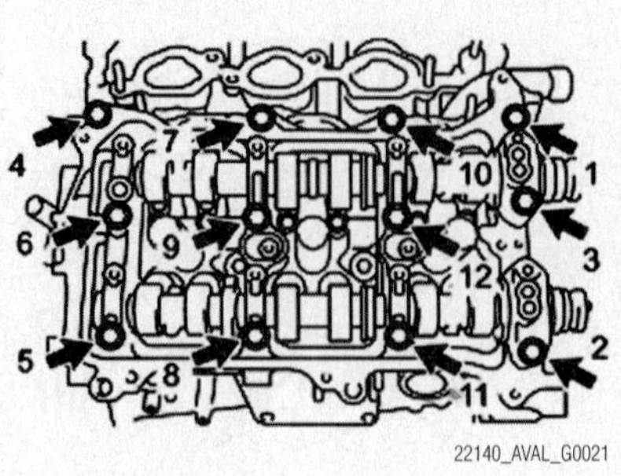

Fig. 36 Bearing cap 12 bolt removal sequence—right-hand bank

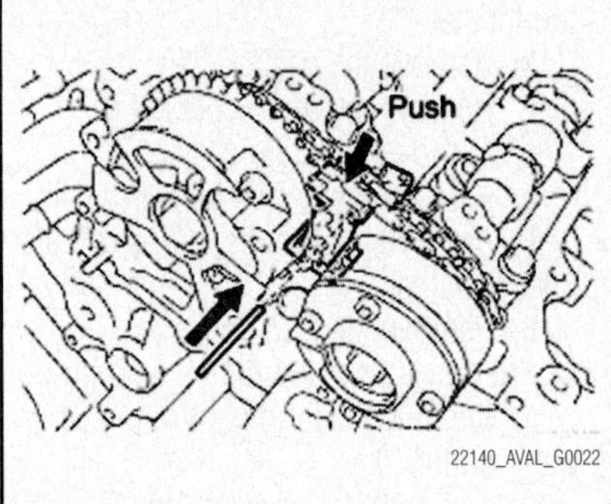

Fig. 37 Pinning No. 3 tensioner—left-hand bank

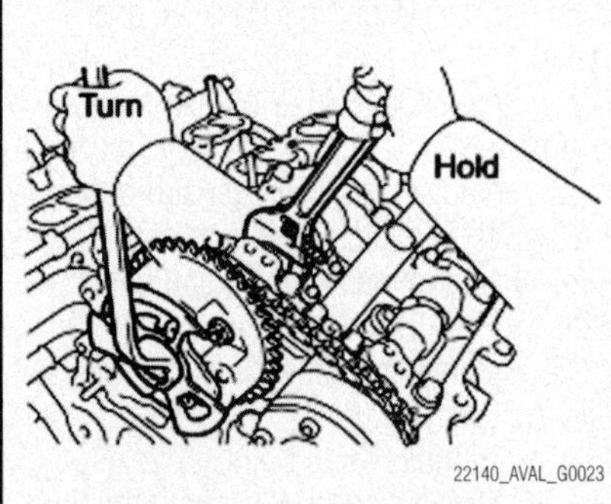

Fig. 38 Removing gear assemblies—left-hand bank

40. Remove the No. 2 camshaft.

41. Remove the right hand camshaft housing sub-assembly by prying between the cylinder head and the camshaft housing with a screwdriver with the tip taped.

➡Be careful not to damage the contact surfaces of the cylinder head and the camshaft housing.

42. Remove the camshaft timing gears and No. 2 chain (for left-hand bank), as follows:

a. While pushing down the No. 3 chain tensioner, insert a pin of 1.0 mm (0.039 in.) into the hole to hold the No. 3 chain tensioner.

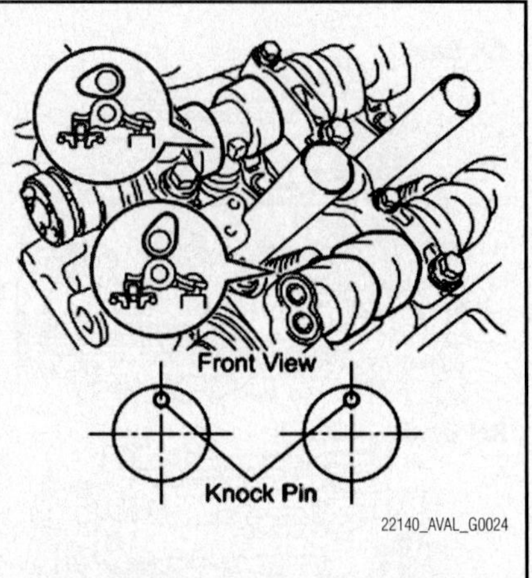

Fig. 39 Knock pin positioning—left-hand bank

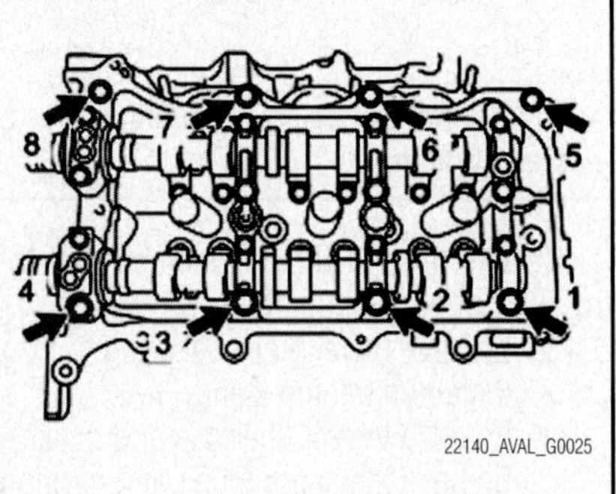

Fig. 40 Bearing cap 8 bolt removal sequence—left-hand bank

b. Hold the hexagonal portion of the camshaft with a wrench, and remove the 2 bolts and 2 camshaft timing gears.

➡Be careful not to damage the cylinder head with the wrench.

➡Do not disassemble the camshaft timing gear assemblies.

c. Remove the No. 2 chain.

43. Remove the bolt and No. 3 chain tensioner.

44. Remove the camshaft bearing cap, as follows:

a. Remove the 3 gaskets.

b. Make sure that the knock pin of the camshaft is positioned as shown in the illustration.

c. Uniformly loosen and remove the 8 bearing cap bolts in the sequence shown in the illustration.

d. Uniformly loosen and remove the 13 bearing cap bolts in the sequence shown in the illustration. Loosen the bolts while keeping the camshaft level

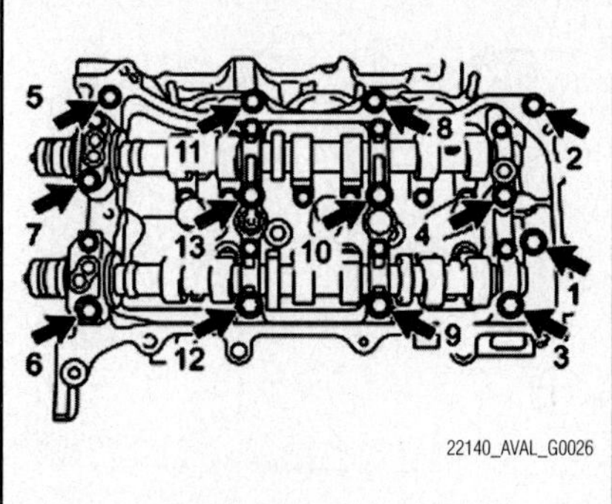

Fig. 41 Bearing cap 12 bolt removal sequence—left-hand bank

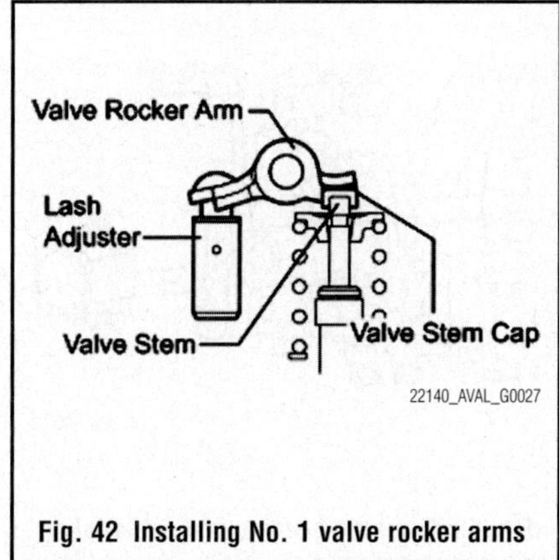

Fig. 42 Installing No. 1 valve rocker arms

e. Remove the 5 camshaft bearing caps.

45. Remove the No. 4 camshaft.

46. Remove the No. 3 camshaft.

47. Remove the left hand camshaft housing sub-assembly by prying between the cylinder head and the camshaft housing with a screwdriver with the tip taped.

➡Be careful not to damage the contact surfaces of the cylinder head and the camshaft housing.

48. Remove the No. 1 valve rocker arm sub-assembly, as follows:

a. Remove the 24 valve rocker arms.

➡Arrange the removed parts in the correct order.

49. Remove the valve lash adjuster assembly, as follows:

a. Remove the 24 valve lash adjusters from the cylinder head.

➡Arrange the removed parts in the correct order.

To install:

50. Install the valve lash adjuster assembly.

51. Install the No. 1 valve rocker arm sub-assembly, as follows:

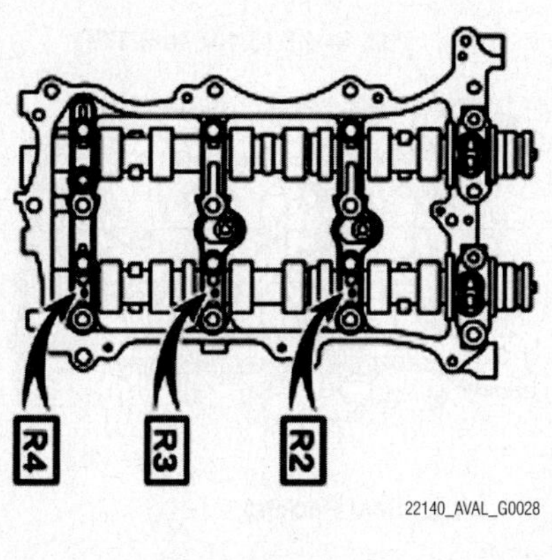

Fig. 43 Camshaft bearing caps placement—right-hand

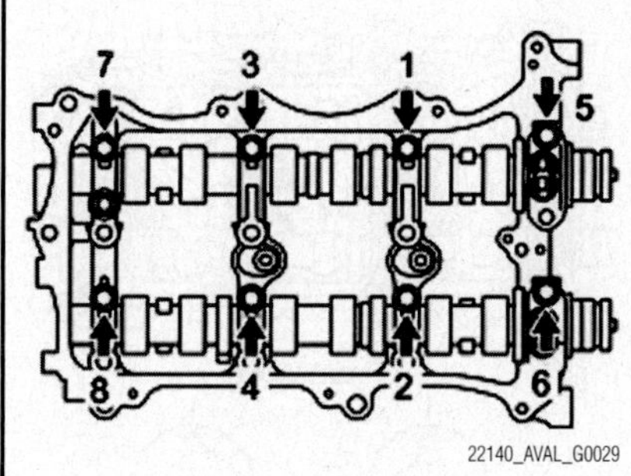

Fig. 44 Camshafts bearing cap tightening sequence—right-hand

a. Apply engine oil to the lash adjuster tips and valve stem cap ends.

b. Make sure that the valve rocker arms are installed as shown in the illustration.

52. Install the right-hand camshaft bearing cap, as follows:

a. Apply engine oil to the camshaft journals, camshaft housing and bearing caps.

b. Install the No. 1camshaft and No. 2 camshaft to the right camshaft housing.

c. Make sure of the marks and numbers on the camshaft bearing caps and place them in each proper position and direction.

d. Temporarily tighten the 8 bearing cap bolts to 7 ft. lbs. (10 Nm) in the order shown in the illustration.

53. Install the right hand camshaft housing sub-assembly, as follows:

a. Apply seal packing in a continuous line as shown in the illustration. Seal packing: Toyota Genuine Seal Packing Black, Three Bond 1207B or equivalent. Seal diameter: 3.5 to 4.5 mm (0.138 to 0.177 in.).

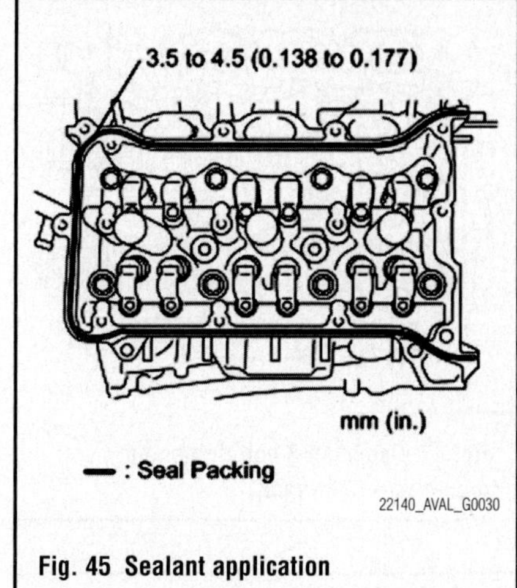

Fig. 45 Sealant application

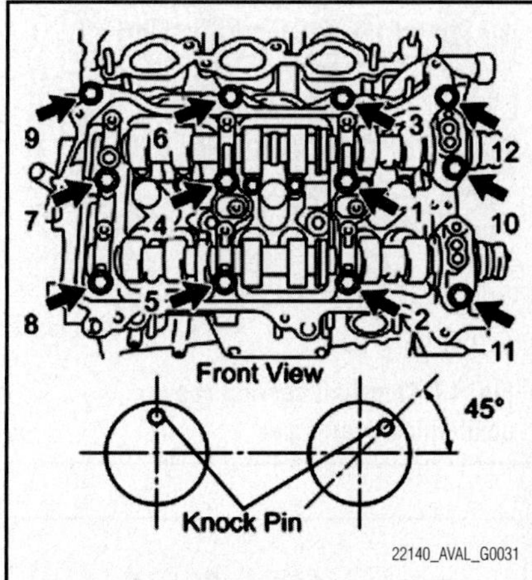

Fig. 46 Camshaft 12 bolt tightening sequence—right-hand

➡Remove any oil from the contact surface. Install the camshaft housing sub-assembly within 3 minutes and tighten the bolts within 15 minutes after applying sealant. Do not start the engine for at least 2 hours after installing.

b. Make sure that the knock pins of the camshafts are positioned as shown. Install the right-hand camshaft housing and tighten the 12 bolts in the order shown in the illustration to 21 ft. lbs. (28 Nm).

c. Tighten the 8 bolts to 12 ft. lbs. (16 Nm) in the order shown in the illustration.

➡Thoroughly wipe clean any seal packing.

d. Install 3 new gaskets.

54. Install the camshaft bearing cap, as follows:

a. Apply engine oil to the camshaft journals, camshaft housing and bearing caps.

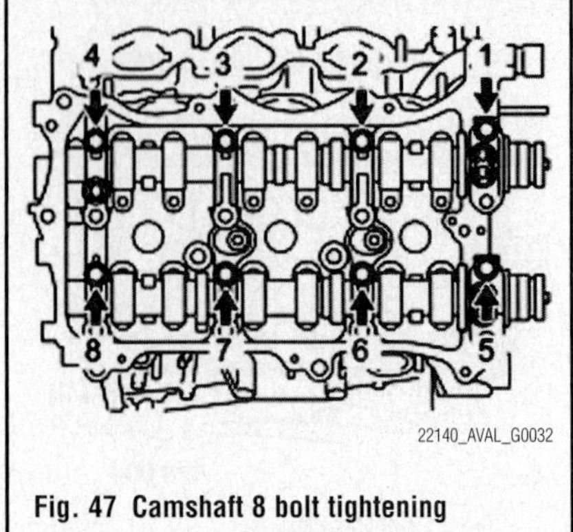

Fig. 47 Camshaft 8 bolt tightening sequence—right-hand

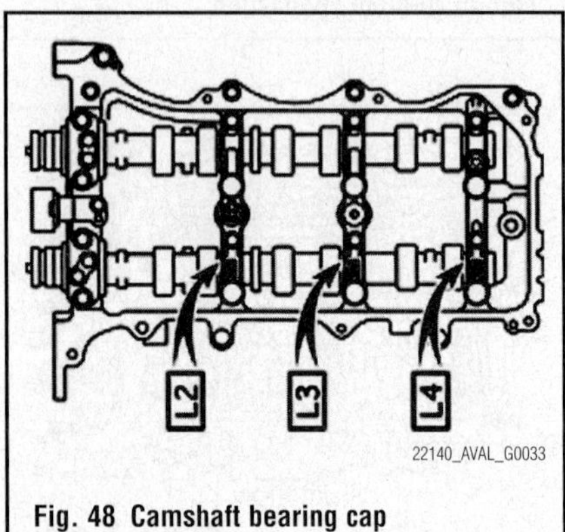

Fig. 48 Camshaft bearing cap positioning—left-hand

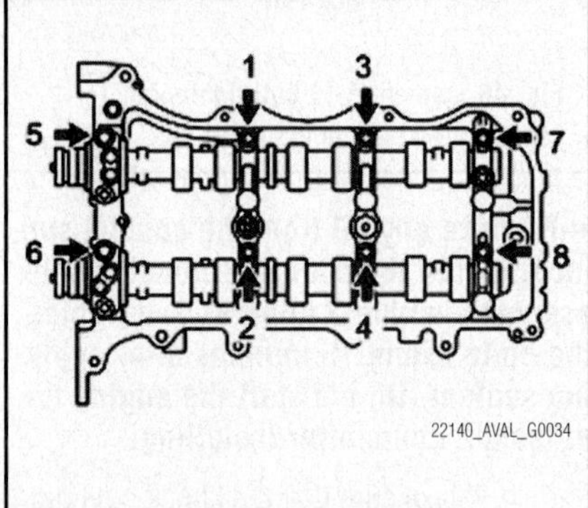

Fig. 49 Camshaft 8 bolt tightening sequence—left-hand

b. Install the No. 3 camshaft and No. 4 camshaft to the left hand camshaft housing.

c. Make sure of the marks and numbers on the camshaft bearing caps and place them in each proper position and direction.

d. Temporarily tighten the 8 bolts in the order shown in the illustration to 7 ft. lbs. (10 Nm).

55. Install the left camshaft housing sub-assembly, as follows:

a. Apply seal packing in a continuous line as shown in the illustration. Seal

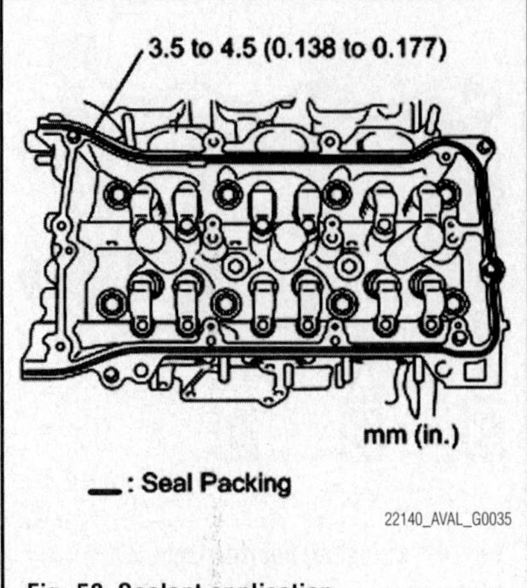

Fig. 50 Sealant application

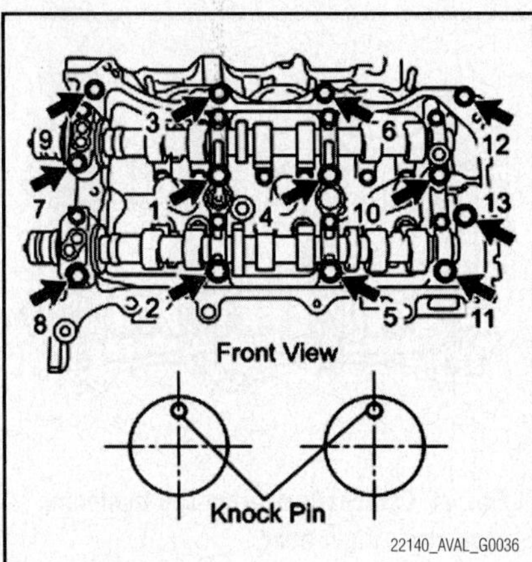

Fig. 51 Camshaft housing 13 bolt tightening sequence—left-hand

packing: Toyota Genuine Seal Packing Black, Three Bond 1207B or equivalent. Seal diameter: 3.5 to 4.5 mm (0.138 to 0.177 in.).

➡Remove any oil from the contact surface. Install the camshaft housing sub-assembly within 3 minutes and tighten the bolts within 15 minutes after applying sealant. Do not start the engine for at least 2 hours after installing.

b. Make sure that the knock pins of the camshafts are positioned as shown. Install the left hand camshaft housing and tighten the 13 bolts in the order shown in the illustration to 21 ft. lbs. (28 Nm).

c. Tighten the 8 bolts to 12 ft. lbs. (16 Nm) in the order shown in the illustration.

➡Thoroughly wipe clean any sealant.

d. Install 3 new gaskets.

56. Install the No. 2 chain tensioner assembly with the bolt and tighten to 15 ft. lbs. (21 Nm).

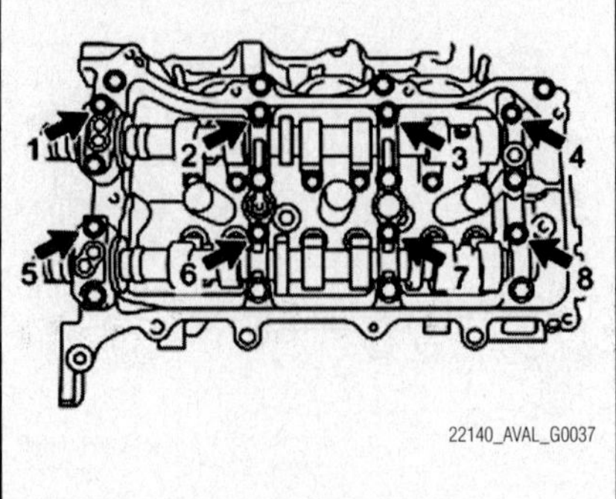

Fig. 52 Camshaft housing 8 bolt tightening sequence—left-hand

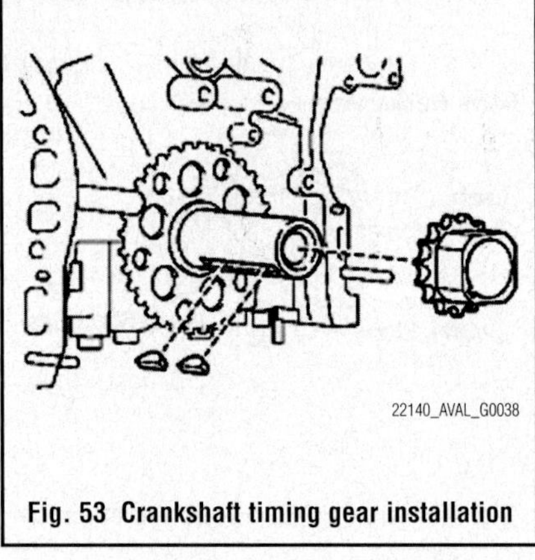

Fig. 53 Crankshaft timing gear installation

57. While pushing in the tensioner, insert a pin of 1.0 mm (0.039 in.) diameter into the hole to hold it.

58. Install the camshaft timing gears and No. 2 chain (for Right-hand Bank).

59. Install the No. 3 chain tensioner assembly with the bolt and tighten to 15 ft. lbs. (21 Nm).

60. While pushing in the tensioner, insert a pin of 1.0 mm (0.039 in.) diameter into the hole to hold it.

61. Install the camshaft timing gears and No. 2 chain (for Left-hand Bank).

62. Install the No. 1 chain vibration damper with the 2 bolts and tighten to 17 ft. lbs. (23 Nm).

63. Install the No. 2 chain vibration damper.

64. Install the timing gear set keys and timing gear as shown in the illustration.

65. Install the idle sprocket assembly.

66. Install the chain sub-assembly.

67. Install the chain tensioner slipper.

68. Install the No. 1 chain tensioner assembly.

69. Install the water pump assembly.

70. Install the timing chain cover sub-assembly.

71. Install the water inlet housing.

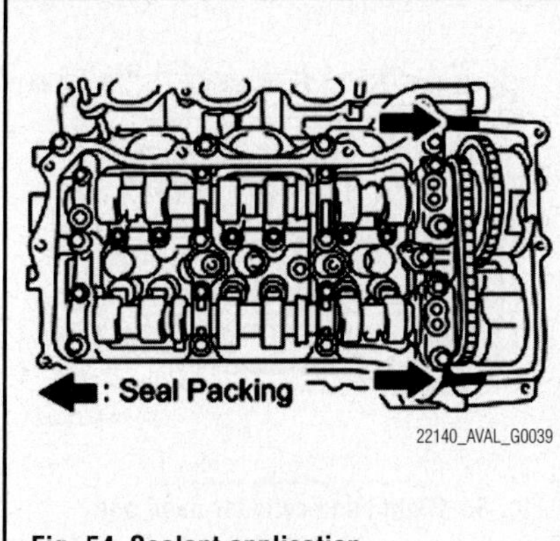

Fig. 54 Sealant application

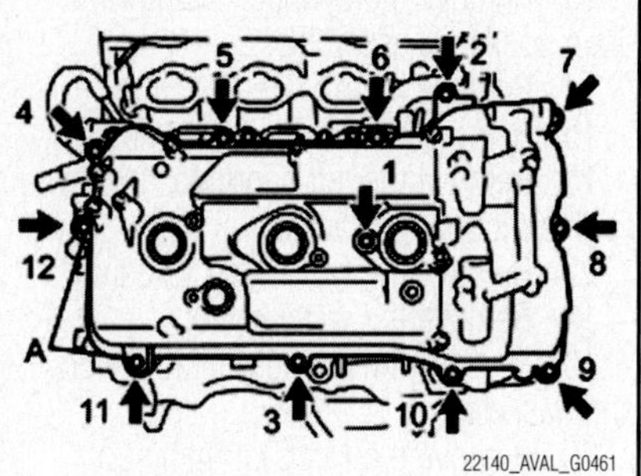

Fig. 55 Cylinder head cover bolt tightening sequence

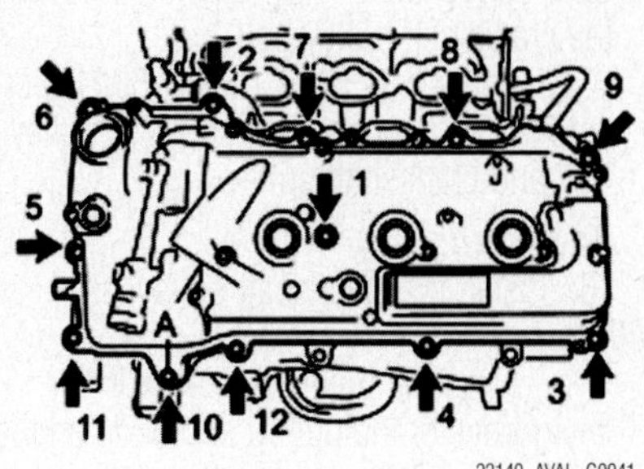

Fig. 57 Left cylinder head cover bolt tightening sequence

72. Install the No. 1 left front engine mounting bracket, as follows:
   a. Install the No. 1 left front engine mounting bracket with the 6 bolts and tighten to 40 ft. lbs. (54 Nm).

➡Install the water inlet and mounting bracket within 15 minutes after installing the chain cover. Do not start the engine for at least 2 hours after installation.

73. Install the No. 1 oil pan baffle plate with the 7 bolts and tighten to 7 ft. lbs. (10 Nm).
74. Install the oil pan sub-assembly.
75. Install the oil strainer sub-assembly.
76. Install the No. 2 oil pan sub-assembly.
77. Install a new gasket and oil pan drain plug and tighten to 30 ft. lbs. (40 Nm).
78. Install the cylinder head cover sub-assembly, as follows:
   a. Apply seal packing (Toyota Genuine Seal Packing Black, Three Bond 1207B or equivalent) as shown in the illustration.

➡Remove any oil from the contact surface. Install the crankcase within 3 minutes after applying seal packing. Do not start the engine for at least 2 hours after installation.

   b. Install the gasket to the head cover.
   c. Install the head cover with the 12 bolts. Tighten bolt A to 15 ft. lbs. (21 Nm), and other bolts to 7 ft. lbs. (10 Nm). Be certain to tighten bolt 1.
79. Install the left-hand cylinder head cover sub-assembly, as follows:
   a. Apply seal packing (Toyota Genuine Seal Packing Black, Three Bond 1207B or equivalent)
   as shown in the illustration.

➡Remove any oil from the contact surface. Install the crankcase within 3 minutes after applying seal packing.

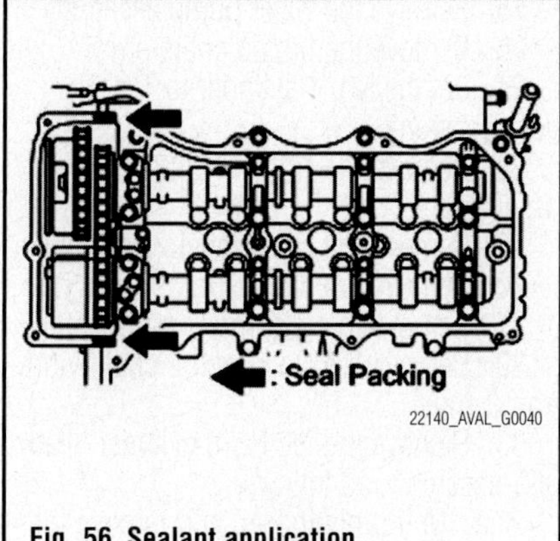

Fig. 56 Sealant application

Do not start the engine for at least 2 hours after installation.

   b. Install the gasket to the head cover.
   c. Install the head cover with the 14 bolts. Tighten bolt A to 15 ft. lbs. (21 Nm), and other bolts to 7 ft. lbs. (10 Nm). Be certain to tighten bolts 1 and 10.
80. Install water outlet.
81. Install the crankshaft pulley.
82. Install the oil filter element.
83. Install the cylinder block water drain cock sub-assembly, as follows:
   a. Apply adhesive around the drain cocks. Adhesive: Toyota Genuine Adhesive 1324, Three Bond 1324 or Equivalent.
   b. Install the water drain cocks and tighten to 18 ft. lbs. (25 Nm). Do not rotate the drain cocks more than 1 revolution (360°) after tightening the drain cocks with the specified torque. Do not loosen after setting correctly.
   c. Install the water drain cock plug to the water drain cocks and tighten to 9 ft. lbs. (13 Nm).
84. Install the No. 1 oil pipe.
85. Install the oil pipe.
86. Install the crankshaft position sensor with the bolt and tighten to 7 ft. lbs. (10 Nm).
87. Install the 4 camshaft timing oil control valves with the 4 bolts and tighten to 7 ft. lbs. (10 Nm).
88. Install the 4 camshaft position sensors with the 4 bolts and tighten to 7 ft. lbs. (10 Nm).
89. Install the ventilation valve sub-assembly, as follows:
   a. Apply adhesive (Toyota Genuine Adhesive 1324, Three Bond 1324 or equivalent) around the ventilation valve.
   b. Install the ventilation valve and tighten to 20 ft. lbs. (27 Nm).
90. Install the 6 spark plugs and the ignition coil assembly.
91. Install the oil filler cap sub-assembly.
92. Remove the engine stand.
93. Install the engine assembly.

CRANKSHAFT DAMPER

REMOVAL & INSTALLATION

1. Before servicing the vehicle, refer to the Precautions Section.
2. Remove the engine assembly.
3. Install on engine stand.
4. Remove the oil filler cap and gasket.
5. Remove the spark plugs and ignition coil assembly.
6. Remove the drain plug and gasket.
7. Remove the ventilation valve.
8. Remove the 4 bolts and 4 camshaft position sensors.
9. Remove the 4 bolts and 4 camshaft timing oil control valves.
10. Remove the bolt and crankshaft position sensor.
11. Remove the No. 1 oil pipe.
12. Remove the oil pipe.
13. Remove the cylinder block water drain cock sub-assembly.
14. Remove the oil filter element.
15. Remove the crankshaft pulley, as follows:
   a. Using SST (SST: 09213-70011,

SST: 09330-00021) or equivalent, loosen the crankshaft pulley bolt.

b. Using SST (SST: 09950-50013) or equivalent, remove the crankshaft pulley bolt and crankshaft pulley.

To install:

16. Install the crankshaft pulley, as follows:

a. Align the pulley set key with the key groove of the pulley, and slide on the pulley.

b. Using SST (SST: 09213-70011, SST: 09330-00021) or equivalent, install the pulley bolt and tighten to 184 ft. lbs. (250 Nm).

17. Install the oil filter element.
18. Install the cylinder block water drain cock sub-assembly.
19. Install the No. 1 oil pipe.
20. Install the oil pipe.
21. Install the crankshaft position sensor.
22. Install the camshaft timing oil control valve assembly.
23. Install the camshaft position sensor.
24. Install the ventilation valve sub-assembly.
25. Install the spark plugs and the ignition coil assembly.
26. Install the oil filler cap sub-assembly.
27. Remove the engine stand.
28. Install the engine assembly.

CRANKSHAFT FRONT SEAL

REMOVAL & INSTALLATION

See Timing Chain Cover and Seal.

CYLINDER HEAD

REMOVAL & INSTALLATION

See Figures 58 through 64.

1. Before servicing the vehicle, refer to the Precautions Section.
2. Remove the engine assembly with transaxle.
3. Secure engine.
4. Remove the oil filler cap sub-assembly.
5. Remove the spark plugs and ignition coil assembly.
6. Remove the oil pan drain plug and gasket.
7. Remove the ventilation valve sub-assembly.
8. Remove the camshaft position sensor.
9. Remove the camshaft timing oil control valve assembly.
10. Remove crankshaft position sensor.
11. Remove the No. 1 oil pipe.
12. Remove the oil pipe.
13. Remove the cylinder block water drain cock sub-assembly.
14. Remove the oil filter.
15. Remove the crankshaft pulley.
16. Remove the left hand No. 1 front engine mounting bracket.
17. Remove the water inlet housing.
18. Remove the water outlet.
19. Remove the cylinder head covers and gaskets.
20. Remove the No. 2 oil pan sub-assembly.
21. Remove the oil strainer sub-assembly.
22. Remove the oil pan sub-assembly.
23. Remove the No. 1 oil pan baffle plate.
24. Remove the engine rear oil seal.
25. Remove the water pump assembly.
26. Remove the timing chain cover.
27. Set the No. 1 cylinder to TDC/compression.
28. Remove the No. 1 chain tensioner assembly.
29. Remove the chain tensioner slipper.
30. Remove the chain sub-assembly.
31. Remove the idle sprocket assembly.
32. Remove the No. 1 and 2 chain vibration damper.
33. Remove the left hand cylinder head sub-assembly, as follows:

a. Uniformly loosen and remove the 2 bolts in the sequence shown in the illustration.

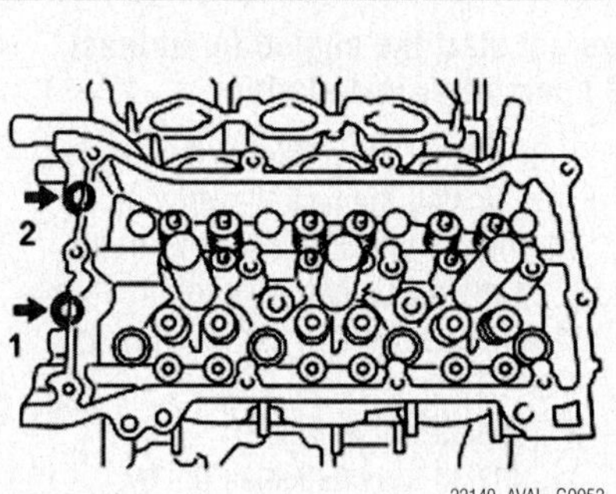

Fig. 58 Left hand cylinder head 2 bolt removal sequence

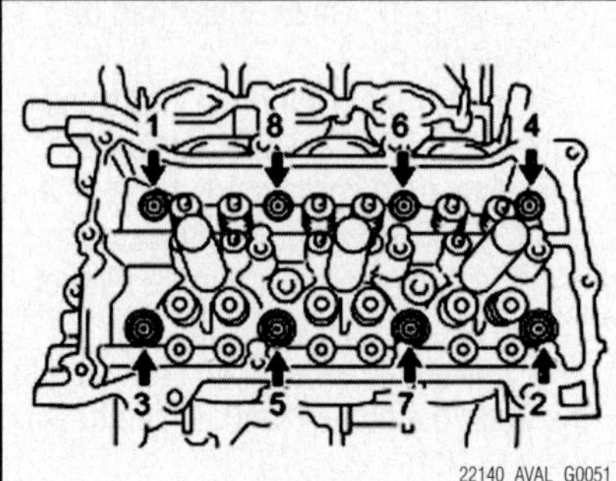

Fig. 59 Left hand cylinder head 8 bolt removal sequence

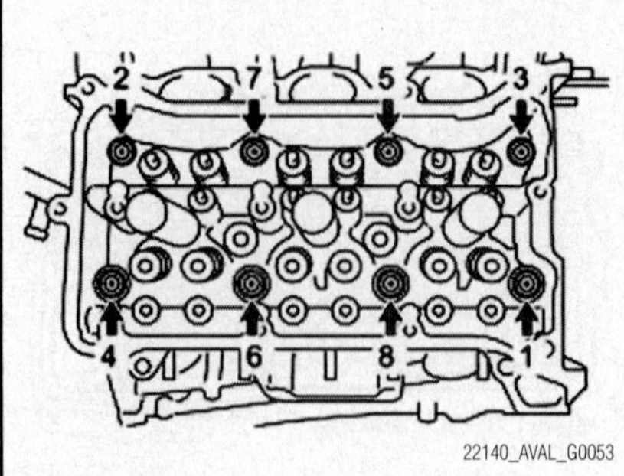

Fig. 60 Right hand cylinder head bolt removal sequence

b. Using a 10 mm bi-hexagon wrench, uniformly loosen the 8 bolts in the sequence shown in the illustration. Remove the 8 cylinder head bolts and plate washers.

WARNING

Be careful not to drop washers into the cylinder head.

WARNING

Cylinder head warpage or cracking could result from removing bolts in an incorrect order.

➡Be sure to keep separate the removed parts for each installation position.

c. Remove the cylinder head and gasket.

34. Remove the right hand cylinder head sub-assembly, as follows:

a. Using a 10 mm bi-hexagon wrench, uniformly loosen the 8 bolts in the sequence shown in the illustration. Remove the 8 cylinder head bolts and plate washers.

WARNING

Be careful not to drop washers into the cylinder head.

WARNING

Cylinder head warpage or cracking could result from removing bolts in an incorrect order.

➡Be sure to keep separate the removed parts for each installation position.

b. Remove the cylinder head and gasket.

To install:

35. Place the right hand cylinder head gasket on the cylinder block surface with the front face of the Lot No. stamp upward.

➡Be careful of the installation direction.

⁂ WARNING

Gently place the cylinder head in order not to damage the gasket with the bottom part of the head.

36. Place the cylinder head on the cylinder block.

➡Do not allow oil to adhere to the mounting surface of the cylinder head.

37. Apply a light coat of engine oil to the threads and under the heads of the cylinder head bolts.

38. The cylinder head bolts are tightened in 3 progressive steps:

a. Step 1: Using a 10 mm bi-hexagon wrench, install and uniformly tighten the 8 cylinder head bolts with the plate washers in several steps and in the sequence shown in the illustration. Tighten to 27 ft. lbs. (36 Nm).

b. Step 2: Mark the cylinder head bolt head with paint as shown in the illustration. Tighten the cylinder head bolts another 90°.

c. Step 3: Tighten the cylinder head bolts an additional 90°. Check that the painted mark is now facing rearward.

d. Seal packing will seep out on the engine's front side. Thoroughly wipe clean any seal packing.

39. Place the left hand cylinder head gasket on the cylinder block surface with the front face of the Lot No. stamp upward.

➡Be careful of the installation direction.

➡Gently place the cylinder head in order not to damage the gasket with the bottom part of the head.

40. Place the cylinder head on the cylinder block.

➡Do not allow oil to adhere to the mounting surface of the cylinder head.

41. Apply a light coat of engine oil to the threads and under the heads of the cylinder head bolts.

42. The cylinder head bolts are tightened in 3 progressive steps:

a. Step 1: Using a 10 mm bi-hexagon wrench, install and uniformly tighten the 8 cylinder head bolts with the plate washers in several steps and in the sequence shown in the illustration. Tighten to 27 ft. lbs. (36 Nm).

b. Step 2: Mark the cylinder head bolt head with paint as shown in the illustration. Tighten the cylinder head bolts another 90°.

c. Step 3: Tighten the cylinder head bolts an additional 90°. Check that the painted mark is now facing rearward.

d. Tighten the 2 bolts in the order shown in the illustration to 22 ft. lbs. (30 Nm). Only use the specifications stated above when tightening the bolts 1 and 2 shown in the illustration.

e. Seal packing will seep out on the engine's front side. Thoroughly wipe clean any seal packing.

43. Install the No. 2 chain tensioner assembly.

44. Install the No. 3 chain tensioner.

45. Install the No. 1 and 2 chain vibration damper.

46. Install the idle sprocket assembly.

47. Install the chain sub-assembly.

48. Install the chain tensioner slipper.

49. Install the No. 1 chain tensioner assembly.

50. Install the water pump assembly.

51. Install the timing chain cover.

52. Install the water inlet housing.

53. Install the left hand No. 1 front engine mounting bracket.

54. Install the No. 1 oil pan baffle plate.

55. Install the oil pan sub-assembly.

56. Install the oil strainer sub-assembly.

57. Install the No. 2 oil pan sub-assembly.

58. Install the oil pan drain plug and gasket.

59. Install the cylinder head covers and gaskets

60. Install the water outlet.

61. Install the crankshaft pulley.

62. Install the oil filter.

63. Install the cylinder block water drain cock sub-assembly.

64. Install the No. 1 oil pipe.

65. Install the oil pipe.

66. Install the crankshaft position sensor.

67. Install the camshaft timing oil control valve assembly.

68. Install the camshaft position sensor.

69. Install the ventilation valve sub-assembly.

70. Install the spark plugs and ignition coil assembly.

71. Install the oil filler cap sub-assembly.

72. Install the engine assembly with transaxle.

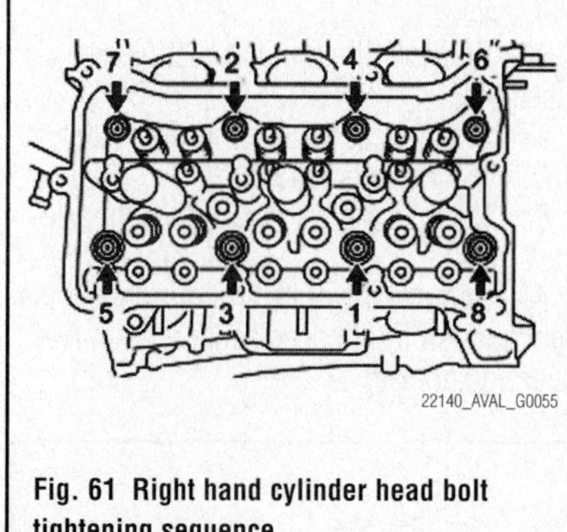

Fig. 61 Right hand cylinder head bolt tightening sequence

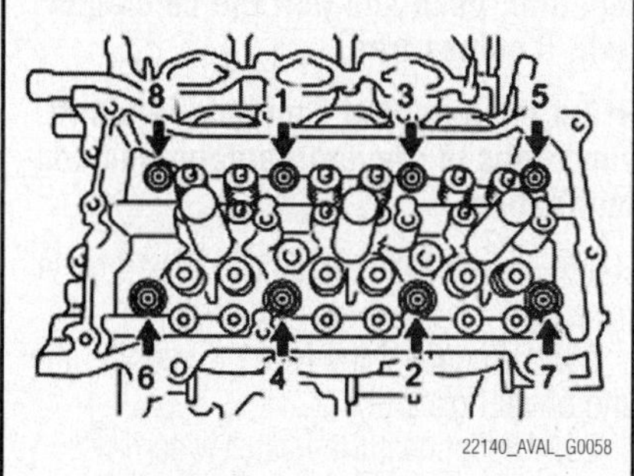

Fig. 63 Left hand cylinder head 8 bolt tightening sequence

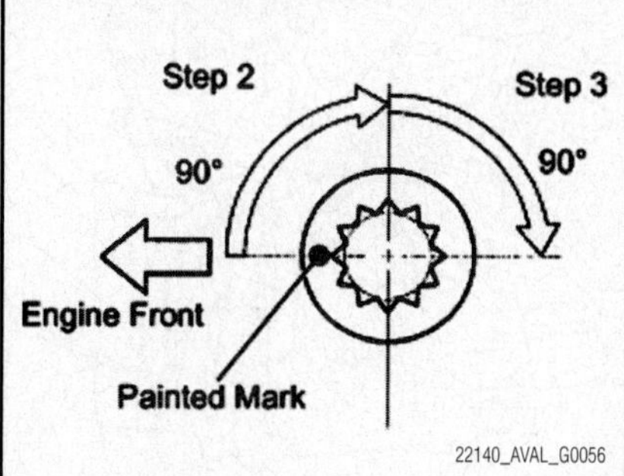

Fig. 62 Mark the cylinder head bolt and tighten another 90°

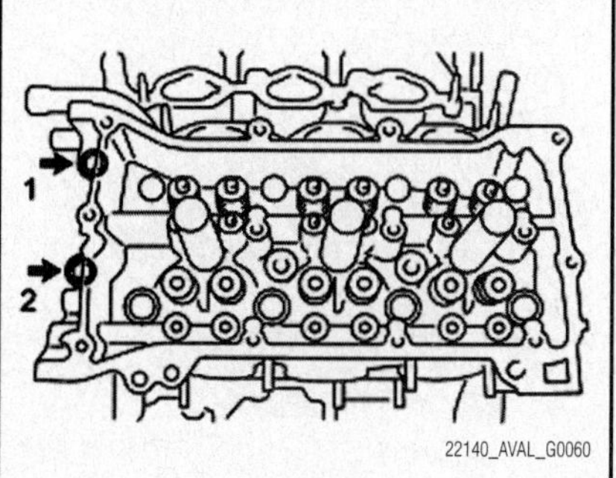
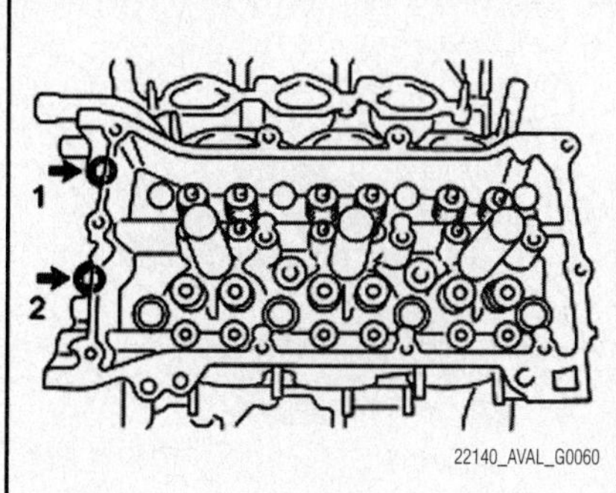

Fig. 64 Left hand cylinder head 2 bolt tightening sequence

ENGINE ASSEMBLY

REMOVAL & INSTALLATION

See Figures 65 through 72.

1. Before servicing the vehicle, refer to the Precautions Section.
2. Discharge fuel system pressure.
3. Remove front wheels.
4. Remove the engine under covers.
5. Remove the right front fender apron seal.
6. Drain engine oil.
7. Drain engine coolant.
8. Drain automatic transaxle fluid.
9. Remove the right and left windshield wiper arm and blade assemblies.
10. Remove the cowl top ventilator louver sub-assembly.
11. Remove the windshield wiper motor and link assembly.
12. Remove the outer cowl top panel.
13. Remove battery.
14. Remove the cool air intake duct seal.
15. Remove the V-bank cover sub-assembly.
16. Remove the V-ribbed belt.
17. Remove the radiator reserve tank assembly.
18. Remove the bolt, 2 nuts, and right No. 2 mounting stay and engine moving control rod.
19. Disconnect the clamp and 2 vacuum hoses, then remove the 2 bolts and the No. 2 air cleaner inlet.
20. Remove the air cleaner cap with hose, as follows:
    a. Disconnect the 3 vacuum hoses and No. 1 ventilation hose.
    b. Disconnect the air flow meter connector.
    c. Separate the No. 1 fuel vapor feed hose.
    d. Loosen the hose clamp bolt.
    e. Remove the 3 clamps and air cleaner cap with air cleaner hose.
21. Remove the air cleaner case, as follows:
    a. Remove the air cleaner filter element from the air cleaner case.
    b. Disconnect the connector, vacuum hose, and hose clamp.
    c. Remove the 3 bolts and air cleaner case.
22. Remove the bolt and No. 1 air cleaner inlet.
23. Disconnect the wire harness clamp, then remove the clip, bolt and intake air resonator.
24. Remove the clamp and disconnect the No. 1 fuel vapor feed hose.
25. Remove the intake air surge tank assembly.
26. Separate the fuel tube sub-assembly, as follows:
    a. Remove the No. 1 fuel pipe clamp.
    b. Disconnect the connector from the tube while pinching.

➡Check for contamination in the pipe and around the connector. Clean if necessary and then disconnect the connector.

➡Disconnect the connector by hand.

➡Do not bend, fold or rotate the nylon tube.

➡If the pipe and connector are stuck together, push and pull the connector until it comes free.

➡Put the pipe and connector ends in vinyl bags to prevent damage and contamination.

27. Remove the radiator inlet and outlet hoses.
28. Disconnect the No. 1 oil cooler inlet and outlet hoses.
29. Disconnect the heater water inlet hose A.
30. Disconnect the heater water outlet hose A.
31. Remove the ECM.
32. Remove the starter assembly.
33. Remove the upper relay block cover.
34. Disconnect the engine wire from the engine room junction block, as follows:
    a. Remove the nut and separate the wire harness.
    b. Using a screwdriver, release the engine room junction block. Pull the junction block upward.
    c. Disconnect the 2 engine wire connectors.
35. Disconnect the engine wire connector and 3 clamps.
36. Remove the 2 bolts and disconnect the 2 ground cables.
37. Remove the bolt and disconnect the ground cable from the wire bracket.
38. Disconnect the engine wire clamp.
39. Remove the 2 bolts and ground cable from the cylinder head.
40. Remove the bolt and disconnect the wire harness.
41. Disconnect the park/neutral position switch connector.
42. Disconnect the transmission control cable assembly.
43. Disconnect the No. 1 oil reservoir to pump hose.
44. Disconnect the steering gear outlet return tube.
45. Remove the No. 1 exhaust pipe support bracket.
46. Remove the front exhaust pipe assembly.
47. Disconnect the left and right front stabilizer link assemblies.
48. Remove the left and right front axle shaft nuts.
49. Remove the bolts and disconnect the right and left front speed sensors from the steering knuckle.

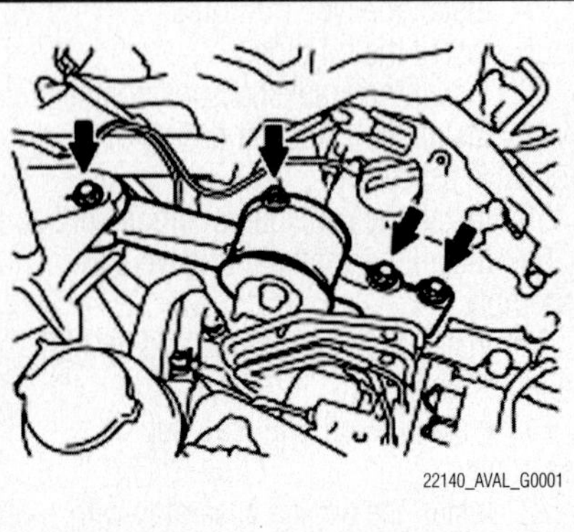

Fig. 65 Engine moving control rod

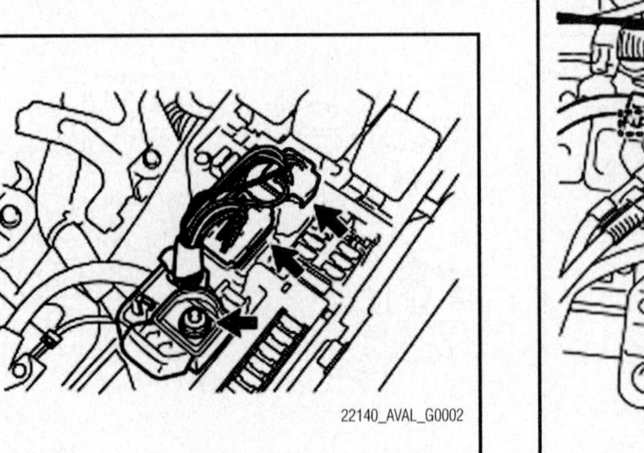

Fig. 66 Remove engine wire from junction block

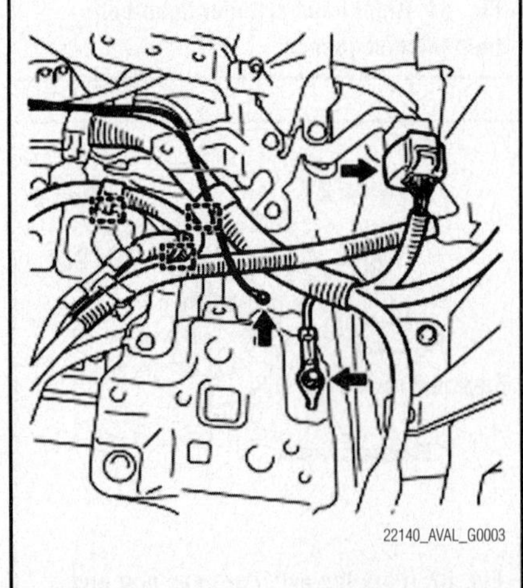

Fig. 67 Engine wire connector and clamps

50. Disconnect the left and right tie rod assemblies.

51. Disconnect the lower No. 1 left and right front suspension arm sub-assemblies.

52. Remove the No. 1 exhaust pipe support bracket.

53. Remove the flywheel housing under cover.

54. Remove the drive plate and torque converter clutch setting bolt, as follows:

a. Using SST (SST: 09213-70011, SST: 09330 00021) or equivalent, hold the crankshaft.

b. Remove the 2 bolts and flywheel housing under cover.

c. Remove the 6 torque converter clutch setting bolts.

55. Disconnect the steering intermediate shaft assembly.

56. Remove the alternator assembly.

57. Separate the cooler compressor assembly.

58. Remove the engine assembly with transaxle, as follows:

a. Set the engine lifter.

b. Remove the 4 bolts, 2 nuts, and right and left frame side rail plates.

c. Remove the 4 bolts, 2 nuts, front suspension member right and left rear brace.

d. Operate the engine lifter, then remove the engine assembly from the vehicle.

➡Make sure the engine is clear of all wiring and hoses.

e. Install the 2 engine hangers with the 4 bolts. Part numbers: Engine hanger No. 1 12281-31120, Engine hanger No. 2 12282-31100, Bolts 90119-A0117. Torque to 24 ft. lbs. (33 Nm).

f. Attach the engine sling device and hang the engine with the chain block.

To install:

59. Set the engine assembly with transaxle on the engine lifter.

60. Install the engine assembly to the vehicle.

61. Install the right and left frame side rail plates and with the 4 bolts and 2 nuts. Tighten A to 63 ft. lbs. (85 Nm) and B to 24 ft. lbs. (32 Nm).

62. Install the right and left front suspension member rear braces with the 4 bolts and 2 nuts. Tighten C to 63 ft. lbs. (85 Nm) and D to 24 ft. lbs. (32 Nm).

63. Install the cooler compressor assembly.

64. Install the alternator assembly.

65. Install the steering intermediate shaft assembly.

66. Install drive plate and torque converter clutch setting bolt, as follows:

a. Using SST (SST: 09213-70011, SST: 09330-00021) or equivalent, hold the crankshaft.

b. Using kerosene or gasoline, clean the bolts thoroughly.

c. Apply adhesive (Toyota Genuine Adhesive 1324, Three Bond 1324 or equivalent) to 2 or 3 threads of the bolt end.

d. Install the 6 torque converter set bolts and tighten 30 ft. lbs. (41 Nm). First tighten the green colored bolt, then the other 5 bolts.

67. Install the flywheel housing under cover with the 2 bolts and tighten 69 inch lbs. (7.8 Nm).

68. Install the No. 1 exhaust pipe support bracket with the 2 bolts and tighten 15 ft. lbs. (21 Nm).

69. Install the left and right front axle assemblies.

70. Install the lower right and left No. 1 suspension arm sub-assemblies.

71. Install the left and right tie rod assemblies.

72. Install the right and left front speed sensors.

73. Install the right and left front axle shaft nuts.

74. Install the right and left front stabilizer link assemblies.

75. Install the front exhaust pipe assembly.

76. Install the front No. 1 exhaust pipe support bracket.

77. Connect the steering gear outlet return tube.

78. Connect the No. 1 oil reservoir to pump hose.

79. Connect the transmission control cable assembly.

80. Connect engine wire, as follows:

a. Install the ground cable to the cylinder head with the 2 bolts and tighten to 9 ft. lbs. (12 Nm).

b. Connect the engine wire clamp.

c. Install the ground cable with the bolt to the wire bracket and tighten to 71 inch lbs. (8 Nm).

d. Install the 2 ground cables with the 2 bolts and tighten bolt A to 74 inch lbs. (8.4 Nm) and bolt B 71 inch lbs. (8 Nm).

e. Connect the engine wire connector and 3 clamps.

f. Connect the wire to the engine room junction block. Then, install it with the nut and 2 connectors and tighten to 74 inch lbs. (8.4 Nm).

81. Install the upper relay block cover.

82. Install the starter assembly.

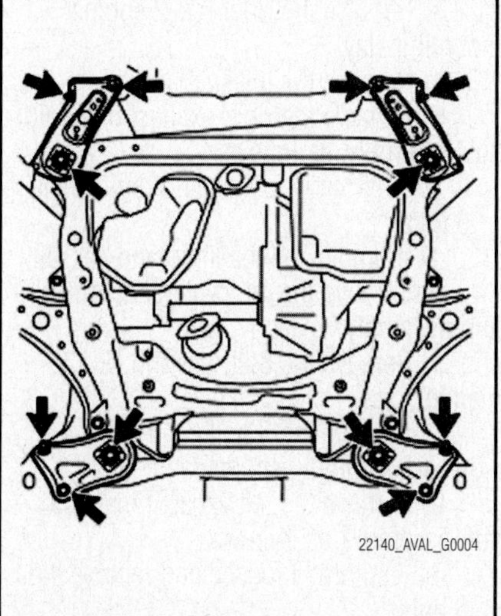

Fig. 68 Engine removal

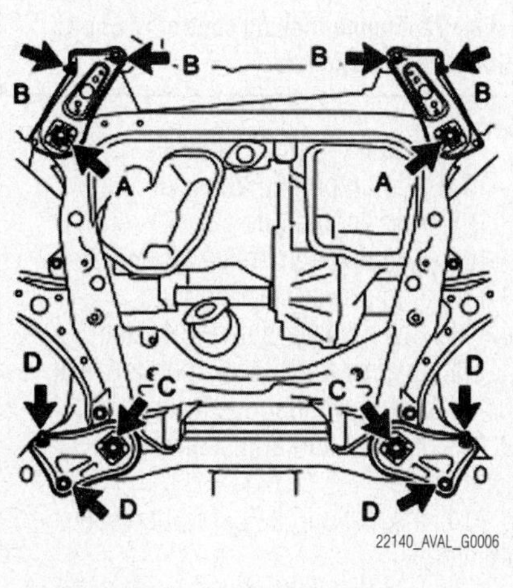

Fig. 69 Engine assembly installation sequence

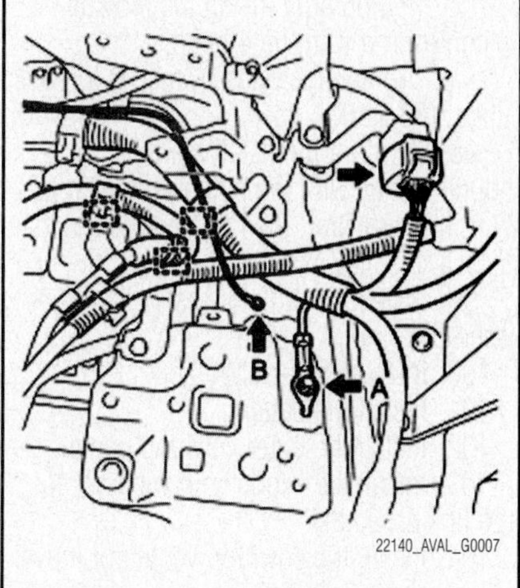

Fig. 70 Ground cable bolt tightening sequence

83. Install the ECM.
84. Connect the heater water outlet hose A.
85. Connect the heater water inlet hose A.
86. Connect the No. 1 oil cooler inlet and outlet hoses.
87. Install the radiator inlet and outlet hoses.
88. Connect the fuel tube sub-assembly, as follows:
 a. Push in the fuel tube connector to the fuel pipe until connector makes a "click" sound.

➡Check for damage or contamination on the connected part of the pipe. Check if the pipe and the connector are securely connected by trying to pull them apart.

 b. Install the No. 1 fuel pipe clamp .
89. Install the intake air surge tank assembly.
90. Install the clamp and connect the No. 1 fuel vapor feed hose.
91. Install the intake air resonator with the bolt and clip and tighten 44 inch lbs. (5 Nm), then connect the wire harness clamp.
92. Install the No. 1 air cleaner inlet with the bolt and tighten 44 inch lbs. (5 Nm).
93. Install the air cleaner case with the 3 bolts and tighten 44 inch lbs. (5 Nm).
94. Install the air cleaner cap with the hose clamp and 3 clips.
95. Install the No. 2 air cleaner inlet with the 2 bolts and tighten 44 inch lbs. (5 Nm), then connect the clamp and 2 vacuum hoses.
96. Connect the vacuum hoses.
97. Install the engine moving control rod, as follows:
 a. Temporarily install the moving control rod with the 4 bolts.
 b. Tighten the bolts A, then B to 28 ft. lbs. (38 Nm).
98. Temporarily install the right No. 2 mounting stay with the bolt and tighten to 28 ft. lbs. (38 Nm), then tighten the 2 nuts to 17 ft. lbs. (23 Nm).
99. Install the radiator reserve tank assembly.
100. Install the V-ribbed belt.
101. Install the battery.
102. Install the outer cowl top panel.
103. Install the windshield wiper motor and link assembly.
104. Install the cowl top ventilator louver sub-assembly.
105. Install the right and left windshield wiper arm and blade assemblies.

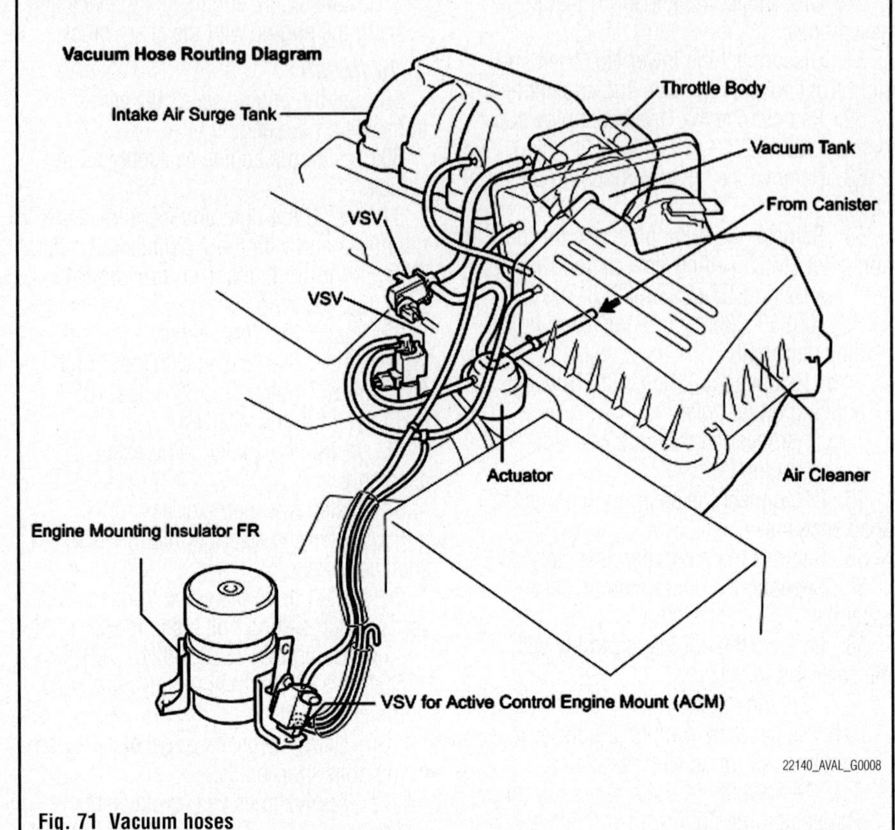

Fig. 71 Vacuum hoses

Fig. 72 Engine moving control rod bolt tightening sequence

106. Install the front wheels.
107. Add automatic transaxle fluid.
108. Add engine oil.
109. Add engine coolant.
110. Add power steering fluid.
111. Bleed power steering system.
112. Inspect for oil leak, coolant leak, fuel leak, and exhaust gas leak.
113. Inspect and adjust shift lever position.
114. Inspect and adjust front wheel alignment.
115. Inspect ignition timing and engine idle speed.
116. Inspect CO/HC.
117. Install the V-bank cover sub-assembly.
118. Install the cool air intake duct seal.
119. Check ABS speed sensor signal.

EXHAUST MANIFOLD

REMOVAL & INSTALLATION

See Figures 73 through 76.

1. Before servicing the vehicle, refer to the Precautions Section.
2. Remove the ignition coil assembly.
3. Remove the right No. 2 engine mounting stay.
4. Remove the intake manifold.
5. Remove the right exhaust manifold sub-assembly, as follows:
 a. Uniformly loosen and remove the 6 nuts.
 b. Remove the manifold and gasket.
6. Remove the oil level gauge guide sub-assembly.
7. Remove the bolt, nut and No. 2 manifold stay.
8. Remove the 3 bolts and No. 2 exhaust manifold heat insulator.
9. Remove the left exhaust manifold sub-assembly, as follows:
 a. Uniformly loosen and remove the 6 nuts.
 b. Remove the manifold and gasket.

Fig. 73 Right-hand exhaust manifold nuts

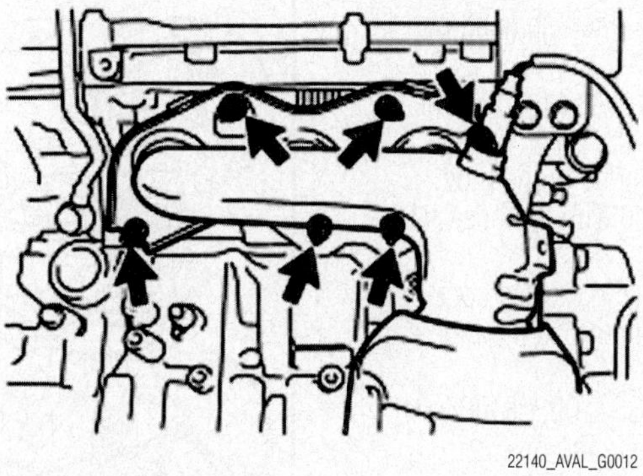

Fig. 74 Left-hand exhaust manifold nuts

Fig. 75 Install a new gasket

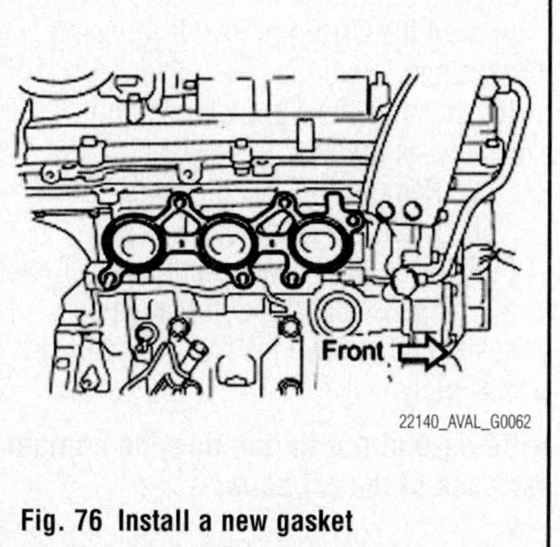

Fig. 76 Install a new gasket

To install:

10. Install the left exhaust manifold sub-assembly, as follows:
    a. Install a new gasket as shown in the illustration.
    b. Install the left exhaust manifold sub-assembly with the 6 nuts and tighten to 15 ft. lbs. (21 Nm).
11. Install the No. 2 exhaust manifold heat insulator with the 3 bolts and tighten to 75 inch lbs. (8.5 Nm).
12. Install the No. 2 manifold stay with the bolt and nut and tighten to 25 ft. lbs. (34 Nm).
13. Install the oil level gauge guide sub-assembly.
14. Install the right exhaust manifold sub-assembly, as follows:
    a. Install a new gasket as shown in the illustration.
    b. Install the right exhaust manifold sub-assembly with the 6 nuts and tighten to 15 ft. lbs. (21 Nm).
15. Install the intake manifold.
16. Install the right No. 2 engine mounting stay.
17. Install the ignition coil assembly.
18. Install the engine assembly with transaxle.

FLEXPLATE

REMOVAL & INSTALLATION

See Figure 77.

1. Remove the automatic transaxle assembly.
2. Remove the drive plate and ring gear sub-assembly, as follows:
    a. Using SST (SST: 09213-70011, SST: 09330-00021) or equivalent tool, hold the crankshaft.
    b. Remove the 8 bolts, front spacer, drive plate and rear spacer.

Fig. 77 Remove the 8 bolts, front spacer, drive plate and rear spacer

To install:

3. Install the drive plate and ring gear sub-assembly, as follows:
    a. Using SST (SST: 09213-70011, SST: 09330-0002) or equivalent, hold the crankshaft.
    b. Apply adhesive (Toyota Genuine Adhesive 1324, Three Bond 1324 or equivalent) to 2 or 3 threads of the mounting bolt end.
    c. Install the front spacer, drive plate and rear spacer on the crankshaft.
    d. Install and tighten the 8 mounting bolts uniformly in several steps. Tighten to 61 ft. lbs (83 Nm).
4. Install automatic transaxle assembly.

INTAKE MANIFOLD

REMOVAL & INSTALLATION

See Figure 78.

1. Before servicing the vehicle, refer to the Precautions Section.
2. Remove the engine assembly with transaxle.
3. Secure the engine.

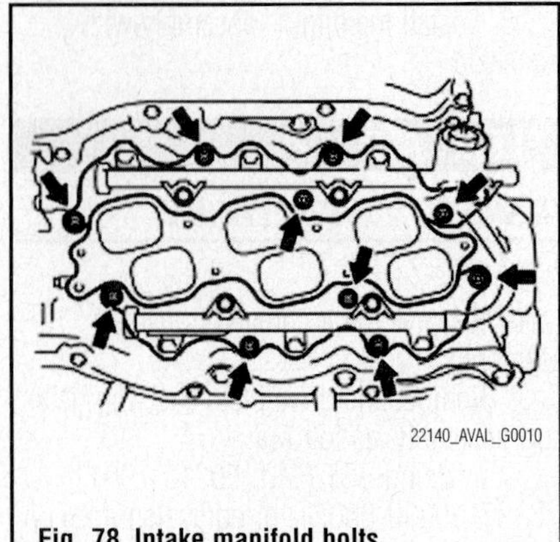

Fig. 78 Intake manifold bolts

4. Remove the ignition coil assembly.
5. Remove the right No. 2 engine mounting stay.
6. Remove the intake manifold, as follows:
   a. Uniformly loosen and remove the 6 bolts and 4 nuts.
   b. Remove the intake manifold and 2 gaskets.

To install:

7. Install the intake manifold, as follows:

⁂ WARNING

DO NOT apply oil to the intake manifold and cylinder head sub-assembly bolts.

   a. Set a new gasket on each cylinder head.

➡Align the port holes of the gasket and cylinder head.

➡Make sure that the gasket is installed in the correct direction.

   b. Set the intake manifold on the cylinder heads.
   c. Install and tighten the 6 bolts and 4 nuts uniformly in several steps to 15 ft. lbs. (21 Nm).
8. Install the right No. 2 engine mounting stay.
9. Install the ignition coil assembly.
10. Install the engine assembly with transaxle.

OIL PAN

REMOVAL & INSTALLATION

See Figures 79 through 87.

1. Before servicing the vehicle, refer to the Precautions Section.
2. Drain the engine oil.
3. Remove the engine assembly with transaxle.
4. Secure the engine.
5. Remove the oil filler cap and gasket.
6. Remove the oil pan drain plug and gasket.
7. Remove the No. 1 oil pipe, as follows:
   a. Remove the 2 oil pipe unions and oil pipe.
   b. Remove the left hand oil control valve filter and gaskets.
8. Remove the oil pipe, as follows:
   a. Remove the bolt.
   b. Remove the 2 oil pipe unions and oil pipe.
   c. Remove the right oil control valve filter and gaskets.
9. Remove the oil filter element, as follows:
   a. Remove the drain plug. Do not remove the O-ring.
   b. Connect the hose to the pipe.
   c. Insert the pipe with the hose into the oil filter cap.
   d. Make sure that the oil is completely drained and remove the pipe and O-ring.
   e. Using SST (SST: 09228-06501) or equivalent, remove the oil filter cap.
   f. Remove the oil filter element and O-ring from the oil filter cap. Do not use any tools when removing the O-ring to prevent the O-ring groove from being damaged.
10. Remove the No. 2 oil pan sub-assembly, as follows:
   a. Remove the 16 bolts and 2 nuts.
   b. Insert the blade of SST (SST: 09032-00100) or equivalent tool between the oil pans. Cut through the applied sealer and remove the No. 2 oil pan sub-assembly.

➡Be careful not to damage the contact surfaces of the oil pans.

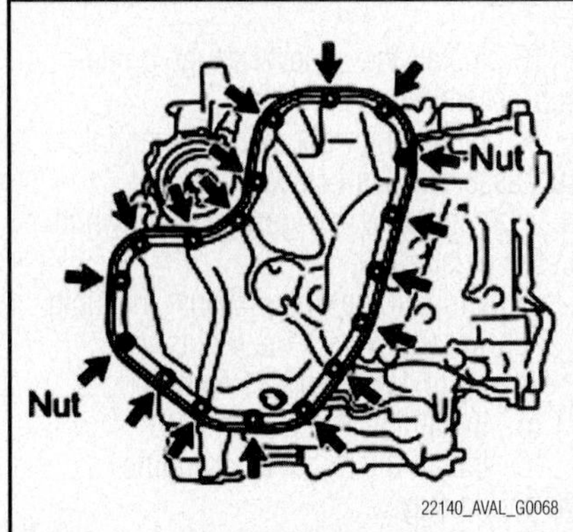

Fig. 79 No. 2 oil pan bolts and nuts

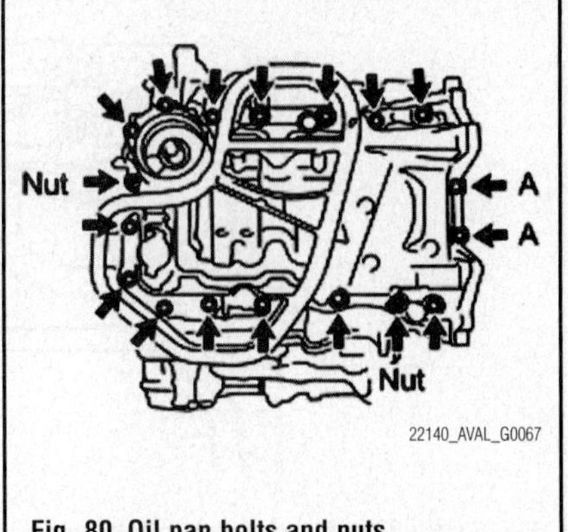

Fig. 80 Oil pan bolts and nuts

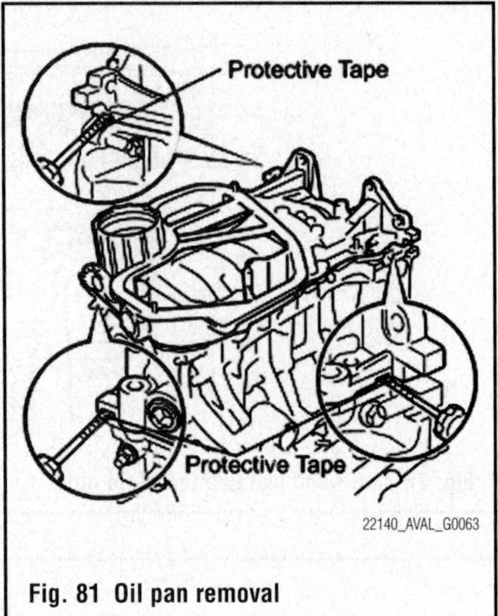

Fig. 81 Oil pan removal

11. Remove the oil pan sub-assembly, as follows:
   a. Remove the 16 bolts and 2 nuts.

➡Be sure to clean the bolts and stud bolts and check the threads for cracks or other damage.

   b. Remove the oil pan by prying between the oil pan and cylinder block with a taped screwdriver.

➡Be careful not to damage the contact surfaces of the cylinder block and oil pan.

   c. Remove the 2 O-rings.

To install:

12. Install the oil pan sub-assembly, as follows:
   a. Using an E8 Torx® socket wrench, install the stud bolts as shown in the illustration. Tighten to 7 ft. lbs (10 Nm).
   b. Apply seal packing (Toyota Genuine Seal Packing Black, Three Bond 1207B or equivalent) in a continuous line as shown in the illustration. Seal diameter: 3.0 to 4.0 mm (0.118 to 0.156 in.).

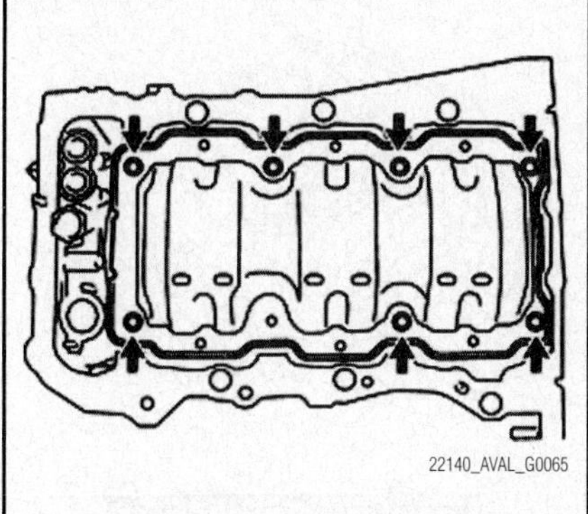

Fig. 82 Oil pan sub-assembly stud bolts

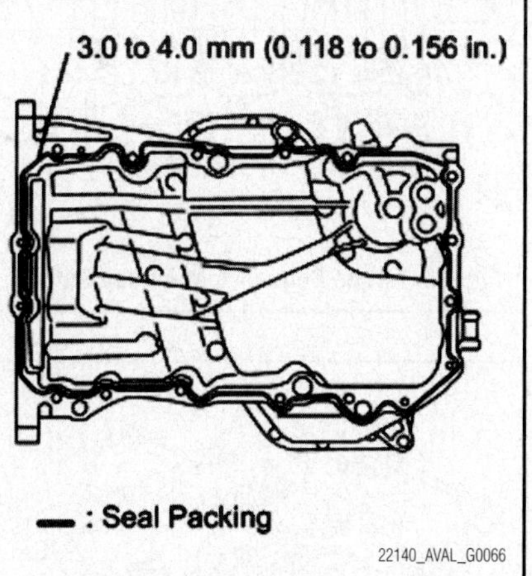

Fig. 83 Sealant application

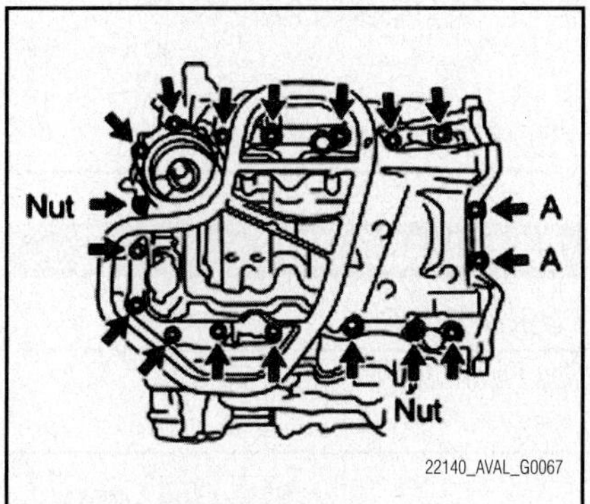

Fig. 84 Oil pan bolts and nuts

➡**Remove any oil from the contact surface.**

➡**Install the oil pan within 3 minutes after applying seal packing.**

➡**Do not start the engine for at least 2 hours after installing.**

c. Install 2 new O-rings.

d. Install the oil pan with the 16 bolts and 2 nuts and tighten to 7 ft. lbs (10 Nm), and 15 ft. lbs (21 Nm).

Fig. 85 Sealant application

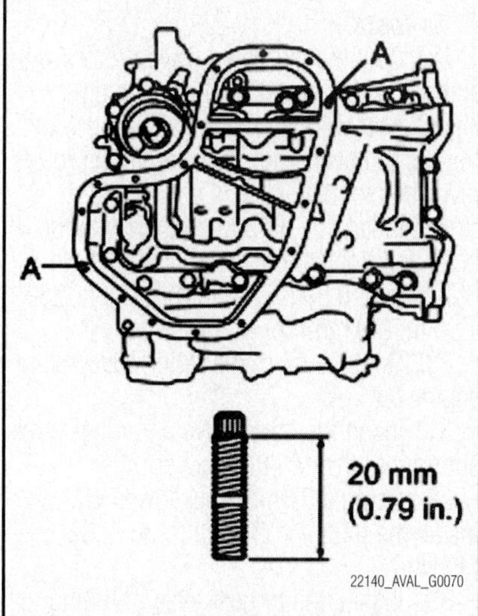

Fig. 86 No. 2 oil pan stud bolt installation

13. Install the No. 2 oil pan sub-assembly, as follows:

a. Apply seal packing (Toyota Genuine Seal Packing Black, Three Bond 1207B or equivalent) in a continuous line as shown in the illustration. Seal diameter: 3.0 to 4.0 mm (0.118 to 0.156 in.).

➡**Remove any oil from the contact surface.**

➡**Install the No. 2 oil pan within 3 minutes after applying seal packing.**

➡**Do not start the engine for at least 2 hours after installing.**

b. Using an E6 Torx® socket wrench, install the stud bolts as shown in the illustration and tighten to 35 inch lbs (4 Nm).

c. Install the No. 2 oil pan with the 16 bolts and 2 nuts and tighten to 7 ft. lbs (10 Nm).

14. Install the oil pan drain plug and a new gasket. Tighten to 30 ft. lbs (40 Nm).

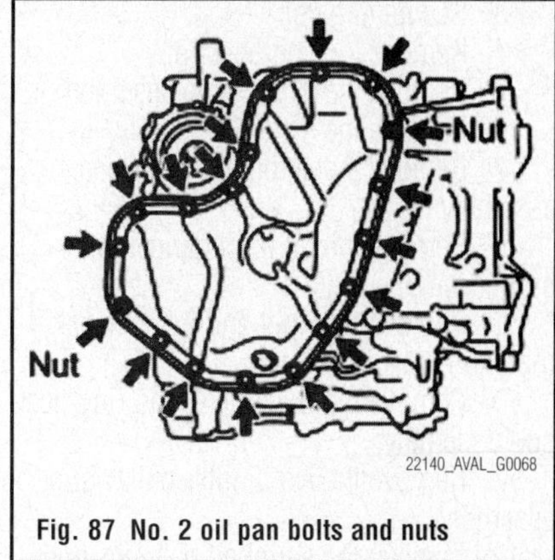

Fig. 87 No. 2 oil pan bolts and nuts

15. Install the oil filter element, as follows:

a. Clean the inside of the oil filter cap, the threads and O-ring groove.

b. Apply a small amount of engine oil to a new O-ring and install it to the oil filter cap.

c. Set a new oil filter element to the oil filter cap.

d. Remove dirt or foreign matter from the installation surface and inside of the engine.

e. Apply a small amount of engine oil to the O-ring again and install the oil filter cap.

➡**Be careful that the O-ring does not get caught between the parts. The O-ring must not be twisted on the groove.**

f. Using SST (SST: 09228-06501) or equivalent, install the oil filter cap and tighten to 18 ft. lbs (25 Nm). Make sure that the oil filter is installed securely as shown in the illustration.

g. Apply a light coat of engine oil to a new O-ring and install it to the oil filter cap. Remove all dirt and foreign matter from the installation surface.

h. Install the oil filter drain plug to the oil filter cap and tighten to 9 ft. lbs (13 Nm). Make sure that the O-ring does not get caught between the parts.

16. Install the oil filler cap sub-assembly.

17. Install the engine assembly with transaxle.

OIL PUMP

REMOVAL & INSTALLATION

See Figures 88 and 89.

1. Before servicing the vehicle, refer to the Precautions Section.

2. Remove the engine assembly with transaxle.

3. Secure engine.
4. Remove the engine wire.
5. Remove the front frame assembly.
6. Remove the starter assembly.
7. Remove the automatic transaxle assembly.
8. Remove the oil level gauge guide sub-assembly.
9. Remove the right and left exhaust manifold sub-assemblies.
10. Remove the drive plate and ring gear sub-assembly.
11. Remove the No. 2 idler pulley sub-assembly.
12. Remove the V-ribbed belt tensioner assembly.
13. Remove the water pump pulley.
14. Remove the water inlet housing.
15. Remove the crankshaft pulley.
16. Remove the No. 2 oil pan sub-assembly.
17. Remove the oil strainer sub-assembly.
18. Remove the oil pan sub-assembly.
19. Remove the intake air surge tank assembly.
20. Remove the ignition coil assembly.
21. Remove the No. 1 and 2 oil pipes.
22. Remove the right and left cylinder head cover sub-assemblies.
23. Remove the timing chain cover sub-assembly. Refer to Timing Chain Cover and Seal procedures for instructions.
24. Using a screwdriver with the tip taped, pry out the timing gear case or timing chain case oil seal.
25. Using a 27 mm socket wrench, remove the relief valve plug.
26. Remove the valve spring and oil pump relief valve.
27. Remove the 8 bolts, oil pump cover, drive rotor and driven rotor.

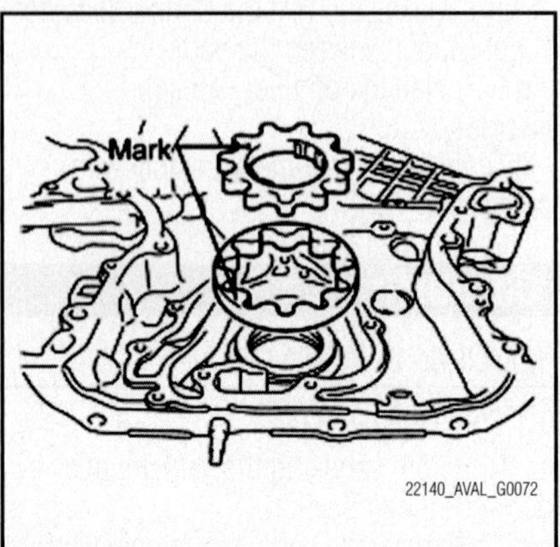

Fig. 88 Oil pump gears

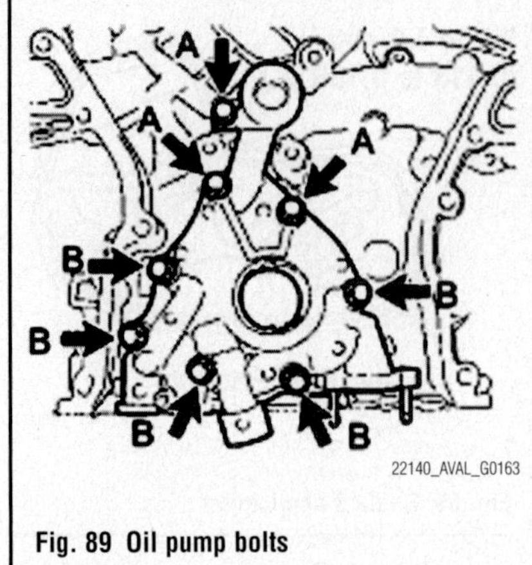

Fig. 89 Oil pump bolts

To install:

28. Coat the drive and driven rotors with engine oil and place them into the timing chain cover with the marks facing outward (oil pump cover side). Check that the rotors revolve smoothly.
29. Install the oil pump cover with the 8 bolts and tighten to 81 inch lbs. (9.1 Nm). Bolt length: 0.87 in. (22 mm) for bolt A, 1.58 in. (40 mm) for bolt B.
30. Coat the oil pump relief valve with engine oil.
31. Insert the relief valve and relief valve spring into the oil pump cover hole.
32. Using a 27 mm socket wrench, install the plug and tighten to 36 ft. lbs. (49 Nm).
33. Install timing gear case or timing chain case oil seal, as follows:
    a. Using SST (SST: 09316-60011) or equivalent tool, tap in a new oil seal until its surface is flush with the timing chain case edge.

➡Keep the lip free from foreign matter.

➡Do not tap on the oil seal at an angle.

➡Make sure that the oil seal edge does not stick out of the timing chain case.

    b. Apply MP grease to the oil seal lip.

34. Install timing chain or belt cover sub-assembly. Refer to Timing Chain Cover and Seal procedures for instructions.
35. The remainder of installation is the reverse of removal.

MAIN BEARING TORQUE SEQUENCE

See Figures 90 and 91.

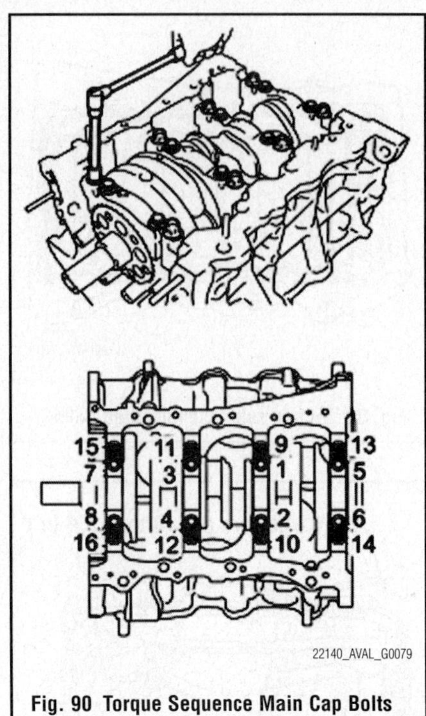

Fig. 90 Torque Sequence Main Cap Bolts

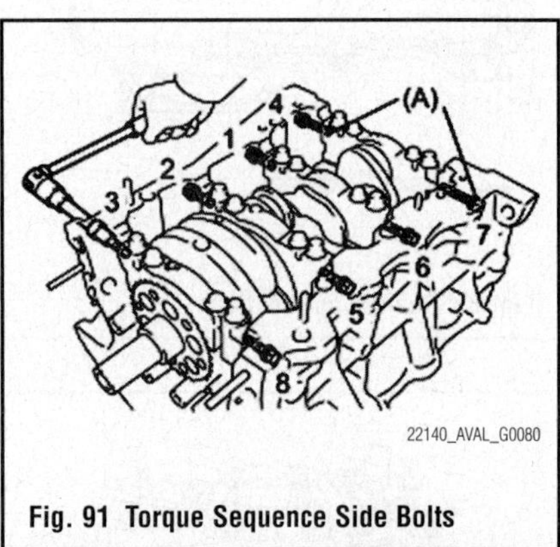

Fig. 91 Torque Sequence Side Bolts

PISTON AND RING

POSITIONING

See Figure 92.

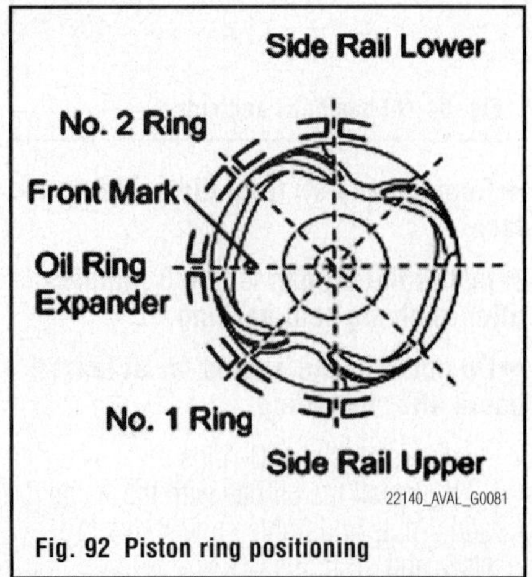

Fig. 92 Piston ring positioning

REAR MAIN SEAL

REMOVAL & INSTALLATION

See Figures 93 and 94.

1. Before servicing the vehicle, refer to the Precautions Section.
2. Remove the automatic transaxle assembly.
3. Remove the drive plate and ring gear sub-assembly.
4. Remove the rear main seal, as follows:
   a. Using a knife, cut off the oil seal lip.
   b. Using a screwdriver with the tip taped, pry out the oil seal.

➡**Be careful not to damage the crankshaft.**

To install:

5. Apply MP grease to a new oil seal lip.
6. Using SST (SST: 09223-15030, SST: 09950-70010) or equivalent and a hammer, tap in the oil seal. Oil seal tap in depth: -0.5 to 0.5 mm (-0.020 to 0.020 in.)

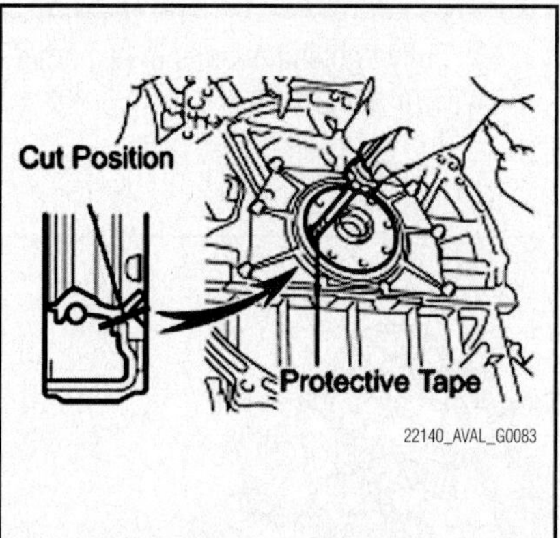

Fig. 93 Cut and pry the oil seal

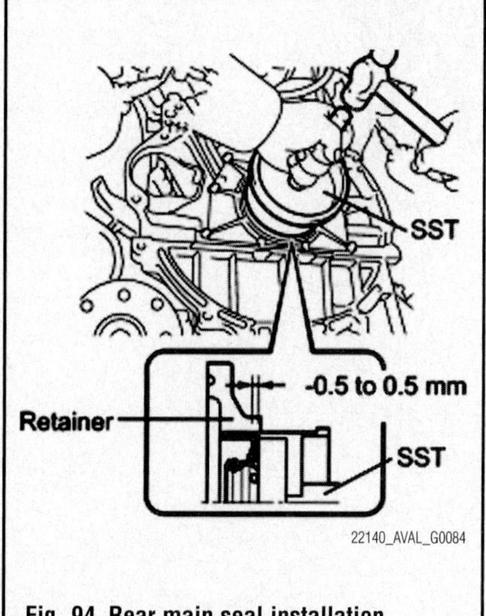

Fig. 94 Rear main seal installation

7. Install the drive plate and ring gear sub-assembly.
8. Install automatic transaxle assembly.

ROCKER ARMS/SHAFTS

REMOVAL & INSTALLATION

See Camshaft and Valve Lifters.

TIMING CHAIN COVER AND SEAL

REMOVAL & INSTALLATION

See Figures 95 through 104.

1. Before servicing the vehicle, refer to the Precautions Section.
2. Remove the engine assembly with transaxle.
3. Secure engine.
4. Remove the oil filler cap sub-assembly.
5. Remove the spark plugs and ignition coil assembly.
6. Remove the oil pan drain plug and gasket.
7. Remove the ventilation valve sub-assembly.
8. Remove the camshaft position sensor.
9. Remove the camshaft timing oil control valve assembly.
10. Remove crankshaft position sensor.
11. Remove the No. 1 oil pipe.
12. Remove the oil pipe.
13. Remove the cylinder block water drain cock sub-assembly.
14. Remove the oil filter.
15. Remove the crankshaft pulley.
16. Remove the left hand No. 1 front engine mounting bracket.
17. Remove the water inlet housing.
18. Remove the water outlet.
19. Remove the left-hand cylinder head cover sub-assembly and gasket.
20. Remove the cylinder head cover sub-assembly and gasket.
21. Remove the No. 2 oil pan sub-assembly.
22. Remove the oil strainer sub-assembly.
23. Remove the oil pan sub-assembly.
24. Remove the water pump assembly.
25. Remove the timing chain cover sub-assembly, as follows:
   a. Remove the 15 bolts and 2 nuts as shown in the illustration.
   b. Remove the timing chain cover by prying between the timing chain cover and cylinder head or cylinder block with a screwdriver with the tip taped.

➡**Be careful not to damage the contact surfaces of the cylinder head, cylinder block and chain cover.**

   c. Remove the 4 bolts, chain cover plate and gasket.
   d. Remove the gasket.

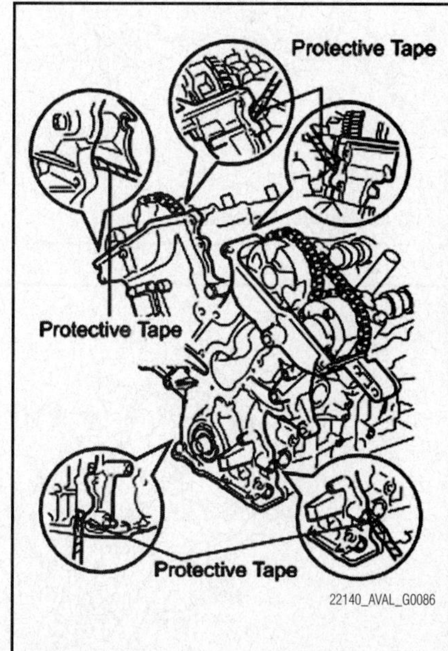

Fig. 96 Timing chain cover removal

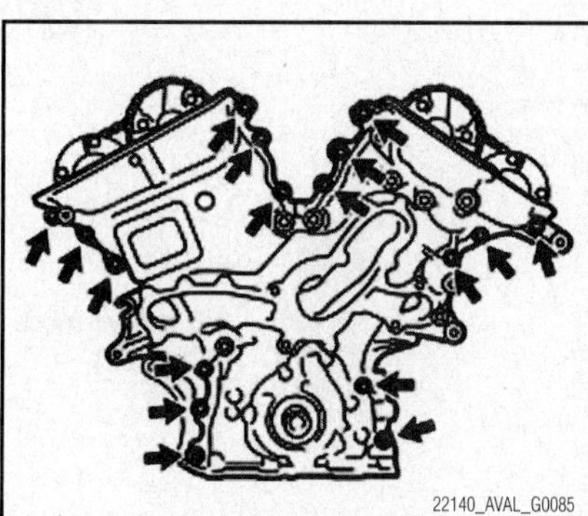

Fig. 95 Timing chain cover bolts and nuts

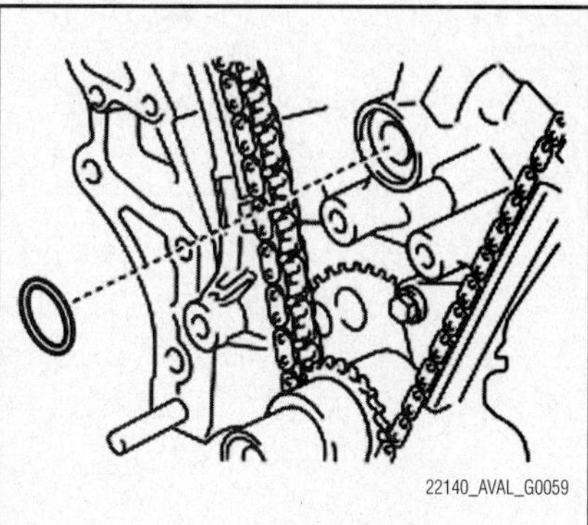

Fig. 97 Timing chain cover gasket

26. Remove the timing chain case oil seal, as follows:

a. Using a screwdriver with the tip taped, pry out the oil seal.

To install:

27. Install timing gear case or timing chain cover oil seal, as follows:

a. Apply MP grease to a new oil seal lip.

b. Using SST (SST: 09316-60011) and a hammer, tap in the oil seal until its surface is flush with the timing chain cover edge.

➡Keep the lip free from foreign matter.

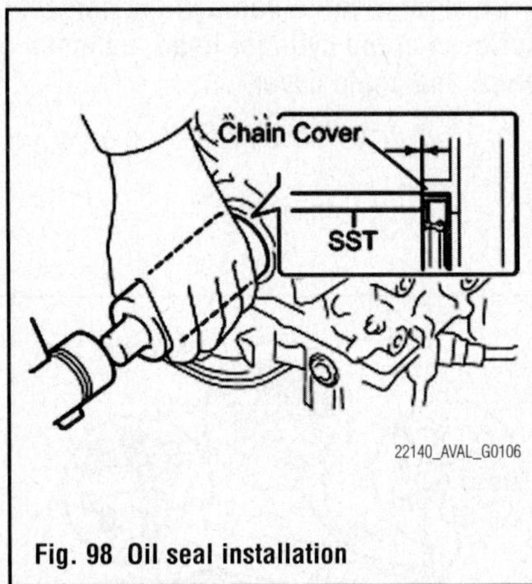

Fig. 98 Oil seal installation

➡Do not tap on the oil seal at an angle.

➡Make sure that the oil seal edge does not stick out of the timing chain cover.

28. Install the timing chain cover sub-assembly, as follows:

a. Install a new gasket and the chain cover plate with the 4 bolts and tighten to 81 inch lbs. (9.1 Nm).

b. Apply seal packing (Toyota Genuine Seal Packing Black, Three Bond 1207B or equivalent) in a continuous line to the engine unit as shown in the illustration. Seal diameter: 3.0 mm (0.118 in.).

➡Be sure to clean and degrease the contact surfaces, especially the surfaces indicated by C in the illustration.

➡When the contact surfaces are wet, wipe them with an oil-free cloth before applying seal packing.

➡Install the chain cover within 3 minutes after applying seal packing.

➡Do not start the engine for at least 2 hours after installing.

c. Apply seal packing in a continuous line to the timing chain cover as shown in the following illustration.

Seal packing: Toyota Genuine Seal Packing Black, Three Bond 1207B or equivalent, Toyota Genuine Seal Packing Black, Three Bond 1282B, Three Bond 1282B or equivalent.

➡When the contact surfaces are wet, wipe them with an oil-free cloth before applying seal packing.

➡Install the crankcase within 3 minutes and tighten the bolts within 15 minutes after applying seal packing.

➡Do not start the engine for at least 2 hours after installing.

d. Install a new gasket.

e. Align the oil pump's drive rotor spline and the crankshaft as shown in the illustration. Install the spline and chain cover to the crankshaft.

f. Loosely install the timing chain cover with the 23 bolts and 2 nuts, but do not tighten the bolts and 2 nuts yet.

WARNING

Make sure that there is no oil on the bolt and nut threads.

g. Fully tighten the bolts in this order: Area 1 and Area 2, tighten to 15 ft. lbs. (21 Nm).

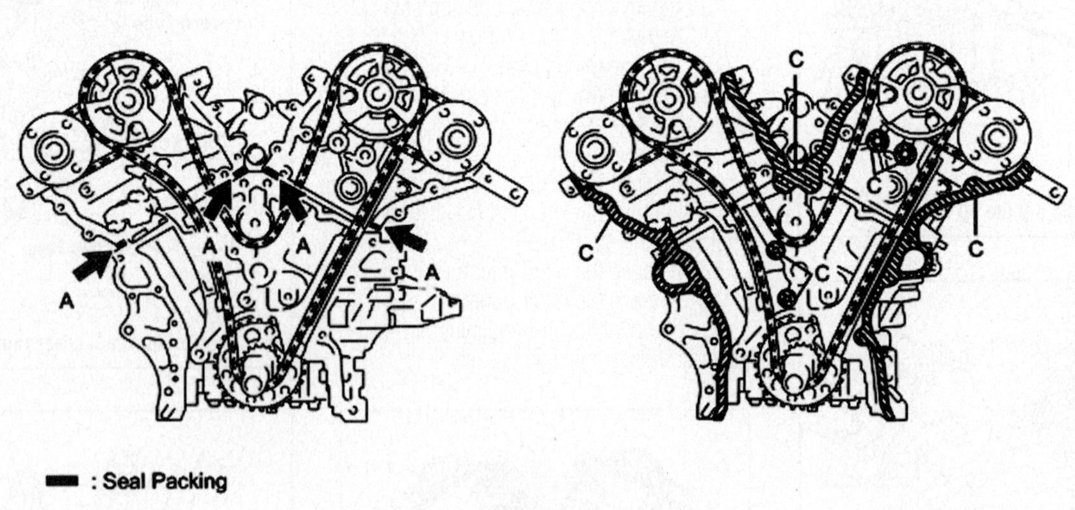

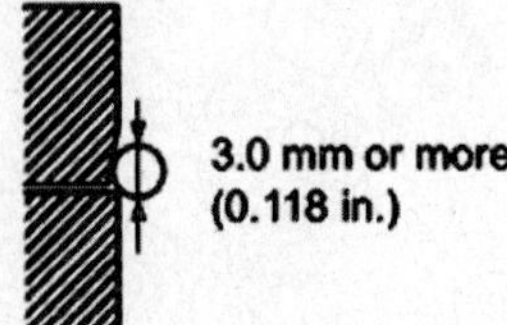

Fig. 99 Sealant application

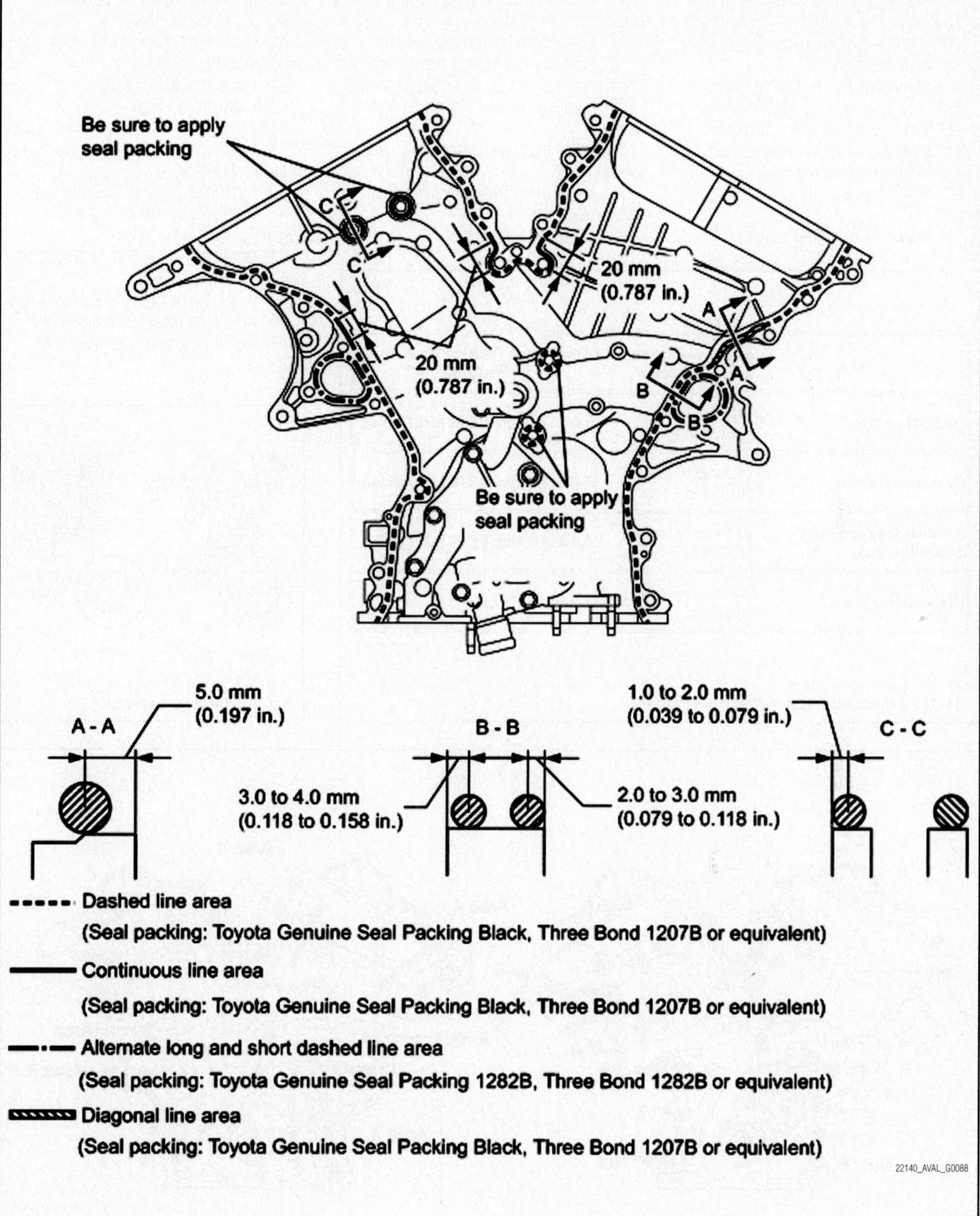

Fig. 100 Timing chain cover sealant application

h. Fully tighten the bolts in Area 3 to 15 ft. lbs. (21 Nm). Tighten the bolts and nuts in the order of upper to lower as shown in the illustration.

i. Fully tighten the bolts in Area 4 to 32 ft. lbs. (43 Nm), and to 15 ft. lbs. (21 Nm). Tighten the bolts and nuts in the order of lower to upper as shown in the illustration.

29. Install the water pump assembly.
30. Install the water inlet housing.
31. Install the left hand No. 1 front engine mounting bracket.
32. Install the oil pan sub-assembly.
33. Install the oil strainer sub-assembly.
34. Install the No. 2 oil pan sub-assembly.
35. Install the oil pan drain plug and gasket.
36. Install the cylinder head cover sub-assembly.
37. Install the left-hand cylinder head cover sub-assembly.
38. Install the water outlet.
39. Install the crankshaft pulley.
40. Install the oil filter.
41. Install the cylinder block water drain cock sub-assembly.
42. Install the No. 1 oil pipe.
43. Install the oil pipe.
44. Install the crankshaft position sensor.
45. Install the camshaft timing oil control valve assembly.
46. Install the camshaft position sensor.
47. Install the ventilation valve sub-assembly.
48. Install the spark plugs and ignition coil assembly.
49. Install the oil filler cap sub-assembly.
50. Install the water pump assembly.
51. Install the engine assembly with transaxle.

Area	Seal Packing Diameter	Application Position from Inside Seal Line
Dashed line area	3.5 mm or more (0.138 in.)	3.0 to 4.0 mm (0.118 to 0.158 in.)
Continuous line area	4.5 mm or more (0.177 in.)	3.0 to 4.0 mm (0.118 to 0.158 in.)
Alternate long and short dashed line area	3.5 mm or more (0.138 in.)	2.0 to 3.0 mm (0.079 to 0.118 in.)
Diagonal line area	6.0 mm or more (0.236 in.)	5.0 mm (0.197 in.)

22140_AVAL_G0089

Fig. 101 Sealant application diameter and position

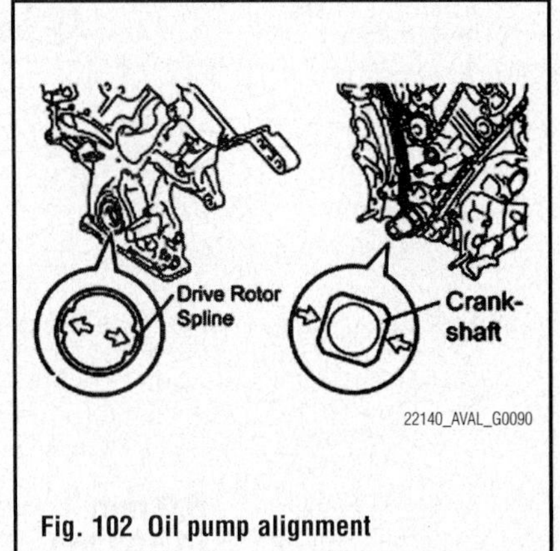

Fig. 102 Oil pump alignment

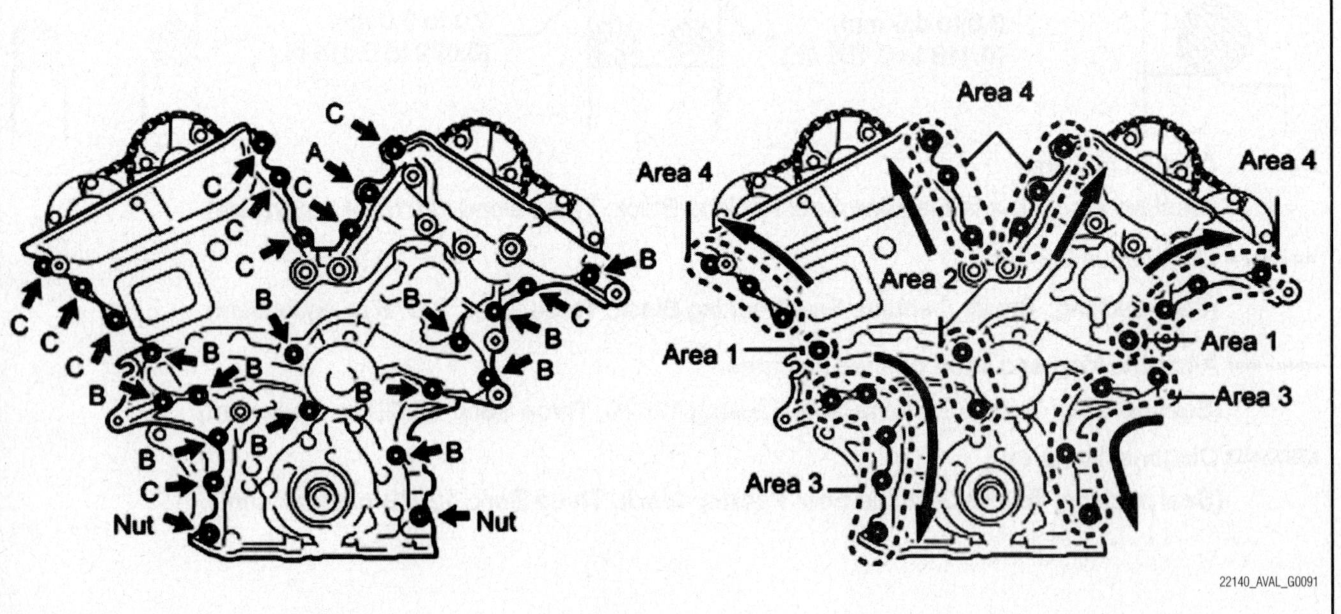

Fig. 103 Timing chain cover bolts and nuts

Item	Length
Bolt A	40 mm (1.57 in.)
Bolt B	55 mm (2.17 in.)
Bolt C	25 mm (0.98 in.)

22140_AVAL_G0092

Fig. 104 Bolt length

TIMING CHAIN AND SPROCKETS

REMOVAL & INSTALLATION

See Figures 105 through 117.

1. Before servicing the vehicle, refer to the Precautions Section.
2. Remove the engine assembly with transaxle.
3. Secure engine.
4. Remove the oil filler cap sub-assembly.
5. Remove the spark plugs and ignition coil assembly.
6. Remove the oil pan drain plug and gasket.
7. Remove the ventilation valve sub-assembly.
8. Remove the camshaft position sensor.
9. Remove the camshaft timing oil control valve assembly.
10. Remove crankshaft position sensor.
11. Remove the No. 1 oil pipe.
12. Remove the oil pipe.
13. Remove the cylinder block water drain cock sub-assembly.
14. Remove the oil filter.
15. Remove the crankshaft pulley.
16. Remove the left hand No. 1 front engine mounting bracket.
17. Remove the water inlet housing.
18. Remove the water outlet.
19. Remove the left-hand cylinder head cover sub-assembly and gasket.
20. Remove the cylinder head cover sub-assembly and gasket.
21. Remove the No. 2 oil pan sub-assembly.
22. Remove the oil strainer sub-assembly.
23. Remove the oil pan sub-assembly.
24. Remove the water pump assembly.
25. Remove the timing chain cover and seal.
26. Remove the No. 1 chain tensioner assembly, as follows:

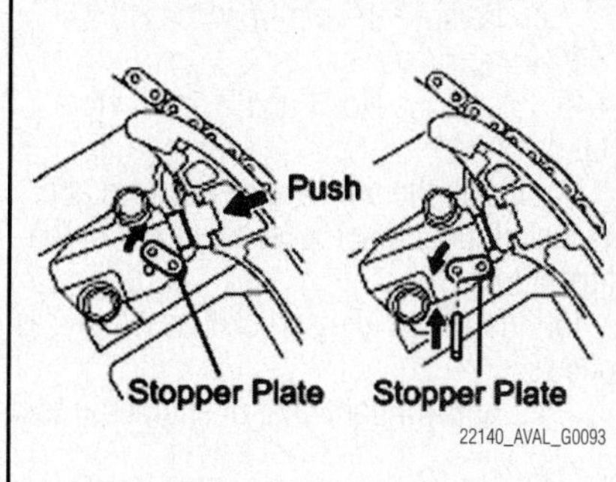

Fig. 105 Remove the No. 1 chain tensioner assembly

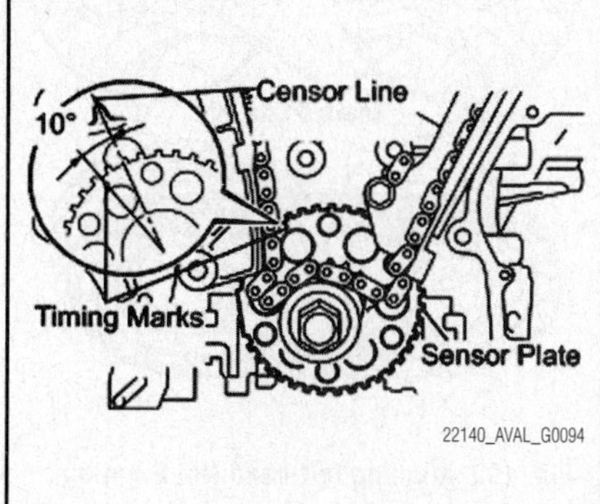

Fig. 106 Crankshaft timing sprocket

a. Move the stopper plate upward to release the lock, and push the plunger deep into the tensioner.

b. Move the stopper plate downward to set the lock, and insert a hexagon wrench into the stopper plate's hole.

c. Remove the 2 bolts and chain tensioner.

27. Remove the chain tensioner slipper.
28. Remove the chain sub-assembly, as follows:

a. Turn the crankshaft counterclockwise 10° to loosen the chain of the crankshaft timing sprocket.

Fig. 107 Camshaft timing gear assembly positioning

b. Remove the chain from the crankshaft timing sprocket and place it on the crankshaft.

c. Turn the camshaft timing gear assembly on the right hand bank clockwise (approximately 60°) and set it as shown in the illustration. Be sure to loosen the chain between the banks.

d. Remove the chain.

29. Remove the idle sprocket assembly, as follows:

a. Using a 10 mm hexagon wrench, remove the No. 2 idle gear shaft, sprocket and No. 1 idle gear shaft.

30. Remove the 2 bolts and the No. 1 chain vibration damper.
31. Remove the No. 2 chain vibration damper.
32. Remove the crankshaft timing sprocket, as follows:

a. Remove the pulley set bolt.

b. Remove the crankshaft timing gear from the crankshaft.

c. Remove the 2 pulley set keys from the crankshaft.

33. Remove the camshaft timing gears and No. 2 chain (for Right-hand Bank), as follows:

a. While raising up the No. 2 chain tensioner, insert a pin of 1.0 mm (0.039 in.) into the hole to hold it.

b. Hold the hexagonal portion of the camshaft with a wrench, and remove the 2 bolts and 2 camshaft timing gears.

➡Be careful not to damage the cylinder head with the wrench.

➡Do not disassemble the camshaft timing gear assemblies.

c. Remove the No. 2 chain.

34. Remove the bolt and No. 2 chain tensioner.
35. Remove the camshaft timing gears and No. 2 chain (for Left-hand Bank), as follows:

a. While pushing down on the No. 3 chain tensioner, insert a pin of 1.0 mm (0.039 in.) into the hole to hold it.

b. Hold the hexagonal portion of the camshaft with a wrench, and remove the 2 bolts and 2 camshaft timing gears.

➡Be careful not to damage the cylinder head with the wrench.

➡Do not disassemble the camshaft timing gear assemblies.

c. Remove the No. 2 chain.

36. Remove the bolt and the No. 3 chain tensioner.

To install:

37. Install the No. 2 chain tensioner assembly with the bolt and tighten to 15 ft. lbs. (21 Nm).

38. While pushing in the tensioner, insert a pin of 1.0 mm (0.039 in.) into the hole to hold it.

39. Install the camshaft timing gears and No. 2 chain (for Right-hand Bank), as follows:

a. Align the mark plate with the timing marks (1-dot mark) of the camshaft timing gears as shown.

b. Apply a light coat of engine oil to the bolt threads and bolt-seating surface.

c. Align the knock pin of the camshaft with pin hole of the camshaft timing gear. Install the camshaft timing gear and the right camshaft timing exhaust gear with the No. 2 chain installed.

d. Hold the hexagonal portion of the camshaft with the wrench and tighten the two bolts to 74 ft. lbs. (100 Nm).

e. Remove the pin from the No. 2 chain tensioner.

40. Install the No. 3 chain tensioner assembly with the bolt and tighten to 15 ft. lbs. (21 Nm).

41. While pushing in the tensioner, insert a pin of 1.0 mm (0.039 in.) into the hole to hold it.

Timing Mark

Mark Plate

22140_AVAL_G0096

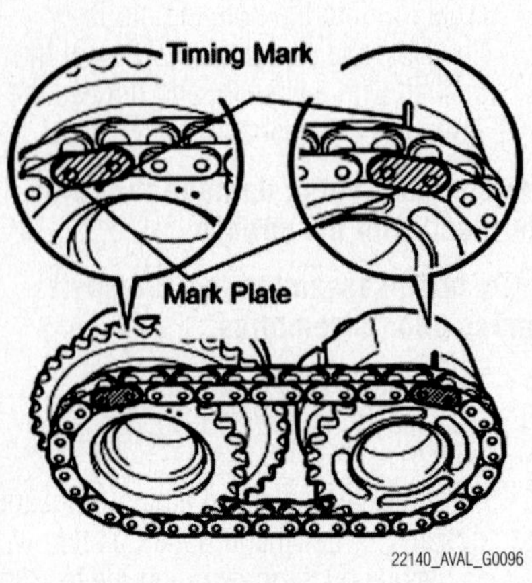

Fig. 108 Aligning No. 2 timing chain

42. Install the camshaft timing gears and No. 2 chain (for Left-hand Bank), as follows:

a. Align the mark plate (yellow) with the timing marks (2-dot mark) of the camshaft timing gears as shown.

b. Apply a light coat of engine oil to the bolt threads and bolts seating surface.

c. Align the knock pin of the camshaft with pin hole of the camshaft timing gear. Install the camshaft timing gear and the left camshaft timing exhaust gear with the No. 2 chain installed.

d. Hold the hexagonal portion of the camshaft with the wrench and tighten the two bolts to 74 ft. lbs. (100 Nm).

e. Remove the pin from the No 2 chain tensioner.

43. Install the No. 1 and 2 chain vibration dampers.

44. Install the timing gear set keys and crankshaft timing sprocket as shown in the illustration.

45. Install the idle sprocket assembly, as follows:

a. Apply a light coat of engine oil to the rotating surface of the No. 1 idle gear shaft.

b. Temporarily install the No. 1 idle gear shaft and idle sprocket with the No. 2 idle gear shaft while aligning the knock pin of the No. 1 idle gear with the knock pin groove of the cylinder block. Be careful of the idle gear direction.

c. Using a 10 mm hexagon wrench, tighten the No. 2 idle gear shaft to 44 ft. lbs. (60 Nm).

d. After installing the idle sprocket assembly, check that the idle sprocket turns smoothly.

46. Install the chain sub-assembly, as follows:

a. Align the mark plate and timing marks as shown in the illustration and install the chain. The camshaft mark plate is orange.

b. Do not pass the chain over the crankshaft, just put it on.

c. Turn the camshaft timing gear assembly on the right bank counterclockwise to tighten the chain between the banks.

➡When the idle sprocket assembly is reused, align the timing chain plate with the mark on the sprocket in order to tighten the chain between the banks.

d. Align the mark plate and timing marks as shown in the illustration and install the chain onto the crankshaft timing sprocket. The crankshaft to mark plate is yellow.

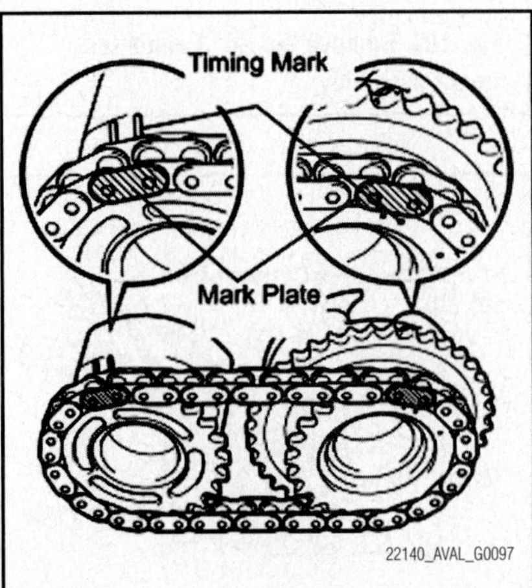

Fig. 109 Aligning left-hand No. 2 timing chain

Fig. 110 Crankshaft timing sprocket

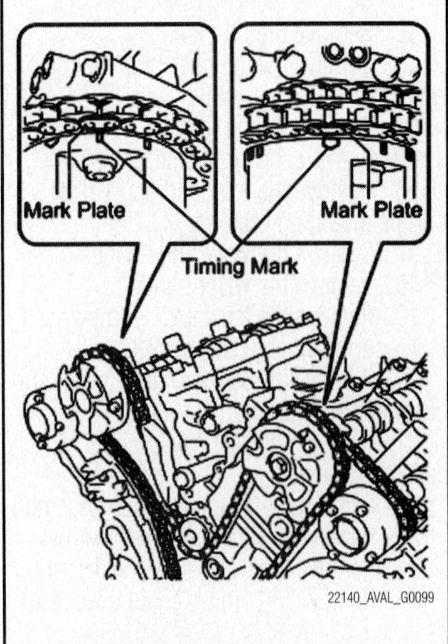

Fig. 111 Aligning timing chain sub-assembly

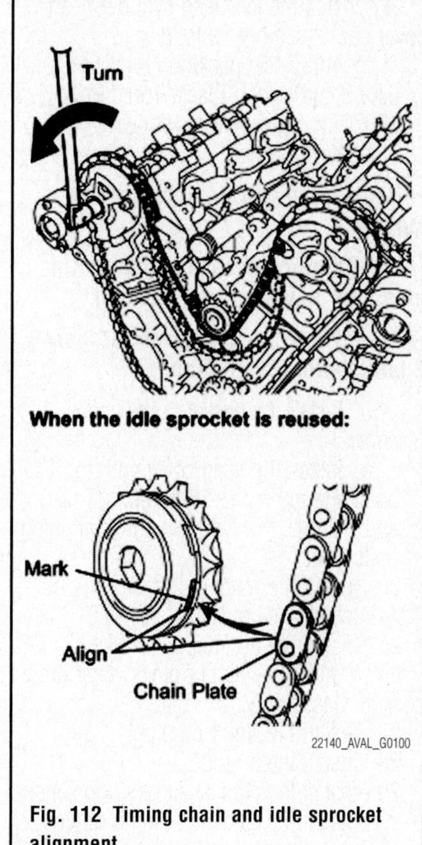

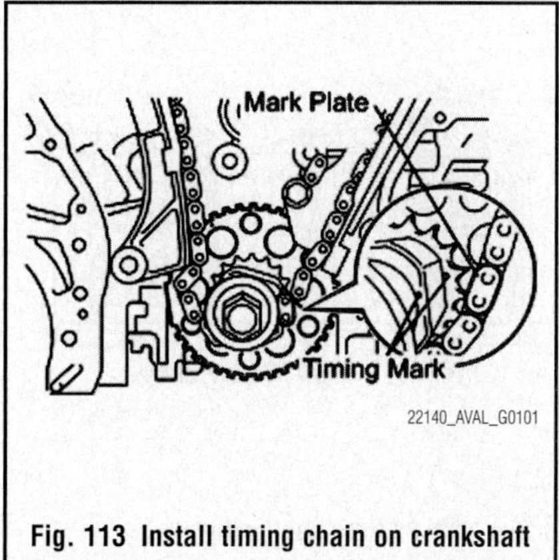

Fig. 112 Timing chain and idle sprocket alignment

Fig. 113 Install timing chain on crankshaft

e. Temporarily tighten the pulley set bolt.

f. Turn the crankshaft clockwise to set it to the right-hand block bore more centerline. (TDC/compression).

47. Install the chain tensioner slipper.

48. Install the No. 1 chain tensioner assembly, as follows:

a. Move the stopper plate upward to release the lock, and push the plunger deep into the tensioner.

b. Move the stopper plate downward to set the lock, and insert a hexagon wrench into the hole of the stopper plate.

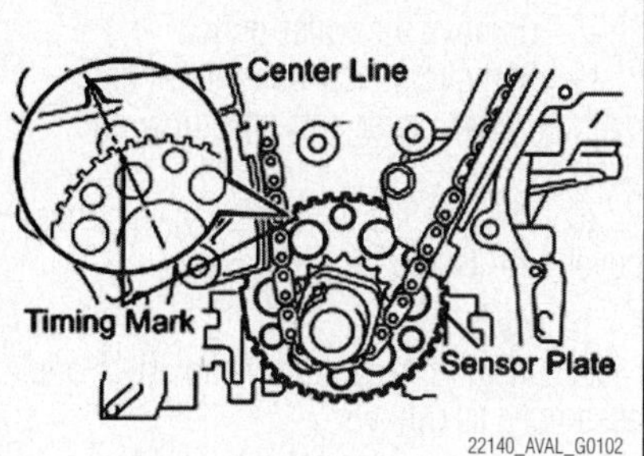

Fig. 114 Timing chain and crankshaft alignment

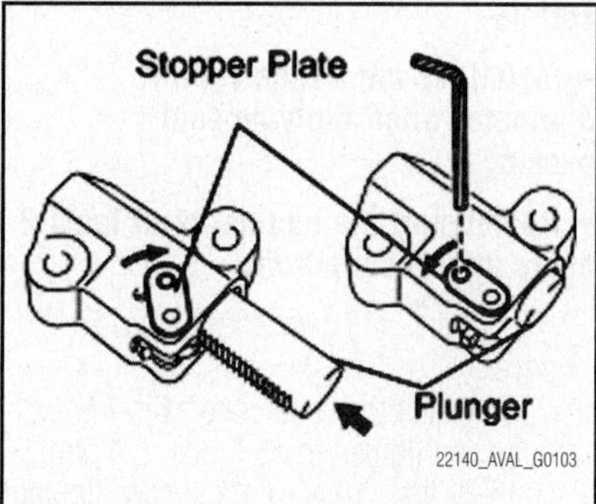

Fig. 115 Set chain tensioner plunger position

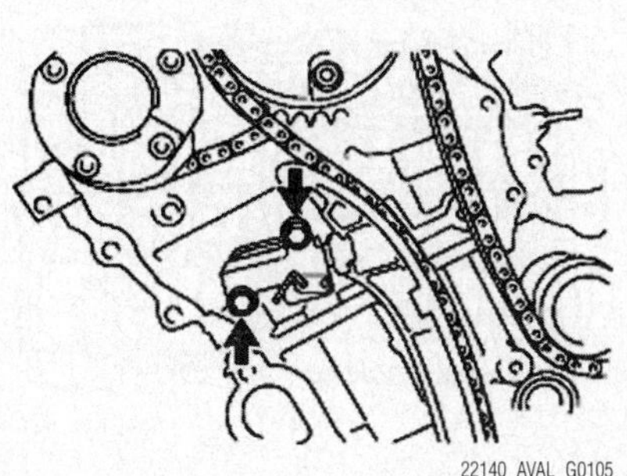

Fig. 116 Install the No. 1 chain tensioner with the 2 bolts

c. Install the No. 1 chain tensioner with the 2 bolts and tighten to 7 ft. lbs. (10 Nm).

d. Remove the lock pin of the No. 1 chain tensioner. Check that each timing mark is aligned with the crankshaft at TDC/compression.

e. Remove the pulley set bolt.

49. Install timing chain cover and seal.

50. Install the water pump assembly.

51. Install the water inlet housing.

52. Install the left hand No. 1 front engine mounting bracket.

53. Install the oil pan sub-assembly.

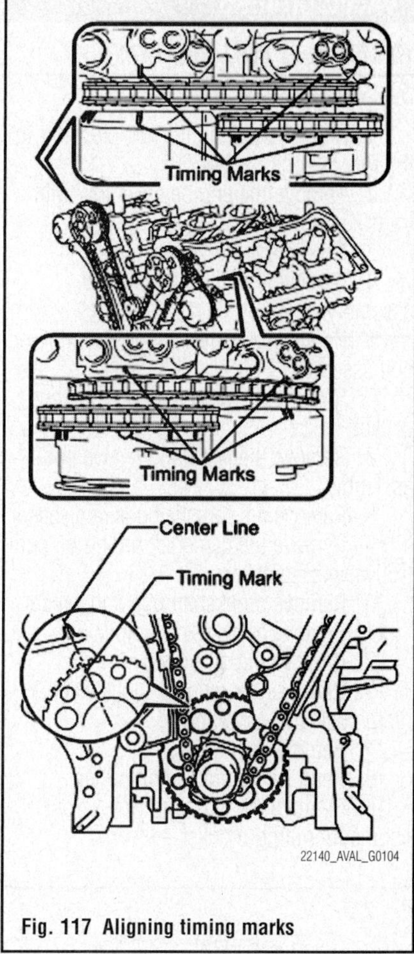

Fig. 117 Aligning timing marks

54. Install the oil strainer sub-assembly.

55. Install the No. 2 oil pan sub-assembly.

56. Install the oil pan drain plug and gasket.

57. Install the cylinder head cover sub-assembly.

58. Install the left-hand cylinder head cover sub-assembly.

59. Install the water outlet.

60. Install the crankshaft pulley.

61. Install the oil filter.

62. Install the cylinder block water drain cock sub-assembly.

63. Install the No. 1 oil pipe.

64. Install the oil pipe.

65. Install the crankshaft position sensor.

66. Install the camshaft timing oil control valve assembly.

67. Install the camshaft position sensor.

68. Install the ventilation valve sub-assembly.

69. Install the spark plugs and ignition coil assembly.

70. Install the oil filler cap sub-assembly.

71. Install the engine assembly with transaxle.

VALVE COVERS

REMOVAL & INSTALLATION

See Figures 118 through 123.

1. Before servicing the vehicle, refer to the Precautions Section.
2. Remove the engine assembly with transaxle.
3. Secure engine.
4. Remove the oil filler cap sub-assembly.
5. Remove the spark plugs and ignition coil assembly.
6. Remove the oil pan drain plug and gasket.
7. Remove the ventilation valve sub-assembly.
8. Remove the camshaft position sensor.
9. Remove the camshaft timing oil control valve assembly.
10. Remove crankshaft position sensor.
11. Remove the No. 1 oil pipe.
12. Remove the oil pipe.
13. Remove the cylinder block water drain cock sub-assembly.
14. Remove the oil filter.
15. Remove the crankshaft pulley.
16. Remove the left hand No. 1 front engine mounting bracket.
17. Remove the water inlet housing.
18. Remove the water outlet.
19. Remove the 12 bolts, left-hand cylinder head cover sub-assembly and gasket.
20. Remove the 12 bolts, cylinder head cover sub-assembly and gasket.

To install:

21. Install the cylinder head cover sub-assembly, as follows:
    a. Apply seal packing (Toyota Genuine Seal Packing Black, Three Bond 1207B or equivalent) as shown in the illustration.

➡Remove any oil from the contact surface.

➡Install the crankcase within 3 minutes after applying seal packing.

➡Do not start the engine for at least 2 hours after installation.

    b. Install the gasket to the head cover.
    c. Install the head cover with the 12 bolts and tighten to 15 ft. lbs. (21 Nm), and 7 ft. lbs. (10 Nm). Be sure to tighten bolt 1.

22. Install the left-hand cylinder head cover sub-assembly, as follows:
    a. Apply seal packing (Toyota Genuine Seal Packing Black, Three Bond 1207B or equivalent) as shown in the illustration.

➡Remove any oil from the contact surface.

➡Install the crankcase within 3 minutes after applying seal packing.

➡Do not start the engine for at least 2 hours after installation.

    b. Install the gasket to the head cover.
    c. Install the head cover with the 14 bolts and tighten to 15 ft. lbs. (21 Nm), and 7 ft. lbs. (10 Nm). Be sure to tighten bolts 1 and 10.

23. Install the water outlet.
24. Install the crankshaft pulley.
25. Install the oil filter.
26. Install the cylinder block water drain cock sub-assembly.
27. Install the No. 1 oil pipe.
28. Install the oil pipe.
29. Install the crankshaft position sensor.
30. Install the camshaft timing oil control valve assembly.

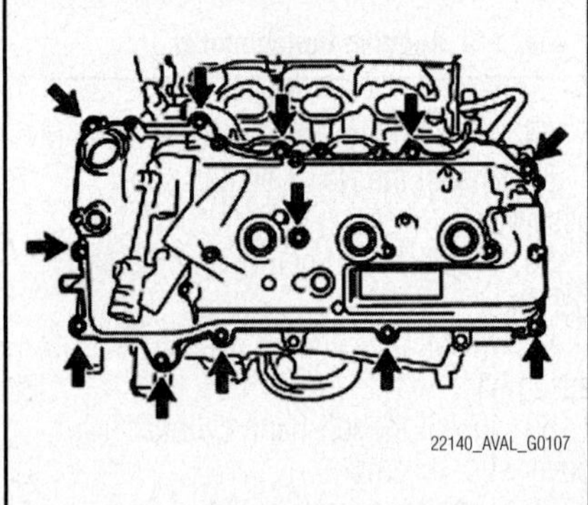

Fig. 118 Left-hand cylinder head cover

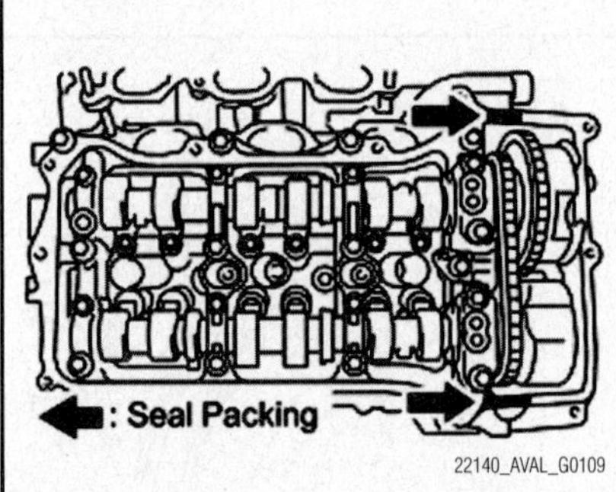

Fig. 120 Sealant application

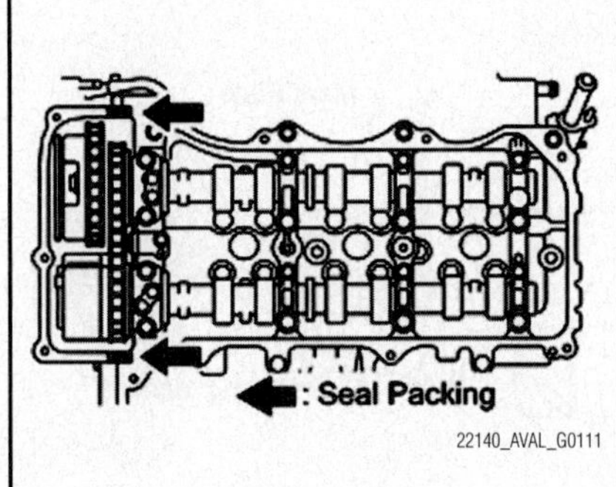

Fig. 122 Sealant application

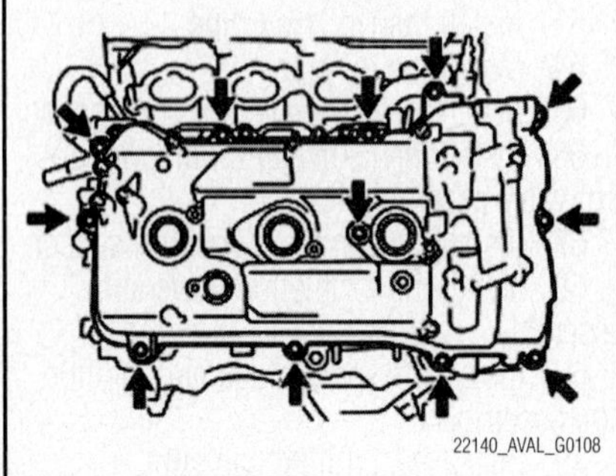

Fig. 119 Right-hand cylinder head cover

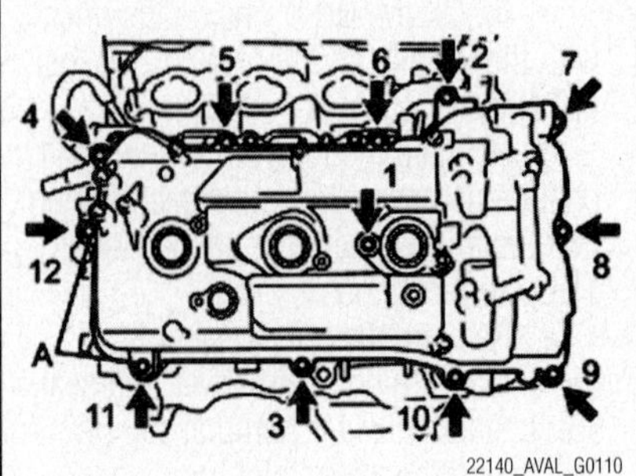

Fig. 121 Right-hand cylinder head cover bolt tightening sequence

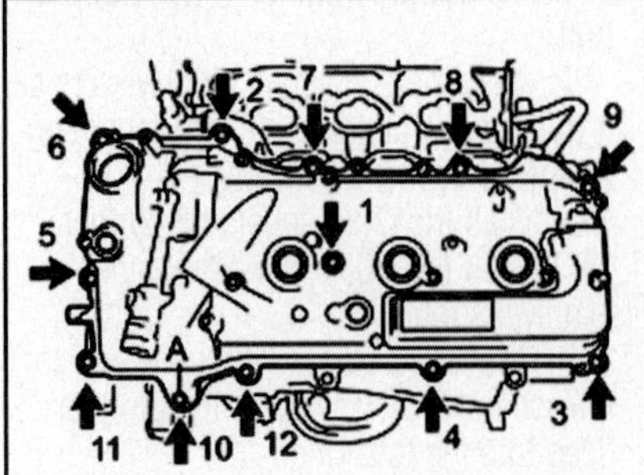

Fig. 123 Left hand cylinder head cover bolt tightening sequence

31. Install the camshaft position sensor.
32. Install the ventilation valve sub-assembly.
33. Install the spark plugs and ignition coil assembly.
34. Install the oil filler cap sub-assembly.
35. Install the water pump assembly.
36. Install the engine assembly with transaxle.

VALVE LASH

ADJUSTMENT

See Figure 124.

➡Keep the lash adjuster free of dirt and foreign objects.

➡Only use clean engine oil.

1. Place the lash adjuster into a container filled with engine oil.
2. Insert the SST's (SST: 09276-75010) tip into the lash adjuster's plunger and use the tip to press down on the check ball inside the plunger.

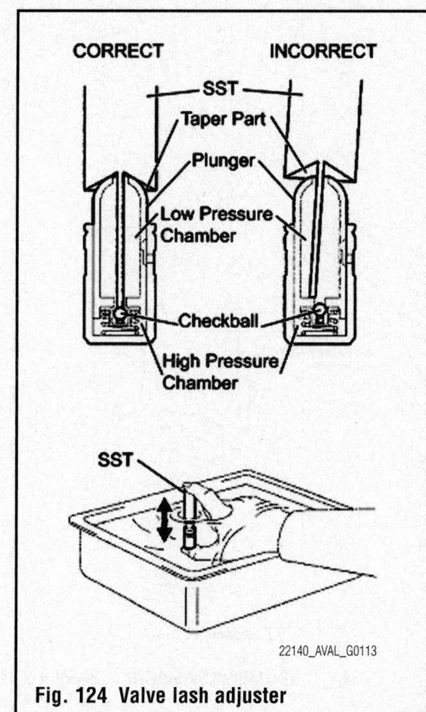

Fig. 124 Valve lash adjuster

3. Squeeze the SST and lash adjuster together to move the plunger up and down 5 to 6 times.
4. Check the movement of the plunger and bleed the air. OK: Plunger moves up and down.

➡When bleeding air from the high-pressure chamber, make sure that the tip of the SST is actually pressing the check ball as shown in the illustration. If the check ball is not pressed, air will not bleed.

5. After bleeding the air, remove the SST. Then, try to press the plunger quickly and firmly with a finger. OK: Plunger is very difficult to move. If the result is not as specified, replace the lash adjuster.
6. Install the lash adjusters.

➡Install the lash adjuster to the same place where it was removed from.

ENGINE PERFORMANCE & EMISSION CONTROLS

ACCELERATOR PEDAL POSITION (APP) SENSOR

LOCATION

See Figure 125.

REMOVAL & INSTALLATION

1. Remove left center floor carpet cover.
2. Disconnect the accelerator pedal connector.
3. Remove the 2 nuts and accelerator pedal assembly.

➡Avoid physical shock to the accelerator pedal assembly.

➡Do not disassemble the accelerator pedal assembly.

To install:

4. Installation is the reverse of the removal procedure. Tighten the accelerator pedal rod nuts to 43 inch lbs. (5.4 Nm).

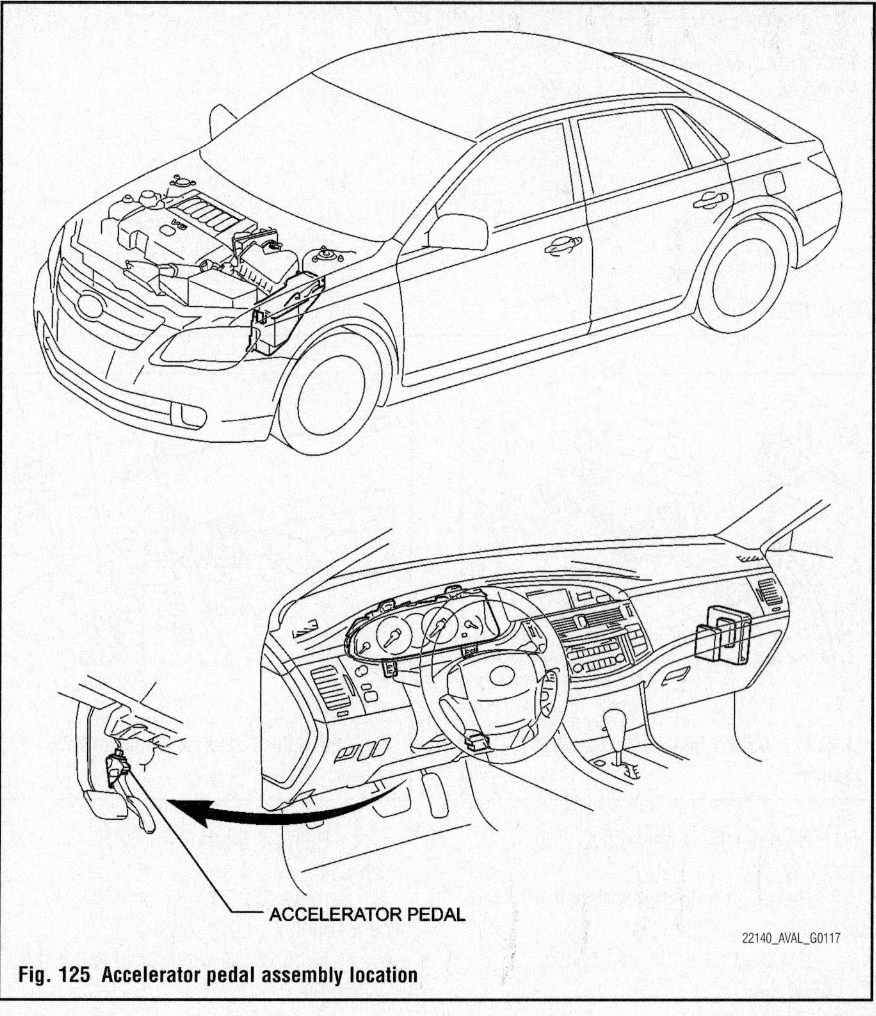

Fig. 125 Accelerator pedal assembly location

CAMSHAFT POSITION (CMP) SENSOR

LOCATION

See Figure 126.

REMOVAL & INSTALLATION

See Figures 127 through 130.

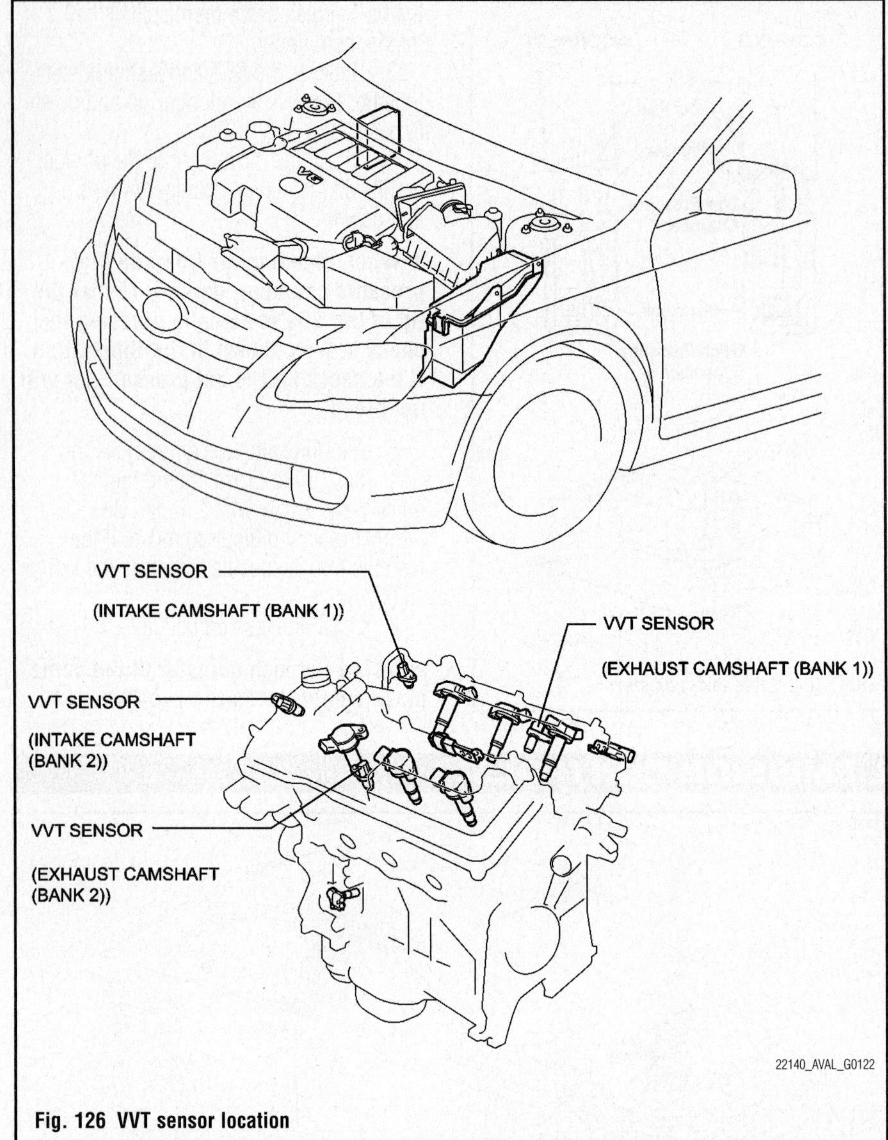

Fig. 126 VVT sensor location

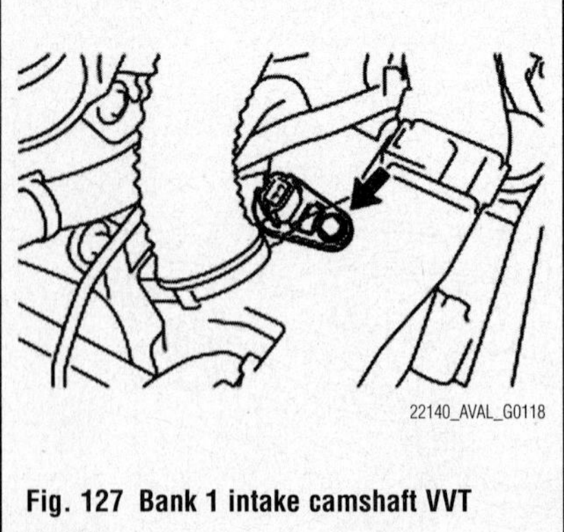

Fig. 127 Bank 1 intake camshaft VVT sensor

Fig. 128 Bank 1 exhaust camshaft VVT sensor

1. Remove the V-bank cover sub-assembly.
2. Remove the Intake camshaft VVT sensor (Bank 1), as follows:
   a. Disconnect the VVT sensor connector.
   b. Remove the bolt and VVT sensor.
3. Remove the Exhaust camshaft VVT sensor (Bank 1), as follows:
   a. Remove the windshield wiper link assembly.

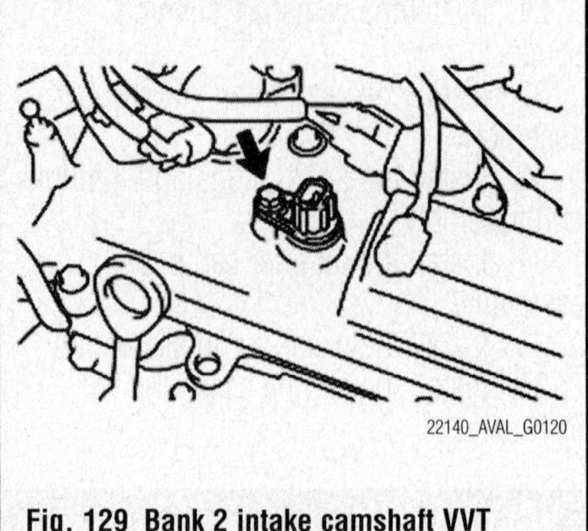

Fig. 129 Bank 2 intake camshaft VVT sensor

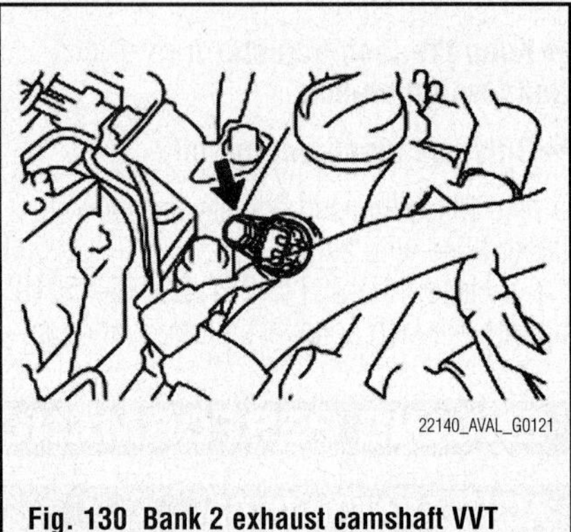

Fig. 130 Bank 2 exhaust camshaft VVT sensor

   b. Remove the front cowl top outside panel.
   c. Disconnect the VVT sensor connector.
   d. Remove the bolt and VVT sensor.
4. Remove the Exhaust camshaft VVT sensor (Bank 2), as follows:
   a. Disconnect the VVT sensor connector.
   b. Remove the bolt and VVT sensor.
5. Remove the Intake camshaft VVT sensor (Bank 2), as follows:
   a. Disconnect the VVT sensor connector.
   b. Remove the bolt and VVT sensor.

CRANKSHAFT POSITION (CKP) SENSOR

LOCATION

See Figure 131.

REMOVAL & INSTALLATION

See Figure 132.

1. Disconnect the negative battery terminal.

Fig. 131 Crankshaft position sensor location

Fig. 132 Crankshaft position sensor

2. Remove the V-ribbed belt.
3. Remove alternator assembly.
4. Disconnect the cooler compressor assembly.
5. Remove the crankshaft position sensor connector.
6. Remove the bolt, and then remove the crankshaft position sensor.

To install:

7. Apply a light coat of engine oil to the O-ring on the crankshaft position sensor.
8. Install the crankshaft position sensor with the bolt and tighten to 7 ft. lbs. (10 Nm).
9. Connect the crankshaft position sensor connector.
10. The remainder of installation is the reverse of the removal procedure.

ELECTRONIC CONTROL MODULE (ECM)

LOCATION

See Figure 133.

REMOVAL & INSTALLATION

See Figure 133.

1. Disconnect the negative battery cable.
2. Remove both windshield wiper arm and blade assemblies.
3. Remove the cowl top ventilator louver sub-assembly.
4. Remove the windshield wiper motor and link assembly.
5. Remove the outer cowl top panel.
6. Remove the ECM, as follows:
   a. Remove the 3 nuts.
   b. Separate the ECM from the body. When separating the ECM, do not apply excessive force to the wire harness.
   c. Raise the 2 levers while pushing the locks on the 2 levers, and disconnect the 2 ECM connectors.

➡After disconnecting the connectors, make sure that dirt, water or other foreign matter does not contact the connections of the connectors.

   d. Remove the ECM.
   e. Remove the 4 screws and 2 ECM brackets.

To install:

7. Install the 2 ECM brackets with the 4 screws, and tighten to 27 inch lbs. (3 Nm).
8. Connect the 2 ECM connectors and lower the 2 levers.

➡Make sure that dirt, water or other foreign matter does not contact the connections of the connectors.

9. Install the ECM to the body.
10. Attach the ECM with the 3 nuts and tighten to 71 inch lbs. (8 Nm).
11. Install the outer cowl top panel.
12. Install the windshield wiper motor and link assembly.
13. Install the cowl top ventilator louver sub-assembly.
14. Install both windshield wiper arm and blade assemblies.
15. Connect the negative battery cable.
16. Register the immobilizer communication ID. If the ECM is replaced, register the ECM communication ID for the immobilizer system (refer to the Service Bulletin for registration).
17. Perform initialization. After replacing the ECM on vehicles with a dynamic laser

cruise control system, it is necessary to initialize the ECM so that the ECM can recognize the dynamic laser cruise control system.

18. Be sure to perform the following procedure after replacing the ECM:
   a. Turn the ignition switch on (IG).
   b. Turn the cruise control main switch on.
   c. With the brake pedal depressed, push the cruise control main switch to RES/ACC 3 times within 3 seconds. Check that the buzzer sounds at this time.

➡Do not turn the headlight dimmer switch on at this time because the optical axis automatic adjustment mode has already started, which may lead to an incorrect optical axis setting. If the headlight dimmer switch is turned on by mistake, readjust the optical axis.

ENGINE COOLANT TEMPERATURE (ECT) SENSOR

LOCATION

See Figure 134.

REMOVAL & INSTALLATION

1. Drain engine coolant.
2. Remove V-bank cover sub-assembly.

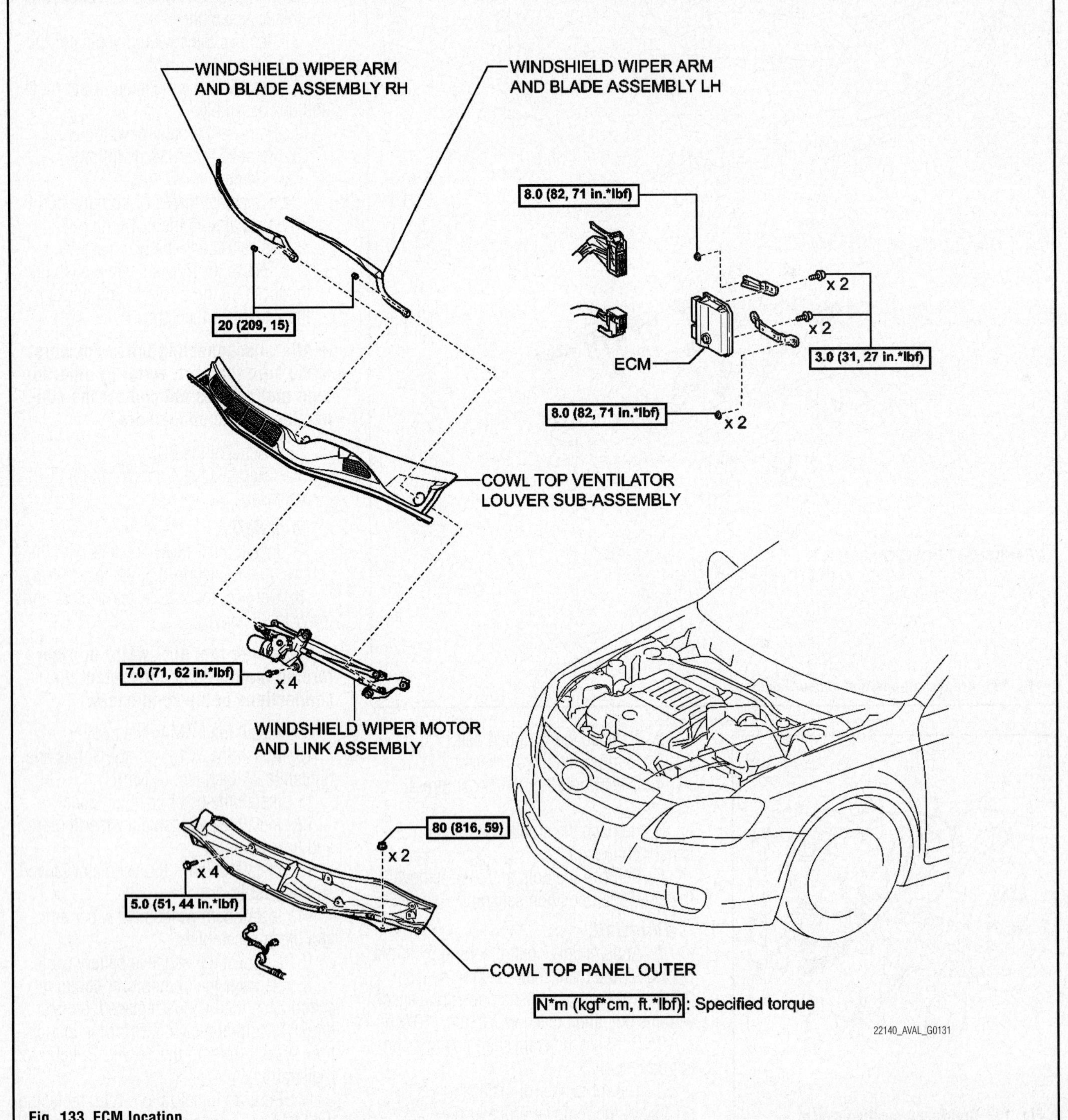

Fig. 133 ECM location

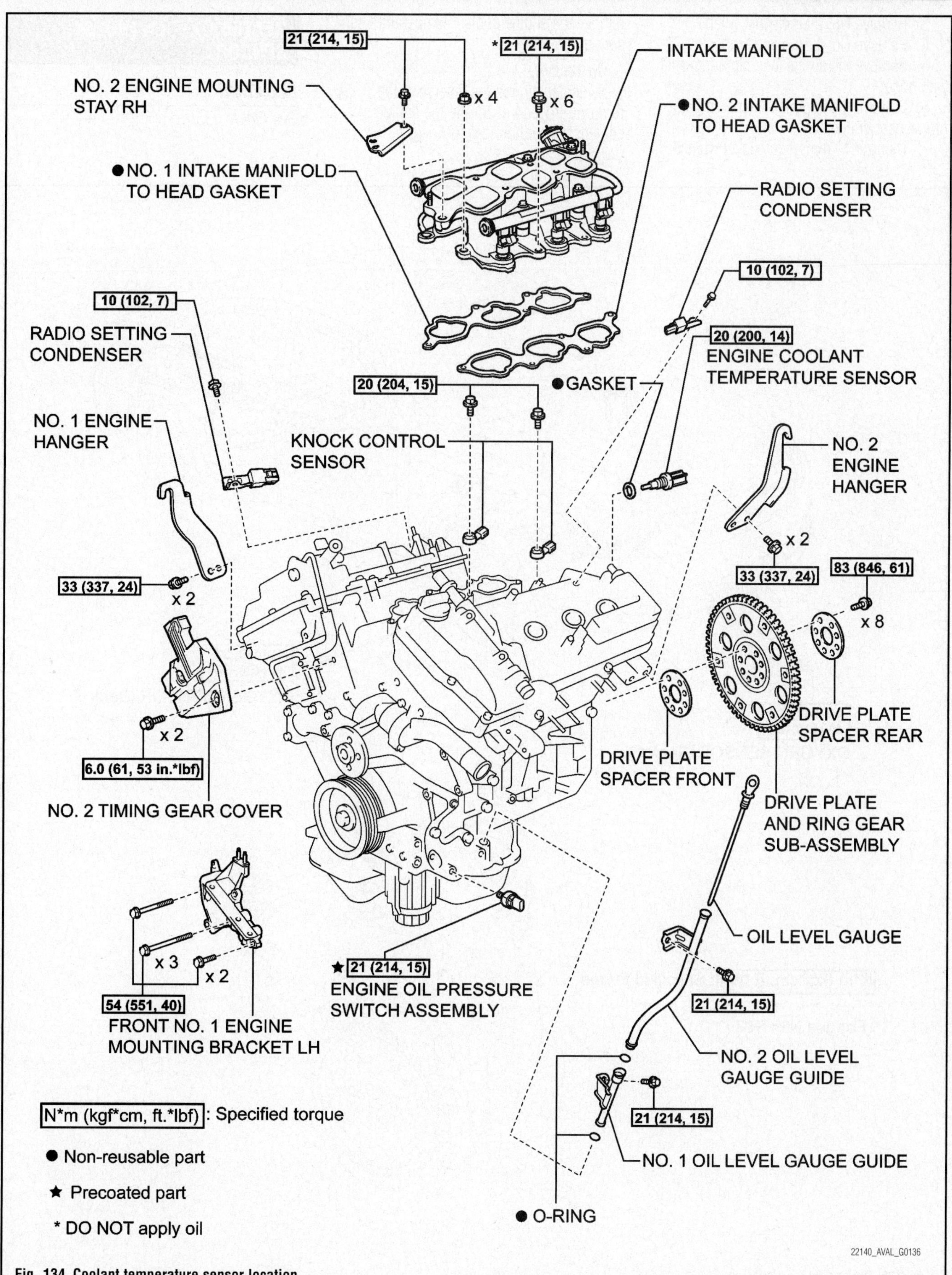

Fig. 134 Coolant temperature sensor location

3. Remove No. 2 air cleaner inlet.
4. Remove No. 1 air cleaner inlet.
5. Remove the air cleaner cap air cleaner with hose.
6. Remove the engine coolant temperature sensor connector.
7. Using a 19 mm deep socket wrench, remove the engine coolant temperature sensor and gasket.

To install:

8. Installation is the reverse of the removal procedure. Torque the engine coolant temperature sensor to 15 ft. lbs. (20 Nm).

HEATED OXYGEN (HO2S) SENSOR

LOCATION

See Figure 135.

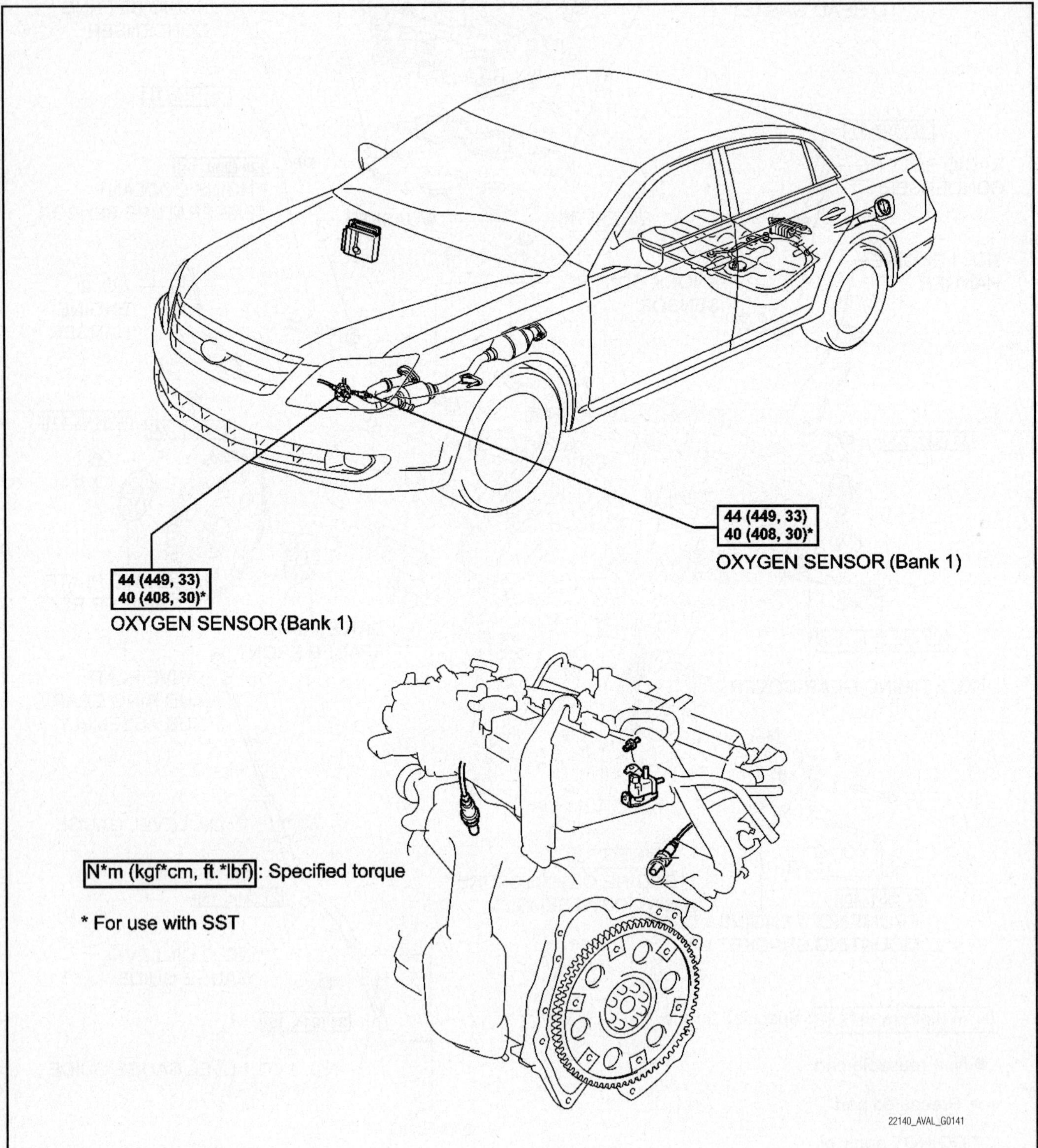

Fig. 135 Oxygen Sensor location

REMOVAL & INSTALLATION

See Figure 135.

1. Disconnect the 2 oxygen sensor connectors.
   a. Using SST (SST: 09224-00010) or equivalent, remove the 2 oxygen sensors from the front pipe assembly.

To install:

2. Install the 2 oxygen sensors to the front pipe assembly. Tighten to 32 ft. lbs. (44 Nm) and 30 ft. lbs. (40 Nm). Use a torque wrench with a fulcrum length of 300 mm (11.81 in.).
3. Connect the 2 oxygen sensor connectors.

INJECTION PRESSURE REGULATOR (IPR)

LOCATION

See Figure 136.

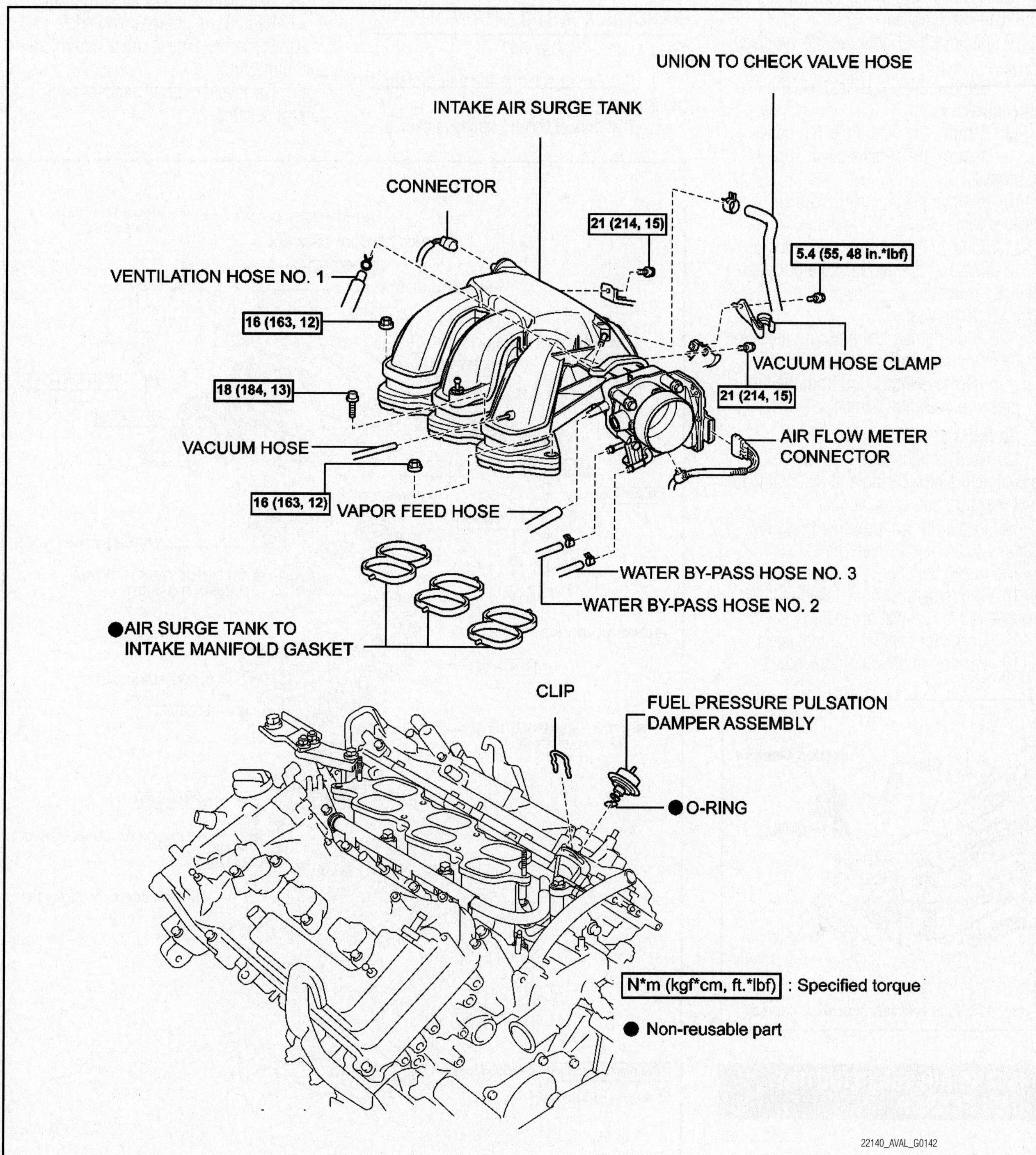

Fig. 136 Fuel pressure pulsation damper (injection pressure regulator) location

REMOVAL & INSTALLATION

See Figure 137.

1. Before servicing the vehicle, refer to the Fuel System Precautions Section.
2. Discharge fuel system pressure.
3. Disconnect the negative battery cable.
4. Drain engine coolant.
5. Remove both windshield wiper arm and blade assemblies.
6. Remove the right cowl top ventilator louver.
7. Remove the windshield wiper motor and link assembly.
8. Remove the outer cowl top panel.
9. Remove the V-bank cover sub-assembly.
10. Remove the air cleaner cap air cleaner with hose.
11. Remove the intake air surge tank.
12. Remove the fuel pressure pulsation damper assembly, as follows:
    a. Remove the clip.
    b. Pull out the fuel pressure pulsation damper from the fuel delivery pipe.
    c. Remove the O-ring from the fuel pressure pulsation damper.

To install:

13. Apply a light coat of spindle oil or gasoline to a new O-ring and install it to the fuel pressure pulsation damper.
14. Install the fuel pressure pulsation damper to the fuel delivery pipe.
15. Install the clip.
16. The remainder of installation is the reverse of the removal procedure.
17. Check for coolant and fuel leaks.
18. Perform initialization procedure.

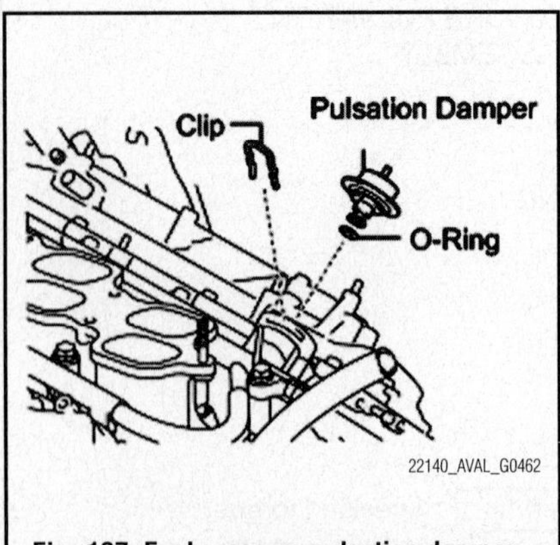

Fig. 137 Fuel pressure pulsation damper

INTAKE AIR TEMPERATURE (IAT) SENSOR

LOCATION

The intake air temperature sensor is mounted on the mass air flow meter.

REMOVAL & INSTALLATION

See Mass Air Flow Meter.

KNOCK SENSOR (KS)

LOCATION

See Figure 138.

REMOVAL & INSTALLATION

See Figures 139 and 140.

1. Properly discharge the fuel system pressure.
2. Disconnect battery negative cable.
3. Drain engine coolant.
4. Remove the windshield wiper motor and link assembly.
5. Remove the outer cowl top panel.
6. Remove the V-bank cover sub-assembly.
7. Remove the air cleaner cap with air cleaner hose.
8. Remove the intake air surge tank.
9. Remove the intake manifold.
10. Disconnect the 2 knock control sensor connectors.
11. Remove the 2 bolts and 2 knock control sensors.

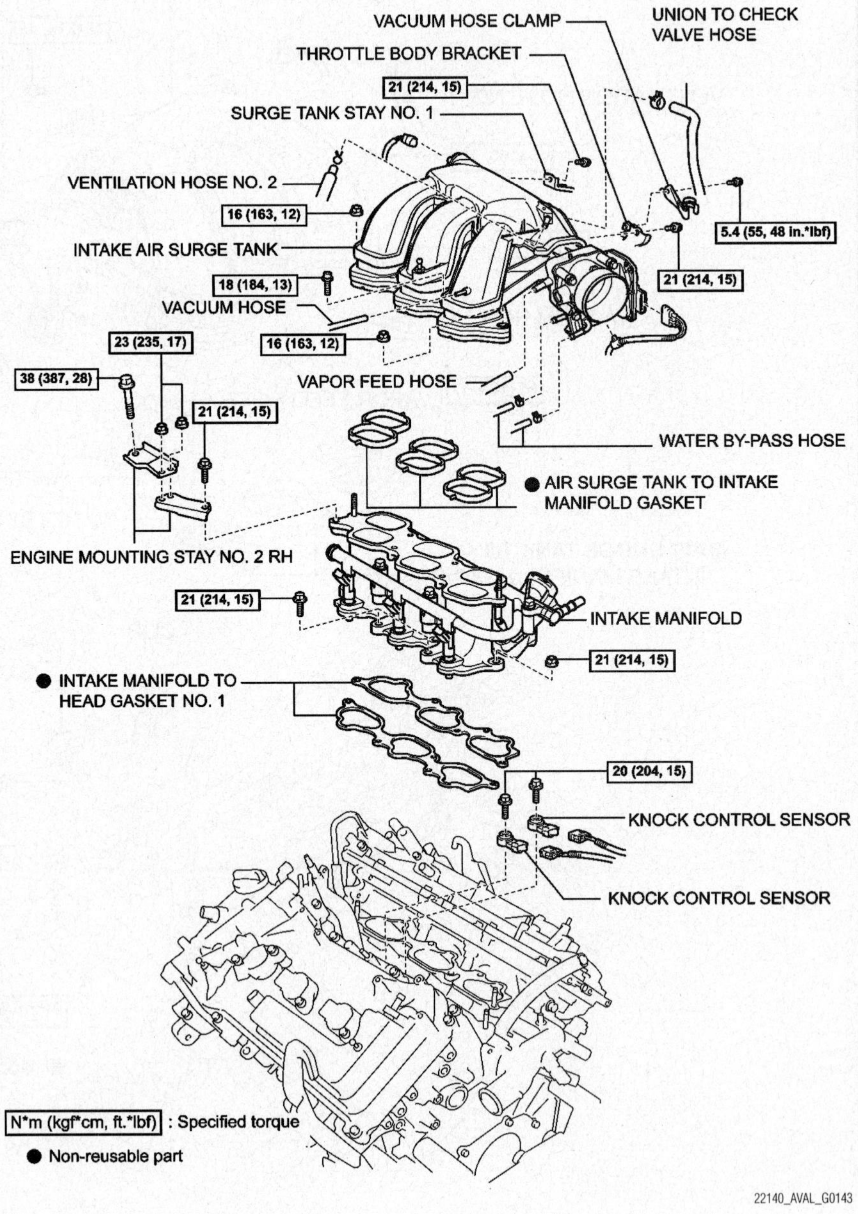

Fig. 138 Knock sensor location

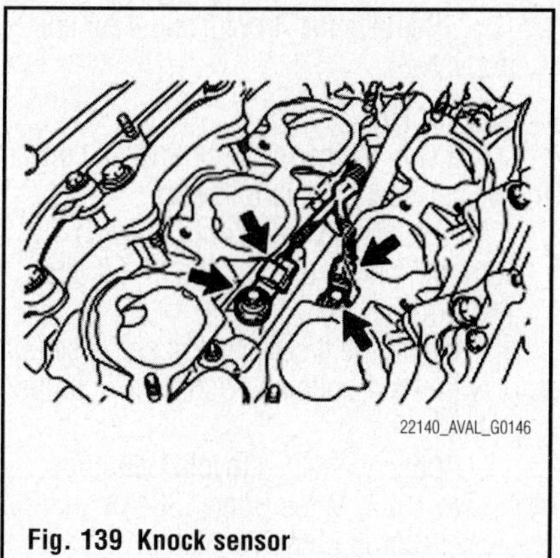

Fig. 139 Knock sensor

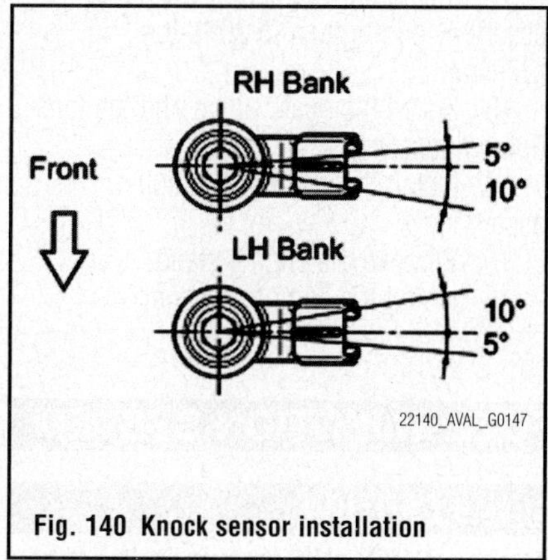

Fig. 140 Knock sensor installation

To install:

12. Install the 2 knock control sensors with the 2 bolts as shown in the illustration and tighten to 15 ft. lbs. (20 Nm).
13. Connect the 2 knock control sensor connectors.
14. The remainder of installation is the reverse of the removal procedure.
15. Inspect for fuel leak and check the function of throttle body.

MALFUNCTION INDICATOR LIGHT (MIL)

RESET PROCEDURES

Clear the DTC codes.

MASS AIR FLOW (MAF) SENSOR

LOCATION

The MAF sensor is between the throttle body and air cleaner housing.

REMOVAL & INSTALLATION

See Figure 141.

1. Disconnect the mass air flow meter connector.

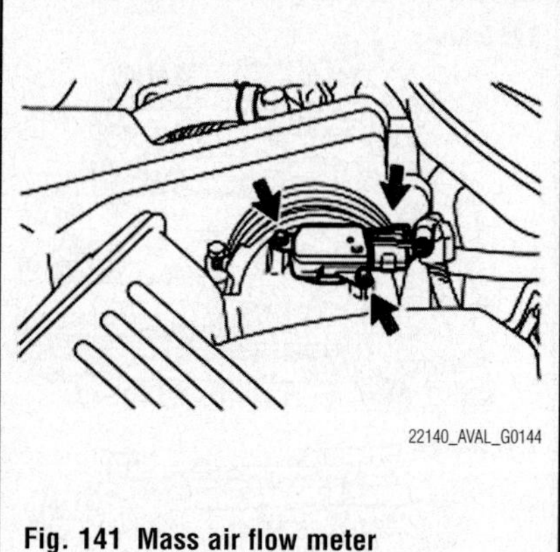

Fig. 141 Mass air flow meter

2. Remove the 2 screws and mass air flow meter.

To install:

3. Installation is the reverse of the removal procedure.

THROTTLE POSITION SENSOR (TPS)

LOCATION

See Figure 142.

The throttle position sensor is located on the throttle body.

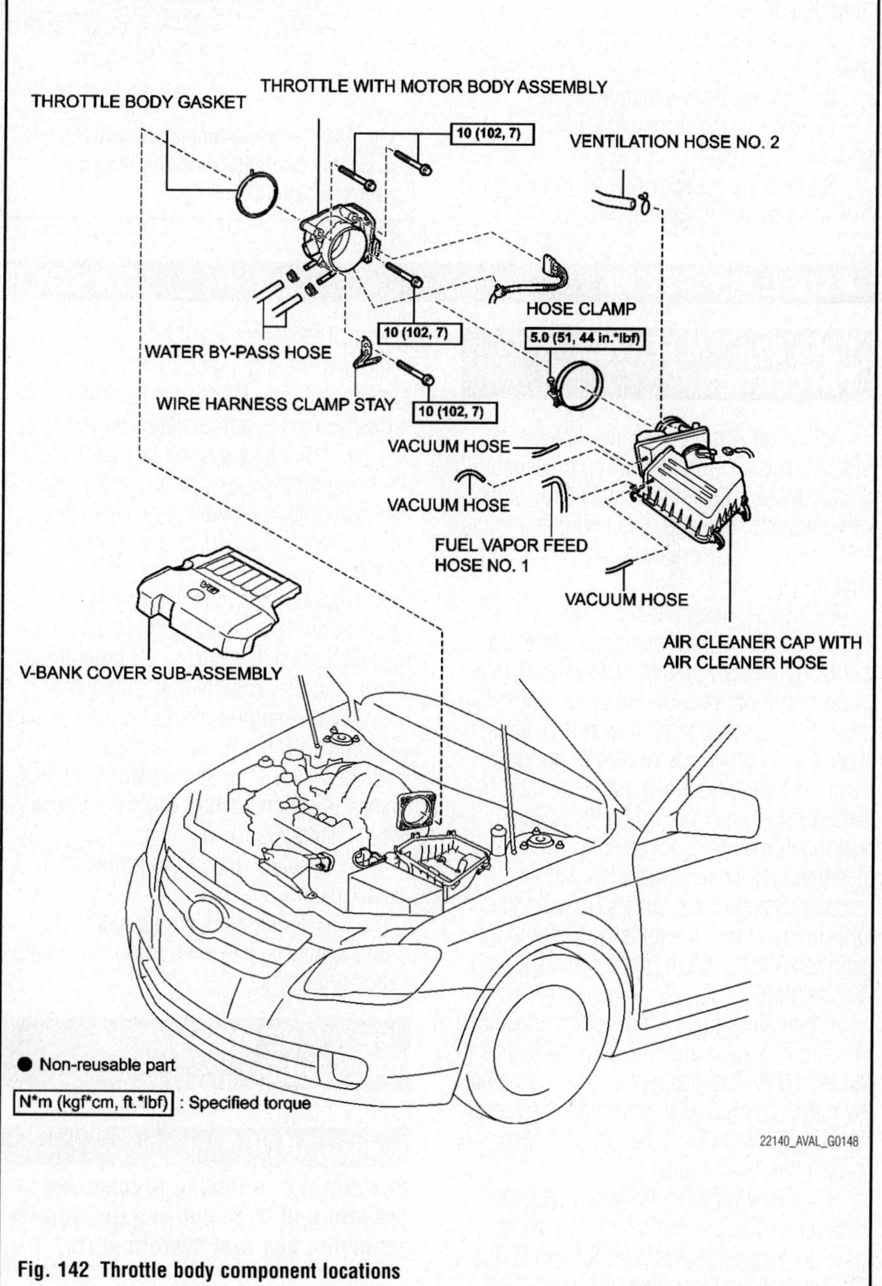

Fig. 142 Throttle body component locations

REMOVAL & INSTALLATION

Refer to the Throttle Body removal and installation procedures.

VARIABLE CAMSHAFT TIMING OIL CONTROL SOLENOID

LOCATION

See Figure 143.

REMOVAL & INSTALLATION

1. Remove the engine assembly.
2. Install on engine stand.
3. Remove the oil filler cap and gasket.
4. Remove the spark plugs and ignition coil assembly.
5. Remove the drain plug and gasket.
6. Remove the ventilation valve.
7. Remove the camshaft position sensor.
8. Remove the camshaft timing oil control valve assembly, as follows:

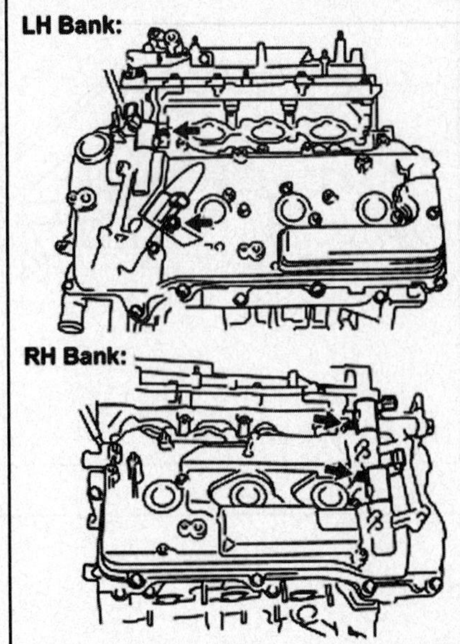

Fig. 143 Camshaft timing oil control valve assembly (Variable camshaft timing oil control solenoid) location

a. Remove the 4 bolts and 4 oil control valves.

To install:

9. Install the camshaft timing oil control valve assembly, as follows:
   a. Install the 4 oil control valves with the 4 bolts and tighten to 7 ft. lbs. (10 Nm).
10. Install the 4 camshaft position sensors with the 4 bolts and tighten to 7 ft. lbs. (10 Nm).
11. Apply adhesive (Toyota Genuine Adhesive 1324, Three Bond 1324 or equivalent) around the ventilation valve.
12. Install the ventilation valve subassembly and tighten to 20 ft. lbs. (27 Nm).
13. Install the spark plugs and the ignition coil assembly.
14. Install the oil filler cap subassembly.
15. Remove the engine stand.
16. Install the engine assembly.

FUEL SYSTEM

GASOLINE FUEL INJECTION SYSTEM

FUEL SYSTEM SERVICE PRECAUTIONS

Before servicing any vehicle, please be sure to read all of the following precautions, which deal with personal safety, prevention of component damage, and important points to take into consideration when servicing a motor vehicle:

- Observe all applicable safety precautions when working around fuel. Whenever servicing the fuel system, always work in a well-ventilated area. Do not allow fuel spray or vapors to come in contact with a spark, open flame, or excessive heat (a hot drop light, for example). Keep a dry chemical fire extinguisher near the work area. Always keep fuel in a container specifically designed for fuel storage; also, always properly seal fuel containers to avoid the possibility of fire or explosion. Refer to the additional fuel system precautions later in this section.
- Fuel injection systems often remain pressurized, even after the engine has been turned **OFF**. The fuel system pressure must be relieved before disconnecting any fuel lines. Failure to do so may result in fire and/or personal injury.
- All new vehicles are now equipped with an air bag system, often referred to as a Supplemental Restraint System (SRS) or Supplemental Inflatable Restraint (SIR) system. The system must be disabled before performing service on or around system components, steering column, instrument panel components, wiring and sensors. Failure to follow safety and disabling procedures could result in accidental air bag deployment, possible personal injury and unnecessary system repairs.
- Disconnecting the negative battery cable on some vehicles may interfere with the functions of the on-board computer system(s) and may require the computer to undergo a relearning process once the negative battery cable is reconnected.

Before inspecting and repairing the fuel system, disconnect the cable from the negative (-) battery terminal.

Keep gasoline away from rubber or leather parts.

1. Check that there are no fuel leaks from the fuel system after doing any maintenance or repairs.

RELIEVING FUEL SYSTEM PRESSURE

CAUTION

Perform the following procedures to prevent fuel from spilling out before removing any fuel system parts.

CAUTION

Pressure will still remain in the fuel line even after performing the following procedures. When disconnecting the fuel line, cover it with a shop rag or a piece of cloth to prevent fuel from spraying or coming out.

1. Disconnect the fuel pump connector:
   a. Remove the rear seat cushion assembly.
   b. Remove the rear floor service hole cover.
   c. Disconnect the fuel pump connector.
   d. Start the engine.
   e. After the engine stops, turn the ignition switch off.

➡DTC P0171/25 (fuel problem) may be detected.

   f. Crank the engine again. Check that the engine does not start.
   g. Remove the fuel tank cap to discharge pressure from the fuel tank.
   h. Disconnect the cable from the negative (-) battery terminal.
   i. Reconnect the fuel pump connector.
   j. Install the rear floor service hole cover.
   k. Install the rear seat.
2. Check that there are no fuel leaks from the fuel system after doing any maintenance or repairs.

FUEL FILTER

REMOVAL & INSTALLATION

See Figure 144.

1. Before servicing the vehicle, refer to the precautions section.
2. Remove the fuel pump from the vehicle.
3. Remove the fuel pump filter, as follows:
   a. Using a screwdriver, pry out the clip.
   b. Pull out the fuel pump filter from the fuel pump.

Fig. 144 Fuel filter removal

To install:

4. Install the fuel pump filter with a new clip.
5. Install the fuel pump.

FUEL PUMP

REMOVAL & INSTALLATION

1. Before servicing the vehicle, refer to the precautions section.
2. Discharge fuel system pressure.
3. Disconnect battery negative cable.
4. Remove the rear seat cushion assembly.
   a. Remove the rear floor service hole cover.
   b. Disconnect the fuel pump connector.
5. Separate the fuel pump tube sub-assembly, as follows:
   a. Remove the tube joint clip, and pull out the fuel pump tube.

➡Check if there is any dirt or mud around the connector before this operation and remove the dirt as necessary.

➡Be careful of mud because the quick connector has an O-ring which seals the pipe and connector that can be contaminated.

➡Do not use any tools in this operation.

➡Do not bend or twist the nylon tube. Cover the fuel tube joint with a plastic bag.

➡When the fuel tube joint and fuel suction plate are stuck, pinch the fuel tank tube between fingers, and turn it carefully to release it. Disconnect the fuel tank tube.

6. Remove fuel tank vent tube set plate, as follows:
   a. Remove the 8 bolts and set plate.
7. Remove fuel suction tube assembly with pump and gauge, as follows:
   a. Pull out the fuel suction tube from the fuel tank.

➡Do not damage the fuel pump filter.

➡Be careful not to bend the arm of the fuel sender gauge.

   b. Remove the gasket from the fuel suction tube.

To install:

8. Install a new gasket to the fuel suction tube.
9. Install the fuel suction tube.

➡Do not damage the fuel pump filter.

➡Be careful not to bend the arm of the fuel sender gauge.

10. Install the fuel tank vent tube set plate, as follows:
    a. Align the mark of the set plate with the fuel suction tube.
    b. Install the set plate with the 8 bolts and tighten to 52 inch lbs. (5.9 Nm).
11. Install the fuel pump tube with the tube joint clip.

➡Check that there is no scratches or foreign objects on the connecting part.

➡Check that the fuel tube joint is inserted securely.

➡Check that the tube joint clip is on the collar of the fuel tube joint.

➡After installing the tube joint clip, check that the fuel tube joint is pulled off.

12. Connect battery negative cable.
13. Inspect for fuel leak.
14. Install the rear floor service hole cover.
15. Install the rear seat cushion assembly.

FUEL RAIL & INJECTORS

REMOVAL & INSTALLATION

See Figure 145.

1. Properly discharge the fuel system pressure.
2. Disconnect battery negative cable.
3. Drain engine coolant.
4. Remove both windshield wiper arm and blade assemblies.
5. Remove the right cowl top ventilator louver.
6. Remove the windshield wiper motor and link assembly.
7. Remove the outer cowl top panel.
8. Remove the V-bank cover sub-assembly.
9. Remove the air cleaner cap with air cleaner hose.
10. Remove the intake air surge tank.
11. Disconnect the fuel tube sub-assembly, as follows:
    a. Remove the No. 2 fuel pipe clamp.
    b. Pinch the tube connector and then pull out the fuel pipe.

➡Check that there is no dirt or other foreign objects around the connector before removing fuel tube, and clean the connector as necessary.

➡It is necessary to prevent mud or dirt from entering the connector. If mud or dirt gets in the connector, the O-rings may not seal properly.

➡Do not use any tools in this operation.

➡Do not bend, kink or twist the nylon tube. Protect the connector by covering it with a plastic bag.

➡When the pipe and connector are stuck, push and pull the connector to release and pull the connector out carefully.

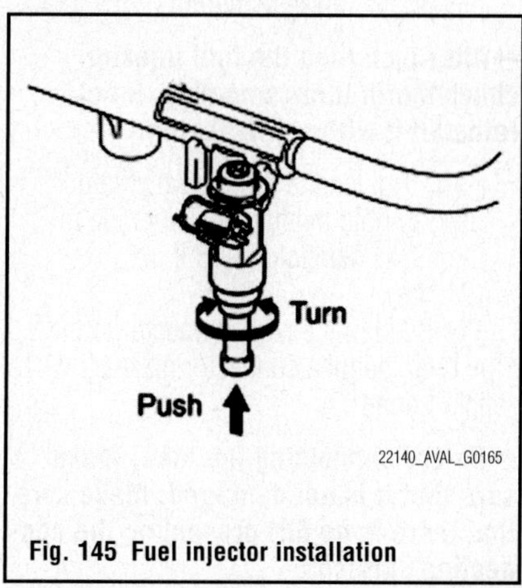

Fig. 145 Fuel injector installation

12. Remove the fuel injector assembly, as follows:
 a. Disconnect the 6 fuel injector connectors.
 b. Remove the 5 bolts and fuel delivery pipe together with the 6 fuel injectors.

➡Be careful not to drop the fuel injectors when removing the fuel delivery pipe.

 c. Remove the 6 insulators from the intake manifold.
 d. Pull out the fuel injector from the fuel delivery pipe.

To install:

13. Install the fuel injector assembly, as follows:
 a. Apply a light coat of spindle oil or gasoline to new O-rings, and install one to each injector.
 b. Apply a light coat of spindle oil or gasoline where the fuel delivery pipe contacts the O-ring.
 c. Push the fuel injector while twisting it back and forth to install it in the fuel delivery pipe.
 d. Position the fuel injector connector outward.

➡Be careful not to twist the O-ring.

➡After installing the fuel injector, check that it turns smoothly. If not, reinstall it with a new O-ring.

 e. Install 6 new insulators to the intake manifold.
 f. Place the fuel delivery pipe and the 6 fuel injectors together to the intake manifold.

➡Be careful not to drop the fuel injectors when installing the fuel delivery pipe.

 g. Temporarily install the 6 bolts which are used to hold the fuel delivery pipe to the intake manifold.

➡After installing the fuel injector, check that it turns smoothly. If not, reinstall it with a new O-ring.

 h. Tighten the 5 bolts which are used to hold the fuel delivery pipe to the intake manifold to 15 ft. lbs. (21 Nm).

14. Push in the tube connector to the pipe until the tube connector makes a "click" sound.

➡Before connecting the tube, make sure that it is not damaged. Make sure that there is no dirt present on the connecting surfaces.

➡After connecting, check if the fuel tube connector and the pipe are securely connected by pulling on them.

15. Install the No. 2 fuel pipe clamp.
16. Te remainder of installation is the reverse of the removal procedure.
17. Check for coolant leak and fuel leak.

FUEL TANK

REMOVAL & INSTALLATION

See Figures 146 and 147.

1. Before servicing the vehicle, refer to the precautions section.
2. Discharge fuel system pressure.
3. Disconnect battery negative cable.
4. Remove rear seat cushion assembly.
5. Remove rear floor service hole cover.
6. Separate fuel pump tube sub-assembly.
7. Remove fuel tank vent tube set plate.
8. Remove fuel suction tube assembly with pump and gauge.
9. Drain fuel.
10. Remove center exhaust pipe assembly.
11. Disconnect no. 2 parking brake cable assembly.
12. Disconnect no. 3 parking brake cable assembly.
13. Remove the 4 bolts and the lower center fuel tank protector.
14. Disconnect the fuel pump tube, as follows:

➡Check that there is no dirt or other foreign objects around the connector before removing fuel tubes, and clean the connector as necessary.

➡It is necessary to prevent mud or dirt from entering the connector. If mud or dirt gets in the connector, the O-rings may not seal properly.

➡Do not use any tools in these operations.

➡Do not bend, kink or twist the nylon tubes. Protect the connector by covering it with a plastic bag.

➡When the pipe and connector are stuck, push and pull the connector to release and pull the connector out carefully.

 a. Pinch the tabs of the retainer to remove the lock claws and pull it down as shown in the illustration.
 b. Pull out the fuel tank main tube.

15. Pinch the tube connector and then pull out the No. 1 fuel tube.
16. Set up a transmission jack underneath the fuel tank.
17. Remove the 2 set bolts of the fuel tank bands.
18. Remove the hose clamp and disconnect the fuel tank to filter pipe hose.
19. Disconnect the fuel tank vent hose from the charcoal canister, as follows:
 a. Push the connector deep into the charcoal canister to release the locking pin.
 b. Pinch portion A.
 c. Pull out the connector.
20. Remove the 2 pins and 2 fuel tank bands as shown in the illustration.
21. Remove the 4 clip nuts.

To install:

22. Install the 4 clip nuts.

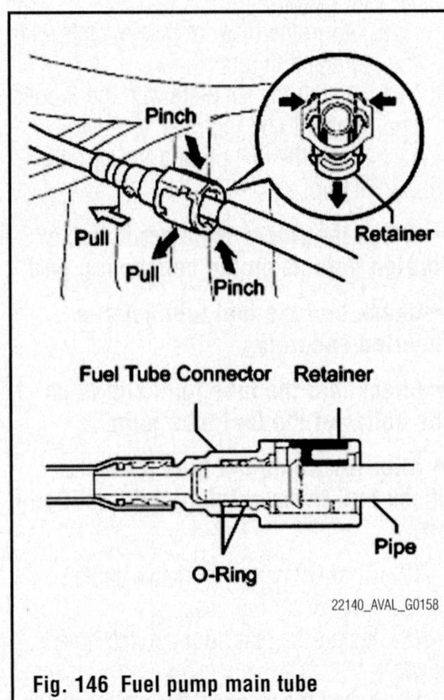

Fig. 146 Fuel pump main tube

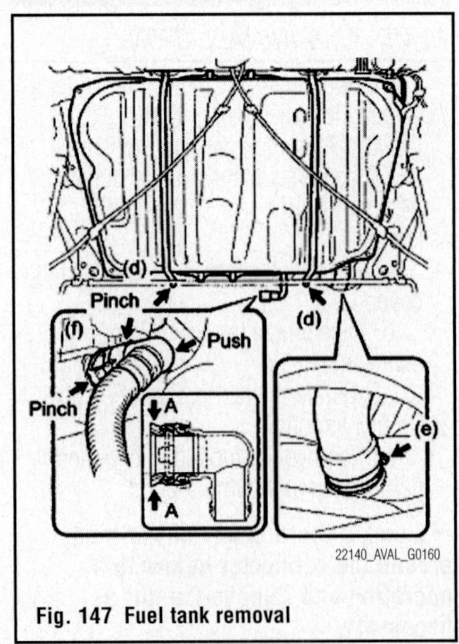

Fig. 147 Fuel tank removal

23. Install the 2 fuel tank bands with the 2 pins.
24. Connect the fuel tank vent hose.
25. Connect the fuel tank inlet pipe with the fuel filter pipe clamp.
26. Tighten the 2 set bolts of the fuel tank bands to 29 ft. lbs. (39 Nm).
27. Connect the No. 1 fuel tube, as follows:
   a. Push the fuel tube connector into the pipe until the fuel tube connector makes a "click" sound.

➡Check that there is no damage or foreign objects on the connected part.

➡After connecting, check if the fuel tube connector and the pipe are securely connected by trying to pull them apart.

28. Connect the fuel pump tube, as follows:
   a. Push in the fuel pump tube connector to the pipe and push up the retainer so that the claws engage.

➡Check that there is no damage or foreign objects on the connected part.

➡After connecting, check if the fuel tube connector and the pipe are securely connected by trying to pull them apart.

29. Install the lower center fuel tank protector and tighten to 48 inch lbs. (5.4 Nm).
30. Install the No. 3 parking brake cable assembly with the bolt and nut and tighten to 53 inch lbs. (6 Nm), and 75 inch lbs. (8.5 Nm).
31. Install the No. 2 parking brake cable assembly with the bolt and nut and tighten to 53 inch lbs. (6 Nm), and 75 inch lbs. (8.5 Nm).
32. Install the center exhaust pipe assembly.
33. Install the fuel suction tube assembly with pump and gauge.
34. Install the fuel tank vent tube set plate.
35. Connect the fuel pump tube sub-assembly.
36. Add fuel.
37. Connect battery negative cable.
38. Inspect for fuel leak and exhaust gas leak.
39. Install the rear floor service hole cover.
40. Install the rear seat cushion assembly.

IDLE SPEED

ADJUSTMENT

Idle speed is maintained by the ECM. No adjustment is necessary or possible.

THROTTLE BODY

REMOVAL & INSTALLATION

1. Before servicing the vehicle, refer to the precautions section.
2. Disconnect battery negative cable.
3. Drain engine coolant.
4. Remove the V-bank cover sub-assembly.
5. Remove the air cleaner cap with air cleaner hose.
6. Disconnect the 2 water by-pass hoses from the throttle w/ motor body assembly.
7. Disconnect the throttle w/ motor body assembly connector.
8. Remove the 4 bolts and throttle w/ motor body assembly from the intake air surge tank.
9. Remove the throttle body gasket from the intake air surge tank.

To install:

10. Install a new throttle body gasket to the intake air surge tank.
11. Install the throttle w/ motor body assembly and wire harness clamp stay to the intake air surge tank with the 4 bolts and tighten to 7 ft. lbs. (10 Nm).
12. Connect the throttle w/ motor body assembly connector.
13. Te remainder of installation is the reverse of the removal procedure.
14. Check for coolant leak.
15. Check the function of the throttle body.

HEATING & AIR CONDITIONING SYSTEM

BLOWER MOTOR

REMOVAL & INSTALLATION

1. Remove the glove box assembly.
2. Disconnect the connector from the blower motor.
3. Remove the 3 screws and the blower motor.

To install:

4. Replace the blower motor and the three screws.
5. Connect the electrical connector.
6. Replace the glove box assembly.

HEATER CORE

REMOVAL & INSTALLATION

See Figures 148 through 155.

1. Before servicing the vehicle, refer to the precautions section.
2. Disconnect battery negative terminal.

⁂ CAUTION

Wait for 90 seconds after disconnecting the cable to prevent the airbag from deploying.

3. Drain engine coolant.
4. Remove both windshield wiper arm and blade assemblies.
5. Remove the right cowl top ventilator louver.
6. Remove the windshield wiper motor and link.
7. Remove the outer front cowl top panel.
8. Discharge refrigerant from refrigeration system.
9. Disconnect the suction hose sub-assembly, as follows:
   a. Remove the bolt, and slide the hook connector.
   b. Disconnect the suction hose sub-assembly.
   c. Remove the O-ring from the suction hose sub-assembly.

➡Seal the openings of the disconnected parts using vinyl tape to prevent moisture and foreign matter from entering.

10. Disconnect the air conditioner tube and accessory.
11. Remove the O-ring from the air conditioner tube and accessory.
12. Slide the clip and disconnect the heater water outlet hose A.
13. Slide the clip and disconnect the heater water inlet hose A.

➡Do not apply excessive force to the heater water hoses

➡Prepare a drain pan or cloth in case the cooling water leaks.

14. Remove the instrument panel safety pad sub-assembly.

➡Refer to the removal procedures for the instrument panel safety pad sub-assembly w/ front passenger airbag assembly.

15. Remove the 3 clips and No. 2 heater to register duct.
16. Disengage the 4 claws and then remove the No. 6 heater to register duct.
17. Remove the clip and No. 1 console box duct.
18. Remove the left and right hand floor carpet brackets, as follows:
   a. Release the clamps (2 from the left and 3 from the right).
   b. Turn back the floor carpet.
   c. Remove the 3 clips.
   d. Remove the floor carpet brackets.
19. Release the 2 claws and remove the rear No. 2 air duct.
20. Release the 2 claws and remove the rear No. 1 air duct.
21. Remove the No. 1 air duct sub-assembly.
22. Separate the steering intermediate shaft assembly.
23. Remove the steering column assembly.
24. Remove instrument panel reinforcement assembly, as follows:
   a. Disconnect each connector and remove each clamp. Disconnect the wire harness.
   b. Remove the 6 nuts and 3 bolts.
   c. Remove the 3 bolts and nut.
   d. Using a Torx® socket wrench (T40), remove the 5 Torx® bolts.

➡The Torx® bolts on the passenger side can be removed with the collar or for adjustment.

   e. Using a hexagon wrench 12 mm, remove the 2 collars and instrument panel reinforcement assembly with the air conditioner unit assembly.
   f. Remove the 3 bolts, 2 screws and instrument panel reinforcement assembly.
25. Remove the No. 2 air duct sub-assembly.
26. Remove the blower assembly.
27. Remove the air outlet control servo motor.
28. Remove the air mix control servo motor.
29. Remove the heater radiator unit sub-assembly, as follows:
   a. Remove the screw and clamp.
   b. Release the 4 claws and remove the clamp.
   c. Remove the heater radiator unit sub-assembly from the air conditioning radiator assembly.

➡Prepare a drain pan or cloth in case the cooling water leaks.

To install:

30. Install the heater radiator unit sub-assembly.
31. Install the air mix control servo motor.
32. Install the air outlet control servo motor.
33. Install the blower assembly.

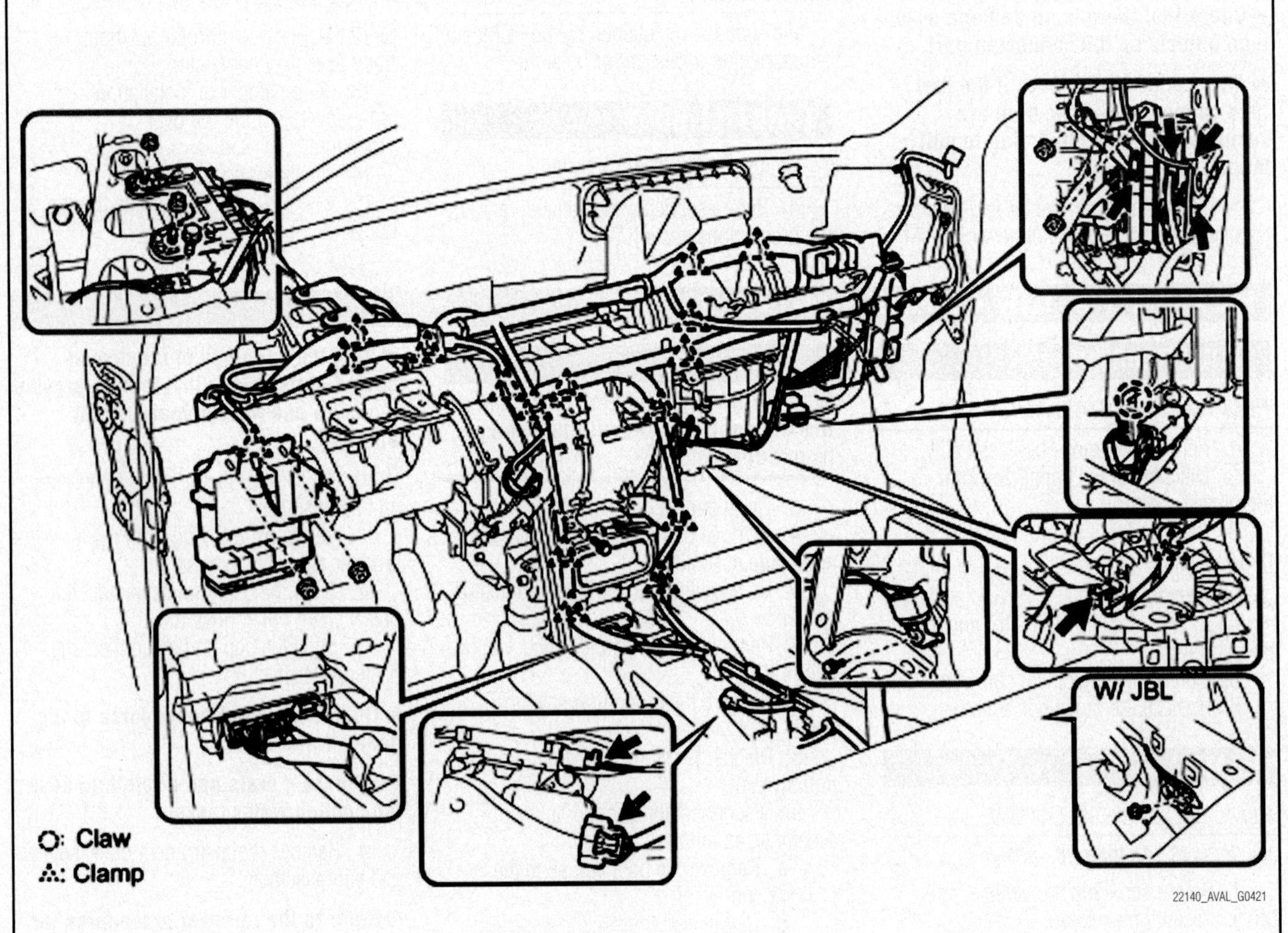

Fig. 148 Instrument panel reinforcement assembly connectors

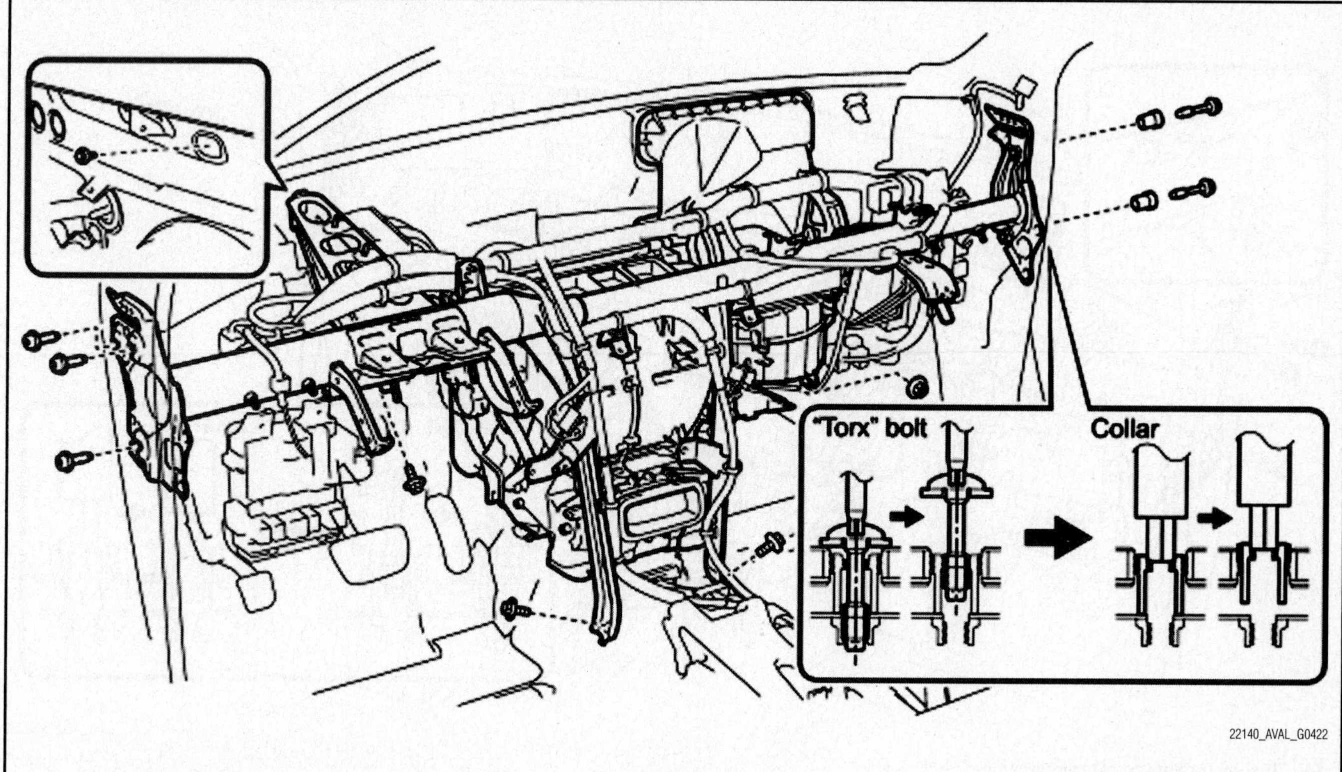

Fig. 149 Instrument panel reinforcement assembly with the air conditioner unit assembly

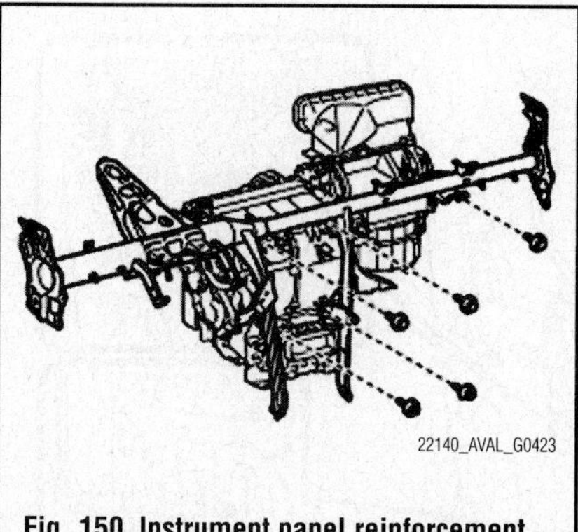

Fig. 150 Instrument panel reinforcement assembly with the air conditioner unit assembly

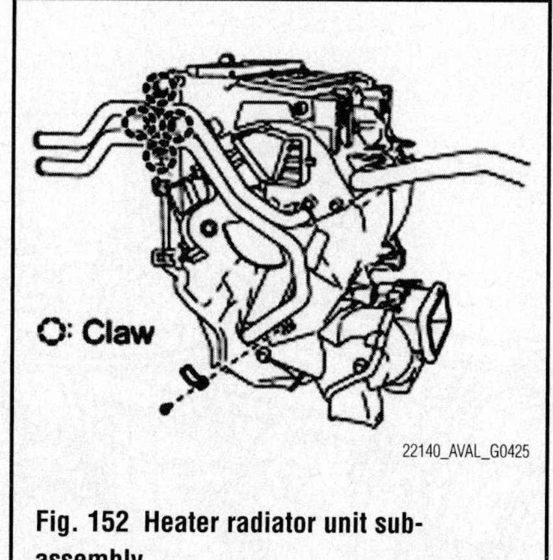

Fig. 152 Heater radiator unit sub-assembly

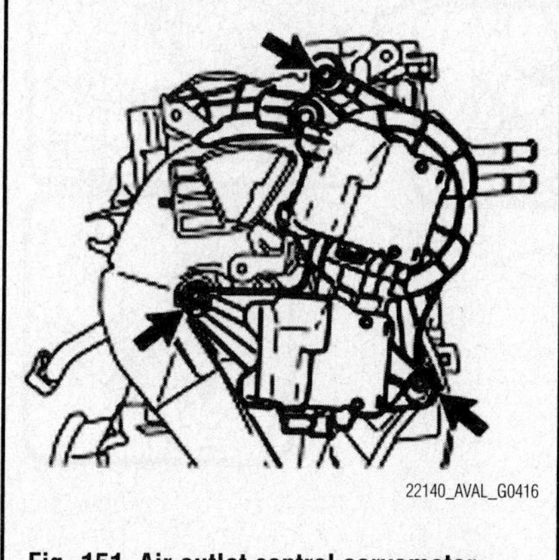

Fig. 151 Air outlet control servomotor

Fig. 153 A/C unit bolt tightening sequence

34. Install the No. 2 air duct sub-assembly.
35. Install the air conditioner unit assembly to the instrument panel reinforcement assembly with the 2 screws and 3 bolts and tighten in the order shown to 87 inch lbs. (9.8 Nm).
36. Driver seat:
    a. Using a Torx® socket wrench (T40), install the instrument panel reinforcement assembly with the 3 Torx® bolts and tighten to 13 ft. lbs. (17 Nm).
37. Passenger seat:
    a. Using a hexagon wrench 12 mm, install the instrument panel reinforcement assembly with the 2 bolts and tighten to 53 inch lbs. (6 Nm).
    b. Using a Torx® socket wrench (T40), install the instrument panel reinforcement assembly with the 2 Torx® bolts and tighten to 15 inch lbs. (20 Nm).
    c. Install the 3 bolts and nut and tighten to 87 inch lbs. (9.8 Nm) and 15 ft. lbs. (20 Nm).
    d. Connect the connectors and clamps.
    e. Install the 6 nuts and 3 bolts.
38. The remainder of installation is the reverse of removal.
39. Initialize system.
40. After adding engine coolant and charging refrigerant, check for coolant and refrigerant leaks.

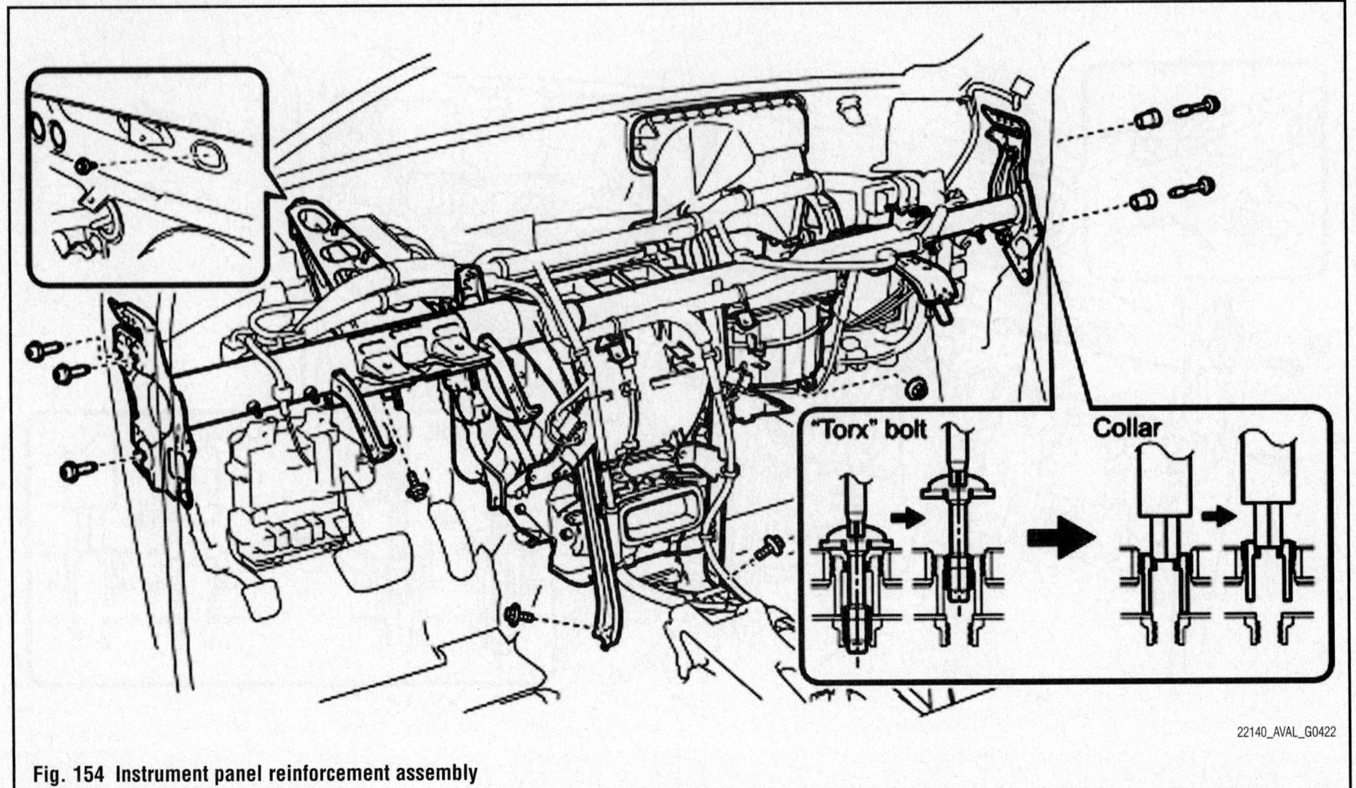

Fig. 154 Instrument panel reinforcement assembly

W/ JBL

○: Claw
∴: Clamp

22140_AVAL_G0421

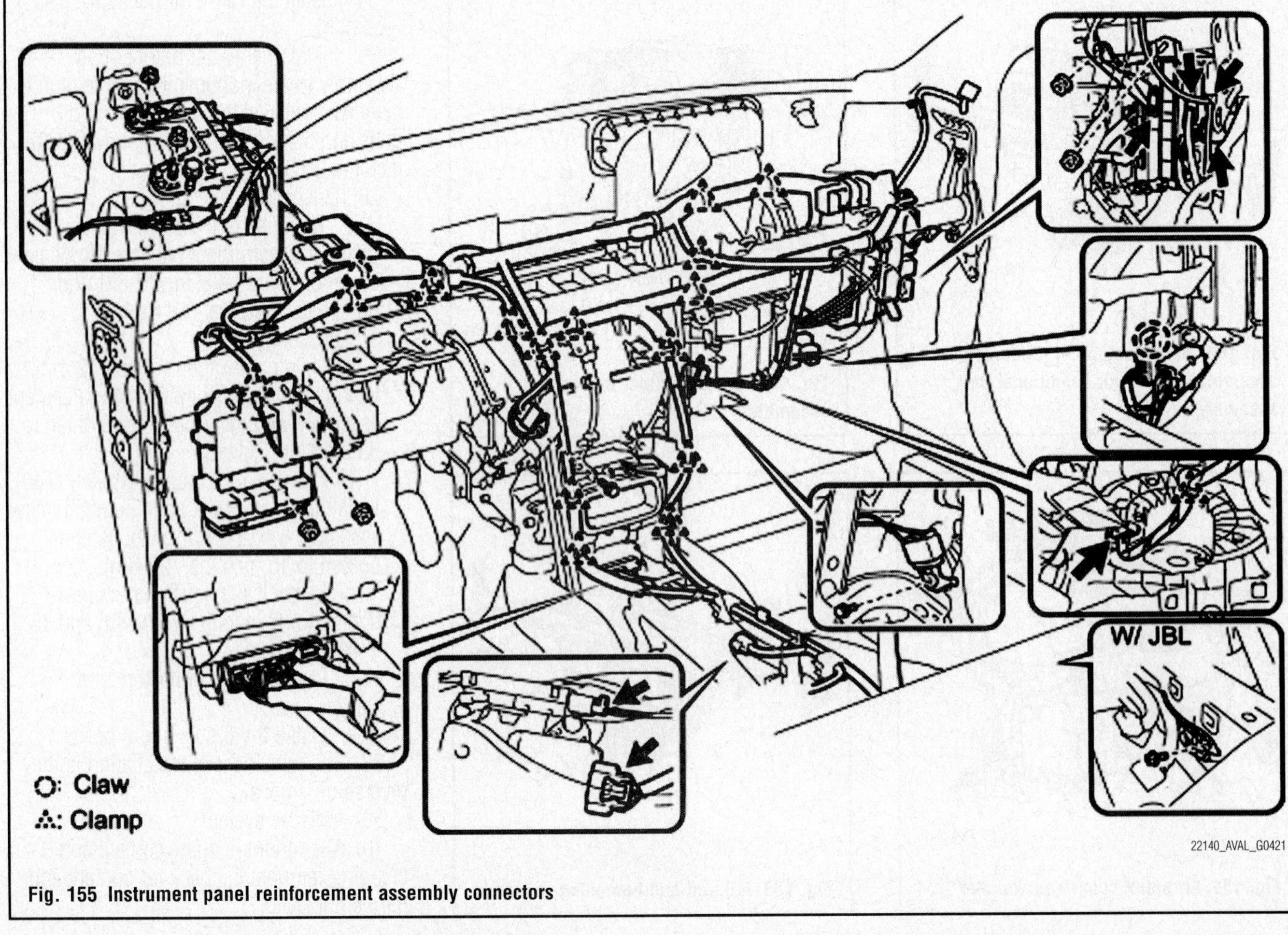

Fig. 155 Instrument panel reinforcement assembly connectors

STEERING

POWER RACK & PINION STEERING GEAR

REMOVAL & INSTALLATION

See Figures 156 through 158.

1. Before servicing the vehicle, refer to the precautions section.
2. Position the front wheels straight ahead.
3. Remove front wheel.
4. Remove right and left front fender apron seals.
5. Separate steering intermediate shaft assembly, as follows:
    a. Secure the steering wheel with the seat belt in order to prevent rotation. This will help prevent damage to the spiral cable.
    b. Remove the bolt.
    c. Put matchmarks on the steering intermediate shaft assembly and the power steering gear assembly.
    d. Separate the intermediate shaft assembly from the steering gear assembly.

Fig. 156 Remove the bolt

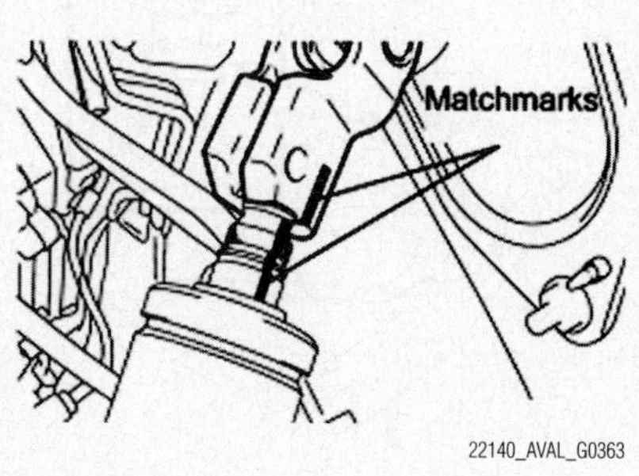

Fig. 157 Matchmarks on the steering intermediate shaft assembly and the power steering gear assembly

6. Separate the left and right tie rod assemblies, as follows:
    a. Remove the cotter pin and castle nut.
    b. Using SST (SST: 09628-62011) or equivalent, separate the tie rod assembly from the steering knuckle.
7. Remove the right and left front stabilizer link assemblies.
8. Remove the 2 bolts and the left and right No. 1 stabilizer bracket and the No. 1 stabilizer bar bush.
9. Disconnect pressure feed tube assembly, as follows:
    a. Using SST (09023-12701) or equivalent, disconnect the return tube assembly from the power steering gear assembly.
    b. Using SST (09023-12701), disconnect the pressure feed tube assembly from the power steering gear assembly.
    c. Remove the 2 bolts and separate the pressure feed tube clamp from the power steering gear assembly.
10. Remove the 2 bolts, nuts and the power steering gear assembly.

➡Do not turn the nut because the nut has its own stopper. Loosen the bolt with the nut secured.

To install:

11. Install the power steering gear assembly with the 2 bolts and 2 nuts and tighten to 52 ft. lbs. (70 Nm).

➡Do not turn the nut because the nut has its own stopper. Tighten the bolt with the nut secured.

➡For the next 2 steps, use a torque wrench with a fulcrum length of 300 mm (11.81 in.). These torque values are effective when SST is parallel to the torque wrench.

12. Using SST (09023-12701) or equivalent, connect the pressure feed tube assembly to the power steering gear assembly and tighten to 16 ft. lbs. (22 Nm).
13. Using SST (09023-12701) or equivalent, connect the return tube assembly to the power steering gear assembly and tighten to 16 ft. lbs. (22 Nm).
14. Install the pressure feed tube clamp with the 2 bolts and tighten to 87 inch lbs. (9.8 Nm).
15. Install the left and right front stabilizer brackets, as follows:
    a. Install the No. 1 stabilizer bar bush to the stabilizer bar.
    b. Install the No. 1 front stabilizer bracket with the 2 bolts and tighten to 20 ft. lbs. (27 Nm).
16. Install the right and left front stabilizer link assemblies.
    a. Connect the left and right tie rod assemblies, as follows:
    b. Connect the tie rod assembly to the steering knuckle with the castle nut and tighten to 36 ft. lbs. (49 Nm).
    c. Install a new cotter pin. If the holes for the cotter pin are not aligned, tighten the nut up to 60° further.
17. Align the matchmarks on the steering intermediate shaft assembly and the power steering gear assembly.
    a. Install the bolt and tighten to 26 ft. lbs. (35 Nm).
18. Bleed power steering fluid.
19. Inspect for power steering fluid.
20. Install the left and right front fender apron seals.
21. Install the front wheel.
22. Inspect the front wheels.
23. Inspect the front wheel alignment.

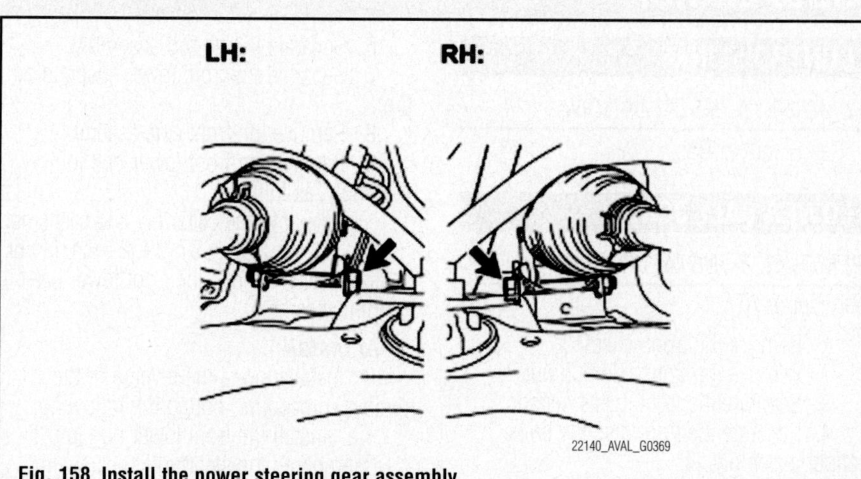

Fig. 158 Install the power steering gear assembly

POWER STEERING PUMP

REMOVAL & INSTALLATION

See Figures 159 and 160.

1. Before servicing the vehicle, refer to the precautions section.

➡Do not over tighten when using a vise.

➡When installing, coat the parts indicated by arrows with power steering fluid. Take care not to spill power steering fluid on the V-belt.

2. Remove the right front wheel.
3. Drain power steering fluid.
4. Remove the right front fender apron seal.
5. Remove fan and generator V-belt.
6. Remove the clip and disconnect the No. 1 oil reservoir to pump hose.

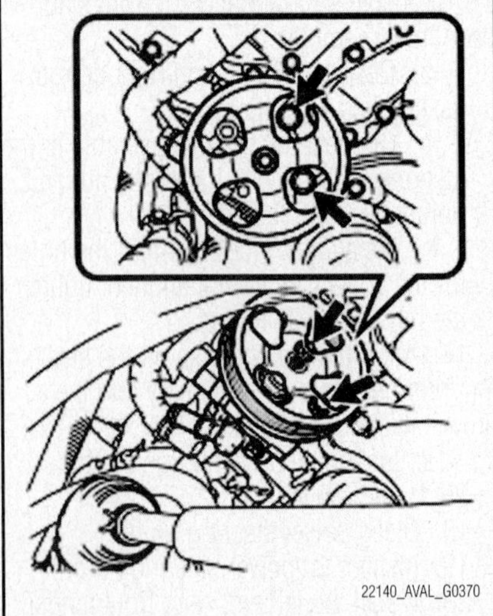

22140_AVAL_G0370

Fig. 159 Remove the 2 bolts and the vane pump assembly

7. Remove the union bolt and disconnect the pressure feed tube assembly from the vane pump assembly.
8. Remove the bolt and separate the pressure feed tube assembly clamp.
9. Remove the gasket from the pressure feed tube assembly.
10. Remove the vane pump assembly, as follows:
    a. Disconnect the connector from the power steering oil pressure switch.
    b. Remove the 2 bolts and the vane pump assembly.

To install:

11. Install the vane pump assembly with the 2 bolts and tighten to 32 ft. lbs. (43 Nm).
12. Connect the connector to the power steering oil pressure switch.
13. Install a new gasket to the pressure feed tube assembly.
14. Connect the pressure feed tube assembly to the vane pump assembly with the union bolt and tighten to 38 ft. lbs. (52 Nm).

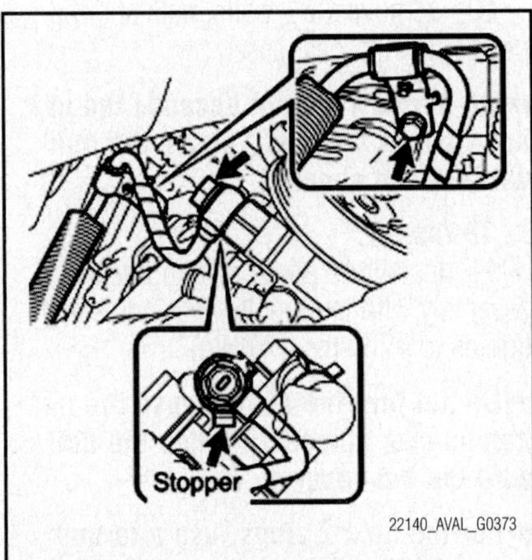

22140_AVAL_G0373

Fig. 160 Connect the pressure feed tube assembly

➡Make sure that the stopper of the pressure feed tube assembly contacts the vane pump assembly as shown in the illustration, then tighten the union bolt.

15. Install the pressure feed tube clamp with the bolt and tighten to 87 inch lbs. (9.8 Nm).
16. Connect the No. 1 oil reservoir to pump hose with the clip.
17. Install the fan and generator V-belt.
18. Install the right front fender apron seal.
19. Install the right front wheel.
20. Bleed the power steering fluid.
21. Inspect for power steering fluid.

BLEEDING

1. Before servicing the vehicle, refer to the precautions section.
2. Check the fluid level.
3. Jack up the front of the vehicle and support it with stands.
4. With the engine stopped, turn the wheel slowly from lock to lock several times.
5. Lower the vehicle.
6. Start the engine.
7. Run the engine at idle for a few minutes.
8. With the engine idling, turn the wheel left or right to the full lock position and keep it there for 2 to 3 seconds, then turn the wheel to the opposite full lock position and keep it there for 2 to 3 seconds.
9. Repeat the above steps several times.
10. Stop the engine.
11. Check for foaming or emulsification. If the system has to be bled twice because of foaming or emulsification, check for fluid leaks in the system.
12. Check the fluid level.

SUSPENSION

CONTROL LINKS

REMOVAL & INSTALLATION

See Stabilizer Bar.

LOWER BALL JOINT

REMOVAL & INSTALLATION

See Figure 161.

1. Remove the front wheel.
2. Remove the front axle hub nut.
3. Separate the front speed sensor.
4. Separate the front disc the brake caliper assembly.
5. Remove front disc.
6. Separate the tie rod assembly.
7. Separate the front lower suspension arm.
8. Remove the front axle assembly.
9. Remove the front lower ball joint assembly, as follows:
    a. Remove the cotter pin and castle nut.
    b. Using SST (SST: 09628-62011) or equivalent, remove the front lower ball joint assembly.

To install:

10. Installation is the reverse of the removal procedure, noting the following:
    a. Install the front lower ball joint assembly to the steering knuckle with the castle nut and tighten to 91 ft. lbs.

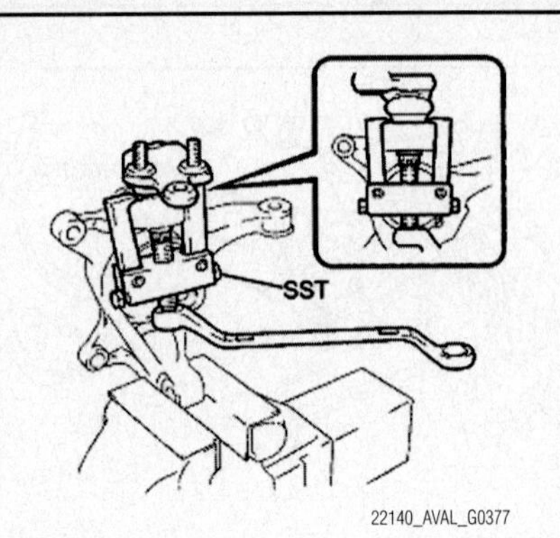

22140_AVAL_G0377

Fig. 161 Remove the front lower ball joint assembly

(123 Nm). Further tighten the nut up to 60° if the holes for the cotter pin are not aligned.

b. Inspect and adjust the front wheel alignment.

c. Inspect the ABS speed sensor signal.

LOWER CONTROL ARM

REMOVAL & INSTALLATION

See Figure 162.

1. Remove the engine assembly with transaxle.
2. Remove the 3 nuts and transverse engine mounting insulator.
3. Remove the 2 bolts on the front side of the lower No. 1 front suspension arm sub-assembly.
4. Remove the bolt and nut on the rear side of the lower No. 1 front suspension arm sub-assembly.
5. Remove the lower No. 1 front suspension arm sub-assembly.
6. Remove the front lower arm bush stopper from the lower No. 1 front suspension arm sub-assembly.

To install:

7. Install the front lower arm bush stopper to the lower No. 1 front suspension arm sub-assembly.
8. Install the 2 bolts on the front side of the lower No. 1 front suspension arm sub-assembly and tighten to 148 ft. lbs. (200 Nm).
9. Install the bolt and nut on the rear side of the lower No. 1 front suspension arm sub-assembly.
10. Install the transverse engine mounting insulator with the 3 nuts.
11. Install the engine assembly with transaxle.

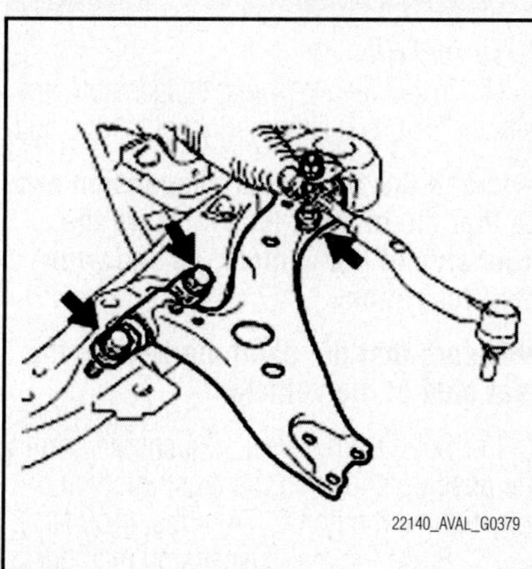

Fig. 162 Lower No. 1 front suspension arm sub-assembly

MACPHERSON STRUT

REMOVAL & INSTALLATION

1. Remove the front wheel.
2. Remove the nut and disconnect the front stabilizer link assembly from the front shock absorber assembly.

➡Use a hexagon (6 mm) wrench to hold the stud if the ball joint turns together with the nut.

3. Remove the front shock absorber with coil spring, as follows:
    a. Loosen the lock nut. If not disassembling the shock absorber it is not necessary to loosen the nut.
    b. Remove the bolt and disconnect the No. 1 front flexible hose and front speed sensor wire harness.
    c. Remove the 2 nuts on the lower side of the front shock absorber with coil spring. Keep the bolts inserted.
    d. Remove the 3 nuts on the upper side of the front shock absorber with coil spring.
    e. Remove the 2 bolts on the lower side of the front shock absorber and front shock absorber with coil spring.

➡Be careful not to drop the collar in the case that there is front suspension upper brace center.

To install:

4. Install the front shock absorber with coil spring.
5. Install the 3 nuts to the upper side of the front shock absorber with coil spring and tighten to 63 ft. lbs. (85 Nm).

➡Be careful not to drop the collar in the case that there is front suspension upper brace center.

6. Install the 2 bolts and 2 nuts to the lower side of the front shock absorber with coil spring and tighten to 155 ft. lbs. (210 Nm).

➡Keep the bolts from rotating and torque the 2 nuts when installing the 2 nuts.

➡Insert the bolts from the front side of the vehicle.

7. Fully tighten the lock nut to 51 ft. lbs. (70 Nm).
8. Install the No. 1 front flexible hose and front speed sensor with the bolt and tighten to 14 ft. lbs. (19 Nm).
9. Install the front stabilizer link assembly with the nut and tighten to 55 ft. lbs. (74 Nm).

➡Use a hexagon (6 mm) wrench to hold the stud if the ball joint turns together with the nut.

10. Install the front wheel.
11. Inspect and adjust front wheel alignment.

STEERING KNUCKLE

REMOVAL & INSTALLATION

See Wheel Hub and Bearing.

STABILIZER BAR

REMOVAL & INSTALLATION

See Figure 163.

1. Remove the front wheel.
2. Remove the nuts and the right and left front stabilizer link assemblies.

➡Use a hexagon (6 mm) wrench to hold the stud if the ball joint turns together with the nut.

3. Remove the engine assembly with transaxle.
4. Remove the bolts and the left and right No. 1 stabilizer brackets.
5. Remove the 2 bushes from the stabilizer.
6. Remove the front stabilizer bar from the vehicle.

To install:

7. Install the stabilizer bar front to the vehicle.

➡Install the bushes as to the outer side of each bush stopper on the stabilizer bar.

➡Place the cutout of the stabilizer bushes as facing the rear side as shown in the illustration.

8. Install the No. 1 left front stabilizer bracket with the 2 bolts and tighten to 20 ft. lbs. (27 Nm).
9. Install the No. 1 right front stabilizer

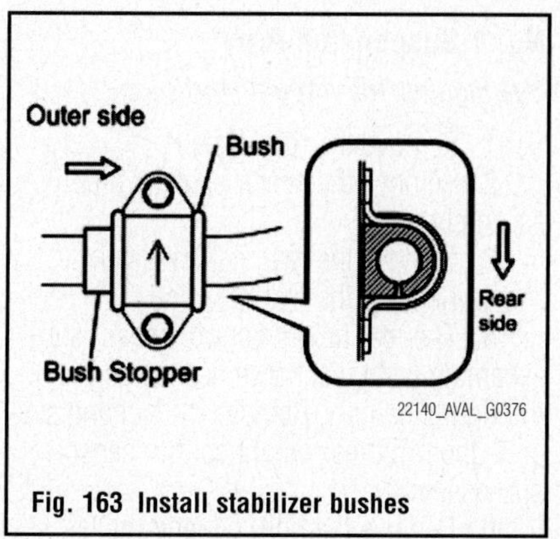

Fig. 163 Install stabilizer bushes

bracket with the 2 bolts and tighten to 20 ft. lbs. (27 Nm).

10. Install the engine assembly with transaxle.

11. Install the left front stabilizer link assembly with the 2 nuts and tighten to 55 ft. lbs. (74 Nm).

12. Install the right front stabilizer link assembly with the 2 nuts and tighten to 55 ft. lbs. (74 Nm).

➡Use a hexagon (6 mm) wrench to hold the stud if the ball joint turns together with the nut.

13. Install the front wheel.

14. Bleed the power steering fluid.

15. Inspect and adjust the front wheel alignment.

WHEEL HUB AND BEARING

REMOVAL & INSTALLATION

See Figure 164.

1. Remove the front wheel.
2. Remove the front axle shaft nut.
3. Separate the front speed sensor.
4. Remove the 2 bolts and separate the front disc brake caliper assembly from the steering knuckle.
5. Remove the front disc.
6. Separate the tie rod end sub-assembly.
7. Separate the No. 1 front suspension arm sub-assembly.
8. Remove the front axle assembly, as follows:
   a. Put matchmarks on the front drive shaft assembly and the front axle hub sub-assembly.
   b. Using a plastic hammer, separate the front drive shaft assembly from the front axle hub sub-assembly.

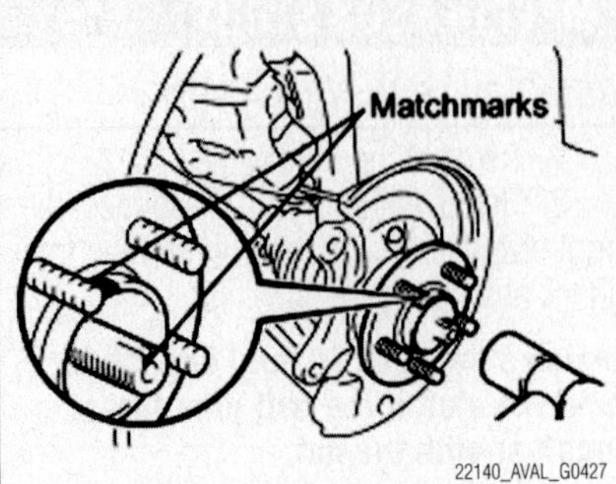

Fig. 164 Matchmarks on the front drive shaft assembly and the front axle hub sub-assembly

⁂ WARNING

Be careful not to damage the boot and ABS speed sensor rotor.

   c. Remove the 2 bolts, nuts and steering knuckle with the front axle hub sub-assembly.

To install:

9. Install the front axle assembly, as follows:
   a. Align the matchmarks and install the front drive shaft assembly to the front axle hub sub-assembly.
   b. Install the steering knuckle with the front axle hub sub-assembly to the front shock absorber assembly with the 2 bolts and 2 nuts and tighten to 155 ft. lbs. (210 Nm).

➡Only when reusing the bolts and nuts, apply the small amount of engine oil to the screw part of the nuts.

⁂ WARNING

Be careful not to damage the drive shaft boot or speed sensor rotor.

10. Install the lower No. 1 front suspension arm sub-assembly.
11. Install the tie rod end sub-assembly.
12. Install the front disc.
13. Install the front disc brake caliper assembly with the 2 bolts to the steering knuckle and tighten to 79 ft. lbs. (107 Nm).
14. Clean the threaded parts on the drive shaft and axle hub nut using a non-residue solvent.

➡Be sure to perform this work for a new drive shaft.

➡Keep the threaded parts free of oil and foreign objects.

15. Using a 30 mm socket wrench, install the front axle hub nut and tighten to 217 ft. lbs. (294 Nm).
16. Remove the 2 bolts and separate the front disc brake caliper assembly from the steering knuckle.
17. Remove the front disc.
18. Inspect front axle hub bearing looseness.
19. Inspect front axle hub runout.
20. Install the front disc.
21. Install the front disc brake caliper assembly with the 2 bolts to the steering knuckle and tighten to 79 ft. lbs. (107 Nm).
22. Install the front speed sensor.
23. Using a chisel and hammer, stake the axle hub nut.
24. Install the front wheel.
25. Inspect and adjust front wheel alignment.
26. Check ABS speed sensor signal.

SUSPENSION

CONTROL ARMS/LINKS

REMOVAL & INSTALLATION

No. 1 Suspension Arm

See Figures 165 through 168.

1. Remove the rear wheel.
2. Remove the center exhaust pipe assembly.
3. Remove the rear stabilizer bar.
4. Remove the rear strut rod.
5. The rear height control sensor sub-assembly (with discharge heard light) is on the right side only. Remove the nut and separate the right rear height control sensor sub-assembly.
6. Remove the bolt, nut and the left No. 2 rear suspension arm (outer side) from the rear axle carrier. When removing the bolt, keep the nut from rotating.
7. Remove the right No. 2 rear suspension arm using the same procedure as for the left side.
8. Remove the bolt, nut and the left and right No. 1 rear suspension arm (outer side) from the rear axle carrier. When removing the bolt, keep the nut from rotating.
9. Support the rear suspension member with a jack.
10. Remove the 4 nuts, 2 bolts and 4 retainers from the rear suspension member.
11. Lower the rear suspension member.
12. Remove the bolt and No. 1 rear suspension arm.

To install:

13. Install the No. 1 rear suspension arm with the bolt, and temporarily tighten the bolt.

➡Install the No. 1 rear suspension arm so that the bracket leans toward the front side of the vehicle, as shown in the illustration.

➡Ensure that the paint mark faces the rear side of the vehicle.

14. Set the No. 1 rear suspension arm in the position shown in the illustration, and fully tighten the bolt to 74 ft. lbs. (100 Nm).
15. Raise the rear suspension member with a jack.
16. Install the rear suspension member with the 4 nuts, 2 bolts and 4 retainers and

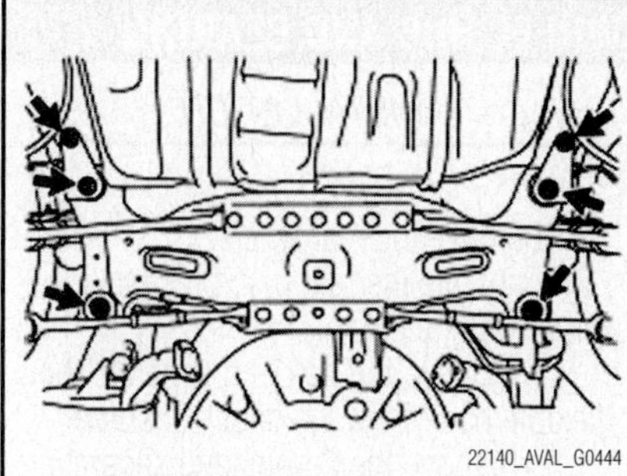
Fig. 165 Remove the 4 nuts, 2 bolts and 4 retainers from the rear suspension member

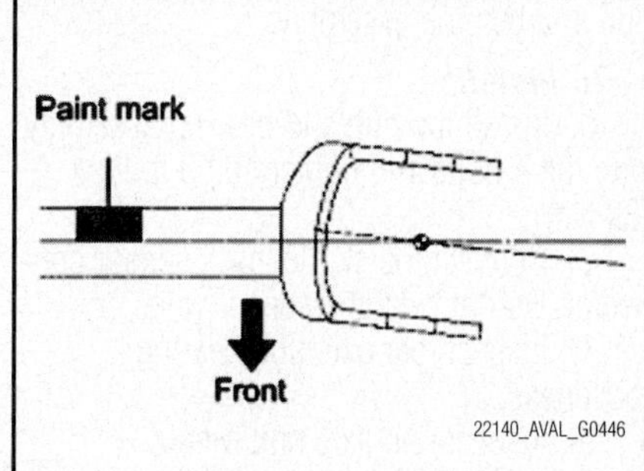

Fig. 166 Ensure that the paint mark faces the rear side of the vehicle

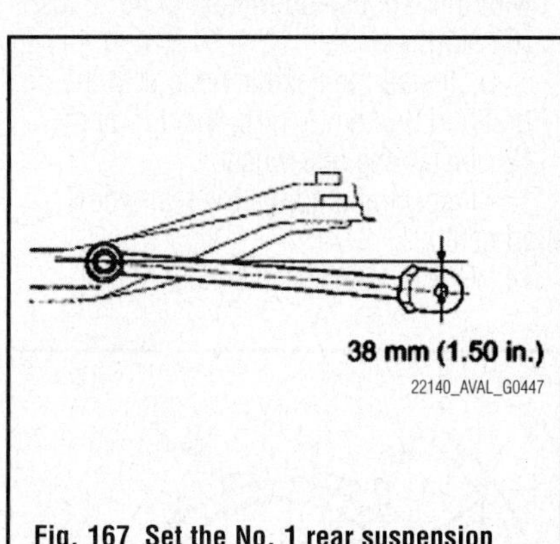

Fig. 167 Set the No. 1 rear suspension arm

Fig. 168 Install the rear suspension member

tighten to 41 ft. lbs. (55 Nm), and 28 ft. lbs. (38 Nm).

17. Connect the left No. 1 rear suspension arm (outer side) to the rear axle carrier with the bolt and nut and temporarily tighten. Insert the bolt from the front side of the vehicle and temporarily install it.

18. Temporarily tighten the right rear No. 1 suspension arm using the same procedure as for the left side.

19. Connect the left No. 2 rear suspension arm (outer side) to the rear axle carrier with the bolt and nut and temporarily tighten the bolt.

20. Temporarily tighten the right rear No. 2 suspension arm using the same procedure as for the left side.

21. Temporarily tighten the rear strut rod.

22. Jack up the rear axle carrier, placing a wooden block to avoid damage. Apply a load to the suspension so that the installed bolt of the No. 1 suspension arm assembly (inner side of the vehicle) is horizontally aligned with the center of the rear axle hub.

23. Fully tighten the left rear No. 1 suspension arm bolt to 74 ft. lbs. (100 Nm). When installing the bolt, hold the nut and tighten the bolt.

24. Fully tighten the right rear No. 1 suspension arm using the same procedure as for the left side.

25. Fully tighten the left rear No. 2 suspension arm bolt to 74 ft. lbs. (100 Nm). When installing the bolt, hold the nut and tighten the bolt.

26. Fully tighten the right rear No. 2 suspension arm using the same procedure as for the left side.

27. Fully tighten the rear strut rod.

28. Install the rear stabilizer bar.

29. Install the right rear height control sensor sub-assembly (with discharge head light) with the nut and tighten to 48 inch lbs. (5.4 Nm).

30. Install the exhaust pipe assembly center.

31. Install the rear wheel.

32. Inspect and adjust the rear wheel alignment.

33. Adjust the headlight aiming.

No. 2 Suspension Arm

See Figure 169.

1. Remove the rear wheel.
2. Remove the exhaust pipe assembly center.
3. Remove the rear stabilizer bar.
4. Remove the rear strut rod.
5. Separate the rear height control sensor sub-assembly.

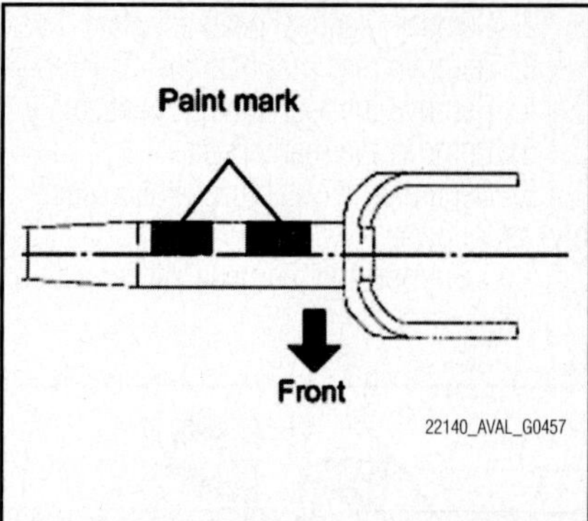

Fig. 169 Ensure that the paint marks face the rear side of the vehicle

6. Separate the left and right rear No. 1 suspension arms.
7. Separate the left and right rear No. 2 suspension arms.
8. Remove the rear suspension member sub-assembly.
9. Remove the bolt, and disconnect the rear No. 2 suspension arm (inner side).

To install:

10. Install the rear No. 2 suspension arm (inner side) with the bolt and tighten to 74 ft. lbs. (100 Nm).

➡Ensure that the paint marks face the rear side of the vehicle.

11. Install the rear suspension member sub-assembly.
12. Temporarily tighten the left and right rear No. 1 suspension arms.
13. Temporarily tighten the left and right rear No. 2 suspension arms.
14. Temporarily tighten the rear strut rod.
15. Stabilize the suspension.
16. Fully tighten the left and right rear No. 1 suspension arms.
17. Fully tighten the left and right rear No. 2 suspension arms.
18. Fully tighten the rear strut rod.
19. Install the rear stabilizer bar.
20. Install the rear height control sensor sub-assembly.
21. Install the exhaust pipe assembly center.
22. Install the rear wheel.
23. Inspect and adjust the rear wheel alignment.
24. Adjust the headlight aiming.

MACPHERSON STRUTS

REMOVAL & INSTALLATION

See Figures 170 through 172.

1. Remove the rear seat cushion assembly.

2. Remove rear seat headrest plate cover.
3. Remove rear seat headrest assembly.
4. Remove the rear seatback assembly.
5. Remove the rear wheel.
6. Separate rear stabilizer link assembly, as follows:
   a. Support the rear axle carrier with a jack.

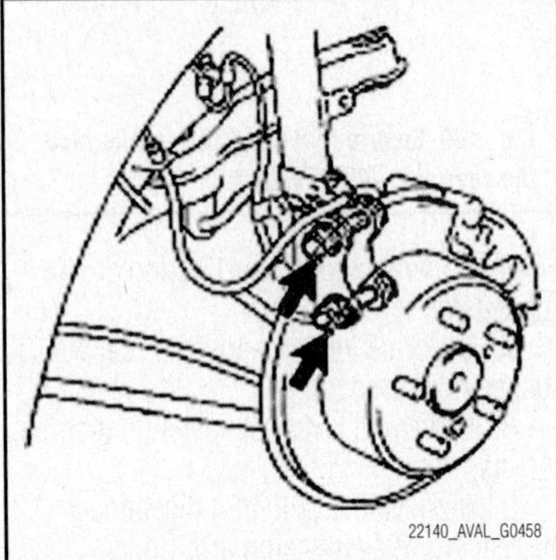

Fig. 170 Loosen the 2 nuts on the lower side of the shock absorber

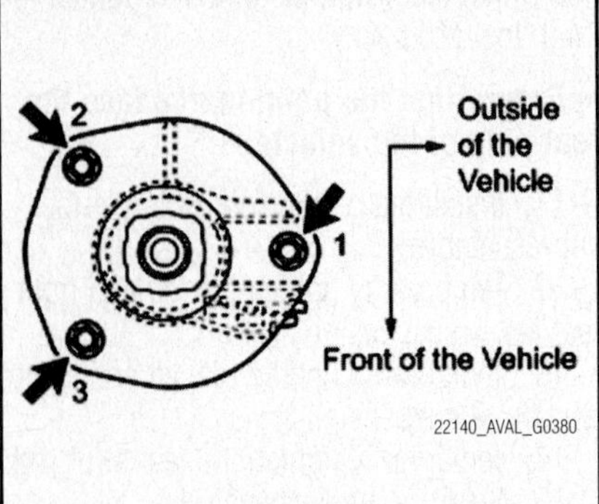

Fig. 171 Tighten the 3 nuts in order from 1 to 3

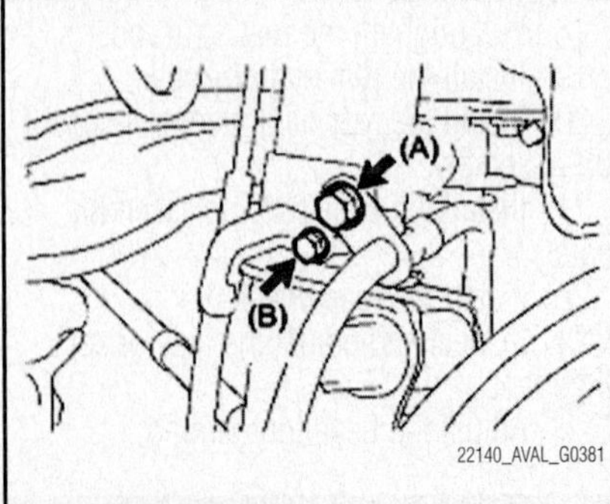

Fig. 172 Flexible hose and skid control sensor wire bolt tightening sequence

   b. Remove the nut, and disconnect the stabilizer link from the shock absorber.

➡If the ball joint turns together with the nut, use a hexagon wrench (5 mm) to hold the stud.

7. Remove the 2 bolts, and disconnect the flexible hose and skid control sensor wire from the shock absorber.
8. Remove the rear shock absorber with coil spring, as follows:
   a. Loosen the 2 nuts on the lower side of the shock absorber. Do not remove the 2 bolts and 2 nuts.
   b. Remove the No. 1 rear suspension support cover.
   c. Loosen the support suspension center nut. Do not remove the nut. It is not necessary to loosen the nut if the shock absorber is not being disassembled.
   d. Remove the 3 nuts.
   e. Lower the rear axle carrier, and remove the 2 nuts and 2 bolts on the lower side of the shock absorber.
   f. Remove the shock absorber with coil spring.

To install:

9. Temporarily install the 3 nuts to the upper side of the rear shock absorber with coil spring.
10. Fully tighten the 3 nuts in order from 1 to 3 to 29 ft. lbs. (39 Nm).
11. Install the 2 bolts and 2 nuts to the shock absorber with coil spring tighten to 133 ft. lbs. (180 Nm).
12. Fully tighten the nut installed on the top of the shock absorber with coil spring and tighten to 41 ft. lbs. (55 Nm). If the shock absorber has not been disassembled, it is not necessary to torque the nut.
13. Install the No. 1 rear suspension support cover.
14. Install the flexible hose and skid control sensor wire with the 2 bolts and tighten to 14 ft. lbs. (19 Nm), 49 inch lbs. (5.5 Nm).
15. Install the stabilizer link to the shock absorber with the nut and tighten to 29 ft. lbs. (39 Nm).

➡If the ball joint turns together with the nut, use a hexagon wrench (5 mm) to hold the stud.

16. Install the front wheel.
17. Inspect and adjust the rear wheel alignment.

WHEEL HUB AND BEARING

REMOVAL & INSTALLATION

See Figure 173.

1. Remove the rear wheel.
2. Separate the rear disc brake caliper assembly, as follows:
   a. Remove the bolt and separate the flexible hose from the shock absorber.
   b. Remove the 2 bolts and separate the rear disc brake caliper assembly.
3. Remove the rear disc.
4. Disconnect the skid control sensor connector.
5. Remove the 4 bolts and the rear axle hub and bearing assembly.

To install:

6. Install the hub and bearing assembly with the 4 bolts and tighten to 59 ft. lbs. (80 Nm).
7. Connect the skid control sensor connector. Do not twist the sensor wire.
8. Inspect rear axle hub bearing looseness.
9. Inspect rear axle hub runout.
10. Install the rear disc.
11. Install the rear disc brake caliper assembly, as follows:
    a. Install the rear disc brake caliper with the 2 bolts and tighten to 46 ft. lbs. (62 Nm).
    b. Install the flexible hose with the bolt and tighten to 14 ft. lbs. (19 Nm).
12. Install the rear wheel.
13. Inspect and adjust the rear wheel alignment.
14. Check ABS speed sensor signal.

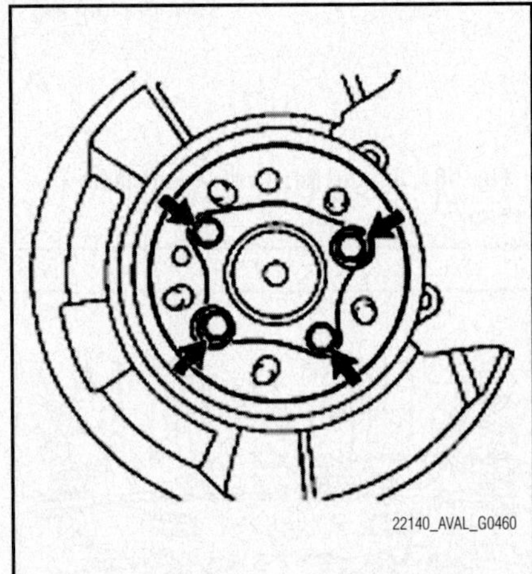

Fig. 173 Remove the 4 bolts and the rear axle hub and bearing assembly

TOYOTA

Camry

11

SPECIFICATIONS AND MAINTENANCE CHARTS

ENGINE AND VEHICLE IDENTIFICATION

Engine							Model Year	
Code ①	Liters (cc)	Cu. In.	Cyl.	Fuel Sys.	Engine Type	Eng. Mfg.	Code ②	Year
2AZ-FE	2.4 (2362)	144	4	SFI	DOHC	Toyota	7	2007
2GR-FE	3.5 (3456)	210	6	SFI	DOHC	Toyota	8	2008

SFI: Sequential Fuel Injection

DOHC: Double Overhead Camshaft

NA: Information not available

① Stamped on the left side of the engine block

② 10th digit of the Vehicle Identification Number (VIN)

22140_CAMR_C0001

GENERAL ENGINE SPECIFICATIONS

All measurements are given in inches.

Year	Model	Engine Displacement Liters	Engine Series VIN	Net Horsepower @ rpm	Net Torque @ rpm (ft. lbs.)	Bore x Stroke (in.)	Com-pression Ratio	Oil Pressure @ rpm
2007	Camry	2.4	2AZ-FE	155@6000	158@4000	3.48x3.78	9.8:1	55@3000
		3.5	2GR-FE	268@6200	248@4700	3.70x3.27	10.8:1	36-78@3000
2008	Camry	2.4	2AZ-FE	155@6000	185@4000	3.48x3.78	9.8:1	55@3000
		3.5	2GR-FE	268@6200	248@4700	3.70x3.27	10.8:1	36-78@3000

22140_CAMR_C0002

GASOLINE ENGINE TUNE-UP SPECIFICATIONS

Year	Engine Displacement Liters	Engine VIN	Spark Plug Gap (in.)	Ignition Timing (deg.)	Fuel Pump (psi)	Idle Speed (rpm)	Valve Clearance (in.)	
							Intake	Exhaust
2007	2.4	2AZ-FE	0.039-0.043	NA	44-50	①	0.0075-0.0114	0.0150-0.0189
	3.5	2GR-FE	0.039-0.043	NA	44-50	650-750	N/A	N/A
2008	2.4	2AZ-FE	0.039-0.043	NA	44-50	①	0.0075-0.0114	0.0150-0.0189
	3.5	2GR-FE	0.039-0.043	NA	44-50	650-750	N/A	N/A

NOTE: The Vehicle Emission Control Information label often reflects specification changes made during production.

The label figures must be used if they differ from those in this chart.

NA: Not available

① Manual transmission: 650 to 750 rpm, Automatic transmission: 610 to 710 rpm

22140_CAMR_C0003

CAPACITIES

Year	Model	Engine Displacement Liters	Engine VIN	Engine Oil with Filter (qts.)	Transmission (pts.) 5-Spd	Transmission (pts.) Auto.	Transfer Case (pts.)	Drive Axle Front (pts.)	Drive Axle Rear (pts.)	Fuel Tank (gal.)	Cooling System (qts.)
2007	Camry	2.4	2AZ-FE	4.5	—	3.7	—	—	—	18.5	6.6
		3.5	2GR-FE	6.4	—	6.8	—	—	—	18.5	8.8
2008	Camry	2.4	2AZ-FE	4.5	—	3.7	—	—	—	18.5	6.6
		3.5	2GR-FE	6.4	—	6.8	—	—	—	18.5	8.8

NOTE: All capacities are approximate. Add fluid gradually and check to be sure a proper fluid level is obtained.

22140_CAMR_C0004

FLUID SPECIFICATIONS

Year	Model	Engine Displacement Liters	Engine ID/VIN	Engine Oil	Auto. Trans.	Drive Axle	Power Steering Fluid	Brake Master Cylinder
2007	Camry	2.4	2AZ-FE	5W-20	NA	—	ATF Dexron II Or III	DOT 3
		3.5	2GR-FE	5W-30	NA	—	ATF Dexron II Or III	DOT 3
2008	Camry	2.4	2AZ-FE	5W-20	NA	—	ATF Dexron II Or III	DOT 3
		3.5	2GR-FE	5W-30	NA	—	ATF Dexron II Or III	DOT 3

DOT: Department Of Transpotation

NA: Not Available

22140_CAMR_C0005

VALVE SPECIFICATIONS

Year	Engine Displacement Liters	Engine VIN	Seat Angle (deg.)	Face Angle (deg.)	Spring Test Pressure (lbs. @ in.)	Spring Installed Height (in.)	Stem-to-Guide Clearance (in.) Intake	Stem-to-Guide Clearance (in.) Exhaust	Stem Diameter (in.) Intake	Stem Diameter (in.) Exhaust
2007	2.4	2AZ-FE	45	44.5	NA	NA	0.0010-0.0031	0.0012-0.0039	0.2154-0.2159	0.2151-0.2157
	3.5	2GR-FE	45	44.5	NA	NA	0.0010-0.0024	0.0012-0.0026	0.2154-0.2159	0.2151-0.2157
2008	2.4	2AZ-FE	45	44.5	NA	NA	0.0010-0.0031	0.0012-0.0039	0.2154-0.2159	0.2151-0.2157
	3.5	2GR-FE	45	44.5	NA	NA	0.0010-0.0024	0.0012-0.0026	0.2154-0.2159	0.2151-0.2157

NA: Not Available

22140_CAMR_C0006

CAMSHAFT AND BEARING SPECIFICATIONS CHART

All measurements are given in inches.

Year	Engine Displ. Liters	Engine ID/VIN	Journal Dia.	Brg. Oil Clearance	Shaft End-play	Runout	Journal Bore	Lobe Height Intake	Lobe Height Exhaust
2007	2.4	2AZ-FE	①	NA	NA	NA	NA	1.8624-1.8664	1.8104-1.1843
	3.5	2GR-FE	②	③	NA	0.0016	NA	1.7447-1.7487	1.7426-1.7465
2008	2.4	2AZ-FE	①	NA	NA	NA	NA	1.8624-1.8664	1.8104-1.1843
	3.5	2GR-FE	②	③	NA	0.0016	NA	1.7447-1.7487	1.7426-1.7465

NA: Not Available

① Mark 1, 2 and 3: 1.4162-1.4167

② No. 1 journal: 1.4152-1.4157
Other Journals: 1.0220-1.0226 in.

③ No. 1 journal: 0.0016-0.0031
Other Journals: 1.0010-1.0024 in.
Maximum No.1 Journal: 0.0039 in.
Other Maximum Journals: 0.0035 in.

22140_CAMR_C0007

CRANKSHAFT AND CONNECTING ROD SPECIFICATIONS

All measurements are given in inches.

Year	Engine Displacement Liters	Engine VIN	Crankshaft Main Brg. Journal Dia.	Crankshaft Main Brg. Oil Clearance	Crankshaft Shaft End-play	Crankshaft Thrust on No.	Connecting Rod Journal Diameter	Connecting Rod Oil Clearance	Connecting Rod Side Clearance
2007	2.4	2AZ-FE	2.0863-2.0866	0.0007-0.0016	0.0016-0.0095	3	1.8894-1.8898	0.0013-0.0025	0.0063-0.0143
	3.5	2GR-FE	2.4011-2.4016	0.0010-0.0019	0.0016-0.0095	2	2.0863-2.0866	0.0018-0.0026	0.0059-0.0157
2008	2.4	2AZ-FE	2.0863-2.0866	0.0007-0.0016	0.0016-0.0095	3	1.8894-1.8898	0.0013-0.0025	0.0063-0.0143
	3.5	2GR-FE	2.4011-2.4016	0.0010-0.0019	0.0016-0.0095	2	2.0863-2.0866	0.0018-0.0026	0.0059-0.0157

22140_CAMR_C0008

PISTON AND RING SPECIFICATIONS

All measurements are given in inches.

Year	Engine Displ. Liters	Engine VIN	Piston Clearance	Ring Gap Top Compression	Ring Gap Bottom Compression	Ring Gap Oil Control	Ring Side Clearance Top Compression	Ring Side Clearance Bottom Compression	Ring Side Clearance Oil Control
2007	2.4	2AZ-FE	0.0020-0.0029	0.0094-0.0122	0.0130-0.0169	0.0040-0.0119	0.0008-0.0028	0.0008-0.0024	0.0008-0.0028
	3.5	2GR-FE	0.0018-0.0020	0.0098-0.0138	0.0197-0.0413	0.0039-0.0157	0.0008-0.0028	0.0008-0.0024	0.0028-0.0059
2008	2.4	2AZ-FE	0.0020-0.0029	0.0094-0.0122	0.0130-0.0169	0.0040-0.0119	0.0008-0.0028	0.0021-0.0037	0.0023-0.0085
	3.5	2GR-FE	0.0018-0.0020	0.0098-0.0138	0.0197-0.0413	0.0039-0.0157	0.0008-0.0028	0.0008-0.0024	0.0028-0.0059

22140_CAMR_C0009

TORQUE SPECIFICATIONS

All readings in ft. lbs.

Year	Engine Displacement Liters	Engine VIN	Cylinder Head Bolts	Main Bearing Bolts	Rod Bearing Bolts	Crankshaft Damper Bolts	Flywheel Bolts	Manifold Intake	Manifold Exhaust	Spark Plugs	Oil Pan Drain Plug
2007	2.4	2AZ-FE	①	②	③	125	④	22	27	13	18
	3.5	2GR-FE	⑤	⑥	⑦	184	61	15	15	13	30
2008	2.4	2AZ-FE	①	②	③	125	④	22	27	13	18
	3.5	2GR-FE	⑤	⑥	⑦	184	61	15	15	13	30

① Step 1: 52 ft. lbs.
Step 2: plus 90 degrees

② Step 1: 15 ft. lbs.
Step 2: 30 ft. lbs.
Step 3: Plus 90 degrees

③ Step 1: Cap bolts to 18 ft. lbs.
Step 2: cap bolts plus 90 degrees

④ Auto driveplate: 72 ft. lbs.
Manual Flywheel : 96 ft. lbs.

⑤ Step 1: 10mm point cao bolts to 27 ft. lbs.
Step 2: 10mm point cap bolts plus 90 degrees
Step 3: 10mm point cap bolts plus 90 degrees
Step 4: Front bolts to 22 ft. lbs.

⑥ Step 1: 18 ft. lbs.
Step 2: Plus 90 degrees

⑦ Step 1: 16 cap bolts to 45 ft. lbs.
Step 2: 16 cap bolts plus 90 degrees
Step 3: 8 side bolts to 38 ft. lbs.

22140_CAMR_C0010

WHEEL ALIGNMENT

Year	Model		Caster Range (+/-Deg.)	Caster Preferred Setting (Deg.)	Camber Range (+/-Deg.)	Camber Preferred Setting (Deg.)	Toe-in (in.)	Steering Axis Inclination (Deg.)
2007	Camry	Front	0.75	2.65	0.75	-0.67	0+/-0.04	12.25+/-0.75
		Rear	—	—	0.75	-1.15	0.16+/-0.08	—
2008	Camry	Front	0.75	2.65	0.75	-0.67	0+/-0.04	12.25+/-0.75
		Rear	—	—	0.75	-1.15	0.16+/-0.08	—

22140_CAMR_C0011

TIRE, WHEEL AND BALL JOINT SPECIFICATIONS

Year	Model	OEM Tires Standard	OEM Tires Optional	Tire Pressures (psi) Front	Tire Pressures (psi) Rear	Wheel Size	Ball Joint Inspection	Lug Nut Torque (ft. lbs.)
2007	Camry	P215/60R16	P215/55R17	31	31	NA	①	76
2008	Camry	P215/60R16	P215/55R17	31	31	NA	①	76

NA: Not Available

OEM: Original Equipment Manufacturer

OPT: Optional

PSI: Pounds Per Square Inch

STD: Standard

① Replace if any measurable movement is found.

22140_CAMR_C0012

BRAKE SPECIFICATIONS

All measurements in inches unless noted

Year	Model		Brake Disc Original Thickness	Brake Disc Minimum Thickness	Brake Disc Maximum Runout	Brake Drum Diameter Original Inside Diameter	Brake Drum Diameter Max. Wear Limit	Brake Drum Diameter Maximum Machine Diameter	Minimum Lining Thickness	Brake Caliper Bracket Bolts (ft. lbs.)	Brake Caliper Mounting Bolts (ft. lbs.)
2007	Camry	F	1.102	0.983	0.0020	—	—	—	0.039	79	25
		R	0.390	0.334	0.0059	—	—	—	0.039	46	32
2008	Camry	F	1.102	0.983	0.0020	—	—	—	0.039	79	25
		R	0.390	0.334	0.0059	—	—	—	0.039	46	32

22140_CAMR_C0013

SCHEDULED MAINTENANCE INTERVALS

TOYOTA—CAMRY

TO BE SERVICED	TYPE OF SERVICE	VEHICLE MILEAGE INTERVAL (x1000)													
		5	10	15	20	25	30	35	40	45	50	55	60	90	120
Engine oil & filter	R	✓	✓	✓	✓	✓	✓	✓	✓	✓	✓	✓	✓	✓	✓
Automatic transmission fluid	S/I			✓			✓			✓			✓	✓	✓
Ball joints & dust covers	S/I			✓			✓			✓			✓	✓	✓
Bolts & nuts on chassis & body	S/I			✓			✓			✓			✓	✓	✓
Brake linings & drums	S/I	✓	✓	✓	✓	✓	✓	✓	✓	✓	✓	✓	✓	✓	✓
Brake line pipes & hoses	S/I			✓			✓			✓			✓	✓	✓
Brake pads & discs (front & rear)	S/I	✓	✓	✓	✓	✓	✓	✓	✓	✓	✓	✓	✓	✓	✓
Brake fluid	R						✓						✓	✓	✓
Rack and pinion assembly	S/I			✓			✓			✓			✓	✓	✓
Steering linkage & boots	S/I			✓			✓			✓			✓	✓	✓
Air cleaner filter	R						✓						✓	✓	✓
Spark plugs ①	R														✓
Drive belts	S/I												✓	✓	✓
Exhaust pipes & mountings	S/I			✓			✓			✓			✓	✓	✓
Fuel lines & connections	S/I						✓						✓	✓	✓
Engine coolant ②	S/I			✓			✓			✓			✓	✓	
Fuel tank cap gasket	S/I						✓						✓	✓	✓
Rotate tires	S/I			✓			✓			✓			✓		✓
Clean air conditioning filter ③	S/I			✓			✓			✓			✓		✓
Axle shaft bolts	S/I			✓			✓			✓			✓	✓	✓
Brake pad thickness and rotor runout	S/I						✓						✓	✓	✓

R: Replace S/I: Service or Inspect

① Spark plugs are replaced at 120,000 miles

② Replace engine coolant at 100,000 miles and then inspect every 15,000 miles

③ Replace air conditioning filter every 30,000 miles

FREQUENT OPERATION MAINTENANCE (SEVERE SERVICE)

If a vehicle is operated under any of the following conditions it is considered severe service:

- Extremely dusty areas.

- 50% or more of the vehicle operation is in 32°C (90°F) or higher temperatures, or constant temperatures below 0°C (32°F).

- Prolonged idling (vehicle operation in stop and go traffic).

- Frequent short running periods (engine does not warm to normal operating temperatures).

- Police, taxi, delivery usage or trailer towing usage.

Air cleaner filter: service or inspect every 5000 miles

Rear differential & transfer case oil: replace every 15,000 miles.

Ball joints & dust covers: service or inspect every 5000 miles.

Bolts & nuts on chassis & body: service or inspect every 5000 miles.

Axle shaft bolts: service or inspect every 5000 miles.

Steering linkage: service or inspect every 5000 miles.

22140_CAMR_C0014

PRECAUTIONS

Before servicing any vehicle, please be sure to read all of the following precautions, which deal with personal safety, prevention of component damage, and important points to take into consideration when servicing a motor vehicle:

- Never open, service or drain the radiator or cooling system when the engine is hot; serious burns can occur from the steam and hot coolant.
- Observe all applicable safety precautions when working around fuel. Whenever servicing the fuel system, always work in a well-ventilated area. Do not allow fuel spray or vapors to come in contact with a spark, open flame, or excessive heat (a hot drop light, for example). Keep a dry chemical fire extinguisher near the work area. Always keep fuel in a container specifically designed for fuel storage; also, always properly seal fuel containers to avoid the possibility of fire or explosion. Refer to the additional fuel system precautions later in this section.
- Fuel injection systems often remain pressurized, even after the engine has been turned **OFF**. The fuel system pressure must be relieved before disconnecting any fuel lines. Failure to do so may result in fire and/or personal injury.
- Brake fluid often contains polyglycol ethers and polyglycols. Avoid contact with the eyes and wash your hands thoroughly after handling brake fluid. If you do get brake fluid in your eyes, flush your eyes with clean, running water for 15 minutes. If eye irritation persists, or if you have taken brake fluid internally, IMMEDIATELY seek medical assistance.
- The EPA warns that prolonged contact with used engine oil may cause a number of skin disorders, including cancer. You should make every effort to minimize your exposure to used engine oil. Protective gloves should be worn when changing oil. Wash your hands and any other exposed skin areas as soon as possible after exposure to used engine oil. Soap and water, or waterless hand cleaner should be used.
- All new vehicles are now equipped with an air bag system, often referred to as a Supplemental Restraint System (SRS) or Supplemental Inflatable Restraint (SIR) system. The system must be disabled before performing service on or around system components, steering column, instrument panel components, wiring and sensors. Failure to follow safety and disabling procedures could result in accidental air bag deployment, possible personal injury and unnecessary system repairs.
- Always wear safety goggles when working with, or around, the air bag system. When carrying a non-deployed air bag, be sure the bag and trim cover are pointed away from your body. When placing a non-deployed air bag on a work surface, always face the bag and trim cover upward, away from the surface. This will reduce the motion of the module if it is accidentally deployed. Refer to the additional air bag system precautions later in this section.
- Clean, high quality brake fluid from a sealed container is essential to the safe and proper operation of the brake system. You should always buy the correct type of brake fluid for your vehicle. If the brake fluid becomes contaminated, completely flush the system with new fluid. Never reuse any brake fluid. Any brake fluid that is removed from the system should be discarded. Also, do not allow any brake fluid to come in contact with a painted surface; it will damage the paint.
- Never operate the engine without the proper amount and type of engine oil; doing so WILL result in severe engine damage.
- Timing belt maintenance is extremely important. Many models utilize an interference-type, non-freewheeling engine. If the timing belt breaks, the valves in the cylinder head may strike the pistons, causing potentially serious (also time-consuming and expensive) engine damage. Refer to the maintenance interval charts for the recommended replacement interval for the timing belt, and to the timing belt section for belt replacement and inspection.
- Disconnecting the negative battery cable on some vehicles may interfere with the functions of the on-board computer system(s) and may require the computer to undergo a relearning process once the negative battery cable is reconnected.
- When servicing drum brakes, only disassemble and assemble one side at a time, leaving the remaining side intact for reference.
- Only an MVAC-trained, EPA-certified automotive technician should service the air conditioning system or its components.

BRAKES

ANTI-LOCK BRAKE SYSTEM (ABS)

GENERAL INFORMATION

The Anti-lock Brake System (ABS) helps prevent wheels from locking when the brake is applied firmly or when braking on a slippery surface.

The ABS warning light comes on when the ABS system is malfunctioning.

The Electronic Brake force Distribution (EBD) control utilizes ABS, realizing proper brake force distribution between front and rear wheels in accordance with driving conditions. In addition, during cornering braking, it also controls the brake force of the right and left wheels, helping to maintain vehicle behavior.

Both of the ABS and BRAKE warning lights come on to indicate a malfunction in the EBD system.

Based on the signals received from the 4 wheel speed sensors, the skid control ECU calculates the speed of each wheel and deceleration, and checks wheel slipping condition. According to the slipping condition, the skid control ECU controls the pressure holding valve and pressure reduction valve in order to adjust the fluid pressure of each wheel cylinder

The primary purpose of the Brake Assist (BA) system is to provide an auxiliary brake force to assist the driver who cannot generate a large enough brake force during emergency braking. This system helps to maximize the brake performance of the vehicle. The ABS warning light comes on to indicate a malfunction in the BA (brake assist) system.

BLEEDING PROCEDURE

Bleeding the Master Cylinder

If the master cylinder is reinstalled or if the reservoir becomes empty, bleed the air from the master cylinder. To prevent brake fluid from adhering, cover nearly painted surfaces with a shop rag or a piece of cloth.

1. Using a union nut wrench (10 mm), disconnect the 2 brake lines from the master cylinder, using a suitable brake line wrench.
2. Have an assistant slowly depress the brake pedal and hold it.
3. Cover the 2 outer holes with your fingers, and have your assistant release the brake pedal.
4. Repeat the previous 2 steps 3 or 4 times.
5. Using a union nut wrench (10 mm), connect the 2 the brake lines to the master cylinder and tighten to 11 ft. lbs. (15 Nm).

➡Use a torque wrench with a fulcrum length of 250 mm (9.84 in.).

➡This torque value is effective when the union nut wrench is parallel to the torque wrench.

Bleeding the Brake Lines

➡Bleed air from the brake line of the wheel farthest from the master cylinder.

1. Raise and safely support the vehicle.
2. Connect a vinyl tube to the bleeder plug.
3. Have an assistant depress the brake pedal several times, then loosen the bleeder plug while the pedal is depressed.
4. When fluid stops coming out, tighten the bleeder plug, then release the brake pedal.
5. Repeat steps 3 and 4 until all the air in the fluid has been bled out.
6. Tighten the brake bleeder plug to 73 inch lbs. (8.3 Nm).
7. Repeat the above steps to bleed the air out of the brake line for each wheel.

Bleeding the Actuator Assembly

➡After bleeding the air from the brake system, if the height or feel of the brake pedal cannot be obtained, perform air bleeding in the brake actuator assembly with a hand-held tester by following the procedures below.

1. Depress the brake pedal more than 20 times with the engine off.
2. Connect the hand-held tester to the DLC3, then turn the ignition switch to the **ON** position, but do NOT start the engine.
3. Select "AIR BLEEDING" on the hand-held tester.

➡Refer to the hand-held tester operator's manual for more details.

4. Bleed the air out of the regular brake line when "Step 1: Increase" appears on the hand-held tester display, as follows:

➡Bleed the air by following the steps displayed on the hand-held tester. Make sure that the brake fluid in the master cylinder reservoir tank does not become empty.

a. Connect the vinyl tube to either one of the bleeder plugs.
b. Have an assistant depress the brake pedal several times, then loosen the bleeder plug connected to the vinyl tube with the pedal depressed.
c. When fluid stops coming out, tighten the bleeder plug and release the brake pedal.
d. Repeat the previous 2 steps until all the air in the fluid is completely bled out.
e. Tighten the bleeder plug completely to 73 inch lbs. (8.3 Nm).
f. Repeat the above procedures for each wheel to bleed the air out of the brake line.

5. Bleed the air out of the suction line when "Step 2: Inhalation" appears on the hand-held tester display, as follows:

➡Bleed the air by following the steps displayed on the hand-held tester. Make sure that the brake fluid in the master cylinder reservoir tank does not become empty.

a. Connect the vinyl tube to the bleeder plug at the right front wheel or the right rear wheel and loosen the bleeder plug.
b. Operate the brake actuator assembly to perform air bleeding from the suction line using the hand-held tester.

➡This operation stops automatically after 4 seconds. At this time, be sure to release the brake pedal.

c. Check if the operation has stopped by referring to the hand-held tester display and tighten the bleeder plug.
d. Repeat the previous 2 steps until all air in the fluid is completely bled out.
e. Tighten the bleeder plug completely to 73 inch lbs. (8.3 Nm).
f. Repeat the above procedures to bleed the air out of the brake line for each wheel.

6. Bleed the air out of the pressure reduction line when "Step 3: Decrease" appears on the hand-held tester display, as follows:

➡Bleed the air by following the steps displayed on the hand-held tester. Make sure that the brake fluid in the master cylinder reservoir tank does not become empty.

a. Connect a vinyl tube to either one of the bleeder plugs.
b. Loosen the bleeder plug.
c. Using the hand-held tester, operate the brake actuator assembly, completely depress the brake pedal and keep it depressed.

➡The operation stops automatically after 4 seconds. When performing this procedure continuously, set an interval of at least 20 seconds. When the operation is complete, the brake pedal goes down slightly. This is a normal phenomenon caused when the solenoid opens. During this procedure, the pedal will feel heavy, but completely depress it so that the brake fluid comes out from the bleeder plug. Be sure to keep depressing the brake pedal. Do not depress and release the pedal repeatedly.

d. Tighten the bleeder plug, then release the brake pedal.
e. Repeat the previous 2 steps until all the air in the fluid is completely bled out.
f. Tighten the bleeder plug completely to 73 inch lbs. (8.3 Nm).
g. Repeat the above procedures for each wheel to bleed the air out of the brake line.

7. Bleed the air out of the brake line again when "Step 4: Increase" appears on the hand-held tester display, as follows:

➡Bleed the air by following the steps displayed on the hand-held tester. Make sure that the brake fluid in the master cylinder reservoir tank does not become empty.

a. Connect the vinyl tube to either one of the bleeder plugs.
b. Depress the brake pedal several times, then loosen the bleeder plug con-

nected to the vinyl tube with the pedal depressed.

8. When fluid stops coming out, tighten the bleeder plug, then release the brake pedal.

a. Repeat the previous 2 steps until all the air in the fluid is completely bled out.

b. Tighten the bleeder plug completely to 73 inch lbs. (8.3 Nm).

c. Repeat the above procedures for each wheel to bleed the air out of the brake line.

d. Finish "AIR BLEEDING" on the hand-held tester and turn the hand-held tester off.

e. Disconnect the hand-held tester from the DLC3 from the DLC3.

f. Turn the ignition switch off.

g. Inspect for fluid leak.

9. Check the fluid level and add fluid if necessary. Use SAE J1703 or FMVSS No. 116 DOT3 Brake fluid.

BRAKES

FRONT DISC BRAKES

BRAKE CALIPER

REMOVAL & INSTALLATION

See Figure 1.

1. Before servicing the vehicle, refer to the Precautions Section.
2. Remove the front wheel.

➡Do not let brake fluid sit on painted surfaces, as it will eat through the paint. Wash it off immediately.

3. Drain brake fluid.
4. Remove the union bolt and gasket from the disc brake cylinder assembly, then disconnect the flexible hose.

➡Remove the disc brake cylinder assembly while holding both of the brake pads or the anti-squeal springs may fall off the brake pads.

5. Hold the front disc brake cylinder slide pin and remove the 2 bolts and disc brake cylinder assembly.

To install:

6. Install the disc brake cylinder assembly with the 2 bolts and tighten to 25 ft. lbs. (34 Nm).
7. Check the installation of the anti-squeal springs. Visually check for any clearance between the brake pad and front disc brake pad support plates.

➡ If the anti-squeal springs are installed correctly, there will be no clearance between the brake pad and the front disc brake pad support plates. If there is a clearance, the anti-squeal springs may not be installed properly.

➡ Check all 4 contact surfaces between the brake pad and the front disc brake pad support plates.

8. Connect the flexible hose with the union bolt and a new gasket and tighten to 21 ft. lbs. (29 Nm).

➡Install the front brake flexible hose lock securely in the lock hole in the disc brake cylinder.

9. Fill reservoir with brake fluid.
10. Bleed master cylinder.
11. Bleed brake line.
12. Bleed brake actuator assembly.
13. Inspect for brake fluid leak.
14. Inspect brake fluid level in reservoir.
15. Install the front wheel and tighten the lug nuts to 76 ft. lbs. (103 Nm).

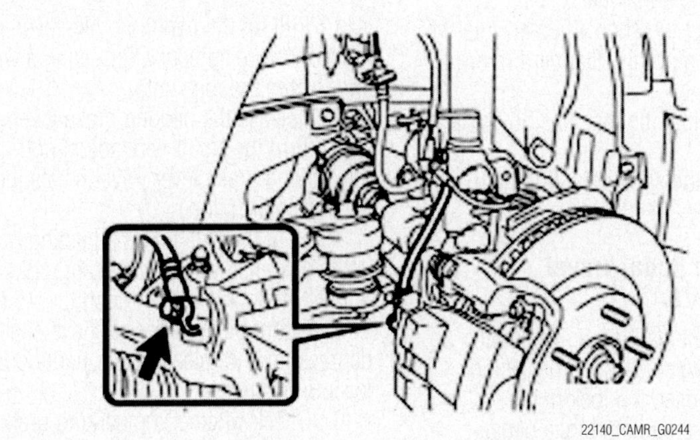

Fig. 1 Removing the union bolt and gasket

DISC BRAKE PADS

REMOVAL AND INSTALLATION

1. Before servicing the vehicle, refer to the Precautions Section.
2. Remove the 2 front disc brake cylinder slide pins (upper and lower) from the front disc brake cylinder mounting.
3. Remove brake cylinder.
4. Remove the 2 anti-squeal springs.
5. Remove the 2 brake pads from the front disc brake cylinder mounting.

To install:

6. Install the 2 brake pads with front anti-squeal shims to the front disc brake cylinder mounting.

➡When replacing worn pads, the front anti-squeal springs must be replaced at the same time.

➡Be sure to install the anti-squeal springs into the front disc brake pad installation holes as far as they will go.

7. Install the 2 front disc brake cylinder slide pins (upper and lower) from the front disc brake cylinder mounting.
8. Install the brake cylinder.

BRAKES — REAR DISC BRAKES

BRAKE CALIPER

REMOVAL & INSTALLATION

1. Before servicing the vehicle, refer to the Precautions Section.
2. Remove the rear wheel.
3. Drain brake fluid.

**** WARNING**

Do not let brake fluid sit on painted surfaces, as it will eat through the paint. Wash it off immediately.

4. Remove the union bolt and the gasket from the rear disc brake cylinder assembly, then disconnect the rear brake flexible hose.
5. Hold the 2 rear disc brake cylinder slide pins and remove the 2 bolts and rear disc brake cylinder assembly.

To install:

6. Install the rear disc brake cylinder assembly with the 2 bolts and tighten to 20 ft. lbs. (77 Nm).
7. Connect the rear brake flexible hose with the union bolt and a new gasket and tighten to 24 ft. lbs. (33 Nm).
8. Fill reservoir with brake fluid.
9. Bleed brake line.
10. Inspect for brake fluid leak.
11. Inspect brake fluid level in reservoir.
12. Install the rear wheel and tighten the lug nuts to 76 ft. lbs. (103 Nm).

DISC BRAKE PADS

REMOVAL & INSTALLATION

1. Before servicing the vehicle, refer to the Precautions Section.
2. Remove the 2 rear disc brake cylinder slide pins (upper and lower) from the rear disc brake cylinder mounting.
3. Remove brake cylinder.
4. Remove the 2 brake pads with the rear anti-squeal shims.

To install:

5. Installation is the reverse of removal procedure.

BRAKES — PARKING BRAKE

PARKING BRAKE CABLES

ADJUSTMENT

1. Inspect the parking brake pedal travel, as follows:
   a. Fully depress the parking brake pedal and release it to engage the parking brake.
   b. Depress the pedal to the floor again, and release it to disengage the parking brake.
   c. Slowly depress the parking brake pedal to the floor, and count the number of clicks. Parking brake pedal travel: 9 to 11 notches at 67.5 lbs. (300 N).
2. Adjust the parking brake pedal travel, as follows:
   a. Depress the parking brake pedal. Hold the wire adjusting nut No. 1 using a wrench and loosen the lock nut.
   b. Release the parking brake pedal.
   c. Turn the No. 1 wire adjusting nut until the parking brake pedal travel meets the above specification.
   d. Hold the No. 1 wire adjusting nut using a wrench or equivalent tool and tighten the lock nut to 48 inch lbs. (5.4 Nm).
   e. Count the number of clicks after depressing and releasing the parking brake pedal 3 or 4 times. Parking brake pedal travel: 9 to 11 notches at 67.5 lbs. (300 N).
   f. Check whether the parking brake drags or not.
   g. When operating the parking brake pedal, check that the parking brake indicator light comes on.

PARKING BRAKE SHOES

ADJUSTMENT

1. Adjust parking brake shoe clearance, as follows:
   a. Temporarily install the hub nuts.
   b. Remove the shoe adjusting hole plug, turn the adjuster and expand the shoes until the disc locks.
   c. Contract the shoe adjuster until the disc rotates smoothly. Standard: returns 8 notches
   d. Check that the disc has no brake drag.
   e. Install the shoe adjusting hole plug.

Parking Brake Pedal Travel Adjustment (A/T)

1. Depress the parking brake pedal. Hold the No. 1 wire adjusting nut using a wrench and loosen the lock nut.
2. Release the parking brake pedal.
3. Turn the No. 1 wire adjusting nut until the parking brake pedal travel meets the above specification.
4. Hold the No. 1 wire adjusting nut using a wrench or an equivalent tool and tighten the lock nut to 48 inch lbs. (5.4 Nm).
5. Count the number of clicks after depressing and releasing the parking brake pedal 3 or 4 times.
6. Check whether the parking brake drags.
7. When operating the parking brake pedal, check that the parking brake indicator light comes on.

Parking Brake Lever Travel Adjustment (M/T)

1. Pull up the parking brake lever. Hold the No. 1 wire adjusting nut using a wrench and loosen the lock nut.
2. Release the parking brake lever.
3. Turn the No. 1 wire adjusting nut until the parking brake lever travel meets the above specification.
4. Hold the No. 1 wire adjusting nut using a wrench or an equivalent tool and tighten the lock nut to 44 inch lbs. (5 Nm).
5. Count the number of clicks after depressing and releasing the parking brake lever 3 or 4 times.
6. Check whether the parking brake drags.
7. When operating the parking brake lever, check that the parking brake indicator light comes on.

GENERAL INFORMATION

SERVICE PRECAUTIONS

The vehicle is equipped with a Supplemental Restraint System (SRS). It consists of a driver airbag, front passenger airbag, driver side knee airbag, front seat side airbag and curtain shield airbag. Failure to carry out service operations in the correct sequence could cause the SRS to unexpectedly deploy during servicing, possibly leading to a serious accident. Further, if a mistake is made in servicing the SRS, it is possible that the SRS may fail to operate when required. Before performing servicing (including removal or installation of parts, inspection or replacement), be sure to read the following items carefully, then follow the correct procedures indicated in the repair manual.

Disconnect and isolate the battery negative cable before beginning any airbag system component diagnosis, testing, removal, or installation procedures. Wait at least 90 seconds after the ignition switch is turned off and the negative (-) terminal cable is disconnected from the battery before starting the operation. The SRS is equipped with a backup power source, so if work is started within 90 seconds after disconnecting the negative (-) terminal cable from the battery, the SRS may be deployed. Failure to disable the airbag system may result in accidental airbag deployment, personal injury, or death.

Do not expose the steering pad, front passenger airbag assembly, driver side knee airbag assembly, center airbag sensor assembly, front airbag sensor, front seat side airbag assembly, side airbag sensor, curtain shield airbag assembly, rear airbag sensor, seat position airbag sensor or occupant classification ECU directly to hot air or flames.

Malfunction symptoms of the SRS are difficult to confirm, so DTCs are the most important source of information when troubleshooting. When troubleshooting the SRS, always inspect DTCs before disconnecting the battery.

Even in the case of a minor collision when the SRS does not deploy, the steering pad, front passenger airbag assembly, driver side knee airbag assembly, center airbag sensor assembly, front airbag sensor, front seat side airbag assembly, side airbag sensor, curtain shield airbag assembly, rear airbag sensor, seat position airbag sensor and occupant classification ECU should be inspected.

Before repair work, remove the airbag sensor if any kind of shock is likely to occur to the airbag sensor during the operation.

Never disassemble or repair any of the following parts in order to reuse them. If any of these parts have been dropped, or a defect is found (e.g. cracks, dents or any other defects) in any of the housings, brackets or connectors, then replace the part with a new one: Steering Pad, Front Passenger Airbag Assembly, Driver Side Knee Airbag Assembly, Front Seat Side Airbag Assembly, Curtain Shield Airbag Assembly, Center Airbag Sensor Assembly, Front Airbag Sensor, Front Seat Inner Belt Assembly, Seat Position Sensor, Occupant Classification ECU, Side Airbag Sensor, Rear Airbag Sensor, Front Seat Outer Belt Assembly.

Use an volt/ohmmeter with high impedance (10 kOhms/V minimum) for troubleshooting the electrical circuits.

Information labels are attached near the SRS components. Follow the instructions in the caution.

Do not place an intact undeployed airbag face down on a solid surface. The airbag will propel into the air if accidentally deployed and may result in personal injury or death.

When carrying or handling an undeployed airbag, the trim side (face) of the airbag should be pointing towards the body to minimize possibility of injury if accidental deployment occurs. Failure to do this may result in personal injury or death.

Replace airbag system components with OEM replacement parts. Substitute parts may appear interchangeable, but internal differences may result in inferior occupant protection. Failure to do so may result in occupant personal injury or death.

Wear safety glasses, rubber gloves, and long sleeved clothing when cleaning powder residue from vehicle after an airbag deployment. Powder residue emitted from a deployed airbag can cause skin irritation. Flush affected area with cool water if irritation is experienced. If nasal or throat irritation is experienced, exit the vehicle for fresh air until the irritation ceases. If irritation continues, see a physician.

Do not use a replacement airbag that is not in the original packaging. This may result in improper deployment, personal injury, or death.

The factory installed fasteners, screws and bolts used to fasten airbag components have a special coating and are specifically designed for the airbag system. Do not use substitute fasteners. Use only original equipment fasteners listed in the parts catalog when fastener replacement is required.

During, and following, any child restraint anchor service, due to impact event or vehicle repair, carefully inspect all mounting hardware, tether straps, and anchors for proper installation, operation, or damage. If a child restraint anchor is found damaged in any way, the anchor must be replaced. Failure to do this may result in personal injury or death.

Deployed and non-deployed airbags may or may not have live pyrotechnic material within the airbag inflator.

Do not dispose of driver/passenger/curtain airbags or seat belt tensioners unless you are sure of complete deployment. Refer to the Hazardous Substance Control System for proper disposal.

Dispose of deployed airbags and tensioners consistent with state, provincial, local, and federal regulations.

After any airbag component testing or service, do not connect the battery negative cable. Personal injury or death may result if the system test is not performed first.

If the vehicle is equipped with the Occupant Classification System (OCS), do not connect the battery negative cable before performing the OCS Verification Test using the scan tool and the appropriate diagnostic information. Personal injury or death may result if the system test is not performed properly.

Never replace both the Occupant Restraint Controller (ORC) and the Occupant Classification Module (OCM) at the same time. If both require replacement, replace one, then perform the Airbag System test before replacing the other.

Both the ORC and the OCM store Occupant Classification System (OCS) calibration data, which they transfer to one another when one of them is replaced. If both are replaced at the same time, an irreversible fault will be set in both modules and the OCS may malfunction and cause personal injury or death.

If equipped with OCS, the Seat Weight Sensor is a sensitive, calibrated unit and must be handled carefully. Do not drop or handle roughly. If dropped or damaged, replace with another sensor. Failure to do so may result in occupant injury or death.

If equipped with OCS, the front passenger seat must be handled carefully as well. When removing the seat, be careful when setting on floor not to drop. If dropped, the sensor may be inoperative, could result in occupant injury, or possibly death.

If equipped with OCS, when the passenger front seat is on the floor, no one should sit in the front passenger seat. This uneven force may damage the sensing ability of the seat weight sensors. If sat on and damaged, the sensor may be inoperative, could result in occupant injury, or possibly death.

DISARMING THE SYSTEM

To avoid personal injury when working on vehicles equipped with an air bag, the negative battery cable must be disconnected and at least 90 seconds must elapse before working on the system. Failure to do so may result in deployment of the air bag.

ARMING THE SYSTEM

To arm the system after service is finished, connect the negative battery cable.

DRIVETRAIN

AUTOMATIC TRANSAXLE ASSEMBLY

REMOVAL & INSTALLATION

U250E Automatic Transaxle

See Figures 2 through 9.

1. Before servicing the vehicle, refer to the Precautions Section.
2. Disconnect the negative battery cable.
3. Place front wheels facing straight ahead.
4. Remove front wheels.
5. Remove both engine under covers.
6. Remove RH front fender apron seal.
7. Drain and recycle the engine coolant.
8. Drain automatic transaxle fluid. Using a 6 mm socket hexagon wrench, remove the drain plug and gasket, and drain ATF. Install a new gasket and drain plug. Tighten the drain plug to 36 ft. lbs. (49 Nm).
9. Remove windshield wiper link assembly.
10. Remove cowl top panel outer sub-assembly.
11. Remove No. 1 engine cover sub-assembly.
12. Remove the drive belt.
13. Remove air cleaner inlet assembly.
14. Remove air cleaner cap sub-assembly.
15. Remove air cleaner case sub-assembly.
16. Remove battery.
17. Remove RH No. 2 engine mounting stay.
18. Remove engine moving control rod sub-assembly.
19. Remove RH No. 2 engine mounting bracket.
20. Disconnect No. 1 vacuum hose connector.
21. Disconnect radiator hose inlet.
22. Disconnect radiator hose outlet.
23. Disconnect oil cooler inlet hose.
24. Disconnect oil cooler outlet hose.
25. Disconnect heater inlet water hose.
26. Disconnect heater outlet water hose.
27. Remove ECM.
28. Disconnect engine wire.
29. Disconnect transmission control cable assembly by performing the following:
    a. Remove the nut from the control shaft lever.
    b. Disconnect the transmission control cable assembly from the control shaft lever.
    c. Remove the clip and disconnect the transmission control cable assembly from the No. 1 control cable bracket.
    d. Disconnect the transmission control cable assembly from the control cable clamp.
30. Disconnect No. 1 oil reservoir to pump hose.
31. Disconnect return tube sub-assembly.
32. Disconnect fuel tube sub-assembly.
33. Remove generator assembly.
34. Separate compressor and magnetic clutch.
35. Remove front exhaust pipe assembly.
36. Remove both front axle hub nut.
37. Remove both front stabilizer link assemblies.
38. Remove both front speed sensors.
39. Disconnect both tie rod assemblies.
40. Disconnect both front suspension lower No. 1 arms.
41. Separate both front axle assemblies
42. Remove drive plate and torque converter clutch setting bolt by performing the following:
    a. Remove the flywheel housing under cover.
    b. Turn the crankshaft to gain access and remove the 6 bolts while holding the crankshaft pulley bolt with a wrench.

➡There will be one green colored bolt.

43. Separate steering sliding yoke.
44. Remove engine assembly with transaxle.
45. Remove vane pump assembly.
46. Remove front frame assembly.
47. Remove both front drive shaft assembly.
48. Remove engine wire.
49. Remove starter assembly.
50. Remove the 3 bolts and engine mounting front bracket.
51. Separate automatic transaxle assembly by performing the following:
    a. Remove the 5 bolts
    b. Remove the 4 lower side mounting bolts.
    c. Separate and remove the automatic transaxle.
52. Remove torque converter clutch assembly.
53. Remove speedometer driven hole cover sub-assembly by performing the following:
    a. Remove the bolt and hole cover from the transaxle case
    b. Remove the O-ring from the hole cover
54. Remove the bolt and No. 2 transmission control cable bracket
55. Disconnect the wire harnesses from the 2 brackets. Remove the 2 bolts and 2 clamps.
56. Remove the bolt and disconnect the wire harness.
57. DISCONNECT transmission electrical CONNECTORS by performing the following:

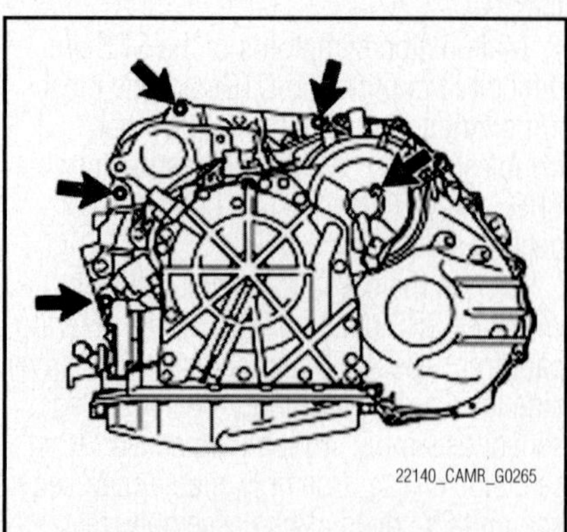

Fig. 2 Removing 5 bolts—U250E A/T

Fig. 3 Removing 4 lower mounting bolts—U250E A/T

a. Disconnect the transmission wire connector.

b. Disconnect the park/neutral position switch connector.

c. Disconnect the 2 speed sensor connectors.

58. Remove no. 1 transmission control cable bracket by performing the following:

a. Remove the bolt and oil cooler tube clamp.

b. Remove the 2 bolts and No. 1 transmission control cable bracket.

59. Remove transmission oil filler tube sub-assembly by performing the following:

a. Remove the bolt and transmission oil filler tube sub-assembly.

b. Remove the O-ring from the oil filler tube sub-assembly.

60. Using SST 09023-12701 and a wrench, disconnect the No. 1 oil cooler inlet tube.

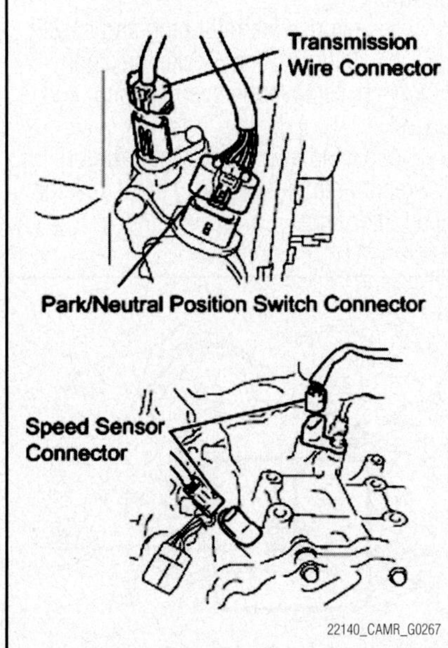

Fig. 4 Disconnecting transmission connectors—U250E A/T

61. Using SST 09023-12701and a wrench, disconnect the No. 1 oil cooler outlet tube.

62. Remove the 3 bolts and transaxle case upper cover.

To install:

63. Install the transmission case upper cover with the 3 bolts and tighten to: 8 ft. lbs. (12 Nm).

64. Install the engine mounting front bracket to the automatic transaxle with the 3 bolts and tighten to 47 ft. lbs. (64 Nm).

65. Coat a new O-ring with ATF, and install it to the oil filler tube. Install the oil filler tube to the automatic transaxle with the bolt and tighten to 49 inch lbs. (6 Nm).

66. Install the ATF level gauge.

67. Install the No. 1 control cable bracket with the 2 bolts and tighten to 9 ft. lbs. (12 Nm).

68. Install No. 1 oil cooler inlet tube by performing the following:

a. Temporarily install the No. 1 oil cooler outlet tube.

b. Temporarily install the No. 1 oil cooler inlet tube.

c. Install the oil cooler tube clamp with the bolt and tighten 48 inch lbs. (5 Nm).

d. Using Special Tool: 09023-12701 and a wrench, tighten the No. 1 oil cooler inlet tube to 25 ft. lbs. (34 Nm).

➡ Use a torque wrench with a fulcrum length of 13.58 inches (345 mm).

69. Using Special Tool: 09023-12701 and a wrench, tighten the No. 1 oil cooler outlet tube to 25 ft. lbs. (34 Nm).

➡ Use a torque wrench with a fulcrum length of 13.58 inches (345 mm).

70. Connect the transmission wire connector. Connect the park/neutral position switch connector. Connect the 2 speed sensor connectors.

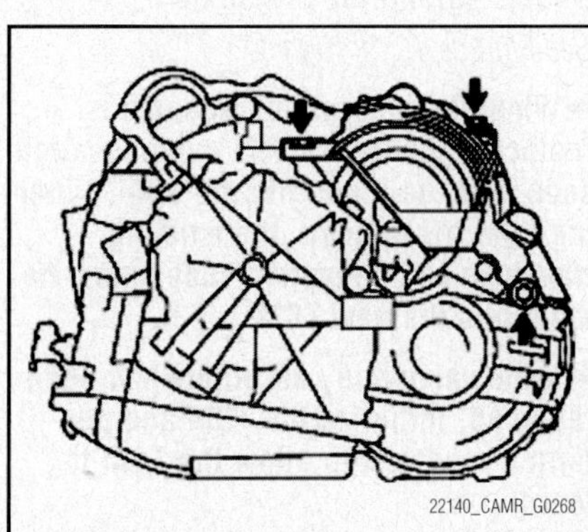

Fig. 5 Removing 3 bolts and transaxle case upper cover —U250E A/T

71. Connect the wire harness with the bolt and tighten to 9 ft. lbs. (12 Nm).

72. Install the wire harness 2 brackets with the 2 bolts and tighten to 74 inch lbs. (8.4 Nm). Connect the wire harnesses to the clamps.

73. Install the No. 2 transmission control cable bracket with the bolt and tighten 9 ft. lbs. (12 Nm).

74. Coat a new O-ring with ATF WS, and install speedometer driven hole cover sub-assembly to the hole cover and tighten to 47 inch lbs. (6 Nm).

75. Install the torque converter clutch assembly, as follows:

a. Using Vernier calipers and a straight edge, measure dimension "A" between the transaxle fitting part of the engine and the converter fitting part of the drive plate.

b. Set the key of the front oil pump drive gear to the top and put a mark on the housing.

c. Put a mark on the torque converter clutch so that its groove can be clearly indicated.

d. Align the 2 marks on the transaxle case and torque converter clutch and fit the splined part of the input shaft to the spline part of the turbine runner.

e. Rotating the torque converter clutch, fit the spline part of the stator shaft to the spline part of the stator. Rotate it about 180°.

⁂ CAUTION

When rotating the torque converter clutch, do not push it with excessive force.

f. Rotating the torque converter clutch, align the 2 marks on the case and the torque converter clutch again and fit the key of the oil pump drive gear into the key way of the torque converter clutch.

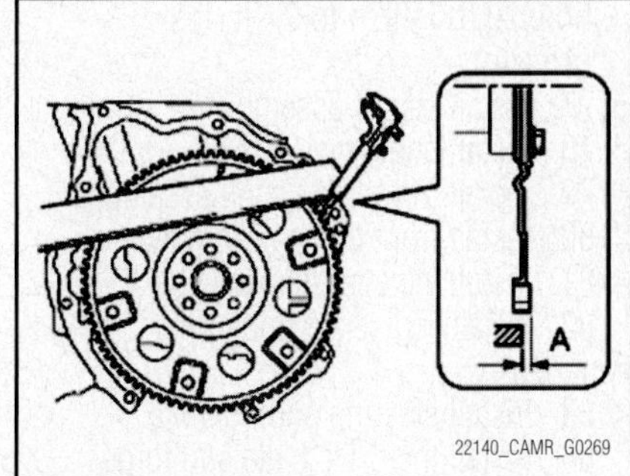

Fig. 6 Measure dimension "A" between the transaxle fitting part of the engine and the converter fitting part of the drive plate

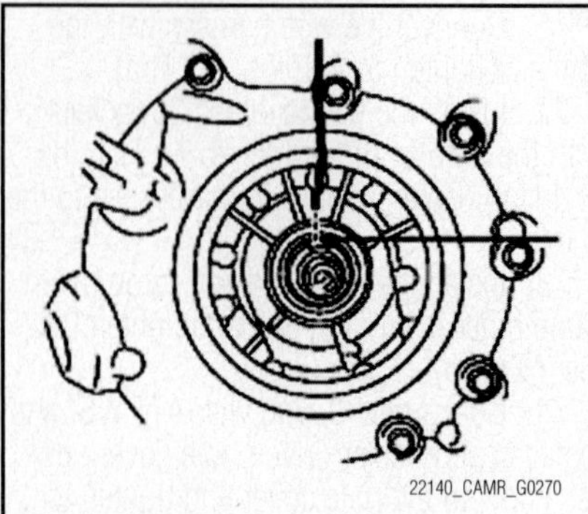

Fig. 7 Setting the key of the front oil pump drive gear

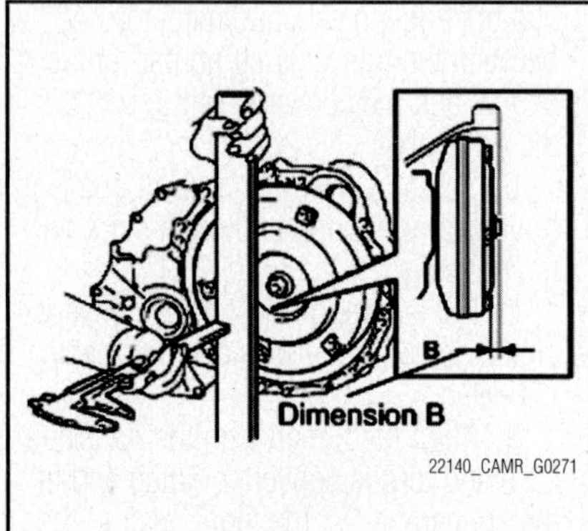

Fig. 8 Measuring dimension "B"

g. Using Vernier calipers and a straight edge, measure dimension "B" shown in the illustration and check that "B" is greater than "A" (measured in previous steps) "A" + 1 (0.04 in.) mm or more. Remember to minus the thickness of the straight edge.

76. Install the automatic transaxle assembly, as follows:

a. Install the automatic transaxle to the engine with the 5 bolts:

- Bolt A: 47 ft. lbs. (64 Nm)
- Bolt B: 34 ft. lbs. (46 Nm)
- Bolt C: 34 ft. lbs. (46 Nm)

b. Install the 4 lower side mounting bolts and tighten to 32 ft. lbs. (44 Nm).

77. Install starter assembly.
78. Install engine wire.
79. Install front drive shaft assembly.
80. Install front frame assembly.
81. Install vane pump assembly.
82. Install engine assembly with transaxle.
83. Install steering sliding yoke.
84. Install the drive plate and torque converter clutch setting bolt, as follows:

a. Apply a few drops of adhesive (Toyota Genuine Adhesive 1324, Three Bond 1324 or equivalent) to 2 threads on the tip of the 6 torque converter clutch mounting bolts.

b. Install the 6 torque converter clutch mounting bolts and tighten to 30 ft. lbs. (41 Nm). First install the black colored bolt, and then the remaining 5 bolts.

c. Install the flywheel housing under cover.

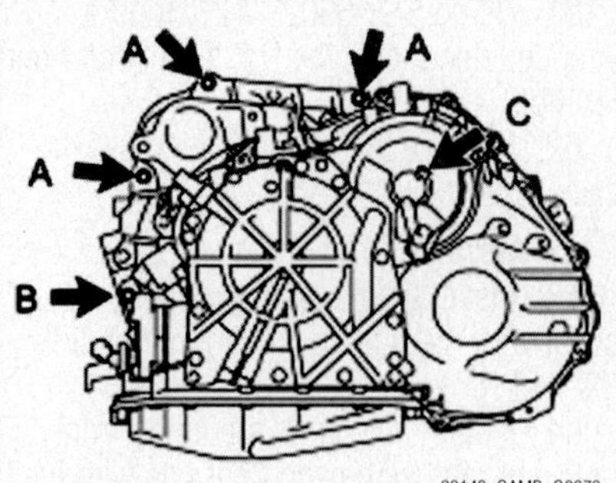

Fig. 9 Automatic transaxle bolt tightening sequence

85. To complete installation, reverse removal procedure.
86. Tighten control shaft lever nut to 11 ft. lbs. (15 Nm).
87. Add engine coolant.
88. Add automatic transaxle fluid (Toyota Genuine ATF WS).
89. Add power steering fluid.
90. Bleed power steering fluid.
91. Inspect automatic transaxle fluid.
92. Check for fuel leaks.
93. Check for engine coolant leaks.
94. Check for exhaust gas leaks.
95. Check for shift lever position.
96. Adjust shift lever position.
97. Inspect and adjust front wheel alignment.
98. Check ignition timing.
99. Check engine idle speed.
100. Inspect CO/HC.
101. Check abs speed sensor signal.

U660E Automatic Transaxle

See Figures 10 through 17.

➡When the automatic transaxle is replaced, the transaxle's compensation code must be input into the TCM. When the TCM is replaced, the existing transaxle compensation codes must be input into the new TCM.

➡If the valve body assembly has been replaced, initialize the TCM and perform a road test to allow the TCM to learn.

➡If the shift solenoid SL1 has been replaced, perform a road test to allow the TCM to learn.

➡If the TCM or transaxle has been replaced, register the transmission compensation code in the TCM.

➡Perform RESET MEMORY (A/T initialization) when replacing the automatic transmission assembly, engine assembly or ECM.

1. Before servicing the vehicle, refer to the Precautions Section.

CAUTION

Observe all applicable safety precautions when working around fuel. Whenever servicing the fuel system, always work in a well ventilated area. Do not allow fuel spray or vapors to come in contact with a spark or open flame. Keep a dry chemical fire extinguisher near the work area. Always keep fuel in a container specifically designed for fuel storage; also, always properly seal fuel containers to avoid the possibility of fire or explosion.

2. Discharge the fuel system pressure.
3. Disconnect cable from negative battery terminal.
4. Place front wheels facing straight ahead.
5. Remove front wheels.
6. Remove the right and left engine under covers.
7. Remove the right front fender apron seal.
8. Drain engine coolant.
9. Drain automatic transaxle fluid, as follows:

a. Remove the refill plug and gasket.

b. Using a 6 mm socket hexagon wrench, remove the overflow plug and gasket.

c. Using a 6 mm socket hexagon wrench, remove the No. 1 transmission oil filler tube.

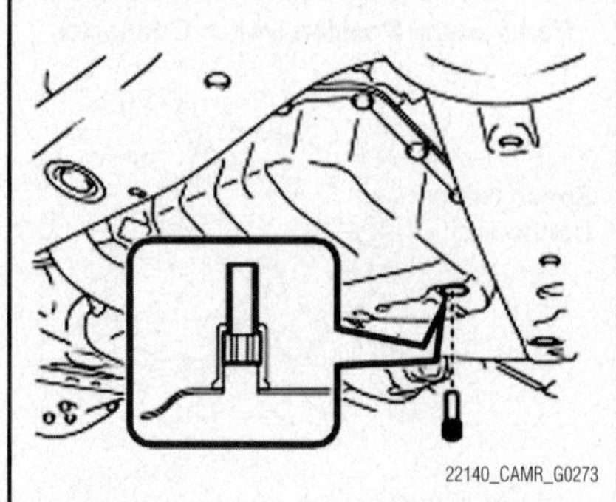

Fig. 10 Remove the No. 1 transmission oil filler tube

d. Drain the automatic transaxle fluid.

e. Using a 6 mm socket hexagon wrench, install the No. 1 transmission oil filler tube and tighten to 15 inch lbs. (1.7 Nm).

f. Using a 6 mm socket hexagon wrench, install the overflow plug with a new gasket and tighten to 30 ft. lbs. (40 Nm).

g. Install the refill plug with a new gasket and tighten to 36 ft. lbs. (49 Nm).

10. Remove windshield wiper link assembly.
11. Remove the cowl top panel outer sub-assembly.
12. Remove the cool air intake duct seal.
13. Remove the V-bank cover sub-assembly.
14. Remove the V-ribbed belt.
15. Remove air cleaner inlet assembly.
16. Remove air cleaner cap sub-assembly.
17. Remove air cleaner case sub-assembly.
18. Remove the No. 1 air cleaner inlet.
19. Remove the battery.
20. Remove the intake air resonator sub-assembly.
21. Remove the right No. 2 engine mounting stay.
22. Remove the engine moving control rod sub-assembly.
23. Disconnect the No. 1 fuel vapor feed hose.
24. Disconnect check valve to brake booster hose.
25. Disconnect the radiator inlet and outlet hoses.
26. Disconnect the No. 1 oil cooler inlet and outlet hoses.
27. Disconnect the heater water inlet hose.
28. Disconnect the heater water outlet hose.
29. Remove the ECM.
30. Remove the upper relay block cover.
31. Disconnect the engine wire.
32. Remove the bolt and disconnect the wire harness.
33. Disconnect the park/neutral position switch connector.
34. Disconnect the transmission control cable assembly, as follows:

a. Remove the nut from the control shaft lever.

b. Disconnect the transmission control cable assembly from the control shaft lever.

c. Remove the clip and disconnect the transmission control cable assembly from the No. 1 transmission control cable bracket.

35. Separate the fuel tube sub-assembly.
36. Disconnect the No. 1 oil reservoir to pump hose.
37. Disconnect return tube sub-assembly.
38. Remove the No. 1 exhaust pipe support bracket.
39. Remove front exhaust pipe assembly.
40. Remove both front axle hub nuts.
41. Disconnect both front stabilizer link assemblies.
42. Disconnect the left and right front speed sensors.
43. Separate the left and right tie rod end sub-assemblies.
44. Separate the left and right front lower No. 1 suspension arm sub-assemblies.
45. Separate the left and right front axle assemblies.
46. Remove the front No. 1 exhaust pipe support bracket.
47. Remove the 2 bolts, exhaust pipe support bracket and flywheel housing under cover.
48. Remove drive plate and torque converter clutch setting bolt, as follows:

a. Turn the crankshaft to gain access and remove the 6 bolts while holding the crankshaft pulley bolt with a wrench.

49. Disconnect steering sliding yoke.
50. Remove generator assembly.
51. Separate cooler compressor assembly.
52. Remove engine assembly with transaxle.
53. Remove vane pump assembly.
54. Install engine hangers.
55. Remove front frame assembly.
56. Remove both front drive shaft assemblies.
57. Remove engine wire.
58. Remove starter assembly.
59. Remove the automatic transaxle assembly, as follows:

a. Remove the bolt and breather bracket from the left camshaft housing sub-assembly.

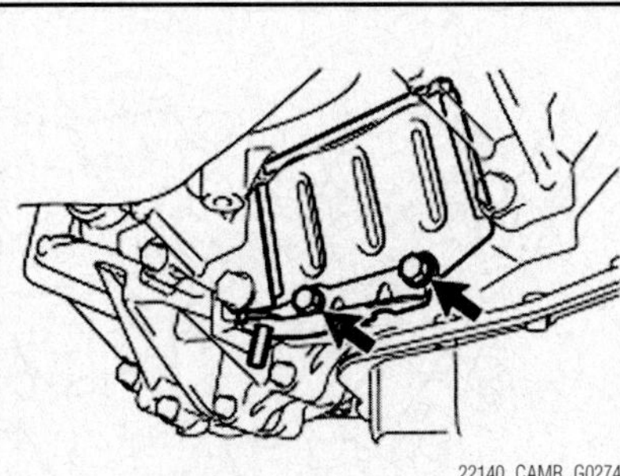

Fig. 11 Removing the 2 bolts, exhaust pipe support bracket and flywheel housing under cover

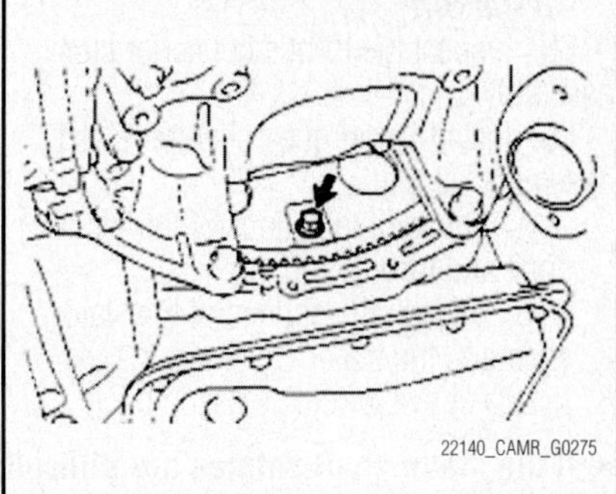

Fig. 12 Removing drive plate and torque converter clutch setting bolt

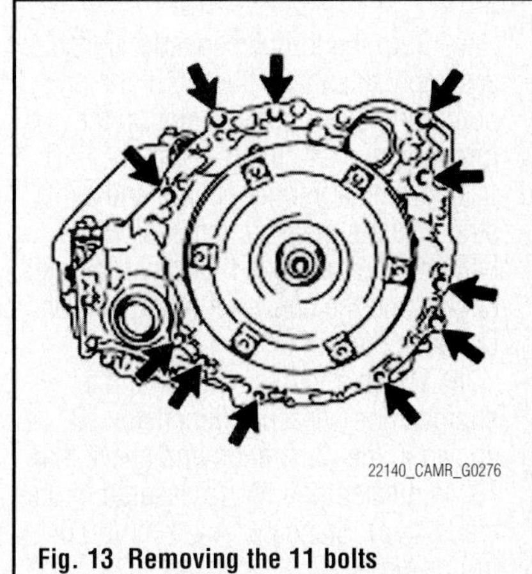

Fig. 13 Removing the 11 bolts

b. Remove the breather hose from the breather bracket.

c. Remove the 11 bolts.

d. Separate and remove the automatic transaxle.

60. Put matchmarks on the transaxle housing and torque converter clutch assembly.
61. Remove the torque converter clutch assembly from the automatic transaxle assembly.
62. Remove the 3 bolts and engine mounting front bracket.
63. Remove the 2 bolts and TCM from the transaxle.
64. Remove the speedometer driven hole cover sub-assembly, as follows:

a. Remove the bolt and hole cover from the transaxle case.

b. Remove the O-ring from the hole cover.

65. Remove the wire harness clamp bracket, as follows:

a. Remove the 4 bolts and 4 clamp brackets.

66. Remove the 2 bolts and No. 1 transmission control cable bracket.
67. Separate the No. 1 oil cooler inlet and outlet hoses.

To install:

68. Inspect the torque converter clutch assembly.

69. Install the torque converter clutch assembly, as follows:

a. Engage the splines of the input shaft and turbine runner.

b. Engage the splines of the stator shaft and the stator while turning the torque converter clutch assembly.

➡If the stator shaft splines are difficult to engage with the stator splines, move the torque converter back approximately 10 mm (0.40 in.) and engage the splines while rotating the torque converter.

c. Turn the torque converter clutch assembly to engage the key of the oil pump drive gear into the slot on the torque converter clutch assembly.

d. Using a Vernier caliper and a straightedge, measure dimension "A" between the transaxle fitting part of the engine and the converter fitting part of the drive plate.

e. Using a Vernier caliper and a straightedge, measure dimension "B" shown in the illustration and check that "B" is greater than "A" (measured in previous step). Standard: A + 1 mm (0.04 in.) or more.

➡Make sure to deduct the thickness of the straightedge.

➡If the transaxle is installed to the engine with the torque converter not sufficiently inserted, the torque converter may be damaged.

➡In the case of where a U660E is used with a 2GR-FE, the standard installation depth is 0.55 inches (14 mm) or more.

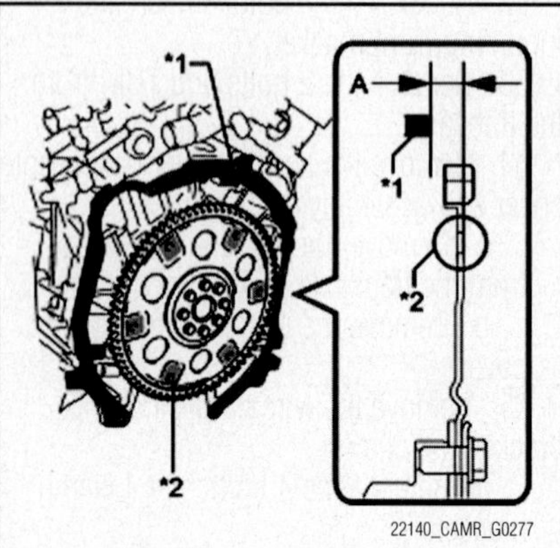

Fig. 14 Measure dimension "A" between the transaxle fitting part of the engine and the converter fitting part of the drive plate

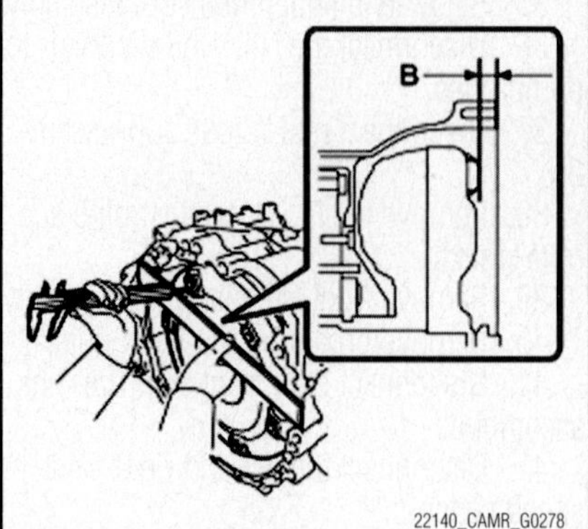

Fig. 15 Measure dimension "B" shown in the illustration and check that "B" is greater than "A"

70. Install the No. 1 oil cooler inlet and outlet hoses.

71. Install the No. 1 transmission control cable bracket with the 2 bolts and tighten to 9 ft. lbs. (12 Nm).

72. Install the 4 clamp brackets and 4 bolts and tighten to 74 inch lbs. (8.4 Nm).

73. Install the speedometer driven hole cover sub-assembly, as follows:

a. Coat a new O-ring with ATF, and install it to the hole cover.

b. Install the hole cover to the transaxle case with the bolt and tighten to 49 inch lbs. (5.5 Nm).

74. Install the TCM, as follows:

a. Install the TCM to the automatic transaxle assembly.

b. Install and tighten the 2 bolts in the order shown in the illustration to 9 ft. lbs. (11 Nm).

75. Install the engine mounting front bracket to the automatic transaxle with the 3 bolts and tighten to 47 ft. lbs. (64 Nm).

76. Install the automatic transaxle assembly, as follows:

a. Install the automatic transaxle to the engine with the 11 bolts and tighten to:

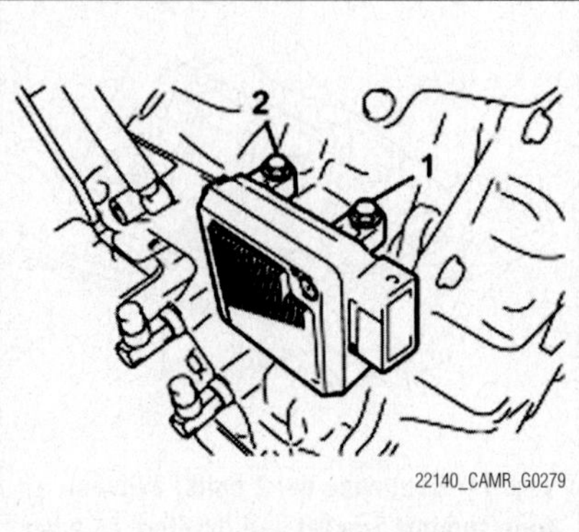

Fig. 16 TCM bolt tightening sequence

- A: 47 ft. lbs. (64 Nm)
- B: 47 ft. lbs. (64 Nm)
- C: 34 ft. lbs. (46 Nm)
- D: 32 ft. lbs. (43 Nm)

➡Make sure that the 2 knock pins are installed on the engine before installing the transaxle.

b. Install the breather bracket to the camshaft housing sub-assembly with the bolt and tighten to 49 inch lbs. (5.5 Nm).

c. Install the breather hose to the breather bracket.

77. Install starter assembly.

78. Connect the park/neutral position switch connector.

79. Connect the wire harness with the bolt and tighten to 9 ft. lbs. (12 Nm).

80. Install the engine wire.

81. Install the left and right front drive shaft assemblies.

82. Install the front frame assembly.

83. Install the vane pump assembly.

84. Install the engine assembly with transaxle.

85. Install the cooler compressor assembly.

86. Install the generator assembly.

87. Install the steering sliding yoke.

88. Install the drive plate and torque converter clutch setting bolt, as follows:

a. Apply a few drops of adhesive (Toyota Genuine Adhesive 1324, Three Bond 1324 or equivalent) to 2 threads on the tip of the 6 torque converter clutch mounting bolts.

b. Install the 6 torque converter clutch mounting bolts and tighten to 30 ft. lbs. (41 Nm). First install the black colored bolt, and then the remaining 5 bolts.

89. Install the flywheel housing under cover and exhaust pipe support bracket to the automatic transaxle with the 2 bolts tighten to 69 inch lbs. (7.8 Nm).

90. Install the No. 1 exhaust pipe support bracket.

91. Install the left and right front axle assemblies.

92. Install the left and right front lower No. 1 suspension arm sub-assemblies.

93. Install the left and right tie rod end sub-assemblies.

94. Install the left and right front speed sensors.

95. Install the left and right front stabilizer link assemblies.

96. Install the left and right front axle shaft nuts.

97. Install the front exhaust pipe assembly.

98. Install the No. 1 exhaust pipe support bracket.

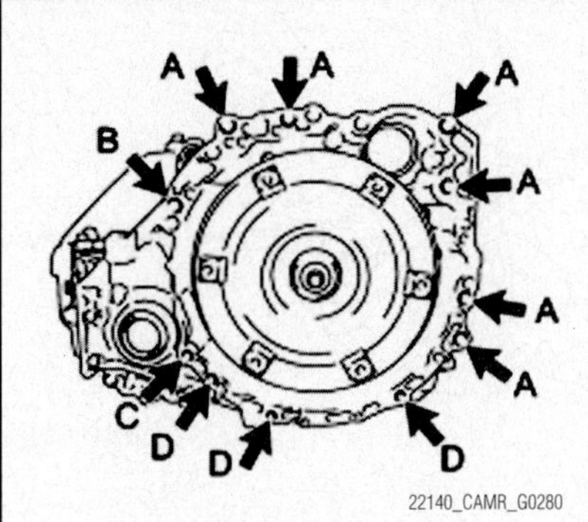

Fig. 17 Automatic transaxle bolt sequence

99. Connect the steering gear outlet return tube.
100. Connect the No. 1 oil reservoir to pump hose.
101. Connect the transmission control cable assembly, as follows:
    a. Install the transmission control cable assembly to the control shaft lever with the nut and tighten to 10 ft. lbs. (13 Nm).
    b. Connect the transmission control cable assembly with a new clip to the No. 1 transmission control cable bracket.
102. To complete installation, reverse remaining removal procedure.
103. Connect the negative battery cable.
104. Add coolant and power steering fluid.
105. Bleed the power steering system.
106. Add automatic transaxle fluid.
107. Inspect and adjust shift lever position.
108. Inspect for coolant leak, fuel leak, and exhaust gas leak.
109. Inspect ignition timing.
110. Inspect the engine idle speed.
111. Check the function of throttle body.
112. Install the V-bank cover sub-assembly.
113. Install the cool air intake duct seal.
114. Inspect CO/HC.
115. Inspect the ABS speed sensor signal.
116. Inspect and adjust front wheel alignment.
117. Install the right front fender apron seal.
118. Install the right and left engine under covers.

MANUAL TRANSAXLE ASSEMBLY

REMOVAL & INSTALLATION

E351 Manual Transaxle

See Figures 18 through 22.

1. Before servicing the vehicle, refer to the Precautions Section.
2. Discharge the fuel system pressure.
3. Place the front wheels facing straight ahead.
4. Remove the front wheels.
5. Remove both engine under covers.
6. Remove RH front fender apron seal.
7. Drain and recycle the engine coolant.
8. Drain the manual transaxle oil.
9. Drain the brake fluid.
10. Remove the windshield wiper motor assembly.
11. Remove the cowl top panel outer sub-assembly.
12. Remove the No. 1 engine cover sub-assembly.
13. Remove the drive belt.
14. Remove the air cleaner inlet assembly.
15. Remove the air cleaner cap sub-assembly.
16. Remove the air cleaner case sub-assembly.
17. Remove the battery.
18. Remove the RH No. 2 engine mounting stay.
19. Remove the engine moving control rod sub-assembly.
20. Remove the RH No. 2 engine mounting bracket.
21. Disconnect the No. 1 vacuum hose connector.
22. Disconnect the radiator inlet and outlet hose.
23. Disconnect the heater inlet and outlet water hose.
24. Remove the ECM.
25. Disconnect the engine wire.
26. Disconnect the transmission control cable assembly.
27. Disconnect the No. 1 oil reservoir to pump hose.
28. Disconnect the return tube sub-assembly.
29. Remove the clutch accumulator assembly.
30. Remove the clutch release cylinder assembly.
31. Disconnect the fuel tube sub-assembly.
32. Remove the generator assembly.
33. Separate the compressor and magnetic clutch.
34. Remove the front exhaust pipe assembly.
35. Remove both front axle hub nuts.
36. Remove both front stabilizer link assemblies.
37. Remove both front speed sensors.
38. Disconnect both tie rod assemblies.
39. Disconnect both front suspension lower no. 1 arms.
40. Separate both front axle assemblies.
41. Separate the steering sliding yoke.
42. Remove the engine assembly with transaxle.
43. Remove the vane pump assembly.
44. Remove front frame assembly.
45. Remove both front drive shaft assemblies.
46. Remove starter assembly.
47. Disconnect wire harness by performing the following:
    a. Remove the bolt and separate the ground cable.
    b. Disconnect the back-up light switch connector.
    c. Disconnect the wire harness.
48. Remove manual transaxle assembly by performing the following:
    a. Remove the 10 bolts.
    b. Separate and remove the transaxle from the engine.
49. Remove the 2 bolts, clutch line and bracket.
50. Remove the 2 bolts and wire harness clamps.
51. Remove the 2 bolts and manual transmission case protector.
52. Remove the 3 bolts and engine mounting front bracket.

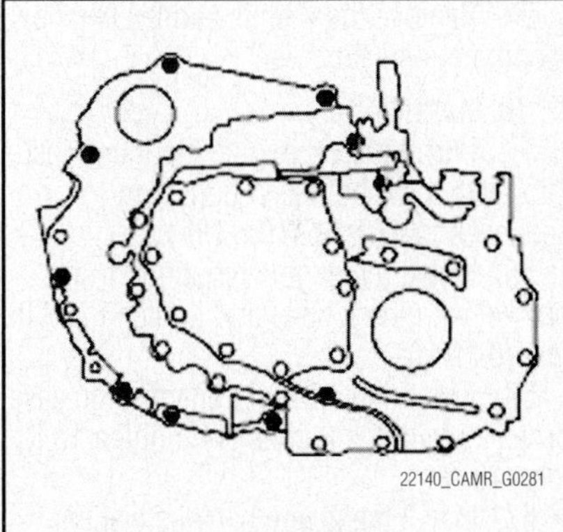

Fig. 18 Removing the 10 bolts

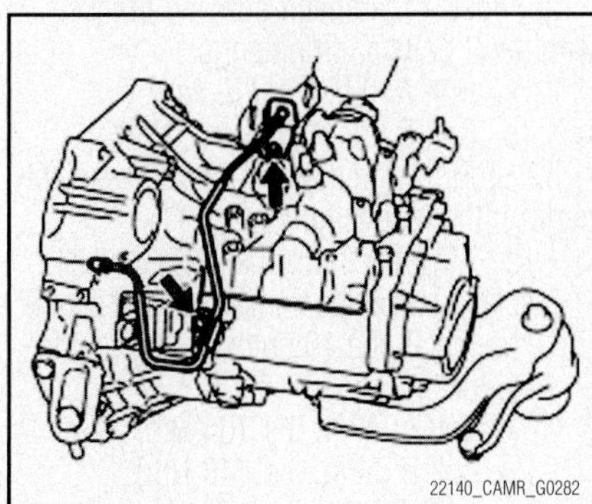

Fig. 19 Removing the 2 bolts, clutch line and bracket

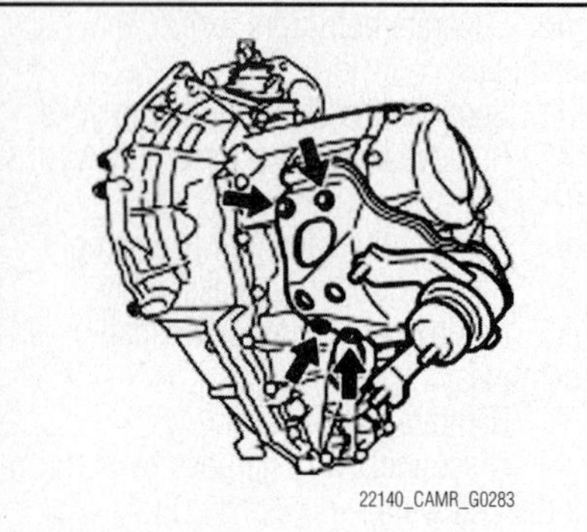

Fig. 20 Removing the 4 bolts and LH engine mounting insulator

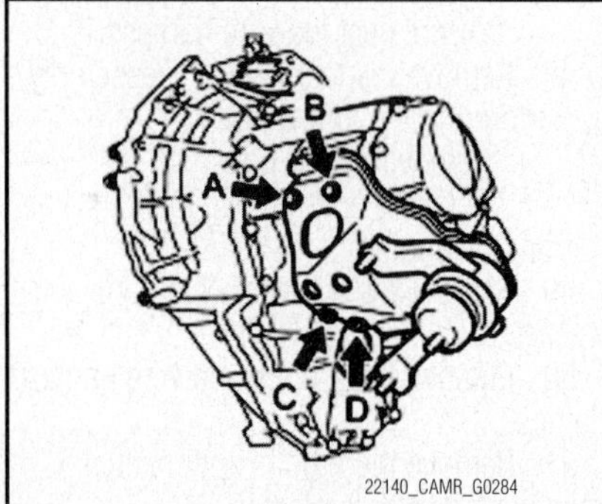

Fig. 21 LH engine mounting insulator tightening sequence

53. Remove the 4 bolts and LH engine mounting insulator.

To install:

54. Install the LH engine mounting insulator with the 4 bolts and tighten in sequence to 47 ft. lbs. (64 Nm).
55. Install the engine mounting front bracket with the 3 bolts and tighten to 47 ft. lbs. (64 Nm).
56. Install the manual transmission case protector with the 2 bolts and tighten 13 ft. lbs. (18 Nm).
57. Install the 2 wire harness clamps with the 2 bolts and tighten to 74 inch lbs. (8 Nm).
58. Install the clutch line and bracket with the 2 bolts and tighten to:
    a. Bolt A: 9 ft. lbs. (12 Nm).
    b. Bolt B: 62 inch lbs. (7 Nm).
59. To install manual transaxle assembly, perform the following:
    a. Align the input shaft with the clutch disc and install the transaxle to the engine.
    b. Install the 10 bolts and tighten in sequence to:
    - Bolt A: 47 ft. lbs. (64 Nm)
    - Bolt B: 34 ft. lbs. (46 Nm)
    - Bolt C: 32 ft. lbs. (44 Nm)
    - Bolt D: 34 ft. lbs. (46 Nm)
    - Bolt E: 47 ft. lbs. (64 Nm)

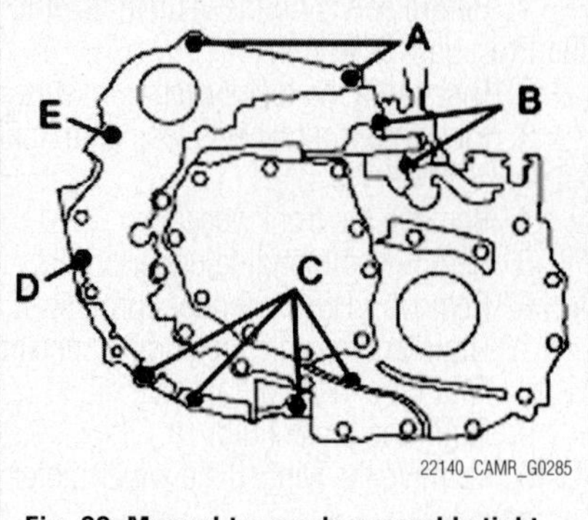

Fig. 22 Manual transaxle assembly tightening sequence

60. Connect wire harness by performing the following:
    a. Install the wire harness.
    b. Connect the back-up light switch connector.
    c. Install the ground cable with the bolt and tighten to 9 ft. lbs. (12 Nm).
61. To complete installation, reverse removal procedure.
62. Bleed clutch line.
63. Add engine coolant.
64. Add manual transaxle oil.
65. Add power steering fluid.
66. Bleed power steering fluid.
67. Check for fuel leaks.
68. Check for engine coolant leaks.
69. Check for exhaust gas leaks.
70. Adjust front wheel alignment.
71. Check ignition timing.
72. Check engine idle speed.
73. Inspect CO/HC.
74. Check abs speed sensor signal.

CLUTCH DRIVEN DISC & PRESSURE PLATE

REMOVAL & INSTALLATION

See Figures 23 through 26.

1. Before servicing the vehicle, refer to the Precautions Section.
2. Remove the manual transaxle assembly.
3. Remove the clutch release fork with the clutch release bearing from the transaxle assembly.
4. Remove the release bearing hub clip and clutch release bearing assembly from the clutch release fork.
5. Remove the release fork support from the manual transaxle assembly.
6. Remove the clutch release fork boot.
7. Remove the clutch cover assembly, as follows:
    a. Put matchmarks on the clutch cover assembly and flywheel sub-assembly.

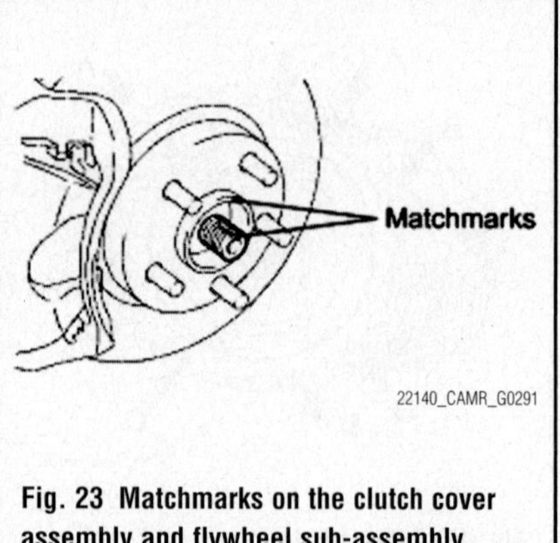

Fig. 23 Matchmarks on the clutch cover assembly and flywheel sub-assembly

    b. Loosen the 6 bolts one turn at a time until spring tension is released.
    c. Remove the 6 bolts and pull off the clutch cover assembly.

➡Do not drop the clutch disc assembly.

8. Remove clutch disc assembly.

➡Keep the lining part of the clutch disc assembly, the pressure plate and surface of the flywheel sub-assembly away from oil and foreign matter.

To install:

9. Insert SST (SST: 09301-00220) into the clutch disc assembly, then insert them into the flywheel sub-assembly.

➡Take care not to insert the clutch disc assembly in the wrong direction.

10. Align the matchmarks on the clutch cover assembly with the one on the flywheel sub-assembly.

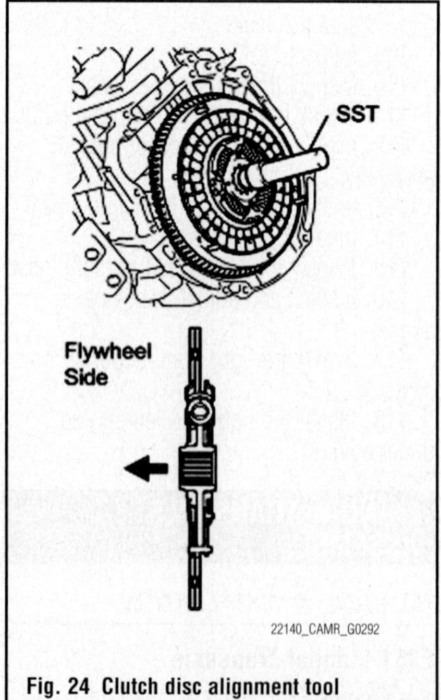

Fig. 24 Clutch disc alignment tool

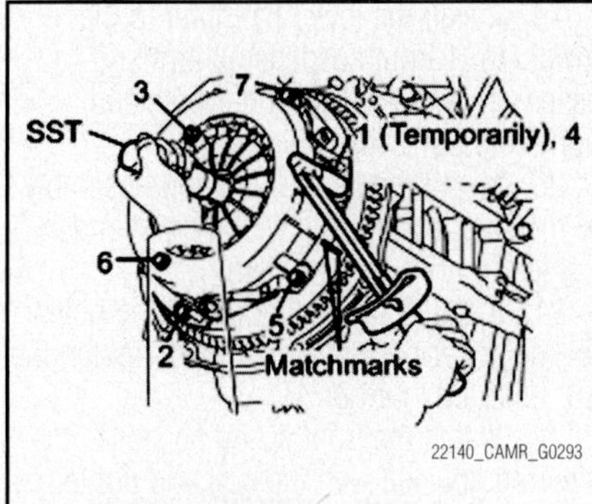

Fig. 25 Clutch cover bolt tightening sequence

11. Following the procedures shown in the illustration, tighten the 6 bolts in order, starting with the bolt located near the knock pin on the top. Tighten to 14 ft. lbs (19 Nm).

➡**Following the order in the illustration, evenly tighten the bolts one turn at a time.**

➡**Lightly move SST (09301-00210) up and down, and right and left after checking that the disc is in the center to tighten the bolts.**

12. Install the release fork support to the manual transaxle assembly and tighten 35 ft. lbs. (47 Nm).
13. Install the release bearing hub clip to the release bearing assembly.
14. Apply release hub grease (Toyota Genuine Release Hub Grease or equivalent) to the contact surfaces between the release fork and the release bearing, the fork and push rod, and the fork and the fork support.
15. Install the release fork to the release bearing assembly.
16. Apply clutch spline grease (Toyota Genuine Clutch Spline Grease or equivalent) to the input shaft spline as shown in the illustration.
17. Install the release fork with the release bearing assembly to the manual transaxle assembly.

➡**After installation, move the fork back and forth to check that the release bearing slides smoothly.**

18. Install the clutch release fork boot.
19. Install the manual transaxle assembly.

HALFSHAFT

REMOVAL & INSTALLATION

See Figures 27 through 32.

1. Before servicing the vehicle, refer to the Precautions Section.
2. Remove the engine under cover.
3. Remove the drain plug and gasket, and then drain the automatic transaxle fluid.
4. Install a new gasket and drain plug and tighten to 36 ft. lbs. (49 Nm).
5. Remove front wheel.
6. Using SST (SST: 09930-00010) and hammer, release the staked part of the front axle hub nut.

➡**Loosen the staked part of the nut completely, otherwise the screw of the drive shaft may be damaged.**

7. While applying the brakes, remove the front axle hub nut.
8. Remove the nut and separate the front stabilizer link assembly.

➡**If the ball joint turns together with the nut, use a hexagon wrench (6mm) to hold the stud.**

9. Remove the bolt and clip, and separate the speed sensor wire and flexible hose from the shock absorber.
10. Remove the bolt and separate the front speed sensor from the steering knuckle.

➡**Do not allow foreign matter to adhere to the speed sensor. Be careful not to damage the speed sensor.**

11. Separate tie rod end sub-assembly, as follows:
    a. Remove the cotter pin and nut.
    b. Using SST (SST: 09628-62011) or equivalent, separate the tie rod end sub-assembly from the steering knuckle.

➡**Do not damage the ball joint dust cover.**

12. Remove the bolt and 2 nuts, and separate the lower No. 1 front suspension arm sub-assembly from the lower ball joint.
13. Put matchmarks on the front drive shaft assembly and the axle hub.
14. Using a plastic hammer, separate the front drive shaft assembly from the front axle hub sub-assembly.

➡**Be careful not to damage the drive shaft boot and speed sensor rotor.**

15. Remove the front drive shaft assembly(s), as follows:
    a. For left front drive shaft, use SST (SST: 09520-01010, SST: 09520-24010) or equivalent, and remove the left front drive shaft assembly.

➡**Be careful not to damage the drive shaft dust cover, boot and oil seal. Be careful not to drop the drive shaft assembly.**

    b. For right front drive shaft, use a screwdriver and remove the bearing bracket hole snap ring.
    c. Remove the bolt and right front drive shaft assembly from the drive shaft bearing bracket.

➡**Do not damage the boot and oil seal.**

16. Fix front axle hub bearing. The hub bearing could be damaged if it is subjected to the vehicle's full weight, such as moving the vehicle with the drive shaft removed. If it is necessary to place the vehicle's weight on the hub bearing, first support it with SST (SST: 09608-16042).

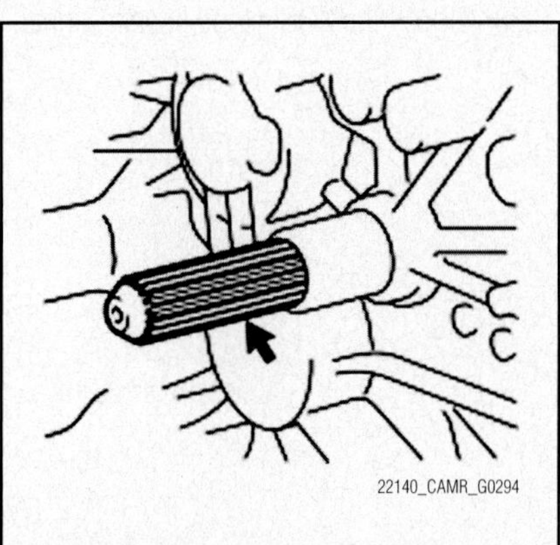

Fig. 26 Clutch spline grease application

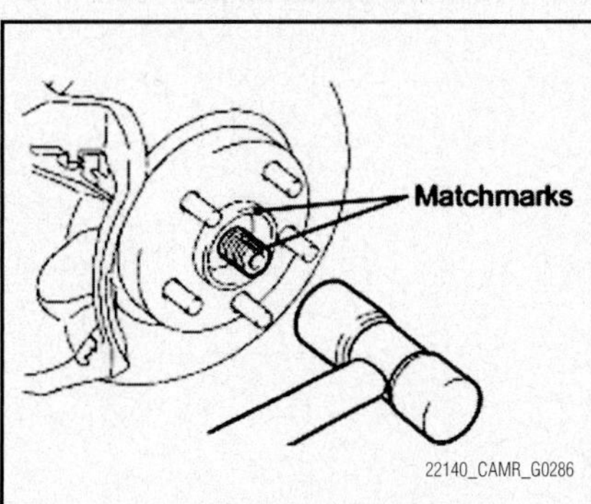

Fig. 27 Put matchmarks on the front drive shaft assembly and the axle hub

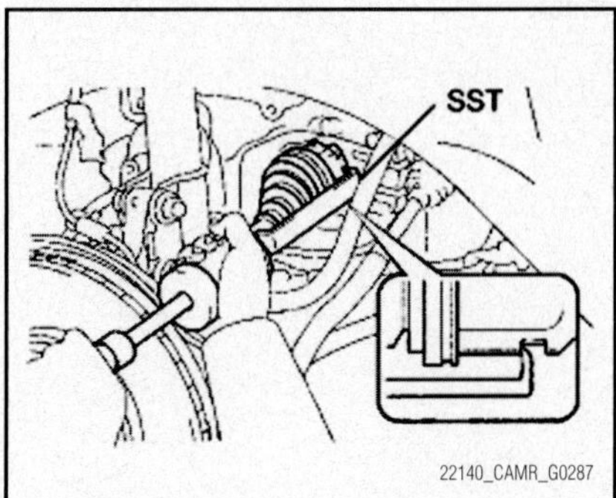

Fig. 28 Remove the left front drive shaft assembly

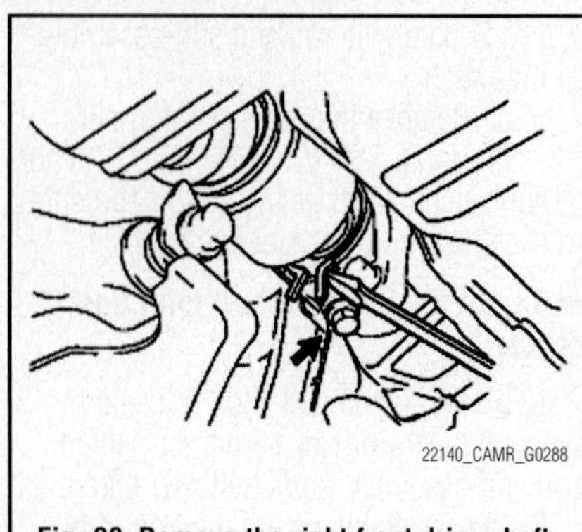

Fig. 29 Remove the right front drive shaft assembly

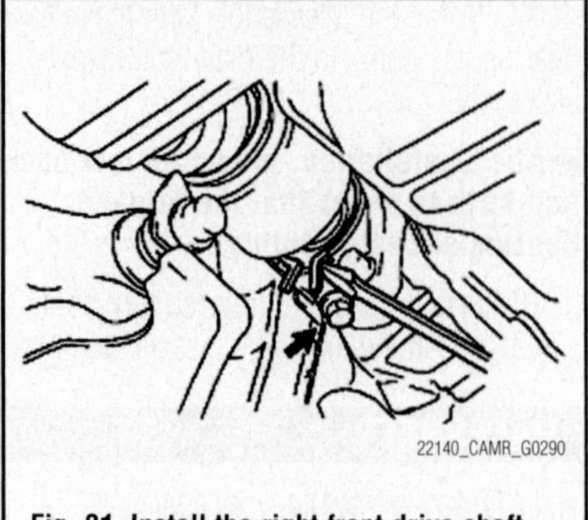

Fig. 31 Install the right front drive shaft assembly

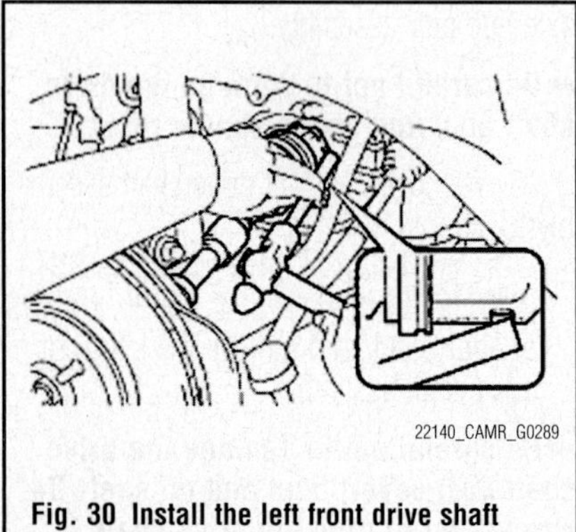

Fig. 30 Install the left front drive shaft assembly

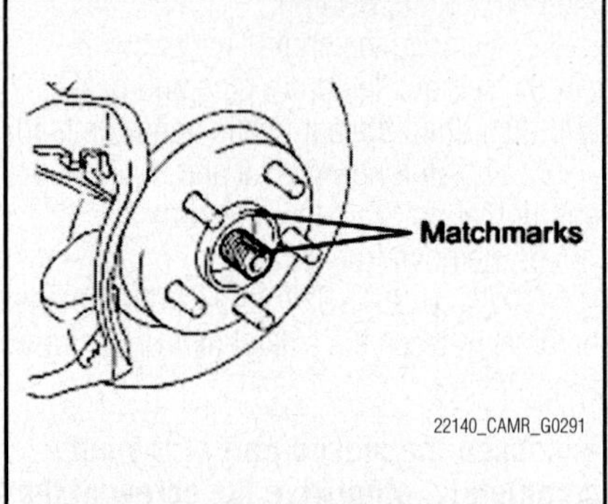

Fig. 32 Align the matchmarks and install the front drive shaft assembly

To install:

17. Install the front drive shaft assembly(s), as follows:
    a. Coat the spline of the inboard joint shaft assembly with automatic transaxle fluid.
    b. For the left front drive shaft, align the shaft splines and install the drive shaft assembly with a brass bar and hammer.

➡Set the shaft snap ring with the opening side facing down. Be careful not to damage the drive shaft dust cover, boot, and oil seal. Move the drive shaft assembly while keeping it level.

    c. For the right front drive shaft, install the drive shaft and use a screwdriver to install a new bearing bracket hole snap ring, and install a new bolt tightened to 24 ft. lbs. (32 Nm).

➡Be careful not to damage the drive shaft dust cover, boot and oil seal.

➡Move the drive shaft assembly while keeping it level.

18. Align the matchmarks and install the front drive shaft assembly to the front axle hub sub-assembly.

➡Be careful not to damage the drive shaft boot and speed sensor rotor.

19. Install the lower ball joint to the lower No. 1 front suspension arm sub-assembly with the bolt and 2 nuts and tighten to 55 ft. lbs. (75 Nm).
20. Install the tie rod end sub-assembly to the steering knuckle with the nut and tighten to 36 ft. lbs. (49 Nm).
21. Install a new cotter pin. If the holes for the cotter pin are not aligned, tighten the nut up to 60° further.
22. Install the front speed sensor to the steering knuckle with the bolt and tighten to 71 inch lbs. (8 Nm).
23. Install the flexible hose and the speed sensor to the shock absorber with the bolt and set the sensor clip on the knuckle and tighten to 14 ft. lbs. (19 Nm).

➡Be careful not to damage the speed sensor. Do not allow foreign matter to adhere to the speed sensor. Do not twist the sensor wire when installing the speed sensor.

24. Install the stabilizer link assembly with the nut and tighten to 55 ft. lbs. (74 Nm).

➡If the ball joint turns together with the nut, use a hexagon wrench (6 mm) to hold the stud.

25. Clean the threaded parts on the drive shaft and axle hub nut using a non-residue solvent.

➡Be sure to perform this work for a new drive shaft. Keep the threaded parts free of oil and foreign objects.

26. Using a socket wrench (30 mm), install a new axle hub nut and tighten to 217 ft. lbs. (294 Nm).
27. Using a chisel and hammer, stake the front axle hub nut.
28. Install front wheel.
29. Add automatic transaxle fluid.
30. Inspect automatic transaxle fluid.
31. Inspect and adjust front wheel alignment.
32. Install the engine under cover.
33. Check the ABS speed sensor signal.

ENGINE COOLING

THERMOSTAT

REMOVAL & INSTALLATION

See Figures 33 through 36.

1. Before servicing the vehicle, refer to the Precautions Section.
2. For 2.4L engines, perform the following:
    a. Remove both front wheel opening extension pads.
    b. Remove both engine under cover.
    c. Drain engine coolant.
3. For 3.5L engines, perform the following:
    a. Drain engine coolant.
    b. Remove the V-bank cover sub-assembly.
    c. Remove the RH front fender apron seal.
    d. Remove the RH No. 2 engine mounting stay.
    e. Remove the drive belt.
    f. Remove No. 2 idler pulley sub-assembly.
4. Separate the radiator hose outlet.
5. Remove the 2 nuts and disconnect the water inlet from the cylinder block.
6. Remove the thermostat and the gasket from the thermostat.

To install:

7. Install a new gasket to the thermostat.
8. Install a new gasket to the thermostat.

➡The jiggle valve may be set within 10° of either side of the prescribed position.

9. Install the thermostat with the jiggle valve facing up.
10. Install the water inlet and tighten to 7 ft. lbs. (10 Nm).
11. The remainder of installation is the reverse of removal.

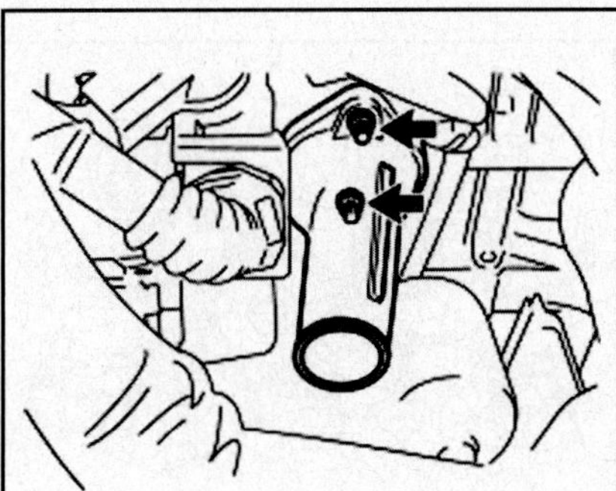

Fig. 33 Removing the 2 thermostat nuts—2.4L engine

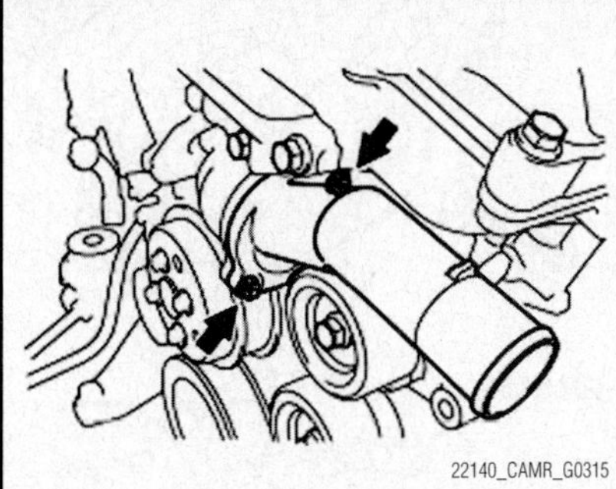

Fig. 34 Removing the 2 thermostat nuts—3.5L engine

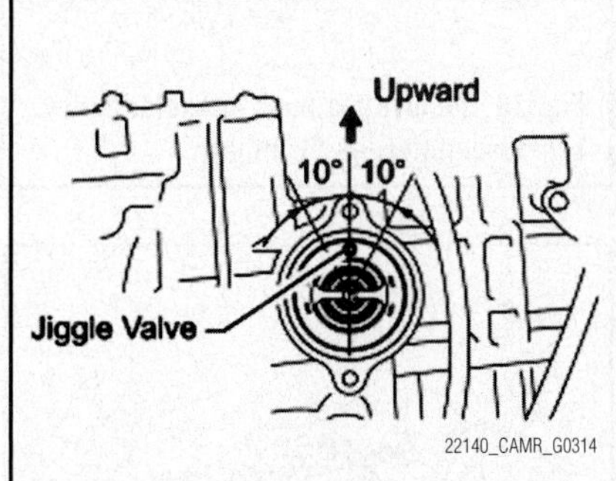

Fig. 35 Radiator jiggle valve—2.4L engine

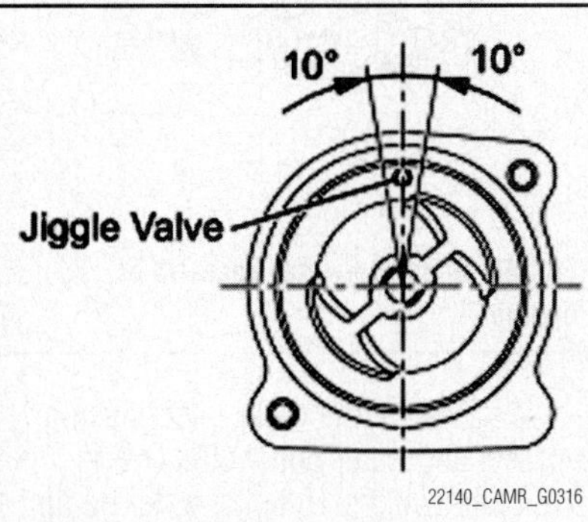

Fig. 36 Radiator jiggle valve—3.5L engine

12. After installation, inspect for coolant leak.

WATER PUMP

REMOVAL & INSTALLATION

2.4L Engine

See Figures 37 and 38.

1. Before servicing the vehicle, refer to the Precautions Section.
2. Disconnect the negative battery cable.
3. Remove both front wheel opening extension pads.
4. Remove both engine under covers.
5. Drain and recycle the engine coolant.
6. Remove RH front fender apron seal.
7. Remove RH No. 2 engine mounting stay.
8. Remove engine moving control rod sub-assembly.
9. Remove RH No. 2 engine mounting bracket
10. Remove the drive belt.
11. Remove generator assembly.
12. Using SST: 09960-10010, remove the 4 bolts and water pump pulley.
13. Remove water pump assembly by performing the following:
    a. Remove the clamp of the crankshaft position sensor from the water pump.
    b. Disconnect the wire of the crankshaft position sensor from the clamp bracket.
    c. Remove the 4 bolts, 2 nuts and clamp bracket.

➡Be careful not to damage the contact surfaces of the water pump and cylinder block.

    d. Using a screwdriver, pry between the water pump and cylinder block, and then remove the water pump. Tape the screwdriver tip before use.

To install:

14. Install water pump assembly:
    a. Remove any old seal packing material from the water pump assembly contact surface.

➡Remove any oil from the contact surface. The parts must be set within 3 minutes after applying seal packing. Otherwise, the material must be removed and reapplied.

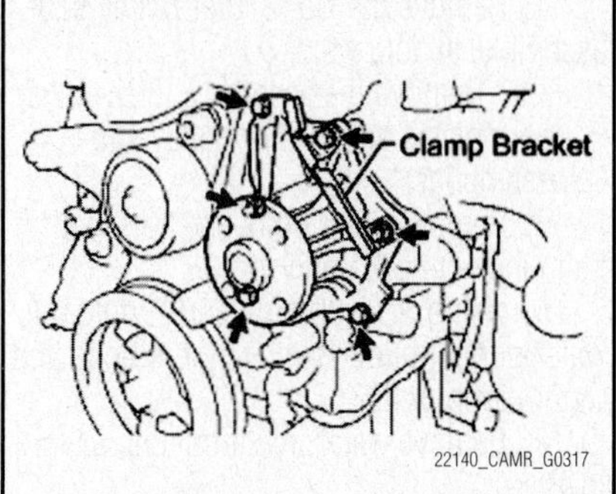

Fig. 37 Removing the 4 bolts, 2 nuts and clamp bracket —2.4L engine

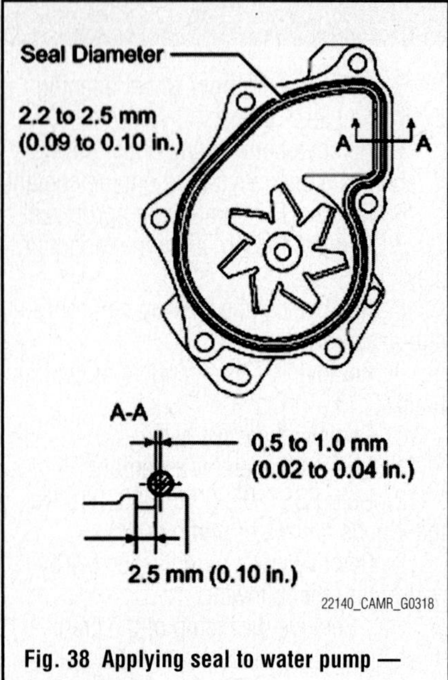

Fig. 38 Applying seal to water pump — 2.4L engine

b. Apply a continuous line of seal packing. Seal packing: Toyota Genuine Seal Packing Black, Three Bond 1207B or Equivalent. Standard seal diameter: 0.09 to 0.10 in. (2.2 to 2.5 mm)

15. Install the water pump and clamp bracket with the 4 bolts and 2 nuts and tighten to 80 inch lbs. (9 Nm).

16. To complete installation, reverse removal procedure.

17. Tighten the water pump bolts to 19 ft. lbs. (26 Nm).

18. Add engine coolant.

19. Inspect for coolant leak.

3.5L Engine

See Figures 39 through 42.

1. Before servicing the vehicle, refer to the Precautions Section.
2. Remove engine assembly and transaxle. secure engine stand.
3. Remove RH front No. 1 engine mounting bracket.
4. Remove the No. 2 idler pulley sub-assembly, as follows:
   a. Remove the 2 bolts, 2 idler pulley cover plates and 2 idler pulley sub-assemblies.
5. Remove the 5 bolts and V-ribbed belt tensioner assembly.
6. Using SST: 09960-10010, hold the water pump pulley. Remove the 4 bolts and water pump pulley.
7. Remove water inlet housing, as follows:
   a. Separate the water hose.
   b. Remove the 2 bolts, nut and water inlet housing.

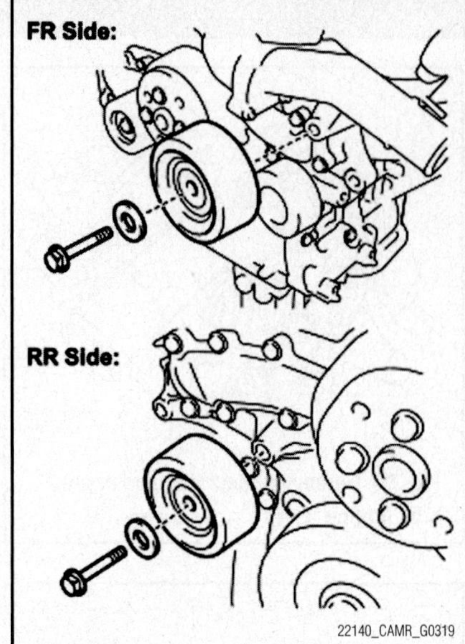

Fig. 39 Remove the bolts and idler pulley sub-assemblies—3.5L engine

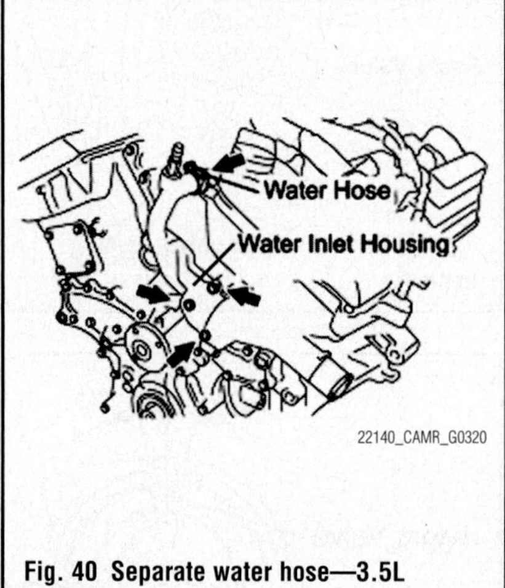

Fig. 40 Separate water hose—3.5L engine

   c. Remove the water inlet housing gasket and water outlet pipe O-ring.
8. Remove the 16 bolts, water pump assembly and water pump gasket.

To install:

➡Make sure that there is no oil on the threads of the A bolts.

➡Be sure to replace the 2 C bolts with new ones or reuse them after applying adhesive (Part No. 08833-00080, three bond 1344 or equivalent).

9. Install a new water pump gasket and the water pump assembly with the 16 bolts and tighten to:
   a. Bolt A: 15 ft. lbs. (21 Nm)
   b. Bolt B: 81 inch lbs. (9.1 Nm).
   c. Bolt C: 81 inch lbs. (9.1 Nm).
10. Install a new water inlet housing No. 1 gasket and water outlet pipe O-ring.

➡Be careful not to allow the O-ring to get caught between the parts.

11. Install water inlet housing, as follows:
   a. Install a new No. 1 water inlet housing gasket and water outlet pipe O-ring.
   b. Install the water inlet with the 2 bolts and nut and tighten to 7 ft. lbs. (10 Nm).
12. Temporarily install the water pump pulley with the 4 bolts.
   a. Using SST (SST: 09960-10010) or equivalent, hold the water pump pulley.
   b. Tighten the 4 bolts to 15 ft. lbs. (21 Nm).
13. Install the V-ribbed belt tensioner assembly with the 5 bolts and tighten to 32 ft. lbs. (43 Nm).
14. Install the 2 idler pulley cover plates and idler pulley sub-assemblies with the 2 bolts and tighten to 32 ft. lbs. (43 Nm).
15. To complete installation, reverse remaining removal procedure.
16. Add engine coolant.
17. Inspect for coolant leak.

Fig. 41 Remove the 16 bolts, water pump assembly—3.5L engine

Fig. 42 Water pump tightening sequence—3.5L engine

ENGINE ELECTRICAL — CHARGING SYSTEM

ALTERNATOR

REMOVAL & INSTALLATION

1. Disconnect the negative battery cable.
2. Remove the V-bank cover sub-assembly.
3. Remove the V-ribbed belt.
4. Remove the alternator assembly, as follows:
   a. Disconnect the wire harness clamp.
   b. Remove the terminal cap.
   c. Remove the nut and disconnect the wire harness from terminal B.
   d. Disconnect the alternator connector from the alternator assembly.
   e. Remove the nut from the cylinder block.
   f. Remove the 2 bolts and alternator assembly.
   g. Remove the bolt and wire harness clamp stay.
   h. Remove the bolt and bracket.

To install:

5. Install the alternator assembly, as follows:
   a. Install the bracket with the bolt and tighten to 15 ft. lbs. (20 Nm).
   b. Install the wire harness clamp stay and tighten to 74 inch lbs. (8.4 Nm).
   c. Install the alternator assembly with the 2 bolts and tighten to 32 ft. lbs. (43 Nm).
   d. Install the nut to the cylinder block and tighten to 15 ft. lbs. (20 Nm).
   e. Connect the alternator connector to the alternator assembly.
   f. Install the alternator wire with the nut and tighten to 87 inch lbs. (9.8 Nm).
   g. Install the terminal cap.
   h. Connect the wire harness clamp.
6. Install the V-ribbed belt.
7. Install the V-bank cover sub-assembly.
8. Connect the negative battery cable.
9. Perform initialization.

ENGINE ELECTRICAL — DISTRIBUTORLESS IGNITION SYSTEM

FIRING ORDER

Firing order for 3.5L engine: 1–2–3–4–5–6

IGNITION COIL

REMOVAL & INSTALLATION

2.4L Engine

See Figure 43.

1. Before servicing the vehicle, refer to the Precautions Section.
2. Disconnect the negative battery cable.
3. Remove engine cover(s).
4. Disconnect the 4 ignition coil connectors. Remove the 4 bolts and 4 ignition coils.
5. To install, reverse removal procedure. Tighten ignition coil bolts to 80 inch lbs. (9 Nm).

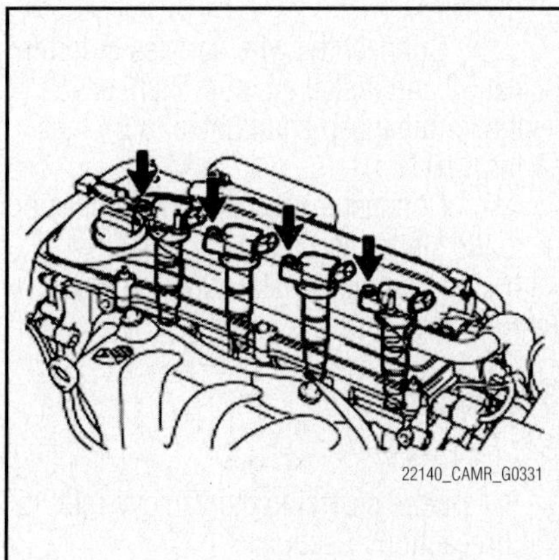

Fig. 43 Removing ignition coils—2.4L engines

3.5L Engine

See Figure 44.

1. Before servicing the vehicle, refer to the Precautions Section.
2. Disconnect the negative battery cable.
3. Drain and recycle the engine coolant.
4. Remove windshield wiper link assembly.
5. Remove cowl top panel outer sub-assembly.
6. Remove v-bank cover sub-assembly.
7. Remove air cleaner cap sub-assembly.

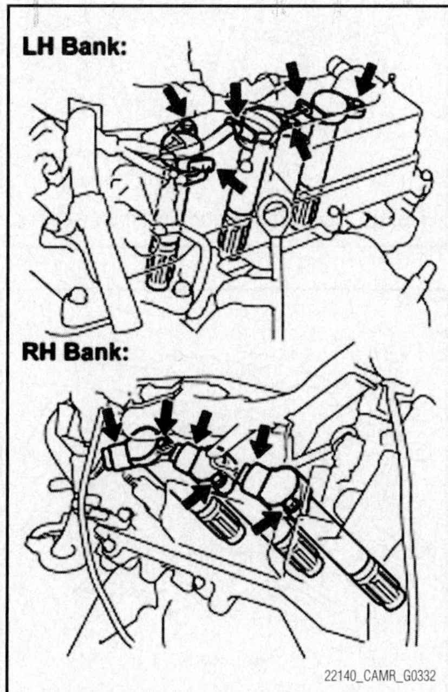

Fig. 44 Removing ignition coils—3.5L engines

8. Remove intake air surge tank assembly.
9. Remove No. 1 surge tank stay by performing the following:
   a. Remove the bolt and disconnect the harness clamp.
   b. Remove the bolt and No. 1 surge tank stay.
10. Disconnect the 6 ignition coil connectors.
11. Remove the 6 bolts and 6 ignition coils.
12. To install, reverse removal procedure.
13. Tighten the following to specification:
    a. 6 ignition coil bolt: 10 ft. lbs. (10 Nm).
    b. No. 1 surge tank stay bolt: 15 ft. lbs. (21 Nm).
    c. No. 1 surge tank stay bolt and clamp: 62 inch lbs. (7 Nm).

IGNITION TIMING

ADJUSTMENT

All engines are equipped with a Distributorless Ignition System (DIS). No timing adjustment is possible.

SPARK PLUGS

REMOVAL & INSTALLATION

2.4L Engine

See Figure 45.

1. Before servicing the vehicle, refer to the Precautions Section.
2. Remove the plastic engine cover.
3. Disconnect the 4 ignition coil con-

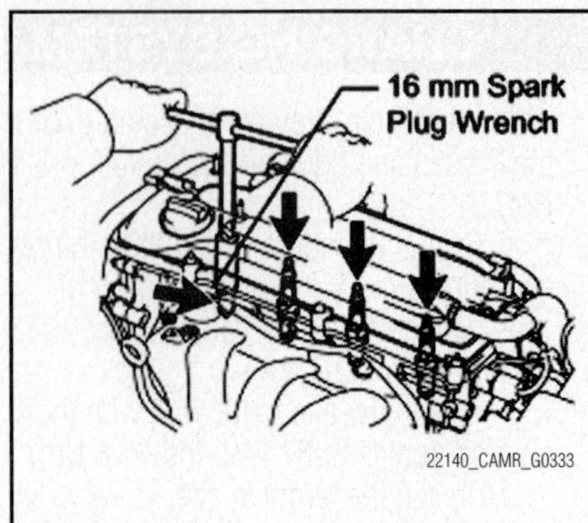

Fig. 45 Removing spark plugs—2.4L engines

nectors and remove the 4 bolts and ignition coils.

4. Using a 16 mm (0.63 in.) spark plug wrench, remove the 4 spark plugs.

5. Installation is the reverse of removal.

a. Torque the ignition coils to 66 inch lbs. (7.5 Nm) and the spark plugs to 13 ft. lbs (18 Nm).

3.5L Engine

See Figure 46.

1. Before servicing the vehicle, refer to the Precautions Section.
2. Remove the V-bank cover.
3. Remove the intake air surge tank.
4. Disconnect the 6 ignition coil connectors.
5. Remove the 6 bolts and 6 ignition coils.
6. Using a 16 mm (0.63 in.) plug wrench, remove the spark plugs.
7. Installation is the reverse of removal, noting the following:

Fig. 46 Removing spark plugs—3.5L engines

a. Torque the ignition coils to 66 inch lbs. (7.5 Nm) and the spark plugs to 13 ft. lbs (18 Nm).

ENGINE ELECTRICAL

STARTING SYSTEM

STARTER

REMOVAL & INSTALLATION

2.4L Engine

See Figures 47 through 49.

1. Before servicing the vehicle, refer to the Precautions Section.
2. Disconnect the negative battery cable.
3. Remove air cleaner inlet assembly.
4. Remove air cleaner cap sub-assembly.
5. Remove air cleaner case sub-assembly.
6. For vehicles equipped with M/T, perform the following:

a. Disconnect the terminal 50 connector from the starter assembly.

b. Remove the nut and disconnect the wire harness from terminal 30.

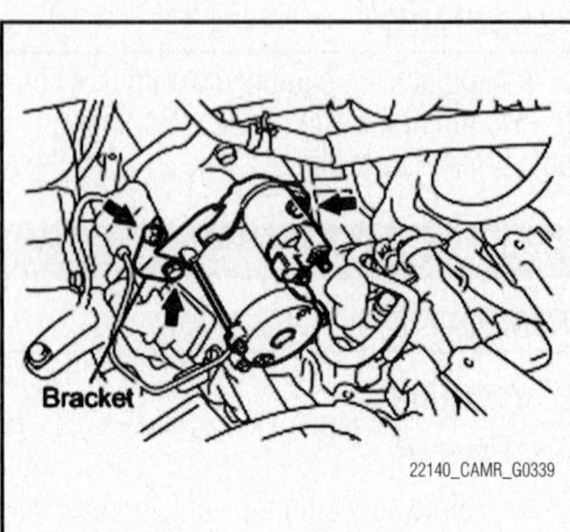

Fig. 47 Removing starter assembly—M/T

c. Remove the 3 bolts, clutch flexible hose bracket and starter assembly.

7. For vehicles equipped with A/T, perform the following:

a. Disconnect the terminal 50 connector from the starter assembly.

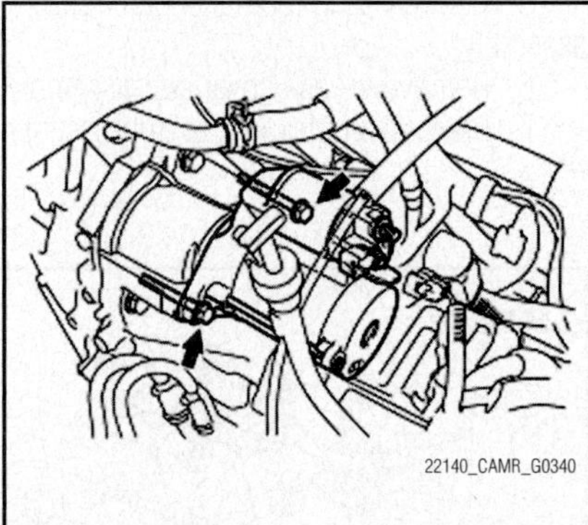
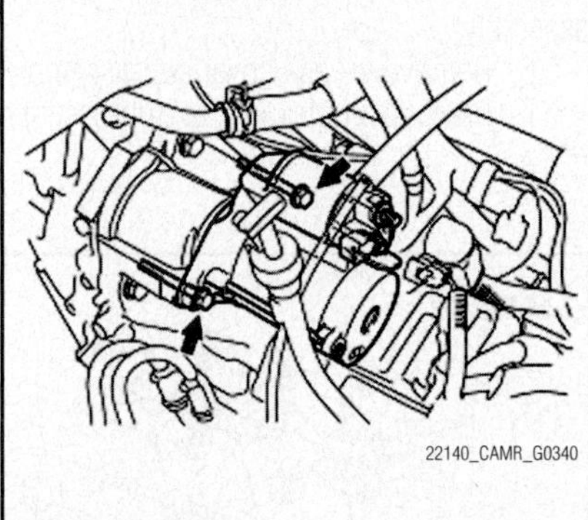

Fig. 48 Removing starter assembly—A/T

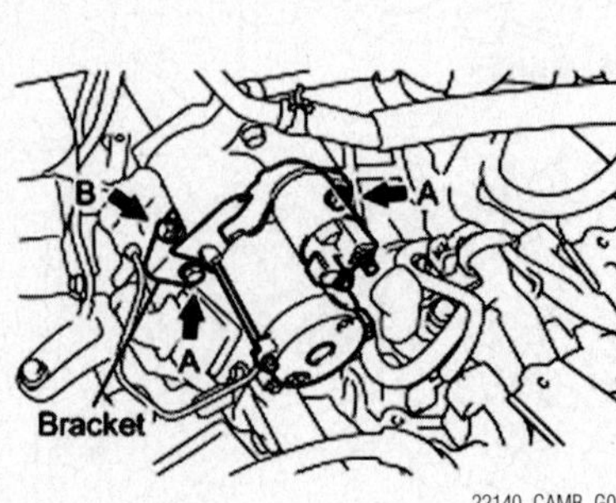

Fig. 49 Installing starter assembly—M/T

b. Remove the nut and disconnect the wire harness from terminal 30.

c. Remove the 2 bolts and starter assembly.

To install:

8. For M/T vehicles:

a. Install the starter assembly and clutch flexible hose bracket with the 3 bolts.

- Bolt A: 28 ft. lbs. (37 Nm).
- Bolt B: 9 ft. lbs. (12 Nm).

b. Connect the wire harness to terminal 30 and install the nut. Then, attach the terminal cap and tighten to 87 inch lbs. (10 Nm).

c. Connect the terminal 50 connector to the starter assembly.

9. For A/T vehicles:

a. Install the starter assembly with the 2 bolts and tighten 28 ft. lbs. (37 Nm).

b. Connect the wire harness to terminal 30 and install the nut. Then, attach the terminal cap and tighten to 87 inch lbs. (10 Nm).

c. Connect the terminal 50 connector to the starter assembly.

10. To complete installation, reverse removal procedure.

3.5L Engine

See Figure 50.

1. Before servicing the vehicle, refer to the Precautions Section.
2. Disconnect the negative battery cable.
3. Remove cool air intake duct seal.

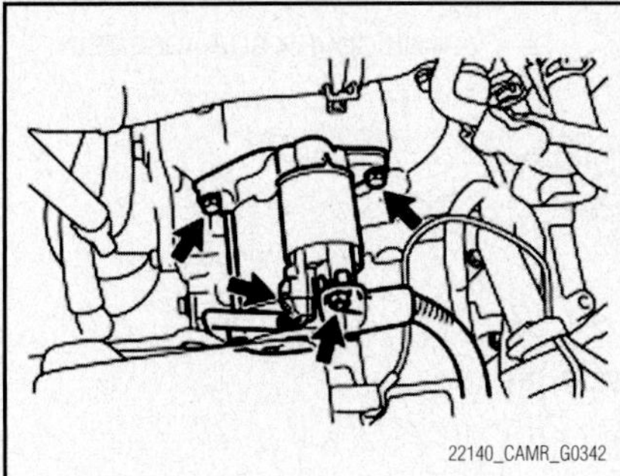
Fig. 50 Removing starter assembly—3.5L engine

4. Remove v-bank cover sub-assembly.
5. Remove air cleaner inlet assembly.
6. Remove air cleaner cap sub-assembly.
7. Remove air cleaner case sub-assembly.
8. Remove No. 1 air cleaner inlet.
9. Disconnect the terminal 50 connector from the starter assembly.
10. Remove the nut and disconnect the wire harness from terminal 30.
11. Remove the 2 bolts and starter assembly.

To install:

12. Install the starter assembly with the 2 bolts and tighten to 26 ft. lbs. (37 Nm).
13. Connect the wire harness to terminal 30 and install the nut and tighten to 87 inch lbs. (9.8 Nm).
14. Cover the nut with the cap.
15. Connect terminal 50 to the starter assembly.
16. To complete installation, reverse removal procedure.

ENGINE MECHANICAL

ACCESSORY DRIVE BELTS

ACCESSORY BELT ROUTING

See Figure 51.

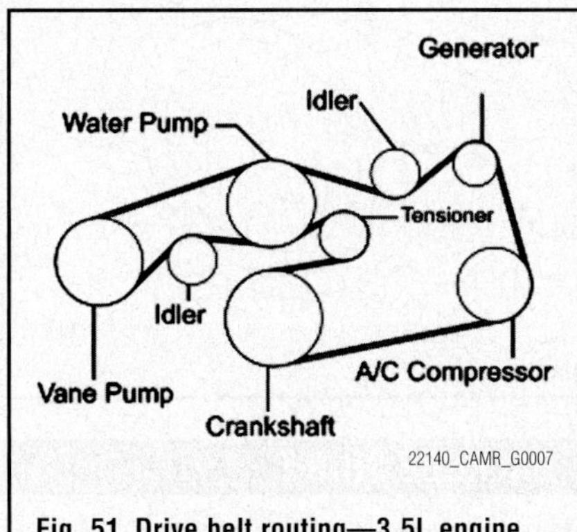

Fig. 51 Drive belt routing—3.5L engine

INSPECTION

See Figure 52.

Visually check the V-ribbed belt for excessive wear, frayed cords, etc. If any defect has been found, replace the V-ribbed belt.

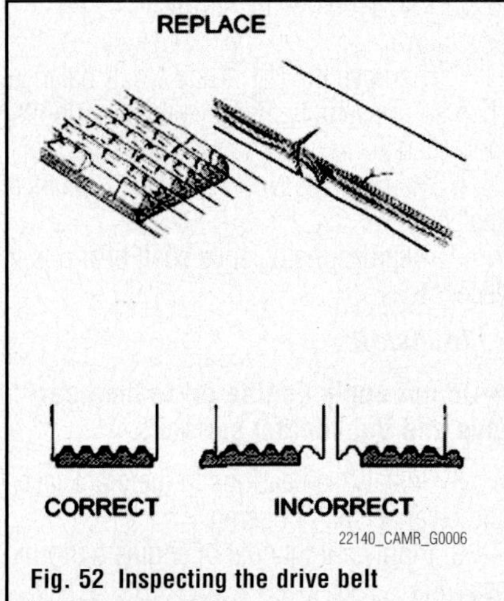

Fig. 52 Inspecting the drive belt

- Cracks on the rib side of a belt are considered acceptable. If the belt has chunks missing from the ribs, it should be replaced
- A "new belt" is a belt which has been used for less than 5 minutes with the engine running
- A "used belt" is a belt which has been used for 5 minutes or more with the engine running

ADJUSTMENT

This vehicle is equipped with an auto-tensioner and cannot be adjusted.

REMOVAL & INSTALLATION

2.4L Engine

See Figure 53.

1. Before servicing the vehicle, refer to the Precautions Section.

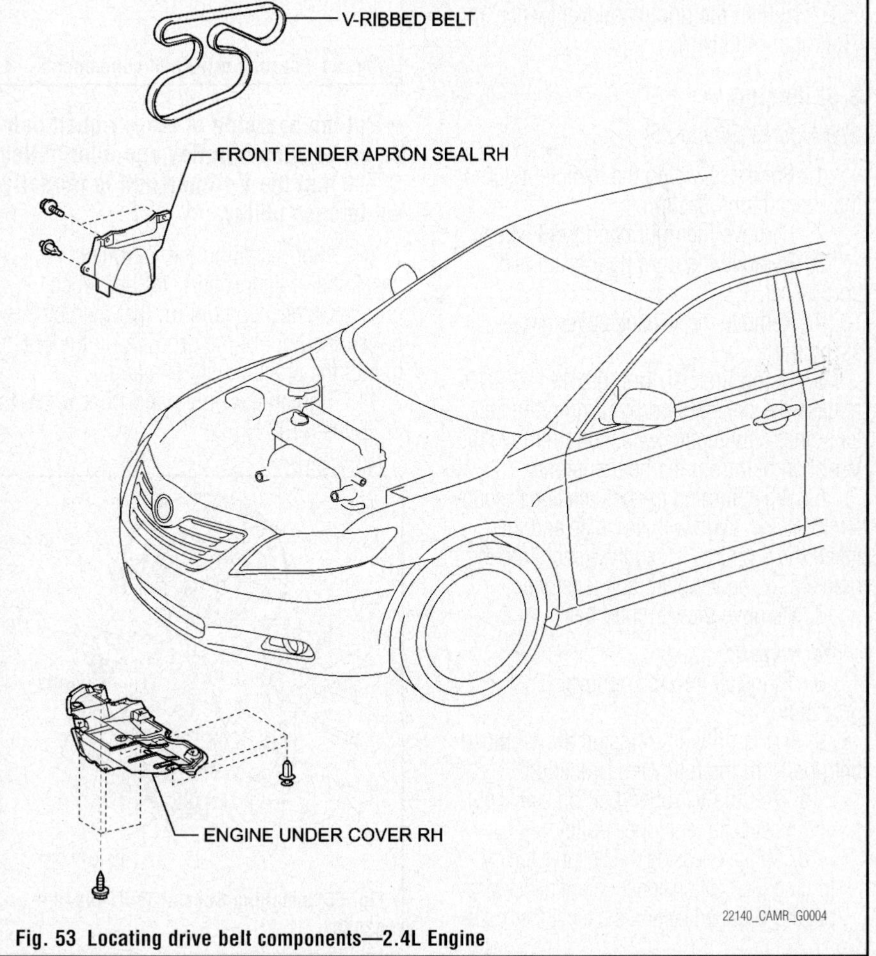

Fig. 53 Locating drive belt components—2.4L Engine

2. Remove the right hand front wheel.
3. Remove the right hand engine under cover.
4. Remove the right hand front fender apron seal.

➡Before removing, take note of the following:

- Be sure to connect Special Tool: 09216—42010 and the tools so that they are in line during use
- When retracting the tensioner, turn it clockwise slowly for 3 seconds or more. Do not apply force rapidly
- After the tensioner is fully retracted, do not apply force any more than necessary

5. Using the Special Tool and a 19 mm socket wrench, loosen the v-ribbed belt tensioner arm clockwise, then remove the v-ribbed belt.
6. Remove the v-ribbed belt.

To install:

7. To install, reverse removal procedure.
8. After installing the V-ribbed belt, check that it fits properly in the ribbed grooves. Check to confirm that the belt has not slipped out of the grooves on the bottom of the crank pulley by hand.
9. Tighten the right hand front wheel to: 76 ft. lbs. (103 Nm).

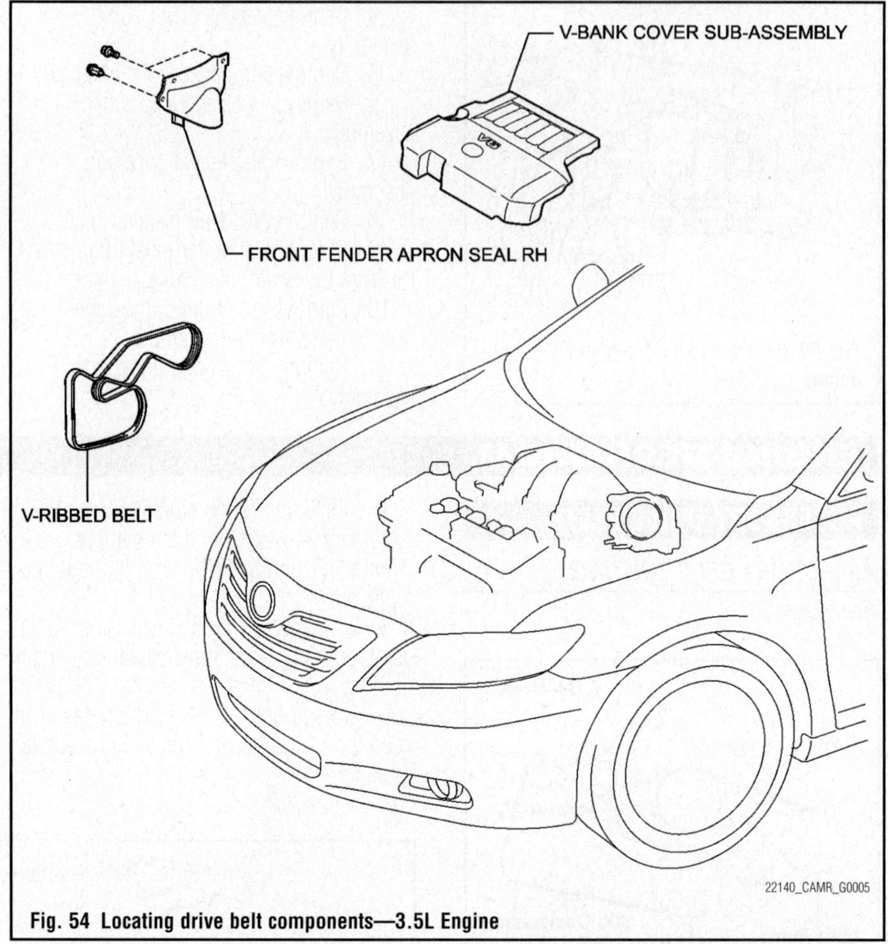

Fig. 54 Locating drive belt components—3.5L Engine

3.5L Engine

See Figures 54 and 55.

1. Before servicing the vehicle, refer to the Precautions Section.
2. Remove the right hand front wheel.
3. Remove the right hand front fender apron seal.
4. Remove the V-bank cover sub-assembly.
5. Using Special Tool: 09249—63010, release the belt tension by turning the belt tensioner counterclockwise, and remove the V-ribbed belt from the belt tensioner.
6. While turning the belt tensioner counterclockwise, align with its holes and then insert the 5 mm bi-hexagon wrench into the holes to fix the V-ribbed belt tensioner.
7. Remove the v-ribbed belt.

To install:

8. To install, reverse removal procedure.
9. If it is difficult to install the V-ribbed belt, perform the following procedure:
   a. Put the V-ribbed belt on every pulley except the tensioner pulley.
   b. While releasing the belt tension by turning the belt tensioner counterclockwise, put the V-ribbed belt on the tensioner pulley.

➡Put the backside of the V-ribbed belt on the tensioner pulley and idler pulley. Check that the V-ribbed belt is properly set to each pulley.

10. After installing the V-ribbed belt, check that it fits properly in the ribbed grooves. Check to confirm that the belt has not slipped out of the grooves on the bottom of the crank pulley by hand.
11. Tighten the right hand front wheel to: 76 ft. lbs. (103 Nm).

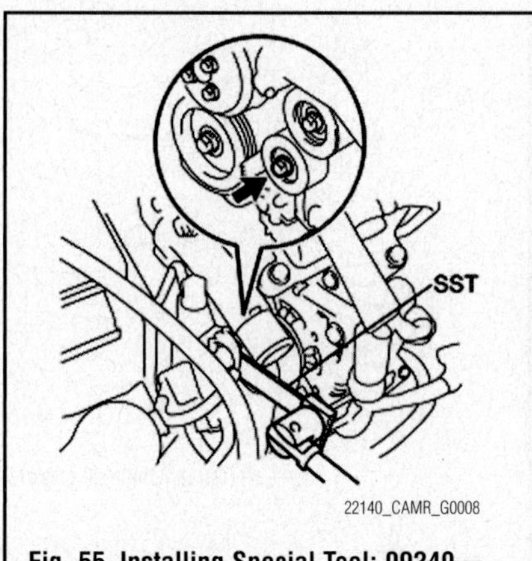

Fig. 55 Installing Special Tool: 09249—63010

BALANCE SHAFT

REMOVAL & INSTALLATION

2.4L Engine

See Figures 56 through 63.

1. Before servicing the vehicle, refer to the Precautions Section.
2. Connect the negative battery cable.
3. Drain the engine oil.
4. Remove the oil pump. Refer to Oil Pump below for removal procedure.
5. Remove the No. 1 and No. 2 balance shaft sub-assembly. Remove the eight bolts in sequence.
6. Remove the No. 1 and No. 2 balance shafts.
7. Remove the balance shaft bearings if necessary.

To install:

➡Do not apply engine oil to the bearings and the contact surfaces.

8. Install the bearings in the crankcase and balance shaft housing.
9. Apply a light coat of engine oil to the bearings.

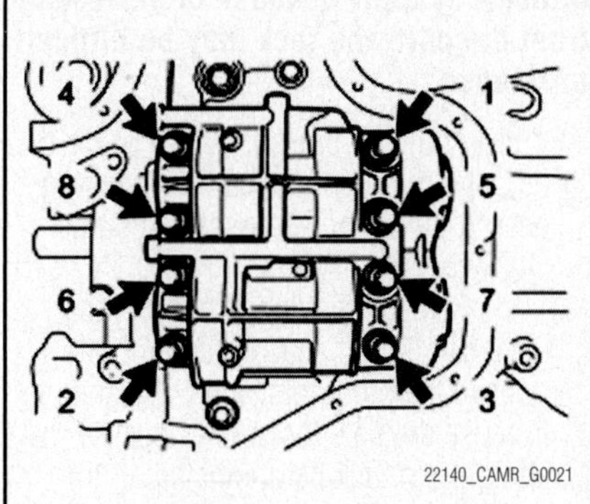

Fig. 56 Sub-assembly bolt removal sequence

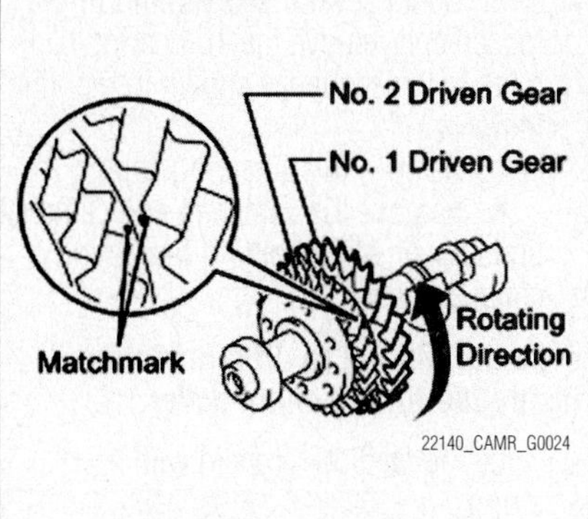

Fig. 59 Rotate the driven gear No. 1 of balance shaft No. 1

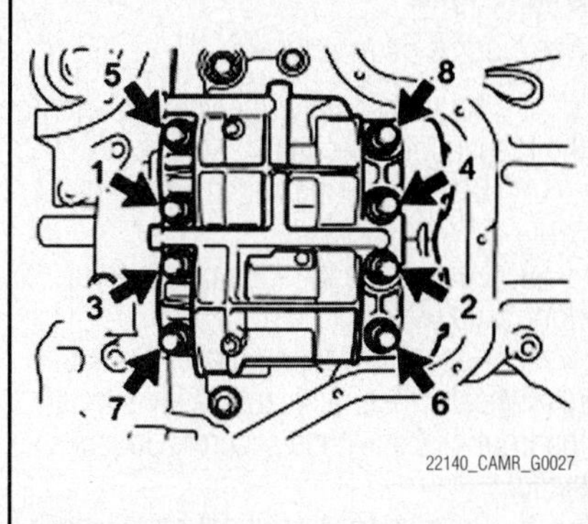

Fig. 62 Balance shaft housing bolt tightening sequence

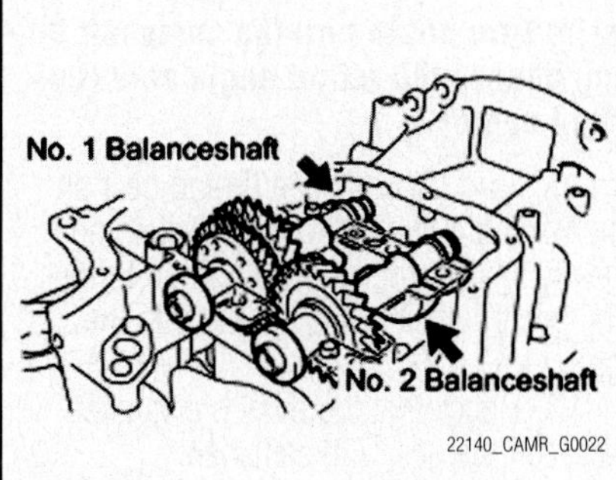

Fig. 57 Removing the No. 1 and No. 2 balance shafts

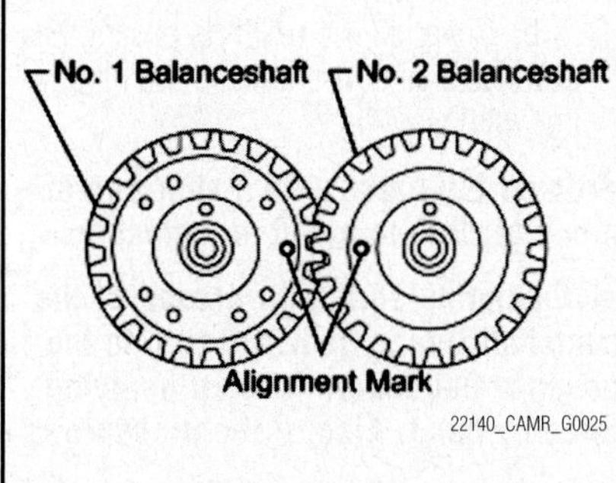

Fig. 60 Aligning marks of the No. 1 and No. 2 balance shafts

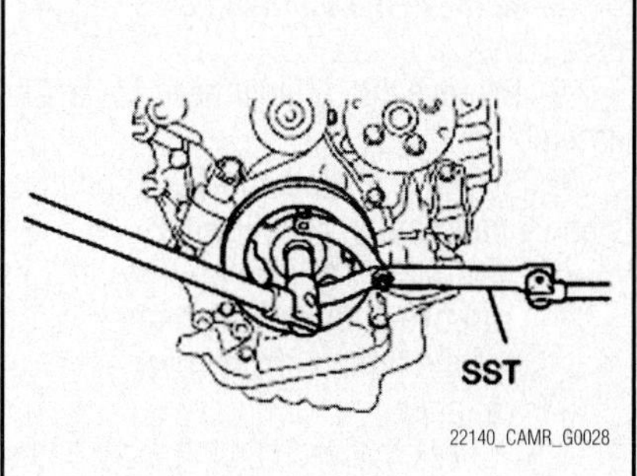

Fig. 63 Marking the bolt head and tightening procedure

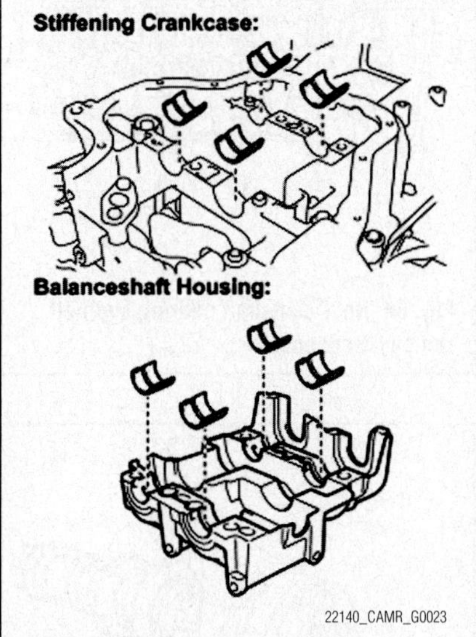

Fig. 58 Identifying and removing the balance shaft bearings

➡Confirm that the match marks on driven gears No. 1 and No. 2 are matched.

10. Install No. 1 and No. 2 balance shaft sub-assembly. Rotate the driven gear No. 1

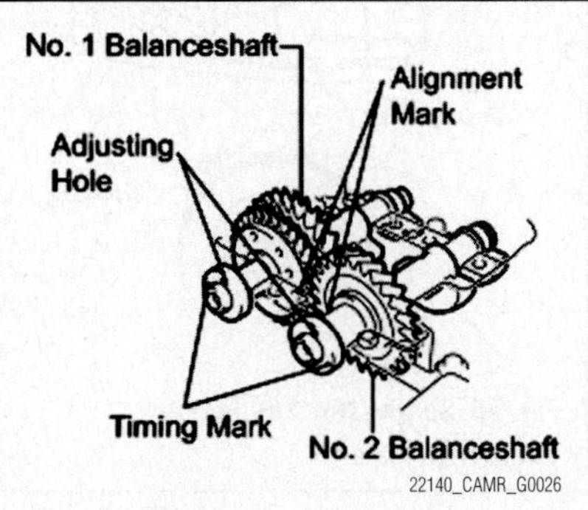

Fig. 61 Placing the No. 1 and No. 2 balance shafts on the crankcase

of balance shaft No. 1 in the rotating direction until it hits the stopper.

11. Align the alignment marks of the No. 1 and No. 2 balance shafts as shown.

12. Place the No. 1 and No. 2 balance shafts on the crankcase.

13. Apply a light coat of engine oil under the heads of the balance shaft housing bolts.

14. Install the balance shaft housing bolts. The balance shaft housing bolts should be tightened in 2 progressive steps as follows:

a. Tighten the eight balance shaft housing bolts in sequence to: 16 ft. lbs. (22 Nm).

b. Mark the front side of each balance shaft housing bolt head with paint. Retighten the bolts by 90°. Check that the paint marks are now at a 90° angle to the front.

15. To complete installation, reverse remaining removal procedure.

16. Check the engine for leaks.

CAMSHAFT AND TIMING GEAR

REMOVAL & INSTALLATION

CAUTION

All models are equipped with a Supplemental Restraint System (SRS), which uses an air bag. Whenever working near any of the SRS components, such as the impact sensors, the air bag module, steering column and instrument panel, disable the SRS.

2.4L Engine

See Figures 64 through 75.

1. Before servicing the vehicle, refer to the Precautions Section.
2. Disconnect the negative battery cable.
3. Loosen the lug nuts on the front right hand wheel.
4. Apply the parking brake, block the rear wheels, then raise and safely support the front of the vehicle securely on jackstands.
5. Remove the front right hand wheel.
6. Remove the left and right hand under cover.
7. Remove the No. 1 engine cover sub-assembly and 2 nuts.
8. Remove the ignition coil assembly.
9. Remove the cylinder head cover sub-assembly
10. Set No. 1 Cylinder to TDC/Compression by performing the following:
    - Turn the crankshaft pulley until its groove and the timing mark "0" of the timing chain cover are aligned
    - Check that each timing mark of the camshaft timing gear and sprocket is aligned with each timing mark located on the No. 1 and No. 2 bearing caps as shown in the illustration. If not, turn the crankshaft by 1 revolution (360°) to align the timing marks

➡Do not turn the crankshaft without the chain tensioner.

11. Remove No. 1 Chain tensioner assembly. Remove the 2 nuts, tensioner and gasket.
12. Remove the No. 2 camshaft by performing the following:
    a. While holding the camshaft with a wrench, loosen the camshaft timing set bolt.
    b. Using several steps, uniformly loosen and remove the 10 bearing cap bolts in the sequence shown in the illustration.
    c. While holding the No. 2 camshaft by hand, remove the camshaft timing sprocket set bolt.
    d. Remove the camshaft timing sprocket from the No. 2 camshaft with the timing chain wrapped on the sprocket.
    e. Remove the camshaft timing sprocket from the timing chain.
13. Remove the No. 1 camshaft by performing the following:
    a. Using several steps, uniformly loosen and remove the 10 bearing cap bolts in the sequence shown in the illustration.
    b. Remove the 5 bearing caps.
    c. Remove the camshaft and camshaft timing gear while holding the timing chain by hand.

➡Be careful not to drop anything inside the timing chain cover.

    d. Tie the timing chain with a string.

14. Remove the camshaft timing gear assembly by performing the following:
    a. Clamp the camshaft in a vise, and make sure that the camshaft timing gear does not rotate.
    b. Cover all the oil ports except the advance side port shown in the illustration with vinyl tape

➡Cover the paths with a shop rag or piece of cloth to avoid oil splashes.

➡Depending on the air pressure, the camshaft timing gear will turn to the advance angle side without applying force by hand. Also, if the pressure is difficult to apply because of air leakage from the port, the lock may be difficult to release.

    c. Apply air pressure of 14 psi to the oil path, then turn the camshaft timing gear in the advance direction (counterclockwise) by hand.
    d. Remove the flange bolt of the camshaft timing gear. Be sure not to remove the other four bolts. If planning to reuse the gear, be sure to release the straight pin lock before installing the gear.

To install:

15. Put the camshaft timing gear and camshaft together with the straight pin and key groove misaligned.

➡Be sure not to turn the camshaft timing gear to the retard angle side (the right angle).

16. Turn the camshaft timing gear as shown in the illustration while pushing it gently against the camshaft. Push further at the position where the pin fits into the groove.
17. Check that there is no clearance between the gear and camshaft.

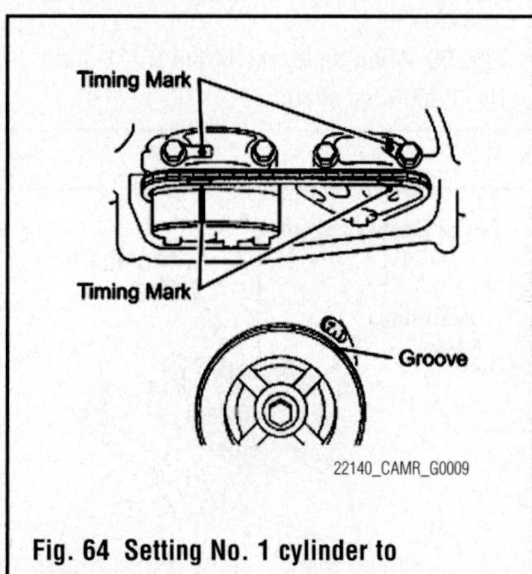

Fig. 64 Setting No. 1 cylinder to TDC/Compression

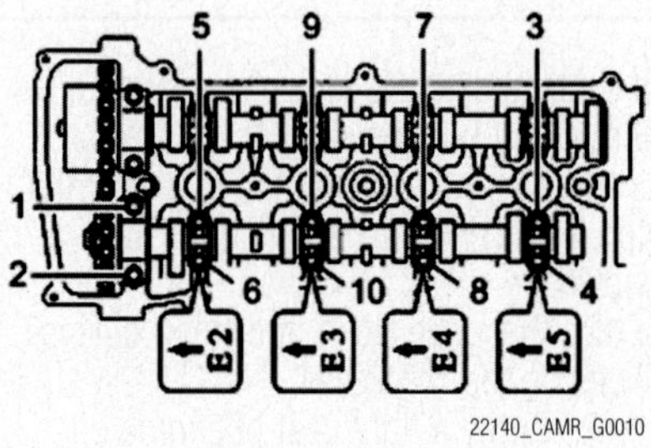

Fig. 65 No. 2 camshaft bearing cap bolt removal sequence

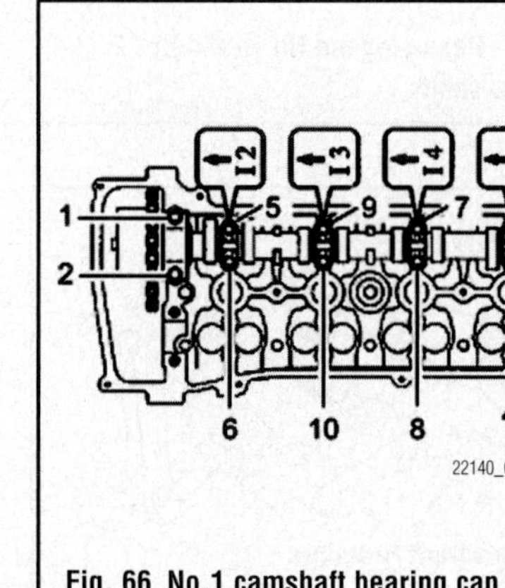

Fig. 66 No.1 camshaft bearing cap bolt removal sequence

Fig. 67 Removing the flange bolt from the camshaft timing gear

18. Tighten the flange bolt with the camshaft timing gear fixed in place and tighten to 40 ft. lbs. (54 Nm).

19. Check that the camshaft timing gear can move to the retard angle side (the right direction) and is locked in the most retarded position.

20. To install the No.1 camshaft, perform the following:

a. Apply a light coat of engine oil to the journal portion of the camshaft.

b. Install the timing chain onto the camshaft timing gear with the paint mark 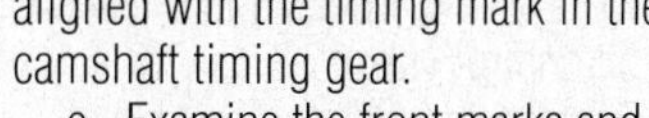 aligned with the timing mark in the camshaft timing gear.

c. Examine the front marks and numbers, and check that the order is as shown in the illustration below. Then install the bearing caps into the cylinder head.

d. Apply a light coat of engine oil to the threads and under the heads of the bearing cap bolts.

e. Using several steps, uniformly tighten the 10 bearing cap bolts in the sequence shown in the illustration. Tighten the following to specification:

- No. 1 Bearing cap: 22 ft. lbs. (30 Nm)
- No. 3 Bearing cap: 80 inch lbs. (9 Nm)

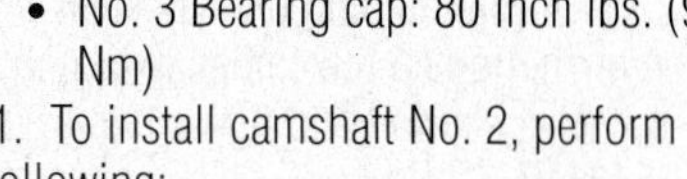

21. To install camshaft No. 2, perform the following:

a. Apply a light coat of engine oil to the journal portion of the No. 2 camshaft.

b. Put the No. 2 camshaft on the cylinder head with the paint mark of the chain aligned with the timing mark on the camshaft timing sprocket.

c. While holding the No. 2 camshaft by hand, temporarily tighten the camshaft timing sprocket set bolt.

d. Examine the front marks and numbers, and check that the order is as shown in the illustration. Then install the bearing caps onto the cylinder head.

e. Apply a light coat of engine oil to the threads and under the heads of the bearing cap bolts.

f. Using several steps, uniformly tighten the 10 bearing cap bolts in the sequence shown in the illustration. Tighten the following to specification:

- No. 1 Bearing cap: 22 ft. lbs. (30 Nm)

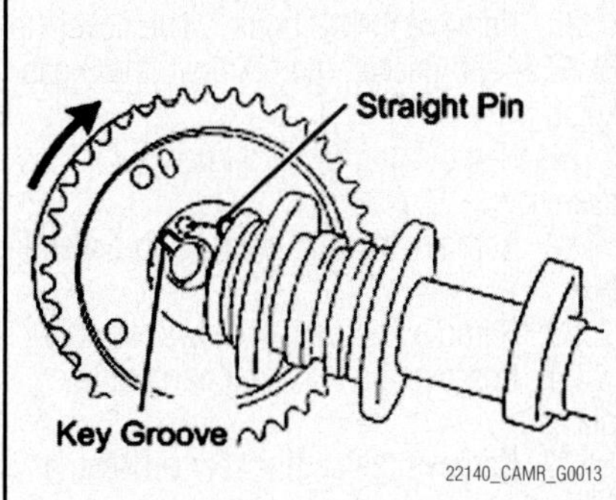

Fig. 68 No. 1 camshaft straight pin and key groove misaligned

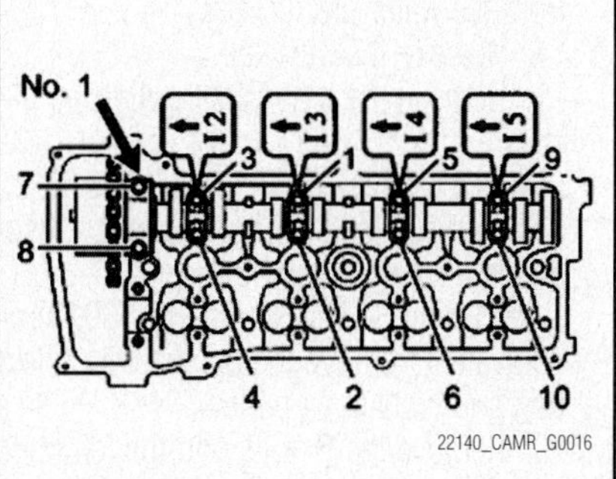

Fig. 71 No. 1 camshaft bearing cap tightening sequence

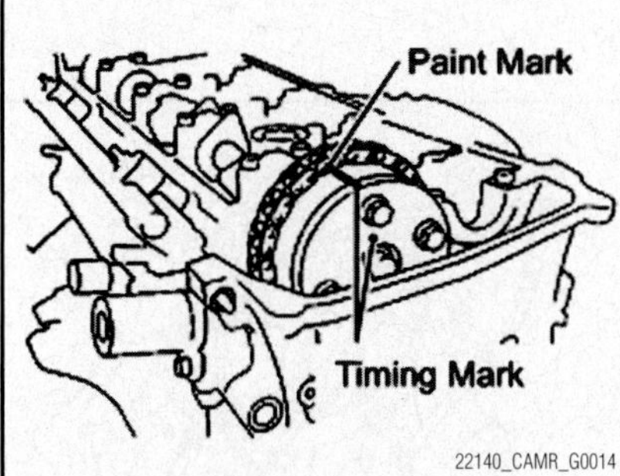

Fig. 69 Aligning No. 1 camshaft paint mark with the timing mark

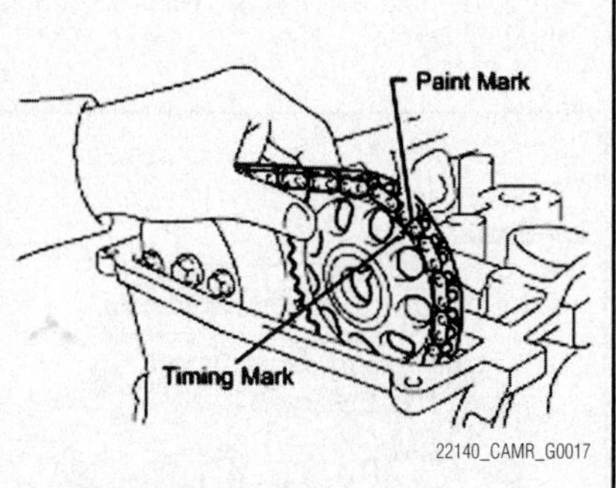

Fig. 72 Aligning No. 2 camshaft paint mark with the timing mark

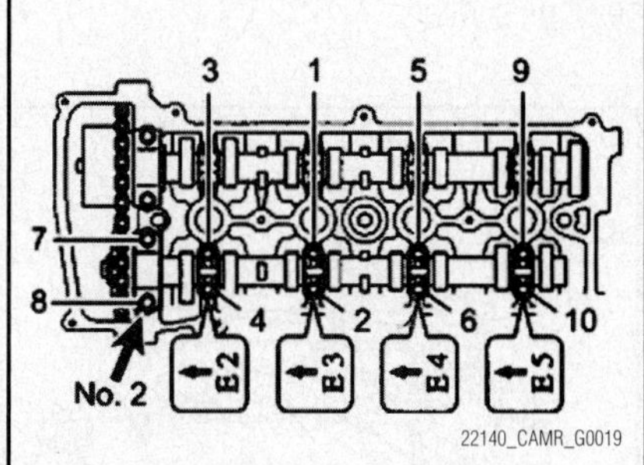

Fig. 74 No. 2 camshaft bearing cap tightening sequence

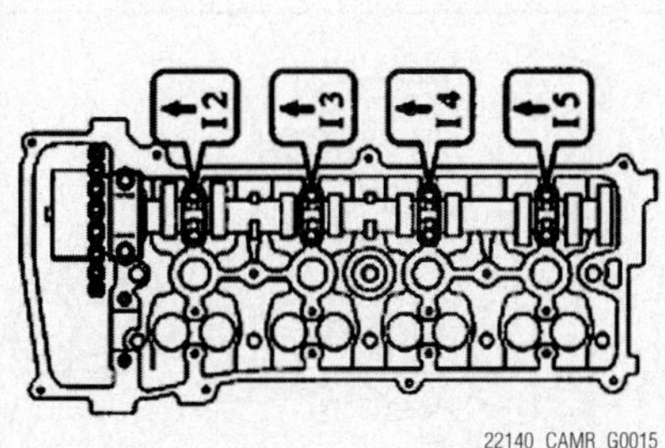

Fig. 70 Checking the No. 1 camshaft bearing cap front marks, numbers and order

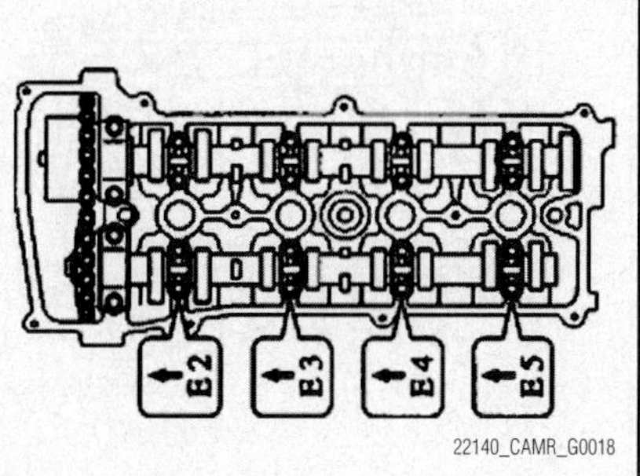

Fig. 73 Checking the No. 2 camshaft bearing cap front marks, numbers and order

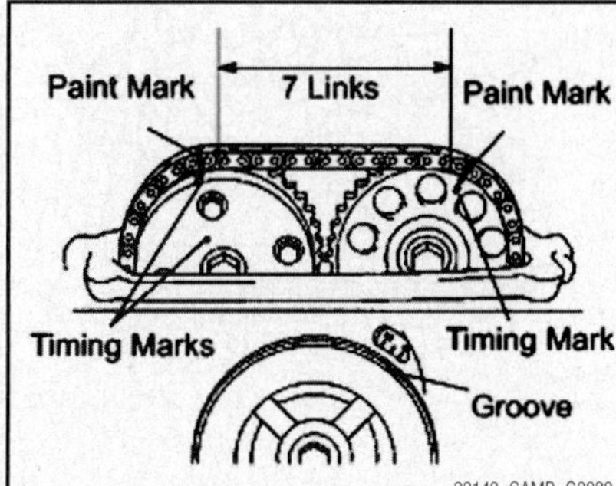

Fig. 75 Aligning paint marks on the chain with the timing marks on the camshaft timing gear and camshaft timing sprocket

- No. 3 Bearing cap: 80 inch lbs. (9 Nm)

g. While holding the camshaft with a wrench, tighten the camshaft timing sprocket set bolt and tighten to 40 ft. lbs. (54 Nm).

h. Check that the paint marks on the chain are aligned with the timing marks on the camshaft timing gear and camshaft timing sprocket. Also, check that the crankshaft pulley groove is aligned with the timing mark "0" of the timing chain cover.

22. To complete installation, reverse remaining removal procedure.
23. Check engine for oil leaks.
24. Connect the negative battery cable.

3.5L Engine

See Figures 76 through 99.

1. Before servicing the vehicle, refer to the Precautions Section.
2. Remove the engine assembly.
3. Install on engine stand.
4. Remove the oil filler cap and gasket.
5. Remove the spark plugs and ignition coil assembly.
6. Remove the drain plug and gasket.
7. Remove the ventilation valve.
8. Remove the 4 bolts and 4 camshaft position sensors.
9. Remove the 4 bolts and 4 camshaft timing oil control valves.
10. Remove the bolt and crankshaft position sensor.
11. Remove the 2 oil pipe unions and oil pipe. Remove the LH oil control valve filter and gaskets.
12. Remove the oil pipe bolt. Remove the 2 oil pipe unions and oil pipe. Remove the RH oil control valve filter and gaskets.
13. Remove the cylinder block water drain cock sub-assembly, as follows:
    a. Remove the water drain cocks from the cylinder block.
    b. Remove the water drain cock plugs from the water drain cocks.
14. Remove the oil filter.
15. Remove the crankshaft pulley, as follows:
    a. Using SST (SST: 09213-70011, SST: 09330-00021) or equivalent, loosen the crankshaft pulley bolt.
    b. Using SST (SST: 09950-50013) or equivalent, remove the crankshaft pulley bolt and crankshaft pulley.
16. Remove the 6 bolts and the left hand No. 1 front engine mounting bracket. Using Torx® socket wrench E8, remove the 2 stud bolts.
17. Remove the water inlet housing, as follows:
    a. Remove the 2 nuts, water inlet and thermostat.
    b. Remove the gasket.
    c. Remove the drain cock plug.
    d. Remove the drain cock.
    e. Remove the 2 stud bolts.
    f. Remove the 2 bolts, nut, and water inlet housing.
    g. Remove the 2 O-rings.
18. Remove the water outlet, as follows:
    a. Remove the 2 bolts, 4 nuts and water outlet.
    b. Remove the 2 gaskets and O-ring.
19. Remove the 12 bolts, valve cover (for Bank 1) and gasket, then remove the 3 gaskets.
20. Remove the 12 bolts, valve cover (for Bank 2) and gasket, then remove the 3 gaskets.
21. Remove the No. 2 oil pan sub-assembly.
22. Remove the oil strainer sub-assembly.
23. Remove the oil pan sub-assembly.
24. Remove the No. 1 oil pan baffle plate.
25. Remove the engine rear oil seal, as follows:
    a. Remove the 6 bolts.
    b. Using a screwdriver with the tip taped, pry out the oil seal retainer. Be careful not to damage the oil seal retainer.

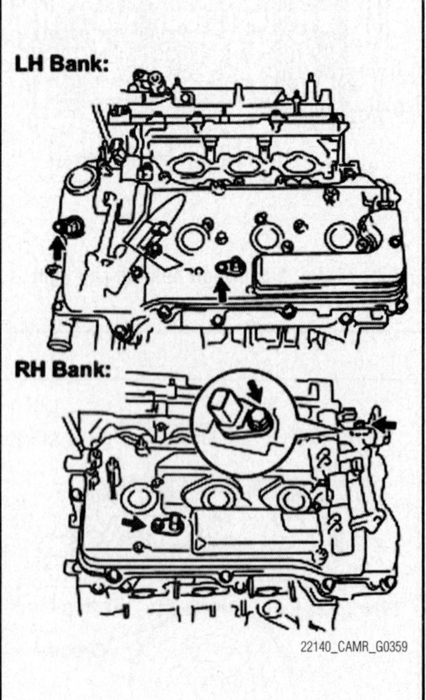

Fig. 76 Camshaft position sensor bolts

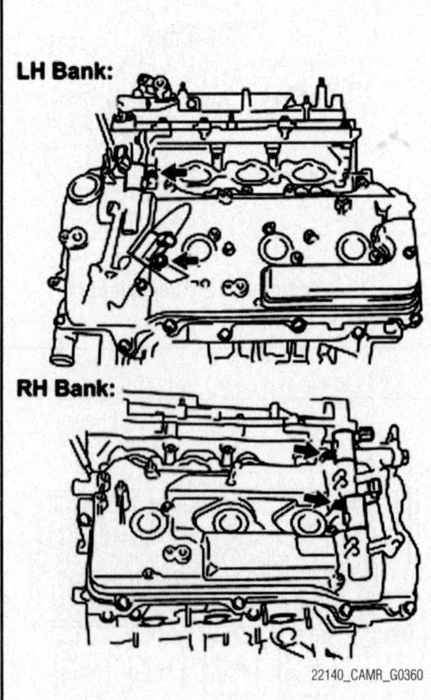

Fig. 77 Camshaft timing oil control valve assembly bolts

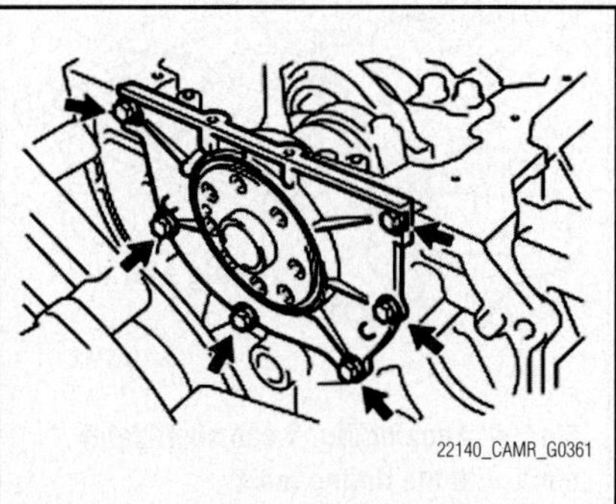

Fig. 78 Engine rear oil seal bolts

Fig. 79 Removing 2 pulley set keys from the crankshaft

➡Be careful not to damage the engine rear oil seal retainer.

➡Tape the screwdriver tip before use.

26. Place the oil seal retainer on wooden blocks. Using a screwdriver and a hammer, tap out the oil seal.
27. Remove the water pump assembly.
28. Remove the timing chain cover sub-assembly.
29. Remove timing chain case oil seal.
30. Set the No. 1 cylinder to TDC/compression.
31. Remove the No. 1 chain tensioner assembly.
32. Remove the chain tensioner slipper.
33. Remove the chain sub-assembly.
34. Remove the idle sprocket assembly.
35. Remove the No. 1 chain vibration damper.
36. Remove crankshaft timing sprocket. Remove the 2 pulley set keys from the crankshaft.
37. Remove camshaft timing gears and No. 2 chain (for Bank 1), as follows:
    a. While raising the No. 2 chain tensioner, insert a pin of 0.039 in (1.0 mm) into the hole to fix the No. 2 chain tensioner.

➡Be careful not to damage the cylinder head with the wrench.

➡Do not disassemble the camshaft timing gear assemblies.

    b. Hold the hexagonal portion of the camshaft with a wrench, and remove the 2 bolts and 2 camshaft timing gears.
    c. Remove the No. 2 chain.
38. Remove the bolt and No. 2 chain tensioner assembly.
39. Remove camshaft bearing cap (for Bank 1), as follows:
    a. Check that the camshafts are positioned as shown in the illustration.
    b. Uniformly loosen and remove the 8 bearing cap bolts in the sequence shown in the illustration.
    c. Uniformly loosen and remove the 12 bearing cap bolts in the sequence shown in the illustration. Uniformly loosen the bolts while keeping the camshaft level.
    d. Remove the 5 bearing caps.
40. Remove the camshaft.
41. Remove the No. 2 camshaft.
42. If necessary, remove the right hand camshaft housing sub-assembly by prying between the cylinder head and the camshaft housing with a screwdriver with the tip taped.

➡Be careful not to damage the contact surfaces of the cylinder head and the camshaft housing.

43. Remove the camshaft timing gears and No. 2 chain (for Left-hand Bank), as follows:
    a. While pushing down the No. 3 chain tensioner, insert a pin of 1.0 mm (0.039 in.) into the hole to fix the No. 3 chain tensioner.
    b. Hold the hexagonal portion of the camshaft with a wrench, and remove the 2 bolts and 2 camshaft timing gears.

➡Be careful not to damage the cylinder head with the wrench.

➡Do not disassemble the camshaft timing gear assemblies.

    c. Remove the No. 2 chain.
44. Remove the bolt and No. 3 chain tensioner.
45. Remove the camshaft bearing cap (bank 2), as follows:
    a. Make sure that the knock pin of the camshaft is positioned as shown in the illustration.

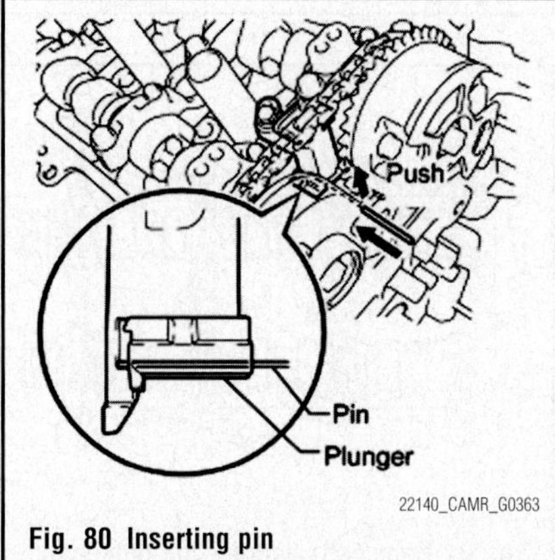

Fig. 80 Inserting pin

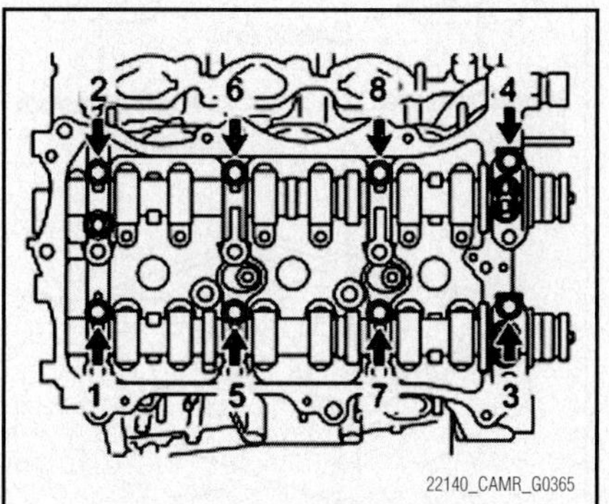

Fig. 82 8 bearing cap bolts removal sequence —bank 1

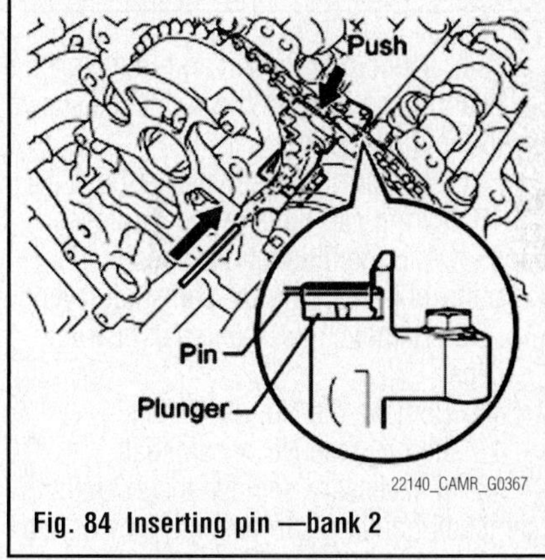

Fig. 84 Inserting pin —bank 2

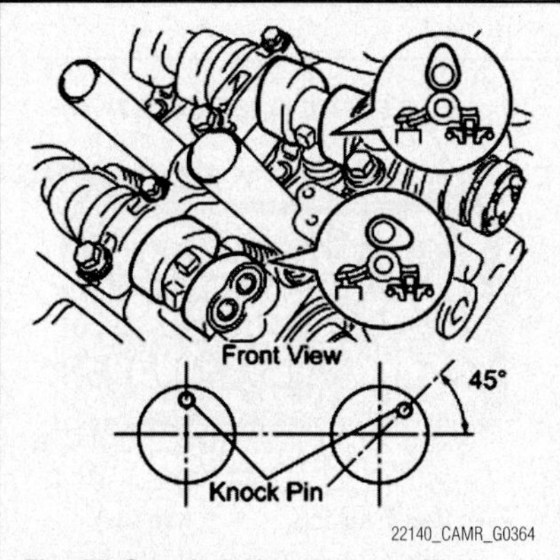

Fig. 81 Camshafts knock pin positioning—bank 1

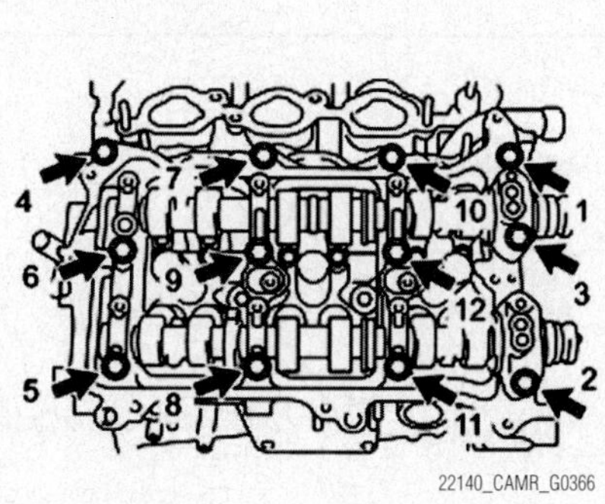

Fig. 83 12 bearing cap bolts removal sequence —bank 1

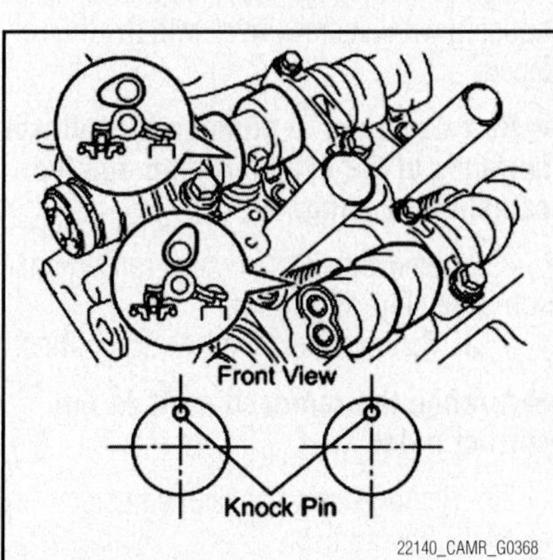

Fig. 85 Camshafts knock pin positioning—bank 2

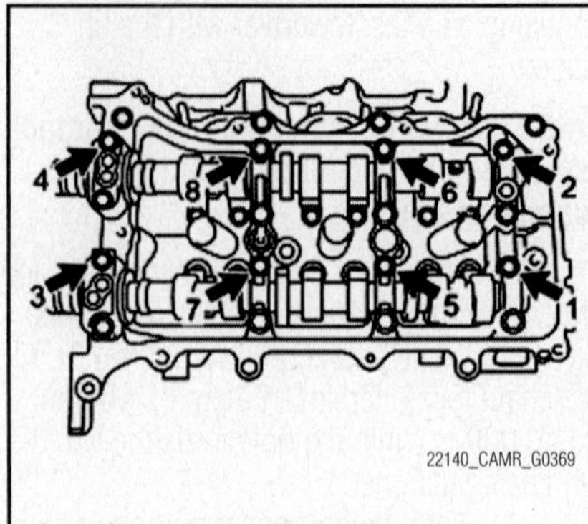

Fig. 86 8 bearing cap bolts removal sequence —bank 1

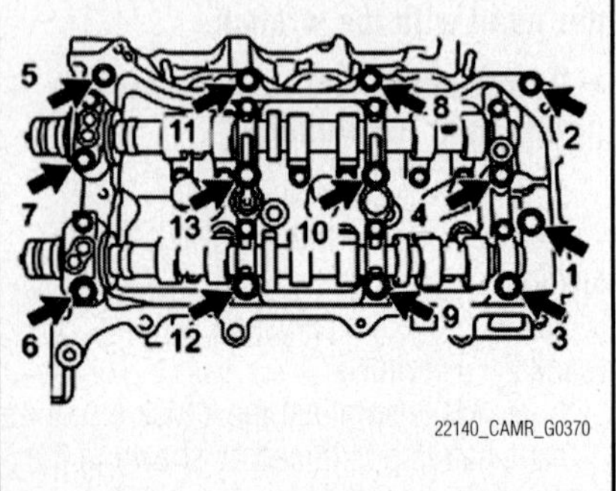

Fig. 87 13 bearing cap bolts removal sequence —bank 1

b. Uniformly loosen and remove the 8 bearing cap bolts in the sequence shown in the illustration.

c. Uniformly loosen and remove the 13 bearing cap bolts in the sequence shown in the illustration. Loosen the bolts while keeping the camshaft level.

d. Remove the 5 camshaft bearing caps.

46. Remove the No. 3 camshaft.

47. Remove the No. 4 camshaft.

48. If necessary, remove the left hand camshaft housing sub-assembly by prying between the cylinder head and the camshaft housing with a screwdriver with the tip taped.

➡Be careful not to damage the contact surfaces of the cylinder head and the camshaft housing.

49. Remove the No. 1 valve rocker arm sub-assembly, as follows:

a. Remove the 24 valve rocker arms.

➡Arrange the removed parts in the correct order.

50. Remove the valve lash adjuster assembly, as follows:

a. Remove the 24 valve lash adjusters from the cylinder head.

➡Arrange the removed parts in the correct order.

To install:

➡Keep the lash adjuster free of dirt and foreign objects. Only use clean engine oil.

51. Install the valve lash adjuster assembly, as follows:

a. Place the lash adjuster into a container filled with engine oil. Insert the SST 09276-75010 tip into the lash adjuster's plunger and use the tip to press down on the check ball inside the plunger.

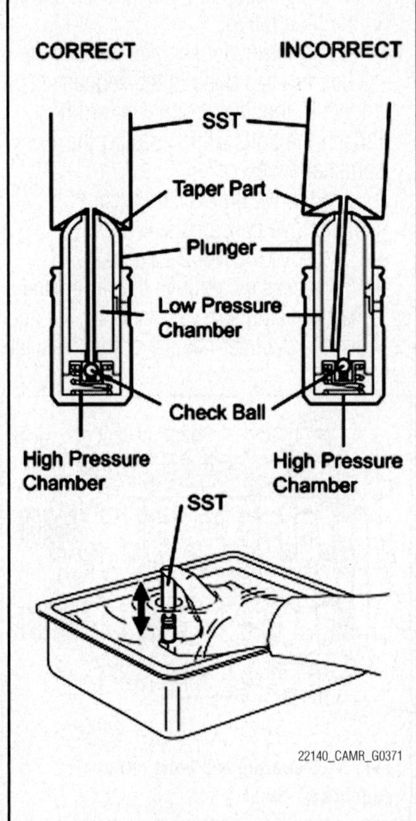
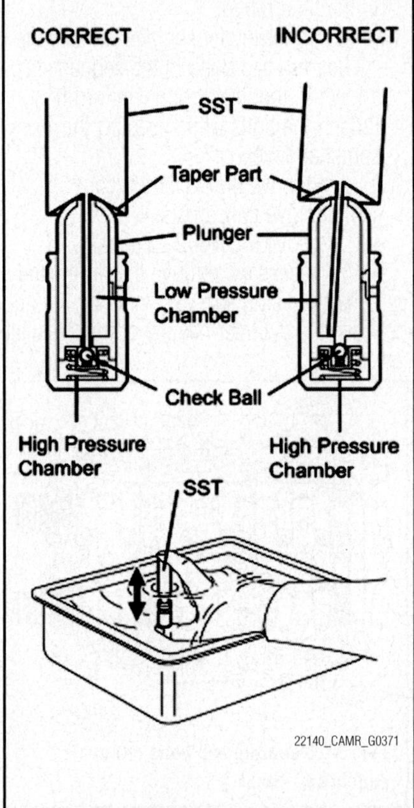

Fig. 88 Installing lash adjuster assembly

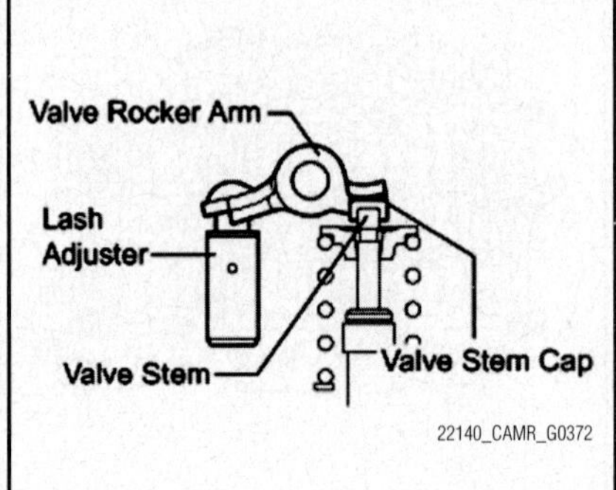

Fig. 89 Installing valve rocker arms

b. Squeeze the SST and lash adjuster together to move the plunger up and down 5 to 6 times.

c. Check the movement of the plunger and bleed the air. Make sure the plunger moves up and down.

➡When bleeding air from the high-pressure chamber, make sure that the tip of the SST is actually pressing the check ball as shown in the illustration. If the check ball is not pressed, air will not bleed.

d. After bleeding the air, remove the SST. Then, try to press the plunger

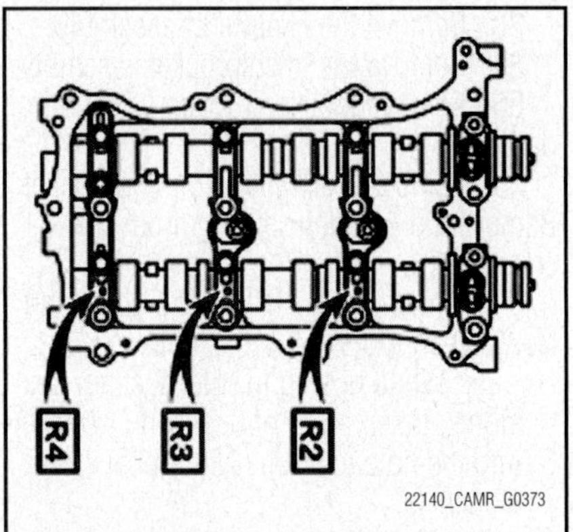

Fig. 90 Camshaft bearing caps placement

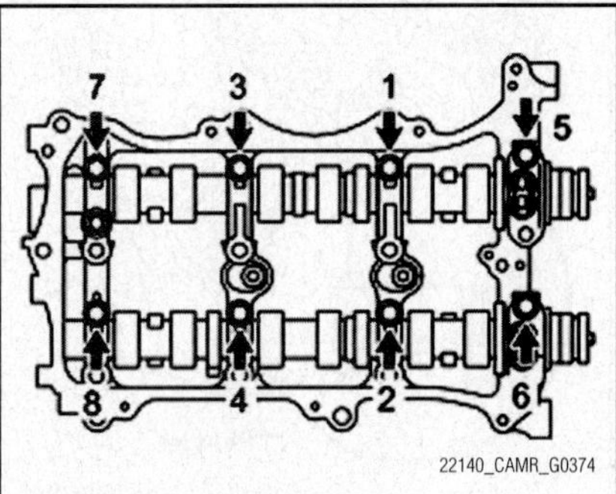

Fig. 91 Camshafts bearing cap tightening sequence

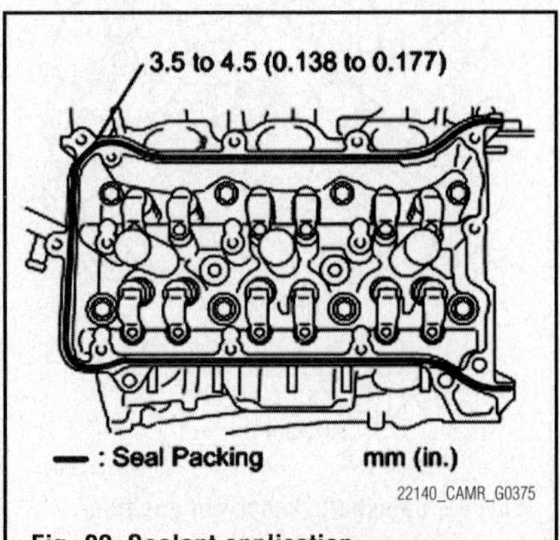

Fig. 92 Sealant application

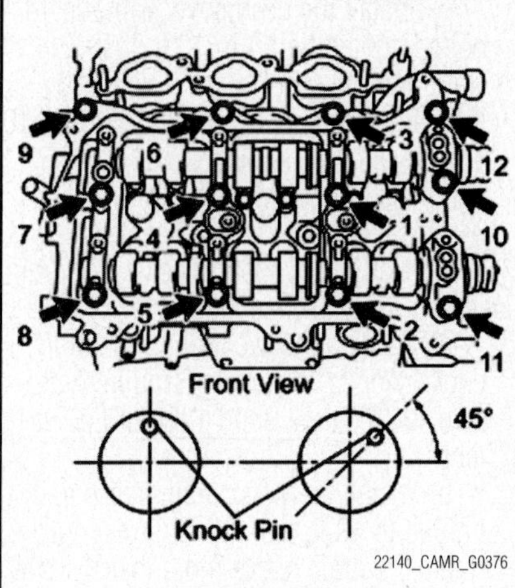

Fig. 93 Camshaft bolt tightening sequence and knock pin position

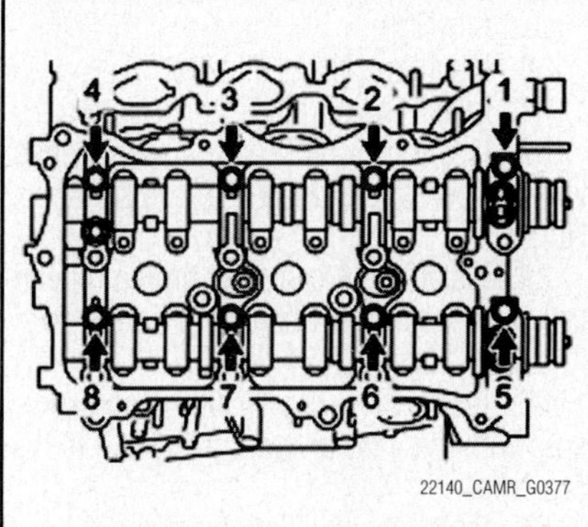

Fig. 94 Camshaft bolt tightening sequence

quickly and firmly with a finger. Make sure the plunger is very difficult to move. If the result is not as specified, replace the lash adjuster.

➡**Install the lash adjuster to the same place it was removed from.**

e. Install the lash adjusters.

52. Install the No. 1 valve rocker arm sub-assembly, as follows:

a. Apply engine oil to the lash adjuster tips and valve stem cap ends.

L2 L3 L4

22140_CAMR_G0378

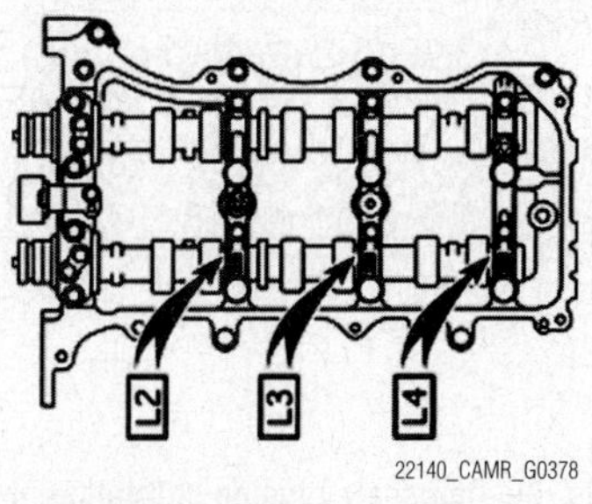

Fig. 95 Camshaft bearing cap positioning

b. Make sure that the valve rocker arms are installed as shown in the illustration.

53. Install the right hand camshaft bearing cap, as follows:

a. Apply engine oil to the camshaft journals, camshaft housing and bearing caps.

b. Install the camshaft and No. 2 camshaft to the right camshaft housing.

c. Make sure of the marks and numbers on the camshaft bearing caps and place them in each proper position and direction.

d. Temporarily tighten the 8 bearing cap bolts to 7 ft. lbs. (10 Nm) in the order shown in the illustration.

54. Install the right hand camshaft housing sub-assembly, as follows:

a. Apply seal packing in a continuous line as shown in the illustration. Seal packing: Toyota Genuine Seal Packing Black, Three Bond 1207B or equivalent. Seal diameter: 3.5 to 4.5 mm (0.138 to 0.177 in.).

➡**Remove any oil from the contact surface. Install the camshaft housing sub-assembly within 3 minutes and tighten the bolts within 15 minutes after applying sealant. Do not start the engine for at least 2 hours after installing.**

b. Make sure that the knock pins of the camshafts are positioned as shown. Install the right hand camshaft housing and tighten the 12 bolts in the order shown in the illustration to 21 ft. lbs. (28 Nm).

➡**When installing the camshaft housing RH, it is necessary to correctly position the camshafts as shown in the illustration. Failure to correctly position these parts may result in damage due to contact between the pistons and valves. If a camshaft is rotated with a piston at TDC, valve contact will occur.**

➡**If any of the bolts are loosened during installation, remove the camshaft housing, clean the installation surfaces, and reapply seal packing.**

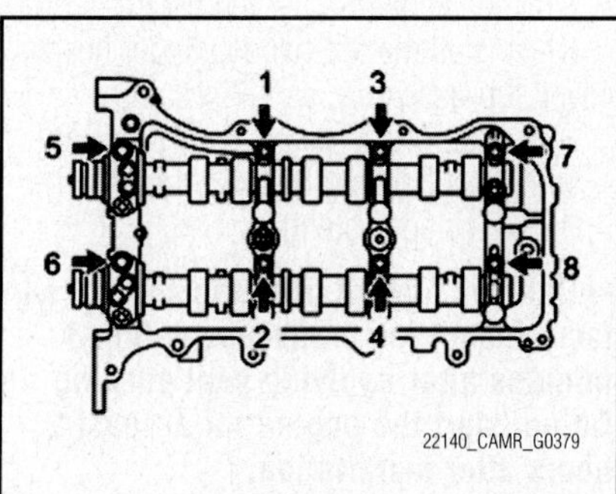

Fig. 96 Camshaft bolt tightening sequence

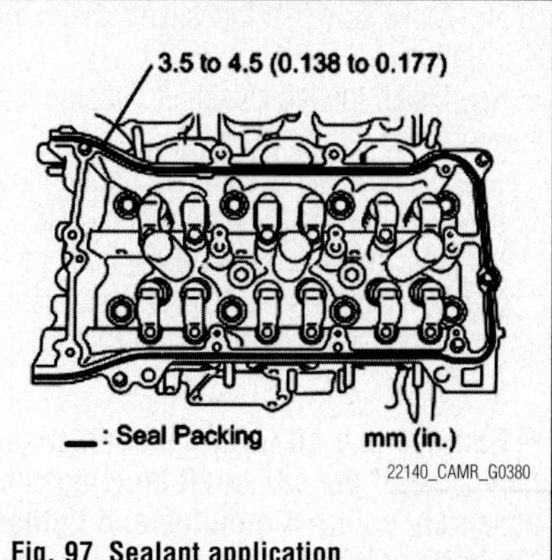

Fig. 97 Sealant application

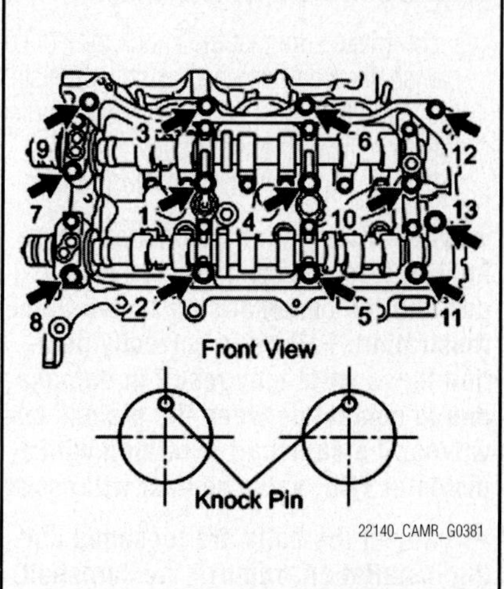

Fig. 98 Camshaft housing bolt tightening sequence and knock pin position

➡**If the camshaft housing is removed because any of the bolts are loosened during installation, make sure that the previously applied seal packing does not enter any oil passages.**

c. Tighten the 8 bolts to 12 ft. lbs. (16 Nm) in the order shown in the illustration.

55. Install the camshaft bearing cap (bank 2), as follows:

a. Apply engine oil to the camshaft journals, camshaft housing and bearing caps.

b. Install the No. 3 camshaft and No. 4 camshaft to the left hand camshaft housing.

c. Make sure of the marks and numbers on the camshaft bearing caps and place them in each proper position and direction.

d. Temporarily tighten the 8 bolts in the order shown in the illustration to 7 ft. lbs. (10 Nm).

56. Make sure that the valve rocker arm is installed.

57. Install the left camshaft housing sub-assembly, as follows:

a. Apply seal packing in a continuous line as shown in the illustration. Seal packing: Toyota Genuine Seal Packing Black, Three Bond 1207B or equivalent. Seal diameter: 3.5 to 4.5 mm (0.138 to 0.177 in.).

➡Remove any oil from the contact surface. Install the camshaft housing sub-assembly within 3 minutes and tighten the bolts within 15 minutes after applying sealant. Do not start the engine for at least 2 hours after installing.

b. Make sure that the knock pins of the camshafts are positioned as shown. Install the left hand camshaft housing and tighten the 13 bolts in the order shown in the illustration to 21 ft. lbs. (28 Nm).

➡When installing the camshaft housing LH, it is necessary to correctly position the camshafts as shown in the illustration. Failure to correctly position these parts may result in damage due to contact between the pistons and valves. If a camshaft is rotated with a piston at TDC, valve contact will occur.

➡If any of the bolts are loosened during installation, remove the camshaft housing, clean the installation surfaces, and reapply seal packing.

➡If the camshaft housing is removed because any of the bolts are loosened during installation, make sure that the previously applied seal packing does not enter any oil passages.

c. Tighten the 8 bolts to 12 ft. lbs. (16 Nm) in the order shown in the illustration.

58. Install the No. 2 chain tensioner assembly with the bolt and tighten to 15 ft. lbs. (21 Nm).

59. While pushing in the tensioner, insert a pin of 1.0 mm (0.039 in.) diameter into the hole to hold it.

60. Install the camshaft timing gears and No. 2 chain (for Right-hand Bank).

61. Install the No. 3 chain tensioner assembly with the bolt and tighten to 15 ft. lbs. (21 Nm).

62. While pushing in the tensioner, insert a pin of 1.0 mm (0.039 in.) diameter into the hole to hold it.

63. Install the camshaft timing gears and No. 2 chain (for Left-hand Bank).

64. Install the No 1 chain vibration damper with the 2 bolts and tighten to 17 ft. lbs. (23 Nm).

65. Install the No 2 chain vibration damper.

66. Install the timing gear set keys and timing gear as shown in the illustration.

67. Install the idle sprocket assembly.

68. Install the chain sub-assembly.

69. Install the chain tensioner slipper.

70. Install the No. 1 chain tensioner assembly.

71. Install the water pump assembly.

72. Install the timing chain cover sub-assembly.

73. Install the water inlet housing.

74. Install the No. 1 left front engine mounting bracket, as follows:

a. Install the No. 1 left front engine mounting bracket with the 6 bolts and tighten to 40 ft. lbs. (54 Nm).

➡Install the water inlet and mounting bracket within 15 minutes after installing the chain cover. Do not start the engine for at least 2 hours after installation.

75. Install the No. 1 oil pan baffle plate with the 7 bolts and tighten to 7 ft. lbs. (10 Nm).

76. Install the oil pan sub-assembly.

77. Install the oil strainer sub-assembly.

78. Install the No. 2 oil pan sub-assembly.

79. Install a new gasket and oil pan drain plug and tighten to 30 ft. lbs. (40 Nm).

80. Install the cylinder head cover sub-assembly, as follows:

a. Apply seal packing (Toyota Genuine Seal Packing Black, Three Bond 1207B or equivalent).

➡Remove any oil from the contact surface. Install the crankcase within 3 minutes after applying seal packing. Do not start the engine for at least 2 hours after installation.

b. Install the gasket to the head cover.

c. Install the head cover with the 12 bolts. Tighten bolt A to 15 ft. lbs. (21 Nm), and other bolts to 7 ft. lbs. (10 Nm). Be certain to tighten bolt 1.

81. Install the left-hand cylinder head cover sub-assembly, as follows:

a. Apply seal packing (Toyota Genuine Seal Packing Black, Three Bond 1207B or equivalent).

➡Remove any oil from the contact surface. Install the crankcase within 3 minutes after applying seal packing. Do not start the engine for at least 2 hours after installation.

b. Install the gasket to the head cover.

c. Install the head cover with the 14 bolts. Tighten bolt A to 15 ft. lbs. (21 Nm), and other bolts to 7 ft. lbs. (10 Nm). Be certain to tighten bolts 1 and 10.

82. Install water outlet.

83. Install the crankshaft pulley.

84. Install the oil filter element.

85. Install the cylinder block water drain cock sub-assembly, as follows:

a. Apply adhesive around the drain cocks. Adhesive: Toyota Genuine Adhesive 1324, Three Bond 1324 or Equivalent.

b. Install the water drain cocks and tighten to 18 ft. lbs. (25 Nm). Do not rotate the drain cocks more than 1 revolution (360°) after tightening the drain cocks with the specified torque. Do not loosen after setting correctly.

c. Install the water drain cock plug to the water drain cocks and tighten to 9 ft. lbs. (13 Nm).

86. Install the No. 1 oil pipe.

87. Install the oil pipe.

88. Install the crankshaft position sensor with the bolt and tighten to 7 ft. lbs. (10 Nm).

89. Install the 4 camshaft timing oil control valves with the 4 bolts and tighten to 7 ft. lbs. (10 Nm).

90. Install the 4 camshaft position sensors with the 4 bolts and tighten to 7 ft. lbs. (10 Nm).

91. Install the ventilation valve sub-assembly, as follows:

a. Apply adhesive (Toyota Genuine Adhesive 1324, Three Bond 1324 or equivalent) around the ventilation valve.

b. Install the ventilation valve and tighten to 20 ft. lbs. (27 Nm).

92. Install the 6 spark plugs and the ignition coil assembly.

93. Install the oil filler cap sub-assembly.

94. Remove the engine stand.

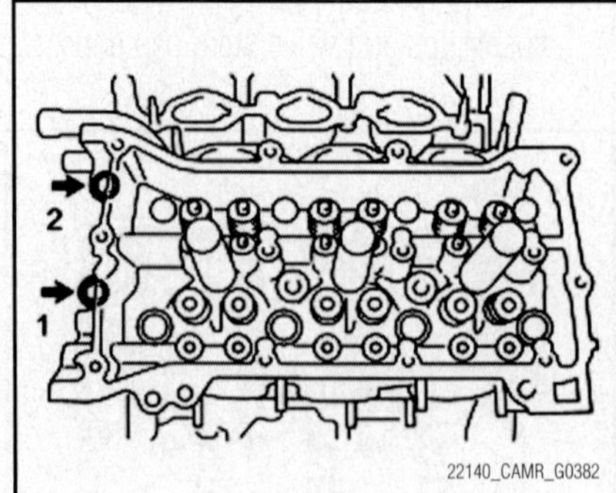

Fig. 99 Camshaft housing bolt tightening sequence

95. Install the engine assembly.

CAMSHAFT BEARING REPLACEMENT

To remove and replace the camshaft bearings, refer to the camshaft removal procedure.

CRANKSHAFT DAMPER

REMOVAL & INSTALLATION

2.4L Engine

1. Before servicing the vehicle, refer to the Precautions Section.
2. Remove the engine assembly.
3. Install on engine stand.
4. Remove the oil filler cap and gasket.
5. Remove the spark plugs and ignition coil assembly.
6. Remove the drain plug and gasket.
7. Remove the ventilation valve.
8. Remove the 4 bolts and 4 camshaft position sensors.
9. Remove the 4 bolts and 4 camshaft timing oil control valves.
10. Remove the bolt and crankshaft position sensor.
11. Remove the No. 1 oil pipe.
12. Remove the oil pipe.
13. Remove the cylinder block water drain cock sub-assembly.
14. Remove the oil filter element.
15. Remove the crankshaft pulley, as follows:
    a. Using SST (SST: 09213-70011, SST: 09330-00021) or equivalent, loosen the crankshaft pulley bolt.
    b. Using SST (SST: 09950-50013) or equivalent, remove the crankshaft pulley bolt and crankshaft pulley.

To install:

16. Install the crankshaft pulley, as follows:
    a. Align the pulley set key with the key groove of the pulley, and slide on the pulley.
    b. Using SST (SST: 09213-70011, SST: 09330-00021) or equivalent, install the pulley bolt and tighten to 184 ft. lbs. (250 Nm).
17. Install the oil filter element.
18. Install the cylinder block water drain cock sub-assembly.
19. Install the No. 1 oil pipe.
20. Install the oil pipe.
21. Install the crankshaft position sensor.
22. Install the camshaft timing oil control valve assembly.
23. Install the camshaft position sensor.
24. Install the ventilation valve sub-assembly.
25. Install the spark plugs and the ignition coil assembly.
26. Install the oil filler cap sub-assembly.
27. Remove the engine stand.
28. Install the engine assembly.

CRANKSHAFT FRONT SEAL

REMOVAL & INSTALLATION

2.4L Engine

See Figures 100 through 104.

1. Before servicing the vehicle, refer to the Precautions Section.
2. Disconnect the negative battery cable.
3. Drain the engine oil
4. Loosen the lug nuts on the left front wheel.
5. Apply the parking brake, block the rear wheels, then raise and safely support the front of the vehicle securely on jack stands.
6. Remove the left front wheel.
7. Remove the right hand front fender apron seal.
8. Remove the left and right hand engine under cover.
9. Remove the drive belt.
10. Remove the crankshaft pulley and take note of the following:
    - For TMMK made crankshaft pulley, use Special Tool 09960-10010 to fix the pulley in place and to loosen the bolt. Use Special Tool 09950-40011 to remove the pulley and bolt.
    - For TMC made crankshaft pulley, use Special Tools 09213-54015 and 09330-00021 to fix the pulley in place and to loosen the bolt. Use Special Tools 09950-50013 and 09950-40011 to remove the pulley and bolt.
11. Using a knife, cut off the oil seal lip. Using a screwdriver with the tip taped, pry out the oil seal.
12. After the removal, check the crankshaft for damage. If it is damaged, smooth the surface with 400-grit sandpaper.

To install:

13. Apply MP grease to a new oil seal lip. Keep the lip free from foreign matter.
14. Using Special Tool 09223-22010 and a hammer, tap in the oil seal until its surface is flush with the rear oil seal retainer edge.
15. For TMMK made pulleys, align the pulley set key with the key groove of the pulley. Using Special Tool 09960-10010,

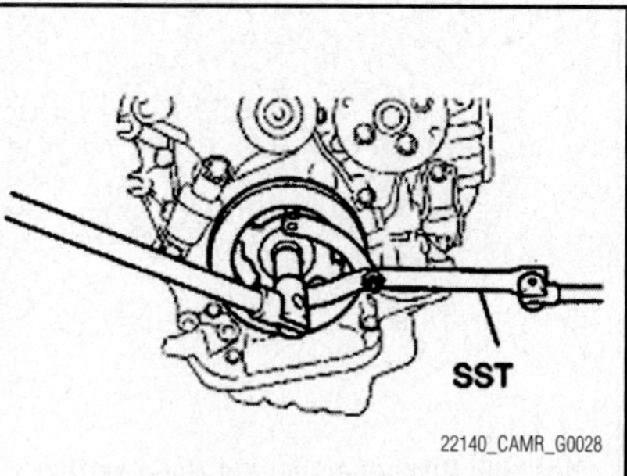

Fig. 100 Using Special Tool 09960-10010, fix the pulley in place and loosen the bolt—TMMK Pulley

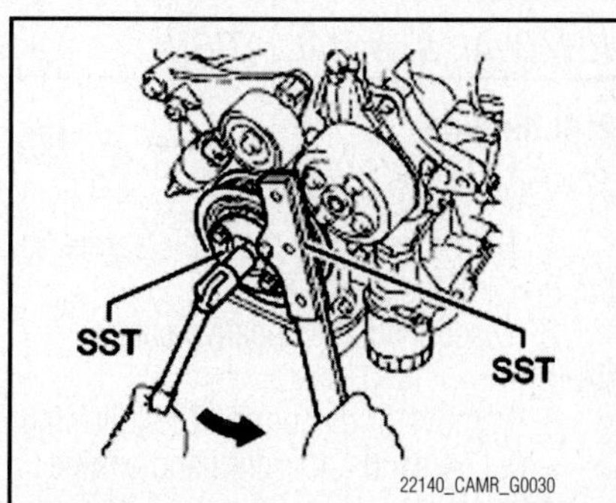

Fig. 102 Using Special Tools 09213-54015 and 09330-00021 to fix the pulley in place and to loosen the bolt—TMC Pulley

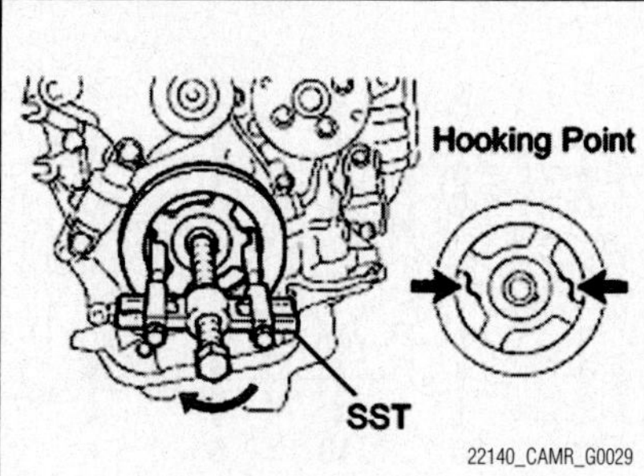

Fig. 101 Using Special Tool 09950-40011 to remove the pulley and bolt—TMMK Pulley

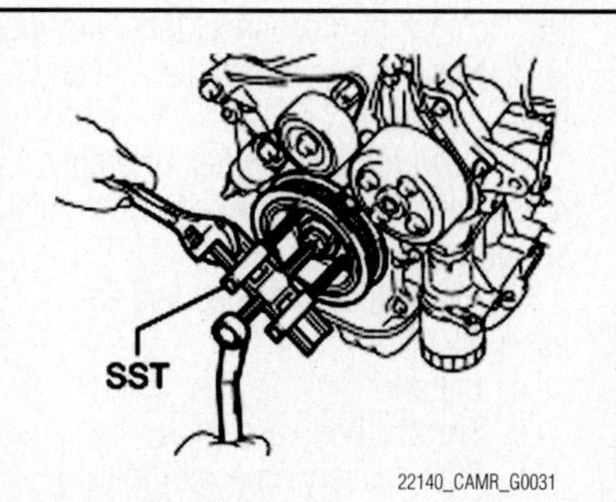

Fig. 103 Using Special Tools 09950-50013 and 09950-40011 to remove the pulley and bolt—TMC Pulley

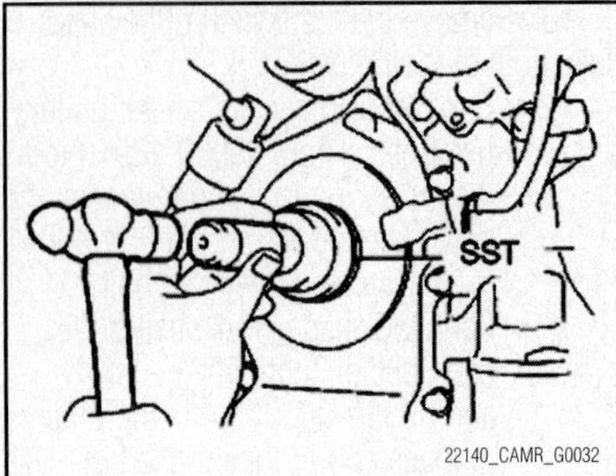

Fig. 104 Using Special Tool 09223-22010 and hammer to install new oil seal

keep the pulley in place and tighten the bolt to: 125 ft. lbs. (170 Nm).

16. For TMC made pulleys, align the pulley set key with the key groove of the pulley. Using Special Tools 09213-54015 and 09330-00021, keep the pulley in place and tighten the bolt to 133 ft. lbs. (180 Nm).

17. To complete installation, reverse remaining removal procedure.

18. Check engine for oil leaks.

3.5L Engine

See Timing Chain Cover and Seal.

CYLINDER HEAD

REMOVAL & INSTALLATION

2.4L Engine

See Figures 105 through 108.

1. Before servicing the vehicle, refer to the Precautions Section.
2. Disconnect the negative battery cable.
3. Remove or disconnect the following:
   a. Left hand and right hand engine under cover.
   b. Right hand front fender apron seal.
   c. Plastic engine cover.
4. Drain and recycle the engine coolant.
5. Drain engine oil
6. Remove or disconnect the following:
   a. Nut and the front wiper arm and blade assembly.
   b. Cowl top panel outer assembly: 4 bolts, 4 nuts and cowl top panel outer sub-assembly.
   c. Air cleaner inlet, cap and case assembly.
   d. Battery.
   e. Throttle body assembly.
   f. Fuel delivery pipe with injector.
   g. Intake manifold.
   h. Intake air control valve (PZEV).
   i. Intake manifold insulator.
   j. Front exhaust pipe assembly.
   k. Right hand engine mounting stay.
   l. Engine mount.
   m. Right hand engine mounting bracket.
   n. Drive belt.
   o. Generator assembly.
   p. Oil level gauge assembly with guide.
   q. Exhaust manifold/catalytic converter stays.
   r. Exhaust manifold converter sub-assembly.
   s. Valve cover.
   t. No. 2 and No. 1 camshafts.
   u. Radio setting condenser connector.
   v. Engine oil pressure switch connector.
   w. Engine coolant temperature sensor connector.
   x. Camshaft position sensor connector.
   y. Bolt and ground cable.
   z. No. 2 camshaft bearing.

➡ Head warpage or cracking could result from removing the bolts in the wrong order.

7. Using several steps, uniformly loosen and remove the 10 cylinder head bolts and 10 plate washers with a 10 mm bi-hexagon wrench in sequence.

➡Be careful not to damage the contact surfaces of the cylinder head and cylinder block.

8. Using a screwdriver with its tip wrapped with tape, pry between the cylinder head and cylinder block, and remove the cylinder head.
9. Remove the cylinder head gasket.

To install:

➡ Remove any oil from the contact surface. Make sure that the gasket is installed in the correct direction.

10. Place a new gasket on the cylinder block surface with the Lot No. stamp facing upward.

➡ The cylinder head bolts are tightened in 2 progressive steps.

➡ Do not drop the washers into the cylinder head.

11. Apply a light coat of engine oil to the

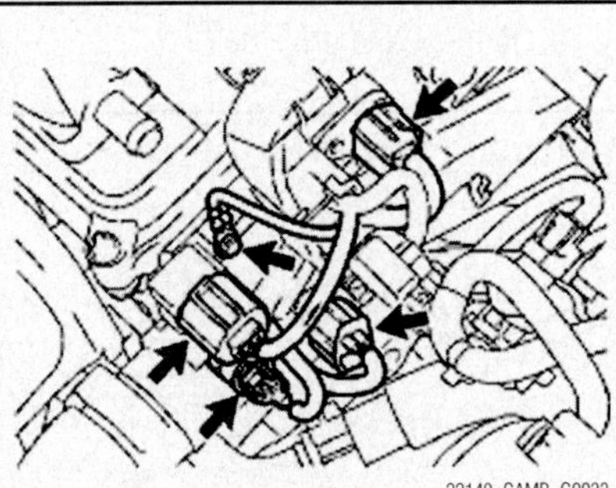

Fig. 105 Disconnecting the radio setting condenser, engine oil pressure switch, ECT and camshaft position sensor connectors—2.4L

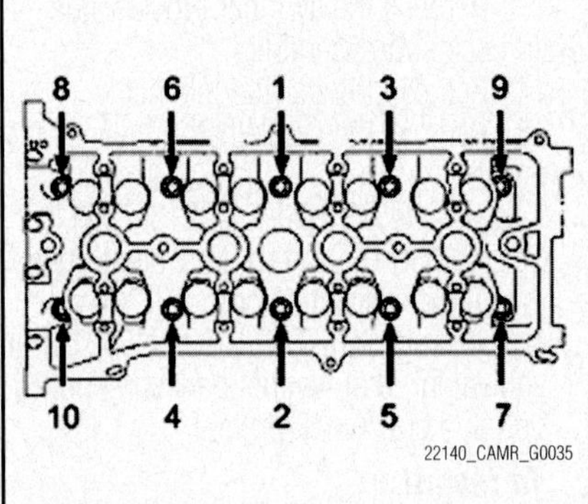

Fig. 107 Installing cylinder head bolts in sequence—2.4L

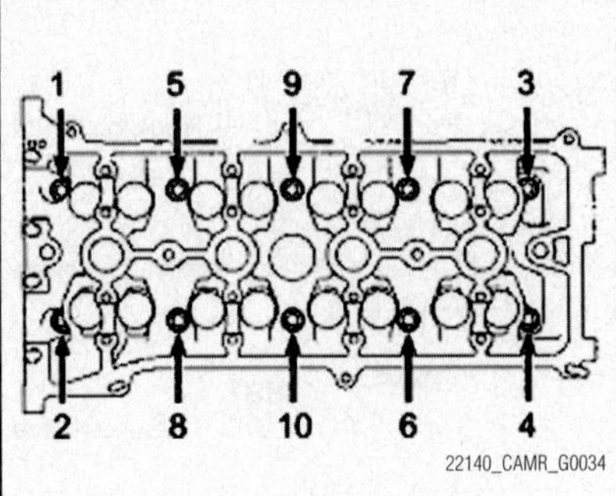

Fig. 106 Removing cylinder head bolts in sequence

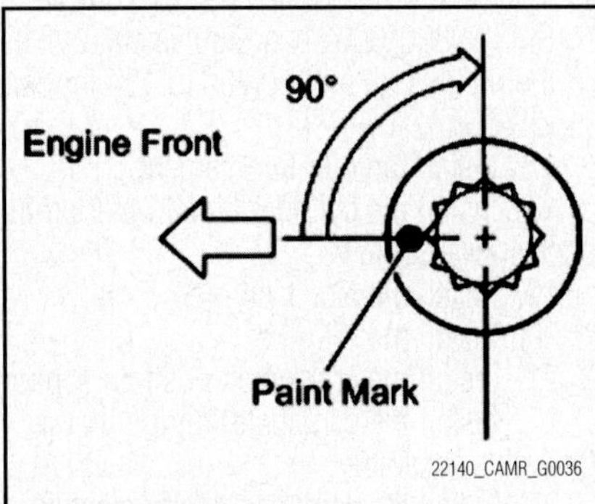

Fig. 108 Cylinder head bolt tightening procedure

bolt threads and the area beneath the bolt heads that come in contact with the washers. Install the bolts and plate washers to the cylinder head.

12. Using several steps, uniformly install and tighten the 10 cylinder head set bolts and plate washers with a 10 mm bi-hexagon wrench in sequence. Tighten the bolts to 52 ft. lbs. (70 Nm).

13. Mark the front side of the cylinder head bolt with paint. Retighten the cylinder head bolts 90° in sequence. Check that the paint mark is now at a 90° angle to the front.

14. To complete installation, reverse removal procedure.

15. Add engine oil and check for leaks.

16. Check for fuel leaks.

17. Add coolant and check for coolant leaks.

18. Check for exhaust gas leaks

19. Inspect ignition timing.

20. Inspect idle speed.

21. Inspect compression.

22. Inspect CO/HC.

2.4L Engine

See Figures 109 through 115.

1. Before servicing the vehicle, refer to the Precautions Section.

2. Remove the engine assembly with transaxle.

3. Secure engine.

4. Remove the oil filler cap sub-assembly.

5. Remove the spark plugs and ignition coil assembly.

6. Remove the oil pan drain plug and gasket.

7. Remove the ventilation valve sub-assembly.

8. Remove the camshaft position sensor.

9. Remove the camshaft timing oil control valve assembly.

10. Remove crankshaft position sensor.

11. Remove the No. 1 oil pipe.

12. Remove the oil pipe.

13. Remove the cylinder block water drain cock sub-assembly.

14. Remove the oil filter.

15. Remove the crankshaft pulley.

16. Remove the left hand No. 1 front engine mounting bracket.

17. Remove the water inlet housing.

18. Remove the water outlet.

19. Remove the cylinder head covers and gaskets.

20. Remove the No. 2 oil pan sub-assembly.

21. Remove the oil strainer sub-assembly.

22. Remove the oil pan sub-assembly.

23. Remove the No. 1 oil pan baffle plate.

24. Remove the engine rear oil seal.

25. Remove the water pump assembly.

26. Remove the timing chain cover.

27. Set the No. 1 cylinder to TDC/compression.

28. Remove the No. 1 chain tensioner assembly.

29. Remove the chain tensioner slipper.

30. Remove the chain sub-assembly.

31. Remove the idle sprocket assembly.

32. Remove the No. 1 and 2 chain vibration damper.

➡Be careful not to drop washers into the cylinder head.

➡Cylinder head warpage or cracking could result from removing bolts in an incorrect order.

➡Be sure to keep separate the removed parts for each installation position.

a. Remove the cylinder head and gasket.

33. Remove the right hand cylinder head sub-assembly, as follows:

a. Using a 10 mm bi-hexagon wrench, uniformly loosen the 8 bolts in the sequence shown in the illustration. Remove the 8 cylinder head bolts and plate washers.

⁂ WARNING

Be careful not to drop washers into the cylinder head.

⁂ WARNING

Cylinder head warpage or cracking could result from removing bolts in an incorrect order.

➡Be sure to keep separate the removed parts for each installation position.

b. Remove the cylinder head and gasket.

34. Remove the left hand cylinder head sub-assembly, as follows:

a. Uniformly loosen and remove the 2 bolts in the sequence shown in the illustration.

➡Be careful not to drop washers into the cylinder head.

➡Cylinder head warpage or cracking could result from removing bolts in an incorrect order.

b. Using a 10 mm bi-hexagon wrench, uniformly loosen the 8 bolts in the sequence shown in the illustration. Remove the 8 cylinder head bolts and plate washers.

c. Remove the cylinder head and gasket

To install:

35. Place the right hand cylinder head gasket on the cylinder block surface with the front face of the Lot No. stamp upward.

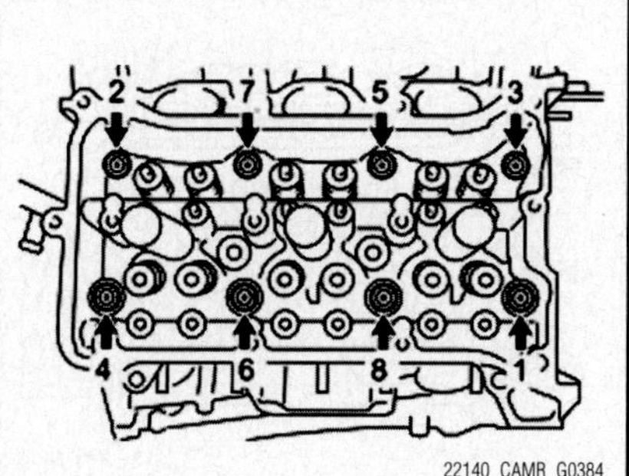

Fig. 109 Right hand cylinder head bolt removal sequence

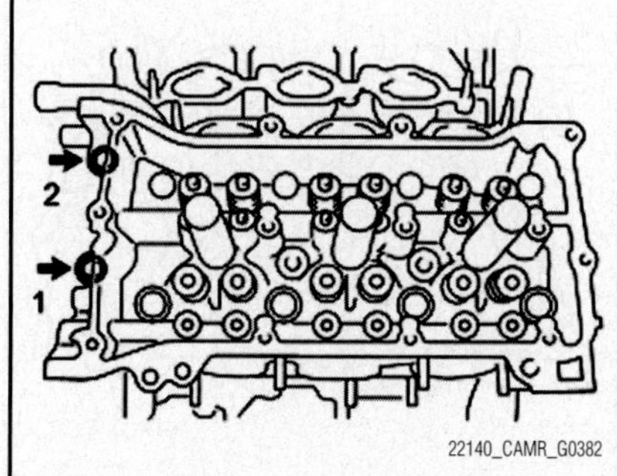

Fig. 110 Left hand cylinder head sub-assembly bolt removal sequence

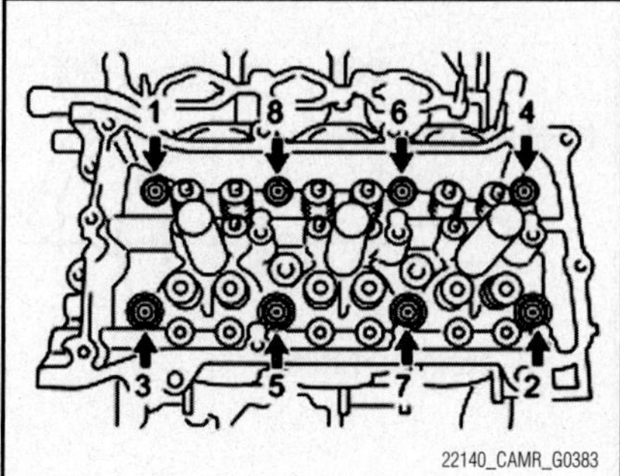

Fig. 111 Left hand cylinder head bolt removal sequence

➡Be careful of the installation direction.

➡Gently place the cylinder head in order not to damage the gasket with the bottom part of the head.

36. Place the cylinder head on the cylinder block.

⁂ CAUTION

Do not allow oil to adhere to the mounting surface of the cylinder head.

37. Apply a light coat of engine oil to the threads and under the heads of the cylinder head bolts.

38. The cylinder head bolts are tightened in 3 progressive steps:

a. Step 1: Using a 10 mm bi-hexagon wrench, install and uniformly tighten the 8 cylinder head bolts with the plate washers in several steps and in the sequence shown in the illustration. Tighten to 27 ft. lbs. (36 Nm).

b. Step 2: Mark the cylinder head bolt head with paint as shown in the illustration. Tighten the cylinder head bolts another 90°.

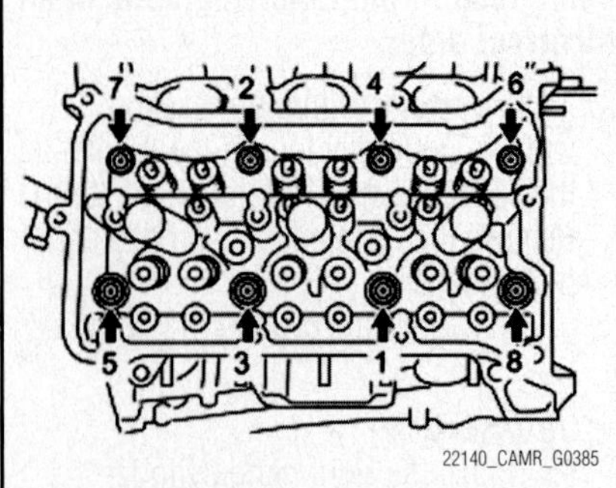

Fig. 112 Right hand cylinder head bolt tightening sequence

c. Step 3: Tighten the cylinder head bolts an additional 90°. Check that the painted mark is now facing rearward.

d. Seal packing will seep out on the engine's front side. Thoroughly wipe clean any seal packing.

39. Place the left hand cylinder head gasket on the cylinder block surface with the front face of the Lot No. stamp upward.

➡Be careful of the installation direction.

➡Gently place the cylinder head in order not to damage the gasket with the bottom part of the head.

40. Place the cylinder head on the cylinder block.

➡Do not allow oil to adhere to the mounting surface of the cylinder head.

41. Apply a light coat of engine oil to the threads and under the heads of the cylinder head bolts.

42. The cylinder head bolts are tightened in 3 progressive steps:

a. Step 1: Using a 10 mm bi-hexagon wrench, install and uniformly tighten the 8 cylinder head bolts with the plate washers in several steps and in the sequence shown in the illustration. Tighten to 27 ft. lbs. (36 Nm).

b. Step 2: Mark the cylinder head bolt head with paint as shown in the illustration. Tighten the cylinder head bolts another 90°.

c. Step 3: Tighten the cylinder head bolts an additional 90°. Check that the painted mark is now facing rearward.

d. Tighten the 2 bolts in the order shown in the illustration to 22 ft. lbs. (30 Nm). Only use the specifications stated above when tightening the bolts 1 and 2 shown in the illustration.

e. Seal packing will seep out on the engine's front side. Thoroughly wipe clean any seal packing.

43. Install the No. 2 chain tensioner assembly.

44. Install the No. 3 chain tensioner.

45. Install the No. 1 and 2 chain vibration damper.

46. Install the idle sprocket assembly.

47. Install the chain sub-assembly.

48. Install the chain tensioner slipper.

49. Install the No. 1 chain tensioner assembly.

50. Install the water pump assembly.

51. Install the timing chain cover.

52. Install the water inlet housing.

53. Install the left hand No. 1 front engine mounting bracket.

54. Install the No. 1 oil pan baffle plate.

55. Install the oil pan sub-assembly.

56. Install the oil strainer sub-assembly.

57. Install the No. 2 oil pan sub-assembly.

58. Install the oil pan drain plug and gasket.

59. Install the cylinder head covers and gaskets

60. Install the water outlet.

61. Install the crankshaft pulley.

62. Install the oil filter.

63. Install the cylinder block water drain cock sub-assembly.

64. Install the No. 1 oil pipe.

65. Install the oil pipe.

66. Install the crankshaft position sensor.

67. Install the camshaft timing oil control valve assembly.

68. Install the camshaft position sensor.

69. Install the ventilation valve sub-assembly.

70. Install the spark plugs and ignition coil assembly.

71. Install the oil filler cap sub-assembly.

72. Install the engine assembly with transaxle.

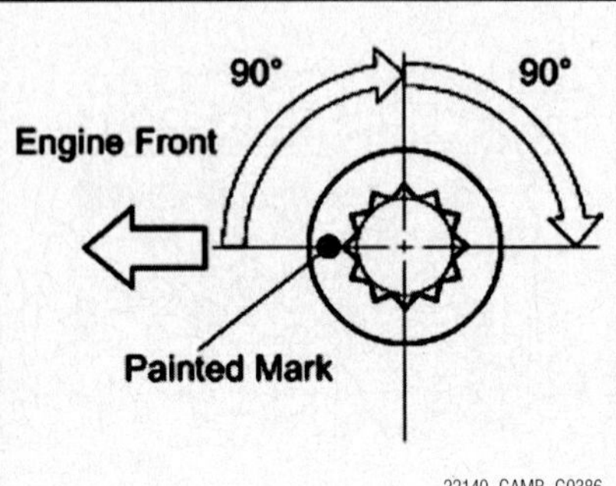

Fig. 113 Mark the cylinder head bolt and tighten another 90°

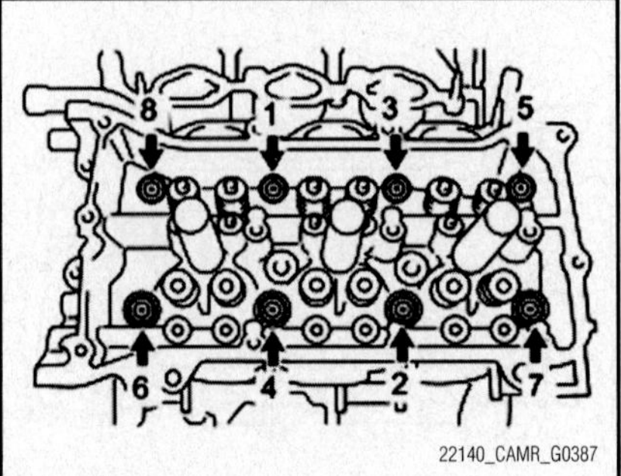

Fig. 114 Left hand cylinder head bolt tightening sequence

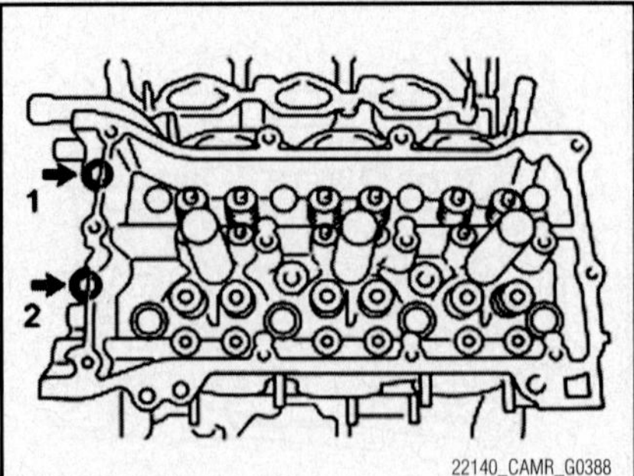

Fig. 115 Left hand cylinder head sub-assembly bolt tightening sequence

ENGINE ASSEMBLY

REMOVAL & INSTALLATION

2.4L Engine

See Figures 116 through 128.

1. Before servicing the vehicle, refer to the Precautions Section.
2. Disconnect the negative battery cable.
3. Make sure wheels are pointing straight ahead.
4. Remove or disconnect the following:
    a. Front wheels.
    b. Right and left hand engine under covers.
    c. Front fender apron seal.
5. Drain engine oil.
6. Drain and recycle the engine coolant.
7. If vehicle is equipped with an automatic transaxle, drain the transaxle fluid.
8. If the vehicle is equipped with a manual transaxle, drain the transaxle fluid.
9. Drain the brake fluid.
10. Remove or disconnect the following:
    a. Windshield wiper link assembly.
    b. Cowl top panel.
    c. Plastic engine cover.
    d. Air cleaner assembly.
    e. Battery.
    f. No. 2 engine mounting stay
    g. Ground cable.
    h. Three bolts and the engine mount
    i. Three bolts and No. 2 right hand mounting bracket.
    j. No. 1 Vacuum hose clamp and disconnect the vacuum hose connector.
    k. Clamp and radiator inlet and outlet hose.
    l. For automatic transaxle, oil cooler inlet and outlet hose.
    m. Heater inlet and outlet hose.
    n. ECM.
    o. Engine wire from the engine room relay block.
    p. Nut and separate the wire harness.
    q. Using a screwdriver, unlock the engine room R/B. Pull the engine room R/B upward.
    r. Engine wire connectors
    s. Clamp from the bracket
    t. Two bolts and clamp from the body
    u. For automatic transaxle vehicles, clip and nut and separate the control lever. Separate the clamp and separate the cable from the transaxle
    v. For manual transaxle vehicles, two clips and two washers. Disconnect the 2 cables from the transaxle. Remove the 2 clips and disconnect the 2 cables from the control cable bracket.
    w. No. 1 oil reservoir to pump hose.
    x. Return tube sub-assembly
    y. For manual transaxle vehicles, clutch accumulator assembly. Clutch release cylinder assembly.

➡ Check for contamination in the pipe and around the connector. Clean if necessary and then disconnect the connector. Disconnect the connector by hand. Do not bend, fold or rotate the nylon tube. If the pipe and connector are stuck together, push and pull the connector until it becomes free. Put the pipe and connector ends in vinyl bags to prevent damage and contamination.

    z. No. 1 fuel pipe clamp. Disconnect the connector from the tube.
    aa. Generator assembly.
    bb. A/C compressor and magnetic clutch.
    cc. Front exhaust pipe assembly.
    dd. Left and right axle hub assembly.
    ee. Left and right front stabilizer link assembly.
    ff. Left and right front speed sensor.
    gg. Left and right tie rod assembly.

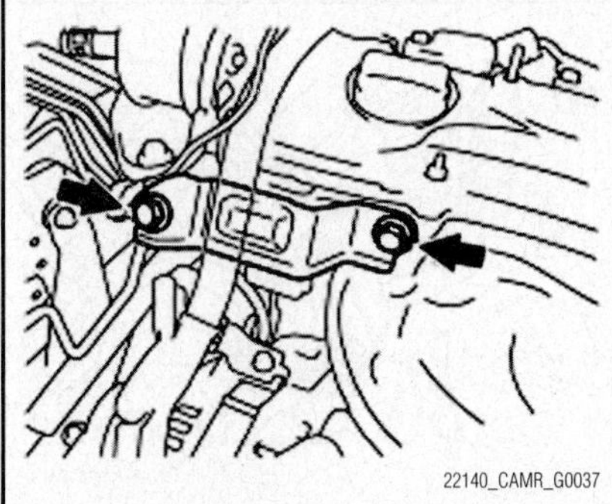

Fig. 116 Removing No. 2 engine mounting stay

Fig. 118 Removing the engine mount and three bolts

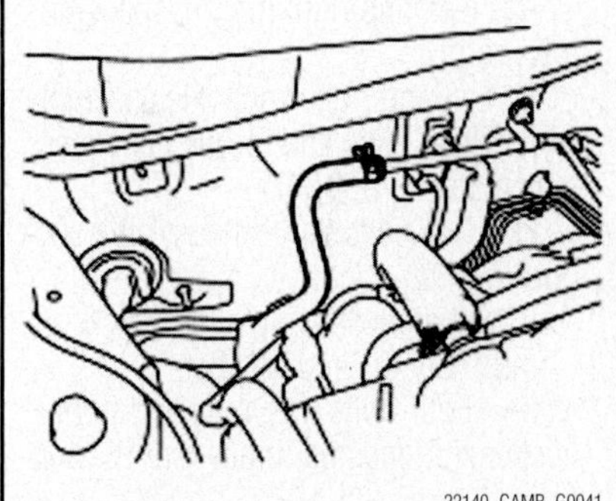

Fig. 120 Removing the No. 1 Vacuum hose clamp and vacuum hose connector

Fig. 117 Disconnecting the ground cable

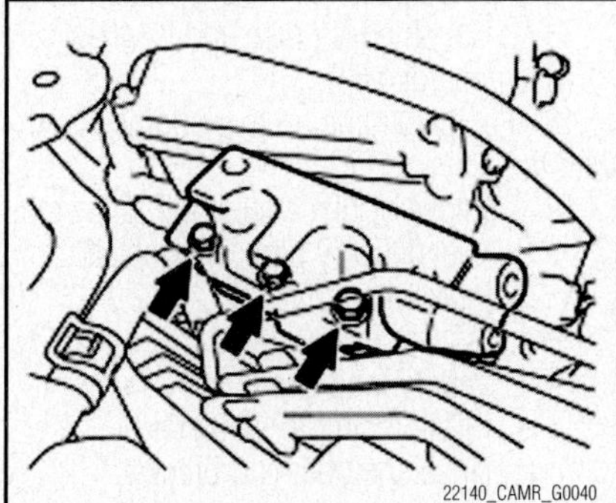

Fig. 119 Removing the No. 2 right hand mounting bracket and three bolts

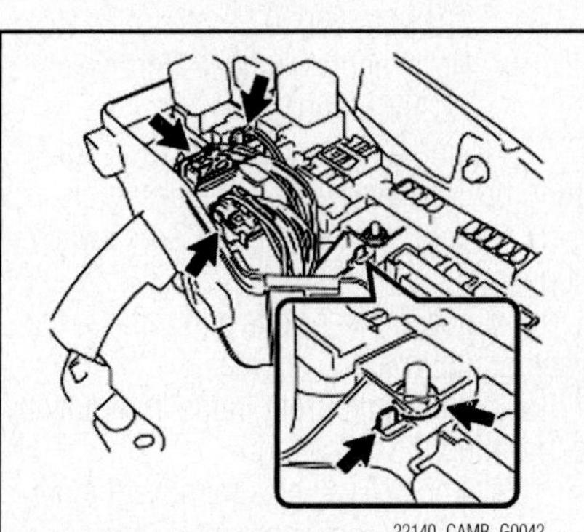

Fig. 121 Removing the Engine wire from the engine room relay block

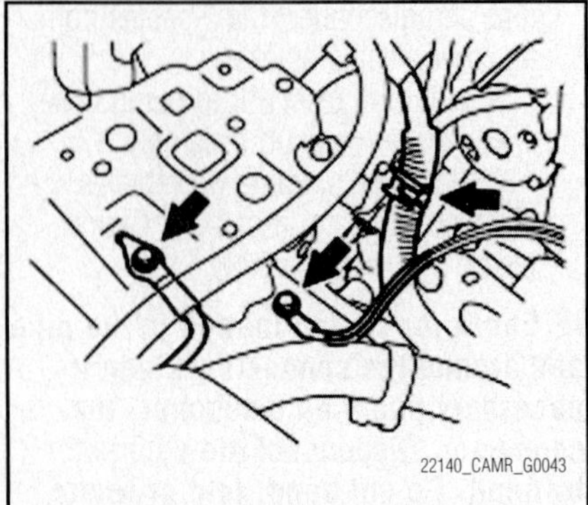

Fig. 122 Removing the two bolts and clamp from the body

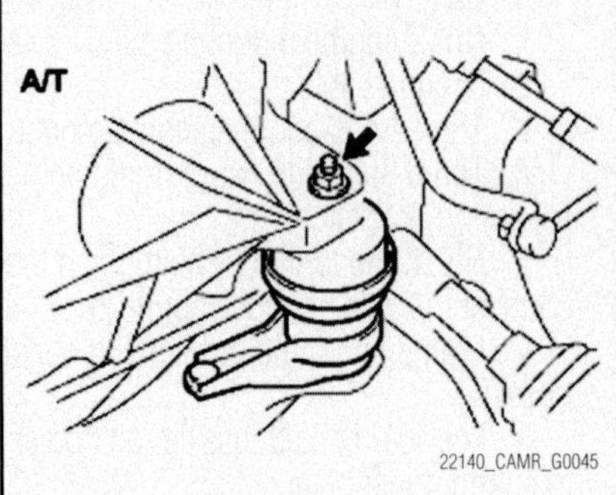

Fig. 124 Removing the nut from the LH engine mounting insulator—A/T

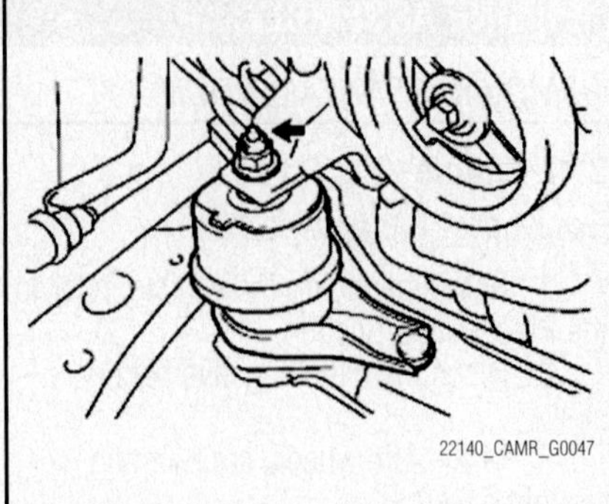

Fig. 126 Removing the nut from the RH engine mounting insulator

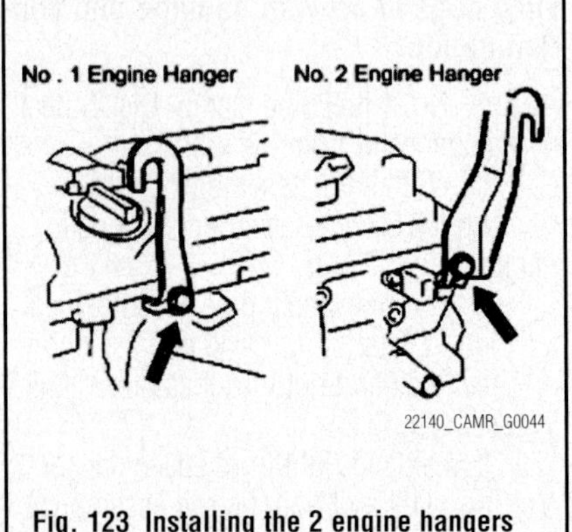

Fig. 123 Installing the 2 engine hangers

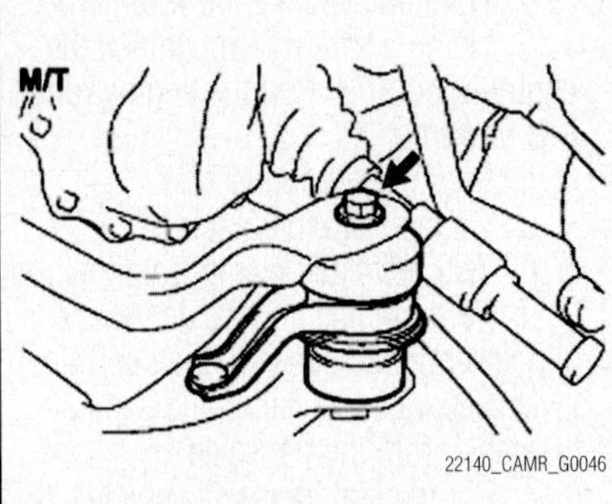

Fig. 125 Removing the nut from the LH engine mounting insulator—M/T

Fig. 127 Removing the nut from the FR engine mounting insulator

hh. Left and right front suspension lower arm.

ii. Left and right front axle assembly

jj. Drive plate and torque converter clutch setting bolt (A/T).

kk. Two bolts and exhaust pipe support bracket.

ll. Separate steering sliding yoke.

mm. Set the engine lifter.

nn. Four bolts, two nuts and front suspension member brace rear RH and LH

11. Install the 2 engine hangers and tighten to 28 ft. lbs. (38 Nm):
- No. 1 engine hanger: Part No. 12281—28010
- No. 2 engine hanger: Part No. 12282—28010

12. Using a chain block and an engine sling device, hang the engine assembly.

13. Disconnect the oil pressure switch connector.

14. Loosen the 2 bolts and remove the vane pump from the engine.

15. Remove the front frame assembly by performing the following:

a. For A/T vehicles, remove the nut from the LH engine mounting insulator.

b. For M/T vehicles, remove the bolt from the LH engine mounting insulator.

c. Remove the nut from the RH engine mounting insulator.

d. Remove the bolt from the FR engine mounting insulator.

e. For M/T vehicles, remove the bolt from the engine lateral control rod

16. Raise the engine assembly and separate the front frame.

17. Remove or disconnect the following:

a. LH and RH front drive shaft assembly.

b. Engine wire harness.

c. Starter assembly.

d. Separate A/T assembly or M/T assembly

e. Drive plate and ring gear for A/T.

f. Clutch cover, disc and flywheel assembly for M/T.

18. Fix the engine on to a engine stand with the bolts.

19. Remove or disconnect the following:

a. Fuel delivery pipe with injector.

b. Intake manifold.

c. IAT (for PZEV).

d. Ventilation hose

e. Intake manifold insulator.

f. Drive shaft bearing bracket.

g. Oil level tube and guide.

h. Manifold/catalytic bolt, nut and stay.

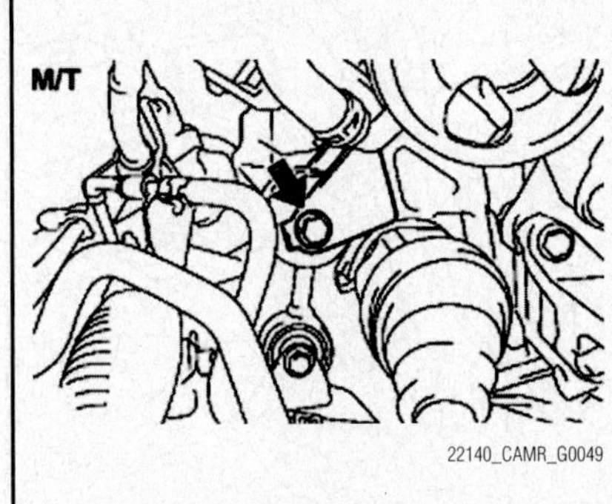

Fig. 128 Removing the bolt from the engine lateral control rod—M/T

i. For PZEV vehicles, air-fuel ratio sensor connector

j. For PZEV vehicles, five nuts, manifold converter and gasket.

k. Except PZEV vehicles, four bolts and No. 1 insulator.

l. Except PZEV vehicles IAT.

m. Except PZEV five nuts, manifold converter and gasket.

n. Water inlet.

o. Thermostat.

p. Two nuts and water by-pass pipe with the gasket.

q. For vehicles with oil cooler pipe;

bolt, 2 nuts and oil cooler pipe with the gasket.

r. Three bolts and RH engine mounting bracket.

s. Bolt, nut and belt tensioner.

t. Four bolts and four ignition coils.

u. Camshaft timing oil control valve assembly.

v. Knock sensor connector, nut and knock sensor.

w. Bolt and radio setting condenser.

x. Engine oil pressure switch assembly, using a 24 mm deep socket wrench.

y. ECT sensor and gasket using Special Tool: 09817—33190.

To install:

20. To install, reverse removal procedure.

21. Tighten the following to specification:

- Drive belt tensioner with the bolt and nut: 44 ft. lbs. (60 Nm)
- RH engine mounting bracket with the 3 bolts: 40 ft. lbs. (54 Nm)
- Oil cooler pipe: 80 inch lbs. (9 Nm)
- Water by-pass pipe bolt, two nuts and install a new gasket: 80 inch lbs. (9 Nm)
- Oil level gauge guide bolt: 80 inch lbs. (9 Nm)
- LH engine mounting insulator nut—A/T: 70 ft. lbs. (95 Nm)
- LH engine mounting insulator bolt—M/T: 105 ft. lbs. (143 Nm)
- RH engine mounting insulator nut: 70 ft. lbs. (95 Nm)
- FR engine mounting insulator bolt: 64 ft. lbs. (87 Nm)
- Engine lateral control rod bolt—M/T: 66 ft. lbs. (89 Nm)
- Engine room junction block nut: 74 inch lbs. (8.4 Nm)
- RH engine mountain bracket bolts: 38 ft. lbs. (52 Nm)
- Engine mount bolts: 47 ft. lbs. (64 Nm)
- Ground cable bolt: 74 inch lbs. (8.4 Nm)
- RH engine mounting stay bolts: 47 ft. lbs. (64 Nm).
- Front wheel: 76 ft. lbs. (103 Nm)

22. Add fluids and check for leaks.

23. Bleed power steering, brakes lines and clutch lines.

24. Check shift lever position.

25. Adjust front wheel alignment.

26. Check ignition timing.

27. Check engine idle speed.

28. Inspect CO/HC.

29. Check ABS speed sensor signal.

3.5L Engine

See Figures 129 through 135.

1. Before servicing the vehicle, refer to the Precautions Section.

2. Prevent gasoline from spilling out.

3. Remove front wheels.

4. Remove the engine under covers.

5. Remove the right front fender apron seal.

6. Drain engine oil.

7. Drain engine coolant.

8. Drain automatic transaxle fluid.

9. Remove the windshield wiper link assembly.

10. Remove the cowl top panel outer sub-assembly.

11. Remove the 7 clips and intake duct seal.

12. Remove the V-bank cover sub-assembly.

13. Remove the drive belt.

14. Remove the 2 bolts, clamp and air cleaner inlet.

15. Remove air cleaner cap sub-assembly.

16. Remove the air cleaner assembly with hose, as follows:

a. Disconnect the vacuum hoses and hose clamp.

b. Remove the 3 bolts and air cleaner case.

17. Remove the bolt and No. 1 air cleaner inlet.

18. Loosen the bolt and nut, and remove the battery clamp.

19. Remove the clip, bolt and intake air resonator.

20. Remove the bolt, 2 nuts, and No. 2 RH mounting stay.

21. Remove the 4 bolts and engine moving control rod

22. Remove the clamp and disconnect the No. 1 fuel vapor feed hose.

23. Remove the clamp and disconnect the check valve to brake booster hose.

24. Remove the clamp and disconnect the radiator inlet and outlet hose.

Fig. 129 Removing the 4 bolts and engine moving control rod

25. Remove the oil cooler inlet and outlet hoses.

26. Disconnect the heater water inlet and outlet hose.

27. Remove the upper relay block cover.

28. Disconnect the engine wire from the engine room junction block, as follows:

a. Remove the nut and separate the wire harness.

b. Using a screwdriver, unlock the engine room junction block. Pull the engine room junction block upward.

c. Disconnect the engine wire connectors.

d. Remove the 2 bolts and 2 clamps from the body.

e. Remove the bolt and clamp from the bracket.

29. Disconnect the transmission control cable assembly. Remove the clip and nut, and separate the cable from the transaxle.

30. Separate the fuel tube sub-assembly, as follows:

a. Remove the No. 1 fuel pipe clamp.

b. Disconnect the connector from the tube while pinching part A with your fingers.

➡Check for contamination in the pipe and around the connector. Clean if necessary and then disconnect the connector.

➡Disconnect the connector by hand.

➡Do not bend, fold or rotate the nylon tube.

➡If the pipe and connector are stuck together, push and pull the connector until it comes free.

➡Put the pipe and connector ends in vinyl bags to prevent damage and contamination.

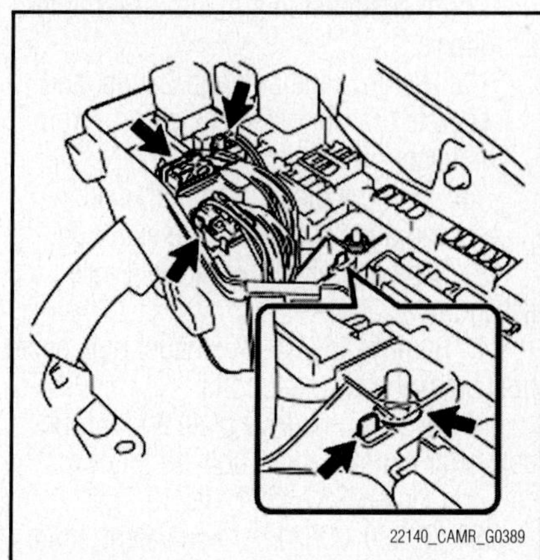

Fig. 130 Disconnecting the engine wire from the engine room junction block

31. Disconnect the oil reservoir to No. 1 pump hose.

32. Disconnect the return tube sub-assembly.

33. Remove the No. 1 exhaust pipe support bracket.

34. Remove the front exhaust pipe assembly.

35. Remove the left front axle shaft nut, as follows:

36. Using SST (SST: 09930-00010) or equivalent and a hammer, strike the lock nut covering to remove it.

a. Set the drive shaft's groove so that it faces up. Then use the SST and a hammer.

b. Remove the covering from the lock nut completely. Otherwise the screw of the drive shaft may be damaged.

c. Do not sharpen the tip of the SST.

d. Using a 30 mm socket wrench, remove the lock nut.

37. Remove the right front axle shaft nut.

38. Disconnect the left front stabilizer link assembly, as follows:

a. Using a 6 mm socket hexagon wrench, hold the ball stud.

b. Remove the nut and disconnect the stabilizer link.

39. Disconnect the right front stabilizer link assembly.

40. Remove the bolts and disconnect the right and left front speed sensors from the steering knuckle.

41. Disconnect the left and right tie rod assemblies, as follows:

a. Remove the cotter pin and nut.

b. Using SST (SST: 09628-62011) or equivalent, disconnect the tie rod ends from the steering knuckle.

➡Do not damage the dust cover of the ball joint.

42. Disconnect the lower No. 1 left and right front suspension arm sub-assemblies, as follows:

a. Remove the bolt and 2 nuts, and separate the front suspension arm from the lower ball joint.

b. Using a plastic hammer, disconnect the drive shaft from the axle hub.

43. Separate left and right front axle assembly.

44. Remove the No. 1 exhaust pipe support bracket with the 2 bolts.

45. Remove the drive plate and torque converter clutch setting bolt, as follows:

a. Using SST (SST: 09213-70011, SST: 09330-00021) or equivalent, hold the crankshaft.

b. Remove the 2 bolts and flywheel housing under cover.

c. Remove the 6 torque converter clutch setting bolts.

46. Disconnect steering sliding yoke.

47. Remove generator assembly.

48. Separate the cooler compressor assembly, as follows:

a. Remove the 2 connector clamps.

b. Remove the 4 bolts and separate the compressor.

➡Hang up the hoses instead of detaching them.

49. Remove the engine assembly with transaxle, as follows:

a. Set the engine lifter.

b. Remove the 4 bolts, 2 nuts, and right and left frame side rail plates.

c. Remove the 4 bolts, 2 nuts, front suspension member right and left rear brace.

d. Operate the engine lifter, then remove the engine assembly from the vehicle.

➡Make sure the engine is clear of all wiring and hoses.

50. Remove the vane pump assembly, as follows:

a. Remove the pressure feed tube clamp bolt.

b. Remove the 2 bolts, nut and vane pump.

51. Install the 2 engine hangers with the 4 bolts as shown in the illustration. Part numbers: Engine hanger No. 1 12281-31120, Engine hanger No. 2 12282-31100, Bolts 91671-10825. Torque to 24 ft. lbs. (33 Nm).

a. Attach the engine sling device and hang the engine with the chain block.

52. Remove the front frame assembly, as follows:

a. Disconnect the connector and clamp.

b. Disconnect the 2 clamps.

c. Remove the 2 nuts and disconnect the right and left engine mounting insulators.

d. Remove the bolt and disconnect the engine mounting insulator FR.

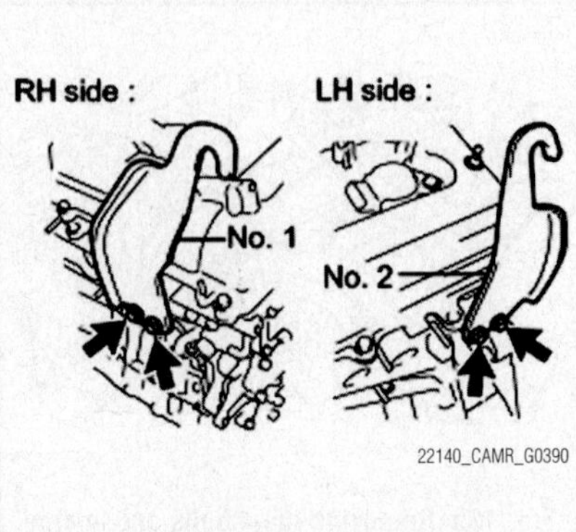

Fig. 131 Install the 2 engine hangers

53. Remove the right and left front drive shaft assemblies.

54. Remove the engine wire.

55. Remove the starter assembly.

56. Remove the automatic transaxle assembly.

57. Remove the drive plate and ring gear sub-assembly.

58. Install engine stand.

59. Remove intake air surge tank assembly, as follows:

a. Disconnect the 2 water by-pass hoses from the throttle with motor body assembly

b. Disconnect the vapor feed hose.

c. Disconnect the throttle with motor body assembly connector and clamp.

d. Disconnect the No. 1 ventilation hose.

e. Remove the bolt and vacuum hose clamp.

f. Disconnect the connector.

g. Remove the 4 bolts, No. 1 surge tank stay and throttle body bracket.

h. Using a 5 mm socket hexagon wrench, remove the 4 bolts. Remove the 2 nuts and intake air surge tank. Remove the gasket from the intake air surge tank.

60. Remove the ignition coil assembly.

61. Remove the bolt and the No. 2 RH engine mounting stay.

62. Remove the intake manifold, as follows:

a. Uniformly loosen and remove the 6 bolts and 4 nuts.

b. Remove the intake manifold and 2 gaskets.

63. Remove the right exhaust manifold sub-assembly, as follows:

a. Disconnect the A/F sensor connector clamp.

b. Uniformly loosen and remove the 6 nuts.

c. Remove the manifold and gasket.

64. Remove the oil level gauge guide sub-assembly, as follows:

a. Remove the oil level gauge.

b. Remove the 2 bolts, and No. 1 and No. 2 oil level gauge guides.

c. Remove the O-rings from the oil level gauge guide.

65. Remove the 2 bolts and the No. 2 manifold stay.

66. Remove the 3 bolts and the No. 2 exhaust manifold heat insulator.

67. Remove the left exhaust manifold sub-assembly, as follows:

a. Uniformly loosen and remove the 6 nuts.

b. Remove the manifold and gasket.

68. Remove the 3 bolts and the right engine mounting bracket.

69. Remove the 3 bolts and drive shaft bearing bracket.

70. Remove the 5 bolts and V-ribbed belt tensioner assembly.

71. Remove the 2 bolts and the No. 2 timing gear cover.

72. Remove the bolt, plate and No. 2 idler pulley sub-assembly.

73. Remove the water pump pulley.

74. Remove the 6 bolts and No. 1 left front engine mounting bracket.

75. Remove the 2 bolts and 2 radio setting condensers.

76. Remove the bolt and No. 1 vacuum switching valve.

77. Using a 24 mm deep socket wrench, remove the engine oil pressure switch assembly.

78. Remove 2 bolts and 2 knock control sensors.

79. Using a 19 mm deep socket wrench, remove the EFI engine coolant temperature sensor and gasket.

80. Replace partial engine assembly as necessary.

To install:

81. Using a 19 mm deep socket wrench, install the EFI engine coolant temperature sensor and a new gasket and tighten to 14 ft. lbs. (20 Nm).

82. Install the knock control sensor.

83. Install the engine oil pressure switch assembly, as follows:

a. Clean the threads of the oil pressure switch. Apply adhesive (Toyota Genuine Adhesive 1344, Three Bond 1344 or equivalent) to 2 or 3 threads of the oil pressure switch.

b. Using a 24 mm deep socket wrench install the oil pressure switch and tighten to 15 ft. lbs. (21 Nm). Install the left front No. 1 engine mounting bracket with the 6 bolts and tighten to 40 ft. lbs. (54 Nm).

84. Install the bolt and No. 1 vacuum switching valve and tighten to 7 ft. lbs. (10 Nm).

85. Install the 2 bolts and 2 radio setting condensers and tighten to 7 ft. lbs. (10 Nm).

86. Install the No. 1 engine mounting bracket front LH with the 6 bolts and tighten to 40 ft. lbs. (54 Nm).

87. Install the water pump pulley.

88. Install the No. 2 idler pulley sub-assembly and cover plate with the bolt and tighten to 32 ft. lbs. (43 Nm).

89. Install the No. 2 timing gear cover with the 2 bolts and tighten to 53 inch lbs. (6 Nm).

90. Install the V-ribbed belt tensioner assembly, as follows:

a. Temporarily install the V-ribbed belt tensioner with the 5 bolts.

b. Install the V-ribbed belt tensioner by tightening the bolt 1 and bolt 2 in the order shown in the illustration to 32 ft. lbs. (43 Nm).

c. Tighten the other bolts to 32 ft. lbs. (43 Nm). Bolt length: A: 70 mm (2.76 in.), B: 33 mm (1.30 in.)

91. Install the drive shaft bearing bracket with the 3 bolts and tighten to 47 ft. lbs. (64 Nm).

92. Install the engine mounting bracket RH with the 3 bolts and tighten to 40 ft. lbs. (54 Nm).

93. Install the left exhaust manifold sub-assembly, as follows:

a. Install a new gasket.

b. Install the exhaust manifold sub-assembly with the 6 nuts and tighten to 15 ft. lbs. (21 Nm).

94. Install the No. 2 exhaust manifold heat insulator with the 3 bolts and tighten to 75 inch lbs. (8.5 Nm).

95. Install the No. 2 manifold stay with the 2 bolts and tighten to 25 ft. lbs. (34 Nm).

96. Install the oil level gauge guide sub-assembly, as follows:

a. Install 2 new O-rings to the oil level gauge guide.

b. Apply a coat of engine oil lightly to the O-rings.

c. Push in the oil level gauge guide end into the guide hole.

d. Install the No. 1 oil level gauge guide with the bolt and tighten to 15 ft. lbs. (21 Nm).

e. Install the No. 2 oil level gauge guide with the bolt and tighten to 15 ft. lbs. (21 Nm).

f. Install the oil level gauge.

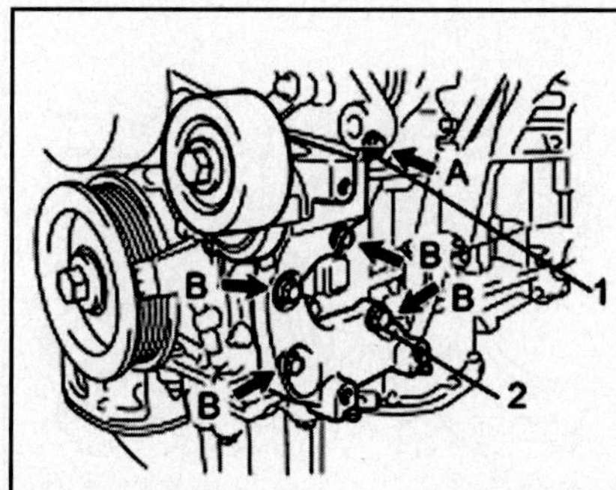

Fig. 132 V-ribbed belt tensioner bolt tightening sequence

97. Install the right exhaust manifold sub-assembly, as follows:

a. Install a new gasket.

b. Install the exhaust manifold sub-assembly with the 6 nuts and tighten to 15 ft. lbs. (21 Nm).

98. Install the intake manifold.

99. Install the right No. 2 mounting stay with the bolt and tighten to 15 ft. lbs. (21 Nm).

100. Install the 6 ignition coil assemblies with the 6 bolts and tighten to 7 ft. lbs. (10 Nm).

CAUTION

DO NOT apply oil to the following bolts: surge tank and intake manifold, No. 1 surge tank stay and cylinder head cover, No. 1 surge tank stay and surge tank, throttle body bracket and cylinder head cover, throttle body bracket and surge tank.

a. Install a new gasket to the intake air surge tank.

b. Using a 5 mm hexagon socket wrench, install the 4 bolts and 2 nuts. Tighten the bolts to 13 ft. lbs. (18 Nm). Tighten the nuts to 12 ft. lbs. (16 Nm).

c. Install the throttle body bracket, No. 1 surge tank stay and 4 bolts and tighten to 15 ft. lbs. (21 Nm).

d. Connect the connector.

e. Install the vacuum hose clamp with the bolt and tighten to 48 inch lbs. (5.4 Nm).

f. Connect the No. 1 ventilation hose.

g. Install the clamp and connect the throttle with motor body assembly connector.

h. Connect the vapor feed hose.

i. Connect the 2 water by-pass hoses to the throttle with motor body assembly.

101. Remove the engine stand.

102. Install the drive plate and ring gear sub-assembly.

103. Install the automatic transaxle assembly.

104. Install the starter assembly.

105. Install the engine wire.

106. Install the right and left front drive shaft assemblies.

107. Install the front frame assembly, as follows:

a. Install the right and left engine mounting insulators and with the 2 nuts and tighten 70 ft. lbs. (95 Nm).

b. Install the engine mounting insulator FR with the bolt and tighten 64 ft. lbs. (87 Nm).

c. Connect the connector and clamp.

d. Connect the 2 clamps.

108. Install the vane pump assembly, as follows:

a. Install the vane pump with the 2 bolts and nut and tighten 32 ft. lbs. (43 Nm).

b. Install the pressure feed tube clamp bolt and tighten 87 inch lbs. (9.8 Nm).

c. Connect the power steering oil pressure switch connector.

109.Install the engine assembly with transaxle, as follows:

a. Set the engine assembly with transaxle on the engine lifter.

b. Install the engine assembly to the vehicle.

c. Install the right and left frame side rail plates and with the 4 bolts and 2 nuts. Tighten A to 63 ft. lbs. (85 Nm) and B to 24 ft. lbs. (32 Nm).

d. Install the right and left front suspension member rear braces with the 4 bolts and 2 nuts. Tighten C to 63 ft. lbs. (85 Nm) and D to 24 ft. lbs. (32 Nm).

110. Install the cooler compressor assembly.

111. Install generator assembly.

112. Install the steering intermediate shaft assembly.

113. Install drive plate and torque converter clutch setting bolt, as follows:

a. Using SST (SST: 09213-70011, SST: 09330-00021) or equivalent, hold the crankshaft.

b. Using kerosene or gasoline, clean the bolts thoroughly.

c. Apply adhesive (Toyota Genuine Adhesive 1324, Three Bond 1324 or equivalent) to 2 or 3 threads of the bolt end.

d. Install the 6 torque converter set bolts and tighten 30 ft. lbs. (41 Nm). First tighten the green colored bolt, then the other 5 bolts.

e. Install the flywheel housing under cover with the 2 bolts and tighten 69 inch lbs. (7.8 Nm).

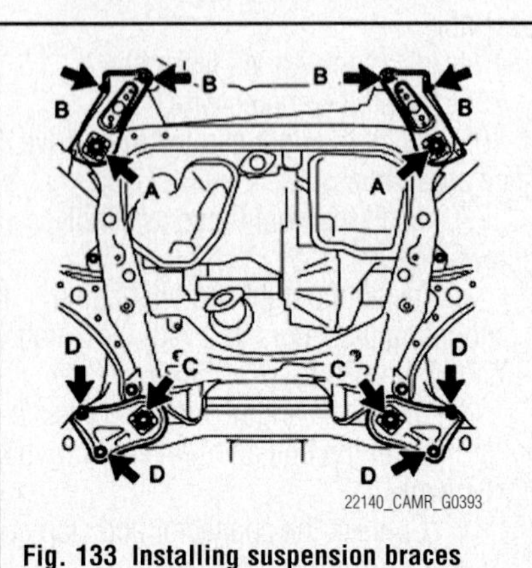

Fig. 133 Installing suspension braces

114. Install the No. 1 exhaust pipe support bracket with the 2 bolts and tighten 15 ft. lbs. (21 Nm).

115. Install both front axle assemblies.

116. Install the lower right and left No. 1 suspension arm sub-assemblies, as follows:

a. Install the drive shaft to the steering knuckle.

b. Install the suspension lower arm with the bolt and 2 nuts and tighten 55 ft. lbs. (75 Nm).

117.Install the left and right tie rod assemblies, as follows:

a. Connect the tie rod end to the steering knuckle with the castle nut and tighten 36 ft. lbs. (49 Nm).

➡Do not allow lubricants to contact the thread and taper portions.

➡After tightening the castle nut, tighten additional 60° so that a cotter pin can be inserted.

b. Insert a new cotter pin.

118. Install the right and left front speed sensors.

119. Install the right and left front stabilizer link assemblies, as follows:

a. Using a 6 mm socket hexagon wrench, hold the ball stud and install the nut and tighten 55 ft. lbs. (74 Nm).

120. Install the right and left front axle shaft hub nuts, as follows:

a. Using a 30 mm socket wrench, install a new hub nut and tighten 217 ft. lbs. (294 Nm).

b. Using a chisel and hammer, tap the hub nut.

121. Install the front exhaust pipe assembly.

122. Install the No. 1 exhaust pipe support bracket.

123. Connect the oil reservoir to the No. 1 pump hose.

Fig. 134 Engine moving control rod bolt tightening sequence

124. Connect the steering gear outlet return tube.

125. Connect the fuel tube sub-assembly, as follows:

a. Push in the fuel tube connector to the fuel pipe until connector makes a "click" sound.

➡Check for damage or contamination on the connected part of the pipe. Check if the pipe and the connector are securely connected by trying to pull them apart.

b. Install the No. 1 fuel pipe clamp .

126. Connect the transmission control cable assembly. Install the clip and nut, and connect the cable to the transaxle. Tighten the nut to 9 ft. lbs. (13 Nm).

127. Connect engine wire, as follows:

a. Install the bolt and clamp to the bracket. Tighten 74 inch lbs. (8.4 Nm).

b. Install the 2 bolts and 2 clamps to the body.

c. Connect the wire to the engine room junction block. Then, install it with the nut and 3 connectors. Tighten to 74 inch lbs. (8.4 Nm).

128. Install upper relay block cover upper.

129. Connect the heater water outlet and inlet hose.

130. Connect the oil cooler inlet and outlet hoses.

131. Install the radiator inlet and outlet hoses.

132. Install the clamp and connect the check valve to brake booster hose.

133. Install the clamp and connect the No. 1 fuel vapor feed hose.

134. Install the engine moving control rod, as follows:

a. Temporarily install the moving control rod with the 4 bolts.

b. Tighten the bolt A to 28 ft. lbs. (38 Nm).

c. Tighten the bolt B in the order shown in the illustration to 28 ft. lbs. (38 Nm).

135. Temporarily install the right No. 2 mounting stay with the bolts and 2 nuts. Tighten to 17 ft. lbs. (23 Nm).

136. Install the intake air resonator with the bolt and clip and tighten to 44 inch lbs. (5 Nm).

137. Install the battery and battery tray.

138. Install the No. 1 air cleaner inlet with the bolt and tighten 44 inch lbs. (5 Nm).

139. Install the air cleaner case sub-assembly, as follows:

a. Install the air cleaner case with the 3 bolts and tighten 44 inch lbs. (5 Nm).

b. Connect the vacuum hose and hose clamp.

140. Install the air cleaner cap sub-assembly.
141. Install the air cleaner inlet with the clamp and 2 bolts and tighten to 44 inch. lbs. (5 Nm).
142. Connect vacuum hoses.
143. Install drive belt.
144. Install the outer cowl top panel.
145. Install the windshield wiper link assembly.
146. Install the front wheels.
147. Add automatic transaxle fluid.
148. Add engine oil.
149. Add engine coolant.
150. Add power steering fluid.
151. Bleed power steering system.
152. Inspect for oil leak, coolant leak, fuel leak, and exhaust gas leak.
153. Inspect and adjust front wheel alignment.
154. Inspect ignition timing and engine idle speed.
155. Check function of throttle body.
156. Install the V-bank cover sub-assembly.
157. Inspect CO/HC.
158. Check ABS speed sensor signal.
159. Perform initialization.

EXHAUST MANIFOLD

REMOVAL & INSTALLATION

2.4L Engine

See Figures 136 through 143.

1. Before servicing the vehicle, refer to the Precautions Section.
2. Disconnect the negative battery cable.
3. Remove engine cover.
4. Remove air cleaner assembly.
5. Remove manifold stays.
6. Disconnect the air-fuel ratio sensor connector.
7. Remove or disconnect remaining components from the exhaust manifold.
8. For PZEV vehicles, perform the following:
   a. Remove the five nuts, manifold converter and gasket.
9. For non-PZEV vehicles, perform the following:
   a. Remove the four bolts and insulator.
   b. Remove the five nuts, manifold converter and gasket.
10. Remove exhaust manifold from catalytic converter.

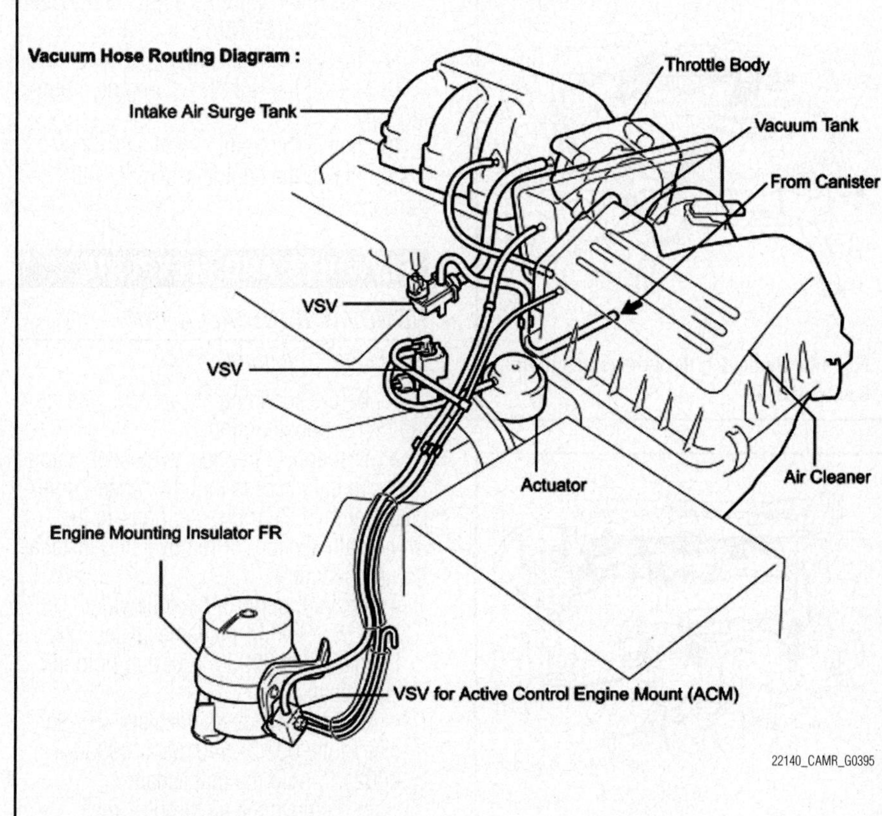

Fig. 135 Connecting vacuum hoses

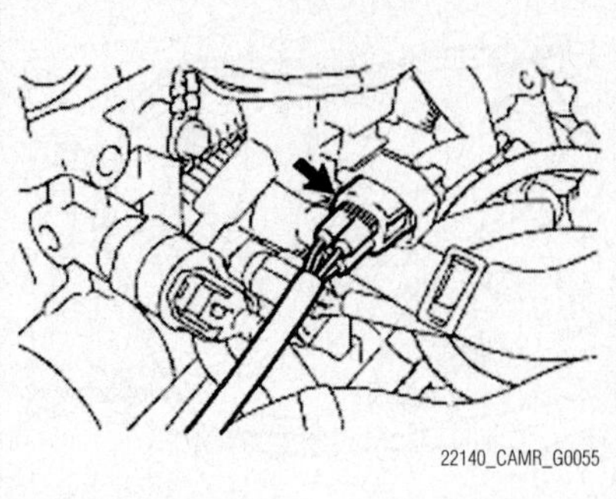

Fig. 136 Disconnecting the air-fuel ratio sensor connector

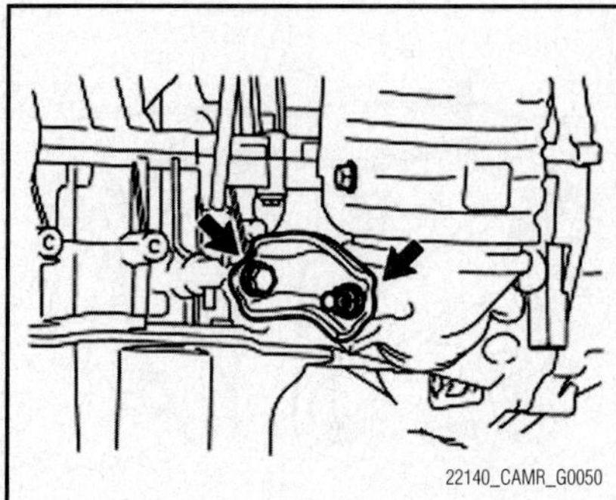

Fig. 137 Removing exhaust manifold stay—1 of 2

Fig. 138 Removing exhaust manifold stay—2 of 2

Fig. 139 Removing five nuts, exhaust manifold and gasket —PZEV vehicles

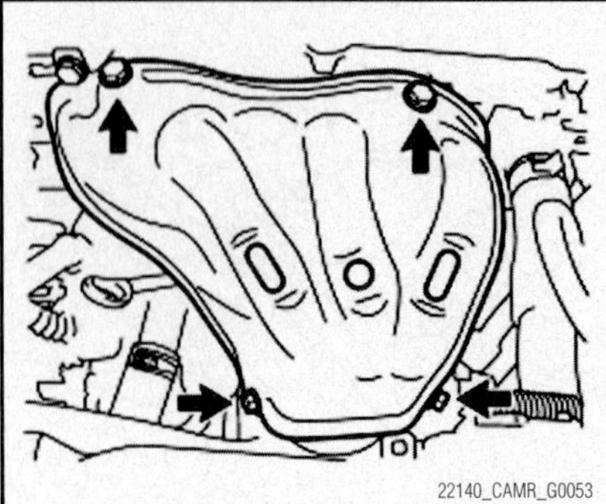
Fig. 140 Removing four bolts and insulator —except PZEV vehicles

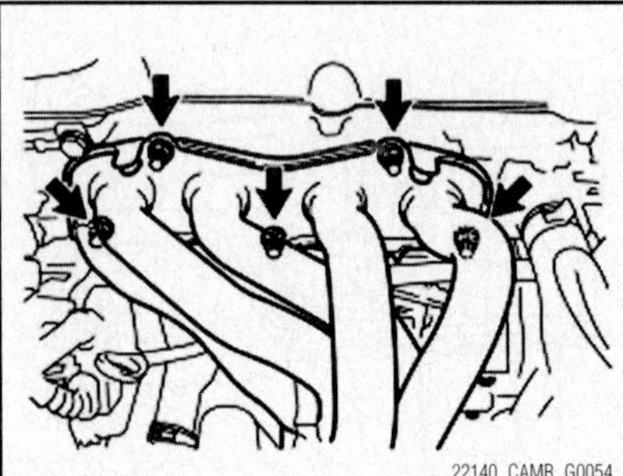
Fig. 141 Removing the five nuts, exhaust manifold and gasket—except PZEV vehicles

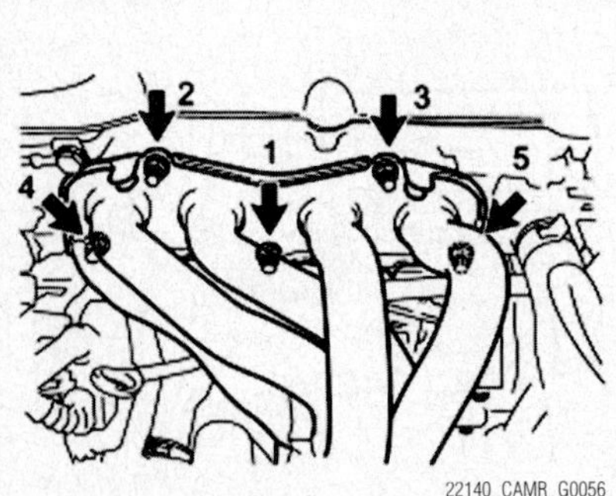

Fig. 142 Exhaust manifold tightening sequence—except PZEV vehicles

Fig. 143 Exhaust manifold tightening sequence—PZEV vehicles

To install:

11. To install, reverse removal procedure.
12. Install new gaskets for exhaust manifolds.
13. For non-PZEV vehicles, tighten exhaust manifold bolts in sequence to: 27 ft. lbs. (37 Nm). Tighten the exhaust manifold heat insulator bolts to: 9 ft. lbs. (12 Nm).
14. For PZEV vehicles, tighten exhaust manifold bolts in sequence to: 27 ft. lbs. (37 Nm).
15. Tighten the exhaust manifold stays to 32 ft. lbs. (44 Nm).

3.5L Engine

See Figures 144 and 145.

1. Before servicing the vehicle, refer to the Precautions Section.
2. Remove the engine assembly with transaxle.
3. Secure the engine.
4. Remove the ignition coil assembly.
5. Remove the right No. 2 engine mounting stay.
6. Remove the intake manifold.
7. Remove the right exhaust manifold sub-assembly, as follows:
   a. Uniformly loosen and remove the 6 nuts.

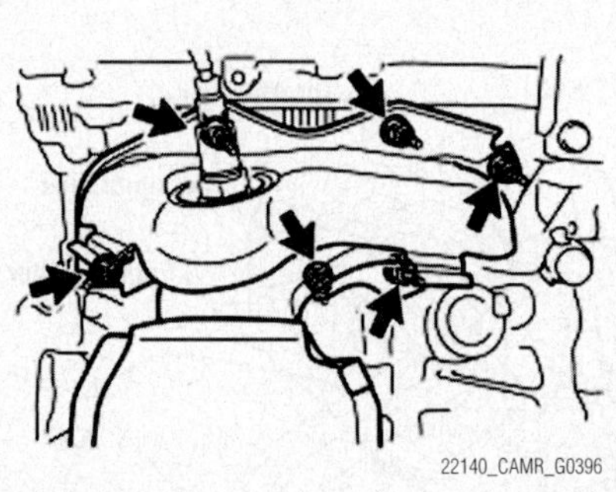
Fig. 144 Removing the exhaust manifold 6 nuts—Right

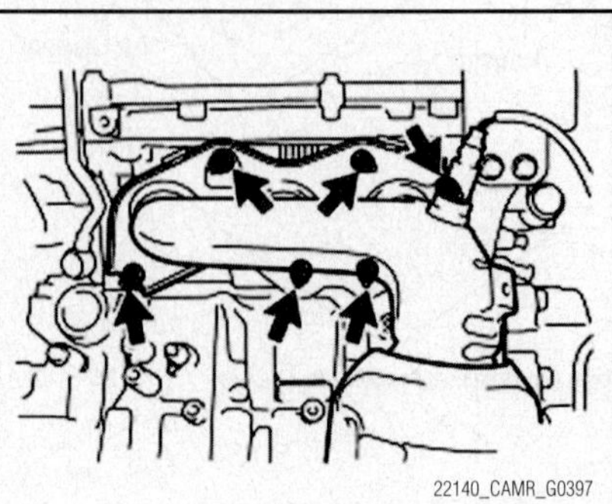
Fig. 145 Removing the exhaust manifold 6 nuts—Left

   b. Remove the manifold and gasket.
8. Remove the oil level gauge guide sub-assembly.
9. Remove the bolt, nut and No. 2 manifold stay.
10. Remove the 3 bolts and No. 2 exhaust manifold heat insulator.
11. Remove the left exhaust manifold sub-assembly, as follows:
   a. Uniformly loosen and remove the 6 nuts.
   b. Remove the manifold and gasket.

To install:

12. Install the left exhaust manifold sub-assembly, as follows:
   a. Install a new gasket.
   b. Install the left exhaust manifold sub-assembly with the 6 nuts and tighten to 15 ft. lbs. (21 Nm).
13. Install the No. 2 exhaust manifold heat insulator with the 3 bolts and tighten to 75 inch lbs. (8.5 Nm).
14. Install the No. 2 manifold stay with the bolt and nut and tighten to 25 ft. lbs. (34 Nm).
15. Install the oil level gauge guide sub-assembly.
16. Install the right exhaust manifold sub-assembly, as follows:
   a. Install a new gasket.
   b. Install the right exhaust manifold sub-assembly with the 6 nuts and tighten to 15 ft. lbs. (21 Nm).
17. Install the intake manifold.
18. Install the right No. 2 engine mounting stay.
19. Install the ignition coil assembly.
20. Install the engine assembly with transaxle.

FLYWHEEL AND FLEXPLATE

REMOVAL & INSTALLATION

See Figures 146 through 149.

1. Before servicing the vehicle, refer to the Precautions Section.
2. Disconnect the negative battery cable.
3. To gain access to the flywheel or flexplate, remove the transaxle. Refer to the Drive Train section for removal and installation procedures.
4. For A/T, perform the following:
   a. For TMMK made flexplates: use Special Tool: 09960-10010 to hold the crankshaft.
   b. For TMC made flexplate: use Special Tools: 09213-54015and 09330-00021 to hold the crankshaft.
   c. Remove the eight bolts, rear spacer, drive plate and front spacer.
5. For M/T, perform the following:

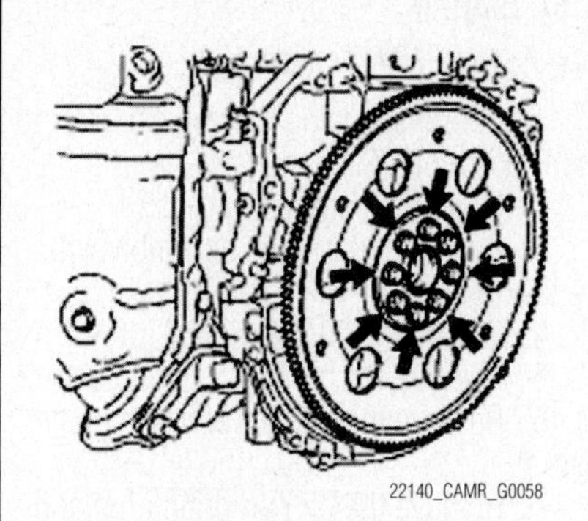

Fig. 146 Removing A/T flexplate bolts

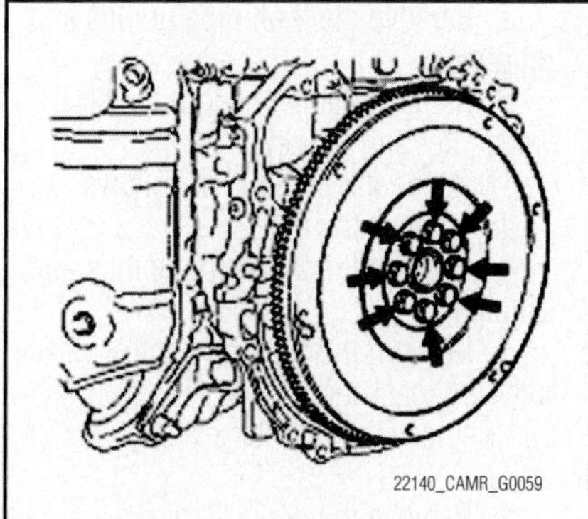

Fig. 147 Removing M/T flywheel bolts

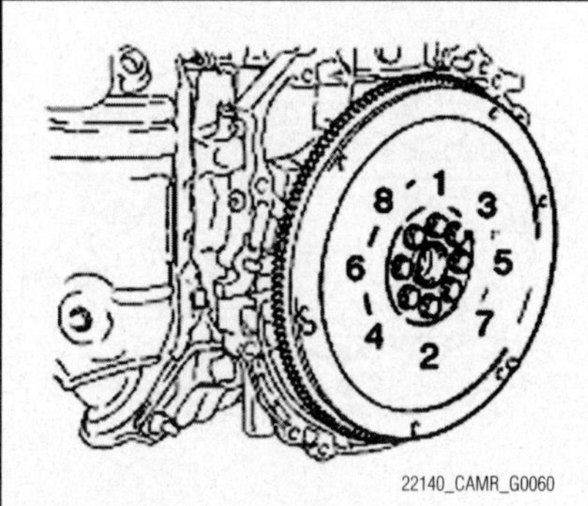

Fig. 148 M/T flywheel bolt tightening sequence

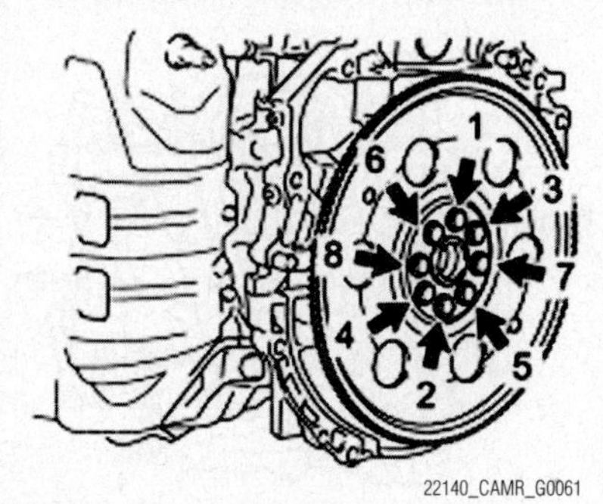

Fig. 149 A/T flywheel bolt tightening sequence

a. Remove the clutch cover assembly and clutch disc assembly.

b. For TMMK made flywheels: use Special Tool: 09960-10010 to hold the crankshaft.

c. For TMC made flywheels: use Special Tools: 09213-54015and 09330-00021 to hold the crankshaft.

d. Remove the eight bolts and flywheel.

To install:

6. For M/T, perform the following:

a. For TMMK made flywheels: use Special Tool: 09960-10010 to hold the crankshaft.

b. For TMC made flywheels: use Special Tools: 09213-54015and 09330-00021 to hold the crankshaft.

c. Clean the bolt and the bolt hole.

d. Apply adhesive to 2 or 3 threads of the bolt end. Adhesive: Part No. 08833-00070, Three Bond or equivalent.

e. Install the flywheel with the eight bolts. Uniformly tighten the 8 bolts in the sequence. Tighten to 96 ft. lbs. (130 Nm).

f. Install clutch cover assembly and clutch disc assembly.

7. For M/T, perform the following:

a. For TMMK made flywheel: use Special Tool: 09960-10010 to hold the crankshaft.

b. For TMC made flywheels: use Special Tools: 09213-54015and 09330-00021 to hold the crankshaft.

c. Clean the bolt and the bolt hole.

d. Apply adhesive to 2 or 3 threads of the bolt end. Adhesive: Part No. 08833-00070, Three Bond or equivalent.

e. Install the front spacer, drive plate and rear spacer with the 8 bolts. Uniformly tighten the eight bolts in the sequence. Tighten 72 ft. lbs. (98 Nm).

INTAKE MANIFOLD

REMOVAL & INSTALLATION

2.4L Engine

See Figure 150.

1. Before servicing the vehicle, refer to the Precautions Section.
2. Drain and recycle the engine coolant.
3. Disconnect the negative battery cable.
4. Remove or disconnect the following:

a. Two nuts and the engine cover.

b. Air cleaner inlet assembly.

c. Mass air flow meter connector.

d. Purge VSV connector.

e. Purge line hose from the clamp.

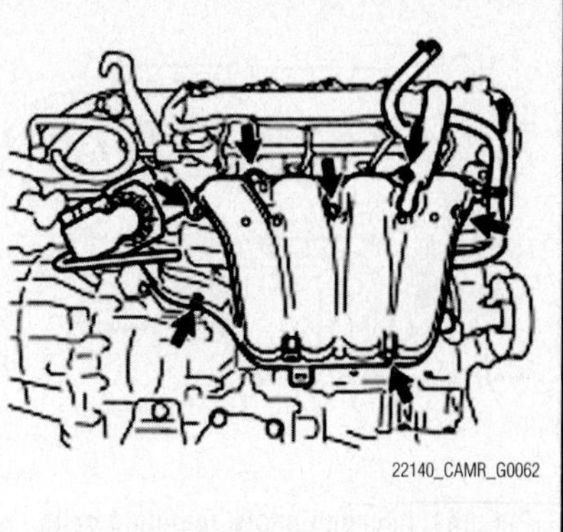

Fig. 150 Removing intake manifold bolts

f. No. 2 ventilation hose from the air cleaner hose.

g. Lock the No. 1 air cleaner hose clamp, and then disconnect the No. 1 air cleaner hose from the throttle body.

h. Two bolts and air cleaner cap.

i. Throttle position sensor connector and wire harness clamp.

j. Four bolts, and then remove the fuel pipe support and throttle body.

k. Purge line hose from the throttle body.

l. Water by-pass hose from the throttle body.

m. N o. 2 water by-pass hose from the throttle body.

n. No. 1 throttle body hose from the throttle body.

o. Gasket from the intake manifold.

p. Five bolts, two nuts, and intake manifold

q. Gasket from the intake manifold.

To install:

5. To install, reverse removal procedure.
6. Install a new throttle body gasket onto the intake manifold.
7. Tighten the throttle body bolts and fuel pipe clamp to: 22 ft. lbs. (30 Nm).
8. Install a new intake manifold gasket.
9. Tighten the intake manifold bolts and nuts to: 22 ft. lbs. (30 Nm).
10. Add coolant and check for leaks.

3.5L Engine

See Figure 151.

1. Before servicing the vehicle, refer to the Precautions Section.
2. Remove the engine assembly with transaxle.
3. Secure the engine.
4. Remove the ignition coil assembly.
5. Remove the right No. 2 engine mounting stay.
6. Remove the intake manifold, as follows:

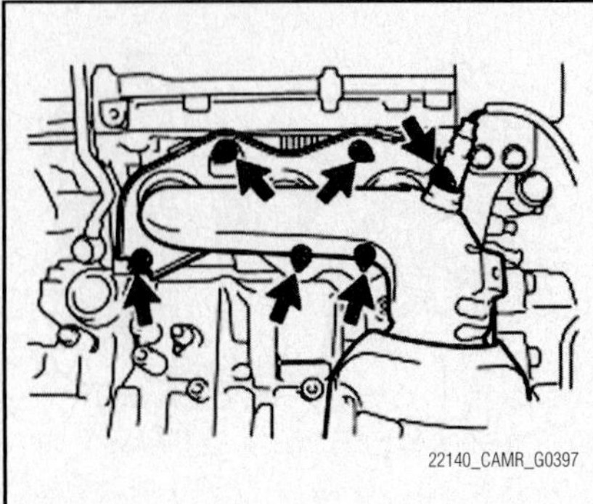
Fig. 151 Locating intake manifold bolts and nuts

a. Uniformly loosen and remove the 6 bolts and 4 nuts.

b. Remove the intake manifold and 2 gaskets.

To install:

7. Install the intake manifold, as follows:

WARNING

DO NOT apply oil to the intake manifold and cylinder head sub-assembly bolts.

a. Set a new gasket on each cylinder head.

Align the port holes of the gasket and cylinder head.

Make sure that the gasket is installed in the correct direction.

b. Set the intake manifold on the cylinder heads.

c. Install and tighten the 6 bolts and 4 nuts uniformly in several steps to 15 ft. lbs. (21 Nm).

8. Install the right No. 2 engine mounting stay.

9. Install the ignition coil assembly.

10. Install the engine assembly with transaxle.

OIL PAN

REMOVAL & INSTALLATION

2.4L Engine

See Figures 152 and 153.

1. Before servicing the vehicle, refer to the Precautions Section.
2. Disconnect the negative battery cable.
3. Drain the engine oil.
4. Remove the oil pan drain plug and gasket.
5. Remove the oil pan 12 bolts and 2 nuts.

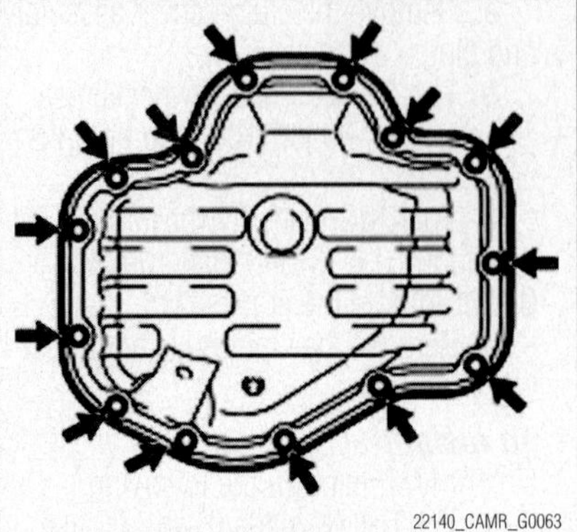
Fig. 152 Removing oil pan bolts

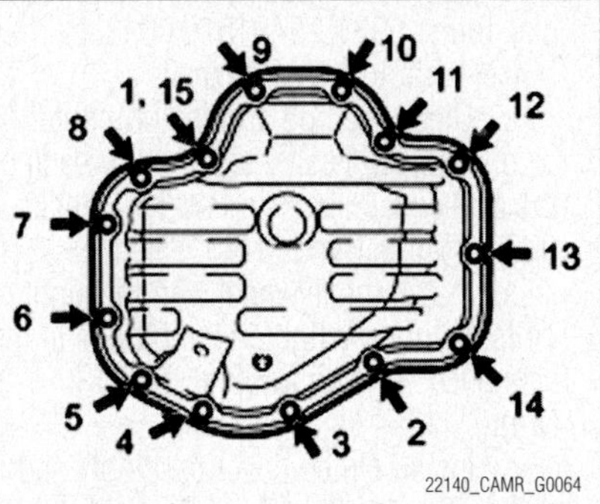

Fig. 153 Oil pan bolt and nut tightening sequence

Be careful not to damage the contact surfaces of the crankcase, chain cover and oil pan.

6. Insert the blade of Special Tool: 09032-00100 between the crankcase and oil pan. Cut through the sealer and remove the oil pan.

To install:

7. Remove any old packing material and be careful not to drop any oil on the contact surfaces of the cylinder block and oil pan.
8. Apply a continuous bead of seal packing (Diameter 0.118 to 0.157 inches (3.0 to 4.0 mm)). Use Toyota Genuine Seal Packing Block, Three Bond 1207B or Equivalent.

Remove any oil from the contact surfaces. Install the oil pan within 3 minutes after applying seal packing. Do not start the engine for at least 2 hours after installing.

9. Uniformly tighten the 12 bolts and 2 nuts in sequence. Tighten the bolts and nuts to 80 inch lbs. (9 Nm).
10. To complete installation, reverse remaining removal procedure.
11. Add oil and check for leaks.

3.5L Engine

See Figures 154 through 158.

1. Before servicing the vehicle, refer to the Precautions Section.
2. Drain the engine oil.
3. Remove the engine assembly with transaxle.
4. Secure the engine.
5. Remove the oil filler cap and gasket.
6. Remove the oil pan drain plug and gasket.
7. Remove the oil pan drain plug and gasket.
8. Remove the No. 1 oil pipe, as follows:
   a. Remove the 2 oil pipe unions and oil pipe.
   b. Remove the left hand oil control valve filter and gaskets.
9. Remove the oil pipe, as follows:
   a. Remove the bolt.
   b. Remove the 2 oil pipe unions and oil pipe.
   c. Remove the right oil control valve filter and gaskets.
10. Remove the oil filter element, as follows:
    a. Remove the drain plug. Do not remove the O-ring.

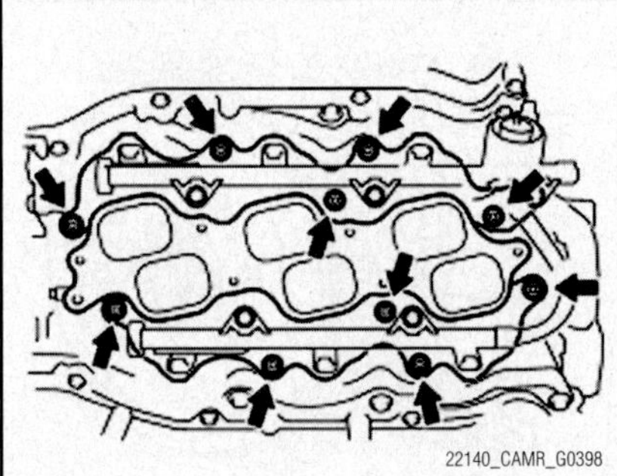
Fig. 154 No. 2 oil pan sub-assembly removal

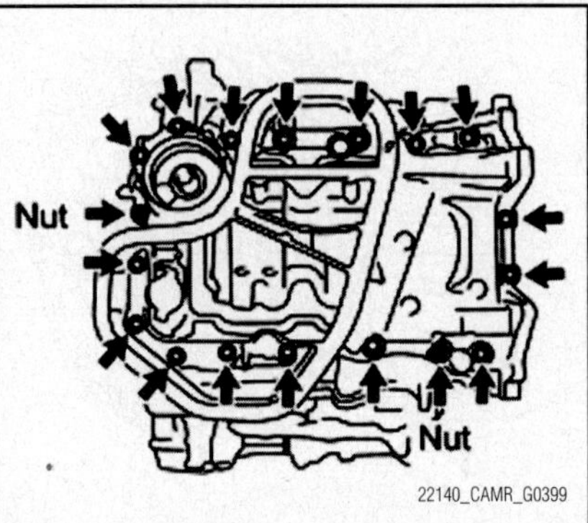

Fig. 155 Oil pan sub-assembly removal

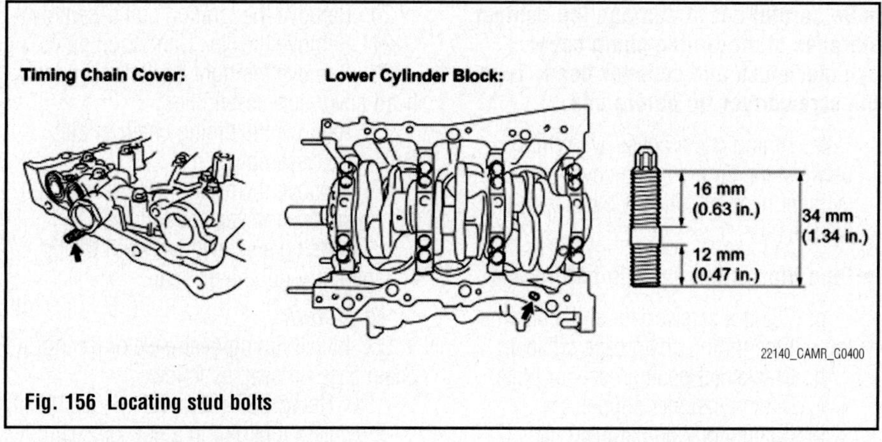

Fig. 156 Locating stud bolts

b. Connect the hose to the pipe.

c. Insert the pipe with the hose into the oil filter cap.

d. Make sure that the oil is completely drained and remove the pipe and O-ring.

e. Using SST (SST: 09228-06501) or equivalent, remove the oil filter cap.

f. Remove the oil filter element and O-ring from the oil filter cap. Do not use any tools when removing the O-ring to prevent the O-ring groove from being damaged.

11. Remove the No. 2 oil pan sub-assembly, as follows:

a. Remove the 16 bolts and 2 nuts.

b. Insert the blade of SST (SST: 09032-00100) or equivalent tool between the oil pans. Cut through the applied sealer and remove the No. 2 oil pan sub-assembly.

➡Be careful not to damage the contact surfaces of the oil pans.

12. Remove the oil pan sub-assembly, as follows:

a. Remove the 16 bolts and 2 nuts.

➡Be sure to clean the bolts and stud bolts and check the threads for cracks or other damage.

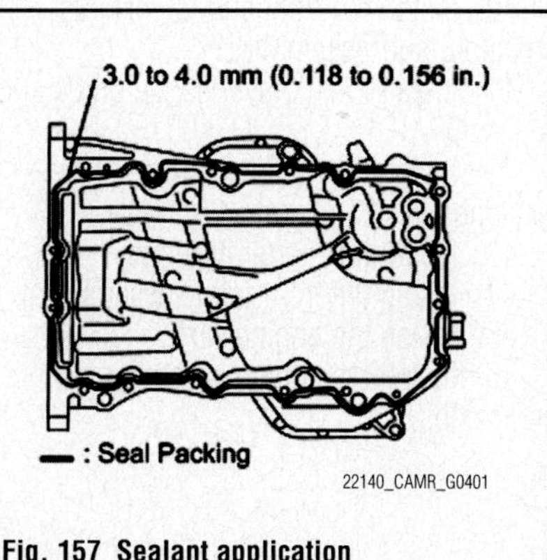

Fig. 157 Sealant application

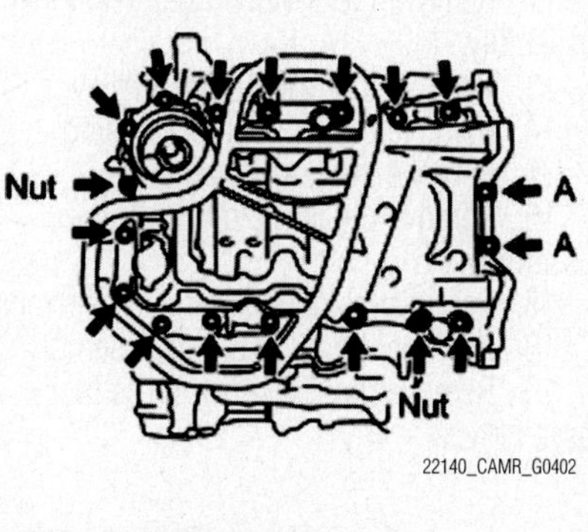

b. Remove the oil pan by prying between the oil pan and cylinder block with a taped screwdriver.

➡Be careful not to damage the contact surfaces of the cylinder block and oil pan.

c. Remove the 2 O-rings.

To install:

13. Install the oil pan sub-assembly, as follows:

a. Using an E8 Torx® socket wrench, install the stud bolts as shown in the illustration. Tighten to 7 ft. lbs (10 Nm).

b. Install 2 new O-rings.

➡Remove any oil from the contact surface.

➡Install the oil pan within 3 minutes after applying seal packing.

➡Do not start the engine for at least 2 hours after installing.

c. Apply seal packing (Toyota Genuine Seal Packing Black, Three Bond 1207B or equivalent) in a continuous line as shown in the illustration. Seal diameter: 3.0 to 4.0 mm (0.118 to 0.156 in.).

d. Install the oil pan with the 16 bolts and 2 nuts and tighten to 7 ft. lbs (10 Nm), and 15 ft. lbs (21 Nm).

Fig. 158 Oil pan bolts and nuts

14. Install the No. 2 oil pan sub-assembly, as follows:

➡Remove any oil from the contact surface.

➡Install the No. 2 oil pan within 3 minutes after applying seal packing.

➡Do not start the engine for at least 2 hours after installing.

a. Using an E6 Torx® socket wrench, install the stud bolts as shown in the illustration and tighten to 35 inch lbs (4 Nm).

b. Apply seal packing (Toyota Genuine Seal Packing Black, Three Bond 1207B or equivalent) in a continuous line as shown in the illustration. Seal diameter: 3.0 to 4.0 mm (0.118 to 0.156 in.).

c. Install the No. 2 oil pan with the 16 bolts and 2 nuts and tighten to 7 ft. lbs (10 Nm).

15. Install the oil pan drain plug and a new gasket. Tighten to 30 ft. lbs (40 Nm).

16. Install the oil filter element, as follows:

a. Clean the inside of the oil filter cap, the threads and O-ring groove.

b. Apply a small amount of engine oil to a new O-ring and install it to the oil filter cap.

c. Set a new oil filter element to the oil filter cap.

d. Remove dirt or foreign matter from the installation surface and inside of the engine.

e. Apply a small amount of engine oil to the O-ring again and install the oil filter cap.

➡Be careful that the O-ring does not get caught between the parts. The O-ring must not be twisted on the groove.

f. Using SST (SST: 09228-06501) or equivalent, install the oil filter cap and tighten to 18 ft. lbs (25 Nm). Make sure that the oil filter is installed securely as shown in the illustration.

17. Install the oil filler cap sub-assembly.

18. Install the engine assembly with transaxle.

19. Check for oil leaks.

OIL PUMP

REMOVAL & INSTALLATION

2.4L Engine

See Figure 159.

1. Before servicing the vehicle, refer to the Precautions Section.

2. Disconnect the negative battery cable.

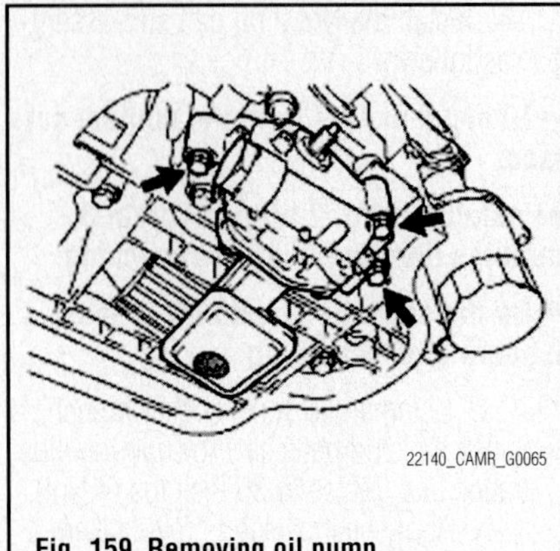

Fig. 159 Removing oil pump

3. Drain the engine oil.
4. Remove or disconnect the following:
   a. Plastic engine cover and two nuts.
   b. Front right wheel.
   c. LH and RH engine under cover.
   d. RH front fender apron seal.
   e. Front exhaust pipe assembly.
   f. RH engine mount stay.
   g. Engine mount
   h. RH engine mounting bracket
   i. Drive belt.
   j. Generator assembly.
   k. Vane pump.
   l. Ignition coil assembly.
   m. Ventilation hoses.
   n. Valve cover.
5. Turn the crankshaft pulley until its groove and the timing mark "0" of the timing chain cover are aligned.
6. Check that each timing mark of the camshaft timing gear and sprocket is aligned with each timing mark located on the intake and exhaust bearing caps. If not, turn the crankshaft by 1 revolution (360°) to align the timing marks.
7. Remove or disconnect the following:
   a. Crankshaft pulley.
   b. Crankshaft position sensor.
   c. Oil pan.
   d. Chain upper tensioner assembly.
8. Install the No. 1 engine hanger (12281-28010) and No. 2 engine hanger (12282-28010) with the bolts (91512-61020) and tighten to: 28 ft. lbs. (38 Nm).
9. Remove or disconnect the following:
   a. Drive belt tensioner.
   b. Engine mount insulator.
   c. RH engine mount bracket.
   d. Using a E10 Torx® socket, remove the stud bolt for the drive belt tensioner from the cylinder block.
   e. Timing chain cover twelve bolts and two nuts.

➡Be careful not to damage the contact surfaces of the timing chain cover, cylinder block and cylinder head. Tape the screwdriver tip before use.

   f. Timing chain cover by prying between the timing chain cover and cylinder head or cylinder block with a screwdriver.

➡Tape the screwdriver tip before use.

   g. Using a screwdriver and a hammer, tap out the timing chain case oil seal.
   h. Crankshaft position sensor plate.
   i. Chain tensioner slipper.
   j. Chain vibration damper.
   k. Timing chain guide.
   l. Upper chain assembly.
   m. Crankshaft timing sprocket.
   n. Lower chain assembly.
   o. Three bolts, oil pump and gasket.

To install:

10. Install a new gasket and the oil pump with the 3 bolts. Tighten the bolts 14 ft. lbs. (19 Nm).
11. To complete installation, reverse removal procedure. Refer to the appropriate sections to install components correctly.

3.5L Engine

1. Before servicing the vehicle, refer to the Precautions Section.
2. Remove the engine assembly with transaxle.
3. Secure engine.
4. Remove the engine wire.
5. Remove the front frame assembly.
6. Remove the starter assembly.
7. Remove the automatic transaxle assembly.
8. Remove the oil level gauge guide sub-assembly.
9. Remove the right and left exhaust manifold sub-assemblies.
10. Remove the drive plate and ring gear sub-assembly.
11. Remove the No. 2 idler pulley sub-assembly.
12. Remove the V-ribbed belt tensioner assembly.
13. Remove the water pump pulley.
14. Remove the water inlet housing.
15. Remove the crankshaft pulley.
16. Remove the No. 2 oil pan sub-assembly.
17. Remove the oil strainer sub-assembly.
18. Remove the oil pan sub-assembly.
19. Remove the intake air surge tank assembly.
20. Remove the ignition coil assembly.
21. Remove the No. 1 and 2 oil pipes.
22. Remove the right and left cylinder head cover sub-assemblies.
23. Remove the timing chain or belt cover sub-assembly.
24. Remove the timing gear case or timing chain case oil seal, as follows:
    a. Using a screwdriver with the tip taped, pry out the oil seal.

To install:

25. Install timing gear case or timing chain case oil seal, as follows:
    a. Using SST (SST: 09316-60011) or equivalent tool, tap in a new oil seal until its surface is flush with the timing chain case edge.

➡Keep the lip free from foreign matter.

➡Do not tap on the oil seal at an angle.

➡Make sure that the oil seal edge does not stick out of the timing chain case.

    b. Apply MP grease to the oil seal lip.
26. Install timing chain or belt cover sub-assembly.
27. Install the right and left cylinder head cover sub-assemblies.
28. Install the No. 1 and 2 oil pipes.
29. Install the ignition coil assembly.
30. Install the intake air surge tank assembly.
31. Install the oil pan sub-assembly.
32. Install the oil strainer sub-assembly.
33. Install the No. 2 oil pan sub-assembly.
34. Install the crankshaft pulley.
35. Install the water inlet housing.
36. Install the water pump pulley.
37. Install the V-ribbed belt tensioner assembly.
38. Install the No. 2 idler pulley sub-assembly.
39. Install the drive plate and ring gear sub-assembly.
40. Install the right and left exhaust manifold sub-assemblies.
41. Install the oil level gauge guide sub-assembly.
42. Install the automatic transaxle assembly.
43. Install the starter assembly.
44. Install the front frame assembly.
45. Install the engine wire.
46. Install the engine assembly with transaxle.

MAIN BEARING TORQUE SEQUENCE

See Figures 160 through 162.

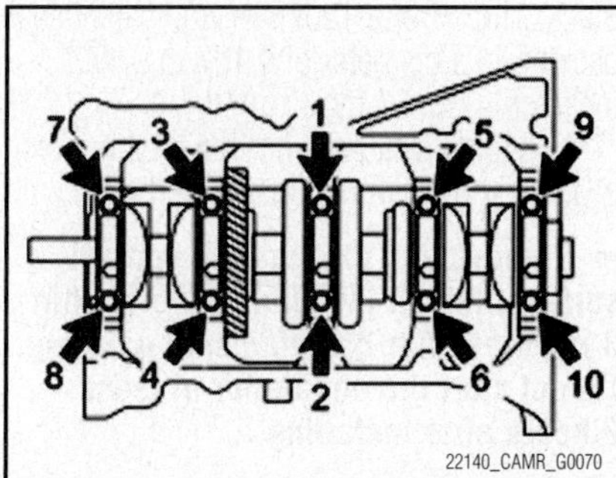

Fig. 160 Main bearing torque sequence—2.4L engine

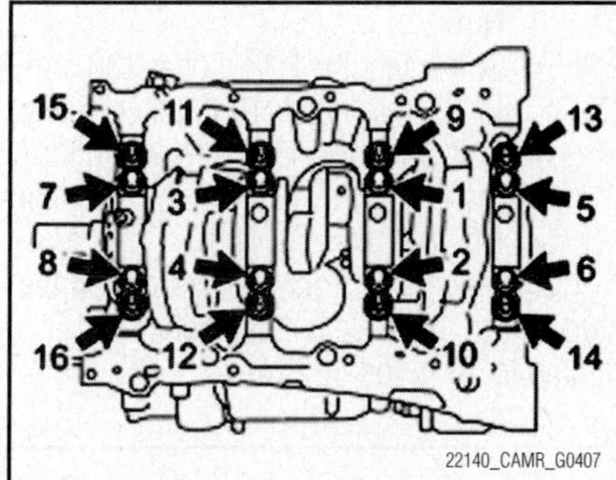

Fig. 161 Main bearing torque sequence main cap bolts—3.5L engine

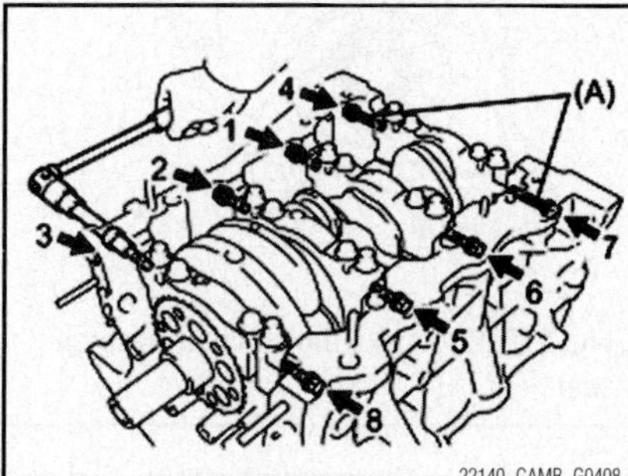

Fig. 162 Main bearing torque sequence side bolts—3.5L engine

PISTON AND RING

POSITIONING

See Figure 163.

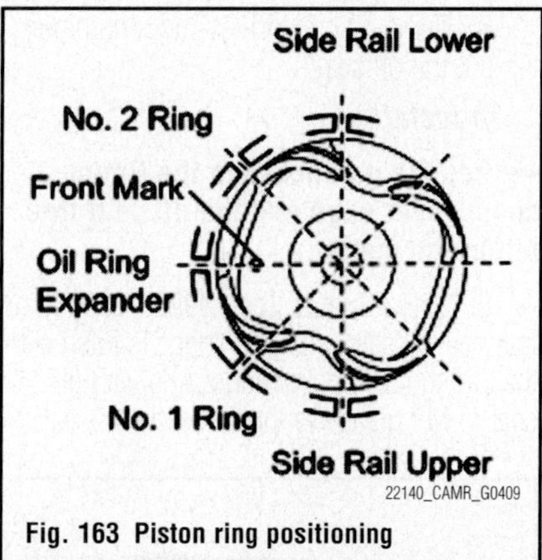

Fig. 163 Piston ring positioning

REAR MAIN SEAL

REMOVAL & INSTALLATION

2.4L Engine

See Figures 164 and 165.

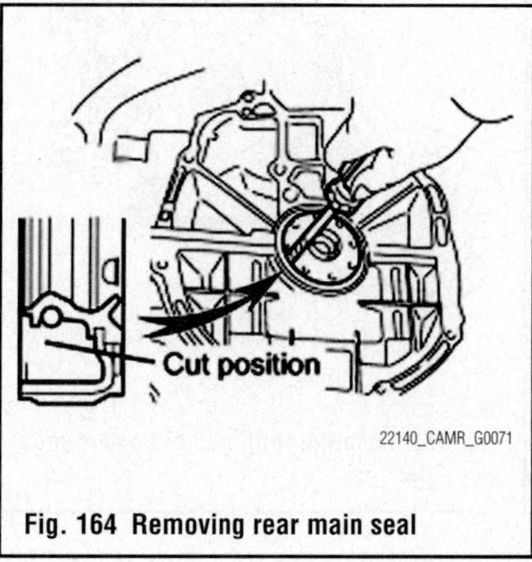

Fig. 164 Removing rear main seal

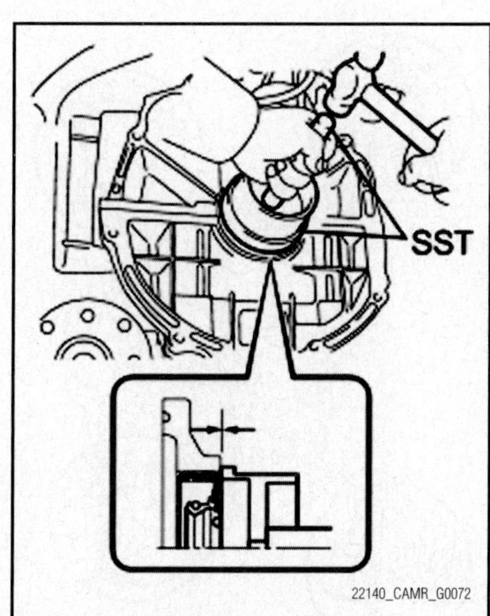

Fig. 165 Installing rear main seal

1. Before servicing the vehicle, refer to the Precautions Section.
2. Disconnect the negative battery cable.
3. For A/T vehicles, separate the transaxle. Refer to the Drive Train section.
4. For M/T vehicles, separate the transaxle. Refer to the Drive Train section.
5. For A/T vehicles, remove the flexplate. Refer to Flexplate and Flywheel above.
6. For M/T vehicles, remove the flywheel. Refer to Flexplate and Flywheel above.
7. Using a knife, cut through the oil seal lip.
8. Using a screwdriver with its tip taped, pry out the oil seal.
9. After the removal, check the crankshaft for damage. If it is damaged, smooth the surface with 400-grit sandpaper.

To install:

➡Keep the lip free from foreign matter.

10. Apply MP grease to a new oil seal lip.
11. Using Special Tools: 09223-15030, 09950-70010 and a hammer, tap in the oil seal until its surface is flush with the rear oil seal retainer edge. Wipe off extra grease from the crankshaft.
12. To complete installation, reverse remaining removal procedure.

3.5L Engine

See Figures 166 and 167.

1. Before servicing the vehicle, refer to the Precautions Section.
2. Remove the automatic transaxle assembly.
3. Remove the drive plate and ring gear sub-assembly.

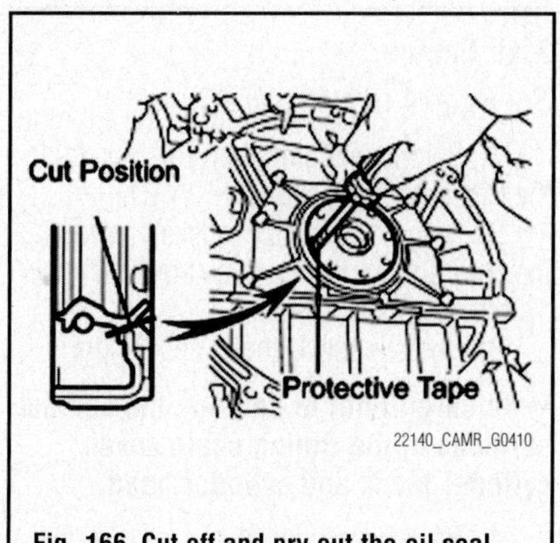

Fig. 166 Cut off and pry out the oil seal

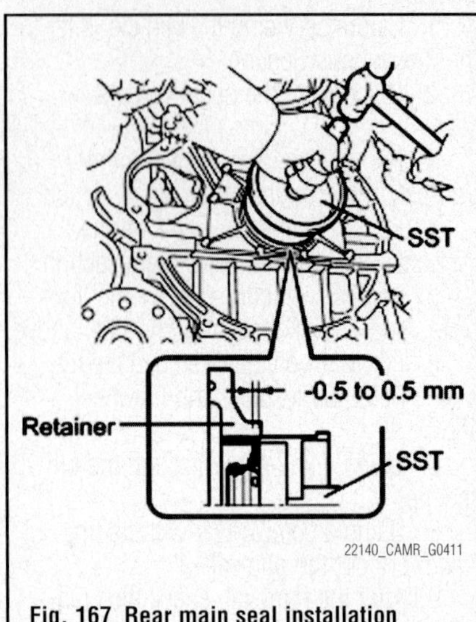

Fig. 167 Rear main seal installation

4. Remove the rear main seal, as follows:

a. Using a knife, cut off the oil seal lip.

To install:

5. Apply MP grease to a new oil seal lip.
6. Using SST (SST: 09223-15030, SST: 09950-70010) or equivalent and a hammer, tap in the oil seal. Oil seal tap in depth: -0.020 to 0.020 in. (-0.5 to 0.5 mm)
7. Install the drive plate and ring gear sub-assembly.
8. Install automatic transaxle assembly.

ROCKER ARMS/SHAFTS

REMOVAL & INSTALLATION

For 3.5L engines, see Camshafts.

TIMING CHAIN COVER AND SEAL

REMOVAL & INSTALLATION

2.4L Engine

See Figures 168 through 173.

1. Before servicing the vehicle, refer to the Precautions Section.
2. Using a E10 Torx® socket, remove the stud bolt for the drive belt tensioner from the cylinder block.
3. Remove the 12 bolts and 2 nuts.

➡Be careful not to damage the contact surfaces of the timing chain cover, cylinder block and cylinder head.

4. Remove the timing chain cover by prying between the timing chain cover and cylinder head or cylinder block with a screwdriver. Tape the screwdriver tip before use.
5. Using a screwdriver and a hammer, tap out the oil seal.

To install:

➡Keep the gap between the timing chain cover edge and the oil seal free of foreign matter.

6. Using Special Tool: 09223-22010, tap in a new oil seal until its surface is flush with the timing chain cover edge. Apply a light coat of MP grease to the lip of the oil seal.

Fig. 168 Locating timing chain cover bolts and nuts

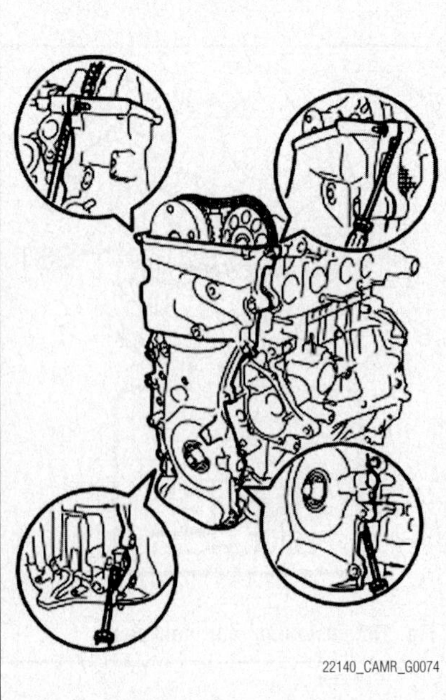

Fig. 169 Removing timing chain cover

7. Remove any old packing (FIPG) material and be careful not to drop any oil on the contact surfaces of the timing chain cover, cylinder head and cylinder block.
8. Apply Toyota Genuine Seal Packing Black, Three Bond 1207B or Equivalent seal packing in a diameter of 0.157 to 0.177 inches (4.0 to 4.5 mm).
9. Apply seal packing in a continuous bead as shown in the illustration below.

➡ Remove any oil from the contact surface. Install the chain cover within 3 minutes after applying seal packing. Do not start the engine for at least 2 hours after installing.

10. Install the timing chain cover with the twelve bolts and two nuts in sequence to the following torque specification:
    - Bolt A length: 1.18 inches (30 mm) for 10 mm head: 80 inch lbs. (9 Nm)
    - Bolt B length: 1.18 inches (30 mm) for 12 mm head: 18 ft. lbs. (25 Nm)
    - Bolt C length: 1.57 inches (40 mm) for 14 mm head: 41 ft. lbs. (55 Nm)
    - Nut: 8 ft. lbs. (11 Nm)
11. Using a E10 Torx® socket, install the stud bolt to the drive belt tensioner and tighten to 16 ft. lbs. (22 Nm).

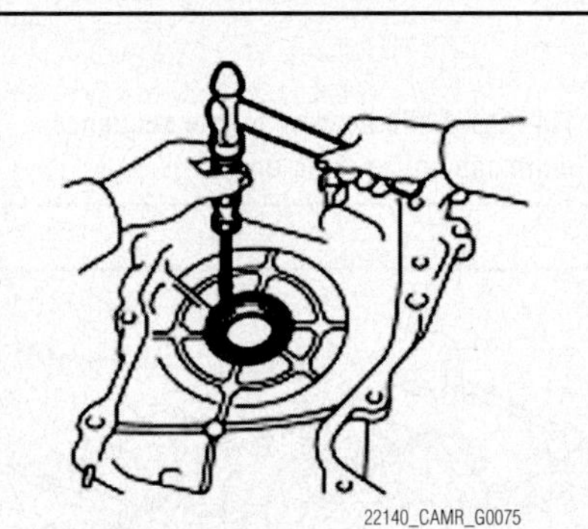

Fig. 170 Removing timing chain case oil seal

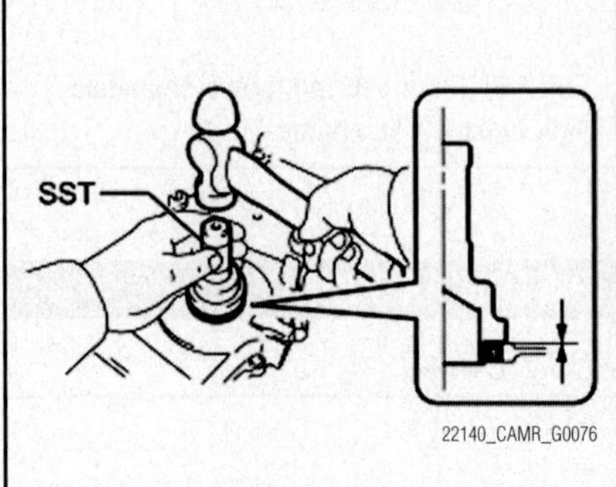

Fig. 171 Installing timing chain case oil seal

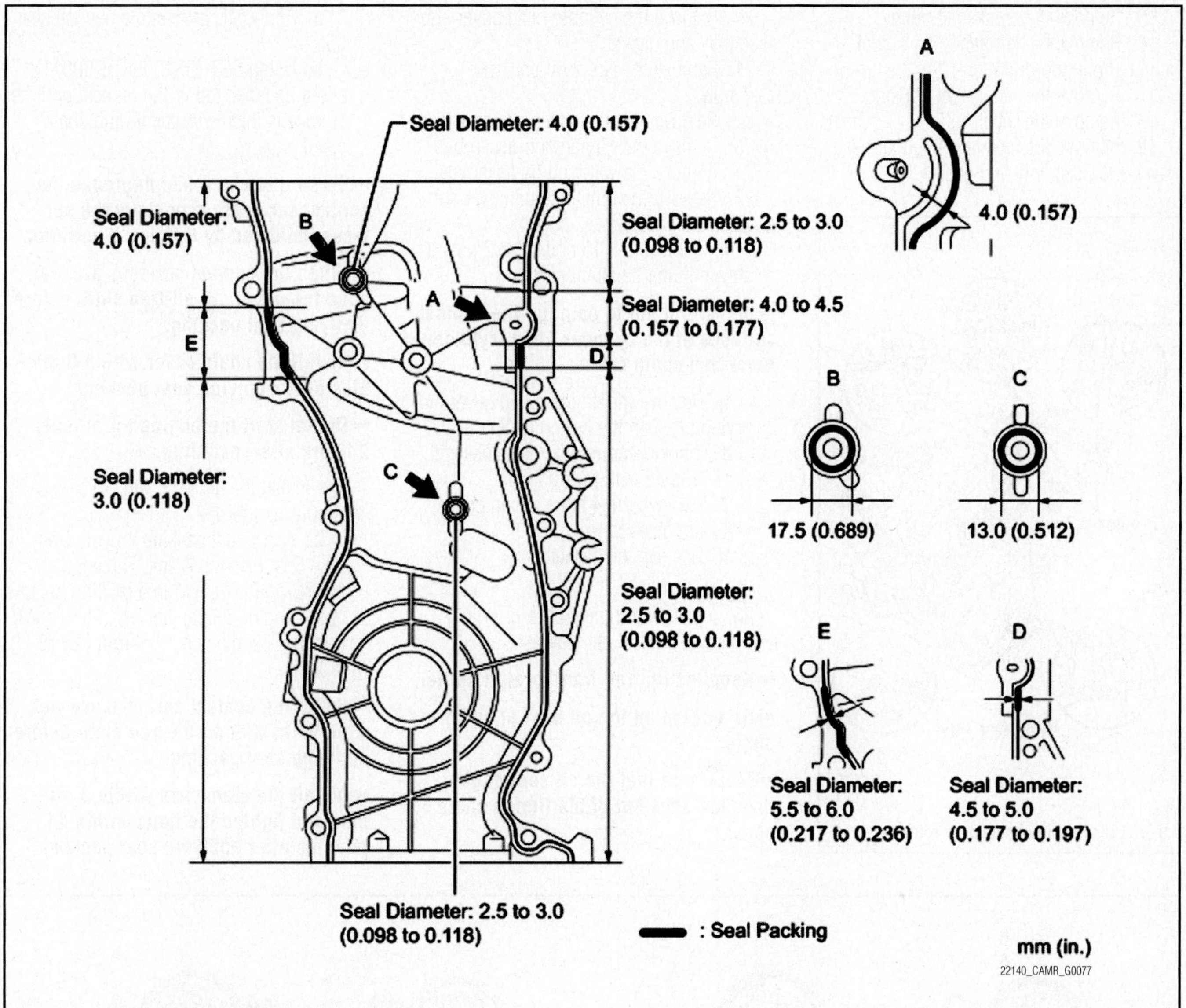

Fig. 172 Applying seal packing in a continuous bead

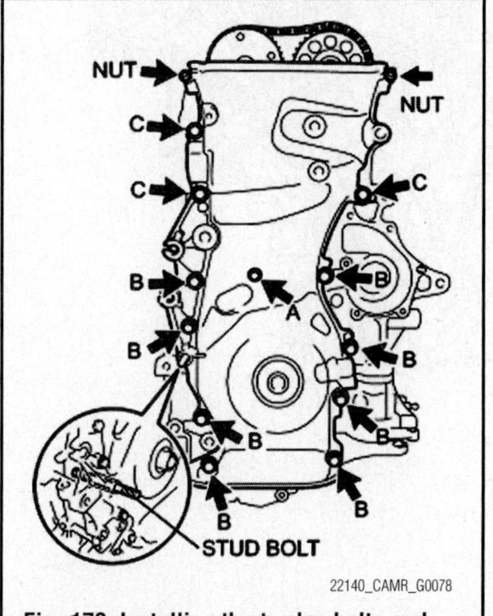

Fig. 173 Installing the twelve bolts and two nuts

3.5L Engine

See Figures 174 through 179.

1. Before servicing the vehicle, refer to the Precautions Section.
2. Remove the engine assembly with transaxle.
3. Secure engine.
4. Remove the oil filler cap sub-assembly.
5. Remove the spark plugs and ignition coil assembly.
6. Remove the oil pan drain plug and gasket.
7. Remove the ventilation valve sub-assembly.
8. Remove the camshaft position sensor.
9. Remove the camshaft timing oil control valve assembly.
10. Remove crankshaft position sensor.
11. Remove the No. 1 oil pipe.
12. Remove the oil pipe.
13. Remove the cylinder block water drain cock sub-assembly.
14. Remove the oil filter.

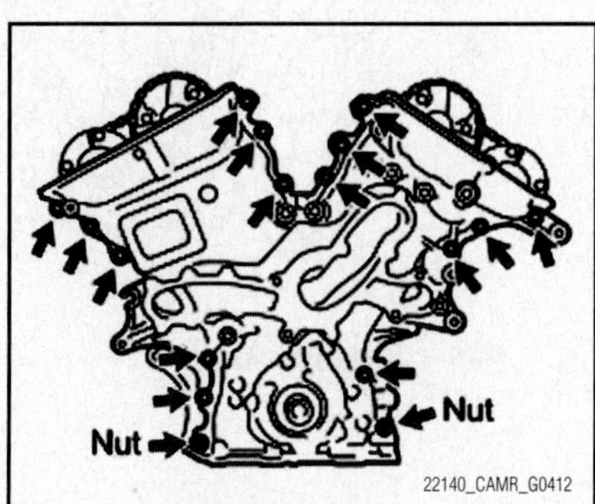

Fig. 174 Locating timing chain cover sub-assembly bolts and nuts

15. Remove the crankshaft pulley.
16. Remove the left hand No. 1 front engine mounting bracket.
17. Remove the water inlet housing.
18. Remove the water outlet.
19. Remove the left-hand cylinder head cover sub-assembly and gasket.

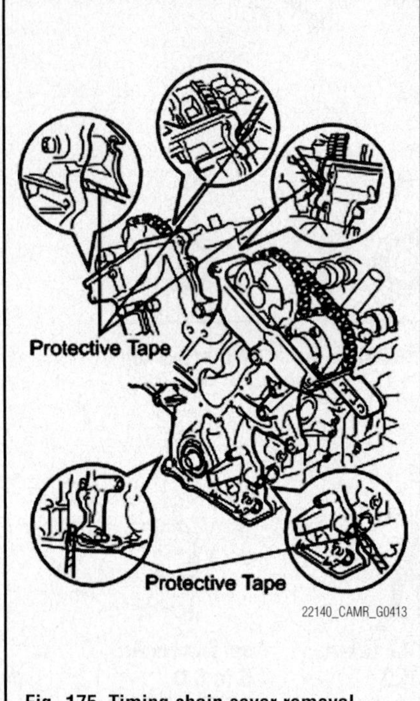

Fig. 175 Timing chain cover removal

20. Remove the cylinder head cover sub-assembly and gasket.
21. Remove the No. 2 oil pan sub-assembly.
22. Remove the oil strainer sub-assembly.
23. Remove the oil pan sub-assembly.
24. Remove the water pump assembly.
25. Remove the timing chain cover sub-assembly, as follows:
   a. Remove the 15 bolts and 2 nuts as shown in the illustration.

➡Be careful not to damage the contact surfaces of the cylinder head, cylinder block and chain cover.

   b. Remove the timing chain cover by prying between the timing chain cover and cylinder head or cylinder block with a screwdriver with the tip taped.
   c. Remove the 4 bolts, chain cover plate and gasket.
   d. Remove the gasket.

To install:

26. Install timing gear case or timing chain cover oil seal, as follows:

➡Keep the lip free from foreign matter.

➡Do not tap on the oil seal at an angle.

➡Make sure that the oil seal edge does not stick out of the timing chain cover.

   a. Apply MP grease to a new oil seal lip.
   b. Using SST (SST: 09316-60011) and a hammer, tap in the oil seal until its surface is flush with the timing chain cover edge.

➡Be sure to clean and degrease the contact surfaces, especially the surfaces indicated by C in the illustration.

➡When the contact surfaces are wet, wipe them with an oil-free cloth before applying seal packing.

➡Install the chain cover within 3 minutes after applying seal packing.

➡Do not start the engine for at least 2 hours after installing.

27. Install the timing chain cover sub-assembly, as follows:
   a. Apply seal packing (Toyota Genuine Seal Packing Black, Three Bond 1207B or equivalent) in a continuous line to the engine unit as shown in the illustration. Seal diameter: 3.0 mm (0.118 in.).

➡When the contact surfaces are wet, wipe them with an oil-free cloth before applying seal packing.

➡Install the crankcase within 3 minutes and tighten the bolts within 15 minutes after applying seal packing.

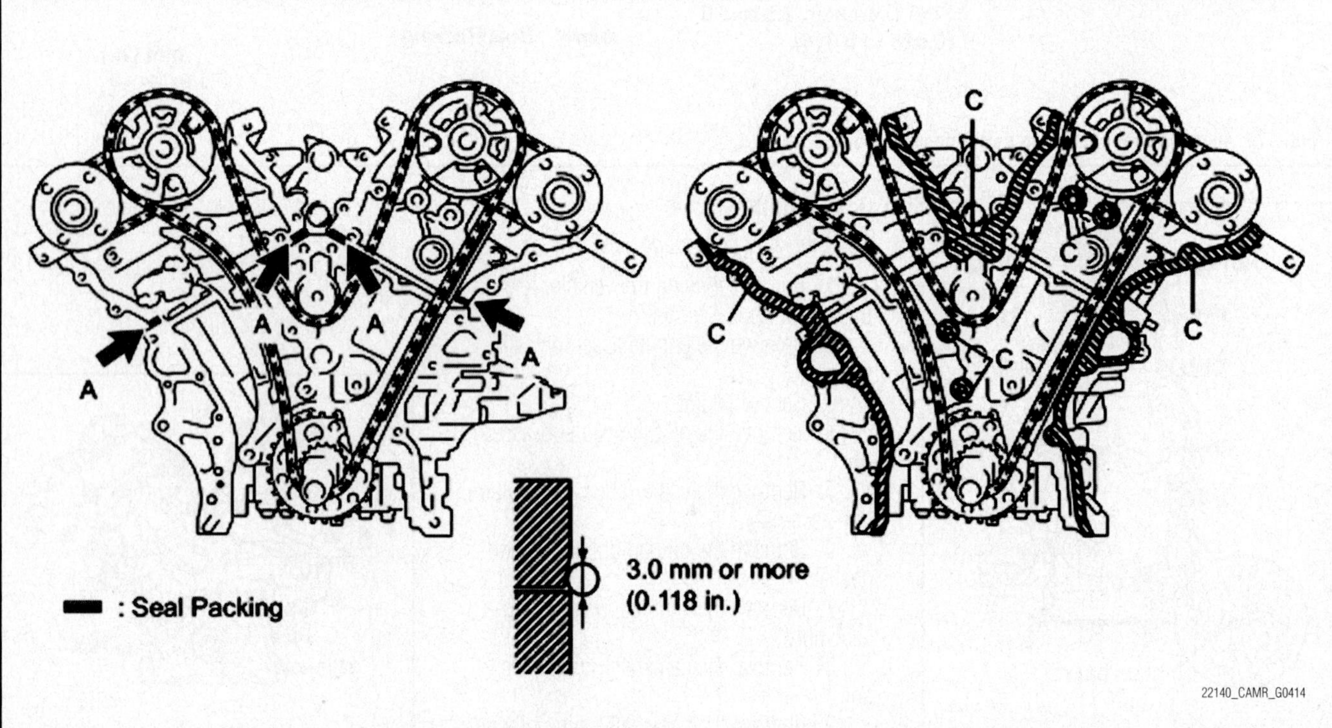

Fig. 176 Applying seal packing to timing chain cover sub-assembly

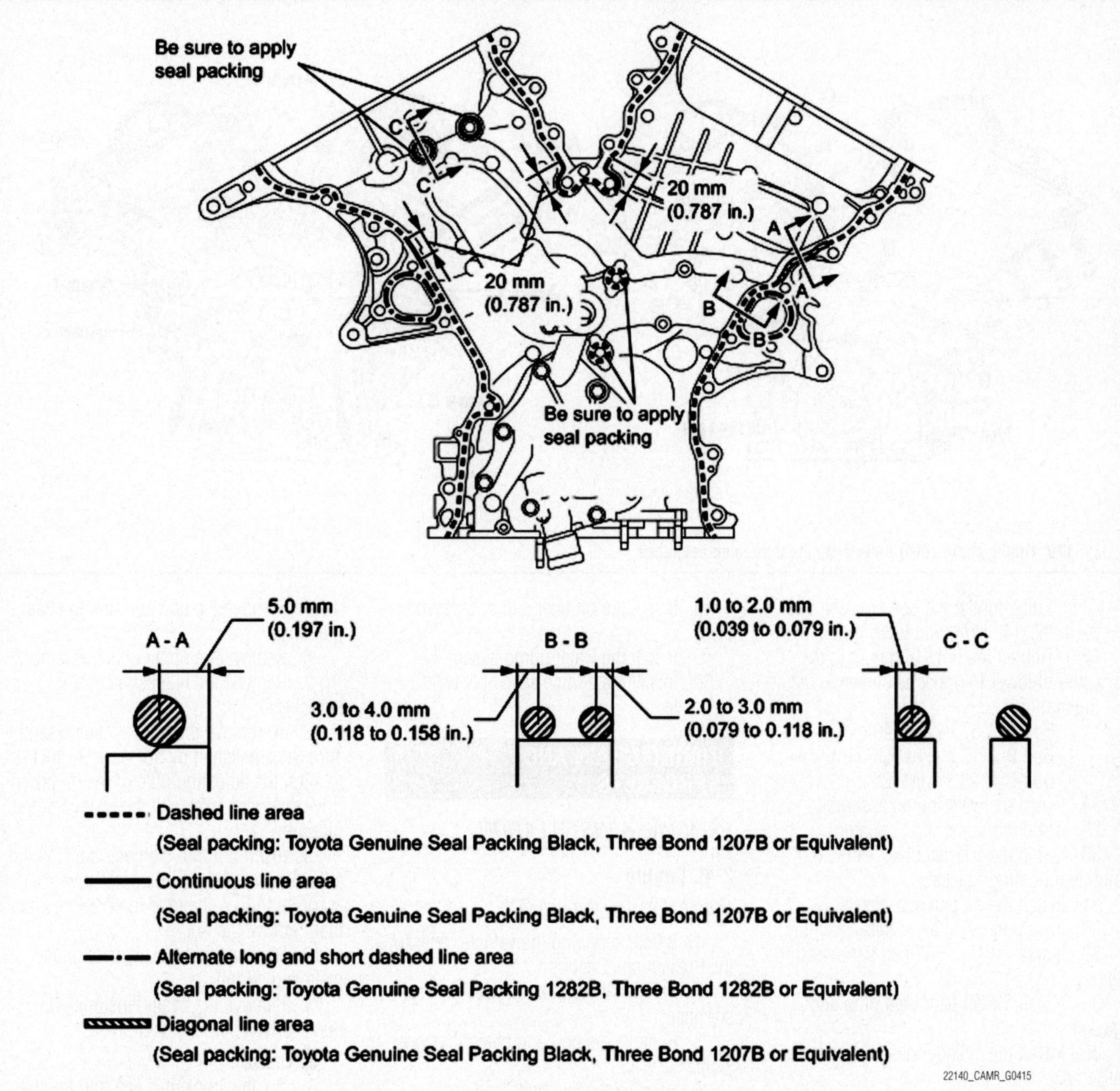

Be sure to apply seal packing

20 mm (0.787 in.)

20 mm (0.787 in.)

Be sure to apply seal packing

A - A

5.0 mm (0.197 in.)

B - B

3.0 to 4.0 mm (0.118 to 0.158 in.)

2.0 to 3.0 mm (0.079 to 0.118 in.)

C - C

1.0 to 2.0 mm (0.039 to 0.079 in.)

Dashed line area (Seal packing: Toyota Genuine Seal Packing Black, Three Bond 1207B or Equivalent)

Continuous line area (Seal packing: Toyota Genuine Seal Packing Black, Three Bond 1207B or Equivalent)

Alternate long and short dashed line area (Seal packing: Toyota Genuine Seal Packing 1282B, Three Bond 1282B or Equivalent)

Diagonal line area (Seal packing: Toyota Genuine Seal Packing Black, Three Bond 1207B or Equivalent)

22140_CAMR_G0415

Fig. 177 Applying seal packing to timing chain cover

➡Do not start the engine for at least 2 hours after installing.

b. Apply seal packing in a continuous line to the timing chain cover as shown in the following illustration. Seal packing: Toyota Genuine Seal Packing Black, Three Bond 1207B or equivalent, Toyota Genuine Seal Packing Black, Three Bond 1282B, Three Bond 1282B or equivalent.

c. Install a new gasket.

d. Align the oil pump's drive rotor spline and the crankshaft as shown in the illustration. Install the spline and chain cover to the crankshaft.

e. Loosely install the timing chain cover with the 23 bolts and 2 nuts, but do not tighten the bolts and 2 nuts yet.

CAUTION

Make sure that there is no oil on the bolt and nut threads.

f. Fully tighten the bolts in this order: Area 1 and Area 2, tighten to 15 ft. lbs. (21 Nm).

g. Fully tighten the bolts in Area 3 to 15 ft. lbs. (21 Nm). Tighten the bolts and nuts in the order of upper to lower as shown in the illustration.

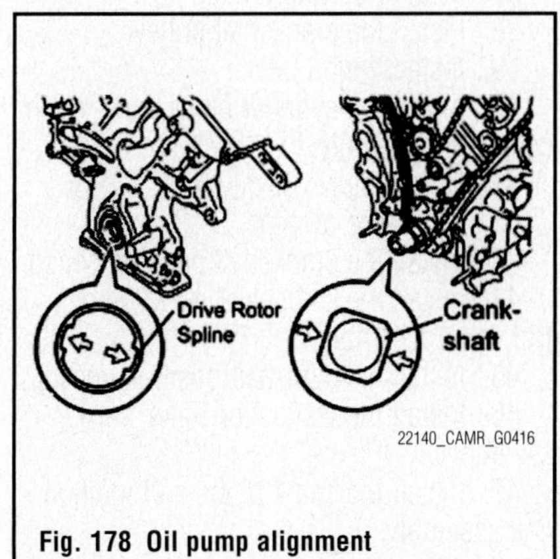

Fig. 178 Oil pump alignment

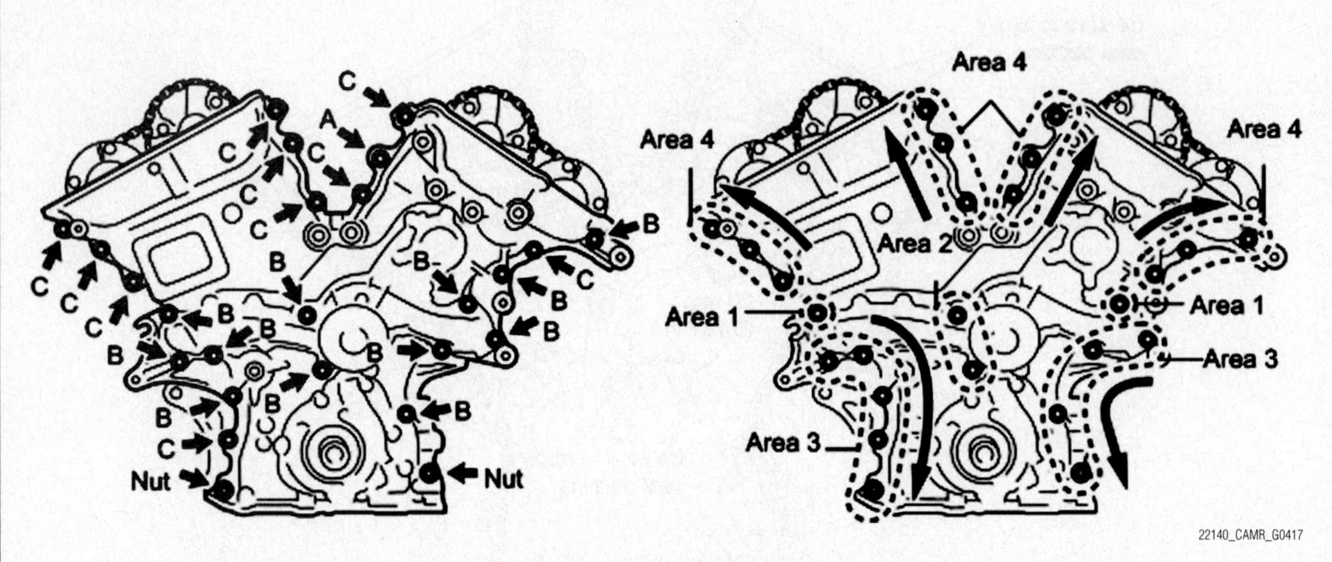

Fig. 179 Timing chain cover bolts and nuts tightening sequence

h. Fully tighten the bolts in Area 4 to 32 ft. lbs. (43 Nm), and to 15 ft. lbs. (21 Nm). Tighten the bolts and nuts in the order of lower to upper as shown in the illustration.

- Bolt A: 1.57 inches (40 mm)
- Bolt B: 2.17 inches (55 mm)
- Bolt C: 0.98 inches (25 mm)

28. Install the water pump assembly.
29. Install the water inlet housing.
30. Install the left hand No. 1 front engine mounting bracket.
31. Install the oil pan sub-assembly.
32. Install the oil strainer sub-assembly.
33. Install the No. 2 oil pan sub-assembly.
34. Install the oil pan drain plug and gasket.
35. Install the cylinder head cover sub-assembly.
36. Install the left-hand cylinder head cover sub-assembly.
37. Install the water outlet.
38. Install the crankshaft pulley.
39. Install the oil filter.
40. Install the cylinder block water drain cock sub-assembly.
41. Install the No. 1 oil pipe.
42. Install the oil pipe.
43. Install the crankshaft position sensor.
44. Install the camshaft timing oil control valve assembly.
45. Install the camshaft position sensor.
46. Install the ventilation valve sub-assembly.
47. Install the spark plugs and ignition coil assembly.
48. Install the oil filler cap sub-assembly.
49. Install the water pump assembly.
50. Install the engine assembly with transaxle.

TIMING CHAIN AND SPROCKETS

REMOVAL & INSTALLATION

2.4L Engine

See Figures 180 through 202.

1. Before servicing the vehicle, refer to the Precautions Section.
2. Remove the crankshaft position sensor plate.
3. Remove the bolt and chain tensioner slipper.
4. Remove the two bolts and chain vibration damper.
5. Remove the bolt and timing chain guide.
6. Remove the upper chain assembly.
7. Remove the crankshaft timing sprocket.
8. To remove the lower chain assembly. Turn the crankshaft by 90° counterclockwise to align the adjusting hole of the oil pump drive shaft sprocket with the groove of the oil pump.
9. Insert a 4 mm diameter bar into the adjusting hole of the oil pump drive shaft sprocket to lock the gear in position, and then remove the nut.
10. Remove the bolt, chain tensioner plate and spring.
11. Remove the chain tensioner, oil pump driven sprocket and chain.

To install:

12. Set the crankshaft key into the left horizontal position.

Fig. 180 Removing the crankshaft position sensor plate

Fig. 181 Removing the bolt and chain tensioner slipper

13. Turn the drive shaft so that the cutout faces upward.

14. Align the yellow mark links with the timing marks of each gear.

15. Install the sprockets onto the crankshaft and oil pump shaft with the chain wrapped on the gears

16. Temporarily tighten the oil pump drive shaft sprocket with the nut.

17. Insert the damper spring into the adjusting hole, and then install the chain tensioner plate with the bolt and tighten to 9 ft. lbs. (12 Nm).

18. Align the adjusting hole of the oil pump drive shaft sprocket with the groove of the oil pump.

19. Insert a 4 mm diameter bar into the adjusting hole of the oil pump drive shaft gear to lock the gear in position, and then tighten the nut to 22 ft. lbs. (30 Nm).

20. Rotate the crankshaft clockwise by 90°, and align the crankshaft key to the top.

21. Install the crankshaft timing sprocket.

22. Install the chain vibration damper with the 2 bolts and tighten to 80 inch lbs. (9 Nm).

23. Set the No. 1 cylinder to TDC/compression.

24. Turn the camshafts with a wrench (using the hexagonal lobe) to align the timing marks of the camshaft timing gear with

Fig. 182 Removing the two bolts and upper chain vibration damper

Fig. 183 Removing the bolt and timing chain guide

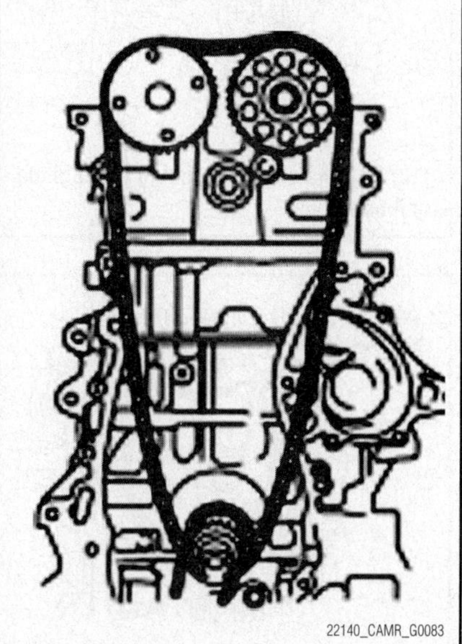

Fig. 184 Removing the upper chain assembly

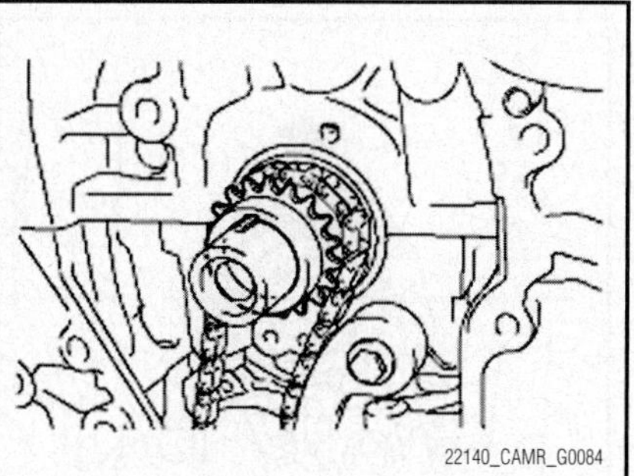

Fig. 185 Removing the crankshaft timing sprocket

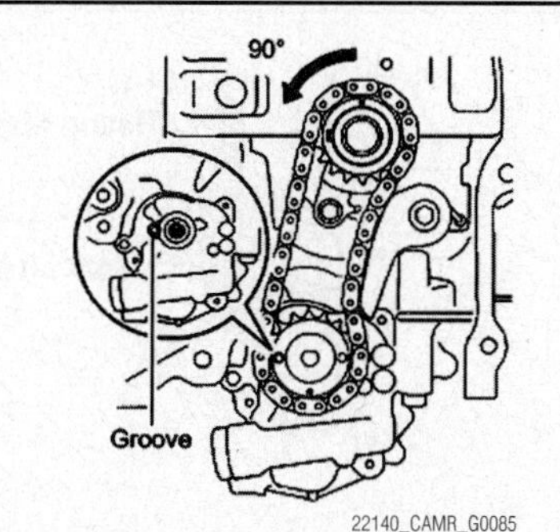

Fig. 186 Turning the crankshaft by 90° counterclockwise to align the adjusting hole

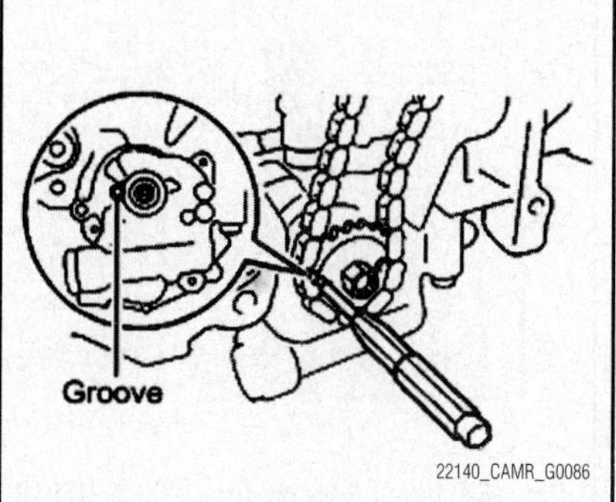

Fig. 187 Locking the gear in position

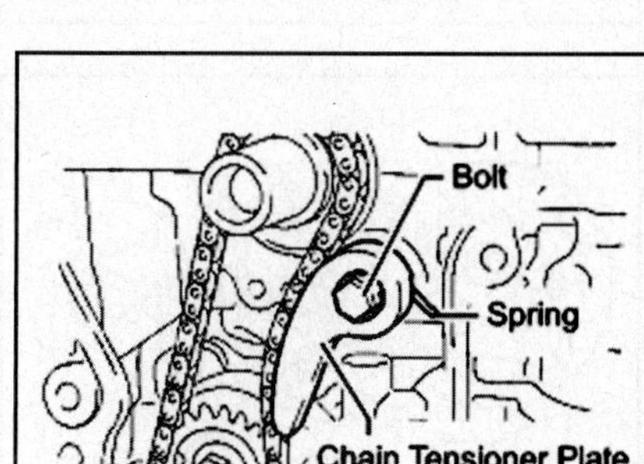

Fig. 188 Removing the bolt, chain tensioner plate and spring

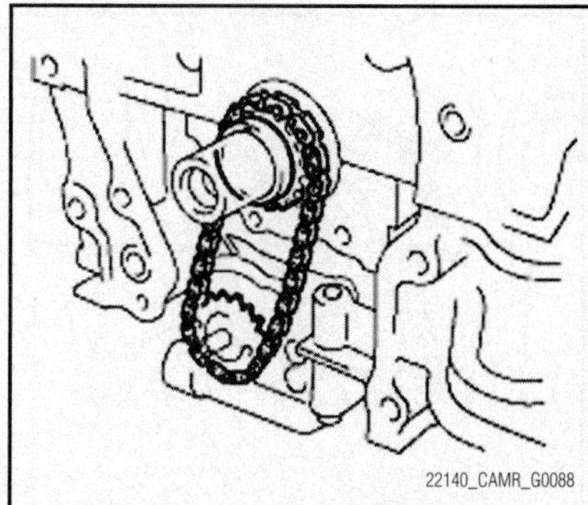

Fig. 189 Removing the chain tensioner, oil pump driven sprocket and chain

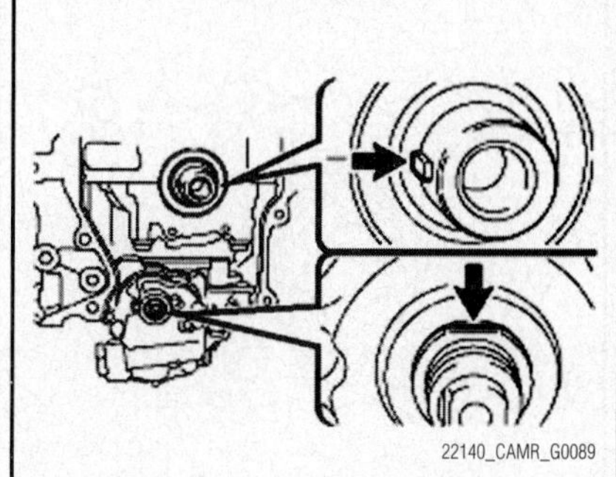

Fig. 190 Setting the crankshaft key and cutout face

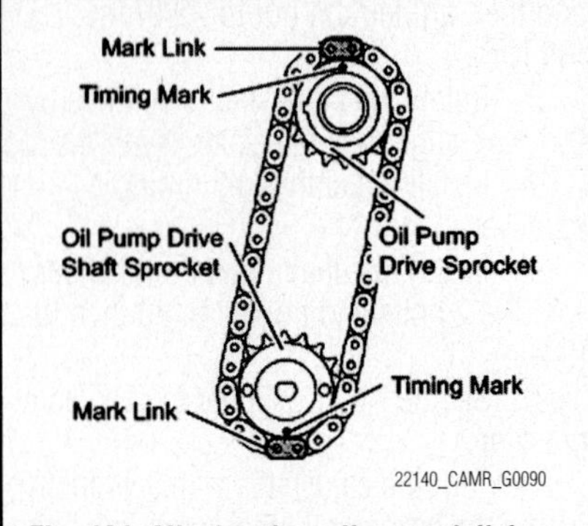

Fig. 191 Aligning the yellow mark links with the timing marks of each gear

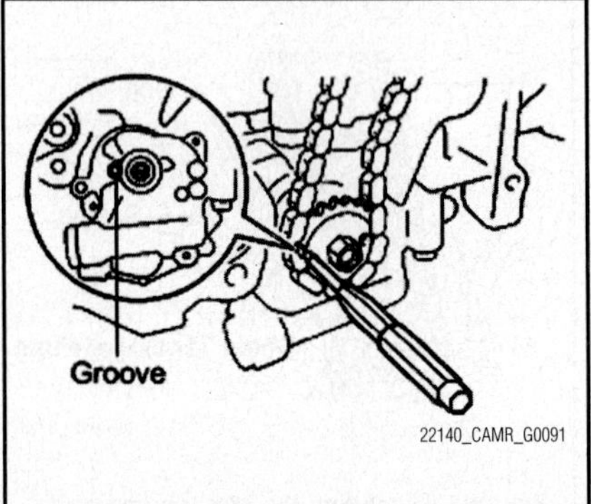

Fig. 192 Aligning the adjusting hole of the oil pump drive shaft sprocket

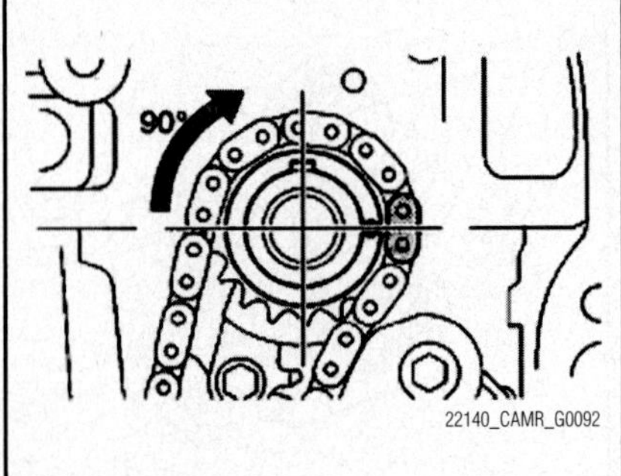

Fig. 193 Rotating the crankshaft clockwise by 90°

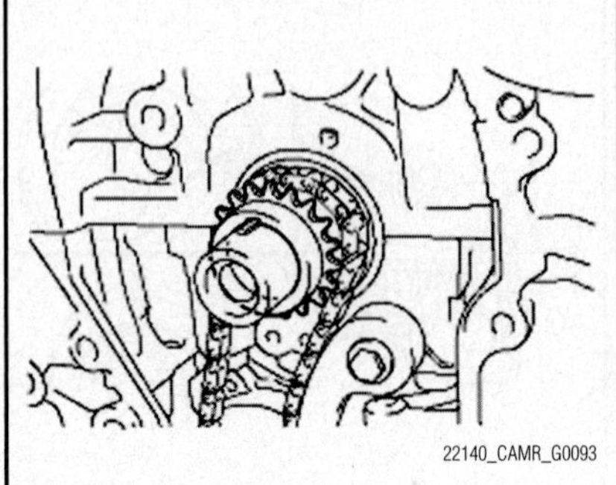

Fig. 194 Installing the crankshaft timing sprocket

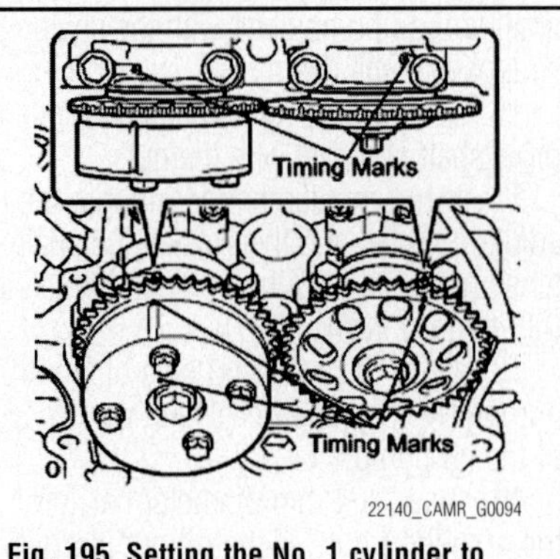

Fig. 195 Setting the No. 1 cylinder to TDC/compression

Fig. 196 Turning the crankshaft to position with the key on the crankshaft upward

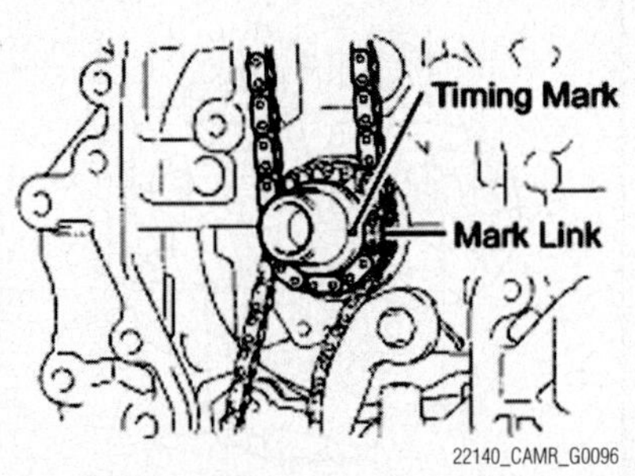

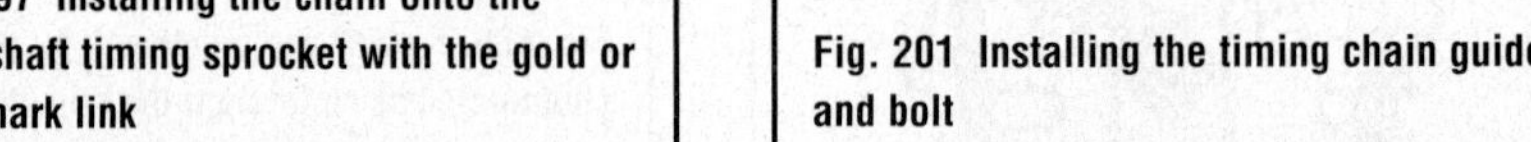

Fig. 197 Installing the chain onto the crankshaft timing sprocket with the gold or pink mark link

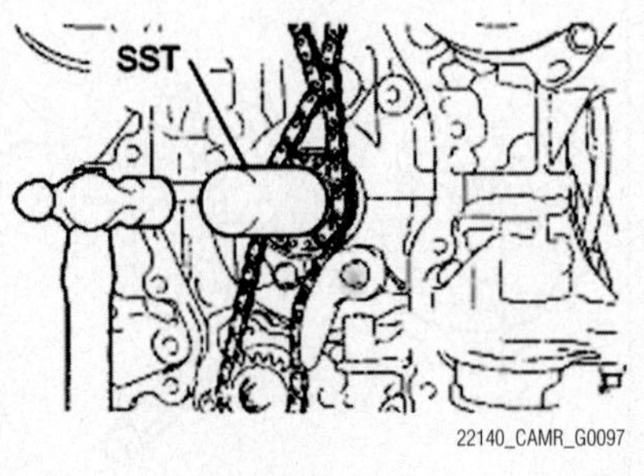

Fig. 198 Using Special Tool: 09309-37010 and a hammer to install the crankshaft timing sprocket

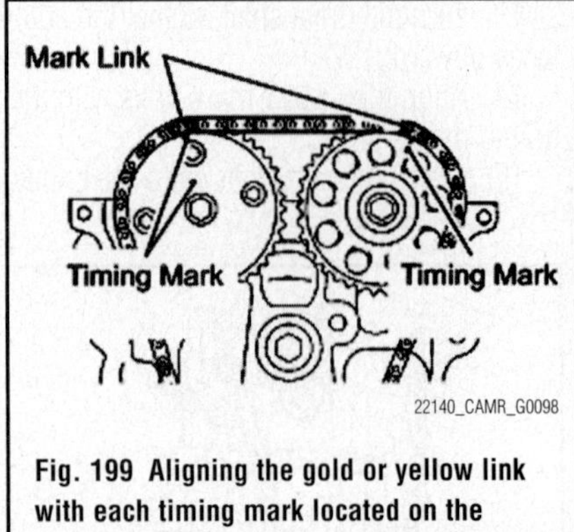

Fig. 199 Aligning the gold or yellow link with each timing mark located on the camshaft timing gear and sprocket

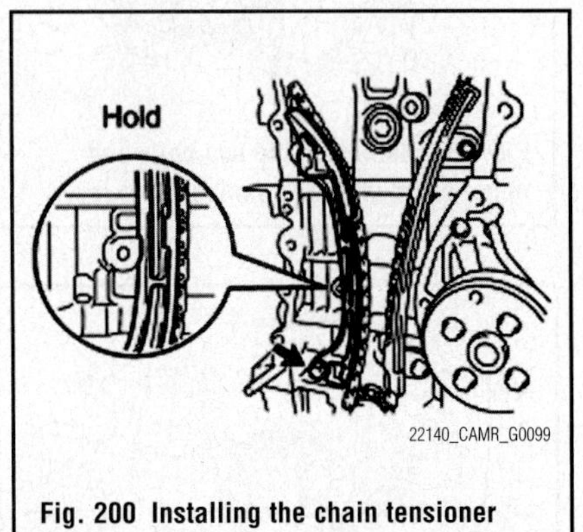

Fig. 200 Installing the chain tensioner slipper and bolt

Fig. 201 Installing the timing chain guide and bolt

Fig. 202 Installing the crankshaft sensor plate with the "F" mark facing forward

each timing mark located on the No. 1 and No. 2 bearing caps.

25. Using the crankshaft pulley bolt, turn the crankshaft to position with the key on the crankshaft upward.

26. Install the chain onto the crankshaft timing sprocket with the gold or pink mark link aligned with the timing mark on the crankshaft.

27. Using Special Tool: 09309-37010 and a hammer, tap in the crankshaft timing sprocket.

28. Align the gold or yellow link with each timing mark located on the camshaft timing gear and sprocket, then install the chain.

29. Install the chain tensioner slipper with the bolt and tighten to: 14 ft. lbs. (19 Nm).

30. Install the timing chain guide with the bolt and tighten to: 80 inch lbs. (9 Nm).

31. Install the crankshaft sensor plate with the "F" mark facing forward.

3.5L Engine

See Figures 203 through 214.

1. Before servicing the vehicle, refer to the Precautions Section.
2. Remove the engine assembly with transaxle.
3. Secure engine.
4. Remove the oil filler cap sub-assembly.
5. Remove the spark plugs and ignition coil assembly.
6. Remove the oil pan drain plug and gasket.
7. Remove the ventilation valve sub-assembly.
8. Remove the camshaft position sensor.
9. Remove the camshaft timing oil control valve assembly.
10. Remove crankshaft position sensor.
11. Remove the No. 1 oil pipe.
12. Remove the oil pipe.
13. Remove the cylinder block water drain cock sub-assembly.
14. Remove the oil filter.
15. Remove the crankshaft pulley.
16. Remove the left hand No. 1 front engine mounting bracket.
17. Remove the water inlet housing.
18. Remove the water outlet.
19. Remove the left-hand cylinder head cover sub-assembly and gasket.
20. Remove the cylinder head cover sub-assembly and gasket.
21. Remove the No. 2 oil pan sub-assembly.
22. Remove the oil strainer sub-assembly.

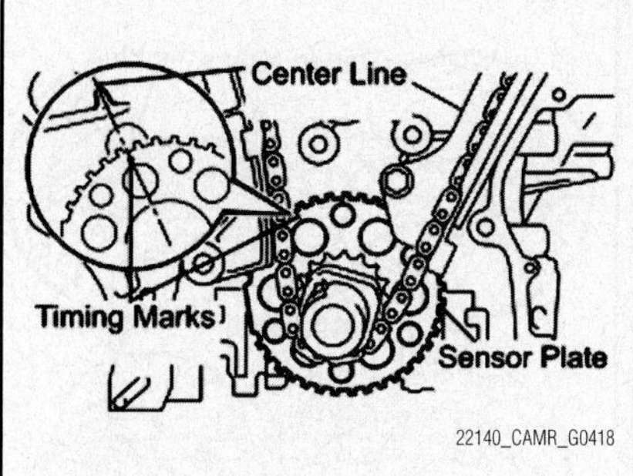

Fig. 203 Set the timing mark on the crank angle sensor plate

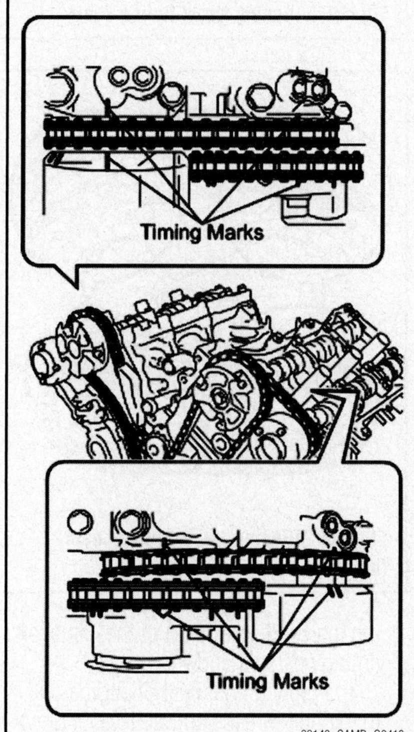

Fig. 204 Check that the timing marks of the camshaft timing gears

23. Remove the oil pan sub-assembly.
24. Remove the water pump assembly.
25. Remove the timing chain cover and seal.
26. Set no. 1 cylinder to TDC/compression, as follows:
    a. Temporarily tighten the pulley set bolt. Set the timing mark on the crank angle sensor plate to the RH block bore center line (TDC/compression).
    b. Check that the timing marks of the camshaft timing gears are aligned with the timing marks of the bearing cap as shown in the illustration. If not, turn the

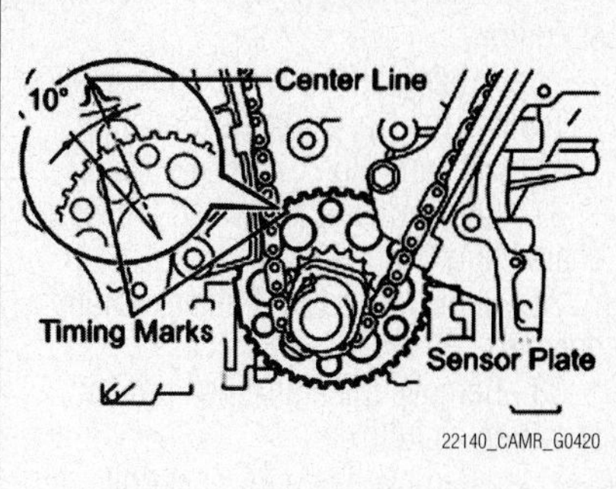

Fig. 205 Turning the crankshaft counter-clockwise 10°

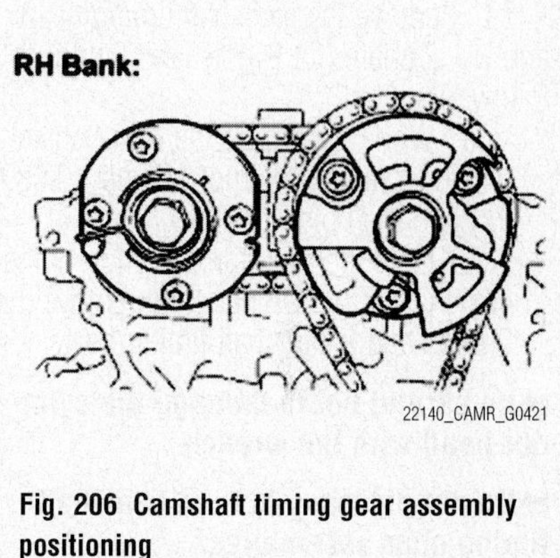

Fig. 206 Camshaft timing gear assembly positioning

crankshaft 1 revolution (360°) and align the timing marks as above.

27. Remove the No. 1 chain tensioner assembly, as follows:
    a. Move the stopper plate upward to release the lock, and push the plunger deep into the tensioner.
    b. Move the stopper plate downward to set the lock, and insert a pin of φ1.27 mm (0.05 in.) into the stopper plate's hole.
    c. Remove the 2 bolts and chain tensioner.
28. Remove the chain tensioner slipper.
29. Remove the chain sub-assembly, as follows:
    a. Turn the crankshaft counterclockwise 10° to loosen the chain of the crankshaft timing sprocket.
    b. Remove the pulley set bolt.
    c. Remove the chain from the crankshaft timing sprocket and place it on the crankshaft.
    d. Turn the camshaft timing gear assembly on the right hand bank clockwise (approximately 60°) and set it as shown in the illustration. Be sure to loosen the chain between the banks.
    e. Remove the chain.

30. Remove the idle sprocket assembly, as follows:

a. Using a 10 mm hexagon wrench, remove the No. 2 idle gear shaft, sprocket and No. 1 idle gear shaft.

31. Remove the 2 bolts and the No. 1 chain vibration damper.

32. Remove the No. 2 chain vibration damper.

33. Remove the crankshaft timing sprocket, as follows:

a. Remove the pulley set bolt.

b. Remove the crankshaft timing gear from the crankshaft.

c. Remove the 2 pulley set keys from the crankshaft.

34. Remove the camshaft timing gears and No. 2 chain (for Right-hand Bank), as follows:

a. While raising up the No. 2 chain tensioner, insert a pin of 1.0 mm (0.039 in.) into the hole to hold it.

b. Hold the hexagonal portion of the camshaft with a wrench, and remove the 2 bolts and 2 camshaft timing gears.

➡Be careful not to damage the cylinder head with the wrench.

➡Do not disassemble the camshaft timing gear assemblies.

c. Remove the No. 2 chain.

35. Remove the bolt and No. 2 chain tensioner.

36. Remove the camshaft timing gears and No. 2 chain (for Left-hand Bank), as follows:

a. While pushing down on the No. 3 chain tensioner, insert a pin of 1.0 mm (0.039 in.) into the hole to hold it.

b. Hold the hexagonal portion of the camshaft with a wrench, and remove the 2 bolts and 2 camshaft timing gears.

➡Be careful not to damage the cylinder head with the wrench.

➡Do not disassemble the camshaft timing gear assemblies.

c. Remove the No. 2 chain.

37. Remove the bolt and the No. 3 chain tensioner.

To install:

38. Install the No. 2 chain tensioner assembly with the bolt and tighten to 15 ft. lbs. (21 Nm).

39. While pushing in the tensioner, insert a pin of 1.0 mm (0.039 in.) into the hole to hold it.

40. Install the camshaft timing gears and No. 2 chain (for Right-hand Bank), as follows:

a. Align the mark plate with the timing marks (1-dot mark) of the camshaft timing gears as shown.

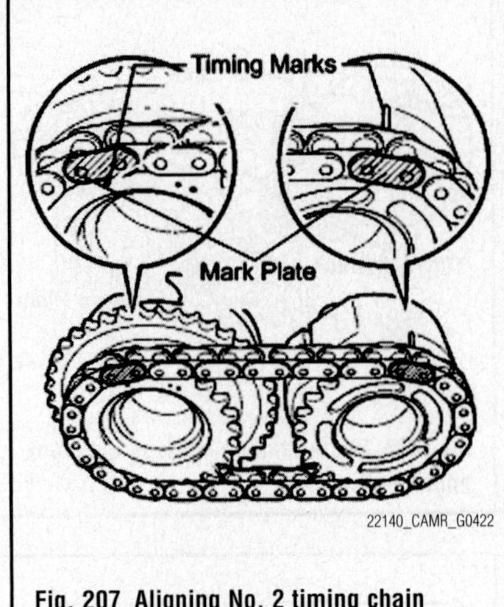

Fig. 207 Aligning No. 2 timing chain

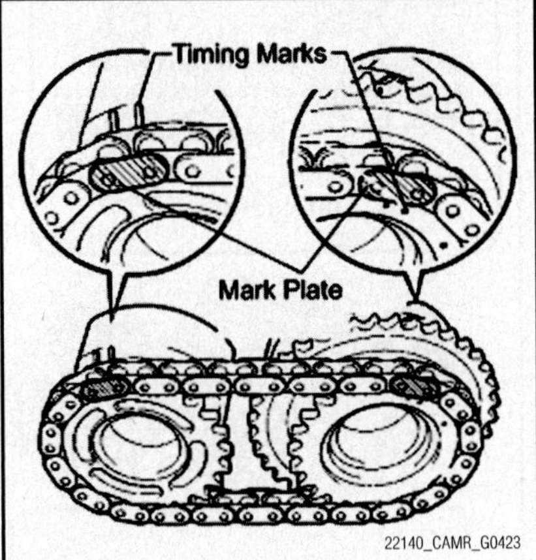

Fig. 208 Aligning No. 2 timing chain (bank 2)

b. Apply a light coat of engine oil to the bolt threads and bolt-seating surface.

c. Align the knock pin of the camshaft with pin hole of the camshaft timing gear. Install the camshaft timing gear and the right camshaft timing exhaust gear with the No. 2 chain installed.

d. Hold the hexagonal portion of the camshaft with the wrench and tighten the two bolts to 74 ft. lbs. (100 Nm).

e. Remove the pin from the No. 2 chain tensioner.

41. Install the No. 3 chain tensioner assembly with the bolt and tighten to 15 ft. lbs. (21 Nm).

42. While pushing in the tensioner, insert a pin of 1.0 mm (0.039 in.) into the hole to hold it.

43. Install the camshaft timing gears and No. 2 chain (for Left-hand Bank), as follows:

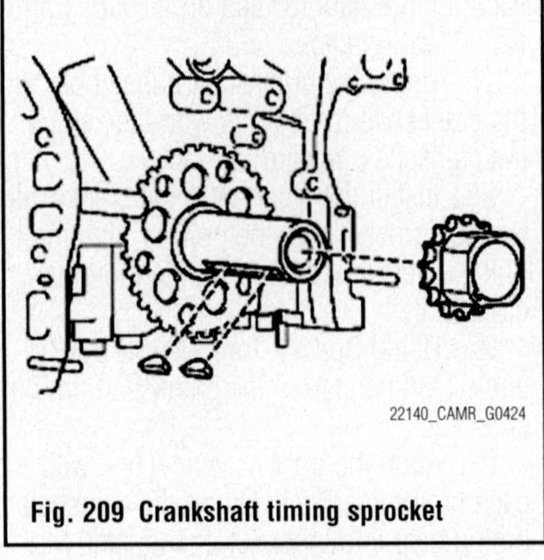

Fig. 209 Crankshaft timing sprocket

a. Align the mark plate (yellow) with the timing marks (2-dot mark) of the camshaft timing gears as shown.

b. Apply a light coat of engine oil to the bolt threads and bolts seating surface.

c. Align the knock pin of the camshaft with pin hole of the camshaft timing gear. Install the camshaft timing gear and the left camshaft timing exhaust gear with the No. 2 chain installed.

d. Hold the hexagonal portion of the camshaft with the wrench and tighten the two bolts to 74 ft. lbs. (100 Nm).

e. Remove the pin from the No. 2 chain tensioner.

44. Install the No. 1 and 2 chain vibration dampers.

45. Install the timing gear set keys and crankshaft timing sprocket as shown in the illustration.

46. Install the idle sprocket assembly, as follows:

a. Apply a light coat of engine oil to the rotating surface of the No. 1 idle gear shaft.

b. Temporarily install the No. 1 idle gear shaft and idle sprocket with the No. 2 idle gear shaft while aligning the knock pin of the No. 1 idle gear with the knock pin groove of the cylinder block. Be careful of the idle gear direction.

c. Using a 10 mm hexagon wrench, tighten the No. 2 idle gear shaft to 44 ft. lbs. (60 Nm).

d. After installing the idle sprocket assembly, check that the idle sprocket turns smoothly.

47. Install the chain sub-assembly, as follows:

a. Align the mark plate and timing marks as shown in the illustration and install the chain. The camshaft mark plate is orange.

b. Do not pass the chain over the crankshaft, just put it on.

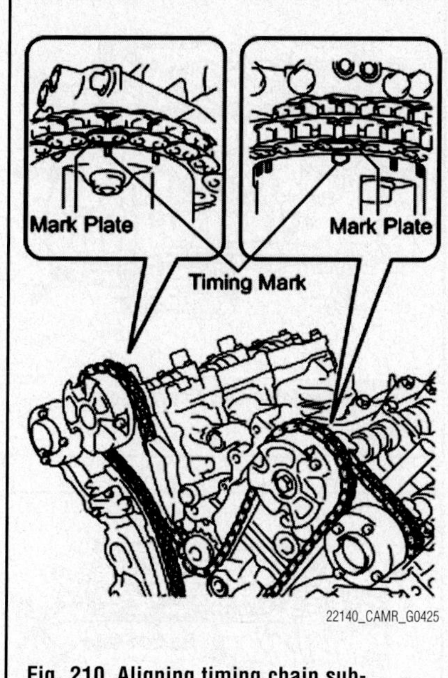

Fig. 210 Aligning timing chain sub-assembly

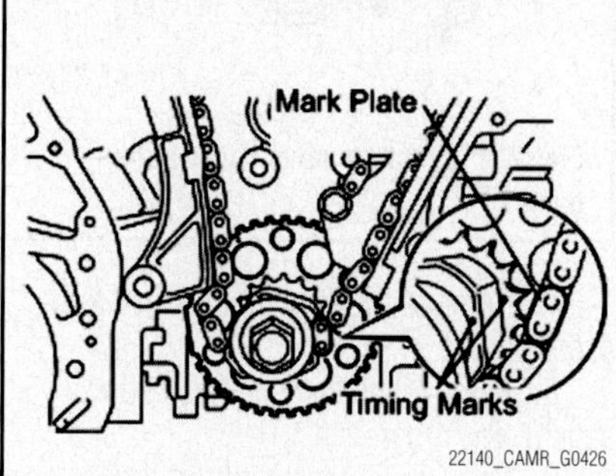

Fig. 211 Aligning the mark plate and timing mark

c. Turn the camshaft timing gear assembly on the right bank counterclockwise to tighten the chain between the banks.

➡When the idle sprocket assembly is reused, align the timing chain plate with the mark on the sprocket in order to tighten the chain between the banks.

d. Align the mark plate and timing marks as shown in the illustration and install the chain onto the crankshaft timing sprocket. The crankshaft to mark plate is yellow.

e. Temporarily tighten the pulley set bolt.

f. Turn the crankshaft clockwise to set it to the right-hand block bore more centerline. (TDC/compression).

48. Install the chain tensioner slipper.

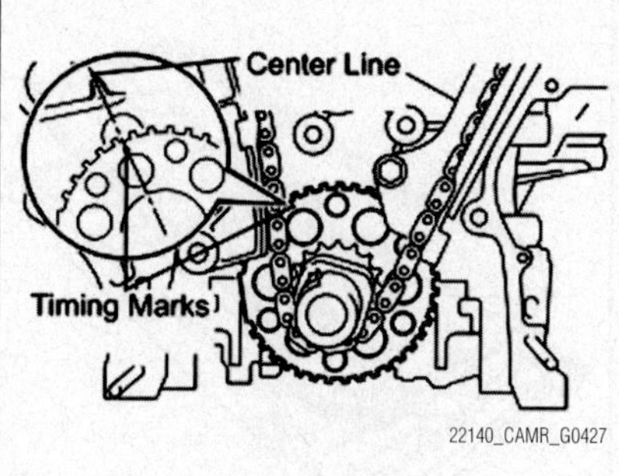

Fig. 212 Turning the crankshaft clockwise (TDC/compression)

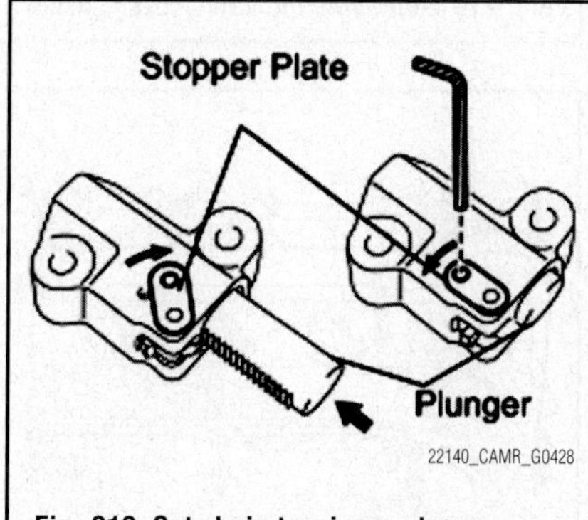

Fig. 213 Set chain tensioner plunger position

49. Install the No. 1 chain tensioner assembly, as follows:

a. Move the stopper plate upward to release the lock, and push the plunger deep into the tensioner.

b. Move the stopper plate downward to set the lock, and insert a hexagon wrench into the hole of the stopper plate.

c. Install the No. 1 chain tensioner with the 2 bolts and tighten to 7 ft. lbs. (10 Nm).

d. Remove the lock pin of the No. 1 chain tensioner. Check that each timing mark is aligned with the crankshaft at TDC/compression.

e. Remove the pulley set bolt.

50. Install timing chain cover and seal.
51. Install the water pump assembly.
52. Install the water inlet housing.
53. Install the left hand No. 1 front engine mounting bracket.
54. Install the oil pan sub-assembly.
55. Install the oil strainer sub-assembly.
56. Install the No. 2 oil pan sub-assembly.
57. Install the oil pan drain plug and gasket.
58. Install the cylinder head cover sub-assembly.

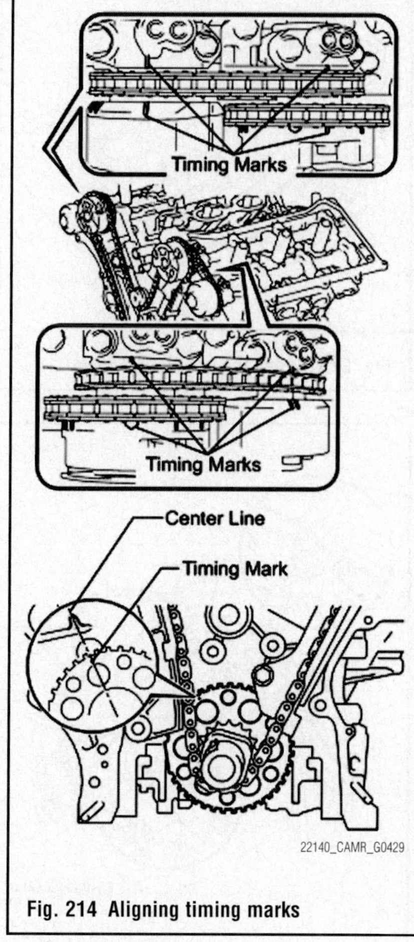

Fig. 214 Aligning timing marks

59. Install the left-hand cylinder head cover sub-assembly.
60. Install the water outlet.
61. Install the crankshaft pulley.
62. Install the oil filter.
63. Install the cylinder block water drain cock sub-assembly.
64. Install the No. 1 oil pipe.
65. Install the oil pipe.
66. Install the crankshaft position sensor.
67. Install the camshaft timing oil control valve assembly.
68. Install the camshaft position sensor.
69. Install the ventilation valve sub-assembly.
70. Install the spark plugs and ignition coil assembly.
71. Install the oil filler cap sub-assembly.
72. Install the engine assembly with transaxle.

VALVE COVERS

REMOVAL & INSTALLATION

2.4L Engine

See Figures 215 through 219.

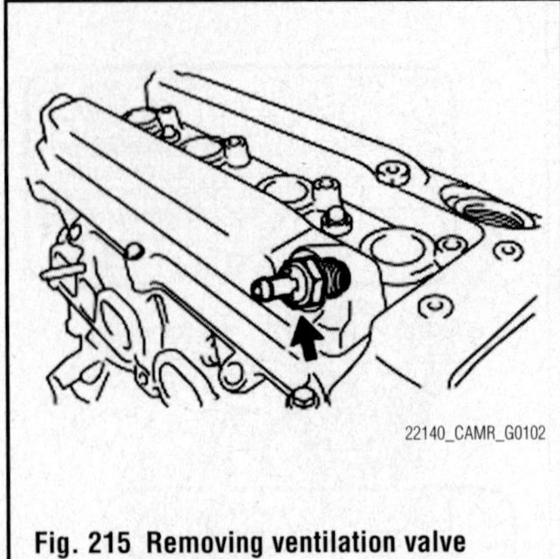

Fig. 215 Removing ventilation valve

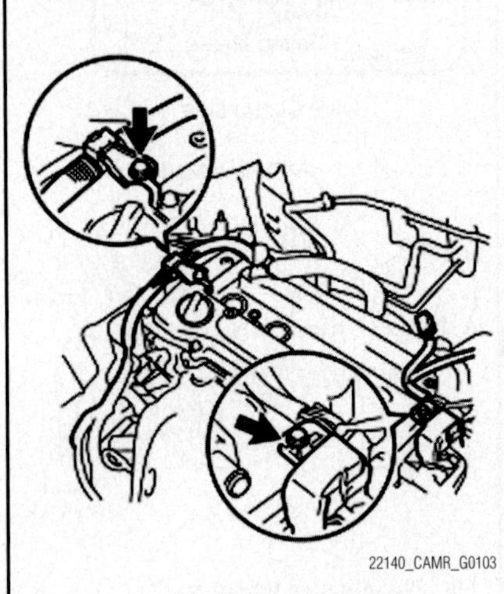

Fig. 216 Remove the two bolts and disconnect the two engine wires

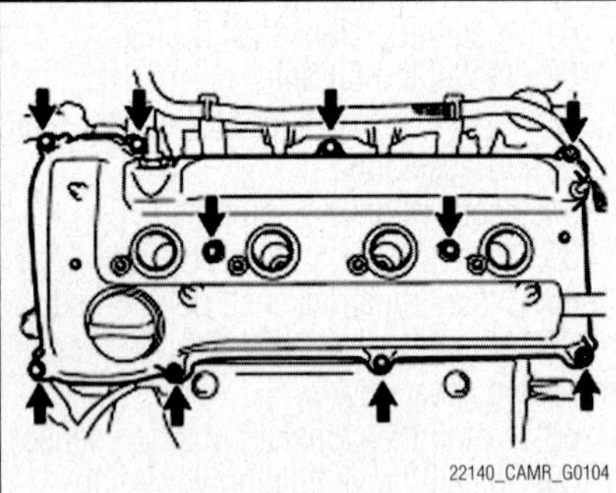

Fig. 217 Removing the eight bolts, two nuts and the valve cover

1. Before servicing the vehicle, refer to the Precautions Section.
2. Remove the oil filler cap.
3. Remove the ventilation valve.
4. Remove the spark plugs.
5. Drain the engine oil and remove the oil filter.
6. Remove the two bolts and disconnect the two engine wires.

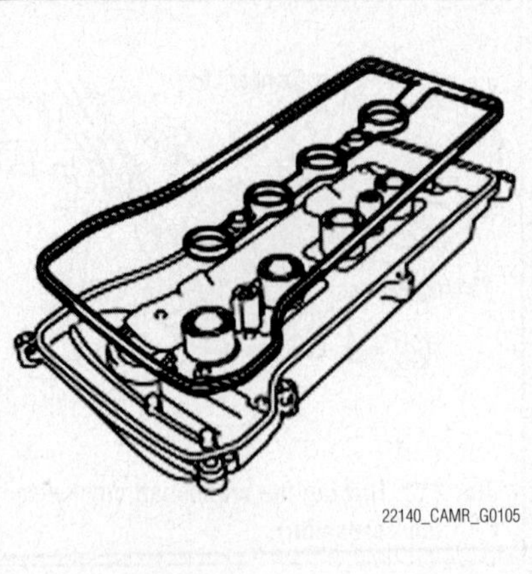

Fig. 218 Removing the valve cover gasket

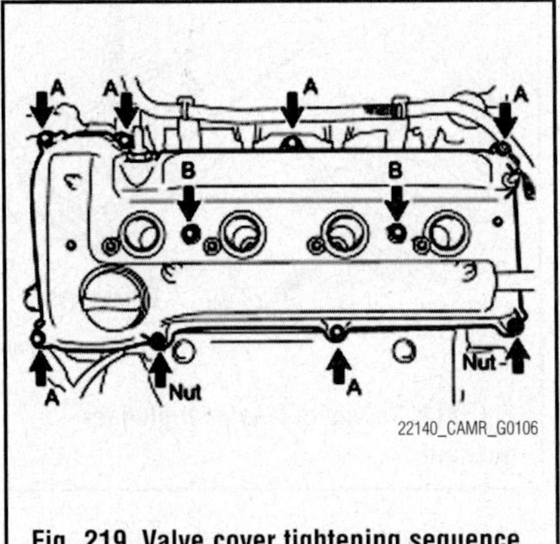

Fig. 219 Valve cover tightening sequence

7. Remove the eight bolts, two nuts and the valve cover.
8. Remove the cylinder head cover gasket.

To install:

9. To install, reverse removal procedure.
10. Install the cylinder head cover with the eight bolts and two nuts . Tighten the following in sequence to specification:
    - Bolt A: 8 ft. lbs. (11 Nm)
    - Bolt B: 10 ft. lbs. (14 Nm)
    - Nut: 8 ft. lbs. (11 Nm)

3.5L Engine

See Figures 220 through 227.

1. Before servicing the vehicle, refer to the Precautions Section.
2. Remove the engine assembly with transaxle.
3. Secure engine.
4. Remove the oil filler cap sub-assembly.
5. Remove the spark plugs and ignition coil assembly.
6. Remove the oil pan drain plug and gasket.

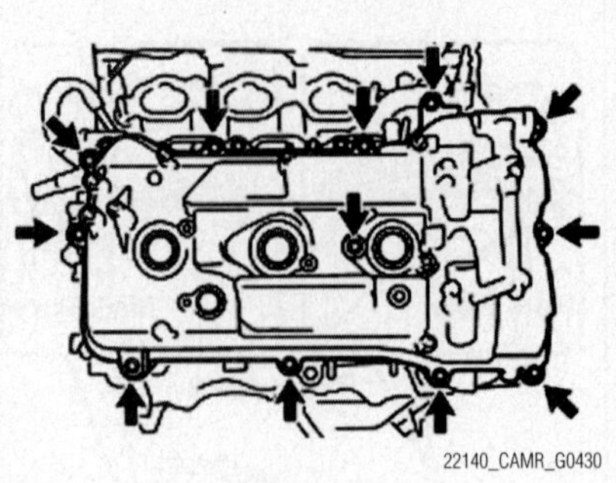

Fig. 220 Locating valve cover bolts—bank 1

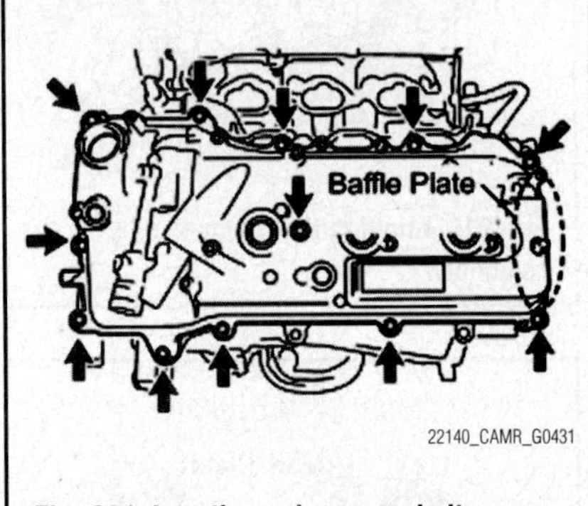

Fig. 221 Locating valve cover bolts—bank 2

7. Remove the ventilation valve sub-assembly.
8. Remove the camshaft position sensor.
9. Remove the camshaft timing oil control valve assembly.
10. Remove crankshaft position sensor.
11. Remove the No. 1 oil pipe.
12. Remove the oil pipe.
13. Remove the cylinder block water drain cock sub-assembly.
14. Remove the oil filter.
15. Remove the crankshaft pulley.
16. Remove the left hand No. 1 front engine mounting bracket.
17. Remove the water inlet housing.
18. Remove the water outlet.
19. Remove the 12 bolts, seal washer, bank 1valve cover and gasket.

➡The baffle plate is located on the back of the portion shown in the illustration. Do not damage the baffle plate when removing the head cover.

20. Remove the 12 bolts, seal washer, bank 2 cover and gasket.

To install:

21. Install the cylinder head cover sub-assembly, as follows:

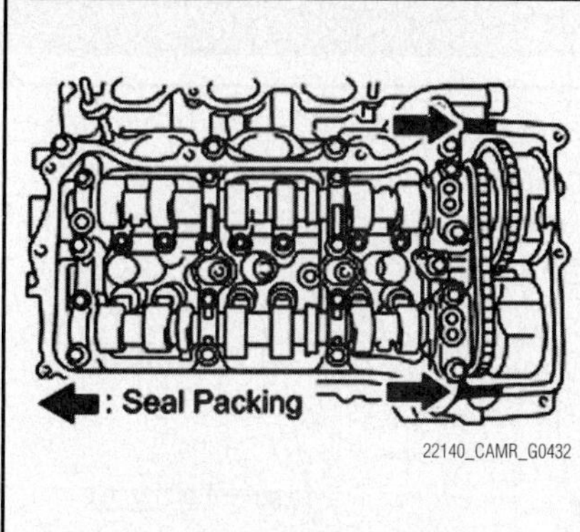

Fig. 222 Sealant application—bank 1

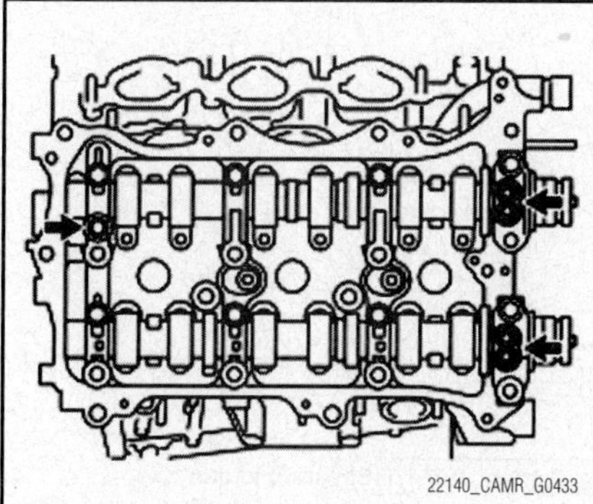

Fig. 223 Installing the 3 gaskets—bank 1

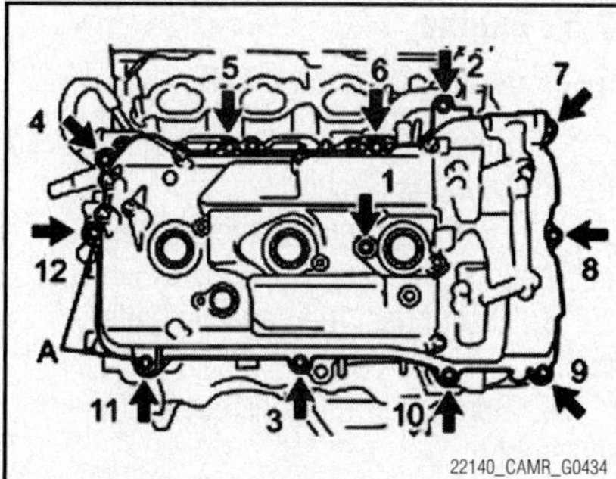

Fig. 224 Valve cover bolt tightening sequence—bank 1

a. Apply seal packing (Toyota Genuine Seal Packing Black, Three Bond 1207B or equivalent) as shown in the illustration.

➡Remove any oil from the contact surface.

➡Install the crankcase within 3 minutes after applying seal packing.

➡Do not start the engine for at least 2 hours after installation.

b. Install 3 new gaskets.

c. Install the gasket to the head cover.

d. Install the head cover with the 12 bolts and a new washer and tighten to 15 ft. lbs. (21 Nm), and 7 ft. lbs. (10 Nm). Be sure to tighten bolt 1.

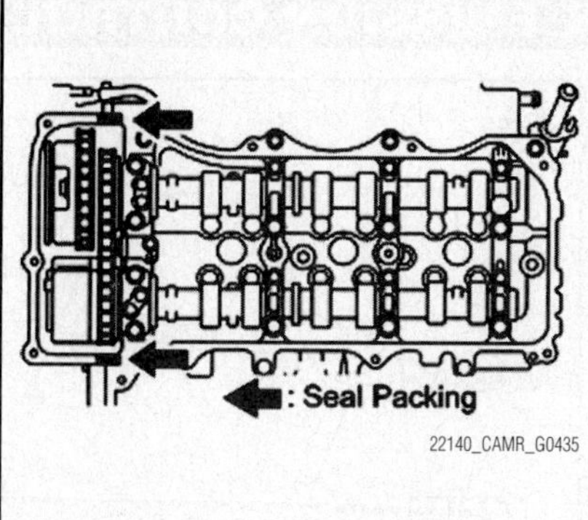

Fig. 225 Sealant application—bank 2

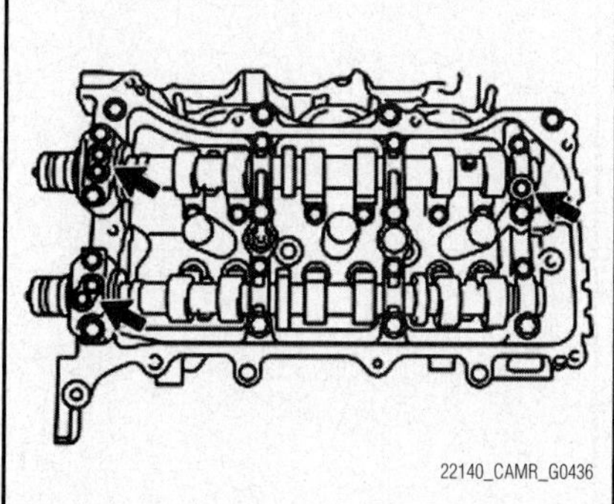

Fig. 226 Installing the 3 gaskets—bank 2

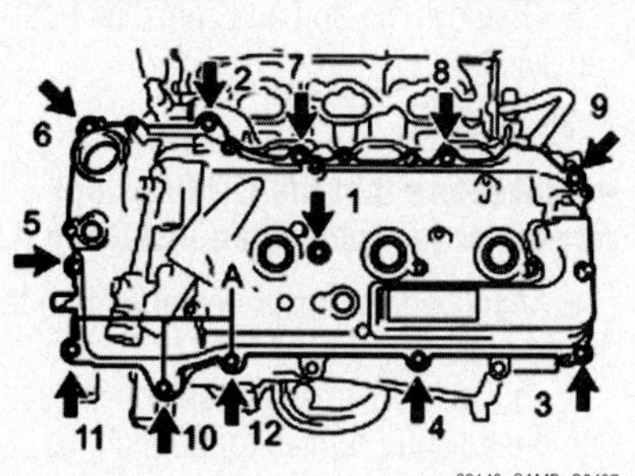

Fig. 227 Valve cover bolt tightening sequence—bank 2

22. Install bank 2 valve cover sub-assembly, as follows:

➡Remove any oil from the contact surface.

➡Install the crankcase within 3 minutes after applying seal packing.

➡Do not start the engine for at least 2 hours after installation.

a. Apply seal packing (Toyota Genuine Seal Packing Black, Three Bond 1207B or equivalent) as shown in the illustration.

b. Install 3 new gaskets as shown in the illustration.

c. Install the gasket to the head cover.

d. Install the head cover with the 14 bolts and tighten to 15 ft. lbs. (21 Nm), and 7 ft. lbs. (10 Nm). Be sure to tighten bolts 1 and 10.

23. Install the water outlet.
24. Install the crankshaft pulley.
25. Install the oil filter.
26. Install the cylinder block water drain cock sub-assembly.
27. Install the No. 1 oil pipe.
28. Install the oil pipe.
29. Install the crankshaft position sensor.
30. Install the camshaft timing oil control valve assembly.
31. Install the camshaft position sensor.
32. Install the ventilation valve sub-assembly.
33. Install the spark plugs and ignition coil assembly.
34. Install the oil filler cap sub-assembly.
35. Install the water pump assembly.
36. Install the engine assembly with transaxle.

VALVE LASH

ADJUSTMENT

➡Keep the lash adjuster free of dirt and foreign objects.

➡Only use clean engine oil.

1. Place the lash adjuster into a container filled with engine oil.
2. Insert the SST's (SST: 09276-75010) tip into the lash adjuster's plunger and use the tip to press down on the check ball inside the plunger.
3. Squeeze the SST and lash adjuster together to move the plunger up and down 5 to 6 times.
4. Check the movement of the plunger and bleed the air. OK: Plunger moves up and down.

➡When bleeding air from the high-pressure chamber, make sure that the tip of the SST is actually pressing the check ball as shown in the illustration. If the check ball is not pressed, air will not bleed.

5. After bleeding the air, remove the SST. Then, try to press the plunger quickly and firmly with a finger. OK: Plunger is very difficult to move. If the result is not as specified, replace the lash adjuster.
6. Install the lash adjusters.

➡Install the lash adjuster to the same place where it was removed from.

ENGINE PERFORMANCE & EMISSION CONTROL

ACCELERATOR PEDAL POSITION (APP) SENSOR

LOCATION

See Figure 228.

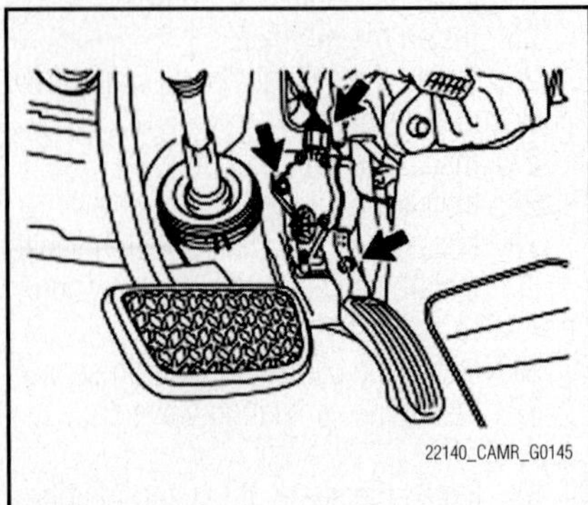

Fig. 228 Accelerator pedal assembly location

REMOVAL & INSTALLATION

See Figure 228.

1. Before servicing the vehicle, refer to the Precautions Section.
2. Remove left center floor carpet cover.
3. Disconnect the accelerator pedal connector.
4. Remove the 2 nuts and accelerator pedal assembly.

➡**Avoid physical shock to the accelerator pedal assembly.**

➡**Do not disassemble the accelerator pedal assembly.**

To install:

5. Installation is the reverse of the removal procedure. Tighten the accelerator pedal rod nuts to 43 inch lbs. (5.4 Nm).

CAMSHAFT POSITION (CMP) SENSOR

LOCATION

See Figure 230.

REMOVAL & INSTALLATION

2.4L Engine

See Figures 231 through 235.

1. Before servicing the vehicle, refer to the Precautions Section.
2. Remove the plastic engine cover.
3. Remove air cleaner cap sub-assembly.
4. Disconnect the camshaft position sensor connector.

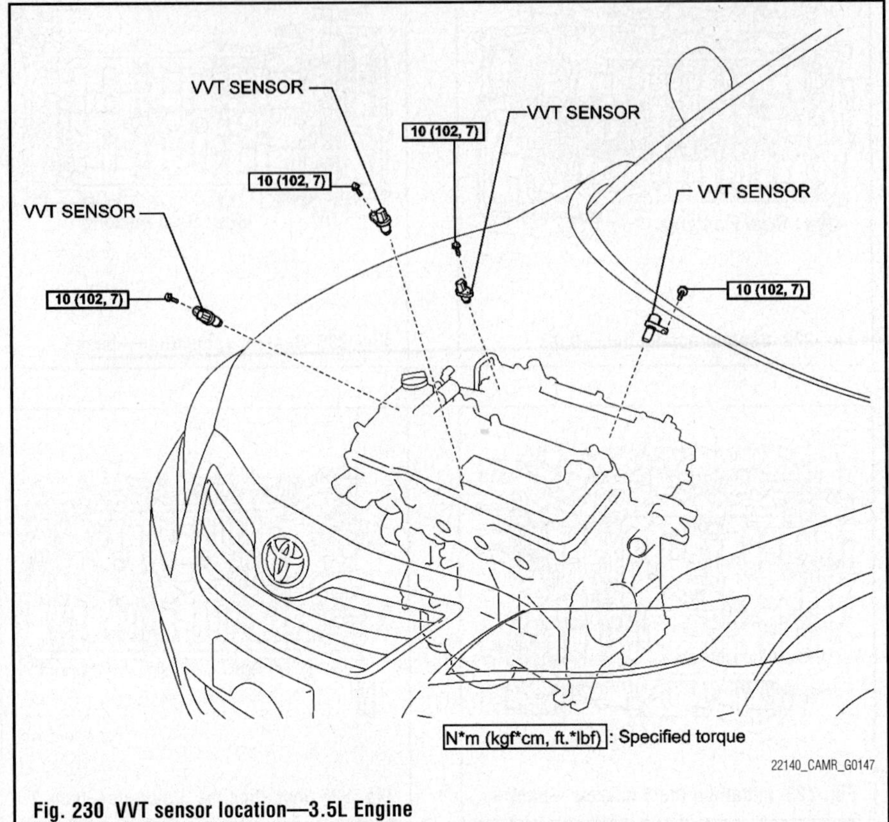

Fig. 230 VVT sensor location—3.5L Engine

5. Remove the bolt and camshaft position sensor.

To install:

➡ **Make sure that the O-ring is not cracked or jammed when installing it.**

6. Apply a light coat of engine oil to the O-ring of the sensor.
7. Install the camshaft position sensor with the bolt and tighten to: 80 inch lbs. (9 Nm).
8. To complete installation, reverse removal procedure.

3.5L Engine

See Figures 232 through 235.

1. Before servicing the vehicle, refer to the Precautions Section.
2. Drain and recycle the engine coolant.
3. Disconnect the negative battery cable.
4. Remove the V-bank cover sub-assembly.
5. Remove the windshield wiper link assembly.

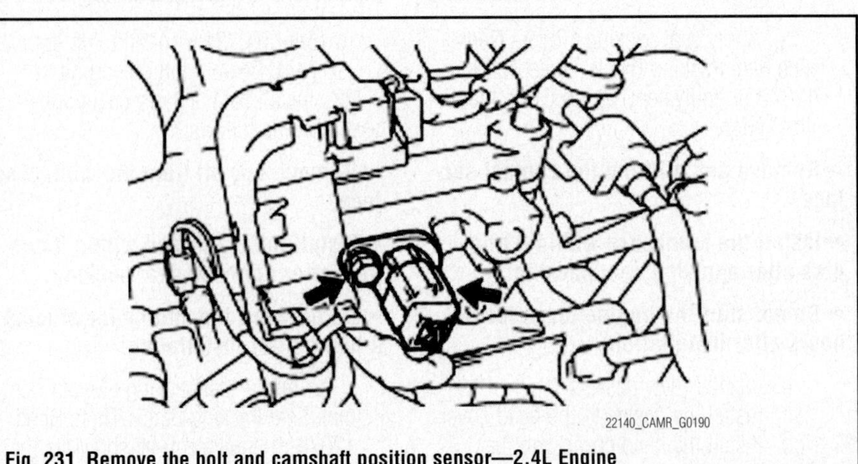

Fig. 231 Remove the bolt and camshaft position sensor—2.4L Engine

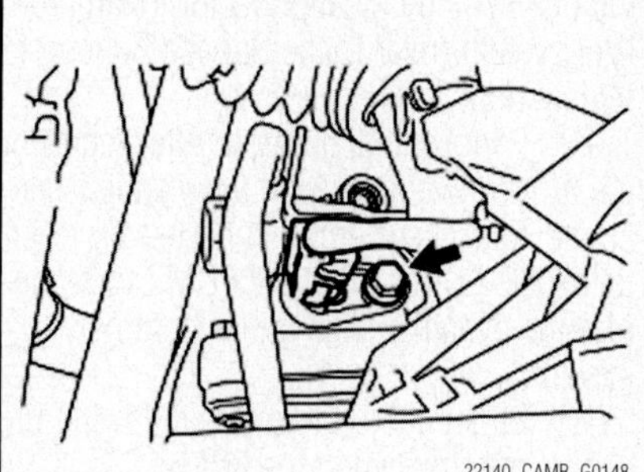

Fig. 232 Bank 1 intake camshaft VVT sensor

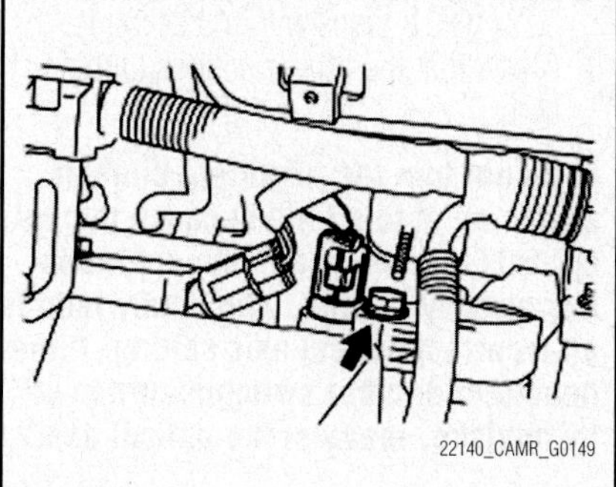

Fig. 233 Bank 1 exhaust camshaft VVT sensor

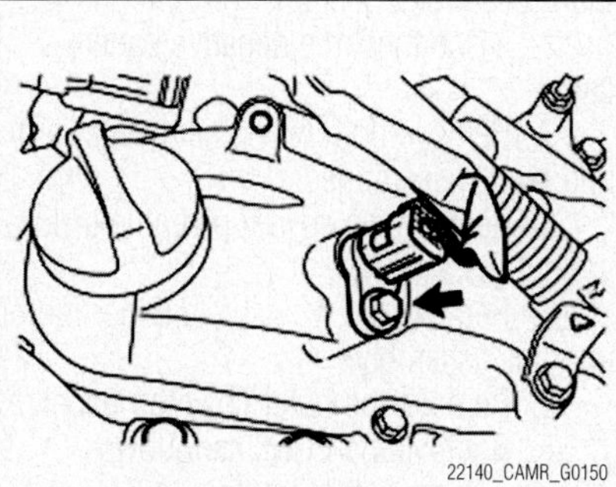

Fig. 234 Bank 2 intake camshaft VVT sensor

Fig. 235 Bank 2 exhaust camshaft VVT sensor

6. Remove the front cowl top outside panel.
7. Remove the Intake camshaft VVT sensor (Bank 1), as follows:
   a. Disconnect the VVT sensor connector.
   b. Remove the bolt and VVT sensor.
8. Remove the Exhaust camshaft VVT sensor (Bank 1), as follows:
   a. Disconnect the VVT sensor connector.
   b. Remove the bolt and VVT sensor.
9. Remove the Exhaust camshaft VVT sensor (Bank 2), as follows:
   a. Disconnect the VVT sensor connector.
   b. Remove the bolt and VVT sensor.
10. Remove the Intake camshaft VVT sensor (Bank 2), as follows:
   a. Disconnect the VVT sensor connector.
   b. Remove the bolt and VVT sensor.

CRANKSHAFT POSITION (CKP) SENSOR

LOCATION

See Figures 236 and 237.

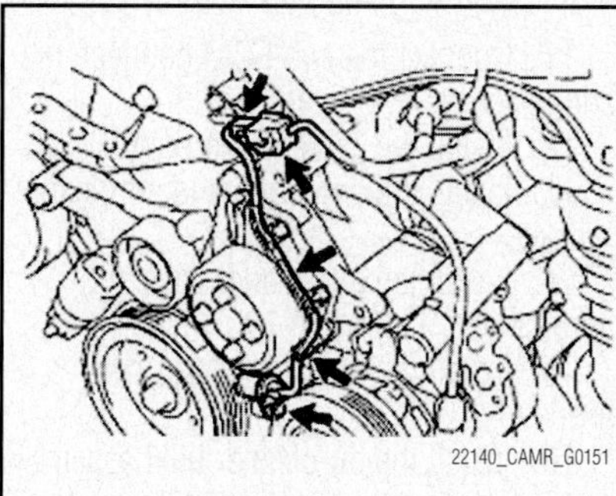

Fig. 236 Crankshaft position sensor location—2.4L engine

Fig. 237 Crankshaft position sensor location—3.5L engine

REMOVAL & INSTALLATION

2.4L Engine

See Figures 238 and 239.

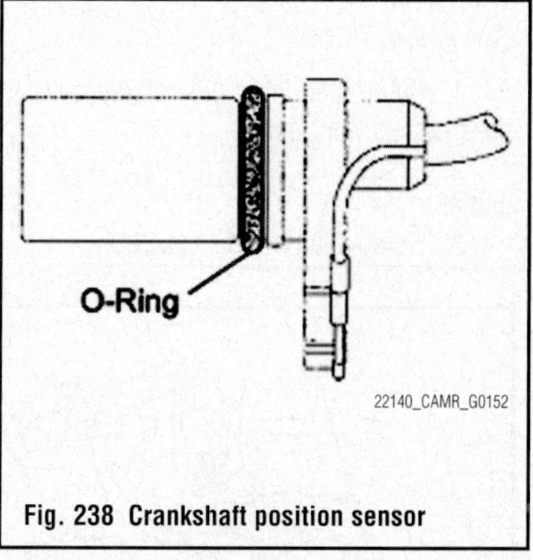

Fig. 238 Crankshaft position sensor

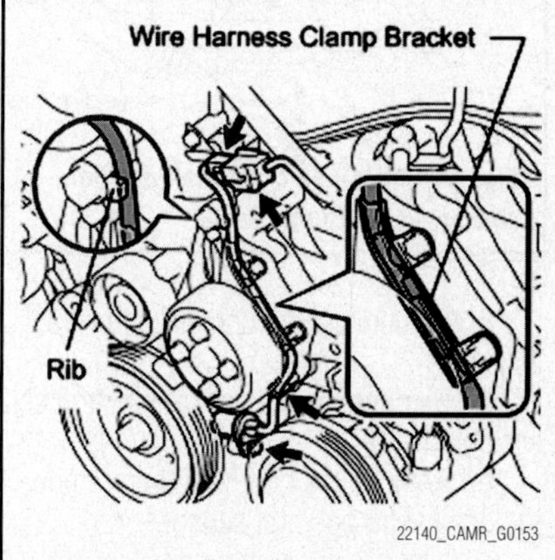

Fig. 239 Installing crankshaft position sensor—2.4L Engine

1. Before servicing the vehicle, refer to the Precautions Section.
2. Disconnect the negative battery cable.
3. Remove the front right wheel.
4. Remove the right hand front fender apron.
5. Remove the drive belt.
6. Remove the generator assembly.
7. Disconnect the crankshaft position sensor connector.
8. Remove the connector clamp and wire harness clamp.
9. Remove the wire harness clamp bracket from the wire harness.
10. Remove the bolt, then remove the crankshaft position sensor.

To install:

11. Apply a light coat of engine oil to the O-ring on the crankshaft position sensor.
12. Install the crankshaft position sensor with the bolt and tighten to 7ft. lbs. (10 Nm).

13. Connect the crankshaft position sensor connector.

14. The remainder of installation is the reverse of the removal procedure.

3.5L Engine

See Figure 240.

1. Before servicing the vehicle, refer to the Precautions Section.
2. Disconnect the negative battery cable.
3. Remove alternator assembly.

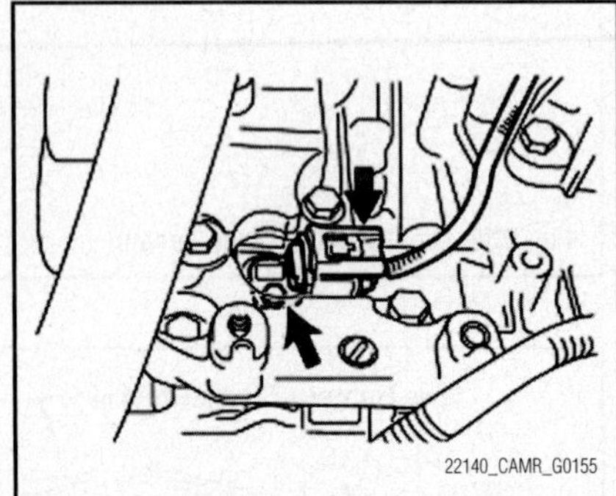

Fig. 240 Installing crankshaft position sensor—3.5L Engine

4. Disconnect the cooler compressor assembly.
5. Remove the crankshaft position sensor connector.
6. Remove the bolt, and then remove the crankshaft position sensor.

To install:

7. Apply a light coat of engine oil to the O-ring on the crankshaft position sensor.
8. Install the crankshaft position sensor with the bolt and tighten to 7ft. lbs. (10 Nm).
9. Connect the crankshaft position sensor connector.
10. The remainder of installation is the reverse of the removal procedure.

ELECTRONIC CONTROL MODULE (ECM)

LOCATION

Refer to Component Locations illustrations at the top of this section to locate the ECM.

REMOVAL & INSTALLATION

2.4L Engine

See Figure 241.

1. Before servicing the vehicle, refer to the Precautions Section.
2. Disconnect the negative battery cable.
3. Remove the plastic engine cover.
4. Remove the air cleaner inlet assembly.
5. Remove the air cleaner cap sub-assembly.
6. Remove the air cleaner case sub-assembly.
7. Remove the two bolts and air cleaner bracket.
8. Disconnect the two ECM connectors by raising the two levers. While pushing the locks on the two levers, disconnect the two ECM connectors.

➡After disconnecting the connectors, make sure that dirt, water or other foreign matter does not contact the connections of the connectors.

a. Remove the ECM with the bracket and three bolts.

b. Remove the four screws and ECM brackets.

To install:

9. Install the bracket to the ECM with the 4 screws, and tighten to 27 inch lbs. (3 Nm).
10. Attach the ECM with the three nuts and tighten to 71 inch lbs. (8 Nm).

➡Make sure that dirt, water or other foreign matter does not contact the connections of the connectors.

11. Connect the two ECM connectors and lower the two levers.
12. Install the ECM to the body.
13. Install the two bolts and air cleaner bracket.
14. Install the air cleaner case sub-assembly.
15. Install the air cleaner cap sub-assembly.
16. Install the air cleaner inlet assembly.
17. Remove the plastic engine cover.
18. Connect the negative battery cable.
19. Register the immobilizer communication ID. If the ECM is replaced, register the ECM communication ID for the immobilizer system (refer to the Service Bulletin for registration).
20. Perform initialization. After replacing the ECM on vehicles with a dynamic laser cruise control system, it is necessary to initialize the ECM so that the ECM can recognize the dynamic laser cruise control system.
21. Be sure to perform the following procedure after replacing the ECM:

a. Turn the ignition switch on (IG).

b. Turn the cruise control main switch on.

c. With the brake pedal depressed, push the cruise control main switch to RES/ACC 3 times within 3 seconds. Check that the buzzer sounds at this time.

➡Do not turn the headlight dimmer switch on at this time because the optical axis automatic adjustment mode has already started, which may lead to an incorrect optical axis setting. If the headlight dimmer switch is turned on by mistake, readjust the optical axis.

3.5L Engine

See Figure 242.

1. Before servicing the vehicle, refer to the Precautions Section.
2. Disconnect the negative battery cable.
3. Remove both windshield wiper arm and blade assemblies.
4. Remove the cowl top ventilator louver sub-assembly.
5. Remove the windshield wiper motor and link assembly.
6. Remove the outer cowl top panel.
7. Remove the ECM, as follows:

a. Remove the 3 nuts.

b. Separate the ECM from the body. When separating the ECM, do not apply excessive force to the wire harness.

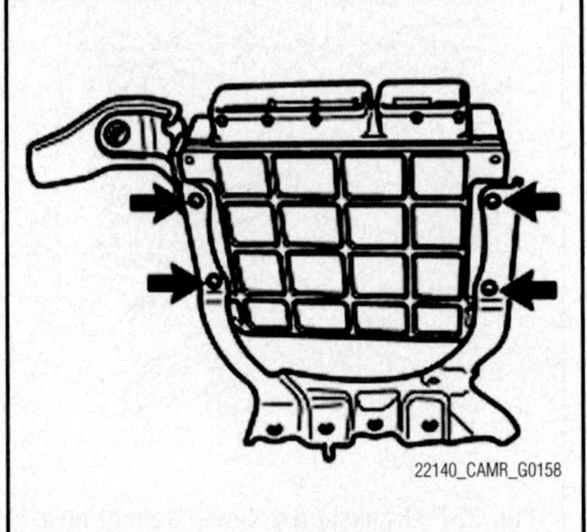

Fig. 241 Removing ECM—2.4L Engine

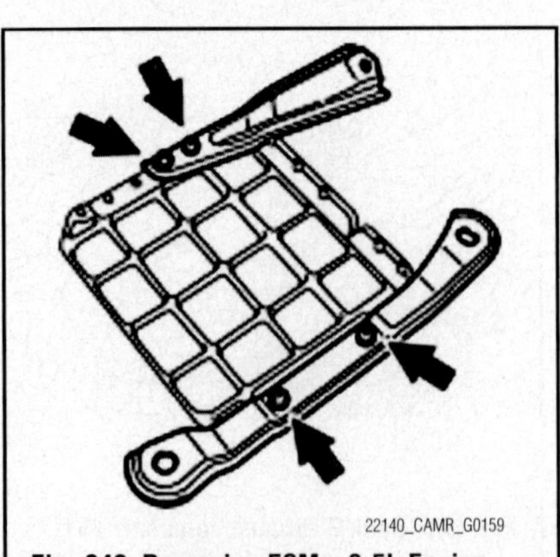

Fig. 242 Removing ECM—3.5L Engine

c. Raise the 2 levers while pushing the locks on the 2 levers, and disconnect the 2 ECM connectors.

➡After disconnecting the connectors, make sure that dirt, water or other foreign matter does not contact the connections of the connectors.

d. Remove the ECM.

e. Remove the 4 screws and 2 ECM brackets.

To install:

8. Install the 2 ECM brackets with the 4 screws, and tighten to 27 inch lbs. (3 Nm).
9. Connect the 2 ECM connectors and lower the 2 levers.

➡Make sure that dirt, water or other foreign matter does not contact the connections of the connectors.

10. Install the ECM to the body.
11. Attach the ECM with the 3 nuts and tighten to 71 inch lbs. (8 Nm).
12. Install the outer cowl top panel.
13. Install the windshield wiper motor and link assembly.
14. Install the cowl top ventilator louver sub-assembly.
15. Install both windshield wiper arm and blade assemblies.
16. Connect the negative battery cable.
17. Register the immobilizer communication ID. If the ECM is replaced, register the ECM communication ID for the immobilizer system (refer to the Service Bulletin for registration).
18. Perform initialization. After replacing the ECM on vehicles with a dynamic laser cruise control system, it is necessary to initialize the ECM so that the ECM can recognize the dynamic laser cruise control system.
19. Be sure to perform the following procedure after replacing the ECM:
    a. Turn the ignition switch on (IG).
    b. Turn the cruise control main switch on.
    c. With the brake pedal depressed, push the cruise control main switch to RES/ACC 3 times within 3 seconds. Check that the buzzer sounds at this time.

➡Do not turn the headlight dimmer switch on at this time because the optical axis automatic adjustment mode has already started, which may lead to an incorrect optical axis setting. If the headlight dimmer switch is turned on by mistake, readjust the optical axis.

ENGINE COOLANT TEMPERATURE (ECT) SENSOR

LOCATION

Refer to Component Locations illustrations at the top of this section to locate the ECT sensor.

REMOVAL & INSTALLATION

2.4L Engine

See Figure 243.

1. Before servicing the vehicle, refer to the Precautions Section.
2. Disconnect the negative battery cable.
3. Drain and recycle the engine coolant.
4. Remove the plastic engine cover.
5. Remove the air cleaner inlet assembly.
6. Remove the air cleaner cap sub-assembly.

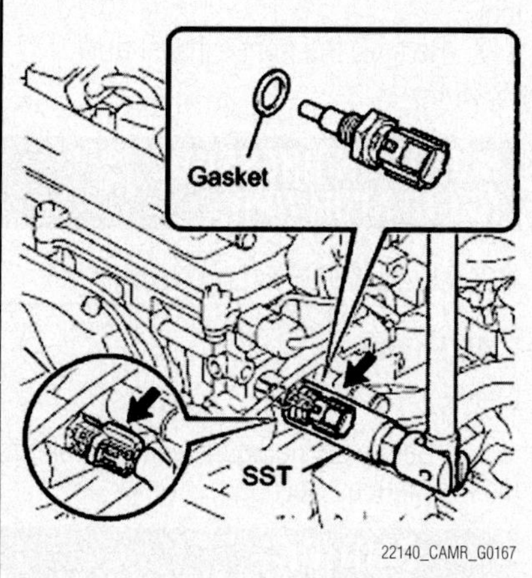

Fig. 243 Removing the ECT sensor—2.4L Engine

7. Remove the air cleaner case sub-assembly.
8. Disconnect the ECT sensor connector.
9. Using Special Tool: 09817—33190, remove the engine coolant temperature sensor and gasket.

To install:

10. Installation is the reverse of the removal procedure. Torque the engine coolant temperature sensor to 15 ft. lbs. (20 Nm).
11. Add coolant and check for leaks.

3.5L Engine

See Figure 244.

1. Before servicing the vehicle, refer to the Precautions Section.
2. Drain engine coolant.

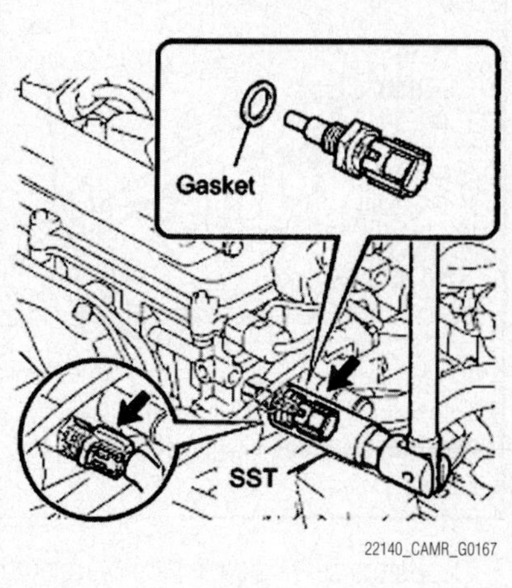

Fig. 244 Removing the ECT sensor—3.5L Engine

3. Remove V-bank cover sub-assembly.
4. Remove the air cleaner inlet assembly.
5. Remove the air cleaner cap sub-assembly.
6. Remove the air cleaner case sub-assembly.
7. Remove No. 1 air cleaner inlet.
8. Remove the engine coolant temperature sensor connector.
9. Using a19 mm deep socket wrench, remove the engine coolant temperature sensor and gasket.

To install:

10. Installation is the reverse of the removal procedure. Torque the engine coolant temperature sensor to 15 ft. lbs. (20 Nm).

HEATED OXYGEN (HO2S) SENSOR

REMOVAL & INSTALLATION

2.4L Engine

See Figures 245 and 246.

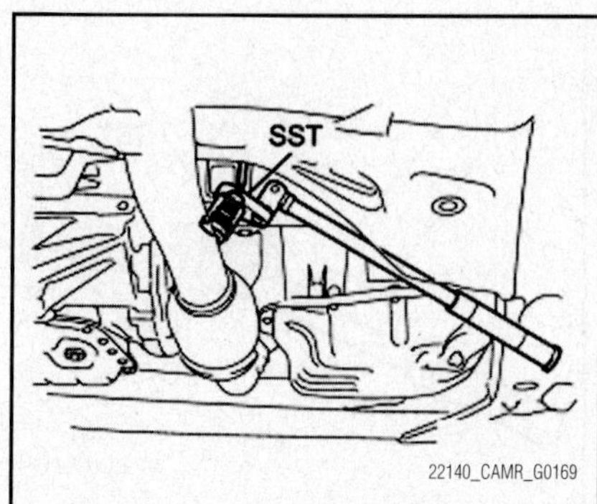

Fig. 245 Removing oxygen sensor—2.4L Engine

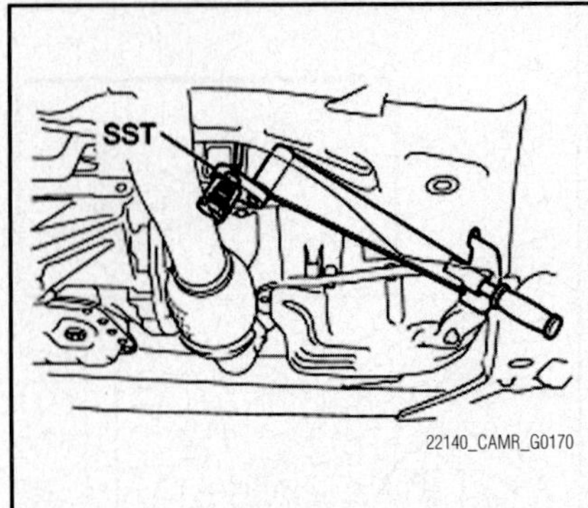

Fig. 246 Installing oxygen sensor—2.4L Engine

1. Before servicing the vehicle, refer to the Precautions Section.
2. Disconnect the oxygen sensor connectors.

➡Do not damage the heated oxygen sensor.

3. Using Special Tool: 09224-00010 or equivalent, remove the two oxygen sensors from the front pipe assembly.

To install:

4. Install the oxygen sensor to the front pipe assembly. Tighten to 32 ft. lbs. (44 Nm). Use a torque wrench with a fulcrum length of 300 mm (11.81 in.).
5. Connect the oxygen sensor connector.

3.5L Engine

See Figures 247 and 248.

1. Before servicing the vehicle, refer to the Precautions Section.
2. Remove the front exhaust pipe assembly.
3. Disconnect the 2 oxygen sensor connectors.
4. Using Special Tool: 09224-00010 or equivalent, remove the 2 oxygen sensors from the front pipe assembly.

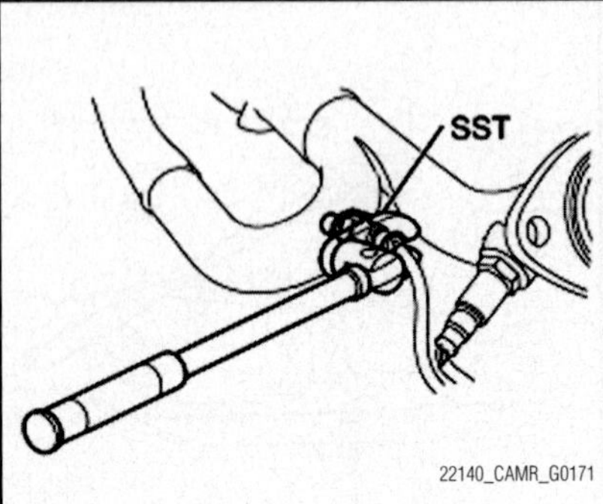

Fig. 247 Removing oxygen sensor—3.5L Engine

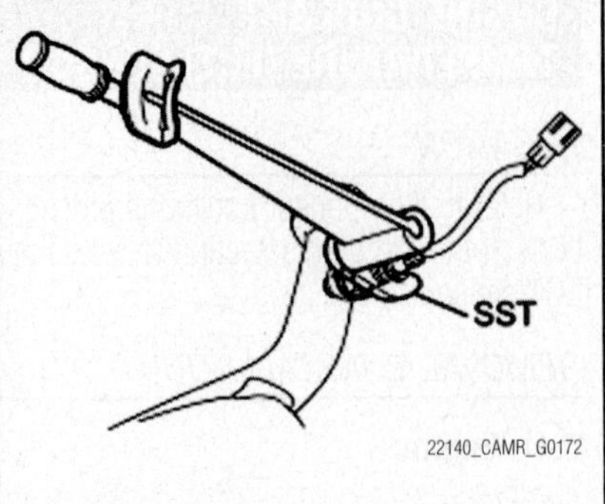

Fig. 248 Installing oxygen sensor—3.5L Engine

To install:

5. Install the 2 oxygen sensors to the front pipe assembly. Tighten to 32 ft. lbs. (44 Nm) and 30 ft. lbs. (40 Nm). Use a torque wrench with a fulcrum length of 300 mm (11.81 in.).
6. Connect the 2 oxygen sensor connectors.
7. Remove the front exhaust pipe assembly.

INJECTION PRESSURE REGULATOR (IPR)

REMOVAL & INSTALLATION

2.4L Engine

See Figure 249.

1. Before servicing the vehicle, refer to the Precautions Section.

⁂ CAUTION

Observe all applicable safety precautions when working around fuel. Whenever servicing the fuel system, always work in a well ventilated area. Do not allow fuel spray or vapors to come in contact with a spark or open flame. Keep a dry chemical fire extinguisher near the work area. Always keep fuel in a container specifically designed for fuel storage; also, always properly seal fuel containers to avoid the possibility of fire or explosion.

2. Disconnect the negative battery cable.
3. Remove the plastic engine cover.
4. Remove the air cleaner cap sub-assembly.
5. Disconnect the fuel tube sub-assembly.
6. Disconnect the metallic type connector.
7. Remove the two bolts and fuel pressure pulsation damper.

Fig. 249 Removing fuel pressure pulsation damper—2.4L Engine

To install:

➡Make sure that the O-ring is not cracked or jammed before installing the damper.

8. Apply a light coat of spindle oil or gasoline to a new O-ring.
9. Install the fuel pressure pulsation damper and tighten the bolts to 80 inch lbs. (9 Nm).
10. The remainder of installation is the reverse of the removal procedure.
11. Check for coolant and fuel leaks.
12. Perform initialization procedure.

3.5L Engine

See Figure 250.

1. Before servicing the vehicle, refer to the Precautions Section.

⁂ CAUTION

Observe all applicable safety precautions when working around fuel. Whenever servicing the fuel system, always work in a well ventilated area. Do not allow fuel spray or vapors to come in contact with a spark or open flame. Keep a dry chemical fire extinguisher near the work area. Always keep fuel in a con-

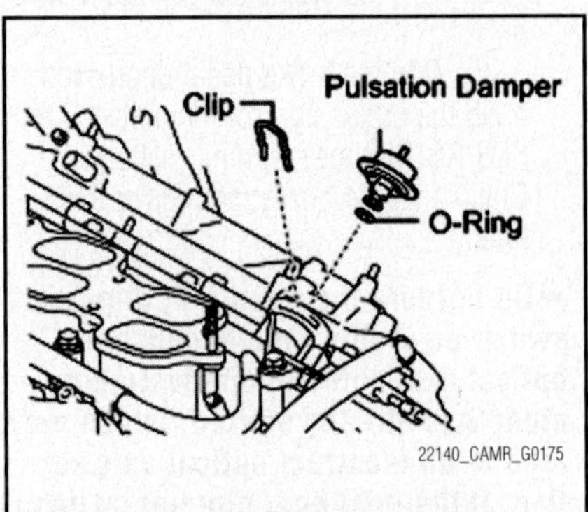

Fig. 250 Removing fuel pressure pulsation damper—3.5L Engine

tainer specifically designed for fuel storage; also, always properly seal fuel containers to avoid the possibility of fire or explosion.

2. Discharge fuel system pressure.
3. Disconnect the negative battery cable.
4. Drain engine coolant.
5. Remove both windshield wiper arm and blade assemblies.
6. Remove the right cowl top ventilator louver.
7. Remove the windshield wiper motor and link assembly.
8. Remove the outer cowl top panel.
9. Remove the V-bank cover sub-assembly.
10. Remove the air cleaner cap air cleaner with hose.
11. Remove the intake air surge tank.
12. Remove the fuel pressure pulsation damper assembly, as follows:
    a. Remove the clip.
    b. Pull out the fuel pressure pulsation damper from the fuel delivery pipe.
    c. Remove the O-ring from the fuel pressure pulsation damper.

To install:

13. Apply a light coat of spindle oil or gasoline to a new O-ring and install it to the fuel pressure pulsation damper.
14. Install the fuel pressure pulsation damper to the fuel delivery pipe.
15. Install the clip.
16. The remainder of installation is the reverse of the removal procedure.
17. Check for coolant and fuel leaks.
18. Perform initialization procedure.

INTAKE AIR TEMPERATURE (IAT) SENSOR

LOCATION

The intake air temperature sensor is mounted on the mass air flow meter.

REMOVAL & INSTALLATION

See Mass Air Flow Meter.

KNOCK SENSOR (KS)

REMOVAL & INSTALLATION

2.4L Engine

See Figures 251 and 252.

CAUTION

Observe all applicable safety precautions when working around fuel. Whenever servicing the fuel system, always work in a well ventilated area. Do not allow fuel spray or vapors to come in contact with a spark or open flame. Keep a dry chemical fire extinguisher near the work area. Always keep fuel in a container specifically designed for fuel storage; also, always properly seal fuel containers to avoid the possibility of fire or explosion.

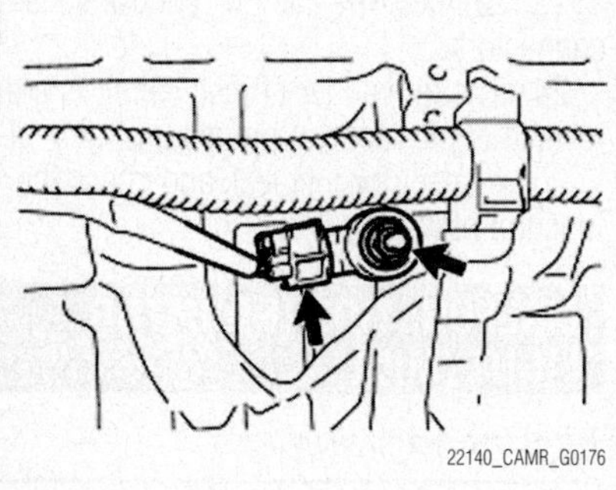

Fig. 251 Removing the nut and knock sensor —2.4L Engine

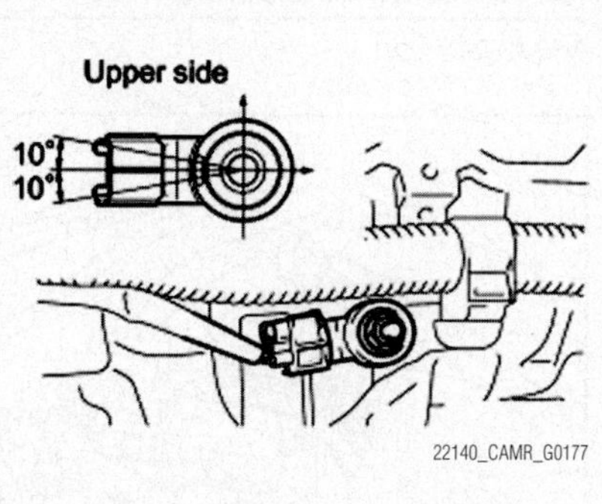

Fig. 252 Installing the nut and knock sensor —2.4L Engine

1. Before servicing the vehicle, refer to the Precautions Section.
2. Properly discharge the fuel system pressure.
3. Disconnect battery negative cable.
4. Remove plastic engine cover.
5. Drain and recycle the engine coolant.
6. Remove windshield wiper arms and blade assemblies.
7. Remove both front fender to cowl side seals.
8. Remove cowl top ventilator louver sub-assembly.
9. Remove windshield wiper motor and link.
10. Remove the four bolts, four nuts and cowl top panel outer sub-assembly.
11. Remove air cleaner cap sub-assembly.
12. Remove air cleaner case sub-assembly.
13. Remove throttle body.
14. Disconnect fuel tube.
15. Remove fuel delivery pipe with injector.
16. Disconnect the union to check valve hose from the brake booster.
17. Disconnect the camshaft timing oil control valve connector.
18. Remove the wire harness clamp.
19. Remove the union to check valve hose from the vacuum hose clamp.
20. Remove the 5 bolts, 2 nuts and intake manifold. Remove the gasket from the intake manifold.
21. Disconnect the knock sensor connector. Remove the nut and knock sensor.

To install:

➡Make sure that the knock sensor is in the correct position.

22. Install the knock control sensor with the nut and tighten to 15 ft. lbs. (20 Nm).
23. Connect the knock control sensor connector.
24. The remainder of installation is the reverse of the removal procedure.
25. Inspect for fuel leak and check the function of throttle body.

3.5L Engine

See Figures 253 and 254.

CAUTION

Observe all applicable safety precautions when working around fuel. Whenever servicing the fuel system, always work in a well ventilated area. Do not allow fuel spray or vapors to come in contact with a spark or open flame. Keep a dry chemical fire extinguisher near the work area. Always keep fuel in a container specifically designed for fuel storage; also, always properly seal fuel containers to avoid the possibility of fire or explosion.

1. Before servicing the vehicle, refer to the Precautions Section.
2. Properly discharge the fuel system pressure.
3. Disconnect battery negative cable.
4. Drain and recycle the engine coolant.
5. Remove both plastic engine under covers.
6. Remove windshield wiper arms and blade assemblies.
7. Remove both of the front fender to cowl side seals.
8. Remove cowl top ventilator louver sub-assembly.
9. Remove the windshield wiper motor and link assembly.

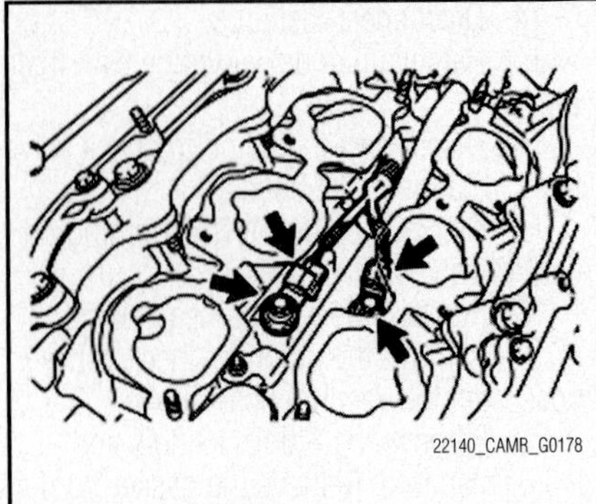

Fig. 253 Removing knock sensor—3.5L Engine

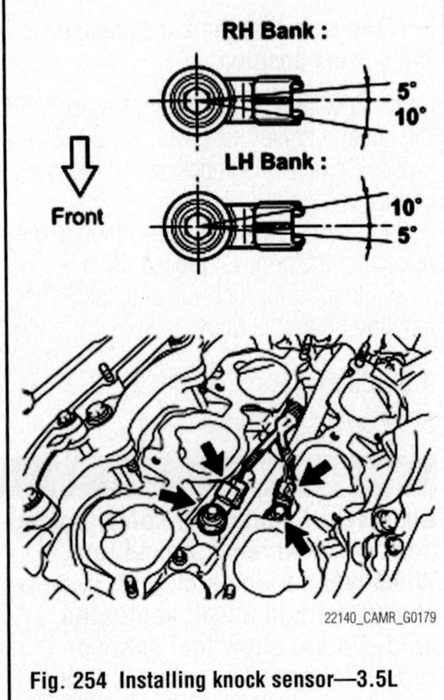

Fig. 254 Installing knock sensor—3.5L Engine

10. Remove the outer cowl top panel.
11. Remove the cool air intake duct seal.
12. Remove the V-bank cover sub-assembly.
13. Remove air cleaner inlet assembly.
14. Remove the air cleaner cap sub-assembly.
15. Remove air cleaner case sub-assembly.
16. Remove the intake air surge tank.
17. Remove no. 1 air cleaner inlet.
18. Separate fuel tube sub-assembly.
19. Remove the intake manifold.
20. Disconnect the 2 knock control sensor connectors.
21. Remove the 2 bolts and 2 knock control sensors.

To install:

22. Install the 2 knock control sensors with the 2 bolts as shown in the illustration and tighten to 15 ft. lbs. (20 Nm).
23. Connect the 2 knock control sensor connectors.
24. The remainder of installation is the reverse of the removal procedure.
25. Inspect for fuel leak and check the function of throttle body.

MALFUNCTION INDICATOR LIGHT (MIL)

RESET PROCEDURES

Clear the DTC codes.

MASS AIR FLOW (MAF) METER

REMOVAL & INSTALLATION

See Figure 255.

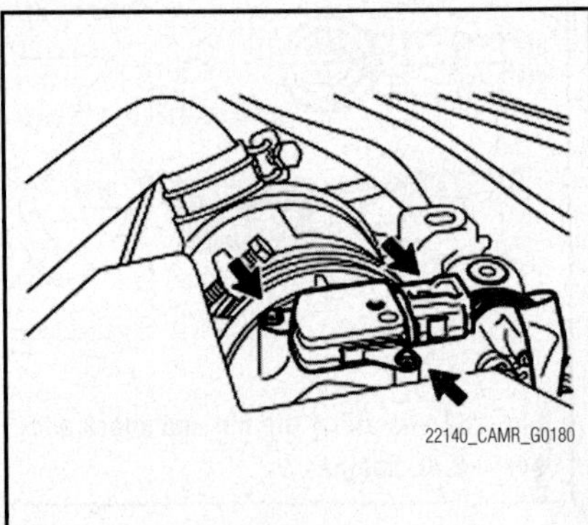

Fig. 255 Mass air flow meter—2.4L shown, 3.5L is similar

1. Before servicing the vehicle, refer to the Precautions Section.
2. Disconnect the mass air flow meter connector.
3. Remove the 2 screws and mass air flow meter.

To install:

4. Installation is the reverse of the removal procedure.

THROTTLE POSITION SENSOR (TPS)

REMOVAL & INSTALLATION

Refer to the Throttle Body removal and installation procedures.

VARIABLE CAMSHAFT TIMING OIL CONTROL SOLENOID

REMOVAL & INSTALLATION

2.4L Engine

See Figure 256.

1. Before servicing the vehicle, refer to the Precautions Section.

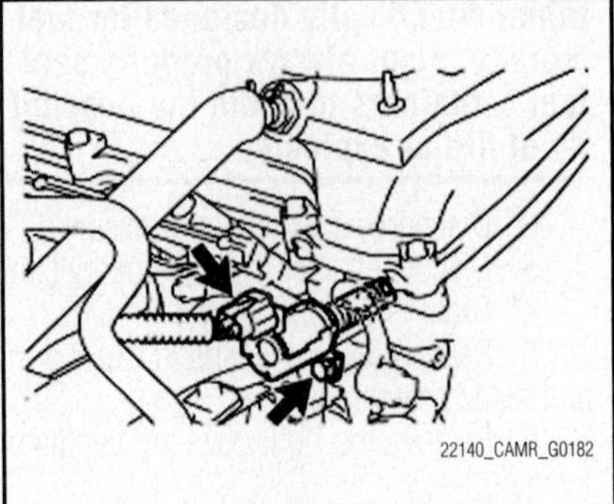

Fig. 256 Removing the camshaft timing oil control valve assembly—2.4L Engine

2. Remove the plastic engine cover.
3. Remove the bolt and disconnect the vacuum hose clamp.
4. Remove the clip nut.
5. Disconnect the camshaft timing oil control valve assembly connector.
6. Remove the bolt and camshaft timing oil control valve assembly.

To install:

7. Apply a light coat of engine oil to an O-ring of the camshaft timing oil control valve assembly sensor.
8. Install the camshaft timing oil control valve assembly with the bolt and tighten to 80 inch lbs. (9 Nm).
9. To complete installation, reverse removal procedure.

3.5L Engine

See Figures 257 through 260.

1. Remove engine under cover.
2. Drain and recycle the engine coolant.
3. Remove windshield wiper arm and blade assembly.
4. Remove front fender to cowl side seal.
5. Remove cowl top ventilator louver sub-assembly.
6. Remove windshield wiper motor and link.

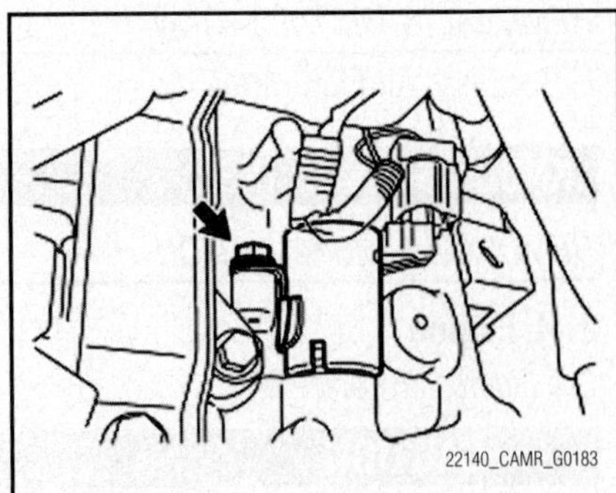

Fig. 257 Removing camshaft timing oil control valve assembly—bank 1, exhaust side—3.5L Engine

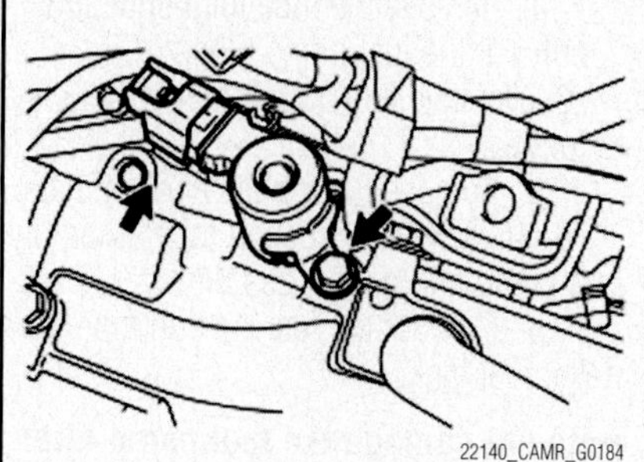

Fig. 258 Removing camshaft timing oil control valve assembly—bank 1, intake side—3.5L Engine

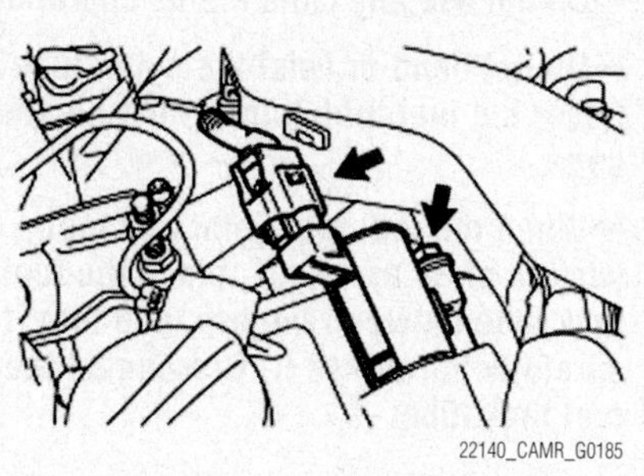

Fig. 259 Removing camshaft timing oil control valve assembly—bank 2, exhaust side—3.5L Engine

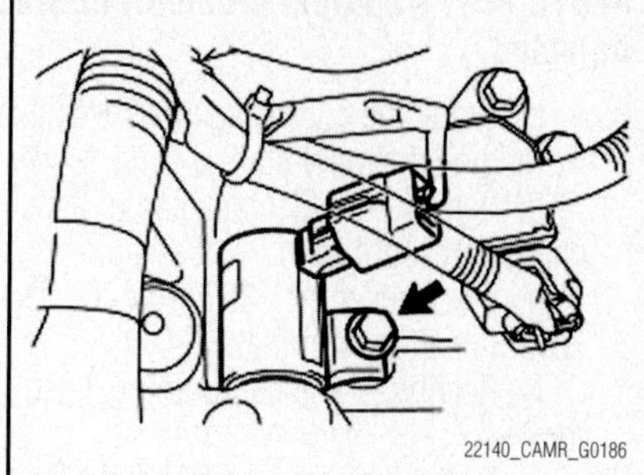

Fig. 260 Removing camshaft timing oil control valve assembly—bank 2, intake side—3.5L Engine

7. Remove the 4 bolts, 4 nuts and cowl top panel outer sub-assembly.
8. Remove cool air intake duct seal.
9. Remove v-bank cover sub-assembly.
10. Remove air cleaner inlet assembly.
11. Remove air cleaner cap sub-assembly.
12. Remove air cleaner case sub-assembly.
13. Remove no. 1 air cleaner inlet.
14. Remove intake air surge tank.
15. Remove the four camshaft timing oil control valves by performing the following:
    a. Disconnect the camshaft timing oil control valve connector.
    b. Remove the bolt and camshaft timing oil control valve.
    c. Remove the O-ring from the camshaft timing oil control valve.

To install:

16. Install the camshaft timing oil control valve assembly, as follows:
    a. Install the 4 oil control valves with the 4 bolts and tighten to 7 ft. lbs. (10 Nm).
17. Install the 4 camshaft position sensors with the 4 bolts and tighten to 7 ft. lbs. (10 Nm).
18. Apply adhesive (Toyota Genuine Adhesive 1324, Three Bond 1324 or equivalent) around the ventilation valve.
19. Install the ventilation valve sub-assembly and tighten to 20 ft. lbs. (27 Nm).
20. Install the spark plugs and the ignition coil assembly.
21. Install the oil filler cap sub-assembly.
22. Remove the engine stand.
23. Install the engine assembly.

FUEL SYSTEM

GASOLINE FUEL INJECTION SYSTEM

FUEL SYSTEM SERVICE PRECAUTIONS

Before servicing any vehicle, please be sure to read all of the following precautions, which deal with personal safety, prevention of component damage, and important points to take into consideration when servicing a motor vehicle:

- Observe all applicable safety precautions when working around fuel. Whenever servicing the fuel system, always work in a well-ventilated area. Do not allow fuel spray or vapors to come in contact with a spark, open flame, or excessive heat (a hot drop light, for example). Keep a dry chemical fire extinguisher near the work area. Always keep fuel in a container specifically designed for fuel storage; also, always properly seal fuel containers to avoid the possibility of fire or explosion. Refer to the additional fuel system precautions later in this section.
- Fuel injection systems often remain pressurized, even after the engine has been turned **OFF**. The fuel system pressure must be relieved before disconnecting any fuel lines. Failure to do so may result in fire and/or personal injury.
- All new vehicles are now equipped with an air bag system, often referred to as a Supplemental Restraint System (SRS) or Supplemental Inflatable Restraint (SIR) system. The system must be disabled before performing service on or around system components, steering column, instrument panel components, wiring and sensors. Failure to follow safety and disabling procedures could result in accidental air bag deployment, possible personal injury and unnecessary system repairs.
- Disconnecting the negative battery cable on some vehicles may interfere with the functions of the on-board computer system(s) and may require the computer to undergo a relearning process once the negative battery cable is reconnected.

Before inspecting and repairing the fuel system, disconnect the cable from the negative (-) battery terminal.

Keep gasoline away from rubber or leather parts.

Check that there are no fuel leaks from the fuel system after doing any maintenance or repairs.

RELIEVING FUEL SYSTEM PRESSURE

✲✲ CAUTION

Perform the following procedures to prevent fuel from spilling out before removing any fuel system parts.

✲✲ CAUTION

Pressure will still remain in the fuel line even after performing the following procedures. When disconnecting the fuel line, cover it with a shop rag or a piece of cloth to prevent fuel from spraying or coming out.

1. Disconnect the fuel pump connector:
    a. Remove the rear seat cushion assembly.
    b. Remove the rear floor service hole cover.
    c. Disconnect the fuel pump connector.
    d. Start the engine.
    e. After the engine stops, turn the ignition switch off.

➡**DTC P0171/25 (fuel problem) may be detected.**

f. Crank the engine again. Check that the engine does not start.

g. Remove the fuel tank cap to discharge pressure from the fuel tank.

h. Disconnect the cable from the negative (-) battery terminal.

i. Reconnect the fuel pump connector.

j. Install the rear floor service hole cover.

k. Install the rear seat.

2. Check that there are no fuel leaks from the fuel system after doing any maintenance or repairs.

FUEL FILTER

REMOVAL & INSTALLATION

1. Before servicing the vehicle, refer to the precautions section.
2. Remove the fuel pump from the vehicle.
3. Remove the fuel pump filter, as follows:

➡**Do not damage the fuel pump filter. Do not remove the suction filter.**

a. Using a screwdriver, pry out the clips.

b. Pull out the fuel pump filter from the fuel pump.

To install:

4. Install the fuel pump filter with a new clip.
5. Install the fuel pump.

FUEL PUMP

REMOVAL & INSTALLATION

See Figures 261 and 262.

1. Before servicing the vehicle, refer to the precautions section.
2. Discharge fuel system pressure.
3. Disconnect battery negative cable.
4. Remove the rear seat cushion assembly.
5. Remove the rear floor service hole cover.
6. Disconnect the fuel pump connector.
7. Separate the fuel pump tube sub-assembly, as follows:

➡**Check if there is any dirt or mud around the connector before this operation and remove the dirt as necessary.**

➡**Be careful of mud because the quick connector has an O-ring which seals the pipe and connector that can be contaminated.**

➡**Do not use any tools in this operation.**

➡**Do not bend or twist the nylon tube. Cover the fuel tube joint with a plastic bag.**

➡**When the fuel tube joint and fuel suction plate are stuck, pinch the fuel tank tube between fingers, and turn it carefully to release it. Disconnect the fuel tank tube.**

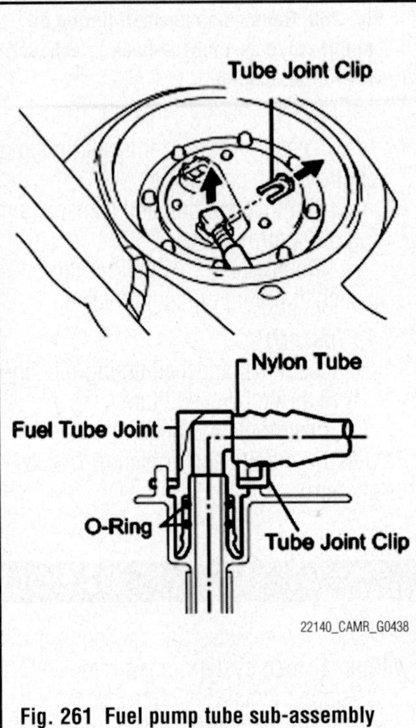

Fig. 261 Fuel pump tube sub-assembly

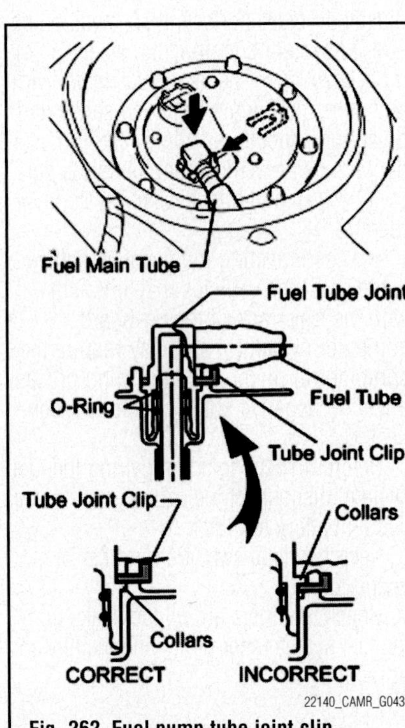

Fig. 262 Fuel pump tube joint clip

a. Remove the tube joint clip, and pull out the fuel pump tube.

8. Remove fuel tank vent tube set plate, as follows:

a. Remove the 8 bolts and set plate.

9. Remove fuel suction tube assembly with pump and gauge, as follows:

a. Pull out the fuel suction tube from the fuel tank.

➡**Do not damage the fuel pump filter.**

➡**Be careful not to bend the arm of the fuel sender gauge.**

b. Remove the gasket from the fuel suction tube.

To install:

10. Install a new gasket to the fuel suction tube.
11. Install the fuel suction tube.

➡**Do not damage the fuel pump filter.**

➡**Be careful not to bend the arm of the fuel sender gauge.**

12. Install the fuel tank vent tube set plate, as follows:

a. Align the mark of the set plate with the fuel suction tube.

b. Install the set plate with the 8 bolts and tighten to 52 inch lbs. (5.9 Nm).

13. Install the fuel pump tube with the tube joint clip.

➡**Check that there is no scratches or foreign objects on the connecting part.**

➡**Check that the fuel tube joint is inserted securely.**

➡**Check that the tube joint clip is on the collar of the fuel tube joint.**

➡**After installing the tube joint clip, check that the fuel tube joint is pulled off.**

14. Connect battery negative cable.
15. Inspect for fuel leak.
16. Install the rear floor service hole cover.
17. Install the rear seat cushion assembly.

FUEL RAIL & INJECTORS

REMOVAL & INSTALLATION

2.4L Engine

See Figures 263 through 266.

1. Before servicing the vehicle, refer to the precautions section.
2. Discharge fuel system pressure.
3. Disconnect battery negative cable.
4. Remove air cleaner cap sub-assembly.

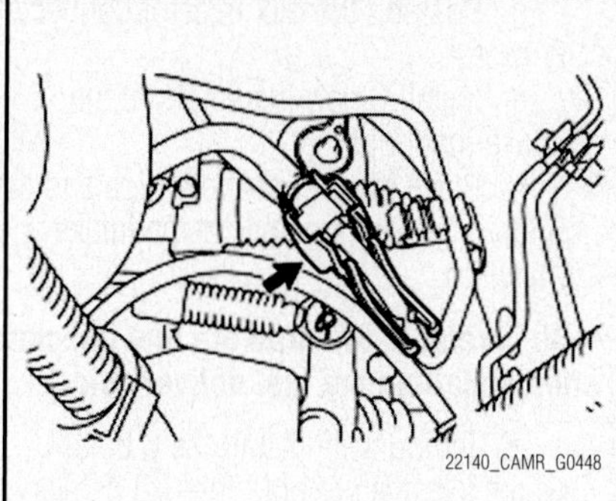
Fig. 263 Removing the No. 1 fuel pipe clamp

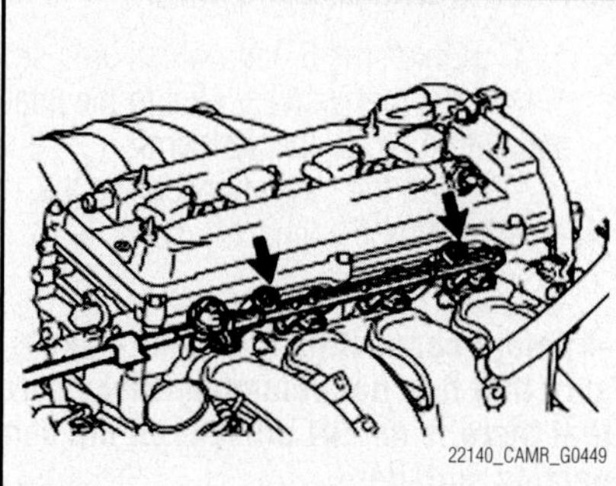
Fig. 264 Removing the fuel delivery pipe together with the 4 fuel injectors

5. Remove No. 1 engine cover sub-assembly.

➡Check for foreign matter on the pipe and around the connector before disconnecting the quick connector. Clean the connector if necessary.

6. Disconnect fuel tube sub-assembly, as follows:
   a. Remove the No. 1 fuel pipe clamp.

⁂ CAUTION

Do not use any tools in this following procedure. Check for foreign matter on the sealing surface of the disconnected pipe. Clean it if necessary.

   b. If the connector and pipe are stuck, pinch the connector, and push and pull the pipe to disconnect them.
   c. Separate the fuel tube from the fuel hose clamp.
7. Disconnect the No. 2 ventilation hose from the ventilation valve.
8. Remove fuel delivery pipe with injector as follows:
   a. Remove the 2 wire harness clamps.

⁂ CAUTION

Be careful not to drop the fuel injectors when removing the fuel delivery pipe.

   b. Remove the 2 bolts, then remove the fuel delivery pipe together with the 4 fuel injectors.
   c. Remove the 2 delivery pipe spacers from the cylinder head.
   d. Remove the 4 insulators from the cylinder head.
9. Pull out the 4 injectors from the delivery pipe.
10. Remove the 4 O-rings from the injectors.

To install:

11. Install the fuel injector assembly, as follows:
   a. Apply a light coat of spindle oil or gasoline to new O-rings, and install one to each injector.
   b. Apply a light coat of spindle oil or gasoline where the fuel delivery pipe contacts the O-ring.
   c. Apply a light coat of gasoline or spindle oil to the O-ring again, then install the right and left fuel injectors onto the fuel delivery pipe.

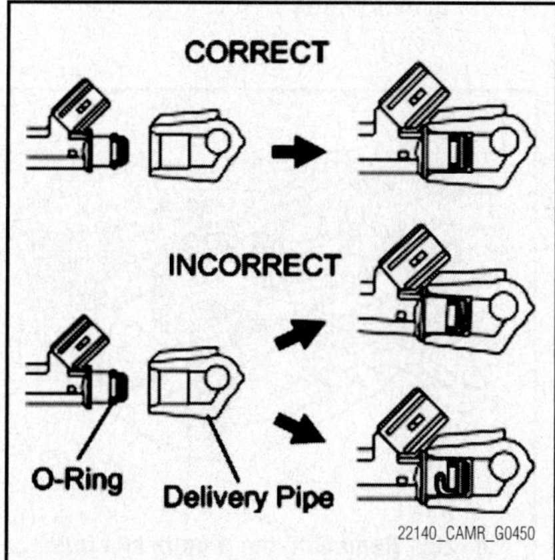

Fig. 265 Installing fuel injector to fuel tube

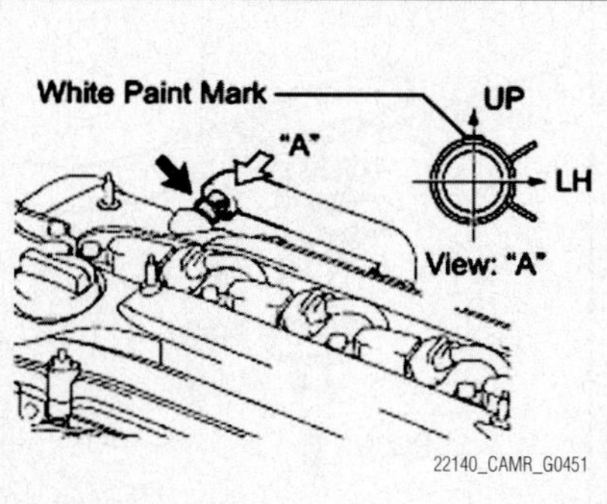

Fig. 266 Positioning paint mark and hose clamp

➡Make sure that the O-ring is not cracked or jammed before installing the injector.

   d. Check that the fuel injector rotates smoothly. If the fuel injector does not rotate, replace the O-ring.
12. Install fuel delivery pipe with injector as follows:
   a. Install 4 new insulators into the cylinder head.
   b. Install the 2 delivery pipe spacers onto the cylinder head.
   c. Install the fuel delivery pipe together with the 4 fuel injectors, then temporarily tighten the 2 bolts.

➡Be careful not to drop the fuel injectors when installing the fuel delivery pipe.

   d. Check that the fuel injector rotates smoothly. If the fuel injector does not rotate smoothly, replace the O-ring.
   e. Tighten the 2 bolts to the specified torque and tighten to 15 ft. lbs. (20 Nm).
   f. Connect the 4 fuel injector connectors.
   g. Install the 2 wire harness clamps.
13. Connect the No. 2 ventilation hose to the ventilation valve.
14. Install the fuel tube to the fuel hose clamp.
15. Push the fuel tube connector until it makes a "click" sound.
16. Install the No. 1 fuel pipe clamp.
17. Install air cleaner cap sub-assembly.
18. Connect cable to negative battery terminal.
19. Check for fuel leaks.
20. Install No. 1 engine cover sub-assembly.

3.5L Engine

See Figures 267 through 269.

1. Properly discharge the fuel system pressure.
2. Disconnect battery negative cable.
3. Drain engine coolant.
4. Remove both windshield wiper arm and blade assemblies.
5. Remove the right cowl top ventilator louver.
6. Remove the windshield wiper motor and link assembly.
7. Remove front fender to cowl side seal.
8. Remove the cowl top ventilator louver sub-assembly.
9. Remove windshield wiper motor and link assembly.

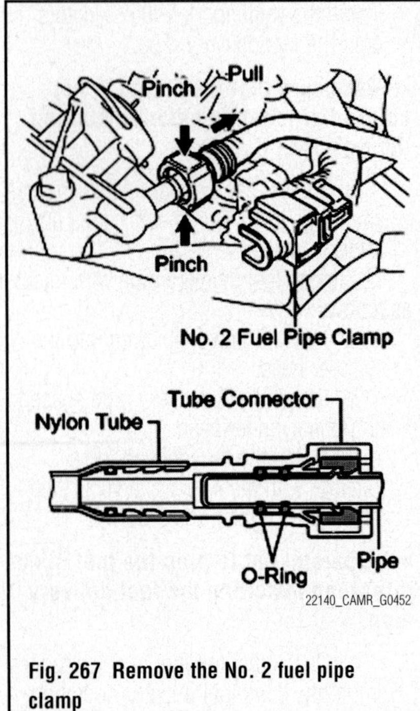

Fig. 267 Remove the No. 2 fuel pipe clamp

10. Remove cowl top panel outer sub-assembly.
11. Remove the V-bank cover sub-assembly.
12. Remove the air cleaner cap with air cleaner hose.
13. Remove the intake air surge tank.
14. Disconnect the fuel tube sub-assembly, as follows:
   a. Remove the No. 2 fuel pipe clamp.
   b. Pinch the tube connector and then pull out the fuel pipe.

➡Check that there is no dirt or other foreign objects around the connector before removing fuel tube, and clean the connector as necessary.

➡It is necessary to prevent mud or dirt from entering the connector. If mud or dirt gets in the connector, the O-rings may not seal properly.

➡Do not use any tools in this operation.

➡Do not bend, kink or twist the nylon tube. Protect the connector by covering it with a plastic bag.

➡When the pipe and connector are stuck, push and pull the connector to release and pull the connector out carefully.

15. Remove the fuel injector assembly, as follows:
   a. Disconnect the 6 fuel injector connectors.
   b. Remove the 5 bolts and fuel delivery pipe together with the 6 fuel injectors.

➡Be careful not to drop the fuel injectors when removing the fuel delivery pipe.

   c. Remove the 6 insulators from the intake manifold.
   d. Pull out the fuel injector from the fuel delivery pipe.
   e. Remove the 6 O-rings from the injectors.

To install:

16. Install the fuel injector assembly, as follows:
   a. Apply a light coat of spindle oil or gasoline to new O-rings, and install one to each injector.
   b. Apply a light coat of spindle oil or gasoline where the fuel delivery pipe contacts the O-ring.

➡Be careful not to twist the O-ring.

➡After installing the fuel injector, check that it turns smoothly. If not, reinstall it with a new O-ring.

   c. Push the fuel injector while twisting it back and forth to install it in the fuel delivery pipe.

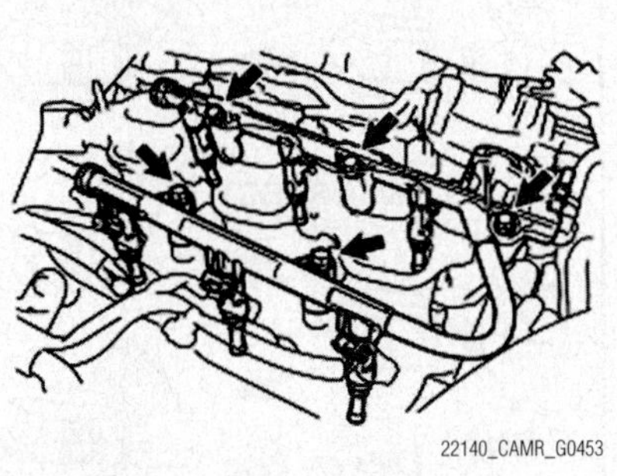

Fig. 268 Removing the 5 bolts and fuel delivery pipe together with the 6 fuel injectors

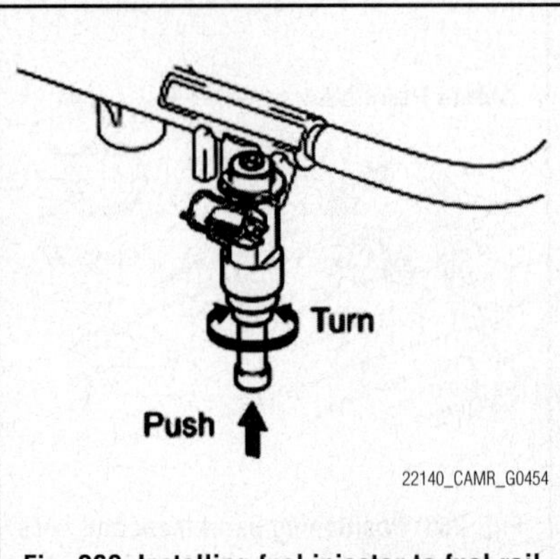

Fig. 269 Installing fuel injector to fuel rail

   d. Position the fuel injector connector outward.
   e. Install 6 new insulators to the intake manifold.
   f. Place the fuel delivery pipe and the 6 fuel injectors together to the intake manifold.

➡Be careful not to drop the fuel injectors when installing the fuel delivery pipe.

   g. Temporarily install the 6 bolts which are used to hold the fuel delivery pipe to the intake manifold.

➡After installing the fuel injector, check that it turns smoothly. If not, reinstall it with a new O-ring.

   h. Tighten the 5 bolts which are used to hold the fuel delivery pipe to the intake manifold to 15 ft. lbs. (21 Nm).
17. Push in the tube connector to the pipe until the tube connector makes a "click" sound.

➡Before connecting the tube, make sure that it is not damaged. Make sure that there is no dirt present on the connecting surfaces.

➡After connecting, check if the fuel tube connector and the pipe are securely connected by pulling on them.

18. Install the No. 2 fuel pipe clamp.
19. Te remainder of installation is the reverse of the removal procedure.
20. Check for coolant leak and fuel leak.

IDLE SPEED

ADJUSTMENT

Idle speed is maintained by the ECM. No adjustment is necessary or possible.

THROTTLE BODY

REMOVAL & INSTALLATION

2.4L Engine

See Figures 270 through 272.

1. Drain engine coolant.
2. Remove No. 1 engine cover sub-assembly.
3. Remove air cleaner cap sub-assembly, as follows:
   a. Disconnect the mass air flow meter connector (1).
   b. Disconnect the purge VSV connector (2).
   c. Disconnect the 2 purge VSV vacuum hoses (3).
   d. Disconnect the purge line hose from the clamp (4).

Fig. 270 Remove air cleaner cap sub-assembly

e. Disconnect the No. 2 ventilation hose from the air cleaner hose.

f. Lock the No. 1 air cleaner hose clamp, and then disconnect the No. 1 air cleaner hose from the throttle body.

g. Remove the 2 bolts and air cleaner cap.

h. Remove the air cleaner filter element from the air cleaner case.

4. Remove air cleaner case sub-assembly.
5. Disconnect the throttle position sensor connector and wire harness clamp.
6. Remove the 4 bolts, and then remove the fuel pipe support and throttle body.
7. Disconnect the purge line hose from the throttle body.

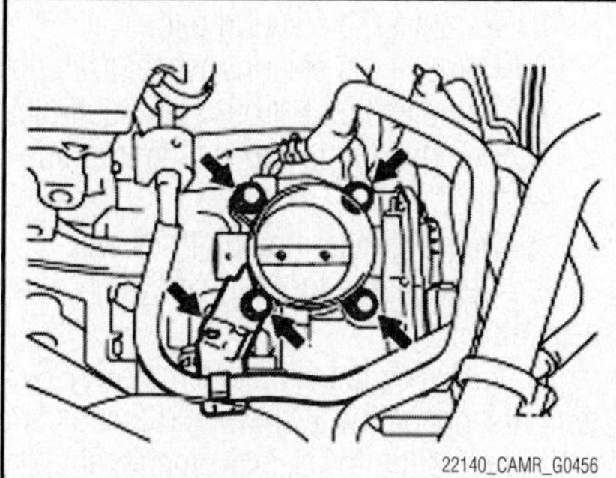

Fig. 271 Removing the 4 bolts and fuel pipe support

Fig. 272 Disconnecting the purge line hose

8. Disconnect the water by-pass hose from the throttle body.
9. Disconnect the No. 2 water by-pass hose from the throttle body.
10. Disconnect the No. 1 throttle body hose from the throttle body.
11. Remove the gasket from the intake manifold.

To install:

12. Install a new gasket onto the intake manifold.
13. Connect the purge line hose to the throttle body.
14. Connect the water by-pass hose to the throttle body.
15. Connect the No. 2 water by-pass hose to the throttle body.
16. Connect the No. 1 throttle body hose to the throttle body.
17. Install the throttle body and fuel pipe clamp with the 4 bolts and tighten to 22 ft. lbs. (30 Nm).
18. Connect the fuel tube into the clamp.
19. Connect the throttle position sensor connector.
20. Connect the wire harness clamp.
21. Install air cleaner case sub-assembly.
22. Install air cleaner cap sub-assembly, as follows:

a. Install the air cleaner filter element onto the air cleaner case.

b. Insert the hinges. Install the air cleaner cap sub-assembly with the 2 bolts.

c. Align the matchmarks of the No. 1 air cleaner hose and throttle body, and then connect the air cleaner hose No. 1 to the throttle body and unfasten the No. 1 air cleaner hose clamp.

➡Make sure that the hose clamp is at the correct angle.

d. Connect the No. 2 ventilation hose to the air cleaner hose.

e. Connect the purge line hose to the clamp.

f. Connect the 2 purge VSV vacuum hoses.

g. Connect the purge VSV connector.

h. Connect the mass air flow meter connector.

23. Install air cleaner inlet assembly.
24. Add engine coolant.
25. Check for engine coolant leaks.
26. Install No. 1 engine cover sub-assembly.

3.5L Engine

See Figures 273 through 276.

1. Before servicing the vehicle, refer to the precautions section.
2. Disconnect battery negative cable.
3. Drain engine coolant.
4. Remove cool air intake duct seal.
5. Remove the V-bank cover sub-assembly.
6. Remove air cleaner inlet assembly.
7. Remove air cleaner cap sub-assembly, as follows:

a. Disconnect the 3 vacuum hoses.

b. Disconnect the mass air flow meter connector (1).

c. Disconnect the No. 2 ventilation hose (2).

d. Disconnect the hose band (3).

e. Disconnect the 3 bands, and remove the air cleaner cap sub-assembly.

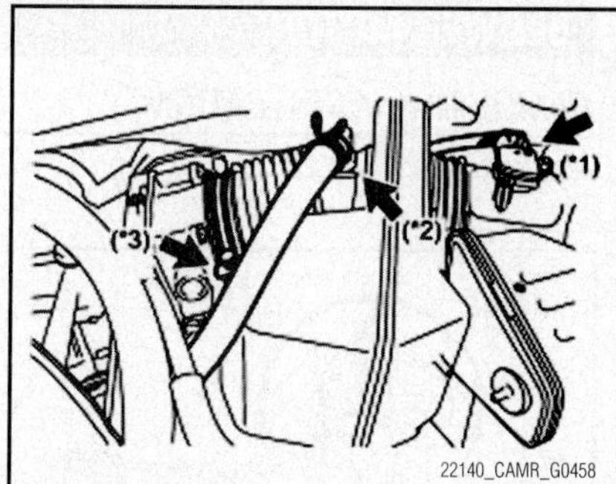

Fig. 273 Removing air cleaner cap sub-assembly

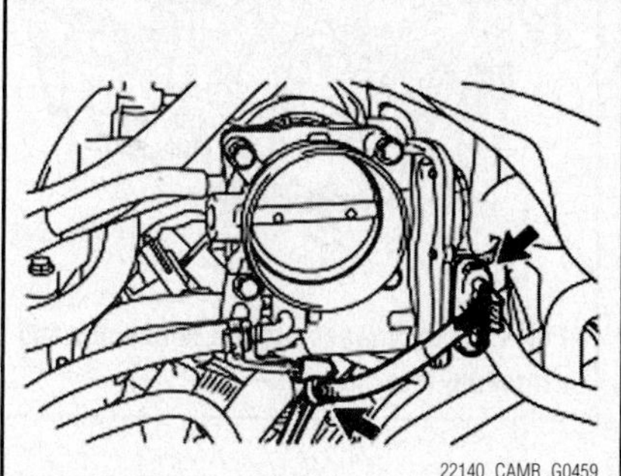

Fig. 274 Disconnecting the throttle body connector

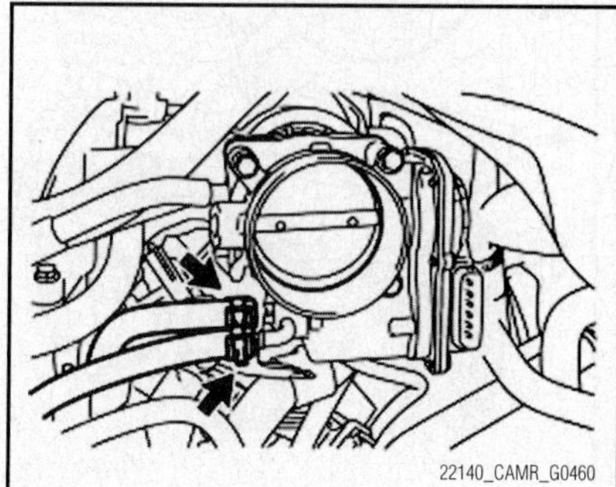

Fig. 275 Disconnecting the 2 water by-pass hoses

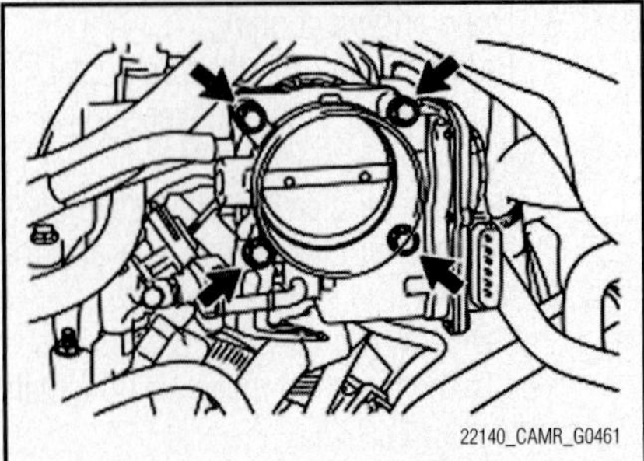

Fig. 276 Removing throttle body bolts

8. Remove air cleaner case sub-assembly.
9. Remove No. 1 air cleaner inlet.
10. Disconnect the throttle body connector and clamp.
11. Disconnect the 2 water by-pass hoses from the throttle body.
12. Remove the 4 bolts and throttle body.
13. Remove the throttle body gasket from the intake air surge tank.

To install:

14. Install a new throttle body gasket to the intake air surge tank.
15. Install the throttle w/ motor body assembly and wire harness clamp stay to the intake air surge tank with the 4 bolts and tighten to 7 ft. lbs. (10 Nm).
16. Connect the throttle w/ motor body assembly connector.
17. Te remainder of installation is the reverse of the removal procedure.
18. Check for coolant leak.
19. Check the function of the throttle body.

HEATING & AIR CONDITIONING SYSTEM

BLOWER MOTOR

REMOVAL & INSTALLATION

See Figures 277 through 279.

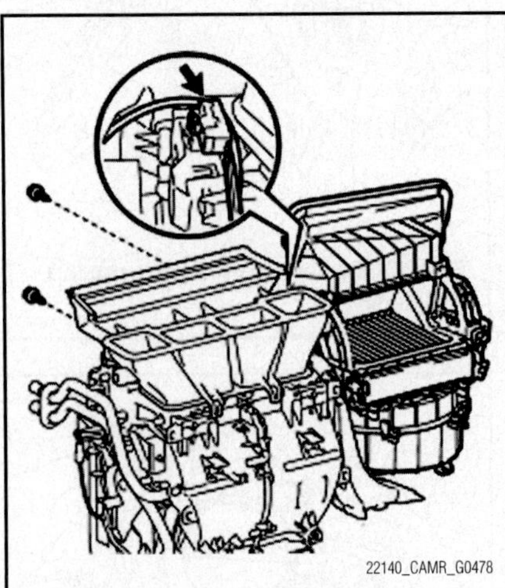

Fig. 277 Disconnecting the connector and removing the 2 screws—TMC

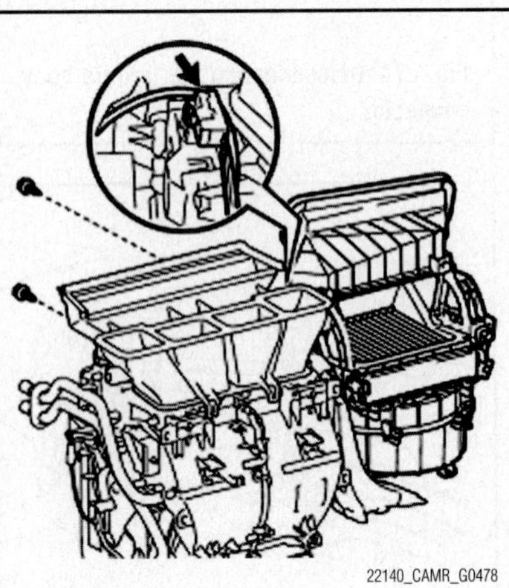

Fig. 278 Removing the connector, clamp and wire harness—TMMK

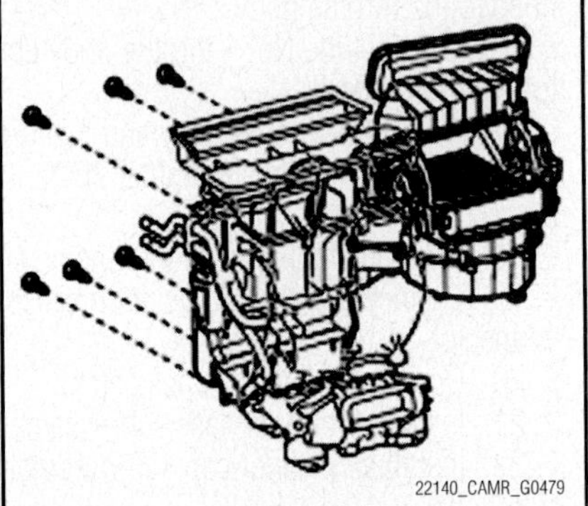

Fig. 279 Removing the 6 screws and blower assembly —TMMK

1. Drain and recycle the engine coolant.
2. Disconnect the negative battery cable.
3. Remove instrument panel.
4. For TMC made:
   a. Disconnect the connector.
   b. Remove the 2 screws and blower assembly.
   c. Remove the 3 screws and blower with fan motor sub-assembly.
5. For TMMK made:
   a. Remove cooler expansion valve.
   b. Remove the connector and clamp, and disconnect the wire harness.
   c. Remove the 6 screws and then the blower assembly with the cooler evaporator sub-assembly.
   d. Remove the 3 screws and blower with fan motor sub-assembly.
6. To install, reverse removal procedure.

HEATER CORE

REMOVAL & INSTALLATION

See Figures 280 through 287.

**** CAUTION**

Models are equipped with a Supplemental Restraint System (SRS), which uses an air bag. Whenever working near any of the SRS components, such as the impact sensors, the air bag module, steering column and instrument panel, disable the SRS.

1. Before servicing the vehicle, refer to the Precautions Section.

**** CAUTION**

Wait for 90 seconds after disconnecting the cable to prevent airbag deployment.

2. Disconnect battery negative terminal.
3. Remove the lower No. 2 and No. 3 steering wheel covers.
4. Remove the steering pad.
5. Remove the steering wheel assembly.
6. Remove the LH front door scuff plate.
7. Remove the LH cowl side trim sub-assembly.
8. Remove steering column cover.
9. Remove the turn signal switch assembly.
10. For vehicles without Smart Key System, disengage the 2 claws and 2 clips and then remove the lower instrument panel finish panel.
11. For vehicles with Smart Key System, disengage the 2 claws and 2 clips. Disconnect the connector and remove the lower instrument panel finish panel.
12. Using a molding remover, disengage the 2 clips. Disengage the guide and 4 claws, and then remove the No. 1 instrument cluster finish panel.
13. Remove the 4 screws. Disconnect each connector and remove the combination meter assembly.
14. Remove the RH front door scuff plate.
15. Remove cowl side trim sub-assembly.
16. Disengage the 4 claws. Disengage the 2 guides and remove the No. 2 under cover sub-assembly.

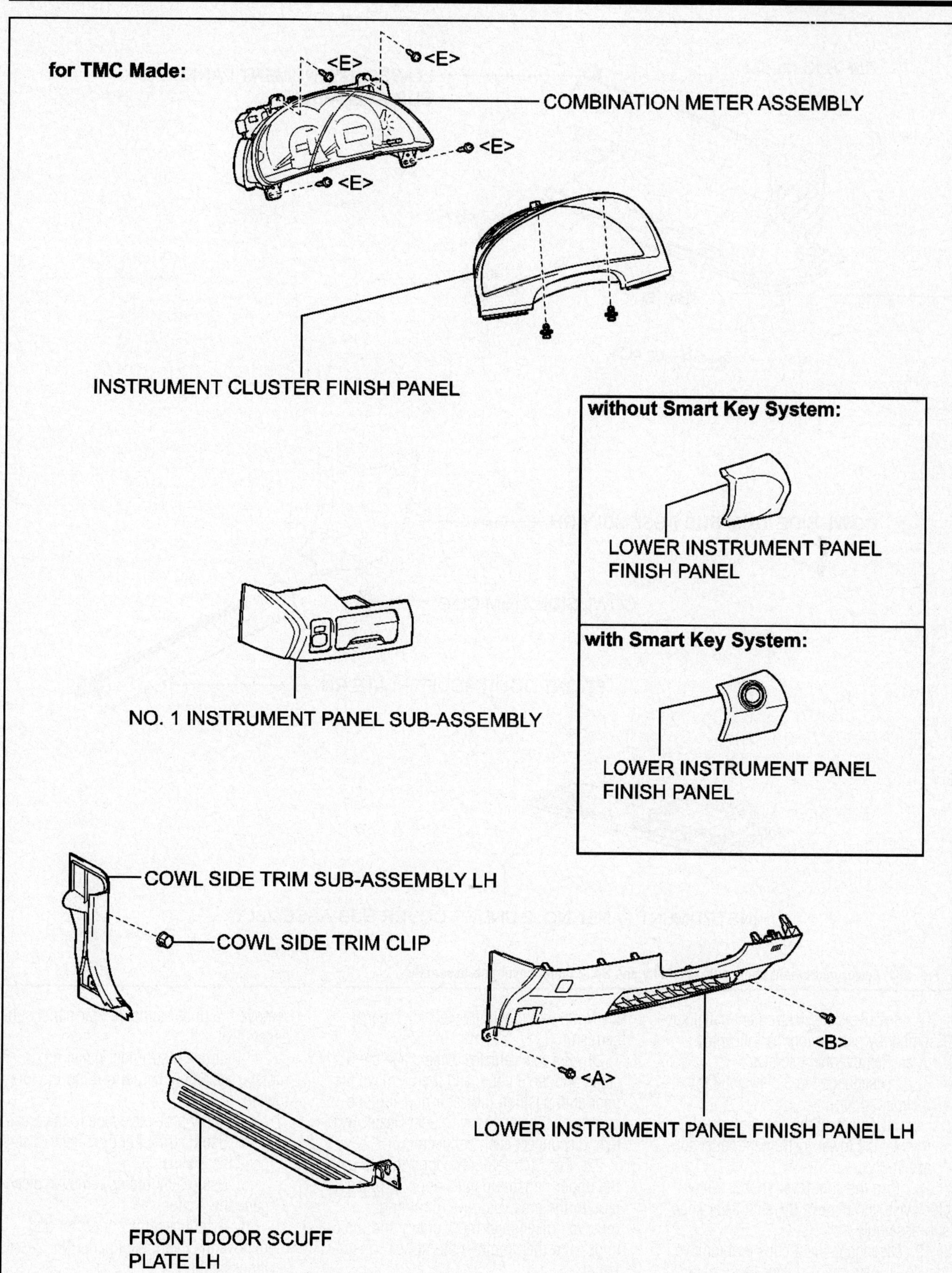

Fig. 280 Instrument cluster, combination meter assembly, instrument panel sub-assembly and lower trip sub-assembly

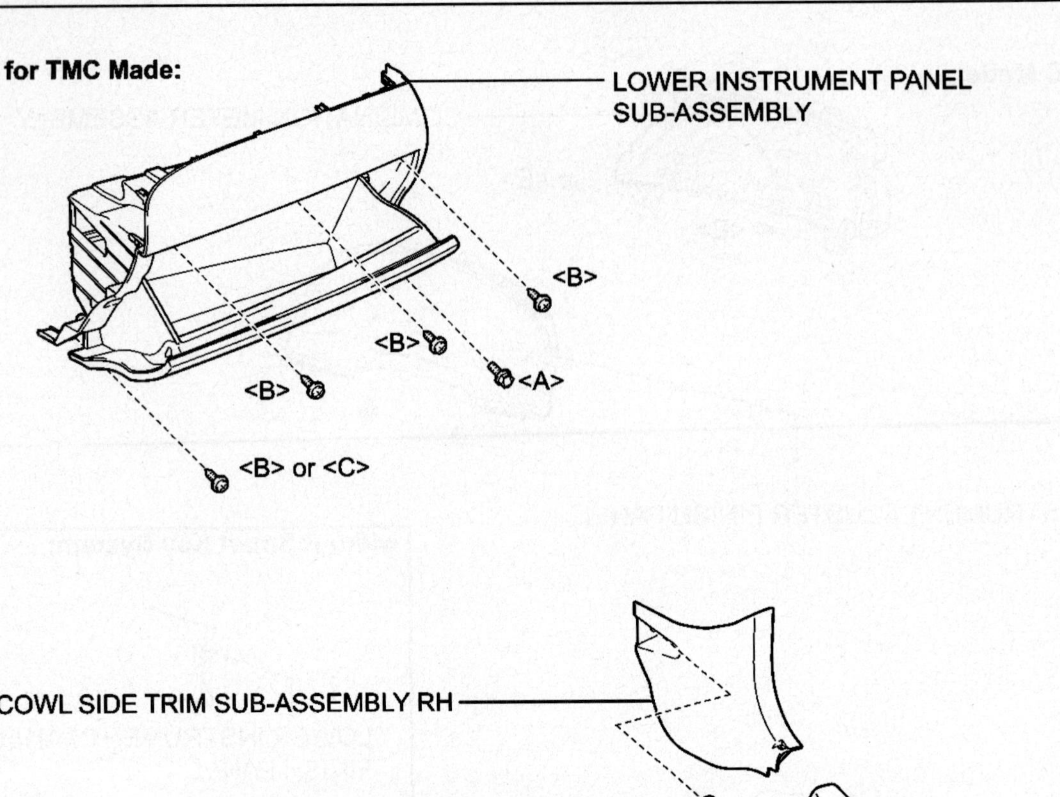

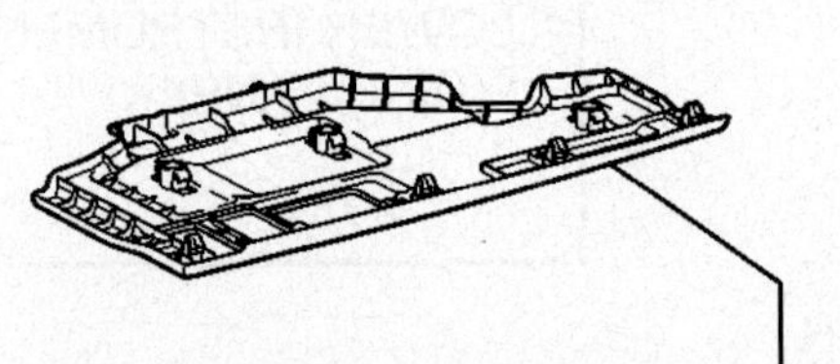

Fig. 281 Lower instrument panel sub-assembly and No. 2 under cover sub-assembly

17. Remove lower instrument panel sub-assembly by performing the following:
   a. Remove the 4 screws.
   b. Disengage the 3 claws and the 3 clips.
   c. Disconnect the connector and remove the lower instrument panel sub-assembly.
18. Turn the shift lever knob counter-clockwise and remove the shift lever knob sub-assembly.
19. Disengage the 2 clips and remove the No. 1 instrument cluster finish panel garnish.
20. Disengage the 2 clips and remove the No. 2 instrument cluster finish panel garnish.
21. For A/T vehicles, Disengage the 6 claws and the 3 clips, and then remove the floor shift position indicator housing sub-assembly. If equipped with Seat Heater System, disconnect each connector.
22. For M/T vehicles, open the lid of the upper console panel. Apply protective tape to the area. Using a moulding remover, disengage the 2 claws and the 5 clips, and then remove the upper console panel.
23. Disengage the 3 claws and the 5 clips. Disconnect the connector and remove the upper console rear panel sub-assembly.
24. Remove instrument panel no. 2 register assembly by performing the following:
   a. Apply protective tape to the areas.
   b. Using a moulding remover, disengage the 3 clips.
   c. Using a moulding remover, disengage the 4 clips.
   d. Disconnect the connector and remove the instrument panel No. 2 register assembly.
25. Remove radio receiver with heater control panel assembly.

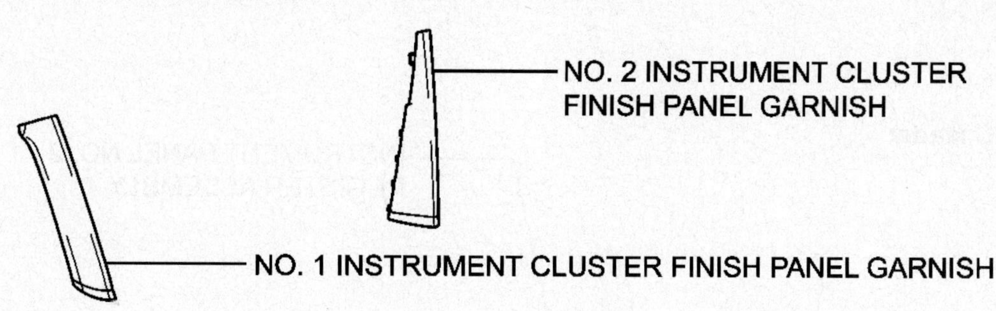

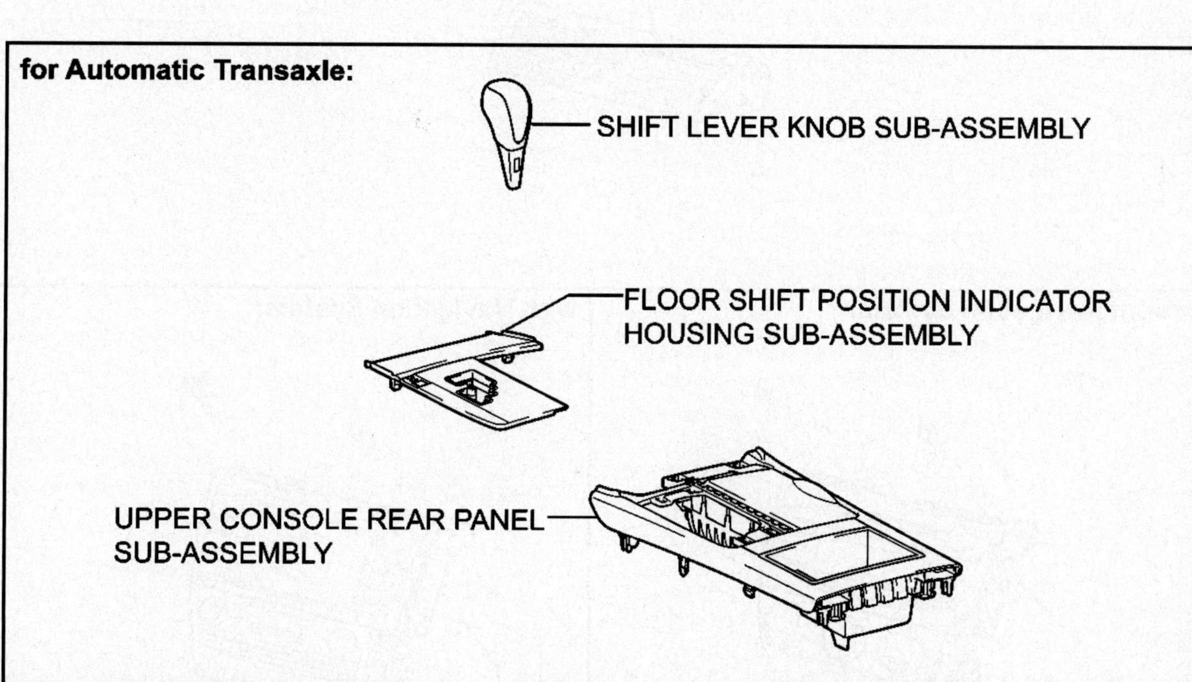

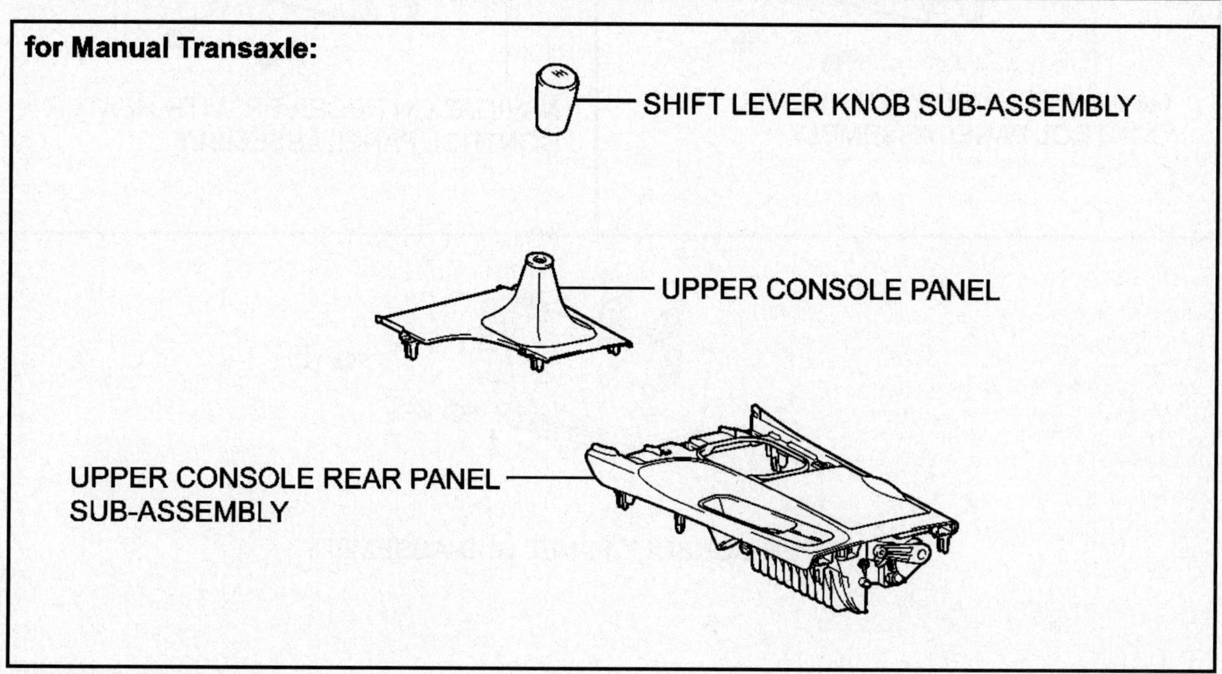

Fig. 282 Instrument cluster panel garnish, floor housing sub-assembly and upper console rear panel sub-assembly

for TMC Made:

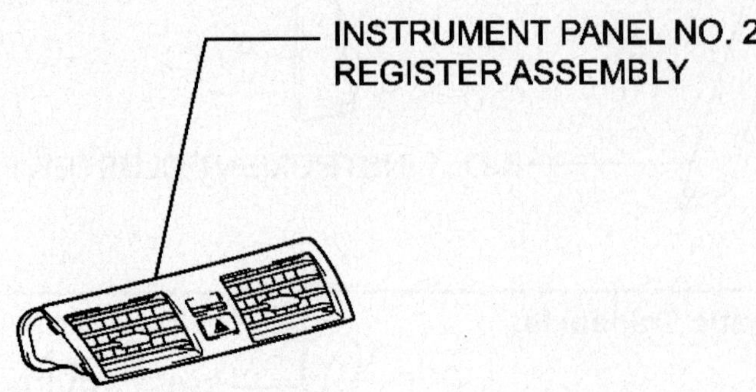

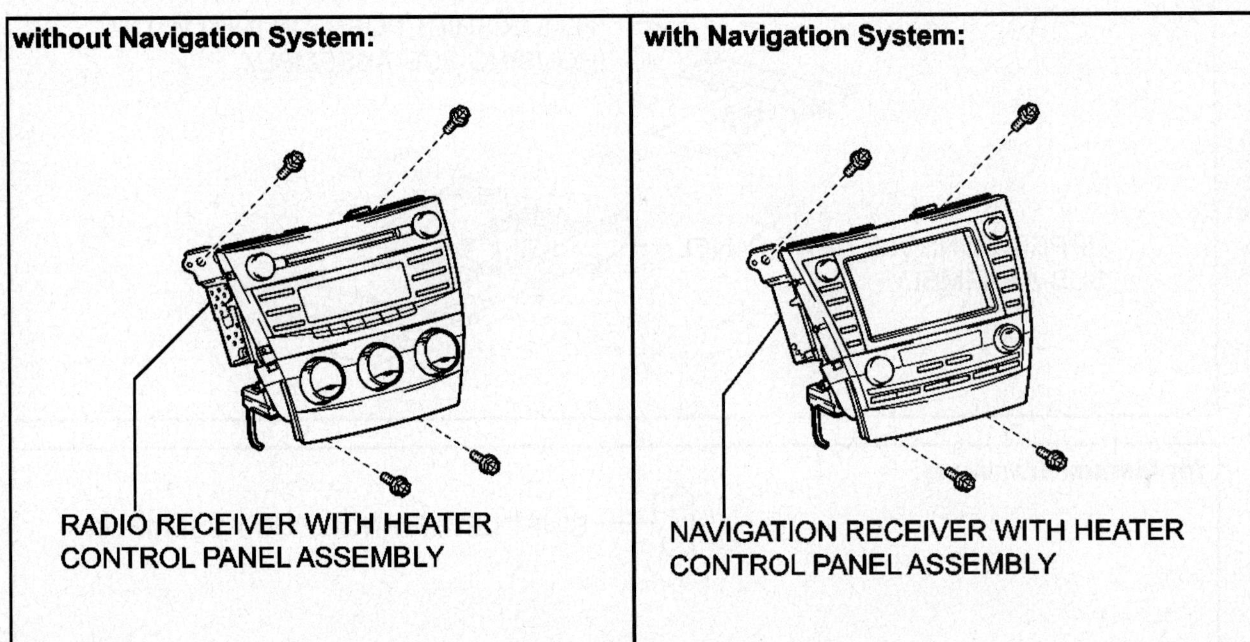

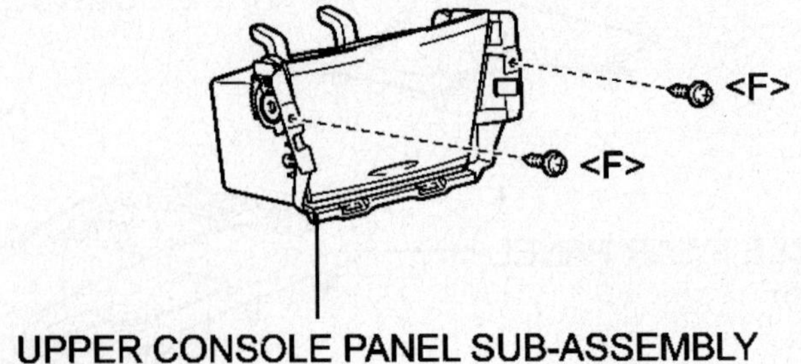

Fig. 283 Instrument panel no. 2 register assembly, control panel assembly and upper console sub-assembly

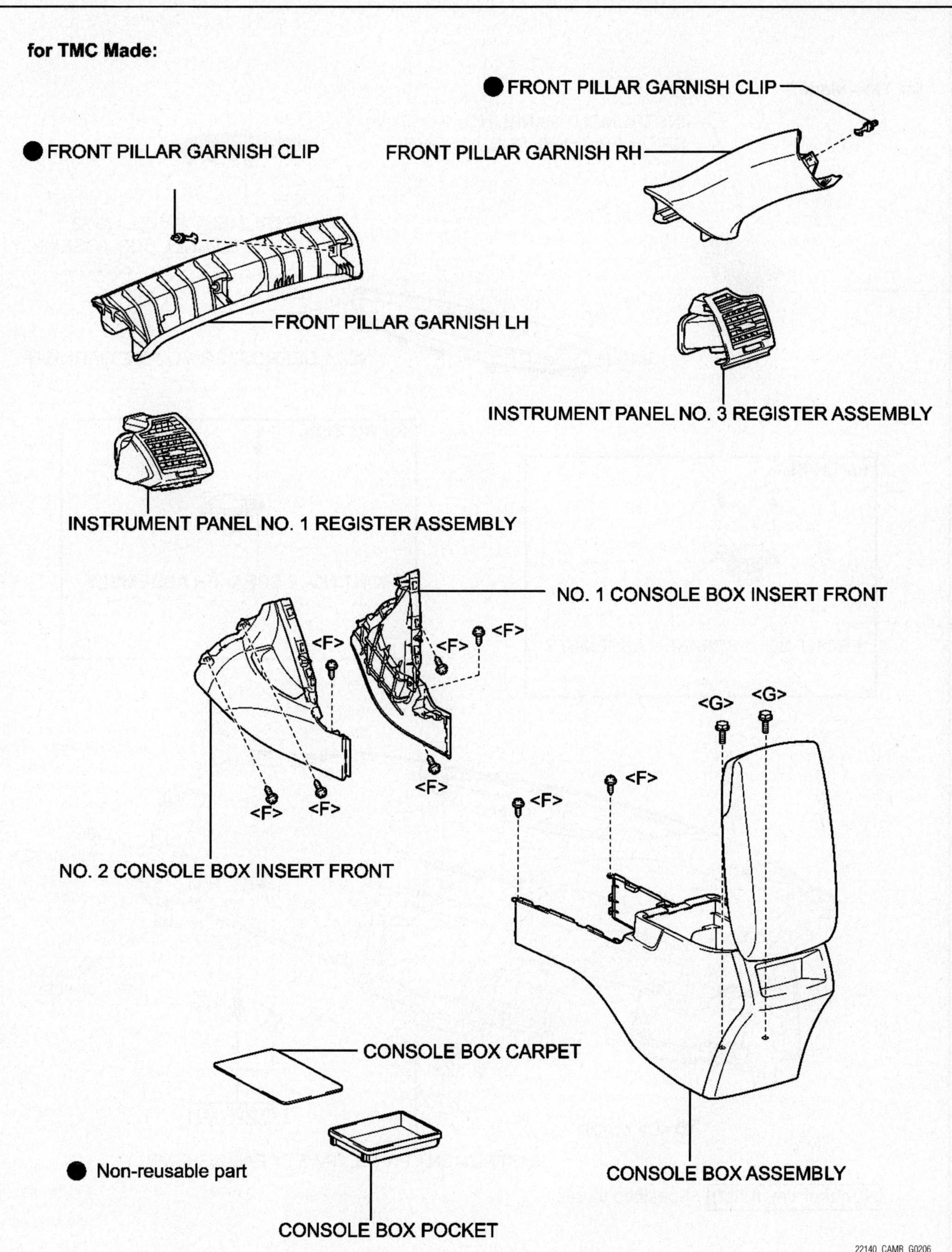

Fig. 284 Console box assembly, front console box inserts, LH front pillar garnish and instrument panel No. 1 and No. 3 register assembly

for TMC Made:

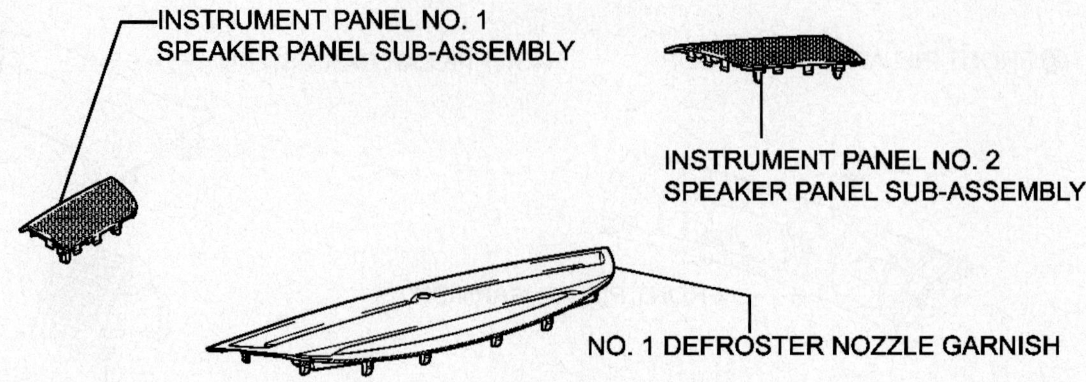

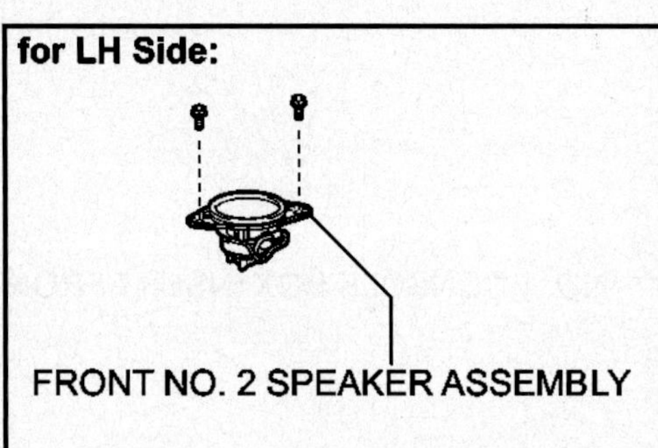

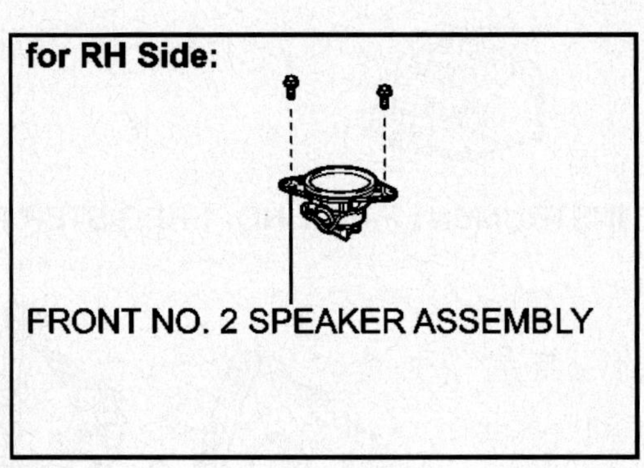

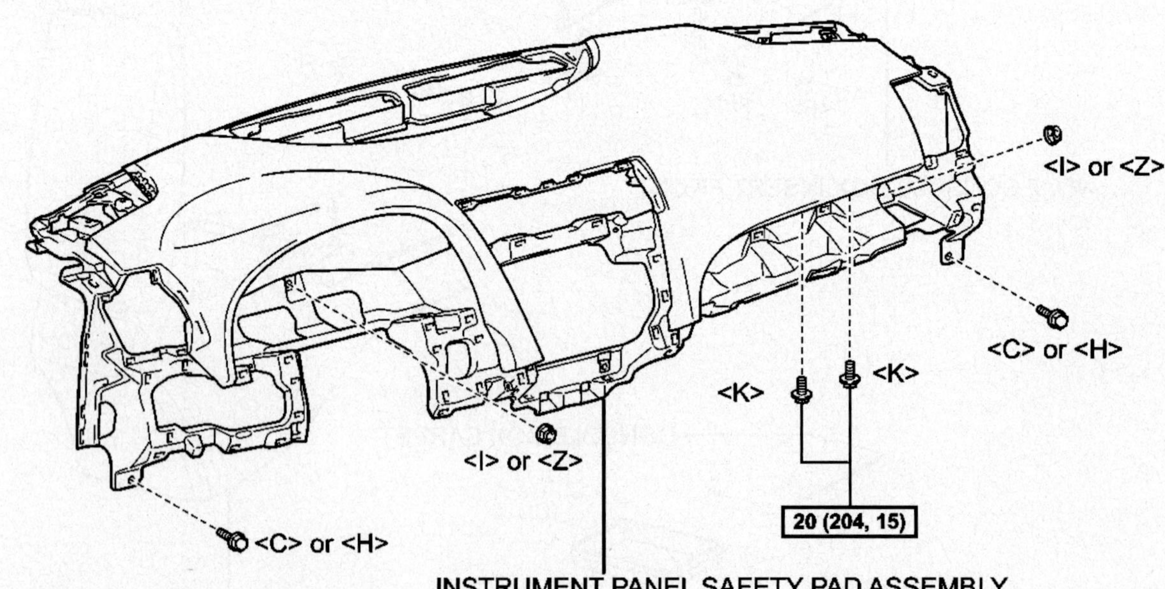

N*m (kgf*cm, ft.*lbf) : Specified torque

22140_CAMR_G0207

Fig. 285 Speakers, speaker panel sub-assembly and No. 1 defroster nozzle garnish

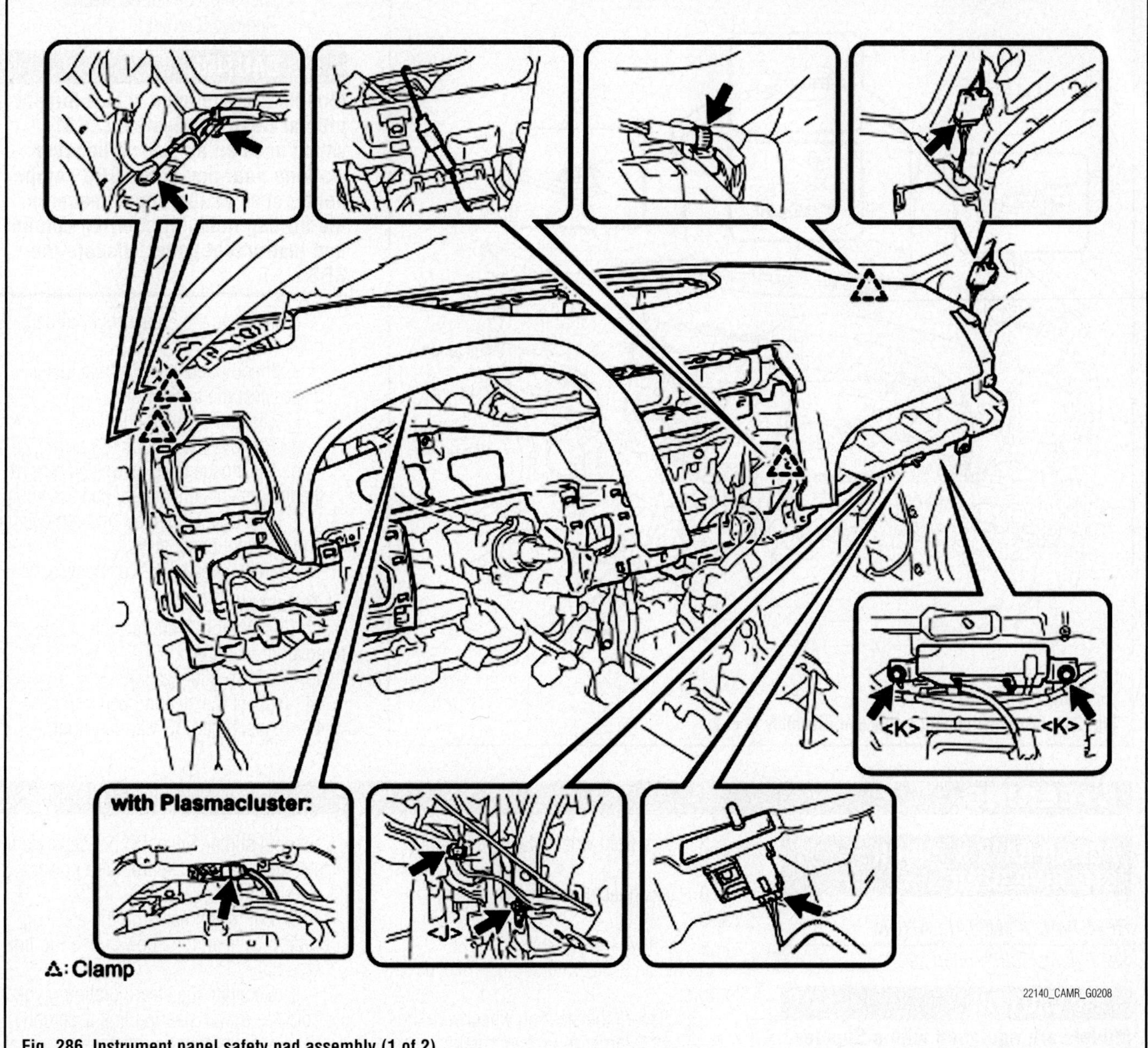

Fig. 286 Instrument panel safety pad assembly (1 of 2)

26. Remove the console box pocket
27. Remove the console box pocket.
28. Remove the console box assembly by performing the following:
    a. Remove the 2 screws.
    b. Disengage the clamp.
    c. Remove the 2 bolts and the console box assembly.
29. Remove both of the front console box inserts by performing the following:
    a. Remove the 3 screws.
    b. Disengage the clip and remove the front console box insert.
30. Remove the LH front pillar garnish.
31. Disengage the 4 clips and remove the instrument panel No. 1 register assembly.
32. Remove instrument panel no. 1 speaker panel sub-assembly by performing the following:
    a. Disengage the 6 claws and the 2 clips.
    b. Disengage the 2 guides and remove the instrument panel No. 1 speaker panel sub-assembly.
33. Remove RH front no. 2 speaker assembly.
34. Remove the RH front pillar garnish.
35. Disengage the 4 clips and remove the instrument panel No. 3 register assembly.
36. Remove instrument panel no. 2 speaker panel sub-assembly by performing the following:
    a. Disengage the 6 claws and the 2 clips.
    b. Disengage the 2 guides and remove the instrument panel No. 2 speaker panel sub-assembly.
37. Remove LH and RH front no. 2 speaker assemblies.
38. Remove no. 1 defroster nozzle garnish by performing the following:
    a. Disengage the 8 clips and the 4 guides.
    b. Disconnect each connector and remove the No. 1 defroster nozzle garnish.
39. Disconnect instrument panel wire assembly.
40. Remove instrument panel safety pad assembly by performing the following:
    a. Disengage each clamp.

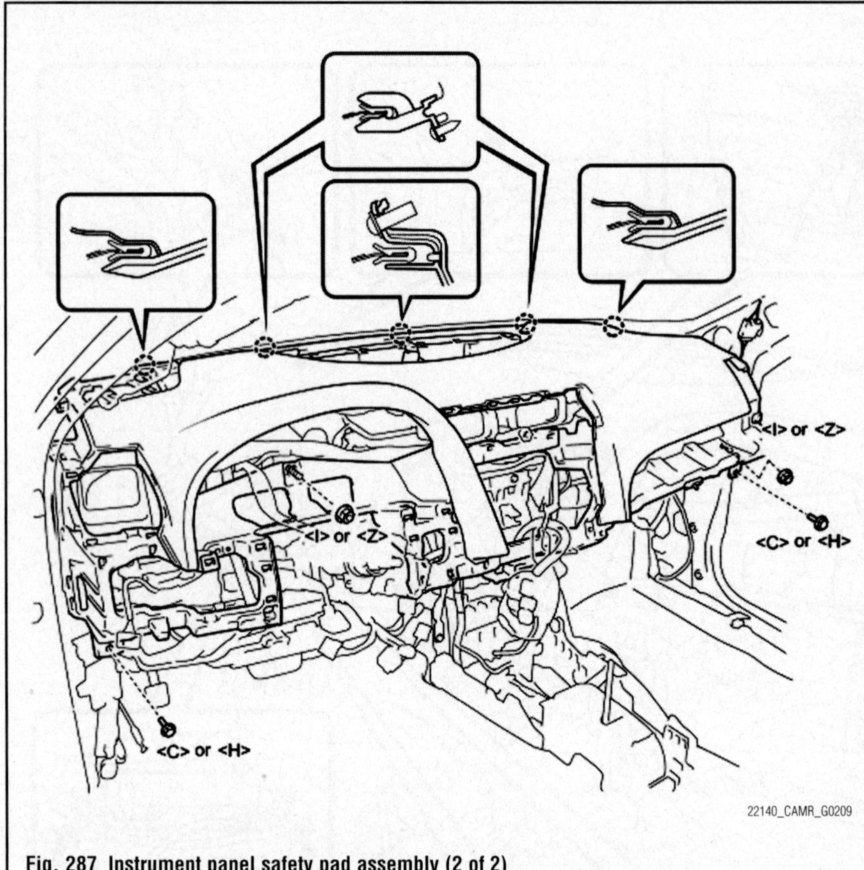

Fig. 287 Instrument panel safety pad assembly (2 of 2)

b. Disconnect each connector.
c. Remove the bolt (J).

CAUTION

Models are equipped with a Supplemental Restraint System (SRS), which uses an air bag. Whenever working near any of the SRS components, such as the impact sensors, the air bag module, steering column and instrument panel, disable the SRS.

d. Remove the 2 passenger airbag bolts (K).
e. If equipped with Plasmacluster, disconnect the connector
f. Disconnect the connector
g. Remove the 2 bolts (C) or (H).
h. Disengage the 5 claws and remove the instrument panel safety pad assembly.
i. Disengage the claw and remove the 5 instrument panel stays.

41. Remove heater core as necessary.

To install:

42. Installation is the reverse of the removal procedure.
43. Perform initialization.
44. Inspect the steering pad.
45. Inspect the SRS warning light.

STEERING

POWER RACK & PINION STEERING GEAR

REMOVAL & INSTALLATION

See Figures 288 through 291.

CAUTION

Models are equipped with a Supplemental Restraint System (SRS), which uses an air bag. Whenever working near any of the SRS components, such as the impact sensors, the air bag module, steering column and instrument panel, disable the SRS.

➡ Be sure to turn the front wheels straight ahead when removing and installing the power steering link assembly.

➡ If disconnecting the steering sliding yoke and the pinion shaft of the power steering link assembly, be sure to put matchmarks before starting the operation.

1. Before servicing the vehicle, refer to the precautions section.
2. Place front wheels facing straight ahead.
3. Disconnect the negative battery cable.
4. Remove front wheels.
5. Separate steering sliding yoke, as follows:
   a. Secure the steering wheel with the seat belt in order to prevent rotation. This operation is useful to prevent damage to the spiral cable.
   b. Remove the bolt and slide the steering sliding yoke. Do not separate the steering sliding yoke from the power steering link assembly.
   c. Put matchmarks on the steering sliding yoke and the power steering link assembly.
   d. Separate the steering sliding yoke from the power steering link assembly.
6. Separate both tie rod assemblies, as follows:
   a. Remove the cotter pin and the nut.
   b. Using SST: 09628-00011, separate

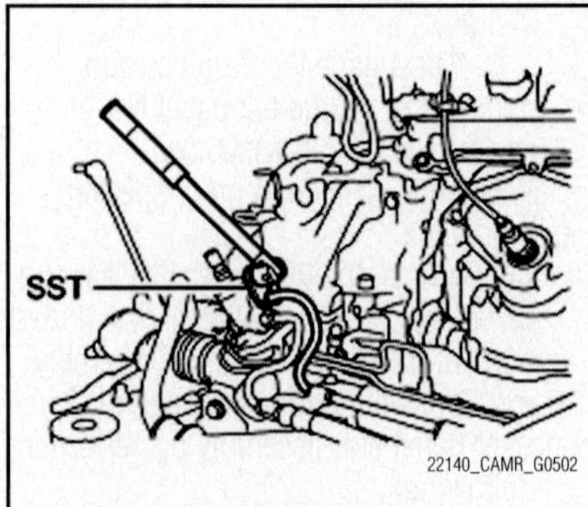

Fig. 288 Disconnect the pressure feed tube assembly —return tube side

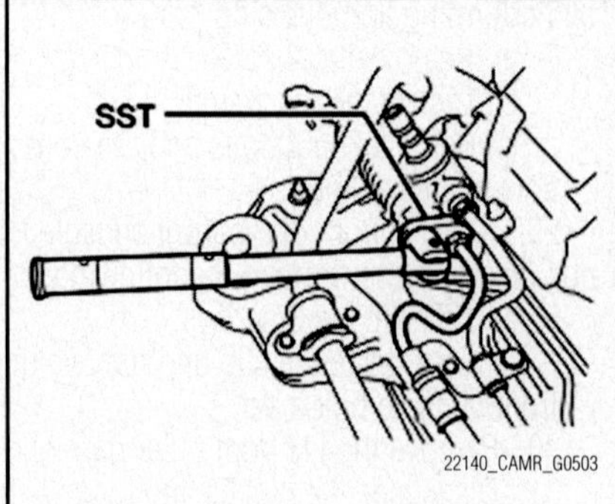

Fig. 289 Disconnect the pressure feed tube assembly — pressure feed tube side

the tie rod assembly LH from the steering knuckle.

7. Remove engine assembly with transaxle.

8. Disconnect pressure feed tube assembly, as follows:

a. Using SST: 09023-12701, disconnect the pressure feed tube assembly (return tube side) from the power steering link assembly.

b. Using SST: 09023-12701, disconnect the pressure feed tube assembly (pressure feed tube side) from the power steering link assembly.

c. Remove the 2 bolts and separate the pressure feed tube clamp.

➡ **Because the nut has its own stopper, do not turn the nut. Loosen the bolt with the nut fixed.**

9. Remove the 2 bolts, 2 nuts, and the power steering link assembly.

10. If equipped, remove the power steering rack housing heat insulator from the power steering link assembly.

To install:

11. If equipped, install the power steering rack housing heat insulator to the power steering link assembly.

12. Install the power steering link assembly with the 2 bolts and 2 nuts. Tighten to 52 ft. lbs. (70 Nm).

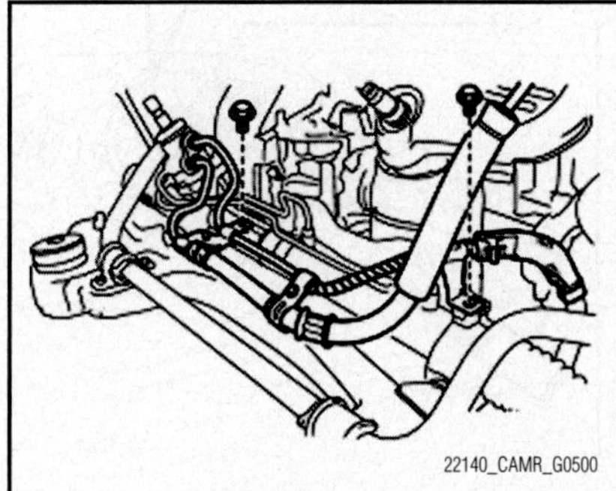

Fig. 290 Removing the 2 bolts and pressure feed tube clamp

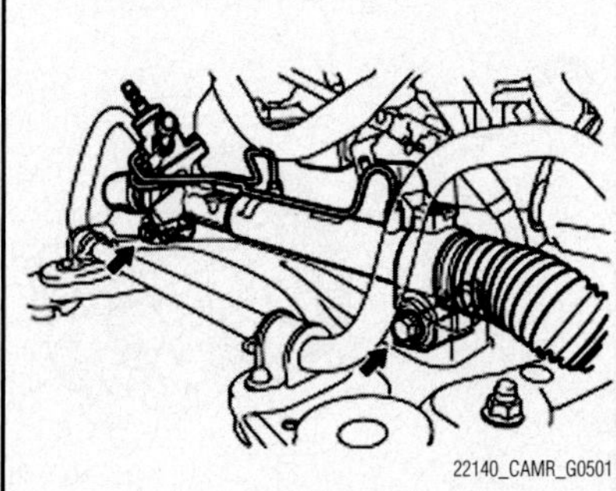

Fig. 291 Removing power steering link assembly

13. Connect pressure feed tube assembly, as follows:

a. Temporarily connect the pressure feed tube assembly to the power steering link assembly.

b. Install the pressure feed tube assembly clamp with the 2 bolts and tighten to 87 inch lbs. (9.8 Nm).

➡ **Use a torque wrench with a fulcrum length of 11.81 inches (300 mm).**

➡ **This torque value is effective when SST is parallel to the torque wrench.**

c. Using SST: 09023-12701, tighten the pressure feed tube assembly (pressure feed tube side) to 16 ft. lbs. (22 Nm).

d. Using SST: 09023-12701, tighten the pressure feed tube assembly (return tube side) to 16 ft. lbs. (22 Nm).

14. Install engine assembly with transaxle.

15. Connect both of the tie rod assemblies to the steering knuckle with the nut. Tighten to 36 ft. lbs. (49 Nm). Install a new cotter pin. Further tighten the nut up to 60° if the holes for the cotter pin are not aligned.

16. Connect steering sliding yoke, as follows:

a. Align the matchmarks on the steering sliding yoke and the steering link assembly.

b. Install the bolt to 26 ft. lbs. (35 Nm).

17. Install front wheels and tighten to 76 ft. lbs. (103 Nm).

18. Connect cable to negative battery terminal.

19. Bleed power steering fluid.

20. Check power steering fluid level.

21. Check for power steering fluid leakage.

22. Check for exhaust gas leaks.

23. Place front wheels facing straight ahead.

24. Inspect and adjust front wheel alignment.

POWER STEERING PUMP

REMOVAL & INSTALLATION

2.4L Engine

See Figures 292 through 295.

1. Before servicing the vehicle, refer to the precautions section.

2. Drain power steering fluid.

3. Remove RH engine under cover.

4. Remove RH front fender apron seal.

5. Remove fan and generator v belt.

6. Slide the clip and disconnect the No. 1 fluid reservoir to pump hose from the vane pump assembly.

7. Disconnect pressure feed tube assembly, as follows:

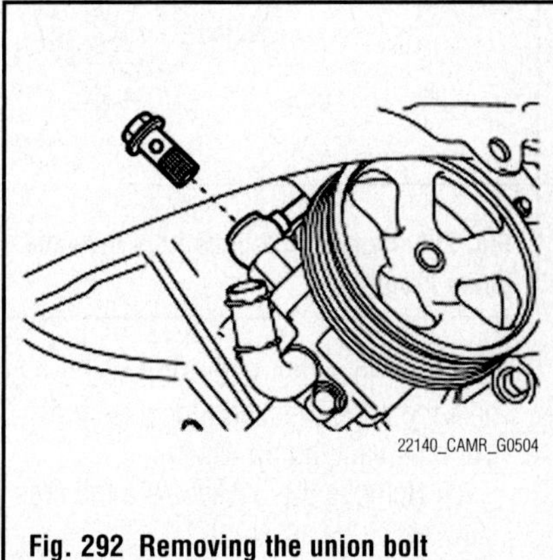

Fig. 292 Removing the union bolt

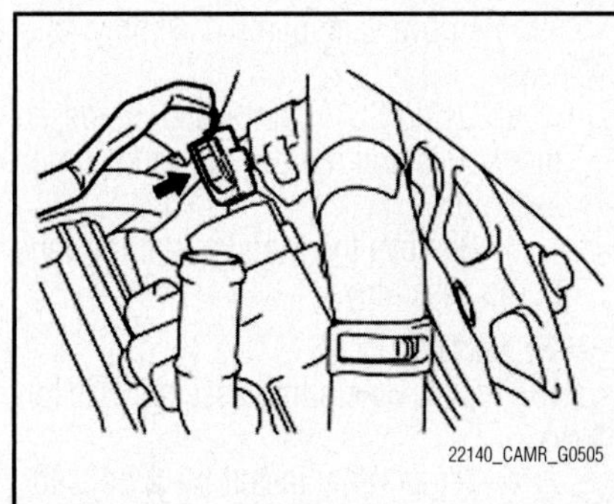

Fig. 293 Disconnecting the power steering fluid pressure switch connector

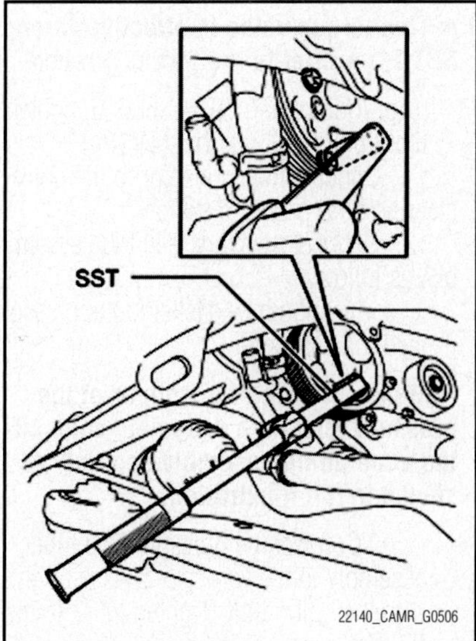

Fig. 294 Removing the vane pump assembly

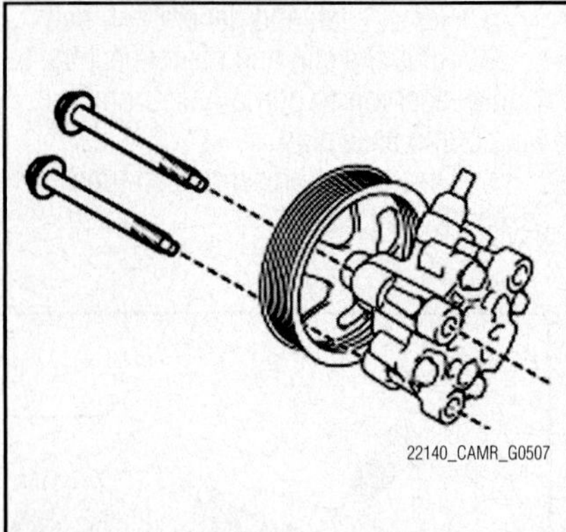

Fig. 295 Removing 2 bolts from the vane pump assembly

a. Remove the union bolt and disconnect the pressure feed tube assembly from the vane pump assembly.

b. Remove the gasket from the pressure feed tube assembly.

8. Disconnect the power steering fluid pressure switch connector.

9. Remove vane pump assembly, as follows:

a. Using SST: 09249-63010, loosen the 2 bolts and remove the vane pump assembly.

b. Remove the 2 bolts from the vane pump assembly.

To install:

10. Install vane pump assembly, as follows:

a. Temporarily install the 2 bolts to the vane pump assembly.

b. Install the vane pump assembly.

➡ Use a torque wrench with a fulcrum length of 11.81 inches (300 mm).

➡This torque value is effective when SST is parallel to the torque wrench.

c. Using SST: 09249-63010, tighten the 2 bolts to 32 ft. lbs. (43 Nm).

11. Connect the connector to the power steering fluid pressure switch.

12. Connect pressure feed tube assembly, as follows:

a. Install a new gasket to the pressure feed tube assembly.

➡Make sure that the stopper of the pressure feed tube assembly contacts the vane pump assembly securely as shown in the illustration.

b. Connect the pressure feed tube assembly to the vane pump assembly with the union bolt. Tighten to 37 ft. lbs. (50 Nm).

13. Connect No. 1 fluid reservoir to pump hose, as follows:

➡Connect the No. 1 oil reservoir to pump hose with the paint mark facing toward the rear of the vehicle.

➡Push the No. 1 oil reservoir to pump hose as far as it will go.

➡Install the clip at the position specified in the illustration.

a. Connect the No. 1 fluid reservoir to pump hose to the vane pump assembly with the clip.

14. To complete installation, reverse removal procedure.

3.5L Engine

See Figures 296 through 299.

1. Before servicing the vehicle, refer to the precautions section.
2. Drain power steering fluid.
3. Remove RH engine under cover.
4. Remove RH front fender apron seal.
5. Remove v-bank cover sub-assembly.
6. Remove fan and generator v belt.
7. Slide the clip and disconnect the No. 1 fluid reservoir to pump hose from the vane pump assembly.
8. Disconnect pressure feed tube assembly, as follows:

a. Remove the union bolt and disconnect the pressure feed tube assembly from the vane pump assembly.

b. Remove the bolt and separate the pressure feed tube clamp.

c. Remove the gasket from the pressure feed tube assembly.

9. Disconnect the power steering fluid pressure switch connector.

10. Using SST: 09249-63010, loosen bolt (A) and remove bolt (B), and then remove the vane pump assembly.

11. Remove the bolt from the vane pump assembly.

To install:

12. Install vane pump assembly, as follows:

a. Temporarily install the bolt to the vane pump assembly.

b. Install the vane pump assembly.

➡ Use a torque wrench with a fulcrum length of 11.81 inches (300 mm).

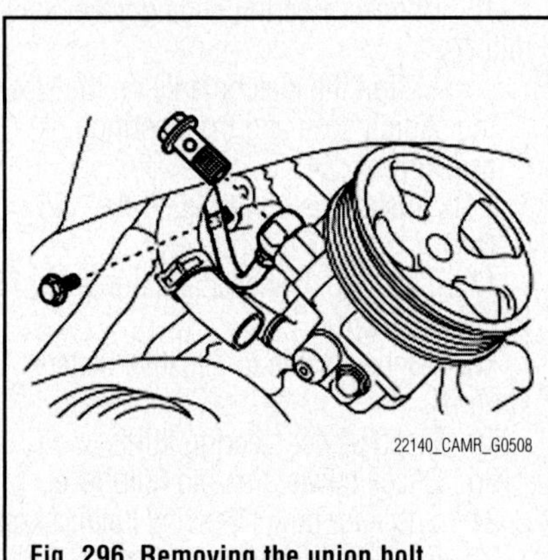

Fig. 296 Removing the union bolt

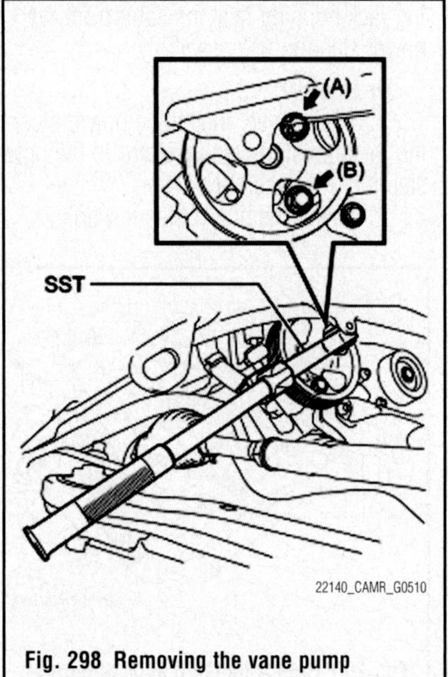

Fig. 298 Removing the vane pump assembly

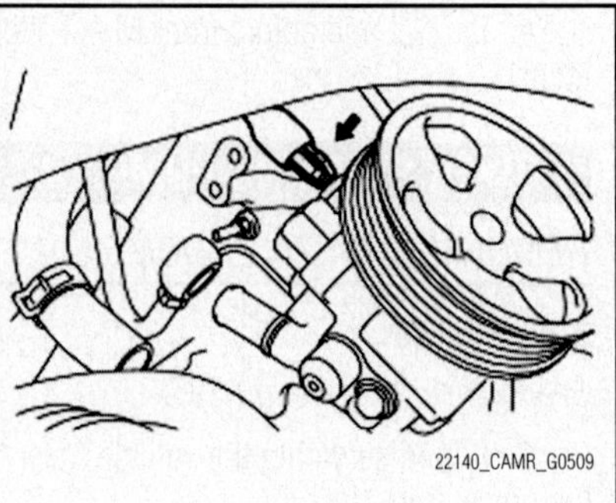

Fig. 297 Disconnecting the power steering fluid pressure switch connector

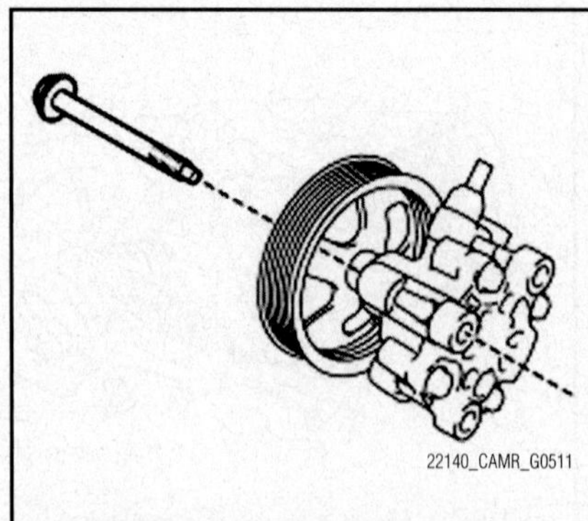

Fig. 299 Removing the bolt from the vane pump assembly

➡**This torque value is effective when SST is parallel to the torque wrench.**

c. Using SST: 09249-63010, tighten the 2 bolts to 32 ft. lbs. (43 Nm).

13. Connect the connector to the power steering fluid pressure switch.

14. Connect pressure feed tube assembly, as follows:

a. Install a new gasket to the pressure feed tube assembly.

b. Temporarily connect the pressure feed tube assembly to the vane pump assembly with the union bolt.

c. Install the pressure feed tube assembly clamp with the bolt. Tighten to 87 ft. lbs. (10 Nm).

d. Fully tighten the union bolt and tighten to 37 ft. lbs. (50 Nm).

➡**Make sure that the stopper of the pressure feed tube assembly contacts the vane pump assembly securely.**

15. Connect No. 1 fluid reservoir to pump hose, as follows:

➡**Connect the No. 1 oil reservoir to pump hose with the paint mark facing toward the rear of the vehicle.**

➡**Push the No. 1 oil reservoir to pump hose as far as it will go as shown in the illustration.**

➡**Install the clip at the position specified in the illustration.**

a. Connect the No. 1 fluid reservoir to pump hose to the vane pump assembly with the clip.

16. To complete installation, reverse removal procedure.

BLEEDING

1. Before servicing the vehicle, refer to the precautions section.

2. Check the fluid level.

3. Jack up the front of the vehicle and support it with stands.

4. With the engine stopped, turn the wheel slowly from lock to lock several times.

5. Lower the vehicle.

6. Start the engine.

7. Run the engine at idle for a few minutes.

8. With the engine idling, turn the wheel left or right to the full lock position and keep it there for 2 to 3 seconds, then turn the wheel to the opposite full lock position and keep it there for 2 to 3 seconds.

9. Repeat the above steps several times.

10. Stop the engine.

11. Check for foaming or emulsification. If the system has to be bled twice because of foaming or emulsification, check for fluid leaks in the system.

12. Check the fluid level.

SUSPENSION

COIL SPRING

REMOVAL & INSTALLATION

See Figures 300 through 305.

1. Before servicing the vehicle, refer to the precautions section.

2. Remove the front shock absorber.

3. As shown in the illustration, secure the front shock absorber with coil spring in a vise using aluminum plates by clamping onto a double nutted bolt affixed to the bracket at the bottom of the absorber.

➡ **Do not use an impact wrench.**

➡ **If the front coil spring is compressed at an angle, using 2 SST will make the work easier.**

4. Using SST: 09727-30021, compress the front coil spring.

5. Remove the front suspension support sub-assembly, front suspension support bearing, front coil spring upper seat, front coil spring upper insulator, front coil spring, front spring bumper, and front coil spring lower insulator from the front shock absorber.

To install:

6. Install front coil spring as follows:

a. Install the front spring bumper to the piston rod.

➡**Align the 2 protrusions of the front coil spring lower insulator and the 2 holes in the front shock absorber.**

➡**Do not use an impact wrench.**

b. Install the front coil spring lower insulator onto the front shock absorber.

c. Using SST: 09727-30021, compress the front coil spring.

➡**The smaller diameter end of the front coil spring must face upward.**

➡**Fit the lower end of the front coil spring into the gap of the insulator.**

d. Install the front coil spring to the front shock absorber.

➡**Any misalignment between the front shock absorber lower bracket and the matchmark must be +/-5°.**

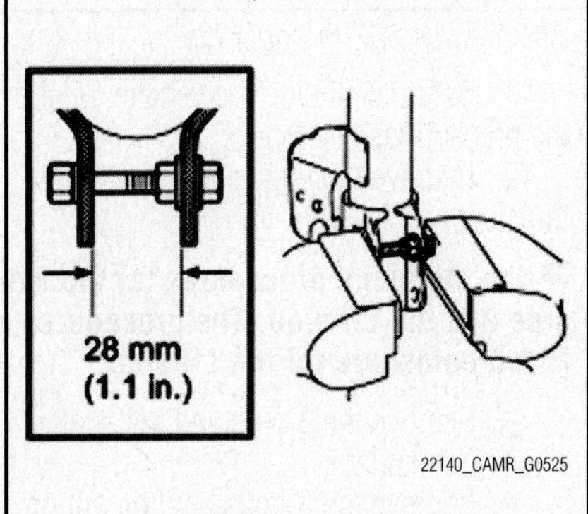

Fig. 300 Secure the front shock absorber

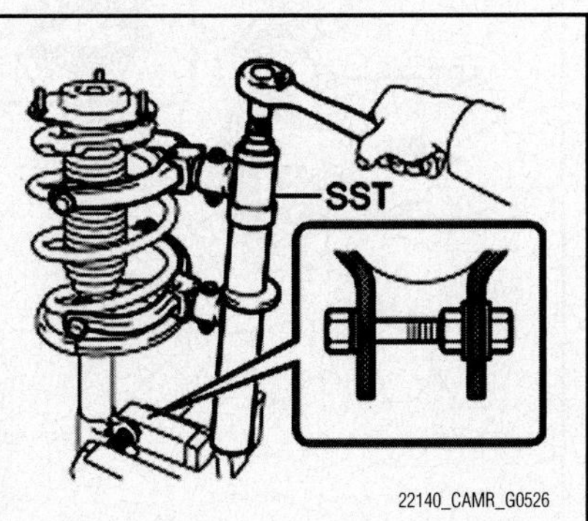

Fig. 301 Removing coil spring components

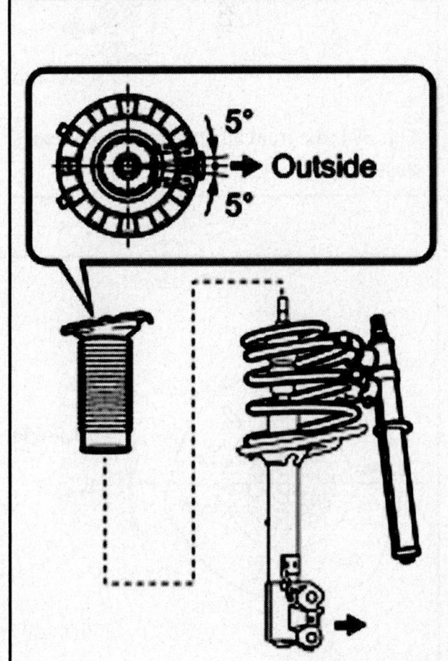

Fig. 302 Installing the front coil spring upper insulator

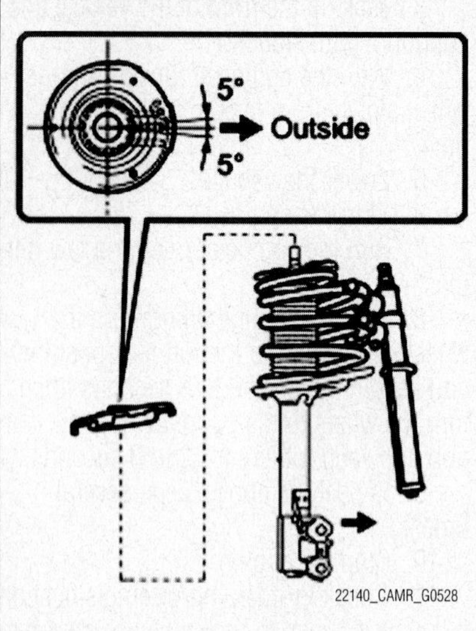

Fig. 303 Installing the front coil spring upper seat

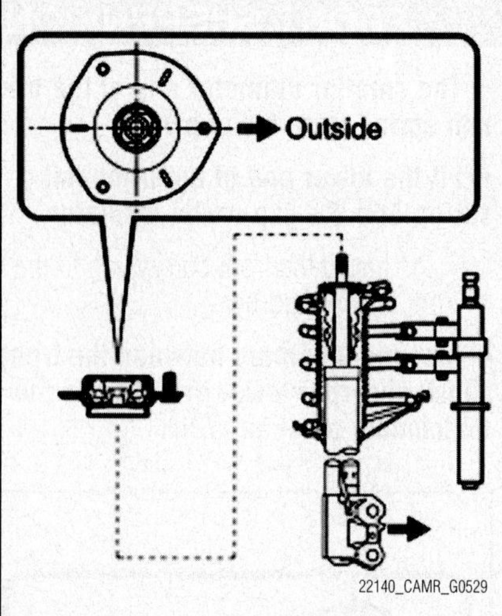

Fig. 304 the front suspension support sub-assembly

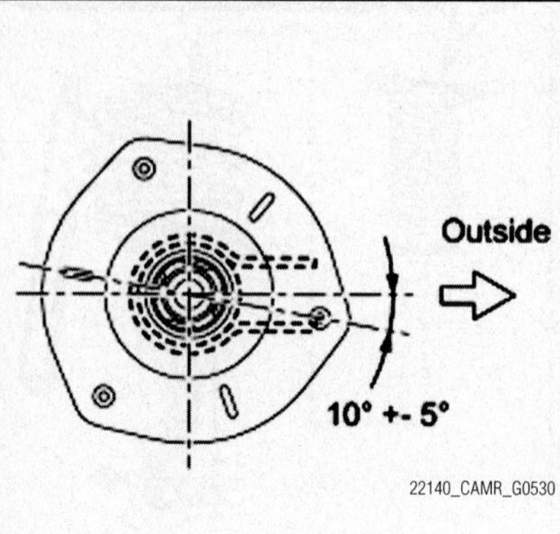

Fig. 305 Aligning the front shock absorber lower bracket and arrows

e. Install the front coil spring upper insulator as shown in the illustration.

➡Any misalignment between the front shock absorber lower bracket and the matchmark must be +/-5°.

7. Install the front coil spring upper seat with the mark facing to the outside of the vehicle.

➡ If there is foreign matter inside the front suspension support bearing, replace it with a new one.

a. Install a new front suspension support bearing.

➡ Check that the flats on the piston rod and the flats on the front suspension support sub-assembly are aligned.

b. Install the front suspension support sub-assembly. Temporarily tighten a new lock nut.

➡ Do not use an impact wrench.

➡Any misalignment between the front shock absorber lower bracket and the matchmark must be +/-5°.

c. Remove the SST slowly in order to release the coil spring.

CONTROL LINKS

REMOVAL & INSTALLATION

See Front Stabilizer Bar.

LOWER BALL JOINT

REMOVAL & INSTALLATION

See Figure 306.

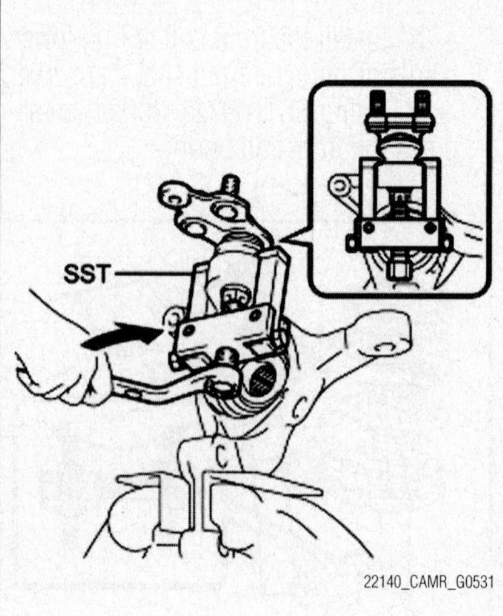

Fig. 306 Remove the front lower ball joint assembly

1. Before servicing the vehicle, refer to the precautions section.
2. Remove the front wheel.
3. Remove the front axle hub nut.
4. Separate the front speed sensor.
5. Separate the front disc the brake caliper assembly.
6. Remove front disc.
7. Separate the tie rod assembly.
8. Separate the No. 1 front lower suspension arm.
9. Remove the front axle assembly.
10. Remove front wheel No. 1 bearing dust deflector.
11. Remove front axle hub hole snap ring.
12. Remove front axle hub.
13. Remove front disc brake dust cover.
14. Remove the front lower ball joint assembly, as follows:
   a. Secure the steering knuckle in a vise using aluminum plates.
   b. Remove the cotter pin and castle nut.

➡Do not damage the dust cover of the ball joint.

➡Do not damage the steering knuckle.

   c. Using SST (SST: 09628-62011) or equivalent, remove the front lower ball joint assembly.

To install:

15. Installation is the reverse of the removal procedure, noting the following:
   a. Install the front lower ball joint assembly to the steering knuckle with the castle nut and tighten to 91 ft. lbs. (123 Nm). Further tighten the nut up to 60° if the holes for the cotter pin are not aligned.
   b. Inspect and adjust the front wheel alignment.
   c. Inspect the ABS speed sensor signal.

LOWER CONTROL ARM

REMOVAL & INSTALLATION

See Figures 307 through 309.

1. Before servicing the vehicle, refer to the precautions section.
2. Remove the engine assembly with transaxle.

➡Use the same procedures for the RH side and the LH side. The procedures listed below are for the LH side.

3. Remove the 3 nuts and the engine mounting insulator.
4. Remove the 3 bolts and the nut on the front suspension lower No. 1 arm and remove it from the front frame assembly.

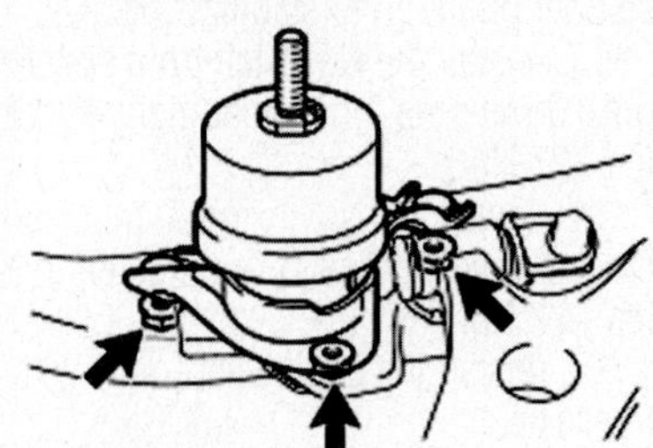

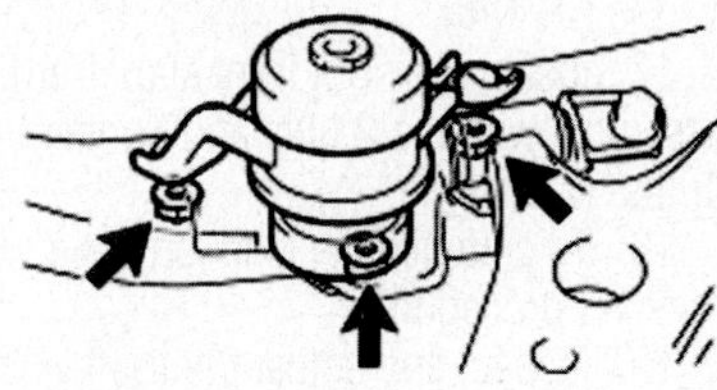

Fig. 307 Removing the 3 nuts and the engine mounting insulator

Fig. 308 Removing the 3 bolts and the nut on the front suspension lower No. 1 arm

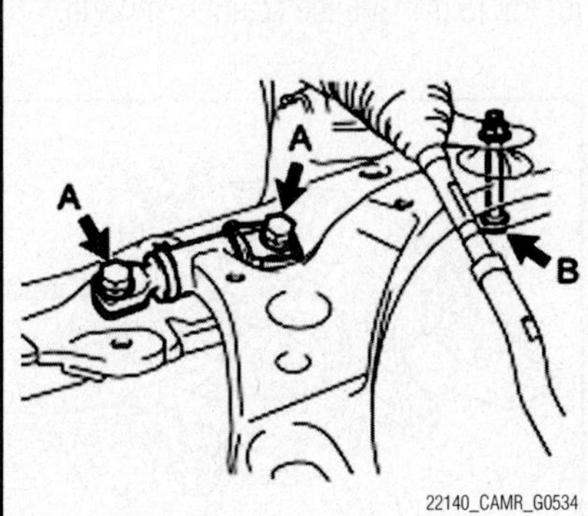

Fig. 309 Tightening the 3 bolts and the nut on the front suspension lower No. 1 arm

5. Remove the front lower arm bushing stopper.

To install:

6. Install the front lower arm bushing stopper.
7. Install the front suspension lower No. 1 arm to the front frame assembly with the 3 bolts and the nut, but do not tighten them yet.
8. Tighten bolts "A" to 148 ft. lbs. (200 Nm). Tighten bolts "B" to 152 ft. lbs. (206 Nm).
9. Install the engine mounting insulator with the 3 nuts and tighten to 64 ft. lbs. (87 Nm).
10. Install the engine assembly with transaxle.

SHOCK ABSORBERS

REMOVAL & INSTALLATION

See Figures 310 through 312.

➡Use the same procedures for the RH side and the LH side. The procedures listed below are for the LH side.

1. Before servicing the vehicle, refer to the precautions section.
2. Remove the front wheel.
3. Remove the nut and disconnect the front stabilizer link assembly from the front shock absorber assembly.
4. Remove front shock absorber with coil spring, as follows:
   a. Loosen the lock nut of the front shock absorber with coil spring.

➡Do not remove the lock nut.

➡Only loosen the nut when disassembling the front shock absorber with coil spring.

   b. Remove the bolt and disconnect the front flexible hose and front speed sensor wire harness from the front shock absorber with coil spring.

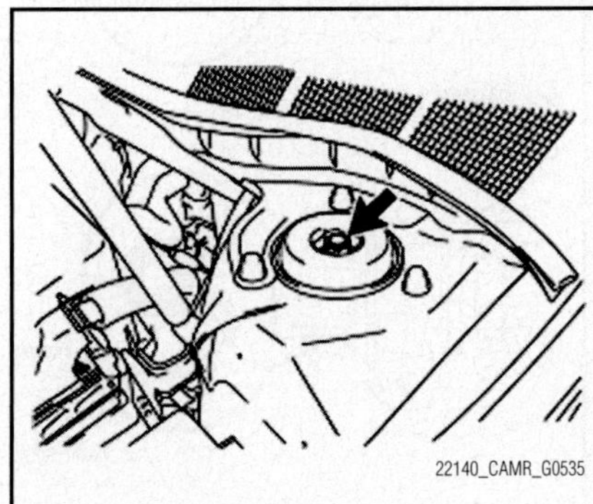

Fig. 310 Loosening the lock nut of the front shock absorber

➡Be sure to remove the front speed sensor from the front shock absorber with coil spring.

   c. Remove the 2 nuts on the lower side of the front shock absorber with coil spring.

➡When removing the nuts, keep the bolts from rotating.

➡Keep the bolts inserted to secure the front axle assembly.

   d. Remove the 3 nuts on the upper side of the front shock absorber with coil spring.
   e. Lower the front axle assembly, and remove the 2 bolts on the lower side of the front shock absorber.

➡Make sure that the front speed sensor is disconnected from the front shock absorber with coil spring.

   f. Remove the front shock absorber with coil spring.

To install:

5. Install front shock absorber with coil spring, as follows:
   a. Install the front shock absorber with coil spring to the front axle assembly and insert the 2 bolts from the front side of the vehicle.

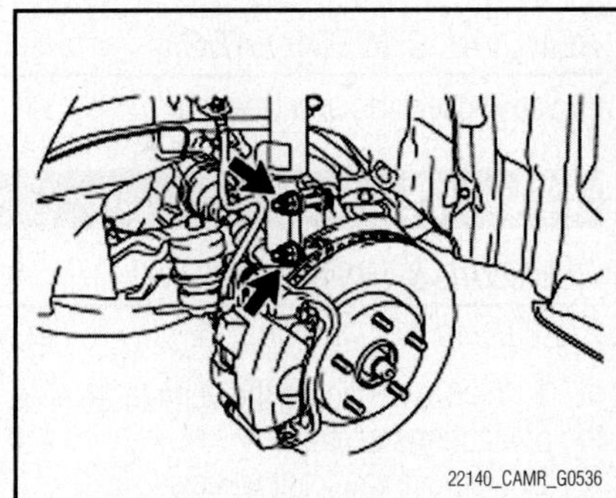

Fig. 311 Remove the 2 nuts on the lower side of the front shock absorber

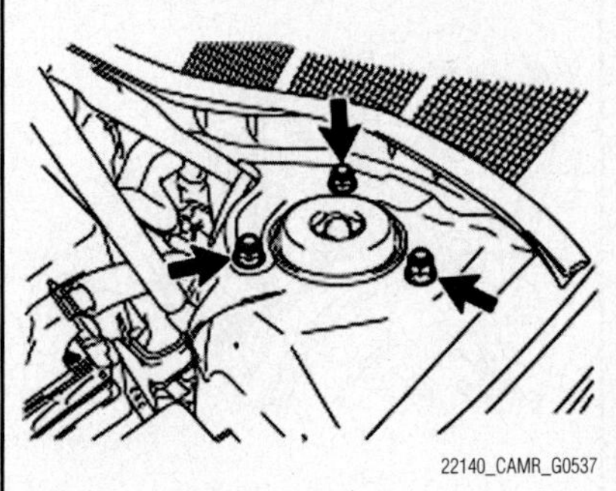

Fig. 312 Removing the 3 nuts on the upper side of the front shock absorber

b. Slowly jack up the vehicle using a wooden block and install the front shock absorber with coil spring (upper side) to the vehicle.

c. Install the 3 nuts to the upper side of the front shock absorber with coil spring and tighten to 63 ft. lbs. (85 Nm).

➡When installing the nuts, keep the bolts from rotating.

d. Install the 2 nuts to the lower side of the front shock absorber with coil spring and tighten to 155 ft. lbs. (210 Nm).

e. Install the front flexible hose and front speed sensor wire harness with the bolt and tighten to 14 ft. lbs. (19 Nm).

f. Fully tighten the lock nut and tighten to 52 ft. lbs. (70 Nm).

➡If the ball joint turns together with the nut, use a hexagon wrench (6 mm) to hold the stud.

6. Install the front stabilizer link assembly with the nut and tighten to 55 ft. lbs. (74 Nm).
7. Install front wheel and tighten to 76 ft. lbs. (103 Nm).
8. Inspect and adjust front wheel alignment.

STEERING KNUCKLE

REMOVAL & INSTALLATION

See Wheel Hub and Bearing.

STABILIZER BAR

REMOVAL & INSTALLATION

See Figures 313 through 315.

1. Before servicing the vehicle, refer to the precautions section.
2. Remove the front wheels.
3. Separate steering intermediate shaft assembly.
4. Separate tie rod end sub-assembly.

Fig. 313 Remove the 2 nuts and front stabilizer link assembly (left hand shown)

➡If the ball joint turns together with the nut, use a hexagon wrench (6 mm) to hold the stud.

5. Remove the 2 nuts and the front stabilizer link assembly
6. Remove the engine assembly with transaxle.
7. Remove the bolts and the left and right No. 1 stabilizer brackets.
8. Remove the engine assembly with transaxle.
9. Remove the bolts and the left and right No. 1 stabilizer brackets.
10. Remove the 2 front No. 1 stabilizer bar bushings from the front stabilizer bar.
11. Remove the front stabilizer bar from the vehicle.

To install:

➡Make sure that the cutout of the front stabilizer bar bushing No. 1 faces the rear side as shown in the illustration.

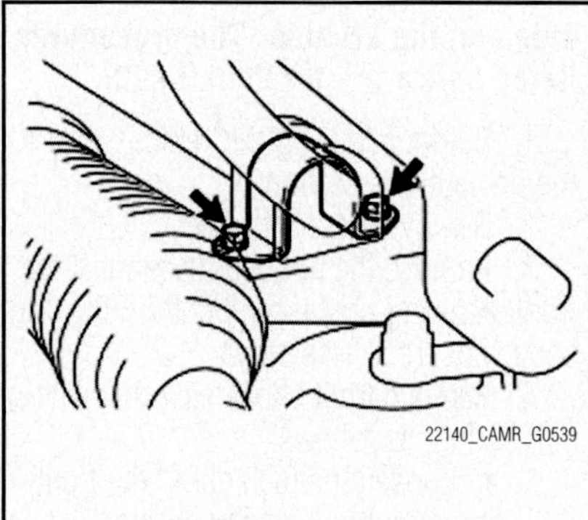

Fig. 314 Remove the 2 bolts and No. 1 stabilizer bracket (left hand shown)

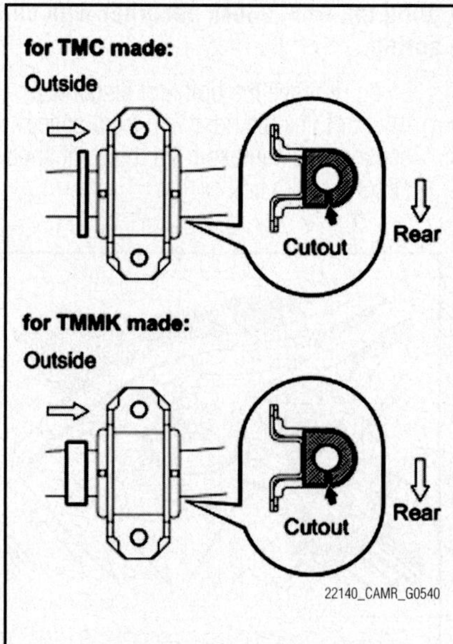

Fig. 315 Installing the 2 front stabilizer bar bushings

12. Install the 2 front stabilizer bar bushings No. 1 to the outside of the bushing stopper on the front stabilizer bar.
13. Install the No. 1 left front stabilizer bracket with the 2 bolts and tighten to 20 ft. lbs. (27 Nm).
14. Install the No. 1 right front stabilizer bracket with the 2 bolts and tighten to 20 ft. lbs. (27 Nm).
15. Install the engine assembly with transaxle.
16. Install the left front stabilizer link assembly with the 2 nuts and tighten to 55 ft. lbs. (74 Nm).
17. Install the right front stabilizer link assembly with the 2 nuts and tighten to 55 ft. lbs. (74 Nm).
18. To complete installation, reverse removal procedure.
19. Inspect and adjust the front wheel alignment.

WHEEL HUB AND BEARING

REMOVAL & INSTALLATION

See Figures 316 through 322.

1. Before servicing the vehicle, refer to the Precautions Section.
2. Remove front wheel.
3. Remove front axle hub nut.
4. Separate front speed sensor.
5. Remove the 2 bolts and separate the front disc brake caliper assembly from the steering knuckle. Use wire or an equivalent tool to keep the brake caliper from hanging down by the flexible hose.
6. Remove front disc.
7. Separate tie rod end sub-assembly.
8. Separate front suspension lower no. 1 arm.
9. Remove front axle assembly.
10. Using a screwdriver with its tip wrapped with vinyl tape, remove the No. 1 front wheel bearing dust deflector. Be careful not to damage the steering knuckle.

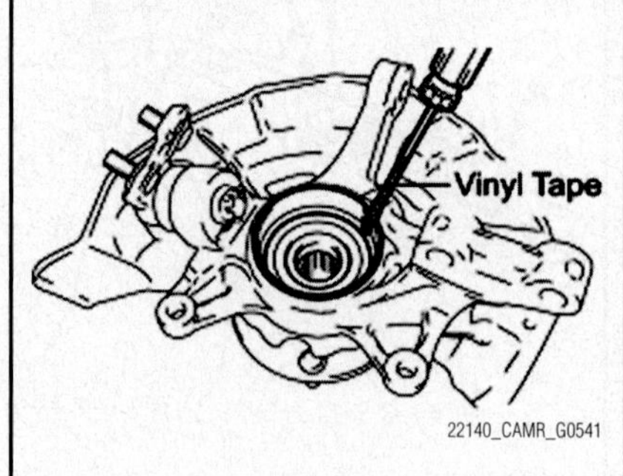

Fig. 316 Remove the No. 1 front wheel bearing dust deflector

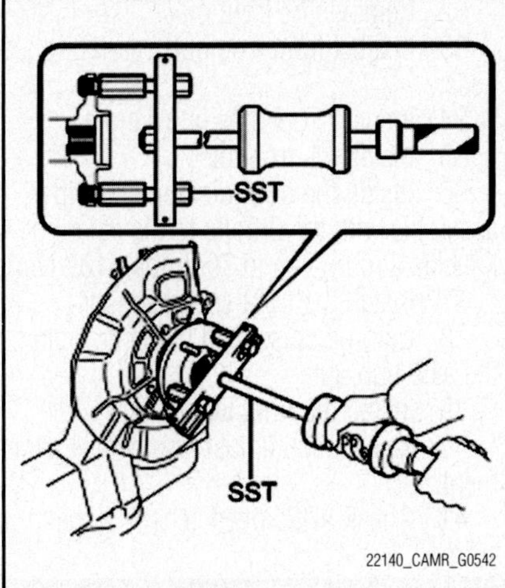

Fig. 317 Remove the front axle hub sub-assembly

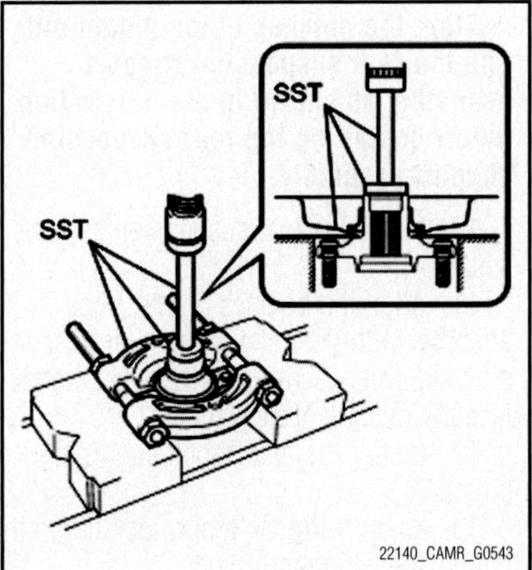

Fig. 318 Remove the bearing inner race (outside) from the front axle hub sub-assembly

11. Using snap ring pliers, remove the front axle hub hole snap ring.

12. Remove front axle hub sub-assembly by performing the following:

a. Hold the front axle assembly between aluminum plates in a vise.

➡Do not overtighten the vise.

b. Using SST 09520-00031, remove the front axle hub sub-assembly.

➡Be careful not to drop the front axle hub sub-assembly.

c. Using SST 09555-55010, SST: 09950-60010 and SST: 09950-70010 and a press, remove the bearing inner race (outside) from the front axle hub sub-assembly.

13. Remove the 4 bolts and disc brake dust cover from the steering knuckle.

14. Remove front lower ball joint assembly.

15. Remove front axle hub bearing by performing the following:

a. Place the bearing inner race (outside) on the front axle hub bearing.

b. Using SST 09527-17011, SST: 09950-60010 and a press, press the front axle hub bearing until it contacts the SST: 09950-70010.

c. Using SST: 09527-20011, SST: 09950-60010 to make the steering knuckle horizontal, fix it to the V-block.

16. Using SST : 09950-70010 and a press, remove the front axle hub bearing from the steering knuckle.

To install:

17. Using SST's: 09950-60020, 09950-70010 and a press, install a new front axle hub bearing to the steering knuckle.

18. Install front lower ball joint assembly.

19. Install the disc brake dust cover to the steering knuckle with the 4 bolts and tighten to 73 inch lbs. (8.3 Nm).

20. Using SST's: 09608-32010, 09950-60020, 09950-70010 and a press, install the front axle hub sub-assembly.

21. Using snap ring pliers, install a new front axle hub hole snap ring.

➡ Align the hole for the speed sensor in the No. 1 front wheel bearing dust deflector with the steering knuckle.

22. Using SST's: 09316-60011, 09608-32010 and a hammer, install a new No. 1 front wheel bearing dust deflector.

➡Only when reusing the bolts and nuts, apply the small amount of engine oil to the screw part of the nuts.

➡Be careful not to damage the drive shaft boot or speed sensor rotor.

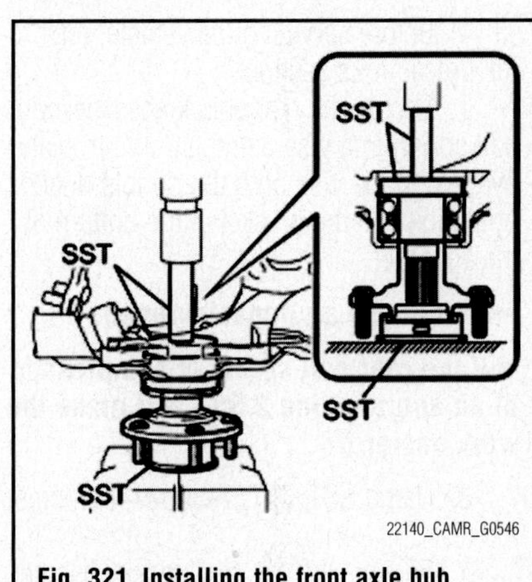

Fig. 321 Installing the front axle hub sub-assembly

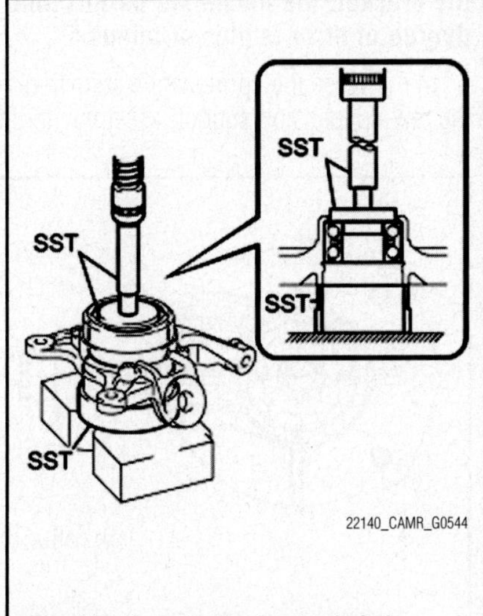

Fig. 319 Pressing the front axle hub bearing

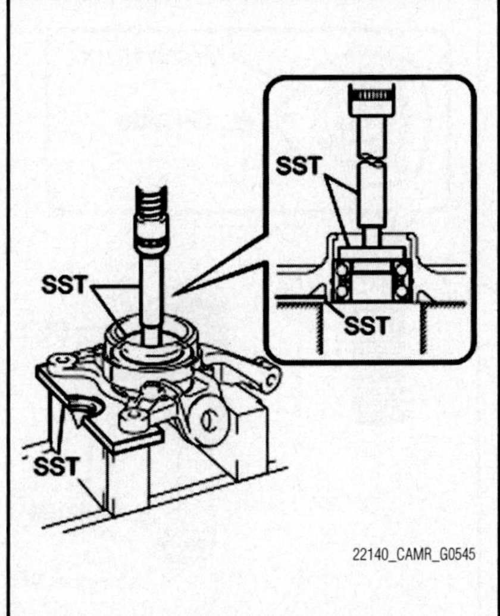

Fig. 320 Removing the front axle hub bearing

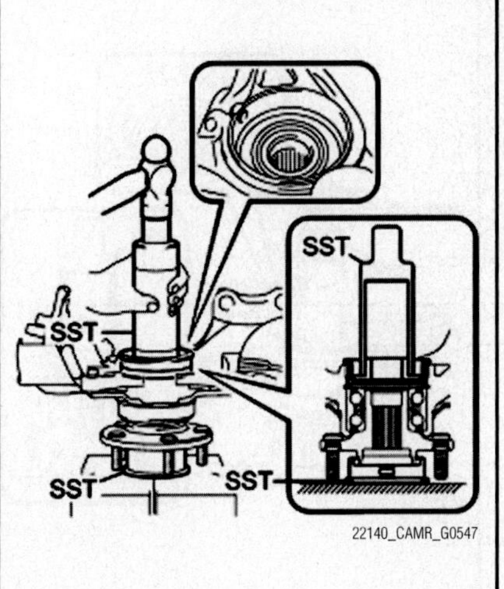

Fig. 322 Installing No. 1 front wheel bearing dust deflector

23. Align the matchmarks and install the front drive shaft assembly to the front axle hub sub-assembly.

24. Install the steering knuckle with the front axle hub sub-assembly to the front shock absorber assembly with the 2 bolts and 2 nuts and tighten to 155 ft. lbs. (210 Nm).

25. Install the lower No. 1 front suspension arm sub-assembly.

26. Install the tie rod end sub-assembly.

27. Install the front disc.

28. Install the front disc brake caliper assembly with the 2 bolts to the steering knuckle and tighten to 79 ft. lbs. (107 Nm).

29. Clean the threaded parts on the drive shaft and axle hub nut using a non-residue solvent.

➡Be sure to perform this work for a new drive shaft.

➡Keep the threaded parts free of oil and foreign objects.

30. Using a 30 mm socket wrench, install the front axle hub nut and tighten to 217 ft. lbs. (294 Nm).

31. Remove the 2 bolts and separate the front disc brake caliper assembly from the steering knuckle.

32. Remove the front disc.

33. Inspect front axle hub bearing looseness.

34. Inspect front axle hub runout.

35. Install the front disc.

36. Install the front disc brake caliper assembly with the 2 bolts to the steering knuckle and tighten to 79 ft. lbs. (107 Nm).

37. Install the front speed sensor.

38. Using a chisel and hammer, stake the axle hub nut.

39. Install the front wheel.

40. Inspect and adjust front wheel alignment.

41. Check ABS speed sensor signal.

SUSPENSION

REAR SUSPENSION

COIL SPRING

REMOVAL & INSTALLATION

See Figures 323 through 325.

1. Before servicing the vehicle, refer to the precautions section.

2. Secure the rear shock absorber with coil spring in a vise using aluminum plates by closing the vise onto the double nutted bolt affixed to the bracket at the bottom of the absorber.

➡Do not use an impact wrench.

➡If the rear coil spring is compressed at an angle, using 2 SST will make the work easier.

3. Using SST: 09727-30021, compress the rear coil spring

4. Remove the nut, rear shock absorber collar and rear suspension support assembly.

5. Remove the rear coil spring, rear No. 1 spring bumper, and rear coil spring lower insulator.

To install:

6. Install the rear No. 1 spring bumper to the piston rod.

7. Install the rear coil spring lower insulator onto the rear shock absorber.

➡Do not use an impact wrench.

8. Using SST: 09727-30021, compress the rear coil spring.

➡The smaller diameter end must face upward.

➡Fit the lower end of the rear coil spring into the gap of the lower seat.

➡If the front coil spring is compressed at an angle, using 2 SST will make the work easier.

9. Install the rear coil spring to the rear shock absorber.

➡Align the notches of the piston rod and the rear suspension support assembly as shown in the illustration before installing the rear suspension support assembly.

10. Install the rear suspension support assembly.

11. Align the notches of the shock absorber with the notch of the rear suspension support assembly so that the notches face the outside of the vehicle.

12. Install the rear shock absorber collar.

13. Loosely tighten a new lock nut to the rear suspension piston rod.

➡ Do not use an impact wrench.

➡ When lining up the rear suspension support assembly's stud bolts at the middle point between the two sides of the bracket, the maximum permissible degree of error is plus or minus 5°.

14. Release the spring while adjusting the rear suspension support assembly to the

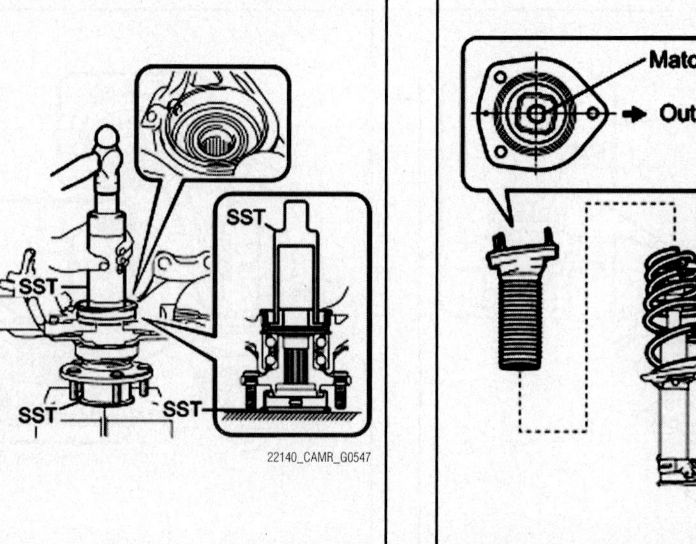

Fig. 323 Remove the rear coil spring

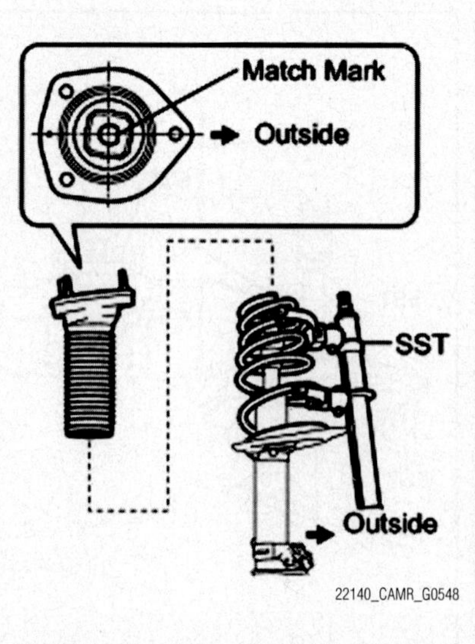

Fig. 324 Align the notches of the shock absorber

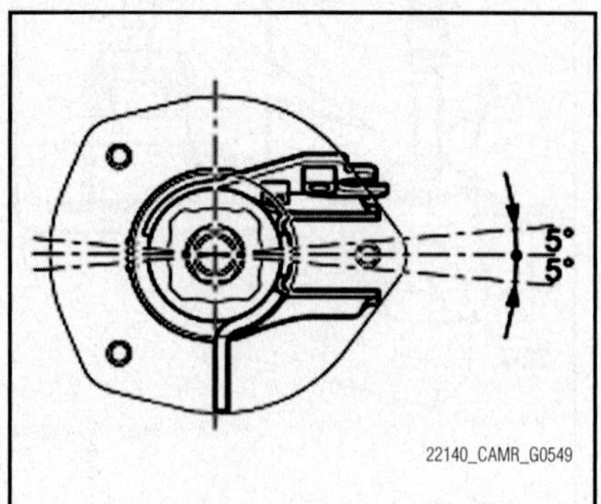

Fig. 325 Lining up the rear suspension support assembly's stud bolts

position shown in the illustration, and remove the SST from the rear coil spring.

CONTROL ARMS/LINKS

REMOVAL & INSTALLATION

No. 1 Suspension Arm

See Figures 326 through 331.

1. Before servicing the vehicle, refer to the precautions section.

➡Check if an old gasket still remains on the pipe. If so, remove it. Also, check if any bolts or nuts are rusted. If so, replace them.

2. Remove rear wheel.
3. Remove center exhaust pipe assembly.
4. Remove tail exhaust pipe assembly.
5. Separate both rear stabilizer link assemblies.
6. Remove rear stabilizer bar no. 2 and no. 1 bracket.
7. Remove rear stabilizer bar.
8. Remove rear stabilizer bushing.
9. Separate rear strut rod.

➡When removing the bolt, keep the nut from rotating.

10. Remove the bolt, nut and separate the rear suspension No. 2 arm (outer side) from the rear axle carrier.

➡ When removing the bolt, keep the nut from rotating.

11. Remove the bolt, nut and the rear No. 1 suspension arm (outer side) from the rear axle carrier.
12. Remove the 2 nuts and the LH rear suspension member lower stopper.
13. Remove the 2 nuts and the RH rear suspension member lower stopper.
14. Support the rear suspension member with a jack.
15. Remove the 2 bolts, and the rear suspension member sub-assembly.
16. Remove the bolt and rear No. 1 suspension arm assembly.

To install:

17. Install the No. 1 rear suspension arm (inner side) with the bolt, and temporarily tighten the bolt.
18. Install the rear No. 1 suspension arm so that the bracket leans toward the front side of the vehicle.
19. Ensure that the paint mark faces the rear side of the vehicle.
20. Set the rear No. 1 suspension arm in the position shown in the illustration, and fully tighten the bolt to 74 ft. lbs. (100 Nm).
21. Raise the rear suspension member with a jack. Install the rear suspension member with the 2 bolts and tighten to 41 ft. lbs. (56 Nm).
22. Install both the rear suspension member lower stoppers with the 2 nuts and tighten to:
    a. Nut A: 41 ft. lbs. (55 Nm).
    b. Nut B: 28 ft. lbs. (38 Nm).

➡Insert the bolt from the front of the vehicle and temporarily install the bolt.

23. Connect the rear No.1 suspension arm (outer side) to the rear axle carrier with the bolt and nut and temporarily tighten the bolt and nut. When temporarily tightening the bolt, keep the nut from rotating.

➡Insert the bolt from the inside of the vehicle and temporarily install the bolt.

24. Connect the strut rod assembly rear to the axle carrier with the bolt and nut and temporarily tighten the bolt. When temporarily tightening the bolt, keep the nut from rotating.
25. Jack up the rear axle carrier, placing a wooden block to avoid damage. Apply load to the suspension so that the installed

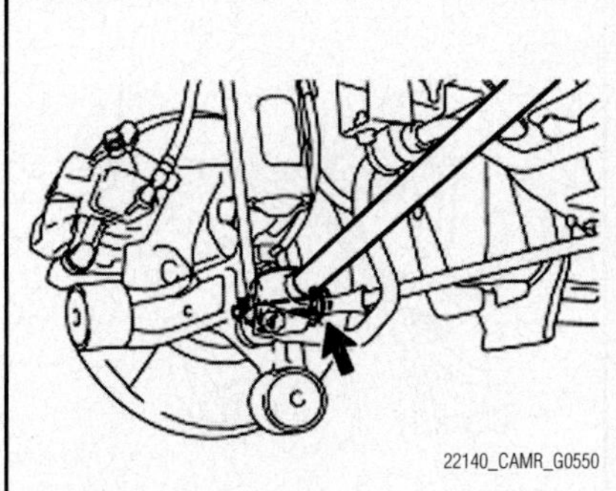

Fig. 326 Removing the bolt, nut and the rear No. 1 suspension arm—LH shown

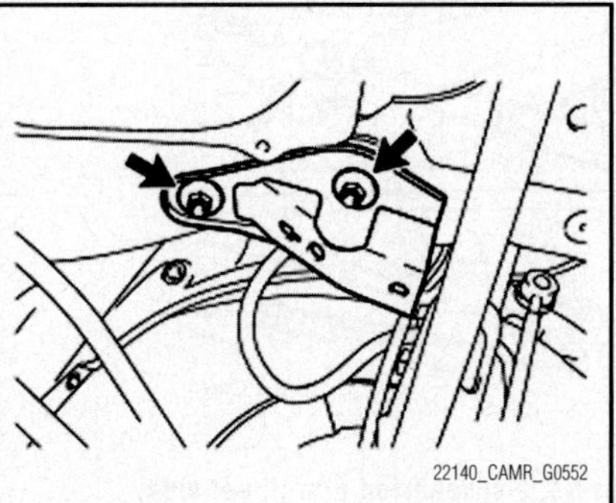

Fig. 328 Removing the 2 nuts and the RH rear suspension member lower stopper

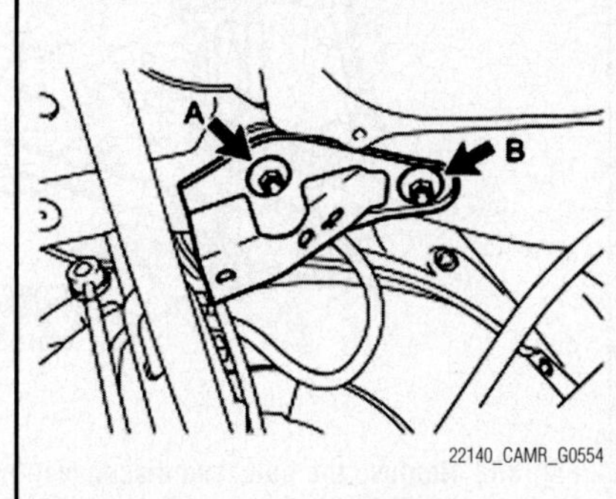

Fig. 330 LH rear suspension member lower stopper tightening sequence

Fig. 327 Removing the 2 nuts and the LH rear suspension member lower stopper

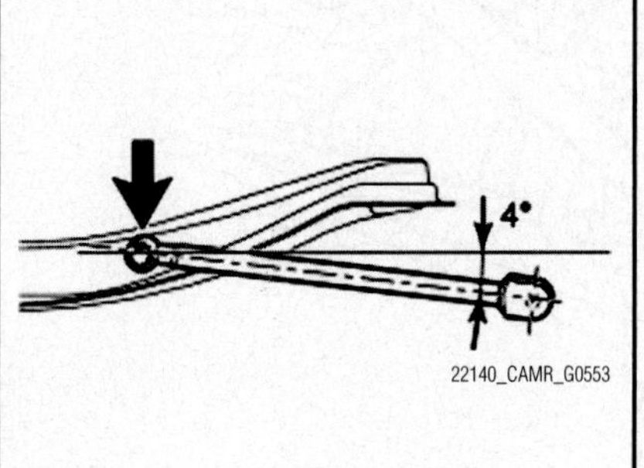

Fig. 329 Set the rear No.1 suspension arm

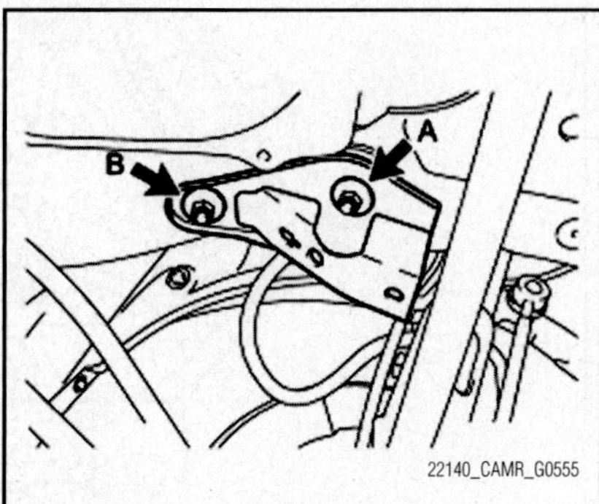

Fig. 331 LH rear suspension member lower stopper tightening sequence

bolt of the rear No. 1 suspension arm (inner side) is horizontally aligned with the center of the rear axle hub.

26. Fully tighten rear No. 1 suspension arm and tighten the bolt to 74 ft. lbs. (100 Nm).

27. Fully tighten rear No. 2 suspension arm and tighten the bolt to 74 ft. lbs. (100 Nm).

28. To complete installation, reverse removal procedure/

No. 2 Suspension Arm

See Figures 332 and 333.

1. Before servicing the vehicle, refer to the precautions section.
2. Remove the rear wheel.
3. Remove the bolt, and disconnect the rear No. 2 suspension arm (inner side).

➡ When removing the bolt, keep the nut from rotating.

4. Remove the bolt, nut and the rear No. 2 suspension arm (outer side) from the rear axle carrier.

To install:

➡Ensure that the paint mark faces to the rear of the vehicle.

5. Install the rear No. 2 suspension arm (inner side) with the bolt, and temporarily tighten the bolt.

➡When temporarily tightening the bolt, keep the nut from rotating.

6. Connect the rear No. 2 suspension arm (outer side) to the rear axle carrier with the bolt and nut, and temporarily tighten the bolt.
7. Stabilize suspension.
8. Fully tighten the rear No. 2 suspension arm bolt (inner side) to 74 ft. lbs. (100 Nm).
9. Fully tighten the rear No. 2 suspension arm bolt (outer side) to 74 ft. lbs. (100 Nm).
10. Install the rear wheel.
11. Inspect and adjust the rear wheel alignment.

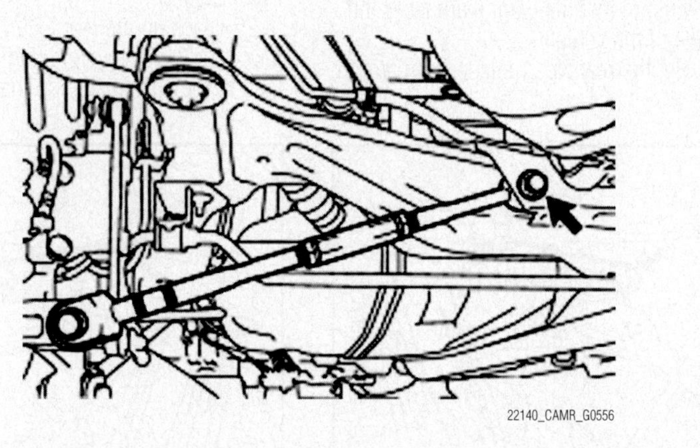

Fig. 332 Remove the bolt, and disconnect the rear No. 2 suspension arm (inner side)

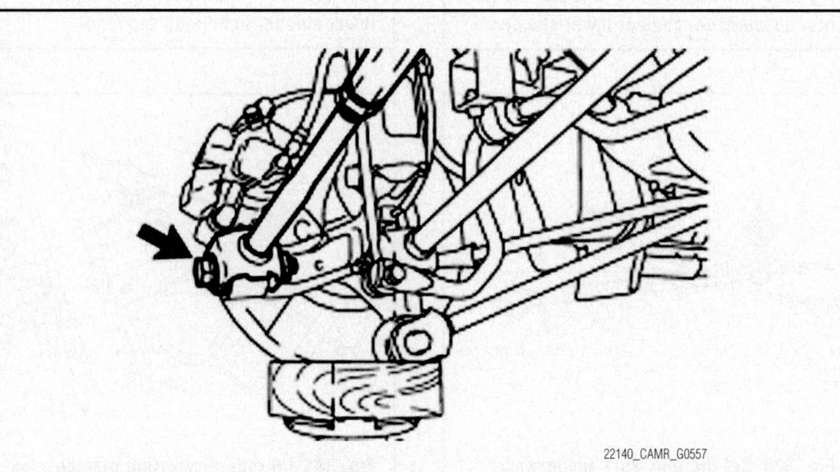

Fig. 333 Removing the bolt, nut and the rear No. 2 suspension arm (outer side)

SHOCK ABSORBER

REMOVAL & INSTALLATION

See Figures 334 and 335.

1. Before servicing the vehicle, refer to the precautions section.
2. Remove the rear seat cushion assembly.
3. Remove rear seat headrest plate cover.
4. Remove rear seat headrest assembly.
5. Remove the rear seatback assembly.
6. Remove the rear wheel.
7. Separate LH rear stabilizer link assembly.
8. Remove the 2 bolts, and disconnect the rear brake flexible hose and rear speed sensor from the rear shock absorber with coil spring and rear axle carrier.
9. Remove the 4 claws and the rear suspension support No. 1 cover.
10. Remove the rear shock absorber with coil spring, as follows:

➡Do not remove the lock nut.

➡Only loosen the nut when disassembling the rear shock absorber with coil spring.

a. Loosen the lock nut of the rear shock absorber with coil spring.

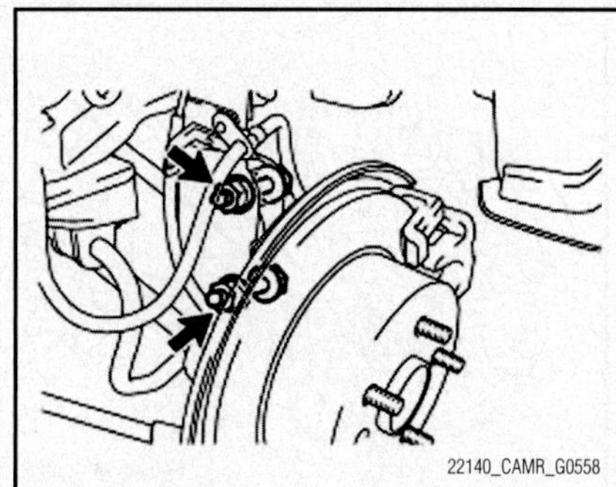

Fig. 334 Loosen the 2 nuts on the lower side of the shock absorber

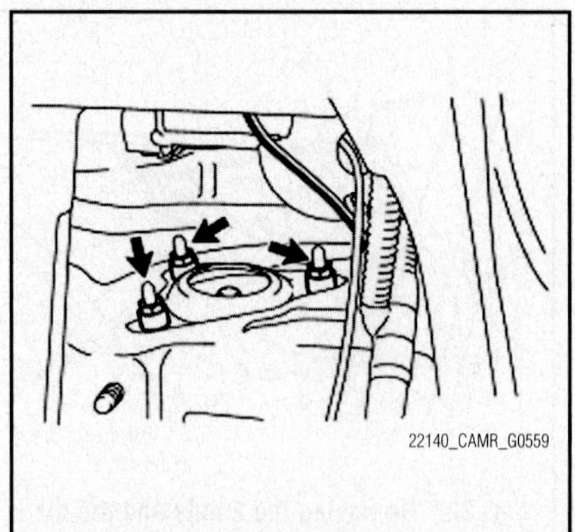

Fig. 335 Remove the 3 nuts

➡**When removing the nuts, keep the bolts from rotating.**

➡**Keep one bolt inserted to secure the hub and disc rotor.**

b. Remove the 2 nuts and 2 bolts on the lower side of the rear shock absorber with coil spring.

c. Remove the 3 nuts on the upper side of the rear shock absorber with coil spring.

➡**Make sure that the rear speed sensor is disconnected from the rear shock absorber with coil spring.**

d. Lower the rear axle carrier, and remove the 2 bolts on the lower side of the rear shock absorber with coil spring.

To install:

11. Install the rear shock absorber with coil spring to the rear axle carrier assembly and insert the 2 bolts from the rear of the vehicle.
12. Slowly jack up the vehicle using a wooden block and install the rear shock absorber with coil spring (upper side) to the vehicle.
13. Install the 3 nuts to the upper side of the rear shock absorber with coil spring and tighten to 29 ft. lbs. (39 Nm).

➡**When installing the nuts, keep the bolts from rotating.**

14. Install the 2 nuts and 2 bolts to the lower side of the rear shock absorber with coil spring and tighten 133 ft. lbs. (180 Nm).
15. Fully tighten the lock nut to 41 ft. lbs. (55 Nm).
16. Connect rear speed sensor.
17. Install the LH rear stabilizer link assembly.
18. Engage the 4 claws and install the rear suspension support No. 1 cover.
19. To complete installation, reverse remaining removal.
20. Check abs speed sensor signal.
21. Inspect and adjust rear wheel alignment.

STABILIZER BAR

REMOVAL & INSTALLATION

See Figures 336 and 337.

1. Before servicing the vehicle, refer to the precautions section.
2. Remove rear wheels.
3. For 2.4L engines, remove center exhaust pipe assembly.
4. For 3.5L engines, remove tail exhaust pipe assembly.
5. For 3.5L engines, center exhaust pipe assembly.
6. Remove rear stabilizer link assembly.
7. Remove the 2 bolts and rear stabilizer bar No. 2 bracket.
8. Remove the 2 bolts and rear stabilizer bar No. 1 bracket.
9. Remove the 2 rear stabilizer bushings from the rear stabilizer bar.
10. Remove rear stabilizer bar.

To install:

11. Install the 2 rear stabilizer bushings to the outside of the stopper ring on the stabilizer bar.
12. Install the rear stabilizer bar No. 2 and No. 1 bracket.
13. Install the rear stabilizer bar with the 2 bolts and tighten to 23 ft. lbs. (31 Nm).
14. Install rear stabilizer link assembly.
15. To complete installation, reverse remaining removal.
16. Check abs speed sensor signal.
17. Inspect and adjust rear wheel alignment.

WHEEL HUB AND BEARING

REMOVAL & INSTALLATION

See Figure 338.

➡**Use the same procedures for the RH side and LH side.**

➡**The procedures listed below are for the LH side.**

1. Before servicing the vehicle, refer to the precautions section.
2. Remove the rear wheel.
3. Separate the rear disc brake caliper assembly, as follows:
   a. Remove the bolt and separate the flexible hose from the shock absorber.
   b. Remove the 2 bolts and separate the rear disc brake caliper assembly.
4. Remove the rear disc.
5. Disconnect the skid control sensor connector.
6. Remove the 4 bolts and the rear axle hub and bearing assembly.
7. Install the hub and bearing assembly with the 4 bolts and tighten to 59 ft. lbs. (80 Nm).

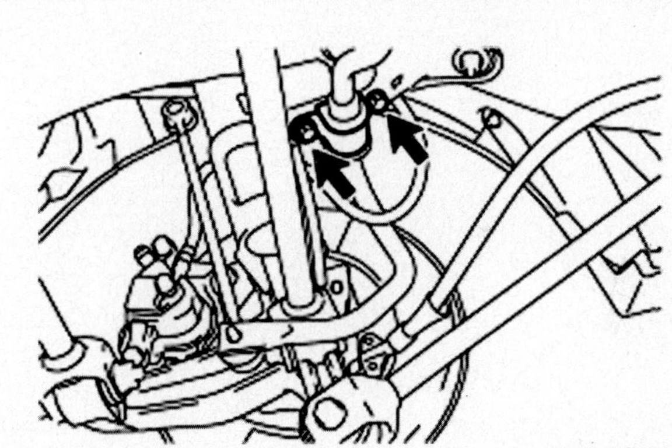
22140_CAMR_G0560

Fig. 336 Removing the 2 bolts and No. 2 bracket

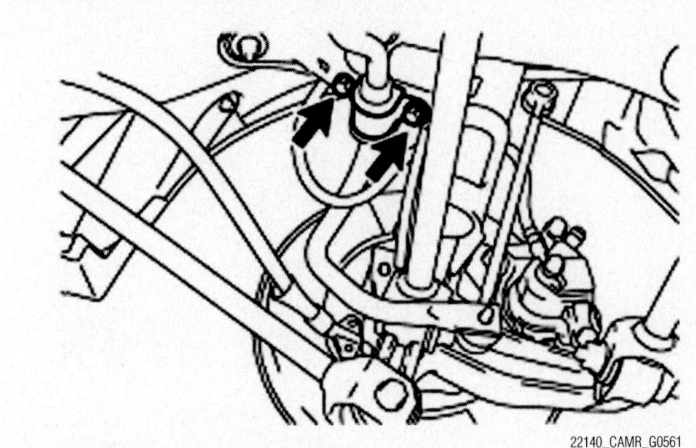
22140_CAMR_G0561

Fig. 337 Removing the 2 bolts and No. 1 bracket

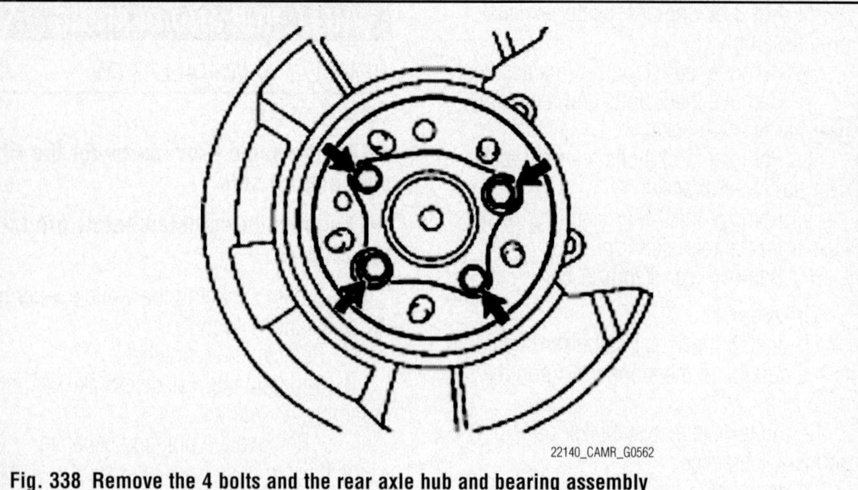

Fig. 338 Remove the 4 bolts and the rear axle hub and bearing assembly

8. Connect the skid control sensor connector. Do not twist the sensor wire.
9. Inspect rear axle hub bearing looseness.
10. Inspect rear axle hub runout.
11. Install the rear disc.
12. Install the rear disc brake caliper assembly, as follows:
    a. Install the rear disc brake caliper with the 2 bolts and tighten to 46 ft. lbs. (62 Nm).
    b. Install the flexible hose with the bolt and tighten to 14 ft. lbs. (19 Nm).
13. Install the rear wheel.
14. Inspect and adjust the rear wheel alignment.
15. Check ABS speed sensor signal.

TOYOTA

Camry Solara

12

SPECIFICATIONS AND MAINTENANCE CHARTS

ENGINE AND VEHICLE IDENTIFICATION

Engine							Model Year	
Code ①	Liters (cc)	Cu. In.	Cyl.	Fuel Sys.	Engine Type	Eng. Mfg.	Code ②	Year
2AZ-FE	2.4 (2362)	144	4	SFI	DOHC	Toyota	7	2007
3MZ-FE	3.3 (3311)	202	6	SFI	DOHC	Toyota	8	2008

SFI: Sequential Fuel Injection

DOHC: Double Overhead Camshaft

NA: Information not available

① Stamped on the left side of the engine block

② 10th digit of the Vehicle Identification Number (VIN)

22140_SOLA_C0001

GENERAL ENGINE SPECIFICATIONS

Year	Model	Engine Displacement Liters	Engine Series ID	Net Horsepower @ rpm	Net Torque @ rpm (ft. lbs.)	Bore x Stroke (in.)	Com-pression Ratio	Oil Pressure @ rpm
2007	Solara	2.4	2AZ-FE	155@6000	158@4000	3.48x3.78	9.8:1	36-78@3000
	Solara	3.3	3MZ-FE	210@5600	220@3600	3.62x3.27	10.8:1	36-78@3000
2008	Solara	2.4	2AZ-FE	155@6000	158@4000	3.48x3.78	9.8:1	36-78@3000
	Solara	3.3	3MZ-FE	210@5600	220@3600	3.62x3.27	10.8:1	36-78@3000

22140_SOLA_C0002

ENGINE TUNE-UP SPECIFICATIONS

Year	Engine Displacement Liters	Engine ID	Spark Plug Gap (in.)	Ignition Timing (deg.)*	Fuel Pump (psi)	Idle Speed (rpm)	Valve Clearance Intake	Valve Clearance Exhaust
2007	2.4	2AZ-FE	0.039-0.043	N/A	44-50	①	0.007-0.011	.0015-0.019
	3.3	3MZ-FE	0.039-0.043	N/A	44-50	630-730	0.006-0.010	0.010-0.014
2008	2.4	2AZ-FE	0.039-0.043	N/A	44-50	①	0.007-0.011	.0015-0.019
	3.3	3MZ-FE	0.039-0.043	N/A	44-50	630-730	0.006-0.010	0.010-0.014

NOTE: The Vehicle Emission Control Information label often reflects specification changes made during production. The label figures must be used if they differ from those in this chart.

① Manual transmission: 650 to 750 rpm, Automatic transmission: 610 to 710 rpm

22140_SOLA_C0003

CAPACITIES

Year	Model	Engine Displacement Liters	Engine ID	Engine Oil with Filter (qts.)	Transmission (qts.) 5-Spd	Transmission (qts.) Auto.*	Transfer Case (pts.)	Drive Axle Front (pts.)	Drive Axle Rear (pts.)	Fuel Tank (gal.)	Cooling System (qts.)
2007	Solara	2.4	2AZ-FE	4.5	—	3.7	—	—	—	18.5	6.6
	Solara	3.3	3MZ-FE	5.0	—	3.7	—	—	—	18.5	9.8
2008	Solara	2.4	2AZ-FE	4.5	—	3.7	—	—	—	18.5	6.6
	Solara	3.3	3MZ-FE	5.0	—	3.7	—	—	—	18.5	9.8

22140_SOLA_C0004

VALVE SPECIFICATIONS

Year	Engine Displacement Liters	Engine ID	Seat Angle (deg.)	Face Angle (deg.)	Spring Test Pressure (lbs. @ in.)	Spring Installed Height (in.)	Stem-to-Guide Clearance (in.) Intake	Stem-to-Guide Clearance (in.) Exhaust	Stem Diameter (in.) Intake	Stem Diameter (in.) Exhaust
2007	2.4	2AZ-FE	45	44.5	NA	NA	0.0010-0.0031	0.0012-0.0039	0.2154-0.2159	0.2151-0.2157
	3.3	3MZ-FE	45	44.5	NA	NA	0.0010-0.0031	0.0012-0.0039	0.2154-0.2159	0.2151-0.2157
2008	2.4	2AZ-FE	45	44.5	NA	NA	0.0010-0.0031	0.0012-0.0039	0.2154-0.2159	0.2151-0.2157
	3.3	3MZ-FE	45	44.5	NA	NA	0.0010-0.0031	0.0012-0.0039	0.2154-0.2159	0.2151-0.2157

NA: Information not available

22140_SOLA_C0005

CRANKSHAFT AND CONNECTING ROD SPECIFICATIONS

All measurements are given in inches.

Year	Engine Displacement Liters	Engine ID	Crankshaft Main Brg. Journal Dia.	Crankshaft Main Brg. Oil Clearance	Crankshaft Shaft End-play	Crankshaft Thrust on No.	Connecting Rod Journal Diameter	Connecting Rod Oil Clearance	Connecting Rod Side Clearance
2007	2.4	2AZ-FE	2.0863-2.0866	0.0007-0.0016	0.0016-0.0095	3	1.8894-1.8898	0.0013-0.0025	0.0063-0.0143
	3.3	3MZ-FE	4.4011-4.4015	①	0.0016 - 0.0118	2	2.0863-2.0866	0.0015-0.0031	0.0059-0.0138
2008	2.4	2AZ-FE	2.0863-2.0866	0.0007-0.0016	0.0016-0.0095	3	1.8894-1.8898	0.0013-0.0025	0.0063-0.0143
	3.3	3MZ-FE	4.4011-4.4015	①	0.0016 - 0.0118	2	2.0863-2.0866	0.0015-0.0031	0.0059-0.0138

① For journals 1 & 4 measure 0.0006-0.0020
For journals 2 & 3 measure 0.0010-0.0024

22140_SOLA_C0006

PISTON AND RING SPECIFICATIONS

All measurements are given in inches.

Year	Engine Displ. Liters	Engine ID	Piston Clearance	Ring Gap Top Comp.	Ring Gap Bottom Comp.	Ring Gap Oil Control	Ring Side Clearance Top Comp.	Ring Side Clearance Bottom Comp.	Ring Side Clearance Oil Control
2007	2.4	2AZ-FE	0.0020-0.0029	0.0094-0.0122	0.0130-0.0169	0.0040-0.0119	0.0008-0.0028	0.0008-0.0024	0.0008-0.0028
	3.3	3MZ-FE	0.0013-0.0051	0.0118-0.0374	0.0197-0.0413	0.0059-0.0394	0.0012-0.0031	0.0008-0.0024	0.0012-0.0043
2008	2.4	2AZ-FE	0.0020-0.0029	0.0094-0.0122	0.0130-0.0169	0.0040-0.0119	0.0008-0.0028	0.0008-0.0024	0.0008-0.0028
	3.3	3MZ-FE	0.0013-0.0051	0.0118-0.0374	0.0197-0.0413	0.0059-0.0394	0.0012-0.0031	0.0008-0.0024	0.0012-0.0043

22140_SOLA_C0007

TORQUE SPECIFICATIONS

All readings in ft. lbs.

Year	Engine Displacement Liters	Engine ID	Cylinder Head Bolts	Main Bearing Bolts	Rod Bearing Bolts	Crankshaft Damper Bolts	Flywheel Bolts	Manifold Intake	Manifold Exhaust	Spark Plugs	Oil Pan Drain Plug
2007	2.4	2AZ-FE	①	②	③	125	④	22	27	13	18
	3.3	3MZ-FE	⑤	⑥	⑦	159	61	11	36	13	33
2008	2.4	2AZ-FE	①	②	③	125	④	22	27	13	18
	3.3	3MZ-FE	⑤	⑥	⑦	159	61	11	36	13	33

① Step 1: 52 ft. lbs.
Step 2: plus 90 degrees

② Step 1: 15 ft. lbs.
Step 2: 30 ft. lbs.
Step 3: Plus 90 degrees

③ Step 1: Cap bolts to 18 ft. lbs.
Step 2: cap bolts plus 90 degrees

④ Auto driveplate to 72 ft. lbs.
Manual Flywheel : 96 ft. lbs.

⑤ Step 1: 40 ft. lbs.
Step 2: Plus 90 degrees
8 mm bolt: 14 ft. lbs.

⑥ 12 point bolt step 1: 15 ft. lbs.
Step 2: Plus 90 degrees
6 point bolt: 20 ft. lbs.

⑦ Step 1: 25 ft. lbs.
Step 2: plus 90 degrees

22140_SOLA_C0008

WHEEL ALIGNMENT

Year	Model		Caster Range (+/-Deg.)	Caster Preferred Setting (Deg.)	Camber Range (+/-Deg.)	Camber Preferred Setting (Deg.)	Toe-in (in.)	Steering Axis Inclination (Deg.)
2007	Solara 2AZ-FE Except Sport	Front	0.75	2.90	0.75	-0.73	0+/-0.08	11.47+/-0.75
		Rear	—	—	0.75	-1.35	0.16+/-0.08	—
	Solara 2AZ-FE Sport	Front	0.75	2.93	0.75	-0.77	0+/-0.08	11.50+/-0.75
		Rear	—	—	0.75	-1.38	0.16+/-0.08	—
	Solara 3MZ-FE Except Sport	Front	0.75	2.83	0.75	-0.75	0+/-0.08	11.52+/-0.75
		Rear	—	—	0.75	-1.35	0.16+/-0.08	—
	Solara 3MZ-FE Sport	Front	0.75	2.88	0.75	-0.77	0+/-0.08	11.55+/-0.75
		Rear	—	—	0.75	-1.37	0.16+/-0.08	—
	Solara Comvertible	Front	0.75	2.85	0.75	-0.73	0+/-0.08	14.52+/-0.75
		Rear	—	—	0.75	-1.33	0.16+/-0.08	—
2008	Solara 2AZ-FE Except Sport	Front	0.75	2.65	0.75	-0.73	0+/-0.08	11.47+/-0.75
		Rear	—	—	0.75	-1.35	0.16+/-0.08	—
	Solara 2AZ-FE Sport	Front	0.75	2.72	0.75	-0.77	0+/-0.08	11.50+/-0.75
		Rear	—	—	0.75	-1.38	0.16+/-0.08	—
	Solara 3MZ-FE Except Sport	Front	0.75	2.70	0.75	-0.75	0+/-0.08	11.52+/-0.75
		Rear	—	—	0.75	-1.35	0.16+/-0.08	—
	Solara 3MZ-FE Sport	Front	0.75	2.80	0.75	-0.77	0+/-0.08	11.55+/-0.75
		Rear	—	—	0.75	-1.37	0.16+/-0.08	—
	Solara Comvertible	Front	0.75	2.85	0.75	-0.73	0+/-0.08	14.52+/-0.75
		Rear	—	—	0.75	-1.33	0.16+/-0.08	—

22140_SOLA_C0009

TIRE, WHEEL AND BALL JOINT SPECIFICATIONS

Year	Model	OEM Tires		Tire Pressures (psi)		Wheel	Ball Joint	Lug Nut Torque
		Standard	Optional	Front	Rear	Size	Inspection	(ft. lbs.)
2007	Solara	P215/60R16	P215/55R17	①	①	6.5-J	②	76
2008	Solara	P215/60R16	P215/55R17	①	①	7-J	②	76

OEM: Original Equipment Manufacturer

PSI: Pounds Per Square Inch

STD: Standard

OPT: Optional

① See manufacturer's recommendations on vehicle

② Replace if any measurable movement is found.

22140_SOLA_C0010

BRAKE SPECIFICATIONS

All measurements in inches unless noted

Year	Model		Brake Disc Original Thickness	Brake Disc Minimum Thickness	Brake Disc Maximum Runout	Minimum Lining Thickness	Brake Caliper Bracket Bolts (ft. lbs.)	Brake Caliper Mounting Bolts (ft. lbs.)
2007	Solara	F	1.102	1.024	0.0020	0.039	79	25
		R	0.472	0.413	0.0059	0.039	46	32
2008	Solara	F	1.102	1.024	0.0020	0.039	79	25
		R	0.472	0.413	0.0059	0.039	46	32

F: Front

R: Rear

22140_SOLA_C0011

SCHEDULED MAINTENANCE INTERVALS

TOYOTA—CAMRY SOLARA

TO BE SERVICED	TYPE OF SERVICE	VEHICLE MILEAGE INTERVAL (x1000)													
		5	10	15	20	25	30	35	40	45	50	55	60	90	120
Engine oil & filter	R	✓	✓	✓	✓	✓	✓	✓	✓	✓	✓	✓	✓	✓	✓
Automatic transmission fluid	S/I			✓			✓			✓			✓	✓	✓
Ball joints & dust covers	S/I			✓			✓			✓			✓	✓	✓
Bolts & nuts on chassis & body	S/I			✓			✓			✓			✓	✓	✓
Brake linings & drums	S/I	✓	✓	✓	✓	✓	✓	✓	✓	✓	✓	✓	✓	✓	✓
Brake line pipes & hoses	S/I			✓			✓			✓			✓	✓	✓
Brake pads & discs (front & rear)	S/I	✓	✓	✓	✓	✓	✓	✓	✓	✓	✓	✓	✓	✓	✓
Brake fluid	R						✓						✓	✓	✓
Rack and pinion assembly	S/I			✓			✓			✓			✓	✓	✓
Steering linkage & boots	S/I			✓			✓			✓			✓	✓	✓
Air cleaner filter	R						✓						✓	✓	✓
Spark plugs ①	R														✓
Drive belts	S/I												✓	✓	✓
Exhaust pipes & mountings	S/I			✓			✓			✓			✓	✓	✓
Fuel lines & connections	S/I						✓						✓	✓	✓
Engine coolant ②	S/I			✓			✓			✓			✓	✓	
Fuel tank cap gasket	S/I						✓						✓	✓	✓
Rotate tires	S/I			✓			✓			✓			✓		✓
Clean air conditioning filter ③	S/I			✓			✓			✓			✓		✓
Axle shaft bolts	S/I			✓			✓			✓			✓	✓	✓
Brake pad thickness and rotor runout	S/I						✓						✓	✓	✓

R: Replace S/I: Service or Inspect

① Spark plugs are replaced at 120,000 miles

② Replace engine coolant at 100,000 miles and then inspect every 15,000 miles

③ Replace air conditioning filter every 30,000 miles

FREQUENT OPERATION MAINTENANCE (SEVERE SERVICE)

If a vehicle is operated under any of the following conditions it is considered severe service:

- Extremely dusty areas.

- 50% or more of the vehicle operation is in 32°C (90°F) or higher temperatures, or constant temperatures below 0°C (32°F).

- Prolonged idling (vehicle operation in stop and go traffic).

- Frequent short running periods (engine does not warm to normal operating temperatures).

- Police, taxi, delivery usage or trailer towing usage.

Air cleaner filter: service or inspect every 5000 miles

Rear differential & transfer case oil: replace every 15,000 miles.

Ball joints & dust covers: service or inspect every 5000 miles.

Bolts & nuts on chassis & body: service or inspect every 5000 miles.

Axle shaft bolts: service or inspect every 5000 miles.

Steering linkage: service or inspect every 5000 miles.

22140_SOLA_C0012

PRECAUTIONS

Before servicing any vehicle, please be sure to read all of the following precautions, which deal with personal safety, prevention of component damage, and important points to take into consideration when servicing a motor vehicle:

- Never open, service or drain the radiator or cooling system when the engine is hot; serious burns can occur from the steam and hot coolant.
- Observe all applicable safety precautions when working around fuel. Whenever servicing the fuel system, always work in a well-ventilated area. Do not allow fuel spray or vapors to come in contact with a spark, open flame, or excessive heat (a hot drop light, for example). Keep a dry chemical fire extinguisher near the work area. Always keep fuel in a container specifically designed for fuel storage; also, always properly seal fuel containers to avoid the possibility of fire or explosion. Refer to the additional fuel system precautions later in this section.
- Fuel injection systems often remain pressurized, even after the engine has been turned **OFF**. The fuel system pressure must be relieved before disconnecting any fuel lines. Failure to do so may result in fire and/or personal injury.
- Brake fluid often contains polyglycol ethers and polyglycols. Avoid contact with the eyes and wash your hands thoroughly after handling brake fluid. If you do get brake fluid in your eyes, flush your eyes with clean, running water for 15 minutes. If eye irritation persists, or if you have taken brake fluid internally, IMMEDIATELY seek medical assistance.
- The EPA warns that prolonged contact with used engine oil may cause a number of skin disorders, including cancer. You should make every effort to minimize your exposure to used engine oil. Protective gloves should be worn when changing oil. Wash your hands and any other exposed skin areas as soon as possible after exposure to used engine oil. Soap and water, or waterless hand cleaner should be used.
- All new vehicles are now equipped with an air bag system, often referred to as a Supplemental Restraint System (SRS) or Supplemental Inflatable Restraint (SIR) system. The system must be disabled before performing service on or around system components, steering column, instrument panel components, wiring and sensors. Failure to follow safety and disabling procedures could result in accidental air bag deployment, possible personal injury and unnecessary system repairs.
- Always wear safety goggles when working with, or around, the air bag system. When carrying a non-deployed air bag, be sure the bag and trim cover are pointed away from your body. When placing a non-deployed air bag on a work surface, always face the bag and trim cover upward, away from the surface. This will reduce the motion of the module if it is accidentally deployed. Refer to the additional air bag system precautions later in this section.
- Clean, high quality brake fluid from a sealed container is essential to the safe and proper operation of the brake system. You should always buy the correct type of brake fluid for your vehicle. If the brake fluid becomes contaminated, completely flush the system with new fluid. Never reuse any brake fluid. Any brake fluid that is removed from the system should be discarded. Also, do not allow any brake fluid to come in contact with a painted surface; it will damage the paint.
- Never operate the engine without the proper amount and type of engine oil; doing so WILL result in severe engine damage.
- Timing belt maintenance is extremely important. Many models utilize an interference-type, non-freewheeling engine. If the timing belt breaks, the valves in the cylinder head may strike the pistons, causing potentially serious (also time-consuming and expensive) engine damage. Refer to the maintenance interval charts for the recommended replacement interval for the timing belt, and to the timing belt section for belt replacement and inspection.
- Disconnecting the negative battery cable on some vehicles may interfere with the functions of the on-board computer system(s) and may require the computer to undergo a relearning process once the negative battery cable is reconnected.
- When servicing drum brakes, only disassemble and assemble one side at a time, leaving the remaining side intact for reference.
- Only an MVAC-trained, EPA-certified automotive technician should service the air conditioning system or its components.

BRAKES — ANTI-LOCK BRAKE SYSTEM (ABS)

GENERAL INFORMATION

The Anti-lock Brake System (ABS) helps prevent wheels from locking when the brake is applied firmly or when braking on a slippery surface.

The ABS warning light comes on when the ABS system is malfunctioning.

The Electronic Brake force Distribution (EBD) control utilizes ABS, realizing proper brake force distribution between front and rear wheels in accordance with driving conditions. In addition, during cornering braking, it also controls the brake force of the right and left wheels, helping to maintain vehicle behavior.

Both of the ABS and BRAKE warning lights come on to indicate a malfunction in the EBD system.

Based on the signals received from the 4 wheel speed sensors, the skid control ECU calculates the speed of each wheel and deceleration, and checks wheel slipping condition. According to the slipping condition, the skid control ECU controls the pressure holding valve and pressure reduction valve in order to adjust the fluid pressure of each wheel cylinder

BLEEDING PROCEDURE

Bleeding the Master Cylinder

If the master cylinder is reinstalled or if the reservoir becomes empty, bleed the air from the master cylinder. To prevent brake fluid from adhering, cover nearly painted surfaces with a shop rag or a piece of cloth.

1. Disconnect the 2 brake lines from the master cylinder, using a suitable brake line wrench.
2. Have an assistant slowly depress the brake pedal and hold it.
3. Cover the 2 outer holes with your fingers, and have your assistant release the brake pedal.
4. Repeat the previous 2 steps 3 or 4 times.
5. Connect the 2 the brake lines to the master cylinder and tighten to 11 ft. lbs. (15 Nm).

Bleeding the Brake Lines

➡Bleed air from the brake line of the wheel farthest from the master cylinder.

1. Raise and safely support the vehicle.
2. Connect a vinyl tube to the bleeder plug.
3. Have an assistant depress the brake pedal several times, then loosen the bleeder plug while the pedal is depressed.
4. When fluid stops coming out, tighten the bleeder plug, then release the brake pedal.
5. Repeat steps 3 and 4 until all the air in the fluid has been bled out.
6. Tighten the brake bleeder plug to 73 inch lbs. (8.3 Nm).
7. Repeat the above steps to bleed the air out of the brake line for each wheel.

Bleeding the Actuator Assembly

➡After bleeding the air from the brake system, if the height or feel of the brake pedal cannot be obtained, perform air bleeding in the brake actuator assembly with a hand-held tester by following the procedures below.

1. Depress the brake pedal more than 20 times with the engine off.
2. Connect the hand-held tester to the DLC3, then turn the ignition switch to the **ON** position, but do NOT start the engine.
3. Select "AIR BLEEDING" on the hand-held tester.

➡Refer to the hand-held tester operator's manual for more details.

4. Bleed the air out of the regular brake line when "Step 1: Increase" appears on the hand-held tester display, as follows:

➡Bleed the air by following the steps displayed on the hand-held tester. Make sure that the brake fluid in the master cylinder reservoir tank does not become empty.

a. Connect the vinyl tube to either one of the bleeder plugs.
b. Have an assistant depress the brake pedal several times, then loosen the bleeder plug connected to the vinyl tube with the pedal depressed.
c. When fluid stops coming out, tighten the bleeder plug and release the brake pedal.
d. Repeat the previous 2 steps until all the air in the fluid is completely bled out.
e. Tighten the bleeder plug completely to 73 inch lbs. (8.3 Nm).
f. Repeat the above procedures for each wheel to bleed the air out of the brake line.

5. Bleed the air out of the suction line when "Step 2: Inhalation" appears on the hand-held tester display, as follows:

➡Bleed the air by following the steps displayed on the hand-held tester. Make sure that the brake fluid in the master cylinder reservoir tank does not become empty.

a. Connect the vinyl tube to the bleeder plug at the right front wheel or the right rear wheel and loosen the bleeder plug.
b. Operate the brake actuator assembly to perform air bleeding from the suction line using the hand-held tester.

➡This operation stops automatically after 4 seconds. At this time, be sure to release the brake pedal.

c. Check if the operation has stopped by referring to the hand-held tester display and tighten the bleeder plug.
d. Repeat the previous 2 steps until all air in the fluid is completely bled out.
e. Tighten the bleeder plug completely to 73 inch lbs. (8.3 Nm).
f. Repeat the above procedures to bleed the air out of the brake line for each wheel.

6. Bleed the air out of the pressure reduction line when "Step 3: Decrease" appears on the hand-held tester display, as follows:

➡Bleed the air by following the steps displayed on the hand-held tester. Make sure that the brake fluid in the master cylinder reservoir tank does not become empty.

a. Connect a vinyl tube to either one of the bleeder plugs.
b. Loosen the bleeder plug.
c. Using the hand-held tester, operate the brake actuator assembly, completely depress the brake pedal and keep it depressed.

➡The operation stops automatically after 4 seconds. When performing this procedure continuously, set an interval of at least 20 seconds. When the operation is complete, the brake pedal goes down slightly. This is a normal phenomenon caused when the solenoid opens. During this procedure, the pedal will feel heavy, but completely depress it so that the brake fluid comes out from the bleeder plug. Be sure to keep depressing the brake pedal. Do not depress and release the pedal repeatedly.

d. Tighten the bleeder plug, then release the brake pedal.
e. Repeat the previous 2 steps until all the air in the fluid is completely bled out.
f. Tighten the bleeder plug completely to 73 inch lbs. (8.3 Nm).
g. Repeat the above procedures for each wheel to bleed the air out of the brake line.

7. Bleed the air out of the brake line again when "Step 4: Increase" appears on the hand-held tester display, as follows:

➡Bleed the air by following the steps displayed on the hand-held tester. Make sure that the brake fluid in the master cylinder reservoir tank does not become empty.

a. Connect the vinyl tube to either one of the bleeder plugs.
b. Depress the brake pedal several times, then loosen the bleeder plug connected to the vinyl tube with the pedal depressed.

8. When fluid stops coming out, tighten the bleeder plug, then release the brake pedal.

a. Repeat the previous 2 steps until all the air in the fluid is completely bled out.

b. Tighten the bleeder plug completely to 73 inch lbs. (8.3 Nm).

c. Repeat the above procedures for each wheel to bleed the air out of the brake line.

d. Finish "AIR BLEEDING" on the hand-held tester and turn the hand-held tester off.

e. Disconnect the hand-held tester from the DLC3 from the DLC3.

f. Turn the ignition switch off.

g. Inspect for fluid leak.

9. Check the fluid level and add fluid if necessary. Use SAE J1703 or FMVSS No. 116 DOT3 Brake fluid.

BRAKES — FRONT DISC BRAKES

BRAKE CALIPER

REMOVAL & INSTALLATION

1. Before servicing the vehicle, refer to the Precautions Section.
2. Remove the front wheel.
3. Drain brake fluid.

WARNING

Do not let brake fluid sit on painted surfaces, as it will eat through the paint. Wash it off immediately.

4. Remove the union bolt and gasket from the disc brake cylinder assembly, then disconnect the front brake flexible hose.
5. Hold the front disc brake cylinder slide pin and remove the 2 bolts and disc brake cylinder assembly.

➡Remove the disc brake cylinder assembly while holding both of the brake pads or the anti-squeal springs may fall off the brake pads.

To install:

6. Install the disc brake cylinder subassembly with the 2 bolts and tighten to 25 ft. lbs. (34 Nm).
7. Connect the flexible hose with the union bolt and a new gasket and tighten to 22 ft. lbs. (29 Nm).

➡Install the front brake flexible hose lock securely in the lock hole in the disc brake cylinder.

8. Fill reservoir with brake fluid.
9. Bleed brake line.
10. For systems with VSC, bleed the brake actuator.
11. Inspect for brake fluid leak.
12. Inspect brake fluid level in reservoir.
13. Install the front wheel and tighten the lug nuts to 76 ft. lbs. (103 Nm).

DISC BRAKE PADS

REMOVAL & INSTALLATION

1. Before servicing the vehicle, refer to the Precautions Section.
2. Remove the 2 front disc brake cylinder slide pins (upper and lower) from the front disc brake cylinder mounting.
3. Remove brake cylinder.
4. Remove the 2 anti-squeal springs.
5. Remove the 2 brake pads from the front disc brake cylinder mounting.

To install:

6. Install the 2 brake pads with front anti-squeal shims to the front disc brake cylinder mounting.

➡When replacing worn pads, the front anti-squeal springs must be replaced at the same time.

➡Be sure to install the anti-squeal springs into the front disc brake pad installation holes as far as they will go.

7. Install the 2 front disc brake cylinder slide pins (upper and lower) from the front disc brake cylinder mounting.
8. Install the brake cylinder.

BRAKES — REAR DISC BRAKES

BRAKE CALIPER

REMOVAL & INSTALLATION

1. Before servicing the vehicle, refer to the Precautions Section.
2. Remove the rear wheel.
3. Drain brake fluid.

WARNING

Do not let brake fluid sit on painted surfaces, as it will eat through the paint. Wash it off immediately.

4. Remove the union bolt and the gasket from the rear disc brake cylinder assembly, then disconnect the rear brake flexible hose.
5. Hold the 2 rear disc brake cylinder slide pins and remove the 2 bolts and rear disc brake cylinder assembly.

To install:

6. Install the rear disc brake cylinder assembly with the 2 bolts and tighten to 20 ft. lbs. (77 Nm).
7. Connect the rear brake flexible hose with the union bolt and a new gasket and tighten to 24 ft. lbs. (33 Nm).
8. Fill reservoir with brake fluid.
9. Bleed brake line.
10. Inspect for brake fluid leak.
11. Inspect brake fluid level in reservoir.
12. Install the rear wheel and tighten the lug nuts to 76 ft. lbs. (103 Nm).

DISC BRAKE PADS

REMOVAL & INSTALLATION

1. Before servicing the vehicle, refer to the Precautions Section.
2. Remove the 2 rear disc brake cylinder slide pins (upper and lower) from the rear disc brake cylinder mounting.
3. Remove brake cylinder.
4. Remove the 2 brake pads with the rear anti-squeal shims.

To install:

5. Installation is the reverse of removal procedure.

PARKING BRAKE SHOES

REMOVAL & INSTALLATION

See Figures 1 through 5.

1. Remove the rear wheel.
2. Remove the 2 bolts and separate the rear disc brake caliper assembly. Do not disconnect the flexible hose from the disc brake caliper assembly.
3. Remove the parking brake shoe adjusting hole plug from the rear disc.
4. Release the parking brake and place the matchmarks on the rear disc and the axle hub.
5. Remove the rear disc.

➡If the disc cannot be removed easily, turn the shoe adjuster until the disc turns freely.

6. Using needle-nose pliers, remove the 2 parking brake shoe return tension No. 1 springs.
7. Remove the parking brake shoe strut and the parking brake shoe strut compression spring.
8. Remove the No. 1 parking brake shoe assembly, as follows:
   a. Release the claw of the parking brake shoe hold down spring No. 2 cup.
   b. Remove the No. 1 parking brake shoe assembly as shown in the illustration.
   c. Remove the parking brake shoe hold down spring No. 1 cup, the parking brake shoe hold down spring, the parking brake shoe hold down spring No. 2 cup, and the parking brake shoe hold down spring No. 1 pin.
9. Remove the parking brake shoe adjusting screw set.
10. Remove the parking brake shoe return tension No. 2 spring.
11. Remove the No. 2 parking brake shoe assembly, as follows:

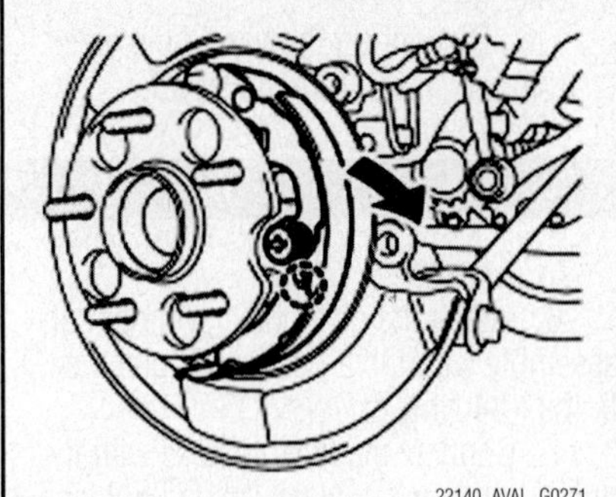

Fig. 2 Remove the No. 2 parking brake shoe assembly

   a. Release the claw of the parking brake shoe hold down spring No. 2 cup.
   b. Remove the No. 2 parking brake shoe assembly as shown in the illustration.
   c. Remove the parking brake shoe hold down spring No. 1 cup, the parking brake shoe hold down spring, the parking brake shoe hold down spring No. 2 cup, and the parking brake shoe hold down spring No. 2 pin.
   d. Using needle-nose pliers, disconnect the No. 3 parking brake cable assembly from the parking brake shoe lever.

➡Be careful not to damage the No. 3 parking brake cable assembly.

12. Using a screwdriver, remove the C-washer, shim and the parking brake shoe lever.
13. Remove the parking brake shoe guide plate set bolt and the parking brake shoe guide plate.

To install:

14. Apply high temperature grease to the backing plate where it contacts the shoe.
15. Apply adhesive (Toyota Genuine Adhesive 1344, Three Bond 1344 or equivalent) to the threads of the parking brake shoe guide plate set bolt.
16. Install the parking brake shoe guide plate with the parking brake shoe guide plate set bolt and tighten to 13 ft. lbs. (18 Nm).
17. Install the parking brake shoe lever and shim to the No. 2 parking brake shoe assembly with a new C-washer.
18. Using a feeler gauge, measure the clearance between the No. 2 parking brake shoe assembly and parking brake shoe lever. Standard clearance: Less than 0.35 mm (0.014 in.). If the clearance is not as specified, replace the shim with one of the correct size.
19. Install the No. 2 parking brake shoe assembly as follows:
   a. Using needle-nose pliers, connect the No. 3 parking brake cable assembly to the parking brake shoe lever.
   b. Install the No. 2 parking brake shoe assembly with the parking brake shoe hold down spring No. 2 pin, the parking brake shoe hold down spring No. 2 cup, the parking brake shoe hold down spring and the parking brake shoe hold down spring No. 1 cup.
   c. Engage the claw of the parking brake shoe hold down spring No. 2 cup to the No. 2 parking brake shoe assembly.
20. Install the parking brake shoe adjusting screw set, as follows:
   a. Apply high temperature grease to the parking brake shoe adjusting screw set as shown in the illustration.
   b. Install the parking brake shoe return tension No. 2 spring to the No. 1

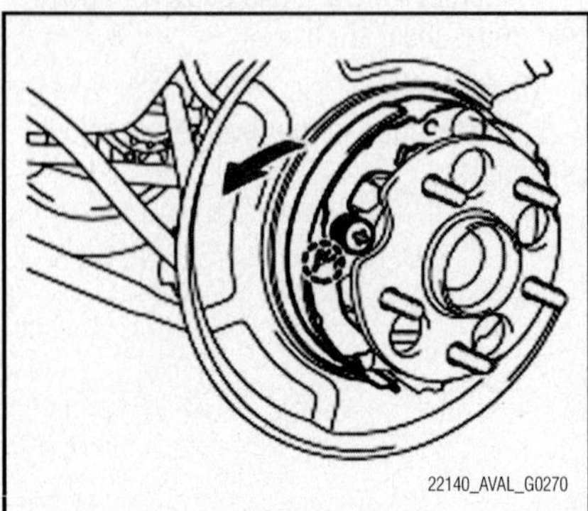

Fig. 1 Remove the No. 1 parking brake shoe assembly

Shim Thickness	Shim Thickness
0.3 mm (0.012 in.)	0.9 mm (0.035 in.)
0.6 mm (0.024 in.)	-

22140_AVAL_G0272

Fig. 3 Shim thickness

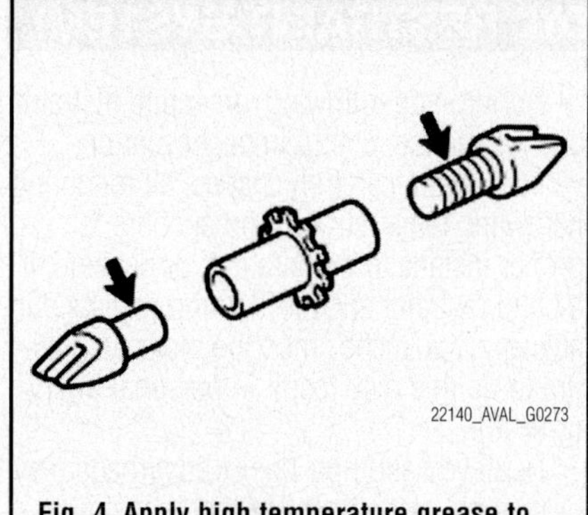

Fig. 4 Apply high temperature grease to the parking brake shoe adjusting screw set

parking brake shoe assembly and the No. 2 parking brake shoe assembly.

c. Install the parking brake shoe adjusting screw set to the No. 1 parking brake shoe assembly and the No. 2 parking brake shoe assembly.

21. Install the No. 1 parking brake shoe assembly as follows:

a. Install the No. 1 parking brake shoe assembly with the parking brake shoe hold down spring No. 1 pin, parking brake shoe hold down spring No. 2 cup, parking brake shoe hold down spring and parking brake shoe hold down spring No. 1 cup.

b. Engage the claw of the parking brake shoe hold down spring No. 2 cup to the No. 1 parking brake shoe assembly.

22. Attach the parking brake shoe strut and the parking brake shoe strut compression spring to the No. 1 parking brake shoe assembly and No. 2 parking brake shoe assembly.

23. Using needle-nose pliers, install the 2 parking brake shoe return tension No. 1 springs. First install the front side spring and then the rear side spring.

24. Inspect parking brake installation and check that each part is installed properly.

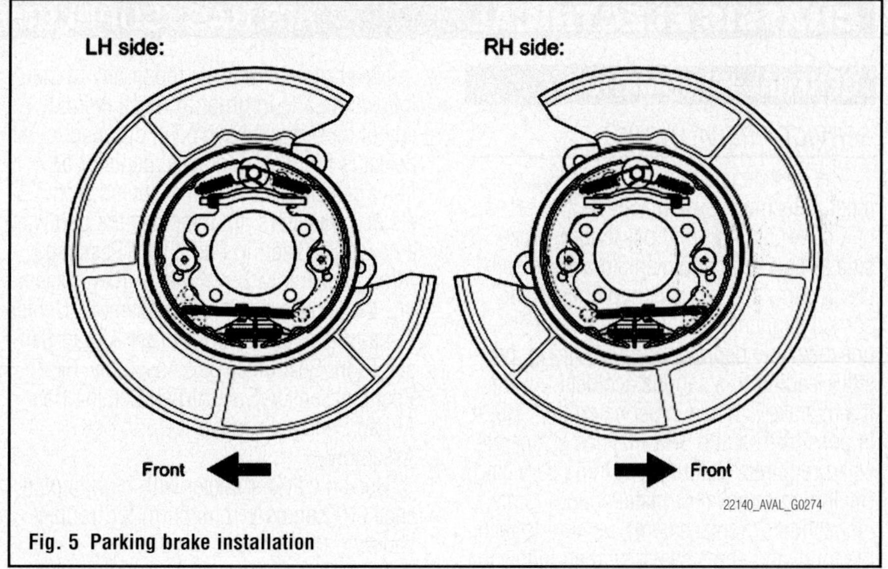

Fig. 5 Parking brake installation

➡There should be no oil or grease on the friction surfaces of the shoe linings and discs.

25. Install the rear disc.

26. Install the parking brake shoe adjusting hole plug.

27. Adjust parking brake shoe clearance.

28. Install the rear disc brake caliper assembly with the 2 bolts and tighten to 46 ft. lbs. (62 Nm).

29. Install the rear wheel.

30. Adjust the parking brake pedal travel.

31. Bed in parking brake shoes to discs, as follows:

a. Drive the vehicle at about 31 mph (50 km/h) on a safe, level and dry road.

b. Depress the parking brake pedal with 34 lbs. (150 N) of force.

32. Drive the vehicle about 0.25 miles (400 m) in this condition.

a. Repeat this procedure 3 times using 5-minute intervals between each procedure to prevent the parking brake assembly from overheating.

33. Remove the rear wheel.

34. Adjust parking brake shoe clearance.

35. Adjust the parking brake pedal travel.

36. Install the rear wheel and tighten the lug nuts to 76 ft. lbs. (103 Nm).

ADJUSTMENT

1. Adjust parking brake shoe clearance, as follows:

a. Temporarily install the hub nuts.

b. Remove the shoe adjusting hole plug, turn the adjuster and expand the shoes until the disc locks.

c. Contract the shoe adjuster until the disc rotates smoothly. Standard: returns 8 notches

d. Check that the disc has no brake drag.

e. Install the shoe adjusting hole plug.

GENERAL INFORMATION

SERVICE PRECAUTIONS

The vehicle is equipped with a Supplemental Restraint System (SRS). It consists of a driver airbag, front passenger airbag, side airbag and curtain shield airbag. Failure to carry out service operations in the correct sequence could cause the SRS to unexpectedly deploy during servicing, possibly leading to a serious accident. Further, if a mistake is made in servicing the SRS, it is possible that the SRS may fail to operate when required. Before performing servicing (including removal or installation of parts, inspection or replacement), be sure to read the following items carefully, then follow the correct procedures indicated in the repair manual.

Disconnect and isolate the battery negative cable before beginning any airbag system component diagnosis, testing, removal, or installation procedures. Wait at least 90 seconds after the ignition switch is turned off and the negative (-) terminal cable is disconnected from the battery before starting the operation. The SRS is equipped with a backup power source, so if work is started within 90 seconds after disconnecting the negative (-) terminal cable from the battery, the SRS may be deployed. Failure to disable the airbag system may result in accidental airbag deployment, personal injury, or death.

Do not expose the steering pad, front passenger airbag assembly, center airbag sensor assembly, front airbag sensor, front seat inner belt assembly, seat position airbag sensor, front seat airbag assembly, side airbag sensor assembly, curtain shield airbag assembly, or rear airbag sensor directly to hot air or flames.

Malfunction symptoms of the SRS are difficult to confirm, so DTCs are the most important source of information when troubleshooting. When troubleshooting the SRS, always inspect DTCs before disconnecting the battery.

Even in the case of a minor collision when the SRS does not deploy, the steering pad, front passenger airbag assembly, center airbag sensor assembly, front airbag sensor, front seat inner belt assembly, seat position airbag sensor, front seat airbag assembly, side airbag sensor assembly, curtain shield airbag assembly, or rear airbag sensor should be inspected.

Before repair work, remove the airbag sensor if any kind of shock is likely to occur to the airbag sensor during the operation.

Never disassemble or repair any of the following parts in order to reuse them. If any of these parts have been dropped, or a defect is found (e.g. cracks, dents or any other defects) in any of the housings, brackets or connectors, then replace the part with a new one: Steering Pad, Front Passenger Airbag Assembly, Front Seat Airbag Assembly, Curtain Shield Airbag Assembly, Center Airbag Sensor Assembly, Front Airbag Sensor, Front Seat Inner Belt Assembly, Seat Position Sensor, Side Airbag Sensor, Rear Airbag Sensor, Front Seat Outer Belt Assembly.

Use an volt/ohmmeter with high impedance (10 Kohms/V minimum) for troubleshooting the electrical circuits.

Information labels are attached near the SRS components. Follow the instructions in the caution.

Do not place an intact undeployed airbag face down on a solid surface. The airbag will propel into the air if accidentally deployed and may result in personal injury or death.

When carrying or handling an undeployed airbag, the trim side (face) of the airbag should be pointing towards the body to minimize possibility of injury if accidental deployment occurs. Failure to do this may result in personal injury or death.

Replace airbag system components with OEM replacement parts. Substitute parts may appear interchangeable, but internal differences may result in inferior occupant protection. Failure to do so may result in occupant personal injury or death.

Wear safety glasses, rubber gloves, and long sleeved clothing when cleaning powder residue from vehicle after an airbag deployment. Powder residue emitted from a deployed airbag can cause skin irritation. Flush affected area with cool water if irritation is experienced. If nasal or throat irritation is experienced, exit the vehicle for fresh air until the irritation ceases. If irritation continues, see a physician.

Do not use a replacement airbag that is not in the original packaging. This may result in improper deployment, personal injury, or death.

The factory installed fasteners, screws and bolts used to fasten airbag components have a special coating and are specifically designed for the airbag system. Do not use substitute fasteners. Use only original equipment fasteners listed in the parts catalog when fastener replacement is required.

During, and following, any child restraint anchor service, due to impact event or vehicle repair, carefully inspect all mounting hardware, tether straps, and anchors for proper installation, operation, or damage. If a child restraint anchor is found damaged in any way, the anchor must be replaced. Failure to do this may result in personal injury or death.

Deployed and non-deployed airbags may or may not have live pyrotechnic material within the airbag inflator.

Do not dispose of driver/passenger/curtain airbags or seat belt tensioners unless you are sure of complete deployment. Refer to the Hazardous Substance Control System for proper disposal.

Dispose of deployed airbags and tensioners consistent with state, provincial, local, and federal regulations.

After any airbag component testing or service, do not connect the battery negative cable. Personal injury or death may result if the system test is not performed first.

If the vehicle is equipped with the Occupant Classification System (OCS), do not connect the battery negative cable before performing the OCS Verification Test using the scan tool and the appropriate diagnostic information. Personal injury or death may result if the system test is not performed properly.

Never replace both the Occupant Restraint Controller (ORC) and the Occupant Classification Module (OCM) at the same time. If both require replacement, replace one, then perform the Airbag System test before replacing the other.

Both the ORC and the OCM store Occupant Classification System (OCS) calibration data, which they transfer to one another when one of them is replaced. If both are replaced at the same time, an irreversible fault will be set in both modules and the OCS may malfunction and cause personal injury or death.

If equipped with OCS, the Seat Weight Sensor is a sensitive, calibrated unit and must be handled carefully. Do not drop or handle roughly. If dropped or damaged, replace with another sensor. Failure to do so may result in occupant injury or death.

If equipped with OCS, the front passenger seat must be handled carefully as well. When removing the seat, be careful when setting on floor not to drop. If dropped, the sensor may be inoperative, could result in occupant injury, or possibly death.

If equipped with OCS, when the passenger front seat is on the floor, no one should sit in the front passenger seat. This uneven force may damage the sensing ability of the seat weight sensors. If sat on and damaged, the sensor may be inoperative, could result in occupant injury, or possibly death.

After work on the SRS is completed, perform the SRS warning light check.

When the negative (-) terminal cable is disconnected from the battery, the memory will be cleared. Because of this, be sure to make a record of the contents memorized in each system before starting work. When work is finished, adjust each system as it was before. Never attempt to avoid erasing vehicle system memories by using a backup power supply from outside the vehicle.

If the vehicle is equipped with a mobile communication system, refer to the precautions.

DISARMING THE SYSTEM

To avoid personal injury when working on vehicles equipped with an air bag, the negative battery cable must be disconnected and at least 90 seconds must elapse before working on the system. Failure to do so may result in deployment of the air bag.

ARMING THE SYSTEM

To arm the system after service is finished, connect the negative battery cable.

DRIVETRAIN

AUTOMATIC TRANSAXLE ASSEMBLY

REMOVAL & INSTALLATION

U151E Automatic Transaxle

See Figures 6 through 10.

1. Remove engine assembly.
2. Remove the right and left front drive shaft assemblies.
3. Remove the bolt and No. 2 transmission control cable bracket.
4. Remove the wire harness clamp, as follows:
   a. Disconnect the wire harnesses from the 2 clamps.
   b. Remove the 2 bolts and 2 clamps.
5. Remove the bolt and disconnect the wire harness.
6. Remove starter assembly.
7. Disconnect the transmission wire connector.
8. Disconnect the park/neutral position switch connector.
9. Disconnect the 2 transmission revolution sensor connectors.
10. Remove the No. 1 transmission control cable bracket, as follows:

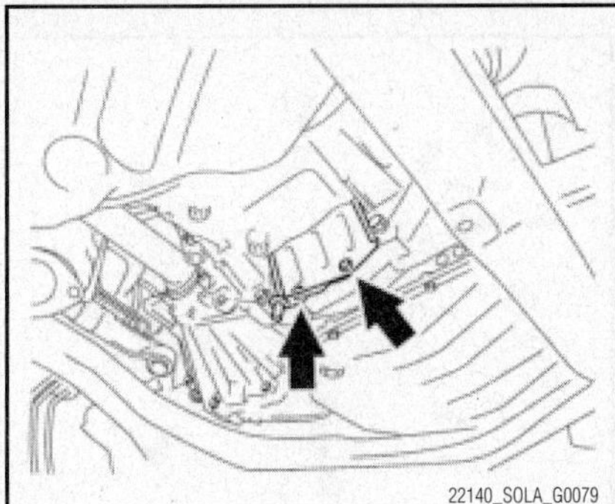

Fig. 7 Transaxle cover bolts—U151E transaxle

   a. Remove the bolt and oil cooler tube clamp.
   b. Remove the 2 bolts and No. 1 transmission control cable bracket.
11. Remove the transmission oil filler tube sub-assembly, as follows:
   a. Remove the ATF level gauge.
   b. Disconnect the breather hose from the wire harness bracket.
   c. Remove the bolt and transmission oil filler tube sub-assembly.
   d. Remove the O-ring from the oil filler tube sub-assembly.
12. Using SST (SST: 09023-12701) and a wrench, disconnect the No. 1 oil cooler inlet and outlet tubes.
13. Remove the 3 bolts and transverse engine mounting bracket.
14. Remove automatic transaxle assembly, as follows:
   a. Remove the 2 bolts.
   b. Remove the exhaust pipe support bracket and the No. 1 exhaust pipe support bracket from the automatic transaxle.
   c. Turn the crankshaft to gain access and remove the 6 bolts while holding the crankshaft pulley bolt with a wrench. There will be one black colored bolt.
   d. Remove the 8 bolts.
   e. Separate and remove the automatic transaxle.

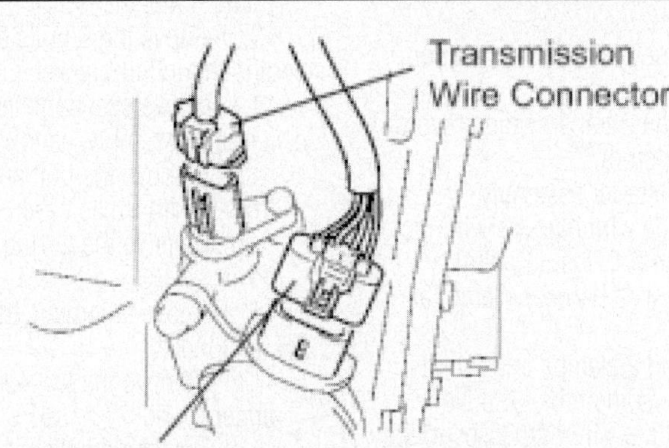

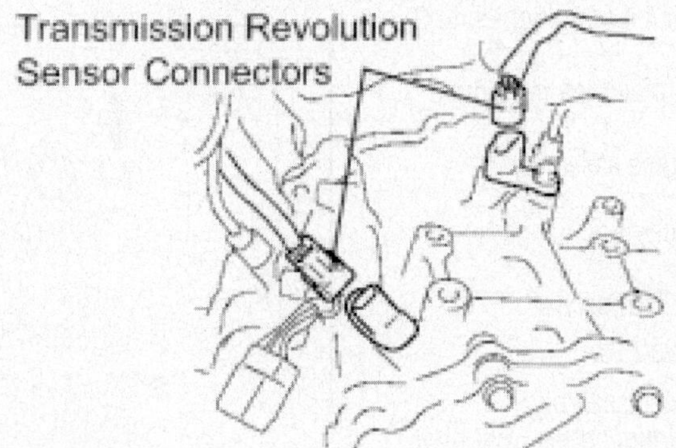

Fig. 6 Transaxle connectors

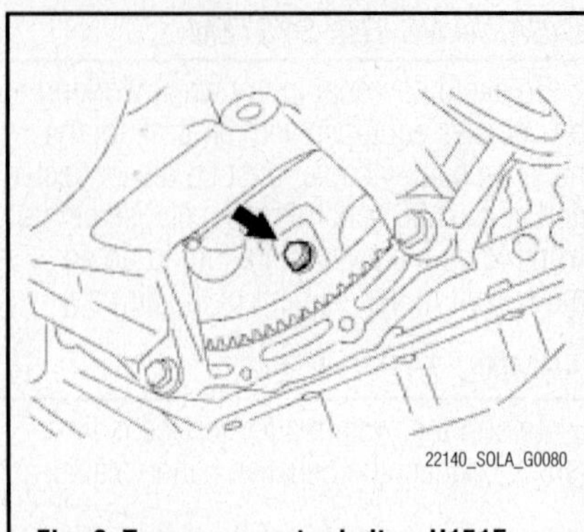

Fig. 8 Torque converter bolts—U151E transaxle

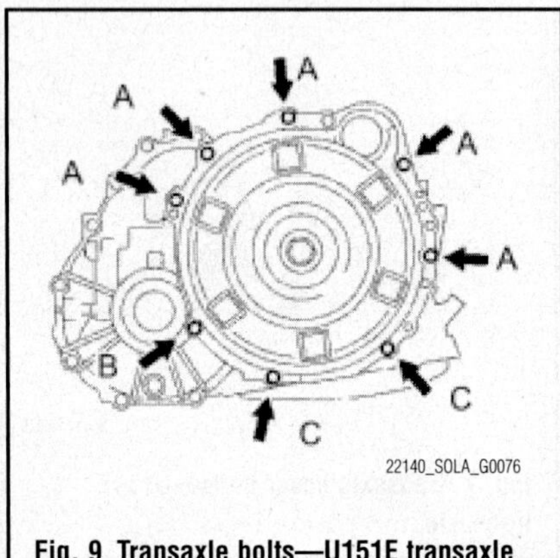

Fig. 9 Transaxle bolts—U151E transaxle

To install:

15. Install the automatic transaxle to the engine with the 8 bolts. Tighten bolt "A" to 47 ft. lbs (64 Nm), bolt "B" to 34 ft lbs. (46 Nm), and bolt "C" to 27 ft. lbs. (37 Nm).
16. Apply a few drops of adhesive (Toyota Genuine Adhesive 1324, Three Bond 1324 or equivalent) to 2 threads on the tip of the 6 torque converter clutch mounting bolts.
17. Install the 6 torque converter clutch mounting bolts and tighten to 30 ft. lbs. (41 Nm). First install the black colored bolt, and then the remaining 5 bolts.
18. Install the No. 1 exhaust pipe support bracket and exhaust pipe support bracket to the automatic transaxle with the 2 bolts and tighten to 69 inch lbs. (7.8 Nm).
19. Install the transverse engine mounting bracket with the 3 bolts to the automatic transaxle and tighten to 47 ft. lbs. (64 Nm).
20. Install transmission oil filler tube sub-assembly, as follows:
    a. Coat a new O-ring with ATF, and install it to the oil filler tube.
    b. Install the oil filler tube to the automatic transaxle with the bolt and tighten to 49 inch lbs. (5.5 Nm).
    c. Connect the breather hose to the wire harness bracket. The breather hose should be on the left side of the vehicle.
    d. Install the ATF level gauge.

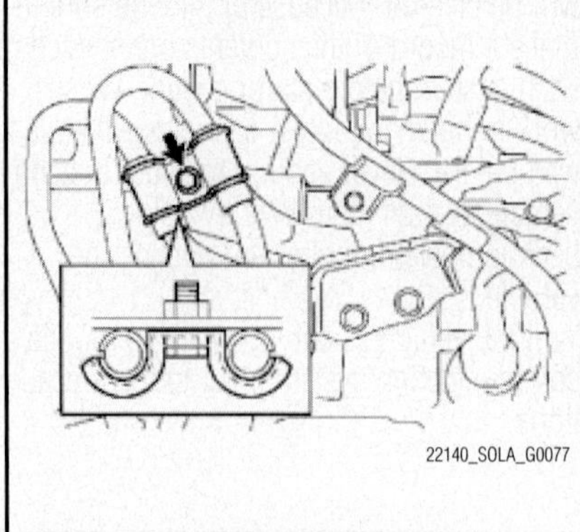

Fig. 10 Oil cooler tube positioning

21. Install the No. 1 control cable bracket with the 2 bolts and tighten to 9 ft. lbs. (12 Nm).
22. Temporarily install the No. 1 oil cooler inlet and outlet tubes.
23. Install the oil cooler tube clamp and bolt and tighten to 48 inch lbs. (5.4 Nm). Install them so that the oil cooler tube cushion is positioned as shown in the illustration.
24. Using SST (SST: 09023-12701) and a wrench, tighten the No. 1 oil cooler inlet and outlet tubes to 25 ft. lbs. (34 Nm).
25. Connect the transmission wire connector.
26. Connect the park/neutral position switch connector.
27. Connect the 2 transmission revolution sensor connectors.
28. Install the starter assembly.
29. Connect the wire harness with the bolt and tighten to 9 ft. lbs. (13 Nm).
30. Install the wire harness clamp, as follows:
    a. Install the 2 clamps and 2 bolts and tighten to 74 inch lbs. (8.4 Nm).
    b. Connect the wire harnesses to the clamps.
31. Install the No. 2 transmission control cable bracket with the bolt and tighten to 9 ft. lbs. (12 Nm).
32. Install the right and left front drive shaft assemblies.
33. Install the engine assembly.
34. Reset memory.
35. Perform initialization.

U250E Automatic Transaxle

See Figures 11 through 16.

1. Remove engine assembly.
2. Remove the right and left front drive shaft assemblies.
3. Remove the bolt and No. 2 transmission control cable bracket.
4. Remove the wire harness clamp, as follows:
    a. Disconnect the wire harnesses from the 2 clamps.
    b. Remove the 2 bolts and 2 clamps.
5. Remove the bolt and disconnect the wire harness.
6. Remove starter assembly.
7. Disconnect the transmission wire connector.
8. Disconnect the park/neutral position switch connector.
9. Disconnect the 2 transmission revolution sensor connectors.
10. Remove the No. 1 transmission control cable bracket, as follows:
    a. Remove the bolt and oil cooler tube clamp.
    b. Remove the 2 bolts and No. 1 transmission control cable bracket.
11. Remove the transmission oil filler tube sub-assembly, as follows:
    a. Remove the ATF level gauge.
    b. Disconnect the breather hose from the wire harness bracket.
    c. Remove the bolt and transmission oil filler tube sub-assembly.
    d. Remove the O-ring from the oil filler tube sub-assembly.
12. Using SST (SST: 09023-12701) and a wrench, disconnect the No. 1 oil cooler inlet and outlet tubes.
13. Remove the 3 bolts and transverse engine mounting bracket.
14. Remove speedometer driven hole cover sub-assembly, as follows:
    a. Remove the bolt and hole cover from the transaxle case.
    b. Remove the O-ring from the hole cover.
15. Remove automatic transaxle assembly, as follows:
    a. Remove the flywheel housing under cover.
    b. Turn the crankshaft to gain access and remove the 6 bolts while holding the

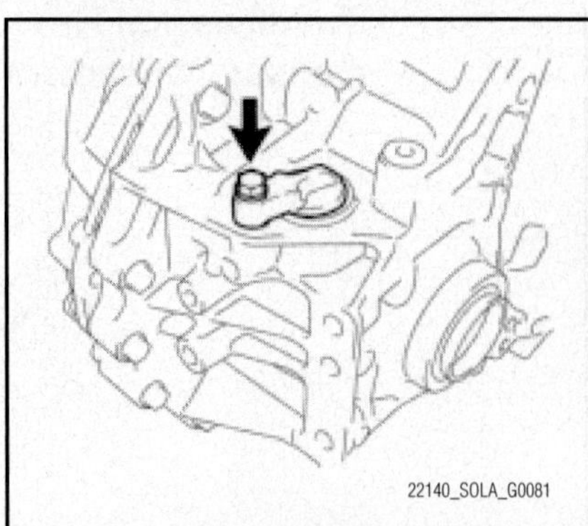

Fig. 11 Speedometer driven hole cover and bolt—U250E transaxle

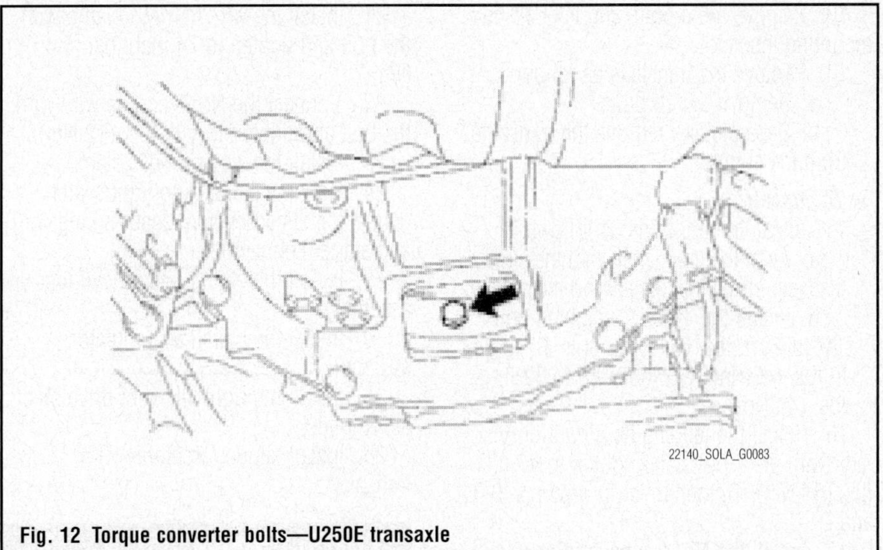

Fig. 12 Torque converter bolts—U250E transaxle

crankshaft pulley bolt with a wrench. There will be one black colored bolt.

c. Remove the 5 bolts.

d. Remove the 4 lower side mounting bolts.

e. Separate and remove the automatic transaxle.

To install:

16. Install the automatic transaxle to the engine with the 5 bolts. Tighten bolt "A" to 47 ft. lbs (64 Nm), bolt "B" to 34 ft lbs. (46 Nm), and bolt "C" to 34 ft. lbs. (46 Nm).

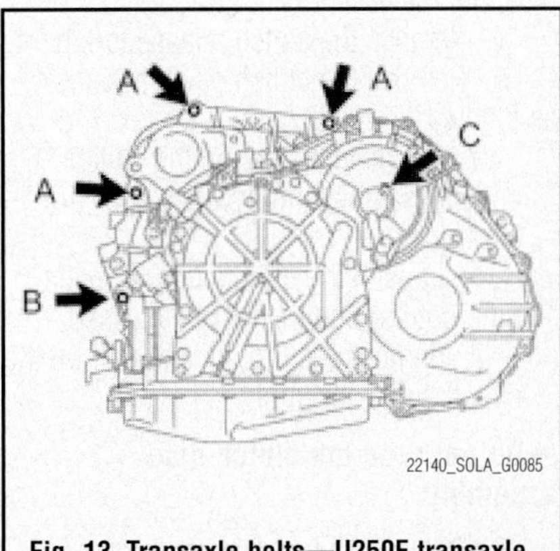

Fig. 13 Transaxle bolts—U250E transaxle

17. Install the 4 lower side mounting bolts and tighten to 32 ft. lbs. (44 Nm).

18. Apply a few drops of adhesive (Toyota Genuine Adhesive 1324, Three Bond 1324 or equivalent) to 2 threads on the tip of the 6 torque converter clutch mounting bolts.

19. Install the 6 torque converter clutch mounting bolts and tighten to 30 ft. lbs. (41 Nm). First install the black colored bolt, and then the remaining 5 bolts.

20. Install the flywheel housing under cover and tighten the bolt to 69 inch lbs. (7.8 Nm) and the nut to 16 ft. lbs. (21 Nm).

21. Install the speedometer driven hole cover to the transaxle case with the bolt and tighten to 47 inch lbs. (5.5 Nm).

22. Install the transverse engine mounting bracket with the 3 bolts to the automatic transaxle and tighten to 47 ft. lbs. (64 Nm).

23. Install transmission oil filler tube sub-assembly, as follows:

a. Coat a new O-ring with ATF, and install it to the oil filler tube.

b. Install the oil filler tube to the automatic transaxle with the bolt and tighten to 49 inch lbs. (5.5 Nm).

c. Connect the breather hose to the wire harness bracket. The breather hose should be on the left side of the vehicle.

d. Install the ATF level gauge.

24. Install the No. 1 control cable bracket with the 2 bolts and tighten to 9 ft. lbs. (12 Nm).

25. Temporarily install the No. 1 oil cooler inlet and outlet tubes.

26. Install the oil cooler tube clamp and bolt and tighten to 48 inch lbs. (5.4 Nm). Install them so that the oil cooler tube cushion is positioned as shown in the illustration.

27. Using SST (SST: 09023-12701) and a wrench, tighten the No. 1 oil cooler inlet and outlet tubes to 25 ft. lbs. (34 Nm).

28. Connect the transmission wire connector.

29. Connect the park/neutral position switch connector.

30. Connect the 2 transmission revolution sensor connectors.

31. Install the starter assembly.

32. Connect the wire harness with the bolt and tighten to 9 ft. lbs. (13 Nm).

33. Install the wire harness clamp, as follows:

a. Install the 2 clamps and 2 bolts and tighten to 74 inch lbs. (8.4 Nm).

b. Connect the wire harnesses to the clamps.

34. Install the No. 2 transmission control cable bracket with the bolt and tighten to 9 ft. lbs. (12 Nm).

35. Install the right and left front drive shaft assemblies.

36. Install the engine assembly.

37. Reset memory.

38. Perform initialization.

39. Check ABS speed sensor signal.

Fig. 14 Transaxle side mounting bolts—U250E transaxle

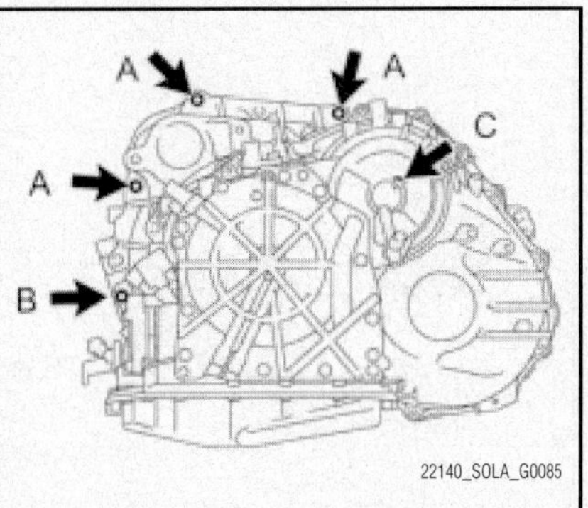

Fig. 15 Transaxle bolt tightening—U250E transaxle

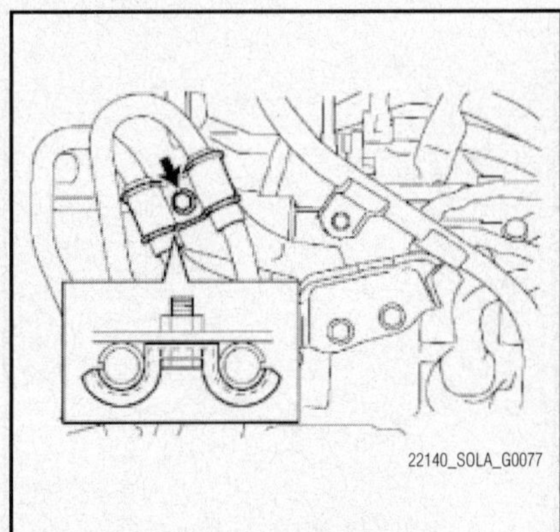

Fig. 16 Oil cooler tube positioning

MANUAL TRANSAXLE ASSEMBLY

REMOVAL & INSTALLATION

See Figures 17 through 20.

1. Remove the engine assembly with transaxle.
2. Remove the right and left front drive shaft assemblies.
3. Remove the clutch accumulator assembly.
4. Remove the clutch release cylinder assembly.
5. Disconnect the wire harness clamp.
6. Disconnect the back-up light switch connector, speedometer sensor connector and starter connector.
7. Remove the bolt and disconnect the No. 3 engine wire.
8. Remove the bolt and wire harness clamp.
9. Remove the 2 bolts and starter assembly.
10. Remove the 3 bolts and front engine mounting bracket.
11. Remove the 2 bolts and manual transmission case protector.
12. Remove the 2 bolts and No. 1 exhaust pipe support bracket.
13. Remove the 4 bolts and left engine mounting insulator.
14. Remove the transaxle as follows:
    a. Remove the 10 bolts.
    b. Separate and remove the transaxle from the engine.

To install:

15. Install the transaxle as follows:
    a. Align the input shaft with the clutch disc and install the transaxle to the engine.
    b. Install the 10 bolts. Tighten bolt "A" to 47 ft. lbs. (64 Nm), bolt "B" to 34 ft. lbs. (46 Nm), and bolt "C" to 32 ft. lbs. (44 Nm).
16. Install the left engine mounting insulator with the 4 bolts and tighten to 47 ft. lbs. (64 Nm). Tighten them in order: A, B, C and D.
17. Install the No. 1 exhaust pipe support bracket with the 2 bolts and tighten to 14 ft. lbs. (19 Nm).
18. Install the manual transmission case protector with the 2 bolts and tighten to 13 ft. lbs. (18 Nm).
19. Install the front engine mounting bracket with the 3 bolts and tighten to 47 ft. lbs. (64 Nm).
20. Install the starter assembly with the 2 bolts and tighten to 29 ft. lbs. (39 Nm).
21. Install the wire harness clamp with the bolt and tighten to 74 inch lbs. (8.4 Nm).
22. Connect the No. 3 engine wire with the bolt and tighten to 9 ft. lbs. (12 Nm).
23. Install the wire harness clamp.
24. Connect the backup light switch connector, speedometer sensor connector and starter connector.
25. Install the clutch release cylinder assembly.
26. Install the clutch accumulator assembly.
27. Install the right left front drive shaft assemblies.
28. Install engine assembly with transaxle.

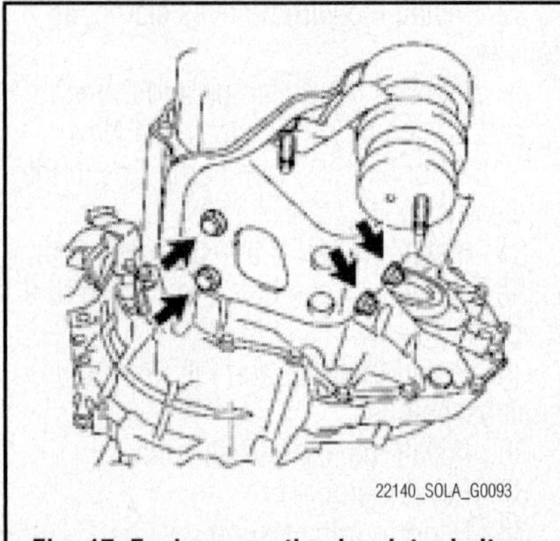

Fig. 17 Engine mounting insulator bolts

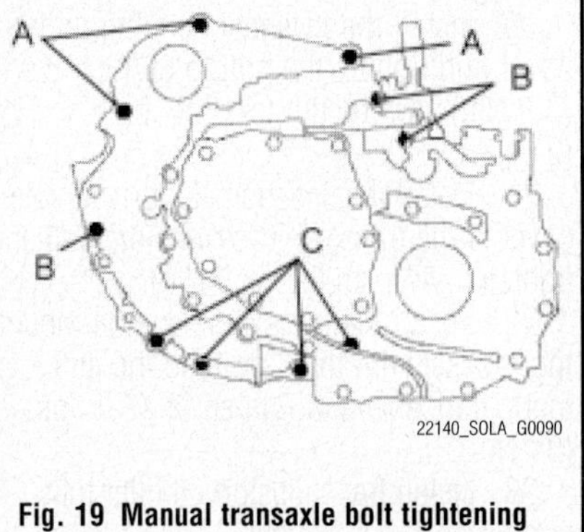

Fig. 19 Manual transaxle bolt tightening

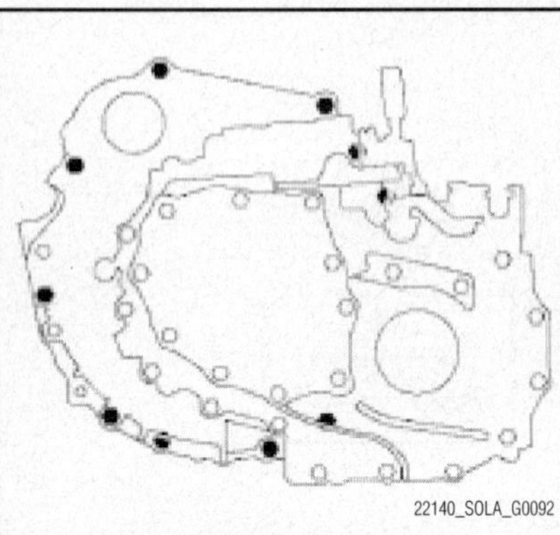

Fig. 18 Manual transaxle bolts

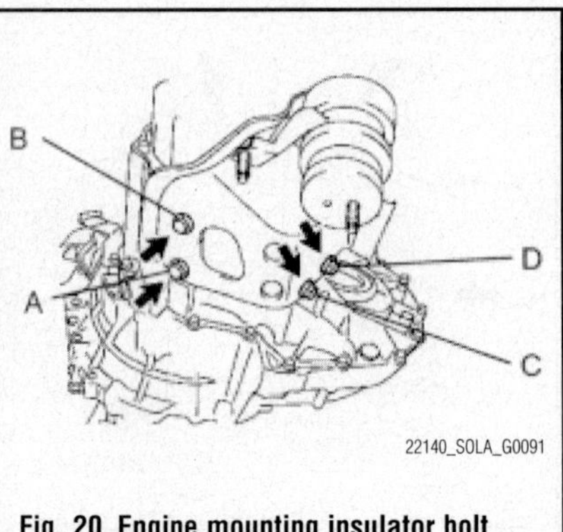

Fig. 20 Engine mounting insulator bolt tightening sequence

CLUTCH DRIVEN DISC & PRESSURE PLATE

REMOVAL & INSTALLATION

See Figures 21 through 24.

1. Remove the manual transaxle assembly.
2. Remove the clutch release fork with the clutch release bearing from the transaxle assembly.
3. Remove the release bearing hub clip and clutch release bearing assembly from the clutch release fork.
4. Remove the clutch release fork boot.
5. Remove the clutch cover assembly, as follows:
    a. Put matchmarks on the clutch cover assembly and flywheel sub-assembly.
    b. Loosen the 6 bolts one turn at a time until spring tension is released.
    c. Remove the 6 bolts and pull off the clutch cover assembly.

➡Do not drop the clutch disc assembly.

6. Remove clutch disc assembly.

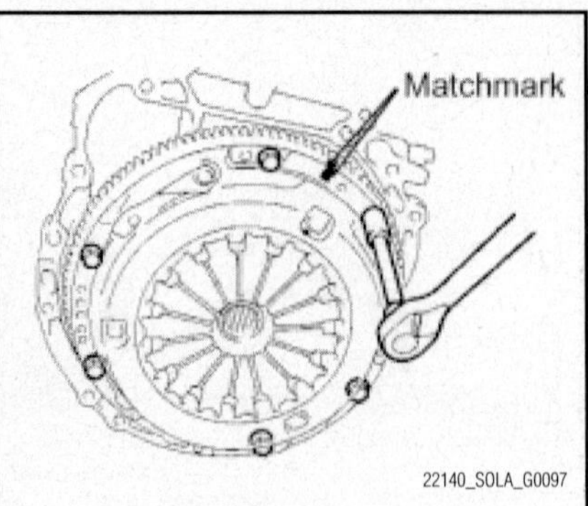

Fig. 21 Matchmarks on the clutch cover assembly and flywheel sub-assembly

➡Keep the lining part of the clutch disc assembly, the pressure plate and surface of the flywheel sub-assembly away from oil and foreign matter.

To install:

7. Insert SST (SST: 09301-00220) into the clutch disc assembly, then insert them into the flywheel sub-assembly.

➡Take care not to insert the clutch disc assembly in the wrong direction.

8. Align the matchmarks on the clutch cover assembly with the one on the flywheel sub-assembly.
9. Following the procedures shown in the illustration, tighten the 6 bolts in order, starting with the bolt located near the knock pin on the top. Tighten to 14 ft. lbs (19 Nm).

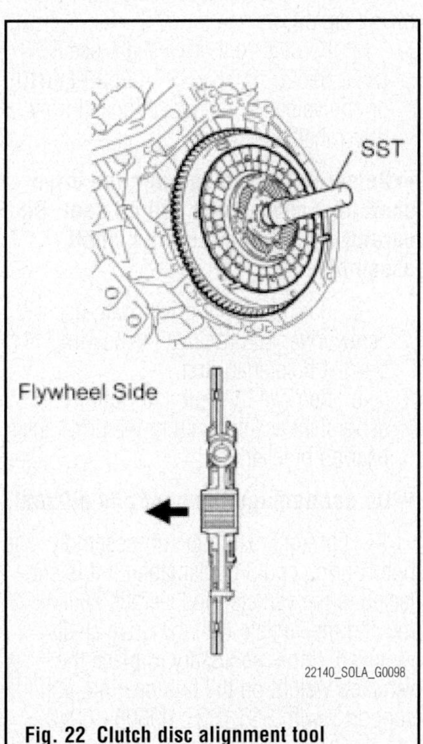

Fig. 22 Clutch disc alignment tool

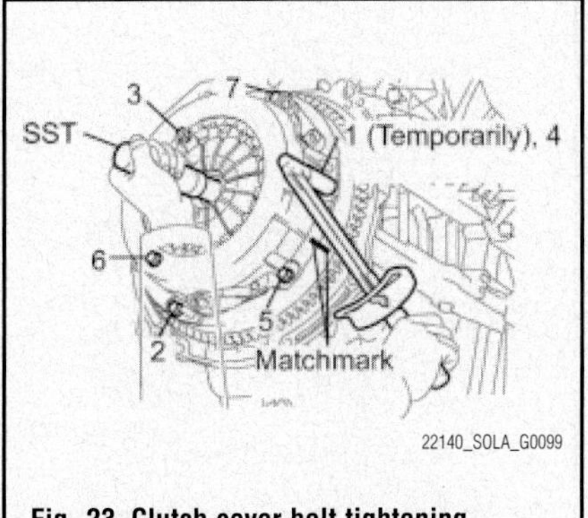

Fig. 23 Clutch cover bolt tightening sequence

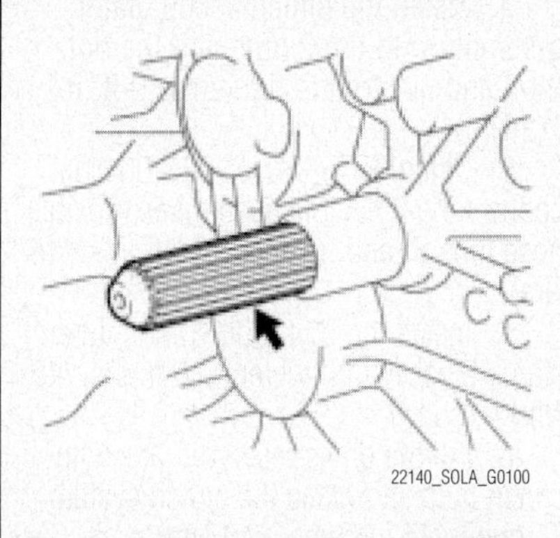

Fig. 24 Clutch spline grease application

➡Following the order in the illustration, evenly tighten the bolts one turn at a time.

➡Lightly move SST (09301-00210) up and down, and right and left after checking that the disc is in the center to tighten the bolts.

10. Install the release bearing hub clip to the release bearing assembly.
11. Apply release hub grease (Toyota Genuine Release Hub Grease or equivalent) to the contact surfaces between the release fork and the release bearing, the fork and push rod, and the fork and the fork support.
12. Install the release fork to the release bearing assembly.
13. Apply clutch spline grease (Toyota Genuine Clutch Spline Grease or equivalent) to the input shaft spline as shown in the illustration.
14. Install the release fork with the release bearing assembly to the manual transaxle assembly.

➡After installation, move the fork back and forth to check that the release bearing slides smoothly.

15. Install the clutch release fork boot.
16. Install the manual transaxle assembly.

CLUTCH MASTER CYLINDER

REMOVAL & INSTALLATION

See Figures 25 and 26.

1. Drain clutch fluid.
2. Remove brake master cylinder sub-assembly.
3. Remove brake booster assembly.
4. Loosen the clip and disconnect the clutch reservoir tube from the clutch master cylinder assembly. Use a container to catch the fluid.
5. Using SST (SST: 09023-00101), disconnect the clutch master cylinder flexible hose tube. Use a container to catch the fluid.

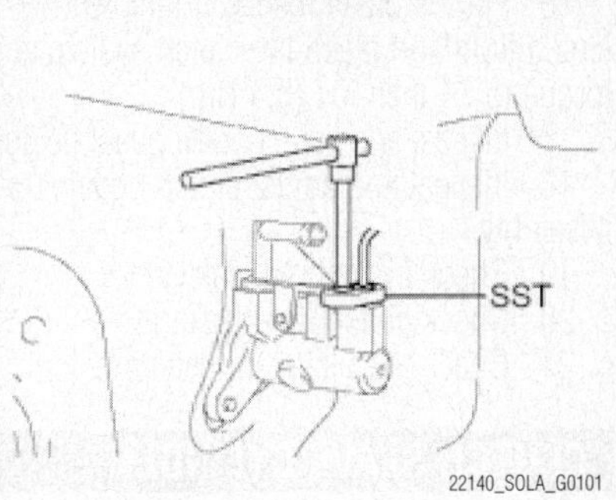

Fig. 25 Disconnect the flexible hose tube

6. Remove clutch master cylinder push rod clevis with hole pin.
7. Remove the 2 nuts and clutch master cylinder assembly.

To install:

8. Install the clutch master cylinder assembly with the 2 nuts and tighten to 9 ft. lbs. (12 Nm).
9. Install clutch master cylinder push rod clevis with hole pin, as follows:
   a. Apply MP grease to the contact surface of the hole pin and bush.
   b. Connect the clevis to the clutch pedal assembly with the hole pin.

➡Install the hole pin from the left side of the vehicle.

   c. Install the clip to the hole pin.
10. Using SST (SST: 09023-00101), connect the clutch master cylinder flexible hose tube and tighten to 11 ft. lbs. (15 Nm).
11. Connect the clutch reservoir tube to the clutch master cylinder assembly with the clip. Connect the clutch reservoir tube so that it will not be twisted.
12. Install the brake booster assembly.
13. Install the brake master cylinder sub-assembly.
14. Bleed the brake master cylinder.
15. Bleed the brake line.

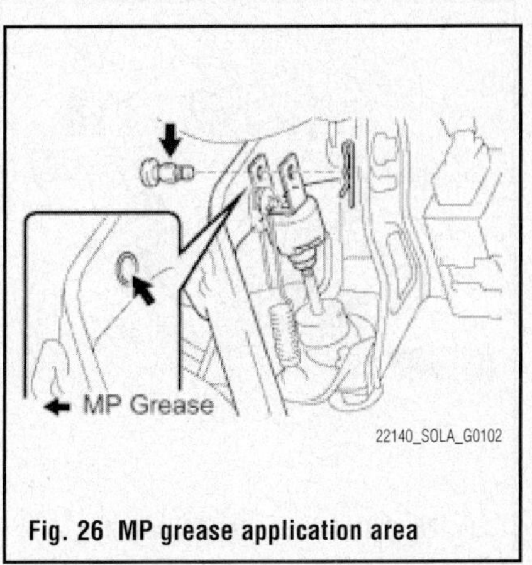

Fig. 26 MP grease application area

16. Fill the brake reservoir tank with clutch fluid and bleed the clutch system and torque to 74 inch lbs (8.4 Nm).
17. Inspect and adjust brake pedal height.
18. Inspect and adjust clutch pedal sub-assembly.
19. Check brake fluid leakage.
20. Check clutch fluid leakage.
21. Check fluid level in reservoir.

CLUTCH SLAVE CYLINDER

REMOVAL & INSTALLATION

See Figures 27 through 28.

1. Disconnect clutch release cylinder to flexible hose tube, as follows:
   a. Remove the bolt and tube clamp bracket.
   b. Using SST (SST: 09023-00101), disconnect the 2 flexible hose tubes. Use a container to catch the fluid.
2. Remove the bolt, 2 nuts and the clutch accumulator assembly.
3. Remove the 2 bolts and clutch release cylinder assembly.

To install:

4. Install the clutch release cylinder assembly with the 2 bolts and tighten to 9 ft. lbs. (12 Nm).

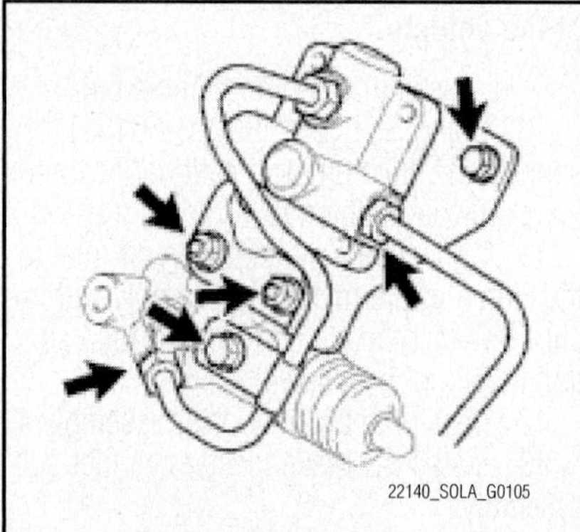

Fig. 27 Clutch release cylinder

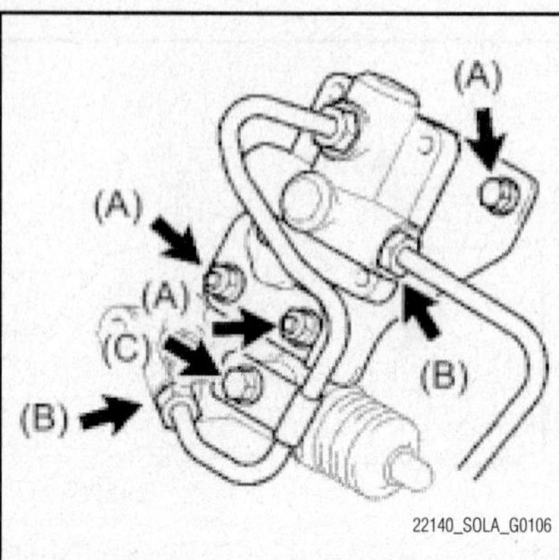

Fig. 28 Clutch accumulator assembly

5. Install the clutch accumulator assembly with the 2 nuts and the bolt (bolt and nut A) and tighten to 9 ft. lbs. (12 Nm).
6. Using SST (SST: 09023-00101), connect the 2 flexible hose tubes (flexible hose tube B) and tighten to 11 ft. lbs. (15 Nm).
7. Install the bolt (bolt C) and tube clamp bracket and tighten to 9 ft. lbs. (12 Nm).
8. Fill the brake reservoir tank with clutch fluid and bleed the clutch system. Torque to 74 inch lbs. (8.4 Nm).
9. Check clutch fluid leakage.
10. Check fluid level in reservoir.

FRONT HALFSHAFT

REMOVAL & INSTALLATION

See Figures 29 through 32.

1. Remove the engine under cover.
2. Drain the automatic transaxle fluid or manual transaxle oil.
3. Remove front wheel.
4. Using SST (SST: 09930-00010) and hammer, release the staked part of the axle hub nut.

➡Loosen the staked part of the nut completely, otherwise the screw of the drive shaft may be damaged.

5. While applying the brakes, remove the lock axle hub nut.
6. Remove the nut and separate the stabilizer link assembly.

➡If the ball joint turns together with the nut, use a hexagon wrench (6 mm) to hold the stud.

7. Remove the bolt and clip, and separate the sensor wire and hose from the shock absorber.
8. Remove the bolt and separate the speed sensor from the steering knuckle.

➡Do not allow foreign matter to adhere to the speed sensor. Be careful not to damage the speed sensor.

9. Separate tie rod end sub-assembly, as follows:
   a. Remove the cotter pin and nut.
   b. Using SST (SST: 09628-62011) or equivalent, separate the tie rod end sub-assembly from the steering knuckle.
10. Remove the bolt and 2 nuts, and separate the lower No. 1 front suspension arm sub-assembly from the lower ball joint.
11. Using a plastic hammer, separate the front drive shaft assembly from the front axle hub.

➡Be careful not to damage the drive shaft boot and speed sensor rotor.

12. Remove the front drive shaft assembly(s), as follows:
   a. For left front drive shaft, use SST (SST: 09520-01010, SST: 09520-24010) or equivalent, and remove the left front drive shaft assembly.

➡Be careful not to damage the drive shaft dust cover, boot and oil seal. Be careful not to drop the drive shaft assembly.

   b. For right front drive shaft, use a screwdriver and remove the bearing bracket hole snap ring.
   c. Remove the bolt and right front drive shaft assembly from the drive shaft bearing bracket.

➡Do not damage the boot and oil seal.

13. Fix front axle hub sub-assembly. The hub bearing could be damaged if it is subjected to the vehicle's full weight, such as moving the vehicle with the drive shaft removed. If it is necessary to place the vehicle's weight on the hub bearing, first support it with SST (SST: 09608-16042).

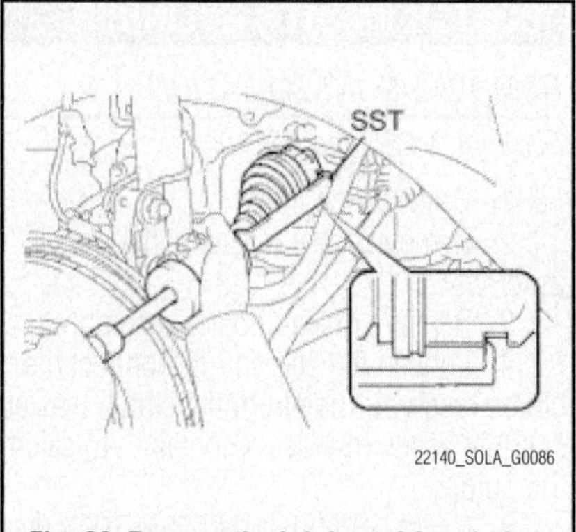

Fig. 29 Remove the left front drive shaft assembly

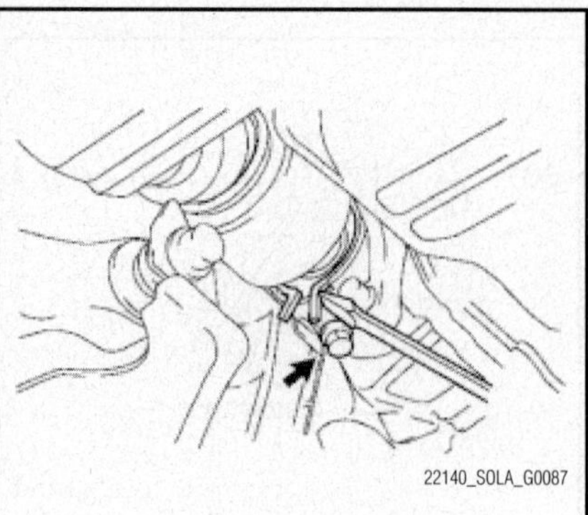

Fig. 30 Remove the right front drive shaft assembly

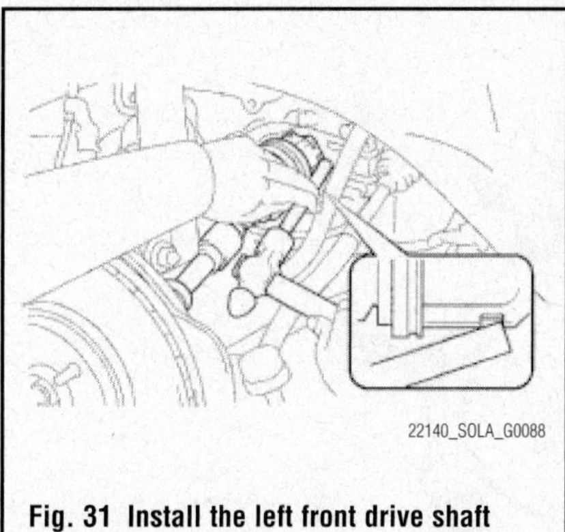

Fig. 31 Install the left front drive shaft assembly

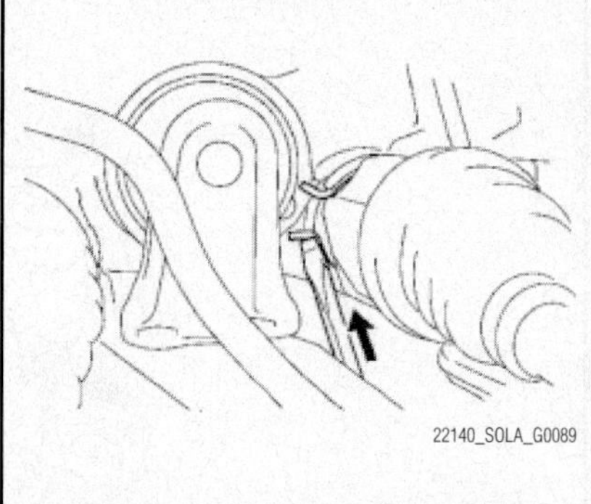

Fig. 32 Install the right front drive shaft assembly

To install:

14. Install the front drive shaft assembly(s), as follows:

a. Coat the spline of the inboard joint shaft assembly with automatic transaxle fluid.

b. For the left front drive shaft, align the shaft splines and install the drive shaft assembly with a brass bar and hammer.

➡Set the shaft snap ring with the opening side facing down.

c. For the right front drive shaft, use a screwdriver to install a new bearing bracket hole snap ring, and install a bolt tightened to 24 ft. lbs. (32 Nm).

➡Be careful not to damage the drive shaft dust cover, boot and oil seal.

➡Move the drive shaft assembly while keeping it level.

15. Install the drive shaft assembly to the front axle assembly.

➡Be careful not to damage the drive shaft boot and speed sensor rotor.

16. Install the lower ball joint to the lower front suspension arm sub-assembly with the bolt and 2 nuts and tighten to 55 ft. lbs. (75 Nm).
17. Install the tie rod end to the steering knuckle with the nut and tighten to 36 ft. lbs. (49 Nm).
18. Install a new cotter pin. If the holes for the cotter pin are not aligned, tighten the nut up to 60° further.
19. Install the front speed sensor to the steering knuckle with the bolt and tighten to 71 inch lbs. (8 Nm).
20. Install the flexible hose and the speed sensor to the shock absorber with the bolt and set the sensor clip on the knuckle and tighten to 14 ft. lbs. (19 Nm).

➡Be careful not to damage the speed sensor. Do not allow foreign matter to adhere to the speed sensor. Do not twist the sensor wire when installing the speed sensor.

21. Install the stabilizer link assembly with the nut and tighten to 55 ft. lbs. (74 Nm).

➡If the ball joint turns together with the nut, use a hexagon wrench (6 mm) to hold the stud.

22. Clean the threaded parts on the drive shaft and axle hub nut using a non-residue solvent.

➡Be sure to perform this work for a new drive shaft. Keep the threaded parts free of oil and foreign objects.

23. Using a socket wrench (30 mm), install a new axle hub nut and tighten to 217 ft. lbs. (294 Nm).
24. Using a chisel and hammer, stake the front axle hub nut.
25. Install front wheel.
26. Add automatic transaxle fluid or manual transaxle oil.
27. Inspect automatic transaxle fluid or manual transaxle oil level.
28. Inspect and adjust front wheel alignment.
29. Install the engine under cover.
30. Check the ABS speed sensor signal.

ENGINE COOLING

THERMOSTAT

REMOVAL & INSTALLATION

2.4L Engine

See Figure 33.

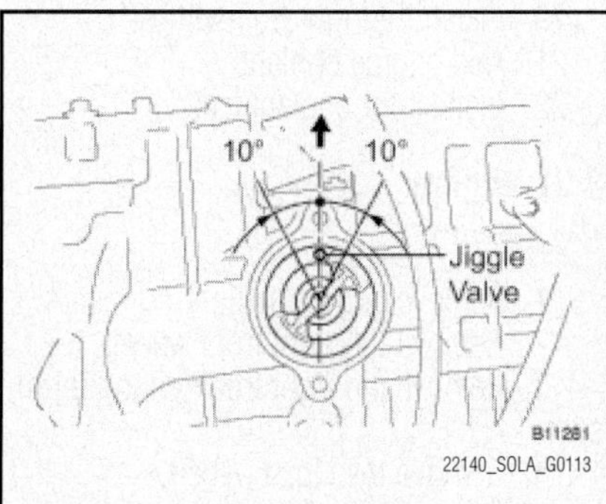

Fig. 33 Jiggle valve positioning—2.4L engine

1. Drain engine coolant.
2. Remove the fan and generator V-belt.
3. Remove the generator assembly.
4. Remove the radiator inlet hose.
5. Remove the 2 nuts and disconnect the water inlet from the cylinder block.
6. Remove the thermostat.

To install:

7. Install a new gasket to the thermostat.
8. Install the thermostat with the jiggle valve facing up.

➡The jiggle valve may be set within 10° on either side of the prescribed position.

9. Install the water inlet with the 2 nuts and tighten to 80 inch lbs. (9 Nm).
10. Install the radiator inlet hose.
11. Install the generator assembly.
12. Install the fan and generator V-belt.
13. Add engine coolant.
14. Inspect for coolant leak.

3.3L Engine

See Figure 34.

1. Drain engine coolant.
2. Remove the air cleaner inlet assembly.
3. Remove the air cleaner assembly.
4. Remove the air cleaner bracket.
5. Remove the No. 1 air cleaner inlet.
6. Remove the radiator outlet hose.
7. Remove the bolt and water inlet pipe.
8. Remove the 3 bolts and water inlet.
9. Remove the thermostat.

To install:

10. Install a new gasket to the thermostat.
11. Align the thermostat jiggle valve with the upper stud bolt, and insert the thermostat in the water inlet housing.

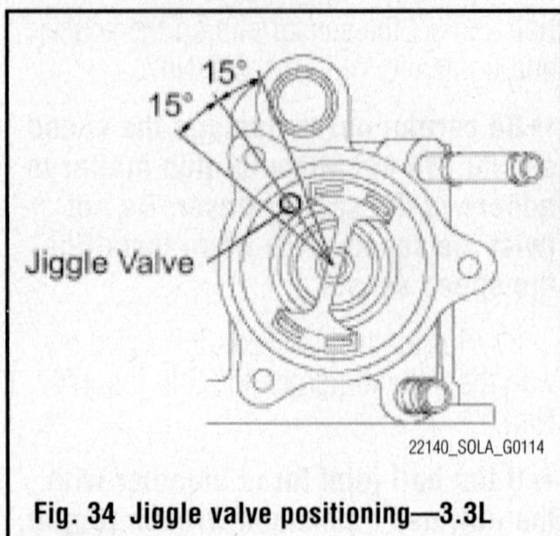

Fig. 34 Jiggle valve positioning—3.3L engine

➡The jiggle valve may be set within 15° of either side of the prescribed position.

12. Install the water inlet and tighten to 71 inch lbs. (8 Nm).
13. Install the water inlet pipe, as follows:
    a. Install a new O-ring to the inlet pipe.
    b. Apply soapy water to the O-ring.
    c. Connect the inlet pipe to the water inlet.
    d. Install the bolt holding the inlet pipe to the cylinder head and tighten to 15 ft. lbs. (20 Nm).
14. Install the air cleaner assembly.
15. Connect the vacuum hoses.
16. Add engine coolant.
17. Inspect for coolant leak.

WATER PUMP

REMOVAL & INSTALLATION

2.4L Engine

See Figures 35 and 36.

1. Drain engine coolant.
2. Remove right front wheel.
3. Remove the right front fender apron seal.
4. Remove the engine moving control rod with bracket.
5. Remove the right No. 2 engine mounting stay.
6. Remove the right No. 2 engine mounting bracket.
7. Remove the fan and generator V-belt.
8. Remove the generator assembly.
9. Using SST (SST: 09960-10010), remove the 4 bolts and water pump pulley.
10. Disconnect the crankshaft position sensor wire harness clamp.
11. Remove the 4 bolts, 2 nuts, bracket and water pump.

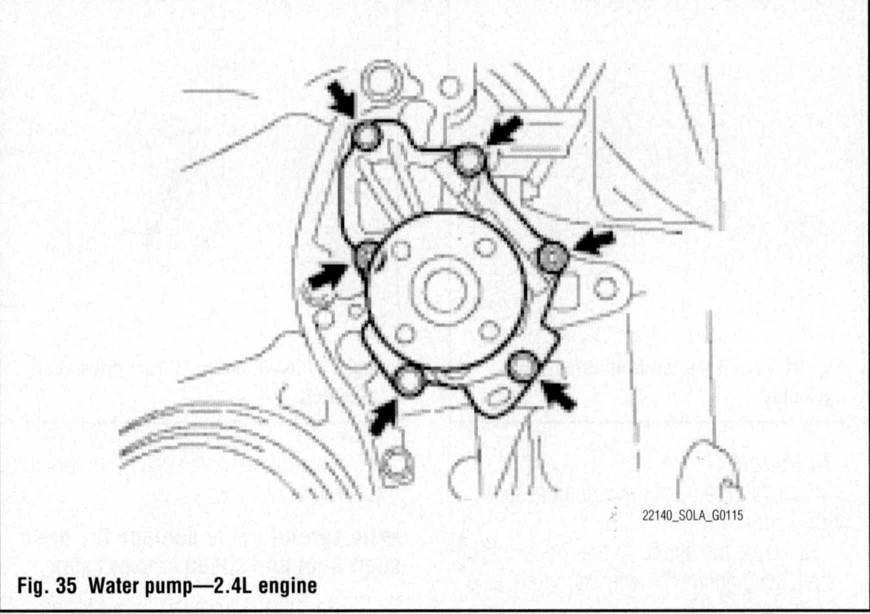

Fig. 35 Water pump—2.4L engine

To install:

12. Install the water pump assembly, as follows:
    a. Remove any old packing (FIPG) material and be careful not to spill any oil on the contacting surfaces of the water pump and cylinder block.

➡Using a razor blade and gasket scraper, remove all the old packing (FIPG) materials from the gasket surfaces and sealing grooves.

➡Thoroughly clean all components to remove all loose material.

➡Using a non-residue solvent, clean both sealing surfaces. Apply seal packing (Part No. 08826-00080 or equivalent) to the water pump as shown in the illustration.

➡Install a nozzle that has been cut to a 2.2 to 2.5 mm (0.09 to 0.10 in.) opening.

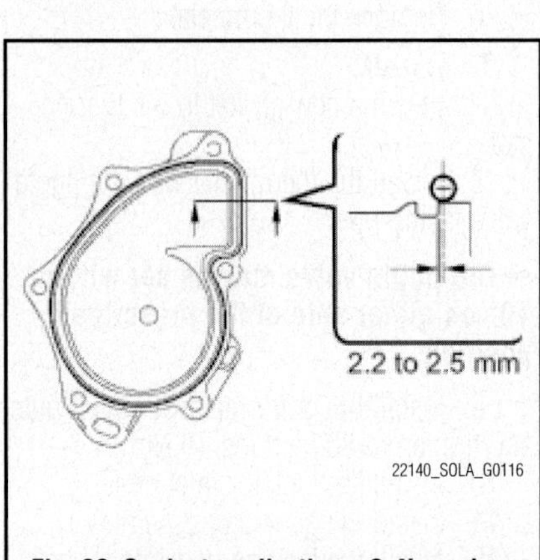

Fig. 36 Sealant application—2.4L engine

➡Parts must be assembled within 5 minutes of application. Otherwise the material must be removed and reapplied.

➡Immediately remove the nozzle from the tube and reinstall cap.

    b. Install the water pump and bracket with the 4 bolts and 2 nuts and tighten to 80 inch lbs. (9 Nm).
    c. Connect the crankshaft position sensor wire harness clamp.
13. Using SST (SST: 09960-10010), install the pump pulley with the 4 bolts and tighten to 19 ft. lbs. (26 Nm).
14. Install the generator assembly.
15. Install the fan and generator V-belt.
16. Install the right No. 2 engine mounting bracket.
17. Install the engine moving control rod with bracket.
18. Install the right No. 2 engine mounting stay.
19. Install the right front fender apron seal.
20. Install the right front wheel.
21. Add engine coolant.
22. Inspect for coolant leak.

3.3L Engine

See Figure 37.

1. Remove engine coolant.
2. Remove the right front wheel.
3. Remove the right front fender apron seal.
4. Remove the No. 1 V-belt.
5. Remove the vane pump V-belt.
6. Remove the engine moving control rod.

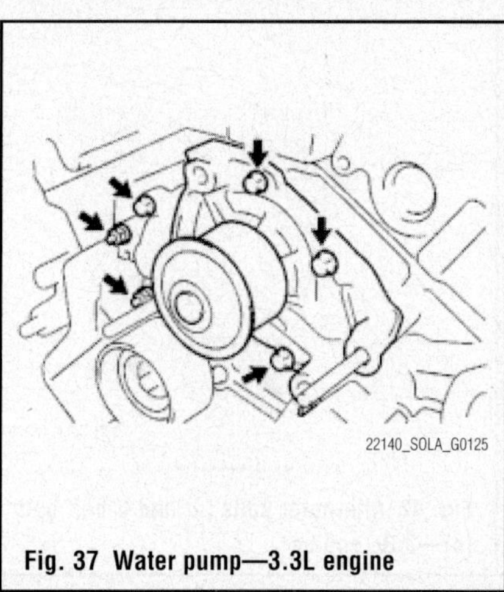
Fig. 37 Water pump—3.3L engine

7. Remove the No. 2 engine mounting stay.
8. Remove the No. 2 generator bracket.
9. Remove the crankshaft pulley.
10. Remove the No. 1 and No. 2 timing belt covers.
11. Remove the right engine mounting bracket.
12. Remove the No. 2 timing belt guide.
13. Remove the timing belt tensioner assembly.
14. Remove the timing belt.
15. Remove the bolt and No. 2 timing belt idler sub-assembly.
16. Remove the camshaft timing pulley.
17. Remove the No. 3 timing belt cover.
18. Remove the 4 bolts, 2 nuts and water pump.

To install:

19. Install a new gasket and the water pump with the 4 bolts and 2 nuts and tighten to 71 inch lbs. (8 Nm).
20. Install the No. 3 timing belt cover.
21. Install the camshaft timing pulley.
22. Install the No. 2 timing belt idler sub-assembly and tighten to 32 ft. lbs (43 Nm).
23. The remainder of installation is the reverse of removal, noting the following:
    a. Inspect the timing belt before installing.
    b. Inspect drive belt tension.
    c. Inspect for coolant leak.

ENGINE ELECTRICAL

CHARGING SYSTEM

ALTERNATOR

REMOVAL & INSTALLATION

2.4L Engine

See Figures 38 through 41.

1. Disconnect cable from negative battery terminal.

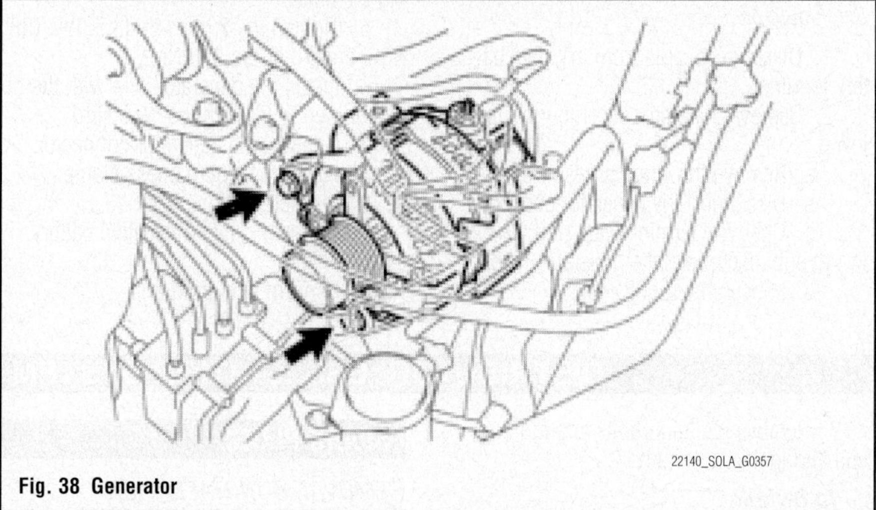
Fig. 38 Generator

2. Remove the right front wheel.
3. Remove the right front fender apron seal.
4. Remove the No. 1 engine cover sub-assembly.
5. Remove the engine moving control rod with bracket.
6. Remove the right No. 2 engine mounting stay.
7. Remove the right No. 2 engine mounting bracket.
8. Remove the fan and generator V-belt.
9. Remove the generator assembly, as follows:
    a. Disconnect the generator connector.
    b. Remove the nut and disconnect the wire harness from terminal B.
    c. Remove the bolt and wire harness clamp bracket.
    d. Remove the wire harness clamps.
    e. Remove the 2 bolts and generator assembly.

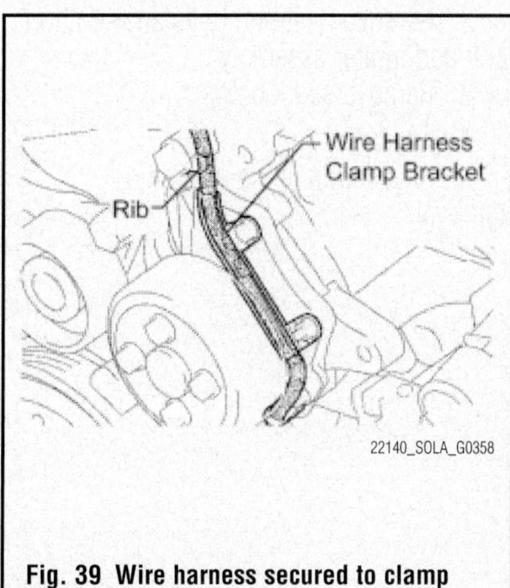

Fig. 39 Wire harness secured to clamp bracket

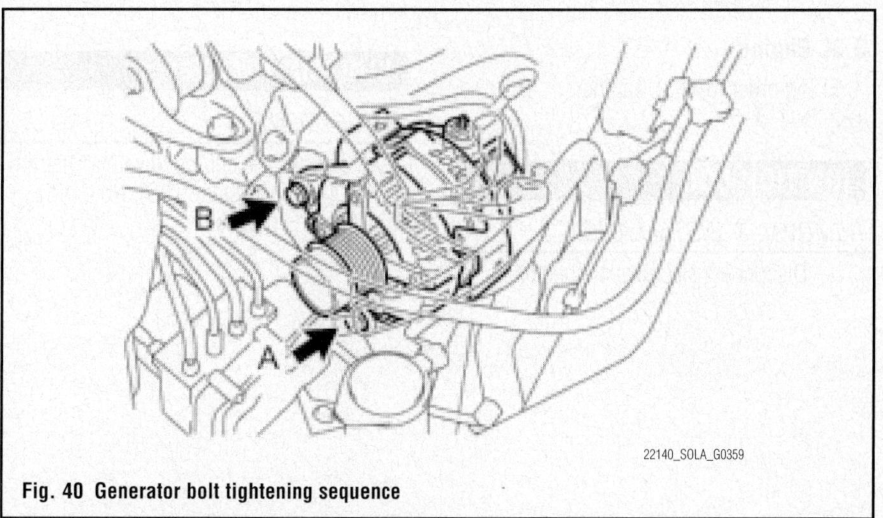

Fig. 40 Generator bolt tightening sequence

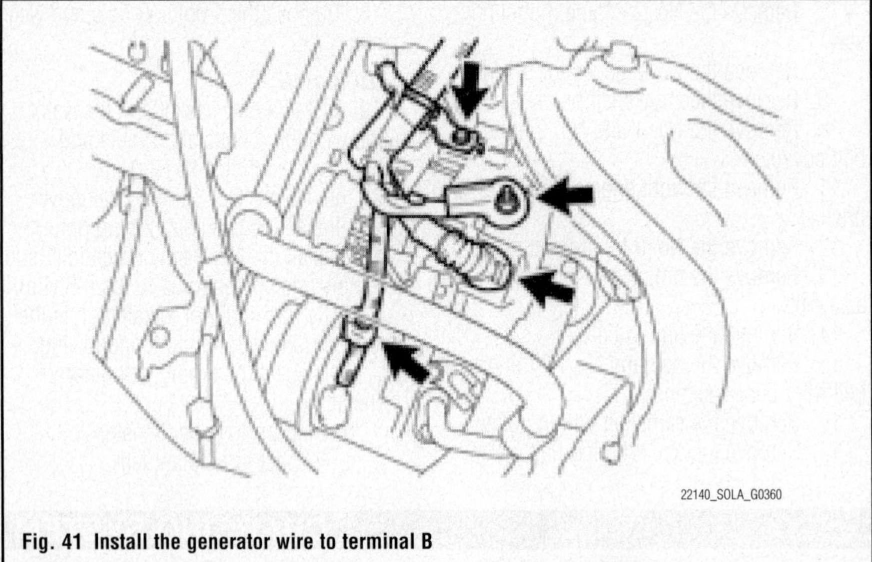

Fig. 41 Install the generator wire to terminal B

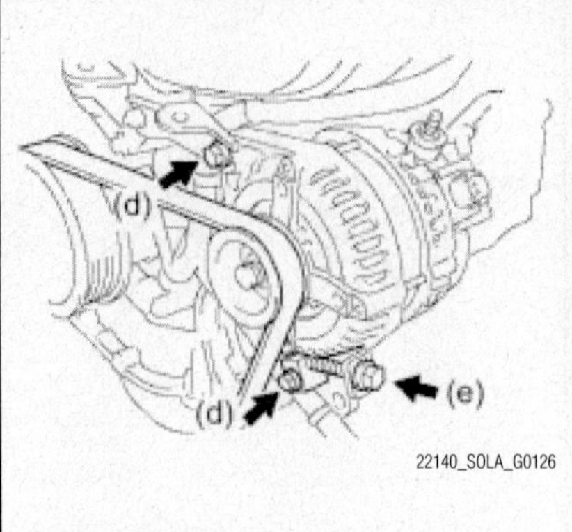

Fig. 42 Alternator bolts (d) and V-belt bolt (e)—3.3L engine

To install:

10. Confirm that the wire harness of the crankshaft position sensor is secured to the wire harness clamp bracket through the back of the rib of the timing chain cover.

11. Install the generator assembly with the 2 bolts. Tighten bolt A to 15 ft. lbs. (21 Nm). And bolt B to 38 ft. lbs. (52 Nm).

12. Install the generator wire to terminal B with the nut and tighten to 87 inch lbs. (9.8 lbs).

13. Install the clamp bracket with the bolt and tighten to 74 inch lbs. (8.4 lbs)

14. Attach the clamp and connect the generator connector to the generator.

15. The remainder of installation is the reverse of removal.

16. Perform initialization.

3.3L Engine

See Figure 42.

1. Disconnect cable from negative battery terminal.

2. Remove generator assembly, as follows:
   a. Remove the wire harness clamp.
   b. Disconnect the generator connector.
   c. Open the terminal cap, remove the nut and disconnect the generator wire.
   d. Loosen the 2 bolts (d).
   e. Loosen the bolt (e) to lessen the tension of the V-belt.
   f. Remove the 2 bolts (d) and generator.

To install:

3. Install the generator with the 2 bolts, but do not fully tighten the bolts.

4. Adjust the V-belt tension by tightening the bolt.

5. Tighten the 2 bolts to 43 ft. lbs. (58 Nm), and 13 ft. lbs. (18 Nm).

6. Install the generator wire with the nut and tighten to 87 inch lbs. (9.8 Nm).

7. Connect the generator connector.

8. Install the wire harness clamp.

9. Inspect the drive belt.

10. Connect cable to negative battery terminal.

11. Perform initialization.

ENGINE ELECTRICAL — DISTRIBUTORLESS IGNITION SYSTEM

FIRING ORDER

2.4L Engine

Firing order for 2.4L engine: 1–3–4–2

3.3L Engine

Firing order for 3.3L engine: 1–2–3–4–5–6

IGNITION COIL

REMOVAL & INSTALLATION

1. Disconnect electrical connectors.

2. Remove retaining bolts and ignition coil and igniter assembly.

To install:

3. Installation is the reverse of removal.

IGNITION TIMING

ADJUSTMENT

All engines are equipped with a Distributorless Ignition System (DIS). No timing adjustment is possible.

SPARK PLUGS

REMOVAL & INSTALLATION

1. Disconnect electrical connectors.

2. Remove retaining bolts and ignition coil and igniter assembly.

3. Remove spark plugs.

To install:

4. Installation is the reverse of removal.

STARTER

REMOVAL & INSTALLATION

2.4L Engine

See Figure 43.

1. Remove battery and tray.
2. Remove air cleaner assembly and bracket.
3. Removc thc 2 bolts and air cleaner inlet.
4. Disconnect the starter connector.
5. Remove the nut and disconnect the starter wire.
6. Remove the 2 bolts and starter.

To install:

7. Install the starter with the 2 bolts and tighten to 29 ft. lbs. (39 Nm).
8. Connect the starter wire with the nut and tighten to 9 ft. lbs. (13 Nm).
9. Connect the starter connector.
10. Install the air cleaner bracket with the 2 bolts and tighten to 9 ft. lbs. (12 Nm).
11. Install the air cleaner assembly and tighten to 44 inch lbs. (5 Nm).
12. Install the air cleaner inlet with the 2 bolts and tighten to 62 inch lbs. (7 Nm).
13. Check vacuum hose connection.
14. Install battery and tray.
15. Perform Initialization.

3.3L Engine

See Figure 44.

1. Remove the battery and battery tray.
2. Remove the air cleaner inlet assembly.
3. Remove the air cleaner assembly.
4. Remove the air cleaner bracket.
5. Remove the No. 1 air cleaner inlet.
6. Disconnect the starter connector.
7. Open the terminal cap, and remove the nut and disconnect the starter wire.
8. Remove the 2 bolts and starter.

To install:

9. Install the starter with the 2 bolts and tighten to 26 ft. lbs (37 Nm).
10. Connect the starter wire with the nut and tighten to 9 ft. lbs (13 Nm).
11. Connect the starter connector.
12. Install the terminal nut and cover the nut with the cap and tighten to 87 inch lbs. (9.8 Nm).
13. The remainder of installation is the reverse of installation.

22140_SOLA_G0355

Fig. 43 Starter

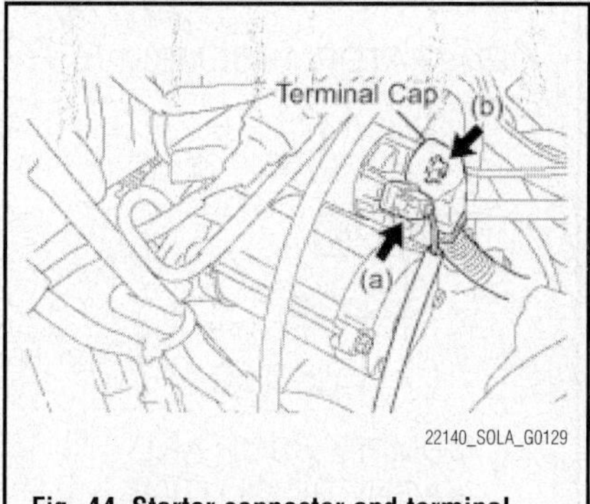

Fig. 44 Starter connector and terminal cap—3.3L engine

ENGINE MECHANICAL

ACCESSORY DRIVE BELTS

INSPECTION

Visually check the V-ribbed belt for excessive wear, frayed cords, etc. All worn or damaged drive belts should be replaced immediately. Cracks on the rib side of a V-ribbed belt are considered acceptable, If the drive belt has chunks missing from its ribs, it should be replaced. After installing the V-ribbed belt, check that it fits properly in the ribbed grooves. Check to confirm that the belt has not slipped out of the grooves on the bottom of the crank pulley by hand.

ADJUSTMENT

Belt tension is maintained by an automatic tensioner and cannot be adjusted.

REMOVAL & INSTALLATION

2.4L Engine

See Figure 45.

1. Before servicing the vehicle, refer to the Precautions Section.
2. Remove the right hand front wheel.
3. Remove the right hand engine under cover.
4. Remove the right hand front fender apron seal.
5. Remove the 3 bolts and the engine moving control rod with bracket.
6. Remove the No. 2 right engine mounting stay.
7. Remove the No. 2 right engine mounting bracket.
8. Using SST, slowly turn the V-ribbed belt tensioner clockwise.
9. Remove the V-ribbed belt from each pulley and slowly return the tensioner.

➡Make sure that SST and other tools are set to the tensioner securely.

➡When compressing the V-ribbed belt tensioner, slowly turn the tensioner.

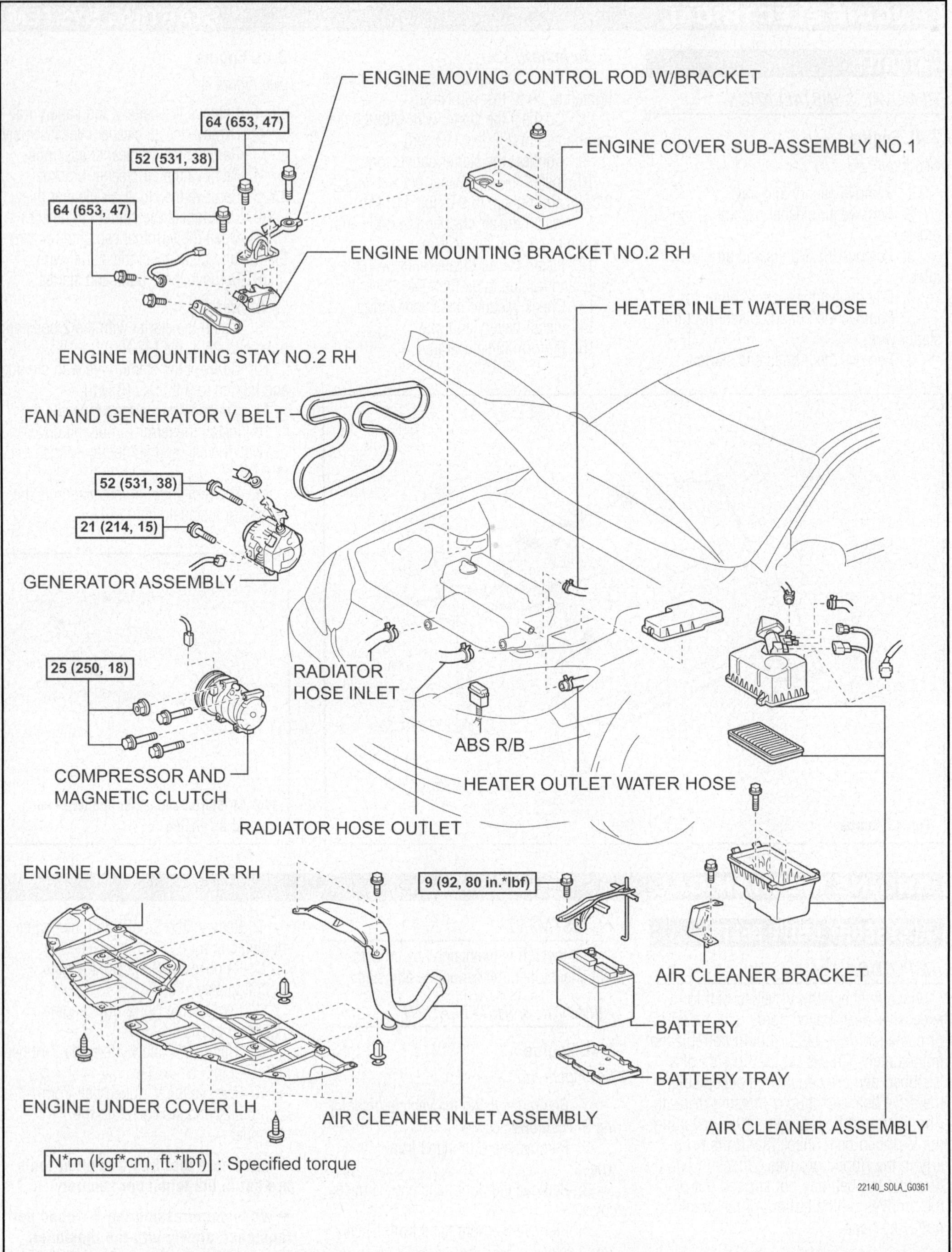

Fig. 45 Locating drive belt components—2.4L Engine

➡Be careful not to pinch your fingers between the parts.

To install:

10. To install, reverse removal procedure.
11. After installing the V-ribbed belt, check that it fits properly in the ribbed grooves. Check to confirm that the belt has not slipped out of the grooves on the bottom of the crank pulley by hand.
12. Tighten the right hand front wheel to: 76 ft. lbs. (103 Nm).

3.3L Engine

See Figures 46 and 47.

1. Remove the right front wheel.
2. Remove the right front fender apron seal.
3. Remove the No. 1 (cooler compressor to crankshaft pulley) V-belt, as follows:
   a. Loosen bolts A and B.
   b. Loosen adjusting bolt C and remove the belt.
4. Loosen bolts A and B, and remove the vane pump V-belt.

To install:

5. Install the vane pump V-belt, as follows:
   a. Install the belt on each pulley.
   b. Using a bar, adjust the belt tension and tighten bolt B to 32 ft. lbs. (43 Nm).
   c. Tighten bolt A to 32 ft. lbs. (43 Nm).
6. Install the No. 1 (cooler compressor to crankshaft pulley) V-belt, as follows:
   a. Install the belt on each pulley.
   b. Tighten adjusting bolt C, and adjust the belt tension.
   c. First tighten bolt A to 43 ft. lbs. (58 Nm), then bolt B to 13 ft. lbs. (18 Nm).
7. Install the right front fender apron seal.
8. Install the right front wheel.

BALANCE SHAFT

REMOVAL & INSTALLATION

See Figures 48 through 55.

1. Before servicing the vehicle, refer to the Precautions Section.
2. Connect the negative battery cable.
3. Drain the engine oil.
4. Remove the oil pump. Refer to Oil Pump below for removal procedure.
5. Remove the No. 1 and No. 2 balance shaft sub-assembly. Remove the eight bolts in sequence.
6. Remove the No. 1 and No. 2 balance shafts.

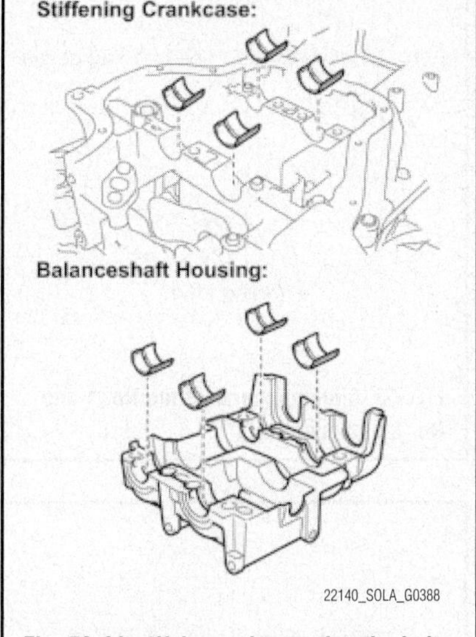

Fig. 50 Identifying and removing the balance shaft bearings

7. Remove the balance shaft bearings if necessary.

To install:

➡Do not apply engine oil to the bearings and the contact surfaces.

8. Install the bearings in the crankcase and balance shaft housing.
9. Apply a light coat of engine oil to the bearings.

➡Confirm that the match marks on driven gears No. 1 and No. 2 are matched.

10. Install No. 1 and No. 2 balance shaft sub-assembly. Rotate the No. 1 driven gear of the No. 1 balance shaft in the rotating direction until it hits the stopper.
11. Align the timing marks of the No. 1 and No. 2 balance shafts as shown.
12. Place the No. 1 and No. 2 balance shafts on the crankcase.
13. Apply a light coat of engine oil under the heads of the balance shaft housing bolts.

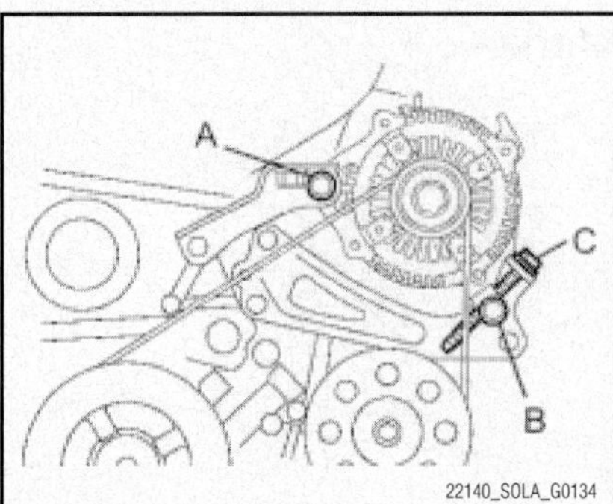

Fig. 46 Cooler compressor to camshaft pulley V-belt and bolts—3.3L engine

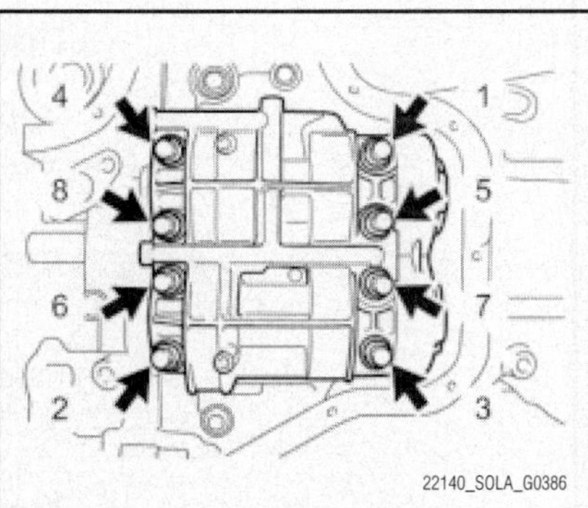

Fig. 48 Sub-assembly bolt removal sequence

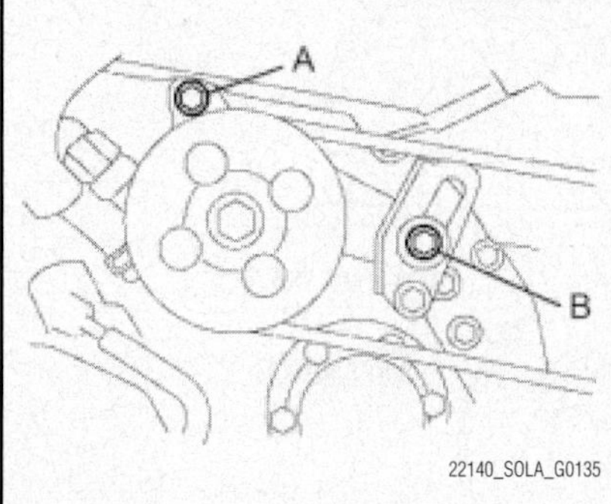

Fig. 47 Vane pump V-belt and bolts—3.3L engine

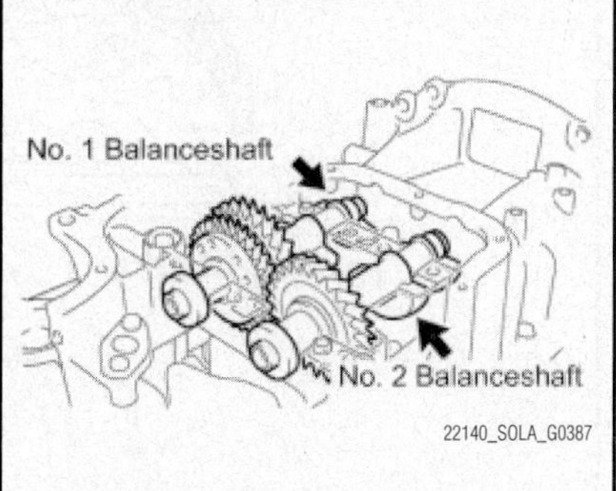

Fig. 49 Removing the No. 1and No. 2 balance shafts

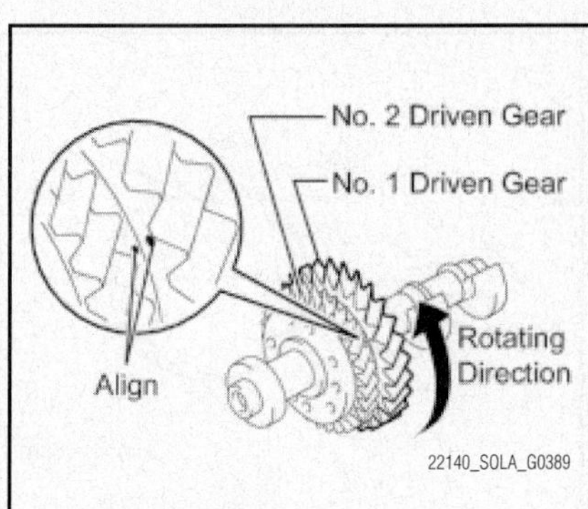

Fig. 51 Rotate the driven gear No. 1 of balance shaft No. 1

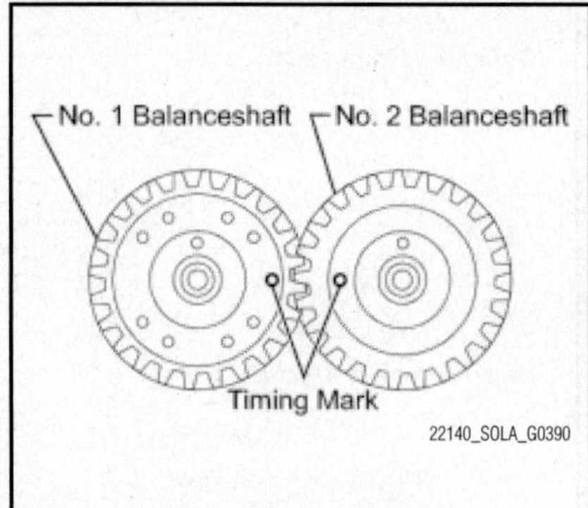

Fig. 52 Aligning marks of the No. 1 and No. 2 balance shafts

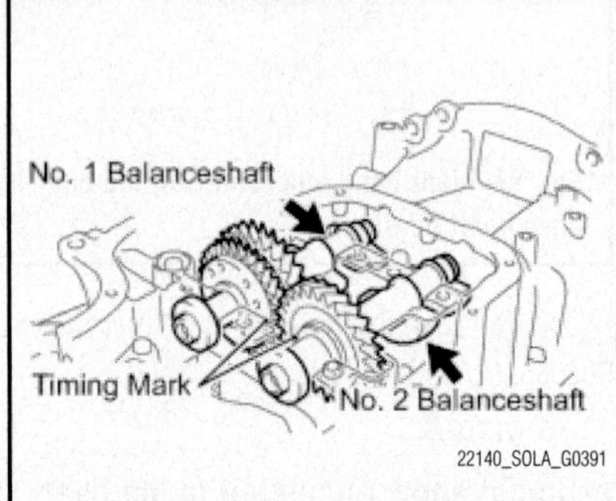

Fig. 53 Placing the No. 1 and No. 2 balance shafts on the crankcase

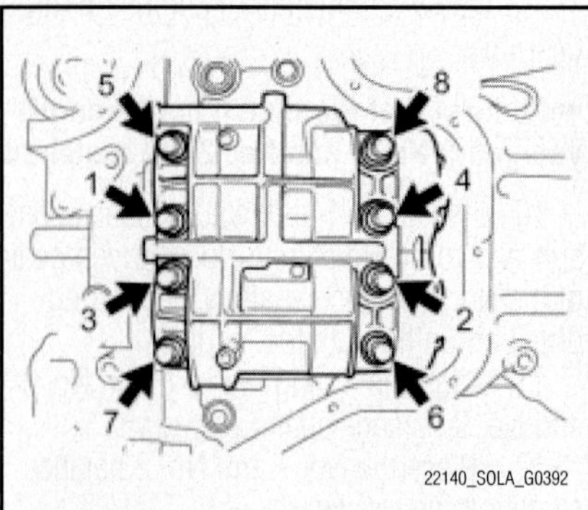

Fig. 54 Balance shaft housing bolt tightening sequence

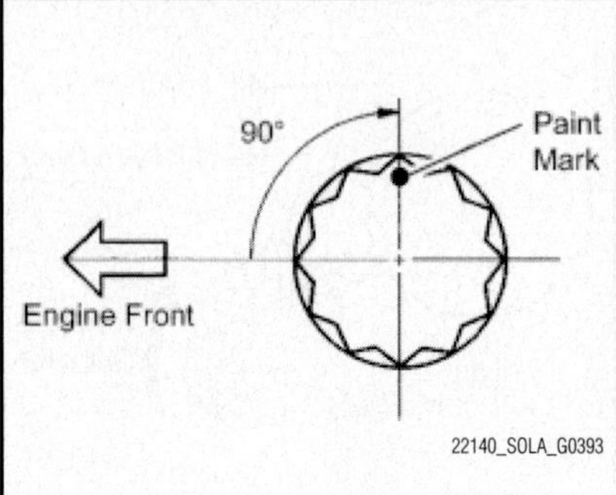

Fig. 55 Marking the bolt head and tightening procedure

14. Install the balance shaft housing bolts. The balance shaft housing bolts should be tightened in 2 progressive steps as follows:
    a. Tighten the 8 balance shaft housing bolts in sequence to: 16 ft. lbs. (22 Nm).
    b. Mark the front side of each balance shaft housing bolt head with paint. Retighten the bolts by 90°. Check that the paint marks are now at a 90° angle to the front.
15. To complete installation, reverse remaining removal procedure.
16. Check the engine for leaks.

CAMSHAFT AND VALVE LIFTERS

REMOVAL & INSTALLATION

2.4L Engine

See Figures 56 through 65.

1. Before servicing the vehicle, refer to the Precautions Section.
2. Disconnect the negative battery cable.
3. Remove the front right hand wheel.
4. Remove the right front fender apron seal.
5. Remove the No. 1 right hand under cover.
6. Remove spark plug.
7. Remove ventilation hose.
8. Remove No. 2 ventilation hose.
9. Remove the engine wire.
10. Remove the cylinder head cover sub-assembly
11. Set No.1 cylinder to TDC/Compression by performing the following:
    a. Turn the crankshaft pulley until its groove and the timing mark "0" of the timing chain cover are aligned
    b. Check that each timing mark of the camshaft timing gear and sprocket is aligned with each timing mark on the bearing caps as shown in the illustration.
    c. Place paint marks on the timing chain.
12. Remove No. 1 chain tensioner assembly. Remove the 2 nuts, tensioner and gasket.
13. Remove the No. 2 camshaft by performing the following:
    a. While holding the camshaft with a wrench, loosen the camshaft timing set bolt.

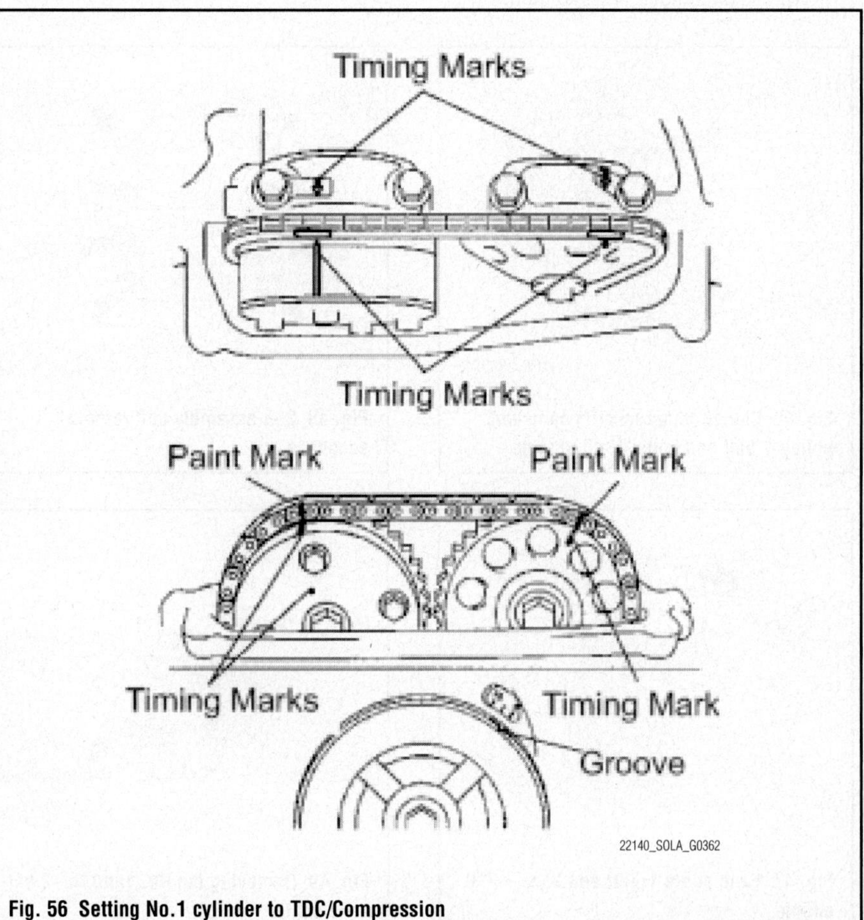

Fig. 56 Setting No.1 cylinder to TDC/Compression

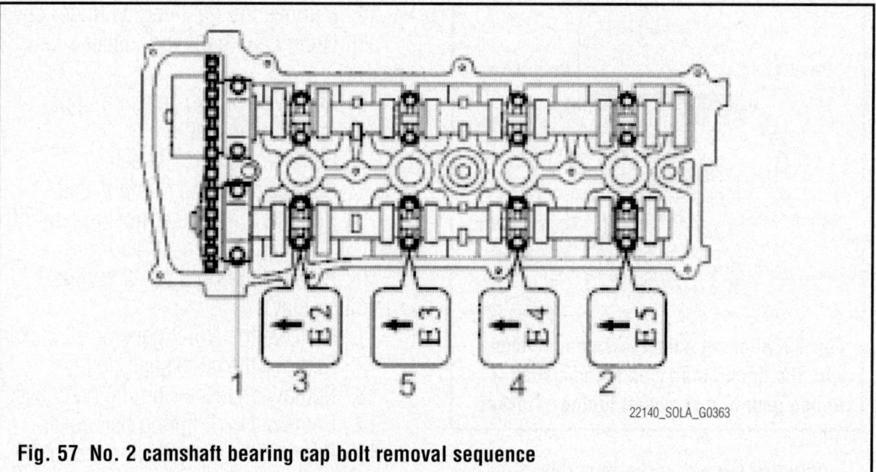

Fig. 57 No. 2 camshaft bearing cap bolt removal sequence

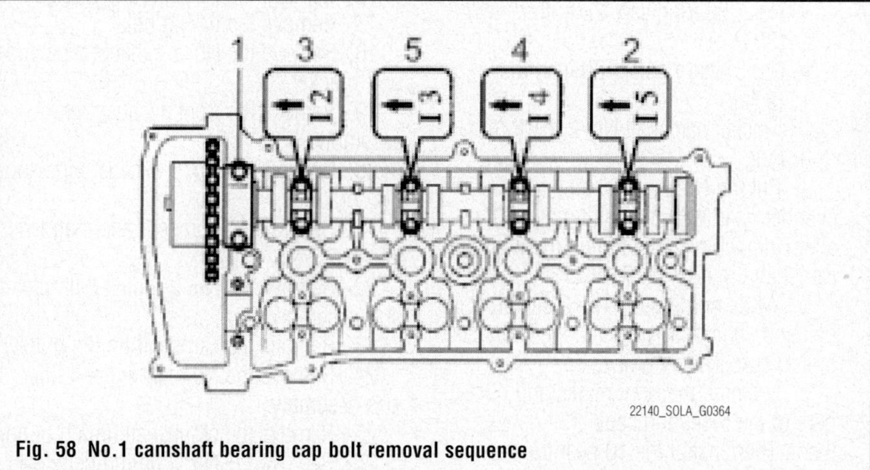

Fig. 58 No.1 camshaft bearing cap bolt removal sequence

b. Loosen and remove the bearing cap bolts on No. 2 camshaft in the sequence shown in the illustration in several passes, and remove the 5 bearing caps.

c. Raise the No. 2 camshaft and remove it. Then remove the sprocket bolt.

d. Remove the camshaft timing sprocket from the No. 2 camshaft with the timing chain wrapped on the sprocket.

e. Remove the camshaft timing sprocket from the timing chain.

14. Remove the No. 1 camshaft by performing the following:

a. Uniformly loosen and remove the 10 bearing cap bolts in the sequence shown in the illustration.

b. Remove the 5 bearing caps.

c. Remove the camshaft and camshaft timing gear while holding the timing chain by hand.

➡Be careful not to drop anything inside the timing chain cover.

d. Tie the timing chain with a string.

15. Remove the camshaft timing gear assembly by performing the following:

a. Clamp the camshaft in a vise, and make sure that the camshaft timing gear does not rotate.

b. Cover all the oil ports except the advance side port with vinyl tape

➡Cover the paths with a shop rag or piece of cloth to avoid oil splashes.

➡Depending on the air pressure, the camshaft timing gear will turn to the advance angle side without applying force by hand. Also, if the pressure is difficult to apply because of air leakage from the port, the lock may be difficult to release.

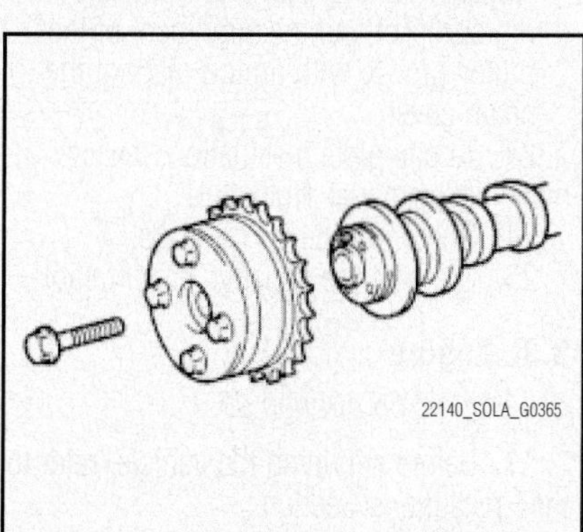

Fig. 59 Removing the flange bolt from the camshaft timing gear

c. Apply air pressure of 14 psi to the oil path, then turn the camshaft timing gear in the advance direction (counterclockwise) by hand.

d. Remove the flange bolt of the camshaft timing gear. Be sure not to remove the other four bolts. If planning to reuse the gear, be sure to release the straight pin lock before installing the gear.

To install:

16. Put the camshaft timing gear against the camshaft.

➡Be sure not to turn the camshaft timing gear to the retard angle side (the right angle).

17. Turn the camshaft timing gear as shown in the illustration while pushing it gently against the camshaft. Push further at the position where the pin fits into the groove.

18. Check that there is no clearance between the gear and camshaft.

19. Tighten the flange bolt with the camshaft timing gear fixed in place and tighten to 40 ft. lbs. (54 Nm).

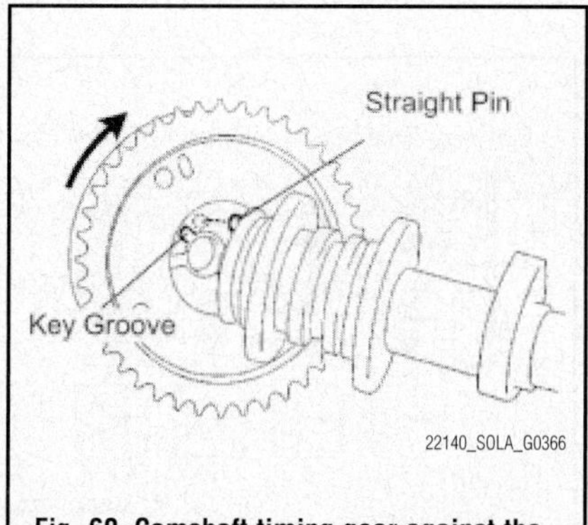

Fig. 60 Camshaft timing gear against the camshaft

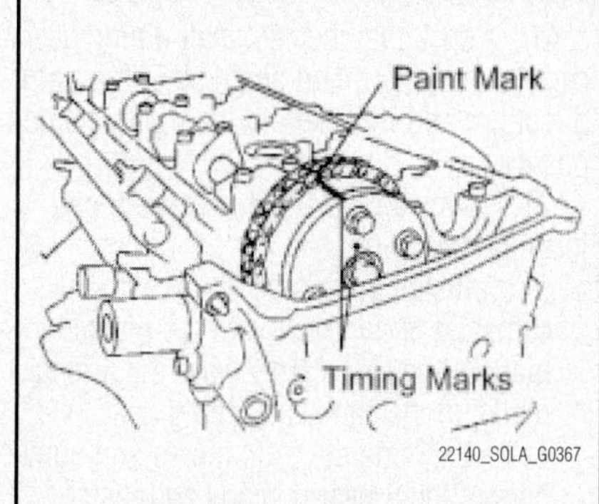

Fig. 61 Aligning No. 1 camshaft paint mark with the timing mark

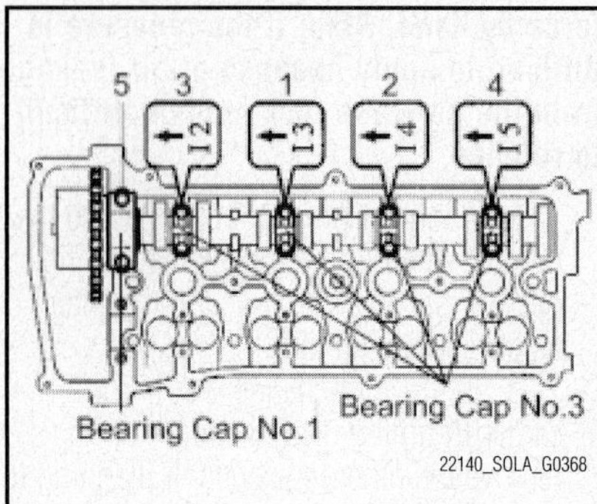

Fig. 62 Checking the No. 1 camshaft bearing cap marks, numbers and order

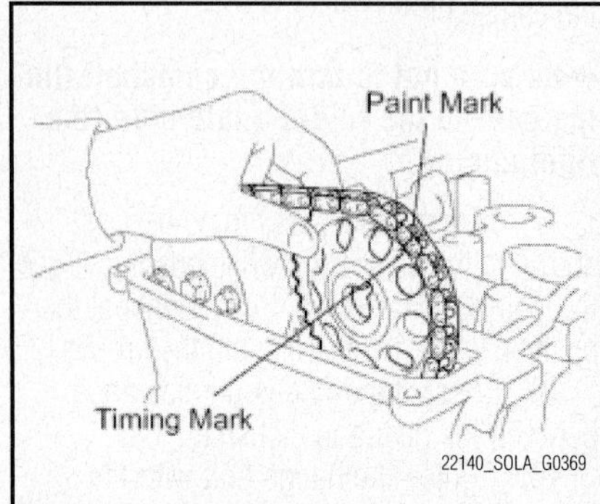

Fig. 63 Aligning No. 2 camshaft paint mark with the timing mark

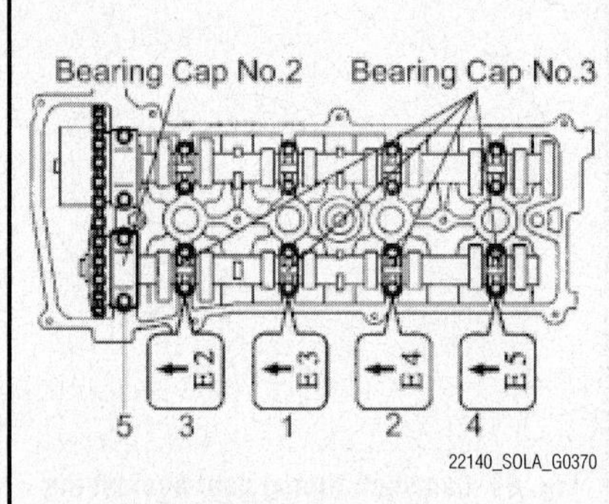

Fig. 64 Checking the No. 2 camshaft bearing cap marks, numbers and order

20. Check that the camshaft timing gear can move to the retard angle side (the right direction) and is locked in the most retarded position.

21. To install the No. 1 camshaft, perform the following:

a. Install the timing chain onto the camshaft timing gear with the painted mark of the link aligned with the timing marks of the camshaft timing gear.

b. Examine the front marks and numbers of the 5 bearing caps and install them. Then install the 10 bearing cap bolts. Uniformly tighten the bolts in the sequence shown in the illustration. Tighten the following to specification:

- No. 1 Bearing cap: 22 ft. lbs. (30 Nm)
- No. 3 Bearing cap: 80 inch lbs. (9 Nm)

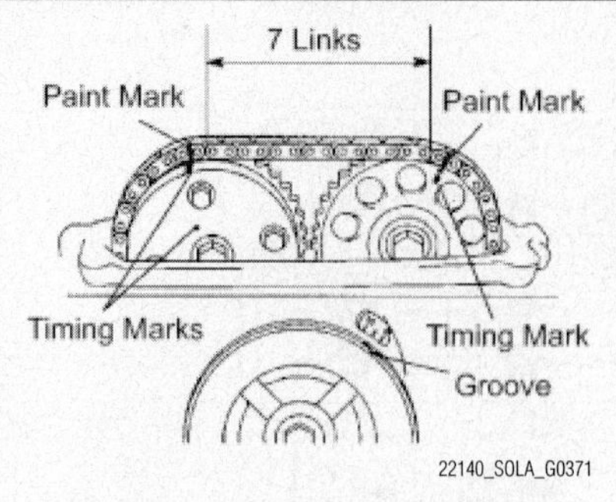

Fig. 65 Aligning paint marks on the chain with the timing marks on the camshaft timing gear and camshaft timing sprocket

22. To install camshaft No. 2, perform the following:

a. Put the No. 2 camshaft on the cylinder head with the paint mark of the chain aligned with the timing mark on the camshaft timing sprocket.

b. While holding the No. 2 camshaft by hand, temporarily tighten the camshaft timing sprocket set bolt.

c. Examine the front marks and numbers of the 5 bearing caps and install them. Then install the 10 bearing cap bolts. Uniformly tighten the bolts in the sequence shown in the illustration. Tighten the following to specification:

- No. 1 Bearing cap: 22 ft. lbs. (30 Nm)
- No. 3 Bearing cap: 80 inch lbs. (9 Nm)

d. While holding the camshaft with a wrench, tighten the camshaft timing sprocket set bolt to 40 ft. lbs. (54 Nm).

e. As shown in the illustration, check the paint marks of the timing chain, camshaft timing gear and camshaft timing sprocket and the alignment of the pulley groove with timing mark of the chain cover.

23. To complete installation, reverse remaining removal procedure.

24. Check engine for oil leaks.

25. Connect the negative battery cable.

3.3L Engine

See Figures 66 through 90.

1. Before servicing the vehicle, refer to the Precautions Section.
2. Remove the right front wheel.
3. Remove the right front fender apron seal.
4. Remove the right engine under cover.
5. Drain engine oil and install a new gasket.
6. Remove front suspension upper brace center, if applicable.
7. Remove the No. 1 V-belt.
8. Remove the vane pump V-belt.
9. Remove the engine moving control rod.
10. Remove the right No. 2 engine mounting stay.
11. Remove the No. 2 generator bracket.
12. Remove the crankshaft pulley.
13. Remove the No. 1 timing belt cover.
14. Remove No. 2 timing belt cover.
15. Remove the right engine mounting bracket.
16. Remove the No. 2 timing belt guide.
17. Remove the timing belt.
18. Remove the No. 1 exhaust pipe support bracket.
19. Remove the front exhaust pipe assembly.
20. Remove the No. 1 exhaust pipe support bracket.
21. Separate compressor and magnetic clutch.
22. Remove the No. 2 timing belt idler sub-assembly.
23. Remove the camshaft timing pulley.
24. Remove the No.1 timing belt idler sub-assembly.
25. Remove the crankshaft timing pulley.
26. Remove the No. 3 timing belt cover.
27. Remove the No. 1 compressor mounting bracket.
28. Remove the oil level gage guide.
29. Separate the engine mounting insulator, as follows:

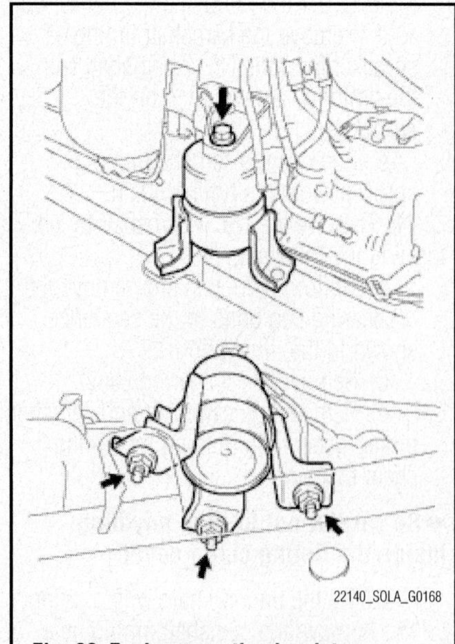

Fig. 66 Engine mounting insulator

a. Remove the 3 nuts and bolt. Then separate the engine mounting insulator. Do not remove the engine mounting insulator.

30. Remove the right engine mounting insulator as follows:

a. Remove the bolt and disconnect the power steering return hose clamp from the frame.

b. Remove the 4 nuts.

c. Prepare a jack. Place a wooden block on the jack and set the jack under the engine. Remove the engine mounting insulator.

d. Raise the jack and lift up the engine. Then remove the right engine mounting insulator. Be careful not to damage the contact surfaces of the oil pan.

31. Remove the right engine mounting bracket.

32. Remove the No. 2 oil pan sub-assembly.

33. Remove the oil strainer sub-assembly.

34. Remove the oil pan sub-assembly.

35. Remove the crankshaft position sensor.

36. Remove intake manifold.

37. Remove ignition coil assembly.

38. Remove cylinder head cover sub-assembly.

39. Remove camshaft, as follows:

WARNING

Since the thrust clearance of the camshaft is small, the camshaft must be kept level while it is being removed. If the camshaft is not kept level, damage to the cylinder head or to the camshaft may result. To avoid this, the following steps should be carried out.

a. Align the camshaft drive and driven gear's timing marks (2 dot marks each) by turning the camshaft with a wrench.

b. Secure the exhaust camshaft sub gear to the main gear with a service bolt.

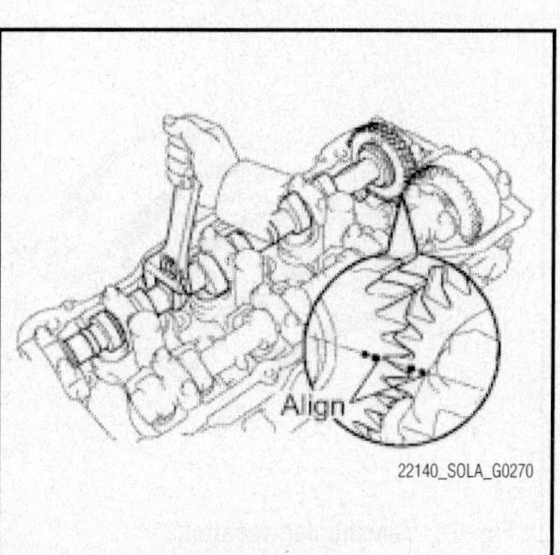

Fig. 67 Camshaft drive alignment

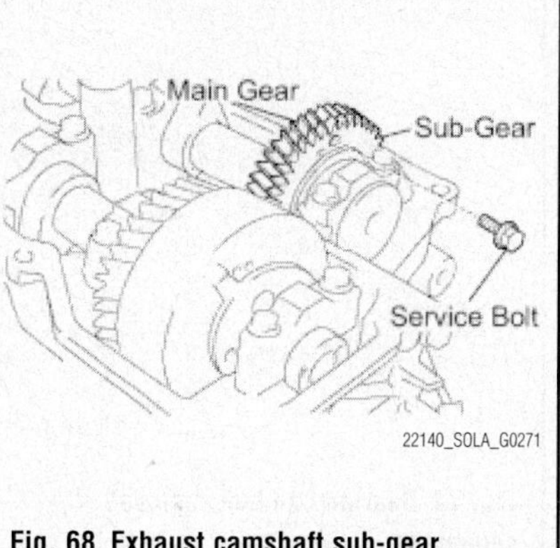

Fig. 68 Exhaust camshaft sub-gear

Recommended service bolt: Thread diameter 6 mm, Thread pitch, 1 mm, Bolt length 16 to 20 mm (0.63 to 0.79 in.). Tighten to 48 inch lbs. (5.4 Nm).

➡**When removing the camshaft, make certain that the torsional spring force of the sub gear has been eliminated by installation of the service bolt.**

c. Uniformly loosen and remove the 10 bearing cap bolts in the sequence shown in the illustration. Remove the 5 bearing caps and the camshaft.

➡**Do not pry out the camshaft.**

➡**Be careful not to damage the portion of the cylinder head receiving the shaft thrust.**

40. Remove the No. 2 camshaft, as follows:

a. Uniformly loosen and remove the 10 bearing cap bolts in the sequence shown in the illustration. Remove the 5 bearing caps and the No. 2 camshaft.

➡**Do not pry out the camshaft.**

➡**Be careful not to damage the portion of the cylinder head receiving the shaft thrust.**

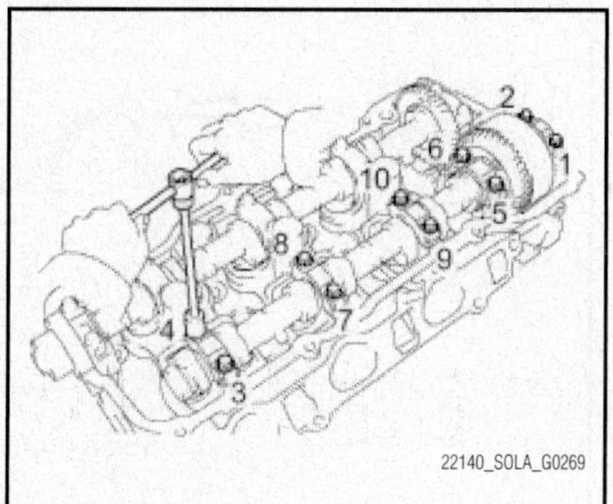

Fig. 69 Bearing cap bolt removal sequence

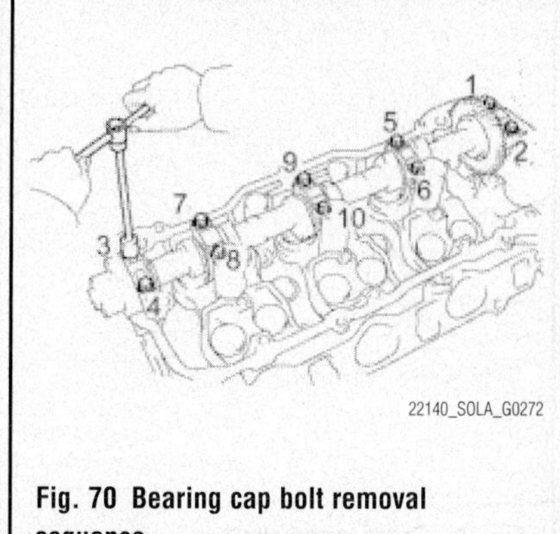

Fig. 70 Bearing cap bolt removal sequence

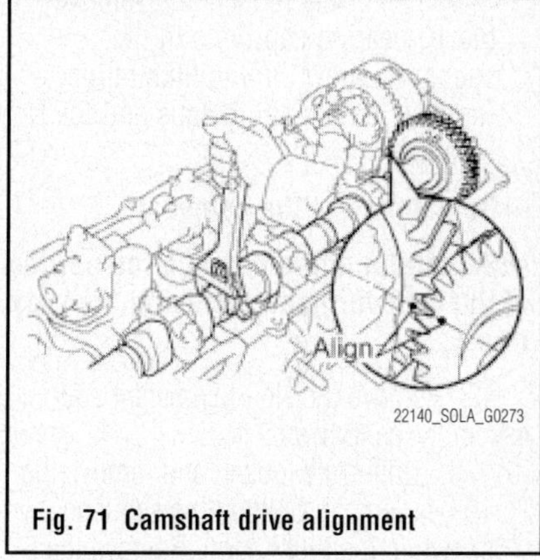

Fig. 71 Camshaft drive alignment

b. Remove the oil seal from the No. 2 camshaft.

41. Remove the No. 3 camshaft sub-assembly, as follows:

WARNING

Since the thrust clearance of the camshaft is small, the camshaft must be kept level while it is being removed. If the camshaft is not kept level, damage to the cylinder head or to the camshaft may result. To avoid this, the following steps should be carried out.

a. Align the camshaft drive and driven gear's timing marks (1 dot mark each) by turning the camshaft with a wrench.

b. Secure the exhaust camshaft sub gear to the main gear with a service bolt. Recommended service bolt: Thread diameter 6 mm, Thread pitch, 1 mm, Bolt length 16 to 20 mm (0.63 to 0.79 in.). Tighten to 48 inch lbs. (5.4 Nm).

➡**When removing the camshaft, make certain that the torsional spring force of the sub gear has been eliminated by installation of the service bolt.**

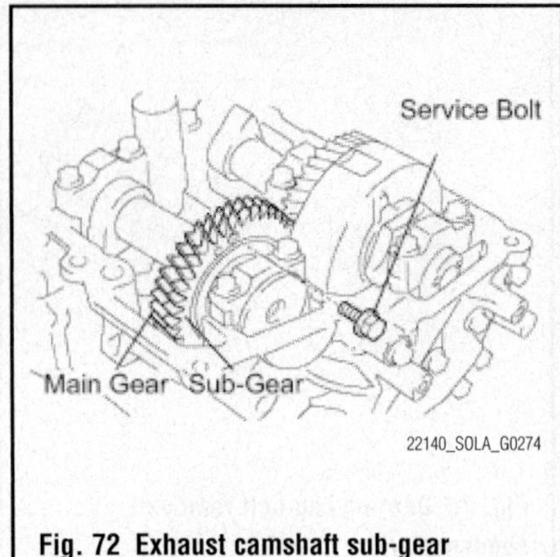

Fig. 72 Exhaust camshaft sub-gear

c. Uniformly loosen and remove the 10 bearing cap bolts in the sequence shown in the illustration. Remove the 5 bearing caps and the No. 3 camshaft.

➡**Do not pry out the camshaft.**

➡**Be careful not to damage the portion of the cylinder head receiving the shaft thrust.**

42. Remove the No. 4 camshaft sub-assembly, as follows:

a. Uniformly loosen and remove the 10 bearing cap bolts in the sequence shown in the illustration. Remove the 5 bearing caps and the No. 4 camshaft.

➡**Do not pry out the camshaft.**

➡**Be careful not to damage the portion of the cylinder head receiving the shaft thrust.**

b. Remove the oil seal from the No. 4 camshaft.

43. Remove valve lifter.

To install:

44. Install valve lifter.

45. Install the No. 4 camshaft sub-assembly, as follows:

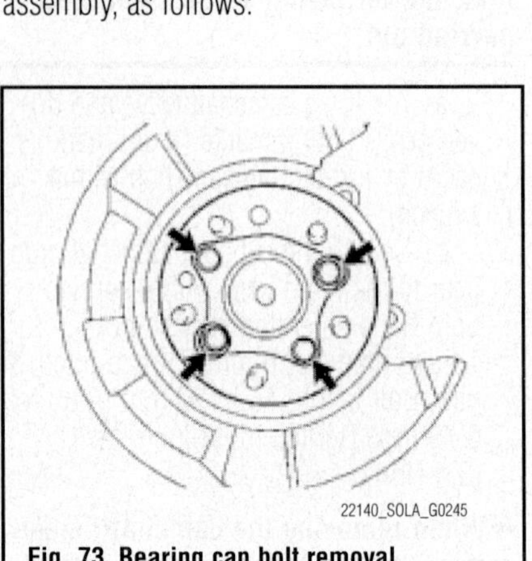

Fig. 73 Bearing cap bolt removal sequence

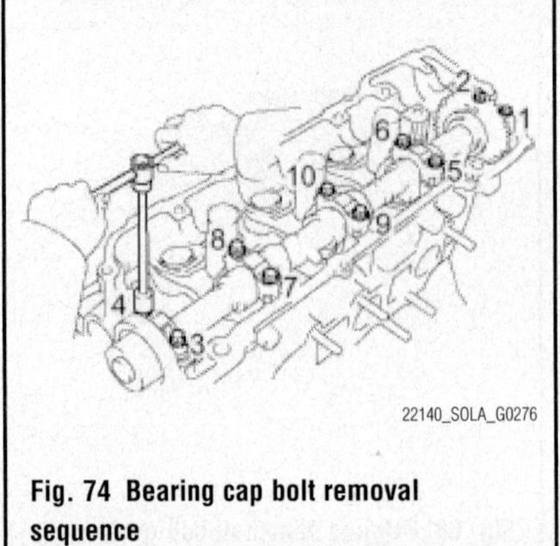

Fig. 74 Bearing cap bolt removal sequence

WARNING

Since the thrust clearance of the camshaft is small, the camshaft must be kept level while it is being removed. If the camshaft is not kept level, damage to the cylinder head or to the camshaft may result. To avoid this, the following steps should be carried out.

a. Apply new engine oil to the thrust portion and journal of the camshaft.

b. Place the No. 4 camshaft at a 90°angle of the timing mark (1 dot mark) on the cylinder head.

c. Apply MP grease to a new oil seal lip.

d. Install the oil seal to the camshaft. Make sure the oil seal is installed facing the correct direction, as shown in the illustration. Insert the oil seal until it cannot be pushed any further.

e. Remove any old packing material from the contact surface.

f. Apply seal packing (Toyota Genuine Seal Packing, Three Bond 1207B or equivalent) to the No. 1 bearing cap as shown in the illustration.

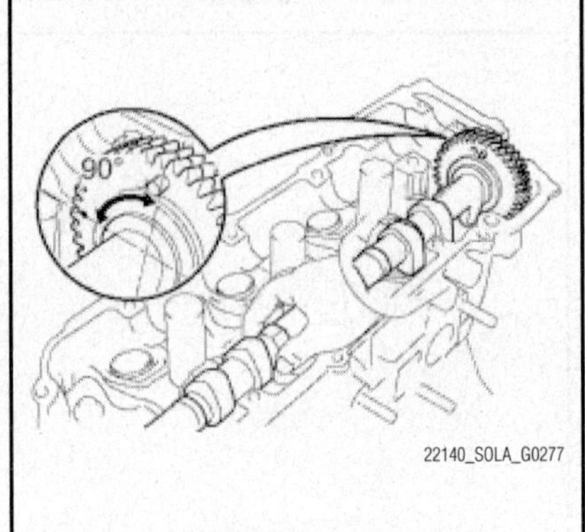

Fig. 75 No. 4 camshaft positioning

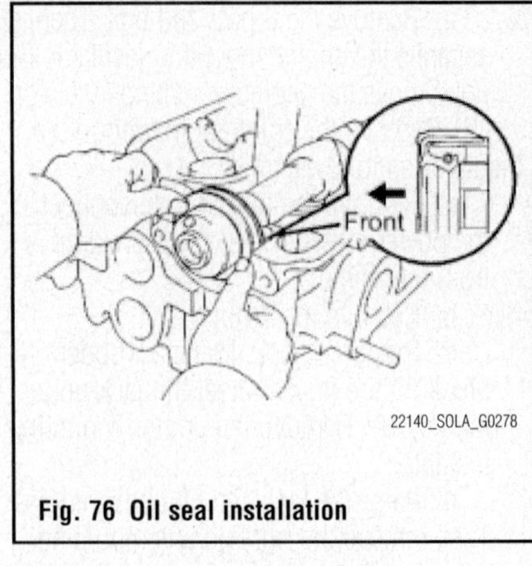

Fig. 76 Oil seal installation

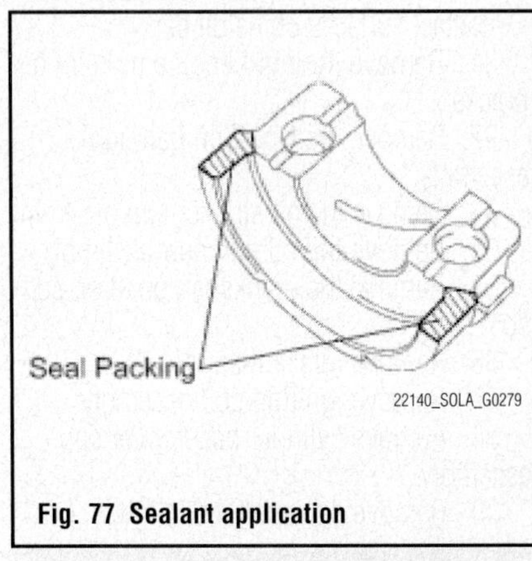

Fig. 77 Sealant application

➡**Install the No. 1 bearing cap within 5 minutes after applying seal packing.**

➡**Do not expose the seal to engine oil for at least 2 hours after installing.**

g. Install the 5 bearing caps in their proper locations.

h. Apply a light coat of engine oil on the threads of the bearing cap bolts.

i. Uniformly tighten the 10 bearing cap bolts in the sequence shown in the illustration to 12 ft. lbs. (16 Nm).

46. Install the No. 3 camshaft sub-assembly, as follows:

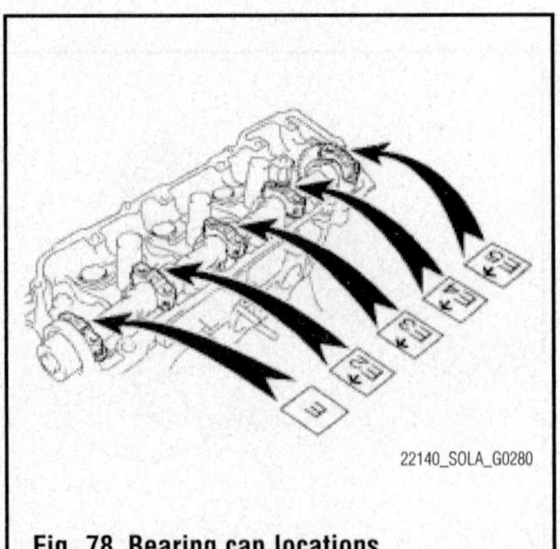

Fig. 78 Bearing cap locations

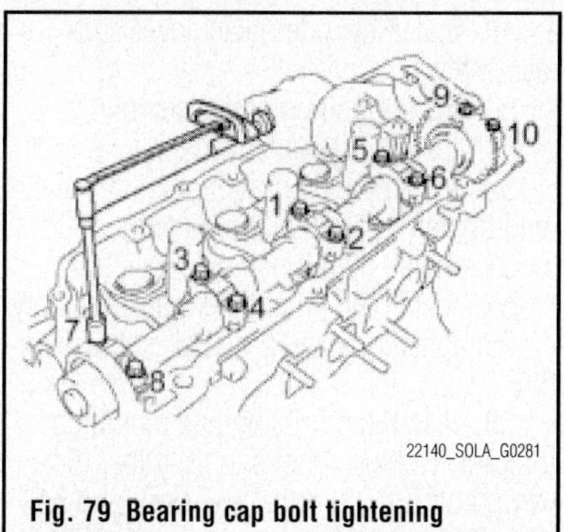

Fig. 79 Bearing cap bolt tightening sequence

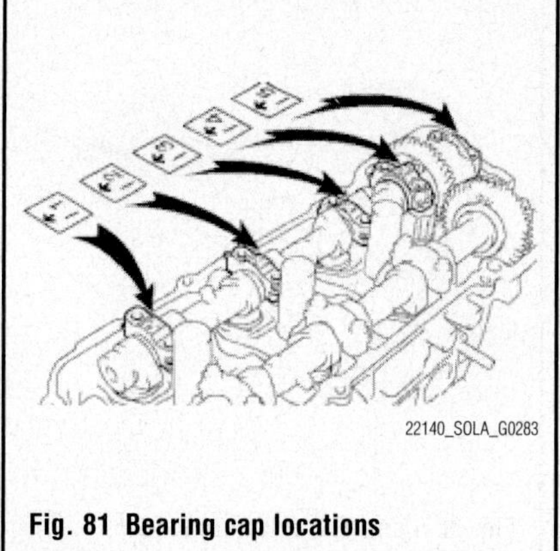

Fig. 81 Bearing cap locations

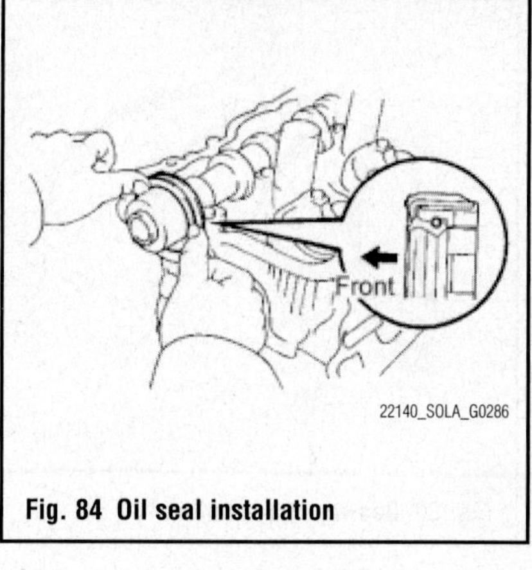

Fig. 84 Oil seal installation

**** WARNING**

Since the thrust clearance of the camshaft is small, the camshaft must be kept level while it is being removed. If the camshaft is not kept level, damage to the cylinder head or to the camshaft may result. To avoid this, the following steps should be carried out.

a. Apply new engine oil to the thrust portion and journal of the camshaft.

b. Align the camshaft drive and driven gear's timing mark (1 dot mark each).

c. Place the camshaft on the cylinder head.

d. Install the 5 bearing caps in their proper locations.

e. Apply a light coat of engine oil on the threads of the bearing cap bolts.

f. Uniformly tighten the 10 bearing cap bolts in the sequence show in the illustration to 12 ft. lbs. (16 Nm).

g. Remove the service bolt.

47. Install the No. 2 camshaft sub-assembly, as follows:

**** WARNING**

Since the thrust clearance of the camshaft is small, the camshaft must be kept level while it is being removed. If the camshaft is not kept level, damage to the cylinder head or to the camshaft may result. To avoid this, the following steps should be carried out.

a. Apply new engine oil to the thrust portion and journal of the camshaft.

b. Place the No. 2 camshaft at a 90° angle of the timing mark (2 dot marks) on the cylinder head.

c. Apply MP grease to a new oil seal lip.

d. Install the oil seal to the camshaft. Make sure the oil seal is installed facing the correct direction, as shown in the illustration. Insert the oil seal until it cannot be pushed any further.

e. Remove any old packing material from the contact surface.

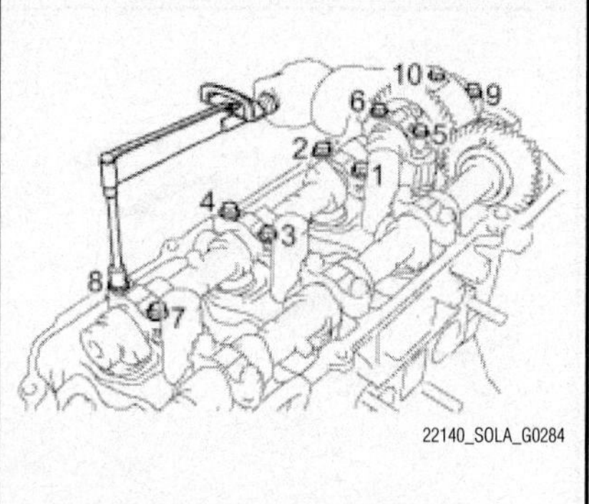

Fig. 82 Bearing cap bolt tightening sequence

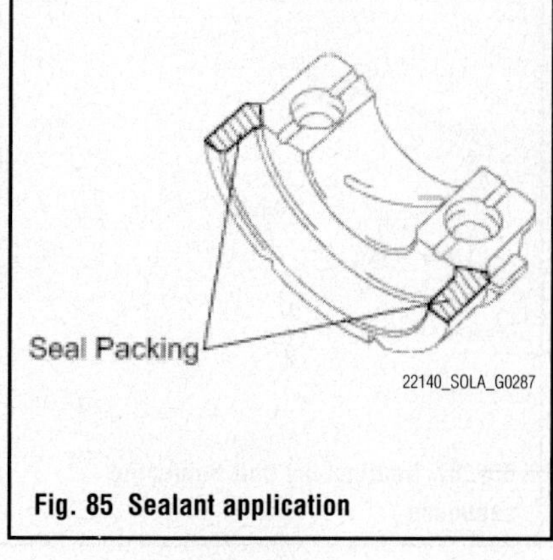

Fig. 85 Sealant application

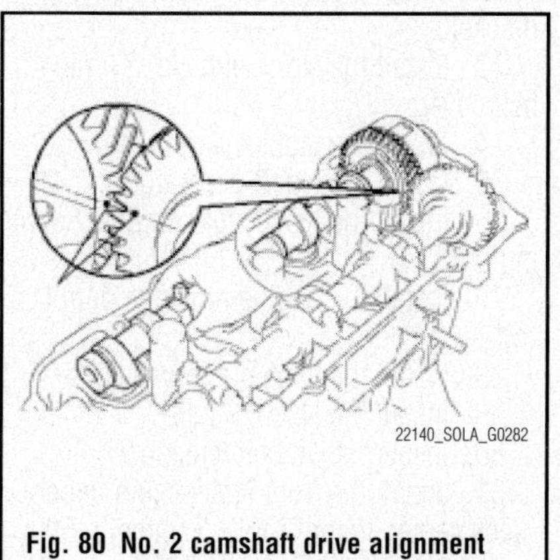

Fig. 80 No. 2 camshaft drive alignment

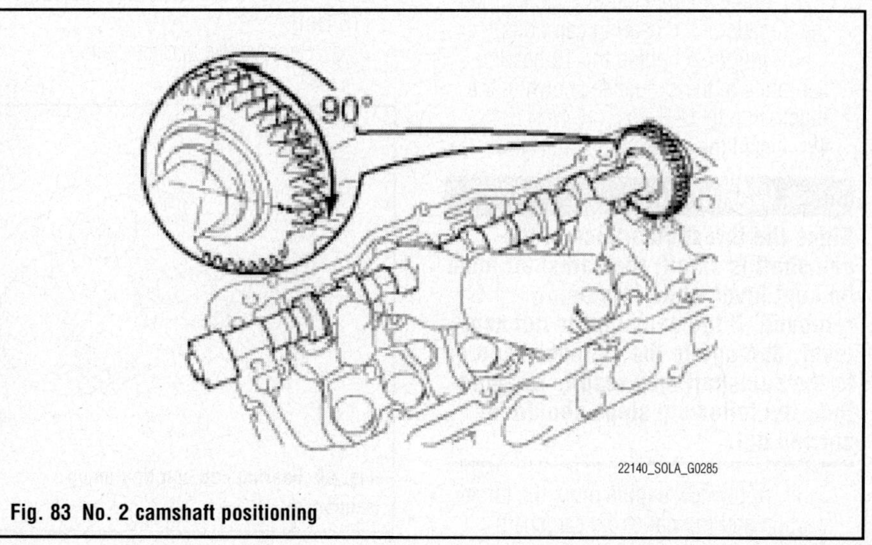

Fig. 83 No. 2 camshaft positioning

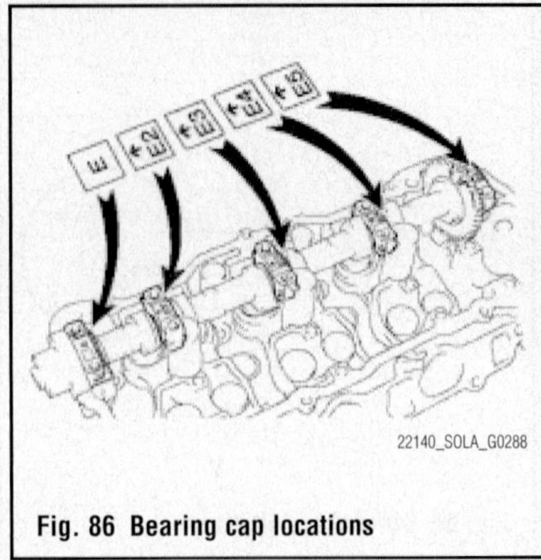

Fig. 86 Bearing cap locations

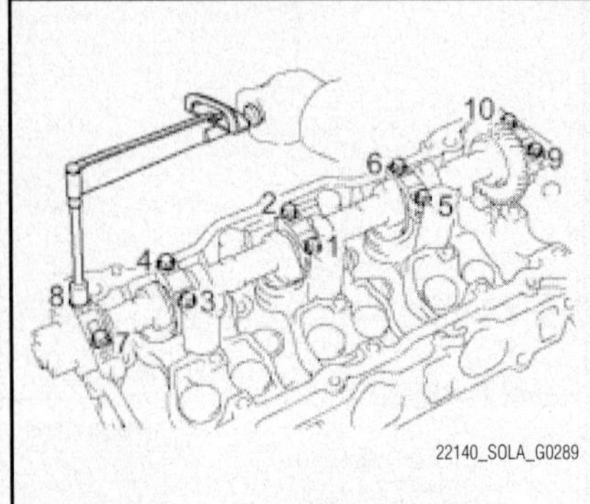

Fig. 87 Bearing cap bolt tightening sequence

f. Apply seal packing (Toyota Genuine Seal Packing, Three Bond 1207B or equivalent) to the No. 1 bearing cap as shown in the illustration.

➡Install the No. 1 bearing cap within 5 minutes after applying seal packing.

➡Do not expose the seal to engine oil for at least 2 hours after installing.

g. Install the 5 bearing caps in their proper locations.

h. Apply a light coat of engine oil on the threads of the bearing cap bolts.

i. Uniformly tighten the 10 bearing cap bolts in the sequence shown in the illustration to 12 ft. lbs. (16 Nm).

48. Install the camshaft, as follows:

⁂ WARNING

Since the thrust clearance of the camshaft is small, the camshaft must be kept level while it is being removed. If the camshaft is not kept level, damage to the cylinder head or to the camshaft may result. To avoid this, the following steps should be carried out.

a. Apply new engine oil to the thrust portion and journal of the camshaft.

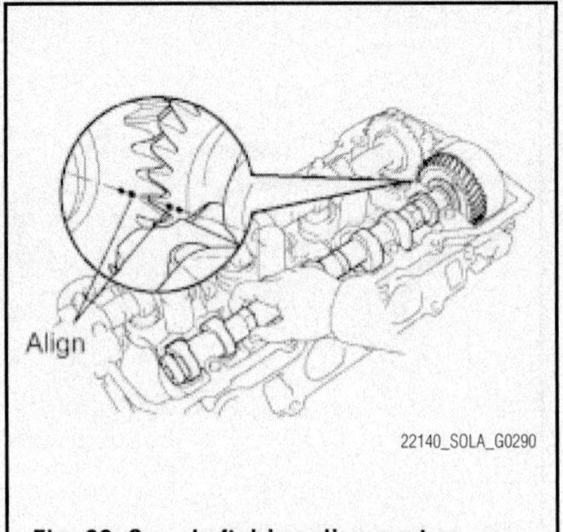

Fig. 88 Camshaft drive alignment

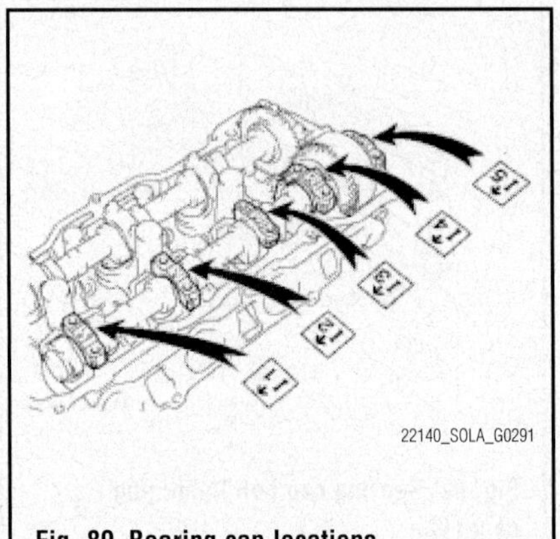

Fig. 89 Bearing cap locations

b. Align the camshaft drive and driven gear's timing mark (2 dot marks each).

c. Place the camshaft on the cylinder head.

d. Install the 5 bearing caps in their proper locations.

e. Apply a light coat of engine oil on the threads of the bearing cap bolts.

f. Uniformly tighten the 10 bearing cap bolts in the sequence show in the illustration to 12 ft. lbs. (16 Nm).

g. Remove the service bolt.

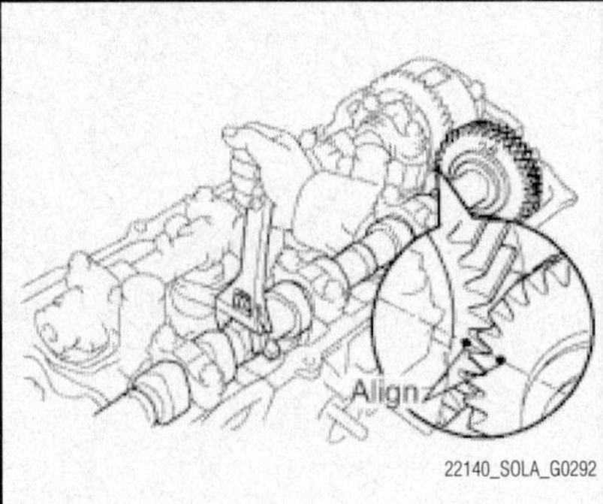
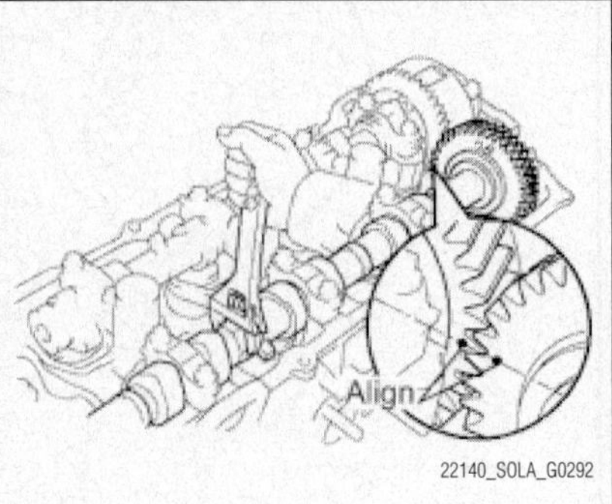

Fig. 90 Bearing cap bolt tightening sequence

49. Install cylinder head cover sub-assembly.
50. Install ignition coil assembly.
51. Install intake manifold.
52. Install the crankshaft position sensor and tighten to 71 inch lbs. (8 Nm).
53. Install the oil pan sub-assembly.
54. Install the oil strainer sub-assembly.
55. Install the No. 2 oil pan sub-assembly.
56. Install the right engine mounting bracket. Tighten A bolts to 40 ft lbs. (54 Nm), bolt B to 40 ft lbs. (54 Nm), and bolt C to 32 ft. lbs. (43 Nm).
57. Install the right engine mounting insulator. Tighten nut A to 70 ft. lbs. (95 Nm), and B nuts to 64 ft. lbs. (87 Nm).
58. Install the engine mounting insulator. Tighten bolt A to 64 ft. lbs. (87 Nm), and B bolts to 38 ft. lbs. (52 Nm).
59. Install the oil level gage guide, as follows:
    a. Install a new O-ring to the gage guide.
    b. Install the gage guide with the bolt and tighten to 71 inch lbs. (8 Nm).
60. Install the No. 1 compressor mounting bracket.
61. Install the No. 3 timing belt cover.
62. Install the No. 1 timing belt idler sub-assembly.
63. Install the No. 2 timing belt idler sub-assembly.
64. Install the camshaft timing pulley.
65. Install the compressor and magnetic clutch.
66. Install the No. 1 exhaust pipe support bracket.
67. Install the front exhaust pipe assembly.
68. Install the No. 1 exhaust pipe support bracket.
69. Inspect and install the timing belt.
70. Install the timing belt tensioner assembly.
71. Install the No. 2 timing belt guide.
72. Install the right engine mounting bracket.
73. Install the No. 1 and No. 2 timing belt cover.
74. Install the crankshaft pulley.
75. Install the No. 2 generator bracket.
76. Install the No. 2 right engine mounting stay.
77. Install the engine moving control rod.
78. Install the vane pump V-belt.
79. Install the No. 1 V-belt.
80. Inspect drive belt tension.
81. Install the front suspension upper brace center, if applicable. Tighten to 59 inch lbs. (80 Nm).

82. Add engine oil and check for leaks.
83. Install the right front wheel.
84. Check for exhaust gas leaks.

CRANKSHAFT DAMPER

REMOVAL & INSTALLATION

3.3L Engine

1. Remove crankshaft pulley.
2. Using SST, loosen the pulley bolt.
3. Using SST and the pulley bolt, remove the pulley.

➡ Before using SST, apply lubricating oil on the threads and tip of the center bolt.

To install:

4. To install, reverse removal procedure.
5. Tighten pulley bolt to 162 ft. lbs. (220 Nm).

CRANKSHAFT FRONT SEAL

REMOVAL & INSTALLATION

2.4L Engine

See Figure 91.

1. Before servicing the vehicle, refer to the Precautions Section.
2. Disconnect the negative battery cable.
3. Drain the engine oil
4. Remove the right front wheel.
5. Remove the right hand front fender apron seal.
6. Remove the left and right hand engine under cover.
7. Remove the drive belt.
8. Remove the crankshaft pulley.
9. Using a knife, cut off the oil seal lip. Using a screwdriver with the tip taped, pry out the oil seal.
10. After the removal, check the crankshaft for damage. If it is damaged, smooth the surface with 400-grit sandpaper.

To install:

11. Apply MP grease to a new oil seal lip. Keep the lip free from foreign matter.
12. Using Special Tool 09223-22010 and a hammer, tap in the oil seal until its surface is flush with the rear oil seal retainer edge.
13. Align the pulley set key with the key groove of the pulley. Using Special Tool 09960-10010, keep the pulley in place and tighten the bolt to: 133 ft. lbs. (180 Nm).
14. To complete installation, reverse remaining removal procedure.
15. Check engine for oil leaks.

3.3L Engine

1. Remove oil pump assembly
2. Using a screwdriver remove oil pump oil seal

➡Be careful not to damage sealing surface when removing.

To install:

3. Using SST and a hammer, install a new oil seal. Tap the surface of the SST with the hammer until the oil seal's surface is flush with oil pump body edge

➡Be careful not to tap the oil seal at an angle. Keep the gap between the oil pump body edge and oil seal free from contamination

4. Install oil pump assembly

CYLINDER HEAD

REMOVAL & INSTALLATION

2.4L Engine

See Figures 92 through 94.

1. Remove hood sub-assembly.
2. Remove right front wheel.
3. Remove engine under cover.
4. Remove front fender apron seal.
5. Drain engine oil.
6. Remove front exhaust pipe assembly.
7. Remove No. 2 right engine mounting stay.
8. Remove No. 2 right engine mounting bracket.
9. Remove fan and generator V-belt.
10. Remove engine wire.
11. Remove generator assembly,
12. Remove vane pump assembly.
13. Remove ignition coil assembly.
14. Disconnect ventilation hose.
15. Remove cylinder head cover sub-assembly.
16. Set No. 1 cylinder to TDC/ compression.
17. Remove crankshaft pulley.
18. Remove crankshaft position sensor.
19. Remove oil pan sub-assembly.
20. Remove No. 1 chain tensioner assembly.
21. Remove V-ribbed belt tensioner assembly.
22. Install engine hangers.
23. Raise the engine and remove the engine mounting insulator.
24. Remove right engine mounting bracket.
25. Remove timing chain cover sub-assembly.

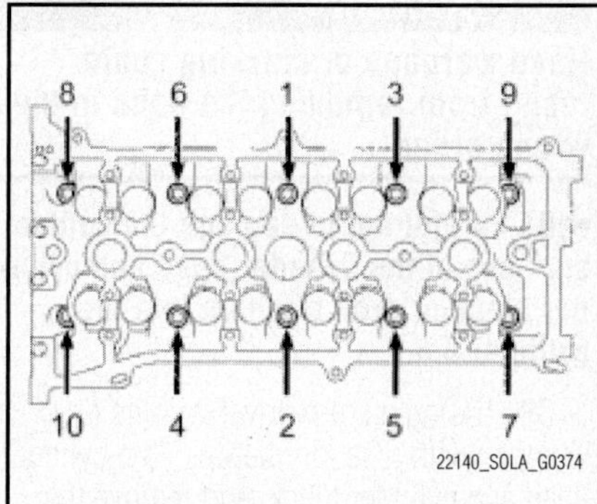

Fig. 93 Cylinder head bolt tightening sequence—2.4L

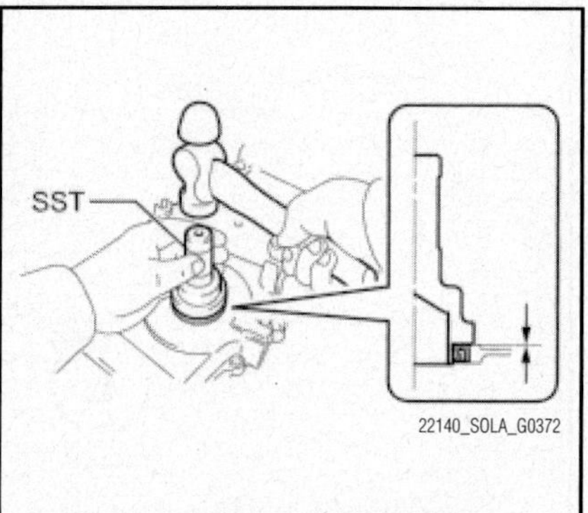

Fig. 91 Using Special Tool 09223-22010 and hammer to install new oil seal

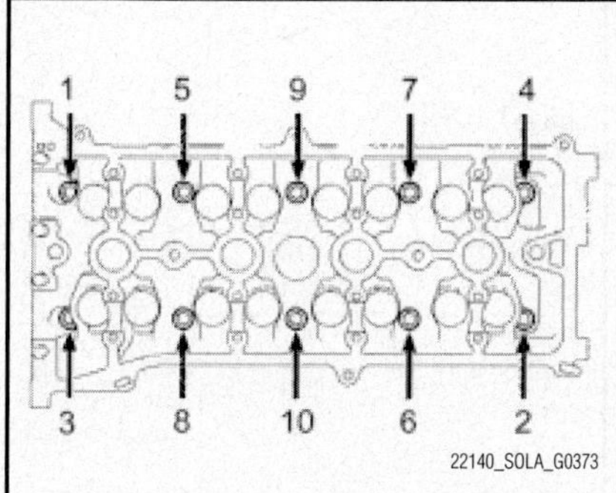

Fig. 92 Removing cylinder head bolts in sequence

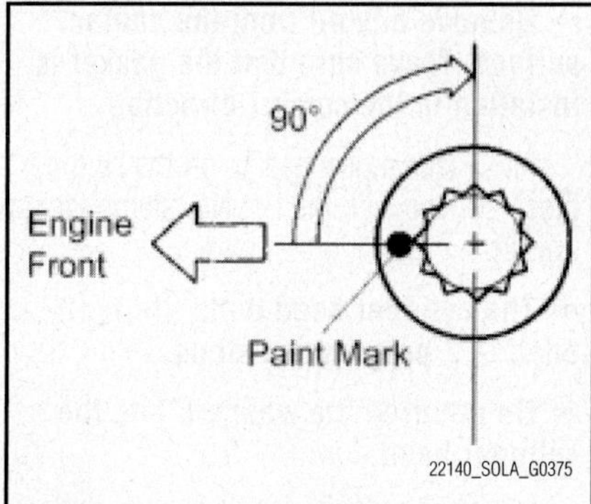

Fig. 94 Cylinder head bolt tightening procedure

26. Remove No. 1 crankshaft position sensor plate.

27. Remove chain tensioner slipper.

28. Remove No. 1 chain vibration damper.

29. Remove timing chain guide.

30. Remove chain sub-assembly.

31. Remove crankshaft timing gear or sprocket.

32. Remove No. 2 chain sub-assembly:

a. Turn the crankshaft counterclockwise by 90°, and align an adjusting hole of the oil pump driven sprocket with the groove of the oil pump.

b. Put a bar (φ 4 mm (0.16 in.) in the adjusting hole of the oil pump driven sprocket to temporarily lock the sprocket in position. Remove the nut.

33. Remove the bolt, chain tensioner plate and spring.

34. Remove the chain tensioner, oil pump driven sprocket and chain.

35. Remove intake manifold.

36. Remove exhaust manifold.

37. Using a 10 mm bi-hexagon wrench, uniformly loosen the 10 bolts in the sequence shown in the illustration. Remove the 10 cylinder head bolts and plate washers.

WARNING

Head warpage or cracking could result from removing the bolts in the wrong order.

Be careful not to damage the contact surfaces of the cylinder head and cylinder block or drop washers into the cylinder head.

38. Using a screwdriver with its tip wrapped with tape, pry between the cylinder head and cylinder block, and remove the cylinder head.

39. Remove the cylinder head gasket.

To install:

Remove any oil from the contact surface. Make sure that the gasket is installed in the correct direction.

40. Place a new gasket on the cylinder block surface with the Lot No. stamp facing upward.

The cylinder head bolts are tightened in 2 progressive steps.

Do not drop the washers into the cylinder head.

41. Apply a light coat of engine oil to the bolt threads and the area beneath the bolt heads that come in contact with the washers. Install the bolts and plate washers to the cylinder head.

42. Using several steps, uniformly install and tighten the 10 cylinder head set bolts and plate washers with a 10 mm bi-hexagon wrench in sequence. Tighten the bolts to 52 ft. lbs. (70 Nm).

43. Mark the front side of the cylinder head bolt with paint. Retighten the cylinder head bolts 90° in sequence. Check that the paint mark is now at a 90° angle to the front.

44. To complete installation, reverse removal procedure.

45. Add engine oil and check for leaks.

46. Check for fuel leaks.

47. Add coolant and check for coolant leaks.

48. Check for exhaust gas leaks

49. Inspect ignition timing.

50. Inspect idle speed.

51. Inspect compression.

52. Inspect CO/HC.

3.3L Engine

See Figures 95 through 98.

1. Before servicing the vehicle, refer to the Precautions Section.

2. Prevent gasoline from spilling out.

3. Drain engine coolant.

4. Drain engine oil.

5. Remove the No. 1 exhaust pipe support bracket.

6. Remove the front exhaust pipe assembly.

7. Remove the right exhaust manifold sub-assembly.

8. Remove the V-bank cover sub-assembly.

9. Remove front suspension upper center brace (w/ front suspension upper center brace).

10. Remove the air cleaner assembly.

11. Remove the intake air surge tank.

12. Remove the No. 1 fuel pipe sub-assembly.

13. Disconnect the heater water inlet hose.

14. Remove the intake manifold.

15. Remove the radiator inlet hose.

16. Remove water outlet.

17. Remove the right front wheel.

18. Remove the right front fender apron seal.

19. Remove the No. 1 (cooler compressor to crankshaft pulley) V-belt.

20. Remove the vane pump V-belt.

21. Remove the engine moving control rod.

22. Remove the right No. 2 engine mounting stay.

23. Remove the No. 2 alternator bracket.

24. Remove crankshaft pulley.

25. Remove the No. 1 and No. 2 timing belt covers.

26. Remove right engine mounting bracket.

27. Remove the No.2 timing belt guide.

28. Remove timing belt.

29. Remove the No. 2 timing belt idler sub-assembly.

30. Remove the camshaft timing pulley.

31. Remove the No. 3 timing belt cover.

32. Remove the vane pump assembly.

33. Remove the ignition coil assembly.

34. Remove the cylinder head cover sub-assembly.

35. Remove camshaft.

36. Remove the No. 2 camshaft.

37. Remove the No. 3 camshaft sub-assembly.

38. Remove the No. 4 camshaft sub-assembly.

39. Remove the camshaft timing gear assembly.

40. Remove the camshaft sub gear.

41. Remove the cylinder head sub-assembly, as follows:

a. Disconnect the VVT sensor connector.

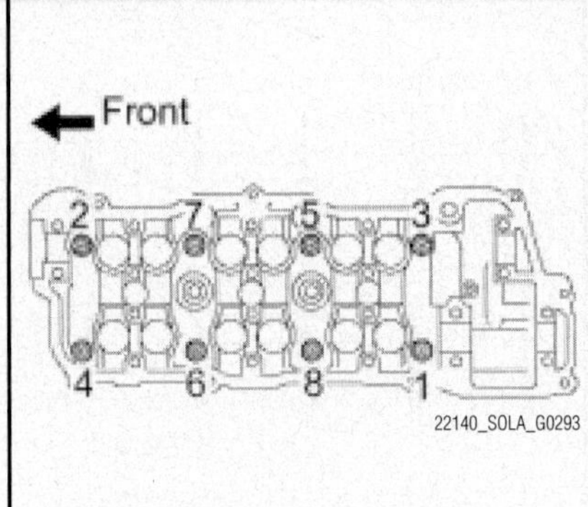

Fig. 95 Cylinder head bolt removal sequence

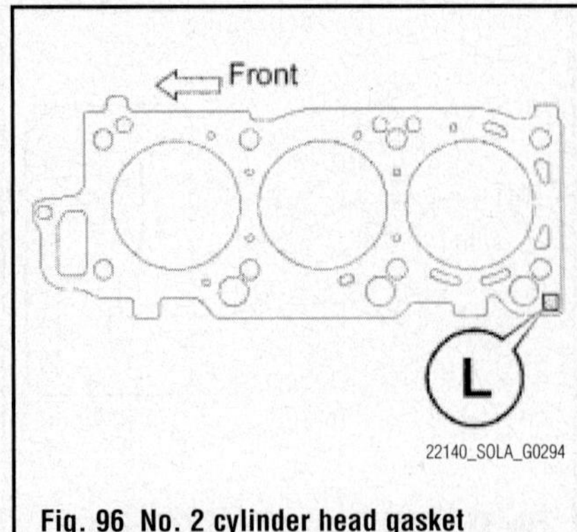

Fig. 96 No. 2 cylinder head gasket positioning

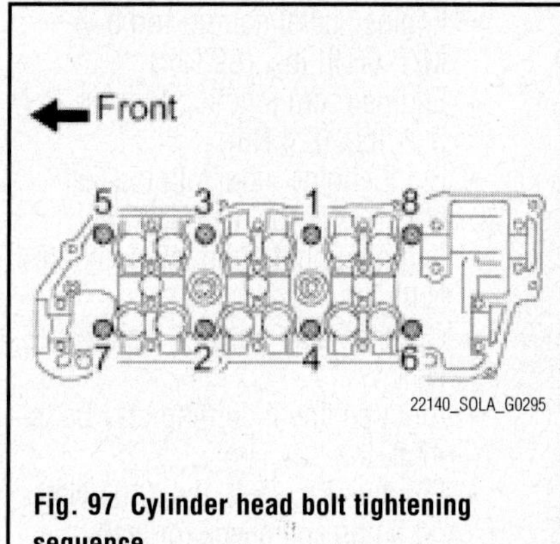

Fig. 97 Cylinder head bolt tightening sequence

b. Disconnect the camshaft timing oil control valve connector.

c. Remove the nut and disconnect the engine wire harness clamp.

d. Using an 8 mm socket hexagon wrench, remove the hexagon bolt.

e. Uniformly loosen the 8 cylinder head bolts in the sequence shown in the illustration. Remove the 8 cylinder head bolts and plate washers.

⁂ WARNING

Be careful not to drop the washers into the cylinder head.

⁂ WARNING

Head warpage or cracking could result from removing bolts in an incorrect order.

42. Remove cylinder head gasket.
43. Remove left cylinder head.
44. Remove No. 2 cylinder head gasket.

To install:

45. Place a new No. 2 cylinder head gasket on the cylinder block with the L mark as shown in the illustration.

➡Remove any oil from the contact surface Make sure the cylinder head gasket is facing in the correct direction.

46. Install cylinder head, as follows:

➡To prevent damage to the block, place the gasket carefully.

➡The cylinder head bolts are tightened in 2 successive steps.

a. Apply a light coat of engine oil on the threads of the cylinder head bolts.

b. Install the plate washer to the cylinder head bolts.

c. Uniformly install and tighten the 8 cylinder head bolts in the sequence shown in the illustration to 40 ft. lbs. (54 Nm).

d. Mark the front side of each cylinder head bolt head with paint as shown in the illustration.

e. Retighten the cylinder head bolts by 90° in the same sequence as shown above.

f. Check that each painted mark is now at a 90° angle to the front.

g. Using an 8 mm socket hexagon wrench, install the hexagon bolt and tighten to 14 ft. lbs. (19 Nm).

h. Install the wire harness clamp bracket with the bolt and tighten to 48 inch lbs. (5.4 Nm).

i. Connect the ground cable with the bolt and tighten to 48 inch lbs. (5.4 Nm).

47. Install cylinder head gasket.
48. Install cylinder head sub-assembly.
49. Install camshaft sub gear.
50. Install camshaft timing gear assembly.
51. Install No. 4 camshaft sub-assembly.
52. Install No. 3 camshaft sub-assembly.
53. Install No. 2 camshaft.
54. Install camshaft.
55. Install cylinder head cover sub-assembly.
56. Install ignition coil assembly.
57. Install vane pump assembly.
58. Install No. 3 timing belt cover.
59. Install camshaft timing pulley.
60. Install No. 2 timing belt idler sub-assembly.
61. Install timing belt.
62. Install timing belt tensioner assembly.
63. Install No. 1 chain vibration damper.
64. Install right engine mounting bracket.
65. Install No. 2 and No. 1 timing belt covers.
66. Install crankshaft pulley.
67. Install No. 2 generator bracket.
68. Install right No. 2 engine mounting stay.
69. Install engine moving control rod.
70. Install vane pump V-belt.
71. Install No. 1 (cooler compressor to crankshaft pulley) V-belt.
72. Inspect drive belt tension.
73. Install water outlet.
74. Install intake manifold.
75. Install No.1 fuel pipe sub-assembly.
76. Install intake air surge tank.
77. Install air cleaner assembly.
78. Connect vacuum hoses.
79. Install front suspension upper center brace.
80. Install V-bank cover sub-assembly.
81. Install right exhaust manifold sub-assembly.
82. Install front exhaust pipe assembly.

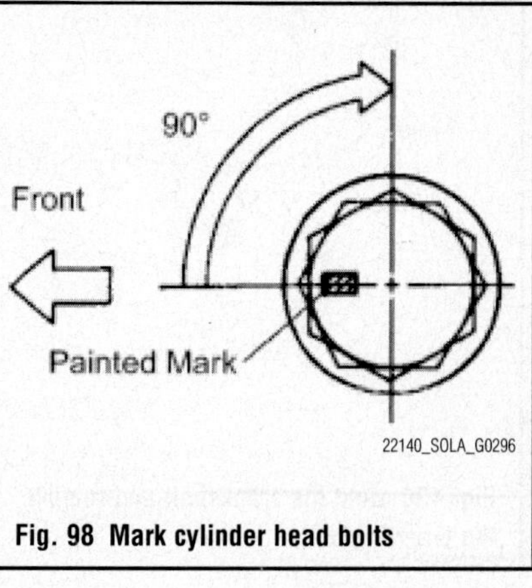

Fig. 98 Mark cylinder head bolts

83. Install No. 1 front exhaust support bracket.
84. Install right front wheel.
85. Add engine oil and coolant.
86. Check for engine oil and coolant leaks.
87. Check for fuel and exhaust gas leaks.
88. Inspect ignition timing.
89. Inspect engine idle speed.
90. Inspect compression.
91. Inspect CO/HC.

ENGINE ASSEMBLY

REMOVAL & INSTALLATION

2.4L Engine

See Figures 99 and 100.

1. Before servicing the vehicle, refer to the Precautions Section.
2. Disconnect the negative battery cable.
3. Prevent gasoline from spilling out.
4. Remove the right front wheel.
5. Remove the engine under covers.
6. Remove the right front fender apron seal.
7. Drain engine oil.
8. Drain engine coolant.
9. Drain transaxle fluid or transaxle oil.

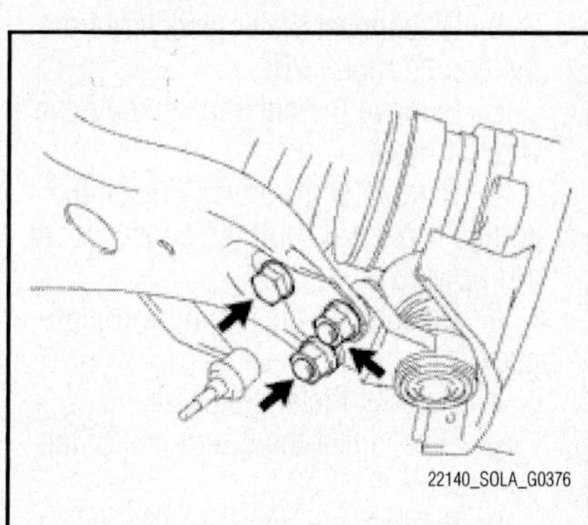

Fig. 99 No. 1 front suspension arm bolts and nuts

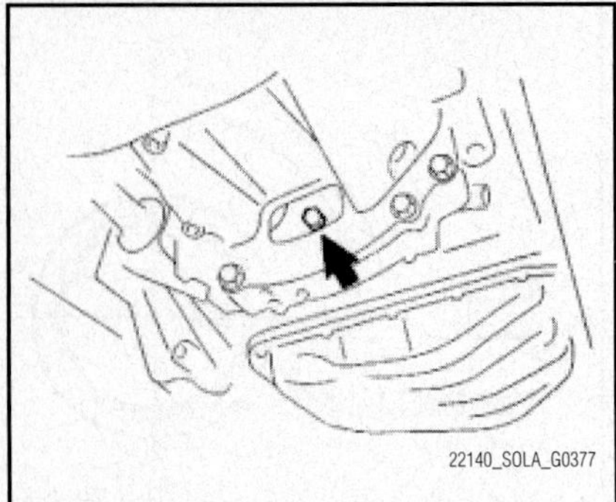

Fig. 100 Hold the crankshaft and remove the 6 setting bolts

10. Remove battery.
11. Remove air cleaner assembly.
12. Remove the No. 1 right hand under cover.
13. Disconnect radiator inlet and outlet hoses.
14. Disconnect the No. 2 and No. 3 oil cooler outlet hoses.
15. Remove the 3 bolts and the engine moving control rod w/bracket.
16. Remove the No. 2 engine mounting stay and No. 2 engine mounting bracket.
17. Remove the fan and generator V-belt.
18. Disconnect the No. 1 oil reservoir to the pump hose.
19. Disconnect the return tube sub-assembly.
20. Disconnect union to connector tube hose.
21. Disconnect floor shift cable transmission control shift.
22. For manual transaxle, disconnect floor shift cable transmission control select.
23. For manual transaxle, remove clutch release cylinder assembly.
24. Disconnect heater inlet and outlet water hoses.
25. Disconnect fuel tube sub-assembly.
26. Disconnect engine wire, as follows:
   a. Disconnect the engine wire from the ECM and passenger side J/B.
   b. Disconnect the engine wire from the engine room J/B.
   c. Remove the nut and separate the wire harness.
   d. Using a screwdriver, unlock the engine room J/B. Pull the engine room J/B upward.
   e. Disconnect the engine wire connectors.
   f. Pull out the engine wire.
   g. Disconnect the B terminal of the generator.
   h. Remove the body ground.
27. Remove the No. 2 engine wire, as follows:
   a. Remove the starter connector.
   b. Remove the nut and disconnect terminal 30 of the starter.
28. Remove generator assembly.
29. Remove compressor and magnetic clutch.
30. Remove front exhaust pipe assembly.
31. Remove the left and right front stabilizer link assemblies.
32. Remove the left and right front axle shaft nuts.
33. Remove the left and right front speed sensors.
34. Disconnect the left and right tie rod assemblies.
35. Remove the left and right front lower No. 1 suspension arm sub-assemblies.
   a. Left and right front axle assembly
36. For automatic transaxle, remove the drive plate and torque converter clutch setting bolt.
37. Remove steering intermediate shaft assembly.
38. Remove engine assembly with transaxle, as follows:
   a. Set the engine lifter.
   b. Remove the 4 bolts, 2 nuts and right and left frame side rail plate.
   c. Remove the 4 bolts, 2 nuts and front right and left suspension member brace rear.
   d. Carefully remove the engine assembly from the vehicle.
   e. Install the 2 engine hangers as shown in the illustration. Engine hanger No.1: 12281-28010,
   f. Engine hanger No.2: 12282-28010, Bolt: 91512-61020. Tighten to 28 ft. lbs. (38 Nm).

To install:

39. To install, reverse removal procedure, tightening to the following to specifications:
   - Drive belt tensioner with the bolt and nut: 44 ft. lbs. (60 Nm)
   - Right engine mounting bracket with the 3 bolts: 40 ft. lbs. (54 Nm)
   - Oil cooler pipe: 80 inch lbs. (9 Nm)
   - Water by-pass pipe bolt, two nuts and install a new gasket: 80 inch lbs. (9 Nm)
   - Oil level gauge guide bolt: 80 inch lbs. (9 Nm)
   - Left engine mounting insulator nut—A/T: 70 ft. lbs. (95 Nm)
   - Left engine mounting insulator bolt—M/T: 105 ft. lbs. (143 Nm)
   - Right engine mounting insulator nut: 70 ft. lbs. (95 Nm)
   - FR engine mounting insulator bolt: 64 ft. lbs. (87 Nm)
   - Engine lateral control rod bolt—M/T: 66 ft. lbs. (89 Nm)
   - Engine room junction block nut: 74 inch lbs. (8.4 Nm)
   - Right engine mountain bracket bolts: 38 ft. lbs. (52 Nm)
   - Engine mount bolts: 47 ft. lbs. (64 Nm)
   - Ground cable bolt: 74 inch lbs. (8.4 Nm)
   - Right engine mounting stay bolts: 47 ft. lbs. (64 Nm).
   - Front wheel: 76 ft. lbs. (103 Nm)
40. Add fluids and check for leaks.
41. Bleed power steering, brakes lines and clutch lines.
42. Check shift lever position.
43. Adjust front wheel alignment.
44. Check ignition timing.
45. Check engine idle speed.
46. Inspect CO/HC.
47. Check ABS speed sensor signal.

3.3L Engine

See Figures 101 through 122.

1. Before servicing the vehicle, refer to the Precautions Section.
2. Prevent gasoline from spilling out.
3. Remove the right front wheel.
4. Remove the engine under covers.
5. Remove the right front fender apron seal.
6. Drain engine oil.
7. Drain engine coolant.
8. Drain automatic transaxle fluid.
9. Remove battery.
10. Remove the No. 1 (cooler compressor to crankshaft pulley) V-belt.
11. Remove the vane pump V-belt.
12. Remove the engine moving control rod.
13. Remove the bolt, right No. 2 mounting stay and right No. 2 mounting bracket.
14. Remove the union to check valve hose.
15. Remove the V-bank cover sub-assembly.
16. Remove the air cleaner inlet assembly.
17. Remove the air cleaner assembly.
18. Remove the air cleaner bracket.
19. Remove the No. 1 air cleaner inlet.
20. Remove the intake air connector sub-assembly.
21. Remove the No. 1 fuel pipe sub-assembly, as follows:
   a. Remove the fuel pipe clamp.
   b. Disconnect the connector from the tube while pinching part A with your fingers as shown in the illustration.

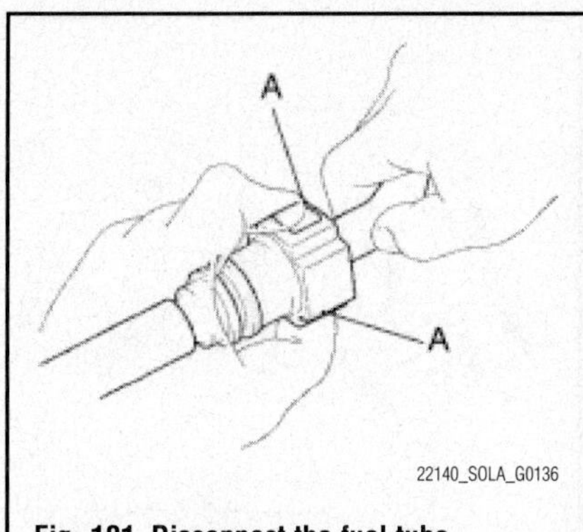

Fig. 101 Disconnect the fuel tube connector

➡Check for contamination in the pipe and around the connector. Clean if necessary and then disconnect the connector.

➡Disconnect the connector by hand.

➡Do not bend, fold or rotate the nylon tube.

➡If the pipe and connector are stuck together, push and pull the connector until it comes free.

➡Put the pipe and connector ends in vinyl bags to prevent damage and contamination.

22. Remove the radiator inlet and outlet hoses.
23. Disconnect the oil cooler inlet and outlet hoses.
24. Disconnect the heater water inlet and outlet hoses.
25. Remove the glove compartment door assembly.
26. Disconnect the engine wire from the ECM and passenger side junction block.

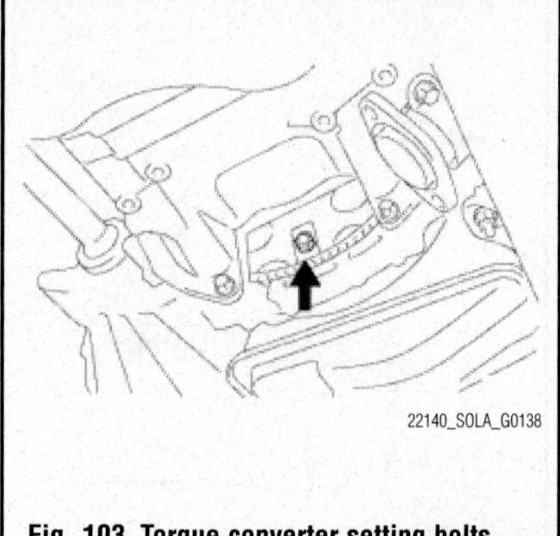

Fig. 103 Torque converter setting bolts

27. Disconnect the engine wire from the engine room junction block, as follows:
    a. Remove the nut and separate the wire harness.
    b. Using a screwdriver, unlock the engine room junction block. Pull the engine room junction block upward.
    c. Disconnect the engine wire connectors.
28. Pull out the engine wire.
29. Remove the body ground.
30. Remove the floor shift cable transmission select control.
31. Disconnect the No. 1 oil reservoir to pump hose.
32. Disconnect the steering gear outlet return tube.
33. Remove the No. 1 exhaust pipe support bracket.
34. Remove the front exhaust pipe assembly.
35. Disconnect the left front stabilizer link assembly, as follows:
    a. Using a 6 mm socket hexagon wrench, hold the ball stud.
    b. Remove the nut and disconnect the stabilizer link.
36. Disconnect the right front stabilizer link assembly.
37. Remove the left and right front axle shaft nuts. Strike the lock nut covering to remove it.
38. Remove the bolts and disconnect the right and left front speed sensors from the steering knuckle.
39. Disconnect the left and right tie rod assemblies, as follows:
    a. Remove the cotter pin and nut.
    b. Using SST (SST: 09628-62011) or equivalent, disconnect the tie rod ends from the steering knuckle.

➡Do not damage the dust cover of the ball joint.

40. Disconnect the lower No. 1 left and right front suspension arm sub-assemblies, as follows:
    a. Remove the bolt and 2 nuts, and separate the front suspension arm from the lower ball joint.
    b. Using a plastic hammer, disconnect the drive shaft from the axle hub.
41. Remove the starter assembly.
42. Remove the No. 1 exhaust pipe support bracket.
43. Remove the drive plate and torque converter clutch setting bolt, as follows:
    a. Using SST (SST: 09213-54015, SST: 09330-00021) or equivalent, hold the crankshaft.
    b. Remove the 2 bolts and flywheel housing under cover.
    c. Remove the 6 torque converter setting bolts.
44. Disconnect the steering intermediate shaft assembly, as follows:
    a. Loosen the sliding yoke bolt.
    b. To prevent the steering wheel from rotating, fix the wheel with the seat belt.

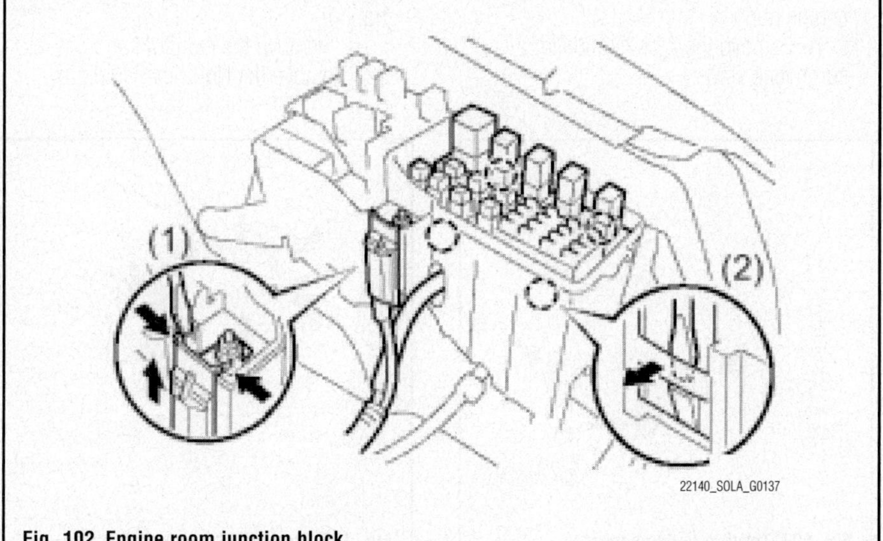

Fig. 102 Engine room junction block

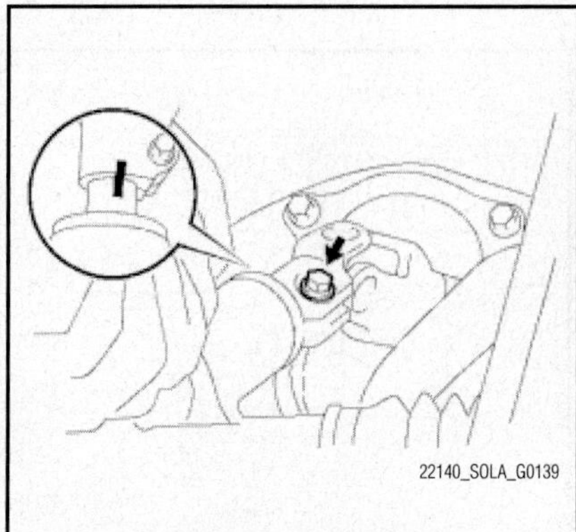

Fig. 104 Matchmarks on the steering intermediate shaft and control valve shaft

⁂ WARNING

If the steering wheel is not fixed, the spiral cable will be damaged.

c. Put matchmarks on the steering intermediate shaft and control valve shaft.

d. Remove the bolt and disconnect the steering intermediate shaft.

45. Remove the alternator assembly.
46. Remove the No. 2 alternator bracket.
47. Remove the alternator belt adjusting bar.
48. Disconnect the compressor and magnetic clutch.
49. Remove the engine assembly with transaxle, as follows:

a. Set the engine lifter.

b. Remove the 4 bolts, 2 nuts, and right and left frame side rail plates.

c. Remove the 4 bolts, 2 nuts, front suspension member right and left rear brace.

d. Operate the engine lifter, then remove the engine assembly from the vehicle.

➡Make sure the engine is clear of all wiring and hoses.

e. Install the engine hanger as shown in the illustration. No. 2 engine hanger part number: 12282-20020, Bolt: 91621-60822. Torque to 15 ft. lbs. (20 Nm).

f. Using a chain block and an engine sling device, hang the engine assembly.

50. Remove the vane pump assembly, as follows:

a. Remove the power steering oil pressure sensor harness.

b. Remove the pressure feed tube clamp.

c. Remove the 2 bolts and vane pump.

51. Remove the front frame assembly, as follows:

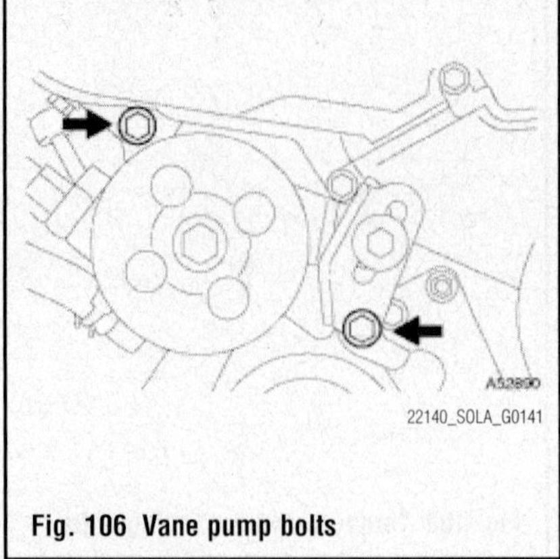

Fig. 106 Vane pump bolts

a. Remove the 2 nuts and disconnect the right and left engine mounting insulators.

b. Remove the bolt and disconnect the engine mounting insulator FR.

52. Remove the right and left front drive shaft assemblies.
53. Remove the engine wire.
54. Remove the automatic transaxle assembly.
55. Remove the drive plate and ring gear sub-assembly.
56. Install engine stand.
57. Remove the intake air surge tank, as follows:

a. Disconnect the throttle body connector.

b. Disconnect the No. 3 water by-pass hose.

c. Disconnect the No. 2 water by-pass hose.

d. Disconnect the union to check valve hose.

e. Disconnect the ventilation hose.

f. Remove the 3 nuts and separate the pressure feed tube.

g. Remove the 2 bolts and No. 1 engine hunger.

h. Remove the 2 bolts and No. 2 surge tank stay.

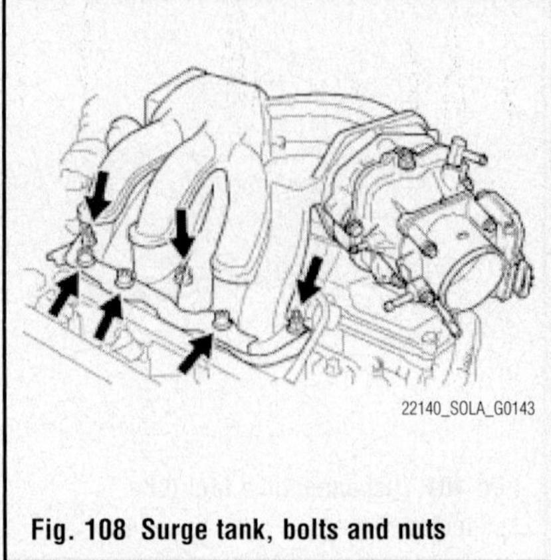

Fig. 108 Surge tank, bolts and nuts

i. Remove the 2 bolts and No. 1 surge tank stay.

j. Using an 8 mm socket hexagon wrench, remove the 4 bolts.

k. Remove the 2 nuts, emission control valve bracket and surge tank.

l. Remove the gasket from the surge tank.

58. Remove the ignition coil assembly.
59. Remove the intake manifold, as follows:

a. Remove the nut and disconnect the ground cable.

b. Disconnect the 6 fuel injector connectors.

c. Uniformly, loosen and remove the 9 bolts and 2 nuts in the sequence shown in the illustration. Remove the intake manifold.

60. Remove the water outlet.
61. Remove the right exhaust manifold sub-assembly, as follows:

a. Disconnect the A/F sensor connector.

b. Remove the 3 bolts and insulator.

c. Uniformly loosen and remove the 6 nuts in the sequence shown in the illustration.

d. Remove the manifold and gasket.

62. Remove the No. 2 manifold stay.

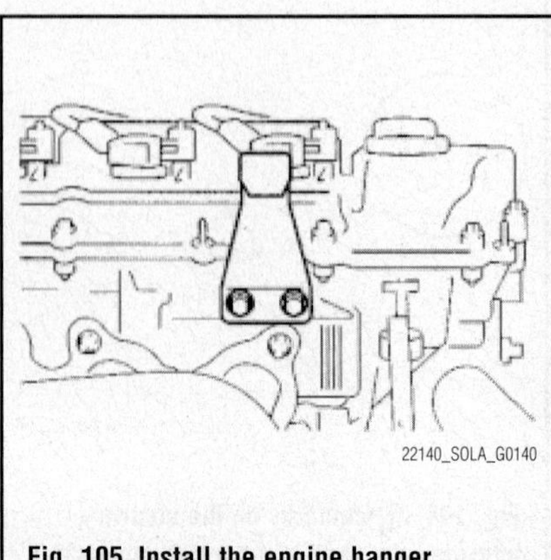

Fig. 105 Install the engine hanger

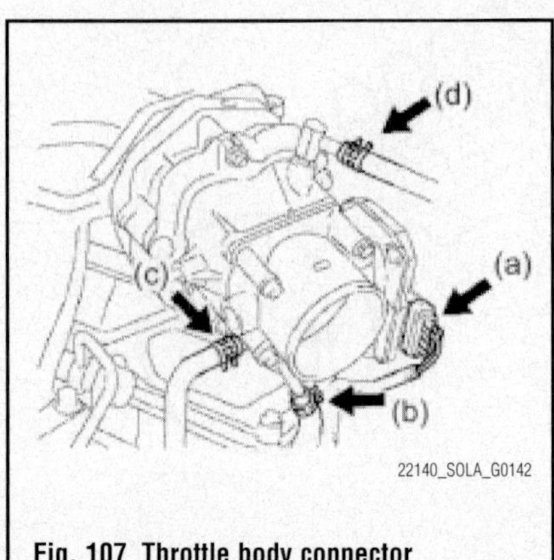

Fig. 107 Throttle body connector

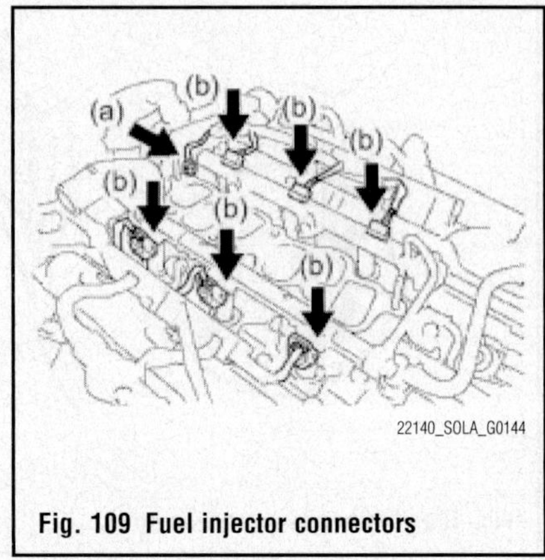

Fig. 109 Fuel injector connectors

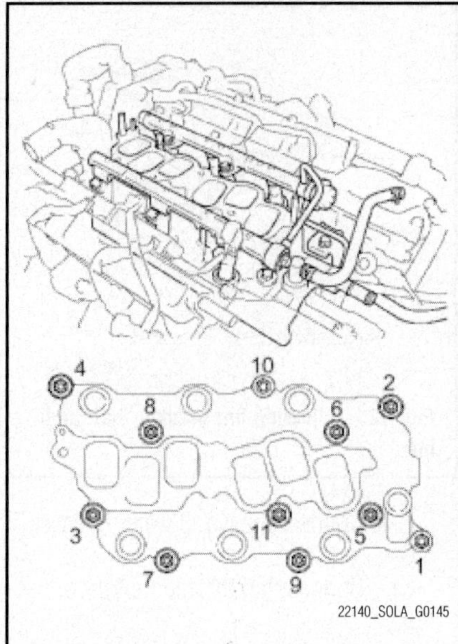

Fig. 110 Intake manifold bolt removal sequence

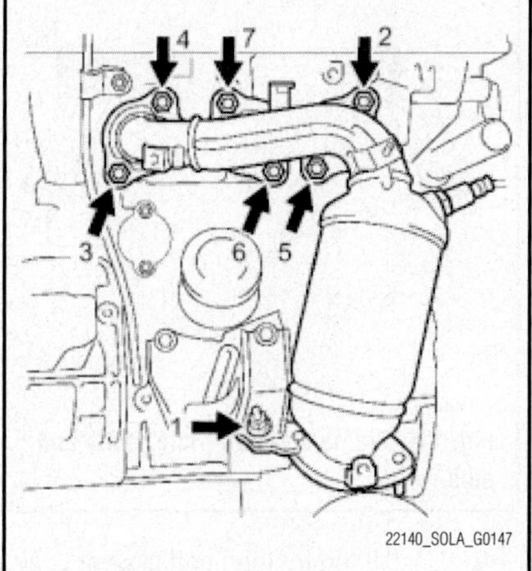

Fig. 112 No. 2 exhaust manifold converter bolt removal sequence

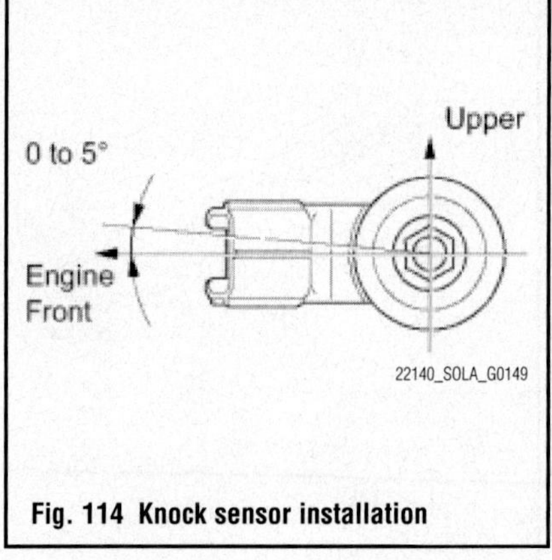

Fig. 114 Knock sensor installation

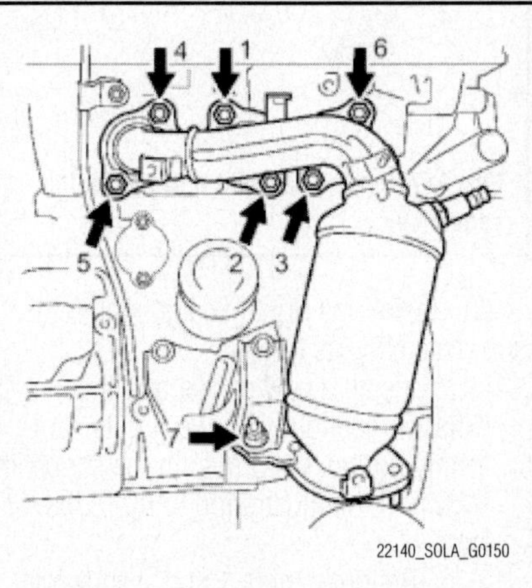

Fig. 115 No. 2 exhaust manifold converter nut tightening sequence

63. Remove the No. 2 exhaust manifold heat insulator, as follows:
   a. Remove the 2 bolts and insulator.
   b. Remove the bolt, nut and insulator.
64. Remove the No. 2 exhaust manifold converter sub-assembly, as follows:
   a. Uniformly loosen and remove the 7 nuts in the sequence shown in the illustration.
   b. Remove the converter and gasket.
65. Remove the drive shaft bearing bracket.
66. Remove the pump bracket.
67. Remove the No. 1 alternator bracket.
68. Remove the No. 1 compressor mounting bracket.
69. Remove the water inlet pipe.
70. Remove the water inlet.
71. Remove the thermostat.
72. Remove the engine oil pressure switch assembly.
73. Remove the knock sensor.
74. Replace partial engine assembly.

To install:

75. Install the 2 knock sensors with the 2 nuts, as shown in the illustration and tighten to 15 ft. lbs. (20 Nm).
76. Clean the threads of the oil pressure switch. Apply adhesive (Toyota Genuine Adhesive 1344, Three Bond 1344 or equivalent) to 2 or 3 threads of the oil pressure switch.
77. Install the oil pressure switch and torque to 11 ft. lbs. (15 Nm).
78. Install the thermostat.
79. Install the water inlet.
80. Install the water inlet pipe.
81. Install the No. 1 compressor mounting bracket and torque to 18 ft. Lbs. (25 nm).
82. Install the No. 1 generator bracket and torque to 43 ft. lbs. (58 Nm).
83. Install the pump bracket and torque to 24 ft. lbs. (32 nm).
84. Install drive shaft bearing bracket and torque to 47 ft. lbs. (64 Nm).
85. Install the No. 2 exhaust manifold converter sub-assembly, as follows:
   a. Install a new gasket and the converter with the 7 nuts. Uniformly, tighten the 7 nuts in the sequence shown in the illustration to 36 ft. lbs. (49 Nm).
   b. Retighten nuts 1 and 2 as shown in the illustration.
86. Install the No. 2 exhaust manifold heat insulator and torque to 75 inch lbs. (8.5 Nm).
87. Install the No. 2 manifold stay and torque to 25 ft. lbs. (34 Nm).

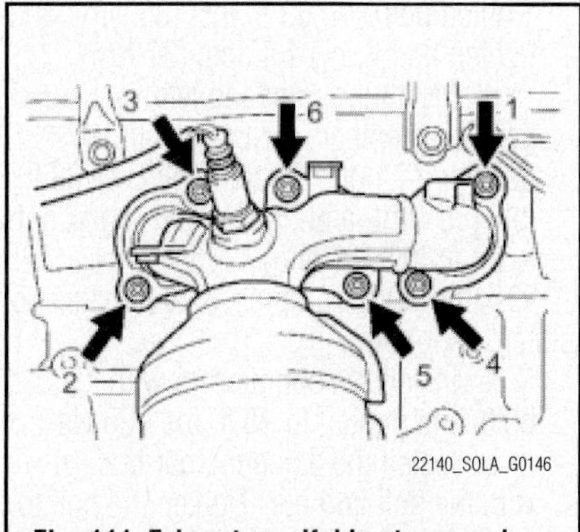

Fig. 111 Exhaust manifold nut removal sequence

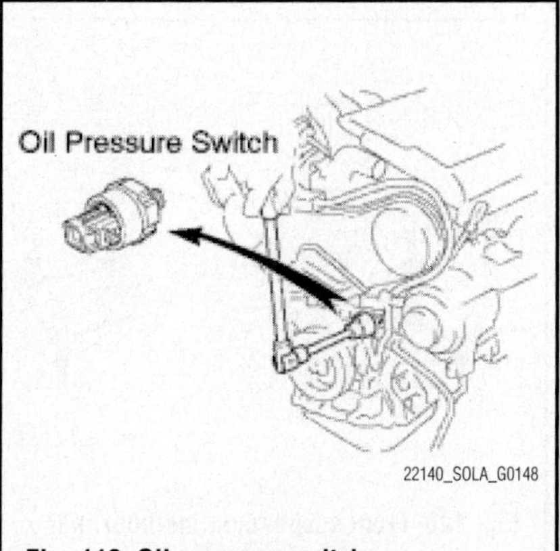

Fig. 113 Oil pressure switch

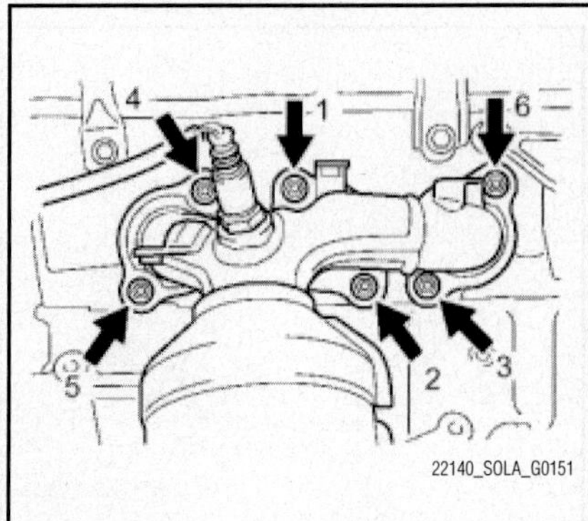

Fig. 116 Exhaust manifold bolt tightening sequence

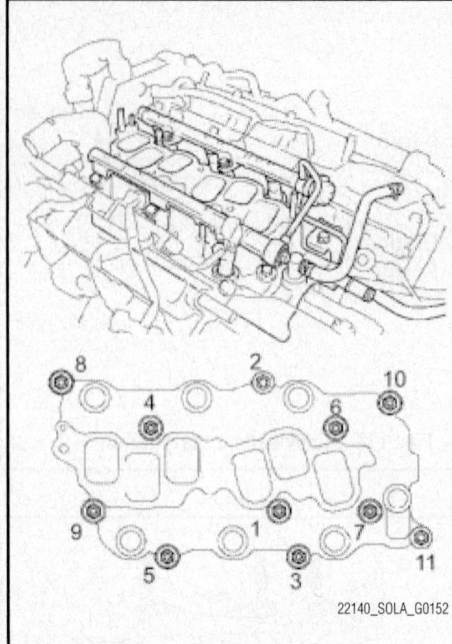

Fig. 117 Intake manifold bolt tightening sequence

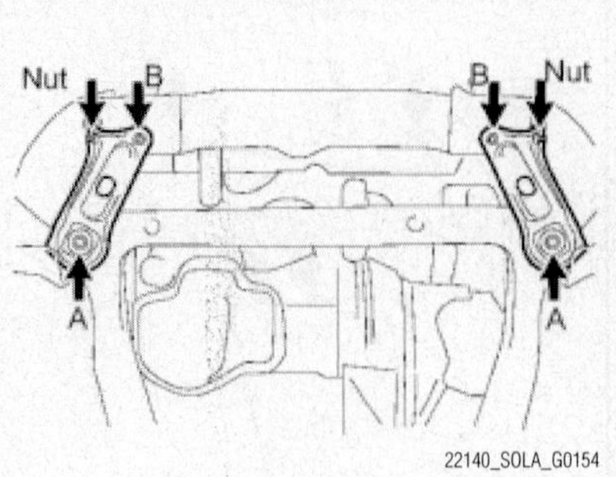

88. Install the right exhaust manifold sub-assembly, as follows:
   a. Install a new gasket and the exhaust manifold with the 5 nuts. Uniformly tighten the 5 nuts in the sequence shown in the illustration to 49 ft. lbs. (49 Nm).
   b. Retighten nuts 1 and 2 shown in the illustration.
89. Install the water outlet.
90. Install the intake manifold as follows:
   a. Install the intake manifold with the 9 bolts, 2 nuts and 2 washers. Uniformly tighten the bolts and nuts in the sequence shown in the illustration to 11 ft. lbs. (15 Nm).
   b. Retighten the water outlet mounting bolts and nuts to 11 ft. lbs. (15 Nm).
   c. Install the ground cable with the nut and tighten to 74 inch lbs. (8.4 Nm).
   d. Connect the heater inlet water hose.

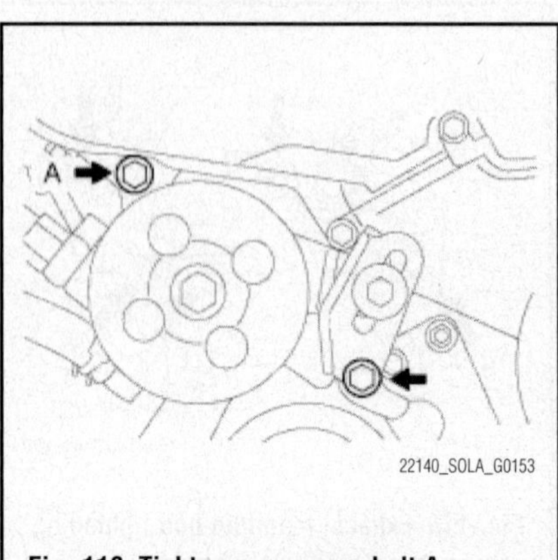

Fig. 118 Tighten vane pump bolt A

Fig. 119 Frame side rail plate, bolts and nuts

91. Install the ignition coil assembly and tighten to 71 inch lbs. (8 Nm).
92. Install the intake air surge tank, as follows:
   a. Install a new gasket to the intake air surge tank.
   b. Install the intake air surge tank and emission control valve bracket with the 2 nuts and tighten to 21 ft. lbs. (28 Nm).
   c. Using an 8 mm socket hexagon wrench, tighten the 4 bolts to 21 ft. lbs. (28 Nm).
   d. Install the No. 2 surge tank stay with the 2 bolts and tighten to 15 ft. lbs. (20 Nm).
   e. Install the No. 1 surge tank stay with the 2 bolts and tighten to 15 ft. lbs. (20 Nm).
   f. Install the No. 1 engine hanger with the 2 bolts and tighten to 15 ft. lbs. (20 Nm).
   g. Install the pressure feed tube with the 3 nuts and tighten to 69 inch lbs. (7.8 Nm).
   h. Connect the ventilation hose.
   i. Connect the union to check valve hose.
   j. Connect the No. 2 water by-pass hose.

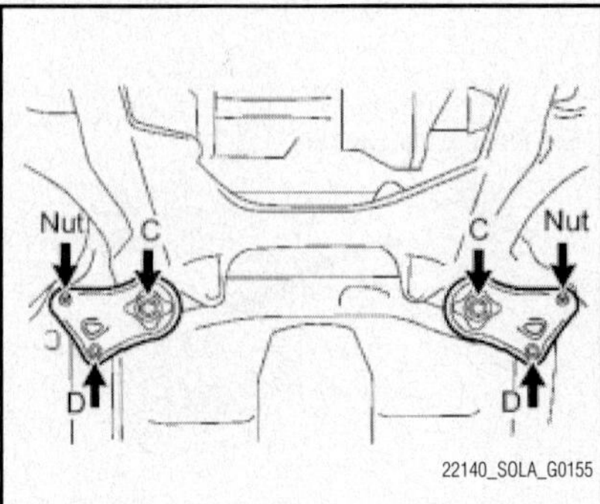

Fig. 120 Front suspension member rear brace, bolts and nuts

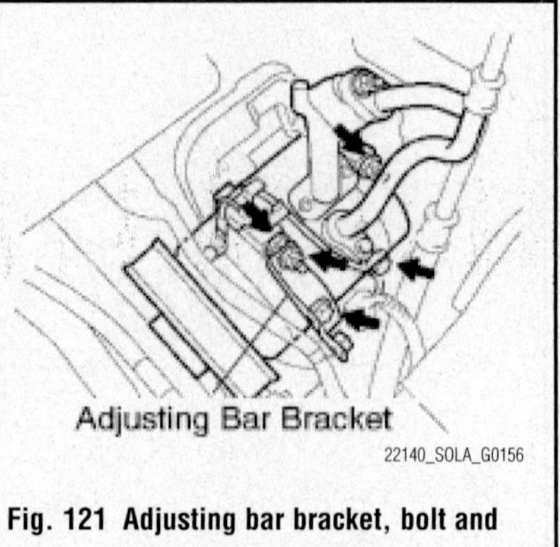

Fig. 121 Adjusting bar bracket, bolt and nut

   k. Connect the No. 3 water by-pass hose.
   l. Connect the throttle motor connector.
93. Install the drive plate and ring gear sub-assembly.
94. Install the automatic transaxle assembly.
95. Install the right and left front drive shaft assemblies.
96. Install the front frame assembly, as follows:
   a. Install the right and left engine mounting insulator and with the 2 nuts and tighten to 70 ft. lbs. (95 Nm).
   b. Install the engine mounting insulator FR with the bolt and tighten to 64 ft. lbs. (87 Nm).
97. Install the vane pump with the 2 bolts and tighten to 32 ft. lbs. (43 Nm).

➡After adjusting the V-ribbed belt, tighten bolt A.

98. Install the engine assembly with transaxle, as follows:
   a. Set the engine assembly with transaxle on the engine lifter.
   b. Install the engine assembly to the vehicle.
   c. Install the right and left frame side rail plate with the 4 bolts and 2 nuts. Tighten bolt A to 63 ft. lbs. (85 Nm), and tighten bolt B to 24 ft. lbs. (32 Nm).
   d. Install the right and left front suspension member rear brace with the 4 bolts and 2 nuts. Tighten bolt C to 63 ft. lbs. (85 Nm), and tighten bolt D and nuts to 24 ft. lbs. (32 Nm).
99. Install the compressor and magnetic clutch, as follows:
   a. Install the compressor with the 3 bolts and tighten to 18 ft. lbs. (25 Nm).
   b. Install the adjusting bar bracket with the bolt and nut. Tighten the bolt to 18 ft. lbs. (25 Nm) and the nut to 19 ft. lbs. (26 Nm).

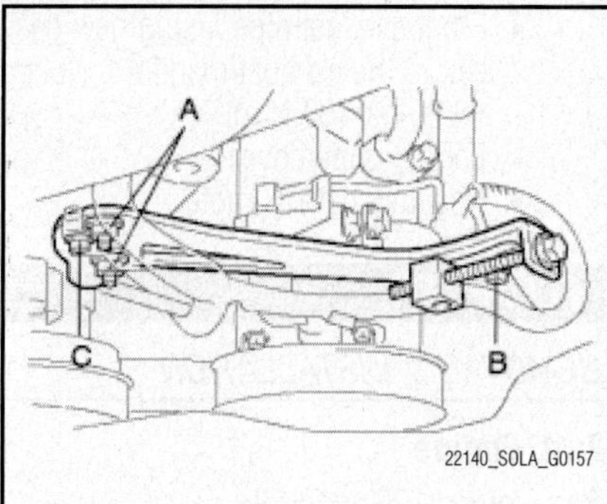

Fig. 122 Alternator belt adjusting bar, bolts and nuts

100.Install the alternator belt adjusting bar with the 2 bolts and 2 nuts. Tighten nut A to 32 ft. lbs. (43 Nm), bolt B to 13 ft. lbs. (18 Nm), and bolt C to 71 inch lbs. (8 Nm).

101.Install the No. 2 alternator bracket and tighten to 21 ft. lbs. (28 Nm).

102.Install the alternator assembly.

103.Install steering intermediate shaft assembly, as follows:

a. Align the matchmarks on the intermediate shaft and control valve shaft, and install the bolt and tighten to 26 ft. lbs. (35 Nm).

b. Tighten the sliding yoke bolt to 26 ft. lbs. (35 Nm).

104.Install the drive plate and torque converter clutch setting bolt, as follows:

a. Using SST (SST: 09213-54015, SST: 09330-00021) or equivalent, hold the crankshaft.

b. Using kerosene or gasoline, clean the bolts thoroughly.

c. Apply adhesive (Toyota Genuine Adhesive 1324, Three Bond 1324 or equivalent) to 2 or 3 threads of the bolt end.

d. Install the 6 torque converter set bolts and tighten to 30 ft. lbs. (41 Nm). First tighten the green colored bolt and then tighten the 5 bolts.

e. Install the flywheel housing under cover with the 2 bolts and tighten to 69 inch lbs. (7.8 Nm).

105.Install the No. 1 exhaust pipe support bracket and tighten to 15 ft. lbs. (20 Nm).

106.Install the starter assembly.

107.Install the left and right No. 1 front lower suspension arm sub-assemblies, as follows:

a. Install the drive shaft to the steering knuckle.

b. Install the lower suspension arm with the bolt and 2 nuts and tighten to 55 ft. lbs. (75 Nm).

108.Install the left and right tie rod assemblies, as follows:

a. Connect the tie rod end to the steering knuckle with the castle nut and tighten to 36 ft. lbs. (49 Nm).

➡Do not allow lubricants from contacting the thread and taper portions.

➡After tightening the castle nut, tighten it an additional 60° so that a cotter pin can be inserted.

b. Insert a new cotter pin.

109.Install the left and right front speed sensors.

110.Install the left and right front axle shaft nut, as follows:

a. Clean the threaded parts on the drive shaft and axle hub nut using a non-residue solvent.

➡Be sure to perform this work for a new drive shaft.

➡Keep the threaded parts free of oil and foreign objects.

b. Using a 30 mm socket wrench, install a new hub nut and tighten to 217 ft. lbs. (294 Nm).

c. Using a chisel and hammer, tapped the hub nut.

111.Install the left and right front stabilizer link assemblies, as follows:

a. Using a 6 mm socket hexagon wrench, hold the ball stud and install the nut and tighten to 55 ft. lbs. (74 Nm).

112.Install the front exhaust pipe assembly.

113.Install the front No. 1 exhaust pipe support bracket.

114.Connect the No. 1 fuel pipe sub-assembly, as follows:

a. Push in the fuel tube connector to the fuel pipe until connector makes a "click" sound.

➡Check for damage or contamination on the connected part of the pipe. Check if the pipe and the connector are securely connected by trying to pull them apart.

b. Install the fuel pipe clamp.

115.Install the air cleaner assembly.

116.Connect the vacuum hoses.

117.Install the V-bank cover sub-assembly.

118.Install the right No. 2 engine mounting stay and mounting bracket with the bolt and tighten to 47 ft. lbs. (64 Nm).

119.Install the engine moving control rod and bracket with the 4 bolts. Tighten bolt A to 47 ft. lbs. (64 Nm), and tighten bolt B to 17 ft. lbs. (23 Nm).

120.Install the vane pump V-belt.

121.Install the No. 1 (cooler compressor to crankshaft pulley) V-belt.

122.Inspect drive belt tension.

123.Install front wheels.

124.Connect cable to negative battery terminal.

125.Add automatic transaxle fluid, engine oil, and engine coolant.

126.Add power steering fluid.

127.Bleed power steering fluid.

128.Inspect for oil, engine coolant, fuel, and exhaust gas leaks.

129.Inspect and adjust front wheel alignment.

130.Inspect ignition timing and engine idle speed.

131.Inspect CO/HC.

132.Check ABS speed sensor signal.

133.Perform initialization.

EXHAUST MANIFOLD

REMOVAL & INSTALLATION

2.4L Engine

See Figures 123 and 124.

1. Before servicing the vehicle, refer to the Precautions Section.
2. Remove or disconnect the following:

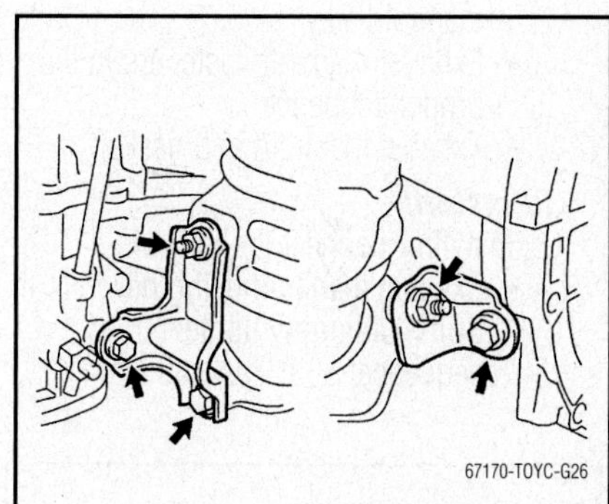

Fig. 123 Exhaust manifold converter stays—2.4L (3AZ-FE) engine

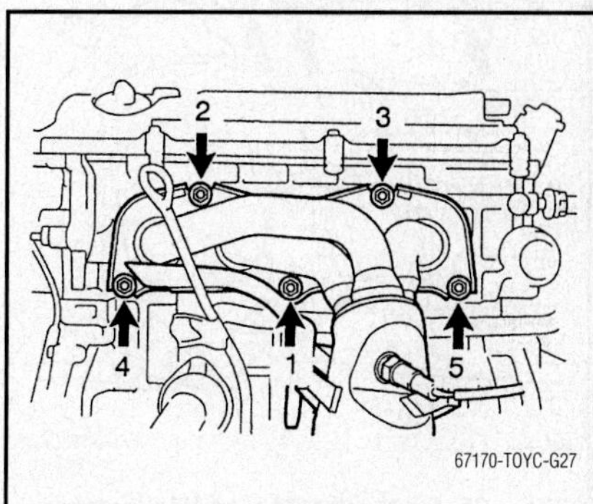

Fig. 124 Exhaust manifold fastener tightening sequence—2.4L (3AZ-FE) engine

- Negative battery terminal
- Exhaust manifold heat insulator
- Oxygen sensor
- Exhaust manifold converter stays
- Exhaust pipe-to-manifold fasteners
- Manifold nuts, manifold and gasket

To install:

3. Place a new gasket and reinstall the exhaust manifold with the nuts. Tighten to 27 ft. lbs. (33 Nm).
4. Install or connect the following:
   - Exhaust manifold converter stays. Tighten to 32 ft. lbs. (44 Nm)
   - Oxygen sensor
   - Exhaust manifold heat insulator
   - Negative battery cable

3.3L Engine

See Figures 125 through 128.

1. Before servicing the vehicle, refer to the Precautions Section.
2. Remove or disconnect the following:
   - Negative battery cable
   - Engine undercover
   - Front exhaust pipes from the exhaust manifold
   - Heated oxygen sensor connector from the exhaust manifold
   - Heat shield insulator
   - For the front manifold, the exhaust manifold stay
   - Exhaust manifold fasteners in the proper sequence
   - Exhaust manifold and gasket

To install:

3. Install or connect the following:
   - Exhaust manifold using new gasket. Bolts, uniformly tightened in sequence: 36 ft. lbs. (49 Nm).

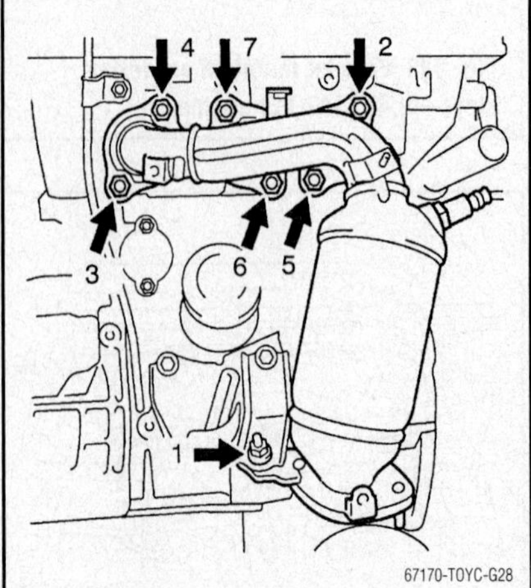

Fig. 125 Front exhaust manifold fastener loosening sequence—3.3L (3MZ-FE) engine

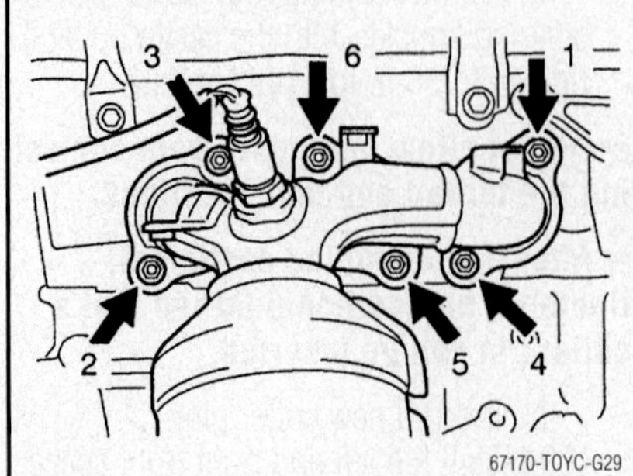

Fig. 126 Rear exhaust manifold fastener loosening sequence—3.3L (3MZ-FE) engine

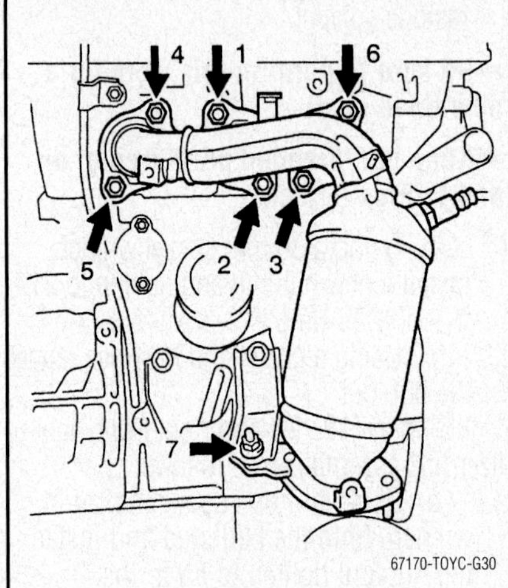

Fig. 127 Front exhaust manifold fastener tightening sequence—3.3L (3MZ-FE) engine

Retighten bolts 1 and 2 to specification again.

- Exhaust manifold stay. Bolt and nut: 25 ft. lbs. (34 Nm).
- Heat shield insulator
- Heated oxygen sensor connector
- Front exhaust pipe, using new gasket, to the exhaust manifold. Nuts: 46 ft. lbs. (62 Nm).
- Engine undercover
- Negative battery cable

FLEXPLATE

REMOVAL & INSTALLATION

2.4L Engine

See Figures 129 and 130.

1. Before servicing the vehicle, refer to the Precautions Section.
2. Disconnect the negative battery cable.
3. To gain access to the flywheel or flexplate, remove the transaxle. Refer to the Drive Train section for removal and installation procedures.
4. For A/T, perform the following:
   a. Use Special Tool: 09960-10010 to hold the crankshaft.
   b. Remove the eight bolts, rear spacer, drive plate and front spacer.
5. For M/T, perform the following:
   a. Remove the clutch cover assembly and clutch disc assembly.
   b. Use Special Tool: 09960-10010 to hold the crankshaft.
   c. Remove the eight bolts, rear spacer and flywheel.

To install:

6. For M/T, perform the following:
   a. Use Special Tool: 09960-10010 to hold the crankshaft.
   b. Clean the bolt and the bolt hole.
   c. Apply adhesive to 2 or 3 threads of the bolt end. Adhesive: Part No. 08833-00070, Three Bond or equivalent.
   d. Install the front spacer, drive plate and rear spacer, and flywheel with the eight bolts. Uniformly tighten the 8 bolts

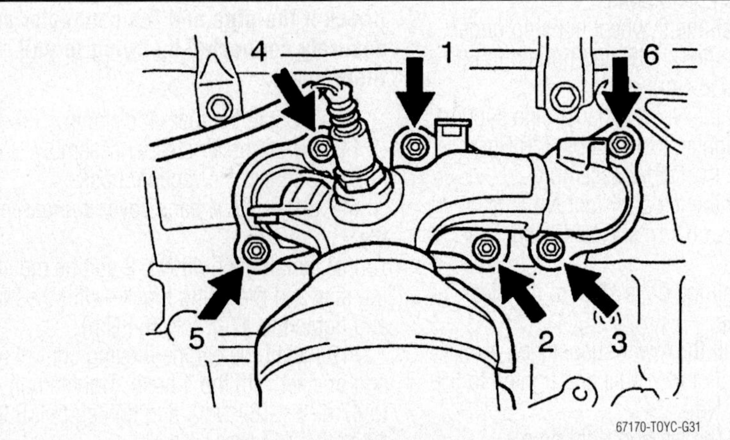

Fig. 128 Rear exhaust manifold fastener tightening sequence—3.3L (3MZ-FE) engine

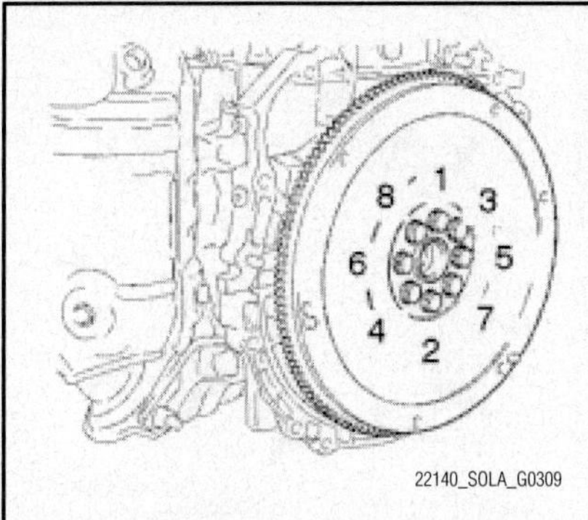

Fig. 129 M/T flywheel bolt tightening sequence

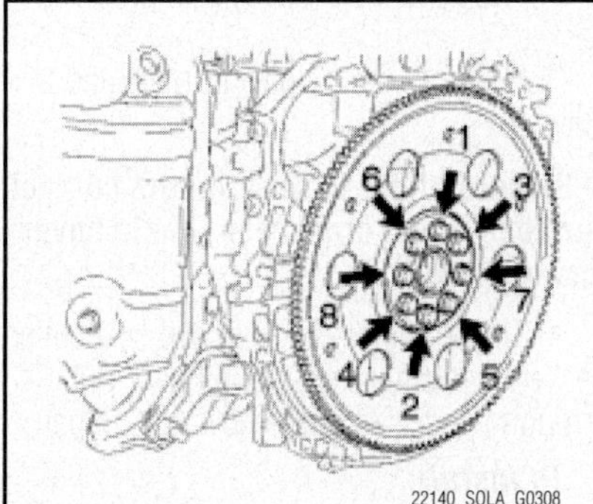

Fig. 130 A/T flywheel bolt tightening sequence

in the sequence. Tighten to 96 ft. lbs. (130 Nm).

e. Install clutch cover assembly and clutch disc assembly.

7. For A/T, perform the following:

a. Use Special Tool: 09960-10010 to hold the crankshaft.

b. Clean the bolt and the bolt hole.

c. Apply adhesive to 2 or 3 threads of the bolt end. Adhesive: Part No. 08833-00070, Three Bond or equivalent.

d. Install the front spacer, drive plate and rear spacer with the 8 bolts. Uniformly tighten the eight bolts in the sequence. Tighten 72 ft. lbs. (98 Nm).

3.3L Engine

See Figure 131.

1. Before servicing the vehicle, refer to the Precautions Section.
2. Disconnect the negative battery cable.
3. Remove the automatic transaxle assembly.
4. Using SST (SST: 09213-54015, SST: 09330-00021) or equivalent tool, hold the crankshaft.
5. Remove the 8 bolts, front spacer, drive plate and rear spacer.

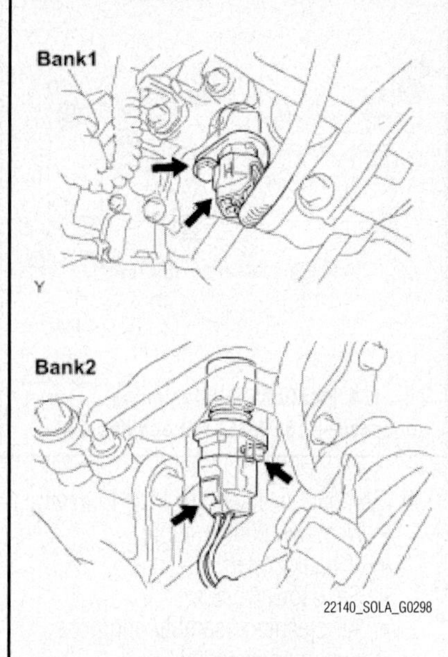

Fig. 131 Flywheel bolt tightening sequence

To install:

6. Using SST (SST: 09213-54015, SST: 09330-00021) or equivalent, hold the crankshaft.
7. Apply adhesive (Toyota Genuine Adhesive 1324, Three Bond 1324 or equivalent) to 2 or 3 threads of the bolts.
8. Install the front spacer, drive plate and rear spacer on the crankshaft.
9. Install and tighten the 8 mounting bolts in the sequence shown in the illustration to 61 ft. lbs (83 Nm).

➡Do not start the engine for at least an hour after installing.

10. Install automatic transaxle assembly.

INTAKE MANIFOLD

REMOVAL & INSTALLATION

2.4L Engine

See Figure 132.

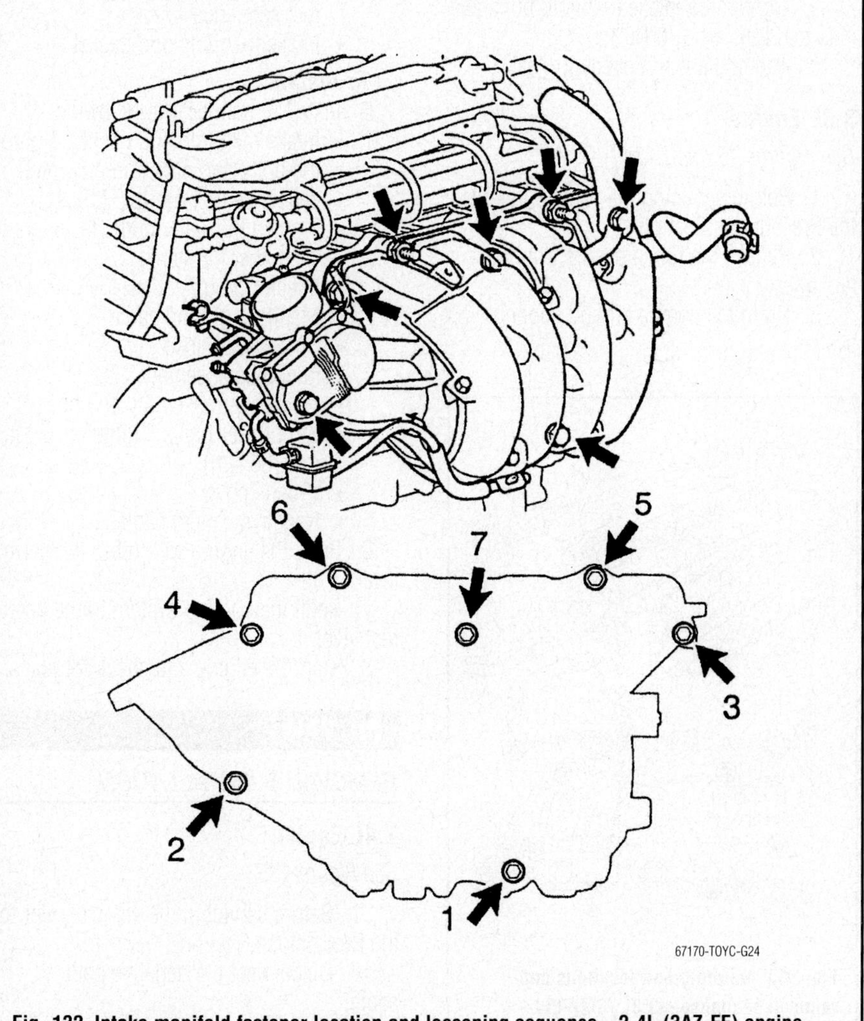

Fig. 132 Intake manifold fastener location and loosening sequence—2.4L (3AZ-FE) engine

1. Before servicing the vehicle, refer to the Precautions Section.
2. Relieve the fuel pressure from the fuel lines.
3. Drain the engine oil and cooling system.
4. Remove or disconnect the following:
- Negative battery cable.
- Strut tower brace
- Radiator hose outlet
- Union to connector tube hose
- Water inlet water hose
- Fuel tube assembly
- Water bypass hoses from the throttle body
- Intake manifold and gasket

To install:

5. To install, reverse removal procedure.
6. Install a new throttle body gasket onto the intake manifold.
7. Tighten the throttle body bolts and fuel pipe clamp to: 22 ft. lbs. (30 Nm).
8. Install a new intake manifold gasket.
9. Tighten the intake manifold bolts and nuts to: 22 ft. lbs. (30 Nm).
10. Add coolant and check for leaks.

3.3L Engine

See Figures 133 and 134.

1. Before servicing the vehicle, refer to the Precautions Section.
2. Relieve the fuel pressure from the fuel lines.
3. Drain the engine oil and cooling system.

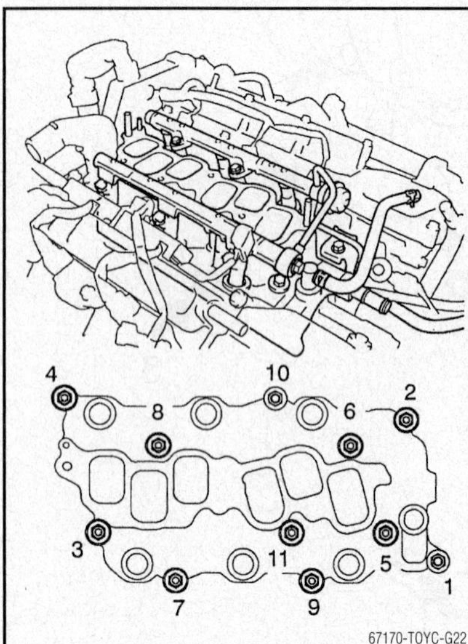

Fig. 133 Manifold bolt locations and removal sequence—3.3L (3MZ-FE) engines

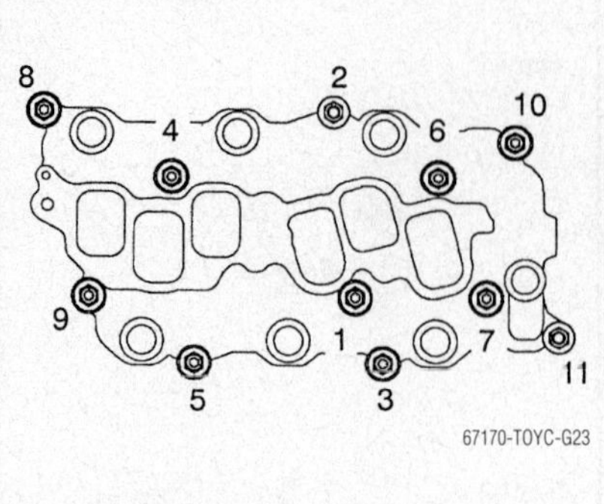

Fig. 134 Manifold bolt tightening sequence—3.3L (3MZ-FE) engines

4. Remove or disconnect the following:
- Negative battery cable.
- V-bank cover
- Strut tower brace
- Air cleaner assembly and hose
- Fuel pipe assembly
- Heater inlet hose
- Manifold ground cable
- Injector plugs
- Manifold bolts and nuts in the correct sequence
- Intake manifold and gasket

To install:

5. Install or connect the following:
- Intake manifold and gasket. Tighten the bolts and nuts in the correct sequence to 11 ft. lbs. (15 Nm)
- Tighten the water outlet fasteners to 11 ft. lbs. (15 Nm)
- Injector plugs
- Manifold ground cable
- Heater inlet hose
- Fuel pipe assembly
- Air cleaner assembly and hose
- Strut tower brace. Tighten to 59 ft. lbs. (80 Nm)
- V-bank cover
- Negative battery cable
6. Refill the engine with oil to the correct level.
7. Refill the cooling system to the correct level.
8. Start the engine and check for leaks.

OIL PAN

REMOVAL & INSTALLATION

2.4L Engine

See Figure 135.

1. Before servicing the vehicle, refer to the Precautions Section.
2. Disconnect the negative battery cable.
3. Drain the engine oil.

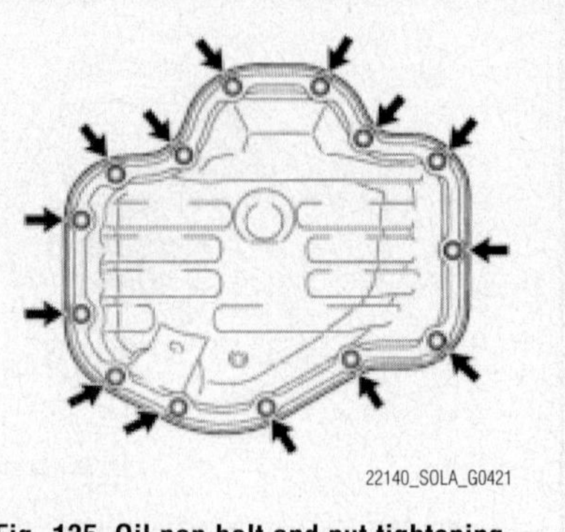

Fig. 135 Oil pan bolt and nut tightening sequence

4. Remove the oil pan drain plug and gasket.
5. Remove the oil pan 12 bolts and 2 nuts.

➡**Be careful not to damage the contact surfaces of the crankcase, chain cover and oil pan.**

6. Insert the blade of oil pan seal cutter between the crankcase and oil pan. Cut through the sealer and remove the oil pan.

To install:

7. Remove any old packing material and be careful not to drop any oil on the contact surfaces of the cylinder block and oil pan.
8. Apply a continuous bead of seal packing (Diameter 0.16 to 0.18 inches (4.0 to 4.5 mm)). Use Toyota Genuine Seal Packing Block, Three Bond 1344 or Equivalent.

➡**Remove any oil from the contact surfaces. Install the oil pan within 3 minutes after applying seal packing. Do not start the engine for at least 2 hours after installing.**

9. Uniformly tighten the 12 bolts and 2 nuts in sequence. Tighten the bolts and nuts to 80 inch lbs. (9 Nm).
10. To complete installation, reverse remaining removal procedure.
11. Add oil and check for leaks.

3.3L Engine

See Figures 136 and 137.

1. Before servicing the vehicle, refer to the Precautions Section.
2. Disconnect the negative battery cable.
3. Drain the engine oil.
4. Remove the oil pan drain plug and gasket.
5. Remove the 10 bolts and 2 nuts from the No. 2 oil pan.

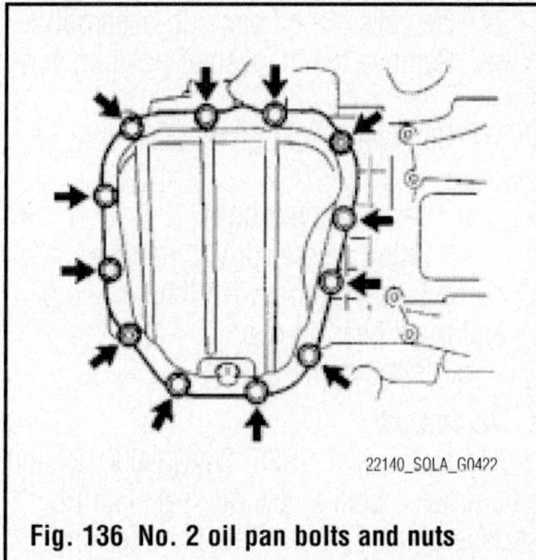

Fig. 136 No. 2 oil pan bolts and nuts

6. Insert the blade of oil pan seal cutter between the No. 1 and oil No. 2 pan oil pan. Cut through the sealer and remove the No. 2 oil pan.

➡Do not damage the contact surface of oil pan or the flange portion of the No. 2 oil pan during removal.

7. Uniformly loosen and remove the 15 bolts and 2 nuts from the oil pan sub-assembly in the sequence shown in the illustration.

8. Using a screwdriver, remove the oil pan by prying between the cylinder block and the oil pan.

➡Be careful not to damage the contact surfaces of the oil pan and cylinder block.

To install:

➡Remove any old seal packing from the contact surface. Remove any oil from the contact surface.

9. Apply a continuous bead of seal packing (Toyota Genuine Seal Packing Black, Three Bond 1207B or equivalent, diameter 3 to 4 mm (0.12 to 0.16 in.)) as shown in the illustration.

10. Apply seal packing to the outer side of the bolt holes in region "X".

11. Apply seal packing to the inner side of the bolt holes in region "Y".

➡Install the oil pan within 3 minutes after applying seal packing. Do not expose the seal to engine oil for at least 2 hours after installing.

12. Install the No. 1 oil pan with the 15 bolts and 2 nuts. Tighten the bolts uniformly in several steps, as follows:

- 10 mm head - Torque: 71 in. lbs. (82 kg. cm., 8.0 Nm)
- 12 mm head - Torque: 14 ft. lbs. (199 kg. cm., 20 Nm)

➡Remove any old seal packing from the contact surface. Remove any oil from the contact surface.

13. Apply a continuous bead of seal packing (Toyota Genuine Seal Packing Black, Three Bond 1207B or equivalent, diameter 3 to 4 mm (0.12 to 0.16 in.)) as shown in the illustration.

➡Install the oil pan within 3 minutes after applying seal packing. Do not expose the seal to engine oil for at least 2 hours after installing.

14. Install the No. 2 oil pan with the 10 bolts and 2 nuts and tighten to 71 inch lbs. (8 Nm).

15. Install the drain plug with a new gasket and tighten to 33 ft. lbs. (45 Nm).

16. Connect battery negative cable.

OIL PUMP

REMOVAL & INSTALLATION

2.4L Engine

See Figure 138.

1. Before servicing the vehicle, refer to the Precautions Section.
2. Disconnect the negative battery cable.
3. Drain the engine oil.
4. Remove right and left engine under covers.
5. Remove front right wheel.
6. Remove right front fender apron seal.
7. Remove front exhaust pipe assembly.
8. Remove engine moving control rod w/bracket.
9. Remove No. 2 right engine mounting stay.
10. Remove No. 2 right engine mounting bracket.
11. Remove fan and generator V-belt.
12. Remove No. 1 engine cover sub-assembly.
13. Disconnect engine wire.
14. Remove generator assembly.
15. Remove vane pump.
16. Remove ignition coil assembly.
17. Disconnect ventilation hoses.
18. Remove valve cover.
19. Turn the crankshaft pulley until its groove and the timing mark "0" of the timing chain cover are aligned.
20. Check that each timing mark of the camshaft timing gear and sprocket is aligned with each timing mark located on the intake and exhaust bearing caps. If not, turn the crankshaft by 1 revolution (360°) to align the timing marks.
21. Remove crankshaft pulley.
22. Remove crankshaft position sensor.
23. Remove oil pan.
24. Remove chain upper tensioner assembly.
25. Install the No. 1 engine hanger (12281-28010) and No. 2 engine hanger (12282-28010) with the bolts (91512-61020) and tighten to: 28 ft. lbs. (38 Nm).
26. Remove drive belt tensioner.

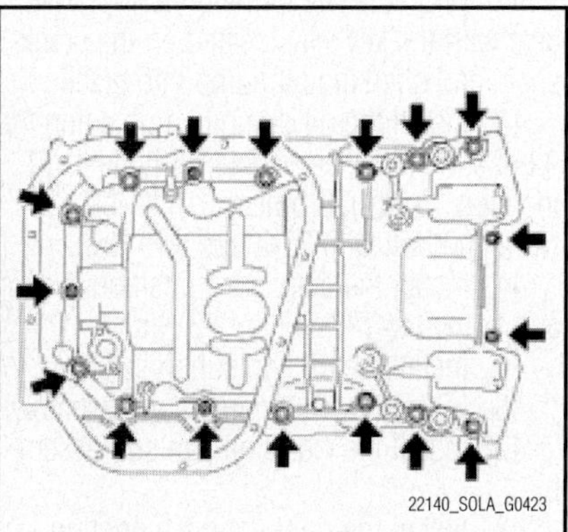

Fig. 137 Oil pan bolt and nut removal sequence

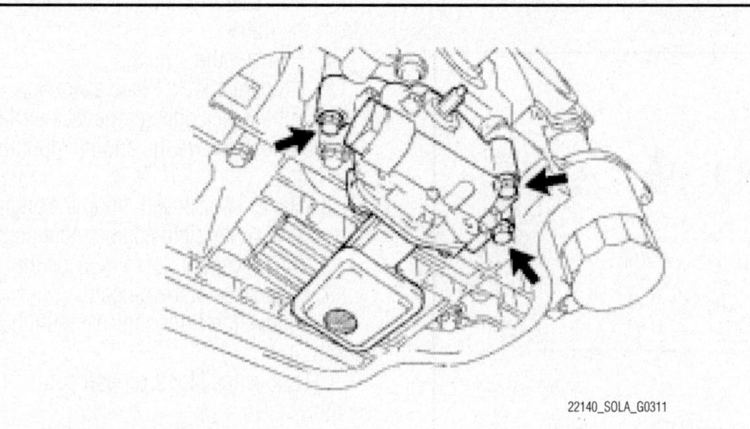

Fig. 138 Removing oil pump

27. Remove right engine mount insulator.
28. Remove right engine mount bracket.
29. Using an E10 Torx®socket, remove the stud bolt for the drive belt tensioner from the cylinder block.
30. Timing chain cover bolts and nuts.

➡Be careful not to damage the contact surfaces of the timing chain cover, cylinder block and cylinder head. Tape the screwdriver tip before use.

31. Remove timing chain cover by prying between the timing chain cover and cylinder head or cylinder block with a screwdriver.

➡Tape the screwdriver tip before use.

32. Using a screwdriver and a hammer, tap out the timing chain case oil seal.
33. Remove No. 1 crankshaft position sensor plate.
34. Remove chain tensioner slipper.
35. Remove No.1 chain vibration damper.
36. Remove chain sub-assembly.
37. Remove crankshaft timing sprocket.
38. Remove No. 2 chain sub-assembly.
    a. Three bolts, oil pump and gasket.

To install:

39. Install a new gasket and the oil pump with the 3 bolts. Tighten the bolts 14 ft. lbs. (19 Nm).
40. To complete installation, reverse removal procedure. Refer to the appropriate sections to install components correctly.

3.3L Engine

See Figures 139 through 144.

1. Before servicing the vehicle, refer to the Precautions Section.
2. Remove the right front wheel.
3. Remove the right front fender apron seal.
4. Remove the right engine under cover.

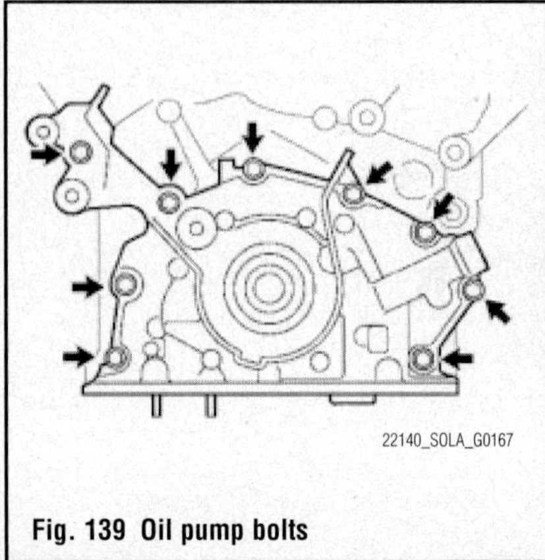

Fig. 139 Oil pump bolts

5. Drain engine oil and install a new gasket.
6. Remove front suspension upper brace center, if applicable.
7. Remove the No. 1 V-belt.
8. Remove the vane pump V-belt.
9. Remove the engine moving control rod.
10. Remove the right No. 2 engine mounting stay.
11. Remove the No. 2 generator bracket.
12. Remove the crankshaft pulley.
13. Remove the No. 1 timing belt cover.
14. Remove No. 2 timing belt cover.
15. Remove the right engine mounting bracket.
16. Remove the No. 2 timing belt guide.
17. Remove the timing belt.
18. Remove the No. 1 exhaust pipe support bracket.
19. Remove the front exhaust pipe assembly.
20. Remove the No. 1 exhaust pipe support bracket.
21. Separate compressor and magnetic clutch.
22. Remove the No. 2 timing belt idler sub-assembly.
23. Remove the camshaft timing pulley.
24. Remove the No.1 timing belt idler sub-assembly.
25. Remove the crankshaft timing pulley.
26. Remove the No. 3 timing belt cover.
27. Remove the No. 1 compressor mounting bracket.
28. Remove the oil level gage guide.
29. Separate the engine mounting insulator, as follows:
    a. Remove the 3 nuts and bolt. Then separate the engine mounting insulator. Do not remove the engine mounting insulator.
30. Remove the right engine mounting insulator as follows:
    a. Remove the bolt and disconnect the power steering return hose clamp from the frame.
    b. Remove the 4 nuts.
    c. Prepare a jack. Place a wooden block on the jack and set the jack under the engine. Remove the engine mounting insulator.
    d. Raise the jack and lift up the engine. Then remove the right engine mounting insulator. Be careful not to damage the contact surfaces of the oil pan.
31. Remove the right engine mounting bracket.
32. Remove the No. 2 oil pan sub-assembly.
33. Remove the oil strainer sub-assembly.
34. Remove the oil pan sub-assembly.
35. Remove the crankshaft position sensor.
36. Remove the oil pump assembly, as follows:
    a. Remove the 9 bolts.
    b. Using a screwdriver, remove the oil pump by prying between the oil pump and main bearing cap.
    c. Remove the O-ring.

To install:

37. Using SST 9SST: 09223-00010) and a hammer, install a new oil seal. Tap the surface of the SST with the hammer until the oil seal's surface is flush with oil pump body edge.

➡Be careful not to tap the oil seal at an angle.

➡Keep the gap between the oil pump body edge and oil seal free from contamination.

38. Apply a small amount of MP grease to the oil seal lip.
39. Remove any old seal packing material from the contact surface.
40. Apply a light coat of engine oil to a new O-ring and place it on the cylinder block.
41. Apply a continuous bead of seal packing (Toyota Genuine Seal Packing Black, Three Bond 1207B or equivalent, diameter: 2 to 3 mm (0.08 to 0.12 in.)) as shown in the illustration.

➡Remove any oil from contact surface.

➡Apply seal packing to the inner side of the bolt holes.

➡Install the oil pump within 3 minutes after applying seal packing.

➡Do not expose the seal to engine oil for at least 2 hours after installing the oil pump.

42. Align the key of the oil pump drive gear with the key way located on the crankshaft, and slide the oil pump into place.
43. Install the oil pump by tightening the 9 bolts in sequence. Tighten A bolts to 71 inch lbs. (8 Nm), B bolts to 15 ft. lbs. (20 Nm), and bolt C to 32 ft. lbs. (43 Nm).
44. Install the crankshaft position sensor and tighten to 71 inch lbs. (8 Nm).
45. Install the oil pan sub-assembly.
46. Install the oil strainer sub-assembly.
47. Install the No. 2 oil pan sub-assembly.
48. Install the right engine mounting bracket. Tighten A bolts to 40 ft lbs. (54 Nm), bolt B to 40 ft lbs. (54 Nm), and bolt C to 32 ft. lbs. (43 Nm).

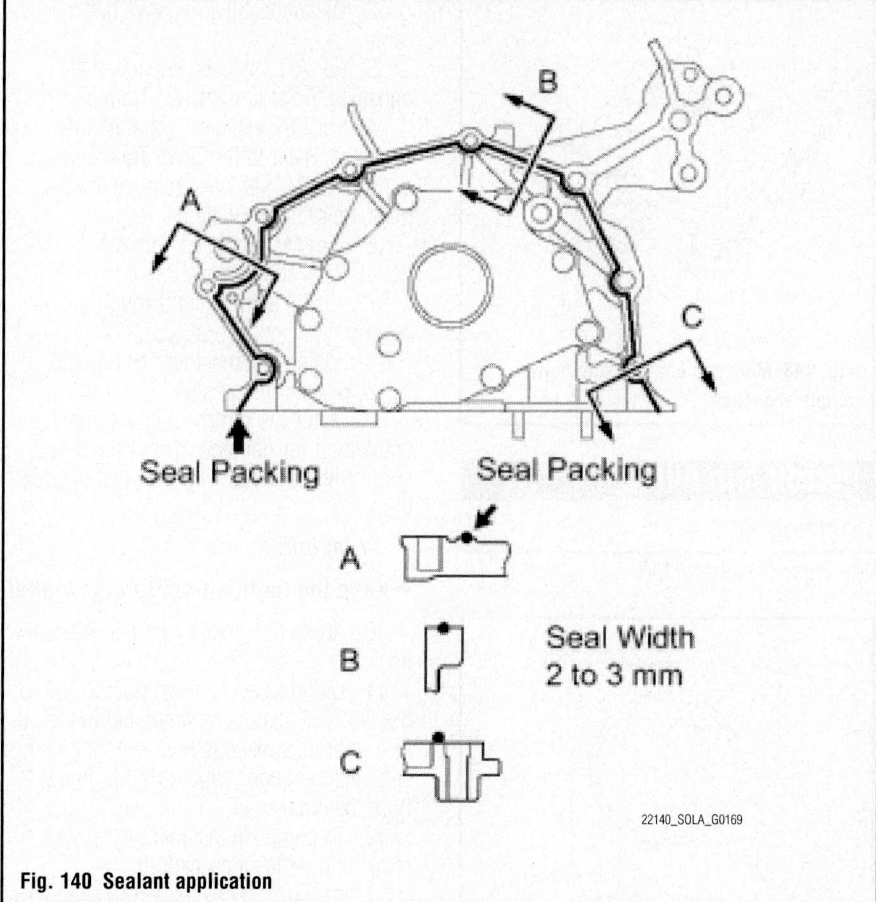

Fig. 140 Sealant application

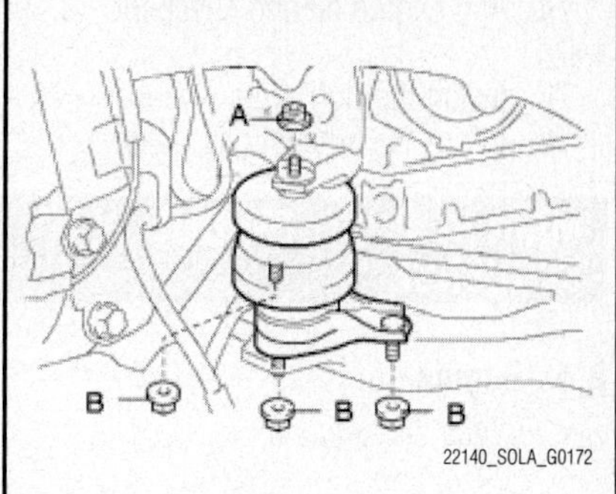

Fig. 143 Right engine mounting insulator nut tightening

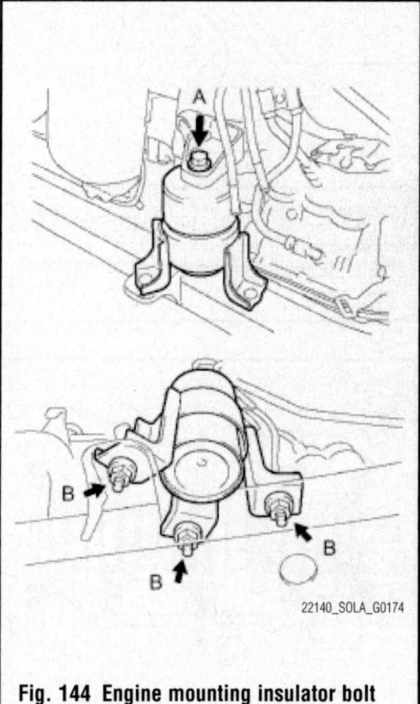

Fig. 144 Engine mounting insulator bolt tightening

49. Install the right engine mounting insulator. Tighten nut A to 70 ft. lbs. (95 Nm), and B nuts to 64 ft. lbs. (87 Nm).

50. Install the engine mounting insulator. Tighten bolt A to 64 ft. lbs. (87 Nm), and B bolts to 38 ft. lbs. (52 Nm).

51. Install the oil level gage guide, as follows:

a. Install a new O-ring to the gage guide.

b. Install the gage guide with the bolt and tighten to 71 inch lbs. (8 Nm).

52. Install the No. 1 compressor mounting bracket.

53. Install the No. 3 timing belt cover.

54. Install the No. 1 timing belt idler sub-assembly.

55. Install the No. 2 timing belt idler sub-assembly.

56. Install the camshaft timing pulley.

57. Install the compressor and magnetic clutch.

58. Install the No. 1 exhaust pipe support bracket.

59. Install the front exhaust pipe assembly.

60. Install the No. 1 exhaust pipe support bracket.

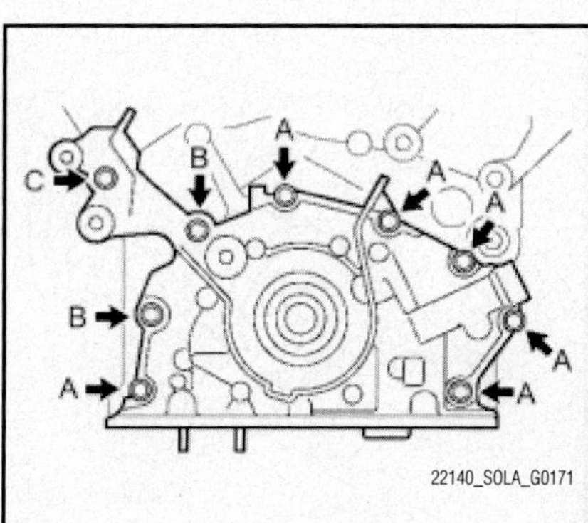

Fig. 141 Oil pump bolt tightening sequence

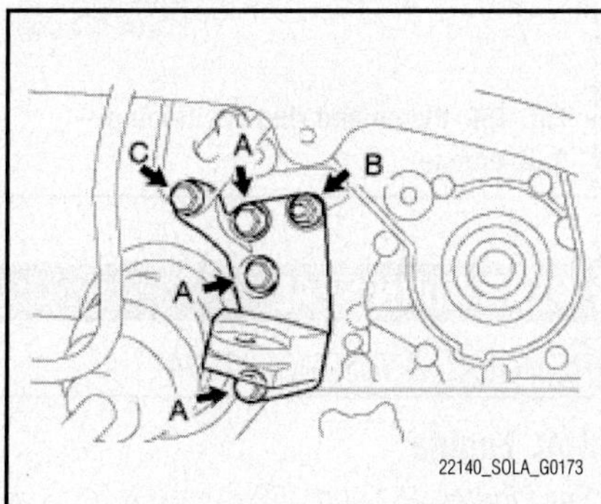

Fig. 142 Engine mounting bracket bolt tightening

61. Inspect and install the timing belt.

62. Install the timing belt tensioner assembly.

63. Install the No. 2 timing belt guide.

64. Install the right engine mounting bracket.

65. Install the No. 1 and No. 2 timing belt cover.

66. Install the crankshaft pulley.

67. Install the No. 2 generator bracket.

68. Install the No. 2 right engine mounting stay.

69. Install the engine moving control rod.

70. Install the vane pump V-belt.

71. Install the No. 1 V-belt.

72. Inspect drive belt tension.

73. Install the front suspension upper brace center, if applicable. Tighten to 59 inch lbs. (80 Nm).

74. Add engine oil and check for leaks.
75. Install the right front wheel.
76. Check for exhaust gas leaks.

MAIN BEARING TORQUE SEQUENCE

2.4L Engine

See Figures 145 and 146.

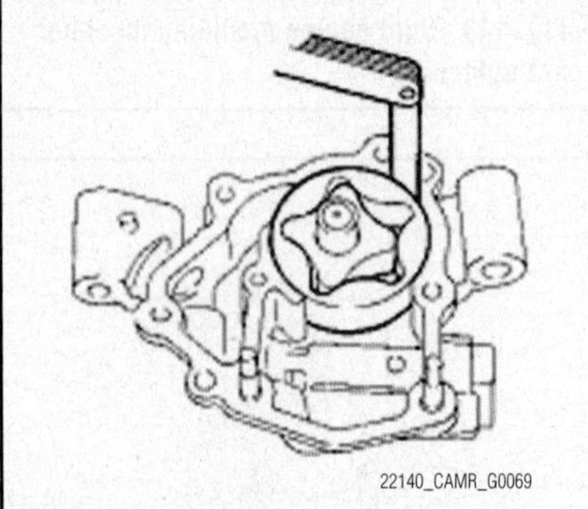

Fig. 145 Main bearing torque sequence—Step 1

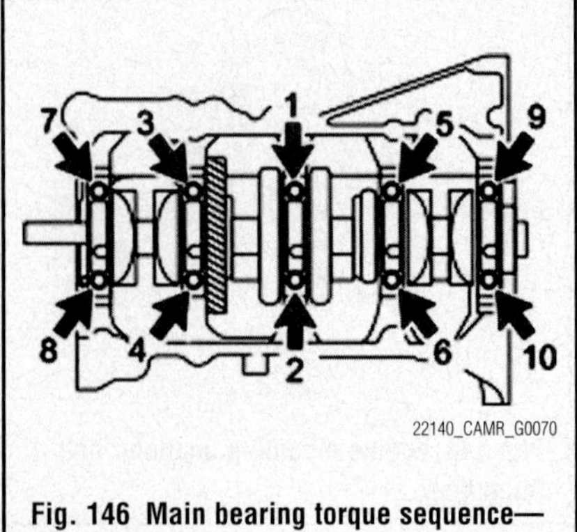

Fig. 146 Main bearing torque sequence—Step 2

3.3L Engine

See Figures 147 and 148.

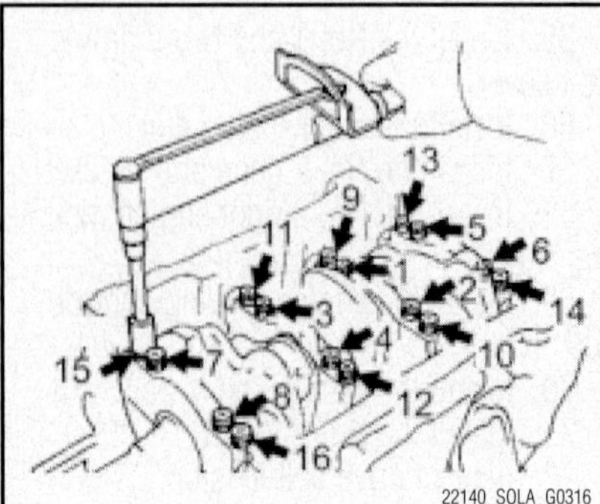

Fig. 147 Main cap bearing bolts torque sequence

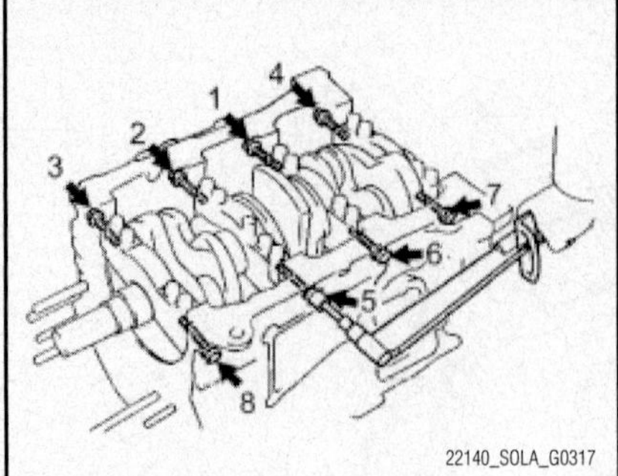

Fig. 148 Main cap bearing side bolts torque sequence

PISTON AND RING

POSITIONING

See Figures 149 and 150.

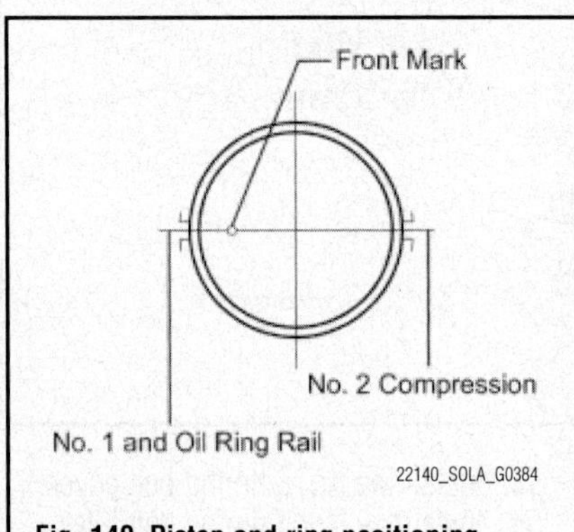

Fig. 149 Piston and ring positioning—2.4L engine

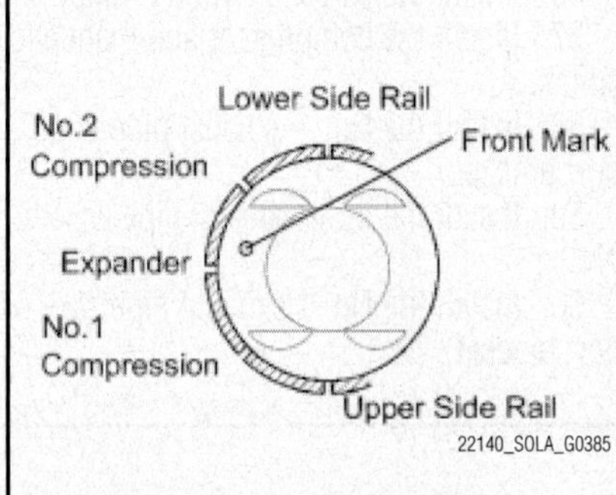

Fig. 150 Piston and ring positioning—3.3L engine

REAR MAIN SEAL

REMOVAL & INSTALLATION

2.4L Engine

See Figures 151 and 152.

1. Before servicing the vehicle, refer to the Precautions Section.
2. Disconnect the negative battery cable.
3. For A/T vehicles, separate the transaxle. Refer to the Drive Train section.
4. For M/T vehicles, separate the transaxle. Refer to the Drive Train section.
5. For A/T vehicles, remove the flexplate. Refer to Flexplate.
6. For M/T vehicles, remove the flywheel. Refer to Flexplate.
7. Using a knife, cut through the oil seal lip.
8. Using a screwdriver with its tip taped, pry out the oil seal.
9. After the removal, check the crankshaft for damage. If it is damaged, smooth the surface with 400 grit sandpaper.

To install:

➡Keep the lip free from foreign matter.

10. Apply MP grease to a new oil seal lip.
11. Using Special Tools: 09223-15030, 09950-70010 and a hammer, tap in the oil seal until its surface is flush with the rear oil seal retainer edge. Wipe off extra grease from the crankshaft.
12. To complete installation, reverse remaining removal procedure.

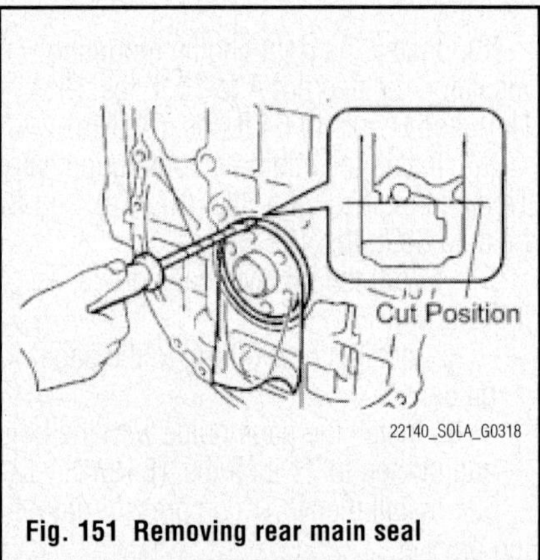

Fig. 151 Removing rear main seal

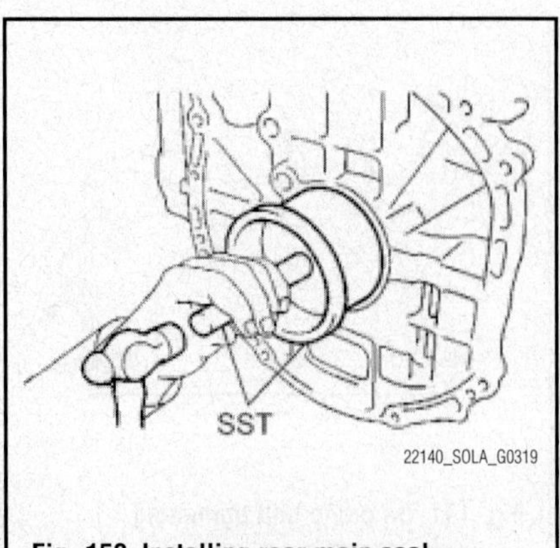

Fig. 152 Installing rear main seal

3.3L Engine

See Figure 153.

1. Remove automatic transaxle assembly.
2. Remove drive plate and ring gear sub-assembly, as follows:
   a.Using SST (SST: 09213-54015, SST: 09330-00021) hold the crankshaft.
   b.Remove the 8 bolts, rear spacer, drive plate and front spacer.
3. Using a knife, cut through the oil seal lip.
4. Using a screwdriver with its tip wrapped in tape, pry out the oil seal.

➡After the removal, check if the crankshaft is not damaged. If it is damaged, smooth the surface with 400-grit sandpaper.

To install:

5. Apply MP grease to a new oil seal lip.

➡Keep the lip free from foreign matter.

6. Using SST (SST: 09223-15030, SST: 09950-70010) and a hammer, tap in the oil seal until its surface is flush with the rear oil seal retainer edge.

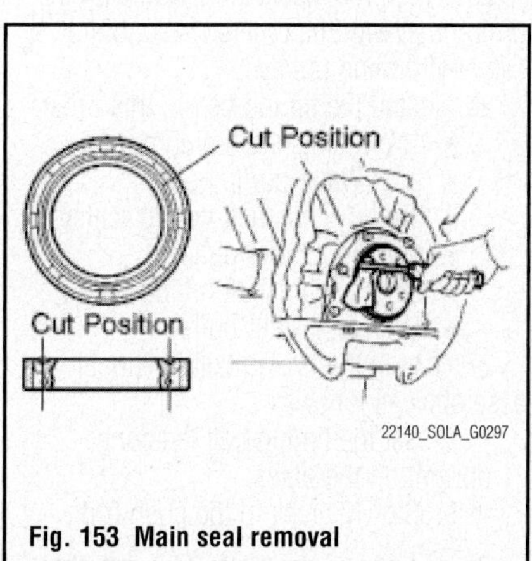

Fig. 153 Main seal removal

➡Be careful not to tap the oil seal at an angle. Wipe off extra grease from the crankshaft.

7. To complete installation, reverse remaining removal procedure.

TIMING BELT AND SPROCKETS

REMOVAL & INSTALLATION

See Figures 154 through 166.

1. Before servicing the vehicle, refer to the Precautions Section.
2. Remove right front wheel.
3. Remove right front fender apron seal.
4. Remove the No. 1 (cooler compressor to crankshaft pulley) V-belt.

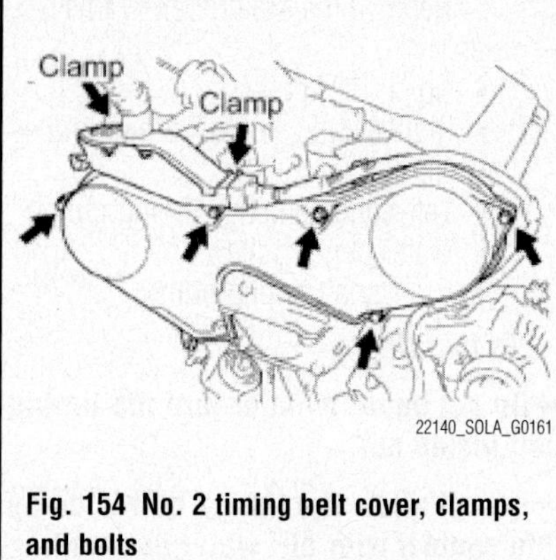

Fig. 154 No. 2 timing belt cover, clamps, and bolts

5. Remove the vane pump V-belt.
6. Remove the engine moving control rod.
7. Remove the right No. 2 mounting stay.
8. Remove the No. 2 alternator bracket.
9. Remove the crankshaft pulley, as follows:
   a. Using SST (SST: 09213-58013, SST: 09330-00021), loosen the pulley bolt.
   b. Using SST (SST: 09950-50013) and the pulley bolt, remove the pulley.

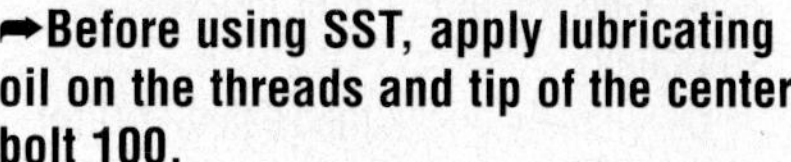

➡Before using SST, apply lubricating oil on the threads and tip of the center bolt 100.

10. Remove the No. 1 timing belt cover.
11. Remove the No. 2 timing belt cover, as follows:
    a. Disconnect the engine wire protector clamps from the timing belt No. 3 cover.
12. Remove the 5 bolts and No 2 timing belt cover.
13. Remove the right engine mounting bracket.

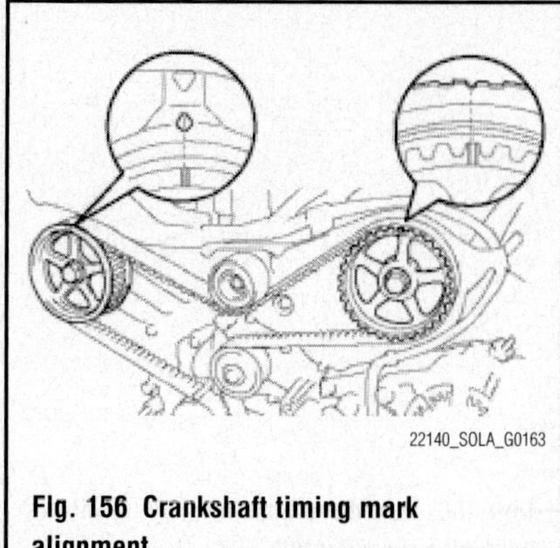

Fig. 156 Crankshaft timing mark alignment

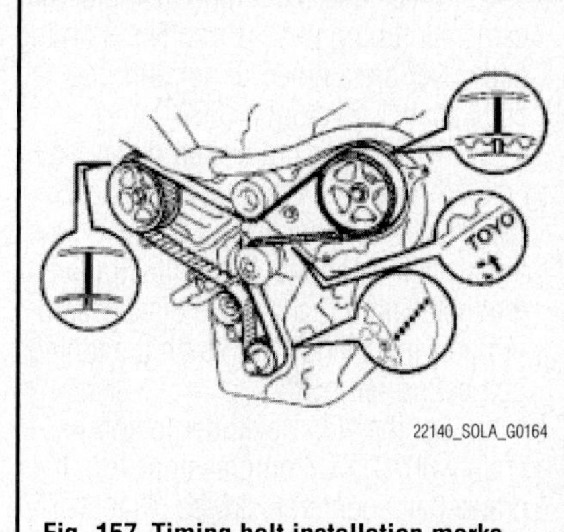

Fig. 157 Timing belt installation marks

14. Remove the No. 2 timing belt guide.
15. Remove timing belt, as follows:
    a. Set the No. 1 cylinder to TDC/compression.
    b. Temporarily install the crankshaft pulley bolt with the washer to the crankshaft.
    c. Turn the crankshaft clockwise, and align the timing marks of the crankshaft timing pulley and oil pump body.

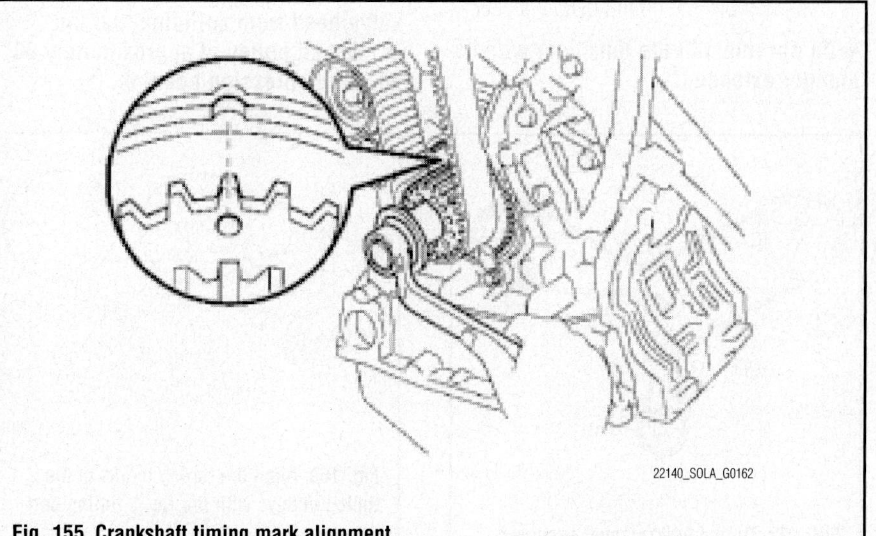

Fig. 155 Crankshaft timing mark alignment

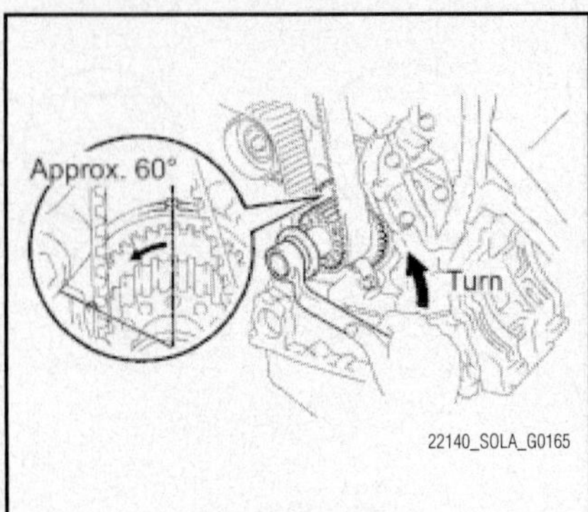

Fig. 158 Turn the crankshaft counterclockwise by approximately 60°

d. Check that the timing marks of the camshaft timing pulleys and No. 3 timing belt cover are aligned. If not, turn the crankshaft 1 revolution (360°).

e. Remove the crankshaft pulley bolt.

f. If reusing the timing belt, check that there are 4 installation marks on the timing belt as shown in the illustration. If the installation marks have disappeared, put new installation marks on the timing belt before removal.

g. Set the No. 1 cylinder to approximately 60°BTDC/compression. Turn the crankshaft counterclockwise by approximately 60°.

WARNING

With timing belt removed: The crankshaft pulley must be at the correct angle to avoid damage in later steps. If the crankshaft pulley is at the wrong angle and then the camshaft timing pulley and the camshaft are removed, the piston head and valve head may come in contact and incur damage.

h. Remove the timing belt tensioner.

➡Do not reinstall the tensioner with its plunger extended.

i. Remove the timing belt in this order:

- No. 1 idler pulley
- Right-hand camshaft timing pulley
- No. 2 idler pulley
- Left-hand camshaft timing pulley
- Water pump pulley
- Crankshaft timing pulley

To install:

➡Do not bend, twist or turn the timing belt inside out.

➡Do not allow the timing belt to come into contact with oil, water or steam.

➡Do not utilize timing belt tension when installing the mounting bolt of the camshaft timing pulley.

16. Remove any oil or water on the pulleys, and keep them clean.

➡If there is a trace of water and/or oil on the timing belt, repair the leakage and install a new timing belt.

➡Only wipe the pulleys. Do not use cleaning agents on the pulleys.

17. Inspect the idler pulleys:

a. Check that the idler pulleys turn smoothly.

b. Visually check the sealed portion of the idler pulleys for oil leakage.

18. Inspect the water pump:

a. Turn the pulley, and check that the water pump bearing moves smoothly without any noise.

b. Visually check the drain hole for coolant leakage.

19. Temporarily install the crankshaft pulley bolt and washer to the crankshaft.

20. Turn the crankshaft counterclockwise by approximately 60°.

➡To prevent the piston head and the valve head from colliding, set the crankshaft pulley at approximately 60° BTDC/compression position.

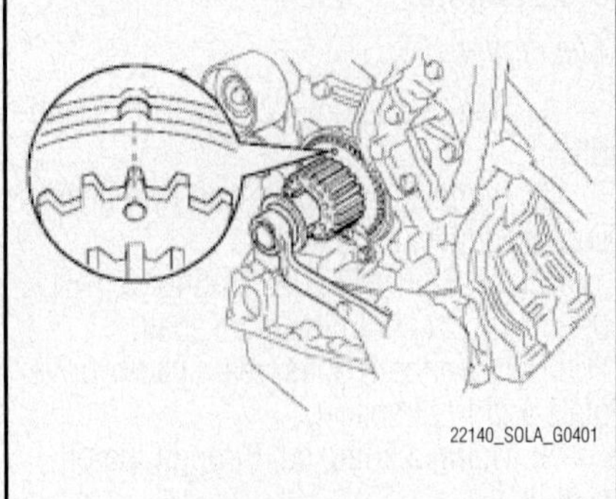

Fig. 161 Align the timing mark of the crankshaft timing pulley with the oil pump

21. Using SST (SST: 09960-10010), turn the timing pulleys, and align the timing marks of the timing pulleys with the No. 3 timing belt cover.

22. Turn the crankshaft, and align the timing mark of the crankshaft timing pulley with the oil pump body.

23. Face the front mark on the timing belt forward.

24. Align the installation mark on the timing belt with the timing mark of the crankshaft timing pulley.

25. Align the installation marks on the timing belt with the timing marks of the camshaft timing pulleys.

26. Install the timing belt in this order:

- 1st: Crankshaft timing pulley
- 2nd: Water pump pulley
- 3rd: LH camshaft timing pulley
- 4th: No. 2 idler pulley
- 5th: RH camshaft timing pulley
- 6th: No. 1 idler pulley

27. Install the timing belt tensioner assembly, as follows:

a. Set the timing belt tensioner upright on the press.

b. Slowly press in the push rod.

➡Do not apply pressure of more than 2,205 lbf (1,000 kgf, 9.8 kN) to the rod.

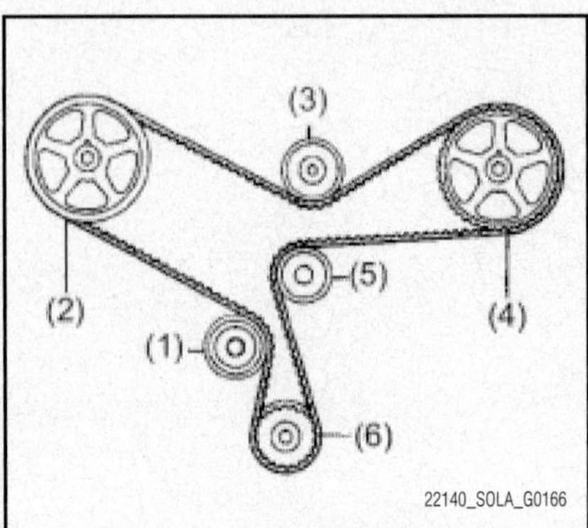

Fig. 159 Timing belt removal sequence

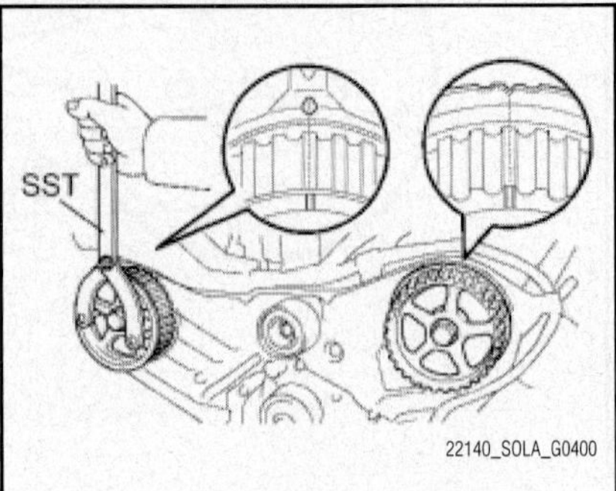

Fig. 160 Align the timing marks of the timing pulleys with the No. 3 timing belt cover

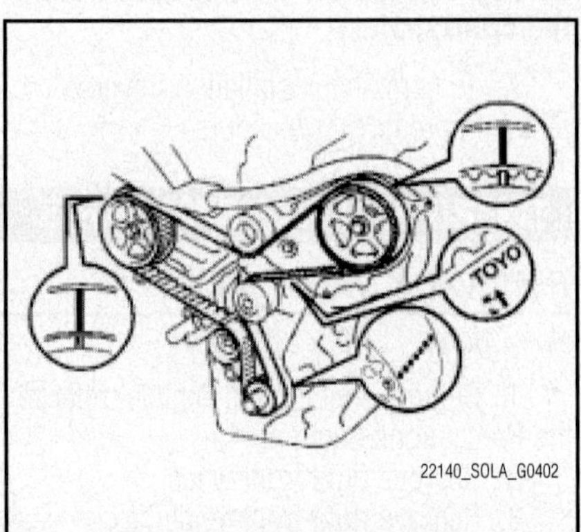

Fig. 162 Timing belt facing forward

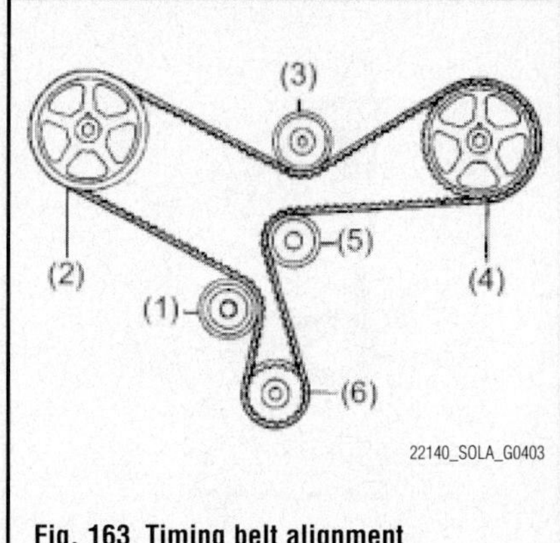

Fig. 163 Timing belt alignment

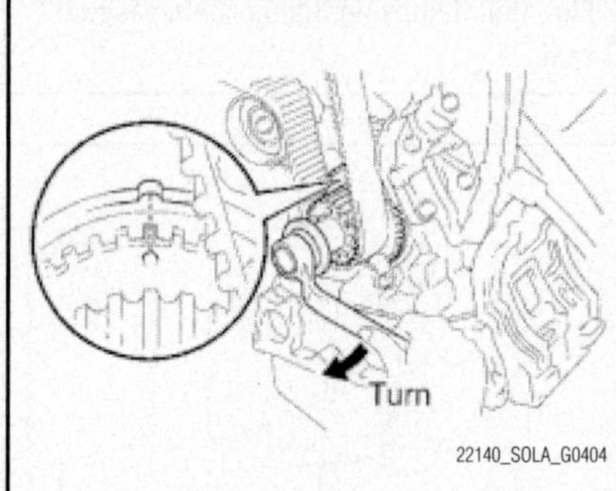

Fig. 164 Align the timing mark of the crankshaft timing pulley with the oil pump

c. Align the holes of the push rod and housing, and pass a 1.5 mm hexagon wrench through the holes to keep the setting position of the push rod.

d. Release the push rod.

e. Temporarily install the tensioner with the 2 bolts and tighten to 20 ft. lbs. (27 Nm).

WARNING

Install the tensioners bolts uniformly and evenly. Installing the tensioner at an angle may cause it to malfunction.

f. Remove the 1.5 mm hexagon wrench from the tensioner.

g. Turn the crankshaft 2 revolutions slowly and align the timing mark of the crankshaft timing pulley with the oil pump body. Always turn the crankshaft clockwise,

h. Check that the timing marks of the right and left timing pulleys are aligned with the timing marks of the No. 3 timing belt cover as shown in the illustration. If the marks do not align, remove the timing belt and reinstall it.

i. Remove the crankshaft pulley bolt.

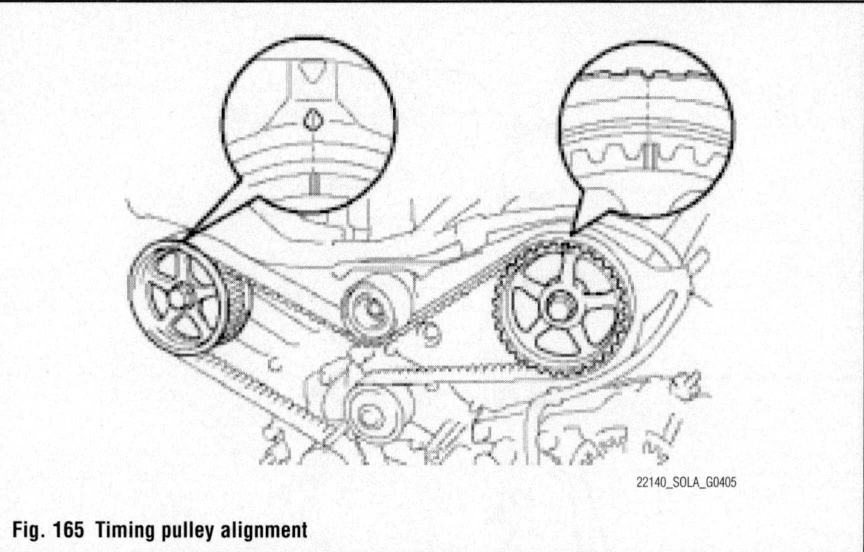

Fig. 165 Timing pulley alignment

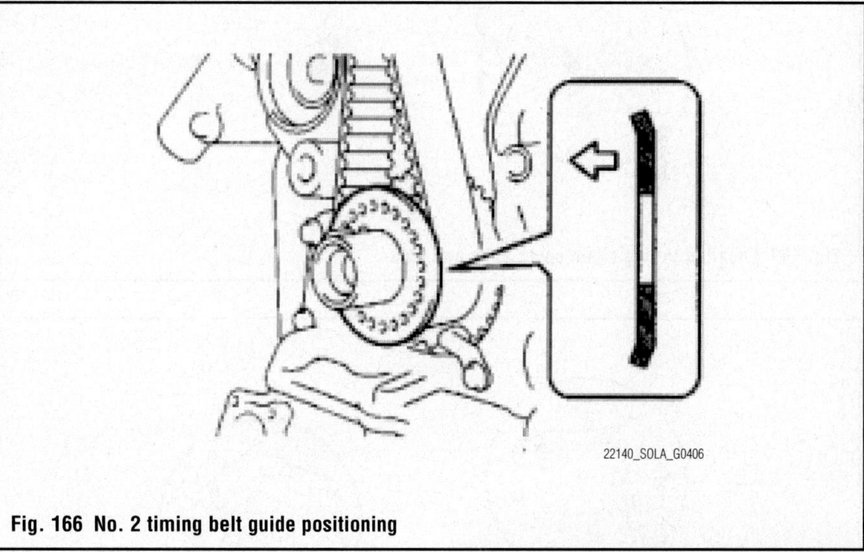

Fig. 166 No. 2 timing belt guide positioning

28. Install the No. 2 timing belt guide, facing the cup side toward the engine front.
29. Install the right engine mounting bracket.
30. Visually check for cracks and breaks on the gasket and install the No 2 timing belt cover. Tighten to 76 inch lbs. (8.5 Nm). If water is entering the gasket, replace the timing belt cover.
31. Visually check for cracks and breaks on the gasket and install the No 1 timing belt cover with the 5 bolts. Tighten to 75 inch lbs. (8.5 Nm). If water is entering the gasket, replace the timing belt cover.
32. The remainder of installation is the reverse of removal.
33. Inspect drive belt tension.

TIMING BELT REAR COVER

REMOVAL & INSTALLATION

See Timing Belt and Sprockets.

TIMING CHAIN COVER AND SEAL

REMOVAL & INSTALLATION

See Figures 167 through 172.

1. Before servicing the vehicle, refer to the Precautions Section.
2. Using an E10 Torx®socket, remove the stud bolt for the drive belt tensioner from the cylinder block.
3. Remove the 14 bolts and 2 nuts.

➡Be careful not to damage the contact surfaces of the timing chain cover, cylinder block and cylinder head.

4. Remove the timing chain cover by prying between the timing chain cover and cylinder head or cylinder block with a screwdriver. Tape the screwdriver tip before use.

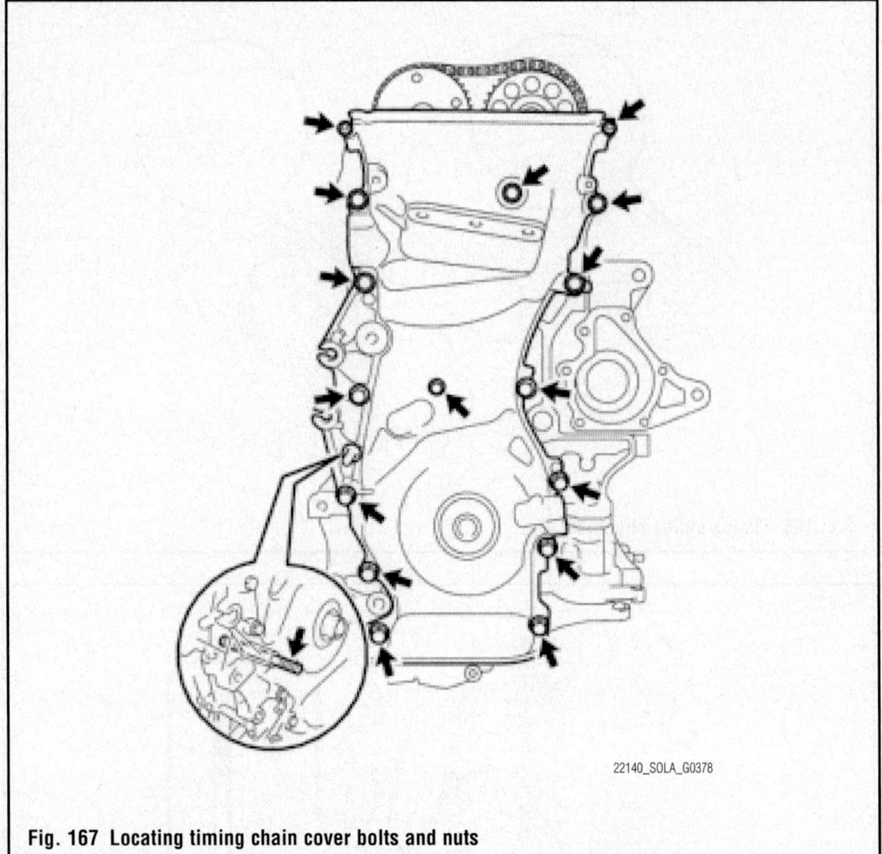

Fig. 167 Locating timing chain cover bolts and nuts

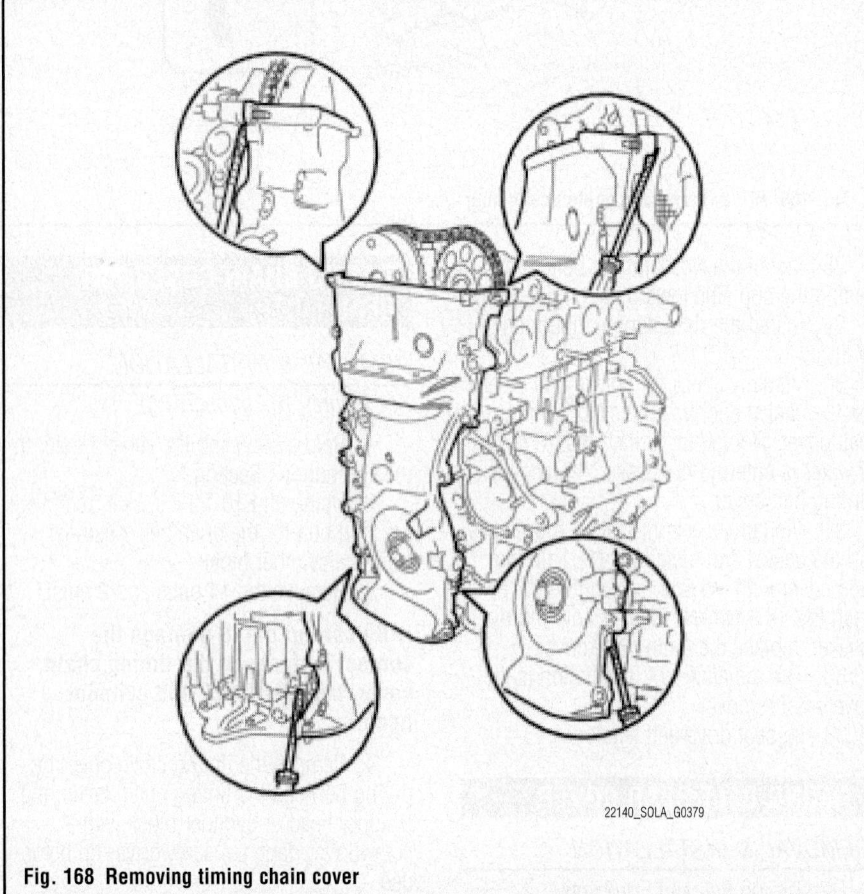

Fig. 168 Removing timing chain cover

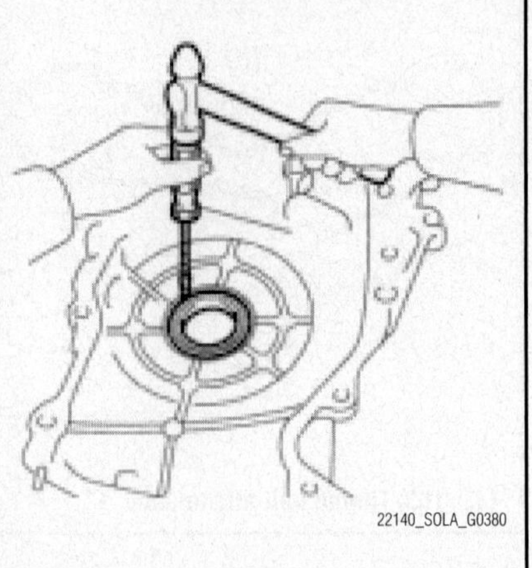

Fig. 169 Removing timing chain case oil seal

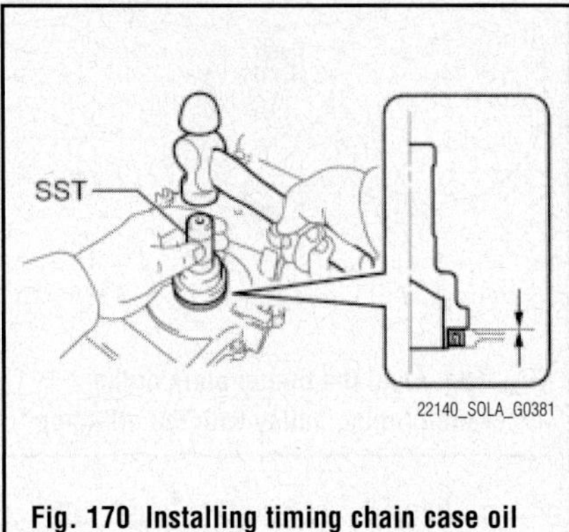

Fig. 170 Installing timing chain case oil seal

5. Using a screwdriver and a hammer, tap out the oil seal.

To install:

➡Keep the gap between the timing chain cover edge and the oil seal free of foreign matter.

6. Using Special Tool: 09223-22010, tap in a new oil seal until its surface is flush with the timing chain cover edge. Apply a light coat of MP grease to the lip of the oil seal.
7. Remove any old packing (FIPG) material and be careful not to drop any oil on the contact surfaces of the timing chain cover, cylinder head and cylinder block.
8. Apply Toyota Genuine Seal Packing Black, Three Bond 1207B or Equivalent seal packing in a diameter of 0.157 to 0.177 inches (4.0 to 4.5 mm).
9. Apply seal packing in a continuous bead as shown in the illustration below.

➡ Remove any oil from the contact surface. Install the chain cover within 3 minutes after applying seal packing.

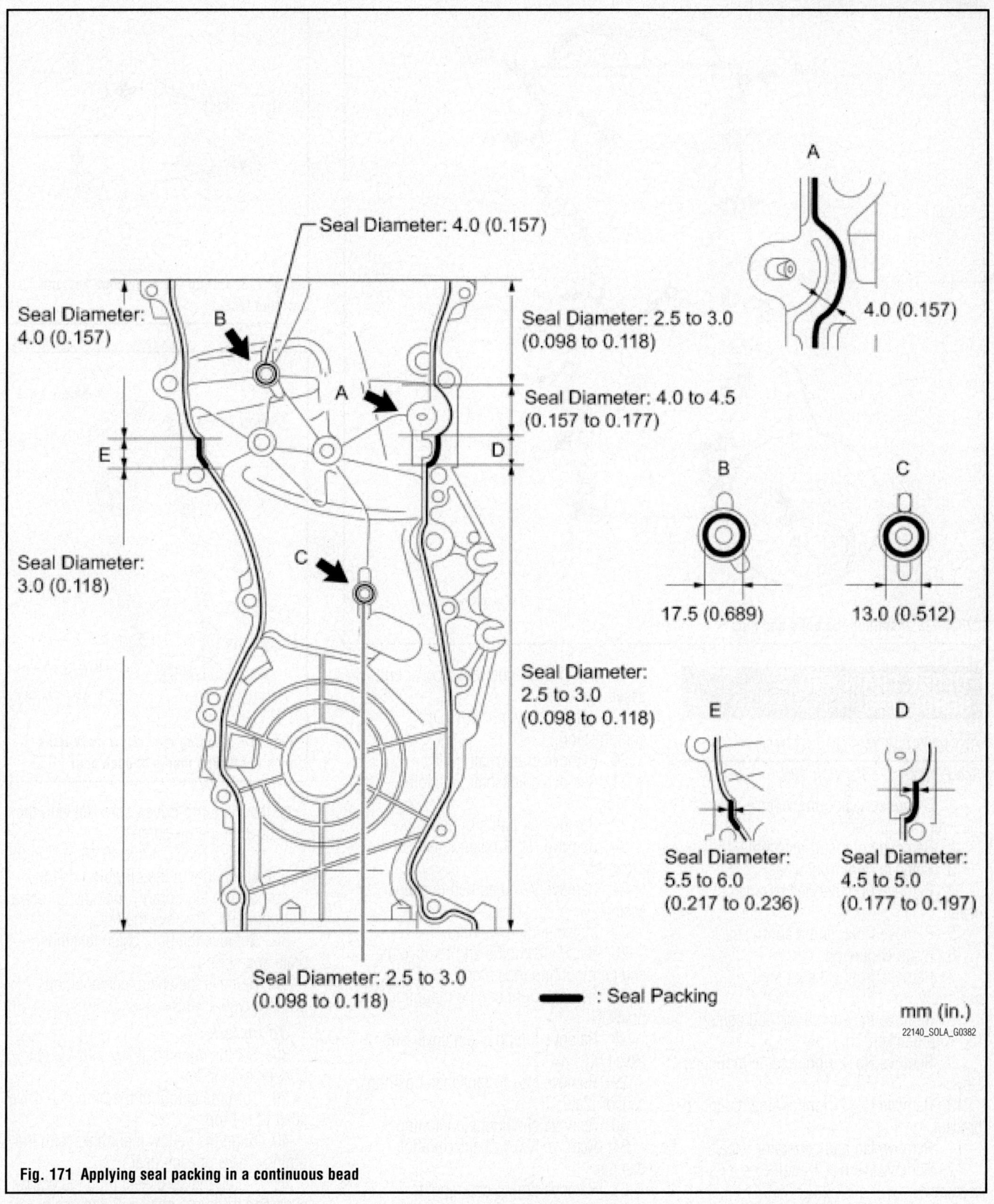

Fig. 171 Applying seal packing in a continuous bead

Do not start the engine for at least 2 hours after installing.

10. Install the timing chain cover with the 14 bolts and 2 nuts in sequence to the following torque specification:

- Bolt A length: 1.18 inches (30 mm) for 10 mm head: 80 inch lbs. (9 Nm)
- Bolt B length: 1.18 inches (30 mm) for 12 mm head: 18 ft. lbs. (25 Nm)
- Bolt C length: 1.57 inches (40 mm) for 14 mm head: 41 ft. lbs. (55 Nm)
- Nut: 8 ft. lbs. (11 Nm)

11. Using a E10 Torx® socket, install the stud bolt to the drive belt tensioner and tighten to 16 ft. lbs. (22 Nm).

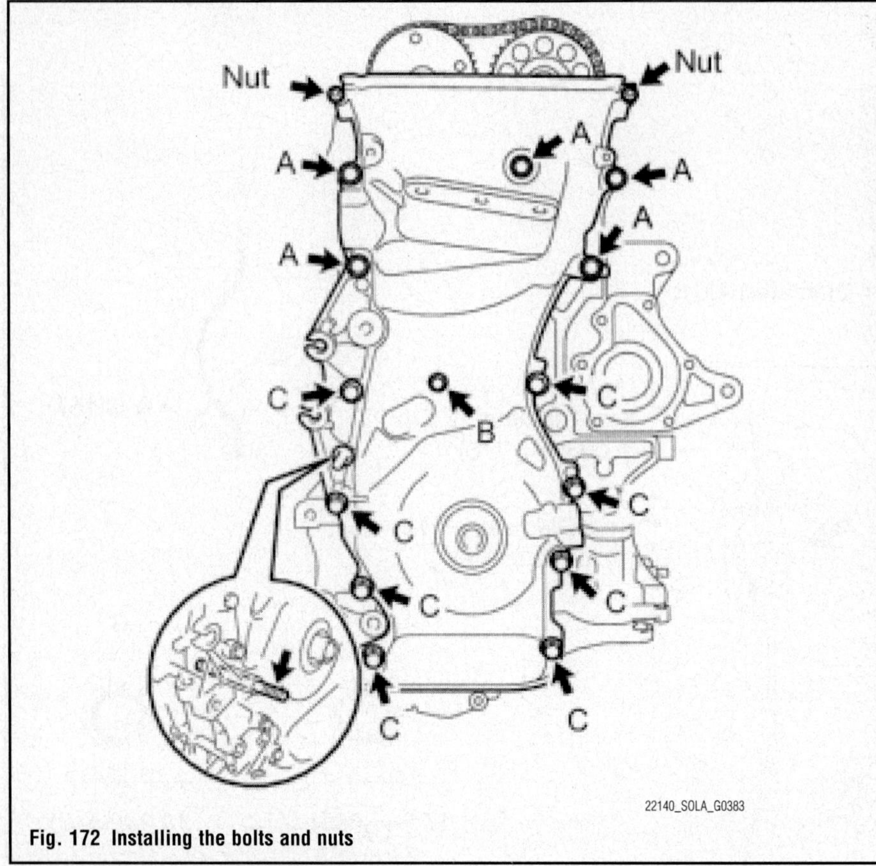

Fig. 172 Installing the bolts and nuts

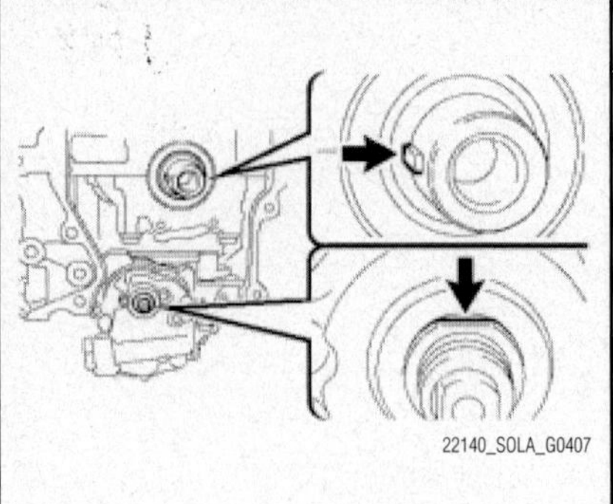

Fig. 173 Setting the crankshaft key and cutout face

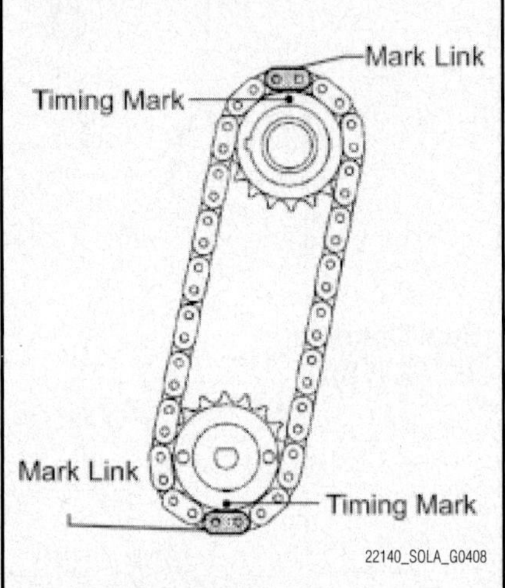

Fig. 174 Aligning the yellow mark links with the timing marks of each gear

TIMING CHAIN AND SPROCKETS

REMOVAL & INSTALLATION

See Figures 173 through 185.

1. Before servicing the vehicle, refer to the Precautions Section.
2. Remove hood sub-assembly.
3. Remove right front wheel.
4. Remove right and left engine under cover.
5. Remove front fender apron seal.
6. Drain engine oil.
7. Remove front exhaust pipe assembly.
8. Remove the engine moving control rod with bracket.
9. Remove No. 2 right engine mounting stay.
10. Remove No. 2 right engine mounting bracket.
11. Remove fan and generator V-belt.
12. Remove the No. 1 engine cover sub-assembly.
13. Remove engine wire.
14. Remove generator assembly,
15. Remove vane pump assembly.
16. Remove ignition coil assembly.
17. Disconnect ventilation hose and No. 2 ventilation hose.
18. Remove cylinder head cover sub-assembly.
19. Set No. 1 cylinder to TDC/ compression.
20. Remove crankshaft pulley.
21. Remove crankshaft position sensor.
22. Remove oil pan sub-assembly.
23. Remove No. 1 chain tensioner assembly.
24. Remove V-ribbed belt tensioner assembly.
25. Install engine hangers.
26. Raise the engine and remove the engine mounting insulator.
27. Remove right engine mounting bracket.
28. Remove timing chain cover sub-assembly.
29. Remove No. 1 crankshaft position sensor plate.
30. Remove chain tensioner slipper.
31. Remove No. 1 chain vibration damper.
32. Remove timing chain guide.
33. Remove chain sub-assembly.
34. Remove crankshaft timing gear or sprocket.
35. Remove No. 2 chain sub-assembly:
    a. Turn the crankshaft counterclockwise by 90°, and align an adjusting hole of the oil pump driven sprocket with the groove of the oil pump.
    b. Put a bar (_ 4 mm (0.16 in.)) in the adjusting hole of the oil pump driven sprocket to temporarily lock the sprocket in position. Remove the nut.
36. Remove the bolt, chain tensioner plate and spring.
37. Remove the chain tensioner, oil pump driven sprocket and chain.

To install:

38. Set the crankshaft key into the left horizontal position.
39. Turn the cutout of the oil pump drive shaft to the top.
40. Align the yellow mark links with the timing marks of each gear.
41. Install the sprockets onto the crankshaft and oil pump shaft with the chain wrapped on the gears.
42. Temporarily tighten the oil pump drive shaft sprocket with the nut.
43. Insert the damper spring into the adjusting hole, and then install the chain

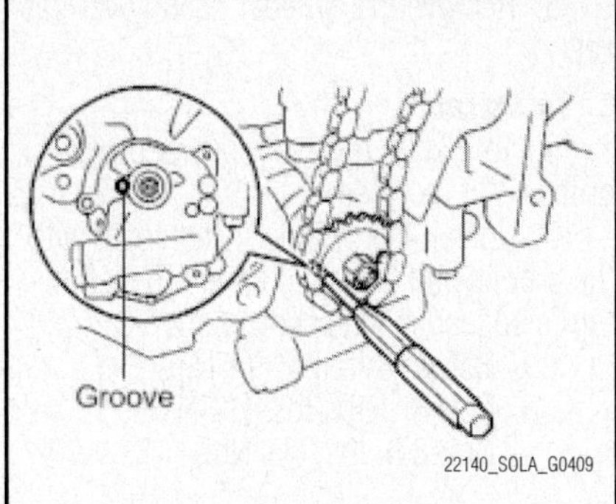

Fig. 175 Aligning the adjusting hole of the oil pump drive shaft sprocket

tensioner plate with the bolt and tighten to 9 ft. lbs. (12 Nm).

44. Align the adjusting hole of the oil pump drive shaft sprocket with the groove of the oil pump.

45. Insert a 4 mm diameter bar into the adjusting hole of the oil pump drive shaft gear to lock the gear in position, and then tighten the nut to 22 ft. lbs. (30 Nm).

46. Rotate the crankshaft clockwise by 90°, and align the crankshaft key to the top.

47. Install the crankshaft timing sprocket.

48. Install the chain vibration damper with the 2 bolts and tighten to 80 inch lbs. (9 Nm).

49. Set the No. 1 cylinder to TDC/compression.

50. Turn the camshafts with a wrench (using the hexagonal lobe) to align the timing marks of the camshaft timing gear with each timing mark located on the No. 1 and No. 2 bearing caps.

51. Using the crankshaft pulley bolt, turn the crankshaft to position with the key on the crankshaft upward.

52. Install the chain onto the crankshaft timing sprocket with the gold or pink mark

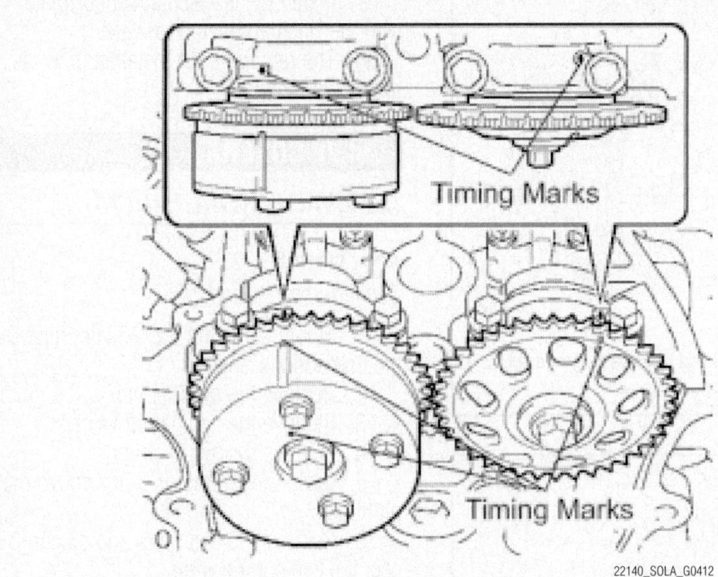

Fig. 178 Setting the No. 1 cylinder to TDC/compression

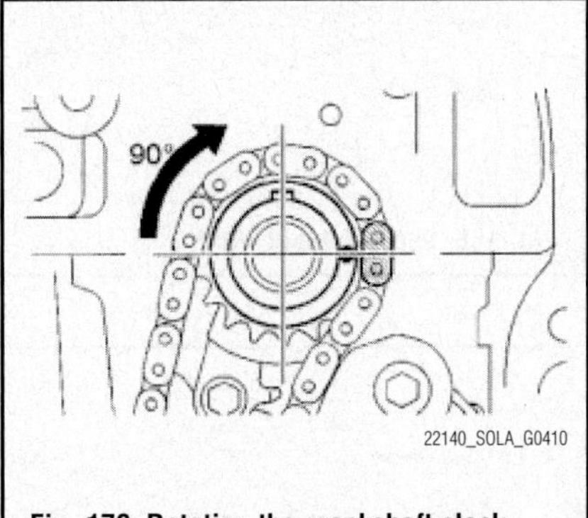

Fig. 176 Rotating the crankshaft clockwise by 90°

Fig. 179 Turning the crankshaft to position with the key on the crankshaft upward

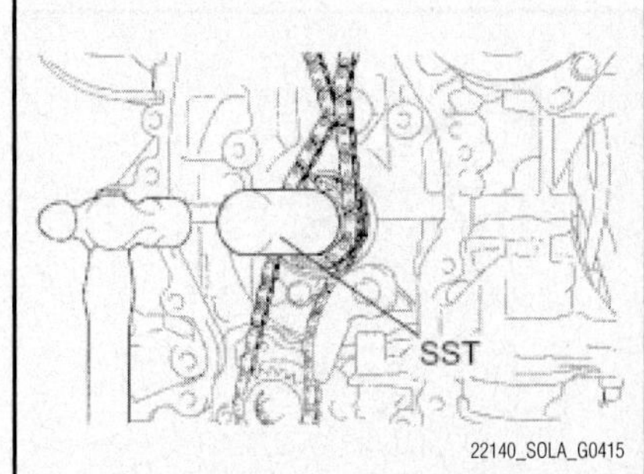

Fig. 181 Using Special Tool: 09309-37010 and a hammer to install the crankshaft timing sprocket

Fig. 177 Installing the crankshaft timing sprocket

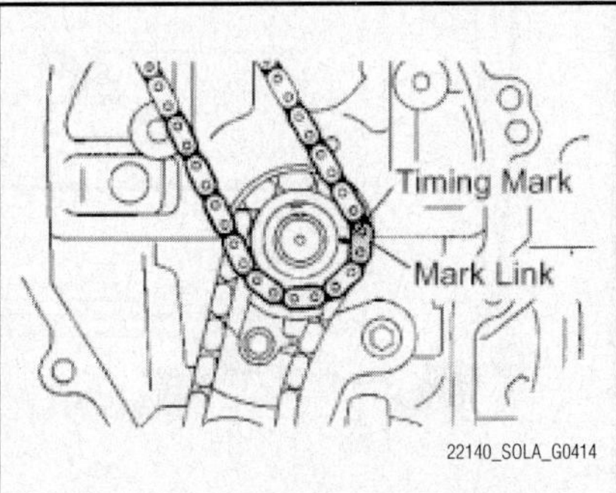

Fig. 180 Installing the chain onto the crankshaft timing sprocket with the gold or pink mark link

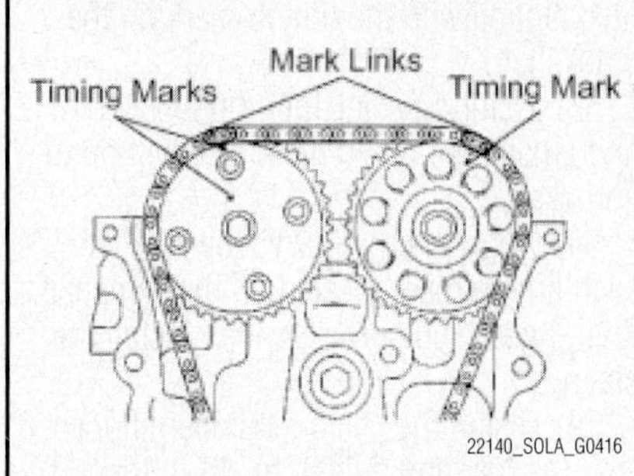

Fig. 182 Aligning the gold or yellow link with each timing mark located on the camshaft timing gear and sprocket

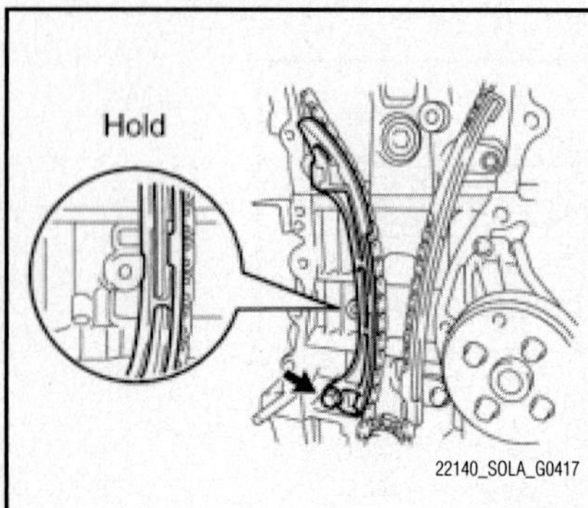

Fig. 183 Installing the chain tensioner slipper and bolt

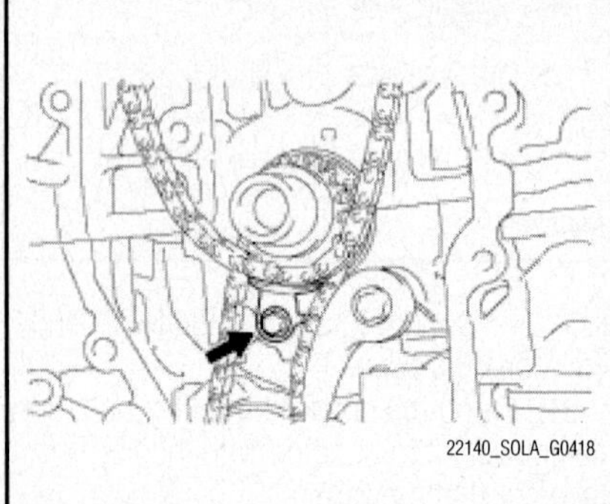

Fig. 184 Installing the timing chain guide and bolt

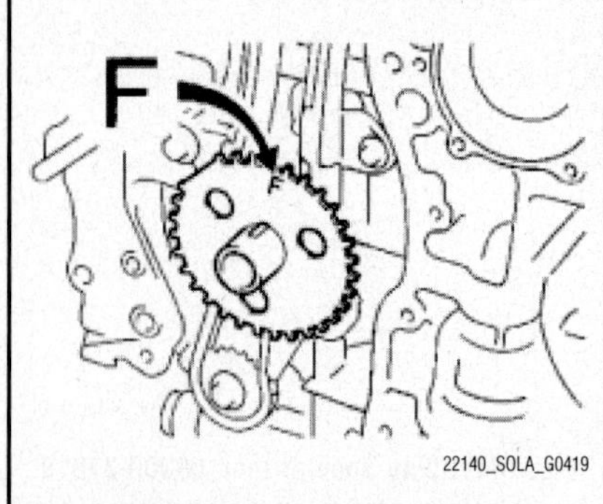

Fig. 185 Installing the crankshaft sensor plate with the "F" mark facing forward

link aligned with the timing mark on the crankshaft.

53. Using Special Tool: 09309-37010 and a hammer, tap in the crankshaft timing sprocket.

54. Align the gold or yellow link with each timing mark located on the camshaft timing gear and sprocket, then install the chain.

55. Install the chain tensioner slipper with the bolt and tighten to: 14 ft. lbs. (19 Nm).

56. Install the timing chain guide with the bolt and tighten to: 80 inch lbs. (9 Nm).

57. Install the crankshaft sensor plate with the "F" mark facing forward.

58. The remainder of installation is the reverse of removal.

VALVE COVERS

REMOVAL & INSTALLATION

2.4L Engine

See Figures 186 and 187.

1. Before servicing the vehicle, refer to the Precautions Section.
2. Remove the oil filler cap.
3. Remove the ventilation valve.
4. Remove the spark plugs.
5. Drain the engine oil and remove the oil filter.
6. Remove the two bolts and disconnect the two engine wires.
7. Remove the 8 bolts, 2 nuts and the valve cover.
8. Remove the cylinder head cover gasket.

To install:

9. To install, reverse removal procedure.
10. Install the cylinder head cover with the 8 bolts and 2 nuts . Tighten the following in sequence to specification:
    - Bolt A: 8 ft. lbs. (11 Nm)
    - Bolt B: 10 ft. lbs. (14 Nm)
    - Nut: 8 ft. lbs. (11 Nm)

3.3L Engine

See Figure 188.

1. Before servicing the vehicle, refer to the Precautions Section.
2. Remove the spark plugs.
3. Remove the ventilation valve sub-assembly.
4. Remove the oil filler cap sub-assembly.

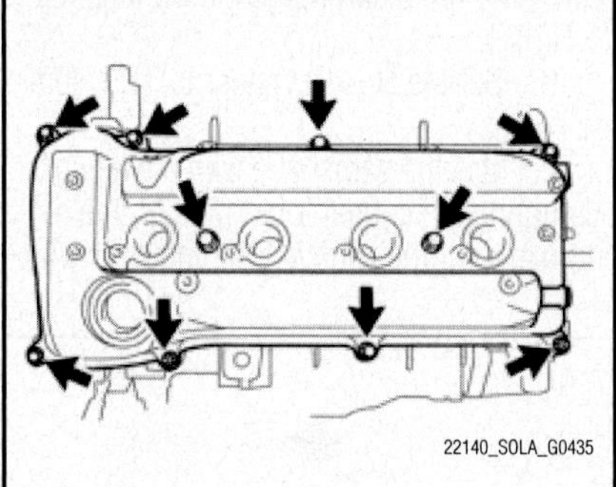

Fig. 186 Removing the 8 bolts, 2 nuts and the valve cover

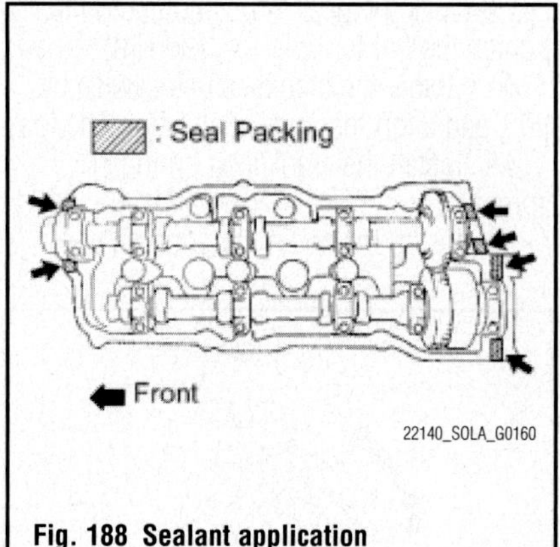

Fig. 188 Sealant application

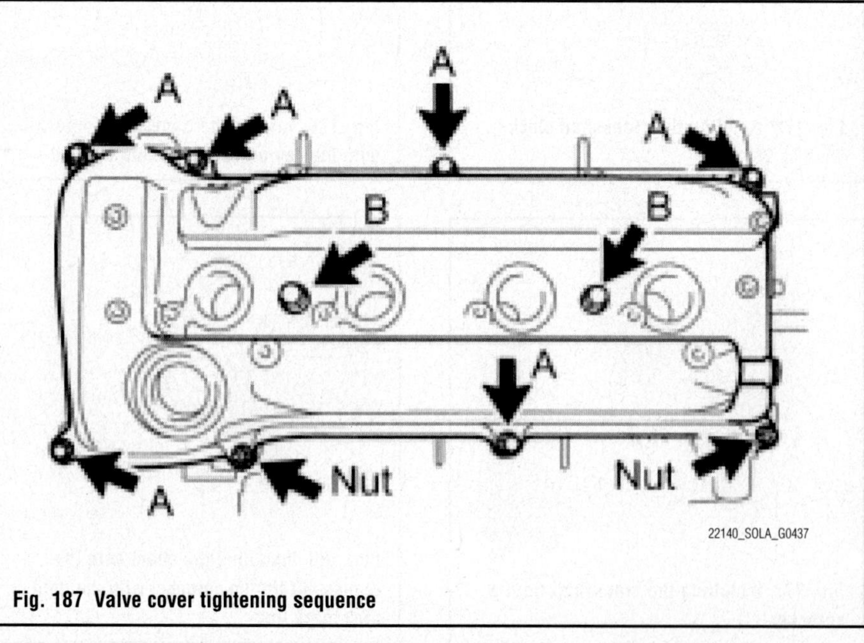

Fig. 187 Valve cover tightening sequence

5. Remove the oil filler cap gasket.
6. Remove the left cylinder head cover sub-assembly and gasket.
7. Remove the right cylinder head cover sub-assembly and gasket.

To install:

8. Install a new gasket and the left cover with the 4 bolts and tighten to 7 ft. lbs. (10 Nm).
9. Install the gasket to the right cylinder head cover.
10. Apply seal packing (Toyota Genuine Seal Packing Black, Three Bond 1207B or equivalent) to the cylinder head as shown in the illustration.

➡Remove any oil from the contact surface.

➡Install the cylinder head cover within 3 minutes after applying seal packing.

➡Do not start the engine for at least 2 hours after installation.

11. Install the cylinder head cover with the 9 bolts. Tighten the bolts uniformly in several steps to 71 inch lbs. (8 Nm).
12. Install the ventilation valve sub-assembly, as follows:
    a. Apply adhesive (Toyota Genuine Adhesive 1324, Three Bond 1324 or equivalent) to 2 or 3 threads.
    b. Install the ventilation valve and tighten to 14 ft. lbs. (19 Nm).
13. Install the oil filler cap gasket.
14. Install the oil filler cap sub-assembly.
15. Install spark plugs.

ENGINE PERFORMANCE & EMISSION CONTROL

ACCELERATOR PEDAL POSITION (APP) SENSOR

LOCATION

See Figure 189.

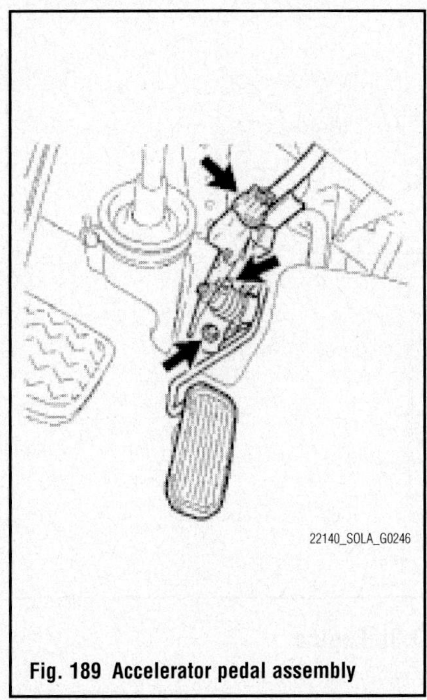

Fig. 189 Accelerator pedal assembly

REMOVAL & INSTALLATION

1. Disconnect the accelerator position sensor connector.

To install:

2. Installation is the reverse of the removal.

CAMSHAFT POSITION (CMP) SENSOR

LOCATION

See Figures 190 and 191.

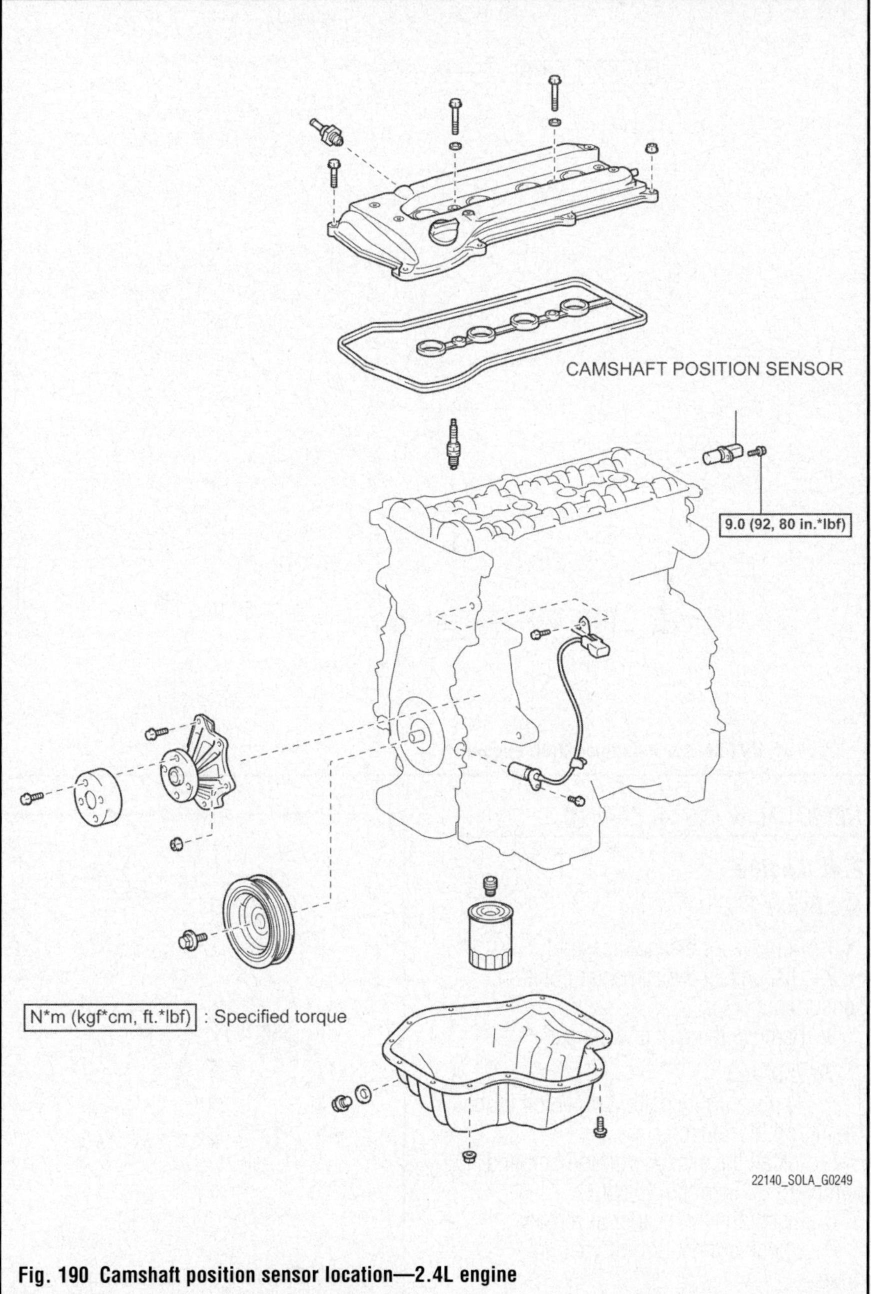

Fig. 190 Camshaft position sensor location—2.4L engine

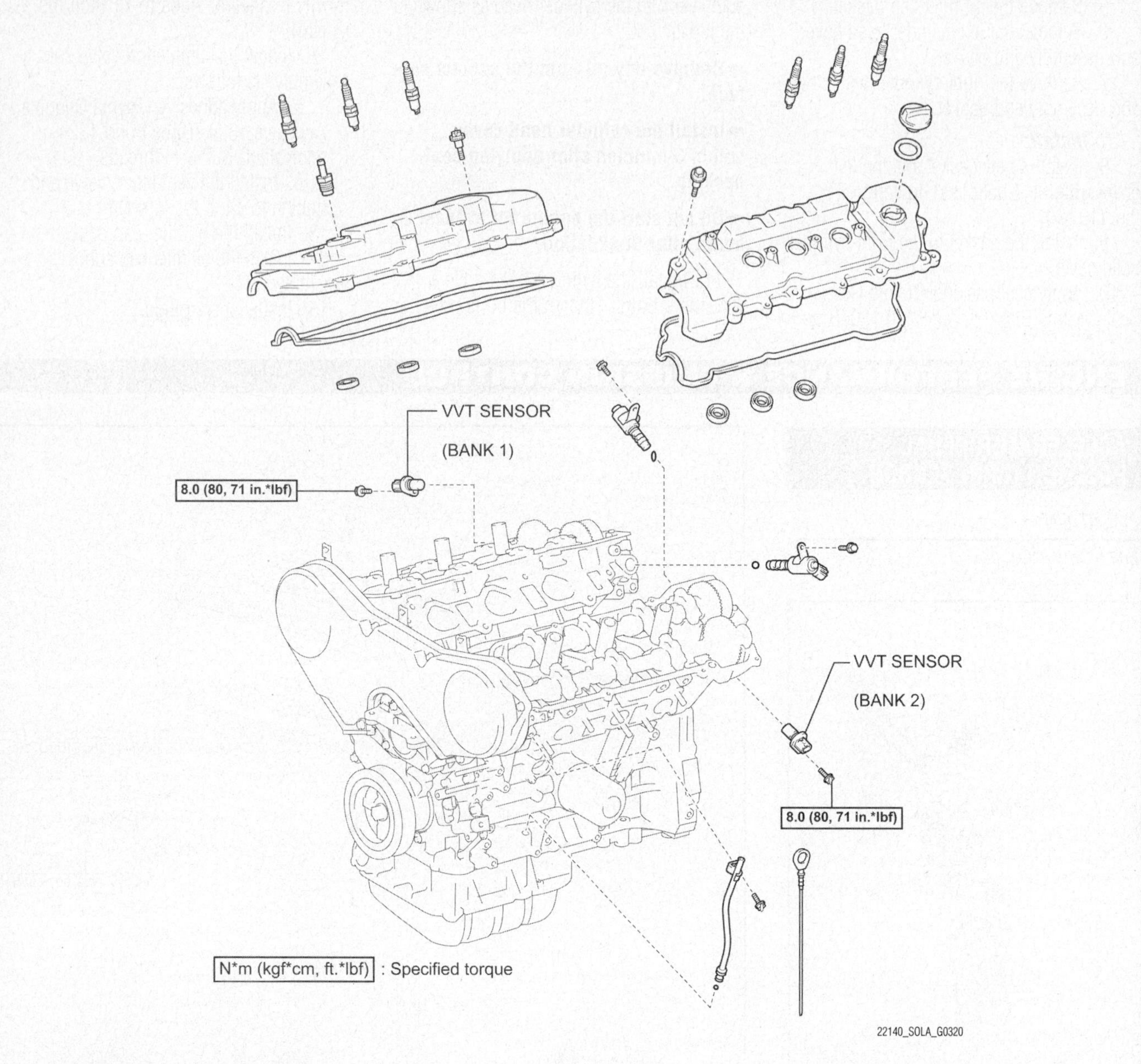

Fig. 191 VVT sensor location—3.3L engine

REMOVAL & INSTALLATION

2.4L Engine

See Figure 192.

1. Remove air cleaner assembly.
2. Disconnect the camshaft position sensor connector.
3. Remove the bolt and sensor.

To install:

4. Apply a light coat of engine oil to the O-ring on the sensor.
5. Install the sensor with the bolt and tighten to 80 inch lbs. (9 Nm).
6. Install the air cleaner assembly.
7. Check connection of vacuum hose.

Fig. 192 Camshaft position sensor

3.3L Engine

See Figure 193.

This engine is equipped with a VVT sensor.

1. Remove air cleaner inlet assembly.
2. Remove air cleaner assembly.
3. Disconnect the VVT sensor connector.
4. Remove the bolt and sensor.

To install:

5. Apply a light coat of engine oil to the O-ring on the sensor.
6. Install the sensor with the bolt and tighten to 71 inch lbs. (8 Nm).
7. Connect the sensor connector.
8. Install air cleaner assembly.
9. Install air cleaner inlet assembly.
10. Check connection of vacuum hose.

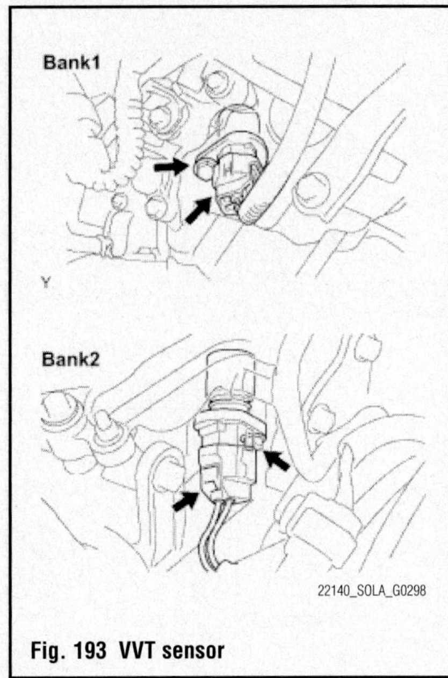

Fig. 193 VVT sensor

CRANKSHAFT POSITION (CKP) SENSOR

LOCATION

See Figure 194.

REMOVAL & INSTALLATION

2.4L Engine

See Figure 195.

1. Remove the 5 screws, 3 clips and right engine under cover.
2. Remove the clip, 2 bolts and front fender apron seal.
3. Disconnect the crankshaft position sensor connector.
4. Remove the bolt, clamp and sensor.

To install:

5. Installation is the reverse of removal.

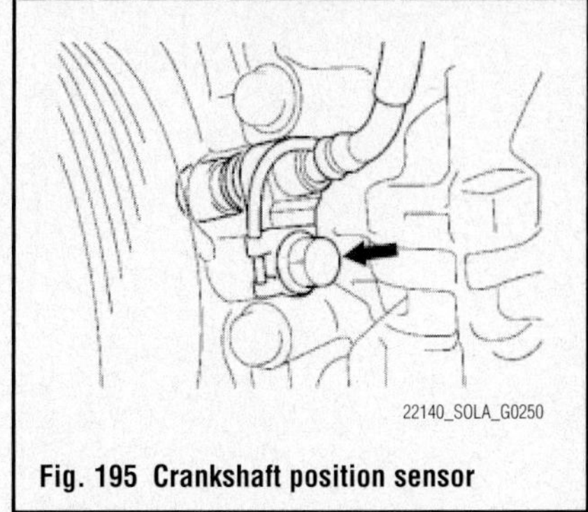

Fig. 195 Crankshaft position sensor

Tighten the crankshaft position sensor to 80 inch lbs. (9 Nm).

ELECTRONIC CONTROL MODULE (ECM)

LOCATION

See Figures 196 and 197.

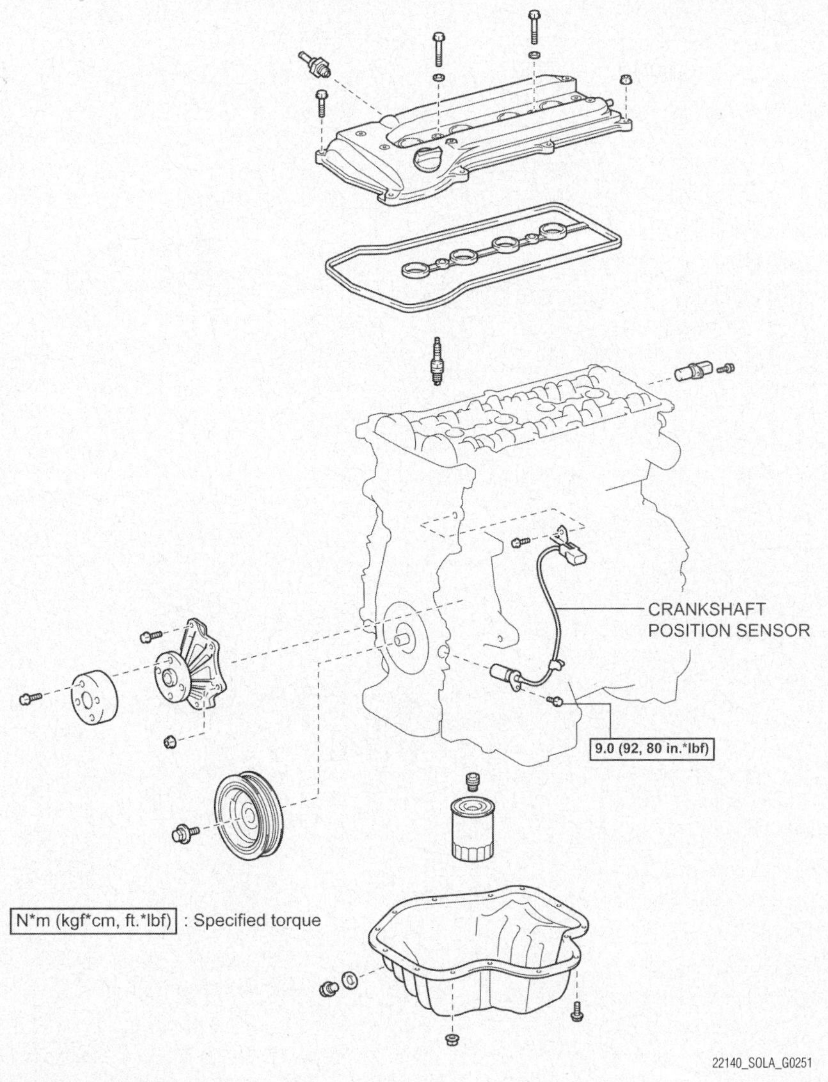

Fig. 194 Crankshaft position sensor—2.4L engine

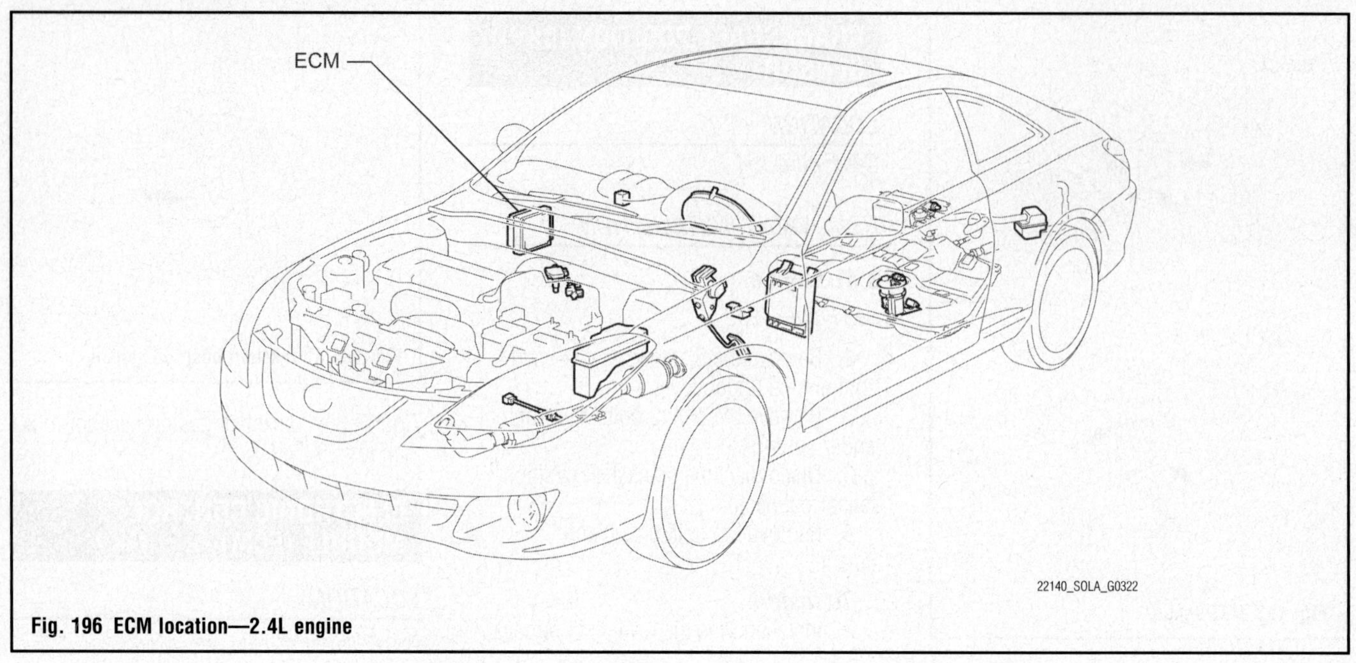

Fig. 196 ECM location—2.4L engine

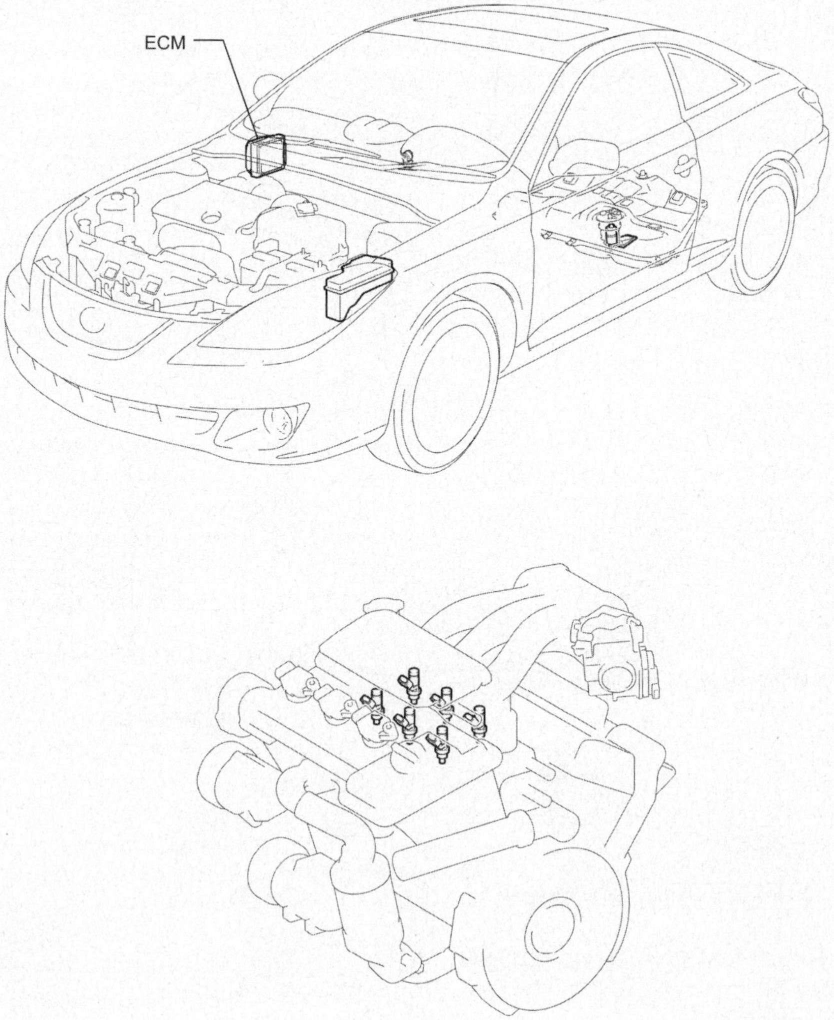

ECM

22140_SOLA_G0323

Fig. 197 ECM location—3.3L engine

REMOVAL & INSTALLATION

1. Disconnect cable from negative battery terminal.
2. Remove the right front door scuff plate.
3. Remove the right cowl side trim sub-assembly
4. Remove the instrument panel No.1 under cover sub-assembly.
5. Remove the instrument panel right finish lower panel.
6. Remove the ECM, as follows:
   a. Remove the 2 wire harness clamps.
   b. Disconnect the 5 ECM connectors.
   c. Remove the 2 nuts and ECM.
   d. Remove the 2 screws and No. 2 ECM bracket.
   e. Remove the 2 screws and ECM bracket.

To install:

7. Installation is the reverse of removal. Tighten the nuts to 49 inch lbs. (5.5 Nm).

ENGINE COOLANT TEMPERATURE (ECT) SENSOR

LOCATION

See Figure 198.

REMOVAL & INSTALLATION

1. Drain engine coolant.
2. Disconnect the ECT sensor connector.
3. Using a 19 mm deep socket wrench, remove the ECT sensor and gasket.

To install:

4. Install a new gasket to the ECT sensor.
5. Using a 19 mm deep socket, install the ECT sensor. Tighten to 14 ft. lbs. (20 Nm).
6. Connect the ECT sensor connector.
7. Refill with engine coolant.

HEATED OXYGEN (HO2S) SENSOR

LOCATION

See Figures 199 and 200.

REMOVAL & INSTALLATION

2.4L Engine

1. Disconnect the heated oxygen sensor (bank 1 sensor 2) connector.

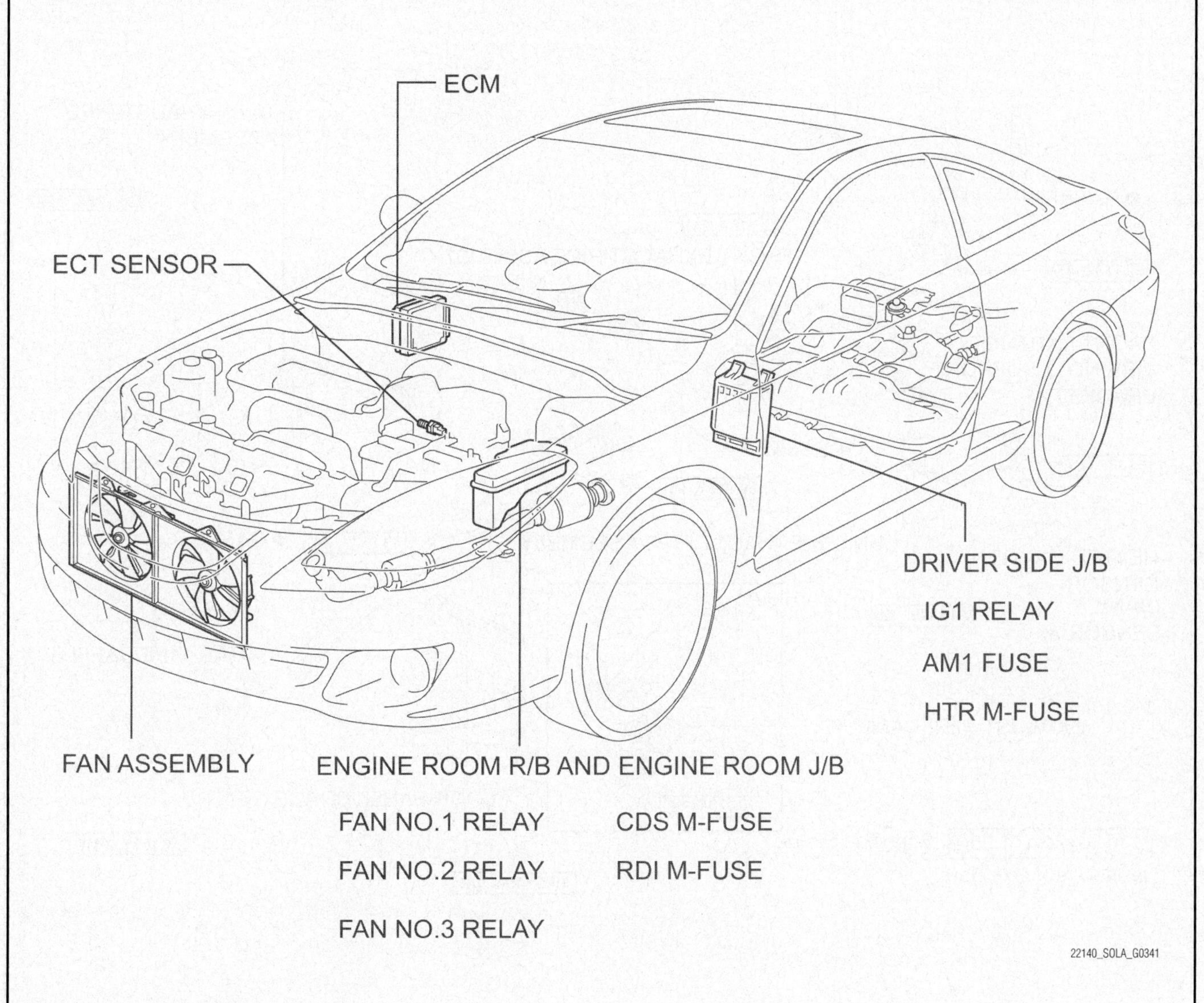

Fig. 198 ECT sensor location—2.4L engine

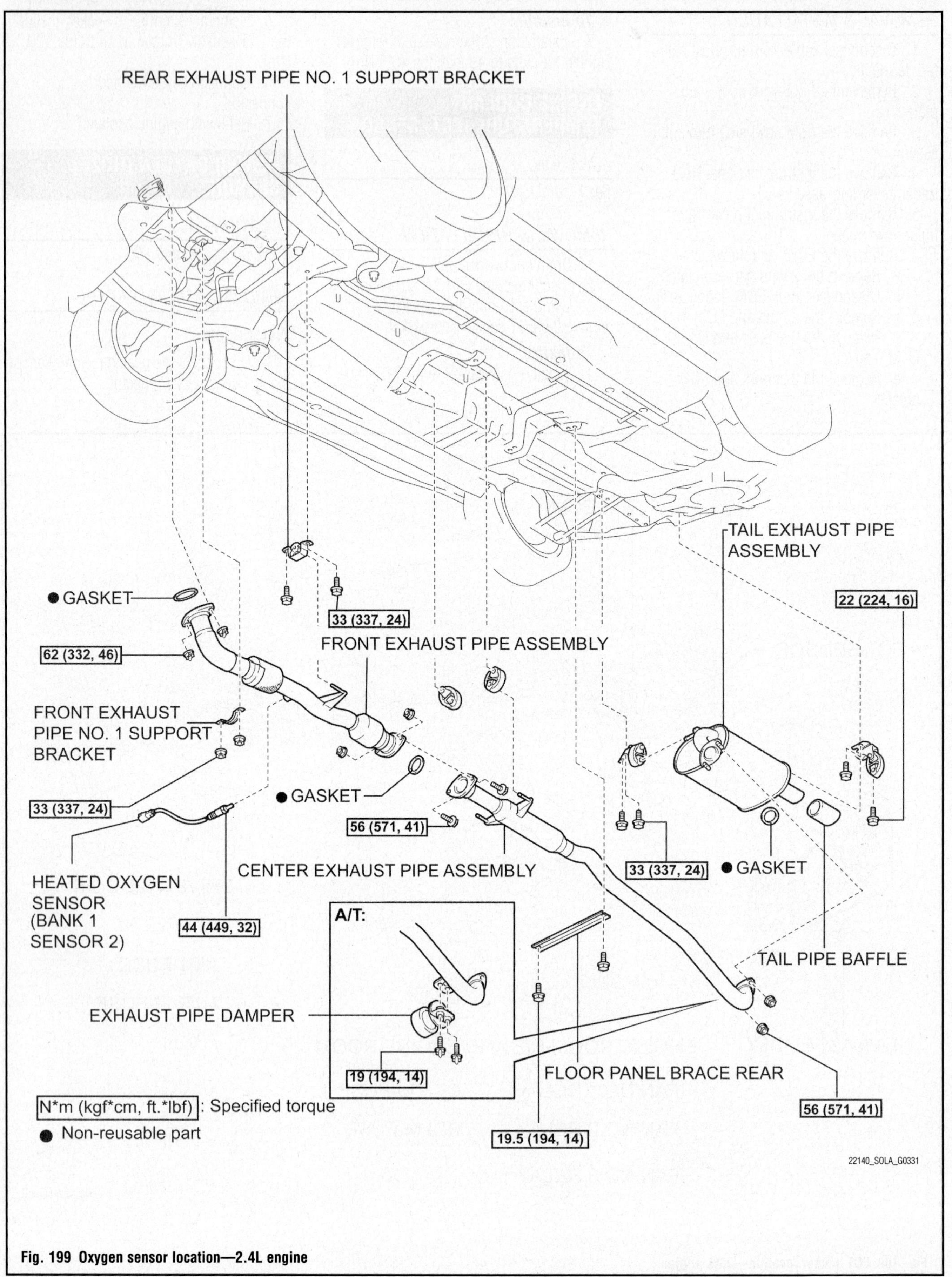

Fig. 199 Oxygen sensor location—2.4L engine

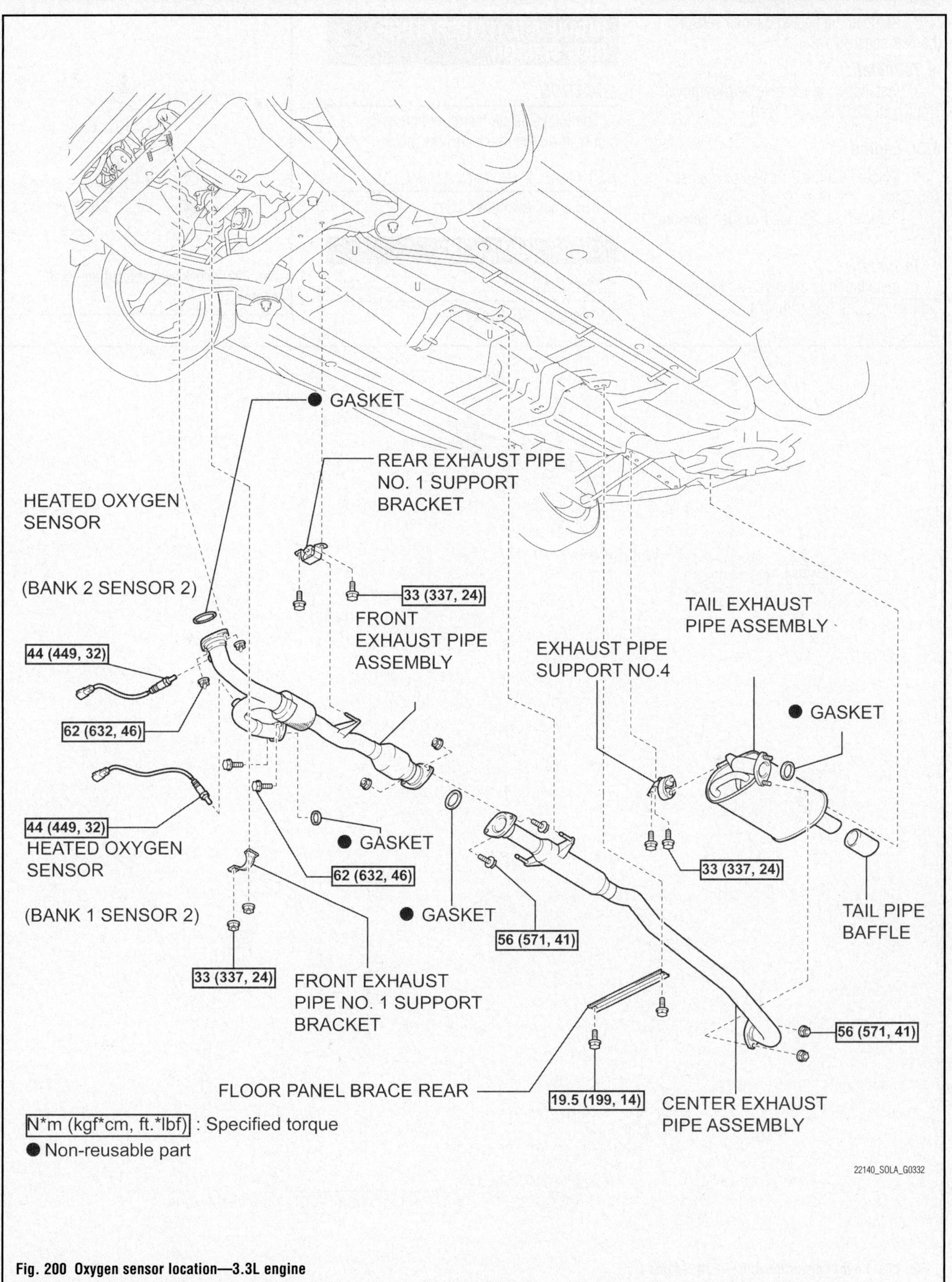

Fig. 200 Oxygen sensor location—3.3L engine

2. Remove the heated oxygen sensor (bank 1 sensor 2).

To install:

3. Installation is the reverse of removal. Tighten to 32 ft. lbs. (44 Nm).

3.3L Engine

1. Disconnect the 2 heated oxygen sensor (bank 1, 2 sensor 2) connectors.
2. Remove the 2 heated oxygen sensors (bank 1, 2 sensor 2).

To install:

3. Installation is the reverse of removal. Tighten to 32 ft. lbs. (44 Nm).

INTAKE AIR TEMPERATURE (IAT) SENSOR

LOCATION

The intake air temperature sensor is mounted on the mass air flow meter.

REMOVAL & INSTALLATION

See Mass Air Flow Meter.

KNOCK SENSOR (KS)

LOCATION

See Figures 201 and 202.

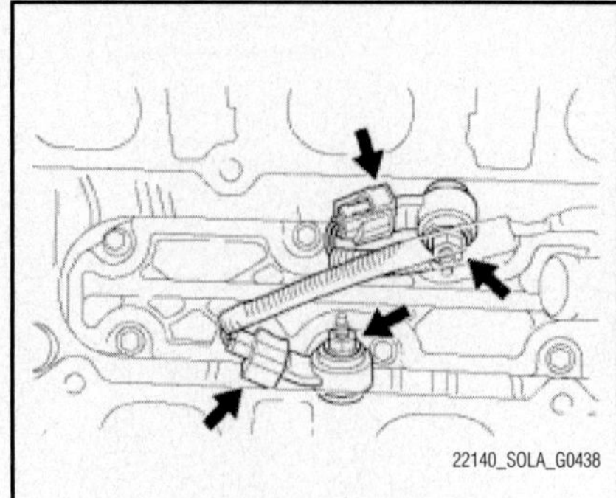

Fig. 202 Knock sensor location—3.3L engine

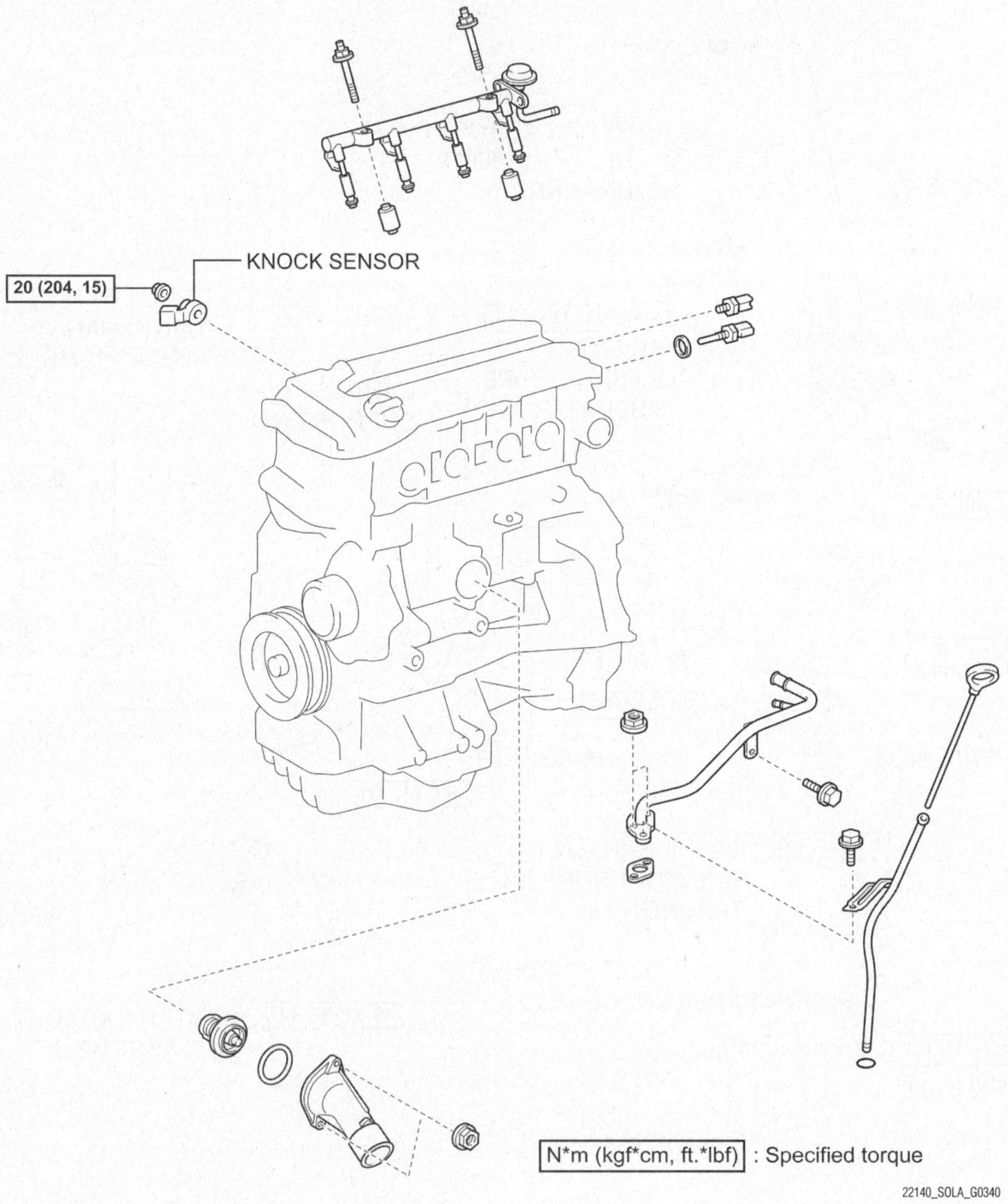

KNOCK SENSOR

20 (204, 15)

N*m (kgf*cm, ft.*lbf) : Specified torque

22140_SOLA_G0340

Fig. 201 Knock sensor location—2.4L engine

REMOVAL & INSTALLATION

2.4L Engine

See Figures 203 and 204.

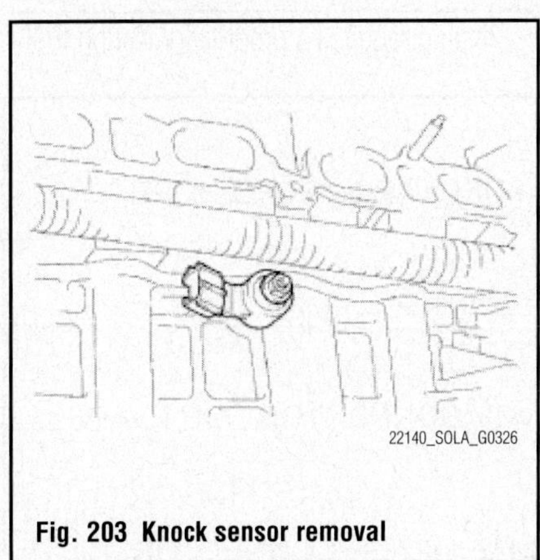

Fig. 203 Knock sensor removal

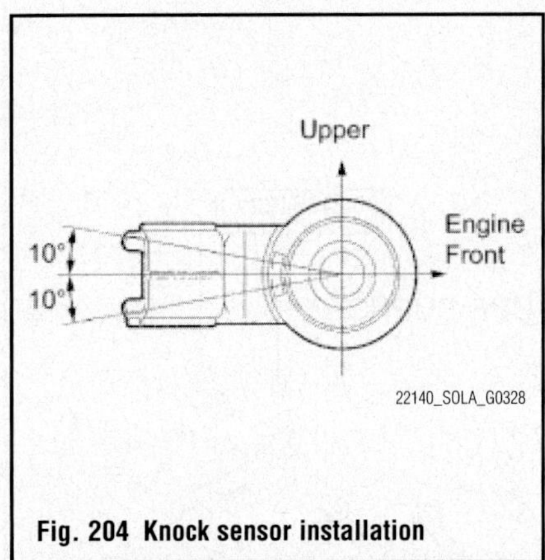

Fig. 204 Knock sensor installation

1. Drain engine coolant.
2. Remove the air cleaner cap with air cleaner hose.
3. Remove the throttle body assembly.
4. Remove the intake manifold.
5. Disconnect the knock sensor connector.
6. Remove the nut and knock sensor.

To install:

7. Install the sensor with the nut tighten to 14 ft. lbs. (20 Nm). Angling the knock sensor below the horizontal (below 0°) is recommended.
8. The remainder of installation is the reverse of the removal procedure.
9. Inspect for coolant leak and check vacuum hose connection.

3.3L Engine

See Figures 205 through 207.

1. Properly discharge the fuel system pressure.
2. Disconnect battery negative cable.
3. Drain engine coolant.

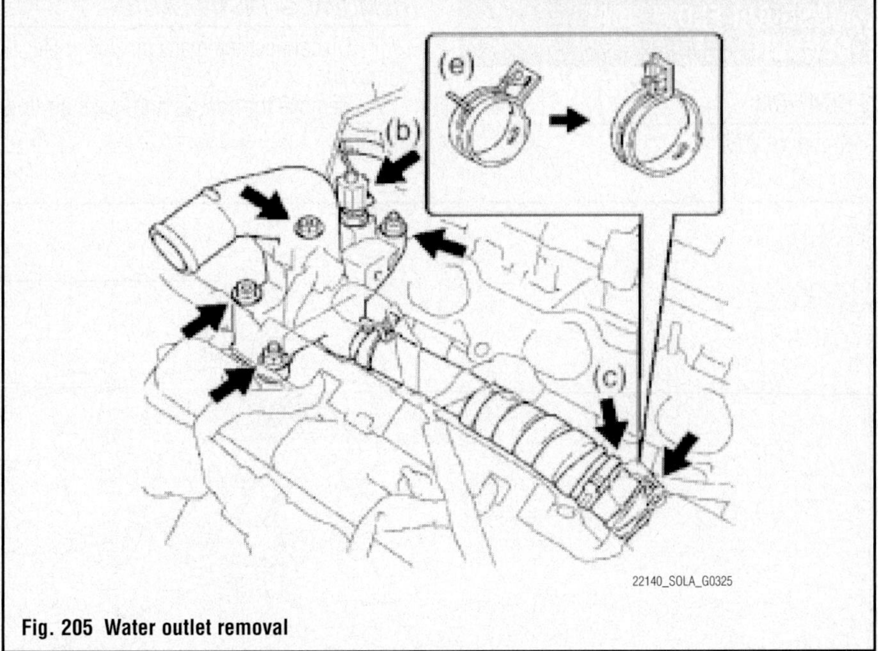

Fig. 205 Water outlet removal

4. Remove the V-bank cover sub-assembly.
5. Remove the air cleaner cap sub-assembly.
6. Remove emission control valve set.
7. Remove the intake air surge tank.
8. Remove the intake manifold.
9. Remove engine moving control rod.
10. Remove water outlet, as follows:
    a. Disconnect the radiator inlet hose.
    b. Disconnect the ECT sensor connector.
    c. Remove the clamp.
11. Remove the 2 bolts, 2 nuts and 2 washers.
    a. Lock the hose clamp as shown in the illustration. Then remove the water outlet together with the No. 1 water by-pass hose.
12. Remove the 2 gaskets from the 2 cylinder heads.

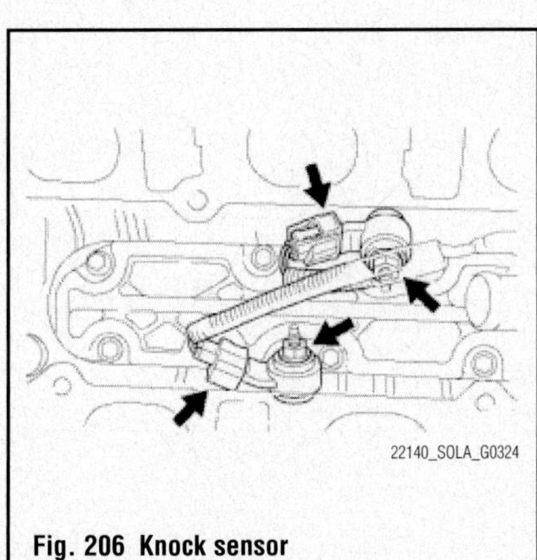

Fig. 206 Knock sensor

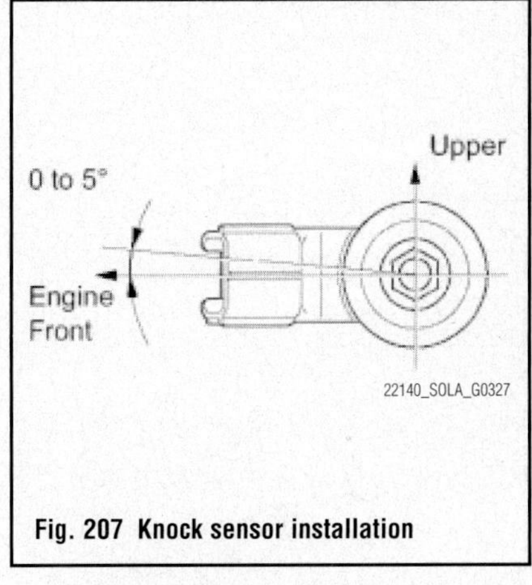

Fig. 207 Knock sensor installation

13. Disconnect the 2 knock control sensor connectors.
14. Remove the 2 bolts and 2 knock control sensors.

To install:

15. Install the 2 knock sensors with the 2 nuts as shown in the illustration and tighten to 14 ft. lbs. (20 Nm).
16. Connect the 2 knock control sensor connectors.
17. The remainder of installation is the reverse of the removal procedure.
18. Inspect for fuel leak and check vacuum hose connection.

MALFUNCTION INDICATOR LIGHT (MIL)

RESET PROCEDURES

Clear the DTC codes.

MASS AIR FLOW (MAF) SENSOR

LOCATION

See Figure 208.

REMOVAL & INSTALLATION

1. Disconnect the mass air flow meter connector.
2. Remove the screws and mass air flow meter.

To install:

3. Installation is the reverse of the removal procedure.

TESTING

Testing requires a proprietary testing tool.

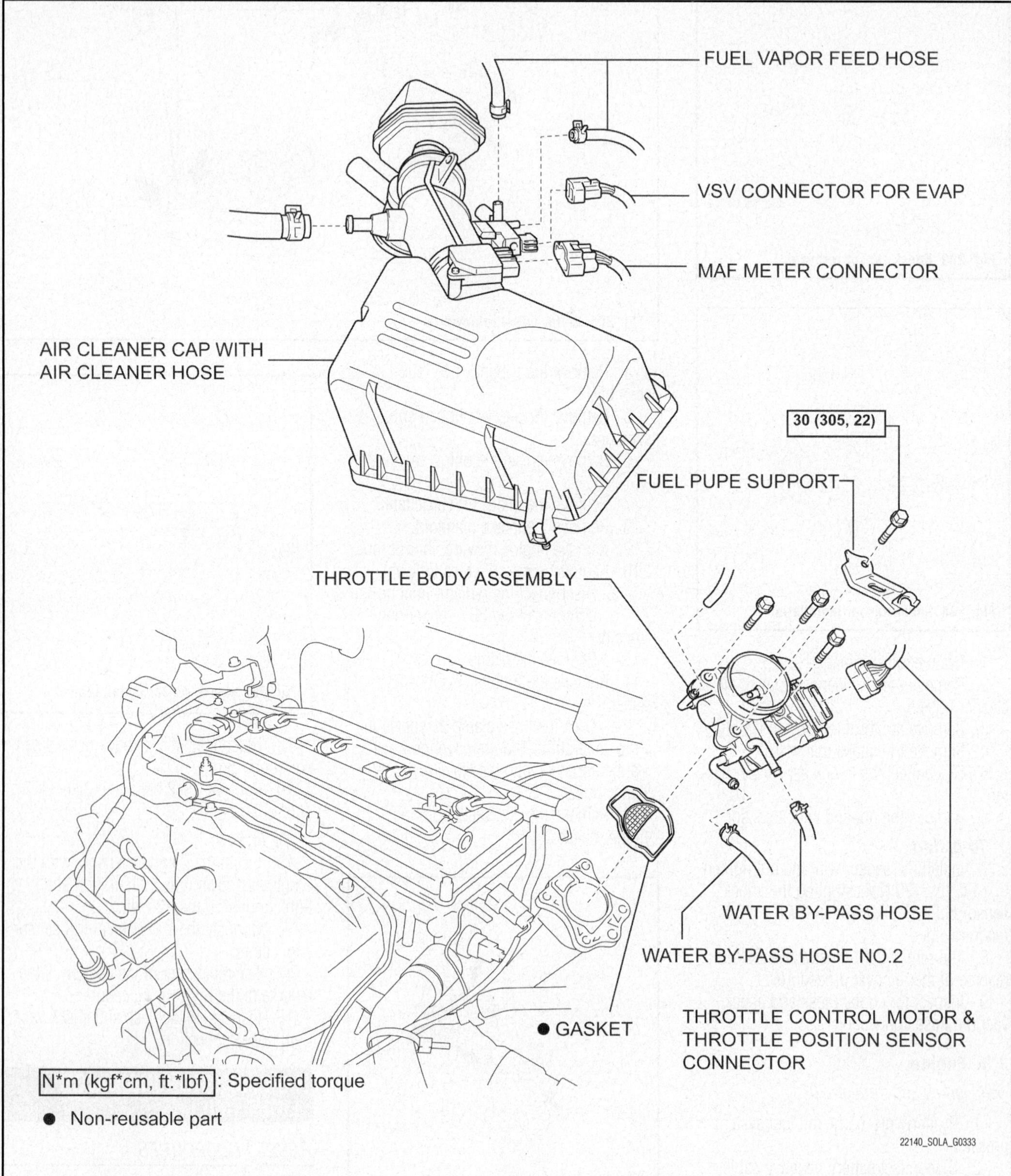

Fig. 208 MAF meter and connector locations

THROTTLE POSITION SENSOR (TPS)

LOCATION

See Figure 209.

The throttle position sensor is located on the throttle body.

REMOVAL & INSTALLATION

Refer to the Throttle Body removal and installation procedures.

VARIABLE CAMSHAFT TIMING OIL CONTROL SOLENOID

LOCATION

See Figures 210 and 211.

REMOVAL & INSTALLATION

1. Remove camshaft timing oil control valve connector.
2. Remove bolt.
3. Remove camshaft timing oil control valve assembly.

To install:

4. Installation is the reverse of removal.

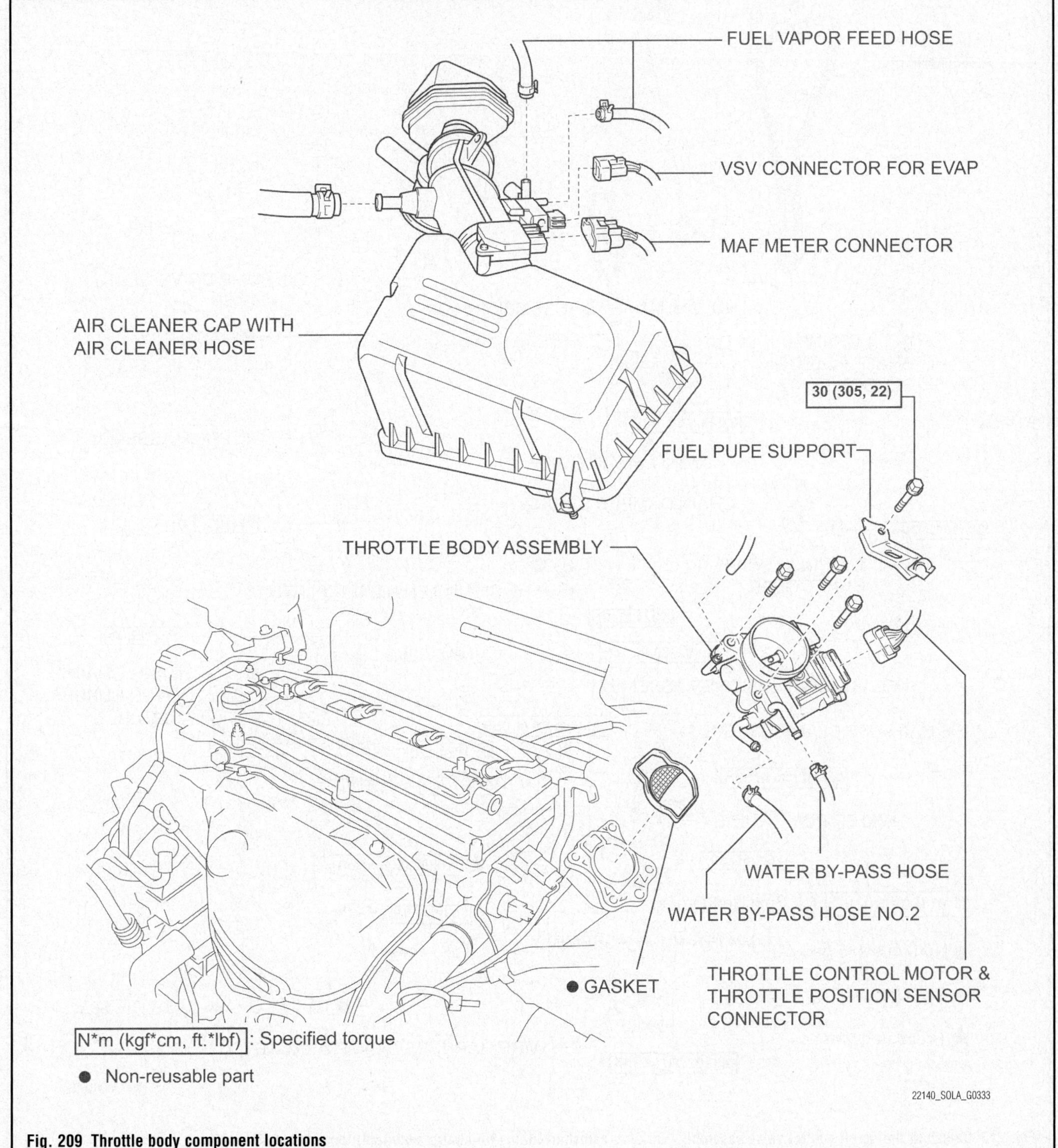

Fig. 209 Throttle body component locations

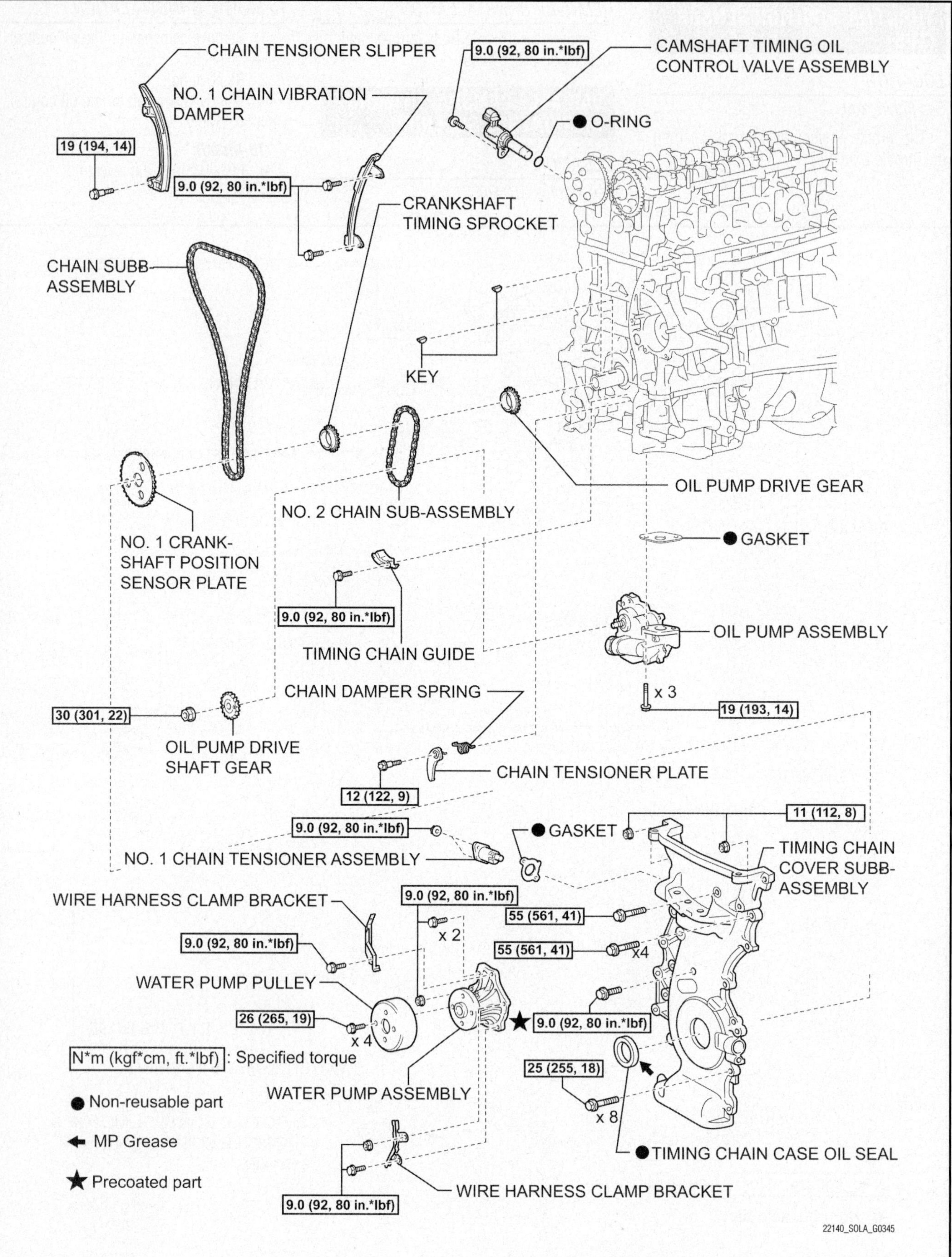

Fig. 210 Camshaft timing oil control valve assembly (Variable camshaft timing oil control solenoid) location—2.4L engine

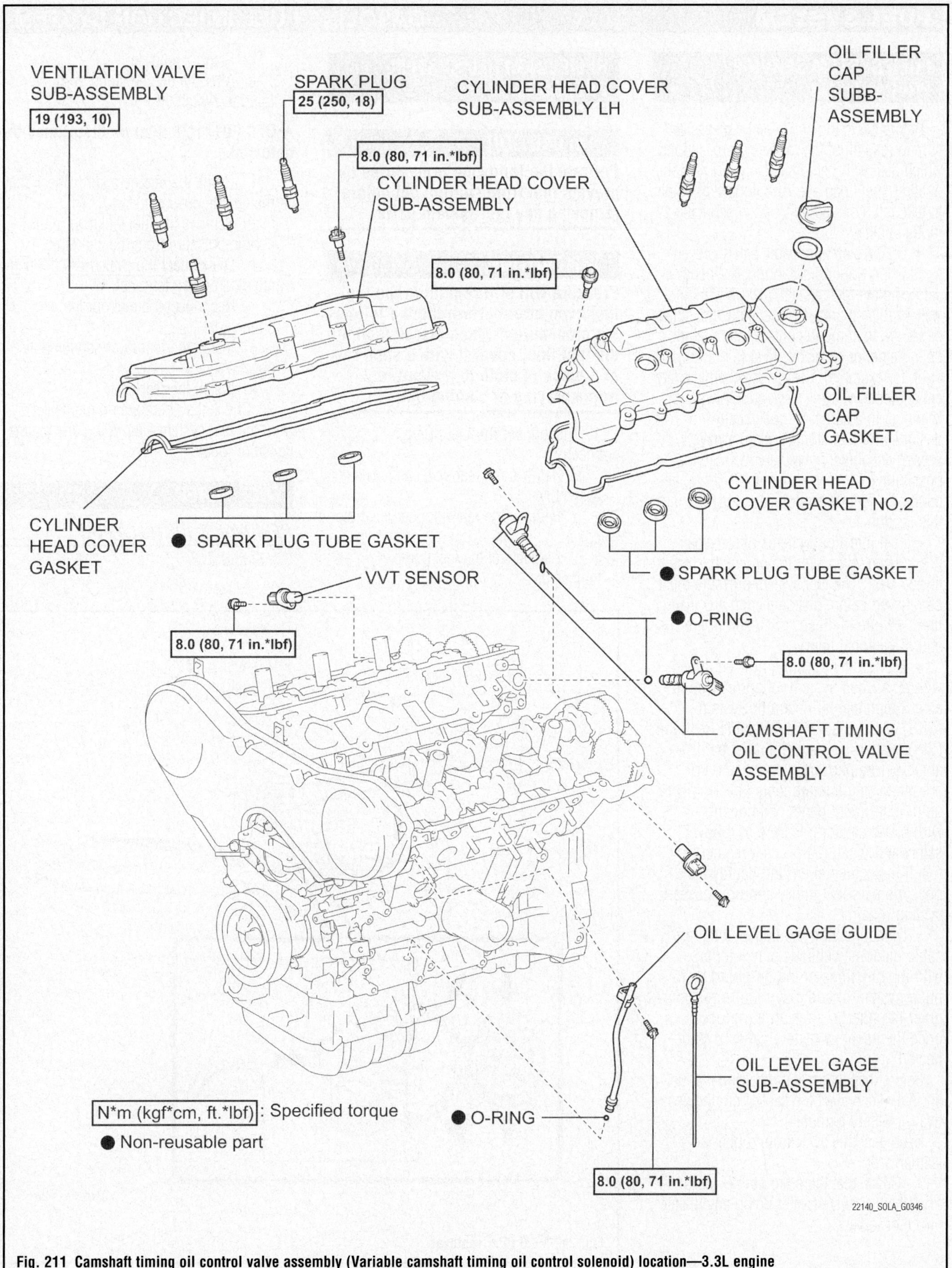

Fig. 211 Camshaft timing oil control valve assembly (Variable camshaft timing oil control solenoid) location—3.3L engine

FUEL SYSTEM SERVICE PRECAUTIONS

Before servicing any vehicle, please be sure to read all of the following precautions, which deal with personal safety, prevention of component damage, and important points to take into consideration when servicing a motor vehicle:

- Observe all applicable safety precautions when working around fuel. Whenever servicing the fuel system, always work in a well-ventilated area. Do not allow fuel spray or vapors to come in contact with a spark, open flame, or excessive heat (a hot drop light, for example). Keep a dry chemical fire extinguisher near the work area. Always keep fuel in a container specifically designed for fuel storage; also, always properly seal fuel containers to avoid the possibility of fire or explosion. Refer to the additional fuel system precautions later in this section.
- Fuel injection systems often remain pressurized, even after the engine has been turned **OFF**. The fuel system pressure must be relieved before disconnecting any fuel lines. Failure to do so may result in fire and/or personal injury.
- All new vehicles are now equipped with an air bag system, often referred to as a Supplemental Restraint System (SRS) or Supplemental Inflatable Restraint (SIR) system. The system must be disabled before performing service on or around system components, steering column, instrument panel components, wiring and sensors. Failure to follow safety and disabling procedures could result in accidental air bag deployment, possible personal injury and unnecessary system repairs.
- Disconnecting the negative battery cable on some vehicles may interfere with the functions of the on-board computer system(s) and may require the computer to undergo a relearning process once the negative battery cable is reconnected.

Before inspecting and repairing the fuel system, disconnect the cable from the negative (-) battery terminal.

Keep gasoline away from rubber or leather parts.

1. Check that there are no fuel leaks from the fuel system after doing any maintenance or repairs.

RELIEVING FUEL SYSTEM PRESSURE

CAUTION

Perform the following procedures to prevent fuel from spilling out before removing any fuel system parts.

CAUTION

Pressure will still remain in the fuel line even after performing the following procedures. When disconnecting the fuel line, cover it with a shop rag or a piece of cloth to prevent fuel from spraying or coming out.

1. Disconnect the fuel pump connector:
   a. Remove the rear seat cushion assembly.
   b. Remove the rear floor service hole cover.
   c. Disconnect the fuel pump connector.
   d. Start the engine.
   e. After the engine stops, turn the ignition switch off.

➡DTC P0171/25 (fuel problem) may be detected.

   f. Crank the engine again. Check that the engine does not start.
   g. Remove the fuel tank cap to discharge pressure from the fuel tank.
   h. Disconnect the cable from the negative (-) battery terminal.
   i. Reconnect the fuel pump connector.
   j. Install the rear floor service hole cover.
   k. Install the rear seat.
2. Check that there are no fuel leaks from the fuel system after doing any maintenance or repairs.

FUEL FILTER

REMOVAL & INSTALLATION

See Figure 212.

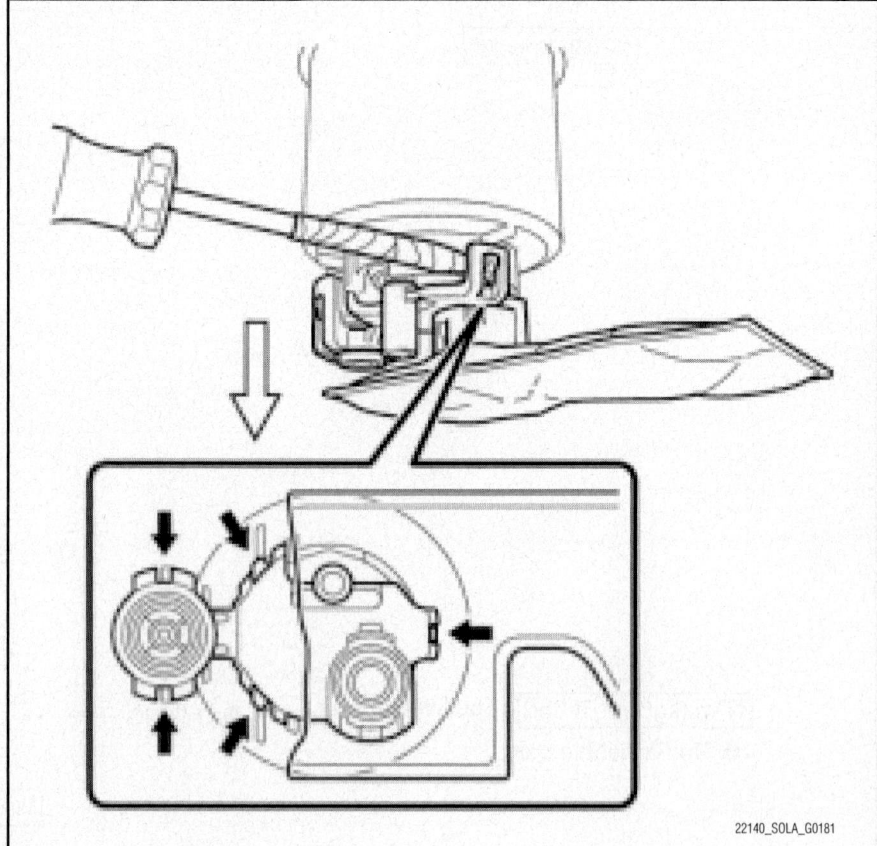

Fig. 212 Fuel filter removal

1. Before servicing the vehicle, refer to the precautions section.
2. Remove the fuel pump from the vehicle.
3. Remove the fuel pump filter, as follows:
   a. Pull out the fuel pump from the fuel filter.

To install:

4. Apply a light coat of gasoline or grease to the O-ring of the fuel pump.
5. Push in the fuel pump to the fuel filter.
6. Install the fuel pump.

FUEL PUMP

REMOVAL & INSTALLATION

See Figures 213 through 215.

1. Before servicing the vehicle, refer to the precautions section.
2. Discharge fuel system pressure.
3. Disconnect battery negative cable.
4. Remove the rear seat cushion assembly.
5. Remove the rear floor service hole cover.

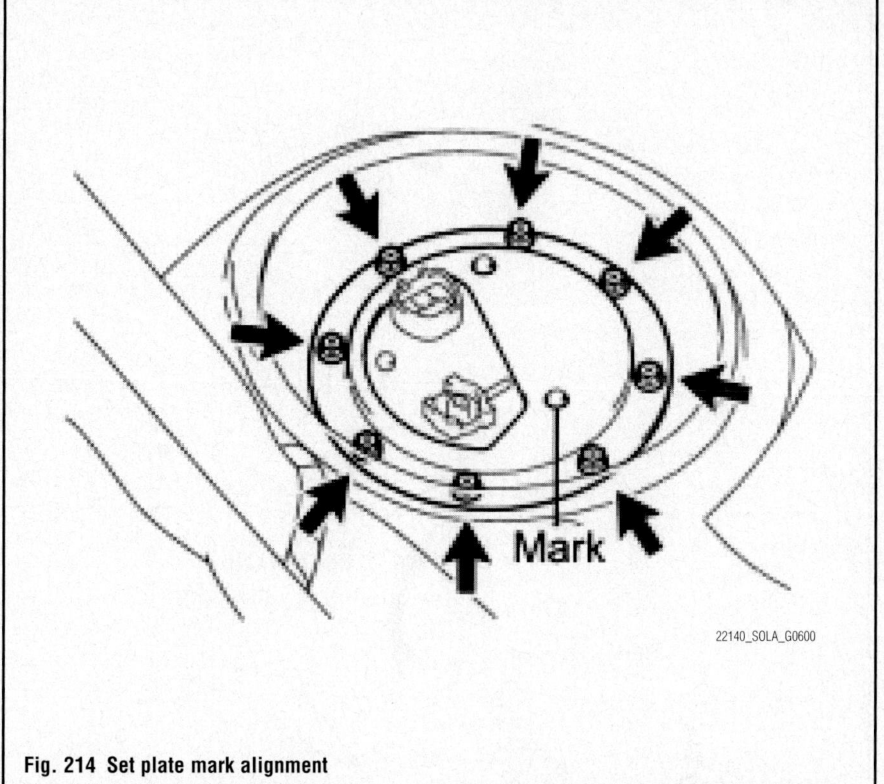

Fig. 214 Set plate mark alignment

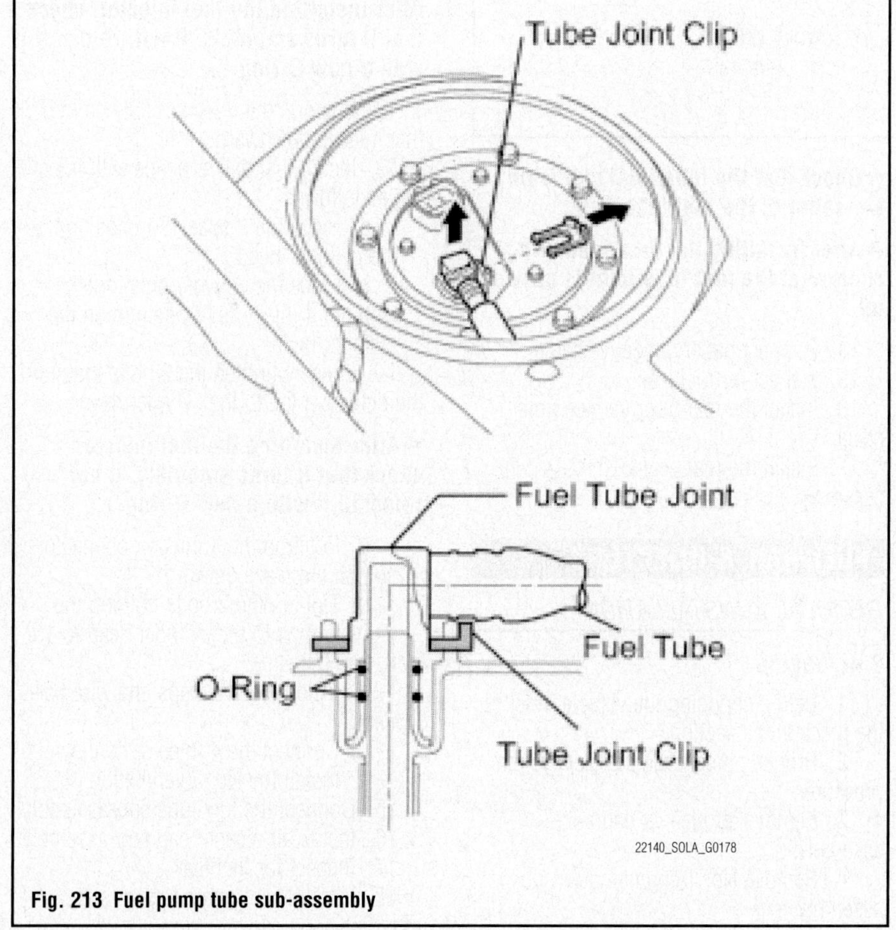

Fig. 213 Fuel pump tube sub-assembly

6. Disconnect the fuel pump connector.
7. Separate the fuel pump tube sub-assembly, as follows:
   a. Remove the tube joint clip, and pull out the fuel pump tube.

➡Check if there is any dirt or mud around the connector before this operation and remove the dirt as necessary.

➡Be careful of mud because the quick connector has an O-ring which seals the pipe and connector that can be contaminated.

➡Do not use any tools in this operation.

➡Do not bend or twist the nylon tube. Cover the fuel tube joint with a plastic bag.

➡When the fuel tube joint and fuel suction plate are stuck, pinch the fuel tank tube between fingers, and turn it carefully to release it. Disconnect the fuel tank tube.

8. Remove fuel tank vent tube set plate, as follows:
   a. Remove the 8 bolts and set plate.
9. Remove fuel suction tube assembly with pump and gauge, as follows:
   a. Pull out the fuel suction tube from the fuel tank.

➡Do not damage the fuel pump filter.

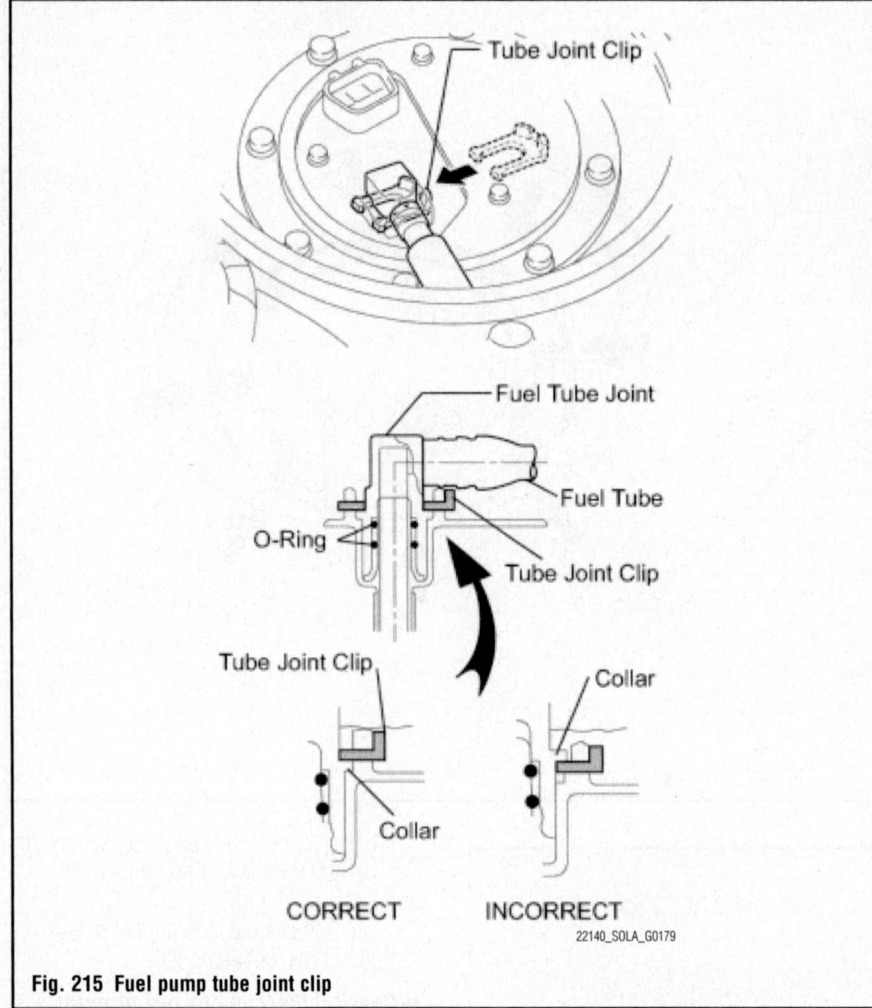

Fig. 215 Fuel pump tube joint clip

➡**Be careful not to bend the arm of the fuel sender gauge.**

b. Remove the gasket from the fuel suction tube.

To install:

10. Install a new gasket to the fuel suction tube.
11. Install the fuel suction tube.

➡**Do not damage the fuel pump filter.**

➡**Be careful not to bend the arm of the fuel sender gauge.**

12. Install the fuel tank vent tube set plate, as follows:
   a. Align the mark of the set plate with the fuel suction tube.
   b. Install the set plate with the 8 bolts and tighten to 52 inch lbs. (5.9 Nm).
13. Install the fuel pump tube with the tube joint clip.

➡**Check that there is no scratches or foreign objects on the connecting part.**

➡**Check that the fuel tube joint is inserted securely.**

➡**Check that the tube joint clip is on the collar of the fuel tube joint.**

➡**After installing the tube joint clip, check that the fuel tube joint is pulled off.**

14. Connect battery negative cable.
15. Inspect for fuel leak.
16. Install the rear floor service hole cover.
17. Install the rear seat cushion assembly.

FUEL RAIL & INJECTORS

REMOVAL & INSTALLATION

2.4L Engine

1. Before servicing the vehicle, refer to the precautions section.
2. Properly discharge the fuel system pressure.
3. Remove air cleaner cap sub-assembly.
4. Remove No. 1 engine cover sub-assembly.
5. Disconnect fuel tube sub-assembly.
6. Remove fuel delivery pipe with injector, as follows:

➡**Be careful not to drop the injectors when removing the delivery pipe.**

   a. Remove the No. 2 ventilation hose.
   b. Disconnect the 4 injector connectors.
   c. Remove the 2 clamps and wire harness from the delivery pipe.
   d. Remove the 2 bolts and delivery pipe together with the 4 injectors.
   e. Remove the 2 spacers from the cylinder head.
7. Pull out the 4 injectors from the delivery pipe.

To install:

8. Install a new insulator to the injector.
9. Apply a light coat of grease or gasoline to a new O-ring and install to the injector.
10. Apply a light coat of grease or gasoline on the place where the delivery pipe touches the O-ring.
11. Install the fuel injector into the fuel delivery pipe by pushing the fuel injector while twisting it back and forth.

➡**Be careful not to twist the O-ring. After installing the fuel injector, check that it turns smoothly. If not, reinstall it with a new O-ring.**

12. Position the injector connector so that it faces downward.
13. Install fuel delivery pipe with injector, as follows:
   a. Place the 2 spacers in position on the cylinder head.
   b. Place the delivery pipe together with the 4 injectors in position on the cylinder head.
14. Loosely tighten the 2 bolts that hold the delivery pipe to the cylinder head.

➡**After installing the fuel injector, check that it turns smoothly. If not, reinstall it with a new O-ring.**

   a. Position the 4 injector connectors so that they face outward.
   b. Tighten the 2 bolts holding the delivery pipe to the cylinder head to 15 ft. lbs. (20 Nm).
   c. Install the 2 clamps and wire harness.
   d. Connect the 4 injector connectors.
   e. Install the No. 2 ventilation hose.
15. Connect the fuel tube sub-assembly.
16. Install air cleaner cap sub-assembly.
17. Inspect for fuel leak.
18. Install No. 1 engine cover sub-assembly.

3.3L Engine

See Figures 216 through 219.

1. Before servicing the vehicle, refer to the precautions section.
2. Properly discharge the fuel system pressure.
3. Disconnect battery negative cable.
4. Drain engine coolant.
5. Remove center front suspension upper brace (w/ performance rod).
6. Remove the V-bank cover sub-assembly.
7. Remove the air cleaner cap sub-assembly.
8. Remove emission control valve set, as follows:
   a. Disconnect the 2 VSV connectors (*1).
   b. Remove the wire harness clamp (*2).
   c. Remove the wire harness clamp (*2).
   d. Disconnect the fuel vapor feed hose No. 1 (*3).

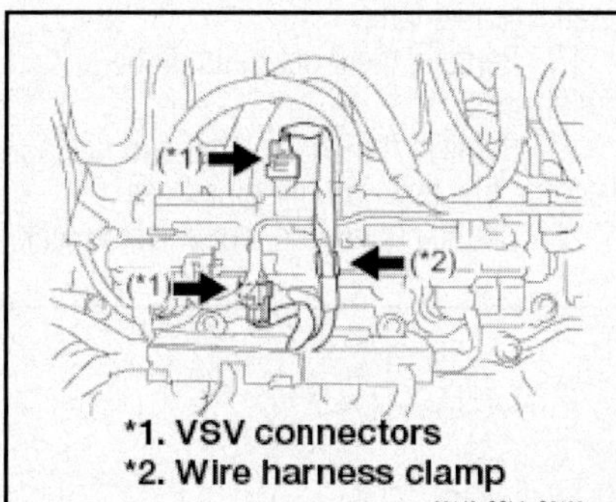

Fig. 216 VSV connectors and wire harness clamp

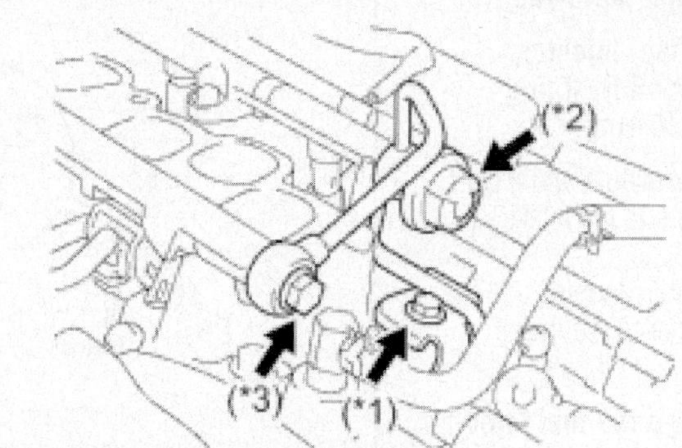

Fig. 218 No. 1 fuel pipe sub-assembly

9. Disconnect the fuel vapor feed hose No. 2 (*4).
   a. Disconnect the 2 vacuum hoses (*5).
   b. Remove the clamp (*6).
   c. Remove the 2 nuts and the emission control valve set.
10. Remove the intake air surge tank.
11. Disconnect the No. 1 fuel pipe sub-assembly, as follows:
    a. Remove the bolt and disconnect the No. 1 fuel pipe (*1).
    b. Remove the fuel pressure pulsation damper and 2 gaskets (*2).
    c. Remove the fuel pipe No. 2 union bolt and 2 gaskets (*3).
12. Remove the fuel injector assembly, as follows:
    a. Disconnect the 6 fuel injector connectors.

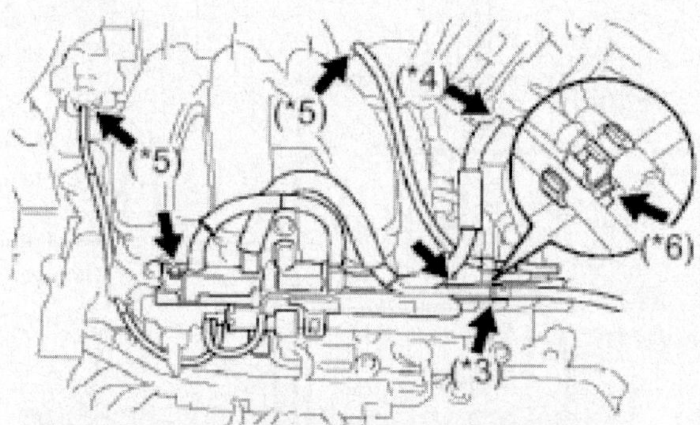

Fig. 217 Emission control valve set hoses

    b. Remove the 4 bolts and 2 fuel delivery pipes together with the 6 fuel injectors.

➡Be careful not to drop the fuel injectors when removing the fuel delivery pipe.

    c. Remove the 4 No. 1 delivery pipe spacers (A) and 6 insulators (B) from the intake manifold.
    d. Pull out the fuel injector from the fuel delivery pipe.

To install:

13. Install the fuel injector assembly, as follows:
    a. Install a new grommet to the injector.
    b. Apply a light coat of spindle oil or gasoline to new O-rings, and install one to each injector.
    c. Apply a light coat of spindle oil or gasoline where the fuel delivery pipe contacts the O-ring.
    d. Push the fuel injector while twisting it back and forth to install it in the fuel delivery pipe.

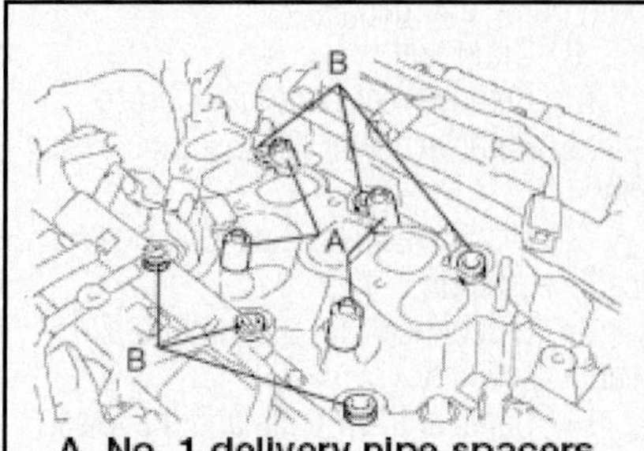

Fig. 219 No. 1 delivery pipe spacers and insulators

➡Be careful not to twist the O-ring.

➡After installing the fuel injector, check that it turns smoothly. If not, reinstall it with a new O-ring.

e. Install 6 new insulators and 4 No. 1 delivery pipe spacers to the intake manifold.

f. Place the fuel delivery pipe and the 6 fuel injectors together to the intake manifold.

➡Be careful not to drop the fuel injectors when installing the fuel delivery pipe.

g. Temporarily install the 4 bolts which are used to hold the fuel delivery pipe to the intake manifold.

➡After installing the fuel injector, check that it turns smoothly. If not, reinstall it with a new O-ring.

h. Tighten the 4 bolts which are used to hold the fuel delivery pipe to the intake manifold to 7 ft. lbs. (10 Nm).

i. Connect the 6 fuel injector connectors.

14. Te remainder of installation is the reverse of the removal procedure.
15. Check vacuum hose connection.
16. Check for coolant leak and fuel leak.

FUEL TANK

REMOVAL & INSTALLATION

See Figures 220 and 222.

1. Before servicing the vehicle, refer to the precautions section.
2. Discharge fuel system pressure.
3. Disconnect battery negative cable.
4. Remove rear seat cushion assembly.
5. Remove rear floor service hole cover.
6. Separate fuel pump tube sub-assembly.
7. Remove fuel tank vent tube set plate.
8. Remove fuel suction tube assembly with pump and gauge.
9. Drain fuel.
10. Remove rear floor panel brace.
11. Remove center exhaust pipe assembly.
12. Disconnect No. 2 parking brake cable assembly.
13. Disconnect No. 3 parking brake cable assembly.
14. Remove the 4 bolts and the lower center fuel tank protector.
15. Disconnect the fuel pump tube, as follows:

➡Check that there is no dirt or other foreign objects around the connector before removing fuel tubes, and clean the connector as necessary.

➡It is necessary to prevent mud or dirt from entering the connector. If mud or dirt gets in the connector, the O-rings may not seal properly.

➡Do not use any tools in these operations.

➡Do not bend, kink or twist the nylon tubes. Protect the connector by covering it with a plastic bag.

➡When the pipe and connector are stuck, push and pull the connector to release and pull the connector out carefully.

a. Pinch the tabs of the retainer to remove the lock claws and pull it down as shown in the illustration.

b. Pull out the fuel tank main tube.

16. Pinch the tube connector and then pull out the No. 1 fuel tube.
17. Set up a transmission jack underneath the fuel tank.
18. Remove the 2 set bolts of the fuel tank bands.
19. Remove the hose clamp and disconnect the fuel tank to filter pipe hose.
20. Slightly lower the transmission jack.

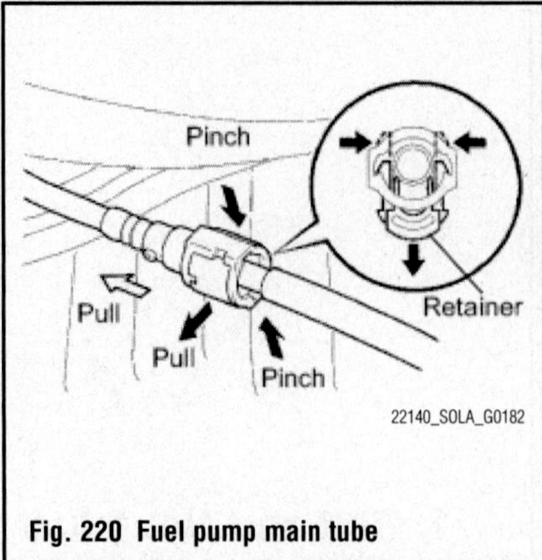

Fig. 220 Fuel pump main tube

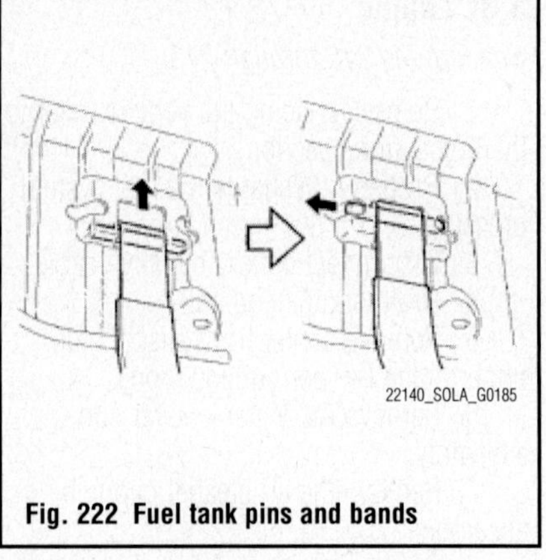

Fig. 222 Fuel tank pins and bands

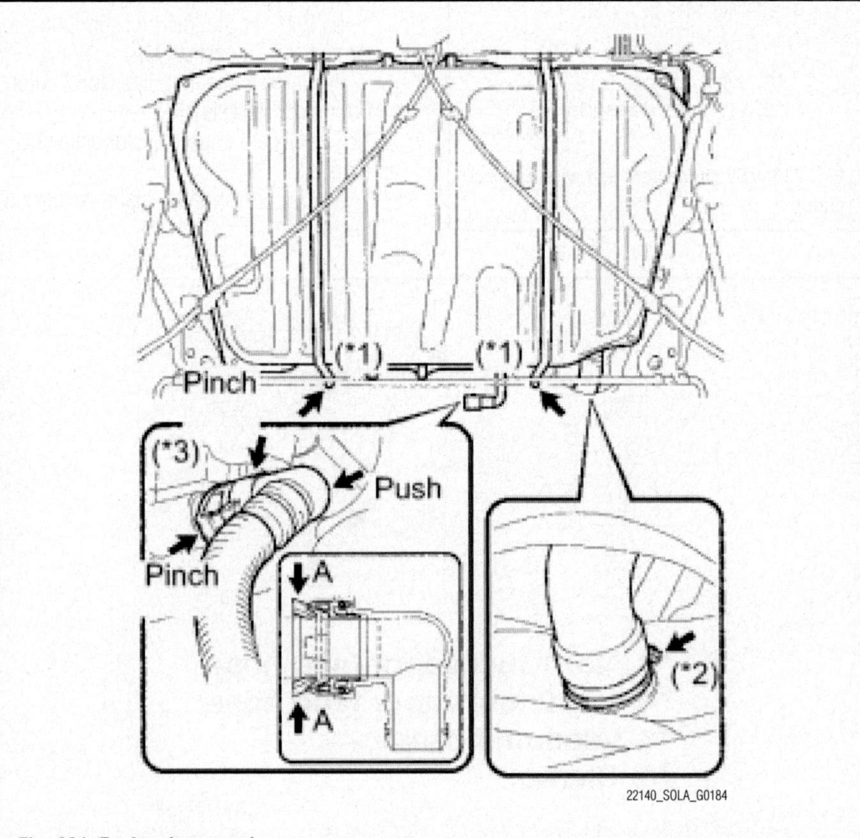

Fig. 221 Fuel tank removal

21. Disconnect the fuel tank vent hose from the charcoal canister, as follows:
  a. Push the connector deep into the charcoal canister to release the locking pin.
  b. Pinch portion A.
  c. Pull out the connector.
22. Remove the 2 pins and 2 fuel tank bands as shown in the illustration.
23. Remove the 4 clip nuts.

To install:

24. Install the 4 clip nuts.
25. Install the 2 fuel tank bands with the 2 pins.
26. Set the fuel tank assembly onto the transmission jack and raise the jack.
27. Connect the fuel tank vent hose.
28. Connect the fuel tank inlet pipe with the fuel filter pipe clamp.
29. Tighten the 2 set bolts of the fuel tank bands to 29 ft. lbs. (39 Nm).
30. Connect the No. 1 fuel tube, as follows:
  a. Push the fuel tube connector into the pipe until the fuel tube connector makes a "click" sound.

➡Check that there is no damage or foreign objects on the connected part.

➡After connecting, check if the fuel tube connector and the pipe are securely connected by trying to pull them apart.

31. Connect the fuel pump tube, as follows:
  a. Push in the fuel pump tube connector to the pipe and push up the retainer so that the claws engage.

➡Check that there is no damage or foreign objects on the connected part.

➡After connecting, check if the fuel tube connector and the pipe are securely connected by trying to pull them apart.

32. Install the lower center fuel tank protector and tighten to 48 inch lbs. (5.4 Nm).
33. Install the No. 3 parking brake cable assembly with the bolt and nut and tighten to 53 inch lbs. (6 Nm), and 48 inch lbs. (5.4 Nm).
34. Install the No. 2 parking brake cable assembly with the bolt and nut and tighten to 53 inch lbs. (6 Nm), and 48 inch lbs. (5.4 Nm).
35. Install the center exhaust pipe assembly.
36. Install rear floor panel brace.
37. Install the fuel suction tube assembly with pump and gauge.
38. Install the fuel tank vent tube set plate.
39. Connect the fuel pump tube sub-assembly.
40. Add fuel.
41. Connect battery negative cable.
42. Inspect for fuel leak and exhaust gas leak.
43. Install the rear floor service hole cover.
44. Install the rear seat cushion assembly.

IDLE SPEED

ADJUSTMENT

Idle speed is maintained by the ECM. No adjustment is necessary or possible.

THROTTLE BODY

REMOVAL & INSTALLATION

2.4L Engine

See Figure 223.

1. Properly discharge the fuel system pressure.
2. Disconnect battery negative cable.
3. Drain engine coolant.
4. Remove the air cleaner cap and hose.
5. Remove throttle body assembly, as follows:
  a. Disconnect the throttle control motor and throttle position sensor connector.
  b. Disconnect the vacuum hose from the throttle body.
  c. Disconnect the 2 water by-pass hoses.
  d. Disconnect the vent hose from the throttle body.
  e. Remove the 4 bolts, fuel pipe support and throttle body.
  f. Remove the gasket.

To install:

6. Install a new gasket on the intake manifold, as shown in the illustration.
7. Install the throttle body and fuel pipe support with the 4 bolts and tighten to 22 ft. lbs. (30 Nm).
8. The remainder of installation is the reverse of removal.

3.3L Engine

See Figure 224.

1. Properly discharge the fuel system pressure.
2. Disconnect battery negative cable.
3. Drain engine coolant.
4. Remove center front suspension upper brace.
5. Remove the V-bank cover sub-assembly.
6. Remove the air cleaner inlet assembly.
7. Remove the air cleaner assembly.
8. Remove the intake air resonator.
9. Remove throttle body assembly, as follows:
  a. Disconnect the throttle control motor and throttle position sensor connector.
  b. Disconnect the No. 2 water by-pass hose.
  c. Disconnect the No. 3 water by-pass hose.
  d. Remove the 2 bolts, 2 nuts and throttle body.
  e. Remove the gasket.

To install:

10. Installation is the reverse of removal, noting the following:
  a. Install a new gasket to the intake air connector.
  b. Tighten the throttle body bolts to 8 ft. lbs. (11 Nm).
  c. Check vacuum hose connection.
  d. After adding coolant, check for coolant leaks.

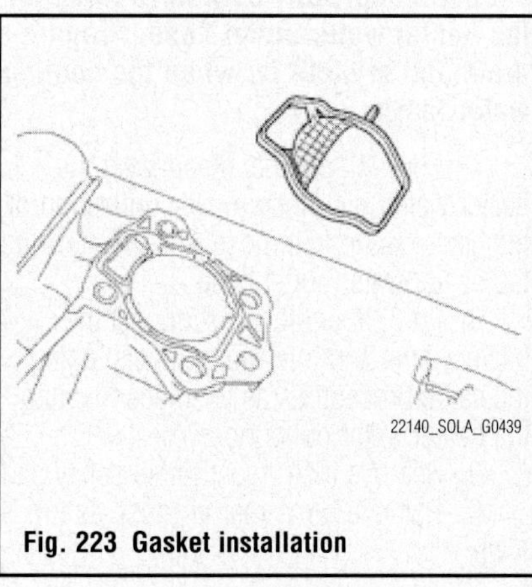

Fig. 223 Gasket installation

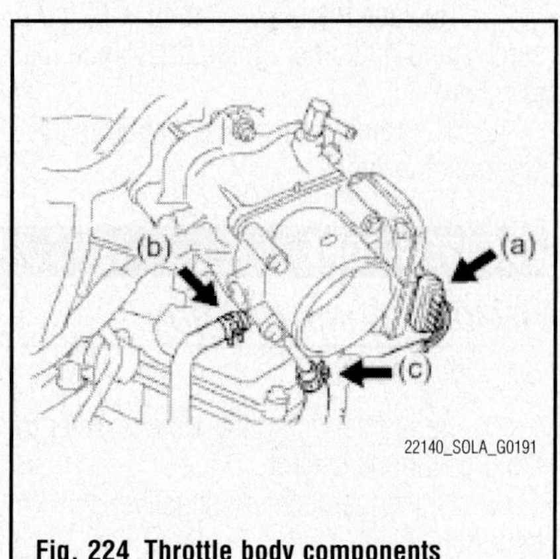

Fig. 224 Throttle body components

HEATING & AIR CONDITIONING SYSTEM

BLOWER MOTOR

REMOVAL & INSTALLATION

See Figure 225.

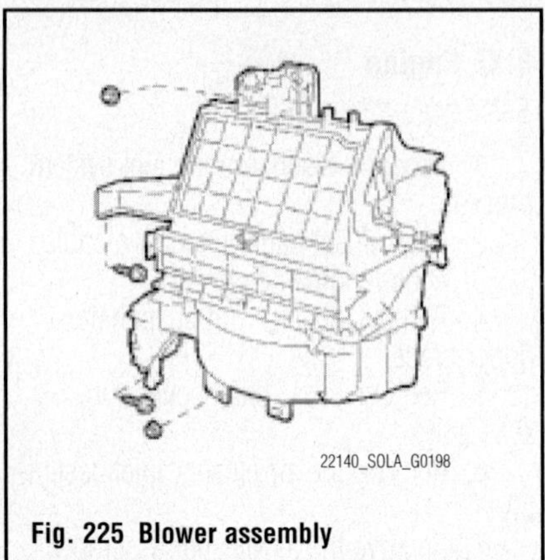

Fig. 225 Blower assembly

1. Remove instrument panel safety pad.
2. Remove the No. 1 heater to foot duct.
3. Remove stereo component amplifier assembly.
4. Remove ECM.
5. Disconnect the connector from the blower motor.
6. Remove the screw, clamp and blower connector holder.
7. Disconnect the 5 wire harness clamps.
8. With stereo component amplifier assembly, disconnect the 2 wire harness clamps.
9. Disconnect the 4 connectors (for manual A/C) or 3 connectors (for Auto A/C) and remove the clamp.
10. Remove the 2 screws and 2 nuts, then remove the blower motor.

To install:

11. Replace the blower motor with the 2 screws and nuts and tighten to 12 inch lbs. (1.5 Nm).
12. The remainder of installation is the reverse of removal.

HEATER CORE

REMOVAL & INSTALLATION

See Figure 226.

1. Before servicing the vehicle, refer to the precautions section.
2. Disconnect battery negative terminal.

CAUTION

Wait for 90 seconds after disconnecting the cable to prevent the airbag from deploying.

3. Discharge refrigerant from refrigeration system.
4. Disconnect the No. 1 cooler refrigerant suction hose, as follows:
   a. Install SST (SST: 09870-00015) to the piping clamp. Check the direction of the piping clamp claw and SST by referring to the illustration shown on the caution label.
   b. Push down on the SST and release the clamp lock. Be careful not to deform the hose when pushing down on the SST.
   c. Pull the SST slightly and push the release lever, then remove the piping clamp with the SST.
   d. Disconnect the No. 1 cooler refrigerant suction hose.

➡Do not use tools such as a screwdriver to remove the hose. Cap the open fittings immediately to keep moisture or dirt out of the system.

   e. Remove the 2 O-rings from the No. 1 cooler refrigerant suction hose.

➡Seal the openings of the disconnected parts using vinyl tape to prevent moisture and foreign matter from entering.

5. Disconnect the cooler refrigerant liquid pipe A. Disconnection of the cooler refrigerant liquid pipe A is the same as the No. 1 cooler refrigerant suction hose.
6. Disconnect the heater water outlet hose A (from the heater unit), as follows:
   a. Using pliers, grip the claws of the clip and slide the clip, then disconnect the heater water outlet hose A.

➡Do not apply any excessive force to the heater water outlet hose. Prepare a drain pan or cloth for when the cooling water leaks.

7. For 2.4L engine, disconnect the heater water inlet hose A. Disconnection of the heater water inlet hose A is the same as the heater water outlet hose A.
8. For3.3L engine, disconnect the heater outlet and inlet water hoses, using the same procedures as for disconnecting the heater water outlet hose A.
9. Remove instrument panel safety pad.
10. Remove no. 1 rear air duct, as follows:
   a. Fold back the floor carpet.
   b. Remove the 2 screws, bolt and nut.
   c. Remove the No. 1 rear air duct.
11. Using the same procedure, remove the No. 2 rear air duct.
12. Remove No. 1 instrument panel brace sub-assembly, as follows:
   a. Remove the 2 bolts and 2 earth wires.
   b. Remove the nut and release the connector holder.
   c. Disconnect the 2 wire harness clamps.
   d. Remove the bolt and screw.
   e. Remove the nut and No. 1 instrument panel brace sub-assembly.
13. Remove the 2 bolts and No. 1 instrument finish panel retainer lower.
14. Remove No. 2 instrument panel brace sub-assembly, as follows:
   a. Remove the bolt and earth wire.
   b. Remove the nut.
   c. Disconnect the passenger side junction block clamp.
   d. Remove the nut and amplifier antenna assembly clamp.
   e. Remove the 2 nuts, bolt and No. 2 instrument panel brace sub-assembly.
15. Remove the clip and No. 3 heater to foot duct.
16. Remove the clip and No. 1 heater to foot duct.
17. Remove the 3 nuts and instrument panel junction block assembly.
18. Remove the nut and center instrument panel bracket sub-assembly.
19. Disconnect steering column assembly, as follows:
   a. Release the wire harness clamp and 3 connectors.
   b. Remove the bolt and 2 nuts and disconnect the steering column assembly.
20. Remove instrument panel reinforcement assembly.
21. Remove stereo component amplifier assembly.
22. Remove ECM.
23. Remove blower assembly, as follows:
   a. Disconnect the connectors.
   b. Remove the screw.
   c. Disconnect the blower connector holder clamp.
   d. Disconnect the 5 wire harness clamps.
   e. Disconnect the 2 wire harness clamps with stereo component amplifier assembly.

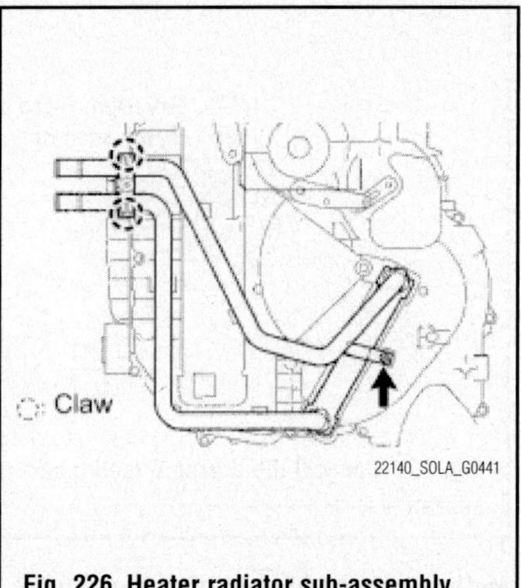

Fig. 226 Heater radiator sub-assembly

f. For manual A/C. disconnect the 4 connectors and remove the clamp.

g. For automatic A/C. disconnect the 3 connectors and remove the clamp.

h. Remove the 2 screws, 2 nuts and blower assembly.

24. Release the 4 fitting claws and remove the lower defroster nozzle assembly.

25. Remove the radiator assembly, as follows:

a. Remove the drain hose.

b. Disconnect the connectors.

c. Remove the 2 nuts and air conditioning radiator assembly.

26. Remove the 3 screws and air outlet control servomotor (mode damper servo sub-assembly.

27. Remove the heater radiator unit sub-assembly, as follows:

a. Remove the screw and plate.

b. Release the 2 fitting claws and remove the piping clamp.

c. Remove the heater radiator unit sub-assembly.

➡Prepare a drain pan or cloth for when the cooling water leaks.

To install:

28. Installation is reverse of removal. Tighten the radiator assembly nuts to 12 inch lbs. (1.5 Nm).

29. After adding coolant and refrigerant, check for leaks.

STEERING

POWER RACK & PINION STEERING GEAR

REMOVAL & INSTALLATION

See Figures 227 through 229.

1. Before servicing the vehicle, refer to the precautions section.

2. Position the front wheels straight ahead.

3. Separate steering intermediate shaft assembly, as follows:

a. Secure the steering wheel with the seat belt in order to prevent rotation. This will help prevent damage to the spiral cable.

b. Loosen bolt A and remove the clamp from the No.1 hole cover.

c. Separate the No. 2 hole cover from the No. 1 hole cover.

d. Loosen bolt B.

e. Put matchmarks on the steering intermediate shaft assembly and the power steering gear assembly.

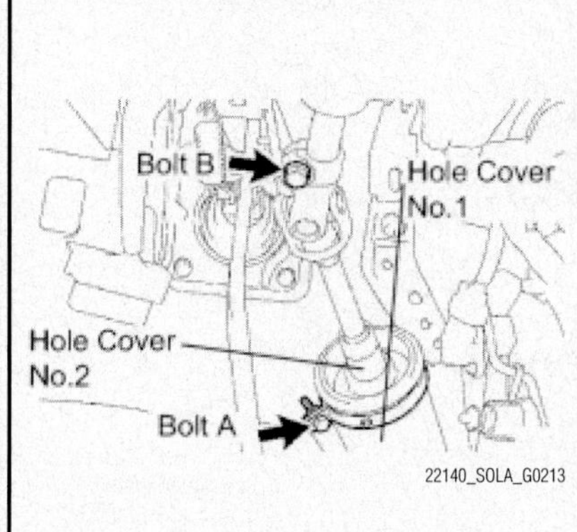

Fig. 227 Loosen the bolts, remove the clamp, and separate the hole cover

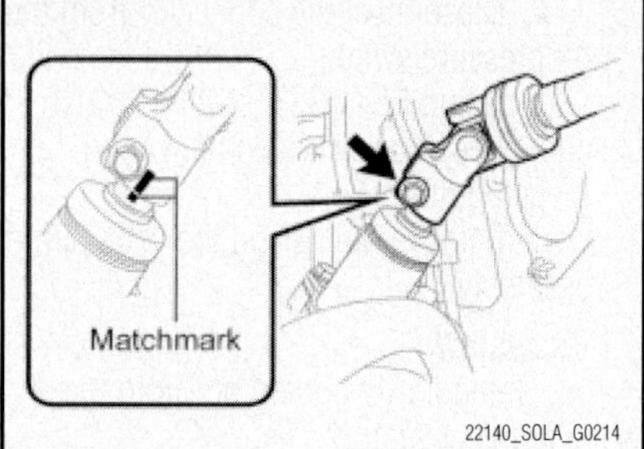

Fig. 228 Matchmarks on the steering intermediate shaft assembly and the power steering gear assembly

f. Remove the bolt and disengage the intermediate shaft assembly.

4. Remove front wheel.

5. Separate the left and right tie rod assemblies.

6. Remove the right and left front stabilizer link assemblies.

7. Remove the 2 bolts and the left and right No. 1 stabilizer bracket and the No. 1 stabilizer bar bush.

8. Disconnect pressure feed tube assembly, as follows:

a. Using SST (09023-12701) or equivalent, disconnect the return tube assembly from the power steering gear assembly.

b. Using SST (09023-12701), disconnect the pressure feed tube assembly from the power steering gear assembly.

c. For 2.4L vehicles, remove the nut and separate the tube clamp.

d. For 3.3L vehicles, remove the nut and separate the tube clamp and the heat insulator.

e. Remove the bolt and separate the tube clamp.

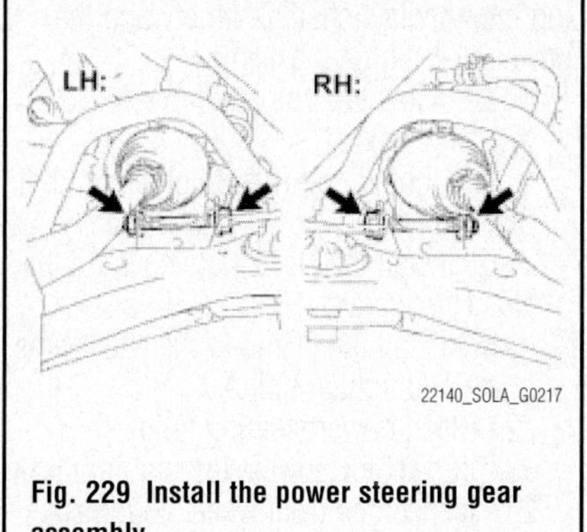

Fig. 229 Install the power steering gear assembly

9. Remove the power steering rack housing heat insulator.

10. Remove the 2 bolts, nuts and the power steering gear assembly.

To install:

11. Install the power steering gear assembly with the 2 bolts and nuts and tighten to 52 ft. lbs. (70 Nm).

12. Install the power steering rack housing heat insulator.

➡For the next 2 steps, use a torque wrench with a fulcrum length of 300 mm (11.81 in.). These torque values are effective when SST is parallel to the torque wrench.

13. Using SST (09023-12701) or equivalent, connect the pressure feed tube assembly to the power steering gear assembly and tighten to 16 ft. lbs. (22 Nm).

14. Using SST (09023-12701) or equivalent, connect the return tube assembly to the power steering gear assembly and tighten to 16 ft. lbs. (22 Nm).

15. For 2.4L vehicles, install the tube clamp with the nut and tighten to 87 inch lbs. (9.8 Nm).

16. For 3.3L vehicles, install the tube clamp and heat insulator with the nut and tighten to 87 inch lbs. (9.8 Nm).

17. Install the tube clamp with the bolt and tighten to 87 inch lbs. (9.8 Nm).

18. Install the left and right front stabilizer brackets, as follows:

  a. Install the No. 1 stabilizer bar bush to the stabilizer bar.

  b. Install the No. 1 front stabilizer bracket with the 2 bolts and tighten to 20 ft. lbs. (27 Nm).

19. Install the right and left front stabilizer link assemblies.

  a. Install the left and right tie rod assemblies.

20. Install the front wheel.

21. Align the matchmarks on the steering intermediate shaft assembly and the power steering gear assembly.

  a. Install the bolt and tighten to 26 ft. lbs. (35 Nm).

  b. Tighten bolt B and tighten to 26 ft. lbs. (35 Nm).

  c. Install the No. 2 hole cover to the No. 1 hole cover.

  d. Install the clamp to the No. 1 hole cover and tighten bolt A.

22. Bleed power steering fluid.

23. Inspect for power steering fluid leak.

24. Inspect the front wheel alignment.

25. Inspect the steering wheel center point.

POWER STEERING PUMP

REMOVAL & INSTALLATION

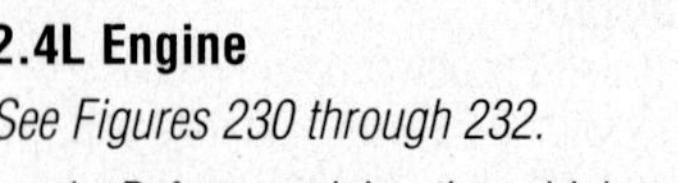

2.4L Engine

See Figures 230 through 232.

1. Before servicing the vehicle, refer to the precautions section.

➡Do not over tighten when using a vise.

➡When installing, coat the parts indicated by arrows with power steering fluid. Take care not to spill power steering fluid on the V-belt.

2. Remove the right front wheel.

3. Drain power steering fluid.

4. Remove the right front fender apron seal.

5. Remove fan and generator V-belt.

6. Remove the clip and disconnect the No. 1 oil reservoir to pump hose.

7. Disconnect the pressure feed tube assembly, as follows:

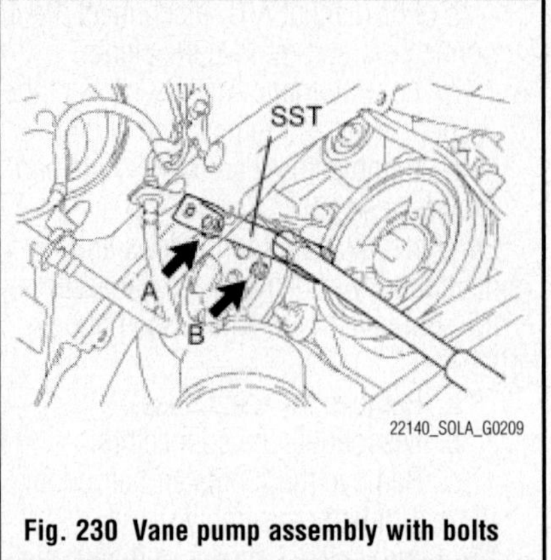

Fig. 230 Vane pump assembly with bolts

  a. Using a wrench (27 mm) to hold the pressure port union, remove the union bolt and the gasket.

8. Remove the vane pump assembly, as follows:

  a. Disconnect the connector from the oil pressure switch.

  b. Using SST (SST: 09249-63010) and a deep socket wrench (14 mm), loosen bolt A.

  c. Remove bolt B and the vane pump assembly.

To install:

9. Temporarily tighten bolt A to the vane pump assembly.

10. Install the vane pump assembly with bolt B and tighten to 27 ft. lbs. (37 Nm).

11. Using SST (SST: 09249-63010) and a deep socket wrench (14 mm), tighten bolt A to 19 ft. lbs. (26 Nm).

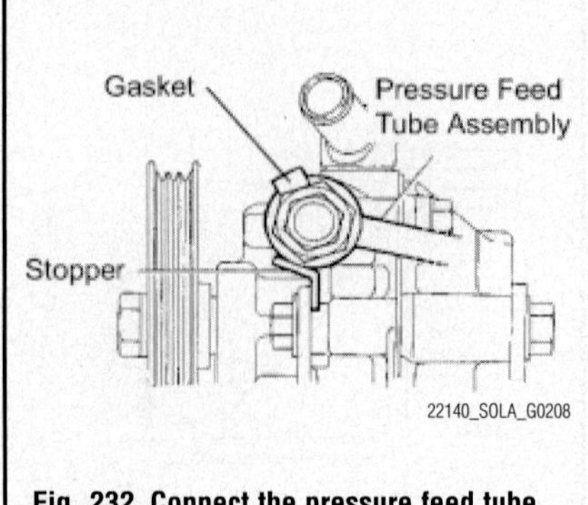

Fig. 232 Connect the pressure feed tube assembly

➡This torque value is effective when SST is parallel to the torque wrench.

➡Use a torque wrench with a fulcrum length of 345 mm (13.58 in.).

12. Connect the connector to the oil pressure switch.

13. Install a new gasket to the pressure feed tube assembly.

14. Connect the pressure feed tube assembly and a new gasket to the vane pump assembly with the union bolt. Using a wrench (27 mm) to hold the pressure port union, torque the union bolt to 38 ft. lbs. (52 Nm).

15. Connect the No. 1 oil reservoir to pump hose with the clip.

16. Install the fan and generator V-belt.

17. Install the right front fender apron seal.

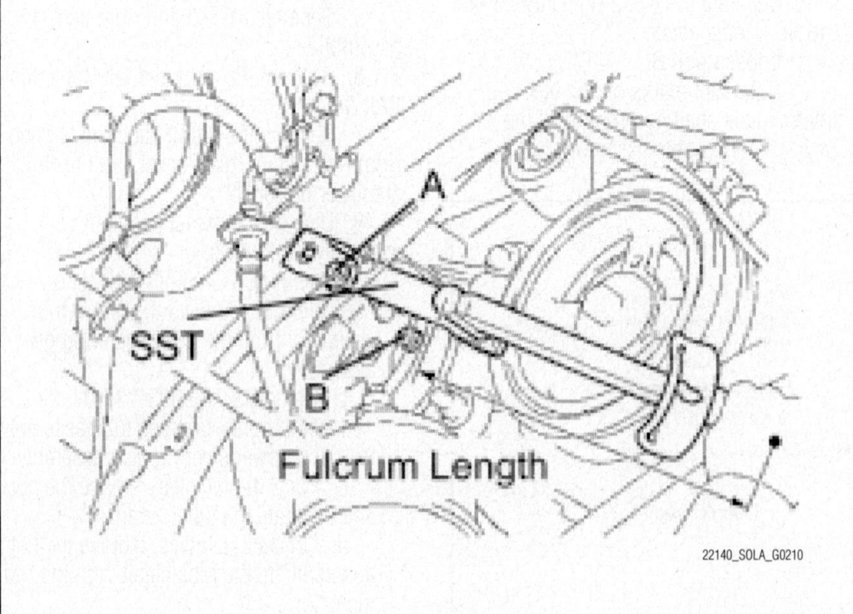

Fig. 231 Install the 2 bolts and the vane pump assembly

18. Install the right front wheel.
19. Bleed the power steering fluid.
20. Inspect for power steering fluid leak.

3.3L Engine

See Figures 233 through 235.

1. Before servicing the vehicle, refer to the precautions section.

➡Do not over tighten when using a vise.

➡When installing, coat the parts indicated by arrows with power steering fluid. Take care not to spill power steering fluid on the V-belt.

2. Remove the right front wheel.
3. Drain power steering fluid.
4. Remove the clip and disconnect the No. 1 oil reservoir to pump hose.

➡Take care not to spill fluid on the V-belt.

5. Remove the oil pressure switch, as follows:
   a. Disconnect the connector.
   b. Remove the oil pressure switch from the union bolt.

➡If the oil pressure switch is dropped or damaged, replace it with a new one.

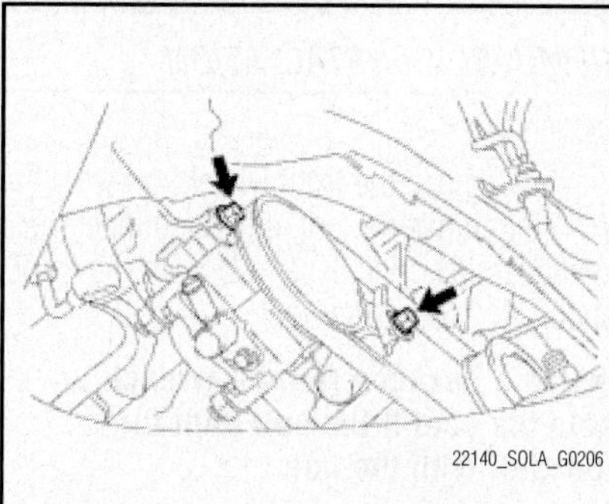

Fig. 233 Remove the 2 bolts and the vane pump assembly

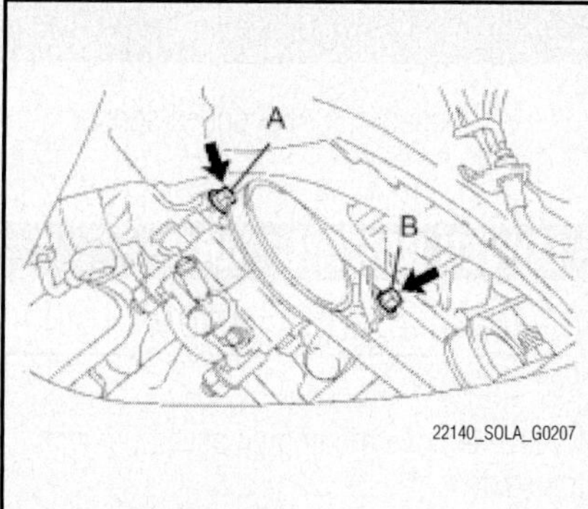

Fig. 234 Install the 2 bolts and the vane pump assembly

6. Using a wrench (24 mm) to keep the pressure port union, remove the union bolt and the gasket and disconnect the pressure feed tube assembly.
7. Loosen the 2 bolts and remove the vane pump V-belt.
8. Remove the 2 bolts and the vane pump assembly.

To install:

9. Temporarily install the vane pump assembly with the 2 bolts.
10. Install the V belt and adjust the tension.
11. Torque bolt A and bolt B to 32 ft. lbs. (43 Nm).
12. Using a wrench (24 mm) to keep the pressure port union, connect the pressure feed tube assembly with the union bolt and a new gasket and tighten to 38 ft. lbs. (52 Nm).

➡Make sure that the stopper of the pressure feed tube assembly contacts the bracket front as shown in the illustration, then tighten the union bolt.

13. Install the oil pressure switch to the union bolt and tighten to 15 ft. lbs. (21 Nm). Be careful to keep oil away from the connector.
14. Connect the connector.
15. Connect the No. 1 oil reservoir to pump hose with the clip.

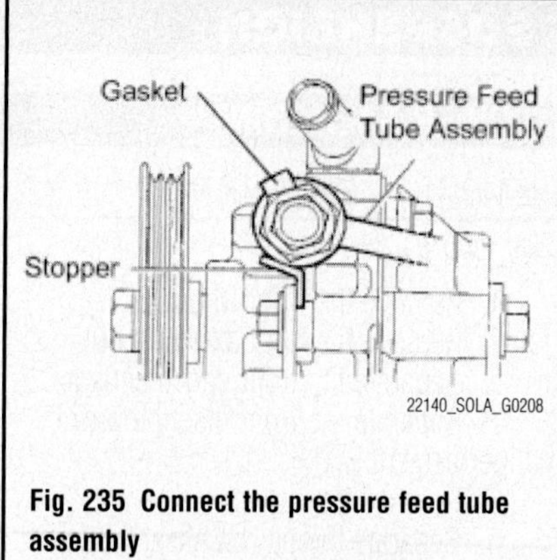

Fig. 235 Connect the pressure feed tube assembly

16. Install the right front wheel.
17. Bleed the power steering fluid.
18. Inspect for fluid leaks.

BLEEDING

1. Before servicing the vehicle, refer to the precautions section.
2. Check the fluid level.
3. Jack up the front of the vehicle and support it with stands.
4. With the engine stopped, turn the wheel slowly from lock to lock several times.
5. Lower the vehicle.
6. Start the engine.
7. Run the engine at idle for a few minutes.
8. With the engine idling, turn the wheel left or right to the full lock position and keep it there for 2 to 3 seconds, then turn the wheel to the opposite full lock position and keep it there for 2 to 3 seconds.
9. Repeat the above steps several times.
10. Stop the engine.
11. Check for foaming or emulsification. If the system has to be bled twice because of foaming or emulsification, check for fluid leaks in the system.
12. Check the fluid level.

LOWER BALL JOINT

REMOVAL & INSTALLATION

See Figure 236.

1. Remove the front wheel.
2. Remove the front axle hub nut.
3. Separate the front speed sensor.
4. Separate the front disc the brake caliper assembly.
5. Remove front disc.
6. Separate the tie rod assembly.
7. Separate the No. 1 front lower suspension arm.
8. Remove the front axle assembly.
9. Remove the front lower ball joint assembly, as follows:
   a. Remove the cotter pin and castle nut.
   b. Using SST (SST: 09628-62011) or equivalent, remove the front lower ball joint assembly.

To install:

10. Installation is the reverse of the removal procedure, noting the following:
   a. Install the front lower ball joint assembly to the steering knuckle with the castle nut and tighten to 91 ft. lbs. (123 Nm). Further tighten the nut up to 60° if the holes for the cotter pin are not aligned.
   b. Inspect and adjust the front wheel alignment.
   c. Inspect the ABS speed sensor signal.

LOWER CONTROL ARM

REMOVAL & INSTALLATION

See Figure 237.

1. Remove the engine assembly with transaxle.
2. Remove the 3 nuts and transverse engine mounting insulator.
3. For manual transaxle, also remove the 4 bolts and bracket from the manual transmission.
4. Remove the 2 bolts on the front side of the lower No. 1 front suspension arm sub-assembly.
5. Remove the bolt and nut on the rear side of the lower No. 1 front suspension arm sub-assembly.
6. Remove the lower No. 1 front suspension arm sub-assembly.
7. Remove the front lower arm bush stopper from the lower No. 1 front suspension arm sub-assembly.

To install:

8. Install the front lower arm bush stopper to the lower No. 1 front suspension arm sub-assembly.
9. Install the 2 bolts on the front side of the lower No. 1 front suspension arm sub-assembly and tighten to 148 ft. lbs. (200 Nm).

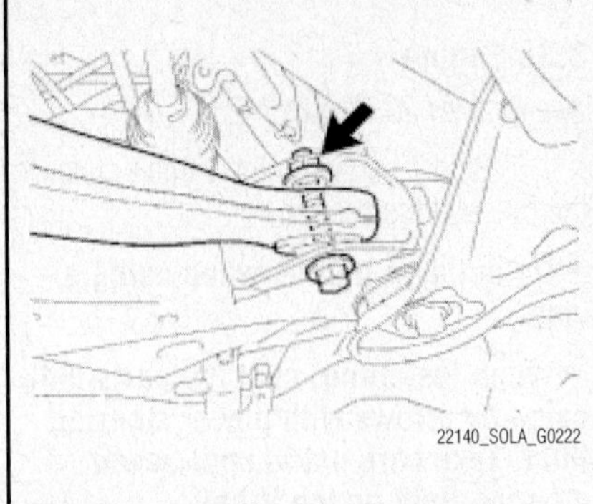

Fig. 237 Lower No. 1 front suspension arm sub-assembly

10. Install the bolt and nut on the rear side of the lower No. 1 front suspension arm sub-assembly and tighten to 152 ft. lbs. (206 Nm).
11. For manual transaxle, install the bracket to the manual transmission with the 4 bolts and tighten to 47 ft. lbs. (64 Nm).
12. Install the transverse engine mounting insulator with the 3 nuts and tighten to 64 ft. lbs. (87 Nm).
13. Install the engine assembly with transaxle.

MACPHERSON STRUT

REMOVAL & INSTALLATION

See Figure 238.

1. Remove the front wheel.
2. Remove the nut and disconnect the front stabilizer link assembly from the front shock absorber assembly.

➡Use a hexagon (6 mm) wrench to hold the stud if the ball joint turns together with the nut.

3. Remove the front shock absorber assembly, as follows:
   a. Loosen the lock nut. If not disassembling the shock absorber it is not necessary to loosen the nut.
   b. Remove the bolt and disconnect the No. 1 front flexible hose and front speed sensor wire harness.
   c. Remove the 2 nuts and 2 bolts on the lower side of the front shock absorber with coil spring. Keep the bolts inserted.
   d. Remove the 3 nuts on the upper side of the front shock absorber with coil spring.
   e. Remove the front shock absorber with coil spring.

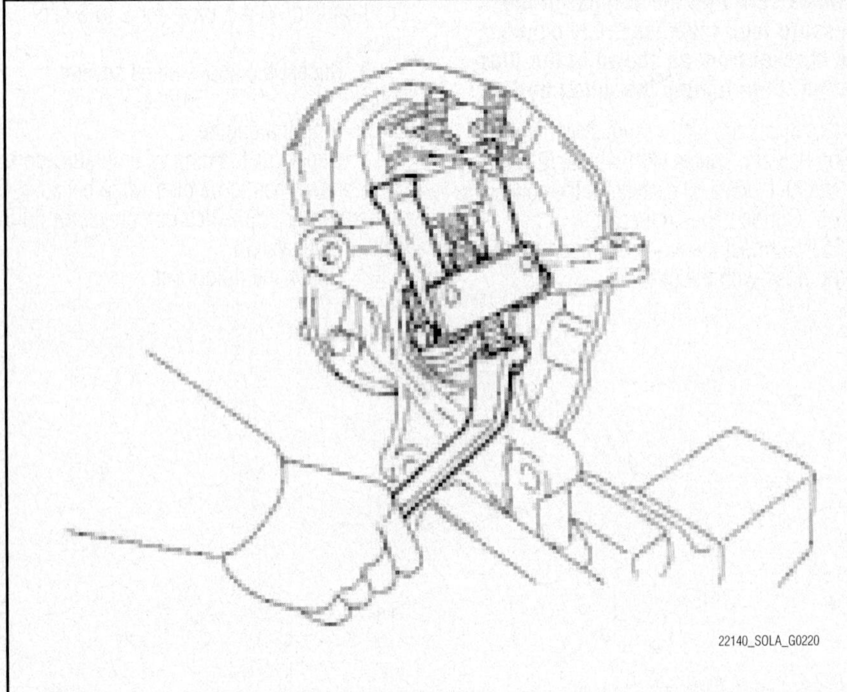

Fig. 236 Remove the front lower ball joint assembly

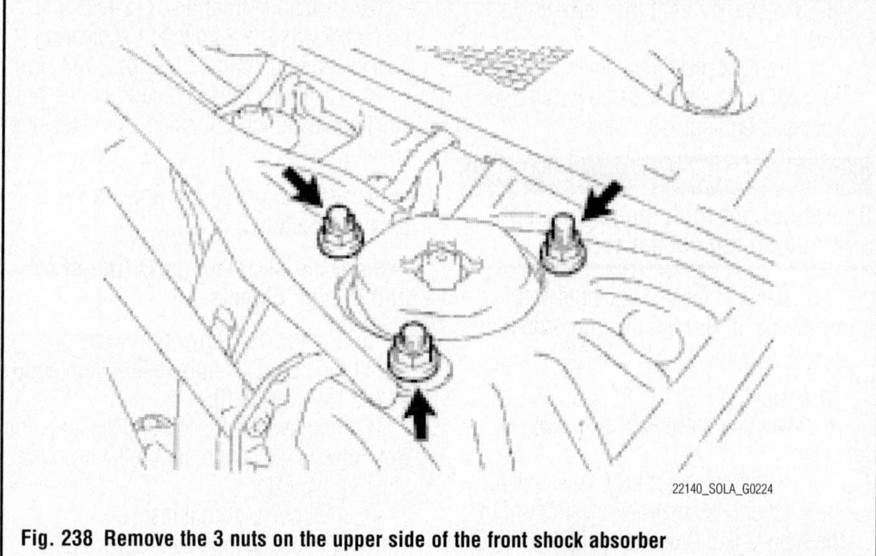

Fig. 238 Remove the 3 nuts on the upper side of the front shock absorber

➡**Be careful not to drop the 2 washers that are installed under the front suspension upper center brace.**

To install:

4. Install the front shock absorber with coil spring.
5. Install the 3 nuts to the upper side of the front shock absorber with coil spring and tighten to 63 ft. lbs. (85 Nm).

➡**Be careful not to drop the collar in the case that there is front suspension upper brace center.**

6. Install the 2 bolts and 2 nuts to the lower side of the front shock absorber with coil spring and tighten to 155 ft. lbs. (210 Nm).

➡**Keep the bolts from rotating and torque the 2 nuts when installing the 2 nuts.**

7. Fully tighten the lock nut to 52 ft. lbs. (70 Nm).
8. Install the No. 1 front flexible hose and front speed sensor with the bolt and tighten to 14 ft. lbs. (19 Nm).
9. Install the front stabilizer link assembly with the nut and tighten to 55 ft. lbs. (74 Nm).

➡**Use a hexagon (6 mm) wrench to hold the stud if the ball joint turns together with the nut.**

10. Install the front wheel.
11. Inspect and adjust front wheel alignment.

STEERING KNUCKLE

REMOVAL & INSTALLATION

See Wheel Hub and Bearing.

STABILIZER BAR

REMOVAL & INSTALLATION

See Figures 239 and 240.

1. Remove the front wheel.
2. Remove the nuts and the right and left front stabilizer link assemblies.

➡**Use a hexagon (6 mm) wrench to hold the stud if the ball joint turns together with the nut.**

3. Remove the bolts and the left and right No. 1 stabilizer brackets.
4. Disconnect tie rod assemblies.
5. Separate steering intermediate shaft assembly.
6. Disconnect steering gear outlet return tube.
7. Disconnect pressure feed tube assembly.

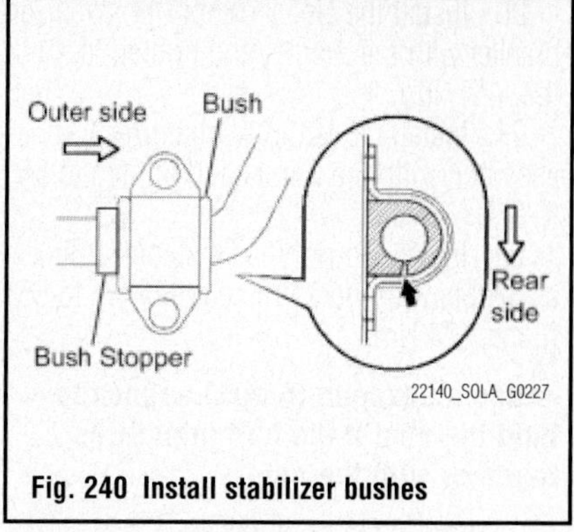

Fig. 240 Install stabilizer bushes

8. Remove rack and pinion power steering gear assembly.
9. Remove the 2 bushes from the stabilizer.
10. Remove the front stabilizer bar from the vehicle.

To install:

11. Install the stabilizer bar front to the vehicle.
12. Install the bushes as to the outer side of each bush stopper on the stabilizer bar.
13. Place the cutout of the stabilizer bushes as facing the rear side as shown in the illustration.
14. Install the rack and pinion power steering gear assembly.
15. Install the pressure feed tube assembly.
16. Install the steering gear outlet return tube.
17. Install the steering intermediate shaft assembly.
18. Install the tie rod assemblies.
19. Install the No. 1 left front stabilizer bracket with the 2 bolts and tighten to 20 ft. lbs. (27 Nm).

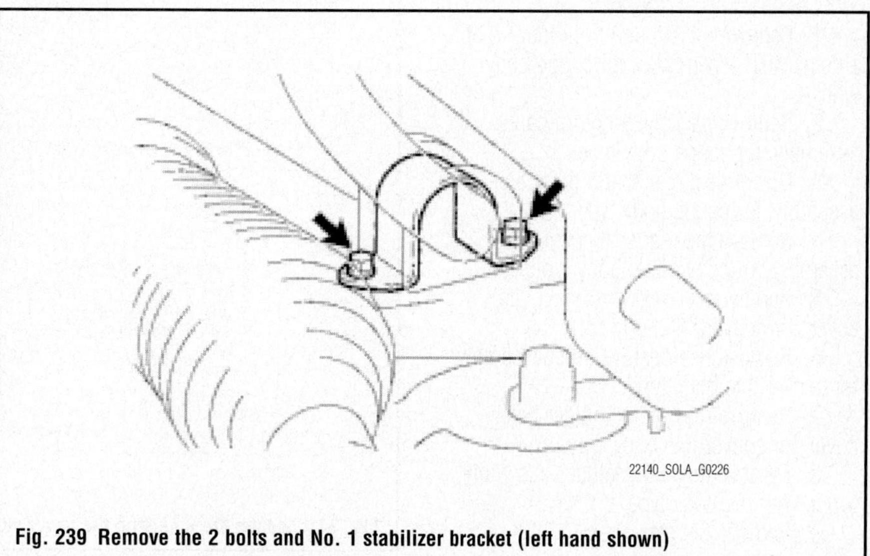

Fig. 239 Remove the 2 bolts and No. 1 stabilizer bracket (left hand shown)

20. Install the No. 1 right front stabilizer bracket with the 2 bolts and tighten to 20 ft. lbs. (27 Nm).
21. Install the left front stabilizer link assembly with the 2 nuts and tighten to 55 ft. lbs. (74 Nm).
22. Install the right front stabilizer link assembly with the 2 nuts and tighten to 55 ft. lbs. (74 Nm).

➡Use a hexagon (6 mm) wrench to hold the stud if the ball joint turns together with the nut.

23. Install the front wheel.
24. Bleed the power steering fluid.
25. Inspect for power steering fluid leak.
26. Inspect and adjust steering wheel center point.
27. Inspect and adjust the front wheel alignment.

WHEEL HUB AND BEARING

REMOVAL & INSTALLATION

1. Remove the front wheel.
2. Remove the front axle shaft nut.
3. Separate the front speed sensor.
4. Remove the 2 bolts and separate the front disc brake caliper assembly from the steering knuckle.
5. Remove the front disc.
6. Separate the tie rod assembly.
7. Separate the lower No. 1 front suspension arm sub-assembly.
8. Remove the front axle assembly, as follows:
   a. Using a plastic hammer, separate the front drive shaft assembly from the front axle hub sub-assembly.

✲✲ WARNING

Be careful not to damage the boot and ABS speed sensor rotor.

   b. Remove the 2 bolts, nuts and steering knuckle with the front axle hub sub-assembly.

To install:

9. Install the front axle assembly, as follows:
   a. Install the front axle assembly to the front shock absorber assembly with the 2 bolts and 2 nuts and tighten to 155 ft. lbs. (210 Nm).

➡Only when reusing the bolts and nuts, apply the small amount of engine oil to the screw part of the nuts. Do not excessively push out the front axle assembly.

✲✲ WARNING

Be careful not to damage the drive shaft boot or speed sensor rotor.

10. Install the lower No. 1 front suspension arm sub-assembly.
11. Install the tie rod assembly.
12. Install the front disc.
13. Install the front disc brake caliper assembly with the 2 bolts to the steering knuckle and tighten to 79 ft. lbs. (107 Nm).
14. Clean the threaded parts on the drive shaft and axle hub nut using a non-residue solvent.

➡Be sure to perform this work for a new drive shaft.

➡Keep the threaded parts free of oil and foreign objects.

15. Using a 30 mm socket wrench, install the front axle hub nut and tighten to 217 ft. lbs. (294 Nm).
16. Remove the 2 bolts and separate the front disc brake caliper assembly from the steering knuckle.
17. Remove the front disc.
18. Inspect bearing backlash.
19. Inspect axle hub deviation.
20. Install the front disc.
21. Install the front disc brake caliper assembly with the 2 bolts to the steering knuckle and tighten to 79 ft. lbs. (107 Nm).
22. Install the front speed sensor.
23. Using a socket wrench (30 mm), install a new axle hub nut and tighten to 217 ft. lbs. (294 Nm).
24. Using a chisel and hammer, stake the axle hub nut.
25. Install the front wheel.
26. Inspect and adjust front wheel alignment.
27. Check ABS speed sensor signal.

SUSPENSION

MACPHERSON STRUTS

REMOVAL & INSTALLATION

See Figures 241 through 243.

1. Remove the left and right inner roof side garnish assemblies, for coupe body type.
2. Remove package tray trim panel assembly, for coupe body type.
3. Disconnect rear seat 3 point type belt assembly, for coupe body type.
4. Remove the rear seat cushion assembly, for convertible body type.
5. Remove rear seat headrest plate cover, for convertible body type.
6. Remove rear seat headrest assembly, for convertible body type.
7. Remove rear seat shoulder belt cover, for convertible body type.
8. Remove the rear seatback assembly, for convertible body type.
9. Remove the rear wheel.

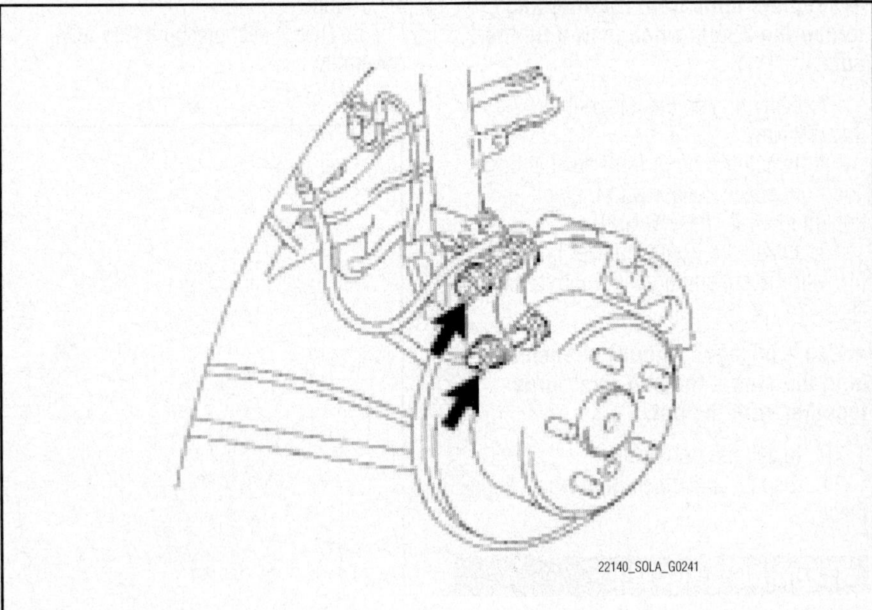

Fig. 241 Loosen the 2 nuts on the lower side of the shock absorber

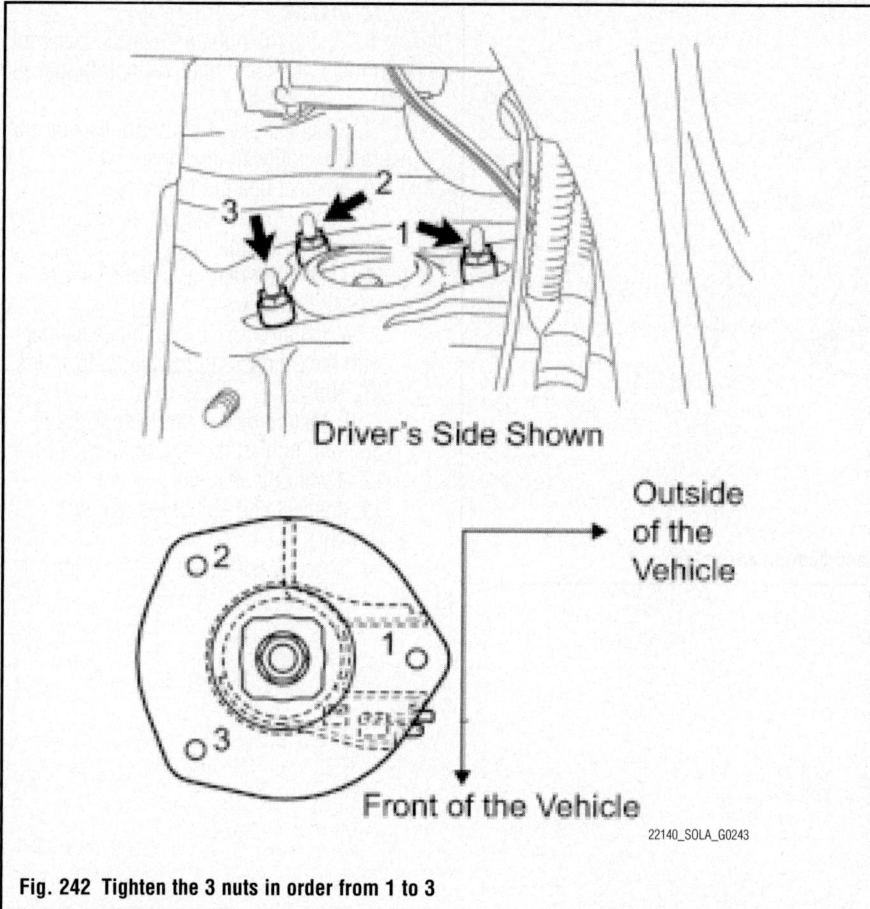

Fig. 242 Tighten the 3 nuts in order from 1 to 3

10. Separate rear stabilizer link assembly, as follows:
   a. Support the rear axle carrier with a jack.
   b. Remove the nut, and disconnect the stabilizer link from the shock absorber.

➡If the ball joint turns together with the nut, use a hexagon wrench (5 mm) to hold the stud.

11. Remove the 2 bolts, and disconnect the flexible hose and skid control sensor wire from the shock absorber.
12. Remove the rear shock absorber with assembly, as follows:
   a. Loosen the 2 nuts on the lower side of the shock absorber. Do not remove the 2 bolts and 2 nuts.
   b. Remove the No. 1 rear suspension support cover.
   c. Loosen the support suspension center nut. Do not remove the nut. It is not necessary to loosen the nut if the shock absorber is not being disassembled.
   d. Remove the 3 nuts.
   e. Lower the rear axle carrier, and remove the 2 nuts and 2 bolts on the lower side of the shock absorber.
   f. Remove the shock absorber with coil spring.

To install:

13. Temporarily install the 3 nuts to the upper side of the rear shock absorber with coil spring.
14. Fully tighten the 3 nuts in order from 1 to 3 to 29 ft. lbs. (39 Nm).
15. Install the 2 bolts and nuts to the shock absorber with coil spring tighten to 133 ft. lbs. (180 Nm).
16. Fully tighten the nut installed on the top of the shock absorber with coil spring and tighten to 41 ft. lbs. (55 Nm). If the shock absorber has not been disassembled, it is not necessary to torque the nut.
17. Install the No. 1 rear suspension support cover.
18. Install the flexible hose and skid control sensor wire with the 2 bolts and tighten bolt A to 14 ft. lbs. (19 Nm), and bolt B to 49 inch lbs. (5.5 Nm).
19. Install the stabilizer link to the shock absorber with the nut and tighten to 29 ft. lbs. (39 Nm).

➡If the ball joint turns together with the nut, use a hexagon wrench (5 mm) to hold the stud.

20. Install the front wheel.
21. Inspect and adjust the rear wheel alignment.
22. Install interior components.

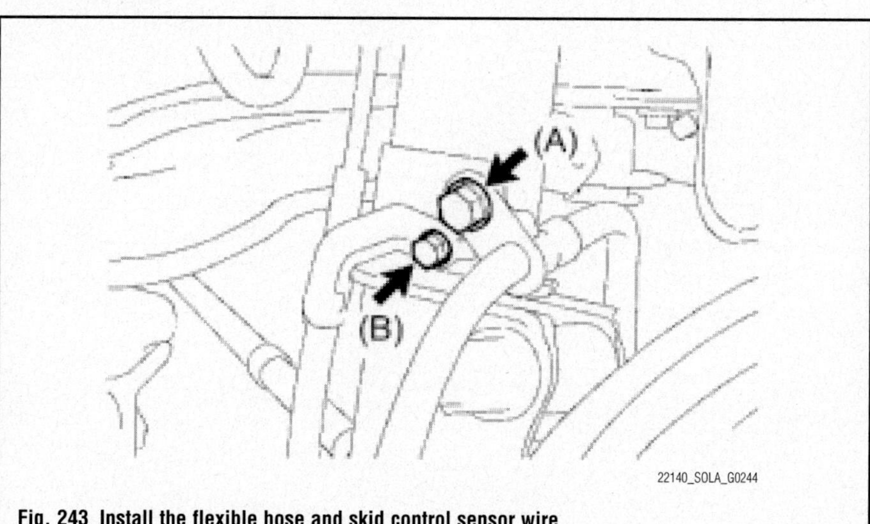

Fig. 243 Install the flexible hose and skid control sensor wire

WHEEL HUB AND BEARING

REMOVAL & INSTALLATION

See Figure 244.

1. Remove the rear wheel.
2. Disconnect the rear disc brake caliper assembly, as follows:
   a. Remove the bolt and separate the flexible hose from the shock absorber.
   b. Remove the 2 bolts and separate the rear disc brake caliper assembly.
3. Remove the rear disc.
4. Disconnect the skid control sensor connector.
5. Remove the 4 bolts and the rear axle hub and bearing assembly.

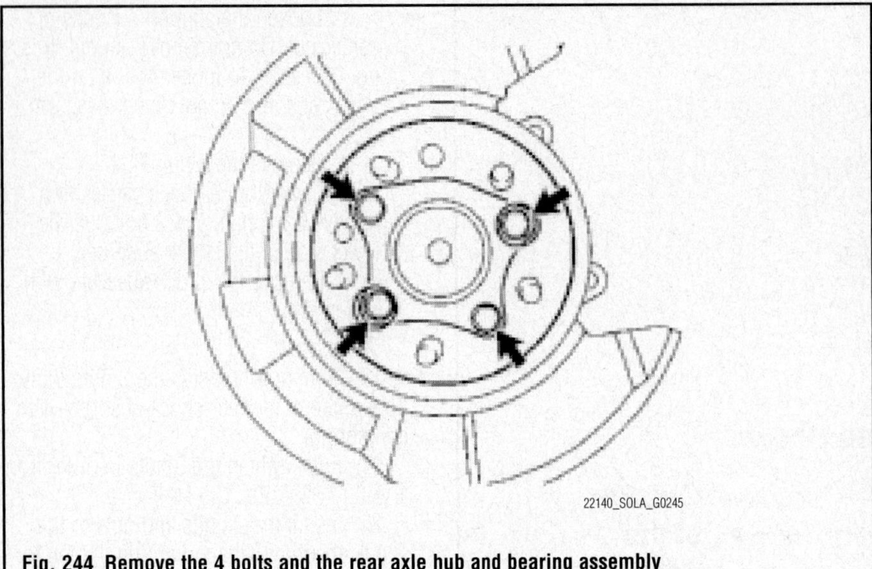

Fig. 244 Remove the 4 bolts and the rear axle hub and bearing assembly

To install:

6. Install the hub and bearing assembly with the 4 bolts and tighten to 59 ft. lbs. (80 Nm).
7. Connect the skid control sensor connector. Do not twist the sensor wire.
8. Inspect bearing backlash.
9. Inspect rear axle hub deviation.
10. Install the rear disc.
11. Install the rear disc brake caliper assembly, as follows:
    a. Install the rear disc brake caliper with the 2 bolts and tighten to 46 ft. lbs. (62 Nm).
    b. Install the flexible hose with the bolt and tighten to 14 ft. lbs. (19 Nm).
12. Install the rear wheel.
13. Inspect and adjust the rear wheel alignment.
14. Check ABS speed sensor signal.

TOYOTA 13

Corolla

SPECIFICATIONS AND MAINTENANCE CHARTS

ENGINE AND VEHICLE IDENTIFICATION

Engine							Model Year	
Code ①	Liters (cc)	Cu. In.	Cyl.	Fuel Sys.	Engine Type	Eng. Mfg.	Code ②	Year
1ZZ-FE	1.8 (1794)	109.5	4	EFI	DOHC	Toyota	7	2008
2ZZ-GE	1.8(1796)	109.6	4	EFI	DOHC	Toyota	8	2008

EFI: Electronic Fuel Injection

DOHC: Double Overhead Camshaft

① 5th digit of VIN

②10th digit of VIN

22140_CORO_C0001

GENERAL ENGINE SPECIFICATIONS

Year	Model	Engine Displacement Liters (cc)	Engine Series (ID/VIN)	Fuel System	Net Horsepower @ rpm	Net Torque @ rpm (ft. lbs.)	Bore x Stroke (in.)	Com-pression Ratio	Oil Pressure @ rpm
2007	Corolla	1.8 (1794)	1ZZ-FE	EFI	126@6000	122@4200	3.11x3.60	10.0:1	4.3
2008	Corolla	1.8 (1794)	1ZZ-FE	EFI	126@6000	122@4200	3.11x3.60	10.0:1	4.3

EFI: Electronic Fuel Injection

22140_CORO_C0002

ENGINE TUNE-UP SPECIFICATIONS

Year	Engine Displacement Liters	Engine ID/VIN	Spark Plug Gap (in.)	Ignition Timing (deg.) ①	Fuel Pump (psi)	Idle Speed (rpm)		Valve Clearance	
						MT	AT	In.	Ex.
2006	1.8	1ZZ-FE	0.043	8-12 BTDC	44-50	650-750	700-800	0.006-0.010	0.010-0.014
	1.8	2ZZ-GE	0.043	8-12 BTDC	44-50	750-850	N/A	0.006-0.010	0.014-0.018
2007	1.8	1ZZ-FE	0.043	8-12 BTDC	44-50	650-750	650-750	0.006-0.010	0.010-0.014
2008	1.8	1ZZ-FE	0.043	8-12 BTDC	44-50	650-750	650-750	0.006-0.010	0.010-0.014

Note: The Vehicle Emission Control Information label often reflects specification changes made during production.

The label figures must be used if they differ from those in this chart.

① With terminal TE1 and E1 connected of DLC1

22140_CORO_C0003

CAPACITIES

Year	Model	Engine Displacement Liters	Engine ID/VIN	Engine Oil with Filter	Transmission (pts.) 5-Spd	6-Spd	Auto.	Drive Axle Front (pts.)	Rear (pts.)	Fuel Tank (gal.)	Cooling System (qts.)
2007	Corolla	1.8	1ZZ-FE	4.4	4.0	NA	3.2	—	—	13.2	7.1
2008	Corolla	1.8	1ZZ-FE	4.4	4.0	NA	3.2	—	—	13.2	7.1

Note: All capacities are approximate. Add fluid gradually and check to be sure a proper fluid level is obtained. Auto trans is a drain and fill capacity.

NA: Not Available

22140_CORO_C0004

VALVE SPECIFICATIONS

Year	Engine Displacement Liters	Engine ID/VIN	Seat Angle (deg.)	Face Angle (deg.)	Spring Test Pressure (lbs. @ in.)	Spring Installed Height (in.)	Stem-to-Guide Clearance (in.) Intake	Exhaust	Stem Diameter (in.) Intake	Exhaust
2007	1.8	1ZZ-FE	45	44.5	35.7-39.5@ 1.323	1.323	0.0010-0.0024	0.0012-0.0025	0.2154-0.2159	0.2152-0.2158
2008	1.8	1ZZ-FE	45	44.5	35.7-39.5@ 1.323	1.323	0.0010-0.0024	0.0012-0.0025	0.2154-0.2159	0.2152-0.2158

① Spring Free length 1.827in. Exhaust 1.821 in.

22140_CORO_C0005

CRANKSHAFT AND CONNECTING ROD SPECIFICATIONS

All measurements are given in inches.

Year	Engine Displacement Liters	Engine ID/VIN	Crankshaft Main Brg. Journal Dia.	Main Brg. Oil Clearance	Shaft End-play	Thrust on No.	Connecting Rod Journal Diameter	Oil Clearance	Side Clearance
2007	1.8	1ZZ-FE	1.8893-1.8898	0.0006-0.0013	0.0008-0.0087	3	1.7320-1.7323	0.0011-0.0024	0.0063-0.0135
2008	1.8	1ZZ-FE	1.8893-1.8898	0.0006-0.0013	0.0008-0.0087	3	1.7320-1.7323	0.0011-0.0024	0.0063-0.0135

22140_CORO_C0007

PISTON AND RING SPECIFICATIONS

All measurements are given in inches.

Year	Engine Displacement Liters	Engine ID/VIN	Piston Clearance	Ring Gap Top Compression	Ring Gap Bottom Compression	Ring Gap Oil Control	Ring Side Clearance Top Compression	Ring Side Clearance Bottom Compression	Ring Side Clearance Oil Control
2007	1.8	1ZZ-FE	0.0033-0.0041	0.0098-0.0138	0.0138-0.0197	0.0059-0.0157	0.0008-0.0028	0.0012-0.0028	0.0012-0.0043
2008	1.8	1ZZ-FE	0.0033-0.0041	0.0098-0.0138	0.0138-0.0197	0.0059-0.0157	0.0008-0.0028	0.0012-0.0028	0.0012-0.0043

22140_CORO_C0006

TORQUE SPECIFICATIONS

All readings in ft. lbs.

Year	Engine Displacement Liters	Engine ID/VIN	Cylinder Head Bolts	Main Bearing Bolts	Rod Bearing Bolts	Crankshaft Damper Bolts	Flywheel Bolts	Manifold Intake	Manifold Exhaust	Spark Plugs	Lug Nuts
2007	1.8	1ZZ-FE	①	②	③	102	65	22	27	13	76
2008	1.8	1ZZ-FE	①	②	③	102	65	22	27	13	76

① Step 1: 36 ft. lbs.
Step 2: 90 degree turn

② Inner 12 point bolts:
Step 1: 16 ft. lbs.
Step 2: 32 ft. lbs.
Step 3: 45 degree turn
Step 4: 45 degree turn
Outer cap bolts: 13 ft. lbs.

③ Step 1: 15 ft. lbs.
Step 2: 90 degree turn

22140_CORO_C0008

WHEEL ALIGNMENT

Year	Model		Caster Range (+/-Deg.)	Caster Preferred Setting (Deg.)	Camber Range (+/-Deg.)	Camber Preferred Setting (Deg.)	Toe-in (Deg.)	Steering Axis Inclination (Deg.)
2007	Corolla	F	0.75	+2.83	0.75	-0.53	0+/-0.2	11.35+/-0.75
		R	—	—	0.5	-1.45	①	—
2008	Corolla	F	0.75	+2.83	0.75	-0.53	0+/-0.2	11.35+/-0.75
		R	—	—	0.5	-1.45	①	—

① For P185/65R15: 0.25+/-0.25 for P195/65R15: 0.26+/-0.26 for 195/55R16: 0.34+/-0.25

22140_CORO_C0009

TIRE, WHEEL AND BALL JOINT SPECIFICATIONS

Year	Model	OEM Tires Standard	OEM Tires Optional	Tire Pressures (psi) Front	Tire Pressures (psi) Rear	Wheel Size	Ball Joint Inspection
2007	Corolla CE	P185/65R15	None	30	30	6.0-JJ	9-43 in. ①
	Corolla S, LE	P195/65R15	None	30	30	6.0-JJ	9-43 in. ①
2008	Corolla CE	P185/65R15	None	30	30	6.0-JJ	9-43 in. ①
	Corolla S, LE	P195/65R15	None	30	30	6.0-JJ	9-43 in. ①

PSI: Pounds Per Square Inch

① Torque required in inch lbs. to rotate ball joint when removed from the knuckle

22140_CORO_C0010

BRAKE SPECIFICATIONS

All measurements in inches unless noted

Year	Model		Brake Disc Original Thickness	Brake Disc Minimum Thickness	Brake Disc Maximum Runout	Brake Drum Diameter Original Inside Diameter	Brake Drum Diameter Max. Wear Limit	Brake Drum Diameter Maximum Machine Diameter	Minimum Lining Thickness	Brake Caliper Bracket Bolts (ft. lbs.)	Brake Caliper Mounting Bolts (ft. lbs.)
2007	Corolla	F	0.984	0.906	0.0020	7.87	—	7.91	0.039	—	25
		R	—	—	—	7.87	—	7.91	0.039	—	—
2008	Corolla	F	0.984	0.906	0.0020	7.87	—	7.91	0.039	—	25
		R	—	—	—	7.87	—	7.91	0.039	—	—

22140_CORO_C0011

SCHEDULED MAINTENANCE INTERVALS
TOYOTA—COROLLA

TO BE SERVICED	TYPE OF SERVICE	VEHICLE MILEAGE INTERVAL (x1000)													
		5	10	15	20	25	30	35	40	45	50	55	60	90	120
Engine oil & filter	R	✓	✓	✓	✓	✓	✓	✓	✓	✓	✓	✓	✓	✓	✓
Drive belts	S/I						✓						✓	✓	✓
Automatic transaxle fluid & filter	S/I						✓						✓	✓	✓
Brake line pipes & hoses	S/I	✓	✓	✓	✓	✓	✓	✓	✓	✓	✓	✓	✓	✓	✓
Brake linings & drums	S/I	✓	✓	✓	✓	✓	✓	✓	✓	✓	✓	✓	✓	✓	✓
Brake pads & discs (front & rear if equipped)	S/I	✓	✓	✓	✓	✓	✓	✓	✓	✓	✓	✓	✓	✓	✓
Cabin air filter	R				✓				✓				✓		✓
Differential oil	S/I						✓						✓	✓	✓
Drive shaft boots	S/I	✓	✓	✓	✓	✓	✓	✓	✓	✓	✓	✓	✓	✓	✓
Drive shaft bolt (tighten)	S/I	✓	✓	✓	✓	✓	✓	✓	✓	✓	✓	✓	✓	✓	✓
Engine coolant	S/I			✓			✓			✓			✓	✓	✓
Manual transaxle oil	S/I						✓						✓	✓	✓
Steering gear housing oil	S/I	✓	✓	✓	✓	✓	✓	✓	✓	✓	✓	✓	✓	✓	✓
Steering linkage	S/I	✓	✓	✓	✓	✓	✓	✓	✓	✓	✓	✓	✓	✓	✓
Air filter	R						✓						✓	✓	✓
Rotate tires	S/I	✓	✓	✓	✓	✓	✓	✓	✓	✓	✓	✓	✓	✓	✓
Spark plugs	R									✓				✓	
Fuel lines & connections	S/I						✓			✓			✓	✓	✓
Fuel tank cap gasket	R									✓				✓	
Charcoal canister	S/I									✓				✓	

R: Replace S/I: Service or Inspect

FREQUENT OPERATION MAINTENANCE (SEVERE SERVICE)

If a vehicle is operated under any of the following conditions it is considered severe service:

- Extremely dusty areas.

- 50% or more of the vehicle operation is in 32°C (90°F) or higher temperatures, or constant operation in temperatures below 0°C (32°F).

- Prolonged idling (vehicle operation in stop and go traffic).

- Frequent short running periods (engine does not warm to normal operating temperatures).

- Police, taxi, delivery usage or trailer towing usage.

Oil & oil filter: change every 5000 miles.

Bolts & nuts on chassis & body: tighten every 5000 miles.

Ball joints & dust covers: service or inspect every 5,000 miles.

Drive shaft boots & except Supra): service or inspect every 12,000 miles.

Steering linkage: service or inspect every 12,000 miles.

Air filter: service or inspect every 5,000 miles.

Exhaust system: service or inspect every 15,000 miles.

Timing belt: replace every 60,000 miles.

22140_CORO_C0012

PRECAUTIONS

Before servicing any vehicle, please be sure to read all of the following precautions, which deal with personal safety, prevention of component damage, and important points to take into consideration when servicing a motor vehicle:

- Never open, service or drain the radiator or cooling system when the engine is hot; serious burns can occur from the steam and hot coolant.
- Observe all applicable safety precautions when working around fuel. Whenever servicing the fuel system, always work in a well-ventilated area. Do not allow fuel spray or vapors to come in contact with a spark, open flame, or excessive heat (a hot drop light, for example). Keep a dry chemical fire extinguisher near the work area. Always keep fuel in a container specifically designed for fuel storage; also, always properly seal fuel containers to avoid the possibility of fire or explosion. Refer to the additional fuel system precautions later in this section.
- Fuel injection systems often remain pressurized, even after the engine has been turned **OFF**. The fuel system pressure must be relieved before disconnecting any fuel lines. Failure to do so may result in fire and/or personal injury.
- Brake fluid often contains polyglycol ethers and polyglycols. Avoid contact with the eyes and wash your hands thoroughly after handling brake fluid. If you do get brake fluid in your eyes, flush your eyes with clean, running water for 15 minutes. If eye irritation persists, or if you have taken brake fluid internally, IMMEDIATELY seek medical assistance.
- The EPA warns that prolonged contact with used engine oil may cause a number of skin disorders, including cancer. You should make every effort to minimize your exposure to used engine oil. Protective gloves should be worn when changing oil. Wash your hands and any other exposed skin areas as soon as possible after exposure to used engine oil. Soap and water, or waterless hand cleaner should be used.
- All new vehicles are now equipped with an air bag system, often referred to as a Supplemental Restraint System (SRS) or Supplemental Inflatable Restraint (SIR) system. The system must be disabled before performing service on or around system components, steering column, instrument panel components, wiring and sensors. Failure to follow safety and disabling procedures could result in accidental air bag deployment, possible personal injury and unnecessary system repairs.
- Always wear safety goggles when working with, or around, the air bag system. When carrying a non-deployed air bag, be sure the bag and trim cover are pointed away from your body. When placing a non-deployed air bag on a work surface, always face the bag and trim cover upward, away from the surface. This will reduce the motion of the module if it is accidentally deployed. Refer to the additional air bag system precautions later in this section.
- Clean, high quality brake fluid from a sealed container is essential to the safe and proper operation of the brake system. You should always buy the correct type of brake fluid for your vehicle. If the brake fluid becomes contaminated, completely flush the system with new fluid. Never reuse any brake fluid. Any brake fluid that is removed from the system should be discarded. Also, do not allow any brake fluid to come in contact with a painted surface; it will damage the paint.
- Never operate the engine without the proper amount and type of engine oil; doing so WILL result in severe engine damage.
- Timing belt maintenance is extremely important. Many models utilize an interference-type, non-freewheeling engine. If the timing belt breaks, the valves in the cylinder head may strike the pistons, causing potentially serious (also time-consuming and expensive) engine damage. Refer to the maintenance interval charts for the recommended replacement interval for the timing belt, and to the timing belt section for belt replacement and inspection.
- Disconnecting the negative battery cable on some vehicles may interfere with the functions of the on-board computer system(s) and may require the computer to undergo a relearning process once the negative battery cable is reconnected.
- When servicing drum brakes, only disassemble and assemble one side at a time, leaving the remaining side intact for reference.
- Only an MVAC-trained, EPA-certified automotive technician should service the air conditioning system or its components.

BRAKES

ANTI-LOCK BRAKE SYSTEM (ABS)

GENERAL INFORMATION

When wheel slip is detected during a brake application, the ABS enters antilock mode. During antilock braking, hydraulic pressure in the individual wheel circuits is controlled to prevent any wheel from slipping. A separate hydraulic line and specific solenoid valves are provided for each wheel. The ABS can decrease, hold, or increase hydraulic pressure to each wheel brake. The ABS cannot, however, increase ==hydraulic pressure above the amount which is transmitted by the master cylinder during braking.

During antilock braking, a series of rapid pulsations is felt in the brake pedal. These pulsations are caused by the rapid changes in position of the individual solenoid valves as the EBCM responds to wheel speed sensor inputs and attempts to prevent wheel slip. These pedal pulsations are present only during antilock braking and stop when normal braking is resumed or when the vehicle comes to a stop. A ticking or popping noise may also be heard as the solenoid valves cycle rapidly. During antilock braking on dry pavement, intermittent chirping noises may be heard as the tires approach slipping. These noises and pedal pulsations are considered normal during antilock operation.

Vehicles equipped with ABS may be stopped by applying normal force to the brake pedal. Brake pedal operation during normal braking is no different than that of previous non-ABS systems. Maintaining a constant force on the brake pedal provides the shortest stopping distance while maintaining vehicle stability.

BLEEDING PROCEDURE

➡ **After bleeding the air from the brake system, if the height or feel of the brake pedal cannot be obtained, perform air bleeding in the brake actuator assembly with the Scan Tool by following the procedures below.**

1. Depress the brake pedal more than 20 times with the engine off.
2. Connect the Scan Tool to the DLC3, and turn the ignition switch to the ON position.
3. Select "AIR BLEEDING" on the Scan Tool.
4. Bleed the air out of the brake line as usual when "Step 1: Increase" appears on the Scan Tool display.

➡ **Bleed the air by following the steps displayed on the Scan Tool. Make sure that the brake fluid in the master cylinder reservoir tank does not become empty.**

a. Connect the vinyl tube to either one of the bleeder plugs.
b. Depress the brake pedal several times, then loosen the bleeder plug connected to the vinyl tube with the pedal depressed.
c. When fluid stops coming out, tighten the bleeder plug and release the brake pedal.
d. Repeat this procedure until all air in the fluid is completely bled out.
e. Tighten the bleeder plug completely.
f. Repeat the above procedures for each wheel to bleed the air out of the brake line.

5. Bleed the air out of the suction line when "Step 2: Inhalation" appears on the Scan Tool display.

➡ **Bleed the air by following the steps displayed on the Scan Tool. Make sure that the brake fluid in the master cylinder reservoir tank does not become empty.**

a. Connect the vinyl tube to the bleeder plug at the right front wheel or the right rear wheel and loosen the bleeder plug.
b. Operate the brake actuator assembly to bleed the air using the Scan Tool.

➡ **At this time, be sure to release the brake pedal.**

➡ **This operation stops automatically after 4 seconds.**

c. Check if the operation has stopped by referring to the Scan Tool display.
d. Repeat this until all air in the fluid is completely bled out.
e. Tighten the bleeder plug.
f. Repeat the above procedures for the other wheels to bleed the air out of the brake line.

6. Bleed the air out of the pressure reduction line when "Step 3: Decrease" appears on the Scan Tool display.

➡ **Bleed the air by following the steps displayed on the Scan Tool. Make sure that the brake fluid in the master cylinder reservoir tank does not become empty.**

a. Connect a vinyl tube to either one of the bleeder plugs.
b. Loosen the bleeder plug.
c. Using the Scan Tool, operate the brake actuator assembly, completely depress the brake pedal and keep it.

➡ **During this procedure, the pedal will feel heavy, but completely depress it so that the brake fluid comes out from the bleeder plug. Be sure to keep depressing the brake pedal. Do not depress and release the pedal repeatedly.**

➡ **The operation stops automatically after 4 seconds. When performing this procedure continuously, set an interval of at least 20 seconds. When the operation is complete, the brake pedal goes down slightly. This is a normal phenomenon caused when the solenoid opens.**

d. Tighten the bleeder plug, then release the brake pedal.
e. Repeat this until all the air in the fluid is completely bled out.
f. Tighten the bleeder plug.
g. Repeat the above procedures for the other wheels to bleed the air out of the brake line.

7. Bleed the air out of the brake line as usual again when "Step 4: Increase" appears on the Scan Tool display.

➡ **Bleed the air by following the steps displayed on the Scan Tool. Make sure that the brake fluid in the master cylinder reservoir tank does not become empty.**

a. Connect the vinyl tube to either one of the bleeder plugs.
b. Depress the brake pedal several times, then loosen the bleeder plug connected to the vinyl tube with the pedal depressed.
c. When fluid stops coming out, tighten the bleeder plug, then release the brake pedal.
d. Repeat this until all the air in the fluid is completely bled out.
e. Tighten the bleeder plug.
f. Repeat the above procedures for the other wheels to bleed the air out of the brake line.
g. Make sure that the air bleeding is complete by referring to the Scan Tool display.
h. Check the fluid level and add fluid if necessary.

BLEEDING THE ABS SYSTEM

➡ **In most circumstances a base brake bleed is all that is required for most component replacements (such as wheel cylinders, calipers, brake tubes, and master cylinder) except for Brake Pressure Modulator Valve (BPMV) replacement.**

➡ **The following automated ABS bleed procedure is required when one of the following occur:**

- Manual bleeding at the wheel cylinders does not achieve the desired pedal height or feel.
- BPMV replacement
- Extreme loss of brake fluid has occurred.
- Air ingestion is suspected.

1. If none of the above conditions apply, use standard bleed procedures.
2. The auto bleed procedure is used on BOSH 5.3 equipped vehicles. This procedure uses a scan tool to cycle the system solenoid valves and run the pump in order to purge the air from the secondary circuits. These secondary circuits are normally closed off, and are only opened during system initialization at vehicle start up and during ABS operation. The automated bleed procedure opens these secondary circuits and allows any air trapped inside the BPMV to flow out toward the wheel cylinders or calipers where it can be purged out of the system.
3. Inspect the battery for full charge, repair the battery and charging system, as necessary.
4. Connect a scan tool to the data link connector (DLC) and select current and history DTCs. Repair any DTCs prior to performing the ABS bleed procedure.

Inspect for visual damage and leaks and repair, as needed.

5. Raise and vehicle on a suitable support.
6. Turn the ignition switch to the **OFF**-position.
7. Remove all 4 tires.
8. Connect the pressure bleeding tool according to the manufacturer's instructions.
9. Turn the ignition switch to **RUN** position, engine off.
10. Connect a scan tool and establish communications with the ABS system.
11. Pressurize the bleeding tool to 30–35 psi (206–241 kPa).
12. Performing the Automated Bleed Procedure

➡The Auto Bleed Procedure may be terminated at any time during the process by pressing the EXIT button. No further Scan Tool prompts pertaining to the Auto Bleed procedure will be given. After exiting the bleed procedure, relieve bleed pressure and disconnect bleed equipment per manufacturer's instructions. Failure to properly relieve pressure may result in spilled brake fluid causing damage to components and painted surfaces.

13. With the pressure bleeding tool at 30–35 psi (206–241 kPa), and all bleeder screws in closed position, select Automated Bleed Procedure on the scan tool and follow the instructions.
14. The first part of the automated bleed procedure will cycle the pump and front release valves for one minute. After the cycling has stopped the scan tool will enter a cool down mode and display a 3 minute timer. The auto bleed will not continue until this timer expires, and cannot be overridden.
15. During the next step, the scan tool will request the technician to open one of the bleeder screws. The scan tool will then cycle the respective release valve and pump motor for one minute.
16. The scan tool will repeat step 3 for the remaining bleeder screws.
17. With the bleeder tool still attached to the vehicle and maintaining 30–35 psi (206–241 kPa), the scan tool will instruct the technician to independently open each bleeder screw for approximately 20 seconds. This should allow any remaining air to be purged from the brake lines.
18. When the automated bleed procedure is completed the scan tool will display the appropriate message.
19. Install all 4 tires.
20. Remove pressure from the pressure bleeding tool and then disconnect the tool from the vehicle.
21. Depress the brake pedal to gage pedal height and feel. Repeat steps 1-8 until the pedal is acceptable.
22. Remove the scan tool from the DLC connector.
23. Lower the vehicle.
24. Inspect the brake fluid level in master cylinder.
25. Road test the vehicle while making sure the brake pedal remains high and firm.
26. If the vehicle is equipped with a traction control system (TCS), the scan tool will cycle both the ABS and the TCS solenoid valves. This bleed procedure is the same as above.

BRAKES — FRONT DISC BRAKES

BRAKE CALIPER

REMOVAL & INSTALLATION

See Figure 1.

1. Before servicing the vehicle, refer to the precautions in the beginning of this section.
2. Remove the wheels.
3. Disconnect the brake hose from the caliper.
4. Remove the bolts that attach the caliper to the torque plate. If applicable, hold the flats of the sliding pin with a wrench while loosening the caliper attaching bolts.
5. Lift up and remove the caliper assembly.

To install:

6. Install the caliper and loosely install the bolts.
7. Hold the flats of the sliding pin with a wrench, then tighten the bolts. Tighten to 25 ft. lbs. (34 Nm).
8. Connect the brake hose to the caliper, using 2 new washers.
9. Fill the brake system to the proper level and bleed the brake system.
10. Add brake fluid to the reservoir to fill to the correct level.
11. Lower the vehicle to the ground.

DISC BRAKE PADS

REMOVAL & INSTALLATION

1. Before servicing the vehicle, refer to the precautions in the beginning of this section.
2. Remove the wheels.
3. Loosen and remove the caliper mounting bolts, then remove the caliper assembly, without disconnecting the brake line. Position it aside.
4. Slide out the old brake pads along with any anti-squeal shims, springs, pad wear indicators and pad support plates.

To install:

5. Install the pad support plates into the torque plate.
6. Install the pad wear indicators onto the pads. Be sure the arrow on the indicator plate is pointing in the direction of rotation.
7. Install the anti-squeal shims on the outside of each pad and then install the pad assemblies into the torque plate.
8. Compress the caliper piston into the bore.
9. Position the caliper back down over the pads.
10. Install and tighten the caliper mounting bolts.
11. Install the wheels. Check the brake fluid level.

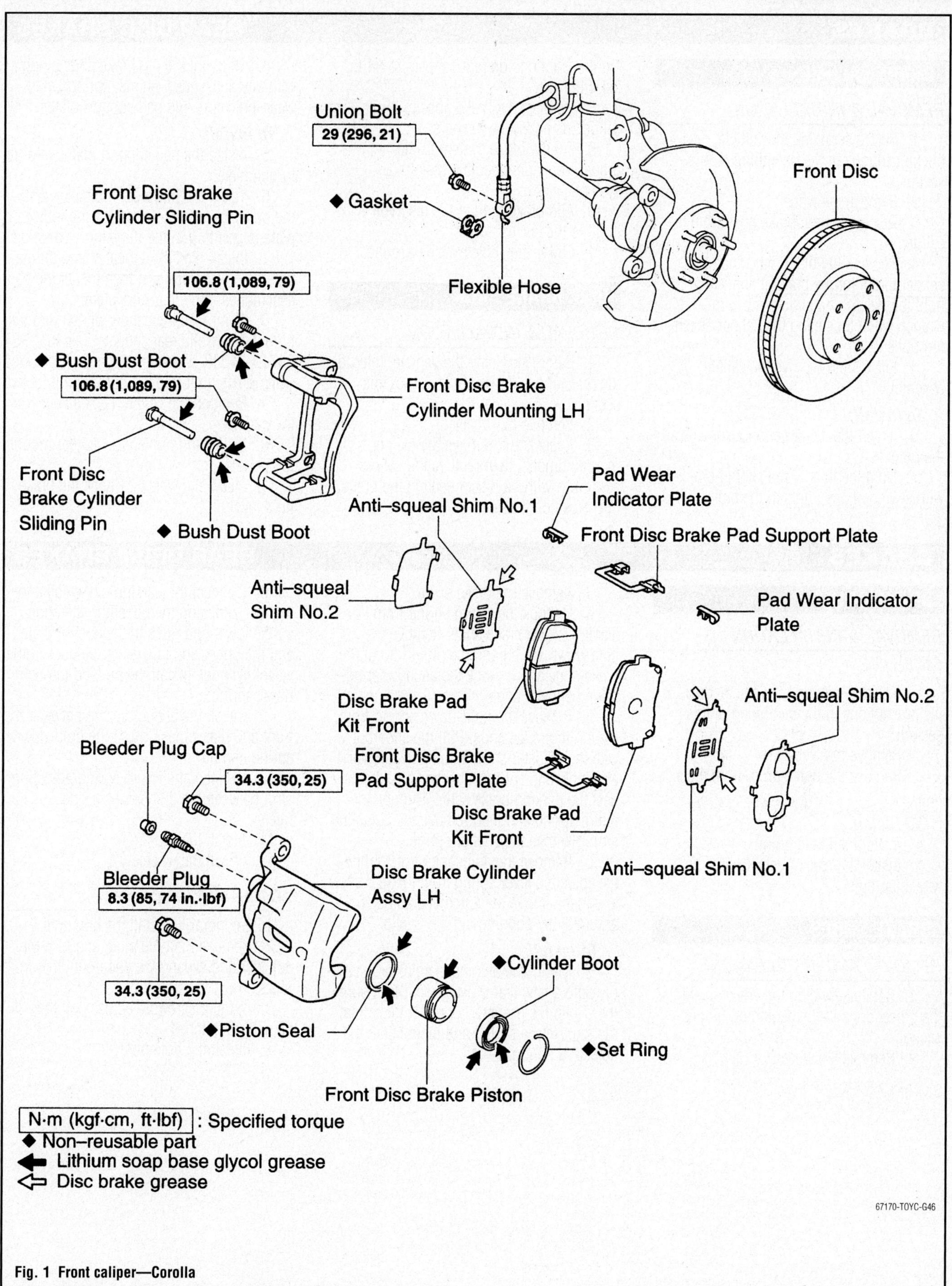

Fig. 1 Front caliper—Corolla

BRAKES — REAR DISC BRAKES

BRAKE CALIPER

REMOVAL & INSTALLATION

1. Before servicing the vehicle, refer to the precautions in the beginning of this section.
2. Remove the wheels.
3. Disconnect the brake hose from the caliper.
4. Remove the bolts that attach the caliper to the torque plate. If applicable, hold the flats of the sliding pin with a wrench while loosening the caliper attaching bolts.
5. Lift up and remove the caliper assembly.

To install:

6. Install the caliper and loosely install the bolts.
7. Hold the flats of the sliding pin with a wrench, then tighten the bolts. Tighten the bolts on rear calipers to 34 ft. lbs. (46 Nm).
8. Connect the brake hose to the caliper, using 2 new washers.
9. Fill the brake system to the proper level and bleed the brake system.
10. Add brake fluid to the reservoir to fill to the correct level.
11. Lower the vehicle to the ground.

DISC BRAKE PADS

REMOVAL & INSTALLATION

1. Before servicing the vehicle, refer to the precautions in the beginning of this section.
2. Remove the wheels.
3. Loosen and remove the caliper mounting bolts, then remove the caliper assembly, without disconnecting the brake line. Position it aside.
4. Slide out the old brake pads along with any anti-squeal shims, springs, pad wear indicators and pad support plates.

To install:

5. Install the pad support plates into the torque plate.
6. Install the pad wear indicators onto the pads. Be sure the arrow on the indicator plate is pointing in the direction of rotation.
7. Install the anti-squeal shims on the outside of each pad and then install the pad assemblies into the torque plate.
8. Compress the caliper piston into the bore. For Corolla rear calipers, use tool SST 09719-14020, to rotate the piston clockwise while pressing it into the bore until it locks.
9. Position the caliper back down over the pads.
10. Install and tighten the caliper mounting bolts.
11. Install the wheels. Check the brake fluid level.

BRAKES — REAR DRUM BRAKES

BRAKE DRUM

REMOVAL & INSTALLATION

See Figure 2.

1. Before servicing the vehicle, refer to the precautions in the beginning of this section.
2. Remove the wheels.
3. Remove the brake drum from the axle hub.

To install:

4. Install the brake drum.
5. Install the rear wheels, tighten the wheel lug nuts.

BRAKE SHOES

REMOVAL AND INSTALLATION

1. Before servicing the vehicle, refer to the precautions in the beginning of this section.
2. Remove the wheels.
3. Remove the brake drum.
4. Unhook the return spring from the leading (front) brake shoe. Remove the hold-down spring and the pin. Pull out the brake shoe and unhook the anchor spring from the lower edge.
5. Remove the hold-down spring from the trailing (rear) shoe. Pull the shoe out with the adjuster strut, automatic adjuster assembly and springs attached and disconnect the parking brake cable. Unhook the return spring and then remove the adjusting strut. Remove the anchor spring.
6. Remove the adjusting strut. Unhook the adjusting lever spring from the rear shoe and then remove the automatic adjuster assembly by popping out the C-clip.

To install:

7. Mount the automatic adjuster assembly onto a new rear brake shoe. Make sure the C-clip fits properly. Connect the adjusting strut/return spring and then install the adjusting spring.
8. Connect the parking brake cable to the rear shoe and then position the shoe so the lower end rides in the anchor plate and the upper end is against the boot in the wheel cylinder. Install the pin and the hold-down spring.
9. Install the anchor spring between the front and rear shoes. Install the hold-down spring and pin.
10. Connect the return spring/adjusting strut between the 2 shoes so it rides freely.
11. Install the drum.
12. Install the wheel.

ADJUSTMENT

1. Temporarily install the hub nuts.
2. Remove the hole plug, and turn the adjuster and expand the shoe until the drum locks.
3. Using a screwdriver, back off the adjuster 8 notches.
4. Install the hole plug.

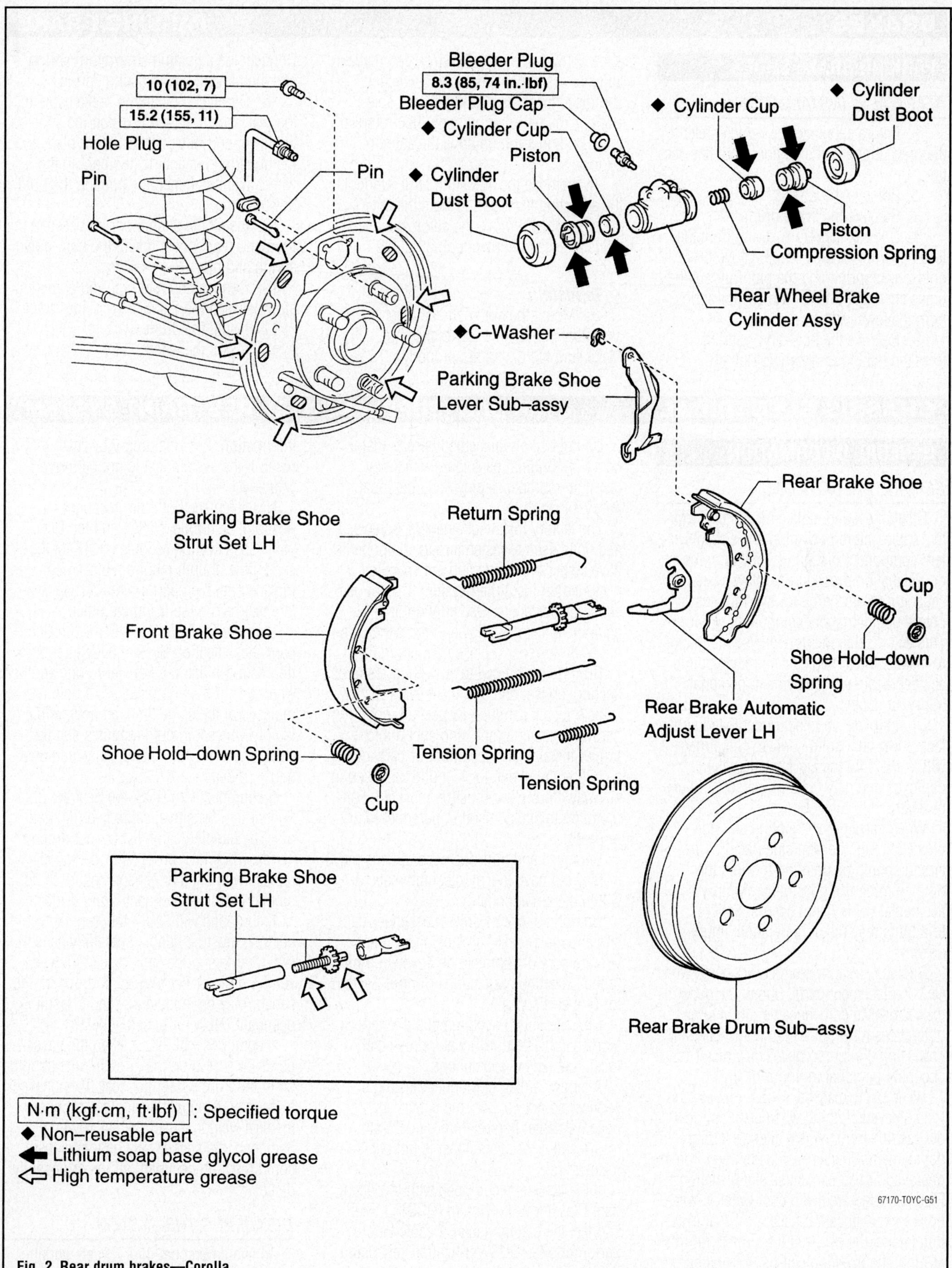

Fig. 2 Rear drum brakes—Corolla

PARKING BRAKE SHOES

REMOVAL & INSTALLATION

1. Before servicing the vehicle, refer to the precautions in the beginning of this section.
2. Remove the wheels.
3. Remove the brake drum.
4. Unhook the return spring from the leading (front) brake shoe. Remove the hold-down spring and the pin. Pull out the brake shoe and unhook the anchor spring from the lower edge.
5. Remove the hold-down spring from the trailing (rear) shoe. Pull the shoe out with the adjuster strut, automatic adjuster assembly and springs attached and disconnect the parking brake cable. Unhook the return spring and then remove the adjusting strut. Remove the anchor spring.
6. Remove the adjusting strut. Unhook the adjusting lever spring from the rear shoe and then remove the automatic adjuster assembly by popping out the C-clip.

To install:

7. Mount the automatic adjuster assembly onto a new rear brake shoe. Make sure the C-clip fits properly. Connect the adjusting strut/return spring and then install the adjusting spring.
8. Connect the parking brake cable to the rear shoe and then position the shoe so the lower end rides in the anchor plate and the upper end is against the boot in the wheel cylinder. Install the pin and the hold-down spring.
9. Install the anchor spring between the front and rear shoes. Install the hold-down spring and pin.
10. Connect the return spring/adjusting strut between the 2 shoes so it rides freely.
11. Install the drum.
12. Install the wheel.

GENERAL INFORMATION

SERVICE PRECAUTIONS

Disconnect and isolate the battery negative cable before beginning any airbag system component diagnosis, testing, removal, or installation procedures. Allow system capacitor to discharge for two minutes before beginning any component service. This will disable the airbag system. Failure to disable the airbag system may result in accidental airbag deployment, personal injury, or death.

Do not place an intact undeployed airbag face down on a solid surface. The airbag will propel into the air if accidentally deployed and may result in personal injury or death.

When carrying or handling an undeployed airbag, the trim side (face) of the airbag should be pointing towards the body to minimize possibility of injury if accidental deployment occurs. Failure to do this may result in personal injury or death.

Replace airbag system components with OEM replacement parts. Substitute parts may appear interchangeable, but internal differences may result in inferior occupant protection. Failure to do so may result in occupant personal injury or death.

Wear safety glasses, rubber gloves, and long sleeved clothing when cleaning powder residue from vehicle after an airbag deployment. Powder residue emitted from a deployed airbag can cause skin irritation. Flush affected area with cool water if irritation is experienced. If nasal or throat irritation is experienced, exit the vehicle for fresh air until the irritation ceases. If irritation continues, see a physician.

Do not use a replacement airbag that is not in the original packaging. This may result in improper deployment, personal injury, or death.

The factory installed fasteners, screws and bolts used to fasten airbag components have a special coating and are specifically designed for the airbag system. Do not use substitute fasteners. Use only original equipment fasteners listed in the parts catalog when fastener replacement is required.

During, and following, any child restraint anchor service, due to impact event or vehicle repair, carefully inspect all mounting hardware, tether straps, and anchors for proper installation, operation, or damage. If a child restraint anchor is found damaged in any way, the anchor must be replaced. Failure to do this may result in personal injury or death.

Deployed and non-deployed airbags may or may not have live pyrotechnic material within the airbag inflator.

Do not dispose of driver/passenger/curtain airbags or seat belt tensioners unless you are sure of complete deployment. Refer to the Hazardous Substance Control System for proper disposal.

Dispose of deployed airbags and tensioners consistent with state, provincial, local, and federal regulations.

After any airbag component testing or service, do not connect the battery negative cable. Personal injury or death may result if the system test is not performed first.

If the vehicle is equipped with the Occupant Classification System (OCS), do not connect the battery negative cable before performing the OCS Verification Test using the scan tool and the appropriate diagnostic information. Personal injury or death may result if the system test is not performed properly.

Never replace both the Occupant Restraint Controller (ORC) and the Occupant Classification Module (OCM) at the same time. If both require replacement, replace one, then perform the Airbag System test before replacing the other.

Both the ORC and the OCM store Occupant Classification System (OCS) calibration data, which they transfer to one another when one of them is replaced. If both are replaced at the same time, an irreversible fault will be set in both modules and the OCS may malfunction and cause personal injury or death.

If equipped with OCS, the Seat Weight Sensor is a sensitive, calibrated unit and must be handled carefully. Do not drop or handle roughly. If dropped or damaged, replace with another sensor. Failure to do so may result in occupant injury or death.

If equipped with OCS, the front passenger seat must be handled carefully as well. When removing the seat, be careful when setting on floor not to drop. If dropped, the sensor may be inoperative, could result in occupant injury, or possibly death.

If equipped with OCS, when the passenger front seat is on the floor, no one should sit in the front passenger seat. This uneven force may damage the sensing ability of the seat weight sensors. If sat on and damaged, the sensor may be inoperative, could result in occupant injury, or possibly death.

DISARMING THE SYSTEM

To avoid personal injury when working on vehicles equipped with an air bag, the negative battery cable must be disconnected and at least 90 seconds must elapse before working on the system. Failure to do so may result in deployment of the air bag.

ARMING THE SYSTEM

After vehicle service is completed, reattach the battery cables (positive cable first!) to rearm the air bag system.

CLOCKSPRING CENTERING

1. Check that the ignition switch is off.
2. Check that the battery negative (-) terminal is disconnected.
3. Rotate the spiral cable counterclockwise slowly by hand until it feels firm.

➡Do not turn the spiral cable by the airbag wire harness.

4. Rotate the spiral cable clockwise approximately 2.5 turns to align the marks.

➡Do not turn the spiral cable by the airbag wire harness. The spiral cable will rotate approximately 2.5 turns to both the left and right from the center.

DRIVETRAIN

AUTOMATIC TRANSAXLE ASSEMBLY

REMOVAL & INSTALLATION

See Figures 3 and 4.

1. Before servicing the vehicle, refer to the precautions in the beginning of this section.
2. Drain the transaxle fluid.
3. Remove or disconnect the following:
   - Negative battery cable. On vehicles equipped with an air bag, wait at least 90 seconds before proceeding
   - Negative battery cable from the transaxle
   - Transaxle level gauge
   - Reservoir tank and air cleaner assembly
   - Throttle cable from the bracket
   - Engine left mounting upper side bolts
   - Engine left mounting stay
   - Ground cable from the transaxle
   - Wiring harness clamp and throttle cable clamp
   - Undercovers
   - Left and right halfshafts
   - Front exhaust pipe
   - Engine support fixture
   - Mounting insulator
   - Starter
   - Vehicle speed sensor connector
   - Solenoid connector and park/neutral position switch connector. Remove the wiring harness clamps
   - Nut from the manual shift lever, then the control cable from the bracket by removing the clip
   - Oil cooler hoses
   - Transaxle filler tube
   - Converter cover
   - Torque converter bolts
   - Transaxle mounting bolts
   - Transaxle

To install:

4. Install or connect the following:
   - Transaxle. Tighten the bolts as follows: Bolt A: 47 ft. lbs. (64 Nm); Bolt B: 34 ft. lbs. (46 Nm); Bolt C: 17 ft. lbs. (23 Nm)
   - Torque converter bolts to the transaxle. Bolts: 20 ft. lbs. (28 Nm)
   - Torque converter cover.
   - Transaxle filler pipe
   - Oil cooler hoses and replace the clips to their original positions
   - Control cable for the transaxle to the bracket and install the clip
   - Control cable to the manual shaft lever by installing the nut
   - Solenoid connector and park/neutral position switch connector. Connect the wiring to the clamps
   - Vehicle speed sensor wiring
   - Starter. Bolt: 29 ft. lbs. (39 Nm)
   - Mounting insulator. Bolt A: 38 ft. lbs. (52 Nm); Bolt and nut B: 59 ft. lbs. (80 Nm)
   - Front exhaust pipe
   - Left and right halfshafts nut: 159 ft. lbs. (49 Nm)
   - Undercovers
   - Transaxle mounting bolts to the transaxle side
   - Wiring harness clamp and throttle cable clamp
   - Ground cable. Bolt: 7 ft. lbs. (10 Nm)
   - Engine left mounting upper side bolts and stay. Bolts: 38 ft. lbs. (52 Nm)
   - Throttle cable
   - Air cleaner and the reservoir tank
   - Transaxle level gauge
   - Negative battery cable
5. Fill the transaxle fluid to the proper level.
6. Start the vehicle, check for leaks and repair if necessary.

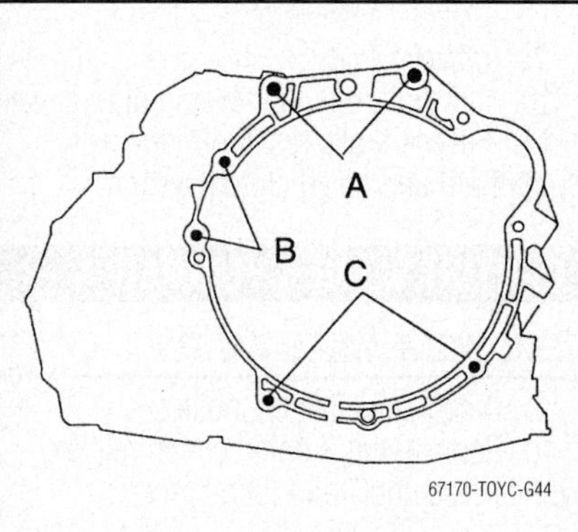

Fig. 3 Transaxle mounting bolts

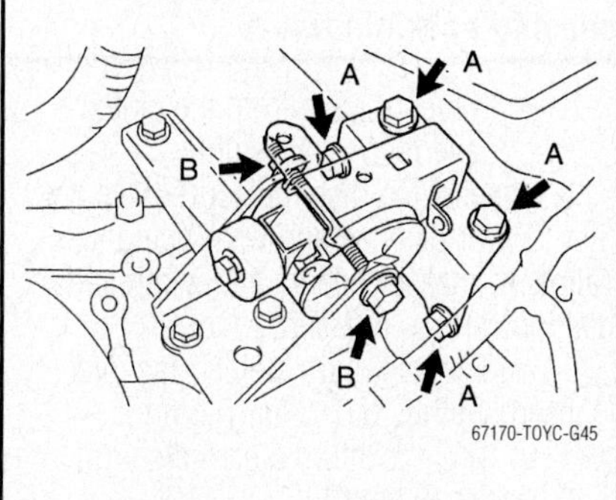

Fig. 4 Mounting insulator fastener identification

MANUAL TRANSAXLE ASSEMBLY

REMOVAL & INSTALLATION

1. Before servicing the vehicle, refer to the precautions in the beginning of this section.
2. Drain the transaxle fluid.
3. Remove or disconnect the following:
   - Negative battery cable. On vehicles equipped with an air bag, wait at least 90 seconds before proceeding.
   - Air cleaner case assembly with hose
   - Coolant reservoir tank
   - Release cylinder tube bracket
   - Clutch release cylinder
   - Back-up light switch connector
   - Ground cable
   - Shift cables from the transaxle
   - Vehicle speed sensor connector or the speedometer cable
   - Engine wire clamps
   - Starter set bolt from the transaxle upper side

- 2 transaxle upper mounting bolts
- Engine left mounting stay
- Engine left mounting set bolt from the rear side
- Under covers
- Lower ball joint from the lower arm.
- Halfshafts
- Front exhaust pipe
- Hole cover
- Engine front mounting set bolts
- Engine rear mounting
- Engine center support member
- Starter
- Transaxle mounting bolts from the engine rear end plate side
- Engine left mounting set bolts from the front side
- Transaxle mounting bolts from the engine front side, then engine rear side
- Transaxle

To install:

4. Align the input shaft with the clutch disc and install the transaxle to the engine. 12mm bolts: 47 ft. lbs. (64 Nm). 10mm bolts: 34 ft. lbs. (46 Nm).
5. Install or connect the following:
    - Left engine mount. Bolts: 41 ft. lbs. (56 Nm).
    - Transaxle mounting bolts to the engine rear end plate side. Bolts: 17 ft. lbs. (23 Nm).
    - Starter, lower bolt and electrical connector to the starter. Bolt: 29 ft. lbs. (39 Nm).
    - Engine center support member. Radiator support bolts: 45 ft. lbs. (61 Nm). Frame bolts: 152 ft. lbs. (206 Nm).
    - Engine rear mounting. Bolts: 35 ft. lbs. (48 Nm).
    - Engine front mounting. Bolts: 47 ft. lbs. (64 Nm).
    - Hole covers
    - Front exhaust pipe
    - Halfshafts Nut: 159 ft. lbs. (49 Nm)
    - Lower ball joint to lower arm. Bolt and nuts: 105 ft. lbs. (142 Nm).
    - Undercovers
    - Engine left mounting set bolt to the rear side. Bolt: 41 ft. lbs. (56 Nm).
    - Engine left mounting stay. Bolt: 15 ft. lbs. (21 Nm).
    - 2 transaxle upper side mounting bolts. Bolts: 29 ft. lbs. (39 Nm).
    - Starter set bolt to the transaxle upper side. Bolt: 29 ft. lbs. (39 Nm).
    - Engine wire clamps
    - Vehicle speed sensor connector or the speedometer cable.
    - Transaxle shift cables and ground cable.
    - Back-up light switch connector.
    - Release cylinder and release cylinder tube bracket. Bolts: 108 inch lbs. (12 Nm).
    - Coolant reservoir tank
    - Air cleaner case assembly
    - Negative battery cable
6. Fill the transaxle fluid to the proper level.
7. Start the vehicle, check for leaks and repair if necessary.

CLUTCH DRIVEN DISC & PRESSURE PLATE

REMOVAL & INSTALLATION

➡Do not allow grease or oil to get on any part of the disc, pressure plate, or flywheel surfaces.

1. Before servicing the vehicle, refer to the precautions in the beginning of this section.
2. Remove or disconnect the following:
    - Negative battery cable. On vehicles equipped with an air bag, wait at least 90 seconds before proceeding
    - Transaxle assembly
3. Make matchmarks on the clutch cover (pressure plate) and flywheel so that the pressure plate can be returned to its original position during installation.
4. Remove or disconnect the following:
    - Release fork bearing clips
    - Release bearing hub, complete with the release bearing
    - Release fork and support

⁂ CAUTION

Slowly unfasten the bolts which attach the pressure plate. Loosen each bolt 1 turn at a time until the spring tension is released. If the bolts are released improperly the clutch assembly could fly apart, causing possible injury.

- Pressure plate from the clutch cover/spring assembly

5. Inspect the disc, pressure plate and flywheel for damage and wear using a caliper to measure depth and width and a dial indicator to measure runout.
    a. The minimum clutch disc rivet head depth is 0.012 in. (0.3mm).
    b. The maximum clutch disc runout is 0.031 in. (0.8mm).
    c. The maximum pressure plate spring depth is 0.024 in. (0.6mm).
    d. The maximum pressure plate spring width is 0.197 in. (5.0mm).
    e. The maximum flywheel runout is 0.004 in. (0.1mm).
6. Replace or machine parts as necessary.

To install:

7. When reassembling, apply a thin coating of multipurpose grease to the release bearing hub and release fork contact points. Also, pack the groove inside the clutch hub with multipurpose grease and lubricate the pivot points of the release fork.
8. Install or connect the following:
    - Clutch disc and pressure plate. The bolts should be tightened in 2 or 3 steps, gradually and evenly. Final bolt torque is 14 ft. lbs. (19 Nm).
    - Release bearing, fork and boot
    - Transaxle assembly
    - Negative battery cable

ADJUSTMENTS

Hydraulic clutch actuating systems used in Toyota vehicles do not require adjustment.

CLUTCH MASTER CYLINDER

REMOVAL & INSTALLATION

1. Drain the clutch fluid.
2. Remove the brake master cylinder subassembly.
3. Remove the brake booster assembly.
4. Loosen the clip and disconnect the clutch reservoir tube from the clutch master cylinder assembly. Use a container to catch the fluid.
5. Disconnect the clutch reservoir tube.
6. Disconnect the flexible hose tube and the clutch master cylinder. Use a container to catch the fluid.
7. Remove the clutch pedal spring.
8. Remove the clutch master cylinder push rod clevis with hole pin.
9. Remove the two nuts and clutch master cylinder assembly.

To install:

10. Installation is the reverse of removal.
11. Fill the brake reservoir tank with clutch fluid and bleed clutch system.

CLUTCH SLAVE CYLINDER

REMOVAL & INSTALLATION

1. Disconnect flexible hose.
2. Remove the 3 bolts, clutch release cylinder assembly and tube bracket.

To install:

3. Installation is the reverse of removal.

4. Fill the brake reservoir tank with clutch fluid and bleed clutch system.

CLUTCH HYDRAULIC SYSTEM BLEEDING

➡If any maintenance on the clutch system was performed or the system is suspected of containing air, bleed the system. Use care; brake fluid will remove the paint from any surface. If the brake fluid spills onto any painted surface, wash it off immediately with soap and water.

1. Before servicing the vehicle, refer to the precautions in the beginning of this section.
2. Fill the clutch reservoir with brake fluid. Check the reservoir level frequently and add fluid as needed.
3. Connect one end of a vinyl tube to the bleeder plug on the slave cylinder and submerge the other end into a clear container half-filled with brake fluid.
4. Slowly pump the clutch pedal several times.
5. Have an assistant hold the clutch pedal down and loosen the bleeder plug until fluid and/or air starts to run out of the bleeder plug. Close the bleeder plug while the pedal is held to the floor.

➡Do not allow the pedal to rise backup while the bleeder is still open. If this happens, it will allow air to reenter the slave cylinder and cause the clutch system not to work properly.

6. Repeat Steps 2 and 3 until all the air bubbles are removed from the system.
7. Tighten the bleeder plug when all the air is gone.
8. Refill the master cylinder to the proper level as required.
9. Check the system for leaks.

HALFSHAFTS

REMOVAL & INSTALLATION

➡The hub bearing could be damaged if subjected to the full weight of the vehicle, such as if the vehicle is moved without the halfshafts. If it is absolutely necessary to place the full vehicle weight on the hub bearing, first support the bearing with SST No. 09608–16041.

1. Before servicing the vehicle, refer to the precautions in the beginning of this section.
2. Drain the transaxle fluid.
3. Remove or disconnect the following:
   - Negative battery cable. On vehicles equipped with an air bag, wait at least 90 seconds before proceeding
   - Cotter pin, locknut cap, and bearing locknut
   - Front wheels
   - Undercovers
   - With ABS, speed sensor
   - Tie rod ball joint from the steering knuckle
   - Lower ball joint from the lower suspension arm
4. Drive the halfshaft from the knuckle.

➡Most halfshafts can be separated from the knuckle using a brass or plastic hammer; some others may require the use of a puller.

5. Remove the halfshaft from the transaxle

To install:

6. Install or connect the following:
   - Snapring, opening side facing downward, to the inboard, oiled, joint tulip
   - Halfshaft into the transaxle. After installing the halfshaft to the transaxle, check that there is 0.08–0.12 in. (2–3mm) of axial play. Check that the halfshaft is making contact with the pinion shaft and that the halfshaft cannot be pulled out
   - Halfshaft into the knuckle
   - Lower suspension arm to the steering knuckle. Nuts and bolts: 105 ft. lbs. (142 Nm).
   - Tie rod end to the steering knuckle. Nut: 36 ft. lbs. (216 Nm).
   - ABS speed sensor
   - Hub locknut and washer Nut: 159 ft. lbs. (49 Nm)
   - Negative battery cable
   - Wheels
   - Locknut cap and NEW cotter pin.
   - Undercovers
7. Fill the transaxle fluid to the proper level.
8. Start the vehicle, check for leaks and repair if necessary.

ENGINE COOLING

THERMOSTAT

REMOVAL & INSTALLATION

1. Remove the right hand cover under the engine.
2. Drain the engine coolant.
3. Remove the fan and generator V belt.
4. Remove the generator assembly
5. Remove the water inlet via the two nuts.
6. Remove the thermostat.

To install:

7. Install a new gasket onto the thermostat.
8. Install the thermostat with the jiggle valve upward. It can be set within 10 degrees on either side of the prescribed position.
9. Install the water inlet via the two nuts. Torque to 8 ft. lbs. (11 Nm).
10. Install the generator assembly.
11. Add engine coolant.
12. Inspect for leaks.
13. Replace the cover under the engine.

WATER PUMP

REMOVAL & INSTALLATION

See Figure 5.

1. Before servicing the vehicle, refer to the precautions in the beginning of this section.
2. Drain the cooling system.
3. Remove or disconnect the following:
   - Negative battery cable
   - Right-hand engine under cover
   - Drive belt
   - Water pump

To install:

4. Install or connect the following:
   - Water pump. Bolts marked **A** (short): 80 inch lbs. (9 Nm). Bolts marked **B** (long): 96 inch lbs. (11 Nm).
   - Drive belt
   - Right engine under cover
   - Negative battery cable
5. Fill the cooling system to the proper level.
6. Start the vehicle, check for leaks and repair if necessary.

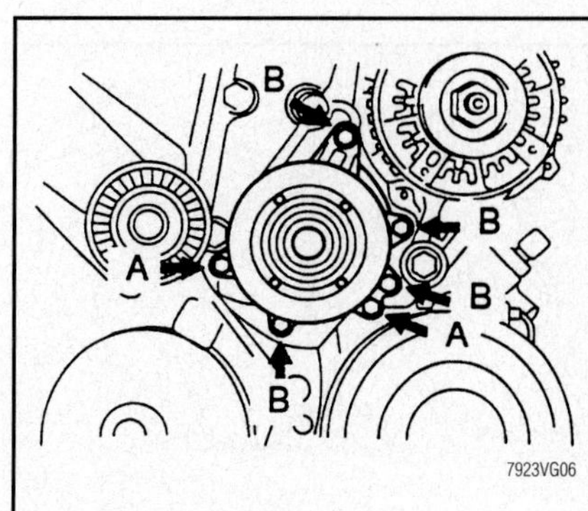

Fig. 5 Water pump bolt identification—1.8L (1ZZ-FE) engine

ENGINE ELECTRICAL — CHARGING SYSTEM

ALTERNATOR

REMOVAL & INSTALLATION

1. Before servicing the vehicle, refer to the precautions in the beginning of this section.

➡**It may be necessary to remove the gravel shield and work from underneath the car in order to gain access to the alternator retaining bolts.**

2. Remove or disconnect the following:
   - Negative battery cable
   - Wiring from the alternator
   - Drive belt
   - Alternator

To install:

3. Install or connect the following:
   - Alternator. Torque the smaller bolt to 18 ft. lbs. (25 Nm) and the larger bolt to 40 ft. lbs. (54 Nm).
   - Alternator connector and wiring
   - Drive belt
   - Negative battery cable

ENGINE ELECTRICAL — DISTRIBUTORLESS IGNITION SYSTEM

FIRING ORDER

See Figure 6.

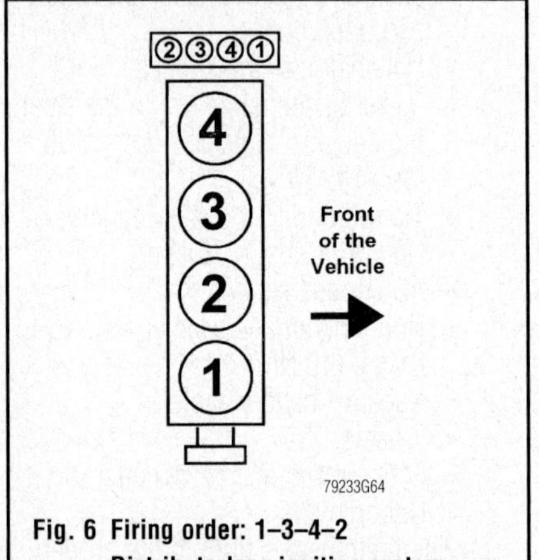

Fig. 6 Firing order: 1–3–4–2 Distributorless ignition system

IGNITION COIL

REMOVAL & INSTALLATION

See Figures 7 and 8.

1. Disconnect the battery's negative terminal.
2. Remove the two nuts and clips and remove the cylinder head cover.
3. Disconnect the four ignition coil connectors.

Fig. 7 The two bolts and two clips of the cylinder head cover

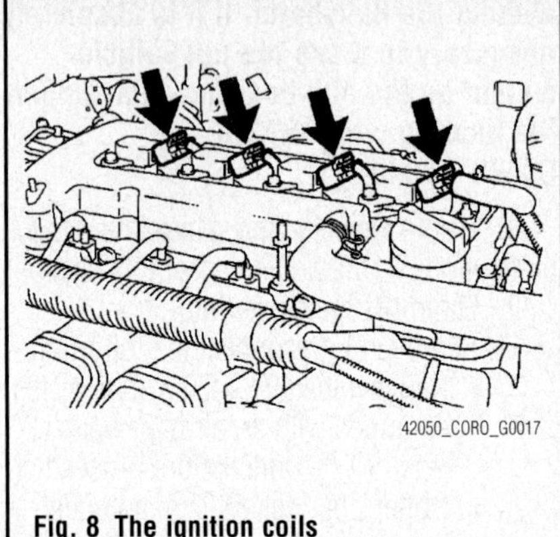

Fig. 8 The ignition coils

4. Remove the four bolts that retain the coils.

To install:

5. Replace the coils and the four bolts.
6. Connect the coil connections.
7. Replace the cylinder head cover.

IGNITION TIMING

ADJUSTMENT

The ignition timing is controlled by the Powertrain Control Module (PCM). No adjustment is necessary or possible.

SPARK PLUGS

REMOVAL & INSTALLATION

➡**Observe the following service precautions:**

- Allow the engine to cool before removing the spark plugs. Attempting to remove spark plugs from a hot engine can cause the spark plugs to seize. This can damage the cylinder head threads.
- Clean the spark plug recess area before removing the spark plug. Failure to do so can result in engine damage due to dirt or foreign material entering the cylinder head, or in contamination of the cylinder head threads. Contaminated threads may prevent proper seating of the new spark plug.
- Use only the spark plugs specified for use in the vehicle. Do not install spark plugs that are either hotter or colder than those specified for the vehicle. Installing spark plugs of another type can severely damage the engine.

1. Turn the ignition **OFF**.
2. If you are replacing the engine right bank (rear) spark plugs, rotate the engine for service access
3. Remove the spark plug wires from the spark plugs.
4. Remove the spark plugs from the engine.

To install:

➡**It is important to check the gap of all new and reconditioned spark plugs before installation. Pre-set gaps may have changed during handling. Use a round wire feeler gauge to be sure of an accurate check, particularly on used plugs. Installing plugs with the wrong gap can cause poor engine performance and may even damage the engine. Gap the spark plugs to the specifications.**

➡**Be sure plug threads smoothly into cylinder head and is fully seated. Use a thread chaser if necessary to clean threads in cylinder head. Cross-threading or failing to fully seat spark plug can cause overheating of plug, exhaust blow-by, or thread damage. Follow the recommended torque specifications carefully. Over or under-tightening can also cause severe damage to engine or spark plug.**

➡**Use the correct fastener in the correct location. Replacement**

fasteners must be the correct part number for that application. Fasteners requiring replacement or fasteners requiring the use of thread locking compound or sealant are identified in the service procedure. Do not use paints, lubricants, or corrosion inhibitors on fasteners or fastener joint surfaces unless specified. These coatings affect fastener torque and joint clamping force and may damage the fastener. Use the correct tightening sequence and specifications when installing fasteners in order to avoid damage to parts and systems.

5. Install the spark plugs.
6. If the spark plugs are installed into a new cylinder head, tighten the spark plugs to 15 lb-ft (20 Nm). If the spark plugs are installed into an existing cylinder head, tighten the spark plugs to 11 ft. lbs. (15 Nm).
7. Install the spark plug wires to the spark plugs.

ENGINE ELECTRICAL

STARTING SYSTEM

STARTER

REMOVAL & INSTALLATION

1. Before servicing the vehicle, refer to the precautions in the beginning of this section.
2. Remove or disconnect the following:
   - Battery and tray. On models equipped with an air bag, work must NOT be started until at least 90 seconds have passed from the time that both the ignition switch is turned to the LOCK position and the negative cable is disconnected from the battery.
   - Coolant reservoir
   - Reservoir hose from the radiator
   - Right side under cover
   - Wiring clamp
   - Wiring from the starter
   - Starter

To install:

3. Install or connect the following:
   - Starter
   - Starter wiring and nut
   - Starter connector and wire clamp
   - Right side engine cover
   - Radiator reservoir and hose
   - Battery and tray
4. Check the cooling system and top off if necessary.

ENGINE MECHANICAL

ACCESSORY DRIVE BELTS

ACCESSORY BELT ROUTING

See Figure 9.

Refer to the accompanying illustration for drive belt routing.

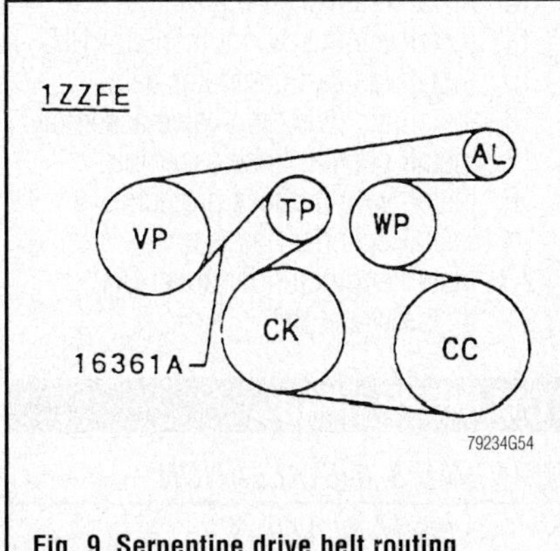

Fig. 9 Serpentine drive belt routing

INSPECTION

Inspect the drive belt for signs of glazing or cracking. A glazed belt will be perfectly smooth from slippage, while a good belt will have a slight texture of fabric visible. Cracks will usually start at the inner edge of the belt and run outward. All worn or damaged drive belts should be replaced immediately.

ADJUSTMENT

1. There is no adjustment necessary for drive belts.

REMOVAL & INSTALLATION

1. Remove the right hand cover under the engine.
2. Turn the drive belt tensioner clockwise to relieve tension on the belt.
3. Remove the fan and generator V belt.
4. Return the tensioner to the unloaded position.

➡ **When retracting the tensioner, turn it clockwise slowly in 3 sec. or more. Be sure not to apply force rapidly. After the tensioner is retracted all the way, do not apply force any more than necessary.**

To install:

5. Turn the drive belt tensioner clockwise
6. Install the belt.
7. Install the right hand cover under the engine

CAMSHAFT AND VALVE LIFTERS

REMOVAL & INSTALLATION

See Figures 10 through 14.

1. Before servicing the vehicle, refer to the precautions in the beginning of this section.
2. Remove or disconnect the following:
   - Negative battery cable. On vehicles equipped with an air bag, wait at least 90 seconds before proceeding.
   - Cylinder head cover
3. Turn the crankshaft so that the No. 1 piston is at TDC on the compression stroke. Check to see that the point marks on the

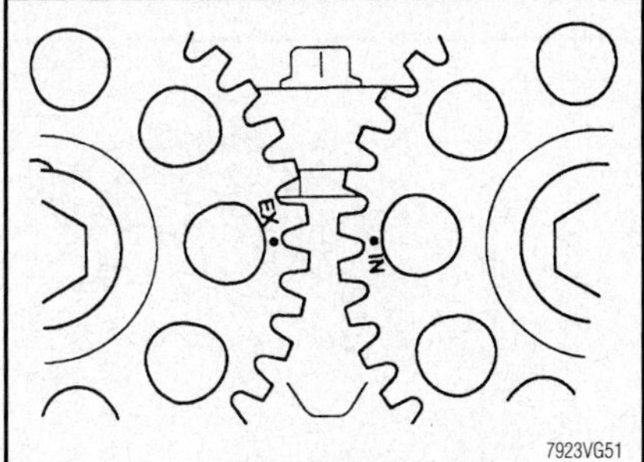

Fig. 10 The sprocket marks will align when the No. 1 piston is at TDC on the compression stroke—1.8L (1ZZ-FE) engine

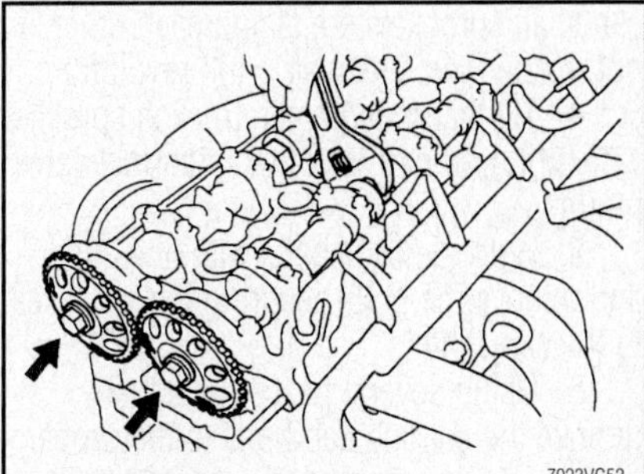

Fig. 11 Hold the camshaft with a wrench while removing the sprocket bolt—1.8L (1ZZ-FE) engine

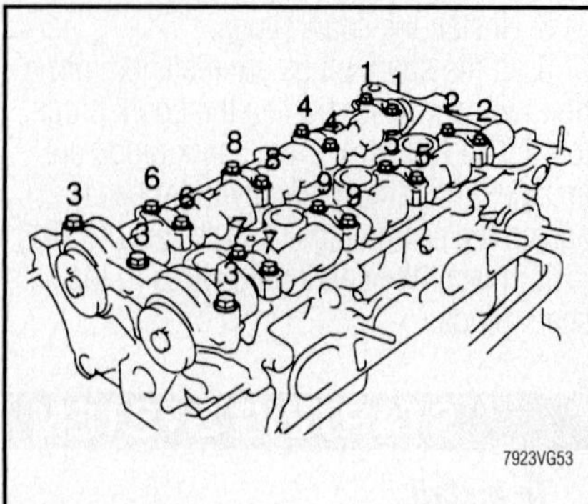

Fig. 12 Camshaft bearing cap bolt removal sequence—1.8L (1ZZ-FE) engine

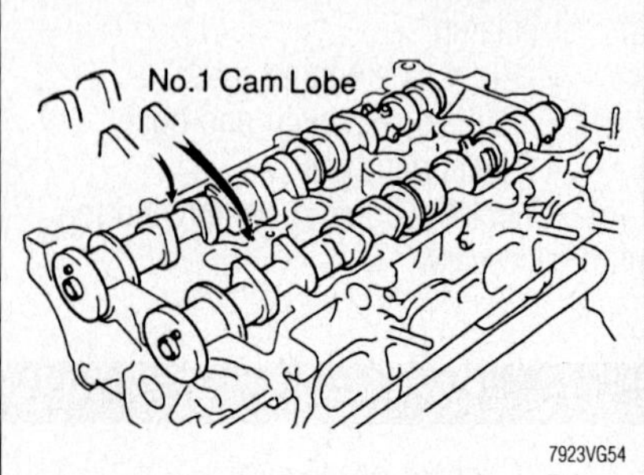

Fig. 13 When installing the camshafts, position the lobes for the No. 1 cylinder as shown—1.8L (1ZZ-FE) engine

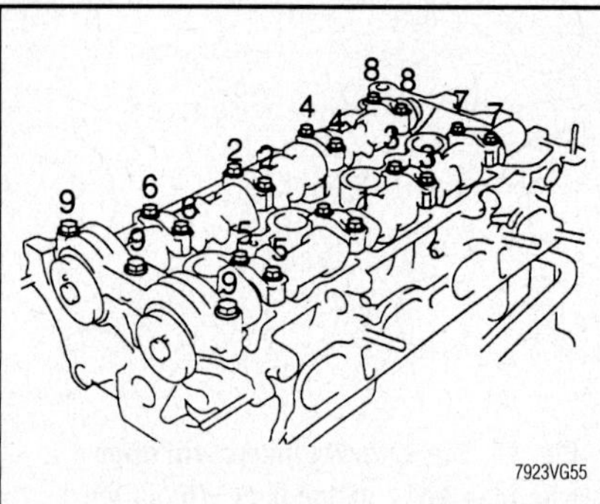

Fig. 14 Camshaft bearing cap bolt tightening sequence—1.8L (1ZZ-FE) engine

camshaft sprockets are facing each other, if not, rotate the crankshaft 1 full revolution.

4. Tie the timing chain to each sprocket with string or wire to maintain correct valve timing.

5. Hold the camshafts with a wrench and remove the bolts securing the sprockets to the camshafts.

6. Using several passes, gradually remove the bearing cap bolts in the proper sequence. Then, remove the camshafts

To install:

7. Lubricate the camshafts with clean engine oil and place them on the cylinder head. Be sure to position the lobes for the No. 1 cylinder as shown in the illustration.

8. Install the bearing caps in their original positions. Apply clean engine oil to the threads and under the heads of the bearing cap bolts. After tightening the bolts on the No. 1 bearing cap to 17 ft. lbs. (23 Nm), tighten the remaining bolts in sequence using several passes to 10 ft. lbs. (13 Nm).

9. Check the valve clearance and make adjustments as needed.

10. Install or connect the following:

- Camshaft sprockets and the chain
- Cylinder head cover
- Negative battery cable

CRANKSHAFT DAMPER

REMOVAL & INSTALLATION

See Figures 15 and 16.

1. Using service tool, remove the pulley bolt

2. Using service tool, remove the pulley.

To install:

3. Align the pulley set key with the key groove of the pulley, then slide on the pulley.

4. Using service tool, install the pulley bolt.

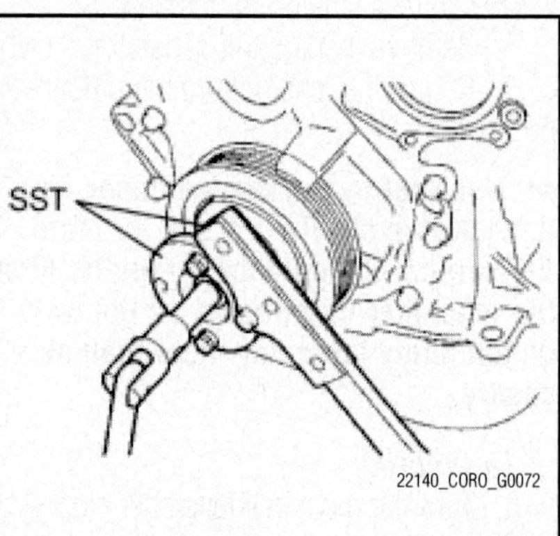

Fig. 15 Removing pulley bolt

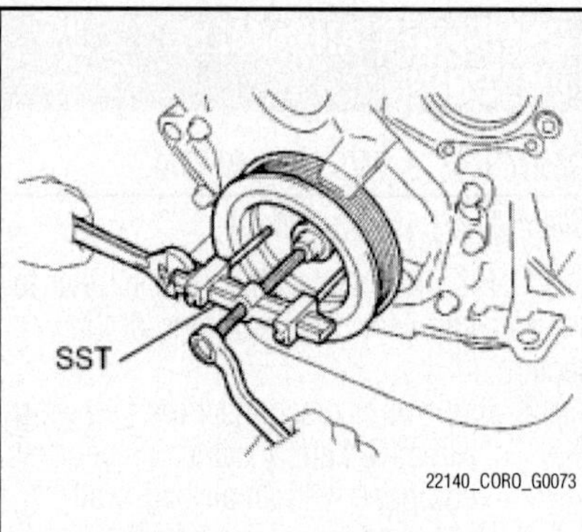

Fig. 16 Removing crankshaft pulley

CRANKSHAFT FRONT SEAL

REMOVAL & INSTALLATION

1. Remove front wheel RH.
2. Remove engine under cover RH.
3. Remove cylinder heads cover.
4. Remove wire harness protector.
5. Separate suction hose assembly.
6. Remove air switching valve assembly.
7. Remove fan and generator belt.
8. Remove engine mount insulator RH.
9. Remove crankshaft pulley.
10. Remove timing chain or belt cover oil seal.
    a. Using a knife, cut off the oil seal lip.
    b. Using a screwdriver with its tip wrapped with tape, pry out the oil seal.

➡After the removal, check if the crankshaft is not damaged. If damaged, smooth the surface with a 400-grit sandpaper.

To install:

11. Apply a light coat of multi-purpose grease to a new oil seal lip.

12. Using service tool and a hammer, tap in the oil seal until its surface is flush with the rear oil seal retainer edge.

➡Wipe any extra grease off the crankshaft.

13. Install crankshaft pulley.
14. Install engine mount insulator RH.
15. Install fan and generator belt.
16. install air switching valve assembly.
17. Install suction hose assembly.
18. Install wire harness protector.
19. Install cylinder heads cover.
20. Install engine under cover RH.
21. Install wheel RH

CYLINDER HEAD

REMOVAL & INSTALLATION

See Figures 17 through 24.

1. Before servicing the vehicle, refer to the precautions in the beginning of this section.

2. Drain the cooling system.

3. Remove or disconnect the following:

- Battery
- ECU box
- Coolant reservoir
- Air cleaner assembly
- Accelerator cable
- Alternator
- Exhaust pipe
- Exhaust manifold

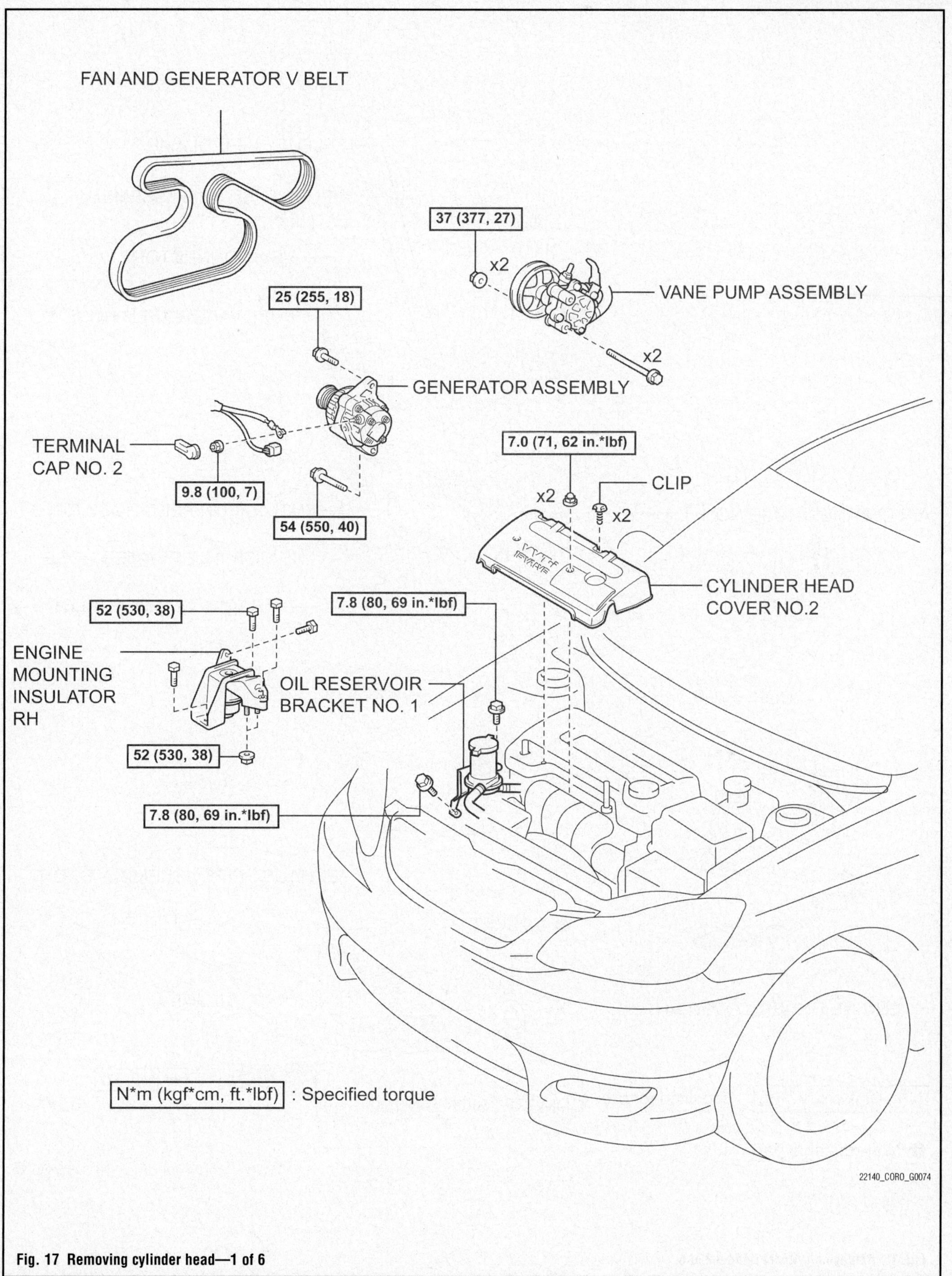

Fig. 17 Removing cylinder head—1 of 6

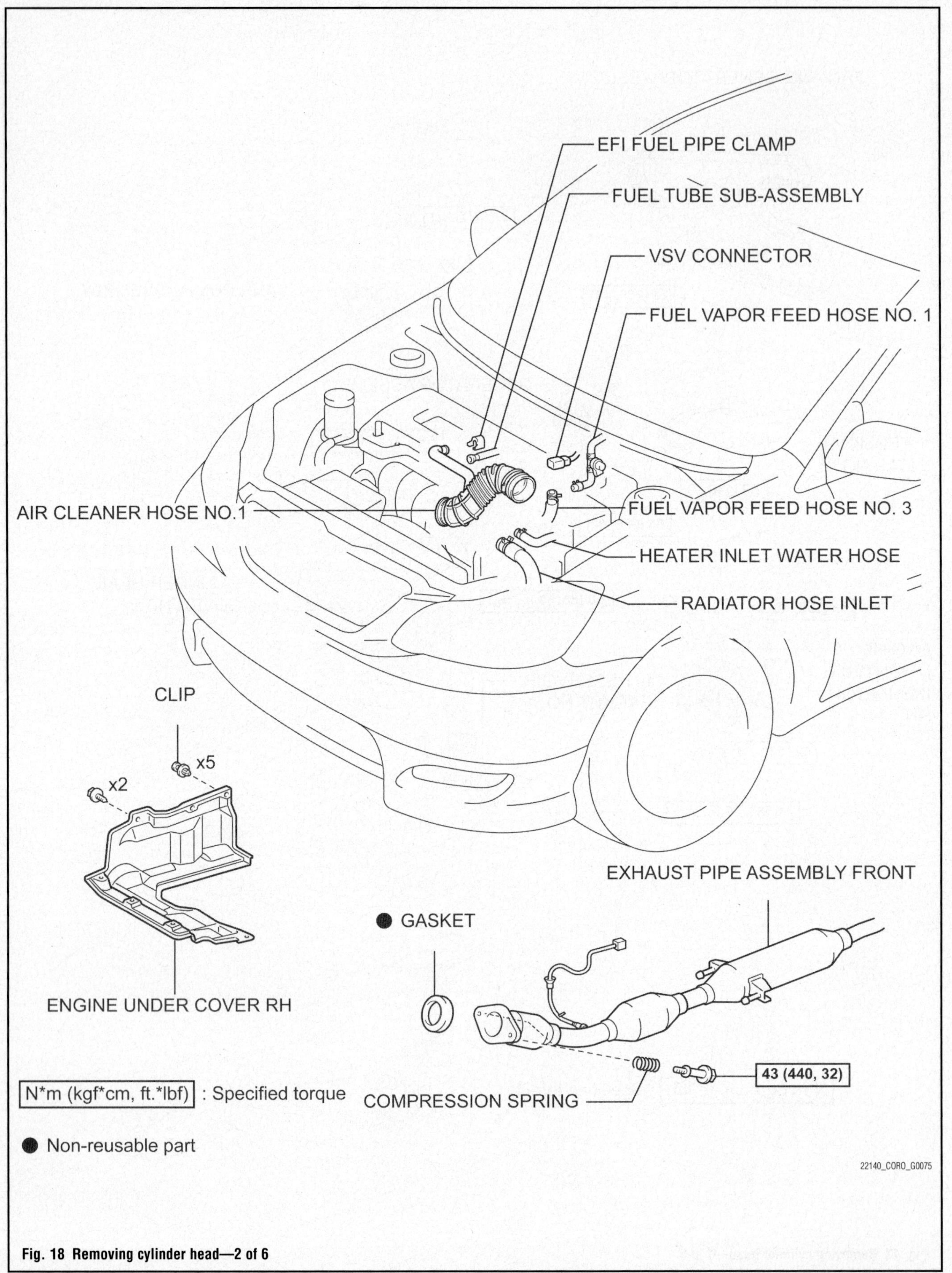

Fig. 18 Removing cylinder head—2 of 6

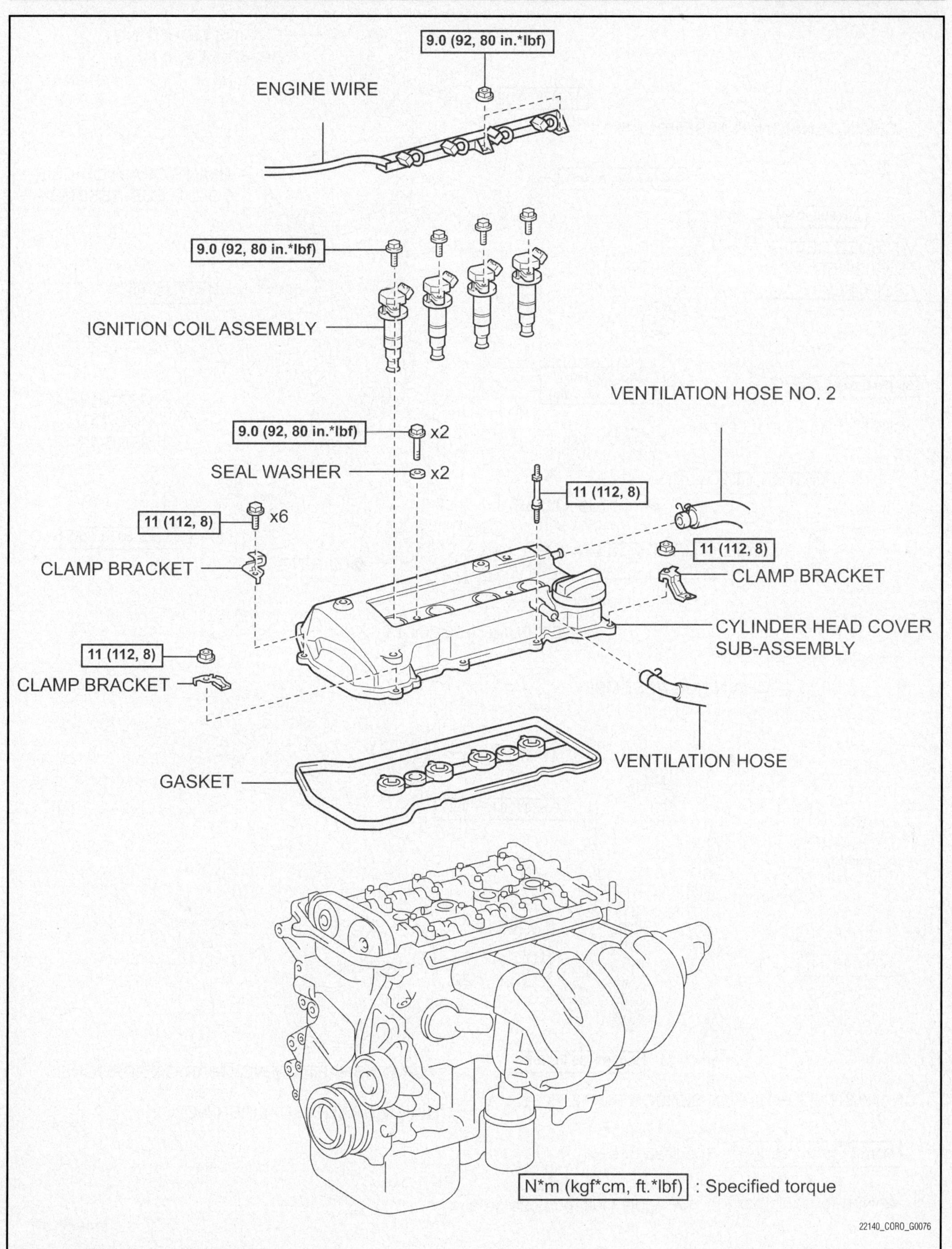

Fig. 19 Removing cylinder head—3 of 6

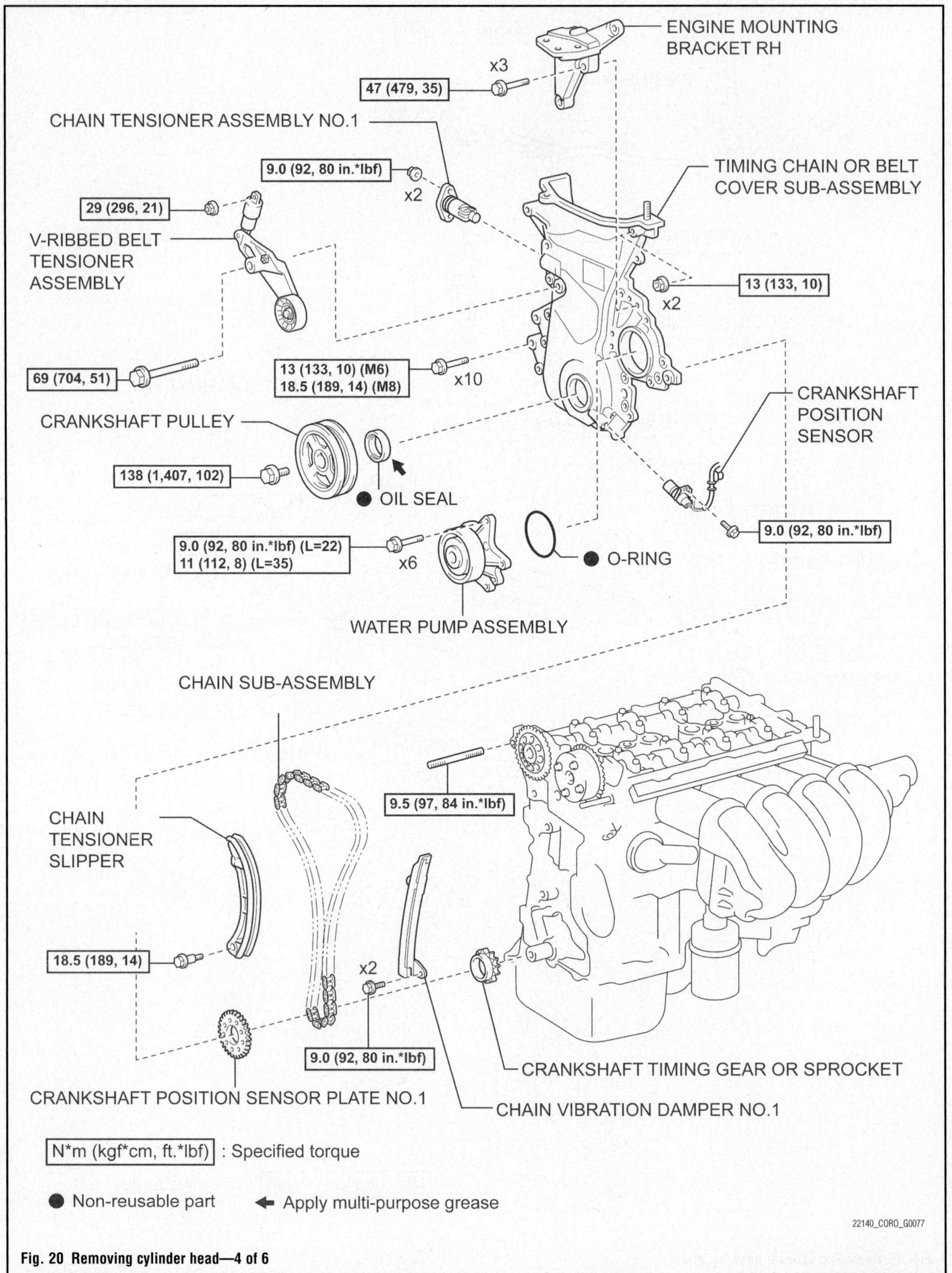

Fig. 20 Removing cylinder head—4 of 6

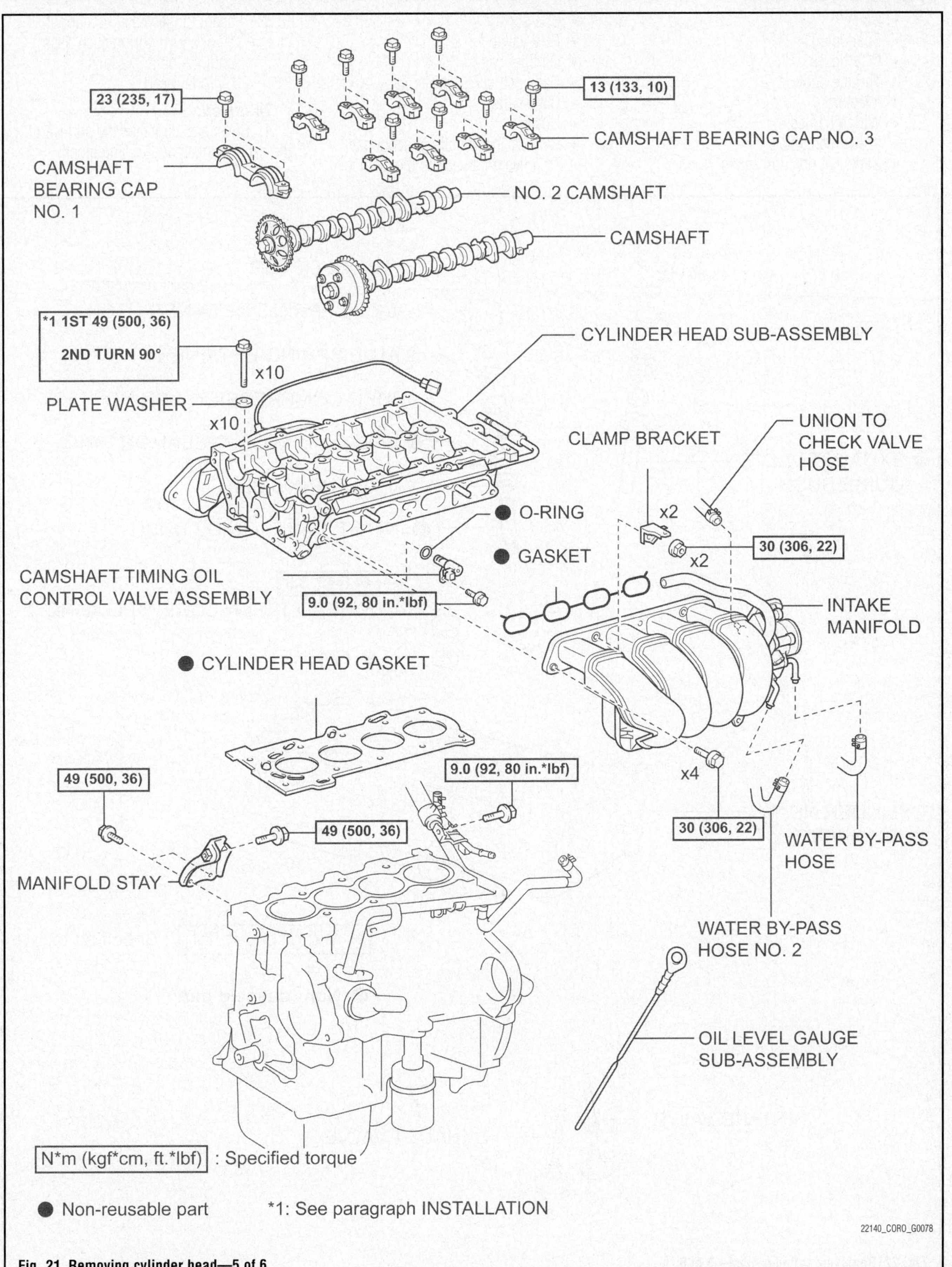

Fig. 21 Removing cylinder head—5 of 6

- Coils
- Spark plugs
- PCV hoses
- Throttle body
- Injectors
- Wiring harness
- Intake manifold
- Camshaft position sensor
- ECT sensor
- PCV valve
- Oil filler cap
- Camshaft sprockets
- Camshafts
- Hoses
- Cylinder head bolts in sequence. To prevent damage to the cylinder head, loosen each bolt about ¼ of a turn during each pass until the bolts are loose.
- Cylinder head

To install:

4. Clean and degrease the surface of the cylinder head and engine block.

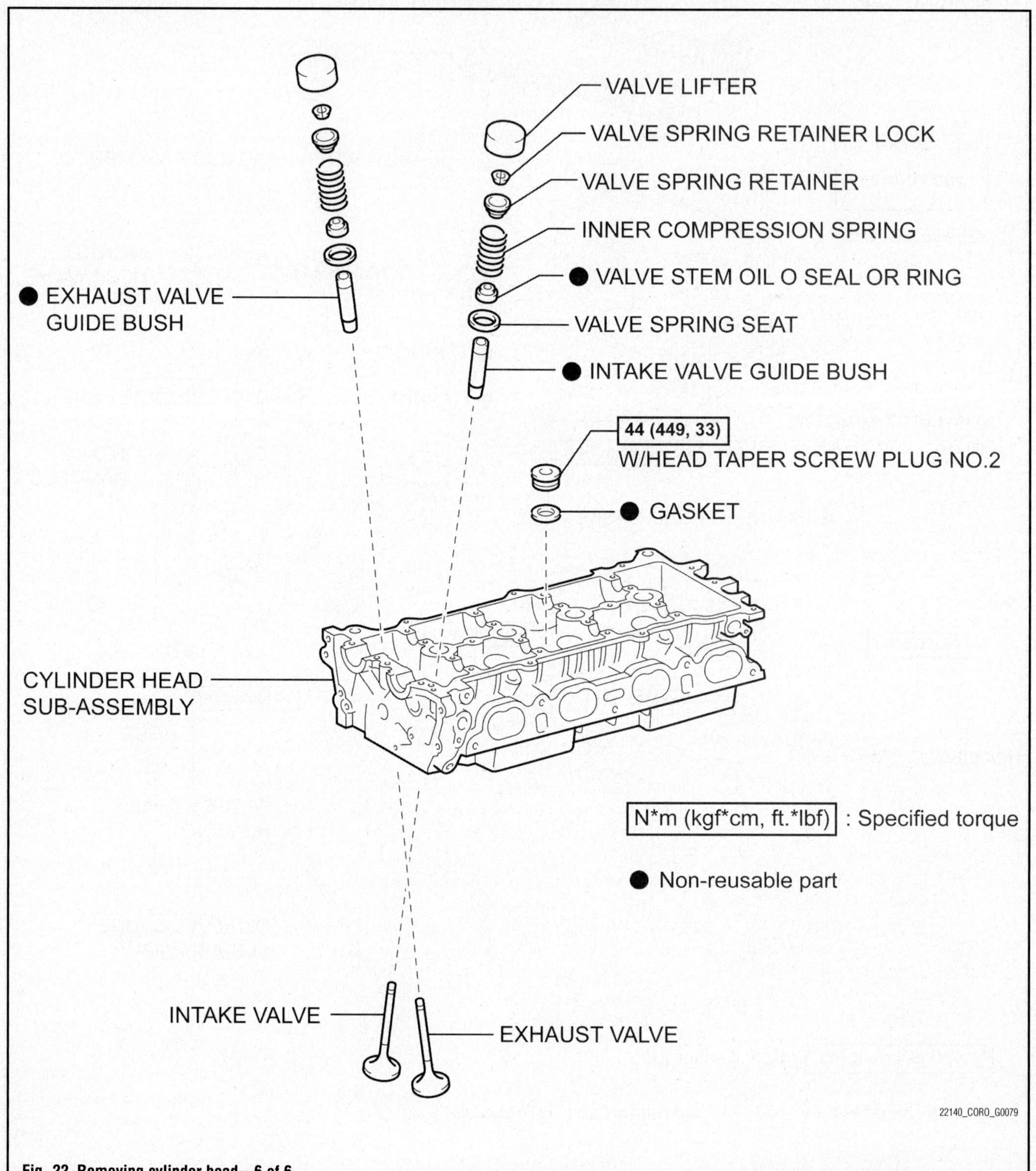

Fig. 22 Removing cylinder head—6 of 6

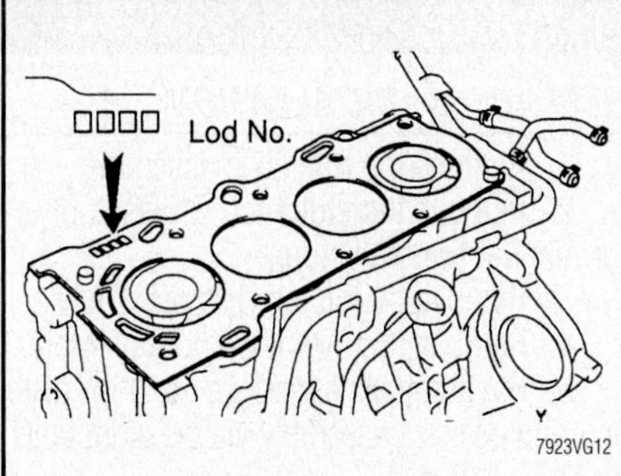

Fig. 23 Position the head gasket correctly on the cylinder head—1.8L (1ZZ-FE) engine

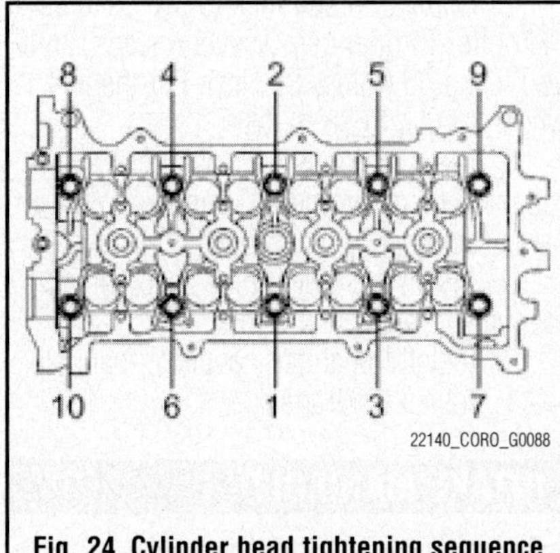

Fig. 24 Cylinder head tightening sequence

5. Install or connect the following:
   - New gasket on the engine block with the Lod No. stamp facing up.
   - Cylinder head
   - Apply a light coat of oil to cylinder head bolt threads and tighten in sequence. Replace any bolt that appears deformed. Bolts: 36 ft. lbs. (49 Nm).
   - Tighten each bolt in sequence an additional 90 degree turn.
   - Camshafts
   - Sprockets
   - Oil filler cap
   - PCV valve
   - ECT sensor
   - Intake manifold
   - Wiring harness
   - Exhaust manifold
   - Exhaust pipe
   - Alternator
   - accelerator cable
   - Air cleaner
   - ECM box
   - Battery
6. Fill the cooling system to the proper level.
7. Start the vehicle, check for leaks and repair if necessary.

ENGINE ASSEMBLY

REMOVAL & INSTALLATION

1. Before servicing the vehicle, refer to the precautions in the beginning of this section.
2. Relieve the fuel system pressure.
3. Drain the cooling system.
4. Drain the engine oil.
5. Drain the transaxle fluid.
6. Remove or disconnect the following:
   - Negative battery cable. On vehicles equipped with an air bag, wait at least 90 seconds before proceeding.
   - Battery, tray & carrier
   - Hood
   - Undercover
   - Accelerator cable
   - With automatic transmission, throttle cable from the accelerator cable.
   - Radiator and cooling fan
   - Air cleaner assembly
   - Coolant reservoir tank stay
   - Electrical connector, the hose, the mounting bolt, and remove the washer tank
   - Cruise control actuator
   - The Manifold Absolute Pressure (MAP) sensor vacuum hose from the gas filter on the intake manifold
   - The brake booster vacuum hose from the intake manifold
   - With air conditioning: the air conditioning vacuum hose from the actuator
   - With power steering: the air hose from the air pipe
   - With air conditioning: the air conditioning actuator connector
7. Disconnect the following wires and connectors from the right-hand fender apron as follows:
   a. The ground strap connector.
   b. The MAP sensor connector.
   c. With air conditioning: the air conditioning pressure switch.
   d. The engine wiring harness from the fender apron.
8. Remove or disconnect the following:
   - Data Link Connector 1 (DLC1) connector and ground strap from the left-hand fender apron.
   - Engine relay box and 4 connectors.
   - Charcoal canister
   - Heater hoses from water inlet housing
   - Fuel inlet and return hoses
   - With manual transmission, clutch release cylinder without disconnecting the pipe
   - Transaxle control cable(s)
9. To disconnect the engine wiring harness, disconnect or remove the following components:
   - Left-hand and right-hand front door scuff plate
   - Lower finish panel
   - Lower panel with the glove compartment
   - Radio and center cluster finish panel
   - Rear console box
   - On manual transmission, shift lever knob
   - On automatic transmission, shifting hole bezel
   - Lower center finish panel
   - Floor carpet bracket
   - The 3 ECM connectors and cowl wire connector
10. Remove or disconnect the remaining components:
    - Air conditioning compressor
    - Front exhaust pipe
    - Halfshafts
    - Power steering pump
    - Engine mounting center member
    - Through-bolt and nut holding the mounting insulator to the mounting bracket
    - Engine and transaxle assembly
    - Front and rear engine mounting bracket
    - Starter
    - Separate the transaxle from the engine

To install:

11. Install or connect the following:
    - Engine to the transaxle
    - Starter
    - Rear engine mounting bracket bolts: 57 ft. lbs. (77 Nm).
    - Front engine mounting bracket. bolts: 57 ft. lbs. (77 Nm).
    - Engine and transaxle assembly into the vehicle
    - Engine mounting center member.
    - Front engine mounting insulator through-bolt and nut. Torque the bolt to 64 ft. lbs. (87 Nm).
    - Halfshafts
    - Front exhaust pipe
    - Power steering pump. Torque the bolts to 29 ft. lbs. (39 Nm).
    - Drive belt
    - Air conditioner compressor. Torque the bolts to 18 ft. lbs. (25 Nm).
    - Drive belt and reconnect the connector.
12. To install and connect the engine wiring harness, perform the following:
    - Push the wire through the cowl

- Connect the 3 ECM connectors
- Attach the cowl wire connector
- Floor carpet bracket
- Center lower finish panel
- With automatic transmission, install the shifting hole bezel, with manual transmission, install the shift lever knob
- Rear console box
- Center cluster finish panel and the radio
- Lower panel with the glove compartment door
- Right and left-hand door scuff plates
- Lower finish panel

13. Install or connect the following:
    - With manual transmission, clutch release cylinder
    - Transaxle control cable(s)
    - Fuel return and inlet hose. Torque the bolt to 22 ft. lbs. (29 Nm).
    - Heater hoses to the water inlet housing
    - Charcoal canister
14. Connect the following wires and connectors on the left-hand fender apron:
    - The 4 connectors to the engine relay box
    - Engine relay box
    - The DLC1 connector
    - The connector on the fender apron
    - The ground strap on the fender apron
15. Install or connect on the right-hand fender apron:
    - The ground strap connector
    - The MAP sensor connector
    - With air conditioning, the air conditioning pressure switch
    - The engine wire from the fender apron
16. Install or connect the following:
    - With A/C, the actuator connector
    - With power steering, the air hoses to the air pipe
    - The vacuum hose from the MAP sensor to the gas filter to the intake chamber
    - The brake booster vacuum hose to the air intake chamber
    - With A/C, the vacuum hose from the actuator
    - With cruise control, actuator, actuator cable and cover
    - Electrical connector and vinyl hose
    - Washer tank with the bolt
    - Coolant reservoir tank stay
    - Air cleaner
    - Radiator and cooling fan
    - With automatic transmission, connect the throttle cable.
    - Accelerator cable
    - All fluids
    - Negative battery cable
    - Undercovers and hood
17. Start the vehicle, check for leaks and repair if necessary.

EXHAUST MANIFOLD

REMOVAL & INSTALLATION

See Figure 25.

1. Before servicing the vehicle, refer to the precautions in the beginning of this section.
2. Remove or disconnect the following:
    - Exhaust pipe from the manifold
    - Exhaust manifold support bracket
    - Heat insulator from the dash panel
    - Upper heat insulator
    - Exhaust manifold and gasket
    - If necessary, the lower heat insulator from the exhaust manifold.

To install:

3. Install or connect the following:
    - Lower heat insulator on the exhaust manifold. Bolts: 108 inch lbs. (12 Nm).
    - Exhaust manifold using a new gasket. Nuts, tightened several passes: 27 ft. lbs. (37 Nm).
    - Upper heat insulator. Bolts: 108 inch lbs. (12 Nm).
    - Heat insulator on the dash panel
    - Exhaust manifold support bracket. Bolts in an alternating pattern: 37 ft. lbs. (49 Nm).
    - Front exhaust pipe to the manifold. Bolts: 46 ft. lbs. (62 Nm).
    - Oxygen sensor, using new gasket and nuts. Nuts: 14 ft. lbs. (20 Nm).
4. Start the vehicle, check for leaks and repair if necessary.

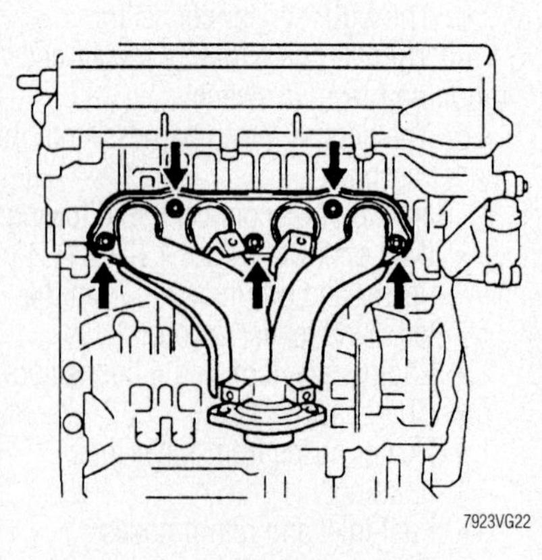

Fig. 25 Exhaust manifold mounting nut locations—1.8L (1ZZ-FE) engine

FLYWHEEL

REMOVAL & INSTALLATION

1. Remove the transaxle assembly.
2. Remove the clutch cover assembly for manual transmissions.
3. Remove the clutch disc assembly.
4. Remove the flywheel sub-assembly.
5. For automatics, remove the drive plate and ring gear subassembly via the eight bolts.

To install:

6. Install the drive plate and ring gear for automatics or the flywheel sub assembly for manuals. Make sure the eight bolts and holes are clean.
7. Apply a thread lock to two or three of the bolts. Torque, using several steps, until you reach 65 ft. lbs (88 Nm). For the flywheel, torque to 36 ft. lbs.
8. Mark the bolts with paint and retighten them another 90 degrees in the same sequence.
9. Install the clutch disc assembly for manuals.
10. Install the clutch cover for manuals.
11. Install the transaxle.

INTAKE MANIFOLD

REMOVAL & INSTALLATION

See Figure 26.

1. Before servicing the vehicle, refer to the precautions in the beginning of this section.
2. Drain the cooling system.
3. Remove or disconnect the following:
    - Negative battery cable
    - Drive belt and alternator
    - Air intake duct
    - Accelerator cable
    - Exhaust pipe from the manifold.
    - Exhaust manifold support bracket
    - Ignition coils
    - Spark plugs
    - PCV hoses
    - Throttle body assembly
    - 2 bolts securing the wiring harness protector
    - Wiring connectors and ground wires
    - Intake manifold support bracket
    - Intake manifold and gasket

To install:

4. Install or connect the following:
    - Intake manifold with a new gasket. Torque the bolts to 14 ft. lbs. (18.5 Nm).
    - Harness wiring to the cylinder head and harness protector
    - Fuel injectors, throttle body and the PCV hoses

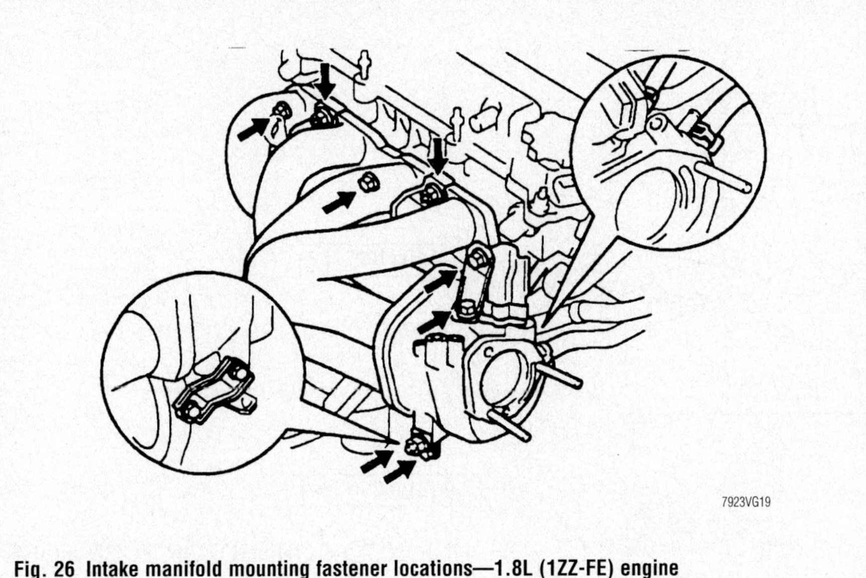

Fig. 26 Intake manifold mounting fastener locations—1.8L (1ZZ-FE) engine

- Spark plugs and ignition coils. Bolts and nuts: 80 inch lbs. (9 Nm).
- Exhaust manifold and support bracket. Bolts: 37 ft. lbs. (49 Nm).
- Front exhaust pipe to the manifold. Bolts: 46 ft. lbs. (62 Nm).
- Oxygen sensor. Nuts: 14 ft. lbs. (20 Nm).
- Accelerator cable and air intake duct
- Alternator and drive belt
- Negative battery cable

5. Fill the cooling system.
6. Start the vehicle, check for leaks and repair if necessary.

OIL PAN

REMOVAL & INSTALLATION

See Figure 27.

1. Before servicing the vehicle, refer to the precautions in the beginning of this section.
2. Drain the engine oil.
3. Remove or disconnect the following:
   - Negative battery cable. On vehicles equipped with an air bag, wait at least 90 seconds before proceeding.
   - Under covers
   - Front exhaust pipe
   - Oil pan mounting bolts and nuts
   - Oil pan, cutting off the applied sealer.

To install:

4. Remove any old sealant from the oil pan flange and thoroughly clean the sealing surface.
5. Install or connect the following:
   - Oil pan. Tighten the bolts and nuts in several passes. Bolts and nuts: 80 inch lbs. (9 Nm).
   - Front exhaust pipe
   - Negative battery cable
   - Under covers
6. Fill the engine with clean oil.
7. Start the vehicle, check for leaks and repair if necessary.

OIL PUMP

REMOVAL & INSTALLATION

See Figures 28 through 32.

1. Before servicing the vehicle, refer to the precautions in the beginning of this section.
2. Drain the engine oil.
3. Remove or disconnect the following:
   - Negative battery cable
   - Timing chain and crankshaft sprocket
   - Oil pump and gasket

To install:

4. Clean the mounting surface.
5. Install or connect the following:
   - Oil pump, with new gasket. Bolts: 80 inch lbs. (9 Nm).
   - Crankshaft sprocket and timing chain
   - Negative battery cable
6. Fill the engine with clean oil.
7. Start the vehicle, check for leaks and repair if necessary.

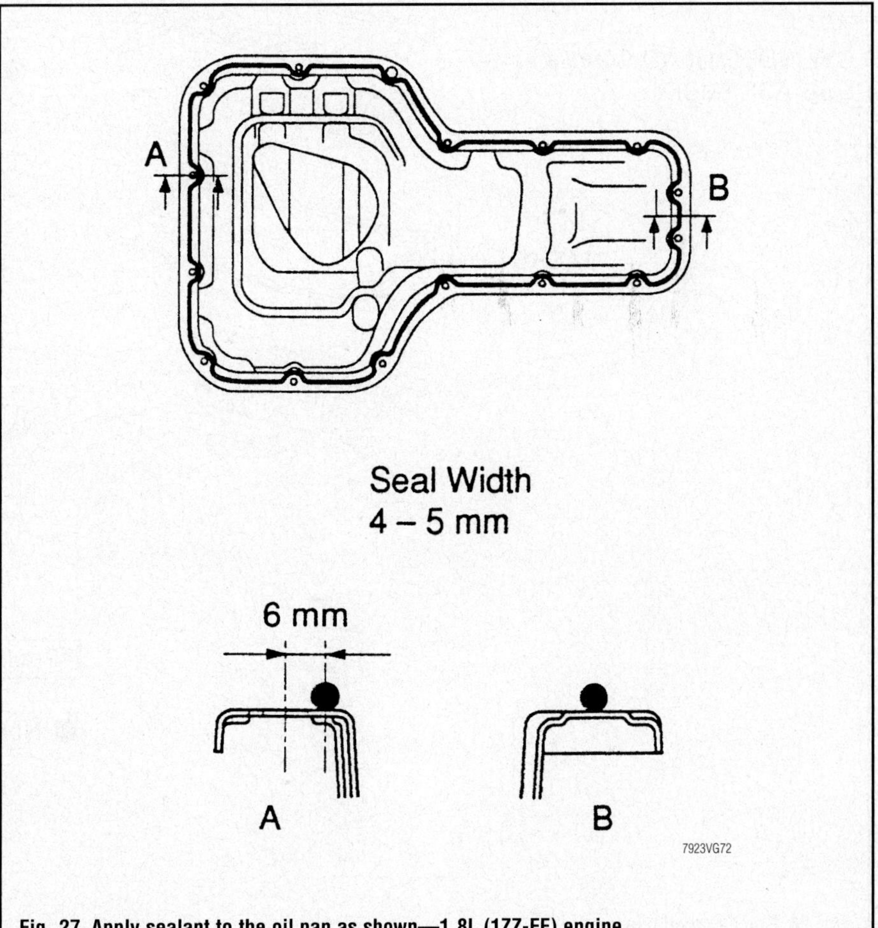

Fig. 27 Apply sealant to the oil pan as shown—1.8L (1ZZ-FE) engine

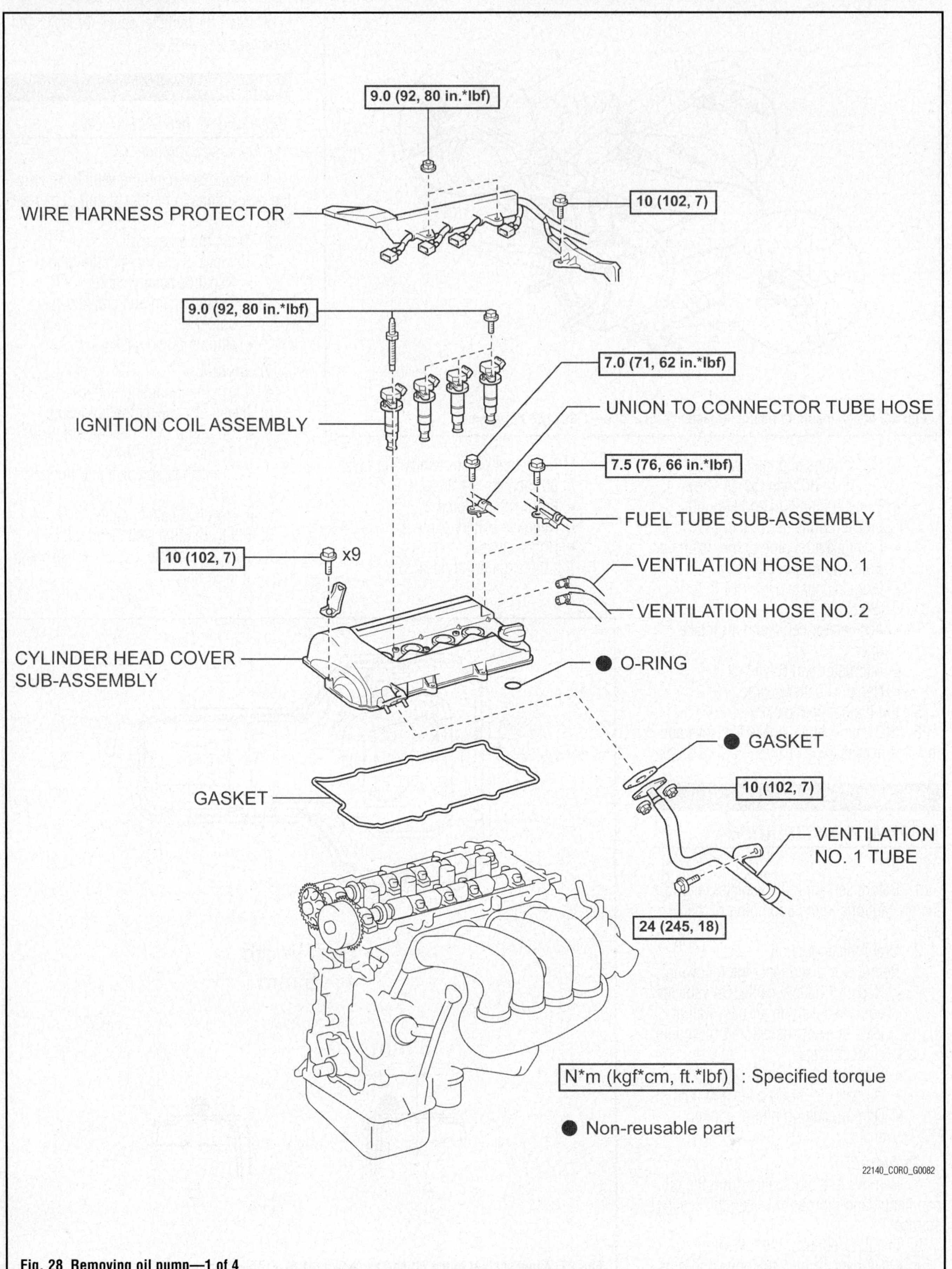

Fig. 28 Removing oil pump—1 of 4

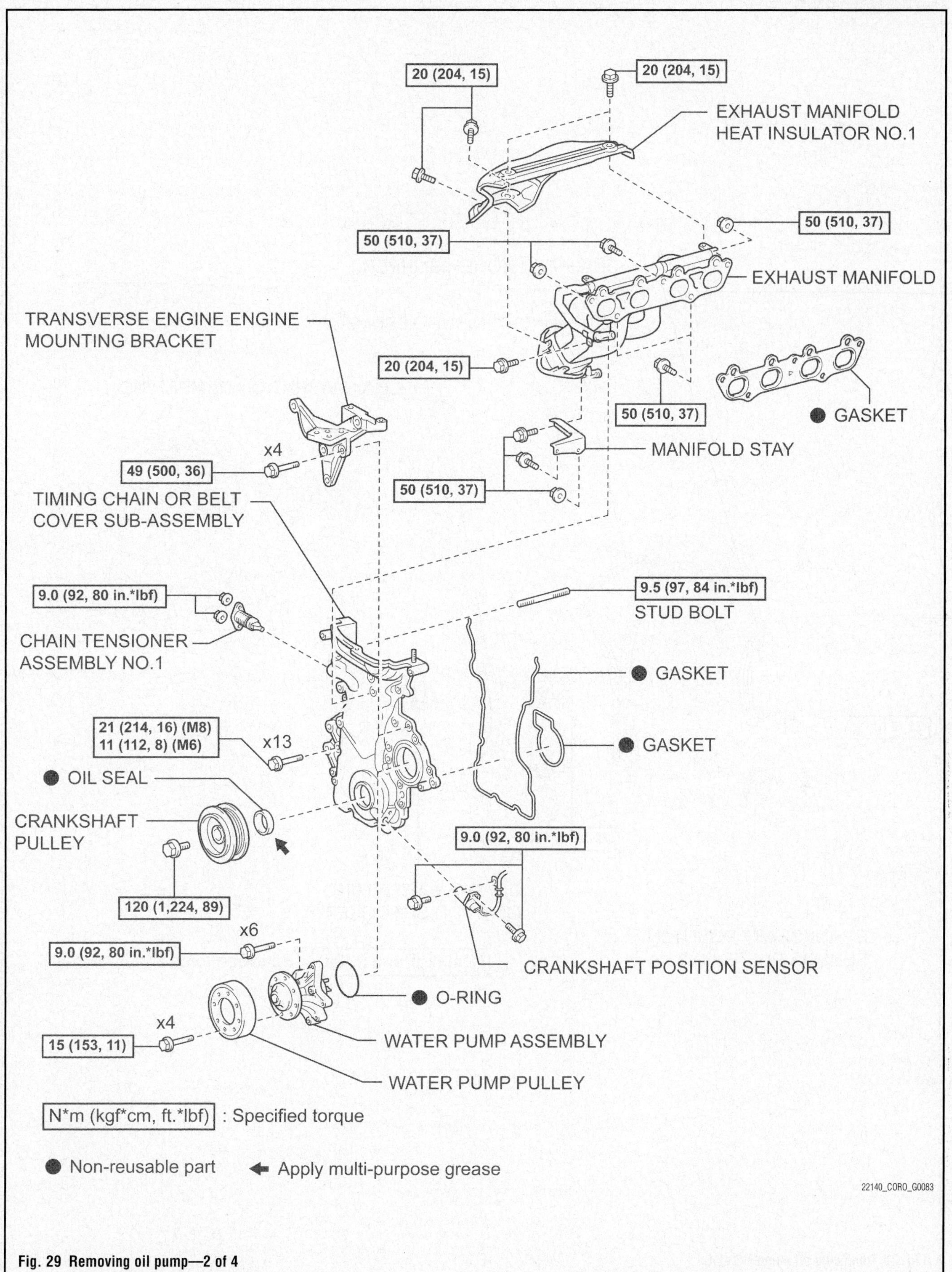

Fig. 29 Removing oil pump—2 of 4

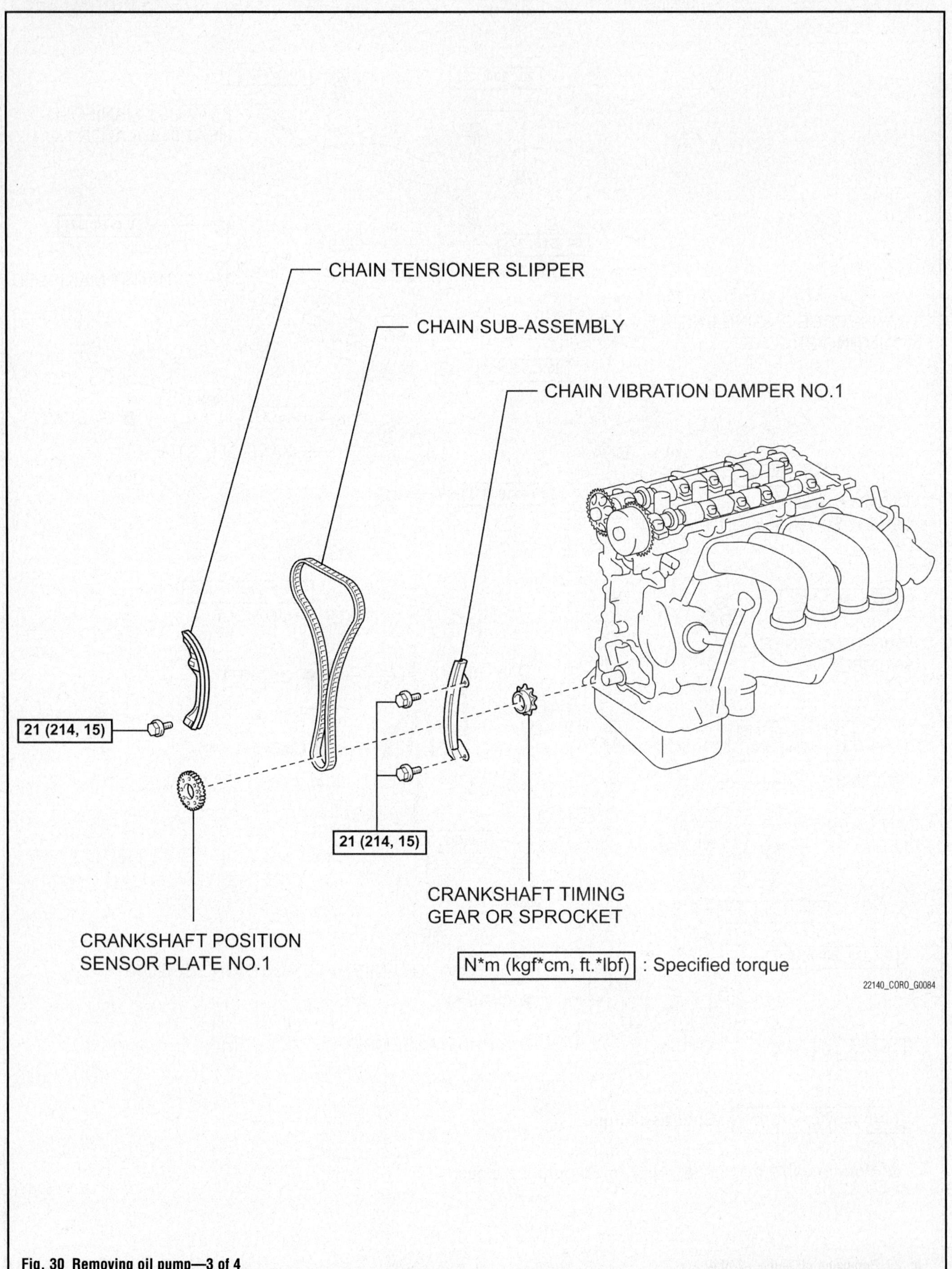

Fig. 30 Removing oil pump—3 of 4

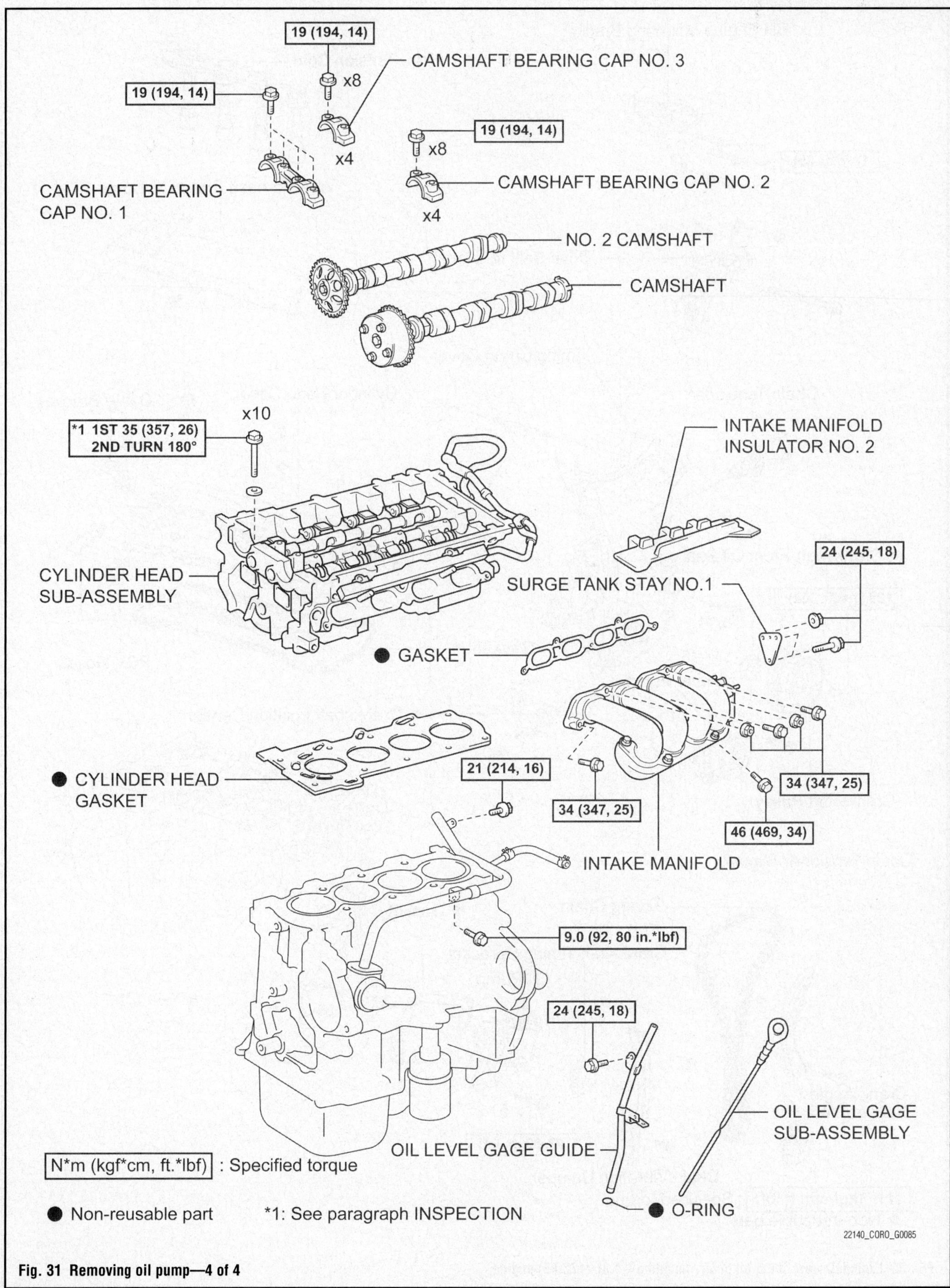

Fig. 31 Removing oil pump—4 of 4

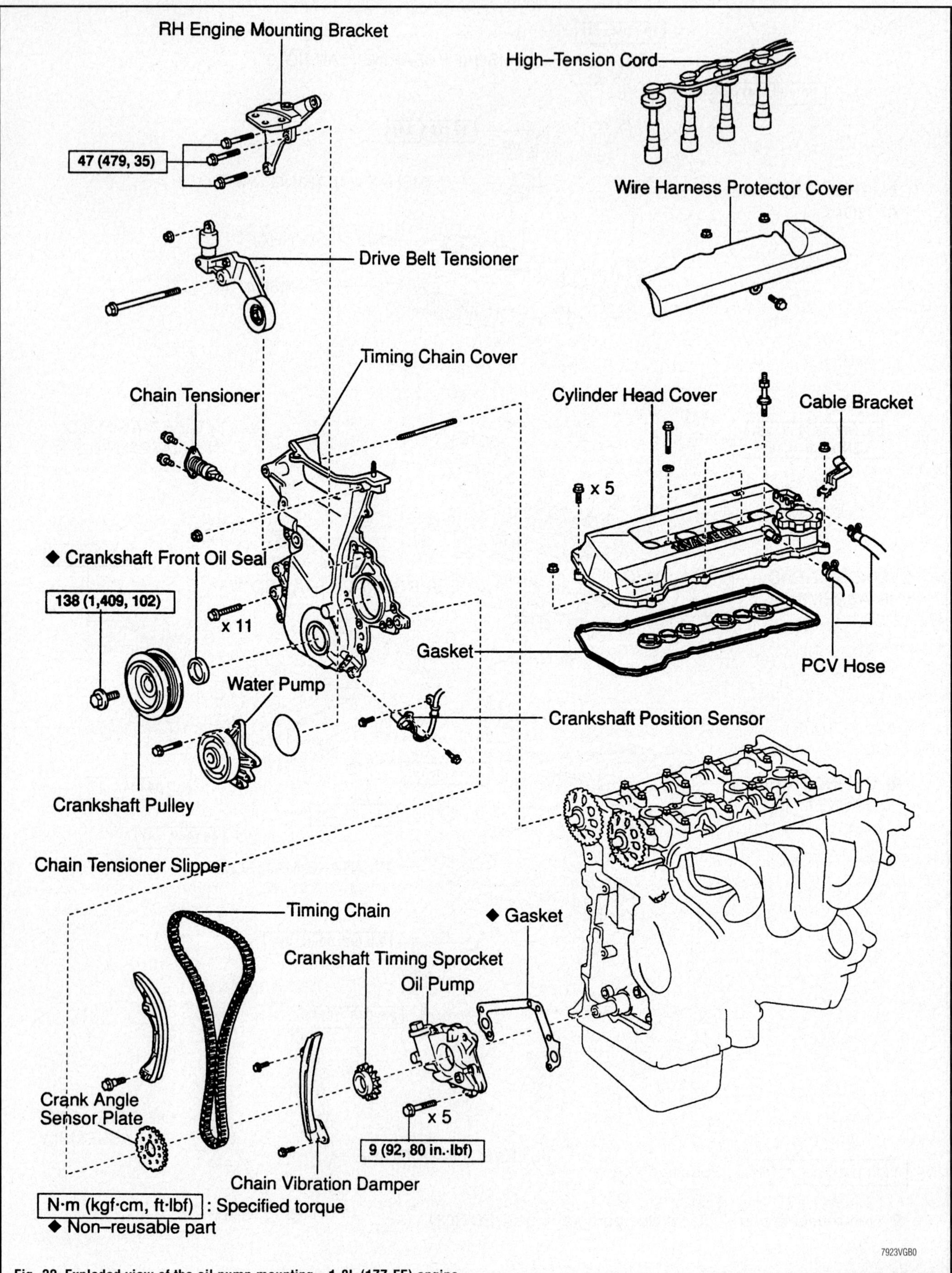

Fig. 32 Exploded view of the oil pump mounting—1.8L (1ZZ-FE) engine

MAIN BEARING TORQUE SEQUENCE

See Figures 33 through 35.

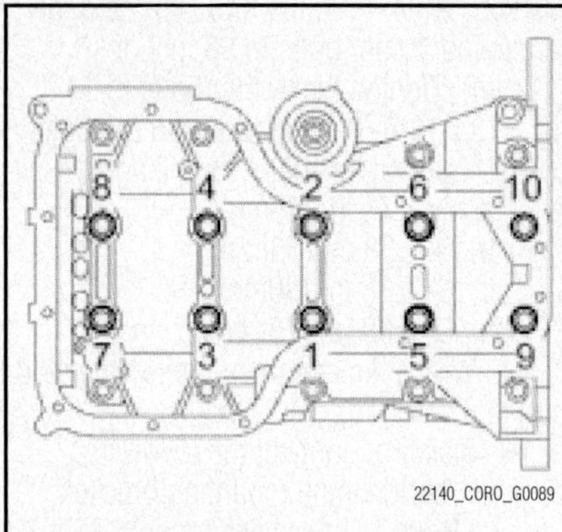

Fig. 33 Main bearing torque sequence

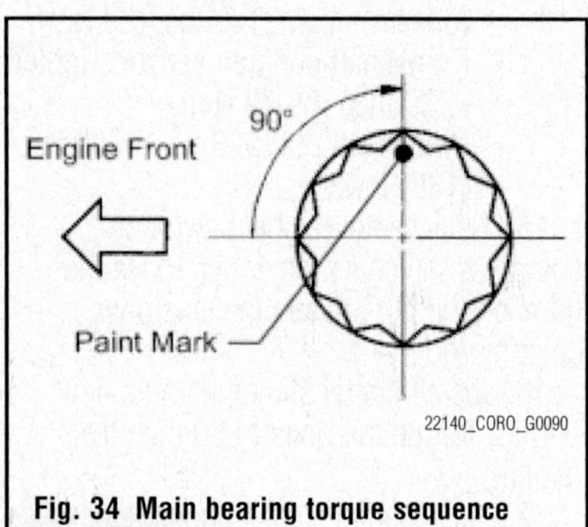

Fig. 34 Main bearing torque sequence

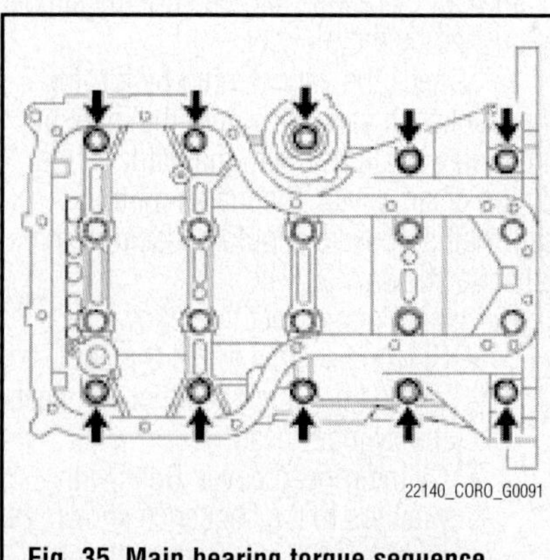

Fig. 35 Main bearing torque sequence

PISTON AND RING

POSITIONING

See Figures 36 and 37.

REAR MAIN SEAL

REMOVAL & INSTALLATION

1. Remove or disconnect the following:

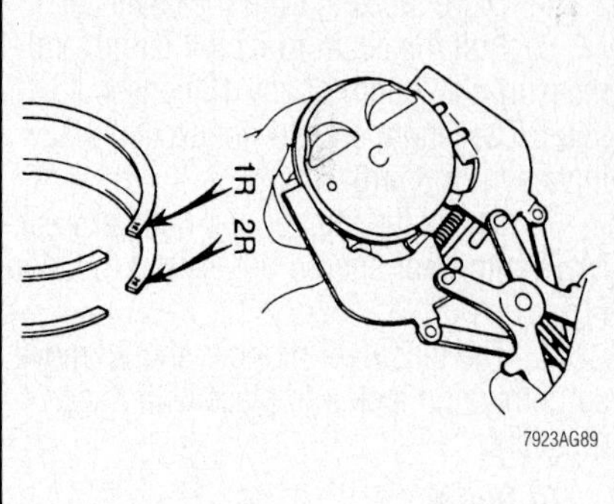

Fig. 36 1ZZ-FE engine—piston ring identification mark locations

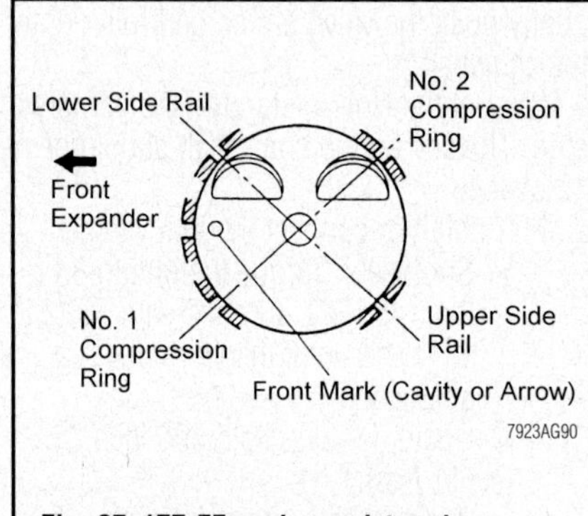

Fig. 37 1ZZ-FE engine—piston ring end-gap spacing

- Transaxle
- Clutch assembly
- Flywheel or flexplate

2. Use a small sharp knife to cut off the lip of the oil seal. Take great care not to score any metal with the knife.
3. Use a small pry tool to pry the old seal from the retaining plate. Be careful not to damage the plate. Protect the tip of the tool with tape and pad the fulcrum point with cloth.
4. Inspect the crankshaft and seal lip contact surfaces for any sign of damage.

To install:

5. Apply a light coat of multi-purpose grease to the lip of a new oil seal. Loosely fit the seal into place by hand, making sure it is not crooked.
6. Use a seal driver of the correct size to install the seal. Tap it into place until the surface of the seal is flush with the edge of the housing.

TIMING CHAIN COVER AND SEAL

REMOVAL & INSTALLATION

1. Before servicing the vehicle, refer to the precautions in the beginning of this section.
2. Drain the cooling system.
3. Remove or disconnect the following:
   - Negative battery cable
   - Right engine cover
   - Accessory drive belt and generator
   - Power steering pump, without disconnecting the hoses.
   - Right engine mount
   - Cylinder head cover
   - Crankshaft pulley
   - Crankshaft position sensor from the timing chain cover.
   - Accessory drive belt tensioner.
   - Right engine mounting bracket
   - Chain tensioner
   - Water pump
   - Timing chain cover

To install:

4. Install or connect the following:
   - New seal in the front cover
   - Silicone sealant to the timing chain cover
   - Timing chain cover
5. Water pump. Tighten the 10mm bolts marked "C" to 80 inch lbs. (9 Nm), those marked "A" to 10 ft. lbs. (13 Nm), and the remaining 10mm bolts to 96 inch lbs. (11 Nm). Tighten the 12mm bolts to 14 ft. lbs. (18.5 Nm). Be sure to install the bolts in their original locations. Bolt lengths:
   a. A: 1.77 in. (45mm).
   b. B: 1.38 in. (35mm).
   c. C: 1.18 in. (30mm).
   d. D: 0.98 in. (25mm).
6. With a Torx wrench, tighten the stud bolt to 82 inch lbs. (9.3 Nm).
   - Right engine mounting bracket and mount. Bolts, with sealant applied: 35 ft. lbs. (47 Nm).
   - Drive belt tensioner. Bolt: 51 ft. lbs. (69 Nm). Nut: 21 ft. lbs. (29 Nm).
   - Crankshaft position sensor. Tighten to 80 inch lbs. (9 Nm).
   - Crankshaft pulley. Bolt: 102 ft. lbs. (138 Nm).
7. Install or connect the following:
   - Silicone sealant to the 2 areas where the timing chain cover meets the cylinder head.
   - Cylinder head cover. Bolts with washers in the sequence shown: 80 inch lbs. (9 Nm). Bolts without washers: 96 inch lbs. (11 Nm).
   - Right engine mount. Bolts and nuts: 38 ft. lbs. (52 Nm).
   - Power steering pump
   - Alternator and drive belt
   - Right engine undercover
   - Negative battery cable
   - Washer tank

8. Fill the cooling system to the proper level.
9. Start the vehicle, check for leaks and repair if necessary.

TIMING CHAIN AND SPROCKETS

REMOVAL & INSTALLATION

See Figure 38.

1. Before servicing the vehicle, refer to the precautions in the beginning of this section.
2. Drain the cooling system.
3. Remove or disconnect the following:
   - Negative battery cable
   - Right engine cover
   - Accessory drive belt and generator
   - Power steering pump, without disconnecting the hoses.
   - Right engine mount
   - Cylinder head cover
   - Turn the crankshaft so the No. 1 piston is at TDC on the compression stroke.
   - Crankshaft pulley
   - Crankshaft position sensor from the timing chain cover.
   - Accessory drive belt tensioner.
   - Right engine mounting bracket
   - Chain tensioner
   - Water pump
   - Timing chain cover
   - Crankshaft angle sensor plate
   - Timing chain tensioner slipper
   - Timing chain and crankshaft timing sprocket.
   - Timing chain vibration damper
   - Valve timing control assembly and camshaft timing sprocket
4. Drive the seal from the cover.
5. Pull the chain to its full length and measure the length of any 16 consecutive links. The length should not exceed 4.827 inches (122.6mm).
6. Check the slipper and damper wear. Maximum wear should not exceed 0.039 in. (1mm).
7. The tensioner plunger should move smoothly and lock into place with finger pressure.

To install:

8. Apply engine oil from the tip of the intake camshaft, back to 16mm.
9. Align the timing mark on the valve timing controller with the knock pin and gently push the valve timing controller onto the camshaft.
10. Set the No.1 piston to TDC compression. The key on the crankshaft should be at 12 o'clock.
11. Install or connect the following:
   - Sprockets. Torque the bolt to 33 ft. lbs. (45 Nm). Turn the camshafts to align the point marks on the sprockets.
   - Chain damper. Bolts: 96 inch lbs. (11 Nm).
   - Timing chain and crankshaft sprocket. Be sure to align the yellow chain link with the mark on the crankshaft sprocket.
   - Timing chain on the camshaft sprockets. Align the yellow links with the marks on the camshaft sprockets.
   - Chain tensioner slipper. Bolt: 14 ft. lbs. (18.5 Nm).
   - Crankshaft angle sensor plate with the **F** mark forward
   - New seal in the front cover
   - Silicone sealant to the timing chain cover as illustrated
   - Timing chain cover
12. Water pump. Tighten the 10mm bolts marked "C" to 80 inch lbs. (9 Nm), those marked "A" to 10 ft. lbs. (13 Nm), and the remaining 10mm bolts to 96 inch lbs. (11 Nm). Tighten the 12mm bolts to 14 ft. lbs. (18.5 Nm). Be sure to install the bolts in their original locations. Bolt lengths:
   a. A: 1.77 in. (45mm).
   b. B: 1.38 in. (35mm).
   c. C: 1.18 in. (30mm).
   d. D: 0.98 in. (25mm).
13. With a Torx wrench, tighten the stud bolt to 82 inch lbs. (9.3 Nm).
14. Install or connect the following:
   - Right engine mounting bracket. Bolts, with sealant applied: 35 ft. lbs. (47 Nm).
   - Drive belt tensioner. Bolt: 51 ft. lbs. (69 Nm). Nut: 21 ft. lbs. (29 Nm).
   - Crankshaft position sensor. Tighten to 80 inch lbs. (9 Nm).
   - Crankshaft pulley. Bolt: 102 ft. lbs. (138 Nm).
15. Release the ratchet pawl and compress the chain tensioner. Place the hook on the pin to keep the tensioner compressed.
16. Install the tensioner, using a new O-ring. Torque the bolts to 80 inch lbs. (9 Nm).
17. Turn the crankshaft counterclockwise and remove the hook from the pin. Turn the crankshaft clockwise and be sure the slipper is pushed by the plunger.
18. Check the valve timing by turning the crankshaft clockwise until the mark of the pulley is aligned with the mark on the timing chain cover. The marks on the camshaft sprockets should be facing each other as shown.
19. Install or connect the following:
   - Silicone sealant to the 2 areas where the timing chain cover meets the cylinder head.
   - Cylinder head cover. Bolts with washers in the sequence shown: 80 inch lbs. (9 Nm). Bolts without washers: 96 inch lbs. (11 Nm).
   - Right engine mount. Bolts and nuts: 38 ft. lbs. (52 Nm).
   - Power steering pump
   - Alternator and drive belt
   - Right engine undercover
   - Negative battery cable
   - Washer tank
20. Fill the cooling system to the proper level.
21. Start the vehicle, check for leaks and repair if necessary.

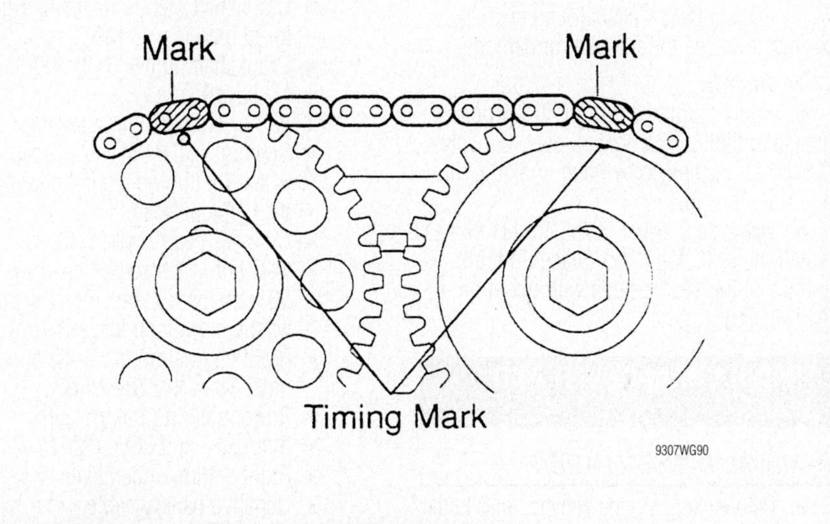

Fig. 38 Timing chain link marks—1ZZ-FE and 2ZZ-GE engines

VALVE COVERS

REMOVAL & INSTALLATION

See Figure 39.

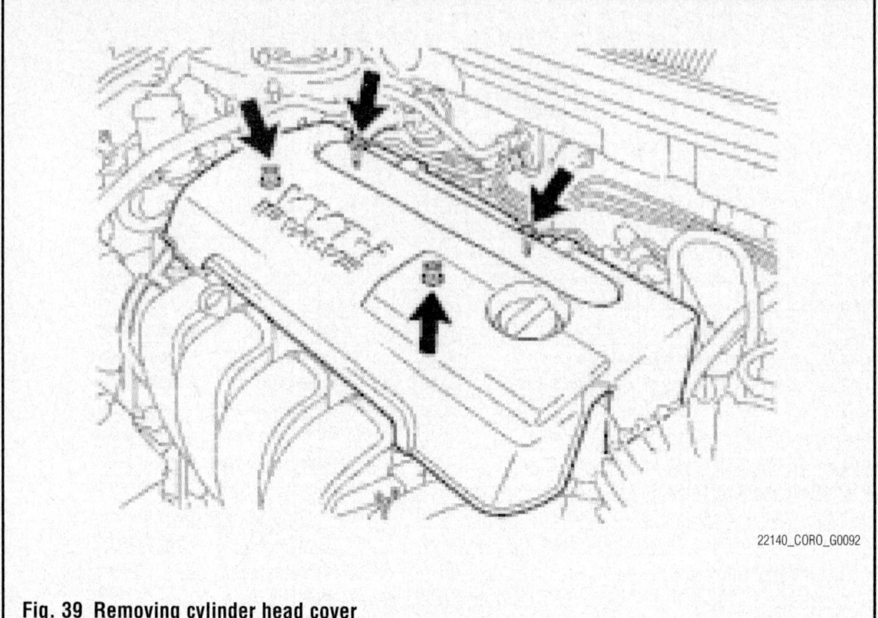

Fig. 39 Removing cylinder head cover

1. Remove the 2 nuts and 2 clips, then remove the cylinder head cover.

To install:

2. Install cylinder head cover.
3. Torque the bolts to 62 inch lbs. (7 Nm)

VALVE LASH

ADJUSTMENT

See Figures 40 through 46.

1. Set the No. 1 cylinder to the TDC/compression.
   a. Check only the valves indicated.
   b. Using a feeler gauge, measure the clearance between the valve lifter and camshaft.
      - Intake: 0.0059 to 0.0098 in. (0.15 to 0.25 mm)
      - Exhaust: 0.0098 to 0.0138 in. (0.25 to 0.35 mm)

   c. Record any out-of-specification valve clearance measurements. They will be used later to determine the required replacement valve lifters.
2. Turn the crankshaft 1 revolution (360°) to set the No. 4 cylinder to the TDC/compression.
   a. Check only the valves indicated.
   b. Using a feeler gauge, measure the clearance between the valve lifter and camshaft
      - Intake: 0.0059 to 0.0098 in. (0.15 to 0.25 mm)
      - Exhaust: 0.0098 to 0.0138 in. (0.25 to 0.35 mm)

   c. Record any out-of-specification valve clearance measurements. They will be used later to determine the required replacement valve lifters.
3. Adjust valve clearance.
   a. Remove the No. 2 camshaft timing chain.
   b. Remove the camshaft.
   c. Remove the valve lifters.

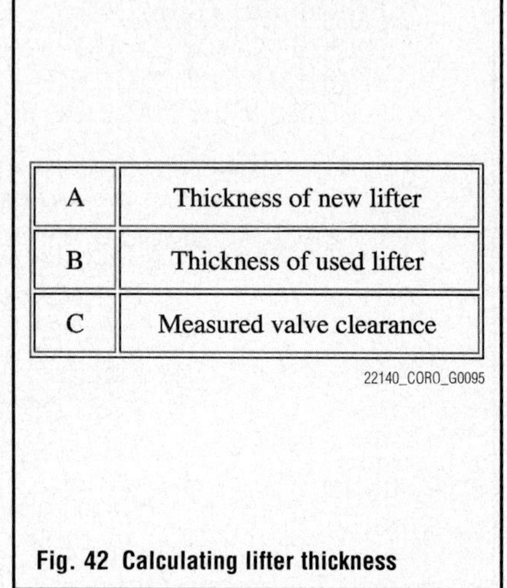

A	Thickness of new lifter
B	Thickness of used lifter
C	Measured valve clearance

22140_CORO_G0095

Fig. 42 Calculating lifter thickness

No. 4 Cylinder TDC/Compression:

22140_CORO_G0094

Fig. 41 Second valve position

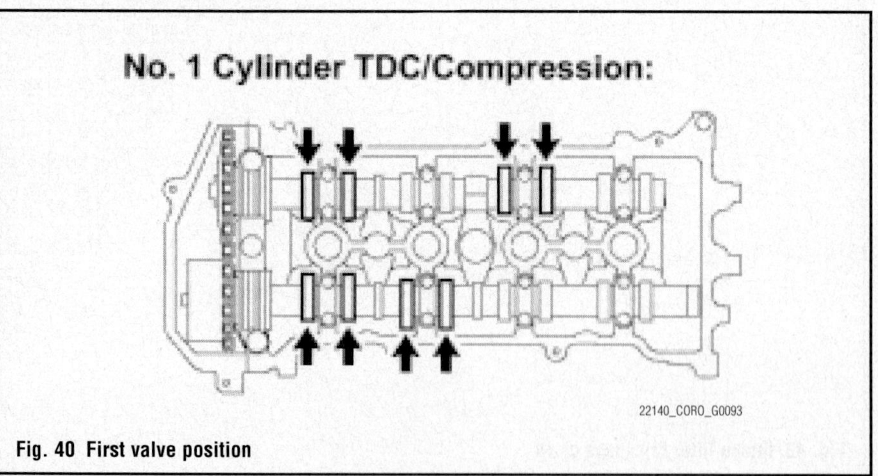

Fig. 40 First valve position

Valve Lifter Selection Chart (Intake)

0.911 - 0.930 (0.0359 - 0.0366)	0.891 - 0.910 (0.0351 - 0.0358)	0.871 - 0.890 (0.0343 - 0.0350)	0.851 - 0.870 (0.0335 - 0.0343)	0.831 - 0.850 (0.0327 - 0.0335)	0.811 - 0.830 (0.0319 - 0.0327)	0.791 - 0.810 (0.0311 - 0.0319)	0.771 - 0.790 (0.0304 - 0.0311)	0.751 - 0.770 (0.0296 - 0.0303)	0.731 - 0.750 (0.0288 - 0.0295)	0.711 - 0.730 (0.0280 - 0.0287)	0.691 - 0.710 (0.0272 - 0.0280)	0.671 - 0.690 (0.0264 - 0.0272)	0.651 - 0.670 (0.0256 - 0.0264)	0.631 - 0.650 (0.0248 - 0.0256)	0.611 - 0.630 (0.0241 - 0.0248)	0.591 - 0.610 (0.0233 - 0.0240)	0.571 - 0.590 (0.0225 - 0.0232)	0.551 - 0.570 (0.0217 - 0.0224)	0.531 - 0.550 (0.0209 - 0.0217)	0.511 - 0.530 (0.0201 - 0.0209)	0.491 - 0.510 (0.0193 - 0.0201)	0.471 - 0.490 (0.0185 - 0.0193)	0.451 - 0.470 (0.0178 - 0.0185)	0.431 - 0.450 (0.0170 - 0.0177)	0.411 - 0.430 (0.0162 - 0.0169)	0.391 - 0.410 (0.0154 - 0.0161)	0.371 - 0.390 (0.0146 - 0.0154)	0.351 - 0.370 (0.0138 - 0.0146)	0.331 - 0.350 (0.0130 - 0.0138)	0.311 - 0.330 (0.0122 - 0.0130)	0.291 - 0.310 (0.0115 - 0.0122)	0.271 - 0.290 (0.0107 - 0.0114)	0.251 - 0.270 (0.0099 - 0.0106)	0.150 - 0.250 (0.0059 - 0.0098)	0.131 - 0.149 (0.0052 - 0.0059)	0.111 - 0.130 (0.0044 - 0.0051)	0.091 - 0.110 (0.0036 - 0.0043)	0.071 - 0.090 (0.0028 - 0.0035)	0.051 - 0.070 (0.0020 - 0.0028)	0.031 - 0.050 (0.0012 - 0.0020)	0.000 - 0.030 (0.0000 - 0.0012)	Measured Clearance mm (in.) / Installed Lifter Thickness mm (in.)
74	74	74	72	70	68	66	64	62	60	58	56	54	52	50	48	46	44	42	40	38	36	34	32	30	28	26	24	22	20	18	16	14	12									5.060 (0.1992)
	74	74	74	72	70	68	66	64	62	60	58	56	54	52	50	48	46	44	42	40	38	36	34	32	30	28	26	24	22	20	18	16	14		06							5.080 (0.2000)
		74	74	74	72	70	68	66	64	62	60	58	56	54	52	50	48	46	44	42	40	38	36	34	32	30	28	26	24	22	20	18	16		06	06						5.100 (0.2008)
			74	74	74	72	70	68	66	64	62	60	58	56	54	52	50	48	46	44	42	40	38	36	34	32	30	28	26	24	22	20	18		06	06	06					5.120 (0.2016)
				74	74	74	72	70	68	66	64	62	60	58	56	54	52	50	48	46	44	42	40	38	36	34	32	30	28	26	24	22	20		08	06	06	06				5.140 (0.2024)
					74	74	74	72	70	68	66	64	62	60	58	56	54	52	50	48	46	44	42	40	38	36	34	32	30	28	26	24	22		10	08	06	06	06			5.160 (0.2031)
						74	74	74	72	70	68	66	64	62	60	58	56	54	52	50	48	46	44	42	40	38	36	34	32	30	28	26	24		12	10	08	06	06	06		5.180 (0.2039)
							74	74	74	72	70	68	66	64	62	60	58	56	54	52	50	48	46	44	42	40	38	36	34	32	30	28	26		14	12	10	08	06	06		5.200 (0.2047)
								74	74	74	72	70	68	66	64	62	60	58	56	54	52	50	48	46	44	42	40	38	36	34	32	30	28		16	14	12	10	08	06	06	5.210 (0.2051)
								74	74	74	72	70	68	66	64	62	60	58	56	54	52	50	48	46	44	42	40	38	36	34	32	30	28		16	14	12	10	08	06	06	5.220 (0.2055)
									74	74	74	72	70	68	66	64	62	60	58	56	54	52	50	48	46	44	42	40	38	36	34	32	30		18	16	14	12	10	08	06	5.230 (0.2059)
									74	74	74	72	70	68	66	64	62	60	58	56	54	52	50	48	46	44	42	40	38	36	34	32	30		18	16	14	12	10	08	06	5.240 (0.2063)
										74	74	74	72	70	68	66	64	62	60	58	56	54	52	50	48	46	44	42	40	38	36	34	32		20	18	16	14	12	10	06	5.250 (0.2067)
										74	74	74	72	70	68	66	64	62	60	58	56	54	52	50	48	46	44	42	40	38	36	34	32		20	18	16	14	12	10	08	5.260 (0.2071)
											74	74	74	72	70	68	66	64	62	60	58	56	54	52	50	48	46	44	42	40	38	36	34		22	20	18	16	14	12	10	5.270 (0.2075)
											74	74	74	72	70	68	66	64	62	60	58	56	54	52	50	48	46	44	42	40	38	36	34		22	20	18	16	14	12	10	5.280 (0.2079)
												74	74	74	72	70	68	66	64	62	60	58	56	54	52	50	48	46	44	42	40	38	36		24	22	20	18	16	14	12	5.290 (0.2083)
												74	74	74	72	70	68	66	64	62	60	58	56	54	52	50	48	46	44	42	40	38	36		24	22	20	18	16	14	12	5.300 (0.2087)
													74	74	74	72	70	68	66	64	62	60	58	56	54	52	50	48	46	44	42	40	38		26	24	22	20	18	16	14	5.310 (0.2091)
													74	74	74	72	70	68	66	64	62	60	58	56	54	52	50	48	46	44	42	40	38		26	24	22	20	18	16	14	5.320 (0.2094)
														74	74	74	72	70	68	66	64	62	60	58	56	54	52	50	48	46	44	42	40		28	26	24	22	20	18	16	5.330 (0.2098)
														74	74	74	72	70	68	66	64	62	60	58	56	54	52	50	48	46	44	42	40		28	26	24	22	20	18	16	5.340 (0.2102)
															74	74	74	72	70	68	66	64	62	60	58	56	54	52	50	48	46	44	42		30	28	26	24	22	20	18	5.350 (0.2106)
															74	74	74	72	70	68	66	64	62	60	58	56	54	52	50	48	46	44	42		30	28	26	24	22	20	18	5.360 (0.2110)
																74	74	74	72	70	68	66	64	62	60	58	56	54	52	50	48	46	44		32	30	28	26	24	22	20	5.370 (0.2114)
																74	74	74	72	70	68	66	64	62	60	58	56	54	52	50	48	46	44		32	30	28	26	24	22	20	5.380 (0.2118)
																	74	74	74	72	70	68	66	64	62	60	58	56	54	52	50	48	46		34	32	30	28	26	24	22	5.390 (0.2122)
																	74	74	74	72	70	68	66	64	62	60	58	56	54	52	50	48	46		34	32	30	28	26	24	22	5.400 (0.2126)
																		74	74	74	72	70	68	66	64	62	60	58	56	54	52	50	48		36	34	32	30	28	26	24	5.410 (0.2130)
																		74	74	74	72	70	68	66	64	62	60	58	56	54	52	50	48		36	34	32	30	28	26	24	5.420 (0.2134)
																			74	74	74	72	70	68	66	64	62	60	58	56	54	52	50		38	36	34	32	30	28	26	5.430 (0.2138)
																			74	74	74	72	70	68	66	64	62	60	58	56	54	52	50		38	36	34	32	30	28	26	5.440 (0.2142)
																				74	74	74	72	70	68	66	64	62	60	58	56	54	52		40	38	36	34	32	30	28	5.450 (0.2146)
																				74	74	74	72	70	68	66	64	62	60	58	56	54	52		40	38	36	34	32	30	28	5.460 (0.2150)
																					74	74	74	72	70	68	66	64	62	60	58	56	54		42	40	38	36	34	32	30	5.470 (0.2154)
																					74	74	74	72	70	68	66	64	62	60	58	56	54		42	40	38	36	34	32	30	5.480 (0.2157)
																						74	74	74	72	70	68	66	64	62	60	58	56		44	42	40	38	36	34	32	5.490 (0.2161)
																						74	74	74	72	70	68	66	64	62	60	58	56		44	42	40	38	36	34	32	5.500 (0.2165)
																							74	74	74	72	70	68	66	64	62	60	58		46	44	42	40	38	36	34	5.510 (0.2169)
																							74	74	74	72	70	68	66	64	62	60	58		46	44	42	40	38	36	34	5.520 (0.2173)
																								74	74	74	72	70	68	66	64	62	60		48	46	44	42	40	38	36	5.530 (0.2177)
																								74	74	74	72	70	68	66	64	62	60		48	46	44	42	40	38	36	5.540 (0.2181)
																									74	74	74	72	70	68	66	64	62		50	48	46	44	42	40	38	5.550 (0.2185)
																									74	74	74	72	70	68	66	64	62		50	48	46	44	42	40	38	5.560 (0.2189)
																										74	74	74	72	70	68	66	64		52	50	48	46	44	42	40	5.570 (0.2193)
																										74	74	74	72	70	68	66	64		52	50	48	46	44	42	40	5.580 (0.2197)
																										74	74	74	74	72	70	68	66		54	52	50	48	46	44	42	5.590 (0.2201)
																											74	74	74	72	70	68	66		54	52	50	48	46	44	42	5.600 (0.2205)
																												74	74	74	72	70	68		56	54	52	50	48	46	44	5.620 (0.2213)
																													74	74	74	72	70		58	56	54	52	50	48	46	5.640 (0.2220)
																														74	74	74	72		60	58	56	54	52	50	48	5.660 (0.2228)
																															74	74	74		62	60	58	56	54	52	50	5.680 (0.2236)
																																74	74		64	62	60	58	56	54	52	5.700 (0.2244)
																																	74		66	64	62	60	58	56	54	5.720 (0.2252)
																																			68	66	64	62	60	58	56	5.740 (0.2260)

22140_CORO_G0096

Fig. 43 Intake lifter thickness chart

d. Using a micrometer, measure the thickness of the removed valve lifters.

e. Calculate the thickness of a new lifter so that the valve clearance comes within the specified values.

4. Valve clearance:
   a. Intake
   b. A = B + (C - 0.20 mm (0.0079 in.))
   c. Exhaust
   d. A = B + (C - 0.30 mm (0.0118 in.))
5. Example:
   a. Measured intake valve clearance = 0.40 mm (0.0158 in.)
   b. 0.40 mm (0.0158 in.) - 0.20 mm (0.0079 in.) = 0.20 mm (0.0079 in.)
   c. (Measured - Specification = Excess clearance)
   d. Used lifter measurement = 5.250 mm (0.2067 in.)
   e. 0.20 mm (0.0079 in.) + 5.250 mm (0.2067 in.) = 5.450 mm (0.2146 in.)
   f. (Excess clearance + Used lifter = Ideal new lifter)
   g. Closest new lifter = 5.460 mm (0.2150 in.)
   h. Select No. 46 lifter

➡Select a new lifter with a thickness as close to the calculated values as possible. Lifters are available in 35 sizes in increments of 0.020mm (0.0008 in.), from 5.060 mm (0.1992 in.) to 5.740 mm (0.2260 in.).

   i. Refer to the new lifter thickness table.
6. Install replacement valve lifters.
   a. Install the camshaft.
   b. Install the No. 2 camshaft.

Lifter No.	Thickness	Lifter No.	Thickness	Lifter No.	Thickness
06	5.060 (0.1992)	30	5.300 (0.2087)	54	5.540 (0.2181)
08	5.080 (0.2000)	32	5.320 (0.2094)	56	5.560 (0.2189)
10	5.100 (0.2008)	34	5.340 (0.2102)	58	5.580 (0.2197)
12	5.120 (0.2016)	36	5.360 (0.2110)	60	5.600 (0.2205)
14	5.140 (0.2024)	38	5.380 (0.2118)	62	5.620 (0.2213)
16	5.160 (0.2031)	40	5.400 (0.2126)	64	5.640 (0.2220)
18	5.180 (0.2039)	42	5.420 (0.2134)	66	5.660 (0.2228)
20	5.200 (0.2047)	44	5.440 (0.2142)	68	5.680 (0.2236)
22	5.220 (0.2055)	46	5.460 (0.2150)	70	5.700 (0.2244)
24	5.240 (0.2063)	48	5.480 (0.2157)	72	5.720 (0.2252)
26	5.260 (0.2071)	50	5.500 (0.2165)	74	5.740 (0.2260)
28	5.280 (0.2079)	52	5.520 (0.2173)		

22140_CORO_G0097

Fig. 44 New Intake lifter thickness chart

Valve Lifter Selection Chart (Exhaust)

1.011 - 1.030 (0.0398 - 0.0406)	0.991 - 1.010 (0.0390 - 0.0398)	0.971 - 0.990 (0.0382 - 0.0390)	0.951 - 0.970 (0.0374 - 0.0382)	0.931 - 0.950 (0.0367 - 0.0374)	0.911 - 0.930 (0.0359 - 0.0366)	0.891 - 0.910 (0.0351 - 0.0358)	0.871 - 0.890 (0.0343 - 0.0350)	0.851 - 0.870 (0.0335 - 0.0343)	0.831 - 0.850 (0.0327 - 0.0335)	0.811 - 0.830 (0.0319 - 0.0327)	0.791 - 0.810 (0.0311 - 0.0319)	0.771 - 0.790 (0.0304 - 0.0311)	0.751 - 0.770 (0.0296 - 0.0303)	0.731 - 0.750 (0.0288 - 0.0295)	0.711 - 0.730 (0.0280 - 0.0287)	0.691 - 0.710 (0.0272 - 0.0280)	0.671 - 0.690 (0.0264 - 0.0272)	0.651 - 0.670 (0.0256 - 0.0264)	0.631 - 0.650 (0.0248 - 0.0256)	0.611 - 0.630 (0.0241 - 0.0248)	0.591 - 0.610 (0.0233 - 0.0240)	0.571 - 0.590 (0.0225 - 0.0232)	0.551 - 0.570 (0.0217 - 0.0224)	0.531 - 0.550 (0.0209 - 0.0217)	0.511 - 0.530 (0.0201 - 0.0209)	0.491 - 0.510 (0.0193 - 0.0201)	0.471 - 0.490 (0.0185 - 0.0193)	0.451 - 0.470 (0.0178 - 0.0185)	0.431 - 0.450 (0.0170 - 0.0177)	0.411 - 0.430 (0.0162 - 0.0169)	0.391 - 0.410 (0.0154 - 0.0161)	0.371 - 0.390 (0.0146 - 0.0154)	0.351 - 0.370 (0.0138 - 0.0146)	0.250 - 0.350 (0.0098 - 0.0138)	0.231 - 0.249 (0.0091 - 0.0098)	0.211 - 0.230 (0.0083 - 0.0091)	0.191 - 0.210 (0.0075 - 0.0083)	0.171 - 0.190 (0.0067 - 0.0075)	0.151 - 0.170 (0.0059 - 0.0067)	0.131 - 0.150 (0.0052 - 0.0059)	0.111 - 0.130 (0.0044 - 0.0051)	0.091 - 0.110 (0.0036 - 0.0043)	0.071 - 0.090 (0.0028 - 0.0035)	0.051 - 0.070 (0.0020 - 0.0028)	0.031 - 0.050 (0.0012 - 0.0020)	0.000 - 0.030 (0.0000 - 0.0012)	Measured Clearance mm (in.) / Installed Shim Thickness mm (in.)
74	74	74	72	70	68	66	64	62	60	58	56	54	52	50	48	46	44	42	40	38	36	34	32	30	28	26	24	22	20	18	16	14	12														5.060 (0.1992)
	74	74	74	72	70	68	66	64	62	60	58	56	54	52	50	48	46	44	42	40	38	36	34	32	30	28	26	24	22	20	18	16	14		06												5.080 (0.2000)
		74	74	74	72	70	68	66	64	62	60	58	56	54	52	50	48	46	44	42	40	38	36	34	32	30	28	26	24	22	20	18	16		06	06											5.100 (0.2008)
			74	74	74	72	70	68	66	64	62	60	58	56	54	52	50	48	46	44	42	40	38	36	34	32	30	28	26	24	22	20	18		06	06	06										5.120 (0.2016)
				74	74	74	72	70	68	66	64	62	60	58	56	54	52	50	48	46	44	42	40	38	36	34	32	30	28	26	24	22	20		08	06	06	06									5.140 (0.2024)
					74	74	74	72	70	68	66	64	62	60	58	56	54	52	50	48	46	44	42	40	38	36	34	32	30	28	26	24	22		10	08	06	06	06								5.160 (0.2031)
						74	74	74	72	70	68	66	64	62	60	58	56	54	52	50	48	46	44	42	40	38	36	34	32	30	28	26	24		12	10	08	06	06								5.180 (0.2039)
							74	74	74	72	70	68	66	64	62	60	58	56	54	52	50	48	46	44	42	40	38	36	34	32	30	28	26		14	12	10	08	06	06							5.200 (0.2047)
								74	74	74	72	70	68	66	64	62	60	58	56	54	52	50	48	46	44	42	40	38	36	34	32	30	28		16	14	12	10	08	06	06						5.210 (0.2051)
								74	74	74	72	70	68	66	64	62	60	58	56	54	52	50	48	46	44	42	40	38	36	34	32	30	28		16	14	12	10	08	06	06						5.220 (0.2055)
									74	74	74	72	70	68	66	64	62	60	58	56	54	52	50	48	46	44	42	40	38	36	34	32	30		18	16	14	12	10	08	06	06					5.230 (0.2059)
									74	74	74	72	70	68	66	64	62	60	58	56	54	52	50	48	46	44	42	40	38	36	34	32	30		18	16	14	12	10	08	06	06	06				5.240 (0.2063)
										74	74	74	72	70	68	66	64	62	60	58	56	54	52	50	48	46	44	42	40	38	36	34	32		20	18	16	14	12	10	08	06	06				5.250 (0.2067)
										74	74	74	72	70	68	66	64	62	60	58	56	54	52	50	48	46	44	42	40	38	36	34	32		20	18	16	14	12	10	08	06	06	06			5.260 (0.2071)
											74	74	74	72	70	68	66	64	62	60	58	56	54	52	50	48	46	44	42	40	38	36	34		22	20	18	16	14	12	10	08	06	06			5.270 (0.2075)
											74	74	74	72	70	68	66	64	62	60	58	56	54	52	50	48	46	44	42	40	38	36	34		22	20	18	16	14	12	10	08	06	06	06		5.280 (0.2079)
												74	74	74	72	70	68	66	64	62	60	58	56	54	52	50	48	46	44	42	40	38	36		24	22	20	18	16	14	12	10	08	06	06		5.290 (0.2083)
												74	74	74	72	70	68	66	64	62	60	58	56	54	52	50	48	46	44	42	40	38	36		24	22	20	18	16	14	12	10	08	06	06		5.300 (0.2087)
													74	74	74	72	70	68	66	64	62	60	58	56	54	52	50	48	46	44	42	40	38		26	24	22	20	18	16	14	12	10	08	06	06	5.310 (0.2091)
													74	74	74	72	70	68	66	64	62	60	58	56	54	52	50	48	46	44	42	40	38		26	24	22	20	18	16	14	12	10	08	06	06	5.320 (0.2094)
														74	74	74	72	70	68	66	64	62	60	58	56	54	52	50	48	46	44	42	40		28	26	24	22	20	18	16	14	12	10	08	06	5.330 (0.2098)
														74	74	74	72	70	68	66	64	62	60	58	56	54	52	50	48	46	44	42	40		28	26	24	22	20	18	16	14	12	10	08	06	5.340 (0.2102)
															74	74	74	72	70	68	66	64	62	60	58	56	54	52	50	48	46	44	42		30	28	26	24	22	20	18	16	14	12	10	08	5.350 (0.2106)
															74	74	74	72	70	68	66	64	62	60	58	56	54	52	50	48	46	44	42		30	28	26	24	22	20	18	16	14	12	10	08	5.360 (0.2110)
																74	74	74	72	70	68	66	64	62	60	58	56	54	52	50	48	46	44		32	30	28	26	24	22	20	18	16	14	12	10	5.370 (0.2114)
																74	74	74	72	70	68	66	64	62	60	58	56	54	52	50	48	46	44		32	30	28	26	24	22	20	18	16	14	12	10	5.380 (0.2118)
																	74	74	74	72	70	68	66	64	62	60	58	56	54	52	50	48	46		34	32	30	28	26	24	22	20	18	16	14	12	5.390 (0.2122)
																	74	74	74	72	70	68	66	64	62	60	58	56	54	52	50	48	46		34	32	30	28	26	24	22	20	18	16	14	12	5.400 (0.2126)
																		74	74	74	72	70	68	66	64	62	60	58	56	54	52	50	48		36	34	32	30	28	26	24	22	20	18	16	14	5.410 (0.2130)
																		74	74	74	72	70	68	66	64	62	60	58	56	54	52	50	48		36	34	32	30	28	26	24	22	20	18	16	14	5.420 (0.2134)
																			74	74	74	72	70	68	66	64	62	60	58	56	54	52	50		38	36	34	32	30	28	26	24	22	20	18	16	5.430 (0.2138)
																			74	74	74	72	70	68	66	64	62	60	58	56	54	52	50		38	36	34	32	30	28	26	24	22	20	18	16	5.440 (0.2142)
																			74	74	74	74	72	70	68	66	64	62	60	58	56	54	52		40	38	36	34	32	30	28	26	24	22	20	18	5.450 (0.2146)
																				74	74	74	72	70	68	66	64	62	60	58	56	54	52		40	38	36	34	32	30	28	26	24	22	20	18	5.460 (0.2150)
																					74	74	74	72	70	68	66	64	62	60	58	56	54		42	40	38	36	34	32	30	28	26	24	22	20	5.470 (0.2154)
																					74	74	74	72	70	68	66	64	62	60	58	56	54		42	40	38	36	34	32	30	28	26	24	22	20	5.480 (0.2157)
																						74	74	74	72	70	68	66	64	62	60	58	56		44	42	40	38	36	34	32	30	28	26	24	22	5.490 (0.2161)
																						74	74	74	72	70	68	66	64	62	60	58	56		44	42	40	38	36	34	32	30	28	26	24	22	5.500 (0.2165)
																							74	74	74	72	70	68	66	64	62	60	58		46	44	42	40	38	36	34	32	30	28	26	24	5.510 (0.2169)
																							74	74	74	72	70	68	68	64	62	60	58		46	44	42	40	38	36	34	32	30	28	26	24	5.520 (0.2173)
																								74	74	74	72	70	68	66	64	62	60		48	46	44	42	40	38	36	34	32	30	28	26	5.530 (0.2177)
																								74	74	74	72	70	70	66	64	62	60		48	46	44	42	40	38	36	34	32	30	28	26	5.540 (0.2181)
																									74	74	74	72	70	68	66	64	62		50	48	46	44	42	40	38	36	34	32	30	28	5.550 (0.2185)
																									74	74	74	72	72	68	66	64	62		50	48	46	44	42	40	38	36	34	32	30	28	5.560 (0.2189)
																										74	74	74	72	70	68	66	64		52	50	48	46	44	42	40	38	36	34	32	30	5.570 (0.2193)
																										74	74	74	74	70	68	66	64		52	50	48	46	44	42	40	38	36	34	32	30	5.580 (0.2197)
																											74	74	74	72	70	68	66		54	52	50	48	46	44	42	40	38	36	34	32	5.590 (0.2201)
																											74	74	74	72	70	68	66		54	52	50	48	46	44	42	40	38	36	34	32	5.600 (0.2205)
																												74	74	74	72	70	68		56	54	52	50	48	46	44	42	40	38	36	34	5.620 (0.2213)
																													74	74	74	72	70		58	56	54	52	50	48	46	44	42	40	38	36	5.640 (0.2220)
																														74	74	74	72		60	58	56	54	52	50	48	46	44	42	40	38	5.660 (0.2228)
																															74	74	74		62	60	58	56	54	52	50	48	46	44	42	40	5.680 (0.2236)
																																74	74		64	62	60	58	56	54	52	50	48	46	44	42	5.700 (0.2244)
																																	74		66	64	62	60	58	56	54	52	50	48	46	44	5.720 (0.2252)
																																			68	66	64	62	60	58	56	54	52	50	48	46	5.740 (0.2260)

22140_CORO_G0098

Fig. 45 Exhaust lifter thickness chart

Lifter No.	Thickness	Lifter No.	Thickness	Lifter No.	Thickness
06	5.060 (0.1992)	30	5.300 (0.2087)	54	5.540 (0.2181)
08	5.080 (0.2000)	32	5.320 (0.2094)	56	5.560 (0.2189)
10	5.100 (0.2008)	34	5.340 (0.2102)	58	5.580 (0.2197)
12	5.120 (0.2016)	36	5.360 (0.2110)	60	5.600 (0.2205)
14	5.140 (0.2024)	38	5.380 (0.2118)	62	5.620 (0.2213)
16	5.160 (0.2031)	40	5.400 (0.2126)	64	5.640 (0.2220)
18	5.180 (0.2039)	42	5.420 (0.2134)	66	5.660 (0.2228)
20	5.200 (0.2047)	44	5.440 (0.2142)	68	5.680 (0.2236)
22	5.220 (0.2055)	46	5.460 (0.2150)	70	5.700 (0.2244)
24	5.240 (0.2063)	48	5.480 (0.2157)	72	5.720 (0.2252)
26	5.260 (0.2071)	50	5.500 (0.2165)	74	5.740 (0.2260)
28	5.280 (0.2079)	52	5.520 (0.2173)		

22140_CORO_G0099

Fig. 46 New Exhaust lifter thickness chart

ENGINE PERFORMANCE & EMISSION CONTROL

ACCELERATOR PEDAL POSITION (APP) SENSOR

LOCATION

The Accelerator Pedal Position (APP) sensor is mounted on the accelerator pedal bracket and has 2 sensor circuits: VPA1 (main) and VPA2 (sub).

REMOVAL & INSTALLATION

1. Disconnect Accelerator Pedal Position Sensor connector.
2. Remove attaching screws.

To install:

3. Replace sensor.
4. Install attaching screws.
5. Connect Accelerator Pedal Position Sensor connector

CAMSHAFT POSITION (CMP) SENSOR

OPERATION

The Camshaft Position (CMP) sensor consists of a magnet and an iron core which is wrapped with copper wire, and is installed onto the cylinder head. When the camshaft rotates, each of 3 teeth on the camshaft passes through the CMP sensor. This activates the internal magnet in the sensor, generating a voltage in the copper wire. The camshaft rotation is synchronized with the crankshaft rotation. When the crankshaft turns twice the voltage is generated 3 times in the CMP sensor. The generated voltage in the sensor acts as a signal, allowing the ECM to locate the camshaft position. This signal is then used to control ignition timing, fuel injection timing, and the VVT system.

REMOVAL & INSTALLATION

1. Remove camshaft position sensor retaining bolt.
2. Disconnect camshaft position sensor connector.
3. Remove sensor.

To install:

4. Reverse removal procedure.

COOLANT TEMPERATURE SENSOR

REMOVAL & INSTALLATION

See Figure 47.

1. Disconnect coolant temperature sensor connector.

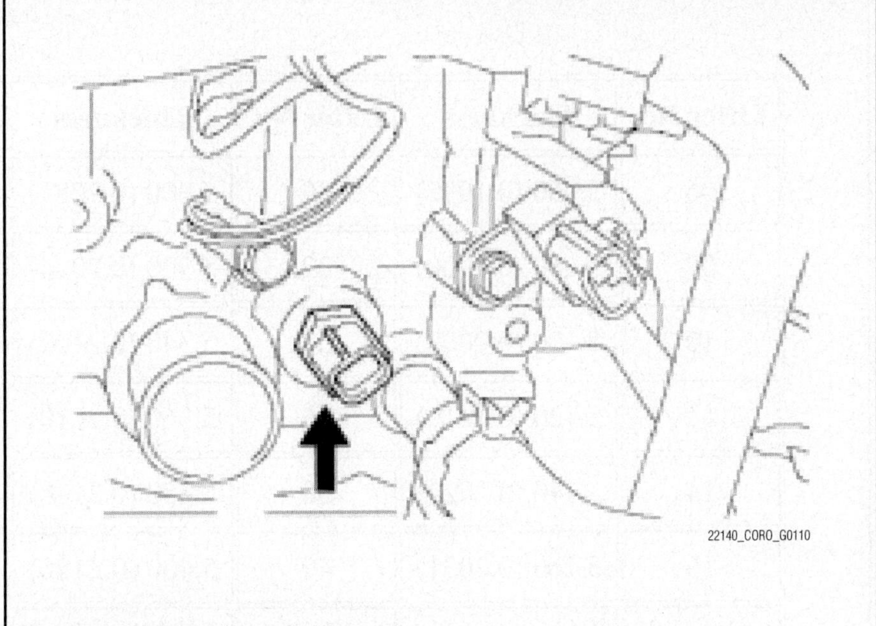

Fig. 47 Removing engine coolant temperature sensor

2. Using appropriate service tool, remove coolant temperature sensor.

To install:

3. Reverse removal procedure.

CRANKSHAFT POSITION (CKP) SENSOR

REMOVAL & INSTALLATION

1. Remove crankshaft position sensor retaining bolt.
2. Disconnect crankshaft position sensor connector.
3. Remove sensor.

To install:

4. Reverse removal procedure.

ELECTRONIC CONTROL MODULE (ECM)

REMOVAL & INSTALLATION

1. Remove the screw from the glove compartment door stoppers.
2. Deform the upper part of the glove compartment door assembly to release the stoppers.
3. Pull the glove compartment door assembly upward to remove it.
4. Remove the 2 clips using a clip remover, then open the cover.
5. Disconnect the 4 ECM connectors.
6. Remove the wire harness from the wire harness clamp.
7. Remove the 2 screws.
8. Unfasten the claw and clip, and then remove the ECM.

To install:

9. Reverse removal procedure

FUEL LEVEL SENDING UNIT

LOCATION

See Figure 48.

REMOVAL & INSTALLATION

See Figure 49.

1. Before servicing the vehicle, refer to the precautions in the beginning of this section.
2. Relieve the fuel system pressure.
3. Remove or disconnect the following:
   - Negative battery cable. On vehicles equipped with an air bag, wait at least 90 seconds before proceeding
   - Rear seat cushion and floor service hole cover
   - Access plate-to-fuel tank bolts, then pull out the plate/fuel pump assembly
   - Fuel pump sender and fuel pump connector
   - Outlet pipe from the fuel pump bracket
   - Return hose from the pump bracket
   - Fuel pump bracket assembly from the fuel tank
   - Lower side of the fuel pump from the pump bracket

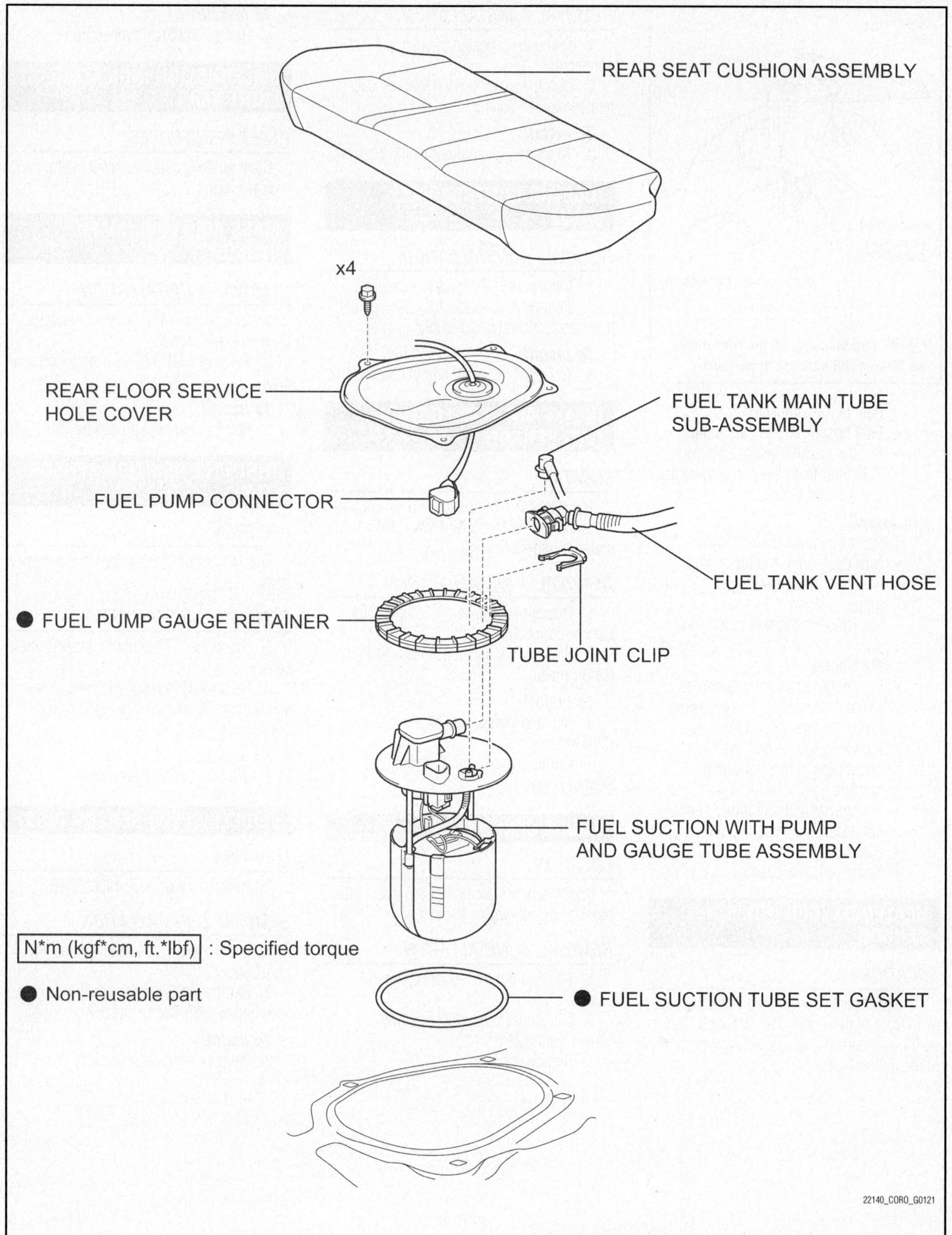

Fig. 48 Fuel system component location

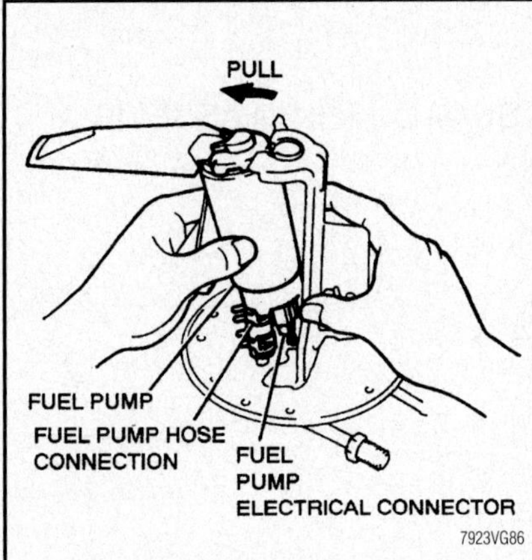

Fig. 49 Pull the pump off the sender unit; the filter is still attached to the pump

- Fuel pump connector
- Fuel hose from the fuel pump
- Rubber cushion from the pump
- Fuel filter from the pump by removing the small clip

To install:

4. Install or connect the following:
   - New cushion to the fuel pump
   - Fuel filter and new clip to the fuel pump
   - Fuel hose to the fuel pump, fuel pump connector and fuel pump to the bracket
   - Fuel pump bracket assembly to the fuel tank using a new gasket. Bolts: 30 inch lbs. (3 Nm).
   - Fuel return hose and the fuel outlet pipe to the fuel pump bracket
   - Fuel pump and fuel pump sender connector
   - Fuel tank
   - Negative battery cable

HEATED OXYGEN (HO2S) SENSOR

LOCATION

The HO2 sensor is located either behind, or in front of the Three Way Catalytic Converter, and detects the oxygen concentration in the exhaust gas.

REMOVAL & INSTALLATION

1. Disconnect heated oxygen sensor connector.
2. Using correct service tool remove heated oxygen sensor.

To install:

3. To install, reverse removal procedure.

IDLE AIR CONTROL (IAC) VALVE

REMOVAL & INSTALLATION

1. Remove idle air control connector.
2. Remove idle air control of assembly from a throttle valve assembly.

To install:

3. To install, reverse removal procedure.

INTAKE AIR TEMPERATURE (IAT) SENSOR

LOCATION

The Intake Air Temperature (IAT) sensor, mounted on the Mass Air Flow (MAF) meter, monitors the IAT.

REMOVAL & INSTALLATION

1. Disconnect Intake Air Temperature Sensor connector.
2. Remove sensor from Mass Air Flow (MAF) meter.

To install:

3. Replace sensor into Mass Air Flow (MAF) meter.
4. Connect Intake Air Temperature Sensor connector.

KNOCK SENSOR (KS)

LOCATION

Knock sensors are fitted onto the engine block to detect engine knocking.

REMOVAL & INSTALLATION

1. Remove knock sensor retaining bolt.
2. Disconnect camshaft position sensor connector.
3. Remove sensor.

To install:

4. Reverse removal procedure.

MALFUNCTION INDICATOR LIGHT (MIL)

RESET PROCEDURES

Clear all DTC codes to reset malfunction indicator light.

MASS AIR FLOW (MAF) SENSOR

REMOVAL & INSTALLATION

1. Disconnect MAF sensor connector from an MAF sensor.
2. Remove MAF sensor from air cleaner case.

To install:

3. Reverse removal procedure

THROTTLE POSITION SENSOR (TPS)

LOCATION

The TP sensor is mounted on the throttle body.

REMOVAL & INSTALLATION

1. Disconnect TP sensor connector from sensor.
2. Remove TP sensor attaching screws and remove TP sensor from the throttle body.

To install:

3. Reverse removal procedure

VEHICLE SPEED SENSOR (VSS)

LOCATION

On transaxle near and output shaft.

REMOVAL & INSTALLATION

1. Disconnect VSS sensor connector from sensor.
2. Remove VSS sensor attaching screws and remove VSS from transaxle assembly.

To install:

3. Reverse removal procedure

FUEL SYSTEM SERVICE PRECAUTIONS

Safety is the most important factor when performing not only fuel system maintenance but any type of maintenance. Failure to conduct maintenance and repairs in a safe manner may result in serious personal injury or death. Maintenance and testing of the vehicle's fuel system components can be accomplished safely and effectively by adhering to the following rules and guidelines.

- To avoid the possibility of fire and personal injury, always disconnect the negative battery cable unless the repair or test procedure requires that battery voltage be applied.
- Always relieve the fuel system pressure prior to disconnecting any fuel system component (injector, fuel rail, pressure regulator, etc.), fitting or fuel line connection. Exercise extreme caution whenever relieving fuel system pressure to avoid exposing skin, face and eyes to fuel spray. Please be advised that fuel under pressure may penetrate the skin or any part of the body that it contacts.
- Always place a shop towel or cloth around the fitting or connection prior to loosening to absorb any excess fuel due to spillage. Ensure that all fuel spillage (should it occur) is quickly removed from engine surfaces. Ensure that all fuel soaked cloths or towels are deposited into a suitable waste container.
- Always keep a dry chemical (Class B) fire extinguisher near the work area.
- Do not allow fuel spray or fuel vapors to come into contact with a spark or open flame.
- Always use a back-up wrench when loosening and tightening fuel line connection fittings. This will prevent unnecessary stress and torsion to fuel line piping.
- Always replace worn fuel fitting O-rings with new. Do not substitute fuel hose or equivalent where fuel pipe is installed.

Before servicing the vehicle, make sure to also refer to the precautions in the beginning of this section as well.

RELIEVING FUEL SYSTEM PRESSURE

✻✻ CAUTION

Failure to relieve fuel pressure before repairs or disassembly can cause serious personal injury and/or property damage. Fuel pressure is maintained within the fuel lines, even if the engine is OFF or has not been run in a period of time. This pressure must be safely relieved before any fuel-bearing line or component is loosened or removed. On vehicles equipped with inflatable restraints or air bag systems, wait at least 90 seconds after disconnecting the battery cable before performing any other work. The back-up power will keep the restraint system energized for a period of time after the battery is disconnected.

1. Before servicing the vehicle, refer to the precautions in the beginning of this section.
2. Perform the following:
   - Remove the fuse for the fuel pump
   - Start the engine until the engine stalls
   - Disconnect the negative battery cable
   - Place a catch-pan under the joint to be disconnected. A large quantity of fuel may be released when the joint is opened
   - Wear eye or full face protection
   - Place a shop towel over the area and slowly release the joint using a wrench of the correct size.
   - Allow the any fuel left in the line to bleed off slowly before fully disconnecting the joint.
   - Plug the opened lines
3. After connecting fuel lines, install the fuse for the fuel pump and start the engine.

FUEL FILTER

REMOVAL & INSTALLATION

1. Before servicing the vehicle, refer to the precautions in the beginning of this section.
2. Remove or disconnect the following:
   - Negative battery cable. On vehicles equipped with an air bag, wait at least 90 seconds before proceeding
   - Protective shield for the fuel filter
   - If necessary, air cleaner hose and cap
   - If necessary, charcoal canister.
   - Slowly loosen the lower flare nut fitting until all the pressure is relieved
   - Banjo fitting and 2 metal gaskets. Discard the gaskets
   - Fuel line with the flared nut from the filter
   - Filter from the mounting bracket

To install:

3. Install or connect the following:
   - New fuel filter
   - Banjo fitting with a new metal gasket on each side and install the union bolt. Bolt: 22 ft. lbs. (30 Nm).
   - Flare nut to the lower connection. Nut: 22 ft. lbs. (30 Nm).
   - Charcoal canister
   - Air cleaner hose and cap
   - Protective shield
   - Negative battery cable

FUEL PUMP

REMOVAL & INSTALLATION

See Figure 50.

1. Before servicing the vehicle, refer to the precautions in the beginning of this section.
2. Relieve the fuel system pressure.
3. Remove or disconnect the following:
   - Negative battery cable. On vehicles equipped with an air bag, wait at least 90 seconds before proceeding
   - Rear seat cushion and floor service hole cover
   - Access plate-to-fuel tank bolts, then pull out the plate/fuel pump assembly
   - Fuel pump sender and fuel pump connector
   - Outlet pipe from the fuel pump bracket
   - Return hose from the pump bracket

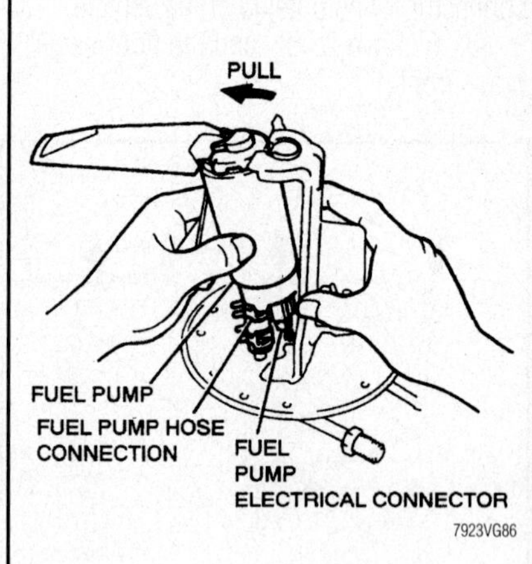

Fig. 50 Pull the pump off the sender unit; the filter is still attached to the pump

- Fuel pump bracket assembly from the fuel tank
- Lower side of the fuel pump from the pump bracket
- Fuel pump connector
- Fuel hose from the fuel pump
- Rubber cushion from the pump
- Fuel filter from the pump by removing the small clip

To install:

4. Install or connect the following:
   - New cushion to the fuel pump
   - Fuel filter and new clip to the fuel pump
   - Fuel hose to the fuel pump, fuel pump connector and fuel pump to the bracket
   - Fuel pump bracket assembly to the fuel tank using a new gasket. Bolts: 30 inch lbs. (3 Nm).
   - Fuel return hose and the fuel outlet pipe to the fuel pump bracket
   - Fuel pump and fuel pump sender connector
   - Negative battery cable
   - Floor service hole cover and rear seat cushion

FUEL TANK

REMOVAL & INSTALLATION

See Figures 51 through 54.

1. Disconnect battery negative terminal.
2. Removed fuel section with pump and gauge tube assembly.
3. Drain fuel
4. Remove front exhaust pipe assembly.
5. Remove front floor carpet.
6. Disconnect the heated oxygen sensor connector.
7. Remove the grommet, then pull the connector to the outside of the vehicle.
8. Remove 2 nuts and the floor panel brace front.
9. Remove the 4 bolts and 4 compression springs.
10. Remove the exhaust pipe front from the 2 exhaust pipe supports.
11. Remove the gasket from the exhaust manifold.
12. Remove the gasket from the exhaust pipe front.
13. Remove the 4 bolts, then remove the fuel tank protector.
14. Remove the 2 bolts, then disconnect the parking brake cable.
15. Disconnect the fuel tank main tube.
    a. Pinch the tab of the retainer of the fuel tube connector to remove the lock claws and push it down as shown in the illustration below.
    b. Pull the fuel tank main tube out of the pipe.
    - Remove any dirt or foreign objects on the fuel tube connector before performing this work.
    - Do not allow any scratches or foreign objects on the parts when disconnecting, as the fuel tube connector has the O-ring that seals the pipe.
    - Perform this work by hand. Do not use any tools.
    - Do not forcibly bend, twist or turn the nylon tube.
    - Protect the disconnected part by covering it with a vinyl bag after disconnecting the fuel tank main tube.
    - If the fuel tube connector and pipe are stuck, push and pull them to release.
16. Disconnect the fuel tank breather tube.
    a. Remove the checker of the fuel tube connector from the pipe.
    b. Pinch the retainer of the fuel tube connector, then pull the fuel tube connector out of the pipe.
    - Remove any dirt or foreign objects on the fuel tube connector before performing this work.
    - Do not allow any scratches or foreign objects on the parts when disconnecting, as the fuel tube connector has the O-ring that seals the pipe.
    - Perform this work by hand. Do not use any tools.
    - Do not forcibly bend, twist or turn the nylon tube.
    - Protect the disconnected part by covering it with a vinyl bag after disconnecting the fuel tank main tube.

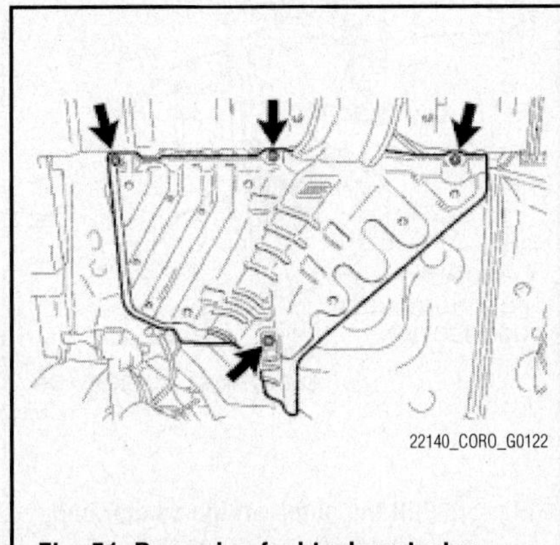

Fig. 51 Removing fuel tank protector

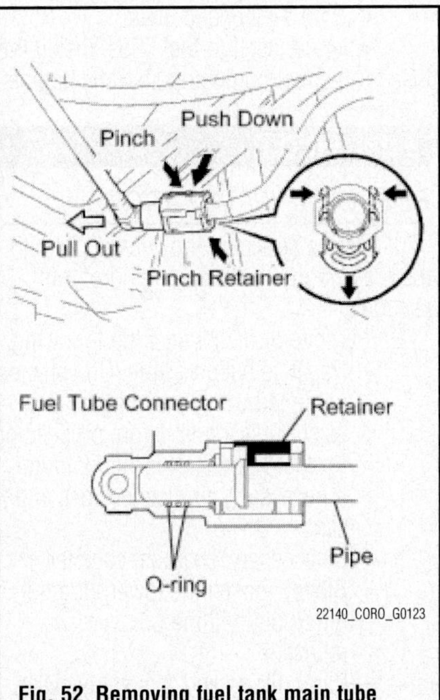

Fig. 52 Removing fuel tank main tube

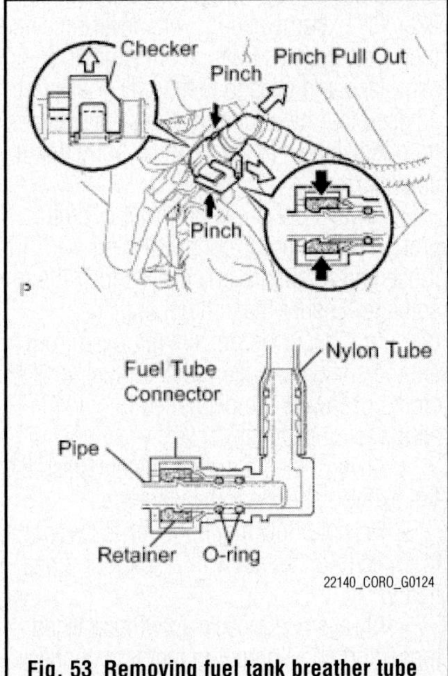

Fig. 53 Removing fuel tank breather tube

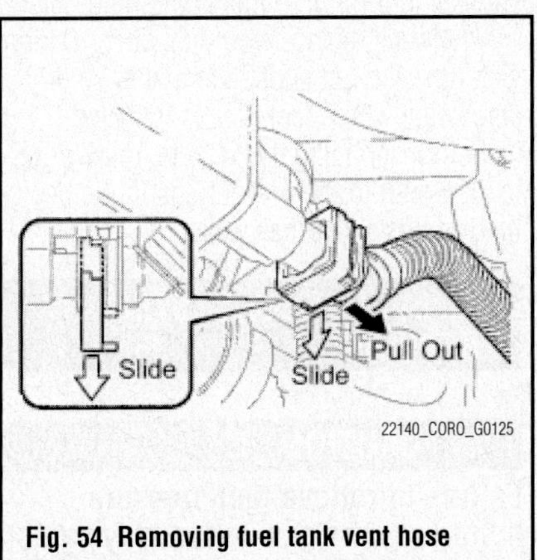

Fig. 54 Removing fuel tank vent hose

- If the fuel tube connector and pipe are stuck, push and pull them to release.

17. Disconnect the fuel tank vent hose.
   a. Pinch the tab of the retainer of the fuel tube connector to remove the lock claws and slide it as shown in the illustration below.
   b. Pull the fuel tank vent hose out of the pipe.
   - Remove any dirt or foreign objects on the fuel tube connector before performing this work.
   - Do not allow any scratches or foreign objects on the parts when disconnecting, as the fuel tube connector has the O-ring that seals the pipe.
   - Perform this work by hand. Do not use any tools.
   - Do not forcibly bend, twist or turn the nylon tube.
   - Protect the disconnected part by covering it with a vinyl bag after disconnecting the fuel tank main tube.
18. Using a screwdriver, unfasten the claw. Then remove the fuel tank filler pipe cover from the fuel tank filler pipe.
19. Loosen the hose clamp bolt, then disconnect the fuel tank filler pipe hose from the fuel tank.
20. Hold and lower the fuel tank using a transmission jack.

To install:

21. To install reverse removal procedure.

FUEL RAIL & INJECTORS

REMOVAL & INSTALLATION

1. Before servicing the vehicle, refer to the precautions in the beginning of this section.
2. Properly relieve the fuel system pressure.
3. Remove or disconnect the following:
   - Negative battery cable
   - Air cleaner
   - Accelerator cable bracket from the throttle body
   - Throttle body from the air intake chamber
   - Engine hanger and air intake chamber stay
   - EGR vacuum modulator if so equipped
   - EGR valve and pipe if so equipped
   - Air intake chamber cover and gasket
   - Injector electrical connections
   - Fuel inlet hose from the delivery pipe
   - Fuel return hose from the fuel pressure regulator
   - Fuel delivery pipe (rail)
   - The 4 insulators and 2 collars from the intake manifold
   - Injectors

To install:

4. Install or connect the following:

➡**Before installing the injectors back into the fuel rail, install a NEW O-ring on each injector, coated with a light coat of gasoline (NEVER use oil of any sort).**

   - Injectors

➡**Make certain each injector can be smoothly rotated. If they do not rotate smoothly, the O-ring is not in its correct position.**

   - Insulators into each injector hole
   - The two spacers on the delivery pipe mounting holes in the intake manifold
5. Place the delivery pipe and injectors on the intake manifold and again check that the injectors rotate smoothly. Position the injector connector upward. Install the two bolts and tighten them to 11 ft. lbs.
6. Install or connect the following:
   - Electrical connectors to each injector
   - Gaskets, the inlet pipe and fuel union bolt. Bolt to 22 ft. lbs.
   - Air intake chamber cover with a NEW gasket. Torque the retaining bolts in steps to 14 ft. lbs.
   - All necessary hoses and electrical connections
   - EGR valve and pipe if so equipped
   - Engine hanger and air intake chamber stay
   - EGR vacuum modulator if so equipped
   - Throttle body. Torque the bolts evenly (in a X-pattern) to 16 ft. lbs.
   - Accelerator cable bracket to the throttle body
   - Air cleaner hose and cap
   - Negative battery cable

IDLE SPEED

ADJUSTMENT

Adjustment is not available or necessary.

THROTTLE BODY

REMOVAL & INSTALLATION

1. Disconnect the battery's negative terminal.
2. Drain the engine coolant.
3. Remove the two nuts and two clips that retains cylinder head cover Number Two.
4. Disc
5. Disconnect the mass air flow meter connector.
6. Disconnect the VSV connector.
7. Remove the 2 wire harness clamps.
8. Disconnect the fuel vapor feed hose No. 1 from the VSV.
9. Disconnect the fuel vapor feed hose No. 3 from the VSV.
10. Disconnect the ventilation hose No. 1 from the cylinder head cover.
11. Loosen the hose clamp bolt and air cleaner hose No. 1.
12. Remove the air cleaner cap together with the air cleaner hose.
13. Disconnect the throttle control motor connector.
14. Disconnect the water by-pass hose from the throttle body.
15. Disconnect the water by-pass hose No. 2 from the throttle body.
16. Remove the 2 bolts and 2 nuts, then remove the throttle body.

To install:

17. Installation is the reverse of the removal process.

HEATING & AIR CONDITIONING SYSTEM

BLOWER MOTOR

REMOVAL & INSTALLATION

See Figure 55.

1. Remove the glove box assembly.
2. Disconnect the connector from the blower motor.
3. Remove the 3 screws and the blower motor.

To install:

4. Replace the blower motor and the three screws.
5. Connect the electrical connector.
6. Replace the glove box assembly.

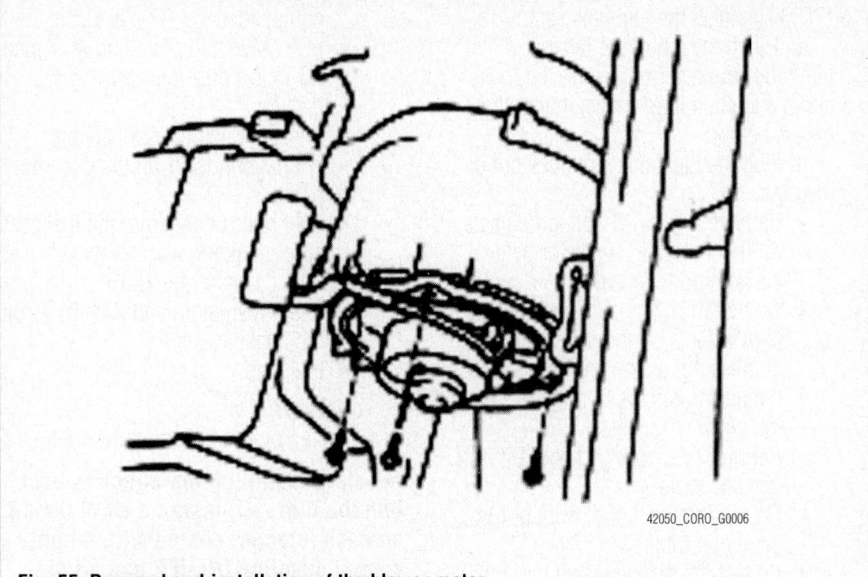

Fig. 55 Removal and installation of the blower motor

HEATER CORE

REMOVAL & INSTALLATION

See Figures 56 through 60.

1. Disconnect the negative battery cable.

⁂ CAUTION

After Connecting the negative battery cable, wait for at least 1½ minutes for the SRS to deplete its energy.

2. Drain the cooling system into a clean container for reuse.
3. Disconnect the heater hoses from the heater core.
4. Discharge and recover the air conditioning system refrigerant.
5. Remove the air bag module and the steering wheel by performing the following procedure:
   - Place the front wheels in the straight-ahead position.
   - At both sides of the steering wheel, remove the screw covers.
   - Using a Torx socket, loosen the 2 air bag module-to-steering wheel Torx screws until the circumference ring catches on the screw case.
   - Carefully, remove the air bag module and disconnect the electrical connector.
   - Remove the steering wheel nut.
   - Using a steering wheel puller, press the steering wheel from the steering column.
6. Remove the passenger's side air bag module by performing the following procedure:
   - Remove the glove box-to-instrument panel 2 bolts and 3 screws; then, pull out the glove box and remove it.
   - Disconnect the passenger's side air bag module electrical connector.
   - Remove the 2 passenger's air bag module-to-instrument panel bolts and nuts. Carefully, remove the air bag module.
7. Remove the following items in the following order:
   - The front door scuff plates
   - The cowl side trims
   - The front pillar garnishes
   - Remove the 2 lower finish panel bolts and the panel.
   - Disconnect the hood lock control cable.
   - Remove the 2 lower insert bolts and the lower insert.
   - Remove the 2 cluster finish panel screws; then, pry out the panel.
   - Remove the steering column cover.
   - Remove the steering column's combination switch.
   - Remove the No. 2 heater-to-register duct.
   - Disconnect and remove the combination meter.
   - Remove the steering column-to-instrument panel fasteners, the steering column shaft pinch bolt and the steering column.
8. Remove the following items in the following order:
   - The lower left side panel
   - The lower right side panel
   - The shifting hole knob (manual transmission)
   - The shifting hole bezel (automatic transmission)
   - The rear console box
   - Remove the center lower cluster finish panel by prying it out, disconnect the electrical connector and remove it.
9. Remove the following items in the following order:
   - The stereo opening cover
   - The lower center finish panel
   - The air conditioning control panel
   - Remove the center cluster finish panel by prying it out.
   - Remove the radio.
   - Disconnect the instrument panel electrical connectors.
   - Remove the 4 instrument panel screws and the instrument panel.
10. Remove the following items in the following order:
    - The heater-to-register duct
    - The lower defroster nozzle
    - The No. 1 and No. 2 braces
    - The reinforcement
11. Remove the evaporator housing by performing the following procedure:
    - Disconnect the cruise control actuator connector.
    - Remove the 3 cruise control actuator set bolts.
    - Using tool No. 09870-00025 (liquid line) and/or 09870-00015 (suction line), disconnect the refrigerant line clamps. Discard the O-rings and plug the openings to prevent contamination.

Cooling Unit
- Expansion Valve
- Evaporator
- Blower Resistor
- Thermistor

Blower Unit
Heater Unit
Relays
Receiver
Condenser Fan
- Condenser Fan Motor

NO. 2 COOLING FAN (A/C)
Condenser
Compressor
- Magnetic Clutch
- Refrigerant Temperature Switch

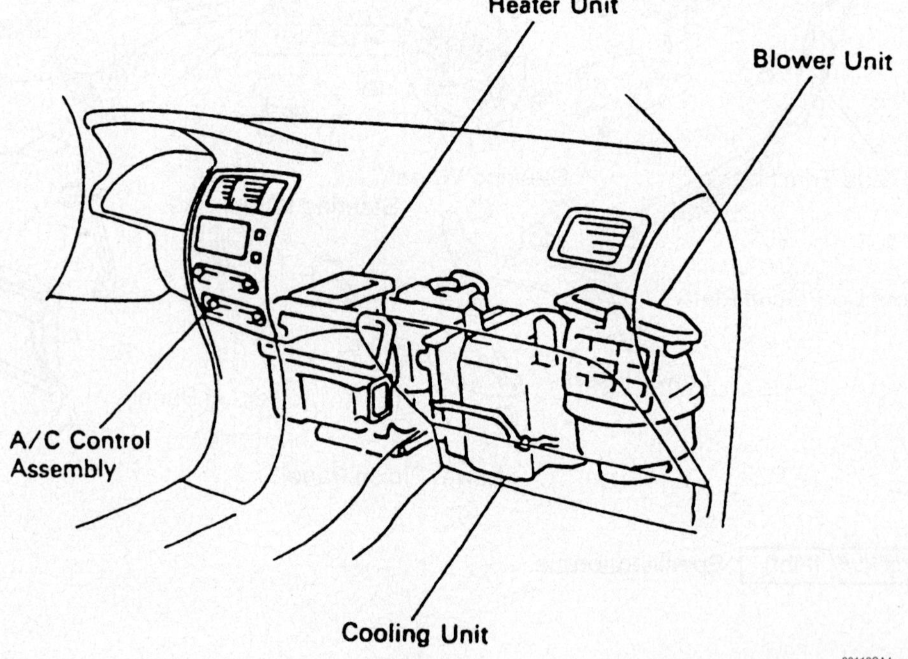

93112GA4

Fig. 56 View of the heater/air conditioning assembly and related components—Corolla

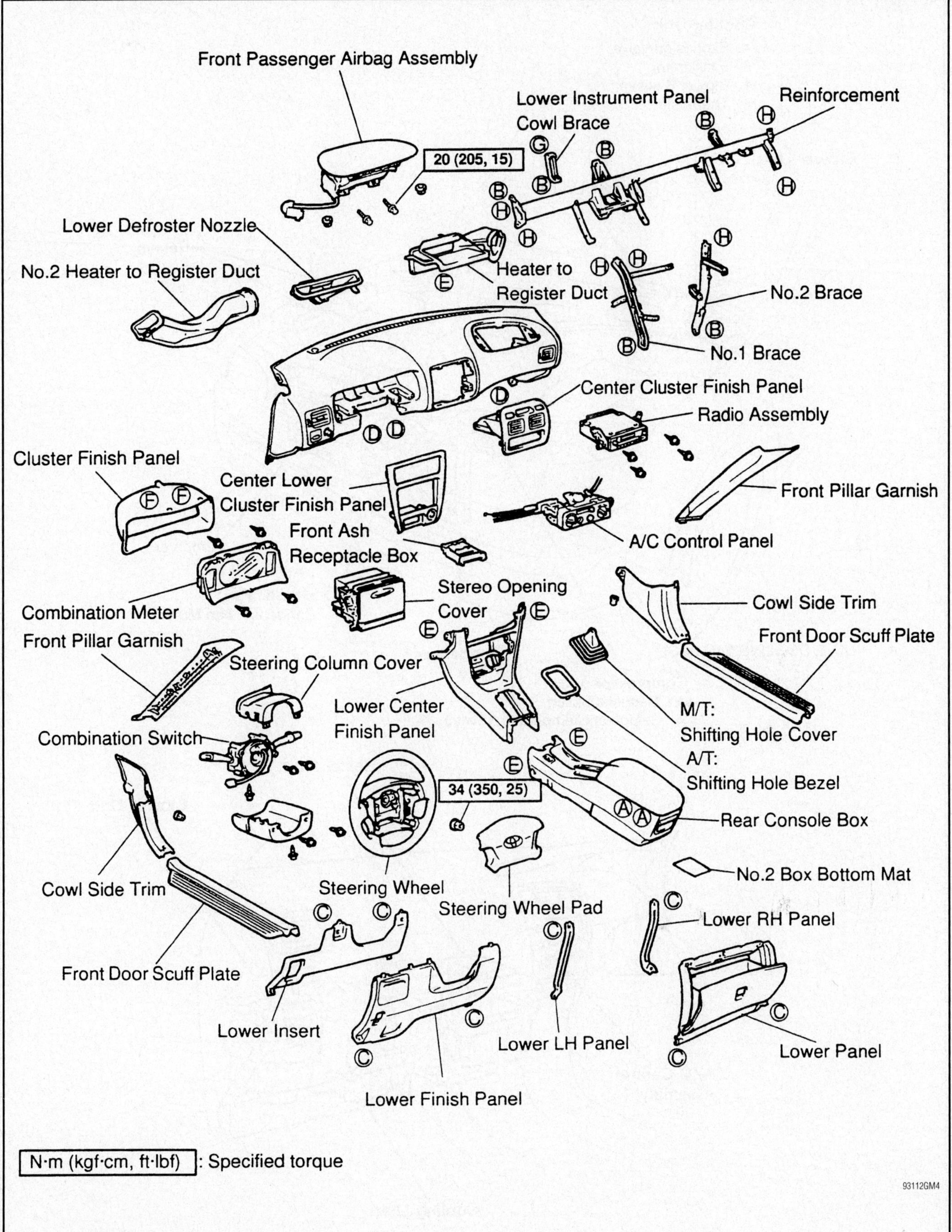

Fig. 57 Exploded view of the instrument panel and related components—Corolla

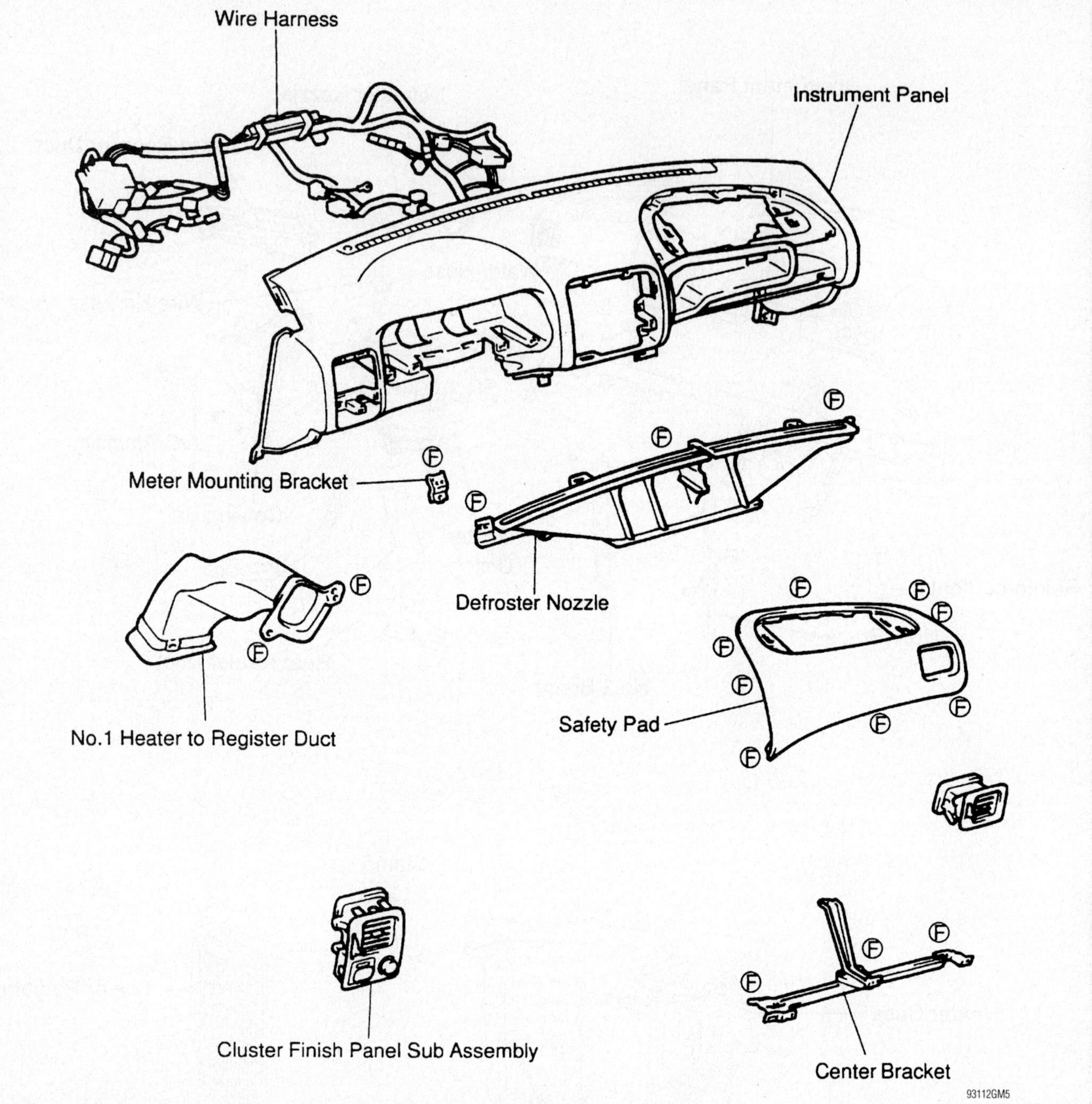

Fig. 58 Exploded view of the ventilation system, wiring harness and related components—Corolla

- Remove the tube grommet and the drain hose grommet.
- Disconnect the air conditioning amplifier electrical connector.
- Remove the 2 air conditioning amplifier nuts and the amplifier.
- Disconnect the evaporator housing's electrical connectors.
- Remove the evaporator housing-to-chassis 3 screws, nut and the housing.

12. Pull back the carpet and remove the rear heater ducts.

13. Disconnect the wiring harness from the heater housing.

14. Remove the 3 heater housing-to-chassis nuts and the heater housing.

15. Remove the air duct-to-heater housing screw, the 5 clips and the duct.

16. Release the 2 heater core cover clips and the cover.

17. Remove the heater core-to-heater housing screw, clamp and the heater core.

To install:

18. Install the heater core, clamp and the heater core-to-heater housing screw.

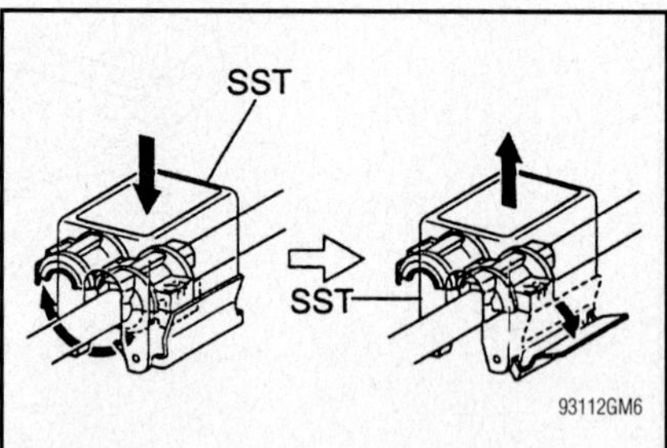

Fig. 59 Using the special tool to remove the air conditioning refrigerant line clamps—Corolla

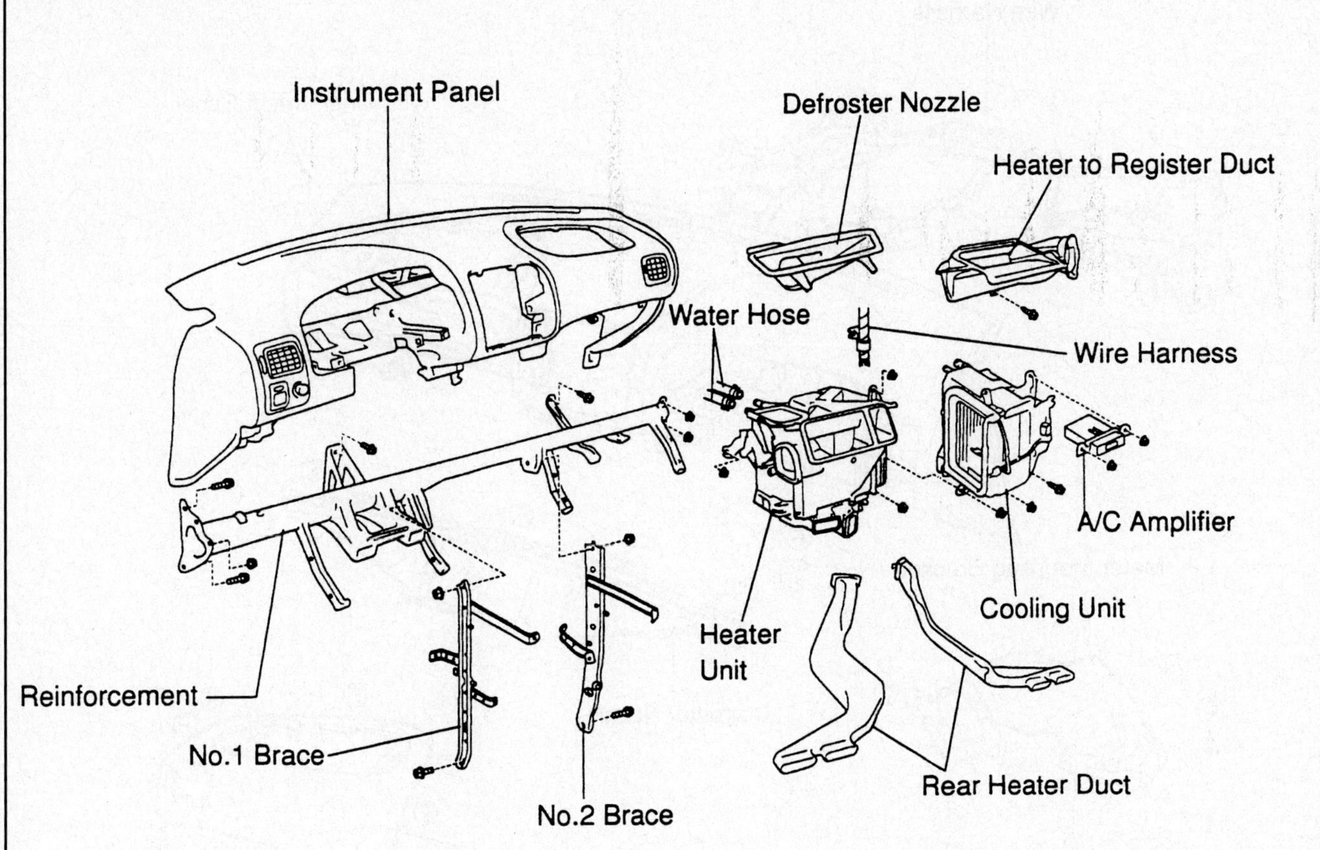

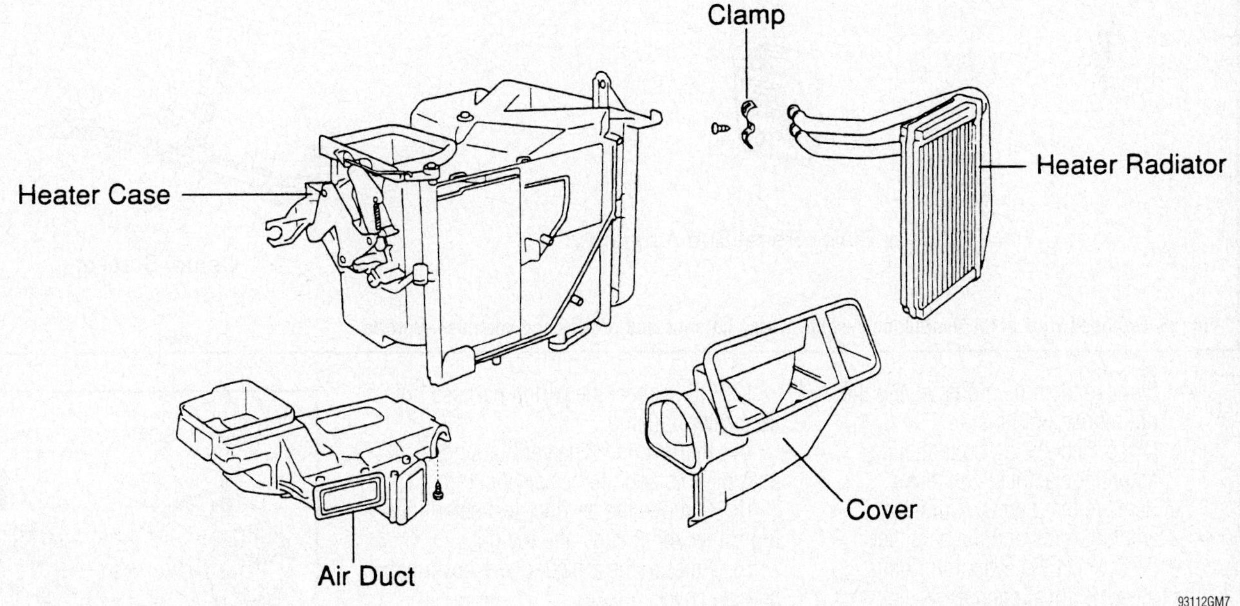

Fig. 60 Exploded view of the heater core, heater housing, evaporator housing and related components—Corolla

19. Install the heater core cover and the 2 cover clips.

20. Install the air duct, the 5 clips and the duct-to-heater housing screw.

21. Install the heater housing and the 3 heater housing-to-chassis nuts.

22. Connect the wiring harness to the heater housing.

23. Install the rear heater ducts and the carpet.

24. Install the evaporator housing by performing the following procedure:

- Install the evaporator housing-to-chassis 3 screws, nut and the housing.
- Connect the evaporator housing's electrical connectors.
- Install the 2 air conditioning amplifier nuts and the amplifier.
- Connect the air conditioning amplifier electrical connector.
- Install the tube grommet and the drain hose grommet.
- Using new O-rings, connect the refrigerant line clamps.
- Install the 3 cruise control actuator set bolts.
- Connect the cruise control actuator connector.

25. Install the following items in the following order:

- The reinforcement
- The No. 1 and No. 2 braces
- The lower defroster nozzle
- The heater-to-register duct
- Install the instrument panel and the 4 instrument panel screws.
- Connect the instrument panel electrical connectors.
- Install the radio.
- Install the center cluster finish panel.

26. Install the following items in the following order:

- The air conditioning control panel
- The lower center finish panel
- The stereo opening cover
- Connect the electrical connector and install the center lower cluster finish panel.

27. Install the following items in the following order:

- The rear console box
- The shifting hole bezel (automatic transmission)
- The shifting hole knob (manual transmission)
- The lower right side panel
- The lower left side panel
- Install the steering column, the steering column shaft pinch bolt and the steering column-to-instrument panel fasteners.
- Connect and install the combination meter.
- Install the No. 2 heater-to-register duct.
- Install the steering column's combination switch.
- Install the steering column cover.
- Install the cluster finish panel and the 2 panel screws.
- Install the lower insert and the 2 lower insert bolts.
- Connect the hood lock control cable.
- Install the lower finish panel and the 2 panel bolts.

28. Install the following items in the following order:

- The front pillar garnishes
- The cowl side trims
- The front door scuff plates

29. Install the passenger's side air bag module by performing the following procedure:

- Carefully, Install the air bag module and the 2 passenger's air bag module-to-instrument panel bolts and nuts.
- Connect the passenger's side air bag module electrical connector.
- Install the glove box and the glove box-to-instrument panel 2 bolts and 3 screws.

30. Install the air bag module and the steering wheel by performing the following procedure:

- Align the matchmarks and install the steering wheel to the steering column.
- Install the steering wheel nut and torque the nut to 26 ft. lbs. (35 Nm).
- Carefully, connect the electrical connector and install the air bag module.
- Using a Torx socket, tighten the 2 air bag module-to-steering wheel Torx screws to 78 inch lbs. (8.8 Nm).
- At both sides of the steering wheel, install the screw covers.

31. Connect the heater hoses to the heater core.

32. Refill the cooling system.

33. Connect the negative battery cable.

34. Evacuate, charge and leak test the air conditioning system refrigerant.

STEERING

POWER RACK & PINION STEERING GEAR

REMOVAL & INSTALLATION

1. Before servicing the vehicle, refer to the precautions in the beginning of this section.

2. Position the front wheels straight ahead.

3. Remove or disconnect the following:

- Negative battery cable. On vehicles equipped with an air bag, wait at least 90 seconds before proceeding

➡If equipped with an air bag, disable the system and secure the steering wheel.

- Steering column hole cover and loosen the upper pinch bolt on the sliding yoke
- Lower pinch bolt at the pinion shaft
- Front wheels
- Left and right engine undercovers.
- Left and right tie rod ends.

4. Install an engine support and tension it to support the engine without raising it.

⁂ CAUTION

The engine hoist is now in place and under tension. Use care when repositioning the vehicle and make necessary adjustments to the engine support.

5. Remove or disconnect the following:

- Lower control arms from the ball joints
- If equipped with a stabilizer bar, stabilizer bar links from both lower control arms
- Right rear control arm bushing retaining bracket. Do this for both lower control arms
- Stabilizer bar
- Grommet in the crossmember
- Bolt and nuts holding in the middle of the crossmember and support the crossmember with a jack
- Bolts from the outer side of the suspension crossmember
- Suspension crossmember with the lower suspension arms
- Exhaust front pipe support
- Engine rear mount insulator
- Engine rear mount bracket
- Pressure feed and return tubes

- Brackets and grommets to the power steering rack

6. Slide the power steering gear assembly to the right side of the vehicle.

To install:

7. Install or connect the following:
   - Power steering assembly
   - Grommets and brackets. Nuts and bolts: 43 ft. lbs. (59 Nm).
   - Pressure feed and return tubes. Nuts: 26 ft. lbs. (36 Nm).
   - Engine rear mount bracket. Bolts: 57 ft. lbs. (77 Nm).
   - Engine rear mount insulator. Bolt: 64 ft. lbs. (87 Nm).
   - Exhaust front pipe support. Bolts: 14 ft. lbs. (19 Nm).
   - Outer 6 bolts to hold the crossmember to the vehicle. Bolts: 152 ft. lbs. (206 Nm).
   - Center crossmember-to-radiator support bolts: 45 ft. lbs. (61 Nm).
   - Lower A frame-to-center bolts: 161 ft. lbs. (218 Nm).
   - Lower A frame-to-outer bolts: 109 ft. lbs. (147 Nm).
   - Front, center and rear mount bolts: 45 ft. lbs. (61 Nm).
   - Grommet to the crossmember
   - Stabilizer bar
   - Lower control arm bushing retaining bracket. Do not tighten the bolts or nut at this time
   - Lower control arm to the lower ball joint. Bolt and nuts: 105 ft. lbs. (142 Nm). Connect both lower control arms to the ball joints
   - Stabilizer bar links to the lower control arms. Nuts: 33 ft. lbs. (44 Nm).
   - Sliding yoke to the pinion shaft. Lower bolt: 26 ft. lbs. (35 Nm). Tighten upper bolt: 20 ft. lbs. (27 Nm).
   - Steering column hole cover. Bolts: 43 inch lbs. (5 Nm).
   - Left and right-hand tie rod ends. Nuts: 36 ft. lbs. (49 Nm).
   - Front wheels
   - Control arm bracket bolts: 108 ft. lbs. (147 Nm).
   - Stabilizer bar bracket bolt: 37 ft. lbs. (50 Nm).
   - Bracket nut: 14 ft. lbs. (19 Nm).
   - Negative battery cable
8. Check and top off the power steering fluid.
9. Check and adjust the alignment, if needed.

POWER STEERING PUMP

REMOVAL & INSTALLATION

See Figure 61.

1. Disconnect the pressure tube from the power steering pump.
2. Loosen the drive belt.
3. Remove the retaining bolts.

To install:

4. Replace the retaining bolts and torque them to 26 ft. lbs.(35 Nm).
5. Adjust the drive belt tension.
6. Connect the pressure tube to the power steering pump.
7. Bleed the system.

BLEEDING

1. Check the fluid level.
2. Jack up the front of the vehicle and support it with the stands.
3. Turn the steering wheel.
   a. With the engine stopped, turn the wheel slowly from lock to lock several times.
4. Lower the vehicle.
5. Start the engine.
   a. Run the engine at idle for a few minutes.
6. Turn the steering wheel.
   a. With the engine idling, turn the wheel to left or right full lock position and keep it there for 2 to 3 seconds, then turn the wheel to the opposite full lock position and keep it there for 2 to 3 seconds.
   b. Repeat procedure several times.
7. Stop the engine.
8. Check for foaming or emulsification. Especially, if the system has to be bled twice because of foaming or emulsification, check for fluid leaks in the system.
9. Check the fluid level.

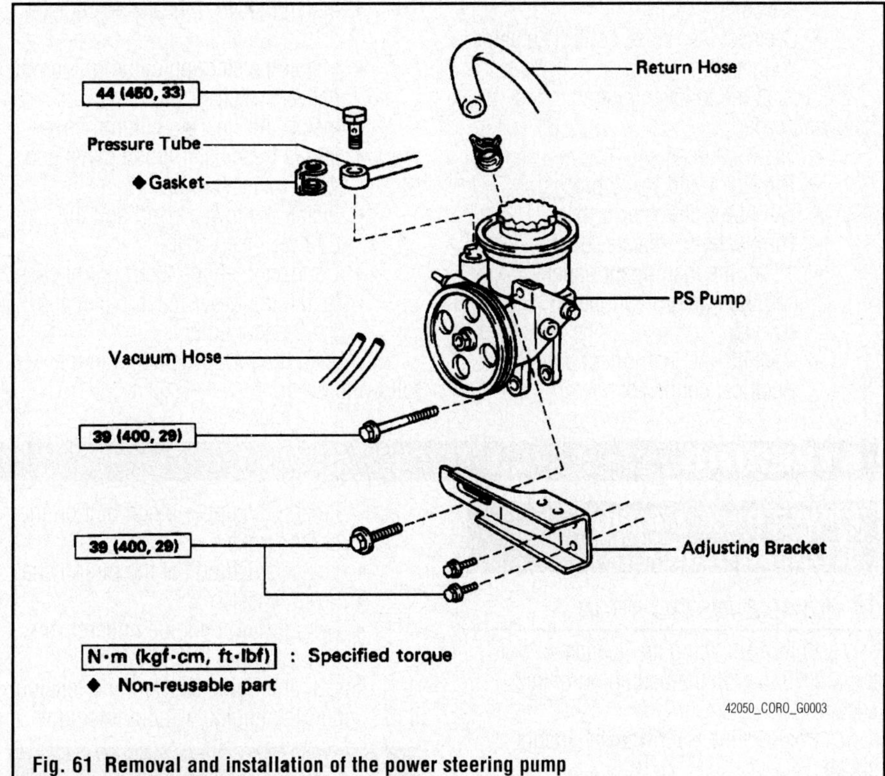

Fig. 61 Removal and installation of the power steering pump

CONTROL LINKS

REMOVAL & INSTALLATION

See Figure 62.

1. Remove front wheel.
2. Remove front stabilizer link assembly.
   a. Remove the 2 nuts and stabilizer bar link.

To install:

3. Installation is reverse of removal.

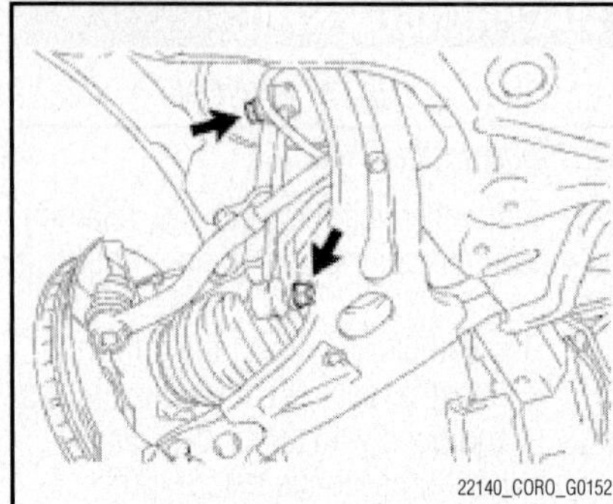

Fig. 62 Removing front stabilizer link assembly

LOWER BALL JOINT

REMOVAL & INSTALLATION

See Figure 63.

1. Before servicing the vehicle, refer to the precautions in the beginning of this section.
2. Remove or disconnect the following:
   - Negative battery cable. On vehicles equipped with an air bag, wait at least 90 seconds before proceeding
   - Front wheels
   - Cotter pin from the bearing locknut cap, then remove the cap
3. Depress the brake pedal and loosen the axle nut
4. Remove or disconnect the following:
   - Brake caliper attaching hardware, position the caliper aside with the hydraulic line still attached and suspend it with a wire
   - ABS speed sensor, if equipped
   - Rotor
5. Loosen the 2 nuts holding the strut to the steering knuckle assembly. Do not remove at this time.
6. Remove or disconnect the following:
   - Cotter pin and nut from the tie rod end. Using a tie rod end removal tool, separate the tie rod end from the steering knuckle

22140_CORO_G0153

Fig. 63 Removing lower ball joint

   - Steering knuckle from the strut assembly
   - Axle nut and grasp the hub and knuckle assembly. With a plastic hammer tap the axle shaft to remove knuckle and hub

➡Cover the halfshaft boot with a shop rag to protect it from any damage.

7. Clamp the steering knuckle in a vise and remove the dust deflector. Remove the nut holding the steering knuckle to the ball joint. Press the ball joint out of the steering knuckle.
8. Remove the ball joint from the arm.

To install:

9. Install the Lower ball joint to the lower arm. Tighten the fasteners to: 66 ft. lbs. (89 Nm).
10. Install the ball joint to the steering knuckle. Tighten the ball joint-to-steering knuckle nut to: 76 ft. lbs. (103 Nm).
11. Install or connect the following:
    - New cotter pin. Drive the deflector shield onto the knuckle
    - Knuckle and hub assembly to the axle and temporarily tighten the axle nut
    - Knuckle assembly to the lower strut bracket. Temporarily insert the mounting bolts from the rear and install the nuts
    - Tie rod end to the knuckle
12. Tighten the bolts on the lower side of the strut assembly.
13. Install or connect the following:
    - ABS speed sensor
    - Brake disc and the caliper
14. Tighten the axle nut.
15. Connect the negative battery cable.
16. Check and adjust the alignment, if needed.

LOWER CONTROL ARM

REMOVAL & INSTALLATION

See Figures 64 through 67.

1. Remove front wheel.
2. Disconnect front stabilizer link assembly.
3. Separate front suspension arm sub assembly lower left and right.
   a. Remove the bolt and 2 nuts, and separate the front suspension arm sub-assembly lower No. 1 LH from the lower ball joint assembly front LH.
4. Separate rack and pinion power steering gear assembly.
   a. Remove the 4 bolts, separate the rack & pinion power steering gear assembly.

➡Loosen the bolt since the nut cannot be rotated.

   b. Suspend the rack & pinion power steering gear assembly.
5. Suspended engine assembly.
6. Separate front suspension cross member subassembly.
   a. Remove the 3 bolts and 3 nuts, disconnect the transverse engine mounting insulator and engine mounting member sub-assembly center from the front suspension cross member sub-assembly.
   b. Remove the 4 bolts.
   c. Lower the transmission jack, remove the front suspension cross member sub-assembly.
7. Remove front suspension arm sub assembly lower.
   a. Remove the 2 bolts, nut and front suspension arm sub- assembly lower No. 1 LH from the front suspension cross member sub-assembly.

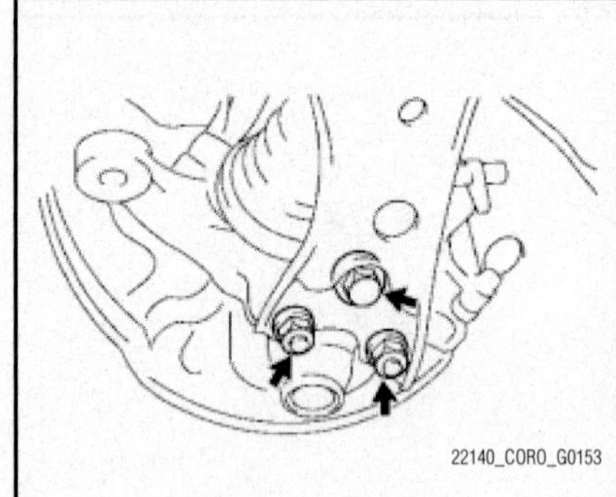

Fig. 64 Removing front suspension arm lower

To install:

8. Temporarily tighten front suspension arm subassembly lower.
   a. Install the front suspension arm sub-assembly lower, temporarily tighten the 2 bolts and nut.
9. Install front suspension cross member subassembly.
   a. Lift the front suspension cross member sub-assembly up with a transmission jack.
   b. Insert service tool to the base hole of the RH side cross member and RH side of the vehicle.
   c. Tighten the bolts temporarily.
   d. Insert service tool to the base hole of the LH side of cross member and LH side of the vehicle.
   e. Tighten the bolts temporarily.
   f. Then tighten the bolt A and B by the specified torque A: 116 ft. lbs. (157 Nm) B: 83 ft. lbs. (113 Nm)
10. Connect the transverse engine mounting insulator and engine mounting member sub-assembly center to the front suspension cross member sub-assembly.
11. Install the 3 bolts and 3 nuts. Tighten to 38 ft. lbs. (52 Nm)
12. Install rack and pinion power steering gear assembly.

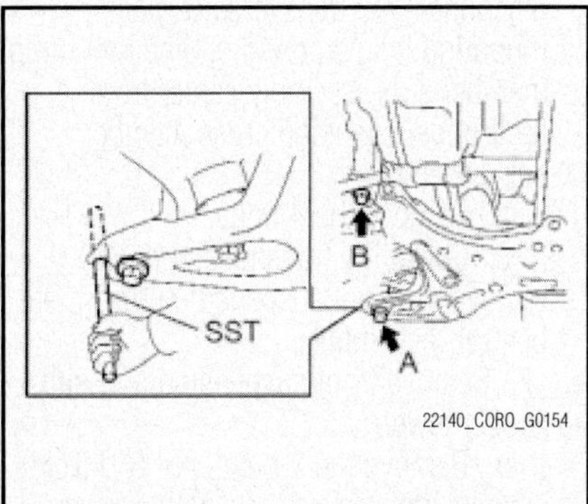

Fig. 65 Installing front suspension arm lower

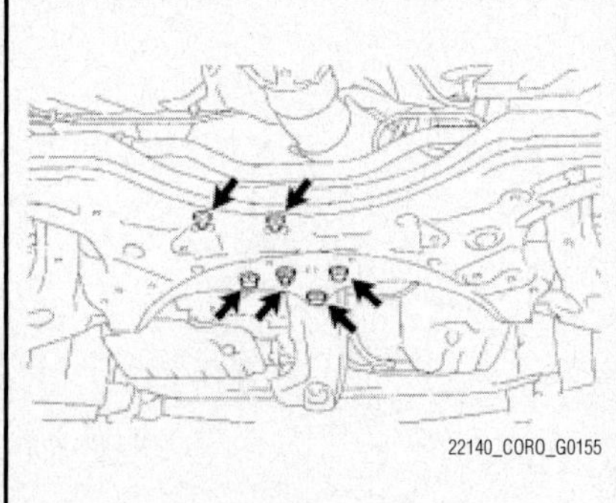

Fig. 66 Installing engine mounting member subassembly

   a. Install the rack & pinion power steering gear assembly with the 4 bolts. Tighten to 66 ft. lbs. (89 Nm)
13. Install front suspension arm sub assembly lowers.
   a. Install the front suspension arm sub-assembly lowers with the 2 nuts and bolt to the lower ball joint assembly front.
14. Install front stabilizer link assemblies.

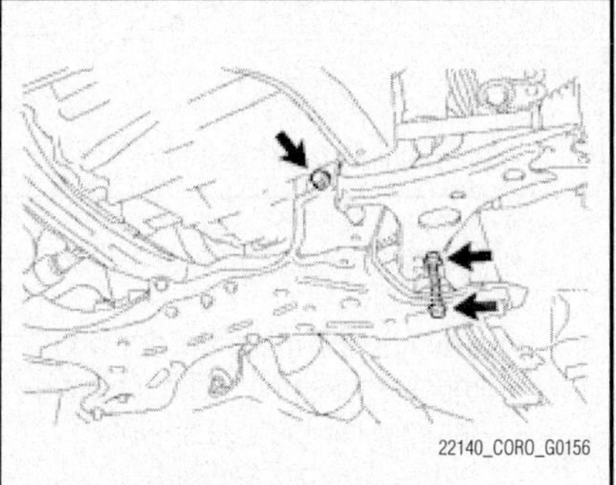

Fig. 67 Tightening front suspension arm lower

15. Stabilizes suspension.
   a. Install the front wheel and jack down the vehicle.
   b. Bounce the vehicle up and down several times to stabilize the suspension.
16. Fully tighten front suspension arm subassembly lower
   a. Fully tighten the 2 bolts and nut.

➡Tighten the bolt since the nut cannot be rotated.

17. Inspect and adjust front wheel alignment.

MACPHERSON STRUT

REMOVAL & INSTALLATION

See Figures 68 and 69.

1. Before servicing the vehicle, refer to the precautions in the beginning of this section.
2. Remove or disconnect the following:
   - Negative battery cable. On vehicles equipped with an air bag, wait at least 90 seconds before proceeding

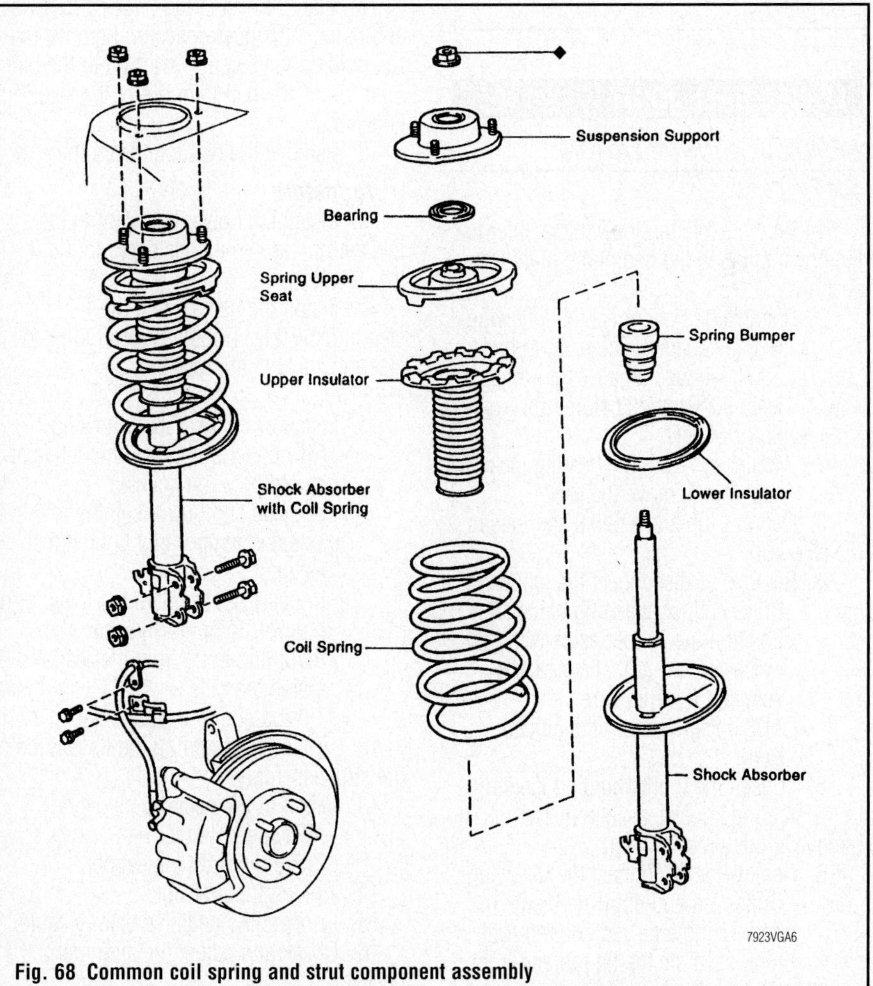

Fig. 68 Common coil spring and strut component assembly

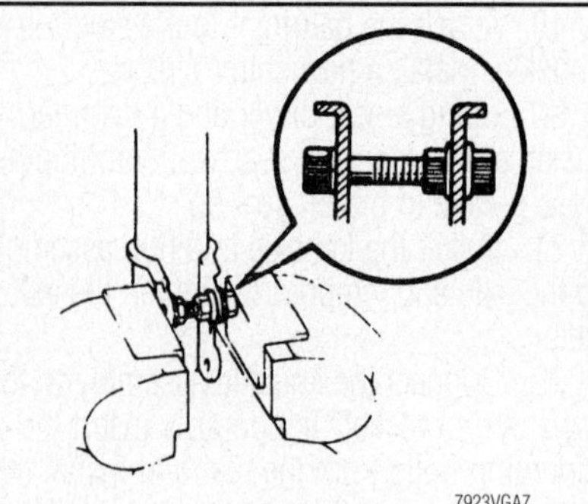

Fig. 69 Proper method of supporting the strut in a vise

** WARNING

Do not support the weight of the vehicle on the suspension arm; the arm will deform under its weight.

- Wheel
- Bolt, and disconnect the brake hose from the strut
- With ABS brakes, wiring harness from the strut
- Bolts and strut from the steering knuckle
- Strut

3. To disassemble the strut:
   - Install a bolt and 2 nuts to the bracket at the lower portion of the strut shell and secure it in a vise
   - Compress the coil spring
   - Remove the dust cover and hold the spring seat so that it will not turn. Remove the nut on the top of the strut
   - Remove the suspension support, bearing, dust seal, spring seat, spring, insulators and bumper

To install:

4. To assemble the strut:
   - Install the spring bumper to piston
   - Using a spring compressor, compress the spring
   - Install the coil spring to the strut. Fit the lower end of the coil spring into the gap of the lower seat
   - Install the spring seat with the insulator
   - Install the dust seal on the spring seat
   - Install the suspension support and tighten 35 ft. lbs. (47 Nm). After the nut has been tighten, release the compressor tool tension
   - Pack multipurpose grease into the suspension support. Install the dust cover

➡Do not use an impact wrench to tighten the nut. Also, check that the bearing fits into the recess in the suspension support.

5. Install or connect the following:
   - Nuts holding the strut to the strut tower. Nuts to 29 ft. lbs. (39 Nm)
   - Steering knuckle to the strut lower bracket
6. Insert the 2 bolts from the rear side and tighten the strut-to-steering knuckle arm bolts. Tighten as follows: 113 ft. lbs. (153 Nm).
7. Install or connect the following:
   - Brake line to the steering knuckle
   - If equipped with ABS, secure the wiring harness
   - Wheel
   - Negative battery cable
8. Check and adjust the alignment, if needed.

STABILIZER BAR

REMOVAL & INSTALLATION

1. Remove the front wheel.
2. Remove the 2 nuts and stabilizer bar link on both sides. If the ball joint turns together with the nut, use a hexagon wrench (6 mm) to hold the stud.
3. Remove the front stabilizer link assembly, left and right sides.
4. Separate the front suspension arm subassembly.
5. Separate the rack and pinion power steering gear assembly.
6. Remove the 4 bolts, the front stabilizer bracket No. 1 on the left side, the front stabilizer bracket No. 1 on the right side, the 2 front stabilizer bar bushes No. 1 and the stabilizer bar front from the front suspension cross-member sub-assembly.

To install:

7. Install the 2 front stabilizer bar bushes No. 1, the front stabilizer bracket No. 1 LH and front stabilizer bracket No. 1 on the right side to the stabilizer bar front. Install the bushing to the inner side of the bushing stopper on the stabilizer bar.
8. Install the stabilizer bar front and the 4 bolts to the front suspension cross-member sub-assembly.
9. Install the rack and pinion power steering gear.
10. Install the front suspension arm sub-assembly on both sides.
11. Install the stabilizer bar link with the 2 nuts and torque them to 55 ft. lbs. (74 Nm).

WHEEL BEARINGS

REMOVAL & INSTALLATION

See Figures 70 through 74.

1. Before servicing the vehicle, refer to the precautions in the beginning of this section.
2. Remove or disconnect the following:
   - Negative battery cable. On vehicles equipped with an air bag, wait at least 90 seconds before proceeding
   - Wheels
   - Axle nut cap
   - Axle nut
   - Caliper. Position the caliper aside with the hydraulic line still attached and suspend it with a wire.
   - ABS speed sensor
   - Rotor
3. Loosen the nuts on the lower side of the strut assembly. Do not remove at this time.
4. Remove or disconnect the following:
   - Tie rod end from the steering knuckle
   - Steering knuckle from the lower control arm
   - Knuckle from the strut assembly
   - Hub

➡Cover the halfshaft boot with a shop rag to protect it from any damage.

5. Clamp the steering knuckle in a vise and remove the dust deflector. Remove the nut holding the steering knuckle to the ball joint. Press the ball joint out of the steering knuckle.
6. Remove the inner axle seal.
7. Using a Torx®wrench, remove the bolts securing the dust cover.
8. Using hub puller, remove the hub and backing plate from the steering knuckle.
9. Using a proper sized driver and a press, remove the inner hub race from the axle hub.
10. Using seal removal tool, remove the outer axle seal.
11. Using snapring pliers, remove the snapring from the inner side of the steering knuckle.
12. Using a proper sized driver and a press, remove the bearing from the steering knuckle. The bearing is pressed from the front of the steering knuckle and is removed through the back of the steering knuckle.

To install:

13. Perform the following:
14. Using a proper sized driver and a press, install a new bearing to the steering knuckle.

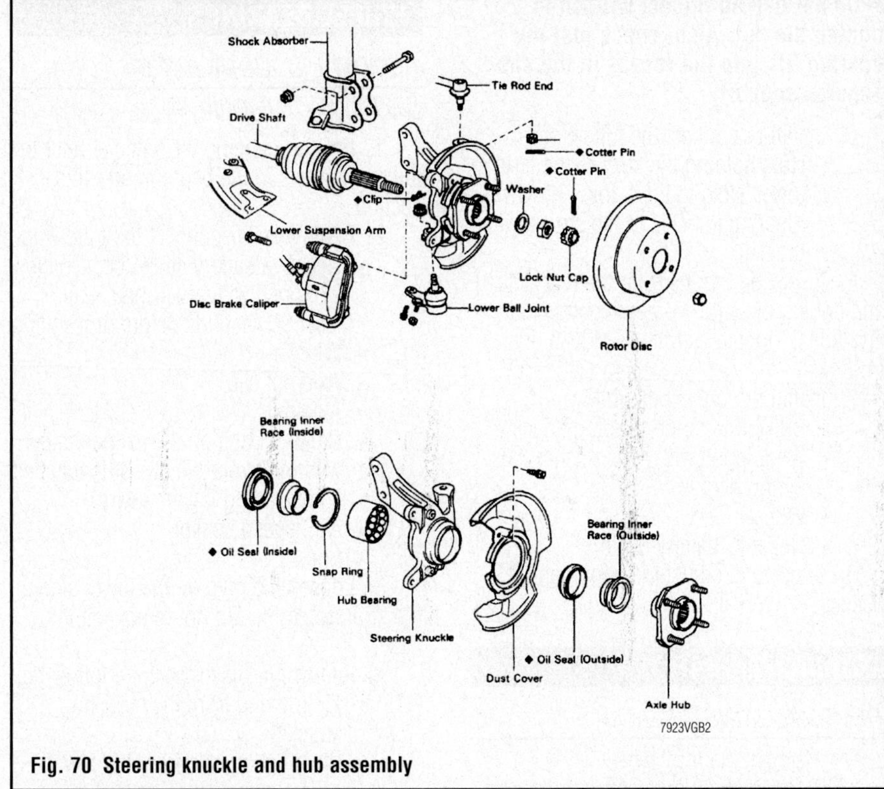

Fig. 70 Steering knuckle and hub assembly

15. Install the snapring to the steering knuckle using snapring pliers.
16. Using a seal driver and a hammer, install a new outer oil seal. Apply multipurpose grease to the oil seal lip.
17. Place the dust cover on the steering knuckle. Bolts: 78 inch lbs. (9 Nm).
18. Using a press and a proper sized driver, install the axle hub to the steering knuckle.

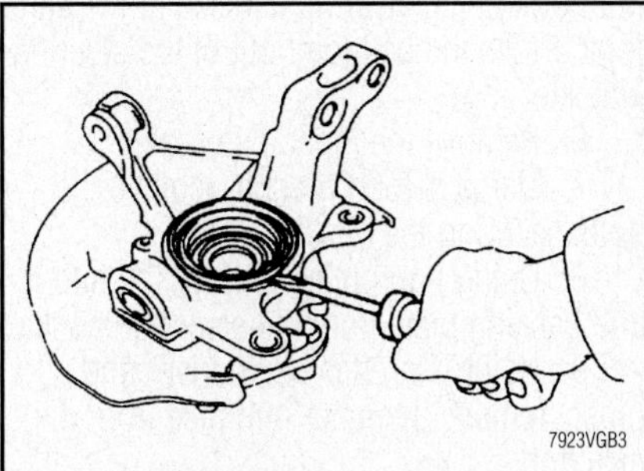

Fig. 71 Removing the inner axle seal from the hub assembly

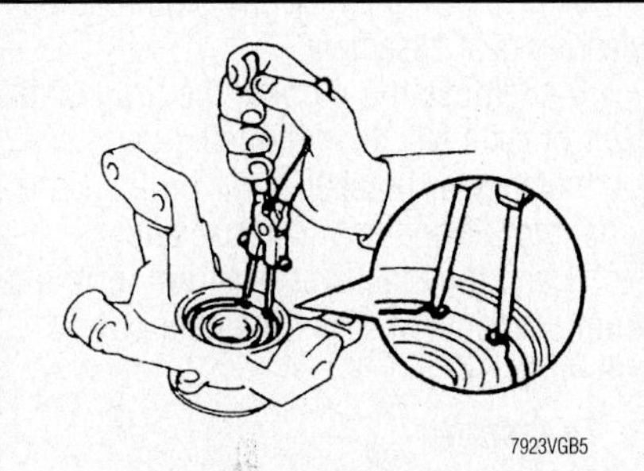

Fig. 73 Removing the snapring from the knuckle before pressing out the bearing

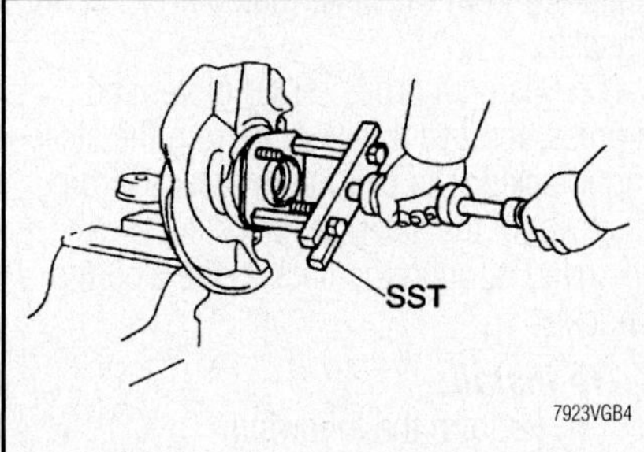

Fig. 72 Removing the axle hub from the knuckle

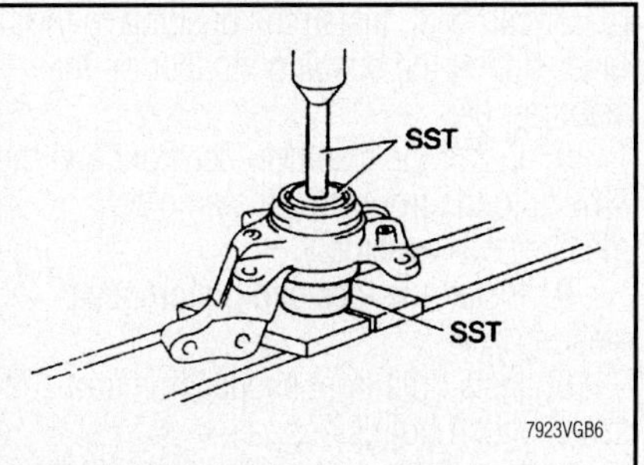

Fig. 74 Removing the bearing from the steering knuckle using a press

19. Attach the ball joint to the steering knuckle. Install a new cotter pin.
20. Using a seal driver and a hammer, install a new inner oil seal. Apply multipurpose grease to the oil seal lip.
21. Install the knuckle and hub assembly to the axle and temporarily tighten the axle nut.
22. Connect the knuckle assembly to the lower strut bracket. Temporarily insert the mounting bolts from the rear and install the nuts making sure the matchmarks made earlier are in alignment.
23. Connect the lower ball joint to lower arm.
24. Connect the tie rod end to the knuckle.
25. Tighten on the lower side of the strut assembly.
26. If equipped, install the ABS speed sensor.
27. Install the brake disc and the caliper.
28. Tighten the axle nut while someone depresses the brake pedal. Install the adjusting nut cap and insert a new cotter pin. Torque the nut to 76 ft. lbs. (103 Nm).
29. Install the wheels to the vehicle. Verify that the wheel turns freely.
30. Connect the negative battery cable to the battery.
31. Check alignment.

ADJUSTMENT

See Figure 75.

1. Before servicing the vehicle, refer to the precautions in the beginning of this section.
2. All models use a non-adjustable wheel bearing. To determine the condition of the wheel bearing, check the backlash in bearing shaft direction and the axle hub deviation. Maximum for backlash should be as follows:
   - 0.0031 in. (0.08mm)
3. Maximum axle hub deviation is:
   - 0.0028 in. (0.07mm)
4. If the wheel bearing is out of specifications, replace the wheel bearing.

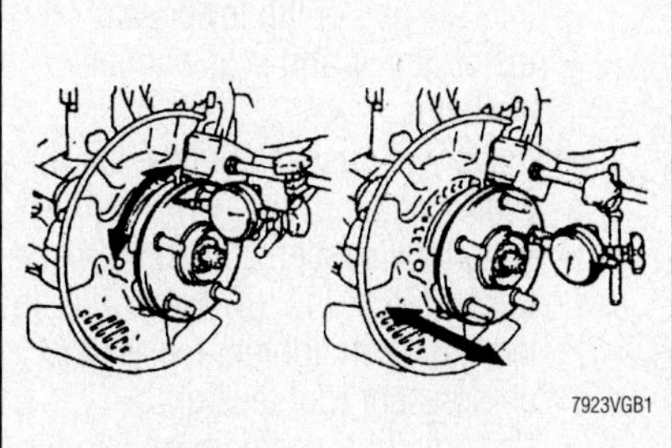

Fig. 75 Checking the wheel bearings for deviation and free-play

MACPHERSON STRUTS

REMOVAL & INSTALLATION

1. Before servicing the vehicle, refer to the precautions in the beginning of this section.
2. Remove or disconnect the following:
   - Negative battery cable from the battery. On vehicles equipped with an air bag, wait at least 90 seconds before proceeding
   - Rear seat cushion and any trim necessary to access the strut towers
3. Support the axle beam with a jack.
4. Remove or disconnect the following:
   - Wheel
   - With ABS, sensor wire from the strut
   - Stabilizer bar
5. Loosen the fasteners securing the strut to the axle carrier. Do not remove the bolts at this time.
6. Support the axle carrier with a jack.
7. Remove or disconnect the following:
   - Strut-to-strut tower nuts

✱✱ CAUTION

Do not loosen the center nut on the top of the strut piston.

   - Strut
8. To disassemble the strut:
   a. Place the strut assembly in a pipe vise or strut vise.

➡Do not attempt to clamp the strut assembly in a flat jaw vise as this will result in damage to the strut tube.

   b. Compress the spring until the upper suspension support is free of any spring tension. Do not over-compress the spring.
   c. Hold the upper support, then remove the nut on the end of the shock piston rod.
   d. Remove the support, coil spring, insulator, and bumper.
9. Inspect the strut as follows:
   a. Check the shock absorber by moving the piston shaft through its full range of travel. It should move smoothly and evenly throughout its entire travel without any trace of binding or notching.
   b. Use a small straightedge to check the piston shaft for any bending or deformation.
   c. Inspect the spring for any sign of deterioration or cracking. The waterproof coating on the coils should be intact to prevent rusting.

To install:

➡Never reuse a self-locking nut. Always replace self-locking nuts and cotter pins as applicable.

10. Assemble the strut as follows:
   a. Loosely assemble all components onto the strut assembly. Be sure the spring end aligns with the hollow in the lower seat.
   b. Align the upper suspension support with the piston rod and install the support.
   c. Align the suspension support with the strut lower bracket. This assures the spring will be properly seated top and bottom.
   d. Compress the spring to expose the strut piston rod threads.
   e. Install a new strut piston nut and tighten to the following:
   - 41 ft. lbs. (56 Nm).
   f. Remove the spring compressor. Be sure the paint mark on the upper support faces the outside of the strut.
11. Place the strut on the vehicle and install the nuts to hold the strut to the strut tower. 59 ft. lbs. (80 Nm).
12. Install or connect the following:
   - Strut to the axle carrier and install the bolt and nut. Do not tighten at this time
   - Stabilizer link to the strut
   - Wheel. Bounce the vehicle up and down to stabilize the suspension
13. With the vehicle weight on the suspension, tighten the bolt holding the strut to the axle carrier as follows:
   - 105 ft. lbs. (142 Nm)
14. Install or connect the following:
   - Rear seat cushion and any trim
   - Negative battery cable

WHEEL BEARINGS

REMOVAL & INSTALLATION

See Figure 76.

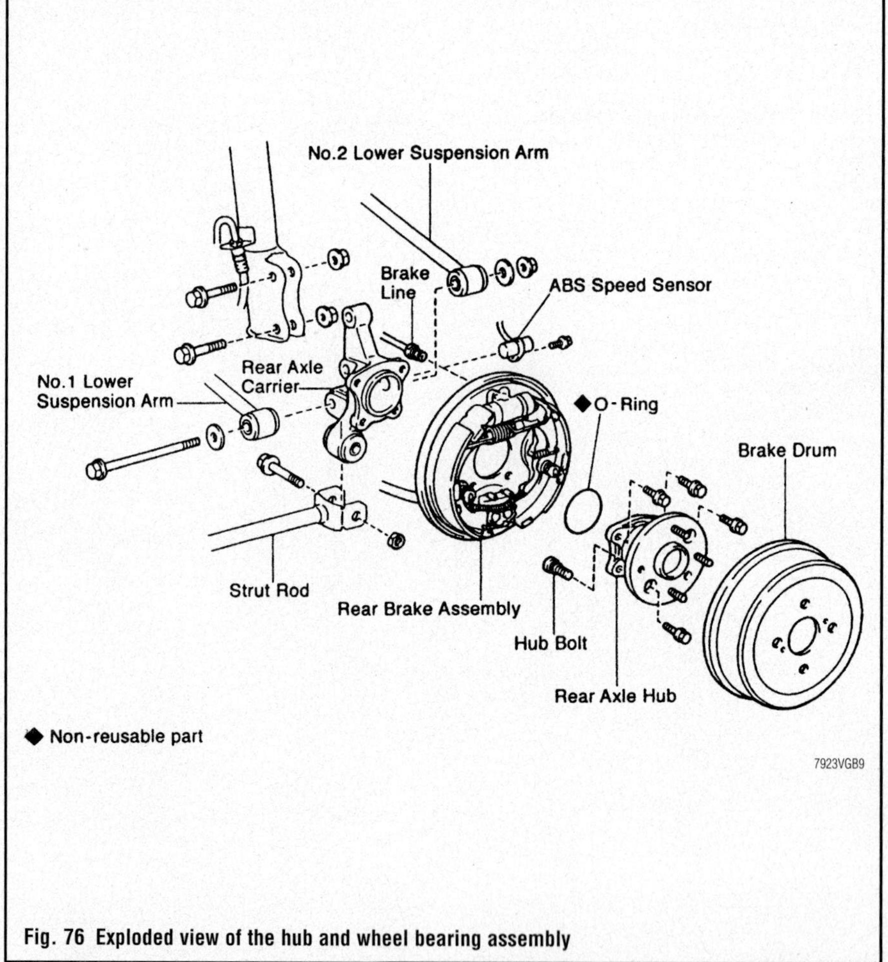

Fig. 76 Exploded view of the hub and wheel bearing assembly

1. Before servicing the vehicle, refer to the precautions in the beginning of this section.
2. Remove or disconnect the following:
   - Negative battery cable. On vehicles equipped with an air bag, wait at least 90 seconds before proceeding
   - Wheel
   - Brake drum or rotor
   - With ABS brakes, ABS wheel speed sensor
   - Hub
   - O-ring from the backing plate

To install:

3. Install or connect the following:
   - New O-ring onto the backing plate. Coat the O-ring with multipurpose grease
   - Hub to the knuckle. Bolts: Except (80 Nm)
   - With ABS brakes, ABS wheel speed sensor
   - Brake drum or rotor
   - Wheel
   - Negative battery cable
4. Check and adjust the alignment, if needed.

ADJUSTMENT

➡The wheel bearing is non-adjustable. If the wheel bearing is out of specifications, replace the wheel bearing.

TOYOTA

FJ Cruiser

14

SPECIFICATIONS AND MAINTENANCE CHARTS

ENGINE AND VEHICLE IDENTIFICATION

Engine							Model Year	
Code/ID ①	Liters (cc)	Cu. In.	Cyl.	Fuel Sys.	Engine Type	Eng. Mfg.	Code ②	Year
1GR-FE/U	4.0 (3956)	241	6	SFI	DOHC	Toyota	7	2007
							8	2008

MFI: Multi-port Fuel Injection

DOHC: Double Overhead Camshaft

① 1GR-FE engine: stamped on the right side of the engine block. Ingine ID is the fifth character of the VIN number.

② 10th digit of the VIN number

22140_FJCR_C0001

GENERAL ENGINE SPECIFICATIONS

Year	Model	Engine Displacement Liters	Engine Series Code/ID	Net Horsepower @ rpm	Net Torque @ rpm (ft. lbs.)	Bore x Stroke (in.)	Com-pression Ratio	Oil Pressure @ rpm
2007	FJ Cruiser	4.0	1GR-FE/U	278@2700	239@5200	3.70x3.74	10.0:1	43-85@3000
2008	FJ Cruiser	4.0	1GR-FE/U	278@2700	239@5200	3.70x3.74	10.0:1	43-85@3000

NA: Not Available

22140_FJCR_C0002

ENGINE TUNE-UP SPECIFICATIONS

Year	Engine Displacement Liters	Engine Code/ID	Spark Plug Gap (in.)	Ignition Timing (deg.)	Fuel Pump (psi)	Idle Speed (rpm)		Valve Clearance	
						MT	AT	Intake	Exhaust
2007	4.0	1GR-FE/U	0.039-0.043	7-24B ①	40.8-41.7	650-750	650-750	0.006-0.010	0.011-0.015
2008	4.0	1GR-FE/U	0.039-0.043	7-24B ①	40.8-41.7	650-750	650-750	0.006-0.010	0.011-0.015

NOTE: The Vehicle Emission Control Information label often reflects specification changes made during production.

The label figures must be used if they differ from those in this chart.

B: Before top dead center

① With terminals TC and CG of the DLC3 connected

22140_FJCR_C0003

CAPACITIES

Year	Model	Engine Displacement Liters	Engine Code/ID	Engine Oil with Filter (qts.)	Transmission (pts.) 6-Spd	Transmission (pts.) Auto.	Transfer Case (pts.)	Drive Axle Front (pts.)	Drive Axle Rear (pts.)	Fuel Tank (gal.)	Cooling System (qts.)
2007	FJ Cruiser	4.0	1GR-FE/U	5.5	3.8	6.0	①	②	6.4 ③	21.1	⑤
2008	FJ Cruiser	4.0	1GR-FE/U	5.5	3.8	6.0	①	④	6.4 ③	21.1	⑤

① MT: 3.0 pts
AT: 2.2 pts

② Full Time 4WD: 3.0 pts
Part Time 4WD: 3.2 pts

③ W/ Locking Differential: 6.2 pts

④ MT: 3.0 pts
AT: 3.4 pts

⑤ MT: 9.9
AT: 10.4

22140_FJCR_C0004

FLUID SPECIFICATIONS

Year	Model	Engine Displ. Liters (VIN)	Engine Oil	Man. Trans.	Auto. Trans.	Drive Axle Front	Drive Axle Rear	Transfer Case	Power Steering Fluid	Brake Master Cylinder	Cooling System
2007	FJ Cruiser	4.0L (U)	5W-30	75W-90	Dexron III	①	80W-90	75W-90	Dexron II or III	DOT 3	S-LLC
2008	FJ Cruiser	4.0L (U)	5W-30	75W-90	Dexron III	②	80W-90	75W-90	Dexron II or III	DOT 3	S-LLC

DOT: Department Of Transportation

S-LLC: Toyota Super Long Life Coolant

① Full Time 4WD: 80W-90
Part Time 4WD: 75W-85

② MT: 80W-90
AT: 75W-85

22140_FJCR_C0014

VALVE SPECIFICATIONS

Year	Engine Displacement Liters	Engine Code/ID	Seat Angle (deg.)	Face Angle (deg.)	Spring Test Pressure (lbs. @ in.)	Spring Installed Height (in.)	Stem-to-Guide Clearance (in.) Intake	Stem-to-Guide Clearance (in.) Exhaust	Stem Diameter (in.) Intake	Stem Diameter (in.) Exhaust
2007	4.0	1GR-FE/U	NA	44.5	41.9-46.3@ 1.311	1.882	0.0010-0.0024	0.0012-0.0026	0.2154-0.2159	0.2152-0.2158
2008	4.0	1GR-FE/U	NA	44.5	41.9-46.3@ 1.311	1.882	0.0010-0.0024	0.0012-0.0026	0.2154-0.2159	0.2152-0.2158

NA: Not Available

22140_FJCR_C0005

CAMSHAFT SPECIFICATIONS

All measurements in inches unless noted

Year	Engine Displacement Liters	Engine Code/ID	Journal Dia.	Brg. Oil Clearance	Shaft End-play	Circle Runout	Lobe Height	
							Intake	Exhaust
2007	4.0	1GR-FE/U	①	②	0.0160-0.0350	0.0024	③	③
2008	4.0	1GR-FE/U	①	②	0.0160-0.0350	0.0024	③	③

① No. 1: 1.4162-1.4167
All others: 0.9039-0.9045

② No. 1: 0.0016-0.0031
All others: 0.0010-0.0024

③ No. 1 camshaft: 1.7389-1.7428
No. 2 camshaft: 1.7551-1.7591

22140_FJCR_C0006

CRANKSHAFT AND CONNECTING ROD SPECIFICATIONS

All measurements are given in inches.

Year	Engine Displacement Liters	Engine Code/ID	Crankshaft				Connecting Rod		
			Main Brg. Journal Dia.	Main Brg. Oil Clearance	Shaft End-play	Thrust on No.	Journal Diameter	Oil Clearance	Side Clearance
2007	4.0	1GR-FE/U	2.8342-2.8346	0.0007-0.0012	0.0016-0.0094	NA	2.2044-2.2047	0.0010-0.0018	0.0059-0.0118
2008	4.0	1GR-FE/U	2.8342-2.8346	0.0007-0.0012	0.0016-0.0094	NA	2.2044-2.2047	0.0010-0.0018	0.0059-0.0118

NA: Not Available

22140_FJCR_C0007

PISTON AND RING SPECIFICATIONS

All measurements are given in inches.

Year	Engine Displacement Liters	Engine Code/ID	Piston Clearance	Ring Gap			Ring Side Clearance		
				Top Compression	Bottom Compression	Oil Control	Top Compression	Bottom Compression	Oil Control
2007	4.0	1GR-FE/U	0.0031-0.0040	0.0118-0.0157	0.0157-0.0197	0.0039-0.0157	0.0008-0.0028	0.0008-0.0024	0.0028-0.0060
2008	4.0	1GR-FE/U	0.0031-0.0040	0.0118-0.0157	0.0157-0.0197	0.0039-0.0157	0.0008-0.0028	0.0008-0.0024	0.0028-0.0060

22140_FJCR_C0008

TORQUE SPECIFICATIONS

All readings in ft. lbs.

Year	Engine Displacement Liters	Engine Code/ID	Cylinder Head Bolts	Main Bearing Bolts	Rod Bearing Bolts	Crankshaft Damper Bolts	Flywheel Bolts	Manifold Intake	Manifold Exhaust	Spark Plugs	Oil Pan Drain Plug
2007	4.0	1GR-FE/U	①	②	③	185	61	19	22	15	30
2008	4.0	1GR-FE/U	①	②	③	185	61	19	22	15	30

① Right side: 27 ft. lbs. plus 180 degrees

Left side (recessed head): 27 ft. lbs. plus 180 degrees

Left side (0.55 inch head): 22 ft. lbs.

② 12 pointed head: 45 ft. lbs. plus 90 degrees

12mm head: 18 ft. lbs.

③ Step 1: 18 ft. lbs.

Step 2: Plus 90 degrees

22140_FJCR_C0009

WHEEL ALIGNMENT

Year	Model	Caster Range (+/-Deg.)	Caster Preferred Setting (Deg.)	Camber Range (+/-Deg.)	Camber Preferred Setting (Deg.)	Toe-in (in.)	Steering Axis Inclination (Deg.)
2007	2WD	0.50	3.57	0.50	-0.57	0.04+/-0.08	12.92+/-0.50
	4WD	0.50	2.82	0.50	0.15	0.04+/-0.08	12.35+/-0.50
2008	2WD	0.50	3.57	0.50	-0.57	0.04+/-0.08	12.92+/-0.50
	4WD	0.50	2.82	0.50	0.15	0.04+/-0.08	12.35+/-0.50

NOTE: All alignment figures based on the following nominal ride heights:

2WD Front: 4.56 in.

Rear: 3.20 in.

4WD Front: 3.43 in.

Rear: 2.43 in.

22140_FJCR_C0010

TIRE, WHEEL AND BALL JOINT SPECIFICATIONS

Year	Model	OEM Tires Standard	OEM Tires Optional	Tire Pressures (psi) Front	Tire Pressures (psi) Rear	Wheel Size	Ball Joint Inspection	Lug Nut Torque (ft. lbs.)
2007	FJ Cruiser	P265/70R17	P265/75R16	32	32	NA	①	82
2008	FJ Cruiser	P265/70R17	P265/75R16	32	32	NA	①	82

① Lower ball joint excessive play, all models: 0.020 inch

22140_FJCR_C0011

BRAKE SPECIFICATIONS

All measurements in inches unless noted

Year		Brake Disc Original Thickness	Brake Disc Minimum Thickness	Brake Disc Maximum Runout	Brake Drum Diameter Original Inside Diameter	Brake Drum Diameter Maximum Machine Diameter	Minimum Lining Thickness Front	Minimum Lining Thickness Rear	Brake Caliper Bracket Bolts (ft. lbs.)	Brake Caliper Mounting Bolts (ft. lbs.)
2007	F	1.102	1.024	0.0020	—	—	0.039	—	—	91
	R	0.709	0.630	0.0079	—	—	—	0.039	—	65
2008	F	1.102	1.024	0.0020	—	—	0.039	—	—	91
	R	0.709	0.630	0.0079	—	—	—	0.039	—	65

22140_FJCR_C0012

SCHEDULED MAINTENANCE INTERVALS

FJ CRUISER

TO BE SERVICED	TYPE OF SERVICE	VEHICLE MILEAGE INTERVAL (x1000)												
		5	10	15	20	25	30	35	40	45	50	55	60	65
Automatic transmission fluid	S/I						✓						✓	
Ball joints & dust covers	S/I			✓			✓			✓			✓	
Brake line pipes & hoses	S/I			✓			✓			✓			✓	
Brake pads & discs/linings & drums (front & rear)	S/I	✓	✓	✓	✓	✓	✓	✓	✓	✓	✓	✓	✓	✓
Cabin air filter	R						✓						✓	
Drive belts	S/I	At 60,000 miles, then every 15,000 miles thereafter												
Driveshaft boots (4WD)	S/I			✓			✓			✓			✓	
Engine air filter	R						✓						✓	
Engine coolant	S/I			✓			✓			✓			✓	
Engine coolant	R	At 100,000 miles, then every 50,000 miles												
Engine oil & filter	R	✓	✓	✓	✓	✓	✓	✓	✓	✓	✓	✓	✓	✓
Exhaust pipes & mountings	S/I			✓			✓			✓			✓	
Front differential oil (4WD)	S/I			✓			✓			✓			✓	
Fuel tank cap gasket	S/I						✓						✓	
Fuel lines & connections	S/I						✓						✓	
Manual transmission oil	S/I						✓						✓	
Propeller shaft (4WD)	L			✓			✓			✓			✓	
Propeller shaft bolt	S/I			✓			✓			✓			✓	
Radiator, condenser and/or intercooler	S/I			✓			✓			✓			✓	
Rear differential oil	S/I			✓			✓			✓			✓	
Rotate tires	S/I	✓	✓	✓	✓	✓	✓	✓	✓	✓	✓	✓	✓	✓
Spark plugs	R						✓						✓	
Steering gear box	S/I			✓			✓			✓			✓	
Steering linkage & boots	S/I			✓			✓			✓			✓	
Transfer case (4WD)	R						✓						✓	

FREQUENT OPERATION MAINTENANCE (SEVERE SERVICE)

Driving on dirt roads or dusty roads:

Air filter: Inspect every 15,000 miles

Ball joints and dust covers: Inspect every 5,000 miles

Drive shaft boots (4WD): Inspect every 5,000 miles

Engine air filter: Inspect every 5,000 miles

Lubricate propeller shaft (4WD): Lubricate every 5,000 miles

Nuts and bolts on chassis: Tighten every 5,000 miles

Propeller shaft bolt: Re-torque every 5,000 miles

Steering linkage and boots: Inspect every 5,000 miles

Driving while towing:

Automatic transmission fluid: Replace every 60,000 miles

Front differential oil (4WD): Replace every 15,000 miles

Manual transmission oil (4WD): Replace every 30,000 miles

Nuts and bolts on chassis: Tighten every 5,000 miles

Propeller shaft (4WD): Lubricate every 5,000 miles

Propeller shaft bolt: Re-torque every 5,000 miles

Rear differential oil : Replace every 15,000 miles

Transfer case oil (4WD): Replace every 60,000 miles

22140_FJCR_C0013

PRECAUTIONS

Before servicing any vehicle, please be sure to read all of the following precautions, which deal with personal safety, prevention of component damage, and important points to take into consideration when servicing a motor vehicle:

- Never open, service or drain the radiator or cooling system when the engine is hot; serious burns can occur from the steam and hot coolant.
- Observe all applicable safety precautions when working around fuel. Whenever servicing the fuel system, always work in a well-ventilated area. Do not allow fuel spray or vapors to come in contact with a spark, open flame, or excessive heat (a hot drop light, for example). Keep a dry chemical fire extinguisher near the work area. Always keep fuel in a container specifically designed for fuel storage; also, always properly seal fuel containers to avoid the possibility of fire or explosion. Refer to the additional fuel system precautions later in this section.
- Fuel injection systems often remain pressurized, even after the engine has been turned **OFF**. The fuel system pressure must be relieved before disconnecting any fuel lines. Failure to do so may result in fire and/or personal injury.
- Brake fluid often contains polyglycol ethers and polyglycols. Avoid contact with the eyes and wash your hands thoroughly after handling brake fluid. If you do get brake fluid in your eyes, flush your eyes with clean, running water for 15 minutes. If eye irritation persists, or if you have taken brake fluid internally, IMMEDIATELY seek medical assistance.
- The EPA warns that prolonged contact with used engine oil may cause a number of skin disorders, including cancer. You should make every effort to minimize your exposure to used engine oil. Protective gloves should be worn when changing oil. Wash your hands and any other exposed skin areas as soon as possible after exposure to used engine oil. Soap and water, or waterless hand cleaner should be used.
- All new vehicles are now equipped with an air bag system, often referred to as a Supplemental Restraint System (SRS) or Supplemental Inflatable Restraint (SIR) system. The system must be disabled before performing service on or around system components, steering column, instrument panel components, wiring and sensors. Failure to follow safety and disabling procedures could result in accidental air bag deployment, possible personal injury and unnecessary system repairs.
- Always wear safety goggles when working with, or around, the air bag system. When carrying a non-deployed air bag, be sure the bag and trim cover are pointed away from your body. When placing a non-deployed air bag on a work surface, always face the bag and trim cover upward, away from the surface. This will reduce the motion of the module if it is accidentally deployed. Refer to the additional air bag system precautions later in this section.
- Clean, high quality brake fluid from a sealed container is essential to the safe and proper operation of the brake system. You should always buy the correct type of brake fluid for your vehicle. If the brake fluid becomes contaminated, completely flush the system with new fluid. Never reuse any brake fluid. Any brake fluid that is removed from the system should be discarded. Also, do not allow any brake fluid to come in contact with a painted surface; it will damage the paint.
- Never operate the engine without the proper amount and type of engine oil; doing so WILL result in severe engine damage.
- Timing belt maintenance is extremely important. Many models utilize an interference-type, non-freewheeling engine. If the timing belt breaks, the valves in the cylinder head may strike the pistons, causing potentially serious (also time-consuming and expensive) engine damage. Refer to the maintenance interval charts for the recommended replacement interval for the timing belt, and to the timing belt section for belt replacement and inspection.
- Disconnecting the negative battery cable on some vehicles may interfere with the functions of the on-board computer system(s) and may require the computer to undergo a relearning process once the negative battery cable is reconnected.
- When servicing drum brakes, only disassemble and assemble one side at a time, leaving the remaining side intact for reference.
- Only an MVAC-trained, EPA-certified automotive technician should service the air conditioning system or its components.

BRAKES

ANTI-LOCK BRAKE SYSTEM (ABS)

GENERAL INFORMATION

This vehicle uses an Antilock Brake System (ABS). If the Antilock Brake Module (ABM) senses impending rear wheel lock-up, it will energize the isolation solenoid. This prevents a further increase of driver induced brake pressure to the rear wheels. If this initial action is not enough to prevent rear wheel lock-up, the ABM will momentarily energize a dump solenoid. This opens the dump valve to vent a small amount of isolated rear brake pressure to an accumulator. The action of fluid moving to the accumulator reduces the isolated brake pressure at the wheel cylinders. The dump (pressure venting) cycle is limited to very short time periods (milliseconds). The ABM will pulse the dump valve until rear wheel deceleration reaches the desired slip rate programmed into the ABM. The system will switch to normal braking once wheel locking tendencies are no longer present.

As part of the anti-lock brake system, this vehicle is equipped with Electronic stability control (ESC). ESC is a computerized technology that improves the safety of a vehicle's handling by detecting and preventing skids. When ESC detects loss of steering control, ESC automatically applies individual brakes to help "steer" the vehicle where the driver wants to go. Braking is automatically applied to individual wheels, such as the outer front wheel to counter oversteer, or the inner rear wheel to counter under steer. Some ESC systems also reduce engine power until control is regained.

ESC compares the driver's intended direction (by measuring steering angle) to the vehicle's actual direction (by measuring lateral acceleration, vehicle rotation (yaw) and individual road wheel speeds). If the vehicle is not going where the driver is steering, ESC then brakes individual front or rear wheels and/or reduces excess engine power as needed to help correct under steer (plowing) or oversteer (fishtailing).

ESC incorporates yaw rate control into the anti-lock braking system (ABS). Yaw is rotation around the vertical axis; i.e. spinning left or right. Anti-lock brakes enable ESC to brake individual wheels. The ESC used on this vehicle also incorporates a Traction Control System (TCS), which senses drive-wheel slip under acceleration and individually brakes the slipping wheel or wheels and/or reduces excess engine power until control is regained.

BLEEDING PROCEDURE

See Figure 1.

➡This procedure is includes vehicles with and without ABS.

1. Before servicing the vehicle, refer to the precautions.

⁂ WARNING

Depressing the brake pedal with the reservoir cap removed will cause the fluid to spray. When bleeding, maintain the amount of fluid in the reservoir between the Min. and Max. lines.

2. Fill reservoir with brake fluid.
3. Bleed brake booster with accumulator pump assembly.
4. If the brake master cylinder is disassembled, the brake line is disconnected from the brake master cylinder or if the reservoir becomes empty, bleed the brake master cylinder.
   a. Turn the ignition switch to **ON**, and wait until the pump motor has stopped. Pump operating sound can be heard.
   b. Turn the ignition switch to **OFF**, and depress the brake pedal more than 20 times.
   c. When pressure in the accumulator is released, the reaction force becomes lighter and the stroke becomes longer.
   d. Repeat the first two steps 5 times.
   e. Turn the ignition switch to **ON**, and check that the pump stops after approximately 8 to 14 seconds.

⁂ WARNING

If the pump does not stop, repeat the procedure again.

5. Bleed brake line.
   a. Turn the ignition switch to **ON**, and wait until the pump motor has stopped. Pump operating sound can be heard.
   b. For the front brake line, connect the vinyl tube to the brake caliper.

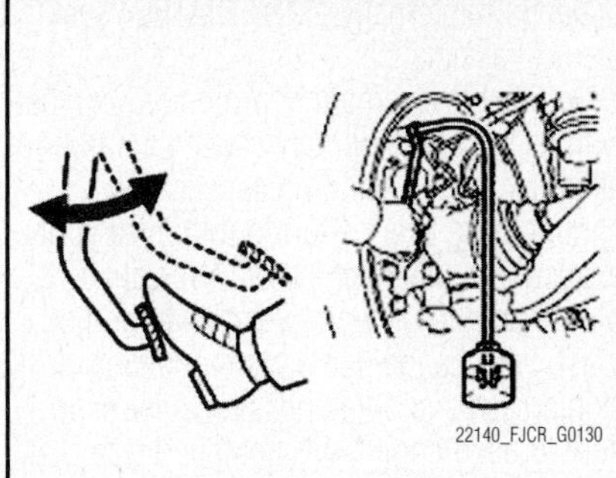

Fig. 1 Connect the vinyl tube to the brake caliper

Depress the brake pedal several times, then loosen the bleeder plug with the pedal held down.
   c. At the point when the fluid stops coming out, tighten the bleeder plug, then release the brake pedal.
   d. Repeat the steps until all the air in the fluid has been bled out. Tighten the bleeder plug to 8 ft. lbs. (11 Nm).
   e. Repeat the above procedures to bleed the other brake line.
6. For the rear brake line, connect the vinyl tube to the brake caliper. Depress the brake pedal, hold it, and then loosen the bleeder plug. Brake fluid is pumped out automatically.
   a. Loosen the bleeder plug and release the air. Keep the brake fluid in the reservoir tank above the **MIN** line during the above procedures.
   b. When the air is completely bled out of the brake fluid through the bleeder plug, tighten the bleeder plug. Tighten the bleeder plug to 8 ft. lbs. (11 Nm).
   c. Repeat the above procedures to bleed the other brake line.
7. Bleed master cylinder solenoid.
8. If the brake master cylinder is disassembled, the brake line is disconnected from the brake master cylinder or if the reservoir becomes empty, bleed the brake master cylinder.
   a. Connect the scan tool to the diagnostic link connector.
   b. Turn the ignition switch to **ON**.
   c. Select **ACTIVE TEST** mode on the scan tool.
   d. Connect the vinyl tube to the rear brake caliper.
   e. Loosen the bleeder plug.
   f. Select **SRMF** to drive the solenoids and bleed air from the rear brake caliper.

⁂ WARNING

Do not depress the brake pedal. Keep the brake fluid in the reservoir tank above the MIN line during the above procedures. Brake fluid is sent through the pump.

➡To protect the solenoids, the Techstream turns OFFautomatically 2 seconds after every solenoid has been turned ON.

   g. Repeat steps until all the air in the brake fluid is bled out.
   h. When the air is completely bled out of the brake fluid through the bleeder plug, tighten the bleeder plug to 8 ft. lbs. (11 Nm).
   i. Repeat the above procedures to bleed the other brake line.
   j. Turn the ignition switch to **OFF**. Turn the ignition switch to **ON**.
   k. Clear DTC.
9. Check fluid level in reservoir
   a. Turn the ignition switch to **OFF**, and depress the brake pedal more than 20 times (until the pedal reaction feels light and pedal stroke becomes longer), and adjust the fluid level to the MAX level.
   b. When the ignition switch is turned to **ON** , brake fluid is sent to the accumulator and the fluid level decreases by approximately 5 mm from the level when the ignition switch is **OFF** (normal).

BRAKE CALIPER

REMOVAL & INSTALLATION

See Figures 2 through 4.

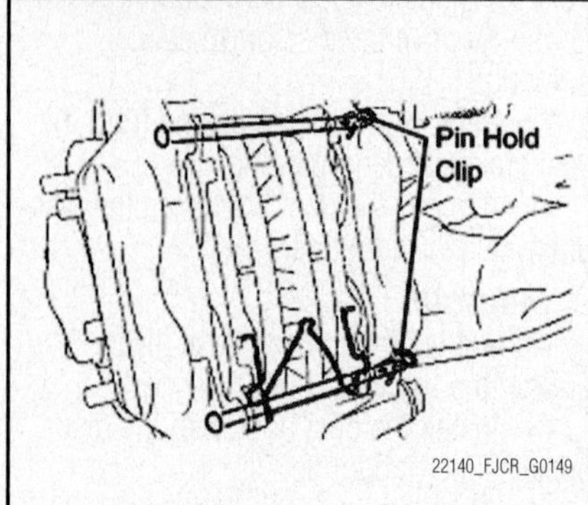

Fig. 2 Connect the vinyl tube to the brake caliper

1. Before servicing the vehicle, refer to the precautions.

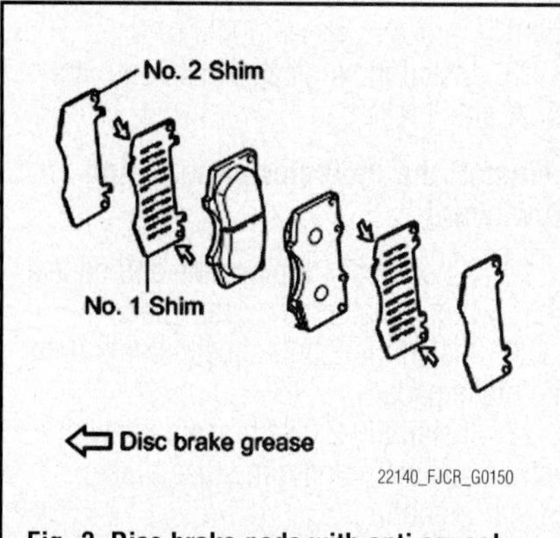

Fig. 3 Disc brake pads with anti squeal shims

2. Raise and safely support the vehicle securely on jackstands.
3. Remove the front wheels.
4. Drain brake fluid.
5. Remove front disc brake anti rattle with hole pin.
   a. Remove the 2 pin hold clips, then remove the 2 hole pins from the disc brake caliper.
   b. Remove the anti rattle spring from the disc brake pad.
6. Remove the disc brake pads with anti squeal shims from the disc brake caliper.

Fig. 4 Disc brake caliper retaining bolts

7. Remove the anti squeal shims from each of the disc brake pads.
8. Using a union nut wrench, separate the brake tube from the disc brake caliper.

➡Use a container to collect the brake fluid as it drains out.

9. Remove the disc brake caliper.

To install:

10. Install the disc brake caliper and tighten the two bolts to 91 ft. lbs. (123 Nm).
11. Install the brake tube onto the disc brake caliper and tighten to 10 ft. lbs. (14 Nm).
12. If necessary, replace the anti squeal shim when replacing the brake pad.
13. Apply disc brake grease to both sides of each No. 1 shim.
14. Install the No. 1 and No. 2 anti squeal shims onto each brake pad.
15. Install the disc brake pads with anti squeal shims onto the disc brake caliper.

**** WARNING**

There should be no oil or grease on the friction surfaces of the disc brake pads or the front disc.

16. Install the anti rattle spring and hole pins onto the disc brake caliper.

➡The anti rattle spring is installed onto the lower hole pin.

17. Install the pin hold clip with its handle oriented in the center of the vehicle.
18. Fill reservoir with brake fluid.
19. Bleed the brakes.
20. Install the front wheel and lower the vehicle.

DISC BRAKE PADS

REMOVAL & INSTALLATION

See Figures 2 and 3.

1. Before servicing the vehicle, refer to the precautions.
2. Raise and safely support the vehicle securely on jackstands.
3. Remove the front wheels.
4. Drain brake fluid.
5. Remove front disc brake anti rattle with hole pin.
   a. Remove the 2 pin hold clips, then remove the 2 hole pins from the disc brake caliper.
   b. Remove the anti rattle spring from the disc brake pad.
6. Remove the disc brake pads with anti squeal shims from the disc brake caliper.
7. Remove the anti squeal shims from each of the disc brake pads.

To install:

8. If necessary, replace the anti squeal shim when replacing the brake pad.
9. Apply disc brake grease to both sides of each No. 1 shim.
10. Install the No. 1 and No. 2 anti squeal shims onto each brake pad.
11. Install the disc brake pads with anti squeal shims onto the disc brake caliper.

**** WARNING**

There should be no oil or grease on the friction surfaces of the disc brake pads or the front disc.

12. Install the anti rattle spring and hole pins onto the disc brake caliper.

➡The anti rattle spring is installed onto the lower hole pin.

13. Install the pin hold clip with its handle oriented in the center of the vehicle.
14. Fill reservoir with brake fluid.
15. Bleed the brakes.
16. Install the front wheel and lower the vehicle.

BRAKE CALIPER

REMOVAL & INSTALLATION

See Figure 5.

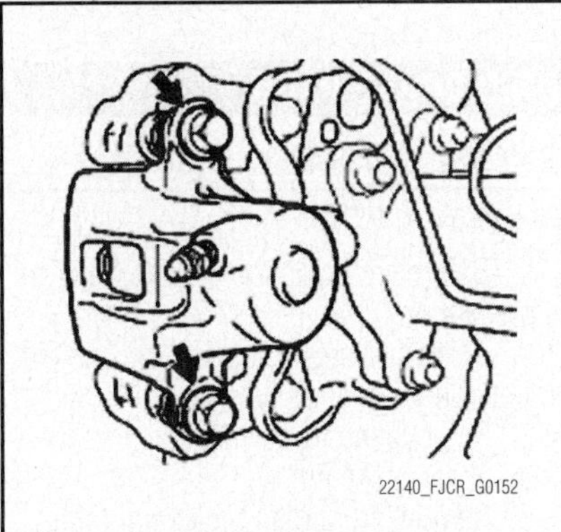

Fig. 5 Disc brake caliper retaining bolts

1. Before servicing the vehicle, refer to the precautions.
2. Raise and safely support the vehicle securely on jackstands.
3. Remove the rear wheels.
4. Drain brake fluid.
5. Using a union nut wrench, separate the brake tube from the disc brake caliper.

➡Use a container to collect the brake fluid as it drains out.

6. Remove the disc brake caliper by remove the 2 slide pins.
7. Remove the 2 disc brake pads with anti squeal shim from the disc brake caliper mounting.
8. Remove the 2 anti squeal shims from each brake.
9. Remove the indicator plate from the inner side of the brake.
10. Remove the 4 pad support plates from the disc brake caliper mounting.
11. Remove the disc brake caliper mounting.
12. Remove the slide bushing from the disc brake caliper mounting.
13. Remove the dust boot from the disc brake caliper mounting.
14. Remove the hole plug from the disc brake caliper mounting.

To install:

15. Install a new hole plug onto the disc brake caliper mounting.
16. Apply lithium soap base glycol grease to a new dust boot.
17. Install the dust boot onto the disc brake caliper mounting.
18. Apply lithium soap base glycol grease to a new slide bushing.
19. Install the slide bushing onto the disc brake caliper mounting.
20. Install the disc brake caliper mounting with the 2 bolts and 2 washers. Tighten to 78 ft. lbs. (105 Nm).
21. Install the 4 pad support plates onto the disc brake caliper mounting.
22. Install the indicator plate onto the inner side brake pad.

➡Install the indicator plate facing downward.

23. If necessary, replace the anti squeal shim when replacing the brake pad.
24. Install the 2 anti squeal shims onto the brake pads.
25. Install the 2 disc brake pads with anti squeal shims onto the disc brake caliper mounting.

⁂ WARNING

There should be no oil or grease on the friction surfaces of the disc brake pads or the rear disc.

26. Apply lithium soap base glycol grease to the 2 slide pins.
27. Install the disc brake caliper onto the disc brake caliper mounting with the 2 slide pins. Tighten to 65 ft. lbs. (88 Nm).
28. Install the flexible hose with the union bolt and a new gasket. Tighten to 23 ft. lbs. (31 Nm).
29. Fill reservoir with brake fluid.
30. Bleed brake system.
31. Check fluid level in reservoir.
32. Check for brake fluid leakage.
33. Install rear wheel.

DISC BRAKE PADS

REMOVAL & INSTALLATION

See Figure 5.

1. Before servicing the vehicle, refer to the precautions.
2. Raise and safely support the vehicle securely on jackstands.
3. Remove the rear wheels.
4. Remove the disc brake caliper and suspend it out of the way.
5. Remove the 2 disc brake pads with anti squeal shim from the disc brake caliper mounting.
6. Remove the 2 anti squeal shims from each brake.
7. Remove the indicator plate from the inner side of the brake.
8. Remove the 4 pad support plates from the disc brake caliper mounting.
9. Remove the disc brake caliper mounting.
10. Remove the slide bushing from the disc brake caliper mounting.
11. Remove the dust boot from the disc brake caliper mounting.

To install:

12. Apply lithium soap base glycol grease to a new dust boot.
13. Install the dust boot onto the disc brake caliper mounting.
14. Apply lithium soap base glycol grease to a new slide bushing.
15. Install the slide bushing onto the disc brake caliper mounting.
16. Install the disc brake caliper mounting with the 2 bolts and 2 washers. Tighten to 78 ft. lbs. (105 Nm).
17. Install the 4 pad support plates onto the disc brake caliper mounting.
18. Install the indicator plate onto the inner side brake pad.

➡Install the indicator plate facing downward.

19. If necessary, replace the anti squeal shim when replacing the brake pad.
20. Install the 2 anti squeal shims onto the brake pads.
21. Install the 2 disc brake pads with anti squeal shims onto the disc brake caliper mounting.

⁂ WARNING

There should be no oil or grease on the friction surfaces of the disc brake pads or the rear disc.

22. Apply lithium soap base glycol grease to the 2 slide pins.
23. Install the disc brake caliper onto the disc brake caliper mounting with the 2 slide pins. Tighten to 65 ft. lbs. (88 Nm).
24. Check the fluid level in the reservoir and correct as necessary.
25. Install the front wheel and lower the vehicle.
26. Pump the brake pedal several times to ensure the brake pads are in contact with the rotors.

PARKING BRAKE SHOES

REMOVAL & INSTALLATION

See Figure 6.

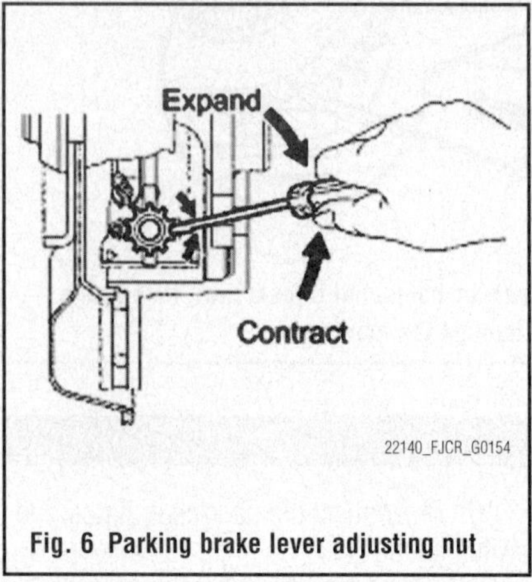

Fig. 6 Parking brake lever adjusting nut

1. Before servicing the vehicle, refer to the precautions.
2. Raise and safely support the vehicle securely on jackstands.
3. Remove rear wheel.
4. Remove the disc brake caliper assembly and suspend it out of the way.
5. Release the parking brake lever.
6. Place matchmarks on the disc and axle hub and remove the disc.

➡If the disc cannot be removed easily, turn the shoe adjuster until the wheel turns freely.

To install:

7. Align the matchmarks of the disc and axle hub and install the disc.
8. Temporarily install 2 hub nuts.
9. Remove the hole plug, and turn the adjuster to expand the shoe until the disc locks.
10. Contract the shoe adjuster until the disc rotates smoothly. Then return 8 notches.
11. Install the hole plug.
12. Install the disc brake caliper.
13. Install rear wheel and lower the vehicle.
14. Inspect parking brake lever travel.
15. Adjust parking brake lever travel.

ADJUSTMENT

See Figure 6.

1. Before servicing the vehicle, refer to the precautions.
2. Raise and safely support the vehicle securely on jackstands.
3. Remove rear wheel.
4. Temporarily install 2 hub nuts.
5. Remove the hole plug, and turn the adjuster to expand the shoe until the disc locks.
6. Contract the shoe adjuster until the disc rotates smoothly. Then return 8 notches.
7. Install rear wheel and lower the vehicle.
8. Inspect parking brake lever travel.
9. Adjust parking brake lever travel.

GENERAL INFORMATION

SERVICE PRECAUTIONS

- The vehicle is equipped with a Supplemental Restraint System (SRS), which consists of a driver airbag, front passenger airbag, side airbags, curtain shield airbags and front seat belt pretensioner. Failure to carry out service operations in the correct sequence could cause the SRS to unexpectedly deploy during servicing, possibly leading to a serious accident. Furthermore, if a mistake is made in servicing the SRS, it is possible that the SRS may fail to operate when required. Before performing servicing (including removal or installation of parts, inspection or replacement), be sure to read the following items carefully, then follow the correct procedures as indicated in the repair manual.
- Wait at least 90 seconds after the ignition switch is turned off and the negative (-) terminal cable is disconnected from the battery before starting the operation. The SRS is equipped with a back-up power source, so if work is started within 90 seconds of disconnecting the negative (-) terminal cable of the battery, the SRS may be deployed.
- Do not directly expose the steering pad, front passenger airbag assembly, center airbag sensor assembly, front airbag sensor, front seat inner belt assembly, seat position sensor, occupant classification ECU, front seat side airbag assembly, side airbag sensor, curtain shield airbag assembly, rear airbag sensor or front seat outer belt assembly to hot air or flames.
- Malfunction symptoms of the SRS are difficult to confirm, so DTCs are the most important source of information when troubleshooting. When troubleshooting the SRS, always inspect DTCs before disconnecting the battery.
- Never use SRS parts from another vehicle. When replacing parts, replace them with new ones.
- Never disassemble or repair any of the following parts in order to reuse them. If any of these parts have been dropped, or a defect is found (e.g. cracks, dents or any other defects) in any of the housings, brackets or connectors, then replace the part with a new one.
- Use a volt/ohmmeter with high impedance (10 kilohms/V minimum) for troubleshooting the electrical circuits.
- Information labels are attached to the periphery of the SRS components. Follow the instructions in the cautions.
- After work on the SRS is completed, perform the SRS warning light check.
- When the negative (-) terminal cable is disconnected from the battery, the memory will be cleared. therefore make a record of the contents stored in each system before starting work. When the work is finished, reset each system as it was before. Never use a back-up power supply from outside the vehicle to avoid erasing the memory in any system.
- When disconnecting the cable from the negative (-) battery terminal, initialize the meter / gauge system after the cable is reconnected.
- In the airbag system, the center airbag sensor assembly, front airbag sensor LH and RH, side airbag sensor LH and RH, rear airbag sensor LH and RH are collectively referred to as the airbag sensors.

DISARMING THE SYSTEM

Wait at least 90 seconds after the ignition switch is turned off and the negative (-) terminal cable is disconnected from the battery before starting the operation. The SRS is equipped with a back-up power source, so if work is started within 90 seconds of disconnecting the negative (-) terminal cable of the battery, the SRS may be deployed.

ARMING THE SYSTEM

Reconnect the negative battery cable and perform an SRS warning light check.

CLOCKSPRING CENTERING

See Figure 7.

1. Check that the ignition switch is turned to **OFF**.
2. Rotate the spiral cable counterclockwise slowly by hand until it feels firm.

**** WARNING**

Do not turn the spiral cable with the airbag wire harness or connector.

3. Rotate the spiral cable clockwise approximately 2.5 turns to align the marks.
4. The spiral cable will rotate approximately 2.5 turns both ways from the center.

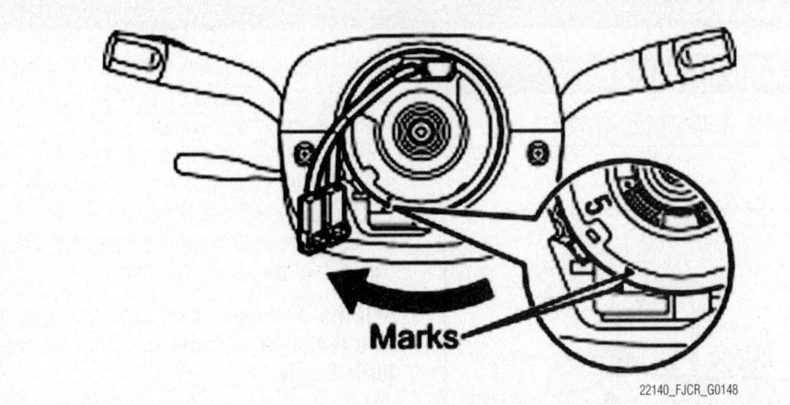

Fig. 7 Rotate the spiral cable counterclockwise slowly by hand until it feels firm, then rotate the spiral cable clockwise approximately 2.5 turns to align the marks

DRIVETRAIN

AUTOMATIC TRANSMISSION ASSEMBLY

REMOVAL & INSTALLATION

See Figures 8 through 15.

1. Before servicing the vehicle, refer to the precautions.
2. On A750F transmissions, perform the following:
   a. Remove shift lever assembly for the automatic transmission.
   b. Remove shift lever knob for the transfer case.
   c. Remove the 4 screws and shift lever boot.
   d. Using needle-nose pliers, remove the snap ring and pull out the shift lever from the shift lever retainer.
3. Disconnect the negative battery cable.
4. Remove No. 1 engine under cover.
5. Remove rear engine under cover assembly.
6. Disconnect oxygen sensor.
7. Remove No. 2 exhaust front pipe assembly.
8. Remove front exhaust pipe assembly.
9. Remove driveshaft assembly.
10. Drain automatic transmission fluid.
    a. Remove the drain plug and gasket, and drain ATF.
    b. Install a new gasket and drain plug. Tighten to 21 ft. lbs. (28 Nm).
11. Remove the 3 bolts and manifold bracket.
12. Remove the 3 bolts and No. 2 manifold bracket.
13. Disconnect oil cooler tube.
    a. Remove the 3 bolts and 2 clamps.
    b. Using SST: 09023-12701, disconnect the No. 1 oil cooler inlet tube.
    c. Using SST: 09023-12701, disconnect the No. 1 oil cooler outlet tube.
14. Disconnect transmission control cable assembly.
    a. Remove the nut and disconnect the transmission control cable.
    b. Remove the 2 bolts and disconnect the transmission control cable.
15. Disconnect the park/neutral position switch connector.
16. Disconnect the transmission wire connector.
17. Disconnect the 2 speed sensor connectors.
18. Disconnect wire harness.
19. Remove starter assembly.
20. Remove the flywheel housing side cover from the cylinder block.
21. Support the automatic transmission assembly with a transmission jack.
22. Remove the 4 bolts and front suspension member bracket.
23. Remove No. 3 frame crossmember.
24. Remove automatic transmission assembly.
    a. Turn the crankshaft to gain access to each bolt.
    b. Hold the crankshaft pulley nut with a wrench and remove the 6 bolts.
    c. Remove the 9 bolts.
    d. Separate and remove the automatic transmission.
25. Remove the 4 bolts and engine

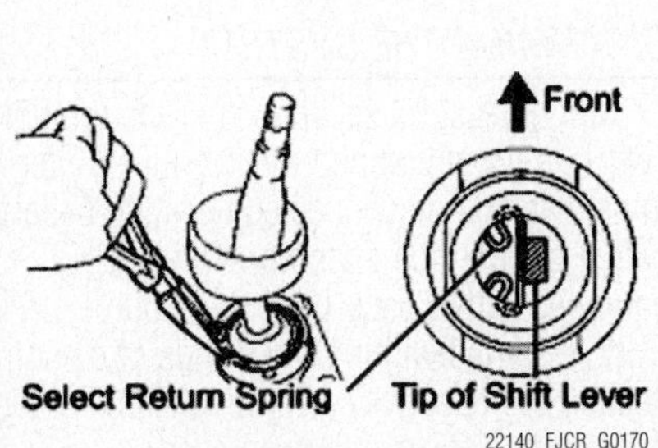

Fig. 8 Using needle-nose pliers, remove the snap ring and pull out the shift lever from the shift lever retainer

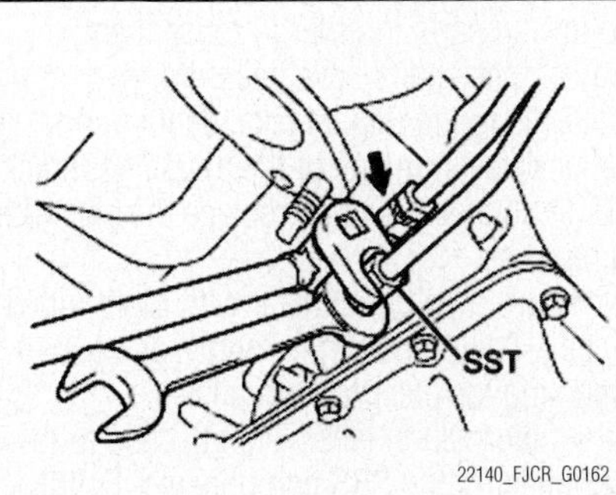

Fig. 9 Using the special tool to disconnect the oil cooler tube

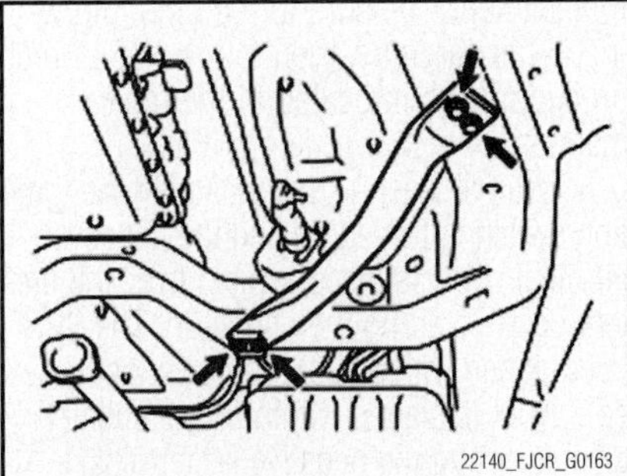

Fig. 10 Front suspension member bracket—left side shown, right side similar

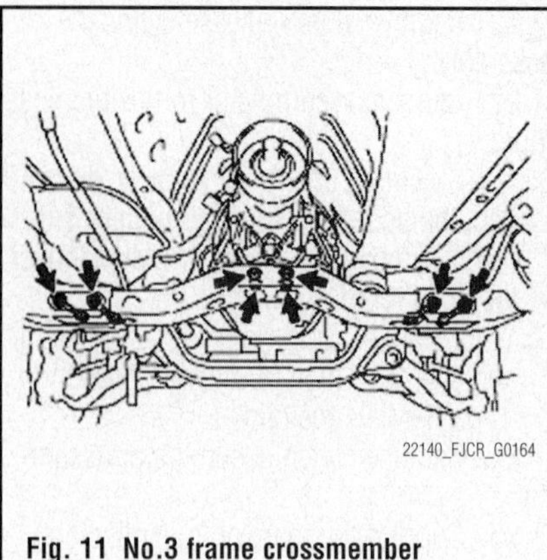

Fig. 11 No.3 frame crossmember

mounting insulator rear from the automatic transmission.

26. Remove torque converter clutch assembly.

27. On A750F transmissions, remove the transfer case assembly.

To install:

28. On A750F transmissions, install the transfer case assembly.

29. Install the torque converter clutch onto the automatic transmission.

30. Using calipers and a straight edge, measure dimension **A**, between the engine and the end surface of the drive plate, as shown. Dimension **A** should be 0.835 in. (21.2 mm).

31. Using calipers and a straight edge, measure dimension **B** shown in the illustration. Check that **B** is greater than **A**. Dimension **B** = **A** + 0.0394 in. (1.0 mm) or more.

32. Install the engine mounting insulator rear onto the automatic transmission with the 4 bolts. Tighten to 48 ft. lbs. (65 Nm).

33. Install the automatic transmission onto the engine with the 9 bolts. Tighten bolt **A** to 53 ft. lbs. (71 Nm) and bolt **B** to 27 ft. lbs. (37 Nm).

34. Install the 6 torque converter clutch mounting bolts. Tighten to 35 ft. lbs. (48 Nm).

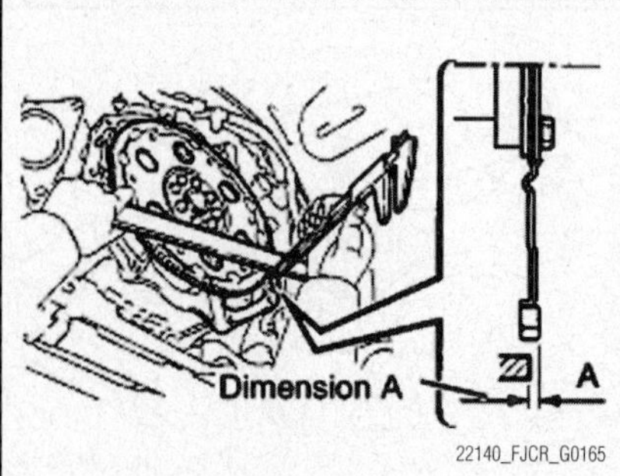

Fig. 12 Dimension A, between the engine and the end surface of the drive plate should be 0.835 in. (21.2 mm)

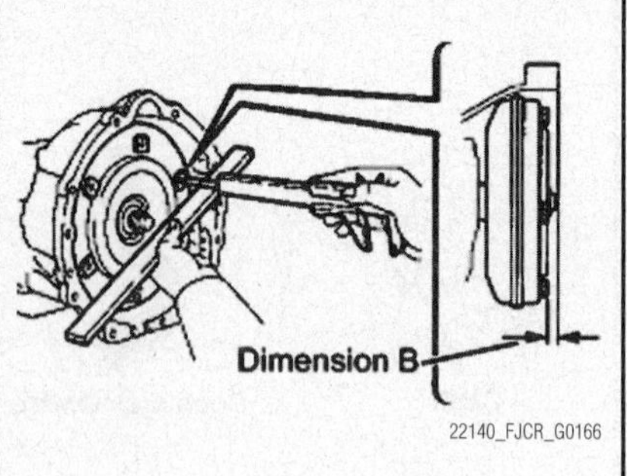

Fig. 13 Dimension B, measured between the torque converter bolt pad and the edge of the transmission should equal dimension A plus 0.0394 in. (1.0 mm) or more

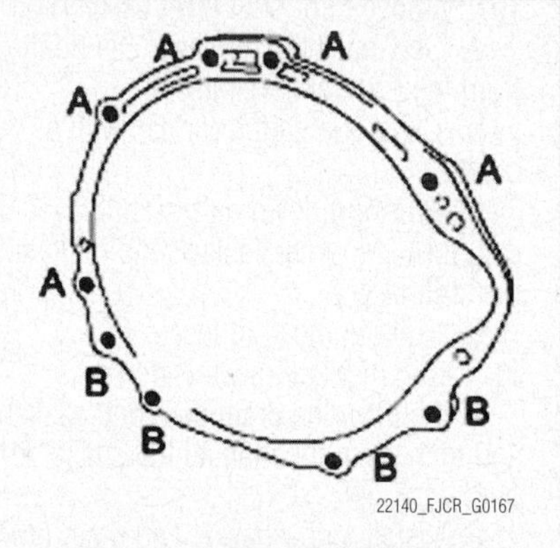

Fig. 14 Automatic transmission bolt tightening torques A to 53 ft. lbs. (71 Nm) and B to 27 ft. lbs. (37 Nm)

➡First install the black bolt and then the other 5 bolts.

35. Install the No. 3 frame crossmember with the 4 bolts and 4 nuts. Tighten 53 ft. lbs. (71 Nm).

36. Install the No. 3 frame crossmember with the 4 bolts. Tighten 14 ft. lbs. (19 Nm).

37. Install the front suspension member bracket with the 4 bolts. Tighten 24 ft. lbs. (33 Nm).

38. Install the flywheel housing side cover onto the cylinder block.

39. Install starter assembly.

40. Connect wire harness.

41. Connect the park/neutral position switch connector.

42. Connect the transmission wire connector.

43. Connect the 2 speed senor connectors.

44. Install the transmission control cable bracket with the 2 bolts. Tighten 10 ft. lbs. (14 Nm).

45. Connect the transmission control cable with the nut. Tighten 10 ft. lbs. (14 Nm).

46. Temporarily install the cooler tubes. Install the 2 clamps and tighten to 10 ft. lbs. (14 Nm). Install the 3 bolts and tighten to 49 inch lbs. (6 Nm).

47. Using SST: 09023-12701, tighten the No. 1 oil cooler tubes to 25 ft. lbs. (34 Nm).

48. Install the No. 2 manifold bracket with the 3 bolts. Tighten to 30 ft. lbs. (40 Nm).

49. Install the manifold bracket with the 3 bolts. Tighten to 30 ft. lbs. (40 Nm).

50. Install driveshaft assembly.

51. Install front exhaust pipe assembly.

52. Install No. 2 exhaust front pipe assembly.

53. Install oxygen sensor.

54. Connect the negative battery cable.

55. Using a scan tool, perform the reset memory (AT initialization).

56. Adjust automatic transmission fluid level.

57. Inspect shift lever position.

a. Check that the shift lever moves smoothly and appropriately from the **P** to the **R** position, only when the ignition switch is turned to **ON** and the brake pedal is depressed.

b. When starting the engine, make sure that the vehicle moves forward when shifting the lever from the **N** to the **D** position and moves rearward when shifting the lever to the **R** position.

c. To adjust the shift lever, push the control shaft fully rearward.

d. Turn the control shaft lever 2 notches counterclockwise to the **N** position.

e. Set the shift lever to the **N** position.

58. Check for exhaust gas leakage.

59. Install rear engine under cover assembly.

60. Install No. 1 engine under cover.

61. On A750F transmissions, perform the following:

a. Install the transfer shift lever onto the shift lever retainer.

Fig. 15 Turn the control shaft lever 2 notches counterclockwise to the N position

b. Using needle-nose pliers, install the snap ring.

c. Install the shift lever boot with the 4 screws.

d. Install shift lever knob for the transfer case.

e. Install shift lever assembly for automatic transmission.

MANUAL TRANSMISSION ASSEMBLY

REMOVAL & INSTALLATION

See Figures 16 through 21.

1. Before servicing the vehicle, refer to the precautions.
2. Disconnect the negative battery cable.
3. Remove parking brake hole cover.
4. Remove console upper rear panel.
5. Remove front console box.
6. Remove shift lever knob for transfer case.
7. Remove shift lever knob for manual transmission.
8. Remove the 4 screws and 2 clips, then remove the shift lever boot.
9. Using needle-nose pliers, remove the snap ring and pull out the shift lever from the shift lever retainer.

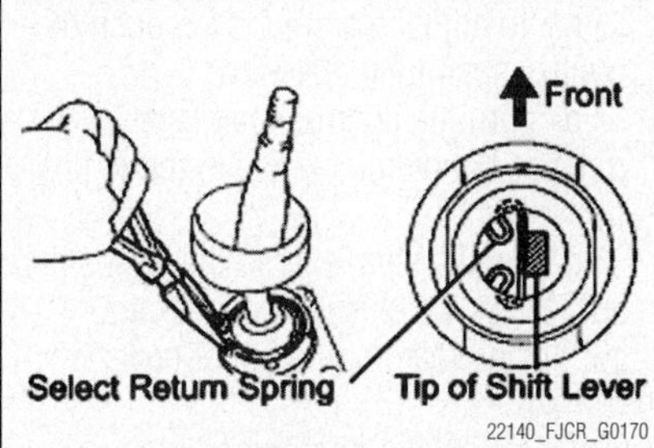

Fig. 16 Using needle-nose pliers, remove the snap ring and pull out the shift lever from the shift lever retainer

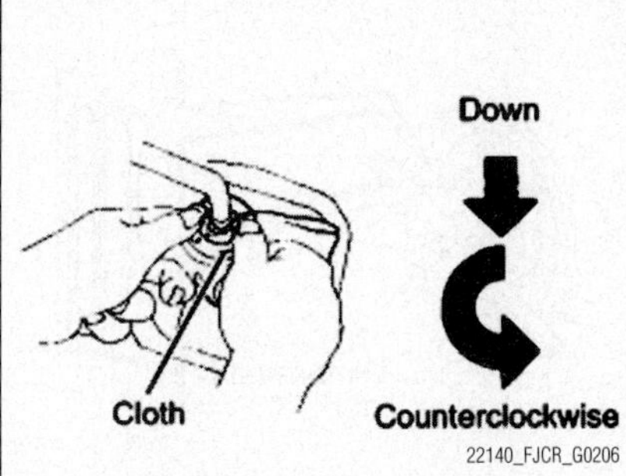

Fig. 18 Cover the shift lever cap with a cloth. Pressing down on the shift lever cap, turn it counterclockwise to remove the shift lever

10. Remove floor shift lever assembly.
    a. Separate the shift lever cap boot from the manual transmission.
    b. Cover the shift lever cap with a cloth.
    c. Pressing down on the shift lever cap, turn it counterclockwise to remove the shift lever.
    d. Pull out the shift lever.
11. Drain manual transmission oil.
    a. Remove the drain plug and gasket and then drain the manual transmission oil.
    b. Install a new gasket and drain plug and tighten to 27 ft. lbs. (37 Nm).
12. Remove No. 2 front exhaust pipe assembly.
13. Remove front exhaust pipe assembly.
14. Remove driveshaft heat insulator.
15. Remove front halfshaft assembly.
16. Remove driveshaft assembly.
17. Remove the 3 bolts and manifold bracket.
18. Remove the 3 bolts and No. 2 manifold bracket.
19. Remove starter assembly.
20. Remove flywheel housing side cover.
21. Remove No. 1 clutch housing cover.
22. Remove clutch release cylinder assembly.
23. Remove accumulator to flexible hose tube.
    a. Using a union nut wrench, disconnect the hose tube for accumulator side.
    b. Remove the 3 nuts and disconnect the hose tube.
    c. Using a union nut wrench, disconnect the hose tube for flexible hose side.
    d. Remove the clip.
24. Remove clutch accumulator assembly.
25. Support the manual transmission with a transmission jack.
26. Remove the 4 bolts and remove the front suspension member brackets.
27. Remove No. 3 frame crossmember.
    a. Remove the 4 bolts from the No. 3 frame crossmember.
    b. Remove the 4 nuts and 4 bolts and remove the No. 3 frame crossmember.
28. Remove the 4 bolts and engine mounting insulator rear from the manual transmission.
29. Tilt down the transmission.
30. Disconnect the 2 transfer indicator switch connectors.
31. Disconnect the speedometer sensor connector.
32. Disconnect the back-up light switch connector.
33. Remove manual transmission with transfer case.
    a. Remove the 9 bolts.
    b. Separate and remove the manual transmission.
34. Remove the 3 bolts and 2 nuts and remove the manual transmission case cover.
35. Remove transfer assembly.

To install:

36. Install transfer case assembly.
37. Install the manual transmission case cover with the 3 bolts and 2 nuts. Tighten to 9 ft. lbs. (12 Nm).

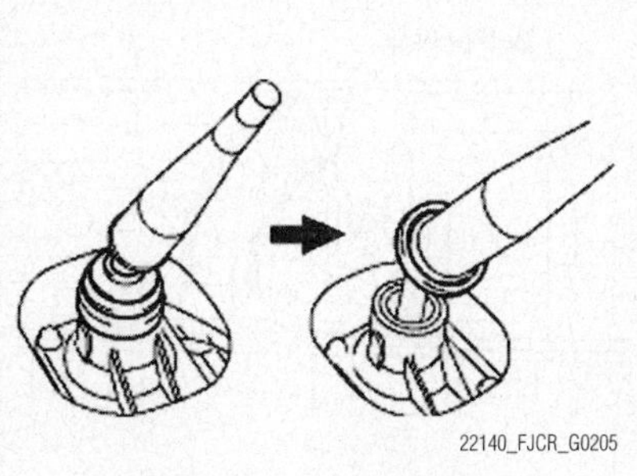

Fig. 17 Separate the shift lever cap boot from the manual transmission

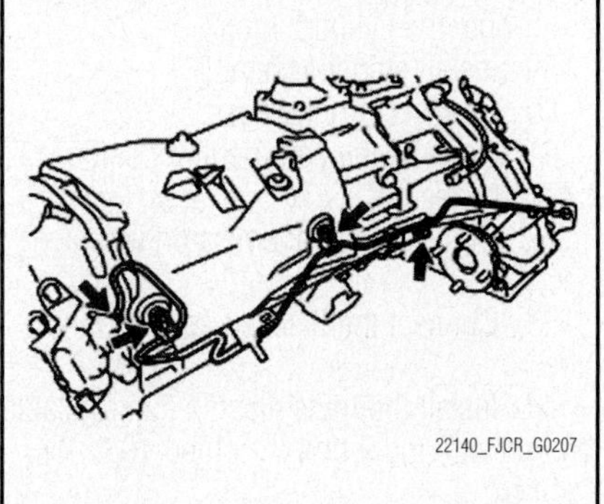

Fig. 19 Accumulator to flexible hose tube

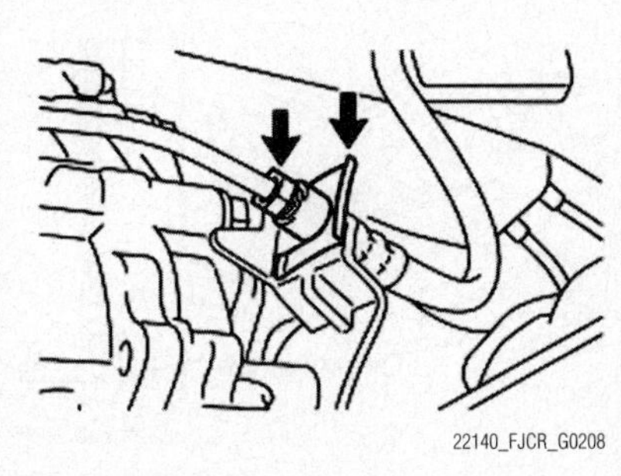

Fig. 20 Using a union nut wrench, disconnect the hose tube for accumulator side

Fig. 21 Manual transmission case cover fastener locations

38. Install the manual transmission with transfer case with the 9 bolts. Tighten bolt **A** to 53 ft. lbs. (71 Nm) and bolt **B** to 27 ft. lbs. (37 Nm).
39. Connect the back-up light switch connector.
40. Connect the speedometer sensor connector.
41. Connect the 2 transfer indicator switch connectors.
42. Install engine mounting insulator rear No. 1 with the 4 bolts. Tighten to 48 ft. lbs. (65 Nm)
43. Install the No. 3 frame crossmember with the 4 bolts and 4 nuts. Tighten to 53 ft. lbs. (72 Nm).
44. Install the No. 3 frame crossmember with the 4 bolts. Tighten to 14 ft. lbs. (19 Nm).
45. Install the front suspension member brackets with the 4 bolts. Tighten to 24 ft. lbs. (33 Nm).
46. Install clutch accumulator assembly.
47. Using a union nut wrench, connect the hose tube for flexible hose side. Tighten to 11 ft. lbs. (15 Nm)
48. Install a new clip.
49. Using a union nut wrench, connect the hose tube for the accumulator side. Tighten to 11 ft. lbs. (15 Nm).
50. Install the 3 nuts and connect the hose tube. Tighten to 14 ft. lbs. (19 Nm).
51. Install clutch release cylinder assembly.
52. Install No. 1 clutch housing cover.
53. Install the flywheel housing side cover onto the cylinder block.
54. Install starter assembly.
55. Install the manifold bracket with the 3 bolts. Tighten to 30 ft. lbs. (40 Nm).
56. Install No. 2 manifold bracket. Tighten to 30 ft. lbs. (40 Nm).
57. Install driveshaft assembly.
58. Install front halfshaft assembly.
59. Install the driveshaft heat insulator with the 2 bolts. Tighten to 12 ft. lbs. (16 Nm).
60. Install front exhaust pipe assembly.
61. Install No. 2 front exhaust pipe assembly.
62. Check and adjust the manual transmission oil level.
63. Apply MP grease to the tip of the shift lever.
64. Cover the shift lever cap with a cloth.
65. Pressing down on the shift lever cap, turn it clockwise to install the shift lever.
66. Install the transfer shift lever onto the shift lever retainer.
67. Using needle-nose pliers, install the snap ring.
68. Install the shift lever boot with the 4 screws and 2 clips.
69. Install shift lever knob (for manual transmission).
70. Install shift lever knob (for transfer).
71. Install front console box.
72. Install console upper rear panel.
73. Install parking brake hole cover.
74. Connect the negative battery cable.
75. Check for exhaust gas leakage.

CLUTCH DRIVEN DISC & PRESSURE PLATE

REMOVAL & INSTALLATION

See Figures 22 through 25.

1. Before servicing the vehicle, refer to the precautions.
2. Remove manual transmission assembly.
3. Remove the clutch release fork together with the clutch release bearing from the transmission assembly.
4. Remove the clutch release bearing assembly from the clutch release fork.
5. Remove the release fork support from the transaxle assembly.
6. Remove release bearing hub clip.
7. Align the matchmark on the clutch cover assembly with the one on the flywheel.
8. Loosen each set bolt one turn at a time until the spring tension is released.
9. Remove the 6 bolts and clutch cover assembly.

✱✱ WARNING

Do not drop the clutch disc assembly.

10. Remove the clutch disc assembly.

✱✱ WARNING

Keep the lining part of the clutch disc assembly, pressure plate and surface of the flywheel free from oil and foreign matter.

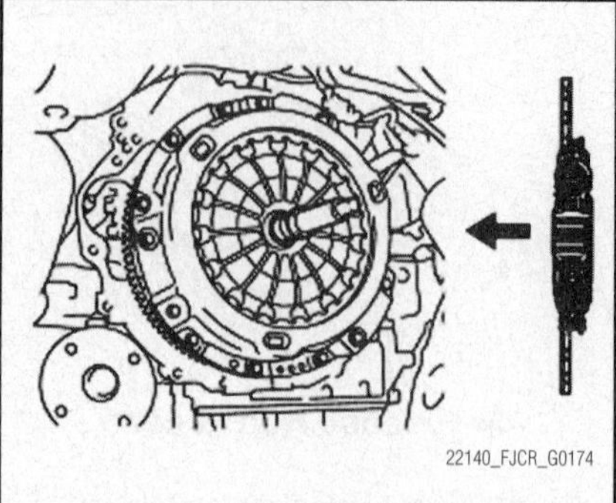

Fig. 22 Insert the clutch disc assembly in the correct orientation

To install:

11. Insert SST: 09301-00220 into the clutch disc assembly, then insert them into the flywheel.

➡Insert the clutch disc assembly in the correct orientation.

12. Align the matchmarks on the clutch cover assembly and flywheel. Move SST up and down, right and left lightly, after checking that the disc is in the center.
13. Following the pattern shown in the illustration, tighten the 6 bolts, in the order starting with the bolt located near the knock pin on the top. Following the order in the illustration, uniformly tighten the bolts to 14 ft. lbs. (19 Nm).
14. Using a dial indicator with roller instrument, check the diaphragm spring tip alignment. Runout should not exceed 0.020 in. (0.5 mm).
15. If the alignment is not as specified, adjust the diaphragm spring tip alignment using SST: 09333-00013.
16. Install the release fork support onto the transaxle assembly and tighten to 35 ft. lbs. (47 Nm).
17. Install release bearing hub clip.

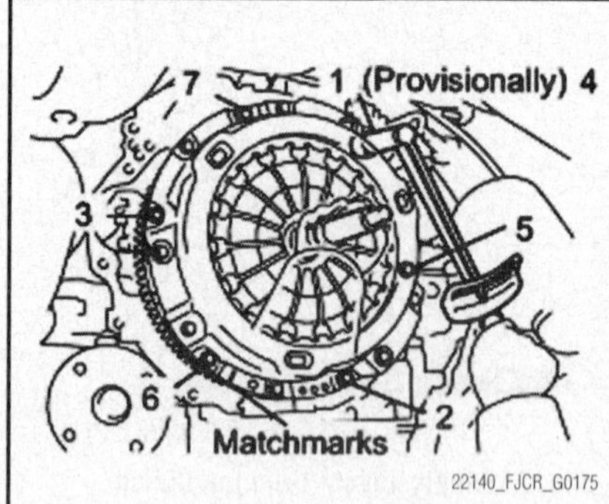

Fig. 23 Clutch disc assembly tightening sequence

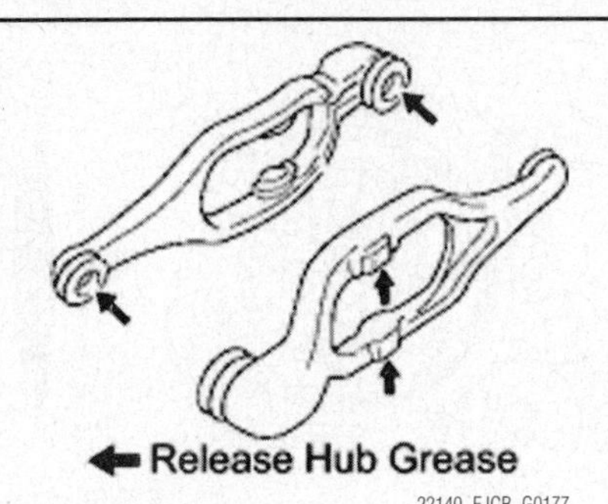

Fig. 24 Apply Toyota Genuine Release Hub Grease or equivalent to the release fork and release bearing assembly contact surfaces

18. Apply Toyota Genuine Release Hub Grease or equivalent to the release fork and release bearing assembly contact surfaces, release fork and push rod contact surface and release fork pivot point.
19. Install the release fork onto the release bearing assembly.
20. Apply Toyota Genuine Clutch Spline Grease or equivalent to the input shaft spline.

WARNING

Do not apply grease to portion A shown in the illustration.

21. Install the bearing onto the release fork, and then install them onto the transaxle assembly.
22. After installation, move the fork forward and backward to check that the release bearing slides smoothly.
23. Install manual transmission assembly.

ADJUSTMENTS

The driven disc and pressure plate used in this vehicle are hydraulically operated and require no adjustment.

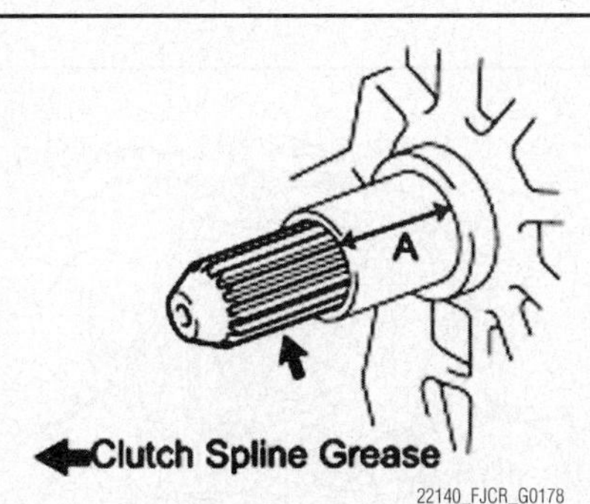

Fig. 25 Apply Toyota Genuine Clutch Spline Grease or equivalent to the input shaft spline. Do not apply grease to portion A shown in the illustration

CLUTCH MASTER CYLINDER

REMOVAL & INSTALLATION

See Figure 26.

1. Before servicing the vehicle, refer to the precautions.
2. Drain clutch fluid.
3. Using a union nut wrench, disconnect the flexible hose tube.
4. Remove front door scuff plate.
5. Remove front floor footrest.
6. Remove footrest clip.
7. Remove cowl side trim board.
8. Remove the instrument panel register.

WARNING

To avoid breakage of the register, remove the instrument panel register with the shut damper of the register open. Pull the instrument panel register up diagonally to remove it.

9. Remove the lower instrument panel finish panel.
10. Remove the lower instrument panel.
11. Separate driver side junction block.
12. Disconnect the clutch start switch connector.
13. On vehicle equipped with cruise control, disconnect the clutch switch connector.
14. Remove the 2 nuts and bolt, then remove the clutch pedal with clutch master cylinder.
15. Disconnect the clutch reservoir tube.
16. Using a union nut wrench, disconnect the 2 way tube from the clutch master cylinder.
17. Remove the nut and clutch start switch from the clutch pedal support.
18. Remove the clip and the pin and separate the push rod clevis from the clutch pedal.
19. Remove the 2 bolts, then remove the clutch master cylinder.

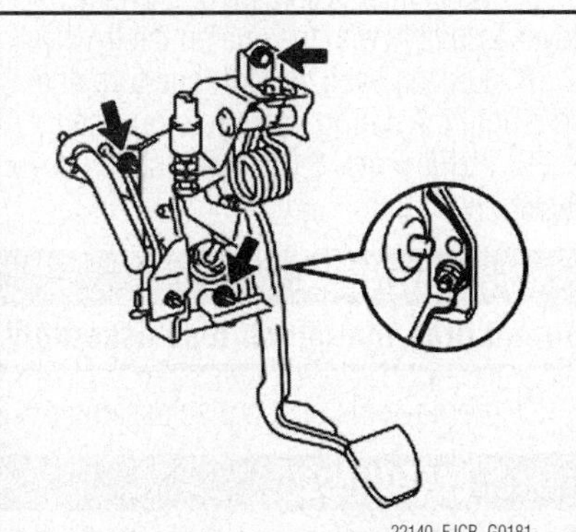

Fig. 26 Remove the 2 nuts and bolt, then remove the clutch pedal with clutch master cylinder

To install:

20. Install the 2 bolts, then install the clutch master cylinder. Tighten to 9 ft. lbs. (12 Nm).
21. Apply MP grease to the pin. Install the clip and the pin and the push rod clevis from the clutch pedal.
22. Install the nut and clutch start switch from the clutch pedal support. Tighten to 12 ft. lbs. (16 Nm).
23. Using a union nut wrench, connect the 2 way tube from the clutch master cylinder. Tighten to 11 ft. lbs. (15 Nm).
24. Connect the clutch reservoir tube.
25. Install the 2 nuts and bolt, then install the clutch pedal with clutch master cylinder. Tighten to 10 ft. lbs. (14 Nm).
26. On vehicle equipped with cruise control, connect the clutch switch connector.
27. Connect the clutch start switch connector.
28. Separate driver side junction block.
29. Install the lower instrument panel.
30. Install the lower instrument panel finish panel.
31. Install the instrument panel register.
32. Install cowl side trim board.
33. Install footrest clip.
34. Install front floor footrest.
35. Install front door scuff plate.
36. Using a union nut wrench, connect the flexible hose tube. Tighten to 11 ft. lbs. (15 Nm).
37. Bleed clutch pipe line. Tighten to 8 ft. lbs. (11 Nm).
38. Inspect and adjust clutch pedal.
39. Check for clutch fluid leakage.
40. Check fluid level in reservoir.

CLUTCH SLAVE CYLINDER

REMOVAL & INSTALLATION

See Figure 27.

1. Before servicing the vehicle, refer to the precautions.
2. Drain clutch fluid.
3. Remove the 3 bolts and No. 1 clutch housing cover.
4. Using a union nut wrench, disconnect the flexible hose tube.
5. Remove the bolt and clutch line bracket.

➡Use a container to catch the fluid.

6. Remove the 2 bolts and pull out the clutch release cylinder.

To install:

7. Install the clutch release cylinder with the 2 bolts. Tighten to 9 ft. lbs. (12 Nm).

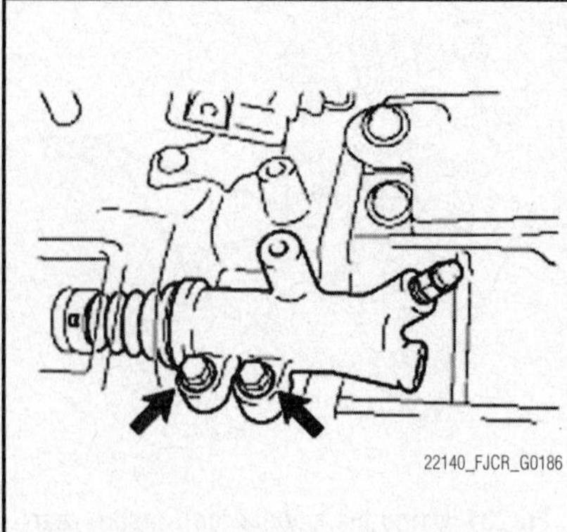

Fig. 27 Clutch release cylinder fasteners

8. Using a union nut wrench, connect the flexible hose tube. Tighten to 11 ft. lbs. (15 Nm).
9. Install the clutch line bracket. Tighten to 9 ft. lbs. (12 Nm).
10. Fill the brake reservoir tank with brake fluid and bleed the clutch system. Tighten to 8 ft. lbs. (11 Nm).
11. Check for clutch fluid leakage.
12. Install the No. 1 clutch housing cover with the 3 bolts. Tighten to 9 ft. lbs. (12 Nm).
13. Check fluid level in reservoir.

CLUTCH HYDRAULIC SYSTEM BLEEDING

➡If any maintenance on the clutch system was performed or the system is suspected of containing air, bleed the system. Use care; brake fluid will remove the paint from any surface. If the brake fluid spills onto any painted surface, wash it off immediately with soap and water.

1. Before servicing the vehicle, refer to the precautions section.
2. Fill the clutch reservoir with brake fluid. Check the reservoir level frequently and add fluid as needed.
3. Connect one end of a vinyl tube to the bleeder plug on the slave cylinder and submerge the other end into a clear container half-filled with brake fluid.
4. Slowly pump the clutch pedal several times.
5. Have an assistant hold the clutch pedal down and loosen the bleeder plug until fluid and/or air starts to run out of the bleeder plug. Close the bleeder plug while the pedal is held to the floor.

➡Do not allow the pedal to rise backup while the bleeder is still open. If this happens, it will allow air to re-enter the slave cylinder and cause the clutch system not to work properly.

6. Repeat Steps 2 and 3 until all the air bubbles are removed from the system.
7. Tighten the bleeder plug when all the air is gone.
8. Refill the master cylinder to the proper level as required.
9. Check the system for leaks.

TRANSFER CASE ASSEMBLY

REMOVAL & INSTALLATION

See Figure 28.

1. Before servicing the vehicle, refer to the precautions.
2. Disconnect the negative battery cable.
3. Drain transfer oil.
4. Remove lower transfer case protector.
5. Remove the transmission assembly.
6. Remove transfer case assembly.

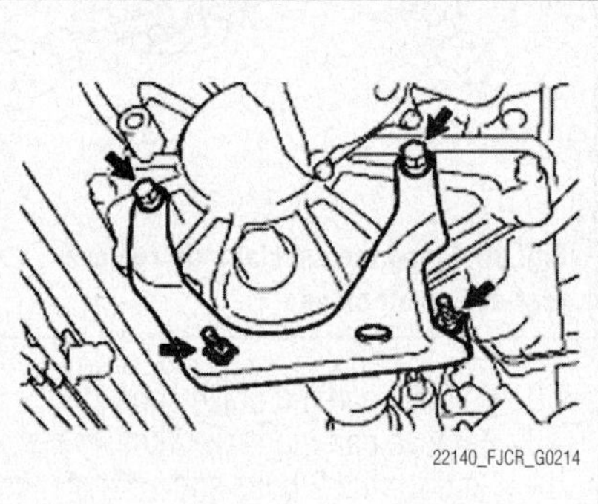

Fig. 28 Lower transfer case protector mounting bolts

To install:

7. Install the transfer case onto the transmission. Tighten the bolts to 17 ft. lbs. (24 Nm).
8. Install the transmission.
9. Install the lower transfer case protector with the 4 bolts. Tighten the bolts to 13 ft. lbs. (18 Nm).
10. Inspect and adjust the transfer case oil.
11. Connect the negative battery cable.
12. Using a scan tool, perform the reset memory procedure to initialize the transmission.
13. Check for transfer oil leakage.

FRONT DIFFERENTIAL CARRIER

REMOVAL & INSTALLATION

1. Before servicing the vehicle, refer to the precautions.
2. Disconnect the negative battery cable.
3. Remove the front wheel.
4. Remove the No.1 engine under cover.
5. Remove the rear engine under cover assembly.
6. Drain the differential oil.
7. Remove the front halfshaft assembly.
8. Disconnect the front speed sensors.
9. Remove front axle hub grease cap.
10. Separate tie rod end.
11. Separate front lower ball joint attachment.
12. Remove front axle hub nut.
13. Remove front drive shaft assembly.
14. Remove the bolt and disconnect the differential breather tube bracket.
15. Support the differential with a jack.
16. Remove the No. 1 differential mounting nut.
17. Remove the 2 mounting bolts and 2 nuts.
18. Lower the jack and remove the front differential.
19. Remove the 3 bolts and front No. 1 differential support.
20. Remove the 2 bolts and front No. 2 differential support.
21. Remove the 2 bolts and front No. 3 differential support.

To install:

22. Install the front No. 3 differential support with the 2 bolts. Tighten to 80 ft. lbs. (108 Nm).
23. Install the front No. 2 differential support with the 2 bolts. Tighten to 118 ft. lbs. (160 Nm).
24. Install the front No. 1 differential support with the 3 bolts. Tighten to 137 ft. lbs. (186 Nm).
25. Support the front differential with a jack.
26. Install the 2 front mounting bolts and 2 nuts. Tighten to 101 ft. lbs. (137 Nm).
27. Install front differential mounting nut No. 1. Tighten to 64 ft. lbs. (87 Nm).
28. Install the front differential breather tube bracket with the bolt. Tighten to 10 ft. lbs. (13 Nm).
29. Lower the jack.
30. Install front halfshaft assembly.
31. Attach front lower ball joint
32. Install tie rod end.
33. Install front axle hub nut.
34. Install front axle hub grease cap.
35. Install front speed sensor.
36. Install front wheel.
37. Inspect differential oil level and correct as necessary.
38. Check for differential oil leakage.

39. Install rear engine under cover assembly.
40. Install No. 1 engine under cover.
41. Connect the negative battery cable.
42. Using a scan tool, check VSC sensor signal.
43. Inspect and adjust front wheel alignment

FRONT HALFSHAFT

REMOVAL & INSTALLATION

1. Before servicing the vehicle, refer to the precautions.
2. Place matchmarks on the driveshaft flange and differential flange.
3. Remove the 4 nuts, 4bolts and 4 washers.
4. Place matchmarks on the driveshaft flange and transfer flange.
5. Remove the 4 nuts, 4 washers and the driveshaft assembly.

To install:

6. Align the matchmarks on the yoke and differential flange.
7. Install the driveshaft assembly with the 4 bolts, 4 nuts and 4 washers. Tighten to 65 ft. lbs. (88 Nm).
8. Align the matchmarks on the yoke and transfer flange.
9. Install the driveshaft assembly with the 4 nuts and 4 washers. Tighten to 65 ft. lbs. (88 Nm).

REAR AXLE SHAFT, BEARING & SEAL

REMOVAL & INSTALLATION

See Figures 29 through 31.

1. Before servicing the vehicle, refer to the precautions.
2. Disconnect the negative battery cable.
3. Raise and safely support the vehicle securely on jackstands.
4. Remove rear wheel.
5. Drain brake fluid.
6. Remove rear disc brake caliper assembly and hang out of the way.
7. Remove rear disc.
8. Remove parking brake shoe return tension spring.

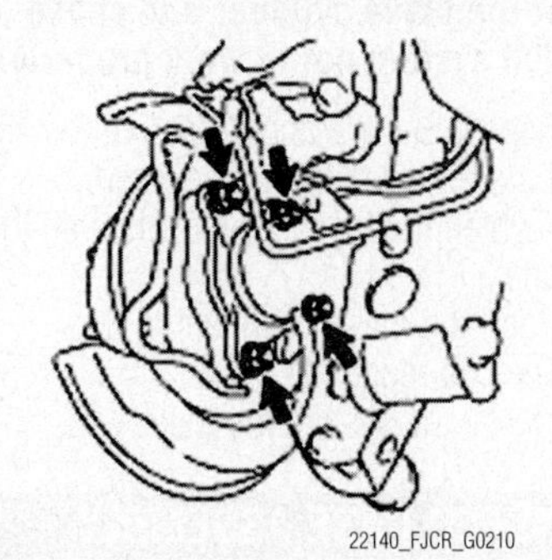

Fig. 29 Remove the 4 nuts and rear axle shaft with parking brake plate

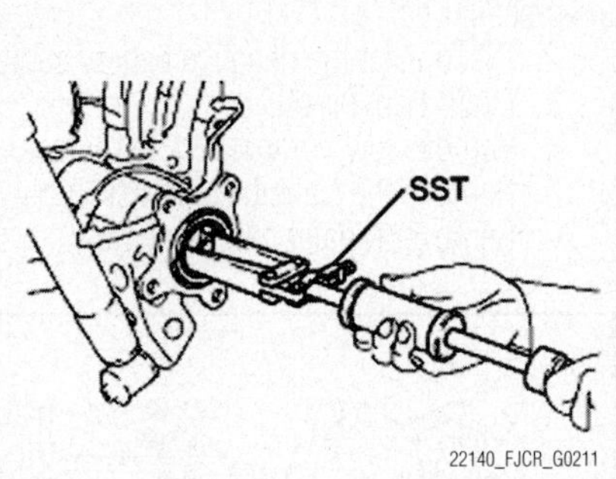

Fig. 30 Using the special tool, remove rear axle shaft oil seal

9. Remove parking brake shoe strut.
10. Remove parking brake shoe.
11. Remove the 2 bolts and separate the parking brake cable from the parking brake plate.
12. Remove rear speed sensor.
13. Remove rear axle shaft.
    a. Remove the 4 nuts and rear axle shaft with parking brake plate.
    b. Remove the O-ring.
14. Using SST: 09308-00010, remove rear axle shaft oil seal.

To install:

15. Using SST: 09950-60020, install rear axle shaft oil seal.
    a. Install the O-ring.
    b. Install the 4 nuts and rear axle shaft with parking brake plate. Tighten to 89 ft. lbs. (120 Nm).
16. Install rear axle shaft.
17. Inspect the axle shaft backlash.

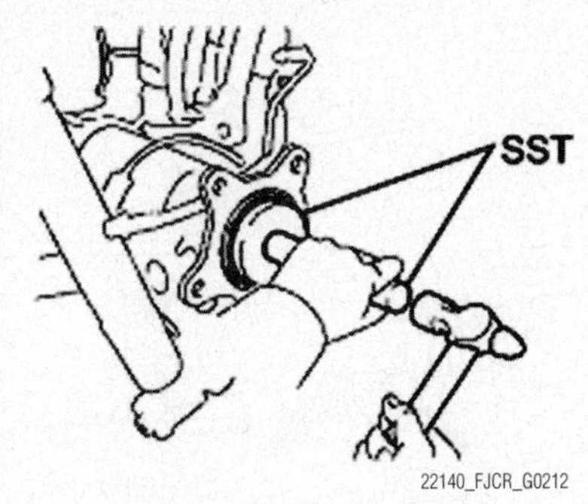

Fig. 31 Using the special tool, install rear axle shaft oil seal

    a. Using a dial indicator, check the backlash near the center of the axle shaft. Maximum backlash is 0.0020 in. (0.05 mm).
    b. If the backlash is greater than the maximum, replace the bearing.
18. Inspect the axle shaft runout.
    a. Using a dial indicator, check the runout of the surface of the axle shaft. Maximum runout is 0.0020 in. (0.05 mm).
    b. If the runout is greater than the maximum, replace the bearing.
19. Install rear speed sensor. Tighten to 71 inch lbs. (8 Nm).
20. Install the 2 bolts and separate the parking brake cable from the parking brake plate.
21. Install parking brake shoe.
22. Install parking brake shoe strut.
23. Install parking brake shoe return tension spring.
24. Install rear disc.
25. Install rear disc brake caliper assembly and hang out of the way.
26. Drain brake fluid.
27. Install rear wheel and lower the vehicle.
28. Inspect differential oil.
29. Check for differential oil leakage.
30. Check and adjust fluid level in reservoir.
31. Check for brake fluid leakage.
32. Inspect and adjust parking brake lever travel.
33. Using a scan tool, check VSC sensor signal.

ENGINE COOLING

THERMOSTAT

REMOVAL & INSTALLATION

See Figure 32.

1. Before servicing the vehicle, refer to the precautions.
2. Drain engine coolant.
3. Remove V-bank cover.

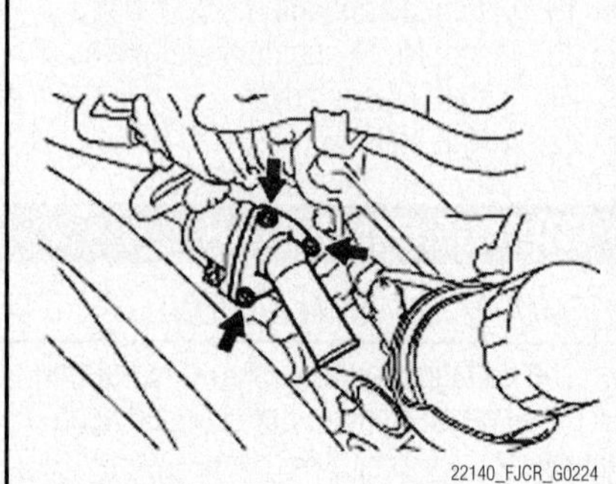

Fig. 32 Remove the 3 nuts, then remove the water inlet

4. Remove No. 2 radiator hose.
5. Remove the 3 nuts, then remove the water inlet with thermostat and gasket.

To install:

6. Install a new gasket onto the water inlet with thermostat.
7. Install the water inlet with thermostat with the 3 nuts. Tighten to 80 inch lbs. (9 Nm)
8. Connect No. 2 radiator hose.
9. Connect the negative battery cable.
10. Add engine coolant.
11. Check for engine coolant leakage.
12. Install V-bank cover.

WATER PUMP

REMOVAL & INSTALLATION

See Figures 33 through 35.

1. Before servicing the vehicle, refer to the precautions.
2. Remove fan.
3. Remove V-bank cover.
4. Remove air cleaner assembly.
5. Remove alternator assembly.
6. Remove water inlet.

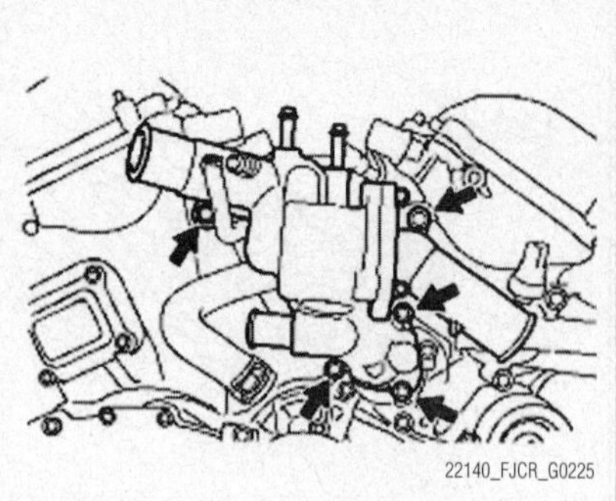

Fig. 33 Water inlet fastener locations

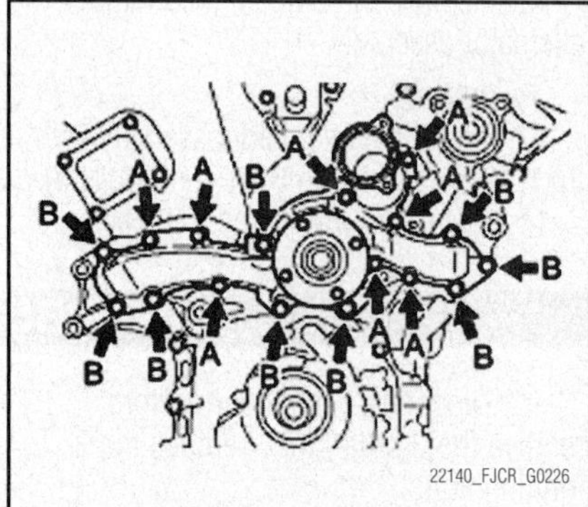

Fig. 34 Water pump bolt tightening torque

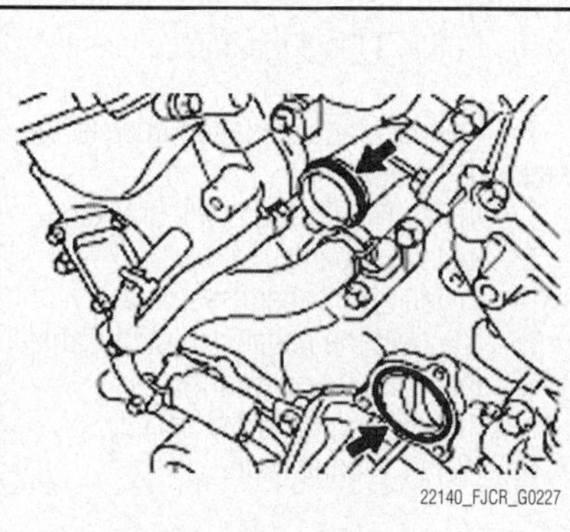

Fig. 35 Water inlet O-rings

a. Disconnect the 2 radiator hoses.
b. Disconnect the 5 water by-pass hoses.
c. Remove the 5 bolts and water inlet.
d. Remove the O-ring from the water outlet pipe.
e. Remove the gasket from the water pump.

7. Remove No. 2 idler pulley.
8. Remove the air conditioning air conditioning compressor assembly and lay it aside.

➡**Do not disconnect the air conditioning hoses from the compressor.**

9. Remove V-ribbed belt tensioner assembly.
10. Separate power steering pump assembly and set it aside.

➡**Do not disconnect the power steering hoses.**

11. Remove the 17 bolts, then remove the water pump and gasket.

To install:

12. Install a new gasket and the water pump with the 17 bolts. Tighten bolts **A** to 80 inch lbs. (9 Nm) and bolts **B** to 17 ft. lbs. (23 Nm).
13. Install power steering pump assembly and set it aside.

➡**Do not disconnect the power steering hoses.**

14. Install V-ribbed belt tensioner assembly.
15. Install air conditioning compressor assembly.
16. Install No. 2 idler pulley.
17. Install water inlet.

a. Install a new O-ring onto the water outlet pipe.
b. Install a new gasket onto the water pump.
c. Apply soapy water to the O-ring.
d. Install the water inlet with the 5 bolts. Tighten bolt to 80 inch lbs. (9 Nm).
e. Connect the 5 water by-pass hoses.
f. Connect the 2 radiator hoses.

18. Install alternator assembly.
19. Install air cleaner assembly.
20. Install fan.
21. Connect the negative battery cable.
22. Install V-bank cover.

ENGINE ELECTRICAL — CHARGING SYSTEM

ALTERNATOR

REMOVAL & INSTALLATION

See Figure 36.

1. Before servicing the vehicle, refer to the precautions.
2. Remove battery.
3. Remove V-bank cover.
4. Remove No. 1 engine under cover.
5. Remove fan and alternator V-belt.
6. Disconnect the wire harness.
7. Remove the bolt and wire harness bracket.
8. Disconnect the connector from the alternator assembly.
9. Remove the terminal cap and nut.
10. Disconnect the wire harness from terminal B.
11. Remove the 2 bolts, then separate the wire harness clamp bracket from the alternator assembly.

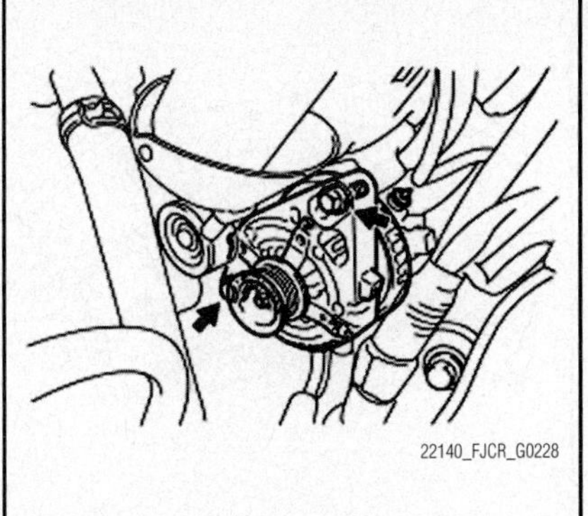

Fig. 36 Alternator mounting bolts

12. Remove the 2 bolts, then remove the alternator assembly.

To install:

13. Install the alternator assembly with the 2 bolts. Tighten to 32 ft. lbs. (43 Nm).
14. Install the wire harness clamp bracket with the 2 bolts. Tighten to 71 inch lbs. (8 Nm).
15. Connect the wire harness.
16. Connect the wire harness to terminal B and install the nut. Tighten to 7 ft. lbs. (10 Nm).
17. Connect the connector to the alternator assembly.
18. Install the wire harness bracket with the bolt. Tighten to 51 inch lbs. (6 Nm).
19. Install fan and alternator V-belt.
20. Install No. 1 engine under cover.
21. Install V-bank cover.
22. Install battery

VOLTAGE REGULATOR

REMOVAL & INSTALLATION

The voltage regulator is an integral part of the alternator and is not serviced separately.

ENGINE ELECTRICAL — DISTRIBUTORLESS IGNITION SYSTEM

FIRING ORDER

See Figure 37.

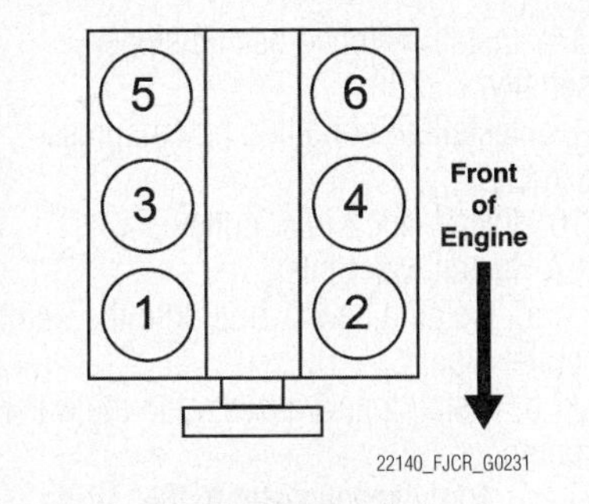

Fig. 37 4.0L V6 DOHC engine
Firing order: 1–2–3–4–5–6
Distributorless ignition system

IGNITION COIL

REMOVAL & INSTALLATION

See Figure 38.

1. Before servicing the vehicle, refer to the precautions.
2. Disconnect the negative battery cable.
3. Remove V-bank cover.
4. Remove air cleaner assembly.
5. Remove oil baffle plate.
6. Remove No. 1 surge tank bracket.
7. Disconnect the ventilation hose from the intake air surge tank.
8. Disconnect the 3 connectors, remove the 3 bolts, then remove the 3 ignition coils.

To install:.

9. Install the 3 ignition coils with the bolts. Tighten the bolts to 7 ft. lbs. (10 Nm).
10. Connect the 3 ignition coil connectors.
11. Connect the ventilation hose to the intake air surge tank.
12. Install No. 1 surge tank bracket.
13. Install oil baffle plate.
14. Install air cleaner assembly.
15. Connect the negative battery cable.
16. Inspect ignition timing.
17. Inspect engine idling speed.
18. Install V-bank cover.

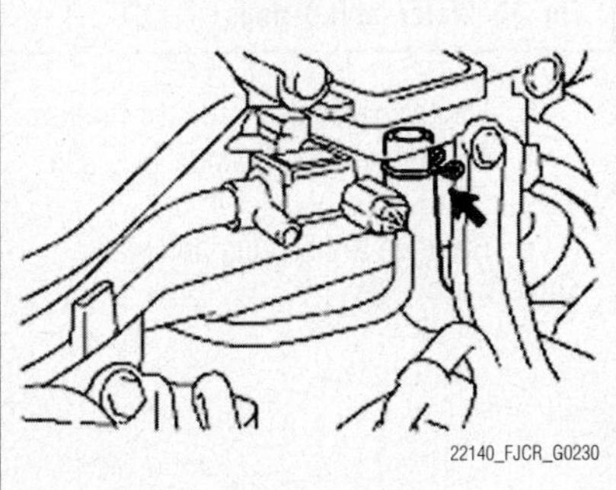

Fig. 38 Ignition coils mounted on the cylinder head

IGNITION TIMING

ADJUSTMENT

The ignition timing is controlled by the Electronic Control Module (ECM). No adjustment is possible.

SPARK PLUGS

REMOVAL & INSTALLATION

1. Before servicing the vehicle, refer to the precautions.
2. Remove the V-bank cover.
3. Disconnect the electrical connector from each ignition coil.
4. Remove the bolt securing each ignition coil from the cylinder head cover.
5. Remove each ignition coil from the cylinder head cover.

➡This engine is equipped with an aluminum cylinder head. Allow the engine to cool before removing spark plugs. Removing the spark plugs from an engine at operating temperature may damage the spark plug threads in the cylinder head. Also be sure to clean any dirt or debris from around spark plug holes prior to removing spark plugs.

6. Remove the spark plugs from the cylinder head.
7. Inspect the spark plugs for electrode wear, carbon deposits and insulator damage.

To install:

⁂ WARNING

Do not touch the tip of the spark plug. Do not damage the iridium surface of the electrode when gapping the plug.

➡Do not adjust the gap on used spark plugs. Replace the spark plug if the gap is greater than specification.

8. Apply a small quantity of anti-seize compound to the plug threads, and screw the plugs into the cylinder head finger-tight.
9. Tighten the spark plugs to 15 ft. lbs. (20 Nm).
10. Install each ignition coil to the cylinder head cover.
11. Secure the ignition coils using the bolts. Tighten the bolts to 80 inch lbs (9 Nm).
12. Connect each ignition coil electrical connector.
13. Install the V-bank cover.

ENGINE ELECTRICAL

STARTER

REMOVAL & INSTALLATION

1. Before servicing the vehicle, refer to the precautions.
2. Disconnect the negative battery cable.
3. Remove the 4 bolts, then remove the No. 2 engine under cover assembly.
4. Remove the 3 bolts, then remove the No. 2 manifold bracket.
5. Remove the bolt, then separate the earth wire harness from the cylinder block.
6. Remove starter assembly.
   a. Disconnect the terminal 50 connector from the starter assembly.
   b. Remove the nut, then disconnect the wire harness from the terminal 30.
   c. Remove the 2 bolts, then remove the starter assembly.

To install:.

7. Install the starter assembly with 2 bolts. Tighten to 27 ft. lbs. (37 Nm).
8. Connect the wire harness to the terminal 30, then install the nut. Tighten to 52 inch lbs. (6 Nm).
9. Connect the terminal 50 connector to the starter assembly.
10. Install the wire harness onto the cylinder block with the bolt. Tighten to 10 ft. lbs. (13 Nm).
11. Install the No. 2 manifold bracket with the 3 bolts. Tighten to 30 ft. lbs. (40 Nm).
12. Install the No. 2 engine under cover with the 4 bolts. Tighten to 21 ft. lbs. (29 Nm).
13. Connect the negative battery cable.

SOLENOID OR RELAY REPLACEMENT

1. Before servicing the vehicle, refer to the precautions.
2. Remove the nut, then disconnect the lead wire from terminal C.
3. Remove the 2 through bolts.
4. Pull the starter yoke and the armature out together.
5. Remove the O-ring from the starter yoke.
6. Remove the 2 screws, then remove the magnet starter switch.

To install:

7. Install the magnet starter switch with the 2 screws. Tighten to 82 inch lbs. (9 Nm).
8. Install a new O-ring onto the groove of the starter yoke.
9. Align the key of the starter yoke with the groove of the magnet starter switch.
10. Install the starter yoke and armature with the 2 through bolts. Tighten to 82 inch lbs. (9 Nm).
11. Connect the lead wire to terminal C with the nut. Tighten to 52 inch lbs. (6 Nm).

ENGINE MECHANICAL

ACCESSORY DRIVE BELTS

ACCESSORY BELT ROUTING

See Figure 39.

Refer to the accompanying illustration for accessory belt routing.

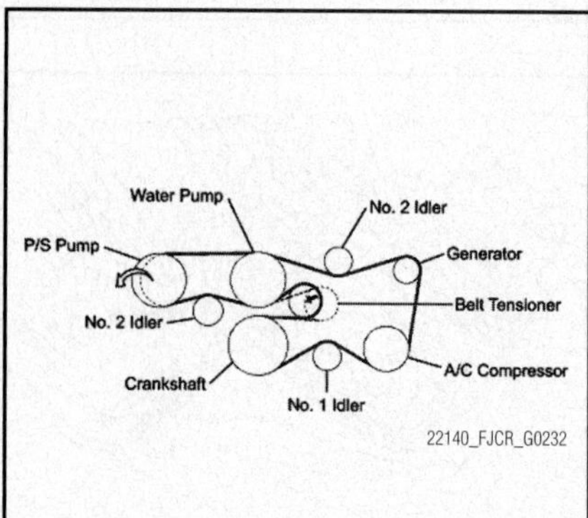

Fig. 39 Accessory belt routing—4.0L DOHC engine

INSPECTION

Inspect the drive belt for signs of glazing or cracking. A glazed belt will be perfectly smooth from slippage, while a good belt will have a slight texture of fabric visible. Cracks will usually start at the inner edge of the belt and run outward. All worn or damaged drive belts should be replaced immediately.

ADJUSTMENT

The belt tension is maintained by an automatic tensioner. No adjustment is possible.

REMOVAL & INSTALLATION

See Figure 40.

1. Before servicing the vehicle, refer to the precautions.
2. Remove the 4 bolts, then remove the No. 1 engine under cover.
3. While releasing the belt tension by turning the belt tensioner counterclockwise,

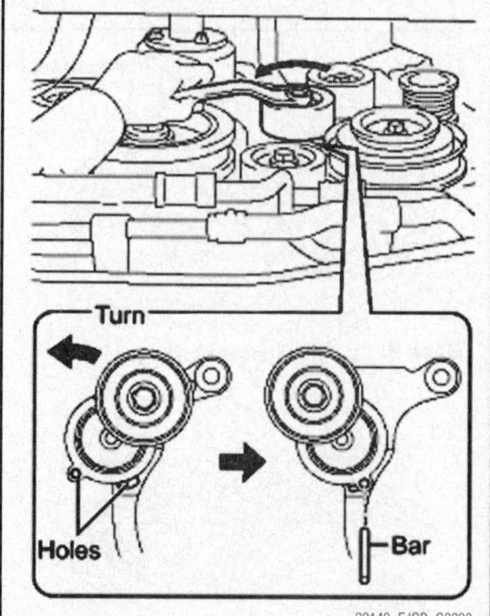

Fig. 40 While turning the belt tensioner counterclockwise, align the holes as shown, and then insert a bar of 0.24 in. (6 mm) into the holes to fix the belt tensioner

remove the V-ribbed belt from the belt tensioner.

To install:

4. Check that nothing gets caught in the tensioner by turning it clockwise and counterclockwise. If a malfunction exists, replace the tensioner.

5. While turning the belt tensioner counterclockwise, align the holes as shown, and then insert a bar of 0.24 in. (6 mm) into the holes to fix the belt tensioner.

6. Install the V-ribbed belt.

7. While turning the belt tensioner counterclockwise, remove the bar.

8. If it is hard to install the V-ribbed belt, perform the following procedure:

   a. Put the V-ribbed belt on all parts except the P/S pump, as shown in the illustration.

   b. While releasing the belt tension by turning the belt tensioner counterclockwise, put the V-ribbed belt on the P/S pump.

9. Install the No. 1 engine under cover with the 4 bolts. Tighten to 21 ft. lbs. (29 Nm).

CAMSHAFT AND VALVE LIFTERS

REMOVAL & INSTALLATION

See Figures 41 through 60.

1. Before servicing the vehicle, refer to the precautions.

2. Disconnect the negative battery cable.

3. Drain engine coolant.

Bank 1: Timing Marks

Bank 2: Timing Marks

Timing Marks

22140_FJCR_G0241

Fig. 41 Check that the timing marks of the camshaft timing gears are aligned with the timing marks of the bearing caps as shown

4. Remove V-bank cover.

5. Remove air cleaner assembly.

6. Remove throttle body bracket.

7. Remove oil baffle plate.

8. Remove surge tank bracket.

9. Remove intake air surge tank.

   a. Disconnect the 2 water by-pass hoses.

   b. Disconnect the fuel vapor feed hose.

   c. Disconnect the ventilation hose.

   d. Disconnect the 2 VSV connectors.

   e. Disconnect the throttle body with motor connector.

   f. Separate the 3 wire harness clamps and the hose clamp.

   g. Remove the 2 nuts.

   h. Using a socket hexagon wrench 8, remove the 4 bolts, intake air surge tank and the gasket.

10. Remove ignition coil assembly.

11. Remove the 10 bolts, 3 seal washers, 2 nuts, cylinder head cover and gasket.

12. Set No. 1 cylinder to TDC on the compression stroke.

   a. Turn the crankshaft pulley until its groove and the "0" timing mark of the timing chain cover are aligned.

   b. Check that the timing marks of the camshaft timing gears are aligned with the timing marks of the bearing caps as shown in the illustration.

13. If not, turn the crankshaft 1 complete revolution (360°) and align the timing marks above.

   a. Place paint marks on the No. 1 chain links corresponding to the timing marks of the camshaft timing gears.

14. Remove No. 1 chain tensioner assembly.

WARNING

Never rotate the crankshaft with the chain tensioner removed. When rotating the camshaft with the timing chain removed, rotate the crankshaft counterclockwise 40°from the TDC first.

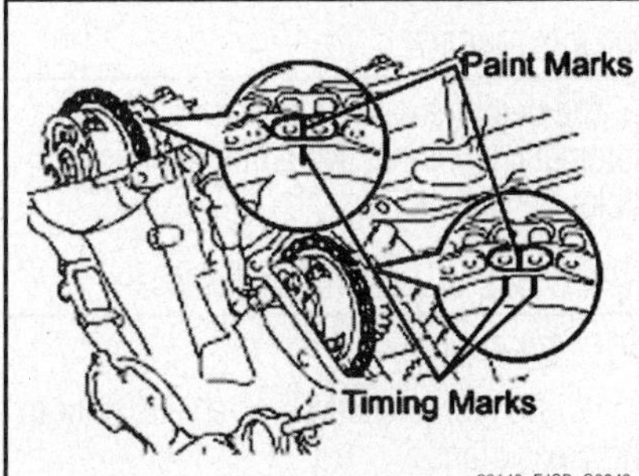

Fig. 42 Place paint marks on the No. 1 chain links corresponding to the timing marks of the camshaft timing gears

   a. Remove the 4 bolts, then remove the timing chain cover plate and gasket.

   b. While turning the stopper plate of the tensioner upward, push in the plunger of the chain tensioner as shown in the illustration.

   c. While turning the stopper plate of the tensioner downward, insert a pin of 0.138 in. (3.5 mm) diameter into the holes in the stopper plate and tensioner to fix the stopper plate.

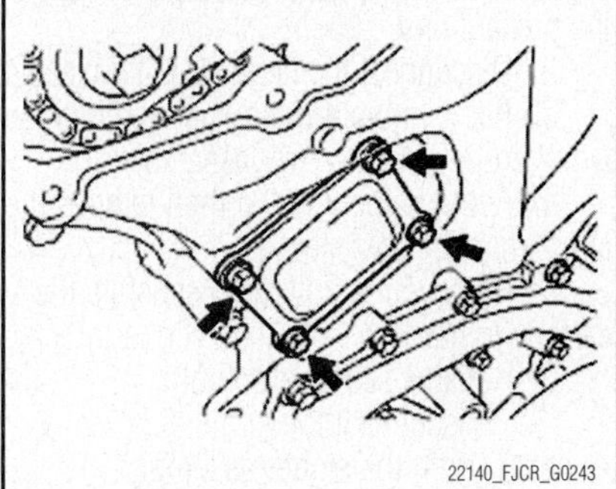

Fig. 43 Timing chain cover plate and gasket

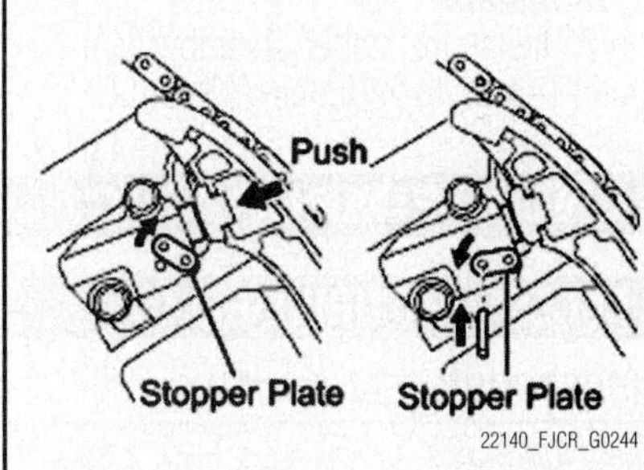

Fig. 44 While turning the stopper plate of the tensioner upward, push in the plunger of the chain tensioner as shown

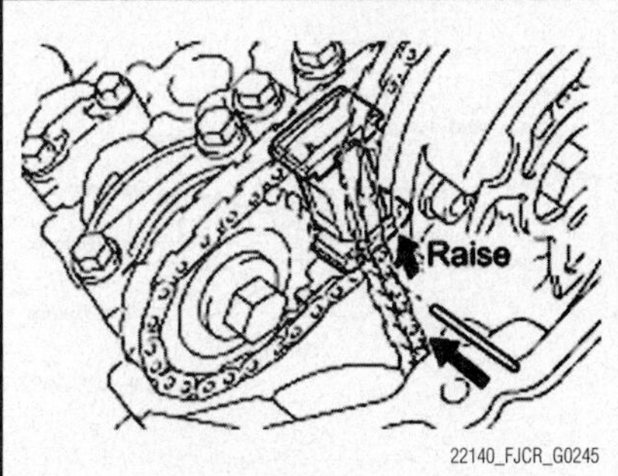

Fig. 45 While raising the chain tensioner No. 2, insert a pin of 0.138 in. (3.5 mm) diameter into the hole to hold it

d. Remove the 2 bolts, then remove the chain tensioner.

15. Remove No. 2 camshaft.

⁂ WARNING

Keep the camshaft level while it is being removed. The camshaft thrust clearance is very small and failing to keep it level could crack or damage the cylinder head journal surface, which receives the thrust force. This could subsequently lead the camshaft to seize or break.

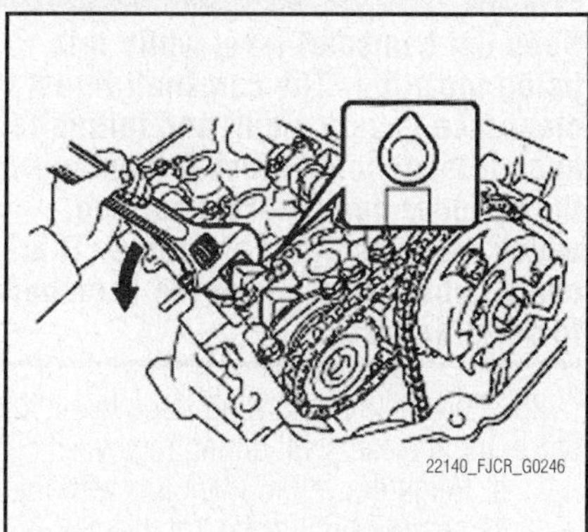

Fig. 46 Cam lobes facing upward

16. Perform the following steps to avoid damaging the camshaft during removal.

a. While raising the chain tensioner No. 2, insert a pin of 0.138 in. (3.5 mm) diameter into the hole to hold it.

b. Hold the hexagonal portion of the No. 2 camshaft with a wrench, and remove the camshaft timing gear set bolt.

⁂ WARNING

Be careful not to damage the cylinder head or valve lifter with the wrench.

c. Separate the camshaft timing gear from the No. 2 camshaft.

d. Rotate the camshaft counterclockwise using the wrench so that the cam lobes of No. 1 cylinder face upward as shown in the illustration.

e. Using several steps, uniformly loosen and remove the 8 bearing cap bolts in the sequence shown in the illustration.

f. Remove the 4 bearing caps and No. 2 camshaft.

17. Remove the No. 2 chain tensioner bolt, then remove the No. 2 chain tensioner and camshaft timing gear.

18. Remove camshaft.

⁂ WARNING

Keep the camshaft level while it is being removed. The camshaft thrust clearance is very small and failing to keep it level could crack or damage the cylinder head journal surface, which receives the thrust force. This could subsequently lead the camshaft to seize or break.

19. Perform the following steps to avoid damaging the camshaft during removal.

a. Hold the hexagonal portion of the No. 1 camshaft with a wrench, and loosen the camshaft timing gear set bolt.

⁂ WARNING

Be careful not to damage the cylinder head or valve lifter with the wrench.

➡Do not disassemble the camshaft timing gear assembly.

b. Slide the camshaft timing gear and separate the No. 1 chain from the camshaft timing gear.

c. Rotate the No. 1 camshaft counterclockwise using the wrench so that the cam lobes of No. 1 cylinder face downward as shown in the illustration.

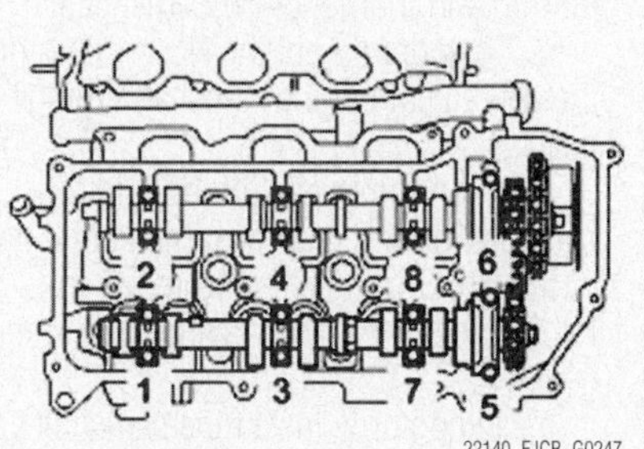

Fig. 47 Using several steps, uniformly loosen and remove the 8 bearing cap bolts in the sequence shown—No. 2 and 4 camshafts

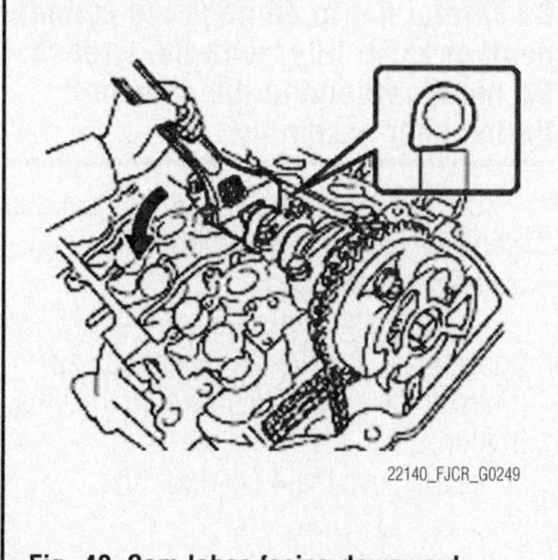

Fig. 49 Cam lobes facing downward

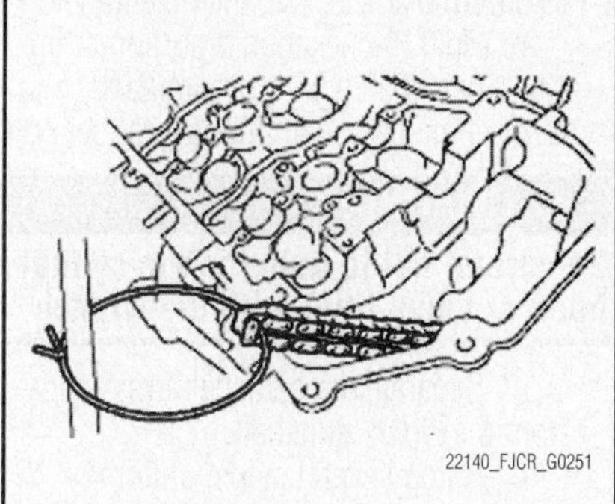

Fig. 51 Tie the No. 1 chain with a piece of string as shown

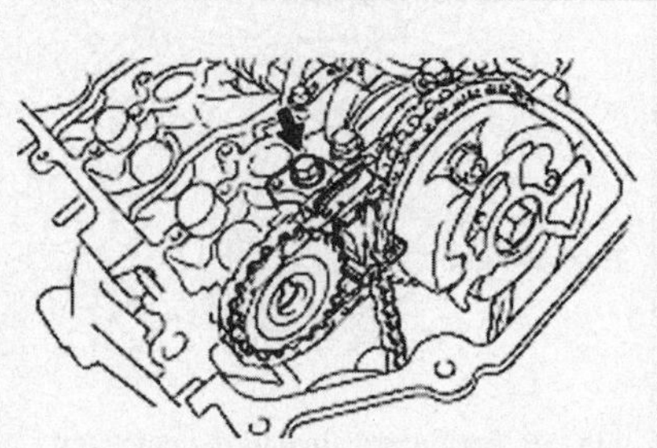

Fig. 48 Remove the No. 2 chain tensioner bolt, then remove the No. 2 chain tensioner and camshaft timing gear

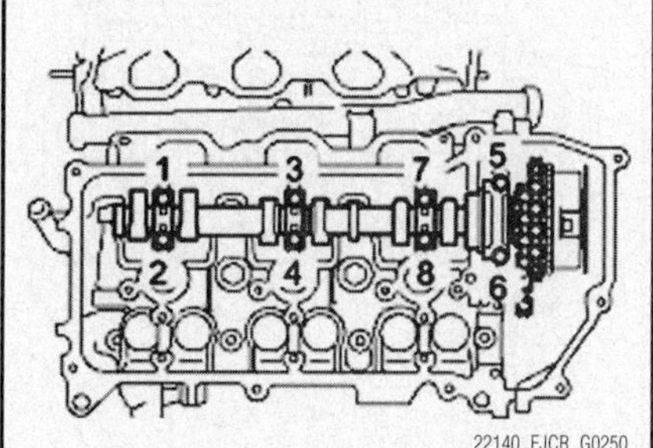

Fig. 50 Using several steps, loosen and remove the 8 bearing cap bolts in the sequence shown—No. 1 and 3 camshafts

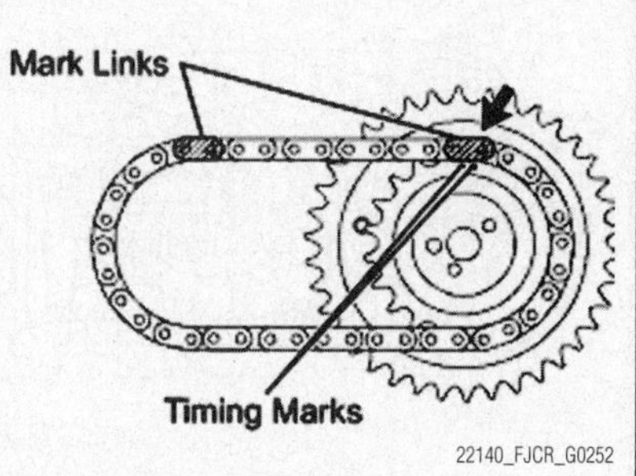

Fig. 52 Align the yellow mark link with the timing mark (2 dot marks) of the camshaft timing gear as shown

d. Using several steps, loosen and remove the 8 bearing cap bolts in the sequence shown in the illustration.

e. Remove the 4 bearing caps.

f. Remove the camshaft timing gear set bolt with the No. 1 camshaft lifted up, then remove the No. 1 camshaft and camshaft timing gear with No. 2 chain.

g. Tie the No. 1 chain with a piece of string as shown in the illustration.

✲✲ WARNING

Be careful not to drop anything inside the timing chain cover.

20. Remove No. 4 camshaft.

✲✲ WARNING

Keep the camshaft level while it is being removed. The camshaft thrust clearance is very small and failing to keep it level could crack or damage the cylinder head journal surface, which receives the thrust force. This could subsequently lead the camshaft to seize or break.

21. Perform the following steps to avoid damaging the camshaft during removal.

a. While pushing down the No. 3 chain tensioner, insert a pin of 0.138 in. (3.5 mm) diameter into the hole to hold it.

b. Hold the hexagonal portion of the No. 4 camshaft with a wrench, and remove the camshaft timing gear set bolt.

✲✲ WARNING

Be careful not to damage the cylinder head or valve lifter with the wrench.

c. Separate the camshaft timing gear from the No. 4 camshaft.

d. Using several steps, uniformly loosen and remove the 8 bearing cap bolts in the sequence shown in the illustration.

e. Remove the 4 bearing caps and No. 4 camshaft.

22. Remove the No. 3 chain tensioner bolt, then remove the No. 3 chain tensioner and camshaft timing gear.

23. Remove No. 3 camshaft.

✲✲ WARNING

Keep the camshaft level while it is being removed. The camshaft thrust clearance is very small and failing to keep it level could crack or damage the cylinder head journal surface, which receives the thrust force. This could subsequently lead the camshaft to seize or break.

24. Perform the following steps to avoid damaging the camshaft during removal.

a. Release the chain tension between the camshaft timing gear on the left cylinder head and crankshaft timing gear by turning the crankshaft pulley counterclockwise slightly.

b. Hold the hexagonal portion of the No. 3 camshaft with a wrench, then loosen the camshaft timing gear set bolt.

✲✲ WARNING

Be careful not to damage the cylinder head or valve lifter with the wrench. Do not disassemble the camshaft timing gear assembly.

c. Slide the camshaft timing gear and separate the No. 1 chain from the camshaft timing gear.

d. Using several steps, uniformly loosen and remove the 8 bearing cap bolts in the sequence shown in the illustration.

e. Remove the 4 bearing caps.

f. Remove the camshaft timing gear set bolt with the No. 3 camshaft lifted up, then remove the No. 3 camshaft and camshaft timing gear with No. 2 chain.

g. Tie the No. 1 chain with a piece of string as shown in the illustration.

✲✲ WARNING

Be careful not to drop anything inside the timing chain cover.

To install:

25. Install No. 3 camshaft.

✲✲ WARNING

Keep the camshaft level while it is being installed. The camshaft thrust clearance is very small and failing to keep it level could crack or damage the cylinder head journal surface, which receives the thrust force. This could subsequently lead the camshaft to seize or break.

26. Perform the following steps to avoid damaging the camshaft during removal.

a. Align the yellow mark link with the timing mark (2 dot marks) of the camshaft timing gear as shown in the illustration.

b. Apply new engine oil to the thrust portion and journal of the camshafts.

c. Temporarily put the No. 1 chain on the No. 2 chain of the camshaft timing gear.

d. Align the knock pin hole in the camshaft timing gear with the knock pin of the No. 3 camshaft, and insert the No. 3 camshaft into the camshaft timing gear.

e. Temporarily install the camshaft timing gear set bolt.

f. Set the No. 3 camshaft onto the left cylinder head with the cam lobes of the

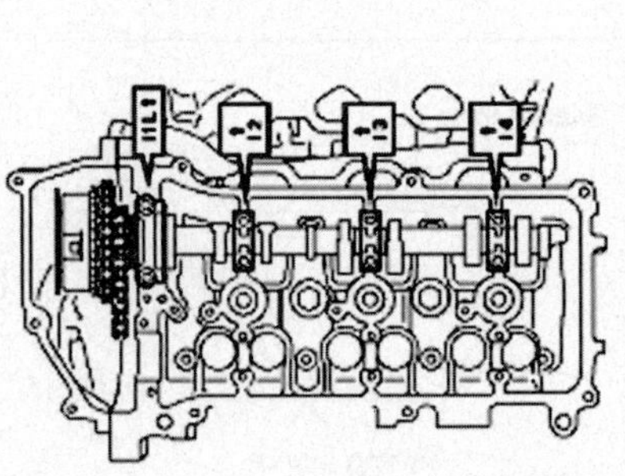

Fig. 53 Install the 4 bearing caps in the proper locations as shown

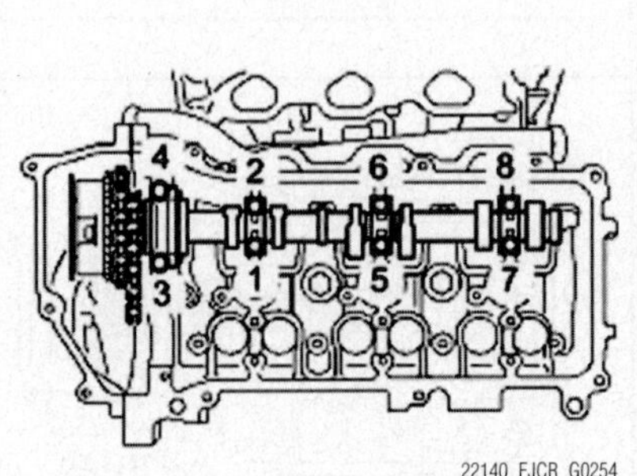

Fig. 54 Using several steps, uniformly install and tighten the 8 bearing cap bolts in the sequence shown—No. 1 and 3 camshafts

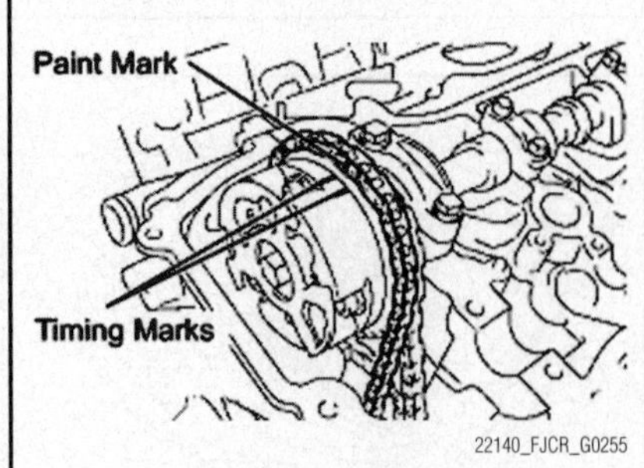

Fig. 55 Temporarily install the camshaft timing gear and No. 3 chain tensioner and align the yellow mark links with the timing marks (1 dot mark and 2 dot marks) of the camshaft timing gears

No. 2 cylinder facing downward as shown in the illustration.

g. Install the 4 bearing caps in the proper locations as shown.

h. Apply a light coat of engine oil to the threads and under the heads of the bearing cap bolts.

i. Using several steps, uniformly install and tighten the 8 bearing cap bolts in the sequence shown in the illustration. Tighten the 0.39 in. (10 mm) head bearing cap bolts to 80 inch lbs. (9 Nm) and the 0.47 in. (12 mm) head bearing cap bolts to 18 ft. lbs. (24 Nm).

j. Align the paint mark of the No. 1 chain with the timing marks of the camshaft timing gear.

k. Hold the hexagonal portion of the No. 3 camshaft with a wrench, and tighten the camshaft timing gear set bolt. Tighten to 74 ft. lbs. (100 Nm).

27. Install No. 3 chain tensioner assembly.

a. While pushing in the tensioner, insert a pin of 0.138 in. (3.5 mm) diameter into the hole to hold it.

b. Temporarily install the camshaft timing gear and No. 3 chain tensioner and align the yellow mark links with the timing marks (1 dot mark and 2 dot marks) of the camshaft timing gears.

c. Tighten the No. 3 chain tensioner bolt. Tighten to 14 ft. lbs. (19 Nm).

28. Install No. 4 camshaft.

⁂ WARNING

Keep the camshaft level while it is being installed. The camshaft thrust clearance is very small and failing to keep it level could crack or damage the cylinder head journal surface, which receives the thrust force. This could subsequently lead the camshaft to seize or break.

29. Perform the following steps to avoid damaging the camshaft during removal.

a. Align the knock pin hole in the camshaft timing gear with the knock pin of the No. 4 camshaft, and insert the No. 4 camshaft into the camshaft timing gear.

b. Temporarily install the camshaft timing gear set bolt.

c. Install the 4 bearing caps in the proper locations as shown.

d. Apply a light coat of engine oil to the threads of the bearing cap bolts.

e. Using several steps, uniformly install and tighten the 8 bearing cap bolts in the sequence shown in the illustration. Tighten the 0.39 in. (10 mm) head bearing cap bolts to 80 inch lbs. (9 Nm) and the 0.47 in. (12 mm) head bearing cap bolts to 18 ft. lbs. (24 Nm).

f. Hold the hexagonal portion of the No. 4 camshaft with a wrench, and tighten the camshaft timing gear set bolt. Tighten to 74 ft. lbs. (100 Nm).

g. Remove the pin from the No. 3 chain tensioner.

h. Release the chain tension between the camshaft timing gear on the right cylinder head and crankshaft timing gear by turning the crankshaft pulley clockwise slightly.

30. Install camshaft.

⁂ WARNING

Keep the camshaft level while it is being installed. The camshaft thrust clearance is very small and failing to keep it level could crack or damage the cylinder head journal surface, which receives the thrust force. This could subsequently lead the camshaft to seize or break.

31. Perform the following steps to avoid damaging the camshaft during removal.

a. Align the yellow mark link with the timing mark (1 dot mark) of the camshaft timing gear as shown in the illustration.

b. Apply new engine oil to the thrust portion and journal of the camshafts.

c. Temporarily install the No. 1 chain onto the No. 2 chain of the camshaft timing gear.

d. Align the knock pin hole in the camshaft timing gear with the knock pin of the No. 1 camshaft, and insert the No. 1 camshaft into the camshaft timing gear.

e. Temporarily install the camshaft timing gear set bolt.

f. Install the No. 1 camshaft onto the right cylinder head with the cam lobes of the No. 1 cylinder facing downward as shown in the illustration.

g. Install the 4 bearing caps in the proper locations as shown.

h. Apply a light coat of engine oil to the threads and under the heads of the bearing cap bolts.

i. Using several steps, uniformly install and tighten the 8 bearing cap bolts in the sequence shown in the illustration. Tighten the 0.39 in. (10 mm) head bearing cap bolts to 80 inch lbs. (9 Nm) and the 0.47 in. (12 mm) head bearing cap bolts to 18 ft. lbs. (24 Nm).

j. Rotate the No. 1 camshaft clockwise using a wrench so that the timing mark of the camshaft timing gear is aligned with the timing mark of the camshaft bearing cap.

k. Align the paint mark of the No. 1 chain with the timing mark of the camshaft timing gear.

l. Hold the hexagonal portion of the No. 1 camshaft with a wrench, and tighten the camshaft timing gear set bolt. Tighten to 74 ft. lbs. (100 Nm).

32. Install No. 2 chain tensioner assembly.

a. While pushing in the tensioner, insert a pin of 0.138 in. (3.5 mm) diameter into the hole to hold it.

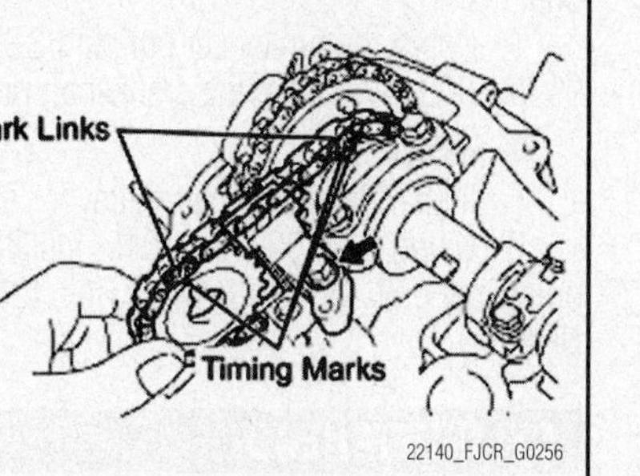

Fig. 56 Using several steps, uniformly install and tighten the 8 bearing cap bolts in the sequence shown—No. 2 and 4 camshafts

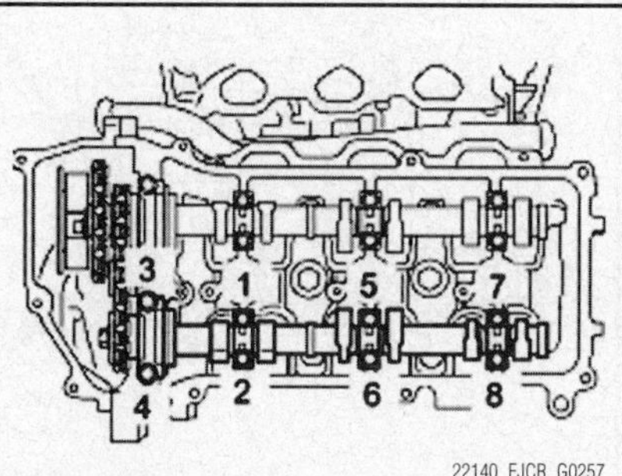

Fig. 57 Using several steps, uniformly install and tighten the 8 bearing cap bolts in the sequence shown—No. 2 and 4 camshafts

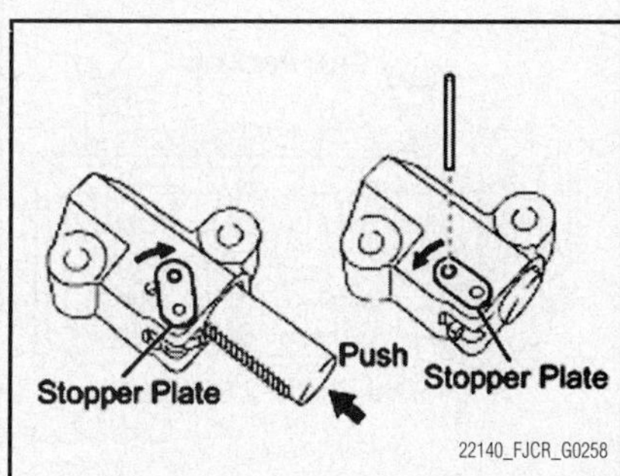

Fig. 58 While turning the stopper plate of the No. 1 chain tensioner clockwise, push in the plunger of the No. 1 chain tensioner as shown

b. Temporarily install the camshaft timing gear and No. 2 chain tensioner and align the yellow mark links with the timing marks (1 dot mark) of the camshaft timing gears.

c. Tighten the No. 2 chain tensioner bolt to 14 ft. lbs. (19 Nm).

33. Install No. 2 camshaft.

WARNING

Keep the camshaft level while it is being installed. The camshaft thrust clearance is very small and failing to keep it level could crack or damage the cylinder head journal surface, which receives the thrust force. This could subsequently lead the camshaft to seize or break.

34. Perform the following steps to avoid damaging the camshaft during removal.

a. Install the No. 2 camshaft onto the right cylinder head with the cam lobes of No. 1 cylinder facing upward as shown in the illustration.

b. Install the 4 bearing caps in the proper locations as shown.

c. Apply a light coat of engine oil to the threads and under the heads of the bearing cap bolts.

d. Using several steps, uniformly install and tighten the 8 bearing cap bolts in the sequence shown in the illustration. Tighten the 0.39 in. (10 mm) head bearing cap bolts to 80 inch lbs. (9 Nm) and the 0.47 in. (12 mm) head bearing cap bolts to 18 ft. lbs. (24 Nm).

e. Rotate the No. 2 camshaft clockwise using a wrench so that the knock pin of the No. 2 camshaft is aligned with the knock pin hole in the camshaft timing gear.

f. Hold the hexagonal portion of the No. 2 camshaft with a wrench, and install the camshaft timing gear set bolt. Tighten to 74 ft. lbs. (100 Nm).

g. Remove the pin from the No. 2 chain tensioner.

35. Install No. 1 chain tensioner assembly.

a. While turning the stopper plate of the No. 1 chain tensioner clockwise, push in the plunger of the No. 1 chain tensioner as shown in the illustration.

b. While turning the stopper plate of the tensioner counterclockwise, insert a pin of 0.138 in. (3.5 mm) diameter into the holes in the stopper plate and No. 1 chain tensioner to fix the stopper plate.

c. Install the No. 1 chain tensioner with the 2 bolts. Tighten to 7 ft. lbs. (10 Nm).

d. Remove the bar from the No. 1 chain tensioner.

e. Install a new gasket and the timing chain cover plate with the 4 bolts. Tighten to 80 inch lbs. (9 Nm).

f. Turn the crankshaft pulley 2 complete revolutions slowly until its groove and the "0" timing mark of the timing chain cover are aligned.

g. Check that the timing marks of the camshaft timing gears are aligned with the timing marks of the bearing cap as shown in the illustration.

36. Set No. 1 cylinder to TDC on the compression stroke.

37. Inspect and adjust valve clearance.

38. Install cylinder head cover.

a. Remove any old sealant.

WARNING

Do not drop any oil on the contact surfaces of the cylinder head, timing chain cover and cylinder head cover.

b. Apply a continuous bead of Three Bond 1207B or the equivalent sealant 0.08–0.12 in. (2–3 mm) diameter to the cylinder head and timing chain cover as shown in the illustration.

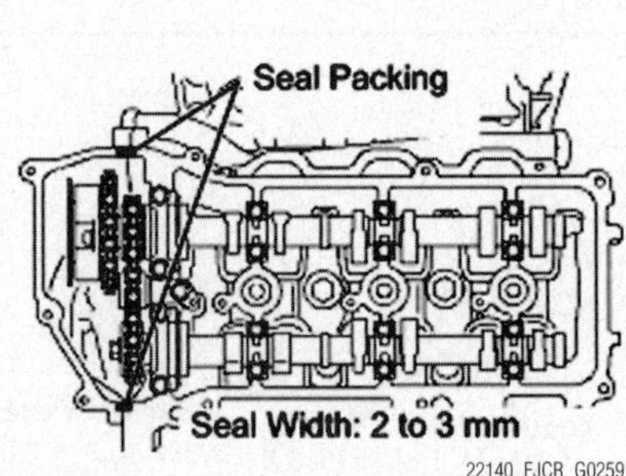

Fig. 59 Apply a continuous bead of Three Bond 1207B or the equivalent sealant 0.08–0.12 in. (2–3 mm) diameter to the cylinder head and timing chain cover as shown

Fig. 60 Tighten the cylinder head cover bolts and nuts uniformly in several steps. Tighten bolts Ato 7 ft. lbs. (10 Nm), bolts Bto 80 inch lbs. (9 Nm) and the nuts to 80 inch lbs. (9 Nm)

WARNING

Install the cylinder head cover within 3 minutes of applying the sealant. Tighten the cylinder head cover bolts and nuts within 15 minutes of installing the cylinder head cover. Otherwise, the sealant must be removed and reapplied.

c. Install the seal washers onto the bolts.

d. Install the cylinder head cover with the 10 bolts and 2 nuts. Tighten the bolts and nuts uniformly in several steps. Tighten bolts **A** to 7 ft. lbs. (10 Nm), bolts **B** to 80 inch lbs. (9 Nm) and the nuts to 80 inch lbs. (9 Nm).

➡Bolt Aare 0.98 in. (25 mm) and bolt B2.36 in. (60 mm).

39. Install ignition coil assembly.
40. Install intake air surge tank.
41. Install No. 2 surge tank bracket.
42. Install No. 1 surge tank bracket.
43. Install oil baffle plate.
44. Install throttle body bracket.
45. Install air cleaner assembly.
46. Connect the negative battery cable.
47. Add engine coolant.
48. Check for engine coolant leakage.
49. Check for engine oil leakage.
50. Inspect ignition timing.
51. Install V-bank cover.

CRANKSHAFT DAMPER

REMOVAL & INSTALLATION

1. Before servicing the vehicle, refer to the precautions.
2. Remove the accessory drive belt.
3. Using SST: 09213-54015, hold the crankshaft pulley and loosen the pulley set bolt.
4. Using the pulley set bolt and SST: 09950-50013, remove the crankshaft pulley.

To install:

5. Install the crankshaft pulley.
6. Using SST: 09213-54015, install the pulley set bolt. Tighten to184 ft. lbs. (250 Nm).

CYLINDER HEAD

REMOVAL & INSTALLATION

See Figures 61 through 68.

1. Before servicing the vehicle, refer to the precautions.

2. Discharge fuel system pressure.
3. Remove timing chain.
4. Remove the 2 bolts and the No. 1 cool air inlet.
5. On 4WD vehicles, remove exhaust pipe stopper bracket.
 a. Remove the 2 bolts, then remove the exhaust pipe stopper bracket.
6. Remove No. 2 front exhaust pipe assembly.
 a. Disconnect the oxygen sensor connector.
 b. Remove the 2 bolts and 2 nuts.
 c. Disengage the support and remove the front exhaust pipe and 2 gaskets.
7. Remove front exhaust pipe assembly.
 a. Disconnect the oxygen sensor connector.
 b. Remove the 2 bolts, 2 springs and 2 nuts, then separate the front exhaust pipe from the exhaust manifold.
8. Remove the 3 bolts and manifold bracket.
9. Remove exhaust manifolds.
 a. Disconnect the oxygen sensor connector.
 b. Remove the 6 nuts and exhaust manifold.
 c. Remove the gasket.
10. Disconnect fuel pipes.
11. Remove intake manifold.
12. Disconnect heater water inlet hose.
13. Remove water by-pass joint.
 a. Disconnect the engine coolant temperature sensor connector.
 b. Remove the 2 bolts and 4 nuts, then remove the water by-pass joint RR and 2 gaskets.
 c. Remove the O-ring from the water outlet hose.
14. Remove camshaft timing gears and No. 2 chain on the right cylinder head.
15. Remove the bolt, then remove the No. 2 chain tensioner.
16. Remove camshafts on the right cylinder head.
17. Remove No. 2 camshaft bearing.
18. Remove right cylinder head.
 a. Remove the bolt and separate the ground cable.
 b. Using several steps, loosen the 8 cylinder head bolts on the cylinder head uniformly with a 10 mm bi-hexagon wrench in the sequence shown in the illustration. Remove the 8 cylinder head bolts and 8 plate washers.

⁂ WARNING

Be careful not to drop the plate washers into the cylinder head. Cylinder head warpage or cracking could result from removing the bolts in the wrong order.

 c. Lift the cylinder head from the dowels on the cylinder block, and place the cylinder head on wooden blocks on a bench.

⁂ WARNING

Be careful not to drop the plate washers into the cylinder head. If the cylinder head is difficult to lift off, pry between the cylinder head and cylinder block with a screwdriver.

19. Remove right cylinder head gasket.
20. Remove the 2 bolts, then remove the No. 1 chain vibration damper.
21. Remove camshaft timing gears and No. 2 chain on the left cylinder head.
22. Remove camshafts on the left cylinder head.
23. Remove left cylinder head.
 a. Remove the bolt, then separate the ground cable.
 b. Remove the bolt, then separate the oxygen connector bracket.
 c. Using several steps, remove the 2 cylinder head bolts from the cylinder head in the sequence shown in the illustration.
 d. Using several steps, uniformly loosen the 8 cylinder head bolts on the cylinder head with a 10 mm bi-hexagon wrench in the sequence shown in the illustration. Remove the 8 cylinder head bolts and 8 plate washers.

⁂ WARNING

Be careful not to drop the plate washers into the cylinder head. Cylinder head warpage or cracking could result from removing the bolts in the wrong order.

 e. Lift the cylinder head from the dowels on the cylinder block, and place the cylinder head on wooden blocks on a bench.

⁂ WARNING

Be careful not to drop the plate washers into the cylinder head. If the cylinder head is difficult to remove, pry between the cylinder head and cylinder block with a screwdriver.

24. Remove right cylinder head gasket.

To install:

25. Install No. 2 cylinder head gasket.
 a. Remove any old sealant.

⁂ WARNING

Do not drop any oil on the contact surface of the cylinder head and cylinder block.

 b. Apply a continuous bead of Three Bond 1207B or the equivalent sealant 0.098–.118 in. (2–3 mm) diameter to a new cylinder head gasket as shown in the illustration.

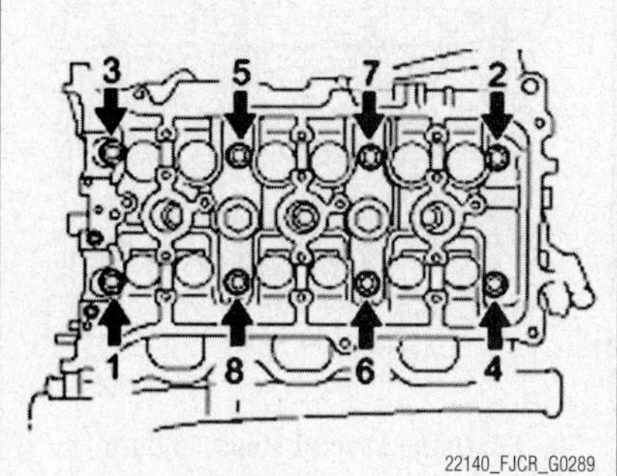

Fig. 61 Right cylinder head bolt loosening sequence

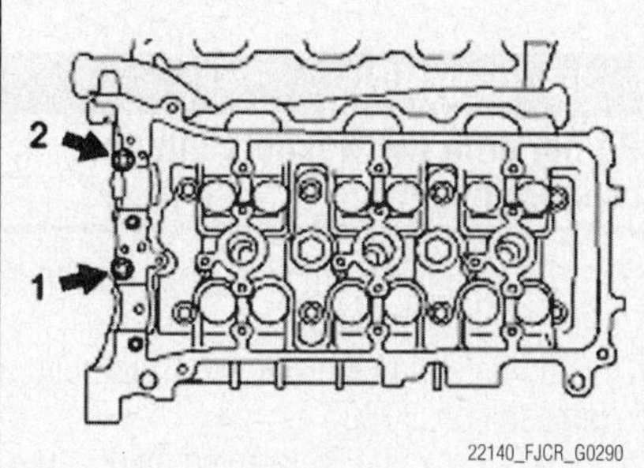

Fig. 62 Left cylinder head front bolt loosening sequence

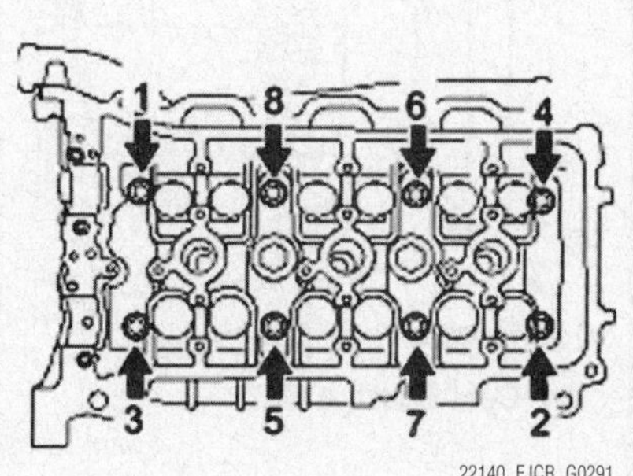

Fig. 63 Left cylinder head bolt loosening sequence

WARNING

Install the cylinder head within 3 minutes of applying the sealant. Tighten the cylinder head bolts within 15 minutes of installing the cylinder head. Otherwise, the sealant must be removed and reapplied.

c. Place the cylinder head gasket on the cylinder block surface with the Lot No. stamp facing upward.

WARNING

Orient the cylinder head gasket correctly. Place the cylinder head carefully in order not to damage the gasket with the bottom part of the head.

26. Install the left cylinder head.

a. Place the cylinder head on the cylinder head gasket.

b. Install the 8 cylinder head bolts.

➡The cylinder head bolts are tightened in 2 successive steps. If any cylinder head bolts are broken or deformed, replace them.

c. Apply a light coat of engine oil to the threads of the cylinder head bolts.

d. Install the plate washer onto the cylinder head bolt.

e. Using several steps, uniformly tighten each bolt with a 10 mm bi-hexagon wrench in the sequence shown in the illustration. Tighten to 27 ft. lbs. (36 Nm).

WARNING

If any cylinder head bolts do not meet the torque specification, replace them.

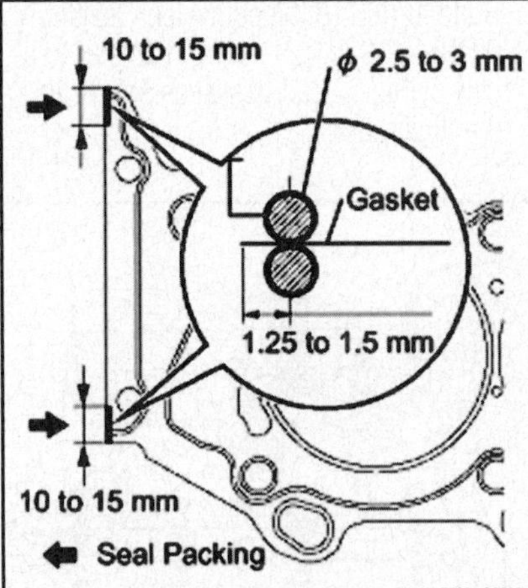

Fig. 64 Apply a continuous bead of Three Bond 1207B or the equivalent sealant 0.098–.118 in. (2–3 mm) diameter

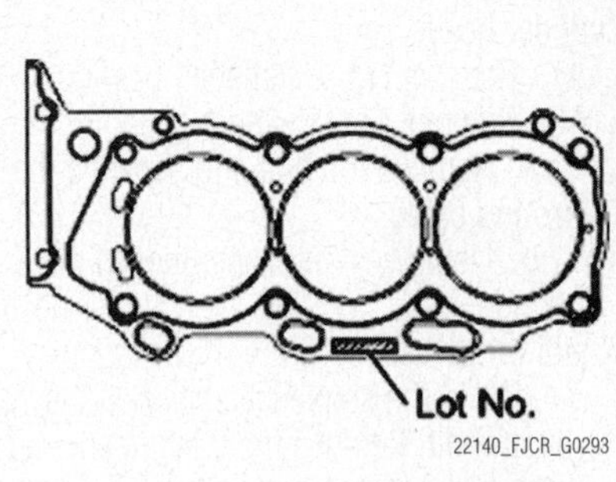

Fig. 65 Apply a continuous bead sealant in the pattern shown

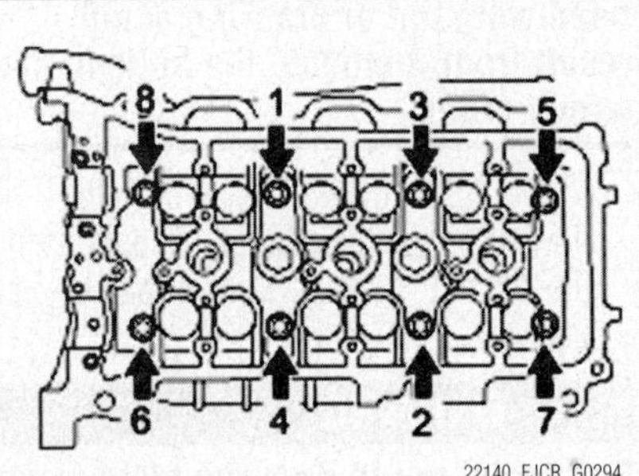

Fig. 66 Using several steps, uniformly tighten each bolt with a 10 mm bi-hexagon wrench in the sequence shown—left cylinder head

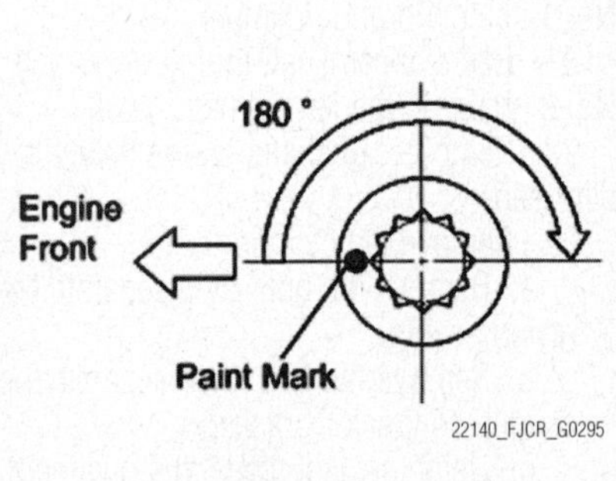

Fig. 67 Check that the painted marks are now at 180° from the engine front

WARNING

Do not drop the washers into the cylinder head.

f. Mark the front side of each cylinder head bolt with paint.

g. Retighten the cylinder head bolts an additional 180°.

h. Check that the painted marks are now at 180°from the engine front.

i. Install the front 2 cylinder head bolts.

j. Apply a light coat of engine oil to the threads of the cylinder head bolts.

k. Using several steps, uniformly install and tighten the 2 cylinder head bolts in the sequence shown in the illustration. Tighten to 22 ft. lbs. (30 Nm).

l. Install the ground cable with the bolt. Tighten to 71 ft. lbs. (96 Nm).

m. Install the oxygen sensor connector bracket with bolt. Tighten to 14 ft. lbs. (19 Nm).

27. Install camshafts on the left cylinder head.

28. Install No. 3 chain tensioner assembly.

29. Install camshaft timing gears and No. 2 chain on the left cylinder head.

30. Install No. 1 chain vibration damper.

a. Install the chain vibration damper with the 2 bolts. Tighten to 14 ft. lbs. (19 Nm).

31. Install the right cylinder head gasket.

a. Remove any old sealant.

WARNING

Do not drop any oil on the contact surfaces of the cylinder head and cylinder block.

b. Apply a continuous bead of Three Bond 1207B or the equivalent sealant 0.098–.118 in. (2–3 mm) diameter to a new cylinder head gasket as shown in the illustration.

WARNING

Install the cylinder head within 3 minutes of applying the sealant. Tighten the cylinder head bolts within 15 minutes of installing the cylinder head. Otherwise, the sealant must be removed and reapplied.

c. Place the cylinder head gasket on the cylinder block surface with the Lot No. stamp facing upward.

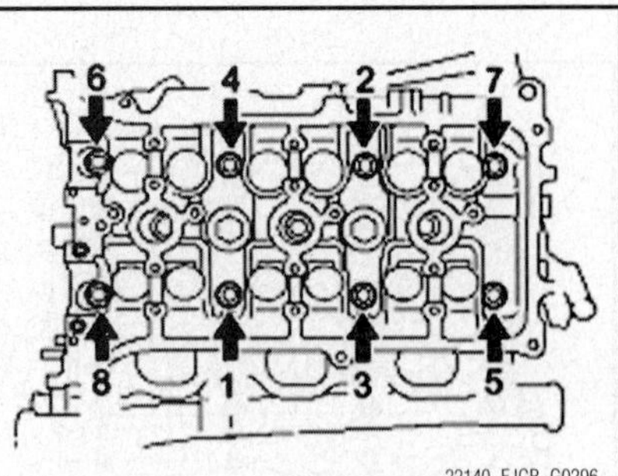

Fig. 68 Using several steps, uniformly tighten each bolt with a 10 mm bi-hexagon wrench in the sequence shown—right cylinder head

⁂ WARNING

Orient the cylinder head gasket correctly. Place the cylinder head carefully in order not to damage the gasket.

32. Install the right cylinder head.
   a. Place the cylinder head on the cylinder head gasket.
   b. Install the 8 cylinder head bolts.

➡The cylinder head bolts are tightened in 2 successive steps. If any cylinder head bolts are broken or deformed, replace them.

   c. Apply a light coat of engine oil to the threads of the cylinder head bolts.
   d. Install the plate washer onto the cylinder head bolt.
   e. Using several steps, uniformly tighten each bolt with a 10 mm bi-hexagon wrench in the sequence shown in the illustration. Tighten to 27 ft. lbs. (36 Nm).

⁂ WARNING

If any cylinder head bolts do not meet the torque specification, replace them.

⁂ WARNING

Do not drop the washers into the cylinder head.

   f. Mark the front side of each cylinder head bolt with paint.
   g. Retighten the cylinder head bolts an additional 180°.
   h. Check that the painted marks are now at 180°from the engine front.
   i. Install the ground cable with the bolt. Tighten to 71 ft. lbs. (96 Nm).
33. Install No. 2 camshaft bearing.
   a. Install the No. 2 camshaft bearing onto the cylinder head.

⁂ WARNING

Clean the installation planes of the back side of the bearing and cylinder head and keep them free of oil.

34. Install camshafts on the right cylinder head.
35. Install No. 2 chain tensioner assembly.
36. Install camshaft timing gears and No. 2 chain on the right cylinder head.
37. Install water by-pass joint.
   a. Install a new O-ring onto the water outlet pipe.
   b. Apply soapy water to the O-ring.
   c. Install 2 new gaskets and water bypass joint rear with the 2 bolts and 4 nuts. Tighten to 80 inch lbs. (9 Nm).
   d. Connect the engine coolant temperature sensor connector.
38. Connect heater water inlet hose.
39. Install intake manifold.
   a. Set a new gasket on each cylinder head.

➡Align the port holes of the gasket and cylinder head. Orient the gasket correctly.

   b. Set the intake manifold on the cylinder heads.
   c. Install and tighten the 10 bolts uniformly in several steps. Tighten to 19 ft. lbs. (26 Nm).
   d. Connect the 6 fuel injector connectors.
40. Connect fuel pipes.
41. Install exhaust manifolds.
   a. Set a new gasket to the cylinder head with the oval shape facing forward.

➡Orient the new gasket correctly.

   b. Install the exhaust manifold with the 6 nuts. Tighten to 22 ft. lbs. (30 Nm).
   c. Connect the oxygen sensor connector.
42. Install the manifold bracket with the 3 bolts. Tighten to 30 ft. lbs. (40 Nm).
43. Install front exhaust pipe assembly.
   a. Check the free length.
   b. Using Vernier calipers, measure the free length of the compression spring. Minimum length should be 0.594 in. (5mm). If the free length is less than the minimum, replace the compression spring.
   c. Install a new gasket onto the right exhaust manifold.
   d. Install a new gasket onto the exhaust front pipe.
   e. Using a wooden block and plastic-faced hammer, tap in the new gasket until it is flush with the exhaust front pipe.

➡Make sure that the gasket is in the correct direction when installing. Do not reuse the removed gasket.

   f. Install the exhaust front pipe with new 2 nuts. Tighten to 40 ft. lbs. (54 Nm).
   g. Install the exhaust center pipe onto the front exhaust pipe with the 2 bolts and 2 springs. Tighten to 32 ft. lbs. (43 Nm).
   h. Connect the oxygen sensor connector.
44. Install No. 2 front exhaust pipe assembly.
   a. Install a new gasket onto the left exhaust manifold.
   b. Install a new gasket onto the No. 2 front exhaust pipe.
   c. Install the No. 2 front exhaust pipe with the 2 new nuts and 2 bolts. Tighten the nut to 40 ft. lbs. (54 Nm) and the bolt to 35 ft. lbs. (48 Nm).
   d. Connect the oxygen sensor connector.
45. On 4WD vehicles, install the exhaust pipe stopper bracket with the 2 bolts. Tighten the nut to 14 ft. lbs. (19 Nm).
46. Install the cool air inlet with the 2 bolts. Tighten to 9 ft. lbs. (12 Nm).
47. Install the timing chain.
48. Check for fuel leakage.
49. Check for exhaust gas leakage.
50. Inspect ignition timing.
51. Inspect engine idling speed.
52. Inspect the cylinder compression.
53. Inspect CO/HC.
54. Inspect and adjust front wheel alignment

ENGINE ASSEMBLY

REMOVAL & INSTALLATION

1. Before servicing the vehicle, refer to the precautions.
2. Discharge fuel system pressure.
3. Remove battery.
4. Drain engine coolant.
5. Drain engine oil.
6. Remove hood support.
7. Remove hood.
   a. disconnect the windshield washer hose.
   b. remove the 4 bolts, then remove the hood.
8. Remove radiator assembly.
9. Remove transmission assembly
10. On manual transmission vehicles, remove clutch disc assembly.
11. Remove flywheel or driveplate.
12. Remove throttle body bracket.
13. Remove oil baffle plate.
14. Remove surge tank brackets.
15. Remove intake air surge tank.
16. Remove alternator assembly.
17. Remove the air conditioner compressor assembly and set it aside.

➡Do not disconnect the air conditioning hoses.

18. Remove power steering pump assembly and set it aside.

➡Do not disconnect the power steering hoses.

19. Disconnect No. 1 fuel pipe.
   a. Remove the No. 2 fuel pipe clamp.
   b. Pinch the retainer as illustrated, then pull the fuel tube connector out of the pipe.

⁂ WARNING

Remove any dirt and foreign matter from the fuel tube connector before performing this work. Do not allow any scratches or foreign matter onto the parts when disconnecting, as the fuel tube connector has the o-rings that seals the pipe. Perform this work by hand. Do not use any tools. Do not forcibly bend, twist or turn the nylon tube. Protect the disconnected part by covering it with a vinyl bag after disconnecting the fuel tube. If the fuel tube connector and pipe are stuck, push and pull to release them.

20. Disconnect No. 2 fuel pipe.
   a. Remove the No. 2 fuel pipe clamp.
   b. Pinch the retainer as illustrated, then pull the fuel tube connector out of the pipe.
21. Disconnect heater inlet water hose.
22. Disconnect heater water outlet hose.
23. Disconnect the connector from the engine room relay block.
24. Remove the nut and separate the engine wire from the engine room relay block.
25. Remove the bolt and separate the ground wire from body.
26. Remove the glove compartment door.
27. On 4WD automatic transmission vehicles, disconnect the connector from the four wheel drive ECU.
28. Disconnect the 3 connectors from the ECM.
29. Disconnect the 2 connectors from the A/C unit.
30. Pull the wire harness into the engine room.
31. Remove the bolt and disengage the clamp, then separate the ground wire from the body.
32. On 4WD automatic transmission vehicles, disconnect the front differential connector.
33. Remove engine assembly.
   a. Install the engine hanger No. 1 12281-310702 and 2 12282-31050 with the 4 bolts.
   b. Attach the engine sling device and hang the engine with a chain block.
   c. Remove the 4 bolts and 4 nuts, and separate the engine mounting brackets from the frame brackets.
   d. Lift the engine out of the vehicle carefully.

⁂ WARNING

Make sure that the engine is clear of all wiring and hoses.

To install:

34. Install engine assembly.
   a. Attach the engine sling device and hang the engine with a chain block.
   b. Lower the engine into the engine compartment carefully.
   c. Attach the engine mounting brackets to the frame brackets.
   d. Install the engine mounting brackets onto the frame brackets with the 4 bolts and 4 nuts. Tighten to 28 ft. lbs. (38 Nm).
   e. Remove the 2 engine hangers.
35. On 4WD automatic transmission vehicles, connect the front differential connector.
36. Install the ground wire with the bolt and engage the clamp. Tighten to 74 inch lbs. (8 Nm).
37. Connect the 2 connectors to the A/C unit.
38. Connect the 3 connectors to the ECM.
39. On 4WD automatic transmission vehicles, connect the connector to the four wheel drive ECU.
40. Install the glove compartment door.
41. Install No. 2 engine wire onto the engine room relay block with the nut. Tighten to 7 ft. lbs. (10 Nm).
42. Install the ground wire onto the frame with the bolt. Tighten to 14 ft. lbs. (19 Nm).
43. Install the ground wire onto the body with the bolt. Tighten to 74 inch lbs. (8 Nm).
44. Connect heater water outlet hose
45. Connect heater inlet water hose
46. Connect the fuel tube connectors and fuel pipe, and install the fuel pipe clamps.
   a. Align the fuel tube connector with the pipe, then push the fuel tube connector in until the retainer makes a click sound. If the connection is tight, apply a small amount of engine oil to the tip of the pipe. After connecting, pull the pipe and connector to make sure that they are securely connected.
47. Install power steering pump assembly.
48. Install air conditioning compressor assembly.
49. Install alternator assembly.
50. Install intake air surge tank.
51. Install surge tank brackets.
52. Install oil baffle plate.
53. Install throttle body bracket.
54. Install flywheel or driveplate.
55. install clutch disc assembly.
56. Install transmission assembly.
57. Install radiator assembly.
58. Install hood.
59. Connect the windshield washer hose.
60. Install hood support.
61. Install battery.
62. Add engine coolant.
63. Add engine oil.
64. On automatic transmission vehicles, using a scan tool, perform the automatic transmission initialization procedure.
65. Check for engine coolant leakage.
66. Check for engine oil leakage.
67. Check for fuel leakage.
68. Check for exhaust gas leakage.
69. Inspect ignition timing.
70. Inspect engine idling speed.
71. Inspect CO/HC.
72. Inspect and adjust front wheel alignment

EXHAUST MANIFOLD

REMOVAL & INSTALLATION

1. Before servicing the vehicle, refer to the precautions.
2. On 4WD vehicles, remove exhaust pipe stopper bracket.
   a. Remove the 2 bolts, then remove the exhaust pipe stopper bracket.
3. Remove No. 2 front exhaust pipe assembly.
   a. Disconnect the oxygen sensor connector.
   b. Remove the 2 bolts and 2 nuts.
   c. Disengage the support and remove the front exhaust pipe and 2 gaskets.
4. Remove front exhaust pipe assembly.
   a. Disconnect the oxygen sensor connector.
   b. Remove the 2 bolts, 2 springs and 2 nuts, then separate the front exhaust pipe from the exhaust manifold.
5. Remove the 3 bolts and manifold bracket.
6. Remove exhaust manifolds.
   a. Disconnect the oxygen sensor connector.
   b. Remove the 6 nuts and exhaust manifold.
   c. Remove the gasket.

To install:

7. Install left exhaust manifold.
   a. Set a new gasket to the left cylinder head with the oval shape facing the correct direction.

➡On the left cylinder head, orient the oval shape to the rear. On the right cylinder head, orient the oval shape to the front.

   b. Install the exhaust manifold with the 6 nuts. Tighten to 22 ft. lbs. (30 Nm).

c. Connect the oxygen sensor connector.

8. Install the No. 2 manifold bracket with the 3 bolts. Tighten to 30 ft. lbs. (40 Nm).

9. Install right exhaust manifold.

a. Set a new gasket to the right cylinder head with the oval shape facing forward.

➡Orient the new gasket correctly.

b. Install the exhaust manifold with the 6 nuts. Tighten to 22 ft. lbs. (30 Nm).

c. Connect the oxygen sensor connector.

10. Install the manifold bracket with the 3 bolts. Tighten to 30 ft. lbs. (40 Nm).

11. Install front exhaust pipe assembly.

a. Check the free length.

b. Using Vernier calipers, measure the free length of the compression spring. Minimum length should be 0.594 in. (5mm). If the free length is less than the minimum, replace the compression spring.

c. Install a new gasket onto the exhaust manifold.

d. Install a new gasket onto the exhaust front pipe.

e. Using a wooden block and plastic-faced hammer, tap in the new gasket until it is flush with the exhaust front pipe.

➡Make sure that the gasket is in the correct direction when installing. Do not reuse the removed gasket.

f. Install the exhaust front pipe with new 2 nuts. Tighten to 40 ft. lbs. (54 Nm).

g. Install the exhaust center pipe onto the front exhaust pipe with the 2 bolts and 2 springs. Tighten to 32 ft. lbs. (43 Nm).

h. Connect the oxygen sensor connector.

12. Install No. 2 front exhaust pipe assembly.

a. Install a new gasket onto the left exhaust manifold.

b. Install a new gasket onto the No. 2 front exhaust pipe.

c. Install the No. 2 front exhaust pipe with the 2 new nuts and 2 bolts. Tighten the nut to 40 ft. lbs. (54 Nm) and the bolt to 35 ft. lbs. (48 Nm).

d. Connect the oxygen sensor connector.

13. On 4WD vehicles, install the exhaust pipe stopper bracket with the 2 bolts. Tighten the nut to 14 ft. lbs. (19 Nm).

14. Check for exhaust gas leakage.

15. Inspect CO/HC.

FLYWHEEL/DRIVEPLATE

REMOVAL & INSTALLATION

See Figure 69.

1. Before servicing the vehicle, refer to the precautions.

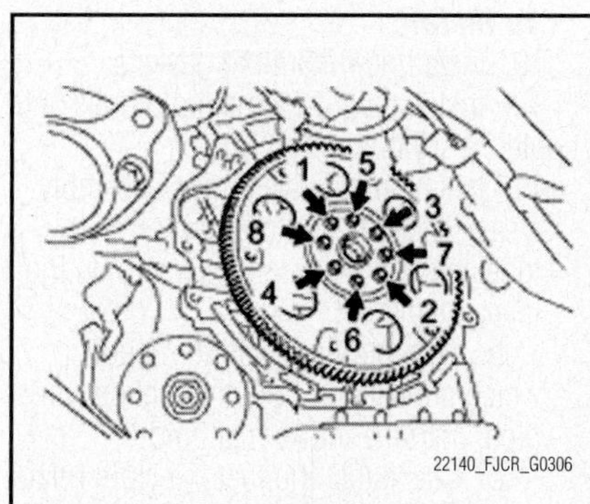

Fig. 69 Install the flywheel with the 6 bolts and tighten in the order shown

2. Remove the transmission assembly.

3. Using SST: 09213-54015, hold the crankshaft.

4. Remove the 8 bolts, then remove the flywheel and 2 spacers.

To install:

5. Apply Three Bond 1324 or the equivalent adhesive to the 2 or 3 end threads of the bolts.

6. Install the flywheel and 2 spacers onto the crankshaft.

7. Install the flywheel with the 6 bolts and tighten in the order shown in the illustration to 61 ft. lbs. (83 Nm).

8. Install the transmission assembly.

INTAKE MANIFOLD

REMOVAL & INSTALLATION

See Figures 70 and 71.

1. Before servicing the vehicle, refer to the precautions.

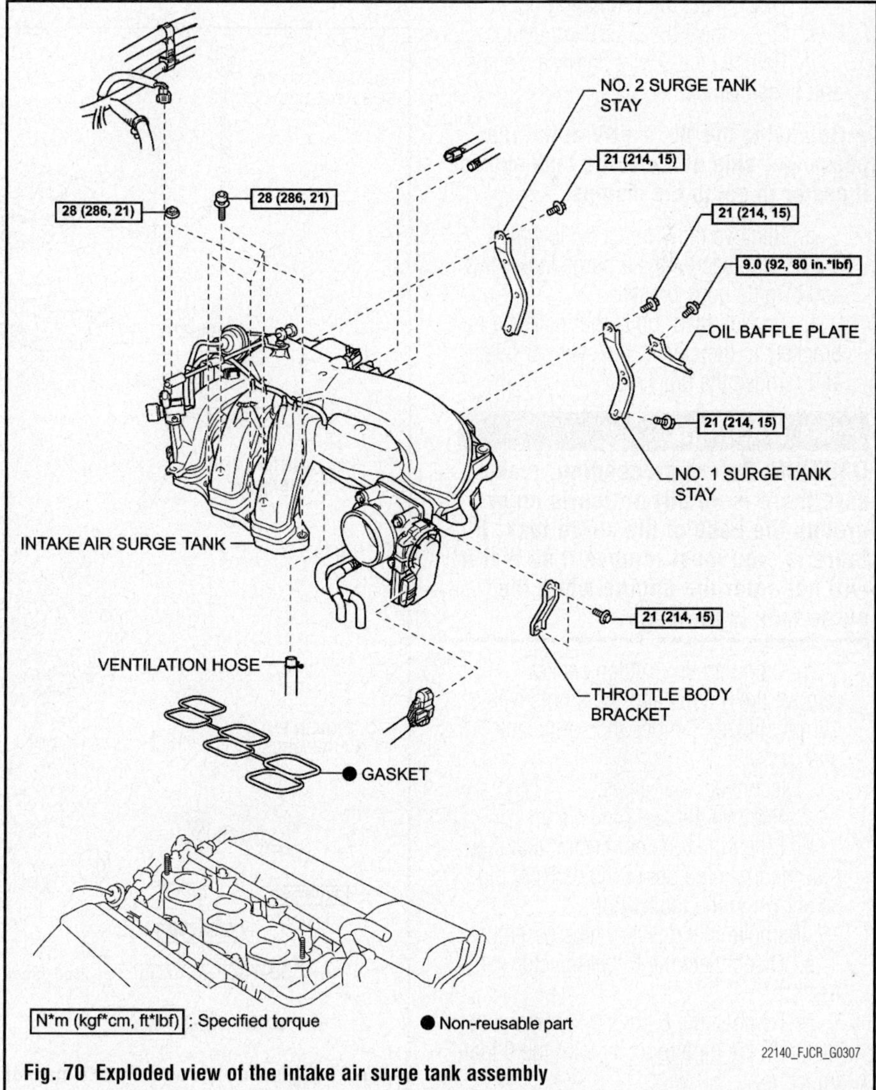

Fig. 70 Exploded view of the intake air surge tank assembly

2. Remove air cleaner assembly.

a. Remove the engine V-bank cover by removing the two acorn nuts.

b. Disconnect the ventilation hose No. 2.

c. Disconnect the vacuum hose.

d. Disconnect the mass air flow meter connector. Remove the 2 wire harness clamps.

e. Loosen the 2 hose clamps. Remove the 2 bolts, then remove the air cleaner.

3. Remove throttle body with motor body assembly.

a. Disconnect the throttle body connector.

b. Remove the 4 bolts, and then remove the throttle w/ motor body and gasket and set it aside

➡Leave the throttle body connected to the coolant hoses.

4. Remove intake air surge tank.

a. Disconnect the fuel vapor feed hose.

b. Disconnect the ventilation hose.

c. Disconnect the 2 VSV connectors.

d. Remove the 3 wire harness clamps and hose clamps.

➡Removing the No. 1 VSV at the rear passenger side of the surge tank makes it easier to get to the clamps.

e. Remove the 3 upper bolts which are used to secure the 2 surge tank stays and throttle body bracket.

f. Loosen the 2 bolts that hold the brackets to the side of the cylinder head and remove the brackets.

WARNING

CAUTION: Before proceeding, make sure there is no dirt or debris on or around the base of the surge tank. If there is, you must remove it so that it will not enter the engine when the surge tank is removed.

g. Using an 8mm Allen socket, remove the 4 bolts and the 2 nuts with a 12mm socket. Remove the surge tank and gasket.

5. Disconnect fuel pipes.

a. Remove the fuel pipe clamp.

b. Pinch the tube connector, and then pull the fuel pipe out of the delivery pipe as shown in the illustration.

6. Remove fuel delivery pipe assembly.

a. Disconnect the 6 fuel injector connectors.

b. Remove the 6 bolts and remove the fuel delivery pipe together with the 6 fuel injectors.

WARNING

Do not drop the injectors when removing the fuel delivery pipe.

7. Remove the intake manifold bolts, then lift the manifold off the engine.

8. Remove the intake manifold gaskets.

To install:

9. Install new manifold gaskets.

10. Install the intake manifold bolts and tighten to 19 ft. lbs. (26 Nm).

11. Install fuel delivery pipe assembly.

a. Place the fuel delivery pipe together with the 6 fuel injectors on the intake manifold.

b. Temporarily install the 6 bolts, which are used to hold the fuel delivery pipe, onto the intake manifold.

c. Check that the fuel injectors rotate smoothly. If the fuel injectors do not rotate smoothly, replace the O-ring.

d. Position each fuel injector connector facing outward.

e. Tighten the 6 fuel delivery pipe attaching bolts to 11 ft. lbs (15 Nm).

f. Connect the 6 fuel injector connectors.

12. Connect fuel pipes.

a. Push the tube connector into the pipe until the tube connector makes a "click" sound.

➡Check if there is any damage or foreign objects on the connected part of the fuel pipe.

b. After connecting, check that the pipe and connector are securely connected by pulling them.

c. Install the fuel pipe clamp.

13. Install intake air surge tank.

a. Install the surge tank using a new gasket. Using an 8mm Allen socket, install the 4 bolts and the 2 nuts with a 12mm socket.

b. Install the brackets and tighten the 2 bolts that hold the brackets to the side of the cylinder head.

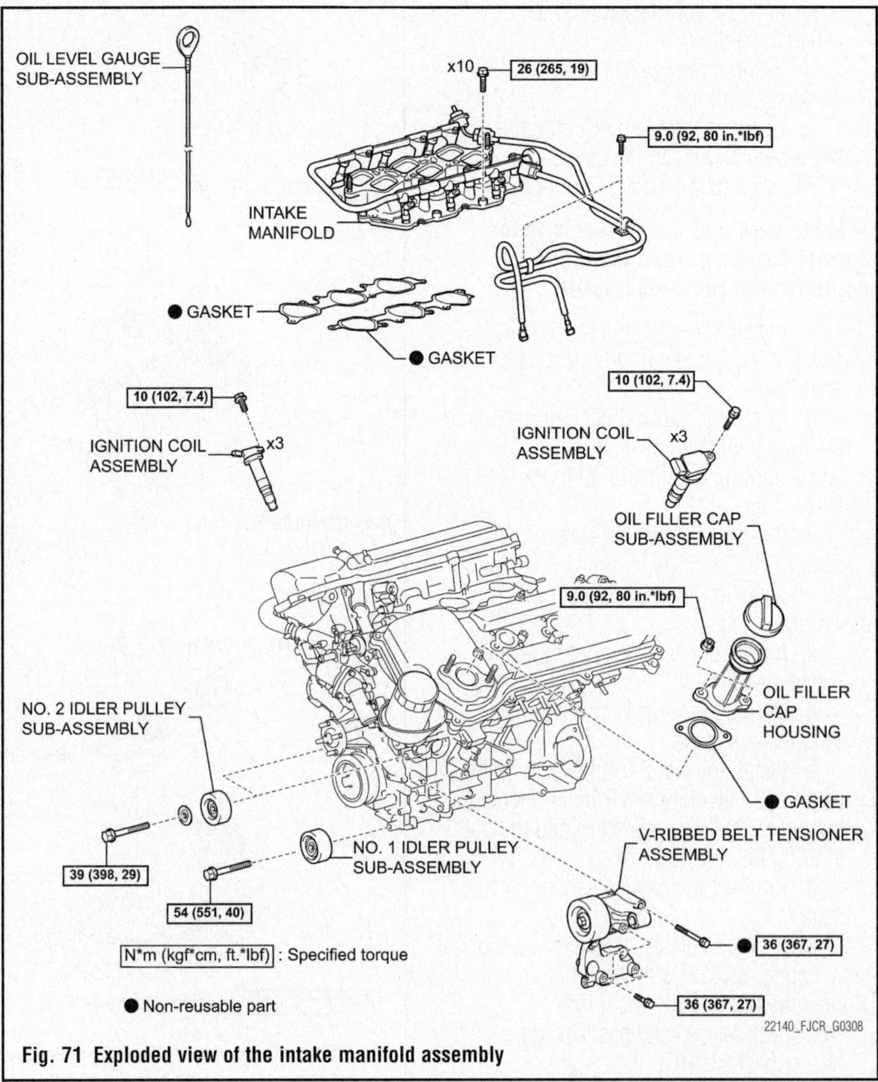

Fig. 71 Exploded view of the intake manifold assembly

c. Install the 3 upper bolts which are used to secure the 2 surge tank stays and throttle body bracket.

d. Install No. 1 VSV at the rear passenger side of the surge tank if removed.

e. Install the 3 wire harness clamps and hose clamps.

f. Connect the 2 VSV connectors.

g. Connect the ventilation hose.

h. Connect the fuel vapor feed hose.

14. Install throttle body with motor body assembly.

a. Install the throttle body with motor using a new gasket Install the 4 bolts, and then and set it aside

b. Connect the throttle body connector.

15. Install air cleaner assembly.

a. Install the air cleaner and 2 bolts. Tighten the 2 hose clamps.

b. Connect the mass air flow meter connector. Install the 2 wire harness clamps.

c. Connect the vacuum hose.

d. Connect the ventilation hose No. 2.

16. Install the engine V-bank cover.

OIL PAN

REMOVAL & INSTALLATION

See Figures 72 through 74.

1. Before servicing the vehicle, refer to the precautions.
2. Remove battery.
3. Drain engine coolant.
4. Drain engine oil.
5. Remove power steering rack assembly.
6. On 4WD vehicles, remove front differential carrier assembly.
7. Remove fan.
8. Remove alternator assembly.
9. Remove the air conditioning compressor and set it aside.

➡Do not disconnect the air conditioning lines.

10. Remove the 5 bolts, then remove the V-ribbed belt tensioner.
11. Remove the oil level gauge.
12. Remove the power steering pump assembly and set it aside.

➡Do not disconnect the power steering lines.

13. Remove idler pulleys.
14. Remove crankshaft pulley.
15. Remove No. 2 oil pan.

a. Remove the 10 bolts and 2 nuts.

b. Insert the blade of SST: 09032-00100 between the oil pan and No. 2 oil pan, cut off applied sealer and remove the No. 2 oil pan.

**** WARNING**

Be careful not to damage the contact surfaces of the oil pan and No. 2 oil pan. Be careful not to damage the No. 2 oil pan flange.

16. Remove the 2 nuts, then remove the oil strainer and gasket.
17. Remove oil pan.

a. Remove the 4 housing bolts.

b. Remove the flywheel housing under cover.

c. Remove the 17 bolts and 2 nuts.

d. Using a screwdriver, remove the oil pan by prying between the oil pan and cylinder block.

**** WARNING**

Be careful not to damage the contact surfaces of the cylinder block and oil pan.

e. Remove the O-ring from the oil pump.

To install:

18. Install oil pan.

a. Remove any old sealant.

**** WARNING**

Do not drop any oil on the contact surfaces of the cylinder block, rear oil seal and oil pan.

b. Install a new O-ring onto the oil pump.

c. Apply a continuous bead of Three Bond or the equivalent sealant 0.12–0.16 in. (3–4 mm) diameter to the oil pan as shown in the illustration.

**** WARNING**

Install the oil pan within 3 minutes of applying the sealant. Tighten the oil pan bolts and nuts within 15 minutes of installing the oil pan. Otherwise, the sealant must be removed and reapplied.

d. Install the oil pan with the 17 bolts and 2 nuts, and tighten the bolts and nuts uniformly in several steps. Tighten the 0.39 in. (10 mm) head oil pan bolt to 7 ft. lbs. (10 Nm), the 0.47 in. (12 mm) head oil pan bolt to 16 ft. lbs. (21 Nm) and the nuts to 16 ft. lbs. (21 Nm).

**** WARNING**

Install bolts in their proper locations. Bolt lengths are as follows: A is 0.98 in. (25 mm) B 1.77 in. (45 mm) and C 0.55 in. (14 mm).

e. Install the 4 housing bolts. Tighten to 27 ft. lbs. (37 Nm).

f. Install the flywheel housing under cover.

19. Install oil strainer.

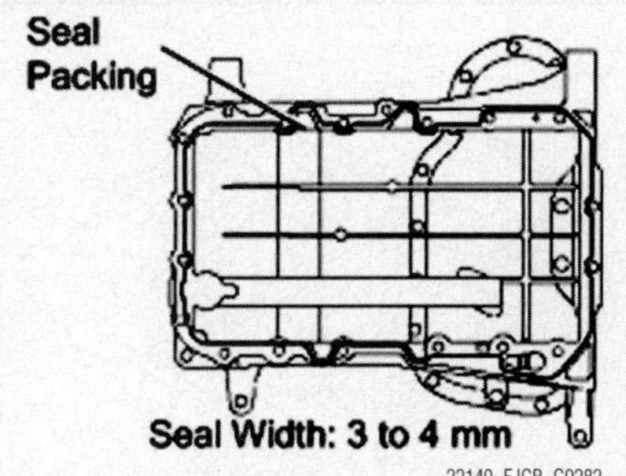

Fig. 72 Apply a continuous bead of Three Bond or the equivalent sealant 0.12–0.16 in. (3–4 mm) diameter to the oil pan as shown

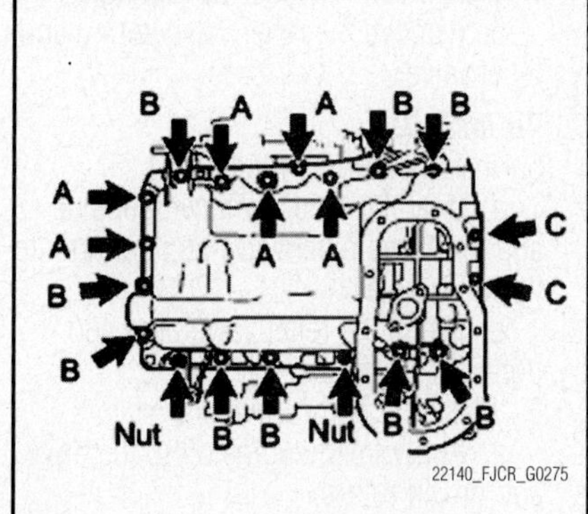

Fig. 73 Oil pan fastener locations

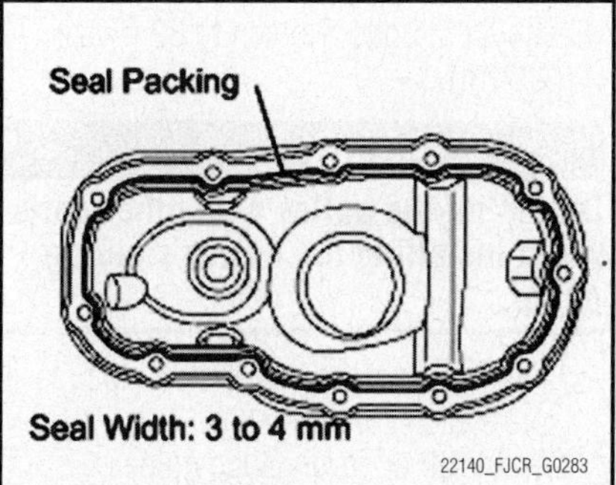

Fig. 74 Apply a continuous bead of Three Bond or the equivalent sealant 0.12–0.16 in. (3–4 mm) diameter to the oil pan

a. Install a new gasket and the oil strainer with the 2 nuts. Tighten to 80 inch lbs. (9 Nm).

20. Install No. 2 oil pan.

a. Remove any old sealant.

WARNING

Do not drop any oil on the contact surfaces of the oil pan and No. 2 oil pan.

b. Apply a continuous bead of Three Bond or the equivalent sealant 0.12–0.16 in. (3–4 mm) diameter to the oil pan as shown in the illustration.

WARNING

Install the oil pan within 3 minutes of applying the sealant. Tighten the oil pan bolts and nuts within 15 minutes of installing the oil pan. Otherwise, the sealant must be removed and reapplied.

c. Install the oil pan with the 17 bolts and 2 nuts, and tighten the bolts and nuts uniformly in several steps. Tighten the bolts to 80 inch lbs. (9 Nm) and the nuts to 7 ft. lbs. (10 Nm).

21. Install crankshaft pulley.

22. Install No. 1idler pulley.

a. Install the idler pulley with the bolt. Tighten to 29 ft. lbs. (39 Nm).

➡DOUBLE is marked on the No. 1 idler pulley to distinguish it from the No. 2 idler pulley.

23. Install No. 2idler pulley.

a. Install the 2 No. 2 idler pulleys with the 2 bolts. Tighten to 40 ft. lbs. (54 Nm).

24. Install power steering pump assembly and set it aside.

➡Do not disconnect the power steering hoses.

a. Install the power steering pump with the 2 bolts. Tighten to 32 ft. lbs. (43 Nm).

WARNING

Do not hit the pulley with other parts when installing the power steering pump.

b. Connect the power steering pressure switch connector.

25. Install oil level gauge guide.

a. Install a new O-ring onto the oil level gauge guide.

b. Apply a light coat of engine oil to the O-ring.

c. Push the oil level gauge guide end into the guide hole in the oil pan.

d. Install the oil level gauge guide with the bolt. Tighten to 80 inch lbs. (9 Nm).

e. Install the oil level gauge guide.

26. Install V-ribbed belt tensioner assembly.

27. Install air conditioner compressor assembly.

28. Install alternator assembly.

29. Install fan.

30. On 4WD vehicles, install front differential carrier assembly.

31. Install power steering link assembly.

32. Install battery.

33. Add engine coolant.

34. Add engine oil.

35. Add power steering fluid and bleed the system, as necessary.

36. On 4WD vehicles, inspect and adjust the differential oil.

37. Check for engine coolant leakage.

38. Check for engine oil leakage.

39. Check for power steering fluid leakage.

40. Check for differential oil leakage.

41. Inspect and adjust front wheel alignment.

OIL PUMP

REMOVAL & INSTALLATION

See Figures 75 and 76.

1. Before servicing the vehicle, refer to the precautions.

2. Remove the timing chain cover.

3. Remove oil pump cover.

a. Remove the 3 bolts, then remove the oil pipe.

b. Remove the 2 O-rings.

c. Remove the 7 bolts, then remove the oil pump cover, drive rotor and driven rotor.

4. Remove oil pump relief valve.

a. Remove the relief valve plug.

b. Remove the relief valve spring and relief valve.

To install:

5. Install oil pump relief valve.

a. Coat the relief valve with engine oil and insert the relief valve and spring into the valve hole.

b. Install the relief valve plug and tighten to 36 ft. lbs. (49 Nm).

6. Install oil pump cover.

a. Apply fresh engine oil to the drive and driven rotors.

b. Place the drive and driven rotors into the timing chain cover with the marks facing the oil pump cover side.

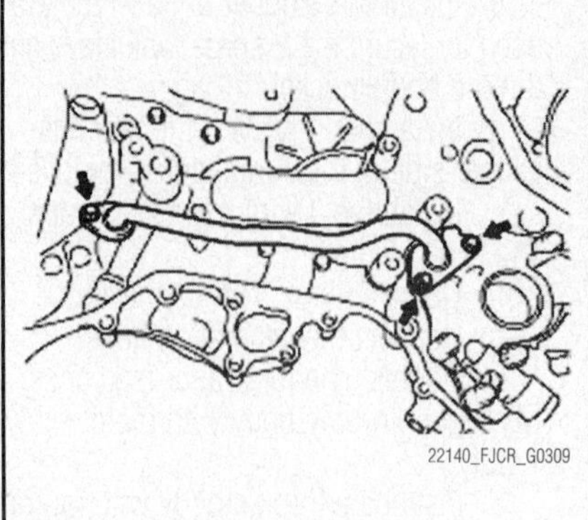

Fig. 75 Oil pipe mounting bolts

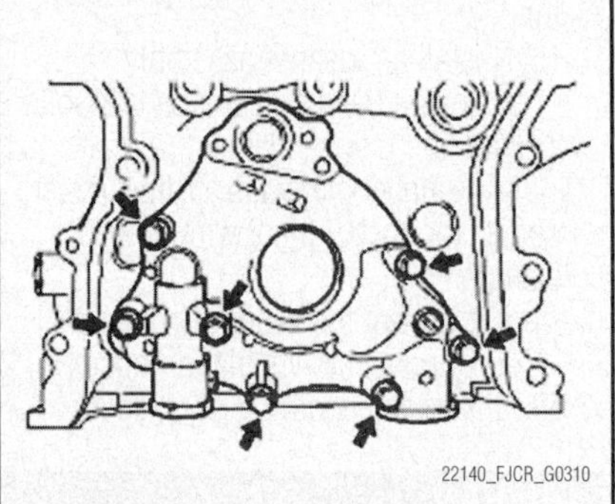

Fig. 76 Oil pump cover mounting bolts

c. Install the oil pump cover with the 7 bolts. Tighten to 80 inch lbs. (9 Nm).

d. Install a new O-ring onto the oil pump cover.

e. Install a new O-ring onto the oil pipe.

f. Install the oil pipe with the 3 bolts. Tighten to 80 inch lbs. (9 Nm).

MAIN BEARING TORQUE SEQUENCE

See Figures 77 through 80.

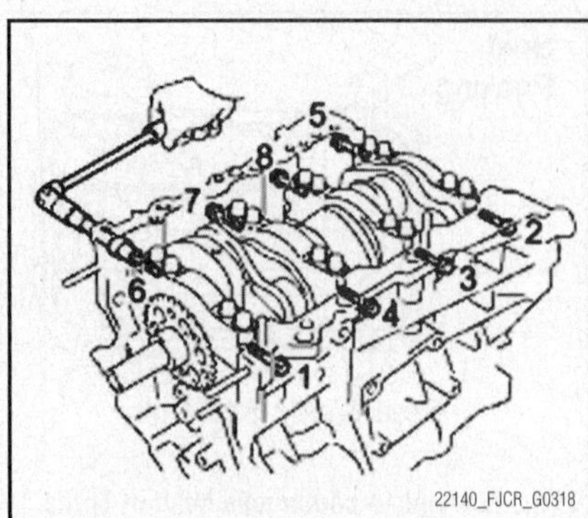

Fig. 77 Main bearing horizontal bolt loosening sequence —4.0L (1GR-FE) engine

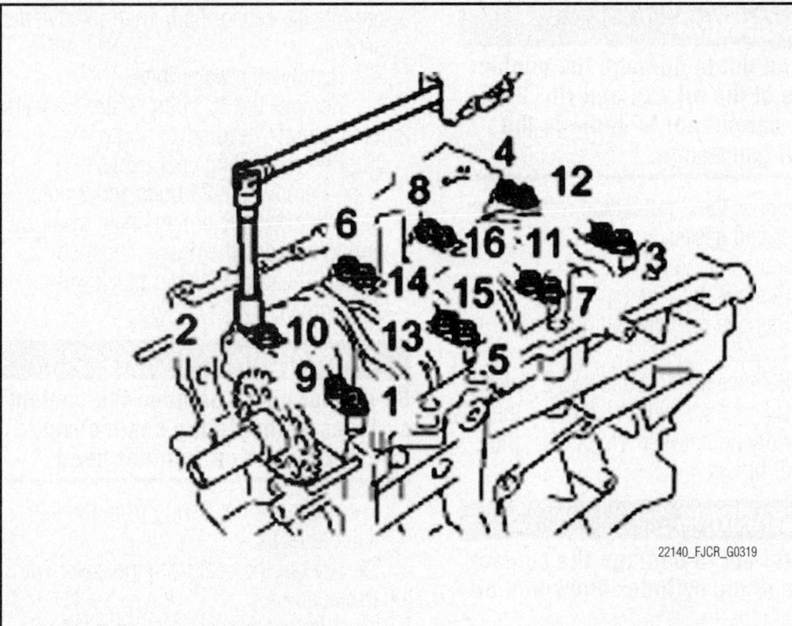

Fig. 78 Main bearing vertical bolt loosening sequence —4.0L (1GR-FE) engine

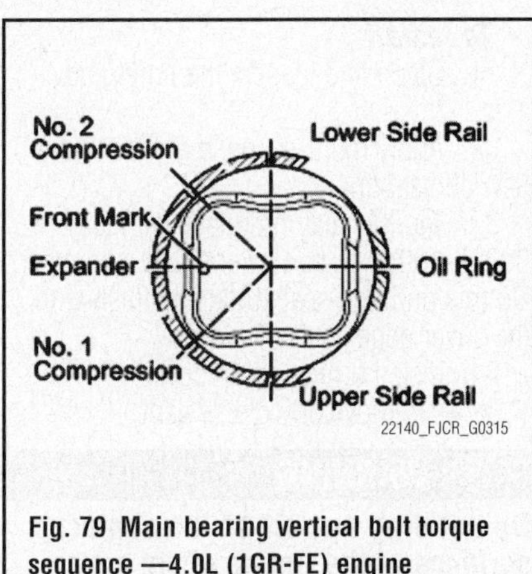

Fig. 79 Main bearing vertical bolt torque sequence —4.0L (1GR-FE) engine

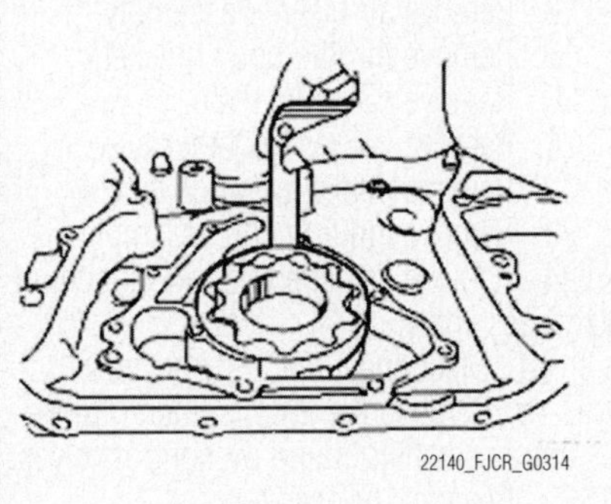

Fig. 81 Piston ring positioning—4.0L (1GR-FE) engine

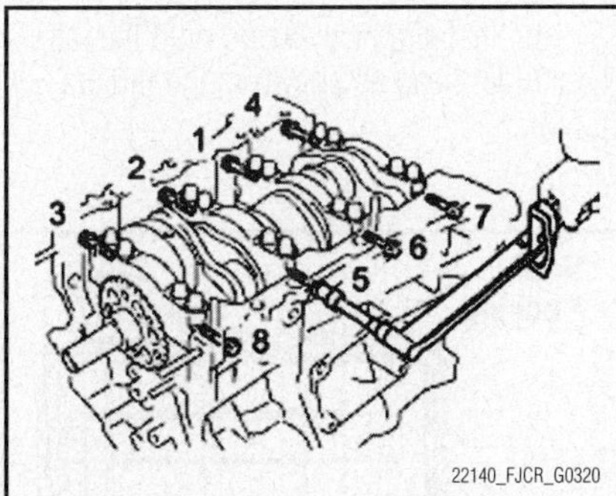

Fig. 80 Main bearing horizontal bolt torque sequence —4.0L (1GR-FE) engine

PISTON AND RING

POSITIONING

See Figure 81.

REAR MAIN SEAL

REMOVAL & INSTALLATION

See Figure 82.

1. Before servicing the vehicle, refer to the precautions.
2. Remove the flywheel/driveplate.
3. Remove the 5 bolts and 2 nuts.
4. Remove rear engine oil seal retainer.
5. Using a screwdriver, remove the oil seal retainer by prying between the oil seal retainer and crankshaft bearing cap.

To install:

6. Remove any old packing material and oil from the contact surfaces of the oil seal retainer and cylinder block.
7. Apply a continuous bead of seal packing (diameter 2 to 3 mm (0.08 to 0.12 in.)) to the oil seal retainer as shown in the illustration.

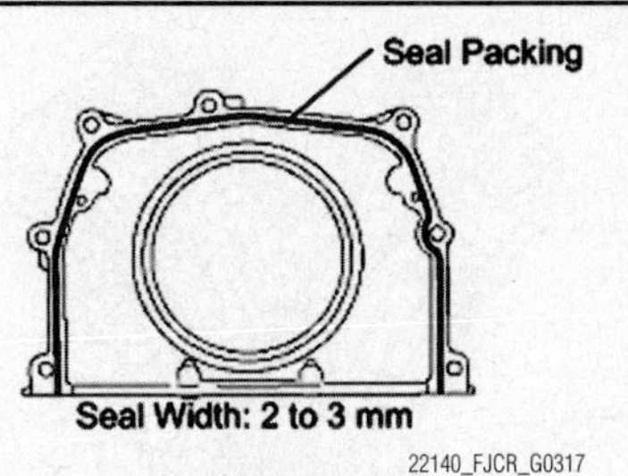

Fig. 82 Apply a continuous bead of Three Bond 1207B or the equivalent sealant 0.08 to 0.12 in. (2 to 3 mm) diameter to the oil seal retainer as shown

✱✱ WARNING

Parts must be assembled within 3 minutes of application. Otherwise, the seal packing must be removed and reapplied.

a. Install the oil seal retainer with the 5 bolts and 2 nuts. Tighten the nuts to 80 inch lbs. (9 Nm) and the bolts to 7 ft. lbs. (10 Nm).

8. Install the flywheel/driveplate.

TIMING CHAIN COVER AND SEAL

REMOVAL & INSTALLATION

See Figures 83 through 88.

1. Before servicing the vehicle, refer to the precautions.
2. Remove battery.
3. Drain engine coolant.
4. Drain engine oil.
5. Remove power steering rack assembly.
6. On 4WD vehicles, remove front differential carrier assembly.
7. Remove fan.
8. Remove alternator assembly.
9. Remove the air conditioning compressor and set it aside.

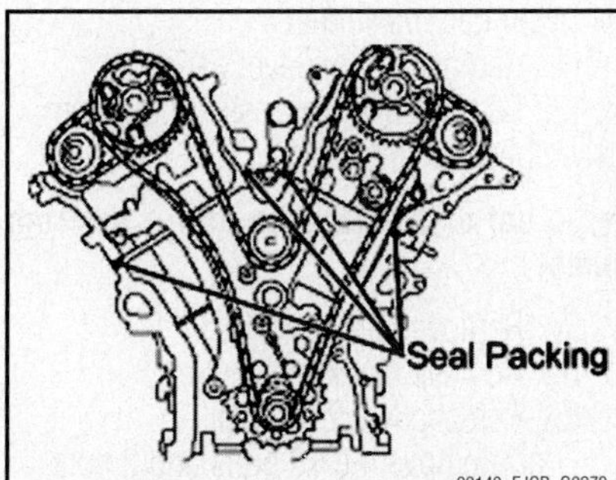

Fig. 83 Apply a continuous bead of Three Bond 1207B or the equivalent sealant 0.12–0.16 in. (3–4 mm) diameter to the 4 locations shown

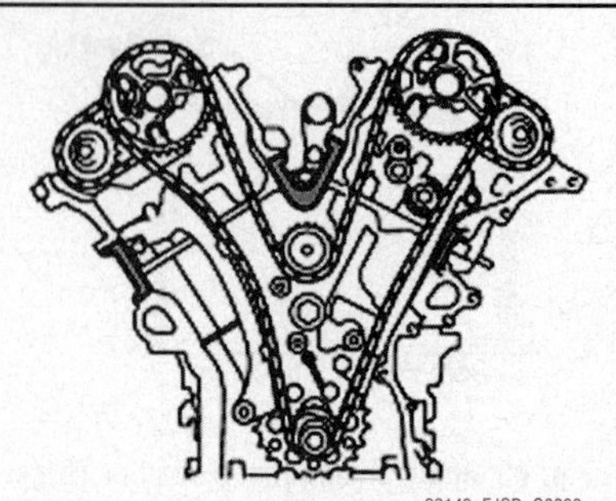

Fig. 84 Keep the seal surface between the cylinder block and the cylinder head shown in the illustration free of oil before installing the chain cover

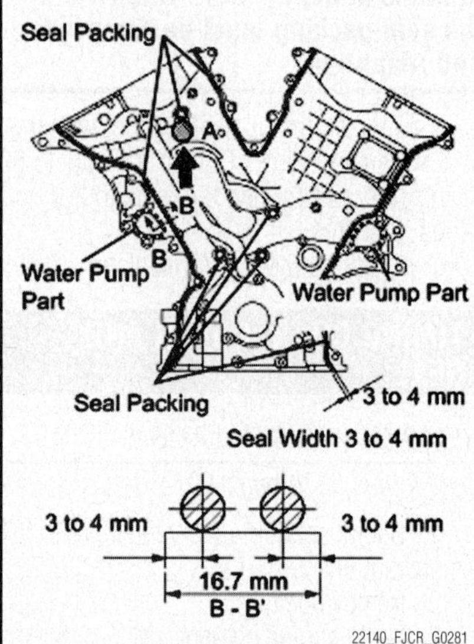

Fig. 85 Apply a continuous bead of Three Bond or the equivalent sealant 0.12–0.16 in. (3–4 mm) diameter to the timing chain cover as shown

➡Do not disconnect the air conditioning lines.

10. Remove the 5 bolts, then remove the V-ribbed belt tensioner.
11. Remove the oil level gauge.
12. Remove the power steering pump assembly and set it aside.

➡Do not disconnect the power steering lines.

13. Remove idler pulleys.
14. Remove crankshaft pulley.
15. Remove No. 2 oil pan.
    a. Remove the 10 bolts and 2 nuts.
    b. Insert the blade of SST: 09032-00100 between the oil pan and No. 2 oil pan, cut off applied sealer and remove the No. 2 oil pan.

⁂ WARNING

Be careful not to damage the contact surfaces of the oil pan and No. 2 oil pan. Be careful not to damage the No. 2 oil pan flange.

16. Remove the 2 nuts, then remove the oil strainer and gasket.
17. Remove oil pan.
    a. Remove the 4 housing bolts.
    b. Remove the flywheel housing under cover.
    c. Remove the 17 bolts and 2 nuts.
    d. Using a screwdriver, remove the oil pan by prying between the oil pan and cylinder block.

⁂ WARNING

Be careful not to damage the contact surfaces of the cylinder block and oil pan.

    e. Remove the O-ring from the oil pump.
18. Remove air cleaner assembly.
19. Remove throttle body bracket.
20. Remove oil baffle plate.
21. Remove surge tank bracket.
22. Remove intake air surge tank.
23. Remove ignition coil assembly.
24. Remove camshaft timing oil control valve assembly.
    a. Disconnect the 2 connectors.
    b. Remove the 2 bolts, then remove the 2 camshaft timing oil control valves.
25. Remove VVT sensor.
    a. For the right cylinder head sensor, disconnect the VVT sensor connector and remove the sensor bolt. Then remove the sensor.
    b. For the left cylinder head sensor, disconnect the No. 4 water by-pass hose and No. 5 water by-pass hose. Disconnect the VVT sensor connector and remove the sensor bolt. Then remove the sensor.
26. Remove the water inlet.
27. Remove the 10 bolts, 3 seal washers, 2 nuts, cylinder head cover and gasket.
28. Remove timing chain cover.
    a. Remove the 24 bolts and 2 nuts.
    b. Remove the timing chain cover by prying between the timing chain cover, cylinder head or cylinder block with a screwdriver.

⁂ WARNING

Be careful not to damage the contact surfaces of the timing chain cover, cylinder block and cylinder head.

    c. Remove the O-ring from the left cylinder head.
29. Pry out the seal from the cover with a flat-bladed tool.
30. It is a good idea to remove the oil pump from the timing cover and replace the O-ring.

To install:

31. Clean and inspect the timing cover area.
32. Apply multi-purpose grease to the new oil seal lip.
33. Tap the seal into place with SST: 09226-10010 or equivalent, and a hammer. Do this until the seal surface is flush with the cover edge.
34. Install timing chain cover.
    a. Remove any old sealant.

⁂ WARNING

Do not drop any oil on the contact surfaces of the timing chain cover, cylinder head and cylinder block.

    b. Install a new O-ring onto the left cylinder head as shown in the illustration.

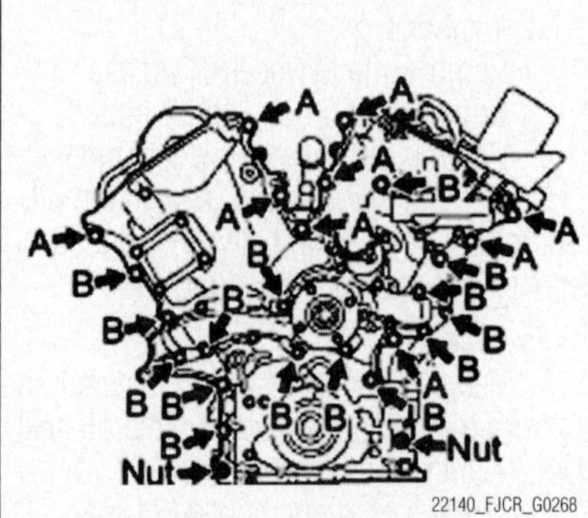

Fig. 86 Timing chain cover fastener locations

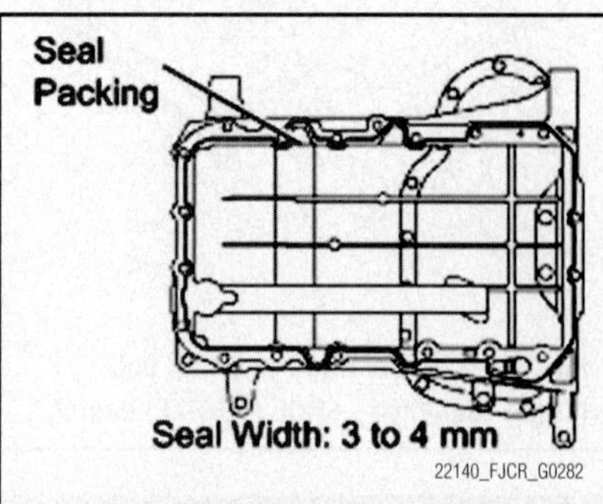

Fig. 87 Apply a continuous bead of Three Bond or the equivalent sealant 0.12–0.16 in. (3–4 mm) diameter to the oil pan as shown

c. Apply a continuous bead of Three Bond 1207B or the equivalent sealant 0.12–0.16 in. (3–4 mm) diameter to the 4 locations shown in the illustration.

d. Keep the seal surface between the cylinder block and the cylinder head shown in the illustration free of oil before installing the chain cover.

e. Apply a continuous bead of Three Bond or the equivalent sealant 0.12–0.16 in. (3–4 mm) diameter to the timing chain cover as shown in the illustration. Use Three Bond 1282B for the water pump part and Three Bond 1207B for the remainder.

⁂ WARNING

Install the timing chain cover within 3 minutes of applying the sealant. The timing chain cover bolts and nuts must be tightened within 15 minutes of the installation. Otherwise the sealant must be removed and reapplied. Do not apply sealant to portion A shown in the illustration.

f. Align the key way of the oil pump drive rotor with the rectangular portion of the crankshaft timing gear, and slide the timing chain cover into place.

g. Install the timing chain cover with the 24 bolts and 2 nuts. Tighten the bolts and nuts to 17 ft. lbs. (23 Nm) uniformly in several steps.

➡Pay attention not to wrap the chain and slipper over the timing chain cover seal line.

⁂ WARNING

Place bolts in their proper positions. Bolt A length is 0.98 in. (25 mm) and bolt B length is 2.17 in. (55 mm).

35. Install cylinder head covers.
36. Install water inlet.

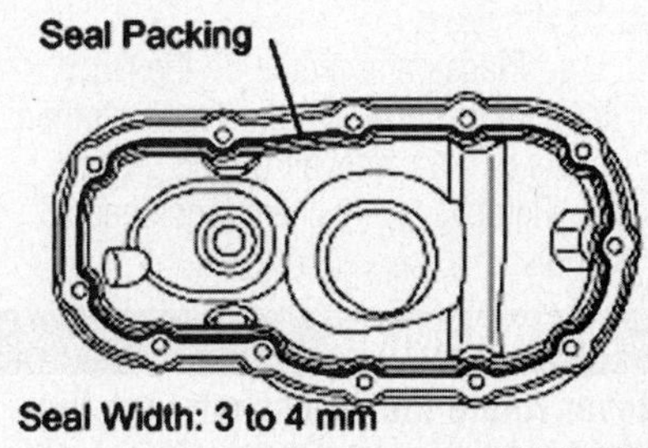

Fig. 88 Apply a continuous bead of Three Bond or the equivalent sealant 0.12–0.16 in. (3–4 mm) diameter to the oil pan

37. Install VVT sensor. Tighten to 71 inch lbs. (8 Nm).
38. Install camshaft timing oil control valve assembly. Tighten to 80 inch lbs. (9 Nm).
39. Install ignition coil assembly.
40. Install intake air surge tank.
41. Install surge tank brackets.
42. Install oil baffle plate.
43. Install throttle body bracket.
44. Install air cleaner assembly.
45. Install oil pan.

a. Remove any old sealant.

⁂ WARNING

Do not drop any oil on the contact surfaces of the cylinder block, rear oil seal and oil pan.

b. Install a new O-ring onto the oil pump.

c. Apply a continuous bead of Three Bond or the equivalent sealant 0.12–0.16 in. (3–4 mm) diameter to the oil pan as shown in the illustration.

⁂ WARNING

Install the oil pan within 3 minutes of applying the sealant. Tighten the oil pan bolts and nuts within 15 minutes of installing the oil pan. Otherwise, the sealant must be removed and reapplied.

d. Install the oil pan with the 17 bolts and 2 nuts, and tighten the bolts and nuts uniformly in several steps. Tighten the 0.39 in. (10 mm) head oil pan bolt to 7 ft. lbs. (10 Nm), the 0.47 in. (12 mm) head oil pan bolt to 16 ft. lbs. (21 Nm) and the nuts to 16 ft. lbs. (21 Nm).

⁂ WARNING

Install bolts in their proper locations. Bolt lengths are as follows: A is 0.98 in. (25 mm) B 1.77 in. (45 mm) and C 0.55 in. (14 mm).

e. Install the 4 housing bolts. Tighten to 27 ft. lbs. (37 Nm).

f. Install the flywheel housing under cover.

46. Install oil strainer.

a. Install a new gasket and the oil strainer with the 2 nuts. Tighten to 80 inch lbs. (9 Nm).

47. Install No. 2 oil pan.

a. Remove any old sealant.

⁂ WARNING

Do not drop any oil on the contact surfaces of the oil pan and No. 2 oil pan.

b. Apply a continuous bead of Three Bond or the equivalent sealant 0.12–0.16 in. (3–4 mm) diameter to the oil pan as shown in the illustration.

⁂ WARNING

Install the oil pan within 3 minutes of applying the sealant. Tighten the oil pan bolts and nuts within 15 minutes of installing the oil pan. Otherwise, the sealant must be removed and reapplied.

c. Install the oil pan with the 17 bolts and 2 nuts, and tighten the bolts and nuts uniformly in several steps. Tighten the bolts to 80 inch lbs. (9 Nm) and the nuts to 7 ft. lbs. (10 Nm).

48. Install crankshaft pulley.
49. Install No. 1idler pulley.

a. Install the idler pulley with the bolt. Tighten to 29 ft. lbs. (39 Nm).

➡DOUBLE is marked on the No. 1 idler pulley to distinguish it from the No. 2 idler pulley.

50. Install No. 2idler pulley.

a. Install the 2 No. 2 idler pulleys with the 2 bolts. Tighten to 40 ft. lbs. (54 Nm).

51. Install power steering pump assembly and set it aside.

➡Do not disconnect the power steering hoses.

a. Install the power steering pump with the 2 bolts. Tighten to 32 ft. lbs. (43 Nm).

⁂ WARNING

Do not hit the pulley with other parts when installing the power steering pump.

b. Connect the power steering pressure switch connector.

52. Install oil level gauge guide.

a. Install a new O-ring onto the oil level gauge guide.

b. Apply a light coat of engine oil to the O-ring.

c. Push the oil level gauge guide end into the guide hole in the oil pan.

d. Install the oil level gauge guide with the bolt. Tighten to 80 inch lbs. (9 Nm).

e. Install the oil level gauge guide.

53. Install V-ribbed belt tensioner assembly.
54. Install air conditioner compressor assembly.
55. Install alternator assembly.

56. Install fan.
57. On 4WD vehicles, install front differential carrier assembly.
58. Install power steering link assembly.
59. Install battery.
60. Add engine coolant.
61. Add engine oil.
62. Add power steering fluid and bleed the system, as necessary.
63. On 4WD vehicles, inspect and adjust the differential oil.
64. Check for engine coolant leakage.
65. Check for engine oil leakage.
66. Check for power steering fluid leakage.
67. Check for differential oil leakage.
68. Inspect and adjust front wheel alignment.

TIMING CHAIN AND SPROCKETS

REMOVAL & INSTALLATION

See Figures 89 through 101.

1. Before servicing the vehicle, refer to the precautions.
2. Remove battery.
3. Drain engine coolant.
4. Drain engine oil.
5. Remove power steering rack assembly
6. On 4WD vehicles, remove front differential carrier assembly.
7. Remove fan.
8. Remove alternator assembly.
9. Remove the air conditioning compressor and set it aside.

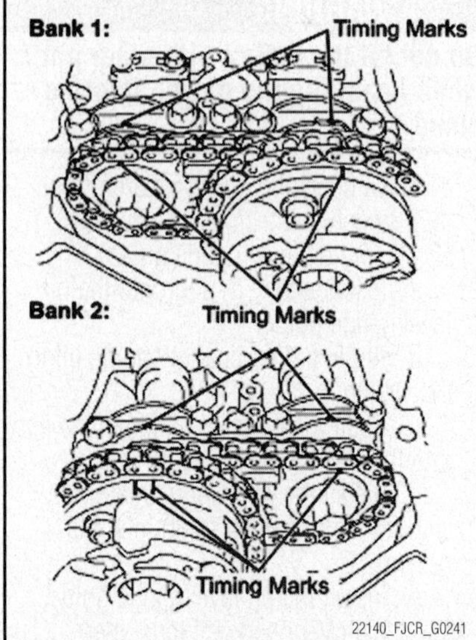

Fig. 89 Check that the timing marks of the camshaft timing gears are aligned with the timing marks of the bearing caps as shown

➡Do not disconnect the air conditioning lines.

10. Remove the 5 bolts, then remove the V-ribbed belt tensioner.
11. Remove the oil level gauge.
12. Remove the power steering pump assembly and set it aside.

➡Do not disconnect the power steering lines.

13. Remove idler pulleys.
14. Remove crankshaft pulley.
15. Remove No. 2 oil pan.
    a. Remove the 10 bolts and 2 nuts.
    b. Insert the blade of SST: 09032-00100 between the oil pan and No. 2 oil pan, cut off applied sealer and remove the No. 2 oil pan.

✱✱ WARNING

Be careful not to damage the contact surfaces of the oil pan and No. 2 oil pan. Be careful not to damage the No. 2 oil pan flange.

16. Remove the 2 nuts, then remove the oil strainer and gasket.
17. Remove oil pan.
    a. Remove the 4 housing bolts.
    b. Remove the flywheel housing under cover.
    c. Remove the 17 bolts and 2 nuts.
    d. Using a screwdriver, remove the oil pan by prying between the oil pan and cylinder block.

✱✱ WARNING

Be careful not to damage the contact surfaces of the cylinder block and oil pan.

    e. Remove the O-ring from the oil pump.
18. Remove air cleaner assembly.
19. Remove throttle body bracket.
20. Remove oil baffle plate.

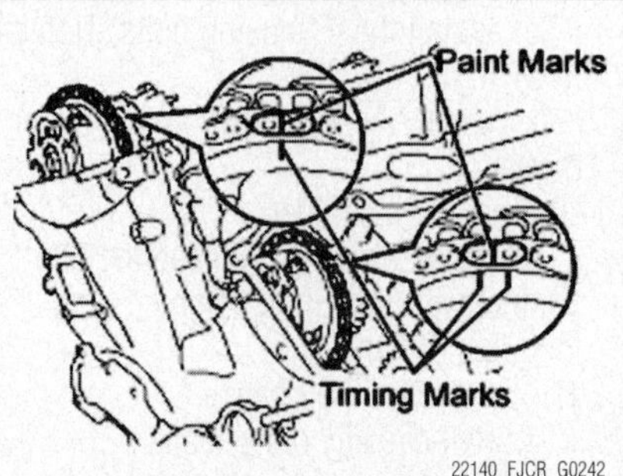

Fig. 90 Place paint marks on the No. 1 chain links corresponding to the timing marks of the camshaft timing gears

21. Remove surge tank bracket.
22. Remove intake air surge tank.
23. Remove ignition coil assembly.
24. Remove camshaft timing oil control valve assembly.
    a. Disconnect the 2 connectors.
    b. Remove the 2 bolts, then remove the 2 camshaft timing oil control valves.
25. Remove VVT sensor.
    a. For the right cylinder head sensor, disconnect the VVT sensor connector and remove the sensor bolt. Then remove the sensor.
    b. For the right cylinder head sensor, disconnect the No. 4 water by-pass hose and No. 5 water by-pass hose. Disconnect the VVT sensor connector and remove the sensor bolt. Then remove the sensor.
26. Remove the water inlet.
27. Remove the 10 bolts, 3 seal washers, 2 nuts, cylinder head cover and gasket.
28. Remove timing chain cover.
    a. Remove the 24 bolts and 2 nuts.
    b. Remove the timing chain cover by prying between the timing chain cover, cylinder head or cylinder block with a screwdriver.

✱✱ WARNING

Be careful not to damage the contact surfaces of the timing chain cover, cylinder block and cylinder head.

    c. Remove the O-ring from the left cylinder head.
29. Remove timing gear case or timing chain case oil seal.
30. Set No. 1 cylinder to TDC on the compression stroke.
    a. Turn the crankshaft pulley until its groove and the "0" timing mark of the timing chain cover are aligned.
    b. Check that the timing marks of the camshaft timing gears are aligned with the timing marks of the bearing caps as shown in the illustration.
31. If not, turn the crankshaft 1 complete revolution (360°) and align the timing marks above.
    a. Place paint marks on the No. 1 chain links corresponding to the timing marks of the camshaft timing gears.
32. Remove No. 1 chain tensioner assembly.

✱✱ WARNING

Never rotate the crankshaft with the chain tensioner removed. When rotating the camshaft with the timing chain removed, rotate the crankshaft counterclockwise 40°from the TDC first.

a. Remove the 4 bolts, then remove the timing chain cover plate and gasket.

b. While turning the stopper plate of the tensioner upward, push in the plunger of the chain tensioner as shown in the illustration.

c. While turning the stopper plate of the tensioner downward, insert a pin of 0.138 in. (3.5 mm) diameter into the holes in the stopper plate and tensioner to fix the stopper plate.

d. Remove the 2 bolts, then remove the chain tensioner.

33. Remove chain tensioner slipper.
34. Remove No. 1 idle gear.

a. Using a 10 mm hexagon wrench, remove the No. 2 idle gear shaft, No. 1 idle gear and No. 1 idle gear shaft.

35. Remove the 2 No. 2 chain vibration dampers.
36. Remove chain.

To install:

37. Install chain tensioner slipper.
38. Install No. 1 chain tensioner assembly.

a. While turning the stopper plate of the No. 1 chain tensioner clockwise, push in the plunger of the No. 1 chain tensioner as shown in the illustration.

b. While turning the stopper plate of the tensioner counterclockwise, insert a pin of 0.138 in. (3.5 mm) diameter into the holes in the stopper plate and No. 1 chain tensioner to fix the stopper plate.

c. Install the No. 1 chain tensioner with the 2 bolts. Tighten to 7 ft. lbs. (10 Nm).

39. Install the timing chain.

a. Set the No. 1 cylinder to TDC on the compression stroke.

b. Align the timing marks of the camshaft timing gears and bearing caps.

c. Using the crankshaft pulley set bolt, turn the crankshaft to align the crankshaft set key with the timing line of the cylinder block.

d. Align the yellow mark link with the timing mark of the crankshaft timing link.

e. Align the orange mark links with the timing marks of the camshaft timing gears, and install the chain.

40. Install the 2 No. 2 chain vibration dampers.
41. Install the idle gear.

a. Apply a light coat of engine oil to rotating surface of the No. 1 idle gear shaft.

b. Temporarily install the No. 1 idle gear shaft together with the No. 2 idle gear shaft with the knock pin of the No. 1 idle gear shaft and the knock pin groove of the cylinder block are aligned.

➡**Orient the idle gear shafts correctly.**

c. Using a 10 mm hexagon wrench, tighten the No. 2 idle gear shaft to 44 ft. lbs. (60 Nm).

d. Remove the bar from the chain tensioner.

42. Install the timing gear case oil seal.
43. Install timing chain cover.

a. Remove any old sealant.

⁂ WARNING

Do not drop any oil on the contact surfaces of the timing chain cover, cylinder head and cylinder block.

b. Install a new O-ring onto the left cylinder head as shown in the illustration.

c. Apply a continuous bead of Three Bond 1207B or the equivalent sealant 0.12–0.16 in. (3–4 mm) diameter to the 4 locations shown in the illustration.

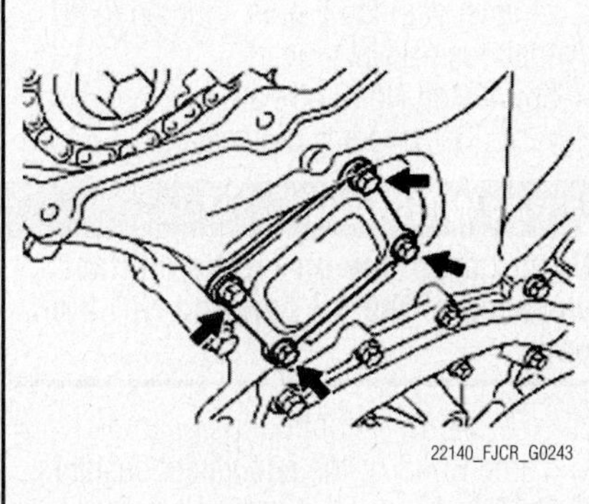

Fig. 91 Timing chain cover plate and gasket

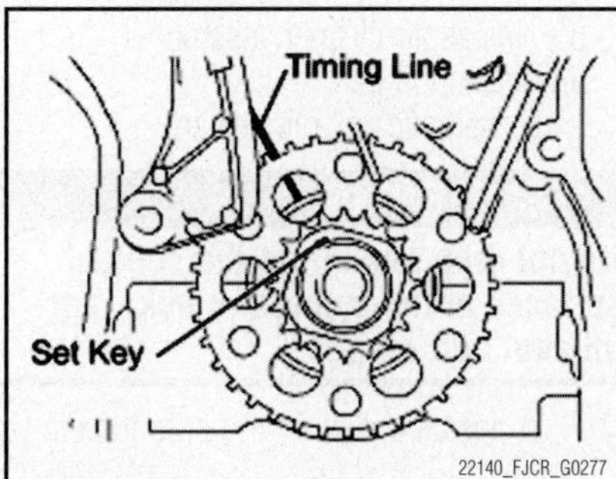

Fig. 93 Using the crankshaft pulley set bolt, turn the crankshaft to align the crankshaft set key with the timing line of the cylinder block

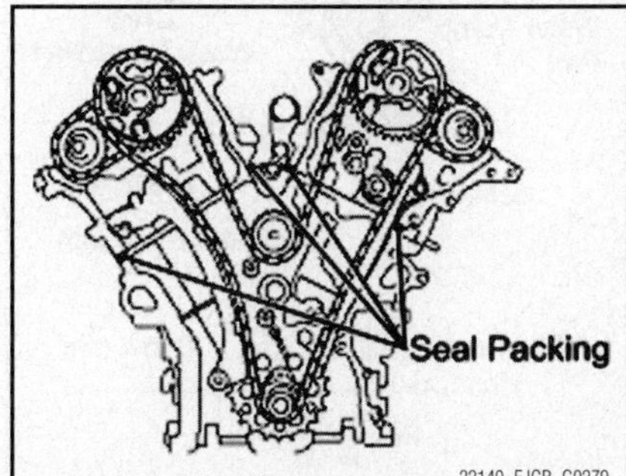

Fig. 95 Apply a continuous bead of Three Bond 1207B or the equivalent sealant 0.12–0.16 in. (3–4 mm) diameter to the 4 locations shown

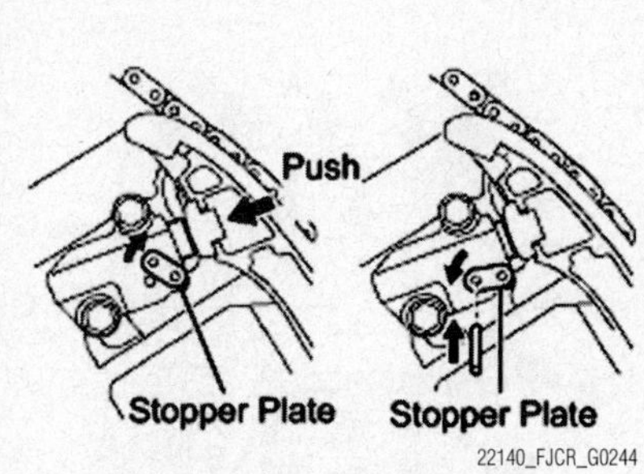

Fig. 92 While turning the stopper plate of the tensioner upward, push in the plunger of the chain tensioner as shown

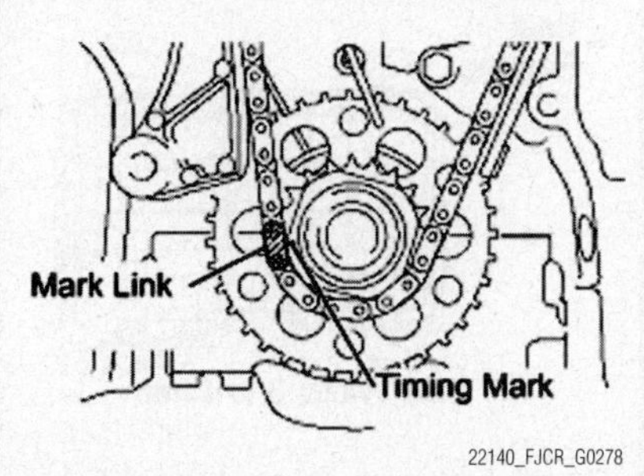

Fig. 94 Align the yellow mark link with the timing mark of the crankshaft timing link

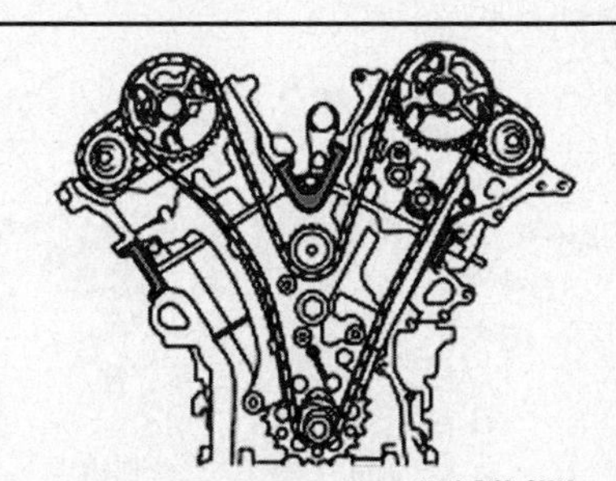

Fig. 96 Keep the seal surface between the cylinder block and the cylinder head shown in the illustration free of oil before installing the chain cover

d. Keep the seal surface between the cylinder block and the cylinder head shown in the illustration free of oil before installing the chain cover.

e. Apply a continuous bead of Three Bond or the equivalent sealant 0.12–0.16 in. (3–4 mm) diameter to the timing chain cover as shown in the illustration. Use Three Bond 1282B for the water pump part and Three Bond 1207B for the remainder.

:: WARNING

Install the timing chain cover within 3 minutes of applying the sealant. The timing chain cover bolts and nuts must be tightened within 15 minutes of the installation. Otherwise the sealant must be removed and reapplied. Do not apply sealant to portion A shown in the illustration.

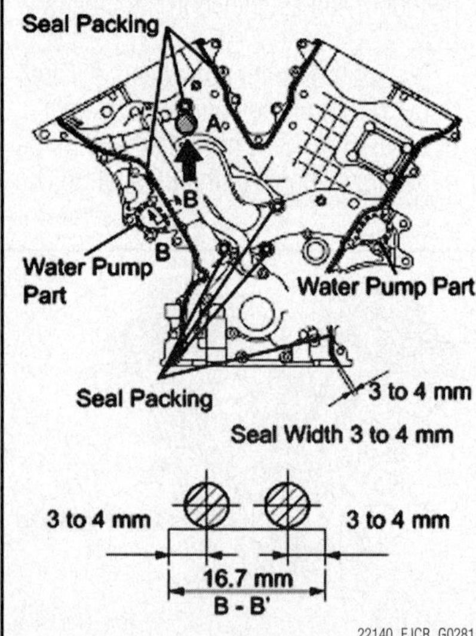

Fig. 97 Apply a continuous bead of Three Bond or the equivalent sealant 0.12–0.16 in. (3–4 mm) diameter to the timing chain cover as shown

f. Align the key way of the oil pump drive rotor with the rectangular portion of the crankshaft timing gear, and slide the timing chain cover into place.

g. Install the timing chain cover with the 24 bolts and 2 nuts. Tighten the bolts and nuts to 17 ft. lbs. (23 Nm) uniformly in several steps.

➡Pay attention not to wrap the chain and slipper over the timing chain cover seal line.

:: WARNING

Place bolts in their proper positions. Bolt A length is 0.98 in. (25 mm) and bolt B length is 2.17 in. (55 mm).

44. Install cylinder head covers.
45. Install water inlet.
46. Install VVT sensor. Tighten to 71 inch lbs. (8 Nm).
47. Install camshaft timing oil control valve assembly. Tighten to 80 inch lbs. (9 Nm).
48. Install ignition coil assembly.
49. Install intake air surge tank.
50. Install surge tank brackets.
51. Install oil baffle plate.
52. Install throttle body bracket.
53. Install air cleaner assembly.
54. Install oil pan.
    a. Remove any old sealant.

:: WARNING

Do not drop any oil on the contact surfaces of the cylinder block, rear oil seal and oil pan.

b. Install a new O-ring onto the oil pump.

c. Apply a continuous bead of Three Bond or the equivalent sealant 0.12–0.16 in. (3–4 mm) diameter to the oil pan as shown in the illustration.

:: WARNING

Install the oil pan within 3 minutes of applying the sealant. Tighten the oil pan bolts and nuts within 15 minutes of installing the oil pan. Otherwise, the sealant must be removed and reapplied.

d. Install the oil pan with the 17 bolts and 2 nuts, and tighten the bolts and nuts uniformly in several steps. Tighten the 0.39 in. (10 mm) head oil pan bolt to 7 ft. lbs. (10 Nm), the 0.47 in. (12 mm) head oil pan bolt to 16 ft. lbs. (21 Nm) and the nuts to 16 ft. lbs. (21 Nm).

:: WARNING

Install bolts in their proper locations. Bolt lengths are as follows: A is 0.98 in. (25 mm) B 1.77 in. (45 mm) and C 0.55 in. (14 mm).

e. Install the 4 housing bolts. Tighten to 27 ft. lbs. (37 Nm).

f. Install the flywheel housing under cover.

55. Install oil strainer.
    a. Install a new gasket and the oil strainer with the 2 nuts. Tighten to 80 inch lbs. (9 Nm).
56. Install No. 2 oil pan.
    a. Remove any old sealant.

:: WARNING

Do not drop any oil on the contact surfaces of the oil pan and No. 2 oil pan.

b. Apply a continuous bead of Three Bond or the equivalent sealant 0.12–0.16 in. (3–4 mm) diameter to the oil pan as shown in the illustration.

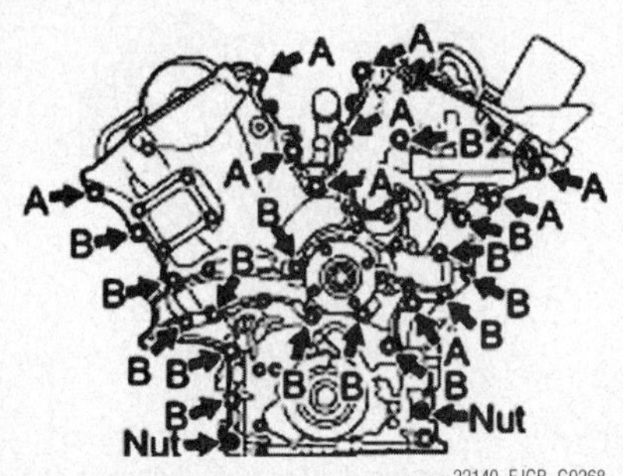

Fig. 98 Timing chain cover fastener locations

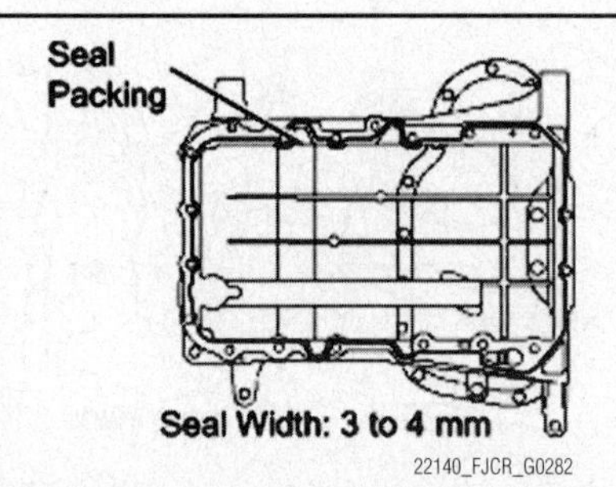

Fig. 99 Apply a continuous bead of Three Bond or the equivalent sealant 0.12–0.16 in. (3–4 mm) diameter to the oil pan as shown

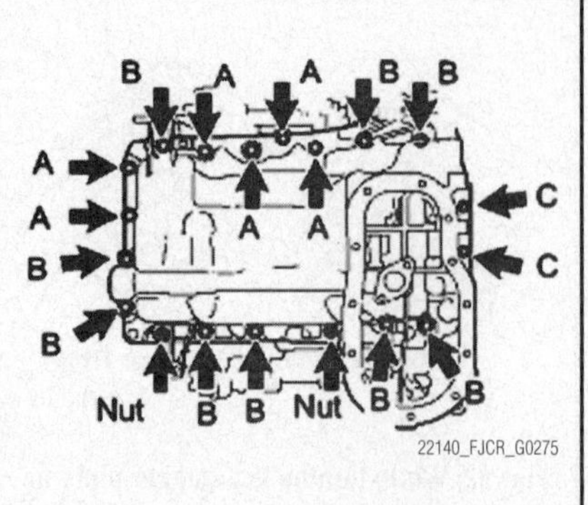

Fig. 100 Oil pan fastener locations

WARNING

Install the oil pan within 3 minutes of applying the sealant. Tighten the oil pan bolts and nuts within 15 minutes of installing the oil pan. Otherwise, the sealant must be removed and reapplied.

c. Install the oil pan with the 17 bolts and 2 nuts, and tighten the bolts and nuts uniformly in several steps. Tighten the bolts to 80 inch lbs. (9 Nm) and the nuts to 7 ft. lbs. (10 Nm).

57. Install crankshaft pulley.

58. Install No. 1idler pulley.

a. Install the idler pulley with the bolt. Tighten to 29 ft. lbs. (39 Nm).

➡DOUBLE is marked on the No. 1 idler pulley to distinguish it from the No. 2 idler pulley.

59. Install No. 2idler pulley.

a. Install the 2 No. 2 idler pulleys with the 2 bolts. Tighten to 40 ft. lbs. (54 Nm).

60. Install power steering pump assembly and set it aside.

➡Do not disconnect the power steering hoses.

a. Install the power steering pump with the 2 bolts. Tighten to 32 ft. lbs. (43 Nm).

WARNING

Do not hit the pulley with other parts when installing the power steering pump.

b. Connect the power steering pressure switch connector.

61. Install oil level gauge guide.

a. Install a new O-ring onto the oil level gauge guide.

b. Apply a light coat of engine oil to the O-ring.

c. Push the oil level gauge guide end into the guide hole in the oil pan.

d. Install the oil level gauge guide with the bolt. Tighten to 80 inch lbs. (9 Nm).

e. Install the oil level gauge guide.

62. Install V-ribbed belt tensioner assembly.

63. Install air conditioner compressor assembly.

64. Install alternator assembly.

65. Install fan.

66. On 4WD vehicles, install front differential carrier assembly.

67. Install power steering link assembly.

68. Install battery.

69. Add engine coolant.

70. Add engine oil.

71. Add power steering fluid and bleed the system, as necessary.

72. On 4WD vehicles, inspect and adjust the differential oil.

73. Check for engine coolant leakage.

74. Check for engine oil leakage.

75. Check for power steering fluid leakage.

76. Check for differential oil leakage.

77. Inspect and adjust front wheel alignment.

VALVE COVERS

REMOVAL & INSTALLATION

See Figures 102 through 104.

1. Before servicing the vehicle, refer to the precautions.
2. Disconnect the negative battery cable.
3. Drain engine coolant.
4. Remove V-bank cover.
5. Remove air cleaner assembly.
6. Remove throttle body bracket.
7. Remove oil baffle plate.
8. Remove surge tank bracket.
9. Remove intake air surge tank.

a. Disconnect the 2 water by-pass hoses.

b. Disconnect the fuel vapor feed hose.

c. Disconnect the ventilation hose.

d. Disconnect the 2 VSV connectors.

e. Disconnect the throttle body with motor connector.

f. Separate the 3 wire harness clamps and the hose clamp.

g. Remove the 2 nuts.

h. Using a socket hexagon wrench 8, remove the 4 bolts, intake air surge tank and the gasket.

10. Remove ignition coil assembly.
11. Remove the 10 bolts, 3 seal washers, 2 nuts, cylinder head cover and gasket.

To install:.

12. Install cylinder head cover.

a. Remove any old sealant.

WARNING

Do not drop any oil on the contact surfaces of the cylinder head, timing chain cover and cylinder head cover.

b. Apply a continuous bead of Three Bond 1207B or the equivalent sealant 0.08–0.12 in. (2–3 mm) diameter to the cylinder head and timing chain cover as shown in the illustration.

WARNING

Install the cylinder head cover within 3 minutes of applying the sealant. Tighten the cylinder head cover bolts and nuts within 15 minutes of installing the cylinder head cover. Otherwise, the sealant must be removed and reapplied.

c. Install the seal washers onto the bolts.

Fig. 101 Apply a continuous bead of Three Bond or the equivalent sealant 0.12–0.16 in. (3–4 mm) diameter to the oil pan

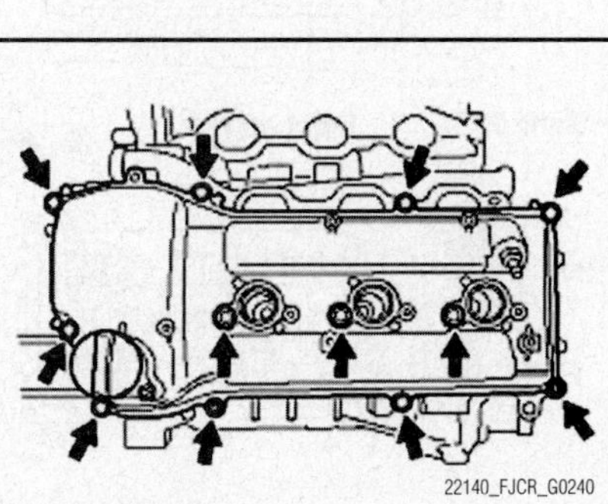

Fig. 102 Cylinder head cover fastener locations—left side shown, right side similar

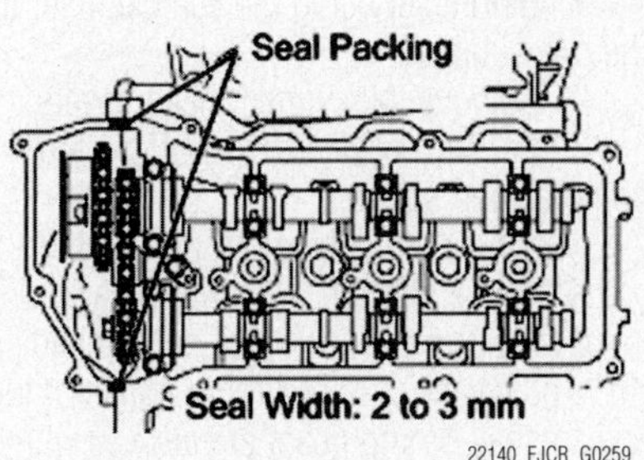

Fig. 103 Apply a continuous bead of Three Bond 1207B or the equivalent sealant 0.08–0.12 in. (2–3 mm) diameter to the cylinder head and timing chain cover as shown

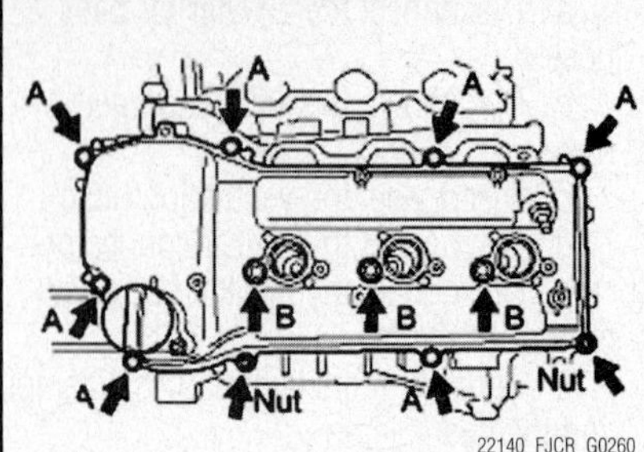

Fig. 104 Tighten the cylinder head cover bolts and nuts uniformly in several steps. Tighten bolts A to 7 ft. lbs. (10 Nm), bolts B to 80 inch lbs. (9 Nm) and the nuts to 80 inch lbs. (9 Nm)

d. Install the cylinder head cover with the 10 bolts and 2 nuts. Tighten the bolts and nuts uniformly in several steps. Tighten bolts **A** to 7 ft. lbs. (10 Nm), bolts **B** to 80 inch lbs. (9 Nm) and the nuts to 80 inch lbs. (9 Nm).

➡Bolt A are 0.98 in. (25 mm) and bolt B 2.36 in. (60 mm).

13. Install ignition coil assembly.
14. Install intake air surge tank.
15. Install No. 2 surge tank bracket.
16. Install No. 1 surge tank bracket.
17. Install oil baffle plate.
18. Install throttle body bracket.
19. Install air cleaner assembly.
20. Connect the negative battery cable.
21. Add engine coolant.
22. Check for engine coolant leakage.
23. Check for engine oil leakage.
24. Inspect ignition timing.
25. Install V-bank cover.

VALVE LASH

ADJUSTMENT

See Figures 105 through 107.

1. Before servicing the vehicle, refer to the precautions.
2. Remove the cylinder head covers.
3. Set No. 1 cylinder to TDC on the compression stroke.

a. Turn the crankshaft pulley until its groove and the "0" timing mark of the timing chain cover are aligned.

b. Check that the timing marks of the camshaft timing gears are aligned with the timing marks of the bearing caps as shown in the illustration.

4. If not, turn the crankshaft 1 complete revolution (360°) and align the timing marks above.
5. Inspect valve clearance.

a. Check the valves indicated in the illustration.

b. Using a feeler gauge, measure the clearance between the valve lifter and camshaft.

c. Record any out-of-specification valve clearance measurements. They will be used later to determine the required replacement valve lifter.

d. Turn the crankshaft 240° clockwise (first time), and check the valves indicated in the illustration.

e. Using a feeler gauge, measure the clearance between the valve lifter and camshaft.

f. Record any out-of-specification valve clearance measurements. They will be used later to determine the required replacement valve lifter.

g. Turn the crankshaft 240° clockwise, and check the valves indicated in the illustration.

h. Using a feeler gauge, measure the clearance between the valve lifter and camshaft.

i. Record any out-of-specification valve clearance measurements. They will be used later to determine the required replacement valve lifter.

6. Adjust valve clearance.

a. Set the No. 1 cylinder to TDC/compression.

b. Turn the crankshaft pulley until its groove and the "0" timing mark of the timing chain cover are aligned.

c. Check that the timing marks of the camshaft timing gears are aligned with the timing marks of the bearing caps as shown in the illustration.

d. If not, turn the crankshaft 1 complete revolution (360°) and align the timing marks as above.

e. Place paint marks on the No. 1 chain links corresponding to the timing marks of the camshaft timing gears.

7. Remove the No. 1 chain tensioner assembly.
8. Remove the No. 2 camshaft.
9. Remove the No. 2 chain tensioner assembly.
10. Remove the camshaft.
11. Remove the No. 4 camshaft.
12. Remove the No. 3 chain tensioner assembly.
13. Remove the No. 3 camshaft.
14. Remove the valve lifters.

To install:

15. Determine the replacement valve lifter size according to the following formulas:

a. Using a micrometer, measure the thickness of the removed lifter.

b. Calculate the thickness of a new lifter so that the valve clearance comes within the specified value.

c. For the intake side: Thickness of new lifter = Thickness of removed lifter + Measured valve clearance - 0.008 in. (0.20 mm).

d. For the exhaust side: Thickness of new lifter = Thickness of removed lifter + Measured valve clearance - 0.013 in. (0.34 mm).

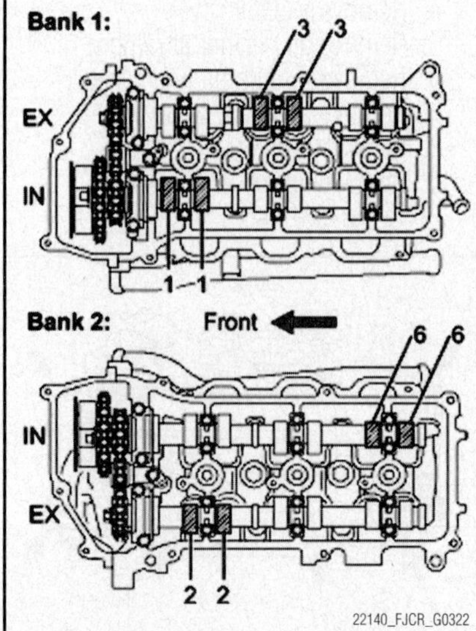

Fig. 105 With the No. 1 cylinder to TDC on the compression stroke check the valves indicated

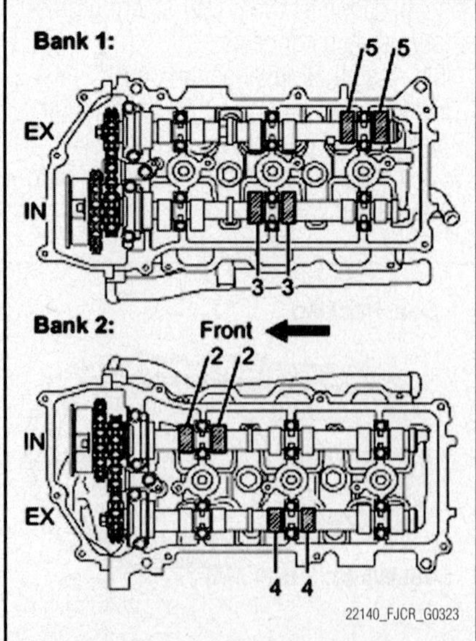

Fig. 106 After rotating the crankshaft 240° clockwise (first time) check the valves indicated

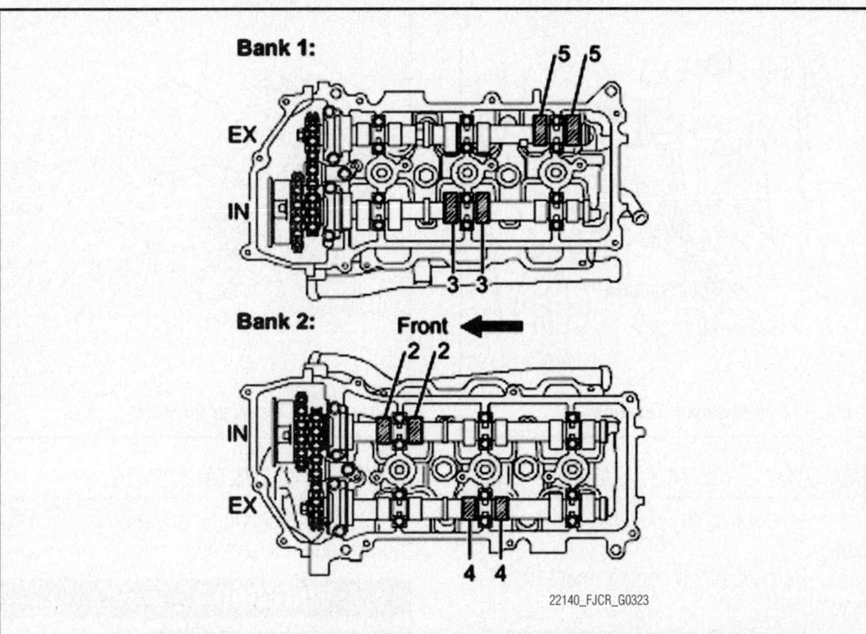

Fig. 107 After rotating the crankshaft 240° clockwise (second time) check the valves indicated

e. Select a new lifter with a thickness as close as possible to the calculated value.

➡Lifters are available in 35 sizes in increments of 0.0008 in. (0.020 mm), from 0.1992 in.–0.2260 in. (5.740–5.060mm).

16. Install the No. 3 camshaft.
17. Install the No. 3 chain tensioner assembly.
18. Install the No. 4 camshaft.
19. Install the camshaft.
20. Install the No. 2 chain tensioner assembly.
21. Install the No. 2 camshaft.
22. Install the No. 1 chain tensioner assembly.
23. Check that the timing marks of the camshaft timing gears are aligned with the timing marks of the bearing cap as shown in the illustration.
24. Install the cylinder head covers.

ENGINE PERFORMANCE & EMISSION CONTROL

ACCELERATOR PEDAL POSITION (APP) SENSOR

LOCATION

See Figure 108.

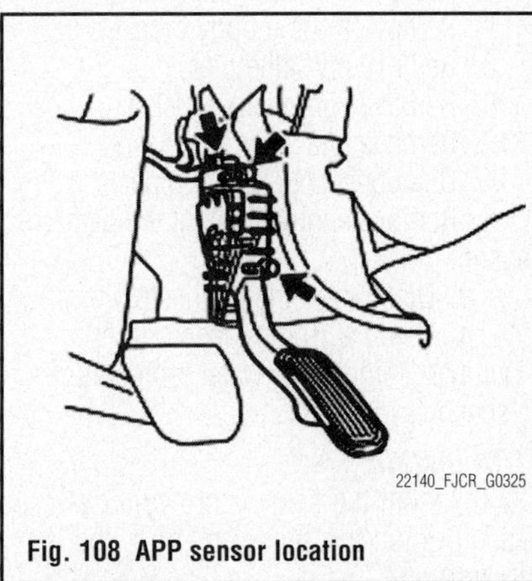

Fig. 108 APP sensor location

REMOVAL & INSTALLATION

1. Before servicing the vehicle, refer to the precautions section.
2. Disconnect the negative battery cable.
3. Disconnect the sensor connector.
4. Remove the sensor attaching bolts, then remove the accelerator pedal.

To install:

5. Install the accelerator pedal. Tighten the pedal attaching bolts to 44 inch lbs. (5 Nm).
6. Connect the sensor connector.
7. Connect the negative battery cable.

CAMSHAFT POSITION (CMP) SENSOR

LOCATION

See Figures 109 and 110.

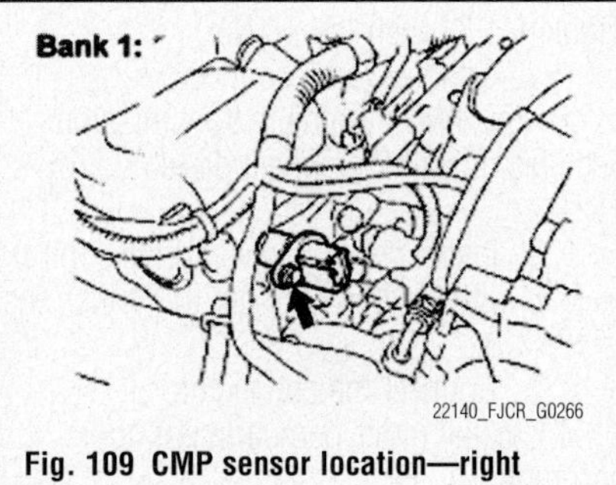

Fig. 109 CMP sensor location—right cylinder head

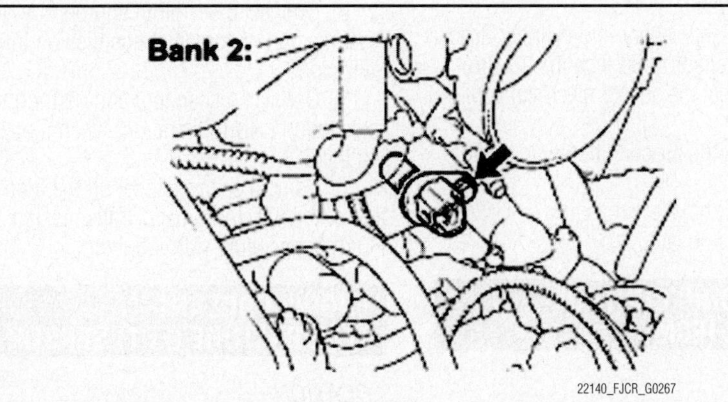

Fig. 110 CMP sensor location—left cylinder head

REMOVAL & INSTALLATION

1. Before servicing the vehicle, refer to the precautions section.
2. Disconnect the negative battery cable.
3. Drain engine coolant.
4. Remove v-bank cover.
5. Remove air cleaner assembly.
6. Remove CMP sensor.
   a. For the right cylinder head sensor, disconnect the CMP sensor connector and remove the sensor bolt. Then remove the sensor.
   b. For the left cylinder head sensor, disconnect the No. 4 water by-pass hose and No. 5 water by-pass hose. Disconnect the CMP sensor connector and remove the sensor bolt. Then remove the sensor.

To install:

7. Install CMP sensor. Tighten to 71 inch lbs. (8 Nm).
8. Install air cleaner assembly.
9. Install V-bank cover.
10. Fill the engine with coolant.
11. Connect the negative battery cable.

CRANKSHAFT POSITION (CKP) SENSOR

LOCATION

See Figure 111.

Fig. 111 CKP sensor location

REMOVAL & INSTALLATION

1. Before servicing the vehicle, refer to the precautions section.
2. Remove the fan.
3. Remove the alternator.
4. Remove the air conditioning compressor and set it aside.

➡Do not disconnect the air conditioning hoses from the compressor.

5. Disconnect the sensor electrical connector.
6. Remove the attaching bolt and sensor.

To install:

7. Install the sensor into cylinder head. Tighten the sensor to 80 inch lbs. (9 Nm).
8. Connect the sensor electrical connector.
9. Install the air conditioning compressor.
10. Install the alternator.
11. Install the fan.

ELECTRONIC CONTROL MODULE (ECM)

LOCATION

See Figure 112.

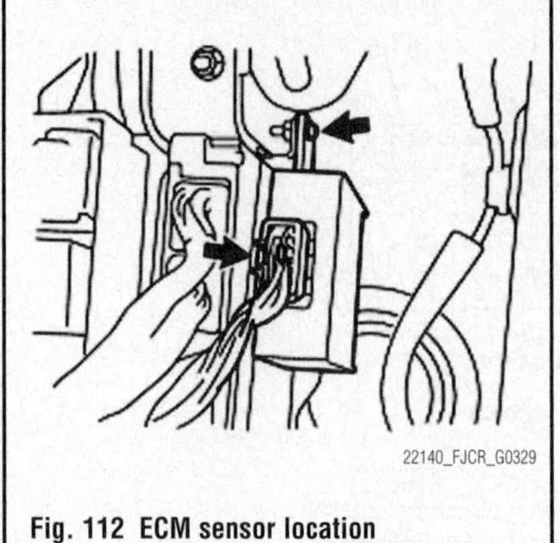

Fig. 112 ECM sensor location

REMOVAL & INSTALLATION

1. Disconnect the negative battery cable.
2. Remove glove compartment door assembly.
3. On 4WD automatic transmission vehicles, remove four wheel drive control ECU.
   a. Disconnect the connector.
   b. Remove the screw and four wheel drive control ECU.
4. Remove ECM.
   a. Disconnect the 5 connectors.
   b. Remove the 3 nuts, then remove the ECM.

To install:

5. Install the ECM with the 3 nuts. Tighten to 71 inch lbs. (8 Nm).
   a. Connect the 5 connectors.
6. On 4WD automatic transmission vehicles, install four wheel drive control ECU.
   a. Install the four wheel drive control ECU with the bolt. Tighten to 71 inch lbs. (8 Nm).
   b. Connect the connector.
7. Install glove compartment door assembly.
8. Connect the negative battery cable.
9. Connect a scan tool and perform the reset memory automatic transmission initialization.
10. Connect a scan tool and perform the registration (VIN registration) when replacing the ECM.
11. Using a scan tool, set up the function of the ATF (Automatic Transmission Fluid) temperature warning lamp.

ENGINE COOLANT TEMPERATURE (ECT) SENSOR

LOCATION

See Figure 113.

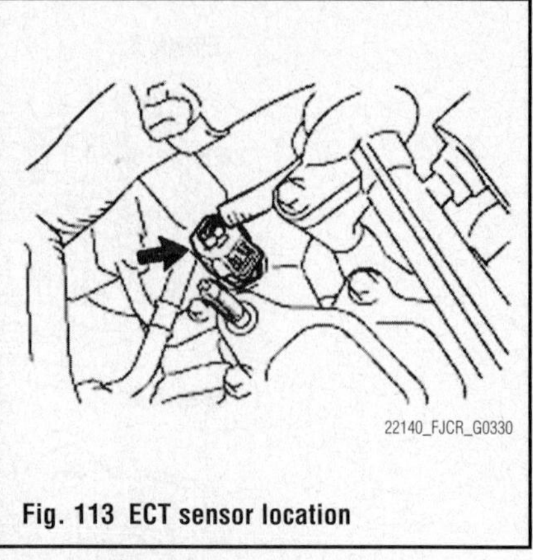

Fig. 113 ECT sensor location

REMOVAL & INSTALLATION

1. Before servicing the vehicle, refer to the precautions section.

✲✲ WARNING

Use care when handling the engine coolant temperature (ECT) sensor. Damage to the ECT sensor will affect proper operation of the fuel injection system.

2. Discharge fuel system pressure.
3. Drain engine coolant.
4. Disconnect the negative battery cable.
5. Remove V-bank cover.
6. Remove air cleaner assembly.
7. Remove throttle body bracket.
8. Remove oil baffle plate.
9. Remove surge tank bracket.
10. Remove intake air surge tank.
11. Disconnect No. 2 fuel pipe.
12. Remove engine coolant temperature sensor.
    a. Disconnect the connector.
    b. Using a 19 mm deep socket wrench, remove the water temperature sensor and gasket.

To install:

13. Install the sensor into cylinder head. Tighten the sensor to 14 ft. lbs. (20 Nm).
    a. Connect the connector.
14. Install engine coolant temperature sensor.
15. Connect No. 2 fuel pipe.
16. Install intake air surge tank.
17. Install surge tank bracket.
18. Install oil baffle plate.
19. Install throttle body bracket.
20. Install air cleaner assembly.
21. Install V-bank cover.
22. Connect the negative battery cable.
23. Fill the engine with coolant.

FUEL LEVEL SENDING UNIT

LOCATION

See Figure 114.

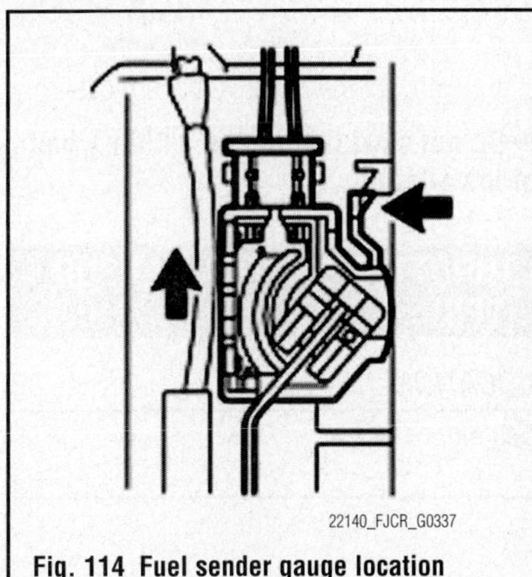

Fig. 114 Fuel sender gauge location

REMOVAL & INSTALLATION

See Figure 115.

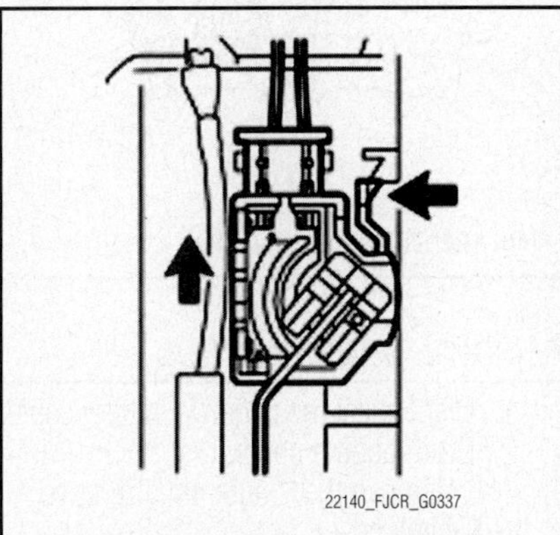

Fig. 115 Unlock the fuel sender gauge, and slide and remove it

1. Before servicing the vehicle, refer to the precautions.
2. Discharge fuel system pressure.
3. Disconnect the negative battery cable.
4. Remove the fuel tank.
5. Disconnect the fuel sender gauge connector.
6. Unlock the fuel sender gauge, and slide and remove it.

To install:

7. Install the fuel sender gauge onto the fuel suction with pump and gauge tube.
8. Connect the fuel sender gauge connector.
9. Install the fuel tank.
10. Connect the negative battery cable.

HEATED OXYGEN (HO2S) SENSOR

LOCATION

See Figure 116.

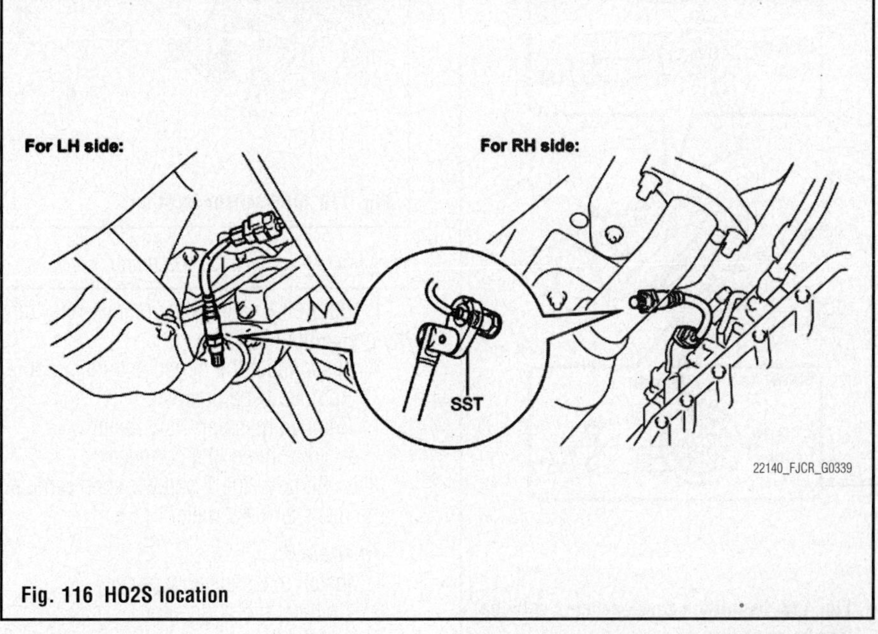

Fig. 116 HO2S location

REMOVAL & INSTALLATION

1. Before servicing the vehicle, refer to the precautions section.
2. Raise and safely support the vehicle, if necessary.
3. Disconnect the sensor electrical connector.
4. Remove the sensor.

To install:

5. Coat the threads of the heated oxygen sensor with the anti-seize compound, if necessary. Some new sensors will come pre-coated.
6. Install the heated oxygen sensor tighten to 35 inch. lbs. (4 Nm).
7. Connect the sensor electrical connector.
8. Lower the vehicle, if raised.
9. Connect the negative battery cable.

INTAKE AIR TEMPERATURE (IAT) SENSOR

The IAT sensor is an integral component of the MAF sensor. Refer to Components & Systems, Mass Air Flow (MAF) Sensor.

KNOCK SENSOR (KS)

LOCATION

See Figure 117.

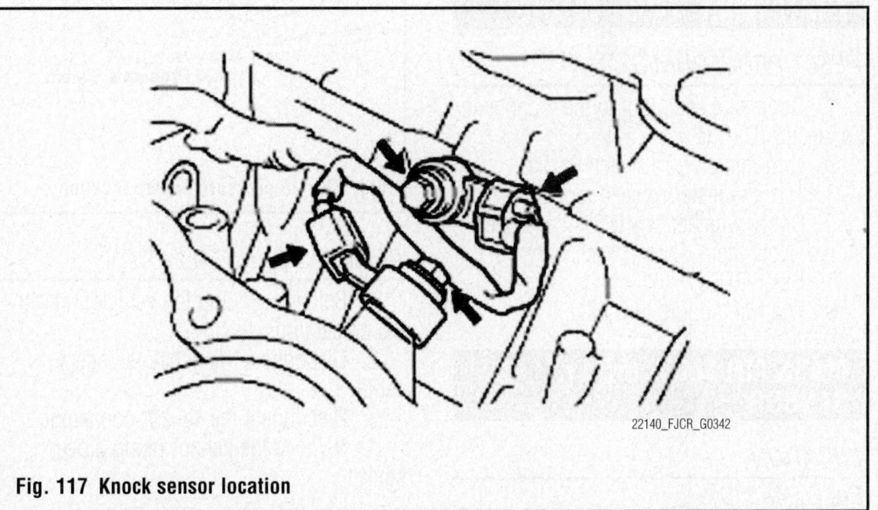

Fig. 117 Knock sensor location

REMOVAL & INSTALLATION

See Figure 118.

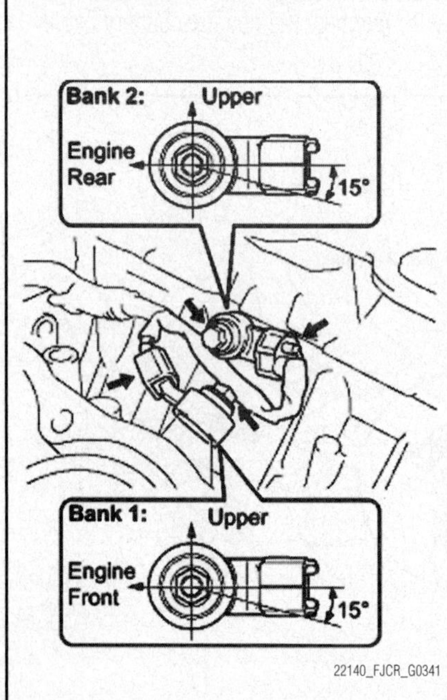

Fig. 118 Install the knock sensors with the 2 bolts as shown

1. Before servicing the vehicle, refer to the precautions section.
2. Remove the cylinder heads.
3. Disconnect the knock sensor connector.
4. Remove the sensor.

To install:

5. Install the sensor and tighten to 15 ft. lbs. (20 Nm).
6. Connect the knock sensor connector.
7. Install the cylinder heads.

MALFUNCTION INDICATOR LIGHT (MIL)

RESET PROCEDURES

1. Connect a scan tool to the diagnostic connector (DLC3).
2. Turn the ignition switch to ON.
3. Turn the tester or scan tool ON
4. Check whether any DTCs have been stored. Note them down if necessary.
5. Clear DTCs.
6. The MIL should turn off.

MASS AIR FLOW (MAF) SENSOR

LOCATION

See Figure 119.

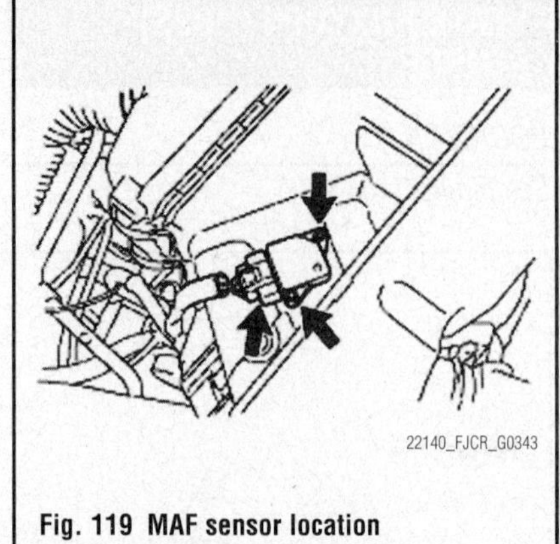

Fig. 119 MAF sensor location

REMOVAL & INSTALLATION

1. Before servicing the vehicle, refer to the precautions section.
2. Disconnect the negative battery cable.
3. Remove V-bank cover.
4. Remove mass air flow meter.
   a. Disconnect the connector.
   b. Remove the 2 screws, then remove the mass air flow meter.

To install:

5. Install mass air flow meter.
6. Connect the connector.
7. Install V-bank cover.
8. Connect the negative battery cable.

OIL PRESSURE SENSOR

LOCATION

See Figure 120.

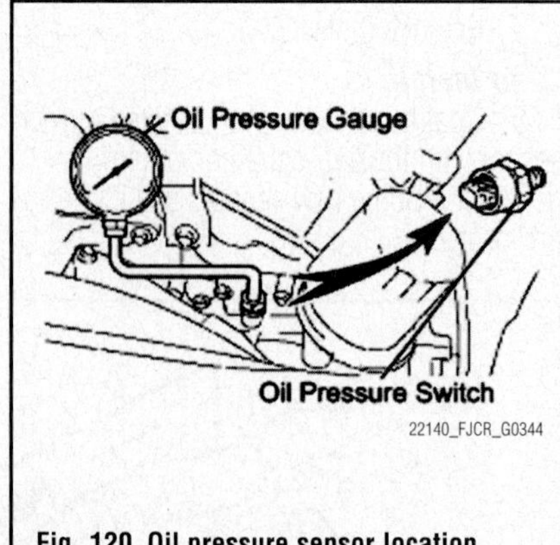

Fig. 120 Oil pressure sensor location

REMOVAL & INSTALLATION

1. Before servicing the vehicle, refer to the precautions section.
2. Disconnect the negative battery cable.
3. Disconnect the sensor connector.
4. Remove the sensor using a deep socket.

To install:

5. Apply a thread locking compound such as Loctite®242 to 2 or 3 threads of the sensor.
6. Install the sensor and tighten to 11 ft. lbs. (15 Nm).
7. Attach the electrical connector.
8. Connect the negative battery cable.

➡Do not start the engine within 1 hour of installation.

THROTTLE POSITION SENSOR (TPS)

LOCATION

See Figure 121.

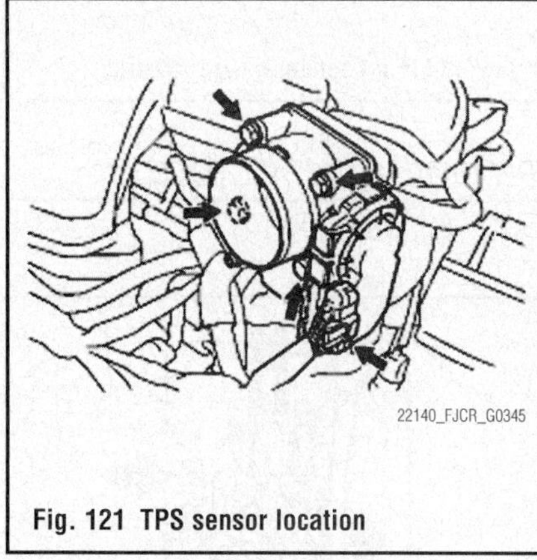

Fig. 121 TPS sensor location

REMOVAL & INSTALLATION

1. Disconnect the negative battery cable.
2. Drain engine coolant.
   a. Remove the 2 nuts, then remove the V-bank cover.
3. Remove air cleaner assembly.
   a. Disconnect the No. 2 ventilation hose.
   b. Disconnect the vacuum hose.
   c. Disconnect the mass air flow meter connector.
   d. Disengage the 2 wire harness clamps.
   e. Loosen the 2 hose clamps.
   f. Remove the 2 bolts.
   g. Disconnect the air cleaner hose and remove the air cleaner.
4. Remove throttle with motor body assembly.
   a. Disconnect the No. 5 water by-pass hose.
   b. Disconnect the No. 4 water by-pass hose.
   c. Disconnect the throttle motor connector.
   d. Remove the 4 bolts, then remove the throttle w/ motor body and gasket.

e. Remove the No. 5 water by-pass hose.

f. Remove the No. 4 water by-pass hose.

To install:

5. Install throttle with motor body assembly.

a. Connect the No. 4 water by-pass hose to the throttle body.

b. Connect the No. 5 water by-pass hose to the throttle body.

c. Install a new gasket and the throttle with motor body with the 4 bolts. Tighten to 9 ft. lbs. (11 Nm).

d. Connect the throttle motor connector.

e. Connect the No. 4 water by-pass hose.

f. Connect the No. 5 water by-pass hose.

6. Install air cleaner assembly.

a. Install the air cleaner onto the throttle body.

b. Connect the air cleaner hose to the air cleaner.

c. Install the air cleaner with the 2 bolts. Tighten to 71 inch lbs. (8 Nm).

d. Connect the mass air flow meter connector.

e. Engage the 2 wire harness clamps.

f. Connect the vacuum hose.

g. Connect the No. 2 ventilation hose.

7. Connect the negative battery cable.

8. Add engine coolant.

9. Check for engine coolant leak.

10. Install v-bank cover.

VARIABLE CAMSHAFT TIMING OIL CONTROL SOLENOID

LOCATION

See Figure 122.

Fig. 122 Camshaft timing oil control valve location

REMOVAL & INSTALLATION

1. Disconnect the negative battery cable.
2. Drain engine coolant.
3. Remove v-bank cover.
4. Remove air cleaner assembly.
5. Remove throttle body bracket.
6. Remove oil baffle plate.
7. Remove surge tank brackets.
8. Remove intake air surge tank.
9. Remove camshaft timing oil control valve assembly.

a. Disconnect the 2 connectors.

b. Remove the 2 bolts, then remove the 2 camshaft timing oil control valves.

To install:

10. Install camshaft timing oil control valve assembly. Tighten to 80 inch lbs. (9 Nm).

a. Connect the 2 connectors.

11. Install intake air surge tank.
12. Install surge tank brackets.
13. Install oil baffle plate.
14. Install throttle body bracket.
15. Install air cleaner assembly.
16. Install v-bank cover.
17. Fill the engine with coolant.
18. Connect the negative battery cable.

FUEL SYSTEM — GASOLINE FUEL INJECTION SYSTEM

FUEL SYSTEM SERVICE PRECAUTIONS

Safety is the most important factor when performing not only fuel system maintenance but any type of maintenance. Failure to conduct maintenance and repairs in a safe manner may result in serious personal injury or death. Maintenance and testing of the vehicle's fuel system components can be accomplished safely and effectively by adhering to the following rules and guidelines.

- To avoid the possibility of fire and personal injury, always disconnect the negative battery cable unless the repair or test procedure requires that battery voltage be applied.
- Always relieve the fuel system pressure prior to disconnecting any fuel system component(injector, fuel rail, pressure regulator, etc.), fitting or fuel line connection. Exercise extreme caution whenever relieving fuel system pressure to avoid exposing skin, face and eyes to fuel spray. Please be advised that fuel under pressure may penetrate the skin or any part of the body that it contacts.
- Always place a shop towel or cloth around the fitting or connection prior to loosening to absorb any excess fuel due to spillage. Ensure that all fuel spillage(should it occur) is quickly removed from engine surfaces. Ensure that all fuel soaked cloths or towels are deposited into a suitable waste container.
- Always keep a dry chemical(Class B) fire extinguisher near the work area.
- Do not allow fuel spray or fuel vapors to come into contact with a spark or open flame.
- Always use a back-up wrench when loosening and tightening fuel line connection fittings. This will prevent unnecessary stress and torsion to fuel line piping.
- Always replace worn fuel fitting O-rings with new. Do not substitute fuel hose or equivalent where fuel pipe is installed.

Before servicing the vehicle, make sure to also refer to the precautions in the beginning of this section as well.

RELIEVING FUEL SYSTEM PRESSURE

See Figure 123.

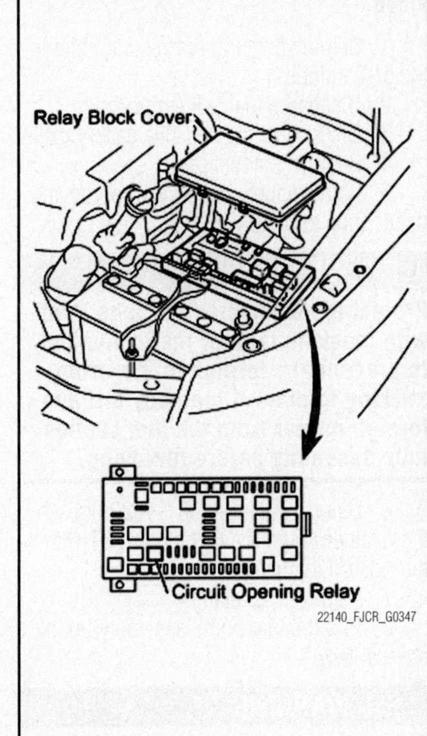

Fig. 123 Circuit opening relay location

⁂ WARNING

Discharge fuel system pressure procedures must be performed before disconnecting any part of the fuel system. As some pressure remains in the fuel line even after taking precautions to prevent gasoline spillage, use a shop rag or piece of cloth to prevent gasoline splashes when disconnecting the fuel line.

1. Disconnect the negative battery cable.
2. Remove the circuit opening relay.
3. Connect the negative battery cable.
4. Start the engine.
5. Turn the ignition switch to **ON** after the engine stops.

➡DTC P0171 (system too lean (Bank 1)) or DTC P0174 (system too learn (Bank 2)) may be set.

6. Crank the engine again and check that the engine stops.
7. Remove the fuel tank cap and discharge the pressure in the fuel tank completely.
8. Install the circuit opening relay.

FUEL FILTER

REMOVAL & INSTALLATION

➡The fuel filter is a integral part of the fuel pump unit and is not normally serviced.

1. Before servicing the vehicle, refer to the precautions.
2. Discharge fuel system pressure.
3. Disconnect the negative battery cable.
4. Remove the fuel tank.
5. Remove fuel suction with pump and gauge tube assembly.

⁂ WARNING

Protect the connector and tube joint with masking tape or the equivalent to prevent any foreign matter from sticking to them. Clean any dirt and foreign matter from the fuel suction tube assembly before removing.

6. Using a SST: 09808-14020, loosen the retainer. Align the tips of the SST with the ribs on the retainer.
7. Remove the retainer.
8. Pull the fuel pump assembly out of the fuel tank.

⁂ WARNING

Do not bend the arm of the sender gauge.

9. Remove the gasket from the fuel tank.
10. Remove fuel sender gauge assembly.
    a. Disconnect the connector.
    b. Disengage the claw and remove the sender gauge by sliding it upward.
11. Remove No. 1 fuel sub-tank.
    a. Disengage the 5 claws and remove the fuel pump tank.
    b. Separate the connector and disengage the clamp.
12. Remove fuel pump assembly.
    a. Disengage the clamp, then disconnect the connector.
    b. Disengage the 5 claws and separate the fuel pump from the fuel pump case.
    c. Disconnect the connector from the fuel pump.
13. Remove the fuel filter from the fuel pump.

To install:

14. Install the fuel filter onto the fuel pump.
15. Install fuel pump assembly.
    a. Connect the fuel filter to the fuel pump.
    b. Engage the 5 claw fittings and install the fuel pump onto the fuel pump case.
    c. Engage the clamp, then connect the connector.
16. Install No. 1 fuel sub-tank.
    a. Install the connector and engage the clamp.
    b. Engage the 5 claws and install the fuel pump tank.
17. Install fuel sender gauge assembly.
    a. Engage the claw and install the sender gauge by sliding it in the direction as shown in the illustration.
    b. Connect the connector.
18. Install the fuel tank.
19. Connect the negative battery cable.

FUEL PUMP

REMOVAL & INSTALLATION

See Figure 124.

1. Before servicing the vehicle, refer to the precautions.
2. Discharge fuel system pressure.
3. Disconnect the negative battery cable.
4. Remove the fuel tank.
5. Remove fuel suction with pump and gauge tube assembly.

⁂ WARNING

Protect the connector and tube joint with masking tape or the equivalent to prevent any foreign matter from sticking to them. Clean any dirt and foreign matter from the fuel suction tube assembly before removing.

6. Using a SST: 09808-14020, loosen the retainer. Align the tips of the SST with the ribs on the retainer.
7. Remove the retainer.
8. Pull the fuel pump assembly out of the fuel tank.

⁂ WARNING

Do not bend the arm of the sender gauge.

9. Remove the gasket from the fuel tank.
10. Remove fuel sender gauge assembly.
    a. Disconnect the connector.
    b. Disengage the claw and remove the sender gauge by sliding it upward.
11. Remove No. 1 fuel sub-tank.
    a. Disengage the 5 claws and remove the fuel pump tank.
    b. Separate the connector and disengage the clamp.
12. Remove fuel pump assembly.
    a. Disengage the clamp, then disconnect the connector.
    b. Disengage the 5 claws and separate the fuel pump from the fuel pump case.
    c. Disconnect the connector from the fuel pump.

To install:

13. Install fuel pump assembly.
    a. Connect the fuel filter to the fuel pump.
    b. Engage the 5 claw fittings and install the fuel pump onto the fuel pump case.
    c. Engage the clamp, then connect the connector.

Fig. 124 Disengage the 5 claws and separate the fuel pump from the fuel pump case

14. Install No. 1 fuel sub-tank.
    a. Install the connector and engage the clamp.
    b. Engage the 5 claws and install the fuel pump tank.
15. Install fuel sender gauge assembly.
    a. Engage the claw and install the sender gauge by sliding it in the direction as shown in the illustration.
    b. Connect the connector.
16. Install the fuel tank.
17. Connect the negative battery cable.

FUEL RAIL & INJECTORS

REMOVAL & INSTALLATION

See Figure 125.

1. Before servicing the vehicle, refer to the precautions.

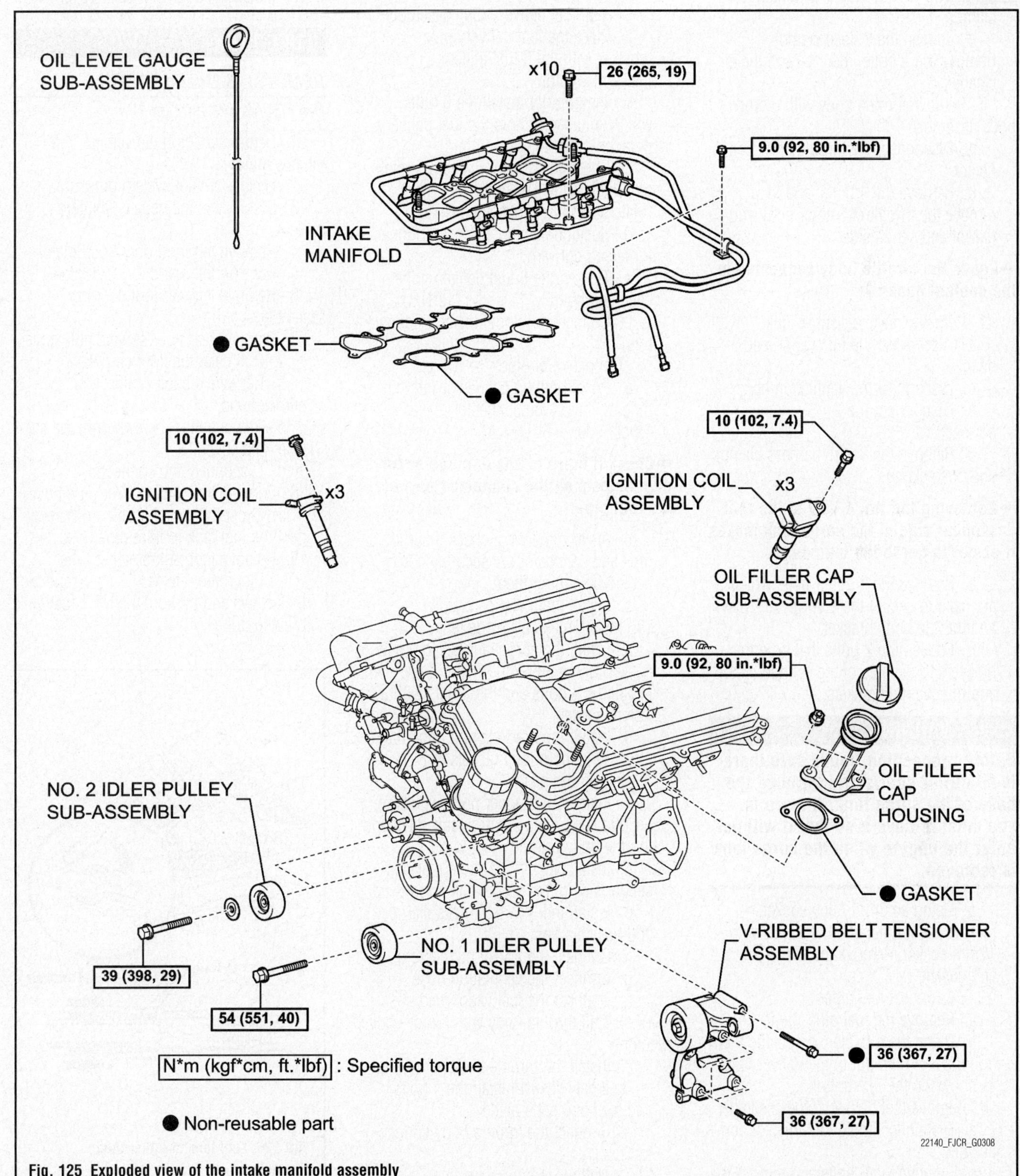

Fig. 125 Exploded view of the intake manifold assembly

2. Remove air cleaner assembly.
a. Remove the engine V-bank cover by removing the two acorn nuts.
b. Disconnect the ventilation hose No. 2.
c. Disconnect the vacuum hose.
d. Disconnect the mass air flow meter connector. Remove the 2 wire harness clamps.
e. Loosen the 2 hose clamps. Remove the 2 bolts, then remove the air cleaner.
3. Remove throttle body with motor body assembly.
a. Disconnect the throttle body connector.
b. Remove the 4 bolts, and then remove the throttle w/ motor body and gasket and set it aside

➡Leave the throttle body connected to the coolant hoses.

4. Remove intake air surge tank.
a. Disconnect the fuel vapor feed hose.
b. Disconnect the ventilation hose.
c. Disconnect the 2 VSV connectors.
d. Remove the 3 wire harness clamps and hose clamps.

➡Removing the No. 1 VSV at the rear passenger side of the surge tank makes it easier to get to the clamps.

e. Remove the 3 upper bolts which are used to secure the 2 surge tank stays and throttle body bracket.
f. Loosen the 2 bolts that hold the brackets to the side of the cylinder head and remove the brackets.

⁂ WARNING

Before proceeding, make sure there is no dirt or debris on or around the base of the surge tank. If there is, you must remove it so that it will not enter the engine when the surge tank is removed.

g. Using an 8mm Allen socket, remove the 4 bolts and the 2 nuts with a 12mm socket. Remove the surge tank and gasket.
5. Disconnect fuel pipes.
a. Remove the fuel pipe clamp.
b. Pinch the tube connector, and then pull the fuel pipe out of the delivery pipe as shown in the illustration.
6. Remove fuel delivery pipe assembly.
a. Disconnect the 6 fuel injector connectors.
b. Remove the 6 bolts and remove the fuel delivery pipe together with the 6 fuel injectors.

⁂ WARNING

Do not drop the injectors when removing the fuel delivery pipe.

To install:

7. Install fuel delivery pipe assembly.
a. Place the fuel delivery pipe together with the 6 fuel injectors on the intake manifold.
b. Temporarily install the 6 bolts, which are used to hold the fuel delivery pipe, onto the intake manifold.
c. Check that the fuel injectors rotate smoothly. If the fuel injectors do not rotate smoothly, replace the O-ring.
d. Position each fuel injector connector facing outward.
e. Tighten the 6 fuel delivery pipe attaching bolts to 11 ft. lbs (15 Nm).
f. Connect the 6 fuel injector connectors.
8. Connect fuel pipes.
a. Push the tube connector into the pipe until the tube connector makes a "click" sound.

➡Check if there is any damage or foreign objects on the connected part of the fuel pipe.

b. After connecting, check that the pipe and connector are securely connected by pulling them.
c. Install the fuel pipe clamp.
9. Install intake air surge tank.
a. Install the surge tank using a new gasket. Using an 8mm Allen socket, install the 4 bolts and the 2 nuts with a 12mm socket.
b. Install the brackets and tighten the 2 bolts that hold the brackets to the side of the cylinder head.
c. Install the 3 upper bolts which are used to secure the 2 surge tank stays and throttle body bracket.
d. Install No. 1 VSV at the rear passenger side of the surge tank if removed.
e. Install the 3 wire harness clamps and hose clamps.
f. Connect the 2 VSV connectors.
g. Connect the ventilation hose.
h. Connect the fuel vapor feed hose.
10. Install throttle body with motor body assembly.
a. Install the throttle body with motor using a new gasket Install the 4 bolts, and then and set it aside
b. Connect the throttle body connector.
11. Install air cleaner assembly.
a. Install the air cleaner and 2 bolts. Tighten the 2 hose clamps.
b. Connect the mass air flow meter connector. Install the 2 wire harness clamps.
c. Connect the vacuum hose.
d. Connect the ventilation hose No. 2.
12. Install the engine V-bank cover.

FUEL TANK

REMOVAL & INSTALLATION

See Figures 126 through 129.

1. Before servicing the vehicle, refer to the precautions.
2. Discharge fuel system pressure.
3. Disconnect the negative battery cable.
4. Remove left front door scuff plate.
5. Remove left rear door scuff plate.
6. Remove left rear seat cushion assembly.
7. Remove rear floor service hole cover.
a. Pull back the floor carpet.
b. Remove the left center floor silencer pad.
c. Remove the 3 screws from the service hole cover.
d. Disconnect the connector.
8. Remove fuel tank to filler pipe hose.
a. Loosen the clamp bolt, and disconnect the fuel tank to filler pipe hose.
9. Remove fuel tank breather tube.
a. Pinch the retainer to disengage the lock claws and pull out the No.1 fuel tank breather tube.

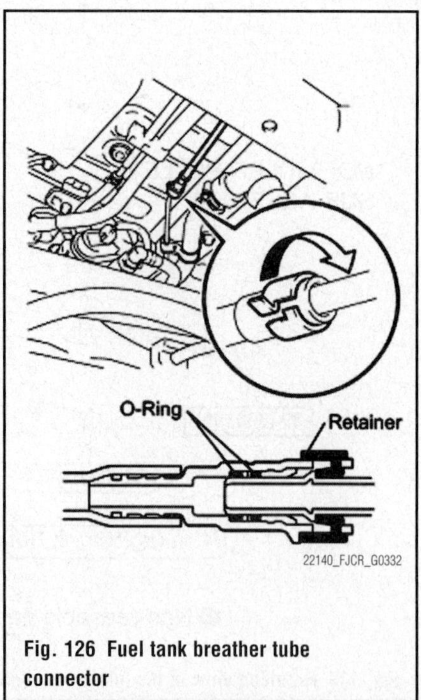

Fig. 126 Fuel tank breather tube connector

※※ WARNING

Check that there is no dirt or mud around the quick connector before performing this work, because the quick connector has an O-ring which seals the pipe and the connector. Clean the connector if necessary.

※※ WARNING

Do not use any tools in this work. Do not bend or twist the nylon tube. To protect the tube, cover it with a vinyl bag after disconnecting it. When the connector and the pipe are stuck, turn the retainer carefully to free and then disconnect the fuel tank tube.

10. Remove No. 1 fuel tank protector.
   a. Remove the 2 nuts and the fuel tank protector bracket. (for automatic transmission)
   b. Remove the nut and the fuel tank protector bracket. (for manual transmission)
   c. Remove the 4 nuts and the fuel tank protector.
11. Disconnect fuel tank main tube and fuel tank return tube.
   a. Pinch the retainer to disengage the lock claws and pull out the 2 fuel tank tubes.
12. Disconnect fuel tank vent hose.
   a. Disconnect the fuel tank breather tube from the fuel tank.
   b. Push the connector deep inside.
   c. Pinch portion A, as shown in the illustration.
   d. Pull out the connector.

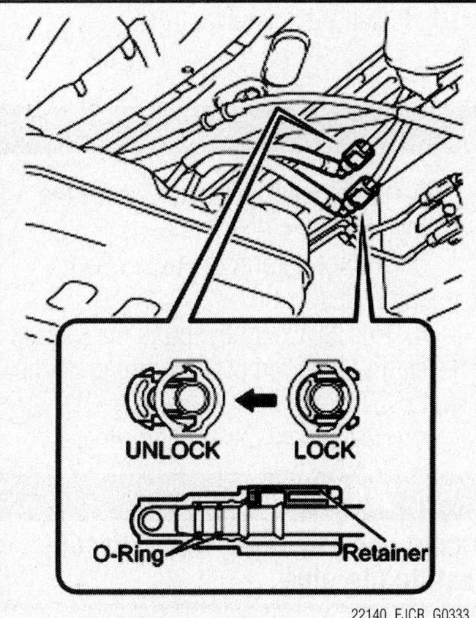

Fig. 127 Fuel tank main and return tube connector

13. Remove fuel tank assembly.
   a. Hold the fuel tank using a mission jack.
   b. Remove the 2 fuel tank bands.
   c. Remove the 2 bolts.
   d. Remove the 2 clips and 2 pins, then remove the 2 fuel tank bands.
14. Remove No. 3 fuel tank protector.
   a. Remove the 2 bolts and disengage the 3 claws, then remove the No. 3 fuel tank protector.
15. Remove fuel tank cushion.
16. Remove fuel tank main tube and fuel tank return tube.
   a. Remove the 2 joint clips, and pull out the 2 fuel tank tubes.

➡After disconnecting, cover the fuel tube joint with a vinyl bag.

※※ WARNING

When the fuel tube joint and fuel suction plate are stuck, turn the fuel tank main tube carefully to free and then disconnect it. Likewise, disconnect the fuel tank return tube.

17. Remove fuel suction with pump and gauge tube assembly.
18. Drain fuel.
19. Remove fuel tank breather tube.
   a. Pinch the retainer to disengage the lock claws and pull out the fuel tank breather tube.
   b. Loosen the clamp bolt, and disconnect the fuel tank to filler pipe hose.

To install:.

20. Install fuel tank to filler pipe hose.
   a. Connect a new fuel tank to filler pipe hose, as shown in the illustration, with the clamp. Install fuel tank breather tube.
   b. Push the tube connector into the pipe until the connector makes a "click" sound and install the retainer.

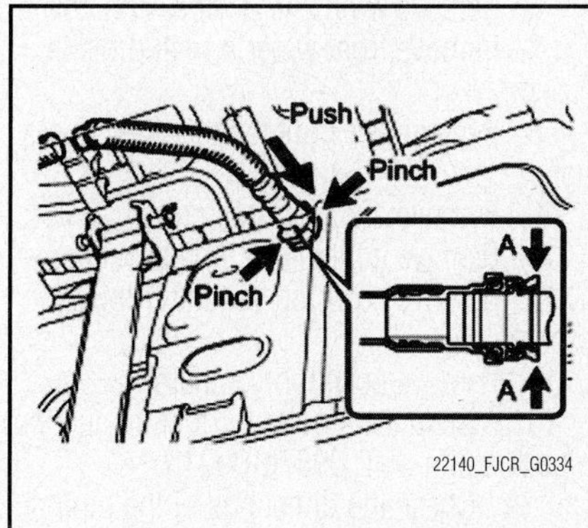

Fig. 128 Fuel tank vent hose connector

➡Check if there is any damage or foreign matter on the connected part of the pipe. After connecting, check if the pipe and the connector are securely connected by pulling them.

21. Install fuel suction with pump and gauge tube assembly.
22. Install fuel tank main tube and fuel tank return tube.
   a. Install the fuel tank main tube and return tube with the 2 joint clips.

➡Check that there are no scratches or foreign matter around the connected part of the fuel tube joint and plug before performing this work. Check that the fuel tube joint is securely inserted into the end. Check that the tube joint clips are on the collar of the fuel tube joint. After installing the tube joint clip, check that the fuel tank main tube and return tube can be pulled out.

23. Install 4 new fuel tank cushions the fuel tank.
24. Install No. 3 fuel tank protector.
   a. Install the No. 3 fuel tank protector with the 2 bolts and 3 claws. Tighten to 44 inch lbs. (5.0 Nm).
25. Install fuel tank assembly.
   a. Set a transmission jack to the fuel tank.

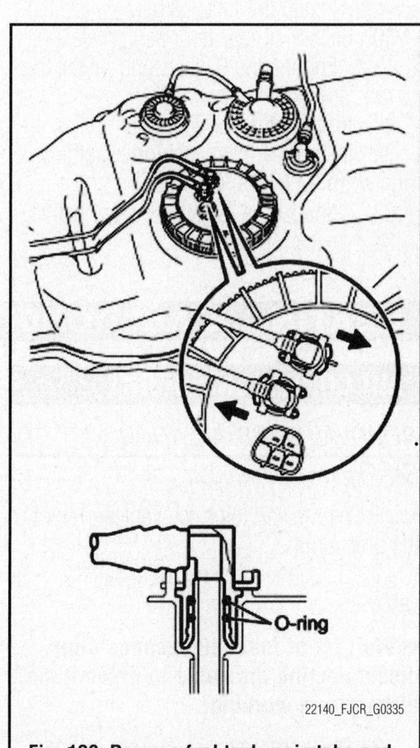

Fig. 129 Remove fuel tank main tube and fuel tank return tube connector at the fuel sender

b. Install the fuel tank and 2 fuel tank bands with the 2 clips, 2 pins and the 2 bolts. Tighten to 30 ft. lbs. (40 Nm).

26. Install fuel tank vent hose.

a. Connect the fuel tank vent hose.

b. Align the fuel tank vent hose connector with the pipe, then push in the fuel tank vent connector until the retainer makes a "click" sound to connect the fuel tank vent hose to the charcoal canister.

27. Connect fuel tank main tube and fuel tank return tube.

a. Push the tube connector into the pipe until the connector makes "click" sound.

b. Push the tube connector into the pipe until the connector makes "click" sound.

c. Connect the fuel tank to filler pipe hose, as shown in the illustration, with the clamp.

28. Install No. 1 fuel tank protector.

a. Install the fuel tank protector with the 4 nuts. Tighten to 15 ft. lbs. (20 Nm).

b. Install the fuel tank protector bracket with the 2 nuts. Tighten to 49 inch lbs. (6 Nm).

29. Install rear floor service hole cover.

a. Connect the connector.

b. Install the 3 screws from the center floor.

c. Install the left center floor silencer pad.

d. Install the floor carpet.

30. Install left rear door scuff plate.

a. Engage the 6 claws and install the rear door scuff plate.

31. Install left front door scuff plate.

32. Install rear seat cushion assembly. Tighten to 27 ft. lbs. (37 Nm).

33. Connect the negative battery cable.

34. Check for fuel leaks.

IDLE SPEED

ADJUSTMENT

The idle speed is controlled by the Electronic Control Module (ECM). No adjustment is possible.

THROTTLE BODY

REMOVAL & INSTALLATION

See Figure 130.

1. Disconnect the negative battery cable.

2. Drain engine coolant.

a. Remove the 2 nuts, then remove the V-bank cover.

3. Remove air cleaner assembly.

a. Disconnect the No. 2 ventilation hose.

b. Disconnect the vacuum hose.

c. Disconnect the mass air flow meter connector.

d. Disengage the 2 wire harness clamps.

e. Loosen the 2 hose clamps.

f. Remove the 2 bolts.

g. Disconnect the air cleaner hose and remove the air cleaner.

4. Remove throttle with motor body assembly.

a. Disconnect the No. 5 water by-pass hose.

b. Disconnect the No. 4 water by-pass hose.

c. Disconnect the throttle motor connector.

d. Remove the 4 bolts, then remove the throttle w/ motor body and gasket.

To install:

5. Install throttle with motor body assembly.

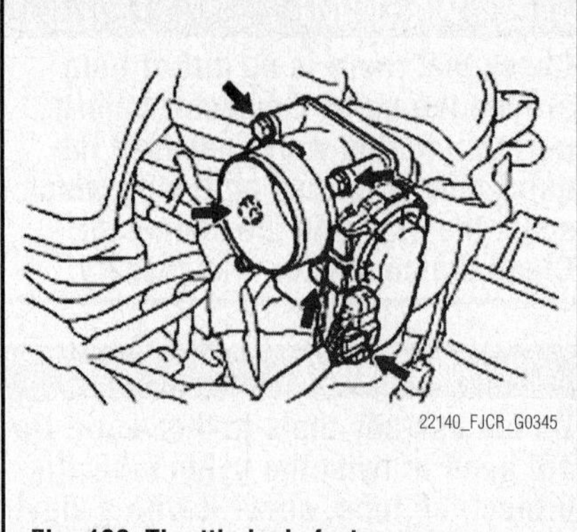

Fig. 130 Throttle body fasteners

a. Install a new gasket and the throttle with motor body with the 4 bolts. Tighten to 9 ft. lbs. (11 Nm).

b. Connect the throttle motor connector.

c. Connect the No. 4 water by-pass hose.

d. Connect the No. 5 water by-pass hose.

6. Install air cleaner assembly.

a. Install the air cleaner onto the throttle body.

b. Connect the air cleaner hose to the air cleaner.

c. Install the air cleaner with the 2 bolts. Tighten to 71 inch lbs. (8 Nm).

d. Connect the mass air flow meter connector.

e. Engage the 2 wire harness clamps.

f. Connect the vacuum hose.

g. Connect the No. 2 ventilation hose.

7. Connect the negative battery cable.

8. Add engine coolant.

9. Check for engine coolant leak.

10. Install v-bank cover.

HEATING & AIR CONDITIONING SYSTEM

AIR CONDITIONING UNIT

REMOVAL & INSTALLATION

See Figures 131 through 133.

1. Before servicing the vehicle, refer to the precautions.

2. Disconnect the negative battery cable.

➡Wait for at least 90 seconds after disconnecting the cable to prevent the airbag from working.

3. Discharge refrigerant from refrigeration system.

4. Drain engine coolant.

5. Remove roof antenna pole.

6. Remove windshield wiper arm cover.

7. Remove front wiper arm and blade assembly.

8. Remove left front fender side panel upper.

9. Remove antenna ornament.

10. Remove front fender side panel upper.

11. Remove cowl top ventilator louver assembly.

12. Remove cowl top ventilator louver.

13. Disconnect cooler refrigerant suction pipe **A** using SST: 09870-00015.

a. Check the directions of the piping clamp and SST by referring to the illustration on the caution label.

b. Push down SST and release the clamp lock.

c. Do not deform the tube when pushing SST.

d. Pull SST slightly, push the release lever, and then remove the piping clamp with SST.

e. Disconnect the suction pipe.

WARNING

Do not use any tools when disconnecting the pipe.

14. Seal the openings of the disconnected parts using vinyl tape to prevent moisture and foreign matter from entering.

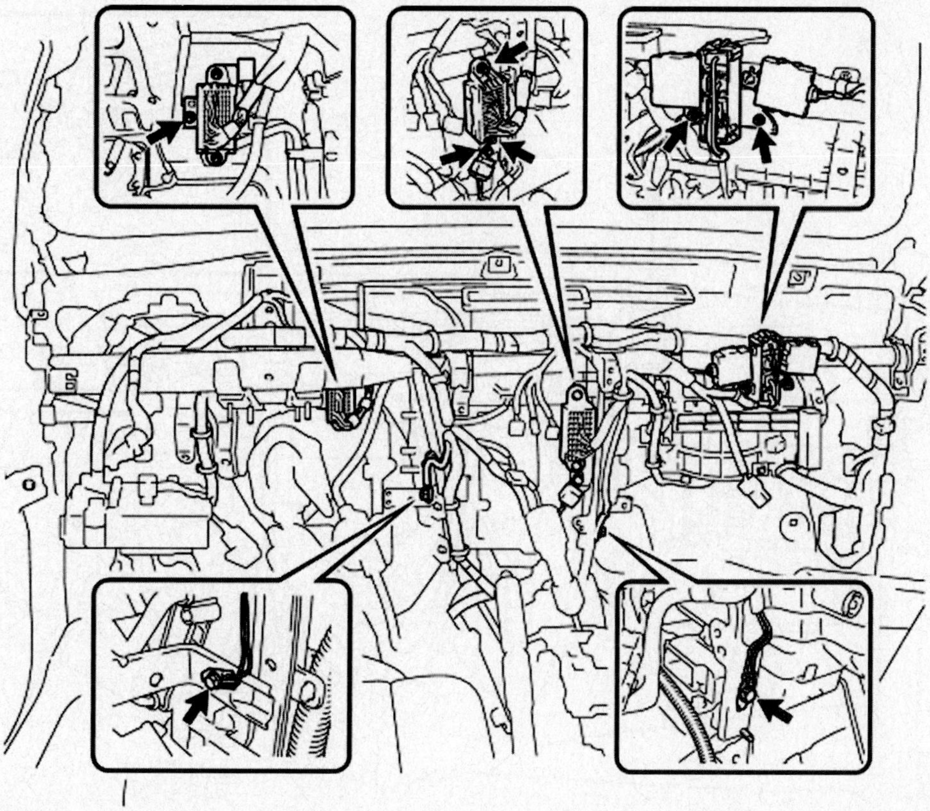

Fig. 131 Instrument panel wiring harness connectors

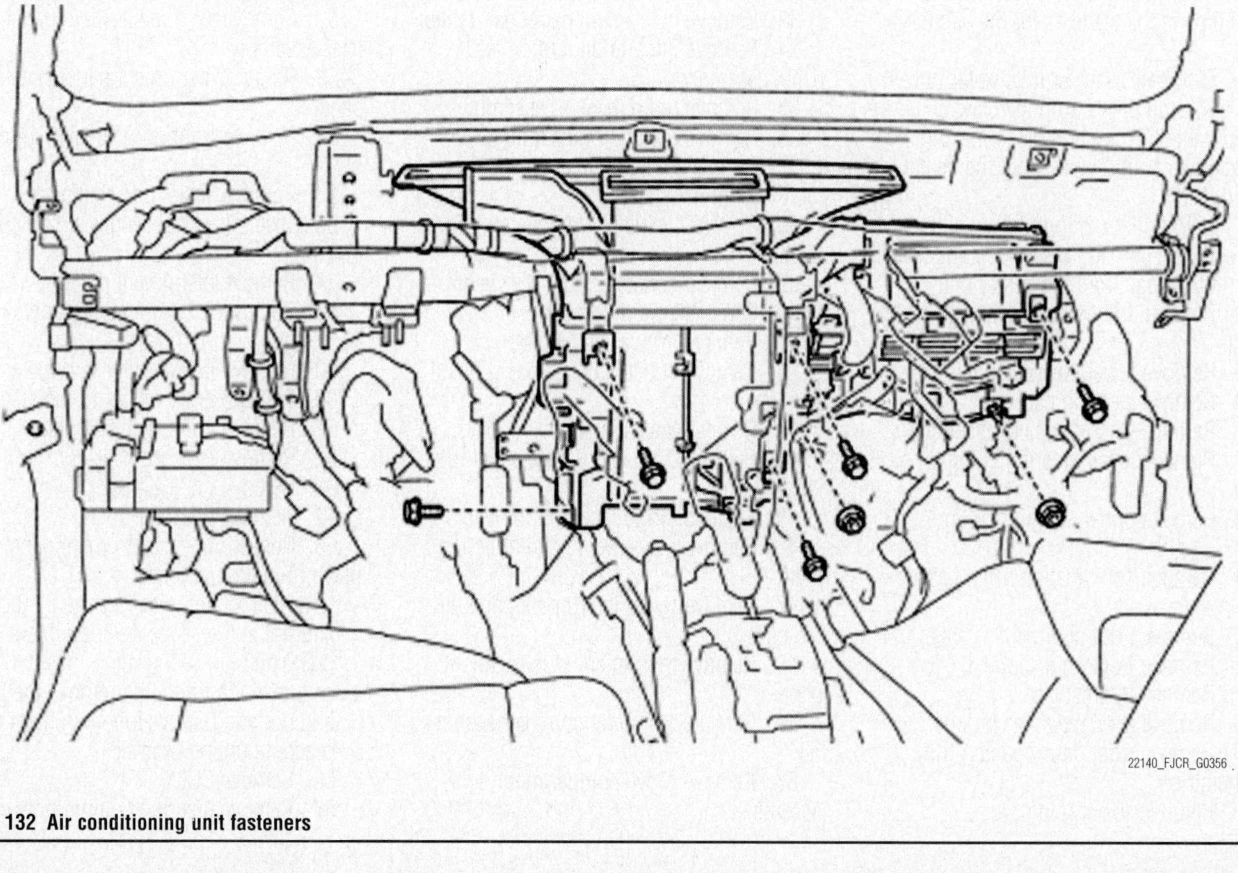

Fig. 132 Air conditioning unit fasteners

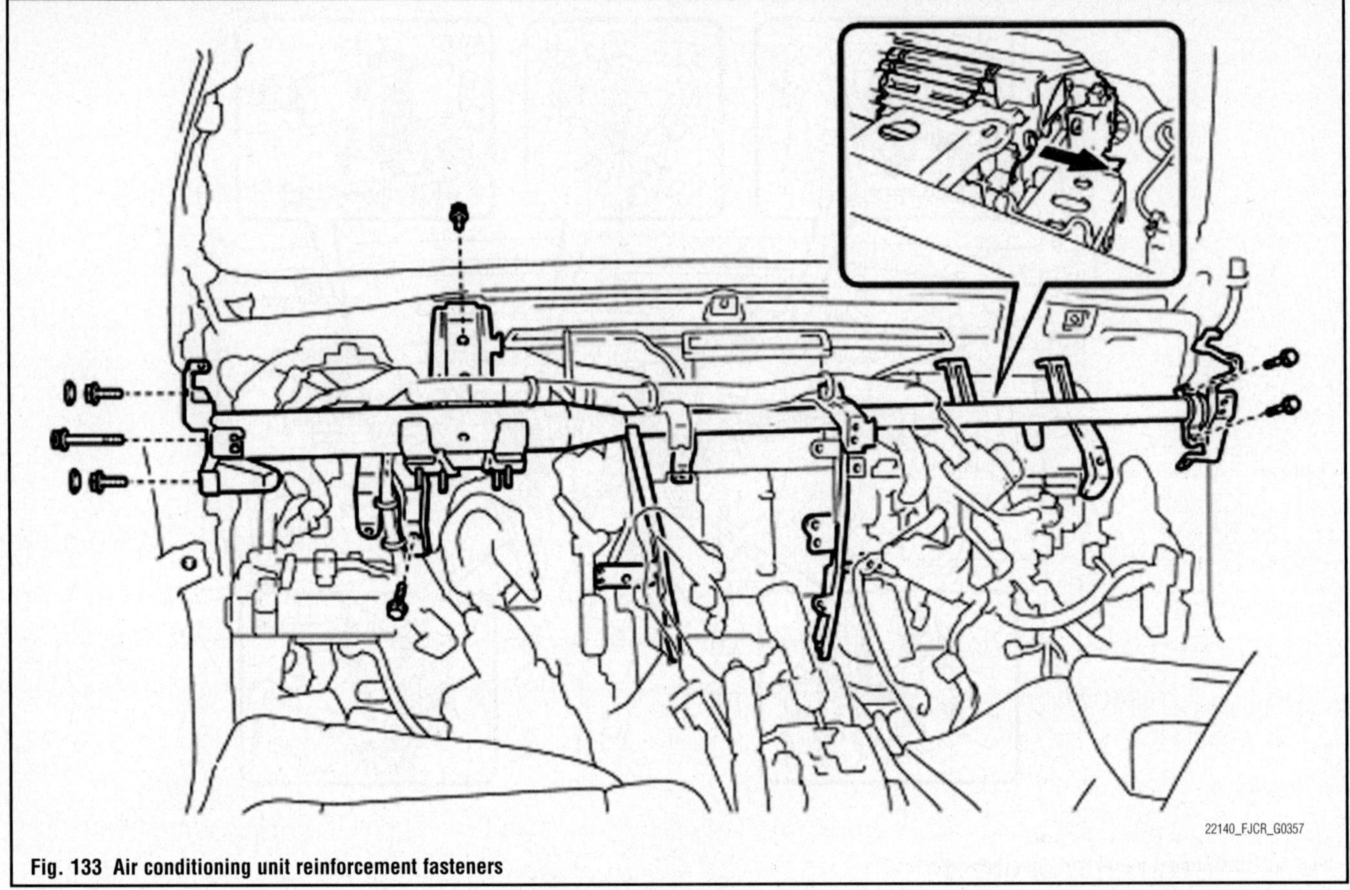

Fig. 133 Air conditioning unit reinforcement fasteners

15. Disconnect cooler refrigerant liquid pipe **A**. The disconnection procedure of the liquid pipe is the same as for the suction pipe.
16. Disconnect heater inlet water hose.
    a. Using pliers, grip the claw of the clip, slide the clip and disconnect the heater water inlet hose from the heater unit.
17. Using pliers, grip the claw of the clip, slide the clip and disconnect the heater water outlet hose from the heater unit.
18. Position front wheels facing straight ahead.
19. Remove lower steering wheel covers.
20. Remove steering pad.
21. Remove steering wheel assembly.
22. Remove lower steering column cover.
23. Remove upper steering column cover.
24. Remove combination switch assembly.
25. Remove front door scuff plates.
26. Remove front floor footrest.
27. Remove footrest clip.
28. Remove cowl side trim boards.
29. Remove front door opening trim weatherstrips.
30. Remove assist grip plug.
31. Remove assist grip assembly.
32. Remove front pillar garnishes.
33. Remove instrument panel garnishes.
34. Remove integration control and panel assembly.
35. Remove radio receiver assembly.
36. Remove parking brake hole cover.
37. On manual transmission vehicles, remove shift lever knob.
38. On 4WD vehicles, remove shift lever knob.
39. Remove console upper rear panel.
40. Remove box bottom mat.
41. Remove front console box.
42. Remove console upper panel no. 1 garnish.
43. Remove instrument lower cover.
44. Remove no 1 instrument panel register assembly.
45. Separate hood lock control lever.
46. Remove instrument panel finish plate.
47. Remove lower instrument panel finish panel.
48. Remove instrument cluster finish panel.
49. Remove combination meter assembly.
50. Remove glove compartment door assembly.
51. Remove instrument panel lower finish panel.
52. Remove no. 2 instrument panel register assembly.
53. Remove instrument panel speaker panel.
54. Remove front no 2 speaker assembly.
55. Remove assist grip retainers.
56. Disconnect passenger airbag connector.
57. Remove instrument panel.
58. Remove left instrument panel finish panel end.
59. Remove no. 1 heater to register duct.
60. Remove no. 2 heater to register duct.
61. Remove rear no. 1 air duct.
62. Remove rear no. 2 air duct.
63. Remove no. 1 air duct.
64. Remove no. 2 air duct.
65. Remove no. 1 instrument panel brace mounting bracket.
    a. On the left side remove the bolt and nut and remove the instrument panel brace mounting bracket.
    b. On the right side remove the bolt and nut and remove the instrument panel brace mounting bracket.
66. Remove ECM.
67. Remove steering column hole cover.

68. Remove the bolt and separate the steering shaft thrust stopper from the steering intermediate shaft assembly.

a. Mark matchmarks on the steering column assembly and steering intermediate shaft.

b. Pull the intermediate shaft assembly and steering shaft thrust stopper out of the steering column assembly.

69. Remove steering column assembly.

70. Remove the bolt and remove the instrument panel side bracket.

71. Separate main body ECU.

a. Remove the 2 nuts and remove the driver side junction block.

72. Remove the cooler unit drain hose.

73. Remove instrument panel reinforcement.

a. Remove the 3 bolts and 5 nuts and disconnect the wire harness.

b. Disconnect the connectors.

c. Disengage the clamps.

d. Remove the 5 bolts and the 2 nuts of the air conditioning unit.

e. Remove the 2 caps and the 7 bolts of the reinforcement.

f. Disengage the reinforcement hook of the air conditioning unit, and remove the reinforcement.

g. Remove the air conditioning unit.

74. Remove air conditioning unit assembly.

a. Remove the 2 screws.

b. Remove the air conditioning unit as shown in the illustration.

75. Remove heater to register duct assembly.

a. Disengage the 4 claws and remove the heater to register duct.

76. Remove lower defroster nozzle assembly.

a. Disengage the 4 claws and remove the lower defroster nozzle.

To install:

77. Install lower defroster nozzle assembly.

78. Install heater to register duct assembly.

79. Install air conditioning unit assembly.

a. Install the air conditioning unit with the 2 screws as shown in the illustration.

80. Install instrument panel reinforcement.

a. Temporarily install the air conditioning unit assembly.

b. Insert the bracket hook into the holes of the reinforcement bracket, and temporarily install the reinforcement.

c. Install the instrument panel reinforcement with the 7 bolts.

d. Install the 2 caps.

e. Install the 5 bolts. Tighten the bolts in the sequence order shown in the illustration to install the air conditioner unit assembly. Tighten to 87 inch lbs. (98 Nm).

f. Install the 2 nuts. Tighten to 48 inch lbs. (54 Nm).

g. Engage the clamps.

h. Connect the connectors.

i. Connect the wire harness with the 3 bolts and 5 nuts. Tighten to 65 inch lbs. (73 Nm).

81. Install cooler unit drain hose.

a. Install the cooler unit drain hose.

82. Install the main body ECU with the 2 nuts. Tighten to 74 inch lbs. (84 Nm).

83. Install the instrument panel side bracket with the bolt. Tighten to 74 inch lbs. 84 Nm).

84. Install steering column assembly.

85. Install steering intermediate shaft assembly.

a. Align the matchmarks on the steering column assembly and the steering intermediate shaft assembly.

b. Install the steering intermediate shaft assembly and thrust stopper onto the steering column assembly with the bolt. Tighten to 27 ft. lbs. (36 Nm).

86. Install steering column hole cover.

87. Install ECM.

88. Install no. 1 instrument panel brace mounting bracket.

a. On the left side, install the instrument panel brace mounting bracket with the bolt and nut.

b. On the right side, install the instrument panel brace mounting bracket with the bolt and nut.

89. Install no. 1 air duct.

90. Install no. 2 air duct.

91. Install rear no. 1 air duct.

92. Install rear no. 2 air duct.

93. Install no. 1 heater to register duct.

94. Install no. 2 heater to register duct.

95. Install instrument panel.

96. Connect passenger airbag connector.

97. Install left instrument panel finish panel end.

98. Install assist grip retainers.

99. Install front no. 2 speaker assembly.

100. Install no. 2 instrument panel speaker panels.

101. Install no. 2 instrument panel register assembly.

102. Install right instrument panel lower finish panel.

103. Install glove compartment door assembly.

104. Install combination meter assembly.

105. Install instrument cluster finish panel.

106. Install lower instrument panel.

107. Install left lower instrument panel finish panel.

108. Install instrument panel finish plate.

109. Connect hood lock control lever.

110. Install no 1 instrument panel register assembly.

111. Install instrument lower cover.

112. Install console upper panel no. 1 garnish.

113. Install front console box.

114. Install box bottom mat.

115. Install console upper rear panel.

116. On manual transmission vehicles, install shift lever knob.

117. On 4WD vehicles, install shift lever knob.

118. Install parking brake hole cover.

119. Install radio receiver assembly.

120. Install integration control and panel assembly.

121. Install instrument panel garnishes.

122. Install front pillar garnishes.

123. Install assist grip assembly.

124. Install assist grip plug.

125. Install front door opening trim weatherstrips.

126. Install cowl side trim boards.

127. Install footrest clip.

128. Install front floor footrest.

129. Install front door scuff plates.

130. Install combination switch assembly.

131. Install upper steering column cover.

132. Install lower steering column cover.

133. Install steering wheel assembly.

134. Install steering pad.

135. Install lower no. 2 steering wheel cover.

136. Install lower no. 3 steering wheel cover.

137. Connect heater water outlet hose to heater unit.

138. Perform the installation with the hose clip and mark at the correct angle as shown in the illustration.

139. Connect heater inlet water hose.

140. Perform the installation with the hose clip and mark at the correct angle as shown in the illustration.

141. Install cooler refrigerant liquid pipe a.

a. Remove the vinyl tape from liquid tube **A** and the connecting portion of the unit.

b. Apply sufficient ND-OIL8 compressor oil to a new O-ring and the connecting part of the liquid pipe.

c. Install the O-ring onto the liquid pipe **A**.

d. Install the liquid and piping clamp.

142. After connection, check the claw fitting of the piping clamp.

143. Install cooler refrigerant suction pipe **A**.

144. Connection procedure of the suction pipe is the same as for the liquid pipe.
145. Install cowl top ventilator louvers.
146. Install cowl top ventilator louver.
147. Install front fender side upper panels.
148. Install antenna ornament.
149. Install roof antenna pole.
150. Install front wiper arm and blade assemblies.
151. Install windshield wiper arm cover.
152. Add engine coolant.
153. Connect cable to negative battery terminal.
154. Check SRS warning light.
155. Charge refrigerant.
156. Warm up engine.
157. Check for engine coolant leak.
158. Check for refrigerant leak.
159. Position front wheels facing straight ahead.

BLOWER MOTOR

REMOVAL & INSTALLATION

1. Before servicing the vehicle, refer to the precautions.
2. Disconnect the connector and the clamp.
3. Remove the 3 screws and the blower motor.

To install:

4. Installation is the reverse of the removal procedure.

HEATER CORE

REMOVAL & INSTALLATION

See Figure 134.

1. Before servicing the vehicle, refer to the precautions.
2. Remove the air conditioning unit.
3. Remove the screw and clamp.
4. Remove the heater radiator unit from the heater case.

To install:

5. Installation is the reverse of the removal procedure.

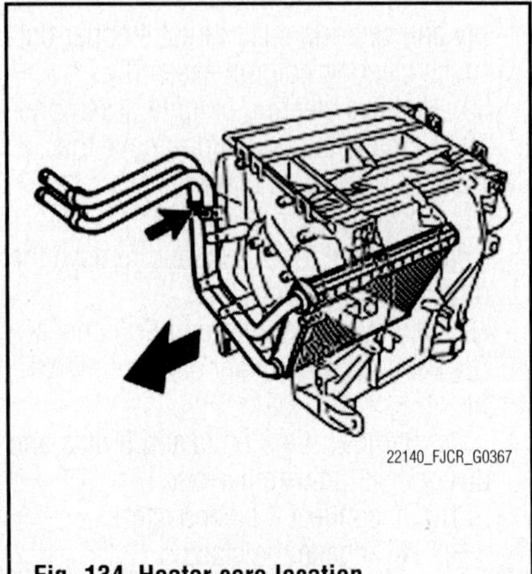

Fig. 134 Heater core location

STEERING

POWER RACK & PINION STEERING GEAR

REMOVAL & INSTALLATION

2WD Models

1. Before servicing the vehicle, refer to the precautions section.
2. Disconnect the negative battery cable.
3. Place front wheels facing straight ahead.
4. Remove no. 1 engine under cover.
5. Remove fan and generator V-belt.
6. Remove front wheels.
7. Remove no. 2 steering intermediate shaft.
   a. Fix the steering wheel with the seat belt in order to prevent it from rotating. This operation is effective for preventing any damage to the spiral cable.
   b. Place matchmarks on the steering sliding yoke, No. 2 steering intermediate shaft and steering intermediate shaft.
   c. Remove bolts from the steering sliding yoke.
   d. Slide the steering sliding yoke up and separate it from the No. 2 steering intermediate shaft.
   e. Pull down the steering sliding yoke from the steering intermediate shaft to remove it.
   f. Place matchmarks on the No. 2 steering intermediate shaft and power steering link.
   g. Remove bolt C from the No. 2 steering intermediate shaft.
   h. Slide the No. 2 steering intermediate shaft up and remove it from the power steering link.
8. Separate tie rod end.
   a. Remove the cotter pin and nut.
   b. Using SST: 09628-62011, separate the tie rod end from the steering knuckle arm.
9. Separate pressure feed tube assembly.
   a. Remove the 2 bolts and separate the tube support brackets.
   b. Loosen the flare nut and separate the pressure feed tube.
   c. Disengage the clip and disconnect the return hose.
10. Remove the 4 bolts and remove the air conditioning compressor and magnetic clutch.
11. Remove power steering link.
    a. Remove the 2 bolts and 2 nuts.

⁂ WARNING

The nut has a detent, so never turn the nut. Be sure to turn only the bolt.

   b. Tilt the transmission and remove the power steering link.

To install:

12. Install the power steering link with the 2 bolts and 2 nuts. Tighten to 74 ft. lbs. (100 Nm).

⁂ WARNING

Never turn the nut since it has a detent. Be sure to turn only the bolt.

13. Install pressure feed tube assembly.
    a. Connect the return hose with the clip.
    b. Tighten the flare nut to 33 ft. lbs. (44 Nm) and connect the pressure feed tube.
    c. Install the tube support brackets with the 2 bolts. Tighten to 21 ft. lbs. (28 Nm).
14. Install the air conditioning compressor and magnetic clutch with the 4 bolts.
15. Install tie rod end.
    a. Install the tie rod end onto the steering knuckle arm.
    b. Install the nut. Tighten to 67 ft. lbs. (91 Nm).
    c. Install a new cotter pin.
16. Install No. 2 steering intermediate shaft.
    a. Align the matchmarks on the No. 2 steering intermediate shaft and power steering link.
    b. Install the No. 2 steering intermediate shaft onto the power steering link. Tighten to 27 ft. lbs. (36 Nm).
    c. Align the matchmarks on the steering intermediate shaft and steering sliding yoke.
    d. Install the steering sliding yoke onto the steering intermediate shaft and slide it upward.

e. Align the matchmarks on the steering sliding yoke and No. 2 steering intermediate shaft.

f. Install the steering sliding yoke. Tighten to 27 ft. lbs. (36 Nm).

17. Install fan and generator V-belt.
18. Install front wheels.
19. Place front wheels facing straight ahead.
20. Inspect steering wheel center point.
21. Connect the negative battery cable.
22. Add power steering fluid.
23. Bleed power steering fluid.
24. Check for power steering fluid leakage.
25. Inspect and adjust front wheel alignment.
26. Install No. 1 engine under cover

4WD Models

1. Before servicing the vehicle, refer to the precautions section.
2. Disconnect the negative battery cable.
3. Place front wheels facing straight ahead.
4. Remove no.1 engine under cover.
5. Remove fan and generator V-belt.
6. Remove front wheels.
7. Remove no. 1 engine under cover.
8. Remove rear engine under cover assembly.
9. Remove no. 2 exhaust front pipe assembly.
10. Remove front exhaust pipe assembly.
11. Remove driveshaft heat insulator.
12. Remove front halfshaft assembly.
13. Remove driveshaft assembly.
14. Remove front suspension member brackets.
15. Remove no. 3 frame crossmember.
16. Remove No. 2 steering intermediate shaft.

a. Fix the steering wheel with the seat belt in order to prevent it from rotating. This operation is effective for preventing any damage to the spiral cable.

b. Place matchmarks on the steering sliding yoke, No. 2 steering intermediate shaft and steering intermediate shaft.

c. Remove bolts from the steering sliding yoke.

d. Slide the steering sliding yoke up and separate it from the No. 2 steering intermediate shaft.

e. Pull down the steering sliding yoke from the steering intermediate shaft to remove it.

f. Place matchmarks on the No. 2 steering intermediate shaft and power steering link.

g. Remove bolt C from the No. 2 steering intermediate shaft.

h. Slide the No. 2 steering intermediate shaft up and remove it from the power steering link.

17. Separate tie rod end.

a. Remove the cotter pin and nut.

b. Using SST: 09628-62011, separate the tie rod end from the steering knuckle arm.

18. Separate pressure feed tube assembly.

a. Remove the 2 bolts and separate the tube support brackets.

b. Loosen the flare nut and separate the pressure feed tube.

c. Disengage the clip and disconnect the return hose.

19. Remove the 4 bolts and remove the air conditioning compressor and magnetic clutch.
20. Remove power steering link.

a. Remove the 2 bolts and 2 nuts.

⁂ WARNING

The nut has a detent, so never turn the nut. Be sure to turn only the bolt.

b. Tilt the transmission and remove the power steering link.

To install:.

21. Install the power steering link with the 2 bolts and 2 nuts. Tighten to 74 ft. lbs. (100 Nm).

⁂ WARNING

Never turn the nut since it has a detent. Be sure to turn only the bolt.

22. Install pressure feed tube assembly.

a. Connect the return hose with the clip.

b. Tighten the flare nut to 33 ft. lbs. (44 Nm) and connect the pressure feed tube.

c. Install the tube support brackets with the 2 bolts. Tighten to 21 ft. lbs. (28 Nm).

23. Install the air conditioning compressor and magnetic clutch with the 4 bolts.
24. Install tie rod end.

a. Install the tie rod end onto the steering knuckle arm.

b. Install the nut. Tighten to 67 ft. lbs. (91 Nm).

c. Install a new cotter pin.

25. Install No. 2 steering intermediate shaft.

a. Align the matchmarks on the No. 2 steering intermediate shaft and power steering link.

b. Install the No. 2 steering intermediate shaft onto the power steering link. Tighten to 27 ft. lbs. (36 Nm).

c. Align the matchmarks on the steering intermediate shaft and steering sliding yoke.

d. Install the steering sliding yoke onto the steering intermediate shaft and slide it upward.

e. Align the matchmarks on the steering sliding yoke and No. 2 steering intermediate shaft.

f. Install the steering sliding yoke. Tighten to 27 ft. lbs. (36 Nm).

26. Install no. 3 frame crossmember.
27. Install front suspension member brackets.
28. Install driveshaft assembly.
29. Install front halfshaft assembly.
30. Install driveshaft heat insulator.
31. Install front exhaust pipe assembly.
32. Install no. 2 exhaust front pipe assembly.
33. Install rear engine under cover assembly.
34. Install no. 1 engine under cover.
35. Install fan and generator V-belt.
36. Install front wheels.
37. Place front wheels facing straight ahead.
38. Inspect steering wheel center point.
39. Connect the negative battery cable.
40. Add power steering fluid.
41. Bleed power steering fluid.
42. Check for power steering fluid leakage.
43. Inspect and adjust front wheel alignment.
44. Install No. 1 engine under cover.

POWER STEERING PUMP

REMOVAL & INSTALLATION

See Figure 135.

1. Before servicing the vehicle, refer to the precautions section.
2. Disconnect the negative battery cable.
3. Remove No. 1 engine under cover.
4. Remove fan and generator v belt.
5. Drain power steering fluid.
6. Disengage the clip and disconnect the No. 1 oil reservoir to pump hose.
7. Disconnect pressure feed tube assembly.

a. Remove the union bolt, then disconnect the pressure feed tube.

b. Remove the gasket from the pressure feed tube.

8. Remove power steering pump.

a. Disconnect the oil pressure switch connector.

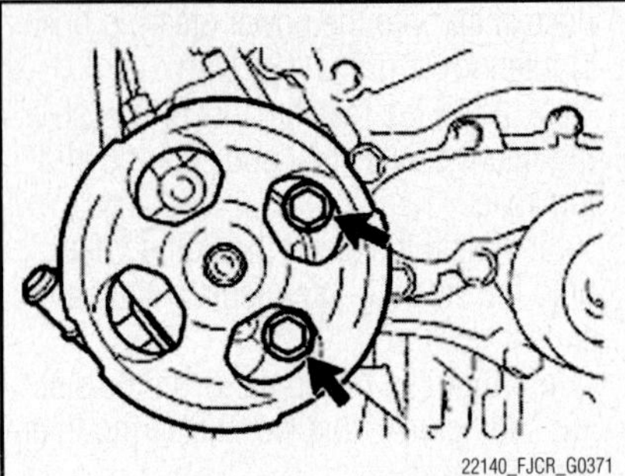

Fig. 135 Rotate the pulley on the power steering pump to access the mounting bolts

b. Remove the bolt and the wire harness clamp bracket.

c. Rotate the pulley on the power steering pump to access the mounting bolts

d. Remove the 2 bolts and the power steering pump assembly.

To install:.

9. Install power steering pump.

a. Install the power steering pump assembly with the 2 bolts. Tighten to 32 ft. lbs. (43 Nm).

b. Install the wire harness clamp bracket with the bolt. Tighten to 71 inch lbs. (8 Nm).

c. Connect the oil pressure switch connector.

**** WARNING**

Make sure that no oil adheres to the connector.

10. Connect pressure feed tube assembly.

a. Install a new gasket onto the pressure feed tube.

b. Install the pressure feed tube with the union bolt. Tighten to 38 ft. lbs. (51 Nm).

11. Connect the No. 1 oil reservoir to pump hose with the clip.

12. Install fan and generator v belt.

13. Connect the negative battery cable.

14. Add power steering fluid.

15. Bleed power steering fluid.

16. Check for power steering fluid leakage.

17. Install No. 1 engine under cover.

BLEEDING

1. Check the fluid level.

2. Raise and safely support the vehicle securely on jackstands.

3. Turn the steering wheel.

4. With the engine stopped, turn the wheel slowly from lock to lock several times.

5. Lower the vehicle.

6. Start the engine.

7. Run the engine at idle for a few minutes.

8. Turn the steering wheel.

9. With the engine idling, turn the wheel to the left or right full lock position and keep it there for 2 to 3 seconds. Then turn the wheel to the opposite full lock position and keep it there for 2 to 3 seconds.

10. Repeat this step several times.

11. Stop the engine.

12. Check for foaming or emulsification.

**** WARNING**

If the system has to be bled twice because of foaming or emulsification, check for fluid leakage in the system.

13. Check the fluid level.

SUSPENSION

FRONT SUSPENSION

COIL SPRING

REMOVAL & INSTALLATION

See Figure 136.

1. Before servicing the vehicle, refer to the precautions section.

2. Remove the shock absorber assembly.

3. Remove front support to front shock absorber nut

4. Using a coil spring compressor, compress the coil spring.

5. While holding the shock absorber rod, remove the nut.

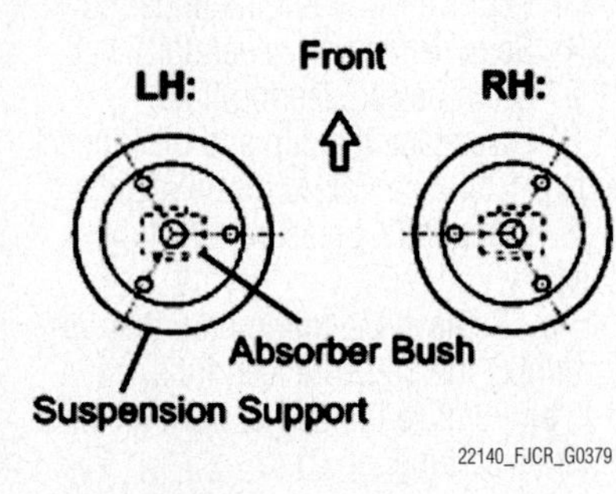

Fig. 136 Align the suspension support and the absorber bushing as shown

**** WARNING**

Do not use an impact wrench. It will damage the shock absorber rod.

6. Remove front shock absorber cushion retainer.

7. Remove front shock absorber no. 1 cushion.

8. Remove front suspension support.

9. Remove front shock absorber cushion retainer.

10. Remove front coil spring.

To install:

11. Using a coil spring compressor, compress the coil spring.

12. Install the coil spring onto the shock absorber.

➡Fit the lower end of the coil spring into the gap of the spring lower seat.

13. Install front coil spring.

14. Install front shock absorber cushion retainer.

15. Install front suspension support.

16. Install front shock absorber no. 1 cushion.

17. Install front shock absorber cushion retainer.

18. Install front support to front shock absorber nut.

19. Align the suspension support and the absorber bushing as shown in the illustration.

20. Fit and tighten a new lock nut. Tighten to 18 ft. lbs. (25 Nm).

21. Release the coil spring while checking the position of the suspension support.

LOWER BALL JOINT

REMOVAL & INSTALLATION

The lower ball joint is an integral part of the lower control arm and is not serviced separately. See Front Suspension, Lower Control Arm, Removal and Installation.

LOWER CONTROL ARM

REMOVAL & INSTALLATION

See Figures 137 and 138.

1. Before servicing the vehicle, refer to the precautions section.

2. Remove front wheel.

3. Inspect front lower suspension arm.

a. Install the hub nuts onto the disc.

b. Using a dial indicator, check the lower ball joint for excessive play when you push the hub nuts up and down with a force of 66 lbs. (294 N). Maximum play is 0.020 in. (0.5 mm).

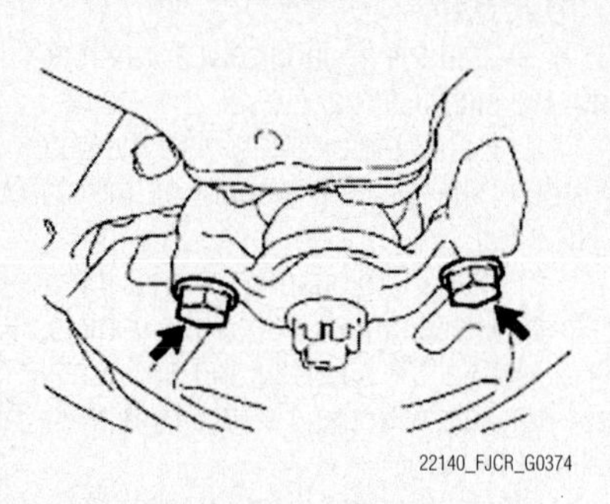

Fig. 137 Lower ball joint attachment to steering knuckle mounting bolts

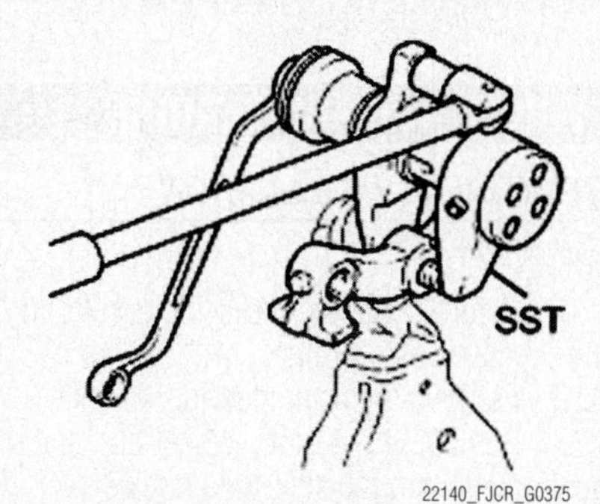

Fig. 138 Using SST: 09628-00011, remove the lower ball joint attachment

c. If it is not within the specification, replace the lower arm.

4. Separate front shock absorber with coil spring.

a. Remove the bolt, nut and washer.

b. Separate the front shock absorber with coil spring from the lower arm.

5. Remove front lower suspension arm.

a. Remove the 2 bolts and separate the lower ball joint attachment from the steering knuckle.

b. Place matchmarks on the No. 2 camber adjust cam and toe adjust cam.

c. Remove the nut, No. 2 camber adjust cam, No. 1 camber adjust cam, bolt, toe adjust cam, toe adjust plate and lower arm.

d. Remove the cotter pin and the nut.

e. Using SST: 09628-00011, remove the lower ball joint attachment.

To install:.

6. Temporarily tighten front lower suspension arm.

a. Align the matchmarks on the No. 2 camber adjust cam and toe adjust cam.

b. Temporarily tighten the bolt and the nut.

c. Install the lower ball joint attachment, a new nut and a new cotter pin. Tighten to 103 ft. lbs. (140 Nm).

d. Install the lower ball joint attachment with the 2 bolts. Tighten to 118 ft. lbs. (160 Nm).

7. Temporarily tighten front shock absorber with coil spring.

a. Install the front shock absorber with coil spring, bolt and washer, and temporarily tighten the nut.

8. Install front wheel and lower the vehicle.

9. Bounce the vehicle up and down several times to stabilize the suspension.

10. Fully tighten front lower suspension arm to 100 ft. lbs. (135 Nm).

11. Fully tighten front shock absorber with coil spring to100 ft. lbs. (135 Nm).

12. Inspect and adjust front wheel alignment.

SHOCK ABSORBERS

REMOVAL & INSTALLATION

See Figures 139 through 141.

1. Before servicing the vehicle, refer to the precautions section.

2. Raise and safely support the vehicle securely on jackstands.

3. Remove front wheels.

4. Remove engine under cover sub-assembly.

5. Remove the nut and separate the stabilizer link from the steering knuckle.

➡If the ball joint turns together with the nut, use a 6 mm hexagon wrench to hold the stud.

6. Remove front stabilizer bar.

7. Separate tie rod end sub-assembly.

8. Remove front shock absorber with coil spring.

a. Remove the bolt, nut and washer.

b. Remove the 3 nuts on the upper side of the front shock absorber with coil spring.

c. Remove the front shock absorber with coil spring.

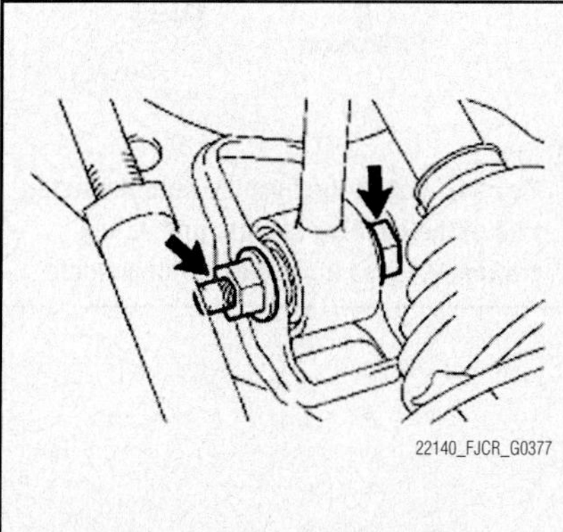

Fig. 139 Lower shock mounting bolts

To install:

9. Temporarily tighten front shock absorber with coil spring.

a. Install the coil spring onto the body with the lower end of the coil spring facing the rear side of the vehicle.

b. Install the 3 nuts onto the upper side of the front shock absorber with coil spring. Tighten to 47 ft. lbs. (64 Nm).

c. Temporarily tighten the bolt, nut and washer as shown in the illustration.

10. Install tie rod end sub-assembly.

11. Install front stabilizer bar.

12. Install front stabilizer link assembly.

a. Install the stabilizer link onto the steering knuckle with the nut. Tighten to 52 ft. lbs. (70 Nm).

13. Install engine under cover sub-assembly. Tighten to 21 ft. lbs. (29 Nm).

14. Install front wheels and lower the vehicle.

15. Bounce the vehicle up and down several times to stabilize the suspension.

16. Fully tighten the nut. Tighten to 100 ft. lbs. (135 Nm).

17. Inspect and adjust front wheel alignment.

Fig. 140 Upper shock mounting nuts

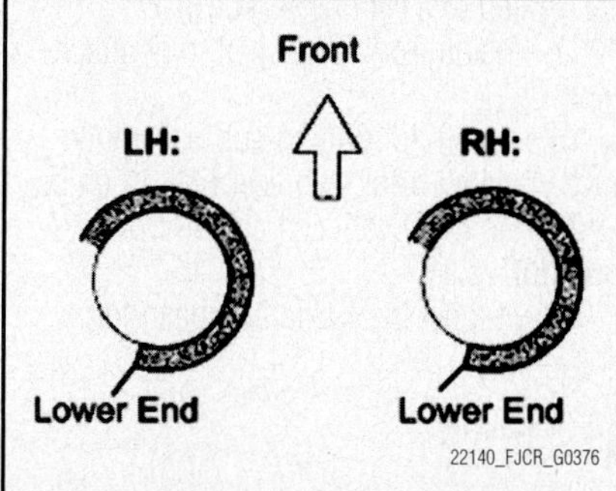

Fig. 141 Install the coil spring onto the body with the lower end of the coil spring facing the rear side of the vehicle

TESTING

Compress and extend the shock absorber rod and check that there is no abnormal resistance or unusual sound during operation. If there is any abnormality, replace the shock absorber with a new one.

STEERING KNUCKLE

REMOVAL & INSTALLATION

1. Before servicing the vehicle, refer to the precautions section.
2. Disconnect the negative battery cable.
3. Raise and safely support the vehicle securely on jackstands.
4. Remove front wheel.
5. Remove front disc brake caliper assembly and suspend it out of the way.
6. Remove front rotor.
7. Separate front speed sensor.
8. On 4WD vehicles, remove front axle hub grease cap.
9. On 4WD vehicles, remove front axle hub nut.
10. Separate front stabilizer link assembly.
11. Separate tie rod end sub-assembly.
12. Separate front lower ball joint attachment.
13. Separate front upper suspension arm.
    a. Support the lower arm with a jack.
    b. Remove the clip and nut.
    c. Using SST: 09628-62011, separate the upper ball joint from the steering knuckle.
14. Remove front steering knuckle.
    a. On 4WD vehicles, use a plastic hammer, separate the front axle hub from the front drive shaft.
    b. Remove the front steering knuckle.

To install:.

15. Install front steering knuckle.
16. Install front upper suspension arm.
    a. Install a new nut and a new clip. Tighten to 81 ft. lbs. (110 Nm).
17. Install front lower ball joint attachment.
18. Install tie rod end sub-assembly.
19. Install front stabilizer link assembly.
20. On 4WD vehicles, install front axle hub nut.
21. Inspect front axle hub bearing.
22. On 4WD vehicles, install front axle hub grease cap.
23. Install front speed sensor.
24. Install front disc.
25. Install front disc brake caliper assembly.
    a. Install the front disc brake caliper assembly with the 2 bolts. Tighten to 91 ft. lbs. (123 Nm).
26. Connect the negative battery cable.
27. Check fluid level in reservoir.
28. Check VSC sensor signal.
29. Inspect and adjust front wheel alignment.

STABILIZER BAR

REMOVAL & INSTALLATION

See Figure 142.

1. Before servicing the vehicle, refer to the precautions section.
2. Raise and safely support the vehicle securely on jackstands.
3. Remove front wheels.
4. Remove engine under cover sub-assembly.
5. Remove the 2 nuts and stabilizer link.

➡If the ball joint turns together with the nut, use a 6 mm hexagon wrench to hold the stud.

6. Remove the 4 bolts and 2 stabilizer brackets and remove the stabilizer bar.
7. Remove the 2 stabilizer bar bushings from the stabilizer bar.

To install:

8. Install the 2 stabilizer bar bushings onto the stabilizer bar.
    a. Install the bushing onto the outer side of the bushing stopper on the stabilizer bar.
    b. Install the bushing so that the protrusion faces inner side of the vehicle.
9. Install the stabilizer bar and 2 stabilizer brackets with the 4 bolts. Tighten to 30 ft. lbs. (40 Nm).
10. Install the stabilizer link with the 2 nuts. Tighten to 52 ft. lbs. (70 Nm).
11. Install engine under cover sub-assembly. Tighten to 21 ft. lbs. (29 Nm).
12. Install front wheels.
13. Lower the vehicle.

Bush Stopper
Outer Side
Front Side
Protrusion
22140_FJCR_G0380

Fig. 142 Install the bushing onto the outer side of the bushing stopper and so the protrusion faces inner side of the vehicle

WHEEL HUB AND BEARING

REMOVAL & INSTALLATION

See Figure 143.

1. Before servicing the vehicle, refer to the precautions section.
2. Remove the front steering knuckle.
3. Remove the 4 bolts, wheel hub and dust cover from the steering knuckle.
4. Remove the O-ring from the wheel hub.

To install:

5. Apply MP grease to a new O-ring.
6. Install the new O-ring onto the axle hub.
7. Install the dust cover and axle hub onto the steering knuckle with the 4 bolts. Tighten to 59 ft. lbs. (80 Nm).
8. Install the front steering knuckle.

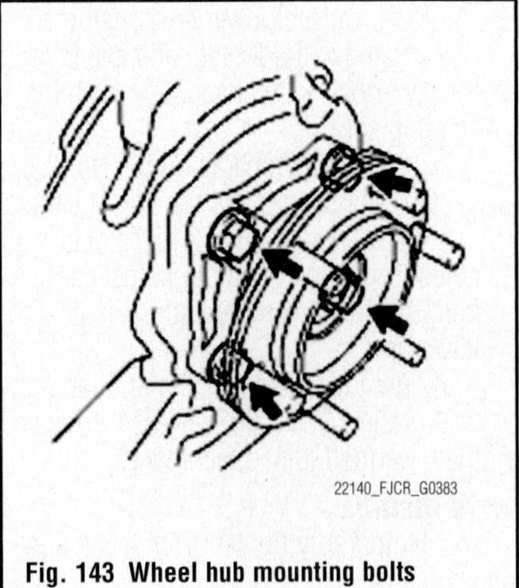

Fig. 143 Wheel hub mounting bolts

COIL SPRING

REMOVAL & INSTALLATION

1. Before servicing the vehicle, refer to the precautions section.
2. Raise and safely support the vehicle securely on jackstands.
3. Remove rear wheel.
4. Separate rear shock absorber.
   a. Support the rear axle housing.
   b. Remove the bolt and separate the shock absorber.
5. Separate rear brake tube flexible hose.
   a. Using a union nut wrench, separate the 2 brake tubes.

➡Use a container to catch the brake fluid.

   b. Remove the 2 clips and disconnect the 2 flexible hoses.
6. Remove rear coil spring.
   a. Start to lower the rear axle housing.

⁂ WARNING

Do not snap the brake line or the parking brake cable.

   b. While lowering the rear axle housing, remove the coil spring.

To install:.

7. Install rear coil spring.
   a. Install the coil spring to the rear axle housing.

➡Fit the lower end of the coil spring into the gap of the spring lower seat.

8. Temporarily tighten rear shock absorber.
   a. Install the shock absorber and temporarily tighten the bolt.
9. Lower the vehicle.
10. Bounce the vehicle up and down several times to stabilize the suspension.
11. Fully tighten rear shock absorber to 72 ft. lbs. (98 Nm).
12. Install rear brake tube flexible hose.
   a. Install the 2 flexible hoses with 2 new clips.
   b. Install the 2 brake tubes onto the flexible hose. Tighten to 11 ft. lbs. (15 Nm).
13. Fill reservoir with brake fluid.
14. Bleed brake line.
15. Check fluid level in reservoir.
16. Check for brake fluid leakage.
17. Install rear wheel.

CONTROL ARMS/LINKS

REMOVAL & INSTALLATION

Lateral Control Rod

1. Before servicing the vehicle, refer to the precautions section.
2. Raise and safely support the vehicle securely on jackstands.
3. Support the rear axle housing.
4. Remove the bolts and nuts and remove the lateral control rod.

➡While holding the nut, turn and remove the bolt.

To install:

5. Install the lateral control rod and temporarily tighten the bolt.
6. Temporarily tighten the bolt and nut.
7. Lower the vehicle.
8. Bounce the vehicle up and down several times to stabilize the suspension.
9. While holding the nut fully tighten the 2 bolts to 96 ft. lbs. (130 Nm).

Lower Control Arm

See Figure 144.

1. Before servicing the vehicle, refer to the precautions section.
2. Raise and safely support the vehicle securely on jackstands.
3. Remove the bolt and separate the parking brake cable.
4. Support the rear axle housing.
5. Remove the bolts and nuts and remove the lateral control rod.

➡While holding the nut, turn and remove the bolt.

To install:

6. Install the lateral control rod and temporarily tighten the bolt.
7. Install the parking brake cable onto the lower control arm. Tighten to 9 ft. lbs. (13 Nm).
8. Lower the vehicle.
9. Bounce the vehicle up and down several times to stabilize the suspension.
10. While holding the nut fully tighten the 2 bolts to 96 ft. lbs. (130 Nm).

Outer Side
Front
22140_FJCR_G0385

Fig. 144 Lateral control rod mounting bolts

SHOCK ABSORBER

REMOVAL & INSTALLATION

See Figures 145 and 146.

1. Before servicing the vehicle, refer to the precautions section.
2. Raise and safely support the vehicle securely on jackstands.
3. Remove rear wheels.
4. Support the rear axle housing.
5. Remove the bolt and separate the bottom of the shock absorber.
6. Remove the nut while keeping the piston rod from rotating.
7. Remove the 3 cushion retainers, the No. 1 cushion, the No. 2 cushion.
8. Remove the rear shock absorber.

To install:

9. Install the rear shock absorber.
10. Install the 3 cushion retainers, the No. 1 cushion, the No. 2 cushion.
11. Temporarily tighten rear shock absorber with coil spring.

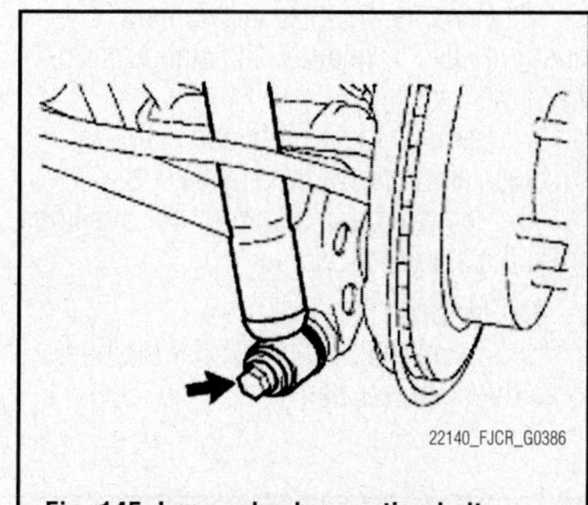

Fig. 145 Lower shock mounting bolts

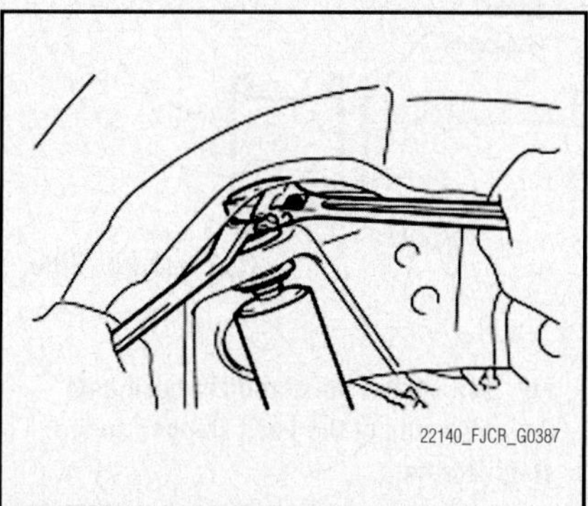

Fig. 146 Upper shock mounting nuts

12. While holding the piston rod, fully tighten a new nut to 18 ft. lbs. (25 Nm).
13. Install rear wheels and lower the vehicle.
14. Bounce the vehicle up and down several times to stabilize the suspension.
15. Fully tighten the nut. Tighten to 72 ft. lbs. (98 Nm).

TESTING

Compress and extend the shock absorber rod and check that there is no abnormal resistance or unusual sound during operation. If there is any abnormality, replace the shock absorber with a new one.

STABILIZER BAR

REMOVAL & INSTALLATION

See Figure 147.

1. Before servicing the vehicle, refer to the precautions section.
2. Raise and safely support the vehicle securely on jackstands.
3. Remove rear wheels.
4. Remove the nut and separate the stabilizer link from the stabilizer bar.

➡If the ball joint turns together with the nut, use a 6 mm hexagon wrench to hold the stud.

5. While holding the stabilizer link with a wrench, remove the nut.
6. Remove the 2 No. 1 retainers, 2 cushions, No. 2 retainer and stabilizer link from the chassis.
7. Remove the 4 bolts and stabilizer brackets and remove the stabilizer bar.
8. Remove the 2 stabilizer bar bushings from the stabilizer bar.

To install:

9. Install the 2 stabilizer bar bushings onto the stabilizer bar.

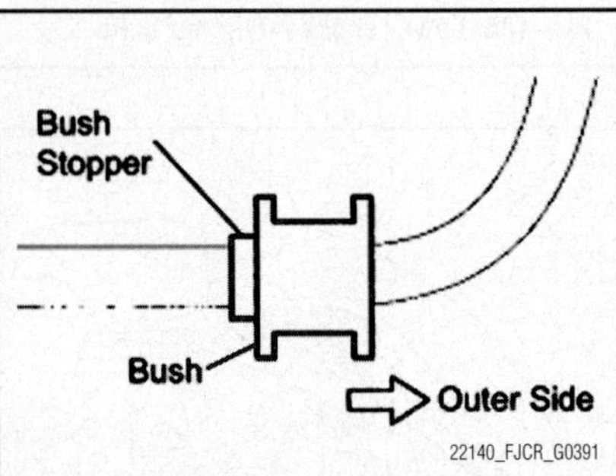

Fig. 147 Install the stabilizer bush onto the outer side of the bush stopper on the stabilizer bar

10. Install the stabilizer bush onto the outer side of the bush stopper on the stabilizer bar.
11. Install the stabilizer bar and 2 stabilizer brackets with the 4 bolts. Tighten to 22 ft. lbs. (30 Nm).
12. Install the stabilizer link with the nuts. Tighten to 52 ft. lbs. (70 Nm).
13. Install the 2 No. 1 retainers, 2 cushions and No. 2 retainer.
14. While holding the stabilizer link with a spanner, install a new nut and tighten to 11 ft. lbs. (15 Nm).
15. Install rear wheels.
16. Lower the vehicle.

WHEEL HUB AND BEARING

REMOVAL & INSTALLATION

See Figures 148 and 149.

1. Remove rear axle shaft from the housing.
2. Using a snap expander, remove the snap ring.

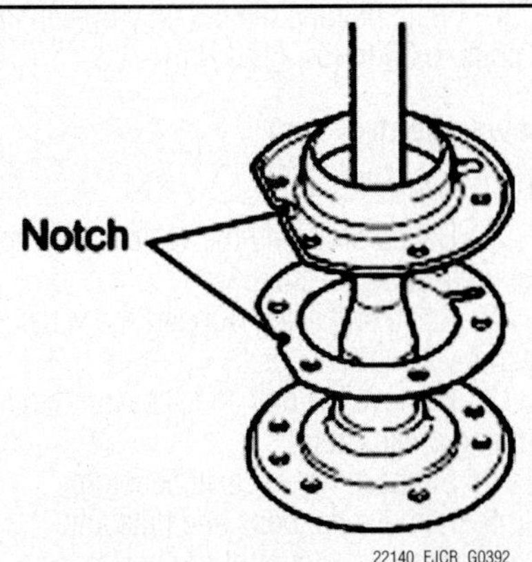

Fig. 148 Install a new deflector gasket and deflector onto the rear axle shaft and align the 2 notches

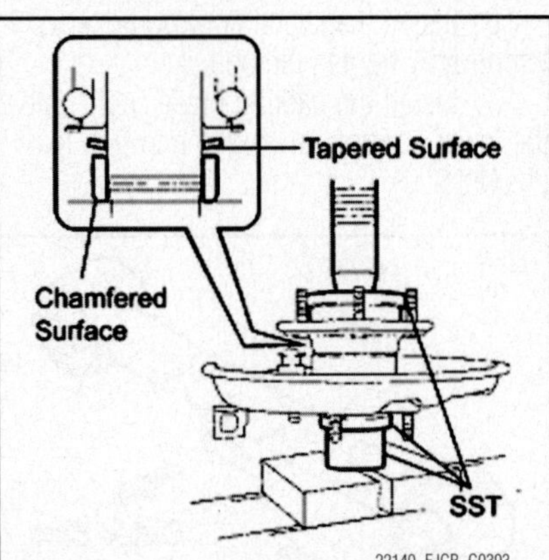

Fig. 149 Install the washer and a new retainer onto the axle hub in the orientations shown

3. Using a press, remove the rear axle shaft.
   a. Remove the rear axle bearing inner retainer from the axle hub.
   b. Remove the rear axle shaft washer from the axle hub.
   c. Grind the rear axle bearing inner race surface using a grinder, then remove it with a chisel.
   d. Remove the rear axle shaft oil seal from the rear axle shaft.
4. Remove rear axle hub and bearing assembly.
   a. Attach the 4 nuts to the housing bolts.
   b. Using a hammer, remove the 4 housing bolts and rear axle hub and bearing assembly.

⁂ WARNING

Do not reuse the nuts previously removed from the vehicle.

5. Remove brake drum oil deflector.
6. Remove the deflector and deflector gasket to the rear axle shaft.

To install:.

7. Install a new deflector gasket and deflector onto the rear axle shaft.
   a. Align the 2 notches.
   b. Install the washer and nut onto a new hub bolt, as shown in the illustration.
   c. Install the hub bolt by tightening the nut.
8. Install rear axle hub and bearing assembly.
   a. Position the parking brake plate on a new rear axle hub and bearing assembly and install the 4 housing bolts using 2 socket wrenches and a press.

⁂ WARNING

The left and right side bearing assemblies have different part numbers and are not interchangeable side to side.

9. Install rear axle shaft.
   a. Install the washer and a new retainer onto the axle hub in the orientations shown in the illustration.

➡Install the washer with its tapered surface facing downward. Install the retainer with its chamfered surface facing downward.

   b. Using a press, install the rear axle shaft onto the rear axle hub and bearing.
10. Using a snap ring expander, install a new snap ring.

TOYOTA

Highlander

15

SPECIFICATIONS AND MAINTENANCE CHARTS

ENGINE AND VEHICLE IDENTIFICATION

Engine							Model Year	
Code ①	Liters (cc)	Cu. In.	Cyl.	Fuel Sys.	Engine Type	Eng. Mfg.	Code ②	Year
2AZ-FE	2.4 (2362)	144	4	SFI	DOHC	Toyota	7	2007
3MZ-FE	3.3 (3311)	202	6	SFI	DOHC	Toyota	8	2008
2GR-FE	3.5 (3456)	210	6	SFI	DOHC	Toyota		

SFI: Sequential Fuel Injection

DOHC: Double Overhead Camshaft

22140_HIGH_C0001

GENERAL ENGINE SPECIFICATIONS

Year	Model	Engine Displacement Liters	Engine Series ID	Net Horsepower @ rpm	Net Torque @ rpm (ft. lbs.)	Bore x Stroke (in.)	Com-pression Ratio	Oil Pressure @ rpm
2007	Highlander	2.4	2AZ-FE	155@5700	163@4000	3.48x3.78	NA	36@3000
		3.3	3MZ-FE	215@5600	222@3600	3.62x3.27	NA	36-78@3000
2008	Highlander	3.5	2GR-FE	270@6200	248@4700	3.70x3.27	NA	36-78@3000

NA: Information not available

22140_HIGH_C0002

ENGINE TUNE-UP SPECIFICATIONS

Year	Engine Displacement Liters	Engine ID	Spark Plug Gap (in.)	Ignition Timing (deg.)*	Fuel Pump (psi)	Idle Speed (rpm)	Valve Clearance	
							Intake	Exhaust
2007	2.4	2AZ-FE	0.041	8-12B	44-50	600-700	0.007-0.011	0.012-0.016
	3.3	3MZ-FE	0.039-0.043	8-12B	44-50	650-750	0.006-0.010	0.010-0.014
2008	3.5	2GR-FE	0.043	N/A	44-50	650-750	N/A	N/A

NOTE: The Vehicle Emission Control Information label often reflects specification changes made during production. The label figures must be used if they differ from those in this chart.

B: Before top dead center

* With terminals TC and CG connected to DLC3

22140_HIGH_C0003

CAPACITIES

Year	Model	Engine Displacement Liters	Engine ID	Engine Oil with Filter (qts.)	Transmission (pts.) 5-Spd	Transmission (pts.) Auto.*	Transfer Case (pts.)	Drive Axle Front (pts.)	Drive Axle Rear (pts.)	Fuel Tank (gal.)	Cooling System (qts.)
2007	Highlander	2.4	2AZ-FE	4.0	—	3.7	2.0	—	2.0	19.1	6.8
		3.3	3MZ-FE	5.0	—	3.7	2.0	—	2.0	19.1	10.7
2008	Highlander	3.5	2GR-FE	6.4	—	3.7	2.0	—	2.0	19.1	①

*After draining, add the following amounts, then, fill to the cold full line.

① Non-towing pkg. 9.3 qt. w/o rear air, 11.6 qts. w/ rear air
Towing pkg. 10 qt. w/o rear air, 12.4 qts. w/ rear air

22140_HIGH_C0004

VALVE SPECIFICATIONS

Year	Engine Displacement Liters	Engine ID	Seat Angle (deg.)	Face Angle (deg.)	Spring Test Pressure (lbs. @ in.)	Spring Installed Height (in.)	Stem-to-Guide Clearance (in.) Intake	Stem-to-Guide Clearance (in.) Exhaust	Stem Diameter (in.) Intake	Stem Diameter (in.) Exhaust
2007	2.4	2AZ-FE	45	44.5	NA	NA	0.0010-0.0024	0.0012-0.0026	0.2154-0.2159	0.2152-0.2157
	3.3	3MZ-FE	45	40.5	41.9-46.3@ 1.437	1.331	0.0010-0.0024	0.0012-0.0026	0.2154-0.2159	0.2152 0.2156
2008	3.5	2GR-FE	45	44.5	NA	NA	0.0010-0.0024	0.0012-0.0026	0.2154-0.2159	0.2151-0.2157

NA: Information not available

22140_HIGH_C0005

CRANKSHAFT AND CONNECTING ROD SPECIFICATIONS

All measurements are given in inches.

Year	Engine Displacement Liters	Engine ID	Crankshaft Main Brg. Journal Dia.	Crankshaft Main Brg. Oil Clearance	Crankshaft Shaft End-play	Crankshaft Thrust on No.	Connecting Rod Journal Diameter	Connecting Rod Oil Clearance	Connecting Rod Side Clearance
2007	2.4	2AZ-FE	2.0654-2.1648	0.0009-0.0019	0.0016-0.0094	2	1.8894-1.8898	0.0009-0.0019	0.0063-0.0143
	3.3	3MZ-FE	2.4011-2.4016	①	0.0016-0.0095	2	2.0863-2.0866	0.0015-0.0026	0.0059-0.0118
2008	3.5	2GR-FE	2.4011-2.4016	0.0010-0.0019	0.0016-0.0095	2	2.0863-2.0866	0.0018-0.0026	0.0059-0.0157

① Journals 1 and 4: 0.0006 - 0.0013 in.
Journals 2 and 3: 0.0010 - 0.0018 in.

22140_HIGH_C0006

PISTON AND RING SPECIFICATIONS

All measurements are given in inches.

Year	Engine Displ. Liters	Engine ID	Piston Clearance	Ring Gap Top Comp.	Ring Gap Bottom Comp.	Ring Gap Oil Control	Ring Side Clearance Top Comp.	Ring Side Clearance Bottom Comp.	Ring Side Clearance Oil Control
2007	2.4	2AZ-FE	0.0020-0.0029	0.0087-0.0126	0.0197-0.0236	0.0039-0.0138	0.0012-0.0028	0.0012-0.0028	SNUG
	3.3	3MZ-FE	0.0013-0.0023	0.0118-0.0138	0.0197-0.0236	0.0059-0.0157	0.0012-0.0031	0.0008-0.0024	0.0012-0.0043
2008	3.5	2GR-FE	0.0018-0.0020	0.0098-0.0138	0.0197-0.0236	0.0039-0.0157	0.0008-0.0028	0.0008-0.0024	0.0028-0.0059

22140_HIGH_C0007

TORQUE SPECIFICATIONS

All readings in ft. lbs.

Year	Engine Displacement Liters	Engine ID	Cylinder Head Bolts	Main Bearing Bolts	Rod Bearing Bolts	Crankshaft Damper Bolts	Flywheel Bolts	Manifold Intake	Manifold Exhaust	Spark Plugs	Oil Pan Drain Plug
2007	2.4	2AZ-FE	①	29	②	125	72	22	27	14	18
	3.3	3MZ-FE	③	④	②	162	61	11	36	18	33
2008	3.5	2GR-FE	⑤	⑥	②	184	61	15	15	18	33

① Step 1: 58
Step 2: plus 90 degrees

② Step 1: 18 ft. lbs.
Step 2: Plus 90 degrees

③ Step 1: 12 point bolts to 40 ft. lbs.
Step 2: 12 point bolts plus 90 degrees
Step 3: Hex head recessed bolt to 13 ft. lbs.

④ Step 1: 12 point cap bolts to 16 ft. lbs.
Step 2: 12 point cap bolts plus 90 degrees
Step 3: Hex head side bolts to 20 ft. lbs.

⑤ Step 1: 10mm bolts to 27 ft. lbs.
Step 2: 10mm point cap bolts plus 90 degrees
Step 3: 10mm point cap bolts plus 90 degrees
Step 4: Front bolts to 22 ft. lbs.

⑥ Step 1: 16 cap bolts to 45 ft. lbs.
Step 2: 16 cap bolts plus 90 degrees
Step 3: 8 side bolts to 38 ft. lbs.

22140_HIGH_C0008

WHEEL ALIGNMENT

Year	Model		Caster Range (+/-Deg.)	Caster Preferred Setting (Deg.)	Camber Range (+/-Deg.)	Camber Preferred Setting (Deg.)	Toe-in (in.)	Steering Axis Inclination (Deg.)
2007	Highlander	2WD F	0.75	+2.75	0.75	-0.67	0+/-0.08	10.75+/-0.75
		4WD F	0.75	+2.75	0.75	-0.58	0+/-0.08	10.58+/-0.75
		2WD R	—	—	0.75	-1.33	0.12+/-0.08	—
		4WD R	—	—	0.75	-0.75	0.12+/-0.08	—
2008	Highlander	Front	0.75	+2.62	0.75	-0.63	0+/-0.08	11.02+/-0.75
		2WD R	—	—	0.75	-1.00	0.12+/-0.08	—
		4WD R	—	—	0.75	-0.60	0.12+/-0.08	—

22140_HIGH_C0009

TIRE, WHEEL AND BALL JOINT SPECIFICATIONS

Year	Model	OEM Tires Standard	OEM Tires Optional	Tire Pressures (psi) Front	Tire Pressures (psi) Rear	Wheel Size	Ball Joint Inspection	Lug Nut Torque (ft. lbs.)
2007	Highlander	P225/70R16	P225/65R17	30	30	6.5-JJ	①	76
2008	Highlander	P245/60R17	P245/50R19	30	30	7.5-J	①	76

OEM: Original Equipment Manufacturer

PSI: Pounds Per Square Inch

STD: Standard

OPT: Optional

① Replace if any measurable movement is found.

22140_HIGH_C0010

BRAKE SPECIFICATIONS

All measurements in inches unless noted

Year	Model		Brake Disc Original Thickness	Brake Disc Minimum Thickness	Brake Disc Maximum Runout	Minimum Lining Thickness	Brake Caliper Bracket Bolts (ft. lbs.)	Brake Caliper Mounting Bolts (ft. lbs.)
2007	Highlander	F	1.102	1.024	0.0020	0.039	77	25
		R	0.394	0.335	0.0059	0.039	58	32
2008	Highlander	F	1.102	1.024	0.0020	0.039	77	25
		R	0.394	0.335	0.0059	0.039	58	32

F: Front

R: Rear

22140_HIGH_C0011

SCHEDULED MAINTENANCE INTERVALS

TOYOTA—HIGHLANDER

TO BE SERVICED	TYPE OF SERVICE	VEHICLE MILEAGE INTERVAL (x1000)												
		7.5	15	22.5	30	37.5	45	52.5	60	67.5	75	82.5	90	97.5
Engine oil & filter	R	✓	✓	✓	✓	✓	✓	✓	✓	✓	✓	✓	✓	✓
Automatic transmission fluid	S/I		✓		✓		✓		✓		✓		✓	
Ball joints & dust covers	S/I		✓		✓		✓		✓		✓		✓	
Bolts & nuts on chassis & body	S/I		✓		✓		✓		✓		✓		✓	
Brake linings & drums	S/I		✓		✓		✓		✓		✓		✓	
Brake line pipes & hoses	S/I		✓		✓		✓		✓		✓		✓	
Brake pads & discs (front & rear)	S/I		✓		✓		✓		✓		✓		✓	
Propeller shaft grease	S/I		✓		✓		✓		✓		✓		✓	
Steering knuckle & chassis grease	S/I		✓		✓		✓		✓		✓		✓	
Steering linkage	S/I		✓		✓		✓		✓		✓		✓	
Air cleaner filter	R				✓				✓				✓	
Spark plugs ①	R				✓				✓				✓	
Drive belts	S/I				✓				✓				✓	
Exhaust pipes & mountings	S/I				✓				✓				✓	
Fuel lines & connections	S/I				✓				✓				✓	
Engine coolant	R						✓				✓			
Charcoal canister	R								✓					
Fuel tank cap gasket	R								✓					
Heated oxygen sensors (except Calif.) ②	R													

R: Replace S/I: Service or Inspect

① Platinum plugs are replaced at 100,000 mile intervals

② Heated oxygen sensors (except Calif.): replace every 80,000 miles.

FREQUENT OPERATION MAINTENANCE (SEVERE SERVICE)

If a vehicle is operated under any of the following conditions it is considered severe service:

- Extremely dusty areas.

- 50% or more of the vehicle operation is in 32°C (90°F) or higher temperatures, or constant temperatures below 0°C (32°F).

- Prolonged idling (vehicle operation in stop and go traffic).

- Frequent short running periods (engine does not warm to normal operating temperatures).

- Police, taxi, delivery usage or trailer towing usage.

Air cleaner filter: service or inspect every 3750 miles

Engine oil & filter: replace every 3750 miles.

Ball joints & dust covers: service or inspect every 7500 miles.

Bolts & nuts on chassis & body: service or inspect every 7500 miles.

Brake pads & discs (front & rear): service or inspect every 7500 miles.

Steering knuckle & chassis grease: service or inspect every 7500 miles.

Steering linkage: service or inspect every 7500 miles.

Exhaust pipes & mountings: service or inspect every 15,000 miles.

22140_HIGH_C0012

PRECAUTIONS

Before servicing any vehicle, please be sure to read all of the following precautions, which deal with personal safety, prevention of component damage, and important points to take into consideration when servicing a motor vehicle:

- Never open, service or drain the radiator or cooling system when the engine is hot; serious burns can occur from the steam and hot coolant.
- Observe all applicable safety precautions when working around fuel. Whenever servicing the fuel system, always work in a well-ventilated area. Do not allow fuel spray or vapors to come in contact with a spark, open flame, or excessive heat (a hot drop light, for example). Keep a dry chemical fire extinguisher near the work area. Always keep fuel in a container specifically designed for fuel storage; also, always properly seal fuel containers to avoid the possibility of fire or explosion. Refer to the additional fuel system precautions later in this section.
- Fuel injection systems often remain pressurized, even after the engine has been turned **OFF**. The fuel system pressure must be relieved before disconnecting any fuel lines. Failure to do so may result in fire and/or personal injury.
- Brake fluid often contains polyglycol ethers and polyglycols. Avoid contact with the eyes and wash your hands thoroughly after handling brake fluid. If you do get brake fluid in your eyes, flush your eyes with clean, running water for 15 minutes. If eye irritation persists, or if you have taken brake fluid internally, IMMEDIATELY seek medical assistance.
- The EPA warns that prolonged contact with used engine oil may cause a number of skin disorders, including cancer. You should make every effort to minimize your exposure to used engine oil. Protective gloves should be worn when changing oil. Wash your hands and any other exposed skin areas as soon as possible after exposure to used engine oil. Soap and water, or waterless hand cleaner should be used.
- All new vehicles are now equipped with an air bag system, often referred to as a Supplemental Restraint System (SRS) or Supplemental Inflatable Restraint (SIR) system. The system must be disabled before performing service on or around system components, steering column, instrument panel components, wiring and sensors. Failure to follow safety and disabling procedures could result in accidental air bag deployment, possible personal injury and unnecessary system repairs.
- Always wear safety goggles when working with, or around, the air bag system. When carrying a non-deployed air bag, be sure the bag and trim cover are pointed away from your body. When placing a non-deployed air bag on a work surface, always face the bag and trim cover upward, away from the surface. This will reduce the motion of the module if it is accidentally deployed. Refer to the additional air bag system precautions later in this section.
- Clean, high quality brake fluid from a sealed container is essential to the safe and proper operation of the brake system. You should always buy the correct type of brake fluid for your vehicle. If the brake fluid becomes contaminated, completely flush the system with new fluid. Never reuse any brake fluid. Any brake fluid that is removed from the system should be discarded. Also, do not allow any brake fluid to come in contact with a painted surface; it will damage the paint.
- Never operate the engine without the proper amount and type of engine oil; doing so WILL result in severe engine damage.
- Timing belt maintenance is extremely important. Many models utilize an interference-type, non-freewheeling engine. If the timing belt breaks, the valves in the cylinder head may strike the pistons, causing potentially serious (also time-consuming and expensive) engine damage. Refer to the maintenance interval charts for the recommended replacement interval for the timing belt, and to the timing belt section for belt replacement and inspection.
- Disconnecting the negative battery cable on some vehicles may interfere with the functions of the on-board computer system(s) and may require the computer to undergo a relearning process once the negative battery cable is reconnected.
- When servicing drum brakes, only disassemble and assemble one side at a time, leaving the remaining side intact for reference.
- Only an MVAC-trained, EPA-certified automotive technician should service the air conditioning system or its components.

BRAKES

ANTI-LOCK BRAKE SYSTEM (ABS)

GENERAL INFORMATION

Anti-lock Brake System (ABS): The ABS helps prevent the wheels from locking when the brakes are applied firmly or when braking on a slippery surface.

➡The skid control ECU is located within the brake actuator assembly.

➡The yaw rate sensor and acceleration sensor are combined in a single unit. This unit communicates with the skid control ECU through CAN communication.

The skid control ECU detects wheel lock based on speed signals received from the wheel speed sensors. Based on this information, the skid control ECU controls the pump motor and solenoid valves. The pump motor and solenoid valves are used to prevent wheel lock by controlling the hydraulic pressure applied to the brakes at each wheel. The ABS warning light will come on when the system is malfunctioning.

➡For vehicles with the multi-information display, the master caution indicator light will come on and a check ABS message will be displayed on the multi-information display if the ABS control cannot be performed.

Electronic Brake force Distribution (EBD): The EBD control utilizes ABS, and performs proper brake force distribution between the front and rear wheels in accordance with driving conditions. When braking while cornering, it also controls the brake force distribution between the right and left wheels, helping to maintain vehicle behavior.

The skid control ECU receives a speed signal from each wheel speed sensor, and uses these signals to detect locking of the wheels. The ECU uses this information in order to determine appropriate control of the solenoid valves. The solenoid valves control the hydraulic pressure applied to the brake cylinder at each wheel. In this way, the solenoid valves are used to control the brake power split between the front and rear, and left and right wheels. The ABS waning and brake warning lights will come on if there is a malfunction in the EBD system.

➡For vehicles with the multi-information display, the master caution indicator light will come on and a brake malfunction message will be displayed on the multi-information display if the EBD system is malfunctioning.

Brake Assist (BA): The primary purpose of the brake assist system is to provide auxiliary brake force to assist a driver who cannot generate a large enough brake force during emergency braking, thus helping to maximize the brake performance of the vehicle.

The skid control ECU receives a speed signal from each speed sensor and a fluid pressure signal from the master cylinder pressure sensor to determine whether brake assist is necessary. If brake assist is necessary, the skid control ECU sends control signals to the pump motor and solenoid. The pump and the solenoid valves then control the pressure applied to each wheel cylinder. The ABS warning light will come on to indicate a malfunction in the BA system.

➡For vehicles with the multi-information display, the master caution indicator light will come on and a check ABS message will be displayed on the multi-information display if the BA system is malfunctioning.

BRAKES — BLEEDING THE BRAKE SYSTEM

BLEEDING PROCEDURE

When any work is done on the brake system that includes disconnecting fluid lines, or if air in the brake lines is suspected, bleed the air from the system.

⁂ WARNING

Do not let brake fluid remain on painted surfaces—it will eat away the paint if left on too long. Wash it off immediately.

Before proceeding, fill the brake fluid reservoir with brake fluid: SAE J1703 or FMVSS no. 116 DOT3

Bleeding the Master Cylinder

See Figure 1.

If the master cylinder has been disassembled or if the reservoir becomes empty, bleed the air from the master cylinder.

1. Remove the air cleaner assembly with hose.
2. Disconnect the brake lines from the master cylinder, using 12mm union nut wrench, or a suitable brake line wrench.
3. Have an assistant slowly depress the brake pedal and hold it.
4. Cover the outer holes with your fingers, and have your assistant release the brake pedal.
5. Repeat steps 3 and 4 several times.

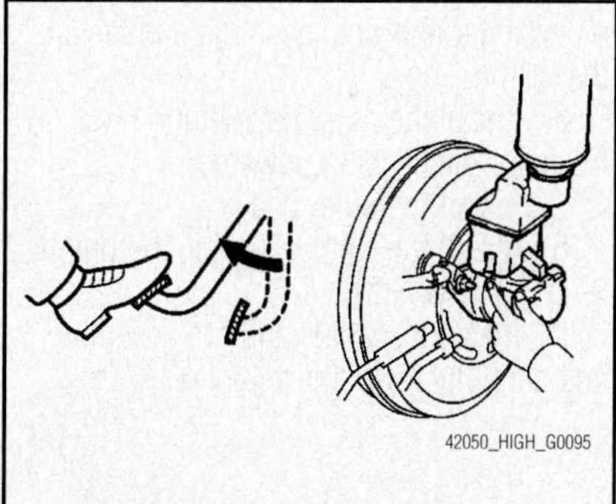

42050_HIGH_G0095

Fig. 1 Use your fingers to cover the outer holes, then release the brake pedal

6. Connect the brake lines. Tighten as follows:
   a. For 2006–07 models, tighten to 11 ft. lbs. (15 Nm).
   b. For 2008 models, use a torque wrench with a fulcrum length of 250 mm (9.84 inches) and tighten to 14 ft. lbs. (19 Nm). This torque value is effective when the union nut wrench is parallel to the torque wrench.
7. Install the air cleaner assembly with hose.

Bleeding the Brake Lines

See Figure 2.

1. Raise and safely support the vehicle.
2. Connect a piece of vinyl tubing to the brake caliper.
3. Have an assistant depress the brake pedal several times, then loosen the bleeder plug while the pedal is depressed.
4. When fluid stops coming out, tighten the bleeder plug, then release the brake pedal.
5. Repeat steps 2 and 3 until all the air in the fluid has been bled out.
6. Tighten the brake bleeder plug to 73 inch lbs. (8.3 Nm).
7. Repeat the above steps to bleed the air out of the brake line for each wheel.

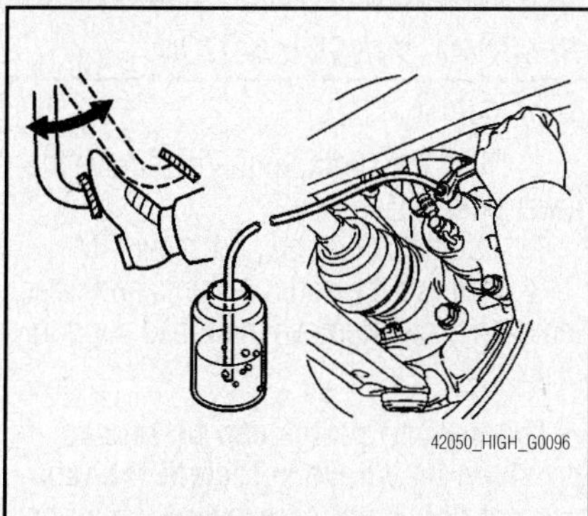

42050_HIGH_G0096

Fig. 2 Bleeding the brake lines at each wheel

Bleeding the Actuator Assembly

See Figure 3.

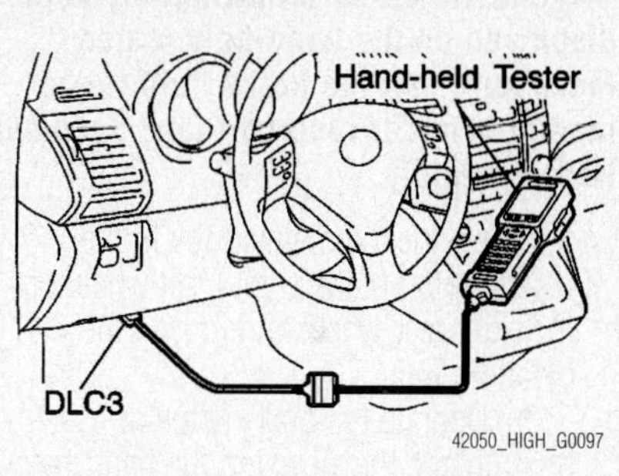

42050_HIGH_G0097

Fig. 3 Connect a suitable hand-held tester to the DLC3 to bleed the actuator assembly

➡After bleeding the air from the brake system, if the height or feel of the brake pedal cannot be obtained, perform air bleeding in the brake actuator assembly with a hand-held tester by following the procedures below.

1. Depress the brake pedal more than 20 times with the engine off.
2. Connect the hand-held tester to the DLC3, then turn the ignition switch to the **ON** position, but do NOT start the engine.
3. Select "AIR BLEEDING" on the hand-held tester.

➡Refer to the hand-held tester operator's manual for more details.

4. Bleed the air out of the regular brake line when "Step 1: Increase" appears on the hand-held tester display, as follows:

➡Bleed the air by following the steps displayed on the hand-held tester. Make sure that the brake fluid in the master cylinder reservoir tank does not become empty.

   a. Connect the vinyl tube to either one of the bleeder plugs.
   b. Have an assistant depress the brake pedal several times, then loosen the bleeder plug connected to the vinyl tube with the pedal depressed.
   c. When fluid stops coming out, tighten the bleeder plug and release the brake pedal.

d. Repeat the previous 2 steps until all the air in the fluid is completely bled out.

e. Tighten the bleeder plug completely to 73 inch lbs. (8.3 Nm).

f. Repeat the above procedures for each wheel to bleed the air out of the brake line.

5. Bleed the air out of the suction line when "Step 2: Inhalation" appears on the hand-held tester display, as follows:

➡Bleed the air by following the steps displayed on the hand-held tester. Make sure that the brake fluid in the master cylinder reservoir tank does not become empty.

a. Connect the vinyl tube to the bleeder plug at the right front wheel or the right rear wheel and loosen the bleeder plug.

b. Operate the brake actuator assembly to bleed the air using the hand-held tester.

➡This operation stops automatically after 4 seconds. At this time, be sure to release the brake pedal.

c. Check if the operation has stopped by referring to the hand-held tester display.

d. Repeat the previous 2 steps until all air in the fluid is completely bled out.

e. Tighten the bleeder plug completely to 73 inch lbs. (8.3 Nm).

f. Repeat the above procedures to bleed the air out of the brake line for each wheel.

6. Bleed the air out of the pressure reduction line when "Step 3: Decrease" appears on the hand-held tester display, as follows:

➡Bleed the air by following the steps displayed on the hand-held tester. Make sure that the brake fluid in the master cylinder reservoir tank does not become empty.

a. Connect a vinyl tube to either one of the bleeder plugs.

b. Loosen the bleeder plug.

c. Using the hand-held tester, operate the brake actuator assembly, completely depress the brake pedal and keep it depressed.

➡The operation stops automatically after 4 seconds. When performing this procedure continuously, set an interval of at least 20 seconds. When the operation is complete, the brake pedal goes down slightly. This is a normal phenomenon caused when the solenoid opens. During this procedure, the pedal will feel heavy, but completely depress it so that the brake fluid comes out from the bleeder plug. Be sure to keep depressing the brake pedal. Do not depress and release the pedal repeatedly.

d. Tighten the bleeder plug, then release the brake pedal.

e. Repeat the previous 3 steps until all the air in the fluid is completely bled out.

f. Tighten the bleeder plug completely to 73 inch lbs. (8.3 Nm).

g. Repeat the above procedures for each wheel to bleed the air out of the brake line.

7. Bleed the air out of the regular brake line again when "Step 4: Increase" appears on the hand-held tester display, as follows:

➡Bleed the air by following the steps displayed on the hand-held tester. Make sure that the brake fluid in the master cylinder reservoir tank does not become empty.

a. Connect the vinyl tube to either one of the bleeder plugs.

b. Depress the brake pedal several times, then loosen the bleeder plug connected to the vinyl tube with the pedal depressed.

8. When fluid stops coming out, tighten the bleeder plug, then release the brake pedal.

a. Repeat the previous 2 steps until all the air in the fluid is completely bled out.

b. Tighten the bleeder plug completely to 73 inch lbs. (8.3 Nm).

c. Repeat the above procedures for each wheel to bleed the air out of the brake line.

9. Check the fluid level and add fluid if necessary. Use SAE J1703 or FMVSS No. 116 DOT3 Brake fluid

BRAKES

FRONT DISC BRAKES

BRAKE CALIPER

REMOVAL & INSTALLATION

See Figure 4.

1. Disconnect the brake line from the caliper and plug it.
2. Hold the caliper slide pins and remove the mounting bolts.
3. Lift off the caliper.
4. Remove the pads and anti-squeal shims.
5. Remove the wear indicator from the inner pad.
6. Installation is the reverse of removal. Grease the caliper slides and bolts with lithium grease or equivalent. Apply disc brake grease to the anti-squeal shims. Torque the caliper bolts to 25 ft. lbs. (34 Nm); the brake line union bolt to 21 ft. lbs. (29 Nm).

DISC BRAKE PADS

REMOVAL & INSTALLATION

See Figure 4.

1. Hold the sliding pin and remove the lower bolt.
2. Lift the caliper up and secure it.
3. Remove the pads, 4 shims and wear indicator plate. Remove the 2 pad support plates.

➡The support plates can be reused, provided they have sufficient rebound, are not deformed or cracked, show no signs of wear and are cleaned of all rust and debris.

To install:

➡Always use new shims and wear indicators, even when re-installing the original pads.

4. Install a wear indicator plate on the inner pad.
5. Apply disc brake grease to both sides of the inner anti-squeal shims and install the shims.
6. Install the inner pad with the wear indicator plate facing upwards.
7. Install the outer pad.
8. Install the caliper. Torque the bolt to 25 ft. lbs. (34 Nm).
9. Install the wheel and tire assembly and carefully lower the vehicle.

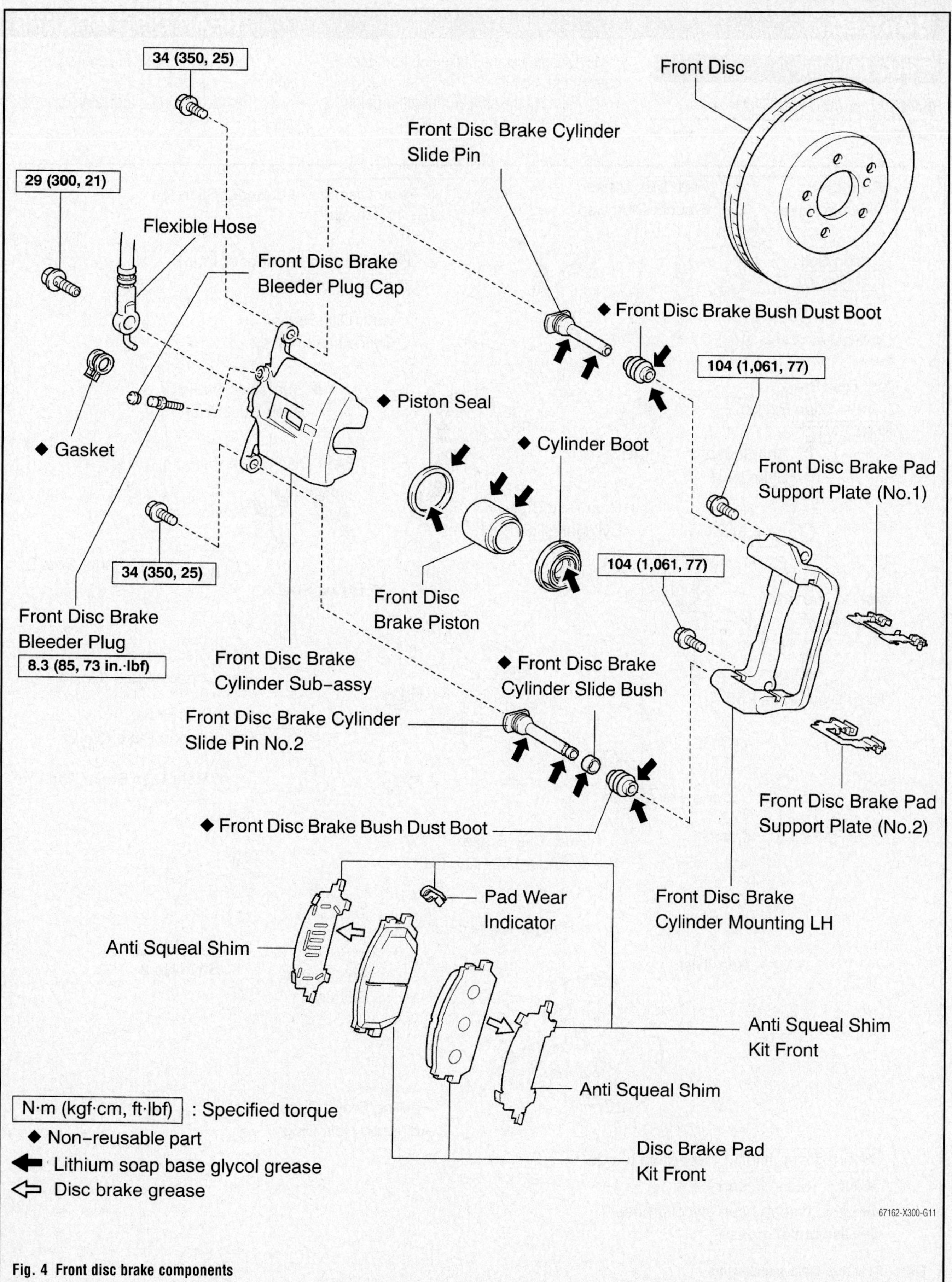

Fig. 4 Front disc brake components

BRAKE CALIPER

REMOVAL & INSTALLATION

See Figure 5.

1. Disconnect the brake line from the caliper and plug it.
2. Remove the caliper mounting bolts.
3. Lift off the caliper.
4. Remove the pads and anti-squeal shims.
5. Remove the wear indicators from each pad.

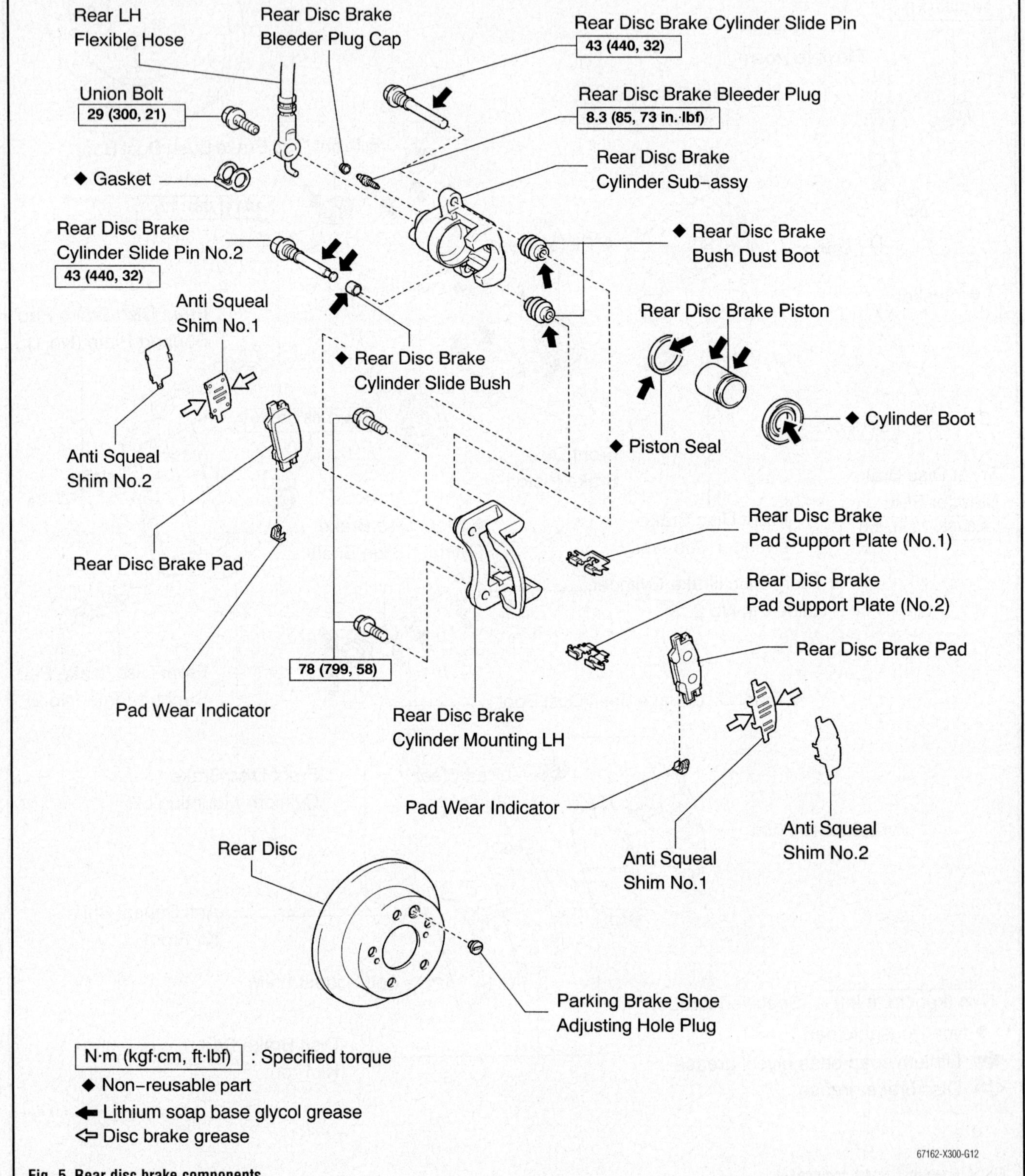

Fig. 5 Rear disc brake components

6. Installation is the reverse of removal. Grease the caliper slides and bolts with lithium grease or equivalent. Apply disc brake grease to the anti-squeal shims. For 2007 models, torque the caliper bolts to 32 ft. lbs. (43 Nm); the brake line union bolt to 21 ft. lbs. (29 Nm). For 2008 models, torque the caliper bolts to 25 ft. lbs. (34 Nm); the brake line union bolt to 24 ft. lbs. (33 Nm).

DISC BRAKE PADS

REMOVAL & INSTALLATION

See Figure 5.

1. Disconnect the brake line from the caliper and plug it.
2. Remove the caliper mounting bolts.
3. Lift off the caliper.
4. Remove the pads and anti-squeal shims.
5. Remove the wear indicators from each pad.
6. Installation is the reverse of removal. Grease the caliper slides and bolts with lithium grease or equivalent. Apply disc brake grease to the anti-squeal shims. For 2007 models, torque the caliper bolts to 32 ft. lbs. (43 Nm); the brake line union bolt to 21 ft. lbs. (29 Nm). For 2008 models, torque the caliper bolts to 25 ft. lbs. (34 Nm); the brake line union bolt to 24 ft. lbs. (33 Nm).

PARKING BRAKE SHOES

REMOVAL & INSTALLATION

See Figures 6 through 16.

1. Raise and safely support the vehicle.
2. Remove the rear wheel and tire assemblies.
3. Unbolt and remove the rear caliper, but do not disconnect the fluid line. Suspend the caliper out of the way with a piece of wire.
4. Matchmark the brake disc (rotor) to the axle hub.
5. Make sure the parking brake is fully released, then remove the rear brake disc (rotor).

➡If the rotor cannot be easily removed, turn the shoe adjuster until the wheel turns freely.

6. Inspect the brake disc (rotor) inside diameter, as follows:
   a. Using a brake drum gauge or equivalent, measure the inside diameter of the disc and compare with the following: Standard inside diameter: 190 mm (7.48 in.). Maximum inside diameter: 191 mm (7.52 in.)
   b. If the inside diameter exceeds the maximum, replace the brake disc.
7. Use needle-nose pliers to remove the 2 parking brake shoe return tension springs.
8. Remove the parking brake shoe strut, as follows:
   a. Remove the parking brake shoe strut and the parking brake shoe strut compression spring.
9. Remove parking brake shoe no. 1, as follows:
   a. Remove the parking brake shoe hold down spring cup No. 1, parking brake shoe hold down spring and parking brake shoe hold down spring cup no. 2.
   b. FWD vehicles, remove the parking brake shoe hold down spring pin No. 1.
   c. Disconnect the parking brake shoe return spring no. 2 and remove the parking brake shoe assembly lh no. 1.

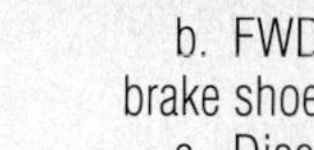

10. Remove parking brake shoe adjusting screw set:
    a. Remove the parking brake shoe adjusting screw set.
    b. Remove the parking brake shoe return tension spring No. 2.
11. Remove parking brake shoe assembly no. 2:
    a. Remove the parking brake shoe hold down spring cup No. 1, parking

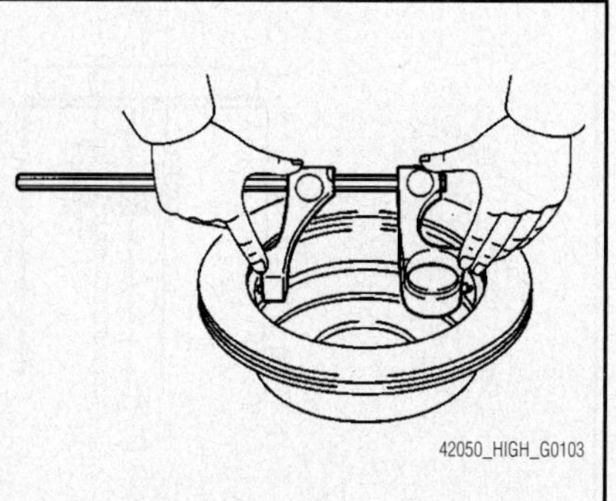

42050_HIGH_G0103

Fig. 7 Use a brake drum gauge to measure the inside of the disc

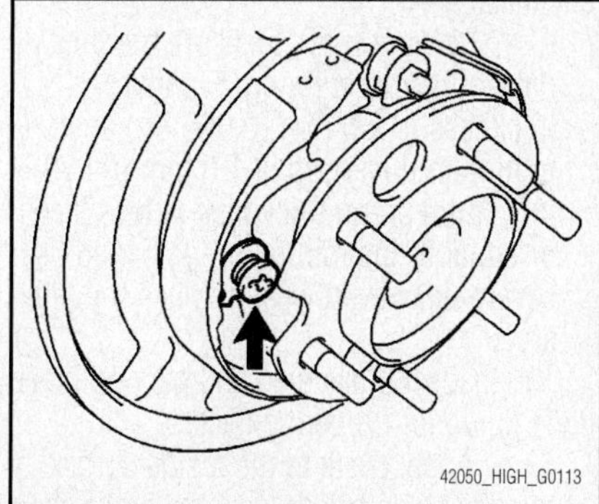

42050_HIGH_G0113

Fig. 9 Remove the parking brake shoe hold-down spring cups

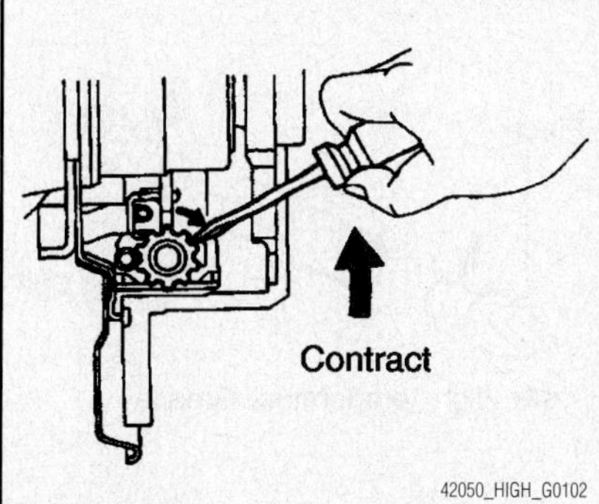

42050_HIGH_G0102

Fig. 6 If you have difficulty removing the rotor, turn the shoe adjuster until the wheel turns freely

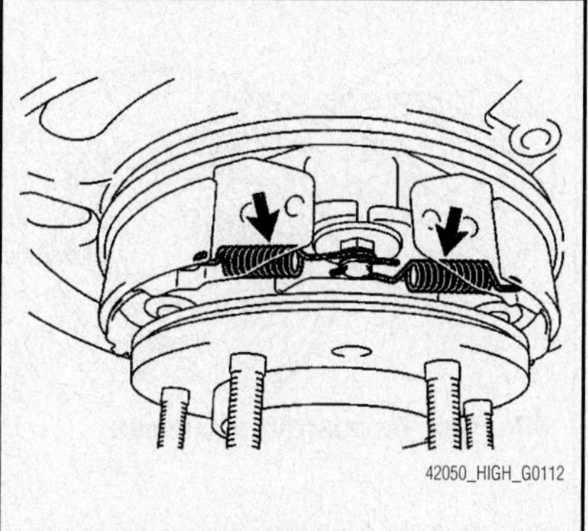

42050_HIGH_G0112

Fig. 8 Use a needle-nose pliers to remove the 2 return tension springs

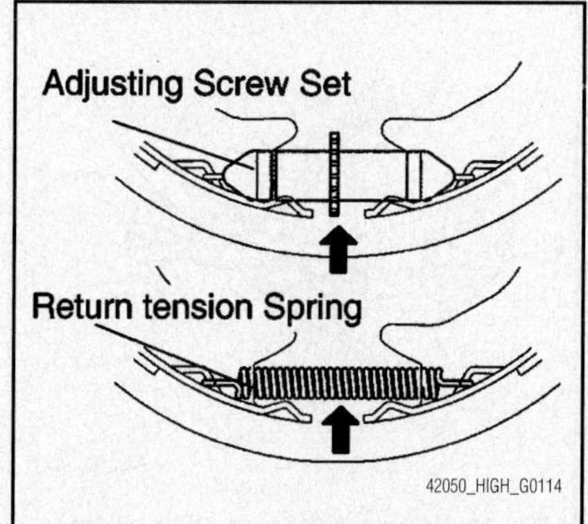

42050_HIGH_G0114

Fig. 10 Remove the parking brake shoe adjusting screw set and shoe return spring

brake shoe hold down spring, parking brake shoe hold down spring cup no. 2 and parking brake shoe hold down spring pin no. 2.

b. Remove the parking brake shoe assembly lh no. 2.

c. Using needle-nose pliers, disconnect the parking brake cable no. 3 from the parking brake cable shoe lever.

WARNING

Be careful not to damage parking brake cable no. 3.

12. On 4WD models, separate the rear speed sensor.
13. On 4WD models, remove the rear axle shaft nut.
14. On 4WD models, remove rear axle hub & bearing assembly
15. On 4WD models, remove parking brake shoe hold down spring pin.
16. Remove parking brake shoe type C-washer, as follows:

a. Using a screwdriver, remove the c-washer.

b. Remove the shim and parking brake shoe lever from the parking brake shoe no. 2.

17. Inspect parking brake shoe lining thickness:

a. Using a ruler, measure the thickness of the shoe lining. Standard thickness is 2.5 mm (0.098 in.) and minimum thickness is 1.0 mm (0.039 in.). If the lining thickness is less than or equal to the minimum, or If there is severe or uneven wear, replace the brake shoe.

18. Inspect brake disc and parking brake shoe lining for proper contact

a. Apply chalk to the inside surface of the disc, then grind down the brake shoe lining to fit disc.

b. If the contact between the brake disc and the shoe lining is improper, repair it using a brake shoe grinder or replace the brake shoe assembly.

To install:

19. Install the parking brake shoe type C-washer, as follows

a. Using a feeler gauge, measure the clearance. Standard clearance: less than 0.35 mm (0.014 in.). If the clearance is not within the specifications, replace the shim with one of the correct size. The shim sizes: 0.3 mm (0.012 in.), 0.9 mm (0.035 in.) or 0.6 mm (0.024 in.).

b. Using pliers, install the parking brake shoe lever and the shim with a new C-washer.

20. Apply high temperature grease to the shaded parts shown in the illustration of the backing plate which make contact with the shoe.
21. On 4WD models, perform the following:

a. Install the parking brake shoe hold down spring pin.

b. Install the rear axle hub & bearing.

c. Install the rear axle shaft nut.

d. Install the rear speed sensor.

Fig. 12 Installing the parking brake shoe type C-washer

22. Install parking brake shoe no. 2, as follows:

a. Using needle-nose pliers, connect the parking brake cable no. 3 to the parking brake cable shoe lever.

➡Be careful not to damage the parking brake cable no.3.

b. Install the parking brake shoe no. 2 with the parking brake shoe hold down spring, parking brake shoe hold down spring cup no. 1, parking brake shoe hold down spring cup no. 2 and parking brake shoe hold down spring pin no. 2.

23. Install the parking brake shoe adjusting screw set, as follows:

a. Apply high temperature grease to the parking brake shoe adjusting bolt and piece.

b. Attach the parking brake shoe return tension spring no. 2 to the parking brake shoe no.1 and parking brake shoe assembly no. 2.

c. Attach the parking brake shoe adjusting screw set to the parking brake shoe no.1 and parking brake shoe no. 2.

24. Install parking brake shoe no. 1:

a. For FWD models, install the parking brake shoe hold down spring pin no. 1.

b. Install the parking brake shoe no. 1 with the parking brake shoe hold down spring, parking brake shoe hold down spring cup no. 2, parking brake shoe hold down spring cup no. 2.

25. attach the parking brake shoe strut and the parking brake shoe strut compression spring to parking brake shoe no. 2 and parking brake shoe no. 1.
26. Install parking brake shoe return tension spring using needle-nose pliers as shown in the illustration.

➡First install the front side spring then the rear side spring.

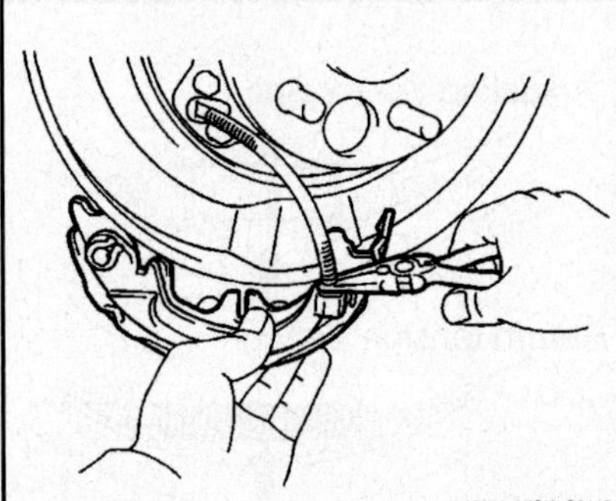

Fig. 11 Using needle-nose pliers, disconnect the parking brake cable no.3 from the parking brake cable shoe lever

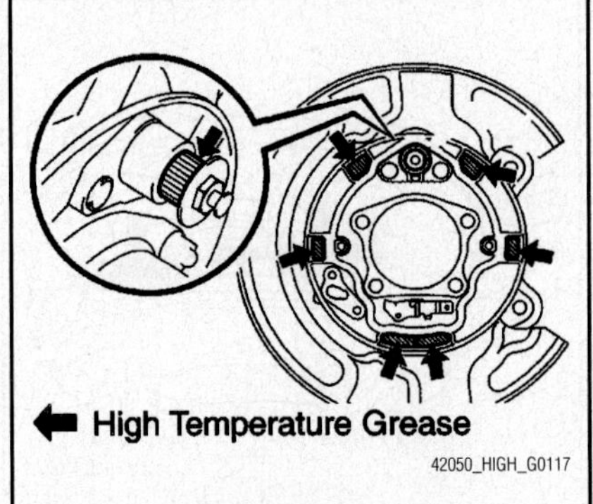

Fig. 13 Apply high temperature grease to the shaded parts of the backing plate which make contact with the shoe

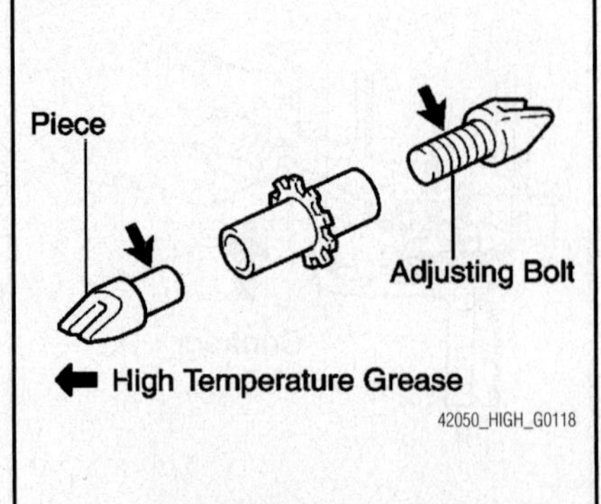

Fig. 14 Apply high temperature grease to the parking brake shoe adjusting bolt and piece

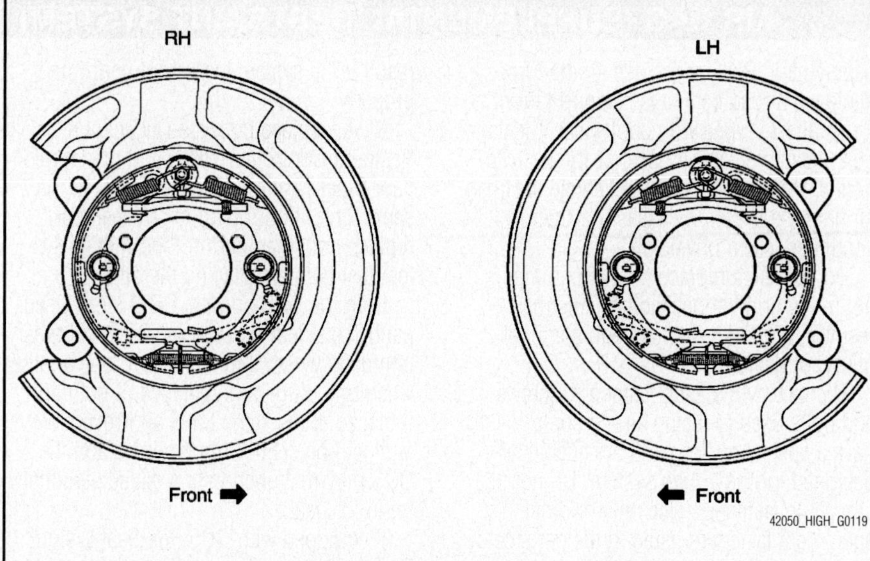

Fig. 15 Check that the parking brake components are properly installed. There should be no oil or grease on the friction surface of the shoe lining and disc

27. Check that the parking brake components are properly installed.

WARNING

There should be no oil or grease on the friction surface of the shoe lining and disc.

28. For 4WD models, inspect the bearing backlash and axle hub deviation.
29. Install the rear disc (rotor), aligning the matchmarks made during removal.
30. Adjust parking brake shoe clearance, as follows:
    a. Temporarily install the hub nuts.
    b. Remove the hole plug, turn the adjuster and expand the shoes until the disc locks.
    c. Contract the shoe adjuster until the disc rotates smoothly. Standard : return 8 notches
    d. Check that the shoe has no brake drag.
    e. Install the hole plug.
31. Install the caliper, as outlined earlier in this section.
32. Install the rear wheel and tire assembly and tighten the lug nuts to 76 ft. lbs. (103 Nm).
33. Inspect and adjust the parking brake pedal travel, as outlined in this section.
34. For 4WD models, check the ABS speed sensor signal.

ADJUSTMENT

See Figures 16 and 17.

1. Raise and safely support the vehicle.
2. Remove the rear wheel and tire assemblies.
3. Adjust parking brake shoe clearance, as follows:
    a. Temporarily install the hub nuts.
    b. Remove the hole plug, turn the adjuster and expand the shoes until the disc locks.
    c. Contract the shoe adjuster until the disc rotates smoothly. Standard : return 8 notches
    d. Check that the shoe has no brake drag.
    e. Install the hole plug.
4. Install the rear wheel and tire assembly and tighten the lug nuts to 76 ft. lbs. (103 Nm).
5. Inspect the parking brake pedal travel, as follows:
    a. Firmly step on the parking brake pedal.
    b. Release the parking brake.
    c. Once more, slowly depress the parking brake pedal all the way, and count the number of clicks. For 2007 models, the parking brake pedal should travel 5 to 7 clicks at 67 lbs. (300 N). For 2008 models, the parking brake pedal should travel 8 to 10 clicks at 67 lbs. (300 N).

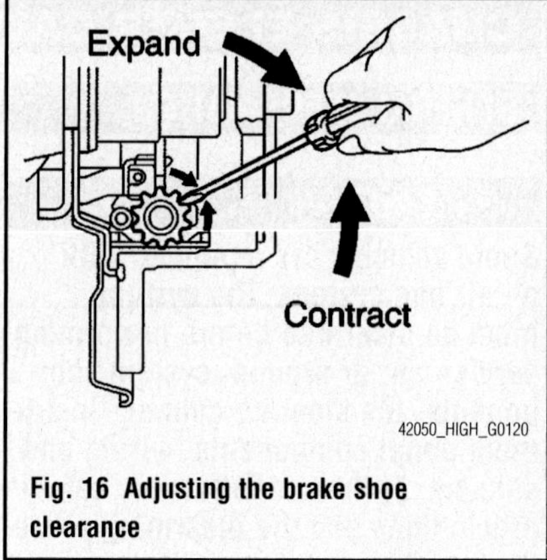

Fig. 16 Adjusting the brake shoe clearance

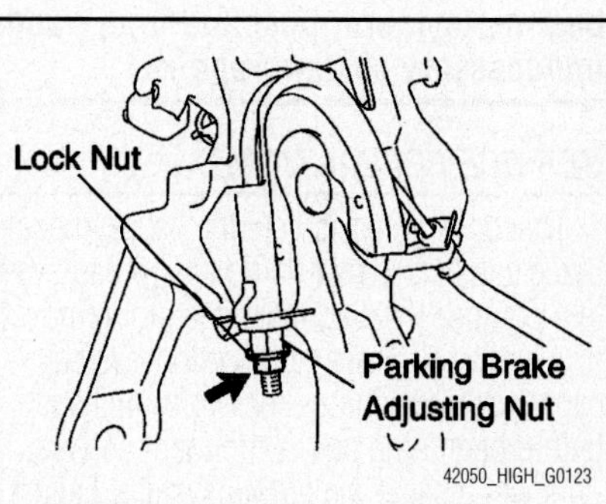

Fig. 17 Use a wrench to hold the parking brake adjusting nut secure while tightening the lock nut

6. If necessary, adjust parking brake pedal travel, as follows:
    a. Remove the lower instrument panel finish panel sub-assembly.
    b. Remove the lower instrument panel insert sub- assembly.
    c. Depress the parking brake pedal 5 clicks to make room for the procedure, and loosen the lock nut with fixing adjusting nut by wrench.
    d. Release the parking brake pedal to the original position.
    e. Turn the parking brake wire adjusting nut until the parking brake pedal travel is correct.
    f. Use a wrench to hold the parking brake adjusting nut, then tighten the lock nut to 62 inch lbs. (7 Nm).
    g. Count the number of clicks after depressing and canceling the parking brake pedal 3 to 4 times.
    h. Check whether the parking brake drags or not.
    i. When operating the parking brake pedal, check that the parking brake pedal indicator light is lit.

GENERAL INFORMATION

** CAUTION

Some vehicles are equipped with an air bag system. The system must be disarmed before performing service on, or around, system components, the steering column, instrument panel components, wiring and sensors. Failure to follow the safety precautions and the disarming procedure could result in accidental air bag deployment, possible injury and unnecessary system repairs.

SERVICE PRECAUTIONS

Disconnect and isolate the battery negative cable before beginning any airbag system component diagnosis, testing, removal, or installation procedures. Allow system capacitor to discharge for two minutes before beginning any component service. This will disable the airbag system. Failure to disable the airbag system may result in accidental airbag deployment, personal injury, or death.

Do not place an intact undeployed airbag face down on a solid surface. The airbag will propel into the air if accidentally deployed and may result in personal injury or death.

When carrying or handling an undeployed airbag, the trim side (face) of the airbag should be pointing towards the body to minimize possibility of injury if accidental deployment occurs. Failure to do this may result in personal injury or death.

Replace airbag system components with OEM replacement parts. Substitute parts may appear interchangeable, but internal differences may result in inferior occupant protection. Failure to do so may result in occupant personal injury or death.

Wear safety glasses, rubber gloves, and long sleeved clothing when cleaning powder residue from vehicle after an airbag deployment. Powder residue emitted from a deployed airbag can cause skin irritation. Flush affected area with cool water if irritation is experienced. If nasal or throat irritation is experienced, exit the vehicle for fresh air until the irritation ceases. If irritation continues, see a physician.

Do not use a replacement airbag that is not in the original packaging. This may result in improper deployment, personal injury, or death.

The factory installed fasteners, screws and bolts used to fasten airbag components have a special coating and are specifically designed for the airbag system. Do not use substitute fasteners. Use only original equipment fasteners listed in the parts catalog when fastener replacement is required.

During, and following, any child restraint anchor service, due to impact event or vehicle repair, carefully inspect all mounting hardware, tether straps, and anchors for proper installation, operation, or damage. If a child restraint anchor is found damaged in any way, the anchor must be replaced. Failure to do this may result in personal injury or death.

Deployed and non-deployed airbags may or may not have live pyrotechnic material within the airbag inflator.

Do not dispose of driver/passenger/curtain airbags or seat belt tensioners unless you are sure of complete deployment. Refer to the Hazardous Substance Control System for proper disposal.

Dispose of deployed airbags and tensioners consistent with state, provincial, local, and federal regulations.

After any airbag component testing or service, do not connect the battery negative cable. Personal injury or death may result if the system test is not performed first.

If the vehicle is equipped with the Occupant Classification System (OCS), do not connect the battery negative cable before performing the OCS Verification Test using the scan tool and the appropriate diagnostic information. Personal injury or death may result if the system test is not performed properly.

Never replace both the Occupant Restraint Controller (ORC) and the Occupant Classification Module (OCM) at the same time. If both require replacement, replace one, then perform the Airbag System test before replacing the other.

Both the ORC and the OCM store Occupant Classification System (OCS) calibration data, which they transfer to one another when one of them is replaced. If both are replaced at the same time, an irreversible fault will be set in both modules and the OCS may malfunction and cause personal injury or death.

If equipped with OCS, the Seat Weight Sensor is a sensitive, calibrated unit and must be handled carefully. Do not drop or handle roughly. If dropped or damaged, replace with another sensor. Failure to do so may result in occupant injury or death.

If equipped with OCS, the front passenger seat must be handled carefully as well. When removing the seat, be careful when setting on floor not to drop. If dropped, the sensor may be inoperative, could result in occupant injury, or possibly death.

If equipped with OCS, when the passenger front seat is on the floor, no one should sit in the front passenger seat. This uneven force may damage the sensing ability of the seat weight sensors. If sat on and damaged, the sensor may be inoperative, could result in occupant injury, or possibly death.

DISARMING THE SYSTEM

To avoid personal injury when working on vehicles equipped with an air bag, the negative battery cable must be disconnected and at least 90 seconds must elapse before working on the system. Failure to do so may result in deployment of the air bag.

ARMING THE SYSTEM

To arm the system after service is finished, connect the negative battery cable.

DRIVETRAIN

AUTOMATIC TRANSAXLE ASSEMBLY

REMOVAL & INSTALLATION

2.4L Engine

See Figure 18.

1. Remove the engine/transaxle assembly.
2. Remove the halfshafts.
3. Remove the engine mounting bracket (4WD).
4. Remove the transfer case (4WD).
5. Disconnect the wiring.
6. Remove the starter.
7. Remove the cables and hoses.
8. Remove the filler tube.
9. Remove the front engine mount bracket.
10. Remove the flywheel housing undercover.
11. Turn the crankshaft to gain access to the torque converter bolts. There is one green bolt.
12. Remove the 9 engine-to-transaxle bolts. Separate the transaxle from the engine.
13. Installation is the reverse of removal. Observe the following torques:
    - Transaxle-to-engine: bolts A 47 ft. lbs (64 Nm); bolts B 34 ft. lbs. (46 Nm); bolts C 27 ft. lbs. (37 Nm)
    - Torque converter bolts (use a thread locking compound such as Three Bond 1324): 30 ft. lbs. (41 Nm)

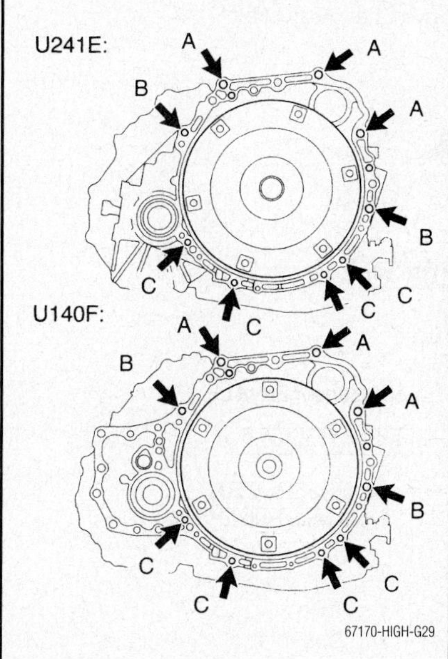

Fig. 18 Engine-to-transaxle bolt identification—2.4L engine

➡Install the green bolt first.

- Undercover: 69 inch lbs. (8 Nm)
- Engine mount bracket: 47 ft. lbs. (64 Nm)
- Transfer case-to-transaxle: 51 ft. lbs. (69 Nm)

3.3L and 3.5L Engines

See Figure 19.

1. Remove the engine/transaxle assembly. See Engine Removal and Installation.
2. Remove the halfshafts.
3. Disconnect the wiring.
4. Remove the starter.
5. Remove the cables and hoses.
6. Remove the filler tube.
7. Remove the front engine mount bracket.
8. Remove the flywheel housing undercover.
9. Turn the crankshaft to gain access to the torque converter bolts. There is one green bolt.
10. Separate the transaxle from the engine.
11. Remove the right stiffener plate.
12. Separate the transfer case from the transaxle.
13. Installation is the reverse of removal. Observe the following torques:
    - Transaxle-to-engine: bolts A 47 ft. lbs (64 Nm); bolts B 34 ft. lbs. (46 Nm); bolts C 27 ft. lbs. (37 Nm)
    - Torque converter bolts (use a thread locking compound such as Three Bond 1324): 30 ft. lbs. (41 Nm)

➡Install the green bolt first.

- Undercover: 69 inch lbs. (8 Nm)
- Stiffener plate: 25 ft. lbs. (34 Nm)
- Engine mount bracket: 47 ft. lbs. (64 Nm)

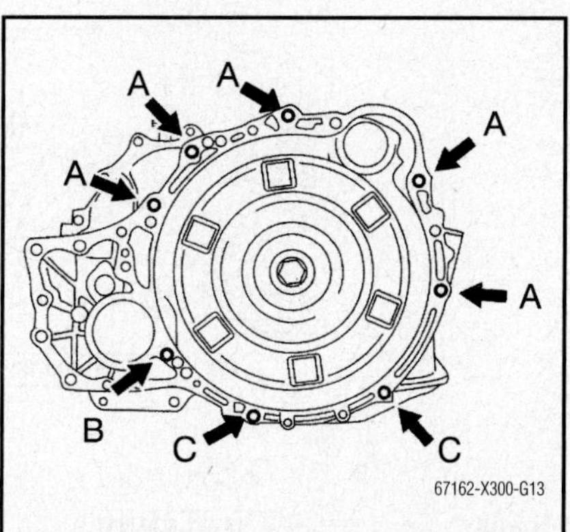

Fig. 19 Transaxle-to-engine bolts—3.3L & 3.5L engines

- Transfer case-to-transaxle: 51 ft. lbs. (69 Nm)

TRANSFER CASE ASSEMBLY

REMOVAL & INSTALLATION

See Figures 20 through 22.

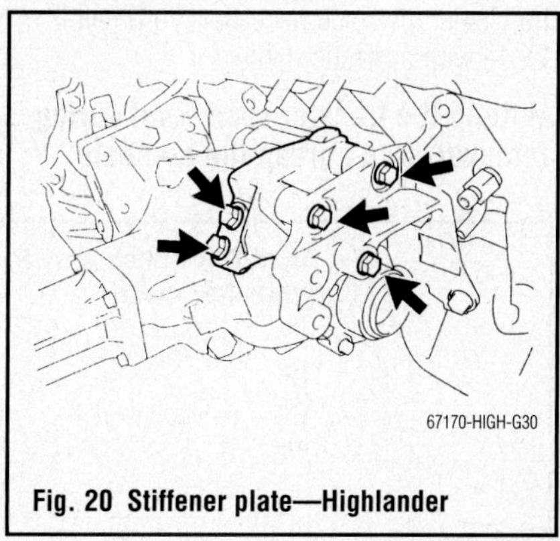

Fig. 20 Stiffener plate—Highlander

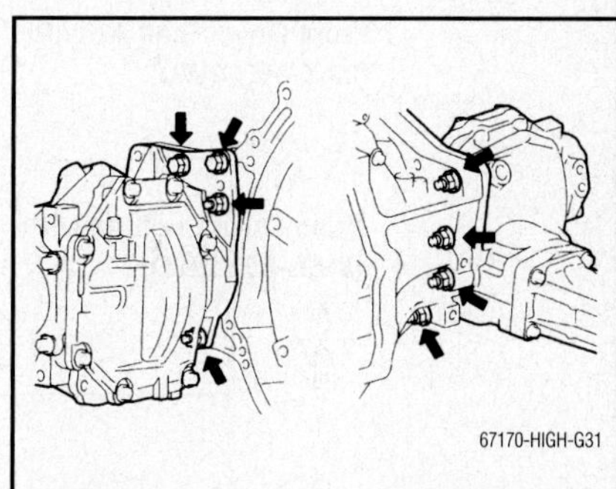

Fig. 21 Transfer case fastener locations—Highlander

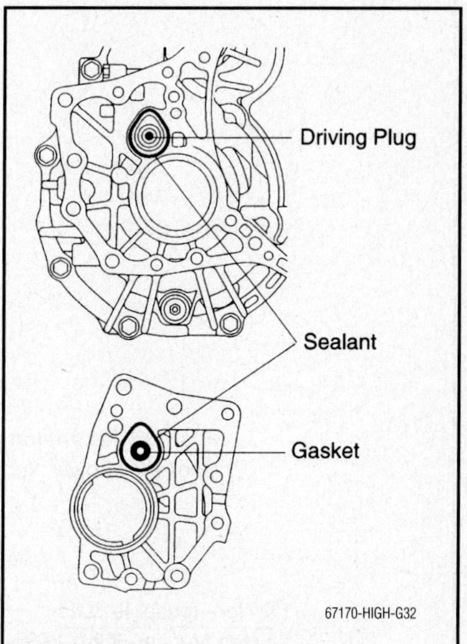

Fig. 22 Gasket material application—Highlander

1. Remove the engine/transaxle assembly.
2. Drain the transaxle.
3. With the 2.4L engine, remove the stiffener plate (5 bolts).
4. With the 2.4L engine, remove the right rear engine mount bracket (3 bolts).
5. Separate the engine and transaxle.
6. Remove the 2 bolts and 6 nuts and separate the transfer case from the transaxle. It will be necessary to break it loose with a plastic mallet.

➡Keep the transfer case level during removal. Don't grasp the oil seals.

To install:

7. Clean all grease from the mating surfaces.
8. Apply a continuous 1.2mm diameter bead of silicone gasket material to the transaxle and transfer case as shown.
9. Join the transfer case to the transaxle within 10 minutes of gasket material application. If not, remove the material and start again.
10. Torque the nuts and bolts to 51 ft. lbs. (69 Nm).
11. The remainder of installation is the reverse of removal. Observe the following torques:
    - Engine mount bracket: 47 ft. lbs. (64 Nm)
    - Stiffener plate: 25 ft. lbs. (34 Nm)
    - Drain plug: 36 ft. lbs. (49 Nm)

FRONT HALFSHAFT

REMOVAL & INSTALLATION

See Figure 23.

1. Before servicing the vehicle, refer to the precautions section.

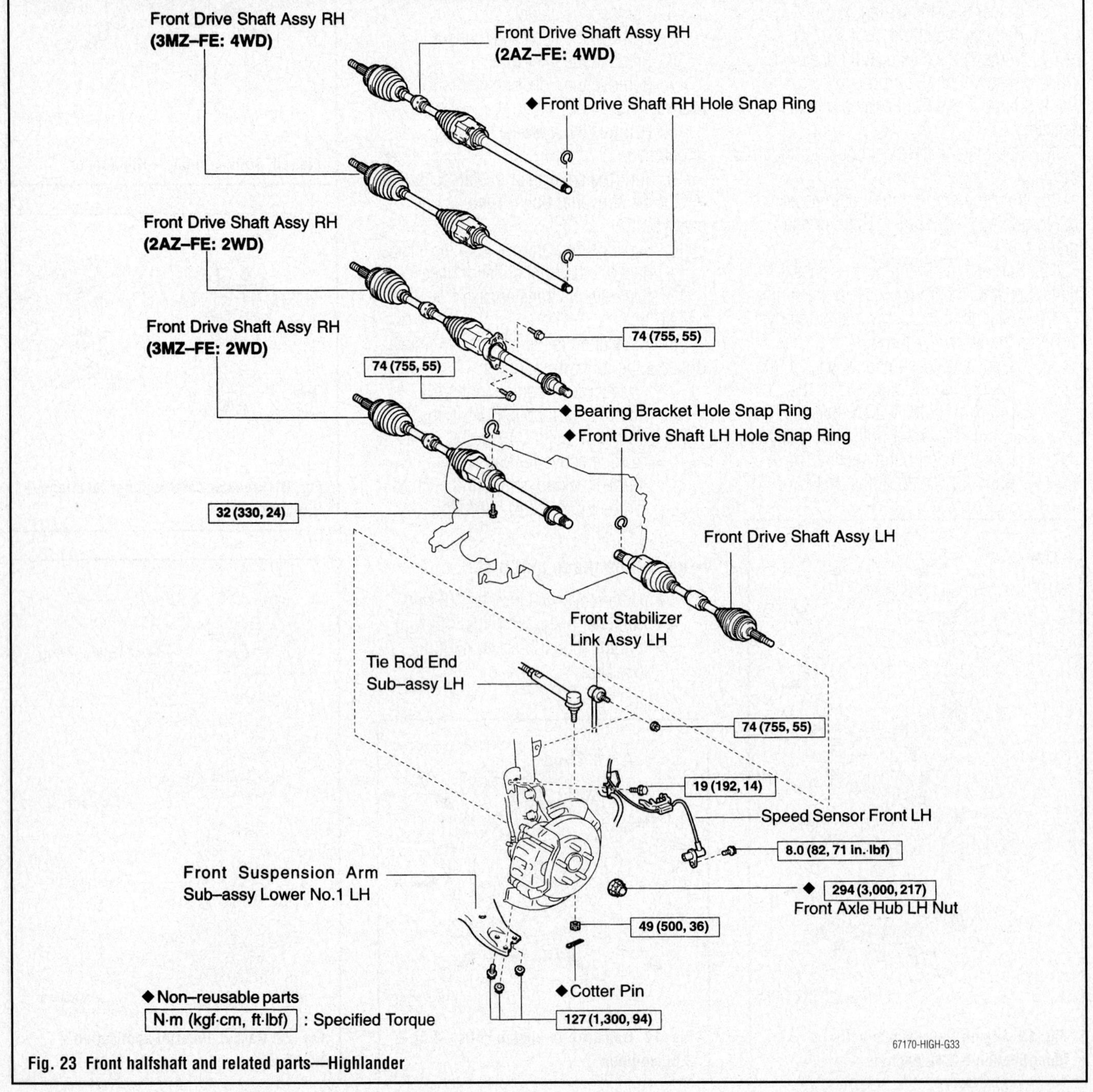

Fig. 23 Front halfshaft and related parts—Highlander

2. Remove or disconnect the following:
   - Front wheels
   - Fender apron seal
   - Transaxle fluid
   - Transfer case oil (4WD)
   - Hub nut
   - Stabilizer bar link
   - Speed sensor
   - Tie rod end
   - Lower arm from the ball joint
3. Slide the halfshaft from the hub, then, carefully, pry the shaft from the transaxle.
4. Installation is the reverse of removal. Torque the hub nut to 217 ft. lbs. (294 Nm).

REAR HALFSHAFT

REMOVAL & INSTALLATION

2007 Models

See Figure 24.

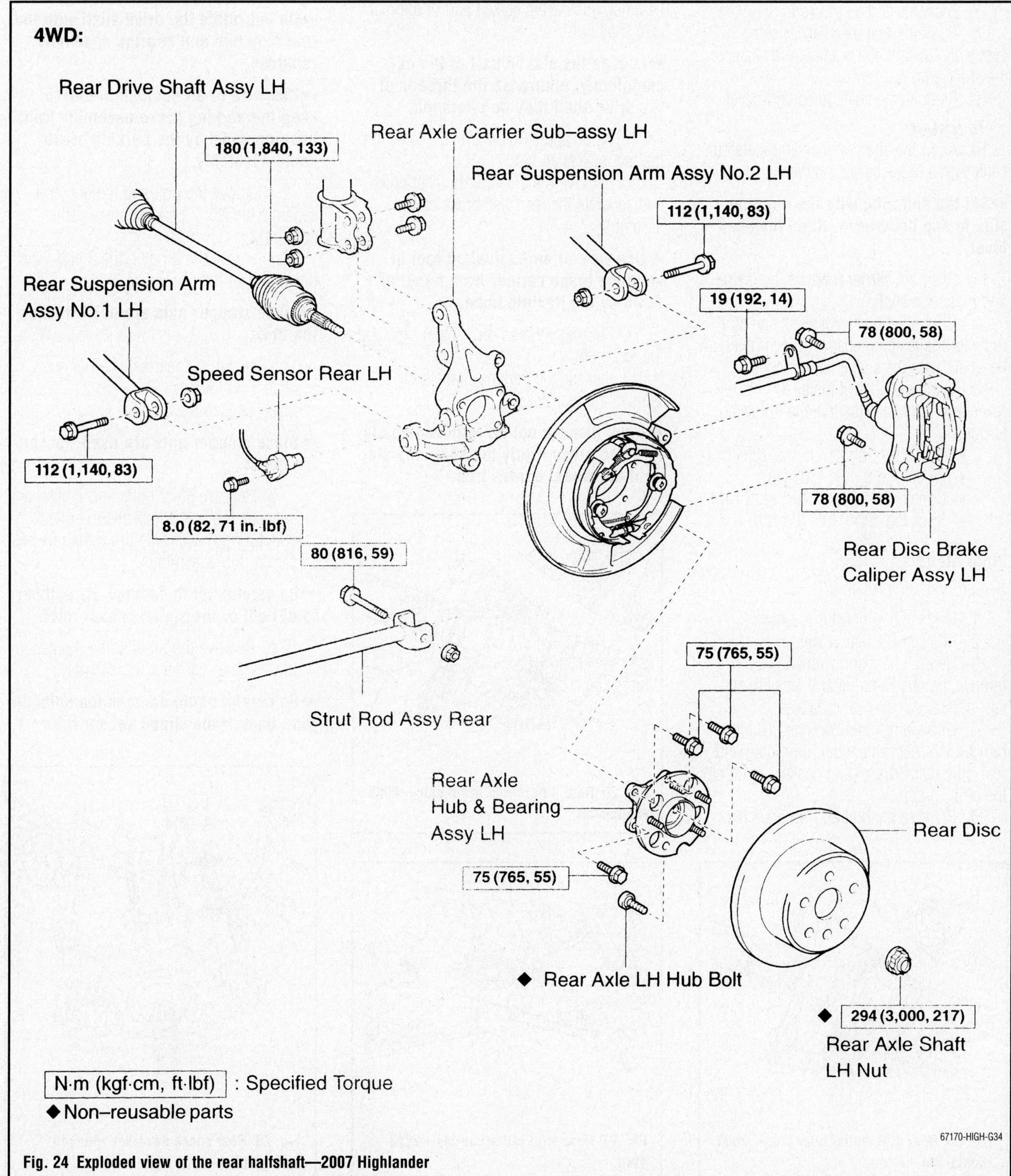

Fig. 24 Exploded view of the rear halfshaft—2007 Highlander

1. Before servicing the vehicle, refer to the precautions section.
2. Remove the rear wheel.
3. Disconnect and remove the speed sensor.
4. Remove the axle shaft nut.
5. Disconnect the height control sensor.
6. Disconnect the rear control arms.
7. Disconnect the strut rod.
8. Push the rear axle carrier sub-assembly towards the outside and separate the shaft from the carrier.
9. Remove the shaft, keeping it level.

To install:

10. Align the shaft splines and install the shaft with a brass bar and hammer.

➡**Set the snapring with the opening side facing downward. Keep the shaft level.**

11. Push the carrier towards the inside and insert the shaft.
12. Connect the control arms and strut rod with the fasteners hand-tight. Tighten all fasteners with the suspension loaded.
13. The remainder of installation is the reverse of removal. Observe the following torques:
    - Axle shaft nut: 217 ft. lbs. (294 Nm)
    - Wheel: 76 ft. lbs. (103 Nm)
    - Control arms: 83 ft. lbs. (112 Nm)
    - Strut rod: 59 ft. lbs. (80 Nm)

2008 Models

See Figures 25 through 30.

1. Raise and support the vehicle.
2. Remove the rear wheel.
3. Using a hexagon wrench (10 mm), remove the rear differential filler plug and rear differential filler plug gasket.
4. Using a hexagon wrench (10 mm), remove the rear differential drain plug and rear differential drain plug gasket, and drain the oil.
5. Remove the bolt and separate the rear speed sensor from the rear axle carrier sub-assembly.

➡**Keep the sensor tip and rear speed sensor installation hole free from foreign matter.**

6. Using Special Service Tool (SST) 09930-00010 or equivalent, along with a hammer, release the staked part of the rear axle shaft nut.

➡**Loosen the staked part of the nut completely, otherwise the threads of the drive shaft may be damaged.**

7. While applying the brakes, remove the rear axle shaft nut.
8. Remove the 2 caliper bracket bolts and separate the rear disc brake caliper assembly.

➡**Use wire or an equivalent tool to keep the brake caliper from hanging down by the flexible hose.**

9. Put matchmarks on the rear disc and the axle hub.
10. Release the parking brake and remove the rear disc.

➡**If the disc cannot be removed easily, turn and press firmly the shoe adjuster until the wheel comes free.**

11. Remove rear axle hub and bearing assembly, as follows:
    a. Put matchmarks on the drive shaft and axle hub.

➡**Do not punch the marks.**

    b. Remove the 4 bolts and the rear axle hub and bearing assembly.

➡**Do not rotate the drive shaft with the rear axle hub and bearing assembly removed.**

➡**Use wire or an equivalent tool to keep the parking brake assembly from hanging down by the parking brake cable assembly.**

12. Remove the bolt and the nut, and separate the No. 3 parking brake cable assembly.
13. Remove the 2 bolts, the 2 nuts, and the rear strut rod assembly.

➡**Since stopper nuts are used, loosen the bolts.**

14. Remove the rear axle carrier sub-assembly, as follows:
    a. Loosen the 2 bolts.

➡**Since stopper nuts are used, loosen the bolts.**

    b. Remove the 2 bolts and 2 nuts, and separate the rear shock absorber with coil spring (lower side) from the rear axle carrier sub-assembly.

➡**Be careful not to damage the outboard joint boot or the speed sensor rotor.**

    c. Remove the 2 bolts, the 2 nuts, and the rear axle carrier sub-assembly.

➡**Be careful not to damage the outboard joint boot or the speed sensor rotor.**

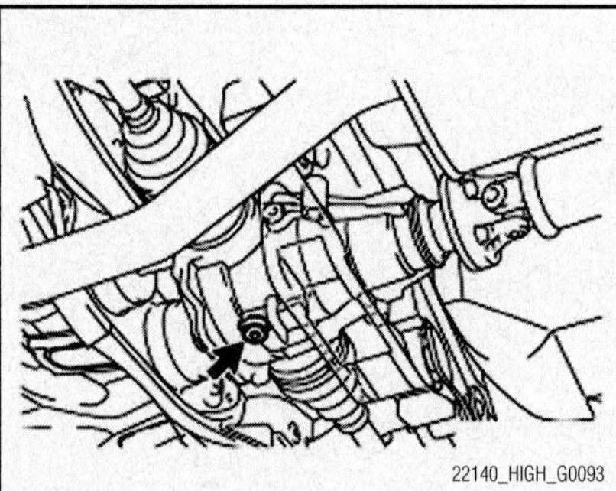

Fig. 26 Rear differential drain plug—2008 Highlander

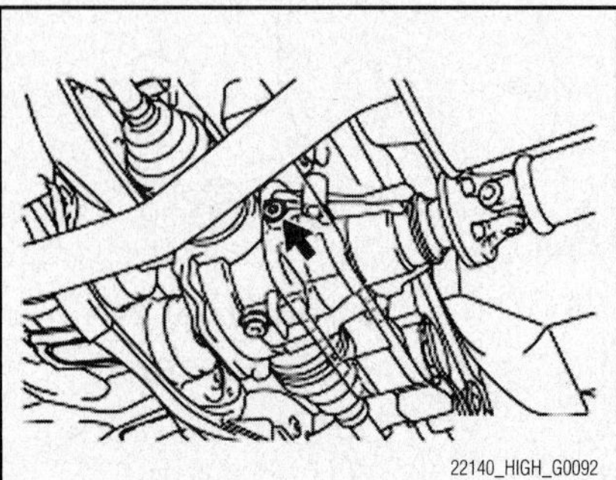

Fig. 25 Rear differential filler plug—2008 Highlander

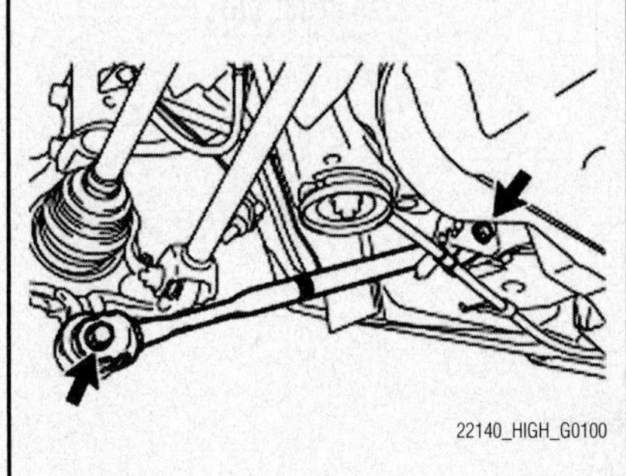

Fig. 27 Rear strut rod assembly—2008 4WD

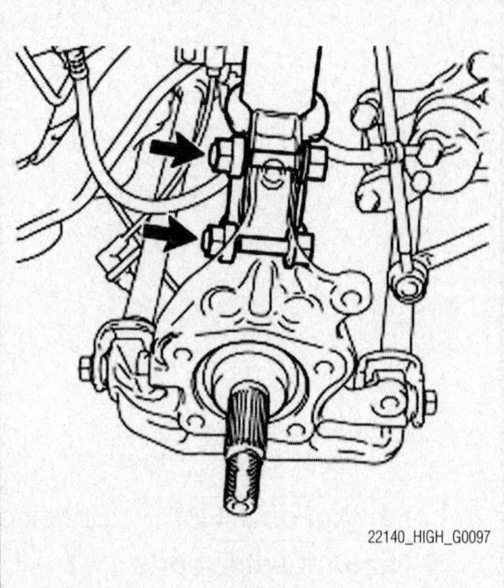

Fig. 28 Rear shock absorber with coil spring bolts—2008 Highlander

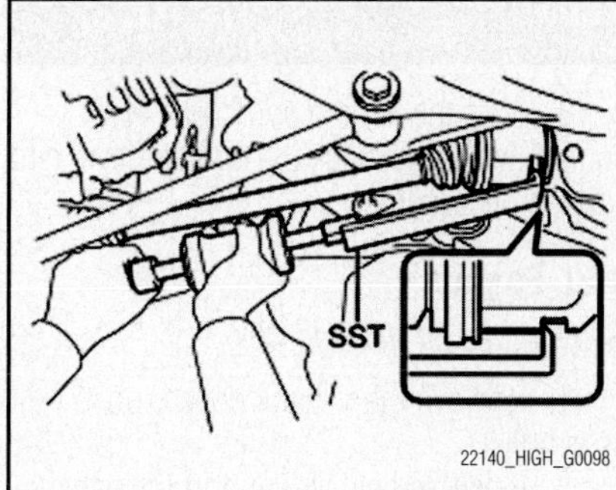

Fig. 29 Remove rear drive shaft assembly—2008 Highlander

15. Using a slide hammer (SST: 09520-01010, SST: 09520-24010 or equivalent), remove the rear drive shaft assembly (halfshaft) as shown in the illustration.

➡**Remove the rear drive shaft assembly while keeping it level.**

To install:

16. Align the shaft splines and install the rear drive shaft assembly (halfshaft) with a brass bar and hammer.

➡**Set the snap ring with the opening facing downward.**

➡**Be careful not to damage the oil seal, boot, or dust cover.**

➡**Install the drive shaft assembly while keeping it level.**

17. Temporarily install the rear axle carrier sub-assembly with the 2 bolts and the 2 nuts.

➡**Be careful not to damage the outboard joint boot.**

➡**Be careful not to damage the speed sensor rotor.**

➡**Prevent foreign matter from adhering to the speed sensor rotor.**

18. Install the rear axle carrier sub-assembly with the 2 bolts and the 2 nuts, and tighten to 213 ft. lbs. (290 Nm).

➡**Do not rotate the drive shaft with the rear axle hub and bearing assembly removed.**

➡**Insert the bolts from the rear side.**

19. Check that the identification mark of the rear strut rod assembly is positioned on the inner side of the vehicle.

20. Temporarily install the rear strut rod assembly with the 2 bolts and the 2 nuts.

➡**Since stopper nuts are used, temporarily tighten the bolts.**

21. Install the rear axle hub and bearing assembly, as follows:

a. Align the matchmarks on the drive shaft and rear axle hub.

➡**Do not rotate the drive shaft.**

b. Install the parking brake assembly and the rear axle hub and bearing assembly with the 4 bolts, and tighten to 55 ft. lbs. (75 Nm).

22. Align the matchmarks and install the rear disc.

➡**When replacing the rear disc with a new one, select the installation position where the rear disc has minimal runout.**

23. Install the rear disc brake caliper assembly with the 2 bolts, and tighten to 57 ft. lbs. (78 Nm).

24. Install the rear speed sensor to the rear axle carrier sub-assembly with the bolt, and tighten to 71 inch lbs. (8 Nm).

➡**Keep the rear speed sensor tip and sensor installation hole free from foreign matter.**

➡**Do not twist the rear speed sensor wire when installing.**

25. Jack up the rear axle carrier sub-assembly, placing a wooden block underneath to avoid damage. Apply load to the suspension so that the rear drive shaft assembly is positioned horizontally.

CAUTION

Do not jack up the rear axle carrier sub-assembly too high as the vehicle may fall.

➡**Do not bend the brake dust cover.**

➡**If the rear drive shaft assembly cannot be positioned horizontally as shown in the illustration even when the rear axle carrier sub-assembly is jacked up, apply additional load to the vehicle such as by having a person sit in the rear seat.**

➡**Use the same procedures for the RH side and LH side.**

26. Fully tighten the rear No. 1 and 2 suspension arm assemblies with the bolts and nuts, and tighten to 82 ft. lbs. (112 Nm).

➡**Since a stopper nut is used, tighten the bolt.**

➡**The final torque must be applied under standard vehicle height conditions.**

27. Complete the installation of the rear strut rod assembly, tightening the bolts to 59 ft. lbs. (80 Nm).

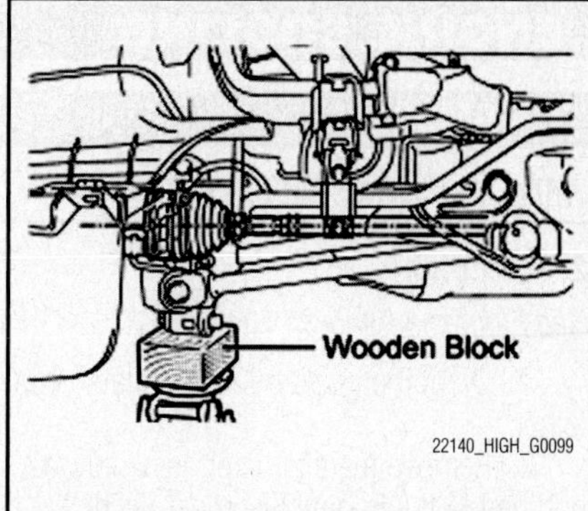

Fig. 30 Stabilizing suspension—2008 Highlander

➡**Since a stopper nut is used, fully tighten the bolt.**

➡**The final torque must be applied under standard vehicle height conditions.**

28. Install the No. 3 parking brake cable assembly with the bolt and the nut, and tighten to 29 ft. lbs. (39 Nm) and 53 inch lbs. (6 Nm).

➡**Do not twist the No. 3 parking brake cable assembly when installing it.**

29. Install the rear axle shaft nut, as follows:

a. Clean the threaded parts on the drive shaft and axle hub nut using a non-residue solvent.

➡**Be sure to perform this work for a new drive shaft.**

➡**Keep the threaded parts free of oil and foreign objects.**

b. Install a new rear axle shaft nut, and tighten to 216 ft. lbs. (294 Nm).

c. Using a chisel and hammer, stake the rear axle shaft nut.

30. Install the rear differential drain plug, as follows:

a. Using a hexagon wrench (10 mm), install the filler plug with a new gasket, and tighten to 36 ft. lbs. (49 Nm).

31. Fill the rear differential carrier assembly with hypoid gear oil.

32. Using a hexagon wrench (10 mm), install the rear differential filler plug with a new gasket, and tighten to 36 ft. lbs. (49 Nm).

33. Install the front wheel and tire assembly and tighten the lug nuts finger-tight.

34. Lower the vehicle, then final tighten the lug nuts to 76 ft. lbs. (103 Nm).

35. Inspect and adjust rear wheel alignment.

36. Check ABS speed sensor signal.

ENGINE COOLING

THERMOSTAT

REMOVAL & INSTALLATION

2.4L Engine

See Figures 31 and 32.

1. Drain the coolant into a suitable container.
2. Remove the alternator assembly, as outlined in the Engine Electrical Section.
3. Remove 2 nuts, and disconnect the water inlet from the cylinder block.
4. Remove the thermostat.
5. Remove and discard the gasket, and thoroughly clean the gasket mating surfaces.

To install:

6. Position a new gasket to the thermostat.
7. Install the thermostat with the jiggle valve upward.

➡The jiggle valve can be placed within 10° on either side of the described position.

8. Install water inlet and secure with the 2 nuts.
9. Install the alternator.
10. Fill the engine with the proper type and amount of engine coolant.
11. Start the engine and check for coolant leaks. Recheck the coolant level and add, as necessary.

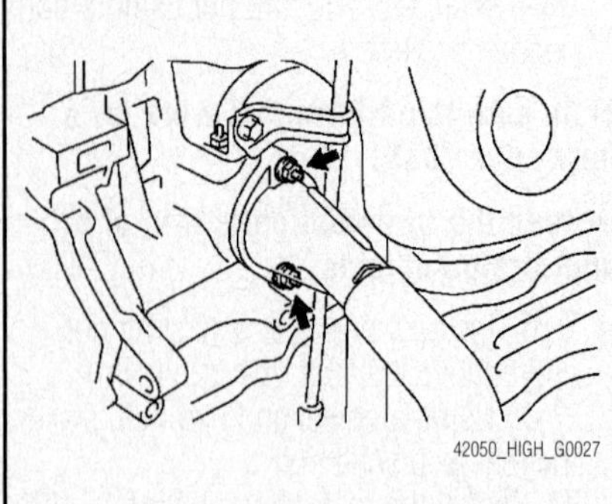

Fig. 31 The water inlet is secured with 2 nuts

3.3L Engine

See Figure 33.

1. Drain the coolant into a suitable container.
2. Disconnect the radiator outlet hose.
3. Remove the bolt and the water inlet pipe.
4. Remove the 3 bolts and water inlet.
5. Remove the thermostat.
6. Remove and discard the gasket, and thoroughly clean the gasket mating surfaces.

To install:

7. Position a new gasket to the thermostat.
8. Align the thermostat jiggle valve with the upper stud bolt, and insert the thermostat into the water inlet housing.

➡The jiggle valve can be placed within 15° of either side of the described position.

9. Install the water inlet and tighten the bolts to 71 inch lbs. (8 Nm).
10. Install the water inlet pipe, as follows:
    a. Install a new O-ring to the inlet pipe.
    b. Apply soapy water to the O-ring.
    c. Connect the inlet pipe to the water inlet.
    d. Install the bolt holding the inlet pipe to the cylinder head and tighten to 15 ft. lbs. (20 Nm).
11. Connect the radiator outlet hose.
12. Fill the engine with the proper type and amount of engine coolant.
13. Start the engine and check for coolant leaks. Recheck the coolant level and add, as necessary.

3.5L Engine

See Figures 34 through 36.

1. Remove the V-bank cover sub-assembly.
2. Raise and safely support the vehicle.
3. Remove the engine under cover assembly.
4. Remove the No. 1 engine under cover.
5. Remove the right front wheel.
6. Remove right front fender molding sub-assembly.
7. Remove the right front fender liner.
8. Remove the 2 bolts, clip and right front fender apron seal.
9. Drain the engine coolant.
10. Using a Special Service Tool (SST: 09961-00950), release the V-ribbed belt tension by turning the V-ribbed belt tensioner assembly counterclockwise, and remove the V-ribbed belt from the V-ribbed belt tensioner assembly.
11. While turning the V-ribbed belt tensioner assembly counterclockwise, align with its holes, and then insert the 5 mm bi-hexagon wrench into the holes to fix the V-ribbed belt tensioner assembly.
12. Disconnect the No. 2 radiator hose from the engine.
13. Remove the bolt, idler pulley cover plate, and idler pulley sub-assembly.
14. Remove the 2 nuts and the water inlet.
15. Remove the thermostat from the water inlet housing.
16. Remove the gasket from the thermostat.

To install:

17. Install a new gasket to the thermostat.

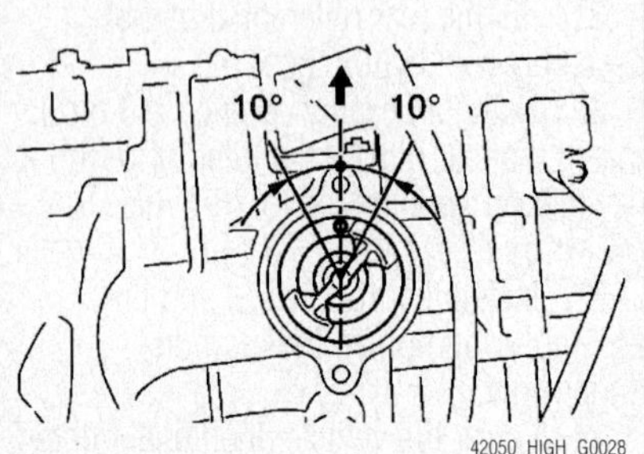

Fig. 32 The jiggle valve may be set within 10° on either side of the described position

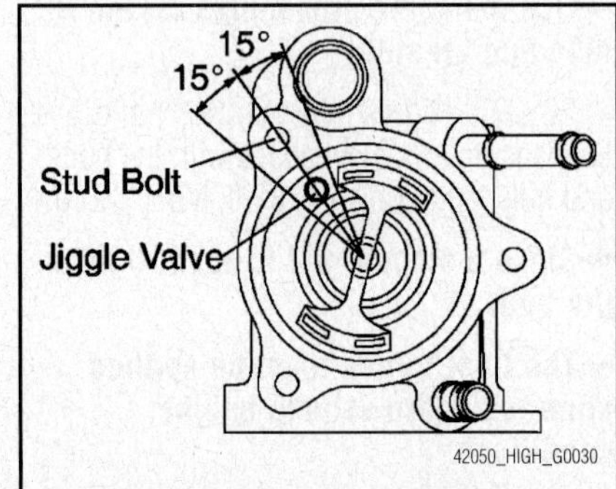

Fig. 33 The jiggle valve may be set within 15° on either side of the described position

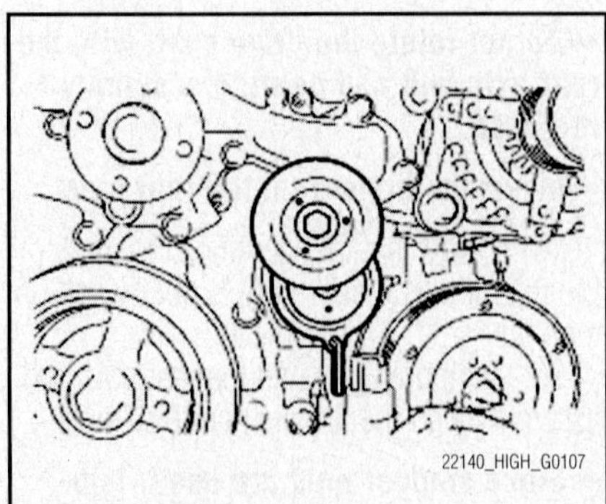

Fig. 34 V-ribbed belt tensioner assembly—3.5L engine

18. Install the thermostat with the jiggle valve facing up.

➡The jiggle valve may be set within 10° on either side of the prescribed position.

19. Install the water inlet with the 2 nuts, and tighten to 7 ft. lbs. (10 Nm).
20. Install the idler pulley cover plate and idler pulley sub-assembly with the bolt, and tighten to 32 ft. lbs. (43 Nm).
21. Connect the No. 2 radiator hose to the engine.
22. Using SST: 09961-00950, turn the V-ribbed belt tensioner assembly counterclockwise and remove the bar.
23. If it is difficult to install the V-ribbed belt, perform the following procedure:
    a. Put the V-ribbed belt on every pulley except the tensioner pulley as shown in the illustration.
    b. Release the V-ribbed belt tension by turning the V-ribbed belt tensioner assembly counterclockwise, and put the V-ribbed belt on the V-ribbed tensioner assembly pulley.

➡Put the backside of the V-ribbed belt on the V-ribbed belt tensioner assembly pulley and the No. 2 idler pulley sub-assembly.

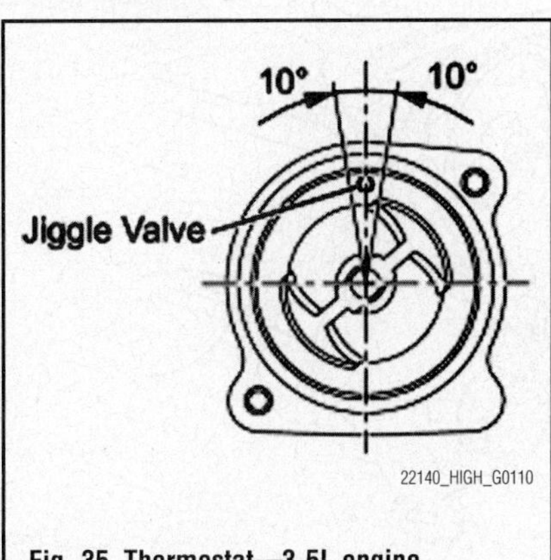

Fig. 35 Thermostat—3.5L engine

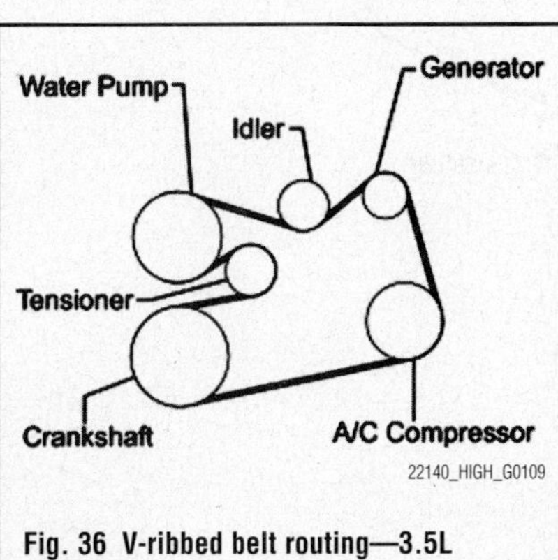

Fig. 36 V-ribbed belt routing—3.5L engine

➡Check that the V-ribbed belt is properly set to each pulley.

    c. After installing the V-ribbed belt, check that it fits properly in the ribbed grooves. Confirm that the belt has not slipped out of the grooves on the bottom of the crankshaft pulley by hand.

24. Install the right front fender apron seal with the 2 bolts and clip.
25. Install the right front fender liner.
26. Install the right front fender molding sub-assembly.
27. Install the right front wheel. Tighten the lug nuts to 76 ft. lbs. (103 Nm).
28. Add engine coolant. Refer to radiator installation procedure for detailed instructions.
29. Inspect for coolant leak. Refer to radiator installation procedure for detailed instructions.
30. Install the No. 1 engine under cover.
31. Install the engine under cover assembly.
32. Install the V-bank cover sub-assembly.

WATER PUMP

REMOVAL & INSTALLATION

2.4L Engine

See Figures 37 and 38.

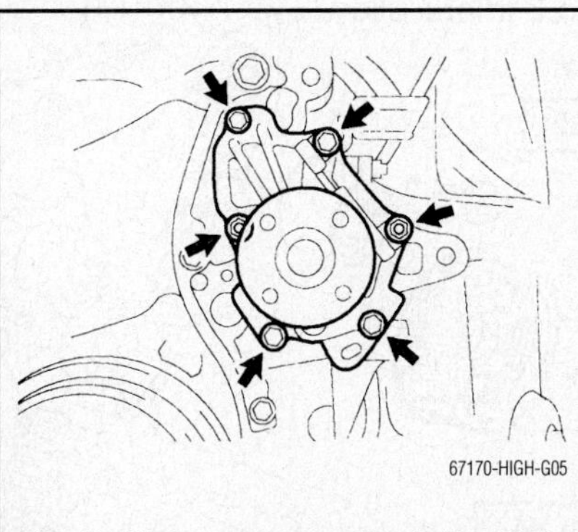

Fig. 37 Water pump mounting bolts—2.4L engine

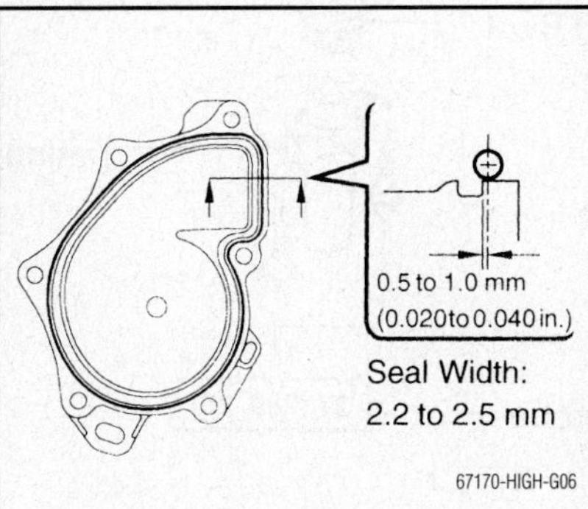

Fig. 38 Sealer application—2.4L engine

1. Before servicing the vehicle, refer to the precautions section.
2. Disconnect the negative battery cable.
3. Drain the engine coolant.
4. Remove or disconnect the following:
    - Alternator
    - Water pump pulley
    - Water pump
5. Installation is the reverse of removal. Torque the pump bolts and nuts to 80 inch lbs. (9 Nm) and the pulley bolts to 19 ft. lbs. (26 Nm).

3.3L Engine

See Figures 39 and 40.

1. Before servicing the vehicle, refer to the precautions section.
2. Disconnect the negative battery cable.
3. Drain the engine coolant.
4. Remove or disconnect the following:
    - Wiper and blade assembly
    - Top cowl seal and panel
    - Window washer hoses, from the ventilator louvers
    - Left and right ventilator louvers
    - Heater air duct
    - Front upper suspension brace
    - Timing belt
5. Mark the left and right camshaft pulleys with a touch of paint.
6. Remove or disconnect the following:
    - Right and left camshaft pulleys bolts
    - Pulleys from the engine

➡Be sure not to mix up the pulleys.

    - No. 2 idler pulley by removing the bolt
    - 3 clamps and engine wire from the rear timing belt cover
    - 6 No. 3 timing belt cover-to-engine bolts
    - Water pump nuts/bolts
    - Water pump and gasket from the engine

To install:

7. Check that the water pump turns smoothly. Also check the air hole for coolant leakage.
8. Apply liquid sealer to the gasket, water pump and engine block.
9. Install or connect the following:
    - Water pump, using a new gasket. Tighten the nuts/bolts to 53 inch lbs. (6 Nm).
    - Rear timing belt cover. Tighten the 6 bolts to 74 inch lbs. (9 Nm).
    - Engine wire with the 3 clamps to the rear timing belt cover
    - No. 2 idler pulley. Tighten the bolt to 32 ft. lbs. (43 Nm).

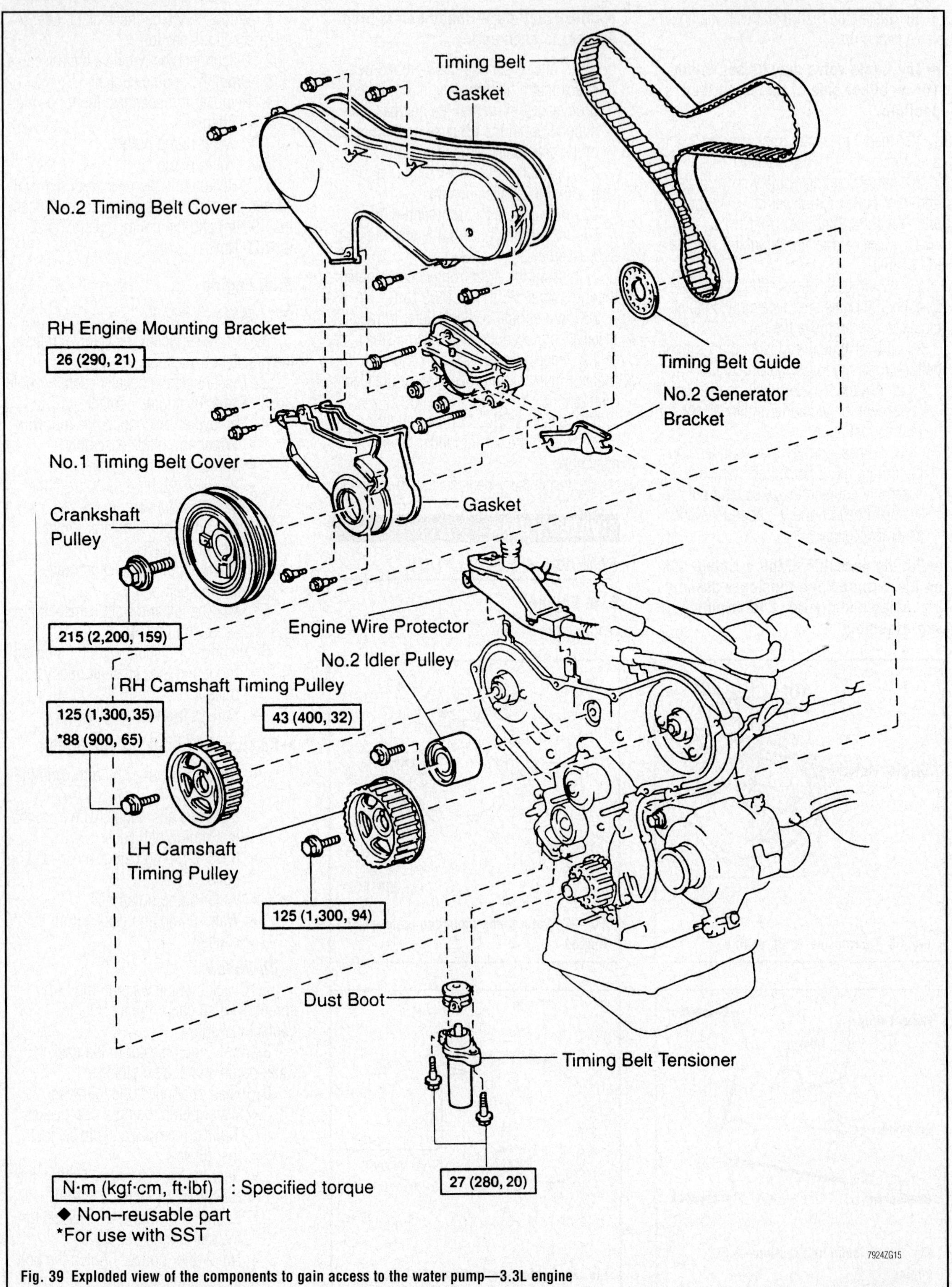

Fig. 39 Exploded view of the components to gain access to the water pump—3.3L engine

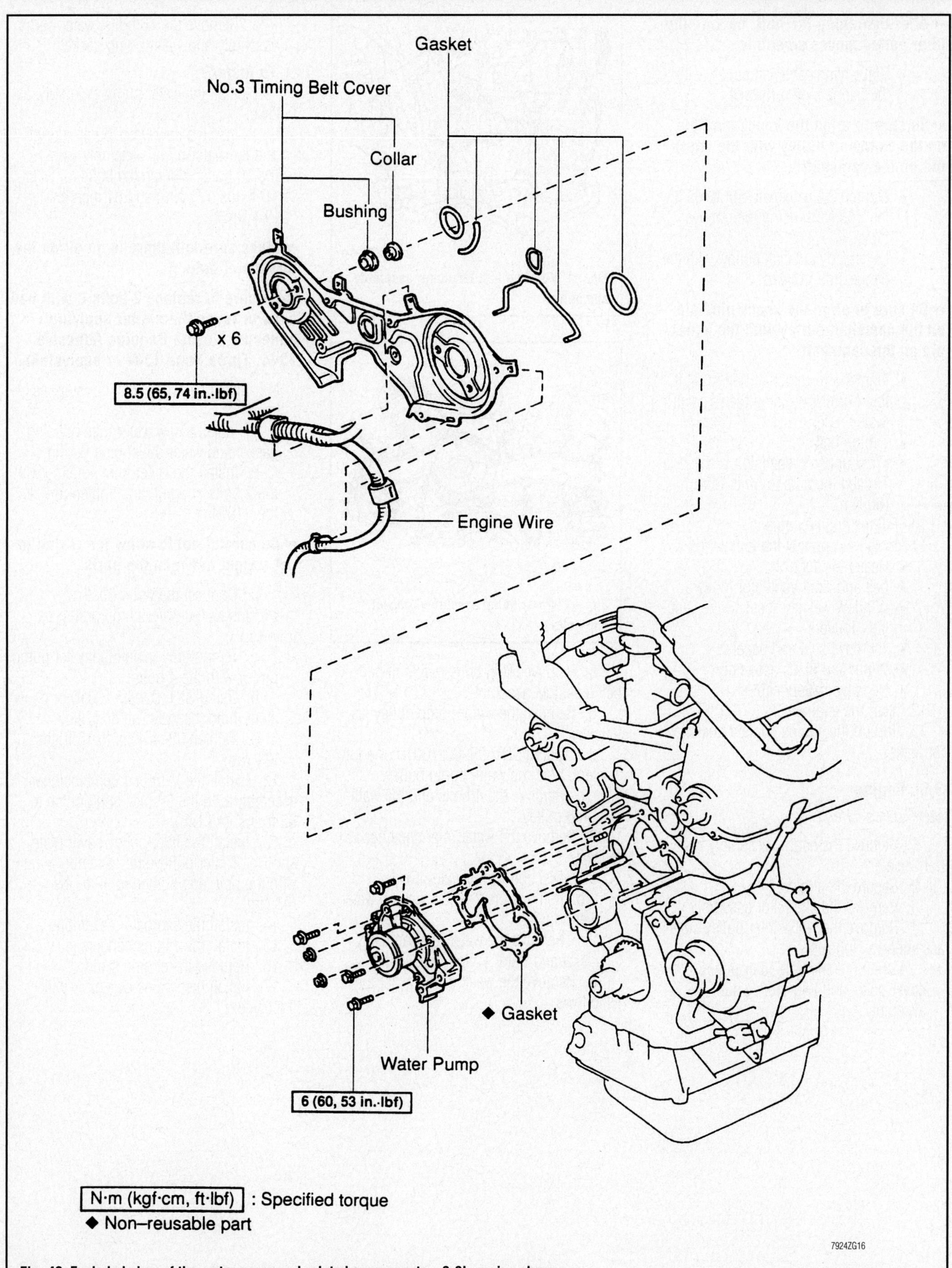

Fig. 40 Exploded view of the water pump and related components—3.3L engine shown

➡**After tightening the bolt, be sure the idler pulley moves smoothly.**

- Right-hand camshaft pulley, with the flange side **outward**.

➡**Be sure to align the knock pin hole on the camshaft pulley with the knock pin on the camshaft.**

- Tighten the camshaft bolt to 65 ft. lbs. (88 Nm), using the removal tools
- Left-hand camshaft pulley, with the flange side **inward**.

➡**Be sure to align the knock pin hole on the camshaft pulley with the knock pin on the camshaft.**

- Tighten the camshaft bolt to 94 ft. lbs. (125 Nm), using the removal tools
- Timing belt
- Front upper suspension brace. Tighten the nuts to 59 ft. lbs. (80 Nm).

10. Fill the engine coolant.
11. Install or connect the following:
    - Heater air duct
    - Left and right ventilator louvers
    - Window washer hoses to the ventilator louvers
    - Top cowl seal and panel
    - Wiper and blade assembly
    - Negative battery cable
12. Start the engine.
13. Top off the engine coolant and check for leaks.

3.5L Engine

See Figures 41 and 42.

1. Remove the engine assembly with transaxle.
2. Secure the engine.
3. Remove the alternator assembly.
4. Remove the No. 2 idler pulley sub-assembly, as follows:
    a. Remove the bolt, idler pulley cover plate, and idler pulley sub-assembly.

Fig. 41 V-ribbed belt tensioner assembly and bolts

Fig. 42 Water pump assembly, gasket, and bolts

5. Remove the 5 bolts and V-ribbed belt tensioner assembly.
6. Remove the water pump pulley, as follows:
    a. Using SST: 09960-10010 or equivalent, hold the water pump pulley.
    b. Remove the 4 bolts and the water pump pulley.
7. Remove the water inlet housing, as follows:
    a. Disconnect the water hose.
    b. Remove the 2 bolts, nut and water inlet housing.
    c. Remove the water inlet housing gasket and water outlet pipe O-ring.
8. Remove the water pump assembly, as follows:
    a. Remove the 16 bolts, water pump assembly and water pump gasket.

To install:

9. Install the water pump assembly, as follows:
    a. Install a new water pump gasket and the water pump assembly with the 16 bolts. Tighten the bolts to 16 ft. lbs. (21 Nm) and 81 inch lbs. (9.1 Nm).

➡**Make sure that there is no oil on the threads of bolts A.**

➡**Be sure to replace 2 bolts C with new ones or reuse them after applying adhesive (Toyota Genuine Adhesive 1344, Three Bond 1344 or equivalent.)**

10. Install the water inlet housing, as follows:
    a. Install a new water inlet housing gasket and water outlet pipe O-ring.
    b. Install the water inlet housing with the 2 bolts and nut, and tighten to 7 ft. lbs. (10 Nm).

➡**Be careful not to allow the O-ring to get caught between the parts.**

    c. Connect the water hose.

11. Install the water pump pulley, as follows:
    a. Temporarily install the water pump pulley with the 4 bolts.
    b. Using SST: 09960-10010 or equivalent, hold the water pump pulley.
    c. Tighten the 4 bolts to 16 ft. lbs. (21 Nm).
12. Install the V-ribbed belt tensioner assembly with the 5 bolts, and tighten to 32 ft. lbs. (43 Nm).
13. Install the idler pulley cover plate and No. 2 idler pulley sub-assembly with the bolt, and tighten to 32 ft. lbs. (43 Nm).
14. Install the alternator assembly.
15. Install the engine hangers.
16. Remove the engine stand.
17. Install the engine assembly with transaxle.

ALTERNATOR

REMOVAL & INSTALLATION

2.4L Engine

See Figure 43.

1. Before servicing the vehicle, refer to the precautions section.
2. Remove or disconnect the following:
   - Electrical wiring from the alternator
   - Drive belt
   - 1 adjusting and 2 mounting bolts
   - Alternator
3. Installation is the reverse of removal. Observe the following torques:
   - M8 bolts: 15 ft. lbs. (21 Nm)
   - M10 bolts: 38 ft. lbs. (52 Nm)

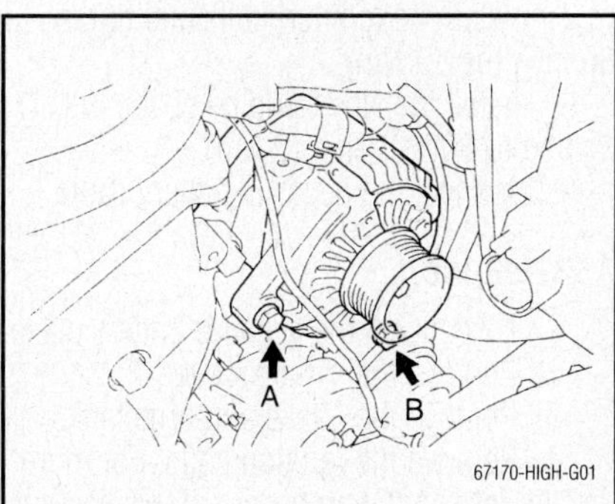

Fig. 43 Alternator bolt locations—2.4L engine

3.3L Engine

See Figure 44.

1. Before servicing the vehicle, refer to the precautions section.
2. Remove or disconnect the following:
   - Alternator electrical connectors
   - Wiring harness from the clip
   - Pivot bolt
   - Plate washer

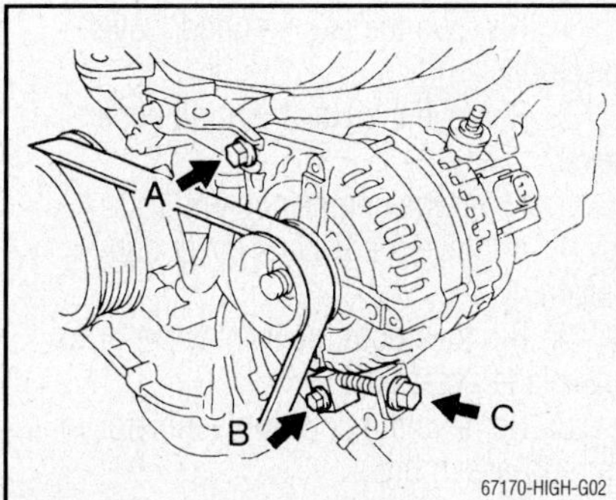

Fig. 44 Alternator bolt locations—3.3L engine

   - Adjusting lockbolt
   - Drive belt
   - Alternator

To install:

3. Install or connect the following:
   - Alternator
   - Drive belt. Tension the belt to 170–180 lbs. for a new belt or 95–135 lbs. for a used belt.
   - Adjusting lockbolt. Tighten the bolt to 13 ft. lbs. (18 Nm).
   - Plate washer
   - Pivot bolt. Tighten the bolt to 41 ft. lbs. (56 Nm) for the 3.0L and 43 ft. lbs. (58 Nm) for the 3.3L.
   - Wiring harness from the clip
   - Alternator electrical connectors

3.5L Engine

1. Raise and safely support the vehicle.
2. Remove the front wheel.
3. Remove the engine under cover assembly.
4. Remove the No. 1 engine under cover.
5. Remove the right front fender molding sub-assembly.
6. Remove the right front fender liner.
7. Remove the right front fender apron seal.
8. Drain engine coolant.
9. Remove the V-bank cover sub-assembly
10. Remove cool air intake duct seal.
11. Remove the battery.
12. Remove the No. 1 and 2 air cleaner inlets, as follows:
    a. Disconnect the 2 vacuum switching valve clamps.
    b. Disconnect the 2 vacuum hoses.
    c. Remove the 2 bolts and No. 2 air cleaner inlet.
    d. Disconnect the 2 vacuum hoses, and remove the 2 bolts and No. 1 air cleaner inlet.
13. Disconnect the No. 1 and 2 radiator hoses.
14. Disconnect the oil cooler hoses.
15. Detach the wire harness clamps from both sides of the fan shroud and disconnect the cooling fan ECU connector.
16. Remove the radiator grill. Refer to radiator removal procedure for detailed instructions.
17. Remove the hood lock assembly.
18. Disconnect the low pitched horn and high pitched horn connectors.
19. Detach the hood lock control cable clamp and remove the 6 bolts and upper radiator support sub-assembly.
20. Remove the 4 bolts and move the cooler condenser assembly.
21. Remove the radiator assembly and fan assembly with motor.
22. Remove the bolt and the No. 2 oil level dipstick guide.
23. Remove the V-ribbed belt.
24. Remove the alternator assembly, as follows:
    a. Remove the terminal cap.
    b. Remove the nut and disconnect the wire harness from terminal B.
    c. Disconnect the alternator connector from the generator assembly.
    d. Disconnect the connector from the compressor and magnetic clutch.
    e. Disconnect the 3 wire harness clamps.
    f. Remove the 2 bolts, and then disconnect the bracket.
    g. Remove the 2 bolts and the alternator assembly.
    h. Disconnect the wire harness clamp, and then remove the alternator bracket.
    i. Remove the bolt and the wire harness clamp stay.

To install:

25. Install the alternator assembly, as follows:
    a. Install the wire harness clamp stay with the bolt, and tighten to 15 ft. lbs. (20 Nm).
    b. Connect the alternator bracket with the wire harness clamp.
    c. Install alternator assembly with the 2 bolts, and tighten to 32 ft. lbs. (43 Nm).
    d. Temporally install the 2 bolts., then fully tighten the 2 bolts to 15 ft. lbs. (20 Nm).
    e. Connect the alternator connector to the alternator assembly.
    f. Install the alternator wire with the nut, and tighten to 87 ft. lbs. (9.8 Nm).
    g. Install the terminal cap.
    h. Connect the 3 wire harness clamps.
    i. Connect the magnetic clutch connector to the compressor and magnetic clutch.
26. Install the V-ribbed belt.
27. Install the No. 2 oil level dipstick guide, as follows:
    a. Install a new O-ring to the No. 2 oil level dipstick guide.
    b. Apply a light coat of engine oil to the O-ring.
    c. Push in the No. 2 oil level dipstick guide end into the No. 1 oil level dipstick guide.

d. Install the No. 2 oil level dipstick guide with the bolt, and tighten to 15 ft. lbs. (20 Nm).

28. Install the radiator assembly and fan assembly with motor.
29. Install the cooler condenser assembly.
30. Install the upper radiator support sub-assembly.
31. Install the hood lock assembly.
32. Install the radiator grill.
33. Connect the cooling fan ECU connector.
34. Connect the oil cooler hose.
35. Connect the No. 1 and 2 radiator hose.
36. Install the No. 1 and 2 air cleaner inlet.
37. Install the battery.
38. Install the cool air intake duct seal.
39. Install the right front fender apron seal.
40. Install the right front fender liner.
41. Install the right front fender molding sub-assembly.
42. Add engine coolant. Refer to radiator installation procedure for detailed instructions.
43. Inspect for coolant leak. Refer to radiator installation procedure for detailed instructions.
44. Inspect automatic transaxle fluid.
45. Inspect for oil leaks.
46. Install the No. 1 engine under cover.
47. Install the engine under cover assembly.
48. Install the right front wheel. Tighten the lug nuts to 76 ft. lbs. (103 Nm).
49. Lower the vehicle.
50. Install the V-bank cover sub-assembly.

ENGINE ELECTRICAL

DISTRIBUTORLESS IGNITION SYSTEM

FIRING ORDER

See Figure 45.

Firing order for 3.5L engine: 1–2–3–4–5–6

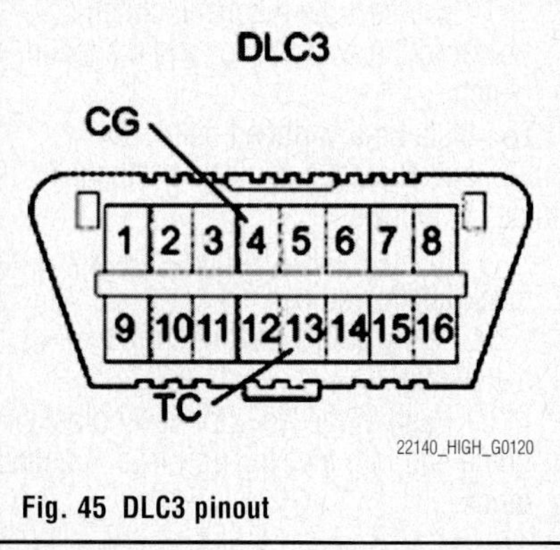

Fig. 45 DLC3 pinout

IGNITION COIL

REMOVAL & INSTALLATION

Refer to Spark Plug procedures.

IGNITION TIMING

INSPECTION

1. Warm up the engine.
2. Using SST: 09843-18040, connect terminals 13 (TC) and 4 (CG) of the DLC3.

➡Confirm the terminal numbers before connecting them. Connecting the wrong terminals can damage the engine.

➡Turn off all electrical systems before connecting the terminals.

➡Perform this inspection after the cooling fan motor is turned off.

3. Remove the V-bank cover sub-assembly.
4. Pull out the red lead wire harness.
5. Connect the tester terminal of the timing light to the red lead wire as shown in the illustration.

➡Use a timing light which detects the No. 1 cylinder ignition signal.

6. Check the ignition timing at idle. Standard ignition timing: 8 to 12° BTDC at idle.

➡When checking the ignition timing, the transmission should be in the neutral position.

➡Run the engine at 1000 to 1300 RPM for 5 seconds, and then check that the engine RPM returns to idle speed.

7. Disconnect terminals 13 (TC) and 4 (CG) of the DLC3.
8. Check the ignition timing at idle. Standard ignition timing: 7 to 24° BTDC at idle.
9. Confirm that the ignition timing advances immediately when the engine RPM is increased.
10. Remove the timing light from the engine.

ADJUSTMENT

All engines are equipped with a Distributorless Ignition System (DIS). No timing adjustment is possible.

SPARK PLUGS

REMOVAL & INSTALLATION

2.4L Engine

1. Disconnect the negative battery cable.
2. Remove the No. 1 engine cover.
3. Remove the ignition coils. Refer to the procedure in this section for more details.
4. Using a 16 mm (0.63 in.) plug wrench, remove the spark plugs.
5. Inspect the condition of the spark plugs, as outlined in this section.

To install:

6. Using a 16 mm (0.63 in.) plug wrench, install the spark plugs and tighten to 14 ft. lbs. (19 Nm).
7. Install the ignition coils and tighten to 80 inch lbs. (9 Nm).
8. Install the No. 1 engine cover and tighten to 80 inch lbs. (9 Nm).
9. Connect the negative battery cable.

3.3L Engine

1. Disconnect the negative battery cable.
2. Remove the V-bank cover.
3. Remove the intake air surge tank.
4. Remove the ignition coils. For more details, refer to the procedure in this section.
5. Using a 16 mm (0.63 in.) plug wrench, remove the spark plugs.

To install:

6. Using a 16 mm (0.63 in.) plug wrench, install the spark plugs and tighten to 18.5 ft. lbs. (25 Nm).
7. Install the ignition coil and tighten the retainers to 71 inch lbs. (8 Nm).
8. Install the intake air surge tank.
9. Install the V-bank cover.
10. Connect the negative battery cable.

3.5L Engine

See Figures 46 through 55.

1. Remove the engine under cover assembly.
2. Remove the No. 1 engine under cover.
3. Drain the engine coolant.
4. Remove the V-bank cover sub-assembly.
5. Remove both the front wiper arms and blade assemblies.
6. Remove the cowl top ventilator louver sub-assembly.
7. Remove the windshield wiper motor and link assembly.
8. Remove the outer cowl top panel sub-assembly.

Fig. 46 Throttle body bracket bolts removal sequence—3.5L engine

Fig. 47 No. 1 surge tank stay bolt removal sequence

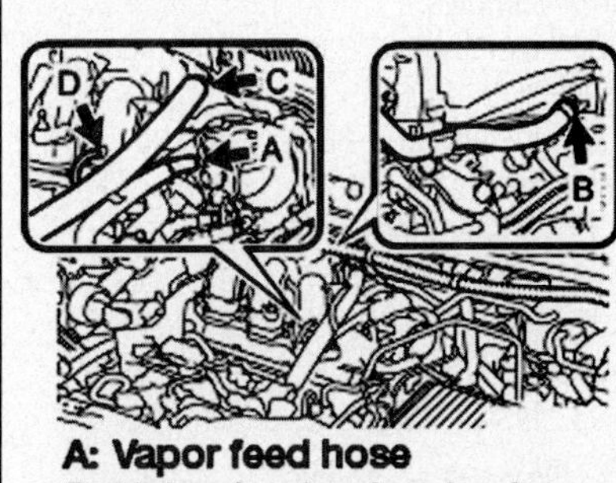

Fig. 48 Intake air surge tank assembly hoses

9. Disconnect the engine room main wire.

10. Remove the 2 bolts in the order shown in the illustration and remove the throttle body bracket.

11. Remove the 2 bolts in the order shown in the illustration and remove the No. 1 surge tank stay.

12. Remove the air cleaner cap sub-assembly.

13. Remove the intake air surge tank assembly, as follows:

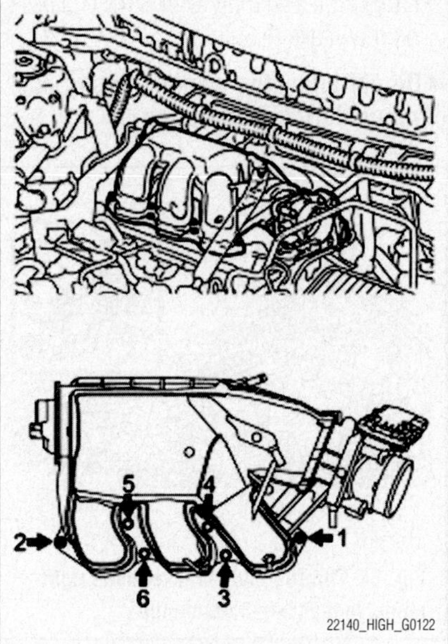

Fig. 49 Intake air surge tank assembly, bolts, and nuts removal sequence

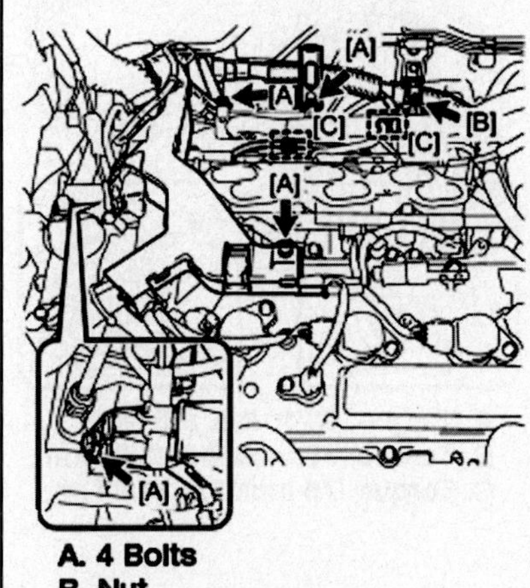

Fig. 50 Ignition coil assembly harness clamps

a. Disconnect the 4 hoses.

b. Disconnect the throttle body connector and clamp.

c. Disconnect the connector.

d. Disconnect the 2 water by-pass hoses from the throttle body.

e. Remove the 4 bolts and 2 nuts in the order shown in the illustration.

➡Use a 5 mm hexagon socket wrench to remove the 4 bolts.

f. Remove the gasket from the intake air surge tank.

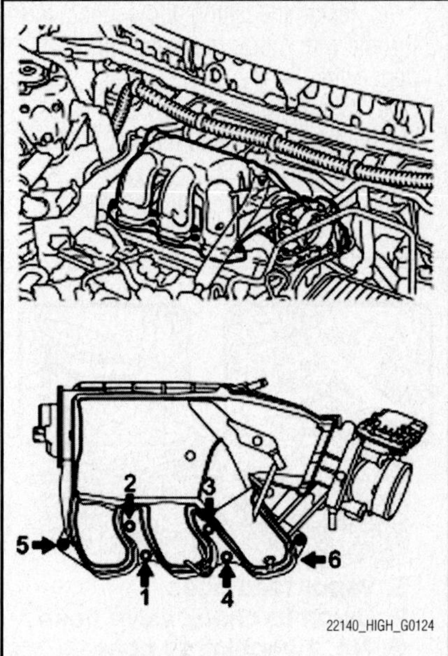

Fig. 51 Intake air surge tank assembly installation sequence

14. Remove the ignition coil assembly, as follows:

a. Remove the 4 bolts.

b. Remove the nut.

c. Disconnect the 2 harness clamps.

d. Disconnect the 6 ignition coil connectors.

e. Remove the 6 bolts and 6 ignition coils

15. Remove the 6 spark plugs.

To install:

16. Install the 6 spark plugs and tighten to 13 ft. lbs. (18 Nm).

17. Install the ignition coil assembly as follows:

a. Install the 6 ignition coils with the 6 bolts and tighten to 7 ft. lbs. (10 Nm).

b. Connect the 6 ignition coil connectors.

c. Install the 4 bolts and tighten to 73 inch lbs. (8.3 Nm).

d. Install the nut and tighten to 73 inch lbs. (8.3 Nm).

e. Install the 2 clamps.

18. Install the intake air surge tank assembly as follows:

a. Install the surge tank with the 4 bolts and 2 nuts in the order shown in the illustration, and tighten to 12 ft. lbs. (16 Nm) and 13 ft. lbs. (18 Nm), using a 5 mm hexagon socket wrench. **DO NOT** apply oil to the bolts.

b. Connect the 2 water by-pass hoses to the throttle with motor body assembly.

c. Connect the connector.

d. Install the clamp and connect the throttle with motor body assembly connector.

e. Connect the 4 hoses.

19. Temporarily install the No. 1 surge tank stay as follows:

a. Temporarily install the intake air surge tank assembly with 3 new gaskets on the intake manifold.

➡Do not allow the gaskets to slip out of place during installation.

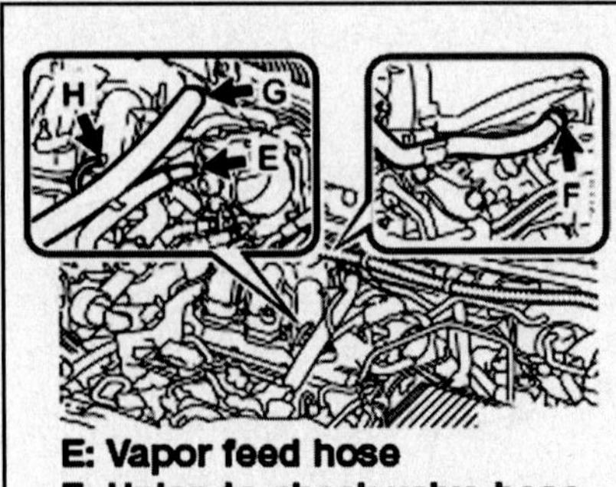

Fig. 52 Intake air surge tank assembly hoses

Fig. 53 No. 1 surge tank stay bolt tightening sequence

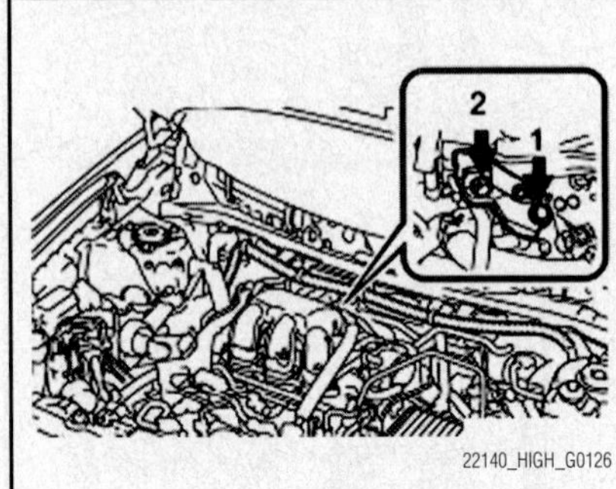

Fig. 54 Throttle body bracket bolts tightening sequence—3.5L engine

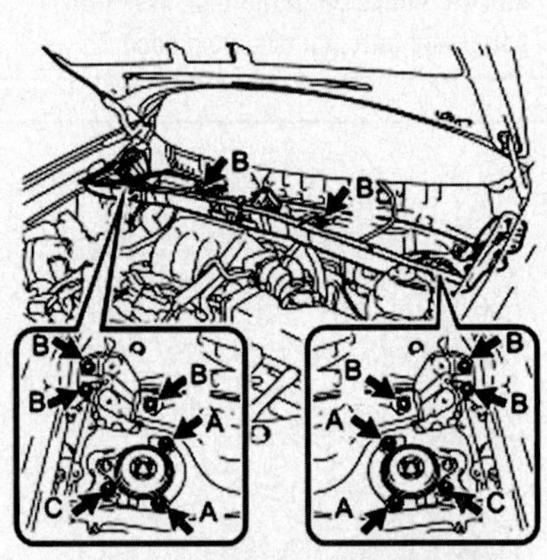

Fig. 55 Outer cowl top panel sub-assembly, bolts, and nuts—3.5L engine

b. Temporarily install the No. 1 surge tank stay with the 2 bolts. **DO NOT** apply oil to the bolts.

20. Temporarily install the throttle body bracket with the 2 bolts. **DO NOT** apply oil to the bolts.

21. Fully tighten the No. 1 surge tank stay, as follows:

a. Fully tighten the 2 bolts in the order shown in the illustration. Tighten to 15 ft. lbs. (21 Nm). **DO NOT** apply oil to the bolts.

22. Fully tighten the throttle body bracket, as follows:

a. Fully tighten the 2 bolts in the order shown in the illustration. Tighten to 15 ft. lbs. (21 Nm). **DO NOT** apply oil to the bolts.

23. Connect the engine room main wire.

24. Install the air cleaner cap sub-assembly.

25. Install the outer cowl top panel sub-assembly, as follows:

a. Install the outer cowl top panel sub-assembly with the 8 bolts and 6 nuts.

b. Engage the 4 clamps.

26. Install the windshield wiper motor and link assembly.

27. Install the cowl top ventilator louver sub-assembly.

28. Install both front wiper arm and blade assemblies.

29. Install the V-bank cover sub-assembly.

30. Add engine coolant.

31. Inspect for engine coolant leak.

32. Install the No. 1 engine under cover.

33. Install the engine under cover assembly.

STARTER

REMOVAL & INSTALLATION

2.4L Engine

See Figure 56.

1. Remove the battery.
2. Remove the battery tray.
3. Disconnect the starter wiring.
4. Remove the mounting bolts.
5. Lift out the starter.
6. Installation is the reverse of removal. Torque the mounting bolts to 27 ft. lbs. (37 Nm).

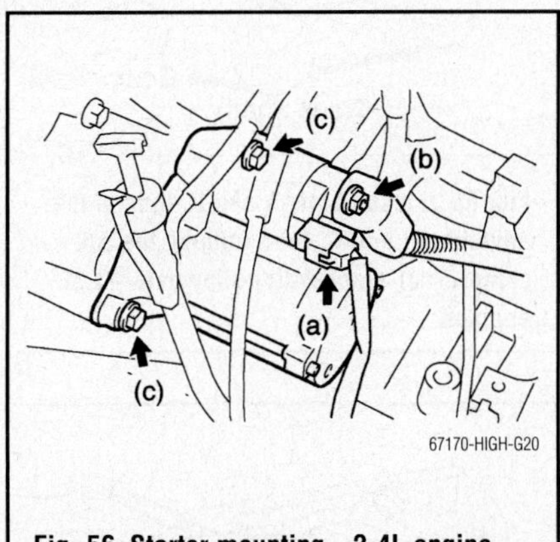

Fig. 56 Starter mounting—2.4L engine

3.3L Engine

See Figure 57.

1. Before servicing the vehicle, refer to the precautions section.
2. Remove the battery.
3. Remove the battery tray.
4. Remove the wiring from the starter.
5. Remove the 2 bolts and lower the starter.
6. Installation is the reverse of removal. Torque the starter bolts to 27 ft. lbs. (37 Nm).

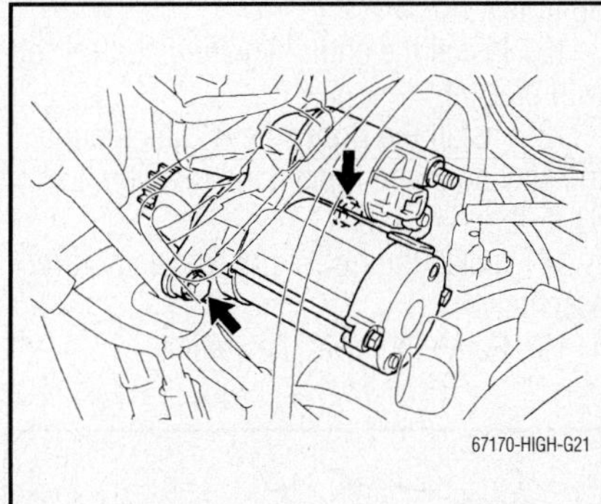

Fig. 57 Starter mounting bolt locations—3.3L engine

3.5L Engine

1. Remove the cool air intake duct seal.
2. Remove the battery.
3. Remove the No.1 and 2 air cleaner inlets.
4. Remove the air cleaner cap sub-assembly.
5. Remove the air cleaner case sub-assembly.
6. Remove the 2 bolts and the air cleaner bracket.
7. Disconnect the starter connector.
8. Open the terminal cap, remove the nut and disconnect the starter wire.
9. Remove the 2 bolts and starter.

To install:

10. Installation is the reverse of removal procedure, noting the following:
    a. After installing the starter assembly with the 2 bolts, tighten to 27 ft. lbs. (37 Nm).
    b. After connecting the starter wire with the nut, tighten to 87 ft. lbs. (9.8 Nm).
    c. After installing the bracket with the 2 bolts, tighten to 9 ft. lbs. (12 Nm).

ENGINE MECHANICAL

ACCESSORY DRIVE BELTS

ACCESSORY BELT ROUTING

See Figures 58 and 59.

Refer to the accompanying illustration for accessory drive belt routing.

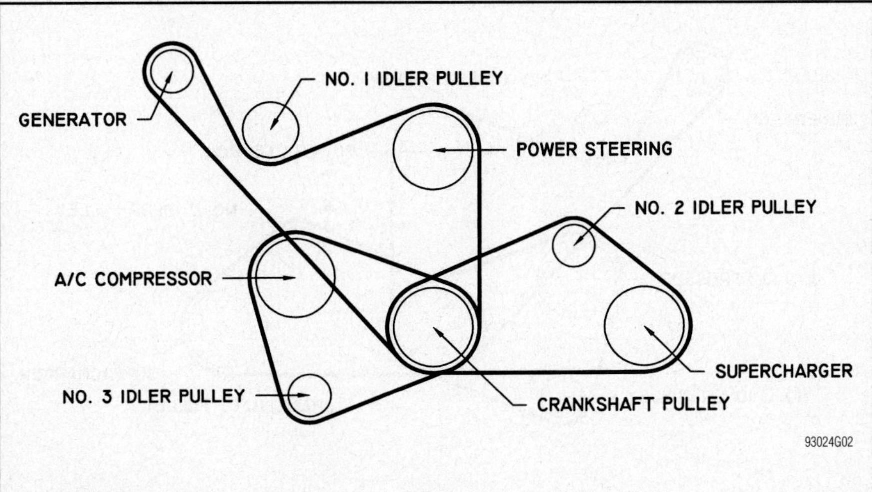

Fig. 58 Accessory drive belt routing —2.4L engine

ADJUSTMENT

2.4L and 3.5L Engines

Belt tension is maintained by an automatic tensioner. No adjustment is necessary of possible.

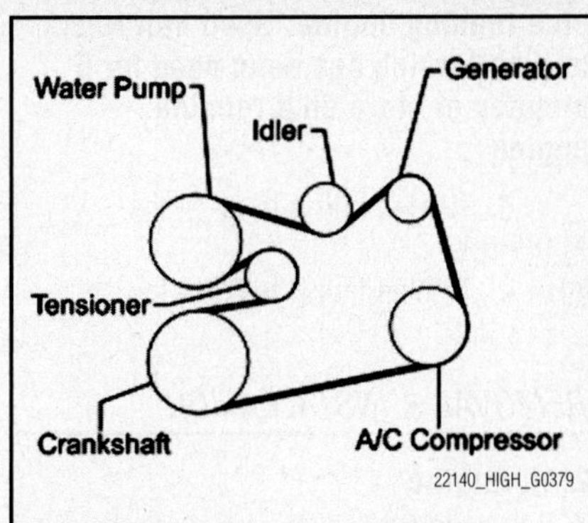

Fig. 59 Accessory drive belt routing — 3.5L engine

3.3L Engine

See Figure 60.

1. To adjust the A/C compressor-crankshaft pulley belt:
    a. Loosen bolt A.
    b. Loosen bolt B.
    c. Apply drive belt tension by turning bolt C. The belt tension should be 160–180 lbs. for a new belt, or 115–135 lbs. for a used belt.

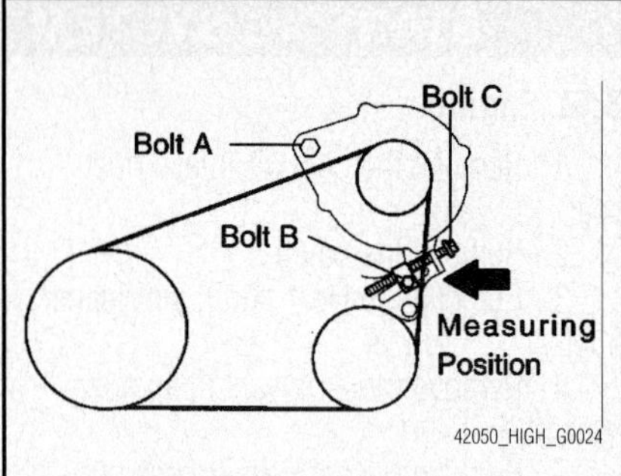

Fig. 60 A/C compressor-crankshaft pulley belt

➡New belt refers to a belt which has been used for less than 5 minutes on a running engine. Used belt refers to a belt which has been used for 5 minutes or more on a running engine.

d. Tighten bolt A to 43 ft. lbs. (58 Nm).

e. Tighten bolt B to 13 ft. lbs. (18 Nm).

2. To adjust the power steering (vane) pump belt:

a. Loosen bolt A.

b. Loosen bolt B.

c. Apply drive belt tension by turning bolt C. The belt tension should be 132–154 lbs. for a new belt, or 55–88 lbs. for a used belt.

➡New belt refers to a belt which has been used for less than 5 minutes on a running engine. Used belt refers to a belt which has been used for 5 minutes or more on a running engine.

d. Tighten bolt A to 32 ft. lbs. (43 Nm).

e. Tighten bolt B to 32 ft. lbs. (43 Nm).

REMOVAL & INSTALLATION

2.4L Engine

See Figures 61 and 62.

1. Raise and safely support the vehicle.
2. Remove the right front wheel and tire assembly.
3. Remove the right front fender splash shield.
4. Remove the right front fender apron seal.
5. Remove the right engine mounting stay no.2.
6. Remove the engine cover sub-assembly no.1.
7. Remove the engine moving control rod with the bracket.
8. Remove the right side engine mounting bracket no. 2.
9. Remove the fan and alternator v belt, as follows:

a. Slowly turn the belt tensioner clockwise for more than 3 seconds using Special Service Tool 09843-18030, and then remove the drive belt.

To install:

10. Slowly turn the belt tensioner clockwise for more than 3 seconds using the special tool, and install the drive belt, then carefully release the tensioner. Make sure the belt is properly routed.
11. Install the right side engine mounting bracket no. 2 and tighten the retainers to 38 ft. lbs. (52 Nm).
12. Install the engine moving control rod with bracket
13. Install the right side engine mounting stay no.2 and tighten the retainers to 47 ft. lbs. (64 Nm).
14. Install the right front wheel and tire assembly.
15. Carefully lower the vehicle.

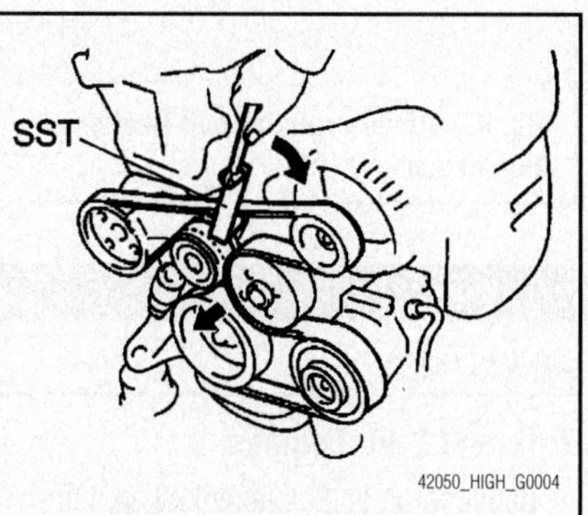

Fig. 61 Belt removal and installation—2.4L engine

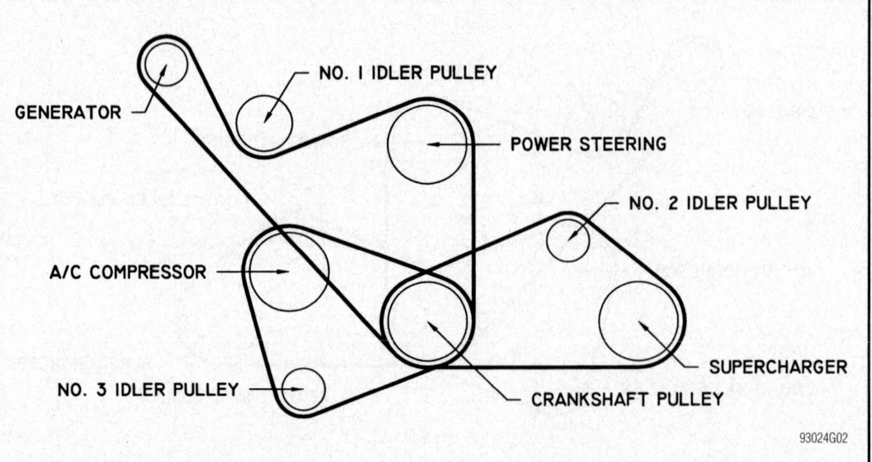

Fig. 62 Accessory drive belt routing—2.4L engine

3.3L Engine

See Figures 63 and 64.

1. Raise and safely support the vehicle.
2. Remove the right front wheel and tire assembly.
3. Remove the engine under cover.
4. Remove the right front fender apron seal
5. To remove the A/C compressor-crankshaft pulley belt :

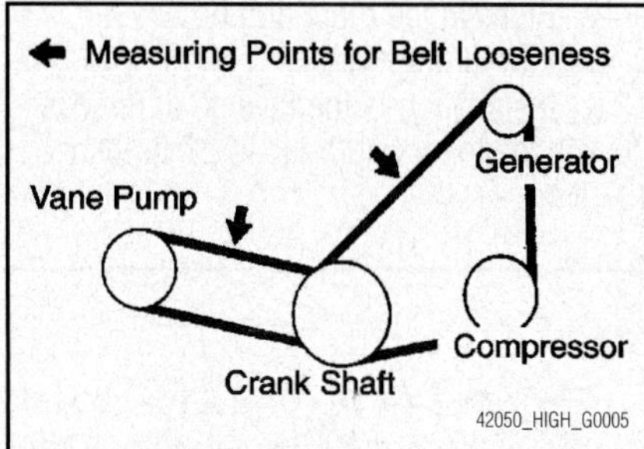

Fig. 63 Loosen bolts A and B, loosen the adjustment bolt C, then remove the A/C compressor-crankshaft pulley belt—3.3L engines

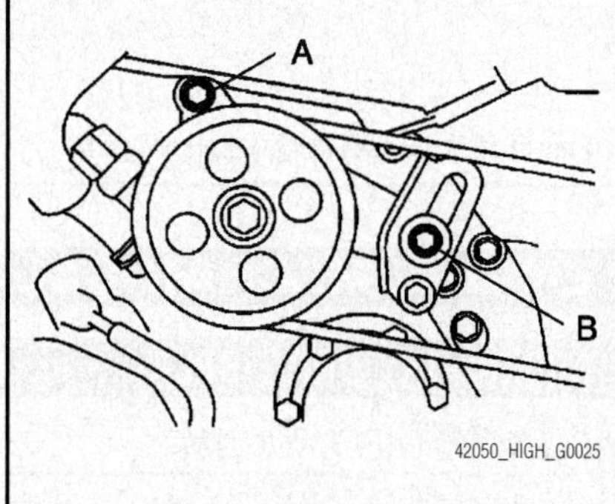

Fig. 64 Power steering (vane) pump removal and installation—3.3L engines

a. Loosen bolts A and B.

b. Loosen adjusting bolt C and remove the belt.

6. To remove the power steering (vane) pump V-belt:

a. Loosen bolts A and B, and remove the belt.

To install:

7. To install the power steering (vane) pump V-belt:

a. Install the belt on each pulley.

b. Using a bar, adjust the belt tension and tighten bolt B to 32 ft. lbs. (43 Nm).

c. Tighten bolt A to 32 ft. lbs. (43 Nm).

8. To install the A/C compressor-crankshaft pulley belt:

a. Install the belt on each pulley.

b. Tighten adjusting bolt C, and adjust the belt tension.

c. First tighten bolt A to 43 ft. lbs. (59 Nm), and then B to 13 ft. lbs. (18 Nm).

9. Check belt tension, as outlined earlier in this section.

10. Install the right front wheel and tire assembly.

11. Carefully lower the vehicle.

3.5L Engine

See Figure 65.

1. Raise and safely support the vehicle.
2. Remove the right front wheel.
3. Remove the engine under cover assembly.
4. Remove the No. 1 engine under cover.
5. Remove the right front fender molding sub-assembly, as follows:

a. Remove the clip.

b. Using a 4 mm hexagon wrench, remove the screw.

c. Peel off the front fender side protector and disengage the 3 clips, and then remove the right front fender molding sub-assembly.

d. Remove the pad from the right front fender molding sub-assembly.

e. Remove the 2 clips No. 4 from the right front fender molding sub-assembly.

f. Remove the front fender side protector from the right front fender molding sub-assembly.

6. Remove the right front fender liner.
7. Remove the right front fender apron seal.
8. Remove the V-ribbed belt as follows:

a. Using SST: 09961-00950 or equivalent, release the V-ribbed belt tension by turning the V-ribbed belt tensioner assembly counterclockwise, and remove the V-ribbed belt from the V-ribbed belt tensioner assembly.

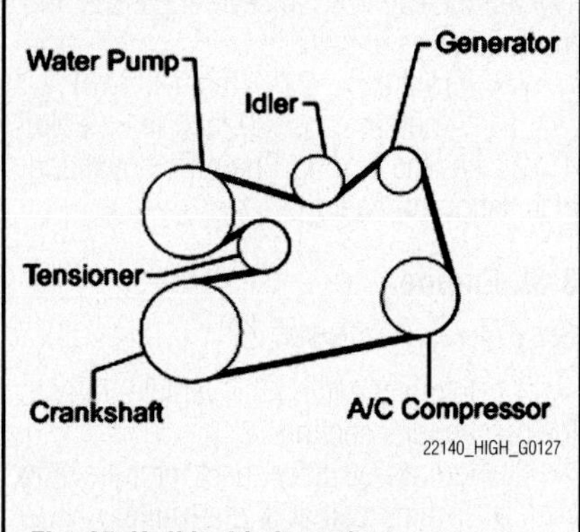

Fig. 65 V-ribbed belt routing

b. While turning the V-ribbed belt tensioner assembly counterclockwise, align with its holes, and then insert the 5 mm bi-hexagon wrench into the holes to fix the V-ribbed belt tensioner assembly.

To install:

9. Install the V-ribbed belt.
10. Using SST: 09961-00950 or equivalent, turn the V-ribbed belt tensioner assembly counterclockwise and remove the bar.
11. If it is difficult to install the V-ribbed belt, perform the following procedure:

a. Put the V-ribbed belt on every pulley except the tensioner pulley as shown in the illustration.

b. Release the V-ribbed belt tension by turning the V-ribbed belt tensioner assembly counterclockwise, and put the V-ribbed belt on the V-ribbed belt tensioner assembly pulley.

➡Put the backside of the V-ribbed belt on the V-ribbed belt tensioner assembly pulley and No. 2 idler pulley sub-assembly.

➡Check that the V-ribbed belt is properly set to each pulley.

c. After installing the V-ribbed belt, check that it fits properly in the ribbed grooves. Confirm that the belt has not slipped out of the grooves on the bottom of the crankshaft pulley by hand.

12. Install the right front fender apron seal.
13. Install the right front fender liner.
14. Install the right front fender molding sub-assembly, as follows:

a. Clean the vehicle body surface by heating the vehicle body surface with a heat light, removing the front fender side protector from the vehicle body, and wiping off any tape adhesive residue with cleaner.

b. If reusing the right front fender molding sub-assembly, heat it with a heat light, remove the front fender side protector, wipe off any tape adhesive residue with cleaner, and install a new front fender side protector to the front fender molding sub-assembly.

c. Using a heat light, heat the vehicle body and the front fender molding sub-assembly.

d. Remove the release paper from the front fender molding sub-assembly.

➡After removing the release paper, keep the exposed adhesive free from foreign matter.

e. Engage the 3 clips and install the right front fender molding sub-assembly.

f. Using a 4 mm hexagon wrench, install the screw.

g. Install the clip.

15. Install the no. 1 engine under cover.
16. Install the engine under cover assembly.
17. Install the right front wheel. Tighten the lug nuts to 76 ft. lbs. (103 Nm).
18. Lower the vehicle.

CAMSHAFT AND VALVE LIFTERS

REMOVAL & INSTALLATION

2.4L Engine

See Figures 66 through 69.

1. Before servicing the vehicle, refer to the precautions section.
2. Disconnect the negative battery cable.
3. Drain the engine coolant.
4. Remove or disconnect the following:

- Right front wheel
- Right fender splash shield
- Right fender apron seal
- No. 1 engine undercover
- Coil pack
- Cylinder head cover

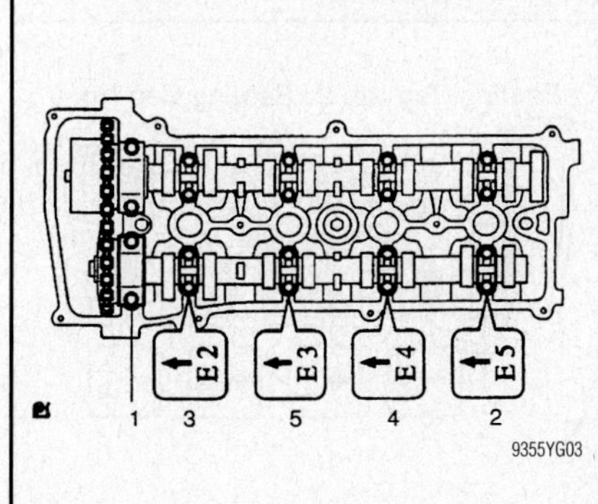

Fig. 66 Exhaust camshaft cap bolt loosening sequence—2.4L engine

5. Set the No. 1 piston at TDC compression.

6. Remove or disconnect the following:
- Timing chain tensioner No. 1

7. Loosen the camshaft timing gear set bolt.

8. Raise the camshaft and remove the set bolt.

9. Remove or disconnect the following:
- Timing gear and chain
- Exhaust camshaft

10. Loosen the intake camshaft cap bolts in several passes, in the sequence shown. Remove the caps. Remove the camshaft.

11. Installation is the reverse of removal.

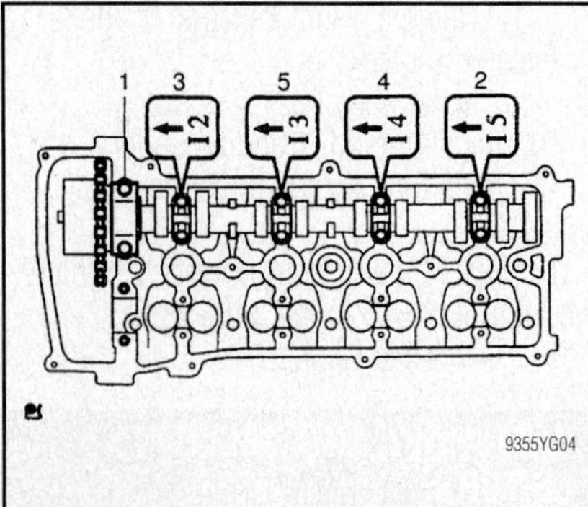

Fig. 67 Intake camshaft cap bolt loosening sequence—2.4L engine

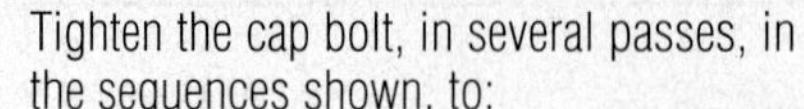

Tighten the cap bolt, in several passes, in the sequences shown, to:
- Front caps: 22 ft. lbs. (30 Nm)
- All other caps: 80 inch lbs. (9 Nm)

12. See the Timing Chain Removal and Installation procedure.

3.3L Engine

See Figures 70 through 79.

1. Before servicing the vehicle, refer to the precautions section.

2. Remove or disconnect the following:
- Timing belt and idler pulley
- Camshaft timing pulleys
- Cylinder head covers

➡The thrust clearance on both the intake and exhaust camshafts is very small; the camshafts must be kept level during removal. If the camshafts are removed without being kept level, the camshaft may be caught in the cylinder head, causing the head to break or the camshaft to seize.

3. Remove the exhaust and intake camshafts from the right side cylinder head, as follows:

a. Turn the camshaft with a wrench until the 2 pointed marks drive and driven gears are aligned. (The right camshaft gears have 2 marks apiece; the left side camshaft gears have 1 mark each.)

b. Secure the exhaust camshaft subgear to the main gear using a service bolt. A bolt 0.63–0.79 in. (16–20mm) long with a 6mm thread diameter and a 1mm pitch is recommended. When removing the exhaust camshaft be sure the sub-gear is not loaded; all the force must be eliminated.

c. Uniformly loosen and remove the exhaust camshaft bearing cap bolts in several passes and in the proper sequence. Remove the 8 bearing cap bolts and remove the caps, keeping them in the correct order.

d. Remove the exhaust camshaft from the engine.

e. Uniformly loosen and remove the 10 bearing cap bolts in several passes, in the proper sequence. Remove the bearing caps, keeping them in order, remove the oil seal, then lift out the intake camshaft.

4. Remove the exhaust and intake camshafts from the left side cylinder head, as follows:

a. Turn the camshaft with a wrench until the pointed marks on the drive

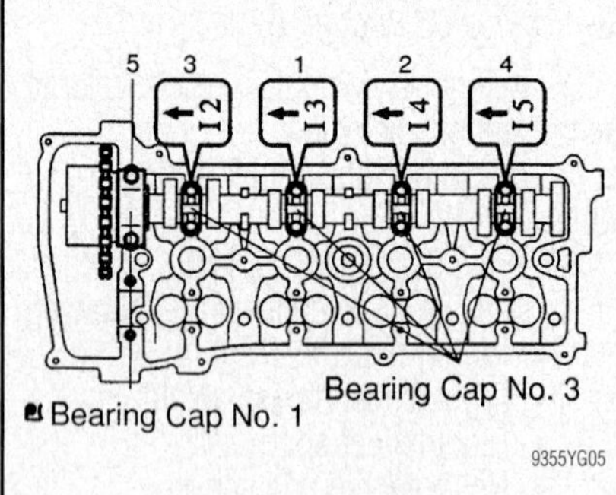

Fig. 68 Intake camshaft cap bolt tightening sequence—2.4L engine

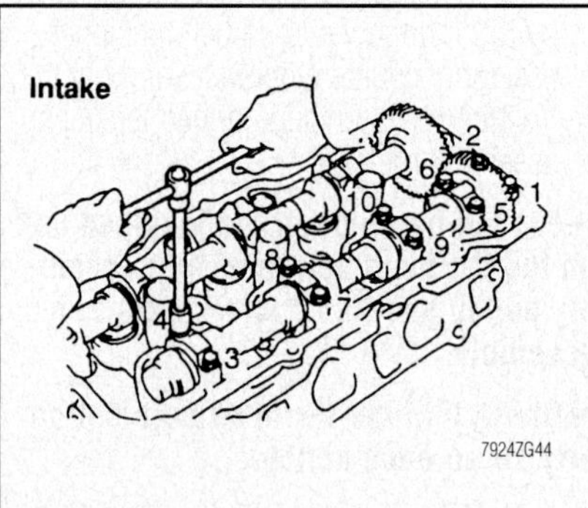

Fig. 70 Right intake camshaft bearing cap bolt loosening sequence—3.3L engine

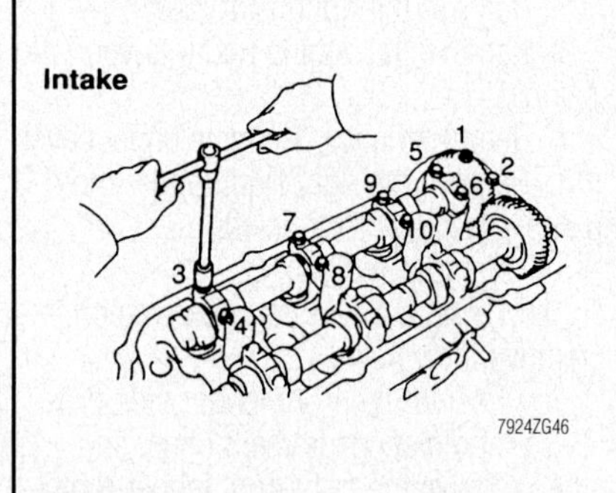

Fig. 72 Left intake camshaft bearing cap bolt loosening sequence—3.3L engine

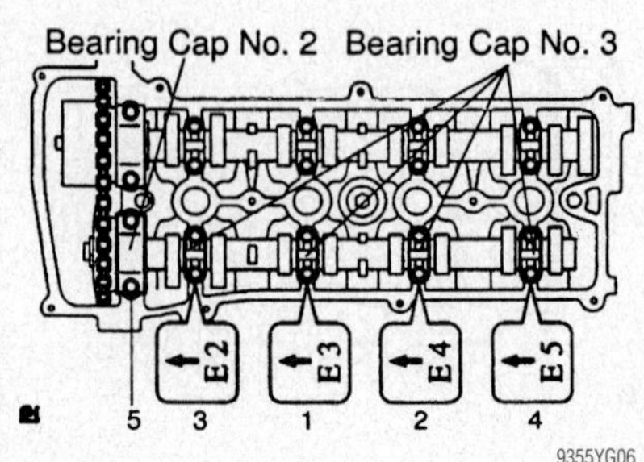

Fig. 69 Exhaust camshaft cap bolt tightening sequence—2.4L engine

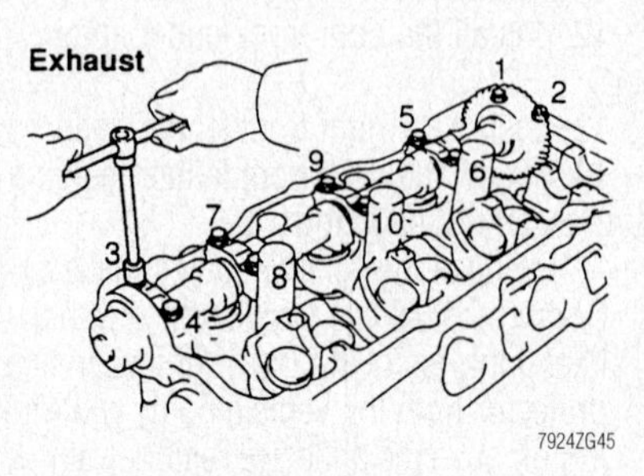

Fig. 71 Right side exhaust camshaft bearing cap bolt loosening sequence—3.3L engine

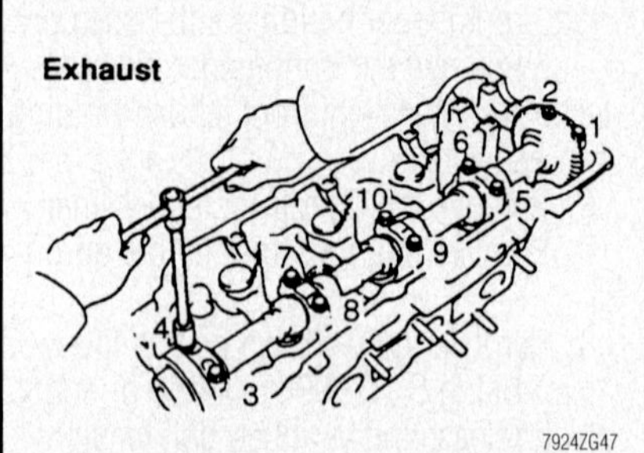

Fig. 73 Left side exhaust camshaft bearing cap bolt loosening sequence—3.3L engine

and driven gears are aligned. (The right camshaft gears have 2 marks apiece; the left side camshaft gears have 1 mark each.)

b. Secure the exhaust camshaft sub-gear to the main gear using a service bolt. A bolt 16–20mm long with a 6mm thread diameter and a 1mm pitch is recommended. When removing the exhaust camshaft be sure the sub-gear is not loaded; all the force must be eliminated.

c. Uniformly loosen and remove the exhaust camshaft bearing cap bolts in several passes and in the proper sequence. Remove the 8 bearing cap bolts and remove the caps. Keep the caps in the correct order.

d. Remove the exhaust camshaft from the engine.

e. Uniformly loosen and remove the 10 bearing cap bolts in several passes, in the reverse order of the installation sequence. Remove the bearing caps, keeping them in order, remove the oil seal, then lift out the intake camshaft.

5. Remove the valve lifter shims and hydraulic lifters. Identify each lifter and shim as it is removed so it can be reinstalled in the same position. If the lifters are to be reused, store them upside down in a sealed container.

To install:

6. Install the valve lifters into their original positions and install the shims. Check valve clearance and replace the shims as necessary.

7. When reinstalling, remember that the camshafts must be handled carefully and kept straight and level to avoid damage.

8. Before installing the camshafts in either cylinder head, apply multi-purpose grease to each camshaft.

9. Install the right camshafts, as follows:

a. Position the intake camshaft on the head so that the alignment marks are at a 90 degrees angle from vertical. The mark should be at the "3 o'clock" position.

b. Apply sealant to the No. 1 bearing cap.

c. Apply a light coat of clean engine oil to the bolt threads and under the bolt head. Install the bearing caps to their proper position. Tighten the bolts evenly and in several passes to 12 ft. lbs. (16 Nm) in the proper sequence.

d. Position the exhaust camshaft on the head so that the alignment marks are at a 90 degrees angle from vertical. The mark should be at the "9 o'clock" position and must align with the marks on the other gear.

e. Apply a light coat of clean engine oil to the bolt threads and under the bolt head. Install the bearing caps to their proper position. Tighten the bolts evenly and in several passes to 12 ft. lbs. (16 Nm) in the proper sequence.

f. Remove the service bolt.

10. Install the left camshafts, as follows:

a. Position the intake camshaft on the head so that the alignment mark is at a 90 degrees angle from vertical. The mark should be at the "9 o'clock" position.

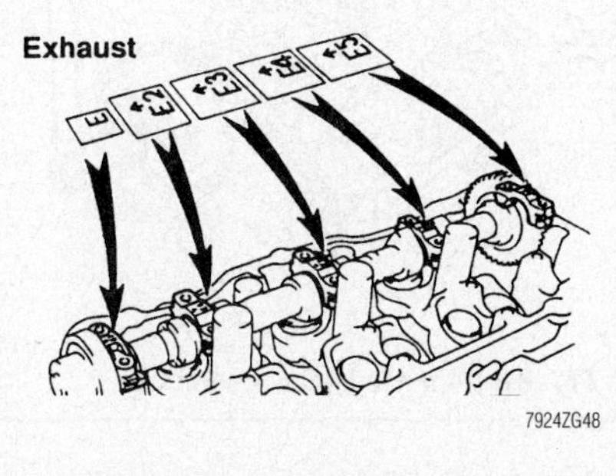

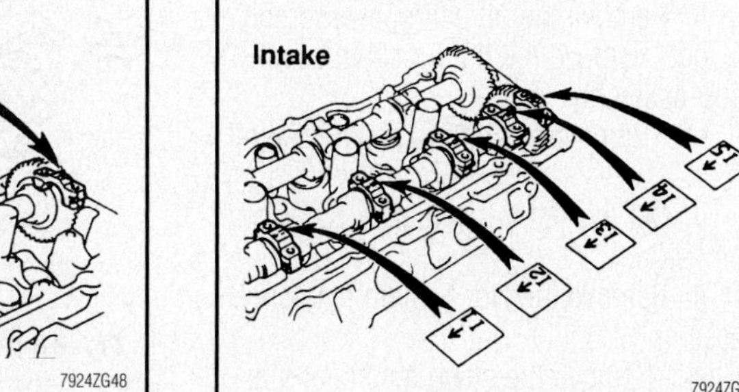

Fig. 74 Right exhaust bearing caps must be placed in their proper locations—3.3L engine

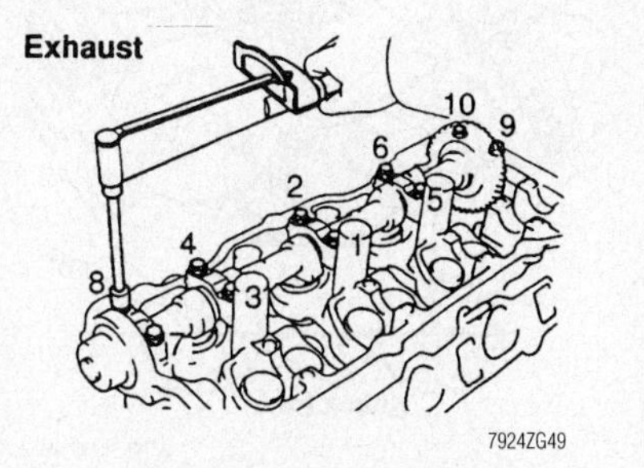

Fig. 75 Right exhaust camshaft bearing cap bolt tightening sequence—3.3L engines

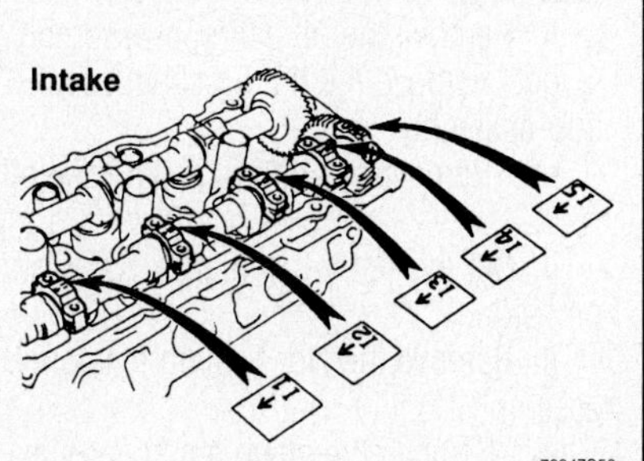

Fig. 76 Right intake bearing caps must be placed in their proper locations—3.3L engine

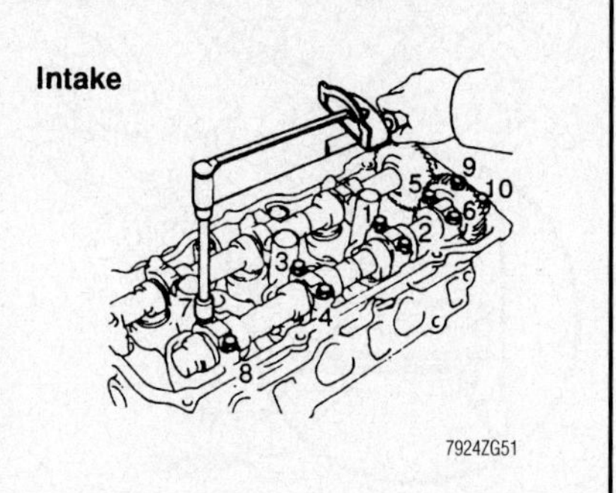

Fig. 77 Right intake camshaft bearing cap bolt tightening sequence—3.3L engine

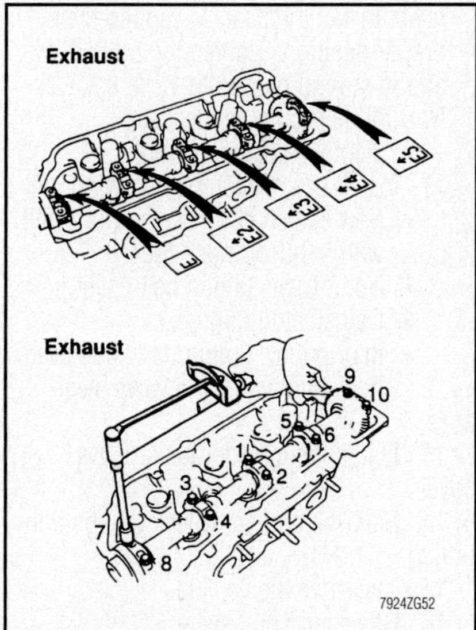

Fig. 78 Left exhaust bearing caps locations and bolt tightening sequence—3.3L engine

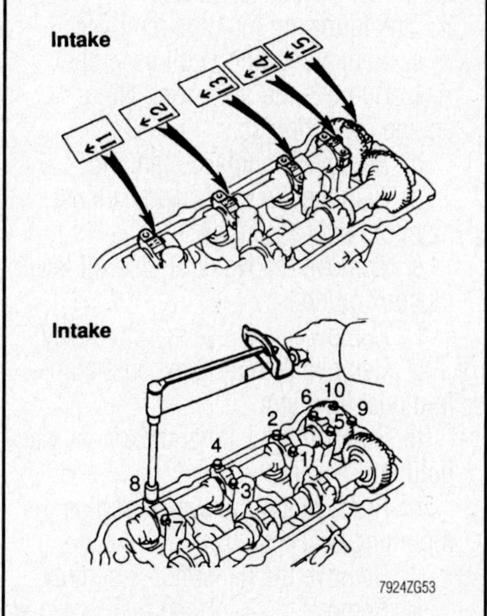

Fig. 79 Left intake camshaft bearing cap locations and bolt tightening sequence—3.3L engine

b. Apply sealant to the No. 1 bearing cap.

c. Apply a light coat of clean engine oil to the bolt threads and under the bolt head. Install the bearing caps to their proper position. Tighten the bolts evenly and in several passes to 12 ft. lbs. (16 Nm) in the proper sequence.

d. Position the exhaust camshaft on the head so that the alignment marks are at a 90 degrees angle from vertical. The mark should be at the "3 o'clock" position and must align with the marks on the other gear.

e. Apply a light coat of clean engine oil to the bolt threads and under the bolt head. Install the bearing caps to their proper position. Tighten the bolts evenly and in several passes to 12 ft. lbs. (16 Nm) in the proper sequence.

f. Remove the service bolt.

11. Install or connect the following:
   - New camshaft oil seals, lubricated with multi-purpose grease
   - No. 3 (rear) timing belt cover
   - Camshaft timing gears
   - Idler pulley, timing belt and covers

12. Check and adjust the valve clearance.

13. Install the cylinder head (valve) covers.

14. Start the engine. Check the ignition timing.

15. Test drive the vehicle.

16. Check all fluid levels.

3.5L Engine

See Figures 80 through 97.

1. Remove the engine assembly.
2. Install on engine stand.
3. The following must be removed:

a. Remove ignition coil assembly.

b. Remove the right hand No. 2 engine mounting stay.

c. Remove the intake manifold.

d. Remove the right hand exhaust manifold sub-assembly.

e. Remove the No. 2 engine oil level dipstick guide.

f. Remove the No. 2 manifold stay.

g. Remove the No. 2 exhaust manifold heat insulator.

h. Remove the left hand exhaust manifold sub-assembly.

i. Remove the transverse engine mounting bracket.

j. Remove the generator assembly.

k. Remove the V-ribbed belt tensioner assembly.

l. Remove the No. 2 timing gear cover.

m. Remove the No. 2 idler pulley sub-assembly.

n. Remove the left hand No. 1 engine front mounting bracket.

o. Remove the left hand 6 bolts and No. 1 front engine mounting bracket.

p. Remove the radio setting condenser.

q. Remove the No. 1 vacuum switching valve.

r. Remove the knock control sensor wire.

s. Remove the knock control sensor.

t. Remove the crankshaft position sensor.

u. Remove the No. 1 oil pipe.

v. Remove the oil pipe.

w. Remove the crankshaft pulley.

x. Remove the oil cooler assembly, if necessary.

y. Remove the No. 1 oil cooler bracket, if necessary.

z. Remove the water inlet housing.

aa. Remove the water outlet.

bb. Remove the cylinder head cover sub-assembly (for Bank 1).

cc. Remove the cylinder head cover sub-assembly (for Bank 2).

dd. Remove the No. 2 oil pan sub-assembly.

ee. Remove the oil strainer sub-assembly.

ff. Remove the oil pan sub-assembly.

gg. Remove the timing chain cover sub-assembly

hh. Remove the timing chain case oil seal.

ii. Set the No. 1 cylinder to TDC/compression.

jj. Remove the No. 1 chain tensioner assembly.

kk. Remove the chain tensioner slipper.

ll. Remove the chain sub-assembly.

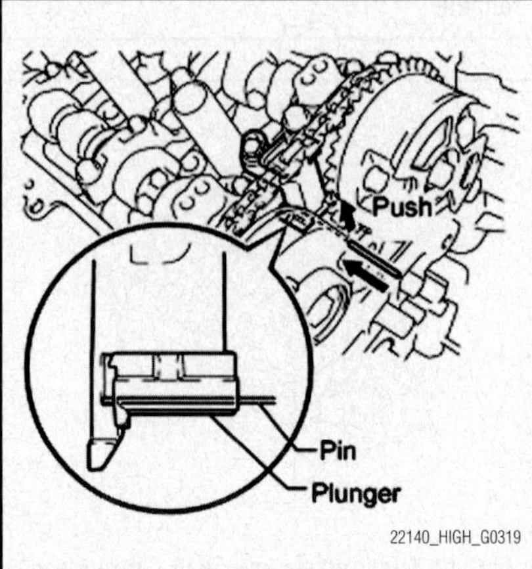

Fig. 80 Pinning tensioner

mm. Remove the idle sprocket assembly.

nn. Remove the camshaft timing gears and No. 2 chain (for Bank 1).

oo. While raising the No. 2 chain tensioner assembly, insert a pin of 1.0 mm (0.039 in.) diameter into the hole to fix the No. 2 chain tensioner assembly.

pp. Hold the hexagonal portion of the camshaft with a wrench, and remove the 2 bolts and 2 camshaft timing gear assemblies.

➡Be careful not to damage the cylinder head with the wrench. Do not disassemble the camshaft timing gear assemblies.

qq. Remove the No. 2 chain assembly.

rr. Remove the bolt and No. 2 chain tensioner assembly.

ss. Check that the camshafts are positioned as shown in the illustration.

tt. Uniformly loosen and remove the 8 bearing cap bolts in several steps and in the sequence shown in the illustration.

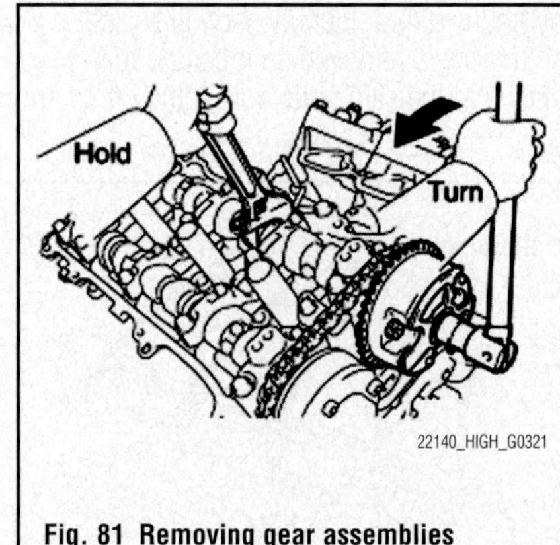

Fig. 81 Removing gear assemblies

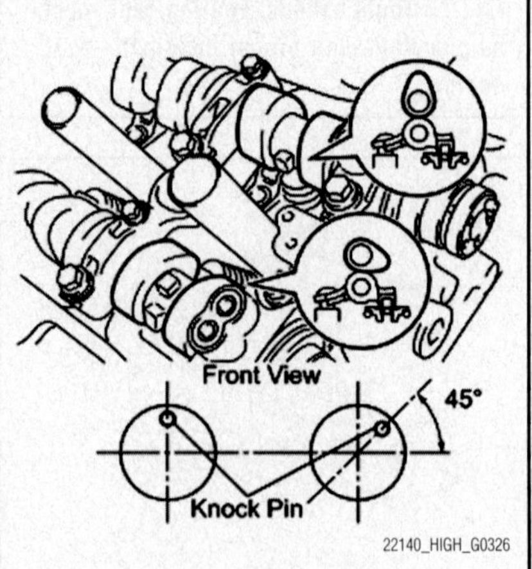

Fig. 82 Positioning camshafts for bearing cap removal

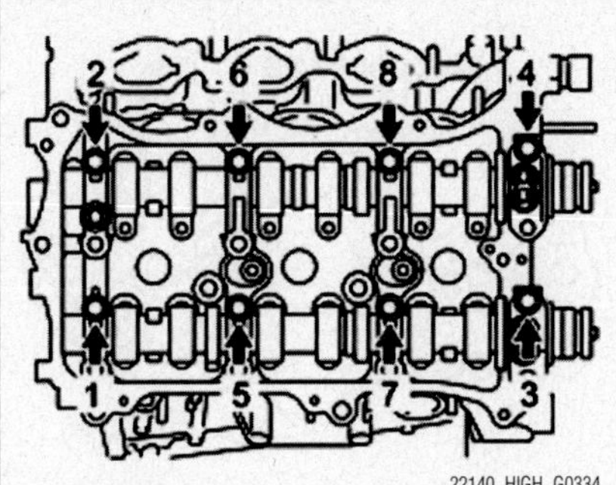
Fig. 83 Camshafts bearing cap removal sequence

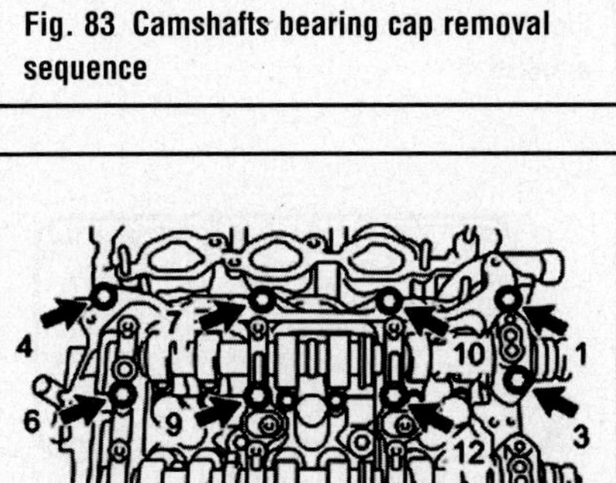
Fig. 84 Positioning camshafts for bearing cap removal

uu. Uniformly loosen and remove the 12 bearing cap bolts in several steps and in the sequence shown in the illustration.

➡Uniformly loosen the bolts while keeping the camshaft level.

vv. Remove the 5 camshaft bearing caps.

ww. Remove the camshaft.

xx. Remove the No. 2 camshaft.

yy. Remove the right hand camshaft housing sub-assembly by prying between the cylinder head and the right hand camshaft housing sub-assembly with a screwdriver.

➡Be careful not to damage the contact surfaces of the cylinder head and the right hand camshaft housing sub-assembly.

To install:

4. Install the following:

a. Apply engine oil to the camshaft journals, camshaft housing sub-assembly RH and camshaft bearing caps.

b. Install the camshaft and No. 2 camshaft to the right hand camshaft housing sub-assembly.

c. Make sure of the marks and numbers on the camshaft bearing caps and

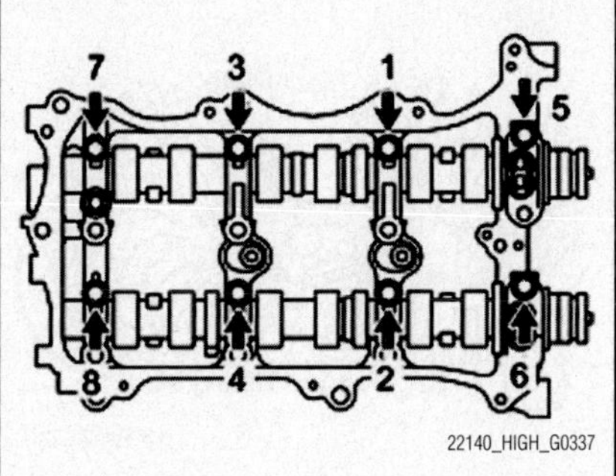
Fig. 85 Camshafts bearing cap tightening sequence

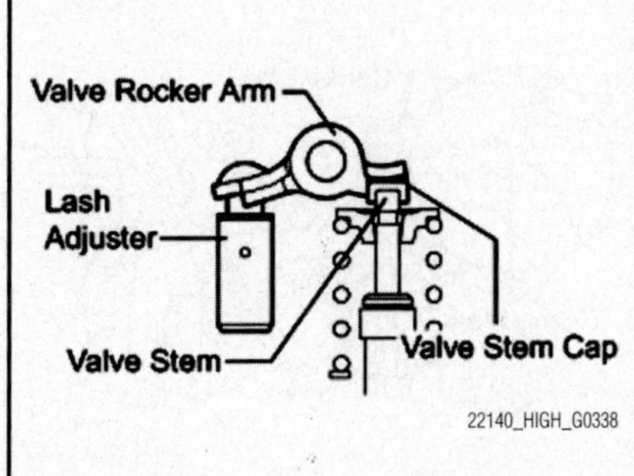

Fig. 86 Valve rocker arm sub-assembly positioning

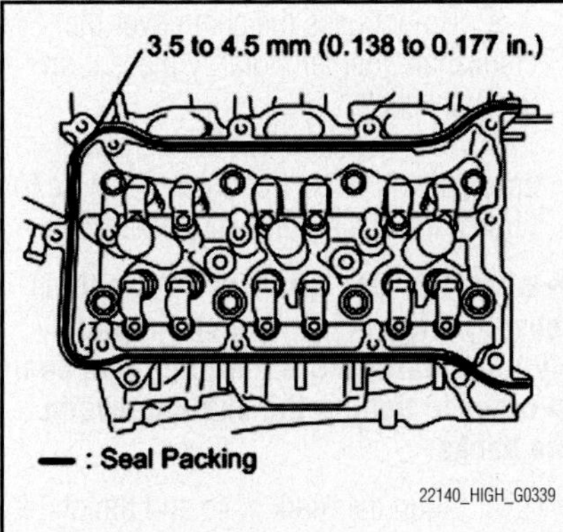

Fig. 87 Applying sealant

place them in each proper position and direction.

d. Temporarily tighten the 8 bearing cap bolts to 7 ft. lbs. (10 Nm) in the order shown in the illustration.

e. Make sure that the No. 1 valve rocker arm sub-assembly is installed as shown in the illustration.

f. Apply seal packing in a continuous line as shown in the illustration.

➡Remove any oil from the contact surface. Install the right hand camshaft housing sub-assembly within 3 minutes. Do not start the engine for at least 2 hours after installing.

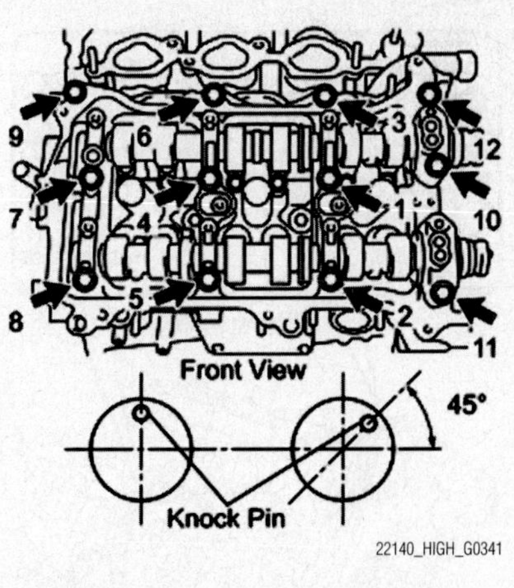

Fig. 88 Camshaft sub-assembly tightening sequence

g. Install the right hand camshaft housing sub-assembly and tighten the 12 bolts to 21 ft. lbs. (28 Nm) in the order shown in the illustration.

➡When installing the right hand camshaft housing, it is necessary to correctly position the camshafts as shown in the removal illustration. If the camshaft housing sub-assembly is removed because any of the bolts are loosened during installation, make sure that the previously applied seal packing does not enter any oil passages.

h. Complete the tightening of the 8 bolts to 12 ft. lbs. (16 Nm) in the order shown above.

i. Install the No. 2 chain tensioner assembly with the bolt and tighten to 15 ft. lbs. (21 Nm).

j. While pushing in the tensioner, insert a pin of 1.0 mm (0.039 in.) diameter into the hole to fix it.

k. Align the mark plate with the timing marks of the camshaft timing gear.

l. Apply a light coat of engine oil to the bolt threads and bolts seating surface.

m. Align the knockpin of the camshaft with pinhole of the camshaft timing gear assembly. Install the camshaft timing gear assembly and camshaft timing exhaust timing gear assembly and camshaft timing exhaust gear with the No. 2 chain sub-assembly installed.

n. Hold the hexagonal portion of the camshaft with the wrench and tighten the two bolts and camshaft timing gear assemblies to 74 ft. lbs. (100 Nm).

o. Remove the pan from the No. 2 chain tensioner assembly.

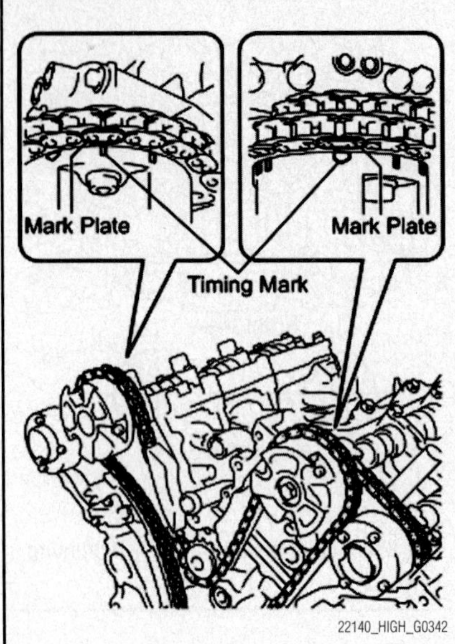

Fig. 89 Aligning timing chain sub-assembly

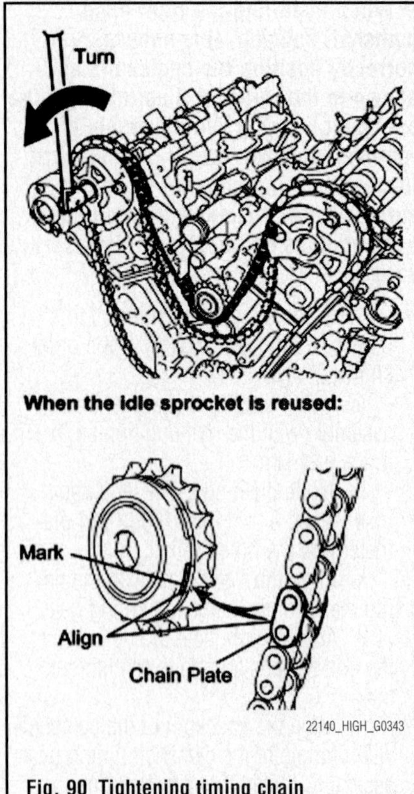

Fig. 90 Tightening timing chain

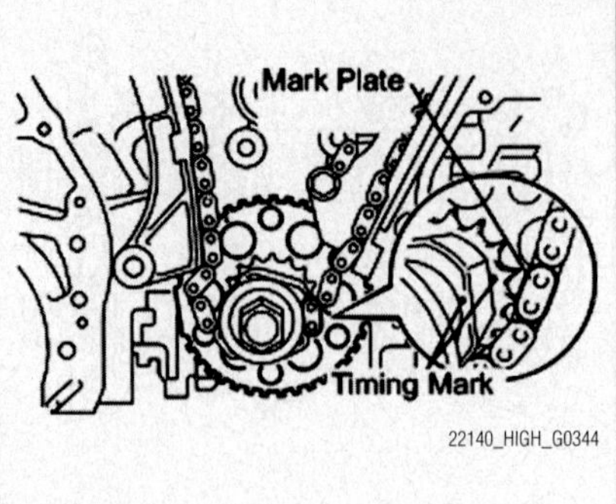

Fig. 91 Installing timing chain on crankshaft

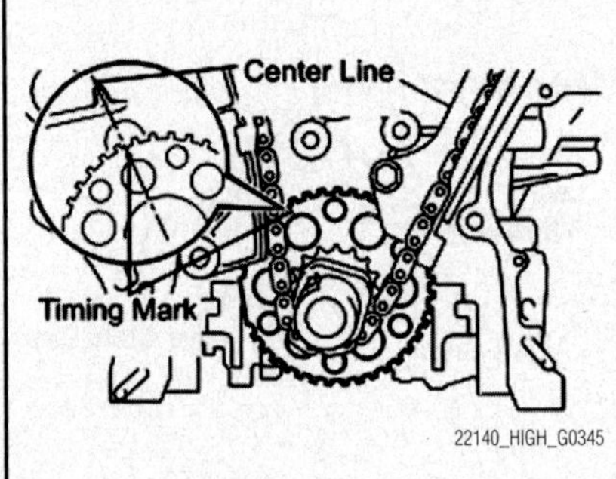

Fig. 92 Aligning timing chain on crankshaft

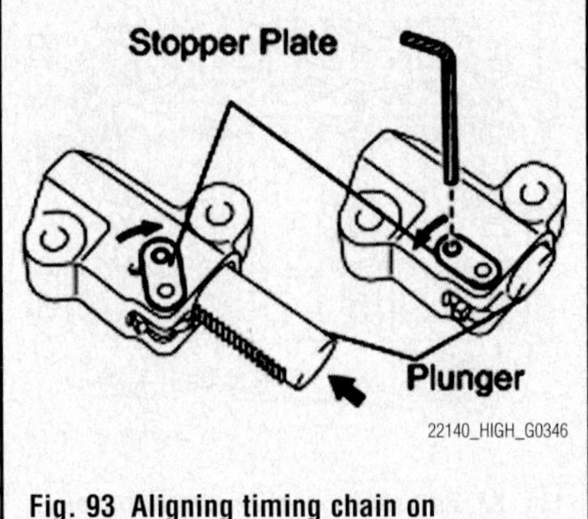

Fig. 93 Aligning timing chain on crankshaft

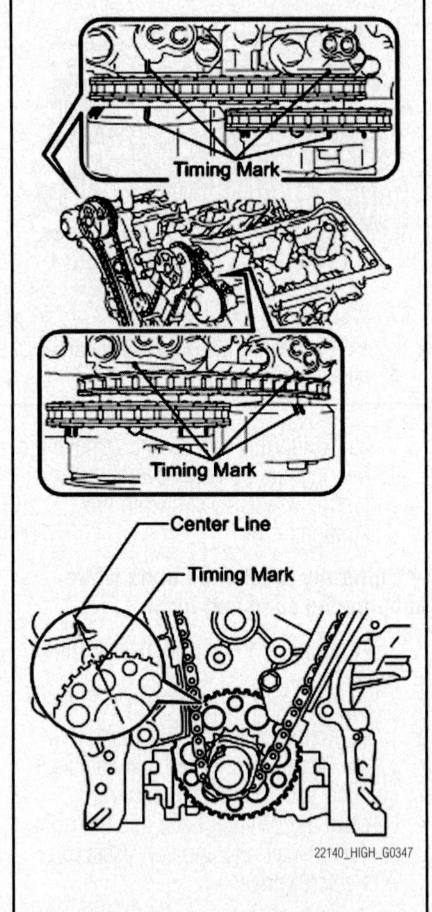

Fig. 94 Aligning complete timing chain

p. Install idle sprocket assembly and tighten to 44 ft. lbs. (60 Nm).

q. Install chain sub-assembly.

r. Align the mark plate and timing marks as shown in the illustration and install the chain.

➡The camshaft mark plates are orange.

s. Do not pass the chain over the crankshaft, just temporarily place it on the crankshaft.

t. Turn the camshaft timing gear assembly on bank 1 counterclockwise to tighten the chain between the banks.

➡When the idle sprocket assembly is reused, align the timing chain plate with the mark where the plate has been in order to tighten the chain between the banks.

u. Align the mark plate and timing marks as shown in the illustration and install the chain onto the crankshaft timing sprocket. The crankshaft to mark plate is yellow.

v. Turn the crankshaft clockwise to set it to the right-hand block bore more centerline. (TDC Compression).

w. Install chain tensioner slipper.

x. Move the stopper plate upward to release the lock, and push the plunger deep into the tensioner.

y. Move the stopper plate downward to set the loss, and insert a hexagon wrench into the hole of the stopper plate.

z. Install No. 1 chain tensioner assembly and tighten bolts to 7 ft. lbs. (10 Nm).

aa. Remove the hexagon wrench from the No. 1 chain tensioner assembly. Check that the each timing mark is aligned with the crankshaft at TDC compression.

bb. Remove the pulley set bolt.

cc. Install timing chain case oil seal.

dd. Install sealant to timing chain cover sub-assembly.

ee. Install new O ring gasket on cylinder block.

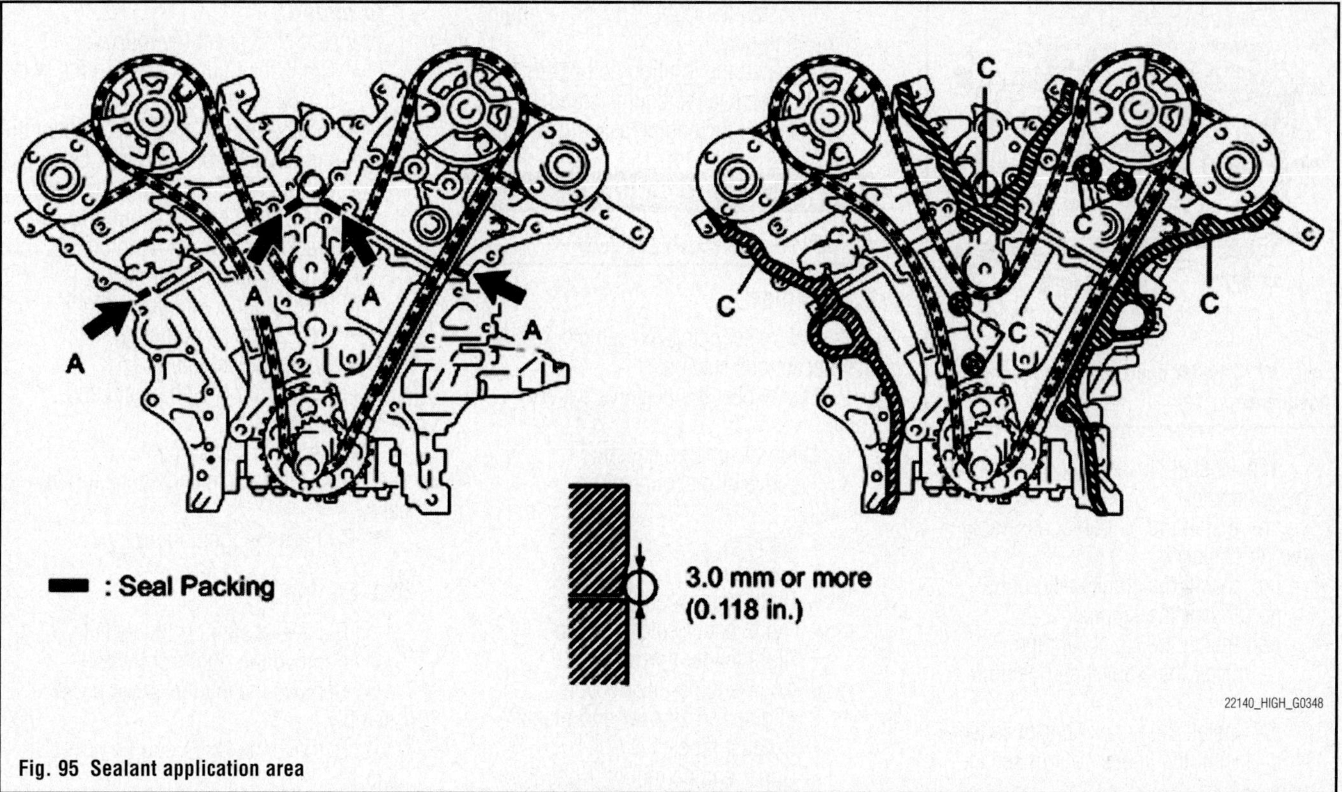

Fig. 95 Sealant application area

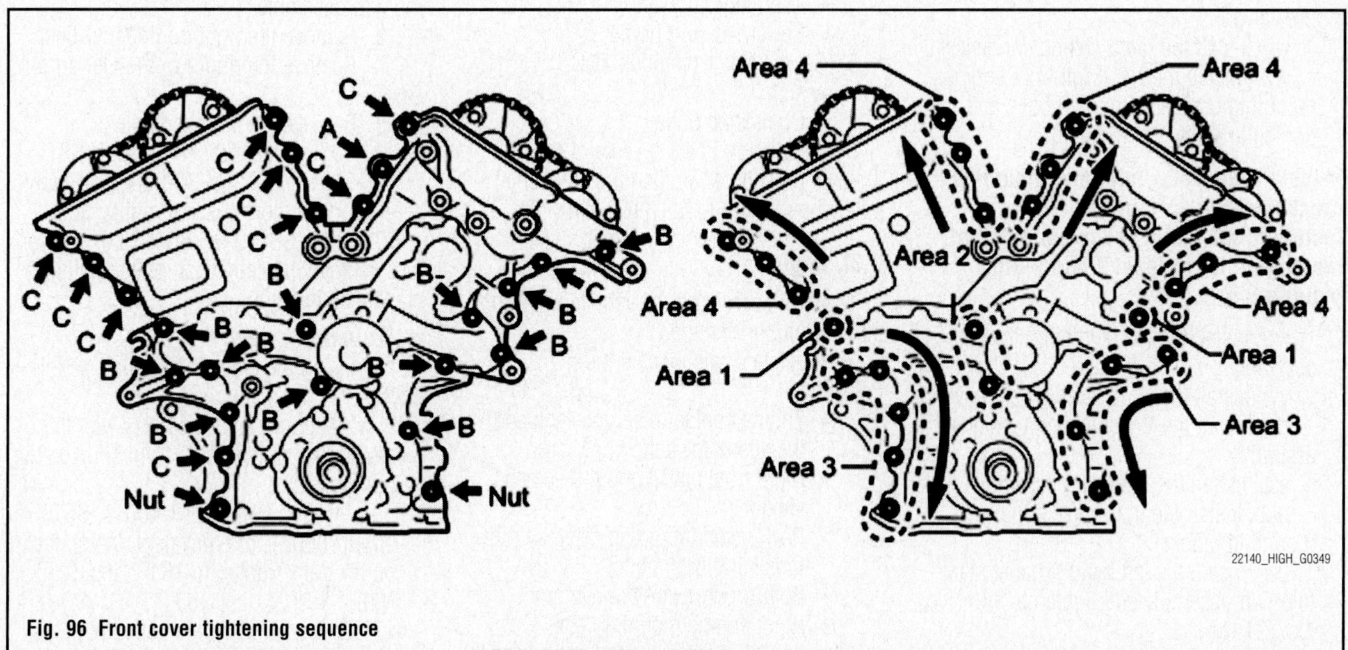

Fig. 96 Front cover tightening sequence

ff. Align the oil pump's drive rotor spline and the crankshaft as shown in the illustration. Install the spline and chain cover to the crankshaft.

gg. Temporarily tighten the timing chain cover with the 23 bolts and 2 nuts.

- Tighten bolts in area 1 and 2: 15 ft. lbs. (21 Nm).
- Tighten bolt in area 3: 15 ft. lbs. (21 Nm).

➡First tighten the upper bolts and nuts, followed by the lower bolts and nuts as shown.

- Tighten bolt in area 4: 32 ft. lbs. (43 Nm)
- Tighten bolt in area 4: 15 ft. lbs. (21 Nm)

hh. Install oil pan sub-assembly and tighten 16 bolts and 2 nuts to 7 ft. lbs. (10 Nm) and 15 ft lbs. (21 Nm).

ii. Install oil strainers sub-assembly and tighten bolts and nuts to 7 ft. lbs. (10 Nm).

jj. Install No.2 oil pan sub-assembly and tighten 16 bolts and 2 nuts to 7 ft. lbs. (10 Nm).

kk. Install cylinder head cover sub-assemblies and tighten to 7 ft. lbs. (10 Nm) and 15 ft. lbs. (21 Nm).

ll. Install the water inlet housing.

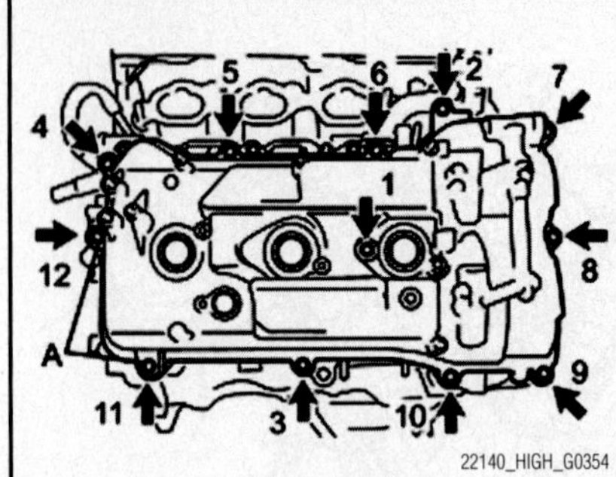

Fig. 97 Cylinder head cover tightening sequence

mm. Install No. 1 oil cooler bracket (w/ oil cooler).

nn. Install the oil cooler assembly (w/ oil cooler).

oo. Install the crankshaft pulley.

pp. Install the oil pipe.

qq. Install the No. 1 oil pipe.

rr. Install the crankshaft position sensor.

ss. Install the knock control sensor.

tt. Install the knock control sensor wire.

uu. Install the No. 1 vacuum switching valve.

vv. Install the radio setting condenser.

ww. Install the No. 1 left front engine mounting bracket with the 6 bolts and tighten to 40 ft. lbs. (54 Nm).

➡Install the water inlet and mounting bracket within 15 minutes after installing the chain cover. Do not start the engine for at least 2 hours after installation.

xx. Install the No. 2 idler pulley sub-assembly.

yy. Install the No. 2 timing gear cover

zz. Install the V-ribbed belt tensioner assembly.

aaa. Install the generator assembly.

bbb. Install the transverse engine mounting bracket.

ccc. Install the left hand exhaust manifold sub-assembly and tighten to 15 ft. lbs. (21 Nm).

ddd. Install the No. 2 exhaust manifold heat insulator.

eee. Install the No. 2 manifold stay.

fff. Install the No. 2 engine oil level dipstick guide.

ggg. Install the right hand exhaust manifold sub-assembly and tighten to 15 ft. lbs. (21 Nm).

hhh. Install the intake manifold and tighten the 6 bolts and 4 nuts uniformly in several steps to 15 ft. lbs. (21 Nm).

iii. Install the right hand No. 2 engine mounting stay.

jjj. Install the ignition coil assembly.

kkk. Remove the engine stand.

lll. Install the engine assembly.

CRANKSHAFT DAMPER

REMOVAL & INSTALLATION

2.4L Engine

1. Before servicing the vehicle, refer to the precautions section.
2. Disconnect the negative battery cable.
3. Drain the engine coolant.
4. Remove or disconnect the following:
   - Hood
   - Engine oil
   - Right front wheel
   - Right fender splash shield
   - Right fender apron seal
   - No. 1 engine undercover
   - Engine roll stopper and bracket
   - Exhaust pipe
   - Upper engine mount, right side
   - Accessory drive belts
   - Alternator
   - Power steering pump
5. Set the No. 1 piston at TDC compression.
   - Crankshaft pulley
6. Installation is the reverse of the removal procedure. Tighten the crankshaft pulley bolt to 125 ft. lbs. (169 Nm).

3.3L Engine

1. Before servicing the vehicle, refer to the precautions section.
2. Remove or disconnect the following:
   - Engine coolant reservoir tank and the alternator belt
   - Right front wheel and the splash shield
   - Power steering pump drive belt, by loosening both bolts
   - Both ground wire connectors
   - Right engine mounting stay
   - Engine moving control rod and the No. 2 right engine mount bracket

➡To extract the engine bracket and control rod, raise the engine slightly.

   - No. 2 alternator bracket
   - Crankshaft pulley bolt, using a pry-bar and wrench or Crankshaft Pulley Holding tool 09213-54015 and Flange Holding tool 09330-00021
   - Crankshaft pulley, using a puller

To install:

3. Install or connect the following:
   - Crankshaft pulley. Tighten the bolt to 162 ft. lbs. (219 Nm).
   - No. 2 alternator bracket. Tighten the nut to 21 ft. lbs. (28 Nm). Do not tighten the pivot bolt at this time.
   - No. 2 right engine mounting bracket and the moving control rod
   - Right engine mount stay
   - Both ground wire connectors
   - Drive belts by adjusting them
   - Coolant reservoir
   - Right front splash shield and wheel
   - Negative battery cable
4. Start the vehicle and check for any leaks.
5. Recheck the ignition timing.

3.5L Engine

1. Raise and support the vehicle.
2. Remove the right front wheel.
3. Remove the engine under cover assembly.
4. Remove the No. 1 engine under cover.
5. Remove the right front fender molding sub-assembly.
6. Remove the right front fender liner.
7. Remove the right front fender apron seal.
8. Remove the V-ribbed belt.
9. Using a special service tool (SST: 09213-70011, SST: 09330-00021 or equivalent), loosen the crankshaft pulley bolt.
10. Using SST: 09950-50013 or equivalent, remove the crankshaft pulley bolt and crankshaft pulley.

To install:

11. Install the crankshaft pulley, as follows:
    a. Align the pulley set key with the key groove of the pulley, and slide on the pulley.
    b. Using SST: 09213-70011, SST: 09330-00021, or equivalent, install the pulley bolt. Tighten to 184 ft. lbs. (250 Nm).
12. The remainder of installation is the reverse of removal. When installing the wheel, tighten the lug nuts to 76 ft. lbs. (103 Nm).

CRANKSHAFT FRONT SEAL

REMOVAL & INSTALLATION

2.4L Engine

For 2.4L engines, see the Timing Chain procedure.

3.3L Engine

1. Before servicing the vehicle, refer to the precautions section.
2. Remove or disconnect the following:
    - Engine coolant reservoir tank and the alternator belt
    - Right front wheel and the splash shield
    - Power steering pump drive belt, by loosening both bolts
    - Both ground wire connectors
    - Right engine mounting stay
    - Engine moving control rod and the No. 2 right engine mount bracket

➡**To extract the engine bracket and control rod, raise the engine slightly.**

- No. 2 alternator bracket
- Crankshaft pulley bolt, using a pry-bar and wrench or Crankshaft Pulley Holding tool 09213-54015 and Flange Holding tool 09330-00021
- Crankshaft pulley, using a puller
- No. 1 timing belt cover

3. Remove the No. 2 timing belt cover, as follows:
    - Engine wire protector from the No. 3 (rear) timing belt cover
    - Engine wire protector clamp from the No. 3 timing belt cover
    - 5 bolts from the No. 2 timing belt cover
    - No. 2 cover

To install:

4. Install or connect the following:
    - No. 2 timing belt cover, using a new gasket

➡**Install it evenly to the part of the belt cover shaded black. After installation, press down on it so that the adhesive sticks to the belt cover firmly.**

- No. 2 timing belt cover. Tighten the 5 bolts to 74 inch lbs. (8 Nm).
- Engine wire protector clamp to the No. 3 timing belt cover
- Engine wire protector to the No. 3 timing belt cover with the bolt
- No. 3 timing belt cover, using a new gasket
- Tighten the 4 No. 1 timing belt cover bolts to 74 inch lbs. (8 Nm).
- Crankshaft pulley. Tighten the bolt to 162 ft. lbs. (219 Nm).
- No. 2 alternator bracket. Tighten the nut to 21 ft. lbs. (28 Nm). Do not tighten the pivot bolt at this time.
- No. 2 right engine mounting bracket and the moving control rod
- Right engine mount stay
- Both ground wire connectors
- Drive belts by adjusting them
- Coolant reservoir
- Right front splash shield and wheel
- Negative battery cable

5. Start the vehicle and check for any leaks.
6. Recheck the ignition timing.

3.5L Engine

See Figures 98 and 99.

1. Raise and support the vehicle.
2. Remove the right front wheel.
3. Remove the engine under cover assembly.
4. Remove the No. 1 engine under cover.
5. Remove the right front fender molding sub-assembly.
6. Remove the right front fender liner.
7. Remove the right front fender apron seal.
8. Remove the V-ribbed belt.
9. Using a special service tool (SST: 09213-70011, SST: 09330-00021 or equivalent), loosen the crankshaft pulley bolt.

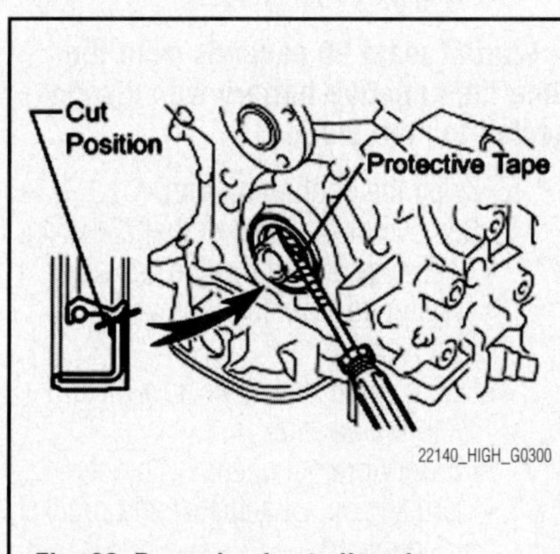

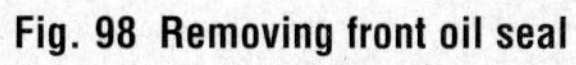

Fig. 98 Removing front oil seal

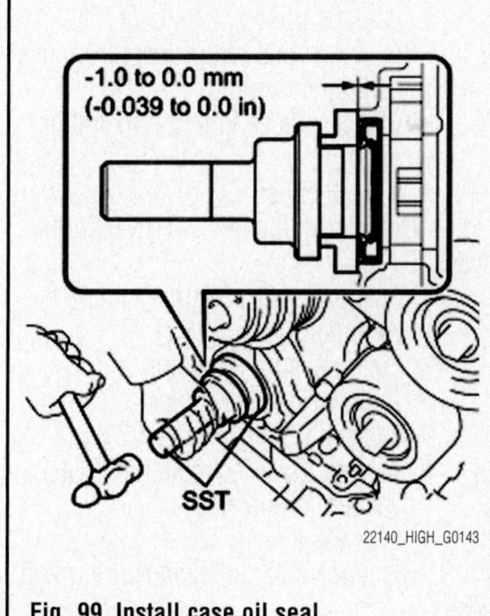

Fig. 99 Install case oil seal

10. Using SST: 09950-50013 or equivalent, remove the crankshaft pulley bolt and crankshaft pulley.
11. Using a screwdriver, pry out the timing chain case oil seal.

➡**Tape the screwdriver tip before use.**

➡**After the removal, check the crankshaft for damage. If it is damaged, smooth the surface with 400-grit sandpaper.**

To install:

12. Install timing chain case oil seal, as follows:
    a. Apply MP grease to a new oil seal lip.
    b. Using a Special Service Tool (SST: 09223-22010, SST: 09506-35010 or equivalent) and a hammer, tap in the oil seal until its surface is flush with the timing chain cover edge.

➡**Keep the lip free of foreign matter.**

➡**Do not tap the oil seal at an angle.**

13. Install the crankshaft pulley, as follows:
    a. Align the pulley set key with the key groove of the pulley, and slide on the pulley.
    b. Using SST: 09213-70011, SST: 09330-00021, or equivalent, install the pulley bolt. Tighten to 184 ft. lbs. (250 Nm).
14. The remainder of installation is the reverse of removal. When installing the wheel, tighten the lug nuts to 76 ft. lbs. (103 Nm).

CYLINDER HEAD

REMOVAL & INSTALLATION

2.4L Engine

See Figures 100 through 102.

1. Before servicing the vehicle, refer to the precautions section.
2. Remove or disconnect the following:
    - Front center suspension brace
    - Timing chain
    - Coolant
    - Transfer case oil
    - Radiator hoses
    - Power steering hoses
    - Heater hoses
    - Fuel rail lines
    - Camshaft timing oil control valve
    - Front driveshaft
    - Rear engine mount insulator (4WD)
    - Transverse engine mount bracket (4WD)

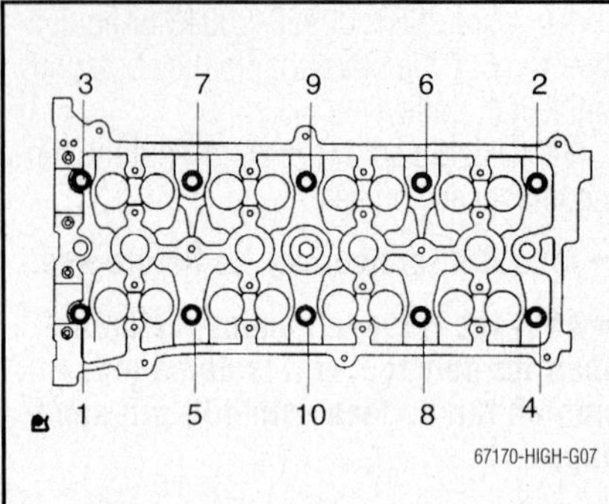

Fig. 100 Cylinder head bolt loosening sequence—2.4L engine

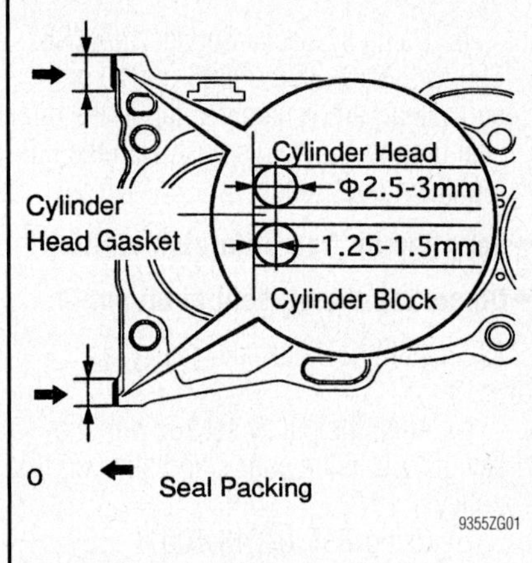

Fig. 101 Apply a bead of RTV sealant as shown—Highlander 2.4L engine cylinder head

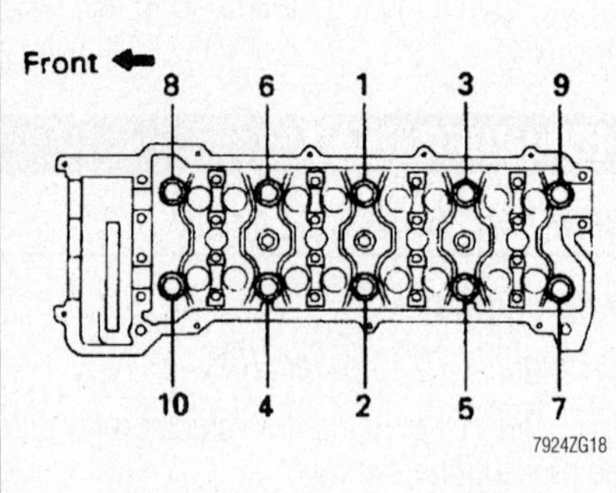

Fig. 102 Cylinder head bolt tightening sequence—2.4L engine

- Intake manifold
- All wires and cables connected to the head
- Exhaust manifold
- Camshafts

3. Loosen the 10 head bolts evenly, a little at a time in several passes and lift off the head. Check the head bolt length. Any bolt longer than 6.465 in. (164.2mm) should be replaced.
4. Installation is the reverse of removal.

Install the head gasket with the lot number stamp upward. Apply a bead of RTV sealer as shown. The head must be installed within 3 minutes of applying the sealer, and the head bolts must be tightened with 15 minutes. The head bolts must be tightened in sequence, in several passes, to 58 ft. lbs. (79 Nm), then, an additional 90 degree turn each.

3.3L Engine

See Figures 103 through 106.

1. Before servicing the vehicle, refer to the precautions section.
2. Remove or disconnect the following:
   - Wiper and blade assembly
   - Top cowl seal and panel
   - Window washer hoses from the ventilator louvers
   - Left and right ventilator louvers
   - Heater air duct
3. Relieve the fuel pressure.
4. Remove or disconnect the following:
   - Turn the ignition key to the **OFF** position
   - Negative battery cable

➡Wait at least 90 seconds from the time the negative battery was disconnected to start work.

5. Drain the cooling system.
6. Remove or disconnect the following:
   - Accelerator and throttle cables, if equipped with an automatic transaxle
   - Air cleaner cover, air flow meter and the air duct
   - Front upper suspension brace
   - Cruise control actuator and bracket, if equipped
   - 2 engine ground straps
   - Right engine mounting support
   - Radiator hoses
   - 2 heater hoses
   - Fuel feed and return lines from the fuel rail assembly
   - Pressure hose from the hydraulic motor
   - V-bank cover
7. Disconnect the following vacuum hoses:
   - Fuel pressure control Vacuum Switching Valve (VSV)
   - Fuel pressure regulator
   - Cylinder head rear plate
   - Intake air control valve VSV
   - Exhaust Gas Recirculation (EGR) vacuum modulator
   - EGR valve
8. Disconnect the following wiring and hoses:
   - Intake air control valve
   - Fuel pressure regulator
   - EGR VSV
9. Remove the 2 nuts and the emission control valve set.
10. Disconnect the following hoses;
    - Brake booster vacuum hose
    - PCV hose
    - Intake air control valve vacuum hose
11. Remove or disconnect the following:
    - Data Link Connector (DLC) from the mounting bracket
    - 2 ground straps from the intake chamber
    - Hydraulic motor pressure hose from the intake chamber
    - Right Oxygen (O_2) sensor connector from the power steering pressure tube
    - 2 nuts and the power steering pressure tube from the intake chamber
    - Both power steering air hoses
    - Engine hanger and the intake chamber support
    - EGR pipe and gaskets
12. Disconnect the following wiring:
    - Throttle Position (TP) sensor connector
    - Idle Air Control (IAC) valve connector
    - EGR gas temperature connector
    - Air conditioning idle up connector
13. Disconnect the following vacuum hoses:
    - 2 vacuum hoses from the Thermal Vacuum Valve (TVV)
    - Vacuum hose from the cylinder head rear plate
    - Vacuum hose from the charcoal canister
14. Remove or disconnect the following:
    - Air assist hose and the 2 water bypass hoses

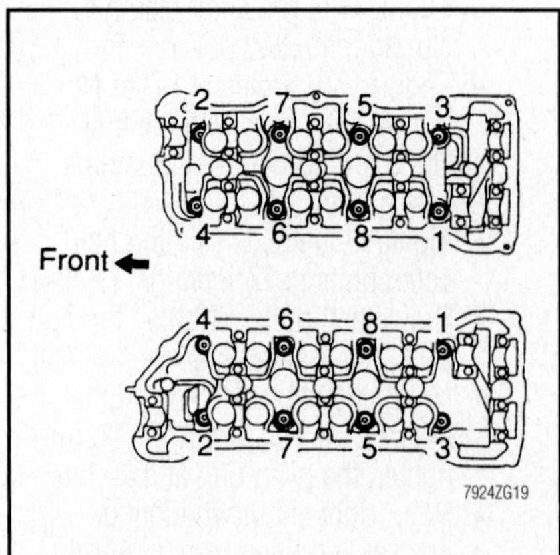

Fig. 103 Cylinder head bolt loosening sequence—3.3L engine

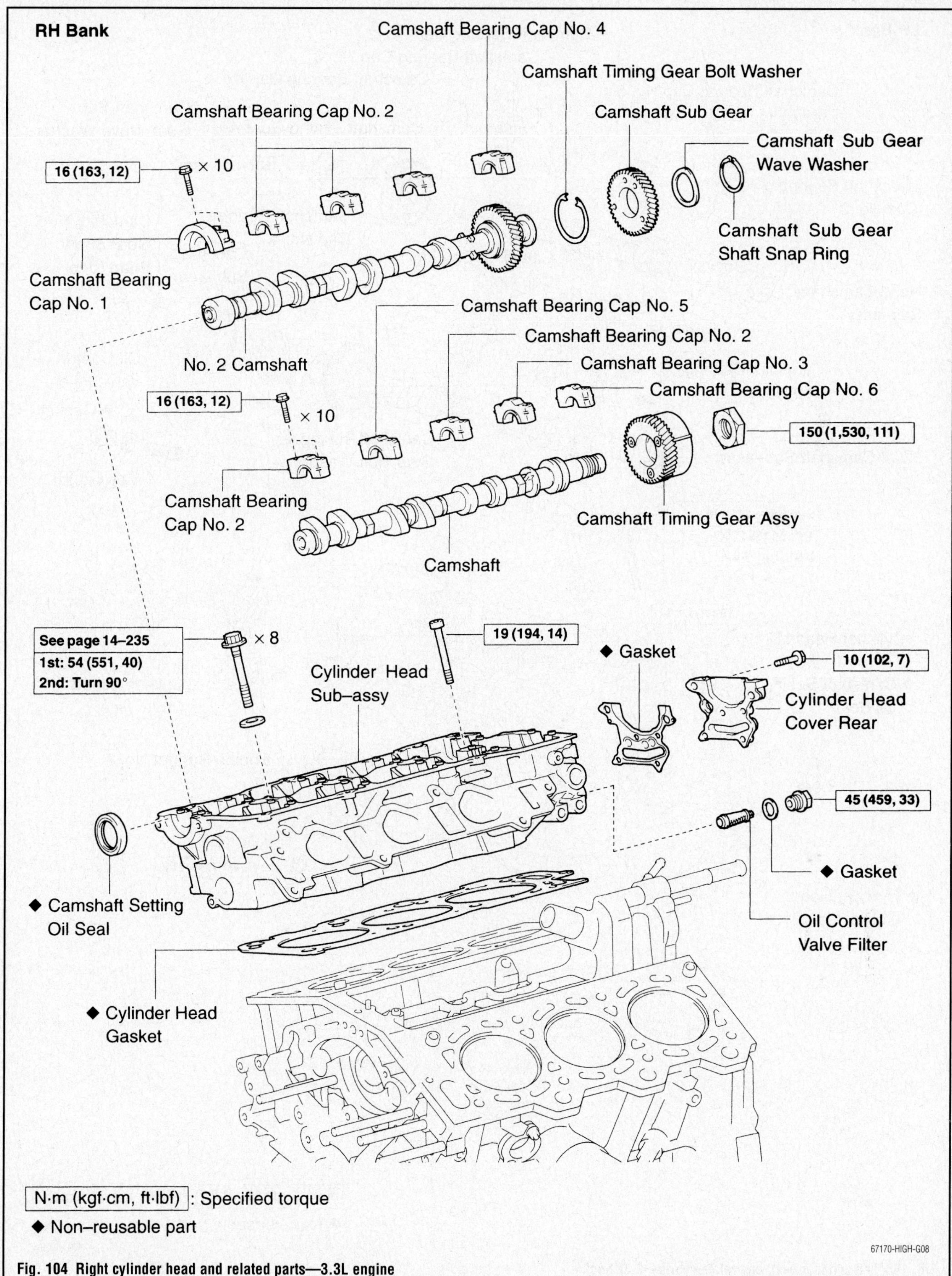

Fig. 104 Right cylinder head and related parts—3.3L engine

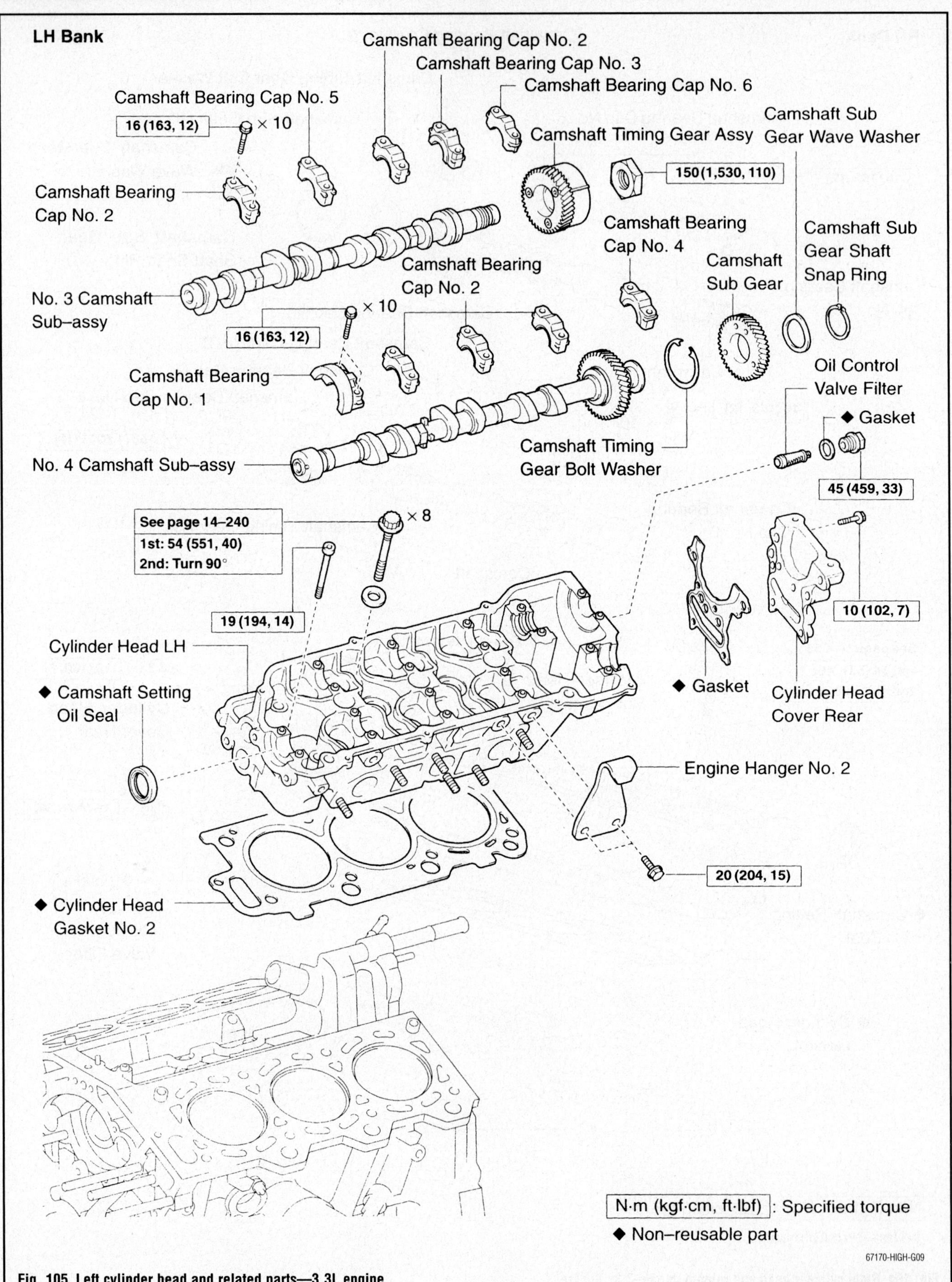

Fig. 105 Left cylinder head and related parts—3.3L engine

- Air intake chamber
- Left engine wiring harness and move it aside
- Wiring harness from the rear of the engine
- Right engine wiring harness and move it aside
- Ignition coils and move them aside
- Timing belt
- Camshaft pulleys and the timing belt rear cover
- Cylinder head rear plate
- Water inlet pipe
- Air assist hose and vacuum hose
- Intake manifold and fuel rail assembly
- Water outlet
- EGR pipe from the right exhaust manifold
- Front exhaust pipe and exhaust manifolds
- Dipstick assembly and the power steering pump bracket
- Valve covers and the Camshaft Position (CMP) sensor
- Camshafts

15. Be sure the engine is at/or near ambient temperature and remove the 2 (1 on each head) 8mm recessed hex bolts. Loosen and remove the 8 head bolts evenly, in 3 passes, in the reverse order of the installation sequence. Carefully lift the head from the engine; if necessary to pry the head loose, take great care not to damage the mating surfaces. Place the head on wood blocks in a clean work area.

➡If the cylinder head bolts are loosened out of sequence, warpage or cracking could result.

16. Remove the cylinder head gasket. With a gasket scraper, carefully remove all the old gasket material from the cylinder head and engine block surfaces.

To install:

17. Place the new cylinder head gasket onto the cylinder block.
18. Install the cylinder head, in sequence, using several steps, as follows:
    - Cylinder head onto the gasket
    - Cylinder head bolts lubricated with clean engine oil
    - Tighten the bolts in sequence in 3 steps to 40 ft. lbs. (54 Nm).

➡If any bolt does not meet the torque, replace it.

- Mark the forward edge of each bolt with paint, then tighten each bolt, in proper sequence, an additional 90 degrees.
- Check that each painted mark is now at a 90 degrees angle to the front

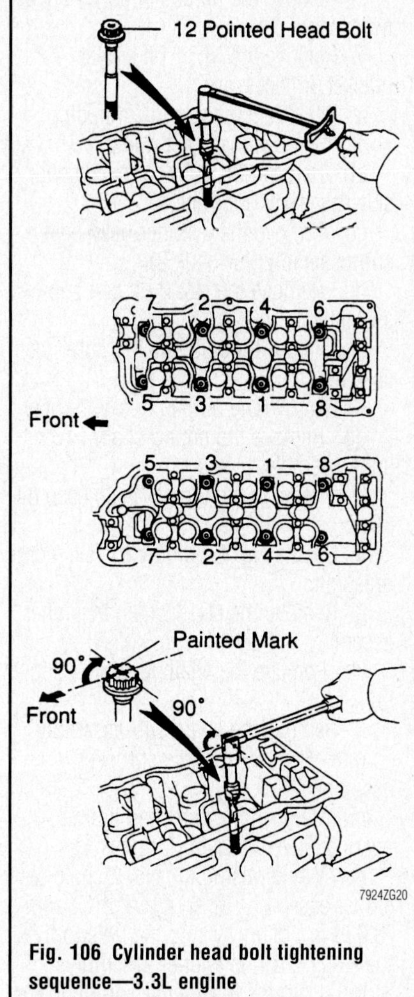

Fig. 106 Cylinder head bolt tightening sequence—3.3L engine

➡The paint mark applied to the bolt in the 9 o'clock position and should now be in the 12 o'clock position.

- Remaining 8mm bolts, lubricated with engine oil. Tighten both bolts to 13 ft. lbs. (18 Nm).

19. Install the camshafts.
20. Check and adjust the valves.
21. Apply sealant to the cylinder heads where the camshaft supports meet the cylinder heads.
22. Install or connect the following:
    - Cylinder head covers, using new gaskets
    - Dipstick and power steering pump bracket
    - Exhaust manifolds. Tighten the nuts to 36 ft. lbs. (49 Nm).
    - EGR pipe to the right exhaust manifold
    - Water outlet
    - Intake manifold and the fuel rail assembly. Tighten the intake manifold nuts/bolts to 11 ft. lbs. (15 Nm).
    - Air assist hose and the 2 water bypass hoses
    - Water inlet pipe and cylinder head rear plate
    - Timing belt rear cover and camshaft pulleys
    - Timing belt
    - Spark plugs and ignition coils
    - Right engine wiring harness
    - Wiring harness to the rear of the engine
    - Left engine wiring harness
    - Air intake chamber
    - EGR pipe, using new gaskets
23. Connect the following vacuum hoses:
    - The 2 TVV vacuum hoses
    - The vacuum hose to the rear cylinder head plate
    - Charcoal canister vacuum hose
24. Connect the following electrical wiring:
    - TP sensor connector
    - IAC valve connector
    - EGR gas temperature connector
    - Air conditioning idle up connector
25. Install or connect the following:
    - Engine hanger and the intake chamber support
    - Both power steering air hoses
    - Power steering pressure tube to the intake chamber
    - O_2 sensor connector to the pressure tube.
    - Both ground straps, to the intake chamber
    - DLC to the bracket
26. Connect the following hoses:
    - Power brake booster vacuum hose
    - PCV hose
    - IAC valve vacuum hose
27. Install or connect the following:
    - Emission control valve set and related vacuum hoses and connectors
    - V-bank cover
    - Pressure hose to the hydraulic motor
    - Fuel lines to the fuel rail assembly
    - Heater and radiator hoses
    - Right engine mounting support
    - both engine ground straps
    - Upper front suspension brace, if removed. Tighten the nuts to 59 ft. lbs. (80 Nm).
    - Cruise control actuator and bracket
    - Air cleaner, air flow meter and air duct assembly
    - Accelerator and throttle cables
28. Fill the cooling system.

29. Install or connect the following:
    - Negative battery cable
    - Heater air duct
    - Left and right ventilator louvers
    - Window washer hoses from the ventilator louvers
    - Top cowl seal and panel
    - Wiper and blade assembly
30. Start the engine and check for leaks.
31. Bleed the air from the cooling system.
32. Road test the vehicle and check for unusual noise, shock, slippage, correct shift points and smooth operation.
33. Recheck the coolant and engine oil levels.

3.5L Engine

See Figures 107 through 117.

1. Remove the engine assembly.
2. Install on engine stand.
3. The following must be removed:

a. Remove the ignition coil assembly.

b. Remove right hand No. 2 engine mounting stay.

c. Remove the intake manifold.

d. Remove the right hand exhaust manifold sub-assembly.

e. Remove the No. 2 engine oil level dipstick guide.

f. Remove the No. 2 manifold stay.

g. Remove the No. 2 exhaust manifold heat insulator.

h. Remove the left exhaust manifold sub-assembly.

i. Remove the transverse engine mounting bracket.

j. Remove the generator assembly.

k. Remove the V-ribbed belt tensioner assembly.

l. Remove the No. 2 timing gear cover.

m. Remove the No. 2 idler pulley sub-assembly.

n. Remove the left No. 1 engine front mounting bracket.

o. Remove the 6 bolts and left hand No. 1 front engine mounting bracket.

p. Remove the radio setting condenser.

q. Remove the No. 1 vacuum switching valve.

r. Remove the knock control sensor wire.

s. Remove the knock control sensor.

t. Remove the crankshaft position sensor.

u. Remove the No. 1 oil pipe.

v. Remove the oil pipe.

w. Remove the crankshaft pulley.

x. Remove the oil cooler assembly, if necessary.

y. Remove the No. 1 oil cooler bracket, if necessary.

z. Remove the water inlet housing.

aa. Remove the water outlet.

bb. Remove the cylinder head cover sub-assembly (for bank 1).

cc. Remove the cylinder head cover sub-assembly (for bank 2).

dd. Remove the No. 2 oil pan sub-assembly.

ee. Remove the oil strainer sub-assembly.

ff. Remove the oil pan sub-assembly.

gg. Remove the timing chain cover sub-assembly.

hh. Remove the timing chain case oil seal.

ii. Set the No. 1 cylinder to TDC/compression.

jj. Remove the No. 1 chain tensioner assembly.

kk. Remove the chain tensioner slipper.

ll. Remove the chain sub-assembly.

mm. Remove the idle sprocket assembly.

nn. Remove the camshaft timing gears and No. 2 chain (for bank 1).

oo. While raising the No. 2 chain tensioner assembly, insert a pin of 1.0 mm (0.039 in.) diameter into the hole to fix the No. 2 chain tensioner assembly.

pp. Hold the hexagonal portion of the camshaft with a wrench, and remove the 2 bolts and 2 camshaft timing gear assemblies.

➡Be careful not to damage the cylinder head with the wrench. Do not disassemble the camshaft timing gear assemblies.

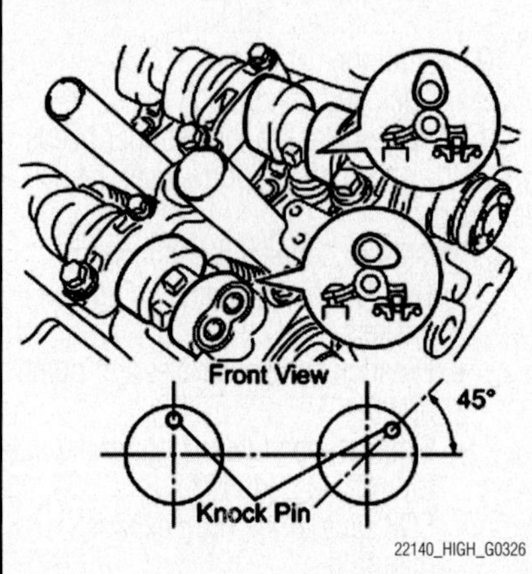

Fig. 107 Positioning camshafts for bearing cap removal

qq. Remove the No. 2 chain assembly.

rr. Remove the bolt and No. 2 chain tensioner assembly.

ss. Check that the camshafts are positioned as shown in the illustration.

tt. Uniformly loosen and remove the 8 bearing cap bolts in several steps and in the sequence shown in the illustration.

uu. Uniformly loosen and remove the 12 bearing cap bolts in several steps and in the sequence shown in the illustration.

➡Uniformly loosen the bolts while keeping the camshaft level.

vv. Remove the 5 camshaft bearing caps.

ww. Remove the camshaft.

xx. Remove the No. 2 camshaft.

yy. Remove the right camshaft housing sub-assembly by prying between the cylinder head and right camshaft housing sub-assembly with a screwdriver.

➡Be careful not to damage the contact surfaces of the cylinder head and the right camshaft housing sub-assembly.

4. Remove the 24 valve lash adjuster assemblies from the cylinder head.

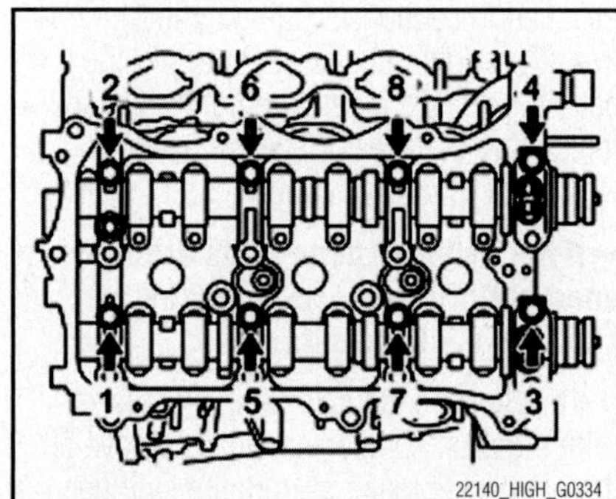

Fig. 108 Camshafts bearing cap removal sequence

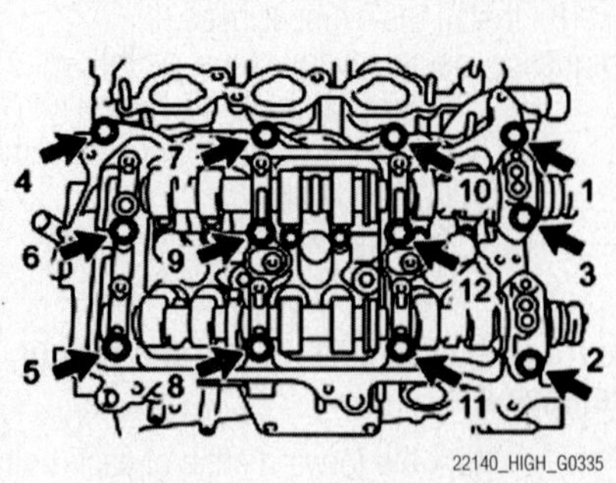

Fig. 109 Camshaft bearing cap loosening sequence

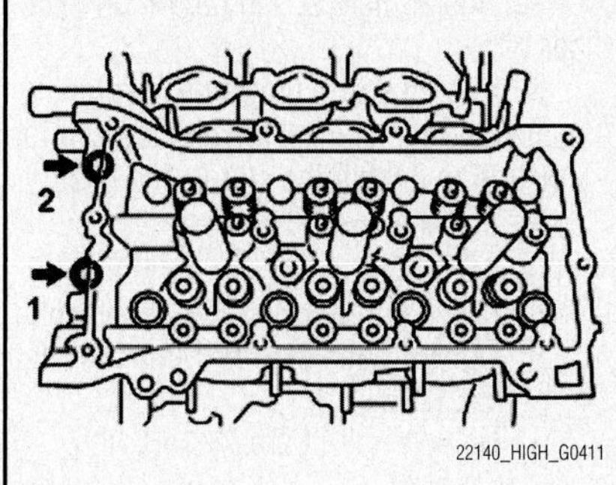

Fig. 110 Loosening cylinder head bolts LH shown

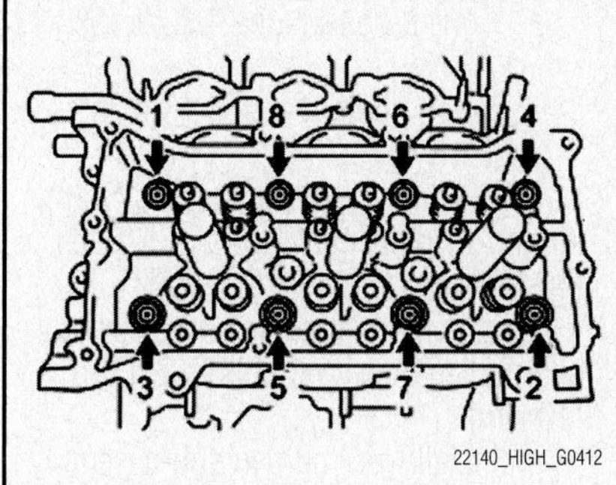
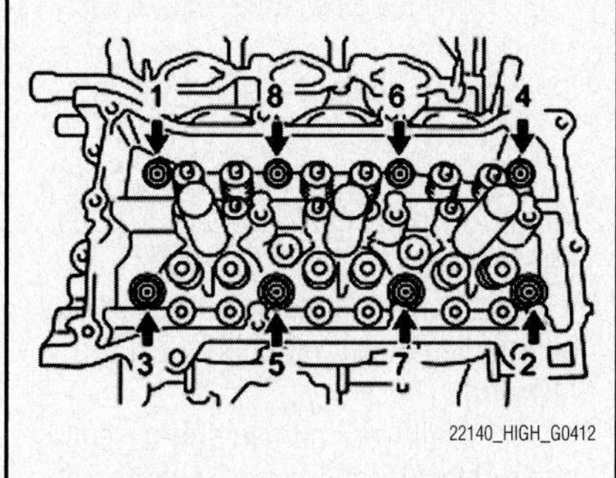

Fig. 111 Cylinder head bolt loosening sequence LH shown

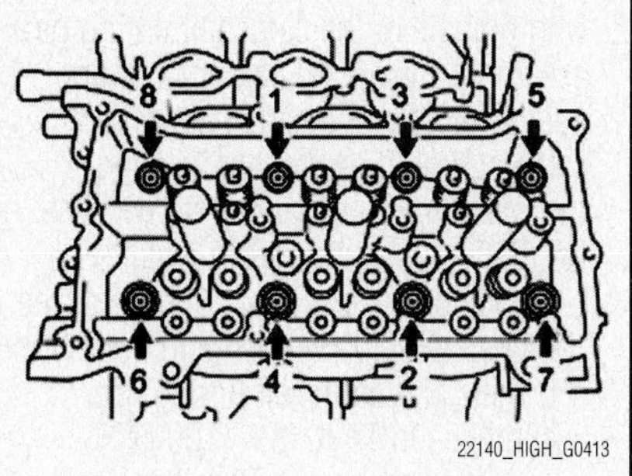

Fig. 112 Cylinder head bolt tightening sequence LH shown

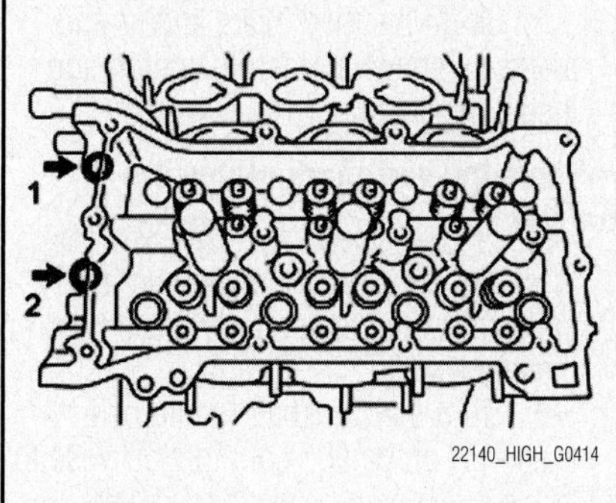

Fig. 113 Tightening cylinder head bolts LH shown

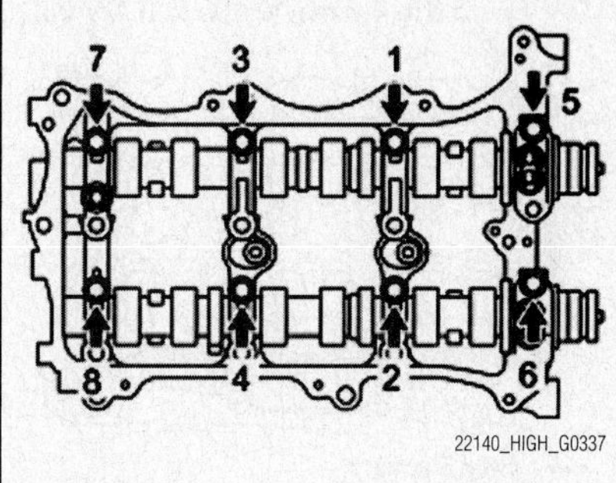

Fig. 114 Camshafts bearing cap tightening sequence

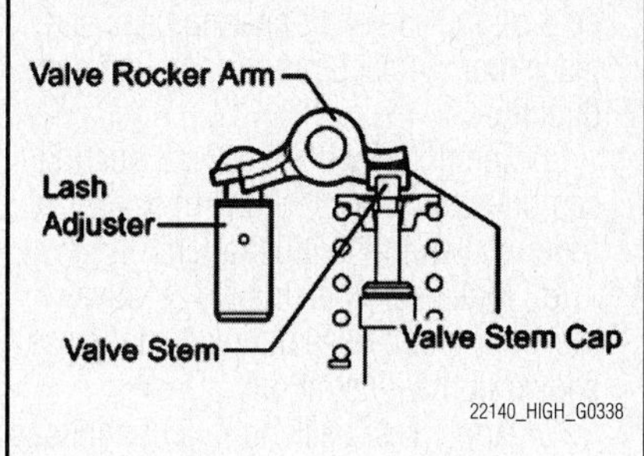

Fig. 115 Valve rocker arm sub-assembly positioning

➡Arrange the removed parts in the correct order.

5. Uniformly loosen and remove the 2 cylinder head set bolts in several steps and in the sequence shown in the illustration.

➡Be careful not to drop washers into the cylinder head. Cylinder head warpage or cracking could result from removing bolts in an incorrect order. Be sure to keep separate the removed parts for each installation position.

6. Using a 10 mm bi-hexagon wrench, uniformly loosen the 8 bolts in the sequence shown in the illustration. Remove the 8 cylinder head bolts and plate washers.
7. Remove the cylinder head sub-assembly.
8. Remove the No. 2 cylinder head gasket.

To install:

9. Place the No. 2 cylinder head gasket on the cylinder block surface with the Lot No. stamp upward.

➡Gently lower the cylinder head in order not to damage the gasket with the bottom part of the head.

10. Place the cylinder head on the cylinder block.

➡Be careful not to allow oil to adhere to the bottom part of the cylinder head.

11. Apply a light coat of engine oil to the threads and under the heads of the cylinder head bolts.

➡The cylinder head bolts are tightened in 3 progressive steps.

a. Step 1: Using a 10 mm bi-hexagon wrench, install and uniformly tighten the 8 cylinder head bolts with the plate washers to 27 ft. lbs. (36 Nm) in several steps in the sequence shown in the illustration.
b. Step 2: Mark the cylinder head bolt head with paint. Tighten the cylinder head bolts another 90°.
c. Step 3: Tighten the cylinder head bolts an additional 90°.

12. Tighten the 2 bolts to 22 ft. lbs. (30 Nm) in the order shown in the illustration.
13. Install the 12 valve stem caps.

➡Keep the lash adjuster free of dirt and foreign objects. Only use clean engine oil.

14. Place the lash adjuster into a container filled with engine oil.
a. Insert Service Tool's tip into the lash adjuster's plunger and use the tip to press down on the check ball inside the plunger.
b. Squeeze Service Tool and lash adjuster together to move the plunger up and down 5 to 6 times.
c. Check the movement of the plunger and bleed the air. Confirm that the plunger moves up and down freely.
d. When bleeding air from the high-pressure chamber, make sure that the tip of SST is actually pressing the check ball as shown in the illustration. If the check ball is not pressed, air will not bleed.
e. After bleeding the air, remove Service Tool. Then, try to press the plunger quickly and firmly with by hand. Confirm that the plunger is very difficult to move.
f. If the results are not as specified, replace the defective lash adjuster.
g. Install the lash adjusters.

➡Install the lash adjuster to the same place where it was removed from.

15. Install the following:
a. Apply engine oil to the camshaft journals, right camshaft housing sub-assembly, and camshaft bearing caps.
b. Install the camshaft and No. 2 camshaft to the right camshaft housing sub-assembly.

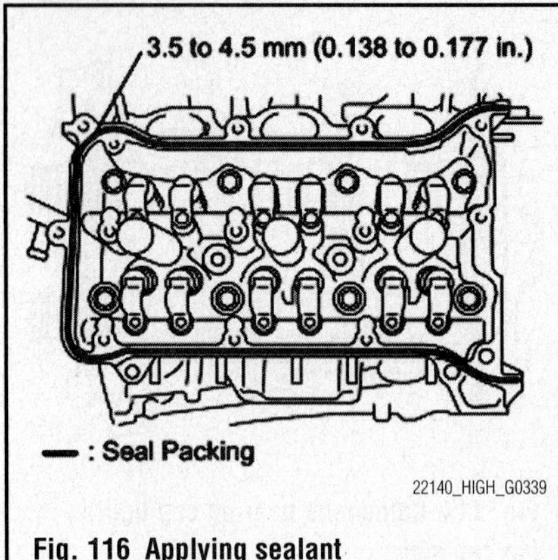

Fig. 116 Applying sealant

c. Make sure of the marks and numbers on the camshaft bearing caps and place them in each proper position and direction.

d. Temporarily tighten the 8 bearing cap bolts to 7 ft. lbs. (10 Nm) in the order shown in the illustration.

e. Make sure that the No. 1 valve rocker arm sub-assembly is installed as shown in the illustration.

f. Apply seal packing in a continuous line as shown in the illustration.

➡Remove any oil from the contact surface. Install the camshaft housing sub-assembly RH within 3 minutes. Do not start the engine for at least 2 hours after installing.

g. Install the right camshaft housing sub-assembly and tighten the 12 bolts to 21 ft. lbs. (28 Nm) in the order shown in the illustration.

➡When installing the camshaft housing RH, it is necessary to correctly position the camshafts as shown in the removal illustration. If the camshaft housing sub-assembly is removed because any of the bolts are loosened during installation, make sure that the previously applied seal packing does not enter any oil passages.

h. Complete the tightening of the 8 bolts to 12 ft. lbs. (16 Nm) in the order shown above.

i. Install the No. 2 chain tensioner assembly with the bolt and tighten to 15 ft. lbs. (21 Nm).

j. While pushing in the tensioner, insert a pin of 1.0 mm (0.039 in.) diameter into the hole to fix it.

k. Align the mark plate with the timing marks of the camshaft timing gear.

l. Apply a light coat of engine oil to the bolt threads and bolts seating surface.

m. Align the knockpin of the camshaft with pinhole of the camshaft timing gear assembly. Install the camshaft timing gear assembly and camshaft timing exhaust timing gear assembly and camshaft timing exhaust gear with the No. 2 chain sub-assembly installed.

n. Hold the hexagonal portion of the camshaft with the wrench and tighten the two bolts and camshaft timing gear assemblies to 74 ft. lbs. (100 Nm).

o. Remove the pan from the No 2 chain tensioner assembly.

p. Install idle sprocket assembly and tighten to 44 ft. lbs. (60 Nm).

q. Install chain sub-assembly.

r. Align the mark plate and timing marks as shown in the illustration and install the chain.

➡The camshaft mark plates are orange.

s. Do not pass the chain over the crankshaft, just temporarily place it on the crankshaft.

t. Turn the camshaft timing gear assembly on bank 1 counterclockwise to tighten the chain between the banks.

➡When the idle sprocket assembly is reused, align the timing chain plate with the mark where the plate has been in order to tighten the chain between the banks.

u. Align the mark plate and timing marks as shown in the illustration and install the chain onto the crankshaft timing sprocket. The crankshaft to mark plate is yellow.

v. Turn the crankshaft clockwise to set it to the right-hand block bore more centerline. (TDC Compression)

w. Install chain tensioner slipper.

x. Move the stopper plate upward to release the lock, and push the plunger deep into the tensioner.

y. Move the stopper plate downward to set the loss, and insert a hexagon wrench into the hole of the stopper plate.

z. Install No. 1 chain tensioner assembly and tighten bolts to 7 ft. lbs. (10 Nm).

aa. Remove the hexagon wrench from the No. 1 chain tensioner assembly. Check that the each timing mark is aligned with the crankshaft at TDC compression.

bb. Remove the pulley set bolt.

cc. Install timing chain case oil seal.

dd. Install sealant to timing chain cover sub-assembly.

ee. Install new O ring gasket on cylinder block.

ff. Align the oil pump's drive rotor spline and the crankshaft as shown in the illustration. Install the spline and chain cover to the crankshaft.

gg. Temporarily tighten the timing chain cover with the 23 bolts and 2 nuts.

- Tighten bolts in area 1 and 2 15 ft. lbs. (21 Nm)
- Tighten bolt in area 3 15 ft. lbs. (21 Nm)

➡First tighten the upper bolts and nuts followed by the lower bolts and nuts as shown.

- Tighten bolt in area 4 32 ft. lbs. (43 Nm)
- Tighten bolt in area 4 15 ft. lbs. (21 Nm)

hh. Install oil pan subassembly and tighten 16 bolts and 2 nuts to 7 ft. lbs. (10 Nm) and 15 ft lbs. (21 Nm).

ii. Install oil strainers sub-assembly and tighten bolts and nuts to 7 ft. lbs. (10 Nm).

jj. Install No.2 oil pan sub-assembly and tighten 16 bolts and 2 nuts to 7 ft. lbs. (10 Nm).

kk. Install cylinder head cover sub-assemblies and tighten to 7 ft. lbs. (10 Nm) and 15 ft. lbs. (21 Nm).

ll. Install water inlet housing

mm. Install the No. 1 oil cooler bracket (w/ oil cooler).

nn. Install the oil cooler assembly (w/ oil cooler).

oo. Install the crankshaft pulley.

pp. Install the oil pipe.

qq. Install the No. 1 oil pipe.

rr. Install the crankshaft position sensor.

ss. Install the knock control sensor.

tt. Install the knock control sensor wire.

uu. Install the No. 1 vacuum switching valve.

vv. Install radio setting condenser.

ww. Install the left hand No. 1 front engine mounting bracket with the 6 bolts and tighten to 40 ft. lbs. (54 Nm).

➡Install the water inlet and mounting bracket within 15 minutes after installing the chain cover. Do not start the engine for at least 2 hours after installation.

xx. Install the No. 2 idler pulley sub-assembly.

yy. Install the No. 2 timing gear cover.

zz. Install the V-ribbed belt tensioner assembly.

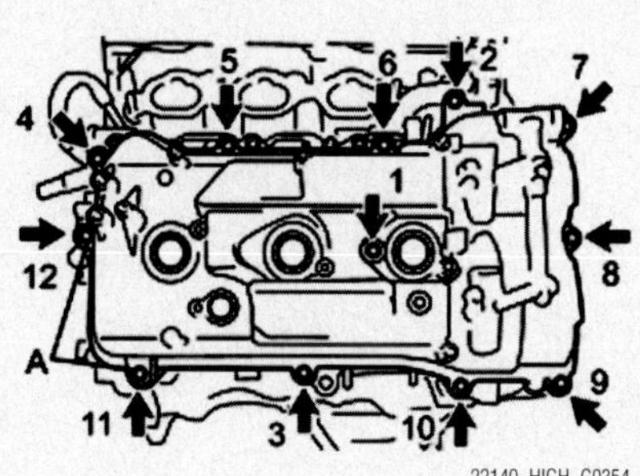

Fig. 117 Cylinder head cover tightening sequence

aaa. Install the generator assembly.

bbb. Install the transverse engine mounting bracket.

ccc. Install left hand exhaust manifold sub-assembly and tighten to 15 ft. lbs. (21 Nm).

ddd. Install the No. 2 exhaust manifold heat insulator.

eee. Install the No. 2 manifold stay.

fff. Install the No. 2 engine oil level dipstick guide.

ggg. Install right hand exhaust manifold sub-assembly and tighten to 15 ft. lbs. (21 Nm).

hhh. Install the intake manifold and tighten the 6 bolts and 4 nuts uniformly in several steps to 15 ft. lbs. (21 Nm).

iii. Install the right hand No. 2 engine mounting stay.

jjj. Install the ignition coil assembly.

kkk. Remove the engine stand.

lll. Install the engine assembly.

ENGINE ASSEMBLY

REMOVAL & INSTALLATION

2.4L Engine

See Figure 118.

1. Before servicing the vehicle, refer to the precautions section.
2. Matchmark the hood position.
3. Remove or disconnect the following:
    - Front wheels
    - No.1 engine undercover
    - Right and left fender splash shields
    - Right fender apron seal
    - Engine oil
    - Coolant
    - Transaxle fluid
    - Transfer case oil
    - Battery
    - Air cleaner
    - Radiator hoses
    - Oil cooler hoses
    - Upper engine stay
    - Upper engine mount bracket
    - Accessory drive belts
    - Steering pump reservoir
    - Steering pump hoses
    - All cables and wires connected to the engine
    - Exhaust pipe
    - Front drive shaft
    - Stabilizer links
    - Left and right axle hub nuts
    - Left and right speed sensors
    - Left and right tie rods
    - Left and right lower control arms
    - Torque converter-to-drive plate bolts
    - Intermediate steering shaft
    - AC compressor
4. Attach a crane, remove the 6 side rail plate subassembly bolts (3 each side) and the front suspension member rear brace.
5. Lift the engine out of the vehicle.
6. Installation is the reverse of removal. Observe the following torques:
    - Frame side plate bolts: Large 63 ft. lbs. (85 Nm); small 24 ft. lbs. (32 Nm)
    - Suspension member rear brace: Large 63 ft. lbs. (85 Nm); small 24 ft. lbs. (32 Nm)
    - Intermediate shaft bolt: 26 ft. lbs. (35 Nm)
    - Torque converter bolts: 30 ft. lbs. (41 Nm)
    - Lower control arms, bolts and nuts: 94 ft. lbs. (127 Nm)
    - Tie rod nuts: 36 ft. lbs. (49 Nm)
    - Speed sensors: 71 inch lbs. (8 Nm)
    - Hub nuts: 217 ft. lbs. (294 Nm)
    - Stabilizer link nuts: 55 ft. lbs. (74 Nm)
    - Driveshaft nuts: 55 ft. lbs. (74 Nm)
    - Engine mount bracket: 15 ft. lbs. (20 Nm)
    - Engine mount stay: 47 ft. lbs. (64 Nm)

3.3L Engine

See Figure 119.

1. Before servicing the vehicle, refer to the precautions section.
2. Drain the coolant, engine oil, transfer case fluid and transmission fluid.
3. Remove the front wheels.
4. Remove the engine undercover assembly.
5. Remove the left and right fender splash shields.
6. Remove the left and right fender apron seals.
7. Remove the wiper arms.
8. Remove the cowl top ventilator louver.
9. Remove the wiper linkage.
10. Remove the cowl top panel outer sub-assembly.
11. Remove the V-bank cover.
12. Remove the battery.
13. Remove the air cleaner assembly.
14. Remove the A/C compressor drive belt.
15. Remove the alternator.
16. Remove the engine roll brace.
17. Remove the front engine mount bracket.
18. Remove the alternator bracket.
19. Remove the alternator belt adjusting bar.
20. Remove the magnetic clutch from the A/C compressor.
21. Remove the transmission control cable.
22. Disconnect the check valve hose.
23. Disconnect the fuel vapor feed hose.
24. Disconnect the fuel pipes.
25. Disconnect the heater hoses.
26. Disconnect the radiator hoses.
27. Disconnect the oil cooler hoses.
28. Disconnect the power steering hoses.
29. Remove the glove compartment door.
30. Disconnect the engine wiring harness.
31. Remove the driveshaft (4WD).
32. Remove the exhaust pipes and brackets.
33. Remove the left and right stabilizer bar links.
34. Remove the left and right front axle hub nuts.
35. Remove the left and right speed sensors.
36. Remove the left and right tie rod ends.
37. Disconnect the left and right lower control arms.
38. Remove the left and right halfshafts.
39. Disconnect the steering intermediate shaft.
40. Disconnect the height control sensor link (air suspension).
41. Attach a lifting crane.
42. Remove the 6 bolts and 2 nuts, then, remove the left and right frame side rail plates.
43. Remove the 6 bolts and 2 nuts, then, remove the left and right front suspension rear braces.
44. Lift the engine/transaxle from the vehicle.
45. Installation is the reverse of removal. Observe the following torques:
    - Engine hanger: 14 ft. lbs. (20 Nm)
    - Alternator bracket: 43 ft. lbs. (58 Nm)

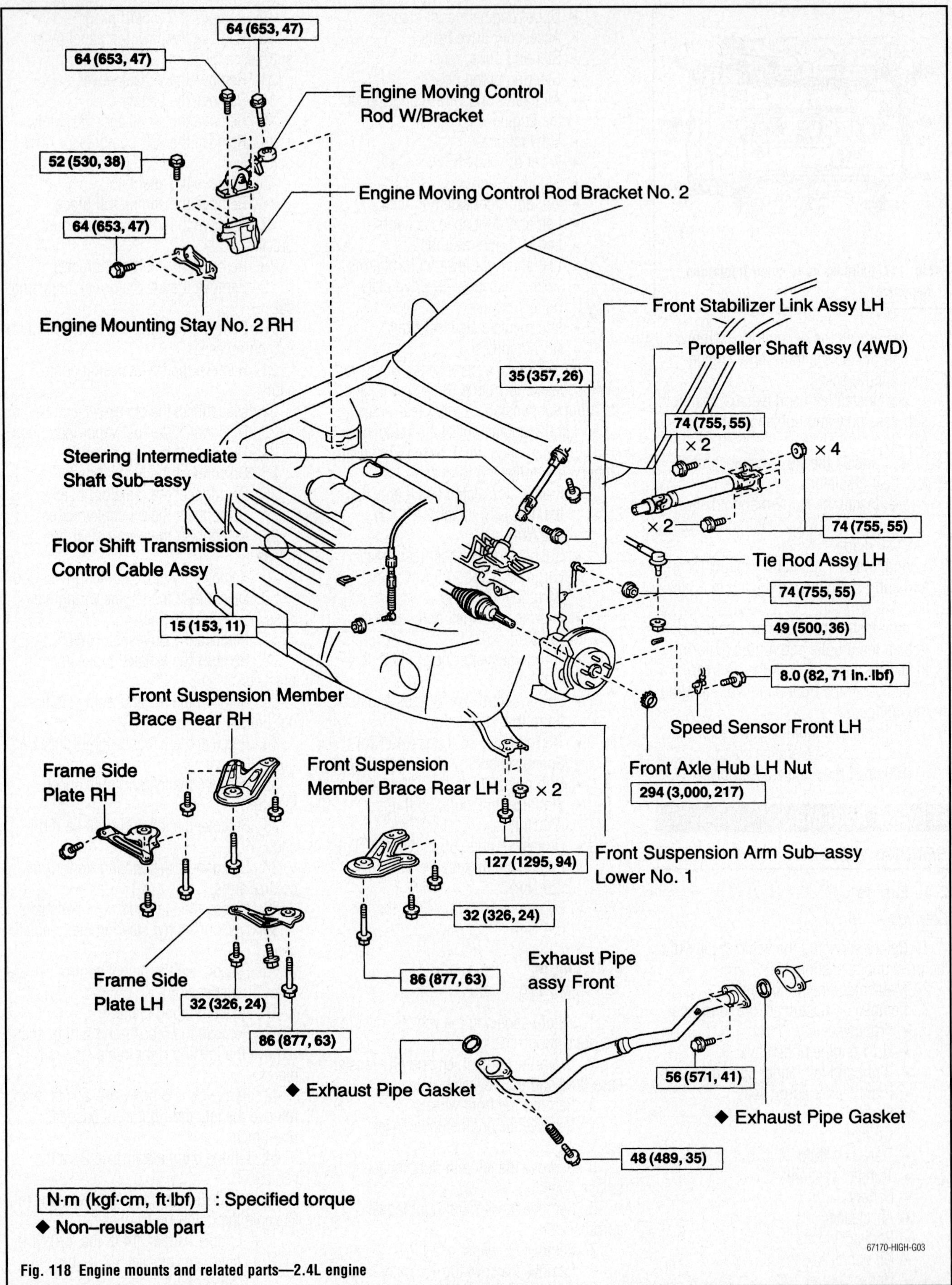

Fig. 118 Engine mounts and related parts—2.4L engine

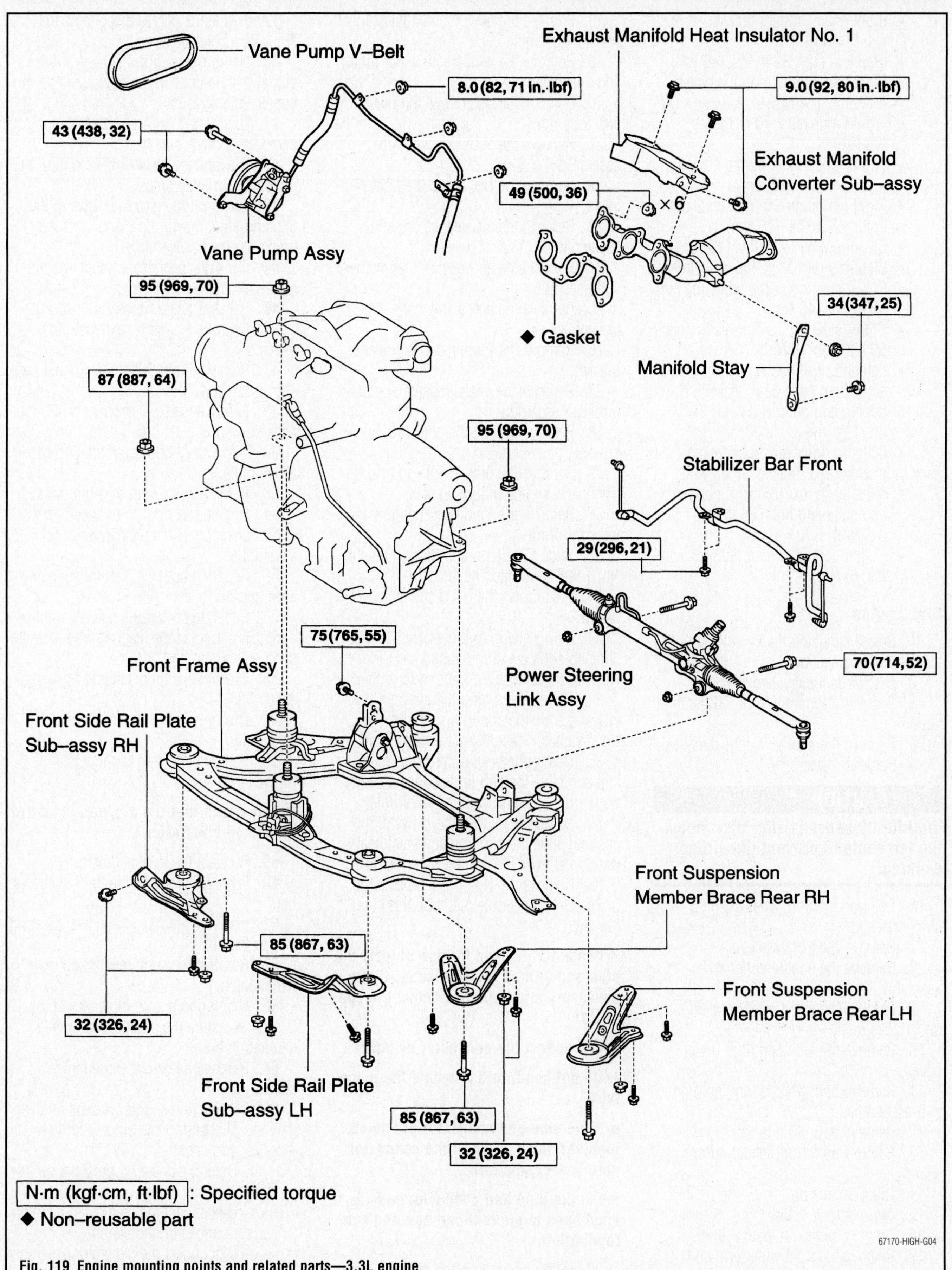

Fig. 119 Engine mounting points and related parts—3.3L engine

- Right engine mount bracket: 40 ft. lbs. (54 Nm)
- Manifold stay: 36 ft. lbs. (49 Nm)
- Right rear engine mount bracket: 47 ft. lbs. (64 Nm)
- Front frame nuts: 70 ft. lbs. (95 Nm)
- Front right engine mount insulator nut: 64 ft. lbs. (87 Nm)
- Right rear engine mount insulator bolts: 55 ft. lbs. (75 Nm)
- Steering link: 52 ft. lbs. (70 Nm)
- Stabilizer bar: 21 ft. lbs. (29 Nm)
- Power steering pump adjusting bar: 32 ft. lbs. (43 Nm)
- Power steering pressure tube nuts: 69 inch lbs. (8 Nm)
- Left and right frame side rail plates: single end bolts 63 ft. lbs. (85 Nm); double end bolts and nuts: 24 ft. lbs. (32 Nm)
- Left and right front suspension member rear braces: single end bolts 63 ft. lbs. (85 Nm); double end bolts and nuts: 24 ft. lbs. (32 Nm)
- Height control sensor link: 48 inch lbs. (5 Nm)

3.5L Engine

1. Before servicing the vehicle, refer to the precautions section.
2. Discharge fuel system pressure.
3. Recover refrigerant from refrigeration system.
4. Remove the cool air intake duct seal.
5. Remove the battery.

CAUTION

Wait for 90 seconds after disconnecting the cable to prevent the airbag working.

6. Position the front wheels facing straight ahead.
7. Remove the front wheels.
8. Remove the engine under cover assembly.
9. Remove the No. 1 and 2 engine under covers.
10. Remove the left hand floor under cover.
11. Remove both front fender molding sub-assemblies.
12. Remove both front fender liners.
13. Remove both front fender apron seals.
14. Drain engine oil.
15. Drain engine coolant.
16. Drain automatic transaxle fluid.
17. Remove the front wiper arm and blade assemblies.
18. Remove the cowl top ventilator louver sub-assembly.
19. Remove the windshield wiper motor and link assembly.
20. Remove the outer cowl top panel sub-assembly.
21. Remove the V-bank cover sub-assembly.
22. Remove the No. 1 and 2 air cleaner inlets.
23. Remove the air cleaner cap sub-assembly.
24. Remove the air cleaner filter element sub-assembly.
25. Remove the air cleaner case sub-assembly.
26. Remove the 2 bolts and air cleaner bracket.
27. Separate the brake master cylinder reservoir assembly.
28. Remove the 2 bolts and reservoir bracket.
29. Remove the bolt, 2 nuts and No. 2 right hand engine mounting stay.
30. Remove the 3 bolts and engine moving control rod.
31. Slide the clamp and disconnect the No. 1 fuel vapor feed hose.
32. Disconnect the No. 1 radiator clamp.
33. Using pliers, grip the claws of the clip and slide the clip to disconnect the No. 1 radiator hose from the water outlet.
34. Using pliers, grip the claws of the clip and slide the clip to disconnect the No. 2 radiator hose from the water inlet.
35. Using pliers, grip the claws of the clip and slide the clip to disconnect heater water outlet hose B from the water inlet.
36. Using pliers, grip the claws of the clip and slide the clip to disconnect heater water inlet hose B from the water outlet.
37. Remove the No. 1 fuel pipe clamp.
38. Disconnect the fuel tube from the fuel pipe.

➡Check for dirt and foreign objects the pipe and around the connector. Clean if necessary and then disconnect the connector.

➡Disconnect the connector by hand.

➡Do not bend, fold or rotate the nylon tube.

➡If the pipe and connector are stuck together, push and pull the connector until it becomes free.

➡Put the pipe and connector ends in vinyl bags to prevent damage and contamination.

39. Using pliers, grip the claws of the clip and slide the clip to disconnect the No. 2 oil cooler hose.
40. Using pliers, grip the claws of the clip and slide the clip to disconnect the oil cooler outlet hose.
41. Disconnect the transmission control cable assembly.
42. Disconnect the engine wire from the engine room relay block.
43. Using pliers, grip the claws of the clip and slide the clip to disconnect the union to check valve hose.
44. For 4WD vehicles, remove the propeller with center bearing shaft assembly.
45. Remove tail exhaust pipe assembly.
46. Remove the center exhaust pipe assembly.
47. Remove the front No. 3 exhaust pipe sub-assembly.
48. Remove the front exhaust pipe assembly.
49. Disconnect both front stabilizer link assemblies.
50. Remove both front axle hub nuts.
51. Disconnect both front speed sensors.
52. Disconnect left and right tie rod assemblies.
53. Separate both No. 1 front suspension lower arms.
54. Separate both front axle assemblies.
55. Disconnect the steering intermediate shaft sub-assembly.
56. Disconnect the discharge hose sub-assembly.
57. Disconnect the suction hose sub-assembly.
58. Remove engine assembly with transaxle.

➡Make sure that the engine is clear of all wiring and hoses.

59. Remove the V-ribbed belt.
60. For 4WD, remove the front stabilizer bar.
61. For 4WD, remove the power steering link assembly.
62. Remove the compressor and magnetic clutch.
63. Using pliers, grip the claws of the 2 clips and slide the 2 clips to remove the ventilation hose.
64. Remove intake air surge tank assembly.
65. Remove the engine wire, as follows:
    a. Disconnect the oil pressure switch cover and clamp.
    b. Disconnect the oil pressure switch assembly connector.
    c. Disconnect the crankshaft position sensor connector and clamp.
    d. Disconnect the generator assembly connector and clamp.

e. Remove the nut and separate the engine wire from the generator assembly.

f. Disconnect the purge VSV connector, air fuel ratio sensor connector and 2 clamps.

g. Remove the bolt.

h. Disconnect the engine coolant temperature sensor connector, knock control sensor wire connector and 2 clamps.

i. Disconnect the air fuel ratio sensor connector and clamp.

j. Remove the bolt and 2 nuts.

k. Disconnect the 2 clamps, radio setting condenser connector, 3 ignition coil assembly connectors, 2 camshaft timing oil control valve assembly connectors, 2 VVT sensor connectors and 3 injector assembly connectors, and separate the engine wire from the cylinder head cover sub-assembly RH.

l. Remove the 3 bolts.

m. Disconnect the 2 clamps, radio setting condenser connector, 3 ignition coil assembly connectors, 2 camshaft timing oil control valve assembly connectors, 2 VVT sensor connectors and 3 injector assembly connectors, and separate the engine wire from the engine assembly.

66. Remove starter assembly.
67. Remove the 3 bolts and transverse engine mounting bracket.
68. Remove the bolt, nut, and manifold stay.
69. Install the 2 engine hangers.
70. Remove front frame assembly.
71. Remove the nut and front engine mounting insulator assembly.
72. Remove the nut and transverse engine mounting insulators.
73. Remove front drive shaft assemblies.
74. For 4WD, remove the right transfer stiffener plate.
75. Remove the automatic transaxle assembly.
76. Remove the drive plate and ring gear sub-assembly.
77. Secure the engine onto an engine stand with the bolts.
78. Remove the engine hangers.
79. Remove transverse engine mounting insulator.

To install:

80. Installation is the reverse of removal. Observe the following torques:
    - Engine hanger: 24 ft. lbs. (33 Nm)
    - Transverse engine mounting insulator: 38 ft. lbs. (52 Nm)
    - Transverse engine mount bracket: 47 ft. lbs. (64 Nm)
    - Manifold stay: 25 ft. lbs. (34 Nm)
    - Engine wire: 74 inch lbs. (8.3 Nm)
    - Ground cable nut: 87 inch lbs. (9.8 Nm)
    - Compressor and magnetic clutch bolts: 18 ft. lbs. (25 Nm)
    - Ground cable bolts: 73 inch lbs. (8.3 Nm)
    - Transmission control cable assembly: 9 ft. lbs. (12 Nm)
    - Left and right frame side rail plates: bolts 63 ft. lbs. (85 Nm); nuts: 24 ft. lbs. (32 Nm)
    - Left and right front suspension member rear braces: bolts 63 ft. lbs. (85 Nm); nuts: 24 ft. lbs. (32 Nm)

EXHAUST MANIFOLD

REMOVAL & INSTALLATION

2.4L Engine

1. Before servicing the vehicle, refer to the precautions section.
2. Remove or disconnect the following:
    - Clamp from the support bracket
    - Support bracket
    - Front exhaust pipe and gaskets from the exhaust manifold
    - Heat insulator
    - Exhaust manifold and gasket

To install:

3. Install or connect the following:
    - Exhaust manifold and gasket. Tighten the nuts to 36 ft. lbs. (49 Nm).
    - Heat insulator. Tighten the bolts and nuts to 48 inch lbs. (5.5 Nm).
    - Front exhaust pipe assembly to the exhaust manifold. Tighten the nuts to 46 ft. lbs. (62 Nm).
    - Support bracket. Tighten the bolts to 29 ft. lbs. (39 Nm).
    - Clamp. Tighten the bolt to 14 ft. lbs. (19 Nm).
4. Start the engine.
5. Check for exhaust leaks.

3.3L Engine

Front Manifold

➡Removing the oil filter helps gain access to a lower bolt in the front exhaust manifold.

1. Before servicing the vehicle, refer to the precautions section.
2. Remove or disconnect the following:
    - Negative battery cable
    - Engine undercovers
    - Front exhaust pipe from the exhaust manifolds, by removing the nuts

➡Check for access to some of the manifold lower bolts, if so remove any possible.

- Heated Oxygen (HO_2) sensor
- Exhaust manifold stay, by removing the bolt and nut
- Remaining exhaust manifold nuts; then, separate the exhaust manifold from the engine

To install:

3. Install or connect the following:
    - Exhaust manifold, using a new gasket. Uniformly, tighten the bolts to 36 ft. lbs. (49 Nm).
    - Exhaust manifold stay. Tighten the nut/bolt to 15 ft. lbs. (20 Nm).
    - Heated Oxygen (HO_2) sensor to the exhaust manifold
    - Front exhaust pipe to the exhaust manifold, using a new gasket. Tighten both nuts to 46 ft. lbs. (62 Nm).
    - Engine undercovers
    - Negative battery cable

Rear Manifold

1. Before servicing the vehicle, refer to the precautions section.
2. Remove or disconnect the following:
    - Negative battery cable
    - Engine undercovers
    - Front exhaust pipe from both exhaust manifolds, from below the engine
    - Exhaust Gas Recirculation (EGR) pipe from the rear exhaust manifold, by removing the 4 nuts
    - Heated Oxygen (HO_2) sensor wiring, from the right exhaust manifold
    - Exhaust manifold stay
    - 6 exhaust manifold nuts and the exhaust manifold

To install:

3. Install or connect the following:
    - Exhaust manifold to the engine, using a new gasket. Tighten the 6 nuts to 36 ft. lbs. (49 Nm).
    - Exhaust manifold stay. Tighten the nut/bolt to 15 ft. lbs. (20 Nm).
    - HO_2 sensor wiring to the exhaust manifold
    - EGR pipe to the exhaust manifold and the engine, using new gaskets. Tighten the 4 nuts to 108 inch lbs. (12 Nm).
    - Front exhaust pipe to the exhaust manifold, use a new gasket. Tighten both nuts to 46 ft. lbs. (62 Nm).
    - Engine undercovers
    - Negative battery cable

3.5L Engine

See Figures 120 through 123.

1. Before servicing the vehicle, refer to the precautions section.
2. Remove the right front wheel.
3. Remove the V-bank cover sub-assembly.
4. Remove the engine under cover assembly.
5. Remove the No. 1 and 2 engine under covers.
6. Drain engine coolant.
7. Disconnect the No. 1 radiator hose.
8. Remove the radiator reserve tank assembly, as follows:
    a. Disconnect the hose.
    b. Remove the 2 bolts and the radiator reserve tank assembly.
9. Remove the No. 2 oil level dipstick guide.
10. Remove the air fuel ratio sensor (for Bank 2 Sensor 1).
11. Remove the 3 bolts and No. 2 exhaust manifold heat insulator.
12. For 4WD vehicles, remove the propeller with center bearing shaft assembly.
13. Remove the tail exhaust pipe assembly.
14. Remove the center exhaust pipe assembly.
15. Remove the front No. 3 exhaust pipe sub-assembly.
16. Remove the front exhaust pipe assembly.
17. Remove the bolt, nut and No. 2 manifold stay.
18. Remove the 6 nuts and left hand exhaust manifold sub-assembly.
19. Remove the gasket.
20. Remove the bolt, nut and manifold stay.
21. Remove the right hand exhaust manifold sub-assembly, as follows:
    a. Disconnect the air fuel ratio sensor (for bank 1 sensor 1) connector and remove the clamp.
    b. Remove the 6 nuts and the right hand exhaust manifold sub-assembly.
    c. Remove the gasket.
22. Remove the air fuel ratio sensor (for Bank 1 Sensor 1).

To install:

23. Install the air fuel ratio sensor (for Bank 1 Sensor 1).
24. Install the right hand exhaust manifold sub-assembly, as follows:
    a. Install a new gasket.
    b. Install the right hand exhaust manifold sub-assembly by tightening the 6 nuts in the order shown to 15 ft. lbs (21 Nm).

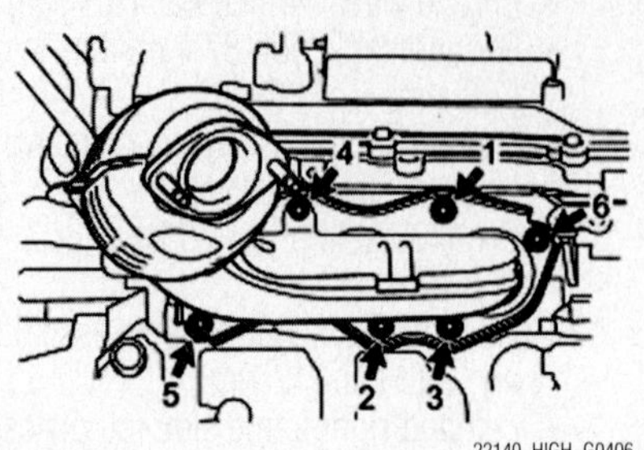

Fig. 120 Right exhaust manifold sub-assembly tightening sequence—3.5L engine

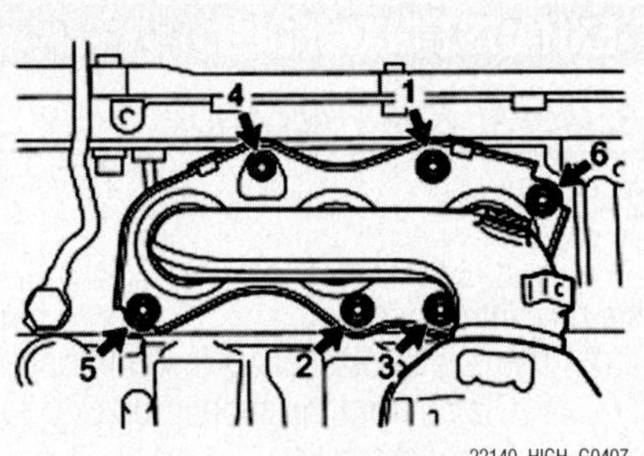

Fig. 121 Left exhaust manifold sub-assembly tightening sequence—3.5L engine

Fig. 122 No. 2 manifold stay tightening sequence—3.5L engine

    c. Connect the air fuel ratio sensor (for Bank 1 Sensor 1) connector and install the clamp.
25. Install the manifold stay with the bolt and nut and tighten to 25 ft. lbs (34 Nm), 26 ft. lbs (35 Nm).
26. Install the left hand exhaust manifold sub-assembly, as follows:
    a. Install a new gasket.
    b. Install the left hand exhaust manifold sub-assembly by tightening the 6 nuts in the order shown to 15 ft. lbs (21 Nm).
27. Install the No. 2 manifold stay by tightening the bolt and nut in the order shown to 25 ft. lbs (34 Nm).

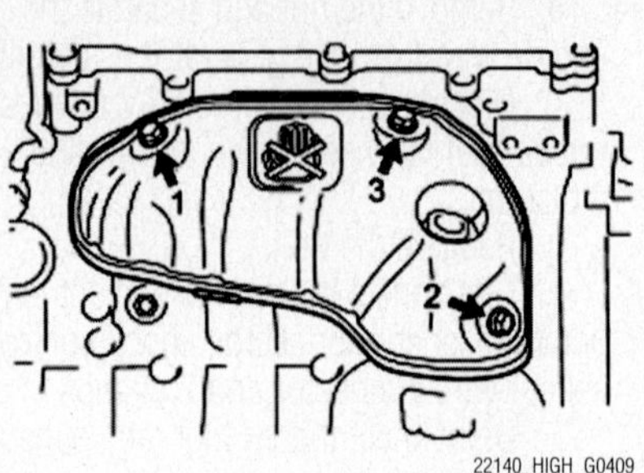

Fig. 123 No. 2 exhaust manifold heat insulator tightening sequence—3.5L engine

28. Install the front exhaust pipe assembly.
29. Install the front No. 3 exhaust pipe sub-assembly.
30. Install the center exhaust pipe assembly.
31. Install the tail exhaust pipe assembly.
32. For 4WD vehicles, temporarily tighten the propeller with center bearing shaft assembly.
33. For 4WD vehicles, fully tighten the propeller with center bearing shaft assembly.
34. Install the No. 2 exhaust manifold heat insulator by tightening the 3 bolts in the order shown to 75 inch lbs (8.5 Nm).
35. Install the air fuel ratio sensor (for Bank 2 Sensor 1).
36. Install the No. 2 oil level dipstick guide.
37. Install the radiator reserve tank assembly with the 2 bolts and tighten to 48 inch lbs (5.4 Nm).
38. Connect the hose.
39. Connect the No. 1 radiator hose.
40. Add engine coolant.
41. Inspect for coolant leak.
42. Inspect for gas leak, and repair as necessary.
43. For 4WD vehicles, inspect and adjust transfer oil.
44. Install the No. 1 and 2 engine under covers.
45. Install the engine under cover assembly.
46. Install the V-bank cover sub-assembly.
47. Install the right front wheel.

FLYWHEEL

REMOVAL & INSTALLATION

See Figures 124 and 125.

1. Remove automatic transaxle assembly

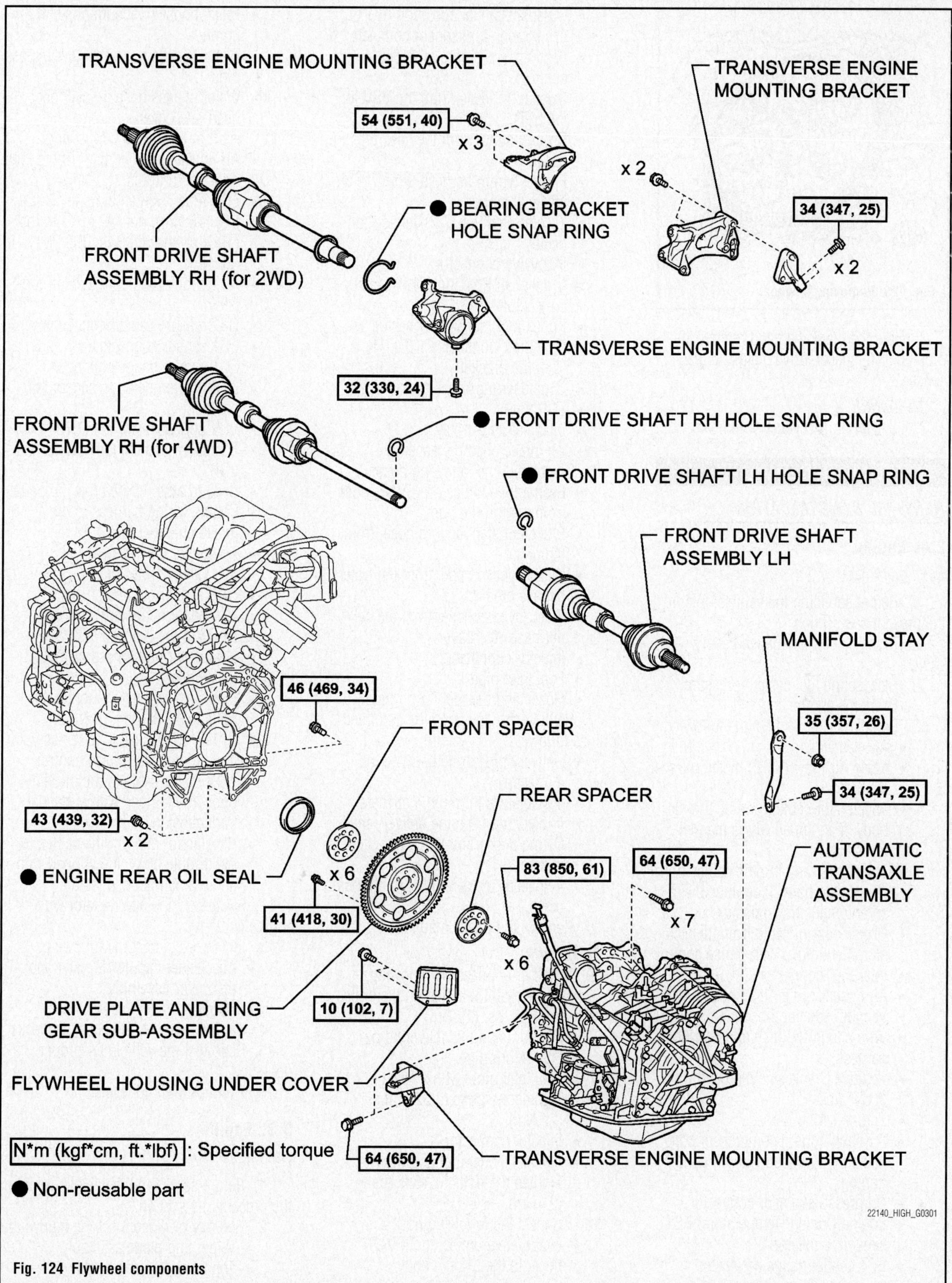

Fig. 124 Flywheel components

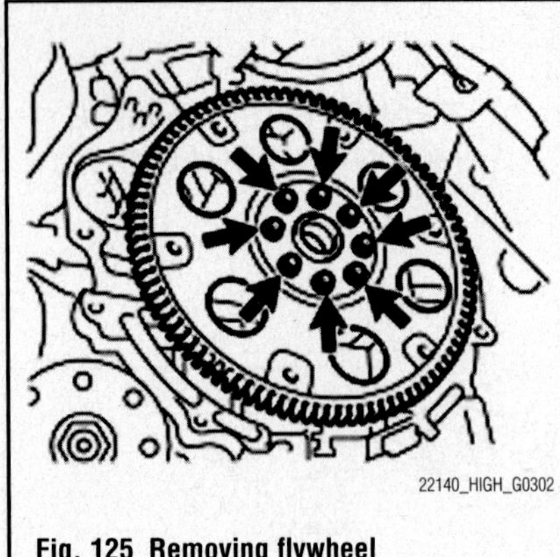

Fig. 125 Removing flywheel

2. Hold the crankshaft and remove the 8 bolts, front spacer, drive plate and rear spacer.

To install:

3. Installation is reverse of removal.

INTAKE MANIFOLD

REMOVAL & INSTALLATION

2.4L Engine

See Figure 126.

1. Before servicing the vehicle, refer to the precautions section.
2. Disconnect the negative battery cable.
3. Release the fuel system pressure.
4. Drain the engine coolant.
5. Remove or disconnect the following:
    - Air cleaner cap
    - Mass Air Flow (MAF) meter and the resonator
    - Accelerator cable from the throttle body, if equipped with a manual transmission
    - Accelerator and throttle cables from the throttle body, if equipped with an automatic transmission
    - Cruise control cable from the actuator, if equipped with cruise control
    - Intake air connector
    - Air hose for Idle Air Control (IAC)
    - Vacuum sensing hose
    - Wire clamp for the engine wiring harness
    - Positive Crankcase Ventilation (PCV) hoses.
    - Engine wiring harness
    - Air conditioning compressor connector, if equipped with air conditioning
    - Oil pressure sensor connector
    - Engine Coolant Temperature (ECT) sensor connector
    - ECT sender gauge connector
    - Exhaust Gas Recirculation (EGR) gas temperature sensor connector
    - Vacuum Switching Valve (VSV) connector, for the EGR
    - 2 vacuum hoses, from the VSV for the EGR
    - Ground strap, from the cowl top panel
    - Engine wiring harness, from the air intake chamber
    - Throttle Position (TP) sensor connector
    - IAC valve connector
    - Crankshaft Position (CKP) sensor connector
    - Knock (KS) sensor connector
    - Data Link Connector 1 (DLC1), from the bracket
    - Engine wiring harness clamp
    - EGR pipe
    - Intake chamber stay
    - Air intake chamber assembly
6. Disconnect the following hoses:
    - Evaporative Emission (EVAP) hose, from the throttle body
    - Brake booster vacuum hose, from the union
    - Water bypass hose, from the water bypass pipe
    - Water bypass hose, from the cylinder head rear cover
    - Injector connectors
    - Fuel inlet pipe
    - Hoses and the fuel return pipe.
7. Remove the delivery pipe and injectors, as follows:
    - Delivery pipe, together with the 4 injectors
    - 4 insulators from the 4 spacers
    - 4 injectors, from the delivery pipe
    - O-ring and grommets, from each injector
    - 4 spacers, by carefully prying them out
8. Remove the intake manifold.

To install:

9. Install or connect the following:
    - Intake manifold. Tighten the bolts to 22 ft. lbs. (29 Nm).
    - Injectors and the delivery pipe
    - Fuel return pipe
    - Fuel inlet pipe, with a new gasket. Tighten the bolts to 22 ft. lbs. (29 Nm).
    - Injector connectors
    - Air intake chamber assembly. Tighten the bolts to 15 ft. lbs. (21 Nm).
10. Connect the following hoses:
    - Evaporative Emissions (EVAP) hose, to the throttle body
    - Brake booster vacuum hose, to the union
    - Water bypass hose, to water bypass pipe
    - Water bypass hose, to cylinder head rear cover
11. Install or connect the following:
    - Air intake chamber stay. Tighten the bolts to 15 ft. lbs. (20 Nm).
    - EGR pipe. Tighten bolts to 13 ft. lbs. (18 Nm), nut "A" to 14 ft. lbs. (19 Nm) and nut B to 15 ft. lbs. (20 Nm).
    - Air conditioning compressor connector
    - Oil pressure sensor connector
    - ECT sensor connector
    - ECT sender gauge connector
    - EGR gas temperature sensor connector
    - VSV connector for the EGR
    - 2 vacuum hose to the VSV for the EGR
    - Ground strap to the cowl top panel
    - Engine wiring harness to the air intake chamber
    - TP sensor connector
    - IAC valve connector
    - CKP sensor connector
    - KS sensor connector
    - DLC1 to the bracket
    - Engine wiring harness clamp
    - PCV hoses
    - Intake air connector. Tighten the bolts to 13 ft. lbs. (18 Nm).
    - Cruise control cable to the actuator, if equipped with cruise control
    - Accelerator cable to the throttle body, if equipped with a manual transmission
    - Accelerator and throttle cables to the throttle body, if equipped with an automatic transmission
12. Fill the engine and radiator with engine coolant.
13. Install or connect the following:
    - Air cleaner cap, MAF meter and resonator assembly
    - Negative battery cable
14. Start the engine and check for leaks.
15. Road test the vehicle for proper operation.
16. Recheck all fluid levels.

3.3L Engine

See Figure 127.

1. Before servicing the vehicle, refer to the precautions section.
2. Remove or disconnect the following:
    - Wiper and blade assembly
    - Top cowl seal and panel

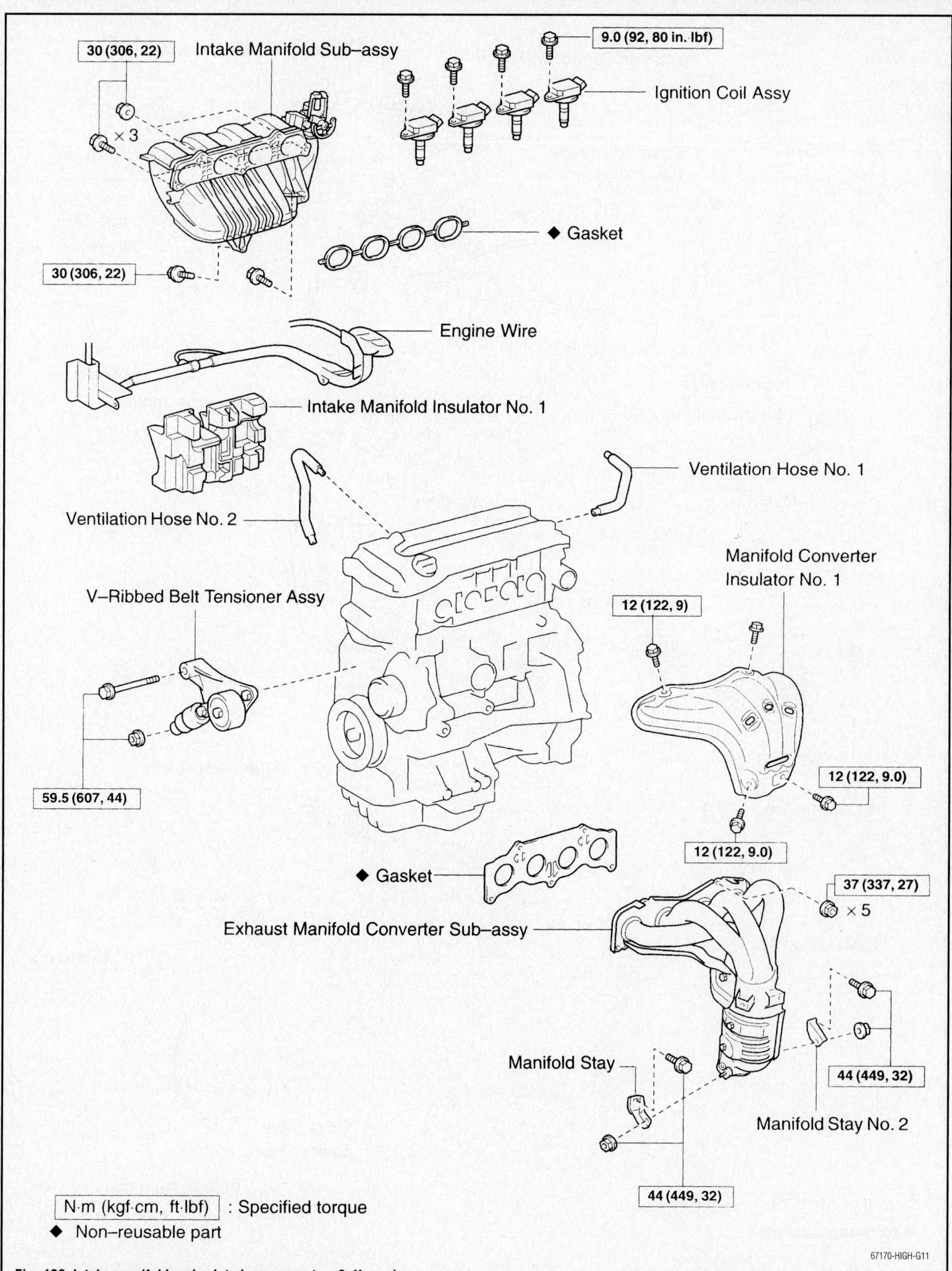

Fig. 126 Intake manifold and related components—2.4L engine

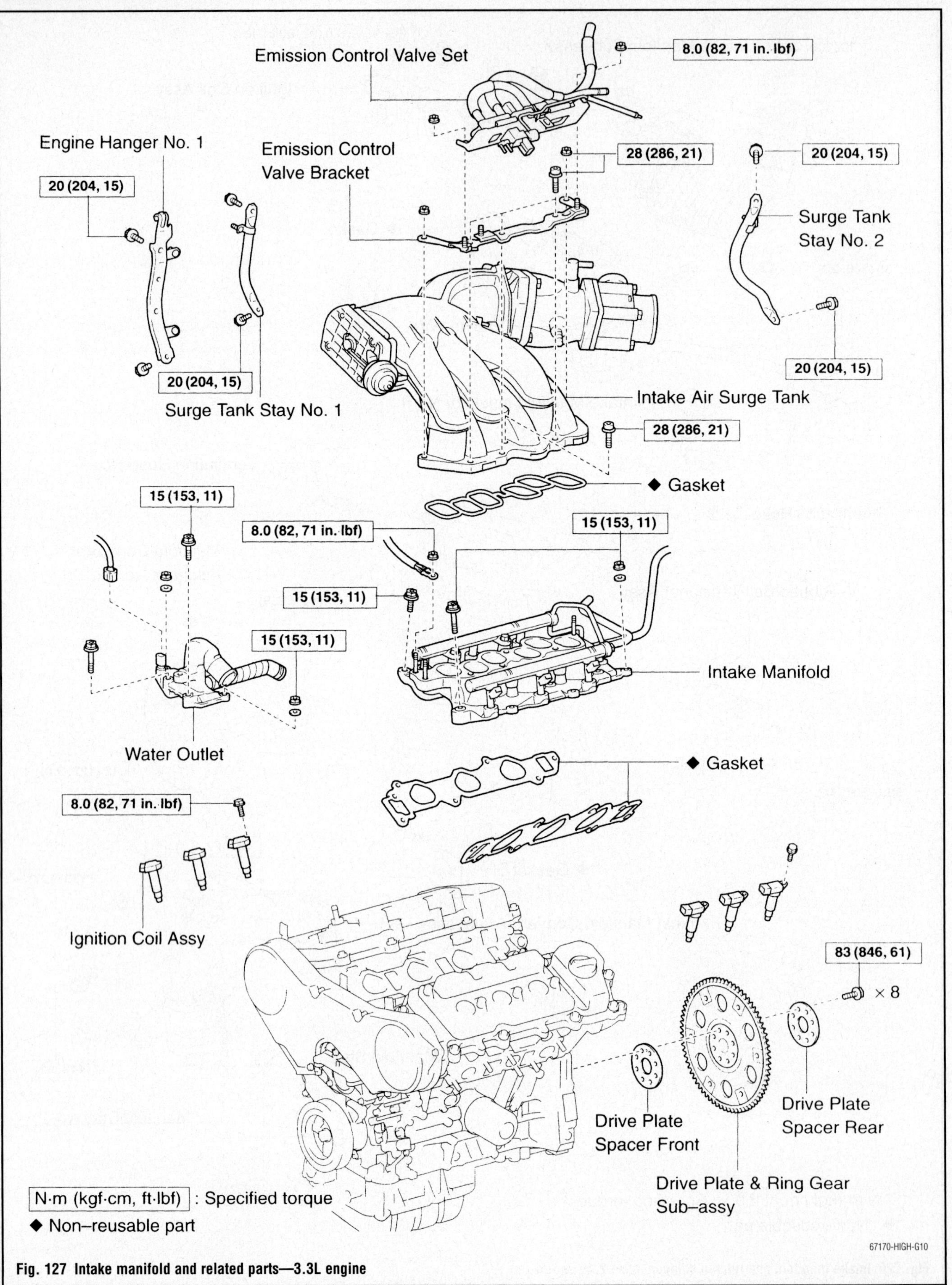

Fig. 127 Intake manifold and related parts—3.3L engine

- Window washer hoses from the ventilator louvers
- Left and right ventilator louvers
- Heater air duct
- Front upper suspension brace

3. Properly relieve the fuel system pressure.
4. Remove the battery and battery tray.
5. Drain and recycle the engine coolant.
6. Remove or disconnect the following:
   - Accelerator cable
   - Throttle cable
   - Air cleaner cap assembly
   - Any wiring or hoses interfering with removal
   - Right side engine mount stay
   - Radiator and heater hoses in the way of the intake manifold removal
   - V-bank cover
   - All the vacuum hose and wiring for the emission control valve set
   - Air intake chamber and discard the gasket
   - Exhaust Gas Recirculation (EGR) pipe and discard the gaskets
   - Hydraulic motor pressure hose from the air intake chamber
   - Engine wiring harnesses from the left side, right side, rear and No. 3 timing belt cover
   - Front exhaust pipe, if necessary
   - Timing belt, camshaft timing pulleys, No. 2 idler pulley and No. 3 timing belt cover
   - Cylinder head rear plate
   - 2 bolts, nuts and plate washers with the intake manifold.

➡The delivery pipes with injectors will be attached to the manifold.

- Other fuel related components such as the No. 2 fuel pipe and pulsation damper, if needed
- Delivery pipes from the intake manifold

7. Clean and inspect the intake manifold mating surfaces. Scrape all old gasket material off.

To install:

8. Install or connect the following:
   - Delivery pipes with injectors to the intake manifold.

➡Be sure to place 4 spacers in position on the manifold. Temporarily install 4 bolts to retain the delivery pipes to the manifold. Inspect the injectors for smooth rotation.

- Tighten the delivery pipes bolts to 84 inch lbs. (10 Nm), once the injectors are properly seated
- No. 2 fuel pipe with union bolts and gaskets. Tighten the bolts to 24 ft. lbs. (32 Nm).
- No. 1 fuel pipe with pulsation damper, using 4 new gaskets. Tighten the damper to 35 ft. lbs. (32 Nm) and the bolt to 11 ft. lbs. (15 Nm).
- Fuel pressure regulator, if removed
- Intake manifold. Tighten the 9 bolts and 2 nuts in a crisscross pattern to 11 ft. lbs. (15 Nm).

➡Be sure the gasket is in place properly prior to tightening.

9. Retighten the water outlet mounting nuts/bolts to 11 ft. lbs. (15 Nm), if loosened.
10. Install or connect the following:
    - Air assist hose and water inlet pipe, using a new O-ring, by applying a small amount of soapy water. Tighten the fastener(s) to 14 ft. lbs. (20 Nm).
    - Ground strap
    - Vacuum hoses removed to the air intake chamber and vacuum tank
    - Any remaining components, using new gaskets. Tighten the air intake chamber nuts/bolts to 32 ft. lbs. (43 Nm), the EGR pipe nuts to 108 inch lbs. (12 Nm) and the emission control valve set to 69 inch lbs. (8 Nm).
    - Air cleaner assembly
    - Heater hoses
    - Battery and tray
    - Throttle cable with bracket onto the throttle body
    - Accelerator cable, by adjusting it, if equipped with an automatic transaxle
    - Front upper suspension brace. Tighten the nuts to 59 ft. lbs. (80 Nm).
11. Refill the cooling system
12. Install or connect the following:
    - Negative battery cable
    - Heater air duct
    - Left and right ventilator louvers
    - Window washer hoses from the ventilator louvers
    - Top cowl seal and panel
    - Wiper and blade assembly
13. Start the engine and inspect for leaks.

3.5L Engine

See Figures 128 through 138.

1. Discharge fuel system pressure.
2. Remove the engine under cover assembly.
3. Remove the No. 1 engine under cover.
4. Drain engine coolant.
5. Remove the V-bank cover sub-assembly.
6. Remove both front wiper arm and blade assemblies.
7. Remove the cowl top ventilator louver sub-assembly.
8. Remove the windshield wiper motor and link assembly.
9. Remove the outer cowl top panel sub-assembly.
10. Remove the air cleaner cap sub-assembly.
11. Disconnect the engine room main wire, as follows:
    a. Disconnect the 5 harness clamps.
12. Remove throttle body bracket:
    a. Remove the 2 bolts in the order shown in the illustration and remove the throttle body bracket.
13. Remove the 2 bolts in the order shown in the illustration and remove the No. 1 surge tank stay.
14. Remove the intake air surge tank assembly, as follows:
    a. Disconnect the 4 hoses.
    b. Disconnect the throttle body connector and clamp.
    c. Disconnect the connector.

Fig. 128 Throttle body bracket bolt removal sequence—3.5L engine

Fig. 129 No. 1 surge tank stay bolt removal sequence—3.5L engine

Fig. 130 Intake air surge tank assembly hoses—3.5L engine

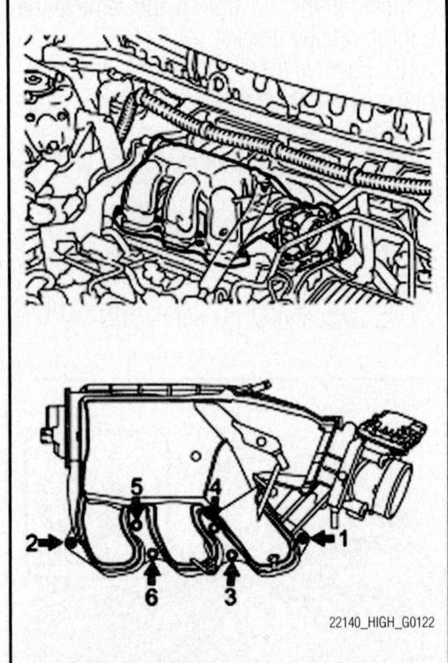

Fig. 131 Intake air surge tank assembly bolt removal sequence—3.5L engine

d. Disconnect the 2 water by-pass hoses from the throttle body.

e. Remove the 4 bolts and 2 nuts in the order shown in the illustration.

➡Use a 5 mm socket hexagon wrench to remove the 4 bolts

f. Remove the gasket from the intake air surge tank.

15. Remove the right hand No. 2 engine mounting stay, as follows:

a. Remove the 5 bolts (E).

b. Remove the 2 nuts (F).

c. Remove the right hand No. 2 engine mounting stay.

16. Disconnect the fuel main tube, as follows:

a. Remove the No. 2 fuel pipe clamp.

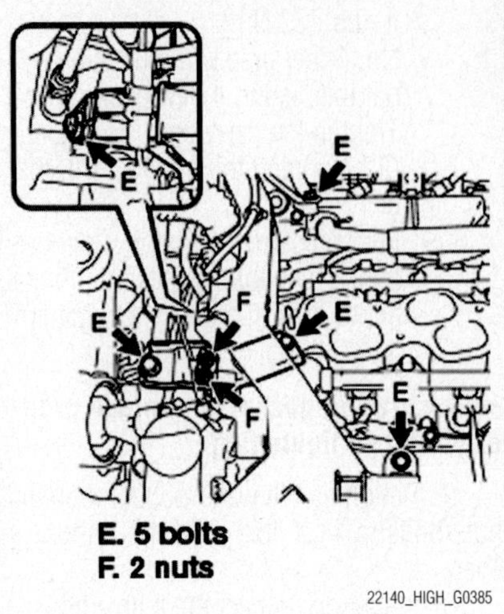

Fig. 132 Right hand No. 2 engine mounting stay bolt and nut removal sequence—3.5L engine

b. Pinch the tube connector and pull out the fuel pipe.

➡Check that there is no dirt or other foreign objects around the connector before disconnecting it. Clean the connector as necessary.

➡It is necessary to prevent dirt or foreign objects from entering the quick connector. If dirt or foreign objects enter the connector, the O-rings may not seal properly.

➡Only disconnect the quick connector by hand.

➡Do not bend, kink or twist the nylon tubes. Protect the connector by covering it with a plastic bag.

➡If the pipe and the connector are stuck, carefully try wiggling or pushing and pulling on the connector to release it. Pull the connector off the pipe carefully.

17. Remove the fuel delivery pipe sub-assembly, as follows:

a. Disconnect the 6 fuel injector connectors.

18. Remove the 5 bolts and fuel delivery pipe sub-assembly together with the 6 fuel injectors.

➡Be careful not to drop the fuel injectors when removing the fuel delivery pipe sub-assembly.

a. Remove the 6 injector vibration insulators from the intake manifold.

19. Remove the intake manifold, as follows:

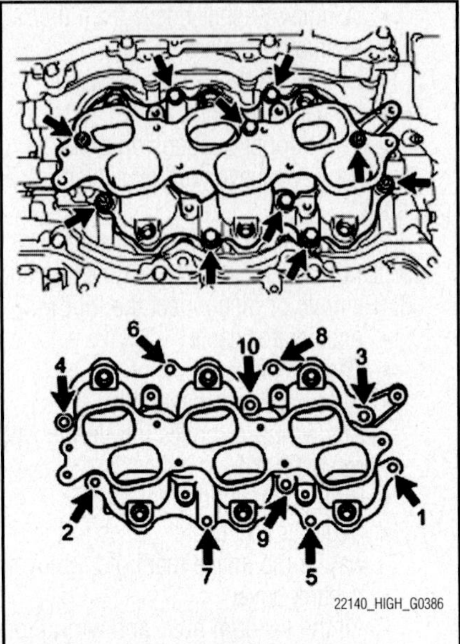

Fig. 133 Intake manifold bolt removal sequence—3.5L engine

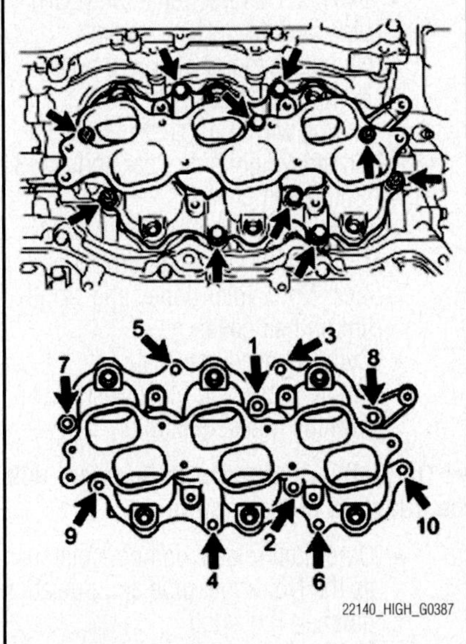

Fig. 134 Intake manifold bolt installation sequence—3.5L engine

a. Remove the 6 bolts and 4 nuts in the order shown in the illustration and remove the intake manifold.

20. Remove the 2 No. 1 intake manifold to head gaskets.

To install:

21. Install the intake manifold, as follows:

a. Set 2 new gaskets on each cylinder head.

➡Align the port holes of the gaskets and cylinder head.

➡Make sure that the gaskets are installed in the correct direction.

b. Set the intake manifold on the cylinder heads.

c. Install the intake manifold with the 6 bolts and 4 nuts in the order shown in the illustration and tighten to 15 ft. lbs. (21 Nm). **DO NOT** apply oil to the bolts.

22. Install the right hand No. 2 engine mounting stay.

23. Install the bolt (A), and tighten to 15 ft. lbs. (21 Nm). **DO NOT** apply oil to the bolt.

24. Install the 2 nuts (B), and tighten to 17 ft. lbs. (23 Nm).

25. Install the bolt (C), and tighten to 28 ft. lbs. (38 Nm). **DO NOT** apply oil to the bolt.

26. Install the 3 bolts (D), and tighten to 73 inch lbs. (8.3 Nm). **DO NOT** apply oil to the bolts.

27. Install the fuel delivery pipe sub-assembly, as follows:

a. Install 6 new insulators to the intake manifold.

b. Place the fuel delivery pipe which has the 6 fuel injectors installed to it in position on the intake manifold.

➡Be careful not to drop the fuel injectors when installing the fuel delivery pipe.

c. Temporarily install the 5 bolts which are used to hold the fuel delivery pipe to the intake manifold. **DO NOT** apply oil to the bolts.

➡After installing the fuel injectors, check that they turn smoothly. If not, reinstall the injectors with new O-rings.

d. Tighten the 5 bolts which are used to hold the fuel delivery pipe to the intake manifold to 15 ft. lbs. (21 Nm). **DO NOT** apply oil to the bolts.

e. Connect the 6 fuel injector connectors.

28. Connect the fuel main tube, as follows:

a. Push in the tube connector onto the pipe until the tube connector clicks.

➡Before connecting the tube, make sure that it is not damaged. Make sure that there is no dirt present on the connecting surfaces.

➡After connecting, check that the fuel tube connector and the pipe are securely connected by pulling on them.

b. Install the No. 2 fuel pipe clamp.

29. Temporarily install the No. 1 surge tank stay, as follows:

a. Temporarily install the intake air surge tank assembly with 3 new gaskets on the intake manifold.

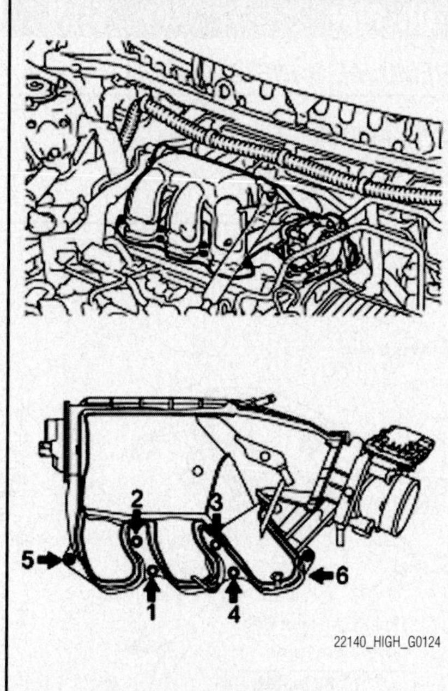

Fig. 135 Surge tank tightening sequence—3.5L engine

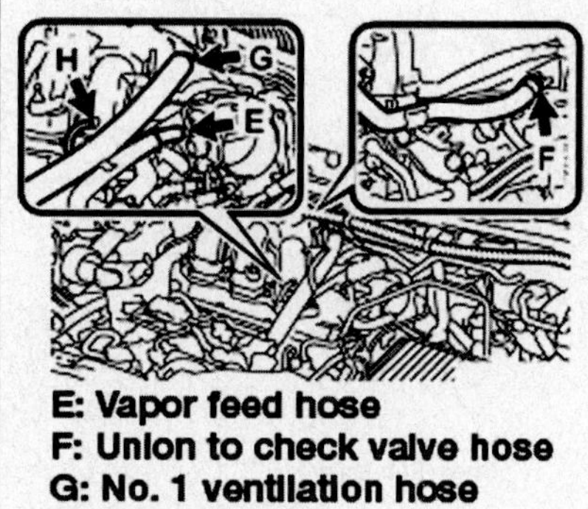

Fig. 136 Intake air surge tank assembly hoses—3.5L engine

➡Do not allow the gaskets to slip out of place during installation.

b. Temporarily install the No. 1 surge tank stay with the 2 bolts. **DO NOT** apply oil to the bolts.

30. Temporarily install the throttle body bracket with the 2 bolts. **DO NOT** apply oil to the bolts.

31. Install the intake air surge tank assembly, as follows:

a. Install the surge tank with the 4 bolts and 2 nuts in the order shown in the illustration and tighten to 12 ft. lbs. (16 Nm) and 13 ft. lbs. (18 Nm). **DO NOT** apply oil to the bolts.

➡Use a 5 mm hexagon socket wrench to tighten the 4 bolts.

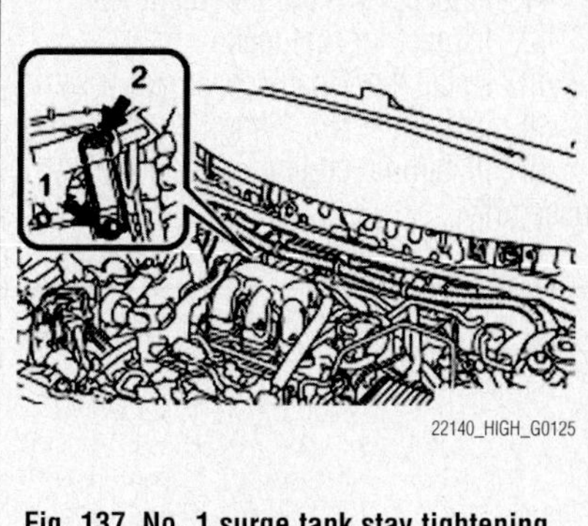

Fig. 137 No. 1 surge tank stay tightening sequence—3.5L engine

Fig. 138 Throttle body bracket tightening sequence—3.5L engine

b. Connect the 2 water by-pass hoses to the throttle with motor body assembly.

c. Connect the connector.

d. Install the clamp and connect the throttle with motor body assembly connector.

e. Connect the 4 hoses.

32. Fully tighten the No. 1 surge tank stay, as follows:

a. Fully tighten the 2 bolts in the order shown in the illustration to 15 ft. lbs. (21 Nm). **DO NOT** apply oil to the bolts.

33. Fully tighten the throttle body bracket, as follows:

a. Fully tighten the 2 bolts in the order shown in the illustration to 15 ft. lbs. (21 Nm). **DO NOT** apply oil to the bolts.

34. Connect the 5 engine room main wire harness clamps.

35. Install the air cleaner cap sub-assembly.

36. Install the outer cowl top panel sub-assembly.

37. Install the windshield wiper motor and link assembly.

38. Install the cowl top ventilator louver sub-assembly.

39. Install the both front wiper arm and blade assemblies.

40. Add engine coolant.

41. Inspect for engine coolant leak.
42. Inspect for fuel leak.
43. Install the V-bank cover sub-assembly.
44. Install the No. 1 engine under cover.
45. Install the engine under cover assembly.

OIL PAN

REMOVAL & INSTALLATION

2.4L Engine

See Figure 139.

1. Before servicing the vehicle, refer to the precautions section.
2. Remove or disconnect the following:
   - Engine undercover
   - Engine oil
   - Oil pan bolts, nuts and pan

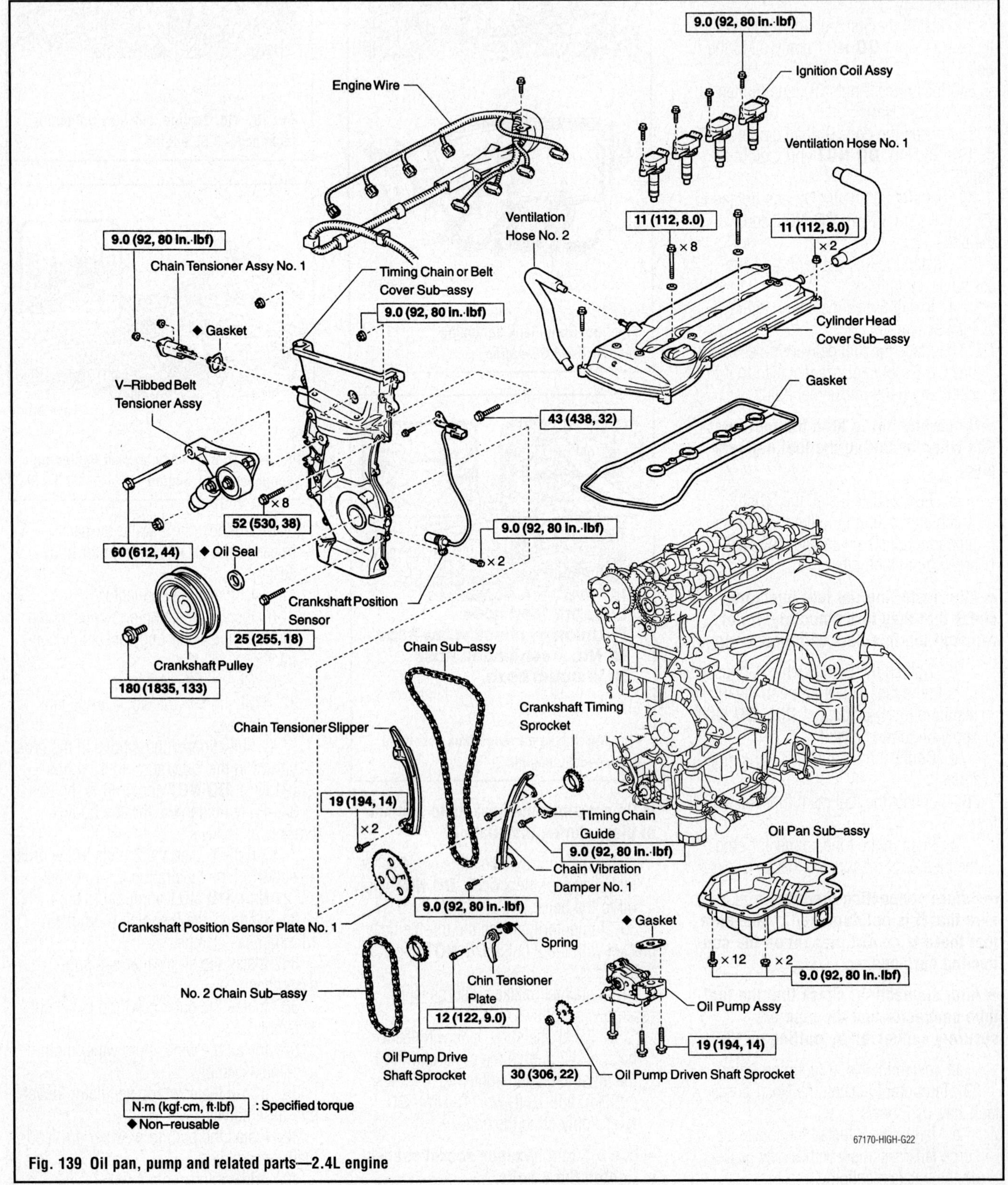

Fig. 139 Oil pan, pump and related parts—2.4L engine

3. Installation is the reverse of removal. Torque the 12 bolts and 2 nuts to 80 inch lbs. (9 Nm).

3.3L Engine

See Figure 140.

1. Before servicing the vehicle, refer to the precautions section.
2. Remove or disconnect the following:
   - Right front wheel
   - Fender apron seal
   - Engine undercover
3. Drain the engine oil from the engine.
4. Remove or disconnect the following:
   - Front exhaust pipe
   - Front exhaust pipe bracket from the No. 1 oil pan
   - Flywheel housing undercover

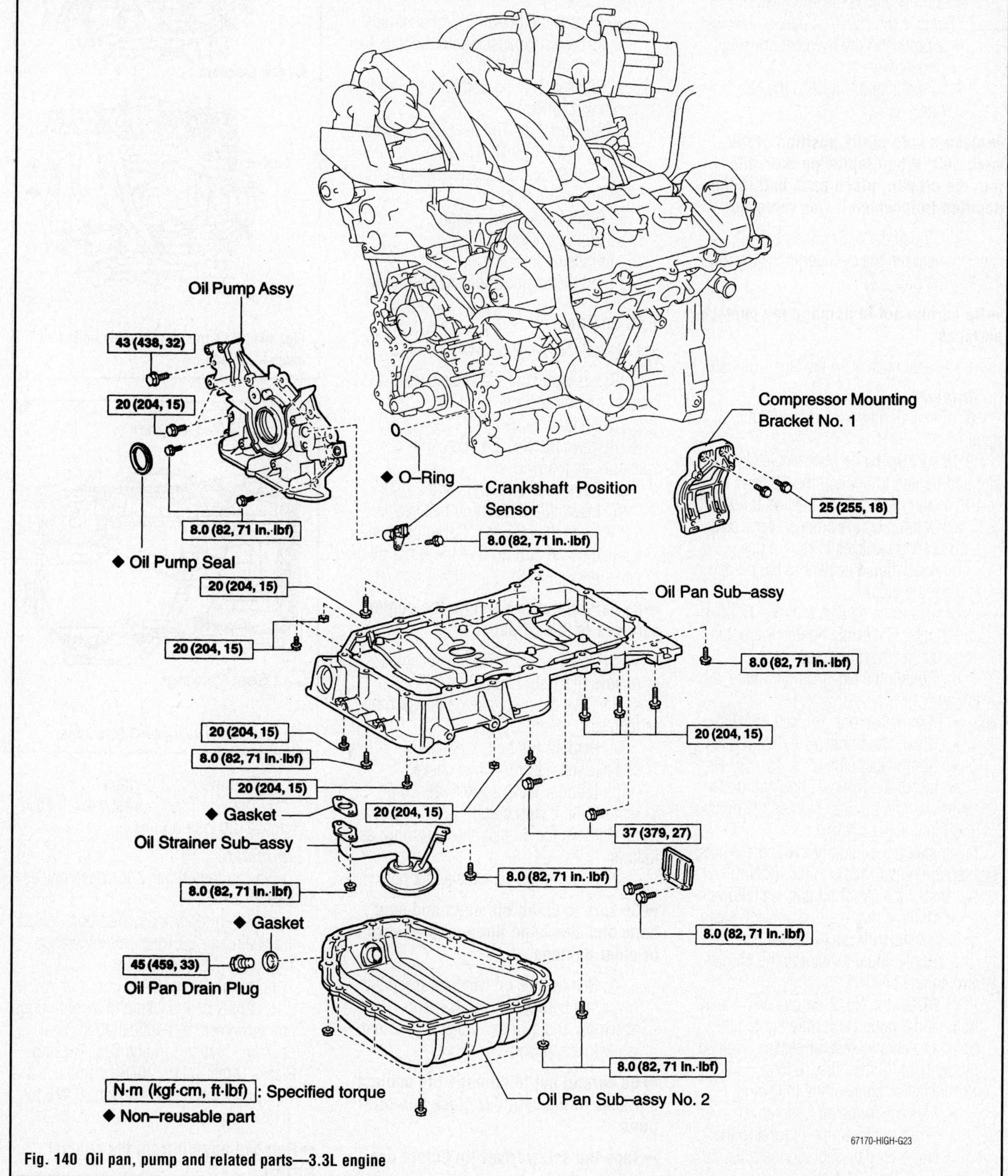

Fig. 140 Oil pan, pump and related parts—3.3L engine

- 10 bolts and 2 nuts to the No. 2 oil pan

5. Insert the blade of the Oil Pan Seal Cutting tool 09032-00100 between the No. 1 and No. 2 oil pans. Clean the surfaces of the oil pans.
6. Remove or disconnect the following:
   - 3 oil strainer nuts and gasket
7. Remove the No. 1 oil pan, as follows:
   - 2 bolts and the flywheel housing undercover
   - 17 bolts and 2 nuts to the No. 1 oil pan

➡Make a note of the position of the each bolt. When replacing the bolts into the oil pan, place each bolt in the position from which it was removed.

- Oil pan, by prying the portions between the cylinder block and the oil pan

➡Be careful not to damage the contact surfaces.

- Baffle plate from the No. 1 oil pan

To install:

8. Clean all mating surfaces of the oil pans.
9. Install the baffle plate to the No. 1 oil pan and tighten to 69 inch lbs. (8 Nm).
10. Install the No. 1 oil pan, as follows:
    a. Using a non residue solvent, clean both sealing surfaces to the oil pan.
    b. Apply liquid sealant to the oil pan and engine block.
    c. Install the oil pan with the 17 bolts and 2 nuts. Uniformly tighten the bolts and nuts in several passes.
    d. Tighten the No. 1 oil pan bolts, as follows:
    - 10mm head bolt: 69 inch lbs. (8 Nm)
    - 12mm head bolt: 14 ft. lbs. (20 Nm)
    - 14mm head bolt: 27 ft. lbs. (37 Nm)

    e. Install the flywheel housing undercover with the 2 bolts. Tighten the bolts to 69 inch lbs. (8 Nm).
11. Install the oil strainer with the 3 nuts. Tighten the nuts to 69 inch lbs. (8 Nm).
12. Install the No. 2 oil pan, as follows:
    a. Using a non residue solvent, clean both sealing surfaces to the oil pan.
    b. Apply liquid sealant to the oil pan and engine block.
    c. Install the No. 2 oil pan with the 10 bolts and 2 nuts. Uniformly tighten the bolts and nuts in several passes. Tighten the bolts to 69 inch lbs. (8 Nm).
13. Install or connect the following:
    - Flywheel housing undercover
    - Front exhaust pipe bracket to the No. 1 oil pan. Tighten the bolts to 15 ft. lbs. (21 Nm).
14. Install the front exhaust pipe, as follows:
    - Temporarily install the 3 new gaskets and the front exhaust pipe with the 2 bolts and 6 nuts
    - Tighten the 4 exhaust manifolds-to-front exhaust pipe nuts to 46 ft. lbs. (62 Nm).
    - Tighten the both front exhaust pipe-to-center exhaust pipe nuts/bolts to 41 ft. lbs. (56 Nm).
    - Bracket. Tighten both bolts to 14 ft. lbs. (19 Nm).
    - Support stay. Tighten both bolts to 22 ft. lbs. (29 Nm).
15. Install or connect the following:
    - Engine undercover
    - Right fender apron seal
    - Right front wheel
16. Fill the engine with oil.
17. Start the engine and check for leaks.

3.5L Engine

See Figures 141 through 145.

1. Before servicing the vehicle, refer to the precautions section.
2. Drain the engine oil.
3. Remove the No. 2 oil pan sub-assembly, as follows:
   a. Remove the 16 bolts and 2 nuts.
   b. Insert the blade of oil pan seal cutter between the oil pans. Cut through the applied sealer and remove the No. 2 oil pan sub-assembly.

➡Be careful not to damage the contact surfaces of the oil pans.

   c. Using a Torx® socket wrench E6, remove the 2 stud bolts.
4. Remove the oil strainer sub-assembly, as follows:
   a. Remove the bolt, 2 nuts, oil strainer sub-assembly and gasket.
   b. Using a Torx® socket wrench E6, remove the 2 stud bolts.
5. Remove the oil pan sub-assembly, as follows:
   a. Remove the 16 bolts and 2 nuts.

➡Be sure to clean the bolts and stud bolts and check the threads for cracks or other damage.

   b. Remove the oil pan sub-assembly by prying between the oil pan sub-assembly and cylinder block sub-assembly with a screwdriver.

➡Be careful not to damage the contact surfaces of the cylinder block and oil pans.

➡Tape the screwdriver tip before use.

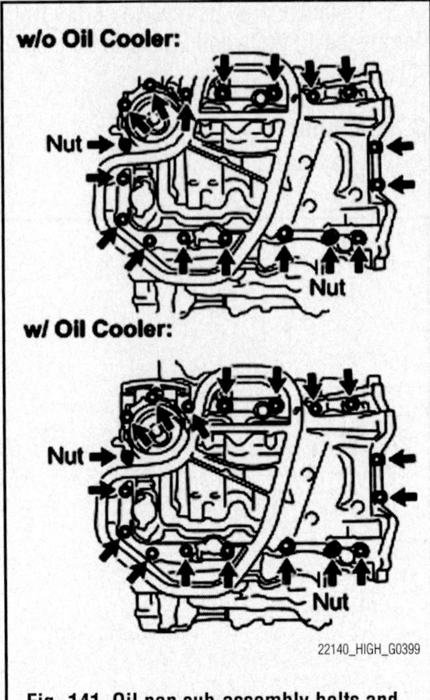

Fig. 141 Oil pan sub-assembly bolts and nuts—3.5L engine

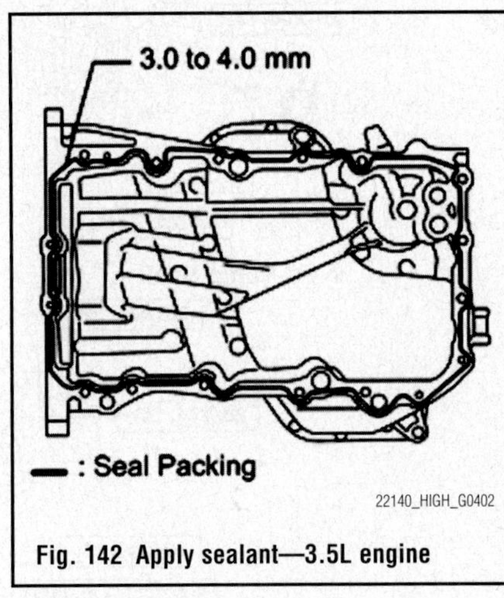

Fig. 142 Apply sealant—3.5L engine

   c. Remove the 2 O-rings.
   d. Using a Torx® socket wrench E8, remove the 2 stud bolts.

To install:

6. Install the oil pan sub-assembly, as follows:
   a. When replacing a stud bolt, install it by using an E8 Torx® socket wrench. Tighten to 7 ft. lbs. (10 Nm).
   b. Install 2 new O-rings.
   c. Apply seal packing in a continuous line as shown in the illustration. Seal packing: Toyota Genuine Seal Packing Black, Three Bond 1207B or equivalent. Seal diameter: 3.0 to 4.0 mm (0.118 to .0156 inches).

➡Remove any oil from the contact surface.

Timing Chain Cover:
16 mm (0.63 in.)
27 mm (1.06 in.)
9 mm (0.35 in.)
22140_HIGH_G0392

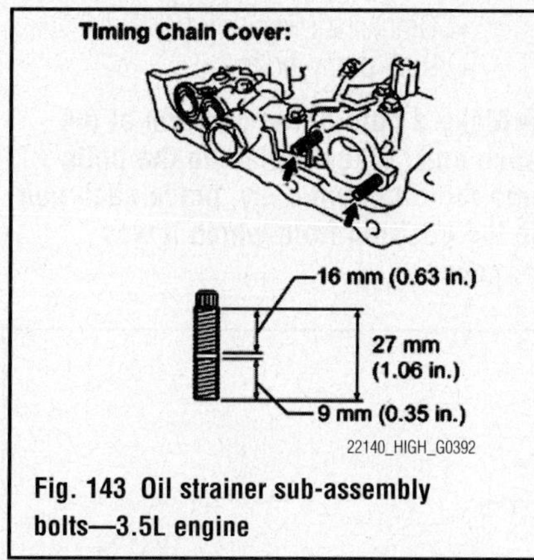

Fig. 143 Oil strainer sub-assembly bolts—3.5L engine

➡Install the oil pan within 3 minutes after applying seal packing.

➡Do not start the engine for at least 2 hours after installing.

d. Install the oil pan with the 16 bolts and 2 nuts and tighten to 7 ft. lbs. (10 Nm), 15 ft. lbs. (21 Nm).

7. Install the oil strainer sub-assembly, as follows:

a. Using an E6 Torx® socket, install the stud bolts as shown in the illustration and tighten to 35 inch lbs. (4 Nm).

b. Install a new gasket and the oil strainer sub-assembly with the bolt and 2 nuts and tighten to 7 ft. lbs. (10 Nm).

9 mm (0.35 in.)
20 mm (0.79 in.)
9 mm (0.35 in.)
E: Vapor feed hose
F: Union to check valve hose
G: No. 1 ventilation hose
H: Vacuum hose
22140_HIGH_G0390

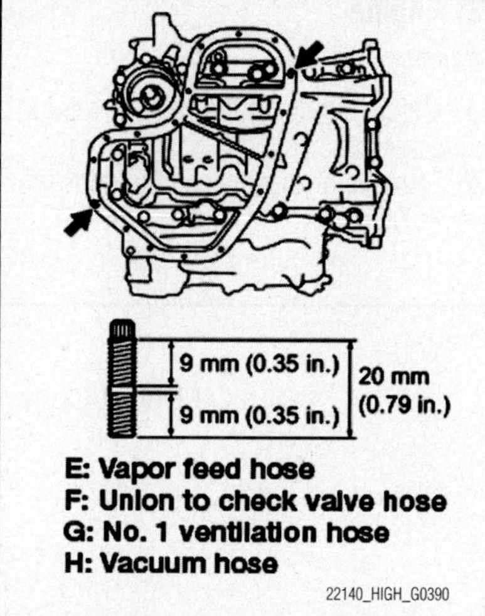

Fig. 144 No. 2 oil pan sub-assembly—3.5L engine

8. Install the No. 2 oil pan sub-assembly, as follows:

a. Using an E6 Torx® socket, install the stud bolts as shown in the illustration and tighten to 35 inch lbs. (4 Nm).

b. Apply seal packing in a continuous line as shown in the illustration. Seal packing: Toyota Genuine Seal Packing Black, Three Bond 1207B or equivalent. Seal diameter: 3.0 to 4.0 mm (0.118 to .0156 inches).

3.0 to 4.0 mm
: Seal Packing
22140_HIGH_G0391

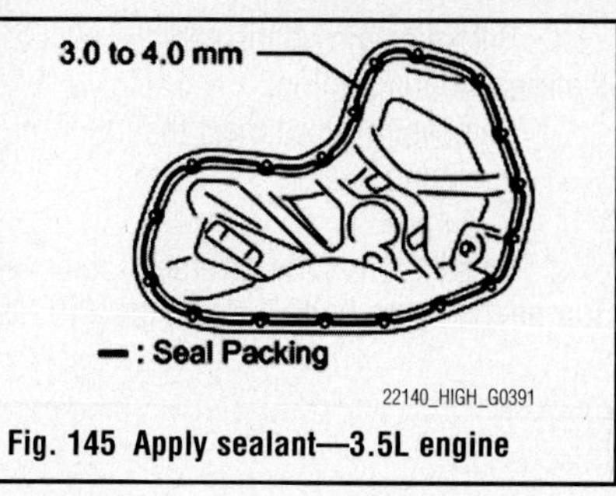

Fig. 145 Apply sealant—3.5L engine

➡Remove any oil from the contact surface.

➡Install the No. 2 oil pan sub-assembly within 3 minutes after applying seal packing.

➡Do not start the engine for at least 2 hours after installing.

c. Install the No. 2 oil pan sub-assembly with the 16 bolts and 2 nuts and tighten to 7 ft. lbs. (10 Nm).

9. Install a new oil pan drain plug gasket and the oil pan drain plug and tighten to 30 ft. lbs. (40 Nm).

OIL PUMP

REMOVAL & INSTALLATION

2.4L Engine

See Figure 146.

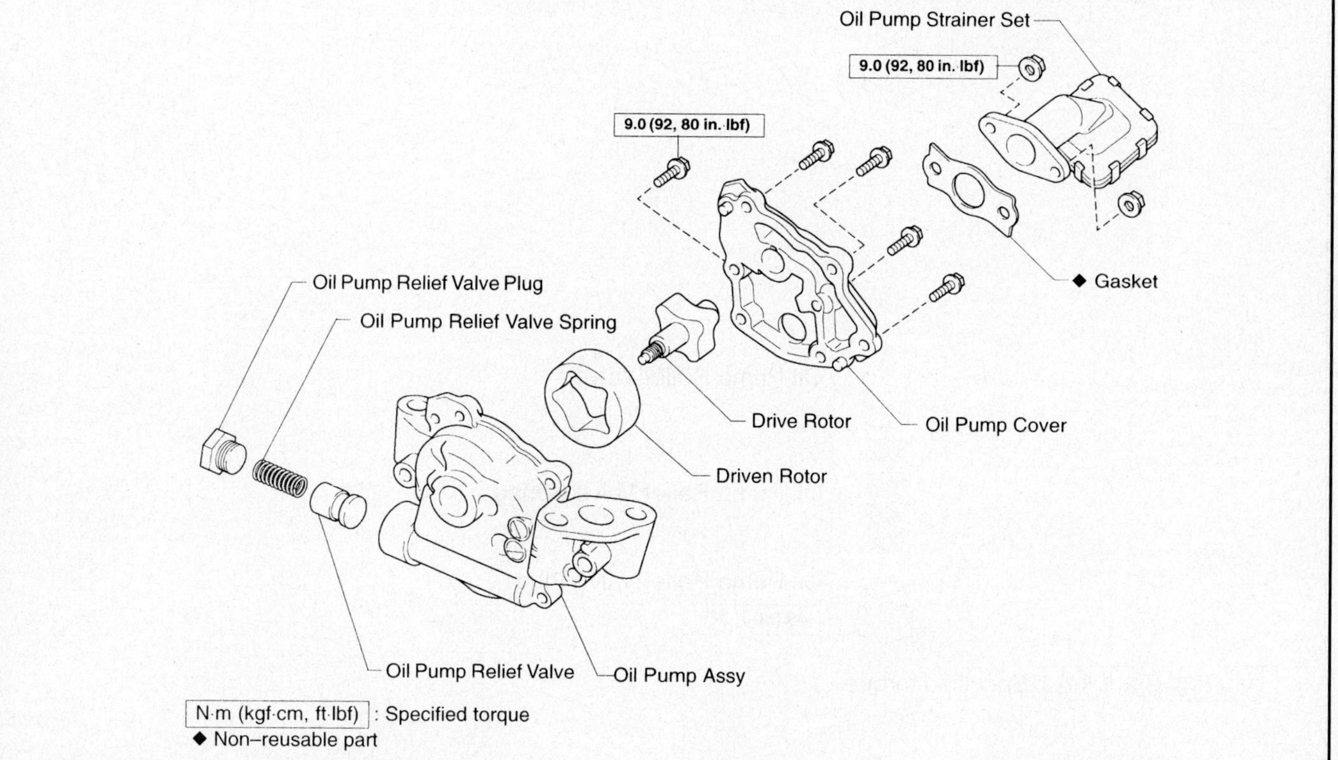

Oil Pump Strainer Set
9.0 (92, 80 in. lbf)
9.0 (92, 80 in. lbf)
◆ Gasket
Oil Pump Relief Valve Plug
Oil Pump Relief Valve Spring
Drive Rotor
Oil Pump Cover
Driven Rotor
Oil Pump Relief Valve
Oil Pump Assy
N·m (kgf·cm, ft·lbf) : Specified torque
◆ Non–reusable part
67170-HIGH-G24

Fig. 146 Oil pump exploded view—2.4L engine

1. Before servicing the vehicle, refer to the precautions section.
2. Remove or disconnect the following:
   - Timing chain
   - Oil pump
3. Installation is the reverse of removal. Torque the pump bolts to 14 ft. lbs. (19 Nm).

3.3L Engine

See Figure 147.

1. Before servicing the vehicle, refer to the precautions section.
2. Remove or disconnect the following:
   - Oil pan
   - Crankshaft Position (CKP) sensor
   - 9 oil pump bolts

➡Make a note of the position of the each bolt. When replacing the bolts into the oil pump body, place each bolt in the position from which it was removed.

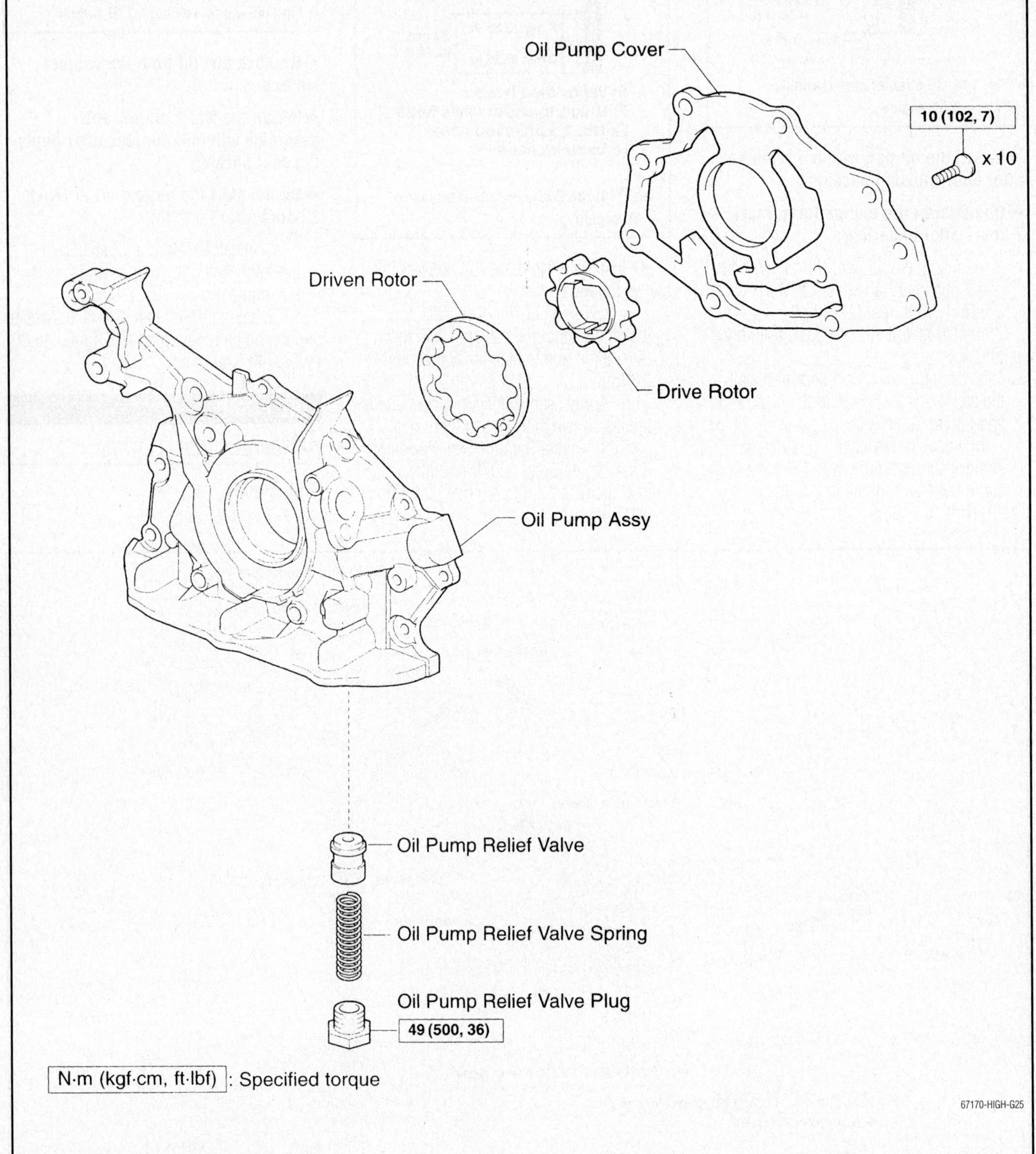

Fig. 147 Oil pump exploded view—3.3L engine

- Oil pump body, by prying between the oil pump and main bearing cap
- O-ring from the cylinder block
- Plug, gasket, spring and relief valve from the oil pump body
- 9 screws, pump body cover, drive and driven rotors

To install:

3. Install or connect the following:
   - Driven rotors, drive, pump body cover, using the 9 screws
   - Oil pump relief valve, spring, gasket and the plug to the oil pump body
   - New O-ring on the cylinder block
4. Using a non residue solvent, clean both sealing surfaces to the oil pump.
5. Apply liquid sealant to the oil pump and engine block.
6. Install or connect the following:
   - Oil pump

➡Be sure to engage the splined teeth of the oil pump drive gear with the large teeth of the crankshaft.

   - 9 oil pump bolts. Tighten the bolts in several passes to 69 inch lbs. (8 Nm), for 10mm; 14 ft. lbs. (20 Nm), for 12mm; 32 ft. lbs. (43 Nm) for 14mm
   - CKP sensor. Tighten the bolt to 69 inch lbs. (8 Nm).
   - Baffle plate to the No. oil pan. Tighten to 69 inch lbs. (8 Nm).
   - No. 1 oil pan, oil strainer and No. 2 oil pan
7. Refill the engine with oil.
8. Start the engine and inspect for leaks.
9. Recheck the engine oil level.

3.5L Engine

See Figures 148 through 157.

1. Remove the engine assembly with transaxle.
2. Secure the engine.

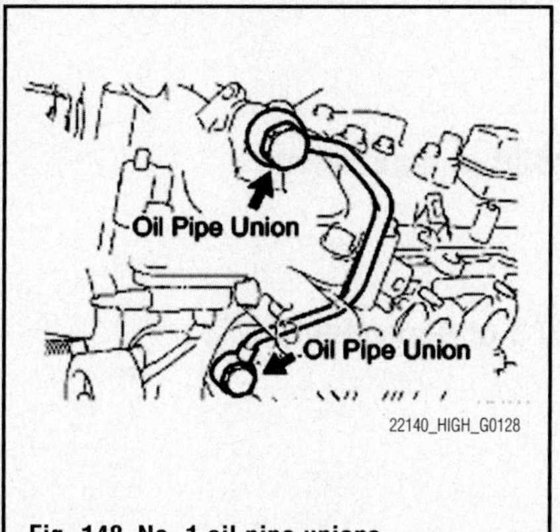

Fig. 148 No. 1 oil pipe unions

3. Remove the No. 1 oil pipe, as follows:
   a. Remove the 2 oil pipe unions, gaskets and No. 1 oil pipe.
   b. Remove the left oil control valve filter and gaskets.
4. Remove the oil pipe, as follows:
   a. Remove the bolt.
   b. Remove the 2 oil pipe unions and oil pipe.

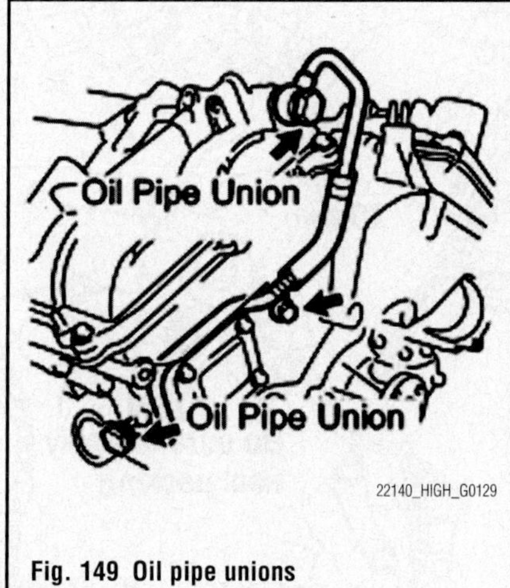

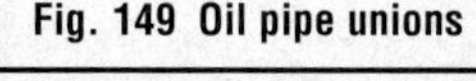

Fig. 149 Oil pipe unions

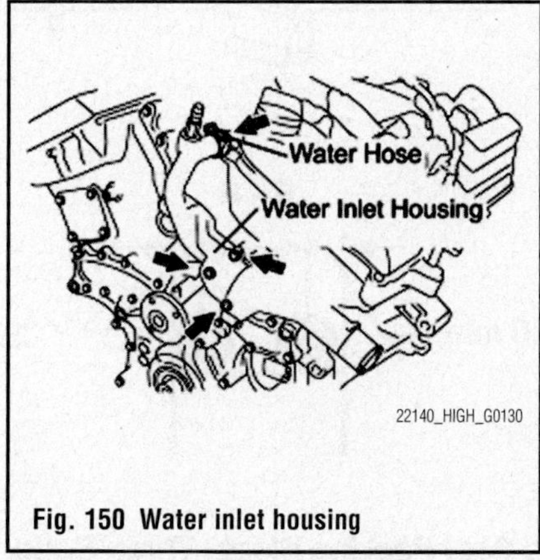

Fig. 150 Water inlet housing

   c. Remove the right oil control valve filter and gaskets.
5. Using a special service tool (SST: 09213-70011, SST: 09213-70011 or equivalent), loosen the crankshaft pulley bolt.
6. Using SST: 09950-50013 or equivalent, remove the crankshaft pulley bolt and crankshaft pulley.
7. Separate the oil cooler pipe (w/ oil cooler), as follows:

Fig. 151 Water inlet housing gasket and water outlet pipe O-ring

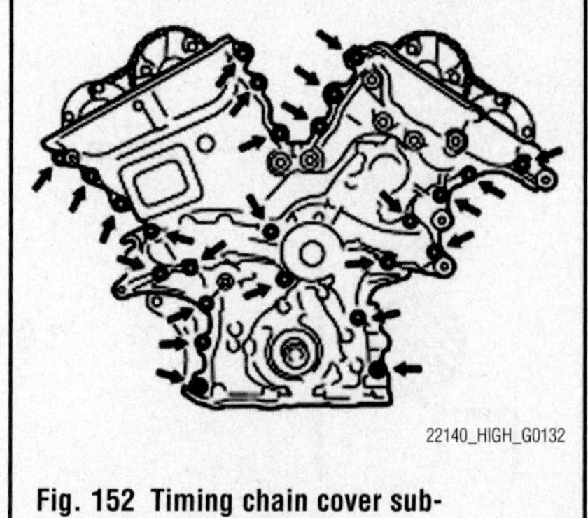

Fig. 152 Timing chain cover sub-assembly bolts and nuts

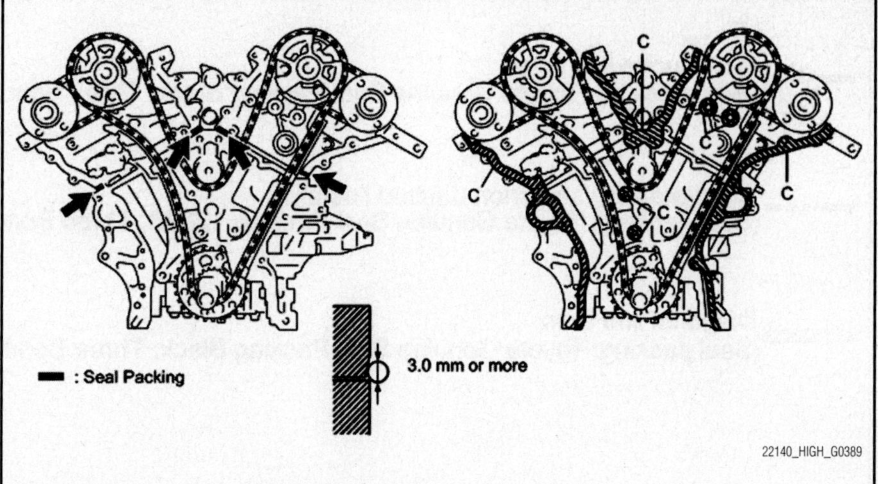

Fig. 153 Engine unit seal packing—3.5L engine

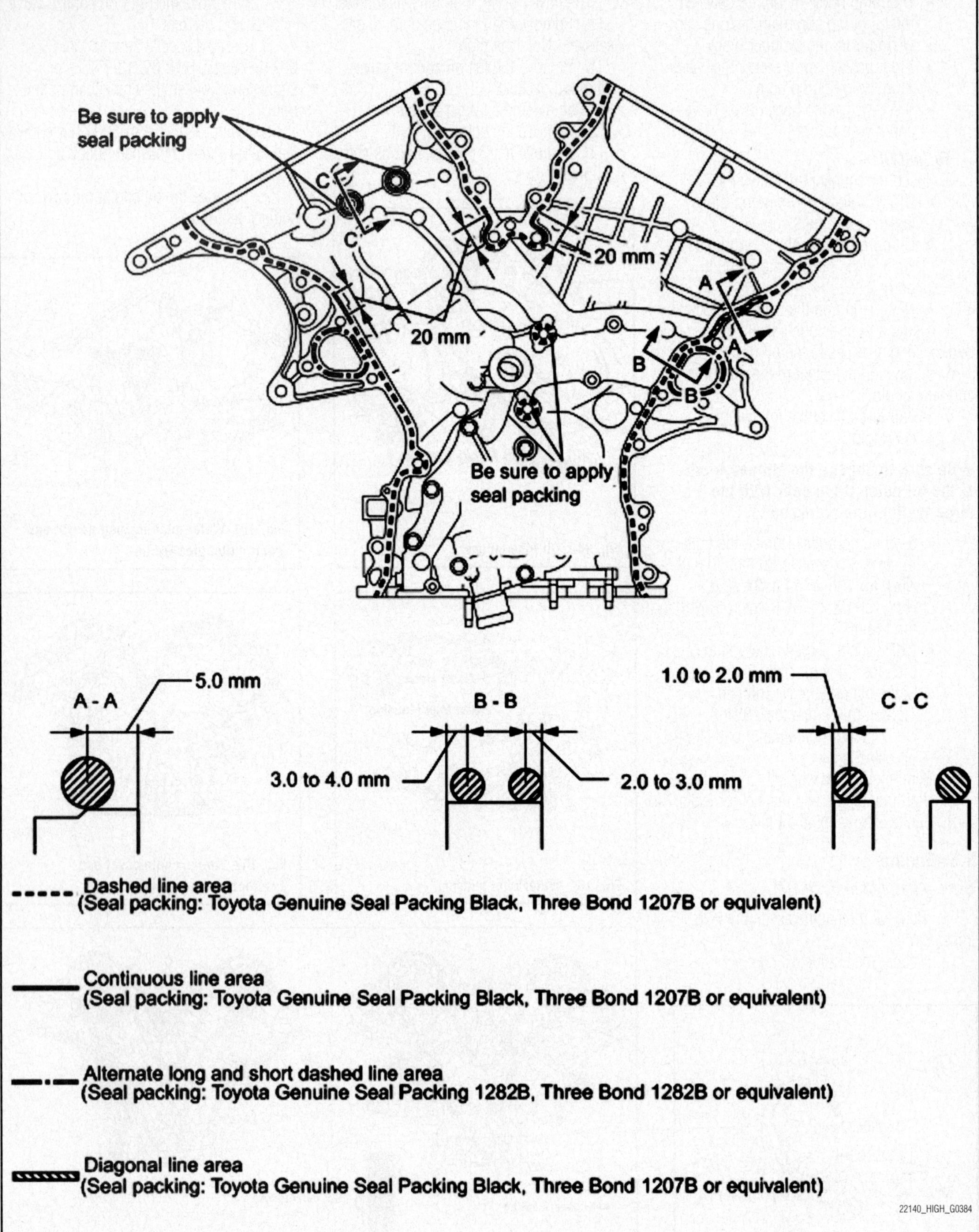

Fig. 154 Timing chain cover seal packing—3.5L engine

a. Remove the bolt and 2 nuts, and disconnect the oil cooler pipe from the oil pan sub-assembly.

b. Remove the gasket from the oil pan sub-assembly.

8. Remove the water inlet housing, as follows:

a. Disconnect the water hose.

b. Remove the 2 bolts, nut and water inlet housing.

c. Remove the water inlet housing gasket and water outlet pipe O-ring.

9. Remove the Bank 1 and Bank 2 cylinder head cover sub-assemblies.

10. Remove the No. 2 oil pan sub-assembly.

11. Remove the oil strainer sub-assembly.

12. Remove the timing chain cover sub-assembly, as follows:

a. Remove the 23 bolts and 2 nuts as shown in the illustration.

b. Remove the timing chain cover by prying between the timing chain cover and cylinder head or cylinder block with a screwdriver.

➡Be careful not to damage the contact surfaces of the cylinder head, cylinder block and chain cover.

➡Tape the screwdriver tip before use.

c. Remove the gasket.

13. Using a screwdriver, pry out and remove the timing chain oil seal.

➡Tape the screwdriver tip before use.

To install:

14. Install the timing chain case oil seal, as follows:

a. Using SST (SST: 09223-22010, SST: 09506-35010) tap in a new oil seal until its surface is flush with the timing chain case edge.

➡Keep the lip free from foreign matter.

➡Make sure that the oil seal edge does not stick out of the timing chain case.

➡Do not tap on the oil seal at an angle.

15. Install the timing chain cover sub-assembly, as follows:

a. Apply seal packing in a continuous line to the engine unit as shown in the illustration. Seal packing: Toyota Seal Packing Black, Three Bond 1207B or equivalent. Seal diameter: 3.0 mm (0.118 in.).

➡Be sure to clean and degrease the contact surfaces, especially the surfaces indicated by C in the illustration.

➡If the contact surfaces are wet, wipe them with an oil-free cloth before applying seal packing.

➡Install the chain cover sub-assembly within 3 minutes after applying seal packing.

➡Do not start the engine for at least 2 hours after installing the chain cover sub-assembly.

b. Apply seal packing in a continuous line to the timing chain cover as shown in the illustration. Seal packing: Toyota Seal Packing Black, Three Bond 1207B, Three Bond 1282B or equivalent.

Area	Seal Packing Diameter	Application Position from Inside Seal Line
Continuous Line Area	4.5 mm (0.177 in.) or more	3.0 to 4.0 mm (0.118 to 0.158 in.)
Alternate Long and Short Dashed Line Area	3.5 mm (0.138 in.) or more	2.0 to 3.0 mm (0.079 to 0.118 in.)
Dashed Line Area	3.5 mm (0.138 in.) or more	3.0 to 4.0 mm (0.118 to 0.158 in.)
Diagonal Line Area	6.0 mm (0.236 in.) or more	5.0 mm (0.197 in.)

22140_HIGH_G0380

Fig. 155 Seal packing specifications—3.5L engine

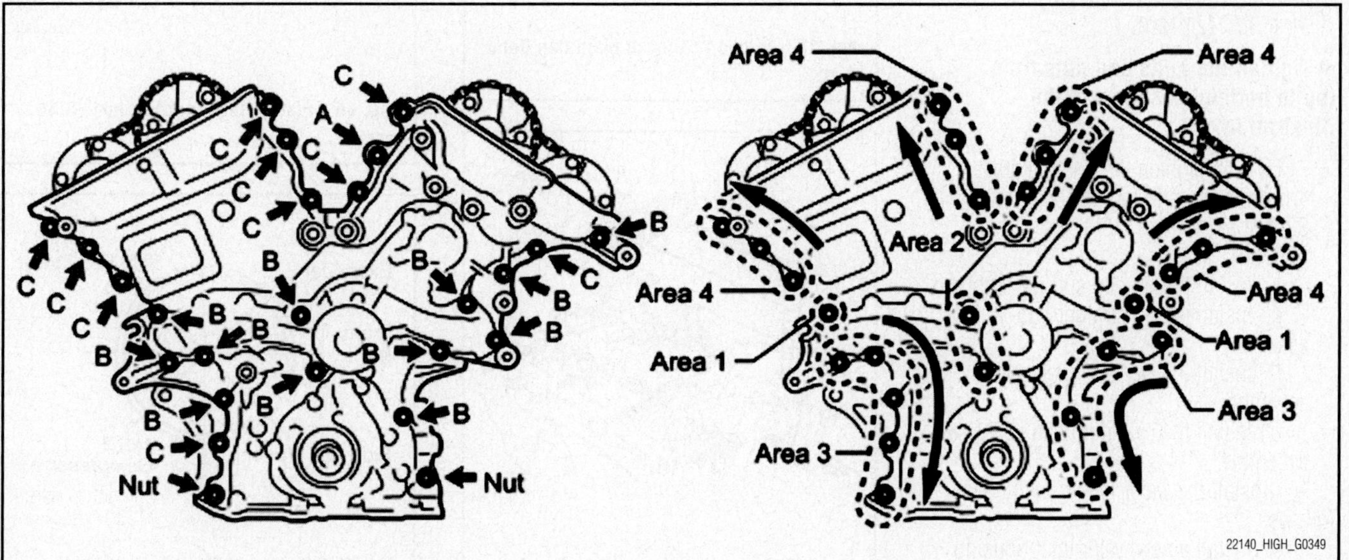

Fig. 156 Timing chain cover tightening sequence

Item	Length
Bolt A	40 mm (1.57 in.)
Bolt B	55 mm (2.17 in.)
Bolt C	25 mm (0.98 in.)

22140_HIGH_G0133

Fig. 157 Timing chain cover bolt length

➡If the contact surfaces are wet, wipe them with an oil-free cloth before applying seal packing.

➡Install the chain cover sub-assembly within 3 minutes and tighten the bolts within 15 minutes after applying seal packing.

➡Do not start the engine for at least 2 hours after installing the chain cover sub-assembly.

c. Apply seal packing as follows:
d. Install a new gasket.
e. Align the oil pump's drive rotor spline and the crankshaft as shown in the illustration. Install the spline and chain cover to the crankshaft.
f. Temporarily tighten the timing chain cover with the 23 bolts and 2 nuts.

➡Make sure that there is no oil on the bolt threads.

g. Fully tighten the bolts in area 1 and area 2 (from top to bottom as shown) to 15 ft. lbs. (21 Nm).
h. Fully tighten the bolts and nuts in area 3 (from top to bottom as shown) to 15 ft. lbs. (21 Nm).

➡Tighten the bolts and nuts from top to bottom as shown in the illustration.

i. Fully tighten the bolts in area 4 (from bottom to top as shown) to 32 ft. lbs. (43 Nm), 15 ft. lbs. (21 Nm).
j. Install the oil pan sub-assembly.
k. Install the oil strainer sub-assembly.
l. Install the No. 2 oil pan sub-assembly.
m. Install the cylinder head cover sub-assembly.

16. Install the water inlet housing, as follows:
a. Install a new water inlet housing No. 1 gasket and water outlet pipe O-ring.
b. Install the water inlet housing with the 2 bolts and nut and tighten to 7 ft. lbs. (10 Nm).

➡Be careful not to allow the O-ring to get caught between the parts.

c. Connect the water hose.

17. Install the oil cooler pipe (w/oil cooler), as follows:
a. Install a new gasket to the oil pan sub-assembly.
b. Install the oil cooler pipe with the bolt and 2 nuts and tighten to 15 ft. lbs. (21 Nm).

18. Install the crankshaft pulley.
19. Install oil pipe.
20. Install the No. 1 oil pipe.
21. Install the engine hangers.
22. Remove engine stand.
23. Install engine assembly with transaxle.

MAIN BEARING TORQUE SEQUENCE

See Figures 158 through 160.

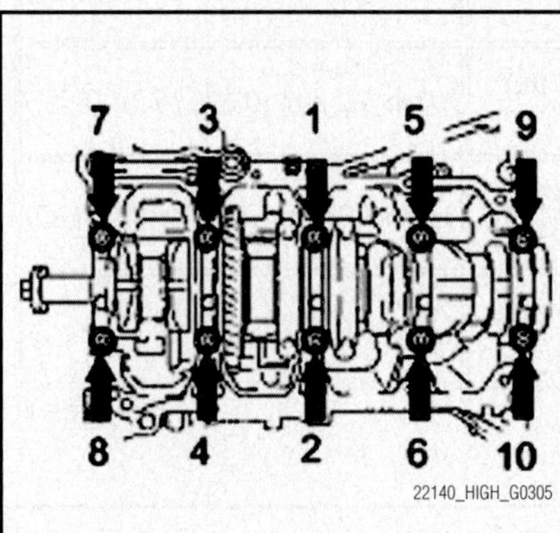

Fig. 158 Torque Sequence Main Cap Bolts 2.4L

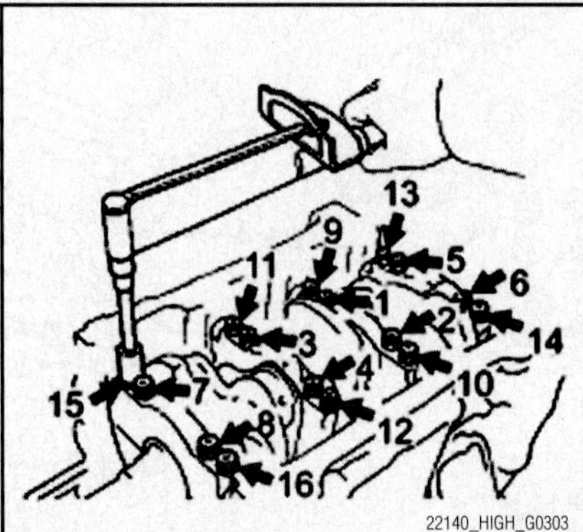

Fig. 159 Torque Sequence Main Cap Bolts 3.3L and 3.5L

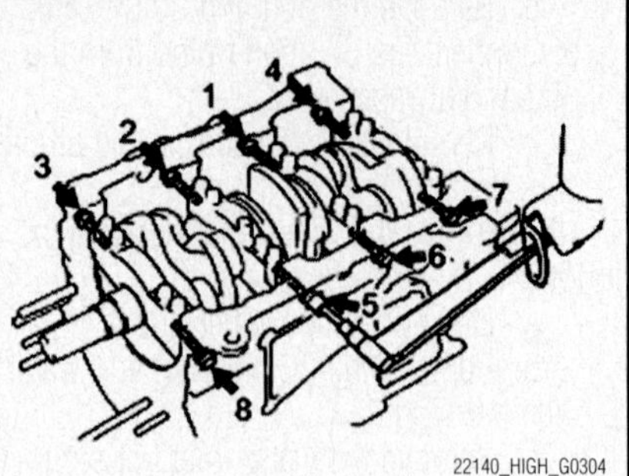

Fig. 160 Torque Sequence Side Bolts 3.3L and 3.5L

PISTON AND RING

POSITIONING

See Figures 161 through 166.

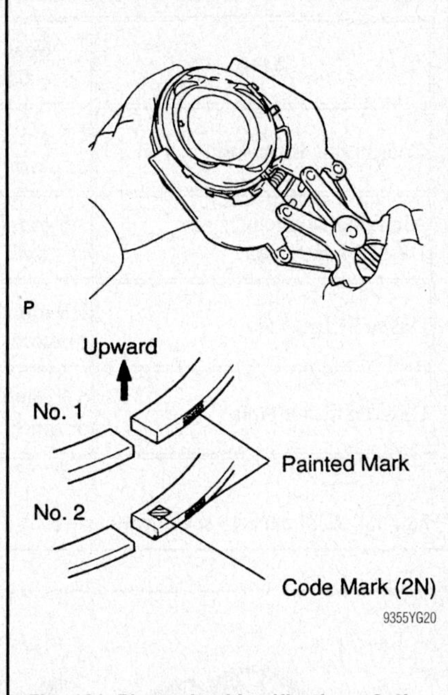

Fig. 161 Piston ring identification—2.4L engine

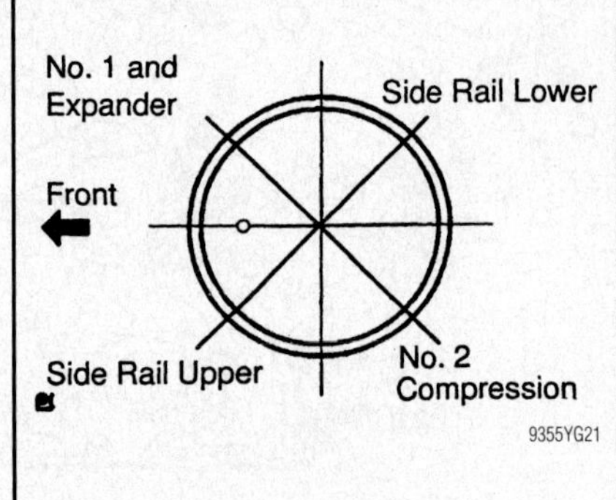

Fig. 162 Piston ring installation—2.4L engine

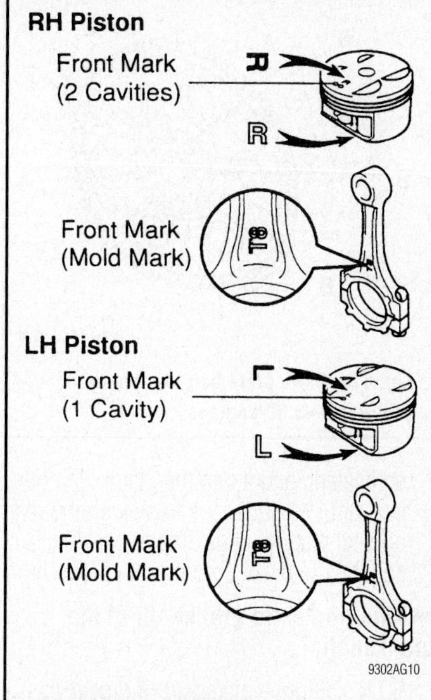

Fig. 163 Piston/connecting rod-to-engine positioning–3.3L engine

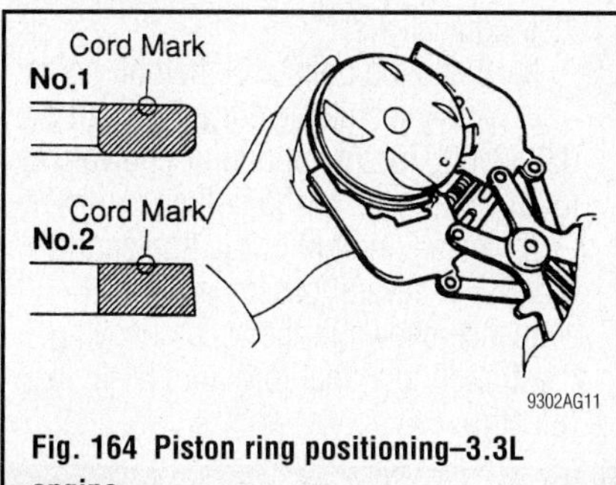

Fig. 164 Piston ring positioning–3.3L engine

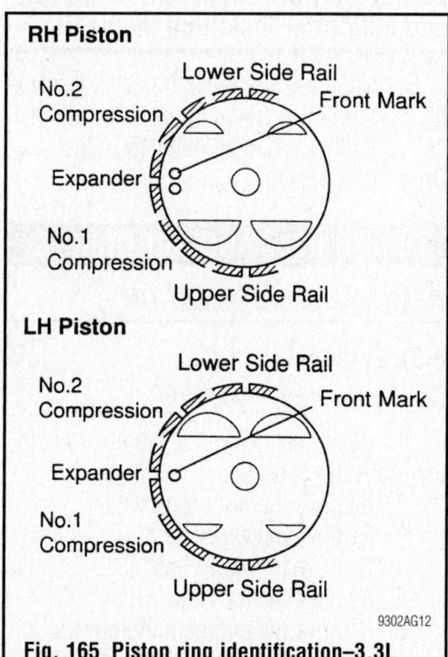

Fig. 165 Piston ring identification–3.3L engine

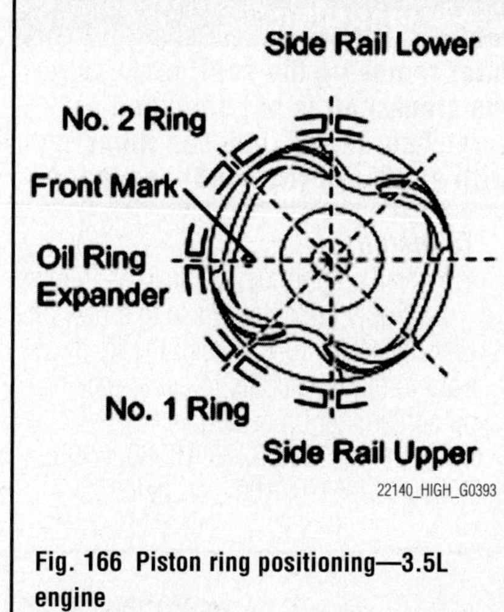

Fig. 166 Piston ring positioning—3.5L engine

REAR MAIN SEAL

REMOVAL & INSTALLATION

2.4L Engine

See Figures 167 through 170.

➡This procedure requires a variety of special tools.

1. Remove the automatic transaxle assembly, as outlined in the Drive Train Section.

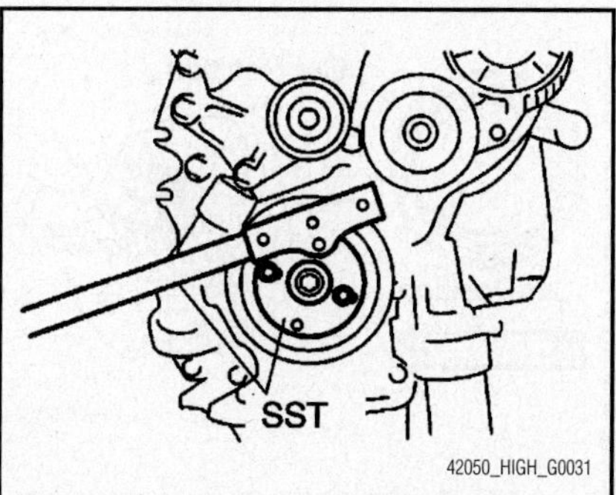

Fig. 167 Hold the crankshaft in place with the special tools as shown

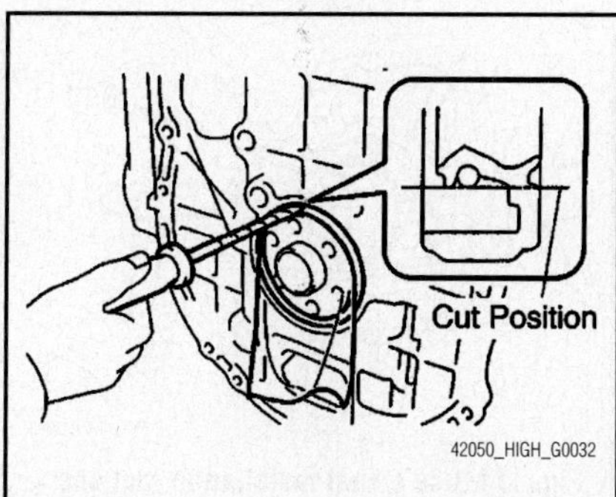

Fig. 168 If using a screwdriver wrapped in tape to remove the rear main seal, be very carefully not to damage the crankshaft

2. Remove the drive plate and ring gear, as follows:
   a. Secure the crankshaft with Special Service Tool (SST) 09213-54015 (91651-60855), 09330-00021 or their equivalents.
   b. Remove the 8 bolts, rear spacer, drive plate and front spacer.
3. Remove the rear main seal, as follows:
   a. Using a knife, carefully cut off the oil seal lip.
   b. Use a suitable prytool, or taped screwdriver, pry out the oil seal.

⁂ WARNING

After removing the seal, make sure the crankshaft is not damaged or scratched. If it is, you can mend it with a fine grit (No. 400) sandpaper.

To install:

4. Install engine rear oil seal, as follows:
   a. Apply suitable grease to a new oil seal lip. Make sure to keep the lip away from foreign materials to avoid picking up contamination or debris.

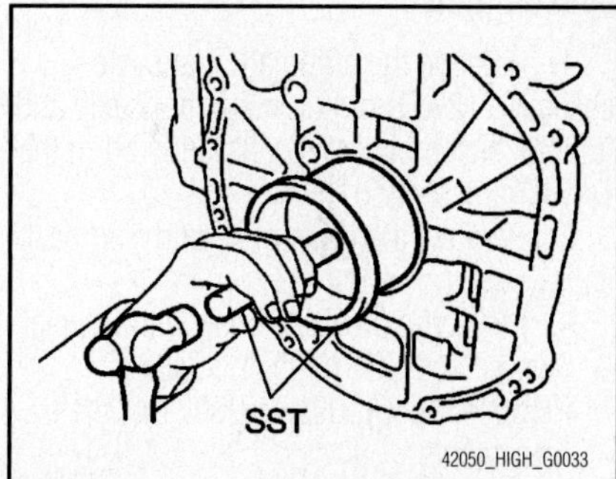

Fig. 169 Use a seal installation tool and hammer to install the rear main seal until its surface is flush with the seal retainer edge

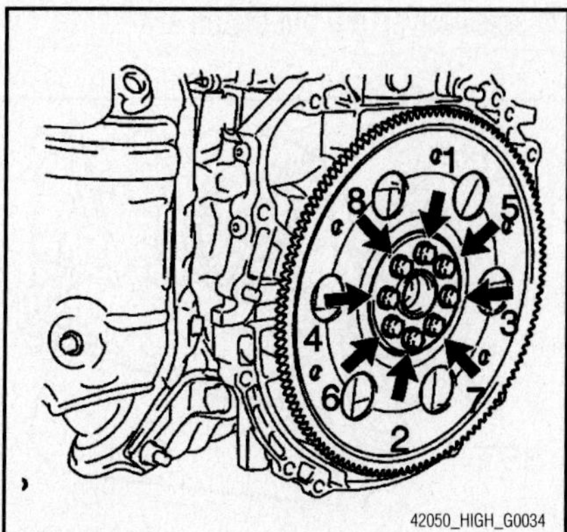

Fig. 170 Drive plate bolt tightening sequence—2.4L engine

b. Using SST 09223-15030, 09950-70010 (09951-07100), or their equivalent, and a hammer, tap in the oil seal until its surface is flush with the rear oil seal retainer edge.

➡**Wipe the extra grease off of the crankshaft.**

5. Install drive plate and ring gear, as follows:

a. Fix the crankshaft with SST 09213-54015 (91651-60855), 09330-00021, or their equivalents.

b. Clean the bolts and the bolt holes.

c. Apply a suitable adhesive (part no. 08833-00070, three bond or equivalent) to 2 or 3 threads of the bolt end.

d. Install and uniformly tighten the 8 bolts, in several passes in the sequence shown in the accompanying illustration to a final torque of 72 ft. lbs. (98 Nm).

6. Install the automatic transaxle, as outlined in the Drive Train Section.

3.3L and 3.5L Engines

See Figures 171 through 175.

➡**This procedure requires a variety of special tools.**

1. Remove the automatic transaxle assembly (2WD) or automatic transaxle and transfer assembly (4WD), as outlined in the Drive Train Section.

2. Remove the drive plate and ring gear, as follows:

a. Secure the crankshaft with Special Service Tool (SST) 09213-54015 (91651-60855), 09330-00021 or their equivalents.

b. Remove the 8 bolts, rear spacer, drive plate and front spacer.

3. Remove the rear main seal, as follows:

a. Using a knife, carefully cut off the oil seal lip.

b. Use a suitable prytool, or taped screwdriver, pry out the oil seal.

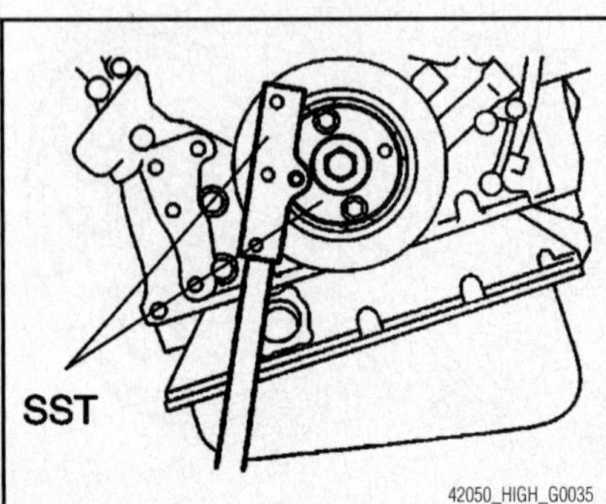

Fig. 171 Hold the crankshaft in place with the special tools as shown

⁂ WARNING

After removing the seal, make sure the crankshaft is not damaged or scratched. If it is, you can mend it with a fine grit (No. 400) sandpaper.

To install:

4. Install engine rear oil seal, as follows:

a. Apply suitable grease to a new oil seal lip. Make sure to keep the lip away from foreign materials to avoid picking up contamination or debris.

b. Using SST 09223-15030, 09950-70010 (09951-07100), or their equivalent, and a hammer, tap in the oil seal until its surface is flush with the rear oil seal retainer edge. Tap the seal in squarely to make sure it seats properly

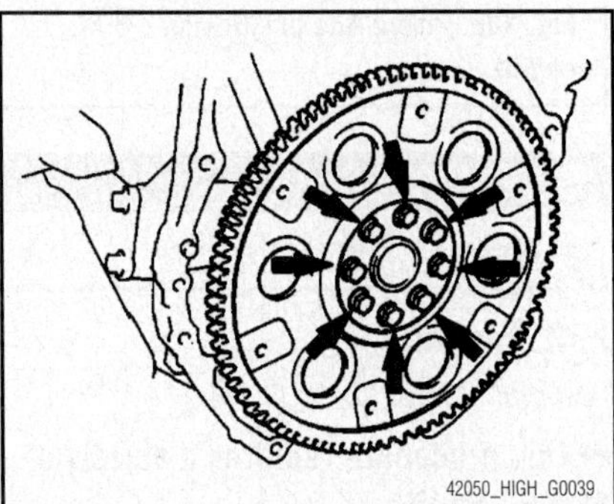

Fig. 172 Remove the 8 bolts, rear spacer, drive plate and front spacer

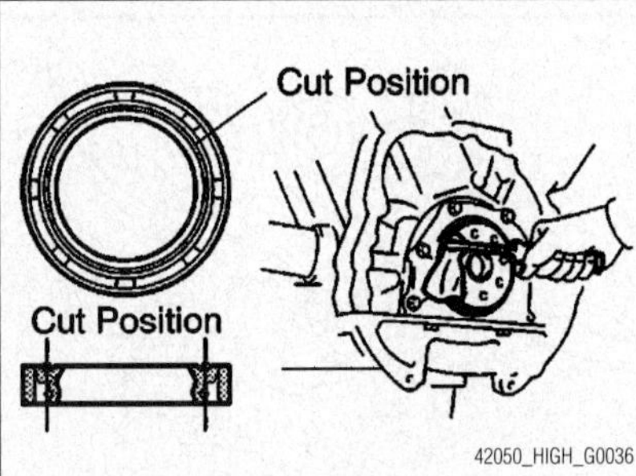

Fig. 173 If using a screwdriver wrapped in tape to remove the rear main seal, be very carefully not to damage the crankshaft

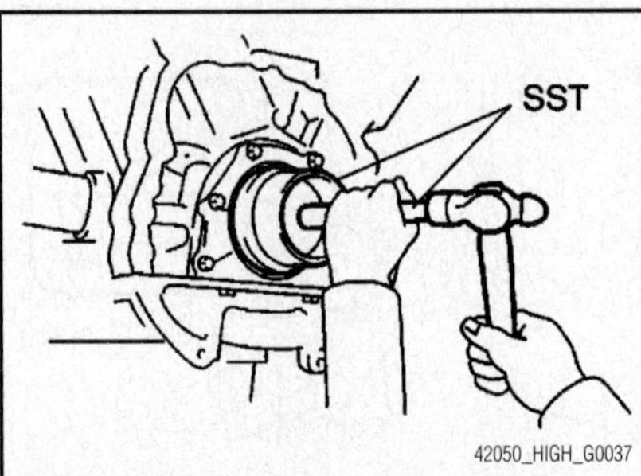
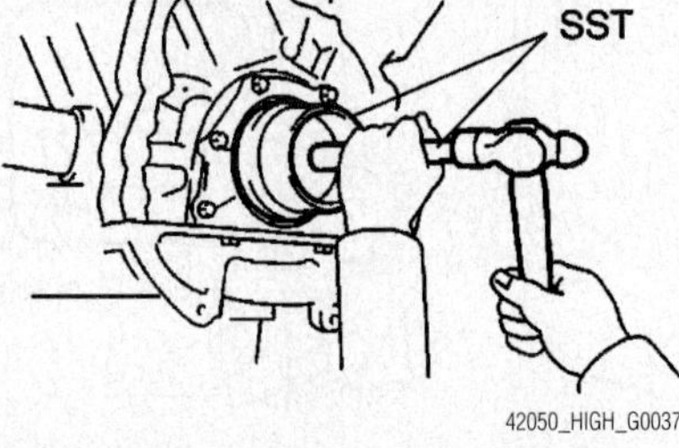

Fig. 174 Use a seal installation tool and hammer to install the rear main seal until its surface is flush with the seal retainer edge

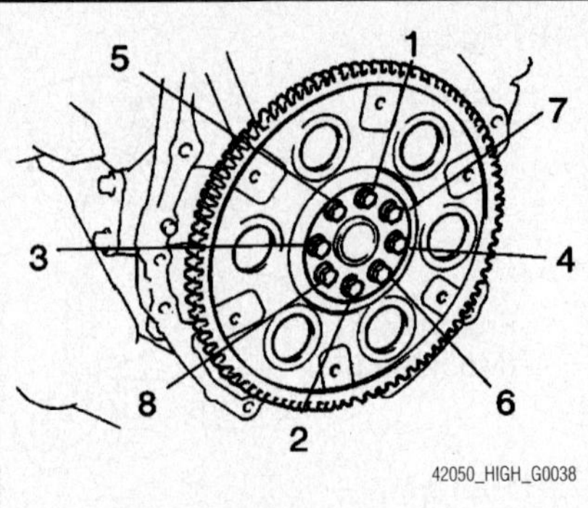

Fig. 175 Drive plate bolt tightening sequence—3.3L engines

➡**Wipe the extra grease off of the crankshaft.**

5. Install drive plate and ring gear, as follows:

a. Fix the crankshaft with SST 09213-54015 (91651-60855), 09330-00021, or their equivalents.

b. Clean the bolts and the bolt holes.

c. Apply a suitable adhesive (part no. 08833-00070, three bond or equivalent) to 2 or 3 threads of the bolt end.

d. Install and uniformly tighten the 8 bolts, in several passes in the sequence shown in the accompanying illustration to a final torque of 61 ft. lbs. (83 Nm).

⁂ WARNING

Do not start the engine for AT LEAST one hour after installing the seal!

6. Install the automatic transaxle assembly (2WD) or automatic transaxle and transfer assembly (4WD), as outlined in the Drive Train Section.

TIMING BELT AND SPROCKETS

REMOVAL & INSTALLATION

3.3L Engine

See Figures 176 through 185.

1. Before servicing the vehicle, refer to the precautions section.
2. Remove the right front wheel.
3. Remove the wiper arms.
4. Remove the wiper linkage.
5. Remove the top cowl panel.
6. Remove the engine undercovers.
7. Remove the right front fender apron seal.

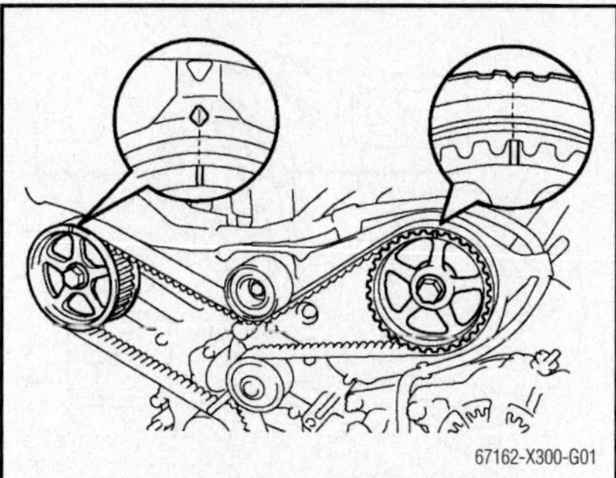

Fig. 176 Check that the timing marks on the camshaft pulleys are aligned with the notches on the inner belt cover

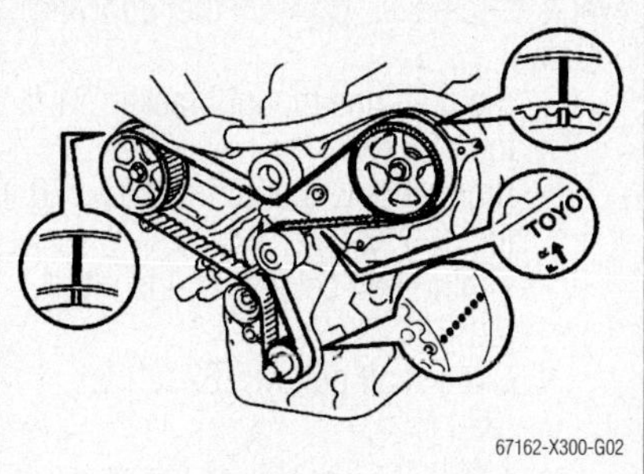

Fig. 178 If the timing belt is re-used, check that the 3 original installation marks are visible on the belt as shown

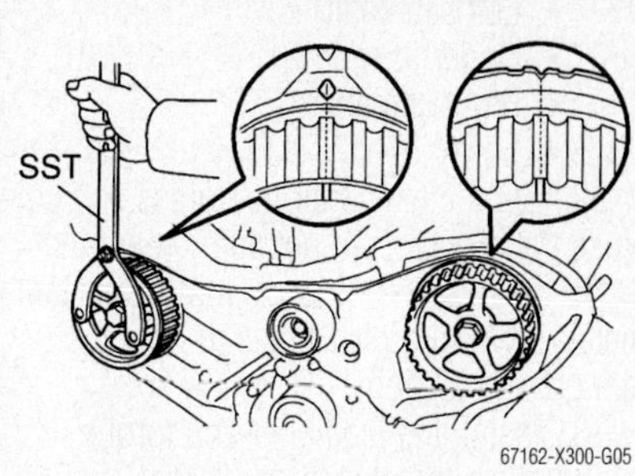

Fig. 181 Turn the camshaft pulleys back into alignment so the marks align with the notches on the inner cover

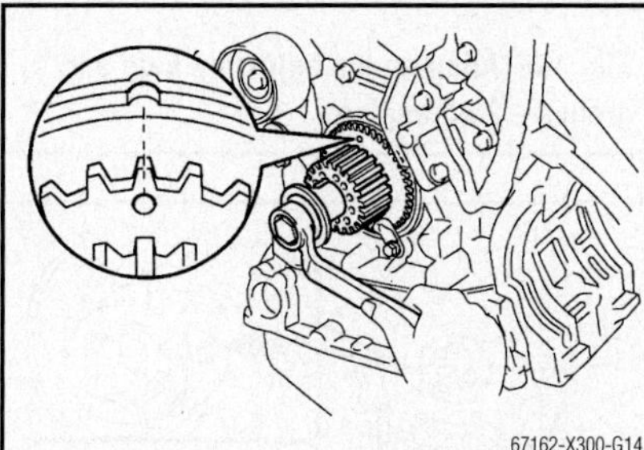

Fig. 177 Turn the crankshaft clockwise to align the timing mark on the crankshaft timing pulley with the notch in the oil pump body

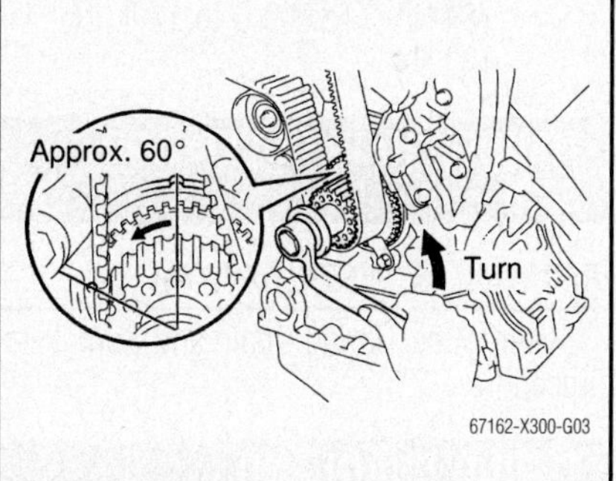

Fig. 179 Turn the crankshaft counterclockwise by 60 degrees

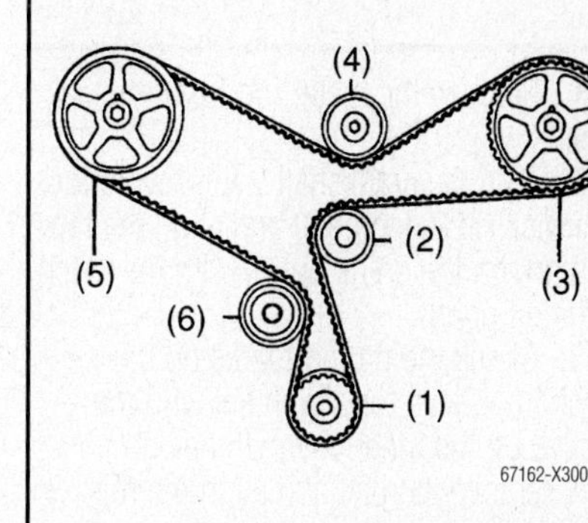

Fig. 182 Install the belt in this order

8. Remove the A/C compressor drive belt.
9. Remove the power steering pump belt.
10. Remove the engine roll control rod.
11. Remove the right side engine mount stay.
12. Remove the alternator bracket.
13. Remove the crankshaft pulley.
14. Remove the upper belt cover.
15. Remove the right engine mount bracket.
16. Remove the no. 2 timing belt guide.
17. Set the no. 1 cylinder to TDC compression.
18. Temporarily install the crank pulley bolt. Turn the crankshaft clockwise to align the timing mark on the crankshaft timing pulley with the notch in the oil pump body. Check that the timing marks on the camshaft pulleys are aligned with the notches on the inner belt cover. If not, rotate the crankshaft 360 degrees clockwise.

➡If the timing belt is re-used, check that the 3 original installation marks are visible on the belt as shown. If not, paint three new marks on the belt.

19. Turn the crankshaft counterclockwise by 60 degrees. Make sure that the belt is still engaged.

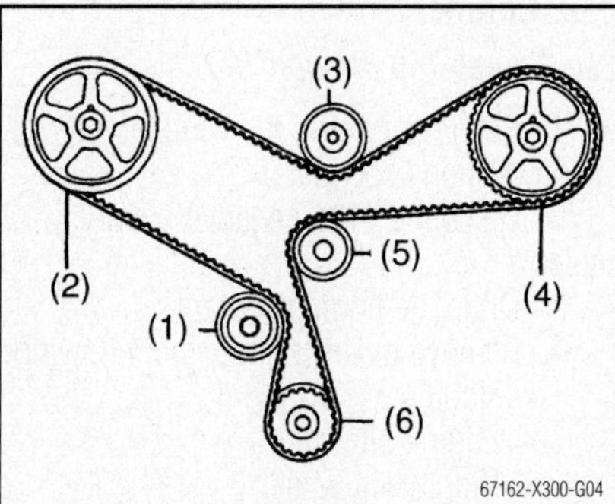

Fig. 180 Remove the belt from the pulleys in this order

20. Remove the tensioner.
21. Remove the belt from the pulleys in this order:
- Lower idler pulley
- Right camshaft pulley
- Upper idler pulley
- Left camshaft pulley
- Water pump pulley
- Crankshaft timing pulley

22. If the belt is being re-used, check it for wear or damage; don't twist it or turn it inside-out. If there is any doubt as to its condition, replace it.

To install:

23. Clean all the pulleys.

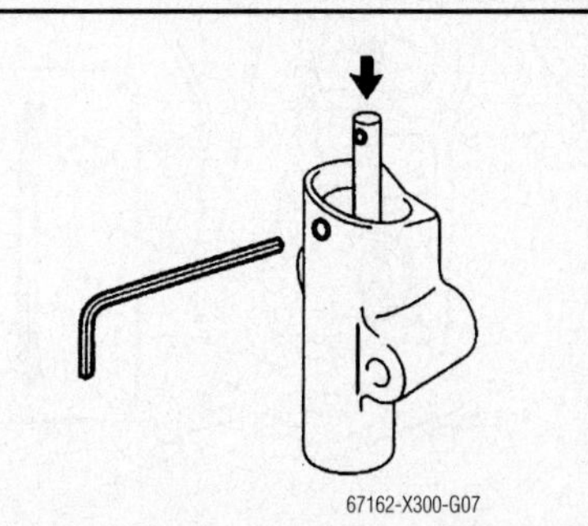

Fig. 183 Set the tensioner in a press and collapse the plunger. Do not apply more that 2,205 lbs (9.8 kN) of force. Insert a suitable metal rod through the holes to hold the plunger in position

24. Turn the crankshaft another 60 degrees counterclockwise.
25. Turn the camshaft pulleys back into alignment so the marks align with the notches on the inner cover.
26. Turn the crankshaft back so that the timing mark aligns with the notch on the oil pump.
27. Align the installation marks on the belt with the timing marks on the pulleys.
28. Install the belt in this order:
- Crankshaft
- Water pump

- Left camshaft
- Upper idler
- Right camshaft
- Lower idler

29. Set the tensioner in a press and collapse the plunger. Do not apply more that 2,205 lbs (9.8 kN) of force. Insert a suitable metal rod through the holes to hold the plunger in position.
30. Install the tensioner and torque the 2 bolts alternately to 20 ft. lbs. (27 Nm).

**** WARNING**

Be sure to tighten to bolts alternately and evenly so the tensioner seats flat.

31. Remove the metal rod from the tensioner.
32. Turn the crankshaft 2 full revolutions clockwise (720 degrees), and align the timing mark on the crank pulley with the notch on the oil pump.
33. Check the timing marks on the camshaft pulleys for alignment with the notches on the inner cover. If they do not align, remove the belt and align the mismatched mark(s).
34. The remainder of installation is the reverse of removal. Observe the following torques:
    - Right engine mount bracket: 21 ft. lbs. (28 Nm)
    - Right engine mount insulator: 70 ft. lbs. (95 Nm)
    - Timing belt covers: 75 inch lbs. (8.5 Nm)
    - Crankshaft pulley: 162 ft. lbs. (220 Nm)
    - Alternator bracket: 21 ft. lbs. (28 Nm)
    - Right engine mount stay: 47 ft. lbs. (64 Nm)
    - Engine roll control rod: tighten first A, then B, then C to 47 ft. lbs. (64 Nm). Torque D to 17 ft. lbs. (23 Nm)

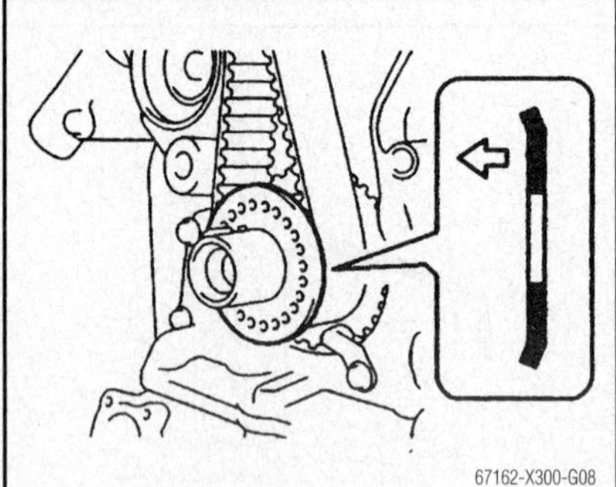

Fig. 184 Install the timing belt guide with the cupped side facing front

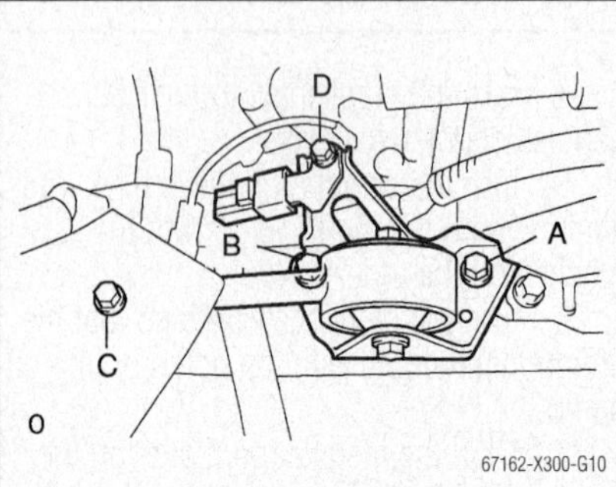

Fig. 185 Tighten the engine roll control rod bolts in this order

TIMING CHAIN COVER AND SEAL

REMOVAL & INSTALLATION

Refer to the Timing Chain and Sprocket procedure.

TIMING CHAIN AND SPROCKETS

REMOVAL & INSTALLATION

2.4L Engine

See Figures 186 through 194.

1. Before servicing the vehicle, refer to the precautions section.
2. Disconnect the negative battery cable.
3. Drain the engine coolant.
4. Remove or disconnect the following:
    - Hood
    - Engine oil
    - Right front wheel
    - Right fender splash shield
    - Right fender apron seal
    - No.1 engine undercover
    - Engine roll stopper and bracket
    - Exhaust pipe
    - Upper engine mount, right side
    - Accessory drive belts
    - Alternator
    - Power steering pump
5. Set the No.1 piston at TDC compression.
    - Crankshaft pulley
    - Oil pan
    - CKP sensor
    - Chain tensioner assembly No.1
    - V-belt tensioner
6. Take up the weight of the engine with a crane.
7. Remove or disconnect the following:

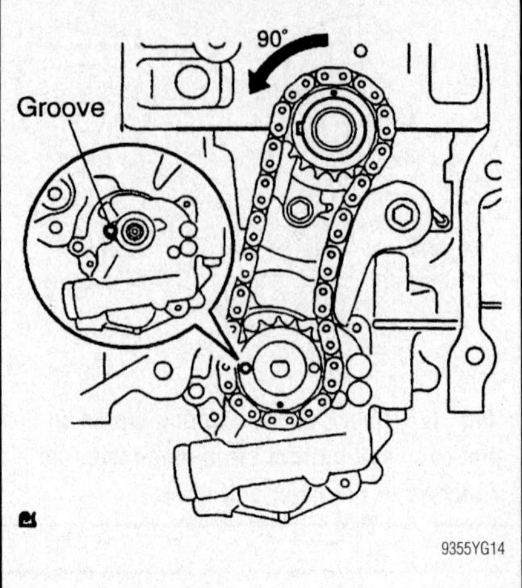

Fig. 186 Aligning the adjusting hole and groove—2.4L engine

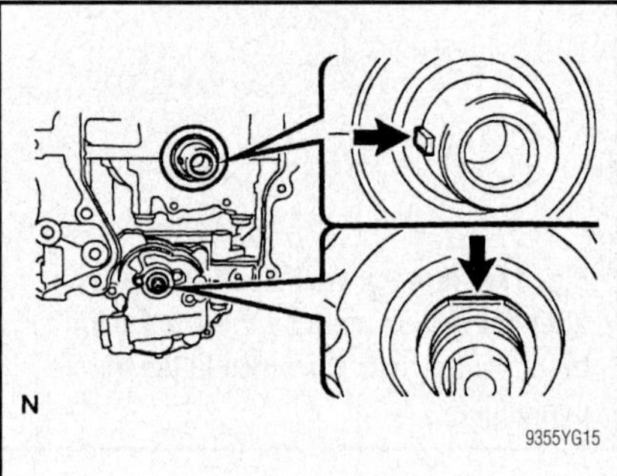

Fig. 187 Aligning the crankshaft with the key in the left horizontal position—2.4L engine

- Transverse engine mount insulator
- Transverse engine mount bracket
- Timing chain cover (14 bolts and 2 nuts
- CKP sensor plate
- Chain tensioner slipper
- Primary chain vibration damper
- Primary chain and crankshaft sprocket

8. Turn the crankshaft 90 degrees counterclockwise and align the adjusting hole of the oil pump drive shaft gear with the groove in the pump.
9. Insert a 4mm diameter bar into the hole to lock the gear in position and remove the nut.
10. Remove the bolt, tensioner plate, spring, tensioner oil pump driveshaft gear and the chain.

To install:

11. Turn the crankshaft so that the key is in the left horizontal position.
12. Install the secondary chain and gears with the timing marks aligned as shown.
13. Install the damper spring and

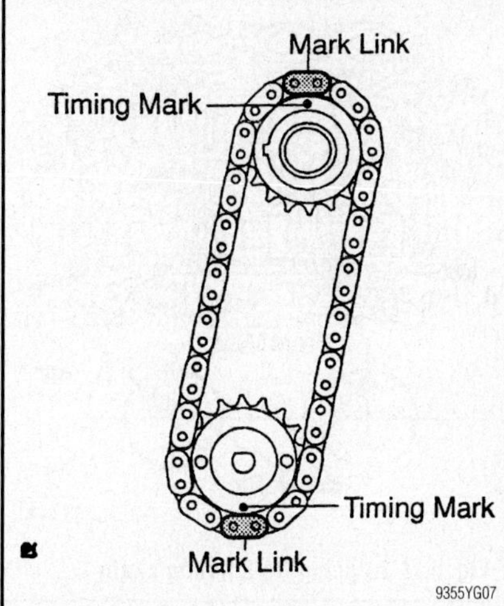

Fig. 188 Install the secondary chain and gears with the timing marks aligned as shown—2.4L engine

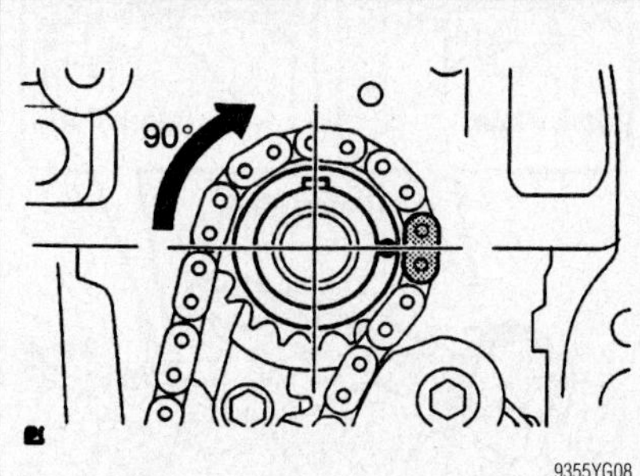

Fig. 189 Aligning the crankshaft with the key in the 12 o'clock position—2.4L engine

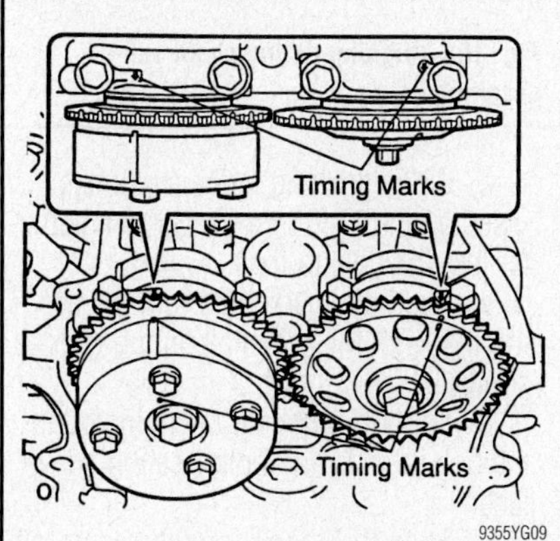

Fig. 190 Aligning the timing marks with the No.1 piston at TDC compression—2.4L engine

tensioner plate. Torque the nut to 10 ft. lbs. (13 Nm).

14. Align the oil pump adjusting hole and groove, lock it with the bar and torque the nut to 22 ft. lbs. (30 Nm).

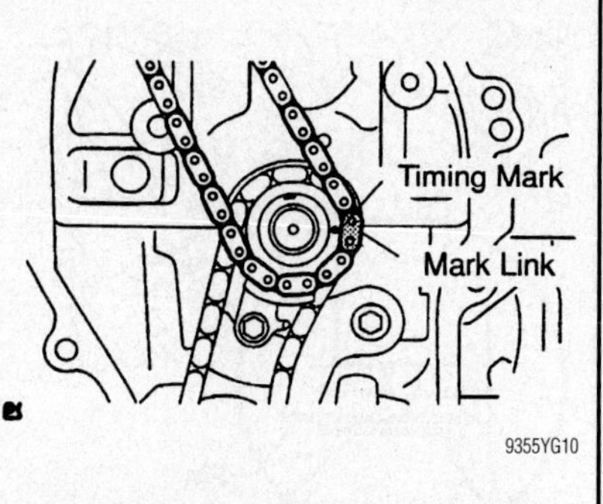

Fig. 191 Aligning the timing chain bottom end marks—2.4L engine

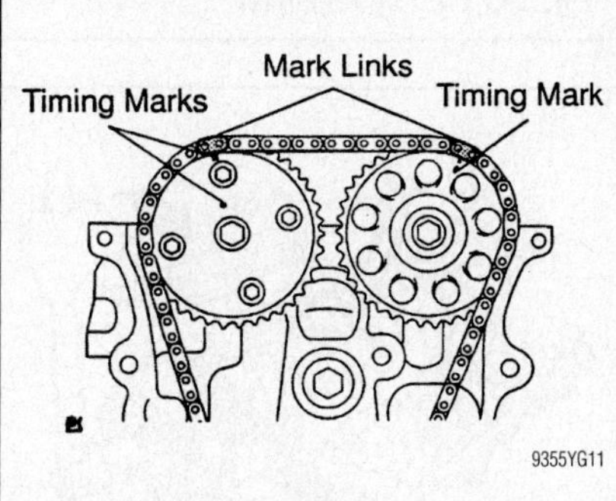

Fig. 192 Aligning the timing chain upper end marks—2.4L engine

15. Rotate the crankshaft counterclockwise 90 degrees so the crankshaft key is at the 12 o'clock position and shown.

16. Install the primary chain damper. Torque the bolts to 80 inch lbs. (9 Nm).

17. Set the No.1 piston at TDC compression with the timing marks aligned as shown.

18. Turn the crankshaft, using the pulley bolt, until the key is at the 12 o'clock position.

19. Install the bottom end of the chain, with sprocket, so that the colored links are aligned as shown.

20. Align the upper end timing marks as shown and install the chain.

21. Install the tensioner slipper. Torque the bolt to 14 ft. lbs. (19 Nm).

22. Install the CKP sensor plate with the **F** mark outwards.

➡When installing the cover, use RTV sealant in the positions shown. The cover must be installed within 3 minutes of seal application. Do not start the engine within 2 hours of seal application.

23. Apply the sealant and install the cover. Torque the cover bolts as follows:

- Bolt A: 80 inch lbs. (9 Nm)
- Bolts B: 15 ft. lbs. (21 Nm)

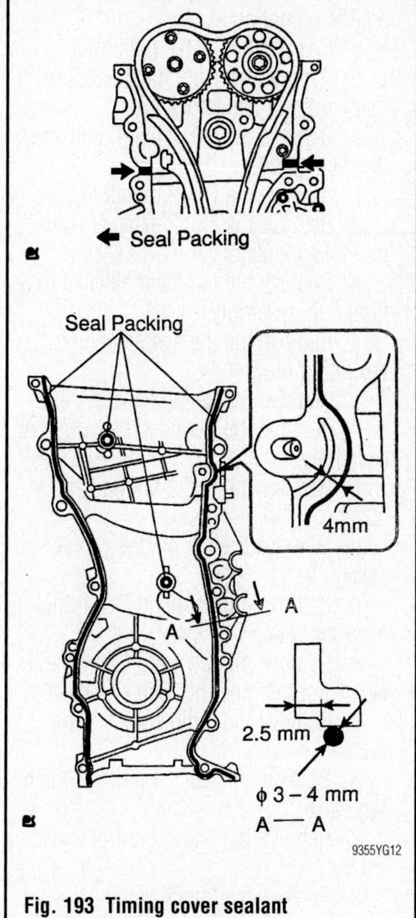

Fig. 193 Timing cover sealant application—2.4L engine

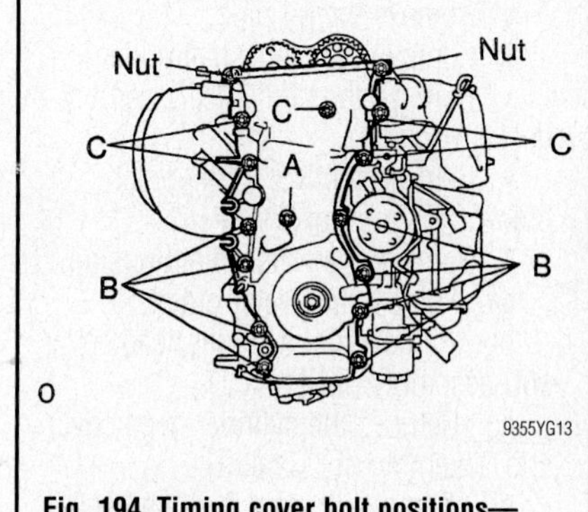

Fig. 194 Timing cover bolt positions—2.4L engine

- Bolts C: 32 ft. lbs. (43 Nm)
- Nuts: 80 inch lbs. (9 Nm)

24. The remainder of installation is the reverse of removal.

3.5L Engine

See Figures 195 through 206.

1. Remove the engine assembly.
2. Install on engine stand.
3. The following must be removed:
   a. Remove ignition coil assembly.

b. Remove the right hand No. 2 engine mounting stay.
c. Remove the intake manifold.
d. Remove the right hand exhaust manifold sub-assembly.
e. Remove the No. 2 engine oil level dipstick guide.
f. Remove the No. 2 manifold stay.
g. Remove the No. 2 exhaust manifold heat insulator.
h. Remove the left hand exhaust manifold sub-assembly.
i. Remove the transverse engine mounting bracket.
j. Remove the generator assembly.
k. Remove the V-ribbed belt tensioner assembly.
l. Remove the No. 2 timing gear cover.
m. Remove the No. 2 idler pulley sub-assembly.
n. Remove the left hand No. 1 engine front mounting bracket.
o. Remove the left hand 6 bolts and No. 1 front engine mounting bracket.
p. Remove the radio setting condenser.
q. Remove the No. 1 vacuum switching valve.
r. Remove the knock control sensor wire.
s. Remove the knock control sensor.
t. Remove the crankshaft position sensor.
u. Remove the No. 1 oil pipe.
v. Remove the oil pipe.
w. Remove the crankshaft pulley.
x. Remove the oil cooler assembly, if necessary.
y. Remove the No. 1 oil cooler bracket, if necessary.
z. Remove the water inlet housing.
aa. Remove the water outlet.
bb. Remove the cylinder head cover sub-assembly (for Bank 1).
cc. Remove the cylinder head cover sub-assembly (for Bank 2).
dd. Remove the No. 2 oil pan sub-assembly.
ee. Remove the oil strainer sub-assembly.
ff. Remove the oil pan sub-assembly.
gg. Remove the timing chain cover sub-assembly
hh. Remove the timing chain case oil seal.
ii. Set the No. 1 cylinder to TDC/compression.
jj. Remove the No. 1 chain tensioner assembly.
kk. Remove the chain tensioner slipper.

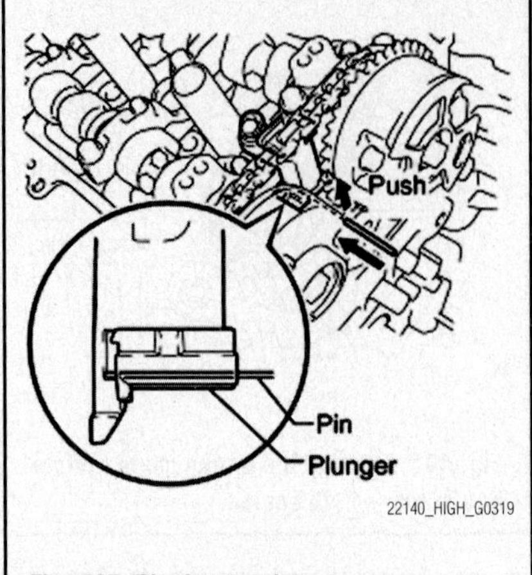

Fig. 195 Pinning tensioner

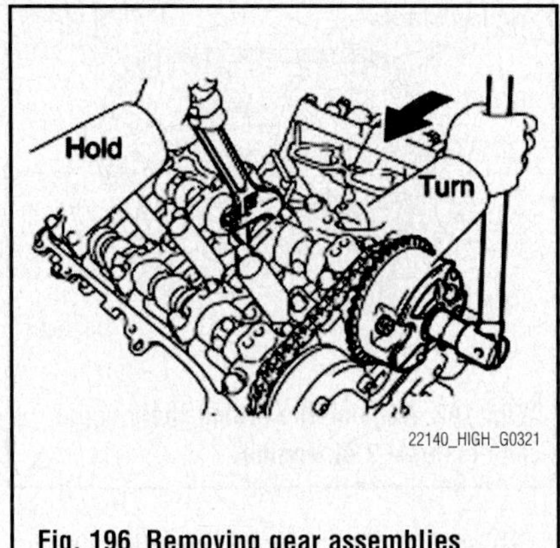

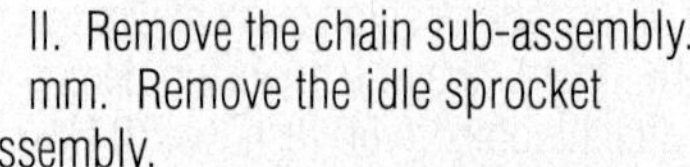
Fig. 196 Removing gear assemblies

ll. Remove the chain sub-assembly.
mm. Remove the idle sprocket assembly.
nn. Remove the camshaft timing gears and No. 2 chain (for Bank 1).
oo. While raising the No. 2 chain tensioner assembly, insert a pin of 1.0 mm (0.039 in.) diameter into the hole to fix the No. 2 chain tensioner assembly.
pp. Hold the hexagonal portion of the camshaft with a wrench, and remove the 2 bolts and 2 camshaft timing gear assemblies.

➡Be careful not to damage the cylinder head with the wrench. Do not disassemble the camshaft timing gear assemblies.

qq. Remove the No. 2 chain assembly.
rr. Remove the bolt and No. 2 chain tensioner assembly.

To install:

4. Install the following:
a. Install the No. 2 chain tensioner assembly with the bolt and tighten to 15 ft. lbs. (21 Nm).

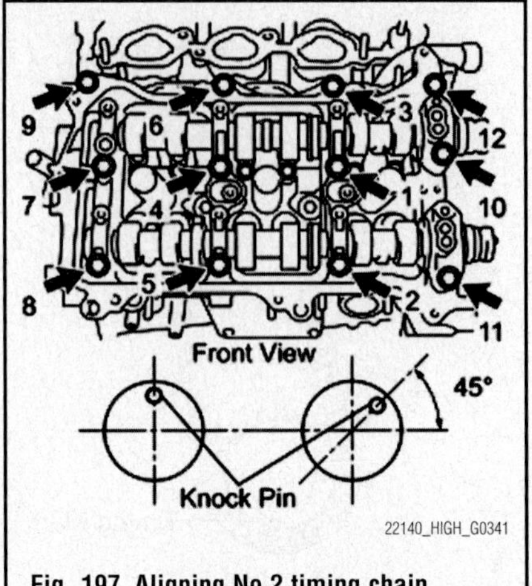

Fig. 197 Aligning No 2 timing chain

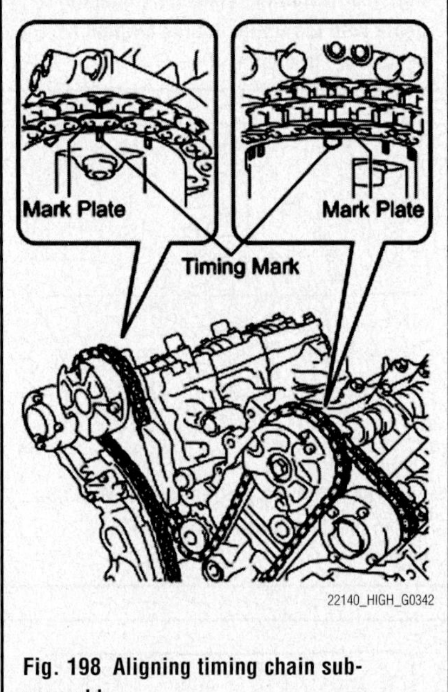

Fig. 198 Aligning timing chain sub-assembly

b. While pushing in the tensioner, insert a pin of 1.0 mm (0.039 in.) diameter into the hole to fix it.
c. Align the mark plate with the timing marks of the camshaft timing gear as shown.
d. Apply a light coat of engine oil to the bolt threads and bolts seating surface.
e. Align the knockpin of the camshaft with pinhole of the camshaft timing gear assembly. Install the camshaft timing gear assembly and camshaft timing exhaust timing gear assembly and camshaft timing exhaust gear with the No. 2 chain sub-assembly installed.
f. Hold the hexagonal portion of the camshaft with the wrench and tighten the two bolts and camshaft timing gear assemblies to 74 ft. lbs. (100 Nm).

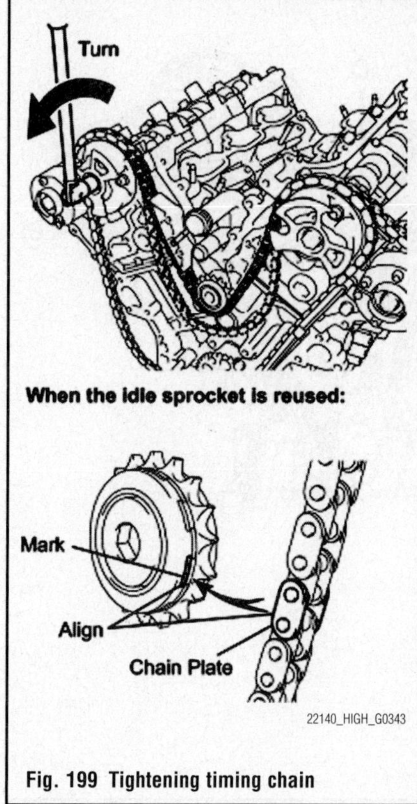

Fig. 199 Tightening timing chain

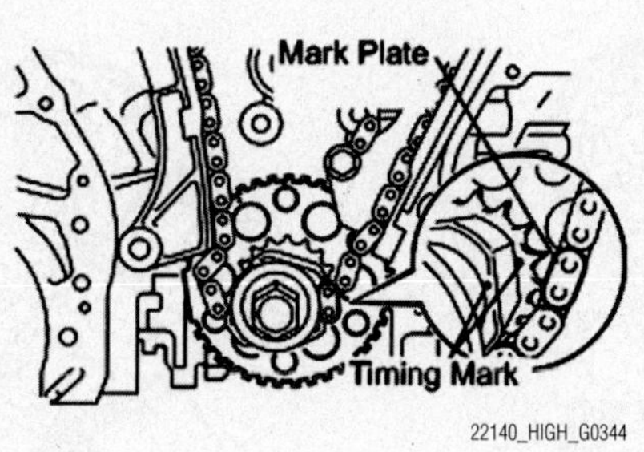

Fig. 200 Installing timing chain on crankshaft

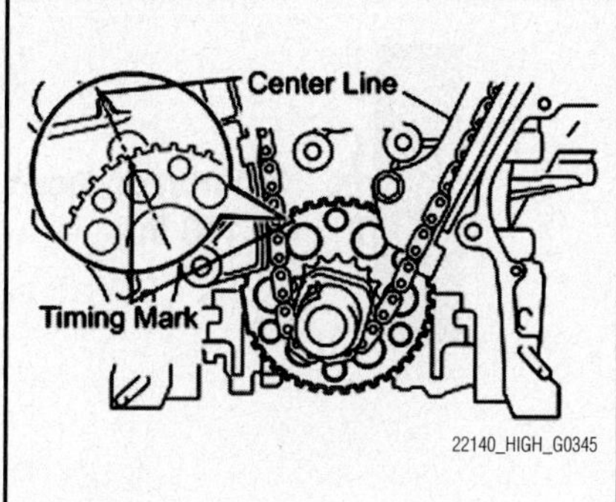

Fig. 201 Aligning timing chain on crankshaft

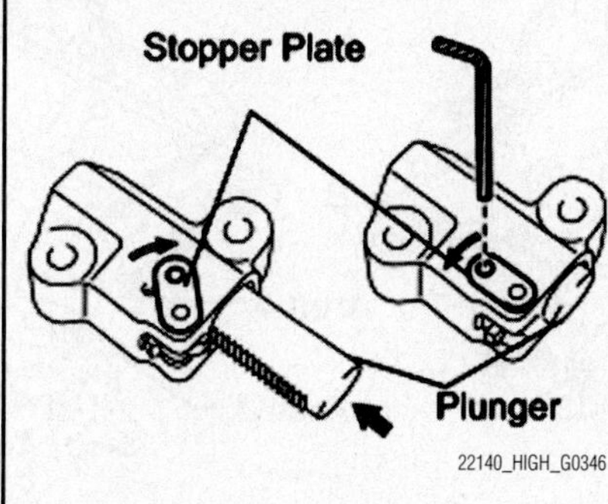

Fig. 202 Aligning timing chain on crankshaft

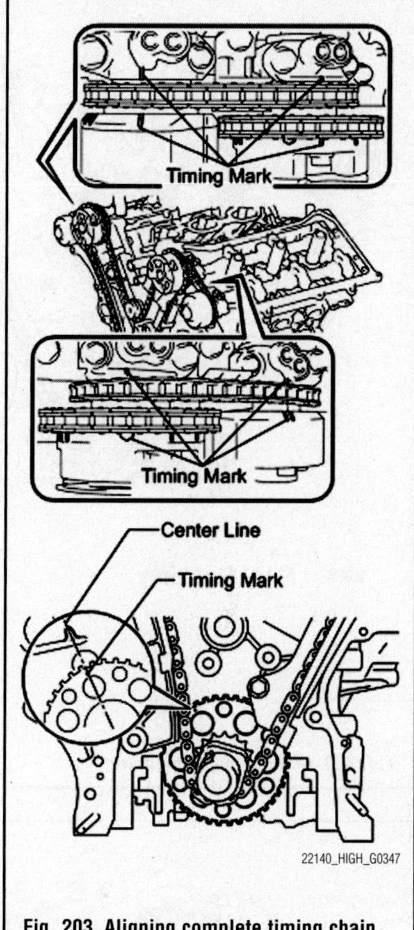

Fig. 203 Aligning complete timing chain

g. Remove the pan from the No. 2 chain tensioner assembly.

h. Install idle sprocket assembly and tighten to 44 ft. lbs. (60 Nm).

i. Install chain sub-assembly.

j. Align the mark plate and timing marks as shown in the illustration and install the chain.

➡The camshaft mark plates are orange.

k. Do not pass the chain over the crankshaft, just temporarily place it on the crankshaft.

l. Turn the camshaft timing gear assembly on bank 1 counterclockwise to tighten the chain between the banks.

➡When the idle sprocket assembly is reused, align the timing chain plate with the mark where the plate has been in order to tighten the chain between the banks.

m. Align the mark plate and timing marks as shown in the illustration and install the chain onto the crankshaft timing sprocket. The crankshaft to mark plate is yellow.

n. Turn the crankshaft clockwise to set it to the right-hand block bore more centerline. (TDC Compression).

o. Install chain tensioner slipper.

p. Move the stopper plate upward to release the lock, and push the plunger deep into the tensioner.

q. Move the stopper plate downward to set the loss, and insert a hexagon wrench into the hole of the stopper plate.

r. Install No. 1 chain tensioner assembly and tighten bolts to 7 ft. lbs. (10 Nm).

s. Remove the hexagon wrench from the No. 1 chain tensioner assembly. Check that the each timing mark is aligned with the crankshaft at TDC compression.

t. Remove the pulley set bolt.

u. Install timing chain case oil seal.

v. Install sealant to timing chain cover sub-assembly.

w. Install new O ring gasket on cylinder block.

x. Align the oil pump's drive rotor spline and the crankshaft as shown in the illustration. Install the spline and chain cover to the crankshaft.

y. Temporarily tighten the timing chain cover with the 23 bolts and 2 nuts.

- Tighten bolts in area 1 and 2: 15 ft. lbs. (21 Nm).
- Tighten bolt in area 3: 15 ft. lbs. (21 Nm).

➡First tighten the upper bolts and nuts, followed by the lower bolts and nuts as shown.

- Tighten bolt in area 4: 32 ft. lbs. (43 Nm)
- Tighten bolt in area 4: 15 ft. lbs. (21 Nm)

z. Install oil pan sub-assembly and tighten 16 bolts and 2 nuts to 7 ft. lbs. (10 Nm) and 15 ft lbs. (21 Nm).

aa. Install oil strainers sub-assembly and tighten bolts and nuts to 7 ft. lbs. (10 Nm).

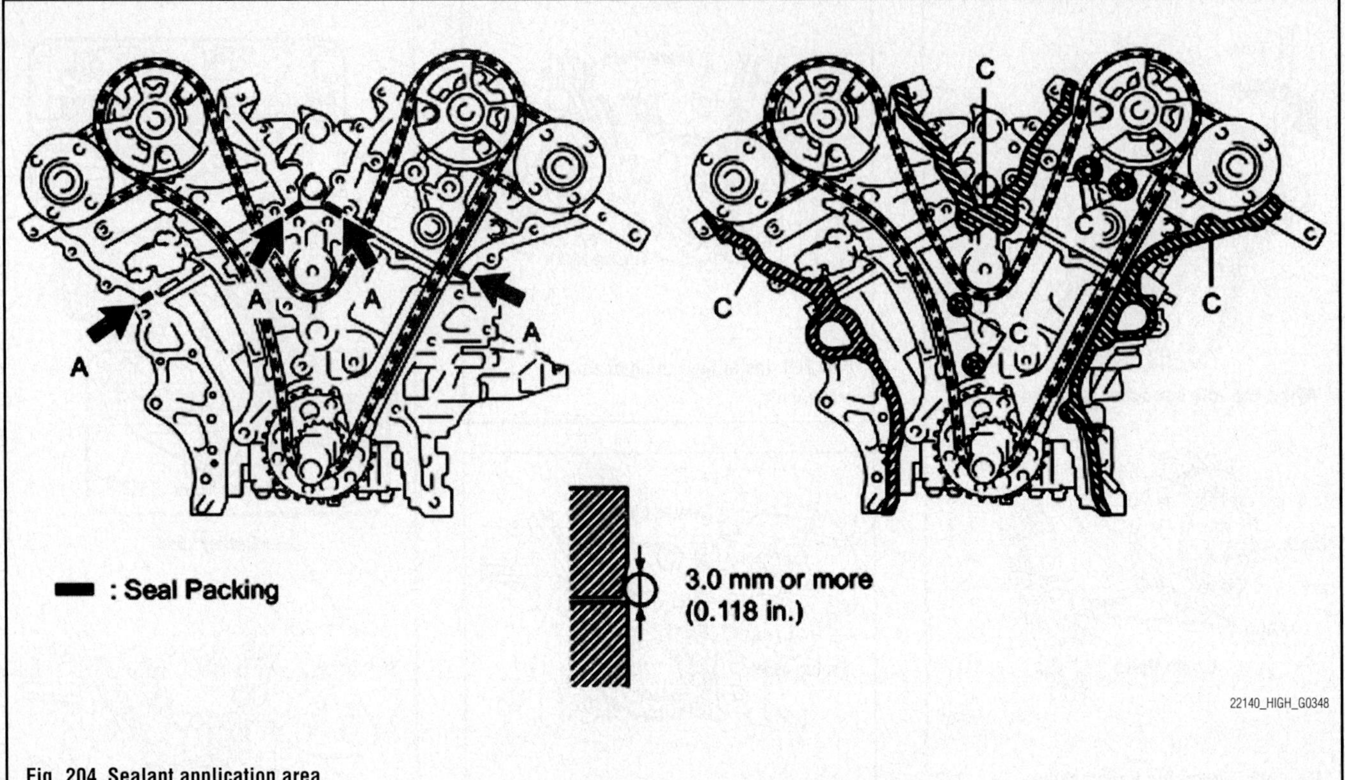

Fig. 204 Sealant application area

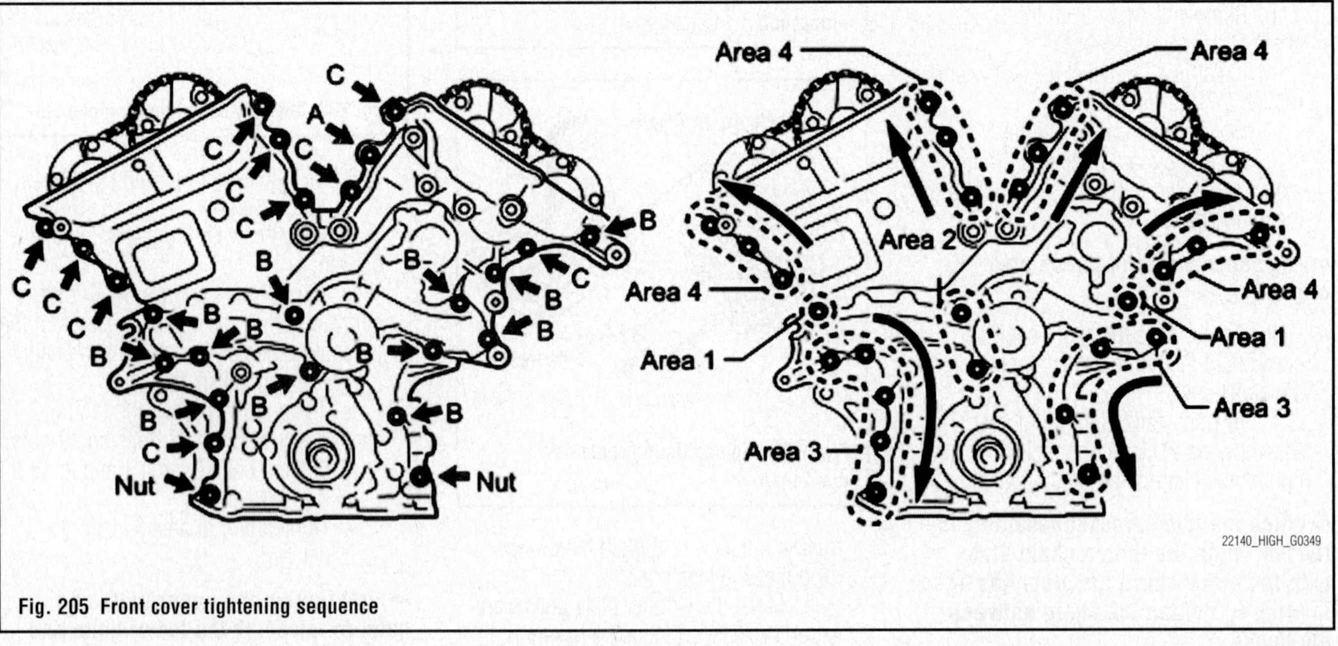

Fig. 205 Front cover tightening sequence

bb. Install No.2 oil pan sub-assembly and tighten 16 bolts and 2 nuts to 7 ft. lbs. (10 Nm).

cc. Install cylinder head cover sub-assemblies and tighten to 7 ft. lbs. (10 Nm) and 15 ft. lbs. (21 Nm).

dd. Install the water inlet housing.

ee. Install No. 1 oil cooler bracket (w/ oil cooler).

ff. Install the oil cooler assembly (w/ oil cooler).

gg. Install the crankshaft pulley.

hh. Install the oil pipe.

ii. Install the No. 1 oil pipe.

jj. Install the crankshaft position sensor.

kk. Install the knock control sensor.

ll. Install the knock control sensor wire.

mm. Install the No. 1 vacuum switching valve.

nn. Install the radio setting condenser.

oo. Install the No. 1 left front engine mounting bracket with the 6 bolts and tighten to 40 ft. lbs. (54 Nm).

➡Install the water inlet and mounting bracket within 15 minutes after installing the chain cover. Do not start the engine for at least 2 hours after installation.

pp. Install the No. 2 idler pulley sub-assembly.

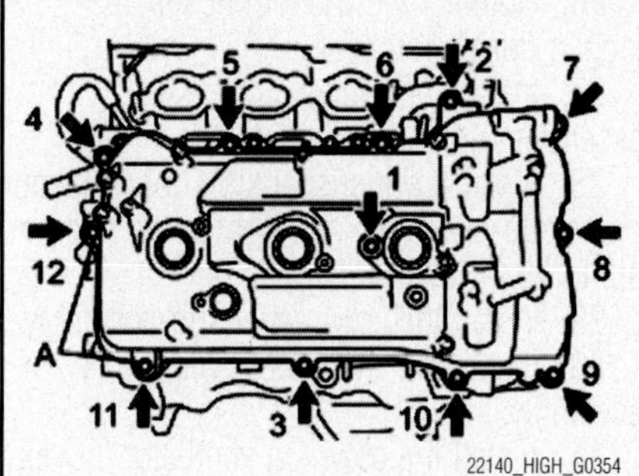

Fig. 206 Cylinder head cover tightening sequence

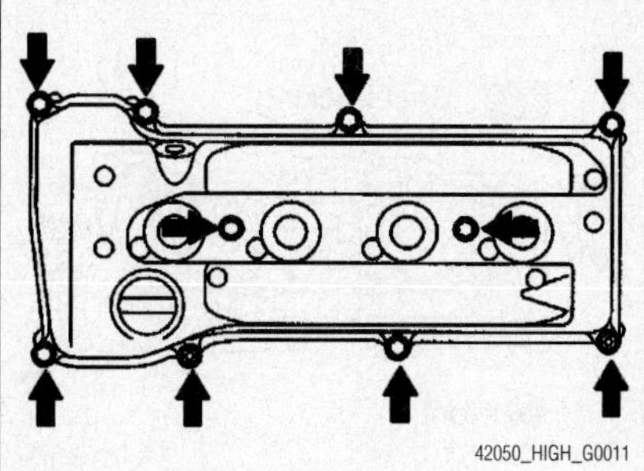

Fig. 207 Location of the 8 bolts and 2 nuts securing the cylinder head (valve) cover—2.4L engine

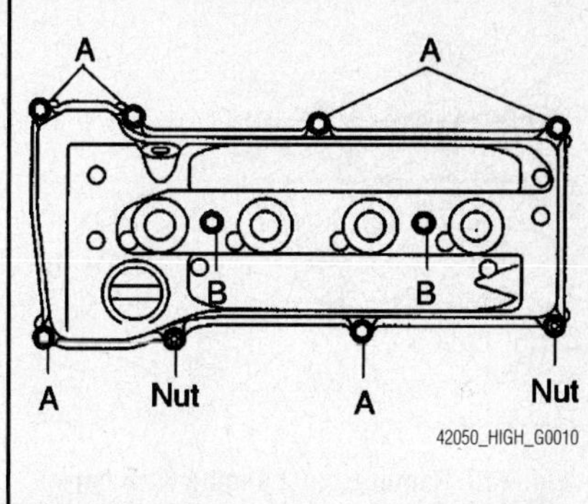

Fig. 209 Cylinder head (valve) cover bolt tightening specifications

qq. Install the No. 2 timing gear cover

rr. Install the V-ribbed belt tensioner assembly.

ss. Install the generator assembly.

tt. Install the transverse engine mounting bracket.

uu. Install the left hand exhaust manifold sub-assembly and tighten to 15 ft. lbs. (21 Nm).

vv. Install the No. 2 exhaust manifold heat insulator.

ww. Install the No. 2 manifold stay.

xx. Install the No. 2 engine oil level dipstick guide.

yy. Install the right hand exhaust manifold sub-assembly and tighten to 15 ft. lbs. (21 Nm).

zz. Install the intake manifold and tighten the 6 bolts and 4 nuts uniformly in several steps to 15 ft. lbs. (21 Nm).

aaa. Install the right hand No. 2 engine mounting stay.

bbb. Install the ignition coil assembly.

ccc. Remove the engine stand.

ddd. Install the engine assembly.

VALVE COVERS

REMOVAL & INSTALLATION

2.4L Engine

See Figures 207 through 209.

1. Raise and safely support the vehicle.
2. Remove the right front wheel and tire assembly.
3. Remove the right front fender splash shield.
4. Remove the right front fender apron seal.
5. Remove engine cover sub-assembly no.1.
6. Remove ignition coil assembly, as outlined in the Engine Electrical.
7. Disconnect the ventilation hose and ventilation hose no. 2.

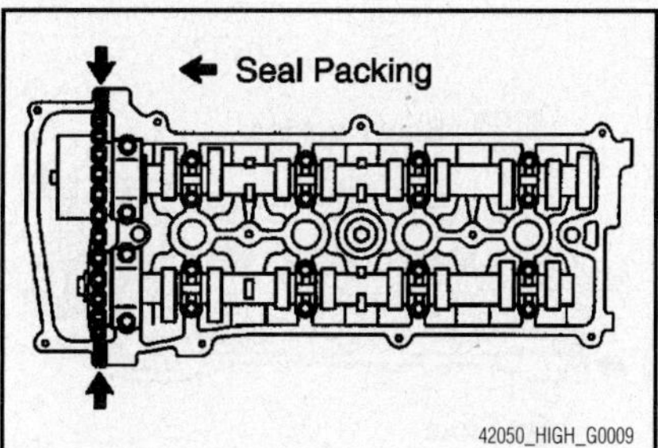

Fig. 208 After removing al of the old packing material, apply suitable seal packing to the 2 locations as shown in the accompanying illustration

8. Remove cylinder head cover sub-assembly, as follows:
   a. Remove the bolt and disconnect the engine wire harness Clamp.
   b. Remove the 8 bolts, 2 nuts, then remove the cylinder head cover and gasket.
9. Thoroughly clean the gasket mating surfaces.

To install:

10. Remove any old packing (fipg) material.
11. Apply suitable seal packing to 2 locations as shown in the illustration. Seal packing: part no. 08826-00080 or equivalent

⁂ WARNING

You must install the cylinder head cover within 5 minutes after applying the seal packing. Also, wait at least 2 hours after installing the cover to fill the engine with oil.

12. Install the cylinder head cover with the 8 bolts and 2 nuts. Tighten the retainers as follows:
    a. Bolt a: 8 ft. lbs. (11 Nm)
    b. Bolt b: 10 ft. lbs. (14 Nm)
    c. Nut: 8 ft. lbs. (11 Nm)
13. Install ignition coil assembly and tighten the retainers to 80 inch lbs. (9 Nm). Refer to the Engine Electrical Section for more information.
14. Install the remaining components in the reverse of the removal procedure. Tighten the lug nuts to 76 ft. lbs. (103 Nm).
15. Wait at least 2 hours, then fill the engine with the proper type and amount of oil. Start the engine and check for leaks.

3.3L Engine

See Figures 210 through 213.

1. Drain the engine coolant into a suitable container.
2. Remove the front wiper arms.
3. Remove the cowl top ventilator louver sub-assembly.
4. Remove the windshield wiper link assembly.
5. Remove the cowl panel sub-assembly.
6. Remove the right side front fender apron seal.
7. Remove the front suspension upper center brace sub-assembly.
8. Remove the v-bank cover sub-assembly.
9. Remove the air cleaner cap sub-assembly.
10. Remove the emission control valve set.
11. Remove the intake air surge tank.
12. Remove radiator inlet hose.
13. Remove ignition coil, as outlined in the Engine Electrical Section.
14. Remove the cylinder head (valve) cover sub-assembly, as follows:
    a. Remove the 2 engine wire harness clamps.
    b. Remove the 3 nuts and disconnect the engine wire harness.
    c. Remove the 9 bolts and cylinder head cover.
15. Remove the cylinder head cover gasket

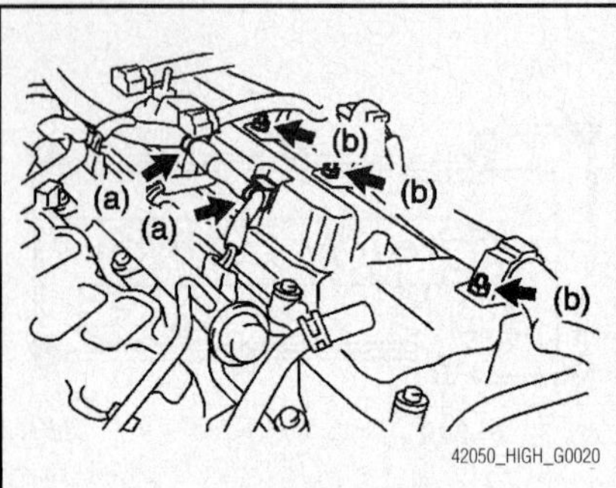

Fig. 210 Remove the 2 engine wire harness clamps (a) and the 3 nuts (b), then disconnect the engine wire harness

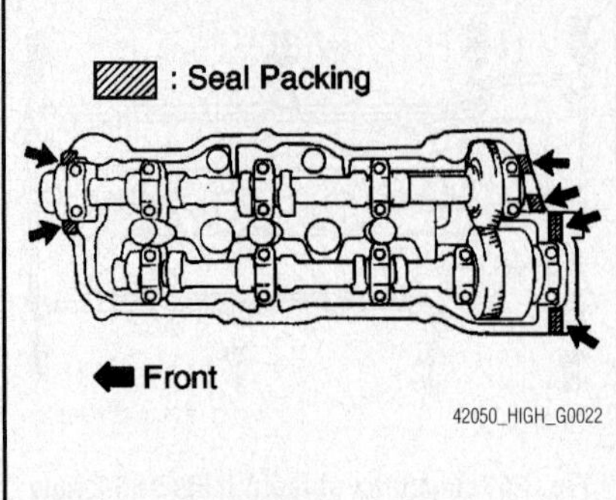

Fig. 212 Applying the seal packing—cylinder head cover sub-assembly

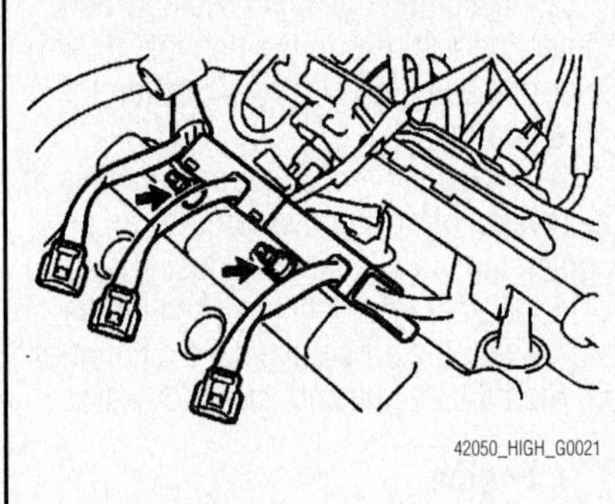

Fig. 211 Remove the 2 bolts, and disconnect the engine wire harness protector

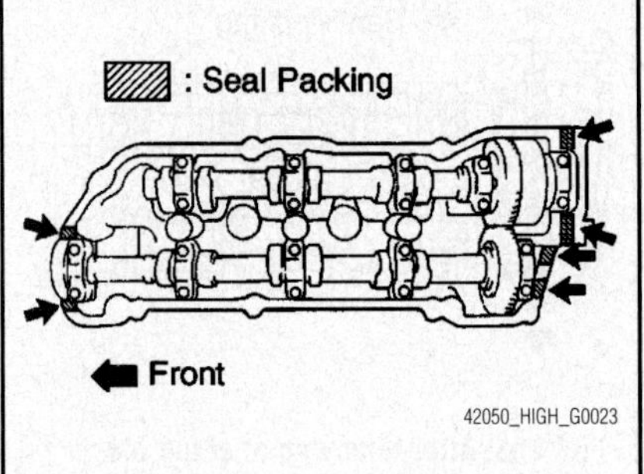

Fig. 213 Applying the seal packing—left side cylinder head cover sub-assembly

16. Remove the left cylinder head cover sub-assembly, as follows:
    a. Using an E6 Torx® socket wrench, remove the 2 bolts and disconnect the engine wire harness protector.
    b. Remove the 9 bolts and cylinder head cover.
17. Remove the no. 2 cylinder head cover gasket.
18. Thoroughly clean the gasket mating surfaces. Remove all traces of oil from the contact surfaces.

To install:

19. Install the cylinder head (valve) cover sub-assembly, as follows:
    a. Install the gasket to the cylinder head cover.
    b. Apply seal packing (part no. 08826-00080, or equivalent) to the cylinder head, as shown in the illustration.

WARNING

You must install the cylinder head cover within 3 minutes after applying seal packing. Also, do NOT start the engine for at least 2 hours after installing.

    c. Install the cylinder head cover with the 9 bolts. Tighten the bolts uniformly in several steps to a final torque of 71 inch lbs. (8 Nm).
20. Install the left cylinder head (valve) cover sub-assembly, as follows:
    a. Install the gasket to the cylinder head cover.
    b. Apply seal packing (part no. 08826-00080, or equivalent) to the cylinder head, as shown in the illustration.

WARNING

You must install the cylinder head cover within 3 minutes after applying seal packing. Also, do NOT start the engine for at least 2 hours after installing.

    c. Install the cylinder head cover with the 9 bolts. Tighten the bolts uniformly in several steps to a final torque of 71 inch lbs. (8 Nm).
21. Install ignition coil, as outlined in the Engine Electrical Section. Tighten the retainers to 71 inch lbs. (8 Nm).
22. Install the inlet radiator hose.
23. Install the intake air surge tank.
24. Install the emission control valve set.
25. Install air cleaner cap sub-assembly.
26. Connect the vacuum hoses, as tagged during removal.
27. Install the v-bank cover sub-assembly.
28. Install front suspension upper center brace sub-assembly and tighten the retainers to 59 ft. lbs. (80 Nm).
29. Install the cowl panel sub-assembly.
30. Install the windshield wiper link assembly.
31. Install the cowl top ventilator louver sub-assembly.
32. Install the front wiper arms.
33. Fill the engine with proper type and amount of engine coolant.
34. After waiting a minimum of 2 hours, start the engine and inspect for engine coolant leaks.

3.5L Engine

See Figures 214 through 221.

1. Remove cylinder head cover sub-assembly (for Bank 1), as follows:
    a. Remove the 12 bolts, seal washer, cylinder head cover sub-assembly and cylinder head cover gasket.
    b. Remove the 3 gaskets.
2. Remove cylinder head cover sub-assembly (for Bank 2), as follows:

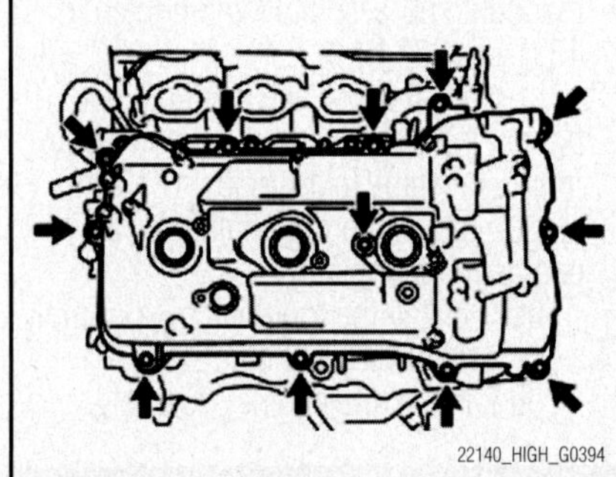

Fig. 214 Cylinder head cover sub-assembly Bank 1—3.5L engine

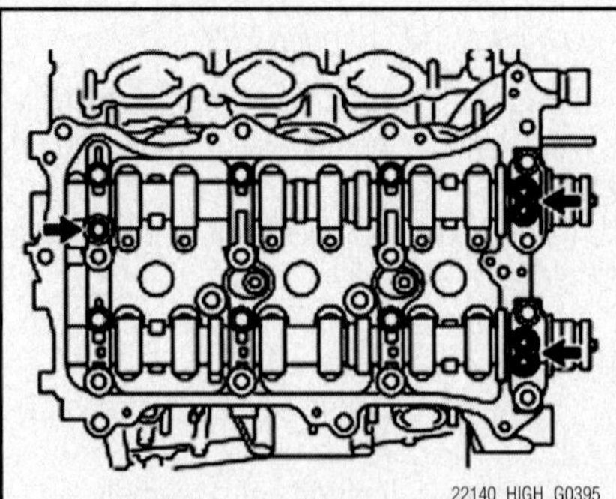

Fig. 215 Cylinder head cover sub-assembly gaskets Bank 1—3.5L engine

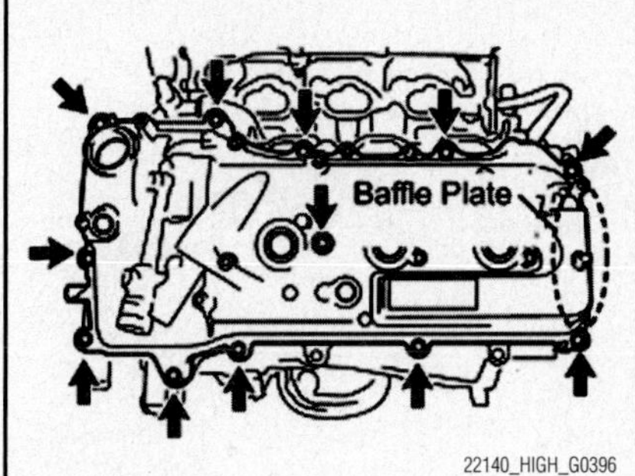

Fig. 216 Cylinder head cover sub-assembly Bank 2—3.5L engine

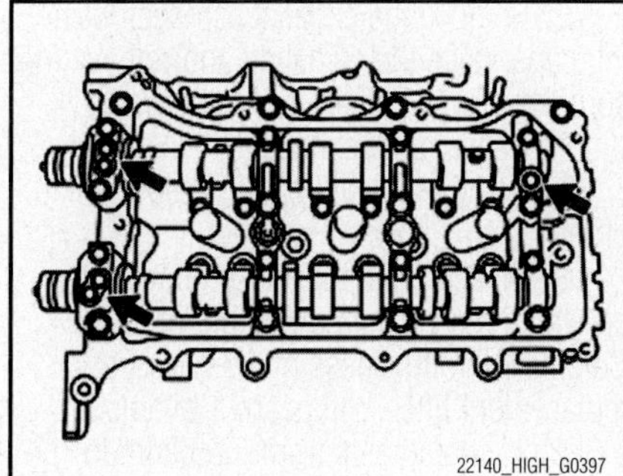

Fig. 217 Cylinder head cover sub-assembly gaskets Bank 2—3.5L engine

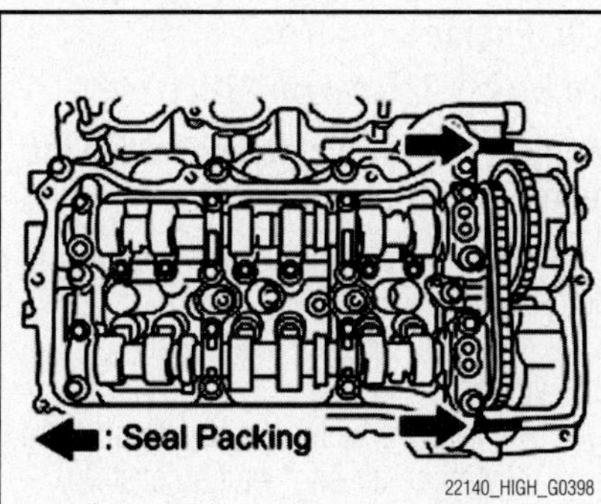

Fig. 218 Apply sealant—3.5L engine

a. Remove the 12 bolts, seal washer, cylinder head cover sub-assembly and cylinder head cover gasket.

➡The baffle plate is located on the back of the portion shown in the illustration. Do not damage the baffle plate when removing the cylinder head cover sub-assembly.

b. Remove the 3 gaskets.

To install:

3. Install cylinder head cover sub-assembly (for Bank 1), as follows:

a. Apply seal packing as shown in the illustration. Seal packing: Toyota Genuine Seal Packing Black, Three Bond 1207B or equivalent.

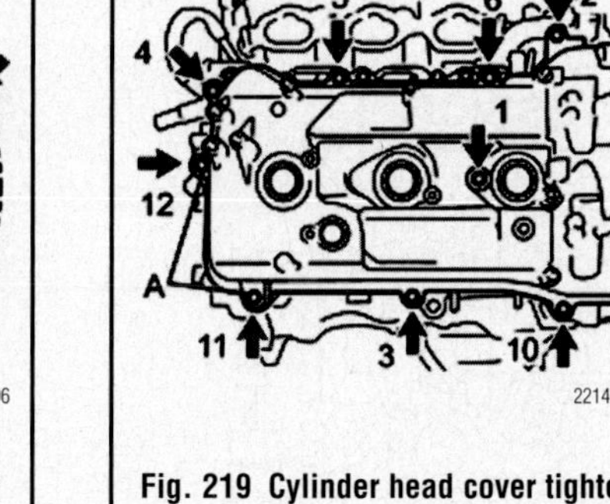

Fig. 219 Cylinder head cover tightening sequence Bank 1—3.5L engine

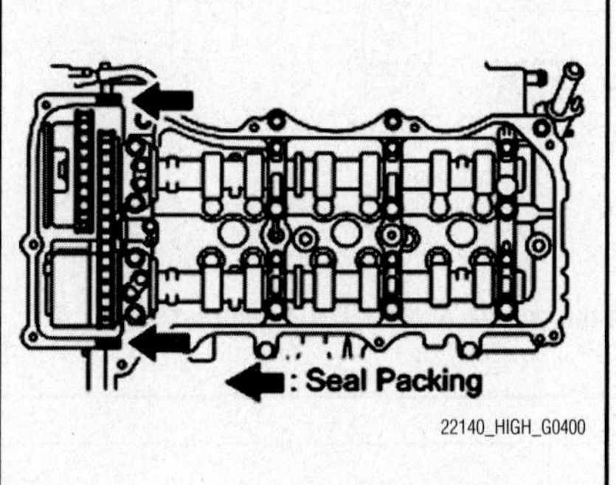

Fig. 220 Apply sealant—3.5L engine

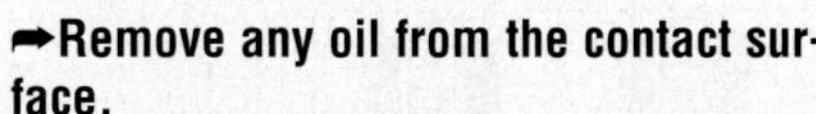

➡Remove any oil from the contact surface.

➡Install the head cover within 3 minutes after applying seal packing.

➡Do not start the engine for at least 2 hours after installing.

b. Install 3 new gaskets as previously removed.

c. Install a new gasket to the head cover.

d. Install a head cover with the 12 bolts and a new seal washer and tighten to 15 ft. lbs. (21 Nm), 7 ft. lbs. (10 Nm).

➡After tightening all bolts, check the tightening torque of 1 and 11. Retighten the bolt if necessary.

4. Install cylinder head cover sub-assembly (for Bank 2), as follows:

a. Apply seal packing as shown in the illustration. Seal packing: Toyota Genuine Seal Packing Black, Three Bond 1207B or equivalent.

➡Remove any oil from the contact surface.

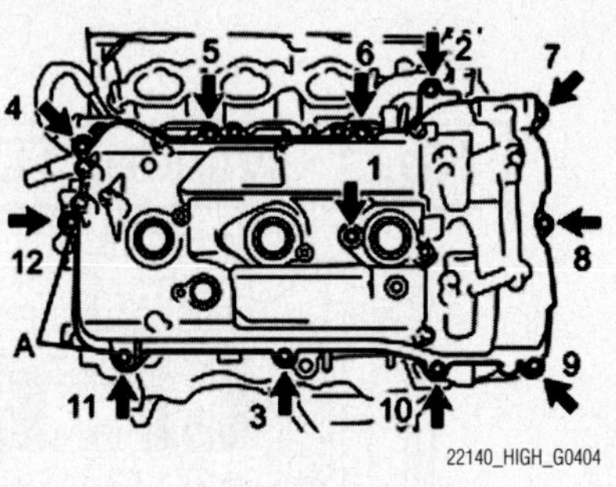

Fig. 221 Cylinder head cover tightening sequence Bank 2—3.5L engine

➡Install the head cover within 3 minutes after applying seal packing.

➡Do not start the engine for at least 2 hours after installing.

b. Install 3 new gaskets as previously removed.

c. Install a new cylinder head cover gasket to the cylinder head cover sub-assembly.

d. Install the cylinder head cover sub-assembly with the 12 bolts and a new seal washer and tighten to 15 ft. lbs. (21 Nm), 7 ft. lbs. (10 Nm).

➡After tightening all bolts, check the tightening torque of 1 and 10. Retighten the bolt if necessary.

VALVE LASH

ADJUSTMENT

2.4L Engine

See Figures 222 through 226.

➡Adjust the valve clearance when the engine is cold.

1. Before servicing the vehicle, refer to the precautions section.
2. Remove or disconnect the following:
   - Negative battery cable. If equipped with an air bag, wait at least 90 seconds before proceeding.
   - Right front wheel, splash shield and apron seal
   - engine undercover
   - Coil pack
   - Air intake hoses
   - Cylinder head cover
3. Place the No. 1 piston at TDC compression. Check only those valves shown. Record the clearance. If out of clearance, the measurement will be used to calculate the adjusting shims.

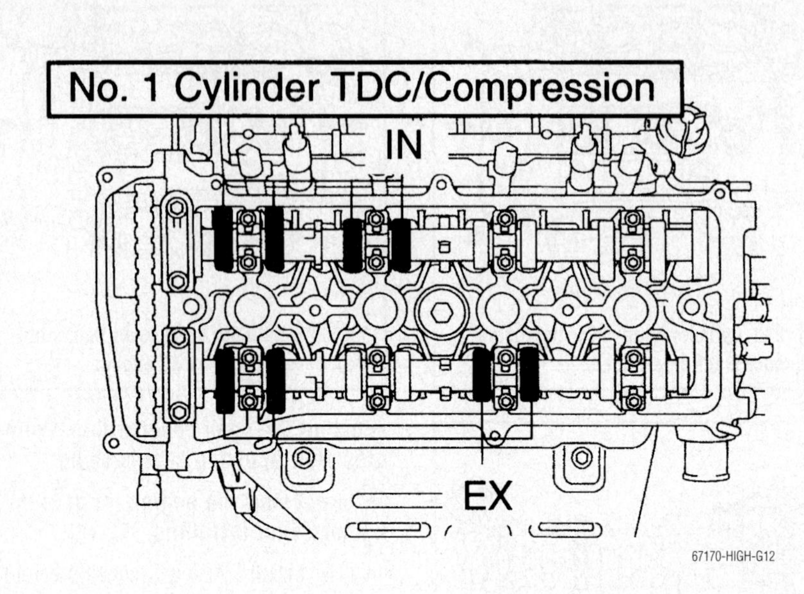

Fig. 222 Check the clearance on these valves with the engine at No.1 TDC compression—2.4L engine

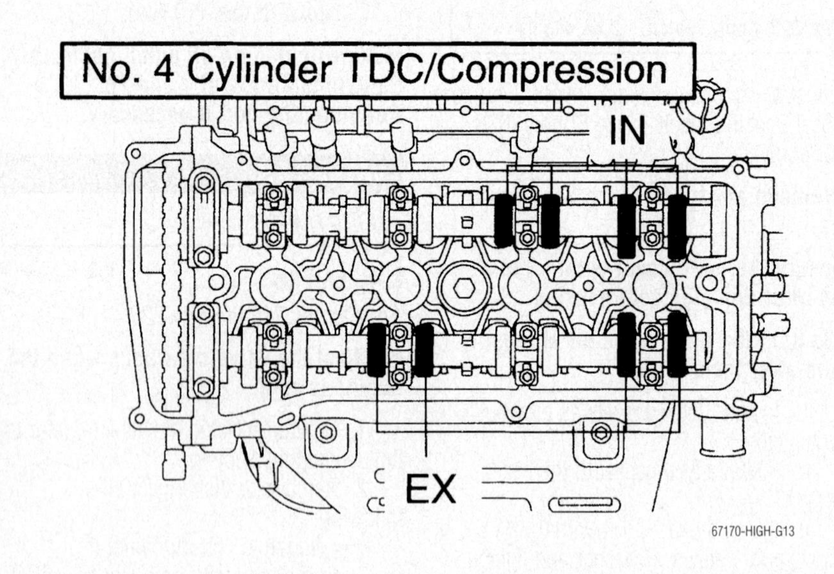

Fig. 223 Check the clearance on these valves with the engine at No.4 TDC compression—2.4L engine

4. Place the No. 4 piston at TDC compression. Check only those valves shown. Record the clearance. If out of clearance, the measurement will be used to calculate the adjusting shims.

Clearance range is:

- Intake: 0.19–0.29mm (0.008–0.011 in.)
- Exhaust: 0.30–0.40mm (0.012–0.016 in.)

To adjust the valves:

5. Turn the crankshaft 1 complete revolution (360 degrees) clockwise and set the No.1 piston at TDC compression. Place matchmarks on the chain and camshaft sprocket.
6. Remove the tensioner.
7. Loosen the camshaft sprocket bolt.

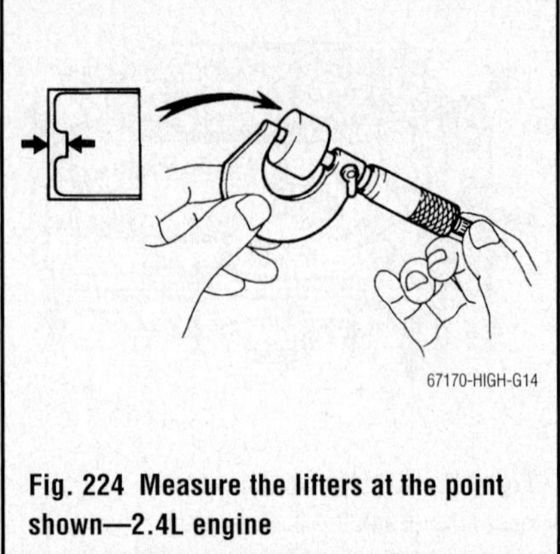

Fig. 224 Measure the lifters at the point shown—2.4L engine

8. Remove the exhaust camshaft bearing caps, raise the camshaft and remove the sprocket. Tie the chain out of the way.
9. Remove the camshaft.
10. Remove the lifters and keep them in order.
11. Measure the thickness of any lifter on which the clearance was out of range. Calculate the thickness of the necessary replacement lifter. Lifters are available in 0.020mm increments from 5.060mm to 5.740mm.
12. For Camshaft and Timing Chain installation, see the respective procedures in this section.

3.3L Engine

See Figures 227 through 232.

➡Adjust the valve clearance when the engine is cold.

1. Before servicing the vehicle, refer to the precautions section.
2. Remove or disconnect the following:
   - Negative battery cable. If equipped with an air bag, wait at least 90 seconds before proceeding.
   - Accelerator/throttle cable from the throttle linkage
   - Air cleaner cover, air flow meter and air duct assembly
   - V-bank cover
   - Emission control valve set
   - Air intake chamber
   - Engine harness from the injectors and the ignition coils
   - Ignition coils and keep them in order for reassembly
   - Spark plugs
   - Cylinder head covers
3. Turn the crankshaft pulley and align its groove with the timing mark **0** of the No. 1 timing cover.
4. Check that the valve lifters on the No. 1 intake are loose and the No. 1 exhaust

Valve Lifter Selection Chart (Intake)

Measure clearance mm (in.) / Installed lifter thickness mm (in.)	5.060 (0.1992)	5.080 (0.2000)	5.100 (0.2008)	5.120 (0.2016)	5.140 (0.2024)	5.160 (0.2031)	5.180 (0.2039)	5.200 (0.2047)	5.210 (0.2051)	5.220 (0.2055)	5.230 (0.2059)	5.240 (0.2063)	5.250 (0.2067)	5.260 (0.2071)	5.270 (0.2075)	5.280 (0.2079)	5.290 (0.2083)	5.300 (0.2087)	5.310 (0.2091)
0.000-0.030 (0.0000-0.0012)													06	06	08	08	10	10	12
0.031-0.050 (0.0012-0.0020)									06	06	06	08	08	10	10	12	12	14	14
0.051-0.070 (0.0020-0.0028)								06	08	08	08	10	10	12	12	14	14	16	16
0.071-0.090 (0.0028-0.0035)							06	08	10	10	10	12	12	14	14	16	16	18	18
0.091-0.110 (0.0036-0.0043)						06	08	10	12	12	12	14	14	16	16	18	18	20	20
0.111-0.130 (0.0044-0.0051)					06	08	10	12	14	14	14	16	16	18	18	20	20	22	22
0.131-0.150 (0.0052-0.0059)				06	08	10	12	14	16	16	16	18	18	20	20	22	22	24	24
0.151-0.170 (0.0059-0.0067)			06	08	10	12	14	16	18	18	18	20	20	22	22	24	24	26	26
0.171-0.189 (0.0067-0.0074)		06	08	10	12	14	16	18	20	20	20	22	22	24	24	26	26	28	28
0.190-0.290 (0.0075-0.0114)																			
0.291-0.310 (0.0115-0.0122)	12	14	16	18	20	22	24	26	28	28	30	30	32	32	34	34	36	36	38
0.311-0.330 (0.0122-0.0130)	14	16	18	20	22	24	26	28	30	30	32	32	34	34	36	36	38	38	40
0.331-0.350 (0.0130-0.0138)	16	18	20	22	24	26	28	30	32	32	34	34	36	36	38	38	40	40	42
0.351-0.370 (0.0138-0.0146)	18	20	22	24	26	28	30	32	34	34	36	36	38	38	40	40	42	42	44
0.371-0.390 (0.0146-0.0154)	20	22	24	26	28	30	32	34	36	36	38	38	40	40	42	42	44	44	46
0.391-0.410 (0.0154-0.0161)	22	24	26	28	30	32	34	36	38	38	40	40	42	42	44	44	46	46	48
0.411-0.430 (0.0162-0.0169)	24	26	28	30	32	34	36	38	40	40	42	42	44	44	46	46	48	48	50
0.431-0.450 (0.0170-0.0177)	26	28	30	32	34	36	38	40	42	42	44	44	46	46	48	48	50	50	52
0.451-0.470 (0.0178-0.0185)	28	30	32	34	36	38	40	42	44	44	46	46	48	48	50	50	52	52	54
0.471-0.490 (0.0185-0.0193)	30	32	34	36	38	40	42	44	46	46	48	48	50	50	52	52	54	54	56
0.491-0.510 (0.0193-0.0201)	32	34	36	38	40	42	44	46	48	48	50	50	52	52	54	54	56	56	58
0.511-0.530 (0.0201-0.0209)	34	36	38	40	42	44	46	48	50	50	52	52	54	54	56	56	58	58	60
0.531-0.550 (0.0209-0.0217)	36	38	40	42	44	46	48	50	52	52	54	54	56	56	58	58	60	60	62
0.551-0.570 (0.0217-0.0224)	38	40	42	44	46	48	50	52	54	54	56	56	58	58	60	60	62	62	64
0.571-0.590 (0.0225-0.0232)	40	42	44	46	48	50	52	54	56	56	58	58	60	60	62	62	64	64	66
0.591-0.610 (0.0233-0.0240)	42	44	46	48	50	52	54	56	58	58	60	60	62	62	64	64	66	66	68
0.611-0.630 (0.0241-0.0248)	44	46	48	50	52	54	56	58	60	60	62	62	64	64	66	66	68	68	70
0.631-0.650 (0.0248-0.0256)	46	48	50	52	54	56	58	60	62	62	64	64	66	66	68	68	70	70	72
0.651-0.670 (0.0256-0.0264)	48	50	52	54	56	58	60	62	64	64	66	66	68	68	70	70	72	72	74
0.671-0.690 (0.0264-0.0272)	50	52	54	56	58	60	62	64	66	66	68	68	70	70	72	72	74	74	74
0.691-0.710 (0.0272-0.0280)	52	54	56	58	60	62	64	66	68	68	70	70	72	72	74	74	74	74	
0.711-0.730 (0.0280-0.0287)	54	56	58	60	62	64	66	68	70	70	72	72	74	74	74	74			
0.731-0.750 (0.0288-0.0295)	56	58	60	62	64	66	68	70	72	72	74	74	74	74					
0.751-0.770 (0.0296-0.0303)	58	60	62	64	66	68	70	72	74	74	74	74							
0.771-0.790 (0.0304-0.0311)	60	62	64	66	68	70	72	74	74	74									
0.791-0.810 (0.0311-0.0319)	62	64	66	68	70	72	74												
0.811-0.830 (0.0319-0.0327)	64	66	68	70	72	74													
0.831-0.850 (0.0327-0.0335)	66	68	70	72	74														
0.851-0.870 (0.0335-0.0343)	68	70	72	74															
0.871-0.890 (0.0343-0.0350)	70	72	74																
0.891-0.910 (0.0351-0.0358)	72	74																	
0.911-0.930 (0.0359-0.0366)	74																		

Measure clearance mm (in.) / Installed lifter thickness mm (in.)	5.320 (0.2094)	5.330 (0.2098)	5.340 (0.2102)	5.350 (0.2106)	5.360 (0.2110)	5.370 (0.2114)	5.380 (0.2118)	5.390 (0.2122)	5.400 (0.2126)	5.410 (0.2130)	5.420 (0.2134)	5.430 (0.2138)	5.440 (0.2142)	5.450 (0.2146)	5.460 (0.2150)	5.470 (0.2154)	5.480 (0.2157)	5.490 (0.2161)
0.000-0.030 (0.0000-0.0012)	12	14	14	16	16	18	18	20	20	22	22	24	24	26	26	28	28	30
0.031-0.050 (0.0012-0.0020)	16	16	18	18	20	20	22	22	24	24	26	26	28	28	30	30	32	32
0.051-0.070 (0.0020-0.0028)	18	18	20	20	22	22	24	24	26	26	28	28	30	30	32	32	34	34
0.071-0.090 (0.0028-0.0035)	20	20	22	22	24	24	26	26	28	28	30	30	32	32	34	34	36	36
0.091-0.110 (0.0036-0.0043)	22	22	24	24	26	26	28	28	30	30	32	32	34	34	36	36	38	38
0.111-0.130 (0.0044-0.0051)	24	24	26	26	28	28	30	30	32	32	34	34	36	36	38	38	40	40
0.131-0.150 (0.0052-0.0059)	26	26	28	28	30	30	32	32	34	34	36	36	38	38	40	40	42	42
0.151-0.170 (0.0059-0.0067)	28	28	30	30	32	32	34	34	36	36	38	38	40	40	42	42	44	44
0.171-0.189 (0.0067-0.0074)	30	30	32	32	34	34	36	36	38	38	40	40	42	42	44	44	46	46
0.190-0.290 (0.0075-0.0114)																		
0.291-0.310 (0.0115-0.0122)	38	40	40	42	42	44	44	46	46	48	48	50	50	52	52	54	54	56
0.311-0.330 (0.0122-0.0130)	40	42	42	44	44	46	46	48	48	50	50	52	52	54	54	56	56	58
0.331-0.350 (0.0130-0.0138)	42	44	44	46	46	48	48	50	50	52	52	54	54	56	56	58	58	60
0.351-0.370 (0.0138-0.0146)	44	46	46	48	48	50	50	52	52	54	54	56	56	58	58	60	60	62
0.371-0.390 (0.0146-0.0154)	46	48	48	50	50	52	52	54	54	56	56	58	58	60	60	62	62	64
0.391-0.410 (0.0154-0.0161)	48	50	50	52	52	54	54	56	56	58	58	60	60	62	62	64	64	66
0.411-0.430 (0.0162-0.0169)	50	52	52	54	54	56	56	58	58	60	60	62	62	64	64	66	66	68
0.431-0.450 (0.0170-0.0177)	52	54	54	56	56	58	58	60	60	62	62	64	64	66	66	68	68	70
0.451-0.470 (0.0178-0.0185)	54	56	56	58	58	60	60	62	62	64	64	66	66	68	68	70	70	72
0.471-0.490 (0.0185-0.0193)	56	58	58	60	60	62	62	64	64	66	66	68	68	70	70	72	72	74
0.491-0.510 (0.0193-0.0201)	58	60	60	62	62	64	64	66	66	68	68	70	70	72	72	74	74	74
0.511-0.530 (0.0201-0.0209)	60	62	62	64	64	66	66	68	68	70	70	72	72	74	74	74	74	
0.531-0.550 (0.0209-0.0217)	62	64	64	66	66	68	68	70	70	72	72	74	74	74	74			
0.551-0.570 (0.0217-0.0224)	64	66	66	68	68	70	70	72	72	74	74	74	74					
0.571-0.590 (0.0225-0.0232)	66	68	68	70	70	72	72	74	74	74	74							
0.591-0.610 (0.0233-0.0240)	68	70	70	72	72	74	74	74	74									
0.611-0.630 (0.0241-0.0248)	70	72	72	74	74	74	74											
0.631-0.650 (0.0248-0.0256)	72	74	74	74	74													
0.651-0.670 (0.0256-0.0264)	74	74	74															
0.671-0.690 (0.0264-0.0272)	74																	

Measure clearance mm (in.) / Installed lifter thickness mm (in.)	5.500 (0.2165)	5.510 (0.2169)	5.520 (0.2173)	5.530 (0.2177)	5.540 (0.2181)	5.550 (0.2185)	5.560 (0.2189)	5.570 (0.2193)	5.580 (0.2197)	5.590 (0.2201)	5.600 (0.2205)	5.620 (0.2213)	5.640 (0.2220)	5.660 (0.2228)	5.680 (0.2236)	5.700 (0.2244)	5.720 (0.2252)	5.740 (0.2260)
0.000-0.030 (0.0000-0.0012)	30	32	32	34	34	36	36	38	38	40	40	42	44	46	48	50	52	54
0.031-0.050 (0.0012-0.0020)	34	34	36	36	38	38	40	40	42	42	44	46	48	50	52	54	56	58
0.051-0.070 (0.0020-0.0028)	36	36	38	38	40	40	42	42	44	44	46	48	50	52	54	56	58	60
0.071-0.090 (0.0028-0.0035)	38	38	40	40	42	42	44	44	46	46	48	50	52	54	56	58	60	62
0.091-0.110 (0.0036-0.0043)	40	40	42	42	44	44	46	46	48	48	50	52	54	56	58	60	62	64
0.111-0.130 (0.0044-0.0051)	42	42	44	44	46	46	48	48	50	50	52	54	56	58	60	62	64	66
0.131-0.150 (0.0052-0.0059)	44	44	46	46	48	48	50	50	52	52	54	56	58	60	62	64	66	68
0.151-0.170 (0.0059-0.0067)	46	46	48	48	50	50	52	52	54	54	56	58	60	62	64	66	68	70
0.171-0.189 (0.0067-0.0074)	48	48	50	50	52	52	54	54	56	56	58	60	62	64	66	68	70	72
0.190-0.290 (0.0075-0.0114)																		
0.291-0.310 (0.0115-0.0122)	56	58	58	60	60	62	62	64	64	66	66	68	70	72	74	74	74	
0.311-0.330 (0.0122-0.0130)	58	60	60	62	62	64	64	66	66	68	68	70	72	74	74	74		
0.331-0.350 (0.0130-0.0138)	60	62	62	64	64	66	66	68	68	70	70	72	74	74	74			
0.351-0.370 (0.0138-0.0146)	62	64	64	66	66	68	68	70	70	72	72	74	74	74				
0.371-0.390 (0.0146-0.0154)	64	66	66	68	68	70	70	72	72	74	74	74	74					
0.391-0.410 (0.0154-0.0161)	66	68	68	70	70	72	72	74	74	74	74							
0.411-0.430 (0.0162-0.0169)	68	70	70	72	72	74	74	74	74									
0.431-0.450 (0.0170-0.0177)	70	72	72	74	74	74	74											
0.451-0.470 (0.0178-0.0185)	72	74	74	74	74													
0.471-0.490 (0.0185-0.0193)	74	74	74															
0.491-0.510 (0.0193-0.0201)	74																	

Intake valve clearance (Cold):
0.19 to 0.29 mm (0.008 to 0.011 in.)

EXAMPLE:

The 5.250 mm (0.2067 in.) lifter is installed, and the measured clearance is 0.400 mm (0.0157 in.). Replace the 5.250 mm (0.2067 in.) lifter with a new No. 42 lifter.

New Lifter Thickness mm (in.)

Lifter No.	Thickness	Lifter No.	Thickness	Lifter No.	Thickness
06	5.060 (0.1992)	30	5.300 (0.2087)	54	5.540 (0.2181)
08	5.080 (0.2000)	32	5.320 (0.2094)	56	5.560 (0.2189)
10	5.100 (0.2008)	34	5.340 (0.2102)	58	5.580 (0.2197)
12	5.120 (0.2016)	36	5.360 (0.2110)	60	5.600 (0.2205)
14	5.140 (0.2024)	38	5.380 (0.2118)	62	5.620 (0.2213)
16	5.160 (0.2031)	40	5.400 (0.2126)	64	5.640 (0.2220)
18	5.180 (0.2039)	42	5.420 (0.2134)	66	5.660 (0.2228)
20	5.200 (0.2047)	44	5.440 (0.2142)	68	5.680 (0.2236)
22	5.220 (0.2055)	46	5.460 (0.2150)	70	5.700 (0.2244)
24	5.240 (0.2063)	48	5.480 (0.2157)	72	5.720 (0.2252)
26	5.260 (0.2071)	50	5.500 (0.2165)	74	5.740 (0.2260)
28	5.280 (0.2079)	52	5.520 (0.2173)		

67170-HIGH-G15

Fig. 225 Intake valve lifter selection chart—2.4L engine

Valve Lifter Selection Chart (Exhaust)

Measure clearance mm (in.) \ Installed lifter thickness mm (in.)	5.060 (0.1992)	5.080 (0.2000)	5.100 (0.2008)	5.120 (0.2016)	5.140 (0.2024)	5.160 (0.2031)	5.180 (0.2039)	5.200 (0.2047)	5.210 (0.2051)	5.220 (0.2055)	5.230 (0.2059)	5.240 (0.2063)	5.250 (0.2067)	5.260 (0.2071)	5.270 (0.2075)	5.280 (0.2079)	5.290 (0.2083)	5.300 (0.2087)	5.310 (0.2091)	5.320 (0.2094)	5.330 (0.2098)	5.340 (0.2102)	5.350 (0.2106)	5.360 (0.2110)	5.370 (0.2114)	5.380 (0.2118)	5.390 (0.2122)	5.400 (0.2126)	5.410 (0.2130)	5.420 (0.2134)	5.430 (0.2138)	5.440 (0.2142)	5.450 (0.2146)	5.460 (0.2150)	5.470 (0.2154)	5.480 (0.2157)	5.490 (0.2161)	5.500 (0.2165)	5.510 (0.2169)	5.520 (0.2173)	5.530 (0.2177)	5.540 (0.2181)	5.550 (0.2185)	5.560 (0.2189)	5.570 (0.2193)	5.580 (0.2197)	5.590 (0.2201)	5.600 (0.2205)	5.620 (0.2213)	5.640 (0.2220)	5.660 (0.2228)	5.680 (0.2236)	5.700 (0.2244)	5.720 (0.2252)	5.740 (0.2260)
0.000–0.030 (0.0000–0.0012)																							06	06	08	08	10	10	12	12	14	14	16	16	18	18	20	20	22	22	24	24	26	26	28	28	30	30	32	34	36	38	40	42	44
0.031–0.050 (0.0012–0.0020)																					06	06	08	08	10	10	12	12	14	14	16	16	18	18	20	20	22	22	24	24	26	26	28	28	30	30	32	32	34	36	38	40	42	44	46
0.051–0.070 (0.0020–0.0028)																			06	06	08	08	10	10	12	12	14	14	16	16	18	18	20	20	22	22	24	24	26	26	28	28	30	30	32	32	34	34	36	38	40	42	44	46	48
0.071–0.090 (0.0028–0.0035)																	06	06	08	08	10	10	12	12	14	14	16	16	18	18	20	20	22	22	24	24	26	26	28	28	30	30	32	32	34	34	36	36	38	40	42	44	46	48	50
0.091–0.110 (0.0036–0.0043)															06	06	08	08	10	10	12	12	14	14	16	16	18	18	20	20	22	22	24	24	26	26	28	28	30	30	32	32	34	34	36	36	38	38	40	42	44	46	48	50	52
0.111–0.130 (0.0044–0.0051)													06	06	08	08	10	10	12	12	14	14	16	16	18	18	20	20	22	22	24	24	26	26	28	28	30	30	32	32	34	34	36	36	38	38	40	40	42	44	46	48	50	52	54
0.131–0.150 (0.0052–0.0059)											06	06	08	08	10	10	12	12	14	14	16	16	18	18	20	20	22	22	24	24	26	26	28	28	30	30	32	32	34	34	36	36	38	38	40	40	42	42	44	46	48	50	52	54	56
0.151–0.170 (0.0059–0.0067)									06	06	08	08	10	10	12	12	14	14	16	16	18	18	20	20	22	22	24	24	26	26	28	28	30	30	32	32	34	34	36	36	38	38	40	40	42	42	44	44	46	48	50	52	54	56	58
0.171–0.190 (0.0067–0.0075)								06	08	08	10	10	12	12	14	14	16	16	18	18	20	20	22	22	24	24	26	26	28	28	30	30	32	32	34	34	36	36	38	38	40	40	42	42	44	44	46	46	48	50	52	54	56	58	60
0.191–0.210 (0.0075–0.0083)							06	08	10	10	12	12	14	14	16	16	18	18	20	20	22	22	24	24	26	26	28	28	30	30	32	32	34	34	36	36	38	38	40	40	42	42	44	44	46	46	48	48	50	52	54	56	58	60	62
0.211–0.230 (0.0083–0.0091)						06	08	10	12	12	14	14	16	16	18	18	20	20	22	22	24	24	26	26	28	28	30	30	32	32	34	34	36	36	38	38	40	40	42	42	44	44	46	46	48	48	50	50	52	54	56	58	60	62	64
0.231–0.250 (0.0091–0.0098)					06	08	10	12	14	14	16	16	18	18	20	20	22	22	24	24	26	26	28	28	30	30	32	32	34	34	36	36	38	38	40	40	42	42	44	44	46	46	48	48	50	50	52	52	54	56	58	60	62	64	66
0.251–0.270 (0.0099–0.0106)				06	08	10	12	14	16	16	18	18	20	20	22	22	24	24	26	26	28	28	30	30	32	32	34	34	36	36	38	38	40	40	42	42	44	44	46	46	48	48	50	50	52	52	54	54	56	58	60	62	64	66	68
0.271–0.290 (0.0107–0.0114)			06	08	10	12	14	16	18	18	20	20	22	22	24	24	26	26	28	28	30	30	32	32	34	34	36	36	38	38	40	40	42	42	44	44	46	46	48	48	50	50	52	52	54	54	56	56	58	60	62	64	66	68	70
0.291–0.299 (0.0115–0.0118)		06	08	10	12	14	16	18	20	20	22	22	24	24	26	26	28	28	30	30	32	32	34	34	36	36	38	38	40	40	42	42	44	44	46	46	48	48	50	50	52	52	54	54	56	56	58	58	60	62	64	66	68	70	72
0.300–0.400 (0.0118–0.0157)																																																							
0.401–0.420 (0.0158–0.0165)	12	14	16	18	20	22	24	26	26	28	28	30	30	32	32	34	34	36	36	38	38	40	40	42	42	44	44	46	46	48	48	50	50	52	52	54	54	56	56	58	58	60	60	62	62	64	64	66	68	70	72	74	74	74	
0.421–0.440 (0.0166–0.0173)	14	16	18	20	22	24	26	28	28	30	30	32	32	34	34	36	36	38	38	40	40	42	42	44	44	46	46	48	48	50	50	52	52	54	54	56	56	58	58	60	60	62	62	64	64	66	66	68	70	72	74	74	74		
0.441–0.460 (0.0174–0.0181)	16	18	20	22	24	26	28	30	30	32	32	34	34	36	36	38	38	40	40	42	42	44	44	46	46	48	48	50	50	52	52	54	54	56	56	58	58	60	60	62	62	64	64	66	66	68	68	70	72	74	74	74			
0.461–0.480 (0.0181–0.0189)	18	20	22	24	26	28	30	32	32	34	34	36	36	38	38	40	40	42	42	44	44	46	46	48	48	50	50	52	52	54	54	56	56	58	58	60	60	62	62	64	64	66	66	68	68	70	70	72	74	74	74				
0.481–0.500 (0.0189–0.0197)	20	22	24	26	28	30	32	34	34	36	36	38	38	40	40	42	42	44	44	46	46	48	48	50	50	52	52	54	54	56	56	58	58	60	60	62	62	64	64	66	66	68	68	70	70	72	72	74	74	74					
0.501–0.520 (0.0197–0.0205)	22	24	26	28	30	32	34	36	36	38	38	40	40	42	42	44	44	46	46	48	48	50	50	52	52	54	54	56	56	58	58	60	60	62	62	64	64	66	66	68	68	70	70	72	72	74	74	74	74						
0.521–0.540 (0.0205–0.0213)	24	26	28	30	32	34	36	38	38	40	40	42	42	44	44	46	46	48	48	50	50	52	52	54	54	56	56	58	58	60	60	62	62	64	64	66	66	68	68	70	70	72	72	74	74	76	76	74							
0.541–0.560 (0.0213–0.0220)	26	28	30	32	34	36	38	40	40	42	42	44	44	46	46	48	48	50	50	52	52	54	54	56	56	58	58	60	60	62	62	64	64	66	66	68	68	70	70	72	72	74	74	74	74	74	74								
0.561–0.580 (0.0221–0.0228)	28	30	32	34	36	38	40	42	42	44	44	46	46	48	48	50	50	52	52	54	54	56	56	58	58	60	60	62	62	64	64	66	66	68	68	70	70	72	72	74	74	74	74	74	74										
0.581–0.600 (0.0229–0.0236)	30	32	34	36	38	40	42	44	44	46	46	48	48	50	50	52	52	54	54	56	56	58	58	60	60	62	62	64	64	66	66	68	68	70	70	72	72	74	74	74	74	74	74												
0.601–0.620 (0.0237–0.0244)	32	34	36	38	40	42	44	46	46	48	48	50	50	52	52	54	54	56	56	58	58	60	60	62	62	64	64	66	66	68	68	70	70	72	72	74	74	74	74	74	74														
0.621–0.640 (0.0244–0.0252)	34	36	38	40	42	44	46	48	48	50	50	52	52	54	54	56	56	58	58	60	60	62	62	64	64	66	66	68	68	70	70	72	72	74	74	74	74	74	74																
0.641–0.660 (0.0252–0.0260)	36	38	40	42	44	46	48	50	50	52	52	54	54	56	56	58	58	60	60	62	62	64	64	66	66	68	68	70	70	72	72	74	74	74	74	74	74																		
0.661–0.680 (0.0260–0.0268)	38	40	42	44	46	48	50	52	52	54	54	56	56	58	58	60	60	62	62	64	64	66	66	68	68	70	70	72	72	74	74	74	74	74	74																				
0.681–0.700 (0.0268–0.0276)	40	42	44	46	48	50	52	54	54	56	56	58	58	60	60	62	62	64	64	66	66	68	68	70	70	72	72	74	74	74	74	74	74																						
0.701–0.720 (0.0276–0.0283)	42	44	46	48	50	52	54	56	56	58	58	60	60	62	62	64	64	66	66	68	68	70	70	72	72	74	74	74	74	74	74																								
0.721–0.740 (0.0284–0.0291)	44	46	48	50	52	54	56	58	58	60	60	62	62	64	64	66	66	68	68	70	70	72	72	74	74	74	74	74	74																										
0.741–0.760 (0.0292–0.0299)	46	48	50	52	54	56	58	60	60	62	62	64	64	66	66	68	68	70	70	72	72	74	74	74	74	74	74																												
0.761–0.780 (0.0300–0.0307)	48	50	52	54	56	58	60	62	62	64	64	66	66	68	68	70	70	72	72	74	74	74	74	74	74																														
0.781–0.800 (0.0307–0.0315)	50	52	54	56	58	60	62	64	64	66	66	68	68	70	70	72	72	74	74	74	74	74	74																																
0.801–0.820 (0.0315–0.0323)	52	54	56	58	60	62	64	66	66	68	68	70	70	72	72	74	74	74	74	74	74																																		
0.821–0.840 (0.0323–0.0331)	54	56	58	60	62	64	66	68	68	70	70	72	72	74	74	74	74	74	74																																				
0.841–0.860 (0.0331–0.0339)	56	58	60	62	64	66	68	70	70	72	72	74	74	74	74	74	74																																						
0.861–0.880 (0.0339–0.0346)	58	60	62	64	66	68	70	72	72	74	74	74	74	74	74																																								
0.881–0.900 (0.0347–0.0354)	60	62	64	66	68	70	72	74	74	74	74	74	74																																										
0.901–0.920 (0.0355–0.0362)	62	64	66	68	70	72	74	74	74	74	74																																												
0.921–0.940 (0.0363–0.0370)	64	66	68	70	72	74	74	74	74																																														
0.941–0.960 (0.0370–0.0378)	66	68	70	72	74	74	74																																																
0.961–0.980 (0.0378–0.0386)	68	70	72	74	74	74																																																	
0.981–1.000 (0.0386–0.0394)	70	72	74	74	74																																																		
1.001–1.020 (0.0394–0.0402)	72	74	74	74																																																			
1.021–1.040 (0.0402–0.0409)	74	74	74																																																				
1.041–1.060 (0.0410–0.0417)	74	74																																																					
1.061–1.080 (0.0418–0.0425)	74																																																						

New lifter thickness mm (in.)

Lifter No.	Thickness	Lifter No.	Thickness	Lifter No.	Thickness
06	5.060 (0.1992)	30	5.300 (0.2087)	54	5.540 (0.2181)
08	5.080 (0.2000)	32	5.320 (0.2094)	56	5.560 (0.2189)
10	5.100 (0.2008)	34	5.340 (0.2102)	58	5.580 (0.2197)
12	5.120 (0.2016)	36	5.360 (0.2110)	60	5.600 (0.2205)
14	5.140 (0.2024)	38	5.380 (0.2118)	62	5.620 (0.2213)
16	5.160 (0.2031)	40	5.400 (0.2126)	64	5.640 (0.2220)
18	5.180 (0.2039)	42	5.420 (0.2134)	66	5.660 (0.2228)
20	5.200 (0.2047)	44	5.440 (0.2142)	68	5.680 (0.2236)
22	5.220 (0.2055)	46	5.460 (0.2150)	70	5.700 (0.2244)
24	5.240 (0.2063)	48	5.480 (0.2157)	72	5.720 (0.2252)
26	5.260 (0.2071)	50	5.500 (0.2165)	74	5.740 (0.2260)
28	5.280 (0.2079)	52	5.520 (0.2173)		

Exhaust valve clearance (Cold):
0.30 to 0.40 mm (0.012 to 0.016 in.)

EXAMPLE:

The 5.340 mm (0.2102 in.) lifter is installed, and the measured clearance is 0.440 mm (0.0173 in.). Replace the 5.340 mm (0.2102 in.) lifter with a new No. 44 lifter.

67170-HIGH-G16

Fig. 226 Exhaust valve lifter selection chart—2.4L engine

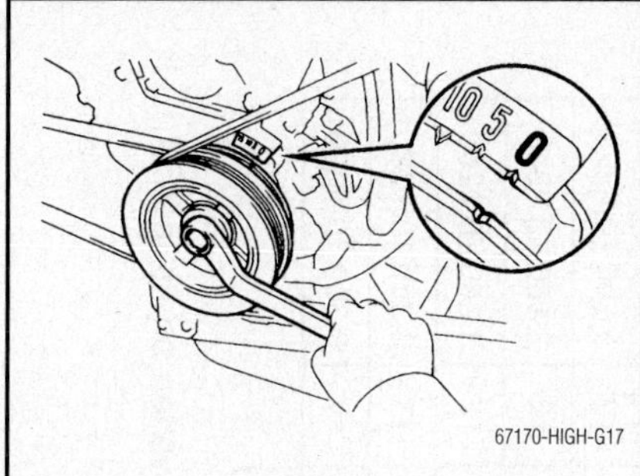

Fig. 227 Turn the crankshaft pulley and align its groove with the timing mark 0 of the No. 1 timing cover—3.3L engine

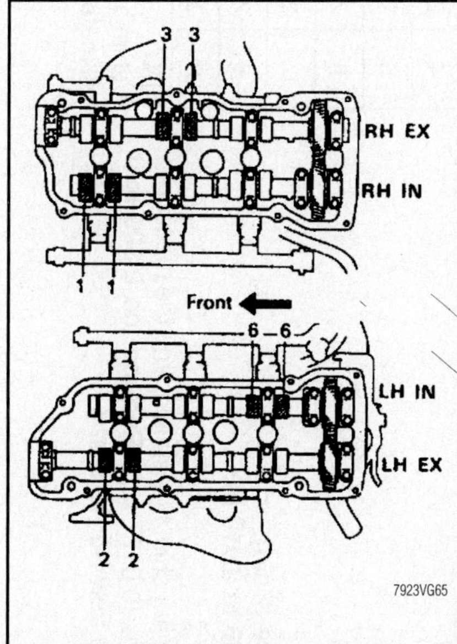

Fig. 228 Adjust these valves during the 1st step—3.3L engine

are tight. If not, turn the crankshaft 1 complete revolution (360 degrees).

➡**All measurements should be written down. These recorded measurements will need to be used in conjunction with a mathematical formula to determine the thickness of the replacement shims.**

5. Measure the clearance between the valve lifters and the camshaft. Record the measurements on valves No. 1 and 6 intake; No. 2 and 3 exhaust.

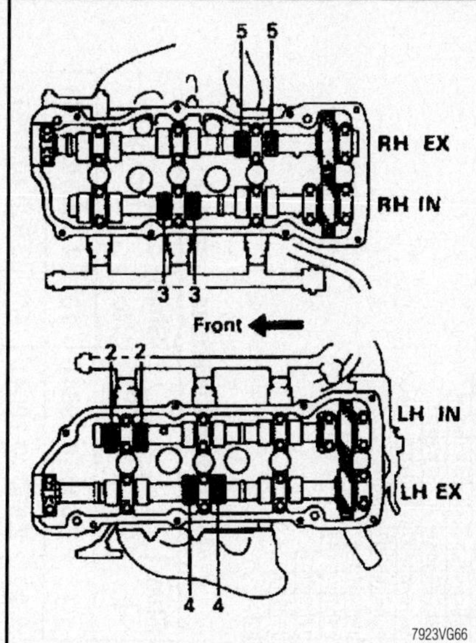

Fig. 229 Adjust these valves during the 2nd step—3.3L engine

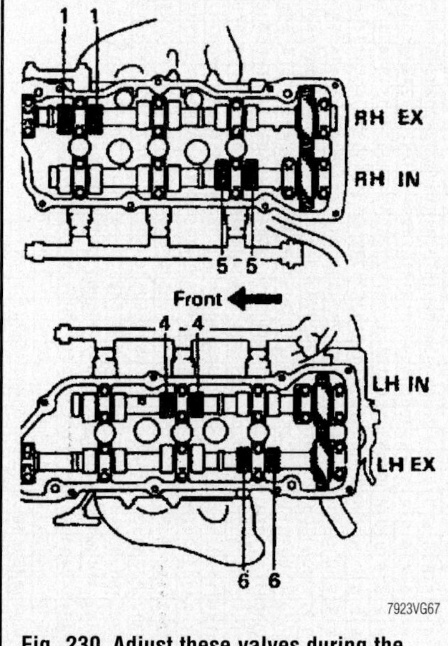

Fig. 230 Adjust these valves during the 3rd step—3.3L engine

a. The intake valve clearance cold is 0.006–0.010 in. (0.15–0.25mm).

b. The exhaust valve clearance cold is 0.010–0.014 in. (0.25–0.35mm).

6. Turn the crankshaft ⅔ of a revolution (240 degrees). Record the measurements on valves No. 2 and 3 intake; No. 4 and 5 exhaust.

7. Turn the crankshaft another ⅔ of a revolution. Record the measurements on valves No. 4 and 5 intake; No. 1 and 6 exhaust.

8. Remove the adjusting shim by turning the crankshaft to position the cam lobe of the camshaft in the up position on the valve to be adjusted. Using a small thin flat bladed tool, turn the valve lifter so that the notches are perpendicular to the camshaft. Press down the valve lifter with tool 09248-55010 part A. Place too 09248-55010 part B between the camshaft and the valve lifter; remove part A.

9. Remove the adjusting shim with a magnet and a small screwdriver.

10. Determine the replacement adjusting shim size by either using the charts or the following formulas:

- Intake: N = T + (A - 0.008 in./0.020mm)
- Exhaust: N = T + (A - 0.012 in./0.30mm)
- T = Thickness of removed shim
- A = Measured valve clearance
- N = Thickness of new shim

11. Select a new shim with a thickness as close as possible to the calculated value. Install the new replacement shim.

➡**Shims are available in 17 sizes in increments of 0.0020 in. (0.050mm), from 0.0984 in. (2.500mm) to 0.1299 in. (3.300mm).**

12. Recheck the valve clearance.

13. Install or connect the following:

- Cylinder head covers
- Spark plugs and the ignition coils
- Engine wiring harness to the injectors and the coils
- Intake chamber
- Emission control valve set
- V-bank cover
- Air flow meter, air duct and air cleaner cover
- Negative battery cable

3.5L Models

No adjustment is necessary on these engines.

Adjusting Shim Selection Chart (Intake)

Measured clearance mm (in.) \ Installed shim thickness mm (in.)	2.500 (0.0984)	2.520 (0.0992)	2.540 (0.1000)	2.550 (0.1004)	2.560 (0.1008)	2.580 (0.1016)	2.600 (0.1024)	2.620 (0.1031)	2.640 (0.1039)	2.650 (0.1043)	2.660 (0.1047)	2.670 (0.1051)	2.680 (0.1055)	2.690 (0.1059)	2.700 (0.1063)	2.710 (0.1067)	2.720 (0.1071)	2.730 (0.1075)	2.740 (0.1079)	2.750 (0.1083)	2.760 (0.1087)	2.770 (0.1091)	2.780 (0.1094)	2.790 (0.1098)	2.800 (0.1102)	2.810 (0.1106)	2.820 (0.1110)	2.830 (0.1114)	2.840 (0.1118)	2.850 (0.1122)	2.860 (0.1126)	2.870 (0.1130)	2.880 (0.1134)
0.000 - 0.020 (0.0000 - 0.0008)									1	1	1	1	1	1	1	1	2	2	2	2	2	3	3	3	3	3	4	4	4	4	4	5	5
0.021 - 0.040 (0.0008 - 0.0016)								1	1	1	1	1	1	1	2	2	2	2	2	3	3	3	3	3	4	4	4	4	4	5	5	5	5
0.041 - 0.060 (0.0016 - 0.0024)							1	1	1	1	1	1	2	2	2	2	2	3	3	3	3	3	4	4	4	4	4	5	5	5	5	5	6
0.061 - 0.080 (0.0024 - 0.0031)						1	1	1	1	1	2	2	2	2	2	3	3	3	3	3	4	4	4	4	4	5	5	5	5	5	6	6	6
0.081 - 0.100 (0.0032 - 0.0039)				1	1	1	1	1	2	2	2	2	2	3	3	3	3	3	4	4	4	4	4	5	5	5	5	5	6	6	6	6	6
0.101 - 0.120 (0.0040 - 0.0047)			1	1	1	1	1	2	2	2	2	3	3	3	3	3	4	4	4	4	4	5	5	5	5	5	6	6	6	6	6	7	7
0.121 - 0.140 (0.0048 - 0.0055)		1	1	1	1	1	2	2	2	3	3	3	3	3	4	4	4	4	4	5	5	5	5	5	6	6	6	6	6	7	7	7	7
0.141 - 0.149 (0.0056 - 0.0059)		1	1	1	1	2	2	2	3	3	3	3	4	4	4	4	4	5	5	5	5	5	6	6	6	6	6	7	7	7	7	7	8
0.150 - 0.250 (0.0059 - 0.0098)																																	
0.251 - 0.260 (0.0099 - 0.0102)	2	3	3	3	3	4	4	5	5	5	5	6	6	6	6	6	7	7	7	7	7	8	8	8	8	8	9	9	9	9	9	10	10
0.261 - 0.280 (0.0103 - 0.0110)	2	3	3	3	4	4	4	5	5	5	6	6	6	6	6	7	7	7	7	7	8	8	8	8	8	9	9	9	9	9	10	10	10
0.281 - 0.300 (0.0111 - 0.0118)	3	3	4	4	4	4	5	5	6	6	6	6	6	7	7	7	7	7	8	8	8	8	8	9	9	9	9	9	10	10	10	10	10
0.301 - 0.320 (0.0119 - 0.0126)	3	4	4	4	4	5	5	6	6	6	6	7	7	7	7	7	8	8	8	8	8	9	9	9	9	9	10	10	10	10	10	11	11
0.321 - 0.340 (0.0126 - 0.0134)	4	4	4	5	5	5	6	6	6	7	7	7	7	7	8	8	8	8	8	9	9	9	9	9	10	10	10	10	10	11	11	11	11
0.341 - 0.360 (0.0134 - 0.0142)	4	4	5	5	5	6	6	6	7	7	7	7	8	8	8	8	8	9	9	9	9	9	10	10	10	10	10	11	11	11	11	11	12
0.361 - 0.380 (0.0142 - 0.0150)	4	5	5	5	6	6	6	7	7	7	8	8	8	8	8	9	9	9	9	9	10	10	10	10	10	11	11	11	11	11	12	12	12
0.381 - 0.400 (0.0150 - 0.0157)	5	5	6	6	6	6	7	7	8	8	8	8	8	9	9	9	9	9	10	10	10	10	10	11	11	11	11	11	12	12	12	12	12
0.401 - 0.420 (0.0158 - 0.0165)	5	6	6	6	6	7	7	8	8	8	8	9	9	9	9	9	10	10	10	10	10	11	11	11	11	11	12	12	12	12	12	13	13
0.421 - 0.440 (0.0166 - 0.0173)	6	6	6	7	7	7	8	8	8	9	9	9	9	9	10	10	10	10	10	11	11	11	11	11	12	12	12	12	12	13	13	13	13
0.441 - 0.460 (0.0174 - 0.0181)	6	6	7	7	7	8	8	8	9	9	9	9	10	10	10	10	10	11	11	11	11	11	12	12	12	12	12	13	13	13	13	13	14
0.461 - 0.480 (0.0181 - 0.0189)	6	7	7	7	8	8	8	9	9	9	10	10	10	10	10	11	11	11	11	11	12	12	12	12	12	13	13	13	13	13	14	14	14
0.481 - 0.500 (0.0189 - 0.0197)	7	7	8	8	8	8	9	9	10	10	10	10	10	11	11	11	11	11	12	12	12	12	12	13	13	13	13	13	14	14	14	14	14
0.501 - 0.520 (0.0197 - 0.0205)	7	8	8	8	8	9	9	10	10	10	10	11	11	11	11	11	12	12	12	12	12	13	13	13	13	13	14	14	14	14	14	15	15
0.521 - 0.540 (0.0205 - 0.0213)	8	8	8	9	9	9	10	10	10	11	11	11	11	11	12	12	12	12	12	13	13	13	13	13	14	14	14	14	14	15	15	15	15
0.541 - 0.560 (0.0213 - 0.0220)	8	8	9	9	9	10	10	10	11	11	11	11	12	12	12	12	12	13	13	13	13	13	14	14	14	14	14	15	15	15	15	15	16
0.561 - 0.580 (0.0221 - 0.0228)	8	9	9	9	10	10	10	11	11	11	12	12	12	12	12	13	13	13	13	13	14	14	14	14	14	15	15	15	15	15	16	16	16
0.581 - 0.600 (0.0229 - 0.0236)	9	9	10	10	10	10	11	11	12	12	12	12	12	13	13	13	13	13	14	14	14	14	14	15	15	15	15	15	16	16	16	16	16
0.601 - 0.620 (0.0237 - 0.0244)	9	10	10	10	10	11	11	12	12	12	12	13	13	13	13	13	14	14	14	14	14	15	15	15	15	15	16	16	16	16	16	17	17
0.621 - 0.640 (0.0244 - 0.0252)	10	10	10	11	11	11	12	12	12	13	13	13	13	13	14	14	14	14	14	15	15	15	15	15	16	16	16	16	16	17	17	17	17
0.641 - 0.660 (0.0252 - 0.0260)	10	10	11	11	11	12	12	12	13	13	13	13	14	14	14	14	14	15	15	15	15	15	16	16	16	16	16	17	17	17	17	17	17
0.661 - 0.680 (0.0260 - 0.0268)	10	11	11	11	12	12	12	13	13	13	14	14	14	14	14	15	15	15	15	15	16	16	16	16	16	17	17	17	17	17	17	17	17
0.681 - 0.700 (0.0268 - 0.0276)	11	11	12	12	12	12	13	13	14	14	14	14	14	15	15	15	15	15	16	16	16	16	16	17	17	17	17	17	17	17	17		
0.701 - 0.720 (0.0276 - 0.0283)	11	12	12	12	12	13	13	14	14	14	14	15	15	15	15	15	16	16	16	16	16	17	17	17	17	17	17	17	17				
0.721 - 0.740 (0.0284 - 0.0291)	12	12	12	13	13	13	14	14	14	15	15	15	15	15	16	16	16	16	16	17	17	17	17	17	17	17	17						
0.741 - 0.760 (0.0292 - 0.0299)	12	12	13	13	13	14	14	14	15	15	15	15	16	16	16	16	16	17	17	17	17	17	17	17	17								
0.761 - 0.780 (0.0300 - 0.0307)	12	13	13	13	14	14	14	15	15	15	16	16	16	16	16	17	17	17	17	17	17	17	17										
0.781 - 0.800 (0.0307 - 0.0315)	13	13	14	14	14	14	15	15	16	16	16	16	16	17	17	17	17	17	17	17	17												
0.801 - 0.820 (0.0315 - 0.0323)	13	14	14	14	14	15	15	16	16	16	16	17	17	17	17	17	17	17	17														
0.821 - 0.840 (0.0323 - 0.0331)	14	14	14	15	15	15	16	16	16	17	17	17	17	17	17	17	17																
0.841 - 0.860 (0.0331 - 0.0339)	14	14	15	15	15	16	16	16	17	17	17	17	17	17	17																		
0.861 - 0.880 (0.0339 - 0.0346)	14	15	15	15	16	16	16	17	17	17	17	17	17																				
0.881 - 0.900 (0.0347 - 0.0354)	15	15	16	16	16	16	17	17	17	17	17																						
0.901 - 0.920 (0.0355 - 0.0362)	15	16	16	16	16	17	17	17	17																								
0.921 - 0.940 (0.0363 - 0.0370)	16	16	16	17	17	17	17	17																									
0.941 - 0.960 (0.0370 - 0.0378)	16	16	17	17	17	17	17																										
0.961 - 0.980 (0.0378 - 0.0386)	16	17	17	17	17	17																											
0.981 - 1.000 (0.0386 - 0.0394)	17	17	17	17	17																												
1.001 - 1.020 (0.0394 - 0.0402)	17	17	17																														
1.021 - 1.040 (0.0402 - 0.0409)	17	17																															
1.041 - 1.050 (0.0410 - 0.0413)	17																																

Measured clearance mm (in.) \ Installed shim thickness mm (in.)	2.890 (0.1138)	2.900 (0.1142)	2.910 (0.1146)	2.920 (0.1150)	2.930 (0.1154)	2.940 (0.1157)	2.950 (0.1161)	2.960 (0.1165)	2.970 (0.1169)	2.980 (0.1173)	2.990 (0.1177)	3.000 (0.1181)	3.010 (0.1185)	3.020 (0.1189)	3.030 (0.1193)	3.040 (0.1197)	3.050 (0.1201)	3.060 (0.1205)	3.080 (0.1213)	3.100 (0.1220)	3.120 (0.1228)	3.140 (0.1236)	3.150 (0.1240)	3.160 (0.1244)	3.180 (0.1252)	3.200 (0.1260)	3.220 (0.1268)	3.240 (0.1276)	3.250 (0.1280)	3.260 (0.1283)	3.280 (0.1291)	3.300 (0.1299)
0.000 - 0.020 (0.0000 - 0.0008)	5	5	5	6	6	6	6	6	7	7	7	7	7	8	8	8	8	8	9	9	10	10	10	10	11	11	12	12	12	12	13	13
0.021 - 0.040 (0.0008 - 0.0016)	5	6	6	6	6	6	7	7	7	7	7	8	8	8	8	8	9	9	9	10	10	10	11	11	11	12	12	12	13	13	13	14
0.041 - 0.060 (0.0016 - 0.0024)	6	6	6	6	7	7	7	7	7	8	8	8	8	8	9	9	9	9	10	10	10	11	11	11	12	12	12	13	13	13	14	14
0.061 - 0.080 (0.0024 - 0.0031)	6	6	7	7	7	7	7	8	8	8	8	8	9	9	9	9	9	10	10	10	11	11	11	12	12	12	13	13	13	14	14	14
0.081 - 0.100 (0.0032 - 0.0039)	7	7	7	7	7	8	8	8	8	8	9	9	9	9	9	10	10	10	10	11	11	12	12	12	12	13	13	14	14	14	14	15
0.101 - 0.120 (0.0040 - 0.0047)	7	7	7	8	8	8	8	8	9	9	9	9	9	10	10	10	10	10	11	11	12	12	12	12	13	13	14	14	14	14	15	15
0.121 - 0.140 (0.0048 - 0.0055)	7	8	8	8	8	8	9	9	9	9	9	10	10	10	10	10	11	11	11	12	12	12	13	13	13	14	14	14	15	15	15	16
0.141 - 0.149 (0.0056 - 0.0059)	8	8	8	8	9	9	9	9	9	10	10	10	10	10	11	11	11	11	12	12	12	13	13	13	14	14	14	15	15	15	16	16
0.150 - 0.250 (0.0059 - 0.0098)																																
0.251 - 0.260 (0.0099 - 0.0102)	10	10	10	11	11	11	11	11	12	12	12	12	12	13	13	13	13	13	14	14	15	15	15	15	16	16	17	17	17	17	17	
0.261 - 0.280 (0.0103 - 0.0110)	10	10	11	11	11	11	11	12	12	12	12	12	13	13	13	13	13	14	14	14	15	15	15	16	16	16	17	17	17	17	17	
0.281 - 0.300 (0.0111 - 0.0118)	11	11	11	11	11	12	12	12	12	12	13	13	13	13	13	14	14	14	14	15	15	16	16	16	16	17	17	17	17	17		
0.301 - 0.320 (0.0119 - 0.0126)	11	11	11	12	12	12	12	12	13	13	13	13	13	14	14	14	14	14	15	15	16	16	16	16	17	17	17	17				
0.321 - 0.340 (0.0126 - 0.0134)	11	12	12	12	12	12	13	13	13	13	13	14	14	14	14	14	15	15	15	16	16	16	17	17	17	17	17					
0.341 - 0.360 (0.0134 - 0.0142)	12	12	12	12	13	13	13	13	13	14	14	14	14	14	15	15	15	15	16	16	16	17	17	17	17	17						
0.361 - 0.380 (0.0142 - 0.0150)	12	12	13	13	13	13	13	14	14	14	14	14	15	15	15	15	15	16	16	16	17	17	17	17	17							
0.381 - 0.400 (0.0150 - 0.0157)	13	13	13	13	13	14	14	14	14	14	15	15	15	15	16	16	16	16	16	17	17	17	17	17								
0.401 - 0.420 (0.0158 - 0.0165)	13	13	13	14	14	14	14	14	15	15	15	15	15	16	16	16	16	16	17	17	17	17										
0.421 - 0.440 (0.0166 - 0.0173)	13	14	14	14	14	14	15	15	15	15	15	16	16	16	16	16	17	17	17	17	17											
0.441 - 0.460 (0.0174 - 0.0181)	14	14	14	14	15	15	15	15	15	16	16	16	16	16	17	17	17	17	17	17												
0.461 - 0.480 (0.0181 - 0.0189)	14	14	15	15	15	15	15	16	16	16	16	16	17	17	17	17	17	17	17													
0.481 - 0.500 (0.0189 - 0.0197)	15	15	15	15	15	16	16	16	16	16	17	17	17	17	17	17	17	17														
0.501 - 0.520 (0.0197 - 0.0205)	15	15	15	16	16	16	16	16	17	17	17	17	17	17	17	17																
0.521 - 0.540 (0.0205 - 0.0213)	15	16	16	16	16	16	17	17	17	17	17	17	17	17																		
0.541 - 0.560 (0.0213 - 0.0220)	16	16	16	16	17	17	17	17	17	17	17	17																				
0.561 - 0.580 (0.0221 - 0.0228)	16	16	17	17	17	17	17	17	17	17																						
0.581 - 0.600 (0.0229 - 0.0236)	17	17	17	17	17	17	17	17																								
0.601 - 0.620 (0.0237 - 0.0244)	17	17	17	17	17	17																										
0.621 - 0.640 (0.0244 - 0.0252)	17	17	17	17																												
0.641 - 0.660 (0.0252 - 0.0260)	17	17																														

Intake valve clearance (Cold):
0.15 to 0.25 mm (0.006 to 0.010 in.)

EXAMPLE:
The 2.800 mm (0.1102 in.) shim is installed, and the measured clearance is 0.450 mm (0.0177 in.). Replace the 2.800 mm (0.1102 in.) shim with a new No. 12 shim.

New shim thickness mm (in.)

Shim No.	Thickness	Shim No.	Thickness
1	2.500 (0.0984)	10	2.950 (0.1161)
2	2.550 (0.1004)	11	3.000 (0.1181)
3	2.600 (0.1024)	12	3.050 (0.1201)
4	2.650 (0.1043)	13	3.100 (0.1220)
5	2.700 (0.1063)	14	3.150 (0.1240)
6	2.750 (0.1083)	15	3.200 (0.1260)
7	2.800 (0.1102)	16	3.250 (0.1280)
8	2.850 (0.1122)	17	3.300 (0.1299)
9	2.900 (0.1142)		

HINT:
A shim's thickness is written on its face in millimeters.

67170-HIGH-G18

Fig. 231 Intake valve shim selection chart—3.3L engine

Adjusting Shim Selection Chart (Exhaust)

Measured clearance mm (in.) / Installed shim thickness mm (in.)	2.500 (0.0984)	2.520 (0.0992)	2.540 (0.1000)	2.550 (0.1004)	2.560 (0.1008)	2.580 (0.1016)	2.600 (0.1024)	2.620 (0.1031)	2.640 (0.1039)	2.650 (0.1043)	2.660 (0.1047)	2.670 (0.1051)	2.680 (0.1055)	2.690 (0.1059)	2.700 (0.1063)	2.710 (0.1067)	2.720 (0.1071)	2.730 (0.1076)	2.740 (0.1079)	2.750 (0.1083)	2.760 (0.1087)	2.770 (0.1091)	2.780 (0.1094)	2.790 (0.1098)	2.800 (0.1102)	2.810 (0.1106)	2.820 (0.1110)	2.830 (0.1114)	2.840 (0.1118)	2.850 (0.1122)	2.860 (0.1126)	2.870 (0.1130)	2.880 (0.1134)	2.890 (0.1138)	2.900 (0.1142)	2.910 (0.1146)	2.920 (0.1150)	2.930 (0.1154)	2.940 (0.1157)	2.950 (0.1161)	2.960 (0.1165)	2.970 (0.1169)	2.980 (0.1173)	2.990 (0.1177)	3.000 (0.1181)	3.010 (0.1185)	3.020 (0.1189)	3.030 (0.1193)	3.040 (0.1197)	3.050 (0.1201)	3.060 (0.1205)	3.080 (0.1213)	3.100 (0.1220)	3.120 (0.1228)	3.140 (0.1236)	3.150 (0.1240)	3.160 (0.1244)	3.180 (0.1252)	3.200 (0.1260)	3.220 (0.1268)	3.240 (0.1276)	3.250 (0.1280)	3.260 (0.1283)	3.280 (0.1291)	3.300 (0.1299)
0.000 – 0.020 (0.0000 – 0.0008)																		1	1	1	1	1	1	1	1	1	2	2	2	2	2	3	3	3	3	3	4	4	4	4	4	5	5	5	5	5	6	6	6	6	6	7	7	8	8	8	8	9	9	10	10	10	10	11	11
0.021 – 0.040 (0.0008 – 0.0016)																1	1	1	1	1	1	1	1	1	2	2	2	2	2	3	3	3	3	3	4	4	4	4	4	5	5	5	5	5	6	6	6	6	6	7	7	7	8	8	8	9	9	9	10	10	10	11	11	11	12
0.041 – 0.060 (0.0016 – 0.0024)														1	1	1	1	1	1	1	1	1	2	2	2	2	2	3	3	3	3	3	4	4	4	4	4	5	5	5	5	5	6	6	6	6	6	7	7	7	7	8	8	8	9	9	9	10	10	10	11	11	11	12	12
0.061 – 0.080 (0.0024 – 0.0031)												1	1	1	1	1	1	1	1	1	2	2	2	2	2	3	3	3	3	3	4	4	4	4	4	5	5	5	5	5	6	6	6	6	6	7	7	7	7	7	8	8	8	9	9	9	10	10	10	11	11	11	12	12	12
0.081 – 0.100 (0.0032 – 0.0039)										1	1	1	1	1	1	1	1	1	2	2	2	2	2	3	3	3	3	3	4	4	4	4	4	5	5	5	5	5	6	6	6	6	6	7	7	7	7	7	8	8	8	8	9	9	10	10	10	10	11	11	12	12	12	12	13
0.101 – 0.120 (0.0040 – 0.0047)									1	1	1	1	1	1	1	1	2	2	2	2	2	3	3	3	3	3	4	4	4	4	4	5	5	5	5	5	6	6	6	6	6	7	7	7	7	7	8	8	8	8	8	9	9	10	10	10	10	11	11	12	12	12	12	13	13
0.121 – 0.140 (0.0048 – 0.0055)								1	1	1	1	1	1	1	2	2	2	2	2	3	3	3	3	3	4	4	4	4	4	5	5	5	5	5	6	6	6	6	6	7	7	7	7	7	8	8	8	8	8	9	9	9	10	10	10	11	11	11	12	12	12	13	13	13	14
0.141 – 0.160 (0.0056 – 0.0063)							1	1	1	1	1	1	2	2	2	2	2	3	3	3	3	3	4	4	4	4	4	5	5	5	5	5	6	6	6	6	6	7	7	7	7	7	8	8	8	8	8	9	9	9	9	10	10	10	11	11	11	12	12	12	13	13	13	14	14
0.161 – 0.180 (0.0063 – 0.0071)						1	1	1	1	1	2	2	2	2	2	3	3	3	3	3	4	4	4	4	4	5	5	5	5	5	6	6	6	6	6	7	7	7	7	7	8	8	8	8	8	9	9	9	9	9	10	10	10	11	11	11	12	12	12	13	13	13	14	14	14
0.181 – 0.200 (0.0071 – 0.0079)				1	1	1	1	1	2	2	2	2	2	3	3	3	3	3	4	4	4	4	4	5	5	5	5	5	6	6	6	6	6	7	7	7	7	7	8	8	8	8	8	9	9	9	9	9	10	10	10	10	11	11	12	12	12	12	13	13	14	14	14	14	15
0.201 – 0.220 (0.0079 – 0.0087)			1	1	1	1	1	2	2	2	2	3	3	3	3	3	4	4	4	4	4	5	5	5	5	5	6	6	6	6	6	7	7	7	7	7	8	8	8	8	8	9	9	9	9	9	10	10	10	10	10	11	11	12	12	12	12	13	13	14	14	14	14	15	15
0.221 – 0.240 (0.0087 – 0.0094)		1	1	1	1	1	2	2	2	3	3	3	3	3	4	4	4	4	4	5	5	5	5	5	6	6	6	6	6	7	7	7	7	7	8	8	8	8	8	9	9	9	9	9	10	10	10	10	10	11	11	11	12	12	12	13	13	13	14	14	14	15	15	15	16
0.241 – 0.249 (0.0095 – 0.0098)		1	1	1	1	2	2	2	3	3	3	3	4	4	4	4	4	5	5	5	5	5	6	6	6	6	6	7	7	7	7	7	8	8	8	8	8	9	9	9	9	9	10	10	10	10	10	11	11	11	11	12	12	12	13	13	13	14	14	14	15	15	15	16	16
0.250 – 0.350 (0.0098 – 0.0138)																																																																	
0.351 – 0.360 (0.0138 – 0.0142)	2	3	3	3	3	4	4	5	5	5	5	6	6	6	6	6	7	7	7	7	7	8	8	8	8	8	9	9	9	9	9	10	10	10	10	10	11	11	11	11	11	12	12	12	12	12	13	13	13	13	13	14	14	15	15	15	15	16	16	17	17	17	17	17	
0.361 – 0.380 (0.0142 – 0.0150)	2	3	3	3	4	4	4	5	5	5	6	6	6	6	6	7	7	7	7	7	8	8	8	8	8	9	9	9	9	9	10	10	10	10	10	11	11	11	11	11	12	12	12	12	12	13	13	13	13	13	14	14	14	15	15	15	16	16	16	17	17	17	17	17	
0.381 – 0.400 (0.0150 – 0.0157)	3	3	4	4	4	4	5	5	6	6	6	6	6	7	7	7	7	7	8	8	8	8	8	9	9	9	9	9	10	10	10	10	10	11	11	11	11	11	12	12	12	12	12	13	13	13	13	13	14	14	14	14	15	15	16	16	16	16	17	17	17	17	17		
0.401 – 0.420 (0.0158 – 0.0165)	3	4	4	4	4	5	5	6	6	6	6	7	7	7	7	7	8	8	8	8	8	9	9	9	9	9	10	10	10	10	10	11	11	11	11	11	12	12	12	12	12	13	13	13	13	13	14	14	14	14	14	15	15	16	16	16	16	17	17	17	17				
0.421 – 0.440 (0.0166 – 0.0173)	4	4	4	5	5	5	6	6	6	7	7	7	7	7	8	8	8	8	8	9	9	9	9	9	10	10	10	10	10	11	11	11	11	11	12	12	12	12	12	13	13	13	13	13	14	14	14	14	14	15	15	15	16	16	16	17	17	17	17	17					
0.441 – 0.460 (0.0174 – 0.0181)	4	4	5	5	5	6	6	6	7	7	7	7	8	8	8	8	8	9	9	9	9	9	10	10	10	10	10	11	11	11	11	11	12	12	12	12	12	13	13	13	13	13	14	14	14	14	14	15	15	15	15	16	16	16	17	17	17	17	17						
0.461 – 0.480 (0.0181 – 0.0189)	4	5	5	5	6	6	6	7	7	7	8	8	8	8	8	9	9	9	9	9	10	10	10	10	10	11	11	11	11	11	12	12	12	12	12	13	13	13	13	13	14	14	14	14	14	15	15	15	15	15	16	16	16	17	17	17	17	17							
0.481 – 0.500 (0.0189 – 0.0197)	5	5	6	6	6	6	7	7	8	8	8	8	8	9	9	9	9	9	10	10	10	10	10	11	11	11	11	11	12	12	12	12	12	13	13	13	13	13	14	14	14	14	14	15	15	15	15	15	16	16	16	16	17	17	17	17	17								
0.501 – 0.520 (0.0197 – 0.0205)	5	6	6	6	6	7	7	8	8	8	8	9	9	9	9	9	10	10	10	10	10	11	11	11	11	11	12	12	12	12	12	13	13	13	13	13	14	14	14	14	14	15	15	15	15	15	16	16	16	16	16	17	17	17	17										
0.521 – 0.540 (0.0205 – 0.0213)	6	6	6	7	7	7	8	8	8	9	9	9	9	9	10	10	10	10	10	11	11	11	11	11	12	12	12	12	12	13	13	13	13	13	14	14	14	14	14	15	15	15	15	15	16	16	16	16	16	17	17	17	17	17											
0.541 – 0.560 (0.0213 – 0.0220)	6	6	7	7	7	8	8	8	9	9	9	9	10	10	10	10	10	11	11	11	11	11	12	12	12	12	12	13	13	13	13	13	14	14	14	14	14	15	15	15	15	15	16	16	16	16	16	17	17	17	17	17	17												
0.561 – 0.580 (0.0221 – 0.0228)	6	7	7	7	8	8	8	9	9	9	10	10	10	10	10	11	11	11	11	11	12	12	12	12	12	13	13	13	13	13	14	14	14	14	14	15	15	15	15	15	16	16	16	16	16	17	17	17	17	17	17	17													
0.581 – 0.600 (0.0229 – 0.0236)	7	7	8	8	8	8	9	9	10	10	10	10	10	11	11	11	11	11	12	12	12	12	12	13	13	13	13	13	14	14	14	14	14	15	15	15	15	15	16	16	16	16	16	17	17	17	17	17	17	17	17														
0.601 – 0.620 (0.0237 – 0.0244)	7	8	8	8	8	9	9	10	10	10	10	11	11	11	11	11	12	12	12	12	12	13	13	13	13	13	14	14	14	14	14	15	15	15	15	15	16	16	16	16	16	17	17	17	17	17	17	17	17																
0.621 – 0.640 (0.0244 – 0.0252)	8	8	8	9	9	9	10	10	10	11	11	11	11	11	12	12	12	12	12	13	13	13	13	13	14	14	14	14	14	15	15	15	15	15	16	16	16	16	16	17	17	17	17	17	17	17	17																		
0.641 – 0.660 (0.0252 – 0.0260)	8	8	9	9	9	10	10	10	11	11	11	11	12	12	12	12	12	13	13	13	13	13	14	14	14	14	14	15	15	15	15	15	16	16	16	16	16	17	17	17	17	17	17	17	17																				
0.661 – 0.680 (0.0260 – 0.0268)	8	9	9	9	10	10	10	11	11	11	12	12	12	12	12	13	13	13	13	13	14	14	14	14	14	15	15	15	15	15	16	16	16	16	16	17	17	17	17	17	17	17	17																						
0.681 – 0.700 (0.0268 – 0.0276)	9	9	10	10	10	10	11	11	12	12	12	12	12	13	13	13	13	13	14	14	14	14	14	15	15	15	15	15	16	16	16	16	16	17	17	17	17	17	17	17	17																								
0.701 – 0.720 (0.0276 – 0.0283)	9	10	10	10	10	11	11	12	12	12	12	13	13	13	13	13	14	14	14	14	14	15	15	15	15	15	16	16	16	16	16	17	17	17	17	17	17	17	17																										
0.721 – 0.740 (0.0284 – 0.0291)	10	10	10	11	11	11	12	12	12	13	13	13	13	13	14	14	14	14	14	15	15	15	15	15	16	16	16	16	16	17	17	17	17	17	17	17	17																												
0.741 – 0.760 (0.0292 – 0.0299)	10	10	11	11	11	12	12	12	13	13	13	13	14	14	14	14	14	15	15	15	15	15	16	16	16	16	16	17	17	17	17	17	17	17	17																														
0.761 – 0.780 (0.0300 – 0.0307)	10	11	11	11	12	12	12	13	13	13	14	14	14	14	14	15	15	15	15	15	16	16	16	16	16	17	17	17	17	17	17	17	17																																
0.781 – 0.800 (0.0307 – 0.0315)	11	11	12	12	12	12	13	13	14	14	14	14	14	15	15	15	15	15	16	16	16	16	16	17	17	17	17	17	17	17	17																																		
0.801 – 0.820 (0.0315 – 0.0323)	11	12	12	12	12	13	13	14	14	14	14	15	15	15	15	15	16	16	16	16	16	17	17	17	17	17	17	17	17																																				
0.821 – 0.840 (0.0323 – 0.0331)	12	12	12	13	13	13	14	14	14	15	15	15	15	15	16	16	16	16	16	17	17	17	17	17	17	17	17																																						
0.841 – 0.860 (0.0331 – 0.0339)	12	12	13	13	13	14	14	14	15	15	15	15	16	16	16	16	16	17	17	17	17	17	17	17	17																																								
0.861 – 0.880 (0.0339 – 0.0346)	12	13	13	13	14	14	14	15	15	15	16	16	16	16	16	17	17	17	17	17	17	17	17																																										
0.881 – 0.900 (0.0347 – 0.0354)	13	13	14	14	14	14	15	15	16	16	16	16	16	17	17	17	17	17	17	17	17																																												
0.901 – 0.920 (0.0355 – 0.0362)	13	14	14	14	14	15	15	16	16	16	16	17	17	17	17	17	17	17	17																																														
0.921 – 0.940 (0.0363 – 0.0370)	14	14	14	15	15	15	16	16	16	17	17	17	17	17	17	17	17																																																
0.941 – 0.960 (0.0370 – 0.0378)	14	14	15	15	15	16	16	16	17	17	17	17	17	17	17																																																		
0.961 – 0.980 (0.0378 – 0.0386)	14	15	15	15	16	16	16	17	17	17	17	17	17																																																				
0.981 – 1.000 (0.0386 – 0.0394)	15	15	16	16	16	16	17	17	17	17	17																																																						
1.001 – 1.020 (0.0394 – 0.0402)	15	16	16	16	16	17	17	17	17																																																								
1.021 – 1.040 (0.0402 – 0.0409)	16	16	16	17	17	17	17	17																																																									
1.041 – 1.060 (0.0410 – 0.0417)	16	16	17	17	17	17	17																																																										
1.061 – 1.080 (0.0418 – 0.0425)	16	17	17	17	17	17																																																											
1.081 – 1.100 (0.0426 – 0.0433)	17	17	17	17	17																																																												
1.101 – 1.120 (0.0433 – 0.0441)	17	17	17																																																														
1.121 – 1.140 (0.0441 – 0.0449)	17	17																																																															
1.141 – 1.150 (0.0449 – 0.0453)	17																																																																

Exhaust valve clearance (Cold):
0.25 to 0.35 mm (0.010 to 0.014 in.)

EXAMPLE:
The 2.800 mm (0.1102 in.) shim is installed, and the measured clearance is 0.450 mm (0.0177 in.). Replace the 2.800 mm (0.1102 in.) shim with a new No. 10 shim.

New shim thickness mm (in.)

Shim No.	Thickness	Shim No.	Thickness
1	2.500 (0.0984)	10	2.950 (0.1161)
2	2.550 (0.1004)	11	3.000 (0.1181)
3	2.600 (0.1024)	12	3.050 (0.1201)
4	2.650 (0.1043)	13	3.100 (0.1220)
5	2.700 (0.1063)	14	3.150 (0.1240)
6	2.750 (0.1083)	15	3.200 (0.1260)
7	2.800 (0.1102)	16	3.250 (0.1280)
8	2.850 (0.1122)	17	3.300 (0.1299)
9	2.900 (0.1142)		

HINT:
A shim's thickness is written on its face in millimeters.

67170-HIGH-G19

Fig. 232 Exhaust valve shim selection chart—3.3L engine

ENGINE PERFORMANCE & EMISSION CONTROLS

ACCELERATOR PEDAL POSITION (APP) SENSOR

LOCATION

See Figure 233.

REMOVAL & INSTALLATION

See Figure 234.

1. Disconnect the accelerator pedal assembly connector.
2. Remove the 2 nuts and accelerator pedal assembly.

To install:

3. Install the accelerator pedal assembly with the and tighten to 48 inch lbs. (5.4 Nm).
4. Connect the accelerator pedal assembly connector.

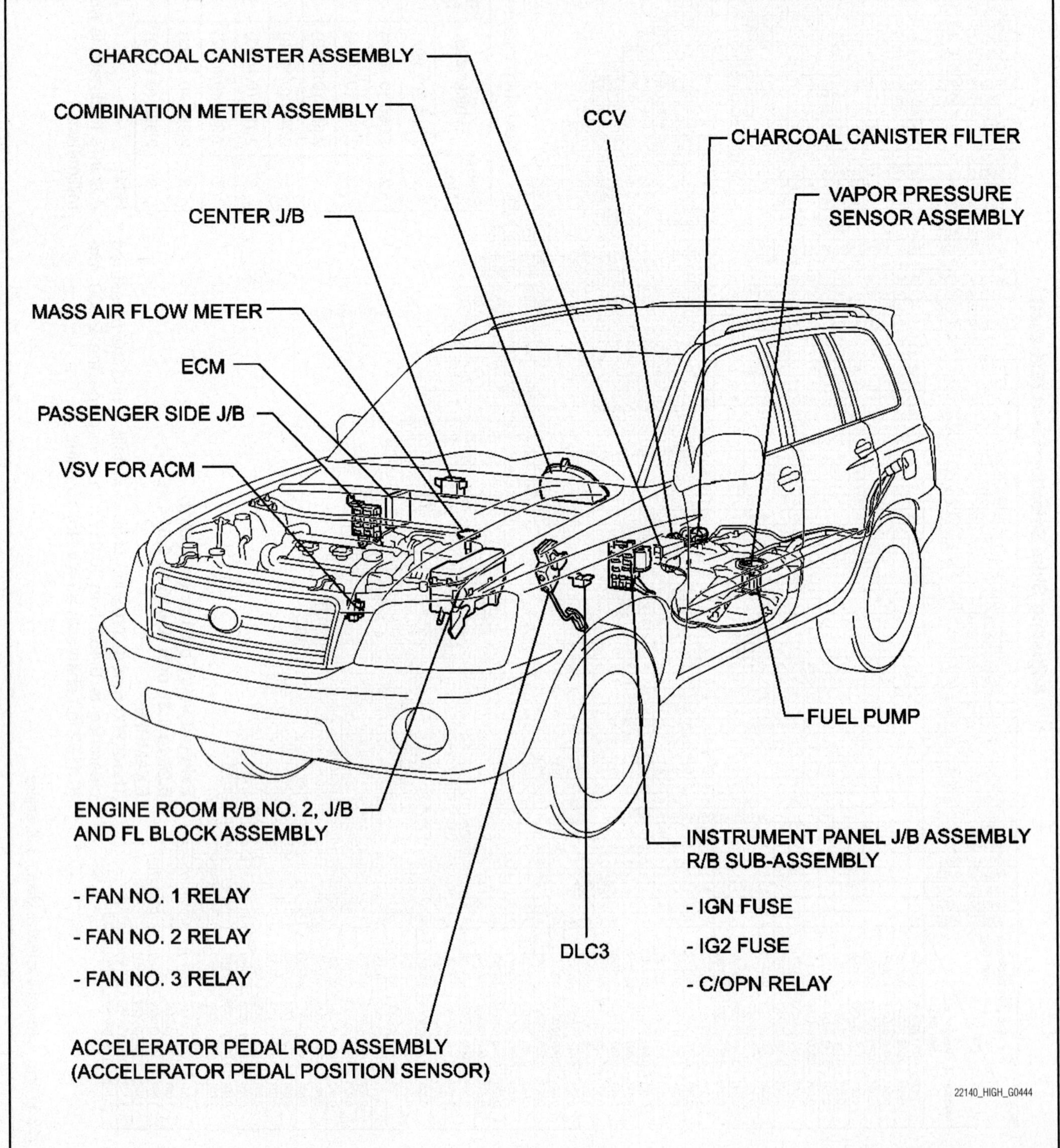

Fig. 233 Accelerator pedal assembly location

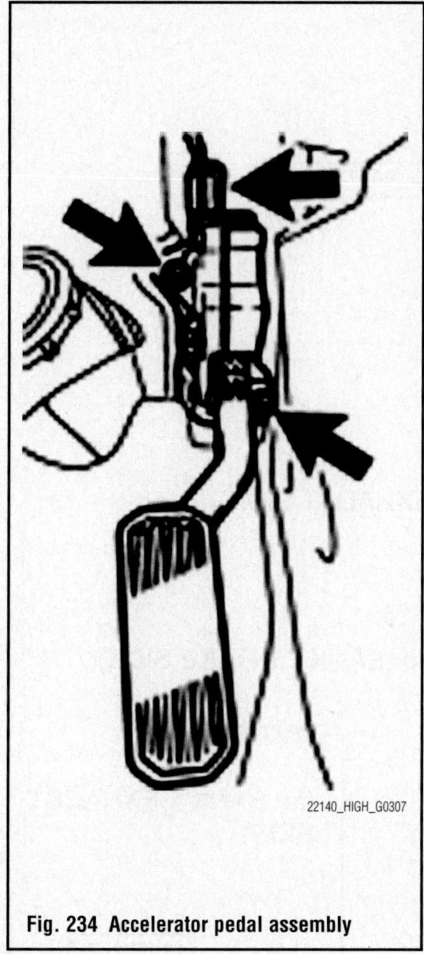

Fig. 234 Accelerator pedal assembly

CAMSHAFT POSITION (CMP) SENSOR

LOCATION

See Figures 235 through 237.

REMOVAL & INSTALLATION

2.4L Engine

See Figure 238.

1. Remove the air cleaner assembly.

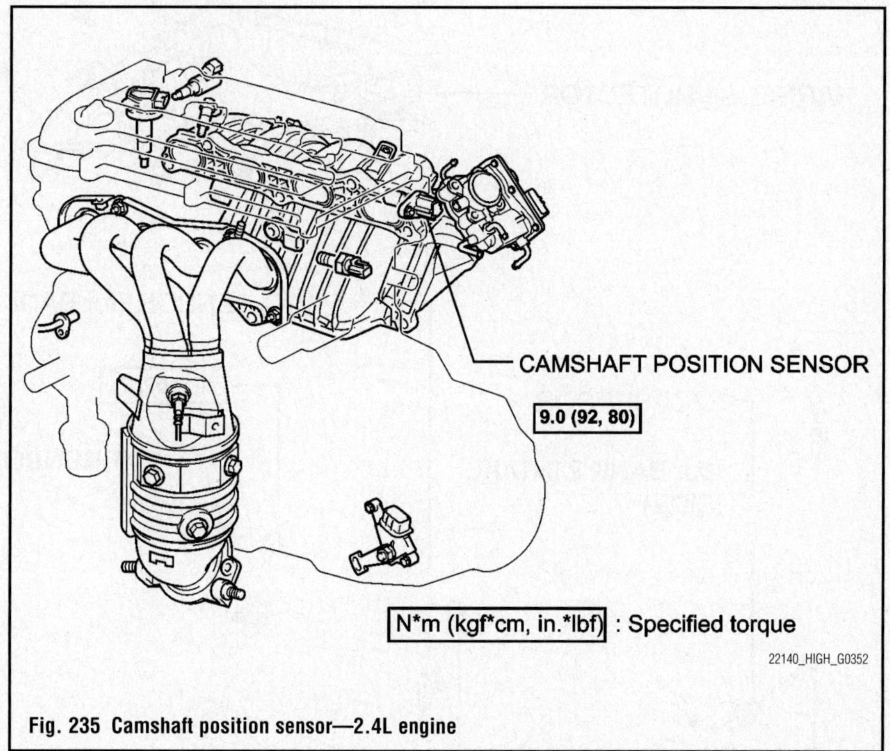

Fig. 235 Camshaft position sensor—2.4L engine

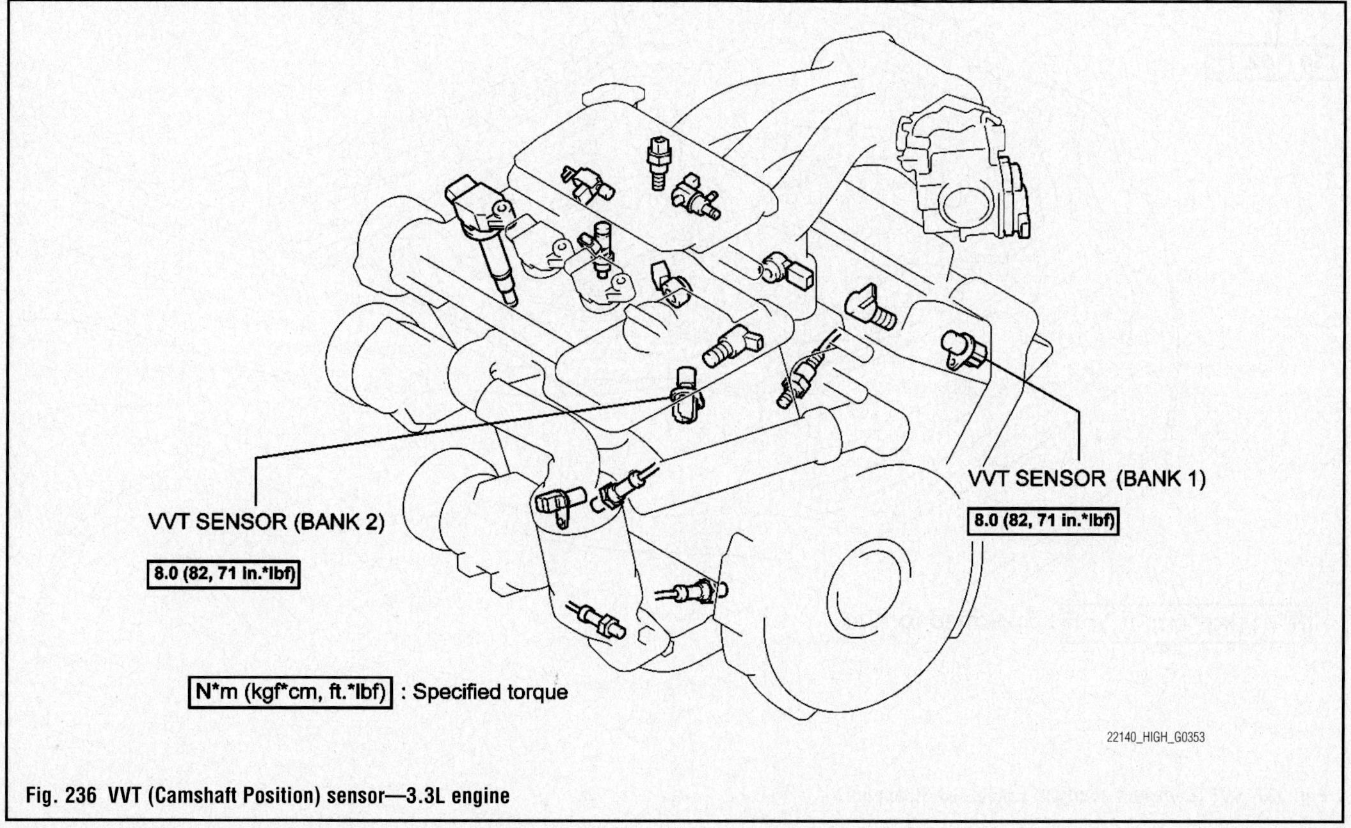

Fig. 236 VVT (Camshaft Position) sensor—3.3L engine

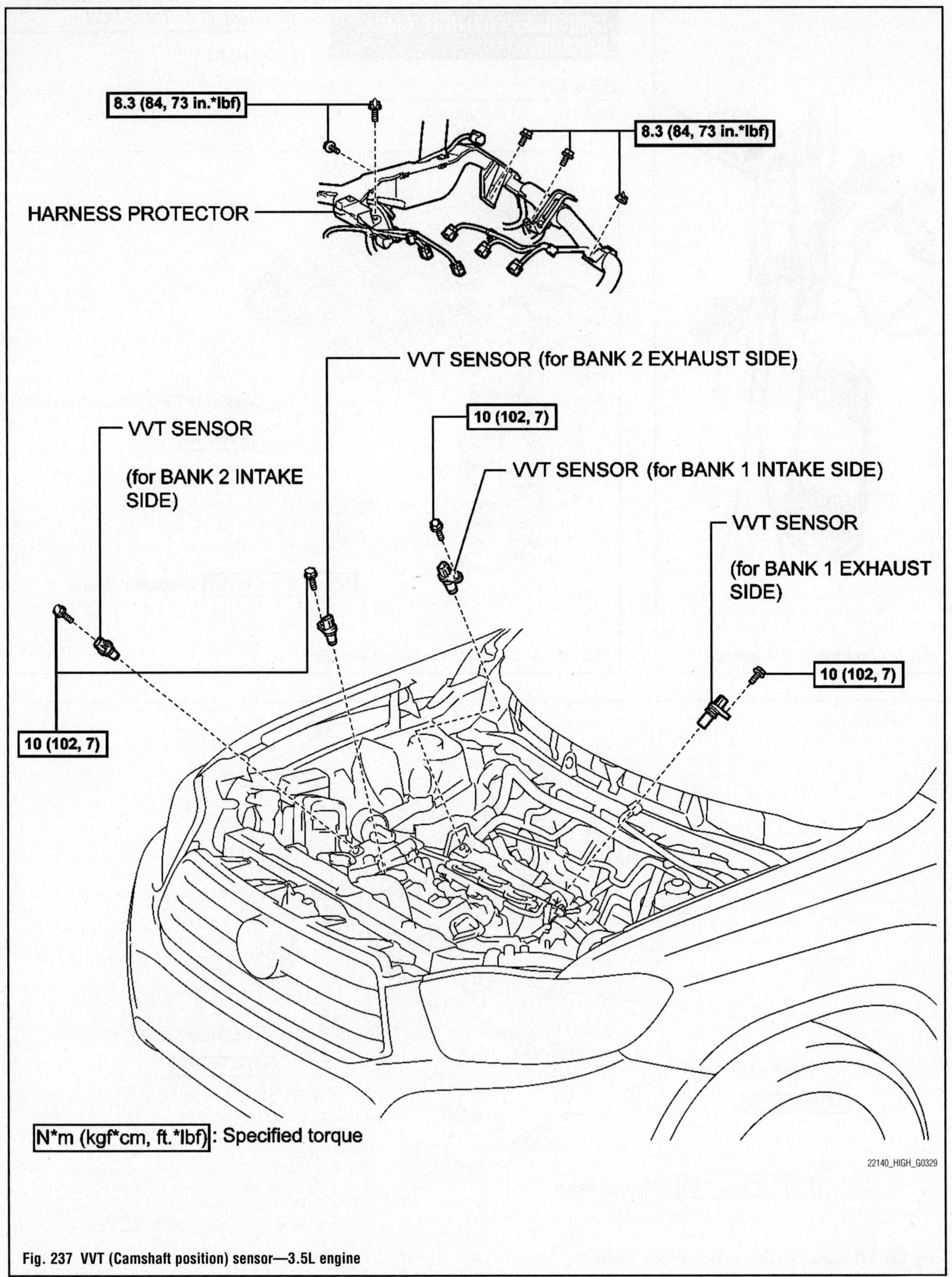

Fig. 237 VVT (Camshaft position) sensor—3.5L engine

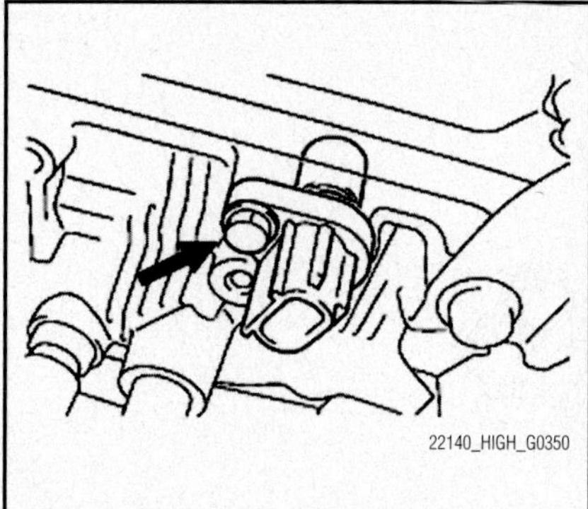

Fig. 238 Camshaft position sensor—2.4L engine

2. Remove the camshaft position sensor by disconnecting the sensor connector and removing the bolt and sensor.

To install:

3. Apply a light coat of engine oil to a new O-ring. Install the O-ring onto the sensor.
4. Install the sensor with the bolt and tighten to 80 inch lbs. (9 Nm).
5. Install the air cleaner assembly.

3.3L Engine

See Figure 239.

1. Remove the air cleaner assembly.
2. Disconnect the VVT (Camshaft Position) sensor connector.
3. Remove the bolt and sensor.

To install:

4. Apply a light coat of engine oil to a new O-ring. Install the O-ring onto the sensor.

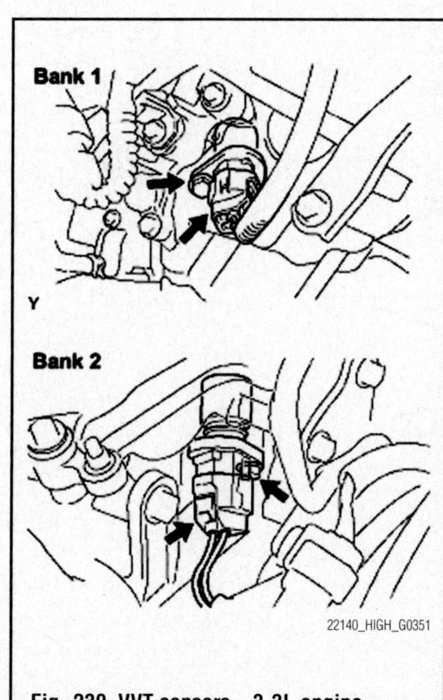

Fig. 239 VVT sensors—3.3L engine

5. Install the sensor with the bolt and tighten to 71 inch lbs. (8 Nm).
6. Connect the sensor connector.
7. Install the air cleaner assembly.

3.5L Engine

See Figures 240 through 243.

1. Remove the engine under cover assembly.
2. Remove the No. 1 engine under cover.
3. Drain engine coolant.
4. Remove both front wiper arm and blade assemblies.

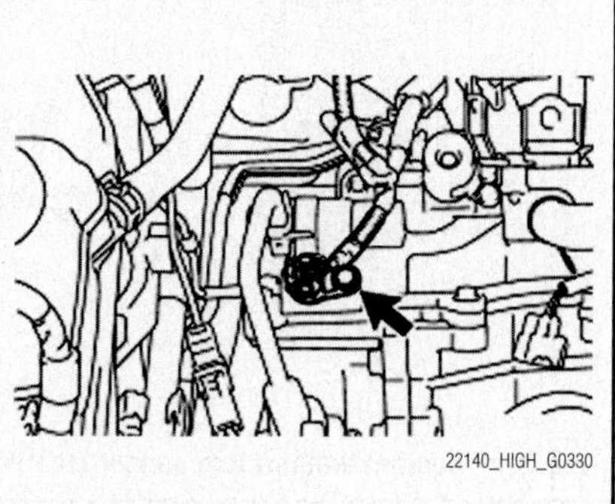

Fig. 240 Bank 1 intake side VVT sensor—3.5L engine

Fig. 241 Bank 1 exhaust side VVT sensor—3.5L engine

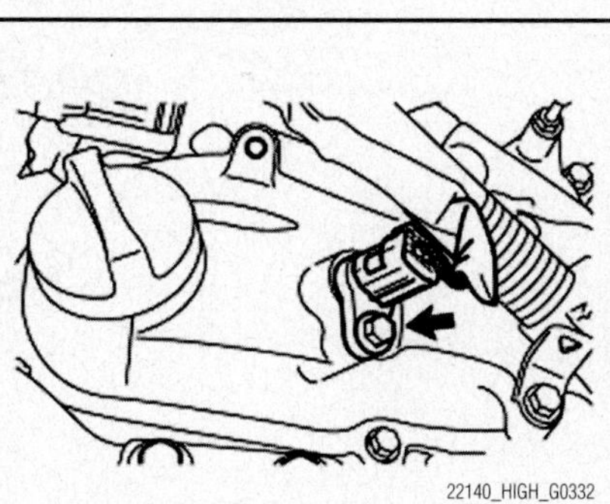

Fig. 242 Bank 2 intake side VVT sensor—3.5L engine

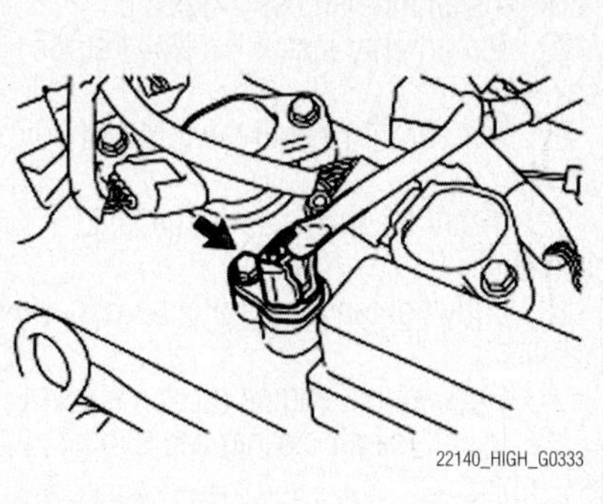

Fig. 243 Bank 2 exhaust side VVT sensor—3.5L engine

5. Remove the cowl top ventilator louver sub-assembly.
6. Remove the windshield wiper motor and link assembly.
7. Remove the outer cowl top panel sub-assembly.
8. Remove the V-bank cover sub-assembly.
9. Remove the air cleaner cap sub-assembly.
10. Disconnect the engine room main wire.
11. Remove the No. 1 surge tank stay.
12. Remove the throttle body bracket.
13. Remove the intake air surge tank assembly.
14. Separate the harness protector.
15. Remove the VVT (Camshaft Position) sensor (for Bank 1), as follows:
    a. Intake side: remove the bolt and VVT sensor.
    b. Exhaust side: remove the bolt and VVT sensor.
16. Remove the VVT (Camshaft Position) sensor (for Bank 2), as follows:
    a. Intake side: remove the bolt and VVT sensor.
    b. Exhaust side: remove the bolt and VVT sensor.

To install:

17. Install the VVT (Camshaft Position) sensor (for Bank 2), as follows:
    a. Exhaust side: install the VVT sensor with the bolt and tighten to 7 ft. lbs. (10 Nm).
    b. Intake side: install the VVT sensor with the bolt and tighten to 7 ft. lbs. (10 Nm).
18. Install the VVT (Camshaft Position) sensor (for Bank 1), as follows:
    a. Exhaust side: install the VVT sensor with the bolt and tighten to 7 ft. lbs. (10 Nm).
    b. Intake side: install the VVT sensor with the bolt and tighten to 7 ft. lbs. (10 Nm).

19. Install the harness protector.
20. Temporarily install the No. 1 surge tank stay.
21. Temporarily install the throttle body bracket.
22. Install the intake air surge tank assembly.
23. Fully tighten the throttle body bracket.
24. Connect the engine room main wire.
25. Install the air cleaner cap sub-assembly.
26. Add engine coolant
27. Inspect for coolant leak.
28. Install the outer cowl top panel sub-assembly.
29. Install the windshield wiper motor and link assembly.
30. Install the cowl top ventilator louver sub-assembly.
31. Install both front wiper arm and blade assemblies.
32. Install the V-bank cover sub-assembly.
33. Install the No. 1 engine under cover.
34. Install the engine under cover assembly.

COOLANT TEMPERATURE SENSOR

LOCATION

See Figures 244 through 246.

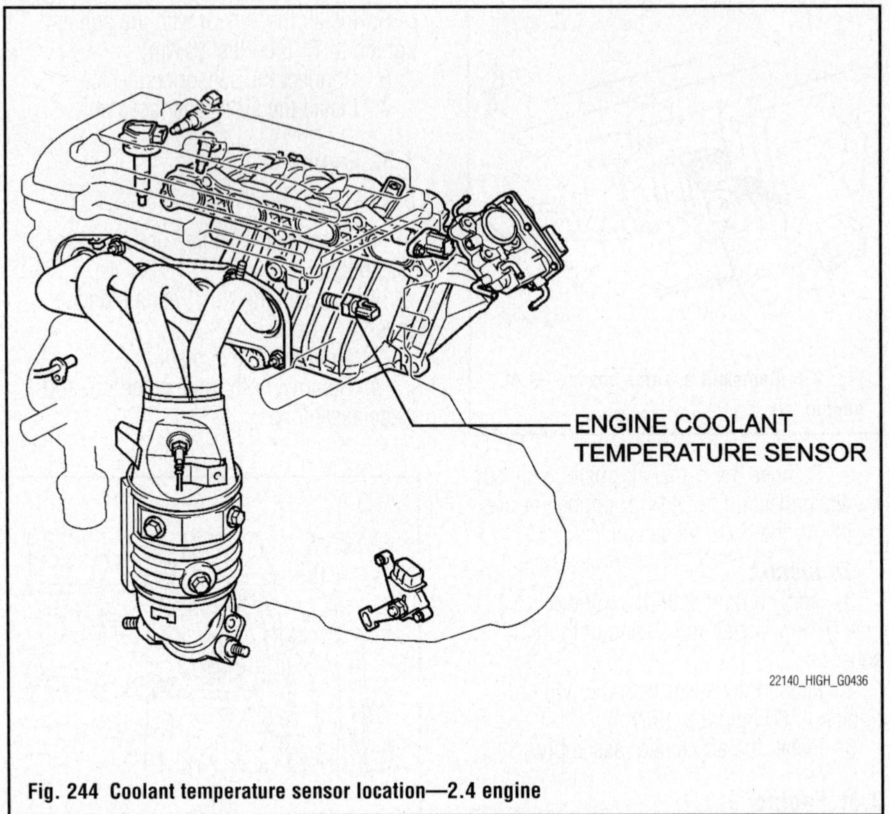

Fig. 244 Coolant temperature sensor location—2.4 engine

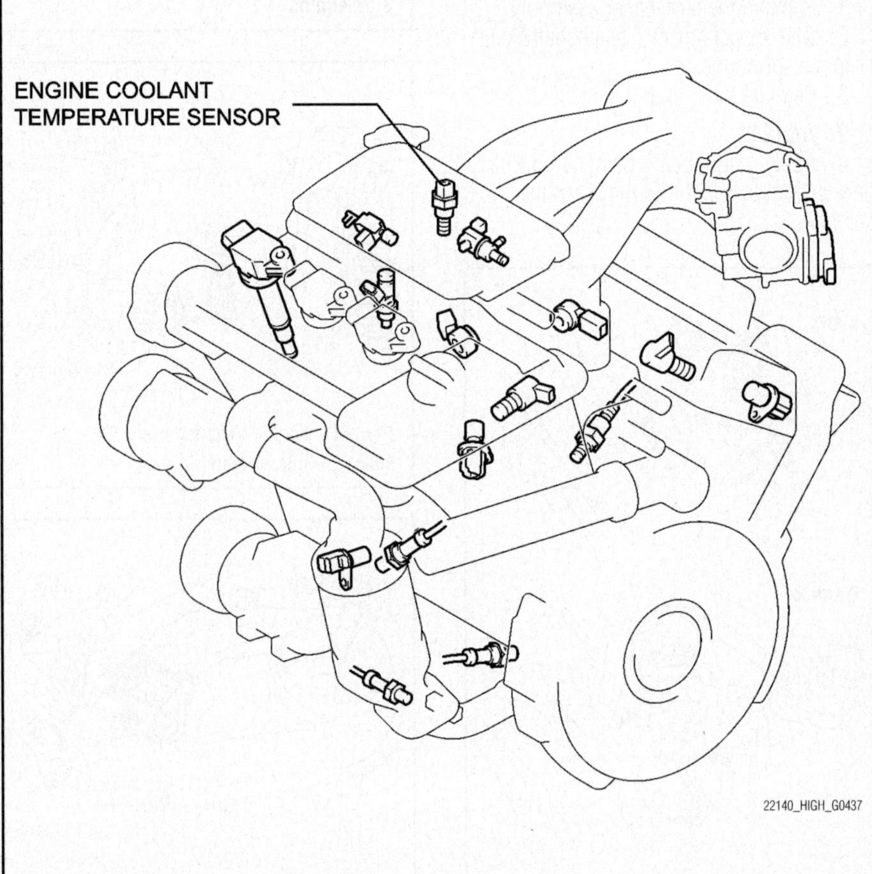

Fig. 245 Coolant temperature sensor location—3.3L engine

REMOVAL & INSTALLATION

3.3L Engine

1. Drain engine coolant.
2. Disconnect the ECT sensor connector.
3. Using a 19 mm deep socket wrench, remove the ECT sensor and gasket.

To install:

4. Install a new gasket to the ECT sensor.
5. Using a 19 mm deep socket, install the ECT sensor. Tighten to 14 ft. lbs. (20 Nm).
6. Connect the ECT sensor connector.
7. Refill with engine coolant.

3.5L Engine

See Figure 247.

1. Remove the V-bank cover sub-assembly.
2. Remove the engine under cover assembly.
3. Remove the No. 1 engine under cover.
4. Drain engine coolant.
5. Remove the cool air intake duct seal.

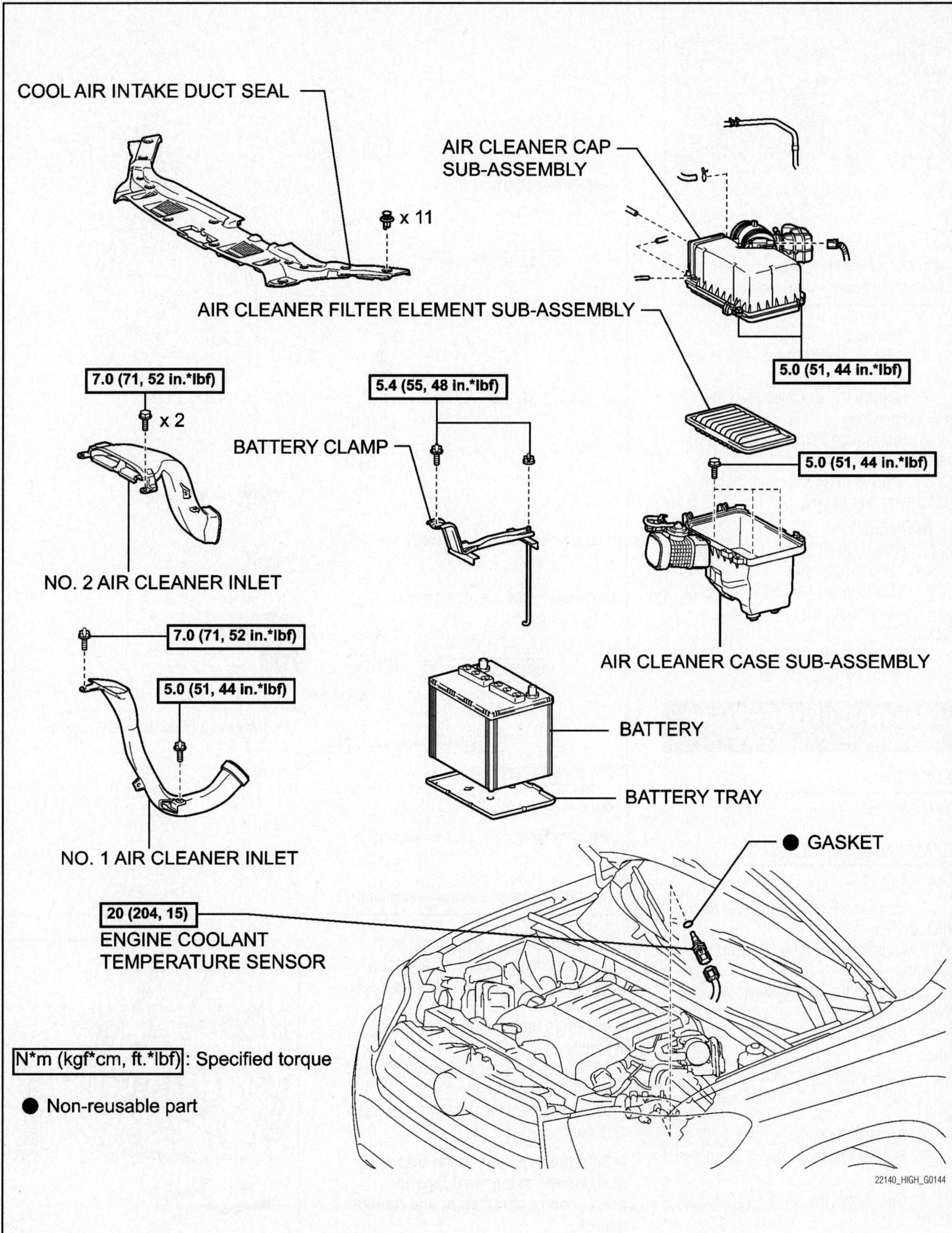

Fig. 246 Coolant temperature sensor location—3.5L engine

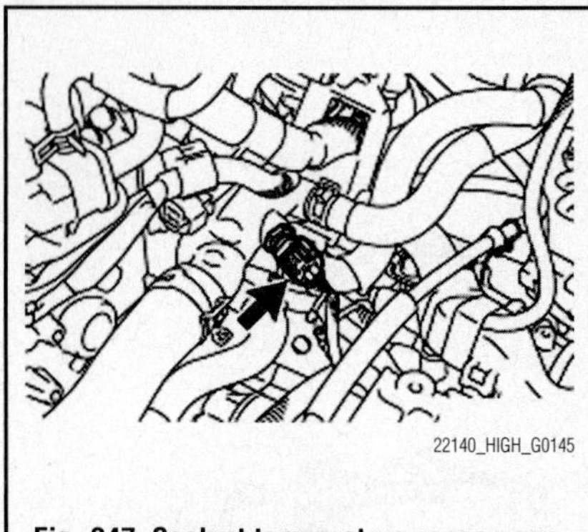

Fig. 247 Coolant temperature sensor, connector, and gasket—3.5L engine

6. Remove the battery.
7. Remove the No. 1 and 2 air cleaner inlets.
8. Remove the air cleaner cap and case sub-assemblies.
9. Disconnect the engine coolant temperature sensor connector.
10. Remove the engine coolant temperature sensor and gasket.

To install:

11. Install a new gasket onto the engine coolant temperature sensor.
12. Install the engine coolant temperature sensor and tighten to 15 ft. lbs. (20 Nm).
13. The remainder of installation is the reverse of the removal procedure.

CRANKSHAFT POSITION (CKP) SENSOR

LOCATION

See Figure 248.

REMOVAL & INSTALLATION

See Figure 249.

1. Recover the refrigerant from refrigeration system.
2. Remove the V-bank cover sub-assembly.
3. Remove the engine under cover assembly.
4. Remove the No. 1 engine under cover.
5. Drain engine coolant.
6. Remove the cool air intake duct seal.
7. Remove the battery.
8. Remove the No. 1 and 2 air cleaner inlets.
9. Disconnect the No. 1 and 2 radiator hoses.
10. Disconnect the oil cooler hose.
11. Disconnect the cooling fan ECU connector.

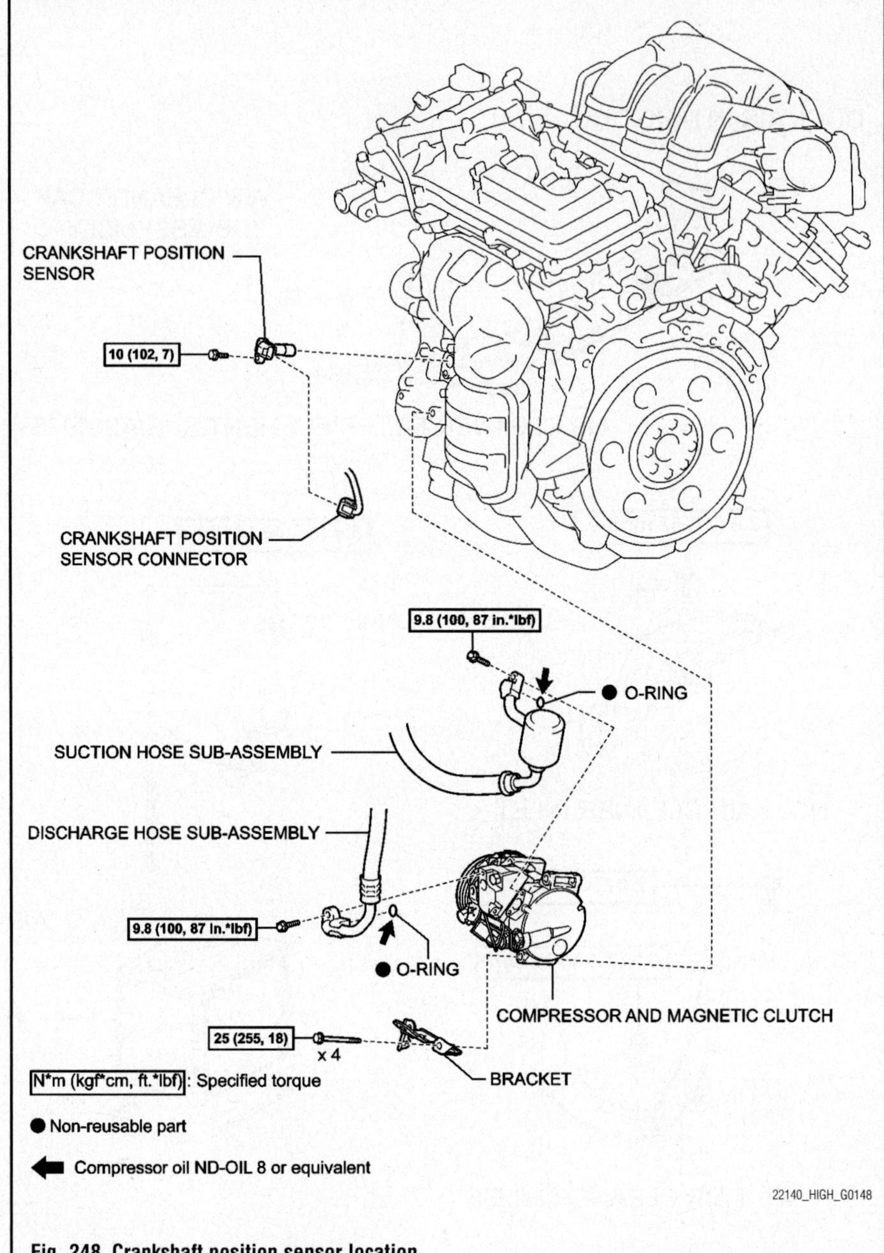

Fig. 248 Crankshaft position sensor location

12. Remove the radiator grill.
13. Remove the hood lock assembly.
14. Remove the upper radiator support sub-assembly.
15. Separate the cooler condenser assembly.
16. Remove the radiator assembly and fan assembly with motor.
17. Disconnect the discharge hose sub-assembly.

➡Seal the openings of the disconnected parts using vinyl tape to prevent entry of moisture and foreign matter.

18. Disconnect the suction hose sub-assembly.

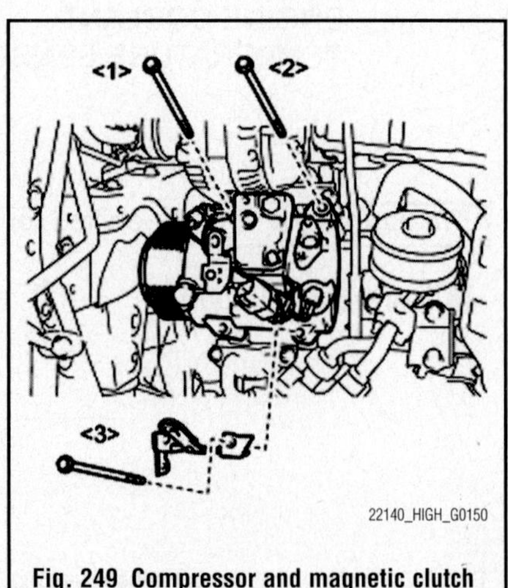

Fig. 249 Compressor and magnetic clutch

19. Remove the compressor and magnetic clutch as follows:
20. Disconnect the crankshaft position sensor connector.
21. Remove the bolt and crankshaft position sensor, as follows:
   a. Disconnect the crankshaft position sensor connector.
   b. Remove the bolt and crankshaft position sensor.

To install:

22. Apply a light coat of engine oil to the O-ring on the crankshaft position sensor.
23. Install the crankshaft position sensor with the bolt and tighten to 7 ft. lbs. (10 Nm).
24. Connect the crankshaft position sensor connector.
25. Temporarily tighten compressor and magnetic clutch with the bolts, in the order shown.
26. Install the compressor and magnetic clutch with the 4 bolts. Tighten the bolts in the order shown above, and torque to 18 ft. lbs. (25 Nm).
27. Connect the suction hose sub-assembly.
28. Connect the discharge hose sub-assembly.
29. Install the radiator assembly and fan assembly with motor.
30. Install the cooler condenser assembly.
31. Install the upper radiator support sub-assembly.
32. Install the hood lock assembly.
33. Install the radiator grill.
34. Connect the cooling fan ECU connector.
35. Connect the oil cooler hose.
36. Connect the No. 1 and 2 radiator hoses.
37. Install the No. 1 and 2 air cleaner inlets.
38. Install the battery.
39. Install the cool air intake duct seal.
40. Charge with refrigerant.
41. Add engine coolant.
42. Inspect for coolant leak.
43. Inspect automatic transaxle fluid.
44. Inspect for oil leak.
45. Warm up the engine.
46. Inspect for refrigerant leak.
47. Install the No. 1 engine under cover.
48. Install the engine under cover assembly.
49. Install the V-bank cover sub-assembly.

ELECTRONIC CONTROL MODULE (ECM)

LOCATION

See Figure 250.

REMOVAL & INSTALLATION

See Figures 251 through 253.

1. Disconnect the negative battery cable.

⁂ CAUTION

Wait at least 90 seconds after disconnecting the cable from the negative (-) battery terminal to prevent airbag and seat belt pretensioner activation.

2. Remove the right front door scuff plate.
3. Remove the right cowl side trim sub-assembly.
4. Remove the No. 2 instrument panel under cover sub-assembly.
5. Remove the lower instrument panel sub-assembly.
6. Remove the certification ECU (Smart Key ECU assembly), if applicable.

⁂ WARNING

Before removing the tire pressure warning ECU, read the registered transmitter IDs of all wheels and write them down to use for re-registration of transmitter IDs.

7. Remove the tire pressure warning ECU.
8. Remove the ECM, as follows:

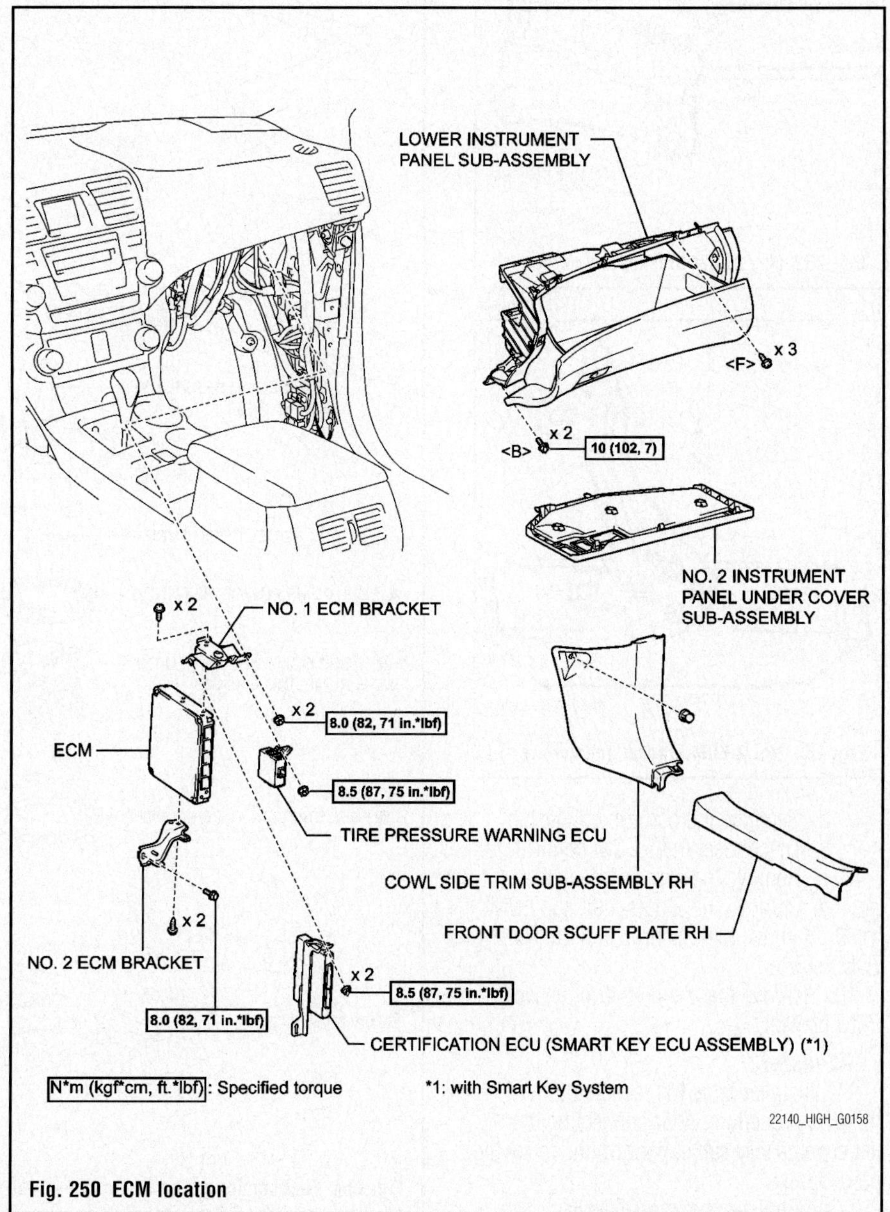

Fig. 250 ECM location

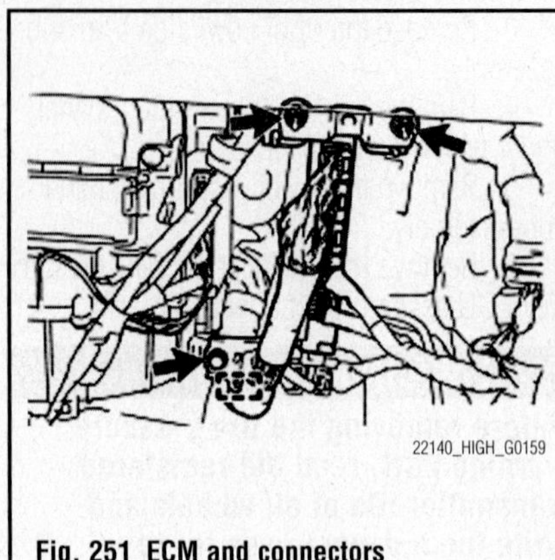

Fig. 251 ECM and connectors

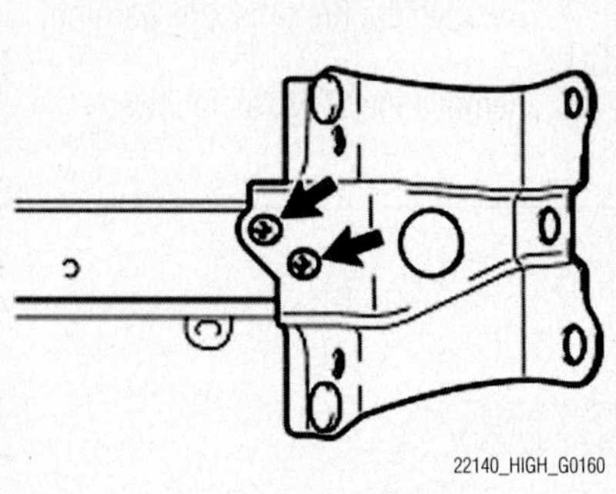

Fig. 252 No. 1 ECM bracket and screws

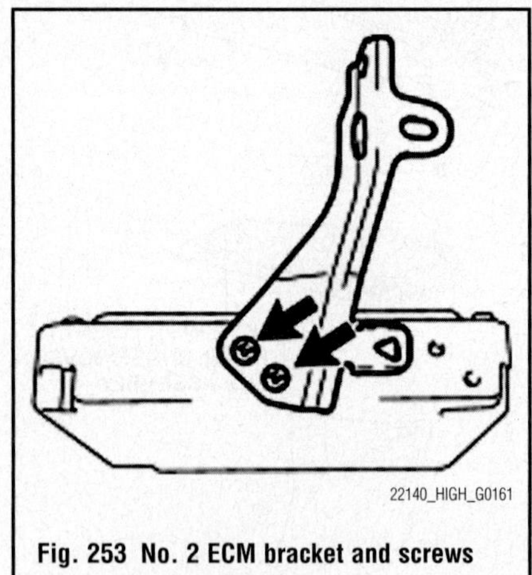

Fig. 253 No. 2 ECM bracket and screws

a. Separate the harness connector.
b. Disconnect the 5 ECM connectors.
c. Remove the wire harness clamp.
d. Remove the 2 nuts, bolt and ECM.

9. Remove the 2 screws and the No. 1 ECM bracket.

10. Remove the 2 screws and the No. 2 ECM bracket.

To install:

11. Installation is the reverse of the removal procedure. After connecting the negative battery cable, perform the following procedures:

a. Register the transmitter ID.
b. Inspect the tire pressure warning system.
c. Initialize tire pressure warning system.

➡Be sure to register the transmitter IDs of all tires in the ECU before initialization.

➡Be sure to inflate all tires to the proper inflation pressure before initialization.

d. Register immobilizer communication ID.

➡If the ECM is replaced, register the ECU communication ID for the immobilizer system.

e. Perform initialization.

➡If the ECM is replaced, perform RESET MEMORY (at initialization).

FUEL LEVEL SENDING UNIT

LOCATION

See Figure 254.

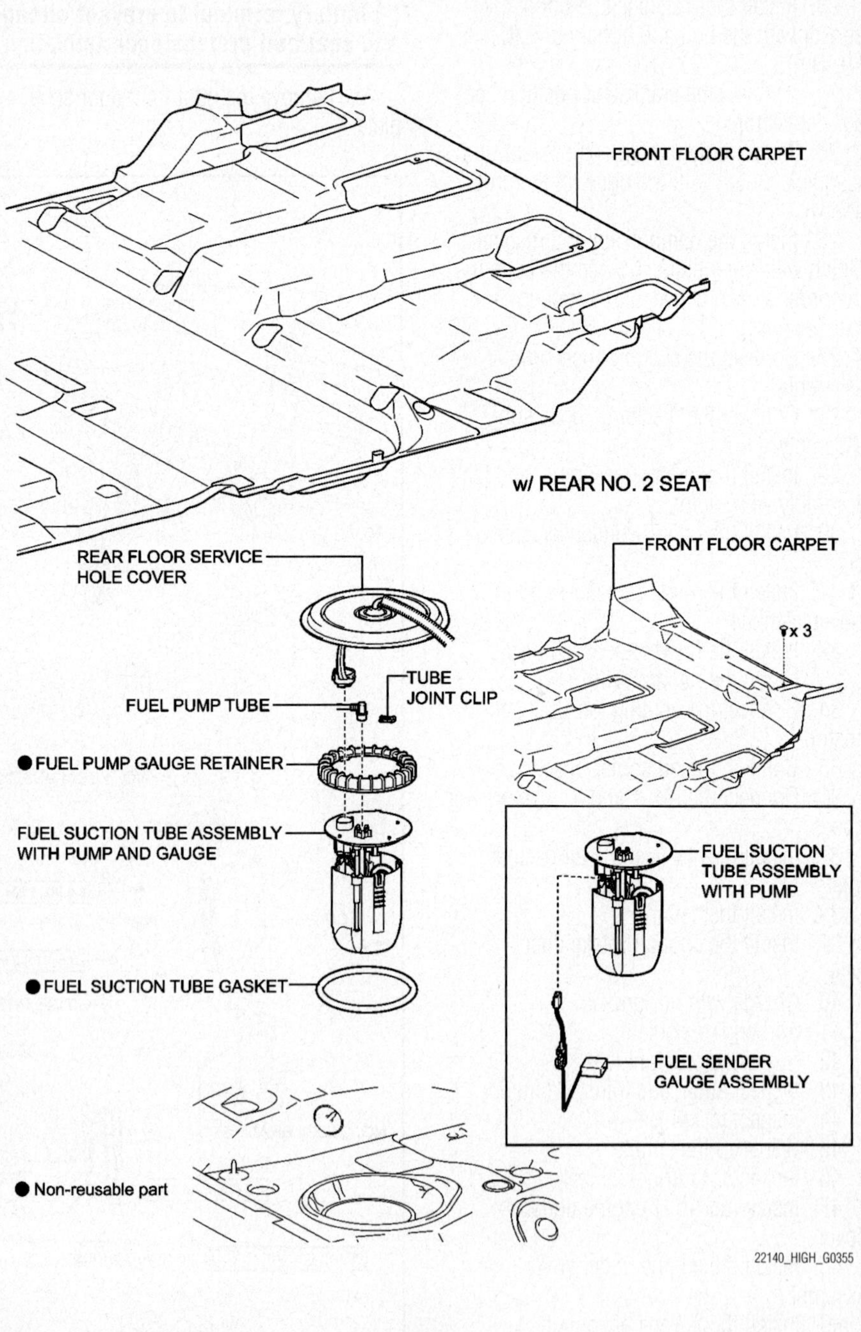

Fig. 254 Fuel sender gauge assembly—2008 models

REMOVAL & INSTALLATION

2007 Models

Refer to the Fuel Pump removal and installation procedures.

2008 Models

See Figures 255 through 257.

1. Before servicing the vehicle, refer to the precautions section.
2. Remove fuel suction tube assembly with pump and gauge. Refer to the Fuel Pump removal procedure for instructions.
3. Disconnect the connector and remove the Fuel Sender Gauge (Fuel Level Sending Unit) from the fuel suction tube.

To install:

4. Install the fuel sender gauge assembly by sliding the fuel sender gauge to fit the claw.
5. Install the fuel suction tube assembly with pump and gauge.
6. Connect the negative battery cable.
7. Inspect for fuel leak.
8. Disconnect the negative battery cable.

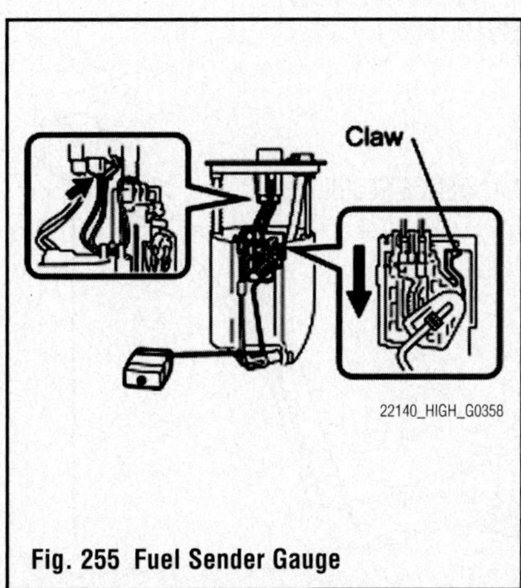

Fig. 255 Fuel Sender Gauge

Float Level	Resistance (Ω)
F	3.0 to 5.0
E	109.0 to 111.0

22140_HIGH_G0439

Fig. 256 Fuel sender gauge assembly standard resistance

Fuel Sender Gauge Assembly:

F

E

Fuel Sender Gauge Assembly Wire Harness View:

1 2 3

22140_HIGH_G0440

Fig. 257 Fuel sender gauge assembly pinout

9. The remainder of installation is the reverse of the removal procedure. Refer to the Fuel Pump installation procedure for instructions.

TESTING

1. Remove the fuel sender gauge assembly.
2. Check the operation of the float moving smoothly between F and E.
3. Measure the resistance between terminals 2 (FE) and 3 (FS) of the connector according to the value(s) in the table below.
4. If the value is not as specified, replace the fuel sender gauge assembly.

HEATED OXYGEN (HO2S) SENSOR

LOCATION

See Figures 258 through 260.

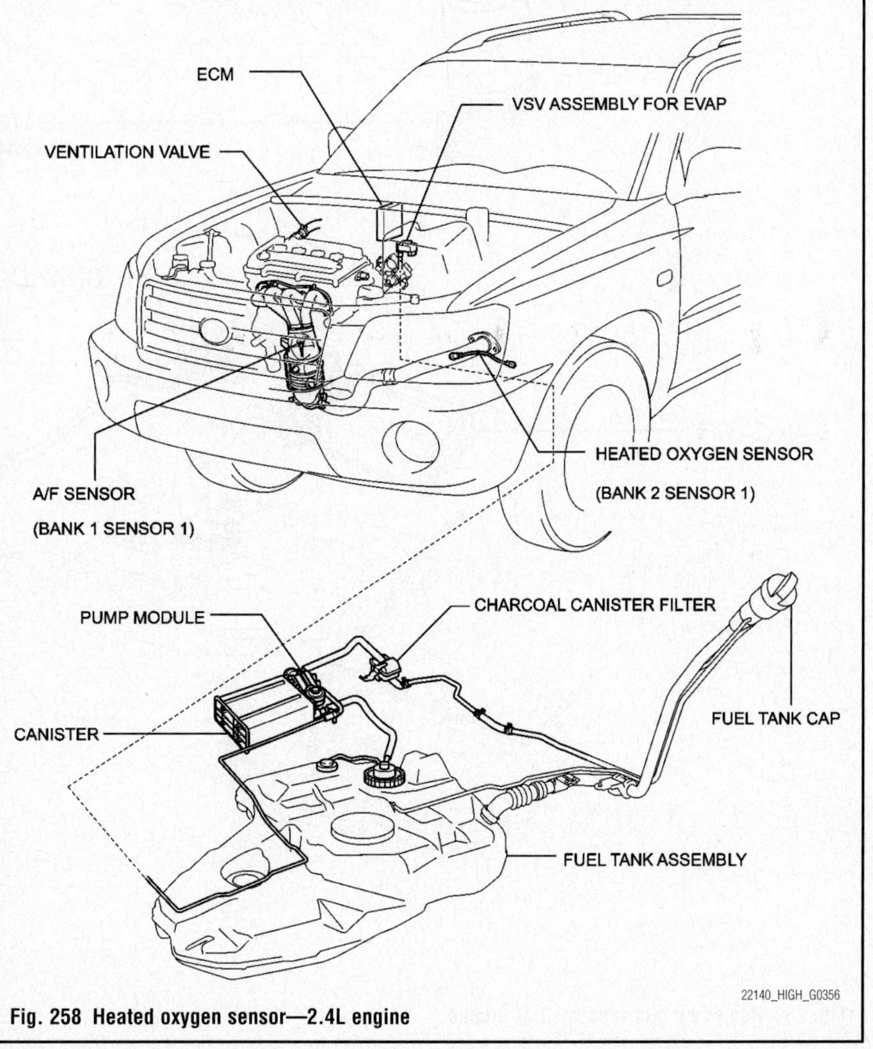

Fig. 258 Heated oxygen sensor—2.4L engine

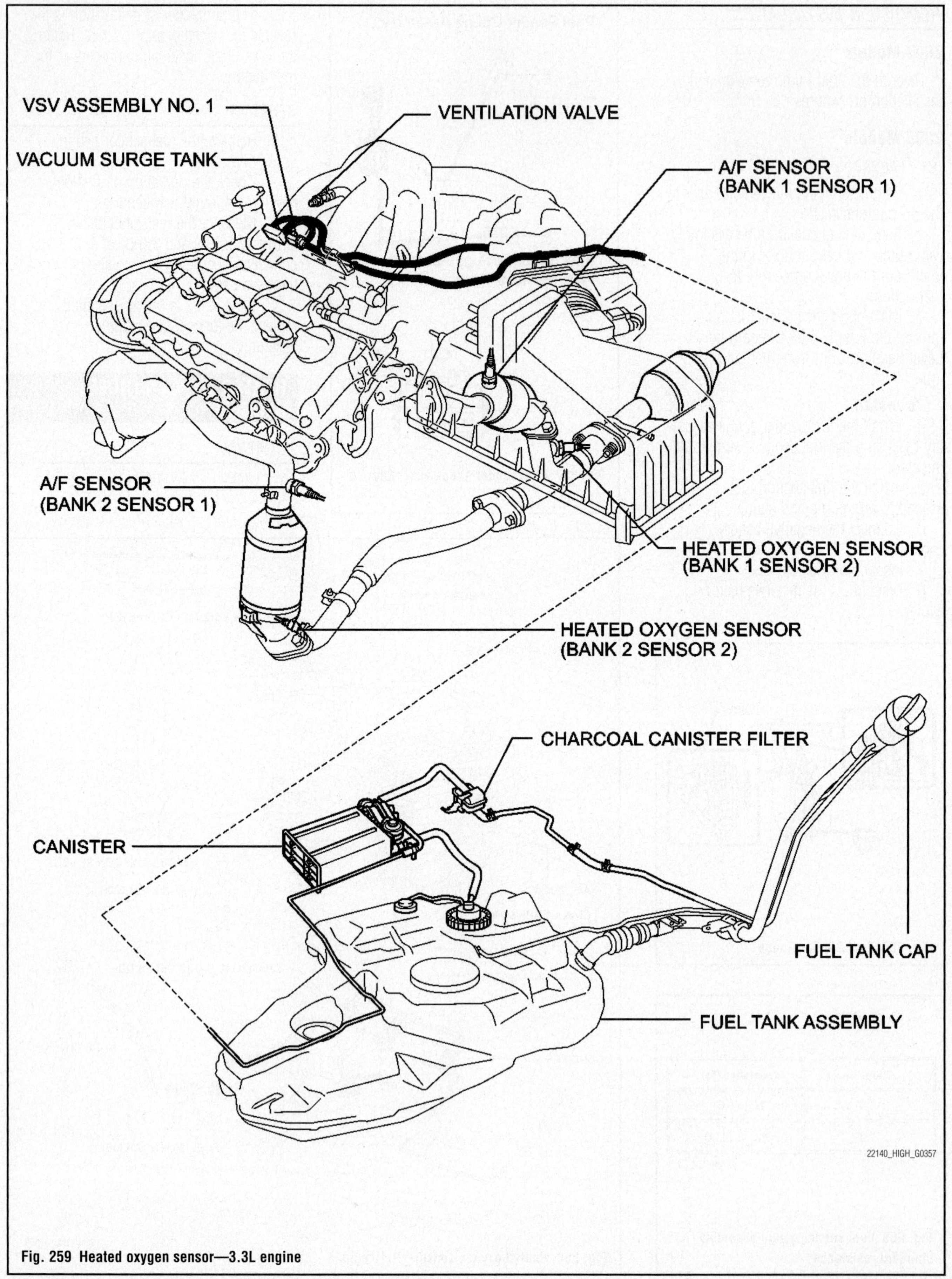

Fig. 259 Heated oxygen sensor—3.3L engine

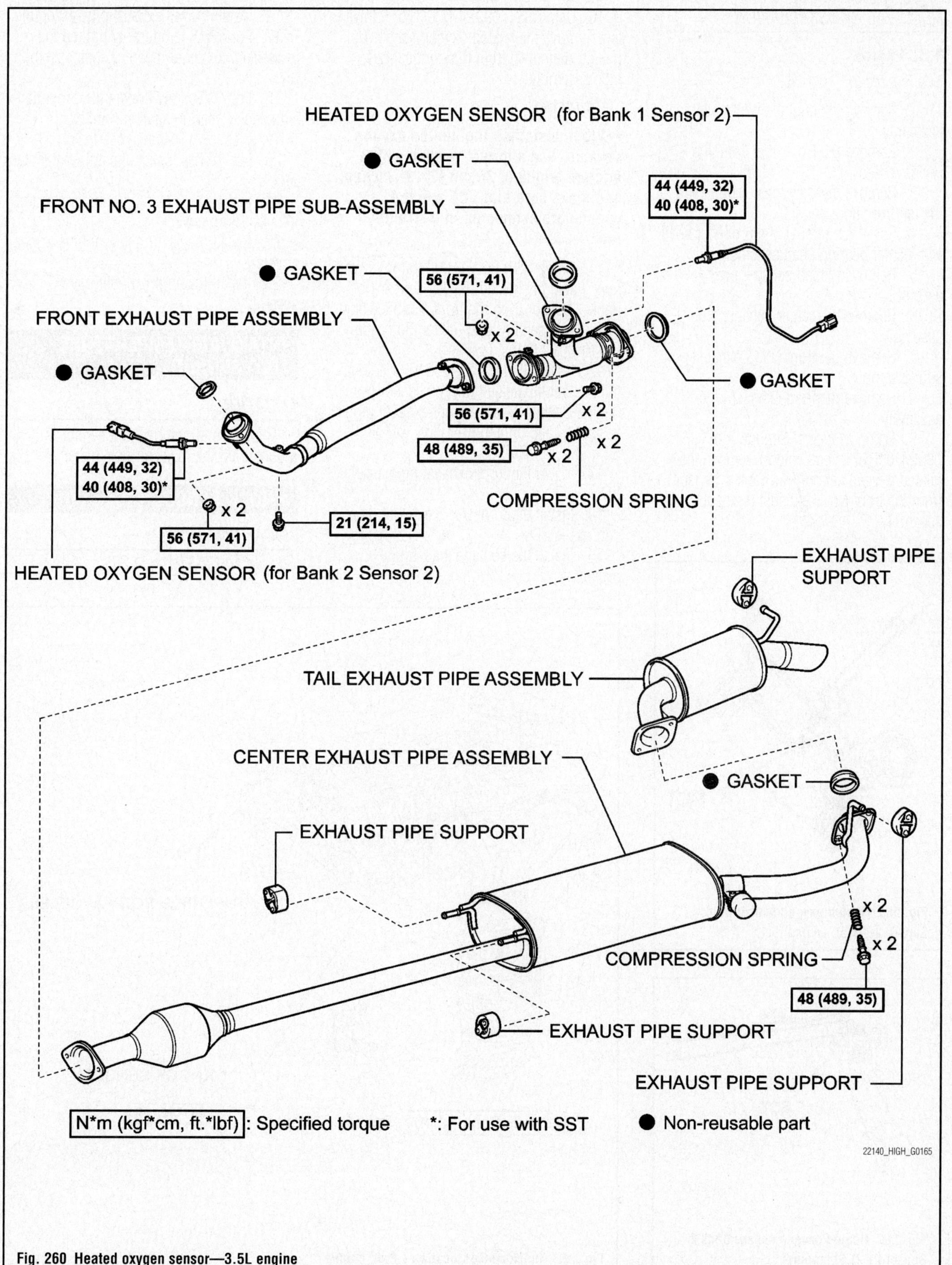

Fig. 260 Heated oxygen sensor—3.5L engine

REMOVAL & INSTALLATION

3.5L Engine

See Figures 261 and 262.

1. Remove the engine under cover assembly.
2. Remove the No. 1 and 2 engine under covers.
3. Remove the 3 clamps, 6 bolts and the left front floor cover.
4. For 4WD, remove the propeller shaft with center bearing shaft assembly.
5. Remove the tail exhaust pipe assembly.
6. Remove the center exhaust pipe assembly.
7. Remove the front No. 3 exhaust pipe sub-assembly.
8. Remove the front exhaust pipe assembly.
9. Using a Special Service Tool (SST: 09224-00010 or equivalent), remove the heated oxygen sensor (for Bank 1 Sensor 2) from the front No. 3 exhaust pipe sub-assembly.
10. Using SST: 09224-00010 or equivalent, remove the heated oxygen sensor (for Bank 2 Sensor 2) from the front exhaust pipe assembly.

To install:

➡When installing the heated oxygen sensors, use a torque wrench with a fulcrum length of 30 cm (11.81 inches), and make sure that SST and the wrench are connected in a straight line.

11. Using SST: 09224-00010 or equivalent, install the heated oxygen sensor (for Bank 1 Sensor 2) to the front No. 3 exhaust pipe sub-assembly, and tighten 30 ft. lbs. (40 Nm), 32 ft. lbs. (44 Nm).
12. Using SST: 09224-00010 or equivalent, install the heated oxygen sensor (for Bank 2 Sensor 2) to the front exhaust pipe assembly, and tighten 30 ft. lbs. (40 Nm), 32 ft. lbs. (44 Nm).
13. Install the front exhaust pipe assembly.
14. Install the front No. 3 exhaust pipe sub-assembly.
15. Install the center exhaust pipe assembly.
16. Install the tail exhaust pipe assembly.
17. For 4WD, temporarily tighten the propeller with center bearing shaft assembly.
18. For 4WD, fully tighten the propeller with center bearing shaft assembly.
19. Inspect for exhaust gas leak.
20. For 4WD, inspect and adjust transfer oil.
21. Install the 3 clamps, 6 bolts and the left front floor cover.
22. Install the No. 1 and 2 engine under covers.
23. Install the engine under cover assembly.

INTAKE AIR TEMPERATURE (IAT) SENSOR

LOCATION

The intake air temperature sensor is mounted on the mass air flow meter.

KNOCK SENSOR (KS)

LOCATION

See Figures 263 through 265.

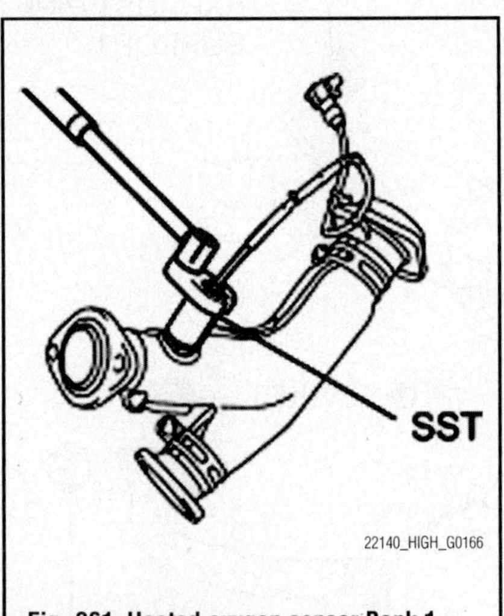

Fig. 261 Heated oxygen sensor Bank 1 Sensor 2—3.5L engine

Fig. 262 Heated oxygen sensor Bank 2 Sensor 2—3.5L engine

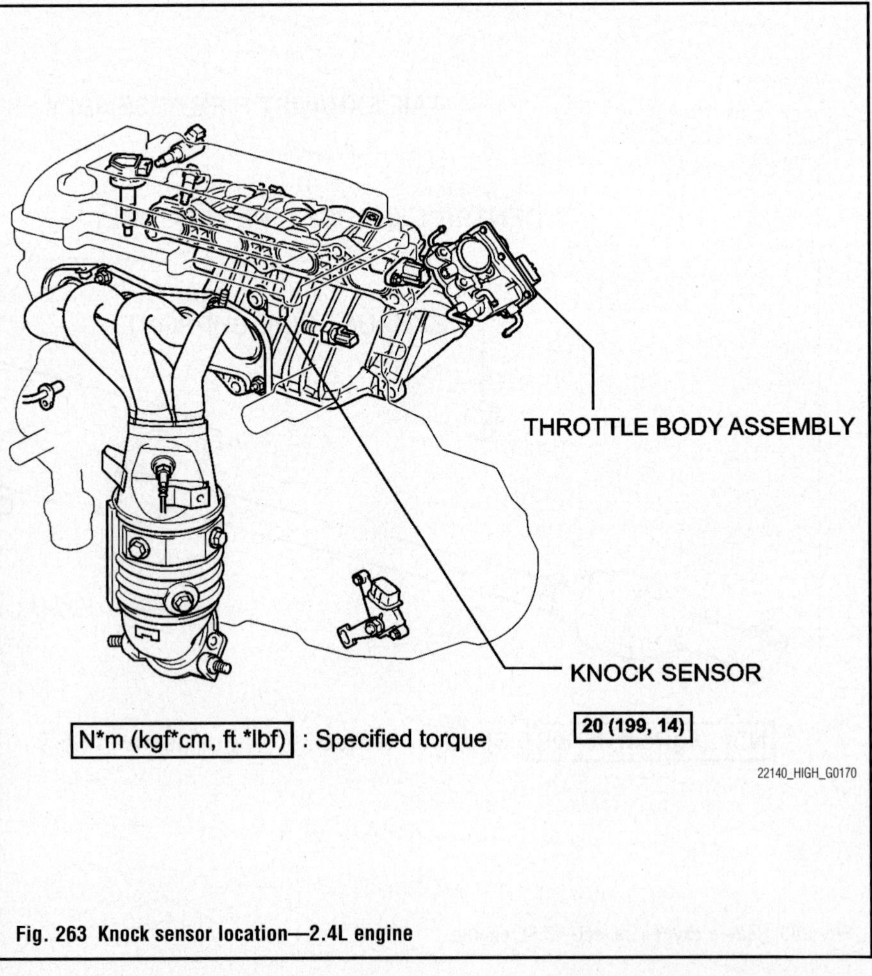

Fig. 263 Knock sensor location—2.4L engine

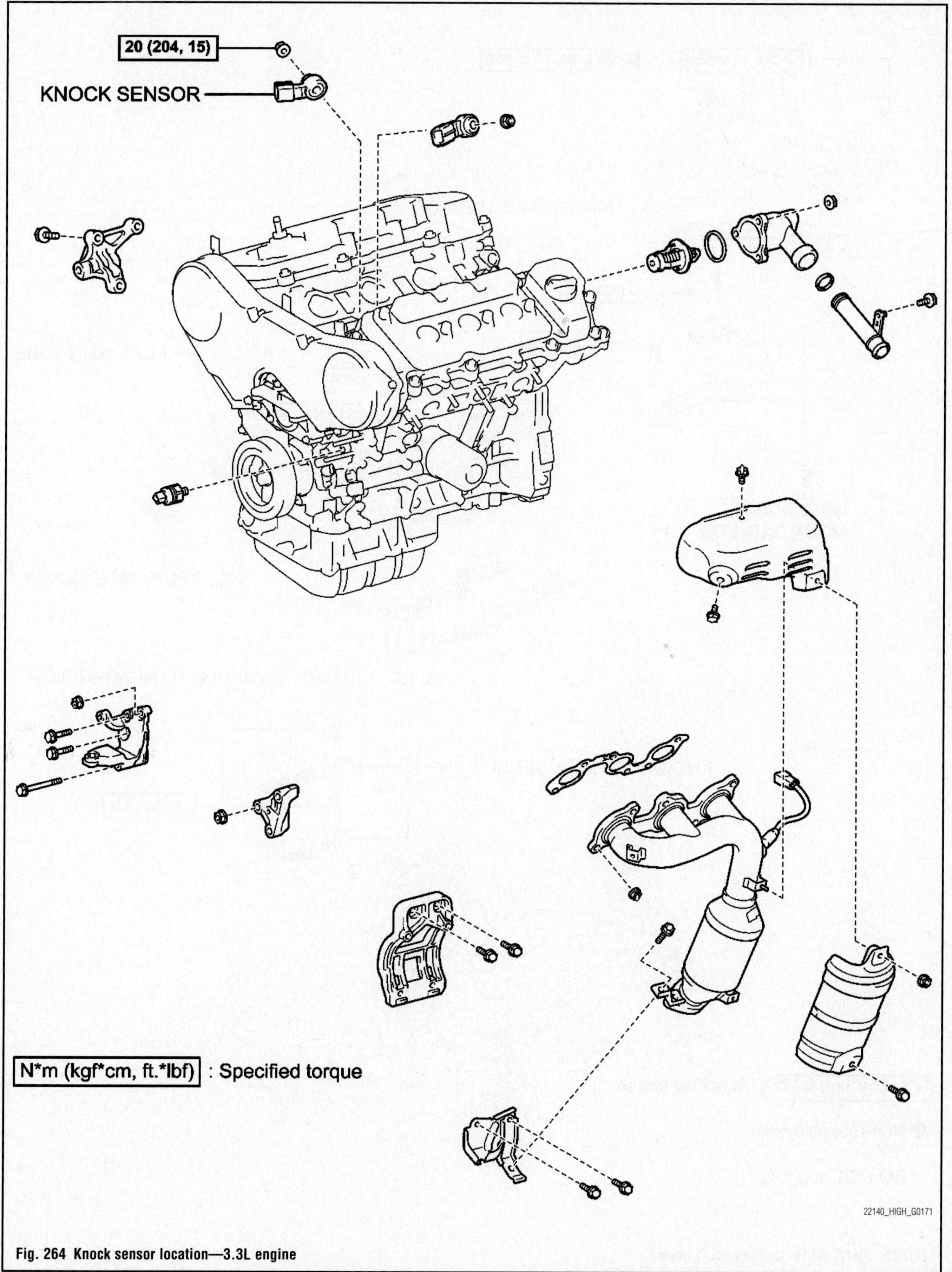

Fig. 264 Knock sensor location—3.3L engine

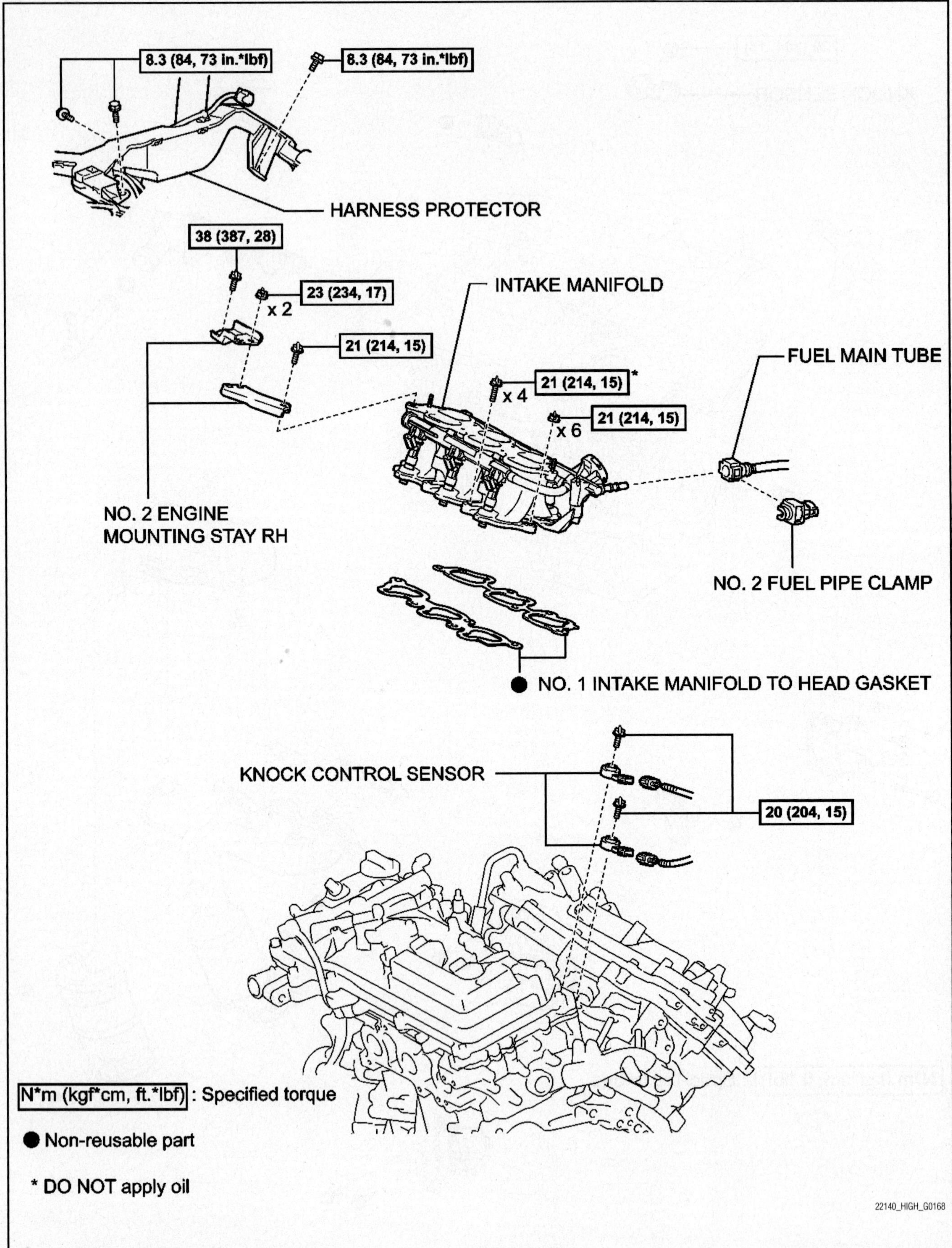

Fig. 265 Knock sensor location—5.7L engine

REMOVAL & INSTALLATION

2.4L Engine

See Figures 266 and 267.

1. Drain engine coolant.
2. Remove the air cleaner cap with air cleaner hose.
3. Remove the throttle body assembly.
4. Remove the 5 bolts, 2 nuts, intake manifold and gasket.
5. Disconnect the knock sensor connector.
6. Remove the nut and sensor.

To install:

7. Install the sensor with the nut and tighten 14 ft. lbs. (20 Nm).

➡Angling the knock sensor below the horizontal (below 0°) is recommended.

8. Connect the sensor connector.
9. Install a new gasket and the intake manifold with the 5 bolts and 2 nuts and tighten 22 ft. lbs. (30 Nm).
10. Install the throttle body assembly.
11. Install the air cleaner cap with air cleaner hose.
12. Add engine coolant.
13. Check for engine coolant leak.

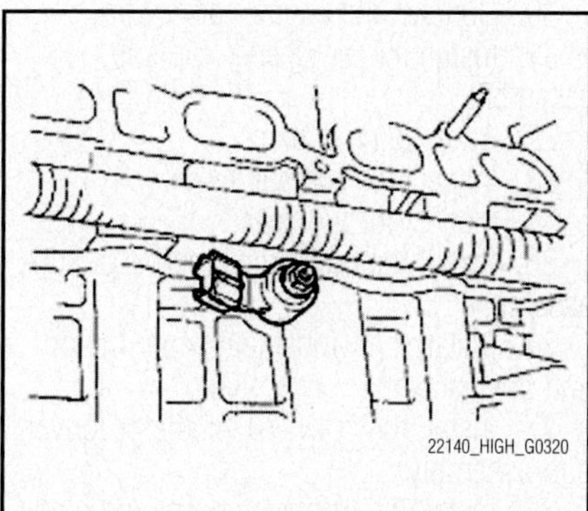

Fig. 266 Knock sensor connector—2.4L engine

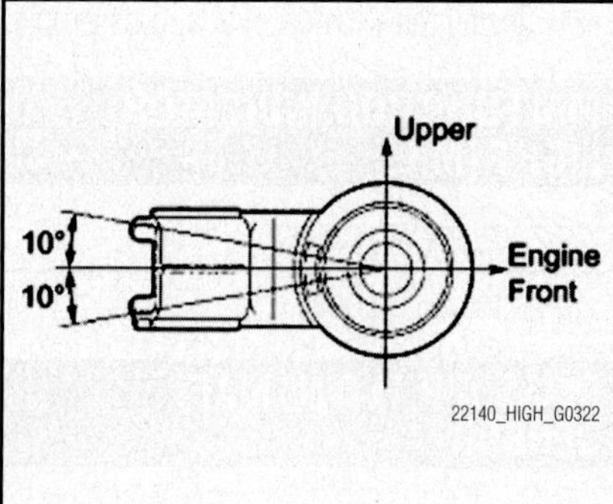

Fig. 267 Positioning knock sensor—2.4L engine

3.3L Engine

See Figures 268 through 271.

1. Discharge fuel system pressure.
2. Drain engine coolant.
3. Remove the V-bank cover sub-assembly.
4. Remove the air cleaner cap sub-assembly.
5. Remove the emission control valve set.
6. Remove the intake air surge tank.
7. Remove the intake manifold, as follows:
   a. Disconnect the fuel pipe No. 1: remove the fuel pipe clamp, and pinch the tube connector and pull out the fuel pipe No. 1.

➡Check the connector for dirt, mud or other contamination. Clean if necessary.

➡Be sure to keep the tube connector, pipe and O-ring clean. They can become contaminated easily.

➡Do not use tools when disconnecting the fuel pipe.

➡Do not bend or twist the nylon tube. Protect the connector by covering it with a vinyl or plastic bag.

➡When the pipe and the connector are stuck, push and pull the connector to release it. Then pull the connector out carefully.

   b. Lock the hose clamp, as shown in the illustration.
   c. Disconnect the heater inlet water hose.

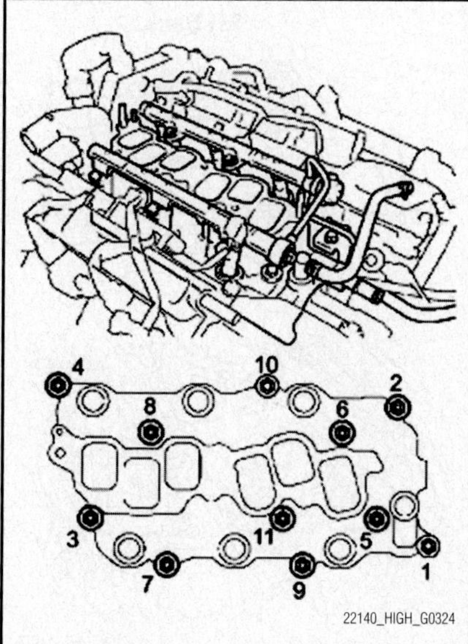

Fig. 268 Intake manifold bolt removal sequence—3.3L engine

   d. Remove the nut and disconnect the ground cable.
   e. Disconnect the 6 fuel injector connectors.
   f. Loosen and remove the intake manifold's 9 bolts and 2 nuts little by little in the numerical order shown in the illustration.
   g. Remove the intake manifold.

➡Fully removing each bolt and nut one by one may damage the intake manifold, bolts and nuts.

8. Remove engine moving control rod.
9. Remove the water outlet, as follows:
   a. Disconnect the radiator hose inlet.
   b. Disconnect the ECT sensor connector.
   c. Remove the clamp.
   d. Remove the 2 bolts, 2 nuts and 2 washers.
   e. Lock the hose clamp as shown in the illustration. Then remove the water outlet together with the water by-pass hose No. 1.
   f. Remove the 2 gaskets from the 2 cylinder heads.
10. Remove the knock sensor, as follows:
   a. Disconnect the 2 sensor connectors.
   b. Remove the 2 nuts and 2 sensors.

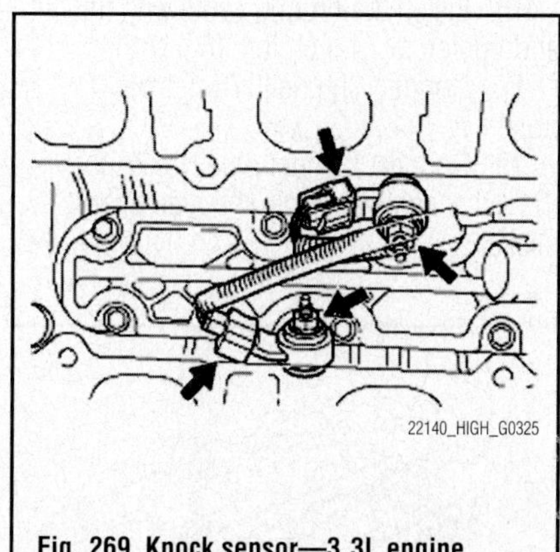

Fig. 269 Knock sensor—3.3L engine

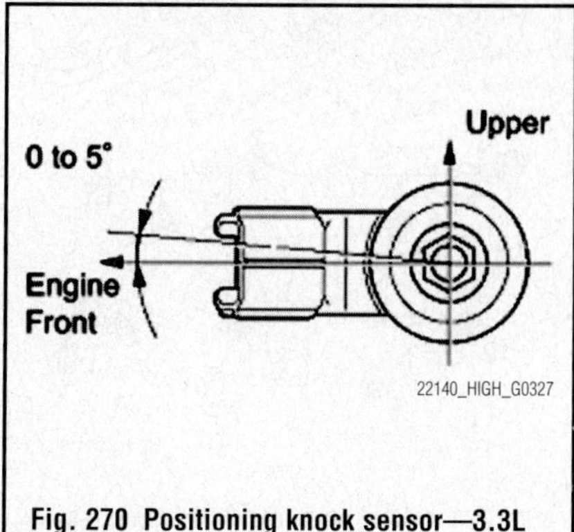

Fig. 270 Positioning knock sensor—3.3L engine

To install:

11. Install the knock sensor, as follows:
    a. Install the 2 sensors with the 2 nuts, as shown in the illustration and tighten to 14 ft. lbs. (20 Nm).
    b. Connect the 2 sensor connectors
12. Install the water outlet, as follows:
    a. Install 2 new gaskets to the 2 cylinder heads.
    b. Install the water outlet together with the water by-pass hose No. 1 and unlock the hose clamp.
    c. Tighten the 2 bolts, 2 nuts and 2 washers to 11 ft. lbs. (15 Nm).
    d. Install the clamp.
    e. Connect the ECT sensor connector.
    f. Connect the radiator hose inlet.
13. Install the engine moving control rod.
14. Tighten the intake manifold's 9 bolts and 2 nuts little by little in the numerical order shown in the illustration to 11 ft. lbs. (15 Nm).

✲✲ WARNING

In this case, fully tightening one of these bolts or nuts without partially tightening the other bolts or nuts in the group may damage the intake manifold, bolts and nuts.

15. Retighten the water outlet mounting bolts and nuts to 11 ft. lbs. (15 Nm).
16. Install the ground cable with the nut and tighten to 74 inch lbs. (8.4 Nm).
17. Connect the heater inlet water hose.
18. Connect the fuel pipe No. 1: push the quick connector into the pipe until it clicks.

➡Check if there is any damage or contamination on the connected part.

➡After connecting, confirm that the connector and pipe are securely connected by trying to pull them apart.

19. Install the fuel pipe clamp.
20. Install the intake air surge tank.
21. Install the emission control valve set.
22. Install the air cleaner cap sub-assembly.
23. Check vacuum hose connection.
24. Add engine coolant.
25. Check for coolant leak.
26. Check for fuel leak.
27. Install the V-bank cover sub-assembly.

3.5L Engine

See Figure 272.

1. Remove the engine under cover assembly.
2. Remove the No. 1 engine under cover.
3. Drain engine coolant.
4. Discharge fuel system pressure.
5. Remove the front wiper arm and blade assemblies.
6. Remove the cowl top ventilator louver sub-assembly.
7. Remove the windshield wiper motor and link assembly.
8. Remove the outer cowl top panel sub-assembly.
9. Remove the V-bank cover sub-assembly.
10. Remove the air cleaner cap sub-assembly.
11. Disconnect the engine room main wire.
12. Remove the throttle body bracket.
13. Remove the No. 1 surge tank stay.
14. Remove the intake air surge tank assembly.
15. Remove the right No. 2 engine mounting stay.
16. Disconnect the fuel main tube.
17. Remove the intake manifold.
18. Disconnect the 2 knock control sensor connectors.
19. Remove the 2 bolts and then remove the 2 knock control sensors.

To install:

20. Install the 2 knock control sensors with the 2 bolts as shown in the illustration, and tighten to 15 ft. lbs. (20 Nm).
21. Connect the 2 knock control sensor connectors.
22. Install the intake manifold.
23. Install the right no. 2 engine mounting stay.
24. Connect the fuel main tube.
25. Temporarily install the No. 1 surge tank stay.
26. Temporarily install the throttle body bracket.
27. Install the intake air surge tank assembly.
28. Fully tighten the No. 1 surge tank stay.
29. Fully tighten the throttle body bracket.
30. Connect the engine room wire.
31. Install the air cleaner cap sub-assembly.
32. Add engine coolant.
33. Inspect for coolant leak.
34. Inspect for fuel leak.
35. Install the outer cowl top panel sub-assembly.
36. Install the windshield wiper motor and link assembly.
37. Install the cowl top ventilator louver sub-assembly.
38. Install the front wiper arm and blade assemblies.
39. Install the No. 1 engine under cover.
40. Install the engine under cover assembly.
41. Install the V-bank cover sub-assembly.

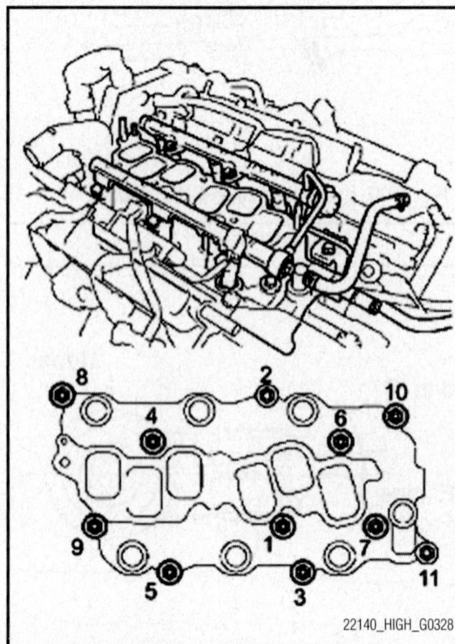

Fig. 271 Intake manifold bolt tightening sequence—3.3L engine

Fig. 272 Knock sensor installation—3.5L engine

MALFUNCTION INDICATOR LIGHT (MIL)

RESET PROCEDURES

Clear the DTC codes.

MASS AIR FLOW (MAF) SENSOR

LOCATION

See Figures 273 through 275.

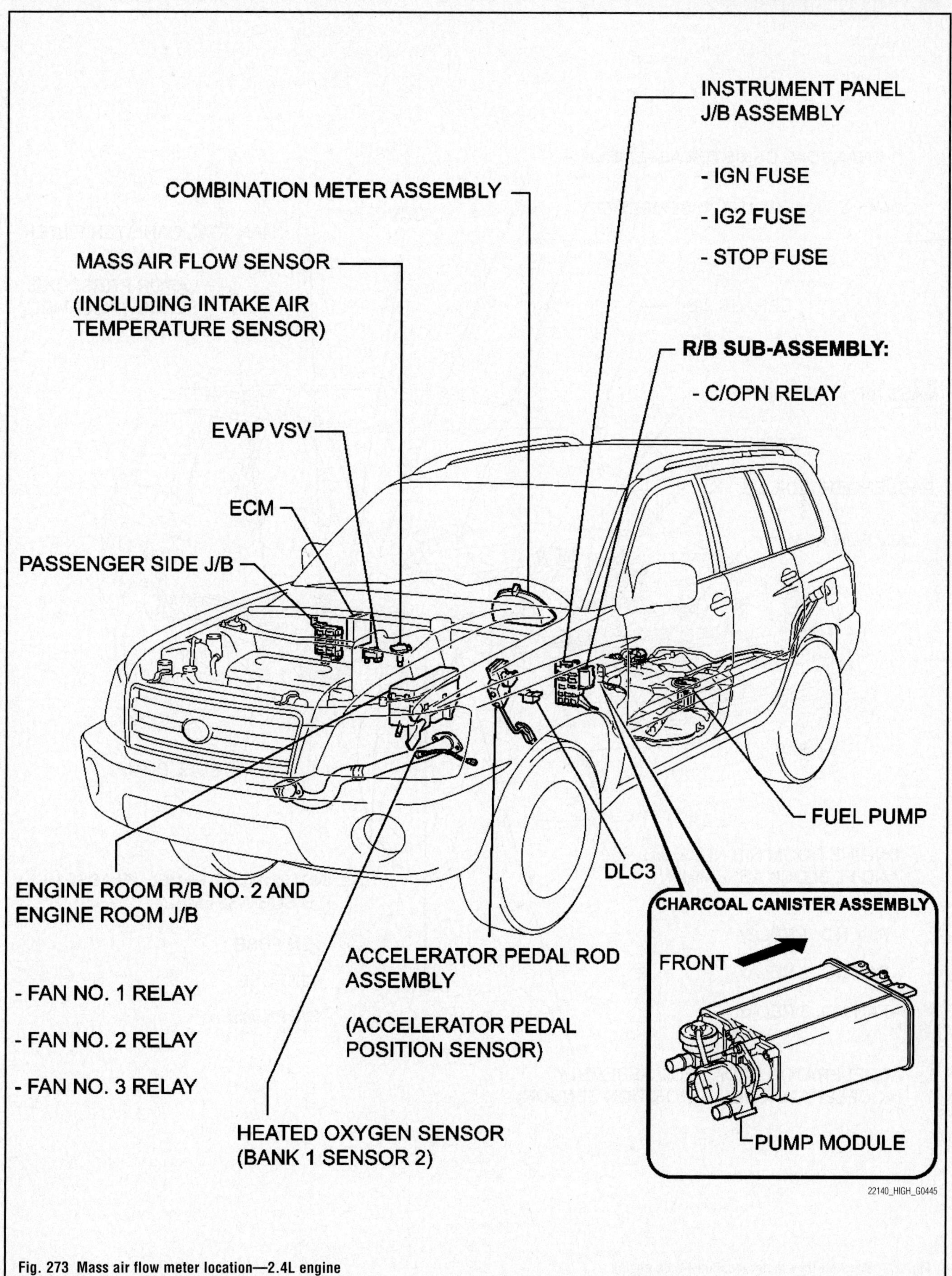

Fig. 273 Mass air flow meter location—2.4L engine

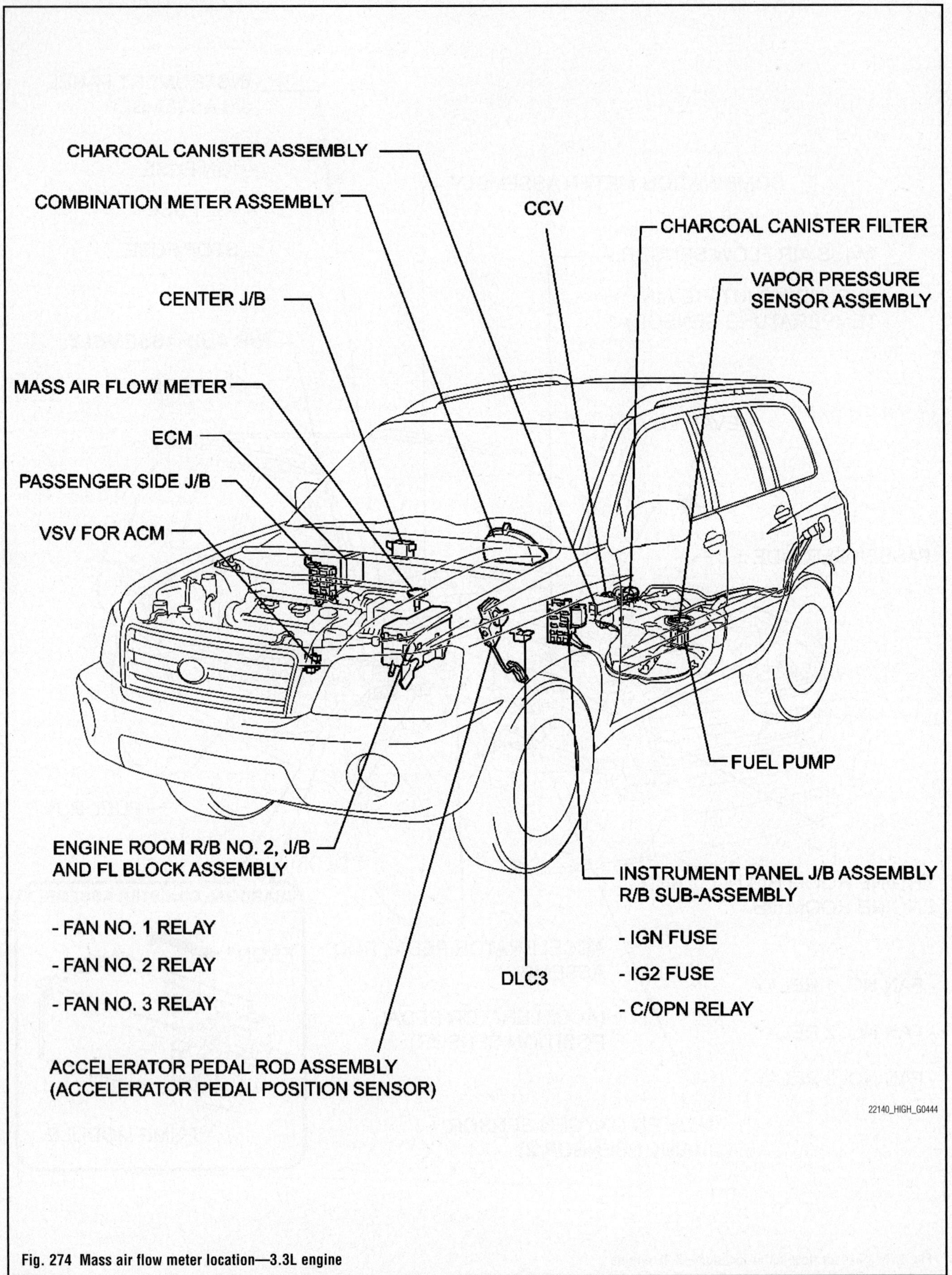

Fig. 274 Mass air flow meter location—3.3L engine

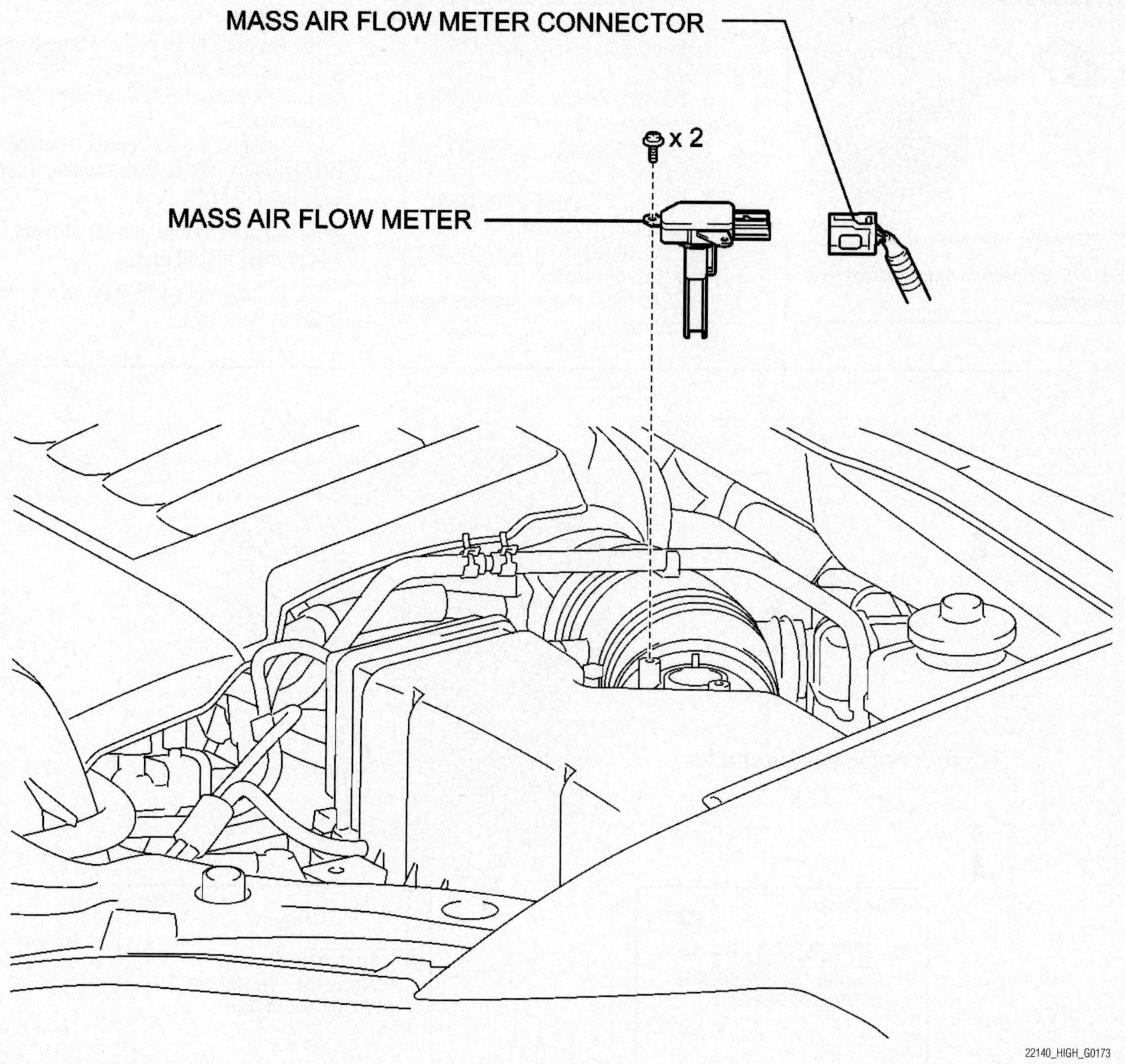

Fig. 275 Mass air flow meter location—3.5L engine

REMOVAL & INSTALLATION

See Figure 276.

1. Disconnect the mass air flow meter connector.
2. Remove the 2 screws and mass air flow meter.

To install:

3. Installation is the reverse of removal.

OIL PRESSURE SENSOR

LOCATION

See Figures 277 through 279.

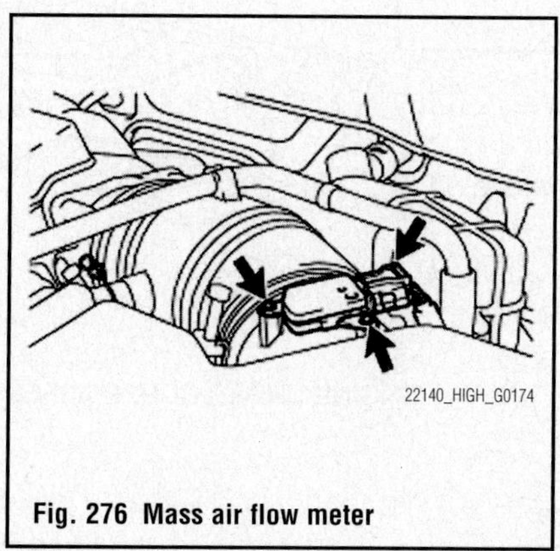

Fig. 276 Mass air flow meter

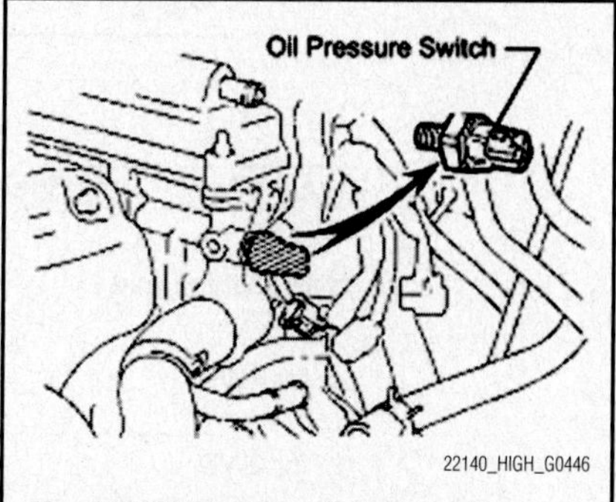

Fig. 277 Oil Pressure Switch location—2.4L engine

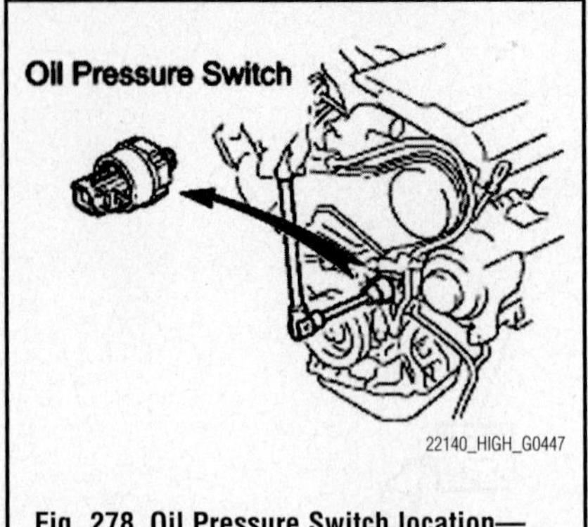

Fig. 278 Oil Pressure Switch location—3.3L engine

REMOVAL & INSTALLATION

1. For 4WD vehicles, remove the right front wheel.
2. Remove the engine under cover assembly.
3. For 4WD vehicles, disconnect the right front fender liner.
4. For 4WD vehicles, remove the right front fender apron seal.
5. Remove the engine oil pressure switch assembly, as follows:
    a. Disconnect the wiring harness protector (with oil cooler).
    b. Disconnect the oil pressure switch assembly connector.
    c. Using a 24 mm deep socket wrench, remove the oil pressure switch.

To install:

6. Apply adhesive to 2 or 3 threads of the oil pressure switch. Adhesive: Toyota Genuine Adhesive 1344, Three Bond 1344 or equivalent.
7. Using a 24 mm deep socket wrench, install the oil pressure switch and tighten to 11 ft. lbs. (15 Nm).

➡Do not start the engine for at least 1 hour after installation.

8. Connect the oil pressure switch assembly connector.

x 2
*1 FRONT FENDER APRON SEAL RH
w/ Oil Cooler:
15 (153, 11)
ENGINE OIL PRESSURE SWITCH ASSEMBLY
HARNESS PROTECTOR
w/o Oil Cooler:
15 (153, 11)
ENGINE OIL PRESSURE SWITCH ASSEMBLY
*2 ENGINE UNDER COVER ASSEMBLY
x 2
x 2
x 5
x 2
*2 ENGINE UNDER COVER ASSEMBLY RR
N*m (kgf*cm, ft.*lbf): Specified torque
★ Precoated part
*1: for 4WD
*2: for 2WD
22140_HIGH_G0435

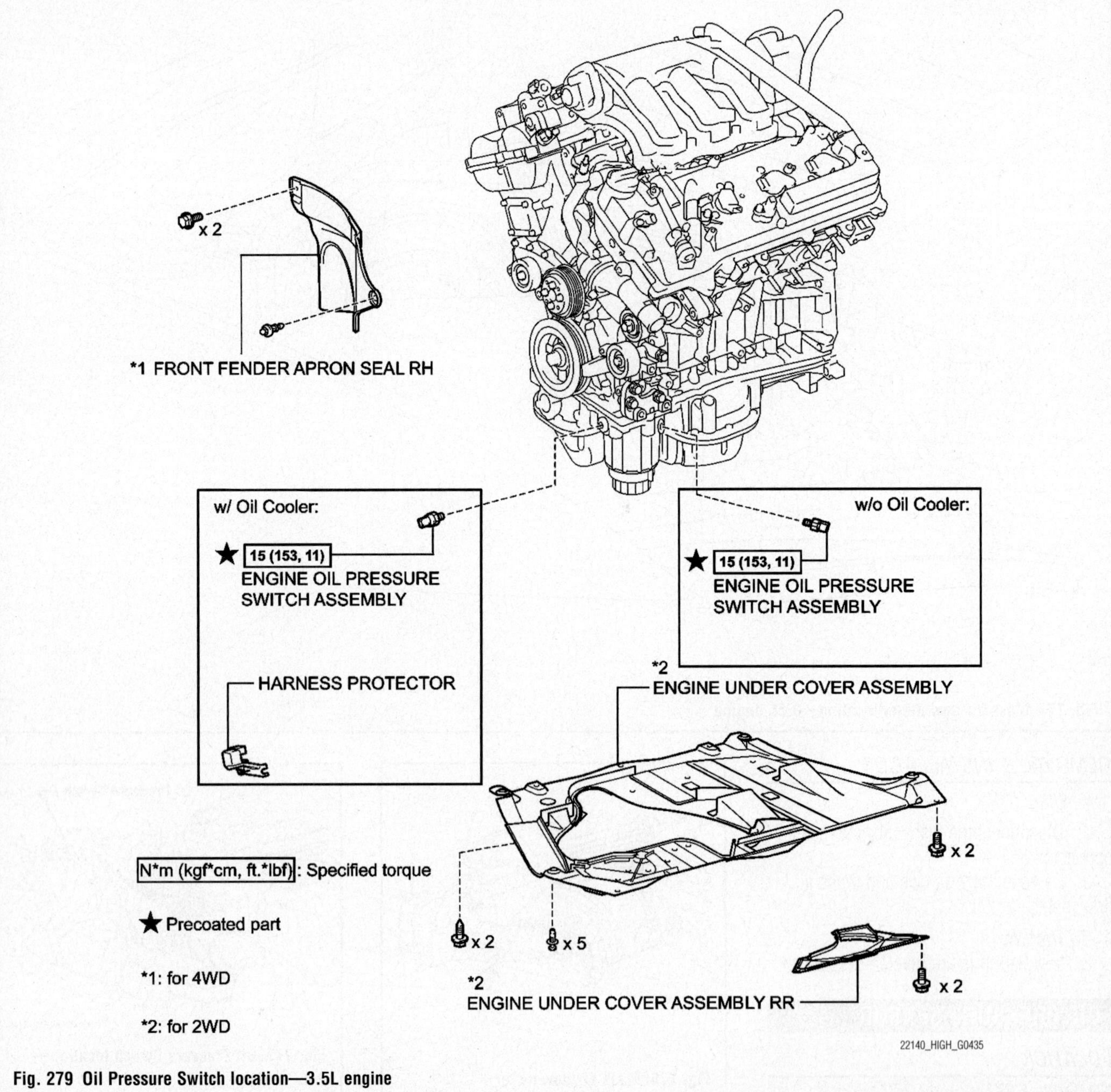

Fig. 279 Oil Pressure Switch location—3.5L engine

9. Connect the wiring harness protector (with oil cooler).

10. Inspect for oil leak. Check for engine oil leaks from the oil pressure switch assembly.

11. For 4WD vehicles, install the right front fender apron seal.

12. For 4WD vehicles, install the right front fender liner.

13. Install the engine under cover assembly.

14. For 4WD vehicles, install the right front wheel.

THROTTLE POSITION SENSOR (TPS)

LOCATION

The throttle position sensor is located on the throttle body.

REMOVAL & INSTALLATION

Refer to the Throttle Body removal and installation procedures.

VARIABLE CAMSHAFT TIMING OIL CONTROL SOLENOID

LOCATION

See Figure 280.

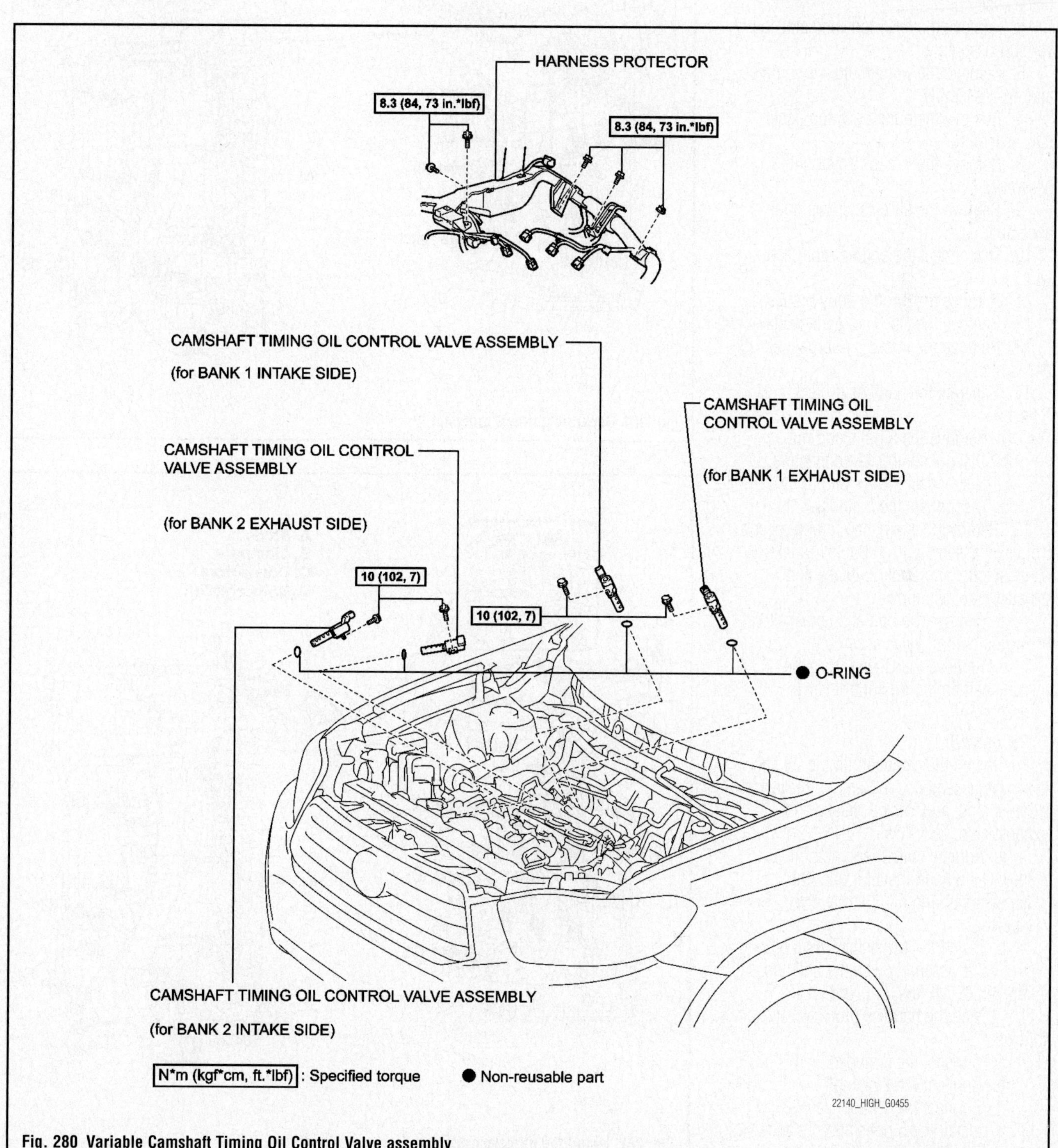

Fig. 280 Variable Camshaft Timing Oil Control Valve assembly

REMOVAL & INSTALLATION

3.5L Engine

See Figures 281 and 282.

1. Remove the engine under cover assembly.
2. Remove the No. 1 engine under cover.
3. Drain engine coolant.
4. Remove both front wiper arm and blade assemblies.
5. Remove the cowl top ventilator louver sub-assembly.
6. Remove the windshield wiper motor and link assembly.
7. Remove the outer cowl top panel sub-assembly.
8. Remove the V-bank cover sub-assembly.
9. Remove the air cleaner cap sub-assembly.
10. Disconnect the engine room main wire.
11. Remove the throttle body bracket.
12. Remove the No. 1 surge tank stay.
13. Remove the intake air surge tank assembly.
14. Separate the harness protector, as follows:
    a. Remove the 5 bolts and nut.
    b. Disconnect the 19 connectors.
    c. Disconnect the 2 clamps.
    d. Disconnect the 2 studs.
15. Remove the camshaft timing oil control valve assembly from Bank 1, intake and exhaust side, and Bank 2, intake and exhaust side, as follows:
    a. Remove the bolt and camshaft timing oil control valve assembly.
    b. Remove the O-ring from the camshaft timing oil control valve assembly.

To install:

16. Install the camshaft timing oil control valve assemblies to Bank 1, intake and exhaust side, and Bank 2, intake and exhaust side, as follows:
    a. Apply a coat of engine oil to a new O-ring and install it onto the camshaft timing oil control valve assembly.
    b. Install the camshaft timing oil control valve assembly with the bolt and tighten to 7 ft. lbs. (10 Nm).
17. Install the harness protector, as follows:
    a. Connect the 2 studs.
    b. Connect the 2 clamps.
    c. Connect the 19 connectors.
    d. Install the 5 bolts and nut and tighten to 73 inch lbs. (8.3 Nm).

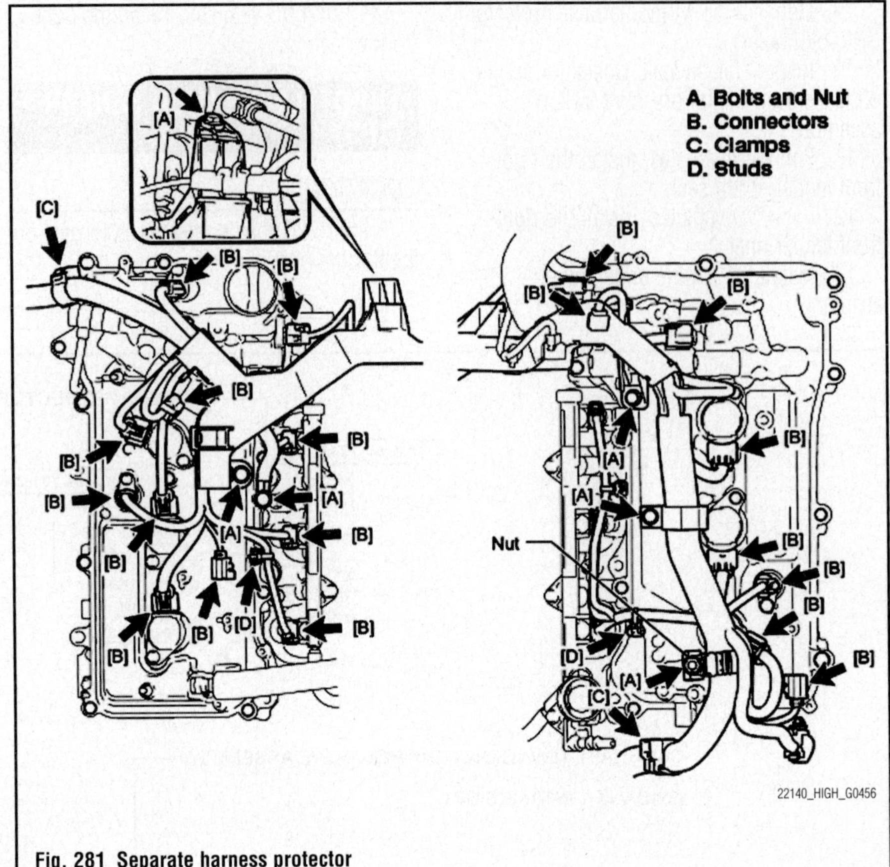

Fig. 281 Separate harness protector

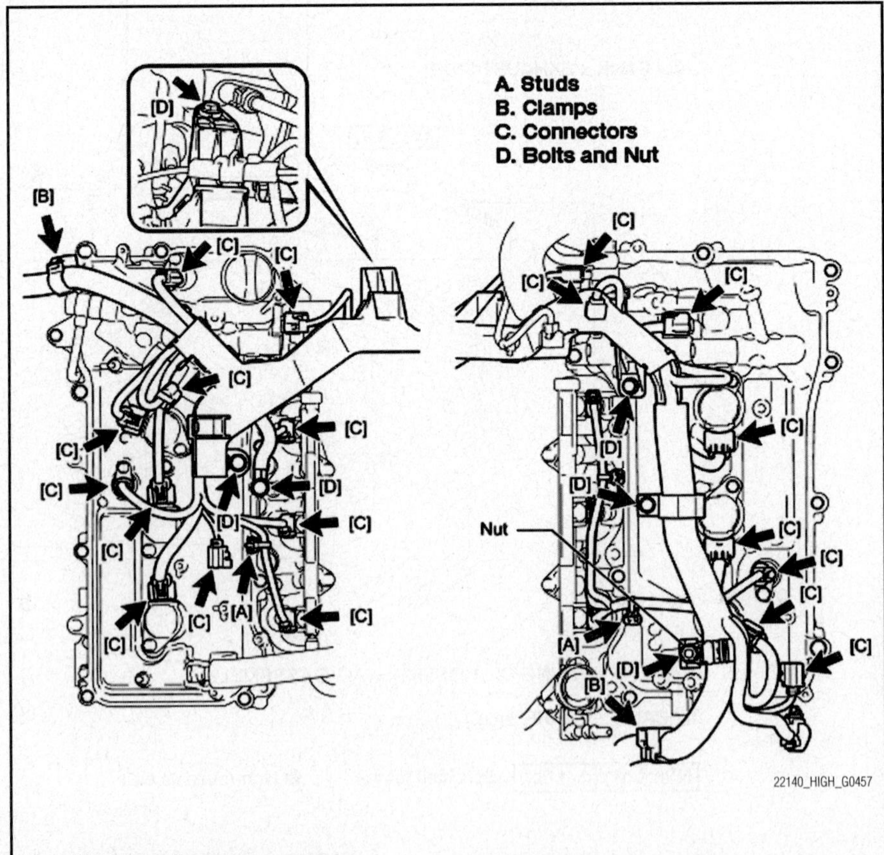

Fig. 282 Install the harness protector

18. Temporarily install the No. 1 surge tank stay.
19. Temporarily install the throttle body bracket.
20. Install the intake air surge tank assembly.
21. Fully tighten the No. 1 surge tank stay.
22. Fully tighten the throttle body bracket.
23. Connect the engine room main wire.
24. Install the air cleaner cap sub-assembly.
25. Add engine coolant.
26. Inspect for coolant leak.
27. Install the outer cowl top panel sub-assembly.
28. Install the windshield wiper motor and link assembly.
29. Install the cowl top ventilator louver sub-assembly.
30. Install both front wiper arm and blade assemblies.
31. Install the V-bank cover sub-assembly.
32. Install the No. 1 engine under cover.
33. Install the engine under cover assembly.

FUEL SYSTEM — GASOLINE FUEL INJECTION SYSTEM

FUEL SYSTEM SERVICE PRECAUTIONS

Before servicing any vehicle, please be sure to read all of the following precautions, which deal with personal safety, prevention of component damage, and important points to take into consideration when servicing a motor vehicle:

- Observe all applicable safety precautions when working around fuel. Whenever servicing the fuel system, always work in a well-ventilated area. Do not allow fuel spray or vapors to come in contact with a spark, open flame, or excessive heat (a hot drop light, for example). Keep a dry chemical fire extinguisher near the work area. Always keep fuel in a container specifically designed for fuel storage; also, always properly seal fuel containers to avoid the possibility of fire or explosion. Refer to the additional fuel system precautions later in this section.
- Fuel injection systems often remain pressurized, even after the engine has been turned **OFF**. The fuel system pressure must be relieved before disconnecting any fuel lines. Failure to do so may result in fire and/or personal injury.
- All new vehicles are now equipped with an air bag system, often referred to as a Supplemental Restraint System (SRS) or Supplemental Inflatable Restraint (SIR) system. The system must be disabled before performing service on or around system components, steering column, instrument panel components, wiring and sensors. Failure to follow safety and disabling procedures could result in accidental air bag deployment, possible personal injury and unnecessary system repairs.
- Disconnecting the negative battery cable on some vehicles may interfere with the functions of the on-board computer system(s) and may require the computer to undergo a relearning process once the negative battery cable is reconnected.

RELIEVING FUEL SYSTEM PRESSURE

1. Before servicing the vehicle, refer to the precautions section.
2. Disconnect the fuel pump wire at the pump.
3. Start the engine. After the engine stops, turn the ignition switch to OFF.
4. Disconnect the negative battery terminal.
5. Connect the fuel pump.

FUEL FILTER

REMOVAL & INSTALLATION

The fuel filter is part of the fuel suction tube/fuel pump assembly and is located in the fuel tank. It is not a normally serviced item.

FUEL PUMP

REMOVAL & INSTALLATION

2007 Models

See Figures 283 through 290.

1. Before servicing the vehicle, refer to the precautions section.
2. Discharge fuel system pressure.
3. Disengage the 5 clips and turn up the front side of the deck board.

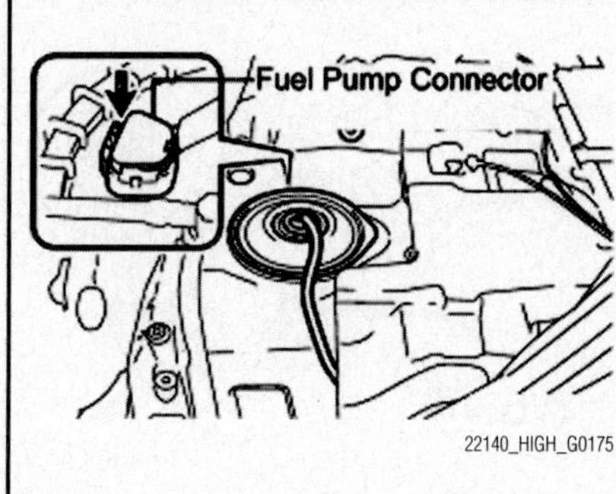

Fig. 283 Rear floor service hole cover and fuel pump connector

4. Remove the rear seat track bracket cover outer.
5. Remove the rear seat assemblies.
6. Remove the left rear door scuff plate.
7. Remove the rear floor service hole cover.
8. Disconnect the fuel pump & sender gauge connector.
9. Remove the tube joint clip, and pull out the fuel pump tube.

➡**Check that there is no dirt or other foreign objects around the connector before disconnecting it. Clean the connector if necessary.**

➡**It is necessary to prevent mud or dirt from entering the quick connector. If mud or dirt gets in the connector, the O-rings may not seal properly.**

➡**Disconnect the quick connector by hand. Do not use any tools.**

➡**Do not bend, kink or twist the nylon tubes. Protect the connector by covering it with a plastic bag.**

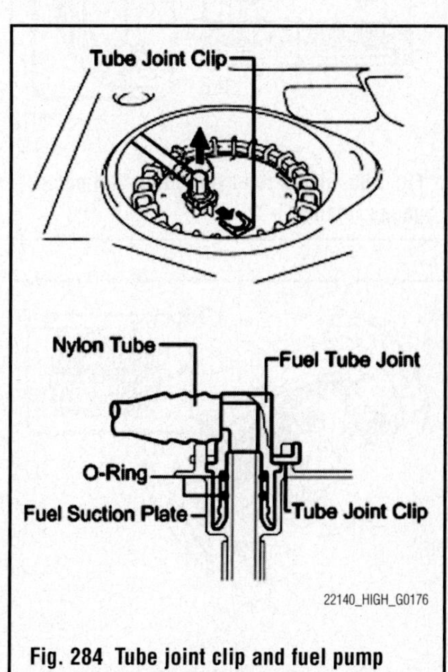

Fig. 284 Tube joint clip and fuel pump tube

➡If the pipe and connector are stuck, carefully try wiggling or pushing and pulling on the connector to release it. Pull the connector off carefully.

10. Using SST: 09808-14020, loosen the fuel pump gauge retainer.

WARNING

Do not use any tools other than specified in this operation. Damage to the fuel pump gauge retainer or the fuel tank may result.

➡Loosen the retainer by turning it counterclockwise while holding SST down. Do not allow the claw of the tank suction tube support to slip out of its groove on the fuel tank.

➡A rib on the fuel pump gauge retainer can be fitted into a tip of the SST.

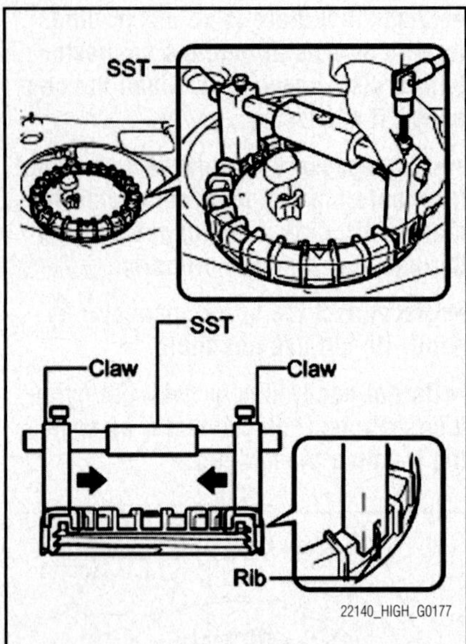

Fig. 285 Install removal tool to fuel pump gauge retainer

11. Remove the fuel pump gauge retainer.

12. Remove the fuel suction tube.

➡Be careful not to bend the arm of the fuel sender gauge.

13. Remove the gasket from the fuel tank.

14. Disconnect the fuel sender gauge connector.

15. Unlock the fuel sender gauge and slide it to remove.

16. Using a screwdriver with the tip wrapped in tape, disconnect the 4 snap-claws from the claw holes and remove the fuel suction plate.

➡Do not damage the fuel suction plate or fuel suction support No. 1.

17. Disconnect the fuel pump connector.

18. Using a screwdriver with its tip wrapped in tape, disconnect the 2 snap-claws from the claw holes and pull out the fuel pump.

➡Do not damage the fuel pump or fuel suction plate.

19. Remove the O-ring and fuel pump spacer from the fuel pump.

To install:

20. Install fuel pump assembly, as follows:

a. Install the fuel pump spacer to the fuel pump.

b. Apply a light coat of gasoline or spindle oil to a new O-ring, and install it to the fuel pump.

c. Apply a light coat of gasoline or spindle oil to the O-ring again, and install the fuel pump to the fuel suction plate.

d. Connect the fuel pump connector.

21. Install fuel suction plate sub-assembly, as follows:

a. Apply a light coat of gasoline or spindle oil to a new O-ring, and install it to the fuel tube joint No. 1.

b. Apply a light coat of gasoline or spindle oil to the O-ring again, and install the fuel suction plate to the fuel suction support No. 1.

22. Install the fuel sender gauge assembly.

23. Install the fuel suction tube assembly with pump and gauge, as follows:

a. Install a new gasket to the fuel tank.

b. Attach the fuel suction tube to the fuel tank.

➡Be careful not to bend the arm of the fuel sender gauge.

c. Align the keyway of the fuel suction tube support with the key of the fuel suction tube.

d. Apply MP grease to the entire interior surface of the fuel pump gage retainer.

e. Align the triangle mark on a new fuel pump gauge retainer with the "S" mark on the fuel tank while pushing down the fuel suction tube, and attach the fuel pump gauge retainer.

f. Rotate the fuel pump gauge retainer by hand, then tighten it one complete turn and half using SST: 09808-14020. The triangle mark on the fuel pump gauge retainer must be positioned between the "A" and "MAX." indications on the fuel tank.

WARNING

Do not use any tools other than specified in this operation. Damage to the fuel pump gauge retainer or the fuel tank may result.

➡A rib on the fuel pump gauge retainer can be fitted into a tip of the SST.

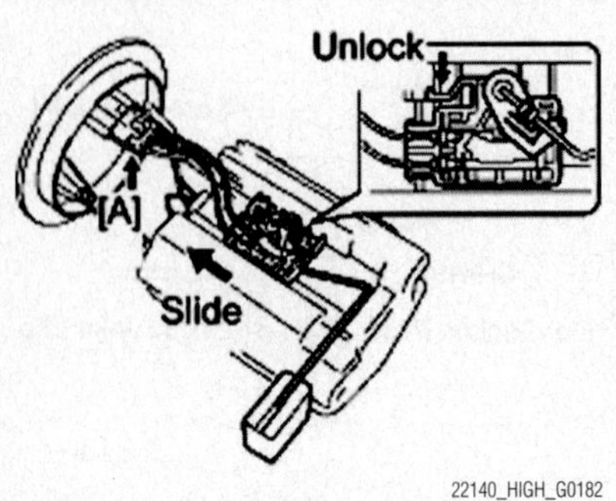

Fig. 286 Unlock and slide the fuel sender gauge

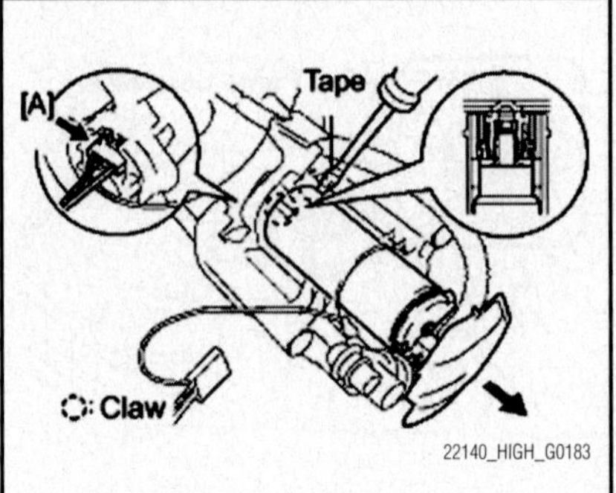

Fig. 287 Disconnect the fuel pump connector (A)

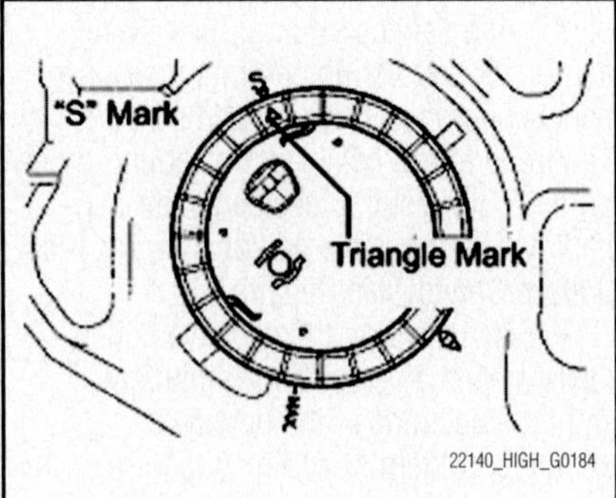

Fig. 288 Align the new fuel pump gauge retainer

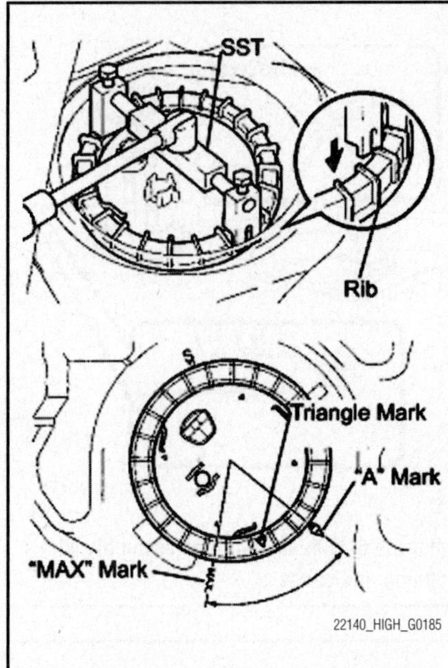

Fig. 289 Installing the fuel pump gauge retainer

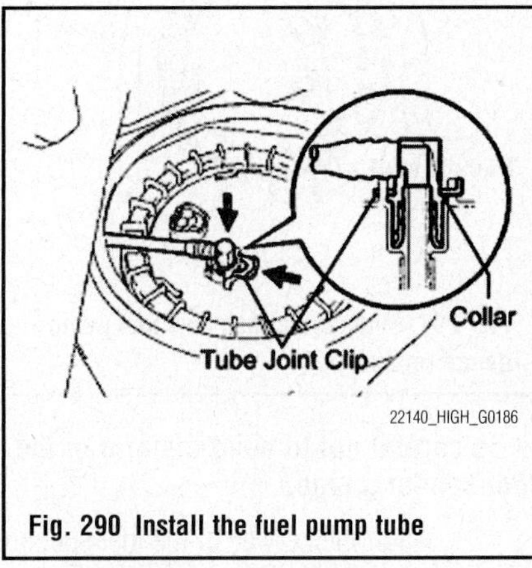

Fig. 290 Install the fuel pump tube

g. Install the fuel pump tube and the tube joint clip.

➡Check that there are no scratches or foreign objects on the connecting part.

➡Check that the fuel tube joint is inserted securely.

➡Check that the tube joint clip is on the collar of the fuel tube joint.

➡After installing the tube joint clip, check that the fuel tube joint has not been pulled off.

24. Check for fuel leaks.
25. Install a new butyl tape to the rear floor service hole cover.
26. Connect the fuel pump & sender gauge connector.
27. Install the rear floor service hole cover.
28. Install the left rear door scuff plate.
29. Install the rear seat assemblies.
30. Install rear seat track bracket cover outer.
31. Install deck board sub-assembly.

2008 Models

See Figures 291 through 299.

1. Before servicing the vehicle, refer to the precautions section.
2. Discharge fuel system pressure.
3. Disconnect cable from negative battery terminal.
4. Remove the rear center seat assembly.
5. Remove the rear seat headrest assemblies.
6. Remove the seat track bracket covers.
7. Remove rear inner and outer track bracket covers.
8. Remove rear seat leg side covers.
9. Remove rear no. 1 seat lock cable assembly, if applicable.
10. Remove rear no. 1 seat assemblies.
11. Remove left rear door scuff plate.
12. Remove left rear door opening trim weatherstrip.
13. Remove deck board assembly.
14. Remove no. 2 and 3 deck board sub-assemblies.
15. Remove tonneau cover assembly, if applicable.
16. Remove rear no. 1 floor board, or rear seat side covers, as necessary.
17. Remove rear seat side covers.
18. Remove deck side trim boxes.
19. Remove jack carrier support.
20. Remove jack carrier cushion.
21. Remove jack assembly.
22. Remove jack carrier assembly.
23. Remove rear mat.
24. Remove deck floor board assembly, if applicable.
25. Remove rear deck floor box, if applicable.
26. Remove rear no. 2 seat inner belt assembly, if applicable.
27. Disconnect rear seat lap type belt assemblies.
28. Remove rear no. 2 seat assembly, if applicable.
29. Remove rear floor finish plate.
30. Remove deck side trim covers.
31. Remove left side trim cover, if applicable.
32. Remove power outlet socket bezel, if applicable.
33. Remove rear combination light service cover.
34. Remove rear power point socket assembly.
35. Remove rear power outlet socket cover

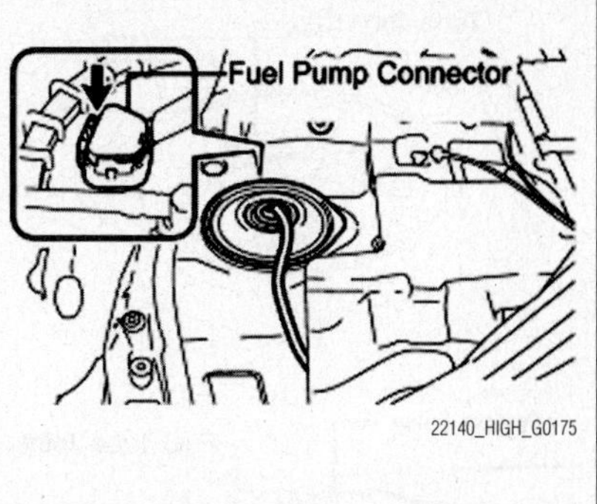

Fig. 291 Rear floor service hole cover and fuel pump connector

36. Remove rear deck trim cover, if applicable.
37. Remove left reclining remote control lever bezel, if applicable.
38. Remove left rope hook assembly.
39. Remove no. 2 deck side trim hook.
40. Remove left front deck side trim cover.
41. Disconnect left rear no. 1 seat outer belt assembly.
42. Remove left deck trim side panel assembly.
43. Remove the rear floor service hole cover, as follows:
    a. Lift the front floor carpet.
    b. Using the appropriate tool, remove the 3 clips and lift up the front floor carpet.
    c. Remove the rear floor service hole cover.
    d. Disconnect the fuel pump connector.
44. Remove the fuel suction tube assembly with pump and gauge, as follows:
    a. Remove the tube joint clip, and pull out the fuel pump tube.

➡Check that there is no dirt or other foreign objects around the connector before disconnecting it. Clean the connector if necessary.

➡It is necessary to prevent mud or dirt from entering the quick connector. If mud or dirt gets in the connector, the O-rings may not seal properly.

➡Disconnect the quick connector by hand. Do not use any tools.

➡Do not bend, kink or twist the nylon tubes. Protect the connector by covering it with a plastic bag.

➡If the pipe and connector are stuck, carefully try wiggling or pushing and pulling on the connector to release it. Pull the connector off carefully.

b. Using a 6 mm socket hexagon wrench, install SST: 09808-14020 to the fuel pump gauge retainer.

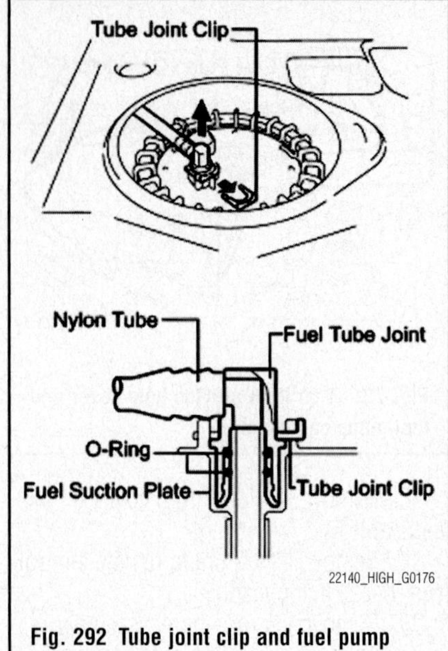

Fig. 292 Tube joint clip and fuel pump tube

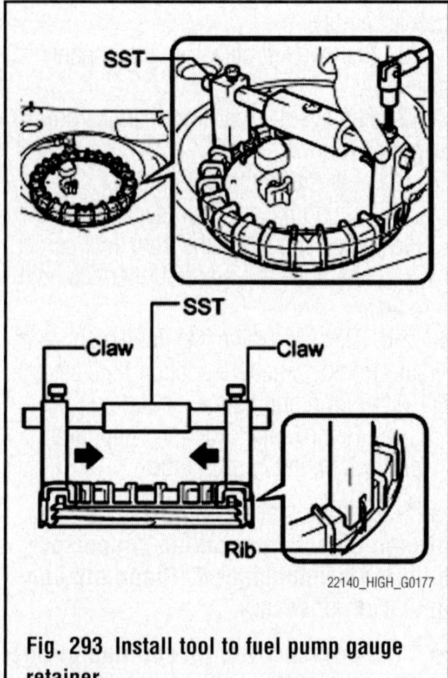

Fig. 293 Install tool to fuel pump gauge retainer

➡Engage the SST claws securely with the fuel pump gauge retainer ribs to secure the SST.

➡Install the SST while pressing the SST claws toward the fuel pump gauge retainer (toward the center of SST).

c. Using SST: 09808-14020, loosen the fuel pump gauge retainer.

d. Remove the fuel pump gauge retainer.

⁂ WARNING

Do not use any tools other than specified in this operation. Damage to the fuel pump gauge retainer or the fuel tank may result.

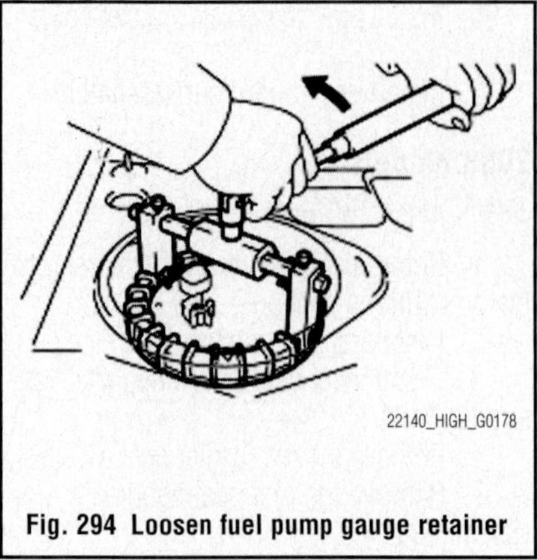

Fig. 294 Loosen fuel pump gauge retainer

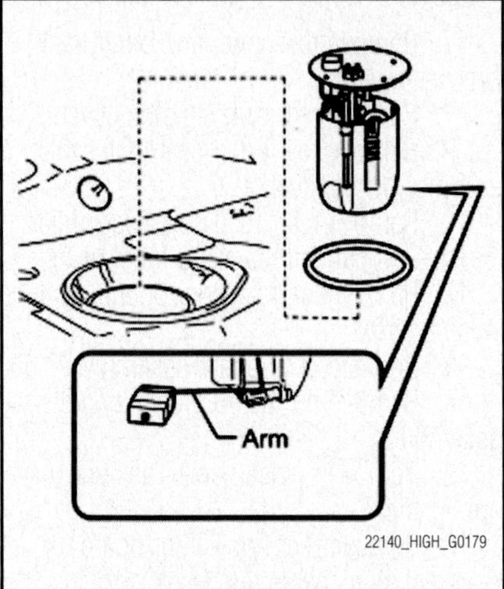

Fig. 295 Remove fuel pump gauge retainer

➡Loosen the retainer by turning it counterclockwise while holding SST down. Do not allow the claw of the tank suction tube support to slip out of its groove on the fuel tank.

➡The ribs on the fuel pump gauge retainer can be fitted into the tips of SST.

e. Remove the fuel suction tube with pump and gauge.

➡Be careful not to bend the arm of the fuel sender gauge.

f. Remove the gasket from the fuel tank.

To install:

45. Install the fuel suction tube assembly with pump and gauge, as follows:

a. Install a new gasket to the fuel tank.

b. Attach the fuel suction tube with pump and gauge to the fuel tank.

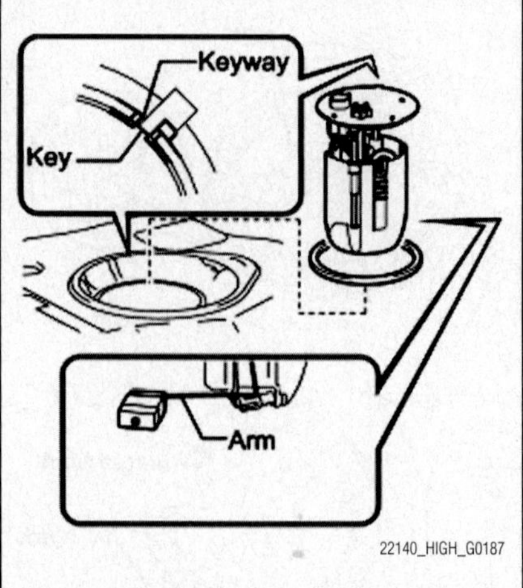

Fig. 296 Fuel suction tube assembly with pump and gauge

Fig. 297 Align and attach new fuel pump gauge retainer

➡Be careful not to bend the arm of the fuel sender gauge.

c. Align the keyway of the fuel suction tube support with the key of the fuel suction tube with pump and gauge.

d. Align the triangle mark on a new fuel pump gauge retainer with the "S" mark on the fuel tank while pushing down the fuel suction tube with pump and gauge, and attach the fuel pump gauge retainer.

46. Using a 6 mm socket hexagon wrench, install SST: 09808-14020 to the fuel pump gauge retainer.

➡Engage the SST claws securely with the fuel pump gauge retainer ribs to secure the SST.

➡Install the SST while pressing the SST claws toward the fuel pump gauge retainer (toward the center of SST).

a. Rotate the fuel pump gauge retainer by hand, then tighten it one complete turn and another half turn using the SST: 09808-14020. The triangle mark on

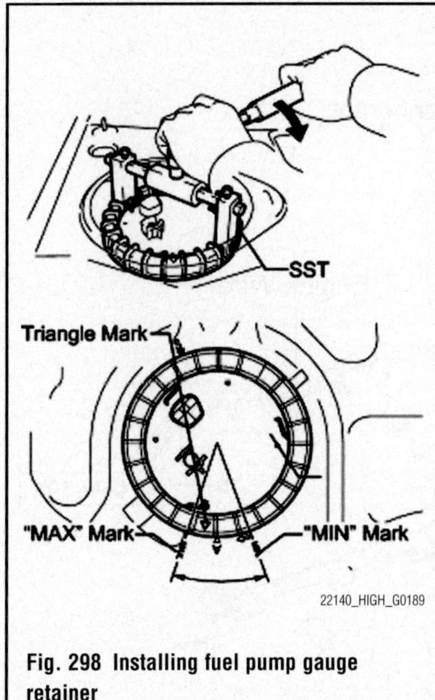

Fig. 298 Installing fuel pump gauge retainer

the fuel pump gauge retainer must be positioned between the "MIN" and "MAX" marks on the fuel tank.

WARNING

Do not use any tools other than specified in this operation. Damage to the fuel pump gauge retainer or the fuel tank may result.

➡Fully tighten the retainer by turning it clockwise while holding the SST down. Do not allow the claw of the tank suction tube support to slip out of its groove on the fuel tank.

➡The ribs on the fuel pump gauge retainer can be fitted into the tips of SST.

b. Install the fuel pump tube and the tube joint clip.

➡Check that there are no scratches or foreign objects on the connecting part.

➡Check that the fuel tube joint is inserted securely.

➡Check that the tube joint clip is on the collar of the fuel tube joint.

➡After installing the tube joint clip, check that the fuel tube joint has not been pulled out of position.

c. Connect the fuel pump connector.

47. Connect the negative battery cable.
48. Inspect for fuel leak.
49. Disconnect the negative battery cable.

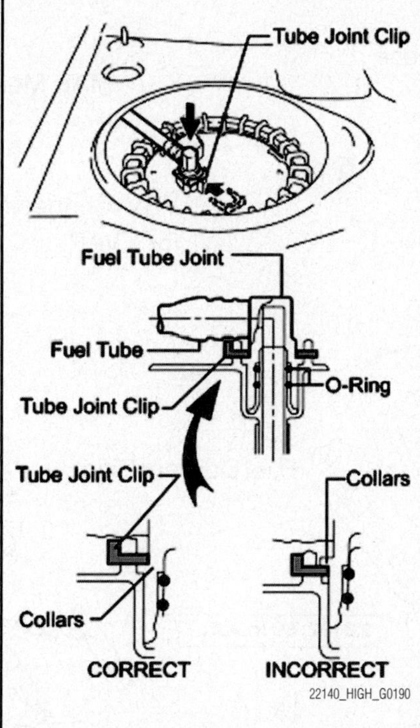

Fig. 299 Fuel pump tube and the tube joint clip

50. Install new butyl tape to the rear floor service hole cover.
51. Install the rear floor service hole cover.
52. Install the front floor carpet with the 3 clips.
53. The remainder of installation is the reverse of the removal procedure.
54. After installing all components and connecting the negative battery cable, inspect the SRS warning light..

FUEL RAIL & INJECTORS

REMOVAL & INSTALLATION

2.4L Engine

See Figure 300.

1. Before servicing the vehicle, refer to the precautions section.
2. Relieve the fuel system pressure.
3. Remove or disconnect the following:
   - Air cleaner and hoses
   - Fuel line from the rail
   - Injector connectors
   - Injector wiring harness
   - Fuel rail with injectors
   - Injector spacers from the head
4. Installation is the reverse of removal. Coat the new o-rings with clean fuel. Before tightening the fuel rail bolts, make sure that each injector rotates smoothly. Tighten the bolts to 15 ft. lbs. (20 Nm).

3.3L Engine

See Figure 301.

1. Before servicing the vehicle, refer to the precautions section.
2. Relieve the fuel system pressure.
3. Drain the coolant.
4. Remove the wiper arms.
5. Remove the wiper linkage.
6. Remove the fender-to-cowl side seals.
7. Remove the rain sensor.
8. Remove the front shock absorber caps (air suspension).
9. Remove the 4 set nuts from the strut (w/o air suspension).
10. Remove the cowl top outer panel.
11. Remove the 6 set nuts from the shock absorber.
12. Remove the V-bank cover.
13. Remove the air cleaner assembly and inlet tubes.
14. Remove the emission control valve set.
15. Remove the upper intake manifold (intake air surge tank). Discard the gasket.
16. Remove the fuel pipe sub-assembly.
17. Disconnect the wiring at the injectors.
18. Remove the 4 bolts and 2 delivery pipe along with the injectors.
19. Remove the delivery pipe spacers and insulators from the manifold.
20. Pull each injector from the pipe.

To install:

21. Install new O-rings on each injector. Apply a light coating of gasoline to the O-rings and mating points on the pipes.
22. Using a twisting motion, install the injectors on the pipes.

➡Be careful to avoid twisting the O-rings. After installation, check that the injectors turn smoothly. If not, use new O-rings.

23. Install the pipes and injectors.
24. Loosely install the bolts and make sure that the injectors still turn freely. If not, replace the O-rings.
25. Torque the bolts to 84 inch lbs. (10 Nm).
26. The remainder of installation is the reverse of removal. Observe the following torques:
    - Fuel line union bolt: 24 ft. lbs. (33 Nm)
    - Pulsation damper: 24 ft. lbs. (33 Nm)
    - Fuel feed pipe: 14 ft. lbs. (20 Nm)
    - Upper intake manifold (air surge tank): 21 ft. lbs. (28 Nm)
    - Upper intake manifold stays: 14 ft. lbs. (20 Nm)

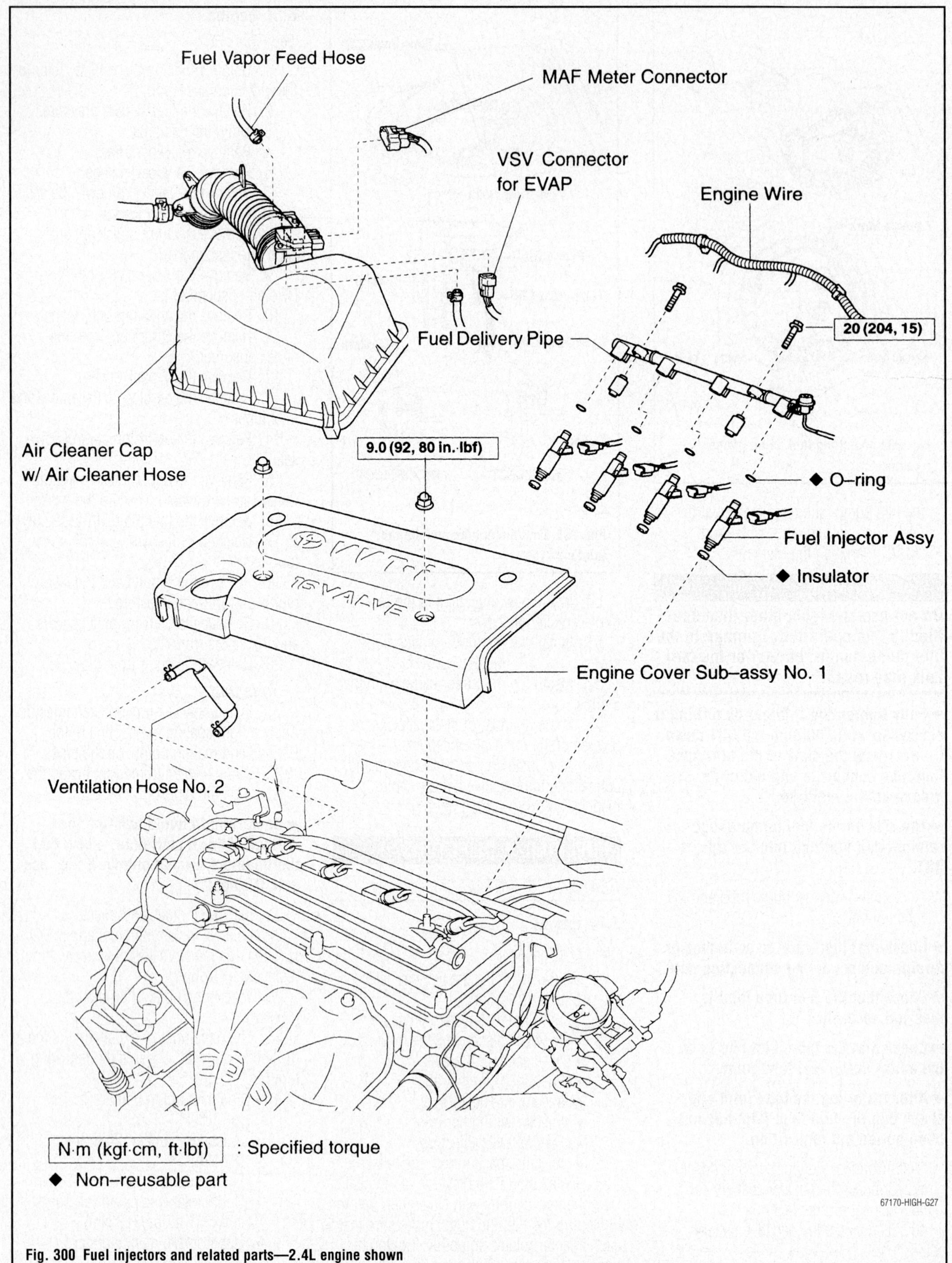

Fig. 300 Fuel injectors and related parts—2.4L engine shown

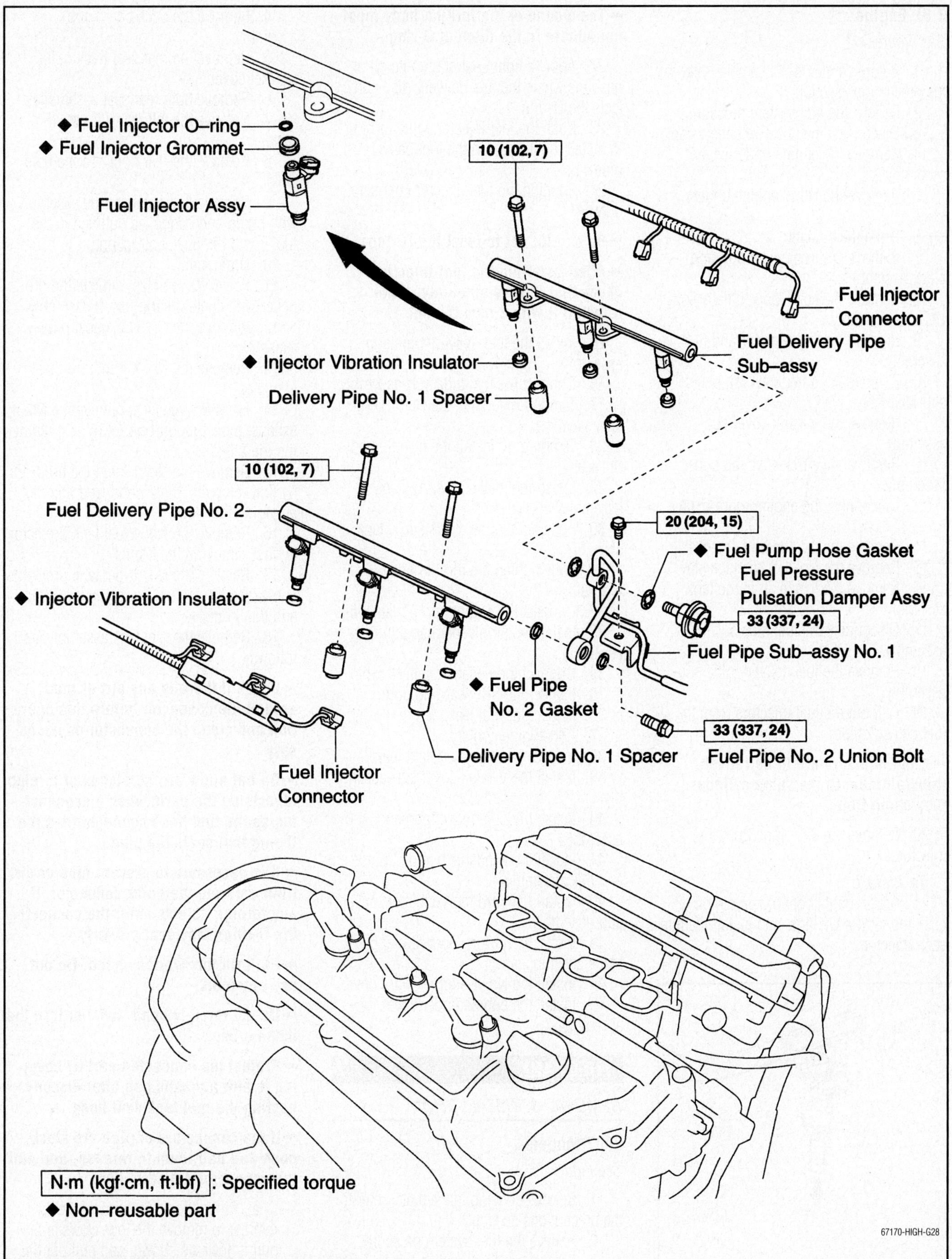

Fig. 301 Fuel injectors and related parts—3.3L engine

3.5L Engine

See Figure 302.

1. Before servicing the vehicle, refer to the precautions section.
2. Relieve the fuel system pressure.
3. Disconnect the negative battery cable.
4. Remove the engine under cover assembly.
5. Remove the No. 1 engine under cover.
6. Drain the coolant.
7. Remove the front wiper arm and blade assemblies.
8. Remove the cowl top ventilator louver sub-assembly.
9. Remove the wiper motor and link assembly.
10. Remove the outer cowl top panel sub-assembly.
11. Remove the V-bank cover sub-assembly.
12. Remove the air cleaner cap sub-assembly.
13. Disconnect the engine room main wire.
14. Remove the throttle body bracket.
15. Remove the No. 1 surge tank stay.
16. Remove the intake air surge tank assembly.
17. Disconnect the fuel tube sub-assembly.
18. Remove the fuel delivery pipe sub-assembly.
19. Pull out the fuel injectors from the fuel delivery pipe.

➡If the injectors are to be reused, reinstall them to the same cylinder they came from.

20. Remove the 6 O-rings from the injectors.

To install:

21. Apply a light coat of spindle oil or gasoline to new O-rings, and install them to each injector.

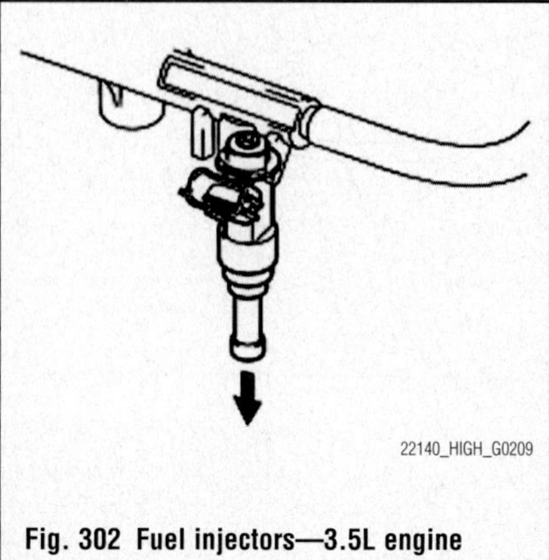

Fig. 302 Fuel injectors—3.5L engine

➡The wound or the foreign body must not adhere in the ditch of O-ring.

22. Apply a light coat of spindle oil or gasoline where the fuel delivery pipe contacts the O-ring.
23. Push the fuel injector while turning it to install the injector in the fuel delivery pipe.
24. Position the fuel injector connector outward.

➡Be careful not to twist the O-ring.

➡After installing the fuel injector, check that it turns smoothly. If not, reinstall it with a new O-ring.

25. Install the fuel delivery pipe sub-assembly.
26. Connect the fuel tube sub-assembly.
27. Temporarily install the No. 1 surge tank stay.
28. Temporarily install the throttle body bracket.
29. Install the intake air surge tank assembly.
30. Fully tighten the No. 1 surge tank stay.
31. Fully tighten the throttle body bracket.
32. Connect the engine room main wire.
33. Install the air cleaner cap sub-assembly.
34. Connect the negative battery cable.
35. Inspect the SRS warning light.
36. Inspect for fuel leak.
37. Add engine coolant.
38. Inspect for engine coolant leak.
39. Install the V-bank cover sub-assembly.
40. Install the outer cowl top panel sub-assembly.
41. Install the windshield wiper motor and link assembly.
42. Install the cowl top ventilator louver sub-assembly.
43. Install both front wiper arm and blade assemblies.
44. Install the No. 1 engine under cover.
45. Install the engine under cover assembly.

FUEL TANK

REMOVAL & INSTALLATION

2007 Models

See Figure 303.

1. Before servicing the vehicle, refer to the precautions section.
2. Relieve the fuel system pressure.
3. Disconnect the negative battery cable.
4. Remove the deck board sub-assembly.
5. Remove the rear seat track outer bracket cover.
6. Remove both rear seat assemblies.
7. Remove the left rear door scuff plate.
8. Remove the rear floor service hole cover.
9. Remove fuel suction tube assembly with pump and gauge, as outlined in the fuel pump removal instructions.
10. Drain fuel.
11. For 4WD vehicles, remove the propeller with center bearing shaft assembly.
12. Remove the center exhaust pipe assembly.
13. Remove the No. 3 front floor heat insulator.
14. For 4WD vehicles, remove the No. 4 exhaust pipe support bracket by first removing the 2 bolts.
15. Remove the No. 1 fuel tube protector by first removing the 2 clips, and then disengaging the claw.
16. Remove the No. 1 fuel tank protector by first removing the 4 nuts.
17. Remove the No. 1 fuel tank protector sub-assembly by first removing the 3 clips and the 6 nuts.
18. Remove the fuel tank assembly, as follows:

➡Check if there is any dirt or mud around the connector before this operation and clean the connector as necessary.

➡Do not allow any scratches or foreign objects on the parts when disconnecting as the fuel hose connector has the O-ring that seals the pipe.

➡It is necessary to prevent mud or dirt from entering the quick connector. If any foreign objects enter the connector, the O-rings may seal properly.

➡Perform this work by hand. Do not use any tools.

➡Do not forcibly bend, twist or turn the nylon tube.

➡Protect the connected part by covering it with a plastic bag after disconnecting the fuel tank vent hose.

➡If the connectors or pipe are stuck, push and pull them to release, and pull the connector out carefully.

a. Pinch the projections of the retainer to remove the lock claws and pull it down as shown, and pull out the fuel pump tube.

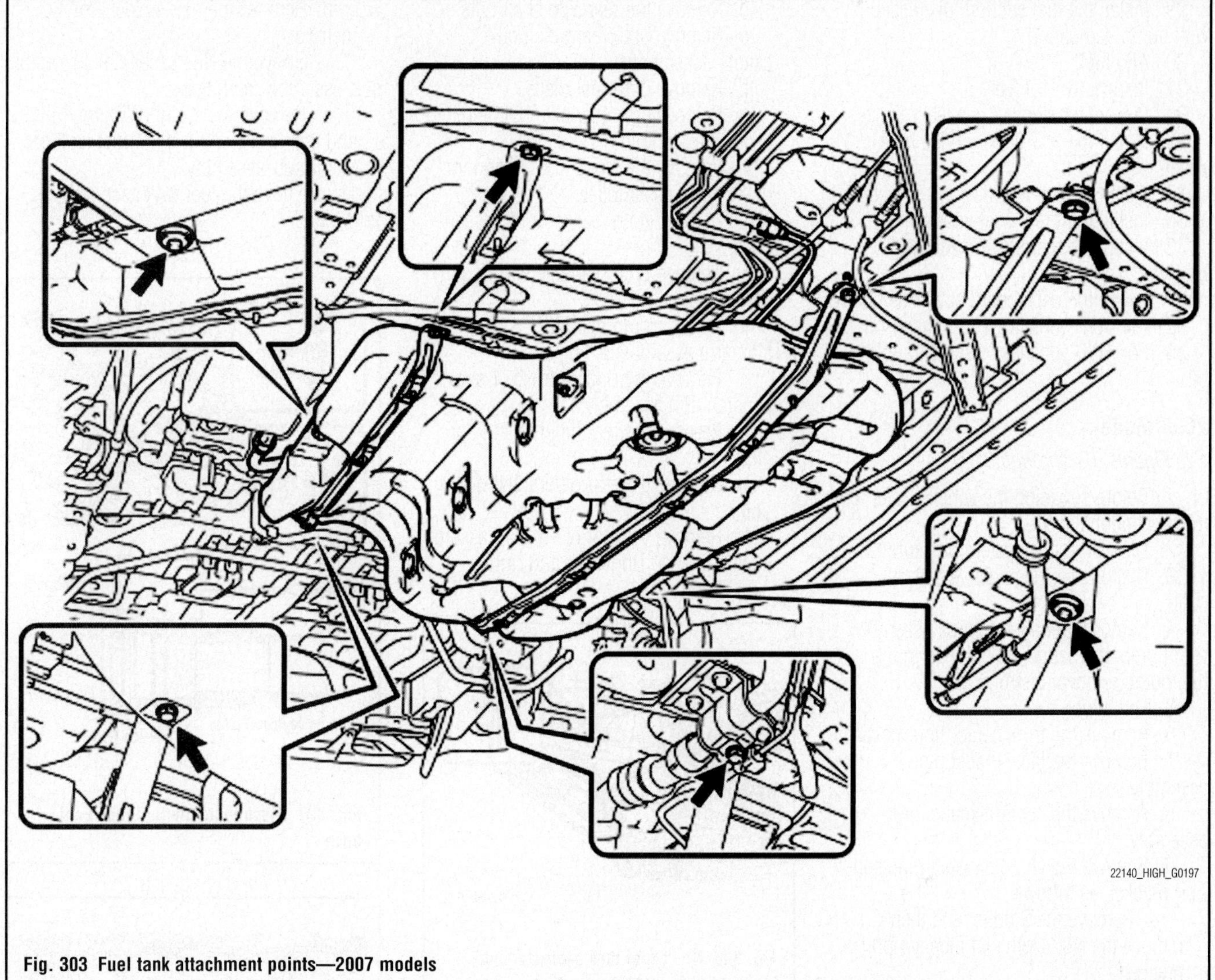

Fig. 303 Fuel tank attachment points—2007 models

b. To disconnect the fuel tank vent hose, deeply push the connector to release the locking tab, pinch portion A, and pull out the connector.

c. Pinch the tube connector and then pull out the No. 3 fuel tank breather tube.

d. Loosen the hose clamp bolt and disconnect the fuel tank to filler pipe hose.

e. For 2006 models, disconnect the fuel tank wire connector and remove the connector clamp.

f. Set up a transmission jack under the fuel tank.

g. Remove the 4 bolts, and then remove the 2 fuel tank bands.

h. Remove the 2 nuts.

i. Operate the transmission jack to remove the fuel tank.

To install:

19. Install fuel tank assembly:

a. Set up the fuel tank to the transmission jack.

b. Operating the transmission jack, install the fuel tank.

c. Tighten the 2 nuts to 14 ft. lbs. (20 Nm).

d. Install the 2 fuel tank bands with the 4 bolts and tighten to 29 ft. lbs. (39 Nm).

e. For 2006 models, install the connector clamp and connect fuel tank wire connector.

f. Connect the fuel tank to filler pipe hose.

g. Connect the No. 3 fuel tank breather tube by pushing the quick connector to the pipe until it clicks.

h. Connect the fuel tank vent hose.

i. Connect the fuel pump tube by pushing the quick connector and push up the retainer to engage the claws.

➡After connecting, check if the quick connector and the pipe are securely connected by pulling on them.

➡Check if there is any damage or foreign objects on the connected part.

20. Install the No. 1 fuel tank protector sub-assembly with the 6 nuts and tighten to 49 inch lbs. (5.5 Nm).

21. Install 3 new clips.

22. Install the No. 1 fuel tank protector with the 4 nuts and tighten to 49 inch lbs. (5.5 Nm).

23. Install the No. 1 fuel tube protector with the 2 clips and claw.

24. For 4WD vehicles, install the No. 4 exhaust pipe support bracket with the 2 bolts and tighten to 16 inch lbs. (22 Nm).

25. Install the No. 3 front floor heat insulator.

26. Install the center exhaust pipe assembly.

27. For 4WD vehicles, install the propeller with center bearing shaft assembly.

28. For 4WD vehicles, fully tighten propeller with center bearing shaft assembly.

29. Install the fuel suction tube assembly with pump and gauge.
30. Add fuel.
31. Inspect for fuel leak.
32. Inspect for exhaust gas leak.
33. Install the rear floor service hole cover.
34. Install the left rear door scuff plate.
35. Install both rear seat assemblies.
36. Install the rear seat track outer bracket cover.
37. Install the deck board sub-assembly.
38. For 4WD vehicles, add transfer oil.
39. For 4WD vehicles, inspect transfer oil.

2008 Models

See Figures 304 through 312.

1. Before servicing the vehicle, refer to the precautions section.
2. Discharge fuel system pressure.
3. Disconnect the negative battery cable.
4. Remove fuel suction tube assembly with pump and gauge, as outlined in the fuel pump removal instructions.
5. Drain fuel.
6. Remove the front center floor cover.
7. Remove the tail exhaust pipe assembly.
8. Remove the center exhaust pipe assembly.
9. Remove the No. 4 exhaust pipe support bracket, as follows:
    a. Remove the 2 bolts, and then remove the No. 4 exhaust pipe support bracket.
10. Remove the No. 3 front floor heat insulator.
11. For 4WD vehicles, remove the following:
    a. Remove the propeller with center bearing shaft assembly.
    b. Remove both rear wheels.
    c. Separate both rear speed sensors.
    d. Remove the rear axle shaft nuts.
    e. Remove both rear disc brake caliper assemblies.
    f. Remove both rear discs.
    g. Remove both rear axle hub and bearing assemblies.
    h. Remove the No. 2 and 3 parking brake cable assemblies.
    i. Remove both rear strut rod assemblies.
    j. Remove both rear axle carrier sub-assemblies.
    k. Remove both rear No. 2 suspension arm assemblies.
    l. Remove the right rear No. 1 suspension arm assembly.
    m. Remove the rear differential filler plug.
    n. Remove the rear differential drain plug.
    o. Remove the rear drive shaft assemblies and snap rings, left and right.
    p. Remove the rear suspension member.
12. Remove the No. 1 fuel tank protector sub-assembly, as follows:
    a. Remove the 3 clips (A) and 7 nuts, and then remove the No. 1 fuel tank protector sub-assembly.
13. Remove the fuel tank assembly, as follows:

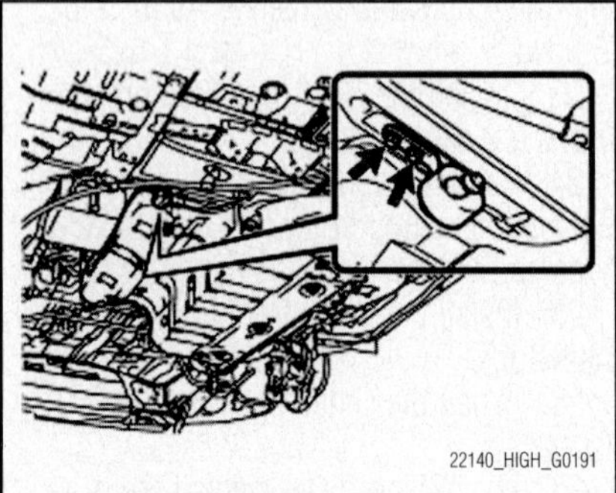

Fig. 304 No. 4 exhaust pipe support bracket

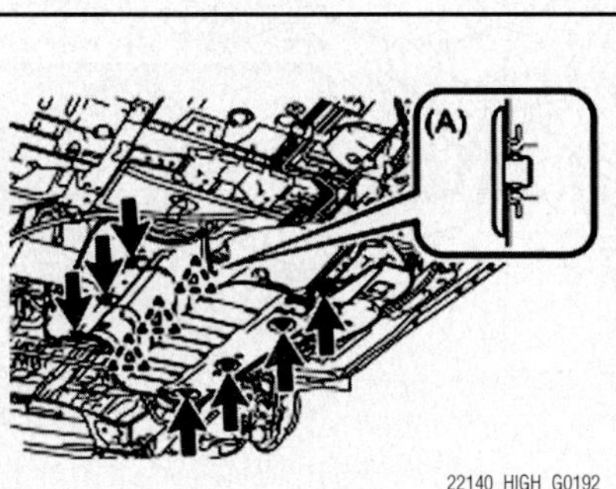

Fig. 305 No. 1 fuel tank protector sub-assembly, clips (A), and nuts

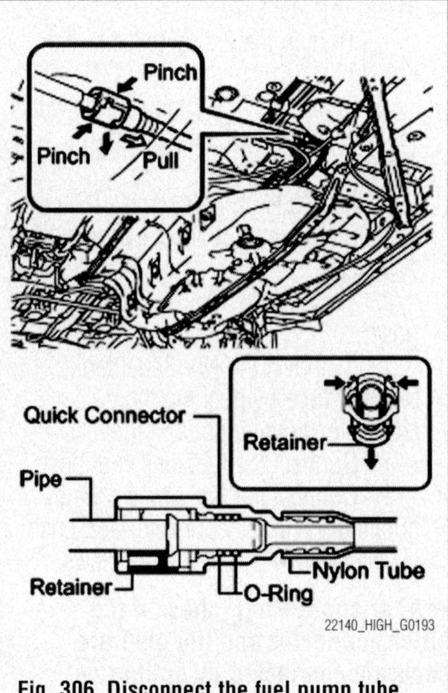

Fig. 306 Disconnect the fuel pump tube

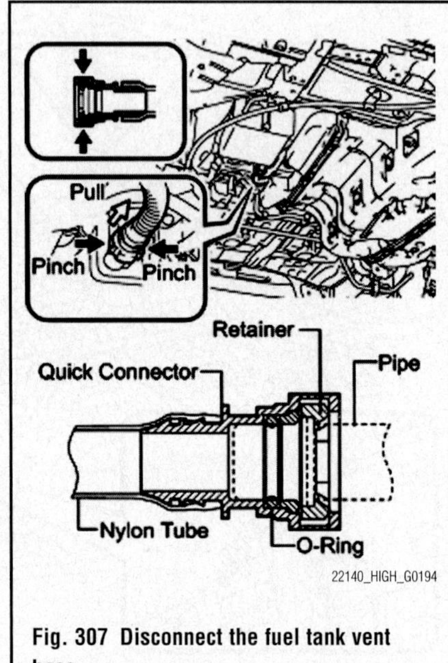

Fig. 307 Disconnect the fuel tank vent hose

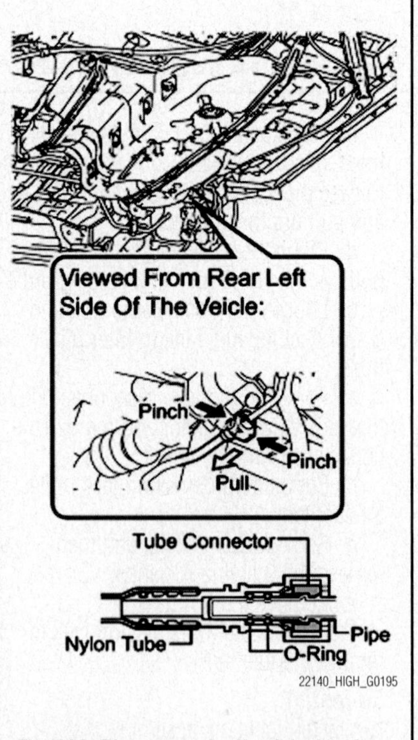

Fig. 308 Disconnect the No. 3 fuel tank breather tube

➡**Check if there is any dirt or mud around the connector before this operation and clean the connector as necessary.**

➡**Do not allow any scratches or foreign objects on the parts when disconnecting as the fuel hose connector has the O-ring that seals the pipe.**

➡**It is necessary to prevent mud or dirt from entering the quick connector. If any foreign objects enter the connector, the O-rings may seal properly.**

➡**Perform this work by hand. Do not use any tools.**

➡**Do not forcibly bend, twist or turn the nylon tube.**

➡**Protect the connected part by covering it with a plastic bag after disconnecting the fuel tank vent hose.**

➡**If the connectors or pipe are stuck, push and pull them to release, and pull the connector out carefully.**

a. Disconnect the fuel pump tube by pinching the tab of the retainer to disengage the lock claws and pull it down as shown.

b. Pull out the fuel pump tube.

c. Pinch the retainer and pull out the quick connector while pushing the quick connector against the pipe to disconnect the fuel tank vent hose from the charcoal canister assembly.

d. Pinch the tube connector and then pull out the No. 3 fuel tank breather tube.

e. Loosen the hose clamp bolt and disconnect the fuel tank to filler pipe hose.

f. Set a transmission jack under the fuel tank.

g. Remove the 4 bolts, and then remove the 2 fuel tank bands.

h. Remove the 2 nuts.

i. Operate the transmission jack to remove the fuel tank.

To install:

14. Set the fuel tank onto a transmission jack.

15. Operating the transmission jack, install the fuel tank.

16. Tighten the 2 nuts to 14 ft. lbs. (20 Nm).

17. Install the 2 fuel tank bands with the 4 bolts and tighten to 29 ft. lbs. (39 Nm).

18. Align the matchmarks and install the fuel tank to filler pipe hose to the fuel tank.

19. Install the hose clamp within the range shown in the illustration.

20. Connect the fuel tank vent hose:

a. Align the quick connector with the pipe, then push in the quick connector until the retainer makes a click sound to connect the charcoal canister fuel hose to the charcoal canister assembly.

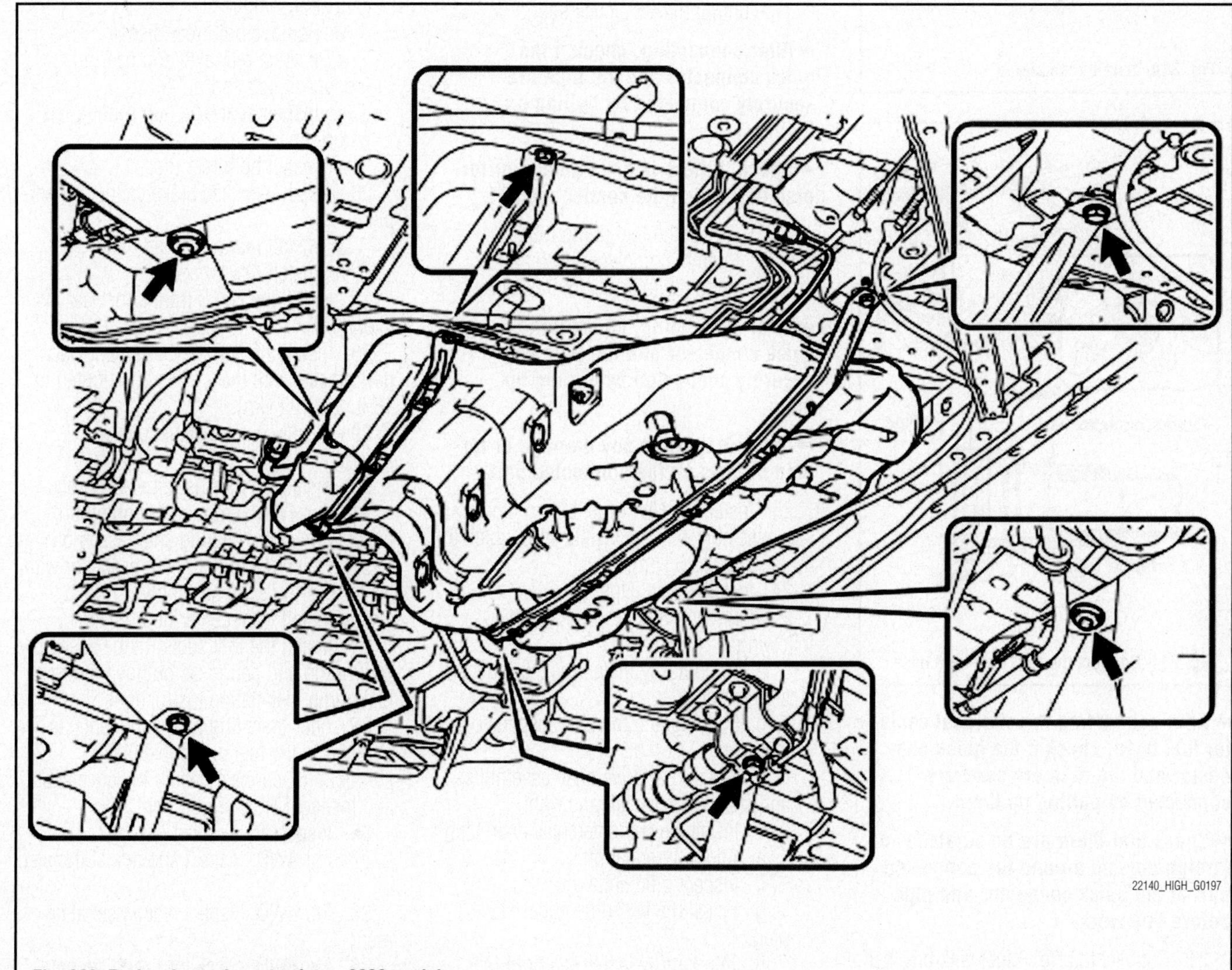

Fig. 309 Fuel tank attachment points—2008 models

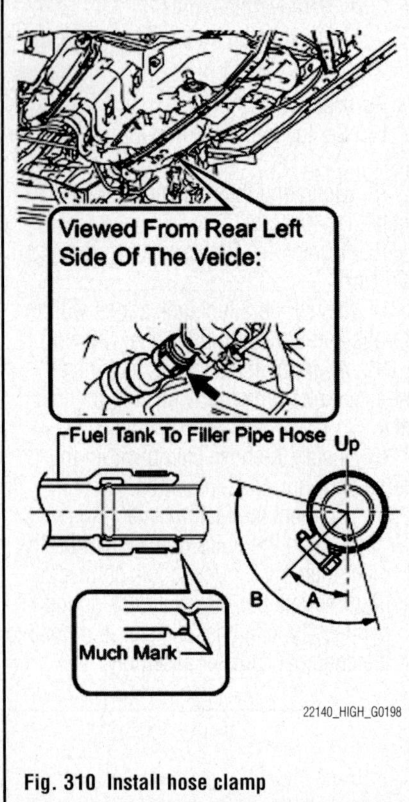

Fig. 310 Install hose clamp

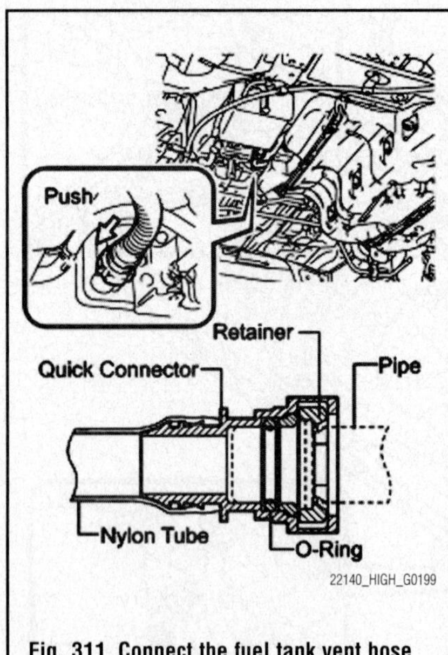

Fig. 311 Connect the fuel tank vent hose

➡After connecting the charcoal canister fuel hose, check if the quick connector and the pipe are securely connected by pulling on them.

➡Check that there are no scratches or foreign objects around the connected part of the quick connector and pipe before this work.

21. Connect the No. 3 fuel tank breather tube:

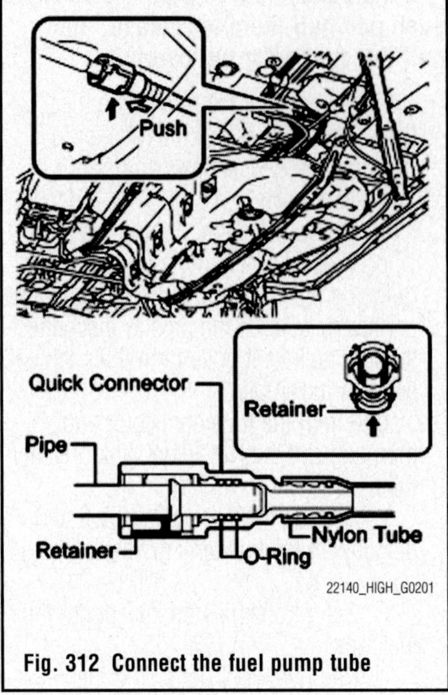

Fig. 312 Connect the fuel pump tube

a. Push the quick connector to the pipe until it makes a click sound.

➡After connecting, check if the quick connector and the pipe are securely connected by pulling on them.

➡Check if there is any damage or foreign objects on the connected part.

22. Connect the fuel pump tube:
   a. Push the quick connector and push in the retainer to lock the claws.

➡After connecting, check if the quick connector and the pipe are securely connected by pulling on them.

➡Check if there is any damage or foreign objects on the connected part.

23. Install the No. 1 fuel tank protector sub-assembly with the 7 nuts and tighten to 49 inch lbs. (5.5 Nm).
24. Install 3 new clips.
25. For 4WD vehicles, install the following:
   a. Temporarily install the rear suspension member.
   b. Fully tighten the rear suspension member.
   c. Install rear drive shaft assemblies and snap rings on left and right.
   d. Install the rear differential drain plug.
   e. Add differential oil.
   f. Inspect differential oil.
   g. Install the rear differential filler plug.
   h. Temporarily install the right rear No. 1 suspension arm assembly.
   i. Temporarily install both rear No. 2 suspension arm assemblies.
   j. Install both rear axle carrier sub-assemblies.
   k. Temporarily install both rear strut rod assemblies.
   l. Install both rear axle hub and bearing assemblies.
   m. Install both rear discs.
   n. Install rear disc brake caliper assemblies.
   o. Temporarily install both rear axle shaft nuts.
   p. Install rear speed sensors.
   q. Stabilize suspension.
   r. Fully tighten both rear no. 1 suspension arm assemblies.
   s. Fully tighten both rear no. 2 suspension arm assemblies.
   t. Fully tighten rear strut rod assemblies.
   u. Install No. 2 and 3 parking brake cable assemblies.
   v. Separate both rear disc brake caliper assemblies.
   w. Remove both rear discs.
   x. Inspect rear axle hub bearing looseness, left and right.
   y. Inspect rear axle hub runout, left and right.
   z. Install both rear discs.
26. Install rear disc brake caliper assemblies.
   a. Install rear axle shaft nuts.
   b. Install rear wheels.
27. Install the No. 3 front floor heat insulator.
28. Install the No. 4 exhaust pipe support bracket with the 2 bolts and tighten to 16 ft. lbs. (22 Nm).
29. Install center exhaust pipe assembly.
30. Install tail exhaust pipe assembly.
31. For 4WD, temporarily tighten propeller with center bearing shaft assembly.
32. For 4WD, fully tighten propeller with center bearing shaft assembly.
33. Install front center floor cover.
34. Install the fuel suction tube assembly with pump and gauge, as outlined in the fuel pump installation instructions.
35. After installing all components as outlined in the fuel pump installation instructions, connect battery negative cable and inspect SRS warning light.
36. Inspect for exhaust gas leak.
37. For 4WD, inspect and adjust transfer oil.
38. For 4WD, inspect rear wheel alignment.
39. For 4WD, check for speed sensor signal.

IDLE SPEED

ADJUSTMENT

Idle speed is maintained by the Powertrain Control Module (PCM). No adjustment is necessary or possible.

THROTTLE BODY

REMOVAL & INSTALLATION

2.4L Engine

See Figures 313 through 317.

1. Disconnect the negative battery cable.
2. Drain the coolant into a suitable container.
3. Remove the air cleaner cap with air cleaner hose.
4. Disconnect the accelerator control cable from the throttle body.
5. Detach the Throttle Position (TP) sensor connector and Idle Speed Control (ISC) valve connector.
6. Disconnect the vacuum hose from the throttle body.
7. Disconnect 2 water by-pass hoses.
8. Remove 3 bolts and throttle body from intake manifold.
9. Remove the throttle body gasket and discard it.
10. If necessary, remove the 2 retaining screws and the TP sensor.
11. If necessary, remove the 3 retaining screws and the ISC valve.

To install:

12. If removed, install the ISC valve with a new gasket and secure with the retaining screws.
13. If removed, install the TP sensor, as follows:
    a. Make sure the throttle valve is fully closed.
    b. Rotate the sensor 45° in the left direction to the throttle valve fully closed position, and insert the TP sensor to the throttle body.
    c. Rotate the sensor 45° in the right direction and install the 2 retaining screws.
14. Position a new throttle body gasket on the intake manifold, as shown in the accompanying illustration.
15. Install the throttle body and secure with the 3 bolts. Tighten to 22 ft. lbs. (30 Nm).
16. Connect the 2 water by-pass hoses to the throttle body.

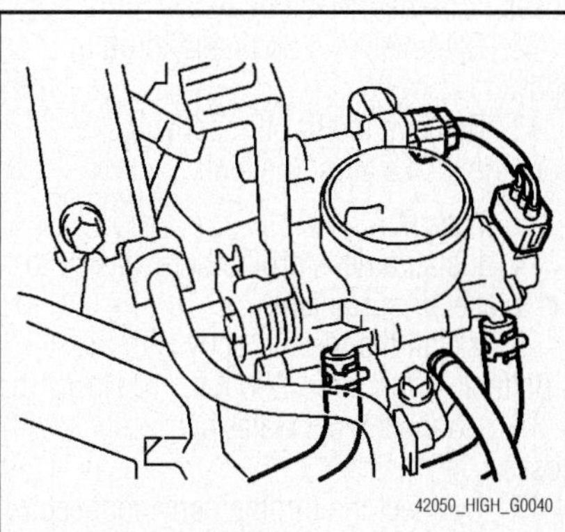

Fig. 313 Disconnect the electrical connectors, vacuum lines and water hoses from the throttle body

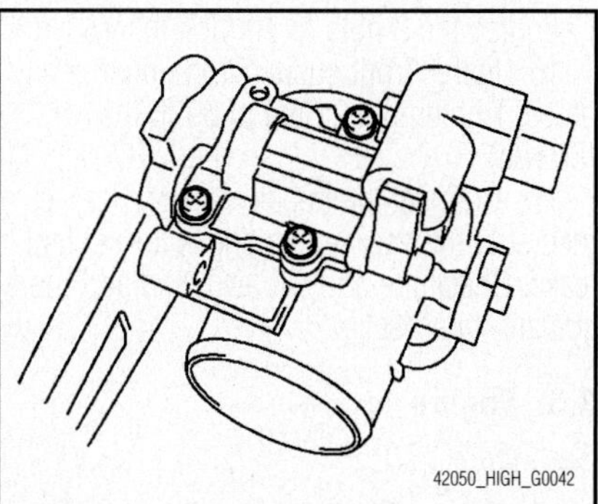

Fig. 315 Remove the 3 retaining screws and the Idle Speed Control (ISC) valve

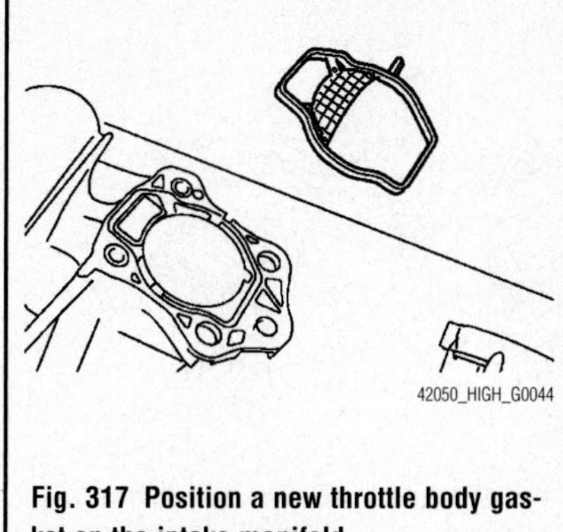

Fig. 317 Position a new throttle body gasket on the intake manifold

17. Connect the vacuum hose to the throttle body.
18. Attach the ISC valve and TP sensor connectors.
19. Connect the accelerator control cable to the throttle body.
20. Install the air cleaner hose and cap.
21. Fill the engine with the proper type and amount of coolant.
22. Connect the negative battery cable, then start the engine and check for coolant leaks. Check the coolant level and add more coolant if necessary.

3.3L Engine

See Figures 318 through 320.

1. Disconnect the negative battery cable.
2. Drain the coolant into a suitable container.
3. Remove the 2 nuts and the front suspension upper center brace
4. Remove v-bank cover, as follows:
    a. Using a socket hexagon wrench 5, remove the 3 nuts.
    b. Remove the v-bank cover.
5. Remove the air cleaner inlet assembly.
6. Remove the air cleaner assembly.

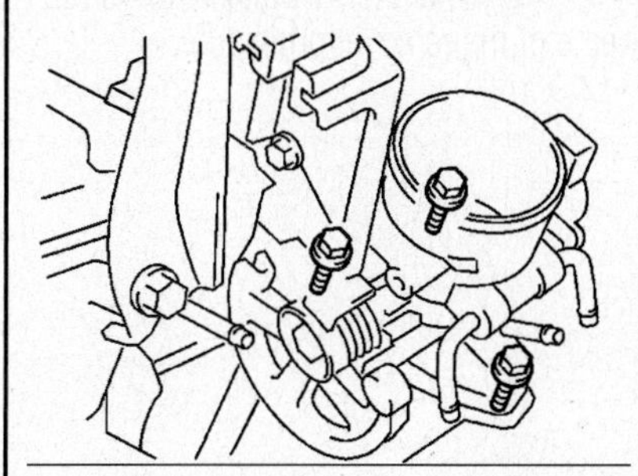

Fig. 314 The throttle body is secured to the intake manifold with 3 bolts

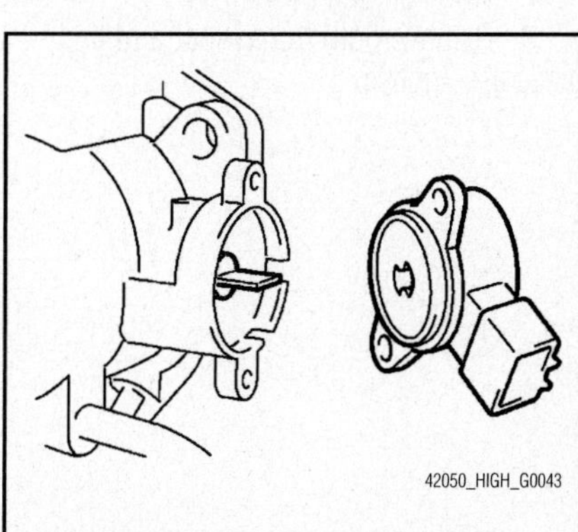

Fig. 316 Installing the Throttle Position (TP) sensor

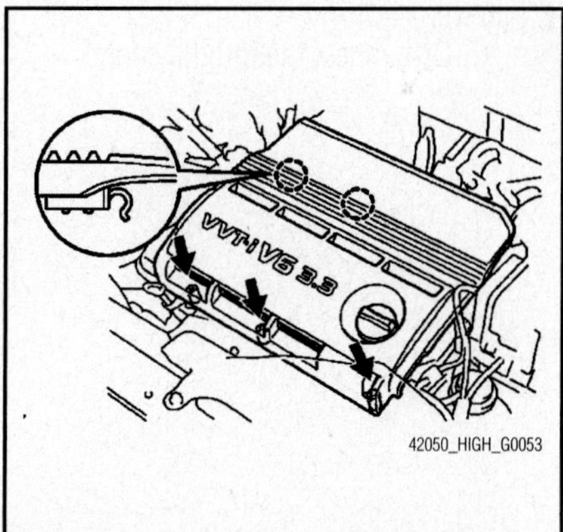

Fig. 318 V-bank cover mounting

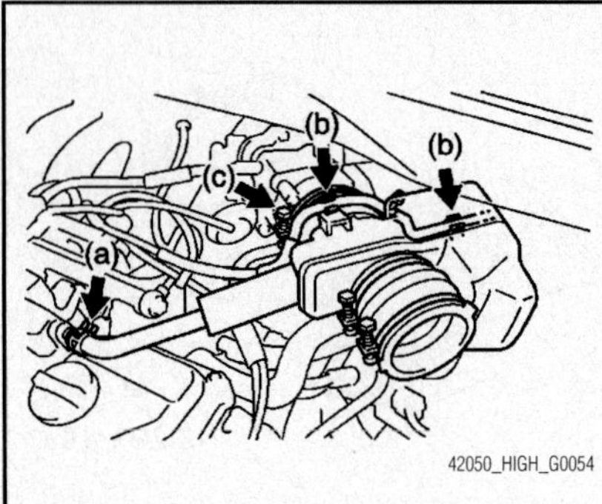

Fig. 319 Intake air resonator removal and installation

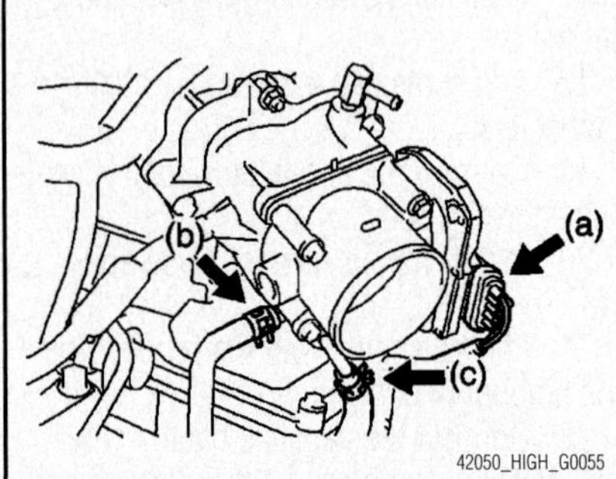

Fig. 320 Detach the throttle control motor and TP sensor connector (A), then remove the no. 2 (b) and no. 3 (c) water by-pass hoses

➡Note that the sub-steps for intake air resonator removal and installation correspond with the letters in the accompanying illustration.

7. Remove the intake air resonator, as follows:
    a. Remove the ventilation hose no. 2.
    b. Remove the fuel vapor feed hose from the 2 hose clamps.
    c. Loosen the air cleaner hose clamp bolt, and then remove the intake air resonator.
8. Remove the throttle body assembly, as follows:
    a. Disconnect the throttle control motor and Throttle Position (TP) sensor connector.
    b. Disconnect the water by-pass hose no. 2.
    c. Disconnect the water by-pass hose no. 3.
    d. Remove the 2 bolts, 2 nuts and throttle body.
    e. Remove and discard the gasket.

To install:

9. Install throttle body assembly, as follows:
    a. Install a new gasket to the intake air connector.
    b. Install the throttle body with the 2 bolts and 2 nuts and tighten to 8 ft. lbs. (11 Nm).
    c. Connect the water by-pass hoses.
    d. Attach the throttle control motor and TP sensor connector.
10. Install the intake air resonator
11. Install the air cleaner assembly.
12. Install the air cleaner inlet assembly.
13. Check the vacuum hose connections.
14. Fill the engine with the proper type and amount of coolant.
15. Check for coolant leaks.
    a. Install v-bank cover, as follows:
    b. Fit the 2 retainers and install the v-bank cover.
    c. Using a socket hexagon wrench 5, tighten the 3 nuts to 70 inch lbs. (8 Nm).
16. Install front suspension upper center brace. Tighten the 2 nuts to 59 ft. lbs. (80 Nm).
17. Connect the negative battery cable, then start the engine and check for coolant leaks. Check the coolant level and add more coolant if necessary.

3.5L Engine

See Figure 321.

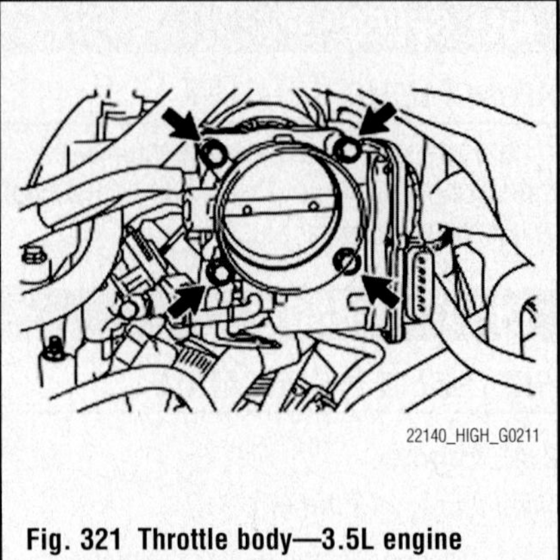

Fig. 321 Throttle body—3.5L engine

1. Remove the engine under cover assembly.
2. Remove the No. 1 engine under cover.
3. Drain engine coolant.
4. Remove both front wiper arm and blade assemblies.
5. Remove the cowl top ventilator louver sub-assembly.
6. Remove the windshield wiper motor and link assembly.
7. Remove the outer cowl top panel sub-assembly.
8. Remove the V-bank cover sub-assembly.
9. Remove the air cleaner cap sub-assembly.
10. Disconnect the throttle body connector and clamp.
11. Disconnect the 2 water by-pass hoses from the throttle body.
12. Remove the 4 bolts and throttle body.
13. Remove the throttle body gasket from the intake air surge tank.

To install:

14. Install a new throttle body gasket to the intake air surge tank.
15. Install the throttle body with the 4 bolts and tighten to 7 ft. lbs. (10 Nm).
16. Connect the 2 water by-pass hoses.
17. Connect the throttle body connector and clamp.
18. Install the air cleaner cap sub-assembly.
19. Add engine coolant.
20. Inspect for coolant leak.
21. The remainder of installation is the reverse of the removal procedure.

HEATING & AIR CONDITIONING SYSTEM

BLOWER MOTOR

REMOVAL & INSTALLATION

2007 Models

1. Remove the glove compartment door assembly.
2. Remove the lower No. 2 instrument panel finish panel.
3. Remove the No. 2 air duct, as follows:
   a. Remove the screw.
   b. Release the 3 claw fittings and remove the No 2 air duct.
4. Remove the ECM, as follows:
   a. Disconnect the connectors.
   b. Remove the 2 nuts and ECM.
5. Remove the blower assembly, as follows:
   a. Disconnect the connector and clamp of the cooling unit motor with fan sub-assembly.
   b. Disconnect the connector of the recirculation damper servo.
   c. Remove the 2 screws and bracket.
   d. Remove the bolt, nut and 2 screws.
   e. Disconnect the part fitted to the air conditioner radiator assembly and remove the blower assembly.
6. Release the 2 claw fittings and remove the No. 1 cooler cover.
7. Remove the clean air filter.
8. Remove the 3 screws and air inlet servo motor.
9. Remove the 3 screws and cooling unit motor with fan sub-assembly.

To install:

10. Install the cooling unit motor with fan sub-assembly.
11. Install the air inlet servo motor.
12. Install the clean air filter.
13. Aligning the part to be fitted with that of the air conditioning radiator assembly, install the blower assembly.
14. Install the blower assembly with the bolt, nut and 2 screws, and tighten to 87 inch lbs. (9.8 Nm).
15. Install the bracket with the 2 screws.
16. Install the connector and clamp.
17. Install the ECM with the 2 nuts and tighten to 49 inch lbs. (5.5 Nm).
18. Install the No. 2 air duct.
19. Install the glove compartment door assembly.

2008 Models

1. Remove the No. 2 instrument panel under cover sub-assembly.
2. Remove front blower motor sub-assembly, as follows:
   a. Disconnect the connector.
   b. Remove the 3 screws and the front blower motor sub-assembly.

To install:

3. Installation is the reverse of the removal procedure.

HEATER CORE

REMOVAL & INSTALLATION

2007 Models

See Figures 322 through 326.

✻✻ CAUTION

Refer to the applicable precautions for this system before performing the following operation. Failure to follow the warnings and cautions could result in possible personal injury or death.

1. Discharge refrigerant from refrigeration system.
2. Disconnect the No. 1 cooler refrigerant suction hose, as follows:
   a. Install SST (SST: 09870-00015, or equivalent) on piping clamp.

➡Make sure the direction of the piping clamp claw and SST by seeing the illustration shown on the caution label.

   b. Push down the SST and release the clamp lock.

➡Be careful not to deform the tube when pushing SST.

   c. Pull SST slightly and push the release lever, and then remove the piping clamp with SST.
   d. Disconnect the No. 1 cooler refrigerant suction hose.

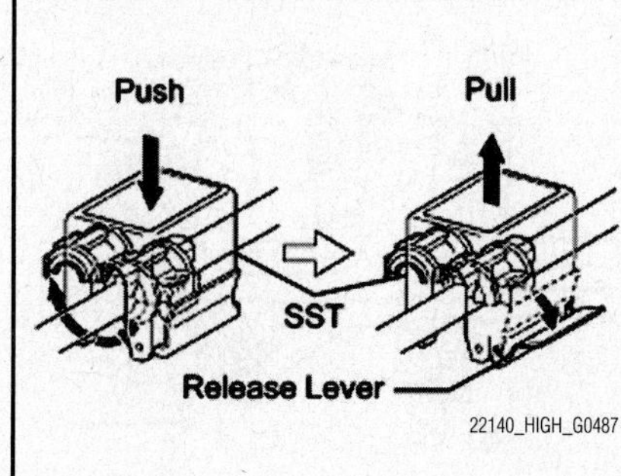

Fig. 322 Push down on Service Tool and release clamp lock

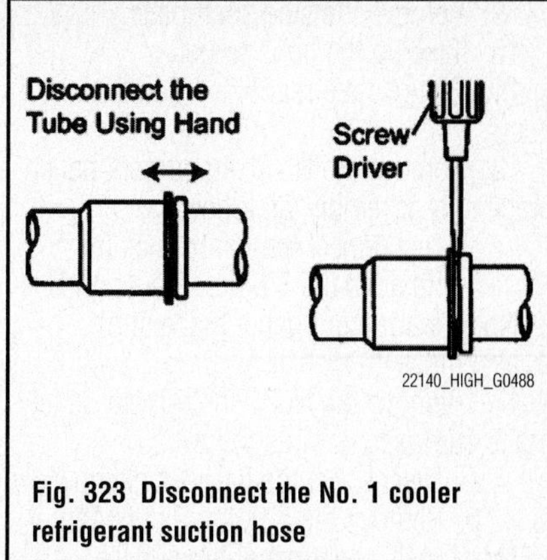

Fig. 323 Disconnect the No. 1 cooler refrigerant suction hose

➡Do not use tools like screwdriver to remove the tube.

➡Cap the open fittings immediately to keep moisture of dirt out of the system.

   e. Remove the 2 O-rings from the No. 1 cooler refrigerant suction hose.
3. Disconnect air conditioning tube and accessory assembly.

➡The procedure to disconnect the air conditioning tube and accessory assembly is same as the No.1 cooler refrigerant suction hose.

4. Using pliers, grip the claws of the clip and slide the clip, and then disconnect the heater water outlet hose.

➡Prepare a drain pan or cloth for when the cooling water leaks.

➡Do not apply any excessive force to the heater water outlet hose.

5. Using pliers, grip the claws of the clip and slide the clip, and then disconnect the heater water inlet hose.

✻✻ CAUTION

Refer to the applicable precautions for this system before performing the following operation. Failure to follow the warnings and cautions could result in possible personal injury or death.

6. Remove instrument panel passenger airbag assembly.
7. Remove the 2 bolts and 2 nuts, and disconnect the steering column assembly form the instrument panel reinforcement assembly.
8. Disconnect the floor shift assembly, as follows:
   a. Remove the 2 nuts and disconnect the shift cable.

b. Remove the 4 nuts and disconnect floor shift assembly.

9. Remove the side scuff plate.

10. Take up the floor carpet.

11. Remove the rear No. 1 and 2 air ducts.

12. Remove the No. 1 instrument panel brace sub-assembly, as follows:

a. Disconnect the harness clamp.

b. Remove the 3 bolts, 2 nuts and No. 1 instrument panel brace sub-assembly.

13. Remove the No. 2 instrument panel brace sub-assembly, as follows:

a. Disconnect the harness clamp.

b. Remove the 2 bolts, 2 nuts and No. 2 instrument panel brace sub-assembly.

14. Remove air conditioner amplifier assembly (w/ Navigation System).

15. Remove the ECM, as follows:

a. Disconnect the connector.

b. Remove the 2 nuts, and ECU.

16. Remove multiplex network body ECU, as follows:

a. Disconnect the connector.

b. Remove the 2 nuts and multiplex network body ECU.

17. Remove the bolt and 2 nuts, disconnect the instrument panel junction block assembly.

18. Remove the skid control ECU assembly, as follows:

a. Disconnect the connector.

b. Remove the 2 nuts and skid control ECU assembly.

19. Remove the 5 bolts, 4 nuts and instrument panel reinforcement assembly.

20. Remove the air conditioner unit assembly, as follows:

a. Disconnect the connector and clamp.

b. Disconnect the connector connecting the wiring air indicator harness sub-assembly and vehicle harness.

c. Remove the 2 bolts, 2 nuts and air conditioner unit assembly.

21. Remove the air conditioning radiator assembly, as follows:

a. Remove the screw, disconnect the connected part of the duct and take off the No. 2 air duct.

b. Remove the 4 screws and bracket.

c. Disconnect the connection between the blower assembly and the air conditioning radiator assembly and remove the air conditioning radiator assembly.

22. Release the 4 claw fittings, remove the heater to register duct center.

23. Remove the 3 screws and air outlet control servomotor.

24. Remove the wiring air conditioning harness sub-assembly from the air conditioner radiator assembly.

25. Remove the 3 screws and air mix control servomotor.

26. Remove the heater radiator unit sub-assembly, as follows:

a. Remove the screw and clamp.

b. Remove the heater radiator unit sub-assembly from the air conditioner radiator assembly.

To install:

27. Install the heater radiator unit sub-assembly.

28. Install the screw and clamp.

29. Install the air mix control servomotor with the 3 screws.

30. Install the wiring air conditioning harness sub-assembly.

31. Engage the connector.

32. Install the air outlet control servomotor with the 3 screws.

33. Engage the 4 claws to install the heater to register duct center.

34. Install the air conditioner unit assembly with the 2 bolts and 2 nuts, and tighten to 87 inch lbs. (9.8 Nm).

35. Connect the connectors connecting the wiring air indicator No. 2 harness and vehicle harness.

36. Install the connector and clamp.

37. Install the instrument panel reinforcement assembly with the 5 bolts and 4 nuts and tighten to 21 inch lbs. (29 Nm).

38. Install the skid control ECU assembly with the 2 nuts and tighten to 43 inch lbs. (5 Nm).

39. Install the instrument panel junction block assembly with the bolt and 2 nuts and tighten to 74 inch lbs. (8.4 Nm).

40. Install the ECM with the 2 nuts and tighten to 49 inch lbs. (5.5 Nm).

41. Install the multiplex network body ECU with the bolt and 2 nuts.

42. Install the air conditioner amplifier assembly (w/ Navigation System).

43. Install the instrument panel brace No.2 sub-assembly with the 2 bolts and 2 nuts and tighten to 21 ft. lbs. (29 Nm).

44. Install the instrument panel brace No.1 sub-assembly with the 3 bolts and 2 nuts and tighten to 21 ft. lbs. (29 Nm).

45. Install the rear No. 1 and 2 air ducts.

46. Install the side scuff plate.

47. Install the floor shift assembly with the 4 nuts and tighten to 9 ft. lbs. (12 Nm).

➡Tighten the nuts in numerical order shown in the illustration to install the floor shift assembly.

48. Install the shift cable with the 2 nuts and tighten to 14 ft. lbs. (20 Nm).

49. Install the steering column assembly with the 2 bolts and 2 nuts and tighten to 15 ft. lbs. (21 Nm).

CAUTION

Refer to the applicable precautions for this system before performing the following operation. Failure to follow the warnings and cautions could result in possible personal injury or death.

50. Install instrument panel passenger airbag assembly.

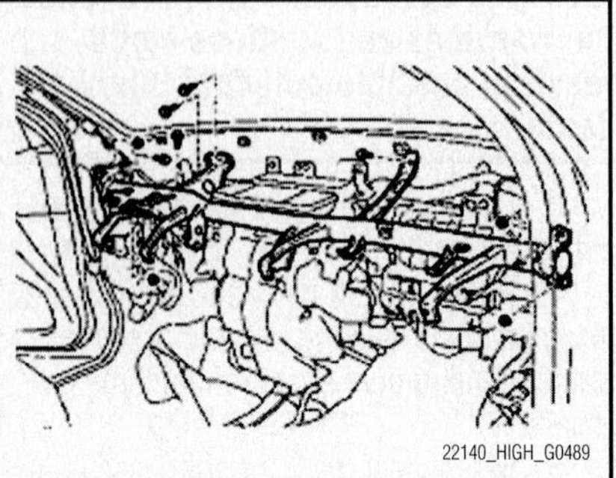

Fig. 324 Remove instrument panel reinforcement assembly

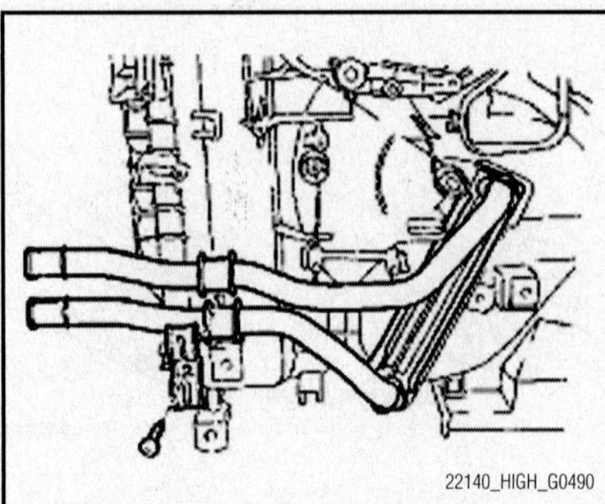

Fig. 325 Remove heater radiator unit sub-assembly

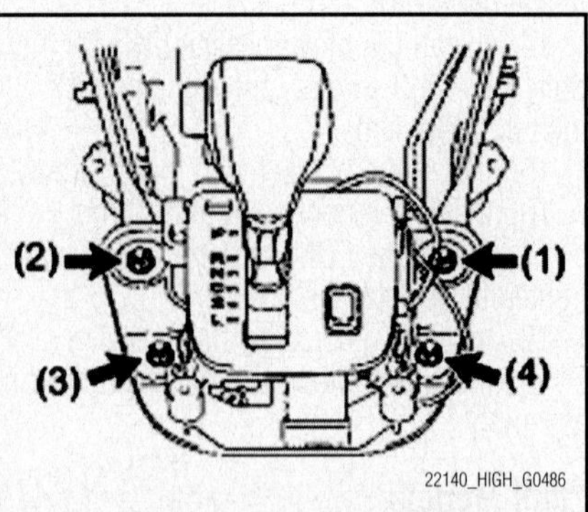

Fig. 326 Floor shift assembly nut tightening sequence

51. Install the heater water inlet and outlet hoses.
52. Install the A/C tube and accessory assembly, as follows:
a. Remove the attached vinyl tape from the tube.
b. Coat a new O-ring with compressor oil (ND-OIL 8 or equivalent) and install it to the hose.
c. Install the air conditioning tube and accessory and piping clamp.

➡Be sure to connect the pipe securely.

➡After connection, check the fitting for claw of the piping clamp.

53. Install the No. 1 cooler refrigerant suction hose, as follows:
54. Remove the attached vinyl tape from the hose.
a. Coat a new O-ring with compressor oil (ND-OIL 8 or equivalent) and install it to the pipe.
b. Install the No. 1 cooler refrigerant suction hose and piping clamp.
55. Add coolant.
56. Charge refrigerant.
57. Inspect for engine coolant leak.
58. Warm up the compressor.
59. Inspect for refrigerant leak.

2008 Models

See Figures 327 and 328.

1. Remove the center instrument panel register assembly.
2. Remove the center instrument cluster finish panel assembly.
3. Remove the heater control and accessory assembly (for manual air conditioning system).
4. Remove the air conditioning control assembly (for automatic air conditioning system).
5. Remove the radio receiver assembly with bracket (w/o navigation system).
6. Remove the navigation receiver assembly with bracket (w/ navigation system).
7. Remove the right front door scuff plate.
8. Remove the right cowl side trim sub-assembly.
9. Remove the No. 2 instrument panel under cover sub-assembly.
10. Remove the lower instrument panel sub-assembly.
11. Remove the upper console panel sub-assembly.
12. Remove the lower rear console box.
13. Remove the console box assembly.
14. Remove the front No. 2 console box insert.
15. Disengage the 3 claws and remove the No. 3 air duct sub-assembly as shown in the illustration.
16. Remove quick heater assembly, as follows:
a. Disconnect the 2 connectors.
b. Remove the 3 screws and the quick heater assembly as shown in the illustration.

To install:

17. Installation is the reverse of the removal procedure.

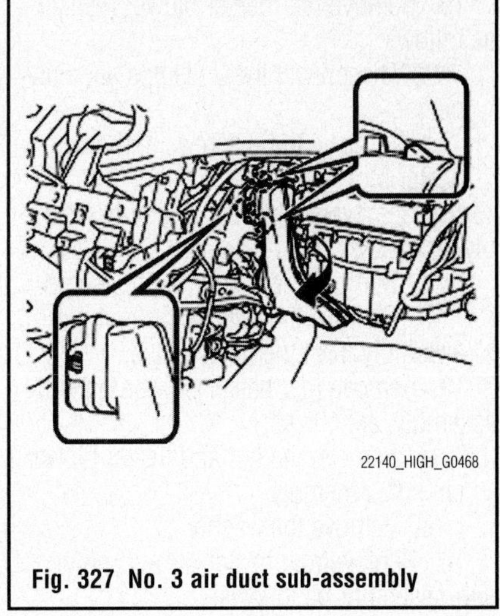

Fig. 327 No. 3 air duct sub-assembly

Fig. 328 Quick heater assembly and screws

AUXILIARY HEATING & AIR CONDITIONING SYSTEM

BLOWER MOTOR

REMOVAL & INSTALLATION

2007 Models

See Figure 329.

1. Remove the deck board No. 2 sub-assembly.
2. Remove the rear floor finish plate.
3. Remove the back door weatherstrip.
4. Remove the right hand deck side trim box.
5. Remove the right hand rear door scuff plate.
6. Remove the right rear door opening trim weatherstrip.
7. Remove the right deck side trim cover.
8. Remove the right deck trim side panel assembly.
9. Using pliers, grip the claws of the clip and slide the clip, and disconnect the heater water outlet hose.

➡When installing or removing the hoses, hold heater water pipes to prevent deformation.

➡Do not apply any excessive force to the heater water outlet hose.

➡Prepare a drain pan or cloth in case the cooling water leaks.

10. Using pliers, grip the claws of the clip and slide the clip, and disconnect the heater water inlet hose.
11. Remove the 2 clips and the rear foot air duct.
12. Remove the rear No. 5 air duct, as follows:
a. Disconnect the wire harness clamp.
b. Remove the 2 clips and rear No. 5 air duct.
13. Remove the rear heater assembly, as follows:
a. Disconnect the washer hose clamp.
b. Remove the 4 bolts and plate.
c. Disconnect the heater relay No. 1 clamp.
d. Disconnect the heater blower resister connector.
e. Remove the 4 bolts and rear heater assembly.
14. Remove water pipe and hose sub-assembly, as follows:
a. Using pliers, grip the claws of the 2 clips and slide the 2 clips, and disconnect the heater water hose

➡Do not apply any excessive force to the heater water hose.

15. Remove rear heater blower resistor, as follows:
   a. Disconnect the 2 clamps and connector.
   b. Remove the 2 screws and rear heater blower resistor.
16. Remove rear heater unit sub-assembly, as follows:
   a. Remove the 9 screws and clamp.
   b. Remove the rear heater unit sub-assembly from the heater case.
17. Remove rear heater blower fan sub-assembly, as follows:
   a. Remove the nut and heater blower fan sub-assembly.
   b. Remove the washer.
18. Remove rear heater blower motor sub-assembly, as follows:
   a. Disconnect the connector clamp.
   b. Remove the 3 screws and heater blower motor sub-assembly.

To install:

19. Install the rear heater blower motor sub-assembly.
20. Install the rear heater blower fan sub-assembly, as follows:
   a. Install the washer.
   b. Install the heater blower fan sub-assembly with the nut and tighten to 36 inch lbs. (4 Nm).
21. Install the rear heater unit sub-assembly.
22. Install the rear heater blower resistor.
23. Install the water pipe and hose sub-assembly.
24. Install the rear heater assembly, as follows:
   a. Install the rear heater assembly with the 4 bolts and tighten to 87 inch lbs. (9.8 Nm).

➡Tighten the bolts in the numerical order shown in the illustration to install the rear heater assembly.

   b. Install the heater blower resistor connector.
   c. Install the heater relay No. 1.
   d. Install the washer hose.
   e. Install the plate with the 4 bolts and tighten to 87 inch lbs. (9.8 Nm).
25. The remainder of installation is the reverse of the removal procedure.

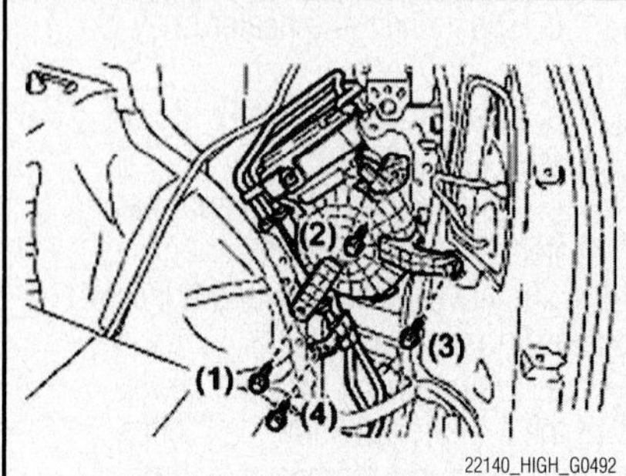

Fig. 329 Rear heater assembly bolt tightening sequence

2008 Models

1. Remove the right rear door scuff plate.
2. Remove the right rear door opening trim weatherstrip.
3. Remove the deck board assembly.
4. Remove the No. 2 and 3 deck board sub-assemblies.
5. Remove the tonneau cover assembly, as applicable.
6. Remove the rear seat side covers.
7. Remove the deck side trim boxes.
8. Remove the jack carrier support.
9. Remove the jack carrier cushion.
10. Remove jack assembly.
11. Remove the jack carrier assembly.
12. Remove rear mat.
13. Remove the deck floor board assembly.
14. Remove the rear No. 2 seat inner belt assembly.
15. Disconnect both rear seat lap type belt assemblies.
16. Remove the rear No. 2 seat assembly.
17. Remove the rear floor finish plate.
18. Remove the rear seat side garnish cap.
19. Remove right deck side trim cover.
20. Remove right deck side trim.
21. Remove right side trim cover (for manual air conditioning system).
22. Remove rear room temperature sensor (for automatic air conditioning system).
23. Remove right rear combination light service cover.
24. Remove right hand rope hook assembly.
25. Remove the No. 1 luggage compartment trim hook
26. Remove the right front deck side trim cover.
27. Remove the right deck trim side panel assembly.
28. Remove the right roof side inner garnish assembly.
29. Remove the rear blower motor sub-assembly, as follows:
30. Disconnect the connector.
31. Remove the 3 screws and the rear blower motor sub-assembly.

To install:

32. Installation is the reverse of the removal procedure.

HEATER CORE

REMOVAL & INSTALLATION

2007 Models

See Figures 330 through 333.

1. Remove the deck board No. 2 sub-assembly.
2. Remove the rear floor finish plate.
3. Remove the back door weatherstrip.
4. Remove the right deck side trim box.
5. Remove the right rear door scuff plate.
6. Remove the right rear door opening trim weatherstrip.
7. Remove the right deck side trim cover.
8. Remove the right deck trim side panel assembly.
9. Using pliers, grip the claws of the clip and slide the clip, and disconnect the heater water inlet and outlet hoses.

➡When installing or removing the hoses, hold heater water pipes to prevent deformation.

➡Do not apply any excessive force to the heater water outlet hose.

➡Prepare a drain pan or cloth in case the cooling water leaks.

10. Remove the 2 clips and the rear foot air duct.
11. Remove the rear No. 5 air duct, as follows:
   a. Disconnect the wire harness clamp.
   b. Remove the 2 clips and the rear No. 5 air duct.
12. Remove the rear heater assembly, as follows:
   a. Disconnect the washer hose clamp.

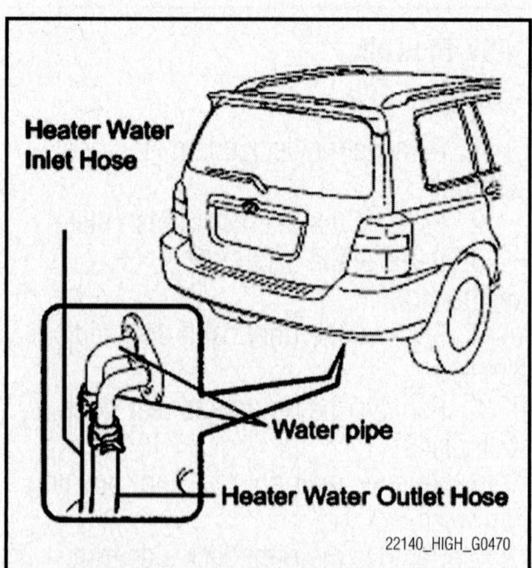

Fig. 330 Rear heater water inlet and outlet hoses

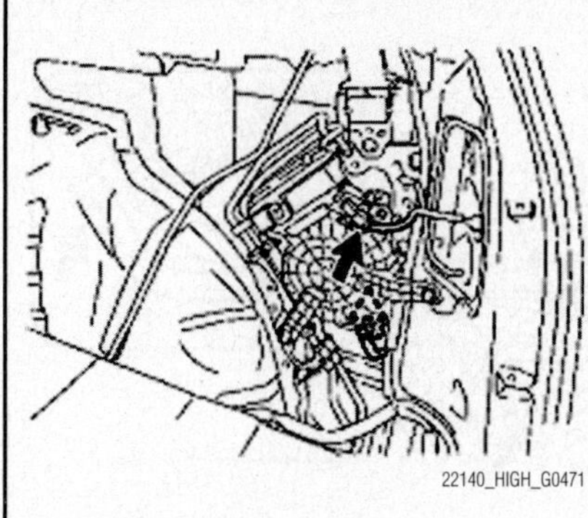
Fig. 331 Rear heater relay No. 1 clamp

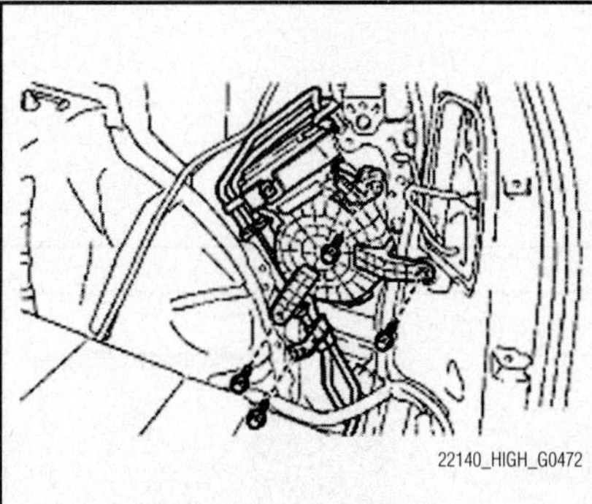
Fig. 332 Rear heater assembly and bolts

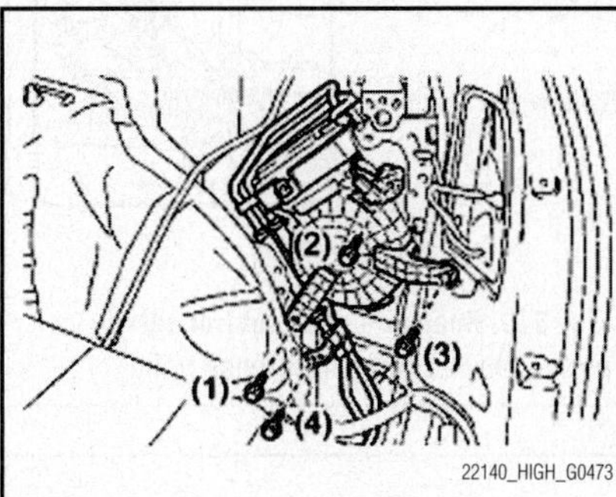

Fig. 333 Rear heater assembly bolt tightening sequence

b. Remove the 4 bolts and plate.
c. Disconnect the heater relay No. 1 clamp.
d. Disconnect the heater blower resister connector.
e. Remove the 4 bolts and rear heater assembly.

To install:

13. Install the rear heater assembly with the 4 bolts and tighten to 87 inch lbs. (9.8 Nm).

➡Tighten the bolts in the numerical order shown in the illustration to install the rear heater assembly.

14. Install the heater blower resistor connector.
15. Install the heater relay No. 1.
16. Install the washer hose.
17. Install the plate with the 4 bolts and tighten to 87 inch lbs. (9.8 Nm).
18. The remainder of installation is the reverse of the removal procedure.

2008 Models

See Figures 334 through 343.

➡Adjust the air outlet mode setting to FOOT.

1. Recover refrigerant from refrigeration system.
2. Disconnect the heater inlet and outlet water hoses.
3. Disconnect the cooler refrigerant liquid pipe C, as follows:
   a. Remove the bolt, and slide the hook connector.
   b. Disconnect the cooler refrigerant liquid pipe C.
   c. Remove the O-ring from the cooler refrigerant liquid pipe C.

➡Seal the openings of the disconnected parts using vinyl tape to prevent entry of moisture and foreign matter.

4. Disconnect the rear cooler refrigerant suction hose.
5. Remove the O-ring from the rear cooler refrigerant suction hose.
6. Remove the right rear door scuff plate.
7. Remove the right rear door opening trim weatherstrip.
8. Remove the deck board assembly.
9. Remove the No. 2 and 3 deck board sub-assemblies.
10. Remove the tonneau cover assembly, as applicable.
11. Remove the rear seat side covers.
12. Remove the deck side trim boxes.
13. Remove the jack carrier support.
14. Remove the jack carrier cushion.
15. Remove jack assembly.
16. Remove the jack carrier assembly.
17. Remove rear mat.
18. Remove the deck floor board assembly.
19. Remove the rear No. 2 seat inner belt assembly.
20. Disconnect both rear seat lap type belt assemblies.
21. Remove the rear No. 2 seat assembly.
22. Remove the rear floor finish plate.
23. Remove the rear seat side garnish cap.
24. Remove right deck side trim cover.

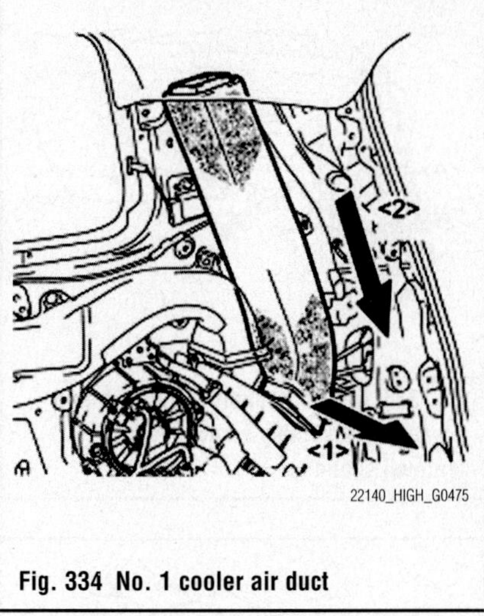

Fig. 334 No. 1 cooler air duct

25. Remove right deck side trim.
26. Remove right side trim cover (for manual air conditioning system).
27. Remove rear room temperature sensor (for automatic air conditioning system).
28. Remove right rear combination light service cover.
29. Remove right hand rope hook assembly.
30. Remove the No. 1 luggage compartment trim hook
31. Remove the right front deck side trim cover.
32. Remove the right deck trim side panel assembly.
33. Remove the right roof side inner garnish assembly.
34. Disengage the 2 claws and remove the cooler plate.
35. Remove the No. 1 cooler air duct as shown in the illustration.
36. Remove the 2 clips and the rear No. 5 air duct.
37. Remove the rear cooling unit assembly, as follows:
   a. Disengage each clamp.
   b. Disconnect each connector.
   c. Remove the 4 bolts and the rear cooling unit assembly.
38. Remove the drain cooler hose.
39. Remove the rear air mix control servo motor sub-assembly, as follows:
   a. Disconnect the connector.
   b. Remove the 2 screws and rear air mix control servo motor sub-assembly.
40. Remove the rear air outlet control servo motor sub-assembly, as follows:
   a. Disconnect the connector.
   b. Remove the 2 screws and rear air outlet control servo motor sub-assembly.
41. Remove the rear air conditioning tube and accessory assembly, as follows:
   a. Remove the packing.

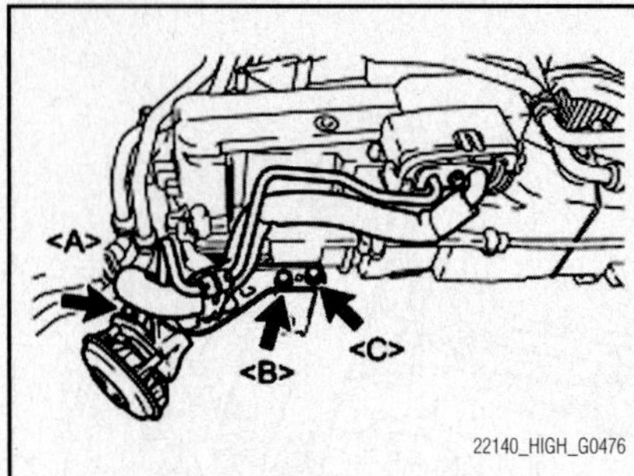

Fig. 335 Rear air conditioning tube and accessory assembly screw and bolt removal sequence

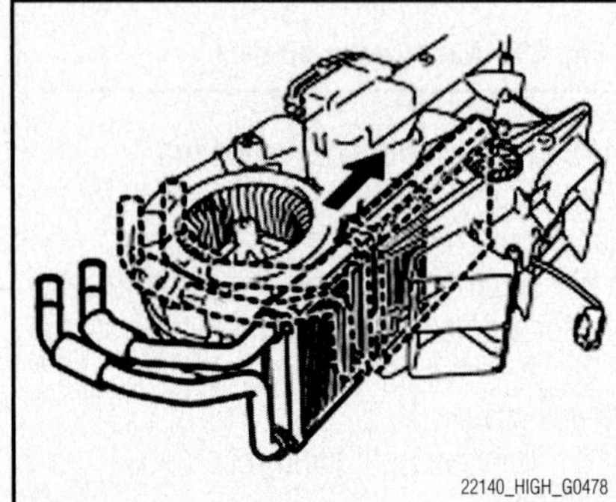

Fig. 336 Install heater radiator unit sub-assembly

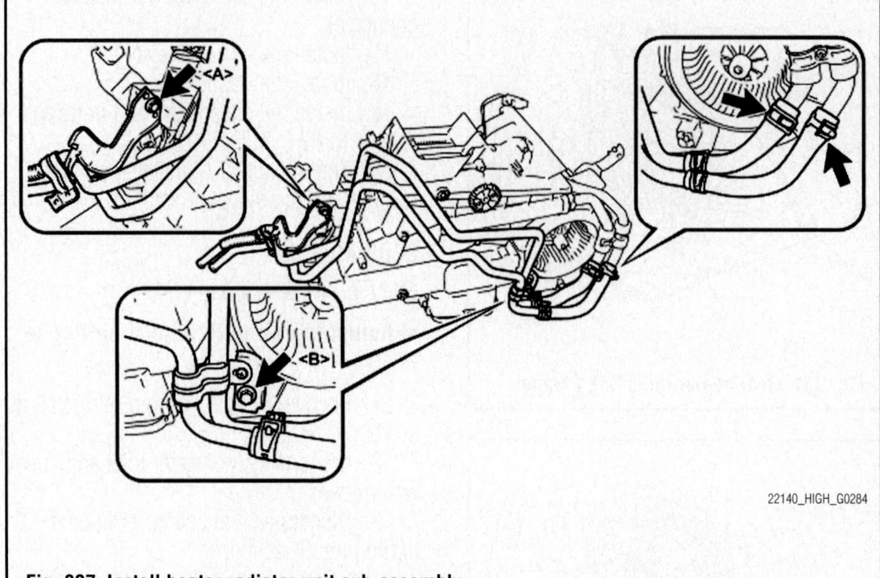

Fig. 337 Install heater radiator unit sub-assembly

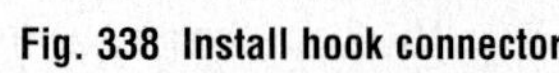

Fig. 338 Install hook connector

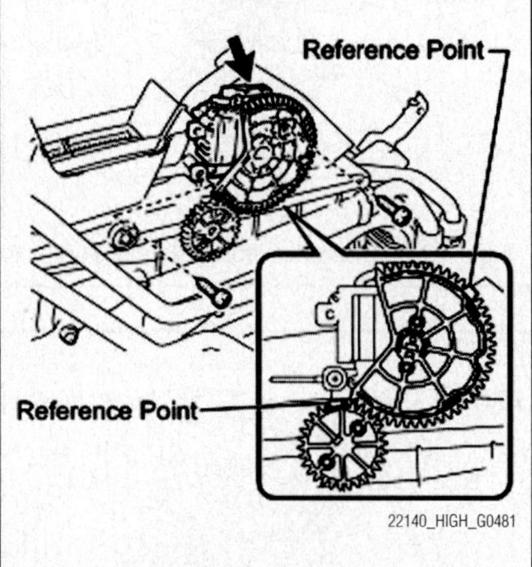

Fig. 340 Rear air outlet control servo motor sub-assembly reference point

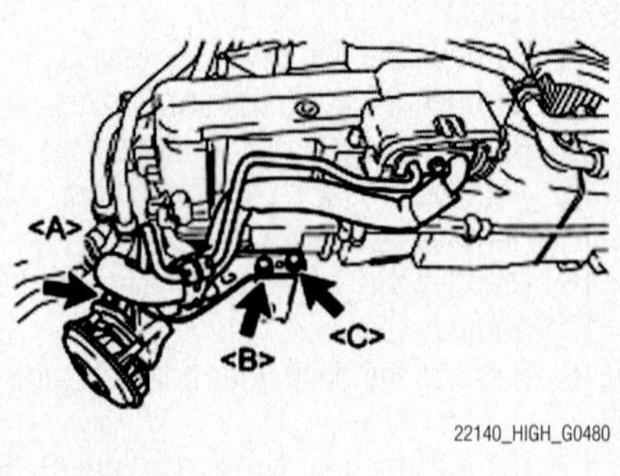

Fig. 339 Install rear air conditioning tube and accessory assembly

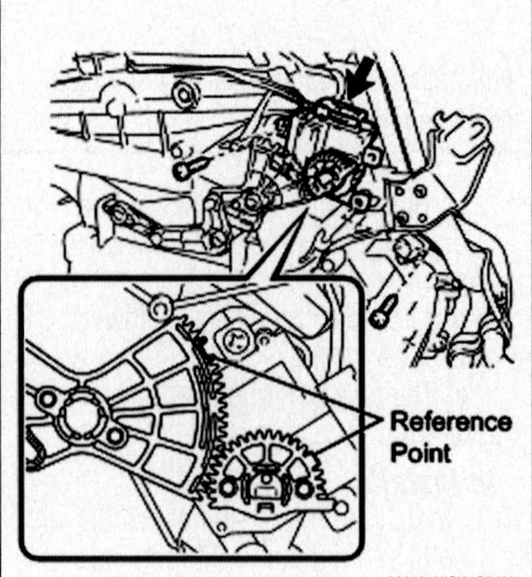

Fig. 341 Rear air mix control servo motor sub-assembly reference point

b. Remove the screws (C then B).

c. Remove the bolt (A).

d. Remove the bolt and slide the hook connector.

e. Remove the rear air conditioning tube and accessory assembly.

f. Remove the 2 O-rings from the rear air conditioning tube and accessory assembly.

42. Using pliers, grip the claws of the 2 clips and slide the clips to disconnect the heater water hose.

43. Remove the 2 screws and the heater water pipe and hose sub-assembly.

44. Remove the heater radiator unit sub-assembly, as follows:

a. Remove the screw and the heater clamp.

b. Remove the heater radiator unit sub-assembly from the rear cooling unit as shown in the illustration.

To install:

45. Install the heater radiator unit sub-assembly as shown in the illustration.

46. Install the clamp with the screw,

47. Install the heater water pipe and hose sub-assembly, as follows:

a. Install screw A, and tighten to 87 inch lbs. (9.8 Nm).

b. Install the heater water pipe and hose sub-assembly with the screw B and tighten to 87 inch lbs. (9.8 Nm).

c. Using pliers, grip the claws of the 2 clips and slide the clips to connect the heater water hose.

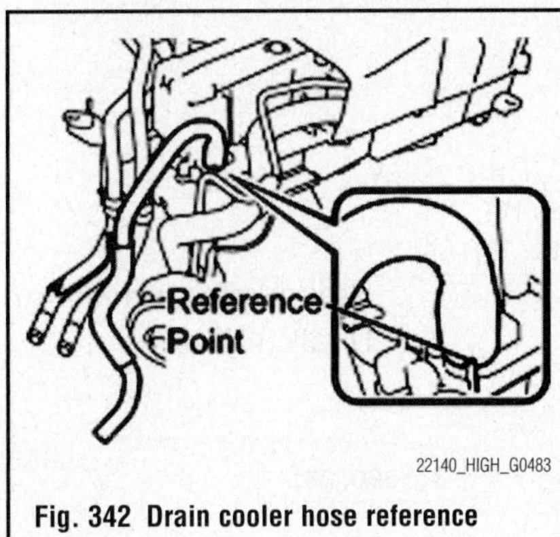

Fig. 342 Drain cooler hose reference point

48. Install the rear air conditioning tube and accessory assembly, as follows:
   a. Sufficiently apply compressor oil (ND-OIL 8 or equivalent) to 2 new O-rings and the fitting surfaces of the rear air conditioning tube and accessory assembly.
   b. Install the 2 new O-rings on the rear air conditioning tube and accessory assembly.

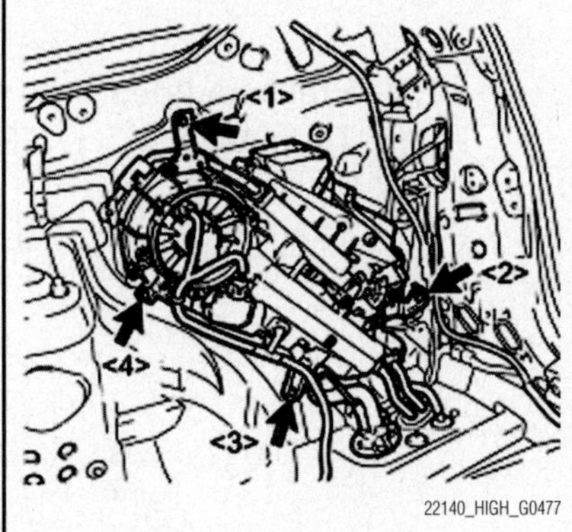

Fig. 343 Rear cooling unit assembly bolts tightening sequence

   c. Move the hook connector in the direction indicated by the arrow in the illustration.
   d. Insert the pipe joint into the fitting hole securely and tighten the bolt to 87 inch lbs. (9.8 Nm).
   e. Install bolt A and tighten to 87 inch lbs. (9.8 Nm).
   f. Install screw B and tighten to 87 inch lbs. (9.8 Nm).
   g. Install the rear air conditioning tube and accessory assembly with screw C and tighten to 87 inch lbs. (9.8 Nm).
   h. Install the packing.
49. Using the reference point, install the rear air outlet control servo motor sub-assembly with the 2 screws.
50. Connect the connector.
51. Using the reference point, install the rear air mix control servo motor sub-assembly with the 2 screws.
52. Connect the connector.
53. Using the reference point, install the drain cooler hose.
54. Install the rear cooling unit assembly with the 4 bolts and tighten in the order shown, to 87 inch lbs. (9.8 Nm).
55. Connect each connector.
56. Engage each clamp.
57. The remainder of installation is the reverse of the removal procedure.

STEERING

POWER RACK & PINION STEERING GEAR

REMOVAL & INSTALLATION

2007 Models

See Figure 344.

1. Before servicing the vehicle, refer to the precautions section.
2. Remove or disconnect the following:
   - Negative battery cable

➡Wait at least 90 seconds before working on the vehicle to allow the Supplemental Restraint System (SRS) system to disarm.

   - Steering wheel
   - Front wheels
   - Tie rod ends
   - Intermediate shaft

➡Matchmark the shaft and gear.

   - Stabilizer bar end links
   - Pressure and return lines
   - Steering gear

To install:

   - Installation is the reverse of removal. Observe the following torques:
   - Rack mounting bolts: 52 ft. lbs. (70 Nm)
   - Stabilizer bar end links: 55 ft. lbs. (74 Nm)
   - Intermediate shaft bolt: 26 ft. lbs. (35 Nm)
   - Tie rod end nuts: 36 ft. lbs. (49 Nm)

2008 Models

See Figures 345 through 347.

1. Before servicing the vehicle, refer to the precautions section.
2. Discharge fuel system pressure.
3. Recover refrigerant from refrigeration system.
4. Remove the cool air intake duct seal.
5. Remove the battery.
6. Place the front wheels facing straight ahead.
7. Secure the steering wheel with the seat belt in order to prevent it from rotating.

✻✻ WARNING

This operation is necessary to prevent damage to the spiral cable.

8. Remove the front wheels.
9. Remove the engine under cover assembly.
10. Remove the No. 1 and 2 engine under covers.
11. Remove the left hand floor under cover.
12. Remove both front fender molding sub-assemblies.
13. Remove both front fender liners.
14. Remove both front fender apron seals.
15. Drain engine oil.
16. Drain engine coolant.
17. Drain automatic transaxle fluid.
18. Remove both front wiper arm and blade assemblies.
19. Remove the cowl top ventilator louver sub-assembly.
20. Remove the windshield wiper motor and link assembly.
21. Remove the outer cowl top panel sub-assembly.
22. Remove the V-bank cover sub-assembly.
23. Remove the No. 1 and 2 air cleaner inlets.
24. Remove the air cleaner cap sub-assembly.
25. Remove the air cleaner filter element sub-assembly.
26. Remove the air cleaner case sub-assembly.
27. Remove the air cleaner bracket.
28. Separate the brake master cylinder reservoir assembly.
29. Remove the reservoir bracket.
30. Remove the right No. 2 engine mounting stay.
31. Remove the engine moving control rod.
32. Disconnect the No. 1 fuel vapor feed hose.

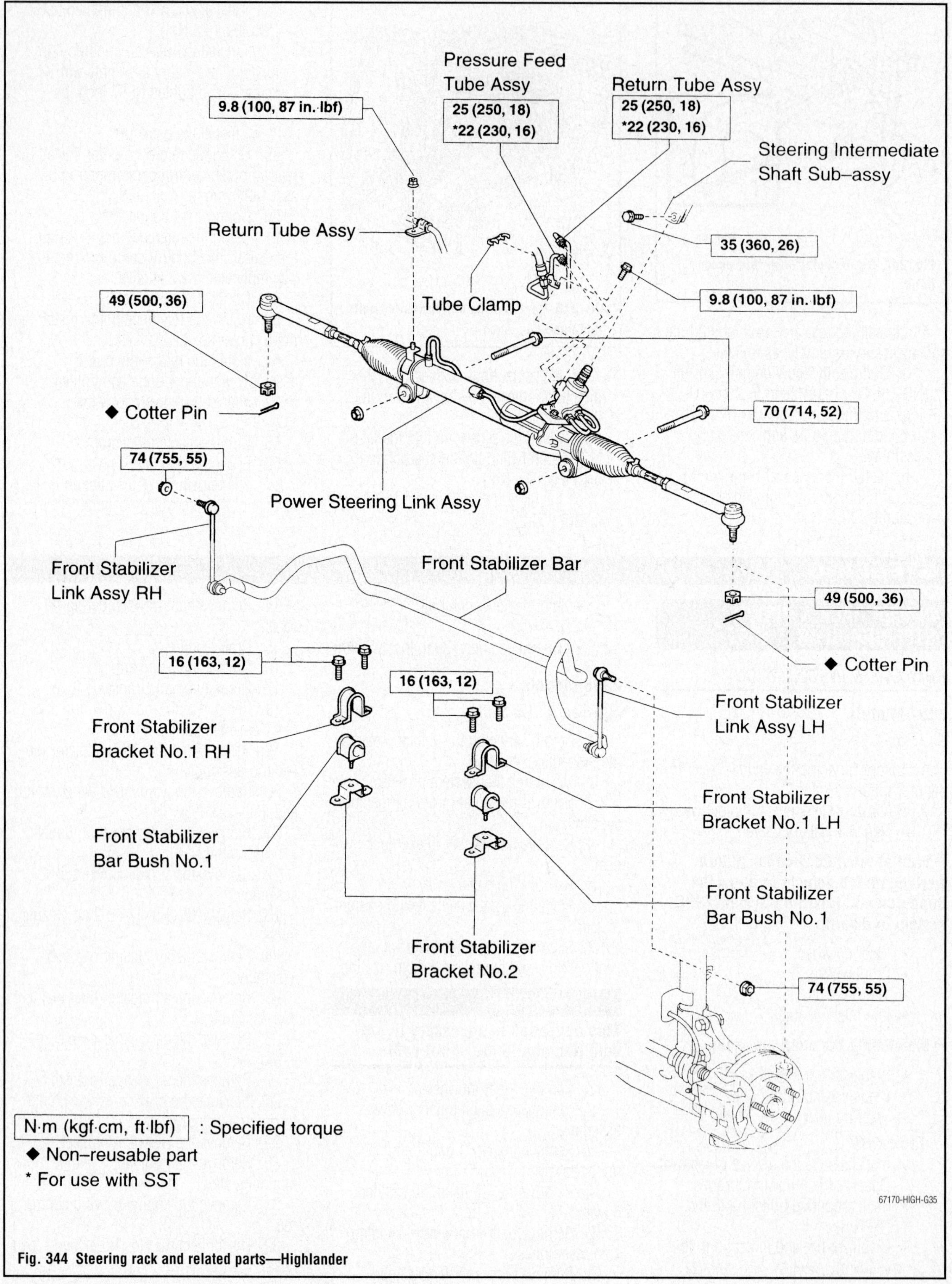

Fig. 344 Steering rack and related parts—Highlander

33. Disconnect the No. 1 and 2 radiator hose.
34. Disconnect the heater water outlet hose B.
35. Disconnect the heater water inlet hose B.
36. Disconnect the fuel tube sub-assembly.
37. Disconnect the oil cooler inlet and outlet hoses.
38. Disconnect the transmission control cable assembly.
39. Disconnect engine wire.
40. Disconnect the union to check valve hose.
41. For 4WD vehicles, remove the propeller with center bearing shaft assembly.

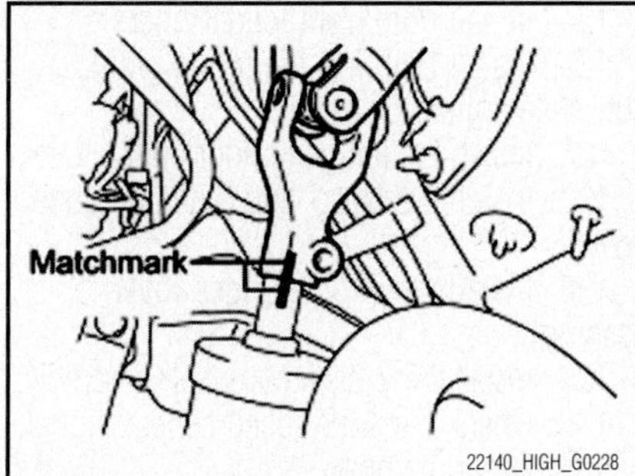

Fig. 345 Matchmarks on the steering intermediate shaft assembly and power steering link assembly

42. Remove the tail exhaust pipe assembly.
43. Remove the center exhaust pipe assembly.
44. Remove the front No. 3 exhaust pipe sub-assembly.
45. Remove the front exhaust pipe assembly.
46. Separate both front stabilizer link assemblies.
47. Remove both front axle hub nuts.
48. Disconnect both front speed sensors.
49. Separate the steering intermediate shaft assembly, as follows:
   a. Remove the bolt and slide the steering intermediate shaft assembly.

➡Do not separate the steering intermediate shaft assembly from the power steering link assembly.

   b. Put matchmarks on the steering intermediate shaft assembly and the power steering link assembly.
   c. Separate the steering intermediate shaft assembly from the power steering link assembly.
50. Separate both tie rod assemblies:
   a. Remove the cotter pin and the nut.
   b. Install SST: 09960-20010 or equivalent to the tie rod end.

➡Make sure that the upper ends of the tie rod end and SST are aligned.

   c. Using SST: 09960-20010, separate the tie rod end from the steering knuckle.

➡When securing SST to the steering knuckle, be sure to tighten the string of SST to prevent it from falling.

➡Install SST so that A and B are parallel.

➡Be sure to place the wrench on the part indicated in the illustration.

➡Do not damage the front disc brake dust cover.

➡Do not damage the ball joint dust cover.

➡Do not damage the steering knuckle.

51. Separate the No. 1 and 2 front suspension lower arms.
52. Separate both front axle assemblies.
53. Disconnect the discharge hose sub-assembly.
54. Disconnect the suction hose sub-assembly.
55. Remove the engine assembly with transaxle.
56. Remove both front No. 1 stabilizer brackets.
57. Remove the front stabilizer bar.
58. Remove the 2 bolts, 2 nuts, and power steering link assembly.

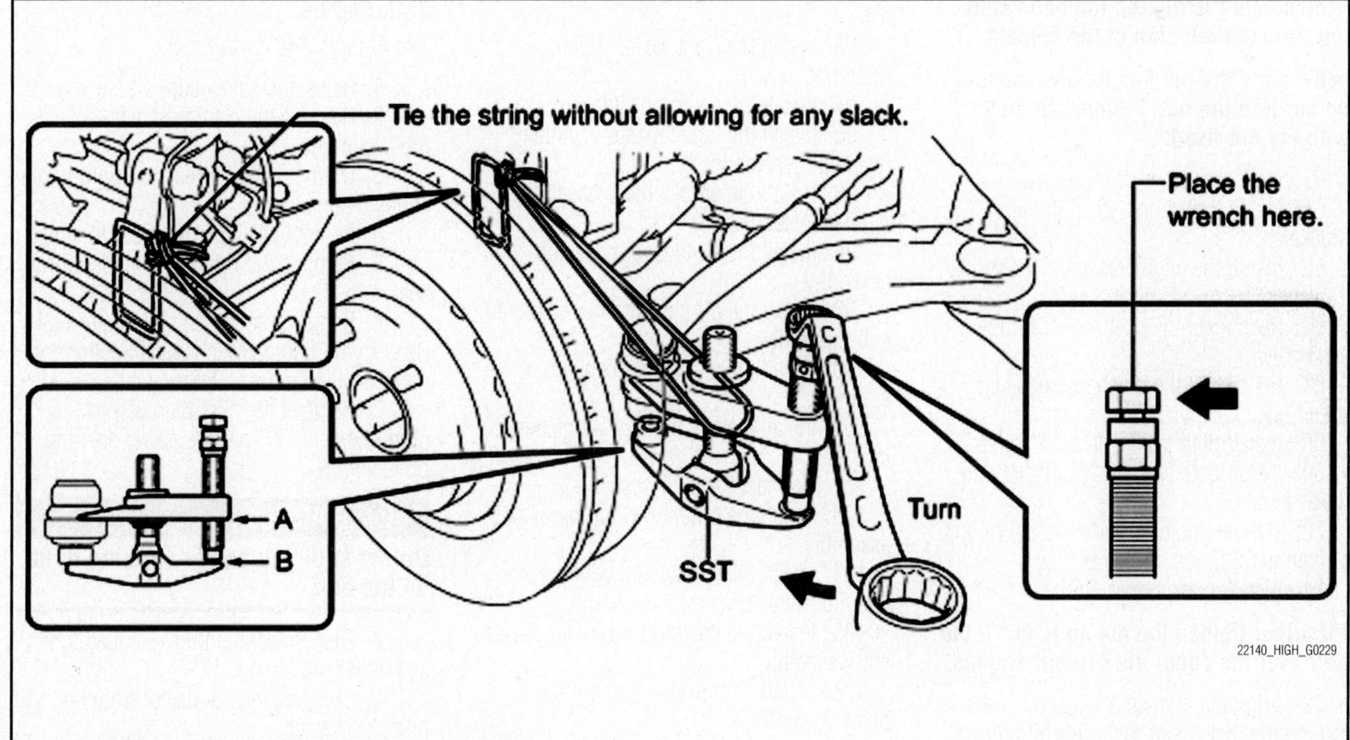

Fig. 346 Separate the tie rod end from the steering knuckle

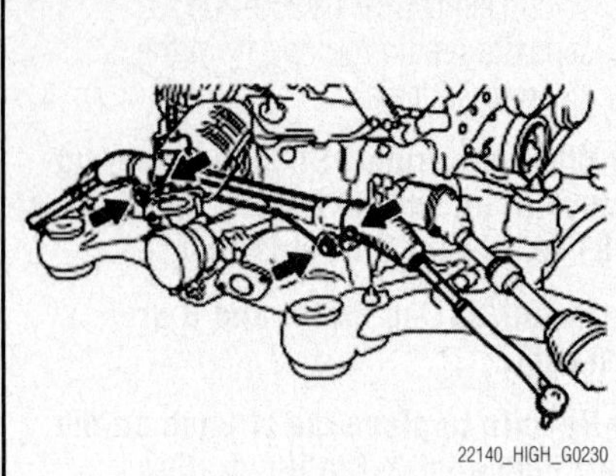

Fig. 347 Power steering link assembly bolts and nuts

➡**Because the nut has its own stopper, do not turn the nut. Loosen the bolt with the nut fixed.**

59. Put matchmarks on the tie rod assemblies and the steering rack end sub-assemblies.
60. Loosen the lock nuts, and remove the tie rod assemblies and the lock nut.

To install:

61. Install the lock nuts and the tie rod assemblies to the steering rack end sub-assembly until the matchmarks are aligned.

➡**After adjusting toe-in, torque the lock nut.**

62. Install the power steering link assembly with the 2 bolts and 2 nuts and tighten to 51 ft. lbs. (70 Nm).

➡**Make sure to tighten the bolts starting from the left side of the vehicle.**

➡**Because the nut has its own stopper, do not turn the nut. Tighten the bolt with the nut fixed.**

63. Install the front stabilizer bar.
64. Install both front No. 1 stabilizer brackets.
65. Install the engine assembly with transaxle.
66. Reconnect the suction hose sub-assembly.
67. Reconnect the discharge hose sub-assembly.
68. Install both front axle assemblies.
69. Install both No. 1 front suspension lower arms.
70. Connect both tie rod assemblies and tighten to 36 ft. lbs. (49 Nm).
71. Install a new cotter pin.

➡**Further tighten the nut up to 60° if the holes for the cotter pin are not aligned.**

72. Align the matchmarks on the steering intermediate shaft assembly and the power steering link assembly.
73. Install the bolt and tighten to 26 ft. lbs. (35 Nm).
74. Install both front speed sensors.
75. Install both front axle hub nuts.
76. Install both front stabilizer link assemblies.
77. Install the front exhaust pipe assembly.
78. Install the front No. 3 exhaust pipe sub-assembly.
79. Install the center exhaust pipe assembly.
80. Install the tail exhaust pipe assembly.
81. For 4WD vehicles, temporarily tighten the propeller with center bearing shaft assembly.
82. For 4WD vehicles, fully tighten the propeller with center bearing shaft assembly.
83. Connect engine wire.
84. Connect the transmission control cable assembly.
85. Connect the fuel tube sub-assembly.
86. Connect the oil cooler inlet and outlet hoses.
87. Connect the heater water inlet hose B.
88. Connect the heater water outlet hose B.
89. Install the No. 1 and 2 radiator hoses.
90. Connect the union to check valve hose.
91. Connect the No. 1 fuel vapor feed hose.
92. Install the engine moving control rod.
93. Install the right No. 2 engine mounting stay.
94. Install the reservoir bracket.
95. Install the brake master cylinder reservoir assembly.
96. Install the air cleaner bracket.
97. Install battery.
98. Install the air cleaner case sub-assembly.
99. Install the air cleaner filter element sub-assembly.
100. Install the air cleaner cap sub-assembly.
101. Install the No. 1 and 2 air cleaner inlets.
102. Connect vacuum hoses.
103. Install the outer cowl top panel sub-assembly.
104. Install the windshield wiper motor and link assembly.
105. Install the cowl top ventilator louver sub-assembly.
106. Install both front wiper arm and blade assemblies.
107. Install front wheels. Tighten lug nuts to 76 ft. lbs. (103 Nm).
108. Add engine oil.
109. Add engine coolant.
110. Add automatic transaxle fluid.
111. Check automatic transaxle fluid.
112. Inspect for fuel leak.
113. Inspect for engine oil leak.
114. Inspect for coolant leak.
115. Inspect for exhaust gas leak.
116. Check shift lever position.
117. Place front wheels facing straight ahead.
118. Inspect and adjust front wheel alignment.
119. Check ignition timing.
120. Check engine idle speed.
121. Check CO/HC.
122. Check function of throttle body assembly.
123. Install both front fender apron seals.
124. Install both front fender liners.
125. Install both front fender molding sub-assemblies.
126. Install the left hand floor under cover.
127. Install the No. 1 and 2 engine under cover.
128. Install the engine under cover assembly.
129. Install the V-bank cover sub-assembly.
130. Check the ABS speed sensor signal.
131. Reset memory.

POWER STEERING PUMP

REMOVAL & INSTALLATION

2.4L Engine

See Figures 348 through 351.

1. Raise and safely support the vehicle.
2. Remove the right front wheel and tire assembly.
3. Drain the power steering fluid into a suitable container.
4. Remove the right front fender liner.
5. Remove the right front fender apron seal.
6. Remove the fan and alternator belt. Refer to the Engine Mechanical Section for more details.
7. Remove the clip, then disconnect pump hose no. 1 from the power steering oil reservoir.

⁂ WARNING

Do not spill any power steering fluid on the belt.

8. Disconnect the pressure feed tube as follows:
9. Separate the connector from the oil pressure switch.
   a. Using a 27mm wrench to hold the

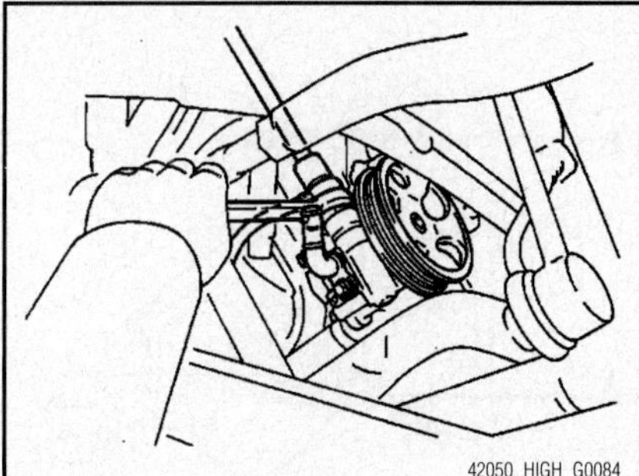

Fig. 348 Using a 27mm wrench to hold the pressure port union, remove the union bolt and gasket

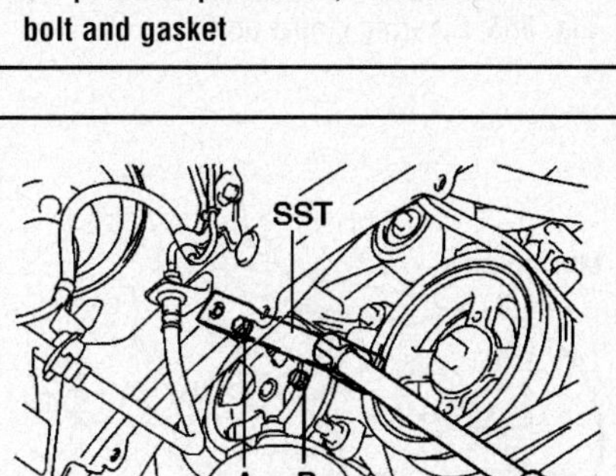

Fig. 349 Use the special tool to loosen bolt A, then remove bolt B and the power steering pump

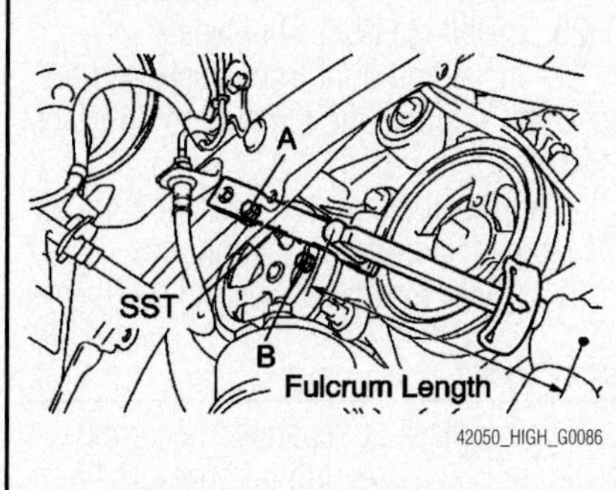

Fig. 350 Installing the power steering pump and bolts

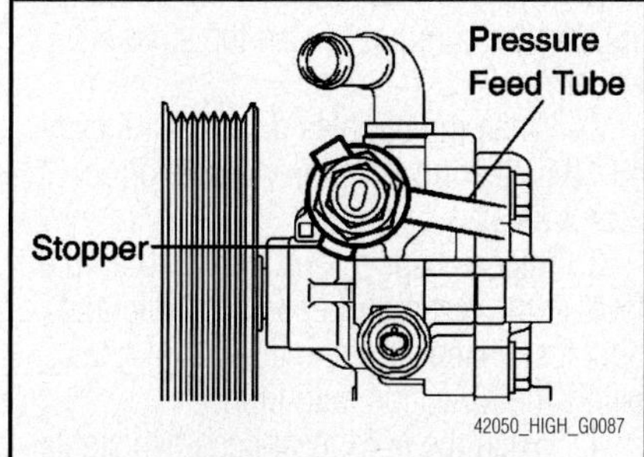

Fig. 351 When installing the pressure feed tube, make sure the stopper of the pressure feed tube touches the pump housing sub-assembly front as shown

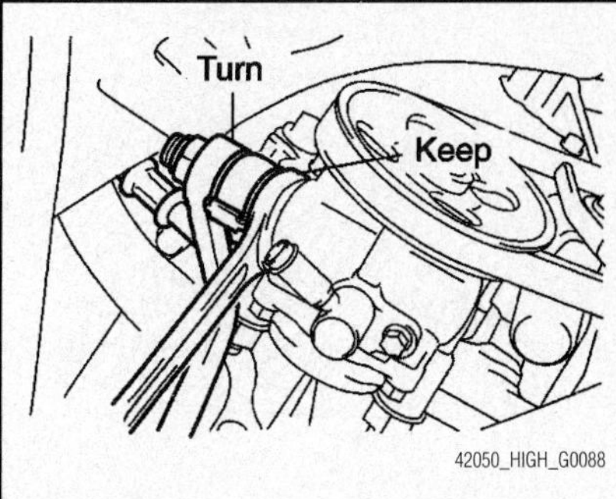

Fig. 352 Using a 24mm wrench to hold the pressure port union, remove the union bolt and gasket

pressure port union, remove the union bolt and gasket.

10. Remove the power steering pump, as follows:

a. Detach the connector from the oil pressure switch.

b. Using Special Service Tool (SST) 09249-63010, or equivalent, and a 14mm deep socket, loosen bolt "A" (as shown in the illustration).

11. Remove bolt "B", as shown in the illustration, then remove the power steering pump.

To install:

12. Install the power steering pump, as follows:

a. Temporarily tighten the bolt "A" to the power steering pump.

b. Install the pump and bolt "B". Tighten the bolt to 32 ft. lbs. (43 Nm).

c. Using the special tool and a 14mm deep socket tighten bolt "A." to 22 ft. lbs. (30 Nm).

➡Use a torque wrench with a fulcrum length of 13.58 in. (345 mm). This torque value is effective in case that SST is parallel to a torque wrench.

d. Attach the connector to the oil pressure switch.

13. Install the pressure feed tube, as follows:

a. Connect pressure feed tube and gasket to the power steering pump with the union bolt.

➡Make sure the stopper of the pressure feed tube touches the pump housing sub-assembly front as shown in the illustration.

b. Using a 27mm wrench to hold the pressure port union, torque the union bolt to 38 ft. lbs. (51 Nm).

14. Connect the pump hose no. 1 to the oil reservoir and install the clip.

15. Install fan and alternator belt.

16. Install the right front fender apron seal.

17. Install the right front fender liner.

18. Install the right front wheel and tire assembly. Tighten the lug nuts to 76 ft. lbs. (103 Nm).

a. Bleed the power steering fluid, as outlined later in this section.

19. Check for power steering fluid leaks.

3.3L Engine

See Figures 352 through 354.

1. Raise and safely support the vehicle.

2. Remove the right front wheel and tire assembly.

3. Drain the power steering fluid into a suitable container.

4. Remove the right front fender liner.

5. Remove the right front fender apron seal.

6. Remove the clip, then disconnect pump hose no. 1 from the power steering oil reservoir.

⁂ WARNING

Do not spill any power steering fluid on the belt.

7. Detach the power steering oil pressure switch connector, then remove the oil pressure switch from the union bolt.

⁂ WARNING

Do not drop the oil pressure switch. If the oil pressure switch is dropped or damaged, it must be replaced with a new one.

8. Separate pressure feed tube, as follows:

a. Using a 22mm (3.0L) or 24mm (3.3L) wrench to hold the pressure port union, remove the union bolt and gasket.

9. Loosen the 2 bolts and remove the power steering pump belt.

10. Remove the 2 mounting bolts and power steering pump assembly.

To install:

11. Temporarily install the power steering pump with the 2 (A and B) bolts.

12. Install the power steering pump belt and adjust the belt tension, as outlined in the Engine Mechanical Section.

13. Torque bolt A to 32 ft. lbs. (43 Nm), then tighten bolt B to 32 ft. lbs. (43 Nm).

14. Connect pressure feed tube, as follows:

a. Using a wrench to hold the pressure port union, connect the pressure

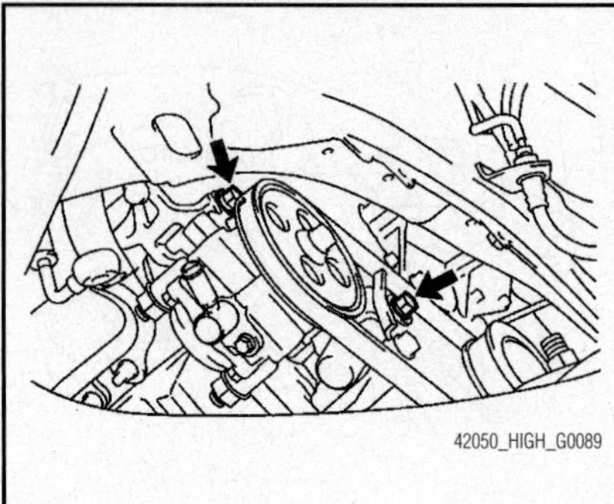
Fig. 353 Loosen the 2 bolts and remove the power steering pump belt

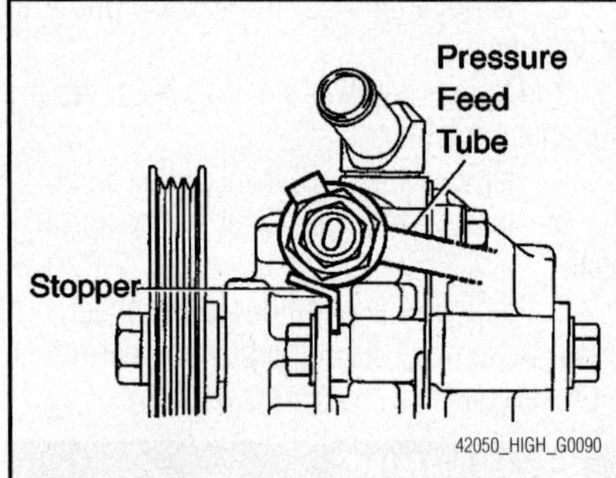

Fig. 354 When installing the pressure tube, make sure the stopper of the pressure feed tube touches the front bracket as shown in the illustration, then install the union bolt

feed tube with the union bolt and a new gasket. Tighten the bolt to 38 ft. lbs. (51 Nm).

➡**Make sure the stopper of the pressure feed tube touches the front bracket as shown in the illustration, then install the union bolt.**

15. Install the oil pressure switch to the union bolt and tighten to 15 ft. lbs. (21 Nm).

➡**Be careful not to prevent oil from getting on the connector.**

16. Attach the oil pressure switch connector.
17. Connect oil reservoir to pump hose no.1 and install the clip.
18. Install the right front fender apron seal.
19. Install the right front fender liner.
20. Install the right front wheel.
21. Install the right front wheel and tire assembly. Tighten the lug nuts to 76 ft. lbs. (103 Nm).
    a. Bleed the power steering fluid, as outlined later in this section.
22. Check for power steering fluid leaks.

BLEEDING

If you replace or separate the components for power steering oil pressure line, you bleed the air in the power steering system after the operation.

1. Start and idle the engine at 1,000 RPM or less until bubbles in the fluid disappear. (Be sure not to turn the steering wheel.)
2. When the bubbles disappear, slowly turn the steering wheel from lock to lock 2 or 3 times.
3. Repeat steps 1. and 2. until the fluid level in the reservoir becomes stable and bubbles disappear. If the fluid level goes below the MAX line, add fluid.
4. When the fluid level becomes stable, increase and decrease oil pressure 2 or 3 times for both left and right by turning the steering wheel to the full lock positions and jiggling it there.
5. Make sure that no bubbles exist in the reservoir. When turning the steering wheel quickly or turning it from lock to lock position, check that the steering wheel is not heavy to turn and there is no abnormal noise or vibration.
6. When there are fluid leaks or abnormalities after bleeding air, repair them and repeat steps 1. through 5. If the abnormalities still exist, replace the related parts

STEERING LINKAGE

REMOVAL & INSTALLATION

See Figures 355 and 356.

1. Inspect center front wheel
2. Remove front wheel
3. Separate tie rod assemblies
4. Fix the steering wheel with the seat belt in order to prevent rotation
5. Loosen bolt A and remove the clamp from the steering column hole cover No. 1

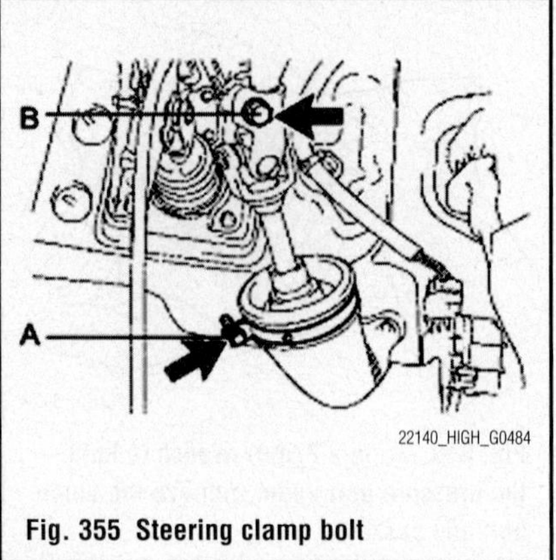

Fig. 355 Steering clamp bolt

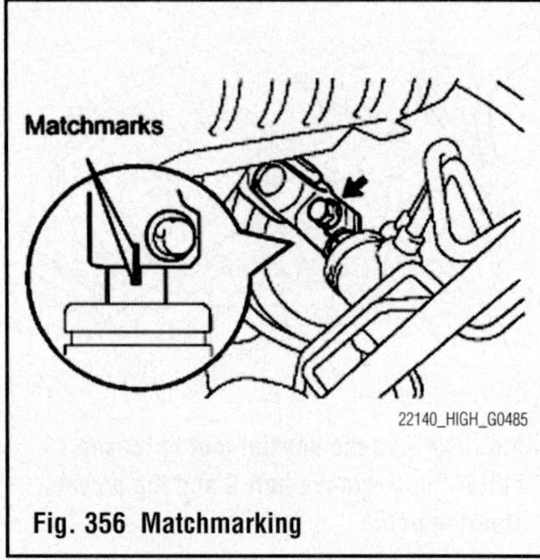

Fig. 356 Matchmarking

6. Separate the steering column hole cover No. 2 from the steering column hole cover No. 1
7. Loosen bolt B
8. Put matchmarks on the steering intermediate shaft sub- assembly and the steering link assembly
9. Remove the bolt and disengage the steering intermediate shaft sub-assembly

To install:

10. Align the matchmarks on the intermediate shaft sub-assembly and the steering link assembly
11. Install bolt A and tighten to 26 ft. lbs. (35 Nm)
12. Install the steering column hole cover No. 2 to the steering hole cover No. 1.
13. Install the clamp to the steering column hole cover No. 1 and tighten bolt B to 26 ft. lbs. (35 Nm).

CONTROL LINKS

REMOVAL & INSTALLATION

➡Perform the same procedure on each side.

1. Remove wheel and tire assembly.
2. Remove the nut and separate the front stabilizer link assembly.

➡If the ball joint turns together with the nut, use a hexagon wrench (6 mm) to hold the stud bolt.

To install:

3. Install the front stabilizer link assembly with the nut and tighten to 55 ft. lbs. (74 Nm).
4. Install the wheel and tire assembly.

LOWER BALL JOINT

REMOVAL & INSTALLATION

2007 Models

1. Remove the front wheel.
2. Remove the front axle shaft nuts.
3. Separate the speed sensor front.
4. Separate the front disc brake caliper assembly.
5. Remove the front disc.
6. Using SST: 09628-62011 or equivalent, separate tie rod assembly.
7. Separate the lower No.1 front suspension arm sub-assembly.
8. Remove front axle assembly.
9. Remove the front lower ball joint assembly:
10. Remove the cotter pin and castle nut.
11. Using SST: 09628-62011 or equivalent, remove the front lower ball joint assembly.

To install:

12. Install the front lower ball joint assembly to the steering knuckle with the nut and tighten to 91 ft. lbs. (123 Nm).

➡Prevent oil from adhering to the screw and tapered parts.

13. Install a new cotter pin to the steering knuckle.

➡If the holes for the cotter pin are not aligned, tighten the nut further up to 60°.

14. Install the front axle assembly.
15. Install the front lower No.1 suspension arm sub-assembly.
16. Install the tie rod assembly.
17. Install the front disc.
18. Install the front disc brake caliper assembly.
19. Install the speed sensor front.
20. Install the front axle shaft nut.
21. Install the front wheel and tighten the lug nuts to 76 ft. lbs. (103 Nm).
22. Inspect and adjust front wheel alignment.
23. Check the ABS speed sensor signal.

2008 Models

See Figures 357 through 360.

1. Remove the front wheel.
2. Remove the front axle hub nut.
3. Remove the bolt and resin clamp, and separate the front speed sensor.

➡Be sure to completely separate the front speed sensor from the front shock absorber with coil spring.

➡Clean the installation hole and the surface for the speed sensor every time the speed sensor is removed.

➡Be careful not to damage the front speed sensor.

4. Put matchmarks on the front drive shaft assembly and the front axle hub sub-assembly.
5. Using a plastic hammer, separate the front drive shaft assembly from the front axle assembly.

➡Loosen the staked part of the front axle hub nut completely, otherwise the threads of the drive shaft may be damaged.

6. Remove the 2 bolts and separate the front disc brake caliper assembly.

➡Use wire or an equivalent tool to keep the brake caliper from hanging down by the flexible hose.

7. Remove the front disc.
8. Separate the tie rod assembly.

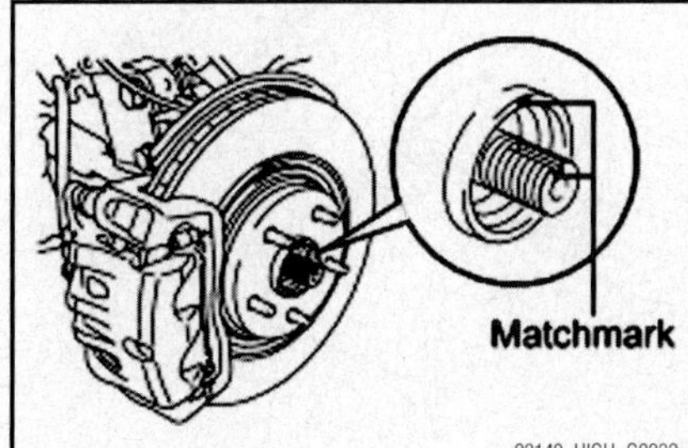

Fig. 357 Matchmark the front drive shaft assembly and the front axle hub sub-assembly

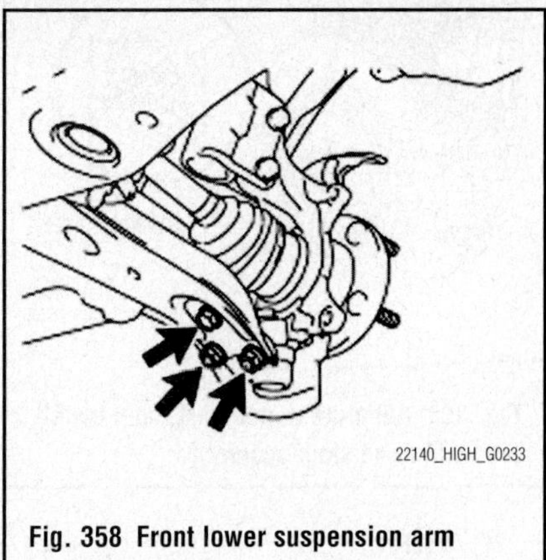

Fig. 358 Front lower suspension arm

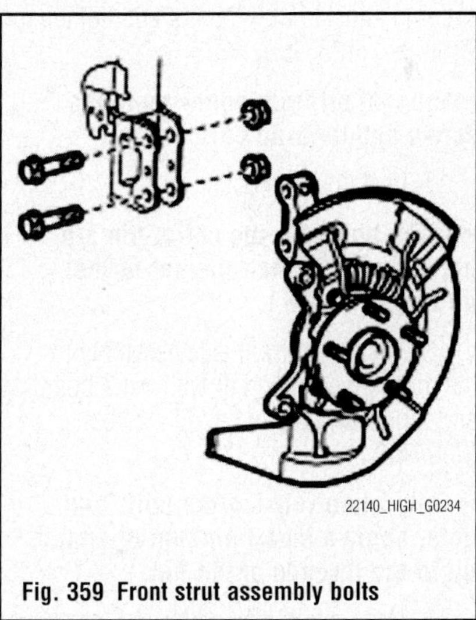

Fig. 359 Front strut assembly bolts

9. Remove the bolt, 2 nuts, and separate the front lower suspension arm from the lower ball joint.
10. Remove the 2 bolts, 2 nuts and front axle assembly.
11. Remove the front lower ball joint, as follows:
12. Secure the front axle assembly in a vise using aluminum plates.

➡When using a vise, do not overtighten it.

13. Remove the cotter pin and nut.
14. Install SST: 09960-20010 or equivalent to the front lower ball joint.
15. Using SST: 09960-20010 or equivalent, remove the front lower ball joint from the front axle assembly.

➡Install SST so that A and B are parallel.

➡Be sure to place a wrench on the part indicated in the illustration.

➡Do not damage the front lower ball joint dust cover.

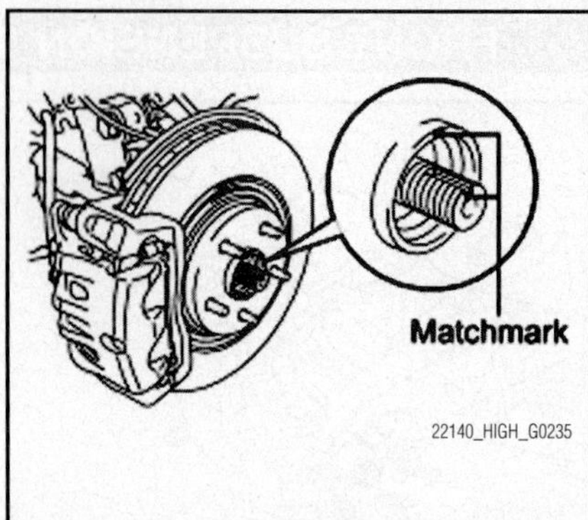

Fig. 360 Align the matchmarks and install the front drive shaft assembly

To install:

16. Install the front lower ball joint to the steering knuckle with the nut and tighten to 91 ft. lbs. (123 Nm).

➡Prevent oil from adhering to the screw and tapered parts.

17. Install a new cotter pin.

➡If the holes for the cotter pin are not aligned, tighten the nut further up to 60°.

18. Install the front axle assembly to the front shock absorber with the 2 bolts and 2 nuts and tighten to 213 ft. lbs. (290 Nm).

➡Only when reusing the bolts and nuts, apply a small amount of engine oil to the threads of the nuts.

19. Align the matchmarks and install the front drive shaft assembly to the front axle hub sub-assembly.
20. Install the front lower suspension arm to the front lower ball joint with the bolt and 2 nuts 68 ft. lbs. (92 Nm).
21. Connect the tie rod assembly.
22. Install the front disc.
23. Install the front disc brake caliper assembly to the steering knuckle with the 2 bolts and tighten to 77 ft. lbs. (104 Nm).
24. Install the clamp and front speed sensor with the bolt and tighten to 71 inch lbs. (8 Nm).

➡Prevent foreign matter from attaching to the sensor tip.

➡Firmly insert the sensor body into the knuckle before tightening the bolt.

➡After installing the sensor to the knuckle, make sure that there is no clearance between the sensor stay and knuckle. Also make sure that no foreign matter is stuck between the parts.

➡To prevent interference between the sensor and magnetic rotor, do not rotate the sensor body during or after the insertion of the sensor body to the knuckle.

25. Install the front axle hub nut.
26. Check the ABS speed sensor signal

➡Check the ABS speed sensor signal.

27. Install front wheel and tighten lug nuts to 76 ft. lbs. (103 Nm).
28. Inspect and adjust front wheel alignment.

LOWER CONTROL ARM

REMOVAL & INSTALLATION

2007 Models

See Figure 361.

1. Remove the engine assembly with transaxle.
2. Remove the transverse engine mounting insulator by first removing the 3 nuts.
3. Remove the lower No. 1 front suspension arm sub-assembly, as follows:
    a. Remove the 3 bolts and the nut from the lower No. 1 suspension arm and remove it from the engine front support member.
    b. Remove the front lower arm bush stopper.

To install:

4. Install the lower No. 1 front suspension arm sub-assembly, as follows:
    a. Install the front lower arm bush stopper.
    b. Install the lower No. 1 suspension lower arm to the engine front support member with the 3 bolts and the nut, but do not tighten.
    c. Tighten the 3 bolts in the numerical order shown. Tighten to 148 ft. lbs. (200 Nm), and 152 ft. lbs. (206 Nm).

➡Start installing the bolts from the front side of the vehicle.

5. Install the transverse engine mounting insulator with the 3 nuts and tighten to 64 inch lbs. (87 Nm).
6. Install the engine assembly with transaxle.

2008 Models

See Figure 362.

1. Remove the engine assembly with transaxle.
2. Remove the No. 1 front stabilizer brackets.
3. Remove the front stabilizer bar with front stabilizer link assembly.
4. Remove the power steering link assembly.
5. Install the engine hangers.
6. Separate the front frame assembly.
7. Remove the front lower suspension arm, as follows:
    a. Remove the 3 bolts, nut, and the front lower suspension arm from the front frame assembly.
    b. Remove the front lower arm bushing stopper from the front lower suspension arm.

To install:

8. Install the front lower suspension arm, as follows:
    a. Install the front lower arm bushing stopper to the front lower suspension arm.
    b. Install the front lower suspension arm to the front frame assembly with the 3 bolts and nut, but do not tighten them yet.
    c. Tighten the 3 bolts in numerical order shown. Tighten to 147 ft. lbs. (200 Nm), and 152 ft. lbs. (206 Nm).

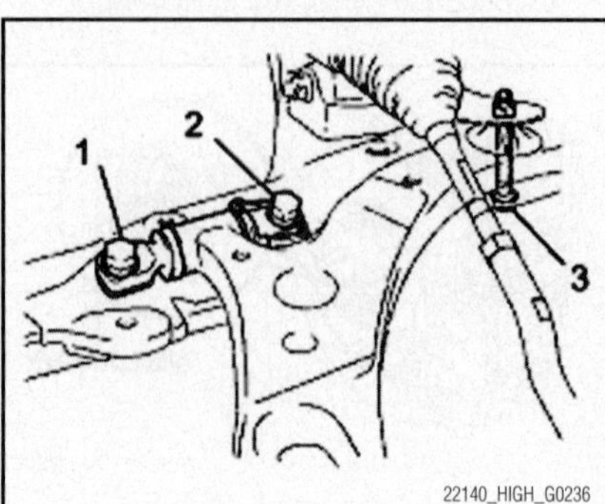

Fig. 361 Front lower control arm bolts tightening sequence–2007 models

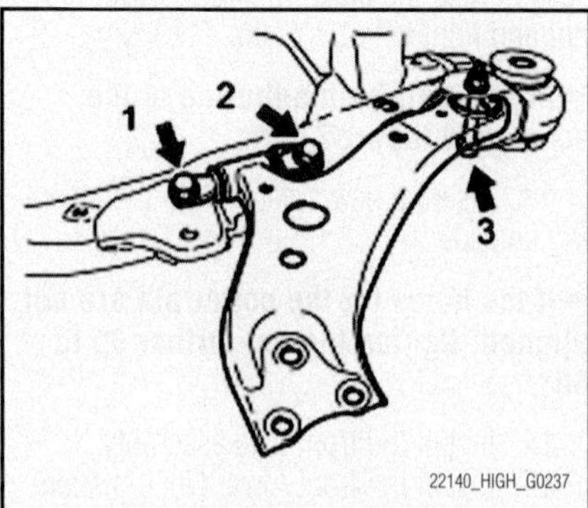

Fig. 362 Front lower control arm bolts tightening sequence–2008 models

➡**Start installing the bolts from the front side of the vehicle.**

9. Connect the front frame assembly.
10. Remove the engine hangers.
11. Install the power steering link assembly.
12. Install the front stabilizer bar with front stabilizer link assembly.
13. Install the No. 1 front stabilizer brackets.
14. Install the engine assembly with transaxle.

MACPHERSON STRUT

REMOVAL & INSTALLATION

2007 Models

See Figures 363 through 366.

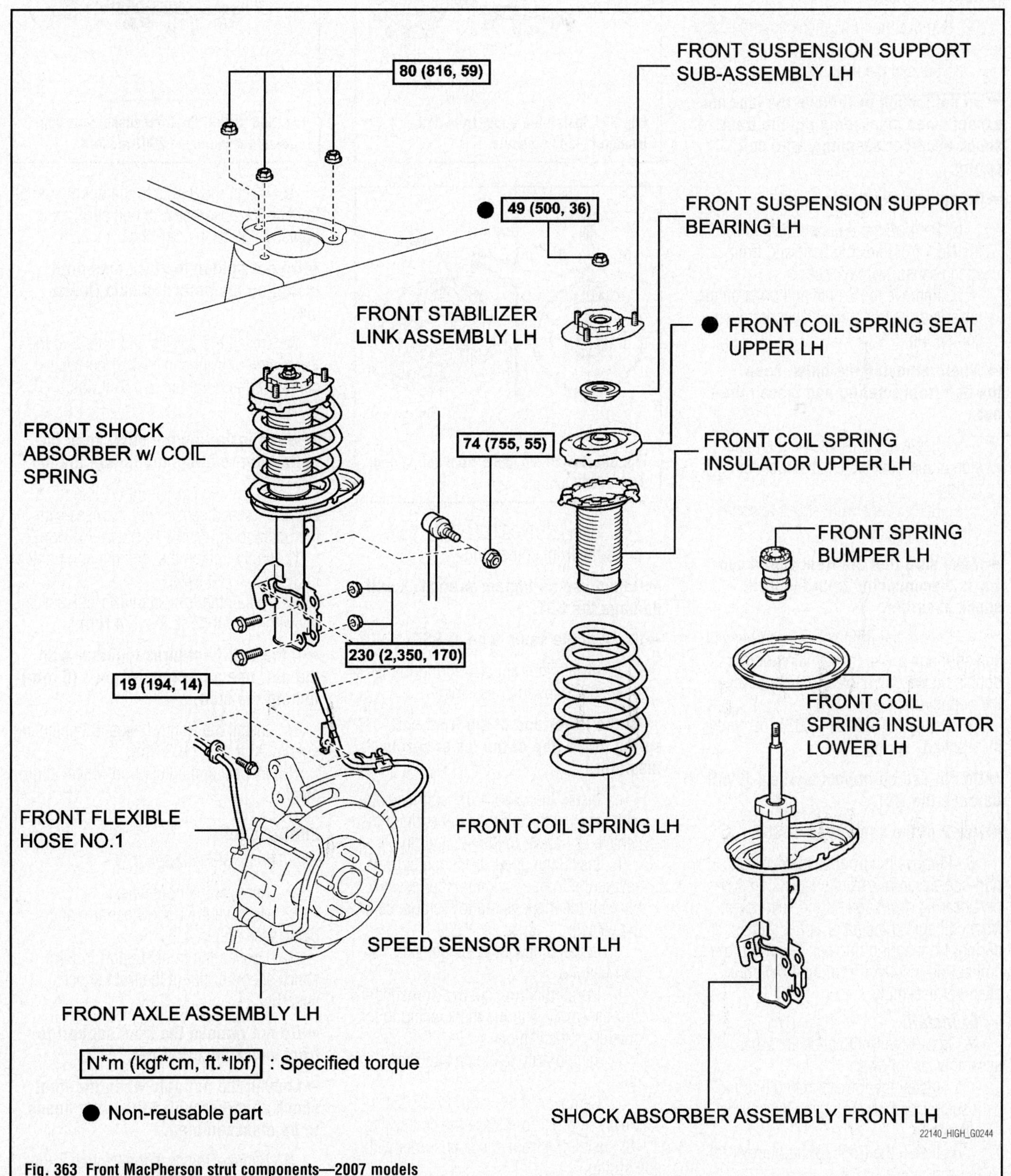

Fig. 363 Front MacPherson strut components—2007 models

1. Remove the front wheel.

2. Remove the nut and separate the front stabilizer link assembly from the front shock absorber assembly.

➡If the ball joint turns together with the nut, use a hexagon wrench (6 mm) to hold the stud.

3. Remove the front shock absorber, as follows:

a. Loosen the lock nut.

➡Do not loosen or remove the lock nut except when disassembling the front shock absorber assembly with coil spring.

➡Do not remove the lock nut.

b. Remove the bolt and disconnect the No.1 front flexible hose and front speed sensor wire harness.

c. Remove the 2 nuts and bolts on the lower side of front shock absorber with coil spring.

➡When removing the bolts, keep the bolt from rotating and loosen the nut.

d. Remove the 3 nuts on the upper side of the front shock absorber with coil spring.

e. Remove the front shock absorber with coil spring.

➡Make sure that the front speed sensor is disconnected from the front shock absorber.

4. Secure the front shock absorber with coil spring in a vise by clamping onto a double nutted bolt affixed to the bracket at the bottom of the absorber.

5. Using SST: 09727-30021, compress the front coil spring.

➡Do not use an impact wrench. It will damage the SST.

➡Use 2 of the same type of SST.

6. Remove the front suspension support sub-assembly, front suspension support bearing, front coil spring seat upper, front coil spring insulator upper, front coil spring, front spring bumper, and front lower coil spring insulator from the front shock absorber assembly.

To install:

7. Install the front shock absorber assembly, as follows:

a. Install the lower front coil spring insulator onto the front shock absorber assembly.

b. Install the front spring bumper to the piston rod.

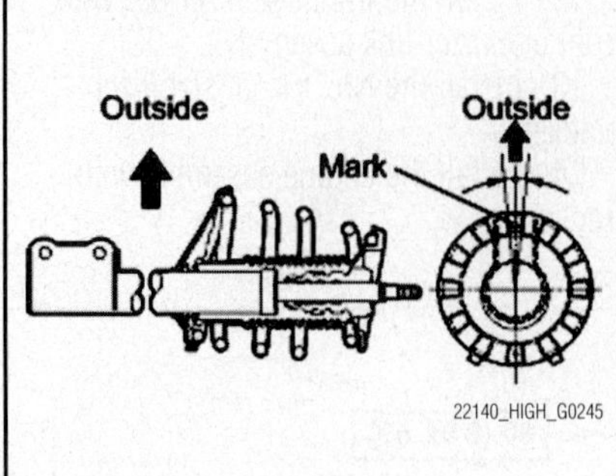

Fig. 364 Install the upper front coil insulator—2007 models

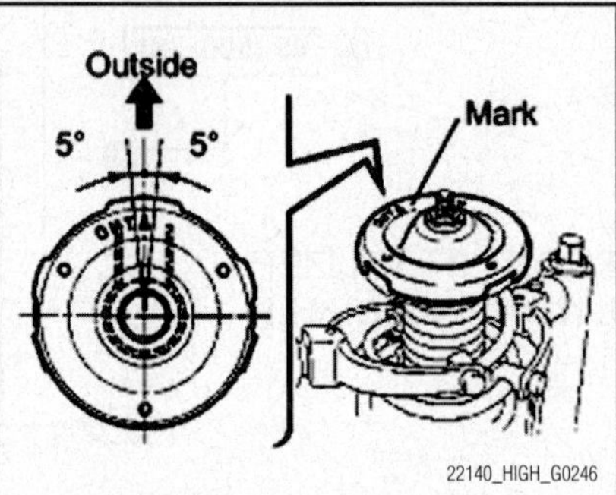

Fig. 365 Install the upper front coil spring seat—2007 models

c. Using SST: 09727-30021, compress the front coil spring.

➡Do not use an impact wrench. It will damage the SST.

➡Use 2 of the same type of SST.

d. Install the front coil spring to the front shock absorber assembly.

➡Fit the lower end of the front coil spring in the gap of the lower spring insulator.

e. Install the upper front coil insulator with the mark facing to the outside of the vehicle as shown in the illustration.

f. Install the upper front coil spring seat to the front shock absorber assembly with the mark facing to the outside of the vehicle.

g. Install a new front suspension support bearing.

h. Install the front suspension support sub-assembly with the mark facing to the outside of the vehicle.

i. Temporarily tighten a new lock nut.

j. Remove the SST: 09727-30021 slowly in order to release the coil spring.

8. Install the front shock absorber with coil spring.

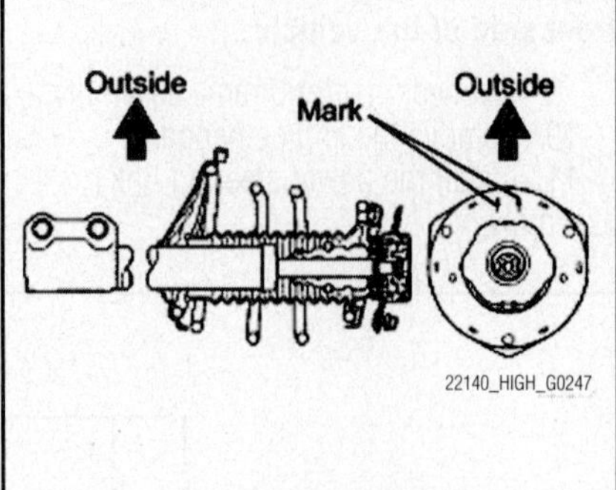

Fig. 366 Install the front suspension support sub-assembly —2007 models

9. Install the 3 nuts to the upper side of front shock absorber with coil spring and tighten to 59 ft. lbs. (80 Nm).

➡Do not tighten the lock nuts until installing the bolts and nuts (lower side).

10. Install the 2 bolts and nuts 2 to the lower side of front shock absorber with coil spring and tighten to 170 ft. lbs. (230 Nm).

➡When installing the nuts, keep the bolts from rotating and torque the nut.

11. Install the No.1 front flexible hose and front speed sensor wire harness with the bolt and tighten to 14 ft. lbs. (19 Nm).

12. Fully tighten the lock nut and tighten to 36 ft. lbs. (49 Nm).

13. Install the front stabilizer link assembly with the nut 55 ft. lbs. (74 Nm).

➡If the ball joint turns together with the nut, use a hexagon wrench (6 mm) to hold the stud.

14. Install the front wheel and tighten lug nuts to 76 ft. lbs. (103 Nm).

15. Inspect and adjust front wheel alignment.

2008 Models

See Figures 367 through 373.

1. Remove the front wheel.

2. Remove the front wiper arm and blade assemblies.

3. Loosen the front support to front shock absorber nut of the front shock absorber.

➡Do not remove the front support to front shock absorber nut.

➡Loosen the nut only when the front shock absorber with coil spring needs to be disassembled.

4. Remove the cowl top ventilator louver sub-assembly.

5. Remove the windshield wiper motor and link.
6. Remove the outer cowl top panel sub-assembly, as follows:
a. Disengage the 4 clamps and separate the wiper wire harness from the outer cowl top panel sub-assembly.
b. Remove the 8 bolts, 6 nuts, and the outer cowl top panel sub-assembly.
7. Remove the bolt and clamp, and separate the front speed sensor and front flexible hose.
8. Remove the nut and separate the front stabilizer link assembly from the front shock absorber.

➡If the ball joint turns together with the nut, use a hexagon wrench (6 mm) to hold the stud.

9. Remove the front shock absorber with coil spring, as follows:
a. Support the front axle using a jack and wooden block.
b. Remove the 2 bolts and 2 nuts, and separate the front shock absorber with coil spring (lower side) from the steering knuckle.'

➡When removing the nuts, keep the bolts from rotating.

c. Remove the nut and 2 spacers on the upper side of the front shock absorber with coil spring.

➡Make sure that the front speed sensor is completely separated from the front shock absorber with coil spring.

10. Secure the front shock absorber with coil spring, as follows:
a. As shown in the illustration, secure the front shock absorber with coil spring in a vise using aluminum plates by clamping onto a double nutted bolt affixed to the bracket at the bottom of the absorber.
11. Remove the front support to front shock absorber nut, as follows:
a. Using A Special Service Tool (SST: 09727-30021, SST: 09727-30021), compress the front coil spring.

➡Do not use an impact wrench. It will damage the SST.

➡If the front coil spring is compressed at an angle, using 2 SST will make the work easier.

b. Check that the front coil spring is fully compressed.
c. Remove the front support to front shock absorber nut.

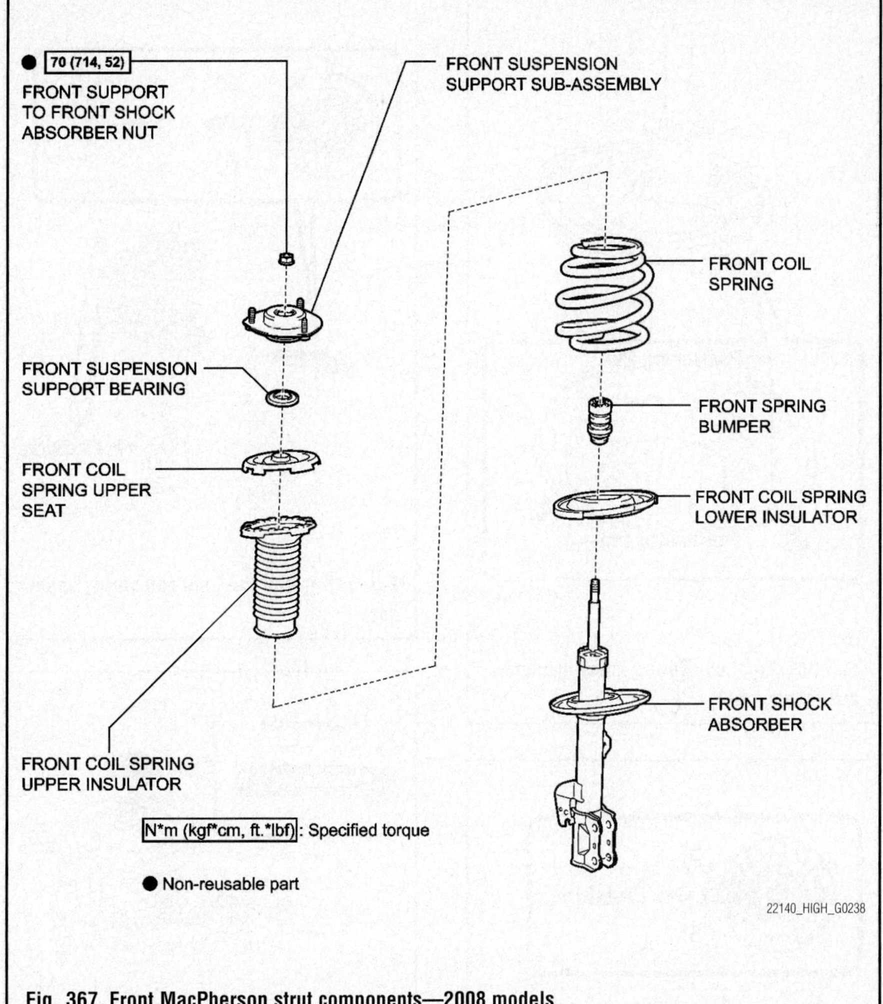

Fig. 367 Front MacPherson strut components—2008 models

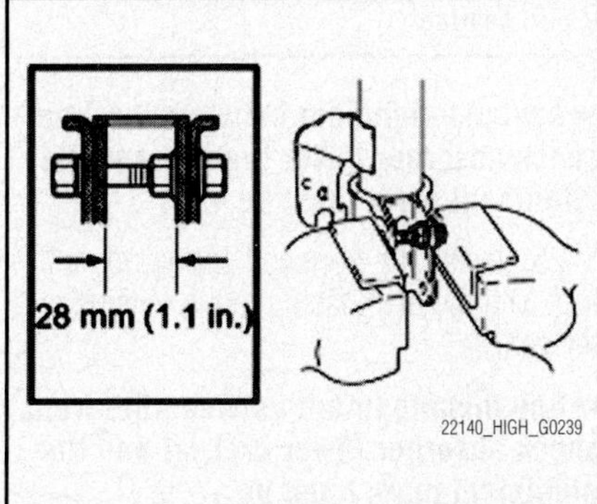

Fig. 368 Secure the front shock absorber with coil spring in a vise

12. Remove the front suspension support sub-assembly.
13. Remove the front suspension support bearing.
14. Remove the front coil spring upper seat.
15. Remove the front coil spring upper insulator.
16. Remove the front coil spring.
17. Remove the front spring bumper.
18. Remove the front coil spring lower insulator.

To install:

19. Secure the front shock absorber assembly, as follows:
a. As shown in the illustration, secure the front shock absorber with coil spring in a vise using aluminum plates by clamping onto a double nutted bolt affixed to the bracket at the bottom of the absorber.
20. Install the front coil spring lower insulator to the front shock absorber.

➡Make sure that the positioning pins on the front coil spring lower insulator are inserted into the holes in the front shock absorber.

21. Install the front spring bumper to the front shock absorber.
22. Using a Special Service Tool (SST: 09727-30021, SST: 09727-00050) or equivalent, compress the front coil spring.

➡Do not use an impact wrench. It will damage the SST.

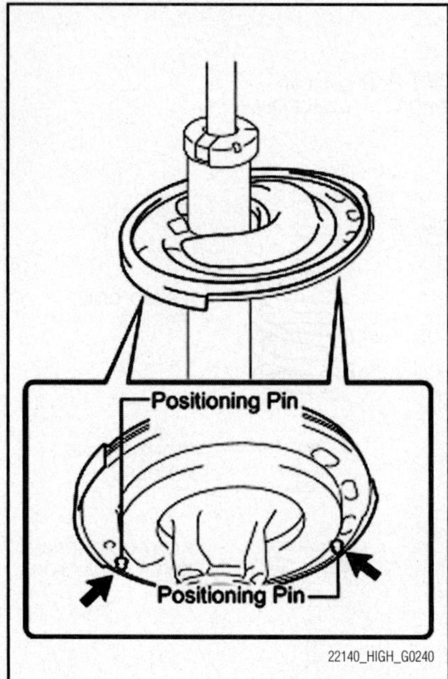

Fig. 369 Front coil spring lower insulator positioning pin

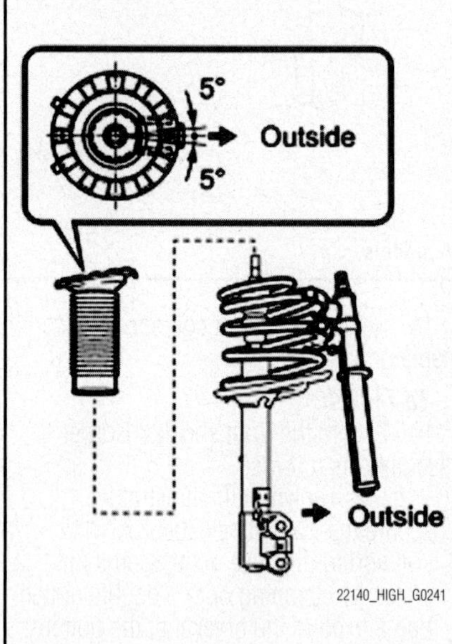

Fig. 370 Install the front coil spring upper insulator

➡**If the front coil spring is compressed at an angle, using 2 SST will make the work easier.**

23. Install the front coil spring to the front shock absorber.

➡**Make sure that the end of the front coil spring is positioned in the depression of the lower spring seat.**

24. Install the front coil spring upper insulator as shown in the illustration.

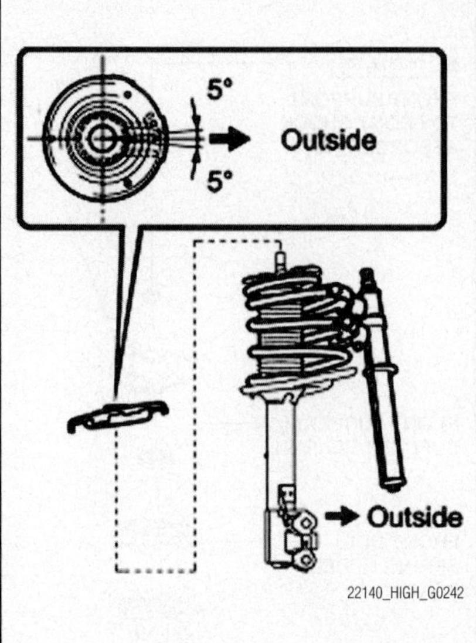

Fig. 371 Install the front coil spring upper seat

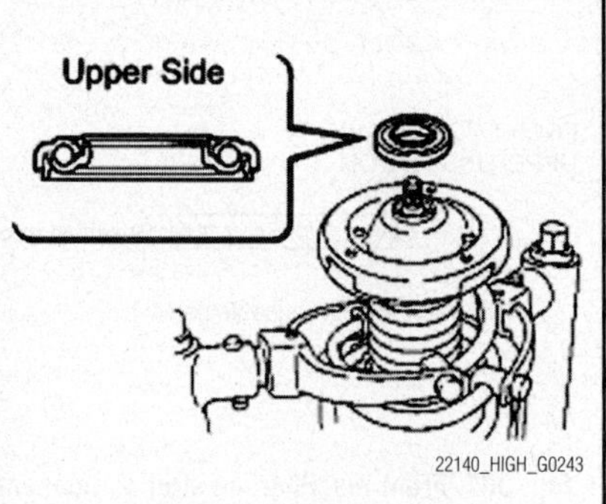

Fig. 372 Install the front suspension support bearing

➡**Any misalignment between the front shock absorber lower bracket and the alignment mark must be +/- 5°.**

25. Install the front coil spring upper seat with the mark facing to the outside of the vehicle.

➡**Any misalignment between the front shock absorber lower bracket and the alignment mark must be +/- 5°.**

26. Install the front suspension support bearing as shown in the illustration.
27. Install the front suspension support sub-assembly as shown in the illustration.

➡**Check that the slot on the piston rod and the slot on the front suspension support sub-assembly are aligned.**

28. Temporarily tighten a new front support to front shock absorber nut.
29. Install the front shock absorber with coil spring (upper side) with the nut and 2 spacers and tighten to 63 ft. lbs. (85 Nm).
30. Install the front shock absorber with coil spring (lower side) to the steering knuckle and insert the 2 bolts and 2 nuts and tighten to 214 ft. lbs. (290 Nm).

➡**When installing the nuts, keep the bolts from rotating.**

31. Install the front stabilizer link assembly to the front shock absorber with the nut and tighten to 55 ft. lbs. (74 Nm).

➡**If the ball joint turns together with the nut, use a hexagon wrench (6 mm) to hold the stud bolt.**

32. Install the front speed sensor and front flexible hose with the bolt and tighten to 14 ft. lbs. (19 Nm).

➡**Do not twist the front speed sensor when installing it.**

33. Install the clamp.
34. Install the outer cowl top panel sub-assembly with the 8 bolts and 6 nuts and tighten to 63 ft. lbs. (85 Nm), 78 inch lbs. (8.8 Nm), 78 inch lbs. (8.8 Nm).
35. Engage the 4 clamps.
36. Fully tighten the front support to front shock absorber nut and tighten to 52 ft. lbs. (70 Nm).
37. Install the windshield wiper motor and link.
38. Install the cowl top ventilator louver sub-assembly.
39. Install the front wiper arm and blade assemblies.
40. Install the front wheel and tighten lug nuts to 76 ft. lbs. (103 Nm).
41. Inspect and adjust front wheel alignment.

STEERING KNUCKLE

REMOVAL & INSTALLATION

1. Remove the front wheel.
2. Remove the front axle hub nut.
3. Separate the front speed sensor.
4. Separate the front disc brake caliper assembly.
5. Remove the front disc.
6. Separate tie rod assembly.
7. Separate the front lower suspension arm.
8. Separate the front drive shaft assembly.
9. Remove the front axle assembly.
10. Remove the front lower ball joint.
11. Remove the No. 1 front wheel bearing dust deflector.
12. Remove the front axle hub hole snap ring.
13. Remove the front axle hub sub-assembly.

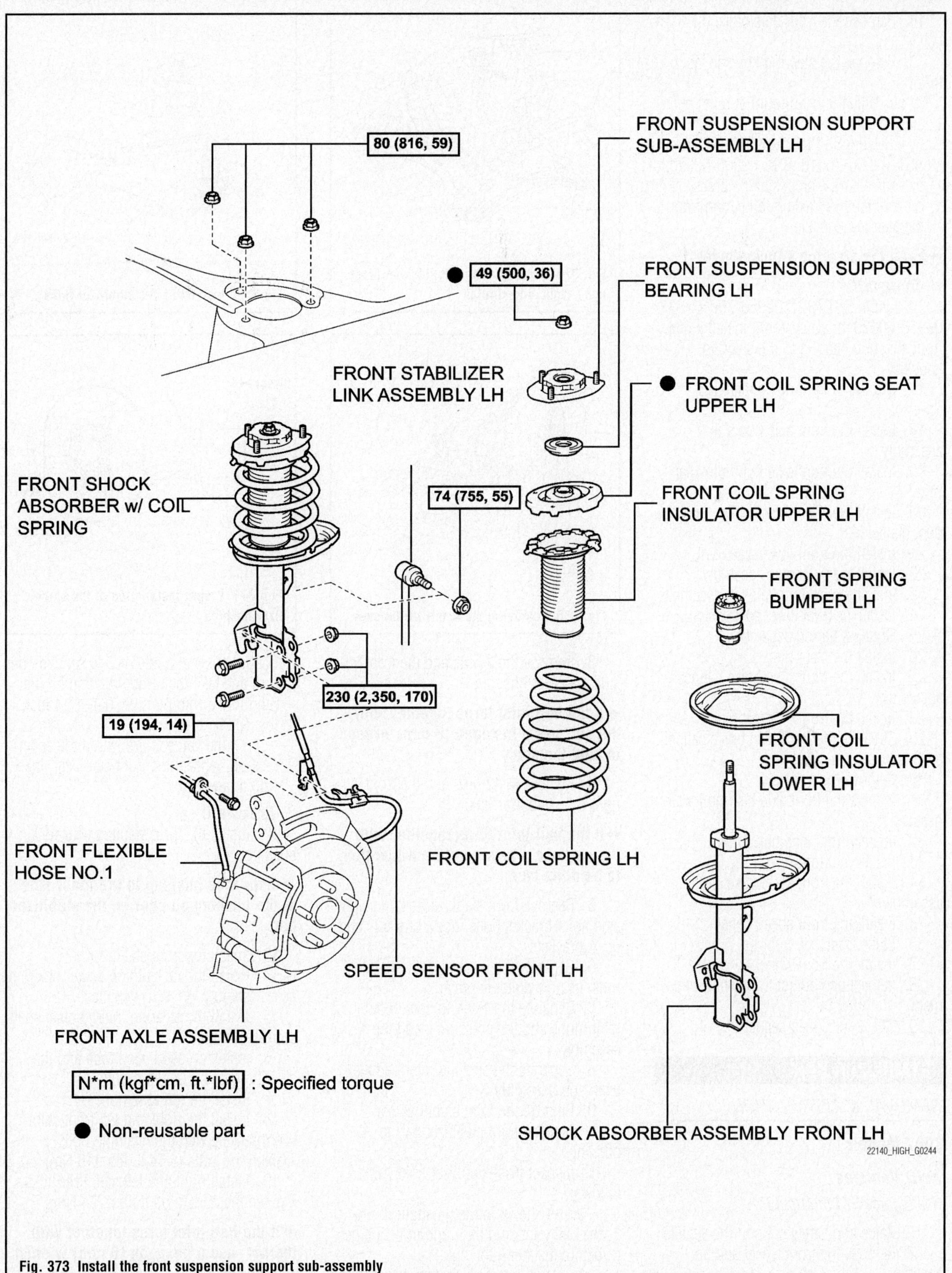

Fig. 373 Install the front suspension support sub-assembly

14. Remove the front disc brake dust cover.
15. Remove the steering knuckle, as follows:
    a. Place the bearing inner race (outside) on the front axle hub bearing.
    b. Using SST (SST: 09950-60010, SST: 09950-70010, SST: 09950-60020, or equivalent),V-blocks and a press, remove the front axle hub bearing from the steering knuckle.

➡Keep the steering knuckle level.

To install:

16. Using SST (SST: 09950-70010, SST: 09950-60020, or equivalent), install a new front axle hub bearing to the steering knuckle.
17. Install the front disc brake dust cover.
18. Install the front axle hub sub-assembly.
19. Install the front axle hub hole snap ring.
20. Install the No. 1 front wheel bearing dust deflector.
21. Install the front lower ball joint.
22. Install the front axle assembly.
23. Install the front drive shaft assembly.
24. Install the front lower suspension arm.
25. Connect tie rod assembly.
26. Install the front disc.
27. Install the front disc brake caliper assembly.
28. Install the front axle hub nut.
29. Separate the front disc brake caliper assembly.
30. Remove the front disc.
31. Inspect the front axle bearing looseness.
32. Inspect front axle hub runout.
33. Install the front disc.
34. Install the front disc brake caliper assembly.
35. Install the front speed sensor.
36. Stake front axle hub nut.
37. Install the front wheel.
38. Inspect and adjust front wheel alignment.
39. Check for speed sensor signal.

STABILIZER BAR

REMOVAL & INSTALLATION

2007 Models

2WD Vehicles

See Figures 374 through 377.

1. Raise and safely support the vehicle.
2. Remove the front wheel and tire assembly.

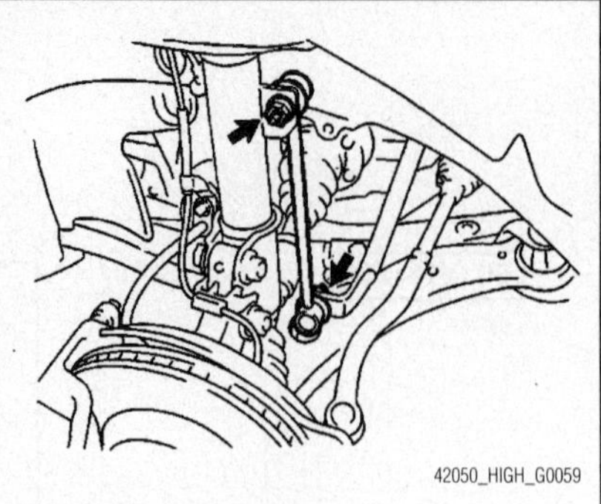

Fig. 374 Removing the left side stabilizer link—right side similar

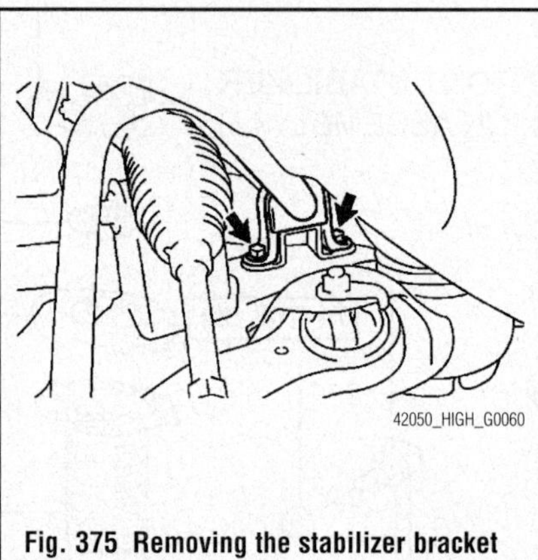

Fig. 375 Removing the stabilizer bracket

3. Remove the 2 nuts and the front left side stabilizer link.

➡If the ball joint turns together with the nut, use a hexagon (6 mm) wrench to hold the stud.

4. Remove the 2 nuts and the front right side stabilizer link.

➡If the ball joint turns together with the nut, use a hexagon (6 mm) wrench to hold the stud.

5. Remove front stabilizer left and right side no. 1 bracket bolts, no. 2 brackets and no. 1 brackets.
6. Separate the left and right side tie rods, using a suitable puller.
7. Separate the steering gear outlet return tube and the pressure feed tube assembly.
8. Separate the steering intermediate shaft sub-assembly.
9. Remove the front stabilizer bar.
10. Remove front stabilizer bar no. 1 bushing
11. Inspect front stabilizer links, as follows:
    a. As shown in the illustration, flip the ball joint stud back and forth 5 times, before installing the nut.
    b. Using a torque wrench, turn the nut continuously at a rate of 2–4 seconds per turn and take the torque reading on the 5th turn. Turning torque: 0.4–17.4 inch lbs. (0.05–1.96 Nm).
    c. If the value is not within the specification, replace the front stabilizer link with a new one.

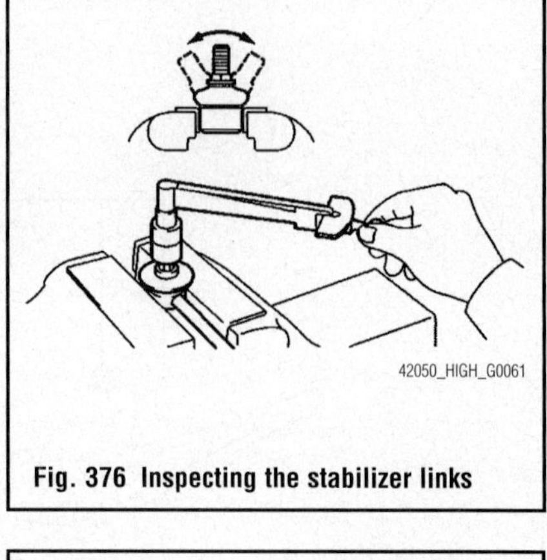

Fig. 376 Inspecting the stabilizer links

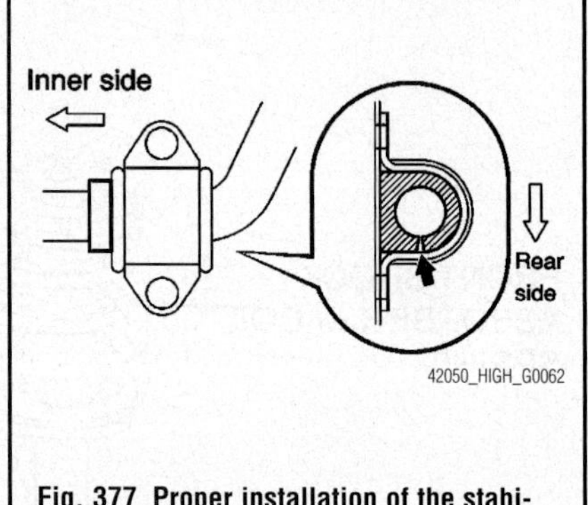

Fig. 377 Proper installation of the stabilizer bushing

To install:

12. Install the front stabilizer bar no. 1 bushing

➡Install the bushing to the inner side of the bushing stopper on the stabilizer bar.

13. Install the front stabilizer bar.
14. Install the rack pinion power steering gear assembly, on 4WD vehicles.
15. Install the steering intermediate shaft sub-assembly.
16. Install pressure feed tube and the steering gear outlet return tube.
17. Install tie rod assemblies
18. Install the right and left front stabilizer brackets with the retaining bolts. Tighten the bolts to 14 ft. lbs. (19 Nm).
19. Install right and left stabilizer links. Tighten the nuts to 55 ft. lbs. (74 Nm).

➡If the ball joint turns together with the nut, use a hexagon (6 mm) wrench to hold the stud.

20. Install the wheel and tire assembly and secure the lug nuts until they are finger-tight.
21. Lower the vehicle, then tighten the lug nuts to 76 ft. lbs. (103 Nm).
22. Bleed the power steering system.
23. Check and adjust the front wheel alignment.

4WD Vehicles

See Figures 378 through 380.

1. Raise and safely support the vehicle.
2. Remove the front wheel and tire assemblies.
3. Remove the 2 nuts and the front left side stabilizer link.

➡If the ball joint turns together with the nut, use a hexagon (6 mm) wrench to hold the stud.

4. Remove the 2 nuts and the front right side stabilizer link.

➡If the ball joint turns together with the nut, use a hexagon (6 mm) wrench to hold the stud.

a. As shown in the illustration, flip the ball joint stud back and forth 5 times, before installing the nut.
b. Using a torque wrench, turn the nut continuously at a rate of 2–4 seconds per turn and take the torque reading on the 5th turn. Turning torque: 0.4–17.4 inch lbs. (0.05–1.96 Nm).

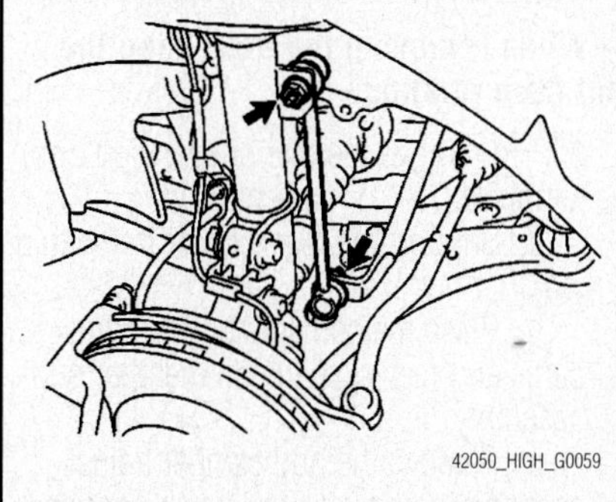

Fig. 378 Removing the left side stabilizer link—right side similar

c. If the value is not within the specification, replace the front stabilizer link with a new one.

5. On 3.3L engines, remove the front exhaust pipe.
6. Remove left and right no. 1 stabilizer brackets bolts and brackets.
7. Remove left and right no. 2 stabilizer bracket, as follows:
   a. Remove the front stabilizer bracket no.2 from the stabilizer bar bushing
8. Separate the left and right side tie rod ends with a suitable puller.
9. Separate steering intermediate shaft sub-assembly
10. Disconnect the return tube and pressure feed tube.
11. Remove the power steering link.
12. Remove front stabilizer bar bushing no. 1:
    a. Remove the 2 bushings from the stabilizer.
13. On 3.3L engines, remove the manifold stay.
    a. On 3.3L engines, remove the exhaust manifold heat insulator no. 1.
14. On 3.3L engines, remove the exhaust manifold converter.
15. Remove the stabilizer bar from the vehicle.

To install:

16. Install the stabilizer bar.
17. On 3.3L engines, install the following:
    a. Exhaust manifold converter
    b. Exhaust manifold heat insulator no. 1
    c. Manifold stay
18. Install the stabilizer bar no. 1 bushing.

➡Install the bushing to the outer side of the bush stopper on the stabilizer bar. Place the cutout of the stabilizer bushing facing the rear side as shown in the illustration.

19. Install the power steering link.
20. Install the pressure feed tube and the return tube.
21. Connect the steering intermediate shaft.
22. Install the right and left tie rod ends
23. Install right and left no. 2 stabilizer brackets to the no. 1 bushings.
24. Install right and left no. 1 stabilizer bracket and secure with the bolts. Tighten the bolts to 12 ft. lbs. (16 Nm).
25. On 3.3L engines, install the front exhaust pipe.
26. Install the right and left stabilizer links and secure with the nuts. Tighten the nuts to 55 ft. lbs. (74 Nm)

➡If the ball joint turns together with the nut, use a hexagon wrench (6 mm) to hold the stud.

27. Install the wheel and tire assembly and secure the lug nuts until they are finger-tight.
28. Lower the vehicle, then tighten the lug nuts to 76 ft. lbs. (103 Nm).
29. Bleed the power steering system.
30. Check and adjust the front wheel alignment.
31. Check and adjust the steering wheel center point

2008 Models

See Figure 381.

1. Remove the engine assembly with transaxle.
2. Remove the nuts and separate both front stabilizer link assemblies.

➡If the ball joint turns together with the nut, use a hexagon wrench (6 mm) to hold the stud.

3. Remove the 2 bolts and both No. 1 front stabilizer brackets from the front frame assembly.
4. Remove the front stabilizer bar.
5. Remove both No. 2 front stabilizer brackets from the front stabilizer bar bushing.
6. Remove the 2 No. 1 front stabilizer bar bushings from the front stabilizer bar.

To install:

7. Install the 2 No. 1 front stabilizer bar bushings to the front stabilizer bar as shown in the illustration.

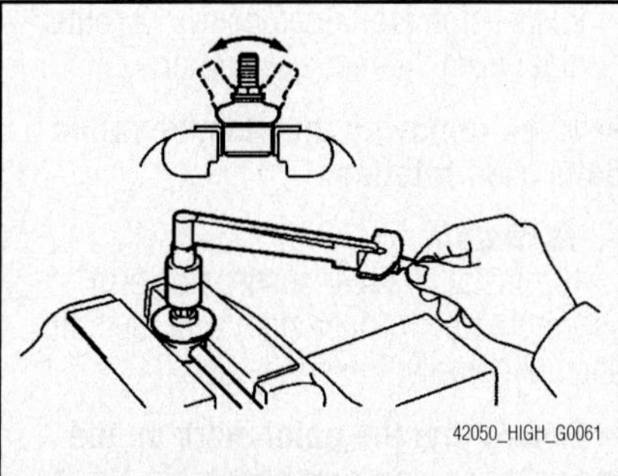

Fig. 379 Inspecting the stabilizer links

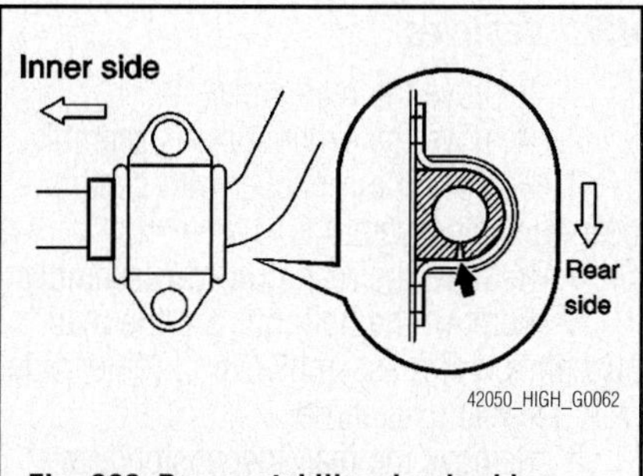

Fig. 380 Proper stabilizer bar bushing installation

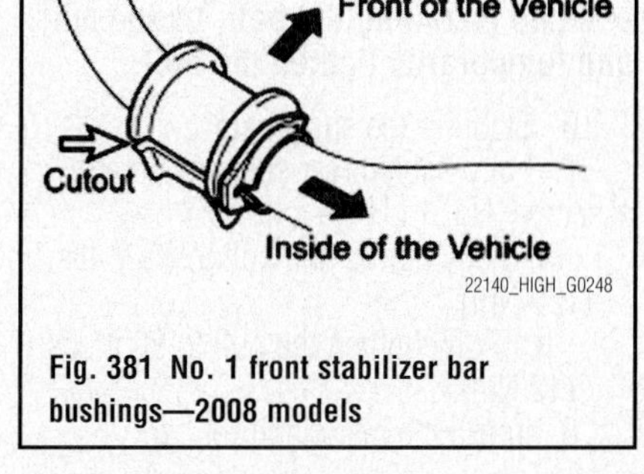

Fig. 381 No. 1 front stabilizer bar bushings—2008 models

➡Install the No. 1 front stabilizer bar bushings so that the cutout faces the rear of the vehicle.

8. Install both No. 2 front stabilizer brackets to the No. 1 front stabilizer bar bushing.
9. Install the front stabilizer bar by inserting it from the right side of the vehicle.
10. Install both No. 1 front stabilizer brackets to the front frame assembly with the 2 bolts and tighten to 21 ft. lbs. (29 Nm).
11. Install both front stabilizer link assemblies with the nuts and tighten to 55 ft. lbs. (74 Nm).

➡If the ball joint turns together with the nut, use a hexagon wrench (6 mm) to hold the stud bolt.

12. Install the engine assembly with transaxle.
13. Inspect and adjust the front wheel alignment.

WHEEL HUB AND BEARING

REMOVAL & INSTALLATION

See Steering Knuckle procedure.

SUSPENSION

REAR SUSPENSION

CONTROL ARMS/LINKS

REMOVAL & INSTALLATION

2007 Models

2WD Vehicles

See Figures 382 and 383.

➡Use the same procedures for the No. 1 or No. 2 rear suspension arm assembly, the following apply to No. 1.

1. Remove the rear wheel.
2. Remove the rear stabilizer bar.
3. Remove the LH No. 1 rear suspension arm assembly, as follows:
   a. Support the rear axle carrier with a jack.
   b. Remove the bolt and disconnect the rear suspension arm assembly No. 1 (inner side).
   c. Remove the bolt, nut and the rear suspension arm assembly No. 1 (outer side) from the rear axle carrier.

➡When removing the bolt, keep the nut from rotating.

To install:

4. Install the rear suspension arm assembly No. 1 (inner side) with the bolt, and temporarily tighten the bolt.

➡Ensure that the paint mark on the rear suspension arm assembly No. 1 faces the rear side of the vehicle.

5. Connect the rear suspension arm assembly No. 1 (outer side) to the rear axle carrier with the bolt and nut and temporarily tighten the bolt and nut.

➡When installing the bolt, fix the nut and temporarily tighten the bolt.

6. Stabilize the suspension.
7. Fully tighten rear suspension arm assembly No. 1 LH, as follows:
   a. Fully tighten the bolt to 89 ft. lbs. (120 Nm).
   b. Fully tighten the bolt to 83 ft. lbs. (112 Nm).
8. Install the rear stabilizer bar.

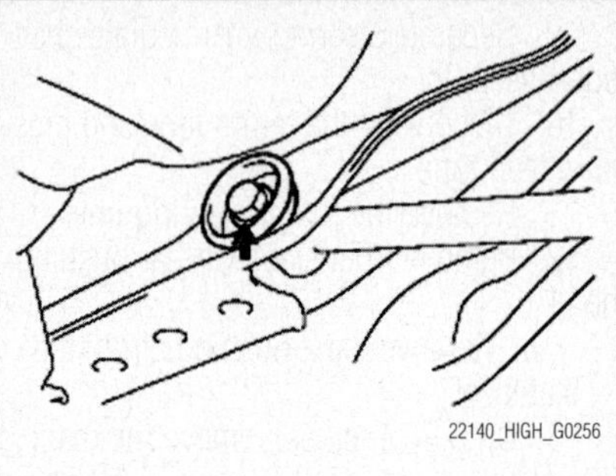

Fig. 382 Rear suspension arm assembly bolt—2007 models

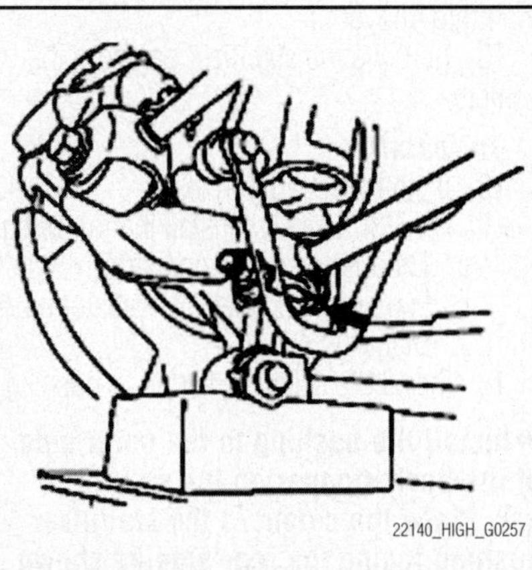

Fig. 383 Rear suspension arm assembly bolt—2007 models

9. Install the wheels and tighten lug nuts to 76 ft. lbs. (103 Nm).
10. Inspect and adjust rear wheel alignment.

4WD Vehicles

1. Remove the rear wheel.
2. Remove the exhaust pipe assembly.
3. Remove the propeller with center bearing shaft assembly.
4. Remove the rear strut rod assembly.
5. Remove the bolt, nut and the rear suspension arm assembly No. 2 (outer side) from the rear axle carrier.
6. Remove the rear suspension arm assembly No. 2 RH.
7. Remove the bolt, nut and the rear suspension arm assembly No. 1 (outer side) from the rear axle carrier.

➡When removing the bolt, keep the nut from rotating.

8. Remove rear suspension arm assembly No. 1 RH
9. Remove the speed sensor rear LH.
10. Remove the speed sensor rear RH.
11. Remove the rear drive shaft assembly LH.
12. Remove the rear drive shaft assembly RH.
13. Remove the rear suspension member sub-assembly, as follows:
   a. Support the rear suspension member with a jack.
   b. Remove the 4 nuts, 2 bolts and 2 retainers from the rear suspension member.
   c. Lower the rear suspension member
14. Remove the rear suspension arm assembly No. 1 LH.

➡When removing the bolt, keep the nut from rotating.

15. To remove the rear suspension arm assembly No. 2 LH, if necessary:
   a. Support the rear axle carrier with a jack.
   b. Place matchmarks on the adjust cams and rear suspension member sub-assembly.
   c. Remove the nut, camber adjust cam and toe adjust cam, and disconnect the rear suspension arm assembly No. 2 (inner side).
   d. Remove the bolt, nut and the rear suspension arm assembly No. 2 (outer side) from the rear axle carrier.

➡When removing the nuts, keep the bolts from rotating.

To install:

16. Install the rear suspension arm assembly No. 1 (inner side) with the bolt, and temporarily tighten the bolt.

➡Ensure that the paint mark on the rear suspension arm assembly No. 1 faces the rear side of the vehicle.

17. Set the rear suspension arm assembly No.1 in the position shown in the illustration, and fully tighten the bolt to 59 ft. lbs. (80 Nm).

18. Raise the rear suspension member with a jack.

19. Install the rear suspension member with the 4 nuts, 2 bolts and 2 retainers. Tighten to 85 ft. lbs. (115 Nm), 134 ft. lbs. (181 Nm).

20. Remove the rear drive shaft assembly LH.

21. Remove the rear drive shaft assembly RH.

22. Install the rear suspension arm assembly No. 1 (outer side) to the rear axle carrier, and temporarily tighten the bolt.

➡When installing the bolt, fix the nut and temporarily tighten the bolt.

23. Temporarily tighten the rear suspension arm assembly No. 1 RH.

24. Install the rear suspension arm No. 2 (outer side) to the rear axle carrier, and temporarily tighten the bolt.

➡When installing the bolt, fix the nut and temporarily tighten the bolt.

25. Temporarily tighten the rear suspension arm assembly No. 2 RH.

26. Temporarily tighten strut rod assembly rear.

27. Rear stabilizer.

28. Fully tighten rear suspension arm assembly No. 1 LH, as follows:

a. Fully tighten the bolt to 83 ft. lbs. (112 Nm).

29. Fully tighten the rear suspension arm assembly No. 1 RH.

30. Fully tighten the rear suspension arm assembly No. 2 LH, as follows:

a. Fully tighten the bolt to 83 ft. lbs. (112 Nm).

31. Fully tighten the rear suspension arm assembly No. 2 RH.

32. Stabilize suspension.

33. To install the rear suspension arm assembly No. 2 LH, if necessary:

a. Install the rear suspension arm No. 2 (inner side) to the rear suspension member sub-assembly with the camber adjust cam and toe adjust cam, and temporarily tighten the nut.

➡Ensure that the paint mark on the rear suspension arm assembly No. 2 faces to the rear side of the vehicle.

b. Connect the rear suspension arm assembly No. 2 (outer side) to the rear axle carrier with the bolt and nut.

➡When installing the bolt, fix the nut and temporarily tighten the bolt.

c. Align the matchmarks on the adjust cams and rear suspension member sub-assembly.

d. Fully tighten the nut to 74 ft. lbs. (100 Nm).

e. Fully tighten the bolt to 83 ft. lbs. (112 Nm).

34. Fully tighten the strut rod assembly rear.

35. Install the wheels and tighten lug nuts to 76 ft. lbs. (103 Nm).

36. Temporarily tighten the propeller w/center bearing shaft assembly.

37. Fully tighten the propeller w/center bearing shaft assembly.

38. Install the exhaust pipe assembly.

39. Check for exhaust gas leak.

40. Inspect and adjust rear wheel alignment.

41. Check the ABS speed sensor signal.

2008 Models

2WD Vehicles

See Figures 384 and 385.

➡Use the same procedures for the RH side and LH side.

1. Remove the deck board assembly.

2. Remove the No. 2 and 3 deck board sub-assemblies, as applicable.

3. Remove the tonneau cover assembly, as applicable.

4. Remove the rear mat.

5. Remove the deck trim service hole cover.

6. Remove the lower spare wheel carrier hinge cover, as applicable.

7. Remove the spare tire.

8. Remove the spare wheel carrier lock cover, as applicable.

9. Remove rear wheel.

10. Remove the nuts and separate the rear stabilizer link assemblies from the rear stabilizer bar.

➡If the ball joint turns together with the nut, use a hexagon wrench (5 mm) to hold the stud bolt.

11. Remove the 4 bolts and rear stabilizer bar.

12. Remove the rear No. 2 suspension arm assembly, as follows:

a. Remove the bolt and the nut, and separate the rear No. 2 suspension arm assembly from the rear axle carrier sub-assembly.

➡Since a stopper nut is used, loosen the bolt.

b. Remove the bolt and the rear No. 2 suspension arm assembly.

13. Remove the rear No. 1 suspension arm assembly, as follows:

a. Remove the bolt and the nut, and separate the rear No. 1 suspension arm assembly from the rear axle carrier sub-assembly.

➡Since a stopper nut is used, loosen the bolt.

b. Remove the bolt and the rear No. 1 suspension arm assembly.

To install:

14. Temporarily install the rear No. 1 and 2 suspension arm assemblies to the rear suspension member with the bolts.

➡Ensure that the identification marks face the rear side of the vehicle.

15. Temporarily install the rear No. 1 and 2 suspension arm assemblies to the rear axle carrier sub-assembly with the bolts and the nuts.

➡Since a stopper nut is used, temporarily tighten the bolts.

16. Jack up the rear axle carrier, placing a wooden block underneath to avoid damage. Apply load to the suspension so that the installed bolt of the rear No. 1 suspension arm (inner side) is horizontally aligned with the center of the rear axle hub.

CAUTION

Do not jack up the rear axle carrier sub-assembly too high as the vehicle may fall.

➡Do not bend the brake dust cover.

➡If the rear drive shaft assembly cannot be positioned horizontally as shown in the illustration even when the rear axle carrier sub-assembly is jacked up, apply additional load to the vehicle such as by having a person sit in the rear seat.

➡Use the same procedures for the RH side and LH side.

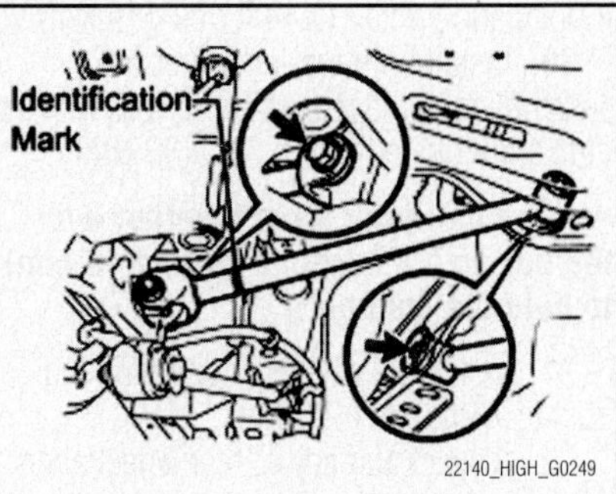

Fig. 384 Temporarily install the rear No. 1 and 2 suspension arm assemblies

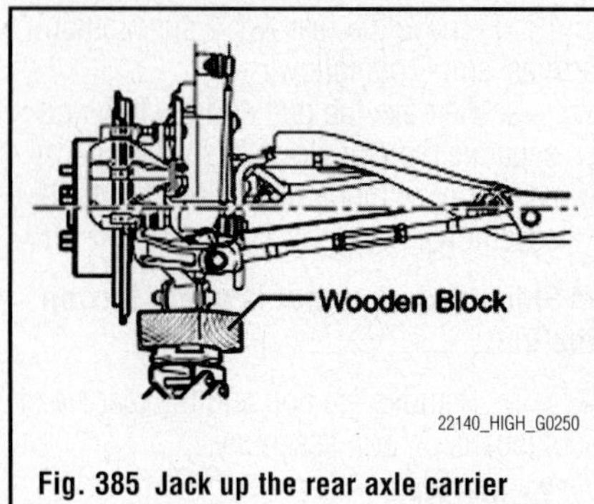

Fig. 385 Jack up the rear axle carrier

17. Fully tighten the rear No. 1 suspension arm assembly, as follows:
   a. Using SST: 09961-00950 or equivalent and a socket wrench (19 mm), fully tighten the bolt to 88 ft. lbs. (120 Nm), 65 ft. lbs. (89 Nm).

➡Use a torque wrench with a fulcrum length of 425 mm (16.73 in.).

➡This torque value is effective when SST is parallel to the torque wrench.

➡The final torque must be applied under standard vehicle height conditions.

   b. Fully tighten the bolt to 82 ft. lbs. (112 Nm).

➡Since a stopper nut is used, fully tighten the bolt.

➡The final torque must be applied under standard vehicle height conditions.

18. Fully tighten the rear No. 2 suspension arm assembly, as follows:
   a. Fully tighten the bolts to 88 ft. lbs. (120 Nm), 82 ft. lbs. (112 Nm).

➡Since a stopper nut is used, fully tighten the bolt.

➡The final torque must be applied under standard vehicle height conditions.

19. Install the rear stabilizer bar with the 4 bolts and tighten to 14 ft. lbs. (19 Nm).
20. Install both rear stabilizer link assemblies to the rear stabilizer bar with the nuts and tighten to 29 ft. lbs. (39 Nm).

➡If the ball joint turns together with the nut, use a hexagon wrench (6 mm) to hold the stud bolt.

21. Install the wheels and tighten lug nuts to 76 ft. lbs. (103 Nm).
22. Inspect and adjust rear wheel alignment.
23. The remainder of installation is the reverse of the removal procedure.

4WD Vehicles

See Figures 386 through 390.

➡The removal procedures for the LH and RH sides are different.

➡When removing RH side components, it is not necessary to follow the steps with (for LH Side).

➡When removing LH side components, it is not necessary to follow the steps with (for RH side).

1. Remove the rear wheel (for RH Side).
2. Remove the rear wheels (for LH Side).
3. Remove the rear no. 2 suspension arm assembly LH (for LH Side), as follows:
   a. Put matchmarks on the adjust cams and the rear suspension member sub-assembly.
   b. Remove the bolt and the nut, and separate the rear No. 2 suspension arm assembly LH from the rear axle carrier sub-assembly LH.

➡Since a stopper nut is used, loosen the bolt.

   c. Remove the nut, the No. 2 camber adjust cam, the rear suspension toe adjust cam sub-assembly, and the rear No. 2 suspension arm assembly LH.

➡When removing the nut, keep the rear suspension toe adjust cam sub-assembly from rotating.

4. Remove the tail exhaust pipe assembly.

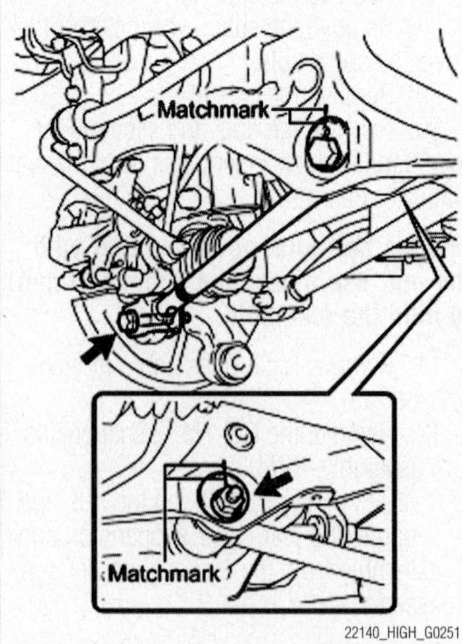

Fig. 386 Put matchmarks on the adjust cams and the rear suspension member sub-assembly

5. Remove the center exhaust pipe assembly.
6. Remove the rear no. 2 suspension arm assembly RH (for RH Side).

➡Perform the same procedure as the rear No. 2 suspension arm assembly LH.

7. Remove the propeller with center bearing shaft assembly (for LH Side).
8. Remove the rear axle shaft nut LH (for LH Side).
9. Remove the rear axle shaft nut RH (for LH Side).
10. Separate the No. 3 parking brake cable assembly (for LH Side).
11. Separate the No. 2 parking brake cable assembly.
12. Remove the rear strut rod assembly LH (for LH Side).
13. Remove the rear strut rod assembly RH.
14. Remove the rear No. 1 suspension arm assembly RH (for RH Side). As follows:
   a. Remove the bolt and the nut, and separate the rear No. 1 suspension arm assembly RH from the rear axle carrier sub-assembly RH.
   b. Remove the bolt, the nut, and the rear No. 1 suspension arm assembly RH from the rear suspension member sub-assembly

➡Since stopper nuts are used, loosen the bolts.

15. Remove the rear speed sensor LH (for LH Side).
16. Remove the rear speed sensor RH (for LH Side).
17. Separate the rear disc brake caliper assembly LH (for LH Side).
18. Separate the rear disc brake caliper assembly RH (for LH Side).
19. Remove the rear disc (for LH Side), as follows:
   a. Remove the rear disc from the rear axle hub and bearing assembly LH.
   b. Remove the rear disc from the rear axle hub and bearing assembly RH.

➡Perform the same procedure as the LH side.

20. Remove the rear axle hub and bearing assembly LH (for LH Side).
21. Remove the rear axle hub and bearing assembly RH (for LH Side).
22. Remove the rear axle carrier sub-assembly LH (for LH Side), as follows:
   a. Loosen the bolt.

➡Since a stopper nut is used, loosen the bolt.

   b. Remove the 2 bolts and 2 nuts, and

separate the rear shock absorber with coil spring from the rear axle carrier sub-assembly LH.

c. Remove the bolt, the nut, and the rear axle carrier sub-assembly LH.

➡Be careful not to damage the outboard joint boot.

➡Be careful not to damage the speed sensor rotor.

➡Use a rope or equivalent to hang the rear drive shaft assembly. Remove rear axle carrier sub-assembly RH (for LH Side), as follows:

d. Loosen the 2 bolts.

e. Remove the 2 bolts and 2 nuts, and separate the rear shock absorber with coil spring from the rear axle carrier sub-assembly RH.

f. Remove the 2 bolts, the 2 nuts, and the rear axle carrier sub-assembly RH.

➡Since stopper nuts are used, loosen the bolts.

➡Be careful not to damage the outboard joint boot.

➡Be careful not to damage the speed sensor rotor.

➡Use a rope or equivalent to hang the rear drive shaft assembly.

23. Remove the rear differential filler plug (for LH Side).

24. Remove the rear differential drain plug (for LH Side).

25. Remove the rear drive shaft assembly LH (for LH Side).

26. Remove the rear drive shaft snap ring LH (for LH Side).

27. Remove the rear drive shaft assembly RH (for LH Side).

28. Remove the rear drive shaft snap ring RH (for LH Side).

29. Remove the rear suspension member (for LH Side).

30. Remove the rear No. 1 suspension arm assembly LH (for LH Side), as follows:

a. Remove the bolt, the nut, and the rear No. 1 suspension arm assembly LH from the rear suspension member sub-assembly.

➡Since a stopper nut is used, loosen the bolt.

To install:

31. Temporarily install rear No. 1 suspension arm assembly LH (for LH Side), as follows:

a. Temporarily install the rear No. 1 suspension arm assembly LH to the rear suspension member sub-assembly with the bolt and the nut.

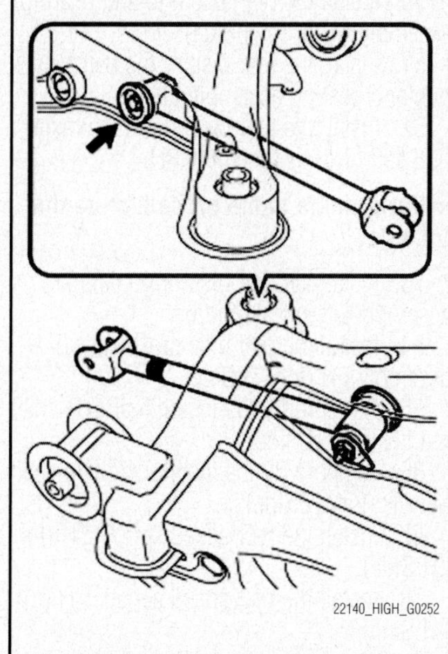

Fig. 387 Rear No. 1 suspension arm assembly LH

➡Ensure that the identification mark faces the rear side of the vehicle.

b. Set the rear No. 1 suspension arm LH in the position shown. Length A: 20 mm (0.787 in.)

c. Fully tighten the bolt to 59 ft. lbs. (80 Nm).

➡Since stopper nuts are used, temporarily tighten the bolts.

32. Install the rear suspension member (for LH Side), as follows:

a. Support the rear suspension member with a jack using a wooden block.

➡Use a properly sized wooden block to keep the jack and suspension member level.

➡Support the suspension member until retightening of the suspension member is complete.

b. Raise the rear suspension member with a jack.

c. Temporarily install the rear suspension member, 2 rear upper suspension member stoppers, and 2 rear lower suspension member stopper retainers with the 4 nuts and 2 bolts.

d. Fully tighten the 2 nuts 85 ft. lbs. (115 Nm).

e. Using SST: 09961-00950 or equivalent and a socket wrench (19 mm), fully tighten the nut (LH side) to 133 ft. lbs. (181 Nm), 98 ft. lbs. (134 Nm).

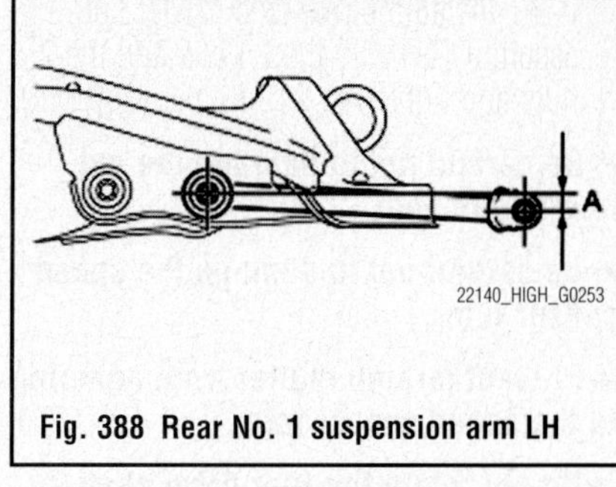

Fig. 388 Rear No. 1 suspension arm LH

f. Using the same tools, fully tighten the nut (RH side) to the same specifications.

➡Use a torque wrench with a fulcrum length of 425 mm (16.73 in.).

➡These torque values are effective when SST is parallel to the torque wrench .

33. Install the rear drive shaft snap ring LH (for LH Side).

34. Install the rear drive shaft assembly LH (for LH Side).

35. Install the rear drive shaft snap ring RH (for LH Side).

36. Install the rear drive shaft assembly RH (for LH Side).

37. Install the rear differential drain plug (for LH Side).

38. Add differential oil (for LH Side).

39. Inspect differential oil (for LH Side).

40. Install the rear differential filler plug (for LH Side).

41. Install the rear axle carrier sub-assembly LH (for LH Side), as follows:

a. Temporarily install the rear axle carrier sub-assembly LH with the bolt and the nut.

b. Install the rear axle carrier sub-assembly LH with the 2 bolts and the 2 nuts and tighten to 213 ft. lbs. (290 Nm).

➡Be careful not to damage the outboard joint boot.

➡Be careful not to damage the speed sensor rotor.

➡Prevent foreign matter from adhering to the speed sensor rotor.

➡Do not rotate the rear drive shaft assembly without the rear axle hub and bearing assembly installed.

➡Insert the bolts from the rear side.

42. Install the rear axle carrier sub-assembly RH (for LH Side), as follows:

a. Temporarily install the rear axle carrier sub-assembly LH with the 2 bolts and the 2 nuts.

b. Install the rear axle carrier sub-assembly LH with the 2 bolts and the 2 nuts and tighten to 213 ft. lbs. (290 Nm).

➡**Be careful not to damage the outboard joint boot.**

➡**Be careful not to damage the speed sensor rotor.**

➡**Prevent foreign matter from adhering to the speed sensor rotor.**

➡**Do not rotate the rear drive shaft assembly without the rear axle hub and bearing assembly installed.**

➡**Insert the bolts from the rear side.**

43. Temporarily install the rear No. 1 suspension arm assembly RH to the rear suspension member sub-assembly with the bolt and the nut.

➡**Ensure that the identification mark faces the rear side of the vehicle.**

44. Temporarily install the rear No. 1 suspension arm assembly RH to the rear axle carrier sub-assembly RH with the bolt and the nut.

➡**Since stopper nuts are used, temporarily tighten the bolts.**

45. Temporarily install the rear strut rod assembly LH (for LH Side).
46. Temporarily install the rear strut rod assembly RH.
47. Temporarily install the rear No. 2 suspension arm assembly LH (for LH Side).
a. Temporarily install the rear No. 2 suspension arm assembly LH to the rear suspension member sub-assembly with the rear suspension toe adjust cam sub-assembly, the No. 2 camber adjust cam and the nut.

➡**Ensure that the identification mark faces the rear side of the vehicle.**

➡**When temporarily tightening the nut, keep the rear suspension toe adjust cam sub-assembly from rotating.**

b. Temporarily install the rear No. 2 suspension arm assembly LH to the rear axle carrier sub-assembly LH with the bolt and the nut.

➡**Since a stopper nut is used, temporarily tighten the bolt.**

48. Temporarily install the rear No. 2 suspension arm assembly RH (for LH Side).

➡**Perform the same procedure as the rear No. 2 suspension arm assembly LH.**

49. Install the rear axle hub and bearing assembly LH (for LH Side).
50. Install the rear axle hub and bearing assembly RH (for LH Side).
51. Install the rear disc to the rear axle hub and bearing assembly LH.
52. Install the rear disc to the rear axle hub and bearing assembly RH.

➡**Perform the same procedure as the LH side.**

53. Install the rear disc brake caliper assembly LH (for LH Side).
54. Install the rear disc brake caliper assembly RH (for LH Side).
55. Temporarily the install rear axle shaft nut LH (for LH Side).
56. Temporarily the install rear axle shaft nut RH (for LH Side).
57. Install the rear speed sensor LH (for LH Side).
58. Install the rear speed sensor RH (for LH Side).
59. Jack up the rear axle carrier sub-assembly, placing a wooden block underneath to avoid damage. Apply load to the suspension so that the rear drive shaft assembly is positioned horizontally.

CAUTION

Do not jack up the rear axle carrier sub-assembly too high as the vehicle may fall.

➡**Do not bend the brake dust cover.**

➡**If the rear drive shaft assembly cannot be positioned horizontally as shown in the illustration even when the rear axle carrier sub-assembly is jacked up, apply additional load to the vehicle such as by having a person sit in the rear seat.**

➡**Use the same procedures for the RH side and LH side.**

60. Fully tighten the rear No. 1 suspension arm assembly LH (for LH Side).
61. Fully tighten the rear No. 1 suspension arm assembly RH (for LH Side).

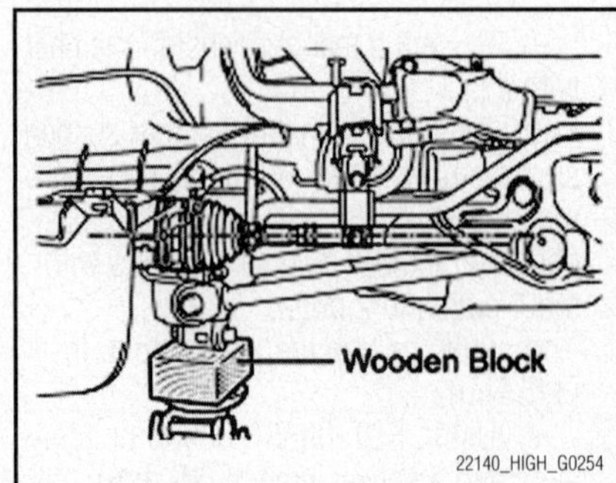

Fig. 389 Jack up the rear axle carrier sub-assembly

➡**Perform the same procedure as the rear No. 1 suspension arm assembly LH.**

62. Fully tighten the rear No. 1 suspension arm assembly RH (for RH Side).
63. Fully tighten the 2 bolts to 59 ft. lbs. (80 Nm), 82 ft. lbs. (112 Nm).

➡**Since a stopper nut is used, temporarily tighten the bolt.**

➡**The final torque must be applied under standard vehicle height conditions.**

64. Fully tighten the rear No. 2 suspension arm assembly LH (for LH Side).
a. Align the matchmarks on the adjust cams and rear suspension member sub-assembly.
b. Fully tighten the nut to 74 ft. lbs. (100 Nm).

➡**The final torque must be applied under standard vehicle height conditions.**

➡**When fully tightening the nut, keep the rear suspension toe adjust cam sub-assembly from rotating.**

c. Fully tighten the bolt to 82 ft. lbs. (112 Nm).

➡**Since a stopper nut is used, temporarily tighten the bolt.**

➡**The final torque must be applied under standard vehicle height conditions.**

65. Fully tighten the rear No. 2 suspension arm assembly RH (for RH Side).

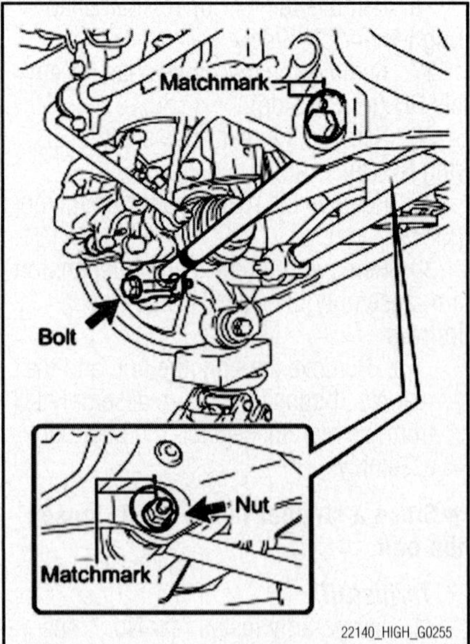

Fig. 390 Align the matchmarks on the adjust cams and rear suspension member sub-assembly

➡**Perform the same procedure as the rear No. 2 suspension arm assembly LH.**

66. Fully tighten rear strut rod assembly LH (for LH Side).
67. Fully tighten rear strut rod assembly RH.
68. Install the No. 3 parking brake cable assembly (for LH Side).
69. Install the No. 2 parking brake cable assembly.
70. Separate the rear disc brake caliper assembly LH (for LH Side).
71. Separate the rear disc brake caliper assembly RH (for LH Side).
72. Remove the rear disc from the rear axle hub and bearing assembly LH.
73. Remove the rear disc from the rear axle hub and bearing assembly RH.

➡**Perform the same procedure as the LH side.**

74. Inspect the rear axle hub bearing looseness.
75. Inspect the rear axle hub bearing runout.

➡**Use the same procedures for the RH side and LH side.**

76. Install the rear disc to the rear axle hub and bearing assembly LH.
77. Install the rear disc to the rear axle hub and bearing assembly RH.

➡**Perform the same procedure as the LH side.**

78. Install the rear disc brake caliper assembly LH (for LH Side).
79. Install the rear disc brake caliper assembly RH (for LH Side).
80. Install the rear axle shaft nut LH (for LH Side).
81. Install the rear axle shaft nut RH (for LH Side).
82. Temporarily tighten the propeller with center bearing shaft assembly (for LH Side).
83. Fully tighten the propeller with center bearing shaft assembly (for LH Side).
84. Inspect and adjust transfer oil (for LH Side).
85. Install the center exhaust pipe assembly.
86. Install the tail exhaust pipe assembly.
87. Inspect for exhaust gas leak
88. Install the wheels and tighten lug nuts to 76 ft. lbs. (103 Nm).
89. Inspect and adjust rear wheel alignment.
90. Check for rear speed sensor signal (for LH Side).

MACPHERSON STRUTS

REMOVAL & INSTALLATION

2007 Models

1. Remove the deck side trim cover.
2. Remove the wheel.
3. Support the rear axle carrier with a jack.
4. Remove the nut and disconnect the stabilizer link from the shock absorber.

➡**If the ball joint turns together with the nut, use a hexagon wrench (5 mm) to hold the stud.**

5. Remove the rear shock absorber with coil spring, as follows:
   a. For 2WD vehicles, disconnect the skid control sensor connector. Remove the 2 bolts, and disconnect the flexible hose and skid control sensor wire from the shock absorber and rear axle carrier.
   b. For 4WD vehicles, remove the 3 bolts, and disconnect the flexible hose and speed sensor from the shock absorber and rear axle carrier.
   c. Loosen the 2 nuts on the lower side of the shock absorber.

➡**Do not remove the 2 bolts and 2 nuts.**

   d. Remove the 3 nuts.
   e. Lower the rear axle carrier, and remove the 2 nuts and 2 bolts on the lower side of the rear shock absorber.
   f. Remove the shock absorber with coil spring.

To install:

6. Install the rear shock absorber with coil spring and 3 nuts and tighten to 43 ft. lbs. (58 Nm).
7. Install the 2 bolts and nuts to the shock absorber with coil spring and tighten to 133 ft. lbs. (180 Nm).

➡**When installing the nuts, keep the bolts from rotating.**

8. For 2WD vehicles, install the flexible hose and skid control sensor wire with the 2 bolts. Connect the skid control sensor connector and tighten to 14 ft. lbs. (19 Nm), 44 inch lbs. (5 Nm).
9. For 4WD vehicles, install the flexible hose and speed sensor wire the 3 bolts and tighten to 44 inch lbs. (5 Nm), 14 ft. lbs. (19 Nm), 71 inch lbs. (8 Nm).
10. Install the stabilizer link to the shock absorber with the nut and tighten to 29 ft. lbs. (39 Nm).

➡**If the ball joint turns together with the nut, use a hexagon wrench (5 mm) to hold the stud.**

11. Install the wheels and tighten lug nuts to 76 ft. lbs. (103 Nm).
12. Inspect and adjust rear wheel alignment.
13. Check the ABS speed sensor signal.

2008 Models

See Figures 391 and 392.

1. Remove the wheel.
2. Remove the deck side trim cover.
3. Remove the deck side trim.
4. For 4WD vehicles, remove the bolt and separate the rear flexible hose from the rear shock absorber with coil spring.
5. Remove the bolt and separate the rear speed sensor from the rear shock absorber with coil spring.
6. Remove the nut and separate the rear stabilizer link assembly from the rear shock absorber with coil spring.

➡**If the ball joint turns together with the nut, use a hexagon wrench (5 mm) to hold the stud bolt.**

7. Disengage the 4 claws and remove the rear No. 1 suspension support cover.
8. Using a jack and wooden block, support the rear axle carrier sub-assembly.

➡**Do not deform the dust cover.**

➡**Support the rear axle carrier sub-assembly until reinstallation of the rear shock absorber with coil spring is complete.**

9. Loosen the rear support to rear shock absorber nut.

➡**Do not remove the rear support to rear shock absorber nut.**

➡**Loosen the nut only when the rear shock absorber with coil spring needs to be disassembled.**

10. Remove the 2 bolts and 2 nuts, and separate the rear shock absorber with coil spring from the rear axle carrier sub-assembly,

➡**When removing the nuts, keep the bolts from rotating.**

11. Remove the 3 nuts and rear shock absorber with coil spring.

➡**Make sure that the rear speed sensor and rear flexible hose are disconnected from the rear shock absorber with coil spring.**

12. Remove the rear support to rear shock absorber nut, as follows:
   a. Using SST: 09727-30021, compress the rear coil spring.

➡Do not use an impact wrench. It will damage SST.

b. Check that the front coil spring is fully compressed.

c. Hold the rear suspension support assembly and remove the rear support to rear shock absorber nut from the rear shock absorber assembly.

13. Remove the rear support to rear shock absorber collar from the rear shock absorber assembly.

14. Remove the rear suspension support assembly from the rear shock absorber assembly.

15. Remove the rear coil spring together with SST from the rear shock absorber assembly.

16. Remove the rear No. 1 spring bumper from the rear shock absorber assembly.

17. Remove the rear lower coil spring insulator from the rear shock absorber assembly.

To install:

18. Install the rear lower coil spring insulator onto the rear shock absorber assembly.

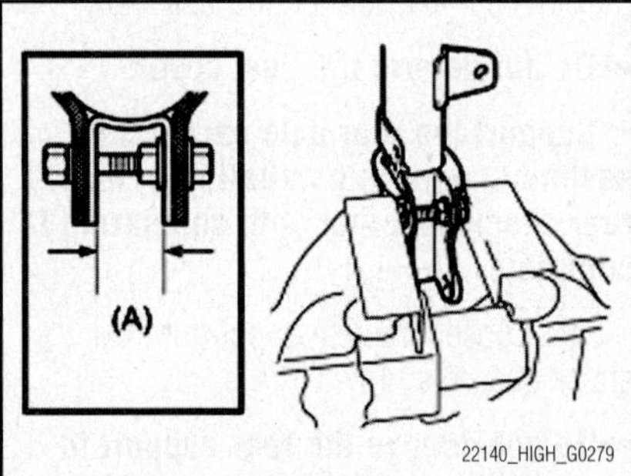

Fig. 391 Install the bolt and nut to the rear shock absorber with coil spring—2WD vehicles

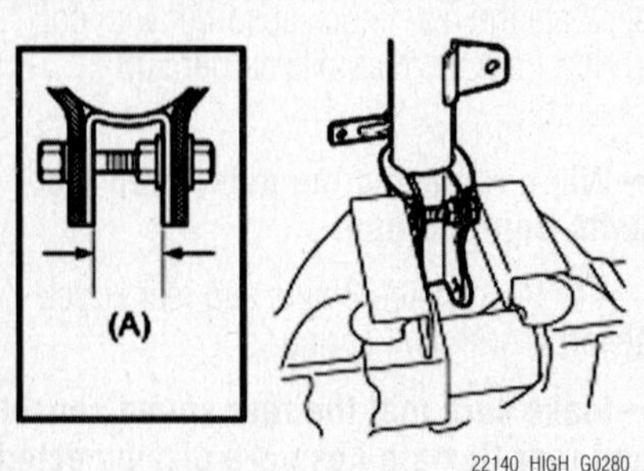

Fig. 392 Install the bolt and nut to the rear shock absorber with coil spring—4WD vehicles

➡Fit the recessed part of the rear lower coil spring insulator into the recession on the shock absorber assembly.

19. Install the rear No. 1 spring bumper to the rear shock absorber assembly.

20. Temporarily install rear coil spring, as follows:

a. Using SST: 09727-30021 or equivalent, compress the rear coil spring.

➡Do not use an impact wrench. It will damage the SST.

b. Temporarily install the rear coil spring together with SST to the rear shock absorber assembly.

21. Install the rear suspension support assembly to the rear shock absorber assembly.

➡Align the cutout on the rear shock absorber assembly with the protrusion on the rear suspension support assembly by referring to the illustration.

22. Install the rear support to rear shock absorber collar to the rear shock absorber assembly.

23. Temporarily install the rear support to rear shock absorber nut to the rear shock absorber assembly.

24. Install the rear coil spring.

➡Do not use an impact wrench. It will damage the SST.

➡Make sure that the end of the rear coil spring is positioned in the depression of the rear lower coil spring insulator.

➡Ensure that the stud bolt is positioned 3.5° to the outside of the vehicle as shown in the illustration. The deviation should be within +-5°.

25. Install the rear shock absorber with coil spring with the 3 nuts and tighten to 43 ft. lbs. (58 Nm).

26. Install the rear shock absorber with coil spring with the 2 bolts and 2 nuts and tighten to 213 ft. lbs. (290 Nm).

➡When installing the nuts, keep the bolts from rotating.

27. Fully tighten the rear support to rear shock absorber nut 40 ft. lbs. (55 Nm).

28. Install the rear No. 1 suspension support cover.

29. Install the rear stabilizer link assembly to the rear shock absorber with coil spring with the nut and tighten to 29 ft. lbs. (39 Nm).

➡If the ball joint turns together with the nut, use a hexagon wrench (5 mm) to hold the stud bolt.

30. Install the rear speed sensor wire to the rear shock absorber with coil spring with the bolt and tighten to 44 inch lbs. (5 Nm).

➡Do not twist the rear speed sensor wire when installing it.

31. Install the rear flexible hose to the rear shock absorber with coil spring with the bolt and tighten to 14 ft. lbs. (19 Nm).

➡Do not twist the rear flexible hose when installing it.

32. Install deck side trim.

33. Install deck side trim cover.

34. Install the wheels and tighten lug nuts to 76 ft. lbs. (103 Nm).

35. Inspect and adjust rear wheel alignment.

WHEEL HUB AND BEARING

REMOVAL & INSTALLATION

1. Disconnect the negative battery cable.
2. Remove the wheel.
3. Separate the rear flexible hose.
4. Remove the 2 bolts and separate the rear disc brake caliper assembly.
5. Remove the rear disc.
6. Using a screwdriver, disconnect the connector from the rear speed sensor.
7. For 4WD vehicles,
8. Remove the 4 bolts and the rear axle hub & bearing assembly.

To install:

9. Install the rear axle hub and bearing assembly with the 4 bolts and tighten to 55 ft. lbs. (75 Nm).
10. Inspect rear axle hub bearing looseness.
11. Inspect rear axle hub runout.
12. Connect the connector to the rear speed sensor.
13. Install the rear disc.
14. Install the rear disc brake caliper assembly with the 2 bolts and tighten to 57 ft. lbs. (78 Nm).
15. Install the rear flexible hose to the shock absorber with coil spring with the bolt and tighten to 14 ft. lbs. (19 Nm).
16. Install the wheel and tighten lug nuts to 76 ft. lbs. (103 Nm).
17. Connect the negative battery cable.
18. Inspect and adjust rear wheel alignment.
19. Check the ABS speed sensor signal.

TOYOTA

Highlander Hybrid

16

SPECIFICATIONS AND MAINTENANCE CHARTS

ENGINE AND VEHICLE IDENTIFICATION

Engine						
Code ①	Liters (cc)	Cu. In.	Cyl.	Fuel Sys.	Engine Type	Eng. Mfg.
3MZ-FE	3.3 (3311)	202.1	6	SFI	DOHC	Toyota

Model Year	
Code ②	Year
6	2006
7	2007
8	2008

SFI: Sequential Fuel Injection

DOHC: Double Overhead Camshaft

① Stamped on the left side of the engine block

② 10th digit of the Vehicle Identification Number (VIN)

22140_HYBR_C0001

GENERAL ENGINE SPECIFICATIONS

Year	Model	Engine Displacement Liters	Engine Series ID	Net Horsepower @ rpm	Net Torque @ rpm (ft. lbs.)	Bore x Stroke (in.)	Com-pression Ratio	Oil Pressure @ rpm
2006	Highlander Hybrid	3.3	3MZ-FE	268@5600	212@4400	3.62x3.27	10.8:1	36-78@3000
2007	Highlander Hybrid	3.3	3MZ-FE	268@5600	212@4400	3.62x3.27	10.8:1	36-78@3000
2008	Highlander Hybrid	3.3	3MZ-FE	268@5600	212@4400	3.62x3.27	10.8:1	36-78@3000

22140_HYBR_C0002

ENGINE TUNE-UP SPECIFICATIONS

Year	Engine Displacement Liters	Engine ID	Spark Plug Gap (in.)	Ignition Timing (deg.)	Fuel Pump (psi)	Idle Speed (rpm)	Valve Clearance	
							Intake	Exhaust
2006	3.3	3MZ-FE	0.039-0.043	8-12B①	44-50	850-950	0.006-0.010	0.010-0.014
2007	3.3	3MZ-FE	0.039-0.043	8-12B①	44-50	850-950	0.006-0.010	0.010-0.014
2008	3.3	3MZ-FE	0.039-0.043	8-12B①	44-50	850-950	0.006-0.010	0.010-0.014

NOTE: The Vehicle Emission Control Information label often reflects specification changes made during production.

The label figures must be used if they differ from those in this chart.

B: Before top dead center

① With terminals TC and CG of DLC3 connected

22140_HYBR_C0003

CAPACITIES

Year	Model	Engine Displacement Liters	Engine ID	Engine Oil with Filter (qts.)	Transaxle (pts)	Rear Transaxle (pts.)	Rear Drive Axle (pts.)	Fuel Tank (gal.)	Cooling System (qts.)
2006	Highlander Hybrid	3.3	3MZ-FE	5.0	①	4.2	N/A	17.2	②
2007	Highlander Hybrid	3.3	3MZ-FE	5.0	①	4.2	N/A	17.2	②
2008	Highlander Hybrid	3.3	3MZ-FE	5.0	①	4.2	N/A	17.2	②

① With towing package: 8.8 pts.
Without towing package: 8.2 pts.

② With Rear Heater: 12.8 qts.
Without Rear Heater: 10.6 qts.

22140_HYBR_C0005

FLUID SPECIFICATIONS

Year	Model	Engine Displacement Liters	Engine ID/VIN	Engine Oil	Auto. Trans. ①	Drive Axle ②	Power Steering Fluid	Brake Master Cylinder	Engine Coolant ③
2006	Highlander Hybrid	3.3	3MZ-FE	5W-30	ATF-WS	75W-90	NA	DOT 3	Toyota coolant
2007	Highlander Hybrid	3.3	3MZ-FE	5W-30	ATF-WS	75W-90	NA	DOT 3	Toyota coolant
2008	Highlander Hybrid	3.3	3MZ-FE	5W-30	ATF-WS	75W-90	NA	DOT 3	Toyota coolant

DOT: Department Of Transpotation

NA: Not Available

① The use of genuine Toyota ATF-WS is recommended

② Synthetic GL-5 (75W-90) or equivalent

③The use of genuine Toyota engine coolant is recommended or similar ethylene glycol based non-silicate, non-amine, non- nitrite, and non- borat coolant

22140_HYBR_C0004

VALVE SPECIFICATIONS

All measurements are given in inches.

Year	Engine Displacement Liters	Engine ID	Seat Angle (deg.)	Face Angle (deg.)	Spring Test Pressure (lbs. @ in.)	Spring Installed Height (in.)	Stem-to-Guide Clearance (in.)		Stem Diameter (in.)	
							Intake	Exhaust	Intake	Exhaust
2006	3.3	3MZ-FE	45	40.5	41.9-46.3@ 1.331	1.331	0.0010-0.0024	0.0012-0.0026	0.2154-0.2159	0.2152 0.2157
2007	3.3	3MZ-FE	45	40.5	41.9-46.3@ 1.331	1.331	0.0010-0.0024	0.0012-0.0026	0.2154-0.2159	0.2152 0.2157
2008	3.3	3MZ-FE	45	40.5	41.9-46.3@ 1.331	1.331	0.0010-0.0024	0.0012-0.0026	0.2154-0.2159	0.2152 0.2157

22140_HYBR_C0006

CAMSHAFT AND BEARING SPECIFICATIONS CHART

All measurements are given in inches.

Year	Engine Displ. Liters	Engine ID/VIN	Journal Dia.	Brg. Oil Clearance	Shaft End-play	Runout	Journal Bore	Lobe Height Intake	Lobe Height Exhaust
2006	3.3	3MZ-FE	1.0614-1.0620	①	0.0016-0.0035	0.0024	NA	1.6981-1.7020	1.6933-1.6972
2007	3.3	3MZ-FE	1.0614-1.0620	①	0.0016-0.0035	0.0024	NA	1.6981-1.7020	1.6933-1.6972
2008	3.3	3MZ-FE	1.0614-1.0620	①	0.0016-0.0035	0.0024	NA	1.6981-1.7020	1.6933-1.6972

NA: Not Available

① Intake Journals 4 and 5: 0.0010 - 0.0022 in.
All Others: 0.0010 - 0.0024 in.

22140_HYBR_C0008

CRANKSHAFT AND CONNECTING ROD SPECIFICATIONS

All measurements are given in inches.

Year	Engine Displacement Liters	Engine ID	Crankshaft Main Brg. Journal Dia.	Crankshaft Main Brg. Oil Clearance	Crankshaft Shaft End-play	Crankshaft Thrust on No.	Connecting Rod Journal Diameter	Connecting Rod Oil Clearance	Connecting Rod Side Clearance
2006	3.3	3MZ-FE	2.4011-2.4016	①	0.0016-0.0094	2	2.0863-2.0866	0.0015-0.0026	0.0059-0.0118
2007	3.3	3MZ-FE	2.4011-2.4016	①	0.0016-0.0094	2	2.0863-2.0866	0.0015-0.0026	0.0059-0.0118
2008	3.3	3MZ-FE	2.4011-2.4016	①	0.0016-0.0094	2	2.0863-2.0866	0.0015-0.0026	0.0059-0.0118

① Journals 1 and 4: 0.0006 - 0.0013 in.
Journals 2 and 3: 0.0010 - 0.0018 in.

22140_HYBR_C0007

PISTON AND RING SPECIFICATIONS

All measurements are given in inches.

Year	Engine Displ. Liters	Engine ID	Piston Clearance	Ring Gap Top Comp.	Ring Gap Bottom Comp.	Ring Gap Oil Control	Ring Side Clearance Top Comp.	Ring Side Clearance Bottom Comp.	Ring Side Clearance Oil Control
2006	3.3	3MZ-FE	0.0013-0.0023	0.0118-0.0157	0.0197-0.0236	0.0059-0.0157	0.0012-0.0031	0.0008-0.0024	0.0012-0.0043
2007	3.3	3MZ-FE	0.0013-0.0023	0.0118-0.0157	0.0197-0.0236	0.0059-0.0157	0.0012-0.0031	0.0008-0.0024	0.0012-0.0043
2008	3.3	3MZ-FE	0.0013-0.0023	0.0118-0.0157	0.0197-0.0236	0.0059-0.0157	0.0012-0.0031	0.0008-0.0024	0.0012-0.0043

22140_HYBR_C0009

TORQUE SPECIFICATIONS

All readings in ft. lbs.

Year	Engine Displacement Liters	Engine ID	Cylinder Head Bolts	Main Bearing Bolts	Rod Bearing Bolts	Crankshaft Damper Bolts	Flywheel Bolts	Manifold Intake	Manifold Exhaust	Spark Plugs	Oil Pan Drain Plug
2006	3.3	3MZ-FE	①	②	③	162	61	11	36	18	33
2007	3.3	3MZ-FE	①	②	③	162	61	11	36	18	33
2008	3.3	3MZ-FE	①	②	③	162	61	11	36	18	33

① Step 1: 12 point bolts to 40 ft. lbs.
Step 2: 12 point bolts plus 90 degrees
Step 3: Hex head recessed bolt to 14 ft. lbs.

② Step 1: 12 point cap bolts to 16 ft. lbs.
Step 2: 12 point cap bolts plus 90 degrees
Step 3: Hex head side bolts to 20 ft. lbs.

③ Step 1: 18 ft. lbs.
Step 2: Plus 90 degrees

22140_HYBR_C0010

WHEEL ALIGNMENT

Year	Model		Caster Range (+/-Deg.)	Caster Preferred Setting (Deg.)	Camber Range (+/-Deg.)	Camber Preferred Setting (Deg.)	Toe-in (in.)
2006	Highlander	2WD F	0.75	+2.75	0.75	-0.58	0+/-0.08
	Hybrid	4WD F	0.75	+2.50	0.75	-0.58	0+/-0.08
		2WD R	—	—	0.75	-1.17	0.12+/-0.08
		4WD R	—	—	0.75	-0.67	0.12+/-0.08
2007	Highlander	2WD F	0.75	+2.75	0.75	-0.58	0+/-0.08
	Hybrid	4WD F	0.75	+2.50	0.75	-0.58	0+/-0.08
		2WD R	—	—	0.75	-1.17	0.12+/-0.08
		4WD R	—	—	0.75	-0.67	0.12+/-0.08
2008	Highlander	2WD F	0.75	+2.75	0.75	-0.58	0+/-0.08
	Hybrid	4WD F	0.75	+2.50	0.75	-0.58	0+/-0.08
		2WD R	—	—	0.75	-1.17	0.12+/-0.08
		4WD R	—	—	0.75	-0.67	0.12+/-0.08

F: Front

R: Rear

22140_HYBR_C0011

TIRE, WHEEL AND BALL JOINT SPECIFICATIONS

Year	Model	OEM Tires		Tire Pressures (psi)		Wheel	Ball Joint	Lug Nut
		Standard	Optional	Front	Rear	Size	Inspection	Torque (ft. lbs.)
2006	Highlander Hybrid	P225/65R17	N/A	32	32	6.5-J	①	76
2007	Highlander Hybrid	P225/65R17	N/A	32	32	6.5-J	①	76
2008	Highlander Hybrid	P225/65R17	N/A	32	32	6.5-J	①	76

OEM: Original Equipment Manufacturer

PSI: Pounds Per Square Inch

STD: Standard

OPT: Optional

① Replace if any measurable movement is found.

22140_HYBR_C0012

BRAKE SPECIFICATIONS

All measurements in inches unless noted

Year	Model		Brake Disc			Minimum	Brake Caliper	
			Original Thickness	Minimum Thickness	Maximum Runout	Lining Thickness	Bracket Bolts (ft. lbs.)	Mounting Bolts (ft. lbs.)
2006	Highlander Hybrid	F	1.102	1.024	0.0020	0.039	78	25
		R	0.394	0.335	0.0059	0.039	56	32
2007	Highlander Hybrid	F	1.102	1.024	0.0020	0.039	78	25
		R	0.394	0.335	0.0059	0.039	56	32
2008	Highlander Hybrid	F	1.102	1.024	0.0020	0.039	78	25
		R	0.394	0.335	0.0059	0.039	56	32

F: Front

R: Rear

22140_HYBR_C0013

SCHEDULED MAINTENANCE INTERVALS

2006-08 TOYOTA Highlander Hybrid

TO BE SERVICED	TYPE OF SERVICE	VEHICLE MILEAGE INTERVAL (x1000)												
		5	10	15	20	25	30	35	40	45	50	55	60	65
Engine oil & filter	R	✓	✓	✓	✓	✓	✓	✓	✓	✓	✓	✓	✓	✓
Automatic transmission fluid	R												✓	
Ball joints & dust covers	S/I	✓	✓	✓	✓	✓	✓	✓	✓	✓	✓	✓	✓	✓
Bolts & nuts on chassis & body	S/I	✓	✓		✓	✓		✓	✓		✓	✓		✓
Brake line pipes & hoses	S/I			✓			✓			✓			✓	
Brake fluid	R						✓						✓	
Brake pads & discs (front & rear)	S/I	✓	✓	✓	✓	✓	✓	✓	✓	✓	✓	✓	✓	✓
Propeller shaft grease	S/I	✓	✓	✓	✓	✓	✓	✓	✓	✓	✓	✓	✓	✓
Steering knuckle & chassis grease	S/I	✓	✓	✓	✓	✓	✓	✓	✓	✓	✓	✓	✓	✓
Steering linkage	S/I	✓	✓	✓	✓	✓	✓	✓	✓	✓	✓	✓	✓	✓
Air cleaner filter	R						✓						✓	
Air conditioner filter	R						✓							
Spark plugs	R	Replace at 120,000 miles												
Exhaust pipes & mountings	S/I			✓			✓			✓			✓	
Fuel lines & connections	S/I												✓	
Engine coolant	R	Replace at 120,000 miles												
Timing belt	R	Replace at 90,000 miles												
Rear differential fluid	R												✓	
Rotate tires	S/I	✓	✓	✓	✓	✓	✓	✓	✓	✓	✓	✓	✓	✓

R: Replace S/I: Service or Inspect

FREQUENT OPERATION MAINTENANCE (SEVERE SERVICE)

If a vehicle is operated under any of the following conditions it is considered severe service:

- Extremely dusty areas.

- 50% or more of the constant operation is in 32°C (90°F) or higher temperatures, or in temperatures below 0°C (32°F).

- Prolonged idling (vehicle operation in stop and go traffic).

- Frequent short running periods (engine does not warm to normal operating temperatures).

- Police, taxi, delivery usage or trailer towing usage.

Air cleaner filter: service or inspect every 3750 miles

Engine oil & filter: replace every 3750 miles.

Ball joints & dust covers: service or inspect every 7500 miles.

Bolts & nuts on chassis & body: service or inspect every 7500 miles.

Brake pads & discs (front & rear): service or inspect every 7500 miles.

Steering knuckle & chassis grease: service or inspect every 7500 miles.

Steering linkage: service or inspect every 7500 miles.

Exhaust pipes & mountings: service or inspect every 15,000 miles.

22140_HYBR_C0014

PRECAUTIONS

CAUTION

The HIGHLANDER HV has a hybrid system that operates at voltages up to 650 volts. Be sure to follow the instructions in this manual to handle the system correctly. Failure to do so may result in serious injury or electrocution. Engineer must undergo special training to be able to perform high-voltage system inspection and servicing.

CAUTION

All high-voltage wire harness connectors are colored orange. The HV battery and other high-voltage components have "High Voltage" caution labels. Do not carelessly touch these wires and components.

CAUTION

Before inspecting or servicing the high-voltage system, be sure to follow safety measures, such as wearing insulated gloves and removing the service plug to prevent electrocution. Carry the removed service plug in your pocket to prevent anyone from reinstalling it while you are servicing the vehicle.

CAUTION

After removing the service plug, wait 5 minutes before touching any of the high-voltage connectors and terminals.

➡Before servicing the vehicle, refer to the Hybrid System (HV) section for all Hybrid related precautions and important system information.

Before servicing any vehicle, please be sure to read all of the following precautions, which deal with personal safety, prevention of component damage, and important points to take into consideration when servicing a motor vehicle:

- Never open, service or drain the radiator or cooling system when the engine is hot; serious burns can occur from the steam and hot coolant.
- Observe all applicable safety precautions when working around fuel. Whenever servicing the fuel system, always work in a well-ventilated area. Do not allow fuel spray or vapors to come in contact with a spark, open flame, or excessive heat (a hot drop light, for example). Keep a dry chemical fire extinguisher near the work area. Always keep fuel in a container specifically designed for fuel storage; also, always properly seal fuel containers to avoid the possibility of fire or explosion. Refer to the additional fuel system precautions later in this section.
- Fuel injection systems often remain pressurized, even after the engine has been turned **OFF**. The fuel system pressure must be relieved before disconnecting any fuel lines. Failure to do so may result in fire and/or personal injury.
- Brake fluid often contains polyglycol ethers and polyglycols. Avoid contact with the eyes and wash your hands thoroughly after handling brake fluid. If you do get brake fluid in your eyes, flush your eyes with clean, running water for 15 minutes. If eye irritation persists, or if you have taken brake fluid internally, IMMEDIATELY seek medical assistance.
- The EPA warns that prolonged contact with used engine oil may cause a number of skin disorders, including cancer. You should make every effort to minimize your exposure to used engine oil. Protective gloves should be worn when changing oil. Wash your hands and any other exposed skin areas as soon as possible after exposure to used engine oil. Soap and water, or waterless hand cleaner should be used.
- All new vehicles are now equipped with an air bag system, often referred to as a Supplemental Restraint System (SRS) or Supplemental Inflatable Restraint (SIR) system. The system must be disabled before performing service on or around system components, steering column, instrument panel components, wiring and sensors. Failure to follow safety and disabling procedures could result in accidental air bag deployment, possible personal injury and unnecessary system repairs.
- Always wear safety goggles when working with, or around, the air bag system. When carrying a non-deployed air bag, be sure the bag and trim cover are pointed away from your body. When placing a non-deployed air bag on a work surface, always face the bag and trim cover upward, away from the surface. This will reduce the motion of the module if it is accidentally deployed. Refer to the additional air bag system precautions later in this section.
- Clean, high quality brake fluid from a sealed container is essential to the safe and proper operation of the brake system. You should always buy the correct type of brake fluid for your vehicle. If the brake fluid becomes contaminated, completely flush the system with new fluid. Never reuse any brake fluid. Any brake fluid that is removed from the system should be discarded. Also, do not allow any brake fluid to come in contact with a painted surface; it will damage the paint.
- Never operate the engine without the proper amount and type of engine oil; doing so WILL result in severe engine damage.
- Timing belt maintenance is extremely important. Many models utilize an interference-type, non-freewheeling engine. If the timing belt breaks, the valves in the cylinder head may strike the pistons, causing potentially serious (also time-consuming and expensive) engine damage. Refer to the maintenance interval charts for the recommended replacement interval for the timing belt, and to the timing belt section for belt replacement and inspection.
- Disconnecting the negative battery cable on some vehicles may interfere with the functions of the on-board computer system(s) and may require the computer to undergo a relearning process once the negative battery cable is reconnected.
- When servicing drum brakes, only disassemble and assemble one side at a time, leaving the remaining side intact for reference.
- Only an MVAC-trained, EPA-certified automotive technician should service the air conditioning system or its components.

The skid control ECU communicates with the hybrid vehicle control ECU and controls braking force according to the motor's regenerative force.

The skid control ECU sends enhanced VSC signals to the hybrid vehicle control ECU and inputs operating signals from the hybrid vehicle control ECU.

The skid control ECU uses CAN communication for communication with the hybrid vehicle control ECU. If a communication malfunction is memorized, the skid control ECU prohibits enhanced VSC operation and a part of electronically controlled brake system control by fail safe function.

GENERAL INFORMATION

Troubleshooting precautions are as follows:

- When there is a malfunction with terminal contact points or part installation problems, removal and installation of the suspected problem parts may return the system to the normal condition either completely or temporarily.
- In order to determine the malfunctioning area, be sure to check the conditions at the time the malfunction occurred, such as DTC output and the freeze frame data, and record it before disconnecting each connector or removing and installing parts.
- Since the system may be influenced by malfunctions in systems other than the brake control system, be sure to check for Daces in other systems.

Handling precautions are as follows:

- Do not remove or install the Electronically Controlled Brake parts such as the steering sensor, yaw rate sensor or brake pedal stroke sensor except when required, as they cannot be adjusted correctly after removal or installation.
- Be sure to perform preparation before work and confirmation after work is completed by following the direction in the repair manual when working on the Electronically Controlled Brake system.
- Be sure to remove and install the ECU, actuator, each sensor, etc. with the ignition switch off unless it is not specified in the inspection procedure.
- Be sure to remove the 2 main relays before removal and installation, or replacement of the Electronically Controlled Brake parts.
- The removal or installation of the actuator, master cylinder or stroke simulator as well as some other procedures can cause the fluid level to drop below the fluid reservoir port. If this happens when performing such work, be sure to remove the 2 motor relays until the bleeding of the air in the pipeline is completed.

➡When the pump motor is operated with the air in the brake actuator hose, bleeding the air becomes difficult due to air in the actuator. The skid control ECU may operate the stroke simulator and drive the pump motor even when the ignition switch is off. With the ignition switch off, the skid control ECU can be operated for 2 minutes after the brake operation is finished.

- Removal of the main relay and motor relay.
- Wait for 2 minutes after turning the ignition switch off, stopping the brake pedal operation and closing the driver door before removing the 2 relays.
- When removing and installing the ECU, actuator and each sensor, be sure to check that the normal display is output in test mode inspection and in DTC output inspection after installing all the parts.

BLEEDING PROCEDURE

See Figures 1 through 3.

➡This procedure requires specialized tools. Please read through the procedure and make sure you have access to the proper equipment before beginning the bleeding procedure.

⁂ CAUTION

Never bleed air from the brake hydraulic system without using the intelligent tester. failure to use the intelligent tester could cause serious injury or an accident.

Note the following before bleeding the brake system:

- Move the shift lever to the P position and apply the parking brake before bleeding.
- Add brake fluid carefully and check that the reservoir level remains between the min and max lines while bleeding the brakes.
- Do not stand the fluid can on the reservoir inlet when bleeding the brake actuator. doing so will cause brake fluid to overflow.
- The actuator pump motor and solenoid can be operated by the driver even if the ignition switch is off.
- If the pump motor operates while air still remains inside the brake actuator hose, air will enter the actuator, making it more difficult to bleed the brakes. If there is concern about air remaining in the actuator hose, remove the two motor relays (skid control relay no.2) until instructed to reinstall them.
- Although a buzzer may sound due to a decline in the accumulator pressure while bleeding, it is not necessary to stop bleeding.
- Daces indicating a malfunction in the motor relays (skid control relay no.2) or the pressure sensor are stored after bleeding. clear the Daces when instructed during or after bleeding.

1. Add SAE J1703 or FMVSS no. 116 DOT 3 brake fluid to the max line in the reservoir.

⁂ CAUTION

Add brake fluid carefully and check that the reservoir level remains between the min and max lines while bleeding the brakes. Do not stand the fluid can on the reservoir inlet when bleeding the brake actuator. Doing so will cause brake fluid to overflow.

2. Disable the brake control (ECB). When using the intelligent tester:

➡When using the intelligent tester , refer to the intelligent tester operator's manual for further details. Bleed the air by following the steps displayed on the intelligent tester.

a. Move the shift lever to the P position and apply the parking brake.

b. Connect the intelligent tester to the DLC3 with the ignition switch **OFF** as shown in the illustration.

c. Turn the ignition switch to the ON position and turn on the intelligent tester.

➡Do not start the engine.

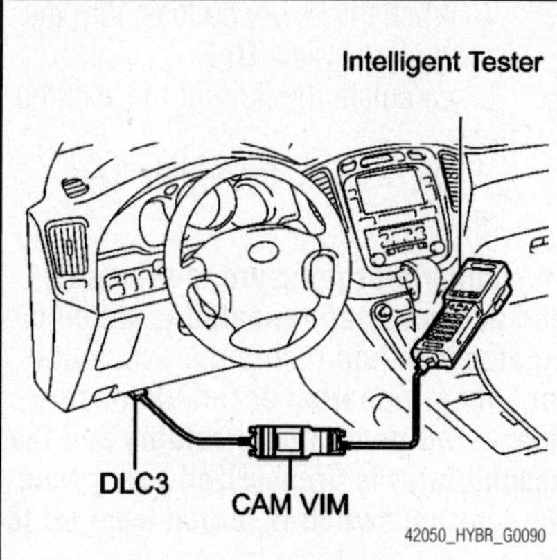

Fig. 1 Connect the intelligent tester to the DLC3 with the ignition switch OFF as shown in the illustration

d. Enter the following menus: DIAGNOSIS / OBD/ MOBD / ABS/ TRAC/VSC / ECB UTILITY / ECB INVALID.

WARNING

If the pump motor operates while air remains inside the brake actuator hose, air will enter the actuator, and this will make bleeding the brakes more difficult.

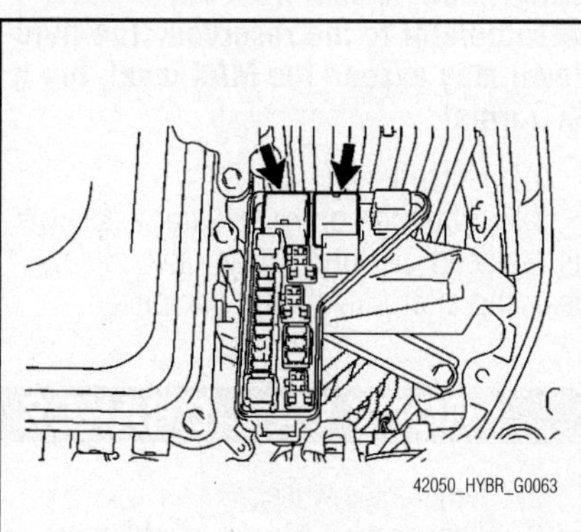

Fig. 2 Location of the 2 ABS motor relays

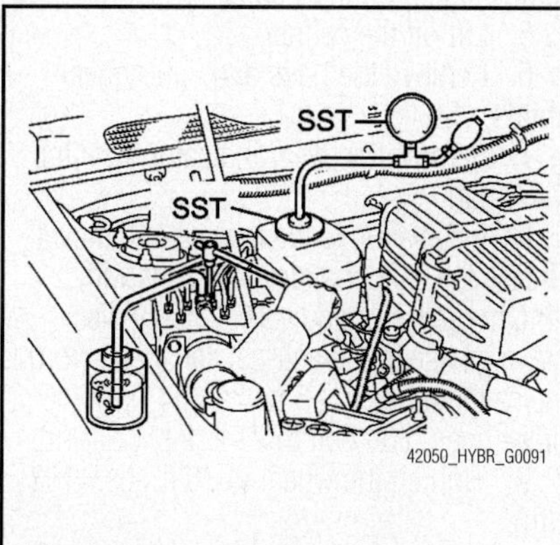

Fig. 3 View of the special tools

e. When removing the ABS motor relay: Remove the 2 ABS motor relays with the ignition switch off in order to disable brake control.

WARNING

If the pump motor operates while air remains inside the brake actuator hose, air will enter the actuator, and this will make bleeding the brakes more difficult.

➡After the brake actuator assembly has been replaced, remove the ABS motor relay before bleeding the brakes.

3. **Bleed the brake actuator hose, as follows:**

a. Connect Special Tool 09992-00242, 0992-00350 or equivalent, to the reservoir with the brake reservoir pressure adapter.

b. Using Special Tool 09023-00101, loosen the bleeder plug of the actuator.

c. Connect a vinyl tube to the bleeder plug of the actuator.

d. Use the SST to boost pressure in the reservoir. Standard pressure is 50 to 80 kPa (0.5 to 0.8 kgf/cm2, 7.3 to 11.6 psi)

e. Drain approximately 100 cc of fluid.

f. Tighten the bleeder plug and boost the pressure in the reservoir again (50 to 80 kPa (0.5 to 0.8 kgf/cm2)). Then, loosen the bleeder plug and bleed the brake actuator hose.

➡Repeat this procedure at least 5 times.

g. When air is completely bled out from the hose between the reservoir and the actuator, tighten the bleeder plug to 74 inch lbs. (8.3 Nm).

4. **Bleed the master cylinder, as follows:**

➡If the master cylinder has been disassembled or if the reservoir becomes empty, bleed the air from the master cylinder.

a. Enter the following menus: DIAGNOSIS / OBD/MOBD / ABS/TRAC/VSC / AIR BLEEDING.

b. Select "USUAL" if the front/rear brakes are removed, installed or disassembled.

c. Select "ACTUATOR" if the actuator is removed, installed or replaced.

d. Select "MASTER CYLINDER" if the brake master cylinder or the brake stroke simulator is removed, installed or replaced.

e. Disconnect the brake lines from the master cylinder.

f. Slowly depress and hold the brake pedal (Procedure A).

g. Cover the outer holes with fingers, and release the brake pedal (Procedure B).

h. Repeat procedure A and B 3 or 4 times.

i. Connect the brake lines to the master cylinder and tighten to 11 ft. lbs. (15 Nm).

5. **Bleed the front brake system, as follows:**

➡Air can be easily bled from the front brake system if air has been bled from the master cylinder when replacing the brake master cylinder assembly.

WARNING

If brake fluid leaks onto any painted surface of the vehicle, wash or otherwise remove it completely.

➡Bleed the air by following the steps displayed on the intelligent tester.
(a) Depress the brake pedal several times and bleed the front brake system from the bleeder plugs on the front brake cylinder RH and LH.

➡Repeat the procedure until air is completely bled from the front brake system.

6. Tighten the bleeder plugs to 74 inch lbs. (8.3 Nm) after bleeding.

7. Cancel brake control (ECB) disable

a. Install the 2 motor relays (skid control relay No.2) if they have been removed.

b. Complete brake control prevention following the prompts on the tester screen. (If brake control has been prevented using the intelligent tester.)

8. Clear the DTC(s).

9. Bleed the rear brake system, as follows:

WARNING

Never bleed air from the brake hydraulic system without using the intelligent tester. Failure to use the intelligent tester could cause serious injury or an accident.

➡Bleed the air by following the steps displayed on the intelligent tester.

a. Connect the intelligent tester to the DLC3 with the ignition switch off.

b. Check that the parking brake is applied and turn the ignition switch to the **ON** position.

c. Enter the following menus: DIAGNOSIS / OBD/ MOBD / ABS/ TRAC/VSC / ECB UTILITY / ECB INVALID.

d. With the brake pedal depressed, bleed the rear brake system from the bleeder plug on the rear disc brake cylinder LH while the pump motor and solenoid are operating.

WARNING

Keep the fluid inside the reservoir above the LOW level by replenishing.

➡Depress and hold the brake pedal. After the solenoid operates for approximately 30 seconds, release the brake pedal to stop the solenoid. Repeat the procedures until air is completely bled from the rear brake system. The ECB warning light comes on and the buzzer sounds while bleeding, but they do not indicate a malfunction.

e. Tighten the bleeder plug to 74 inch lbs. (8.3 Nm) after bleeding.

f. Enter the following menus: DIAGNOSIS / OBD/ MOBD / ABS/ TRAC/VSC / ECB UTILITY / ECB INVALID.

g. With the brake pedal depressed, bleed the rear brake system from the bleeder plug on the rear disc brake cylinder RH while the pump motor and solenoid are operating.

WARNING

Keep the fluid inside the reservoir above the LOW level by replenishing.

➡Depress and hold the brake pedal. After the solenoid operates for approximately 30 seconds, release the brake pedal to stop the solenoid. Repeat the procedures until air is completely bled from the rear brake system. The ECB warning light comes on and the buzzer sounds while bleeding, but they do not indicate a malfunction.

h. Tighten the bleeder plug to 74 inch lbs. (8.3 Nm) after bleeding.

10. Perform the accumulator zero down:

WARNING

Never bleed air from the brake hydraulic system without using the intelligent tester. Failure to use the intelligent tester could cause serious injury or an accident. Be sure to perform this procedure before replacement, removal, or installation of the actuator.

➡Perform accumulator zero down by following the steps displayed on the intelligent tester.

a. Connect the intelligent tester to the DLC3 with the ignition switch **OFF**.

b. Depressurize the accumulator:

- Check that the parking brake is applied and turn the ignition switch to the **ON** position.
- Enter the following menus: DIAGNOSIS / OBD/ MOBD / ABS/ TRAC/VSC / ECB UTILITY / ZERO DOWN.
- When the buzzer sounds, turn the ignition switch **OFF**.

c. Circulate the fluid in the accumulator.

d. Depressurize the accumulator 5 times.

➡Accumulator pressure is released and accumulated repeatedly, which circulates the fluid inside the accumulator, when repeating accumulator zero down. The pump motor rotates and the accumulator is pressurized every time the ignition switch is turned from off to on.

11. Check the brake fluid level:

a. After performing accumulator zero down (accumulator depressurizing), return the fluid in the accumulator back to the reservoir and then adjust the fluid level in the master cylinder reservoir to the MAX level.

➡After performing accumulator zero down (accumulator depressurizing), fluid is built up in the accumulator by turning the ignition switch to the ON position and the fluid level of the reservoir lowers. If the fluid level is adjusted without performing accumulator zero down (accumulator depressurizing), fluid is sent from the accumulator to the reservoir. The fluid level may exceed the MAX level, but it is normal.

12. Clear the DTC(s).

13. When the brake actuator assembly is replaced, perform linear valve offset learning after bleeding is completed.

BRAKES

FRONT DISC BRAKES

CAUTION

Dust and dirt accumulating on brake parts during normal use may contain asbestos fibers from production or aftermarket brake linings. Breathing excessive concentrations of asbestos fibers can cause serious bodily harm. Exercise care when servicing brake parts. Do not sand or grind brake lining unless equipment used is designed to contain the dust residue. Do not clean brake parts with compressed air or by dry brushing. Cleaning should be done by dampening the brake components with a fine mist of water, then wiping the brake components clean with a dampened cloth. Dispose of cloth and all residue containing asbestos fibers in an impermeable container with the appropriate label. Follow practices prescribed by the Occupational Safety and Health Administration (OSHA) and the Environmental Protection Agency (EPA) for the handling, processing, and disposing of dust or debris that may contain asbestos fibers.

BRAKE CALIPER

REMOVAL & INSTALLATION

2006–07 Models

See Figure 4.

1. Carefully raise the vehicle.
2. Remove the wheel.
3. Disconnect the brake line from the caliper and plug it.
4. Hold the caliper slide pins and remove the mounting bolts.
5. Lift off the caliper.
6. Remove the pads and anti-squeal shims.
7. Remove the wear indicator from the inner pad.
8. Installation is the reverse of removal. Grease the caliper slides and bolts with lithium grease or equivalent. Apply disc brake grease to the anti-squeal shims. Torque the caliper bolts to 25 ft. lbs. (34 Nm); the brake line union bolt to 21 ft. lbs. (29 Nm).
9. Tighten the wheel to 76 ft. lbs. (103 Nm).

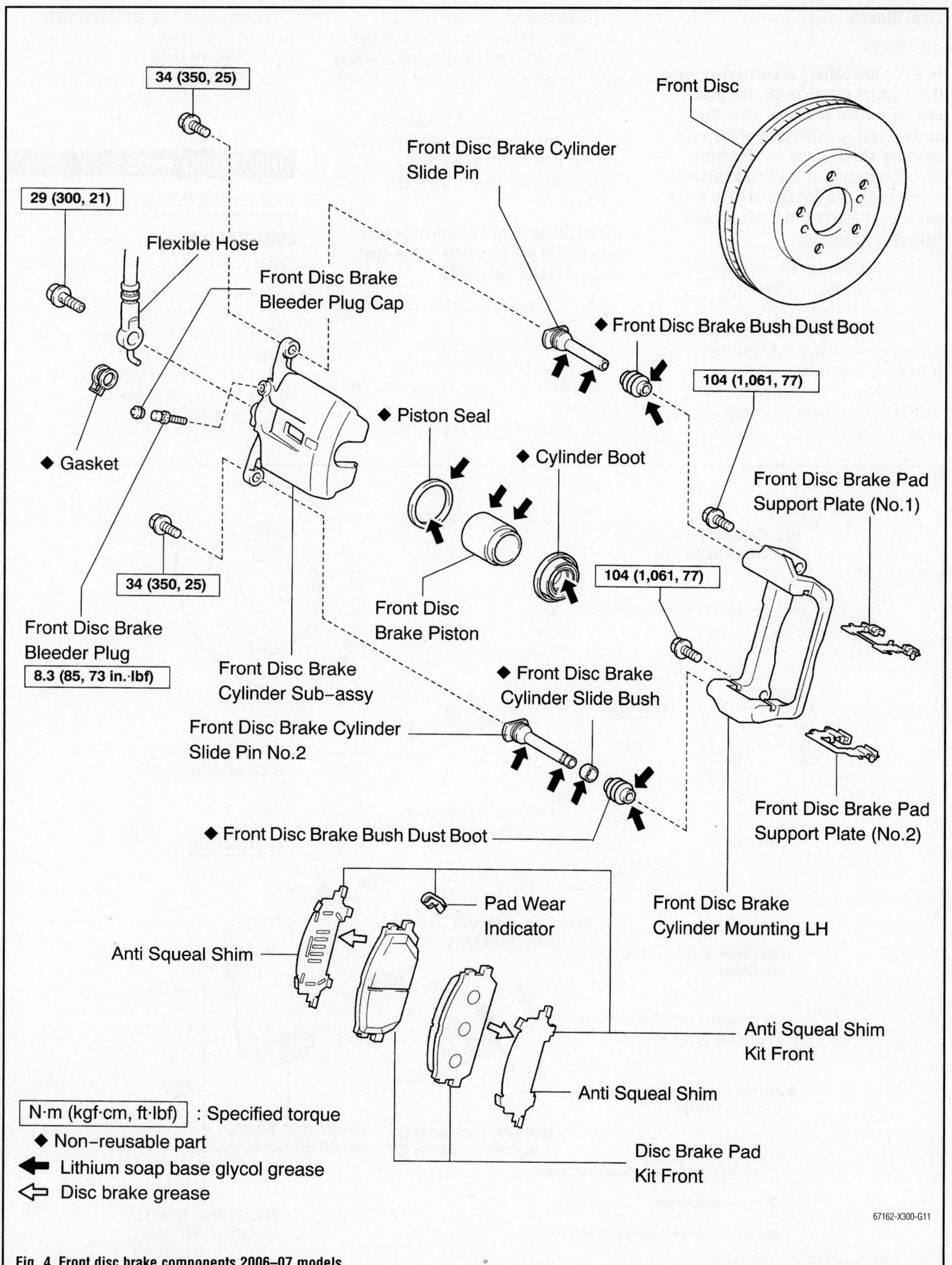

Fig. 4 Front disc brake components 2006–07 models

2008 Models

See Figure 5.

➡While the battery is connected, even if the power switch is off, the brake control system activates when the brake pedal is depressed or the door courtesy switch turns on. Therefore, during servicing of the brake system components, do not operate the brake pedal or open/close the doors while the battery is connected.

1. Carefully raise the vehicle.
2. Remove the front wheel.
3. Drain the brake fluid.
4. Remove the bolt and gasket, and disconnect the front flexible hose from the disc brake cylinder assembly.
5. Hold the front disc brake cylinder slide pins and remove the 2 bolts and disc brake cylinder assembly.

To install:

6. Hold the front disc brake cylinder slide pins and install the disc brake cylinder assembly with the 2 bolts. Tighten the bolts to 25 ft. lbs. 34 (Nm).
7. Connect the front flexible hose with the union bolt and a new gasket to the disc brake cylinder assembly. Tighten the banjo bolt to 20 ft. lbs. (30 Nm).

➡Install the front flexible hose lock securely in the lock hole in the disc brake cylinder assembly.

8. Fill the master cylinder reservoir with brake fluid.
9. Bleed the brake system.
10. Inspect for brake fluid leaks.
11. Perform the accumulator zero down procedure.
12. Inspect the brake fluid level.
13. Install the front wheel and tighten to 76 ft. lbs. (103 Nm).
14. Clear the Daces
15. Check for Daces. If any DTC is output, perform the troubleshooting for that DTC

DISC BRAKE PADS

REMOVAL & INSTALLATION

2006–07 Models

See Figure 6.

1. Raise the vehicle.
2. Remove the wheel.
3. Disconnect the brake line from the caliper and plug it.
4. Hold the caliper slide pins and remove the mounting bolts.
5. Lift off the caliper.

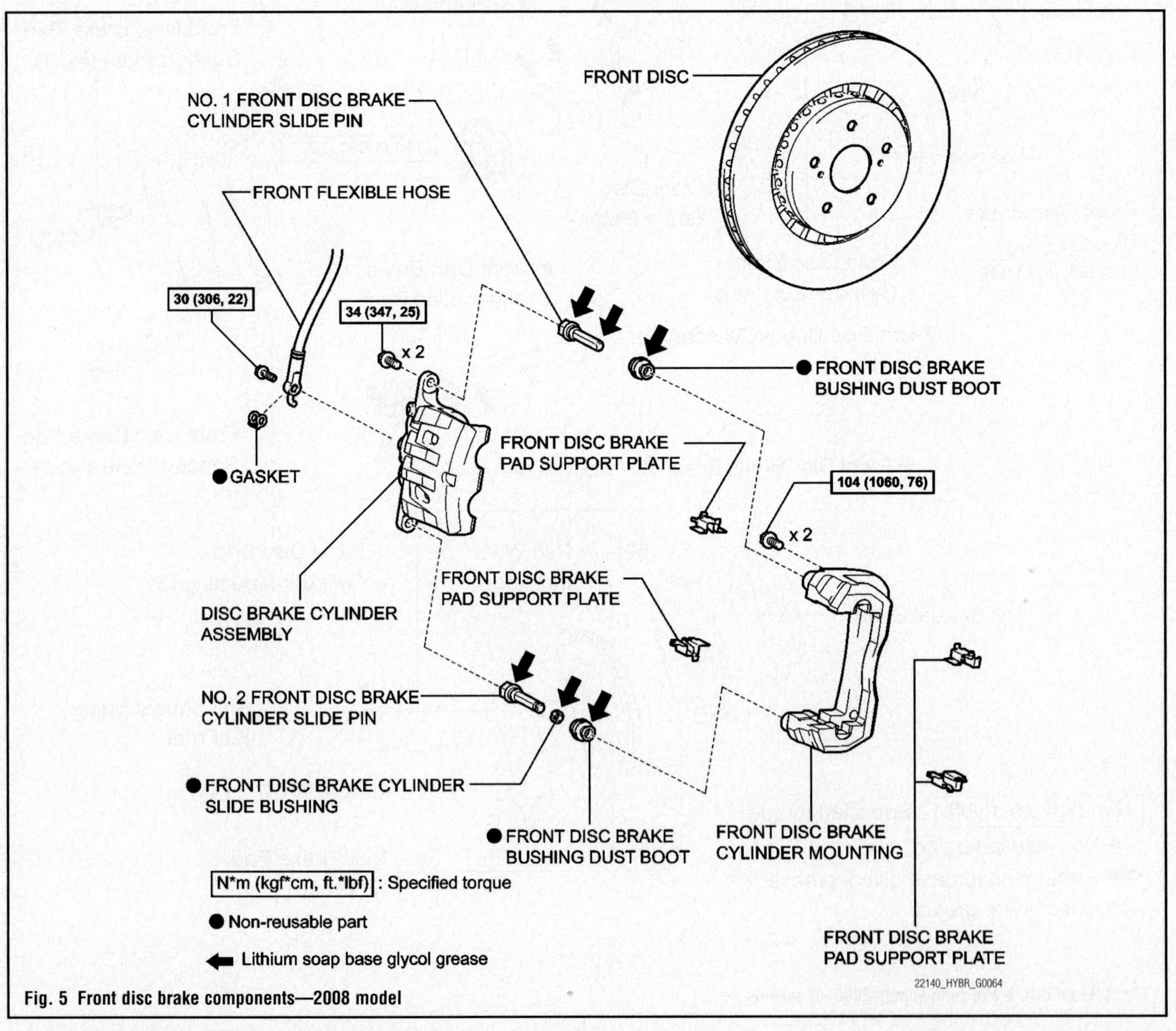

Fig. 5 Front disc brake components—2008 model

6. Remove the pads and anti-squeal shims.
7. Remove the wear indicator from the inner pad.
8. Installation is the reverse of removal. Grease the caliper slides and bolts with lithium grease or equivalent. Apply disc brake grease to the anti-squeal shims. Torque the caliper bolts to 25 ft. lbs. (34 Nm); the brake line union bolt to 21 ft. lbs. (29 Nm).
9. Tighten the wheel to 76 ft. lbs. (103 Nm).

2006–07 Models

See Figures 7 and 8.

1. Carefully raise the vehicle.
2. Remove the front wheel.

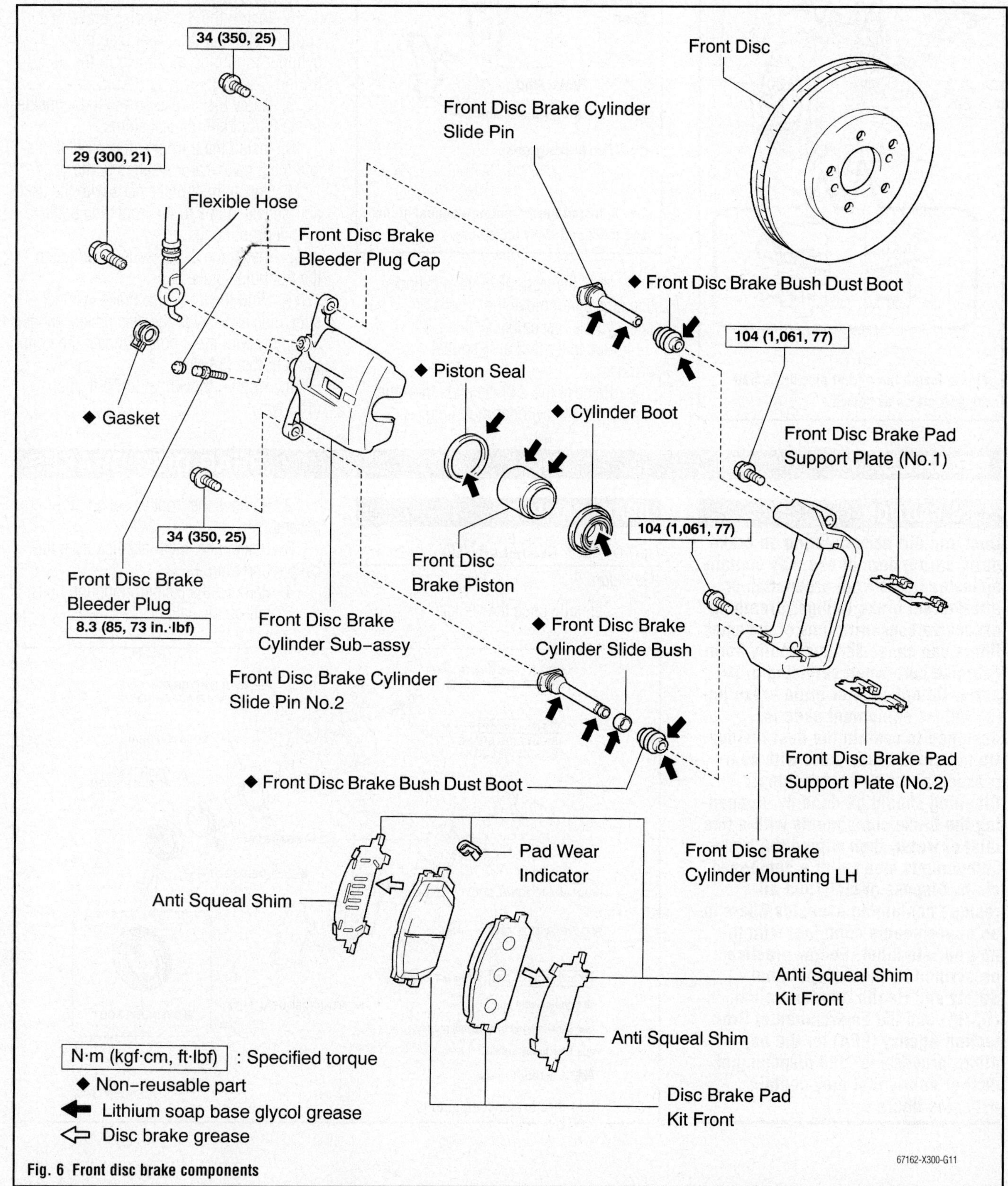

Fig. 6 Front disc brake components

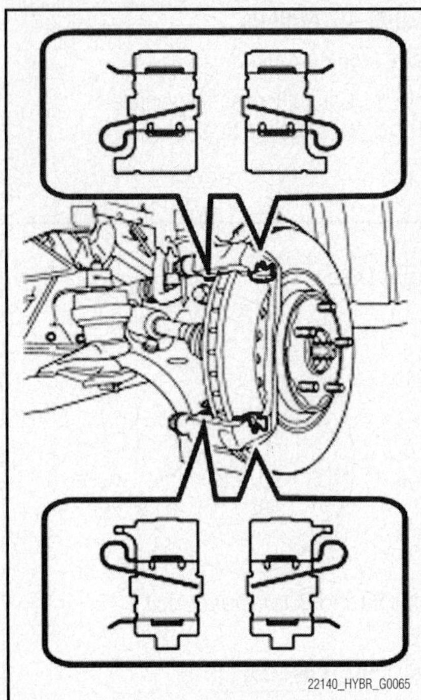

Fig. 7 Install the 4 front disc brake pad support plates as shown

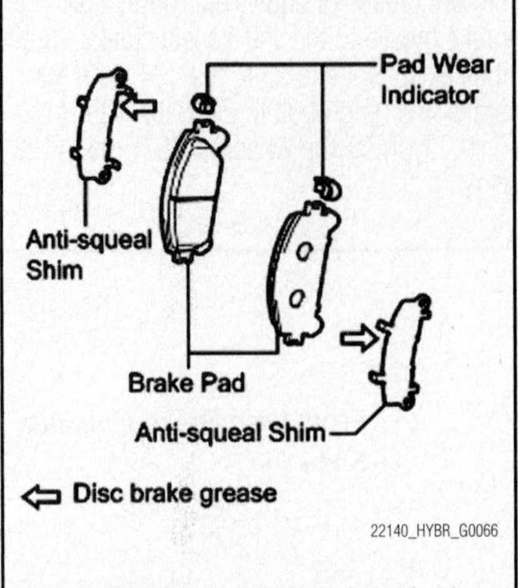

Fig. 8 Install the 2 front anti-squeal shims and the 2 pad wear indicators

3. Hold the front disc brake cylinder slide pins and remove the 2 bolts and disc brake cylinder assembly.

4. Remove the 2 anti-squeal springs.

5. Remove the 2 brake pads from the front disc brake cylinder mounting.

6. Remove the 2 anti-squeal shims and the 2 pad wear indicators from the pads.

7. Remove the 4 front disc brake pad support plates from the front disc brake cylinder mounting.

To install:

8. Install the 4 front disc brake pad support plates to the front disc brake cylinder mounting as shown in the illustration.

9. Apply disc brake grease to the inside of the 2 front anti-squeal shims.

10. Install the 2 front anti-squeal shims and the 2 pad wear indicators to the pads.

11. Install the 2 brake pads with the front anti-squeal shims to the front disc brake cylinder mounting.

12. Install the 2 anti-squeal springs to the front disc brake pads.

13. Hold the front disc brake cylinder slide pins and install the disc brake cylinder assembly with the 2 bolts. Tighten the bolts to 25 ft. lbs. (34 Nm).

14. Tighten the wheel to 76 ft. lbs. (103 Nm).

BRAKES

REAR DISC BRAKES

CAUTION

Dust and dirt accumulating on brake parts during normal use may contain asbestos fibers from production or aftermarket brake linings. Breathing excessive concentrations of asbestos fibers can cause serious bodily harm. Exercise care when servicing brake parts. Do not sand or grind brake lining unless equipment used is designed to contain the dust residue. Do not clean brake parts with compressed air or by dry brushing. Cleaning should be done by dampening the brake components with a fine mist of water, then wiping the brake components clean with a dampened cloth. Dispose of cloth and all residue containing asbestos fibers in an impermeable container with the appropriate label. Follow practices prescribed by the Occupational Safety and Health Administration (OSHA) and the Environmental Protection Agency (EPA) for the handling, processing, and disposing of dust or debris that may contain asbestos fibers.

BRAKE CALIPER

REMOVAL & INSTALLATION

See Figure 9.

1. Carefully raise the vehicle.

2. Remove the front wheel and tire assembly.

3. Disconnect the brake line from the caliper and plug it.

4. Remove the caliper mounting bolts.

5. Lift off the caliper.

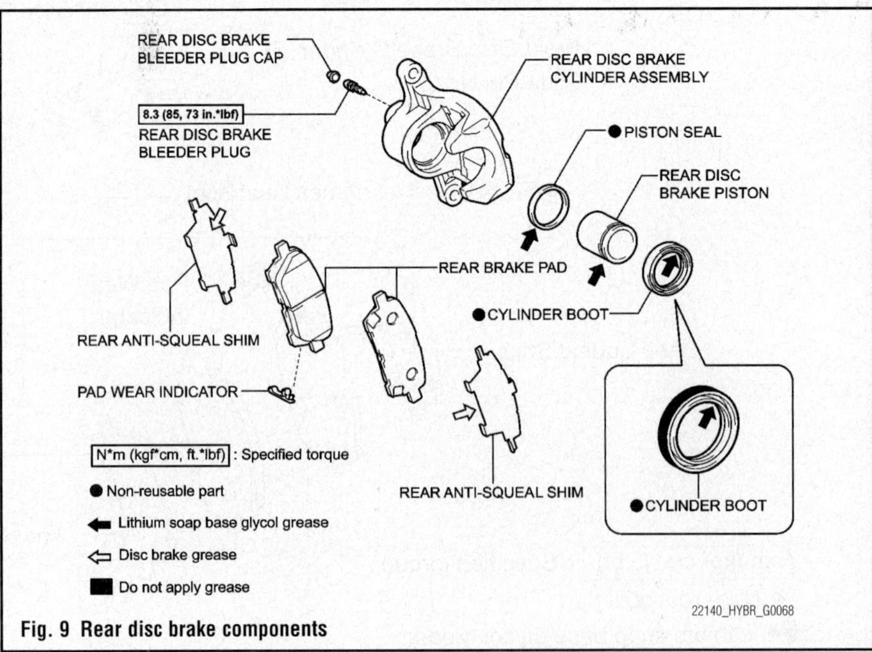

Fig. 9 Rear disc brake components

6. Remove the pads and anti-squeal shims.

7. Remove the wear indicators from each pad.

To install:

8. Install the 2 rear brake pads with rear anti-squeal shims to the rear disc brake cylinder mounting.

9. Hold the rear disc brake cylinder slide pins and install the disc brake cylinder assembly with the 2 bolts. Tighten the mounting bolts to 25 ft. lbs. (34 Nm).

10. Install the brake hose to the caliper, tighten the banjo bolt to 24 ft. lbs. (33 Nm).

11. Tighten the wheel to 76 ft. lbs. (103 Nm).

12. Fill the master cylinder reservoir with brake fluid.

13. Bleed the brake system.

14. Inspect for brake fluid leaks.

15. Perform the accumulator zero down procedure.

16. Inspect the brake fluid level.

17. Install the front wheel and tighten to 76 ft. lbs. (103 Nm).

18. Clear the Daces

DISC BRAKE PADS

REMOVAL & INSTALLATION

1. Disconnect the brake line from the caliper and plug it.

2. Remove the caliper mounting bolts.

3. Lift off the caliper.

4. Remove the pads and anti-squeal shims.

5. Remove the wear indicators from each pad.

6. Installation is the reverse of removal. Grease the caliper slides and bolts with lithium grease or equivalent. Apply disc brake grease to the anti-squeal shims. Tighten the caliper bolts to 32 ft. lbs. (43 Nm); the brake line union bolt to 24 ft. lbs. (33 Nm).

BRAKES — PARKING BRAKE

PARKING BRAKE CABLES

ADJUSTMENT

2006–07 Models

See Figures 10 and 11.

1. Raise and safely support the vehicle.

2. Remove the rear wheel and tire assemblies.

3. Adjust parking brake shoe clearance, as follows:

a. Temporarily install the hub nuts.

b. Remove the hole plug, turn the adjuster and expand the shoes until the disc locks.

c. Contract the shoe adjuster until the disc rotates smoothly. Standard : return 8 notches

d. Check that the shoe has no brake drag.

e. Install the hole plug.

4. Install the rotor and caliper.

5. Install the rear wheel and tire assembly and tighten the lug nuts to 76 ft. lbs. (103 Nm).

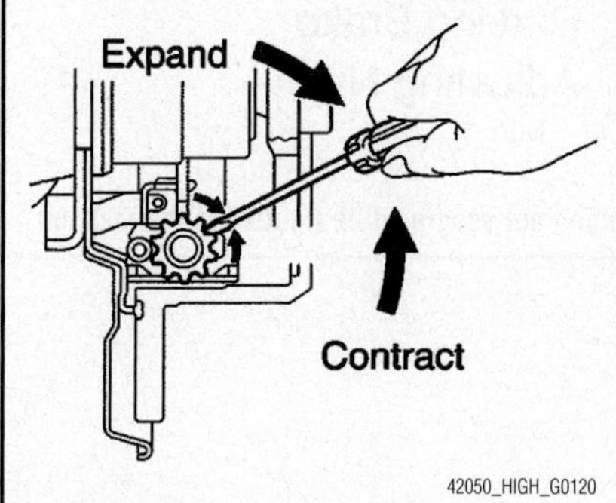

Fig. 10 Adjusting the brake shoe clearance

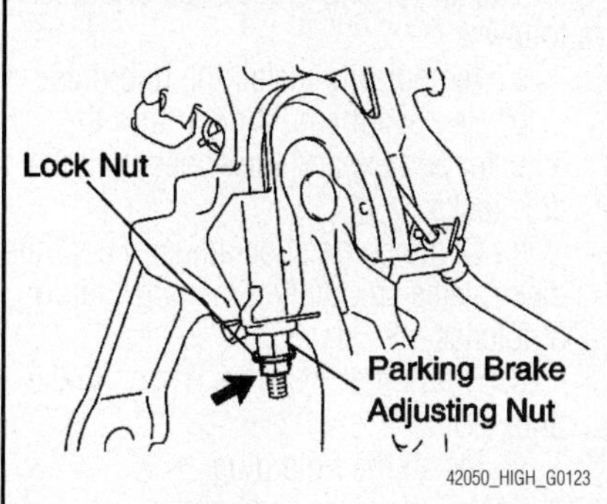

Fig. 11 Use a wrench to hold the parking brake adjusting nut secure while tightening the lock nut

6. Inspect the parking brake pedal travel, as follows:

a. Firmly step on the parking brake pedal.

b. Release the parking brake.

c. Once more, slowly depress the parking brake pedal all the way, and count the number of clicks. The parking brake pedal should travel 5 to 7 clicks at 68 lbs. (300 N).

7. If necessary, adjust parking brake pedal travel, as follows:

a. Remove the lower instrument panel finish panel sub-assembly.

b. Remove the lower instrument panel insert sub- assembly.

c. Depress the parking brake pedal 5 clicks to make room for the procedure, and loosen the lock nut with fixing adjusting nut by wrench.

d. Release the parking brake pedal to the original position.

e. Turn the parking brake wire adjusting nut until the parking brake pedal travel is correct.

f. Use a wrench to hold the parking brake adjusting nut, then tighten the lock nut to 53 inch lbs. (6 Nm).

g. Count the number of clicks after depressing and canceling the parking brake pedal 3 to 4 times.

h. Check whether the parking brake drags or not.

i. When operating the parking brake pedal, check that the parking brake pedal indicator light is lit.

2008 Models

See Figures 12 and 13.

1. Inspect parking brake pedal travel, as follows:

a. Fully depress the parking brake pedal to engage the parking brake.

b. Depress the pedal again to disengage the parking brake.

c. Slowly depress the parking brake pedal using the specified force, and

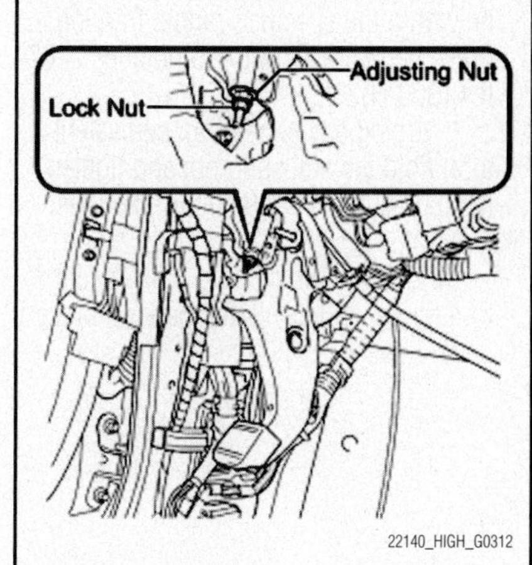

Fig. 12 Parking brake lock nut and adjusting nut

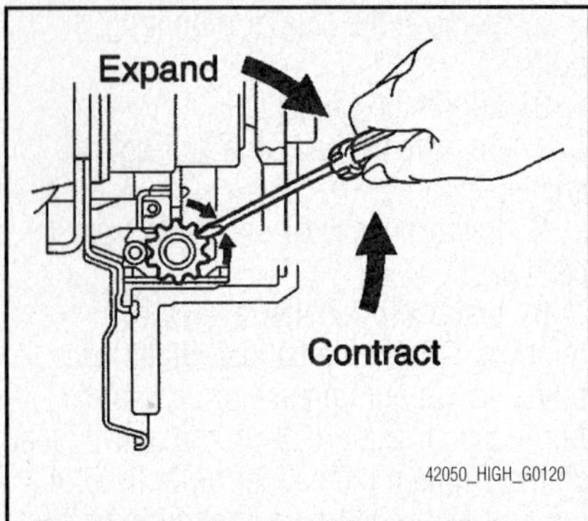

Fig. 13 Adjusting the brake shoe clearance

count the number of clicks. Parking brake pedal travel: 8 to 10 notches at 67 lbs (300 N). If the parking brake pedal travel is not as specified, adjust the parking brake shoe clearance and parking brake pedal travel.

2. Adjust parking brake shoe clearance and parking brake pedal travel, as follows:
   a. Remove the driver side knee airbag.
   b. Completely release the parking brake pedal.
   c. Loosen the lock nut and the adjusting nut to completely release the parking brake cable.
   d. Remove the rear wheel.
   e. Temporarily install the hub nuts.
   f. Remove the shoe adjusting hole plug.
   g. Turn the shoe adjuster and expand the shoe until the disc locks.
   h. Turn and contract the shoe adjuster until the disc can rotate smoothly. Standard: Return 8 notches.
   i. Check that there is no brake drag against the shoe.
   j. Install the shoe adjusting hole plug.
   k. Turn the adjusting nut until the parking brake pedal travel is corrected to be within the specified range. Parking brake pedal travel: 8 to 10 notches at 67 lbs (300 N).
   l. Using a wrench or an equivalent tool, hold the adjusting nut and tighten the lock nut and tighten to 62 inch lbs. (7 Nm).
   m. Operate the parking brake pedal 3 to 4 times, and check the parking brake pedal travel.
   n. Check that there is no brake drag against the shoe.
   o. Remove the hub nuts.
   p. Install the rear wheel and tighten the lug nuts to 76 ft. lbs. (103 Nm).
   q. Install the driver side knee airbag.
3. When operating the parking brake pedal, check that the brake warning light illuminates. Standard: the brake warning light always illuminates at the first click.

PARKING BRAKE SHOES

ADJUSTMENT

See Figures 13 and 14.

1. Raise and safely support the vehicle.
2. Remove the rear wheel and tire assemblies.
3. Adjust parking brake shoe clearance, as follows:
   a. Temporarily install the hub nuts.
   b. Remove the hole plug, turn the adjuster and expand the shoes until the disc locks.
   c. Contract the shoe adjuster until the disc rotates smoothly. Standard : return 8 notches
   d. Check that the shoe has no brake drag.
   e. Install the hole plug.
4. Install the rotor and caliper.
5. Install the rear wheel and tire assembly and tighten the lug nuts to 76 ft. lbs. (103 Nm).
6. Inspect the parking brake pedal travel, as follows:
   a. Firmly step on the parking brake pedal.
   b. release the parking brake.
   c. Once more, slowly depress the parking brake pedal all the way, and count the number of clicks. The parking brake pedal should travel 5 to 7 clicks at 67 lbs. (300 N).
7. If necessary, adjust parking brake pedal travel, as follows:
   a. Remove the lower instrument panel finish panel sub-assembly.
   b. Remove the lower instrument panel insert sub- assembly.
   c. Depress the parking brake pedal 5 clicks to make room for the procedure, and loosen the lock nut with fixing adjusting nut by wrench.
   d. Release the parking brake pedal to the original position.
   e. Turn the parking brake wire adjusting nut until the parking brake pedal travel is correct.
   f. Use a wrench to hold the parking brake adjusting nut, then tighten the lock nut to 53 inch lbs. (6 Nm).
   g. Count the number of clicks after depressing and canceling the parking brake pedal 3 to 4 times.
   h. Check whether the parking brake drags or not.
   i. When operating the parking brake pedal, check that the parking brake pedal indicator light is lit.

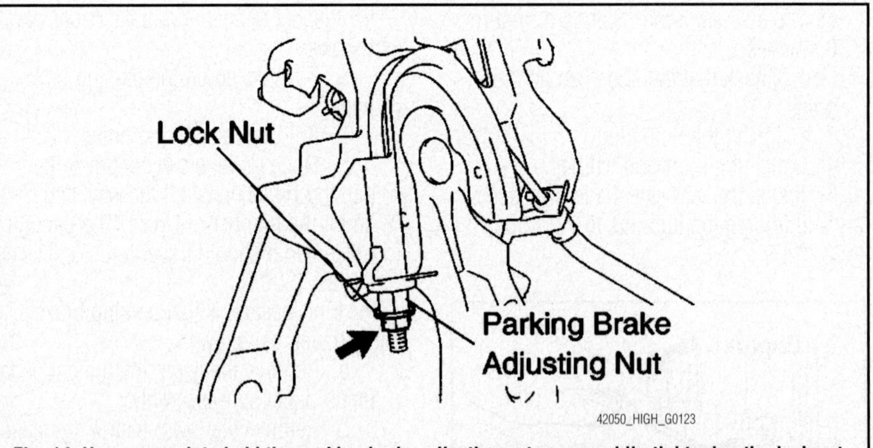

Fig. 14 Use a wrench to hold the parking brake adjusting nut secure while tightening the lock nut

GENERAL INFORMATION

See Figures 15 through 17.

** CAUTION

The vehicle is equipped with a Supplemental Restraint System (SRS). It consists of a driver airbag, front passenger airbag, side airbag, curtain shield airbag, and front seat belt pretensioner. Failure to carry out service operations in the correct sequence could cause the SRS to unexpectedly deploy during servicing, possibly leading to a serious accident. Further, if a mistake is made in servicing the SRS, it is possible that the SRS may fail to operate when required. Before performing servicing (including removal or installation of parts, inspection or replacement), be sure to read the following items carefully, then follow the correct procedures indicated in the repair manual.

** CAUTION

Wait at least 90 seconds after the ignition switch is turned to the "LOCK" position and the negative (-) terminal cable is disconnected from the battery before starting the operation. (The SRS is equipped with a backup power source, so that if work is started within 90 seconds after disconnecting the negative (-) terminal cable of the battery, the SRS may be deployed).

** CAUTION

Do not expose the steering pad, front passenger airbag assembly, center airbag sensor assembly, front airbag sensor, front seat inner belt assembly, seat position sensor, occupant classification ECU, front seat side airbag assembly, side airbag sensor, curtain shield airbag assembly, rear airbag sensor, or front seat outer belt assembly directly to hot air or flames.

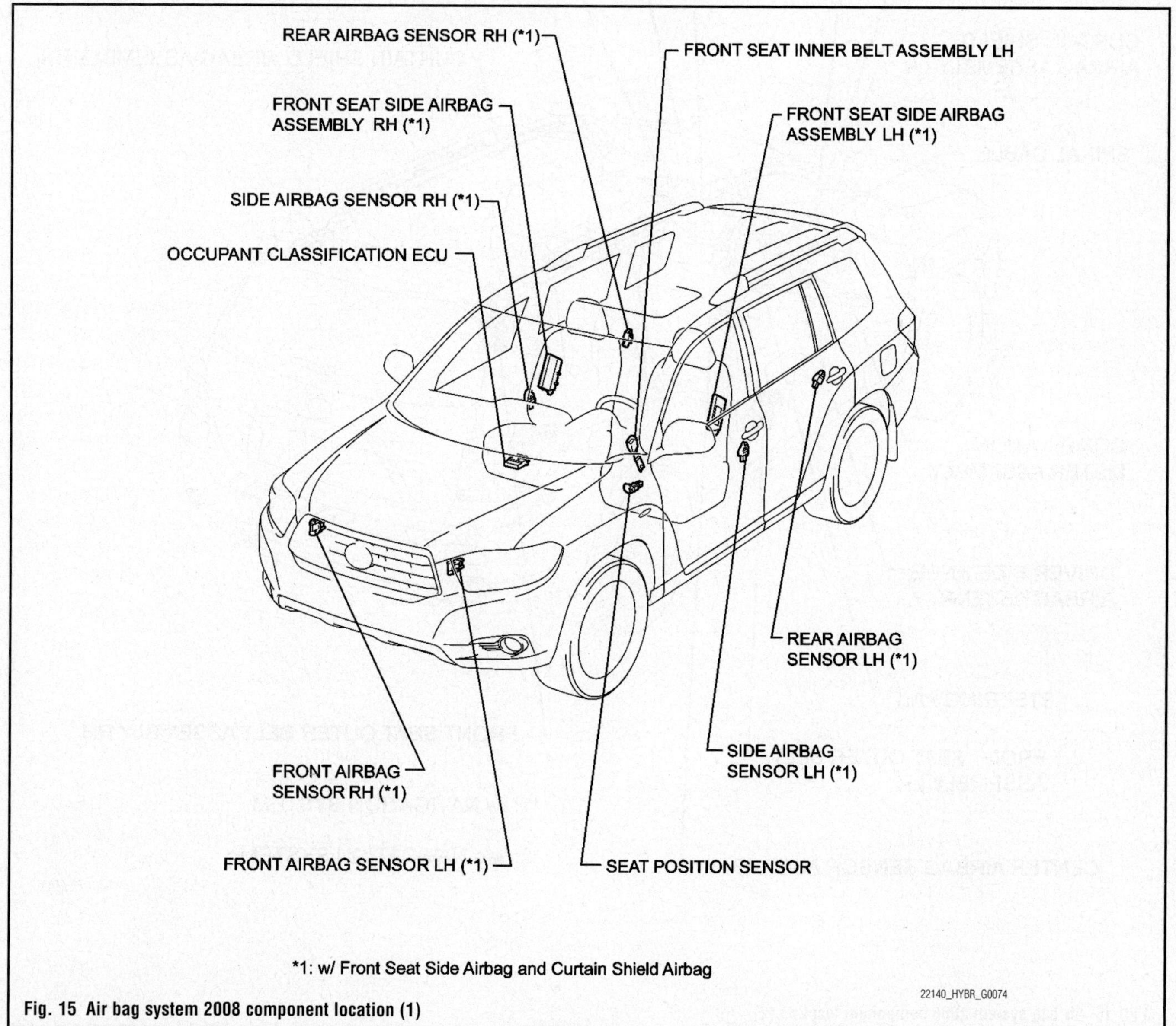

Fig. 15 Air bag system 2008 component location (1)

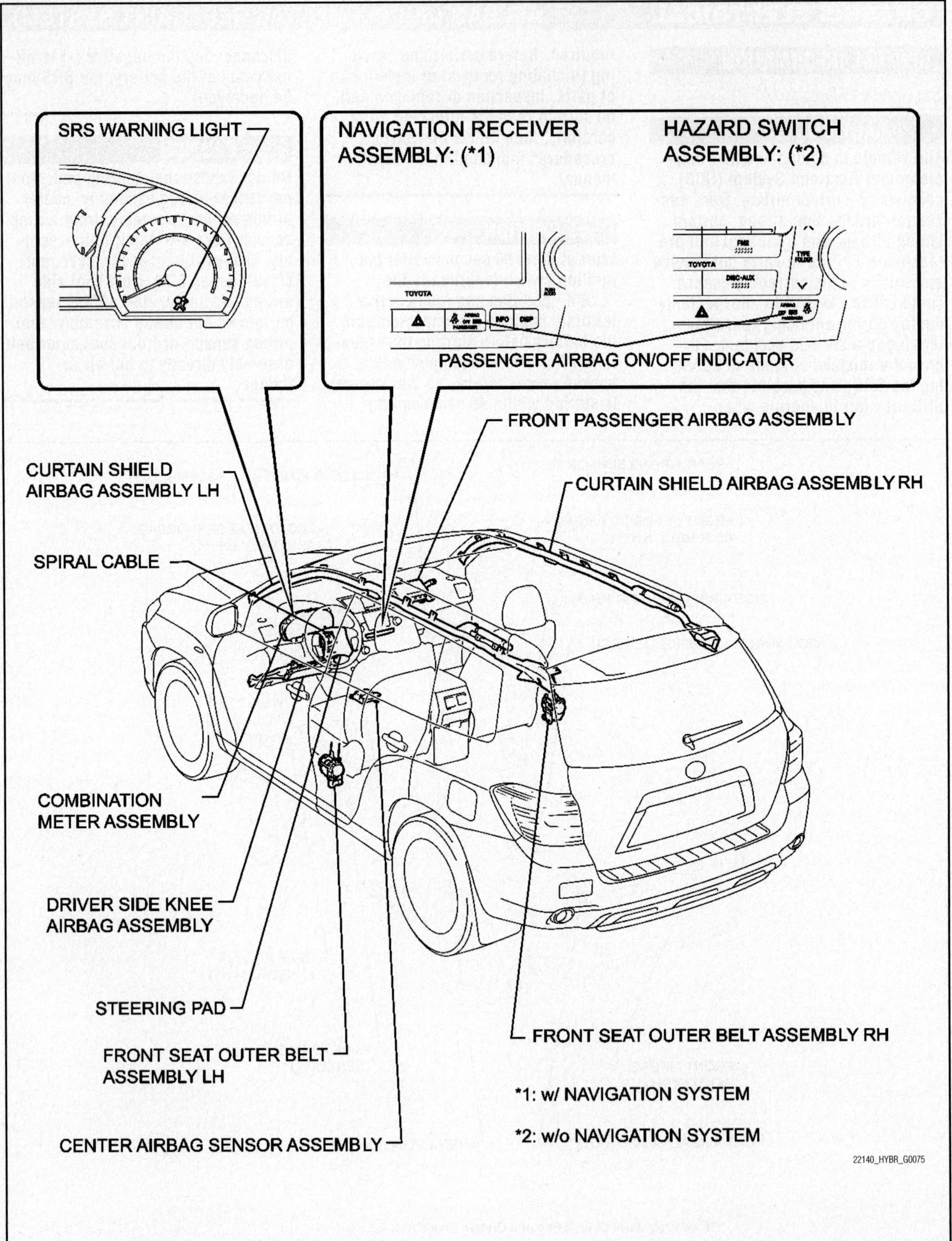

Fig. 16 Air bag system 2008 component location (2)

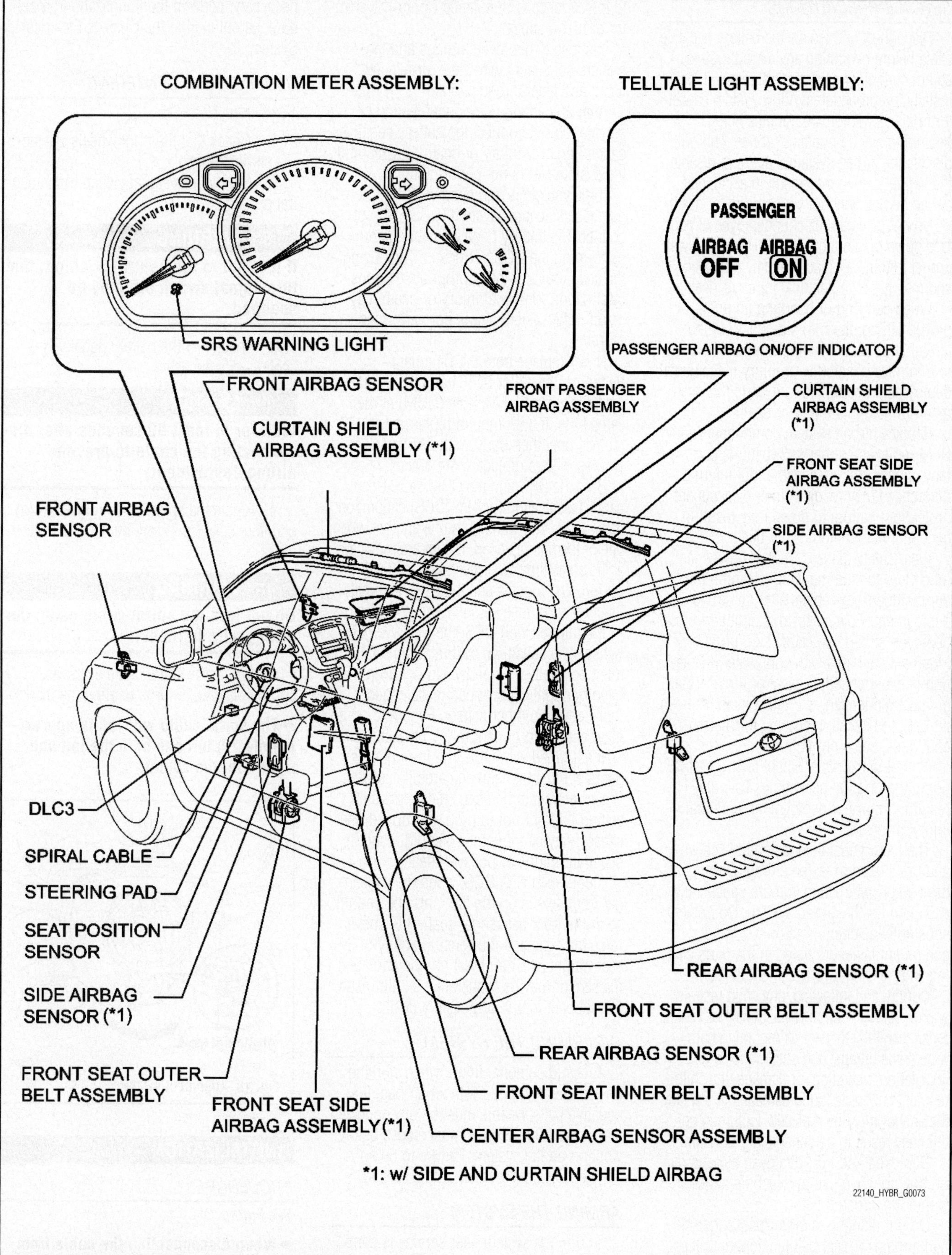

Fig. 17 Air bag system 2006–07 component location

SERVICE PRECAUTIONS

Disconnect and isolate the battery negative cable before beginning any airbag system component diagnosis, testing, removal, or installation procedures. Allow system capacitor to discharge for two minutes before beginning any component service. This will disable the airbag system. Failure to disable the airbag system may result in accidental airbag deployment, personal injury, or death.

Do not place an intact undeployed airbag face down on a solid surface. The airbag will propel into the air if accidentally deployed and may result in personal injury or death.

When carrying or handling an undeployed airbag, the trim side (face) of the airbag should be pointing towards the body to minimize possibility of injury if accidental deployment occurs. Failure to do this may result in personal injury or death.

Replace airbag system components with OEM replacement parts. Substitute parts may appear interchangeable, but internal differences may result in inferior occupant protection. Failure to do so may result in occupant personal injury or death.

Wear safety glasses, rubber gloves, and long sleeved clothing when cleaning powder residue from vehicle after an airbag deployment. Powder residue emitted from a deployed airbag can cause skin irritation. Flush affected area with cool water if irritation is experienced. If nasal or throat irritation is experienced, exit the vehicle for fresh air until the irritation ceases. If irritation continues, see a physician.

Do not use a replacement airbag that is not in the original packaging. This may result in improper deployment, personal injury, or death.

The factory installed fasteners, screws and bolts used to fasten airbag components have a special coating and are specifically designed for the airbag system. Do not use substitute fasteners. Use only original equipment fasteners listed in the parts catalog when fastener replacement is required.

During, and following, any child restraint anchor service, due to impact event or vehicle repair, carefully inspect all mounting hardware, tether straps, and anchors for proper installation, operation, or damage. If a child restraint anchor is found damaged in any way, the anchor must be replaced. Failure to do this may result in personal injury or death.

Deployed and non-deployed airbags may or may not have live pyrotechnic material within the airbag inflator.

Do not dispose of driver/passenger/curtain airbags or seat belt tensioners unless you are sure of complete deployment. Refer to the Hazardous Substance Control System for proper disposal.

Dispose of deployed airbags and tensioners consistent with state, provincial, local, and federal regulations.

After any airbag component testing or service, do not connect the battery negative cable. Personal injury or death may result if the system test is not performed first.

If the vehicle is equipped with the Occupant Classification System (OCS), do not connect the battery negative cable before performing the OCS Verification Test using the scan tool and the appropriate diagnostic information. Personal injury or death may result if the system test is not performed properly.

Never replace both the Occupant Restraint Controller (ORC) and the Occupant Classification Module (OCM) at the same time. If both require replacement, replace one, then perform the Airbag System test before replacing the other.

Both the ORC and the OCM store Occupant Classification System (OCS) calibration data, which they transfer to one another when one of them is replaced. If both are replaced at the same time, an irreversible fault will be set in both modules and the OCS may malfunction and cause personal injury or death.

If equipped with OCS, the Seat Weight Sensor is a sensitive, calibrated unit and must be handled carefully. Do not drop or handle roughly. If dropped or damaged, replace with another sensor. Failure to do so may result in occupant injury or death.

If equipped with OCS, the front passenger seat must be handled carefully as well. When removing the seat, be careful when setting on floor not to drop. If dropped, the sensor may be inoperative, could result in occupant injury, or possibly death.

If equipped with OCS, when the passenger front seat is on the floor, no one should sit in the front passenger seat. This uneven force may damage the sensing ability of the seat weight sensors. If sat on and damaged, the sensor may be inoperative, could result in occupant injury, or possibly death.

DISARMING THE SYSTEM

To avoid personal injury when working on vehicles equipped with an air bag, the negative battery cable must be disconnected and at least 90 seconds must elapse before working on the system. Failure to do so may result in deployment of the air bag.

ARMING THE SYSTEM

To arm the system after service is completed, connect the negative battery cable. If necessary, perform the initialization procedure, as outlined in the Chassis Electrical System.

CLOCKSPRING CENTERING

See Figure 18.

1. Check that the front wheels are facing straight ahead.
2. Set the turn signal switch to the neutral position.

**** WARNING**

If it is not in the neutral position, the turn signal switch pin may be snapped.

3. Check that the battery negative (-) cable is disconnected.

**** CAUTION**

Wait for at least 90 seconds after disconnecting the cable to prevent airbag deployment.

4. Rotate the spiral cable (clockspring) counterclockwise slowly by hand until it stops.

**** WARNING**

Do not turn the spiral cable using the airbag wire harness.

5. Rotate the spiral cable clockwise approximately 2.5 turns to align the marks.

➡The spiral cable will rotate approximately 2.5 turns to both the left and right from the center.

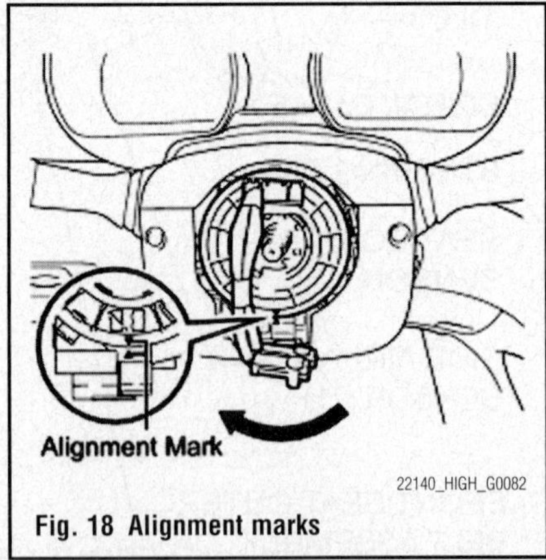

Fig. 18 Alignment marks

INITIALIZATION

PROCEDURE

See Figure 19.

➡When disconnecting the cable from negative (-) battery terminal, initialize

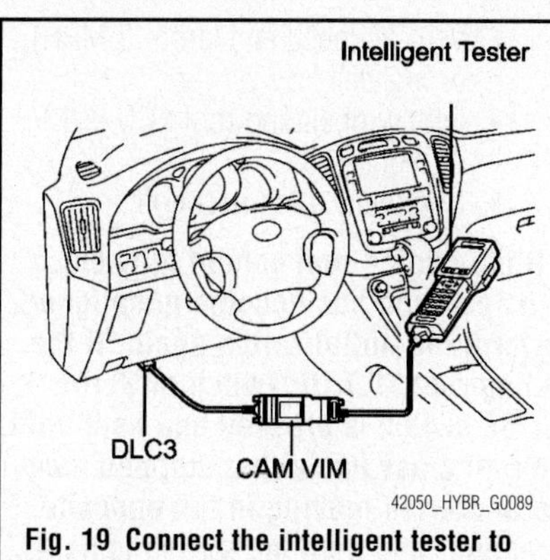

Fig. 19 Connect the intelligent tester to the DLC3

the following system(s) after the terminal is reconnected:

- SFI (fuel injection) system
- Power window control system
- Sunroof system

1. PERFORM INITIALIZATION

a. Apply the parking brake and chock the vehicle.

b. Connect the intelligent tester to the DLC3.

c. Put the engine in the inspection mode.

d. Select the following menu items: DIAGNOSIS / ENHANCED OBD II / DATA LIST / PRIMARY / COOLANT TEMP.

e. Warm up the engine with the A/C switch off until the coolant temperature is 83°C or more.

f. Select the following menu items: DIAGNOSIS /HV CONTROL ECU / DATALIST / SOC.

g. Move the shift lever to the N position, set the A/C switch to MAX cool or turn on the front defroster, and leave the vehicle until the SOC value is below 40 percent.

➡If the SOC is substantially below 40 percent, the auxiliary battery may go dead. Therefore, do not allow the SOC to reach substantially below 40 percent.

h. Move the shift lever to the P position and turn the ignition switch **OFF**.

i. Disconnect the negative terminal from the auxiliary battery and leave it for 1 minute or more.

j. Reconnect the negative terminal to the auxiliary battery and turn the ignition switch **ON**.

k. Select the following menu items: DIAGNOSIS / ENHANCED OBD II / DATALIST / PRIMARY / COOLANT TEMP.

l. Check that the engine coolant temperature is 83°C or more.

➡If the engine coolant temperature is below 83°C, start the engine by depressing the accelerator pedal with the shift lever in the P position to increase the coolant temperature.

m. Select the following menu items: DIAGNOSIS / ENHANCED OBDII / DATA LIST / CALCLOAD.

n. Move the shift lever to the D position while depressing the brake pedal with the left foot.

o. Depress the accelerator pedal with the right foot while firmly depressing the brake pedal with the left foot, and maintain an engine load value of 45 percent or more (60 to 70 percent of full acceleration) for approximately 30 seconds.

⁂ WARNING

Do not perform this step for 40 seconds or more.

p. Shift the shift lever to the P position. Turn the ignition switch off and then wait for 5 seconds before putting the vehicle into the READY-on state again.

q. Select the following menu items: DIAGNOSIS / ENHANCED OBDII / DATA LIST / ISC LEARNING.

r. Check that the air conditioner is off. Lightly depress the accelerator pedal and release it when the engine starts.

s. Check that "COMPLETE" is displayed on the intelligent tester screen.

➡If "ISC LEARNING" is not completed within 1 minute, repeat steps (o) through (r). The engine will usually stop when "ISC LEARNING" is completed. However, the engine will not stop even when "ISC LEARNING" is completed in such cases when the charge level of the auxiliary battery is decreased.

2. RESET (INITIALIZE) POWER WINDOW REGULATOR MOTOR

➡Resetting the power window motor (initializing the pulse sensor) is necessary when the battery terminal is disconnected; when the power window regulator master switch, power window regulator switch, wire harness, power window regulator and power window motor are replaced or removed/installed; or when the P FR DOOR fuse, ECU-B fuse and IG1 fuse are replaced. AUTO operation function, jam protection function and remote operation function via the power window regulator master switch do not operate without resetting.

➡Be sure to reset the power window regulator motors using the power window switch on each door because the motors cannot be reset by remote operation using the power window master switch.

⁂ WARNING

In order to prevent a large amount of current from flowing in the wires, do not simultaneously reset 2 or more power window regulator motors.

➡If the power window switch has been continuously operated for a long time, the power window regulator motor will stop so that no additional load will be applied to the motor. Continuing to operate the switch even after the motor has stopped will cause the switch to blink while restricting the AUTO UP/DOWN function and the remote UP/DOWN function. If this happens, reset the motor again after several minutes have passed.

➡Whenever disconnecting the battery terminal, reset all the other systems besides the power window control system.

a. Turn the ignition switch **ON**.

b. Halfway open the power window by pressing the power window switch.

c. Fully pull up the switch until the power window is fully closed and continue to hold the switch for approx. 1 second after the power window is fully closed.

d. Check that the AUTO UP/DOWN function operates normally. If the AUTO UP/DOWN function operates normally, reset operations have been completed at this time. If not normal, proceed with the steps.

e. Disconnect the negative battery terminal for 10 seconds.

f. Connect the battery terminal.

g. Turn the ignition switch **ON**.

h. Halfway open the power window by pressing the power window switch.

i. If the AUTO UP/DOWN function operates normally, reset operations have been completed at this time. If not normal, follow steps below.

j. Turn the ignition switch **ON**.

k. Halfway open the power window by pressing the power window switch.

l. Fully pull up the switch until the power window is fully closed and continue to hold the switch for approx. 12 seconds after the power window is fully closed.

m. Check that the AUTO UP/DOWN function operates normally.

3. RESET SLIDING ROOF DRIVE GEAR SUBASSEMBLY

➡When disconnecting the cable from the negative (-) battery terminal, or disconnecting the sliding roof drive gear or connector, sliding roof drive gear requires initialization of the pulse sensor. If a reset is not executed, the following functions do not operate: auto operation, jam protection function, key off operation, key-linked open and close, transmitter-linked open and close.

a. Turn the ignition switch **ON**.

b. If the sliding roof is opened, close it fully.

c. Push the open switch of the slide switch or the up switch of the tilt switch on the personal light, making the sliding roof operation as follows: Tilt up _ Approximately 1 second _ tilt down _ slide open _ slide close

d. Sliding roof stops at the fully closed position.

e. Finish the initialization.

f. Check that the operation works normally with AUTO operation.

➡If the following conditions occur while operating, initialization will fail.

- Ignition switch is turned off.
- Pushed switch is released while sliding roof is operating.
- Vehicle speed is 5 km/h (3 MPH) or higher.
- Battery of sliding roof ECU is 1 V or higher.
- Communication is cut off.

➡If the sliding roof cannot fully close or its position has become misaligned, perform the initialization again. If the sliding roof TILT UP switch or SLIDE CLOSE switch is pressed and held until the roof glass has either stopped moving or started moving in the opposite direction, and then the switch is held for another 10 seconds or more, perform the initialization again. If the AUTO operation function and jam protection function do not operate after the drive gear has been reset, replace the sliding roof drive gear (sliding roof ECU).

DRIVETRAIN

AUTOMATIC TRANSAXLE

Refer to Hybrid Transaxle assembly, in the Hybrid System (HV) Section.

FRONT HALFSHAFT

REMOVAL & INSTALLATION

See Figures 20 and 21.

1. Before servicing the vehicle, refer to the Precautions Section.
2. Remove or disconnect the following:
   - Engine under covers
   - Front wheels
   - Drain HV transaxle fluid
   - Cotter pin and hub nut
   - Front speed sensor
   - Brake caliper
   - Brake disc

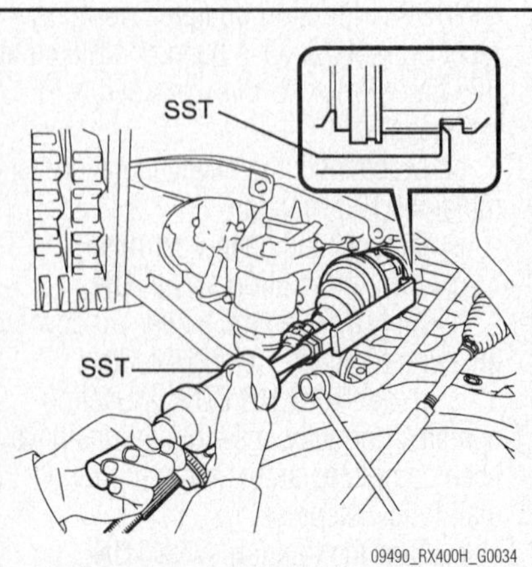

Fig. 20 Use the Special Tool to remove the halfshaft from the transaxle.

Fig. 21 Bearing bracket hole snap ring and bolt

   - Tie rod end, from the steering knuckle
   - Steering knuckle, from the lower control arm
   - Halfshaft from the axle hub, using a plastic hammer
   - Stabilizer link
3. Using Special Tool 095020-01010, remove the halfshaft from the transaxle.
4. For RH axle remove the bearing bracket hole snap ring from the drive shaft bearing bracket.
5. Remove the bolt and front drive shaft assembly RH from the drive shaft bearing bracket.

To install:

6. Install a new halfshaft hole snapring.
7. Coat the splines of the inboard joint shaft assembly with ATF.
8. Align the shaft splines and install the halfshaft assembly with a brass drift and hammer.
9. For RH side axle install the bracket hole snap ring and bolt. Tighten the bolt to 24 ft. lbs. (32 Nm).
10. Tighten the axle shaft nut and tighten to 216 ft. lbs. (294 Nm).
11. The remainder of installation is the reverse order of removal.
12. Fill the HV transaxle with gear oil, install the engine under covers, check front end alignment and test drive.

➡If the cotter pin holes do not align, always correct by tightening the nut until the next hole aligns.

13. Install a new cotter pin.
14. Tighten the front wheels to 76 ft. lbs. (103 Nm).

REAR HALFSHAFT

REMOVAL & INSTALLATION

See Figure 22.

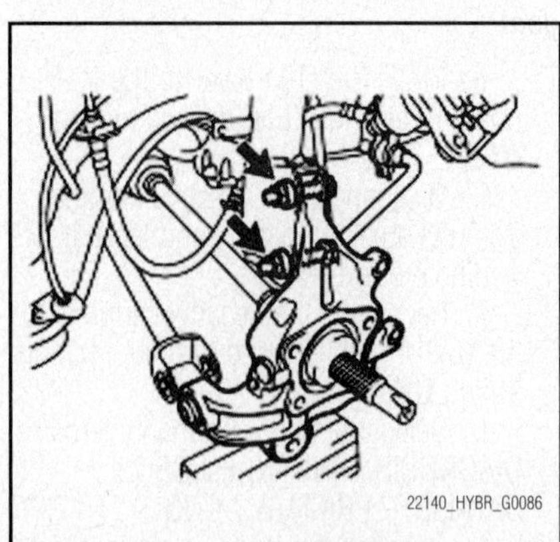

Fig. 22 Rear carrier sub assembly

1. Before servicing the vehicle, refer to the Precautions Section.
2. Remove the rear wheel.
3. Disconnect and remove the speed sensor.
4. Remove the brake disc and brake caliper assembly.
5. Remove the axle shaft nut.
6. Separate the rear axle hub and bearing assembly.
7. Separate the parking brake cable.
8. Disconnect the strut rod.
9. Separate the rear suspension arms.
10. Remove the rear carrier sub assembly.
11. Put matchmarks on the rear drive shaft assembly and differential side gear shaft.
12. Remove the 4 nuts, washers and rear drive shaft assembly.

To install

13. Align the matchmarks.
14. Install the rear drive shaft assembly with the 4 nuts and washers. Tighten to 41 ft. lbs. (56 Nm).
15. Install the rear carrier sub assembly. Tighten the 2 mounting strut bolts to 133 ft. lbs. (180 Nm).
16. Temporarily tighten the rear suspension arm assembly No.2 with the bolt and nut.
17. Temporarily tighten the rear suspension arm assembly No.1 with the bolt and nut.
18. Temporarily tighten the strut rod assembly rear with the bolt and nut.
19. Install the hub and bearing assembly with the 4 bolts and tighten to 55 ft. lbs. (75 Nm).
20. Install the rear brake disc.
21. Install the rear disc brake caliper assembly with the 2 bolts to the rear axle carrier. Tighten the bolts to 58 ft. lbs. (78 Nm).
22. Install the speed sensor and tighten the mounting bolt to 71 inch. lbs. (8 Nm).
23. Using a socket wrench (30mm), install a new rear axle shaft nut. Tighten the axle nut to 217 ft. lbs. (294 Nm).
24. Using a chisel and hammer, stake the rear axle shaft nut.
25. Stabilize the suspension. Then tighten the control arms to 83 ft. lbs. (112 Nm). Tighten the strut rod to 133 ft. lbs. (180 Nm).
26. Install the parking brake cable assembly No.3 with the nut. Tighten the nut to 53 inch. lbs. (6 Nm)
27. Install the rear wheel and tighten to 76 ft. lbs. (103 Nm)
28. Check rear wheel alignment.
29. Verify speeds sensor operation.

ENGINE COOLING

THERMOSTAT

REMOVAL & INSTALLATION

2006–07 Models

See Figures 23 through 28.

1. Before servicing the vehicle, refer to the Precautions Section.

CAUTION

The HIGHLANDER HV has a hybrid system that operates at voltages up to 650 volts. Be sure to follow the instructions in this manual to handle the system correctly. Failure to do so may result in serious injury or electrocution. Engineer must undergo special training to be able to perform high-voltage system inspection and servicing.

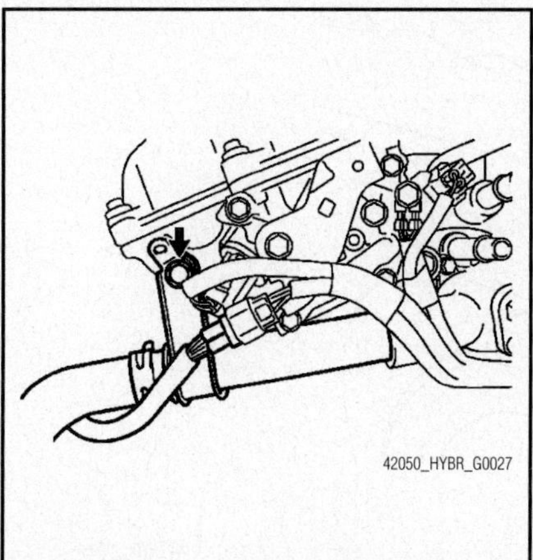

Fig. 23 Remove the water inlet pipe

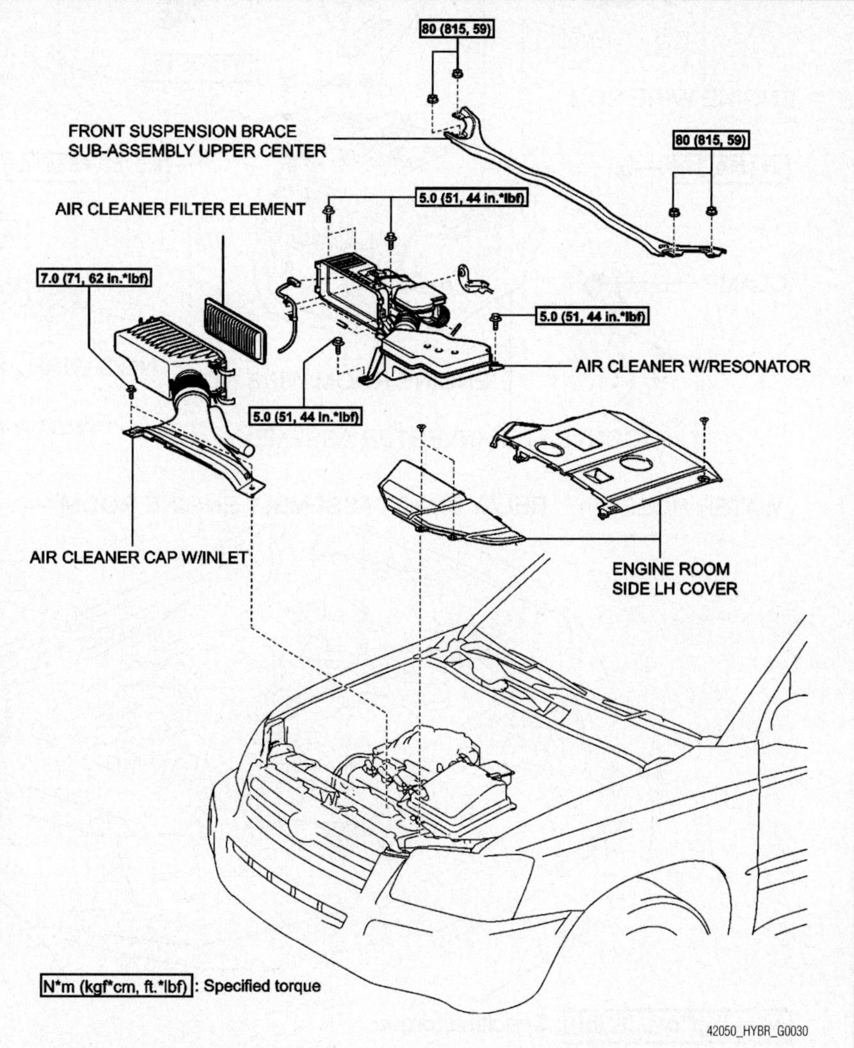

Fig. 24 Exploded view of the components you need to remove to access the thermostat (1 of 3)—all models

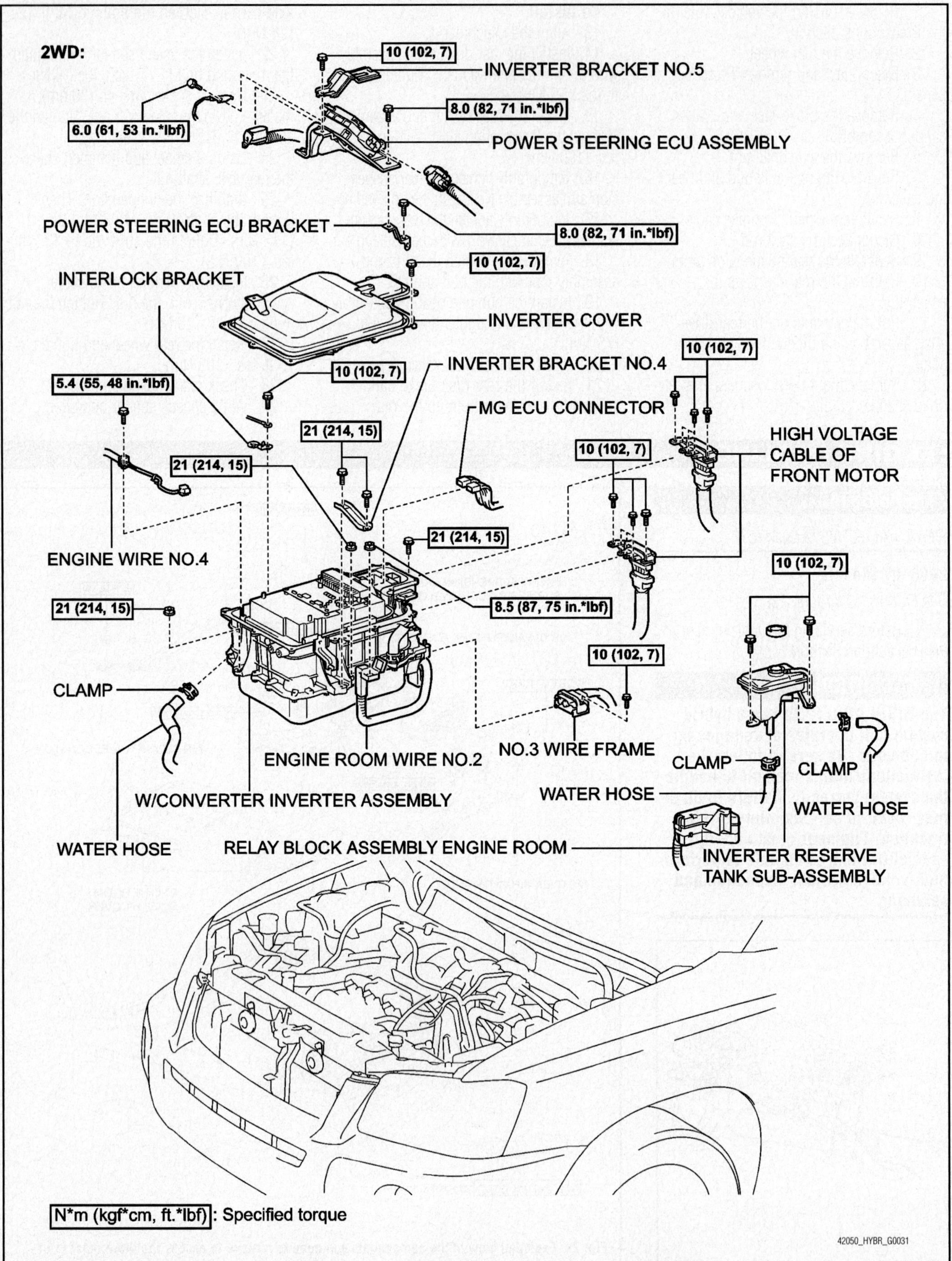

Fig. 25 Exploded view of the components you need to remove to access the thermostat (2 of 3)—2WD models

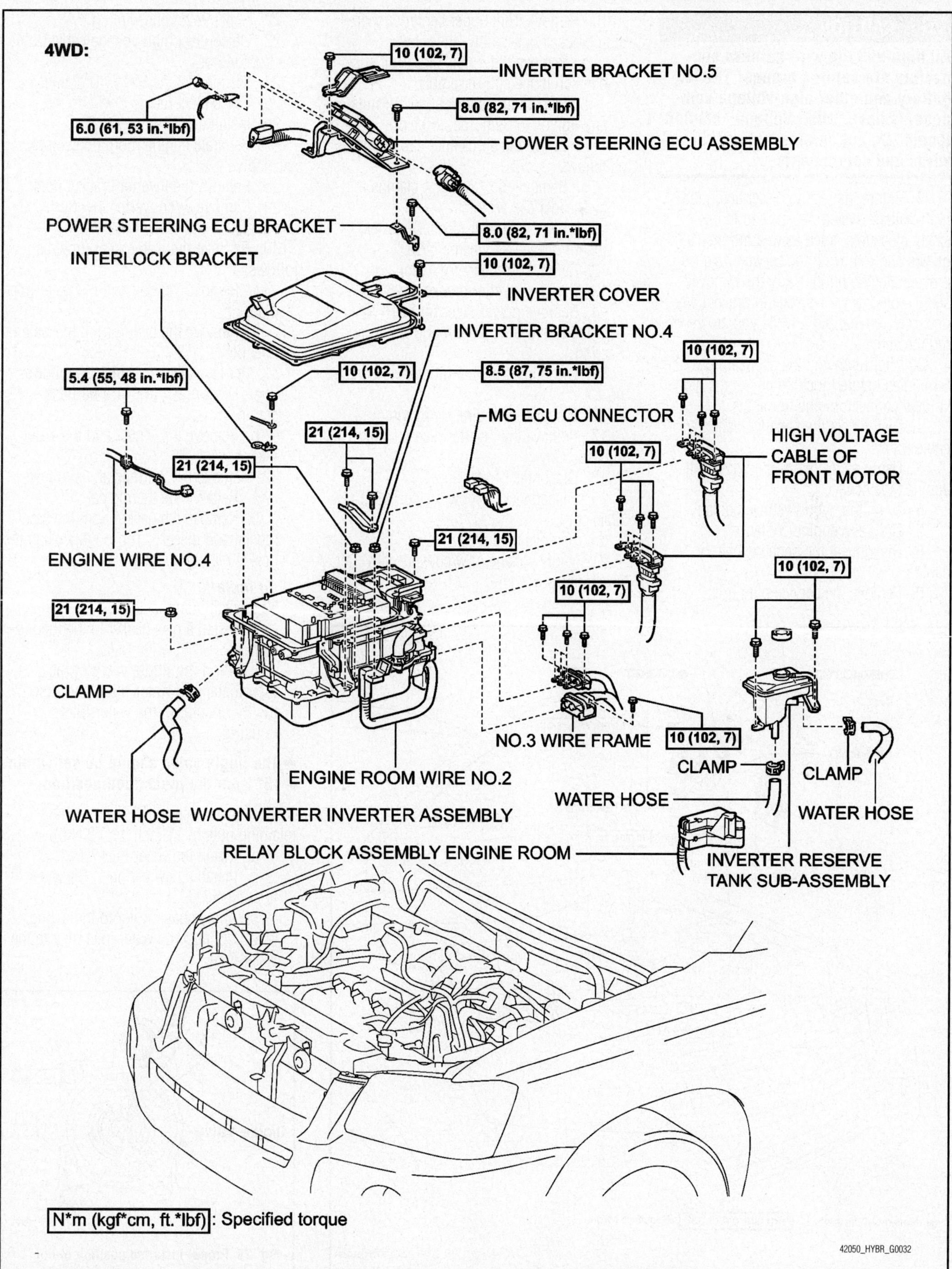

Fig. 26 Exploded view of the components you need to remove to access the thermostat (3 of 3)—4WD models

CAUTION

All high-voltage wire harness connectors are colored orange. The HV battery and other high-voltage components have "High Voltage" caution labels. Do not carelessly touch these wires and components.

2. Before inspecting or servicing the high-voltage system, be sure to follow safety measures, such as wearing insulated gloves and removing the service plug to prevent electrocution. Carry the removed service plug in your pocket to prevent anyone from reinstalling it while you are servicing the vehicle.
3. After removing the service plug, wait 5 minutes before touching any of the high-voltage connectors and terminals.
4. Disconnect the negative battery cable.
5. Using a clip remover, remove the engine side cover.
6. Drain the hybrid vehicle coolant.
7. Drain the engine coolant.
8. Disconnect the negative battery cable.
9. Remove the service plug grip.
10. Remove front upper center suspension brace subassembly, as follows:
    a. Remove the 4 nuts and front suspension brace sub-assembly.
    b. Install the 4 shock absorber nuts and tighten to 59 ft .lbs. (80 Nm).
11. Remove the air cleaner cap w/ inlet, as follows:
    a. Remove the 2 bolts, 4 clamps and air cleaner cap w/ inlet.
    b. Remove the air cleaner filter element from the air cleaner case.
12. Remove air cleaner w/resonator.
13. Remove inverter bracket no.5.
14. Remove power steering ECU, as outlined in the Suspension and Steering Section.
15. Disconnect the circuit breaker sensor no.1.
16. Disconnect engine room wire no.2.
17. Remove the inverter reserve tank subassembly.
18. Disconnect water hose.
19. Disconnect the power steering ECU bracket.
20. Remove the inverter cover.
21. Verify voltage of w/converter inverter assembly is 0 v.
22. Separate engine wire no.4.
23. Disconnect high voltage cable from the front motor.
24. Disconnect the MG ECU connector.
25. Disconnect the no.3 wire frame.
26. Install the inverter cover.
27. Separate engine room relay block assembly
28. Remove the inverter bracket no.4.
29. Remove w/converter inverter assembly.
30. Remove the water inlet pipe, as follows:
    a. Remove the bolt and the water inlet pipe.
    b. Remove the o-ring from the water inlet pipe.
31. Remove the water inlet, as follows:
    a. Disconnect the wire harness clamp.
    b. Remove the 3 nuts and the water inlet.
32. Remove the thermostat, as follows:
    a. Remove the thermostat.
    b. Remove the gasket from the thermostat and discard. Thoroughly clean the gasket mating surfaces.

To install:

33. Install thermostat:
    a. Install a new gasket to the thermostat.
    b. Align the jiggle valve of the thermostat and water inlet, and insert the thermostat in the water inlet housing.

➡The jiggle valve should be set within +-15° from the prescribed position.

34. Install water inlet. Tighten the 3 retaining nuts to 71 inch lbs. (8 Nm).
    a. Install the water inlet pipe:
    b. Install a new o-ring to the water inlet pipe.
    c. Apply soapy water to the o-ring.
    d. Connect the water inlet pipe to the water inlet.

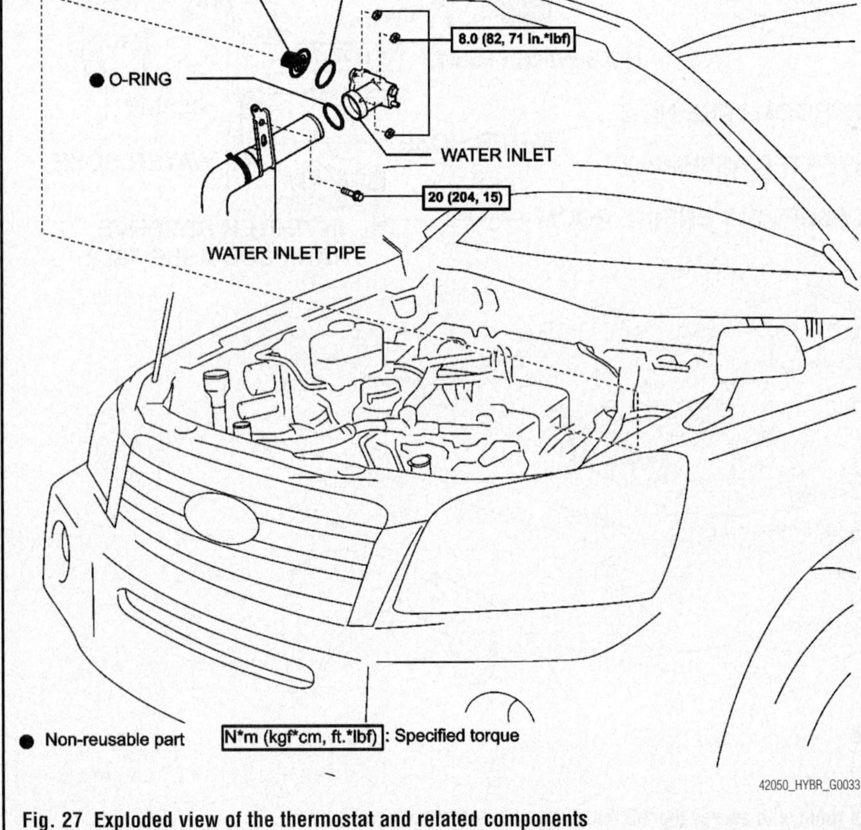

Fig. 27 Exploded view of the thermostat and related components

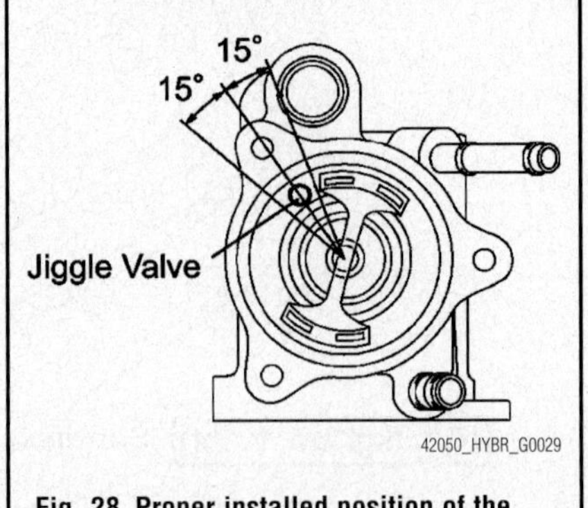

Fig. 28 Proper installed position of the jiggle valve

e. Install the bolt which is used to fix the water inlet pipe to the cylinder head with the bolt. Tighten to 15 ft. lbs. (20 Nm).

35. Install w/converter inverter assembly.
36. Install inverter bracket NO. 4.
37. Install engine compartment relay block assembly.
38. Remove inverter cover.
39. Connect no. 3 wire frame.
40. Connect MG ECU connector.
41. Connect high voltage cable of front motor
42. Connect engine wire NO. 4.
43. Check high voltage cable connection.
44. Install inverter cover.
45. Install the power steering ECU bracket.
46. Connect the water hose.
47. Install inverter reserve tank sub-assembly.
48. Connect engine room wire no. 2.
49. Connect circuit breaker sensor no.1.
50. Install the power steering ECU assembly.
51. Install the invertcr bracket No. 5.
52. Install air cleaner w/resonator.
53. Install air cleaner cap w/inlet.
54. Install the front suspension upper center brace sub-assembly.
55. Install service plug grip.
56. Connect the negative battery cable.
57. Add hybrid vehicle coolant.
58. Add engine coolant.
59. Check hybrid vehicle coolant leaks
60. Check for engine coolant leaks
61. Install engine side cover.
62. If necessary, perform the initialization procedure, as outlined in the Chassis Electrical System.

2008 Models

See Figures 29 and 30.

1. Before servicing the vehicle, refer to the Precautions Section.

⁂ CAUTION

The HIGHLANDER HV has a hybrid system that operates at voltages up to 650 volts. Be sure to follow the instructions in this manual to handle the system correctly. Failure to do so may result in serious injury or electrocution. Engineer must undergo special training to be able to perform high-voltage system inspection and servicing.

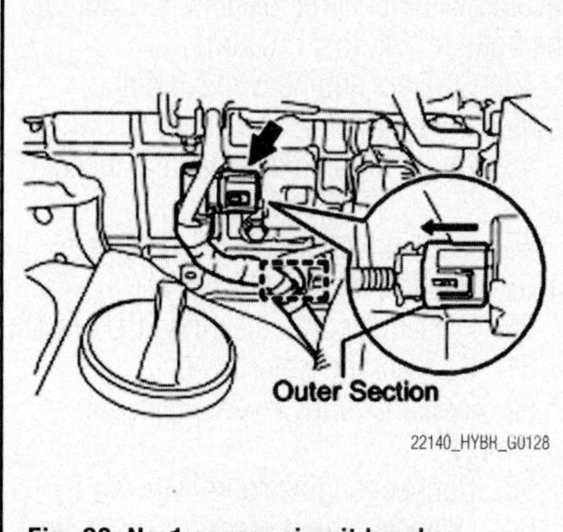

Fig. 29 No.1 sensor circuit breaker

⁂ CAUTION

All high-voltage wire harness connectors are colored orange. The HV battery and other high-voltage components have "High Voltage" caution labels. Do not carelessly touch these wires and components.

2. Before inspecting or servicing the high-voltage system, be sure to follow safety measures, such as wearing insulated gloves and removing the service plug to prevent electrocution. Carry the removed service plug in your pocket to prevent anyone from reinstalling it while you are servicing the vehicle.
3. After removing the service plug, wait 5 minutes before touching any of the high-voltage connectors and terminals.
4. Disconnect the negative battery cable.
5. Remove the engine room left side cover.
6. Remove the engine room cover.
7. Drain the coolant for the inverter.
8. Remove the wiper arm assembly LH and RH.
9. Remove the cowl top ventilator top louver sub-assembly.
10. Remove the wiper motor and link assembly.
11. Remove the cowl top outer panel sub-assembly.
12. Remove the cool air intake duct seal.
13. Remove the air cleaner cap with inlet.
14. Remove the air cleaner with resonator.
15. Remove the bolt and inverter bracket No. 5.
16. Remove bolt and ground cable terminal from power steering ECU assembly.
17. Move the outer section to the wire harness side as illustrated, then disconnect the circuit breaker sensor No. 1.
18. Remove the inverter reserve tank subassembly.
19. Disconnect water hose.
20. Disconnect the power steering ECU bracket.

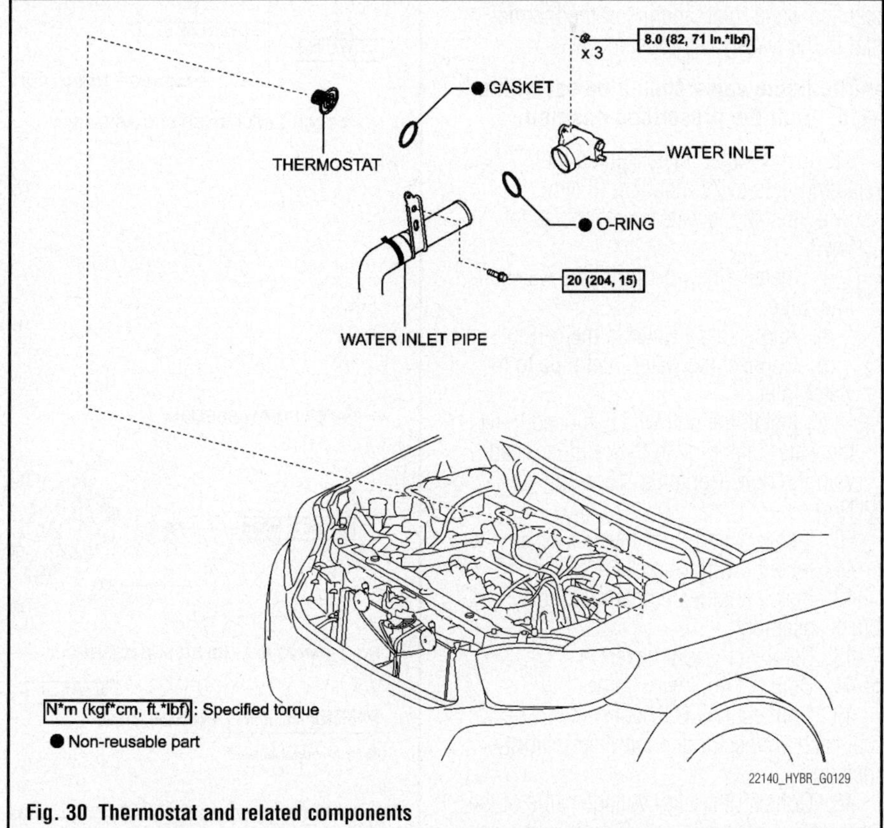

Fig. 30 Thermostat and related components

21. Remove the inverter cover.
22. Verify voltage of w/converter inverter assembly is 0 v.
23. Separate engine wire no.4.
24. Disconnect high voltage cable from the front motor.
25. Disconnect the MG ECU connector.
26. Disconnect the no.3 wire frame.
27. Install the inverter cover.
28. Separate engine room relay block assembly
29. Remove the inverter bracket no.4.
30. Remove w/converter inverter assembly. Since the inverter with converter assembly is very heavy, 2 people are needed to remove the inverter with converter assembly. When removing the inverter with converter assembly, do not damage the parts around it.
31. Remove the water inlet pipe, as follows:
32. Remove the bolt and the water inlet pipe.
33. Remove the o-ring from the water inlet pipe.
34. Disconnect the wire harness clamp.
35. Remove the 3 nuts and the water inlet.
36. Remove the thermostat and gasket.

To install:

37. Install a new gasket and thermostat.
38. Align the jiggle valve of the thermostat and water inlet, and insert the thermostat in the water inlet housing.

➡The jiggle valve should be set within +-15° from the prescribed position.

39. Install water inlet. Tighten the 3 retaining nuts to 71 inch lbs. (8 Nm).
    a. Install the water inlet pipe as follows:
    b. Install a new o-ring to the water inlet pipe.
    c. Apply soapy water to the o-ring.
    d. Connect the water inlet pipe to the water inlet.
    e. Install the bolt which is used to fix the water inlet pipe to the cylinder head with the bolt. Tighten to 15 ft. lbs. (20 Nm).
40. Install w/converter inverter assembly.
41. Install inverter bracket No.4.
42. Install engine compartment relay block assembly.
43. Remove inverter cover.
44. Connect no.3 wire frame.
45. Connect MG ECU connector. Check that each connector and terminal is firmly installed.
46. Connect the high voltage cable of the generator (MG1) with new 5 bolts to the inverter with converter assembly. Tighten the bolts to 7 ft. lbs. (10 Nm).
47. Connect engine wire No.4 and tighten the bolts to 48 inch lbs. (5.4 Nm).
48. Check high voltage cable connection.
49. Install inverter cover. Tighten the cover bolts to 7 ft. lbs. (10 Nm).
50. Install the power steering ECU bracket.
51. Connect the water hose.
52. Install inverter reserve tank sub-assembly.
53. Connect engine room wire No.2.
54. Connect circuit breaker sensor No.1.
55. Install the inverter bracket No.5.
56. Install air cleaner w/resonator.
57. Install air cleaner cap w/inlet.
58. Install air cleaner w/resonator.
59. Install air cleaner cap w/inlet.
60. Install the cool air intake duct seal.
61. Install the outer cowl top panel.
62. Install wiper motor and link assembly.
63. Install cowl top ventilator louver sub-assembly.
64. Install the LH and RH wiper arm assembly.
65. Install the service plug grip.
66. Connect the negative battery cable.
67. Refill Coolant for inverter and engine.
68. Check for leaks.
69. If necessary, perform the initialization procedure, as outlined in the Chassis Electrical System.

WATER PUMP

REMOVAL & INSTALLATION

See Figure 31.

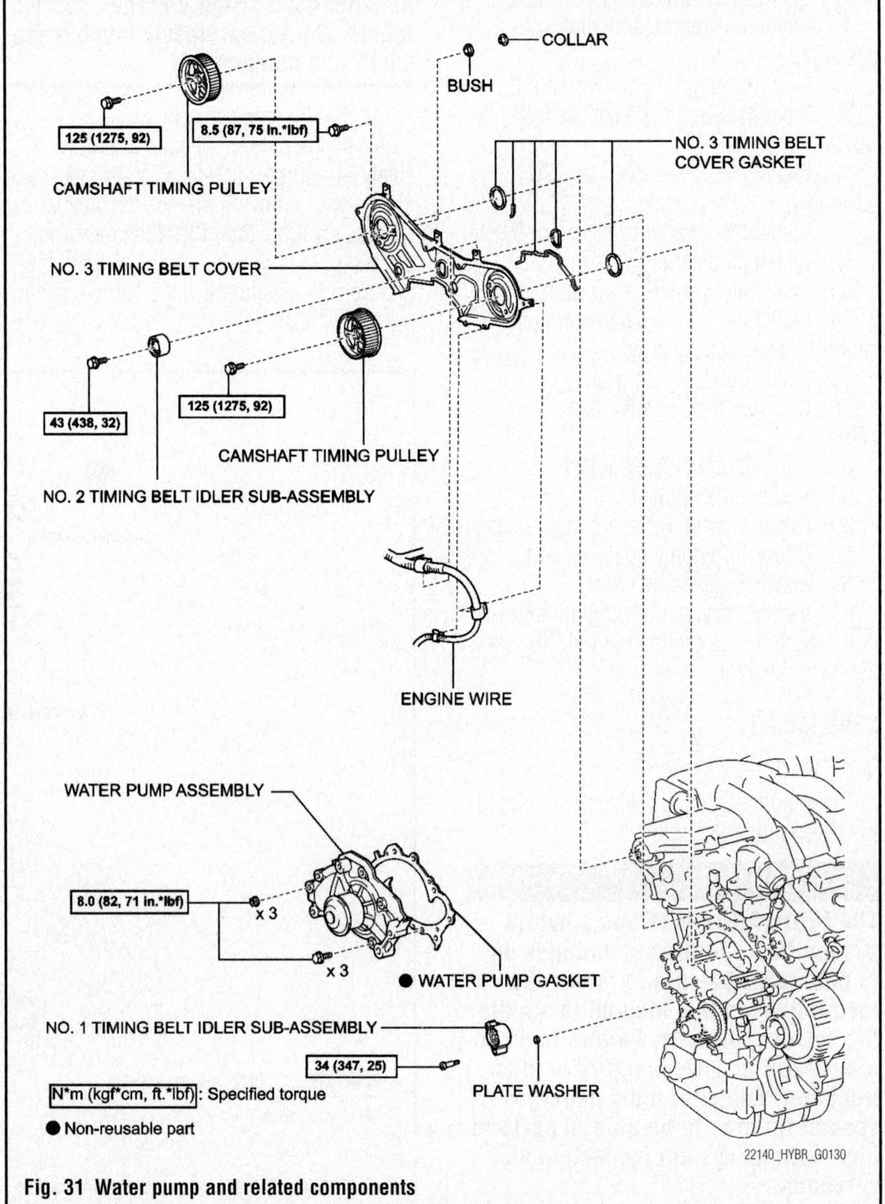

Fig. 31 Water pump and related components

1. Before servicing the vehicle, refer to the Precautions Section.
2. Drain the engine coolant.
3. Remove or disconnect the following:
   - Negative battery cable
   - Engine side covers
   - Right-hand front wheel
   - Engine splash shield
   - Right-hand front fender apron seal
   - Wiper arm and blade assembly
   - Top cowl ventilator louver assembly
   - Wiper motor and link assembly
   - Battery and battery tray
   - Air intake assembly
   - Brake master cylinder reservoir
   - Reservoir support bracket
   - Air cleaner support bracket
   - Engine moving control rod
   - Right-hand engine mount
   - Crankshaft pulley
   - Timing belt cover No.1 and 2
   - Timing belt guide
   - Timing belt
   - Timing belt idler sub-assembly No.1
   - Camshaft timing pulley
   - Timing belt cover No.3
   - Timing belt idler sub-assembly No.2
   - Water pump
4. Installation is the reverse order of removal. Tighten the water pump with a new gasket to 71 inch lbs. (8 Nm).
5. Refill the coolant to the correct level.
6. Start the engine and check for leaks.
7. If necessary, perform the initialization procedure, as outlined in the Chassis Electrical System.

ENGINE ELECTRICAL — CHARGING SYSTEM

The HV transaxle assembly consists of the planetary gear unit, Motor, and Generator. The gear unit uses the planetary gear to split engine output in accordance with a driving request while the vehicle is driven or the HV battery is charged. The Motor assists engine output while increasing vehicle driving force. The Motor also converts the energy, which is consumed in the form of heat during normal braking, into electrical energy and recover it into the HV battery to effect regenerative braking. The Generator supplies power, which is used for charging the HV battery or driving the Motor. It also controls the stepless transmission function of the transaxle by regulating the amount of electricity generated to change Generator speed. In addition, the Generator is used as a starter Motor to start the engine. The transmission input damper absorbs the shock generated when the driving force from the engine is transmitted.

**** CAUTION**

The HIGHLANDER HV has a hybrid system that operates at voltages up to 650 volts. Be sure to follow the instructions in this manual to handle the system correctly. Failure to do so may result in serious injury or electrocution. Engineer must undergo special training to be able to perform high-voltage system inspection and servicing.

**** CAUTION**

All high-voltage wire harness connectors are colored orange. The HV battery and other high-voltage components have "High Voltage" caution labels. Do not carelessly touch these wires and components.

**** CAUTION**

Before inspecting or servicing the high-voltage system, be sure to follow safety measures, such as wearing insulated gloves and removing the service plug to prevent electrocution. Carry the removed service plug in your pocket to prevent anyone from reinstalling it while you are servicing the vehicle.

**** CAUTION**

After removing the service plug, wait 5 minutes before touching any of the high-voltage connectors and terminals.

ALTERNATOR

REMOVAL & INSTALLATION

The 3.3L engine has a DC electric converter and therefore does not use a standard alternator.

ENGINE ELECTRICAL — DISTRIBUTORLESS IGNITION SYSTEM

FIRING ORDERS

See Figure 32.

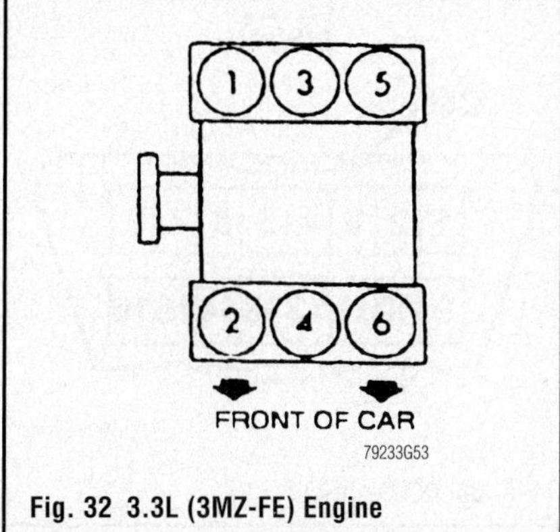

Fig. 32 3.3L (3MZ-FE) Engine

IGNITION COIL

REMOVAL & INSTALLATION

See Figures 33 through 35.

1. Disconnect the negative battery cable.
2. Remove the RH and LH wiper arm assembly.
3. Remove the cowl and seal.
4. Remove the wiper motor and link assembly.
5. Remove the outer cowl top panel.
6. Remove the cool air intake duct seal.
7. Remove the LH engine side cover.
8. Drain the engine coolant.
9. Remove air cleaner cap and case assembly.
10. Remove the air cleaner bracket.
11. Remove the emission control valve set.
12. Remove the upper intake plenum.
13. Disconnect the 6 ignition coil connectors.
14. Remove the 6 bolts and 6 ignition coils.
15. Remove the 6 spark plugs.

To install:

16. Install the spark plug and tighten to 13 ft. lbs. (18 Nm).
17. Install the ignition coils and tighten retaining bolts to 71 inch. lbs. (8 Nm).
18. Connect the ignition coil connectors.
19. Install intake plenum with a new gasket

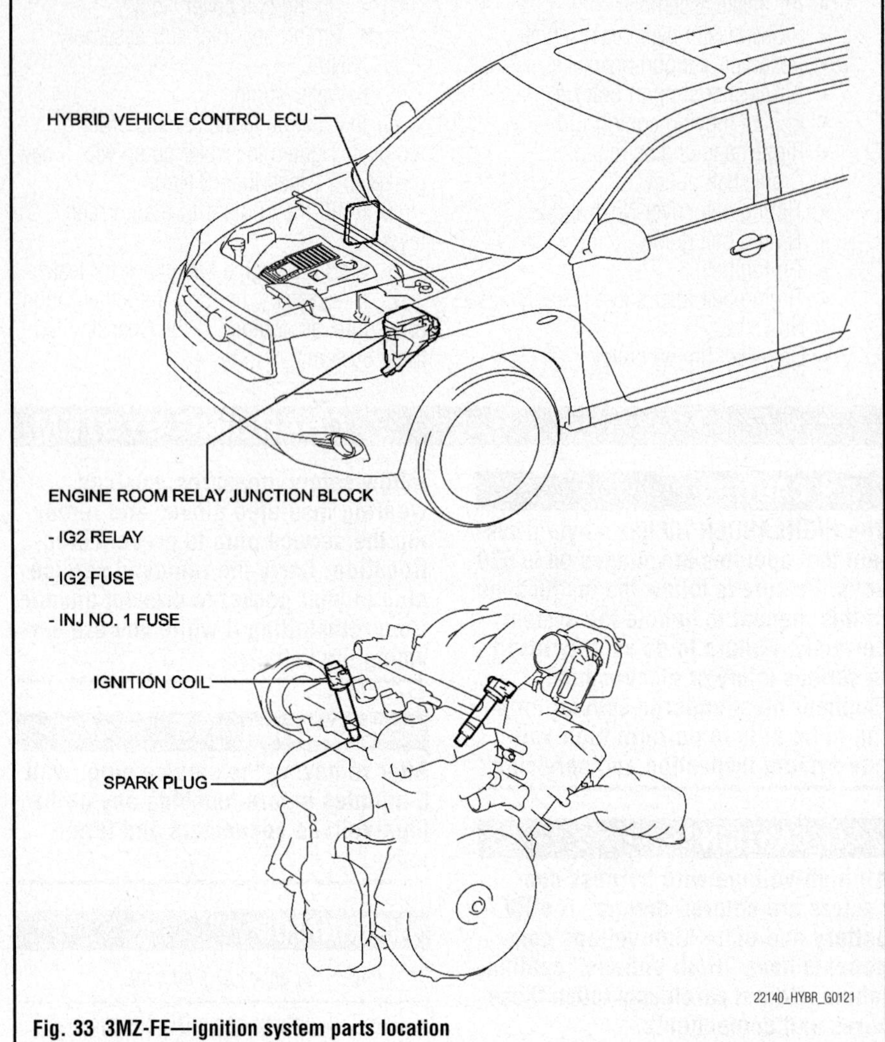

Fig. 33 3MZ-FE—ignition system parts location

20. Using a socket hexagon wrench 8 mm, install the upper intake plenum with the 4 bolts and 2 nuts.
21. Using several steps, tighten the bolts and nuts uniformly to 21 ft. lbs. (28 Nm) follow the sequence shown in the illustration.
22. Install emission control vale set.
23. Install air cleaner bracket, case and cap.

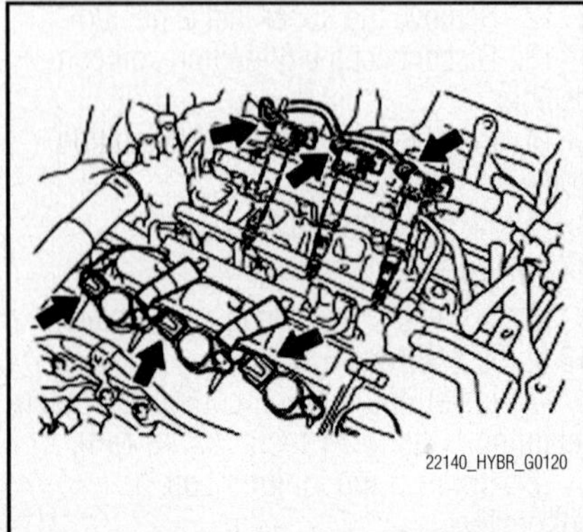

Fig. 34 Ignition coil and connector view

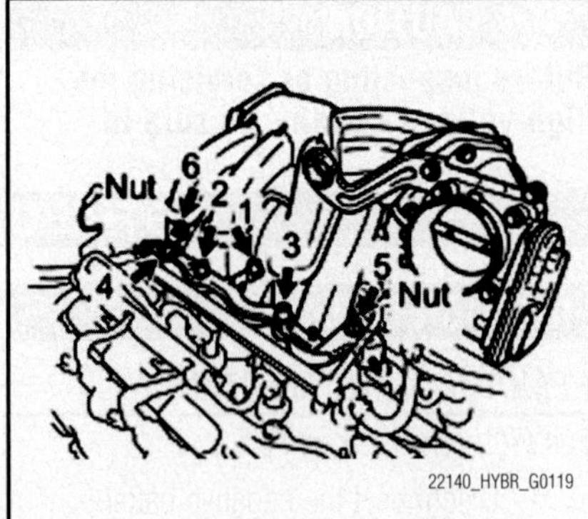

Fig. 35 Upper intake tightening sequence

24. Add engine coolant.
25. Install engine room side cover.
26. Install cool air intake duct seal.
27. Install the outer cowl top panel.
28. Install the windshield wiper motor and link assembly.
29. Install the top cowl and hood seal.
30. Install the LH and RH wiper arm assembly.
31. Connect the negative battery cable.
32. Check for coolant leaks.

➡**When disconnecting the cable, some systems need to be initialized after the cable is reconnected.**

IGNITION TIMING

INSPECTION

See Figure 36.

1. Warm up the engine.
2. Using SST: 09843-18040, connect terminals 13 (TC) and 4 (CG) of the DLC3.

➡**Confirm the terminal numbers before connecting them. Connecting the wrong terminals can damage the engine.**

➡**Turn off all electrical systems before connecting the terminals.**

➡**Perform this inspection after the cooling fan motor is turned off.**

3. Remove the V-bank cover sub-assembly.
4. Pull out the red lead wire harness.
5. Connect the tester terminal of the timing light to the red lead wire as shown in the illustration.

➡**Use a timing light which detects the No. 1 cylinder ignition signal.**

6. Check the ignition timing at idle. Standard ignition timing: 8 to 12° BTDC at idle.

➡**When checking the ignition timing, the transmission should be in the neutral position.**

➡**Run the engine at 1000 to 1300 RPM for 5 seconds, and then check that the engine RPM returns to idle speed.**

7. Disconnect terminals 13 (TC) and 4 (CG) of the DLC3.
8. Check the ignition timing at idle. Standard ignition timing: 7 to 24° BTDC at idle.

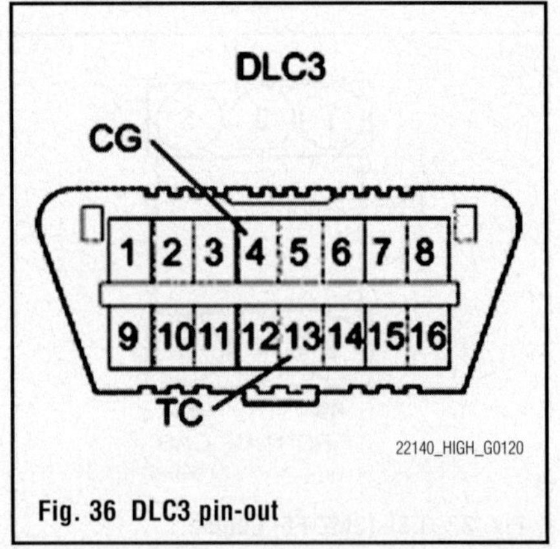

Fig. 36 DLC3 pin-out

9. Confirm that the ignition timing advances immediately when the engine RPM is increased.

10. Remove the timing light from the engine.

ADJUSTMENT

All engines are equipped with a Distributorless Ignition System (DIS). No timing adjustment is possible.

SPARK PLUGS

REMOVAL & INSTALLATION

See Figures 37 and 38.

1. Disconnect the negative battery cable.
2. Remove the RH and LH wiper arm assembly.
3. Remove the cowl and seal.
4. Remove the wiper motor and link assembly.
5. Remove the outer cowl top panel.
6. Remove the cool air intake duct seal.
7. Remove the LH engine side cover.
8. Drain the engine coolant.
9. Remove air cleaner cap and case assembly.
10. Remove the air cleaner bracket.
11. Remove the emission control valve set.
12. Remove the upper intake plenum.
13. Disconnect the 6 ignition coil connectors.

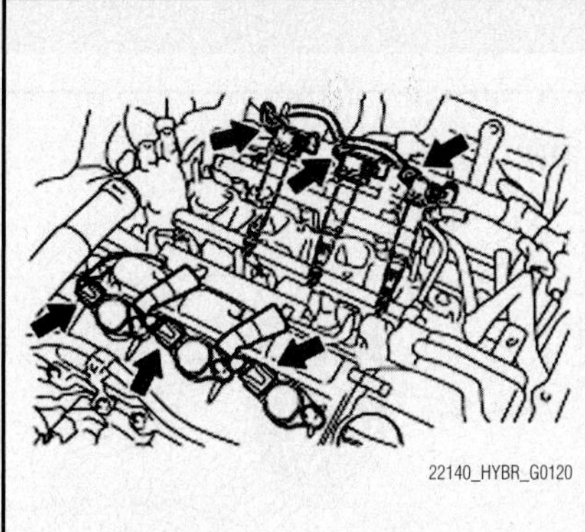

Fig. 37 Ignition coil and connector view

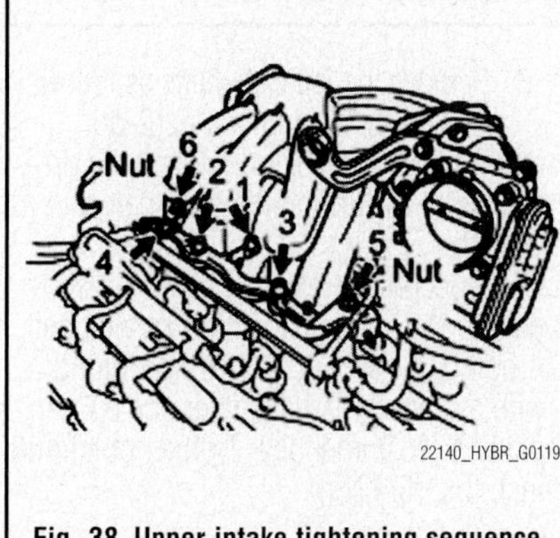

Fig. 38 Upper intake tightening sequence

14. Remove the 6 bolts and 6 ignition coils.
15. Remove the 6 spark plugs.

To install:

16. Install the spark plug and tighten to 13 ft. lbs. (18 Nm).
17. Install the ignition coils and tighten retaining bolts to 71 inch. lbs. (8 Nm).
18. Connect the ignition coil connectors.
19. Install intake plenum with a new gasket
20. Using a socket hexagon wrench 8 mm, install the upper intake plenum with the 4 bolts and 2 nuts.
21. Using several steps, tighten the bolts and nuts uniformly to 21 ft. lbs. (28 Nm) follow the sequence shown in the illustration.
22. Install emission control vale set.
23. Install air cleaner bracket, case and cap.
24. Add engine coolant.
25. Install engine room side cover.
26. Install cool air intake duct seal.
27. Install the outer cowl top panel.
28. Install the windshield wiper motor and link assembly.
29. Install the top cowl and hood seal.
30. Install the LH and RH wiper arm assembly.
31. Connect the negative battery cable.
32. Check for coolant leaks.

➡When disconnecting the cable, some systems need to be initialized after the cable is reconnected.

ENGINE ELECTRICAL — STARTING SYSTEM

STARTER

REMOVAL & INSTALLATION

The 3.3L engine has a DC electric converter and therefore does not use a standard starter.

ENGINE MECHANICAL

CAMSHAFT AND VALVE LIFTERS

REMOVAL & INSTALLATION

See Figures 39 through 54.

⁂ WARNING

The thrust clearance on both the intake and exhaust camshafts is very small; the camshafts must be kept level during removal. If the camshafts are removed without being kept level, the camshaft may be caught in the cylinder head, causing the head to break or the camshaft to seize.

1. Before servicing the vehicle, refer to the Precautions Section.
2. Relieve the fuel system pressure.
3. Drain the engine oil.
4. Drain the coolant from the engine radiator and hybrid transaxle.
5. Remove or disconnect the following:
   - Negative battery cable
   - Engine cover
   - Right-hand front wheel
   - Engine splash shields
   - Right-hand front fender apron seal
   - Wiper and blade assembly
   - Top cowl ventilator louver assembly
   - Wiper motor and link assembly
   - Battery and battery tray
   - Air intake assembly
   - Emission control valve hoses
   - Air intake surge tank
   - Radiator intake hose
   - Brake master cylinder reservoir and bracket
   - Air cleaner bracket
   - Engine moving control rod
   - Right-hand engine No. 2 mounting stay
   - Ignition coil
   - Valve covers
   - Crankshaft pulley
   - Timing belt No.1 and No. 2 covers
   - Right-hand engine mounting bracket
   - No. 2 timing belt guide
   - Timing belt
   - Timing belt idler
6. Using Special Tool 09960-10010, remove the camshaft timing pulleys.

➡Keep all valvetrain components in order for reassembly.

7. Disconnect the engine wiring harness clamps from the No.3 timing belt cover and remove the cover.

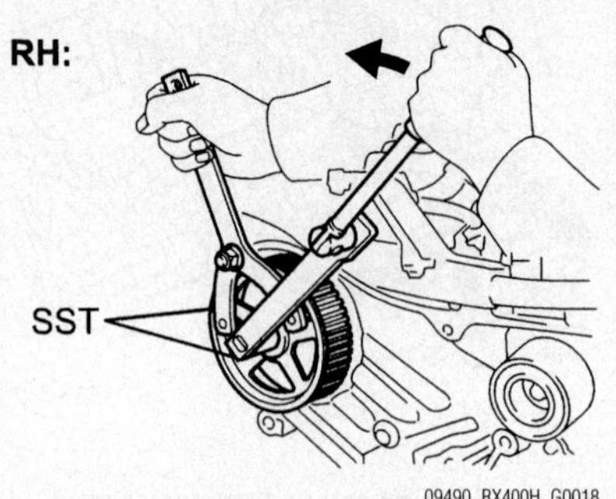

Fig. 39 Removing the right-hand camshaft timing pulley, left-hand similar.

8. Remove the left camshafts as follows:
   a. Align the timing marks (2-dot mark) of the camshaft drive and the driven gears by turning the camshaft with a wrench.
   b. Secure the exhaust camshaft sub-gear to the main gear with a service bolt. A bolt 0.63–0.79 in. (16–20mm) long with a 6mm thread diameter and a 1mm pitch is recommended. Tighten bolt to 48 inch lbs. (5.4 Nm).

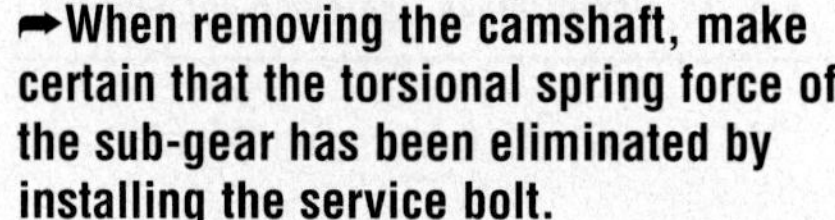

➡When removing the camshaft, make certain that the torsional spring force of the sub-gear has been eliminated by installing the service bolt.

   c. Using several steps, loosen and remove the 10 bearing cap bolts uniformly in the sequence shown.
   d. Remove the 5 bearing caps and the exhaust camshaft.
   e. Using several steps, loosen and remove the 10 bearing cap bolts uniformly in the sequence shown.
   f. Remove the 5 bearing caps and the intake camshaft.
   g. Remove the oil seal from the intake camshaft.

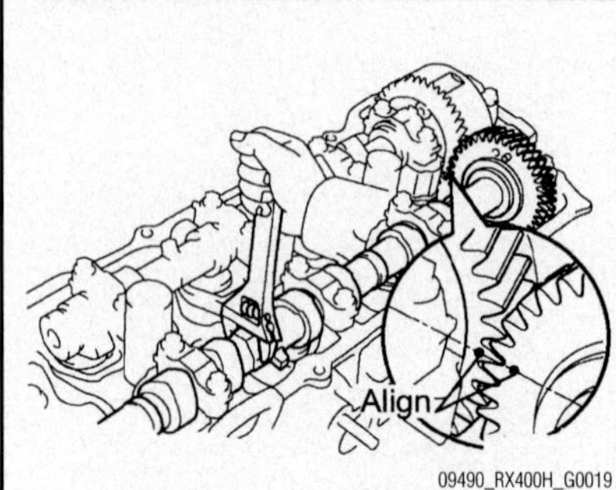

Fig. 40 Align the timing marks of the camshaft gears.

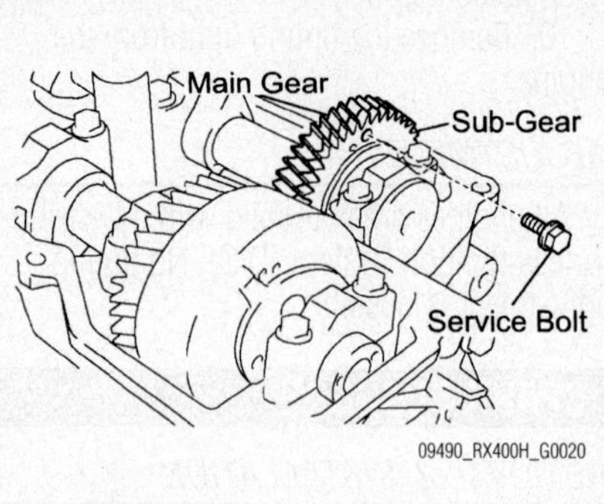

Fig. 41 Install a service bolt to secure the camshaft gears.

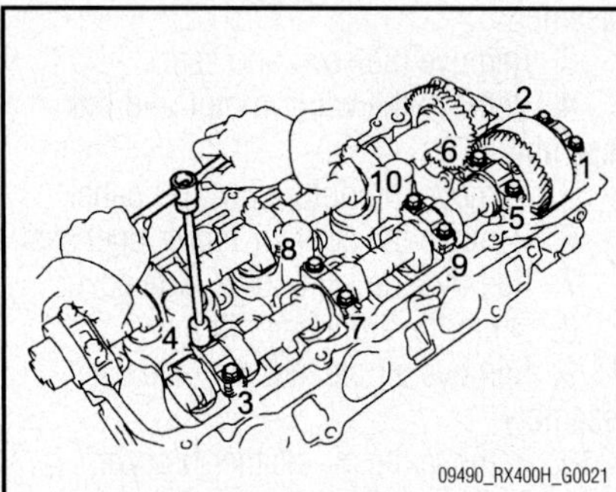

Fig. 42 Intake camshaft bearing cap loosening sequence

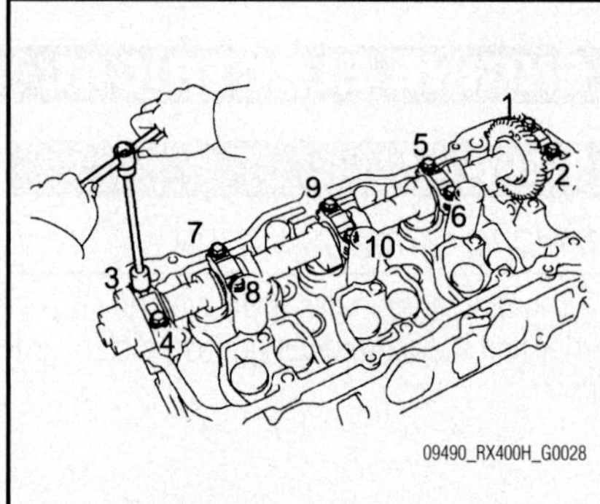

Fig. 43 Exhaust camshaft bearing cap loosening sequence

9. Repeat the same process to remove the right-side camshafts, beginning with the intake camshaft.
10. Remove the valve lifter shims and hydraulic lifters. Identify each lifter and shim as it is removed so it can be reinstalled in the same position. If the lifters are to be reused, store them upside down in a sealed container.

To install:

11. Install the valve lifters into their original positions and install the shims. Check

valve clearance and replace the shims as necessary.

12. When reinstalling, remember that the camshafts must be handled carefully and kept straight and level to avoid damage.

13. Install the right camshafts, as follows:

a. Apply new engine oil to the thrust portion and journal of the camshaft.

b. Position the exhaust camshaft on the head so that the alignment marks are at a 90 degrees angle from vertical.

c. Apply multi-purpose grease to the lip of a new oil seal.

d. Install the oil seal to the camshaft.

e. Apply sealant to the No. 1 bearing cap.

f. Apply a light coat of clean engine oil to the bolt threads and under the bolt head. Install the bearing caps to their proper position. Tighten the bolts evenly and in several passes to 12 ft. lbs. (16 Nm) in the proper sequence.

g. Position the intake camshaft on the head so that the alignment marks are at a 90 degrees angle from vertical. The mark should be at the "9 o'clock" position and must align with the marks on the other gear.

h. Apply a light coat of clean engine oil to the bolt threads and under the bolt head. Install the bearing caps to their proper position. Tighten the bolts evenly and in several passes to 12 ft. lbs. (16 Nm) in the proper sequence.

i. Remove the service bolt.

14. Install the left camshafts, as follows:

a. Apply new engine oil to the thrust portion and journal of the camshaft

b. Position the exhaust camshaft on the head so that the alignment mark is at a 90 degrees angle from vertical. The mark should be at the "9 o'clock" position.

c. Apply multi-purpose grease to the oil seal lip and install the new oil seal to the camshaft.

d. Apply sealant to the No. 1 bearing cap.

e. Apply a light coat of clean engine oil to the bolt threads and under the bolt head. Install the bearing caps to their proper position. Tighten the bolts evenly and in several passes to 12 ft. lbs. (16 Nm) in the proper sequence.

f. Position the intake camshaft on the head so that the alignment marks are at a 90 degrees angle from vertical. The mark should be at the "3 o'clock" position and must align with the marks on the exhaust camshaft gear.

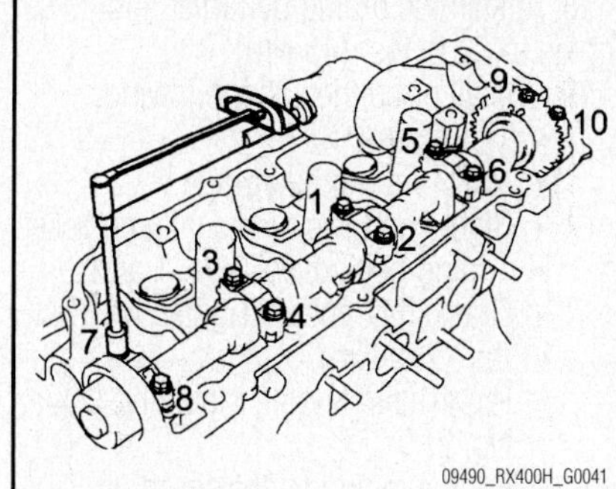

Fig. 46 Right exhaust camshaft bearing cap torque sequence

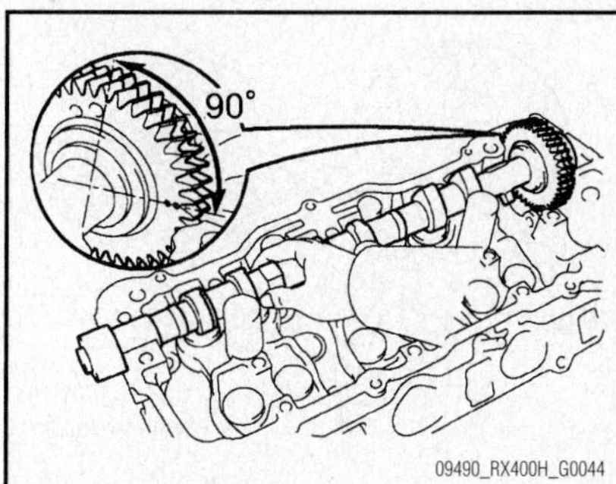

Fig. 49 Install the left exhaust camshaft with the alignment mark in the correct position.

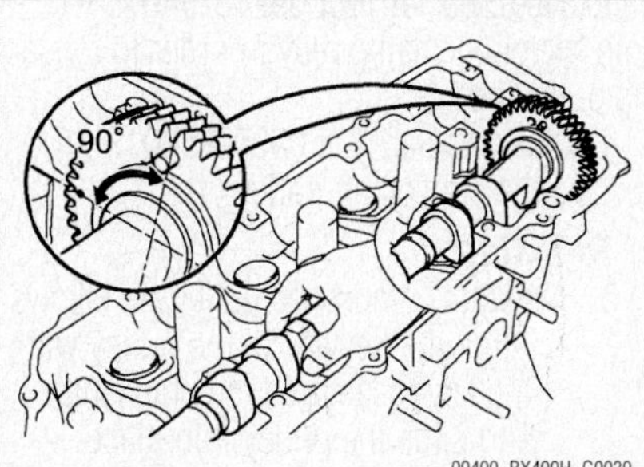

Fig. 44 Install the right exhaust camshaft with the alignment marks in the correct position.

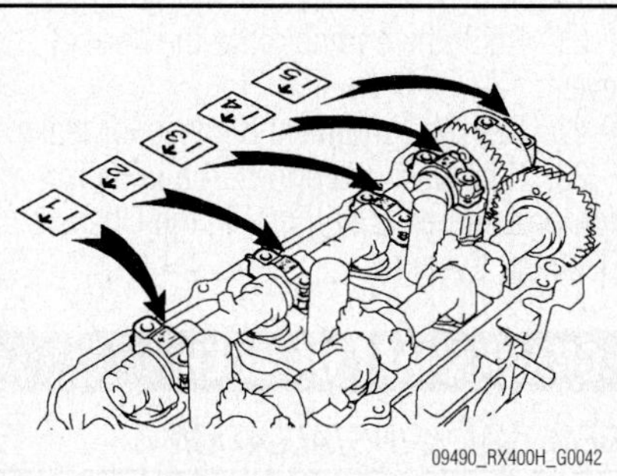

Fig. 47 Right intake camshaft bearing caps must be placed in their proper locations

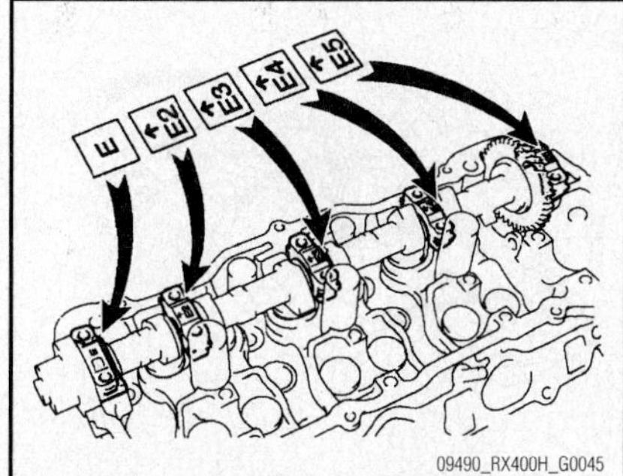

Fig. 50 Left exhaust camshaft bearing caps must be placed in their proper locations

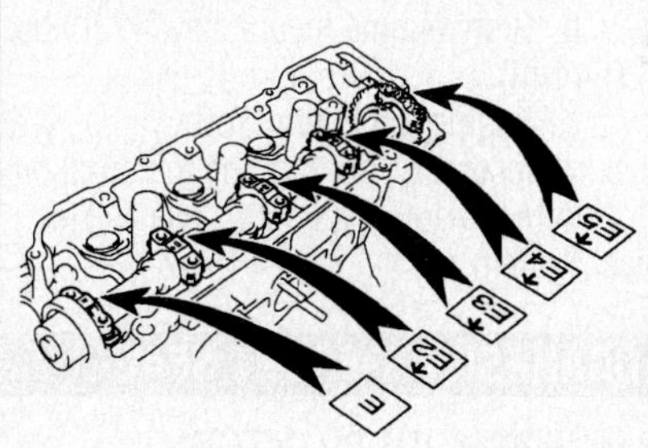

Fig. 45 Right exhaust camshaft bearing caps must be placed in their proper locations.

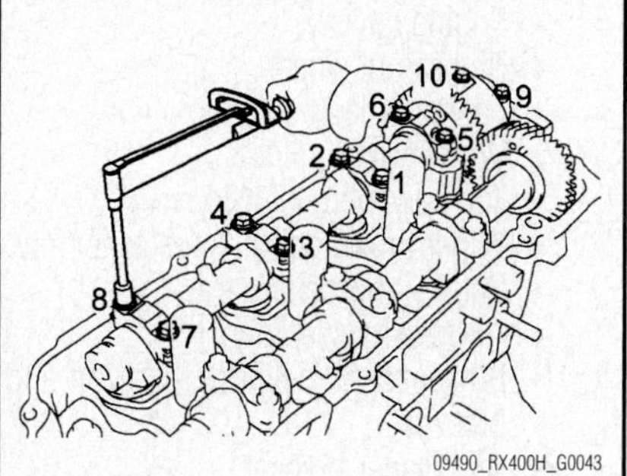

Fig. 48 Right intake camshaft bearing cap bolt tightening sequence

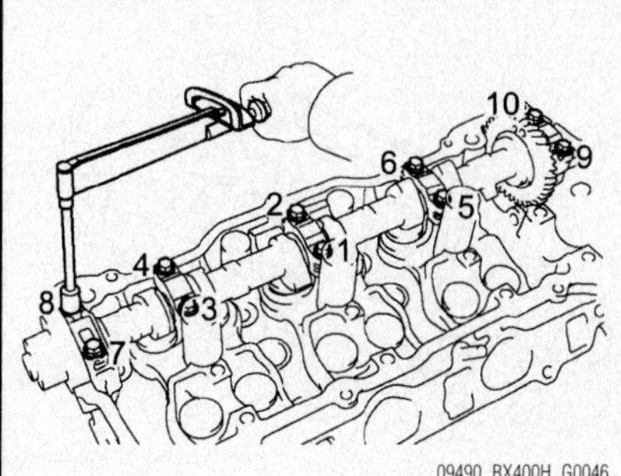

Fig. 51 Left exhaust camshaft bearing cap torque sequence

g. Apply a light coat of clean engine oil to the bolt threads and under the bolt head. Install the bearing caps to their proper position. Tighten the bolts evenly and in several passes to 12 ft. lbs. (16 Nm) in the proper sequence.

h. Remove the service bolt.

15. Install or connect the following:

16. Install the timing belt cover No. 3. Tighten to 76 inch lbs. (8.5 Nm)

17. Using Special Tool 09960-10010, install the camshaft timing pulleys. Tighten to 92 ft. lbs. (125 Nm).

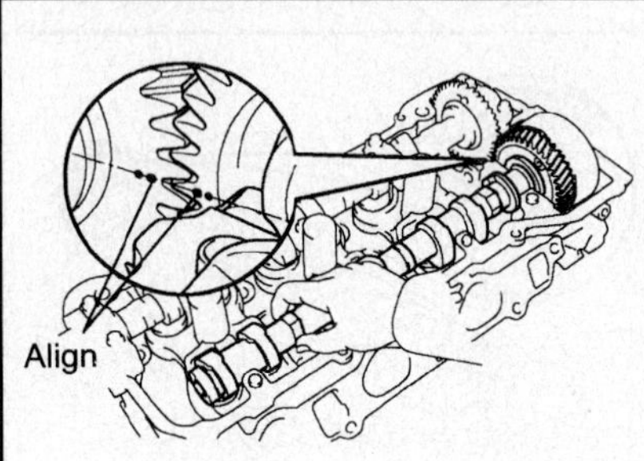

Fig. 52 Install the left intake camshaft with the alignment mark in the correct position.

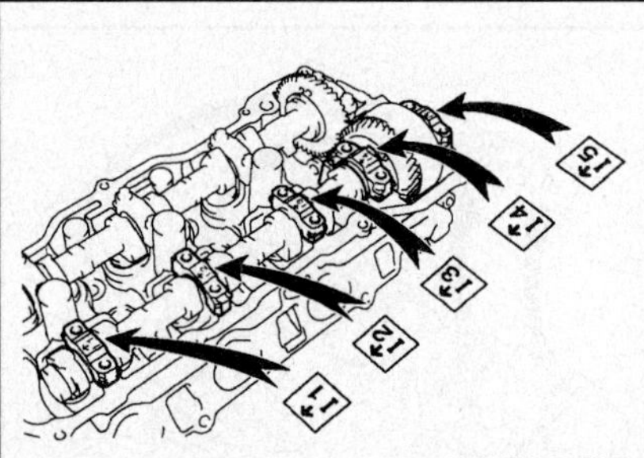

Fig. 53 Left exhaust camshaft bearing caps must be placed in their proper locations.

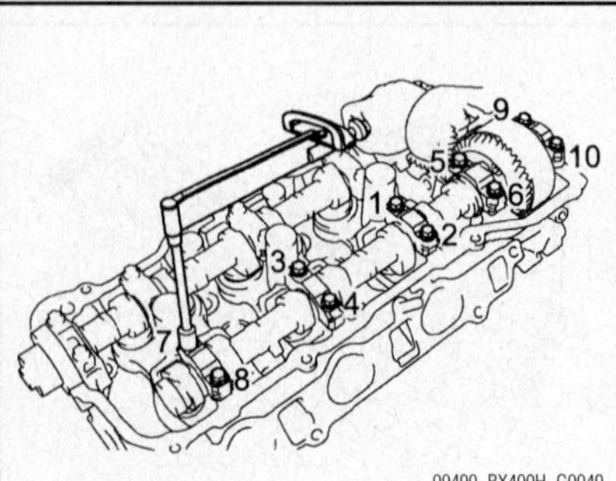

Fig. 54 Left exhaust bearing cap torque sequence

18. Install the timing belt idler and tighten to 32 ft. lbs. (43 Nm).

19. Install or connect the following:

- Timing belt
- No. 2 Timing belt guide
- Right-hand engine mounting bracket
- Timing belt covers Nos. 1 and 2
- Crankshaft pulley. Tighten to 162 ft. lbs. (220 Nm).
- Right-hand engine mounting stay No. 2
- Engine moving control rod
- Air cleaner bracket
- Brake master cylinder reservoir and bracket
- Valve covers
- Ignition coil
- Radiator intake hose
- Air intake surge tank
- Emission control valve hoses
- Air intake assembly
- Battery and battery tray
- Wiper motor and link assembly
- Top cowl ventilator louver assembly
- Wiper and blade assembly
- Right-hand front fender apron seal
- Engine splash shields
- Right-hand front wheel
- Engine covers
- Negative battery cable

20. Refill the cooling system to the correct level.

21. Refill the engine oil to the correct level.

22. Start the engine and check for leaks.

23. If necessary, perform the initialization procedure, as outlined in the Chassis Electrical System.

CRANKSHAFT DAMPER

REMOVAL & INSTALLATION

See Figures 55 and 56.

1. Before servicing the vehicle, refer to the precautions section.

2. Remove or disconnect the following:

- Negative battery cable
- Engine covers
- Right front wheel
- Fender splash shields
- Wiper arms
- Top cowl ventilator louver
- Wiper motor and linkage assembly
- Battery and battery tray
- Air intake assembly
- Brake master cylinder reservoir and bracket
- Air cleaner bracket
- Engine moving control rod
- Right-hand engine mounting stay No. 2

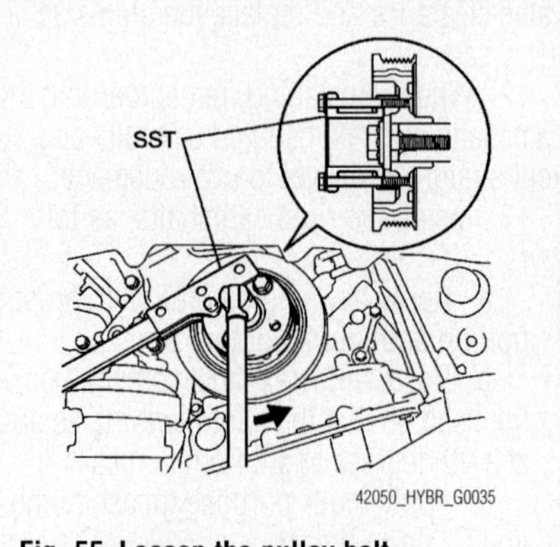

Fig. 55 Loosen the pulley bolt

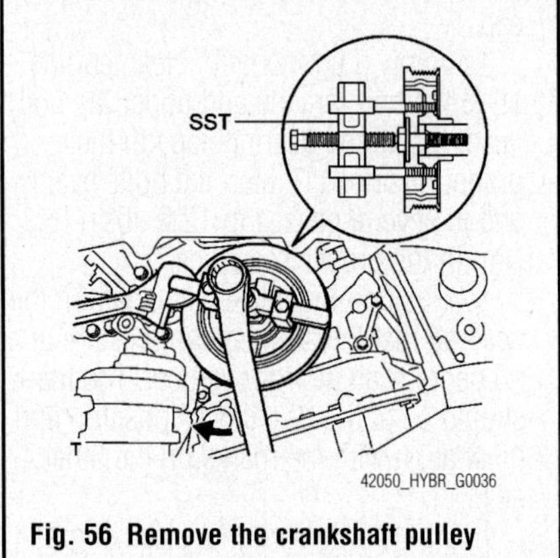

Fig. 56 Remove the crankshaft pulley

3. Use Special Tool 09213-54015 to hold the crankshaft pulley in order to loosen the pulley bolt.

4. Use Special Tool 09950-50013 to remove the crankshaft pulley.

To install:

5. Install the crankshaft pulley, as follows

- Align the keyway of the pulley with the key located on the crankshaft and slide the pulley into place
- Crankshaft pulley. Tighten the bolt to 162 ft. lbs. (219 Nm).

6. The remainder of installation is the reverse of the removal procedure.

7. Observe the following torques:

a. Drive belts by adjusting them

b. Right engine mount stay: 47 ft. lbs. (64 Nm).

c. Engine roll control rod.

8. If necessary, perform the initialization procedure, as outlined in the Chassis Electrical System

CRANKSHAFT FRONT SEAL

REMOVAL & INSTALLATION

See Figures 57 and 58.

1. Before servicing the vehicle, refer to the precautions section.

2. Remove or disconnect the following:
   - Wiper motor and link assembly
   - Top cowl ventilator louver assembly
   - Wiper and blade assembly
   - Right-hand front fender apron seal
   - Engine splash shields
   - Right-hand front wheel
   - Engine covers
   - Battery and battery tray
   - Air cleaner cap and case assembly
   - Master cylinder Sub-assembly
   - Air cleaner bracket
   - Engine moving control rod
   - RH engine mounting stay

3. Use Special Tool 09213-54015 to hold the crankshaft pulley in order to loosen the pulley bolt.

4. Use Special Tool 09950-50013 to remove the crankshaft pulley.

5. Remove the timing belt covers No.1 and No.2.

6. Remove the RH engine mounting bracket.

7. Remove the timing belt as follows:
   - Set No. 1 cylinder to TDC/compression.
   - Temporarily install the crankshaft pulley bolt and washer to the crankshaft.
   - Turn the crankshaft clockwise, and align the timing mark of the crankshaft timing pulley with the oil pump body.
   - Check that the timing marks of the camshaft timing pulleys and No. 3 timing belt cover are aligned.
   - If not, turn the crankshaft 1 revolution (360°).

8. Remove the bolt and the timing belt plate.

9. Install the pulley bolt to the crankshaft.

10. Using SST, remove the crankshaft timing pulley.

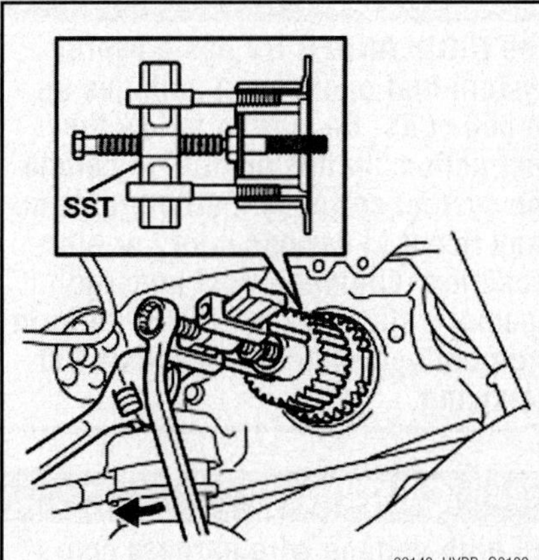

Fig. 57 Remove the timing crankshaft pulley

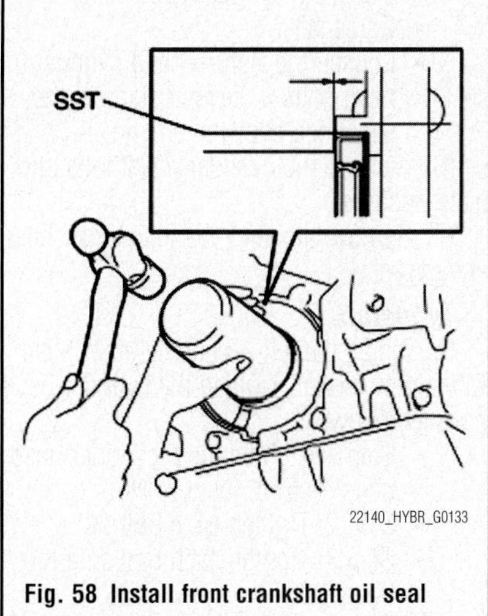

Fig. 58 Install front crankshaft oil seal

11. Using a knife, cut off the oil seal lip.

12. Using a screwdriver with its tip taped, pry out the oil seal.

To install:

13. Apply MP grease to a new oil seal lip.

14. Using SST and a hammer, tap in a new oil seal until its surface is flush with the oil pump edge.

15. Reverse the removal procedure for installation and note the following:
   - Install new cover gaskets
   - Tighten timing belt covers to 74 inch lbs. (8 Nm).
   - Crankshaft pulley. Tighten the bolt to 162 ft. lbs. (219 Nm).

16. Start the vehicle and check for any leaks.

17. Recheck the ignition timing.

18. If necessary, perform the initialization procedure, as outlined in the Chassis Electrical System

CYLINDER HEAD

REMOVAL & INSTALLATION

See Figures 59 through 63.

1. Before servicing the vehicle, refer to the Precautions Section.

⁂ CAUTION

The HIGHLANDER HV has a hybrid system that operates at voltages up to 650 volts. Be sure to follow the instructions in this manual to handle the system correctly. Failure to do so may result in serious injury or electrocution. Engineer must undergo special training to be able to perform high-voltage system inspection and servicing.

⁂ CAUTION

All high-voltage wire harness connectors are colored orange. The HV battery and other high-voltage components have "High Voltage" caution labels. Do not carelessly touch these wires and components.

2. Before inspecting or servicing the high-voltage system, be sure to follow safety measures, such as wearing insulated gloves and removing the service plug to prevent electrocution. Carry the removed service plug in your pocket to prevent anyone from reinstalling it while you are servicing the vehicle.

3. After removing the service plug, wait 5 minutes before touching any of the high-voltage connectors and terminals.

➡Wear insulating gloves and protective glasses.

4. Relieve the fuel system pressure.

5. Drain the engine oil.

6. Drain the coolant from the engine radiator and hybrid transaxle.

7. Remove the service plug grip, found underneath the Battery Service cover on the rear seat. Wait 5 minutes to discharge the high voltage capacitor.

8. Remove or disconnect the following:
   - Negative battery cable
   - Engine cover
   - Right-hand front wheel
   - Engine splash shields
   - Right-hand front fender apron seal
   - Wiper and blade assembly
   - Top cowl ventilator louver assembly
   - Wiper motor and link assembly
   - Battery and battery tray
   - Air intake assembly
   - Converter with Inverter assembly
   - Emission control valve set
   - Intake air surge tank
   - Fuel supply hose
   - Heater inlet hose
   - Intake manifold
   - Radiator hoses
   - Water outlet from the cylinder heads
   - Brake master cylinder reservoir
   - Air cleaner bracket
   - Engine moving control rod
   - Right-hand No.2 engine mounting stay
   - Crankshaft pulley
   - Timing belt No.1 and No.2 covers
   - Right-hand engine mounting bracket
   - Timing belt guide No.2
   - Timing belt

- Timing belt No.2 idler
- Camshaft timing pulley
- Timing belt No.3 cover
- Front exhaust pipe assembly
- Exhaust manifold heat insulator
- Exhaust manifold stay
- Right-hand exhaust manifold and gasket
- Ignition coil
- Right-hand cylinder head cover
- Camshaft
- VVT sensor connector
- Camshaft timing oil control valve connecter

9. Loosen the right-hand cylinder head bolts in several steps in the sequence shown.

10. Remove the cylinder head bolts and plate washers.

11. Remove the right-hand cylinder head and gasket

12. Remove the manifold converter No.3 insulator.

13. Remove the exhaust manifold No.2 heat insulator.

14. Separate the cooling fan ECU and hang securely with mechanic's wire.

15. Remove or disconnect the following:

- Left-hand exhaust manifold
- Oil level gauge guide
- Water inlet pipe

Fig. 59 Left—hand cylinder head loosening sequence

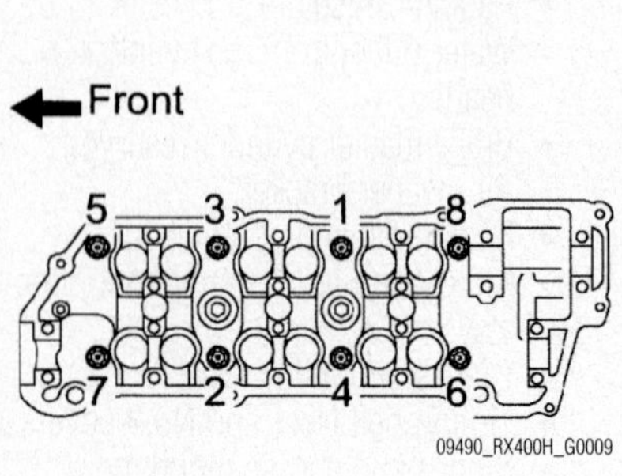

Fig. 60 Right—hand cylinder head loosening sequence

- Left-hand cylinder head cover
- Camshaft
- Loosen the right—hand cylinder head bolts in several steps in the sequence shown.

16. Remove the cylinder head bolts and plate washers.

17. Remove the left—hand cylinder head and gasket.

To install:

18. Install the left—hand cylinder head with a new gasket. Tighten the cylinder head bolts as follows:

- Step 1: Tighten the 8 cylinder head bolts to 40 ft. lbs. (54 Nm)
- Step 2: Tighten each bolt 90°
- Step 3: Tighten each bolt an additional 90°
- Step 4: Tighten the single 8mm hexagon bolt to 14 ft. lbs. (19 Nm)

19. Install the or connect the following:

- Camshaft
- Left-hand cylinder head cover. Tighten to 71 inch lbs. (8 Nm).
- Water inlet pipe
- Oil level gauge guide. Tighten to 71 inch lbs. (8 Nm).
- Cooling fan ECU
- Exhaust manifold No.2 heat insulator

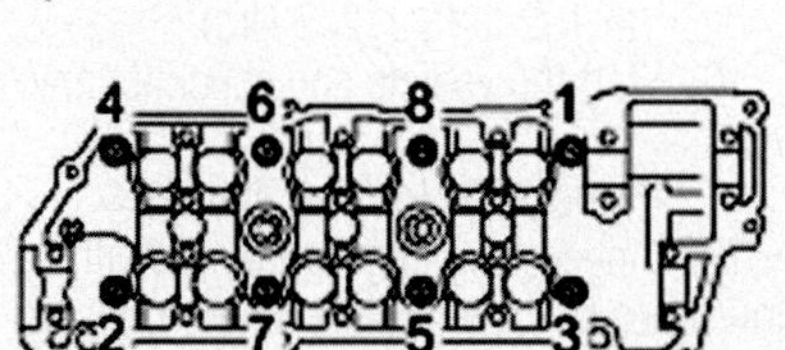

Fig. 61 Left—hand cylinder head tightening sequence

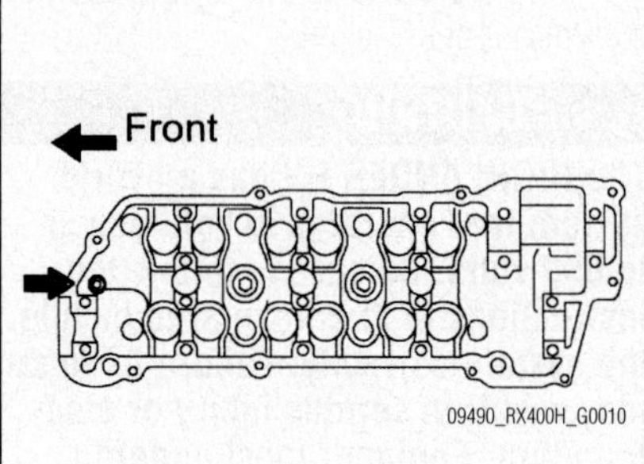

Fig. 62 8mm hexagon bolt on the cylinder head

Fig. 63 Right—hand cylinder head tightening sequence

- Manifold converter No.2 insulator

20. Install the right—hand cylinder head with a new gasket. Tighten the cylinder head bolts as follows:

- Step 1: Tighten the 8 cylinder head bolts to 40 ft. lbs. (54 Nm)
- Step 2: Tighten each bolt 90°
- Step 3: Tighten each bolt an additional 90°
- Step 4: Tighten the single 8mm hexagon bolt to 14 ft. lbs. (19 Nm)

21. The remainder of installation is the reverse order of removal.

22. Refill the engine oil to the correct level.

23. Refill the coolant to the engine radiator and hybrid transaxle to the correct level.

24. Replace the service plug grip.

25. Start the engine and check for leaks.

26. If necessary, perform the initialization procedure, as outlined in the Chassis Electrical System

DRIVE PLATE

REMOVAL & INSTALLATION

See Figures 64 through 66.

CAUTION

The HIGHLANDER HV has a hybrid system that operates at voltages up to 650 volts. Be sure to follow the instructions in this manual to handle the system correctly. Failure to do so may result in serious injury or electrocution. Engineer must undergo special training to be able to perform high-voltage system inspection and servicing.

CAUTION

All high-voltage wire harness connectors are colored orange. The HV battery and other high-voltage components have "High Voltage" caution

labels. Do not carelessly touch these wires and components.

1. Before inspecting or servicing the high-voltage system, be sure to follow safety measures, such as wearing insulated gloves and removing the service plug to prevent electrocution. Carry the removed service plug in your pocket to prevent anyone from reinstalling it while you are servicing the vehicle.
2. After removing the service plug, wait 5 minutes before touching any of the high-voltage connectors and terminals.

➡Wear insulating gloves and protective glasses.

➡This procedure requires a variety of special tools.

3. Remove the automatic transaxle assembly (2WD) or automatic transaxle and transfer assembly (4WD), as outlined in the Drive Train Section.
4. Remove the drive plate and ring gear, as follows:
    a. Secure the crankshaft with Special Service Tool (SST) 09213-54015 (91651-60855), 09330-00021 or their equivalents.
    b. Remove the 8 bolts, rear spacer, drive plate and front spacer.

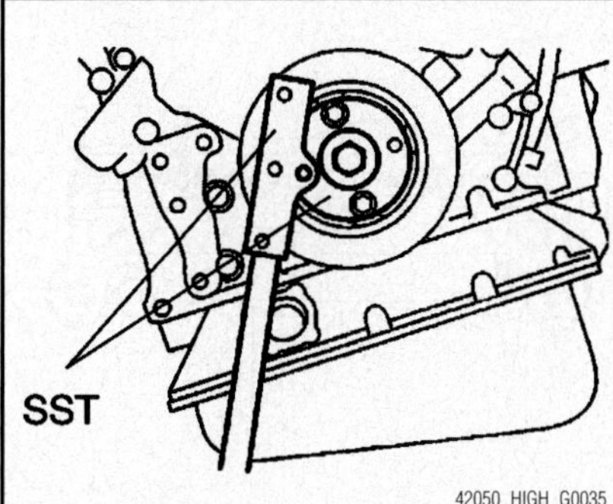

Fig. 64 Hold the crankshaft in place with the special tools as shown

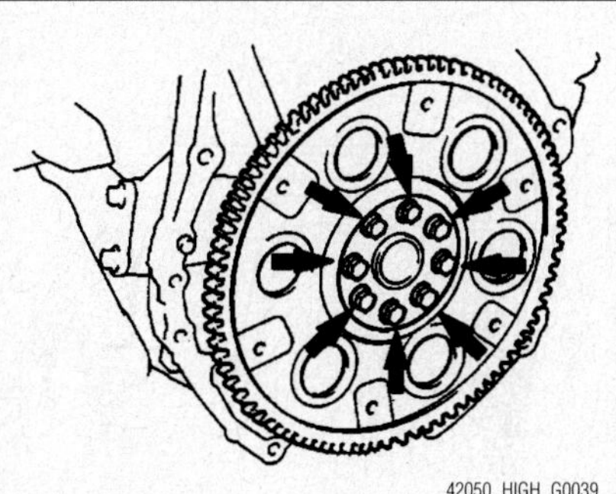

Fig. 65 Remove the 8 bolts, rear spacer, drive plate and front spacer

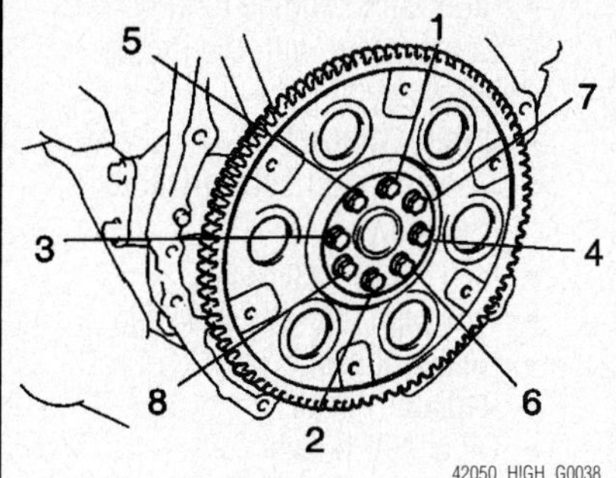

Fig. 66 Drive plate bolt tightening sequence

To install:

5. Install drive plate and ring gear, as follows:
    a. Fix the crankshaft with SST 09213-54015 (91651-60855), 09330-00021, or their equivalents.
    b. Clean the bolts and the bolt holes.
    c. Apply a suitable adhesive (part no. 08833-00070, three bond or equivalent) to 2 or 3 threads of the bolt end.
    d. Install and uniformly tighten the 8 bolts, in several passes in the sequence shown in the accompanying illustration to a final torque of 61 ft. lbs. (83 Nm).

⁂ WARNING

Do not start the engine for AT LEAST one hour after installing the seal!

6. Install the automatic transaxle assembly (2WD) or automatic transaxle and transfer assembly (4WD), as outlined in the Drive Train Section.
7. If necessary, perform the initialization procedure, as outlined in the Chassis Electrical System.

ENGINE ASSEMBLY

REMOVAL & INSTALLATION

See Figures 67 through 69.

1. Before servicing the vehicle, refer to the Precautions Section.

⁂ CAUTION

The HIGHLANDER HV has a hybrid system that operates at voltages up to 650 volts. Be sure to follow the instructions in this manual to handle the system correctly. Failure to do so may result in serious injury or electrocution. Engineer must undergo special training to be able to perform high-voltage system inspection and servicing.

⁂ CAUTION

All high-voltage wire harness connectors are colored orange. The HV battery and other high-voltage components have "High Voltage" caution labels. Do not carelessly touch these wires and components.

2. Before inspecting or servicing the high-voltage system, be sure to follow safety measures, such as wearing insulated gloves and removing the service plug to prevent electrocution. Carry the removed service plug in your pocket to prevent anyone from reinstalling it while you are servicing the vehicle.
3. After removing the service plug, wait 5 minutes before touching any of the high-voltage connectors and terminals.

➡Wear insulating gloves and protective glasses.

4. Drain the coolant from the engine and hybrid assemblies.
5. Drain the engine oil.
6. Drain the transfer case fluid, if equipped.
7. Drain the hybrid transaxle fluid.
8. Discharge the fuel system pressure.
9. Remove the service plug grip, found underneath the Battery Service cover on the rear seat. Wait 5 minutes to discharge the high voltage capacitor.
10. Remove or disconnect the following:
    - Engine room side covers
    - Front wheels
    - Engine under cover assembly
    - Left and right fender splash shields
    - Left and right fender apron seals
    - Wiper arms
    - Cowl top ventilator louver
    - Wiper linkage
    - Cowl top panel outer sub-assembly
    - Battery and battery tray
    - Air intake assembly
    - Inverter support bracket No. 5
    - Power steering ECU assembly
11. Remove the inverter with converter assembly as follows:
    a. Disconnect circuit breaker sensor No. 1.
    b. Disconnect engine room wire No. 2.
    c. Remove the inverter reserve tank sub-assembly.
    d. Disconnect the coolant hose.
    e. Disconnect the power steering ECU bracket.
    f. Remove the inverter cover.
    g. Verify the voltage of w/ converter inverter assembly is 0 volts.
    h. Disconnect engine wire No. 4.

i. Disconnect the high voltage cable of the front motor.

j. Disconnect the Motor Generator ECU connector.

k. Disconnect No. 3 wire frame.

l. Install the inverter cover.

m. Separate the engine room relay block assembly.

n. Remove the inverter bracket No. 4.

o. Remove the w/ converter inverter assembly.

12. Remove or disconnect the following:
- Brake master cylinder reservoir sub-assembly
- Master cylinder reservoir bracket
- Air intake assembly bracket
- Engine moving control rod
- Right-hand engine mount
- Compressor with the motor assembly
- Inverter bracket No. 1
- Transmission control cable assembly
- Fuel vapor feed hose
- Fuel pipe sub-assembly No. 1
- Heater hoses
- Radiator hoses
- Oil cooler hoses
- Water pump assembly
- Glove compartment door assembly
- Engine wire harnesses from the ECU
- Front exhaust pipe assembly
- Front stabilizer link assembly
- Front axle hub nut, both sides
- Wheel speed sensors
- Tie rod end sub-assemblies
- Lower control arms

13. Separate the halfshafts from the axle hub.

14. Separate the steering intermediate shaft sub-assembly.

15. Attach a lifting crane.

16. Remove the left and right hand frame side rail plates.

17. Remove the front suspension member rear braces.

18. Lift the engine/transaxle from the vehicle.

19. Installation is the reverse of removal. Observe the following torques:
- Frame side rail plates—Bolt A: 63 ft. lbs. (85 Nm). Bolt B: 24 ft. lbs. (32 Nm). Nut C: 24 ft. lbs. (32 Nm).
- Front suspension member brace—Bolt A: 63 ft. lbs. (85 Nm). Bolt B: 24 ft. lbs. (32 Nm). Nut C: 24 ft. lbs. (32 Nm).
- Front exhaust pipe assembly: 41 ft. lbs. (56 Nm).
- Front stabilizer link assembly: 55 ft. lbs. (74 Nm).
- Steering link: 26 ft. lbs. (35 Nm)

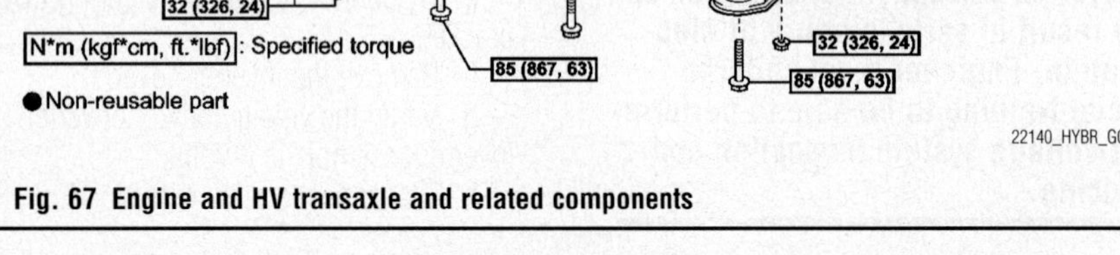

Fig. 67 Engine and HV transaxle and related components

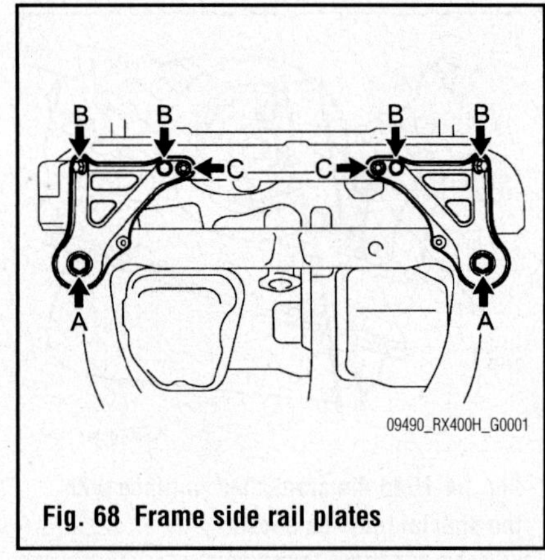

Fig. 68 Frame side rail plates

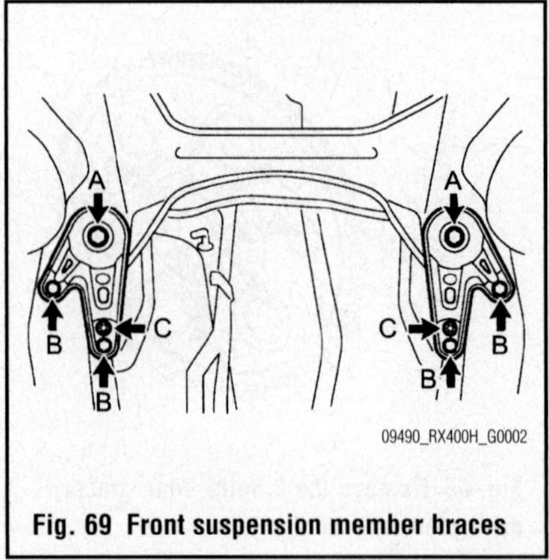

Fig. 69 Front suspension member braces

20. If necessary, perform the initialization procedure, as outlined in the Chassis Electrical System.

EXHAUST MANIFOLD

REMOVAL & INSTALLATION

Front

See Figure 70.

➡Removing the oil filter helps gain access to a lower bolt in the front exhaust manifold.

1. Before servicing the vehicle, refer to the Precautions Section.
2. Remove or disconnect the following:
    - Negative battery cable
    - Engine under covers
    - Front exhaust pipe from the exhaust manifolds, by removing the nuts

➡Check for access to some of the manifold lower bolts, if so remove any possible.

- Heated Oxygen (HO_2) sensor
- Exhaust manifold stay, by removing the bolt and nut
- Remaining exhaust manifold nuts; then, separate the exhaust manifold from the engine

To install:

3. Install or connect the following:
    - Exhaust manifold, using a new gasket. Uniformly, tighten the bolts to 36 ft. lbs. (49 Nm).
    - Exhaust manifold stay. Tighten the nut/bolt to 15 ft. lbs. (20 Nm).
    - Heated Oxygen (HO_2) sensor to the exhaust manifold
    - Front exhaust pipe to the exhaust manifold, using a new gasket. Tighten both nuts to 41 ft. lbs. (56 Nm).

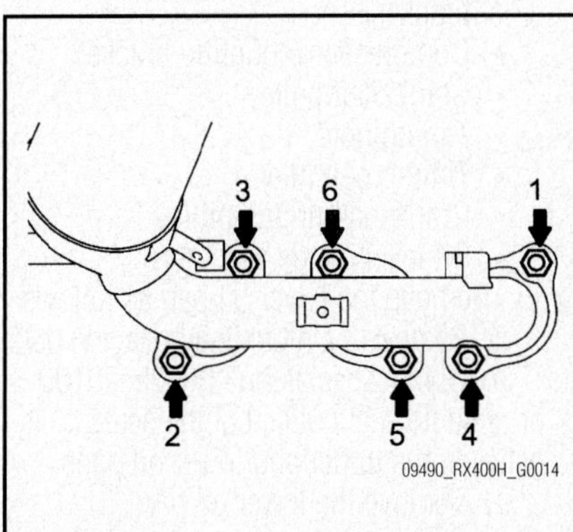

Fig. 70 Front manifold nut locations

- Engine under covers
- Negative battery cable

4. If necessary, perform the initialization procedure, as outlined in the Chassis Electrical System.

Rear

See Figures 71 and 72.

1. Before servicing the vehicle, refer to the Precautions Section.
2. Remove or disconnect the following:
    - Negative battery cable
    - Engine under covers

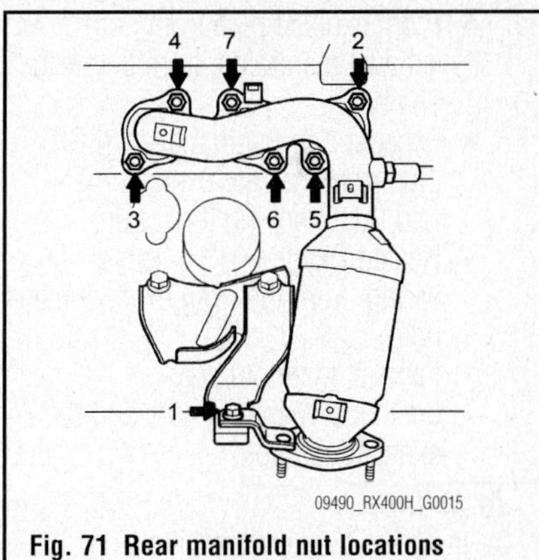

Fig. 71 Rear manifold nut locations

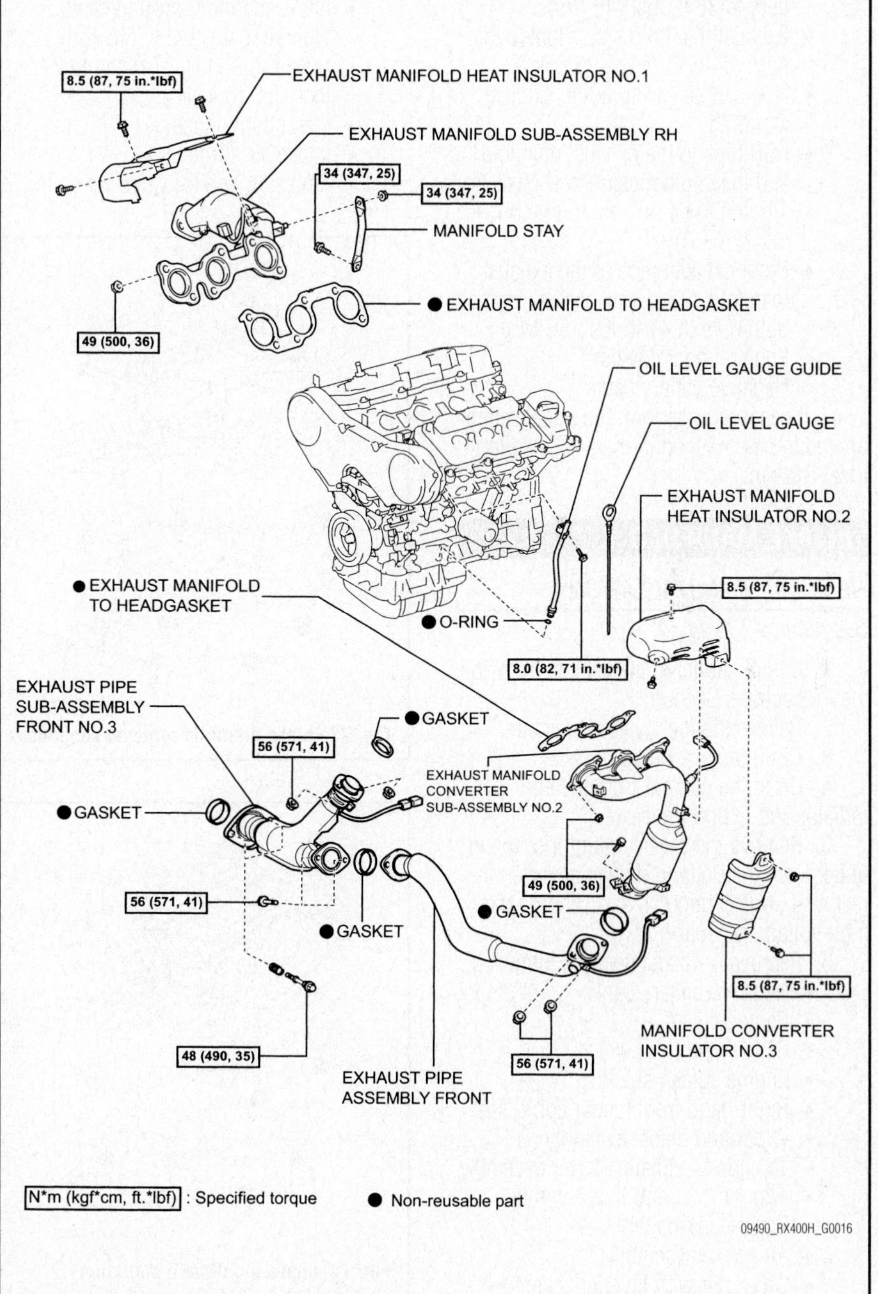

Fig. 72 Exploded view of the exhaust system

- Front exhaust pipe from both exhaust manifolds, from below the engine
- Exhaust Gas Recirculation (EGR) pipe from the rear exhaust manifold, by removing the 4 nuts
- Heated Oxygen (HO_2) sensor wiring, from the right exhaust manifold
- Exhaust manifold stay
- 6 exhaust manifold nuts and the exhaust manifold

To install:

3. Install or connect the following:
   - Exhaust manifold to the engine, using a new gasket. Tighten the 6 nuts to 36 ft. lbs. (49 Nm).
   - Exhaust manifold stay. Tighten the nut/bolt to 25 ft. lbs. (34 Nm).
   - HO_2 sensor wiring to the exhaust manifold
   - EGR pipe to the exhaust manifold and the engine, using new gaskets. Tighten the 4 nuts to 108 inch lbs. (12 Nm).
   - Front exhaust pipe to the exhaust manifold, use a new gasket. Tighten both nuts to 41 ft. lbs. (56 Nm).
   - Engine under covers
   - Negative battery cable
4. If necessary, perform the initialization procedure, as outlined in the Chassis Electrical System.

INTAKE MANIFOLD

REMOVAL & INSTALLATION

See Figures 73 and 74.

1. Before servicing the vehicle, refer to the Precautions Section.
2. Relieve the fuel system pressure.
3. Drain the engine oil.
4. Drain the coolant from the engine radiator and hybrid transaxle.
5. Remove the service plug grip, found underneath the Battery Service cover on the rear seat. Wait 5 minutes to discharge the high voltage capacitor.
6. Remove or disconnect the following:
   - Negative battery cable
   - Engine cover
   - Right-hand front wheel
   - Engine splash shields
   - Right-hand front fender apron seal
   - Wiper and blade assembly
   - Top cowl ventilator louver assembly
   - Wiper motor and link assembly
   - Battery and battery tray
   - Air intake assembly
   - Converter with Inverter assembly
   - Emission control valve set
   - Intake air surge tank
   - Fuel supply hose
   - Heater inlet hose
   - Intake manifold ground cable
   - Fuel injector connectors
7. Loosen the intake manifold mounting bolts in several steps, in sequence as shown.
8. Remove the intake manifold and gaskets.

To install:

9. Install the intake manifold and gaskets. Tighten the bolts in sequence to 11 ft. lbs. (15 Nm).
10. Install or connect the following:
    - Fuel injector connectors
    - Intake manifold ground cable. Tighten to 11 ft. lbs. (15 Nm).
    - Heater inlet hose. Tighten to 74 inch lbs. (8.4 Nm).
    - Fuel supply hose
    - Intake air surge tank
    - Emission control valve set

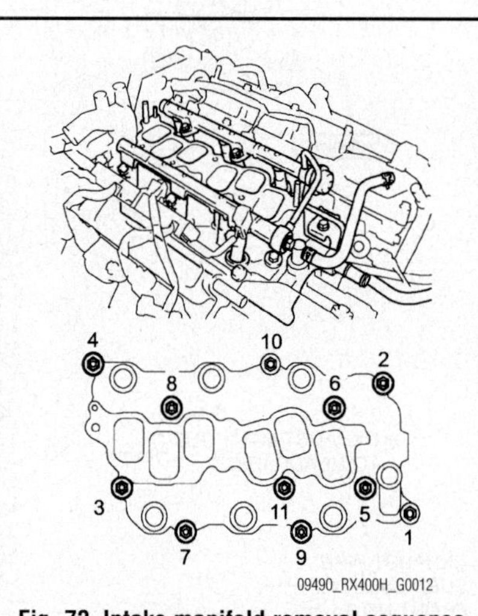

Fig. 73 Intake manifold removal sequence

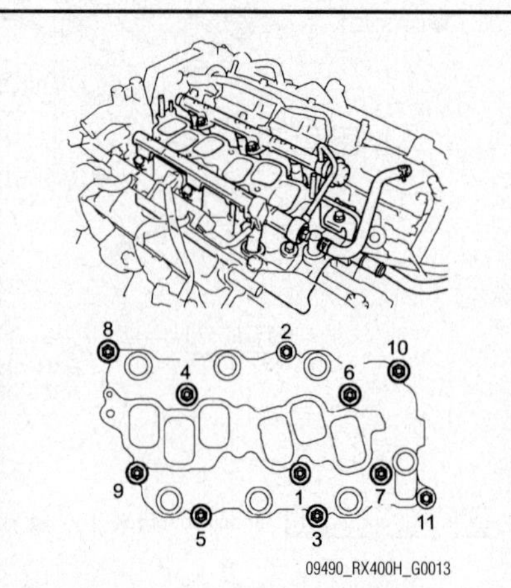

Fig. 74 Intake manifold installation sequence

- Converter with Inverter assembly
- Air intake assembly
- Battery and battery tray
- Wiper motor and link assembly
- Top cowl ventilator louver assembly
- Wiper and blade assembly
- Right-hand front fender apron seal
- Engine splash shields
- Right-hand front wheel
- Engine cover
- Negative battery cable

11. Refill the engine oil to the correct level.
12. Refill the coolant to the engine radiator and hybrid transaxle to the correct level.
13. Replace the service plug grip.
14. If necessary, perform the initialization procedure, as outlined in the Chassis Electrical System.
15. Start the engine and check for leaks.

OIL PAN

REMOVAL & INSTALLATION

See Figures 75 through 78.

1. Before servicing the vehicle, refer to the Precautions Section.
2. Remove or disconnect the following:
   - Engine/transaxle assembly from the vehicle
   - Right-hand exhaust manifold
   - Transaxle mass damper
   - Front frame assembly
   - Halfshafts
   - Flywheel housing undercover
   - Front engine mounting bracket
   - Transaxle assembly from the engine
   - Transmission input damper assembly
   - Flywheel
3. Install the engine to a suitable engine stand.
4. Remove or disconnect the following:
   - Remaining exhaust manifold heat shields
   - Right-hand engine mounting bracket
   - Compressor mounting bracket
   - Crankshaft pulley
   - Timing belt
   - Timing belt idler
   - Crankshaft timing pulley
   - Oil level gauge assembly
5. Remove the lower oil pan as follows:
   a. Remove the mounting bolts and nuts
   b. Using Special Tool 09032-00100 or suitable seal cutter, cut the sealant between the upper and lower oil pans.
   c. Remove the lower oil pan.
6. Remove the oil strainer and gasket.
7. Remove the upper oil pan as follows:

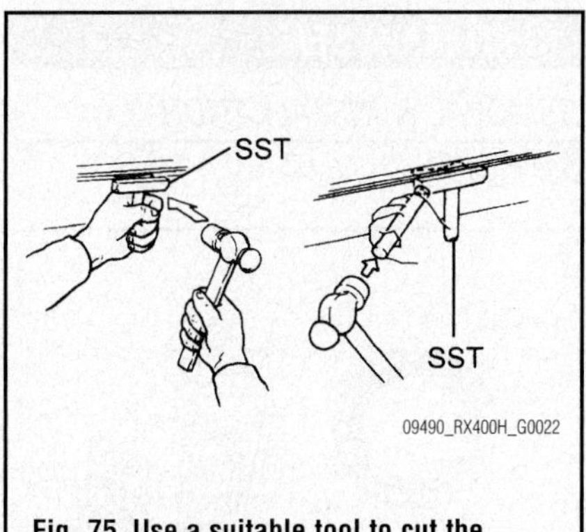

Fig. 75 Use a suitable tool to cut the sealant between the oil pans.

a. Uniformly loosen and remove the mounting bolts.

b. Using a suitable pry tool, pry the upper oil pan from the cylinder block.

To install:

8. Remove any old sealant from the mating surface of the oil pans.

9. Install the upper oil pan as follows:

a. Apply a 0.12–0.16 inch (3–4 mm) wide continuous bead of sealant to the mating surface as shown in the illustration.

b. Install the upper oil pan mounting bolts and tighten in several steps. Tighten bolts 'A' to 71 inch lbs. (8 Nm) and bolts 'B' to 14 ft. lbs. (20 Nm).

10. Install the oil strainer with a new gasket. Tighten to 71 inch lbs. (8 Nm).

11. Install the lower oil pan as follows:

a. Apply a 0.16–0.20 inch (4–5 mm) wide continuous bead of sealant to the mating surface as shown in the illustration.

b. Install the lower oil pan mounting bolts and nuts. Tighten to 71 inch lbs. (8 Nm).

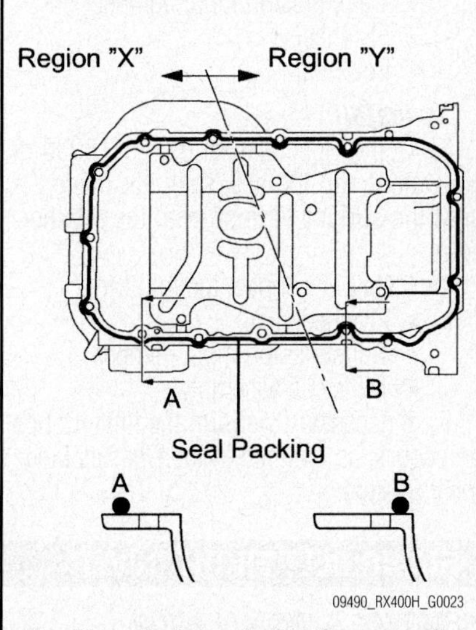

Fig. 76 Apply the sealant to the upper oil pan as shown.

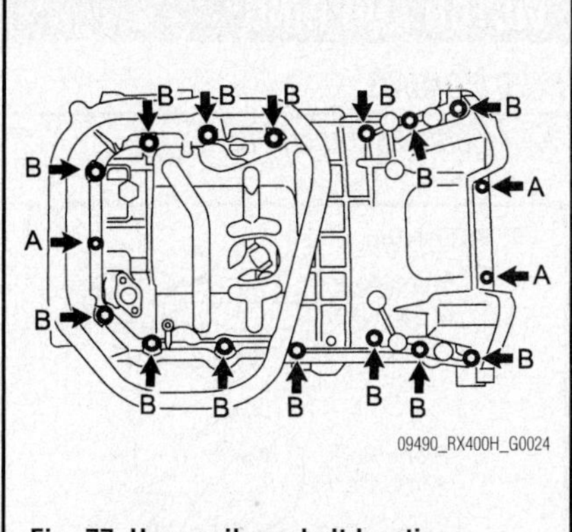

Fig. 77 Upper oil pan bolt locations

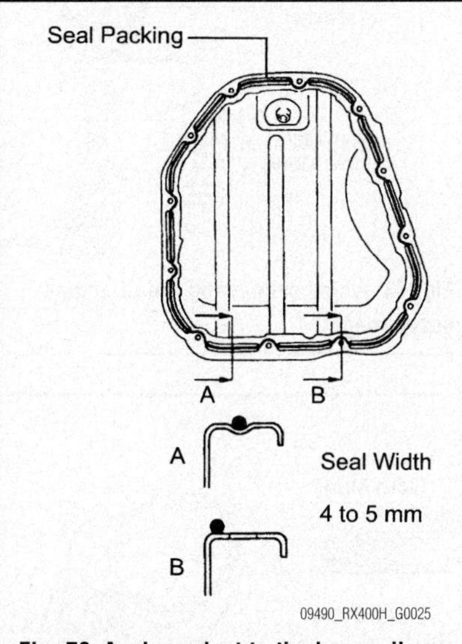

Fig. 78 Apply sealant to the lower oil pan as shown.

12. The remainder of the installation is the reverse order of removal.

13. Refill the engine with oil to the correct level.

14. Start the engine and check for leaks.

15. If necessary, perform the initialization procedure, as outlined in the Chassis Electrical System.

OIL PUMP

REMOVAL & INSTALLATION

See Figures 79 and 80.

1. Before servicing the vehicle, refer to the Precautions Section.
2. Remove the engine with transaxle.
3. Remove the timing belt.
4. Remove the crankshaft gear.
5. Remove or disconnect the following:
   - Upper and lower oil pans
   - Crankshaft Position (CKP) sensor
   - 9 oil pump bolts

➡Make a note of the position of the each bolt. When replacing the bolts into the oil pump body, place each bolt in the position from which it was removed.

- Oil pump body, by prying between the oil pump and main bearing cap
- O-ring from the cylinder block
- Plug, gasket, spring and relief valve from the oil pump body
- 9 screws, pump body cover, drive and driven rotors

To install:

6. Install or connect the following:
   - Driven rotors, drive, pump body cover, using the 9 screws
   - Oil pump relief valve, spring, gasket and the plug to the oil pump body
   - New O-ring on the cylinder block
7. Using a non residue solvent, clean both sealing surfaces to the oil pump.
8. Apply liquid sealant to the oil pump and engine block.
9. Install or connect the following:
   - Oil pump

➡Be sure to engage the splined teeth of the oil pump drive gear with the large teeth of the crankshaft.

- 9 oil pump bolts. Tighten the bolts in several passes to 71 inch lbs. (8 Nm) for bolt 'A'; 14 ft. lbs. (20 Nm), for bolts 'B'; 32 ft. lbs. (43 Nm) for bolt 'C'
- CKP sensor. Tighten the bolt to 71 inch lbs. (8 Nm).
- Upper and lower oil pans
- crankshaft gear

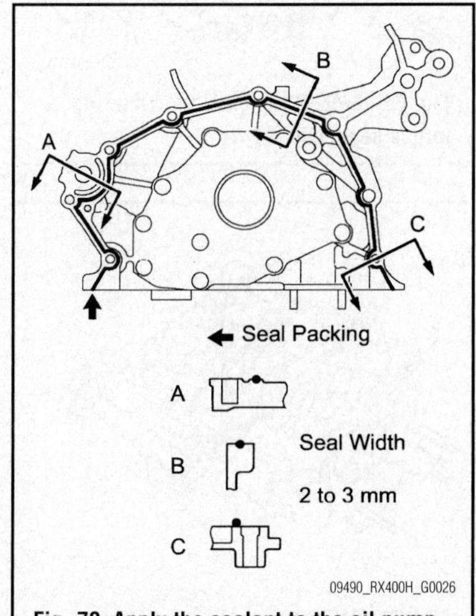

Fig. 79 Apply the sealant to the oil pump as shown

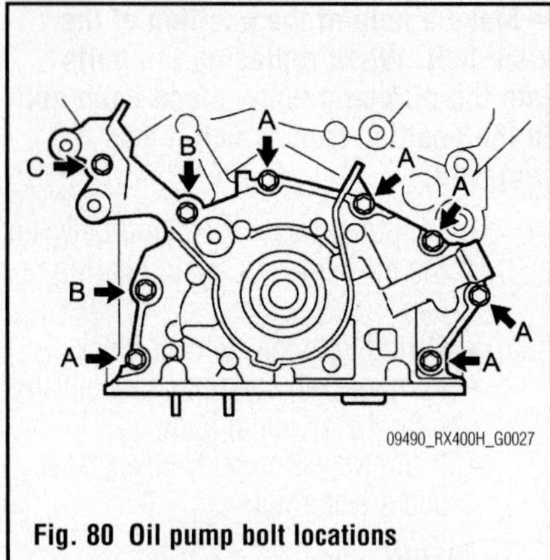

Fig. 80 Oil pump bolt locations

- Timing belt, and covers
- Engine with the transaxle.

10. Refill the engine with oil to the correct level.
11. Start the engine and inspect for leaks.
12. Recheck the engine oil level.
13. If necessary, perform the initialization procedure, as outlined in the Chassis Electrical System.

MAIN BEARING TORQUE SEQUENCE

See Figures 81 and 82.

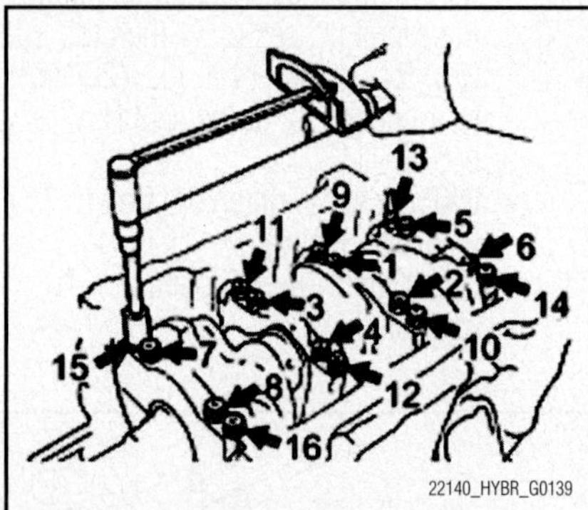

Fig. 81 (3MZ-FE) engine main bearing torque sequence (1)

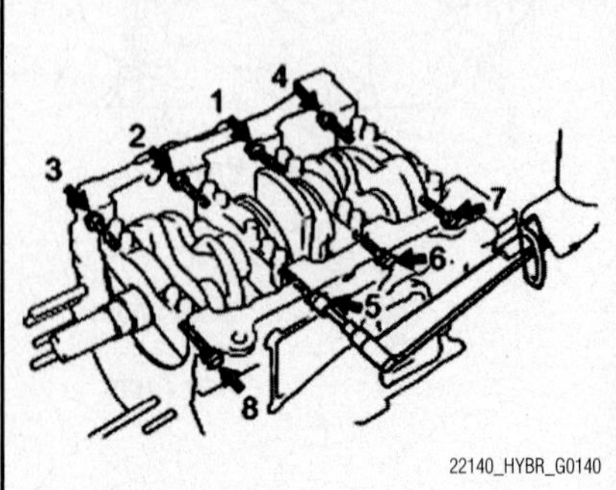

Fig. 82 (3MZ-FE) engine main bearing torque sequence (2)

PISTON AND RING

POSITIONING

See Figures 83 through 85.

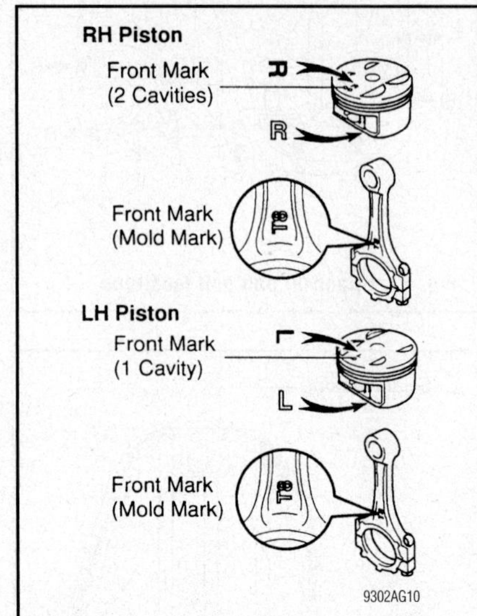

Fig. 83 Piston/connecting rod-to-engine positioning

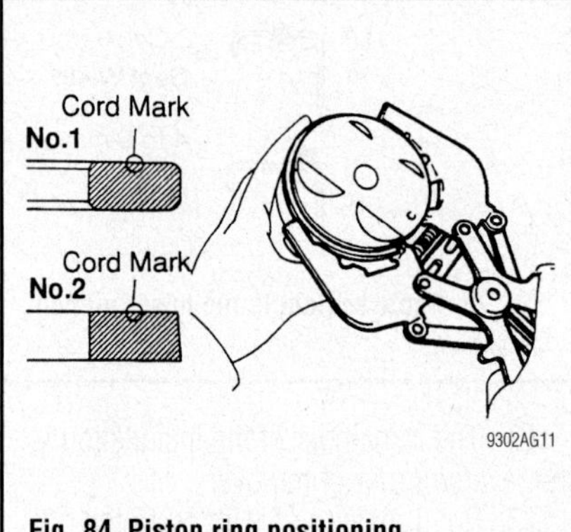

Fig. 84 Piston ring positioning

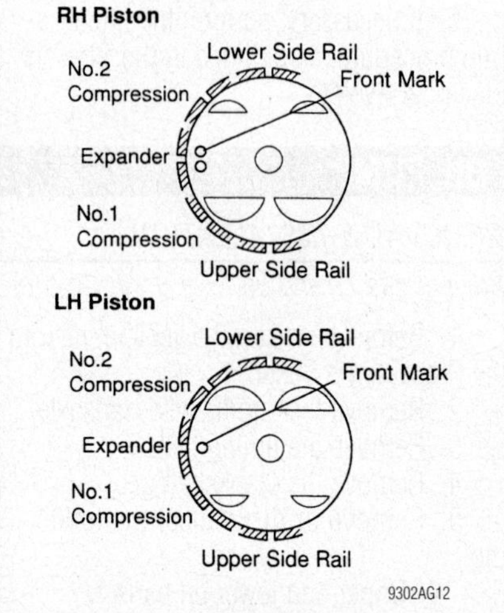

Fig. 85 Piston ring identification

REAR MAIN SEAL

REMOVAL & INSTALLATION

See Figures 86 and 87.

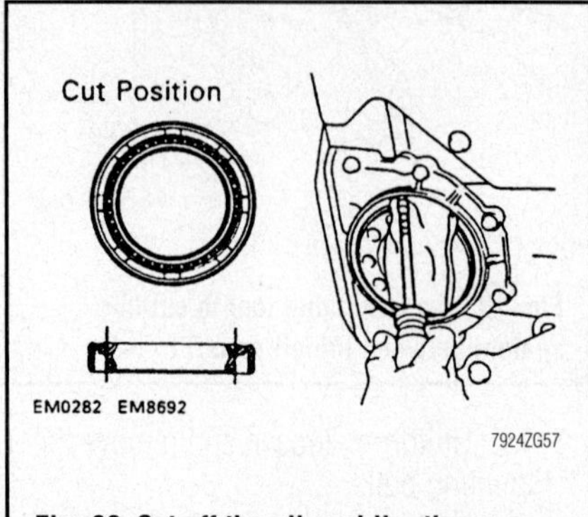

Fig. 86 Cut off the oil seal lip, then pry the seal out of the retaining plate

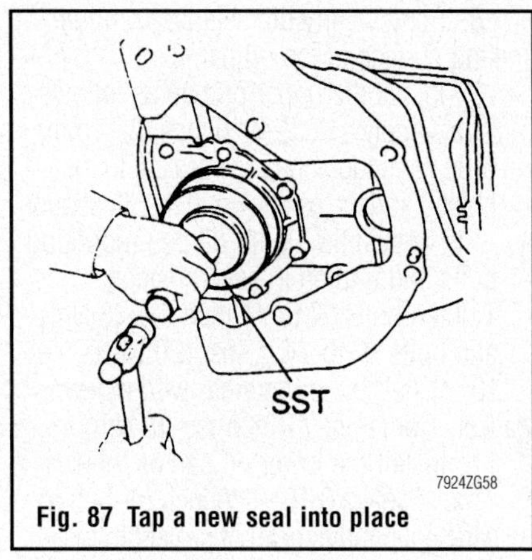

Fig. 87 Tap a new seal into place

1. Before servicing the vehicle, refer to the Precautions Section.
2. Remove or disconnect the following:
   - Transaxle assembly
   - Transmission input damper
   - Flywheel
   - Rear main seal

To install:

3. Using Special Tool 09223-15030 or equivalent, tap the new seal into place until the surface is flush with the retainer edge.
4. Install or connect the following:
   - Flywheel
   - Transmission input damper
   - Transaxle assembly
5. If necessary, perform the initialization procedure, as outlined in the Chassis Electrical System.

TIMING BELT FRONT COVER

REMOVAL & INSTALLATION

Refer to Timing Belt and Sprockets.

TIMING BELT AND SPROCKETS

REMOVAL & INSTALLATION

See Figures 88 through 100.

1. Before servicing the vehicle, refer to the Precautions Section.
2. Remove or disconnect the following:
   - Negative battery cable
   - Engine covers
   - Right front wheel
   - Fender splash shields
   - Wiper arms
   - Top cowl ventilator louver
   - Wiper motor and linkage assembly
   - Battery and battery tray
   - Air intake assembly
   - Brake master cylinder reservoir and bracket
   - Air cleaner bracket
   - Engine moving control rod
   - Right-hand engine mounting stay No. 2
3. Use Special Tool 09213-54015 to hold the crankshaft pulley in order to loosen the pulley bolt.
4. Use Special Tool 09950-50013 to remove the crankshaft pulley.
5. Remove or disconnect the following:

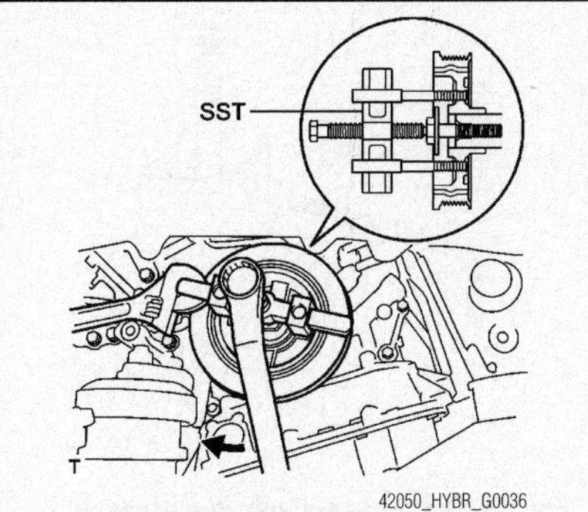

Fig. 90 Use the special tool to remove the crankshaft pulley

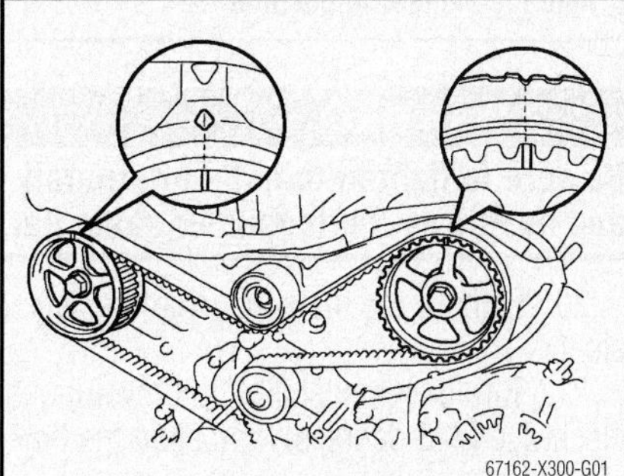

Fig. 91 Check that the timing marks on the camshaft pulleys are aligned with the notches on the inner belt cover

   - Timing belt cover No. 1
   - Timing belt cover No. 2
   - Right-hand engine mounting bracket
   - Timing belt guide No. 2
6. Temporarily install the crank pulley bolt. Turn the crankshaft clockwise to align the timing mark on the crankshaft timing pulley with the notch in the oil pump body.
7. Check that the timing marks on the camshaft pulleys are aligned with the notches on the inner belt cover. If not, rotate the crankshaft 360 degrees clockwise.

➡If the timing belt is re-used, check that the 3 original installation marks are visible on the belt as shown. If not, paint three new marks on the belt.

8. Turn the crankshaft counterclockwise by 60 degrees. Make sure that the belt is still engaged.
9. Remove the timing belt tensioner.
10. Remove the belt from the pulleys in this order:
    - Lower idler pulley
    - Right camshaft pulley
    - Upper idler pulley
    - Left camshaft pulley
    - Water pump pulley
    - Crankshaft timing pulley

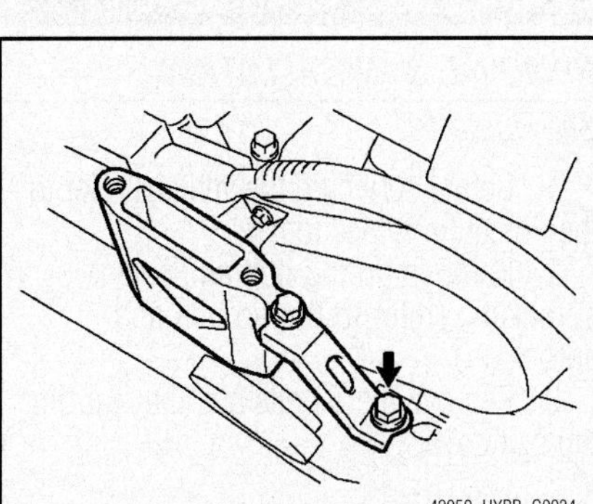

Fig. 88 Remove the right-hand engine mounting stay No. 2

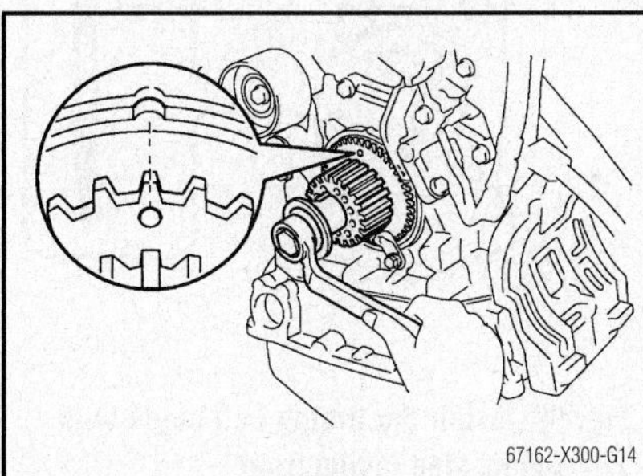

Fig. 92 Turn the crankshaft clockwise to align the timing mark on the crankshaft timing pulley with the notch in the oil pump body

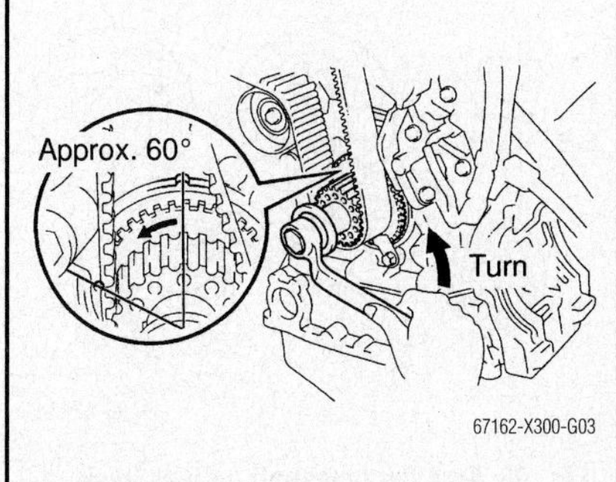

Fig. 94 Turn the crankshaft counterclockwise by 60 degrees

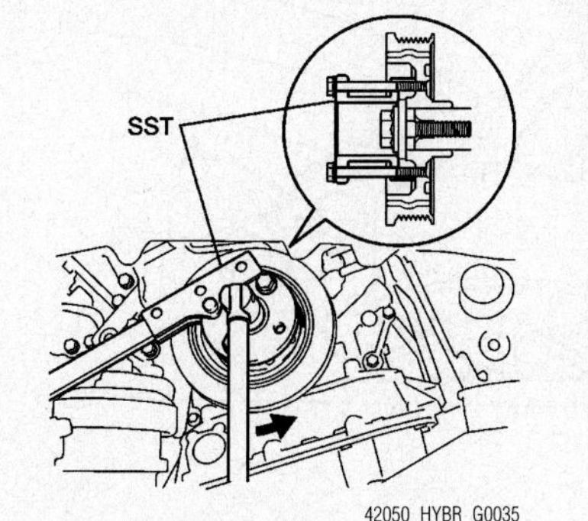

Fig. 89 Use the special tool to hold the pulley in order to loosen the pulley bolt

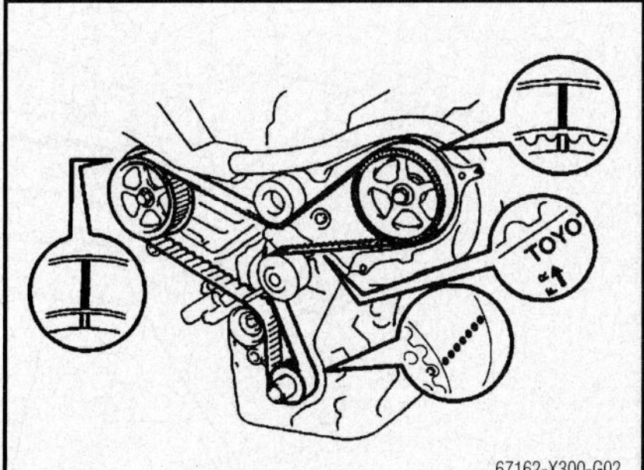

Fig. 93 If the timing belt is re-used, check that the 3 original installation marks are visible on the belt as shown

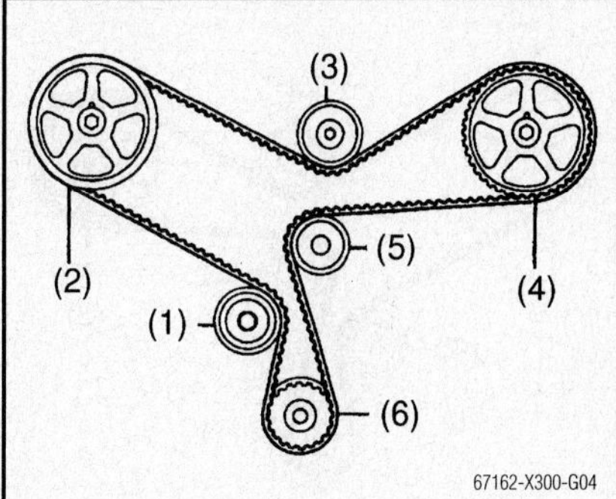

Fig. 95 Remove the belt from the pulleys in this order

11. If the belt is being re-used, check it for wear or damage; don't twist it or turn it inside-out. If there is any doubt as to its condition, replace it.

To install:

12. Clean all the pulleys.
13. Turn the crankshaft another 60 degrees counterclockwise.
14. Turn the camshaft pulleys back into alignment so the marks align with the notches on the inner cover.
15. Turn the crankshaft back so that the timing mark aligns with the notch on the oil pump.
16. Align the installation marks on the belt with the timing marks on the pulleys.
17. Install the belt in this order:
    - Crankshaft
    - Water pump
    - Left camshaft
    - Upper idler
    - Right camshaft
    - Lower idler
18. Set the tensioner in a press and collapse the plunger. Do not apply more that 2,205 lbs (9.8 kn) of force. Insert a suitable metal rod through the holes to hold the plunger in position.
19. Install the tensioner and torque the 2 bolts alternately to 20 ft. lbs. (27 Nm).

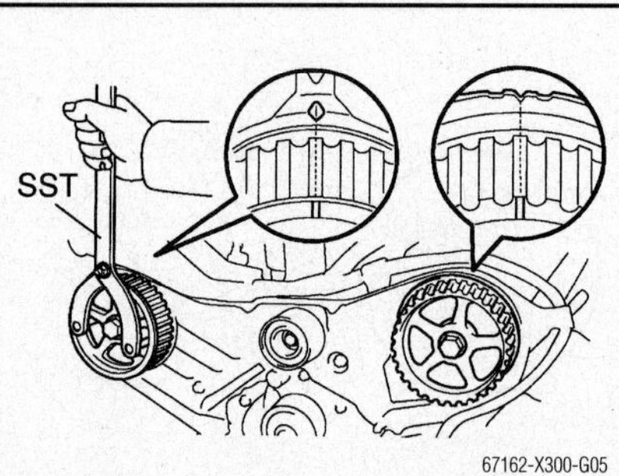

Fig. 96 Turn the camshaft pulleys back into alignment so the marks align with the notches on the inner cover

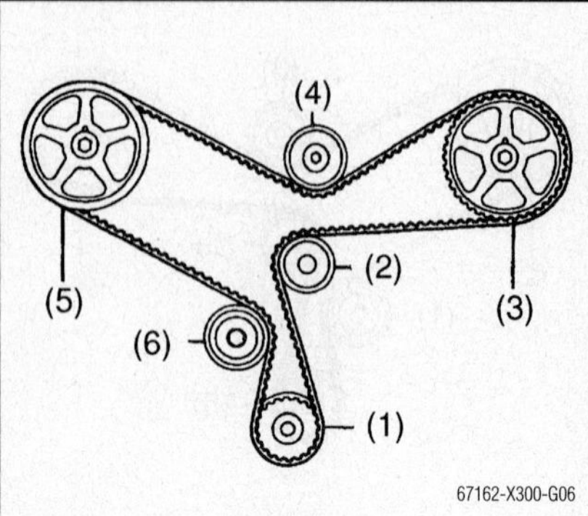

Fig. 97 Install the belt in this order

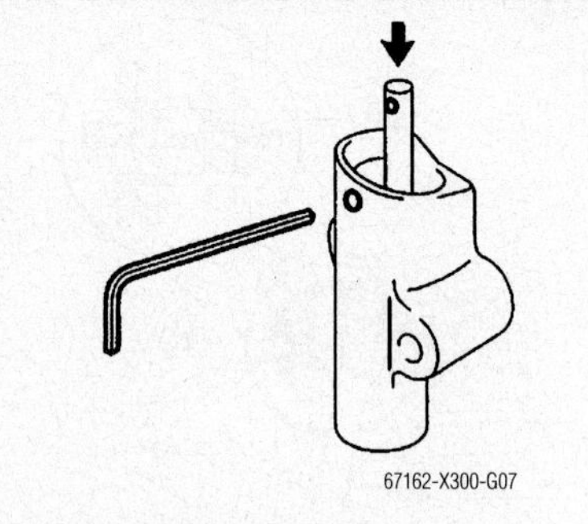

Fig. 98 Set the tensioner in a press and collapse the plunger. Do not apply more that 2,205 lbs (9.8 kn) of force. Insert a suitable metal rod through the holes to hold the plunger in position

WARNING

Be sure to tighten to bolts alternately and evenly so the tensioner seats flat.

20. Remove the metal rod from the tensioner.
21. Turn the crankshaft 2 full revolutions clockwise (720 degrees), and align the timing mark on the crank pulley with the notch on the oil pump.

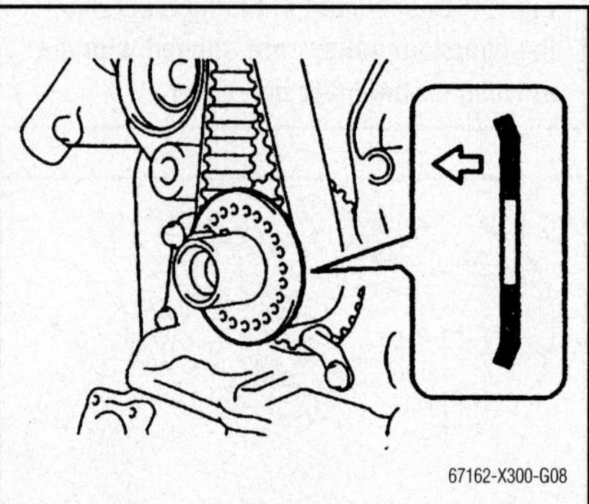

Fig. 99 Install the timing belt guide with the cupped side facing front

22. Check the timing marks on the camshaft pulleys for alignment with the notches on the inner cover. If they do not align, remove the belt and align the mismatched mark(s).
23. The remainder of installation is the reverse of removal. Observe the following torques:
    - Right engine mount bracket: 21 ft. lbs. (28 Nm)
    - Right engine mount insulator: 70 ft. lbs. (95 Nm)
    - Timing belt covers: 75 inch lbs. (8.5 Nm)
    - Crankshaft pulley: 162 ft. lbs. (220Nm)
    - Alternator bracket: 21 ft. lbs. (28 Nm)
    - Right engine mount stay: 47 ft. lbs. (64 Nm)
    - Engine roll control rod: tighten first A, then B, and then C to 47 ft. lbs. (64 Nm). Torque D to 17 ft. lbs. (23 Nm)
24. If necessary, perform the initialization procedure, as outlined in the Chassis Electrical System.

TIMING BELT REAR COVER

REMOVAL & INSTALLATION

See Figure 101.

1. Before servicing the vehicle, refer to the Precautions Section.
2. Remove timing belt and cam sprockets. (refer to timing belt and sprockets)
3. Remove the 6 bolts and the rear timing belt cover.

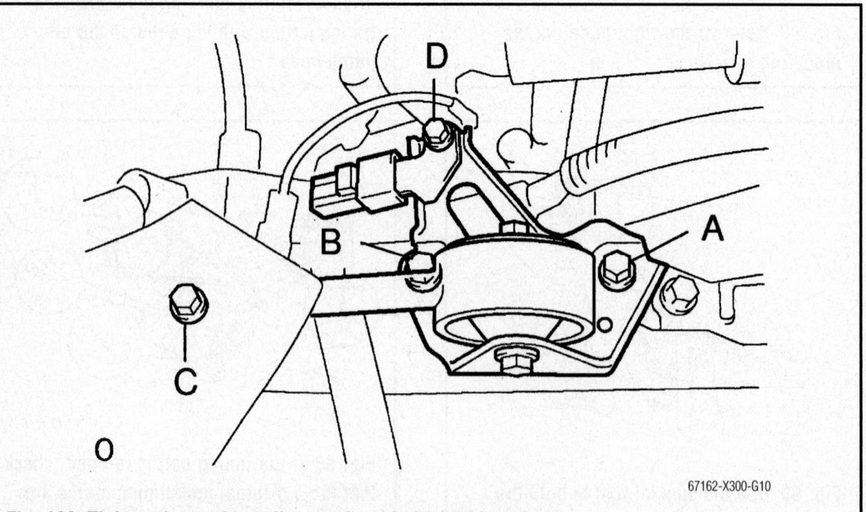

Fig. 100 Tighten the engine roll control rod bolts in this order

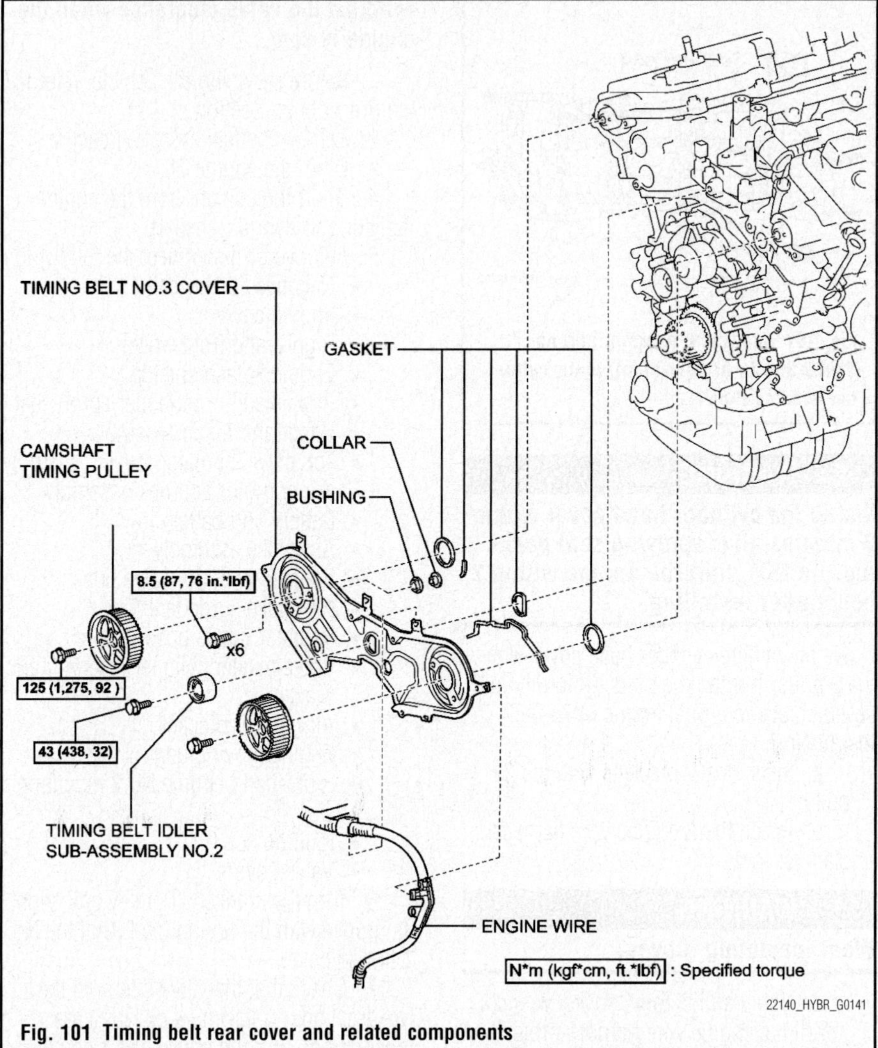

Fig. 101 Timing belt rear cover and related components

To install:

4. Install the rear timing belt cover No.3 with the 6 bolts. Tighten the bolts to 76 inch. lbs. (8.5 Nm).
5. Install the camshaft sprockets and tighten to 92 ft. lbs. (125 Nm).
6. Install the timing belt and front cover. (refer to timing belt and sprockets)
7. Some system need initialization when reconnecting the battery cable.

VALVE COVERS

REMOVAL & INSTALLATION

See Figures 102 through 107.

1. Disconnect the negative battery cable.
2. Using a clip remover, remove the engine side cover.
3. Drain the coolant and engine oil.
4. Remove front upper center suspension brace subassembly, as follows:
   a. Remove the 4 nuts and front suspension brace subassembly.
   b. Install the 4 shock absorber nuts and tighten to 59 ft .lbs. (80 Nm).
5. Remove the air cleaner cap w/ inlet, as follows:
   a. Remove the 2 bolts, 4 clamps and air cleaner cap w/ inlet.
   b. Remove the air cleaner filter element from the air cleaner case.
6. Remove the air cleaner case w/ resonator.
7. Remove the emission control valve set.
8. Remove the intake air surge tank.
9. Disconnect the inlet radiator hose.
10. Remove the ignition coil assembly
11. Remove the cylinder head cover subassembly, as follows:
   a. Remove the 2 engine wire harness clamps.
   b. Remove the 3 nuts and disconnect the engine wire harness.
   c. Remove the 9 bolts and the cylinder head cover.
12. Remove left cylinder head cover subassembly:
   a. Using an E6 Torx® socket wrench, remove the 2 bolts and disconnect the engine wire harness protector.
   b. Put on insulating gloves, then remove the 2 engine wire harness clamps.
   c. Remove the 2 bolts and 2 brackets.
   d. Remove the 9 bolts and the cylinder head cover.
13. Thoroughly clean the gasket mating surfaces.

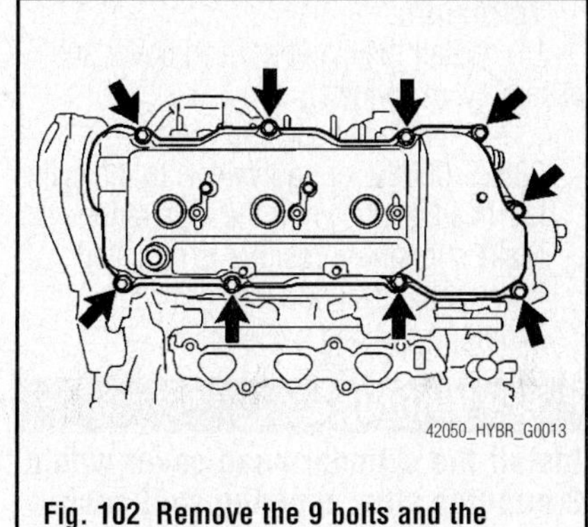

Fig. 102 Remove the 9 bolts and the cylinder head cover

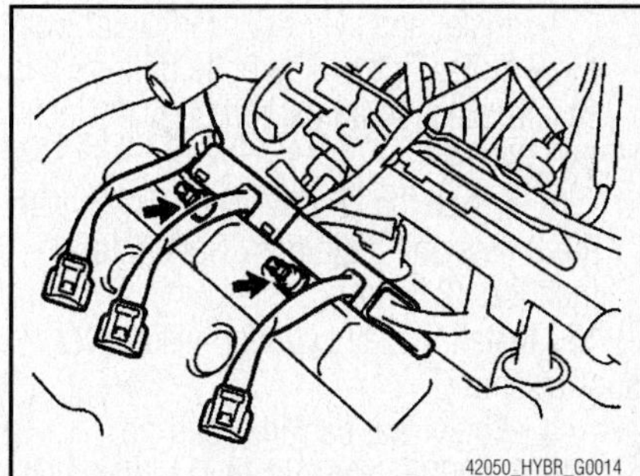

Fig. 103 Use a Torx® wrench to remove the 2 bolts, then disconnect the engine wire harness protector

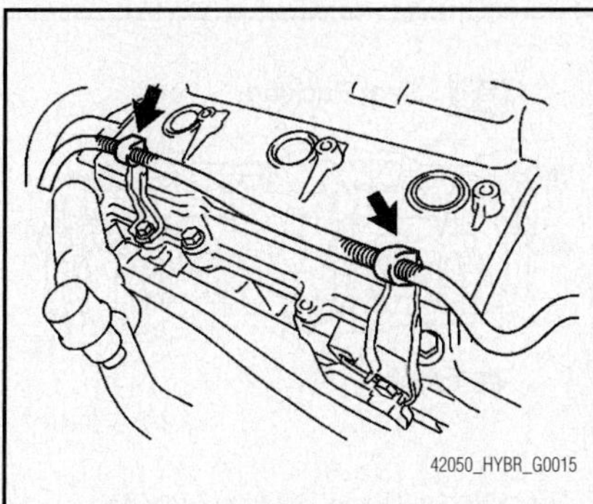

Fig. 104 Put on insulating gloves, then remove the 2 engine wire harness clamps

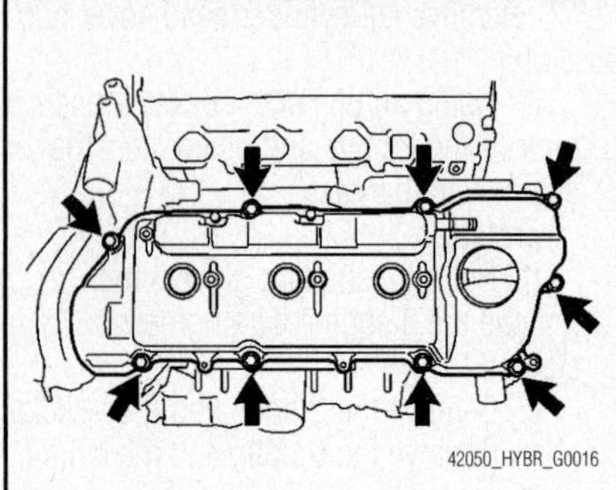

Fig. 105 Remove the 9 bolts and the cylinder head cover

To install:

14. Install the cylinder head cover sub-assembly, as follows:
   a. Apply seal packing (part no. 08826-00080, or equivalent) to the cylinder head as shown in the illustration. Make sure all oil residue is removed from the contact surfaces prior to installing the seal packing.

⁂ WARNING

Install the cylinder head cover within 3 minutes after applying seal packing. Do NOT start the engine within 2 hours after installing.

   b. Install the cylinder head cover with the 9 bolts. Tighten the bolts uniformly in several steps to a final torque of 71 inch lbs. (8 Nm).
   c. Install the engine wire harness with the 3 nuts and 2 clamps. Tighten to 74 inch lbs. (8.4 Nm).
15. Install the left cylinder head cover sub-assembly:
   a. Apply seal packing (part no. 08826-00080, or equivalent) to the cylinder head as shown in the illustration. Make sure all oil residue is removed from the contact surfaces prior to installing the seal packing.

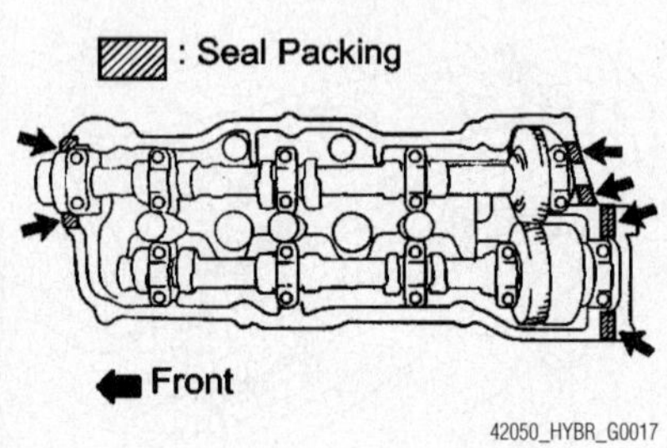

Fig. 106 Apply seal packing (part no. 08826-00080, or equivalent) to the cylinder head as shown

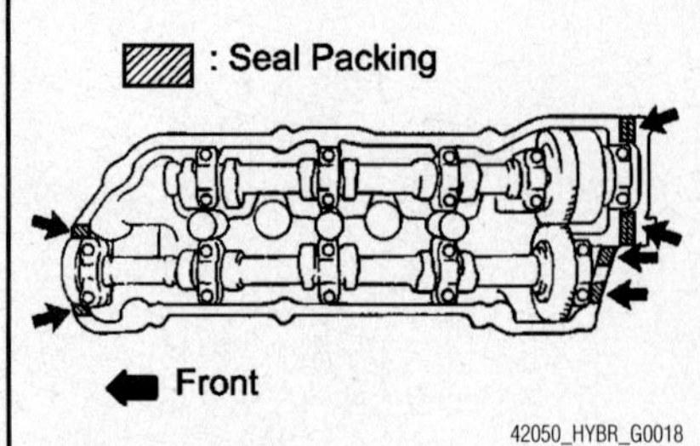

Fig. 107 Apply seal packing (part no. 08826-00080, or equivalent) to the cylinder head as shown

⁂ WARNING

Install the cylinder head cover within 3 minutes after applying seal packing. Do NOT start the engine within 2 hours after installing.

16. Install the cylinder head cover with the 9 bolts. Tighten the bolts uniformly in several steps to a final torque of 71 inch lbs. (8 Nm).
   a. Install the 2 brackets with the 2 bolts.
   b. Install the 2 engine wire harness clamps.

⁂ WARNING

Wear insulating gloves.

   c. Using an E6 Torx® socket wrench, install the engine wire harness protector with the 2 bolts. Tighten to 74 inch lbs. (8.4 Nm).
17. Install the ignition coil assembly. Tighten the retainers to 71 inch lbs. (8 Nm).
18. Connect the radiator inlet hose.
19. Install the intake air surge tank.
20. Install the emission control valve set.
21. Install the air cleaner case w/ resonator.
22. Install the air cleaner cap w/ inlet.
23. Install the front upper center suspension brace subassembly.
24. Add the proper and amount of engine oil and coolant to the engine.
25. Check for engine coolant and/or oil leaks.
26. Install the engine side cover:
27. If necessary, perform the initialization procedure, as outlined in the Chassis Electrical System.

VALVE LASH

ADJUSTMENT

See Figures 108 through 110.

➡Adjust the valve clearance when the engine is cold.

1. Before servicing the vehicle, refer to the Precautions Section.
2. Relieve the fuel system pressure.
3. Drain the engine oil.
4. Drain the coolant from the engine radiator and hybrid transaxle.
5. Remove or disconnect the following:
   - Negative battery cable
   - Engine cover
   - Right-hand front wheel
   - Engine splash shields
   - Right-hand front fender apron seal
   - Wiper and blade assembly
   - Top cowl ventilator louver assembly
   - Wiper motor and link assembly
   - Battery and battery tray
   - Air intake assembly
   - Emission control valve hoses
   - Air intake surge tank
   - Radiator intake hose
   - Brake master cylinder reservoir and bracket
   - Air cleaner bracket
   - Engine moving control rod
   - Right-hand engine No.2 mounting stay
   - Ignition coil
   - Valve covers
6. Turn the crankshaft pulley and align its groove with the timing mark **0** of the No. 1 timing cover.
7. Check that the valve lifters on the No. 1 cylinder (intake and exhaust) are loose. If not, turn the crankshaft 1 complete revolution (360 degrees).

➡All measurements should be written down. These recorded measurements will need to be used in conjunction

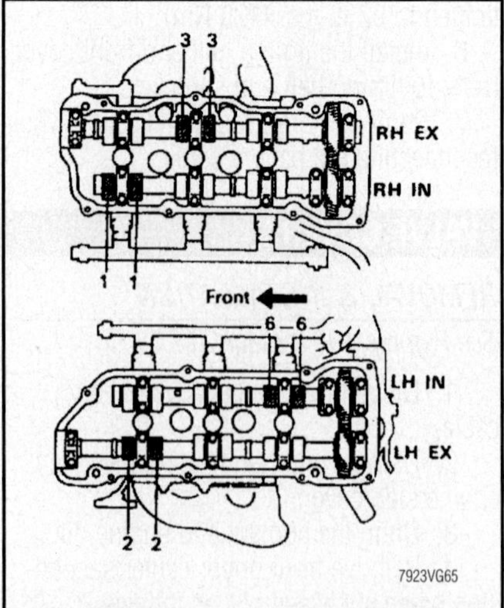

Fig. 108 Adjust these valves during the 1st step

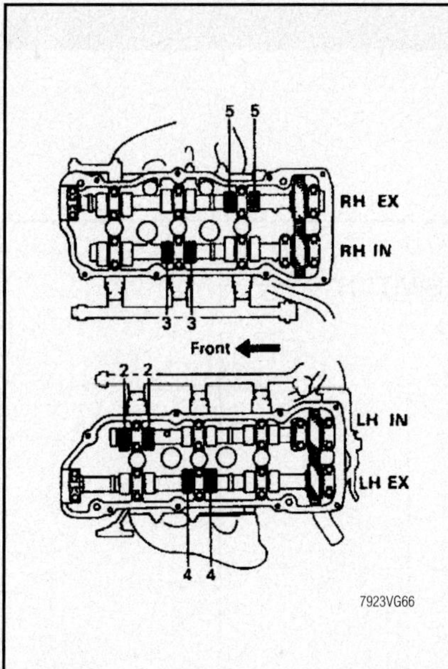

Fig. 109 Adjust these valves during the 2nd step

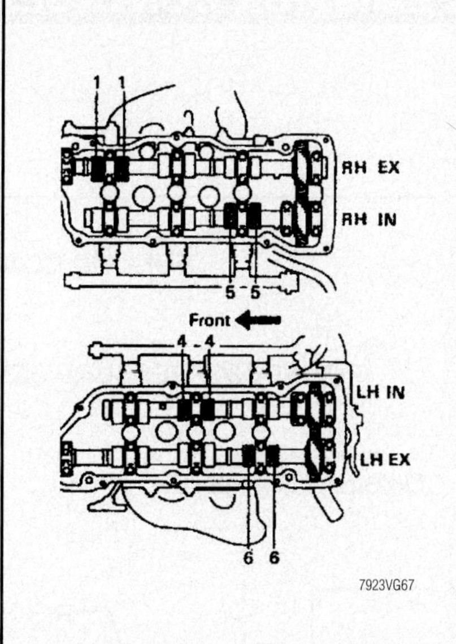

Fig. 110 Adjust these valves during the 3rd step

with a mathematical formula to determine the thickness of the replacement shims.

8. Measure the clearance between the valve lifters and the camshaft. Record the measurements on valves No. 1 and 6 intake; No. 2 and 3 exhaust.
   a. The intake valve clearance cold is 0.006–0.010 in. (0.15–0.25mm).
   b. The exhaust valve clearance cold is 0.010–0.014 in. (0.25–0.35mm).
9. Turn the crankshaft ⅔ of a revolution (240 degrees). Record the measurements on valves No. 2 and 3 intake; No. 4 and 5 exhaust.
10. Turn the crankshaft another ⅔ of a revolution (240 degrees). Record the measurements on valves No. 4 and 5 intake; No. 1 and 6 exhaust.
11. Remove the adjusting shim by turning the crankshaft to position the cam lobe of the camshaft in the up position on the valve to be adjusted. Using a small thin flat bladed tool, turn the valve lifter so that the notches are perpendicular to the camshaft. Press down the valve lifter with tool 09248-55010 part A. Place too 09248-55010 part B between the camshaft and the valve lifter; remove part A.
12. Remove the adjusting shim with a magnet and a small screwdriver.
13. Determine the replacement adjusting shim size by either using the charts or the following formulas:
   - Intake: N = T + (A - 0.008 in./0.020mm)
   - Exhaust: N = T + (A - 0.012 in./0.30mm)
   - T = Thickness of removed shim
   - A = Measured valve clearance
   - N = Thickness of new shim
14. Select a new shim with a thickness as close as possible to the calculated value. Install the new replacement shim.

➡Shims are available in 17 sizes in increments of 0.0020 in. (0.050mm), from 0.0984 in. (2.500mm) to 0.1299 in. (3.300mm).

15. Recheck the valve clearance.
16. Install or connect the following:
   - Valve covers
   - Ignition coil
   - Radiator intake hose
   - Air intake surge tank
   - Emission control valve hoses
   - Air intake assembly
   - Battery and battery tray
   - Wiper motor and link assembly
   - Top cowl ventilator louver assembly
   - Wiper and blade assembly
   - Right-hand front fender apron seal
   - Engine splash shields
   - Right-hand front wheel
   - Engine covers
   - Negative battery cable
17. Refill the cooling system to the correct level.
18. Refill the engine oil to the correct level.
19. Start the engine and check for leaks.
20. If necessary, perform the initialization procedure, as outlined in the Chassis Electrical System.

ENGINE PERFORMANCE & EMISSION CONTROLS

COMPONENT LOCATIONS

See Figures 111 and 112.

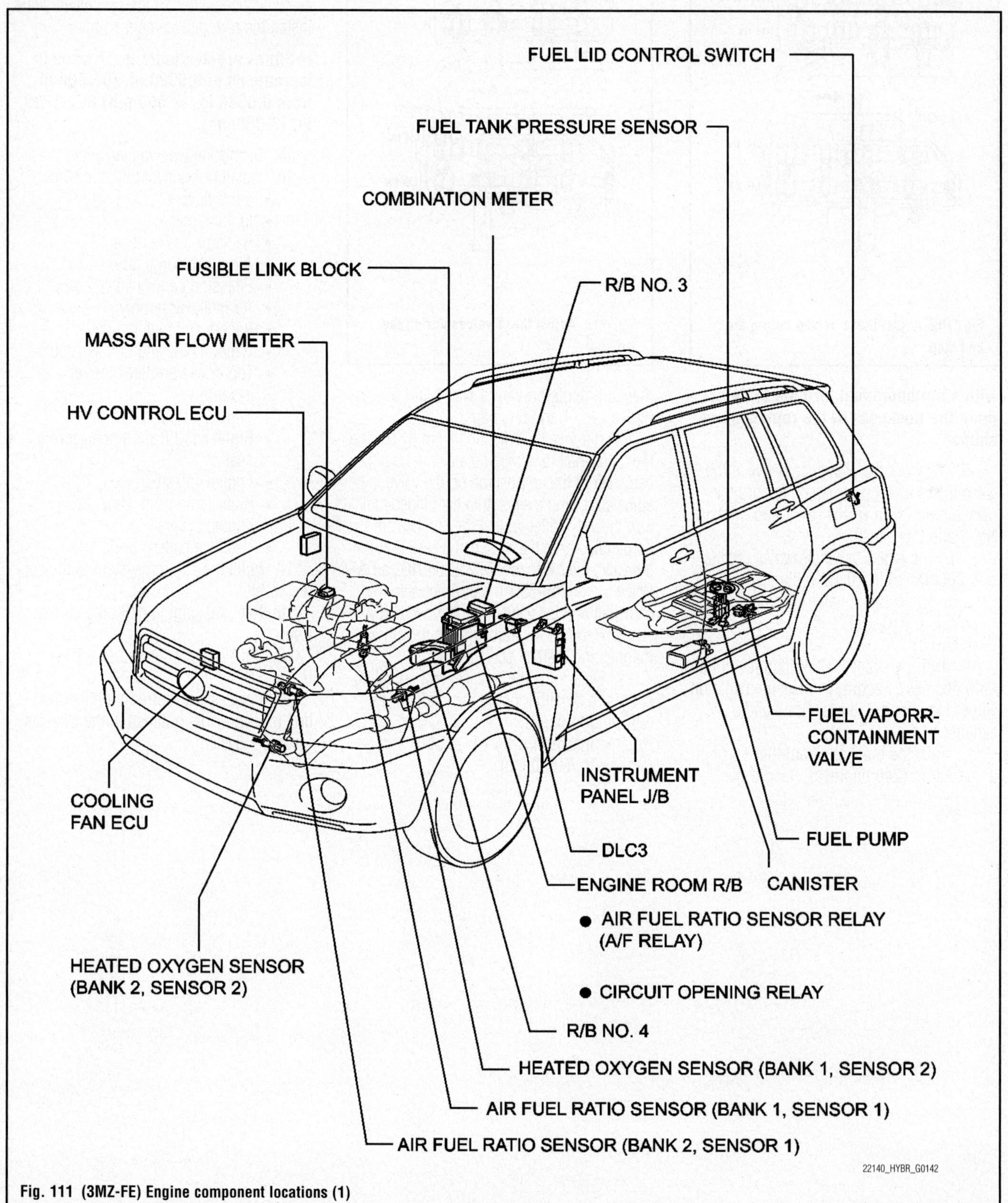

Fig. 111 (3MZ-FE) Engine component locations (1)

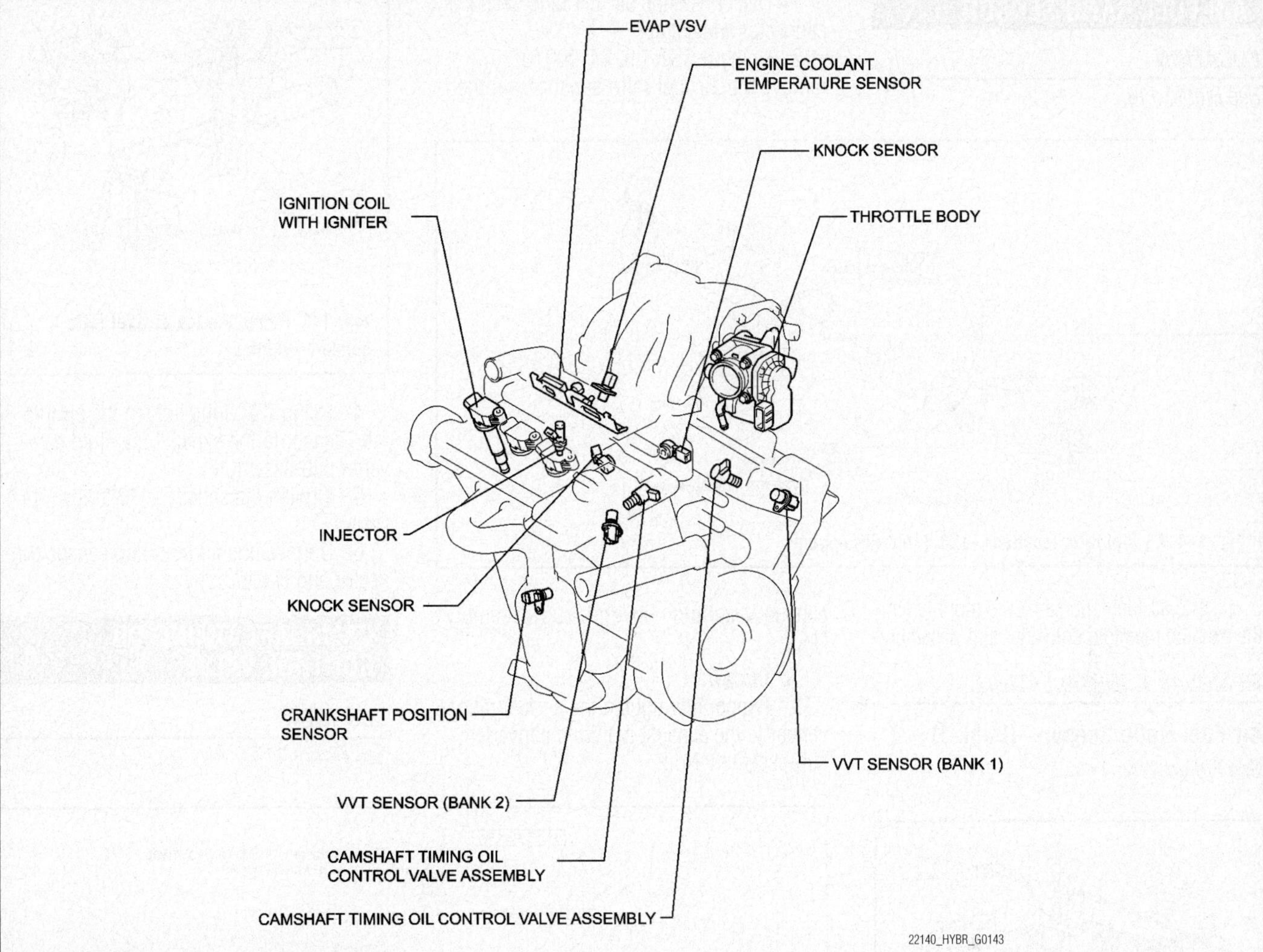

Fig. 112 (3MZ-FE) Engine component locations (2)

ACCELERATOR PEDAL POSITION (APP) SENSOR

LOCATION

See Figure 113.

The Accelerator Pedal Position (APP) sensor is located inside the vehicle and is part of the accelerator pedal. It is referred to as the accelerator rod assembly.

REMOVAL & INSTALLATION

1. Disconnect the connector from the accelerator pedal rod assembly.
2. Remove the 2 nuts and accelerator pedal rod assembly.

To install:

3. Install the accelerator pedal rod assembly with the 2 nuts.
4. Tighten the mounting nuts to 43 inch. lbs. (4.9 Nm).
5. Connect the connector to the accelerator pedal rod assembly.

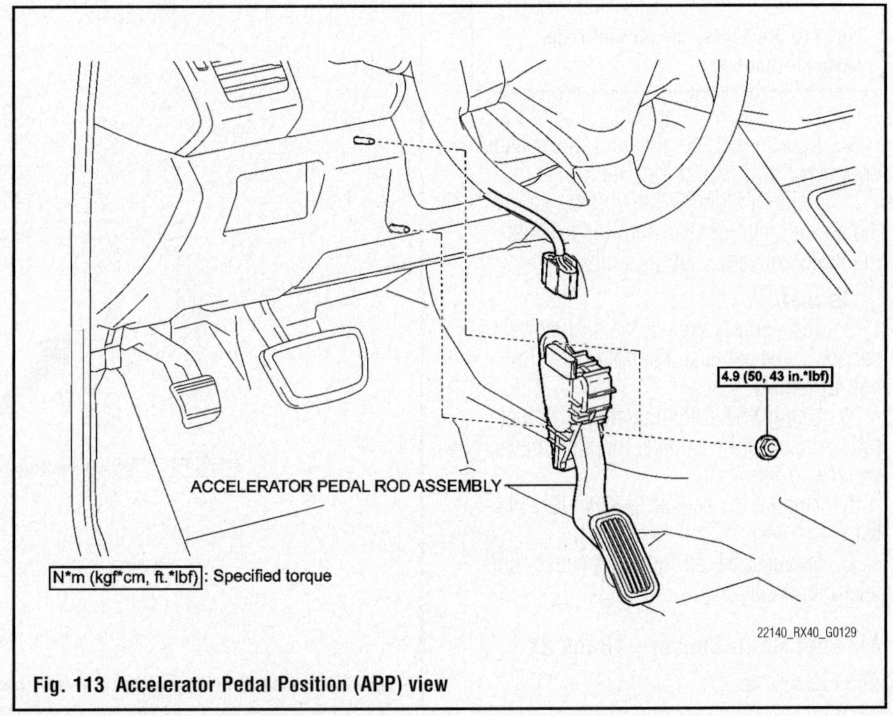

Fig. 113 Accelerator Pedal Position (APP) view

AIR FUEL RATIO SENSOR

LOCATION

See Figure 114.

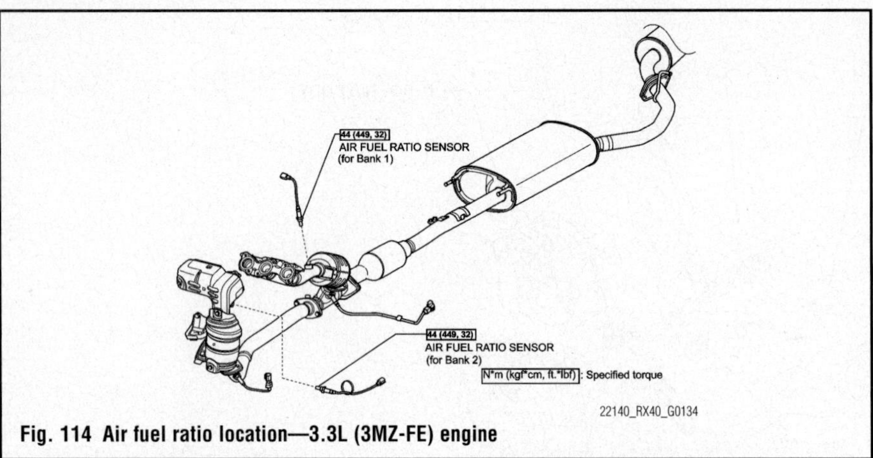

Fig. 114 Air fuel ratio location—3.3L (3MZ-FE) engine

1. The air fuel ratio sensor is mounted in the exhaust manifold converter sub-assembly.

REMOVAL & INSTALLATION

Air Fuel Ratio Sensor—(Bank 1)

See Figure 115.

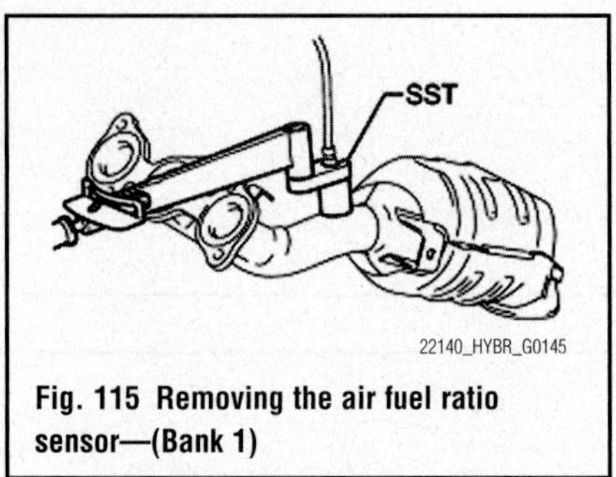

Fig. 115 Removing the air fuel ratio sensor—(Bank 1)

1. Disconnect the air fuel ratio sensor connector and clamp for (Bank 1).
2. Using a SST: 09224-00010, remove the air fuel ratio sensor from the exhaust manifold converter sub-assembly.

To install:

3. Temporarily tighten the air fuel ratio sensor to the exhaust manifold converter sub-assembly.
4. Using SST, fully tighten the air fuel ratio sensor to the exhaust manifold converter sub-assembly.
5. Tighten the sensor to 32 ft. lbs. (44 Nm).
6. Connect the air fuel ratio sensor connector and clamp.

Air Fuel Ratio Sensor—(Bank 2)

See Figure 116.

1. Disconnect the air fuel ratio sensor connector and clamp.
2. Using a SST: 09224-00010, remove the air fuel ratio sensor from the exhaust manifold converter sub-assembly No.2.

To install:

3. Temporarily tighten the air fuel ratio sensor to the exhaust manifold converter sub-assembly No.2.

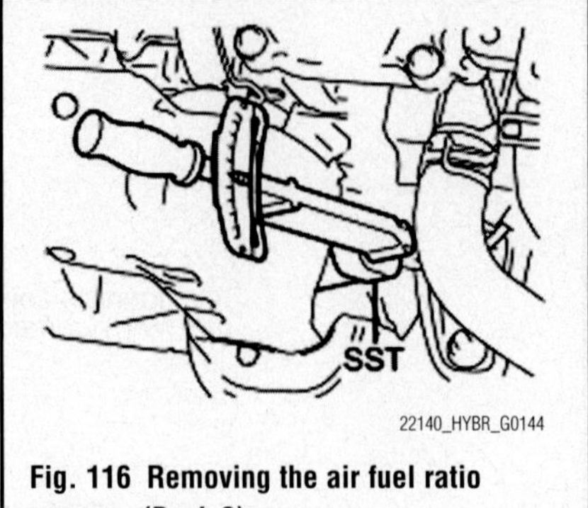

Fig. 116 Removing the air fuel ratio sensor—(Bank 2)

4. Using SST, fully tighten the air fuel ratio sensor to the exhaust manifold converter sub-assembly.
5. Tighten the sensor to 32 ft. lbs. (44 Nm).
6. Connect the air fuel ratio sensor connector and clamp.

CAMSHAFT TIMING OIL CONTROL VALVE

LOCATION

See Figure 117.

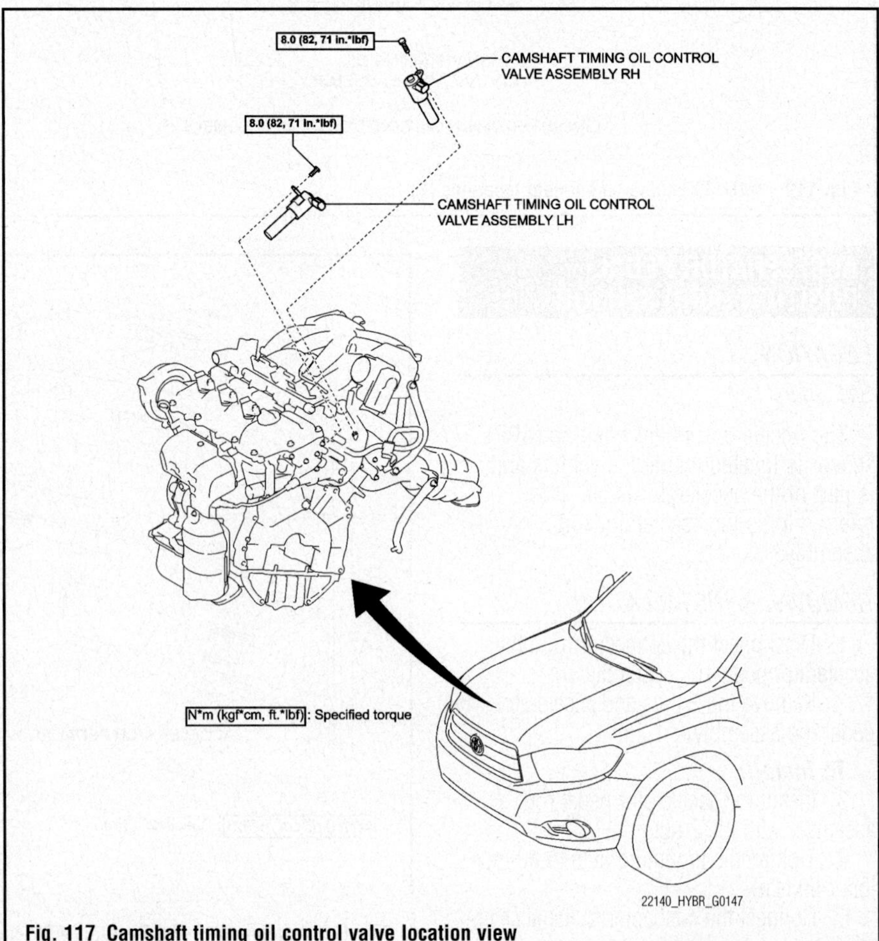

Fig. 117 Camshaft timing oil control valve location view

The camshaft timing oil control valves are mounted in the rear of the cylinder heads.

REMOVAL & INSTALLATION

1. Remove the engine room covers.
2. Remove the RH and LH wiper arm assembly.
3. Remove the cowl and seal.
4. Remove the wiper motor and link assembly.
5. Remove the cool air intake duct seal.
6. Remove the air cleaner cap with inlet.
7. Remove the air cleaner case with resonator.
8. Disconnect the oil control valve connector.
9. Remove the bolt and oil control valve.

To install:

10. Apply a coat of engine oil to an O-ring of the sensor.
11. Install the camshaft timing oil control valve with the bolt.
12. Tighten the mounting bolt to 71 inch. (8 Nm).
13. Connect the camshaft timing oil control valve connector.
14. Install the air cleaner case with resonator.
15. Install the air cleaner cap with inlet.
16. Install the cool air intake duct seal.
17. Install the wiper motor and link assembly.
18. Install the cowl and seal.
19. Install the RH and LH wiper arm assembly and engine room cover.

CRANKSHAFT POSITION (CKP) SENSOR

LOCATION

See Figure 118.

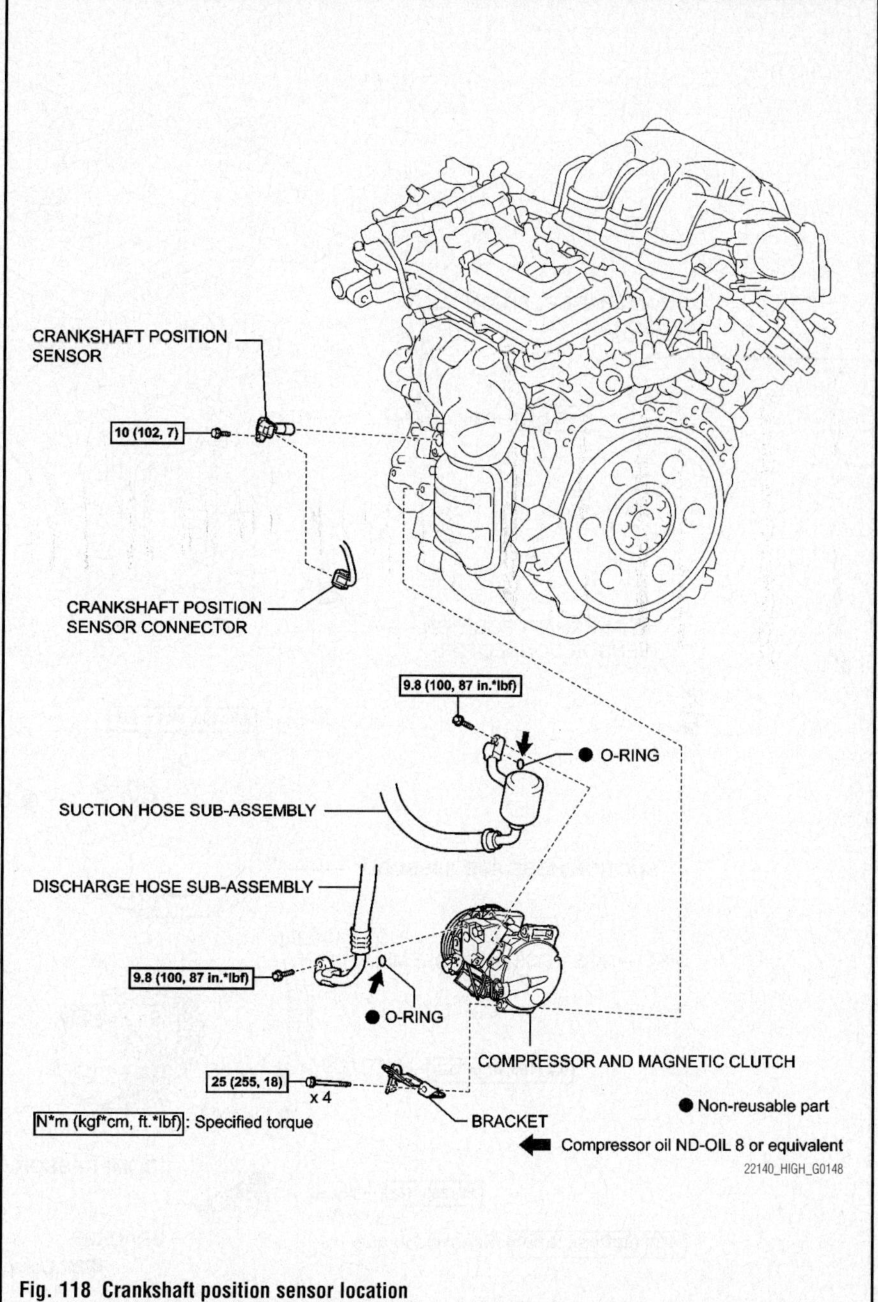

Fig. 118 Crankshaft position sensor location

REMOVAL & INSTALLATION

See Figures 119 and 120.

1. Recover the refrigerant from refrigeration system.
2. Remove the V-bank cover sub-assembly.
3. Remove the engine under cover assembly.
4. Remove the No. 1 engine under cover.
5. Drain engine coolant.
6. Remove the cool air intake duct seal.
7. Remove the battery.
8. Remove the No. 1 and 2 air cleaner inlets.
9. Disconnect the No. 1 and 2 radiator hoses.
10. Disconnect the oil cooler hose.
11. Disconnect the cooling fan ECU connector.
12. Remove the radiator grill.
13. Remove the hood lock assembly.
14. Remove the upper radiator support sub-assembly.
15. Separate the cooler condenser assembly.
16. Remove the radiator assembly and fan assembly with motor.
17. Disconnect the discharge hose sub-assembly.

➡Seal the openings of the disconnected parts using vinyl tape to prevent entry of moisture and foreign matter.

18. Disconnect the suction hose sub-assembly.
19. Remove the compressor and magnetic clutch as follows:
20. Disconnect the crankshaft position sensor connector.
21. Remove the bolt and crankshaft position sensor, as follows:
    a. Disconnect the crankshaft position sensor connector.
    b. Remove the bolt and crankshaft position sensor.

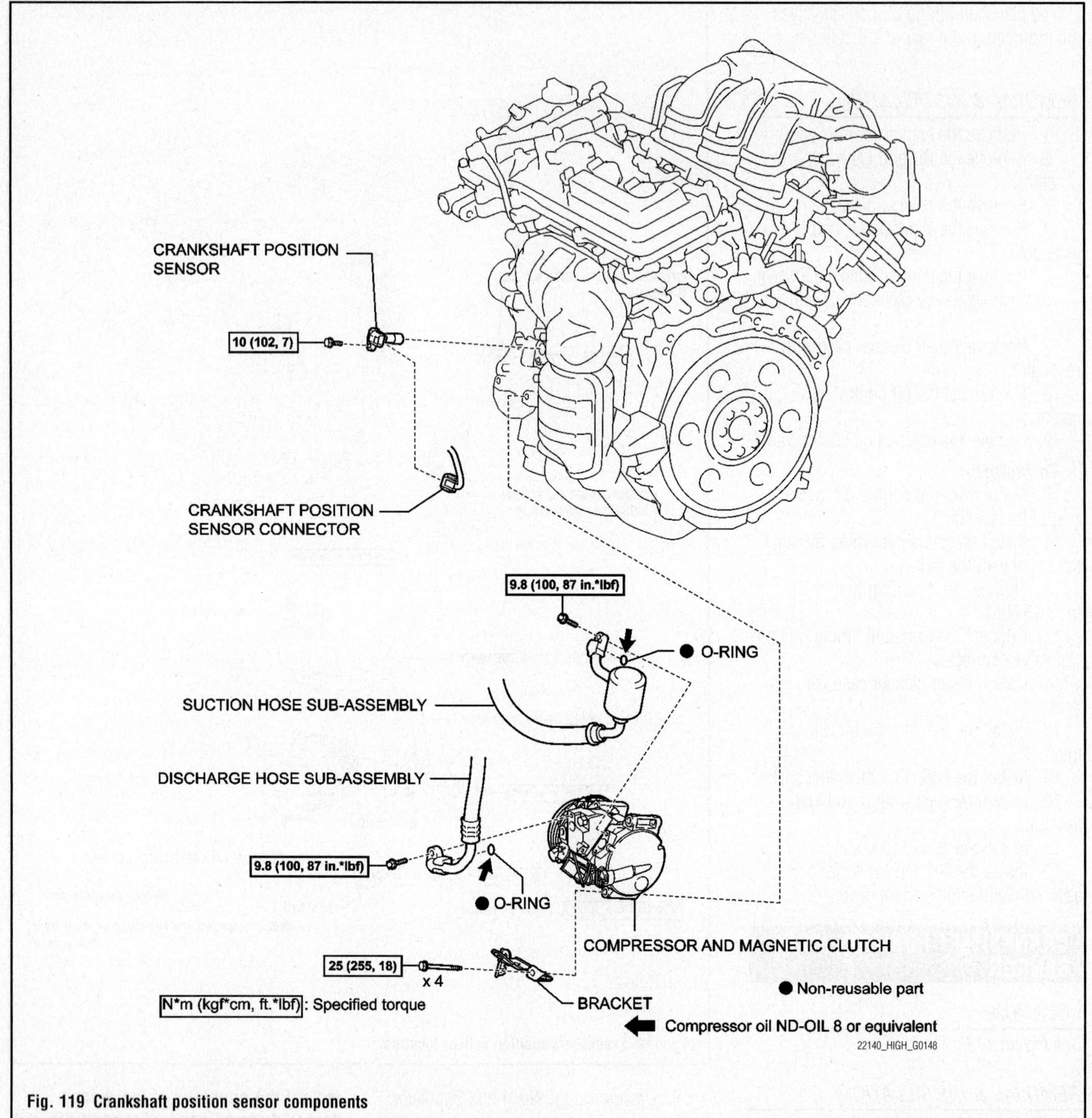

Fig. 119 Crankshaft position sensor components

To install:

22. Apply a light coat of engine oil to the O-ring on the crankshaft position sensor.
23. Install the crankshaft position sensor with the bolt and tighten to 7 ft. lbs. (10 Nm).
24. Connect the crankshaft position sensor connector.
25. Temporarily tighten compressor and magnetic clutch with the bolts, in the order shown.
26. Install the compressor and magnetic clutch with the 4 bolts. Tighten the bolts in the order shown above, and torque to 18 ft. lbs. (25 Nm).
27. Connect the suction hose sub-assembly.
28. Connect the discharge hose sub-assembly.
29. Install the radiator assembly and fan assembly with motor.
30. Install the cooler condenser assembly.
31. Install the upper radiator support sub-assembly.
32. Install the hood lock assembly.
33. Install the radiator grill.
34. Connect the cooling fan ECU connector.
35. Connect the oil cooler hose.
36. Connect the No. 1 and 2 radiator hoses.
37. Install the No. 1 and 2 air cleaner inlets.
38. Install the battery.
39. Install the cool air intake duct seal.
40. Charge with refrigerant.
41. Add engine coolant.
42. Inspect for coolant leak.

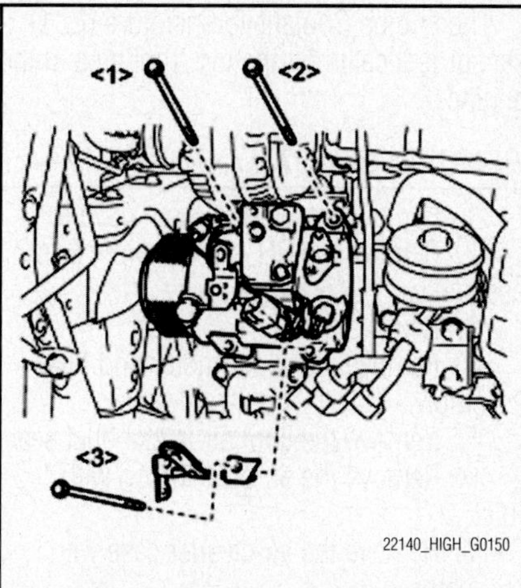

Fig. 120 Compressor and magnetic clutch

43. Inspect automatic transaxle fluid.
44. Inspect for oil leak.
45. Warm up the engine.
46. Inspect for refrigerant leak.
47. Install the No. 1 engine under cover.
48. Install the engine under cover assembly.
49. Install the V-bank cover sub-assembly.

ELECTRONIC CONTROL MODULE (ECM)

LOCATION

See Figure 121.

The Hybrid ECU is located behind the glove compartment and to the right of the blower unit.

REMOVAL & INSTALLATION

1. Disconnect the negative battery cable.
2. Remove the instrument under cover.
3. Remove the right door scuff plate.
4. Remove the cowl side trim.
5. Remove the glove compartment door assembly.
6. Disconnect the 3 wire harness clamps and 6 connectors from the hybrid vehicle control ECU
7. Remove the 2 nuts and hybrid vehicle control ECU.
8. Remove the ECU.
9. If ECU is to be changed remove the brackets.

To install:

10. Install brackets to the ECU if previously removed.
11. Install the wire harness clamp bracket with the 2 screws.
12. Install the hybrid vehicle control ECU with the 2 nuts. Tighten the mounting nuts to 49 inch. lbs. (5.5 Nm).
13. Connect the 6 connectors to the hybrid vehicle control ECU.

HYBRID VEHICLE CONTROL ECU
STOP LIGHT SWITCH
COMBINATION METER
DLC3
ACCELERATOR PEDAL SENSOR ASSEMBLY
INSTRUMENT PANEL JUNCTION BLOCK (MAIN BODY ECU)
22140_HYBR_G0153

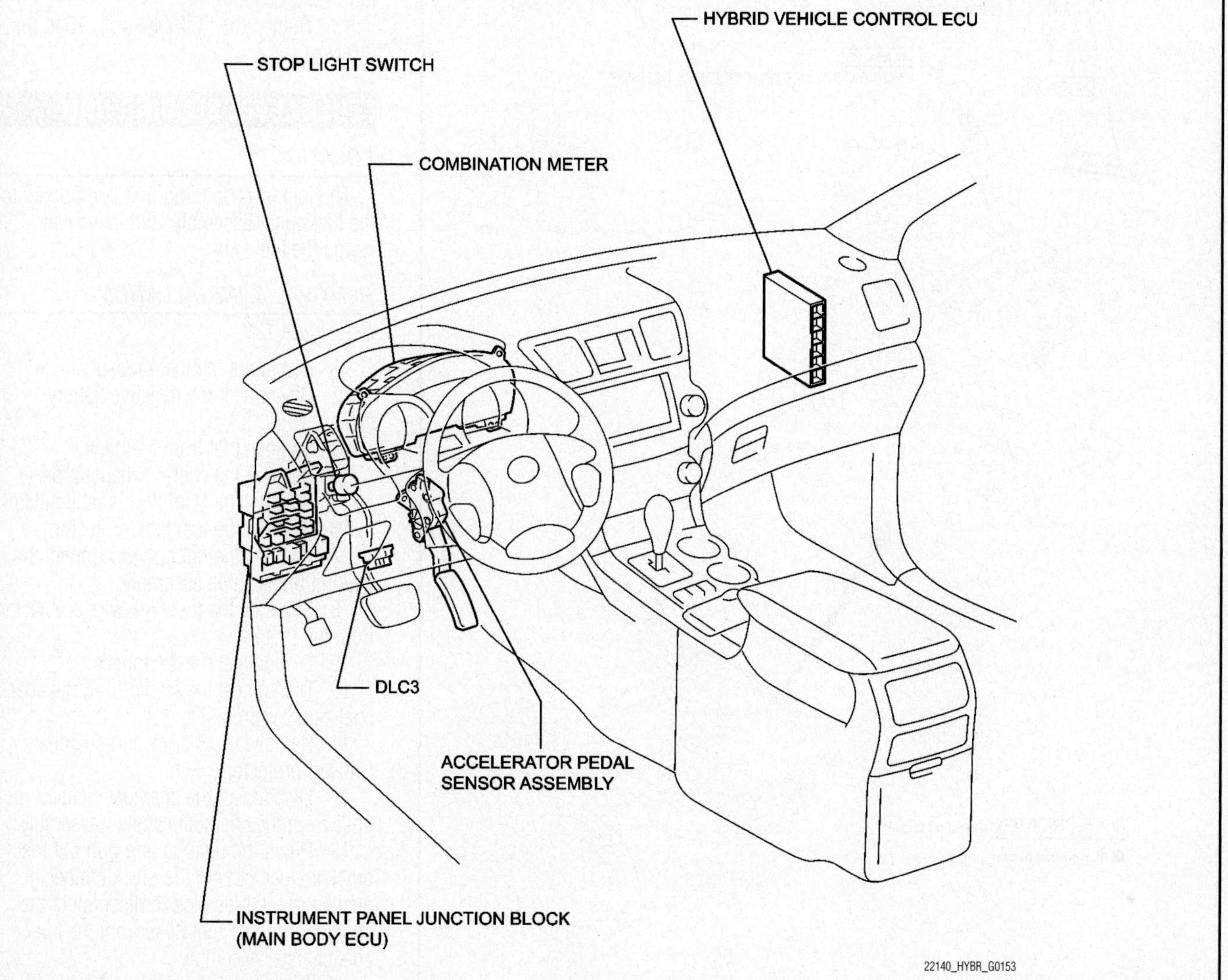

Fig. 121 Hybrid ECU location view

14. Connect the 3 wire harness clamps.
15. Install the glove compartment door assembly.
16. Install the undercover.
17. Install the cowl side trim.
18. Install the scuff plate.
19. Connect the negative battery cable.

➡After replacing the hybrid vehicle control ECU on vehicles with a dynamic laser cruise control system, it is necessary to initialize the hybrid vehicle control ECU so that the ECU can recognize the dynamic laser cruise control system.

Be sure to perform the following procedures after replacing the ECU.

- Turn the ignition switch to the on position.
- Turn the cruise main switch on.
- With the brake pedal depressed, push the cruise control main switch to RES/ACC 3 times within 3 seconds. Check that the buzzer sounds at this time.

➡Do not turn the headlight dimmer switch on at this time because the optical axis automatic adjustment mode has already started, which may lead to an incorrect optical axis setting. If the headlight dimmer switch is turned on by mistake, readjust the optical axis.

20. Some systems need initialization when disconnecting the cable from the negative battery terminal.

ENGINE COOLANT TEMPERATURE (ECT) SENSOR

LOCATION

See Figure 122.

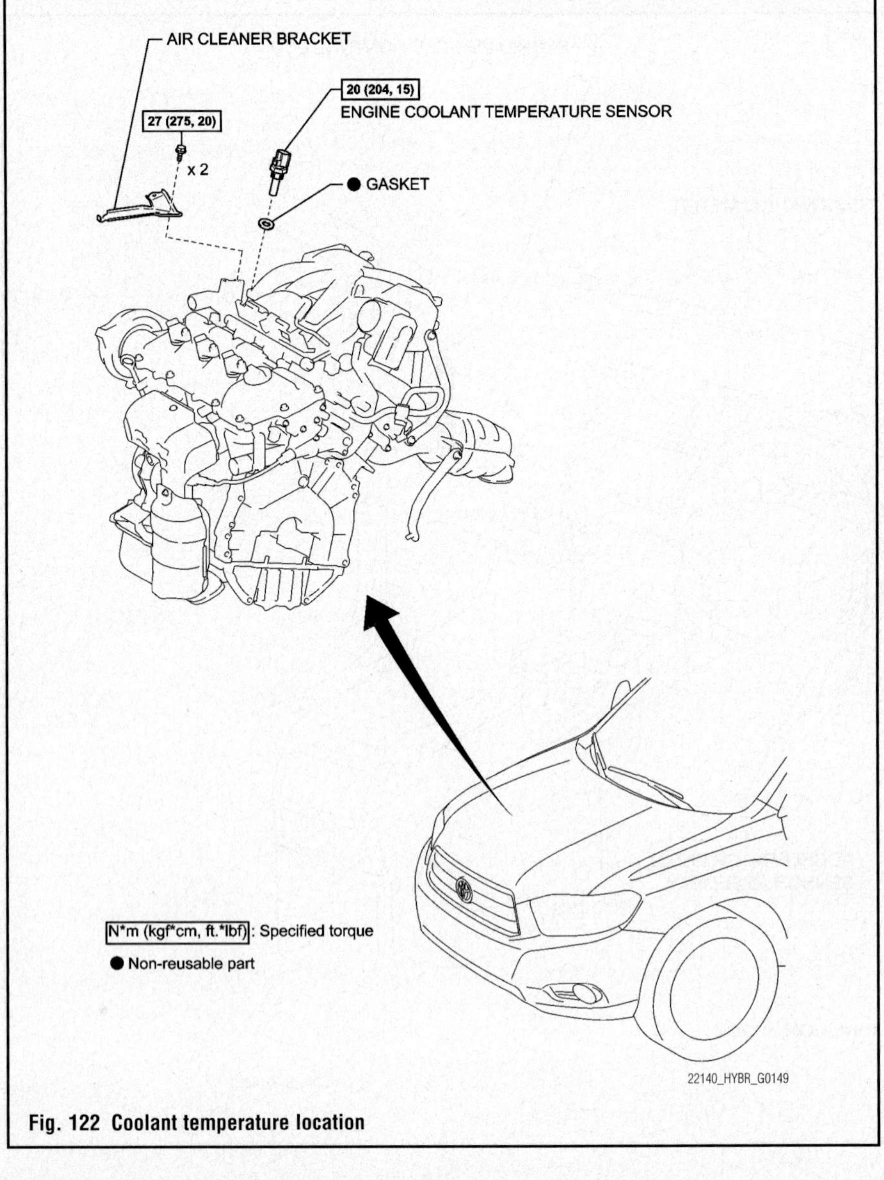

Fig. 122 Coolant temperature location

The Engine Coolant Temperature (ECT) sensor is located on the top front area of the engine.

REMOVAL & INSTALLATION

1. Remove the engine room covers.
2. Remove the RH and LH wiper arm assembly.
3. Remove the cowl and seal.
4. Remove the wiper motor and link assembly.
5. Remove the cool air intake duct seal.
6. Remove the air cleaner cap with inlet.
7. Remove the air cleaner case with resonator.
8. Air cleaner bracket.
9. Disconnect the ECT sensor connector.
10. Using a deep socket, remove the sensor from the top of the engine.
11. Remove the gasket from the sensor.
12. Installation is the reverse of the removal procedure.
13. Tighten the ECT sensor to 15 ft. lbs. (20 Nm).

FUEL LEVEL SENDING UNIT

LOCATION

The fuel level sending unit is mounted to the fuel pump assembly, and is located inside the fuel tank.

REMOVAL & INSTALLATION

See Figure 123.

1. Discharge the fuel pressure.
2. Disconnect the negative battery cable.
3. Remove fuel from fuel tank.
4. Remove the center exhaust pipe.
5. Remove the front floor heat insulator.
6. Remove the fuel tank protector.
7. Remove the nut and disconnect the parking brake cable assembly.
8. Separate the fuel tank wire connector from the bracket.
9. Disconnect the 2 connectors.
10. Disconnect the clamp from the charcoal canister protector.
11. Remove the 3 bolts and charcoal canister protector.
12. Disconnect the charcoal canister fuel hose from the charcoal canister assembly.
13. Pinch the retainer and pull out the quick connector with the quick connector pushed to the pipe side to disconnect the charcoal canister fuel hose from the charcoal canister assembly.
14. Disconnect the main fuel supply line. Pinch the tabs of the retainer to disengage

the lock claws and pull the retainer down as shown in the illustration.

15. Pinch the retainer of the breather lower tube connector, and pull out the breather lower tube connector to disconnect the breather lower tube from the fuel tank.

16. Set a transmission jack to the fuel tank.

17. Remove the fuel filler pipe clamp and fuel tube connector from the fuel tank inlet pipe.

18. Remove the 4 bolts and the fuel tank bands.

19. Operate the transmission jack, and disconnect the fuel inlet pipe.

20. Operate the transmission jack, and remove the fuel tank.

21. Disconnect the fuel pump connector.

22. Disconnect the clamp and vapor pressure sensor connector.

23. Remove the tube joint clip and clamp, and pull out the fuel pump tube.

24. Remove the 8 bolts, fuel tank vent tube set plate and fuel pump connector plate.

25. Pull out the fuel pump assembly from the fuel tank.

26. Remove the sending unit from the fuel pump assembly if applicable.

To install:

27. Install new fuel tank cushions.

28. Install the 2 fuel tube clamps of the fuel tank main tube to the fuel tank.

29. Install the fuel tank main fuel assembly.

30. Install the fuel tank wire.

31. Connect the fuel tank vent hose to the fuel tank

32. Install the 2 nuts to the fuel vapor containment valve, and tighten the nuts to 71 inch. lbs. (8 Nm).

33. Connect the connector and clamp.

34. Install a new gasket on the fuel pump assembly.

35. Install the fuel pump assembly.

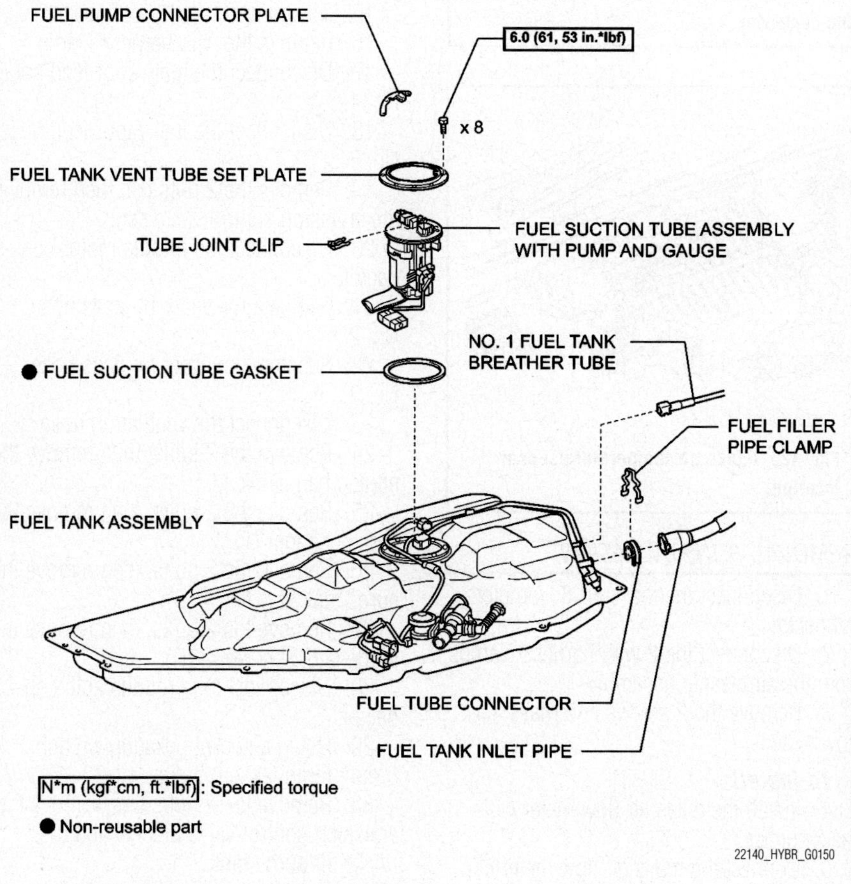

36. Install the fuel tank vent tube set plate and fuel pump connector plate with the 8 bolts. Tighten the bolts to 53 inch. lbs. (6 Nm).

37. Connect the fuel tank main tube with the tube joint clip and clamp.

38. Connect the fuel tank wire connector.

39. Connect the vapor pressure sensor connector and clamp.

40. Set the fuel tank on a transmission jack.

41. Operate the transmission jack, and install the fuel tank to the vehicle.

42. Operate the transmission jack, and connect the fuel tank inlet pipe with the tube connector and clamp.

43. Install the 2 fuel tank bands with the 4 bolts and tighten to 29 ft. lbs. (39 Nm).

44. Connect the breather lower tube. Push in the tube connector to the pipe until the tube connector makes a "click" sound.

45. Install the checker to the pipe.

46. Connect the fuel tank main supply line. Push in the fuel tube connector to the pipe and push up the retainer to engage the claws.

47. Align the quick connector with the pipe, then push in the quick connector until the retainer makes a "click" sound to connect the charcoal canister fuel hose to the charcoal canister assembly.

48. Connect the clamp to the charcoal canister protector.

49. Install the charcoal canister protector with the 3 bolts. Tighten the bolts to 48 inch. lbs. (5.4 Nm).

50. Connect the fuel tank wire connectors, and install them to the bracket.

51. Install the parking brake cable assembly No.3 with the nut.

52. Install the fuel tank protector.

53. Install the front floor heat insulator and tighten the retaining nuts to 43 inch. lbs. (4.9 Nm).

54. Install the center pipe exhaust assembly.

55. Connect the negative battery cable.

56. Add fuel and inspect for fuel leaks.

57. Inspect for exhaust leaks.

58. Some systems need initialization after reconnecting the cable to the negative battery terminal.

Fig. 123 Fuel level sending unit and related parts

HEATED OXYGEN (HO2S) SENSOR

LOCATION

See Figure 124.

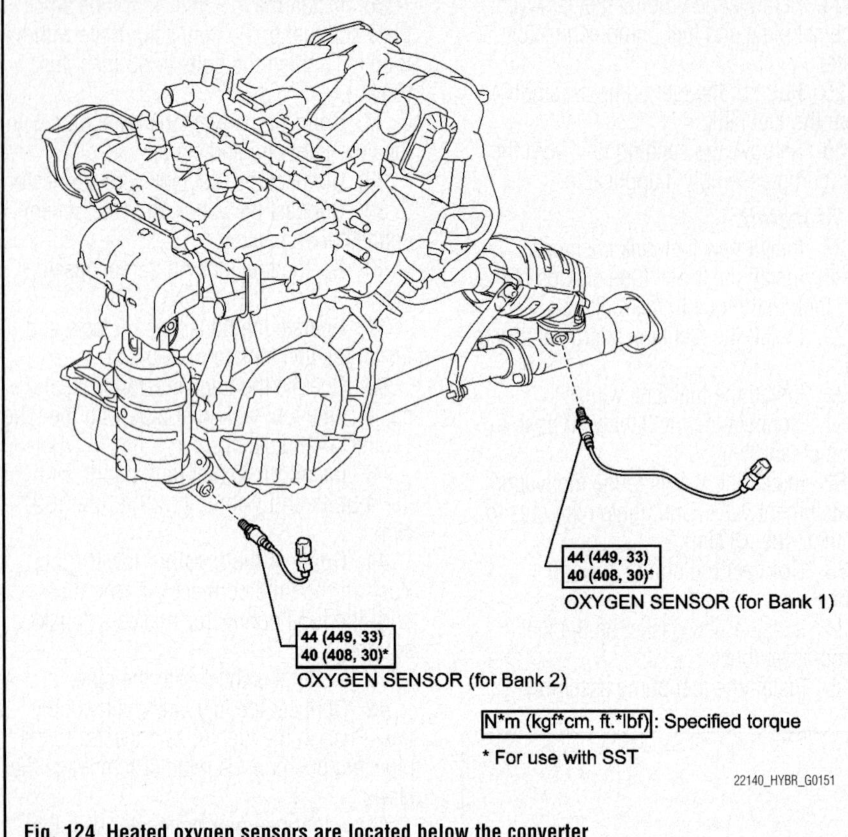

Fig. 124 Heated oxygen sensors are located below the converter

The heated oxygen sensors are located below the converter. They are mounted in the front exhaust pipes.

REMOVAL & INSTALLATION

1. Disconnect the oxygen sensor connector and clamp.
2. Using SST 09224-00010, remove the oxygen sensor from the exhaust pipe assembly front.

To install:

3. Temporarily tighten the oxygen sensor to the exhaust pipe sub-assembly front.
4. Using SST, fully tighten the oxygen sensor to the exhaust pipe sub-assembly front.
5. Tighten the sensor to 33 ft. lbs. (44 Nm),
6. Connect the oxygen sensor connector.

INTAKE AIR TEMPERATURE (IAT) SENSOR

LOCATION

See Figure 125.

The intake air temperature sensor is integral to the mass air flow sensor The intake air temperature sensor is located in the air cleaner housing.

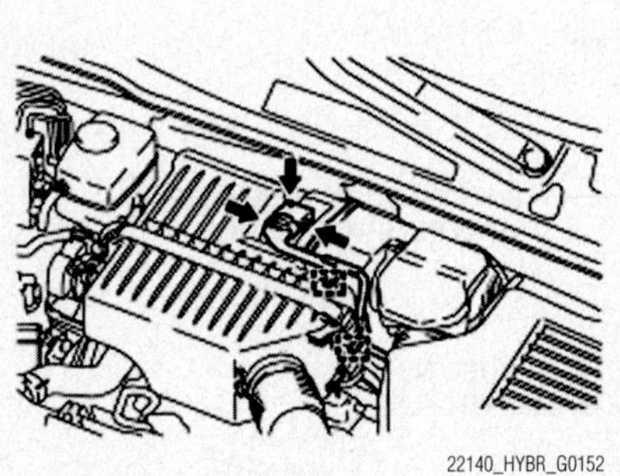

Fig. 125 Intake air temperature sensor location.

REMOVAL & INSTALLATION

1. Disconnect the mass air flow meter connector.
2. Disconnect the 2 wire harness clamps from the air cleaner assembly.
3. Remove the 2 screws and mass air flow meter.

To install:

4. Install the mass air flow meter with the 2 screws.
5. Connect the mass air flow meter connector.
6. Connect the 2 wire harness clamps to the air cleaner assembly.

KNOCK SENSOR (KS)

LOCATION

The knock sensors are located under the intake manifold.

REMOVAL & INSTALLATION

See Figures 126 through 129.

1. Discharge the fuel system.
2. Remove the engine room covers.
3. Disconnect the negative battery cable.
4. Drain the engine coolant.
5. Remove LH and RH wiper arm assembly.
6. Remove the cowl top ventilator louver.
7. Remove the wiper motor and link assembly.
8. Remove the cowl top panel.
9. Remove the cool air intake duct seal.
10. Remove the air cleaner assembly.
11. Remove the brake master cylinder reservoir.
12. Remove the air cleaner bracket.
13. Remove the air filter bracket.
14. Remove engine moving control rod.
15. Disconnect the VSV connector.
16. Remove the wire harness clamp.
17. Disconnect the fuel vapor feed hose No. 1.
18. Disconnect the fuel vapor feed hose No. 2.
19. Remove the 2 nuts (E), then remove the emission control valve set.
20. Disconnect the throttle motor connector.
21. Separate the water by-pass hose No. 2.
22. Separate the water by-pass hose No. 3.
23. Disconnect the ventilation hose.
24. Remove the 2 bolts, then remove the engine hanger No. 1
25. Remove the 2 bolts, then remove the engine hanger No. 1
26. Remove the 2 bolts, then remove the surge tank stay No. 1 (B).
27. Remove the 2 bolts, then remove the surge tank stay No. 2 (C).
28. Disconnect the ground cable connector.
29. Using a socket hexagon wrench 8 mm, remove the 4 bolts.
30. Remove the 2 nuts, then remove the emission control valve bracket and the intake air surge tank.
31. Remove the gasket from the intake air surge tank.
32. Remove the EFI fuel pipe clamp No. 1.

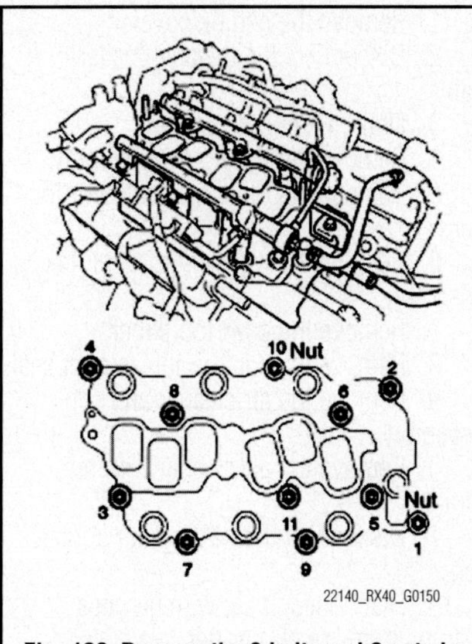

Fig. 126 Remove the 9 bolts and 2 nuts in the sequence shown

33. Pinch the quick connector and then pull out the fuel pipe No. 1.
34. Disconnect the heater water inlet hose.
35. Disconnect the radiator hose inlet.
36. Remove the nut and ground cable.
37. Disconnect the 6 fuel injector connectors.
38. In order to remove the intake manifold, using several steps, remove the 9 bolts and 2 nuts in the sequence shown in the illustration.
39. Remove the water outlet.
40. Disconnect the 2 knock sensor connectors.
41. Remove the 2 nuts, and then remove the 2 knock sensors.

To install:

42. Install the 2 knock sensors so that it is horizontal as shown in the illustration. Then install the 2 bolts and tighten to 15 ft. lbs. (20 Nm).
43. Connect the 2 knock sensor connectors.
44. Install the intake manifold with the 9 bolts, 2 nuts and 2 washers. Using several steps, tighten the bolts and nuts uniformly in the removal sequence. Tighten to 11 ft. lbs. (1 Nm).
45. Connect the 6 fuel injector connectors.
46. Install the ground cable with the nut.
47. Connect the radiator hose inlet.
48. Connect the heater water inlet hose.
49. Align the quick connector with the pipe, then push in the quick connector until the retainer makes a "click" sound to connect the fuel hose to the fuel pipe.
50. Using a socket hexagon wrench 8 mm, install the intake manifold with the 4 bolts and 2 nuts . Using several steps, tighten the bolts and nuts uniformly in the sequence shown in the illustration. Tighten the nuts and bolts to 21 ft. lbs. (28 Nm).
51. Connect the ground cable connector.
52. Install the surge tank stay brackets and tighten to 14 ft. lbs. 20 (Nm).
53. Connect the ventilation hose.
54. Connect the fuel vapor feed hose.
55. Connect both of the water by-pass hoses.
56. Reconnect the throttle motor connector.
57. Install the emission control valve set with the 2 nuts and tighten to 80 inch. lbs (8 Nm).
58. Reconnect both of the fuel vapor hose.
59. Connect the wire harness clamp.
60. Connect the VSV connector.
61. Install the engine moving control rod.
62. Install the reservoir bracket and tighten to 71 inch. lbs. (9 Nm).
63. Install air filter assembly bracket and tighten to 14 ft. lbs. 20 (Nm).
64. Install the 2 bolts and brake master cylinder reservoir to the bracket. Tighten to 80 inch. lbs. (9 Nm).
65. Install the air cleaner assembly.

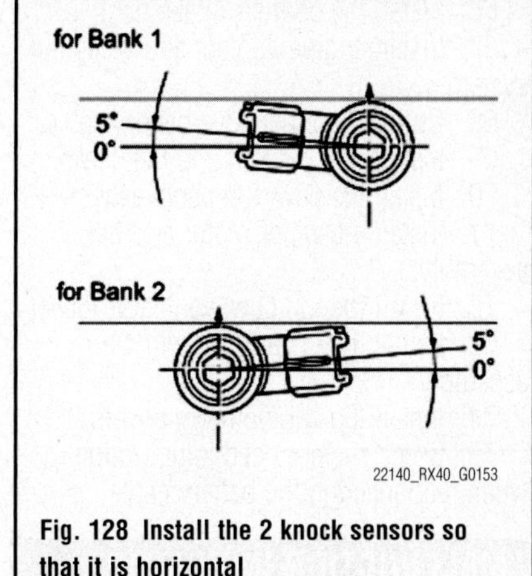

Fig. 128 Install the 2 knock sensors so that it is horizontal

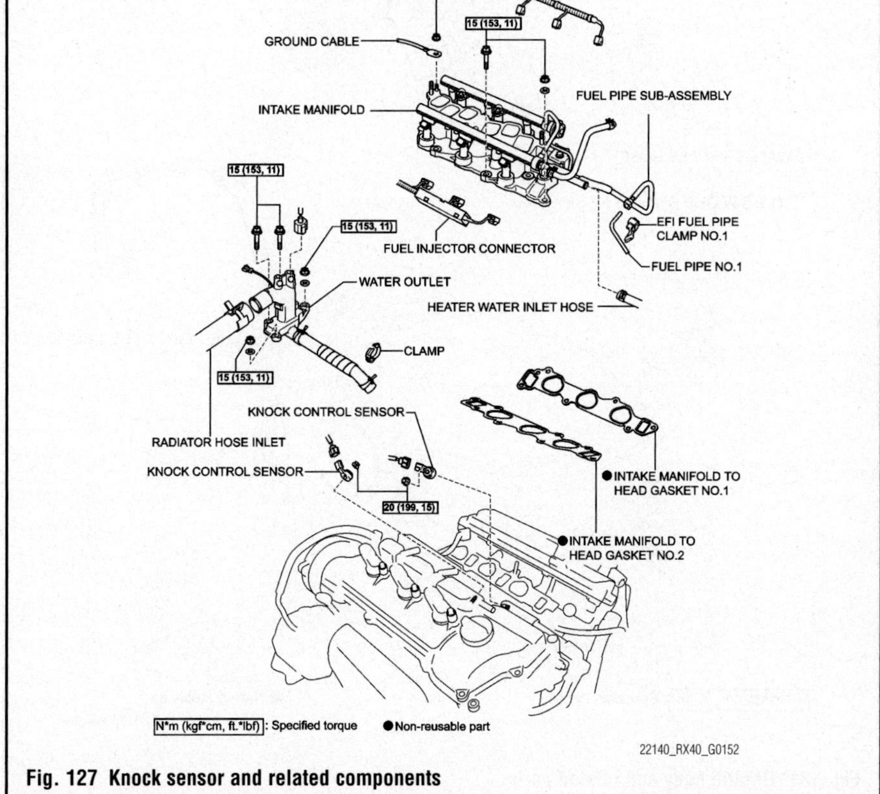

Fig. 127 Knock sensor and related components

Fig. 129 Using several steps, tighten the bolts and nuts uniformly in the sequence shown

66. Install the cool air intake duct.
67. Install engine coolant and bleed the system.
68. Connect the negative battery cable.
69. Inspect for coolant and fuel leaks.
70. Install the cowl top panel assembly.
71. Install the wiper motor and link assembly.
72. Install the top cowl ventilator louver.
73. Install the LH and RH wiper arm assembly.
74. Install the engine room covers.
75. Some system need initialization when reconnecting the battery cable

MALFUNCTION INDICATOR LIGHT (MIL)

RESET PROCEDURES

1. Clear DTC (Using the Techstream) as follows:
    a. Connect the Techstream to the DLC3.
    b. Turn the ignition switch ON.
    c. Enter the following menus: Powertrain / Engine and ECT / Trouble Codes.
    d. Press the YES button.
2. Clear DTC (Without using the Techstream). Perform either one of the following operations.
    a. Disconnect the negative (-) battery cable for more than 1 minute.
    b. Remove the EFI and ETCS fuses from the engine room No. 2 relay block located inside the engine compartment for more than 1 minute.

MASS AIR FLOW (MAF) SENSOR

LOCATION

See Figure 130.

The mass air flow sensor is located in the air cleaner housing.

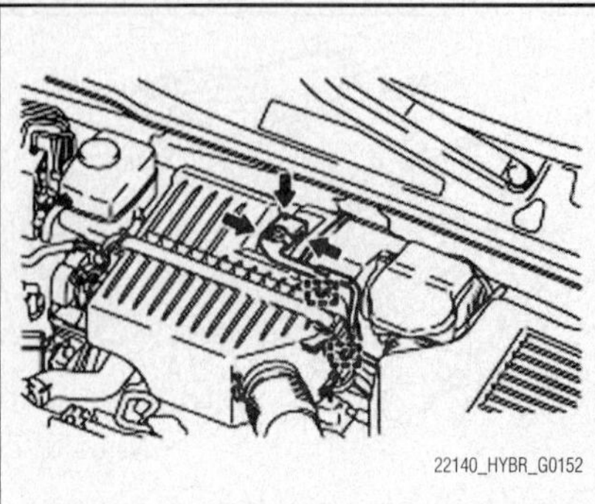

Fig. 130 Intake air temperature sensor location.

REMOVAL & INSTALLATION

1. Disconnect the mass air flow meter connector.
2. Disconnect the 2 wire harness clamps from the air cleaner assembly.
3. Remove the 2 screws and mass air flow meter.

To install:

4. Install the mass air flow meter with the 2 screws.
5. Connect the mass air flow meter connector.
6. Connect the 2 wire harness clamps to the air cleaner assembly.

THROTTLE POSITION SENSOR (TPS)

LOCATION

The throttle position sensor is integral to the electric throttle body assembly.

REMOVAL & INSTALLATION

See Figure 131.

1. Remove the engine covers.
2. Disconnect the negative battery cable.
3. Drain the engine coolant.
4. Remove the wiper arm assembly.
5. Remove the top cowl ventilator louver.
6. Remove the wiper motor and link assembly.
7. Remove the cowl top panel.
8. Remove the cool air intake duct seal.
9. Remove the air cleaner and cap assembly.
10. Remove the air cleaner with resonator.
11. Disconnect the MAF meter connector.
12. Disconnect the 2 wire harness clamps from the air cleaner.
13. Remove the bolts from the air cleaner case w/ resonator.
14. Remove the hose clamp, and separate the air cleaner hose No. 1.
15. Disconnect the throttle motor connector.

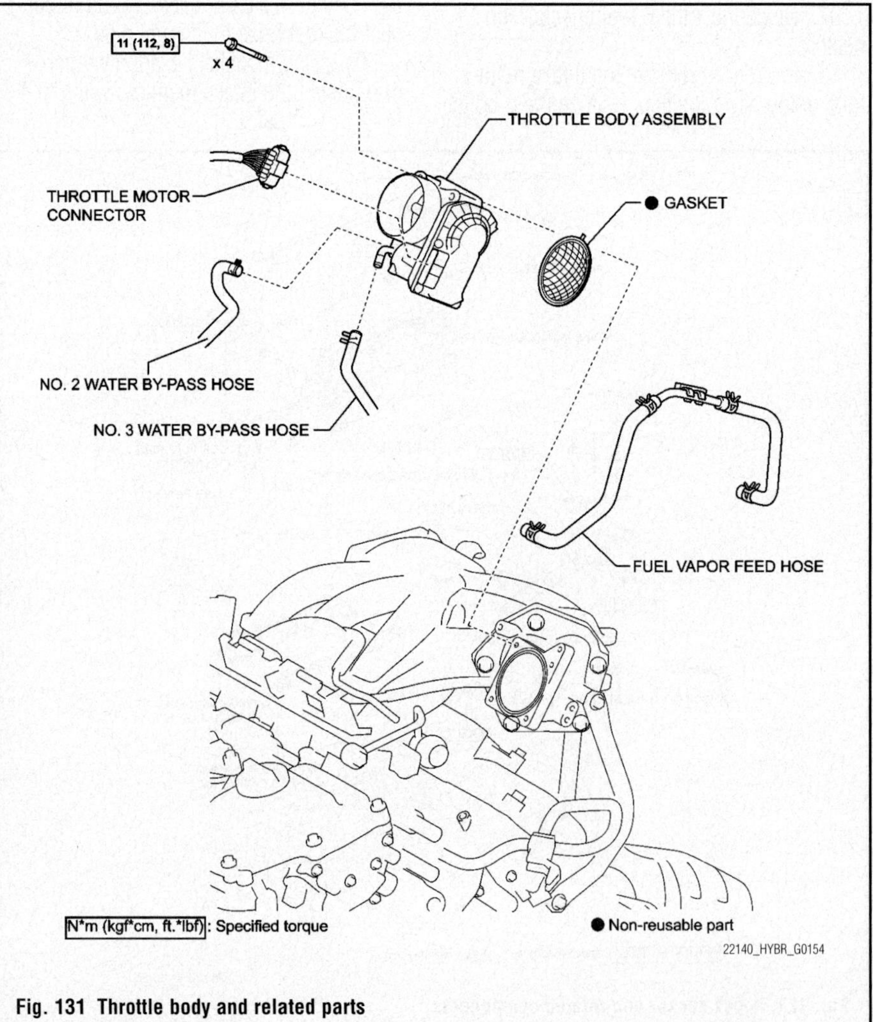

Fig. 131 Throttle body and related parts

16. Separate both of the water by-pass hoses.
17. Separate the fuel vapor feed hose.
18. Remove the 4 bolts and the throttle body assembly.
19. Remove the throttle body gasket from the intake air connector.

To install:

20. Install a new throttle body gasket to the intake air connector.
21. Install the throttle body assembly with the 4 bolts. Tighten the bolts to 8 ft. lbs. (11 Nm).
22. Connect the fuel vapor feed hose.
23. Connect the water by-pass hoses.
24. Reconnect the throttle body connector.
25. Install the air cleaner hose No.1 to the throttle body assembly with the hose clamp.
26. Install the air cleaner case w/ resonator with the 5 bolts and tighten to 44 inch. lbs. (5 Nm).
27. Connect the MAF meter connector.
28. Connect the 2 wire harness clamps to the air cleaner.
29. Connect the ventilation hose.
30. Install a new air cleaner element to the air cleaner case.
31. Install the air cleaner cap with inlet and tighten.
32. Install cool air intake duct seal.
33. Install engine coolant.
34. Check for coolant leaks.
35. Install cowl top panel assembly.
36. Install the wiper motor and link assembly.
37. Install the cowl top ventilator louver assembly.
38. Install the LH and RH wiper arm assembly.
39. Connect the negative battery cable.
40. Install engine room covers.
41. Some system need initialization when reconnecting the battery cable.

VARIABLE VALVE TIMING (VVT) SENSOR

REMOVAL & INSTALLATION

See Figures 132 through 136.

1. Disconnect the negative battery cable.
2. Remove air cleaner cap w/ inlet.
3. Remove the bank 1 VVT sensor, as follows:
   a. Disconnect the VVT sensor (bank 1) connector.
   b. Remove the bolt, and then remove the bank 1 VVT sensor.
4. Remove the bank 2 VVT sensor, as follows:

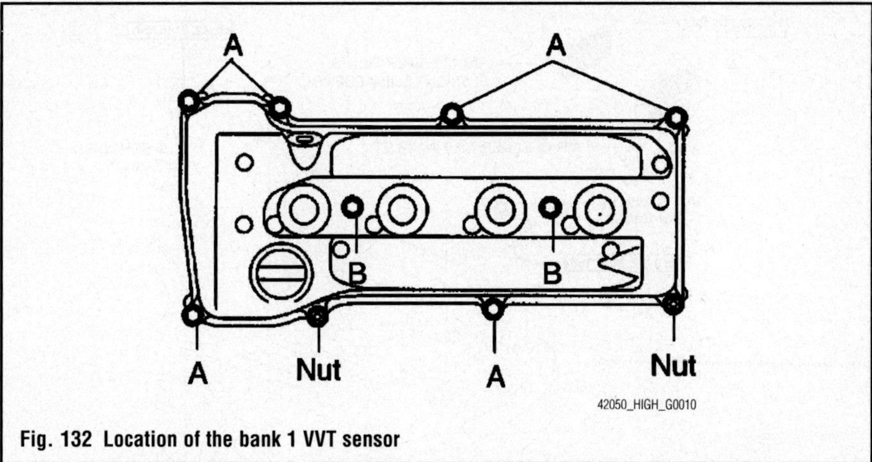

Fig. 132 Location of the bank 1 VVT sensor

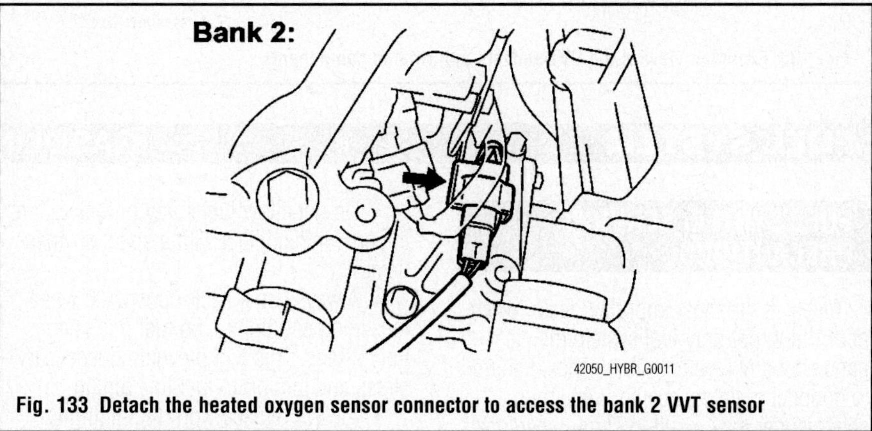

Fig. 133 Detach the heated oxygen sensor connector to access the bank 2 VVT sensor

   a. Disconnect the heated oxygen sensor connector (bank 2) from the wire harness bracket.
   b. Disconnect the bank 2 VVT sensor connector.

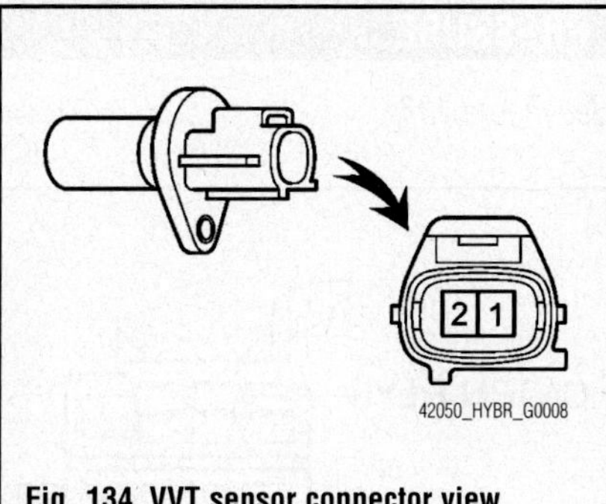

Fig. 134 VVT sensor connector view

   c. Remove the bolt, and then remove the bank 2 VVT sensor.
5. Inspect the VVT sensor:
   a. Using an ohmmeter, measure the resistance according to the value(s) in the accompanying chart. Standard resistance cold is 14–122°f (-10–50°c) and hot is 122–212°f (50–100°c. If the resistance is not within specifications, replace the sensor.

➡Cold and hot refer to the temperature of the coils themselves.

To install:

6. Install the bank 1 VVT sensor:
   a. Apply a light coat of engine oil to the o-ring on the VVT sensor (bank 1).
   b. Install the VVT sensor (bank 1)

Condition	Specified Condition
Cold	835 to 1,400 Ω
Hot	1,060 to 1,645 Ω

42050_HYBR_G0009

Fig. 135 VVT sensor resistance chart

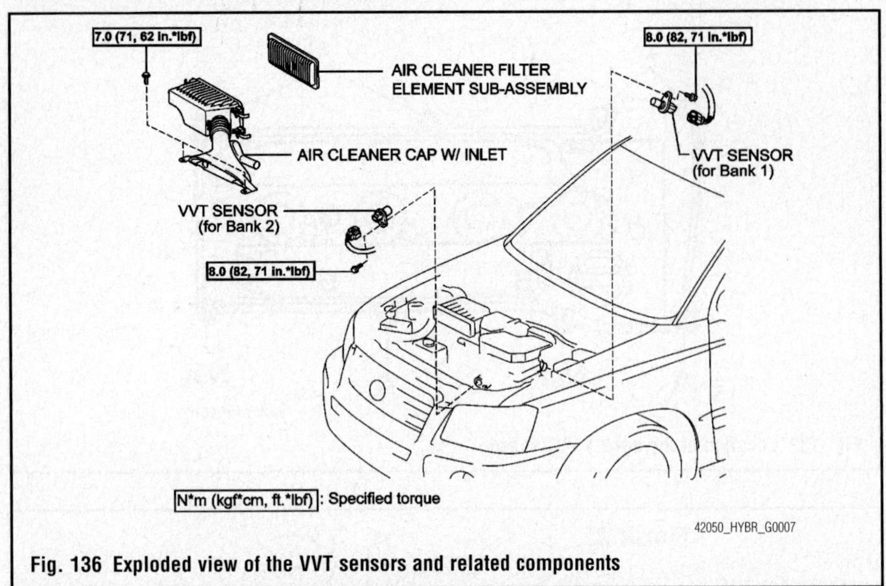

Fig. 136 Exploded view of the VVT sensors and related components

with the bolt an tighten to 71 inch lbs. (8 Nm).

c. Connect the VVT sensor (bank 1) connector.

7. Install the bank 2 VVT sensor:

a. Apply a light coat of engine oil to the o-ring on the VVT sensor (bank 2).

b. Install the VVT sensor (bank 2) with the bolt and tighten to 71 inch lbs. (8 Nm).

c. Connect the VVT sensor (bank 2) connector.

d. Install the heated oxygen sensor connector (bank 2) to the wire harness bracket.

8. Install air cleaner cap w/ inlet.

9. If necessary, perform the initialization procedure, as outlined in the Chassis Electrical System.

FUEL SYSTEM — GASOLINE FUEL INJECTION SYSTEM

FUEL SYSTEM SERVICE PRECAUTIONS

Safety is the most important factor when performing not only fuel system maintenance but any type of maintenance. Failure to conduct maintenance and repairs in a safe manner may result in serious personal injury or death. Maintenance and testing of the vehicle's fuel system components can be accomplished safely and effectively by adhering to the following rules and guidelines.

- To avoid the possibility of fire and personal injury, always disconnect the negative battery cable unless the repair or test procedure requires that battery voltage be applied.
- Always relieve the fuel system pressure prior to disconnecting any fuel system component (injector, fuel rail, pressure regulator, etc.), fitting or fuel line connection. Exercise extreme caution whenever relieving fuel system pressure to avoid exposing skin, face and eyes to fuel spray. Please be advised that fuel under pressure may penetrate the skin or any part of the body that it contacts.
- Always place a shop towel or cloth around the fitting or connection prior to loosening to absorb any excess fuel due to spillage. Ensure that all fuel spillage (should it occur) is quickly removed from engine surfaces. Ensure that all fuel soaked cloths or towels are deposited into a suitable waste container.
- Always keep a dry chemical (Class B) fire extinguisher near the work area.
- Do not allow fuel spray or fuel vapors to come into contact with a spark or open flame.
- Always use a back-up wrench when loosening and tightening fuel line connection fittings. This will prevent unnecessary stress and torsion to fuel line piping.
- Always replace worn fuel fitting O-rings with new. Do not substitute fuel hose or equivalent where fuel pipe is installed.

Before servicing the vehicle, make sure to also refer to the precautions in the beginning of this section as well.

RELIEVING FUEL SYSTEM PRESSURE

See Figure 137.

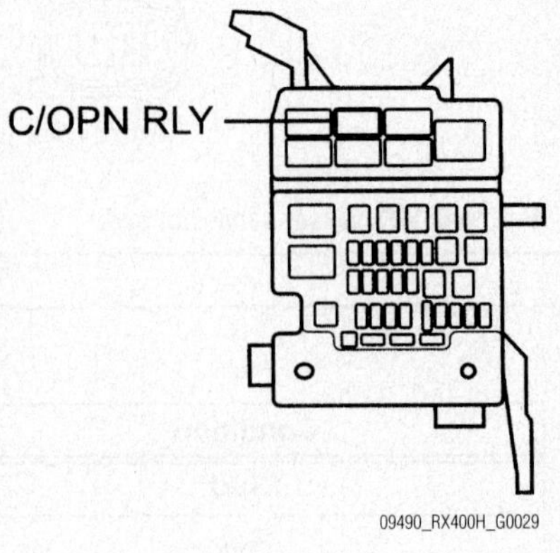

Fig. 137 Location of the C/OPN relay in the junction box.

1. Before servicing the vehicle, refer to the Precautions Section.
2. Disconnect the No. 3 relay block.
3. Remove the No. 2 junction block cover.
4. Remove the C/OPN RLY.
5. Put the vehicle in Inspection Mode and start the engine.
6. Turn the ignition switch to **OFF** immediately after the engine comes to 'rough idle state'.

➡The hybrid system has a complicated process from an 'out of gas' to 'engine stall' condition. Therefore, 'rough idle' is regarded as 'stop'.

7. Disconnect the negative battery cable.
8. Reinstall the C/OPN RLY.

FUEL FILTER

REMOVAL & INSTALLATION

The fuel filter is part of the fuel suction tube/fuel pump assembly and is located in the fuel tank.

FUEL PUMP

REMOVAL & INSTALLATION

See Figure 138.

1. Before servicing the vehicle, refer to the Precautions Section.
2. Relieve the fuel system pressure, as outlined in this section.
3. Disconnect the negative battery cable.
4. Remove the front floor heat insulator no. 3.

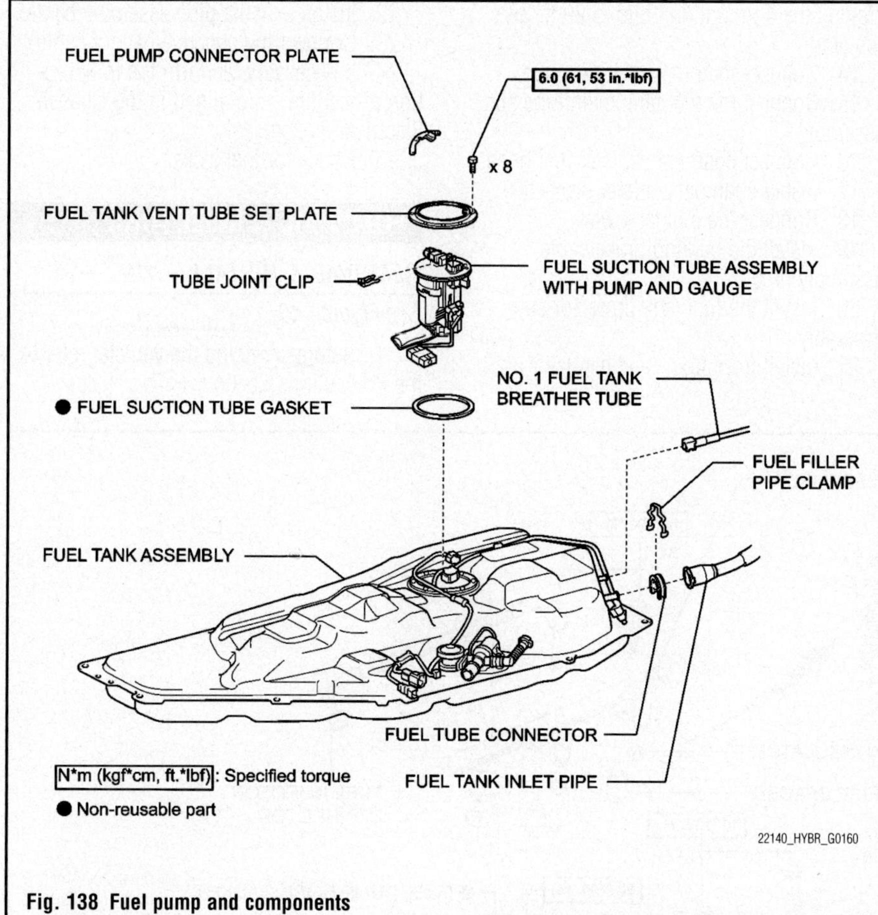

Fig. 138 Fuel pump and components

5. Remove the fuel tank protector subassembly no.1.
6. Separate parking brake cable assembly no.3.
7. Disconnect the fuel tank wire.
8. Remove the charcoal canister protector.
9. Disconnect the charcoal canister fuel hose.
10. Disconnect the fuel tank main tube subassembly.
11. Disconnect the breather lower tube.
12. Remove the fuel tank assembly.
13. Remove the fuel pump and gauge with the suction tube, as follows:
   a. Disconnect the fuel pump connector.
   b. Disconnect the clamp and vapor pressure sensor connector.
14. Remove the tube joint clip and clamp, and pull out the fuel pump tube.

⁂ WARNING

Check that there is no dirt or foreign matter around the fuel tube joint and after cleaning off any excess dirt, proceed with the work. Be careful not to allow dirt or foreign matter to scratch or come into contact with the fuel tube connector and fuel suction plate that are sealed by o-rings. Do not use any tools for this work. Do not bend or twist the nylon tube by force. Cover the disconnected fuel tube joint with a plastic bag. When the fuel tube joint and fuel suction plate are stuck, pinch the fuel tank tube with the fingers, and turn it carefully to release. Disconnect the fuel tank tube.

   a. Remove the 8 bolts, fuel tank vent tube set plate and fuel pump connector plate.
   b. Pull out the fuel pump assembly from the fuel tank.

To install:

   c. Install the fuel pump assembly.
   d. Install the fuel tank vent tube set plate by aligning it with the cutout on the fuel pump assembly.
   e. Install the fuel tank vent tube set plate and fuel pump connector plate with the 8 bolts. Tighten to 53 inch lbs. (6.0 Nm).
   f. Connect the fuel tank main tube with the tube joint clip and clamp.

➡Check that there are no scratches or foreign objects on the connecting part. Check that the fuel tube joint is inserted securely. Check that the tube joint clip is on the collar of the fuel tube joint. after installing the tube joint clip, check that the fuel tube joint is not pulled off.

   g. Connect the fuel tank wire connector.
   h. Connect the vapor pressure sensor connector and clamp.
15. Install the fuel tank assembly.
16. Connect the breather lower tube,
17. Connect the fuel tank main tube subassembly.
18. Connect charcoal canister fuel hose.
19. Install charcoal canister protector.
20. Connect the fuel tank wire.
21. Install the parking brake cable assembly no. 3.
22. Install the fuel tank protector subassembly no. 1.
23. Install front floor heat insulator no. 3.
24. Install exhaust pipe assembly center.
25. Connect the negative battery cable.
26. If necessary, perform the initialization procedure, as outlined in the Chassis Electrical System.
27. Check for fuel leaks.

FUEL TANK

REMOVAL & INSTALLATION

See Figure 139.

1. Before servicing the vehicle, refer to the Precautions Section.
2. Relieve the fuel system pressure, as outlined in this section.
3. Disconnect the negative battery cable.
4. Remove the front floor heat insulator no.3.
5. Remove the fuel tank protector subassembly no. 1.

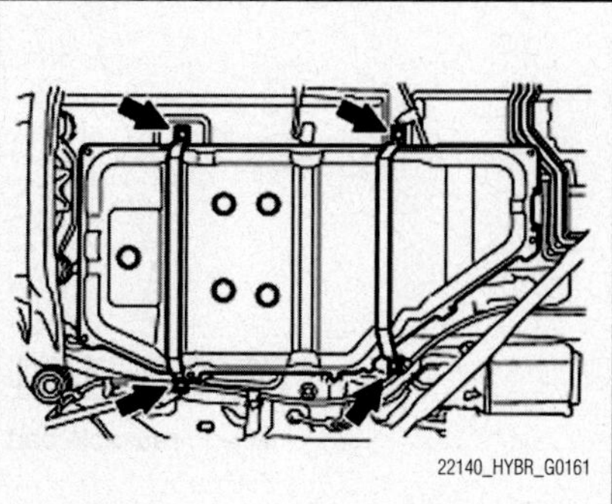

Fig. 139 Fuel tank straps and mounting bolts

6. Separate parking brake cable assembly no.3.
7. Disconnect the fuel tank wire.
8. Remove the charcoal canister protector.
9. Disconnect the charcoal canister fuel hose.
10. Disconnect the fuel tank main tube subassembly.
11. Disconnect the breather lower tube.
12. Remove the fuel tank assembly.

To install:

13. Install the fuel tank assembly. Tighten the 4 mounting bolts to 29 ft. lbs. (39 Nm).
14. Connect the breather lower tube,
15. Connect the fuel tank main tube sub-assembly.
16. Connect charcoal canister fuel hose.
17. Install charcoal canister protector.
18. Connect the fuel tank wire.
19. Install the parking brake cable assembly no.3.
20. Install the fuel tank protector sub-assembly no.1.
21. Install front floor heat insulator no.3.
22. Install exhaust pipe assembly center.
23. Connect the negative battery cable.
24. If necessary, perform the initialization procedure, as outlined in the Chassis Electrical System.
25. Check for fuel leaks.

FUEL RAIL & INJECTORS

REMOVAL & INSTALLATION

See Figure 140.

1. Before servicing the vehicle, refer to the Precautions Section.

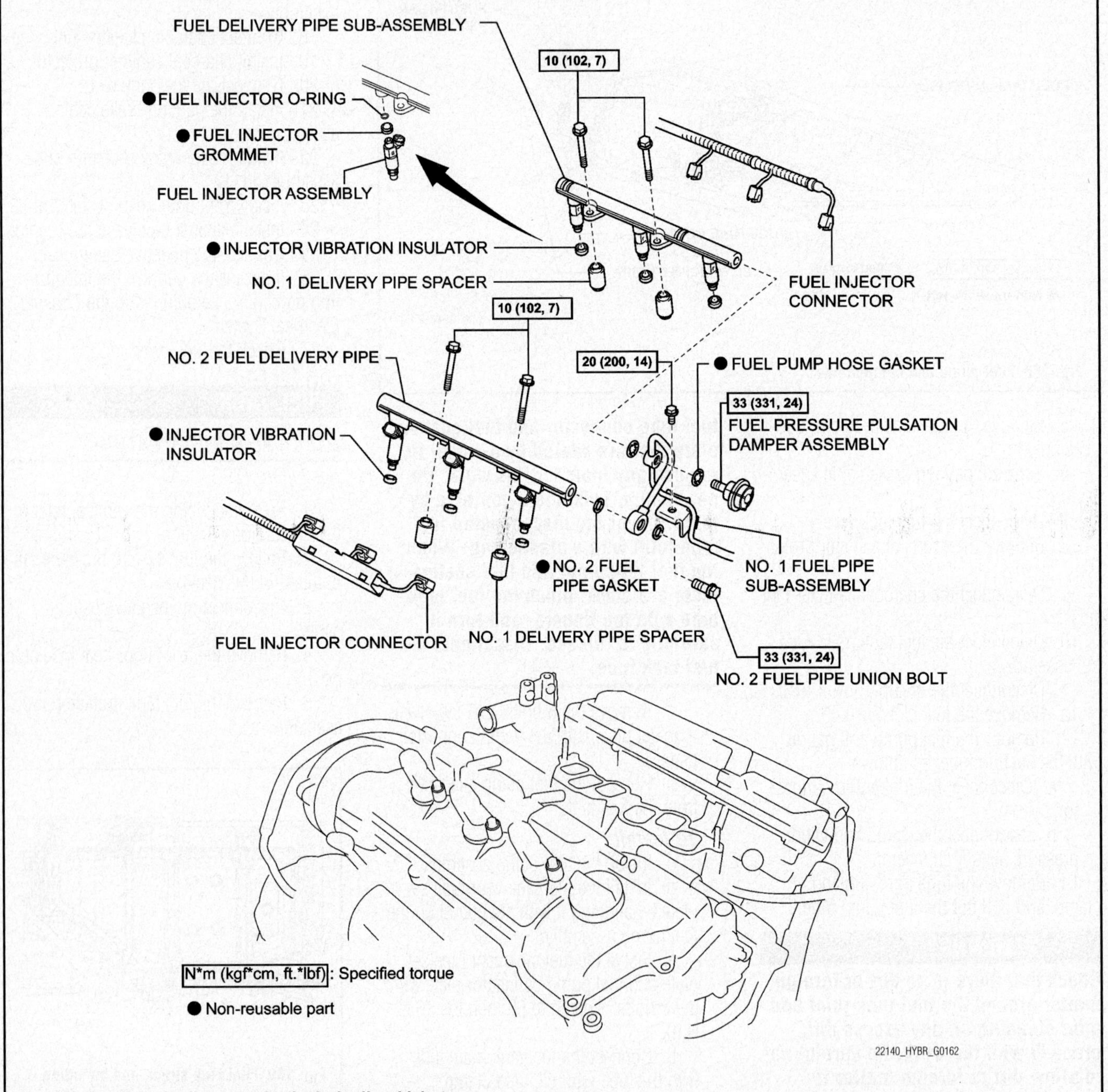

Fig. 140 Location of components for the fuel rail and injectors

2. Relieve the fuel system pressure.
3. Drain the cooling system.
4. Remove or disconnect the following:
- Negative battery cable
- Engine cover
- Wiper arm and blade assembly
- Top cowl ventilator louver assembly
- Wiper motor and linkage assembly
- Air intake assembly
- Emission control valve hoses
- Air intake surge tank

5. Remove the mounting bolt and separate fuel hose No. 1
6. Remove the fuel pressure pulsation damper and gaskets.
7. Remove the fuel hose No. 2 union bolt and gaskets.
8. Disconnect the wiring at the injectors.
9. Remove the 4 bolts and each fuel rail with the injectors still attached.
10. Pull each injector from the fuel rail..

To install:

11. Install new O-rings on each injector. Apply a light coating of gasoline to the O-rings and mating points on the pipes.
12. Using a twisting motion, install the injectors on the pipes.

➡Be careful to avoid twisting the O-rings. After installation, check that the injectors turn smoothly. If not, use new O-rings.

13. Install the pipes and injectors.
14. Loosely install the bolts and make sure that the injectors still turn freely. If not, replace the O-rings.
15. Torque the bolts to 84 inch lbs. (10 Nm).
16. The remainder of installation is the reverse of removal. Observe the following torques:
- Fuel hose No. 2 union bolt: 24 ft. lbs. (33 Nm)
- Pulsation damper: 24 ft. lbs. (33 Nm)
- Fuel hose No.1: 14 ft. lbs. (20 Nm)

17. If necessary, perform the initialization procedure, as outlined in the Chassis Electrical System.

IDLE SPEED

ADJUSTMENT

Idle speed is maintained by the Engine Control Module (ECM). No adjustment is necessary or possible.

THROTTLE BODY

REMOVAL & INSTALLATION

See Figure 141.

1. Before servicing the vehicle, refer to the Precautions Section.
2. Relieve the fuel system pressure.
3. Drain the cooling system.
4. Remove or disconnect the following:
- Negative battery cable
- Engine cover
- Wiper arm and blade assembly
- Top cowl ventilator louver assembly
- Wiper motor and linkage assembly

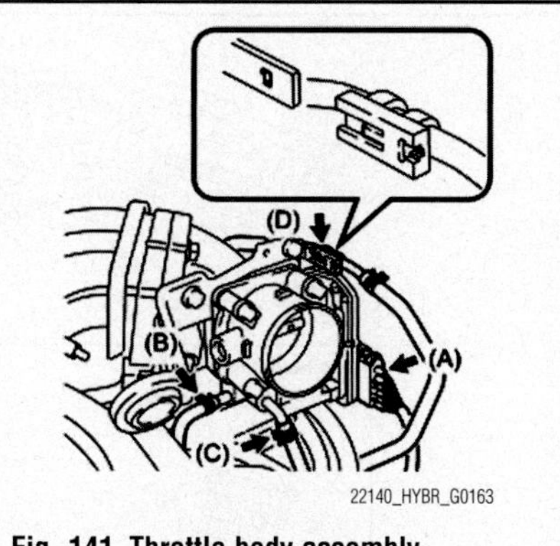

Fig. 141 Throttle body assembly

- Cool air intake duct seal
- Air cleaner cap and case assembly

5. Remove the throttle body as follows:
   a. Disconnect the throttle motor connector (A).
   b. Separate the No. 2 water by-pass hose (B).
   c. Separate the No. 3 water by-pass hose (C).
   d. Separate the fuel vapor feed hose (D).
   e. Remove the 4 bolts and the throttle body assembly.
   f. Remove the throttle body gasket from the intake air surge tank.

To install:

6. Install the throttle body assembly as follows:
7. Install a new throttle body gasket to the intake air surge tank.
   a. Tighten the 4 mounting bolts to 8 ft. lbs. (11 Nm).
   b. Connect the fuel vapor feed hose (A).
   c. Connect the No. 3 water by-pass hose (B).
   d. Connect the No. 2 water by-pass hose (C).
   e. Connect the throttle motor connector (D).
8. Install air cleaner case and cap assembly.
9. Add the engine coolant.
10. Inspect for any coolant leaks.
11. Connect the negative battery cable.
12. Install or connect the following:
- Air cleaner cap and case assembly
- Cool air intake duct seal
- Wiper motor and linkage assembly
- Top cowl ventilator louver assembly
- Wiper arm and blade assembly
- Engine cover
- Negative battery cable

HEATING & AIR CONDITIONING SYSTEM

BLOWER MOTOR

REMOVAL & INSTALLATION

2006–07 Models

See Figure 142.

1. Before servicing the vehicle, refer to the Precautions Section.
2. Remove the glove compartment door assembly.
3. Remove the lower No. 2 instrument panel finish panel.
4. Remove the No. 2 air duct, as follows:
    a. Remove the screw.
    b. Release the 3 claw fittings and remove the No 2 air duct.
5. Remove the ECM, as follows:
    a. Disconnect the connectors.
    b. Remove the 2 nuts and ECM.
6. Remove the blower assembly, as follows:
    a. Disconnect the connector and clamp of the cooling unit motor with fan sub-assembly.
    b. Disconnect the connector of the recirculation damper servo.
    c. Remove the 2 screws and bracket.
    d. Remove the bolt, nut and 2 screws.
    e. Disconnect the part fitted to the air conditioner radiator assembly and remove the blower assembly.
7. Release the 2 claw fittings and remove the No. 1 cooler cover.
8. Remove the clean air filter.
9. Remove the 3 screws and air inlet servo motor.
10. Remove the 3 screws and cooling unit motor with fan sub-assembly.

To install:

11. Install the cooling unit motor with fan sub-assembly.
12. Install the air inlet servo motor.

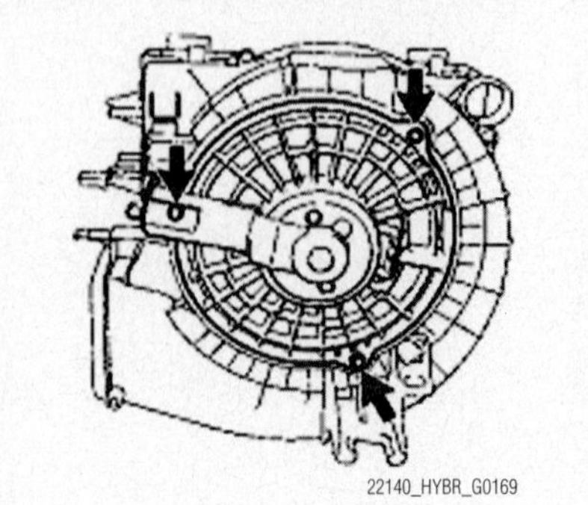

Fig. 142 Blower motor and retaining screws

13. Install the clean air filter.
14. Aligning the part to be fitted with that of the air conditioning radiator assembly, install the blower assembly.
15. Install the blower assembly with the bolt, nut and 2 screws, and tighten to 87 inch lbs. (9.8 Nm).
16. Install the bracket with the 2 screws.
17. Install the connector and clamp.
18. Install the ECM with the 2 nuts and tighten to 49 inch lbs. (5.5 Nm).
19. Install the No. 2 air duct.
20. Install the glove compartment door assembly.

2008 Models

See Figure 143.

1. Before servicing the vehicle, refer to the Precautions Section.
2. Remove the No. 2 instrument panel under cover sub-assembly.
3. Remove front blower motor sub-assembly, as follows:
    a. Disconnect the connector.
    b. Remove the 3 screws and the front blower motor sub-assembly.

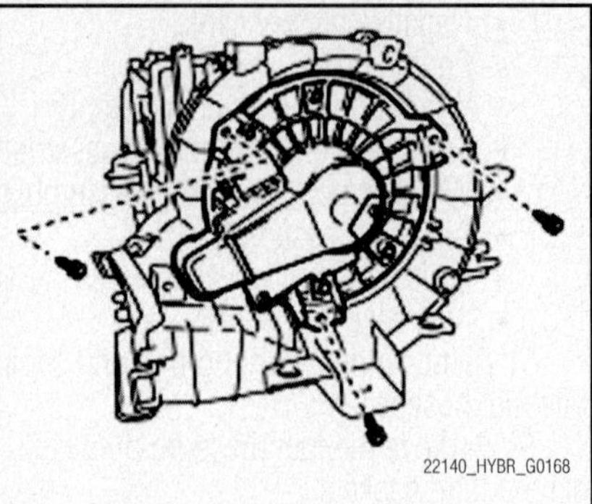

Fig. 143 Blower motor and retaining screws

To install:

4. Installation is the reverse of the removal procedure.

HEATER CORE

REMOVAL & INSTALLATION

2006–07 Models

See Figures 144 through 149.

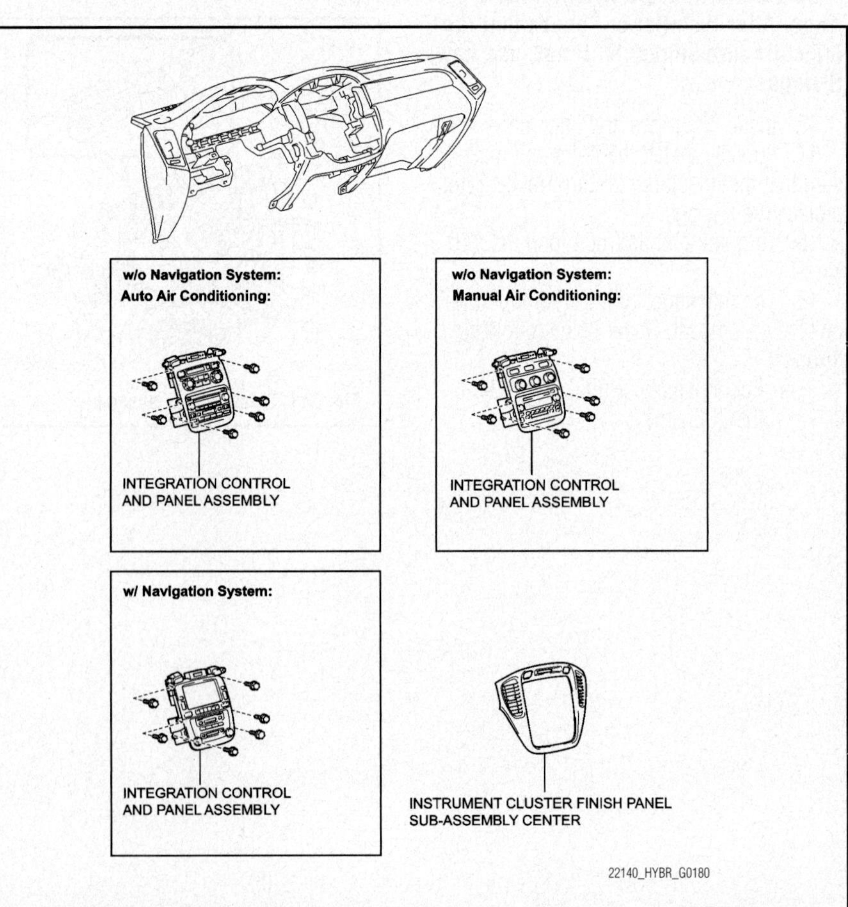

Fig. 144 Integration control panel

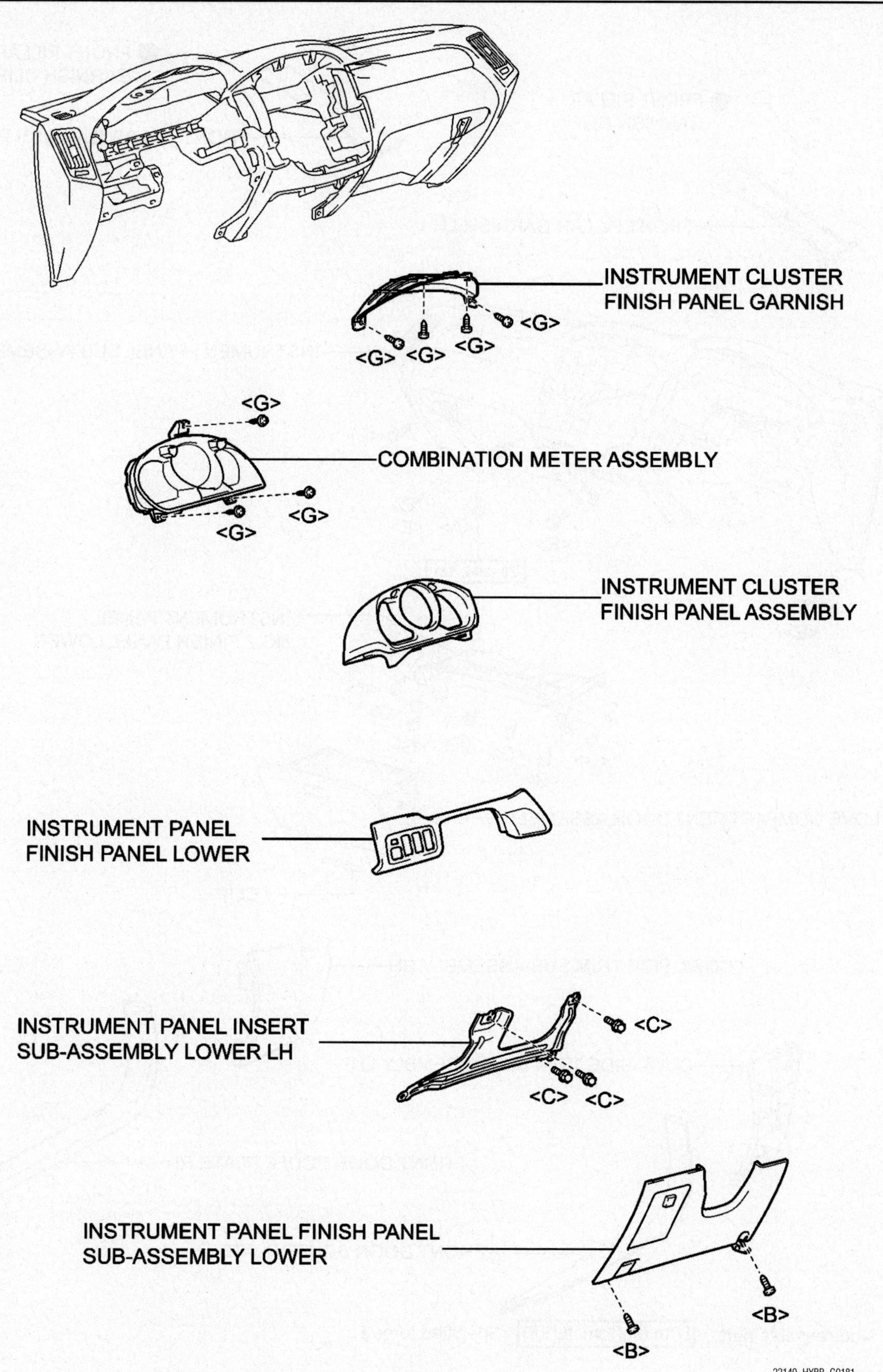

Fig. 145 Combination meter and related components

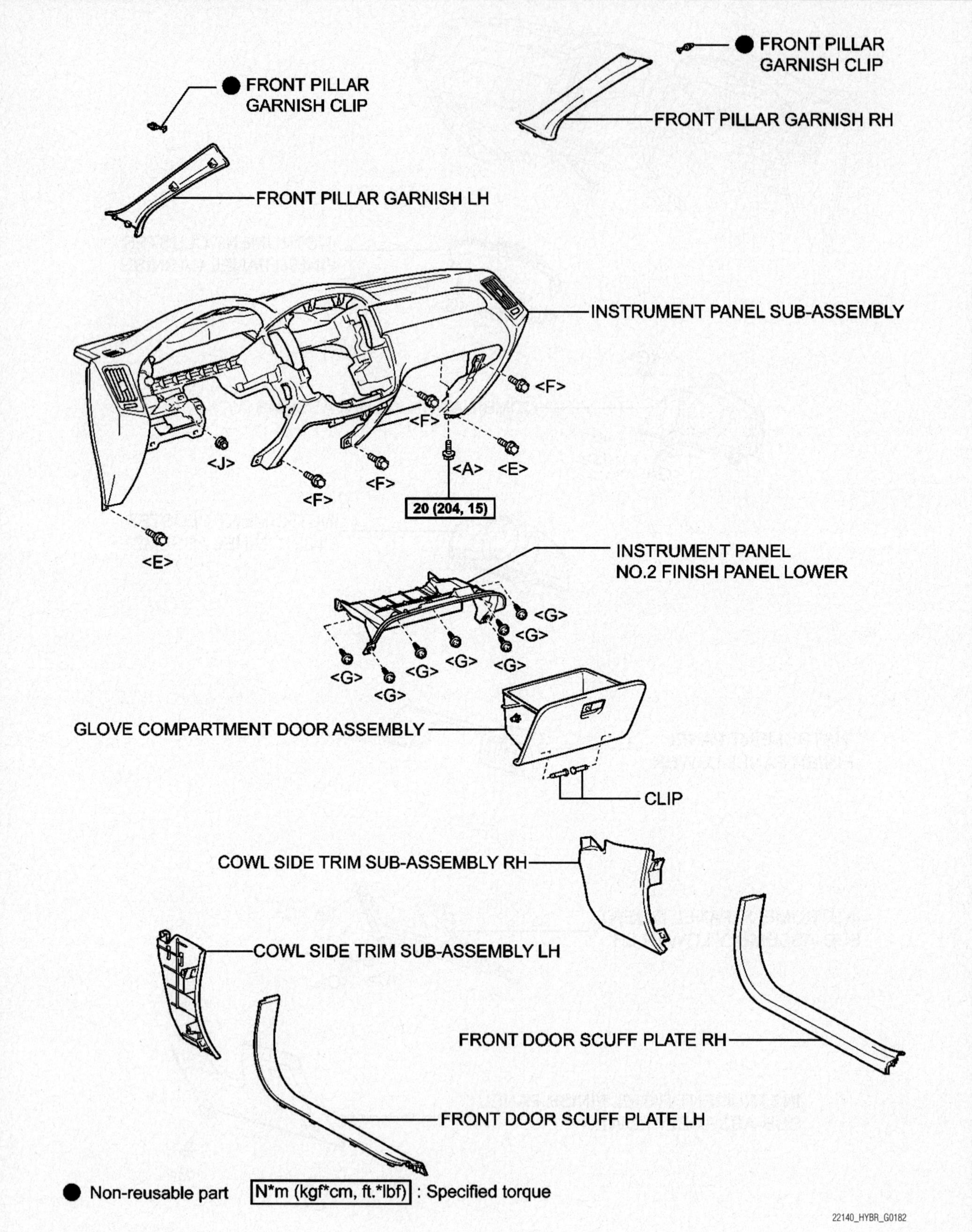

Fig. 146 Instrument panel sub-assembly and related components.

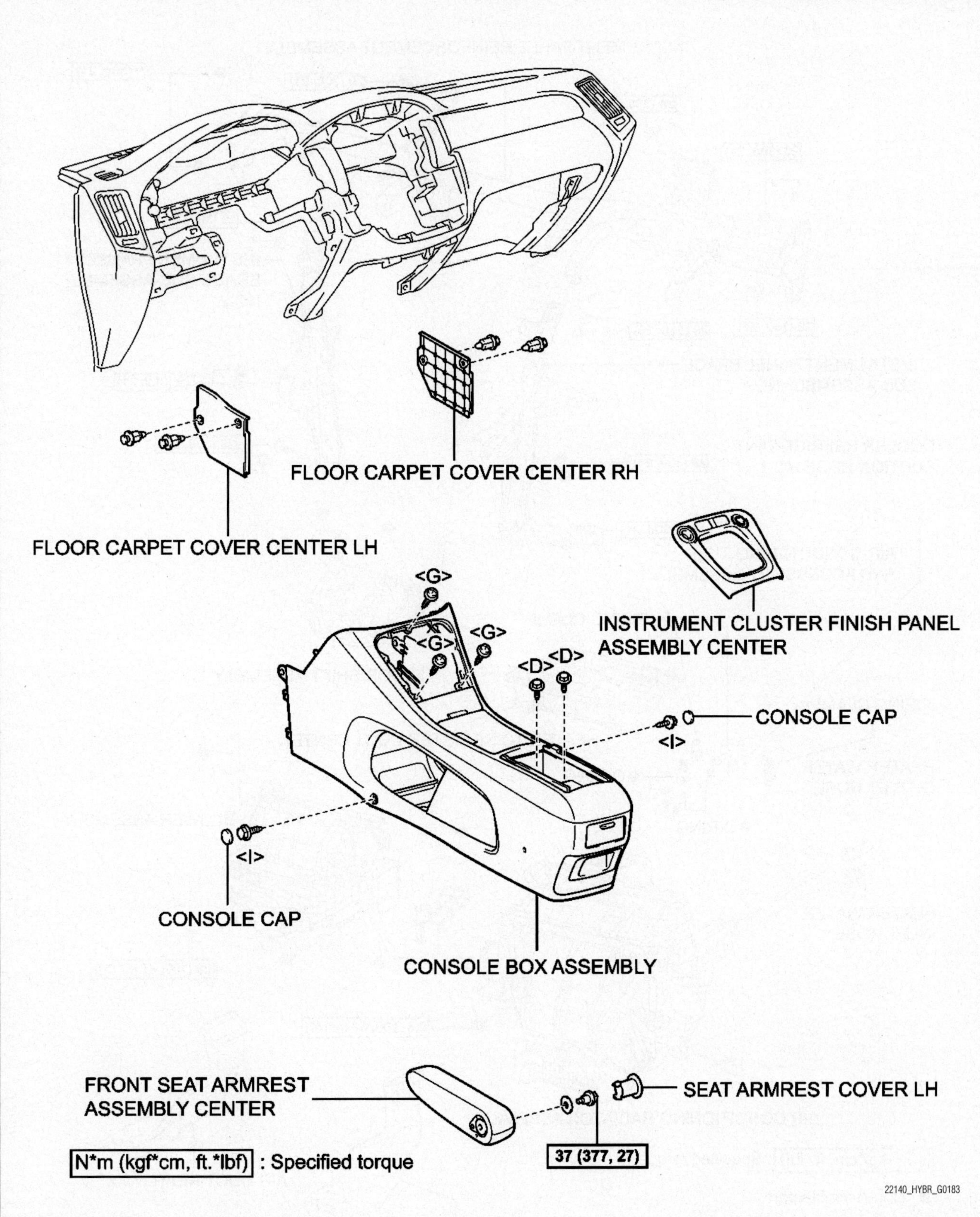

Fig. 147 Console box assembly and related components

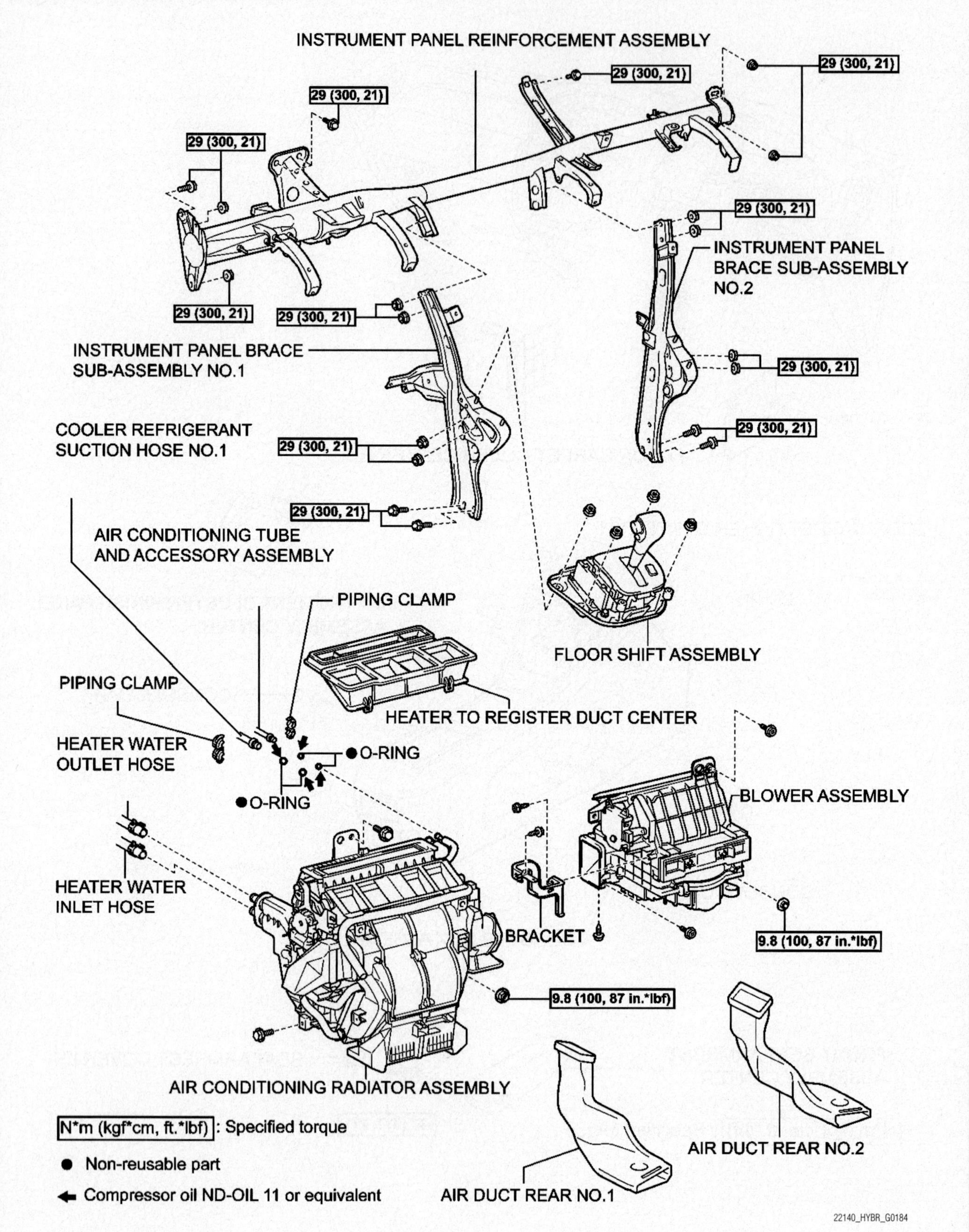

Fig. 148 Instrument panel reinforcement assembly and related components

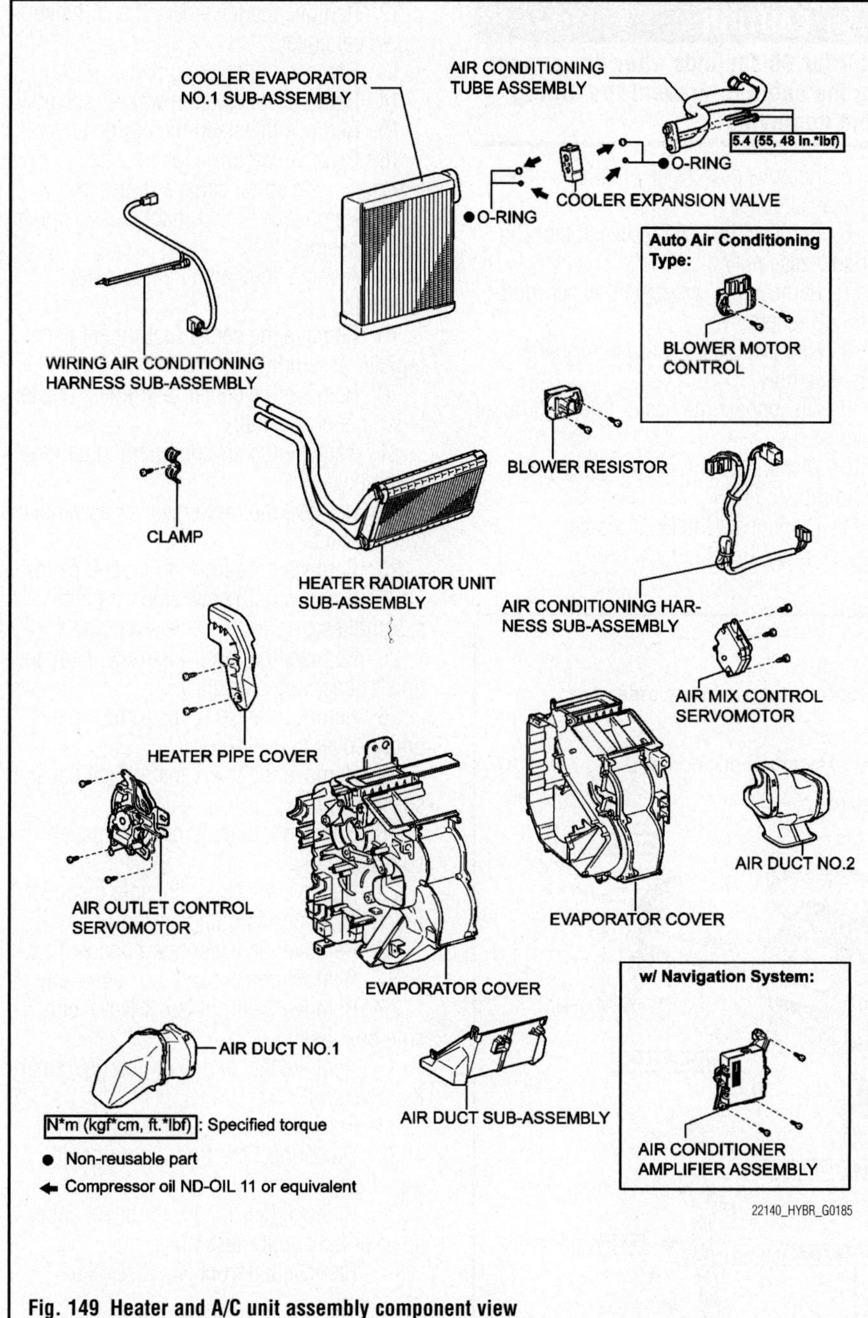

Fig. 149 Heater and A/C unit assembly component view

1. Before servicing the vehicle, refer to the Precautions Section.
2. Recover the refrigerant from the A/C system using a refrigerant recovery unit.
3. Disconnect the A/C lines.
4. Disconnect the heater hoses.
5. Install instrument panel air bag assembly.
6. Disconnect the steering column assembly.
7. Disconnect the floor shift assembly.
8. Remove the rear air ducts No. 1 and No. 2.
9. Remove the 3 bolts, 2 nuts and instrument panel brace sub-assembly No.1.
10. Remove the 2 bolts, 2 nuts and instrument panel brace sub-assembly No.2.
11. Remove the air conditioner amplifier assembly. (w/ navigation system)
12. Remove the 2 nuts and multiplex network body ECU.
13. Remove the bolt and 2 nuts, disconnect the instrument panel junction block assembly.
14. Remove the 5 bolts, 4 nuts and instrument panel reinforcement assembly.
15. Disconnect the heater and A/C unit assembly as follows:
    - Disconnect the connector and clamp.
    - Disconnect the connector connecting the wiring air indicator harness sub-assembly and vehicle harness.
    - Remove the 2 bolts, 2 nuts and air conditioner unit assembly.
    - Remove the screw, disconnect the connected part of the duct and take off the air duct No.2.
    - Remove the 4 screws and bracket.
    - Disconnect the connection between the blower assembly and the heater and A/C unit assembly and remove the unit.
16. Remove the 2 screws and heater pipe cover.
17. Remove the heater core.

To install:

18. Install the heater core.
19. Install the heater pipe cover with the 2 screws.
20. Install the air conditioner unit assembly with the 2 bolts and 2 nuts. Tighten to 87 inch. lbs. (9.8 Nm).
21. Connect the connectors connecting the wiring air indicator No.2 harness and vehicle harness.
22. Install the connector and clamp.
23. Install the instrument panel reinforcement assembly with the 5 bolts and 4 nuts. Tighten to 21 ft. lbs. (29 Nm).
24. Install the instrument panel junction block assembly with the bolt and 2 nuts. Tighten to 74 inch. lbs. (8.4 Nm).
25. Install the multiplex network body ECU with the bolt and 2 nuts.
26. Install the air conditioner amplifier assembly. (w/ navigation system)
27. Install both instrument panel braces. Tighten the bolts and nuts to Tighten to 21 ft. lbs. (29 Nm).
28. Install both rear air ducts.
29. Install the floor shift assembly with the 4 nuts and tighten to 9 ft. lbs. (12 Nm).
30. Install the shift cable with the 2 nuts.
31. Install the steering column assembly with the 2 bolts and 2 nuts. Tighten the nuts to 15 ft. lbs. (21 Nm).
32. Install instrument panel air bag assembly.
33. Install the A/C lines with new o-rings and coat with oil. (ND-OIL 11 or equivalent)
34. Install both heater hoses.
35. Refill the engine coolant.
36. Vacuum and recharge the A/C system.
37. Start the engine and check for coolant and refrigerant leaks.

2008 Models

See Figures 150 through 152.

1. Before servicing the vehicle, refer to the Precautions Section.

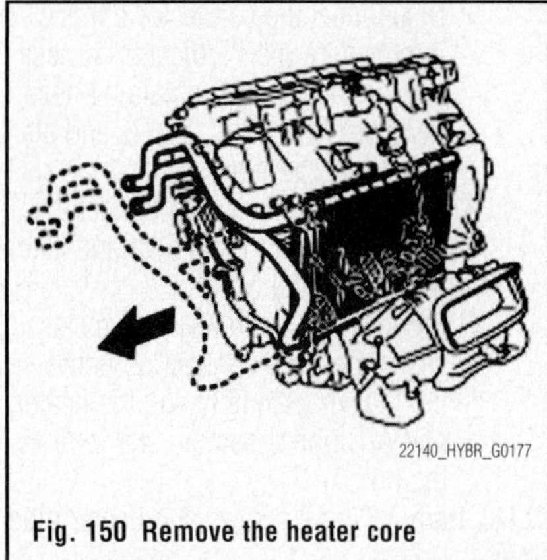

Fig. 150 Remove the heater core

2. Recover refrigerant from refrigeration system.
3. Position front wheels straight ahead.
4. Remove both front wheels.

CAUTION

Wait for 90 seconds after disconnecting the cable to prevent the airbag from deploying.

5. Remove both front wiper arm and blade assemblies.
6. Remove the cowl top ventilator louver sub-assembly.
7. Remove the windshield wiper motor and link assembly.
8. Remove the cowl top outer panel sub-assembly.
9. Disconnect the heater inlet and outlet water hoses.
10. Disconnect the cooler refrigerant liquid pipe.
11. Disconnect the No. 1 cooler refrigerant suction pipe.

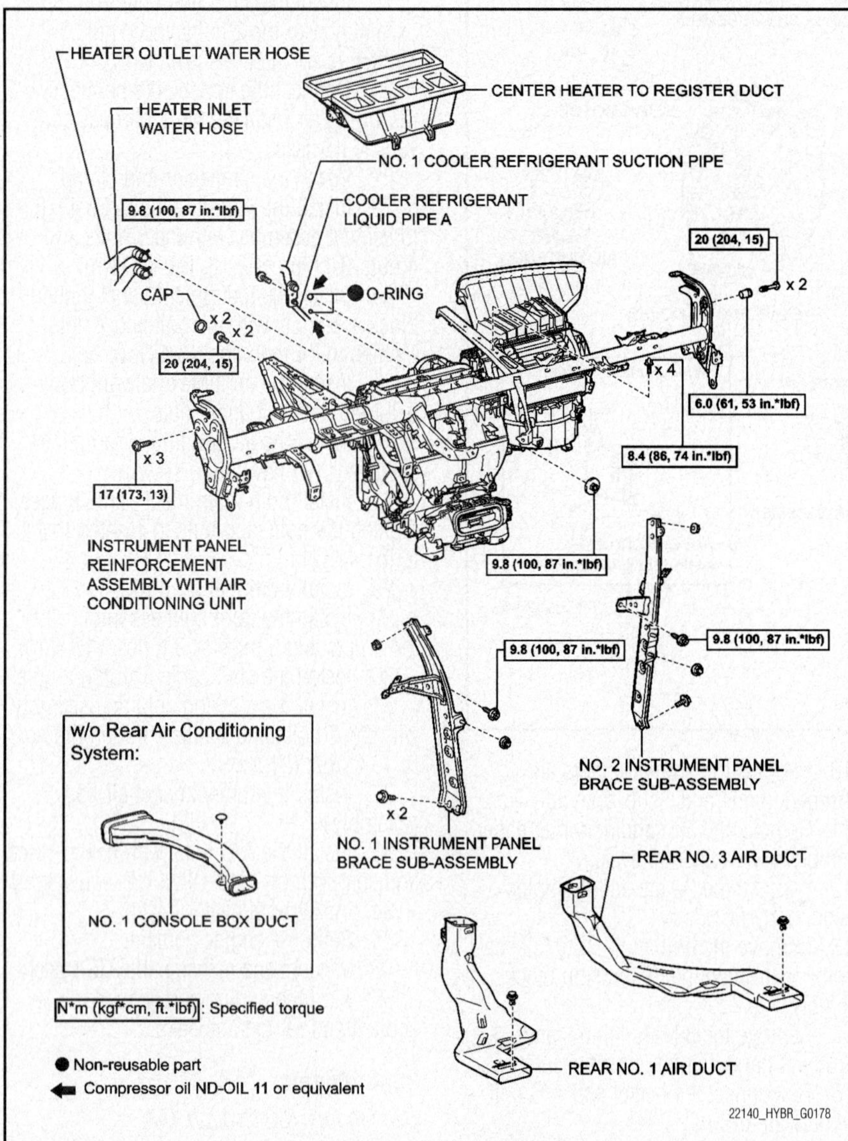

Fig. 151 Heater and A/C assembly, reinforcement assembly and related components

12. Remove the lower No. 2 and 3 steering wheel covers.
13. Remove the steering pad.
14. Remove the steering wheel assembly.
15. Remove the steering column cover.
16. Remove the turn signal switch assembly with spiral cable sub-assembly.
17. Remove the instrument cluster finish panel assembly.
18. Remove the combination meter assembly.
19. Remove the center instrument panel register assembly.
20. Remove the center instrument cluster finish panel assembly.
21. Remove the air conditioning control assembly.
22. Remove the radio receiver assembly with bracket.
23. Remove both front door scuff plates.
24. Remove both cowl side trim sub-assemblies.
25. Remove the lower instrument panel finish panel sub-assembly.
26. Remove the No. 2 instrument panel under cover sub-assembly.
27. Remove the lower instrument panel sub-assembly.
28. Remove the upper console panel sub-assembly.
29. Remove the No. 2 console box duct (w/o rear air conditioning system).
30. Remove the lower rear console box.
31. Remove the console box assembly.
32. Remove the front No. 1 No. 2 console box insert.
33. Remove the engine switch (W/ Smart Key System).
34. Remove the front pillar garnish.
35. Disconnect the front door opening trim weather strip.
36. Remove the No. 1 instrument panel speaker panel sub-assembly.
37. Remove the front No. 2 speaker assembly.
38. Disconnect the instrument panel wire assembly.
39. Remove the instrument panel safety pad assembly.
40. Remove the brake pedal return spring.
41. Remove the stop light switch assembly.
42. Separate the brake master cylinder push rod clevis.
43. Remove the brake pedal support sub-assembly.
44. Remove the driver side knee airbag assembly.
45. Remove the No. 1 air duct sub-assembly.
46. Separate the steering intermediate shaft sub-assembly.

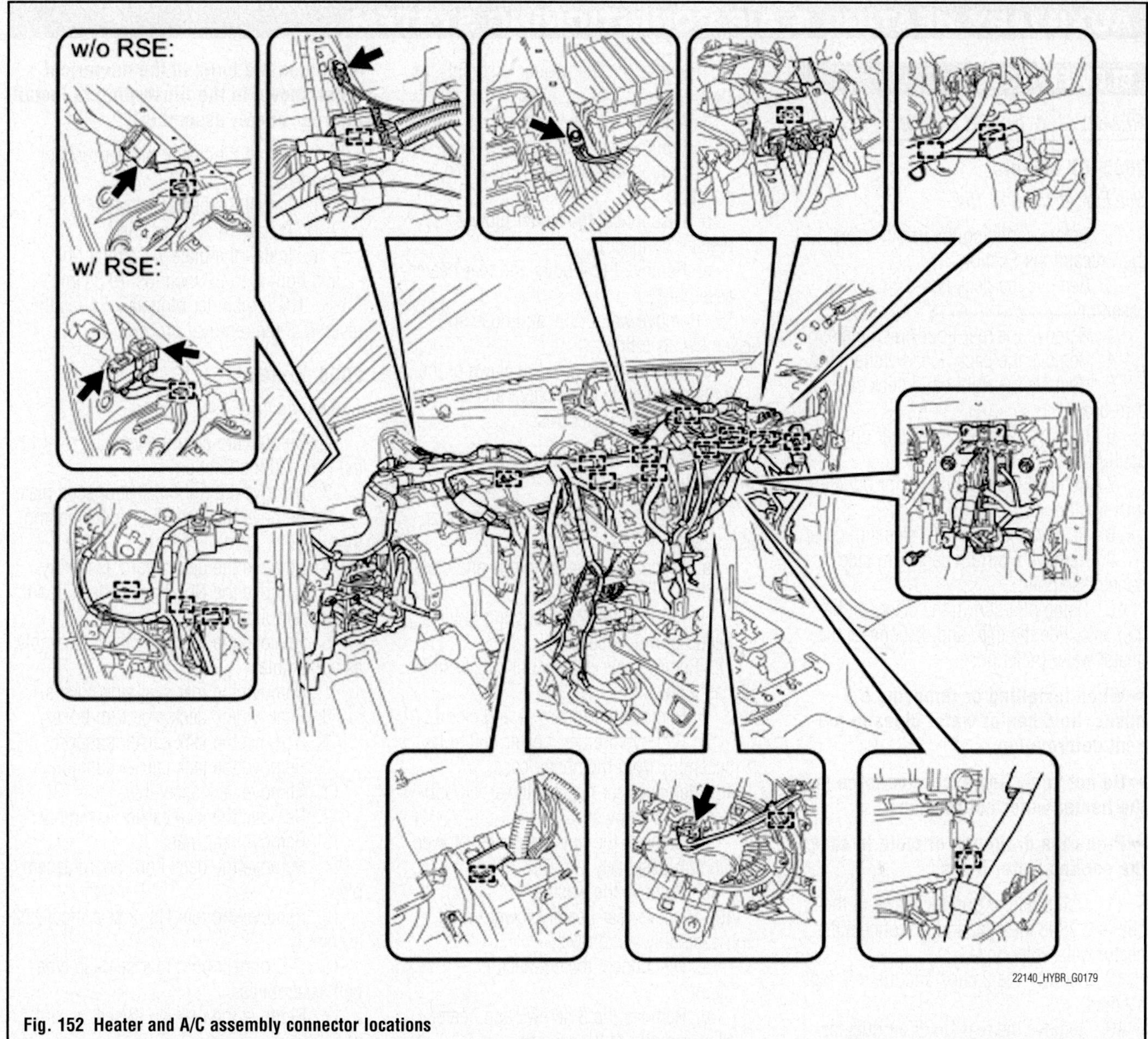

Fig. 152 Heater and A/C assembly connector locations

47. Remove the steering column assembly.

48. Remove the certification ECU (smart key ECU assembly) (W/ smart key system).

49. Remove the air conditioning amplifier assembly.

50. Remove the rear No. 1 air duct.

51. Remove the rear No. 3 air duct.

52. Remove the No. 1 console box duct (w/o rear air conditioning system).

53. Remove the center heater to register duct.

54. Remove the No. 1 and 2 instrument panel brace sub-assemblies.

55. Remove instrument panel reinforcement assembly with air conditioning unit.

56. Remove the 2 caps and the 2 bolts from the engine compartment side.

57. Remove the heater core tube clamps.

58. Remove the heater core as shown in the illustration

To install:

59. Install heater core and tighten the tube brackets.

60. The remainder of installation is reverse of the removal procedure.

61. Check and note the following:

   a. Inspect the SRS warning light.
   b. Refill the engine coolant.
   c. Vacuum and recharge A/C system.
   d. Warm up the engine and check for leaks.

AUXILIARY HEATING & AIR CONDITIONING SYSTEM

BLOWER MOTOR

REMOVAL & INSTALLATION

2006–07 Models

See Figure 153 and 154.

1. Before servicing the vehicle, refer to the Precautions Section.
2. Remove the deck board No.2 sub-assembly.
3. Remove the rear floor finish plate.
4. Remove the back door weather strip.
5. Remove the right hand deck side trim box.
6. Remove the right hand rear door scuff plate.
7. Remove the right rear door opening trim weather strip.
8. Remove the right deck side trim cover.
9. Remove the right deck trim side panel assembly.
10. Using pliers, grip the claws of the clip and slide the clip, and disconnect the heater water outlet hose.

➡When installing or removing the hoses, hold heater water pipes to prevent deformation.

➡Do not apply any excessive force to the heater water outlet hose.

➡Prepare a drain pan or cloth in case the cooling water leaks.

11. Using pliers, grip the claws of the clip and slide the clip, and disconnect the heater water inlet hose.
12. Remove the 2 clips and the rear foot air duct.
13. Remove the rear No. 5 air duct, as follows:
    a. Disconnect the wire harness clamp.
    b. Remove the 2 clips and rear No. 5 air duct.

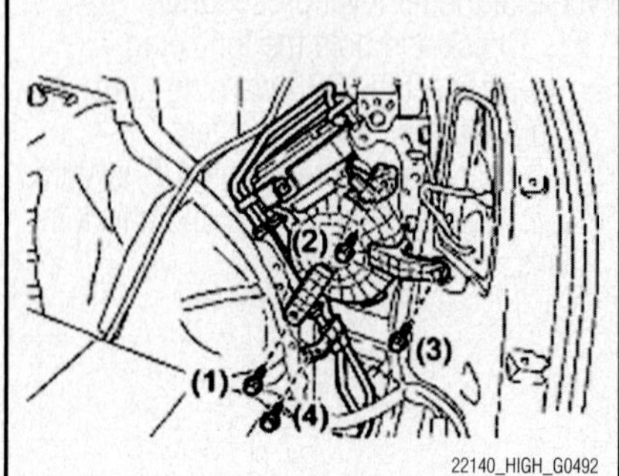

Fig. 153 Rear heater assembly bolt tightening sequence

14. Remove the rear heater assembly, as follows:
    a. Disconnect the washer hose clamp.
    b. Remove the 4 bolts and plate.
    c. Disconnect the heater relay No.1 clamp.
    d. Disconnect the heater blower resister connector.
    e. Remove the 4 bolts and rear heater assembly.
15. Remove water pipe and hose sub-assembly, as follows:
    a. Using pliers, grip the claws of the 2 clips and slide the 2 clips, and disconnect the heater water hose

➡Do not apply any excessive force to the heater water hose.

16. Remove rear heater blower resistor, as follows:
    a. Disconnect the 2 clamps and connector.
    b. Remove the 2 screws and rear heater blower resistor.
17. Remove rear heater unit sub-assembly, as follows:
    a. Remove the 9 screws and clamp.
    b. Remove the rear heater unit sub-assembly from the heater case.
18. Remove rear heater blower fan sub-assembly, as follows:
    a. Remove the nut and heater blower fan sub-assembly.
    b. Remove the washer.
19. Remove rear heater blower motor sub-assembly, as follows:
    a. Disconnect the connector clamp.
    b. Remove the 3 screws and heater blower motor sub-assembly.

To install:

20. Install the rear heater blower motor sub-assembly.
21. Install the rear heater blower fan sub-assembly, as follows:
    a. Install the washer.
    b. Install the heater blower fan sub-assembly with the nut and tighten to 36 inch lbs. (4 Nm).
22. Install the rear heater unit sub-assembly.
23. Install the rear heater blower resistor.
24. Install the water pipe and hose sub-assembly.
25. Install the rear heater assembly, as follows:
    a. Install the rear heater assembly with the 4 bolts and tighten to 87 inch lbs. (9.8 Nm).

➡Tighten the bolts in the numerical order shown in the illustration to install the rear heater assembly.

    b. Install the heater blower resistor connector.
    c. Install the heater relay No.1.
    d. Install the washer hose.
    e. Install the plate with the 4 bolts and tighten to 87 inch lbs. (9.8 Nm).
26. The remainder of installation is the reverse of the removal procedure.

2008 Models

See Figure 155.

1. Before servicing the vehicle, refer to the Precautions Section.
2. Remove the right rear door scuff plate.
3. Remove the right rear door opening trim weather strip.
4. Remove the deck board assembly.
5. Remove the No. 2 and 3 deck board sub-assemblies.
6. Remove the tonneau cover assembly, as applicable.
7. Remove the rear seat side covers.
8. Remove the deck side trim boxes.
9. Remove the jack carrier support.
10. Remove the jack carrier cushion.
11. Remove jack assembly.
12. Remove the jack carrier assembly.
13. Remove rear mat.
14. Remove the deck floor board assembly.
15. Remove the rear No. 2 seat inner belt assembly.
16. Disconnect both rear seat lap type belt assemblies.
17. Remove the rear No. 2 seat assembly.
18. Remove the rear floor finish plate.
19. Remove the rear seat side garnish cap.
20. Remove right deck side trim cover.
21. Remove right deck side trim.
22. Remove right side trim cover (for manual air conditioning system).
23. Remove rear room temperature sensor (for automatic air conditioning system).
24. Remove right rear combination light service cover.
25. Remove right hand rope hook assembly.
26. Remove the No. 1 luggage compartment trim hook
27. Remove the right front deck side trim cover.
28. Remove the right deck trim side panel assembly.

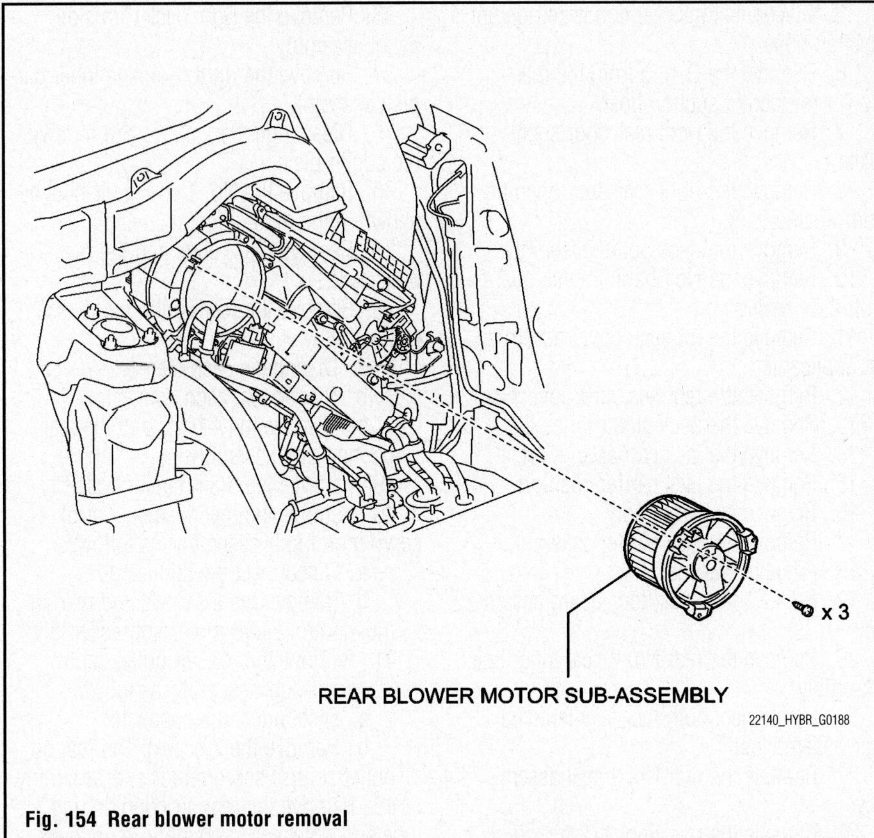

Fig. 154 Rear blower motor removal

29. Remove the right roof side inner garnish assembly.
30. Remove the rear blower motor sub-assembly, as follows:
    a. Disconnect the connector.
    b. Remove the 3 screws and the rear blower motor sub-assembly.

To install:

31. Install the 3 screws and the rear blower motor sub-assembly
32. Reconnect the connector.
33. Install the right roof side inner garnish assembly.
34. Install the right deck trim side panel assembly.
35. Install the right front deck side trim cover.
36. Install the No. 1 luggage compartment trim hook
37. Install the right hand rope hook assembly.
38. Install the right rear combination light service cover.
39. Install rear room temperature sensor (for automatic air conditioning system).
40. Install right deck side trim.
41. Install right deck side trim cover.
42. Install the rear seat side garnish cap.
43. Install the rear floor finish plate.
44. Install the rear No. 2 seat assembly.
45. Install both rear seat lap type belt assemblies.
46. Install the rear No. 2 seat inner belt assembly.
47. Install the deck floor board assembly.
48. Install the rear mat.
49. Install the jack carrier cushion.
50. Install the jack carrier support.
51. Install the deck side trim boxes.
52. Install the rear seat side covers.
53. Install the tonneau cover assembly, as applicable.
54. Install the No. 2 and 3 deck board sub-assemblies.
55. Install the deck board assembly.
56. Install the right rear door opening trim weather strip.
57. Install the right rear door scuff plate.

HEATER CORE

REMOVAL & INSTALLATION

2006–07 Models

See Figures 155 through 157.

1. Before servicing the vehicle, refer to the Precautions Section.
2. Remove the deck board No. 2 sub-assembly.
3. Remove the rear floor finish plate.
4. Remove the back door weather strip.
5. Remove the right deck side trim box.

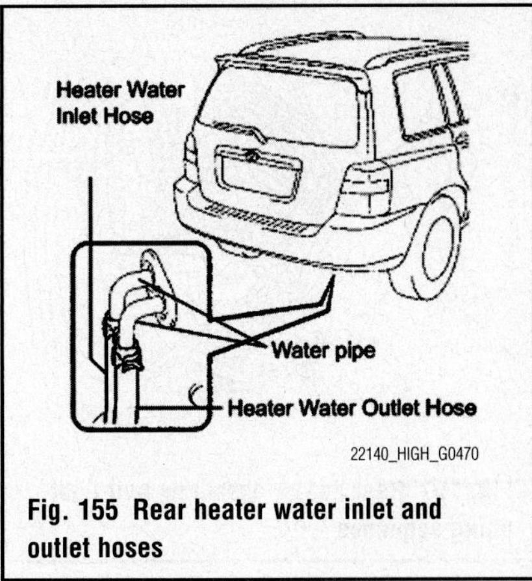

Fig. 155 Rear heater water inlet and outlet hoses

6. Remove the right rear door scuff plate.
7. Remove the right rear door opening trim weather strip.
8. Remove the right deck side trim cover.
9. Remove the right deck trim side panel assembly.
10. Using pliers, grip the claws of the clip and slide the clip, and disconnect the heater water inlet and outlet hoses.

➡When installing or removing the hoses, hold heater water pipes to prevent deformation.

➡Do not apply any excessive force to the heater water outlet hose.

➡Prepare a drain pan or cloth in case the cooling water leaks.

11. Remove the 2 clips and the rear foot air duct.
12. Remove the rear No. 5 air duct, as follows:
    a. Disconnect the wire harness clamp.
    b. Remove the 2 clips and the rear No. 5 air duct.
13. Remove the rear heater assembly, as follows:
    a. Disconnect the washer hose clamp.
    b. Remove the 4 bolts and plate.

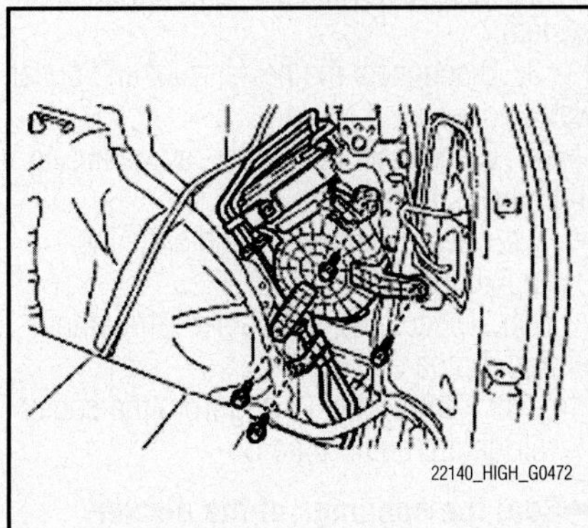

Fig. 156 Rear heater assembly and bolts

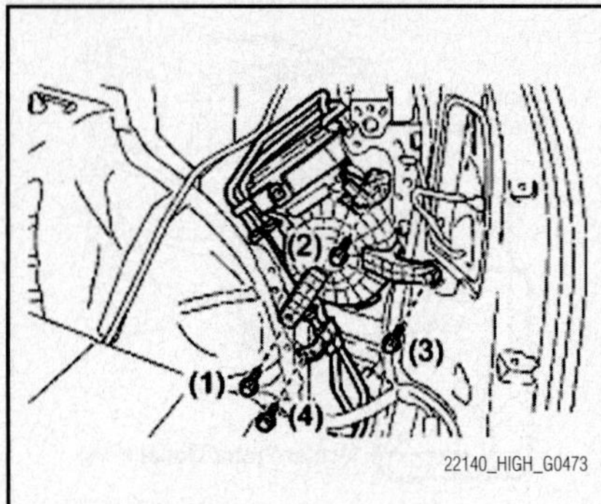

Fig. 157 Rear heater assembly bolt tightening sequence

c. Disconnect the heater relay No.1 clamp.

d. Disconnect the heater blower resister connector.

e. Remove the 4 bolts and rear heater assembly.

To install:

14. Install the rear heater assembly with the 4 bolts and tighten to 87 inch lbs. (9.8 Nm).

➡Tighten the bolts in the numerical order shown in the illustration to install the rear heater assembly.

15. Install the heater blower resistor connector.
16. Install the heater relay No.1.
17. Install the washer hose.
18. Install the plate with the 4 bolts and tighten to 87 inch lbs. (9.8 Nm).
19. The remainder of installation is the reverse of the removal procedure.

2008 Models

See Figures 158 through 167.

1. Before servicing the vehicle, refer to the Precautions Section.

➡Adjust the air outlet mode setting to FOOT.

2. Recover refrigerant from refrigeration system.
3. Disconnect the heater inlet and outlet water hoses.
4. Disconnect the cooler refrigerant liquid pipe C, as follows:
   a. Remove the bolt, and slide the hook connector.
   b. Disconnect the cooler refrigerant liquid pipe C.
   c. Remove the O-ring from the cooler refrigerant liquid pipe C.

➡Seal the openings of the disconnected parts using vinyl tape to prevent entry of moisture and foreign matter.

5. Disconnect the rear cooler refrigerant suction hose.
6. Remove the O-ring from the rear cooler refrigerant suction hose.
7. Remove the right rear door scuff plate.
8. Remove the right rear door opening trim weather strip.
9. Remove the deck board assembly.
10. Remove the No. 2 and 3 deck board sub-assemblies.
11. Remove the tonneau cover assembly, as applicable.
12. Remove the rear seat side covers.
13. Remove the deck side trim boxes.
14. Remove the jack carrier support.
15. Remove the jack carrier cushion.
16. Remove jack assembly.
17. Remove the jack carrier assembly.
18. Remove rear mat.
19. Remove the deck floor board assembly.
20. Remove the rear No. 2 seat inner belt assembly.
21. Disconnect both rear seat lap type belt assemblies.
22. Remove the rear No. 2 seat assembly.
23. Remove the rear floor finish plate.
24. Remove the rear seat side garnish cap.
25. Remove right deck side trim cover.
26. Remove right deck side trim.
27. Remove right side trim cover (for manual air conditioning system).
28. Remove rear room temperature sensor (for automatic air conditioning system).
29. Remove right rear combination light service cover.
30. Remove right hand rope hook assembly.
31. Remove the No. 1 luggage compartment trim hook
32. Remove the right front deck side trim cover.
33. Remove the right deck trim side panel assembly.
34. Remove the right roof side inner garnish assembly.
35. Disengage the 2 claws and remove the cooler plate.
36. Remove the No. 1 cooler air duct as shown in the illustration.
37. Remove the 2 clips and the rear No. 5 air duct.
38. Remove the rear cooling unit assembly, as follows:
   a. Disengage each clamp.
   b. Disconnect each connector.
   c. Remove the 4 bolts and the rear cooling unit assembly.
39. Remove the drain cooler hose.
40. Remove the rear air mix control servo motor sub-assembly, as follows:
   a. Disconnect the connector.
   b. Remove the 2 screws and rear air mix control servo motor sub-assembly.
41. Remove the rear air outlet control servo motor sub-assembly, as follows:
   a. Disconnect the connector.
   b. Remove the 2 screws and rear air outlet control servo motor sub-assembly.
42. Remove the rear air conditioning tube and accessory assembly, as follows:
   a. Remove the packing.
   b. Remove the screws (C then B).
   c. Remove the bolt (A).
   d. Remove the bolt and slide the hook connector.
   e. Remove the rear air conditioning tube and accessory assembly.
   f. Remove the 2 O-rings from the rear air conditioning tube and accessory assembly.
43. Using pliers, grip the claws of the 2 clips and slide the clips to disconnect the heater water hose.
44. Remove the 2 screws and the heater water pipe and hose sub-assembly.
45. Remove the heater radiator unit sub-assembly, as follows:

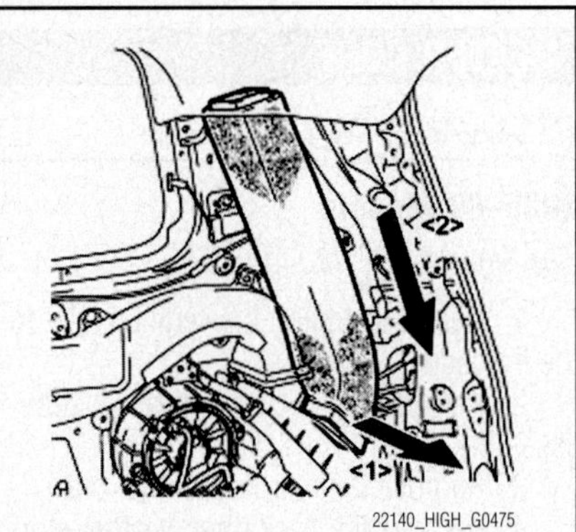

Fig. 158 No. 1 cooler air duct

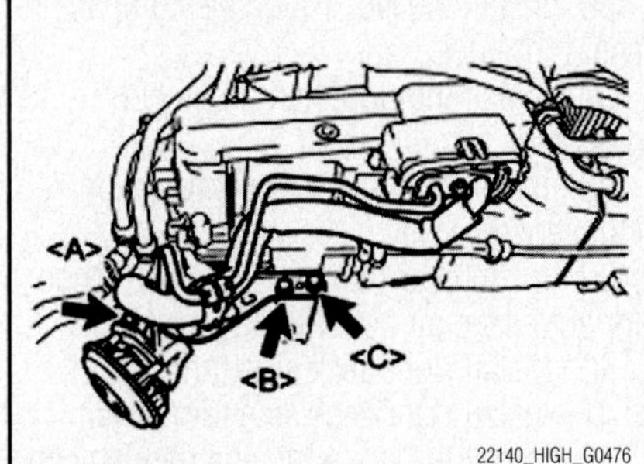

Fig. 159 Rear air conditioning tube and accessory assembly screw and bolt removal sequence

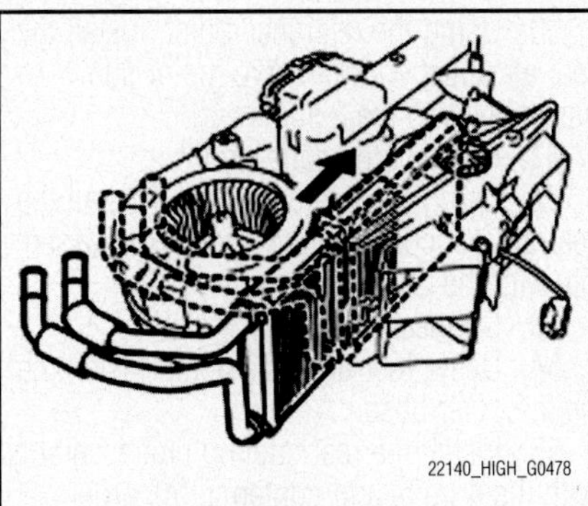

Fig. 160 Install heater radiator unit sub-assembly

a. Remove the screw and the heater clamp.

b. Remove the heater radiator unit sub-assembly from the rear cooling unit as shown in the illustration.

To install:

46. Install the heater radiator unit sub-assembly as shown in the illustration.

47. Install the clamp with the screw,

48. Install the heater water pipe and hose sub-assembly, as follows:

a. Install screw A, and tighten to 87 inch lbs. (9.8 Nm).

b. Install the heater water pipe and hose sub-assembly with the screw B and tighten to 87 inch lbs. (9.8 Nm).

c. Using pliers, grip the claws of the 2 clips and slide the clips to connect the heater water hose.

49. Install the rear air conditioning tube and accessory assembly, as follows:

a. Sufficiently apply compressor oil (ND-OIL 8 or equivalent) to 2 new O-rings and the fitting surfaces of the rear air conditioning tube and accessory assembly.

b. Install the 2 new O-rings on the rear air conditioning tube and accessory assembly.

c. Move the hook connector in the direction indicated by the arrow in the illustration.

d. Insert the pipe joint into the fitting hole securely and tighten the bolt to 87 inch lbs. (9.8 Nm).

e. Install bolt A and tighten to 87 inch lbs. (9.8 Nm).

f. Install screw B and tighten to 87 inch lbs. (9.8 Nm).

g. Install the rear air conditioning tube and accessory assembly with screw C and tighten to 87 inch lbs. (9.8 Nm).

h. Install the packing.

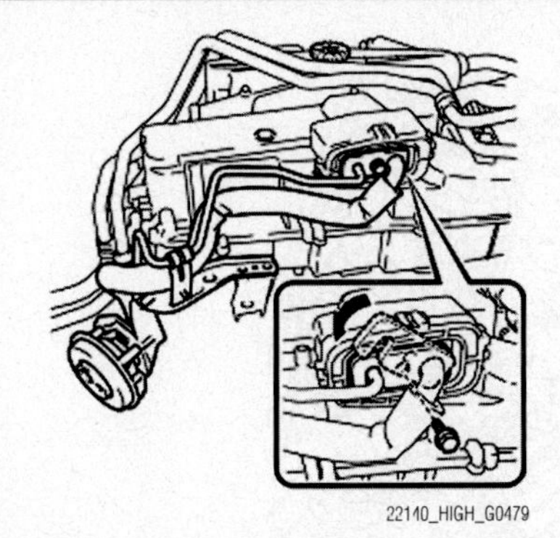

Fig. 162 Install hook connector

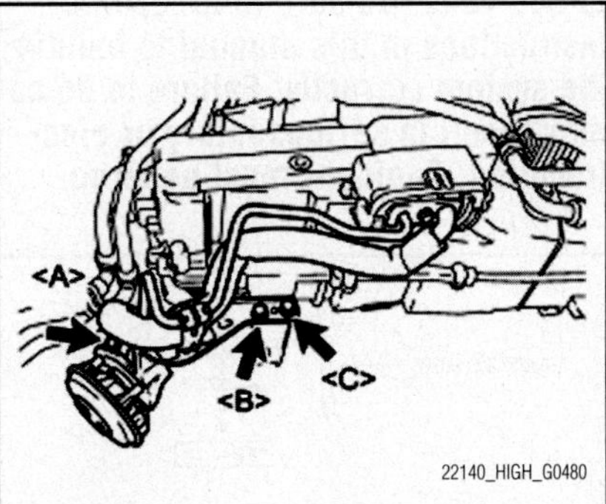

Fig. 163 Install rear air conditioning tube and accessory assembly

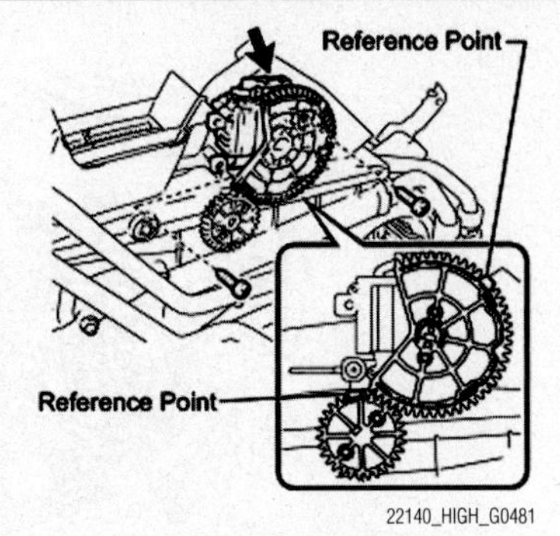

Fig. 164 Rear air outlet control servo motor sub-assembly reference point

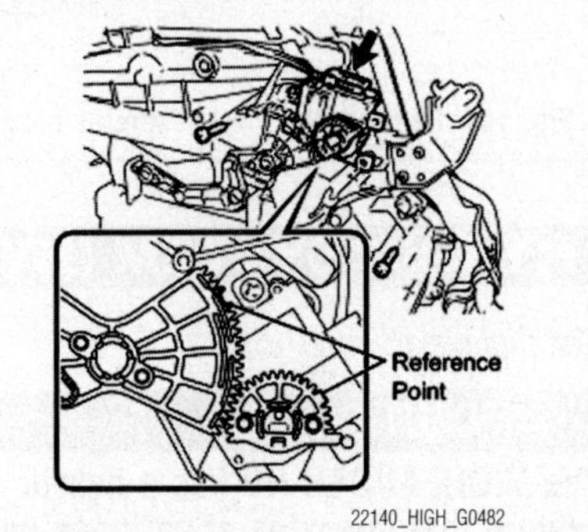

Fig. 165 Rear air mix control servo motor sub-assembly reference point

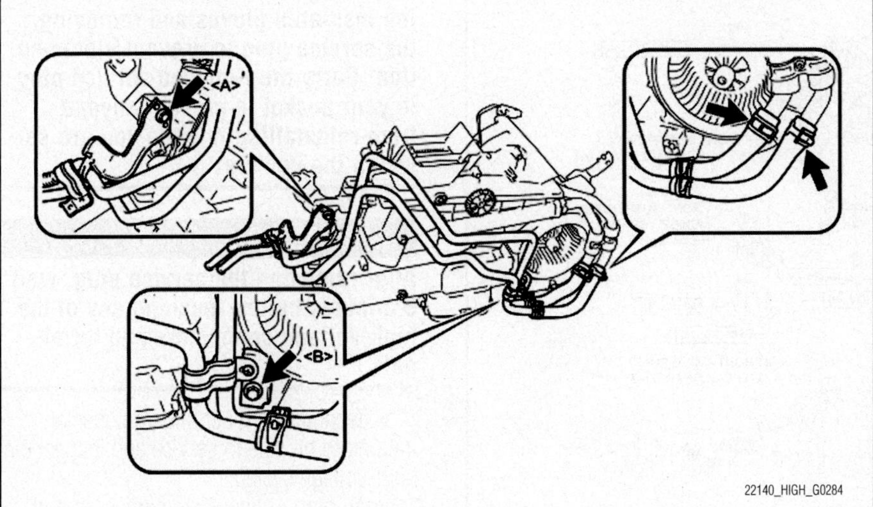

Fig. 161 Install heater radiator unit sub-assembly

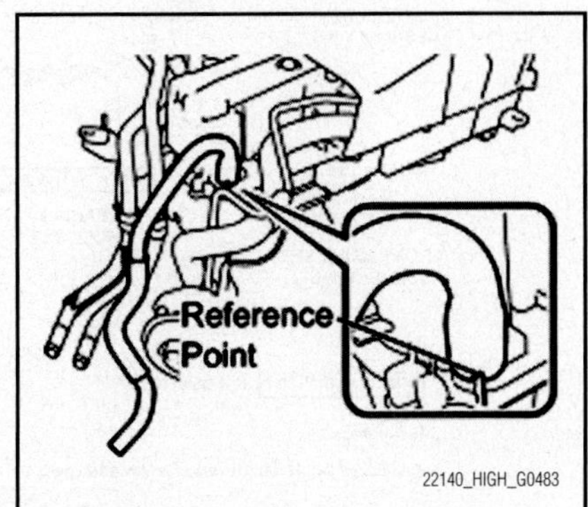

Fig. 166 Drain cooler hose reference point

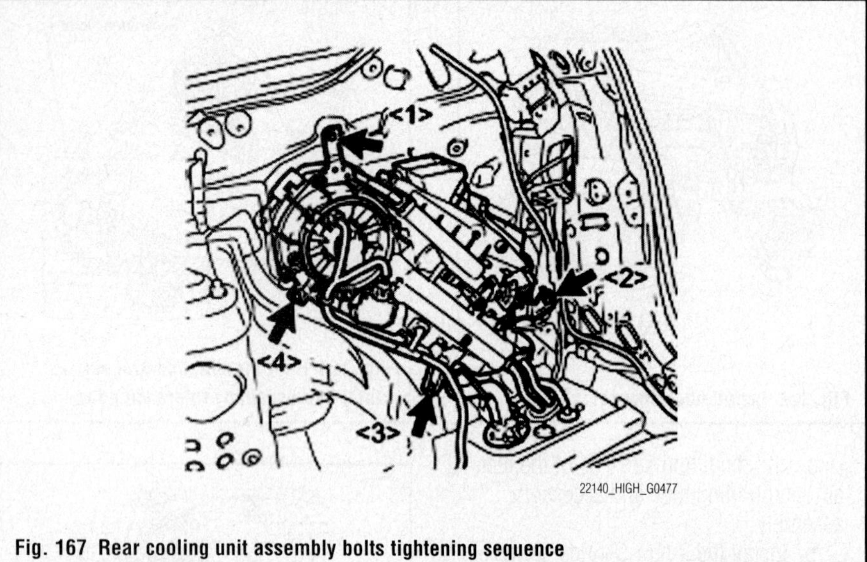

Fig. 167 Rear cooling unit assembly bolts tightening sequence

50. Using the reference point, install the rear air outlet control servo motor sub-assembly with the 2 screws.
51. Connect the connector.
52. Using the reference point, install the rear air mix control servo motor sub-assembly with the 2 screws.
53. Connect the connector.
54. Using the reference point, install the drain cooler hose.
55. Install the rear cooling unit assembly with the 4 bolts and tighten in the order shown, to 87 inch lbs. (9.8 Nm).
56. Connect each connector.
57. Engage each clamp.
58. The remainder of installation is the reverse of the removal procedure.
59. Install engine coolant.
60. Vacuum and recharge A/C system.
61. Start the engine and check for leaks.

HYBRID SYSTEM (HV)

PRECAUTIONS

See Figures 168 and 169.

CAUTION

The HIGHLANDER HV has a hybrid system that operates at voltages up to 650 volts. Be sure to follow the instructions in this manual to handle the system correctly. Failure to do so may result in serious injury or electrocution. Engineer must undergo special training to be able to perform high-voltage system inspection and servicing.

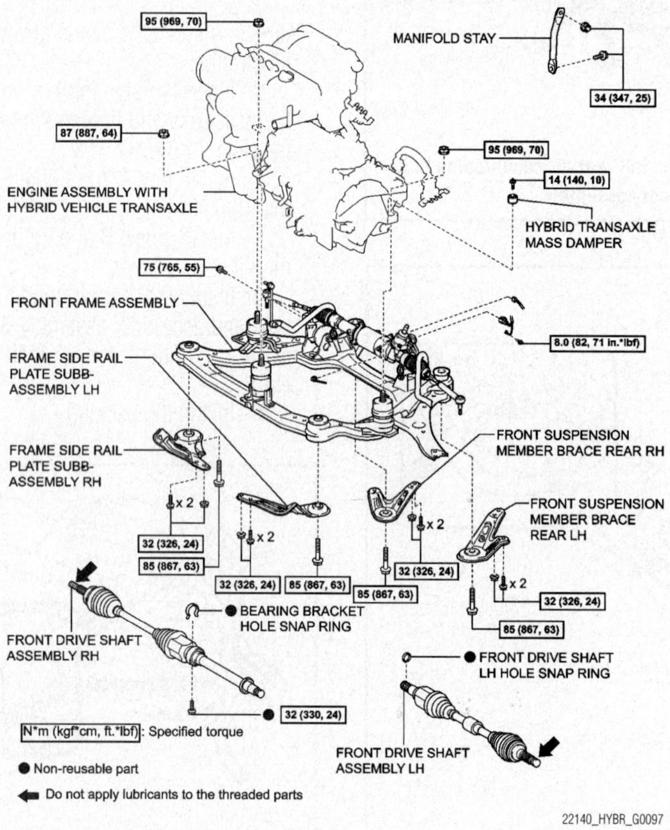

Fig. 168 Hybrid transaxle assembly and related parts

CAUTION

All high-voltage wire harness connectors are colored orange. The HV battery and other high-voltage components have "High Voltage" caution labels. Do not carelessly touch these wires and components.

CAUTION

Before inspecting or servicing the high-voltage system, be sure to follow safety measures, such as wearing insulated gloves and removing the service plug to prevent electrocution. Carry the removed service plug in your pocket to prevent anyone from reinstalling it while you are servicing the vehicle.

CAUTION

After removing the service plug, wait 5 minutes before touching any of the high-voltage connectors and terminals.

- Technicians must undergo special training to be able to service and inspect the high-voltage system.
- All high-voltage wire harnesses and connectors are colored orange. The HV battery and other high-voltage components have "High Voltage" caution labels. Do not

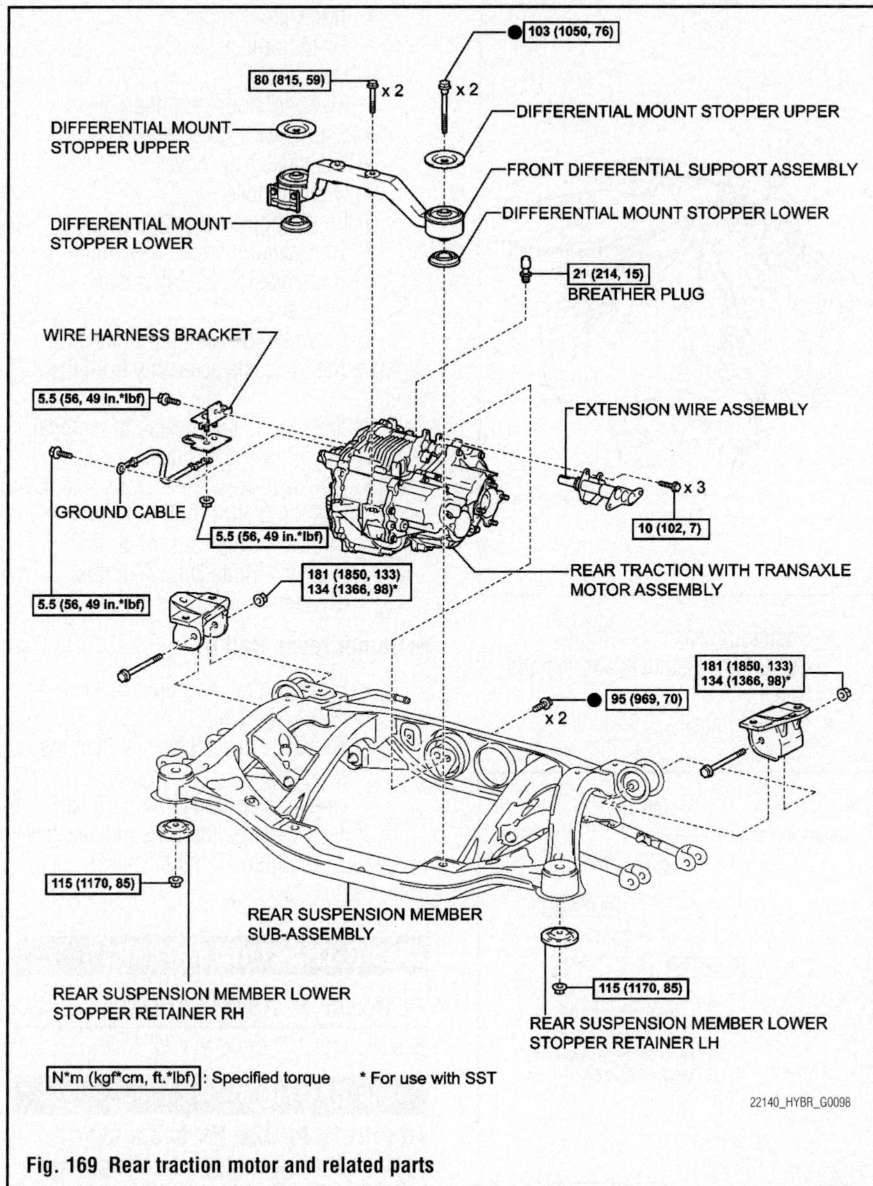

Fig. 169 Rear traction motor and related parts

carelessly touch these wires or components.

• When there is a problem with the wire harness or connector of a high-voltage circuit, repairs to the harness or connector should not be attempted. Replace damaged or malfunctioning high voltage cables or connectors.

• Before inspecting or servicing the high-voltage system, be sure to follow all safety measures, such as wearing insulated gloves and removing the service plug to prevent electrocution. Carry the removed service plug in your pocket to prevent other technicians from accidentally reconnecting it while you are servicing the vehicle.

• After removing the service plug, wait 5 minutes before touching any of the high-voltage connectors and terminals.

• Before wearing insulated gloves, make sure that they are not cracked, ruptured, torn, or damaged in any way. Do not wear wet insulated gloves

• When servicing the vehicle, do not carry metal objects like mechanical pencils or scales that can be dropped accidentally and cause a short circuit.

• Before touching a bare high-voltage terminal, wear insulated gloves and use a tester to make sure that the terminal voltage is 0 V.

• After disconnecting or exposing a high-voltage connector or terminal, insulate it immediately using insulating tape.

• Bolts and nuts high-voltage terminals should be tightened firmly to the specified torque. Both insufficient and excessive torque can cause failure

• After servicing the high-voltage system and before reinstalling the service plug, check again that you have not left a part or tool inside, that the high-voltage terminals are firmly tightened, and that the connectors are correctly connected.

• When installing hybrid system components such as the HV battery, make sure that the polarity of all connections is correct.

HYBRID TRANSAXLE ASSEMBLY

REMOVAL & INSTALLATION

See Figures 170 through 172.

✲✲ CAUTION

The HIGHLANDER HV has a hybrid system that operates at voltages up to 650 volts. Be sure to follow the instructions in this manual to handle the system correctly. Failure to do so may result in serious injury or electrocution. Engineer must undergo special training to be able to perform high-voltage system inspection and servicing.

✲✲ CAUTION

All high-voltage wire harness connectors are colored orange. The HV battery and other high-voltage components have "High Voltage" caution labels. Do not carelessly touch these wires and components.

1. Before inspecting or servicing the high-voltage system, be sure to follow safety measures, such as wearing insulated gloves and removing the service plug to prevent electrocution. Carry the removed service plug in your pocket to prevent anyone from reinstalling it while you are servicing the vehicle.
2. After removing the service plug, wait 5 minutes before touching any of the high-voltage connectors and terminals.

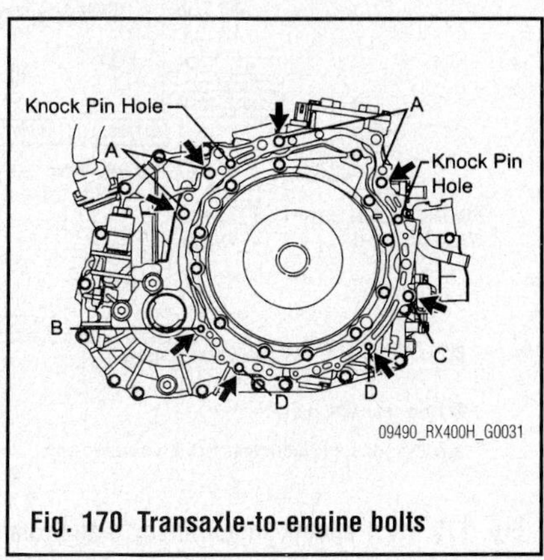

Fig. 170 Transaxle-to-engine bolts

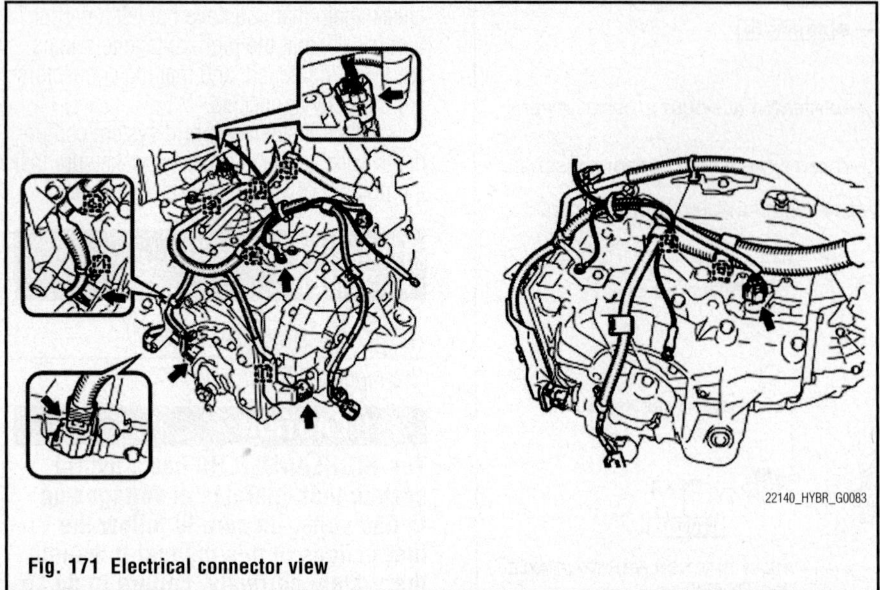

Fig. 171 Electrical connector view

3. Before servicing the vehicle, refer to the Precautions Section.

4. Remove or disconnect the following:

- Engine/transaxle assembly. See Engine Removal and Installation.
- Manifold stay
- Transaxle damper
- Front frame assembly
- Halfshafts
- Flywheel housing undercover
- Engine wiring harnesses
- Transaxle case cover
- Coolant hose
- Front engine mounting bracket
- Transaxle oil cooler assembly
- Transmission control cable bracket

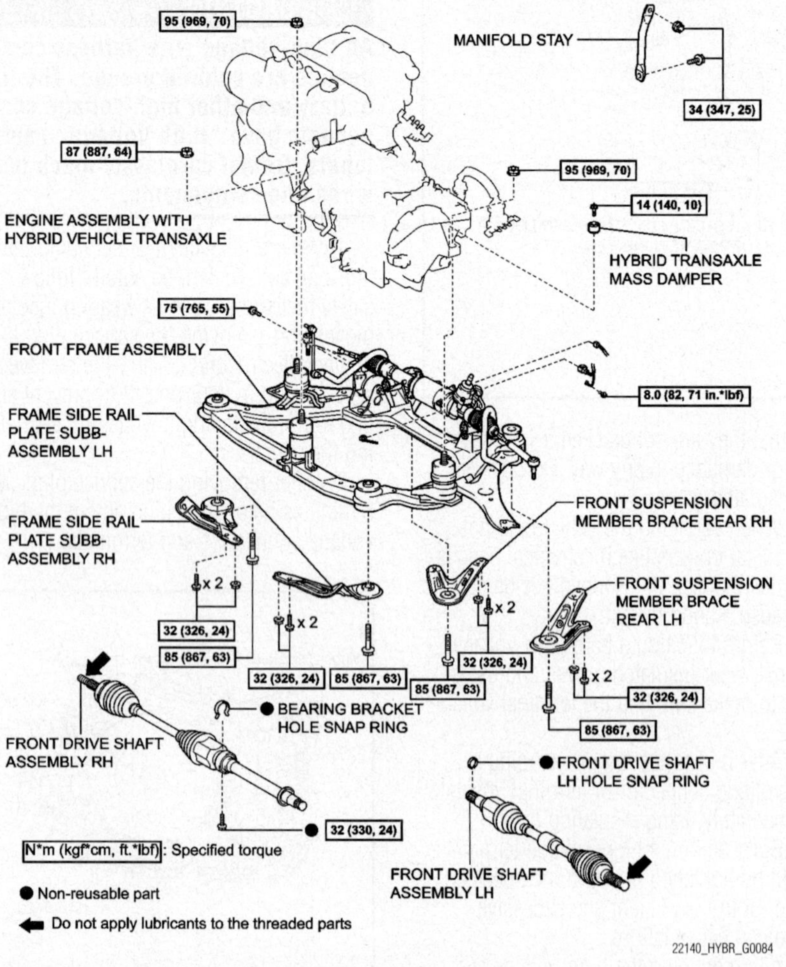

5. Remove the 8 mounting bolts and separate the transaxle assembly from the vehicle.

6. Installation is the reverse of removal. Observe the following torques:

- Transaxle-to-engine: Bolts A to 47 ft. lbs (64 Nm); Bolt B to 34 ft. lbs. (46 Nm); Bolts C to 47 ft. lbs. (64 Nm); Bolts D to 27 ft. lbs. (37 Nm)

➡**Do not reuse Bolt B.**

- Front engine mounting bracket: 47 ft. lbs. (64 Nm)
- Transaxle case cover: 74 inch lbs. (8.4 Nm)
- Undercover: 69 inch lbs. (8 Nm)

7. If necessary, perform the initialization procedure, as outlined in the Chassis Electrical System.

95 (969, 70)
MANIFOLD STAY
34 (347, 25)
87 (887, 64)
95 (969, 70)
14 (140, 10)
ENGINE ASSEMBLY WITH HYBRID VEHICLE TRANSAXLE
HYBRID TRANSAXLE MASS DAMPER
75 (765, 55)
FRONT FRAME ASSEMBLY
8.0 (82, 71 in.*lbf)
FRAME SIDE RAIL PLATE SUBB-ASSEMBLY LH
FRONT SUSPENSION MEMBER BRACE REAR RH
FRAME SIDE RAIL PLATE SUBB-ASSEMBLY RH
FRONT SUSPENSION MEMBER BRACE REAR LH
x 2
32 (326, 24)
85 (867, 63)
x 2
32 (326, 24)
85 (867, 63)
32 (326, 24)
85 (867, 63)
x 2
x 2
32 (326, 24)
85 (867, 63)
● BEARING BRACKET HOLE SNAP RING
FRONT DRIVE SHAFT ASSEMBLY RH
● FRONT DRIVE SHAFT LH HOLE SNAP RING
32 (330, 24)
N*m (kgf*cm, ft.*lbf): Specified torque
● Non-reusable part
Do not apply lubricants to the threaded parts
FRONT DRIVE SHAFT ASSEMBLY LH
22140_HYBR_G0084

Fig. 172 HV Transmission/Transaxle and related parts

REAR TRACTION MOTOR

REMOVAL & INSTALLATION

See Figures 173 through 176.

⁂ CAUTION

The HIGHLANDER HV has a hybrid system that operates at voltages up to 650 volts. Be sure to follow the instructions in this manual to handle the system correctly. Failure to do so may result in serious injury or electrocution. Engineer must undergo

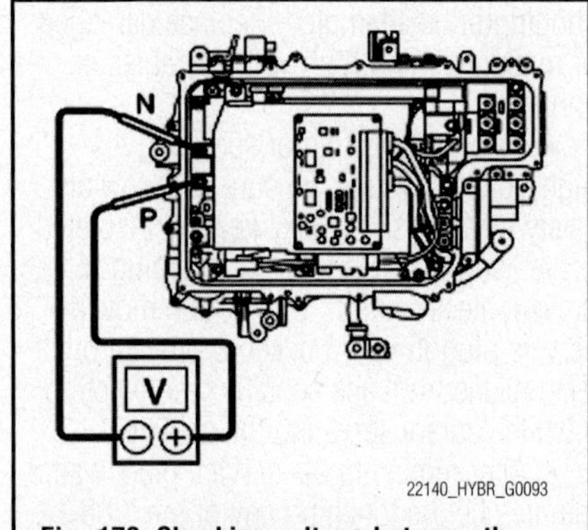

Fig. 173 Checking voltage between the terminals of the 2 phase connectors

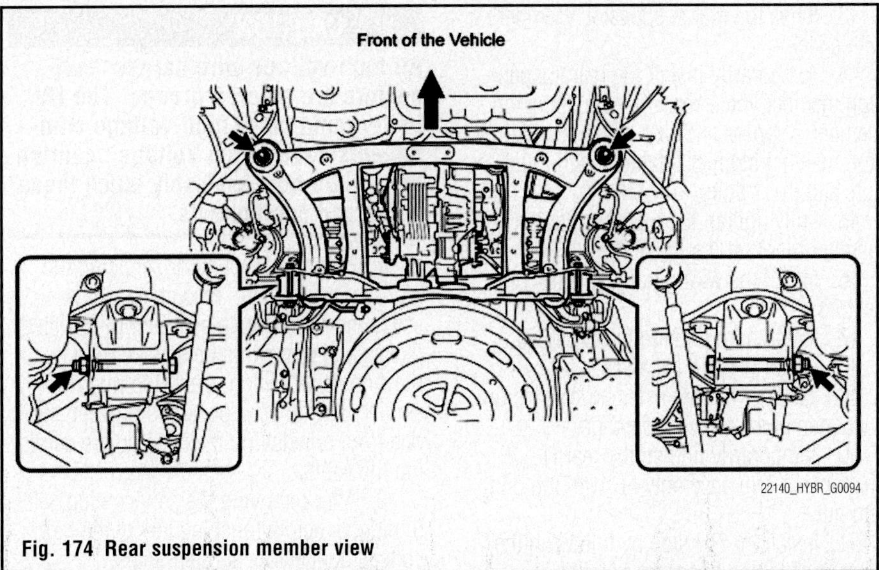

Fig. 174 Rear suspension member view

special training to be able to perform high-voltage system inspection and servicing.

⁂ CAUTION

All high-voltage wire harness connectors are colored orange. The HV battery and other high-voltage components have "High Voltage" caution labels. Do not carelessly touch these wires and components.

1. Before inspecting or servicing the high-voltage system, be sure to follow safety measures, such as wearing insulated gloves and removing the service plug to prevent electrocution. Carry the removed service plug in your pocket to prevent anyone from reinstalling it while you are servicing the vehicle.
2. After removing the service plug, wait 5 minutes before touching any of the high-voltage connectors and terminals.
3. Before servicing the vehicle, refer to the Precautions Section.
4. Disconnect the negative battery cable.
5. When disconnecting the cable, some systems need to be initialized after the cable is reconnected.
6. Remove the service plug grip, found underneath the Battery Service cover on the rear seat. Wait 5 minutes to discharge the high voltage capacitor.
7. Remove the LH and RH wiper arm assembly.
8. Remove the cowl top ventilator louver sub-assembly.
9. Remove the wiper motor and link assembly.
10. Remove the outer cowl top panel.
11. Remove the LH room side cover.
12. Remove the cool air intake duct seal.
13. Remove the air cleaner cap sub assembly.
14. Remove the air cleaner case sub-assembly.
15. Remove the inverter reserve tank sub assembly.
16. Remove the inverter cover.
17. Using the voltmeter, measure the voltage between the terminals of the 2 phase connectors (N-P). Standard voltage: 0 volts.
18. Install the inverter cover and tighten bolts to 7 ft. lbs. (10 Nm).
19. Install the inverter reserve tank sub-assembly with the 2 bolts. Tighten the bolts to 7 ft. lbs. (10 Nm).
20. Install the air cleaner case and cap assembly.
21. Install outer panel top cowl assembly.

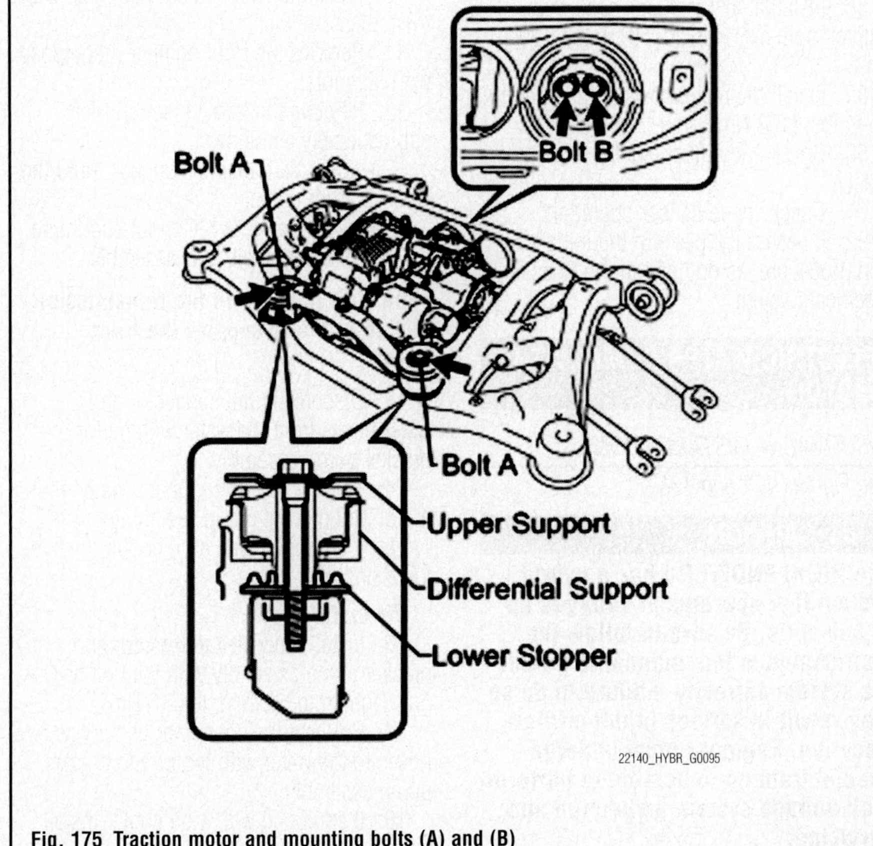

Fig. 175 Traction motor and mounting bolts (A) and (B)

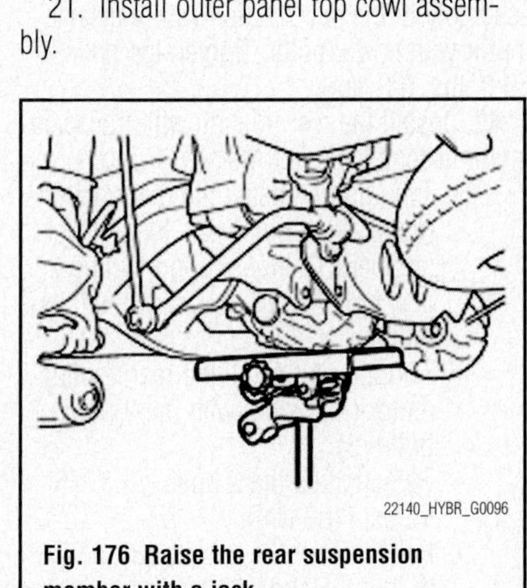

Fig. 176 Raise the rear suspension member with a jack

22. Install wiper motor and link assembly.
23. Install the cowl top ventilator assembly.
24. Install the LH and RH wiper arm assembly.
25. Install the cool air duct seal.
26. Install the engine room side cover.
27. Drain the rear traction motor fluid.
28. Remove the rear wheels.
29. Remove exhaust pipe assembly.
30. Remove the RH and LH axle shaft nuts.
31. Remove the LH and RH strut rod assembly.
32. Remove the LH and RH suspension arm assembly. No.1 and No. 2.
33. Remove axle assembly for RH and LH side.
34. Remove the nuts, and separate the both parking brake cables.
35. Remove the nut, and separate the ground cable. And all wiring harness clamps.
36. Wear insulated gloves. Remove the 2 nuts, and separate the No. 3 frame wire from the rear traction motor.
37. Support the rear suspension member with a jack.
38. Remove the rear suspension member as follows:
   - Remove the 4 nuts, 2 bolts and 2 rear lower suspension member stopper retainers.
   - Lower the rear suspension member.
   - Remove the 2 rear upper suspension member stoppers.
39. Remove the 4 bolts and rear traction with transaxle motor assembly.
40. Remove the front differential support assembly.

To install:

41. Install the front differential support assembly to the rear traction with transaxle motor with new 2 bolts. Tighten the bolts to 59 ft. lbs. (80 Nm).
42. Install the rear traction with transaxle motor assembly as follows :
   - Temporarily install the rear traction motor (front side) with the 2 lower stoppers, 2 upper supports, and 2 new bolts (A) as shown in the illustration.
   - Temporarily install the rear traction motor (rear side) with the 2 new bolts (B).
   - Fully tighten the 2 bolts (A) to 76 ft. lbs. (103 Nm).
   - Fully tighten the 2 bolts (B) to 77 ft. lbs. (95 Nm).
43. Raise the rear suspension member with a jack.
44. Temporarily install the rear suspension member, the 2 rear upper suspension member stoppers and rear lower suspension member stopper retainers with the 4 nuts and the 2 bolts.
45. Fully tighten the rear suspension member to 133 ft lbs. (181 Nm).
46. Install the wiring harness and connectors.
47. Install both parking brake cables.
48. Install RH and LH axles.
49. Jack up the rear axle carrier, placing a wooden block to avoid damage.
50. Temporarily tighten the rear No. 2 suspension arm assembly LH with the bolt and nut.
51. Install the RH side by following the same procedures as for the LH side
52. Temporarily tighten the rear No. 1 suspension arm assembly LH with the bolt and nut.
53. Install the RH side by following the same procedures as for the LH side
54. Temporarily tighten the rear strut rod assembly with the bolt and nut.
55. Stabilize the suspension.
56. Fully tighten all suspension arm bolts to 82 ft. lbs. (112 Nm).
57. Fully tighten both strut rods to 59 ft. lbs. (80 Nm).
58. Install a new rear axle shaft nuts. Tighten both axle nuts to 216 ft. lbs. (294 Nm).
59. Install the rear wheels and tighten to 76 ft. lbs. (103 Nm).
60. Check and adjust rear wheel alignment.
61. Check speed sensor operation.
62. If necessary, perform the initialization procedure, as outlined in the Chassis Electrical System.

ELECTRIC WATER PUMP WITH MOTOR

REMOVAL & INSTALLATION

See Figures 177 and 178.

CAUTION

The HIGHLANDER HV has a hybrid system that operates at voltages up to 650 volts. Be sure to follow the instructions in this manual to handle the system correctly. Failure to do so may result in serious injury or electrocution. Engineer must undergo special training to be able to perform high-voltage system inspection and servicing.

CAUTION

All high-voltage wire harness connectors are colored orange. The HV battery and other high-voltage components have "High Voltage" caution labels. Do not carelessly touch these wires and components.

1. Before inspecting or servicing the high-voltage system, be sure to follow safety measures, such as wearing insulated gloves and removing the service plug to prevent electrocution. Carry the removed service plug in your pocket to prevent anyone from reinstalling it while you are servicing the vehicle.
2. After removing the service plug, wait 5 minutes before touching any of the high-voltage connectors and terminals.
3. Before servicing the vehicle, refer to the Precautions Section.
4. Disconnect the negative battery cable.
5. Remove the left engine room side cover.
6. Remove the transaxle side reserve tank.
7. Loosen the bleeder plug and drain the coolant from inverter cooler.
8. Loosen the bleeder plug and drain the coolant from inverter.
9. Remove the engine under cover.
10. Drain transaxle fluid if equipped with oil cooler.
11. Remove the front bumper if equipped with oil cooler.
12. Remove the frame side rail Plate sub-assembly as follows:
   - Using a transmission jack, hold the front frame.
   - Remove the 3 bolts, nut and frame side rail plate sub-assembly.

➡Be sure to position the transmission jack to properly support the front frame.

13. Disconnect the connector and 2 water hoses from the water with motor and bracket pump assembly.
14. Remove the bolt, nut and water with motor and bracket pump assembly.
15. If equipped with a oil cooler remove the 2 hoses.

To install:

16. Install the water with motor and bracket pump assembly with the bolt and nut. Tighten to 53 inch. lbs. (6 Nm).
17. Connect the connector and 2 water hoses to the water with motor and bracket pump assembly.
18. If equipped with a oil cooler install the 2 hoses.

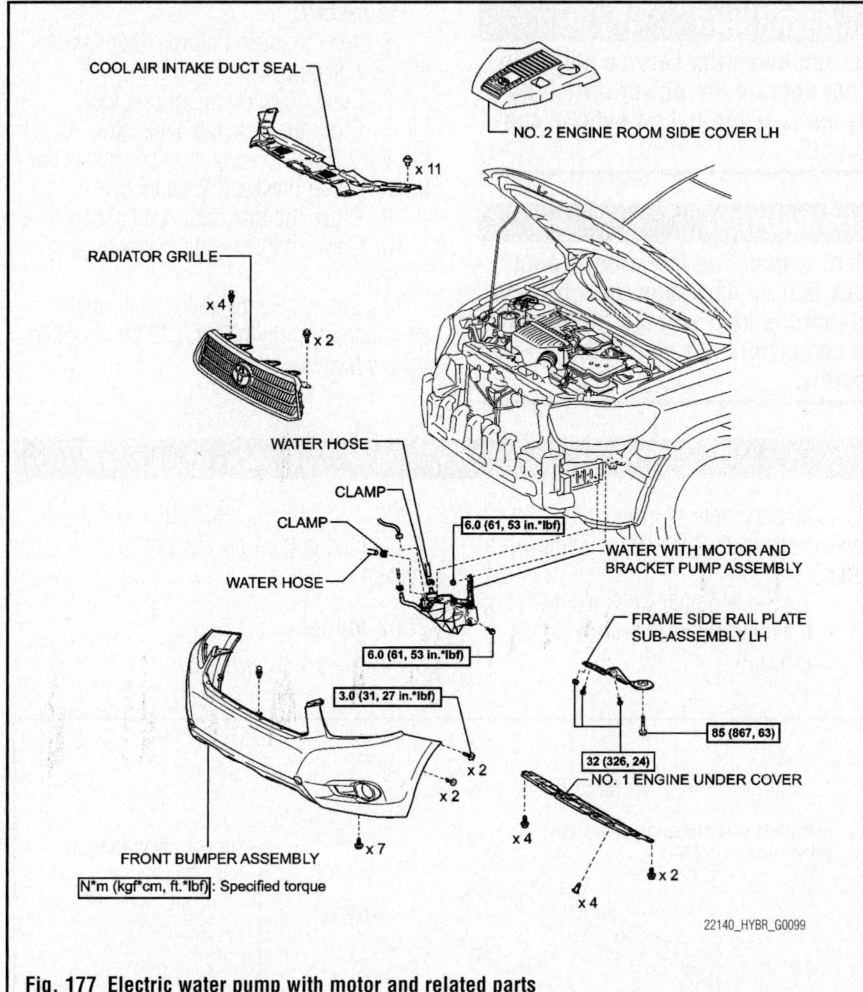

Fig. 177 Electric water pump with motor and related parts

19. Install the frame side rail plate sub-assembly with the 3 bolts and nut.
20. Tighten the bolt (A) to 63 ft. lbs. (85 Nm).
21. Tighten the bolts (B) and nut to 24 ft. lbs. (32 Nm).
22. Install bumper if removed for oil cooler. Inspect fluid level for hybrid transaxle.
23. Add engine coolant to inverter.

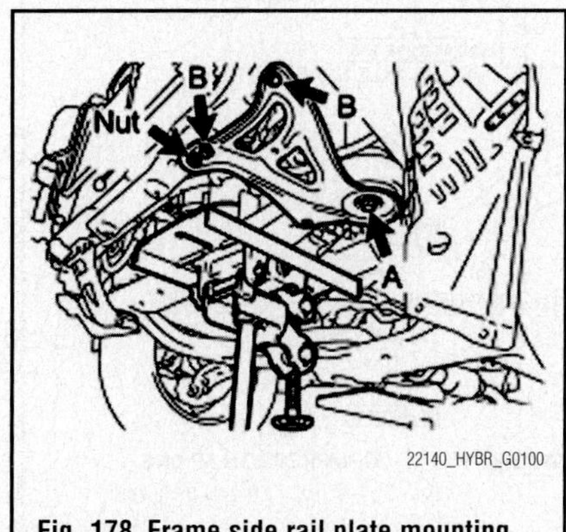

Fig. 178 Frame side rail plate mounting bolts (A) and nut (B)

24. Connect the negative battery cable.
25. Check for coolant leaks.
26. Check oil cooler lines if removed.
27. Install the engine undercover.
28. Install the engine room left side cover.

SERVICE PLUG GRIP

REMOVAL & INSTALLATION

See Figure 179.

CAUTION

The hybrid system contains a 288V high-voltage system with a strong alkali solution of potassium hydroxide. Be sure to follow the instructions in this manual to handle the system correctly. Failure to do so may result in serious injury or electrocution. Engineer must undergo special training to be able to perform high-voltage system inspection and servicing.

CAUTION

All high-voltage wire harness connectors are colored orange. The HV battery and other high-voltage components have "High Voltage" caution labels. Do not carelessly touch these wires and components.

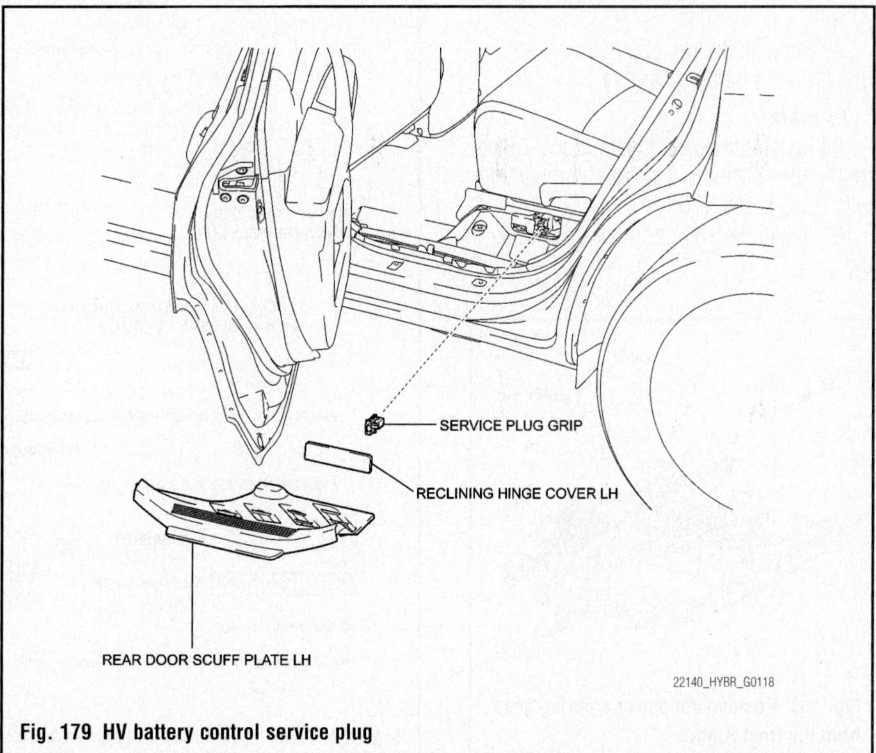

Fig. 179 HV battery control service plug

1. Check for Daces. Check for Daces and confirm that P0AA6 (High voltage insulation is unusual) is not output before doing removal or installation inside the battery. If the DTC is output, perform troubleshooting first.
2. Disconnect the negative battery cable.
3. Remove the 5 clips and door scuff plate LH.
4. Remove the 2 clips and reclining hinge cover.
5. Wear insulated glove and remove the service plug grip, after sliding up the lever of the service plug grip.

WARNING

After removing the service plug grip, do not operate the power switch as it may damage the hybrid vehicle control ECU.

CAUTION

Before connecting the service plug, check that no parts and tools remain and that the high voltage terminals and connectors are connected securely.

To install:

6. Wear insulated gloves, then insert the service plug.
7. Push down on the grip to lock.
8. Close the reclining hinge cover. Check that the 2 clips are securely connected to the battery carrier bracket (click sound).
9. Close the rear door scuff plate LH.
10. Connect the negative battery cable.
11. Some systems need initialization after reconnecting the cable to the negative battery terminal

STEERING

POWER RACK & PINION STEERING GEAR

REMOVAL & INSTALLATION

2006–07 Models

See Figure 180.

1. Before servicing the vehicle, refer to the Precautions Section.
2. Center the steering wheel to place the front wheels facing straight ahead.
3. Disconnect the negative battery cable.
4. Matchmark and disconnect the intermediate shaft.
5. Remove the wheels.
6. Separate the tie rods.
7. Remove the engine/transaxle assembly.
8. Remove the power steering gear from the front frame assembly.

To install:

9. Install the power steering gear to the front frame assembly. Tighten the mounting bolts to 52 ft. lbs. (70 Nm).
10. Install the engine/transaxle assembly.

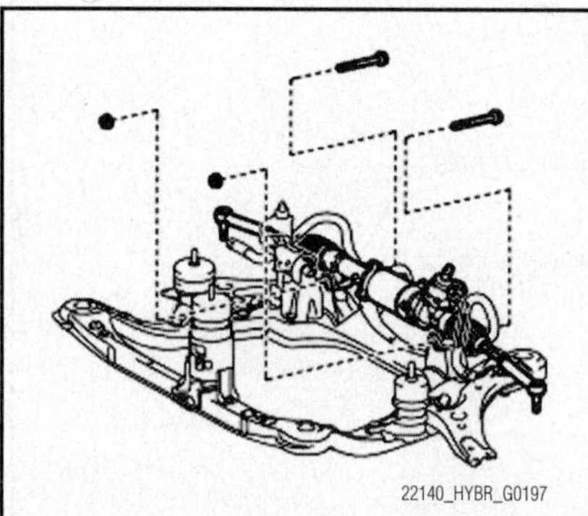

Fig. 180 Remove the power steering gear from the front frame

11. The remainder of installation is the reverse of removal. Observe the following torques:
    - Tie rod end nuts: 36 ft. lbs. (49 Nm)
    - Intermediate shaft bolt: 26 ft. lbs. (35 Nm)

12. Perform the initialization procedure, as outlined in the Chassis Electrical System.

2008 Models

See Figures 181 and 182.

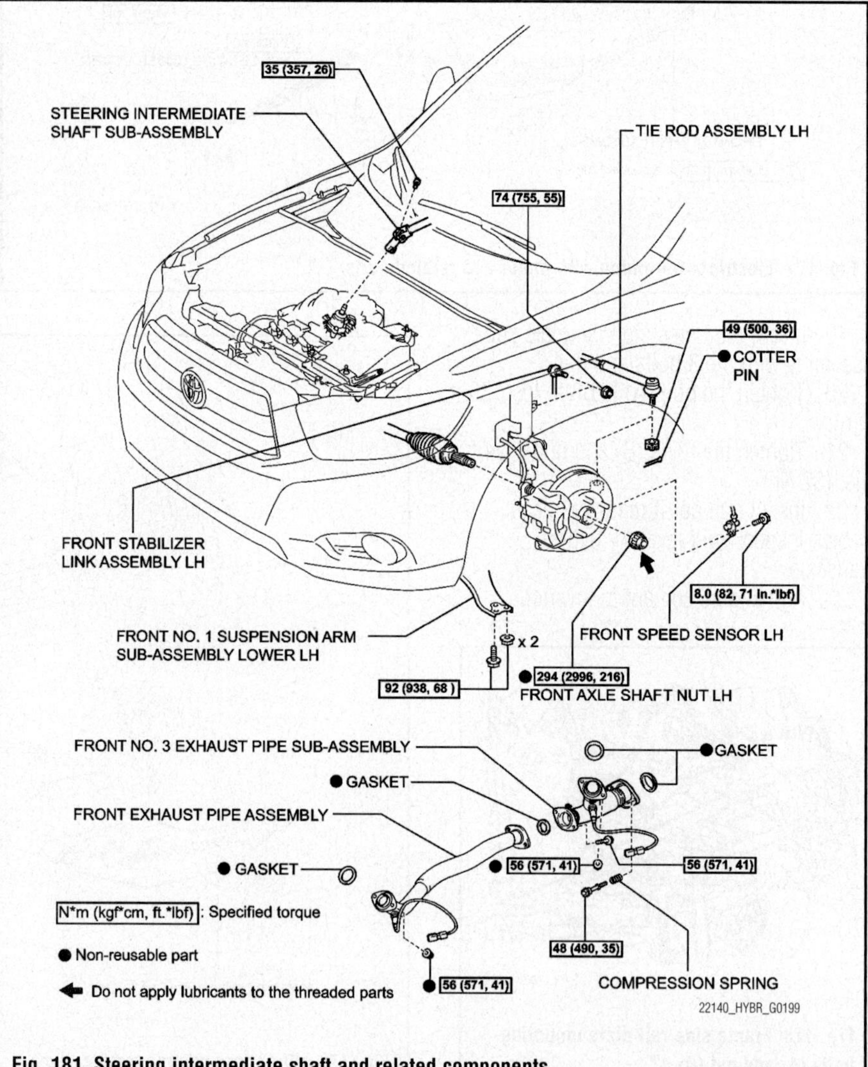

Fig. 181 Steering intermediate shaft and related components

1. Check for Daces
2. Remove the right and left front wiper arm assembly.
3. Remove the hood cowl top seal.
4. Remove the cowl top ventilator louver sub-assembly
5. Remove the wiper motor and link assembly.
6. Remove the outer cowl top panel.
7. Discharge the fuel system.
8. Place front wheel in straight ahead position.
9. Secure the steering wheel. This operation is to prevent damage to the spiral cable.
10. Remove the service plug grip.
11. Remove the LH engine room side cover.
12. Remove the cool air intake duct seal.
13. Raise the vehicle and remove the front wheels.
14. Remove the No.1 engine cover.
15. Separate LH and RH fender liners.
16. Remove the front fender apron seals.
17. Remove the No.2 engine under cover.
18. Drain the coolant for the engine and inverter.
19. Drain the engine oil.
20. Drain the HV transaxle fluid.
21. Remove the primary batter, tray and carrier.
22. Remove the air cleaner cap and case assembly.
23. Move the outer section to the wire harness side, then disconnect the No. 1 circuit breaker sensor.
24. Disconnect No.3 frame wire.
25. Remove the inverter reserve tank assembly.
26. Slide the clamp, and disconnect the water hose from the inverter with converter assembly.
27. Remove the inverter cover.
28. Using the voltmeter, measure the voltage between the terminals of the 2 phase connectors (N-P). It should rear 0 volts.
29. Separate engine room block relay assembly.
30. Remove the No. 4 inverter bracket.
31. Remove the inverter with converter assembly.
32. Remove the brake master cylinder Reservoir.
33. Remove the air cleaner bracket.
34. Remove the No.2 engine stay.
35. Remove the engine moving control rod.
36. Remove the engine mounting control rod.
37. Remove the RH No.2 engine mounting bracket.

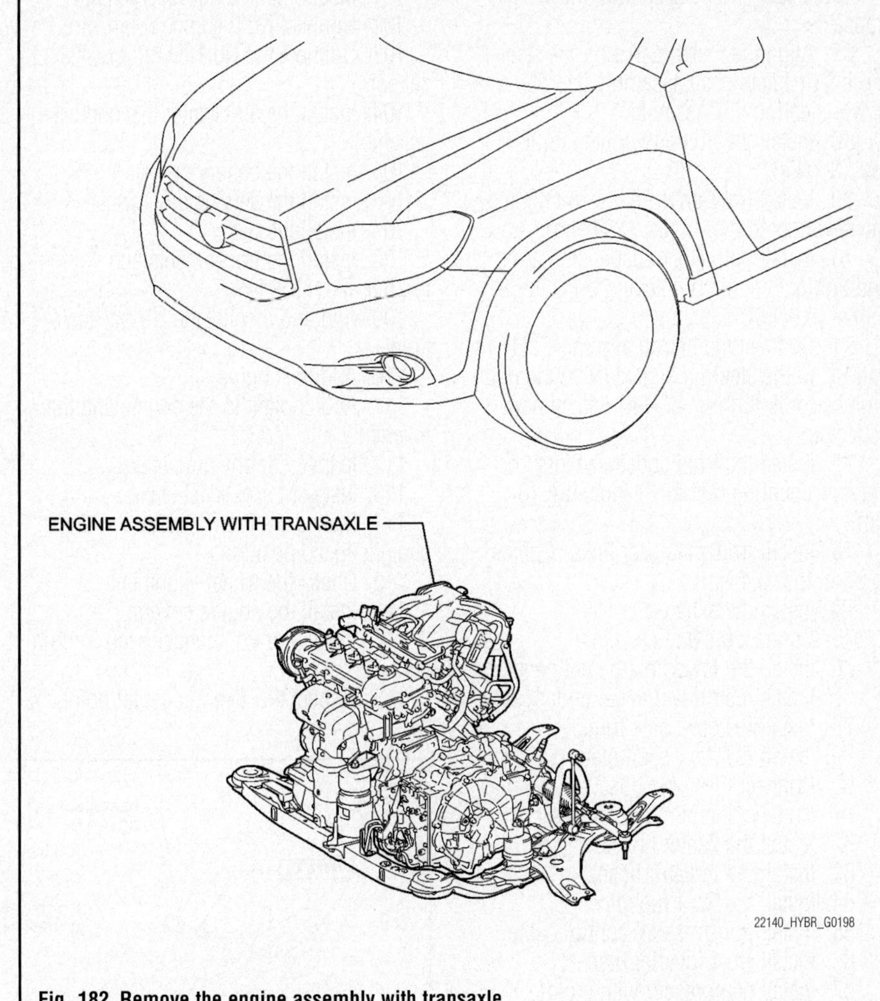

Fig. 182 Remove the engine assembly with transaxle

38. Separate compressor with motor assembly.
39. Remove the No.1 inverter bracket.
40. Separate transmission control cable assembly.
41. Disconnect the fuel vapor feed hose.
42. Disconnect the No.1 fuel pipe.
43. Disconnect heater hoses and radiator hoses.
44. Disconnect the oil cooler lines.
45. Disconnect the water hose.
46. Remove the RH front door scuff plate.
47. Remove the RH side cowl trim.
48. Remove the No.2 instrument under cover.
49. Remove the lower instrument panel.
50. Separate the engine wire.
51. Remove the exhaust pipes.
52. Remove the RH and LH stabilizer links.
53. Remove the front axle shaft nuts.
54. Remove the bolt and resin clamp, and separate the front speed sensors.
55. Separate the RH and LH tie rod ends from steering knuckle.
56. Remove the bolts and nuts, and separate the front lower suspension arms from the front lower ball joints.
57. Remove the LH and RH drive shafts.
58. Install matchmark and separate the steering intermediate shaft.
59. Remove the engine assembly with transaxle.
60. Remove the RH and LH stabilizer brackets.
61. Remove the stabilizer bar.
62. Remove the power steering link assembly.

To install:

63. Install the power steering link assembly with the 2 bolts and 2 nuts. Tighten to 51 ft. lbs. (70 Nm).
64. Install front stabilizer and brackets. Tighten bracket bolts to 21 ft. lbs. (29 Nm).
65. Install the front stabilizer links and tighten to 55 ft. lbs. (74 Nm).

66. Install engine assembly with transaxle.
67. Align the matchmarks on the steering intermediate shaft assembly and the power steering link assembly.
68. Install the bolt and tighten to 26 ft. lbs. (35 Nm).
69. Install front drive shafts and tighten the new nuts to 216 ft. lbs. (294 Nm).
70. Install both No.1 suspension arms and tighten mounting nuts and bolts to 68 ft. lbs. (92 Nm).
71. Connect the tie rod assembly LH and RH to the steering knuckle with the nut. Tighten to 36 ft. lbs. (49 Nm). Install new cotter pin.
72. Install the front speed sensors and tighten mounting bolt to 71 inch. lbs. (8 Nm).
73. Install front stabilizer links. Tighten to 55 ft. lbs. (74 Nm).
74. Install the exhaust.
75. Connect the engine wire.
76. Install the lower instrument panel.
77. Install instrument panel under cover.
78. Install RH cowl side trim.
79. Install RH door scuff plate.
80. Connect the water hose.
81. Connect the oil cooler hoses.
82. Install the heater hoses.
83. Install the radiator hoses.
84. Install the No.1 fuel pipe.
85. Install transmission control cable.
86. Install No.1 inverter bracket.
87. Install compressor with motor assembly.
88. Install RH No.2 engine mounting bracket.
89. Install engine mounting control bracket.
90. Install the engine moving control rod.
91. Install the RH engine mounting stay.
92. Install the air cleaner bracket.
93. Install the brake master cylinder reservoir.
94. Install inverter with the converter assembly.
95. Remove the inverter cover.
96. Connect the No. 3 frame wire (high voltage cable of the rear motor (MGR) with new 5 bolts to the inverter with converter assembly.
97. Connect the No. 4 engine wire with the bolt to the inverter with converter assembly.
98. Check that each connector and terminal is firmly installed.
99. Install the inverter cover with the 12 bolts to the inverter with converter assembly.
100. Connect the water hose.
101. Install the inverter reserve tank.
102. Connect No.2 engine room wire,
103. Connect the No.1 circuit breaker sensor.
104. Install the air cleaner cap and case assembly.
105. Install the battery carrier.
106. Install the service plug grip.
107. Install the battery.
108. Install the front wheels and tighten to 76 ft. lbs. (103 Nm).
109. Add and inspect the HV transaxle fluid.
110. Add the engine oil.
111. Add coolant to the engine and the inverter.
112. Inspect for any fluid leaks.
113. Inspect for exhaust leaks.
114. Make sure the wheels are in the straight ahead position.
115. Check the wheel alignment.
116. Install the engine covers.
117. Install the front fender seals and liners.
118. Inspect the steering center point.
119. Check ignition timing and engine speed.
120. Check speed sensor operation.
121. Install the front wiper system in the reverse of the removal procedure.

POWER STEERING DC-DC CONVERTER

REMOVAL & INSTALLATION

2006–07 Models

See Figure 183.

1. Before servicing the vehicle, refer to the Precautions Section.
2. Disconnect the negative battery cable.
3. Remove the service plug grip.
4. Remove the center instrument cluster finish panel.
5. Remove the left and right center floor carpet covers.
6. Remove the front seat center armrest assembly.
7. Remove the console box assembly.

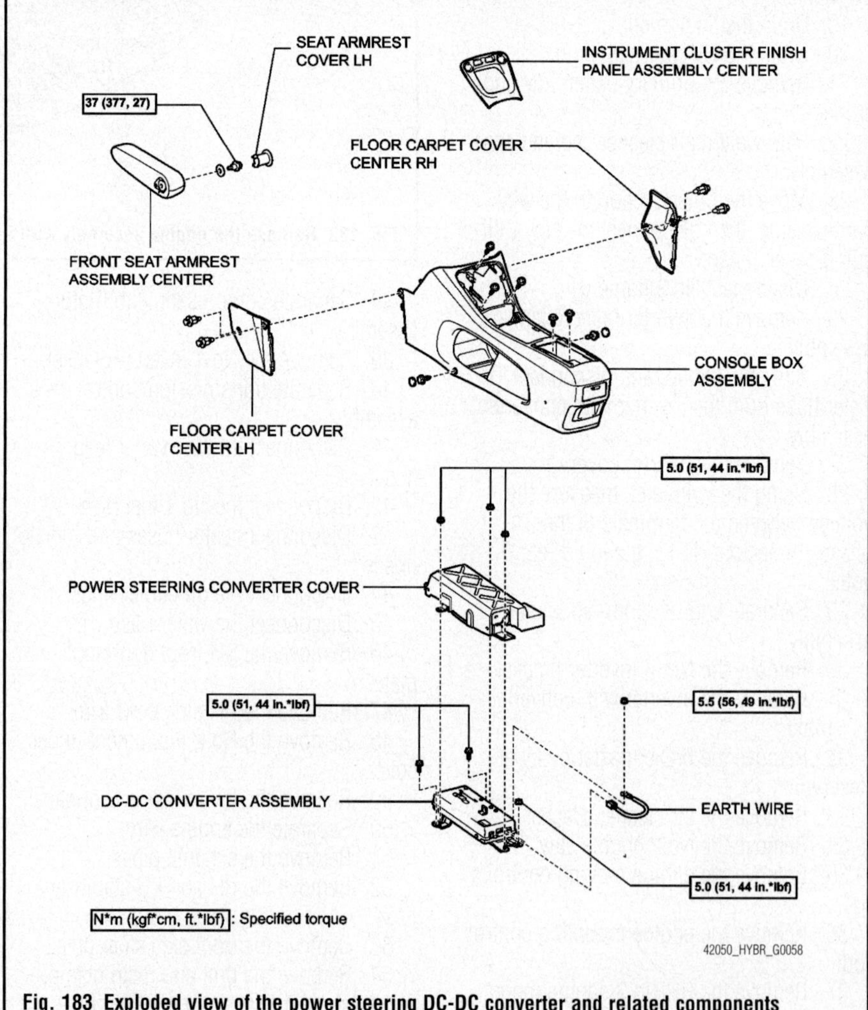

Fig. 183 Exploded view of the power steering DC-DC converter and related components

8. Remove power steering converter cover:

a. Separate wire harness clamps from the power steering converter cover.

b. Remove the 3 nuts and the power steering converter cover.

9. Remove the DC-DC converter assembly:

✲✲ CAUTION

Be sure to wear insulated gloves.

a. Disconnect the 3 connectors from the DC-DC converter assembly.

b. Remove the nut and separate the wire harness clamp.

c. Remove the 2 bolts, nut and the DC-DC converter assembly.

10. Remove the ground wire:

a. Disconnect the ground (earth) wire connector from the DC-DC converter assembly.

b. Remove the nut and the ground wire from the DC-DC converter assembly.

To install:

11. Install the ground wire:

a. Install the ground wire to the DC-DC converter assembly with the nut. Tighten to 49 inch lbs. (5.5 Nm).

b. Connect the ground wire connector to the DC-DC converter assembly.

12. Install the DC-DC converter assembly

✲✲ CAUTION

Be sure to wear insulated gloves.

a. Install the DC-DC converter assembly with the 2 bolts and the nut and tighten to 44 inch lbs. (5 Nm).

b. Attach the 3 connectors to the DC-DC converter assembly.

c. Connect the wire harness clamp to the Dc-DC converter assembly.

d. Install the nut and tighten to 49 inch lbs. (5.5 Nm).

✲✲ CAUTION

Be sure to wear insulated gloves when checking the high voltage cable connector.

13. Check that the high voltage cable connector is securely connected to the power steering converter assembly.

14. Install power steering converter cover, as follows:

a. Install the power steering converter cover with the 3 nuts and tighten to 44 inch lbs. (5 Nm).

b. Install the wire harness clamps to the power steering ECU bracket.

15. Install the remaining components in the reverse of the removal procedure.

16. Perform the initialization procedure, as outlined in the Chassis Electrical System.

POWER STEERING ECU

REMOVAL & INSTALLATION

2007–07 Models

See Figure 184.

1. Before servicing the vehicle, refer to the Precautions Section.
2. Disconnect the negative battery cable.
3. Remove the left-hand engine compartment side cover.
4. Remove the bolt and inverter bracket no.5.
5. Remove the power steering ECU assembly, as follows:

a. Remove the bolt and ground cable terminal from the power steering ECU assembly.

b. Release the locks of the 2 power steering ECU assembly connectors and disconnect the connectors.

c. Separate the 2 wirc harness clamps from the power steering ECU assembly.

d. Remove the 2 bolts and the power steering ECU assembly.

To install:

6. Install power steering ECU, as follows:

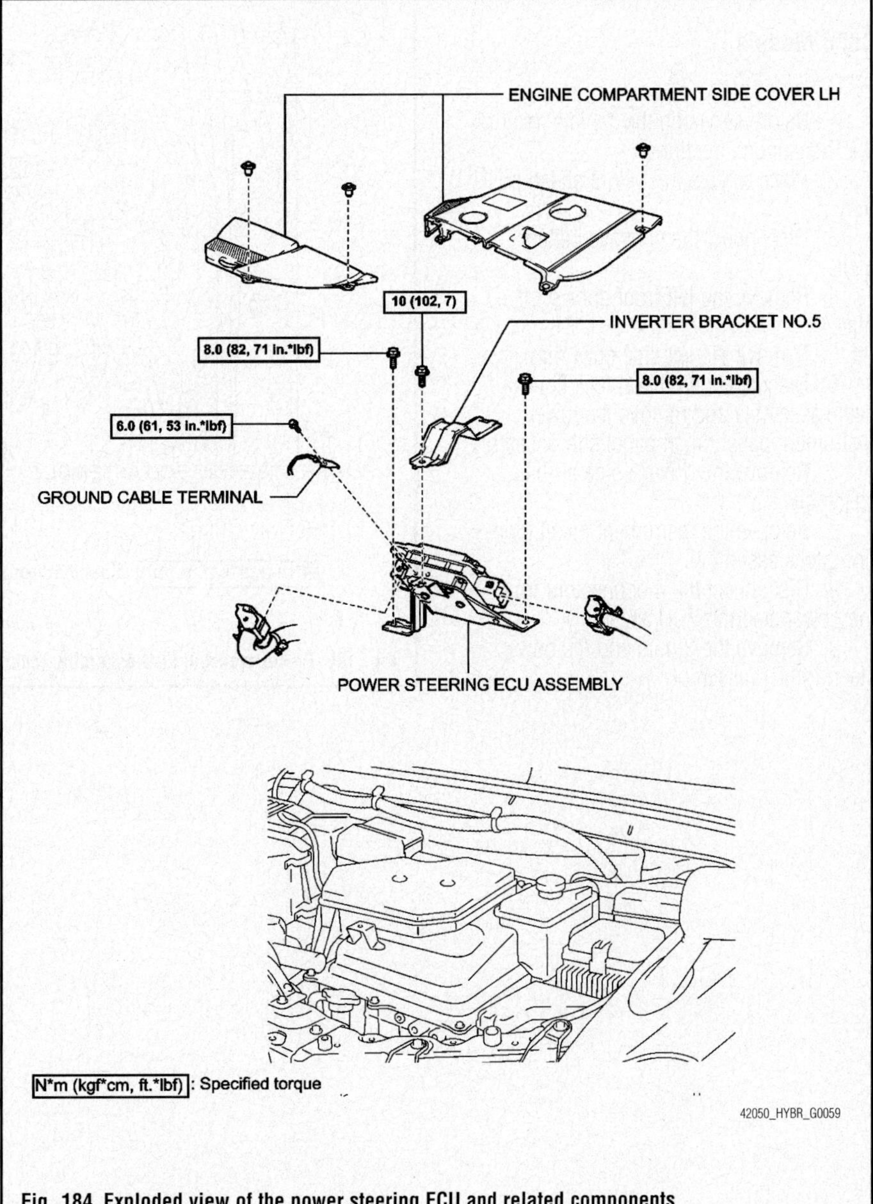

Fig. 184 Exploded view of the power steering ECU and related components

a. Install the power steering ECU assembly with the 2 bolts and tighten to 71 inch lbs. (8 Nm).

b. Connect the 2 wire harness clamps to the power steering ECU assembly.

c. Connect the 2 power steering ECU assembly connectors and securely lock the connectors.

d. Install the ground cable terminal to the power steering ECU assembly with the bolt and tighten to 53 inch lbs. (6.0 Nm).

7. Install the inverter bracket no.5 to the power steering ECU assembly with the bolt and tighten to 7 ft. lbs. (10 Nm).

8. Install the left-hand engine compartment side cover.

9. Connect the negative battery cable.

10. Perform the initialization procedure, as outlined in the Chassis Electrical System.

2008 Models

See Figure 185.

1. Before servicing the vehicle, refer to the Precautions Section.

2. Place wheels in straight ahead position.

3. Disconnect the negative battery cable.

4. Remove the left front door scuff plate.

5. Remove the left side cowl trim.

6. Disconnect the hood lock control cable assembly and remove the lower instrument panel finish panel sub-assembly.

7. Remove the driver's side air bag assembly.

8. Remove the instrument panel junction block assembly.

9. Disconnect the 4 connectors from the power steering ECU assembly.

10. Remove the 3 nuts and the power steering ECU assembly.

To install:

11. Install the power steering ECU assembly with the 3 nuts, tighten the nuts to 10 ft. lbs. (14 Nm).

12. Check that the connector lever is at the fully unlocked position before installation.

13. Connect the 4 connectors to the power steering ECU assembly.

14. Connect the connectors to the back of the instrument panel junction block assembly.

15. Engage the wire harness clamp onto the instrument panel junction block assembly.

16. Install the instrument panel junction block assembly with the 3 nuts, tighten the nuts to 74 inch. lbs. (8.4 Nm).

17. Connect the connectors to the instrument panel junction block assembly.

18. Engage the wire harness clamp onto the instrument panel junction block assembly.

19. Install the driver side knee airbag assembly with the 4 bolts. Tighten the bolts to 7 ft. lbs. (10 Nm).

20. Install the left cowl side trim.

21. Install the left front door scuff plate.

22. Connect the negative battery cable.

➡When disconnecting the cable, some systems need to be initialized after the cable is reconnected

23. Inspect the SRS warning light.

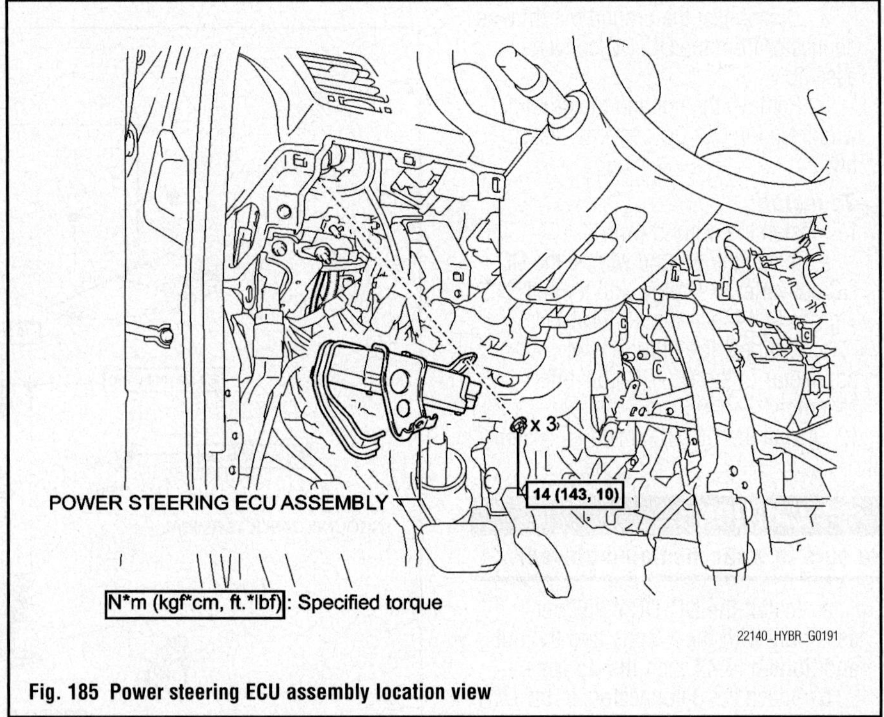

Fig. 185 Power steering ECU assembly location view

COIL SPRING

REMOVAL & INSTALLATION

Refer to the front strut removal.

CONTROL LINKS

REMOVAL & INSTALLATION

See Figure 186.

➡Perform the same procedure on each side.

1. Remove wheel and tire assembly.
2. Remove the nut and separate the front stabilizer link assembly.

➡If the ball joint turns together with the nut, use a hexagon wrench (6 mm) to hold the stud bolt.

To install:

3. Install the front stabilizer link assembly with the nut and tighten to 55 ft. lbs. (74 Nm).
4. Install the wheel and tire assembly.

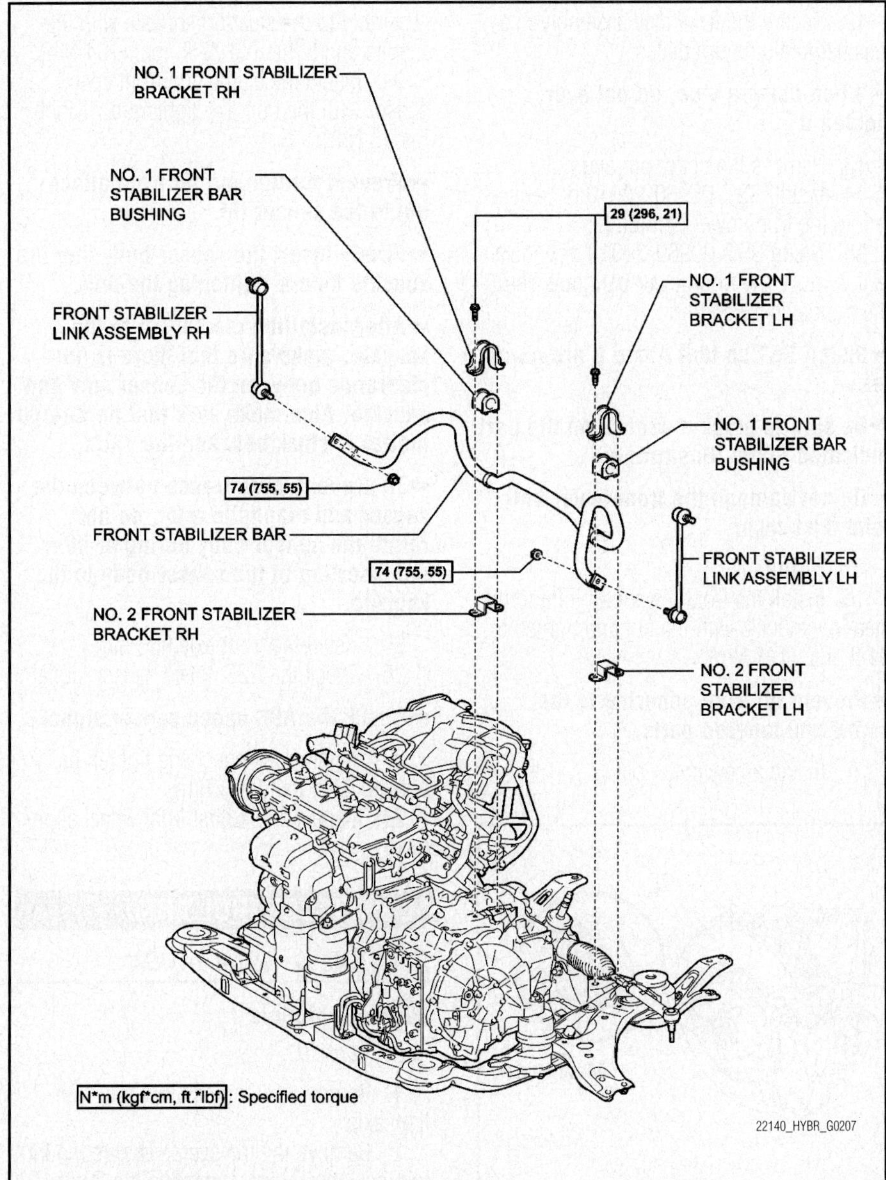

Fig. 186 Front stabilizer control links and related parts.

LOWER BALL JOINT

REMOVAL & INSTALLATION

2006–07 Models

See Figure 187.

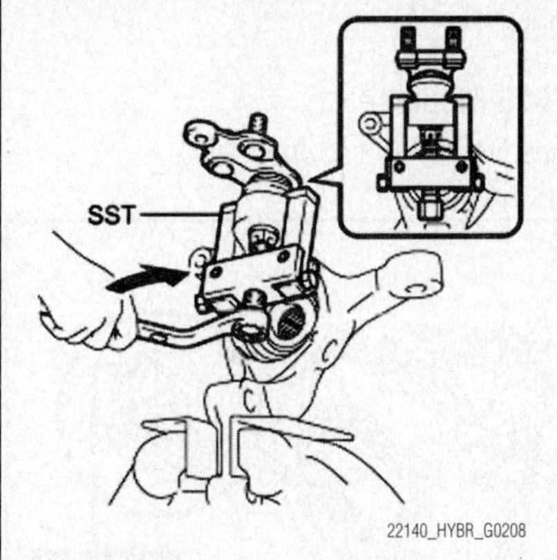

Fig. 187 Ball joint removal with tool 09628-62011

1. Remove the front wheel.
2. Remove the front axle shaft nuts.
3. Separate the speed sensor front.
4. Separate the front disc brake caliper assembly.
5. Remove the front disc.
6. Using SST: 09628-62011 or equivalent, separate tie rod assembly.
7. Separate the lower No.1 front suspension arm sub-assembly.
8. Remove front axle assembly.
9. Remove the front lower ball joint assembly:
10. Remove the cotter pin and castle nut.
11. Using SST: 09628-62011 or equivalent, remove the front lower ball joint assembly.

To install:

12. Install the front lower ball joint assembly to the steering knuckle with the nut and tighten to 91 ft. lbs. (123 Nm).

➡Prevent oil from adhering to the screw and tapered parts.

13. Install a new cotter pin to the steering knuckle.

➡If the holes for the cotter pin are not aligned, tighten the nut further up to 60°.

14. Install the front axle assembly.
15. Install the front lower No.1 suspension arm sub-assembly.
16. Install the tie rod assembly.
17. Install the front disc.
18. Install the front disc brake caliper assembly.
19. Install the speed sensor front.
20. Install the front axle shaft nut.
21. Install the front wheel and tighten the lug nuts to 76 ft. lbs. (103 Nm).

22. Inspect and adjust front wheel alignment.
23. Check the ABS speed sensor signal.

2008 Models

See Figures 188 through 190.

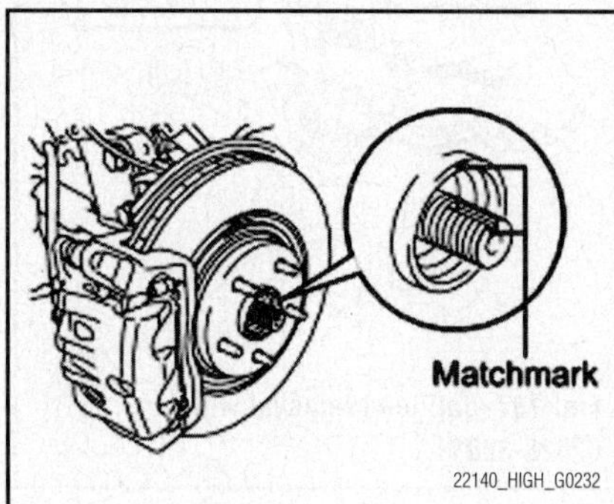

Fig. 188 Matchmark the front drive shaft assembly and the front axle hub sub-assembly

1. Remove the front wheel.
2. Remove the front axle hub nut.
3. Remove the bolt and resin clamp, and separate the front speed sensor.

➡Be sure to completely separate the front speed sensor from the front shock absorber with coil spring.

➡Clean the installation hole and the surface for the speed sensor every time the speed sensor is removed.

➡Be careful not to damage the front speed sensor.

4. Put matchmarks on the front drive shaft assembly and the front axle hub sub-assembly.
5. Using a plastic hammer, separate the front drive shaft assembly from the front axle assembly.

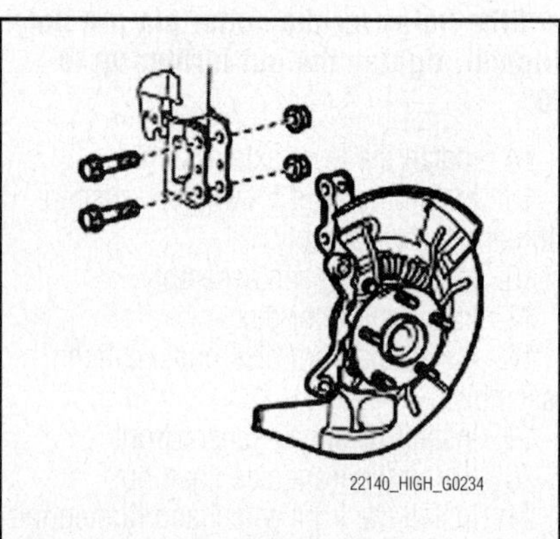

Fig. 189 Front strut assembly bolts

➡Loosen the staked part of the front axle hub nut completely, otherwise the threads of the drive shaft may be damaged.

6. Remove the 2 bolts and separate the front disc brake caliper assembly.

➡Use wire or an equivalent tool to keep the brake caliper from hanging down by the flexible hose.

7. Remove the front disc.
8. Separate the tie rod assembly.
9. Remove the bolt, 2 nuts, and separate the front lower suspension arm from the lower ball joint.
10. Remove the 2 bolts, 2 nuts and front axle assembly.
11. Remove the front lower ball joint, as follows:
12. Secure the front axle assembly in a vise using aluminum plates.

➡When using a vise, do not over tighten it.

13. Remove the cotter pin and nut.
14. Install SST: 09960-20010 or equivalent to the front lower ball joint.
15. Using SST: 09960-20010 or equivalent, remove the front lower ball joint from the front axle assembly.

➡Install SST so that A and B are parallel.

➡Be sure to place a wrench on the part indicated in the illustration.

➡Do not damage the front lower ball joint dust cover.

To install:

16. Install the front lower ball joint to the steering knuckle with the nut and tighten to 91 ft. lbs. (123 Nm).

➡Prevent oil from adhering to the screw and tapered parts.

17. Install a new cotter pin.

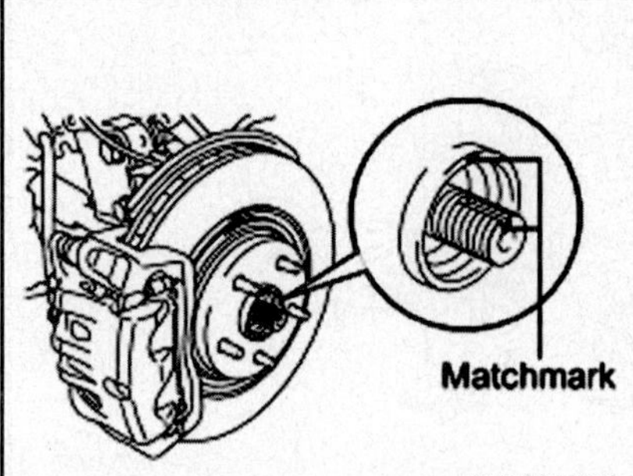

Fig. 190 Align the matchmarks and install the front drive shaft assembly

➡If the holes for the cotter pin are not aligned, tighten the nut further up to 60°.

18. Install the front axle assembly to the front shock absorber with the 2 bolts and 2 nuts and tighten to 213 ft. lbs. (290 Nm).

➡Only when reusing the bolts and nuts, apply a small amount of engine oil to the threads of the nuts.

19. Align the matchmarks and install the front drive shaft assembly to the front axle hub sub-assembly.
20. Install the front lower suspension arm to the front lower ball joint with the bolt and 2 nuts 68 ft. lbs. (92 Nm).
21. Connect the tie rod assembly.
22. Install the front disc.
23. Install the front disc brake caliper assembly to the steering knuckle with the 2 bolts and tighten to 77 ft. lbs. (104 Nm).
24. Install the clamp and front speed sensor with the bolt and tighten to 71 inch lbs. (8 Nm).

➡Prevent foreign matter from attaching to the sensor tip.

➡Firmly insert the sensor body into the knuckle before tightening the bolt.

➡After installing the sensor to the knuckle, make sure that there is no clearance between the sensor stay and knuckle. Also make sure that no foreign matter is stuck between the parts.

➡To prevent interference between the sensor and magnetic rotor, do not rotate the sensor body during or after the insertion of the sensor body to the knuckle.

25. Install the front axle hub nut.
26. Check the ABS speed sensor signal

➡Check the ABS speed sensor signal.

27. Install front wheel and tighten lug nuts to 76 ft. lbs. (103 Nm).
28. Inspect and adjust front wheel alignment.

LOWER CONTROL ARM

REMOVAL & INSTALLATION

2006–07 Models

See Figure 191.

1. Remove the engine assembly with transaxle.
2. Remove the transverse engine mounting insulator by first removing the 3 nuts.
3. Remove the lower No. 1 front suspension arm sub-assembly, as follows:

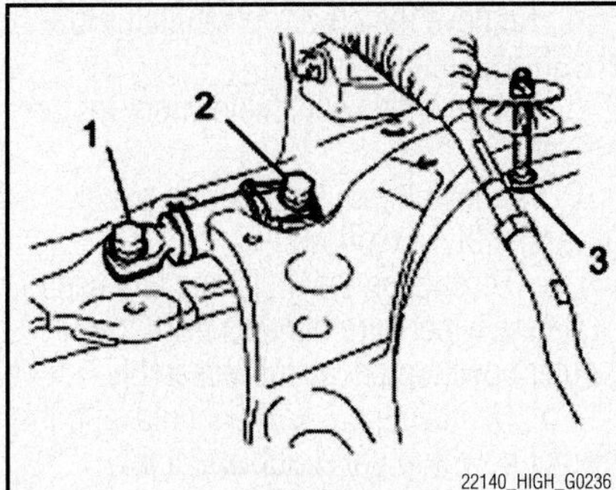

Fig. 191 Front lower control arm bolts tightening sequence–2006–07 models

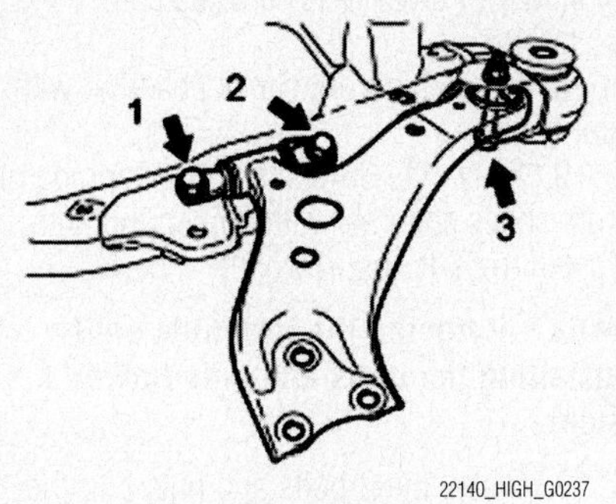

Fig. 192 Front lower control arm bolts tightening sequence–2008 models

a. Remove the 3 bolts and the nut from the lower No. 1 suspension arm and remove it from the engine front support member.

b. Remove the front lower arm bush stopper.

To install:

4. Install the lower No. 1 front suspension arm sub-assembly, as follows:

a. Install the front lower arm bush stopper.

b. Install the lower No. 1 suspension lower arm to the engine front support member with the 3 bolts and the nut, but do not tighten.

c. Tighten the 3 bolts in the numerical order shown. Tighten to 148 ft. lbs. (200 Nm), and 152 ft. lbs. (206 Nm).

➡Start installing the bolts from the front side of the vehicle.

5. Install the transverse engine mounting insulator with the 3 nuts and tighten to 64 inch lbs. (87 Nm).

6. Install the engine assembly with transaxle.

2008 Models

See Figure 193.

1. Remove the engine assembly with transaxle.

2. Remove the No. 1 front stabilizer brackets.

3. Remove the front stabilizer bar with front stabilizer link assembly.

4. Remove the power steering link assembly.

5. Install the engine hangers.

6. Separate the front frame assembly.

7. Remove the front lower suspension arm, as follows:

a. Remove the 3 bolts, nut, and the front lower suspension arm from the front frame assembly.

b. Remove the front lower arm bushing stopper from the front lower suspension arm.

To install:

8. Install the front lower suspension arm, as follows:

a. Install the front lower arm bushing stopper to the front lower suspension arm.

b. Install the front lower suspension arm to the front frame assembly with the 3 bolts and nut, but do not tighten them yet.

c. Tighten the 3 bolts in numerical order shown. Tighten to 147 ft. lbs. (200 Nm), and 152 ft. lbs. (206 Nm).

➡Start installing the bolts from the front side of the vehicle.

9. Connect the front frame assembly.

10. Remove the engine hangers.

11. Install the power steering link assembly.

12. Install the front stabilizer bar with front stabilizer link assembly.

13. Install the No. 1 front stabilizer brackets.

14. Install the engine assembly with transaxle.

MACPHERSON STRUT

REMOVAL & INSTALLATION

2006–07 Models

See Figures 193 through 195.

1. Remove the front wheel.

2. Remove the nut and separate the front stabilizer link assembly from the front shock absorber assembly.

➡If the ball joint turns together with the nut, use a hexagon wrench (6 mm) to hold the stud.

3. Remove the front shock absorber, as follows:

a. Loosen the lock nut.

➡Do not loosen or remove the lock nut except when disassembling the front shock absorber assembly with coil spring.

➡Do not remove the lock nut.

b. Remove the bolt and disconnect the No.1 front flexible hose and front speed sensor wire harness.

c. Remove the 2 nuts and bolts on the lower side of front shock absorber with coil spring.

➡When removing the bolts, keep the bolt from rotating and loosen the nut.

d. Remove the 3 nuts on the upper side of the front shock absorber with coil spring.

e. Remove the front shock absorber with coil spring.

➡Make sure that the front speed sensor is disconnected from the front shock absorber.

4. Secure the front shock absorber with coil spring in a vise by clamping onto a double nut bolt affixed to the bracket at the bottom of the absorber.

5. Using SST: 09727-30021, compress the front coil spring.

➡Do not use an impact wrench. It will damage the SST.

➡Use 2 of the same type of SST.

6. Remove the front suspension support sub-assembly, front suspension support bearing, front coil spring seat upper, front coil spring insulator upper, front coil spring, front spring bumper, and front lower coil spring insulator from the front shock absorber assembly.

To install:

7. Install the front shock absorber assembly, as follows:

a. Install the lower front coil spring insulator onto the front shock absorber assembly.

b. Install the front spring bumper to the piston rod.

c. Using SST: 09727-30021, compress the front coil spring.

➡Do not use an impact wrench. It will damage the SST.

➡Use 2 of the same type of SST.

d. Install the front coil spring to the front shock absorber assembly.

➡Fit the lower end of the front coil spring in the gap of the lower spring insulator.

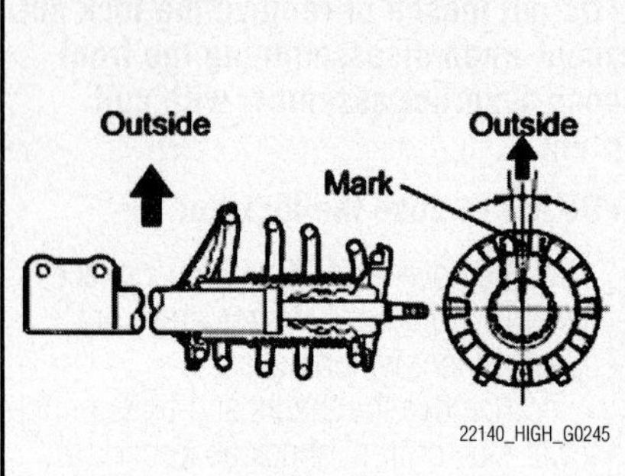

Fig. 193 Install the upper front coil insulator—2006-07 models

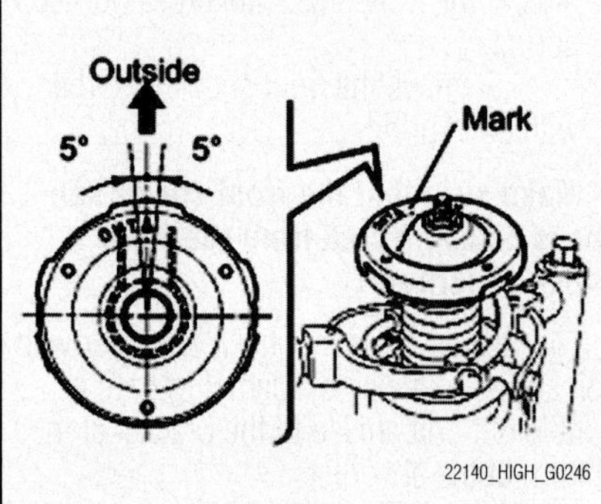

Fig. 194 Install the upper front coil spring seat—2006-07 models

e. Install the upper front coil insulator with the mark facing to the outside of the vehicle as shown in the illustration.

f. Install the upper front coil spring seat to the front shock absorber assembly with the mark facing to the outside of the vehicle.

g. Install a new front suspension support bearing.

h. Install the front suspension support sub-assembly with the mark facing to the outside of the vehicle.

i. Temporarily tighten a new lock nut.

j. Remove the SST: 09727-30021

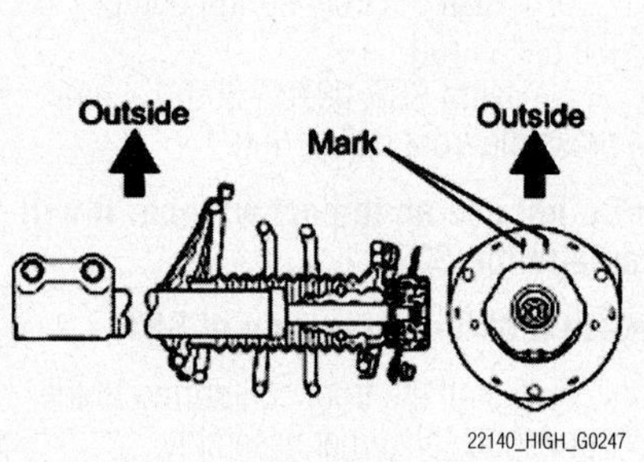

Outside

Outside

Mark

22140_HIGH_G0247

Fig. 195 Install the front suspension support sub-assembly —2006-07 models

slowly in order to release the coil spring.

8. Install the front shock absorber with coil spring.

9. Install the 3 nuts to the upper side of front shock absorber with coil spring and tighten to 59 ft. lbs. (80 Nm).

➡Do not tighten the lock nuts until installing the bolts and nuts (lower side).

10. Install the 2 bolts and nuts 2 to the lower side of front shock absorber with coil spring and tighten to 170 ft. lbs. (230 Nm).

➡When installing the nuts, keep the bolts from rotating and torque the nut.

11. Install the No.1 front flexible hose and front speed sensor wire harness with the bolt and tighten to 14 ft. lbs. (19 Nm).

12. Fully tighten the lock nut and tighten to 36 ft. lbs. (49 Nm).

13. Install the front stabilizer link assembly with the nut 55 ft. lbs. (74 Nm).

➡If the ball joint turns together with the nut, use a hexagon wrench (6 mm) to hold the stud.

14. Install the front wheel and tighten lug nuts to 76 ft. lbs. (103 Nm).

15. Inspect and adjust front wheel alignment.

2008 Models

See Figures 197 through 204.

1. Remove the front wheel.

2. Remove the front wiper arm and blade assemblies.

3. Loosen the front support to front shock absorber nut of the front shock absorber.

➡Do not remove the front support to front shock absorber nut.

➡Loosen the nut only when the front shock absorber with coil spring needs to be disassembled.

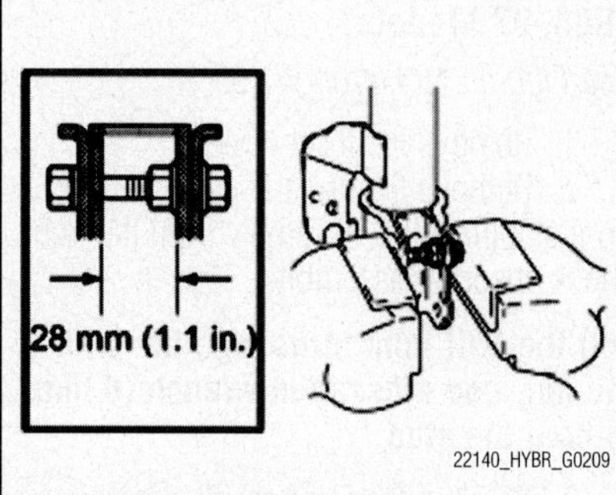

Fig. 196 Secure the front shock absorber with coil spring in a vise

4. Remove the cowl top ventilator louver sub-assembly.

5. Remove the windshield wiper motor and link.

6. Remove the outer cowl top panel sub-assembly, as follows:

a. Disengage the 4 clamps and separate the wiper wire harness from the outer cowl top panel sub-assembly.

b. Remove the 8 bolts, 6 nuts, and the outer cowl top panel sub-assembly.

7. Remove the bolt and clamp, and separate the front speed sensor and front flexible hose.

8. Remove the nut and separate the front stabilizer link assembly from the front shock absorber.

➡If the ball joint turns together with the nut, use a hexagon wrench (6 mm) to hold the stud.

9. Remove the front shock absorber with coil spring, as follows:

a. Support the front axle using a jack and wooden block.

b. Remove the 2 bolts and 2 nuts, and separate the front shock absorber with coil spring (lower side) from the steering knuckle.'

➡When removing the nuts, keep the bolts from rotating.

c. Remove the nut and 2 spacers on the upper side of the front shock absorber with coil spring.

➡Make sure that the front speed sensor is completely separated from the front shock absorber with coil spring.

10. Secure the front shock absorber with coil spring, as follows:

a. As shown in the illustration, secure the front shock absorber with coil spring in a vise using aluminum plates by clamping onto a double nut bolt affixed to the bracket at the bottom of the absorber.

11. Remove the front support to front shock absorber nut, as follows:

a. Using A Special Service Tool (SST: 09727-30021, SST: 09727-30021), compress the front coil spring.

➡Do not use an impact wrench. It will damage the SST.

➡If the front coil spring is compressed at an angle, using 2 SST will make the work easier.

b. Check that the front coil spring is fully compressed.

c. Remove the front support to front shock absorber nut.

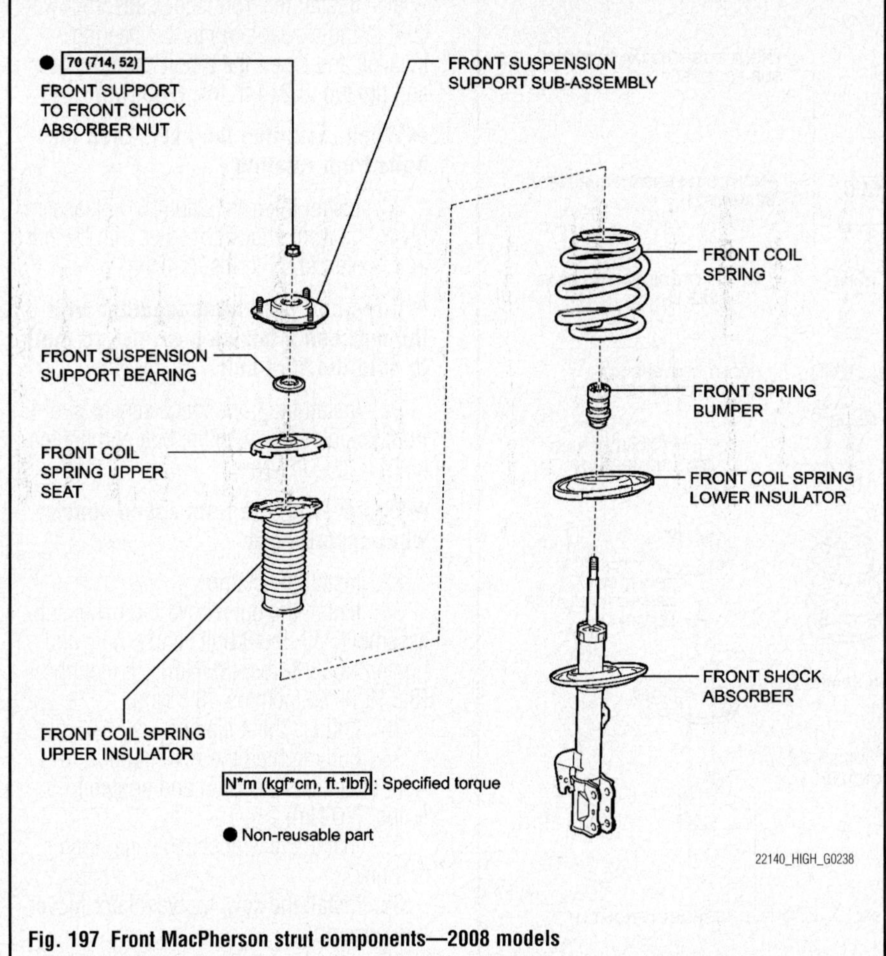

Fig. 197 Front MacPherson strut components—2008 models

12. Remove the front suspension support sub-assembly.
13. Remove the front suspension support bearing.
14. Remove the front coil spring upper seat.
15. Remove the front coil spring upper insulator.
16. Remove the front coil spring.
17. Remove the front spring bumper.
18. Remove the front coil spring lower insulator.

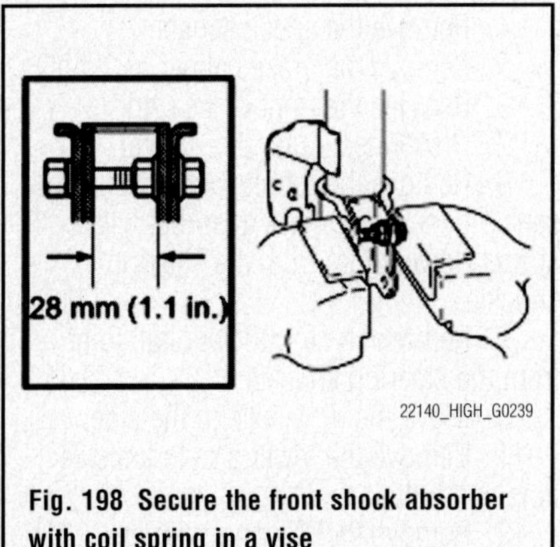

Fig. 198 Secure the front shock absorber with coil spring in a vise

To install:

19. Secure the front shock absorber assembly, as follows:
    a. As shown in the illustration, secure the front shock absorber with coil spring in a vise using aluminum plates by clamping onto a double nut bolt affixed to the bracket at the bottom of the absorber.
20. Install the front coil spring lower insulator to the front shock absorber.

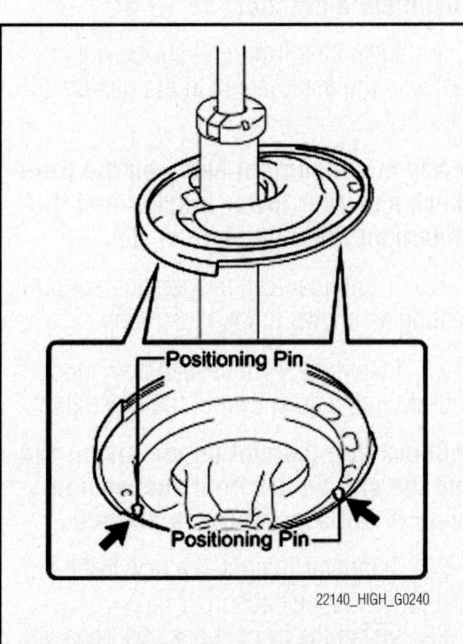

Fig. 199 Front coil spring lower insulator positioning pin

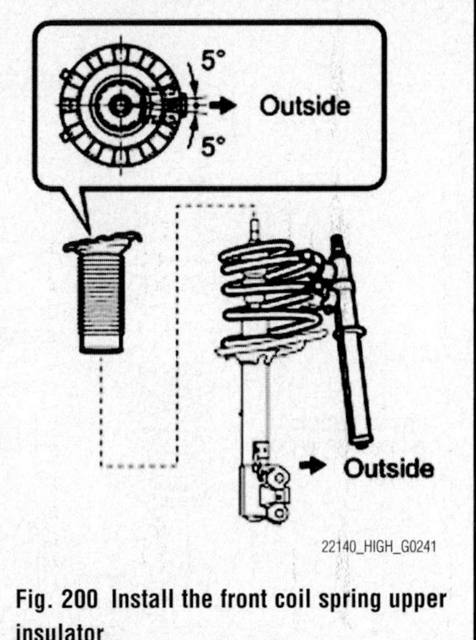

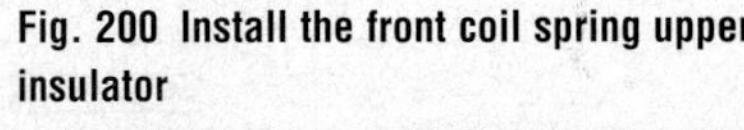

Fig. 200 Install the front coil spring upper insulator

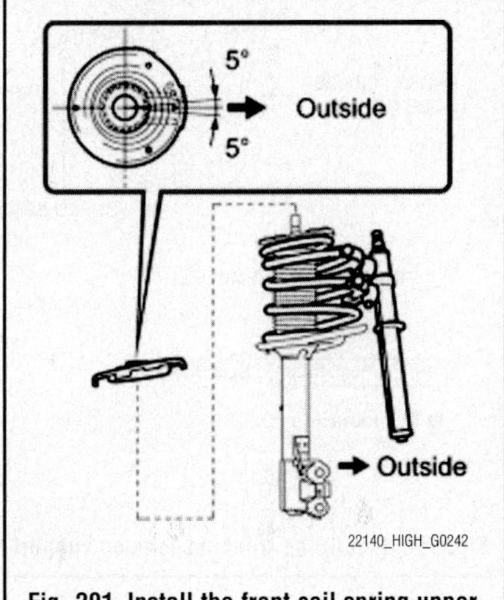

Fig. 201 Install the front coil spring upper seat

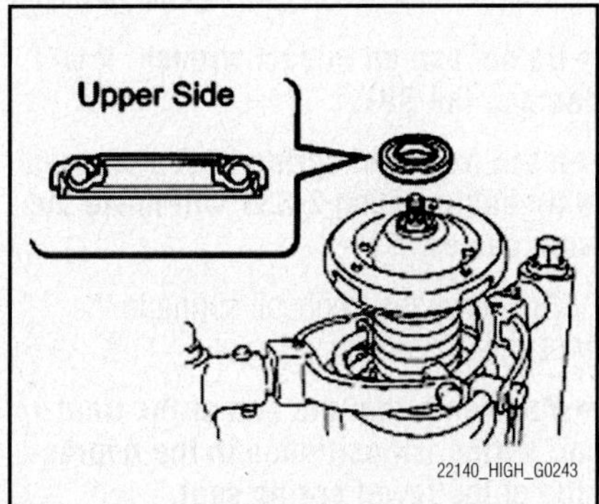

Fig. 202 Install the front suspension support bearing

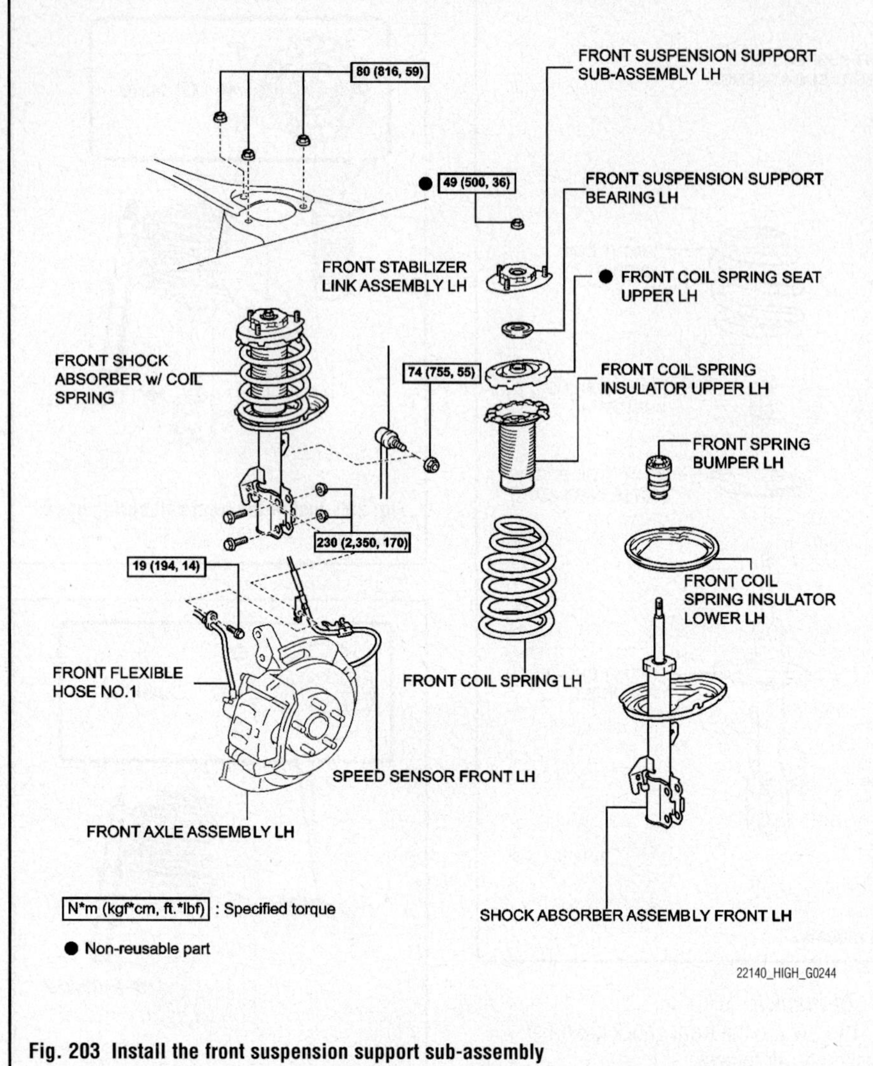

Fig. 203 Install the front suspension support sub-assembly

➡Make sure that the positioning pins on the front coil spring lower insulator are inserted into the holes in the front shock absorber.

21. Install the front spring bumper to the front shock absorber.
22. Using a Special Service Tool (SST: 09727-30021, SST: 09727-00050) or equivalent, compress the front coil spring.

➡Do not use an impact wrench. It will damage the SST.

➡If the front coil spring is compressed at an angle, using 2 SST will make the work easier.

23. Install the front coil spring to the front shock absorber.

➡Make sure that the end of the front coil spring is positioned in the depression of the lower spring seat.

24. Install the front coil spring upper insulator as shown in the illustration.

➡Any misalignment between the front shock absorber lower bracket and the alignment mark must be +/- 5°.

25. Install the front coil spring upper seat with the mark facing to the outside of the vehicle.

➡Any misalignment between the front shock absorber lower bracket and the alignment mark must be +/- 5°.

26. Install the front suspension support bearing as shown in the illustration.

27. Install the front suspension support sub-assembly as shown in the illustration.

➡Check that the slot on the piston rod and the slot on the front suspension support sub-assembly are aligned.

28. Temporarily tighten a new front support to front shock absorber nut.
29. Install the front shock absorber with coil spring (upper side) with the nut and 2 spacers and tighten to 63 ft. lbs. (85 Nm).
30. Install the front shock absorber with coil spring (lower side) to the steering knuckle and insert the 2 bolts and 2 nuts and tighten to 214 ft. lbs. (290 Nm).

➡When installing the nuts, keep the bolts from rotating.

31. Install the front stabilizer link assembly to the front shock absorber with the nut and tighten to 55 ft. lbs. (74 Nm).

➡If the ball joint turns together with the nut, use a hexagon wrench (6 mm) to hold the stud bolt.

32. Install the front speed sensor and front flexible hose with the bolt and tighten to 14 ft. lbs. (19 Nm).

➡Do not twist the front speed sensor when installing it.

33. Install the clamp.
34. Install the outer cowl top panel sub-assembly with the 8 bolts and 6 nuts and tighten to 63 ft. lbs. (85 Nm), 78 inch lbs. (8.8 Nm), 78 inch lbs. (8.8 Nm).
35. Engage the 4 clamps.
36. Fully tighten the front support to front shock absorber nut and tighten to 52 ft. lbs. (70 Nm).
37. Install the windshield wiper motor and link.
38. Install the cowl top ventilator louver sub-assembly.
39. Install the front wiper arm and blade assemblies.
40. Install the front wheel and tighten lug nuts to 76 ft. lbs.

STEERING KNUCKLE

REMOVAL & INSTALLATION

See Figures 204 and 205.

1. Raise and support the vehicle.
2. Remove the tire and wheel assembly.
3. Remove the drive axle retaining nut.
4. Remove the speed sensor.
5. Remove the brake caliper assembly.
6. Remove the front brake rotor.
7. Disconnect outer tie rod end.
8. Remove the 2 bolts and 2 nuts, and separate the front shock absorber with coil spring (lower side) from the steering knuckle.
9. Remove the front lower ball joint from the steering knuckle.
10. Move the drive axle to the side.
11. Remove the lower arm to knuckle nuts and bolts.
12. Remove the steering knuckle.

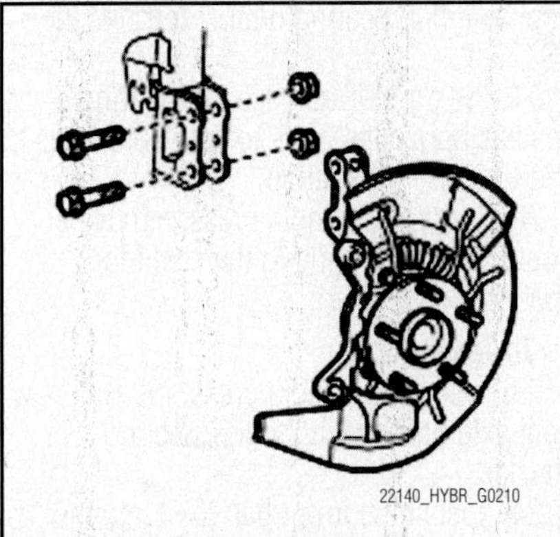

Fig. 204 Remove the 2 bolts and 2 nuts

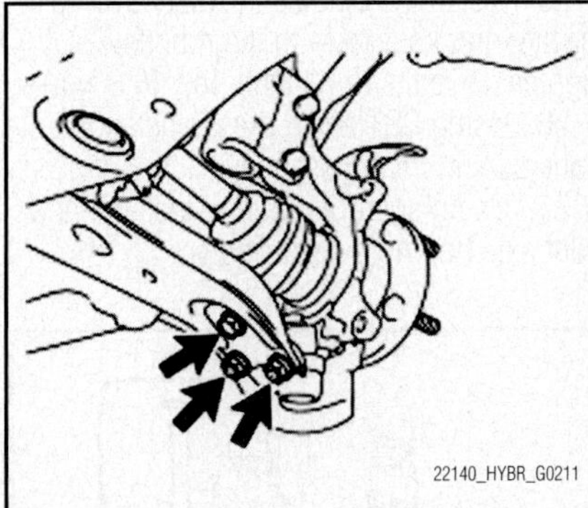

Fig. 205 Remove front lower suspension arm to the front lower ball joint

To install:

13. Install steering knuckle and drive axle.
14. Install ball joint retaining nut and tighten to 36 ft. lbs. (49 Nm). Install cotter pin.
15. Install the lower arm to knuckle nuts and bolts. Tighten to 68 ft. lbs. (92 Nm).
16. Install the 2 bolts and 2 nuts, and tighten the front strut with coil spring (lower side) to the steering knuckle. Tighten to 213 ft. lbs. (290 Nm).
17. Connect the outer tie rod end and tighten the castle nut to 36 ft. lbs. (49 Nm). Install a new cotter pin.
18. Install the front brake rotor.
19. Install front brake caliper assembly and tighten the mounting bolts to 77 ft. lbs. (104 Nm).
20. Install the speed sensor and tighten mounting bolt to 71 inch. lbs. (8 Nm).
21. Install the new drive axle retaining nut and tighten to 216 ft. lbs. (294 Nm). Using a chisel and hammer, stake the front axle hub nut.
22. Install and tighten the front wheel assembly to 76 ft. lbs. (103 Nm).
23. Check and adjust wheel alignment.
24. Check speed sensor operation.

STABILIZER BAR

REMOVAL & INSTALLATION

See Figure 206.

1. Make sure the vehicle's front wheels are in the straight-ahead position.
2. Disconnect the negative battery cable.
3. Raise and safely support the vehicle.
4. Remove the remove front wheel and tire assemblies.
5. Remove the engine under cover.
6. Separate the steering intermediate shaft subassembly.
7. Separate the tie rod.
8. Remove the 2 nuts, then remove the front left stabilizer link assembly:

➡If the ball joint turns together with the nut, use a hexagon wrench (6 mm) to hold the stud.

9. Remove the 2 nuts, then remove the front right stabilizer link assembly:

➡If the ball joint turns together with the nut, use a hexagon wrench (6 mm) to hold the stud.

10. Remove the 2 bolts, then remove the left stabilizer bracket no. 1 from the front frame assembly.
11. Remove the 2 bolts, then remove the right stabilizer bracket no. 1 from the front frame assembly.
12. Remove the left and right no. 2 stabilizer brackets from the bushings.
13. Remove the 2 stabilizer bar bushings no.1 from the stabilizer bar.
14. Remove the front no. 3 exhaust pipe subassembly
15. Remove the front exhaust pipe/
16. Remove the bolt, nut and the manifold stay.
17. Remove the power steering link.
18. Remove the front stabilizer bar from the right side of the vehicle.
19. Inspect the turning of the stabilizer link ball joint:
    a. secure the front stabilizer link assembly in a vise using aluminum plates.
    b. Install the nut to the front stabilizer link assembly stud.
    c. If using a torque wrench, turn the nut continuously at a rate of 3 to 5 seconds per turn and take the torque reading on the 5th turn. The turning torque should be 18 inch lbs. (2.0 Nm) or less.

➡If the turning torque is not within the specified range, replace the front stabilizer link assembly with a new one.

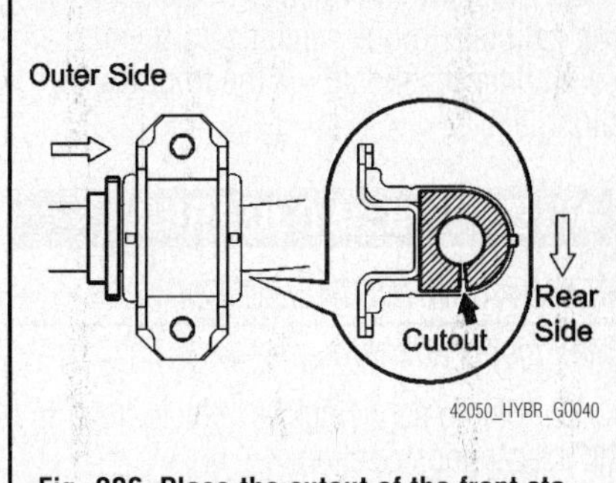

Fig. 206 Place the cutout of the front stabilizer bar bushing no.1 as facing the rear side as shown

20. Inspect the dust cover:
    a. Check that the dust cover is not cracked and that there is no grease on it.

To install:

21. Install the front stabilizer bar by inserting it from the right side of the vehicle.
22. Install the power steering link.
23. Install the manifold stay with the bolt and nut. Tighten to 25 ft. lbs. (34 Nm).
24. Install the front exhaust pipe.
25. Install the front no. 3 exhaust pipe subassembly.
26. Install the 2 front stabilizer bar bushings no.1 to the outer side of the bushing stopper on the front stabilizer bar.

➡Place the cutout of the front stabilizer bar bushing no.1 as facing the rear side as shown in the illustration.

27. Install the right and left no.2 stabilizer brackets to the stabilizer bushings.
28. Install the right and left no.1 stabilizer bracket to the front frame assembly with the bolts. Tighten to 12 ft. lbs. (16 Nm).
29. Install the right and left stabilizer link assemblies with the nuts and tighten to 55 ft. lbs. (75 Nm).

➡If the ball joint turns together with the nut, use a hexagon (6 mm) wrench to hold the stud.

30. Connect the tie rod.
31. Connect steering intermediate shaft subassembly.
32. Make sure the front wheels are facing straight-ahead.
33. Install the engine under cover.
34. Install the front wheels.
35. Connect the negative battery cable.
36. If necessary, perform the initialization procedure, as outlined in the Chassis Electrical System.

37. Initialize the rotation angle sensor and calibrate torque sensor zero point.

38. Inspect and adjust the front wheel alignment.

WHEEL HUB AND BEARING

REMOVAL & INSTALLATION

See Figures 207 through 213.

1. Before servicing the vehicle, refer to the Precautions Section.
2. Remove the front wheel.
3. Remove the front axle hub nut.
4. Separate the front speed sensor.
5. Separate the front disc brake caliper assembly.
6. Remove the front brake rotor.
7. Separate and remove the tie rod end.
8. Remove the bolt and 2 nuts, and separate the front suspension arm sub-assembly lower No.1 from the lower ball joint.
9. Using a plastic hammer, separate the drive shaft from the axle hub.
10. Remove the 2 bolts, nuts and steering knuckle.

WARNING

Be careful not to damage the boot and speed sensor rotor.

11. Using a screwdriver with its tip wrapped with vinyl tape, remove the bearing dust deflector No.1.
12. Using snap ring pliers, remove the front axle hub hole snap ring.
13. Hold the front axle assembly between aluminum plates in a vise.
14. Using SST 09520-00031, remove the front axle hub sub-assembly.
15. Using SST and a press, remove the bearing inner race (outside) from the front axle hub sub-assembly.
16. Remove the 4 bolts and disc brake dust cover from steering knuckle.

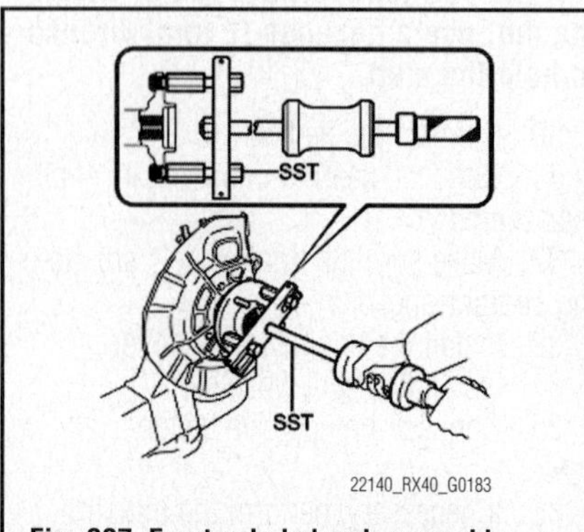

Fig. 207 Front axle hub sub-assembly removal

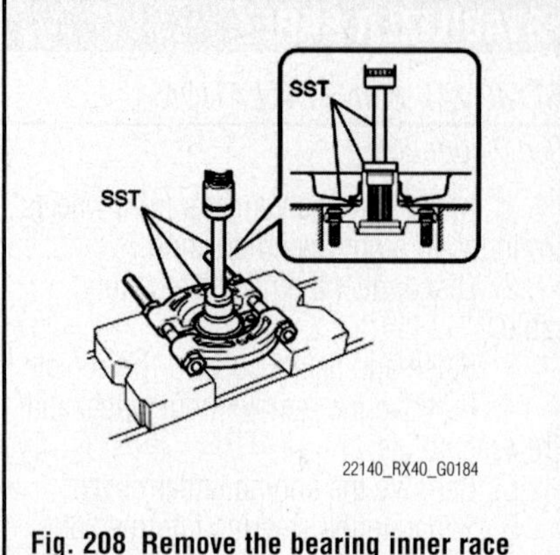

Fig. 208 Remove the bearing inner race

17. Remove the lower ball joint from the hub assembly.
18. Remove the front axle hub bearing as follows:
19. Place the bearing inner race (outside) on the front axle hub bearing.
20. Using SST and a press, press the front axle hub bearing until it contacts the SST.
21. Using SST to make the steering knuckle horizontal, fix it to the V-block, as shown in the illustration.
22. Using SST and a press, remove a front axle hub bearing to the steering knuckle.

To install:

23. Using SST and a press, install a new front axle hub bearing to the steering knuckle.
24. Install the lower ball joint into the steering knuckle and tighten castle nut to 91 ft. lbs. (93 Nm).
25. Install the disc brake dust cover to the steering knuckle with the 4 bolts. Tighten the bolts to 73 inch. lbs. (8.3 Nm).
26. Using SST and a press, install the front axle hub sub-assembly.
27. Using snap ring pliers, install a new front axle hub hole snap ring.

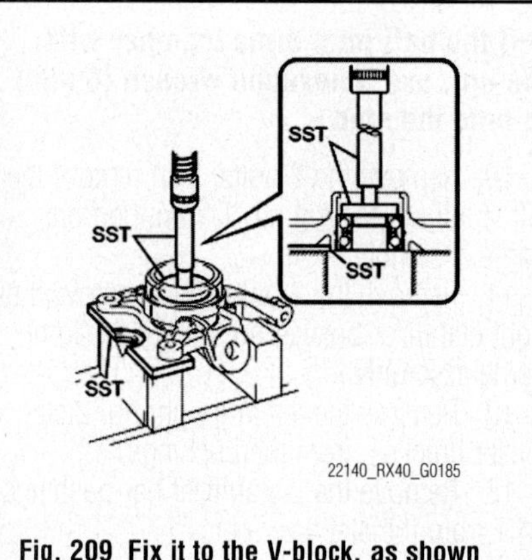

Fig. 209 Fix it to the V-block, as shown

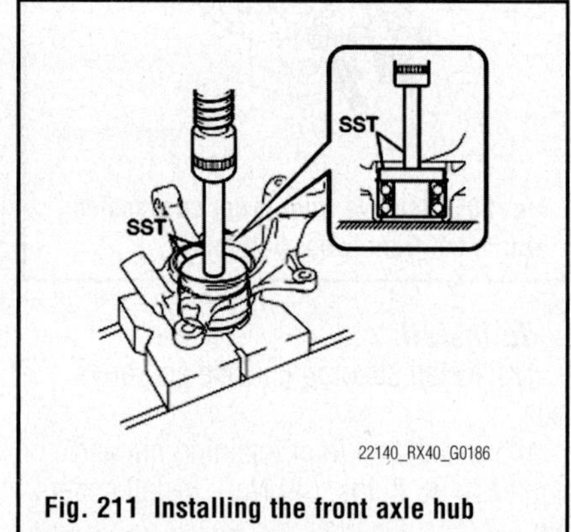

Fig. 211 Installing the front axle hub bearing to the steering knuckle

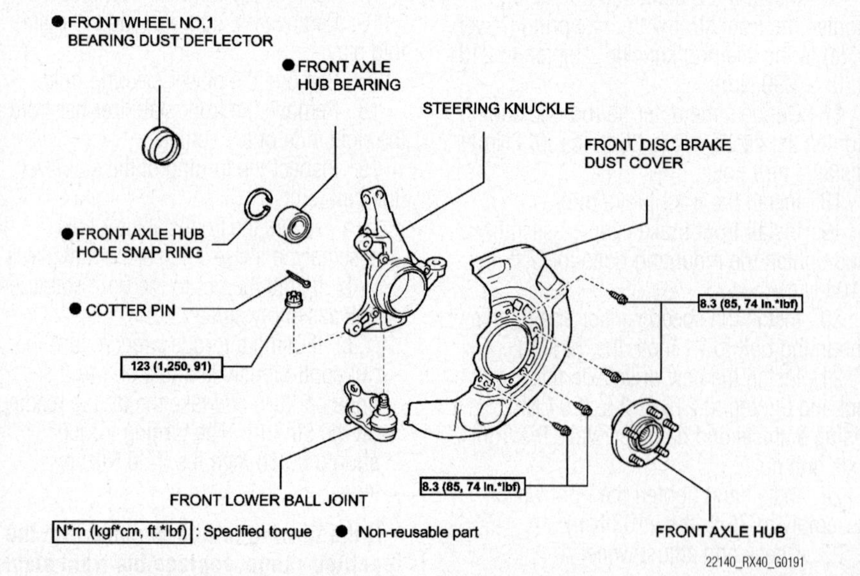

Fig. 210 Hub and bearing assembly view

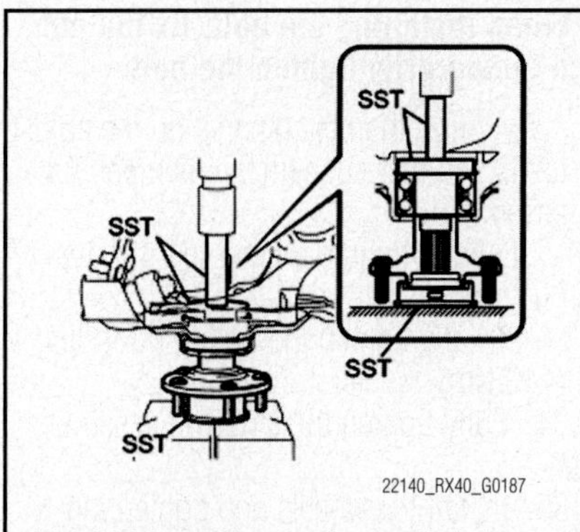

Fig. 212 Installing the front axle hub sub-assembly

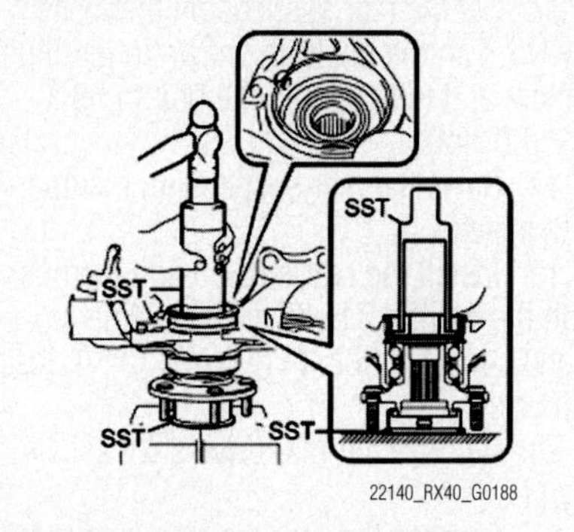

Fig. 213 Installing the bearing dust deflector No.1.

28. Using SST and a hammer, install the bearing dust deflector No.1.

➡Align the hole for the speed sensor in the bearing dust deflector No.1 with the steering knuckle.

29. Install the front axle assembly to the front drive shaft assembly.

30. Install the front axle assembly to the front shock absorber assembly with the 2 bolts and nuts. Tighten the bolts and nuts to 217 ft. lbs. (290 Nm).

31. Install the lower ball joint to the front suspension arm sub-assembly lower with the bolt and 2 nuts. Tighten the nuts and bolts to 94 ft. lbs. (127 Nm).

32. Install the tie rod end to the steering knuckle with the nut. Tighten the nut to 36 ft. lbs. (49 Nm).

33. Install the front brake rotor.

34. Install the front brake caliper assembly.

35. Install the front axle hub nut and tighten to 217 ft. lbs. (294 Nm). Using a chisel and hammer, stake the axle hub nut.

36. Install the speed sensor to the steering knuckle with the bolt and tighten to 71 inch. lbs. (8 Nm).

37. Install the front wheel and tighten to 76 ft. lbs. (103 Nm).

38. Adjust the front wheel alignment.

39. Check the speed sensor operation.

SUSPENSION

REAR SUSPENSION

COIL SPRING

REMOVAL & INSTALLATION

Refer to rear strut removal.

CONTROL ARMS/LINKS

REMOVAL & INSTALLATION

2006–07 Models

2WD Vehicles

See Figures 214 and 215.

➡Use the same procedures for the No. 1 or No. 2 rear suspension arm assembly, the following apply to No. 1.

1. Remove the rear wheel.
2. Remove the rear stabilizer bar.
3. Remove the LH No 1. rear suspension arm assembly, as follows:
   a. Support the rear axle carrier with a jack.
   b. Remove the bolt and disconnect the rear suspension arm assembly No.1 (inner side).
   c. Remove the bolt, nut and the rear suspension arm assembly No.1 (outer side) from the rear axle carrier.

➡When removing the bolt, keep the nut from rotating.

To install:

4. Install the rear suspension arm assembly No.1 (inner side) with the bolt, and temporarily tighten the bolt.

➡Ensure that the paint mark on the rear suspension arm assembly No.1 faces the rear side of the vehicle.

5. Connect the rear suspension arm assembly No.1 (outer side) to the rear axle carrier with the bolt and nut and temporarily tighten the bolt and nut.

➡When installing the bolt, fix the nut and temporarily tighten the bolt.

6. Stabilize the suspension.
7. Fully tighten rear suspension arm assembly No.1 LH, as follows:
   a. Fully tighten the bolt to 89 ft. lbs. (120 Nm).
   b. Fully tighten the bolt to 83 ft. lbs. (112 Nm).
8. Install the rear stabilizer bar.
9. Install the wheels and tighten lug nuts to 76 ft. lbs. (103 Nm).

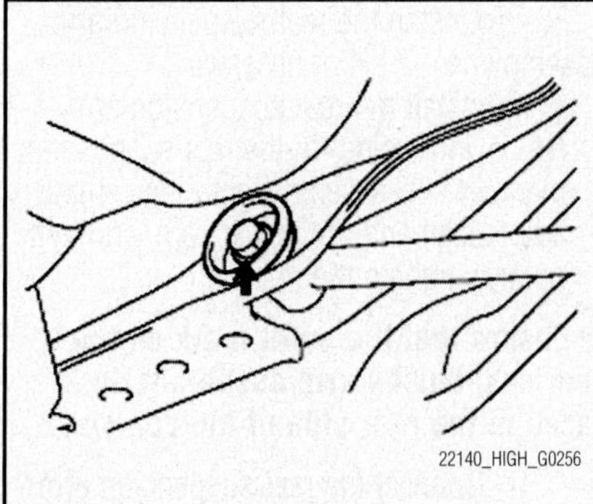

Fig. 214 Rear suspension arm assembly bolt—2006-07 models

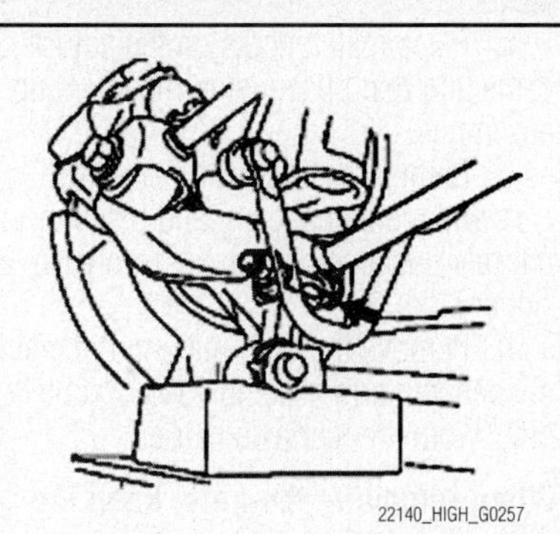

Fig. 215 Rear suspension arm assembly bolt—2006-07 models

10. Inspect and adjust rear wheel alignment.

4WD Vehicles

See Figure 216.

1. Remove the rear wheel.
2. Remove the exhaust pipe assembly.
3. Remove the propeller with center bearing shaft assembly.
4. Remove the rear strut rod assembly.
5. Remove the bolt, nut and the rear suspension arm assembly No.2 (outer side) from the rear axle carrier.
6. Remove the rear suspension arm assembly No. 2 RH.
7. Remove the bolt, nut and the rear suspension arm assembly No.1 (outer side) from the rear axle carrier.

➡When removing the bolt, keep the nut from rotating.

8. Remove rear suspension arm assembly No.1 RH
9. Remove the speed sensor rear LH.
10. Remove the speed sensor rear RH.
11. Remove the rear drive shaft assembly LH.
12. Remove the rear drive shaft assembly RH.
13. Remove the rear suspension member sub-assembly, as follows:
   a. Support the rear suspension member with a jack.
   b. Remove the 4 nuts, 2 bolts and 2 retainers from the rear suspension member.
   c. Lower the rear suspension member
14. Remove the rear suspension arm assembly No. 1 LH.

➡When removing the bolt, keep the nut from rotating.

15. To remove the rear suspension arm assembly No. 2 LH, if necessary:
   a. Support the rear axle carrier with a jack.
   b. Place matchmarks on the adjust cams and rear suspension member sub-assembly.
   c. Remove the nut, camber adjust cam and toe adjust cam, and disconnect the rear suspension arm assembly No. 2 (inner side).
   d. Remove the bolt, nut and the rear suspension arm assembly No. 2 (outer side) from the rear axle carrier.

➡When removing the nuts, keep the bolts from rotating.

To install:

16. Install the rear suspension arm assembly No. 1 (inner side) with the bolt, and temporarily tighten the bolt.

➡Ensure that the paint mark on the rear suspension arm assembly No.1 faces the rear side of the vehicle.

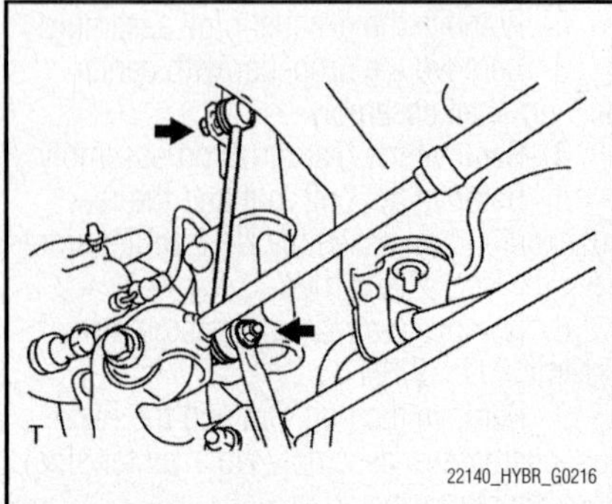

Fig. 216 Stabilizer links and related components

17. Set the rear suspension arm assembly No. 1 in the position shown in the illustration, and fully tighten the bolt to 59 ft. lbs. (80 Nm).
18. Raise the rear suspension member with a jack.
19. Install the rear suspension member with the 4 nuts, 2 bolts and 2 retainers. Tighten to 85 ft. lbs. (115 Nm), 134 ft. lbs. (181 Nm).
20. Remove the rear drive shaft assembly LH.
21. Remove the rear drive shaft assembly RH.
22. Install the rear suspension arm assembly No. 1 (outer side) to the rear axle carrier, and temporarily tighten the bolt.

➡When installing the bolt, fix the nut and temporarily tighten the bolt.

23. Temporarily tighten the rear suspension arm assembly No.1 RH.
24. Install the rear suspension arm No. 2 (outer side) to the rear axle carrier, and temporarily tighten the bolt.

➡When installing the bolt, fix the nut and temporarily tighten the bolt.

25. Temporarily tighten the rear suspension arm assembly No. 2 RH.
26. Temporarily tighten strut rod assembly rear.
27. Rear stabilizer.
28. Fully tighten rear suspension arm assembly No. 1 LH, as follows:
   a. Fully tighten the bolt to 83 ft. lbs. (112 Nm).
29. Fully tighten the rear suspension arm assembly No. 1 RH.
30. Fully tighten the rear suspension arm assembly No. 2 LH, as follows:
   a. Fully tighten the bolt to 83 ft. lbs. (112 Nm).
31. Fully tighten the rear suspension arm assembly No. 2 RH.
32. Stabilize suspension.
33. To install the rear suspension arm assembly No. 2 LH, if necessary:
   a. Install the rear suspension arm No. 2 (inner side) to the rear suspension member sub-assembly with the camber adjust cam and toe adjust cam, and temporarily tighten the nut.

➡Ensure that the paint mark on the rear suspension arm assembly No.2 faces to the rear side of the vehicle.

   b. Connect the rear suspension arm assembly No. 2 (outer side) to the rear axle carrier with the bolt and nut.

➡When installing the bolt, fix the nut and temporarily tighten the bolt.

   c. Align the matchmarks on the adjust cams and rear suspension member sub-assembly.
   d. Fully tighten the nut to 74 ft. lbs. (100 Nm).
   e. Fully tighten the bolt to 83 ft. lbs. (112 Nm).
34. Fully tighten the strut rod assembly rear.
35. Install the wheels and tighten lug nuts to 76 ft. lbs. (103 Nm).
36. Temporarily tighten the propeller w/center bearing shaft assembly.
37. Fully tighten the propeller w/center bearing shaft assembly.
38. Install the exhaust pipe assembly.
39. Check for exhaust gas leak.
40. Inspect and adjust rear wheel alignment.
41. Check the ABS speed sensor signal.

2008 Models

Rear Lower Arm

See Figure 217.

1. Remove the rear wheel.
2. Remove the rear exhaust.
3. Separate the No. 3 parking brake cable.
4. Remove the 2 bolts, the 2 nuts, and the rear strut rod assembly.
5. Put matchmarks on the adjust cams and the rear suspension member sub-assembly.
6. Remove the bolt and nut, and separate the rear No. 2 suspension arm assembly from the rear axle carrier sub-assembly.
7. Remove the nut, the No. 2 camber adjust cam, the rear suspension toe adjust cam sub-assembly, and the rear No. 2 suspension arm assembly.
8. Remove the 2 bolts, the 2 nuts, and the rear No. 1 suspension arm assembly.

To install:

9. Temporarily install the rear No. 1 suspension arm assembly with the 2 bolts and the 2 nuts.
10. Ensure that the identification mark faces the rear side of the vehicle.
11. Temporarily install the rear No. 2 suspension arm assembly to the rear suspension member sub-assembly with the rear suspension toe adjust cam sub-assembly, the No. 2 camber adjust cam and the nut.
12. Temporarily install the rear No. 2 suspension arm assembly to the rear axle carrier sub-assembly with the bolt and the nut.

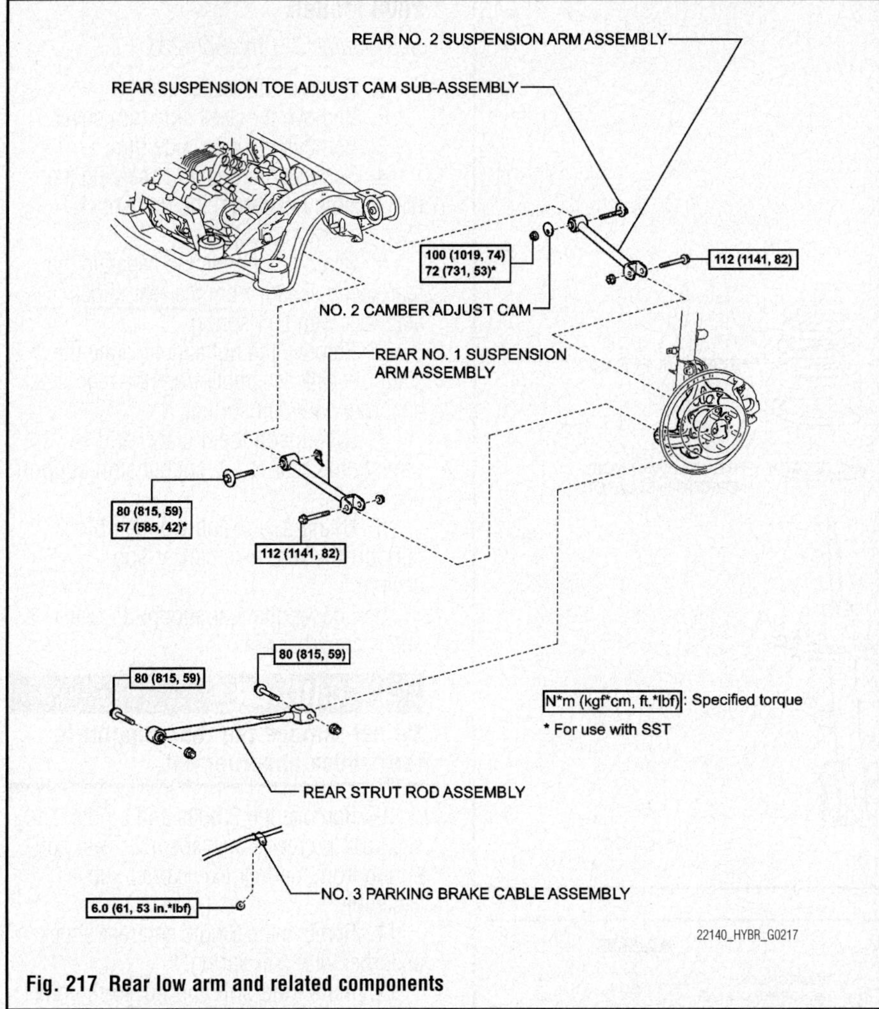

Fig. 217 Rear low arm and related components

13. Temporarily install the rear strut rod assembly with the 2 bolts and the 2 nuts.
14. Stabilize the suspension.
15. Fully tighten the rear strut rod assembly to 59 ft. lbs. (80 Nm).
16. Install the No.3 parking brake cable.
17. Install the rear wheel and tighten to 76 ft. lbs. (103 Nm).

Rear Strut Rod Assembly

See Figure 218.

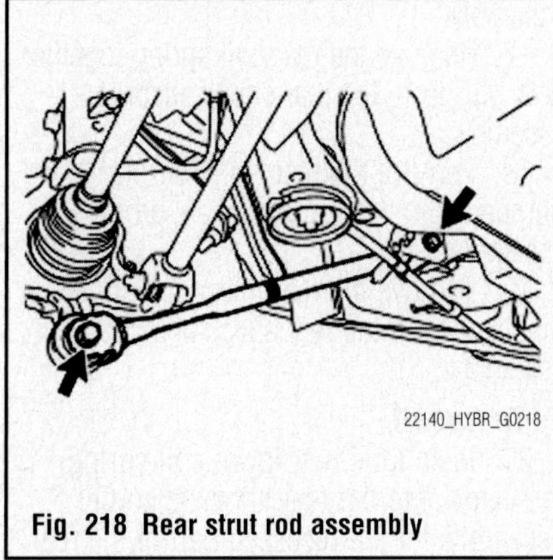

Fig. 218 Rear strut rod assembly

1. Remove the rear wheel.
2. Separate the No. 3 parking brake cable.
3. Remove the 2 bolts, the 2 nuts, and the rear strut rod assembly.

To install:

4. Check that the identification mark of the rear strut rod assembly is positioned on the inner side of the vehicle.
5. Temporarily install the rear strut rod assembly with the 2 bolts and the 2 nuts.
6. Stabilize the suspension.
7. Fully tighten the rear strut rod assembly to 59 ft. lbs. (80 Nm).
8. Install the No. 3 parking brake cable.
9. Install the rear wheel and tighten to 76 ft. lbs. (103 Nm).

MACPHERSON STRUTS

REMOVAL & INSTALLATION

2006–07 Models

See Figure 219.

1. Remove the deck side trim cover.
2. Remove the wheel.
3. Support the rear axle carrier with a jack.
4. Remove the nut and disconnect the stabilizer link from the shock absorber.

➡If the ball joint turns together with the nut, use a hexagon wrench (5 mm) to hold the stud.

5. Remove the rear shock absorber with coil spring, as follows:
   a. For 2WD vehicles, disconnect the skid control sensor connector. Remove the 2 bolts, and disconnect the flexible hose and skid control sensor wire from the shock absorber and rear axle carrier.
   b. For 4WD vehicles, remove the 3 bolts, and disconnect the flexible hose and speed sensor from the shock absorber and rear axle carrier.
   c. Loosen the 2 nuts on the lower side of the shock absorber.

➡Do not remove the 2 bolts and 2 nuts.

   d. Remove the 3 nuts.
   e. Lower the rear axle carrier, and remove the 2 nuts and 2 bolts on the lower side of the rear shock absorber.
   f. Remove the shock absorber with coil spring.

To install:

6. Install the rear shock absorber with coil spring and 3 nuts and tighten to 43 ft. lbs. (58 Nm).
7. Install the 2 bolts and nuts to the shock absorber with coil spring and tighten to 133 ft. lbs. (180 Nm).

➡When installing the nuts, keep the bolts from rotating.

8. For 2WD vehicles, install the flexible hose and skid control sensor wire with the 2 bolts. Connect the skid control sensor connector and tighten to 14 ft. lbs. (19 Nm), 44 inch lbs. (5 Nm).
9. For 4WD vehicles, install the flexible hose and speed sensor wire the 3 bolts and tighten to 44 inch lbs. (5 Nm), 14 ft. lbs. (19 Nm), 71 inch lbs. (8 Nm).
10. Install the stabilizer link to the shock absorber with the nut and tighten to 29 ft. lbs. (39 Nm).

➡If the ball joint turns together with the nut, use a hexagon wrench (5 mm) to hold the stud.

11. Install the wheels and tighten lug nuts to 76 ft. lbs. (103 Nm).
12. Inspect and adjust rear wheel alignment.
13. Check the ABS speed sensor signal.

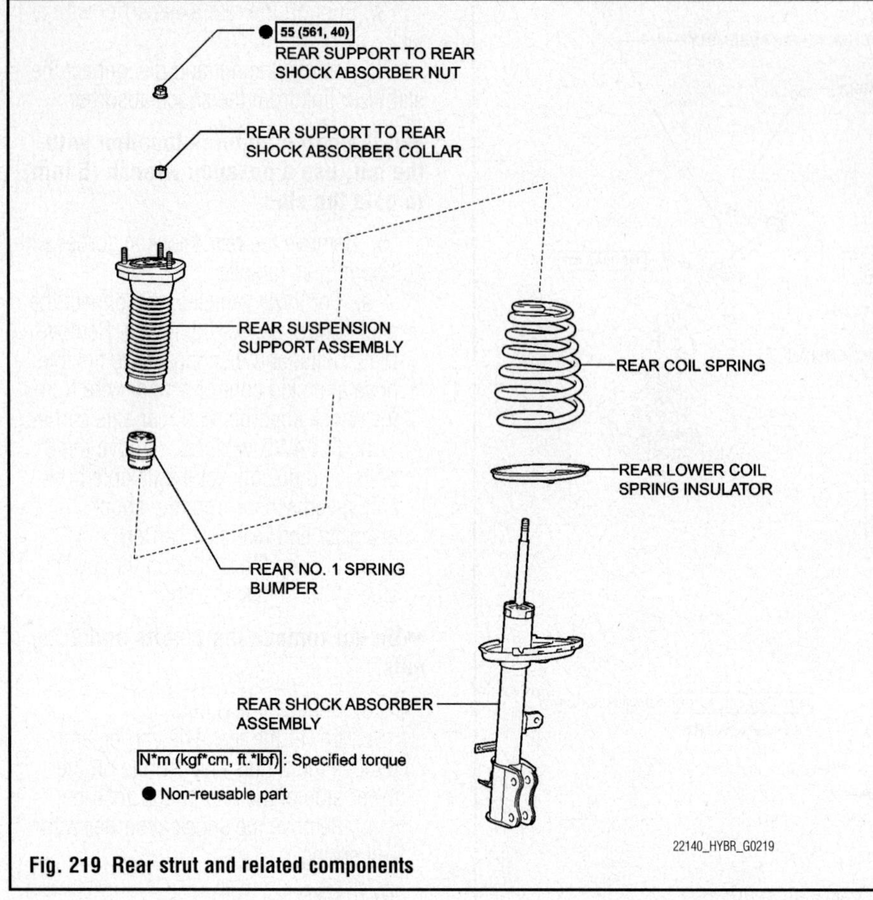

Fig. 219 Rear strut and related components

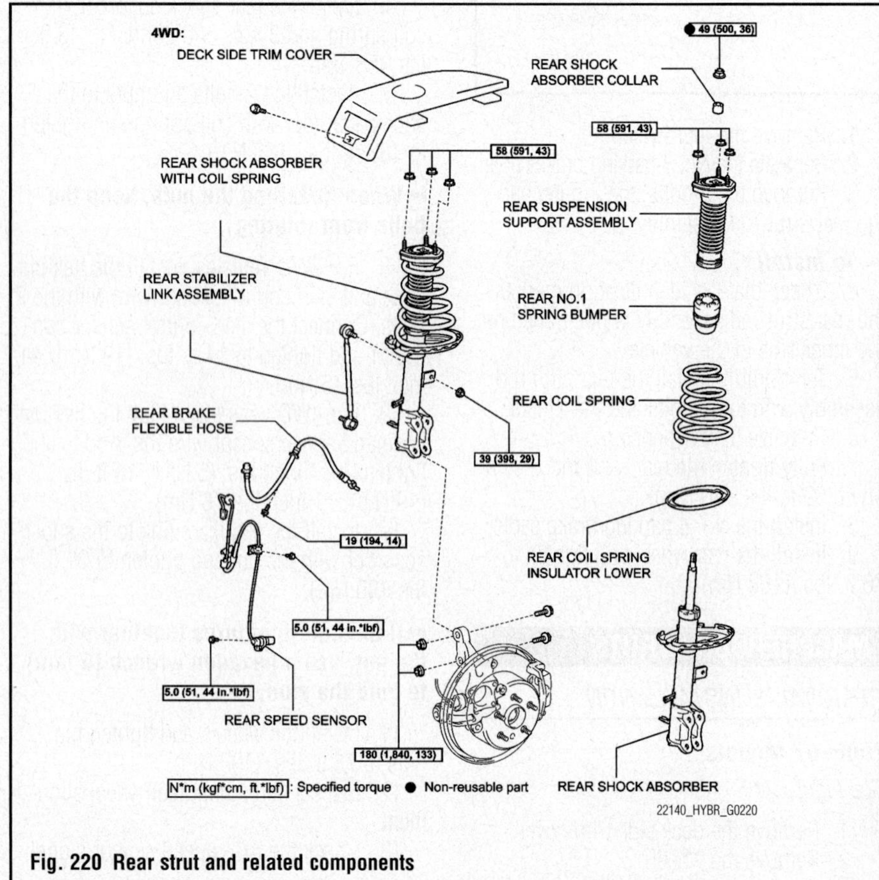

Fig. 220 Rear strut and related components

2008 Models

See Figures 221 through 226.

1. Remove the rear wheel.
2. Remove the deck side trim cover.
3. Remove the deck side trim.
4. Remove the bolt and separate the rear flexible hose from the rear shock absorber with coil spring.
5. Remove the bolt and separate the rear speed sensor from the rear shock absorber with coil spring.
6. Remove the nut and separate the rear stabilizer link assembly from the rear shock absorber with coil spring.
7. Disengage the 4 claws and remove the rear No. 1 suspension support cover.
8. Using a jack and wooden block, support the rear axle carrier sub-assembly.
9. Loosen the rear support to rear shock absorber nut.

CAUTION

Do not remove the rear support to rear shock absorber nut.

10. Remove the 2 bolts and 2 nuts, and separate the rear shock absorber with coil spring from the rear axle carrier sub-assembly.
11. Remove the 3 nuts and rear shock absorber with coil spring.
12. Install the bolt and nut to the rear shock absorber with coil spring as shown in the illustration and secure the rear shock absorber with coil spring in a vise.
13. Using a spring compressor, compress the rear coil spring.
14. Check that the front coil spring is fully compressed.
15. Remove the rear support to rear shock absorber collar from the rear shock absorber assembly.
16. Remove the rear suspension support assembly from the rear shock absorber assembly.
17. Remove the rear coil spring together with SST from the rear shock absorber assembly.
18. Remove the rear No. 1 spring bumper from the rear shock absorber assembly.
19. Remove the rear lower coil spring insulator from the rear shock absorber assembly.

To install:

20. Install the rear lower coil spring insulator onto the rear shock absorber assembly. Fit the recessed part of the rear

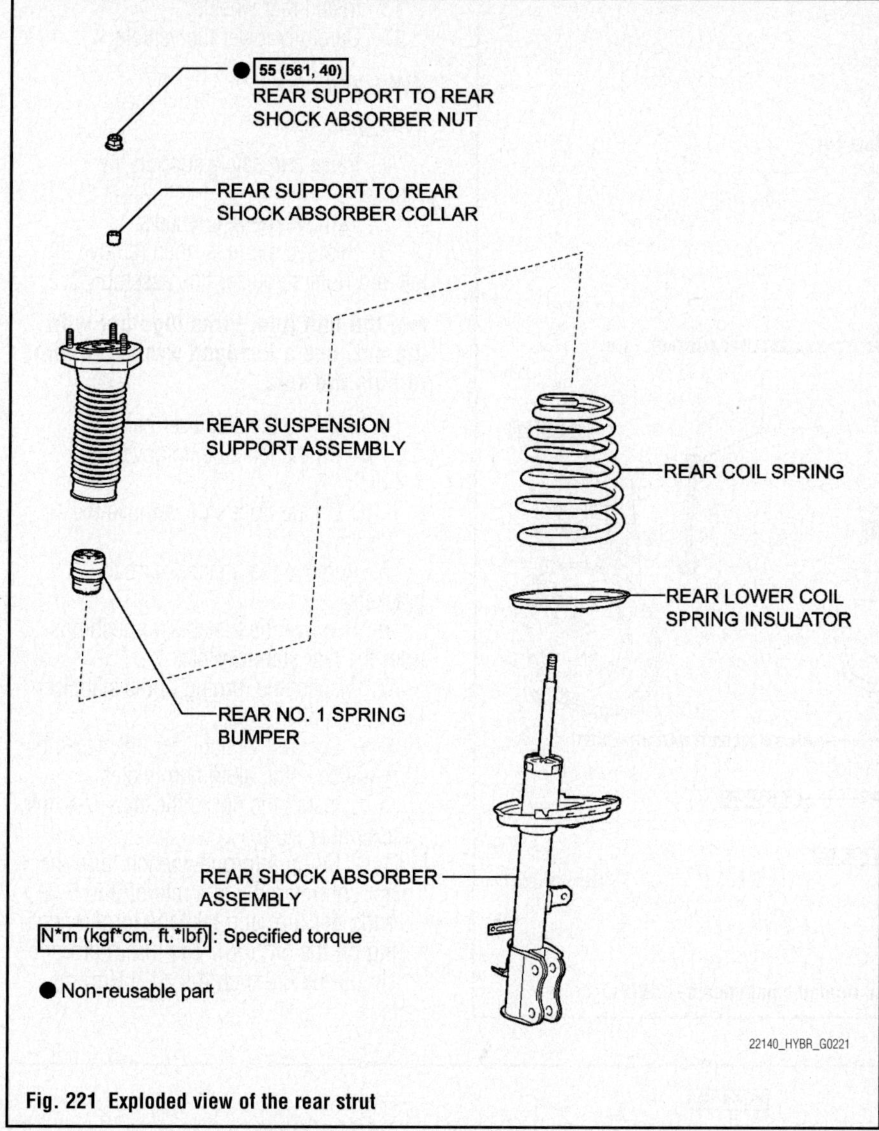

Fig. 221 Exploded view of the rear strut

lower coil spring insulator into the recession on the shock absorber assembly.

21. Install the rear No. 1 spring bumper to the rear shock absorber assembly.
22. Using spring, compress the rear coil spring.
23. Temporarily install the rear coil spring together with SST to the rear shock absorber assembly.
24. Install the rear suspension support assembly to the rear shock absorber assembly. Align the cutout on the rear shock absorber assembly with the protrusion on the rear suspension support assembly by referring to the illustration.
25. Install the rear support to rear shock absorber collar to the rear shock absorber assembly.
26. Temporarily install the rear support to rear shock absorber nut to the rear shock absorber assembly.
27. Install the rear coil spring. Ensure that the stud bolt is positioned 3.5° to the outside of the vehicle as shown in the illustration. The deviation should be within plus or minus 5°.
28. Install the rear shock absorber with coil spring with the 3 nuts. Tighten the upper mounting nuts to 43 ft. lbs. (58 Nm).
29. Install the rear shock absorber with coil spring with the 2 bolts and 2 nuts. Tighten to 213 ft. lbs. (290 Nm).
30. Fully tighten the rear support to rear shock absorber nut. Tighten the nut to 40 ft. lbs. (55 Nm).
31. Install the rear No. 1 suspension support cover.
32. Install the rear stabilizer link assembly to the rear shock absorber with coil spring with the nut.
33. Tighten the link assembly to 29 ft. lbs. (39 Nm).
34. Install the rear speed sensor to the rear shock absorber, and tighten the bolt to 54 inch. lbs. (5 Nm).
35. Install the rear speed sensor to the rear shock absorber and tighten the bolt 14 ft. lbs. (19 Nm).
36. Install the deck side trim and cover.
37. Install the wheels and tighten lug nuts to 76 ft. lbs. (103 Nm).
38. Inspect and adjust rear wheel alignment.
39. Check the ABS speed sensor signal.

STABILIZER BAR

REMOVAL & INSTALLATION

2006–07 Models

2WD Vehicles

See Figures 222 and 223.

1. Raise and safely support the vehicle.
2. Remove the rear wheels.
3. Remove the nuts, then remove the left and right stabilizer link assemblies.

➡If the ball joint turns together with the nut, use a hexagon wrench (5 mm) to hold the stud.

4. Remove the 4 bolts and the stabilizer bar.
5. Remove the 2 stabilizer bar brackets.
6. Remove the 2 stabilizer bushings from the rear stabilizer bar.
7. Inspect the turning of the stabilizer link ball joint:
   a. Secure the stabilizer link assembly in a vise using aluminum plates.
   b. Install the nut to the stabilizer link assembly stud.
   c. Using a torque wrench, turn the nut continuously at a rate of 3 to 5 seconds per turn and take the torque

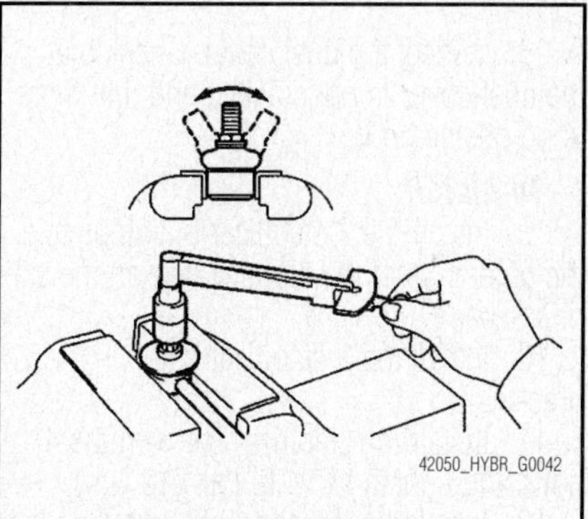

Fig. 222 Using a torque wrench, turn the nut continuously at a rate of 3 to 5 seconds per turn and take the torque reading on the 5th turn. The turning torque should be 8.9 inch lbs. (1.0 Nm) or less

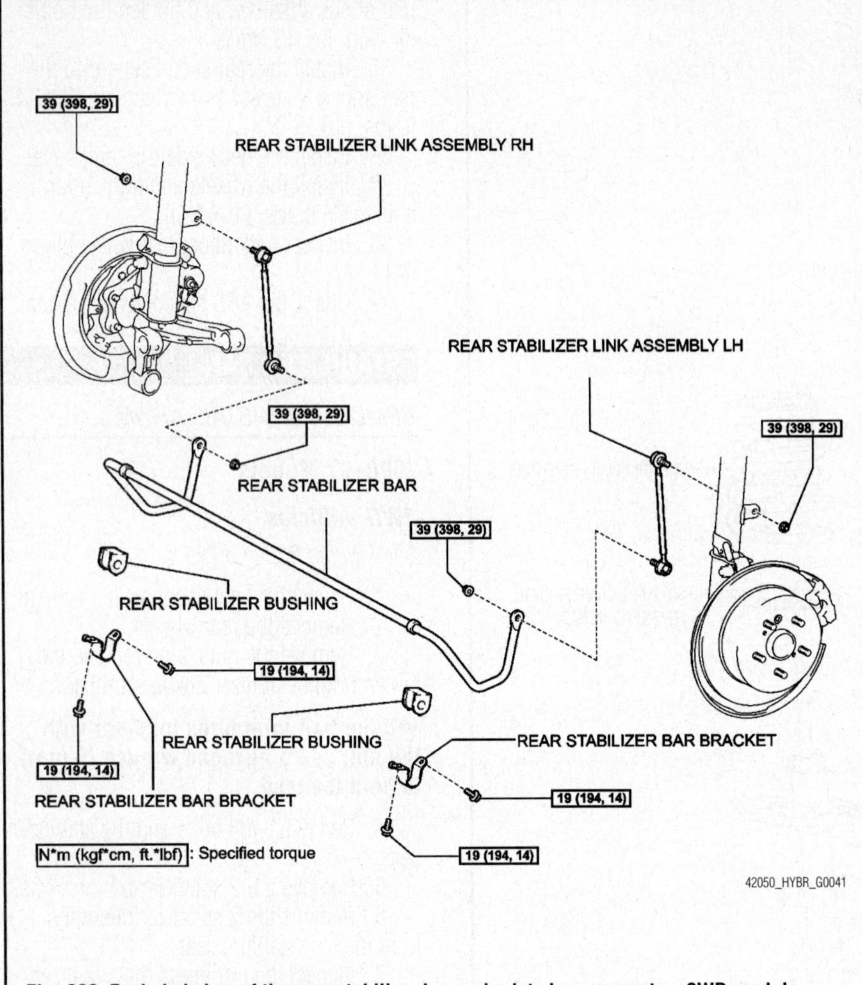

Fig. 223 Exploded view of the rear stabilizer bar and related components—2WD models

reading on the 5th turn. The turning torque should be 8.9 inch lbs. (1.0 Nm) or less.

➡If the turning torque is not within the specified range, replace the rear stabilizer link assembly with a new one.

8. Inspect the dust cover. Check that the dust cover is not cracked and that there is no grease on it.

To install:

9. Install the 2 stabilizer bushings to the outer side of the stopper ring on the stabilizer bar.
10. Install the 2 stabilizer bar brackets.
11. Install the stabilizer bar with the 4 bolts and tighten to 14 ft. lbs. (19 Nm).
12. Install the left and right stabilizer link assemblies. Tighten the nuts to 29 ft. lbs. (39 Nm).

➡If the ball joint turns together with the nut, use a hexagon wrench (5 mm) to hold the stud.

13. Install rear wheels.
14. Carefully lower the vehicle.

4WD Vehicles

See Figure 225.

1. Raise and safely support the vehicle.
2. Remove the rear wheels.
3. Remove the nuts, then remove the left and right stabilizer link assemblies.

➡If the ball joint turns together with the nut, use a hexagon wrench (5 mm) to hold the stud.

4. Remove the stabilizer bar:
   a. On the left side, remove the bolt.
   b. On the right side, remove the 2 bolts.
5. Remove the 2 stabilizer bar brackets.
6. Remove the 2 stabilizer bushings from the rear stabilizer bar.
7. Inspect the turning of the stabilizer link ball joint:
   a. Secure the stabilizer link assembly in a vise using aluminum plates.
   b. Install the nut to the stabilizer link assembly stud.
   c. Using a torque wrench, turn the nut continuously at a rate of 3 to 5 seconds per turn and take the torque reading on the 5th turn. The turning torque should be 8.9 inch lbs. (1.0 Nm) or less.

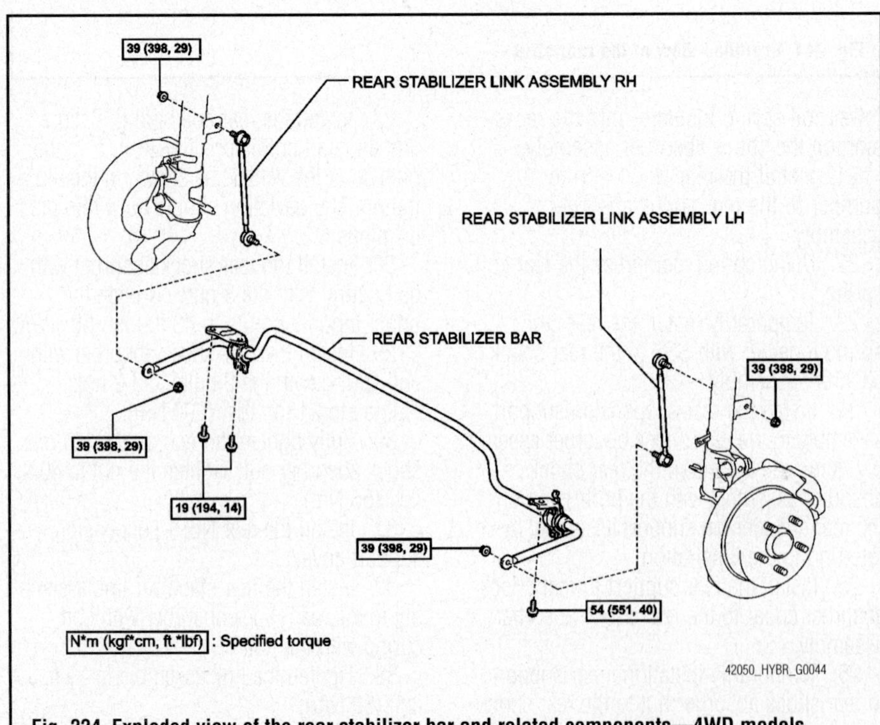

Fig. 224 Exploded view of the rear stabilizer bar and related components—4WD models

To install:

8. Install the stabilizer bar:
   a. Install the stabilizer bracket.
   b. On the left side, install the stabilizer bar with the bolt and tighten to 40 ft. lbs. (54 Nm).
   c. On the right side, install the stabilizer bar with the 2 bolts and tighten to 14 ft. lbs. (19 Nm).
9. Install the right and left stabilizer links and tighten the nuts to 29 ft. lbs. (39 Nm).

➡If the ball joint turns together with the nut, use a hexagon wrench (5 mm) to hold the stud.

10. Install the rear wheels
11. Carefully lower the vehicle.

2008 Models

See Figures 225 and 226.

1. Remove the deck board assembly.
2. Remove the No. 3 and No. 2 deck board assembly. (w/ tonneau cover)
3. Remove the rear mat.

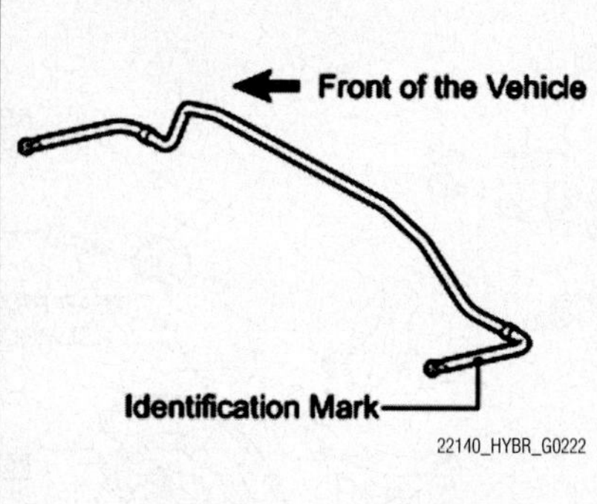
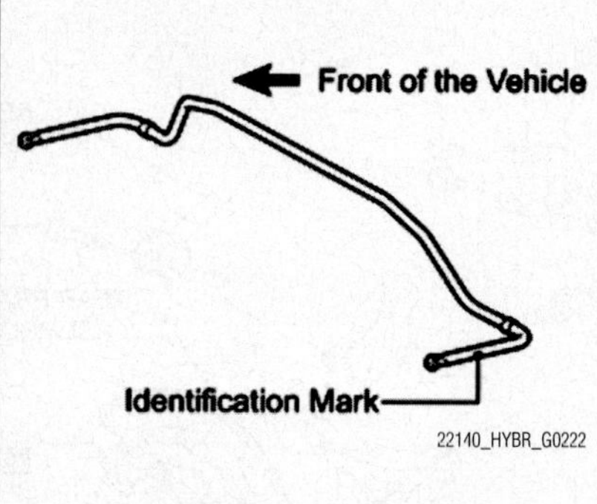

Fig. 226 Stabilizer bar and identification mark

4. Remove the deck trim service hole cover.
5. Remove the spare tire.
6. Remove the rear wheels.
7. Remove the rear exhaust.
8. Remove the LH and RH rear stabilizer link assembly.
9. Remove both rear stabilizer bar brackets.
10. Remove stabilizer bar.

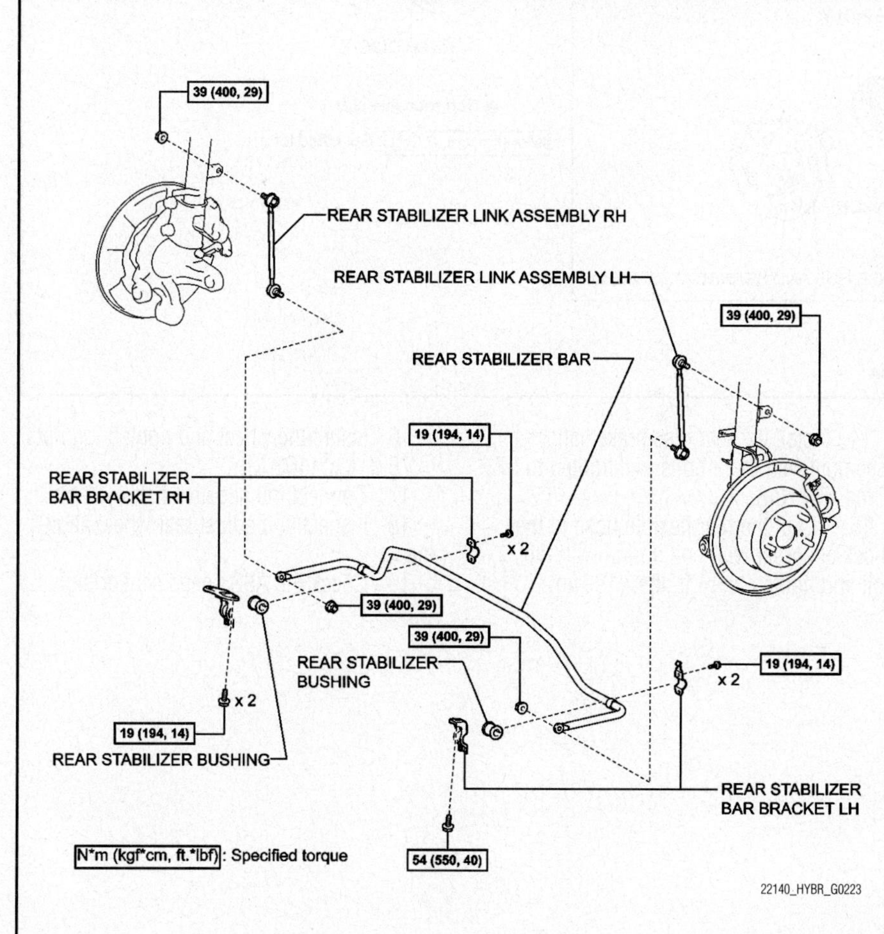

Fig. 225 Stabilizer bar and related components

To install:

11. Temporarily install the rear stabilizer bar with the identification mark positioned on the left side of the vehicle.
12. Temporarily install the rear stabilizer bar bracket LH (front side) with the bolt. Loosely tighten the bolt so that the bracket can be moved by hand.
13. Temporarily install the rear stabilizer bar bracket RH (front side) with the 2 bolts. Loosely tighten the bolt so that the bracket can be moved by hand.
14. Install the rear stabilizer bushing (RH side) to the rear stabilizer bar.
15. Install the rear stabilizer bar bracket RH (rear side) with the 2 bolts.
16. Install the rear stabilizer bushing (LH side) to the rear stabilizer bar.
17. Install the rear stabilizer bar bracket LH (rear side) with the 2 bolts.
18. Tighten the stabilizer bracket bolts to 14 ft. lbs. (19 Nm).
19. Install the rear stabilizer link assembly LH and RH to the rear shock absorber. Tighten to 29 ft. lbs. (39 Nm).
20. Install the rear exhaust assembly.
21. Inspect for exhaust leaks.
22. Install and tighten the rear wheels to 76 ft. lbs. (103 Nm).
23. Install the rear spare.
24. Install the rear mat.
25. Install the tonneau cover if equipped.
26. Install No.3 and No.2 deck board assembly. (if equipped with a tonneau cover)
27. Install the deck board assembly.

WHEEL HUB AND BEARING

REMOVAL & INSTALLATION

See Figure 227.

1. Disconnect the negative battery cable.
2. Remove the wheel.
3. Separate the rear flexible hose.
4. Remove the 2 bolts and separate the rear disc brake caliper assembly.
5. Remove the rear disc.
6. Using a screwdriver, disconnect the connector from the rear speed sensor.
7. For 4WD vehicles,
8. Remove the 4 bolts and the rear axle hub & bearing assembly.

To install:

9. Install the rear axle hub and bearing assembly with the 4 bolts and tighten to 55 ft. lbs. (75 Nm).

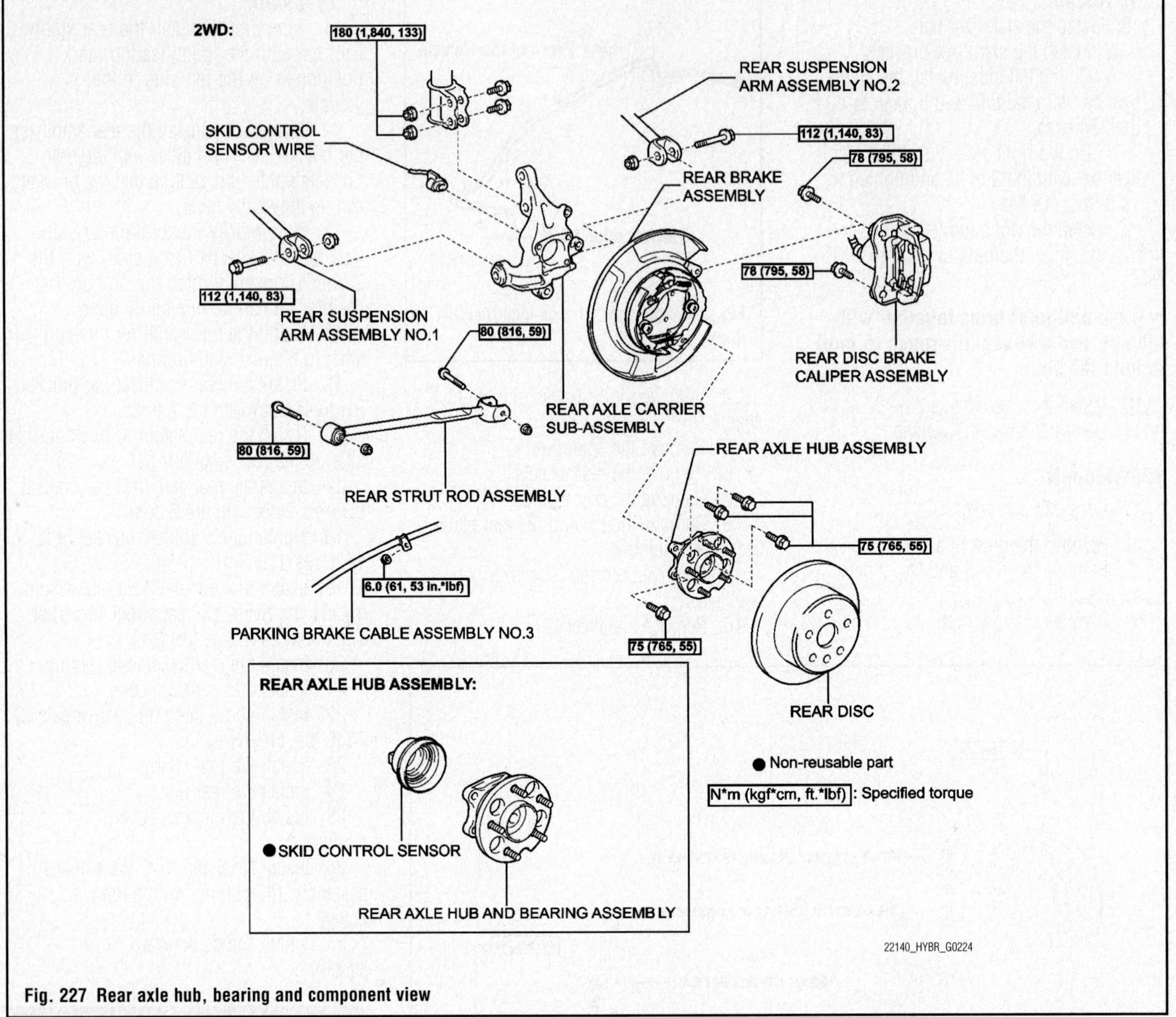

Fig. 227 Rear axle hub, bearing and component view

10. Inspect rear axle hub bearing looseness.
11. Inspect rear axle hub runout.
12. Connect the connector to the rear speed sensor.
13. Install the rear disc.
14. Install the rear disc brake caliper assembly with the 2 bolts and tighten to 57 ft. lbs. (78 Nm).
15. Install the rear flexible hose to the shock absorber with coil spring with the bolt and tighten to 14 ft. lbs. (19 Nm).
16. Install the wheel and tighten lug nuts to 76 ft. lbs. (103 Nm).
17. Connect the negative battery cable.
18. Inspect and adjust rear wheel alignment.
19. Check the ABS speed sensor signal.

LEXUS AND TOYOTA

Land Cruiser • LX 470

17

SPECIFICATIONS AND MAINTENANCE CHARTS

ENGINE AND VEHICLE IDENTIFICATION

Engine						
Code ①	Liters (cc)	Cu. In.	Cyl.	Fuel Sys.	Engine Type	Eng. Mfg.
2UZ-FE	4.7 (4664)	285	8	SFI	DOHC	Toyota

Model Year	
Code ②	Year
7	2007
8	2008

SFI: Sequential Fuel Injection

DOHC: Double Overhead Camshaft

① Stamped on the left side of the engine block

22140_LAND_C0001

GENERAL ENGINE SPECIFICATIONS

Year	Model	Engine Displacement Liters	Engine Series ID	Net Horsepower @ rpm	Net Torque @ rpm (ft. lbs.)	Bore x Stroke (in.)	Com-pression Ratio	Oil Pressure @ rpm
2006	LX470	4.7	2UZ-FE	245@4800	315@3400	3.70x3.31	9.6:1	43-85@3000
	Land Cruiser	4.7	2UZ-FE	245@4800	320@3400	3.70x3.31	9.6:1	43-85@3000
2007	LX470	4.7	2UZ-FE	245@4800	315@3400	3.70x3.31	9.6:1	43-85@3000
	Land Cruiser	4.7	2UZ-FE	245@4800	315@3400	3.70x3.31	9.6:1	43-85@3000

22140_LAND_C0002

ENGINE TUNE-UP SPECIFICATIONS

Year	Engine Displacement Liters	Engine ID	Spark Plug Gap (in.)	Ignition Timing (deg.)*	Fuel Pump (psi)	Idle Speed (rpm)		Valve Clearance	
						MT	AT	Intake	Exhaust
2006	4.7	2UZ-FE	0.043	N/A	38-44	—	650-750	0.006-0.010	0.010-0.014
2007	4.7	2UZ-FE	0.043	N/A	38-44	—	650-750	0.006-0.010	0.010-0.014

NOTE: The Vehicle Emission Control Information label often reflects specification changes made during production.
The label figures must be used if they differ from those in this chart.

B: Before top dead center

* With terminals TC and E1 connected to DLC1 or for 5.7L Terminal TC and CG of DLC3 connected

22140_LAND_C0003

CAPACITIES

Year	Model	Engine Displacement Liters	Engine ID	Engine Oil with Filter (qts.)	Transmission (pts.) 5-Spd	Transmission (pts.) Auto.*	Transfer Case (pts.)	Drive Axle Front (pts.)	Drive Axle Rear (pts.)	Fuel Tank (gal.)	Cooling System (qts.)
2006	LX470	4.7	2UZ-FE	7.2	—	6.4	2.8	3.4	7.0	25.4	①
	Land Cruiser	4.7	2UZ-FE	6.0	—	6.4	2.8	3.4	7.0	25.4	13.6
2007	LX470	4.7	2UZ-FE	7.2	—	6.4	2.8	3.4	7.0	25.4	①
	Land Cruiser	4.7	2UZ-FE	6.0	—	6.4	2.8	3.4	7.0	25.4	13.6

*After draining, add the following amounts, then fill to the cold full line

① Without rear heater: 15.6

With rear heater: 16.2

22140_LAND_C0004

VALVE SPECIFICATIONS

Year	Engine Displacement Liters	Engine ID	Seat Angle (deg.)	Face Angle (deg.)	Spring Test Pressure (lbs. @ in.)	Spring Installed Height (in.)	Stem-to-Guide Clearance (in.) Intake	Stem-to-Guide Clearance (in.) Exhaust	Stem Diameter (in.) Intake	Stem Diameter (in.) Exhaust
2006	4.7	2UZ-FE	45	44.5	45.9-50.7@ 1.378	1.380	0.0010-0.0024	0.0012-0.0026	0.2154-0.2159	0.2152-0.2157
2007	4.7	2UZ-FE	45	44.5	45.9-50.7@ 1.378	1.380	0.0010-0.0024	0.0012-0.0026	0.2154-0.2159	0.2152-0.2157

22140_LAND_C0005

CRANKSHAFT AND CONNECTING ROD SPECIFICATIONS

All measurements are given in inches.

Year	Engine Displacement Liters	Engine ID	Crankshaft Main Brg. Journal Dia.	Crankshaft Main Brg. Clearance	Crankshaft Shaft End-play	Crankshaft Thrust on No.	Connecting Rod Journal Diameter	Connecting Rod Oil Clearance	Connecting Rod Side Clearance
2006	4.7	2UZ-FE	2.6373-2.6378	①	0.0008-0.0087	3	2.0465-2.0472	0.0011-0.0021	0.0063-0.0138
2007	4.7	2UZ-FE	2.6373-2.6378	①	0.0008-0.0087	3	2.0465-2.0472	0.0011-0.0021	0.0063-0.0138

① Nos. 1 and 2: 0.0011-0.0018

All others: 0.0016-0.0023

22140_LAND_C0006

PISTON AND RING SPECIFICATIONS

All measurements are given in inches.

Year	Engine Displacement Liters	Engine ID	Piston Clearance	Ring Gap Top Comp.	Ring Gap Bottom Comp.	Ring Gap Oil Control	Ring Side Clearance Top Comp.	Ring Side Clearance Bottom Comp.	Ring Side Clearance Oil Control
2006	4.7	2UZ-FE	0.0035-0.0044	0.0118-0.0157	0.0157-0.0217	0.0051-0.0150	0.0012-0.0031	0.0012-0.0028	SNUG
2007	4.7	2UZ-FE	0.0035-0.0044	0.0118-0.0157	0.0157-0.0217	0.0051-0.0150	0.0012-0.0031	0.0012-0.0028	SNUG

22140_LAND_C0007

TORQUE SPECIFICATIONS

All readings in ft. lbs.

Year	Engine Displacement Liters	Engine ID	Cylinder Head Bolts	Main Bearing Bolts	Rod Bearing Bolts	Crankshaft Damper Bolts	Flywheel Bolts	Manifold Intake	Manifold Exhaust	Spark Plugs	Oil Pan Drain Plug
2006	4.7	2UZ-FE	①	②	③	181	④	13	33	13	29
2007	4.7	2UZ-FE	①	②	③	181	④	13	33	13	29

① Step 1: 30
Step 2: Plus 90 degrees
Step 3: Plus 90 degrees

② Step 1: 20 ft. lbs.
Step 2: Plus 90 degrees

③ Step 1: 18 ft. lbs.
Step 2: Plus 90 degrees

④ Step 1: 35 ft. lbs.
Step 2: Plus 90 degrees

22140_LAND_C0008

WHEEL ALIGNMENT

Year	Model	Caster Range (+/-Deg.)	Caster Preferred Setting (Deg.)	Camber Range (+/-Deg.)	Camber Preferred Setting (Deg.)	Toe-in (in.)	Steering Axis Inclination (Deg.)
2006	LX470	0.75	+3.08	0.75	0	0+/-0.08	12.25+/-0.75
	Land Cruiser	0.75	+2.50	0.75	-0.08	0.06+/-0.12	12.17+/-0.75
2007	LX470	0.75	+3.08	0.75	0	0+/-0.08	12.25+/-0.75
	Land Cruiser	0.75	+2.50	0.75	-0.08	0.06+/-0.12	12.17+/-0.75

Note: All alignment specifications are based on nominal ride height and standard tires

22140_LAND_C0009

TIRE, WHEEL AND BALL JOINT SPECIFICATIONS

Year	Model	OEM Tires Standard	OEM Tires Optional	Tire Pressures (psi) Front	Tire Pressures (psi) Rear	Wheel Size	Ball Joint Inspection	Lug Nut Torque (ft. lbs.)
2006	LX470	P275/60HR18	None	①	①	8-JJ	②	97
	Land Cruiser	P275/60R18	None	①	①	8-JJ	②	97
2007	LX470	P275/60HR18	None	①	①	8-JJ	②	97
	Land Cruiser	P275/60R18	None	①	①	8-JJ	②	97

OEM: Original Equipment Manufacturer

PSI: Pounds Per Square Inch

STD: Standard

OPT: Optional

NA: Not Available

① See placard on vehicle

② Upper arm ball joint turning torque: 9-39 inch lbs.
Lower arm ball joint turning torque: 2.6-26 inch lbs.

22140_LAND_C0010

BRAKE SPECIFICATIONS

All measurements in inches unless noted

Year	Model		Brake Disc Original Thickness	Brake Disc Minimum Thickness	Brake Disc Maximum Runout	Minimum Lining Thickness	Brake Caliper Bracket Bolts (ft. lbs.)	Brake Caliper Mounting Bolts (ft. lbs.)
2006	LX470	F	1.260	1.181	0.0028	0.039	—	90
		R	0.709	0.611	0.0040	0.039	76	20
	Land Cruiser	F	1.260	1.181	0.0028	0.039	—	90
		R	0.709	0.630	0.0040	0.039	76	20
2007	LX470	F	1.260	1.181	0.0028	0.039	—	90
		R	0.709	0.611	0.0040	0.039	76	20
	Land Cruiser	F	1.260	1.181	0.0028	0.039	—	90
		R	0.709	0.630	0.0040	0.039	76	20

F: Front

R: Rear

22140_LAND_C0011

SCHEDULED MAINTENANCE INTERVALS

Lexus/Toyota - Land Cruiser/LX470

TO BE SERVICED	TYPE OF SERVICE	VEHICLE MILEAGE INTERVAL (x1000)												
		7.5	15	22.5	30	37.5	45	52.5	60	67.5	75	82.5	90	97.5
Engine oil & filter	R	✓	✓	✓	✓	✓	✓	✓	✓	✓	✓	✓	✓	✓
Automatic transmission fluid & filter	S/I		✓		✓		✓		✓		✓		✓	
Ball joints & dust covers	S/I		✓		✓		✓		✓		✓		✓	
Bolts & nuts on chassis & body	S/I		✓		✓		✓		✓		✓		✓	
Brake line pipes & hoses	S/I		✓		✓		✓		✓		✓		✓	
Brake pads & discs	S/I		✓		✓		✓		✓		✓		✓	
Propeller shaft grease	S/I		✓		✓		✓		✓		✓		✓	
Steering knuckle & chassis grease	S/I		✓		✓		✓		✓		✓		✓	
Steering linkage	S/I		✓		✓		✓		✓		✓		✓	
Transfer and differential oil	S/I		✓		✓		✓		✓		✓		✓	
Air cleaner filter	R				✓				✓				✓	
Spark plugs ①	R				✓				✓				✓	
Drive belts	S/I				✓				✓				✓	
Exhaust pipes & mountings	S/I				✓				✓				✓	
Fuel lines & connections	S/I				✓				✓				✓	
Engine coolant	R						✓				✓			
Charcoal canister	R								✓					
Fuel tank cap gasket	R								✓					
Heated oxygen sensors (exc. Cal.) ②	R													

R: Replace S/I: Service or Inspect

① Platinum plugs, replace every 100,000 miles

② Heated oxygen sensors (except Calif.): replace every 80,000 miles.

FREQUENT OPERATION MAINTENANCE (SEVERE SERVICE)

If a vehicle is operated under any of the following conditions it is considered severe service:

- Extremely dusty areas.

- 50% or more of the vehicle operation is in 32°C (90°F) or higher temperatures, or constant operation in temperatures below 0°C (32°F).

- Prolonged idling (vehicle operation in stop and go traffic).

- Frequent short running periods (engine does not warm to normal operating temperatures).

- Police, taxi, delivery usage or trailer towing usage.

Air cleaner filter: service or inspect every 3750 miles

Engine oil & filter: replace every 3750 miles.

Ball joints & dust covers: service or inspect every 7500 miles.

Bolts & nuts on chassis & body: service or inspect every 7500 miles.

Brake pads & discs (front & rear): service or inspect every 7500 miles.

Steering knuckle & chassis grease: service or inspect every 7500 miles.

Steering linkage: service or inspect every 7500 miles.

Propeller shaft grease: service or inspect every 7500 miles.

Exhaust pipes & mountings: service or inspect every 15,000 miles.

22140_LAND_C0012

PRECAUTIONS

Before servicing any vehicle, please be sure to read all of the following precautions, which deal with personal safety, prevention of component damage, and important points to take into consideration when servicing a motor vehicle:

- Never open, service or drain the radiator or cooling system when the engine is hot; serious burns can occur from the steam and hot coolant.
- Observe all applicable safety precautions when working around fuel. Whenever servicing the fuel system, always work in a well-ventilated area. Do not allow fuel spray or vapors to come in contact with a spark, open flame, or excessive heat (a hot drop light, for example). Keep a dry chemical fire extinguisher near the work area. Always keep fuel in a container specifically designed for fuel storage; also, always properly seal fuel containers to avoid the possibility of fire or explosion. Refer to the additional fuel system precautions later in this section.
- Fuel injection systems often remain pressurized, even after the engine has been turned **OFF**. The fuel system pressure must be relieved before disconnecting any fuel lines. Failure to do so may result in fire and/or personal injury.
- Brake fluid often contains polyglycol ethers and polyglycols. Avoid contact with the eyes and wash your hands thoroughly after handling brake fluid. If you do get brake fluid in your eyes, flush your eyes with clean, running water for 15 minutes. If eye irritation persists, or if you have taken brake fluid internally, IMMEDIATELY seek medical assistance.
- The EPA warns that prolonged contact with used engine oil may cause a number of skin disorders, including cancer. You should make every effort to minimize your exposure to used engine oil. Protective gloves should be worn when changing oil. Wash your hands and any other exposed skin areas as soon as possible after exposure to used engine oil. Soap and water, or waterless hand cleaner should be used.
- All new vehicles are now equipped with an air bag system, often referred to as a Supplemental Restraint System (SRS) or Supplemental Inflatable Restraint (SIR) system. The system must be disabled before performing service on or around system components, steering column, instrument panel components, wiring and sensors. Failure to follow safety and disabling procedures could result in accidental air bag deployment, possible personal injury and unnecessary system repairs.
- Always wear safety goggles when working with, or around, the air bag system. When carrying a non-deployed air bag, be sure the bag and trim cover are pointed away from your body. When placing a non-deployed air bag on a work surface, always face the bag and trim cover upward, away from the surface. This will reduce the motion of the module if it is accidentally deployed. Refer to the additional air bag system precautions later in this section.
- Clean, high quality brake fluid from a sealed container is essential to the safe and proper operation of the brake system. You should always buy the correct type of brake fluid for your vehicle. If the brake fluid becomes contaminated, completely flush the system with new fluid. Never reuse any brake fluid. Any brake fluid that is removed from the system should be discarded. Also, do not allow any brake fluid to come in contact with a painted surface; it will damage the paint.
- Never operate the engine without the proper amount and type of engine oil; doing so WILL result in severe engine damage.
- Timing belt maintenance is extremely important. Many models utilize an interference-type, non-freewheeling engine. If the timing belt breaks, the valves in the cylinder head may strike the pistons, causing potentially serious (also time-consuming and expensive) engine damage. Refer to the maintenance interval charts for the recommended replacement interval for the timing belt, and to the timing belt section for belt replacement and inspection.
- Disconnecting the negative battery cable on some vehicles may interfere with the functions of the on-board computer system(s) and may require the computer to undergo a relearning process once the negative battery cable is reconnected.
- When servicing drum brakes, only disassemble and assemble one side at a time, leaving the remaining side intact for reference.
- Only an MVAC-trained, EPA-certified automotive technician should service the air conditioning system or its components.

BRAKES — ANTI-LOCK BRAKE SYSTEM (ABS)

GENERAL INFORMATION

The ABS helps prevent the wheels from locking when the brakes are applied firmly or when braking on a slippery surface. The skid control ECU detects wheel lock condition by receiving vehicle speed signals from each speed sensor, and sends control signals to the pump motor and solenoid valve. The pump motor and solenoid valve avoid wheel lock by controlling the oil pressure of each wheel cylinder. The EBD control utilizes ABS, realizing proper brake force distribution between the front and rear wheels in accordance with driving conditions. In addition, when braking while cornering, it also controls the brake forces of the right and left wheels, helping to maintain vehicle behavior.

BRAKES — BLEEDING THE BRAKE SYSTEM

BLEEDING PROCEDURE

➡If any work is done on the brake system or if air is suspected in the brake lines, bleed the air from the system. Bleeding Hydraulic Brake Booster is only possible with a Toyota proprietary scan system.

➡ When bleeding, keep the amount of the fluid within the line of reservoir between Min. and Max. Do not let brake fluid remain on a painted surface. Wash it off immediately.

1. Before servicing the vehicle, refer to the precautions in the beginning of this section.
2. Check the fluid level in the reservoir after bleeding each wheel. Add DOT3 fluid, if necessary.
3. If the hydraulic brake booster was disassembled or if the reservoir becomes empty, bleeding Hydraulic Brake Booster is only possible with a Toyota proprietary scan system.
4. Bleeding the air from the hydraulic brake lines can be performed as follows:
   a. Turn the ignition switch OFF, depress the brake pedal 40 times or more.
   b. Turn the ignition switch to the ON position and start the brake booster pump. The pump stops after approximately 30 to 40 seconds.

➡If the pump does not operate as specified, repeat the above and recheck the operating time.

   c. Holding the brake pedal depressed, bleed the right and left rear brake cylinders.
   d. Turn the ignition switch ON, depress the brake pedal 20 times or more.
5. Bleeding Front Brake Lines:
   a. Turn the ignition switch to the ON position and wait until the pump motor has stopped.
   b. Connect the vinyl tube to the brake caliper.
   c. Depress the brake pedal several times, then loosen the bleeder plug with the pedal held down.
   d. At the point when the fluid stops coming out, tighten the bleeder plug,, 8 ft. lbs. (11 Nm) then release the brake pedal.
   e. Repeat procedure until all the air in the fluid has been bled out.
   f. Repeat the above procedures to bleed the other brake line.
6. Bleeding Rear Brake Lines:
   a. Turn the ignition switch to the ON position and depress the brake pedal.
   b. Connect the vinyl tube to the brake caliper.
   c. Loosen the bleeder plug and release air.

➡Brake fluid is sent through the pump, so keep the brake pedal depressed until the air is completely bled out.

   d. When the air is completely bled out of the brake fluid through the bleeder plug, tighten the bleeder plug to 8 ft. lbs. (11 Nm) then release.
   e. Repeat the above procedures to bleed the other brake line.
7. Bleeding Hydraulic Brake Booster is only possible with a Toyota proprietary scan system.

BRAKES — FRONT DISC BRAKES

⁂ CAUTION

Dust and dirt accumulating on brake parts during normal use may contain asbestos fibers from production or aftermarket brake linings. Breathing excessive concentrations of asbestos fibers can cause serious bodily harm. Exercise care when servicing brake parts. Do not sand or grind brake lining unless equipment used is designed to contain the dust residue. Do not clean brake parts with compressed air or by dry brushing. Cleaning should be done by dampening the brake components with a fine mist of water, then wiping the brake components clean with a dampened cloth. Dispose of cloth and all residue containing asbestos fibers in an impermeable container with the appropriate label. Follow practices prescribed by the Occupational Safety and Health Administration (OSHA) and the Environmental Protection Agency (EPA) for the handling, processing, and disposing of dust or debris that may contain asbestos fibers.

BRAKE CALIPER

REMOVAL & INSTALLATION

1. Disconnect the negative battery cable from the battery.
2. Raise and support the vehicle safely.
3. Remove the wheels.
4. Disconnect the brake hose from the caliper by removing the union bolt and 2 gaskets. Plug the end of the hose to prevent loss of fluid.
5. Remove the bolts that attach the caliper to the torque plate.
6. Lift the bottom of the caliper up and remove the caliper assembly.

To install:

7. Grease the caliper slides and bolts with lithium grease or equivalent. Install the caliper and secure with the bolts. Torque the bolts to 90 ft. lbs. (123 Nm).
8. Connect the brake hose to the caliper, using 2 new washers. Make sure the flexible hose lock is securely in the lock hole of the caliper. Torque the union bolt to 22 ft. lbs. (30 Nm).
9. Fill the brake system to the proper level and bleed the brake system.
10. Install the tire and wheel assembly.
11. Top off the brake fluid level in the master cylinder. Check for leaks and proper brake operation.
12. Connect the negative battery cable to the battery.

DISC BRAKE PADS

REMOVAL & INSTALLATION

1. Raise the vehicle and support it safely.
2. Remove the wheels.
3. Remove the clip, pins and anti-rattle spring.
4. Withdraw the pads and remove the anti-squeal shims.

To install:

5. Before installing the new pads, check the disc thickness and disc runout.
6. Siphon out a small amount of brake fluid from the reservoir.
7. Press in the pistons with a hammer handle or equivalent.

8. Apply disc brake grease to both sides of the inner anti-squeal shim. Install the anti-squeal shims to the new pads.

9. Install the pads.

10. Install the anti-rattle springs and pins. Install the clip.

11. Install the wheels.

12. Check and adjust the fluid level. Apply the brake pedal several times.

13. Road-test the vehicle for proper operation.

BRAKES — REAR DISC BRAKES

⁂ CAUTION

Dust and dirt accumulating on brake parts during normal use may contain asbestos fibers from production or aftermarket brake linings. Breathing excessive concentrations of asbestos fibers can cause serious bodily harm. Exercise care when servicing brake parts. Do not sand or grind brake lining unless equipment used is designed to contain the dust residue. Do not clean brake parts with compressed air or by dry brushing. Cleaning should be done by dampening the brake components with a fine mist of water, then wiping the brake components clean with a dampened cloth. Dispose of cloth and all residue containing asbestos fibers in an impermeable container with the appropriate label. Follow practices prescribed by the Occupational Safety and Health Administration (OSHA) and the Environmental Protection Agency (EPA) for the handling, processing, and disposing of dust or debris that may contain asbestos fibers.

BRAKE CALIPER

REMOVAL & INSTALLATION

1. Disconnect the negative battery cable from the battery.
2. Raise and support the vehicle safely.
3. Remove the wheels.
4. Disconnect the brake hose from the caliper by removing the union bolt and 2 gaskets. Plug the end of the hose to prevent loss of fluid.
5. Remove the bolts that attach the caliper to the torque plate.
6. Lift the bottom of the caliper up and remove the caliper assembly.

To install:

7. Grease the caliper slides and bolts with lithium grease or equivalent. Install the caliper and secure with the bolts. Torque the bolts to 20 ft. lbs. (27 Nm).
8. Connect the brake hose to the caliper, using 2 new washers. Make sure the flexible hose lock is securely in the lock hole of the caliper. Torque the union bolt to 22 ft. lbs. (30 Nm).
9. Fill the brake system to the proper level and bleed the brake system.
10. Install the tire and wheel assembly.
11. Top off the brake fluid level in the master cylinder. Check for leaks and proper brake operation.
12. Connect the negative battery cable to the battery.

DISC BRAKE PADS

REMOVAL & INSTALLATION

1. Raise the vehicle and support it safely.
2. Remove the wheels.
3. Remove the brake caliper and suspend it so the hose is not stretched.
4. Remove the brake pads, anti-squeal shim, pad support plates and wear indicators.

To install:

5. Before installing the new pads, check the disc thickness and disc runout.
6. Install the pad support plates.
7. Install the pad wear indicator plates on each pad.
8. Install the anti-squeal shim to the outer pad. Install the pads.
9. Install the brake caliper.
10. Install the wheels.
11. Apply the brake pedal several times.
12. Road-test the vehicle for proper operation.

BRAKES — PARKING BRAKE

PARKING BRAKE CABLES

ADJUSTMENT

See Figure 1.

1. Remove rear wheel.
2. Adjust parking brake shoe clearance.
3. Install rear wheel an tighten to 82 ft. lbs. (112 Nm)
4. Inspect parking brake lever travel.
5. Slowly depress the parking brake lever all the way, and count the number of clicks.
   a. Parking brake lever travel at 44 ft. lbs. (196 Nm) 4 to 6 clicks.
   b. Loosen the lock nut and turn the turn buckle until the lever travel is correct.
   c. Tighten the lock nut to 48 inch lbs. (5.4 Nm)
6. Check whether parking brake drags or not.

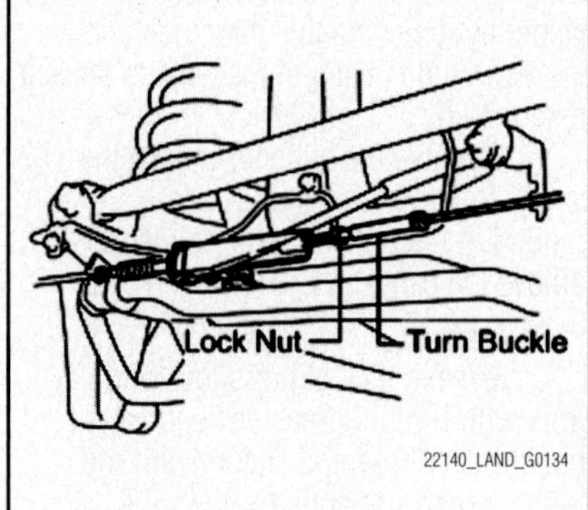

Fig. 1 Park brake cable adjustment

7. When operating the parking brake lever, check that the parking brake lever indicator light comes on.

PARKING BRAKE SHOES

REMOVAL & INSTALLATION

See Figures 2 through 4.

1. Before servicing the vehicle, refer to the precautions in the beginning of this section.
2. Raise and safely support the vehicle.
3. Remove the rear wheel.
4. Remove the 2 mounting bolts and remove the disc brake cylinder assembly.
5. Suspend the disc brake securely and so the hose is not stretched.
6. Release the parking brake lever.
7. Place matchmarks on the disc and rear axle hub.
8. Remove the disc.

➡If the disc cannot be removed easily, turn the shoe adjuster until the wheel turns freely.

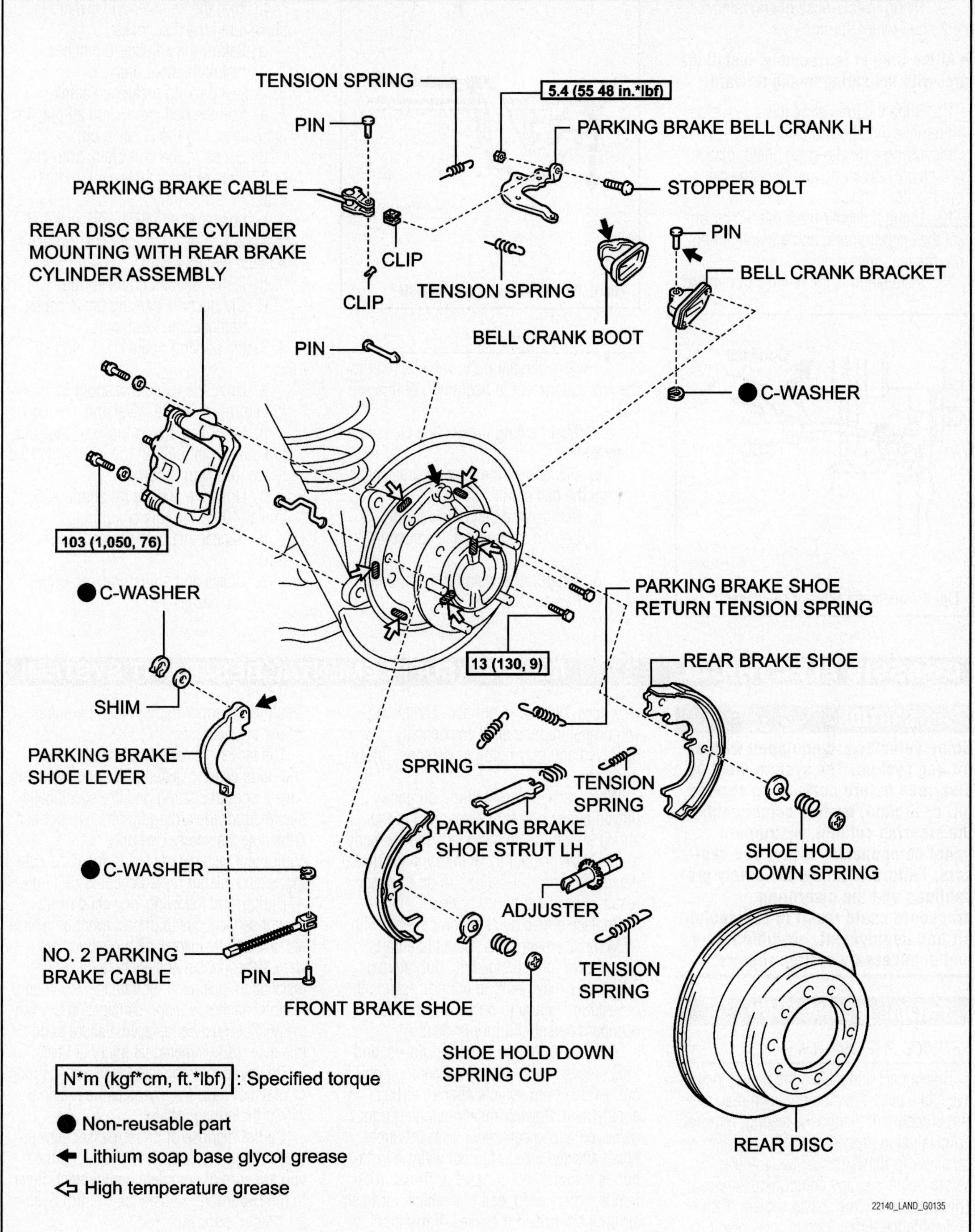

Fig. 2 Parking brake components

9. Using needle-nose pliers, remove the 2 shoe return springs.

➡At the time of reassembly, install the strut with the spring facing forward.

10. Slide the front shoe toward outside and remove the shoe adjuster.
11. Using a needle-nose pliers, disconnect the anchor spring and tension spring from the front shoe.
12. Using a needle-nose pliers, disconnect the anchor spring and tension spring from the rear shoe.
13. Installation is the reverse of removal.

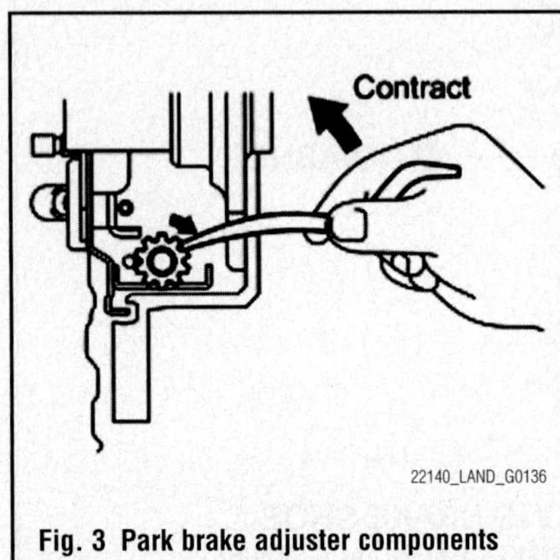

Fig. 3 Park brake adjuster components

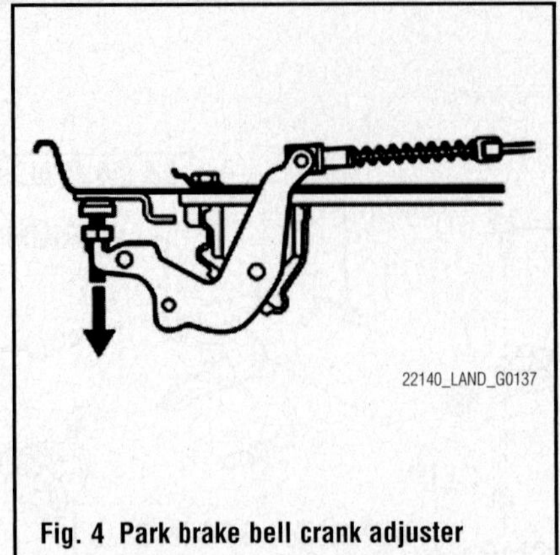

Fig. 4 Park brake bell crank adjuster

ADJUSTMENT

1. Before servicing the vehicle, refer to the precautions in the beginning of this section.
2. Adjust parking brake shoe clearance as follows:
   a. Disconnect the parking brake cable from the bell crank.
   b. Remove the bell crank spring.
   c. Loosen the bell crank adjusting bolt.
   d. Temporarily install the 3 hub nuts.
   e. Remove the hole plug.
   f. Turn the adjuster and expand the shoes until the disc locks.
   g. Return the adjuster 8 notches.
   h. Install the hole plug.
3. Adjust parking brake bell crank
   a. Pull the bell crank until all play in the interior linkage is taken up.
   b. Screw in the bell crank adjusting bolt to where it contacts on the dust seal.
   c. Loosen it one turn, and lock it at that position with the lock nut. Tighten to 48 inch lbs. (5.4 Nm).
   d. Install the bell crank spring.
   e. Connect the parking brake cable.
   f. Remove the 3 hub nuts.
4. Settle parking brake shoe and rear disc.
   a. Drive the vehicle at about 31 mph (50 km/h) on a safe, level and dry road.
   b. With the parking brake release button pushed in, pull on the lever with 19.8 ft. lbs. (88 Nm) of force.
   c. Drive the vehicle for about 0.25 miles (400 m) in this condition.
   d. Repeat this procedures 2 or 3 times.
   e. Check and adjust Parking Brake Cable, as necessary.

CHASSIS ELECTRICAL

AIR BAG (SUPPLEMENTAL RESTRAINT SYSTEM)

✲✲ CAUTION

Some vehicles are equipped with an air bag system. The system must be disarmed before performing service on, or around, system components, the steering column, instrument panel components, wiring and sensors. Failure to follow the safety precautions and the disarming procedure could result in accidental air bag deployment, possible injury and unnecessary system repairs.

GENERAL INFORMATION

SERVICE PRECAUTIONS

Disconnect and isolate the battery negative cable before beginning any airbag system component diagnosis, testing, removal, or installation procedures. Allow system capacitor to discharge for two minutes before beginning any component service. This will disable the airbag system. Failure to disable the airbag system may result in accidental airbag deployment, personal injury, or death.

Do not place an intact undeployed airbag face down on a solid surface. The airbag will propel into the air if accidentally deployed and may result in personal injury or death.

When carrying or handling an undeployed airbag, the trim side (face) of the airbag should be pointing towards the body to minimize possibility of injury if accidental deployment occurs. Failure to do this may result in personal injury or death.

Replace airbag system components with OEM replacement parts. Substitute parts may appear interchangeable, but internal differences may result in inferior occupant protection. Failure to do so may result in occupant personal injury or death.

Wear safety glasses, rubber gloves, and long sleeved clothing when cleaning powder residue from vehicle after an airbag deployment. Powder residue emitted from a deployed airbag can cause skin irritation. Flush affected area with cool water if irritation is experienced. If nasal or throat irritation is experienced, exit the vehicle for fresh air until the irritation ceases. If irritation continues, see a physician.

Do not use a replacement airbag that is not in the original packaging. This may result in improper deployment, personal injury, or death.

The factory installed fasteners, screws and bolts used to fasten airbag components have a special coating and are specifically designed for the airbag system. Do not use substitute fasteners. Use only original equipment fasteners listed in the parts catalog when fastener replacement is required.

During, and following, any child restraint anchor service, due to impact event or vehicle repair, carefully inspect all mounting hardware, tether straps, and anchors for proper installation, operation, or damage. If a child restraint anchor is found damaged in any way, the anchor must be replaced. Failure to do this may result in personal injury or death.

Deployed and non-deployed airbags may or may not have live pyrotechnic material within the airbag inflator.

Do not dispose of driver/passenger/curtain airbags or seat belt tensioners unless you are sure of complete deployment. Refer to the Hazardous Substance Control System for proper disposal.

Dispose of deployed airbags and tensioners consistent with state, provincial, local, and federal regulations.

After any airbag component testing or service, do not connect the battery negative cable. Personal injury or death may result if the system test is not performed first.

If the vehicle is equipped with the Occupant Classification System (OCS), do not connect the battery negative cable before performing the OCS Verification Test using the scan tool and the appropriate diagnostic information. Personal injury or death may result if the system test is not performed properly.

Never replace both the Occupant Restraint Controller (ORC) and the Occupant Classification Module (OCM) at the same time. If both require replacement, replace one, then perform the Airbag System test before replacing the other.

Both the ORC and the OCM store Occupant Classification System (OCS) calibration data, which they transfer to one another when one of them is replaced. If both are replaced at the same time, an irreversible fault will be set in both modules and the OCS may malfunction and cause personal injury or death.

If equipped with OCS, the Seat Weight Sensor is a sensitive, calibrated unit and must be handled carefully. Do not drop or handle roughly. If dropped or damaged, replace with another sensor. Failure to do so may result in occupant injury or death.

If equipped with OCS, the front passenger seat must be handled carefully as well. When removing the seat, be careful when setting on floor not to drop. If dropped, the sensor may be inoperative, could result in occupant injury, or possibly death.

If equipped with OCS, when the passenger front seat is on the floor, no one should sit in the front passenger seat. This uneven force may damage the sensing ability of the seat weight sensors. If sat on and damaged, the sensor may be inoperative, could result in occupant injury, or possibly death.

DISARMING THE SYSTEM

To avoid personal injury when working on vehicles equipped with an air bag, the negative battery cable must be disconnected and at least 90 seconds must elapse before working on the system. Failure to do so may result in deployment of the air bag.

ARMING THE SYSTEM

The system is self-arming when it is operating properly.

DRIVETRAIN

AUTOMATIC TRANSMISSION ASSEMBLY

REMOVAL & INSTALLATION

1. Before servicing the vehicle, refer to the precautions section.
2. Remove or disconnect the following:
   - Battery and tray
   - Air intake assembly
   - Cooling fan and shroud
   - Coolant recovery reservoir
   - Transmission dipstick tube
   - Center console
   - Transmission gear select lever and rod
   - Transfer case shift lever and rod
   - Wire harness
   - Engine under covers
   - Exhaust front pipes
   - Front and rear driveshafts
   - Vehicle Speed (VSS) sensor connectors
   - Overdrive clutch speed sensor connector
   - Solenoid harness connector
   - Transmission fluid temperature sensor connector
   - Park/Neutral Position (PNP) switch connector
   - Center differential lock indicator switch connector
   - L4 solenoid valve position switch connector
   - Motor actuator connector
   - Torque converter
   - Transmission oil cooler lines
   - Ground cable
   - Transmission mount crossmember. Support the transmission with a jack.
   - Engine mounting insulator
   - Transmission flange bolts
   - Transmission
   - Transfer assembly

To install:

3. Install or connect the following:
   - Transfer assembly. Tighten adaptor bolts to 51 ft. lbs. (69 Nm)
   - Transmission. Tighten the flange upper bolts to 53 ft. lbs. (72 Nm). Lower bolts to 27 ft. lbs. (37 Nm).
   - Engine mounting insulator. Tighten bolts to 43 ft. lbs. (59 Nm)
   - Transmission mount crossmember. Tighten the bolts to 37 ft. lbs. (50 Nm) and the nuts to 54 ft. lbs. (74 Nm).
   - Transmission oil cooler lines. Tighten with service tool to 23 ft. lbs. (31 Nm)
   - Ground cable
   - Torque converter. Tighten the bolts to 35 ft. lbs. (48 Nm).
   - Motor actuator connector
   - L4 solenoid valve position switch connector
   - Center differential lock indicator switch connector
   - PNP switch connector
   - Transmission fluid temperature sensor connector
   - Solenoid harness connector
   - Overdrive clutch speed sensor connector
   - VSS sensor connectors
   - Front driveshaft. Tighten the fasteners to 59 ft. lbs. (80 Nm).
   - Rear driveshaft. Tighten the fasteners to 78 ft. lbs. (106 Nm).
   - Exhaust front pipes
   - Engine under covers
   - Wire harness
   - Transfer case shift lever and rod
   - Transmission gear select lever and rod
   - Center console
   - Transmission dipstick tube
   - Coolant recovery reservoir
   - Cooling fan and shroud
   - Air intake assembly
   - Battery and tray
4. Check the transmission and transfer case fluid levels and adjust as necessary.

TRANSFER CASE ASSEMBLY

REMOVAL & INSTALLATION

1. Before servicing the vehicle, refer to the precautions section.
2. Drain the transfer case oil.
3. Remove or disconnect the following:
   - Transfer case protector
   - Front and rear driveshafts
   - Transfer case shift lever rod
   - Ground cable
   - Transmission mount crossmember. Support the transmission with a jack.
   - Transfer case vent hose
   - Vehicle Speed (VSS) sensor connector
   - Center differential lock indicator switch connector

- Motor actuator connectors
- Transfer case adapter bolts
- Transfer case

To install:

4. Install or connect the following:
   - Transfer case. Tighten the adapter bolts to 51 ft. lbs. (69 Nm).
   - Motor actuator connectors
   - Center differential lock indicator switch connector
   - VSS sensor connector
   - Transfer case vent hose
   - Transmission mount crossmember. Tighten the bolts to 37 ft. lbs. (50 Nm) and the nuts to 54 ft. lbs. (74 Nm).
   - Ground cable
   - Transfer case shift lever rod
   - Front driveshaft. Tighten the fasteners to 59 ft. lbs. (80 Nm).
   - Rear driveshaft. Tighten the fasteners to 78 ft. lbs. (106 Nm).
   - Transfer case protector
5. Fill the transfer case to the correct level.

FRONT HALFSHAFTS

REMOVAL & INSTALLATION

See Figures 5 and 6.

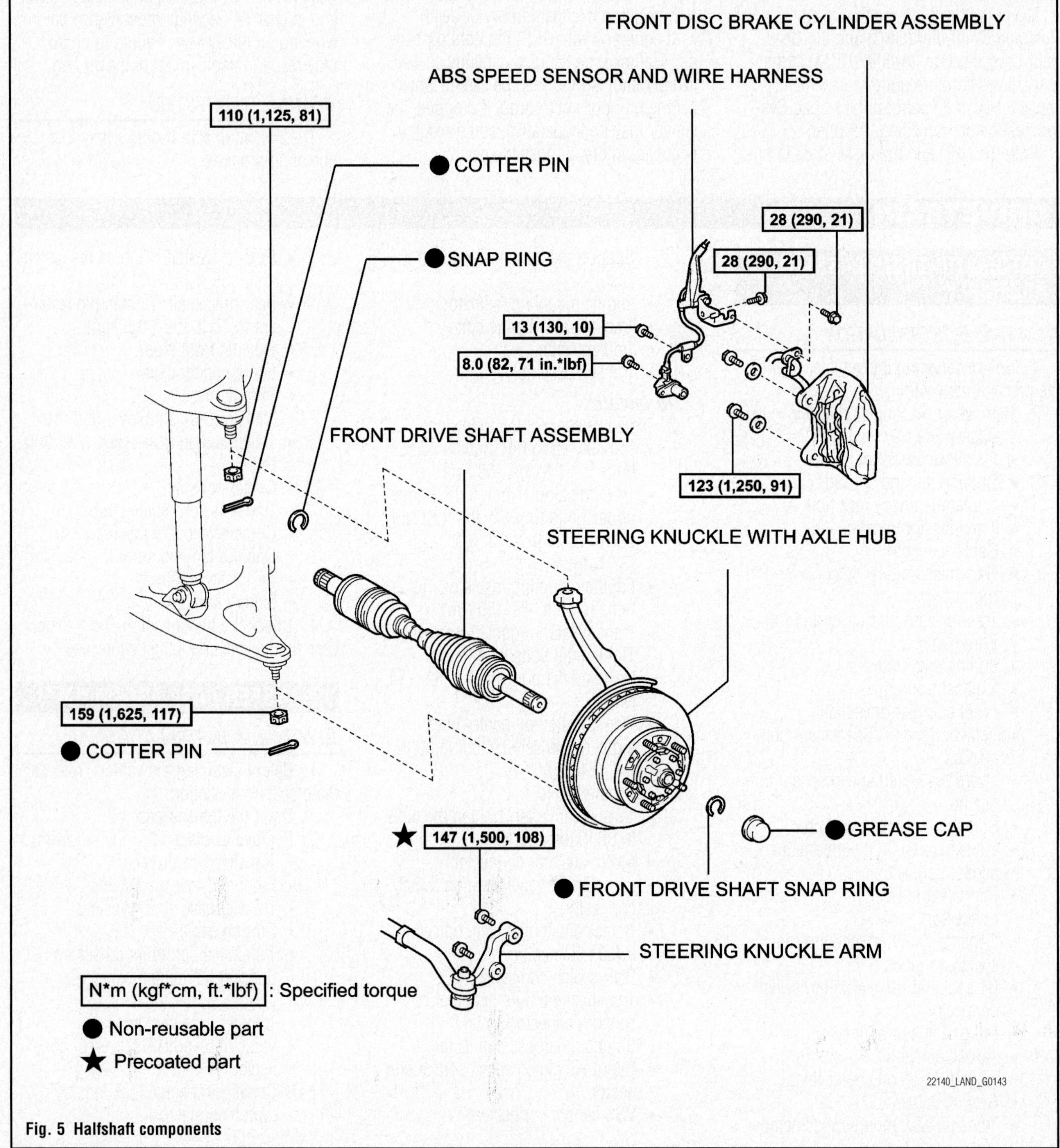

Fig. 5 Halfshaft components

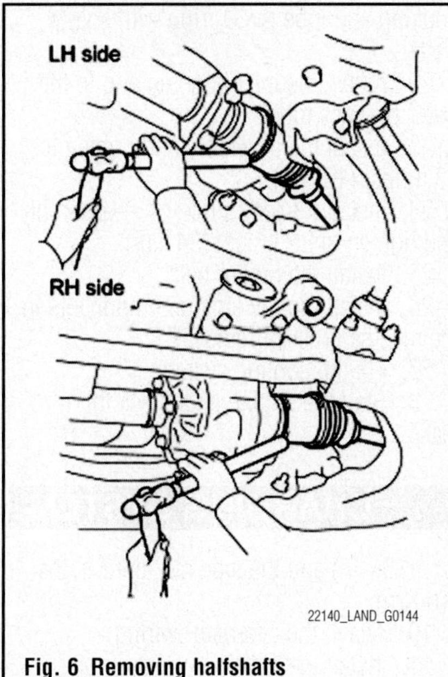

Fig. 6 Removing halfshafts

1. Remove front wheel
2. Remove front disc brake cylinder assembly
3. Remove the bolt and disconnect the flexible hose from the steering knuckle
4. Remove the 2 bolts, washers and brake cylinder
5. Support the brake cylinder securely
6. Using a screwdriver and hammer, remove the grease cap from the flange.
7. Using snap ring pliers, remove the snap ring
8. Remove the 3 bolts and disconnect the ABS speed sensor and wire harness
9. Remove the 2 bolts and disconnect the steering knuckle arm
10. Remove the cotter pin and nut
11. Disconnect the steering knuckle from the lower suspension arm
12. Temporarily install the nut to the lower suspension arm
13. Support the lower suspension arm with a jack
14. Remove the cotter pin and nut.
15. Disconnect the steering knuckle from the upper suspension arm
16. Remove the nut and steering knuckle with axle hub
17. Using a brass bar and hammer, tap out the drive shaft

To install:

18. Installation is reverse of removal
19. Tighten:
    - Brake cylinder to 91 ft. lbs. (123 Nm)
    - Steering knuckle arm to 108 ft. lbs. (147 Nm)
    - Steering knuckle 117 ft. lbs. (159 Nm)
    - Front wheel 97 ft. lbs. (131 Nm)

REAR AXLE SHAFT, BEARING & SEAL

REMOVAL & INSTALLATION

1. Before servicing the vehicle, refer to the precautions section.
2. Remove or disconnect the following:
    - Rear wheel
    - Brake caliper and rotor
    - Parking brake shoes and hardware
    - Bearing case nuts
    - Axle shaft and bearing assembly
3. Separate the backing plate from the bearing case by removing the serrated bolts.
4. Grind a flat spot on the wheel speed sensor rotor and retainer, then split them with a hammer and chisel.
5. Remove the axle snapring.
6. Press the axle bearing case, bearing and retainer off of the axle.
7. Press the axle bearing from the bearing case.
8. Remove or disconnect the following:
    - Backing plate
    - Axle housing oil seal
    - Bearing case oil seal

To install:

9. Press the wheel bearing into the bearing case.
10. Install the bearing case to the backing plate with the serrated bolts.
11. Install or connect the following:
    - Bearing case oil seal
    - Axle housing oil seal
    - Axle shaft to backing plate and bearing assembly
    - Bearing retainer
    - Axle snapring
    - Wheel speed sensor rotor and retainer
    - Axle shaft and bearing assembly to the axle housing. Tighten the nuts to 91 ft. lbs. (123 Nm).
    - Parking brake shoes and hardware
    - Brake caliper and rotor
    - Rear wheel

ENGINE COOLING

THERMOSTAT

REMOVAL & INSTALLATION

See Figure 7.

1. Before servicing the vehicle, refer to the precautions in the beginning of this section.
2. Drain engine coolant.
3. Remove the 3 nuts and disconnect the water inlet from the water inlet housing.
4. Remove the thermostat.
5. Remove the gasket from the thermostat.

To install:

6. Install a new gasket to the thermostat.
7. Insert the thermostat into the water inlet housing with the jiggle valve facing straight upward.

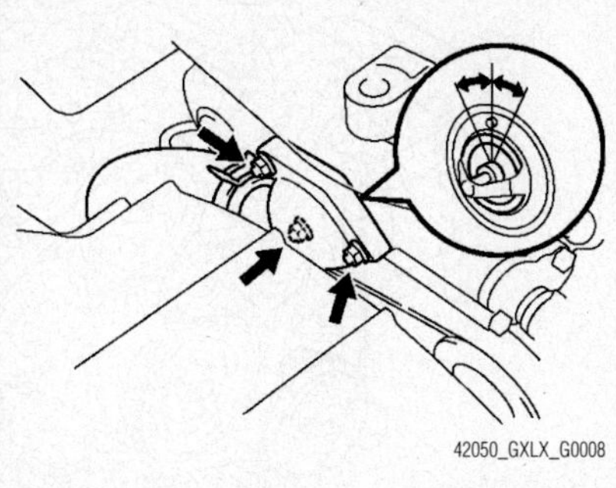

Fig. 7 Thermostat positioning and installation

➡The jiggle valve may be set within 30° of either side of the prescribed position.

8. Install the water inlet with the 3 nuts and tighten to 19 Nm (14 ft. lbs.).
9. Fill with engine coolant
10. Start engine and check for coolant leaks
11. Recheck engine coolant level

WATER PUMP

REMOVAL & INSTALLATION

1. Before servicing the vehicle, refer to the precautions section.
2. Disconnect the negative battery cable.
3. Remove the engine undercover.
4. Drain the cooling system.
5. Remove the v-bank cover.
6. Remove the intake air pipe.
7. Remove the A/C compressor and set aside with the lines attached.
8. Remove the drive belt.
9. Remove the radiator.
10. Remove the oil cooler pipe.

11. Remove the timing belt.
12. Remove the number 2 timing belt idler pulley sub assembly.
13. Disconnect the water bypass hose from the water inlet housing.
14. Remove the bolts, nuts and stud attaching the inlet housing to the water pump.
15. Disconnect the hose from the inlet housing front joint and remove the inlet housing.
16. Remove the water pump bolts, gasket and the O–ring from the by-pass pipe.

To install:

17. Coat a new O–ring with soapy water, install the O–ring on the pipe end and connect the water pump to the by-pass pipe.
18. Using a new gasket, install the water pump. Tighten the bolts to 15 ft. lbs. (21 Nm), the nuts and stud to 24 ft. lbs. (32 Nm).
19. Clean all packing material from the water inlet housing gasket surfaces and sealing groove.
20. Apply a 0.078–0.118 inch (2–3 mm) bead of sealant to the inlet housing.
21. Install a new O–ring to the inlet housing and lube the O–ring with soapy water.
22. Attach the inlet housing end to the water by-pass tube.
23. Install the assembly and tighten to 13 ft. lbs. (18 Nm).
24. Install the idler pulley sub-assembly and tighten to 25 ft. lbs. (34 Nm).
25. Install the timing belt.
26. Install the remaining components in the reverse order of removal.
27. Fill the cooling system.
28. Start the engine and check for leaks.

ENGINE ELECTRICAL — CHARGING SYSTEM

ALTERNATOR

REMOVAL & INSTALLATION

1. Before servicing the vehicle, refer to the precautions section.
2. Disconnect the negative battery cable.
3. Remove the drive belt.
4. Disconnect the 2 oil cooler lines from the fan shroud, remove the fan shroud.
5. Remove the 4 nuts and remove the fan with the fluid coupling.
6. Disconnect the vane pump assembly.
7. Disconnect the alternator wiring.
8. Remove the nuts and bolts and remove the alternator.

To install:

9. Install the alternator. Tighten the bolt to 29 ft. lbs. (39 Nm), the upper nut to 29 ft. lbs. (39 Nm) and the side nut to 12 ft. lbs. (16 Nm).
10. Attach the alternator wiring.
11. Install the vane pump assembly.
12. Install the shroud and fluid coupling together and tighten shroud bolts. Tighten the fan coupling nuts to 21 ft. lbs. (29 Nm).

ENGINE ELECTRICAL — DISTRIBUTORLESS IGNITION SYSTEM

IGNITION COIL

REMOVAL & INSTALLATION

1. Disconnect cable from negative battery terminal.

CAUTION

Wait at least 90 seconds after disconnecting the cable from the negative (-) battery terminal to prevent airbag and seat belt pretensioner activation.

2. Remove the V-bank cover.
3. Remove intake air connector pipe
   a. Disconnect the 3 hoses.
   b. Remove the 2 bolts.
   c. Loosen the 2 clamp bolts and remove the intake air connector.
4. Disconnect the 2 engine wire clamps from the cylinder head cover LH side.
5. Disconnect the connector and remove ignition coil assembly.

To install:

6. Install the ignition coil with the bolt tighten bolt to 66 inch lbs. (7.5 Nm)
7. Connect the connector.
8. Connect the 2 engine wire clamps to the cylinder head cover LH side.
9. Install intake air connector pipe.
10. Install v-bank cover sub-assembly.
11. Connect cable to negative battery terminal.
12. Perform initialization if necessary.

FIRING ORDER

See Figure 8.

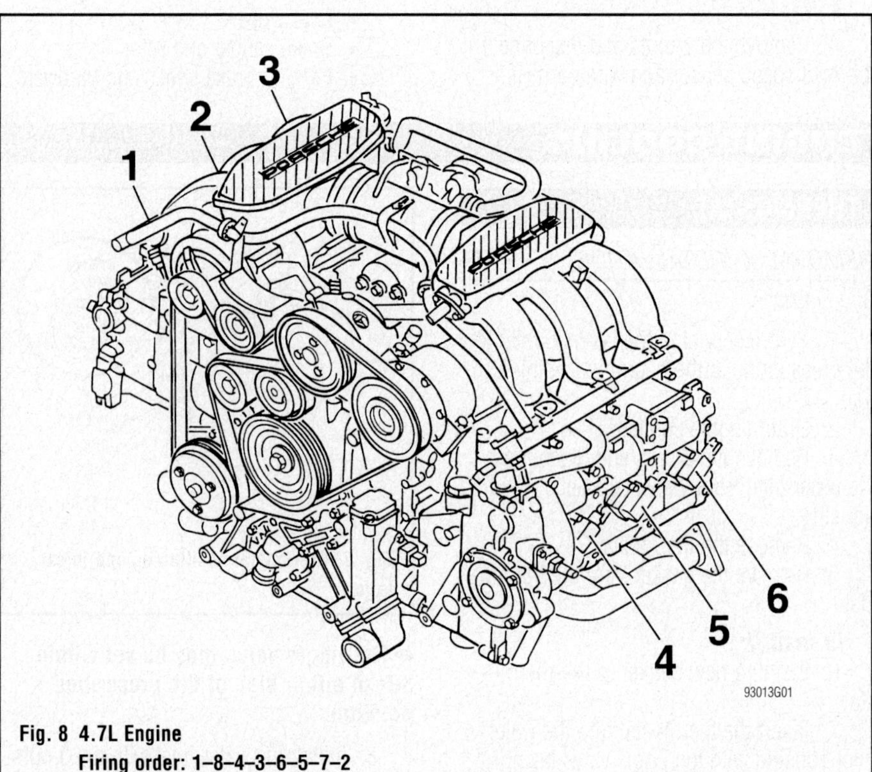

Fig. 8 4.7L Engine
Firing order: 1–8–4–3–6–5–7–2
Distributorless ignition system

IGNITION TIMING

ADJUSTMENT

The ignition timing is controlled by the Powertrain Control Module (PCM). No adjustment is necessary or possible.

SPARK PLUGS

REMOVAL & INSTALLATION

1. Remove the ignition coils.
2. Using a 16 mm plug wrench, remove the spark plugs.
3. Clean the spark plugs.

To install:

4. Adjust the spark plug electrode gap. Electrode gap for new spark plug is 1.0 to 1.1 mm (0.039 to 0.043 in.).
5. Using a 16 mm plug wrench, install the spark plugs and tighten to 17.5 Nm (13 ft lbs)
6. Reinstall the ignition coils.

ENGINE ELECTRICAL

STARTING SYSTEM

STARTER

REMOVAL & INSTALLATION

1. Before servicing the vehicle, refer to the precautions section.
2. Drain the cooling system.
3. Relieve the fuel system pressure.
4. Remove or disconnect the following:
   - Negative battery cable
   - Engine appearance cover
   - Air intake tube
   - No1 & No2 fuel hoses
   - Ventilation hose
   - Purge VSV
   - Vacuum control valve
   - Engine wire harness
   - Throttle body water bypass hose
   - Intake manifold
   - Air pump and switching valve
   - Starter motor mounting bolts
   - Starter wiring connectors
   - Starter motor

To install:

5. Install or connect the following:
   - Starter motor
   - Starter wiring connectors. Tighten the cable nut to 86 inch lbs. (10 Nm).
   - Starter motor mounting bolts. Tighten the bolts to 29 ft. lbs. (39 Nm).
   - Air pump and switching valve
   - Intake manifold
   - Throttle body water bypass hose
   - Engine wire harness
   - Vacuum control valve
   - Purge VSV
   - Ventilation hose
   - No1 & No2 fuel hoses
   - Air intake tube
   - Engine appearance cover
   - Negative battery cable
6. Fill the cooling system.
7. Start the engine and check for leaks.
8. Perform initialization if necessary

ENGINE MECHANICAL

ACCESSORY DRIVE BELTS

ACCESSORY BELT ROUTING

See Figure 9.

INSPECTION

Inspect the drive belt for signs of glazing or cracking. A glazed belt will be perfectly smooth from slippage, while a good belt will have a slight texture of fabric visible. Cracks will usually start at the inner edge of the belt and run outward. All worn or damaged drive belts should be replaced immediately.

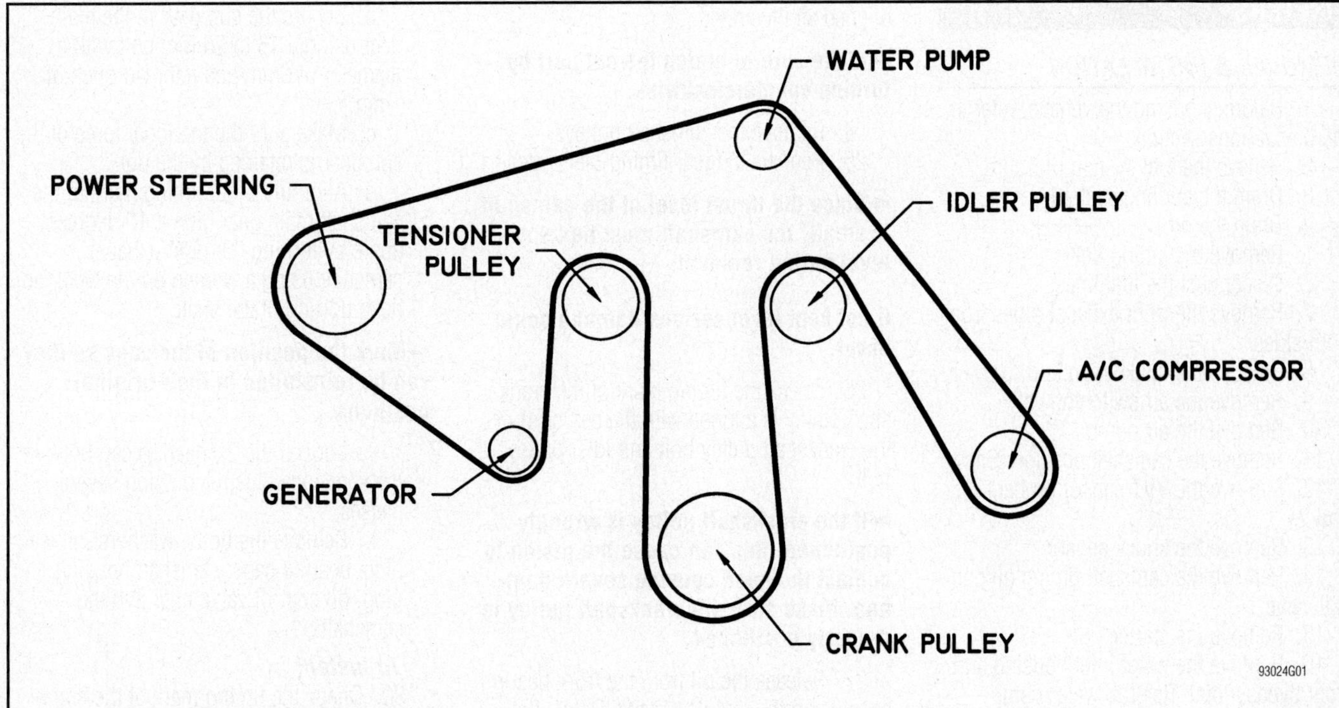

Fig. 9 Accessory drive belt routing—4.7L engine

REMOVAL & INSTALLATION

See Figure 10.

1. Before servicing the vehicle, refer to the precautions in the beginning of this section.
2. Loosen the drive belt tension by turning the drive belt tensioner counterclockwise, and remove the drive belt.
3. Installation is the reverse of removal.

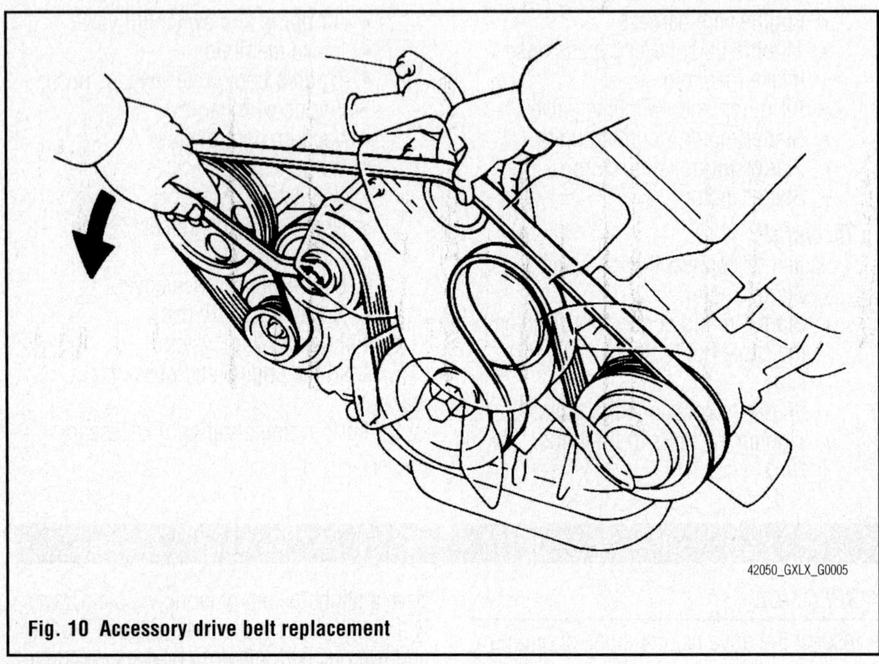

Fig. 10 Accessory drive belt replacement

CAMSHAFT AND VALVE LIFTERS

REMOVAL & INSTALLATION

1. Before servicing the vehicle, refer to the precautions section.
2. Relieve the fuel system pressure.
3. Drain the cooling system.
4. Drain the oil.
5. Remove the timing belt.
6. Disconnect the fuel lines.
7. Remove the front exhaust pipes from both sides.
8. Remove the intake manifold.
9. Remove the air switching valve.
10. Remove the air pump.
11. Remove the camshaft position sensor.
12. Remove the VVT sensor on banks 1 and 2.
13. Remove the knock sensor.
14. Remove the camshaft timing oil control valve.
15. Remove the starter.
16. Remove the water inlet housing and front bypass joint. Refer to water pump removal.
17. Disconnect the heater hose and remove the water bypass pipe sub assembly.
18. Remove the rear water bypass joint.
19. Remove the oil dipstick and tube.
20. Remove the ignition coils.
21. Remove the spark plugs.
22. Remove the valve covers.
23. Turn the crankshaft pulley counterclockwise by 5 degrees until the marks are aligned as illustrated.

➡Make sure to match the cut part by turning counterclockwise.

24. Remove the camshaft pulleys.
25. Remove the rear timing belt plates.

➡Since the thrust level of the camshaft is small, the camshaft must be kept level during removal.

If not kept level serious damage could occur.

26. Check the timing mark of the crankshaft pulley is aligned with the center(s) of the crankshaft pulley bolt and idler pulley bolt.

➡If the crankshaft pulley is wrongly positioned, this can cause the piston to contact the head causing severe damage. Make sure the crankshaft pulley is properly positioned.

27. Release the oil from the front bearing caps using the tool illustrated. Rotate the camshaft timing tube from left to right 2 to 3

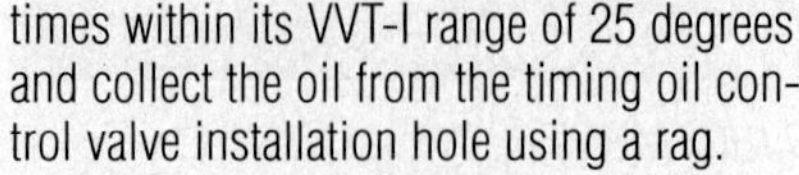

times within its VVT-I range of 25 degrees and collect the oil from the timing oil control valve installation hole using a rag.

28. Remove the left hand camshafts as follows:
   a. Bring the service bolt of the sub gear up by turning the left exhaust camshaft using a wrench on the hexagon head portion of the shaft.
   b. Secure the sub gear to the main gear using a 16 to 20 mm bolt with a diameter of 6mm and a thread pitch of 1mm.
   c. Make sure the torsional force of the sub gear is retained by the bolt.
   d. Align the 2 dot timing mark of the left side camshaft by turning the left exhaust camshaft using a wrench on the hexagon head portion of the shaft.

➡Mark the position of the caps so they can be reinstalled in their original positions.

   e. Loosen the 22 bearing cap bolts in the sequence illustrated using several passes.
   f. Remove the bolts, washers, oil feed pipe, bearing caps, camshaft housing plug, oil control valve filter and the camshafts.

29. Remove the right hand camshafts as follows:
   a. Bring the service bolt of the sub gear up by turning the right exhaust camshaft using a wrench on the hexagon head portion of the shaft.
   b. Secure the sub gear to the main gear using a 16 to 20 mm bolt with a diameter of 6mm and a thread pitch of 1mm.
   c. Make sure the torsional force of the sub gear is retained by the bolt.
   d. Align the 1 dot timing mark of the camshaft main gear (about 10 degrees) angle by turning the right exhaust camshaft using a wrench on the hexagon head portion of the shaft.

➡Mark the position of the caps so they can be reinstalled in their original positions.

   e. Loosen the 22 bearing cap bolts in the sequence illustrated using several passes.
   f. Remove the bolts, washers, oil feed pipe, bearing caps, camshaft housing plug, oil control valve filter and the camshafts.

To install:

30. Check the timing mark of the crankshaft pulley is aligned with the center(s) of

the crankshaft pulley bolt and idler pulley bolt.

➡If the crankshaft pulley is wrongly positioned, this can cause the piston to contact the head causing severe damage. Make sure the crankshaft pulley is properly positioned.

31. Install the left side camshafts as follows:
    a. Apply multipurpose grease to the thrust portion of the camshafts.
    b. Align the 2 dot timing mark of the camshaft drive and driven main gears and install the camshafts.
    c. Apply seal packing to the camshaft housing plug.
    d. Install the camshaft housing plug on the cylinder head as illustrated. Install the strainer on the head being careful it is properly positioned.
    e. Apply seal packing to the front bearing cap.
    f. Install the front bearing cap and then the other caps in the sequence illustrated.
    g. Push in the camshaft oil seal.
    h. Install 4 new seal washers to the bearing cap bolts A and B, refer to the illustration.
    i. Apply a light coating of clean oil to the threads and underside of the bolt heads D and E. make sure no oil gets under the heads of bolts A, B and C.
    j. The bolt lengths and positions are as follows. refer to the illustration for bolt location:
    - 94mm bolts A
    - 72mm bolts B
    - 25mm bolts C
    - 52mm bolts D
    - 38mm bolts E

    k. Tighten the cap bolts using several passes. Tighten bolt C to 66 inch lbs. (7.5 Nm) an the remaining bolts to 12 ft. lbs. (16 Nm).
    l. Remove the service bolt.
32. Install the right side camshafts as follows:
    a. Apply multipurpose grease to the thrust portion of the camshafts.
    b. Align the 1 dot timing mark of the camshaft drive and driven main gears and install the camshafts.
    c. Set the 1 dot timing mark of the camshaft drive and driven gears at a 10 degree angle.
    d. Apply seal packing to the camshaft housing plug.
    e. Install the camshaft housing plug on the cylinder head as illustrated. Install the strainer on the head being careful it is properly positioned.
    f. Apply seal packing to the front bearing cap.
    g. Install the front bearing cap and then the other caps in the sequence illustrated.
    h. Push in the camshaft oil seal.
    i. Install 4 new seal washers to the bearing cap bolts A and B, refer to the illustration.
    j. Apply a light coating of clean oil to the threads and underside of the bolt heads D and E. make sure no oil gets under the heads of bolts A, B and C.
    k. The bolt lengths and positions are as follows. refer to the illustration for bolt location:
    - 94mm bolts A
    - 72mm bolts B
    - 25mm bolts C
    - 52mm bolts D
    - 38mm bolts E

    l. Tighten the cap bolts using several passes. Tighten bolt C to 66 inch lbs. (7.5 Nm) an the remaining bolts to 12 ft. lbs. (16 Nm).
    m. Remove the service bolt.
33. Check and adjust the valve clearance.
34. Install the camshaft timing control valve.
35. Install the 4 half moon plugs onto the cylinder heads.
36. Install the valve covers and tighten to 53 inch lbs. (6 Nm).
37. Install the spark plugs.
38. Install the ignition coils.
39. Install the engine hangers and tighten to 27 ft. lbs. (37 Nm).
40. Install the VVT sensors.
41. Install the oil dipstick tube and dipstick.
42. Install the water bypass joint and tighten the retainers to 13 ft. lbs. (18 Nm).
43. Install the water inlet and housing assembly.
44. Install new intake manifold gaskets and the manifold. Tighten the bolts to 13 ft. lbs. (18 Nm) in several passes.
45. Install the throttle cover bracket, wire bracket and wire to the engine hanger bracket.
46. Install the wire to the timing belt rear plate.
47. Attach the wire protector to the intake manifold.
48. Attach the 2 ground cables the cylinder heads.
49. Connect the water bypass hoses to the throttle body.
50. Connect the wire clamps to the bracket on the right delivery pipe.
51. Attach the hoses to the intake manifold.
52. Attach the electrical connectors to the intake manifold.
53. Connect the fuel hose.
54. Install the timing belt rear plates, right plates first, then left plates. Tighten the retainers to 66 inch lbs. (7 Nm).
55. Install the throttle body cover.
56. Install the front exhaust pipe, power steering pump.
57. Install the camshaft position sensor and camshaft timing pulleys, tighten to 25 ft. lbs. (34 Nm).
58. Install the timing belt.
59. Fill the cooling system and perform an oil change.
60. Start the vehicle and check for leaks.

CRANKSHAFT DAMPER

REMOVAL & INSTALLATION

1. Before servicing the vehicle, refer to the precautions in the beginning of this section.
2. Drain the cooling system.
3. Remove or disconnect the following:
    - Negative battery cable
    - Engine under cover
    - Engine appearance cover
    - Air intake assembly
    - Accessory drive belt
    - Cooling fan and pulley
    - Radiator
    - Drive belt idler pulley
    - Camshaft Position (CMP) sensor connector
    - Upper timing covers
    - Oil cooler pipe
    - Center timing cover
    - A/C compressor
    - Cooling fan bracket
    - Crankshaft pulley

To install:

- Crankshaft pulley. Tighten the bolt to 181 ft. lbs. (245 Nm).
- Cooling fan bracket. Tighten the 12mm bolts to 12 ft. lbs. (16 Nm) and the 14mm bolts to 24 ft. lbs. (32 Nm).
- A/C compressor
- Center timing cover
- Oil cooler pipe
- Upper timing covers
- CMP sensor connector
- Drive belt idler pulley. Tighten the bolt to 27 ft. lbs. (37 Nm).
- Radiator
- Cooling fan and pulley. Tighten the nuts to 16 ft. lbs. (21 Nm).
- Accessory drive belt
- Air intake assembly
- Engine appearance cover

- Engine under cover
- Negative battery cable

4. Fill the cooling system.
5. Start the engine and check for leaks.

CRANKSHAFT FRONT SEAL

REMOVAL & INSTALLATION

1. Before servicing the vehicle, refer to the precautions in the beginning of this section.
2. Drain the cooling system.
3. Remove or disconnect the following:
   - Negative battery cable
   - Engine under cover
   - Engine appearance cover
   - Air intake assembly
   - Accessory drive belt
   - Cooling fan and pulley
   - Radiator
   - Drive belt idler pulley
   - Camshaft Position (CMP) sensor connector
   - Upper timing covers
   - Oil cooler pipe
   - Center timing cover
   - A/C compressor
   - Cooling fan bracket
   - Crankshaft pulley
   - Lower timing cover
   - Timing belt.
   - Crankshaft timing sprocket
   - Front crankshaft seal

To install:

4. Install the oil seal so that it is flush with the oil pump housing.
5. Install or connect the following:
   - Crankshaft timing sprocket
   - Timing belt
   - Lower timing cover
   - Crankshaft pulley. Tighten the bolt to 181 ft. lbs. (245 Nm).
   - Cooling fan bracket. Tighten the 12mm bolts to 12 ft. lbs. (16 Nm) and the 14mm bolts to 24 ft. lbs. (32 Nm).
   - A/C compressor
   - Center timing cover
   - Oil cooler pipe
   - Upper timing covers
   - CMP sensor connector
   - Drive belt idler pulley. Tighten the bolt to 27 ft. lbs. (37 Nm).
   - Radiator
   - Cooling fan and pulley. Tighten the nuts to 16 ft. lbs. (21 Nm).
   - Accessory drive belt
   - Air intake assembly
   - Engine appearance cover
   - Engine under cover
   - Negative battery cable

6. Fill the cooling system.
7. Start the engine and check for leaks.

CYLINDER HEAD

REMOVAL & INSTALLATION

See Figure 11.

1. Before servicing the vehicle, refer to the precautions section.
2. Relieve the fuel system pressure.
3. Drain the cooling system.
4. Drain the oil.
5. Remove the timing belt. See Timing Belt and Sprockets
6. Disconnect the fuel lines.
7. Remove the front exhaust pipes from both sides.
8. Remove the intake manifold.
9. Remove the air switching valve.
10. Remove the air pump.
11. Remove the camshaft position sensor.
12. Remove the VVT sensor on banks 1 and 2.
13. Remove the knock sensor.
14. Remove the camshaft timing oil control valve.
15. Remove the starter.
16. Remove the water inlet housing and front bypass joint. Refer to water pump removal.
17. Disconnect the heater hose and remove the water bypass pipe sub assembly.
18. Remove the rear water bypass joint.
19. Remove the oil dipstick and tube.
20. Remove the ignition coils.
21. Remove the spark plugs.
22. Remove the valve covers.
23. Turn the crankshaft pulley counterclockwise by 5 degrees until the marks are aligned as illustrated.

➡Make sure to match the cut part by turning counterclockwise.

24. Remove the camshaft pulleys.
25. Remove the rear timing belt plates.

➡Since the thrust level of the camshaft is small, the camshaft must be kept level during removal.

If not kept level serious damage could occur.

26. Check the timing mark of the crankshaft pulley is aligned with the center(s) of the crankshaft pulley bolt and idler pulley bolt.

➡If the crankshaft pulley is wrongly positioned, this can cause the piston to contact the head causing severe damage. Make sure the crankshaft pulley is properly positioned.

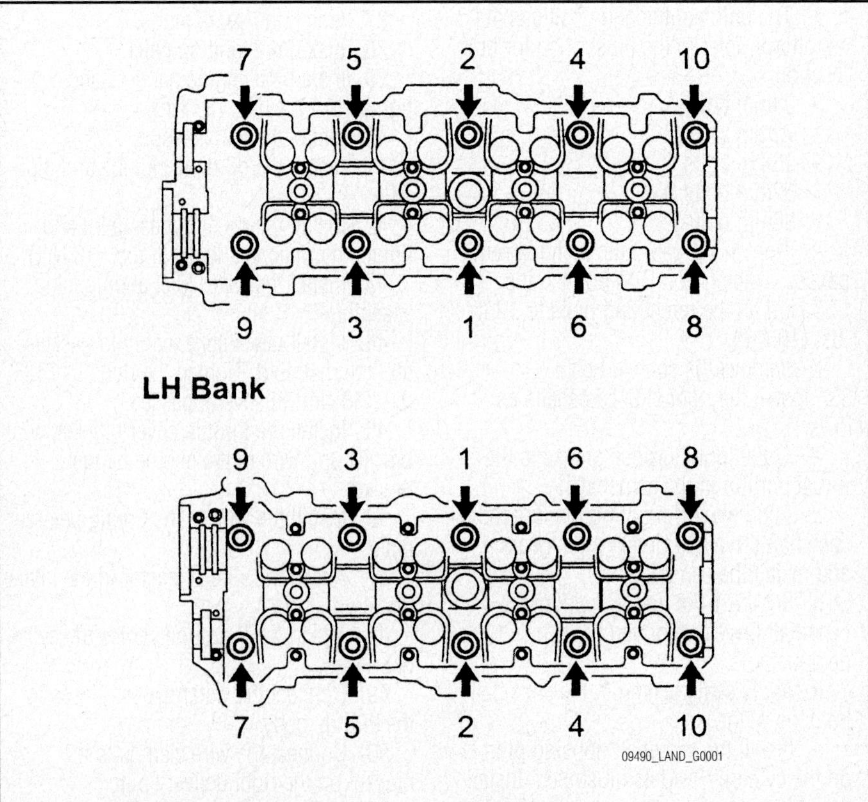

Fig. 11 Cylinder head loosening sequence—4.7L 2UZ-FE engine

27. Release the oil from the front bearing caps using the tool illustrated. Rotate the camshaft timing tube from left to right 2 to 3 times within its VVT-I range of 25 degrees and collect the oil from the timing oil control valve installation hole using a rag.

28. Remove the left hand camshafts as follows:

a. Bring the service bolt of the sub gear up by turning the left exhaust camshaft using a wrench on the hexagon head portion of the shaft.

b. Secure the sub gear to the main gear using a 16 to 20 mm bolt with a diameter of 6mm and a thread pitch of 1mm.

c. Make sure the torsional force of the sub gear is retained by the bolt.

d. Align the 2 dot timing mark of the left side camshaft by turning the left exhaust camshaft using a wrench on the hexagon head portion of the shaft.

➡Mark the position of the caps so they can be reinstalled in their original positions.

e. Loosen the 22 bearing cap bolts in the sequence illustrated using several passes.

f. Remove the bolts, washers, oil feed pipe, bearing caps, camshaft housing plug, oil control valve filter and the camshafts.

29. Remove the right hand camshafts as follows:

a. Bring the service bolt of the sub gear up by turning the right exhaust camshaft using a wrench on the hexagon head portion of the shaft.

b. Secure the sub gear to the main gear using a 16 to 20 mm bolt with a diameter of 6mm and a thread pitch of 1mm.

c. Make sure the torsional force of the sub gear is retained by the bolt.

d. Align the 1 dot timing mark of the camshaft main gear (about 10 degrees) angle by turning the right exhaust camshaft using a wrench on the hexagon head portion of the shaft.

➡Mark the position of the caps so they can be reinstalled in their original positions.

e. Loosen the 22 bearing cap bolts in the sequence illustrated using several passes.

f. Remove the bolts, washers, oil feed pipe, bearing caps, camshaft housing plug, oil control valve filter and the camshafts.

30. Remove the engine hangers.

31. Loosen the cylinder head bolts in the sequence shown, using several passes.

32. Remove the cylinder heads and exhaust manifolds together as an assembly.

To install:

33. Install new gaskets and the cylinder heads

34. Tighten the bolts in sequence as follows:

a. Step 1: 30 ft. lbs. (40 Nm).

b. Step 2: Plus 90 degrees.

c. Step 3: Plus 90 degrees.

35. Check the timing mark of the crankshaft pulley is aligned with the center(s) of the crankshaft pulley bolt and idler pulley bolt.

➡If the crankshaft pulley is wrongly positioned, this can cause the piston to contact the head causing severe damage. Make sure the crankshaft pulley is properly positioned.

36. Install the left side camshafts as follows:

a. Apply multipurpose grease to the thrust portion of the camshafts.

b. Align the 2 dot timing mark of the camshaft drive and driven main gears and install the camshafts.

c. Apply seal packing to the camshaft housing plug.

d. Install the camshaft housing plug on the cylinder head as illustrated. Install the strainer on the head being careful it is properly positioned.

e. Apply seal packing to the front bearing cap.

f. Install the front bearing cap and then the other caps in the sequence illustrated.

g. Push in the camshaft oil seal.

h. Install 4 new seal washers to the bearing cap bolts A and B, refer to the illustration.

i. Apply a light coating of clean oil to the threads and underside of the bolt heads D and E. make sure no oil gets under the heads of bolts A, B and C.

j. The bolt lengths and positions are as follows. refer to the illustration for bolt location:

- 94mm bolts A
- 72mm bolts B
- 25mm bolts C
- 52mm bolts D
- 38mm bolts E

k. Tighten the cap bolts using several passes. Tighten bolt C to 66 inch lbs. (7.5 Nm) an the remaining bolts to 12 ft. lbs. (16 Nm).

l. Remove the service bolt.

37. Install the right side camshafts as follows:

a. Apply multipurpose grease to the thrust portion of the camshafts.

b. Align the 1 dot timing mark of the camshaft drive and driven main gears and install the camshafts.

c. Set the 1 dot timing mark of the camshaft drive and driven gears at a 10 degree angle.

d. Apply seal packing to the camshaft housing plug.

e. Install the camshaft housing plug on the cylinder head as illustrated. Install the strainer on the head being careful it is properly positioned.

f. Apply seal packing to the front bearing cap.

g. Install the front bearing cap and then the other caps in the sequence illustrated.

h. Push in the camshaft oil seal.

i. Install 4 new seal washers to the bearing cap bolts A and B, refer to the illustration.

j. Apply a light coating of clean oil to the threads and underside of the bolt heads D and E. make sure no oil gets under the heads of bolts A, B and C.

k. The bolt lengths and positions are as follows. refer to the illustration for bolt location:

- 94mm bolts A
- 72mm bolts B
- 25mm bolts C
- 52mm bolts D
- 38mm bolts E

l. Tighten the cap bolts using several passes. Tighten bolt C to 66 inch lbs. (7.5 Nm) an the remaining bolts to 12 ft. lbs. (16 Nm).

m. Remove the service bolt.

38. Check and adjust the valve clearance.

39. Install the camshaft timing control valve.

40. Install the 4 half moon plugs onto the cylinder heads.

41. Install the valve covers and tighten to 53 inch lbs. (6 Nm).

42. Install the spark plugs.

43. Install the ignition coils.

44. Install the engine hangers and tighten to 27 ft. lbs. (37 Nm).

45. Install the VVT sensors.

46. Install the oil dipstick tube and dipstick.

47. Install the water bypass joint and tighten the retainers to 13 ft. lbs. (18 Nm).

48. Install the water inlet and housing assembly.

49. Install new intake manifold gaskets and the manifold. Tighten the bolts to 13 ft. lbs. (18 Nm) in several passes.

50. Install the throttle cover bracket, wire bracket and wire to the engine hanger bracket.

51. Install the wire to the timing belt rear plate.

52. Attach the wire protector to the intake manifold.

53. Attach the 2 ground cables the cylinder heads.

54. Connect the water bypass hoses to the throttle body.

55. Connect the wire clamps to the bracket on the right delivery pipe.

56. Attach the hoses to the intake manifold.

57. Attach the electrical connectors to the intake manifold.

58. Connect the fuel hose.

59. Install the timing belt rear plates, right plates first, then left plates. Tighten the retainers to 66 inch lbs. (7 Nm).

60. Install the throttle body cover.

61. Install the front exhaust pipe, power steering pump.

62. Install the camshaft position sensor and camshaft timing pulleys, tighten to 25 ft. lbs. (34 Nm).

63. Install the timing belt.

64. Fill the cooling system and perform an oil change.

65. Start the vehicle and check for leaks.

ENGINE ASSEMBLY

REMOVAL & INSTALLATION

1. Before servicing the vehicle, refer to the precautions section.
2. Properly relieve the fuel system pressure.
3. Disconnect the negative battery cable.
4. Remove the hood.
5. Remove the under-covers.
6. Drain the engine oil.
7. Drain and recycle the engine coolant.
8. Drain the transmission fluid.
9. Remove the V-bank cover.
10. Remove the air cleaner assembly.
11. Remove the drive belt.
12. Disconnect the radiator hoses and the cooler tubes.
13. Remove the fan shroud.
14. Remove the radiator.
15. Tag and disconnect all hoses, pipes and wires necessary for engine removal.
16. Remove the front exhaust pipes.
17. Remove the compressor and secure it out of the way.
18. Disconnect the vane pump.
19. Remove the rear and front drive shafts.
20. Attach a crane and equalizer to the engine.
21. Support the weight of the engine with the crane and remove the mount bolts.
22. Remove the engine and transmission assembly.
23. Installation is the reverse of removal. Observe the following torques:
    - Frame cross member bolts to 37 ft. lbs. (50 Nm) and the nuts to 55 ft. lbs. (74 Nm)
    - Engine mount bolts: 22 ft. lbs. (30 Nm)
    - Transmission bolts: 14mm 27 ft. lbs. (37 Nm) and 17mm (53 ft. lbs. (72 Nm)
    - Exhaust pipes: 46 ft. lbs. (62 Nm)
    - Rear drive shaft: 78 ft. lbs. (106 Nm)
    - Front drive shaft: 59 ft. lbs. (80 Nm)
    - Compressor: bolt, 36 ft. lbs. (49 Nm); and stay, 45 ft. lbs. (61 Nm)

EXHAUST MANIFOLD

REMOVAL & INSTALLATION

1. Before servicing the vehicle, refer to the precautions section.
2. Attach a hoist to the engine lifting eyes.
3. Remove or disconnect the following:
    - Negative battery cable
    - Heated Oxygen (HO_2S) sensor connectors
    - Exhaust manifold heat shield
    - Exhaust front pipe
    - Motor mount
    - Motor mount bracket
    - Exhaust manifold

To install:

➡Use new exhaust manifold nuts for assembly.

4. Install or connect the following:
    - Exhaust manifold. Tighten the nuts to 32 ft. lbs. (44 Nm).
    - Motor mount bracket. Tighten the bolts to 27 ft. lbs. (36 Nm).
    - Motor mount. Tighten the fasteners to 22 ft. lbs. (30 Nm).
    - Exhaust front pipe. Tighten the nuts to 46 ft. lbs. (62 Nm).
    - Exhaust manifold heat shield
    - HO_2S sensor connectors
    - Negative battery cable
5. Start the engine and check for leaks.

FLEXPLATE

REMOVAL & INSTALLATION

See Figure 12.

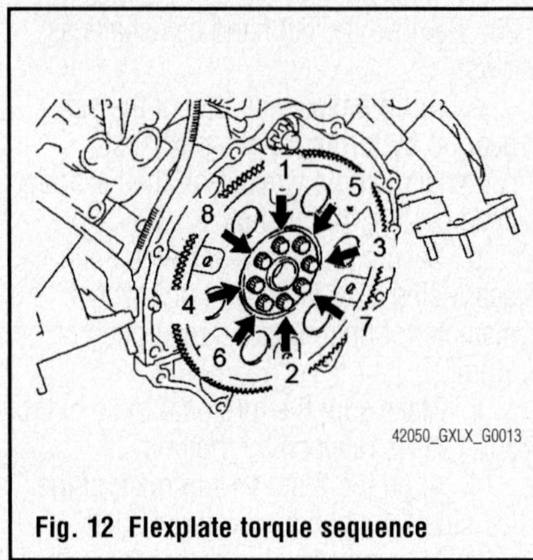

Fig. 12 Flexplate torque sequence

1. Before servicing the vehicle, refer to the precautions in the beginning of this section.
2. Remove the transmission. See transmission removal.
3. Remove the 8 bolts and the Flexplate.

To install:

4. Install the flexplate.
5. Coat two or three threads of each mounting bolt with Part No. 008833-00070 Three Bond 1324 or equivalent.
6. Tighten the bolts in sequence as follows:
    a. 49 Nm (36 ft. lbs.)
    b. Plus 90°.
7. Install the transmission.

INTAKE MANIFOLD

REMOVAL & INSTALLATION

1. Before servicing the vehicle, refer to the precautions section.
2. Relieve the fuel system pressure.
3. Drain the cooling system.
4. Drain the oil.
5. Remove the timing belt.
6. Disconnect the fuel lines.
7. Remove the front exhaust pipes from both sides.
8. Remove the intake manifold as follows:
    a. Remove the V bank cover.
    b. Disconnect the all electrical connectors from the manifold.
    c. Disconnect all hoses from the intake manifold.
    d. Disconnect the water bypass hoses from the throttle body.
    e. Disconnect the wire clamp bracket on the right hand delivery pipe.

f. Remove the engine wire harness protector from the water bypass joint and right hand cylinder head.

g. Remove the two ground cables from the left and right head.

h. Remove the throttle body cover.

i. Remove the engine wire from the engine hanger and bracket.

j. Remove the bolts and nuts that attach the upper manifold to the lower manifold.

To install:

9. Install new intake manifold gaskets and the manifold. Tighten the bolts to 13 ft. lbs. (18 Nm) in several passes.

10. Install the throttle cover bracket, wire bracket and wire to the engine hanger bracket.

11. Install the wire to the timing belt rear plate.

12. Attach the wire protector to the intake manifold.

13. Attach the 2 ground cables the cylinder heads.

14. Connect the water bypass hoses to the throttle body.

15. Connect the wire clamps to the bracket on the right delivery pipe.

16. Attach the hoses to the intake manifold.

17. Attach the electrical connectors to the intake manifold.

18. Connect the fuel hose.

19. Install the throttle body cover.

20. Install the front exhaust pipe, power steering pump.

21. Install the camshaft position sensor and camshaft timing pulleys, tighten to 25 ft. lbs. (34 Nm).

22. Install the timing belt.

23. Fill the cooling system and perform an oil change.

24. Start the vehicle and check for leaks.

OIL PAN

REMOVAL & INSTALLATION

See Figures 13 through 15.

1. Before servicing the vehicle, refer to the precautions section.

2. Remove the engine from the vehicle and mount it on a stand.

3. Remove or disconnect the following:

- Oil dipstick tube
- Lower oil pan
- Oil pan baffle
- Upper oil pan

To install:

4. The upper oil pan bolts are different lengths and are identified as follows:

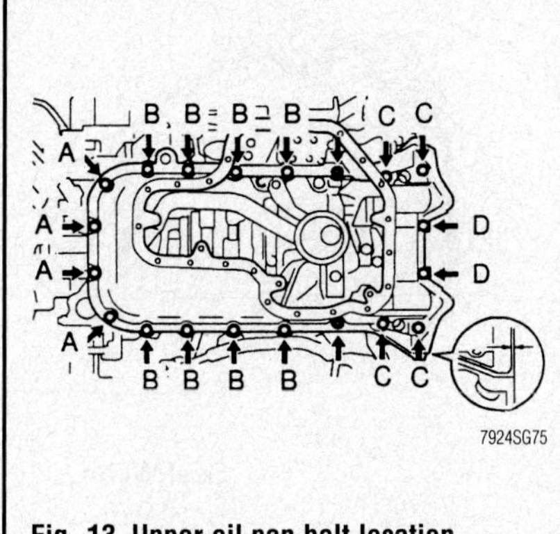

Fig. 13 Upper oil pan bolt location

- A: 0.79 inch (20mm) w/10mm head
- B: 0.98 inch (25mm) w/12mm head
- C: 2.36 inch (60mm) w/12mm head
- D: 1.38 inch (35mm) w/10mm head

5. Apply silicone sealant to the upper oil pan as shown.

6. Install the upper oil pan and tighten the fasteners in several passes to the following specifications:

- 10mm: 66 inch lbs. (7.5 Nm)
- 12mm: 21 ft. lbs. (28 Nm)

7. Install or connect the following:

- Oil pan baffle. Tighten the fasteners to 66 inch lbs. (7.5 Nm).

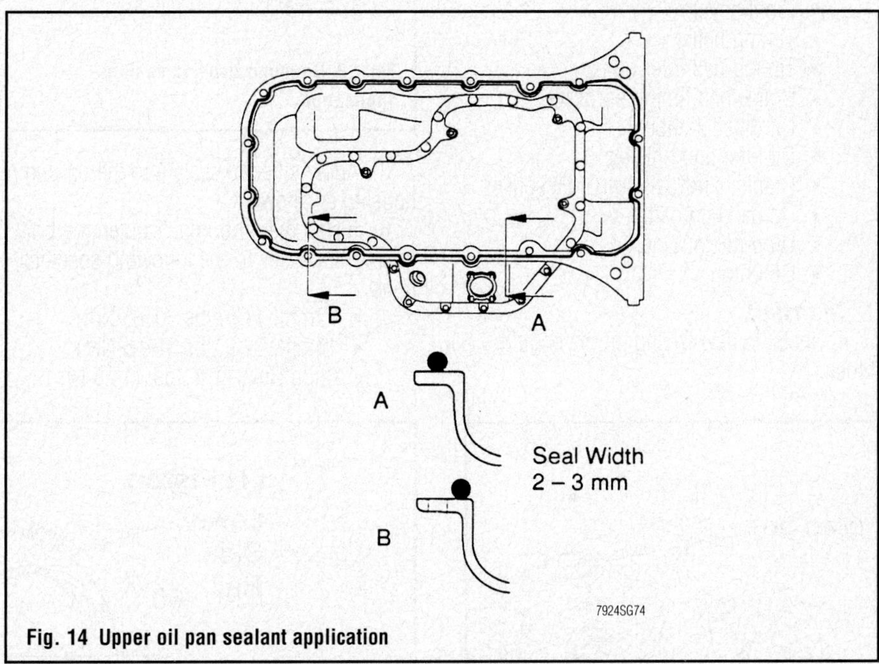

Fig. 14 Upper oil pan sealant application

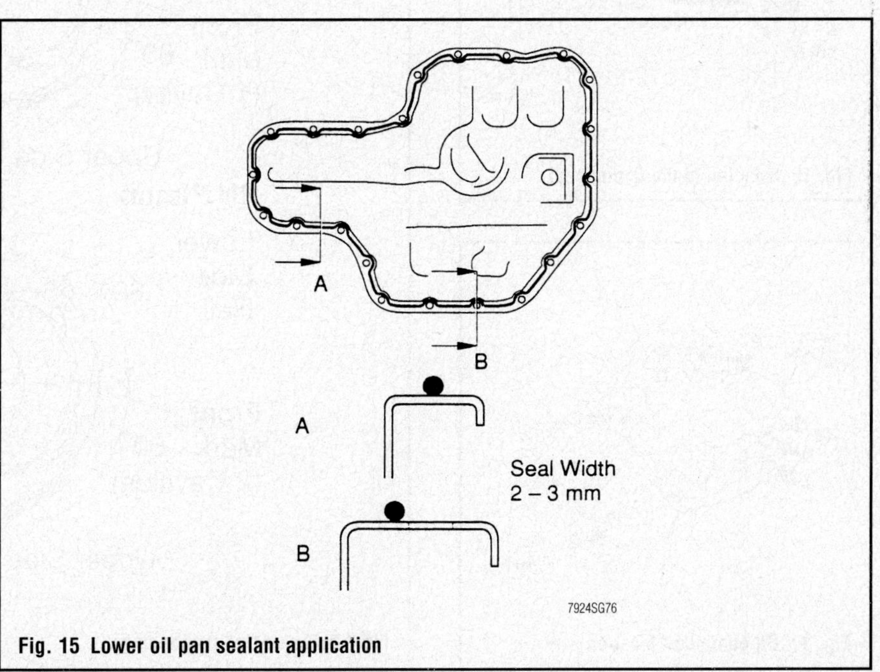

Fig. 15 Lower oil pan sealant application

- Lower oil pan. Tighten the fasteners in several passes to 66 inch lbs. (7.5 Nm).
- Oil dipstick tube

8. Install the engine.

OIL PUMP

REMOVAL & INSTALLATION

See Figures 16 through 18.

1. Before servicing the vehicle, refer to the precautions section.
2. Remove the engine from the vehicle and mount it on a stand.
3. Remove or disconnect the following:
   - Front cover
   - Timing belt.
   - Timing belt idler pulleys
   - Crankshaft timing sprocket
   - Oil dipstick tube
   - Oil filter and bracket
   - Crankshaft Position (CKP) sensor
   - Oil pan and baffle
   - Oil pump strainer
   - Oil pump

To install:

4. Install a new O-ring on the engine block.

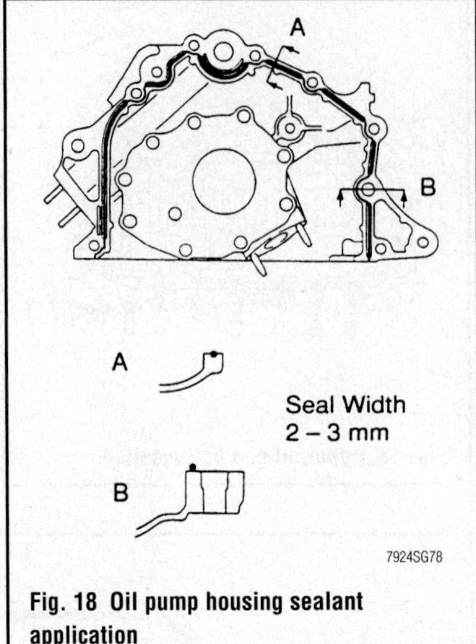

Fig. 18 Oil pump housing sealant application

5. Apply silicone sealant to the oil pump housing as shown.
6. Install the oil pump. Tighten the bolts in several passes to the following specifications:
   - 12mm: 11 ft. lbs. (15.5 Nm)
   - 14mm: 22 ft. lbs. (30.5 Nm)
   - 6mm Hex: 11 ft. lbs. (15.5 Nm)
7. The upper oil pan bolts are different lengths and are identified as follows:
   - A: 1.38 inch (35mm) w/12mm head
   - B: 1.97 inch (50mm) w/12mm head
   - C: 4.17 inch (106mm) w/12mm head
   - D: 1.57 inch (40mm) w/14mm head
   - E: 1.18 inch (30mm) w/6mm hex head
8. Install or connect the following:
   - Oil pump pickup tube. Tighten the bolts to 66 inch lbs. (7.5 Nm).
   - Oil pan and baffle
   - CKP sensor
   - Oil filter and bracket. Tighten the bolts to 13 ft. lbs. (18 Nm).
   - Oil dipstick tube
   - Crankshaft timing sprocket
   - Timing belt idler pulleys
   - Timing belt
   - Front cover
9. Install the engine.

PISTON AND RING

POSITIONING

See Figures 19 through 21.

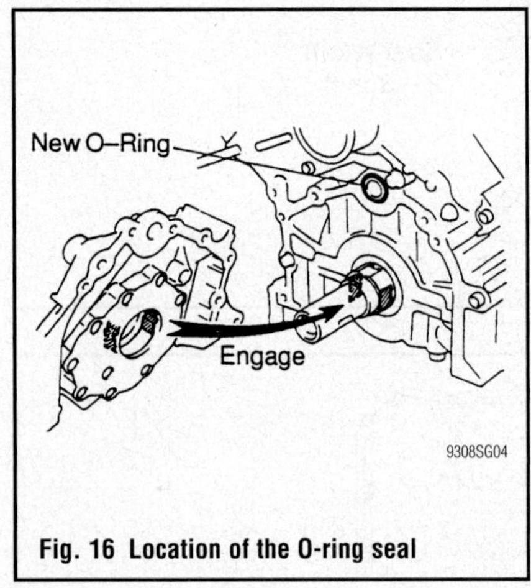

Fig. 16 Location of the O-ring seal

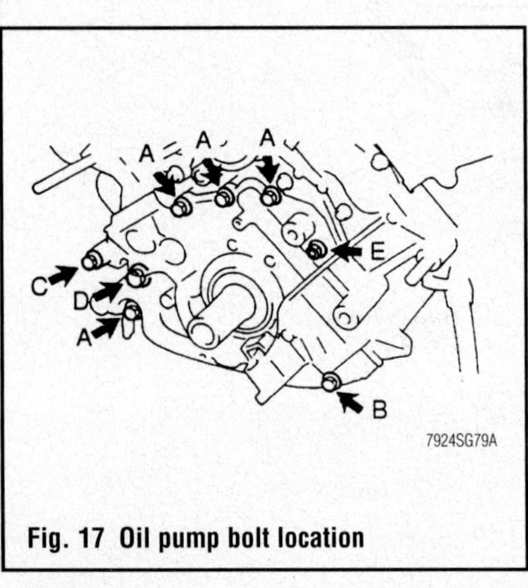

Fig. 17 Oil pump bolt location

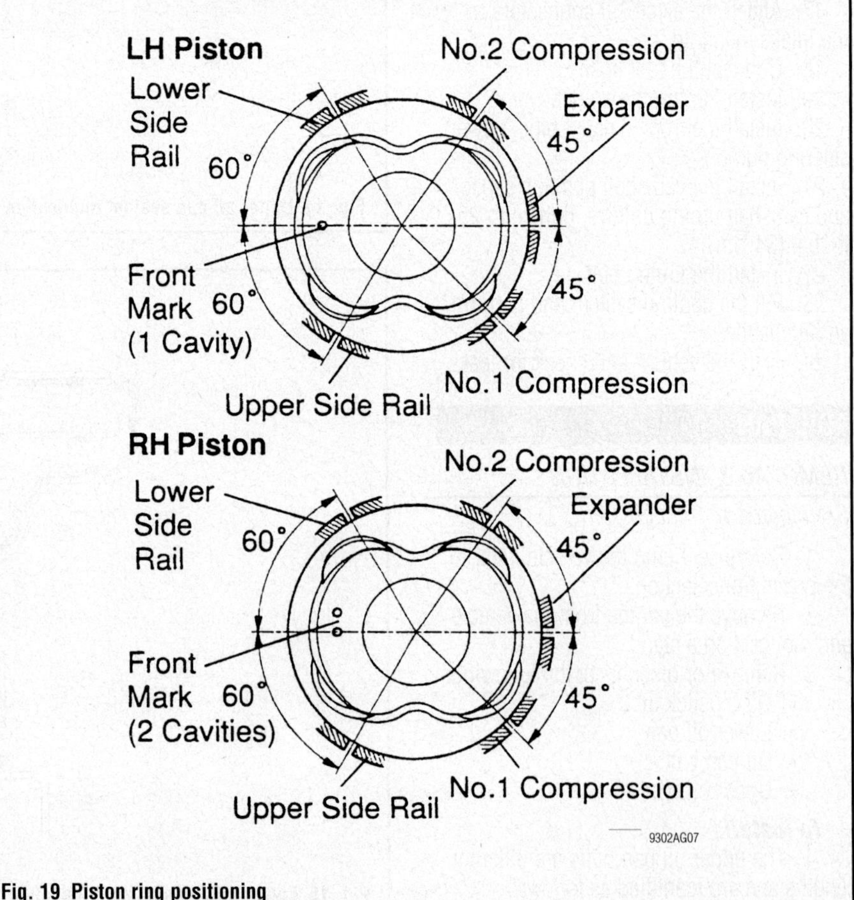

Fig. 19 Piston ring positioning

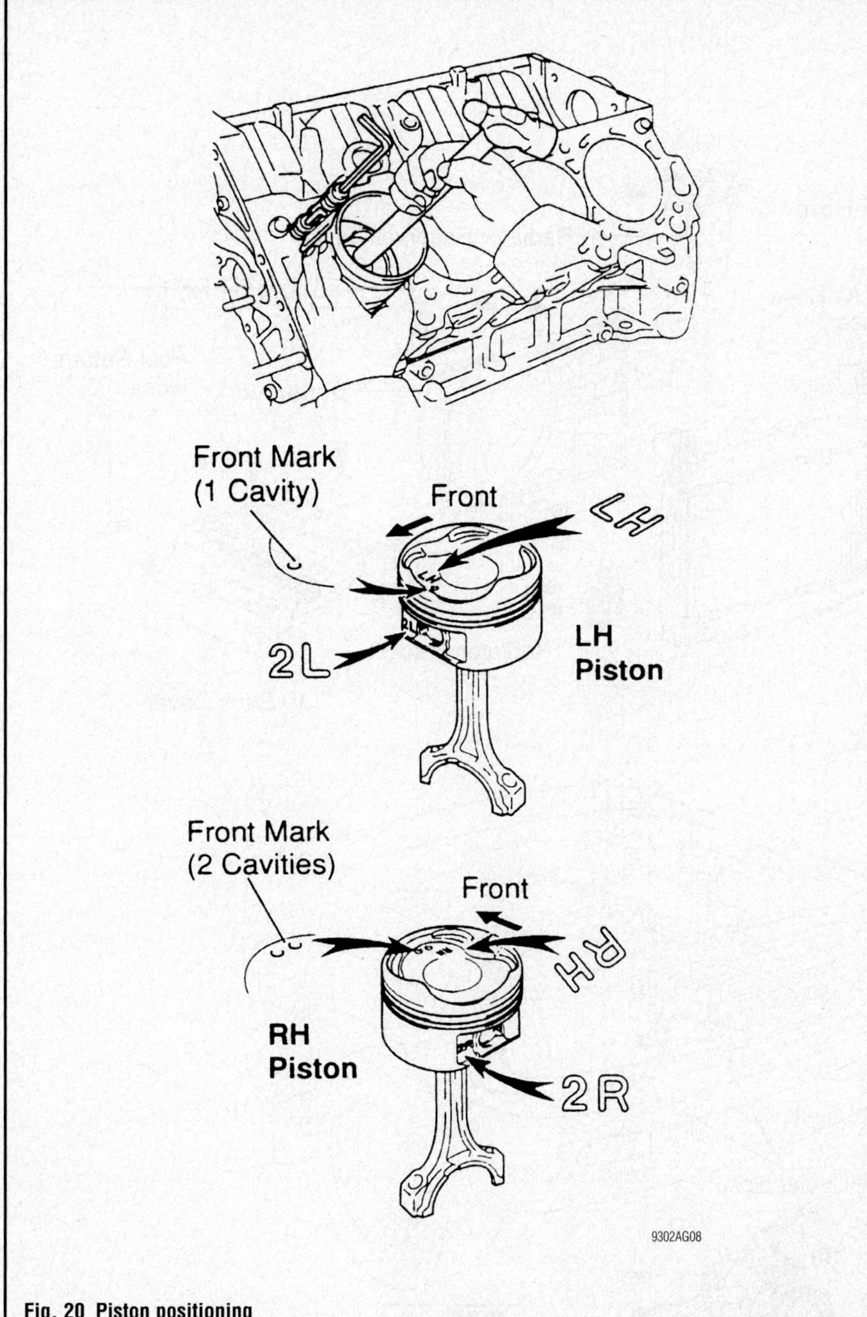

Fig. 20 Piston positioning

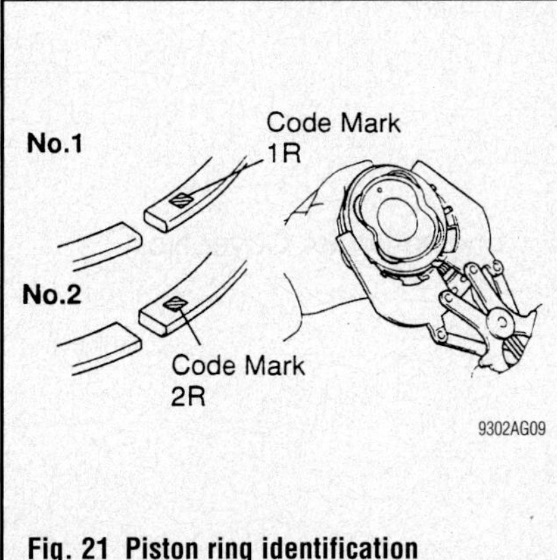

Fig. 21 Piston ring identification

REAR MAIN SEAL

REMOVAL & INSTALLATION

1. Before servicing the vehicle, refer to the precautions section.
2. Remove the transmission and flywheel from the vehicle.
3. Cut off the rubber lip portion of the seal with a sharp knife.
4. Pry out the oil seal.

To install:

5. Install the rear main seal so that it is flush with the seal retainer housing.
6. Install or connect the following:
   - Flywheel/driveplate. Tighten the bolts to 35 ft. lbs. (48 Nm) plus a 90 degree turn.
   - Transmission

REMOVAL & INSTALLATION

See Timing Belt and Sprockets

TIMING BELT AND SPROCKETS

REMOVAL & INSTALLATION

See Figures 22 through 31.

1. Disconnect the negative battery cable.
2. Raise and safely support the vehicle.
3. Remove the oil pan protector and the engine under cover.
4. Drain the cooling system and store the coolant for refilling purposes.
5. Lower the vehicle and remove the battery clamp cover.
6. From the top of the engine, remove the fuel return hose, the engine cover nuts/bolts and the cover.
7. Remove the air cleaner and the intake air connector assembly.
8. Remove the cooling fan pulley by performing the following procedures:
   a. Loosen the 4 fan clutch-to-fan pulley nuts.
   b. Using a box-end wrench on the serpentine drive belt tensioner bolt, rotate the tensioner counterclockwise and remove the drive belt.

➡ The serpentine drive belt tensioner bolt is a left-hand thread.

   c. Remove the fan clutch-to-fan pulley nuts, the fan, the clutch assembly and the fan pulley.
9. Remove the radiator by performing the following procedures:
   a. Disconnect the upper, lower and reservoir hoses from the radiator.
   b. Disconnect and plug the automatic transmission oil cooler at the radiator. Disconnect the automatic transmission oil cooler hoses from the fan shroud clamp.
   c. Remove the radiator reservoir tank.
   d. Remove the fan shroud-to-radiator bolts and the shroud.
   e. Remove the 2 upper radiator-to-chassis nuts.
   f. Remove the middle radiator-to-chassis nut/bolts and brackets.
   g. Carefully, lift the radiator from the vehicle.
10. Remove the serpentine drive belt idler pulley bolt, cover plate and pulley.

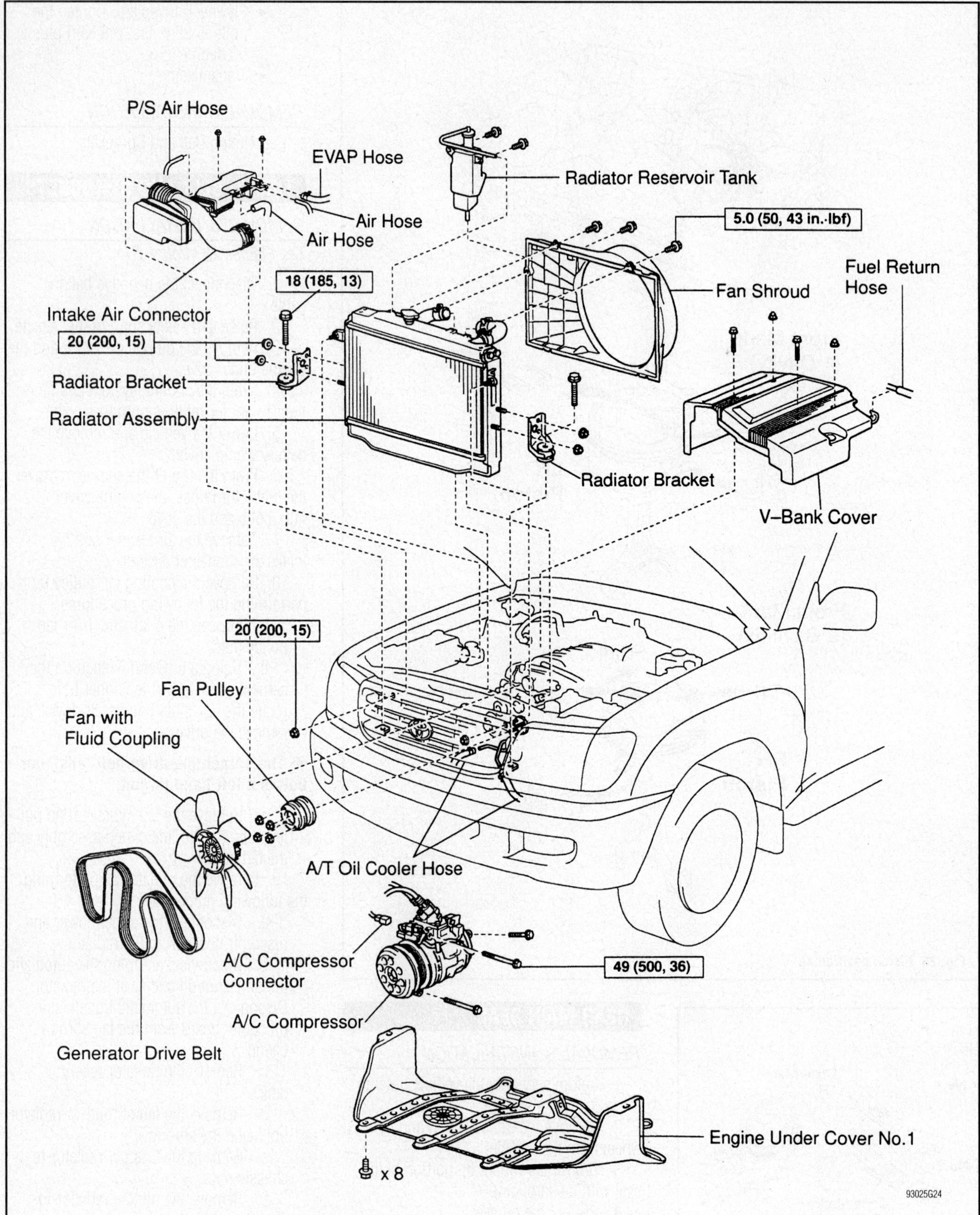

Fig. 22 Exploded view of vehicle components for timing belt replacement

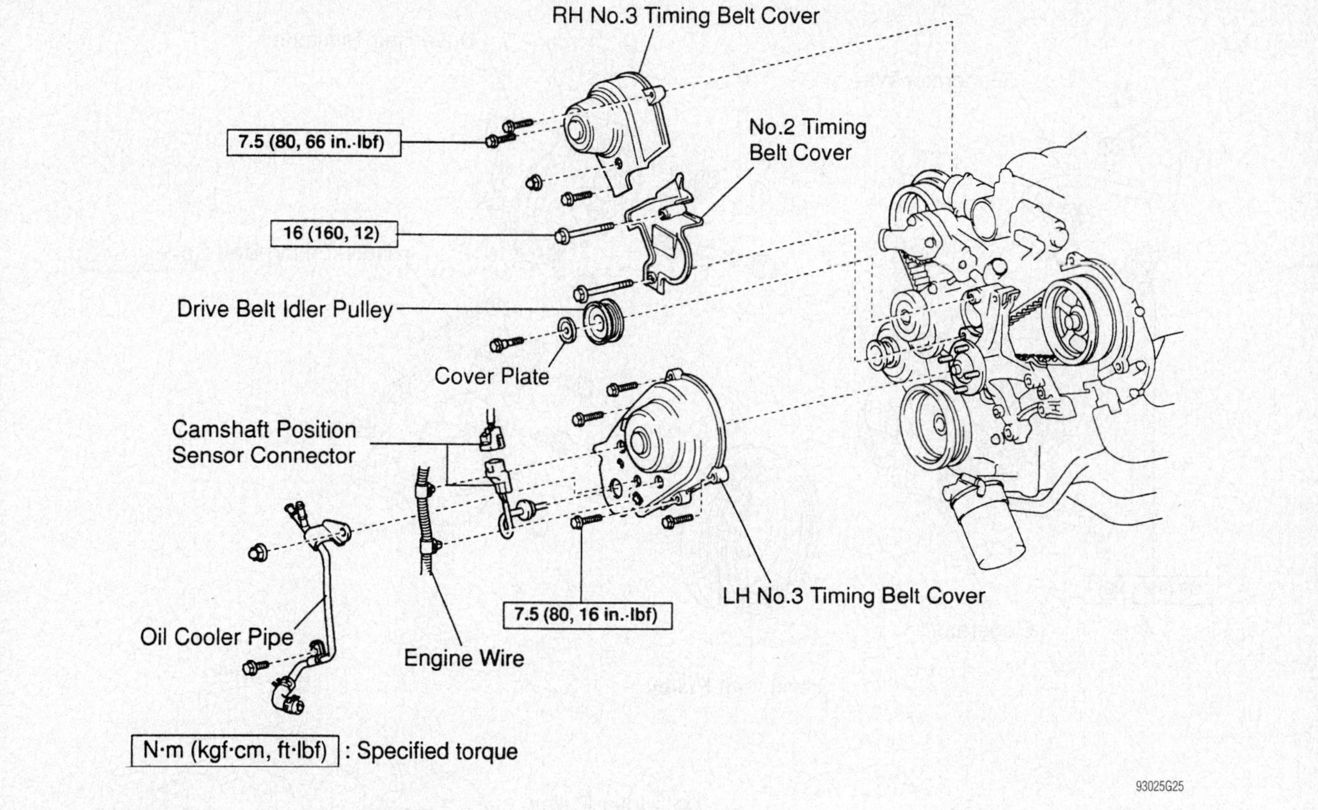

Fig. 23 Exploded view of upper timing belt covers

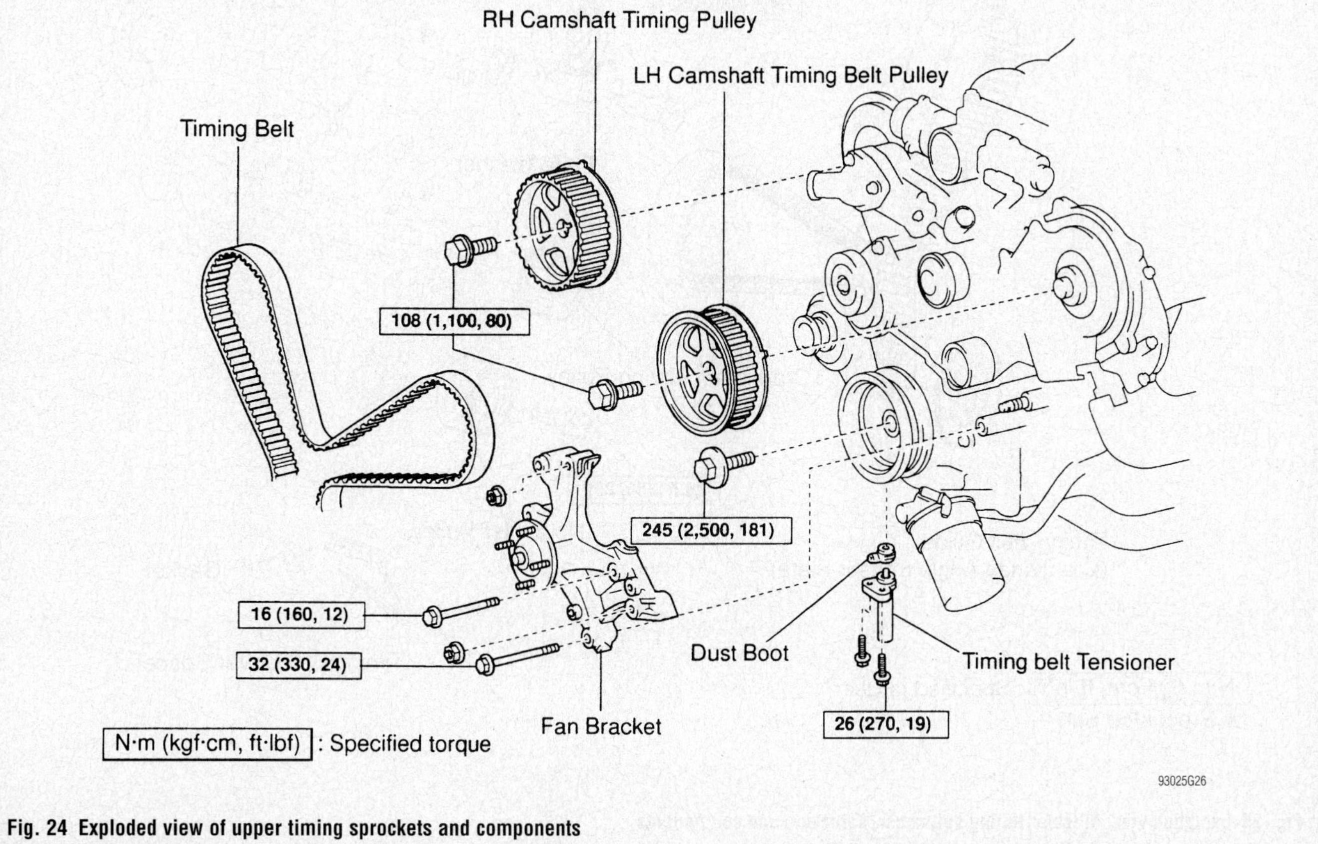

Fig. 24 Exploded view of upper timing sprockets and components

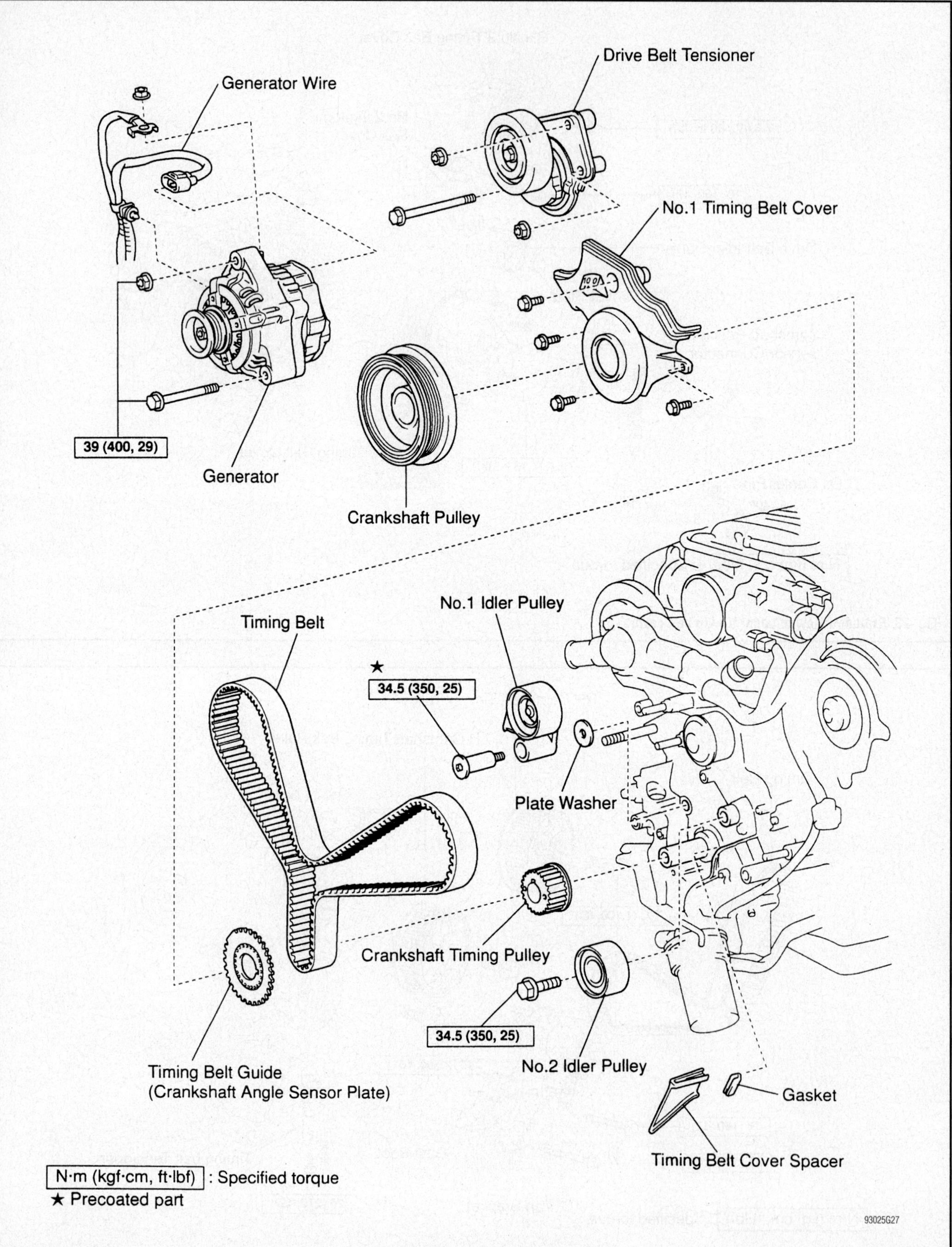

Fig. 25 Exploded view of lower timing belt cover, sprockets and components

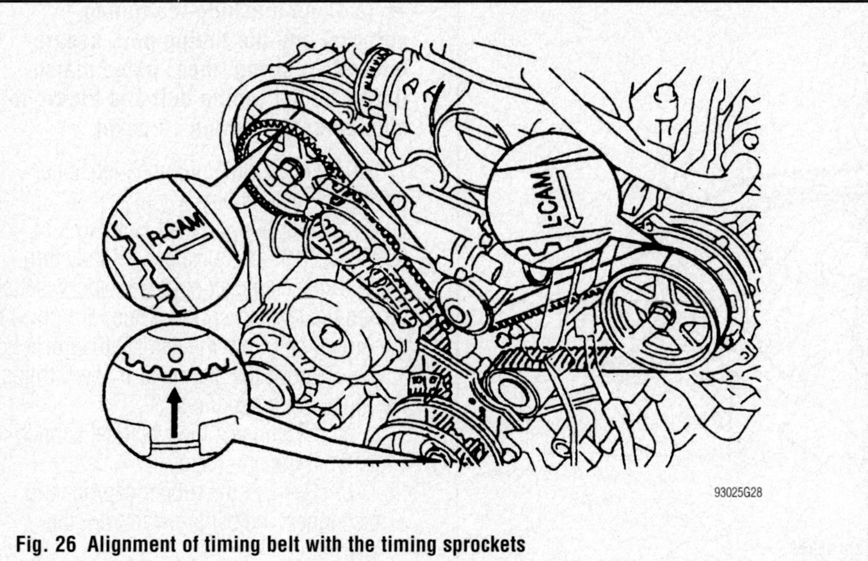

Fig. 26 Alignment of timing belt with the timing sprockets

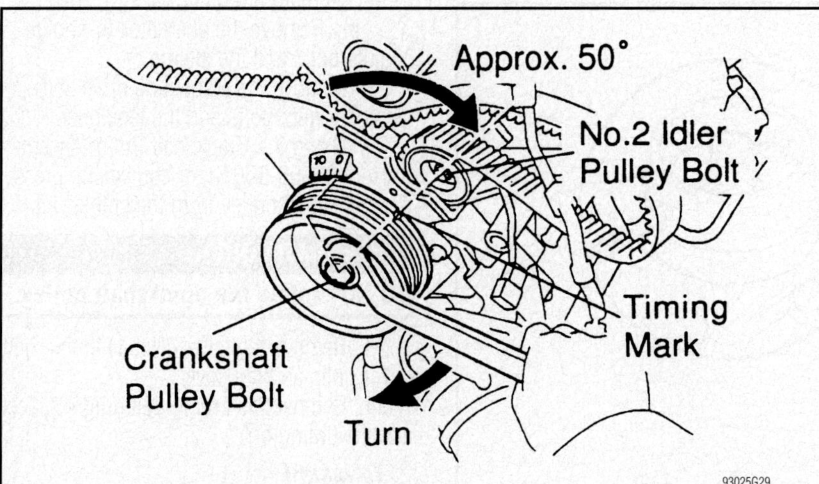

Fig. 27 Aligning of crankshaft pulley timing mark with the center line of the crankshaft pulley bolt and the idler pulley bolt

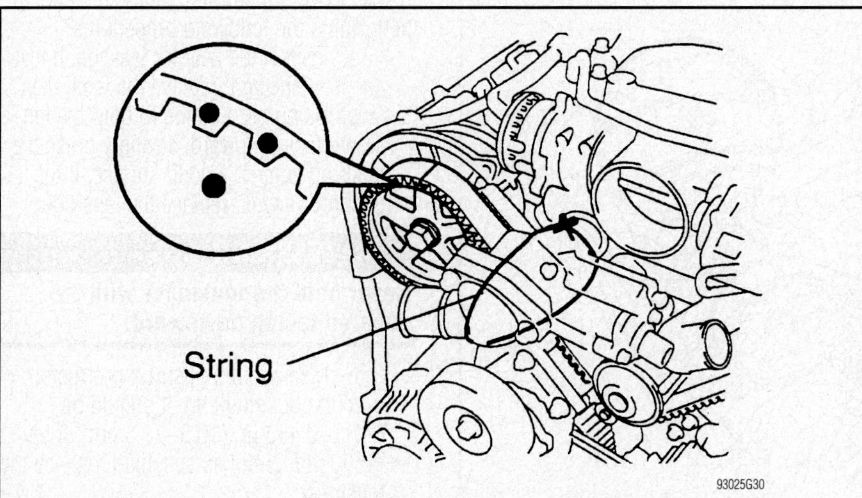

Fig. 28 Securing the timing belt with string and matchmarking the camshaft with the timing belt

11. Remove the right side (No. 3) timing belt cover.
12. Remove the left side (No. 3) timing belt cover by performing the following procedures:
    a. Disconnect the engine wire from both wire clamps.
    b. Disconnect the camshaft position sensor wire from the wire clamp on the left-side (No.3) timing belt cover.
    c. Disconnect the sensor connector from the connector bracket.
    d. Disconnect the sensor connector.
    e. Remove the wire grommet from the left-side (No. 3) timing belt cover.
    f. Remove the oil cooler tube bolts and tube.
13. Remove the middle (No. 2) timing belt cover bolts and cover.
14. Remove the cooling fan bracket nuts/bolts and bracket.

➡ If reusing the timing belt, make sure that there are 3 installation marks on the belt; if there are none, install them.

15. Using the Crankshaft Pulley Holding tool 09213-70010, Bolt tool 90105-08076 and Companion Flange Holding tool 09330-00021, or equivalent, loosen the crankshaft pulley bolt.
16. Position the No. 1 cylinder to approximately 50 degrees After Top Dead Center (ATDC) of the compression stroke by performing the following procedures:
    a. Rotate the crankshaft pulley (CLOCKWISE) to align its groove with the timing mark "0" on the lower (No. 1) timing belt cover.
    b. Check that the camshaft sprocket timing marks are aligned with the rear timing belt plate marks; if not, rotate the crankshaft 1 revolution (360 degrees).
    c. Rotate the crankshaft pulley approximately 50 degrees (CLOCKWISE) and align the crankshaft pulley timing mark between the centers of the crankshaft pulley bolt and the idler pulley bolt.

⁂ WARNING

If the timing belt is disengaged, having the crankshaft pulley in the wrong angle can cause the valve to come into contact with the piston when removing the camshaft pulley.

17. Remove the crankshaft pulley bolt.

➡ If reusing the timing belt and the installation marks have disappeared, place new installation marks on the timing belt to match the camshaft timing sprocket marks.

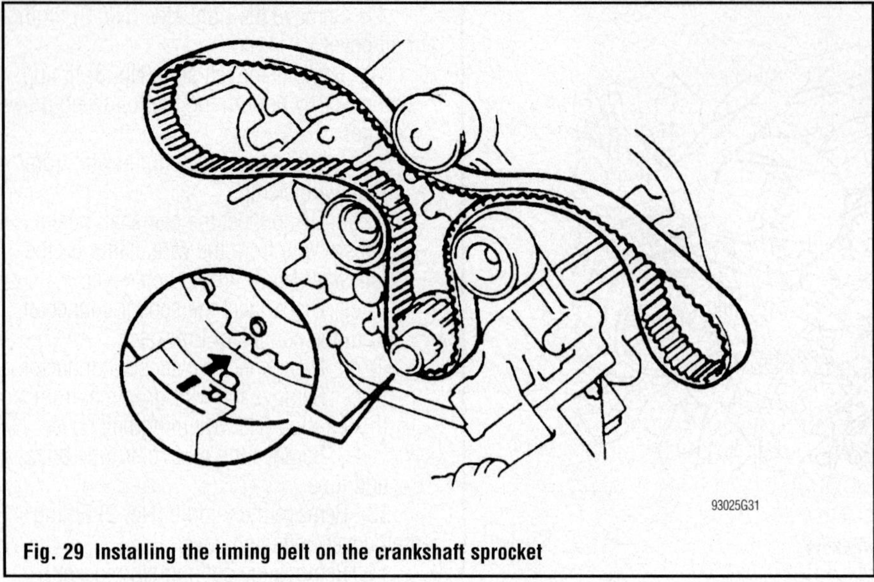

Fig. 29 Installing the timing belt on the crankshaft sprocket

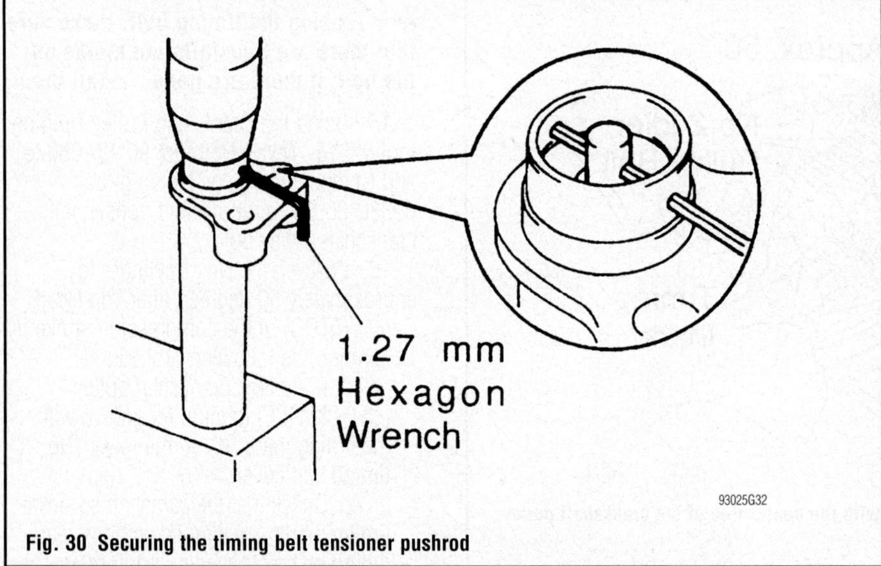

Fig. 30 Securing the timing belt tensioner pushrod

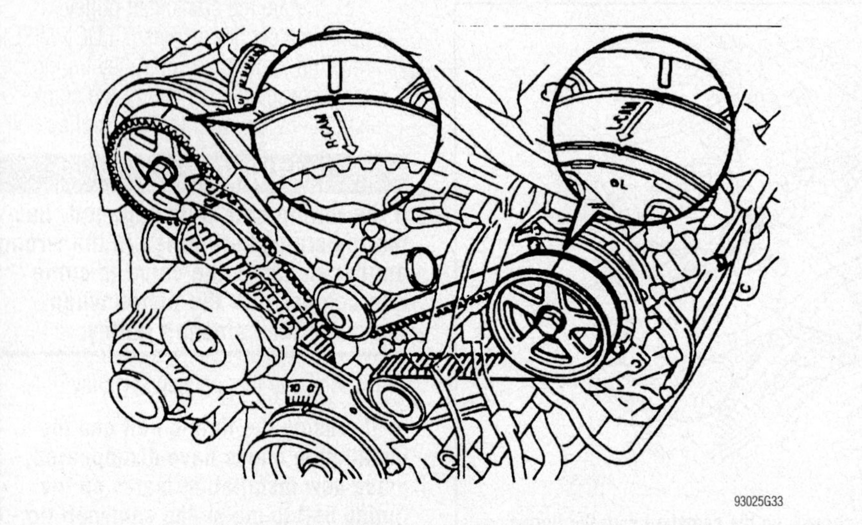

Fig. 31 Checking the TDC alignment marks after rotating the crankshaft 2 revolutions

➡ To avoid meshing the timing sprocket and the timing belt, secure one with a string; then, place matchmarks on the timing belt and the right-side camshaft timing sprocket.

18. Remove the timing belt tensioner bolts and the tensioner.
19. Using the Camshaft Holding tool 09960-10010, or equivalent, slightly turn the left-side camshaft sprocket clockwise to loosen the tension spring. Then, disconnect the timing belt from the camshaft sprockets.
20. Remove the alternator by performing the following procedures:
    a. Disconnect the electrical connector from the alternator.
    b. Remove the rubber cap/nut and disconnect the battery wire from the alternator.
    c. Disconnect the wire clamp from the alternator cord clip.
    d. Remove the alternator-to-engine nuts/bolts and the alternator.
21. Remove the serpentine drive belt tensioner nuts/bolts and the tensioner.
22. Using the Crankshaft Puller Assembly tool 09950-50012, or equivalent, press the crankshaft pulley from the crankshaft.

✱✱ WARNING

DO NOT rotate the crankshaft pulley.

23. Remove the lower (No. 1) timing belt cover bolts and the cover.
24. Remove the timing belt guide, spacer and the timing belt.

To install:

➡With the timing belt removed, this is a perfect opportunity to inspect and/or replace the water pump.

25. Inspect the timing belt tensioner by performing the following procedures:
    a. Inspect the seal for leakage; if leakage is suspected, replace the tensioner.
    b. Using both hands to hold the tensioner facing upward, strongly press the pushrod against a solid surface. If the pushrod moves, replace the tensioner.

✱✱ WARNING

Never hold the tensioner with the pushrod facing downward.

    c. Measure the pushrod protrusion from the housing end, it should be 0.413–0.453 in. (10.5–11.5mm). If the protrusion is not as specified, replace the tensioner.
26. Temporarily install the timing belt by performing the following procedures:

a. Align the timing belt's installation mark with the crankshaft timing sprocket.

b. Install the timing belt on the crankshaft timing sprocket, the No. 1 idler pulley and the No. 2 idler pulley.

27. Install the gasket to the timing belt cover spacer and install the cover spacer.

28. Install the timing belt guide with the cup side facing outward.

29. Install the lower (No. 1) timing belt cover.

30. Install the crankshaft pulley by performing the following procedures:

a. Align the crankshaft pulley with the crankshaft key.

b. Using the Crankshaft Installer tool 09223-46011, or equivalent, and a hammer, tap the crankshaft pulley into position.

31. Install the serpentine drive belt tensioner and torque the tensioner-to-engine bolts to 12 ft. lbs. (16 Nm).

➡To install the serpentine drive belt tensioner, use a bolt 4.18 in. (106mm) in length.

32. Check that the crankshaft pulley's timing mark is aligned with the centers of the idler pulley and crankshaft pulley bolts.

33. Install the alternator and torque the alternator-to-engine nuts/bolts to 29 ft. lbs. (39 Nm). Connect the alternator's electrical connectors and clip.

34. Install the timing belt to the left-side camshaft by performing the following procedures:

a. Rotate the left-side camshaft pulley to align the timing belt installation mark with the camshaft sprocket's timing mark and slide the belt onto the camshaft timing sprocket.

b. Using the Camshaft Holding tool 09960-10010, or equivalent, slightly turn the left-side camshaft sprocket counterclockwise to place tension on the timing belt between the crankshaft sprocket and the camshaft sprocket.

35. Rotate the right-side camshaft pulley to align the timing belt installation mark with the camshaft sprocket's timing mark and slide the belt onto the camshaft timing sprocket.

36. Using a vertical press, slowly press the pushrod into the housing using 200–2205 lbs. (981–9807 N) until the holes align, then, install a 1.27mm Allen® wrench to secure the pushrod and release the press. Install the dust boot on the tensioner housing.

37. Install the timing belt tensioner and torque the bolts to 19 ft. lbs. (26 Nm).

38. Using a pair of pliers, remove the Allen® wrench from the tensioner housing.

39. Check the valve timing by performing the following procedure:

a. Temporarily install the crankshaft pulley bolt.

b. Slowly, rotate the crankshaft pulley 2 revolutions (CLOCKWISE) and realign the TDC marks.

➡If the pulley/sprocket timing marks do not realign, remove the timing belt and reinstall it.

40. Using the Crankshaft Pulley Holding tool 09213-70010, Bolt tool 90105-08076 and Companion Flange Holding tool 09330-00021, or equivalent, torque the crankshaft pulley bolt to 181 ft. lbs. (245 Nm).

41. Install the cooling fan bracket and torque the 12mm (head size) bolt to 12 ft. lbs. (16 Nm) and the 14mm (head size) bolt to 24 ft. lbs. (32 Nm).

42. Install the air conditioning compressor.

43. Install the middle (No. 2) timing belt cover and torque the bolts to 12 ft. lbs. (16 Nm).

44. Install the upper right-side (No. 3) timing belt cover and torque the bolts to 66 inch lbs. (7.5 Nm).

45. Install the upper left-side (No. 3) timing belt cover by performing the following procedures:

a. Install the oil cooler tube and bolt.

b. Feed the Camshaft Position Sensor (CPS) through the left-side (No. 3) timing belt cover hole.

c. Install the left-side (No. 3) timing belt cover and torque the bolts to 66 inch lbs. (7.5 Nm).

d. Install the wire grommet to the left-side (No. 3) timing belt cover.

e. Install the sensor connector to the connector bracket and connect the sensor connector.

f. Install the sensor wire and the engine wire to the clamps on the left-side (No. 3) timing belt cover.

46. Install the drive belt idler pulley and cover plate; then, torque the pulley bolt to 27 ft. lbs. (37 Nm).

47. To complete the installation, reverse the removal procedures.

48. Refill the cooling system and connect the negative battery cable.

VALVE COVERS

REMOVAL & INSTALLATION

1. Before servicing the vehicle, refer to the precautions in the beginning of this section.

2. Remove battery clamp cover.

3. Remove throttle body cover.

4. Remove air cleaner and intake air connector assembly.

5. Remove the upper timing belt covers.

6. Remove ignition coils.

7. Remove the 9 bolts, seal washers and the right cylinder head cover.

8. Remove the oil dipstick for the transmission.

9. Disconnect the PCV hose.

10. Disconnect the engine wire clamp from the wire bracket on the cylinder head cover.

11. Remove the 9 bolts, 9 seal washers and the left cylinder head cover.

To install:

12. Install the left cylinder head cover. Tighten bolts to 53 inch lbs. (6.0 Nm)

13. Connect the engine wire clamp to the wire bracket on the cylinder head cover.

14. Connect the PCV hose.

15. Replace the oil dipstick for the transmission.

16. Install the right cylinder head cover. Tighten bolts to 53 inch lbs. (6.0 Nm)

17. Install the ignition coils.

18. Install the upper timing belt covers. Tighten bolts to 66 inch lbs. (7.5 Nm)

19. Install the air cleaner and intake air connector assembly.

20. Install the throttle body cover.

21. Install the battery clamp cover.

VALVE LASH

ADJUSTMENT

See Figures 32 and 33.

➡Measure valve clearance with the engine cold.

1. Before servicing the vehicle, refer to the precautions section.

2. Drain the cooling system.

3. Remove or disconnect the following:

- Negative battery cable
- Ignition coils
- Valve covers

4. Set the engine to the top of the compression stroke with the valves closed for the cylinder to be measured.

5. Check the valve clearance. The valve clearance specifications are as follows:

- Intake: 0.006–0.010 in. (0.15–0.25mm)
- Exhaust: 0.010–0.014 in. (0.25–0.35mm)

6. Record the measurements for each valve.

Installed shim thickness mm (in.) / Measured clearance mm (in.)	2.000 (0.0787)	2.020 (0.0795)	2.040 (0.0803)	2.060 (0.0811)	2.080 (0.0819)	2.100 (0.0827)	2.120 (0.0835)	2.140 (0.0843)	2.160 (0.0850)	2.180 (0.0858)	2.200 (0.0866)	2.210 (0.0870)	2.220 (0.0874)	2.230 (0.0878)	2.240 (0.0882)	2.250 (0.0886)	2.260 (0.0890)	2.270 (0.0894)	2.280 (0.0898)	2.290 (0.0902)	2.300 (0.0906)	2.310 (0.0909)	2.320 (0.0913)	2.330 (0.0917)	2.340 (0.0921)	2.350 (0.0925)	2.360 (0.0929)	2.370 (0.0933)	2.380 (0.0937)	2.390 (0.0941)	2.400 (0.0945)	2.410 (0.0949)	2.420 (0.0953)	2.430 (0.0957)	2.440 (0.0961)	2.450 (0.0965)	2.460 (0.0969)	2.470 (0.0972)	2.480 (0.0976)	2.490 (0.0980)	2.500 (0.0984)	2.510 (0.0988)	2.520 (0.0992)	2.530 (0.0996)	2.540 (0.1000)	2.550 (0.1004)	2.560 (0.1008)	2.570 (0.1012)	2.580 (0.1016)	2.590 (0.1020)	2.600 (0.1024)	2.620 (0.1031)	2.640 (0.1039)	2.660 (0.1047)	2.680 (0.1055)	2.700 (0.1063)	2.720 (0.1071)	2.740 (0.1079)	2.760 (0.1087)	2.780 (0.1094)	2.800 (0.1102)
0.000-0.030 (0.0000-0.0012)							00	00	00	00	02	02	04	04	06	06	08	08	10	10	12	12	14	14	16	16	18	18	20	20	22	22	24	24	26	26	28	28	30	30	32	32	34	34	36	36	38	38	40	40	42	44	46	48	50	52	54	56	58	60	62
0.031-0.050 (0.0012-0.0020)						00	00	00	00	02	04	06	06	08	08	10	10	12	12	14	14	16	16	18	18	20	20	22	22	24	24	26	26	28	28	30	30	32	32	34	34	36	36	38	38	40	40	42	42	44	44	46	48	50	52	54	56	58	60	62	64
0.051-0.070 (0.0020-0.0028)					00	00	00	00	02	04	06	08	08	10	10	12	12	14	14	16	16	18	18	20	20	22	22	24	24	26	26	28	28	30	30	32	32	34	34	36	36	38	38	40	40	42	42	44	44	46	46	48	50	52	54	56	58	60	62	64	66
0.071-0.090 (0.0028-0.0035)				00	00	00	00	02	04	06	08	10	10	12	12	14	14	16	16	18	18	20	20	22	22	24	24	26	26	28	28	30	30	32	32	34	34	36	36	38	38	40	40	42	42	44	44	46	46	48	48	50	52	54	56	58	60	62	64	66	68
0.091-0.110 (0.0036-0.0043)			00	00	00	00	02	04	06	08	10	12	12	14	14	16	16	18	18	20	20	22	22	24	24	26	26	28	28	30	30	32	32	34	34	36	36	38	38	40	40	42	42	44	44	46	46	48	48	50	50	52	54	56	58	60	62	64	66	68	70
0.111-0.130 (0.0044-0.0051)		00	00	00	00	02	04	06	08	10	12	14	14	16	16	18	18	20	20	22	22	24	24	26	26	28	28	30	30	32	32	34	34	36	36	38	38	40	40	42	42	44	44	46	46	48	48	50	50	52	52	54	56	58	60	62	64	66	68	70	72
0.131-0.149 (0.0052-0.0059)		00	00	00	02	04	06	08	10	12	14	16	16	18	18	20	20	22	22	24	24	26	26	28	28	30	30	32	32	34	34	36	36	38	38	40	40	42	42	44	44	46	46	48	48	50	50	52	52	54	54	56	58	60	62	64	66	68	70	72	74
0.150-0.250 (0.0059-0.0098)																																																													
0.251-0.270 (0.0099-0.0106)	06	08	10	12	14	16	18	20	22	24	26	28	28	30	30	32	32	34	34	36	36	38	38	40	40	42	42	44	44	46	46	48	48	50	50	52	52	54	54	56	56	58	58	60	60	62	62	64	64	66	66	68	70	72	74	76	78	80	80	80	
0.271-0.290 (0.0107-0.0114)	08	10	12	14	16	18	20	22	24	26	28	30	30	32	32	34	34	36	36	38	38	40	40	42	42	44	44	46	46	48	48	50	50	52	52	54	54	56	56	58	58	60	60	62	62	64	64	66	66	68	68	70	72	74	76	78	80	80	80		
0.291-0.310 (0.0115-0.0122)	10	12	14	16	18	20	22	24	26	28	30	32	32	34	34	36	36	38	38	40	40	42	42	44	44	46	46	48	48	50	50	52	52	54	54	56	56	58	58	60	60	62	62	64	64	66	66	68	68	70	70	72	74	76	78	80	80	80			
0.311-0.330 (0.0122-0.0130)	12	14	16	18	20	22	24	26	28	30	32	34	34	36	36	38	38	40	40	42	42	44	44	46	46	48	48	50	50	52	52	54	54	56	56	58	58	60	60	62	62	64	64	66	66	68	68	70	70	72	72	74	76	78	80	80	80				
0.331-0.350 (0.0130-0.0138)	14	16	18	20	22	24	26	28	30	32	34	36	36	38	38	40	40	42	42	44	44	46	46	48	48	50	50	52	52	54	54	56	56	58	58	60	60	62	62	64	64	66	66	68	68	70	70	72	72	74	74	76	78	80	80	80					
0.351-0.370 (0.0138-0.0146)	16	18	20	22	24	26	28	30	32	34	36	38	38	40	40	42	42	44	44	46	46	48	48	50	50	52	52	54	54	56	56	58	58	60	60	62	62	64	64	66	66	68	68	70	70	72	72	74	74	76	76	78	80	80	80						
0.371-0.390 (0.0146-0.0154)	18	20	22	24	26	28	30	32	34	36	38	40	40	42	42	44	44	46	46	48	48	50	50	52	52	54	54	56	56	58	58	60	60	62	62	64	64	66	66	68	68	70	70	72	72	74	74	76	76	78	78	80	80	80							
0.391-0.410 (0.0154-0.0161)	20	22	24	26	28	30	32	34	36	38	40	42	42	44	44	46	46	48	48	50	50	52	52	54	54	56	56	58	58	60	60	62	62	64	64	66	66	68	68	70	70	72	72	74	74	76	76	78	78	80	80	80	80								
0.411-0.430 (0.0162-0.0169)	22	24	26	28	30	32	34	36	38	40	42	44	44	46	46	48	48	50	50	52	52	54	54	56	56	58	58	60	60	62	62	64	64	66	66	68	68	70	70	72	72	74	74	76	76	78	78	80	80	80	80	80									
0.431-0.450 (0.0170-0.0177)	24	26	28	30	32	34	36	38	40	42	44	46	46	48	48	50	50	52	52	54	54	56	56	58	58	60	60	62	62	64	64	66	66	68	68	70	70	72	72	74	74	76	76	78	78	80	80	80	80	80	80										
0.451-0.470 (0.0178-0.0185)	26	28	30	32	34	36	38	40	42	44	46	48	48	50	50	52	52	54	54	56	56	58	58	60	60	62	62	64	64	66	66	68	68	70	70	72	72	74	74	76	76	78	78	80	80	80	80	80	80	80											
0.471-0.490 (0.0185-0.0193)	28	30	32	34	36	38	40	42	44	46	48	50	50	52	52	54	54	56	56	58	58	60	60	62	62	64	64	66	66	68	68	70	70	72	72	74	74	76	76	78	78	80	80	80	80	80	80	80													
0.491-0.510 (0.0193-0.0201)	30	32	34	36	38	40	42	44	46	48	50	52	52	54	54	56	56	58	58	60	60	62	62	64	64	66	66	68	68	70	70	72	72	74	74	76	76	78	78	80	80	80	80	80	80	80															
0.511-0.530 (0.0201-0.0209)	32	34	36	38	40	42	44	46	48	50	52	54	54	56	56	58	58	60	60	62	62	64	64	66	66	68	68	70	70	72	72	74	74	76	76	78	78	80	80	80	80	80	80	80																	
0.531-0.550 (0.0209-0.0217)	34	36	38	40	42	44	46	48	50	52	54	56	56	58	58	60	60	62	62	64	64	66	66	68	68	70	70	72	72	74	74	76	76	78	78	80	80	80	80	80	80	80																			
0.551-0.570 (0.0217-0.0224)	36	38	40	42	44	46	48	50	52	54	56	58	58	60	60	62	62	64	64	66	66	68	68	70	70	72	72	74	74	76	76	78	78	80	80	80	80	80	80	80																					
0.571-0.590 (0.0225-0.0232)	38	40	42	44	46	48	50	52	54	56	58	60	60	62	62	64	64	66	66	68	68	70	70	72	72	74	74	76	76	78	78	80	80	80	80	80	80	80																							
0.591-0.610 (0.0233-0.0240)	40	42	44	46	48	50	52	54	56	58	60	62	62	64	64	66	66	68	68	70	70	72	72	74	74	76	76	78	78	80	80	80	80	80	80	80																									
0.611-0.630 (0.0241-0.0248)	42	44	46	48	50	52	54	56	58	60	62	64	64	66	66	68	68	70	70	72	72	74	74	76	76	78	78	80	80	80	80	80	80	80																											
0.631-0.650 (0.0248-0.0256)	44	46	48	50	52	54	56	58	60	62	64	66	66	68	68	70	70	72	72	74	74	76	76	78	78	80	80	80	80	80	80	80																													
0.651-0.670 (0.0256-0.0264)	46	48	50	52	54	56	58	60	62	64	66	68	68	70	70	72	72	74	74	76	76	78	78	80	80	80	80	80	80	80																															
0.671-0.690 (0.0264-0.0272)	48	50	52	54	56	58	60	62	64	66	68	70	70	72	72	74	74	76	76	78	78	80	80	80	80	80	80	80																																	
0.691-0.710 (0.0272-0.0280)	50	52	54	56	58	60	62	64	66	68	70	72	72	74	74	76	76	78	78	80	80	80	80	80	80	80																																			
0.711-0.730 (0.0280-0.0287)	52	54	56	58	60	62	64	66	68	70	72	74	74	76	76	78	78	80	80	80	80	80	80	80																																					
0.731-0.750 (0.0288-0.0295)	54	56	58	60	62	64	66	68	70	72	74	76	76	78	78	80	80	80	80	80	80	80																																							
0.751-0.770 (0.0296-0.0303)	56	58	60	62	64	66	68	70	72	74	76	78	78	80	80	80	80	80	80	80																																									
0.771-0.790 (0.0304-0.0311)	58	60	62	64	66	68	70	72	74	76	78	80	80	80	80	80	80	80																																											
0.791-0.810 (0.0311-0.0319)	60	62	64	66	68	70	72	74	76	78	80	80	80	80	80	80																																													
0.811-0.830 (0.0319-0.0327)	62	64	66	68	70	72	74	76	78	80	80	80	80	80																																															
0.831-0.850 (0.0327-0.0335)	64	66	68	70	72	74	76	78	80	80	80	80																																																	
0.851-0.870 (0.0335-0.0343)	66	68	70	72	74	76	78	80	80	80																																																			
0.871-0.890 (0.0343-0.0350)	68	70	72	74	76	78	80	80	80																																																				
0.891-0.910 (0.0351-0.0358)	70	72	74	76	78	80	80	80																																																				
0.911-0.930 (0.0359-0.0366)	72	74	76	78	80	80	80																																																						
0.931-0.950 (0.0367-0.0374)	74	76	78	80	80	80																																																							
0.951-0.970 (0.0374-0.0382)	76	78	80	80	80																																																								
0.971-0.990 (0.0382-0.0390)	78	80	80	80																																																									
0.991-1.010 (0.0390-0.0398)	80	80	80																																																										
1.011-1.030 (0.0398-0.0406)	80	80																																																												
1.031-1.050 (0.0406-0.0413)	80																																																													

Intake valve clearance (Cold):
0.15 – 0.25 mm (0.006 – 0.010 in.)

EXAMPLE:
The 2.300 mm (0.0906 in.) shim is installed, and the measured clearance is 0.440 mm (0.0173 in.). Replace the 2.300 mm (0.0906 in.) shim with a No. 54 shim.

New shim thickness mm (in.)

Shim No.	Thickness	Shim No.	Thickness	Shim No.	Thickness
00	2.000 (0.0787)	28	2.280 (0.0898)	56	2.560 (0.1008)
02	2.020 (0.0795)	30	2.300 (0.0906)	58	2.580 (0.1016)
04	2.040 (0.0803)	32	2.320 (0.0913)	60	2.600 (0.1024)
06	2.060 (0.0811)	34	2.340 (0.0921)	62	2.620 (0.1031)
08	2.080 (0.0819)	36	2.360 (0.0929)	64	2.640 (0.1039)
10	2.100 (0.0827)	38	2.380 (0.0937)	66	2.660 (0.1047)
12	2.120 (0.0835)	40	2.400 (0.0945)	68	2.680 (0.1055)
14	2.140 (0.0843)	42	2.420 (0.0953)	70	2.700 (0.1063)
16	2.160 (0.0850)	44	2.440 (0.0961)	72	2.720 (0.1071)
18	2.180 (0.0858)	46	2.460 (0.0969)	74	2.740 (0.1079)
20	2.200 (0.0866)	48	2.480 (0.0976)	76	2.760 (0.1087)
22	2.220 (0.0874)	50	2.500 (0.0984)	78	2.780 (0.1094)
24	2.240 (0.0882)	52	2.520 (0.0992)	80	2.800 (0.1102)
26	2.260 (0.0890)	54	2.540 (0.1000)		

7924SG71

Fig. 32 Intake valve clearance shim selection chart

Measured clearance mm (in.) \ Installed shim thickness mm (in.)	2.000 (0.0787)	2.020 (0.0795)	2.040 (0.0803)	2.060 (0.0811)	2.080 (0.0819)	2.100 (0.0827)	2.120 (0.0835)	2.140 (0.0843)	2.160 (0.0850)	2.180 (0.0858)	2.200 (0.0866)	2.210 (0.0870)	2.220 (0.0874)	2.230 (0.0878)	2.240 (0.0882)	2.250 (0.0886)	2.260 (0.0890)	2.270 (0.0894)	2.280 (0.0898)	2.290 (0.0902)	2.300 (0.0906)	2.310 (0.0909)	2.320 (0.0913)	2.330 (0.0917)	2.340 (0.0921)	2.350 (0.0925)	2.360 (0.0929)	2.370 (0.0933)	2.380 (0.0937)	2.390 (0.0941)	2.400 (0.0945)	2.410 (0.0949)	2.420 (0.0953)	2.430 (0.0957)	2.440 (0.0961)	2.450 (0.0965)	2.460 (0.0969)	2.470 (0.0972)	2.480 (0.0976)	2.490 (0.0980)	2.500 (0.0984)	2.510 (0.0988)	2.520 (0.0992)	2.530 (0.0996)	2.540 (0.1000)	2.550 (0.1004)	2.560 (0.1008)	2.570 (0.1012)	2.580 (0.1016)	2.590 (0.1020)	2.600 (0.1024)	2.620 (0.1031)	2.640 (0.1039)	2.660 (0.1047)	2.680 (0.1055)	2.700 (0.1063)	2.720 (0.1071)	2.740 (0.1079)	2.760 (0.1087)	2.780 (0.1094)	2.800 (0.1102)
0.000–0.030 (0.0000–0.0012)												00	00	00	00	00	00	00	00	00	02	02	04	04	06	06	08	08	10	10	12	12	14	14	16	16	18	18	20	20	22	22	24	24	26	26	28	28	30	30	32	34	36	38	40	42	44	46	48	50	52
0.031–0.050 (0.0012–0.0020)											00	00	00	00	00	00	00	02	02	04	04	06	06	08	08	10	10	12	12	14	14	16	16	18	18	20	20	22	22	24	24	26	26	28	28	30	30	32	32	34	34	36	38	40	42	44	46	48	50	52	54
0.051–0.070 (0.0020–0.0028)										00	00	00	00	00	00	02	02	04	04	06	06	08	08	10	10	12	12	14	14	16	16	18	18	20	20	22	22	24	24	26	26	28	28	30	30	32	32	34	34	36	36	38	40	42	44	46	48	50	52	54	56
0.071–0.090 (0.0028–0.0035)									00	00	00	00	00	02	02	04	04	06	06	08	08	10	10	12	12	14	14	16	16	18	18	20	20	22	22	24	24	26	26	28	28	30	30	32	32	34	34	36	36	38	38	40	42	44	46	48	50	52	54	56	58
0.091–0.110 (0.0036–0.0043)								00	00	00	00	02	02	04	04	06	06	08	08	10	10	12	12	14	14	16	16	18	18	20	20	22	22	24	24	26	26	28	28	30	30	32	32	34	34	36	36	38	38	40	40	42	44	46	48	50	52	54	56	58	60
0.111–0.130 (0.0044–0.0051)							00	00	00	00	02	04	04	06	06	08	08	10	10	12	12	14	14	16	16	18	18	20	20	22	22	24	24	26	26	28	28	30	30	32	32	34	34	36	36	38	38	40	40	42	42	44	46	48	50	52	54	56	58	60	62
0.131–0.150 (0.0052–0.0059)						00	00	00	00	02	04	06	06	08	08	10	10	12	12	14	14	16	16	18	18	20	20	22	22	24	24	26	26	28	28	30	30	32	32	34	34	36	36	38	38	40	40	42	42	44	44	46	48	50	52	54	56	58	60	62	64
0.151–0.170 (0.0059–0.0067)					00	00	00	00	02	04	06	08	08	10	10	12	12	14	14	16	16	18	18	20	20	22	22	24	24	26	26	28	28	30	30	32	32	34	34	36	36	38	38	40	40	42	42	44	44	46	46	48	50	52	54	56	58	60	62	64	66
0.171–0.190 (0.0067–0.0075)				00	00	00	00	02	04	06	08	10	10	12	12	14	14	16	16	18	18	20	20	22	22	24	24	26	26	28	28	30	30	32	32	34	34	36	36	38	38	40	40	42	42	44	44	46	46	48	48	50	52	54	56	58	60	62	64	66	68
0.191–0.210 (0.0075–0.0083)			00	00	00	00	02	04	06	08	10	12	12	14	14	16	16	18	18	20	20	22	22	24	24	26	26	28	28	30	30	32	32	34	34	36	36	38	38	40	40	42	42	44	44	46	46	48	48	50	50	52	54	56	58	60	62	64	66	68	70
0.211–0.230 (0.0083–0.0091)		00	00	00	00	02	04	06	08	10	12	14	14	16	16	18	18	20	20	22	22	24	24	26	26	28	28	30	30	32	32	34	34	36	36	38	38	40	40	42	42	44	44	46	46	48	48	50	50	52	52	54	56	58	60	62	64	66	68	70	72
0.231–0.249 (0.0091–0.0098)		00	00	00	02	04	06	08	10	12	14	16	16	18	18	20	20	22	22	24	24	26	26	28	28	30	30	32	32	34	34	36	36	38	38	40	40	42	42	44	44	46	46	48	48	50	50	52	52	54	54	56	58	60	62	64	66	68	70	72	74
0.250–0.350 (0.0098–0.0138)																																																														
0.351–0.370 (0.0138–0.0146)	06	08	10	12	14	16	18	20	22	24	26	28	28	30	30	32	32	34	34	36	36	38	38	40	40	42	42	44	44	46	46	48	48	50	50	52	52	54	54	56	56	58	58	60	60	62	62	64	64	66	66	68	70	72	74	76	78	80	80	80	
0.371–0.390 (0.0146–0.0154)	08	10	12	14	16	18	20	22	24	26	28	30	30	32	32	34	34	36	36	38	38	40	40	42	42	44	44	46	46	48	48	50	50	52	52	54	54	56	56	58	58	60	60	62	62	64	64	66	66	68	68	70	72	74	76	78	80	80	80		
0.391–0.410 (0.0154–0.0161)	10	12	14	16	18	20	22	24	26	28	30	32	32	34	34	36	36	38	38	40	40	42	42	44	44	46	46	48	48	50	50	52	52	54	54	56	56	58	58	60	60	62	62	64	64	66	66	68	68	70	70	72	74	76	78	80	80	80			
0.411–0.430 (0.0162–0.0169)	12	14	16	18	20	22	24	26	28	30	32	34	34	36	36	38	38	40	40	42	42	44	44	46	46	48	48	50	50	52	52	54	54	56	56	58	58	60	60	62	62	64	64	66	66	68	68	70	70	72	72	74	76	78	80	80	80				
0.431–0.450 (0.0170–0.0177)	14	16	18	20	22	24	26	28	30	32	34	36	36	38	38	40	40	42	42	44	44	46	46	48	48	50	50	52	52	54	54	56	56	58	58	60	60	62	62	64	64	66	66	68	68	70	70	72	72	74	74	76	78	80	80	80					
0.451–0.470 (0.0178–0.0185)	16	18	20	22	24	26	28	30	32	34	36	38	38	40	40	42	42	44	44	46	46	48	48	50	50	52	52	54	54	56	56	58	58	60	60	62	62	64	64	66	66	68	68	70	70	72	72	74	74	76	76	78	80	80	80						
0.471–0.490 (0.0185–0.0193)	18	20	22	24	26	28	30	32	34	36	38	40	40	42	42	44	44	46	46	48	48	50	50	52	52	54	54	56	56	58	58	60	60	62	62	64	64	66	66	68	68	70	70	72	72	74	74	76	76	78	78	80	80	80							
0.491–0.510 (0.0193–0.0201)	20	22	24	26	28	30	32	34	36	38	40	42	42	44	44	46	46	48	48	50	50	52	52	54	54	56	56	58	58	60	60	62	62	64	64	66	66	68	68	70	70	72	72	74	74	76	76	78	78	80	80	80	80								
0.511–0.530 (0.0201–0.0209)	22	24	26	28	30	32	34	36	38	40	42	44	44	46	46	48	48	50	50	52	52	54	54	56	56	58	58	60	60	62	62	64	64	66	66	68	68	70	70	72	72	74	74	76	76	78	78	80	80	80	80	80									
0.531–0.550 (0.0209–0.0217)	24	26	28	30	32	34	36	38	40	42	44	46	46	48	48	50	50	52	52	54	54	56	56	58	58	60	60	62	62	64	64	66	66	68	68	70	70	72	72	74	74	76	76	78	78	80	80	80	80	80	80										
0.551–0.570 (0.0217–0.0224)	26	28	30	32	34	36	38	40	42	44	46	48	48	50	50	52	52	54	54	56	56	58	58	60	60	62	62	64	64	66	66	68	68	70	70	72	72	74	74	76	76	78	78	80	80	80	80	80	80	80											
0.571–0.590 (0.0225–0.0232)	28	30	32	34	36	38	40	42	44	46	48	50	50	52	52	54	54	56	56	58	58	60	60	62	62	64	64	66	66	68	68	70	70	72	72	74	74	76	76	78	78	80	80	80	80	80	80	80													
0.591–0.610 (0.0233–0.0240)	30	32	34	36	38	40	42	44	46	48	50	52	52	54	54	56	56	58	58	60	60	62	62	64	64	66	66	68	68	70	70	72	72	74	74	76	76	78	78	80	80	80	80	80	80	80															
0.611–0.630 (0.0241–0.0248)	32	34	36	38	40	42	44	46	48	50	52	54	54	56	56	58	58	60	60	62	62	64	64	66	66	68	68	70	70	72	72	74	74	76	76	78	78	80	80	80	80	80	80	80																	
0.631–0.650 (0.0248–0.0256)	34	36	38	40	42	44	46	48	50	52	54	56	56	58	58	60	60	62	62	64	64	66	66	68	68	70	70	72	72	74	74	76	76	78	78	80	80	80	80	80	80	80																			
0.651–0.670 (0.0256–0.0264)	36	38	40	42	44	46	48	50	52	54	56	58	58	60	60	62	62	64	64	66	66	68	68	70	70	72	72	74	74	76	76	78	78	80	80	80	80	80	80	80																					
0.671–0.690 (0.0264–0.0272)	38	40	42	44	46	48	50	52	54	56	58	60	60	62	62	64	64	66	66	68	68	70	70	72	72	74	74	76	76	78	78	80	80	80	80	80	80	80																							
0.691–0.710 (0.0272–0.0280)	40	42	44	46	48	50	52	54	56	58	60	62	62	64	64	66	66	68	68	70	70	72	72	74	74	76	76	78	78	80	80	80	80	80	80	80																									
0.711–0.730 (0.0280–0.0287)	42	44	46	48	50	52	54	56	58	60	62	64	64	66	66	68	68	70	70	72	72	74	74	76	76	78	78	80	80	80	80	80	80	80																											
0.731–0.750 (0.0288–0.0295)	44	46	48	50	52	54	56	58	60	62	64	66	66	68	68	70	70	72	72	74	74	76	76	78	78	80	80	80	80	80	80	80																													
0.751–0.770 (0.0296–0.0303)	46	48	50	52	54	56	58	60	62	64	66	68	68	70	70	72	72	74	74	76	76	78	78	80	80	80	80	80	80	80																															
0.771–0.790 (0.0304–0.0311)	48	50	52	54	56	58	60	62	64	66	68	70	70	72	72	74	74	76	76	78	78	80	80	80	80	80	80	80																																	
0.791–0.810 (0.0311–0.0319)	50	52	54	56	58	60	62	64	66	68	70	72	72	74	74	76	76	78	78	80	80	80	80	80	80	80																																			
0.811–0.830 (0.0319–0.0327)	52	54	56	58	60	62	64	66	68	70	72	74	74	76	76	78	78	80	80	80	80	80	80	80																																					
0.831–0.850 (0.0327–0.0335)	54	56	58	60	62	64	66	68	70	72	74	76	76	78	78	80	80	80	80	80	80	80																																							
0.851–0.870 (0.0335–0.0343)	56	58	60	62	64	66	68	70	72	74	76	78	78	80	80	80	80	80	80	80																																									
0.871–0.890 (0.0343–0.0350)	58	60	62	64	66	68	70	72	74	76	78	80	80	80	80	80	80	80																																											
0.891–0.910 (0.0351–0.0358)	60	62	64	66	68	70	72	74	76	78	80	80	80	80	80	80																																													
0.911–0.930 (0.0359–0.0366)	62	64	66	68	70	72	74	76	78	80	80	80	80	80																																															
0.931–0.950 (0.0367–0.0374)	64	66	68	70	72	74	76	78	80	80	80	80																																																	
0.951–0.970 (0.0374–0.0382)	66	68	70	72	74	76	78	80	80	80																																																			
0.971–0.990 (0.0382–0.0390)	68	70	72	74	76	78	80	80	80																																																					
0.991–1.010 (0.0390–0.0398)	70	72	74	76	78	80	80	80																																																						
1.011–1.030 (0.0398–0.0406)	72	74	76	78	80	80	80																																																							
1.031–1.050 (0.0406–0.0413)	74	76	78	80	80	80																																																								
1.051–1.070 (0.0414–0.0421)	76	78	80	80	80																																																									
1.071–1.090 (0.0422–0.0429)	78	80	80	80																																																										
1.091–1.110 (0.0430–0.0437)	80	80	80																																																											
1.111–1.130 (0.0437–0.0445)	80	80																																																												
1.131–1.150 (0.0445–0.0453)	80																																																													

Exhaust valve clearance (Cold):
0.25 – 0.35 mm (0.010 – 0.014 in.)

EXAMPLE:
The 2.300 mm (0.0906 in.) shim is installed, and the measured clearance is 0.440 mm (0.0173 in.). Replace the 2.300 mm (0.0906 in.) shim with a No. 44 shim.

New shim thickness mm (in.)

Shim No.	Thickness	Shim No.	Thickness	Shim No.	Thickness
00	2.000 (0.0787)	28	2.280 (0.0898)	56	2.560 (0.1008)
02	2.020 (0.0795)	30	2.300 (0.0906)	58	2.580 (0.1016)
04	2.040 (0.0803)	32	2.320 (0.0913)	60	2.600 (0.1024)
06	2.060 (0.0811)	34	2.340 (0.0921)	62	2.620 (0.1031)
08	2.080 (0.0819)	36	2.360 (0.0929)	64	2.640 (0.1039)
10	2.100 (0.0827)	38	2.380 (0.0937)	66	2.660 (0.1047)
12	2.120 (0.0835)	40	2.400 (0.0945)	68	2.680 (0.1055)
14	2.140 (0.0843)	42	2.420 (0.0953)	70	2.700 (0.1063)
16	2.160 (0.0850)	44	2.440 (0.0961)	72	2.720 (0.1071)
18	2.180 (0.0858)	46	2.460 (0.0969)	74	2.740 (0.1079)
20	2.200 (0.0866)	48	2.480 (0.0976)	76	2.760 (0.1087)
22	2.220 (0.0874)	50	2.500 (0.0984)	78	2.780 (0.1094)
24	2.240 (0.0882)	52	2.520 (0.0992)	80	2.800 (0.1102)
26	2.260 (0.0890)	54	2.540 (0.1000)		

7924SG72

Fig. 33 Exhaust valve clearance shim selection chart

7. When all valve clearances have been measured, remove the camshafts.

8. Remove the valve shims and measure them. Note this measurement along with the clearance measurement recorded earlier.

9. Using the valve clearance and shim thickness measurements, find replacement shims in the Adjusting Shim Selection charts.

To install:

10. Install or connect the following:
- Replacement valve shims
- Camshafts
- Valve covers
- Ignition coils
- Negative battery cable

11. Fill the cooling system.

12. Start the engine and check for leaks.

ENGINE PERFORMANCE & EMISSION CONTROL COMPONENT LOCATIONS

See Figures 34 and 35.

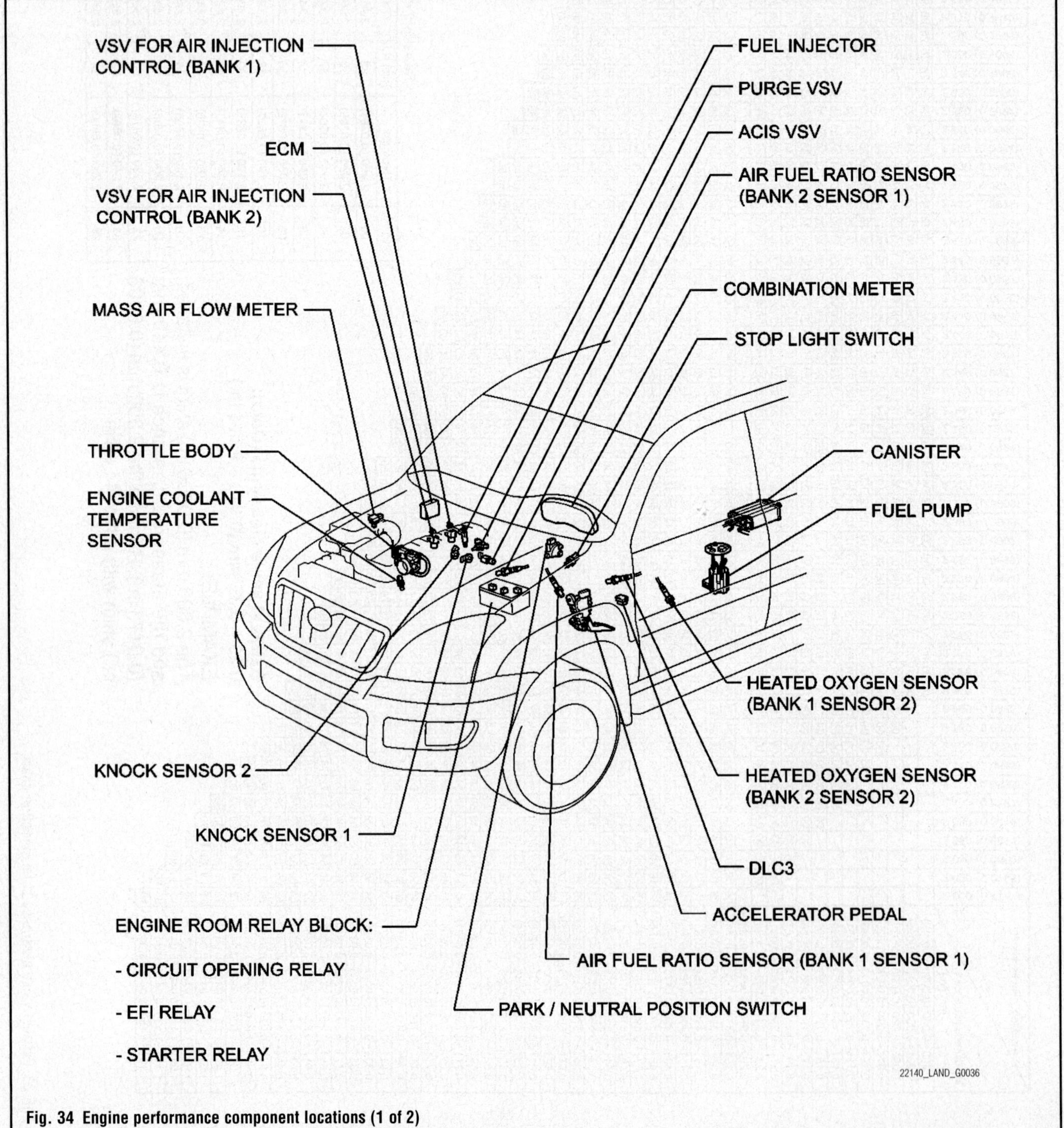

Fig. 34 Engine performance component locations (1 of 2)

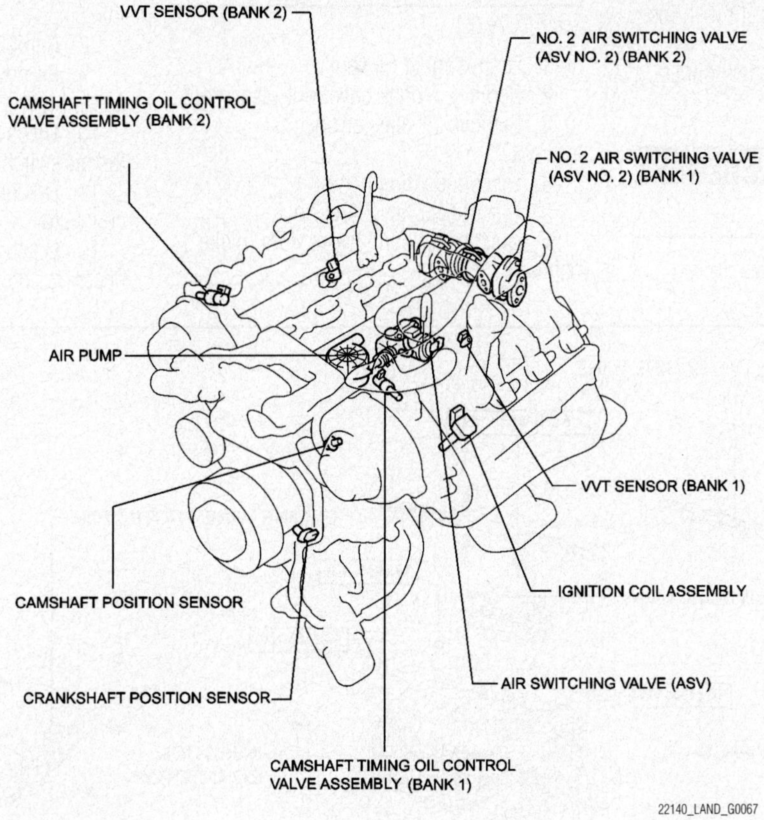

VVT SENSOR (BANK 2)
NO. 2 AIR SWITCHING VALVE (ASV NO. 2) (BANK 2)
CAMSHAFT TIMING OIL CONTROL VALVE ASSEMBLY (BANK 2)
NO. 2 AIR SWITCHING VALVE (ASV NO. 2) (BANK 1)
AIR PUMP
VVT SENSOR (BANK 1)
CAMSHAFT POSITION SENSOR
IGNITION COIL ASSEMBLY
AIR SWITCHING VALVE (ASV)
CRANKSHAFT POSITION SENSOR
CAMSHAFT TIMING OIL CONTROL VALVE ASSEMBLY (BANK 1)
22140_LAND_G0067

Fig. 35 Engine performance component locations (2 of 2)

ACCELERATOR PEDAL POSITION (APP) SENSOR

LOCATION

See Figure 36.

REMOVAL & INSTALLATION

1. Disconnect a accelerator pedal connector.
2. Remove the 2 nuts and accelerator pedal assembly.

To install:

➡Be care not to give a shock to the accelerator pedal assembly. Be care not to disassemble the accelerator pedal assembly.

ACCELERATOR PEDAL ASSEMBLY
(ACCELERATOR PEDAL POSITION SENSOR)
22140_LAND_G0035

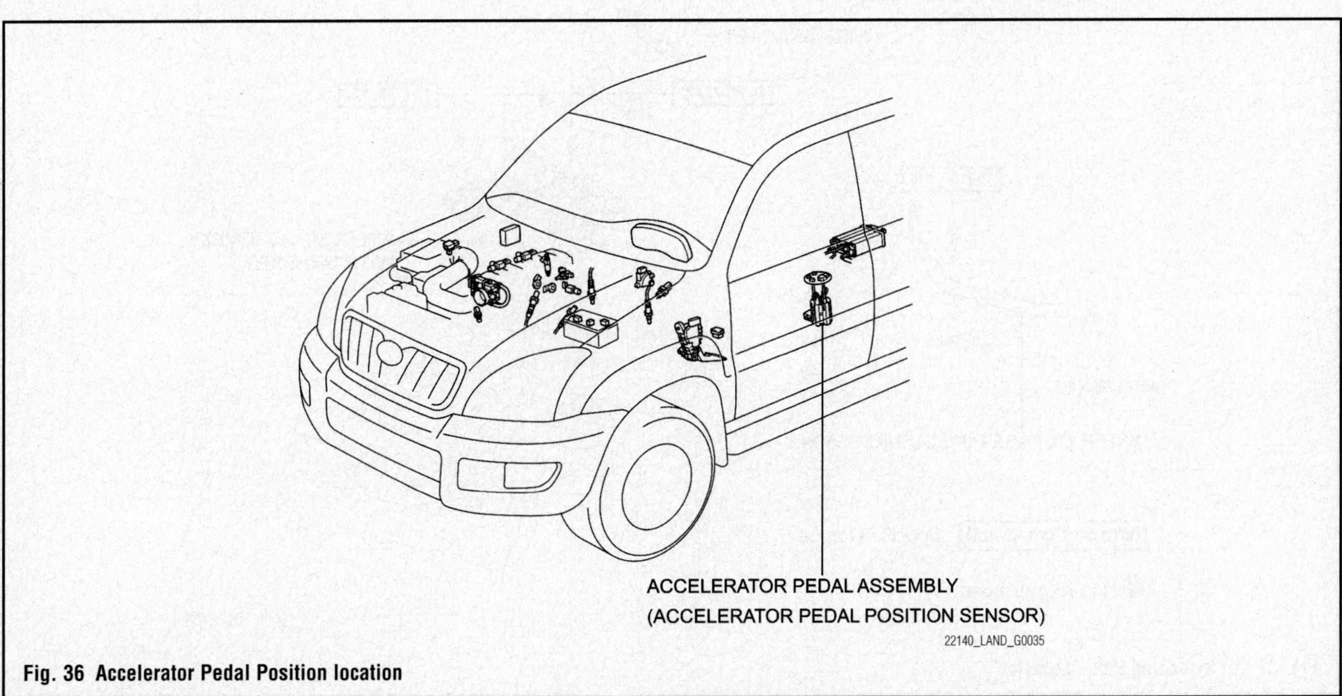

Fig. 36 Accelerator Pedal Position location

3. Install the accelerator pedal assembly with the 2 nuts. Tighten nuts to 44 inch lbs. (5.0 Nm)
4. Connect a accelerator pedal connector.

AIR BYPASS VALVE (ABV)

LOCATION

See Figure 37.

REMOVAL & INSTALLATION

See Figure 38.

1. Drain engine coolant
2. Remove v-bank cover sub-assembly
3. Remove air cleaner hose assembly
4. Disconnect fuel hose
5. Disconnect no. 2 fuel hose
6. Disconnect connectors from intake manifold
7. Disconnect hoses from intake manifold
8. Remove intake manifold assembly
9. Remove water by-pass pipe sub-assembly
10. Disconnect the no.2 air hose from the air switching valve.
11. Disconnect the air switching valve connector.
12. Disconnect the pressure sensor connector for the air injection system.

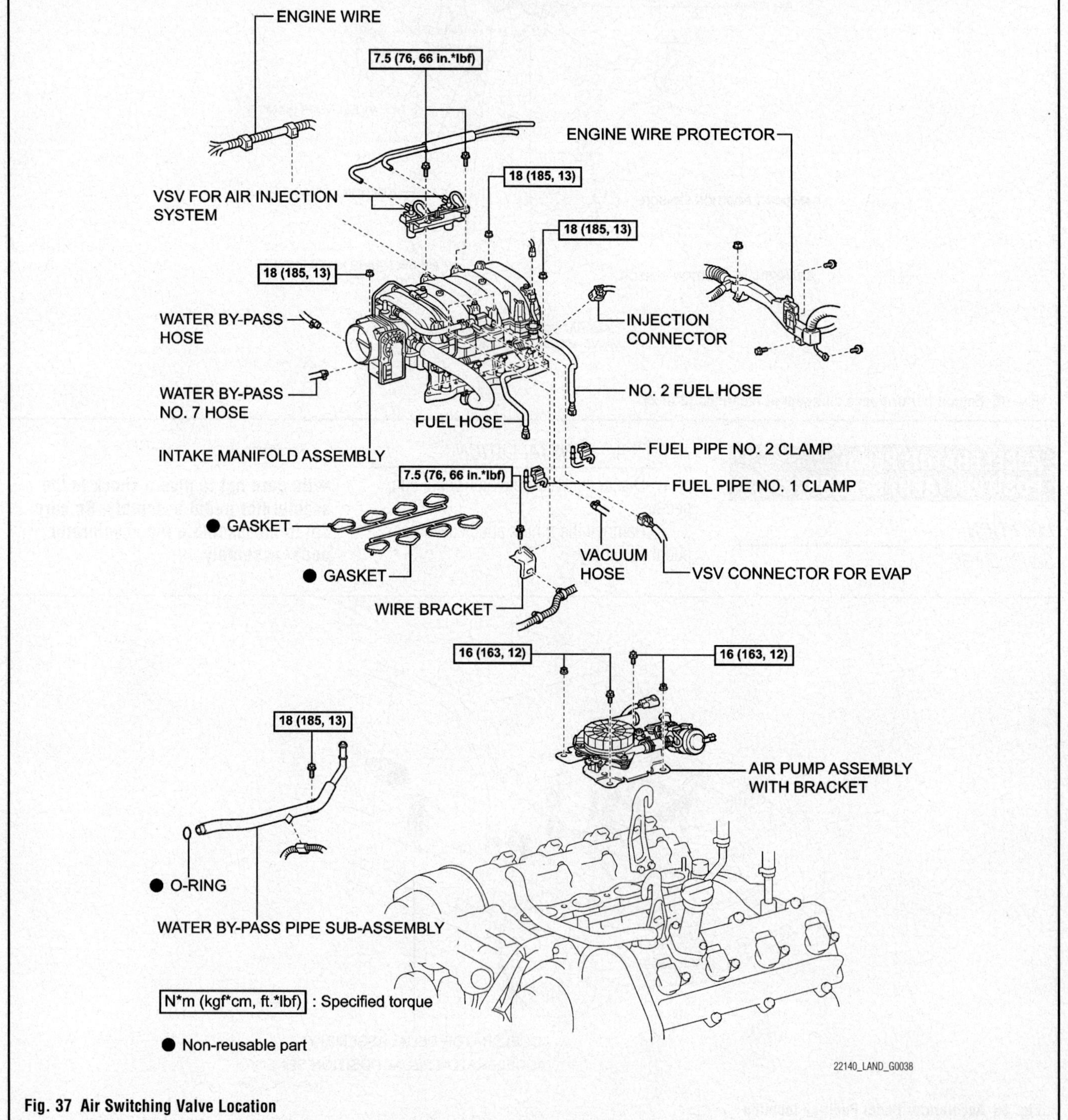

Fig. 37 Air Switching Valve Location

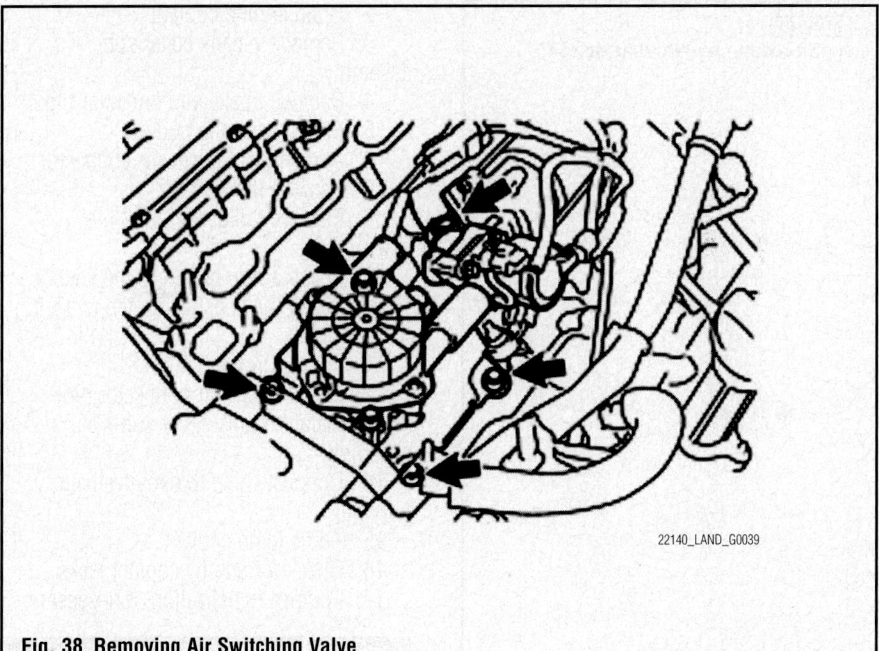

Fig. 38 Removing Air Switching Valve

13. Remove the 4 bolts and air pump assembly with bracket.

To install:

14. Install the air pump assembly with the 4 bolts. Tighten bolts to 7 ft. lbs. (10 Nm)
15. Connect the pressure sensor connector for the air injection system.
16. Connect the air switching valve connector.
17. Connect the No. 2 air hose to the air switching valve.
18. Install water by-pass pipe sub-assembly.
19. Install intake manifold assembly.
20. Connect hoses to intake manifold
21. Connect connectors to intake manifold
22. Connect no. 2 fuel hose
23. Connect fuel hose
24. Install air cleaner hose assembly
25. Install v-bank cover sub-assembly
26. Add engine coolant
27. Check for engine coolant leak
28. Check for fuel leak

CAMSHAFT POSITION (CMP) SENSOR

LOCATION

See Figure 39.

REMOVAL & INSTALLATION

1. Drain engine coolant
2. Remove v-bank cover sub-assembly
3. Remove fan and generator v belt
4. Remove oil cooler pipe
5. Remove timing belt cover sub-assembly no. 3 LH
6. Disconnect the camshaft position sensor connector.
7. Remove the bolt, stud bolt and camshaft position sensor.

To install:

8. Install the camshaft position sensor with the bolt and stud bolt. Tighten bolt to 66 inch lbs. (7.5 Nm)
9. Reconnect the camshaft position sensor connector.
10. Install timing belt cover sub-assembly no. 3 LH
11. Install oil cooler pipe
12. Install fan and generator v belt
13. Add engine coolant
14. Check for engine coolant leaks
15. Install v-bank cover sub-assembly

COOLANT TEMPERATURE SENSOR

LOCATION

See Figure 40.

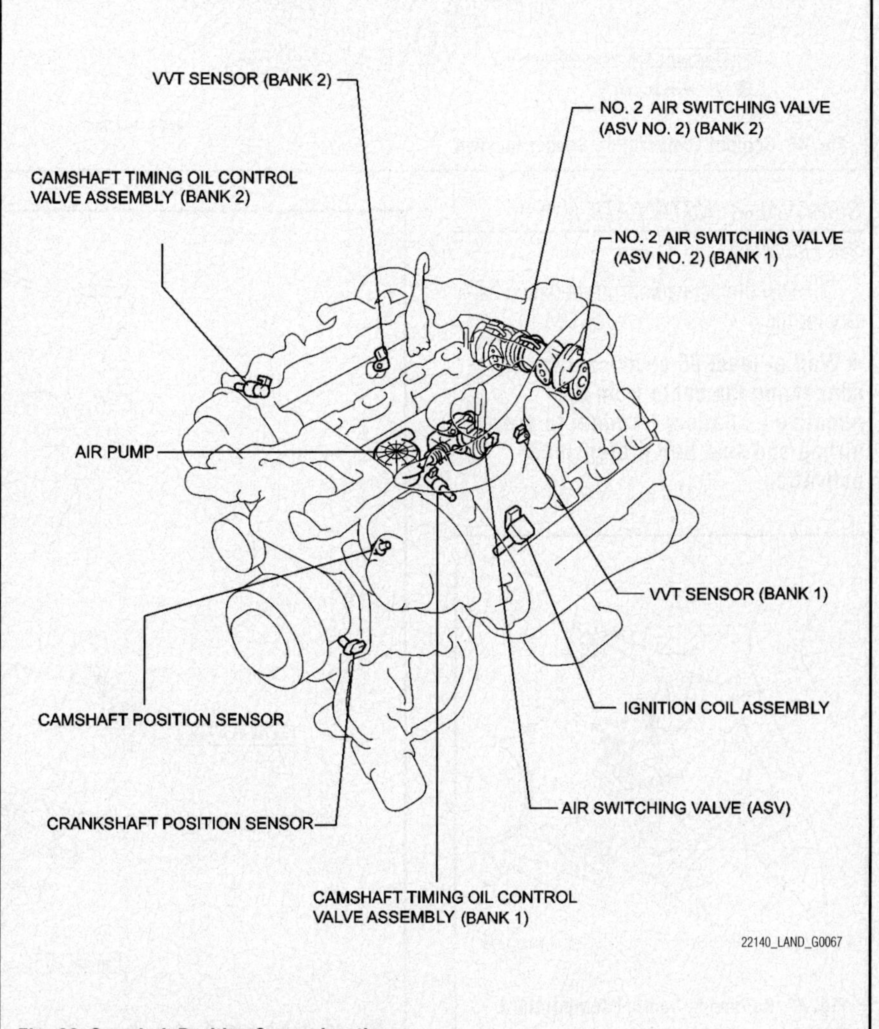

Fig. 39 Camshaft Position Sensor location

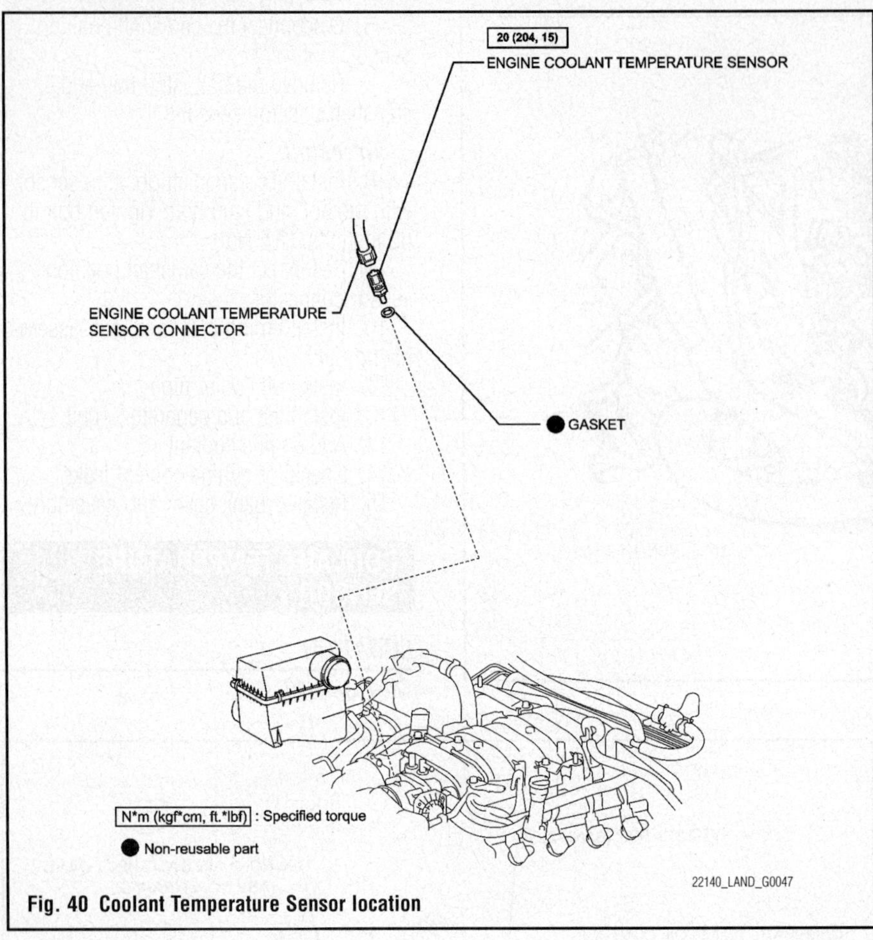

Fig. 40 Coolant Temperature Sensor location

REMOVAL & INSTALLATION

See Figure 41.

1. Disconnect cable from negative battery terminal

➡**Wait at least 90 seconds after disconnecting the cable from the negative (-) battery terminal to prevent airbag and seat belt pretensioner activation**

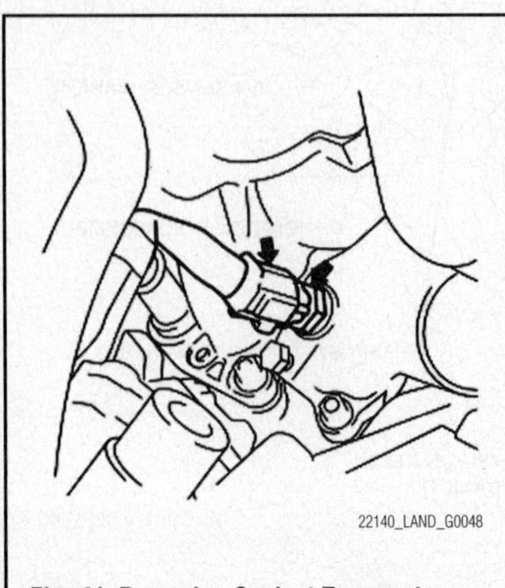

Fig. 41 Removing Coolant Temperature Sensor

2. Drain engine coolant
3. Remove v-bank cover sub-assembly
4. Remove intake air connector pipe
5. Remove throttle body
6. Disconnect the sensor connector
7. Remove the sensor
8. Remove the gasket from the sensor
9. Install a new gasket to the sensor
10. Install the sensor. Tighten to 15 ft. lbs. (20 Nm)
11. Install throttle body
12. Install intake air connector pipe
13. Install v-bank cover sub-assembly
14. Connect cable to negative battery terminal
15. Add engine coolant
16. Check for engine coolant leaks
17. Perform initialization, if necessary.

CRANKSHAFT POSITION (CKP) SENSOR

LOCATION

See Figure 42.

REMOVAL & INSTALLATION

1. Disconnect cable from negative battery terminal

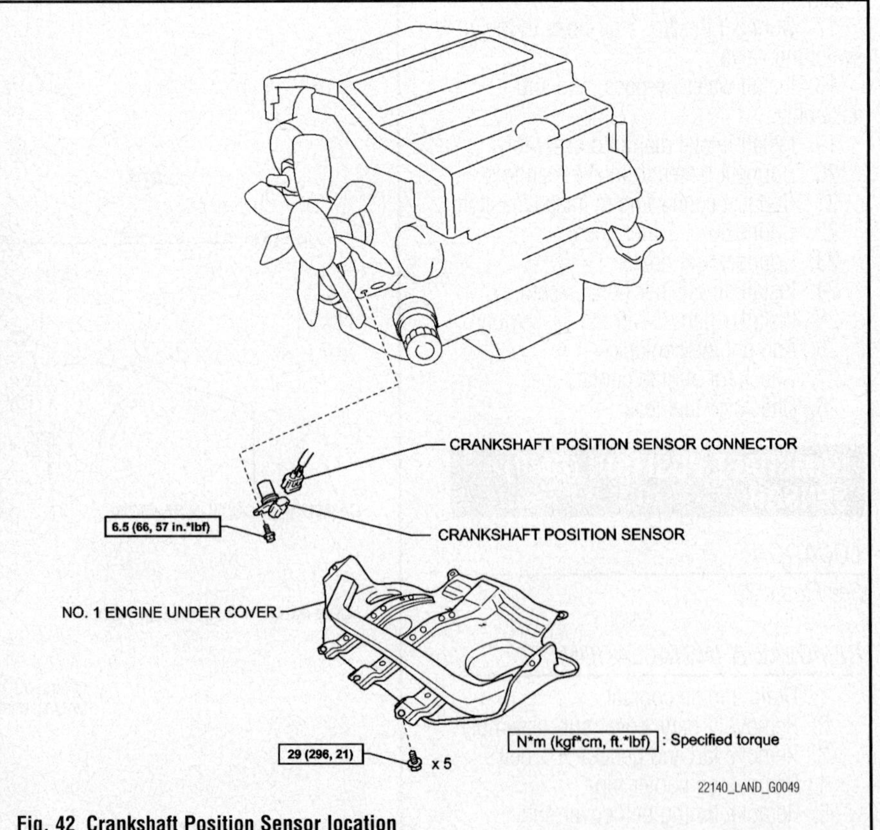

Fig. 42 Crankshaft Position Sensor location

➡Wait at least 90 seconds after disconnecting the cable from the negative (-) battery terminal to prevent airbag and seat belt pretensioner activation.

2. Remove no. 1 engine under cover
3. Disconnect the sensor connector.
4. Remove the bolt and sensor

To install:

5. Install crankshaft position sensor
6. Install the sensor with the bolt. Tighten bolt to 57 inch lbs. (6.5 Nm)
7. Connect the sensor connector.
8. Install no. 1 engine under cover
9. Connect cable to negative battery terminal
10. Perform initialization, if necessary.

TESTING

See Figure 43.

1. Measure the resistance between the terminals.
2. If the result is not as specified, replace the sensor.

➡In the table below, the terms "cold" and "hot" refer to the temperature of the coils. "Cold" means approximately -10 to 50°C (14 to 122°F). "Hot" means approximately 50 to 100°C (122 to 212°F

Temperature	Specified Condition
Cold	1,630 to 2,740 Ω
Hot	2,065 to 3,225 Ω

22140_LAND_G0050

Fig. 43 Crankshaft Position Sensor standard resistance

ELECTRONIC CONTROL MODULE (ECM)

LOCATION

See Figure 44.

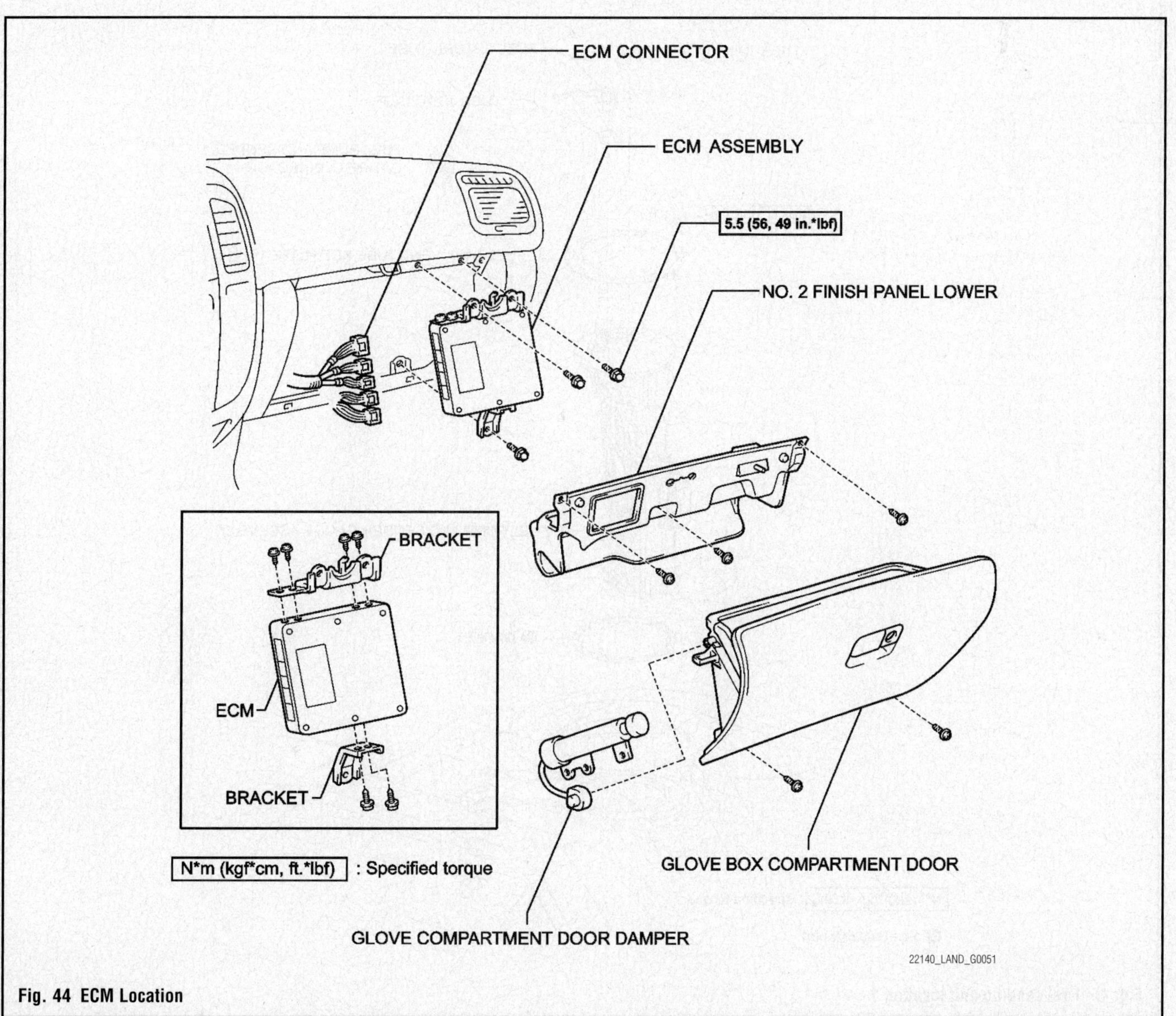

Fig. 44 ECM Location

REMOVAL & INSTALLATION

1. Disconnect cable from negative battery terminal.

➡**Wait at least 90 seconds after disconnecting the cable from the negative (-) battery terminal to prevent airbag and seat belt pretensioner activation.**

2. Remove glove box compartment door.
3. Remove the 2 screws and glove compartment door.
4. Remove no. 2 finish panel lower.
5. Remove the 3 screws and no. 2 finish panel lower.
6. Disconnect the 5 ECM connectors.
7. Remove the 3 screws and ECM.

To install:

8. Install the ECM with the 3 screws, tighten to 49 inch lbs. (5.5 Nm).
9. Connect the 5 ECM connectors.

➡**Be sure to securely connect the connectors.**

10. Install the No. 2 finish panel lower with the 3 screws
11. Install the glove compartment door with the 2 screws
12. Connect cable to negative battery terminal.
13. Perform initialization, if necessary.

➡**Certain systems need to be initialized after disconnecting and reconnecting the cable from the negative (-) battery terminal.**

FUEL LEVEL SENDING UNIT

LOCATION

See Figures 45 and 46.

REMOVAL & INSTALLATION

See Figures 47 through 49.

⁂ CAUTION

Do not smoke or work near an open flame when working on the fuel pump.

1. Discharge fuel system pressure.
2. Disconnect cable from negative battery terminal

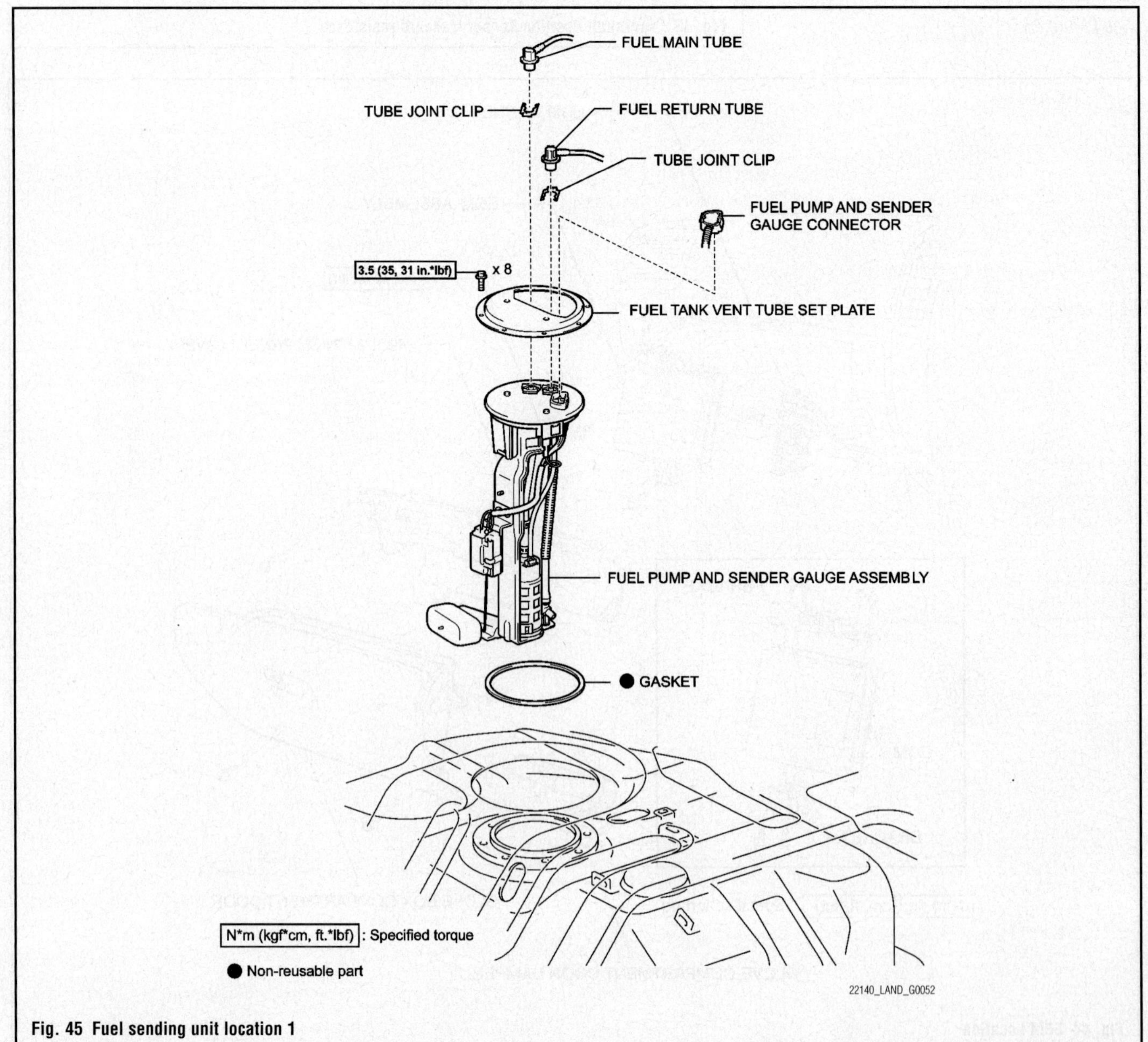

Fig. 45 Fuel sending unit location 1

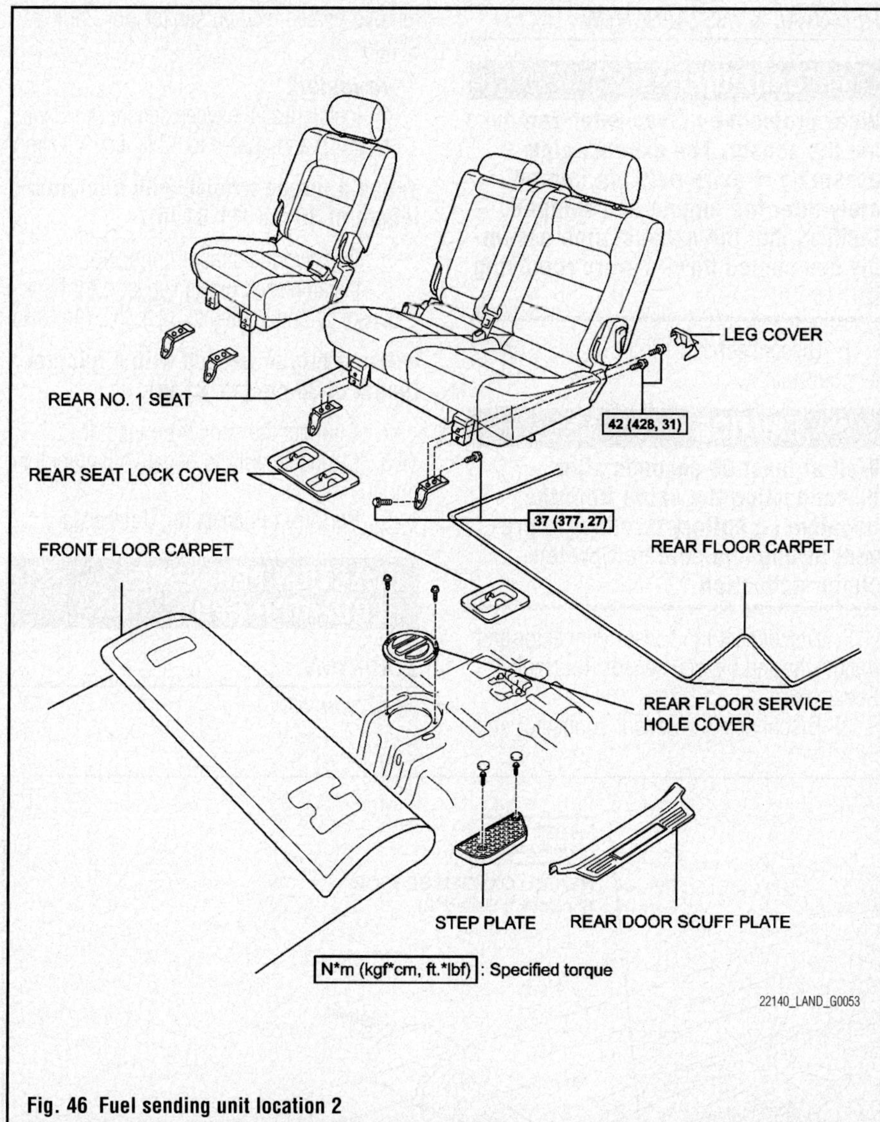

Fig. 46 Fuel sending unit location 2

CAUTION

Wait at least 90 seconds after disconnecting the cable from the negative (-) battery terminal to prevent airbag and seat belt pretensioner activation.

3. Remove the 2 rear seats.
4. Remove rear door scuff plate.
5. Remove step plate.
6. Remove rear seat lock cover.
7. Take off the front and rear floor carpets.
8. Remove the 2 screws and floor service hole cover.

CAUTION

Prevent the retained pressure in the fuel line from splashing inside the vehicle compartment. When sealing the tube and suction plates with the O-ring of the quick connector, be careful not to damage any contact surfaces or allow foreign matter to contact any surface. Be sure to perform the disconnection by hand. Do not use tools. Do not bend or turn the nylon tube by force.

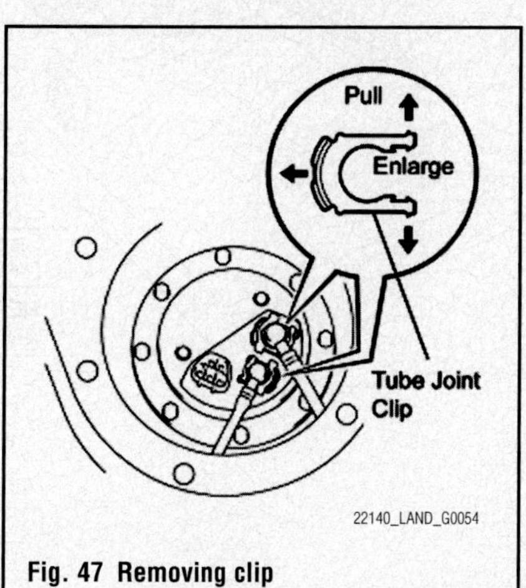

Fig. 47 Removing clip

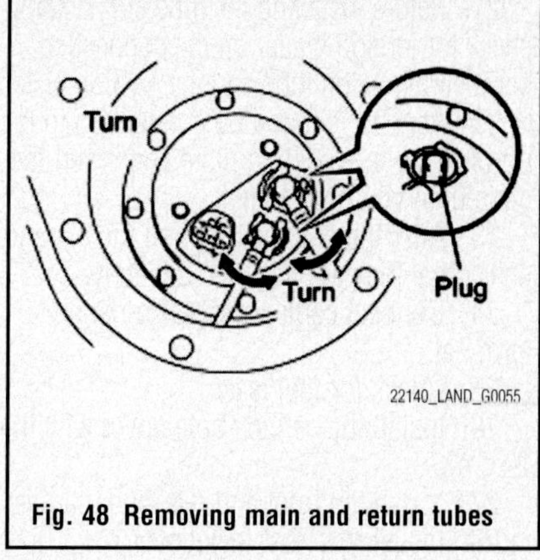

Fig. 48 Removing main and return tubes

9. Disconnect fuel main tube and return tube.
10. Before the operation, remove foreign matter or dirt sticking to the tube joint clips.
11. Widen the tip of the clips with your fingers and pull them out for disconnection.
12. Pull out the fuel main tube and the return tube. If the nylon tube and the suction plate stick together, turn the nylon tube with your fingers and pull it out for disconnection.
13. After the disconnection, protect the connector with a plastic bag.
14. Remove the 8 bolts.
15. Pull out the fuel pump and sender gauge assembly.

➡Do not damage the fuel pump filter. Be careful that the arm of the sender gauge is not bent.

16. Remove sending unit.

To install:

17. Install sending unit.
18. Install a new gasket to the fuel suction plate.
19. Insert the fuel pump and sender gauge assembly into the fuel tank.
20. Install the fuel tank vent tube set plate with the 8 bolts tighten bolts to 31 inch lbs. (3.5 Nm).

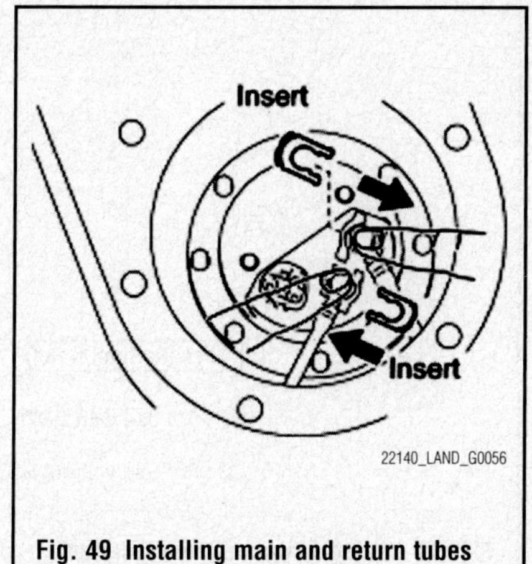

Fig. 49 Installing main and return tubes

21. Before installing the tube connectors, check for foreign matter on the connection between the nylon tube and the suction plate.
22. Attach the fuel tube connectors to the ports of the fuel suction plate and insert the clips until you hear a click.
23. After the connection, pull the clips to check that they are installed securely.
24. Connect cable to negative battery terminal.
25. Check for fuel leaks.
26. Install the service hole cover with the 2 screws.
27. Install the front and rear floor carpets.
28. Install rear seat lock cover.
29. Install step plate.
30. Install rear door scuff plate.
31. Install the 2 rear seats.
32. Perform initialization, if necessary.

HEATED OXYGEN (HO2S) SENSOR

LOCATION

See Figure 50.

REMOVAL & INSTALLATION

CAUTION

Wear protective gloves when removing the sensor. The exhaust pipe assembly is extremely hot immediately after the engine has stopped. Confirm that the exhaust pipe assembly has cooled down before removing it.

1. Disconnect cable from negative battery terminal

CAUTION

Wait at least 90 seconds after disconnecting the cable from the negative (-) battery terminal to prevent airbag and seat belt pretensioner activation.

2. Disconnect the sensor connector and remove heated oxygen sensor (for Bank 1 Sensor 2).
3. Disconnect the sensor connector and remove heated oxygen sensor (for Bank 2 Sensor 2).

To install:

4. Install heated oxygen sensor (for Bank 1 Sensor 2) and tighten to 32 ft. lbs. (44 Nm.).

➡Use a torque wrench with a fulcrum length of 30 cm (11.81 in).

5. Connect the sensor connector.
6. Install heated oxygen sensor (for Bank 2 Sensor 2) and tighten to 32 ft. lbs. (44 Nm.).

➡Use a torque wrench with a fulcrum length of 30 cm (11.81 in).

7. Connect the sensor connector.
8. Connect cable to negative battery terminal.
9. Perform initialization, if necessary.

INJECTION PRESSURE REGULATOR (IPR)

LOCATION

See Figure 51.

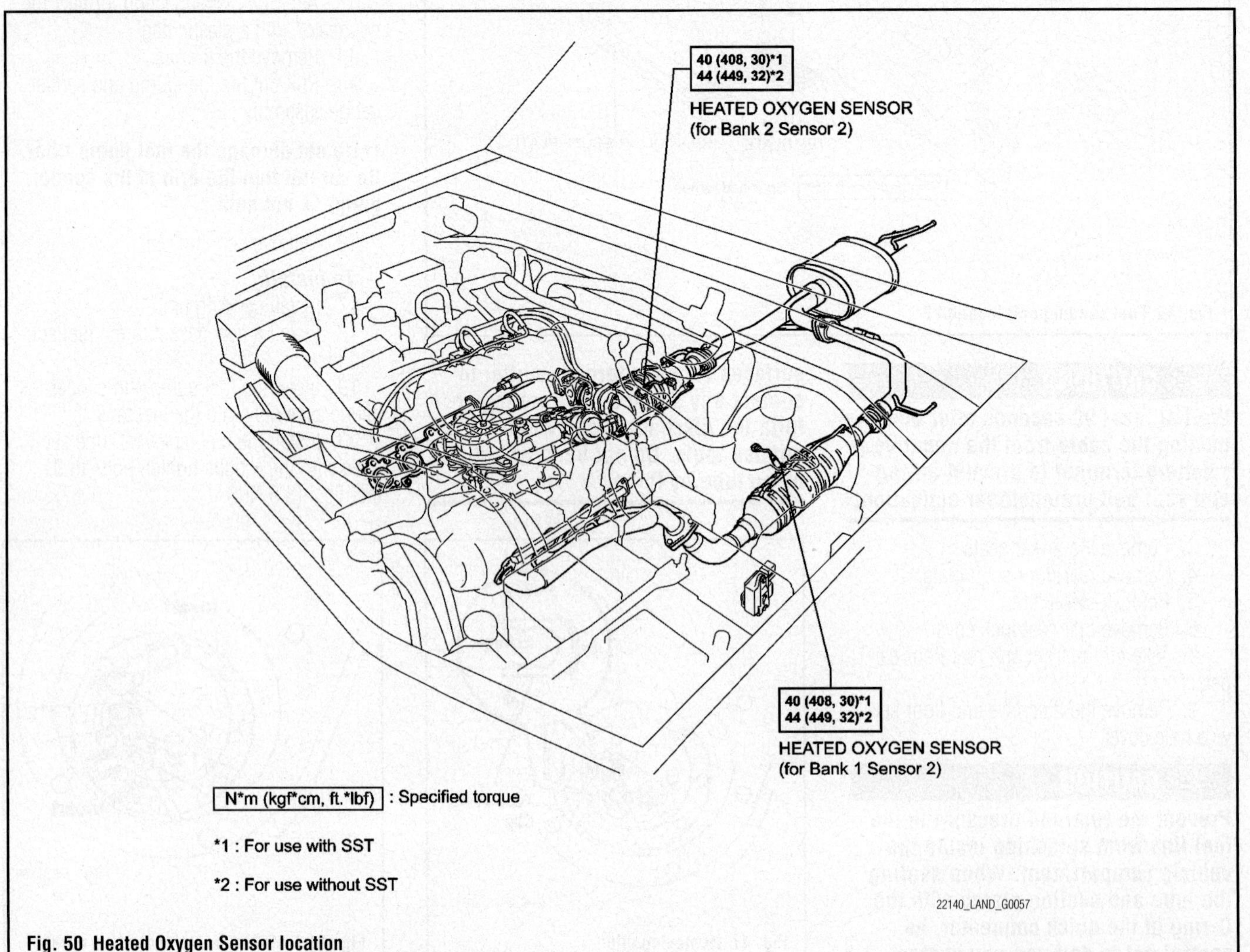

Fig. 50 Heated Oxygen Sensor location

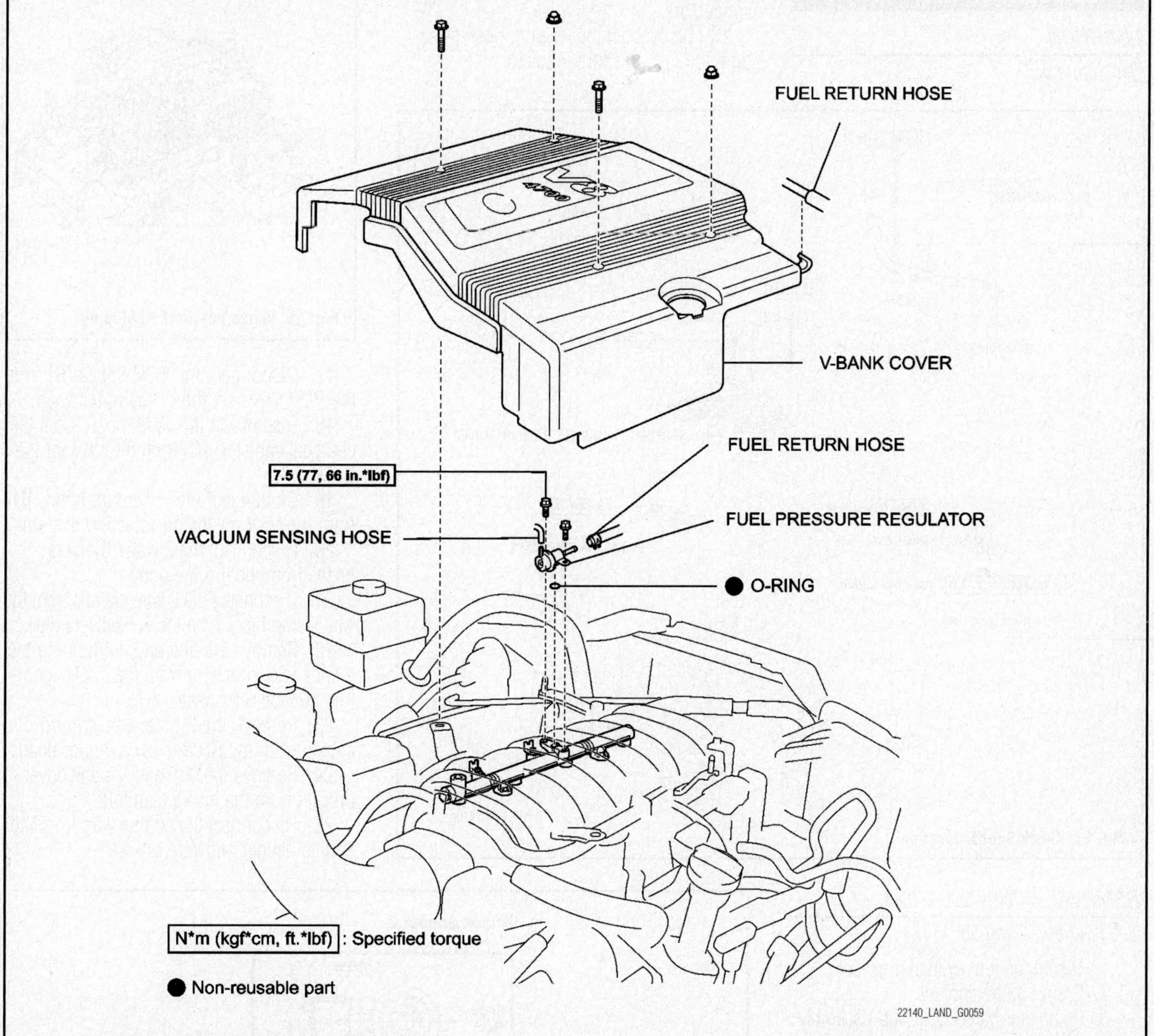

FUEL RETURN HOSE
V-BANK COVER
FUEL RETURN HOSE
7.5 (77, 66 in.*lbf)
VACUUM SENSING HOSE
FUEL PRESSURE REGULATOR
● O-RING
N*m (kgf*cm, ft.*lbf) : Specified torque
● Non-reusable part
22140_LAND_G0059

Fig. 51 Injection Pressure Regulator location

REMOVAL & INSTALLATION

See Figure 52.

1. Discharge fuel system pressure.
2. Disconnect cable from negative battery terminal.
3. Remove v-bank cover.
4. Disconnect the vacuum sensing hose from the pressure regulator.
5. Disconnect the fuel return hose from the pressure regulator
6. Remove the 2 bolts, and pull out the pressure regulator.
7. Remove the O-ring from the pressure regulator.

To install:

8. Apply a light coat of gasoline to a new O-ring, and install it to the pressure regulator.

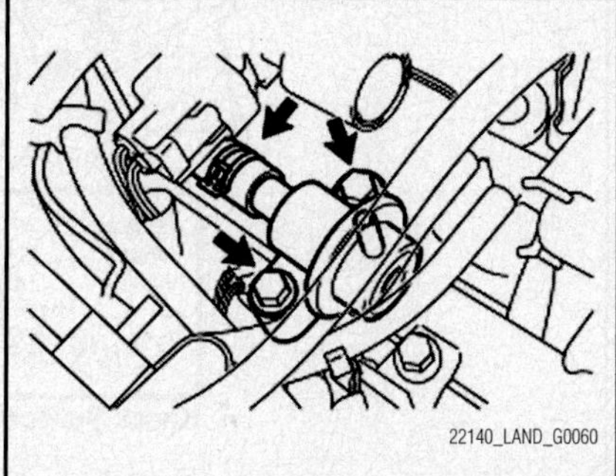

Fig. 52 Removing Injection Pressure Regulator

9. While turning the pressure regulator left and right, install it to the delivery pipe.
10. Install the pressure regulator with the 2 bolts and tighten bolts to 66 inch lbs. (7.5 Nm).

INTAKE AIR TEMPERATURE (IAT) SENSOR

LOCATION

The Intake Air Temperature (IAT) sensor, built into the Mass Air Flow (MAF) meter.

REMOVAL & INSTALLATION

See Mass Air Flow Meter

KNOCK SENSOR (KS)

LOCATION

See Figure 53.

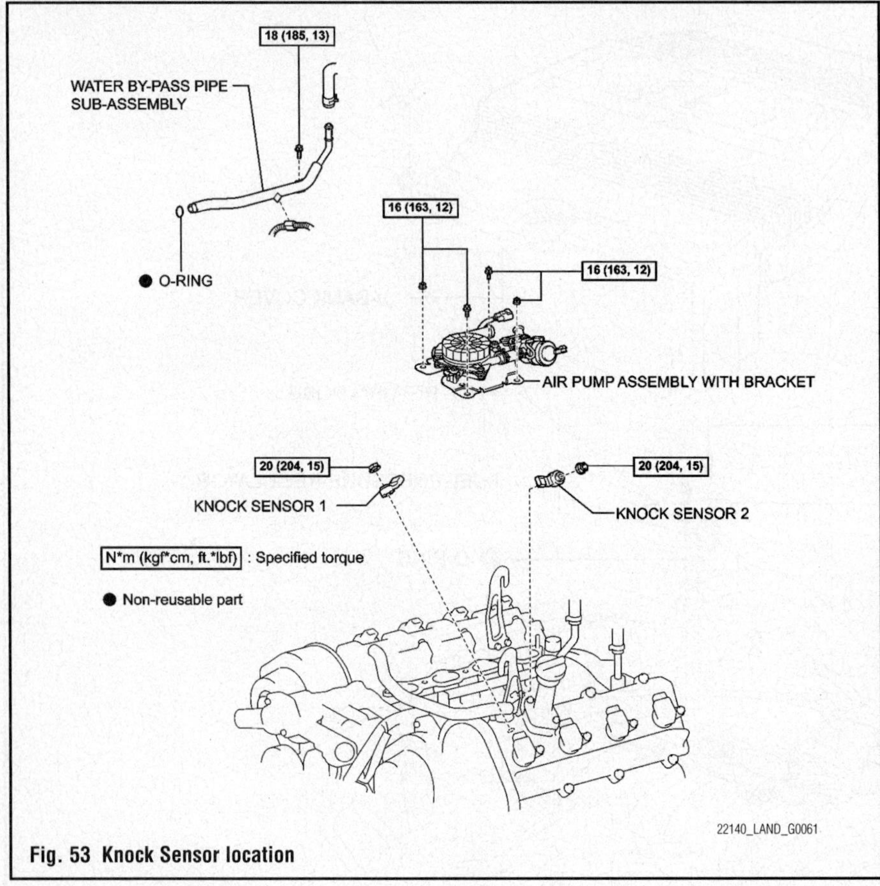

Fig. 53 Knock Sensor location

REMOVAL & INSTALLATION

See Figures 54 and 55.

1. Discharge fuel system pressure
2. Drain engine coolant
3. Remove v-bank cover sub-assembly
4. Disconnect the vacuum hoses (for the power steering idle-up and fuel pressure regulator) and ventilation hose.
5. Remove the air cleaner hose assembly.
6. Disconnect fuel hose.
7. Disconnect fuel hose no.2.
8. Disconnect the throttle control connector.
9. Disconnect the purge VSV connector.
10. Disconnect the 8 injector connectors.
11. Disconnect the ECT sensor connector.
12. Disconnect the 8 ignition coil connectors.
13. Disconnect the 2 VSV connectors for the air injection system.
14. Disconnect the 8 ignition coil connectors.
15. Disconnect the 2 air fuel ratio sensor connectors.
16. Disconnect the vacuum hose [A] from the fuel pressure regulator.

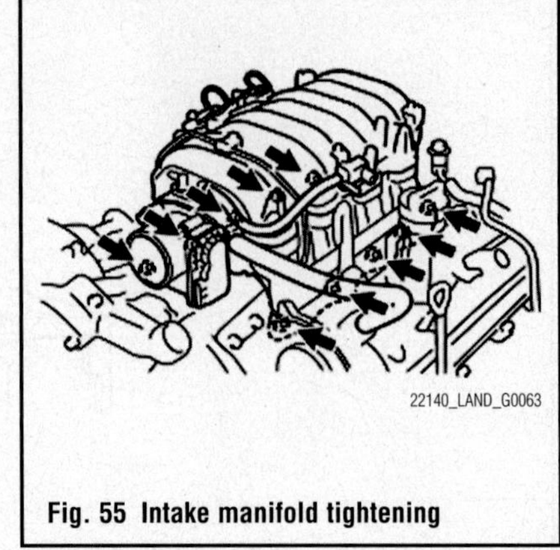

Fig. 55 Intake manifold tightening

17. Disconnect the PCV hoses [B] from the PCV valve on the LH cylinder head.
18. Disconnect the EVAP hose (from the charcoal canister) [C] from the VSV for the EVAP.
19. Disconnect the 2 vacuum hoses [D] from the VSV for the air injection system.
20. Disconnect the 2 water by-pass hoses from the throttle body.
21. Disconnect the 2 wire clamps from the wire clamp bracket on the RH delivery pipe.
22. Remove the bolt and nut holding the engine wire protector from the intake manifold and cylinder head.
23. Remove the 2 bolts and ground cables from the RH and LH cylinder heads.
24. Remove the bolt and V-bank cover bracket from the intake manifold.
25. Disconnect the engine wire from the engine hanger and wire bracket.

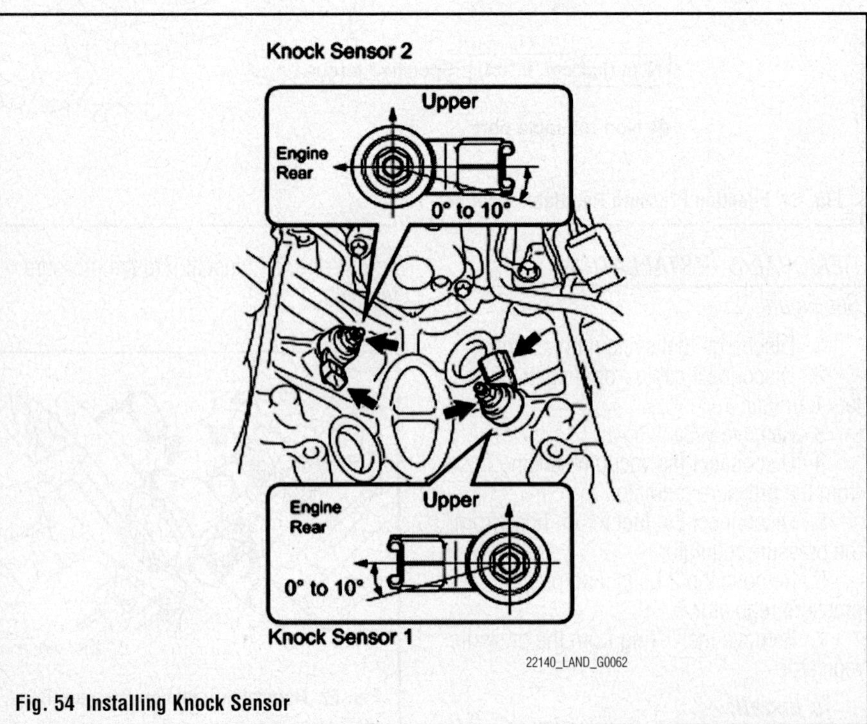

Fig. 54 Installing Knock Sensor

26. Remove the bolt and wire bracket from the intake manifold.
27. Remove the 6 bolts, 4 nuts, intake manifold assembly and 2 gaskets.
28. Remove air pump assembly with bracket.
29. Remove knock sensor.
30. Disconnect the 2 knock sensor connectors.

To install:

31. Install the 2 knock sensors with the 2 nuts as shown in the illustration and tighten nuts to 15 ft. lbs. (20 Nm).
32. Connect the 2 knock sensor connectors.
33. Place 2 new gaskets on the intake manifold.
34. Place the intake manifold on the cylinder heads.
35. Install and uniformly tighten the 6 bolts and 4 nuts in several steps to 13 ft. lbs. (18 Nm).
36. Install the V-bank cover bracket to the intake manifold.
37. Install the wire bracket to the intake manifold with the bolt.
38. Connect the engine wire to the engine hanger and wire bracket.
39. Connect the wire protector to the intake manifold and cylinder heads with the bolt and nut.
40. Install the 2 ground cables with the 2 bolts to the RH and LH cylinder heads.
41. Connect the 2 water by-pass hoses to the throttle body.
42. Connect the 2 wire clamps to the wire clamp bracket on the RH delivery pipe.
43. Connect the vacuum hose to the fuel pressure regulator.
44. Connect the PCV hose to the PCV valve on the LH cylinder head.
45. Connect the EVAP hose (from the charcoal canister) to the purge VSV.
46. Connect the 2 vacuum hoses to the VSV for the air injection system.
47. Connect the throttle control connector.
48. Connect the 2 VSV connectors for the air injection system.
49. Connect the purge VSV connector.
50. Connect the 8 injector connectors.
51. Connect the ECT sensor connector.
52. Connect the 8 ignition coil connectors.
53. Connect the 2 air fuel ratio sensor connectors.
54. Install fuel hose no.2.
55. Install fuel hose.
56. Install v-bank cover sub-assembly.
57. Add engine coolant.
58. Check for engine coolant leaks.
59. Check for fuel leaks.

MALFUNCTION INDICATOR LIGHT (MIL)

RESET PROCEDURES

Clearing DTC codes resets MIL.

1. To clear codes without intelligent tester:

a. Disconnect the negative (-) battery cable for more than 1 minute, or Remove the EFI OR ECD NO. 1 and ETCS fuses from the engine room junction block located inside the engine compartment for more than 1 minute.

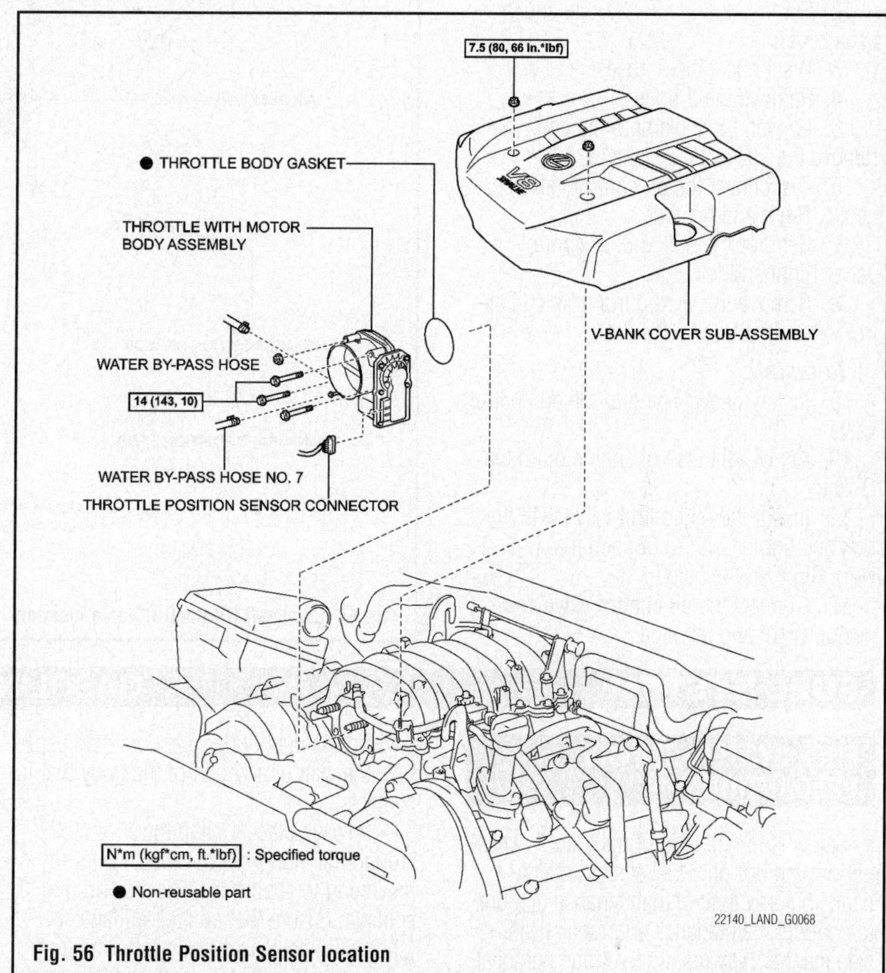

Fig. 56 Throttle Position Sensor location

MASS AIR FLOW (MAF) SENSOR

LOCATION

The MAF is located in the air intake snorkel.

REMOVAL & INSTALLATION

1. Disconnect connector.
2. Remove attaching screws an remove MAF.

To install:

3. Reverse removal procedure.

THROTTLE POSITION SENSOR (TPS)

LOCATION

See Figure 56.

OPERATION

The Throttle Position (TP) sensor is mounted on the throttle body and it detects the opening angle of the throttle valve. This sensor is electronically controlled and uses Hall-effect elements so that accurate control and reliability can be obtained.

REMOVAL & INSTALLATION

1. Disconnect connector.
2. Remove attaching screws and remove TPS.

To install:

3. Reverse removal procedure.

VARIABLE CAMSHAFT TIMING OIL CONTROL SOLENOID

LOCATION

See Figure 57.

REMOVAL & INSTALLATION

1. Disconnect cable from negative battery terminal.
2. Remove the 2 nuts, 2 bolts and V-bank cover.
3. Disconnect the 3 hoses.
4. Remove the 2 bolts.
5. Loosen the 2 clamp bolts and remove the intake air connector.
6. Disconnect the oil control valve connector. Right and left side.
7. Remove the bolt and oil control valve. Right and left side.
8. Remove the O-ring from the oil control valve. Right and left side.

To install:

9. Install a new O-ring to the oil control valve
10. Apply light coat of engine oil to the O-ring
11. Install the oil control valve with the bolt and tighten bolt to 66 inch lbs. (7.5 Nm). Right and left side
12. Connect the oil control valve connector. Right and left side.

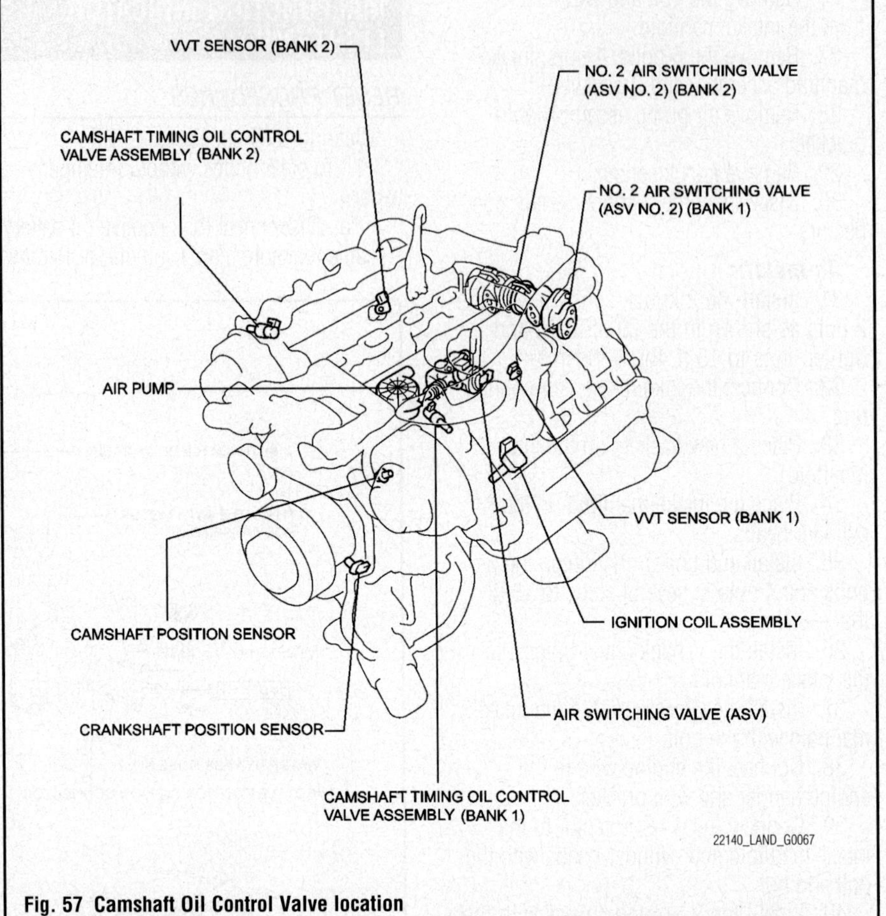

Fig. 57 Camshaft Oil Control Valve location

FUEL SYSTEM

GASOLINE FUEL INJECTION SYSTEM

FUEL SYSTEM SERVICE PRECAUTIONS

Safety is the most important factor when performing not only fuel system maintenance but any type of maintenance. Failure to conduct maintenance and repairs in a safe manner may result in serious personal injury or death. Maintenance and testing of the vehicle's fuel system components can be accomplished safely and effectively by adhering to the following rules and guidelines.

- To avoid the possibility of fire and personal injury, always disconnect the negative battery cable unless the repair or test procedure requires that battery voltage be applied.
- Always relieve the fuel system pressure prior to disconnecting any fuel system component (injector, fuel rail, pressure regulator, etc.), fitting or fuel line connection. Exercise extreme caution whenever relieving fuel system pressure to avoid exposing skin, face and eyes to fuel spray. Please be advised that fuel under pressure may penetrate the skin or any part of the body that it contacts.
- Always place a shop towel or cloth around the fitting or connection prior to loosening to absorb any excess fuel due to spillage. Ensure that all fuel spillage (should it occur) is quickly removed from engine surfaces. Ensure that all fuel soaked cloths or towels are deposited into a suitable waste container.
- Always keep a dry chemical (Class B) fire extinguisher near the work area.
- Do not allow fuel spray or fuel vapors to come into contact with a spark or open flame.
- Always use a back-up wrench when loosening and tightening fuel line connection fittings. This will prevent unnecessary stress and torsion to fuel line piping.
- Always replace worn fuel fitting O-rings with new. Do not substitute fuel hose or equivalent where fuel pipe is installed.

Before servicing the vehicle, make sure to also refer to the precautions in the beginning of this section as well.

RELIEVING FUEL SYSTEM PRESSURE

1. Remove the fuel pump relay from the engine compartment relay block.
2. Start the engine and let it run until it shuts off.
3. Turn the ignition to OFF.
4. Try to start the engine and make sure it won't start.
5. Disconnect the negative battery cable.
6. Install the relay.

FUEL FILTER

REMOVAL & INSTALLATION

See Figure 58.

1. Before servicing the vehicle, refer to the precautions section.
2. Relieve the fuel system pressure.
3. Remove or disconnect the following:
   - Negative battery cable
   - Fuel lines
   - Fuel filter

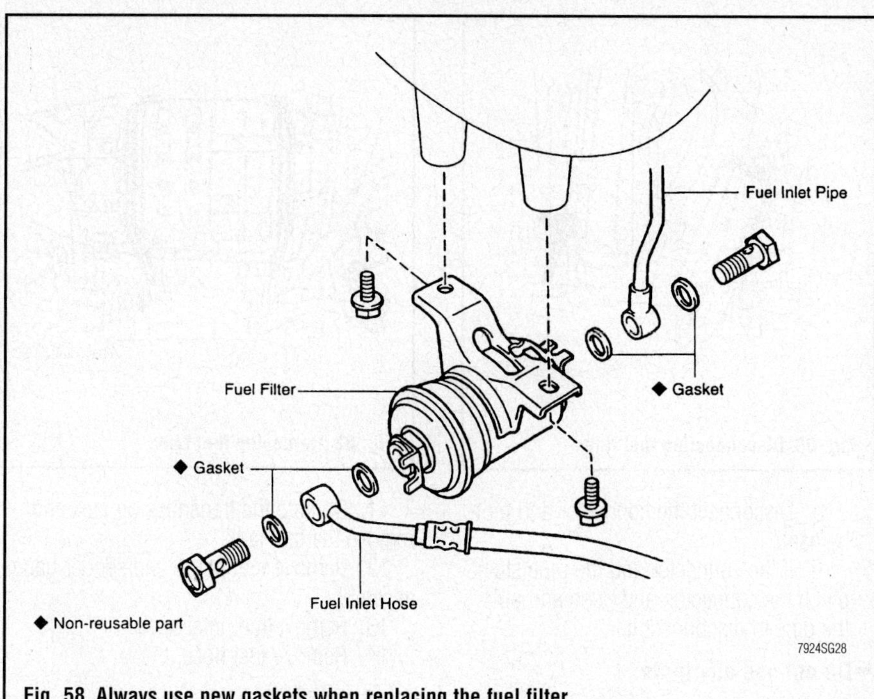

Fig. 58 Always use new gaskets when replacing the fuel filter

To install:

4. Install the fuel filter.
5. Use new washers and tighten the fuel line bolts to the following specifications:
   - Banjo bolt fittings: 21 ft. lbs. (29 Nm)
   - Flare nut fitting: 28 ft. lbs. (38 Nm)
6. Connect the negative battery cable.
7. Start the engine and check for leaks.

FUEL PUMP

REMOVAL & INSTALLATION

See Figure 59.

1. Before servicing the vehicle, refer to the precautions section.
2. Relieve the fuel system pressure.
3. Remove or disconnect the following:
   - Negative battery cable
   - Rear seats
   - Door sill trim plates
   - Carpeting and floor mats
   - Access panel
   - Fuel pump harness connector
   - Fuel lines
   - Fuel pump module

To install:

4. Install or connect the following:
   - Fuel pump module. Tighten the bolts to 35 inch lbs. (4 Nm).
   - Fuel lines
   - Fuel pump harness connector
   - Access panel
   - Carpeting and floor mats
   - Door sill trim plates
   - Rear seats
   - Negative battery cable
5. Start the engine and check for leaks.

FUEL RAIL & INJECTORS

REMOVAL & INSTALLATION

1. Before servicing the vehicle, refer to the precautions section.
2. Relieve the fuel system pressure.
3. Remove or disconnect the following:
   - Negative battery cable
   - Engine appearance cover
   - Air intake tube
   - Fuel lines
   - Fuel pulsation damper
   - Fuel pressure regulator vacuum line
   - Accelerator cable and bracket
   - Positive Crankcase Ventilation (PCV) valve and hose

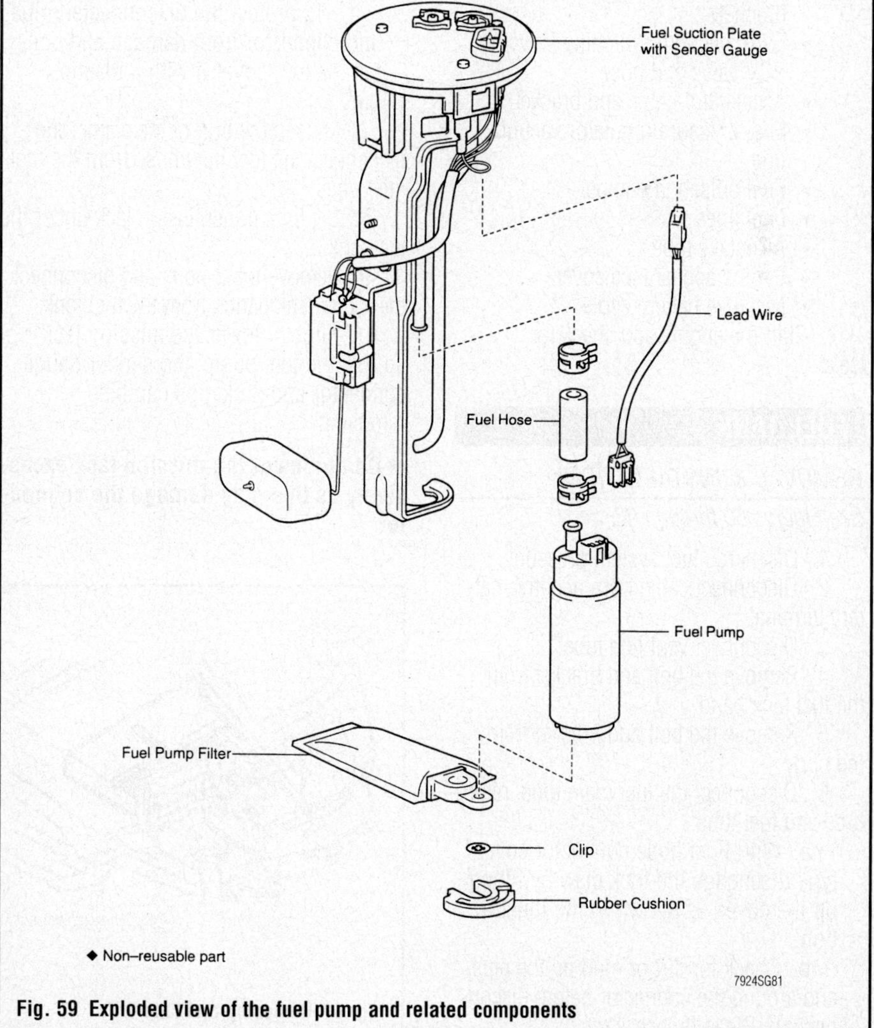

Fig. 59 Exploded view of the fuel pump and related components

- Evaporative Emissions (EVAP) vacuum switching valve
- Engine appearance cover brackets
- Fuel injector harness connectors
- Engine harness protector
- Fuel supply manifold crossover pipe
- Fuel supply manifolds with injectors attached
- Fuel injectors

To install:

4. Install the fuel injectors to the supply manifold with new O-ring seals and new grommets.
5. Install new injector insulators to the intake manifold.
6. Install or connect the following:
   - Fuel supply manifolds with injectors attached. Tighten the bolts to 66 inch lbs. (7.5 Nm).
   - Fuel supply manifold crossover pipe. Tighten the bolts to 29 ft. lbs. (39 Nm).
   - Engine harness protector
   - Fuel injector harness connectors
   - Engine appearance cover brackets
   - EVAP vacuum switching valve
   - PCV valve and hose
   - Accelerator cable and bracket
   - Fuel pressure regulator vacuum line
   - Fuel pulsation damper
   - Fuel lines
   - Air intake tube
   - Engine appearance cover
   - Negative battery cable
7. Start the engine and check for leaks.

FUEL TANK

REMOVAL & INSTALLATION

See Figures 60 through 62.

1. Discharge fuel system pressure
2. Disconnect cable from negative battery terminal
3. Disconnect vent line tube
4. Remove the bolt and bracket from the fuel tank band
5. Remove the bolt and bracket from the body
6. Disconnect the fuel main tube, return tube and fuel tube.
   a. With Fuel hose connector cover type disengage the lock claw by lifting up the cover, as shown in the illustration.
   b. Check for dirt or mud on the pipe and around the connector before disconnection. Clean if necessary

Fig. 60 Disconnecting fuel lines

   c. Disconnect the connector and pipe by hand.
   d. If the connector and the pipe stuck, pinch the connector, and push and pull the pipe to disconnect it.

➡Do not use any tools.

   e. Check for dirt or mud on the seal surface of the disconnected pipe. Clean if necessary.
   f. To protect the disconnected pipe and connector from damage and contamination, cover it with a plastic bag.
7. Loosen the bolt of the clamp and disconnect the fuel inlet hose from the fuel inlet pipe.
8. Set up a transmission jack under the fuel tank.
9. Remove the 2 bolts and disconnect the 2 fuel tank bands from the fuel tank.
10. Slightly lower the mission jack so that the fuel pump and sender gauge connector and 2 clamps can be removed.

➡Do not lower the mission jack excessively as this may damage the connector.

Fig. 61 Removing fuel tank

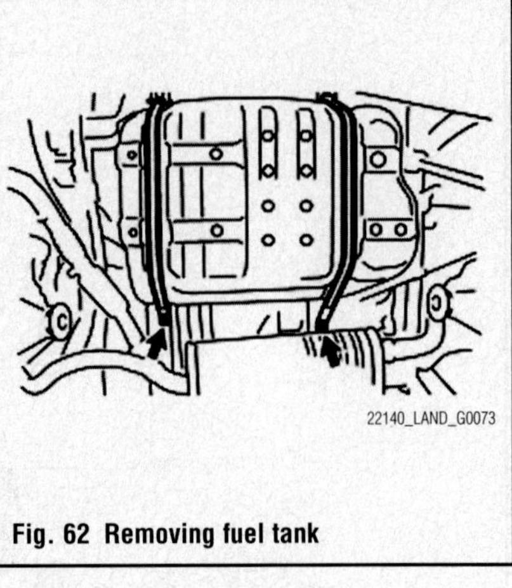

Fig. 62 Removing fuel tank

11. Operate the transmission jack and remove the fuel tank.
12. Remove fuel pump and sender gauge assembly.
13. Remove fuel inlet hose.
14. Remove fuel hose.

To install:

15. Install fuel hose
16. Install fuel inlet hose
17. Install fuel pump and sender gauge assembly
18. Install fuel tank assembly
19. Operate the transmission jack so that the fuel pump and sender gauge connector and 2 clamps can be installed. Then raise the transmission jack again to install the fuel tank.
20. Install the 2 fuel tank bands with the 2 bolts and tighten to 30 ft. lbs. (40 Nm).
21. Connect the fuel main tube, return tube and fuel hose.
   a. Check that there is no damage or contamination in the connected part of the pipe.
   b. Align the axis of the connector with the axis of the pipe. Push the pipe into the connector until the connector makes a "click" sound. If the connection is tight, apply a little amount of fresh engine oil on the tip of the pipe.
   c. After having finished the connection, try to pull apart the pipe and the connector and confirm that they are securely connected.
   d. With Fuel hose connector cover type attach the lock claw by lifting up the cover, as shown in the illustration.
22. Connect the fuel inlet hose to the fuel inlet pipe and tighten the bolt of the clamp to 66 inch lbs. (7.5 Nm).
23. Install the bracket to the body with the bolt and tighten bolt to 11 ft. lbs. (15 Nm).

24. Install the bracket to the fuel tank band with the bolt and tighten bolt to 11 ft. lbs. (15 Nm).
25. Connect the vent line tube to the fuel tank.
26. Check for fuel leaks.
27. Connect cable to negative battery terminal.
28. Perform initialization, if necessary.

IDLE SPEED

ADJUSTMENT

Idle speed is maintained by the Powertrain Control Module (PCM). No adjustment is necessary or possible.

THROTTLE BODY

REMOVAL & INSTALLATION

See Figures 63 and 64.

1. Before servicing the vehicle, refer to the precautions in the beginning of this section.
2. Remove throttle body cover.
3. Drain engine coolant.

Fig. 63 Throttle body coolant hoses

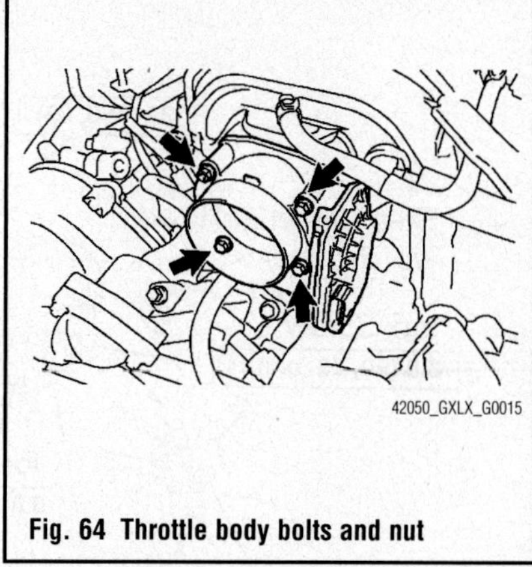

Fig. 64 Throttle body bolts and nut

4. Remove intake air connector.
5. Disconnect the throttle control connector.
6. Disconnect the 2 water bypass hoses from the throttle body.
7. Remove the nut and 3 bolts, and remove the throttle body from the intake manifold.

To install:

8. Install the throttle body with the nut and 3 bolts. Tighten them to 14 Nm (10 ft. lbs.).
9. Connect the 2 water bypass hoses to the throttle body.
10. Connect the throttle control connector.
11. Install intake air connector.
12. Fill with engine coolant.
13. Start engine and check for engine coolant leaks.
14. Install throttle body cover.

HEATING & AIR CONDITIONING SYSTEM

BLOWER MOTOR

REMOVAL & INSTALLATION

1. Before servicing the vehicle, refer to the precautions in the beginning of this section.
2. Disconnect negative battery cable.
3. Disconnect the connector.
4. Remove the three screws and the blower motor.

To install:

5. Install the blower motor with three screws.
6. Connect the connector.
7. Perform initialization, if necessary.

HEATER CORE

REMOVAL & INSTALLATION

See Figures 65 through 71.

1. Disconnect the negative battery cable.
2. Drain the cooling system into a clean container for reuse.
3. Disconnect the heater hoses from the heater core.
4. Remove the steering wheel by performing the following procedure:
   a. Position the front wheels facing straight-ahead.
   b. Remove the steering wheel side covers.
   c. Using a Torx® wrench, loosen the 2 screws located at each side of the steering wheel until the screw's circumference groove catches on the screw case.
   d. Pull the air bag module from the steering wheel and disconnect the electrical connector.

✱✱ CAUTION

Place the air bag module in a safe place with the front side facing upward.

   e. Remove the steering wheel nut.
   f. Place alignment marks on the steering wheel and the main shaft.
   g. Using a steering wheel puller, press the steering wheel from the steering column.
5. Remove the instrument panel and reinforcement by performing the following procedure:
   a. Remove the front door scuff plates, the cowl side trim and the front door opening trim.
   b. At the driver's side, remove the 2 assist grip plugs, the 2 screws and assist grip and the front pillar garnish.
   c. At the passenger's side, remove the 4 assist grip plugs, the 4 screws, the 2 assist grips and the front pillar garnish.
   d. Remove the instrument cluster finish panel.
   e. Remove the 2 screws and the hood lock control cable.
   f. Remove the 2 screws and the fuel lid control cable lever.
   g. Remove the lower No. 1 panel screw and the panel.
   h. Remove the lower left side panel.
   i. Remove the 3 steering column cover screws and the covers.

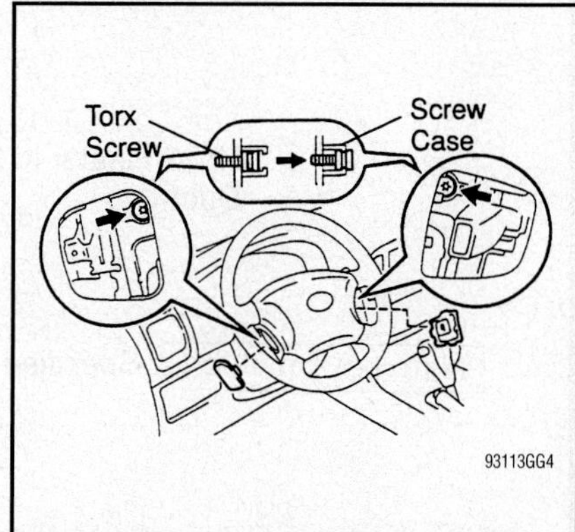

Fig. 65 View of the steering wheels Torx® bolts

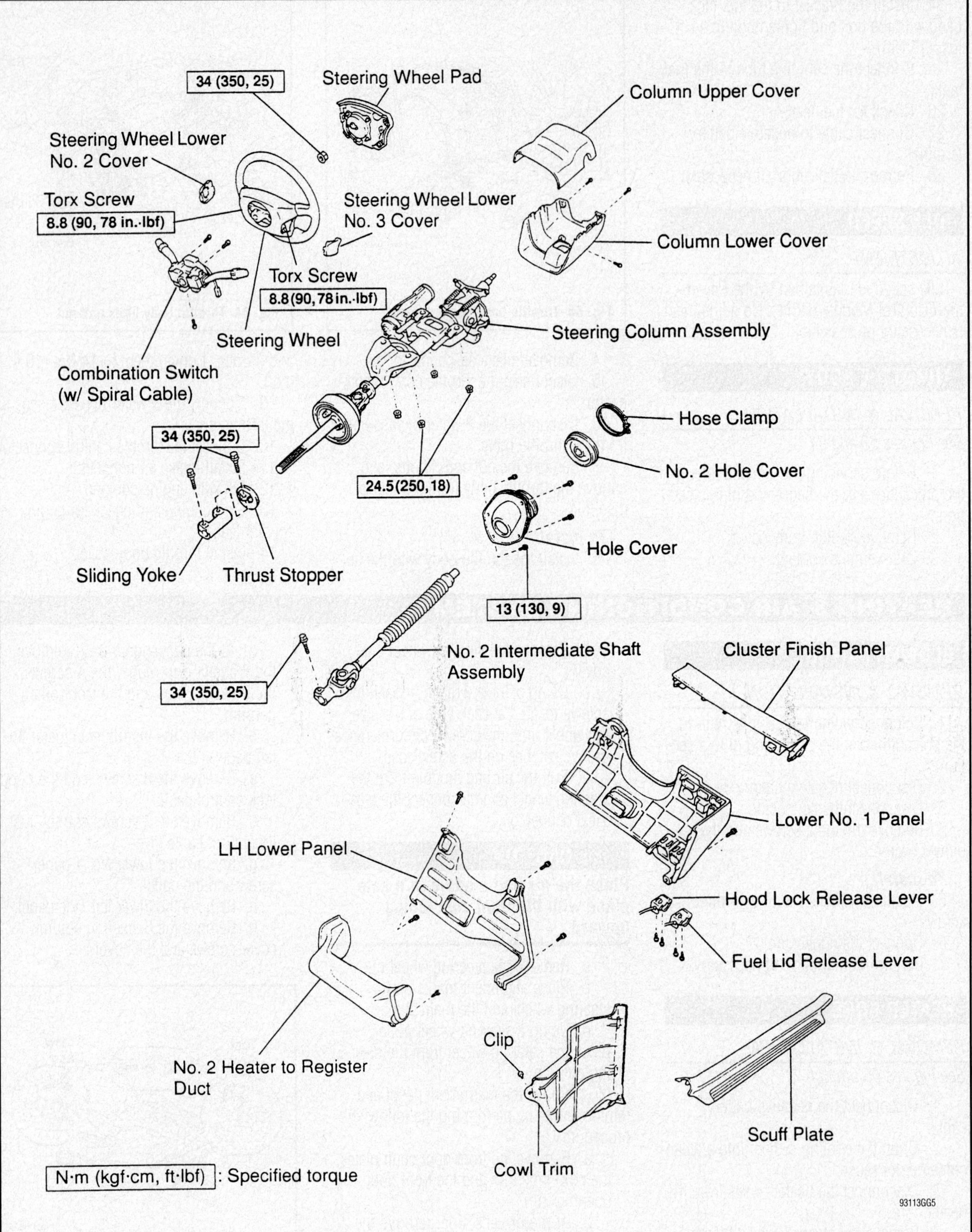

Fig. 66 Exploded view the steering column—Part 1 of 2

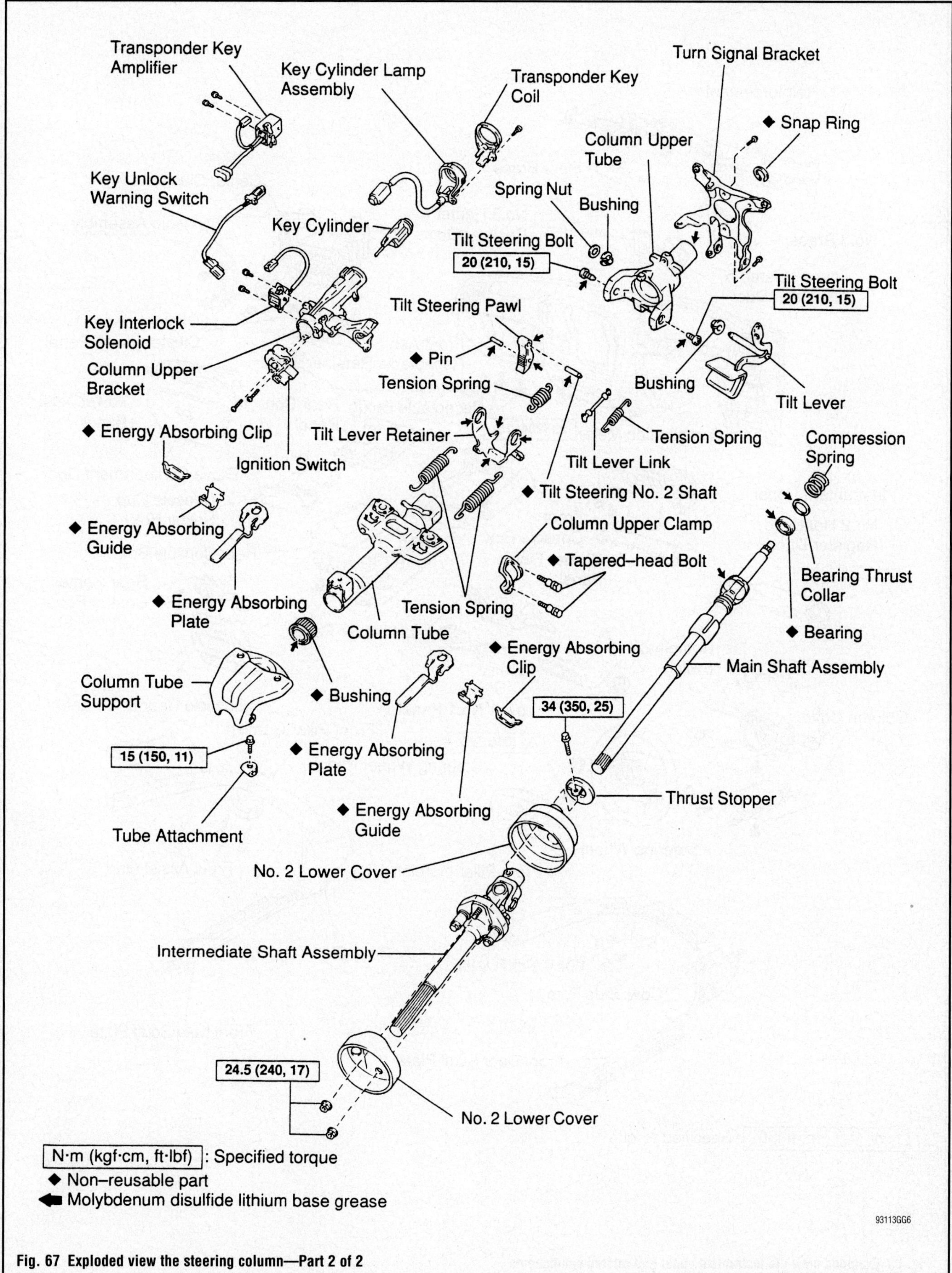

Fig. 67 Exploded view the steering column—Part 2 of 2

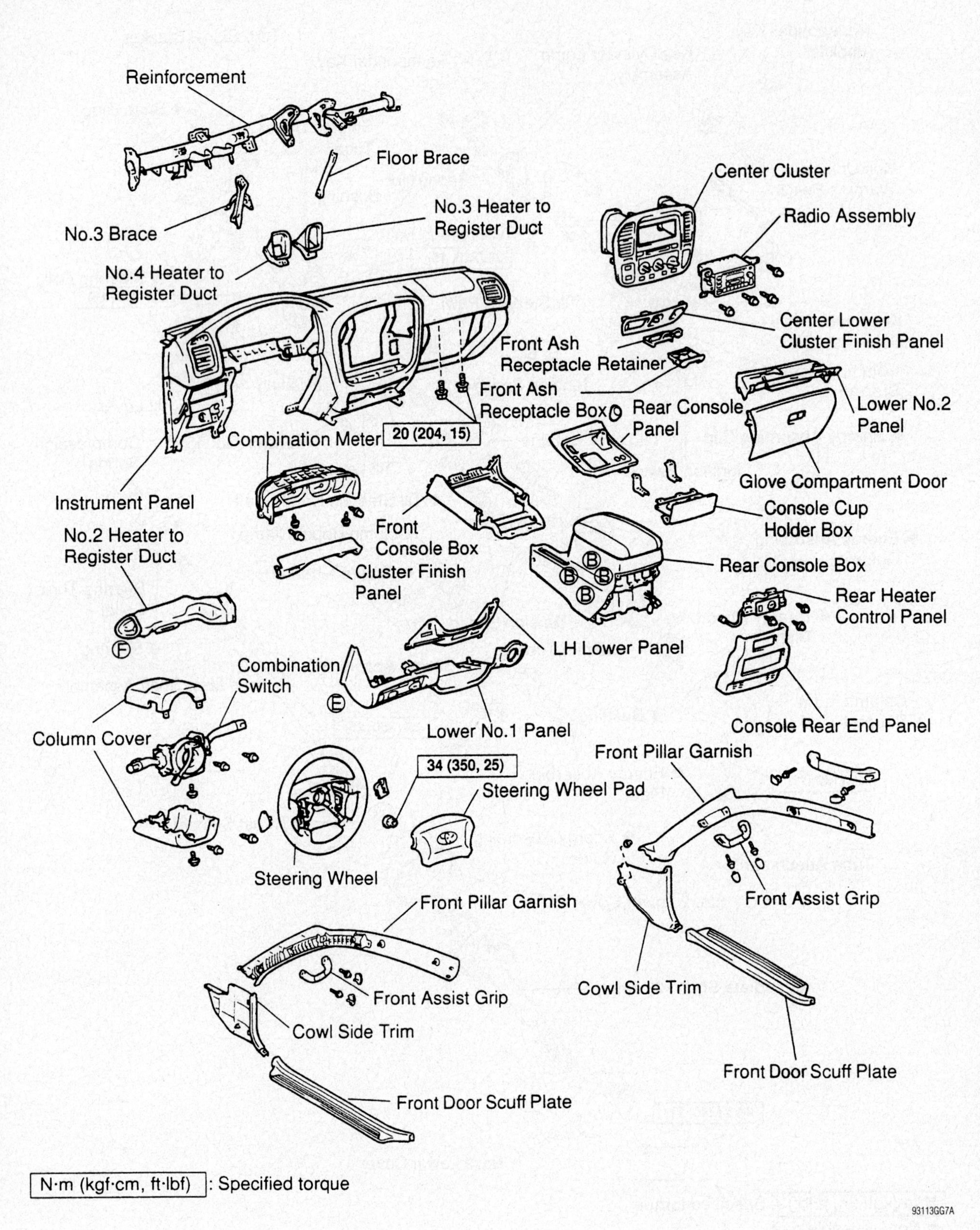

Fig. 68 Exploded view the instrument panel and related components

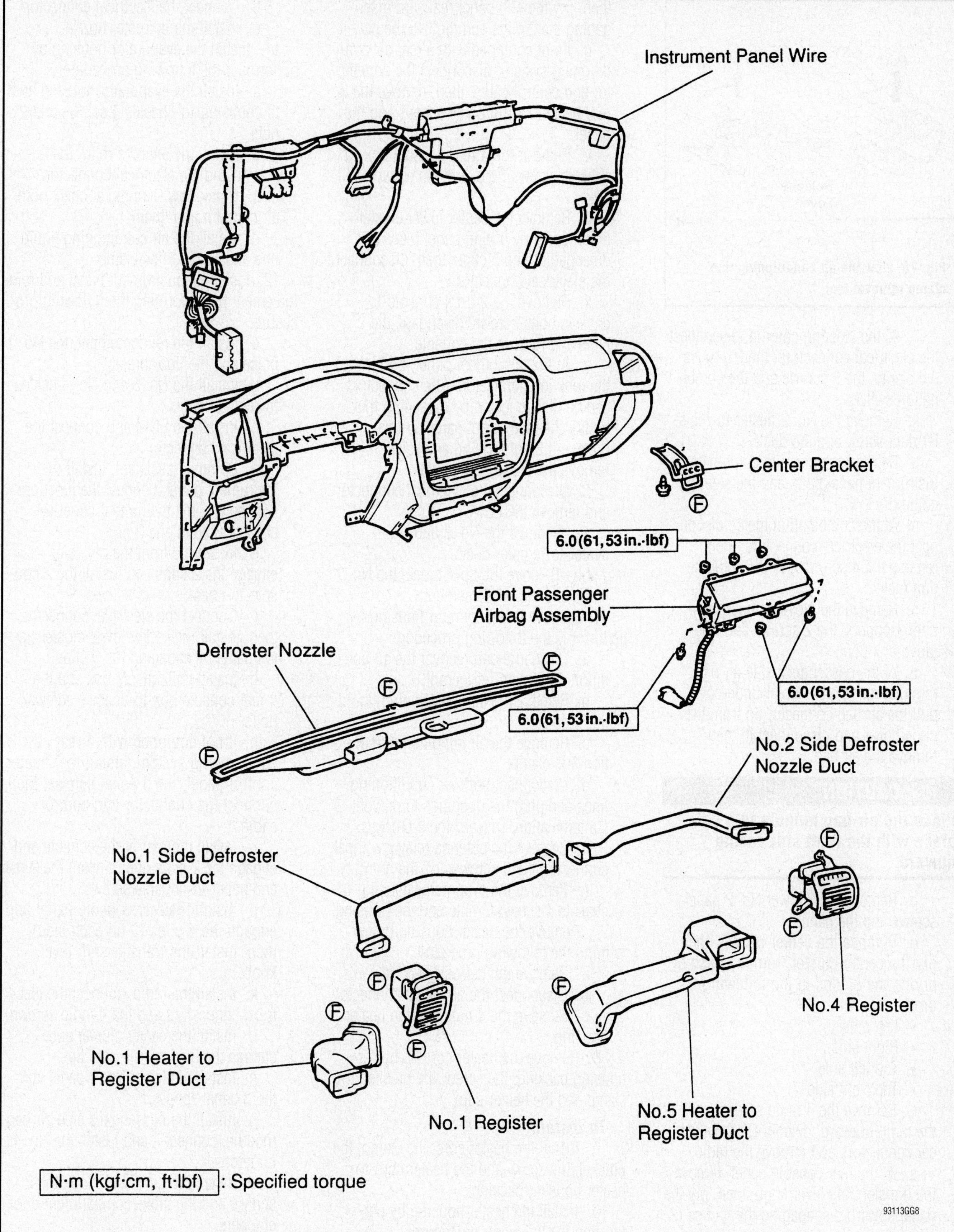

Fig. 69 Exploded view the front ventilation ducts and related components

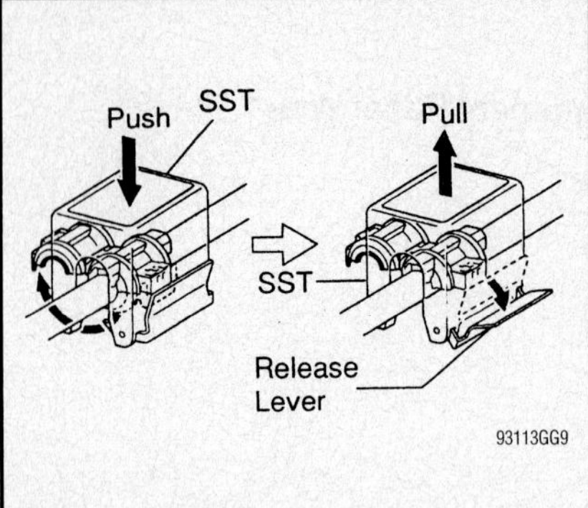

Fig. 70 View the air conditioning line clamp removal tool

j. At the steering column, disconnect the electrical connectors; then, remove the clamp, the 3 screws and the combination switch.

k. Remove the No. 2 heater-to-register duct screw and the duct.

l. Remove the steering column-to-instrument panel bolts and the steering column.

m. At the combination meter, disconnect the electrical connectors; then, remove the 4 screws and the combination meter.

n. Remove the glove compartment door stoppers, the 2 screws and the glove box door.

o. At the passenger's side air bag module, remove the No. 1 undercover, pull the air bag connector up from the undercover and disconnect it; then, remove the air bag.

CAUTION

Place the air bag module in a safe place with the front side facing upward.

p. Remove the 3 lower No. 2 panel screws and the panel.

q. Remove the center cluster; then, pry the center cluster from the dash by prying the 8 clips in the following order:

- Left side
- Right side
- Top left side
- Top right side

r. Remove the 4 radio screws, pull the radio outward, disconnect the electrical connectors and remove the radio.

s. At the rear console panel, remove the transfer shift lever knob; then, pry the panel upward disengaging the 4 clips (2 on each side) and remove the panel.

t. At the rear of the console, remove the 2 rear end panel-to-console screws; then, pry the end panel rearward disengaging the 2 clips and remove the panel.

u. If not equipped with a rear air conditioning system, disconnect the connector and control cable; then, remove the 3 rear heater control panel screws and the panel.

v. Remove the 4 rear console box-to-chassis screws/bolts and the console box.

w. Remove the center lower cluster finish panel by prying panel rearward disengaging the 5 clips; then, disconnect the electrical connector.

x. Remove the 2 front console-to-chassis bolts/screws, disengage the 2 clips and remove the console.

y. At the instrument panel, disconnect the junction connectors (the connectors can be disconnected by loosening the bolts), the instrument panel-to-chassis 8 bolts and 2 nuts. Using an assistant, remove the instrument panel.

z. Disconnect the electrical connector and remove the ECM.

aa. Remove the No. 3 and No. 4 heater-to-register ducts.

bb. Remove the floor brace, the No. 1 brace and the reinforcement.

6. Remove the evaporator housing by performing the following procedure:

a. Discharge and recover the air conditioning system refrigerant.

b. Remove the air conditioning liquid line clamp.

c. Remove the air conditioning suction line clamp.

d. Disconnect both air conditioning lines and plug the openings to prevent contamination. Discard the 4 O-rings.

e. Remove the antenna relay electrical connector, the 2 screws and the relay.

f. Remove the evaporator housing-to-chassis 4 screws/2 nuts and the housing.

7. Remove the heater housing by performing the following procedure:

a. Remove the defroster nozzle.

b. Disconnect the electrical connector.

c. Remove the 4 nuts and the heater housing.

8. Remove the heater core-to-heater housing packing, the screw, the bracket, the clamp and the heater core.

To install:

9. Install the heater core, the clamp, the bracket, the screw and the heater core-to-heater housing packing.

10. Install the heater housing by performing the following procedure:

a. Install the heater housing and the 4 nuts.

b. Connect the electrical connector.

c. Install the defroster nozzle.

11. Install the evaporator housing by performing the following procedure:

a. Install the evaporator housing and the housing-to-chassis 4 screws and 2 nuts.

b. Install the antenna relay, the 2 screws and the electrical connector.

c. Using new O-rings, connect both air conditioning lines.

d. Install the air conditioning liquid line and suction line clamp.

12. Install the instrument panel and reinforcement by performing the following procedure:

a. Install the reinforcement, the No. 1 brace and the floor brace.

b. Install the No. 3 and No. 4 heater-to-register ducts.

c. Install the ECM and connect the electrical connector.

d. Using an assistant, install the instrument panel, connect the junction connectors, the instrument panel-to-chassis 8 bolts and 2 nuts.

e. Install the front the console, engage the 2 clips and install the 2 console-to-chassis bolts/screws.

f. Connect the electrical connector; then, install the center lower cluster finish panel by engaging the 5 clips.

g. Install the console box and the 4 rear console box-to-chassis screws/bolts.

h. If not equipped with a rear air conditioning system, install rear heater control panel, the 3 panel screws; then, connect the connector and control cable.

i. Install the rear of the console and engage the 2 clips; then, install the 2 rear end panel-to-console screws.

j. Install the rear console panel and engage the 4 clips (2 on each side); then, install the transfer shift lever knob.

k. Install the radio, connect the electrical connectors and the 4 radio screws.

l. Install the center cluster and engage the 8 center cluster clips.

m. Install the lower No. 2 panel and the 3 panel screws.

n. Install the passenger's side air bag module, connect it and install the No. 1 undercover.

o. Install the glove box door, the 2 screws and the glove compartment door stoppers.

p. Install the combination meter and the 4 screws; then, connect the electrical connectors.

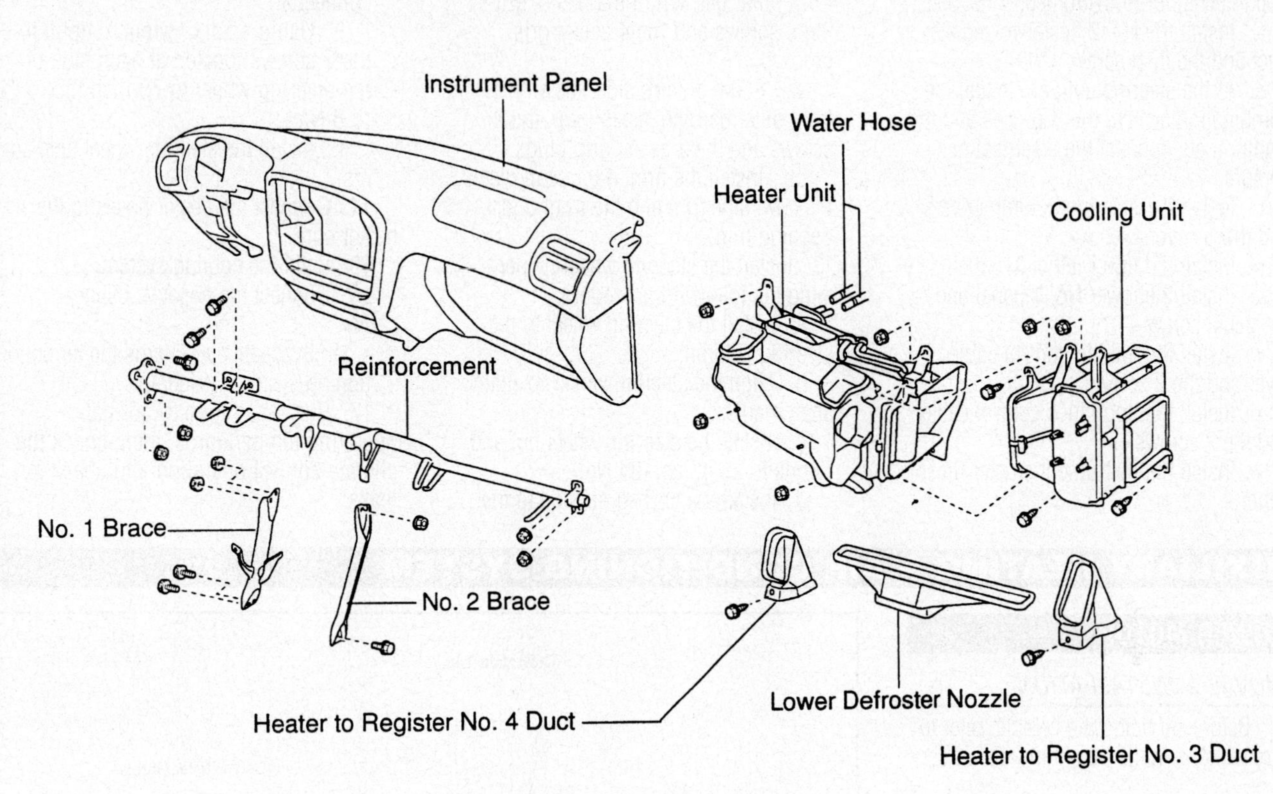

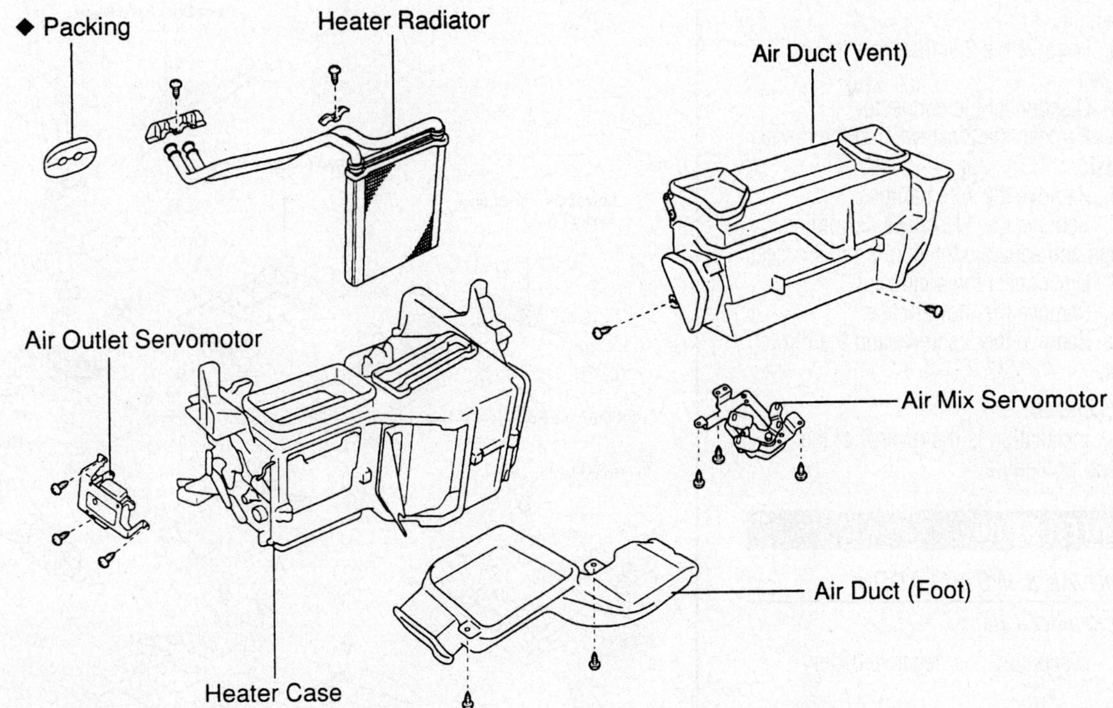

Fig. 71 Exploded view the front heater core, heater housing, evaporator housing and related components

q. Install the steering column and the steering column-to-instrument panel bolts.

r. Install the No. 2 heater-to-register duct and the duct screw.

s. At the steering column, install the combination switch, the 3 screws and the clamp; then, connect the electrical connectors.

t. Install the steering column covers and the 3 covers screws.

u. Install the lower left side panel.

v. Install the lower No. 1 panel and the panel screw.

w. Install the fuel lid control cable lever and the 2 screws.

x. Install the hood lock control cable and the 2 screws.

y. Install the instrument cluster finish panel.

z. At the passenger's side, install the front pillar garnish, the 2 assist grips, the 4 screws and the 4 assist grip plugs.

aa. At the driver's side, install the front pillar garnish, assist grip, the 2 screws and the 2 assist grip plugs.

bb. Install the front door scuff plates, the cowl side trim and the front door opening trim.

13. Install the steering wheel by performing the following procedure:

a. Install the steering wheel to the steering column.

b. Align the steering wheel-to-main shaft marks.

c. Install the steering wheel nut and torque to 25 ft. lbs. (34 Nm).

d. Install the air bag module to the steering wheel and connect the electrical connector.

e. Using a Torx® wrench, tighten the 2 screws located at each side of the steering wheel to 78 inch lbs. (8.8 Nm).

f. Install the steering wheel side covers.

14. Connect the heater hoses to the heater core.

15. Refill the cooling system.

16. Connect the negative battery cable.

a. Evacuate and charge the air conditioning system refrigerant.

17. Run the engine to normal operating temperatures; then, check the climate control operation and check for leaks.

AUXILIARY HEATING & AIR CONDITIONING SYSTEM

BLOWER MOTOR

REMOVAL & INSTALLATION

1. Before servicing the vehicle, refer to the precautions in the beginning of this section.
2. Remove the rear cooling unit from the vehicle.
3. Remove the 2 screws and the A/C amplifier.
4. Disconnect the connector.
5. Remove the 2 screws and the power transistor.
6. Remove the wire harness.
7. Remove the 11 screws, 3 holding springs and separate the cooling unit cases.
8. Disconnect the connector.
9. Remove the blower case.
10. Remove the 3 screws and the blower motor.

To install:

11. Installation is the reverse of the removal procedure.

HEATER CORE

REMOVAL & INSTALLATION

See Figures 72 and 73.

1. Disconnect the negative battery cable.
2. Drain the cooling system into a clean container for reuse.
3. Disconnect the heater hoses from the rear heater core.
4. Remove the front seats.
5. Remove the rear heater control assembly.
6. Remove the rear console box.

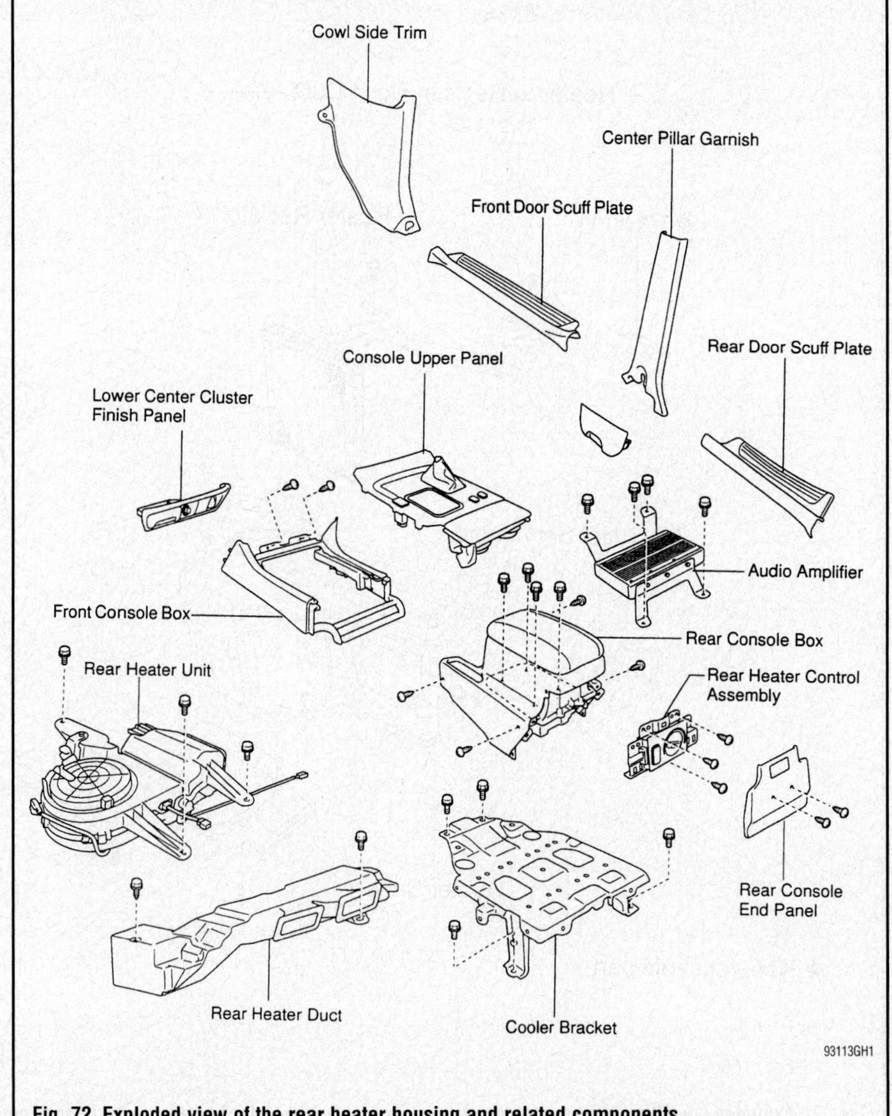

Fig. 72 Exploded view of the rear heater housing and related components

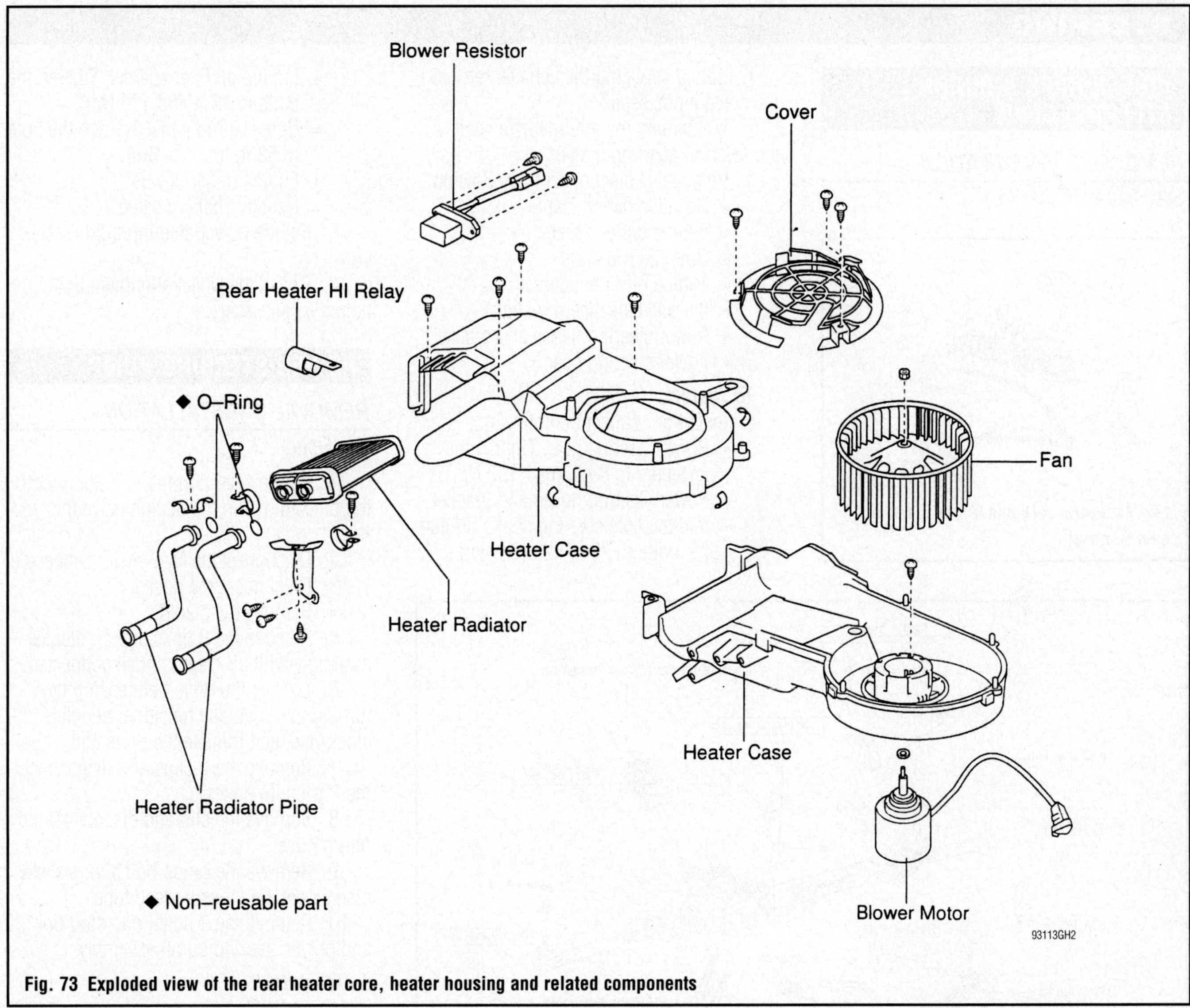

Fig. 73 Exploded view of the rear heater core, heater housing and related components

7. Remove the front console box cover.
8. Remove the lower center cluster finish panel.
9. Remove the front door scuff plates.
10. Remove the cowl side trim.
11. Remove the rear door scuff plates.
12. Remove the center pillar garnishes.
13. Slide the carpet rearward.
14. Remove the cooler bracket bolts and the bracket.
15. Remove the rear heater duct bolt/screw and the duct.
16. Disconnect the rear heater housing electrical connector.
17. Remove the 3 rear heater housing-to-chassis bolts and the heater housing.
18. Remove the heater core-to-heater housing 3 screws and 2 clamps.
19. Remove the heater core from the heater housing.

To install:

20. Install the heater core to the heater housing.
21. Install the heater core-to-heater housing 3 screws and 2 clamps.
22. Install the heater housing and the 3 rear heater housing-to-chassis bolts.
23. Connect the rear heater housing electrical connector.
24. Install the rear heater duct and the duct bolt/screw.
25. Install the cooler bracket and the bracket bolts.
26. Slide the carpet rearward.
27. Install the center pillar garnishes.
28. Install the rear door scuff plates.
29. Install the cowl side trim.
30. Install the front door scuff plates.
31. Install the lower center cluster finish panel.
32. Install the front console box cover.
33. Install the rear console box.
34. Install the rear heater control assembly.
35. Install the front seats.
36. Connect the heater hoses to the rear heater core.
37. Refill the cooling system.
38. Connect the negative battery cable.

STEERING

POWER RACK & PINION STEERING GEAR

REMOVAL & INSTALLATION

See Figures 74 and 75.

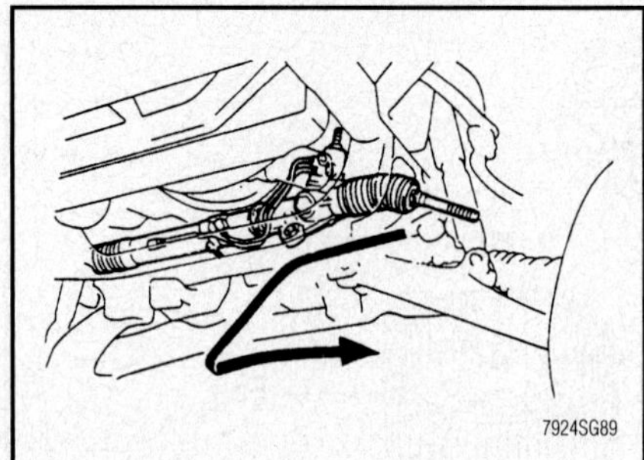

Fig. 74 Power rack and pinion steering gear removal

1. Before servicing the vehicle, refer to the precautions section.
2. Matchmark the intermediate shaft to the steering gear input shaft.
3. Remove or disconnect the following:
   - Negative battery cable
   - Engine under covers
   - Outer tie rod ends
   - Engine oil filter adapter
   - Intermediate steering shaft
   - Power steering hoses and bracket
   - Power steering gear

To install:

4. Install or connect the following:
   - Power steering gear. Tighten the fasteners to 89 ft. lbs. (120 Nm).
   - Power steering hoses and bracket
   - Intermediate steering shaft. Tighten the bolts to 25 ft. lbs. (34 Nm).
   - Engine oil filter adapter. Tighten the bolts to 13 ft. lbs. (18 Nm).
   - Outer tie rod ends. Tighten the nuts to 53 ft. lbs. (72 Nm).
   - Engine under covers
   - Negative battery cable
5. Fill the power steering fluid reservoir.
6. Check the wheel alignment and adjust as necessary.

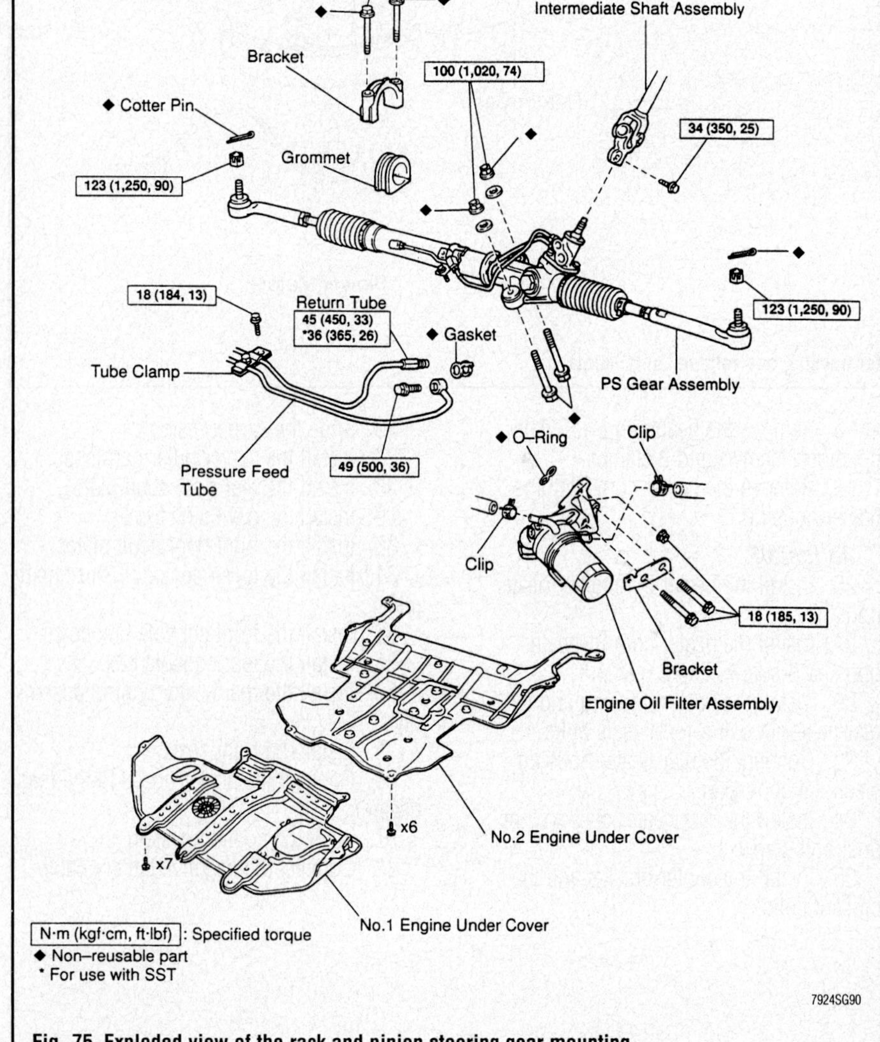

Fig. 75 Exploded view of the rack and pinion steering gear mounting

POWER STEERING PUMP

REMOVAL & INSTALLATION

See Figure 76.

1. Before servicing the vehicle, refer to the precautions in the beginning of this section.
2. Disconnect the MAF meter connector.
3. Disconnect the hoses.
4. Remove the clamp.
5. Remove the 3 bolts and air cleaner assembly with air cleaner hose connected.
6. Loosen the drive belt tension by turning the drive belt tensioner counterclockwise, and remove the drive belt.
7. Remove the 2 clips and disconnect the 2 vacuum hoses.
8. Remove the clip and disconnect the return hose.
9. Remove the union bolt and gasket, disconnect the pressure feed tube.
10. Remove the 2 bolts, nut, stud bolt and power steering pump assembly.

To install:

11. Install the power steering pump assembly with the stud bolt.
12. Tighten the stud bolt to 22 Nm (16 ft. lbs.)
13. Install the 2 bolts and nut and tighten them to 44 Nm (33 ft. lbs.)
14. Install a new gasket and the union bolt on the pressure feed tube.

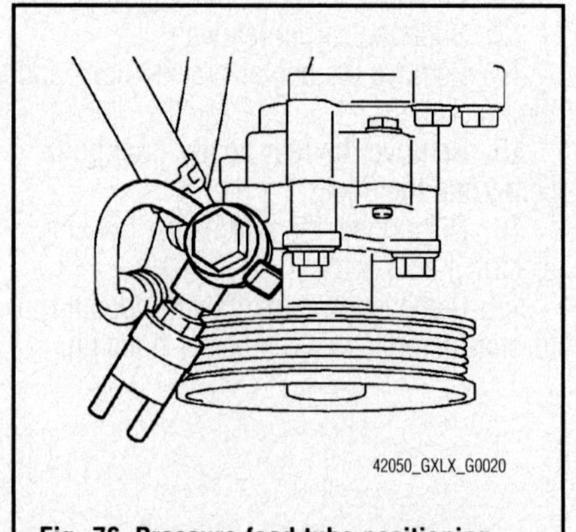

Fig. 76 Pressure feed tube positioning

➡Make sure that the stopper of the pressure feed tube contacts the power steering pump body as shown in the illustration.

15. Tighten the union bolt to 46.5 Nm (34 ft. lbs.)
16. Connect the return hose with the clip.
17. Connect the 2 vacuum hoses and install the 2 clips.
18. Loosen the drive belt tension by turning the drive belt tensioner counterclockwise, and install the belt.
19. Install the air cleaner assembly with air cleaner hose and the 3 bolts.
20. Install the clamp.
21. Connect the MAF meter connector.
22. Fill with power steering fluid and bleed the system.

BLEEDING

1. Before servicing the vehicle, refer to the precautions in the beginning of this section.
2. Check fluid level
3. Jack up front of vehicle and support it with stands
4. With the engine stopped, turn the wheel slowly from lock to lock several times.
5. Lower the vehicle.
6. Start the engine and run at idle for a few minutes.
7. With the engine idling, turn the wheel left or right to the full lock position and keep it there for 2 to 3 seconds, then turn the wheel to the opposite full lock position and keep it there for 2 to 3 seconds. Repeat several times.
8. Stop the engine.
9. Check for foaming or emulsification of the power steering fluid.
10. If the system has to be bled twice specifically because of foaming or emulsification, check for fluid leaks in the system.
11. Check fluid level.

SUSPENSION

FRONT SUSPENSION

CONTROL LINKS

REMOVAL & INSTALLATION

See Figure 77.

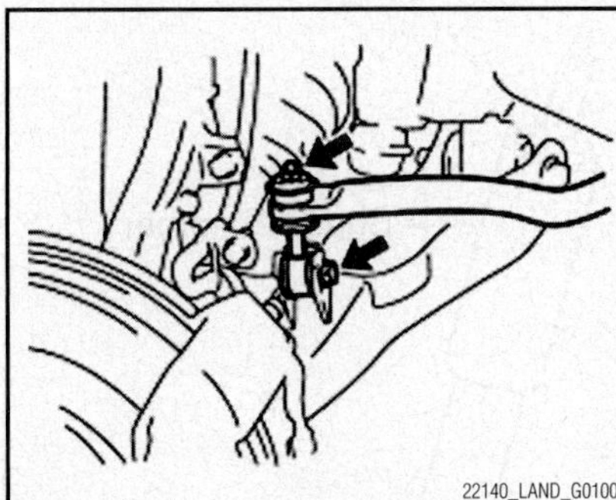

Fig. 77 Removing control link

1. Remove engine under cover
2. Remove the bolt, nut, 2 retainers, 2 cushions and stabilizer bar link
3. Remove front stabilizer link assembly

To install:

4. Installation is reverse of removal.
   a. Tighten bolt to 38 ft. lbs. (52 Nm)
   b. Tighten nut to 18 ft. lbs. (25 Nm)

LOWER BALL JOINT

REMOVAL & INSTALLATION

The lower ball joint is serviced with the lower control arm as an assembly.

LOWER CONTROL ARM

REMOVAL & INSTALLATION

1. Before servicing the vehicle, refer to the precautions section.
2. Remove or disconnect the following:
   - Front wheel
   - Engine under cover
   - Torsion bar
   - Stabilizer bar link
   - Shock absorber
   - Lower ball joint
   - Lower control arm

To install:

3. Install or connect the following:
   - Lower control arm. Tighten the bolts to 123 ft. lbs. (167 Nm).
   - Lower ball joint. Tighten the nut to 117 ft. lbs. (159 Nm).
   - Shock absorber. Tighten nut to 100 ft. lbs. (135 Nm)
   - Stabilizer bar link. Tighten the bolt to 38 ft. lbs. (52 Nm).
   - Torsion bar
   - Engine under cover
   - Front wheel
4. Check the wheel alignment and adjust as necessary.

SHOCK ABSORBERS

REMOVAL & INSTALLATION

Without Active Height Control

1. Before servicing the vehicle, refer to the precautions section.
2. Support the axle with a jackstand.
3. Remove or disconnect the following:
   - Front wheel
   - Shock absorber

To install:

4. Install or connect the following:
   - Shock absorber. Tighten the upper nut to 51 ft. lbs. (68 Nm) and lower nut to 100 ft. lbs. (135 Nm).
   - Front wheel

With Active Height Control

See Figures 78 and 79.

⁂ CAUTION

The vehicle ride height may change suddenly when relieving system pressure.

1. Before servicing the vehicle, refer to the precautions section.
2. Relieve the Active Height Control (AHC) hydraulic pressure as follows:
   a. Connect a hose to the control actuator bleed screw and place the other end in a container.
   b. Open the bleed screw.
   c. When the fluid pressure has dropped and oil stops flowing, close the bleed screw.
3. Remove or disconnect the following:
   - Front wheel
   - Inner fender liner
   - Lower shock absorber mounting bolt
   - AHC pressure hose
   - Upper shock absorber mounting nut
   - Shock absorber

To install:

4. Install or connect the following:
   - Shock absorber. Tighten the upper nut to 51 ft. lbs. (68 Nm) and the lower bolt to 100 ft. lbs. (135 Nm).
   - AHC pressure hose with new O-ring seals. Tighten the bolts to 13 ft. lbs. (18 Nm).
   - Inner fender liner
   - Front wheel

➡Do not let the AHC reservoir run empty during this procedure.

5. Bleed the AHC system as follows:

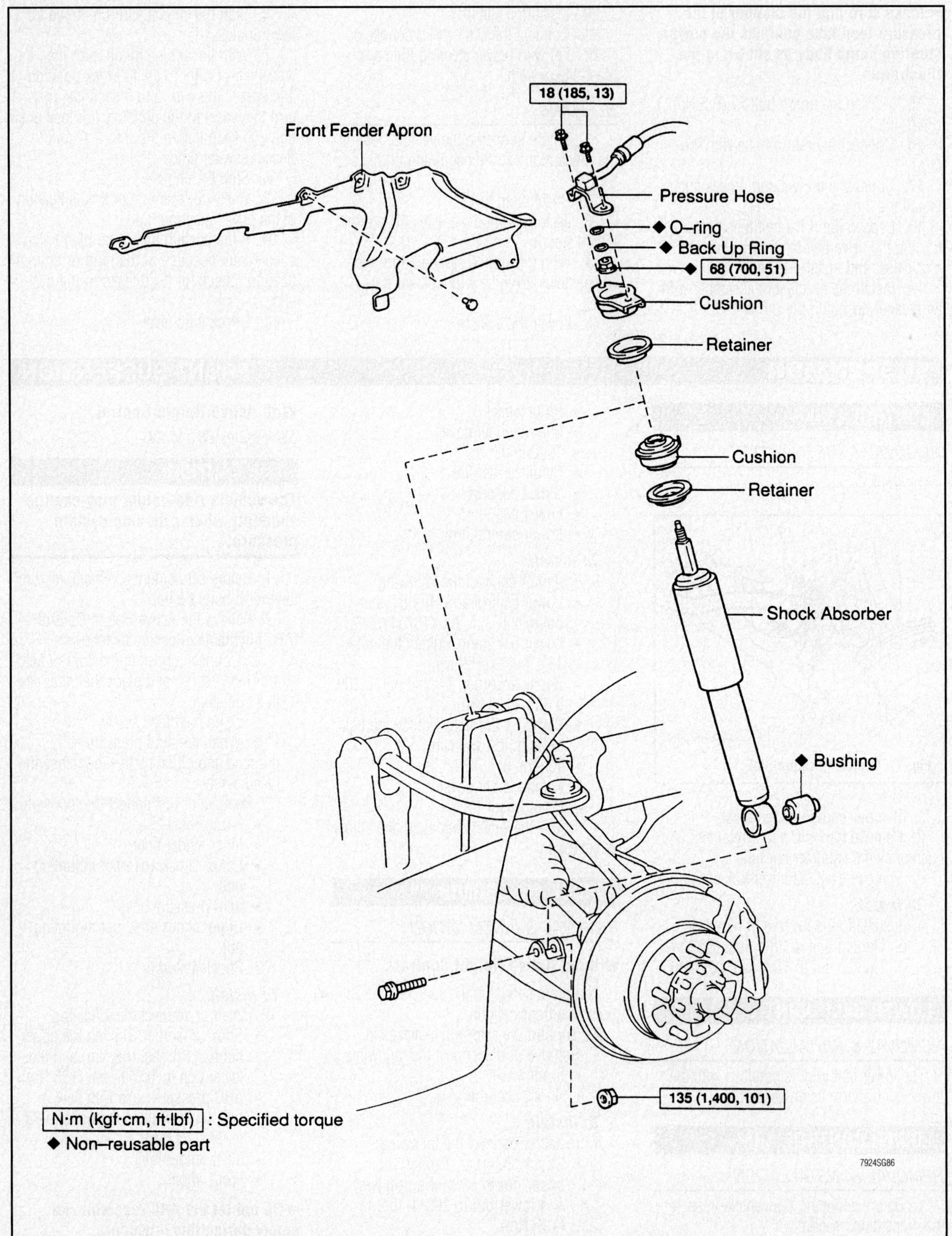

Fig. 78 Exploded view of the front shock absorber mounting—LX470 models with Active Height Control (AHC)

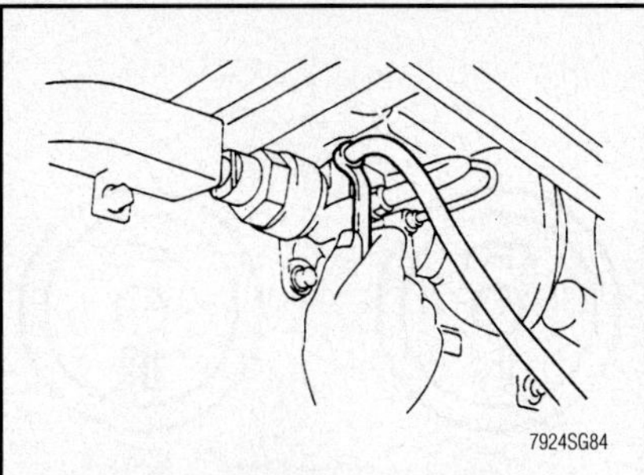

Fig. 79 Relieving system pressure—LX470 with Active Height Control (AHC)

a. Fill the AHC system reservoir with AHC fluid 08886-01805.

b. Start the engine and push **N** on the vehicle height select switch.

c. When the AHC pump stops, turn the engine **OFF**.

d. Open the bleed screw and allow any air in the system to escape.

e. Repeat until no air is expelled from the bleed screw.

f. Fill the AHC reservoir to the correct level.

STEERING KNUCKLE

REMOVAL & INSTALLATION

See Figure 80.

1. Remove front wheel
2. Remove axle hub and disc. See Wheel Bearing in this section.
3. Remove the bolt and disconnect the flexible hose from the steering knuckle
4. Remove the 4 bolts, oil seal, gasket and dust cover

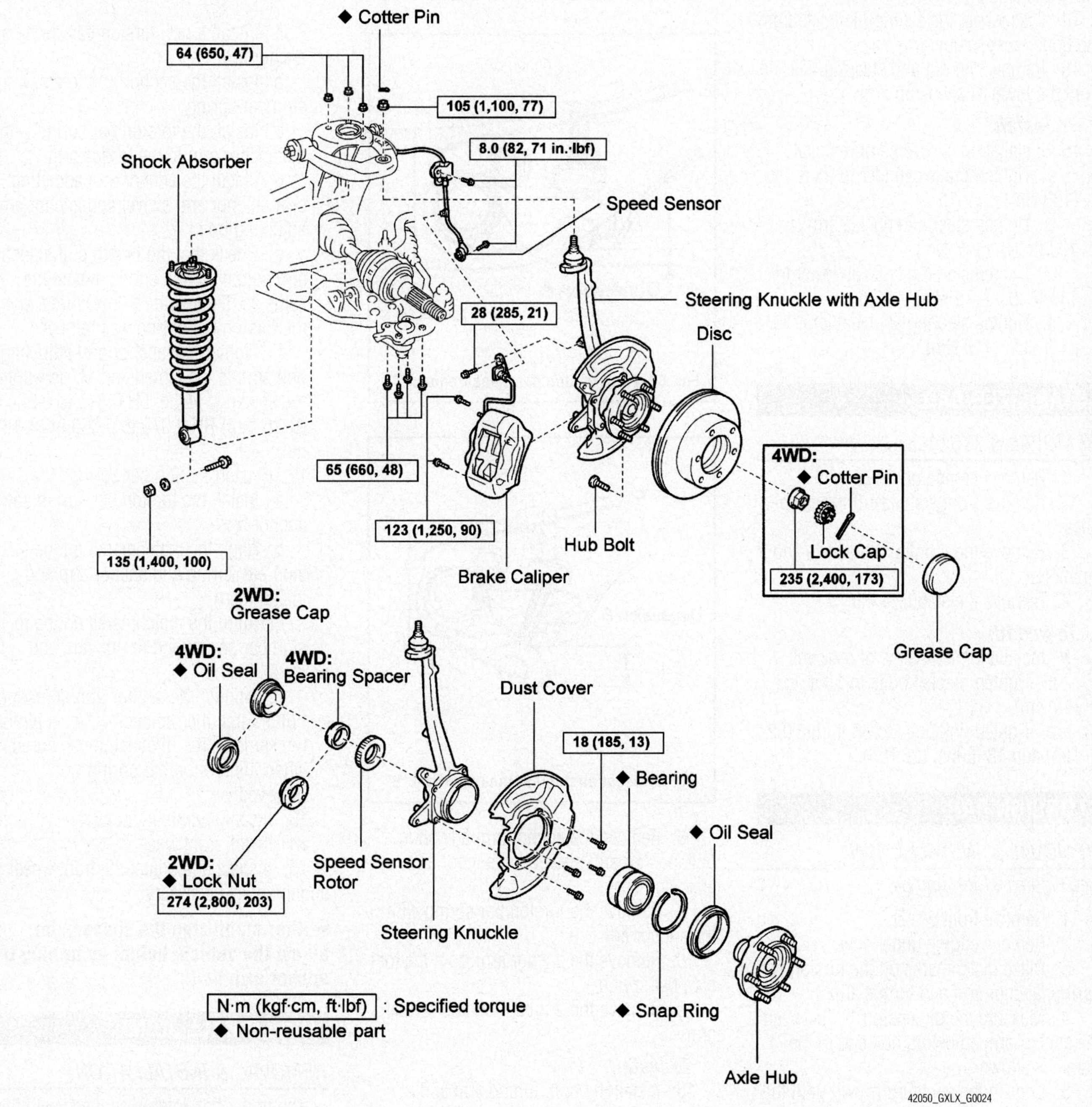

◆ Cotter Pin
64 (650, 47)
105 (1,100, 77)
Shock Absorber
8.0 (82, 71 in.·lbf)
Speed Sensor
Steering Knuckle with Axle Hub
Disc
28 (285, 21)
65 (660, 48)
123 (1,250, 90)
Hub Bolt
Brake Caliper
135 (1,400, 100)
4WD:
◆ Cotter Pin
Lock Cap
235 (2,400, 173)
Grease Cap
2WD:
Grease Cap
4WD:
◆ Oil Seal
4WD:
Bearing Spacer
Dust Cover
18 (185, 13)
◆ Bearing
◆ Oil Seal
2WD:
◆ Lock Nut
274 (2,800, 203)
Speed Sensor Rotor
Steering Knuckle
N·m (kgf·cm, ft·lbf) : Specified torque
◆ Non-reusable part
◆ Snap Ring
Axle Hub
42050_GXLX_G0024

Fig. 80 Front suspension exploded view

5. Remove the 2 bolts, washers and brake cylinder
6. Support the brake cylinder securely
7. Remove ABS speed sensor and harness
8. Remove the cotter pin and nut
9. Disconnect the tie rod end from the steering knuckle arm
10. Remove the cotter pin and nut
11. Disconnect the lower suspension arm from the steering knuckle
12. Remove the cotter pin and nut
13. Temporarily install the nut to the lower suspension arm
14. Disconnect the steering knuckle from the upper suspension arm
15. Remove the nut and steering knuckle from the lower suspension arm

To install:

16. Installation is reverse of removal.
   a. Tighten brake cylinder to 91 ft. lbs. (123 Nm)
   b. Tighten steering knuckle arm to 108 ft. lbs. (147 Nm)
   c. Tighten steering knuckle lower to 117 ft. lbs. (159 Nm)
   d. Tighten steering knuckle upper to 81 ft. lbs. (110 Nm)

STABILIZER BAR

REMOVAL & INSTALLATION

1. Remove engine under cover
2. Remove front stabilizer link assemblies
3. Remove the 4 bolts, 2 brackets and 2 cushions
4. Remove the stabilizer bar

To install:

5. Installation is reverse of removal
   a. Tighten bracket bolts to 13 ft. lbs. (18 Nm)
   b. Tighten link bolts to 38 ft. lbs. (52 Nm) and 18 ft. lbs. (25 Nm)

TORSION BAR

REMOVAL & INSTALLATION

See Figures 81 through 84.

1. Remove front wheel
2. Remove engine under cover
3. Place matchmarks on the torsion bar spring, anchor arm and torque arm
4. Measure the dimension "A" between the anchor arm adjusting bolt end of the frame as shown
5. Loosen the anchor arm bolt until the spring tension is free and measurement the anchor arm bolt dimension "B"

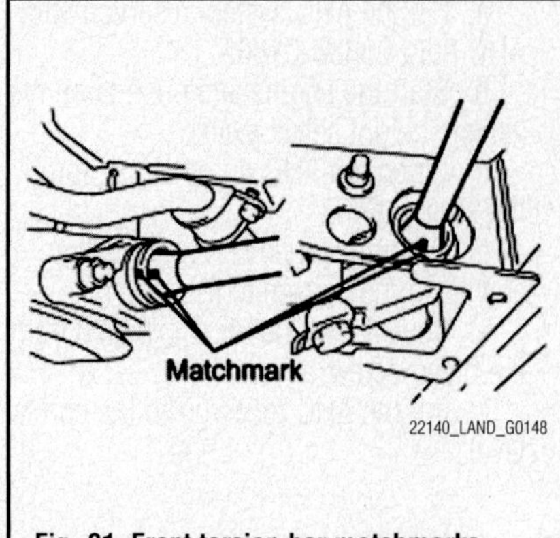

Fig. 81 Front torsion bar matchmarks

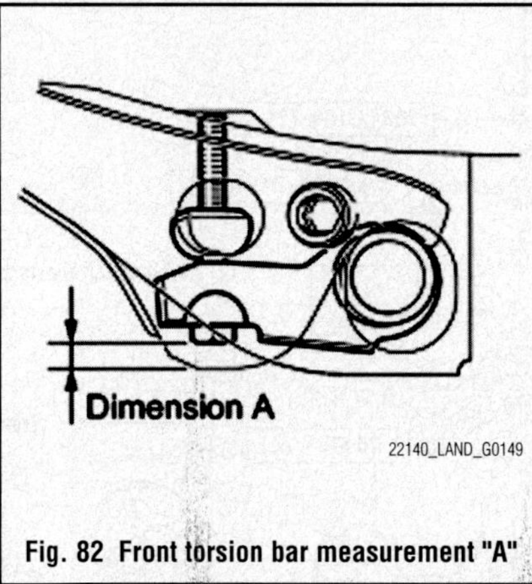

Fig. 82 Front torsion bar measurement "A"

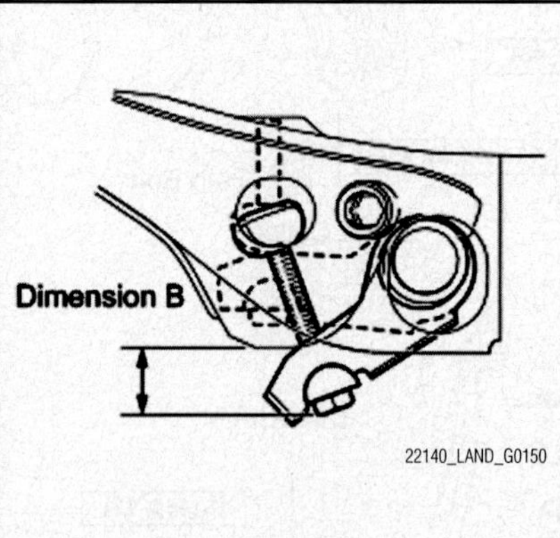

Fig. 83 Front torsion bar measurement "B"

6. Remove the anchor arm adjusting bolt, anchor arm swivel and anchor arm adjusting seat
7. Remove the torsion bar spring with the anchor arm
8. Remove the anchor arm from the torsion bar spring
9. Remove the 2 nuts, bolt and torque arm

To install:

10. Install the bolt, torque arm and 2 nuts and tighten to 16 ft. lbs.(225 Nm)
11. For a new torsion bar spring:

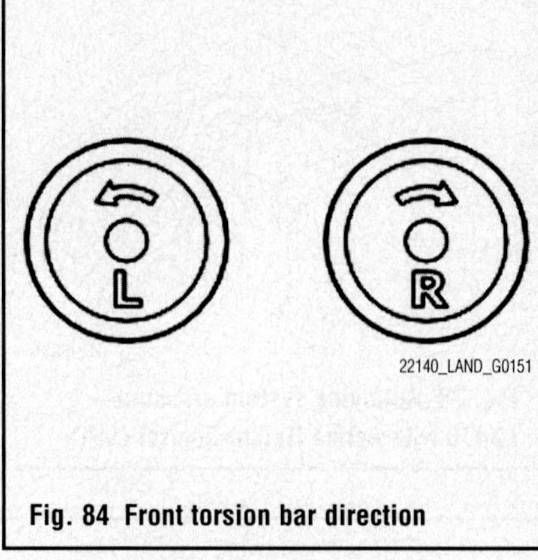

Fig. 84 Front torsion bar direction

   a. Install a new torsion bar spring and anchor arm
   b. Install the anchor arm to a new torsion bar spring
   c. Install the torsion bar spring with the anchor arm to the torque arm
   d. Install the anchor arm adjusting seat, anchor arm swivel and anchor arm adjusting bolt
   e. Check that the length of the anchor arm adjusting bolt end is almost the same as dimension "B" measured when the torsion bar spring was removed
   f. Tighten the anchor arm adjusting bolt so that the dimension "A" is within the specified value: LH 0.315 to 0.984 in (8-25 mm) RH 0.079 to 0.709 in (2-18 mm).
12. Reused torsion bar spring:
   a. Install the torsion bar spring and anchor arm
   b. Align the matchmarks on the torsion bar spring and anchor arm and install them
   c. Align the matchmarks on the torsion bar spring and torque arm and install them
   d. Tighten the anchor arm adjusting bolt so that the dimension "A" is almost the same as the dimension measured when the torsion bar spring was removed
13. Install engine under cover
14. Install front wheel
15. Inspect and adjust the front wheel alignment as necessary

➡After stabilizing the suspension, adjust the vehicle height by turning the anchor arm bolt.

UPPER BALL JOINT

REMOVAL & INSTALLATION

The upper ball joint is serviced with the upper control arm as an assembly.

UPPER CONTROL ARM

REMOVAL & INSTALLATION

See Figure 85.

1. Before servicing the vehicle, refer to the precautions section.
2. Remove or disconnect the following:
   - Front wheel
   - Inner fender liner
   - Place matchmarks on the height control sensor link and upper suspension arm
   - Remove the nut and disconnect the height control sensor link.
   - Wheel speed sensor harness
   - Upper ball joint
   - Adjustment cam bolts
   - Upper control arm

To install:

3. Install or connect the following:
   - Upper control arm. Tighten the adjustment cam bolts to 72 ft. lbs. (98 Nm).
   - Upper ball joint. Tighten the nut to 81 ft. lbs. (110 Nm).
   - Wheel speed sensor harness. Tighten the bolts to 10 ft. lbs. (13 Nm).
   - Inner fender liner
   - Front wheel
   - Aligning matchmarks, install the height control sensor link and connect the nut.
4. Check the wheel alignment and adjust as necessary.

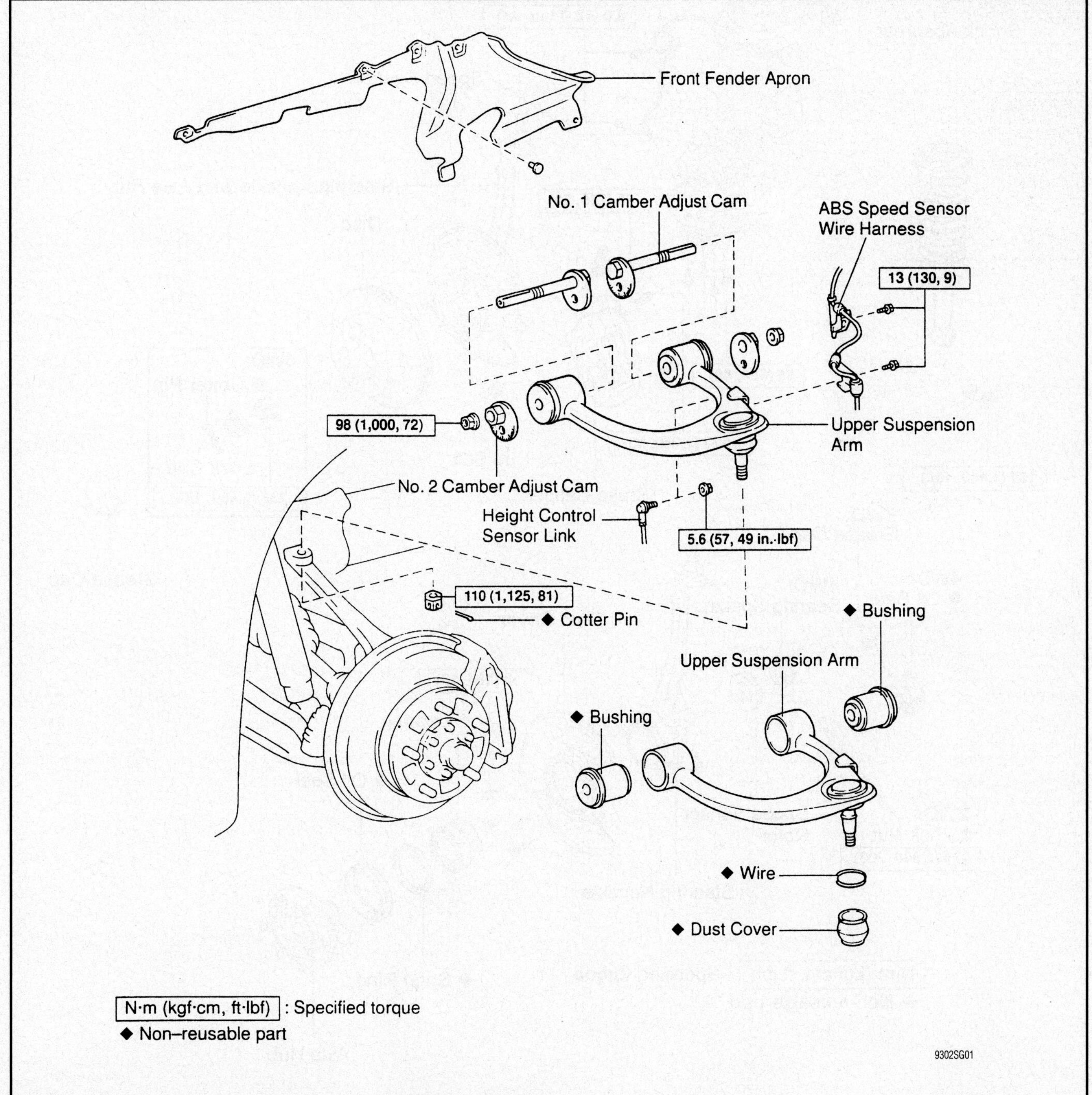

Fig. 85 Exploded view of the upper control arm and related components

WHEEL BEARINGS

REMOVAL & INSTALLATION

See Figures 86.

1. Before servicing the vehicle, refer to the precautions section.
2. Remove front wheel
3. Remove the bolt and disconnect the flexible hose from the steering knuckle
4. Remove the 2 bolts, washers and brake cylinder
5. Support the brake cylinder securely
6. Using a screwdriver and hammer, remove the grease cap from the flange

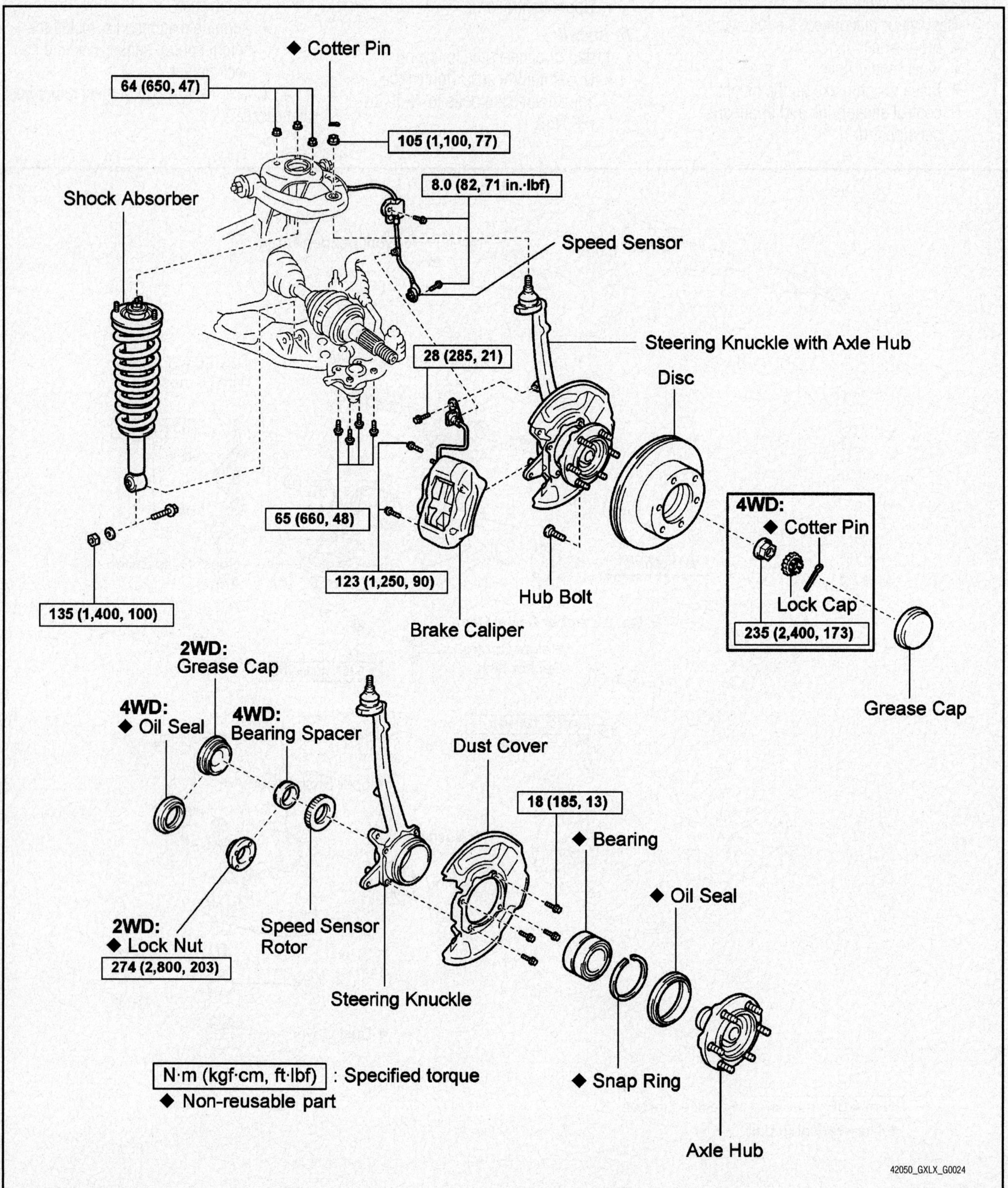

Fig. 86 Front suspension exploded view

7. Using a snap ring expander, remove the snap ring
8. Remove the 6 nuts and 6 washers
9. Install the 6 nuts temporarily to protect the threads of the stud bolts
10. Using a brass bar and hammer, tap on the bolt heads and remove the 6 nuts and cone washers
11. Remove the lock nut
12. Remove the flange and gasket
13. Using a screwdriver, release the lock washer
14. Remove the lock nut
15. Remove the lock washer
16. Remove the adjusting nut.
17. Remove the axle hub with disc
18. Remove the claw washer and bearing from the axle hub

To install:

19. Place the axle hub with disc on the steering knuckle
20. Install the outer bearing
21. Install the claw washer
22. Adjust preload by tightening adjusting nut to 43 ft. lbs. (59 Nm)
    a. Turn the axle hub several times to settle down the bearing
    b. Loosen the adjusting nut until it can be rotated by hand
    c. Retighten the adjusting nut to 38-57 inch lbs. (4.3 to 6.5 Nm)
    d. Using a spring tension gauge, measure the preload. Standard preload is 9.5-15 ft. lbs. (42-67 Nm)
23. Install a new lock washer and the lock nut tighten to 47 ft. lbs.(64 Nm)
    a. Check that the axle hub rotates smoothly and there is no looseness on the bearing
    b. Using a spring tension gauge, measure the preload. Standard preload is 9.5-15 ft. lbs. (42-67 Nm)

➡Make sure to check the preload in the direction of rotation. If the preload is not within the specified value, adjust it again with the adjusting nut

24. Secure the lock nut by bending one of the lock washer teeth inward and the other lock washer teeth outward
25. Place a new gasket in position on the axle hub
26. Install the flange to the axle hub.
27. Install the 6 cone washers, 6 washers and 6 new nuts. Tighten nuts to 24 ft. lbs. (33 Nm)
28. Pull out the drive shaft to the outside of the vehicle and select the snap ring which ensures the clearance between the tip of the flange and the snap ring is less than 0.008 in. (0.2 mm)
29. Using a snap ring expander, install a new snap ring to the drive shaft
30. Install a new grease cap to the flange
31. Install the brake caliper with the 2 washers and 2 bolts and tighten brake cylinder to 91 ft. lbs. (123 Nm)
32. Install the flexible hose and bolt to the steering knuckle
33. Install front wheel and tighten to 97 ft. lbs. (131 Nm)
34. Check ABS speed sensor signal.

ADJUSTMENT

1. Adjust preload by tightening adjusting nut to 43 ft. lbs. (59 Nm)
    a. Turn the axle hub several times to settle down the bearing
    b. Loosen the adjusting nut until it can be rotated by hand
    c. Retighten the adjusting nut to 38-57 inch lbs. (4.3 to 6.5 Nm)
    d. Using a spring tension gauge, measure the preload. Standard preload is 9.5-15 ft. lbs. (42-67 Nm)
2. Install a new lock washer and the lock nut tighten to 47 ft. lbs.(64 Nm)
    a. Check that the axle hub rotates smoothly and there is no looseness on the bearing
    b. Using a spring tension gauge, measure the preload. Standard preload is 9.5-15 ft. lbs. (42-67 Nm)

➡Make sure to check the preload in the direction of rotation. If the preload is not within the specified value, adjust it again with the adjusting nut

3. Secure the lock nut by bending one of the lock washer teeth inward and the other lock washer teeth outward

SUSPENSION

REAR SUSPENSION

COIL SPRING

REMOVAL & INSTALLATION

1. Before servicing the vehicle, refer to the precautions section.
2. Support the vehicle at the frame.
3. Support the axle with a floor jack.
4. Remove or disconnect the following:
    - Rear wheel
    - Shock absorber
    - Stabilizer bar brackets
    - Lateral control rod
    - Coil spring

To install:

5. Install or connect the following:
    - Coil spring
    - Lateral control rod. Tighten the axle housing bolt to 110 ft. lbs. (149 Nm).
    - Stabilizer bar brackets. Tighten the bolts to 13 ft. lbs. (18 Nm)
    - Shock absorber
    - Rear wheel

CONTROL ARMS/LINKS

REMOVAL & INSTALLATION

Without Active Height Control

1. Remove rear wheel
2. Remove the nut and disconnect the stabilizer bar from the stabilizer link

➡If the ball joint turns together with the nut, use a hexagon wrench (6 mm) to hold the stud

3. Hold the stabilizer bar link with a wrench and remove the nut, retainer, cushion and link.
4. Remove the 2 retainers and cushion from the stabilizer link.

To install:

5. Install the stabilizer link with the nut and tighten nut to 52 ft. lbs. (70 Nm).
6. Install the 2 retainers and the cushion to the stabilizer link.
7. Hold the stabilizer bar link with a wrench, and install the nut, retainer, cushion and link and tighten to 11 ft. lbs. (15 Nm).
8. Install rear wheel and tighten to 83 ft. lbs. (112 Nm)

With Active Height Control

See Figures 88 and 89.

1. Remove the bolt, nut and rear stabilizer link assembly.

To install:

2. Install the rear stabilizer link assembly with the bolt and nut, tighten to 96 ft. lbs. (130 Nm)

SHOCK ABSORBER

REMOVAL & INSTALLATION

Without Active Height Control

1. Before servicing the vehicle, refer to the precautions section.

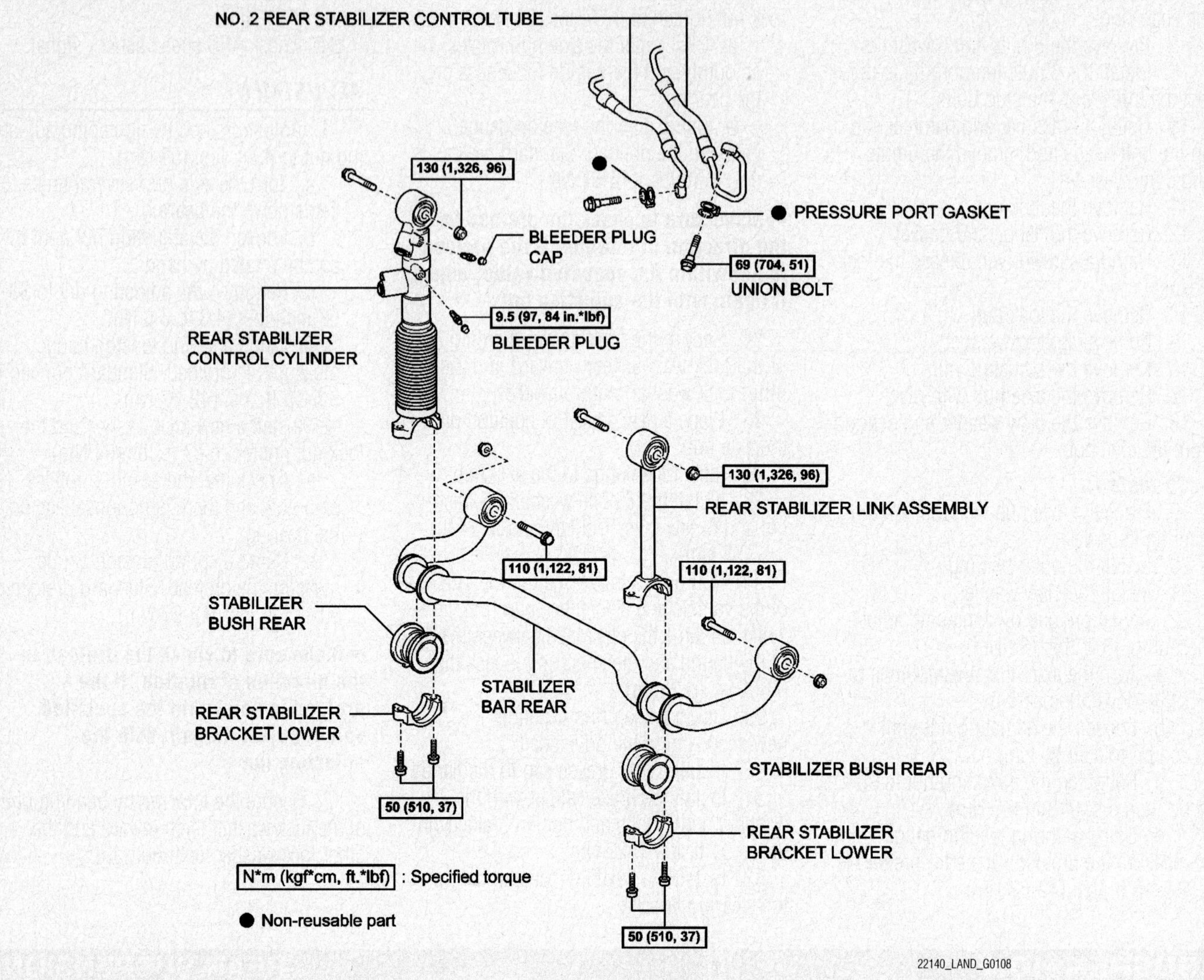

Fig. 88 Rear control link components

2. Support the axle with a jackstand.
3. Remove or disconnect the following:
   - Rear wheel
   - Shock absorber

To install:

4. Install or connect the following:

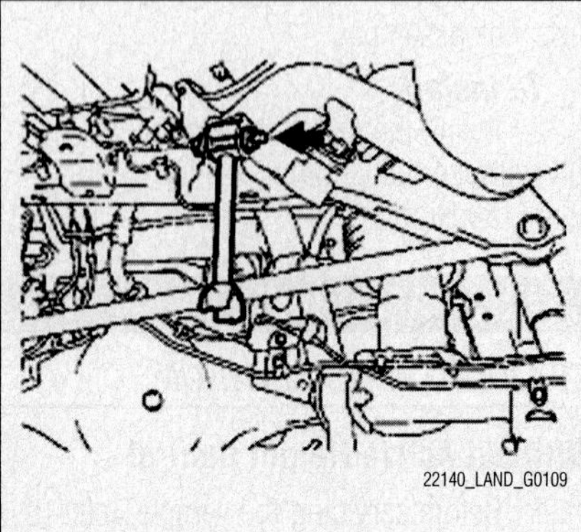

Fig. 89 Rear control link removal

   - Shock absorber. Tighten the nut to 51 ft. lbs. (69 Nm) and the bolt to 72 ft. lbs. (98 Nm).
   - Rear wheel

With Active Height Control

See Figure 90.

CAUTION

The vehicle ride height may change suddenly when relieving system pressure.

1. Before servicing the vehicle, refer to the precautions section.
2. Support the rear axle with a jack or stands.
3. Relieve the Active Height Control (AHC) hydraulic pressure as follows:
   a. Connect a hose to the control actuator bleed screw and place the other end in a container.
   b. Open the bleed screw.
   c. When the fluid pressure has dropped and oil stops flowing, close the bleed screw.
4. Remove or disconnect the following:
   - Rear wheel
   - Lower shock absorber mounting bolt
   - AHC pressure hose
   - Upper shock absorber mounting nut
   - Shock absorber

To install:

5. Install or connect the following:
   - Shock absorber. Tighten the upper nut to 51 ft. lbs. (68 Nm) and the lower bolt to 72 ft. lbs. (98 Nm).
   - AHC pressure hose with new O-ring seals. Tighten the bolts to 13 ft. lbs. (18 Nm).
   - Rear wheel

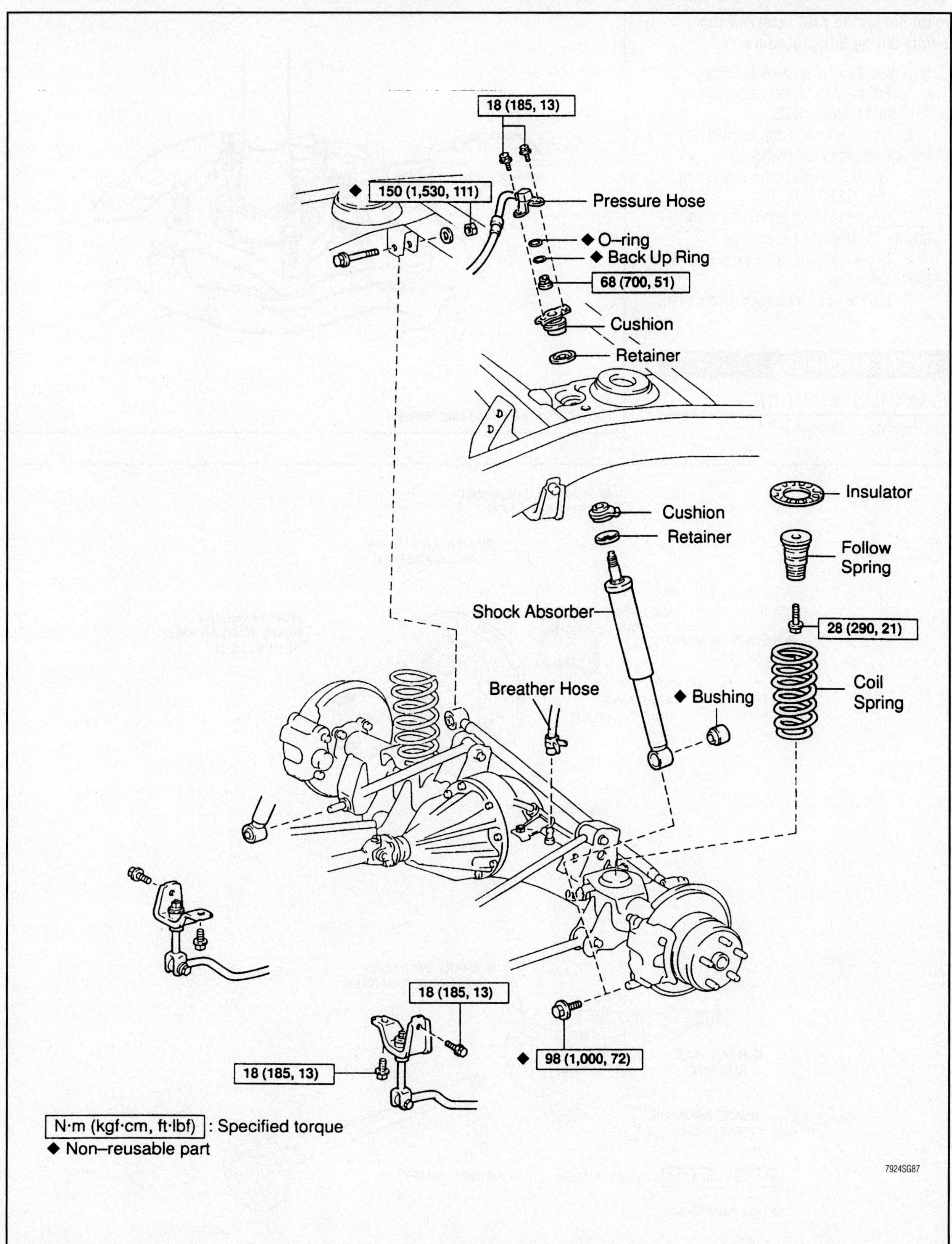

Fig. 90 Exploded view of the rear shock absorber mounting—LX470 with Active Height Control (AHC)

➡Do not let the AHC reservoir run empty during this procedure.

6. Bleed the AHC system as follows:
   a. Fill the AHC system reservoir with AHC fluid 08886-01805.
   b. Start the engine and push **N** on the vehicle height select switch.
   c. When the AHC pump stops, turn the engine **OFF**.
   d. Open the bleed screw and allow any air in the system to escape.
   e. Repeat until no air is expelled from the bleed screw.
   f. Fill the AHC reservoir to the correct level.

WHEEL BEARINGS

REMOVAL & INSTALLATION

See Figures 91 through 97.

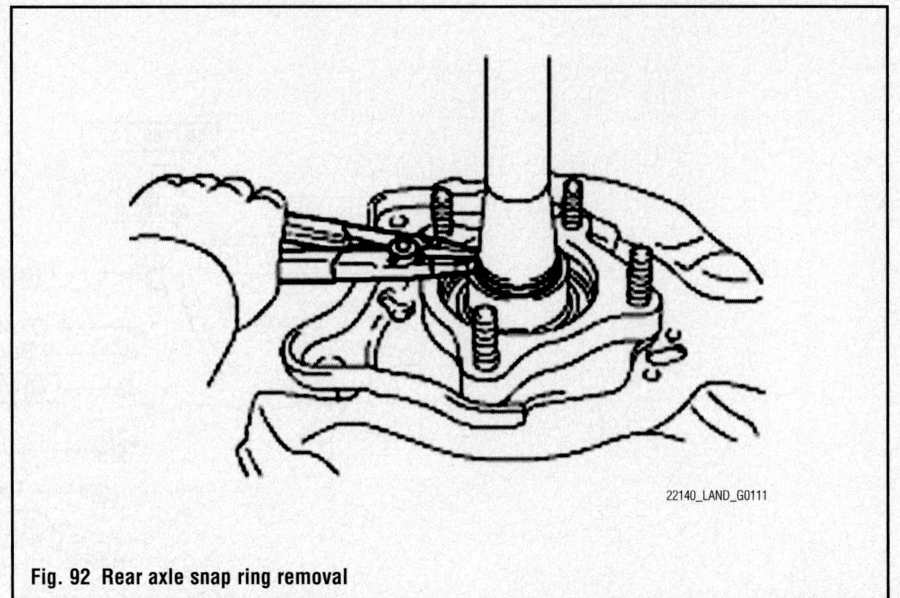

Fig. 92 Rear axle snap ring removal

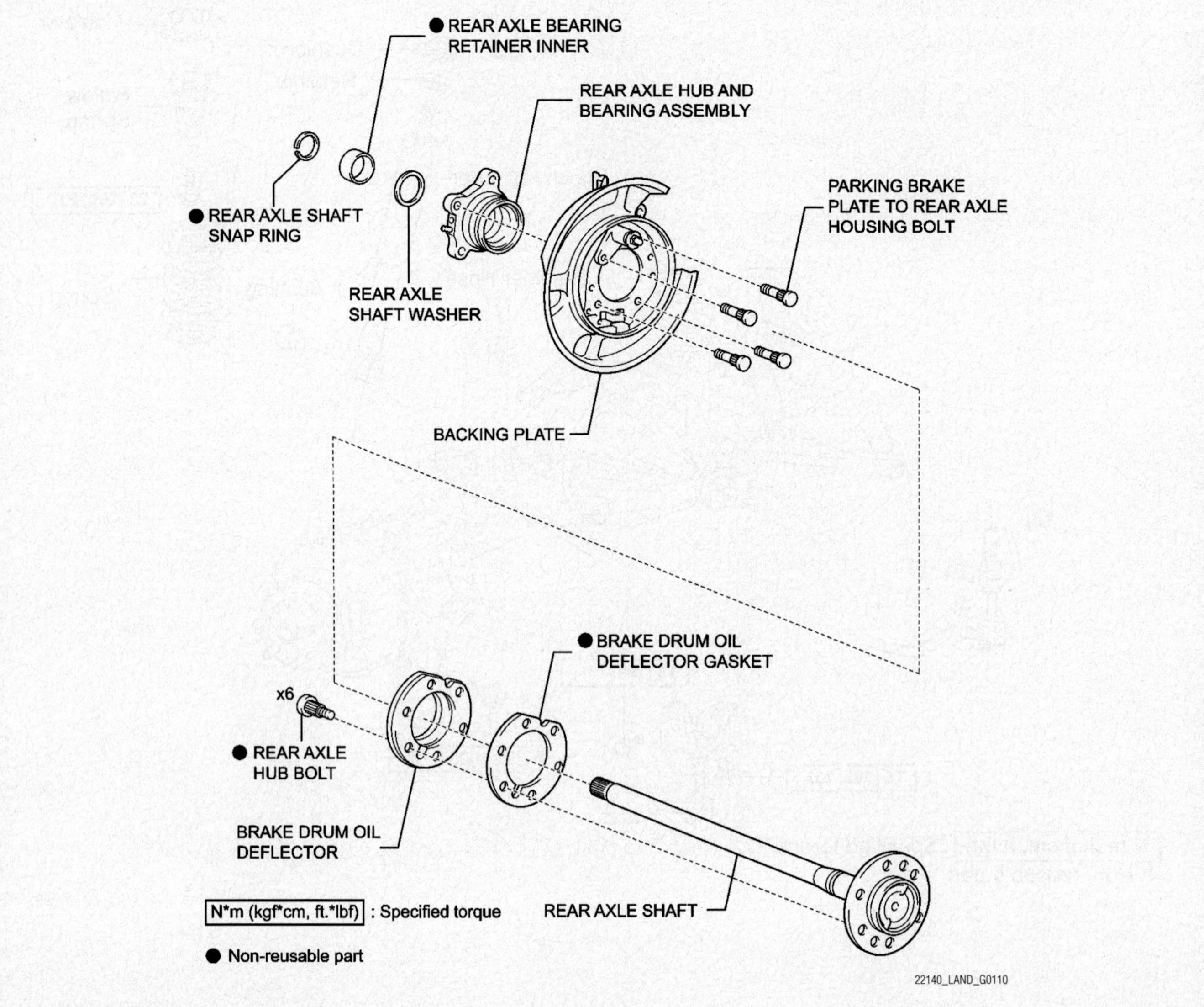

● REAR AXLE BEARING RETAINER INNER
REAR AXLE HUB AND BEARING ASSEMBLY
PARKING BRAKE PLATE TO REAR AXLE HOUSING BOLT
● REAR AXLE SHAFT SNAP RING
REAR AXLE SHAFT WASHER
BACKING PLATE
● BRAKE DRUM OIL DEFLECTOR GASKET
x6
● REAR AXLE HUB BOLT
BRAKE DRUM OIL DEFLECTOR
N*m (kgf*cm, ft.*lbf) : Specified torque
REAR AXLE SHAFT
● Non-reusable part
22140_LAND_G0110

Fig. 91 Rear axle assembly components

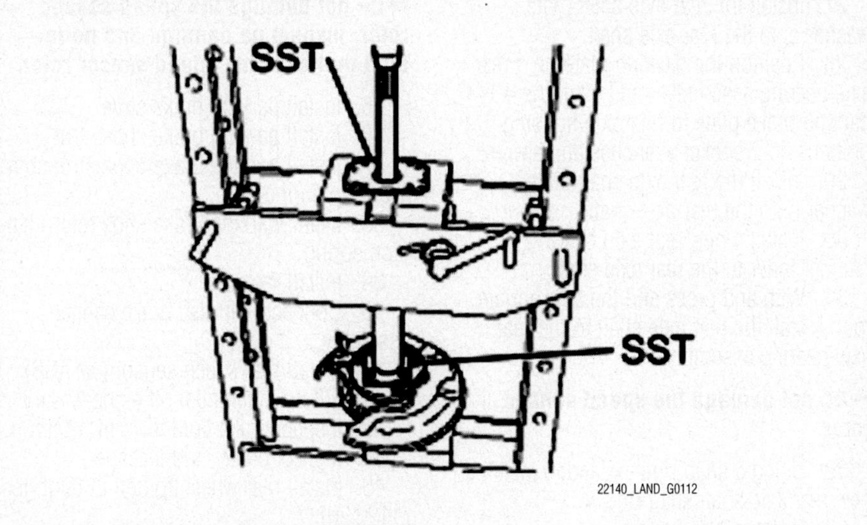

Fig. 93 Removing rear axle shaft from bearing

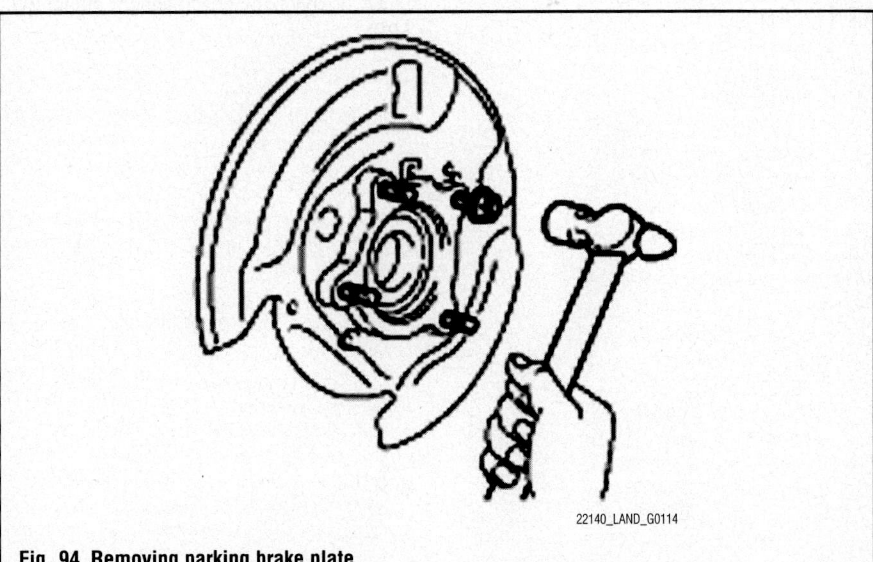

Fig. 94 Removing parking brake plate

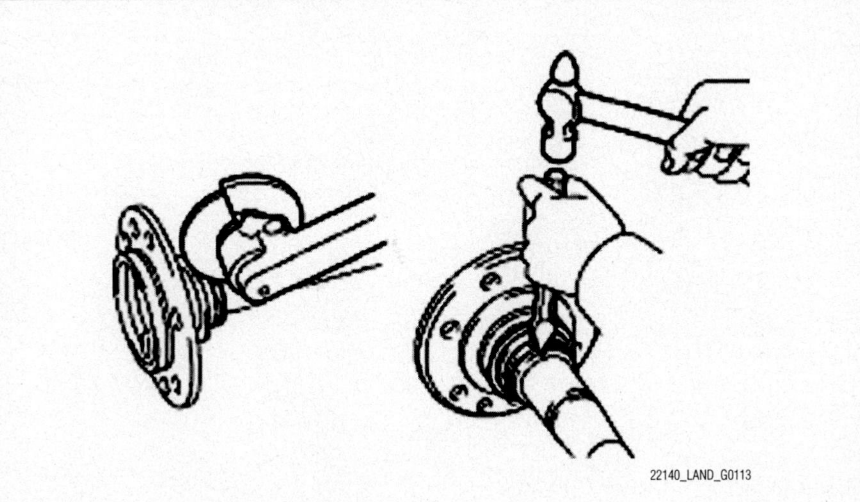

Fig. 95 Removing rear axle bearing inner race

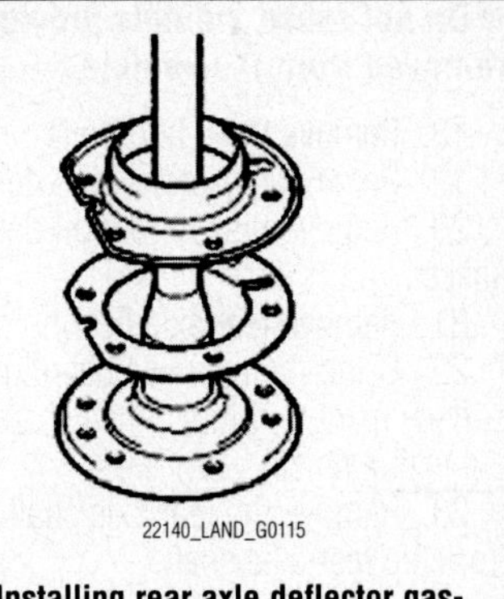

Fig. 96 Installing rear axle deflector gaskets

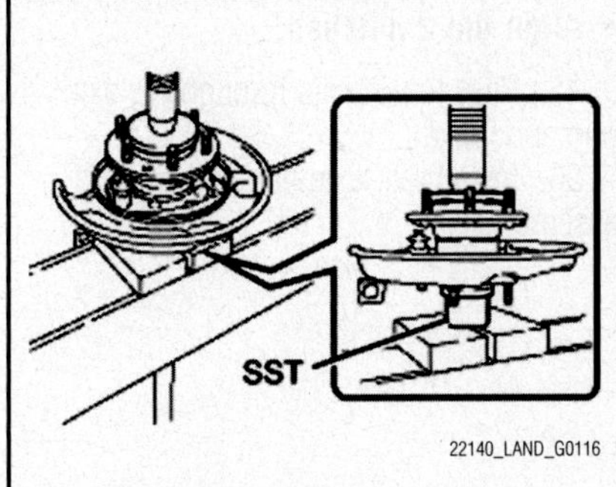

Fig. 97 Assembling rear axle shaft-to-axle bearing assembly

1. Remove rear wheel.
2. Remove rear speed sensor (w/ ABS).
3. Separate rear disc brake caliper assembly.
4. Remove rear disc.
5. Remove parking brake shoe return tension spring.
6. Remove parking brake shoe strut compression spring.
7. Remove parking brake shoe strut.
8. Remove parking brake shoe.
9. Remove rear axle shaft with backing plate.
10. Remove the 4 nuts and rear axle shaft with backing plate.
11. Remove the O-ring.
12. Using a snap ring expander, remove the snap ring.
13. Remove the rear axle shaft from bearing.
14. Remove the rear axle bearing retainer inner from the rear axle bearing assembly.
15. Remove the rear axle shaft washer from the rear axle bearing assembly.
16. Attach the 4 nuts to the parking brake plate to rear axle housing bolts.
17. Using a hammer, remove the 4 parking brake plate to rear axle housing bolts and rear axle bearing assembly.

➡**Do not reuse the nuts previously removed from the vehicle.**

18. Remove the 6 hub bolts.
19. Remove brake drum oil deflector.
20. Remove brake drum oil deflector gasket.
21. Remove rear axle bearing oil seal.
22. Grind the rear axle bearing inner race surface using a grinder, then chisel them out with a chisel.
23. Remove the rear axle shaft oil seal from the rear axle shaft.

To install:

24. Install a new deflector gasket and deflector to the rear axle shaft.

➡**Align the 2 notches.**

25. Pass the 6 bolts through the axle hub and install.
26. Install rear axle hub and bearing assembly.
27. Install the rear axle shaft plate washer onto the rear axle shaft.
28. Position the backing plate on a rear axle bearing assembly, and install the 4 parking brake plate to rear axle housing bolts using 2 socket wrenches and a press.
29. Install the rear axle shaft plate washer onto the rear axle shaft.
30. Install a new rear axle bearing retainer inner to the rear axle shaft.
31. With and press and the appropriate tool, install the rear axle shaft to the rear axle bearing assembly.

➡**Do not damage the speed sensor rotor.**

32. Using a snap ring expander, install a new rear axle shaft snap ring.
33. Install a new O-ring.
34. Install the rear axle shaft with backing plate with the 4 nuts, tighten to 89 ft. lbs. (120 Nm).

➡**Do not damage the speed sensor rotor. Inspect no damage and no foreign matter at the speed sensor rotor.**

35. Install parking brake shoe
36. Install parking brake shoe strut
37. Install parking brake shoe strut compression spring
38. Install parking brake shoe return tension spring
39. Install rear disc
40. Connect rear disc brake caliper assembly
41. Install rear speed sensor (w/ ABS)
42. Fill up differential oil as necessary
43. Inspect brake fluid level in reservoir
44. Inspect brake fluid leakage
45. Install rear wheel tighten to 83 ft. lbs. (112 Nm)
46. Inspect and adjust parking brake lever travel
47. Inspect abs speed sensor signal (w/ ABS)

TOYOTA

Matrix

18

SPECIFICATIONS AND MAINTENANCE CHARTS

ENGINE AND VEHICLE IDENTIFICATION

Engine						
Code ①	Liters (cc)	Cu. In.	Cyl.	Fuel Sys.	Engine Type	Eng. Mfg.
1ZZ-FE	1.8 (1794)	109.5	4	EFI	DOHC	Toyota
2ZZ-GE	1.8 (1796)	109.5	4	EFI	DOHC	Toyota

Model Year	
Code ②	Year
6	2006
7	2007
8	2008

EFI: Electronic Fuel Injection

DOHC: Double Overhead Camshaft

① 8th digit of VIN

② 10th digit of VIN

22140_MATR_C0001

GENERAL ENGINE SPECIFICATIONS

Year	Model	Engine Displacement Liters (VIN)	Net Horsepower @ rpm	Net Torque @ rpm (ft. lbs.)	Bore x Stroke (in.)	Com-pression Ratio	Oil Pressure @ idle
2006	Matrix	1.8 (1ZZ-FE)	①	②	3.11x3.60	10.0:1	27
	Matrix	1.8 (2ZZ-GE)	164@7600	125@4400	3.23x3.35	11.5:1	27
2007	Matrix	1.8 (1ZZ-FE)	126@6000	122@4200	3.11x3.60	10.0:1	27
2008	Matrix	1.8 (1ZZ-FE)	126@6000	122@4200	3.11x3.60	10.0:1	27

EFI: Electronic Fuel Injection

① 2WD models: 126@6000

4WD models: 118@6000

② 2WD models: 122@4200

4WD models: 115@4200

22140_MATR_C0002

ENGINE TUNE-UP SPECIFICATIONS

Year	Engine Displacement Liters (VIN)	Spark Plug Gap (in.)	Ignition Timing (deg.)	Fuel Pump (psi)	Idle Speed (rpm)		Valve Clearance	
					MT	AT	In.	Ex.
2006	1.8 (1ZZ-FE)	0.039-0.051	①	44-50	650-750	650-750	0.0059-0.0098	0.0098-0.0138
	1.8 (2ZZ-GE)	0.039-0.051	②	44-50	750-850	700-800	0.0031-0.0071	0.0087-0.0126
2006	1.8 (1ZZ-FE)	0.039-0.051	①	44-50	650-750	650-750	0.0059-0.0098	0.0098-0.0138
2007	1.8 (1ZZ-FE)	0.039-0.051	①	44-50	650-750	650-750	0.0059-0.0098	0.0098-0.0138
2008	1.8 (1ZZ-FE)	0.039-0.051	①	44-50	650-750	650-750	0.0059-0.0098	0.0098-0.0138

Note: The Vehicle Emission Control Information label often reflects specification changes made during production. The label figures must be used if they differ from those in this chart.

① With terminal TC and CG of DLC3 connected: 8-12 degrees BTDC

With terminal TC and CG of DLC3 disconnected: 10-18 degrees BTDC

② With terminal TC and CG of DLC3 connected: 8-12 degrees BTDC

With terminal TC and CG of DLC3 disconnected:

A/T: 10-18 degrees BTDC

M/T: 4-12 degrees BTDC

22140_MATR_C0003

CAPACITIES

Year	Model	Engine Displacement Liters (VIN)	Engine Oil with Filter	Transmission (pts.) 5-Spd	6-Spd	Auto.	Drive Axle Front (pts.)	Rear (pts.)	Fuel Tank (gal.)	Cooling System (qts.)
2006	Matrix	1.8 (1ZZ-FE)	4.4	4.0	—	5.2	NA	1.04	①	6.9
	Matrix	1.8 (2ZZ-GE)	4.7	4.8	4.8	6.1	NA	—	13.2	7.1
2007	Matrix	1.8 (1ZZ-FE)	4.4	4.0	—	5.2	NA	—	13.2	6.9
2008	Matrix	1.8 (1ZZ-FE)	4.4	4.0	—	5.2	NA	—	13.2	6.9

Note: All capacities are approximate. Add fluid gradually and check to be sure a proper fluid level is obtained.

NA - Not available

① 2WD: 13.2 gallons
4WD: 11.9 gallons

22140_MATR_C0004

FLUID SPECIFICATIONS

Year	Model	Engine Displacement Liters	Engine ID/VIN	Engine Oil	Auto. Trans.	Manual Trans.	Drive Axle	Transfer Case	Power Steering Fluid	Brake Master Cylinder
2006	Matrix	1.8	1ZZ-FE	5W-30	ATF Type T-IV	75W-90	①	①	Dexron II or III	DOT 3
	Matrix	1.8	2ZZ-GE	5W-30	ATF Type T-IV	75W-90	①	①	Dexron II or III	DOT 3
2007	Matrix	1.8	1ZZ-FE	5W-30	ATF Type T-IV	75W-90	①	①	Dexron II or III	DOT 3
2008	Matrix	1.8	1ZZ-FE	5W-30	ATF Type T-IV	75W-90	①	①	Dexron II or III	DOT 3

DOT: Department Of Transpotation

① Above O degrees F - API GL-5 SAE 90
Below O degrees F - API GL-5 SAE 85 or 85W-90

22140_MATR_C0014

VALVE SPECIFICATIONS

Year	Engine Displacement Liters (VIN)	Seat Angle (deg.)	Face Angle (deg.)	Spring Test Pressure (lbs. @ in.)	Spring Installed Height (in.)	Stem-to-Guide Clearance (in.) Intake	Exhaust	Stem Diameter (in.) Intake	Exhaust
2006	1.8 (1ZZ-FE)	45	44.5	31.3-34.8@ 1.252	1.323	0.0010-0.0024	0.0012-0.0026	0.2154-0.2159	0.2152-0.2158
	1.8 (2ZZ-GE)	45	44.5	①	1.516	0.0010-0.0023	0.0012-0.0025	0.2145-0.2156	0.2144-0.2154
2007	1.8 (1ZZ-FE)	45	44.5	31.3-34.8@ 1.252	1.323	0.0010-0.0024	0.0012-0.0026	0.2154-0.2159	0.2152-0.2158
2008	1.8 (1ZZ-FE)	45	44.5	31.3-34.8@ 1.252	1.323	0.0010-0.0024	0.0012-0.0026	0.2154-0.2159	0.2152-0.2158

① Intake: 49.6-55.5@1.516
Exhaust: 47.6-52.6@1.516

22140_MATR_C0005

CAMSHAFT AND BEARING SPECIFICATIONS CHART

All measurements are given in inches.

Year	Engine Displ. Liters	Engine VIN	Journal Dia.	Brg. Oil Clearance	Shaft End-play	Runout	Journal Bore	Lobe Lift	
								Intake	Exhaust
2006	1.8	1ZZ-FE	①	0.0001-0.0024	0.0016-0.0037	0.0012	NA	1.7454-1.7493	1.7229-1.7268
	1.8	2ZZ-GE	②	0.0001-0.0024	0.0016-0.0055	0.0012	NA	③	④
2007	1.8	1ZZ-FE	①	0.0001-0.0024	0.0016-0.0037	0.0012	NA	1.7454-1.7493	1.7229-1.7268
2008	1.8	1ZZ-FE	①	0.0001-0.0024	0.0016-0.0037	0.0012	NA	1.7454-1.7493	1.7229-1.7268

NA - Not available

① No 1: 1.3563-1.3569
All others: 0.9035-0.9041

② No 1: 1.3563-1.3569
All others: 1.1004-1.1010

③ No 1 camshaft intake: 1.6030-1.6069
No 2 camshaft intake: 1.755-1.5795

④ No 1 camshaft exhaust: 1.5236-1.5303
No 2 camshaft exhaust: 1.5300-1.5340

22140_MATR_C0008

CRANKSHAFT AND CONNECTING ROD SPECIFICATIONS

All measurements are given in inches.

Year	Engine Displacement Liters	Engine ID/VIN	Crankshaft				Connecting Rod		
			Main Brg. Journal Dia.	Main Brg. Oil Clearance	Shaft End-play	Thrust on No.	Journal Diameter	Oil Clearance	Side Clearance
2006	1.8	1ZZ-FE	1.8893-1.8898	0.0006-0.0013	0.0008-0.0087	3	1.7320-1.7323	0.0011-0.0024	0.0063-0.0135
	1.8	2ZZ-GE	1.8893-1.8898	0.0006-0.0013	0.0016-0.0094	3	1.7713-1.7717	0.0011-0.0020	0.0063-0.0135
2007	1.8	1ZZ-FE	1.8893-1.8898	0.0006-0.0013	0.0008-0.0087	3	1.7320-1.7323	0.0011-0.0024	0.0063-0.0135
2008	1.8	1ZZ-FE	1.8893-1.8898	0.0006-0.0013	0.0008-0.0087	3	1.7320-1.7323	0.0011-0.0024	0.0063-0.0135

22140_MATR_C0006

PISTON AND RING SPECIFICATIONS

All measurements are given in inches.

Year	Engine Displacement Liters	Engine ID/VIN	Piston Clearance	Ring Gap			Ring Side Clearance		
				Top Compression	Bottom Compression	Oil Control	Top Compression	Bottom Compression	Oil Control
2006	1.8	1ZZ-FE	0.0026-0.0035	0.0098-0.0138	0.0138-0.0197	0.0059-0.0197	0.0008-0.0028	0.0012-0.0028	0.0012-0.0043
	1.8	2ZZ-GE	0.0003-0.0015	0.0098-0.0138	0.0138-0.0197	NA	0.0009-0.0028	0.0012-0.0028	NA
2007	1.8	1ZZ-FE	0.0026-0.0035	0.0098-0.0138	0.0138-0.0197	0.0059-0.0197	0.0008-0.0028	0.0012-0.0028	0.0012-0.0043
2008	1.8	1ZZ-FE	0.0026-0.0035	0.0098-0.0138	0.0138-0.0197	0.0059-0.0197	0.0008-0.0028	0.0012-0.0028	0.0012-0.0043

NA - Not available

22140_MATR_C0007

TORQUE SPECIFICATIONS

All readings in ft. lbs.

Year	Engine Displacement Liters (VIN)	Cylinder Head Bolts	Main Bearing Bolts	Rod Bearing Bolts	Crankshaft Damper Bolts	Flywheel Bolts	Manifold Intake	Manifold Exhaust	Spark Plugs	Oil Pan Drain Plug
2006	1.8 (1ZZ-FE)	①	②	③	102	①	22	27	18	76
	1.8 (2ZZ-GE)	④	⑤	⑥	87	①	⑦	37	13	76
2007	1.8 (1ZZ-FE)	①	②	③	102	①	22	27	18	76
2008	1.8 (1ZZ-FE)	①	②	③	102	①	22	27	18	76

① Step 1: 36 ft. lbs.
Step 2: 90 degree turn

② 12 pointed bolts:
Step 1: 33 ft. lbs.
Step 2: 90 degree turn
Hex head bolts: 14 ft. lbs.

③ Step 1: 15 ft. lbs.
Step 2: 90 degree turn

④ Step 1: 26 ft. lbs.
Step 2: 180 degree turn

⑤ 12 pointed bolts:
Step 1: 16 ft. lbs.
Step 2: 32 ft. lbs.
Step 3: 45 degree turn
Step 4: 45 degree turn
Hex head bolts: 13 ft. lbs.

⑥ Step 1: 22 ft. lbs.
Step 2: 90 degree turn

⑦ Bolt A: 25 ft. lbs.
Bolt B: 34 ft. lbs.

22140_MATR_C0009

WHEEL ALIGNMENT

Year	Model		Caster Range (+/-Deg.)	Caster Preferred Setting (Deg.)	Camber Range (+/-Deg.)	Camber Preferred Setting (Deg.)	Toe-in (in.)
2006	Matrix - 2WD	F	0.75	+2.78	0.75	-0.77	0+/-0.08
		R	—	—	0.50	-1.45	0.11+/-0.11
	Matrix - 4WD	F	0.75	+2.77	0.75	-0.48	0+/-0.08
		R	—	—	0.75	-0.73	0.08+/-0.08
2007	Matrix	F	0.75	+2.78	0.75	-0.77	0+/-0.08
		R	—	—	0.50	-1.45	0.11+/-0.11
2008	Matrix	F	0.75	+2.78	0.75	-0.77	0+/-0.08
		R	—	—	0.50	-1.45	0.11+/-0.11

22140_MATR_C0010

TIRE, WHEEL AND BALL JOINT SPECIFICATIONS

		OEM Tires		Tire Pressures (psi)		Wheel	Ball Joint	Lug Nuts
Year	Model	Standard	Optional	Front	Rear	Size	Inspection	(ft. lbs.)
2006	Matrix	205/55R16	—	②	31	6.5-JJ	9-26 in. ①	76
2007	Matrix	205/55R16	—	②	31	6.5-JJ	9-26 in. ①	76
2008	Matrix	205/55R16	—	②	31	6.5-JJ	9-26 in. ①	76

OEM: Original Equipment Manufacturer

PSI: Pounds Per Square Inch

STD: Standard

OPT: Optional

① Torque required in inch lbs. to rotate ball joint when removed from the knuckle

② FWD: 31 PSI , AWD 34 PSI.

22140_MATR_C0011

BRAKE SPECIFICATIONS

All measurements in inches unless noted

			Brake Disc			Brake Drum Diameter			Minimum	Brake Caliper	
			Original	Minimum	Maximum	Original Inside	Max. Wear	Maximum Machine	Lining	Bracket Bolts	Mounting Bolts
Year	Model		Thickness	Thickness	Runout	Diameter	Limit	Diameter	Thickness	(ft. lbs.)	(ft. lbs.)
2006	Matrix	F	0.984	0.906	0.0020	—	—	—	0.039	79	25
		R	0.354	0.295	0.0059	9.00	—	9.04	0.039	—	34
2007	Matrix	F	0.984	0.906	0.0020	—	—	—	0.039	79	25
		R	0.354	0.295	0.0059	9.00	—	9.04	0.039	—	34
2008	Matrix	F	0.984	0.906	0.0020	—	—	—	0.039	79	25
		R	0.354	0.295	0.0059	9.00	—	9.04	0.039	—	34

F: Front

R: Rear

22140_MATR_C0012

SCHEDULED MAINTENANCE INTERVALS

TOYOTA—MATRIX

TO BE SERVICED	TYPE OF SERVICE	VEHICLE MILEAGE INTERVAL (x1000)												
		7.5	15	22.5	30	37.5	45	52.5	60	67.5	75	82.5	90	97.5
Engine oil & filter	R	✓	✓	✓	✓	✓	✓	✓	✓	✓	✓	✓	✓	✓
Drive belts	S/I								✓	✓	✓	✓	✓	✓
Automatic transaxle fluid &	S/I		✓		✓		✓		✓		✓		✓	
Ball joints & dust covers	S/I		✓		✓		✓		✓		✓		✓	
Bolts & nuts on body & chassis	S/I		✓		✓		✓		✓		✓		✓	
Brake line pipes & hoses	S/I		✓		✓		✓		✓		✓		✓	
Brake linings & drums	S/I		✓		✓		✓		✓		✓		✓	
Brake pads & discs (front & rear if equipped)	S/I		✓		✓		✓		✓		✓		✓	
Differential oil	S/I		✓		✓		✓		✓		✓		✓	
Drive shaft boots	S/I		✓		✓		✓		✓		✓		✓	
Manual transaxle oil	S/I		✓		✓		✓		✓		✓		✓	
Steering gear housing oil	S/I		✓		✓		✓		✓		✓		✓	
Steering linkage	S/I		✓		✓		✓		✓		✓		✓	
Air filter	R				✓				✓				✓	
Spark plugs	R				✓				✓				✓	
Spark plugs (platinum tip)	R								✓					
Exhaust system	S/I				✓				✓				✓	
Fuel lines & connections	S/I				✓				✓				✓	
Valve clearance	S/I				✓				✓				✓	
Engine coolant	R						✓				✓			
Fuel tank cap gasket	R								✓					
Charcoal canister	S/I								✓					

R: Replace I: Inspect A: Adjust

FREQUENT OPERATION MAINTENANCE (SEVERE SERVICE)

If a vehicle is operated under any of the following conditions it is considered severe service:

- Extremely dusty areas.

- 50% or more of the vehicle operation is in 32°C (90°F) or higher temperatures, or constant operation in temperatures below 0°C (32°F).

- Prolonged idling (vehicle operation in stop and go traffic).

- Frequent short running periods (engine does not warm to normal operating temperatures).

- Police, taxi, delivery usage or trailer towing usage.

Oil & oil filter: change every 6000 miles.

Bolts & nuts on chassis & body: tighten every 7500 miles.

Ball joints & dust covers: service or inspect every 12,000 miles.

Brake linings & drums: service or inspect ever 12,000 miles.

Brake pads & discs (front & rear if equipped): service or inspect every 12,000 miles.

Drive shaft boots & except Supra): service or inspect every 12,000 miles.

Steering linkage: service or inspect every 12,000 miles.

Air filter: service or inspect every 15,000 miles.

Exhaust system: service or inspect every 15,000 miles.

Timing belt: replace every 60,000 miles.

22140_MATR_C0013

PRECAUTIONS

Before servicing any vehicle, please be sure to read all of the following precautions, which deal with personal safety, prevention of component damage, and important points to take into consideration when servicing a motor vehicle:

- Never open, service or drain the radiator or cooling system when the engine is hot; serious burns can occur from the steam and hot coolant.
- Observe all applicable safety precautions when working around fuel. Whenever servicing the fuel system, always work in a well-ventilated area. Do not allow fuel spray or vapors to come in contact with a spark, open flame, or excessive heat (a hot drop light, for example). Keep a dry chemical fire extinguisher near the work area. Always keep fuel in a container specifically designed for fuel storage; also, always properly seal fuel containers to avoid the possibility of fire or explosion. Refer to the additional fuel system precautions later in this section.
- Fuel injection systems often remain pressurized, even after the engine has been turned **OFF**. The fuel system pressure must be relieved before disconnecting any fuel lines. Failure to do so may result in fire and/or personal injury.
- Brake fluid often contains polyglycol ethers and polyglycols. Avoid contact with the eyes and wash your hands thoroughly after handling brake fluid. If you do get brake fluid in your eyes, flush your eyes with clean, running water for 15 minutes. If eye irritation persists, or if you have taken brake fluid internally, IMMEDIATELY seek medical assistance.
- The EPA warns that prolonged contact with used engine oil may cause a number of skin disorders, including cancer. You should make every effort to minimize your exposure to used engine oil. Protective gloves should be worn when changing oil. Wash your hands and any other exposed skin areas as soon as possible after exposure to used engine oil. Soap and water, or waterless hand cleaner should be used.
- All new vehicles are now equipped with an air bag system, often referred to as a Supplemental Restraint System (SRS) or Supplemental Inflatable Restraint (SIR) system. The system must be disabled before performing service on or around system components, steering column, instrument panel components, wiring and sensors. Failure to follow safety and disabling procedures could result in accidental air bag deployment, possible personal injury and unnecessary system repairs.
- Always wear safety goggles when working with, or around, the air bag system. When carrying a non-deployed air bag, be sure the bag and trim cover are pointed away from your body. When placing a non-deployed air bag on a work surface, always face the bag and trim cover upward, away from the surface. This will reduce the motion of the module if it is accidentally deployed. Refer to the additional air bag system precautions later in this section.
- Clean, high quality brake fluid from a sealed container is essential to the safe and proper operation of the brake system. You should always buy the correct type of brake fluid for your vehicle. If the brake fluid becomes contaminated, completely flush the system with new fluid. Never reuse any brake fluid. Any brake fluid that is removed from the system should be discarded. Also, do not allow any brake fluid to come in contact with a painted surface; it will damage the paint.
- Never operate the engine without the proper amount and type of engine oil; doing so WILL result in severe engine damage.
- Timing belt maintenance is extremely important. Many models utilize an interference-type, non-freewheeling engine. If the timing belt breaks, the valves in the cylinder head may strike the pistons, causing potentially serious (also time-consuming and expensive) engine damage. Refer to the maintenance interval charts for the recommended replacement interval for the timing belt, and to the timing belt section for belt replacement and inspection.
- Disconnecting the negative battery cable on some vehicles may interfere with the functions of the on-board computer system(s) and may require the computer to undergo a relearning process once the negative battery cable is reconnected.
- When servicing drum brakes, only disassemble and assemble one side at a time, leaving the remaining side intact for reference.
- Only an MVAC-trained, EPA-certified automotive technician should service the air conditioning system or its components.

BRAKES — ANTI-LOCK BRAKE SYSTEM (ABS)

GENERAL INFORMATION

The anti-lock braking system or ABS has been developed to obtain maximum braking effect with maintained stability in the most varying conditions. Many factors have a bearing on the final braking distance, e.g. weather conditions, condition of the road surface, current traffic situation and the braking force employed. ABS is a modulating function that utilizes maximum braking capacity in critical situations in varying road conditions. The anti-lock braking system or ABS has been developed to obtain maximum braking effect with maintained stability in the most varying conditions. Many factors have a bearing on the final braking distance, e.g. weather conditions, condition of the road surface, current traffic situation and the braking force employed. ABS is a modulating function that utilizes maximum braking capacity in critical situations in varying road conditions.

The ABS system used on this vehicle includes a Vehicle Stability Control (VSC) system. VSC helps prevent skidding or spinning while cornering. It does this by controlling the ABS and engine power output to help keep the vehicle traveling in the intended direction, even if the driver enters a turn too fast or steers the vehicle too sharply for road conditions.

BLEEDING PROCEDURE

➡If the master cylinder has been disassembled or if the reservoir becomes empty, bleed the air out of the master cylinder first, then bleed the remainder of the brake system.

1. Bleeding master cylinder.
   a. Disconnect the brake lines from the master cylinder.
   b. Slowly depress and hold the brake pedal.
   c. Cover the outer holes with your fingers, and release the brake pedal.
   d. Repeat the procedure 3 or 4 times.
   e. Connect the brake lines to the master cylinder and tighten to 11 ft. lbs. (15 Nm).
2. Bleed brake line.
   a. Connect the vinyl tube to the bleeder plug.
   b. Depress the brake pedal several times, then loosen the bleeder plug with the pedal depressed.
   c. When fluid stops coming out, tighten the bleeder plug, then release the brake pedal.
   d. Repeat the procedure until all the air in the fluid is completely bled out.
   e. Tighten the bleeder plug completely.
   f. Repeat the above procedures for each wheel to bleed the air out of the brake line.

BLEEDING THE ABS SYSTEM

The bleeding procedure for the ABS System is the same as the conventional bleeding procedure. Refer to Bleeding the Brake System.

If after bleeding the air from the brake system using the conventional procedure the height or feel of the brake pedal cannot be obtained, perform air bleeding in the brake actuator assembly with Techstream or equivalent scan tool by following the procedures below.

1. Let the vacuum pressure out from the brake booster assembly.
2. Depress the brake pedal more than 20 times with the engine off.
3. Connect Techstream or equivalent scan tool to the diagnostic connector, and turn the ignition switch to the **ON** position but do not start the engine.
4. Select "Air Bleeding" and bleed the air out of the brake line as usual when "Step 1: Increase" appears on the Techstream or equivalent scan tool display. Bleed the air by following the steps displayed on Techstream or equivalent scan tool.

➡Make sure that the brake fluid in the master cylinder reservoir tank does not become empty.

   a. Connect the vinyl tube to either one of the bleeder plugs.
   b. Depress the brake pedal several times, then loosen the bleeder plug connected to the vinyl tube with the pedal depressed.
   c. When fluid stops coming out, tighten the bleeder plug and release the brake pedal.
   d. Repeat the procedure until all air in the fluid is completely bled out.
   e. Tighten the bleeder plug completely.
5. Repeat the above procedures for each wheel to bleed the air out of the brake line.
6. Bleed the air out of the suction line when "Step 2: Inhalation" appears on the Techstream or equivalent scan tool display.
   a. Connect the vinyl tube to the bleeder plug at the right front wheel or the right rear wheel and loosen the bleeder plug.
   b. Operate the brake actuator assembly to bleed the air using Techstream or equivalent scan tool.

➡Make sure to release the brake pedal. This operation stops automatically after 4 seconds.

   c. Check if the operation has stopped by referring to the Techstream or equivalent scan tool display.
   d. Repeat the procedure until all air in the fluid is completely bled out.
7. Repeat the above procedures for the other wheels to bleed the air out of the brake line.
8. Bleed the air out of the pressure reduction line when "Step 3: Decrease" appears on the Techstream or equivalent scan tool display.
   a. Connect a vinyl tube to either one of the bleeder plugs.
   b. Loosen the bleeder plug.
   c. Using Techstream or equivalent scan tool, operate the brake actuator assembly, completely depress the brake pedal and keep it.

➡During this procedure, the pedal will feel heavy, but completely depress it so that the brake fluid comes out from the bleeder plug. Be sure to keep depressing the brake pedal. Do not depress and release the pedal repeatedly. The operation stops automatically after 4 seconds. When performing this procedure continuously, set an interval of at least 20 seconds.

   d. When the operation is complete, the brake pedal goes down slightly. This is a normal phenomenon caused when the solenoid opens.
   e. Tighten the bleeder plug, then release the brake pedal.
   f. Repeat the procedure until all the air in the fluid is completely bled out.
9. Repeat the above procedures for the other wheels to bleed the air out of the brake line.
10. Bleed the air out of the brake line as usual again when "Step 4: Increase" appears on the Techstream or equivalent scan tool display.
   a. Connect the vinyl tube to either one of the bleeder plugs.
   b. Depress the brake pedal several times, then loosen the bleeder plug connected to the vinyl tube with the pedal depressed.
   c. When fluid stops coming out, tighten the bleeder plug, then release the brake pedal.
   d. Repeat procedure F until all the air in the fluid is completely bled out.
   e. Tighten the bleeder plug.
11. Repeat the above procedures for the other wheels to bleed the air out of the brake line.
12. Make sure that the air bleeding is complete by referring to the Techstream or equivalent scan tool display.

BRAKE CALIPER

REMOVAL & INSTALLATION

See Figure 1.

1. Before servicing the vehicle, refer to the precautions section.
2. Remove some fluid from the reservoir with a suction pump.
3. Raise and safely support the vehicle securely on jackstands.
4. Remove the front wheels.
5. Remove the banjo bolt and disconnect the brake hose from the caliper. Plug the hose to prevent fluid loss and contamination.
6. Remove the mounting bolts while holding the slide pin.
7. Remove the caliper.

To install:

8. Compress the caliper piston using a C-clamp or other suitable tool.
9. Install the caliper. Tighten the mounting bolts to 25 ft. lbs. (34 Nm)
10. Install the brake hose to the caliper using new sealing washers. Carefully torque the banjo bolt to 21 ft. lbs. (29 Nm).

➡There are two types of sealing washers, a 2-piece type and a 1-piece type. Make sure to install the correct type.

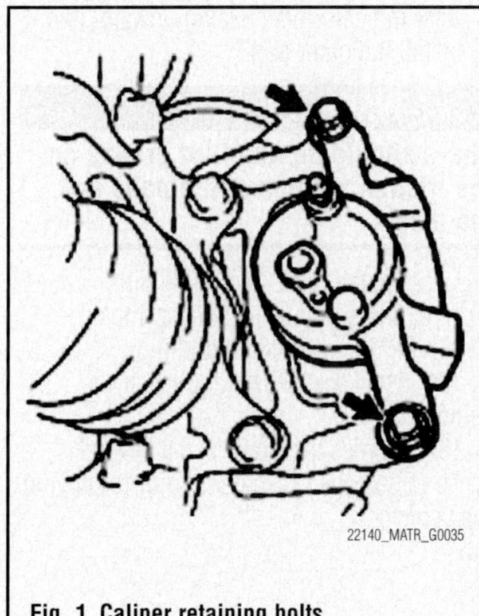

Fig. 1 Caliper retaining bolts

11. Fill the reservoir with fluid and bleed the brakes.
12. Install the front wheels and lower the vehicle.
13. Depress the brake pedal several times to seat the brake pads prior to moving the vehicle.

DISC BRAKE PADS

REMOVAL & INSTALLATION

See Figure 2.

1. Before servicing the vehicle, refer to the precautions section.
2. Remove some fluid from the reservoir with a suction pump.
3. Raise and safely support the vehicle securely on jackstands.
4. Remove the brake caliper.
5. Remove disc brake pad kit from the caliper
   a. Remove the 2 brake pads with anti-squeal shims.
   b. Remove the anti-squeal shims from each pad.

1ZZ-FE(FF) Type:
Anti-squeal Shim No.1
Front Disc Brake Pad Wear Indicator Plate
Anti-squeal Shim No.1
Anti-squeal Shim No.2
Disc Brake Pad Kit
Except 1ZZ-FE(FF) Type:
Anti-squeal Shim No.2
Disc Brake Pad Kit
Anti-squeal Shim No.2
Anti-squeal Shim No.2
Anti-squeal Shim No.1
Anti-squeal Shim No.1
42356-TMAT-G03

Fig. 2 Front pads and related components

   c. Remove the pad wear indicator plates from each pad.
   d. Remove the 2 front disc brake pad support plates from the cylinder mounting.
6. Remove front disc brake cylinder slide pin.
   a. Remove the 2 cylinder slide pins from the disc brake cylinder mounting.

To install:

7. Compress the caliper piston using a C–clamp or other suitable tool.
8. Install front disc brake cylinder slide pin.
   a. Apply the lithium soap base glycol grease to the sliding part and the seal surface of the 2 cylinder slide pins.
   b. Install the 2 cylinder slide pins to the disc brake cylinder mounting.
9. Install front disc brake pad support plate.
   a. Install the 2 front disc brake pad support plates to the cylinder mounting.
10. Install front disc brake pad support plate.
11. Install disc brake pad kit on the caliper.

➡If necessary, replace the anti-squeal shim kit when replacing the brake pad.

   a. Apply disc brake grease to each anti-squeal shim No. 1.
   b. Install anti-squeal shims on each pad.
   c. Install the pad wear indicator plate facing upward, and install each pad.

⁂ WARNING

There should be no oil or grease on the friction surfaces of the pads and the disc.

12. Install the caliper. Tighten the mounting bolts to 25 ft. lbs. (34 Nm).
13. Install the front wheels and lower the vehicle.
14. Fill the reservoir with fluid.
15. Depress the brake pedal several times to seat the brake pads prior to moving the vehicle.

BRAKE CALIPER

REMOVAL & INSTALLATION

See Figures 3 and 4.

1. Before servicing the vehicle, refer to the precautions section.
2. Remove some fluid from the reservoir with a suction pump.
3. Raise and safely support the vehicle securely on jackstands.
4. Remove the wheels.
5. Remove disc brake pad kit.
    a. Remove the clip and the anti-rattle springs.

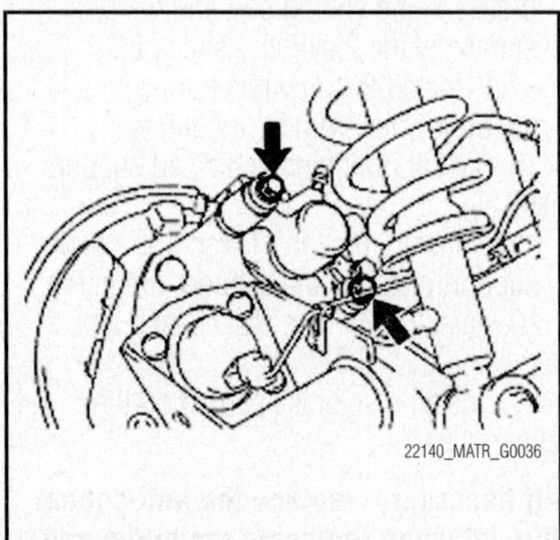

Fig. 3 Caliper retaining bolts

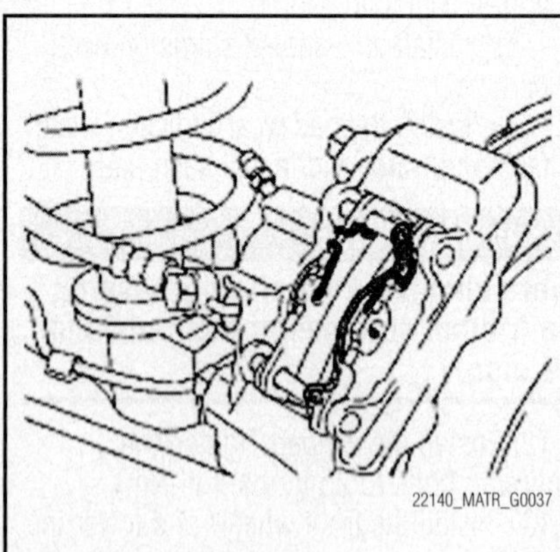

Fig. 4 Anti-rattle springs

    b. Remove the 2 pad guide pins and disc brake pads with anti-squeal shims.
    c. Remove the anti-squeal shims from each pad.
6. Remove the banjo bolt and disconnect the brake hose from the caliper. Plug the hose to prevent fluid loss and contamination.
7. Remove the mounting bolts while holding the slide pin.
8. Remove the caliper.

To install:

9. Compress the caliper piston using a C-clamp or other suitable tool.
10. Install the caliper. Tighten the caliper bolts to 34 ft. lbs. (46 Nm).
11. Install the brake hose to the caliper using new sealing washers. Carefully torque the banjo bolt to 21 ft. lbs. (29 Nm).

➡There are two types of sealing washers, a 2-piece type and a 1-piece type. Make sure to install the correct type.

12. Install disc brake pad kit.

➡If necessary, replace the anti-squeal shim kit when replacing the brake pad.

    a. Apply disc brake grease to inside of anti-squeal shims, and install the anti-squeal shims on each pad.
    b. Install the inner pad with the pad wear indicator plate facing upward, and install the outer pad.

✲✲ WARNING

There should be no oil or grease on the friction surface of the pads and the disc.

    c. Install the 2 pad guide pins.
    d. Install the anti-rattle springs.
13. Fill the reservoir with fluid and bleed the brakes.
14. Install the wheels and lower the vehicle.
15. Depress the brake pedal several times to seat the brake pads prior to moving the vehicle.

DISC BRAKE PADS

REMOVAL & INSTALLATION

1. Before servicing the vehicle, refer to the precautions section.
2. Remove some fluid from the reservoir with a suction pump.
3. Raise and safely support the vehicle securely on jackstands.
4. Remove the wheels.
5. Remove disc brake pad kit.
    a. Remove the clip and the anti-rattle springs.
    b. Remove the 2 pad guide pins and disc brake pads with anti-squeal shims.
    c. Remove the anti-squeal shims from each pad.

To install:

6. Compress the caliper piston using a C-clamp or other suitable tool.
7. Install disc brake pad kit.

➡If necessary, replace the anti-squeal shim kit when replacing the brake pad.

    a. Apply disc brake grease to inside of anti-squeal shims, and install the anti-squeal shims on each pad.
    b. Install the inner pad with the pad wear indicator plate facing upward, and install the outer pad.

✲✲ WARNING

There should be no oil or grease on the friction surface of the pads and the disc.

    c. Install the 2 pad guide pins.
    d. Install the anti-rattle springs.
8. Fill the reservoir with fluid.
9. Install the wheels and lower the vehicle.
10. Depress the brake pedal several times to seat the brake pads prior to moving the vehicle.

BRAKE DRUM

REMOVAL & INSTALLATION

See Figure 5.

1. Before servicing the vehicle, refer to the precautions section.
2. Release the parking brake.
3. Remove the rear wheel.
4. Remove the inspection plug from the brake drum.
5. Insert a suitable regular screwdriver through the plug hole in the drum and hold the adjusting lever away from the adjuster.
6. Insert a second regular screwdriver and turn the brake shoe adjuster to reduce the size of the adjuster.
7. Remove the brake drum.

To install:

8. Measure the brake drum inside diameter and the diameter of the installed brake shoe assembly.
9. Turn the brake shoe adjuster to either retract or expand the brake shoe assembly until the diameter of the brake she assembly is slightly less than that of the drum.
10. Install the drum, and turn the wheel/drum assembly. There should be a hardly noticeable drag as the drum is turned.
11. Install the wheel and pump the brake pedal several times to fully seat the brakes.
12. Start the vehicle and very slowly drive back and forwards. Driving backwards and applying the brake pedal causes the rear brake adjuster to self adjust.

➡Do not drive backwards at high speeds, for long distances or apply the brakes too hard. A slow speed under 1 mile per hour for a short distance and plying the brake will help the adjuster to self adjust.

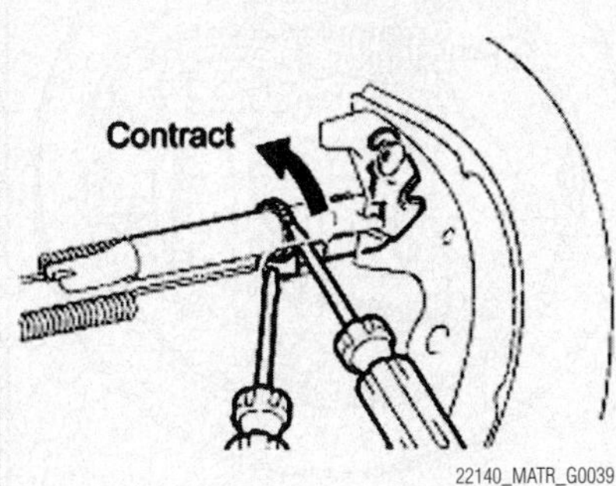

Fig. 5 Insert a suitable regular screwdriver through the plug hole in the drum and hold the adjusting lever away from the adjuster.

BRAKE SHOES

REMOVAL & INSTALLATION

See Figures 6 through 15.

1. Before servicing the vehicle, refer to the precautions section.
2. Remove rear wheel.
3. Remove rear drum brake.
   a. Release the parking brake lever, and remove the brake drum.
4. If the brake drum cannot be removed easily, do the following steps.
   a. Remove the hole plug and insert a screwdriver through the hole in the backing plate, and hold the automatic adjusting lever away from the adjuster.
   b. Using an another screwdriver, reduce the brake shoe adjuster by turning the adjusting wheel.
5. Remove front brake shoe.
   a. Using a brake tool, remove the upper side tension spring with spacer.
   b. Using a needle-nose pliers, remove the anchor side return spring.
   c. Using a tool remove the cup, shoe hold-down sprig and pin.
   d. Disconnect the upper side return spring from the front shoe, and remove the front brake shoe.
6. Remove parking brake shoe strut.
7. Remove rear brake automatic adjust lever.

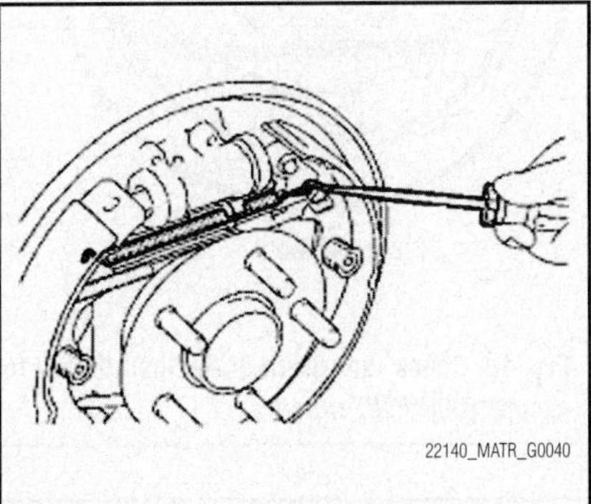

Fig. 6 Using a brake tool, remove the upper side tension spring with spacer

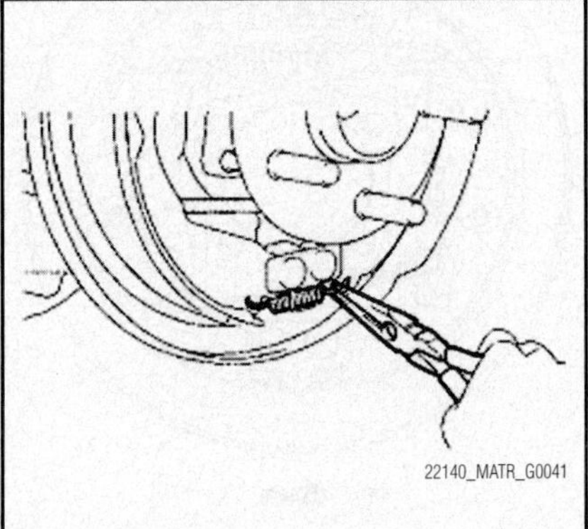

Fig. 7 Using a needle-nose pliers, remove the anchor side return spring

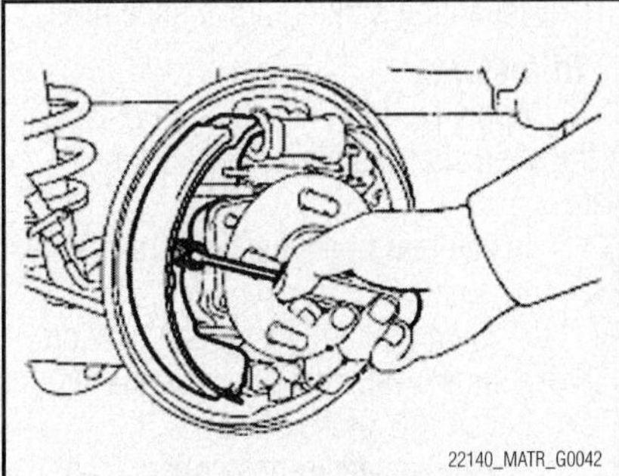

Fig. 8 Using a brake tool, remove the cup, shoe hold-down sprig and pin from the front brake shoe

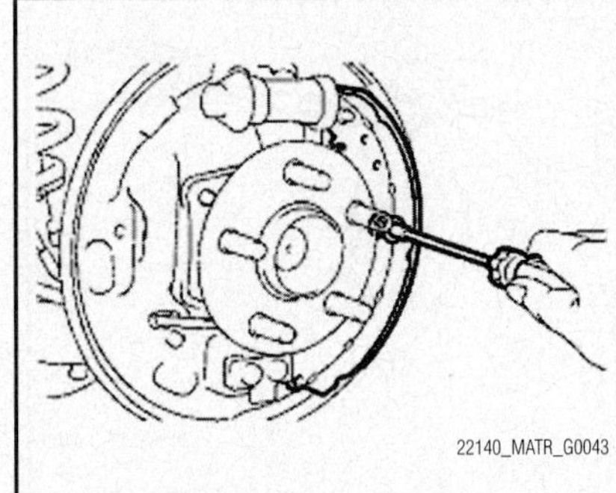

Fig. 9 Using a brake tool, remove the cup, shoe hold-down spring and pin from the rear brake shoe

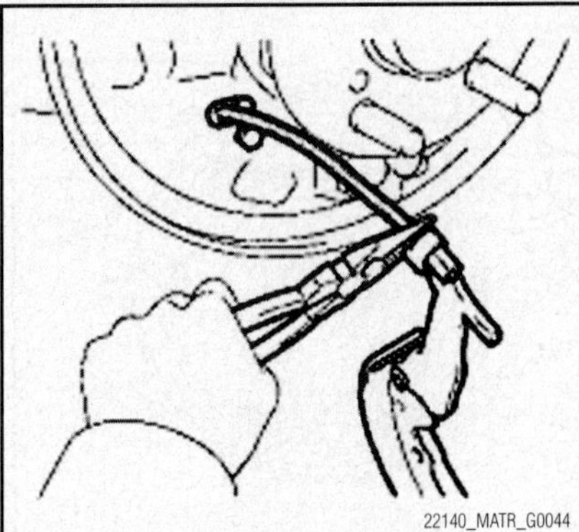

Fig. 10 Using a needle-nose pliers, disconnect the parking brake cable and remove the rear brake shoe

8. Remove rear brake shoe.
    a. Using a brake tool, remove the upper side tension spring with spacer.
    b. Using a needle-nose pliers, remove the anchor side return spring.
    c. Using a tool remove the cup, shoe hold-down sprig and pin.
    d. Using a needle-nose pliers, disconnect the parking brake cable and remove the rear brake shoe.
9. Remove rear brake parking brake shoe lever.
    a. Using a screwdriver, remove the C-washer and parking brake shoe lever.

To install:

10. Apply the high temperature grease to the shoe attached surface of backing plate.
11. Install rear brake parking brake shoe lever.
    a. Using a pliers, install the parking brake shoe lever with a new C-washer.
12. Install rear brake shoe.
    a. Using a needle-nose pliers, connect the parking brake cable No. 3 to the parking brake shoe lever.

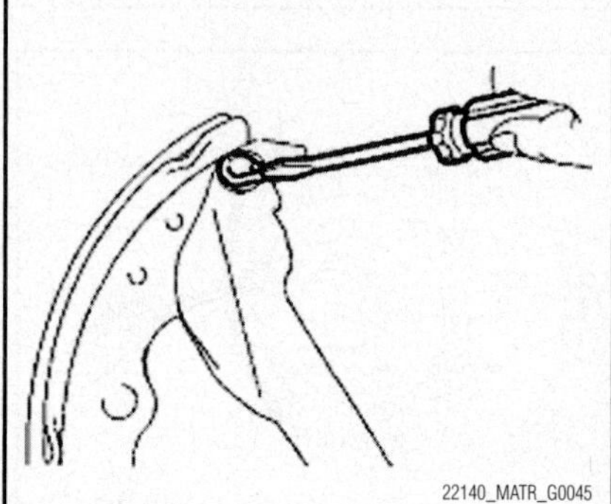

Fig. 11 Using a screwdriver, remove the C-washer and parking brake shoe lever

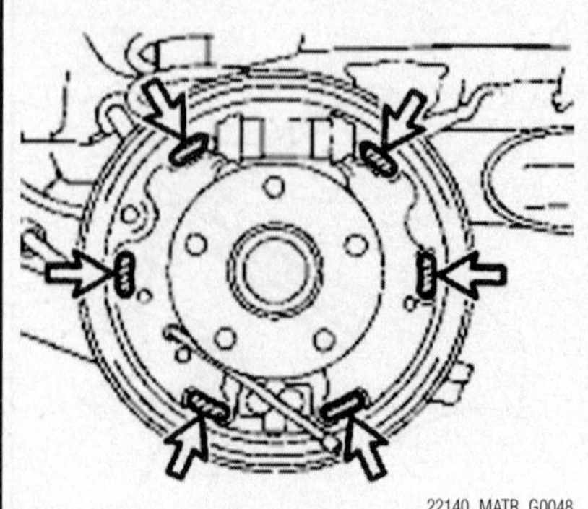

Fig. 12 Apply the high temperature grease to the shoe attached surface of backing plate

    b. Connect the upper side return spring to the rear brake shoe.
    c. Using a brake tool, install the rear brake shoe, pin, shoe hold-down spring and cup.

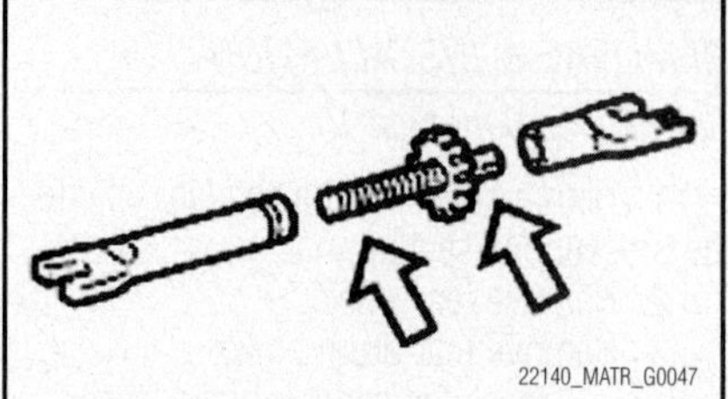

Fig. 13 Apply the high temperature grease to the adjusting bolt.

13. Install rear brake automatic adjust lever .
14. Install parking brake shoe strut.
    a. Apply the high temperature grease to the adjusting bolt.
    b. Install the parking brake shoe strut.
15. Install front brake shoe.
    a. Install the upper side return spring to the front brake shoe.
    b. Using a brake tool, install the front brake shoe, pin, shoe hold-down spring and cup.
16. Using a needle-nose pliers, install the anchor side return spring.
17. Install the spacer to the upper side tension spring.
    a. Using a brake tool, install the upper side tension spring with spacer.

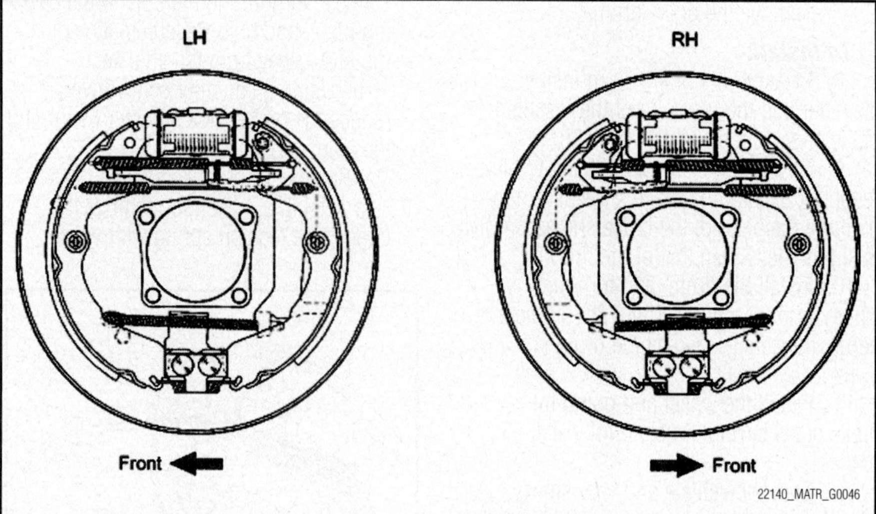

Fig. 14 Check rear drum brake installation to ensure that each part is installed properly—except 2006 AWD models

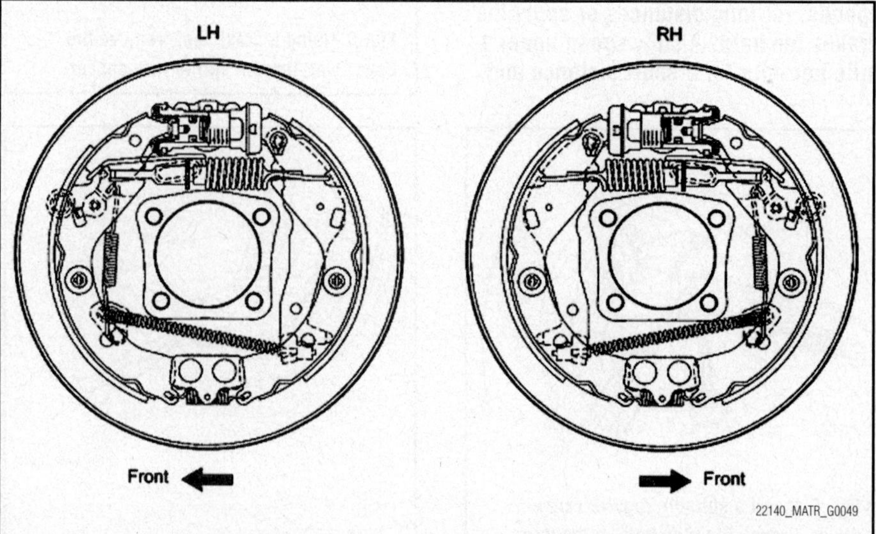

Fig. 15 Check rear drum brake installation to ensure that each part is installed properly—2006 AWD models

18. Check rear drum brake installation against the illustration to ensure that each part is installed properly.
19. Adjust the rear brakes.
20. Install the rear drum.
    a. Readjust the parking brake cable as necessary.
21. Install the wheels.
22. Refill the master cylinder and pump pedal to attain full brake pedal before road-testing the vehicle.

ADJUSTMENT

See Figure 16.

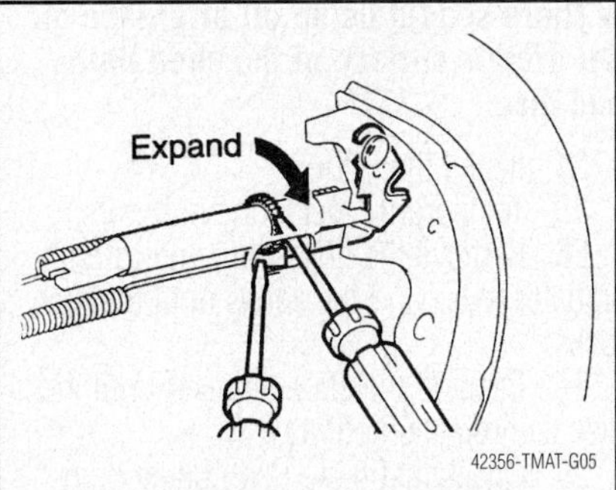

Fig. 16 Turn the adjuster to expand the shoe until the drum locks

1. Temporarily install the drum and hub nuts.
2. Remove the hole plug from the backing plate.
3. Turn the adjuster to expand the shoe until the drum locks.
4. Back off the adjuster eight notches using a suitable adjustment tool.
5. Install the hole plug. into the backing plate to prevent dirt and moisture from entering.
6. Readjust the parking brake cable as necessary.

BRAKES — PARKING BRAKE

PARKING BRAKE SHOES

REMOVAL & INSTALLATION

With Rear Drum Brakes

On models with rear drum brakes, the rear brakes serve as the parking brakes. Refer to Rear Drum Brakes, Brake Shoes, Removal & Installation.

With Rear Disc Brakes

See Figures 17 through 19.

1. Before servicing the vehicle, refer to the precautions section.
2. Raise and safely support the vehicle securely on jackstands.
3. Remove the rear wheel
4. Remove the rear disc brake caliper.
5. Remove the rotor.
6. Using needle-nose pliers, remove the 2 upper side tension springs.
7. Remove the parking brake shoe strut.
8. Using a needle-nose pliers, remove the anchor side tension spring.
9. Remove the shoe adjusting screw set.
10. Using a brake tool, remove the parking brake shoe hold-down compression spring, parking brake shoe hold-down spring pin and parking brake shoe assembly.
11. Using a needle-nose pliers, disconnect the parking brake cable assembly and remove the parking brake shoe lever.

To install:

12. Using a ruler, measure the thickness of the shoe lining. The standard thickness is 0.138 in. (3.5 mm). and min thickness is 0.039 in. (1mm). If shoe lining is not within specification, replace shoe lining.

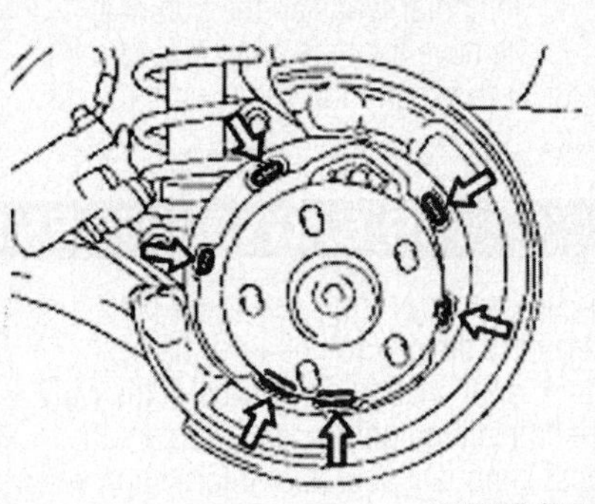

Fig. 17 Apply high temperature grease to the shoe contact surfaces of the backing plate

13. Apply high temperature grease to the shoe contact surfaces of the backing plate.

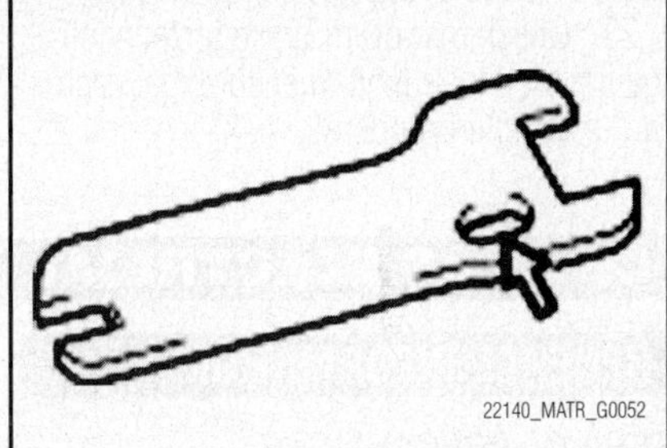

Fig. 18 Apply high temperature grease to the shoe contact surfaces of the backing plate

14. Using a needle-nose pliers, connect the parking brake cable assembly to the parking brake shoe lever.
15. Apply the high temperature grease to the contact part of the parking brake shoe and parking brake shoe lever.

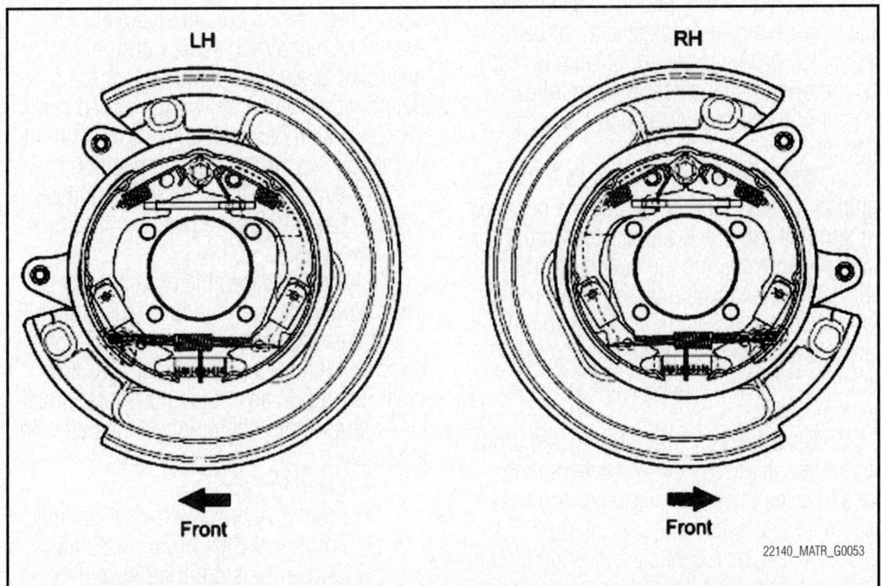

Fig. 19 Check rear drum brake installation to ensure that each part is installed properly.

16. Using a brake tool, install the parking brake shoe hold-down compression spring, parking brake shoe hold-down spring pin and parking brake shoe assembly.
17. Apply the high temperature grease to the adjusting bolt and support piece.
18. Install the parking brake shoe adjusting screw set
19. Using a needle-nose pliers, install the tension spring.
20. Apply the high temperature grease to the contact part of the shoe strut and tension spring.
21. Install the parking brake shoe strut.
22. Using a needle-nose pliers, install the 2 upper side tension springs securely to the grooves of the shoe strut.
23. Check rear drum brake installation against the illustration to ensure that each part is installed properly.

➡There should be no oil or grease on the friction surface of the shoe lining and disc.

24. Install the rotor.
25. Temporarily install the hub nuts.
26. Remove the hole plug, and turn the adjuster to expand the shoes until the disc locks.
27. Back off the shoe adjuster until the disc can rotate smoothly.
28. Check that there is no brake drag and install the hole plug.
29. Install the caliper and rear wheel.
30. Adjust parking brake lever travel.

ADJUSTMENT

See Figure 20.

1. Temporarily install the hub nuts.
2. Remove the hole plug, and turn the adjuster to expand the shoes until the disc locks.

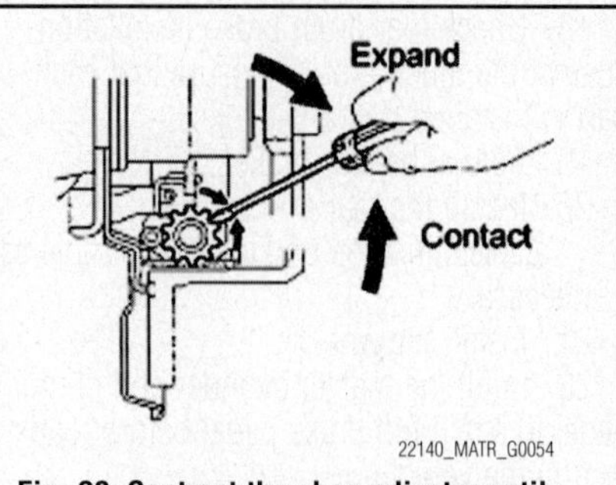

Fig. 20 Contract the shoe adjuster until the disc can rotate smoothly

3. Contract the shoe adjuster until the disc can rotate smoothly.
4. Check that there is no brake drag and install the hole plug.
5. Install the caliper and rear wheel.
6. Adjust parking brake lever travel. Refer to Parking Brake, Parking Brake Cables, Adjustment.

CHASSIS ELECTRICAL

AIR BAG (SUPPLEMENTAL RESTRAINT SYSTEM)

GENERAL INFORMATION

SERVICE PRECAUTIONS

1. The vehicle is equipped with a Supplemental Restraint System (SRS). It consists of a driver airbag, front passenger airbag, side airbag, curtain shield airbag and front seat belt pretensioner. Failure to carry out service operations in the correct sequence could cause the SRS to unexpectedly deploy during servicing, possibly leading to a serious accident. Further, if a mistake is made in servicing the SRS, it is possible that the SRS may fail to operate when required. Before performing servicing (including removal or installation of parts, inspection or replacement), be sure to read the following items carefully, then follow the correct procedures indicated in the repair manual.
2. Wait at least 90 seconds after the ignition switch is turned off and the negative (-) terminal cable is disconnected from the battery before starting the operation. The SRS is equipped with a backup power source, so that if work is started within 90 seconds after disconnecting the negative (-) terminal cable of the battery, the SRS may be deployed.
3. Do not expose the steering pad, front passenger airbag assembly, center airbag sensor assembly, front airbag sensor, front seat inner belt assembly, seat position airbag sensor, occupant classification ECU, front seat side airbag assembly, side airbag sensor, curtain shield airbag assembly, rear airbag sensor or front seat outer belt assembly directly to hot air or flames.
4. Malfunction symptoms of the SRS are difficult to confirm, so DTCs are the most important source of information when troubleshooting. When troubleshooting the SRS, always inspect DTCs before disconnecting the battery.
5. Even in the case of a minor collision when the SRS does not deploy, the steering pad, front passenger airbag assembly, center airbag sensor assembly, front airbag sensor, front seat inner belt assembly, seat position airbag sensor, occupant classification ECU, front seat side airbag assembly, side airbag sensor, curtain shield airbag assembly, rear airbag sensor or front seat outer belt assembly should be inspected. Before repair work, remove the airbag sensor if any kind of shock is likely to occur to the airbag sensor during the operation.
6. Never use SRS parts from another vehicle. When replacing parts, replace them with new ones.
7. Never disassemble or repair any of the following parts in order to reuse them. If any of these parts have been dropped, or a defect is found (e.g. cracks, dents or any other defects) in any of the housings, brackets or connectors, then replace the part with a new one.
   - Steering Pad
   - Front Passenger Airbag Assembly
   - Front Seat Side Airbag Assembly
   - Curtain Shield Airbag Assembly
   - Center Airbag Sensor Assembly
   - Front Airbag Sensor
   - Front Seat Inner Belt Assembly
   - Seat Position Airbag Sensor
   - Occupant Classification ECU
   - Side Airbag Sensor
   - Rear Airbag Sensor
   - Front Seat Outer Belt Assembly
8. Use a volt/ohmmeter with high impedance (10 kilohms/V minimum) for troubleshooting the electrical circuits.
9. Information labels are attached near the SRS components. Follow the instructions in the caution.
10. After work on the SRS is completed, perform the SRS warning light check .
11. When the negative (-) terminal cable is disconnected from the battery, the memory will be cleared. Because of this, be sure to make a record of the contents memorized in each system before starting work.
12. When work is finished, adjust each system as it was before. Never attempt to avoid erasing vehicle system memories by using a backup power supply from outside the vehicle.

DISARMING THE SYSTEM

To avoid personal injury when working on vehicles equipped with an air bag, the negative battery cable must be disconnected and at least 90 seconds must elapse before working on the system. Failure to do so may result in deployment of the air bag.

ARMING THE SYSTEM

After vehicle service is completed, reattach the battery cables (positive cable first!) to rearm the air bag system.

DRIVETRAIN

AUTOMATIC TRANSAXLE ASSEMBLY

REMOVAL & INSTALLATION

FWD—A246E & U240E Transaxles

See Figures 21 and 22.

1. Before servicing the vehicle, refer to the precautions section.
2. Drain the transaxle fluid.
3. Remove or disconnect the following:
   - Negative battery cable
   - Hood
   - No. 2 cylinder head cover
   - Battery and battery carrier
   - Air cleaner assembly with hose
   - Floor shift cable transmission control shift
   - Transmission control cable support
   - No. 1 transmission control cable bracket
   - Wiring harness and brackets
   - Transmission wire connector
   - Park/neutral position switch connector, with Anti-lock Brake System (ABS)
   - Speedometer sensor connector, without ABS
   - Transmission revolution sensor connectors, if equipped
   - Transmission fluid filler tube
   - No. 1 oil cooler inlet and outlet tubes
   - Foot rest
   - Floor carpet
   - Oxygen (O_2) sensor connector
4. Suspend the engine.
   a. Disconnect the 2 Positive Crankcase Ventilation (PCV) hoses.
   b. Install the No. 1 and No. 2 engine hangers in the correct direction.
   c. No. 1 engine hanger: P/N 12281-22021 (A246E) or 12281-88600 (U240E).
   d. No. 2 engine hanger: P/N 12281-15040 (A246E) or 12281-88600 (U240E)
   e. Bolt: P/N 91512-B1016.
   f. Torque the bolt to 28 ft. lbs. (38 Nm).
   g. Attach an engine chain hoist to the engine hangers.
   - Front wheels
   - Right and left engine undercovers
   - Front floor panel brace, U240E transaxle
   - Front exhaust pipe
   - Front halfshafts
   - Automatic transmission case protector
   - Starter
5. Support the transaxle with a floor jack
   - Left side transverse engine mounting insulator and bracket
   - Right side front and rear engine mount insulators
   - 4 bolts, dynamic damper and member
   - Front and rear right side transverse engine mounting brackets
   - Flywheel housing undercover
   - Automatic transaxle. Turn the crankshaft for access to the bolts while holding the crankshaft pulley bolt with a wrench.
   - Torque converter clutch

To install:

6. Installation is the reverse of the removal procedure, noting the following specifications:
   a. Automatic transaxle: Bolt "A" to 47 ft. lbs. (64 Nm), bolt "B" to 34 ft. lbs. (47 Nm) and bolt "C" to 17 ft. lbs. (23 Nm).
   b. Torque converter bolts: 20 ft. lbs. (28 Nm).

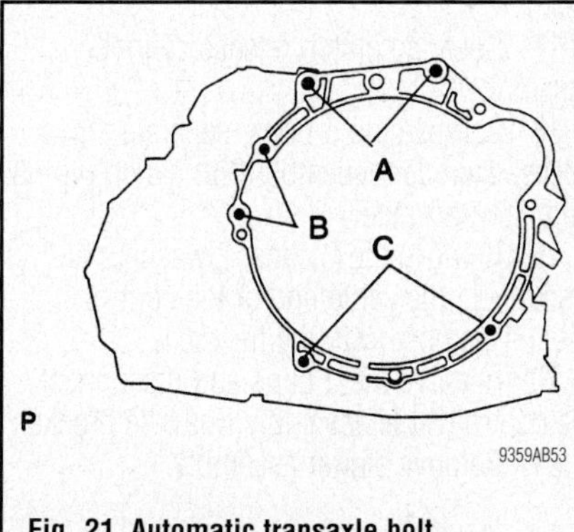

Fig. 21 Automatic transaxle bolt locations—A246E & U240E transaxles

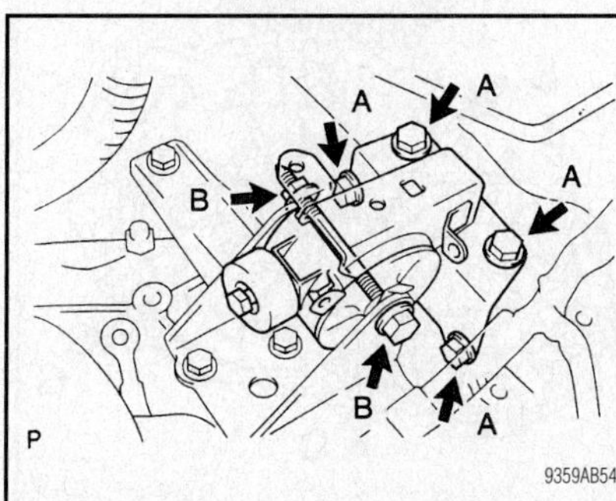

Fig. 22 Left side engine mount insulator bolt and nut locations—A246E & U240E transaxles

   c. Front and rear right transverse engine mounting bracket bolts: 47 ft. lbs. (64 Nm).
   d. Member center bolts: "A" bolts to 29 ft. lbs. (39 Nm) and "B" bolts to 38 ft. lbs. (52 Nm).
   e. Right rear engine mounting insulator-to-engine mounting bracket bolt: 64 ft. lbs. (87 Nm).
   f. Right rear engine mount insulator nuts and bolt: 38 ft. lbs. (52 Nm).
   g. Left side engine mounting bracket-to-transaxle bolts: 38 ft. lbs. (52 Nm).
   h. Left side engine mounting insulator bolts and nut: Bolt "A" to 38 ft. lbs. (52 Nm), Bolt "B" and Nut "B" to 59 ft. lbs. (80 Nm).
   i. Front right engine mount insulator-to-mounting bracket bolt and nut: 38 ft. lbs. (52 Nm).
   j. Starter bolts: 29 ft. lbs. (39 Nm).
   k. Automatic transmission case protector bolts: 14 ft. lbs. (18 Nm).
   l. Wheel lug nuts: 76 ft. lbs. (103 Nm).
   m. Oil cooler clamp bolts: 49 inch lbs. (5.5 Nm).
   n. Oil cooler inlet and outlet tubes: 25 ft. lbs. (34 Nm).
   o. Wire harness bracket bolt: 9 ft. lbs. (13 Nm).
   p. Transmission control cable bracket bolts: 9 ft. lbs. (12 Nm).
   q. Transmission control cable support: 9 ft. lbs. (12 Nm).
   r. Battery carrier: 10 ft. lbs. (13 Nm).
   s. Air cleaner assembly: 62 inch lbs. (7 Nm).
   t. Cylinder head cover bolts: 62 inch lbs. (7 Nm).
   u. Hood bolts: 10 ft. lbs. (13 Nm).
7. Fill the transaxle fluid to the proper level.
8. Start the vehicle, check for leaks and repair if necessary.

AWD—U341F Transaxle

See Figure 23.

1. Before servicing the vehicle, refer to the precautions section.
2. Drain the transaxle fluid.
3. Remove or disconnect the following:
   - Negative battery cable
   - Engine and transaxle assembly
   - Transfer case
   - Automatic transmission case protector
   - Front left side halfshaft

- Transmission control cable support and bracket
- Wire harness clamp bracket, bolts and 2 wire harnesses
- Transmission wire connector
- Park/neutral position switch connector
- Transmission revolution sensor connectors, if equipped
- Transmission fluid filler tube
- Oil cooler inlet and outlet tubes
- Transverse engine mounting brackets
- Flywheel housing undercover
- Automatic transaxle. Turn the crankshaft for access to the bolts while holding the crankshaft pulley bolt with a wrench.
- Torque converter clutch

To install:

4. Installation is the reverse of the removal procedure, noting the following specifications:
    a. Automatic transaxle: Bolt "A" to 47 ft. lbs. (64 Nm), bolt "B" to 34 ft. lbs. (47 Nm) and bolt "C" to 17 ft. lbs. (23 Nm).
    b. Oil cooler clamp bolts: 8 ft. lbs. (11 Nm) for the top bolt and 49 inch lbs. (5.5 Nm) for the bottom bolt
    c. Oil cooler inlet and outlet tube bolts: 25 ft. lbs. (34 Nm).
    d. Wire harness clamp bracket bolt: 48 inch lbs. (5 Nm).
    e. Transmission control cable bracket and support bolts: 9 ft. lbs. (12 Nm).
    f. Automatic transmission case protector bolts: 17 ft. lbs. (23 Nm).
5. Fill the transaxle fluid to the proper level.
6. Start the vehicle, check for leaks and repair if necessary.

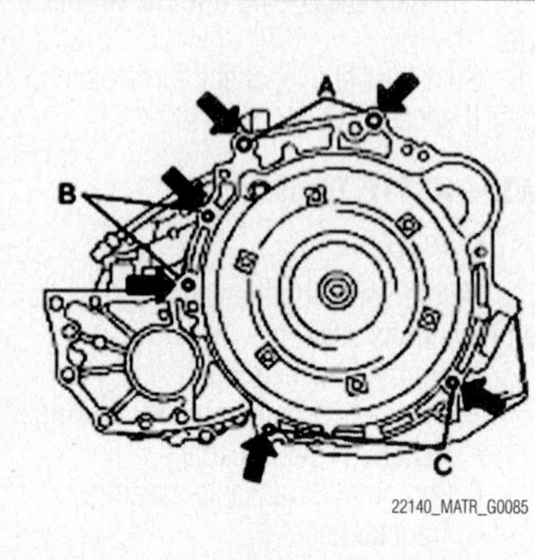

Fig. 23 Automatic transaxle bolt locations—U341F transaxle

MANUAL TRANSAXLE ASSEMBLY

REMOVAL & INSTALLATION

See Figures 24 through 26.

1. Before servicing the vehicle, refer to the precautions section.
2. Place front wheels facing straight ahead.
3. Separate steering intermediate shaft.
4. Raise and safely support the vehicle securely on jackstands.
5. Remove front wheels.
6. Remove engine undercover.
7. Remove exhaust pipe assembly.
8. Drain transaxle oil.
9. Remove hood.
10. Remove the nuts caps, clips and cylinder head cover.
11. Remove air cleaner assembly.
12. Remove battery clamp.
13. Remove battery, battery tray and battery carrier.
14. Disconnect the wire harness clamp. Remove the bolts and disconnect the 2 wire harness brackets. Remove the bolts and 2 ground cables.
15. Disconnect the back-up light switch connector.
16. Disconnect the speed sensor connector.
17. Separate clutch release cylinder assembly.
18. Remove the 5 bolts, separate the release cylinder assembly with clutch piping from the transaxle.
19. Remove the 2 clips, 2 washers and disconnect the cable end of the control cable assembly from the transaxle.
20. Remove the 2 clips and disconnect the control cable assembly from the bracket.
21. Remove starter assembly.

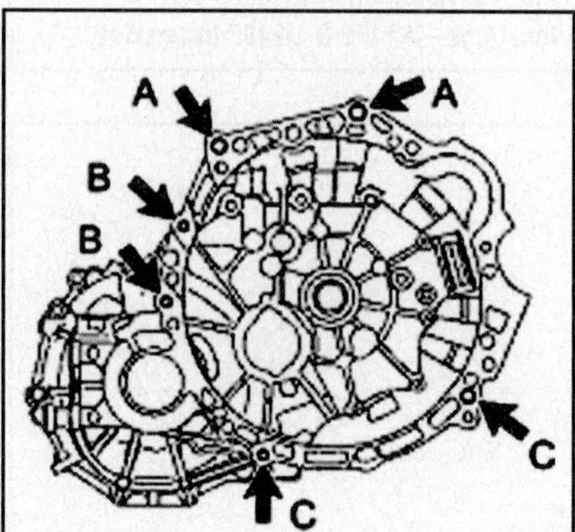

Fig. 24 Tighten manual transaxle assembly bolts as specified

22. Disconnect tie rod ends.
23. Disconnect fuel system pressure feed and return tubes.
24. Remove front drive shaft assembly.
25. Suspend engine assembly.

**** WARNING**

Do not attempt to hang the engine by hooking the chain to any other part except the engine hangers.

26. Remove front suspension crossmember.
27. Support the transaxle with a transmission jack.
28. Remove transverse engine mounting insulator.
29. Remove transverse engine mounting bracket.
30. Remove manual transaxle assembly.

To install:

31. Align the input shaft with the clutch disc and install the transaxle to the engine. Tighten bolts "A" to 47 ft. lbs. (64 Nm), bolts "B" to 35 ft. lbs. (64 Nm) and bolts "C" to 17 ft. lbs. (23 Nm).
32. Install the engine mounting bracket to the transaxle. Tighten bolts to 38 ft. lbs. (52 Nm).

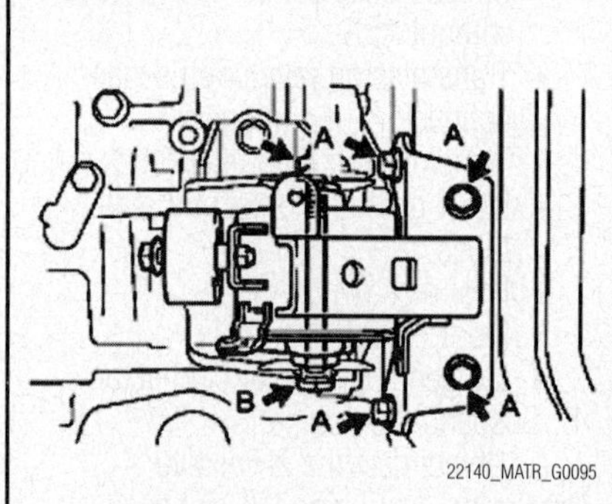

Fig. 25 Tighten engine mounting insulator bolts as specified

Fig. 26 Tighten engine clutch release cylinder bolts as specified

33. Install the engine mounting insulator. Tighten bolts "A" to 38 ft. lbs. (52 Nm), bolts "B" to 59 ft. lbs. (80 Nm).
34. Install front suspension crossmember.
35. Install front drive shaft
36. Connect the fuel system return and feed tubes.
37. Connect tie rod end.
38. Install starter assembly. Tighten bolts to 27 ft. lbs. (37 Nm).
39. Connect the starter connector. Install the wire and nut to starter assembly and tighten to 8 ft. lbs. (9 Nm).
40. Connect transmission control cable assembly.
41. Connect clutch release cylinder. Tighten bolts "A" to 19 ft. lbs. (25 Nm), bolts "B" to 9 ft. lbs. (12 Nm) and bolts "C" to 44 in. lbs.(5 Nm).
42. Connect the back-up light switch connector.
43. Connect wire harness.
44. Install battery carrier. Tighten bolts to 10 ft. Lbs. (13 nm).
45. Install battery tray and battery.
46. Install air cleaner.
47. Install cylinder head cover.
48. Install hood.
49. Add transaxle oil.
50. Install front wheels.
51. Install steering column cover.
52. Connect steering intermediate shaft.
53. Place front wheels facing straight ahead.
54. Center spiral cable.
55. Install exhaust pipe.
56. Install engine undercover.
57. Inspect front wheel alignment.
58. Check the ABS speed sensor signal.

CLUTCH DRIVEN DISC & PRESSURE PLATE

REMOVAL & INSTALLATION

See Figures 27 through 30.

1. Before servicing the vehicle, refer to the precautions section.

➡Do not allow grease or oil to get on any part of the disc, pressure plate, or flywheel surfaces.

2. Remove the manual transaxle assembly
3. Make matchmarks on the clutch cover (pressure plate) and flywheel so that the pressure plate can be returned to its original position during installation.
4. Remove the release fork bearing clips.
5. Remove the release bearing hub, complete with the release bearing.
6. Remove the release fork and support.

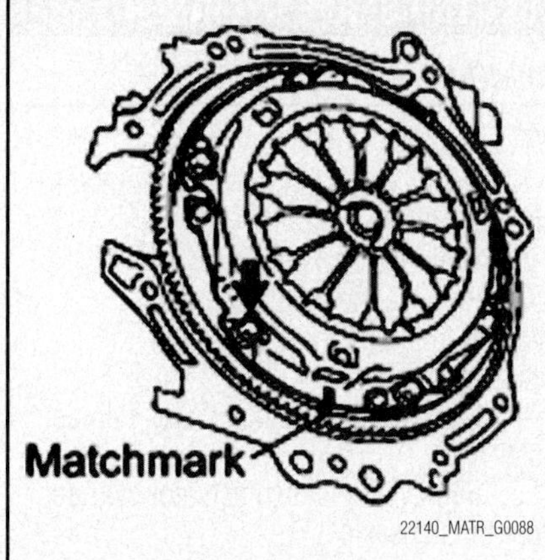

Fig. 27 Matchmarks on the clutch cover (pressure plate) and flywheel

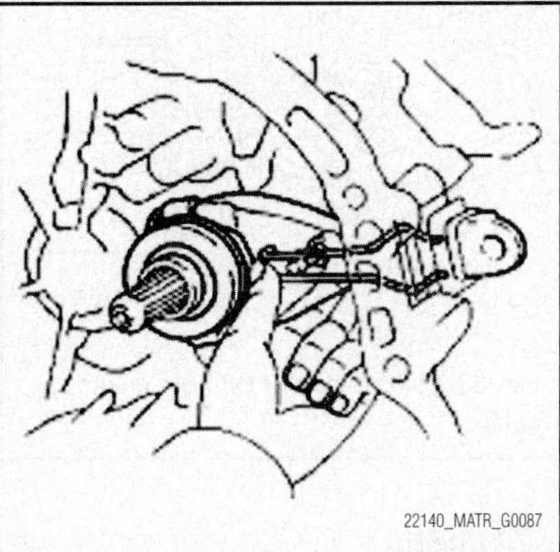

Fig. 28 Remove the release fork bearing clips

✻✻ CAUTION

Slowly unfasten the bolts which attach the pressure plate. Loosen each bolt 1 turn at a time until the spring tension is released. If the bolts are released improperly the clutch assembly could fly apart, causing possible injury.

7. Remove the pressure plate from the clutch cover/spring assembly.
8. Replace or machine parts as necessary.

To install:

9. Pack the groove inside the clutch hub with multipurpose grease.
10. Install the clutch disc and pressure plate. The bolts should be tightened in sequence using 2 or 3 steps, gradually and evenly. Final bolt torque is 14 ft. lbs. (19 Nm).
11. Apply a thin coating of multipurpose grease to the release bearing hub and release fork contact points and lubricate the pivot points of the release fork.

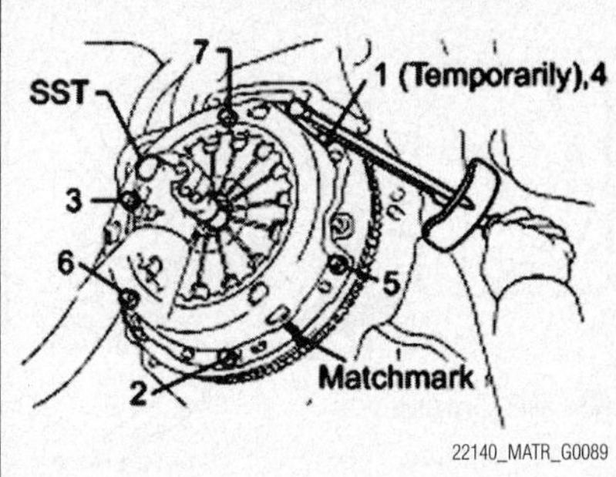

Fig. 29 Bolts should be tightened in sequence using 2 or 3 steps, gradually and evenly

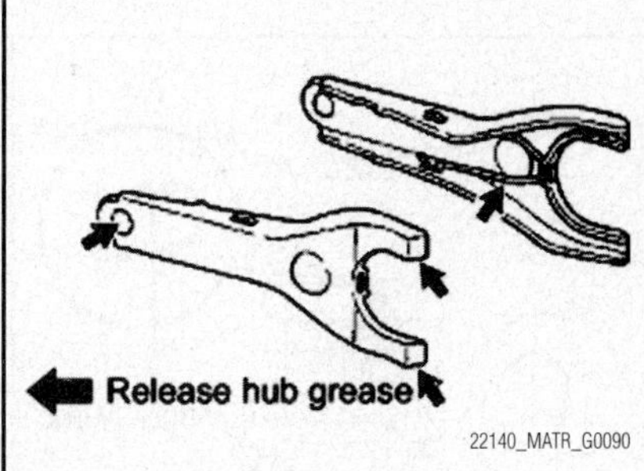

Fig. 30 Grease the contact points and lubricate the pivot points of the release fork

12. Install the release bearing, fork and boot. After the installation, move the folk forward and backward to check that the release bearing slides smoothly.
13. Install the transaxle assembly

ADJUSTMENTS

Hydraulic clutch actuating systems used in Toyota vehicles do not require adjustment.

CLUTCH MASTER CYLINDER

REMOVAL & INSTALLATION

See Figures 31 and 32.

1. Before servicing the vehicle, refer to the precautions section.
2. Drain the brake fluid.
3. Remove the brake master cylinder and brake booster.
4. Disconnect the clutch reservoir tube from the clutch master cylinder assembly.
5. Disconnect the clutch master cylinder flexible hose and the clutch reservoir tube.
6. Remove the clutch pedal spring.
7. Remove the clutch master cylinder push rod clevis pin and clip.

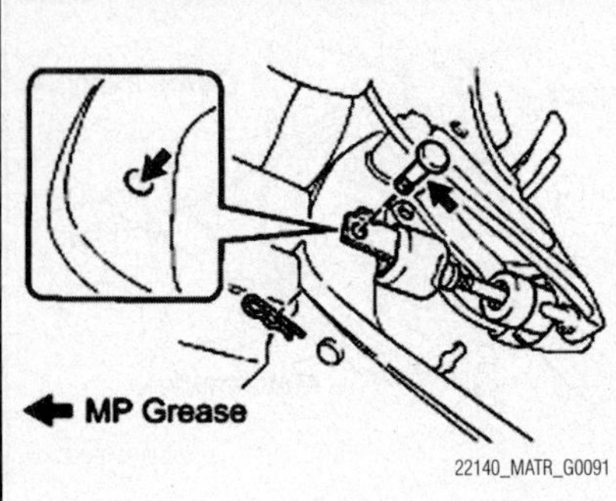

Fig. 31 Apply multi-purpose grease to the contacting surfaces of the pin and clevis bushing

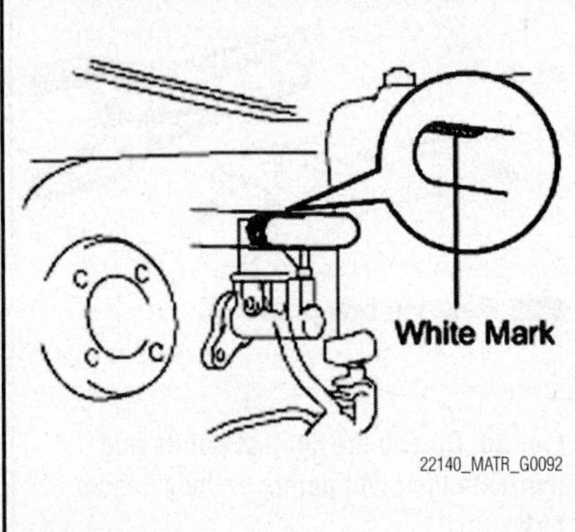

Fig. 32 Install the clutch reservoir tube with the white marking facing upward

8. Remove the clutch master cylinder retaining nuts.
9. Remove the clutch master cylinder from the vehicle.

To install:

10. Install the clutch master cylinder in the vehicle.
11. Install the clutch master cylinder retaining nuts and tighten to 9 ft. lbs. (12 Nm).
12. Apply multi-purpose grease to the contacting surfaces of the pin and clevis bush.
13. Install the clevis pin and clip. The pin is easier to install from the right side.
14. Install the clutch pedal spring.
15. Install the hydraulic pipe to the clutch master cylinder. Tighten to 11 ft. lbs. (15 Nm).
16. Install the clutch reservoir tube with the white marking facing upward.
17. Install the brake booster assembly in the vehicle.
18. Install the brake master cylinder.
19. Connect the reservoir tube to the clutch master cylinder.
20. Bleed the clutch.

CLUTCH SLAVE CYLINDER

REMOVAL & INSTALLATION

See Figure 33.

1. Before servicing the vehicle, refer to the precautions section.
2. Drain the brake fluid.
3. Remove the clutch actuator cylinder fluid line from the clutch actuator cylinder.
4. Remove the clutch actuator cylinder bolts.
5. Remove the clutch actuator cylinder from the transaxle.

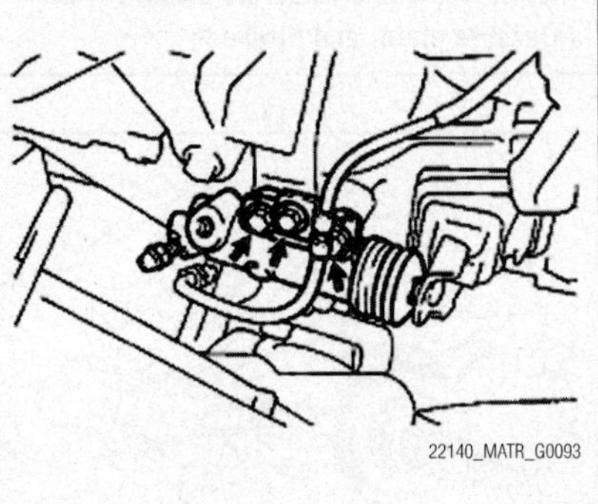

Fig. 33 Clutch actuator cylinder mounting bolts

To install:

6. Install the clutch actuator cylinder to the transaxle.
7. Install the clutch actuator cylinder bolts and tighten to 9 ft. lbs. (12 Nm).
8. Install the clutch actuator cylinder fluid line to the clutch actuator cylinder.
9. Bleed the hydraulic clutch system.

CLUTCH HYDRAULIC SYSTEM BLEEDING

➡If any maintenance on the clutch system was performed or the system is suspected of containing air, bleed the system. Use care; brake fluid will remove the paint from any surface. If the brake fluid spills onto any painted surface, wash it off immediately with soap and water.

1. Before servicing the vehicle, refer to the precautions section.
2. Fill the clutch reservoir with brake fluid. Check the reservoir level frequently and add fluid as needed.
3. Connect one end of a vinyl tube to the bleeder plug on the slave cylinder and submerge the other end into a clear container half-filled with brake fluid.
4. Slowly pump the clutch pedal several times.
5. Have an assistant hold the clutch pedal down and loosen the bleeder plug until fluid and/or air starts to run out of the bleeder plug. Close the bleeder plug while the pedal is held to the floor.

➡Do not allow the pedal to rise back-up while the bleeder is still open. If this happens, it will allow air to re-enter the slave cylinder and cause the clutch system not to work properly.

6. Repeat Steps 2 and 3 until all the air bubbles are removed from the system.
7. Tighten the bleeder plug when all the air is gone.
8. Refill the master cylinder to the proper level as required.
9. Check the system for leaks.

TRANSFER CASE ASSEMBLY

REMOVAL & INSTALLATION

See Figure 34.

1. Before servicing the vehicle, refer to the precautions section.
2. Drain the transfer case fluid.
3. Remove the engine and transaxle assembly.
4. Remove the vane pump.
5. Remove the steering gear.
6. Remove the crossmember.
7. Remove the manifold stays
8. Remove the oxygen sensor.
9. Remove the exhaust manifold heat shield.
10. Remove the exhaust manifold.
11. Remove the starter.
12. Remove the right side halfshaft.
13. Remove the transverse engine mounting bracket.

Fig. 34 Transfer case stiffener plate bolt locations

14. Remove the center and right side transfer case stiffener plates.

✲✲ WARNING

When removing the transfer case, DO NOT touch the oil seal.

15. Remove the transfer case bolts, and transfer case assembly, using a mallet to dislodge it from the transaxle.

To install:

16. Install the transfer case assembly and transfer case bolts. Tighten to 51 ft. lbs. (69 Nm).

✲✲ WARNING

When installing the transfer case, DO NOT touch the oil seal.

17. Install the center and right side transfer case stiffener plates. Tighten to 25 ft. lbs. (34 Nm).
18. Install the transverse engine mounting bracket. Tighten to 47 ft. lbs. (64 Nm).
19. Install the right side halfshaft.
20. Install the starter.
21. Install the exhaust manifold.
22. Install the exhaust manifold heat shield.
23. Install the oxygen sensor.
24. Install the manifold stays
25. Install the crossmember.
26. Install the steering gear.
27. Install the vane pump.
28. Install the engine and transaxle assembly.
29. Add fluid to the transfer case, and check for leaks.

FRONT HALFSHAFT

REMOVAL & INSTALLATION

See Figure 35.

✲✲ WARNING

The hub bearing could be damaged if subjected to the full weight of the vehicle, such as if the vehicle is moved without the halfshafts. If it is absolutely necessary to place the full vehicle weight on the hub bearing, first support the bearing with special tool No. 09608–16041.

1. Before servicing the vehicle, refer to the precautions section.
2. Drain the transaxle fluid and as applicable the transfer case oil.
3. Remove the front wheels.
4. Remove the engine undercover.

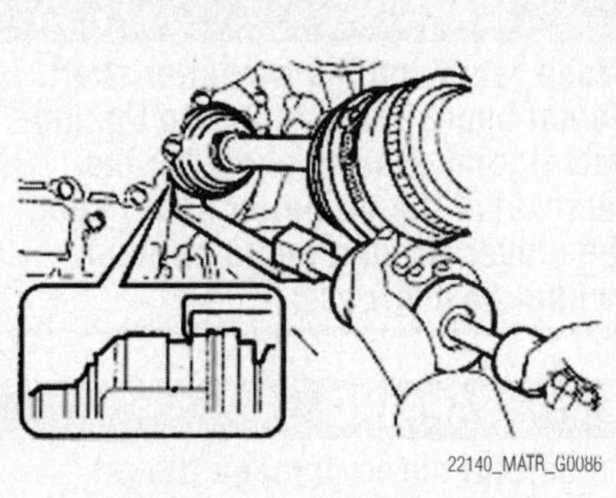

Fig. 35 Separate the halfshaft from the transaxle using a puller

5. Remove the cotter pin, locknut cap, and the hub nut.
6. Remove the speed sensors.
7. Remove the tie rod ball joint from the steering knuckle.
8. Remove the stabilizer bar link from the lower suspension arm.
9. Remove the bolt and nuts, separate the front suspension arm from the lower ball joint.
10. Using a plastic hammer, separate the halfshaft from the steering knuckle.

➡Be careful not to damage the inner oil seal or the ABS sensor rotor on the halfshaft.

11. To remove the left side halfshaft, separate the halfshaft from the transaxle using a puller.
12. To remove the right side halfshaft perform the following steps:
    a. Remove the bolts of the center bearing bracket.
    b. Pull the halfshaft out together with the center bearing case and the center halfshaft.

➡Do not damage the oil seal lip.

To install:

13. Install the snap ring opening side facing downward, on the oiled inboard joint tulip
14. Install the left side halfshaft into the transaxle.
15. Install the right side halfshaft, with the bearing case and center shaft, into the transaxle.
16. After installing either halfshaft, check that there is 0.08–0.12 in. (2–3mm) of axial play. Check that the halfshaft is making contact with the pinion shaft and that the halfshaft cannot be pulled out.
17. Install the halfshaft into the knuckle
18. Connect the lower suspension arm to the lower ball joint. Torque the bolt and nuts to 66 ft. lbs. (89 Nm).
19. Install the tie rod end to the steering knuckle. Tighten the nut to 36 ft. lbs. (49 Nm).
20. Install the stabilizer bar link to the lower suspension arm. Torque the nuts to 55 ft. lbs. (74 Nm).
21. Install the front wheels.
22. Install the hub nut and washer and tighten to 159 ft. lbs. (216 Nm)
23. Install the locknut cap and a new cotter pin.
24. Install the speed sensors.
25. Install the engine undercover.
26. Fill the transaxle fluid to the proper level.
27. Start the vehicle, check for leaks and repair if necessary.

REAR DIFFERENTIAL CARRIER

REMOVAL & INSTALLATION

1. Before servicing the vehicle, refer to the precautions section.
2. Raise and safely support the vehicle securely on jackstands.
3. Remove rear wheels.
4. Remove the rear differential carrier cover plug (filler plug) and gasket.
5. Remove the rear differential drain plug and gasket, drain the oil.
6. Install the rear differential drain plug and a new gasket.
7. Remove driveshaft.
8. Remove exhaust pipe assembly.
9. Remove rear stabilizer link assembly.
10. Remove rear axle shaft nut.
11. Remove rear brake drum, shoes, parking brake and brake tube.
12. Separate rear speed sensor.
13. Separate upper control arm.
14. Separate upper control arm.
15. Remove rear axle assembly.
16. Remove rear suspension member.
17. Remove rear halfshaft assembly.
18. Remove rear differential carrier assembly.

To install:

19. Install rear differential carrier assembly. Tighten bolts to 87 ft. lbs. (118 Nm).
20. Install the bolts to the rear suspension member rear side and tighten to 101 ft. lbs. (137 Nm).
21. Install rear suspension member.
22. Install halfshaft assembly.
23. Temporarily tighten upper control arm.
24. Install rear axle.
25. Temporarily tighten the upper control arm.
26. Install the brake tube, parking brake, shoes and brake drum.

➡**Apply high temperature grease to the contact points .**

27. Adjust rear drum brake shoe clearance .
28. Install rear speed sensor.
29. Install rear stabilizer link.
30. Install driveshaft.
31. Install exhaust pipe.
32. Bleed the brakes.
33. Inspect and adjust differential oil level .
34. Install rear wheel.
35. Lower the vehicle.
36. Stabilize the suspension .
37. Fully tighten upper control arm assembly.
38. Fully tighten upper control arm.
39. Inspect fluid level in reservoir.
40. Inspect and adjust rear wheel alignment.

REAR DRIVESHAFT (PROPELLER SHAFT)

REMOVAL & INSTALLATION

See Figures 36 through 38.

1. Before servicing the vehicle, refer to the precautions section.
2. Depress the brake pedal and hold it. Loosen the cross groove joint set bolts 1/2 turn.
3. Put a piece of cloth inside the cross groove joint cover so that the bolt does not touch the inside of the cross groove joint cover.
4. Place matchmarks on the propeller shaft and differential carrier assembly.
5. Separate the propeller shaft.
6. Separate the propeller with center bearing shaft assembly.
7. Pull out the propeller with center bearing shaft assembly from the transfer.

WARNING

When removing the propeller shaft, do not apply a large force to the universal joint. During and after the removal of the propeller shaft, keep the universal joint angle straight (within 15 degrees).

WARNING

Be careful not to damage the oil seal.

8. Plug the end of the transfer case to prevent oil leakage.

To install:

9. Remove plug from the transfer case.
10. Insert the propeller with center bearing shaft assembly.
11. Align matchmarks on the propeller shaft and differential carrier assembly, install the bolts and washers temporarily.
12. Connect the center support bearing assembly. Use the removed washers during installation. Tighten the bolts to 54 ft. lbs. (74 Nm).
13. Remove the cloth from the universal joint.
14. Depress the brake pedal and hold it. Tighten the driveshaft bolts to 20 ft. lbs. (27 Nm).
15. With the vehicle in an unladen condition, adjust the dimension between the rear side if cover and the shaft, as shown in the illustration.
16. With the same condition, adjust the dimension between the rear side of center bearing housing and the rear side of cushion, as shown in the illustration.
17. Check that the center line of the bracket is at right angles at the shaft axial direction.
18. Tighten the bolts to 27 ft. lbs. (37 Nm).

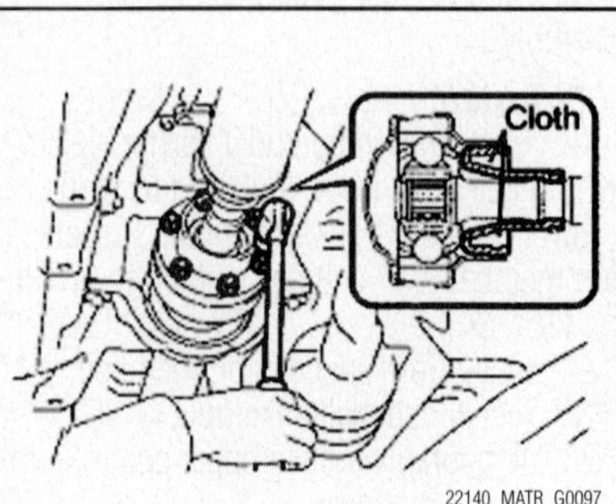

Fig. 36 Loosen the cross groove joint set bolts, then place a cloth inside the joint cover

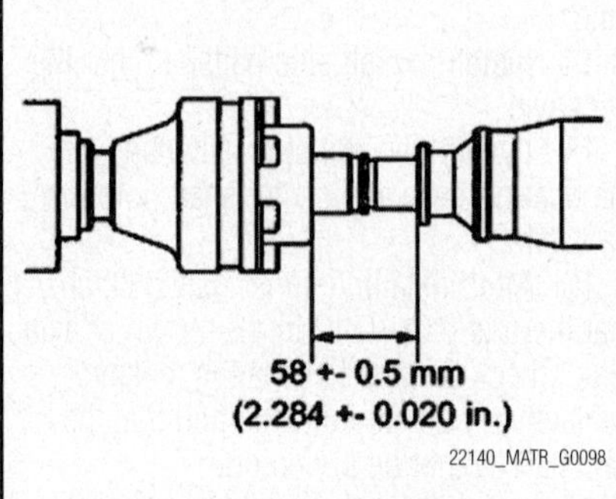

Fig. 37 Adjust the dimension between the rear side if cover and the shaft

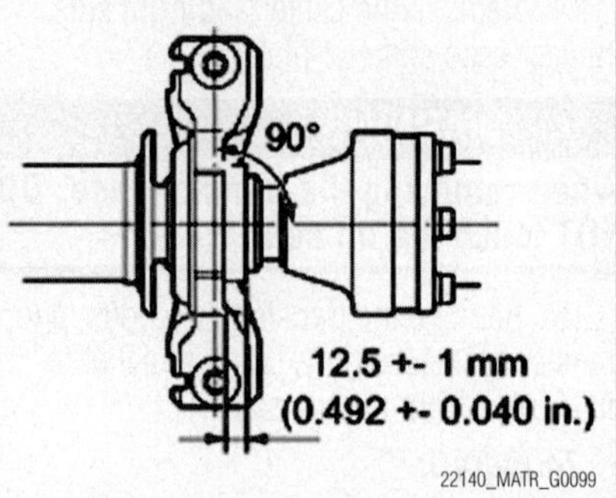

Fig. 38 Adjust the dimension between the rear side of center bearing housing and the rear side of cushion

REAR HALFSHAFT

REMOVAL & INSTALLATION

See Figures 39 and 40.

1. Raise and safely support the vehicle securely on jackstands.
2. Remove the rear wheel
3. Separate the rear stabilizer link assembly.
4. Remove rear axle shaft nut.
5. Remove rear brake drum assembly.
6. Separate the rear speed sensor.
7. Remove the parking brake shoe strut set.
8. Remove the brake shoe assembly.
9. Separate the parking brake cable assembly No: 3.
10. Separate the rear brake tube no.4.
11. Separate upper control arm assembly no.1.
12. Separate the upper control arm assembly.
13. Using a hammer and a hub nut wrench, remove the rear drive shaft assembly on the left side.

➡**Be careful not to damage the dust cover, boot and oil seal.**

14. Using a brass bar and a hammer, remove the rear drive shaft assembly on the right side.

To install:

15. Align the shaft splines and install the rear drive shaft assembly using a brass bar and hammer.

➡**Install the snap ring with opening side facing downward.**

16. Temporarily tighten the upper control arm assembly.
17. Install the rear axle assembly.
18. Temporarily tighten upper control arm assembly no.1.

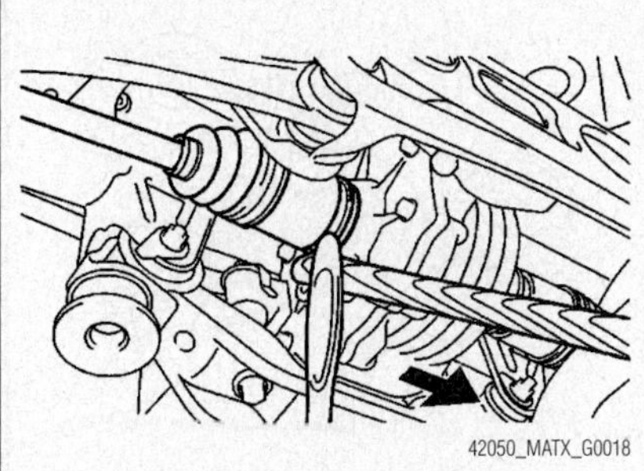

Fig. 39 Using a hammer and a hub nut wrench, remove the rear halfshaft assembly on the left side

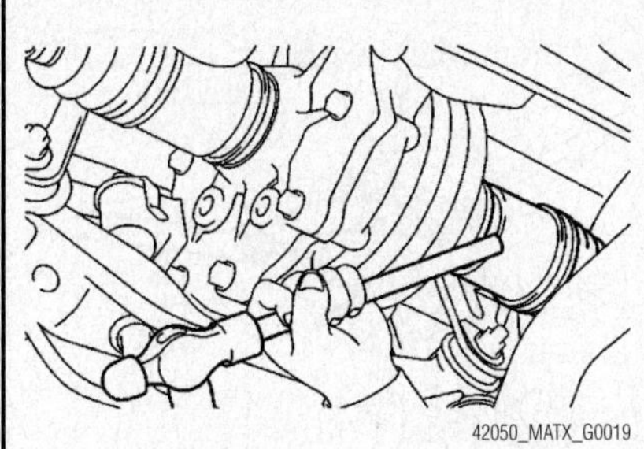

Fig. 40 Using a brass bar and a hammer, remove the rear halfshaft assembly on the right side

19. Install parking brake cable and rear brake assembly.
20. Install the rear speed sensor.
21. Install the rear axle shaft nut. Tighten the rear axle shaft nut to 159 ft. lbs. (216 Nm).
22. Install the rear stabilizer link assembly.
23. Bleed the brake line.
24. Install the rear wheel
25. Fully tighten the upper control arm assembly.
26. Fully tighten the upper control arm assembly no.1.
27. Inspect and adjust the rear wheel alignment.

ENGINE COOLING

THERMOSTAT

REMOVAL & INSTALLATION

1ZZ-FE Engine

See Figure 41.

1. Before servicing the vehicle, refer to the precautions section.
2. Remove the right hand engine undercover.
3. Drain the cooling system.
4. Remove the alternator.
5. Remove the nuts securing the water inlet.
6. Remove the thermostat and gasket.
7. Clean the gasket mating surfaces.

To install:

➡Position the air bleed valve facing upward.

8. Install the thermostat. The jiggle valve may be 10 degrees on either side of center.

Fig. 41 The thermostat jiggle valve may be 10 degrees on either side of center

9. Install a new thermostat gasket.
10. Install the thermostat housing to the cylinder block.
11. Use the nuts in order to secure the water inlet.
    - On low output engine, tighten the thermostat cap nuts to 8 ft. lbs. (11 Nm).
    - On high output engine, tighten the thermostat cap nuts to 7 ft. lbs. (10 Nm).
12. Install the alternator.
13. Fill the radiator using approved coolant.
14. Start the engine.
15. Run the engine until the coolant is at operating temperature.
16. Inspect for leaks in the cooling system.
17. Turn off the engine.
18. Install the radiator cap.

2ZZ-GE Engine

1. Before servicing the vehicle, refer to the precautions section.
2. Drain engine coolant.
3. Remove engine undercover.
4. Remove cylinder head cover.
5. Disconnect negative battery cable.
6. Remove wire harness protector.
7. Separate suction hose.
8. Remove air switching valve assembly.
9. Remove fan and alternator belt.
10. Remove alternator bracket and assembly.
11. Remove water inlet.
12. Remove thermostat.

To install:

13. Install thermostat using a new gasket. Install the thermostat with the jiggle valve upward as shown in the illustration.
14. Install water inlet. Tighten to 7 ft. lbs. (10 Nm)
15. Install alternator assembly.
16. Install fan and alternator belt.
17. Install air switching valve assembly.
18. Install suction hose assembly. Tighten to 87 inch lbs. (10 Nm).
19. Install wire harness protector.
20. Connect negative battery cable.
21. Add engine coolant and check for leaks.
22. Install cylinder head cover and tighten to 62 inch lbs. (7 Nm).
23. Install engine undercover.

WATER PUMP

REMOVAL & INSTALLATION

1ZZ-FE Engine

See Figure 42.

1. Before servicing the vehicle, refer to the precautions section.
2. Drain the cooling system.
3. Disconnect the negative battery cable.
4. Remove the right-hand engine undercover.
5. Remove the drive belt.
6. Remove the alternator.
7. Remove the water pump.

To install:

8. Install the water pump. Torque bolts marked "A" (short) to 80 inch lbs. (9 Nm) and bolts marked "B" (long) to 96 inch lbs. (11 Nm).
9. Install the alternator.
10. Install the drive belt.
11. Install the right engine undercover.
12. Connect the negative battery cable.
13. Fill the cooling system to the proper level.
14. Start the vehicle and inspect for leaks.

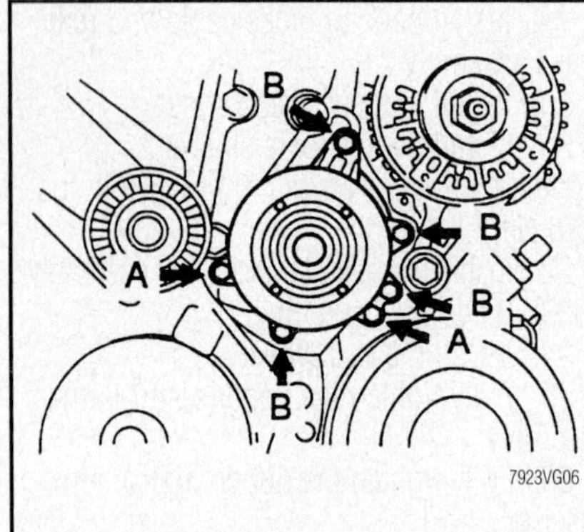

Fig. 42 Water pump bolt identification—1ZZ-FE engine

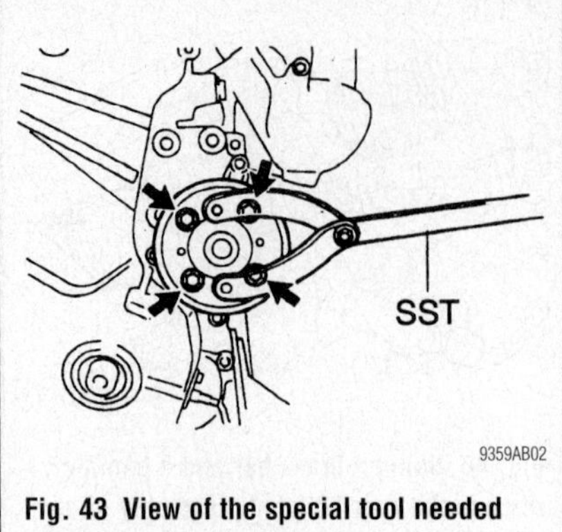

Fig. 43 View of the special tool needed to remove and install the water pump pulley—2ZZ-GE engine

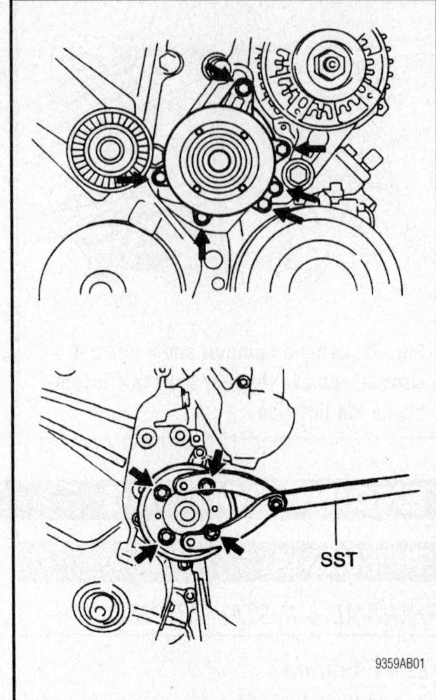

Fig. 44 Water pump mounting and bolt locations—2ZZ-GE engine

2ZZ-GE Engine

See Figures 43 and 44.

1. Before servicing the vehicle, refer to the precautions section.
2. Drain the cooling system.
3. Disconnect the negative battery cable.
4. Remove the right-hand engine undercover.
5. Remove the drive belt.
6. Remove the alternator.
7. Remove the water pump pulley, using special tool 09960-10010
8. Remove the water pump and O-ring

To install:

9. Install or connect the following:
10. Install the water pump with new O-ring. Torque the bolts to 80 inch lbs. (9 Nm).
11. Install the water pump pulley, using special tool 09960-100Torque the bolts to 11 ft. lbs. (15 Nm).
12. Install the alternator.
13. Install the drive belt.
14. Install the right engine undercover.
15. Connect the negative battery cable.
16. Fill the cooling system to the proper level.
17. Start the vehicle and inspect for leaks.

ENGINE ELECTRICAL

CHARGING SYSTEM

ALTERNATOR

REMOVAL & INSTALLATION

1ZZ-FE Engine

1. Before servicing the vehicle, refer to the precautions section.
2. Disconnect negative battery cable.
3. Remove engine undercover .
4. Remove fan and alternator belt.
5. Disconnect the alternator wiring harness.
6. Remove alternator assembly.

To install:

7. Install alternator assembly. Tighten the upper bolt to 18 ft. lbs. (25 Nm) and the lower bolt to 40 ft. lbs. (54 Nm).
8. Install the alternator wire with the nut and tighten to 7 ft. lbs. (10 Nm).
9. Connect the alternator wiring harness.
10. Install fan and alternator belt.
11. Install engine undercover .
12. Connect negative battery cable.

2ZZ-GE Engine

1. Before servicing the vehicle, refer to the precautions section.
2. Disconnect negative battery cable.
3. Remove cylinder head cover.
4. Remove air switching valve assembly.
5. Remove engine undercover .
6. Remove fan and alternator belt.
7. Disconnect the alternator wiring harness.
8. Remove air pump outlet pipe.
9. Remove alternator assembly.

To install:

10. Install alternator assembly. Tighten the upper bolt to 18 ft. lbs. (25 Nm) and the lower bolt to 43 ft. lbs. (58 Nm).
11. Install the alternator wire harness clam bracket and tighten to 48 inch lbs. (5 Nm).
12. Connect the alternator wiring harness.
13. Install the alternator wire with the nut and tighten to 7 ft. lbs. (10 Nm).
14. Install the alternator bracket and tighten to 21 ft. lbs. (29 Nm).
15. Install air pump outlet pipe.
16. Connect the alternator wiring harness.
17. Install fan and alternator belt.
18. Install engine undercover .
19. Install air switching valve assembly.
20. Install cylinder head cover.
21. Connect negative battery cable.

ADJUSTMENT

The ignition timing is controlled by the Powertrain Control Module (PCM). No adjustment is possible.

IGNITION COIL

REMOVAL & INSTALLATION

1ZZ-FE Engine

1. Remove the 2 fasteners.
2. Remove the 2 plastic retainers.
3. Remove the engine cover from the engine.
4. Disconnect the 4 electrical connectors from the ignition coils (3).
5. Remove the fasteners (4) from the ignition coils.
6. Remove the fasteners from the electrical harness .
7. Remove the electrical harness package.
8. Remove the ignition coils (3) from the cylinder head.

To install:

9. Install the ignition coils to the engine.

➡Use the correct fastener in the correct location. Replacement fasteners must be the correct part number for that application. Fasteners requiring replacement or fasteners requiring the use of thread locking compound or sealant are identified in the service procedure. Do not use paints, lubricants, or corrosion inhibitors on fasteners or fastener joint surfaces unless specified. These coatings affect fastener torque and joint clamping force and may damage the fastener. Use the correct tightening sequence and specifications when installing fasteners in order to avoid damage to parts and systems.

10. Secure the ignition coils with the fasteners. Tighten the ignition coil fasteners to
11. Install the electrical harness and secure with the 2 fasteners. Tighten the fasteners to 78 inch lbs (8.8 Nm).
12. Connect the electrical connectors to the ignition coils.
13. Install the engine cover. Secure the engine cover with the 2 fasteners. Tighten the 2 fasteners to 80 inch lbs (9 Nm).
14. Install the 2 plastic retainers.

2ZZ-GE Engine

1. Remove the engine cover from the engine.
2. Disconnect the 4 electrical connectors from the ignition coils.
3. Remove the fasteners from the ignition coils.
4. Remove the fasteners from the electrical harness.
5. Remove the electrical harness package.
6. Remove the ignition coils from the cylinder head.

To install:

7. Install the ignition coils to the engine.

➡Use the correct fastener in the correct location. Replacement fasteners must be the correct part number for that application. Fasteners requiring replacement or fasteners requiring the use of thread locking compound or sealant are identified in the service procedure. Do not use paints, lubricants, or corrosion inhibitors on fasteners or fastener joint surfaces unless specified. These coatings affect fastener torque and joint clamping force and may damage the fastener. Use the correct tightening sequence and specifications when installing fasteners in order to avoid damage to parts and systems.

8. Secure the ignition coils with the fasteners. Tighten the ignition coil fasteners to 80 inch lbs (9 Nm).
9. Install the electrical harness and secure with the 2 fasteners. Tighten the fasteners to78 inch lbs (8.8 Nm).
10. Connect the electrical connectors to the ignition coils.
11. Install the engine cover to the engine.
12. Secure the engine cover with the four fasteners. Tighten the fasteners to 62 inch lbs (7 Nm).

SPARK PLUGS

REMOVAL & INSTALLATION

1. Remove the engine cover from the engine.
2. Disconnect the electrical connector from each ignition coil.
3. Remove the bolt securing each ignition coil from the cylinder head cover.
4. Remove each ignition coil from the cylinder head cover.

➡This engine is equipped with an aluminum cylinder head. Allow the engine to cool before removing spark plugs. Removing the spark plugs from an engine at operating temperature may damage the spark plug threads in the cylinder head. Also be sure to clean any dirt or debris from around spark plug holes prior to removing spark plugs.

5. Remove the spark plugs from the cylinder head.
6. Inspect the spark plugs for electrode wear, carbon deposits and insulator damage.

To install:

⁂ WARNING

Do not touch the tip of the spark plug. Do not damage the iridium surface of the electrode when gapping the plug.

➡Do not adjust the gap on used spark plugs. Replace the spark plug if the gap is greater than specification.

7. Set the spark plug gap to specification.
8. Apply a small quantity of anti-seize compound to the plug threads, and screw the plugs into the cylinder head finger-tight.
9. Tighten the spark plugs to 18 ft. lbs. (25 Nm) for 1ZZ-FE engines or 13 ft. lbs. (18 Nm) for 2ZZ-GE engines.
10. Install each ignition coil to the cylinder head cover.
11. Secure the ignition coils using the bolts. Tighten the bolts to 80 inch lbs (9 Nm).
12. Connect each ignition coil electrical connector.
13. Install the engine cover on the engine.

ENGINE ELECTRICAL

STARTER

REMOVAL & INSTALLATION

See Figure 45.

1. Before servicing the vehicle, refer to the precautions section.
2. Disconnect negative battery cable.
3. Remove right side engine undercover.
4. Disconnect starter wiring.
5. Remove starter.

To install:

6. Installation is the reverse of removal.
7. Torque the bolts to 27 ft. lbs. (37 Nm) and the nut to 7 ft. lbs. (10 Nm).

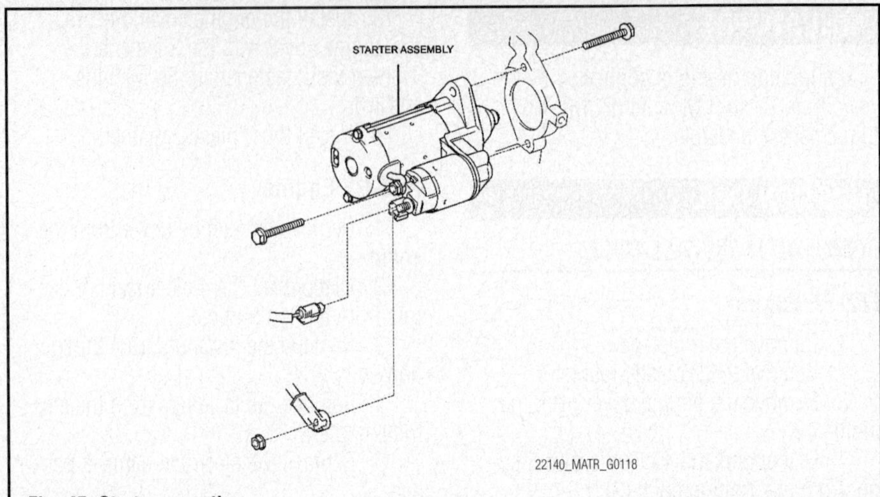

Fig. 45 Starter mounting

ENGINE MECHANICAL

ACCESSORY DRIVE BELTS

ACCESSORY BELT ROUTING

See Figures 46 and 47.

INSPECTION

Inspect the drive belt for signs of glazing or cracking. A glazed belt will be perfectly smooth from slippage, while a good belt will have a slight texture of fabric visible. Cracks will usually start at the inner edge of the belt and run outward. All worn or damaged drive belts should be replaced immediately.

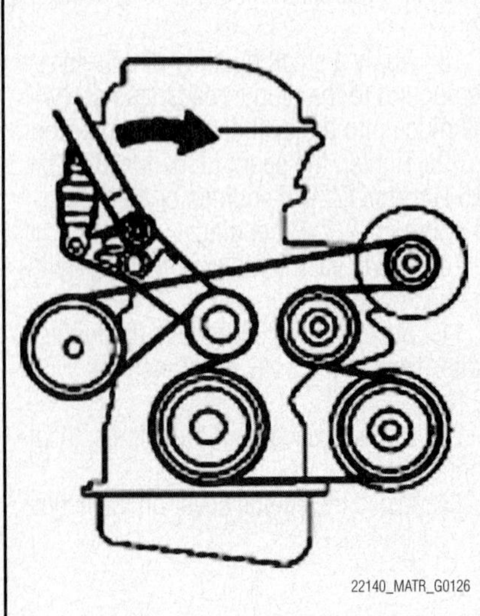

Fig. 46 Accessory belt routing—1ZZ-FE engine

Fig. 47 Accessory belt routing—2ZZ-GE engine

ADJUSTMENT

The belt tension is maintained by an automatic tensioner. No adjustment is possible.

REMOVAL & INSTALLATION

1ZZ-FE Engine

See Figures 48 and 49.

1. Before servicing the vehicle, refer to the precautions section.
2. Disconnect the negative battery cable.
3. Use a wrench and rotate the belt tensioner clockwise.
4. With pressure applied to the wrench, and tension relieved from the drive belt, remove the accessory drive belt.

To install:

5. Raise and safely support the vehicle.

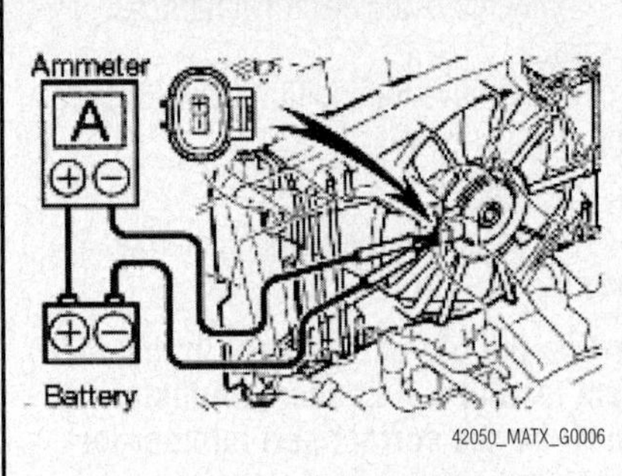

Fig. 48 Use a wrench and rotate the belt tensioner clockwise—1ZZ-FE engine

Fig. 49 Use a wrench and rotate the belt tensioner clockwise.

6. Remove the right side lower engine splash shield.
7. Properly route and install the accessory drive belt onto the drive pulleys.
8. Use a wrench and rotate the belt tensioner clockwise.

9. With pressure applied to the wrench, install the accessory drive belt.

10. With the accessory drive belt installed properly, release the belt tensioner.

11. Install the right side lower engine splash shield.

12. Connect the negative battery cable. Tighten the battery cable bolt to 11 ft. lbs. (15 Nm).

2ZZ-GE Engine

See Figure 50.

1. Before servicing the vehicle, refer to the precautions section.
2. Disconnect the negative battery cable.
3. Remove the engine cover.
4. Remove the cylinder head cover.
5. Unfasten the clips and remove the wire harness connector, then disconnect the pressure sensor connector.
6. Remove the nuts, then reposition the A/C hose aside.
7. Remove the air switching valve.
8. Use a wrench and rotate the belt tensioner clockwise.
9. With pressure applied to the wrench, and tension relieved from the drive belt, remove the accessory drive belt.

To install:

10. Properly route and install the accessory drive belt onto the drive pulleys.
11. Use a wrench and rotate the belt tensioner clockwise.
12. With pressure applied to the wrench, install the accessory drive belt.
13. With the accessory drive belt installed properly, release the belt tensioner.
14. Install the remaining components in the reverse order of removal.

Fig. 50 Use a wrench and rotate the belt tensioner clockwise—2ZZ-GE engine

CAMSHAFT AND VALVE LIFTERS

REMOVAL & INSTALLATION

1ZZ-FE Engine

See Figures 51 through 69.

1. Before servicing the vehicle, refer to the precautions section.
2. Disconnect battery negative terminal.
3. Remove engine undercover.
4. Remove the upper cylinder head cover.
5. Remove the ignition coil assembly.
6. Remove the lower cylinder head cover.
7. Set the No. 1 cylinder to TDC on the compression stroke.
   a. Turn the crankshaft pulley until its groove and timing mark "0" of the timing chain cover are aligned.
   b. Check that the point marks of the camshaft timing gears are in a straight line on the timing chain cover surface as shown in the illustration.
   c. If not, turn the crankshaft 1 revolution (360°) to align the marks as above.
8. Remove camshaft.
   a. Place matchmarks on the timing chain and camshaft timing gears as illustrated.
   b. Remove the nuts, then remove the chain tensioner.

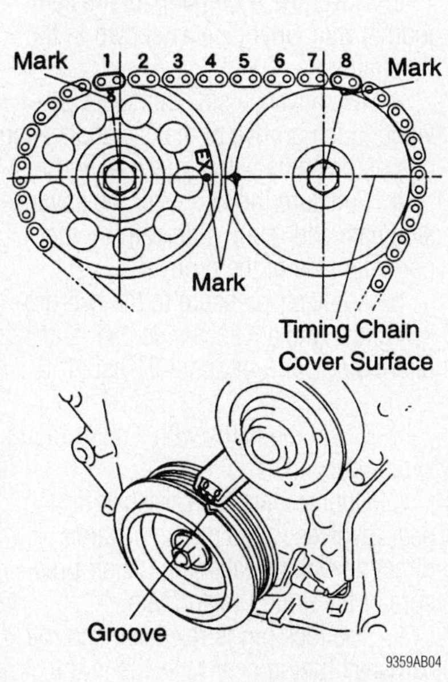

Fig. 51 Proper timing mark alignment for TDC

**** WARNING**

Do not turn the crankshaft without the chain tensioner.

   c. While holding the camshaft with a wrench, loosen the camshaft timing gear set bolt.

**** WARNING**

Be careful not to damage the valve lifter.

   d. Using several steps, uniformly loosen and remove the bolts in the

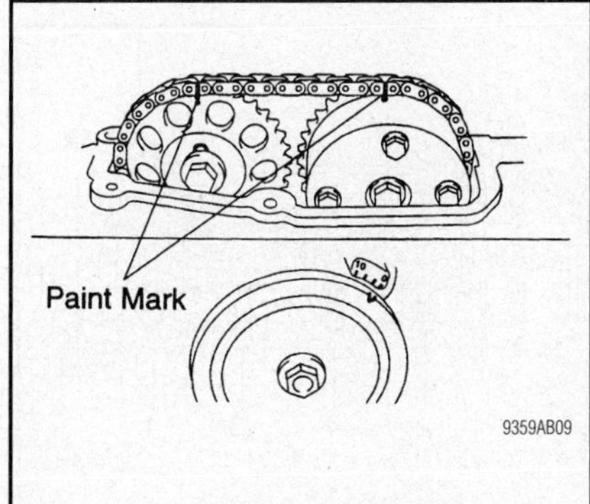

Fig. 52 Matchmark the timing chain and cam sprockets—1ZZ-FE engine

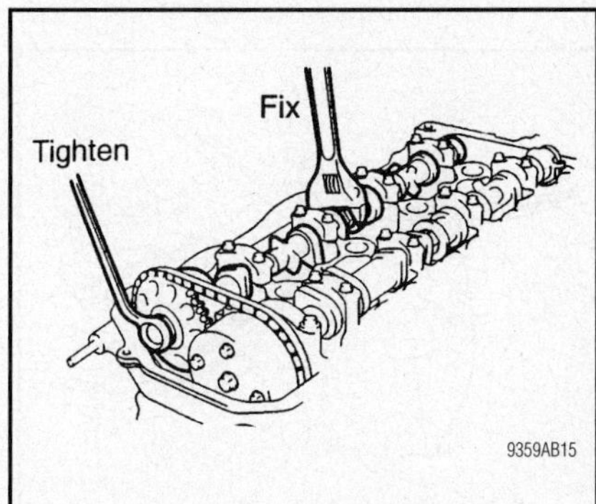

Fig. 53 Hold the camshaft with a wrench while removing the set bolt

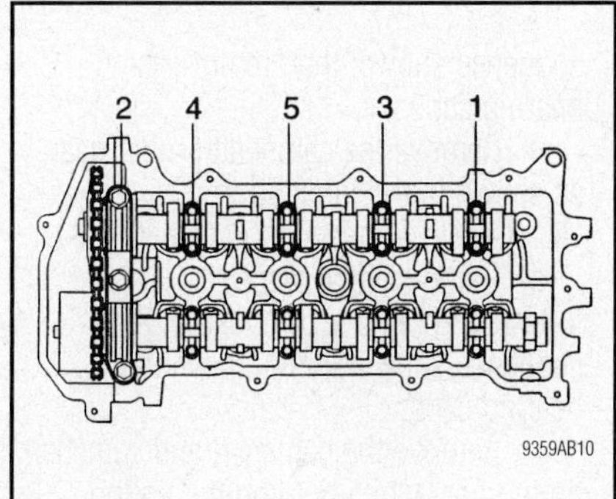

Fig. 54 Camshaft bearing cap bolt removal sequence—1ZZ-FE engine

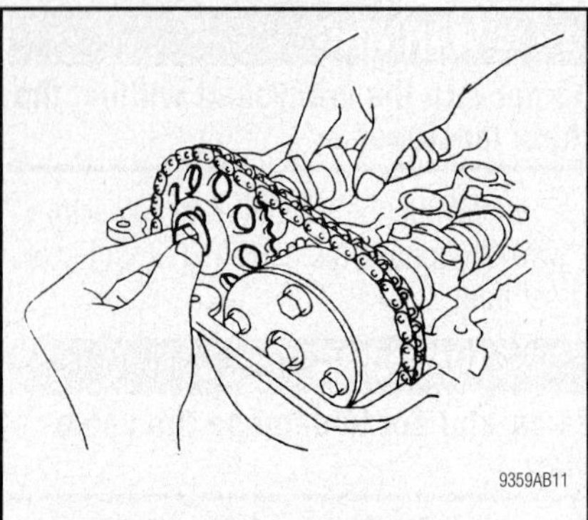

Fig. 55 Carefully remove the cam and timing gear

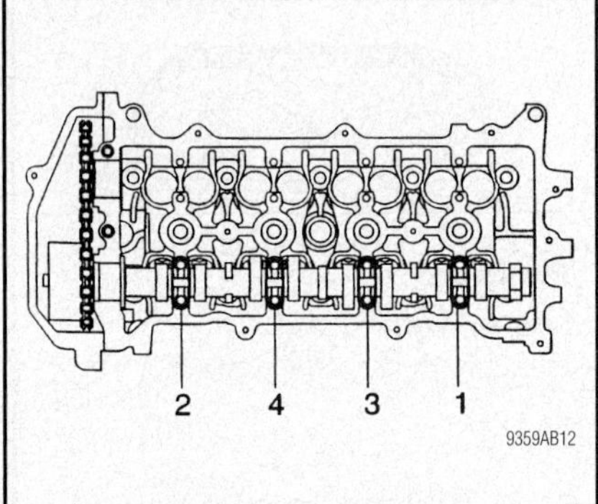

Fig. 56 Camshaft bearing cap bolt removal sequence—1ZZ-FE engine

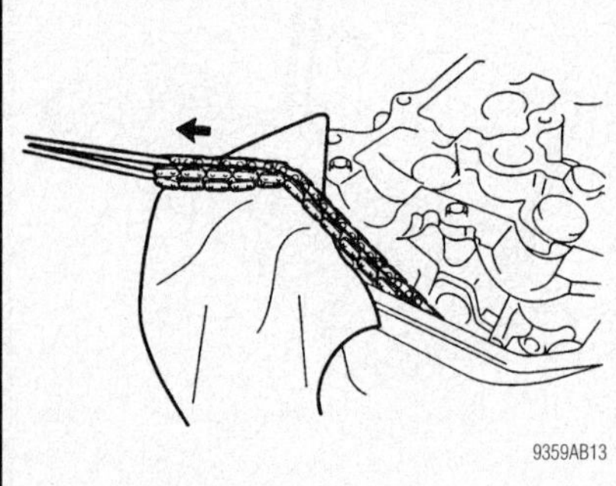

Fig. 57 Secure the timing chain with string to prevent it from slipping down into the timing chain cover

sequence shown, then remove the 5 bearing caps.

e. Remove the camshaft timing gear as shown in the illustration.

9. Remove the second camshaft.

a. Using several steps, uniformly loosen and remove the bolts in the sequence shown, then remove the 4 bearing caps.

b. Remove the camshaft and camshaft timing gear while holding the timing chain by hand.

10. Tie the timing chain with a string as shown in the illustration.

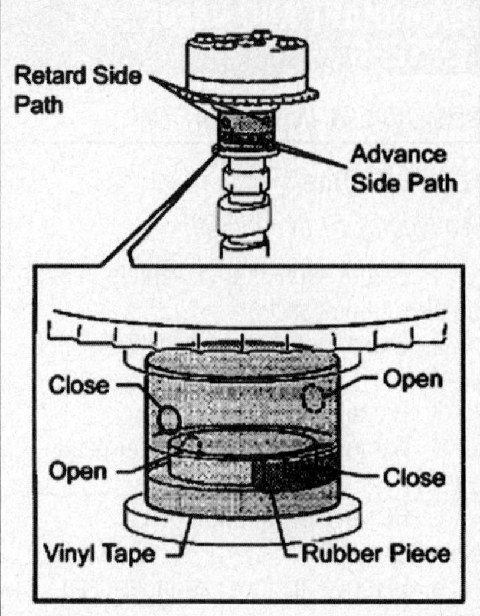

Fig. 58 Inspecting the camshaft timing gear

⁂ WARNING

Be careful not to drop anything inside the timing chain cover.

To install:

11. Inspect camshaft timing gear.

a. Check the lock of the camshaft timing gear.

b. Clamp the camshaft in a vise, then confirm that the camshaft timing gear is locked.

c. Be careful not to damage the camshaft.

d. Release the lock pin.

e. Cover the 4 oil paths of the cam journal with vinyl tape as shown in the illustration.

f. Two advance side paths are provided in the groove of the camshaft. Plug one of the paths with a piece of rubber.

g. Puncture the tape over the advance side path and retard side path on the opposite side of the groove.

h. Apply air pressure to the two broken paths (the advance side path and retard side path) of about 21 psi (150 kPa).

i. Cover the paths with a shop rag to avoid oil splashes.

j. Confirm that the camshaft timing gear revolves in the timing advance direction when weakening the air pressure of the timing retard path.

k. The lock pin is released, and the camshaft timing gear revolves in the advance direction.

l. When the camshaft timing gear comes to the most advanced position,

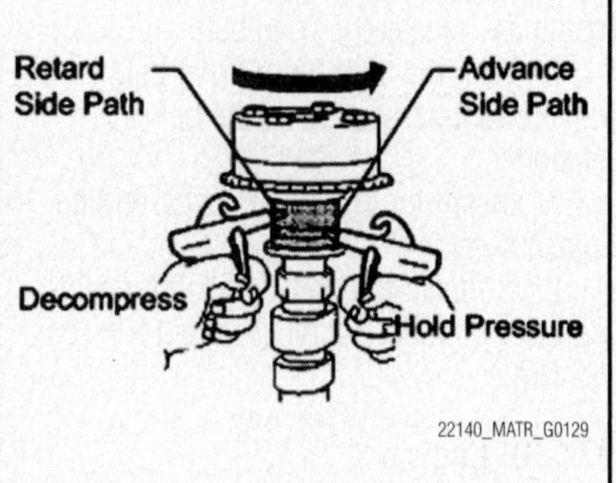

Fig. 59 Apply air pressure to the two broken paths and confirm that the camshaft timing gear revolves in the timing advance direction when weakening the air pressure of the timing retard path

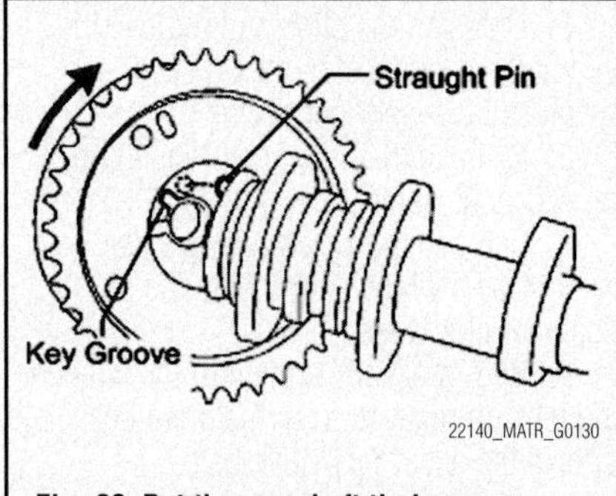

Fig. 60 Put the camshaft timing gear together with the camshaft with the straight pin off the key groove

release the air pressure of the timing retard side path, then release that of the timing advance side path.

m. The camshaft timing gear occasionally shifts to the retard side abruptly if the air compression of the advance side path is released before the retard side path. This often causes the breakage of the lock pin.

n. Check the smooth revolution.

o. Revolve the camshaft timing gear several times within the movable range except for the most retarded position and check that it turns smoothly.

p. Be sure to perform this check by hand, instead of using air pressure.

q. Check that the lock is in the most retarded position.

r. Confirm that the camshaft timing gear is locked in the most retarded position.

12. Install camshaft timing gear assembly.

a. Put the camshaft timing gear together with the camshaft with the straight pin off the key groove.

b. Turn the camshaft timing gear in the left as shown in the illustration while

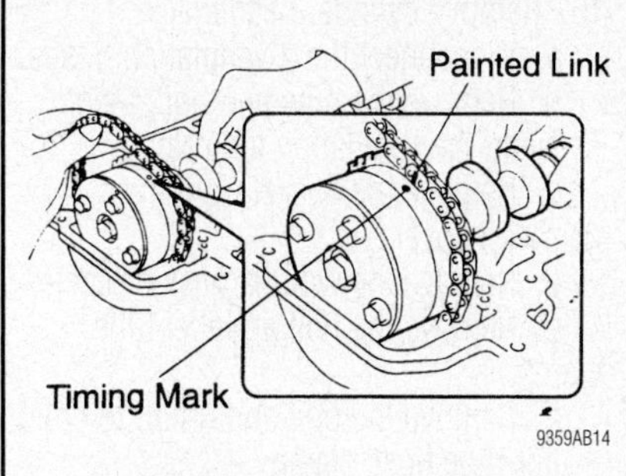

Fig. 61 Make sure the alignment marks on the timing chain and camshaft gear match up—1ZZ-FE engine

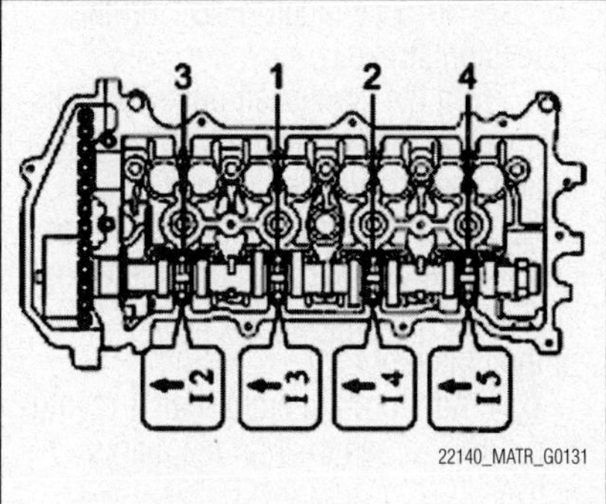

Fig. 62 Camshaft cap bolt tightening sequence—1ZZ-FE engine

pushing it gently against the camshaft. Push further at the position where the pin fits into the groove.

c. Be sure not to turn the camshaft timing gear to the retard angle side (the right angle).

d. Check that there is no clearance between the gear fringe and camshaft.

e. Tighten the fringe bolt with the camshaft timing gear fixed to 40 ft. lbs (54 Nm).

f. Check that the camshaft timing gear can move to the retard angle side (the right angle), and is locked in the most retarded position.

13. Install camshaft.

a. Install the timing chain onto the camshaft timing gear with the paint mark aligned with the timing mark on the camshaft timing gear as shown in the illustration.

b. Examine the front marks and numbers and check that the sequence order is as shown in the illustration. Then tighten the bolts to 10 ft. lbs (13 Nm).

14. Install second camshaft.

a. Put the second camshaft on the cylinder head with the paint mark of the chain aligned with the timing mark on the camshaft timing gear.

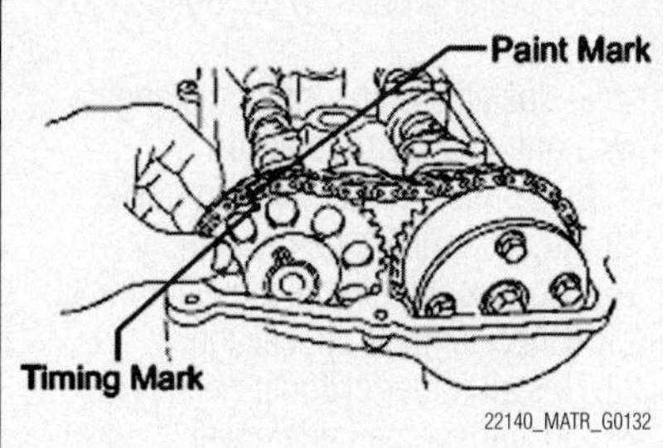

Fig. 63 Put the second camshaft on the cylinder head with the paint mark of the chain aligned with the timing mark on the camshaft timing gear

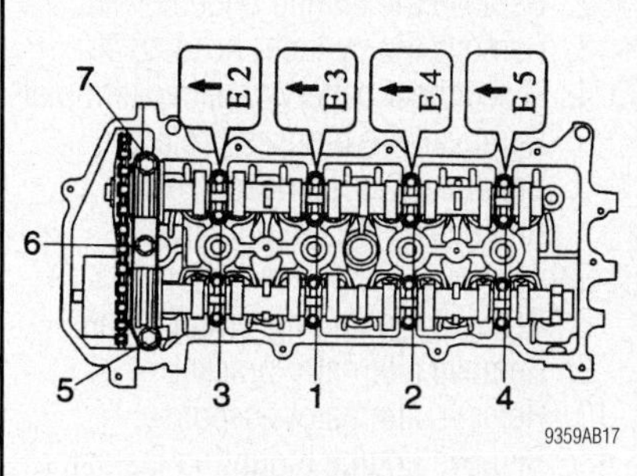
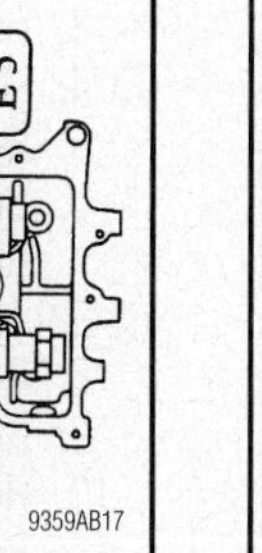

Fig. 64 Camshaft cap bolt tightening sequence—1ZZ-FE engine

b. Temporarily install the camshaft timing gear set bolt .

c. Examine the front marks and numbers and check that the sequence order is as shown in the illustration. Then tighten the bolts to 10 ft. lbs (13 Nm).

d. Install the bearing cap No. 1 and tighten to 17 ft. lbs (23 Nm).

e. While holding the camshaft with a wrench, tighten the camshaft timing gear set bolt to 40 ft. lbs (54 Nm).

**** WARNING**

Be careful not to damage the valve lifter.

f. Check that the 2 point marks are aligned on the timing chain cover surface. Also, check that each timing mark is aligned with its corresponding painted link on the timing chain as shown in the illustration. In addition, check that the groove is aligned with timing mark "0" of the chain cover.

15. Install the chain tensioner.

a. Check that the O-ring is clean, and set the hook as shown in the illustration.

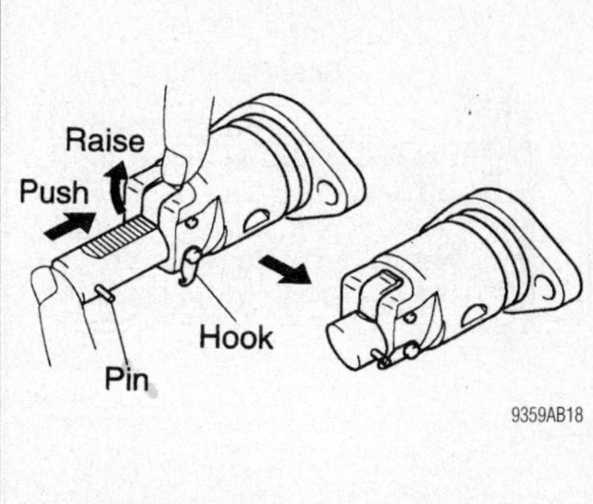

Fig. 65 Set the timing chain tensioner hook properly

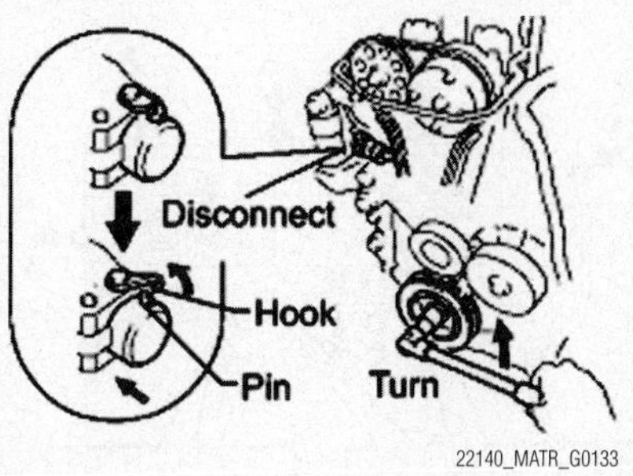

Fig. 66 Turn the crankshaft counterclockwise to disconnect the plunger knock pin from the hook

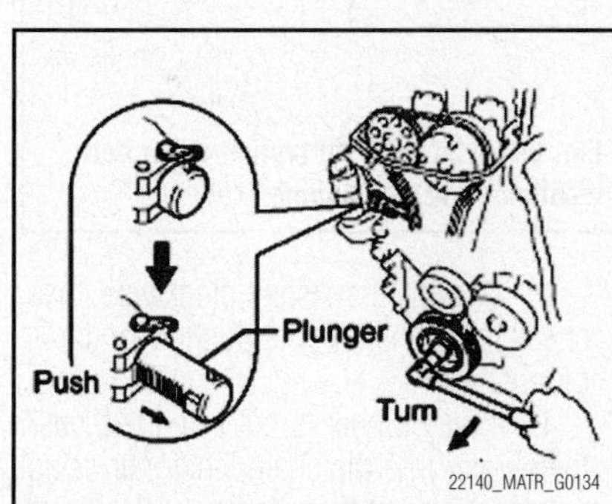

Fig. 67 Turn the crankshaft clockwise, then check that the slipper is pushed by the plunger

b. Apply a light coat of engine oil to the O-ring.

c. Install the chain tensioner and tighten the nuts to 40 inch lbs. (9 Nm).

**** WARNING**

Be careful not to twist the O-ring. When installing the chain tensioner, set the hook again if the hook releases the plunger.

d. Turn the crankshaft counterclockwise to disconnect the plunger knock pin from the hook.

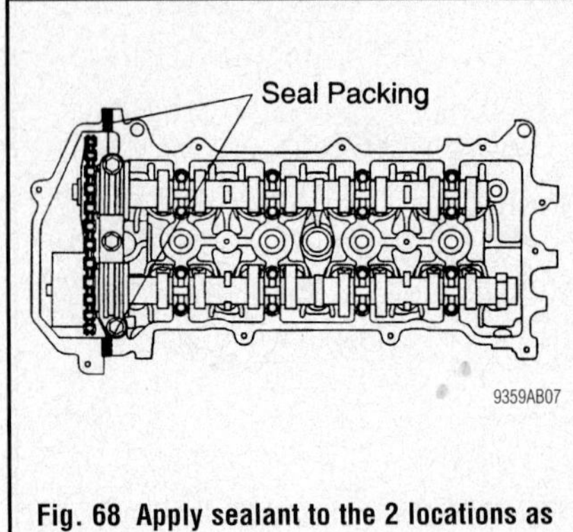

Fig. 68 Apply sealant to the 2 locations as shown

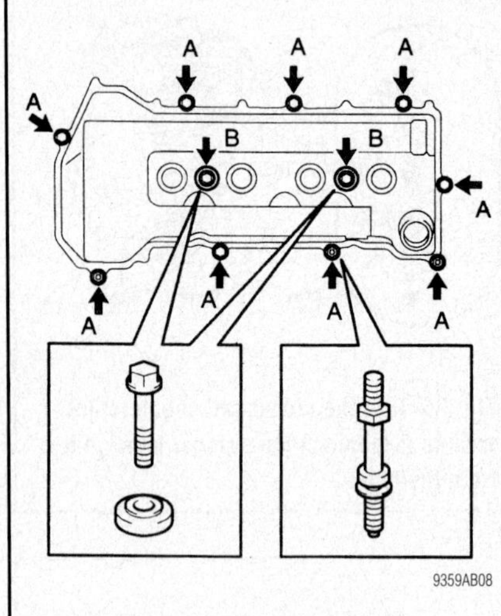

Fig. 69 Cylinder head (valve) cover bolt locations—1ZZ-FE engine

e. Turn the crankshaft clockwise, then check that the slipper is pushed by the plunger.

f. If the plunger is not extended, push the slipper into the chain tensioner using a screwdriver so that the hook is released from the knock pin and the plunger can be extended.

16. Check and adjust valve clearance, as necessary.

17. Install cylinder head cover.

a. Remove any old packing material from the contact surface.

b. Apply sealant to the 2 locations as shown in the illustration.

c. Remove any oil from the contact surface.

➡**Install the cylinder head cover within 3 minutes of applying sealant. Do not put in engine oil for at least 2 hours after installation.**

18. Install the cylinder head cover and clamp brackets. Tighten bolts "A" to 8 ft. lbs. (11 Nm) and bolts "B" to 80 inch lbs. (9 Nm).

a. Install the clamps of the engine wire onto the clamp bracket.

b. Connect the ventilation hoses.

19. Install ignition coil assembly.
20. Connect battery negative terminal.
21. Check engine oil leakage.
22. Install upper cylinder head cover.
23. Install engine undercover.
24. Lower the vehicle

2ZZ-GE Engine

See Figures 70 through 79.

1. Before servicing the vehicle, refer to the precautions section.
2. Remove the engine undercover.
3. Remove the cylinder head cover.
4. Disconnect battery negative terminal.
5. Remove wire harness protector.
6. Separate suction hose.
7. Remove air switching valve assembly.
8. Remove fan and generator v belt.
9. Remove alternator bracket.
10. Remove alternator assembly.
11. Remove engine mounting insulator.

a. Place a wooden block between the jack and engine and jack up the engine.

b. Remove the bolts and nuts, then remove the engine mounting insulator .

12. Remove ribbed belt tensioner.

➡**Jack the engine up and down to remove the bolt.**

13. Disconnect the 4 ignition coil connectors.
14. Remove the ignition coils.

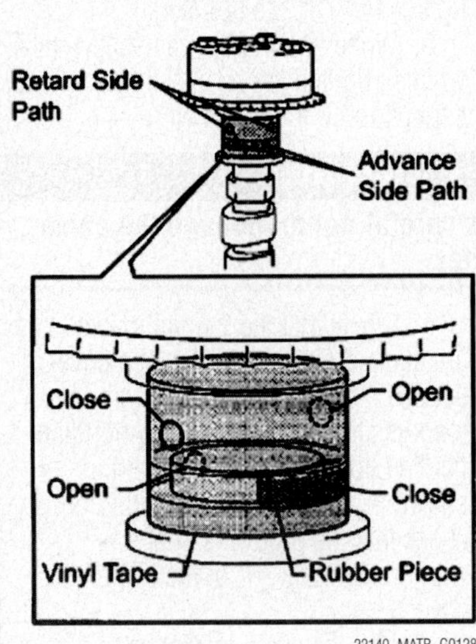

Fig. 70 Engine mount insulator fastener locations—2ZZ-GE engine

15. Remove cylinder head cover.

a. Disconnect the 2 ventilation hoses.

b. Remove the nuts and bolt, then separate the ventilation tube No. 1.

c. Remove the gasket from the cylinder head cover.

d. Remove the wire harness protector.

e. Remove the bolt and fuel tube clamp.

f. Remove the bolt and union to connector tube hose clamp.

g. Remove the bolts, bracket, wire harness protector, and cylinder head cover.

h. Remove the o-ring from the cylinder head cover.

16. Set No. 1 cylinder to TDC on the compression stroke.

a. Turn the crankshaft pulley until its groove and the timing mark "0" of the timing chain cover are aligned.

b. Check that the point marks of the camshaft timing sprocket and the variable valve timing sprocket are as shown in the illustration.

c. If not, turn the crankshaft 1 revolution (360°) so as to align the marks above.

17. Remove chain tensioner.

⁂ WARNING

Do not to turn the crankshaft without the chain tensioner.

18. Remove camshaft.

a. Hold the hexagonal lobe of the camshaft, then loosen the camshaft timing gear set bolt.

b. Do not to turn the crankshaft without the chain tensioner.

c. Using several steps, uniformly loosen and remove the camshaft bearing cap bolts in the sequence shown, then remove the caps.

d. Lift up the camshaft slightly and remove the bolt.

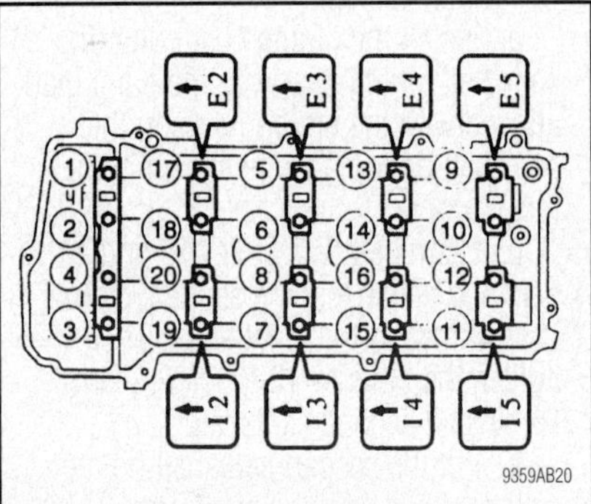

Fig. 71 Camshaft cap bolt removal sequence—2ZZ-GE Engine

e. Hold the timing chain by hand, then remove the camshaft.

f. Tie the timing chain with a string as shown in the illustration.

WARNING

Be careful not to drop anything inside the timing chain cover.

19. Remove camshaft timing gear.

a. Check the lock of camshaft timing gear.

b. Clamp the camshaft in a vise, then confirm the camshaft timing gear is locked.

WARNING

Be careful not to damage the camshaft.

c. Check that the camshaft timing gear assembly does not turn.

WARNING

Do not damage the camshaft by clamping it in a vise too tightly.

d. Cover the 4 oil paths of the cam journal with vinyl tape. One of the 2 grooves located on the cam journal is for retarding cam timing (upper) and the other is for advance cam timing (lower). Each groove has 2 oil paths. Plug one of the 2 oil paths for each groove with a pieces of rubber before wrapping the cam journal with the tape.

e. Puncture the tape which covers the advance side path and retard side path on the opposite side.

f. Apply approximately 21 psi (150 kpa) of air pressure to the retard side oil path. When applying air pressure, cover the paths with a shop rag to prevent oil splashes.

g. Apply approximately 21 psi (150 kpa) of air pressure to the advance side oil path.

h. Confirm that the camshaft timing gear assembly revolves in the advance direction when reducing the air pressure on the retard side path.

i. If the timing gear assembly does not revolve in the advance direction by air pressure, turn it forcibly by hand after air pressure is applied to both oil paths. When revolving, first turn it in the retard direction, then in the advance direction.

j. The lock pin is released, and the camshaft timing gear revolves in the advance direction.

k. When the camshaft timing gear assembly comes to the most advanced position, release the air pressure of the retard side path, then release the air pressure of the advance side path.

l. If the air pressure of the advance side path is released first, the camshaft timing gear assembly occasionally shifts abruptly in the retard direction. It may cause the breakage of the lock pin. Be sure to release the air pressure on the retard side path first.

m. Remove the bolt and camshaft timing gear assembly.

n. Do not remove the other 4 bolts.

o. When reusing the camshaft timing gear assembly, unlock the lock pin inside the camshaft timing gear first.

20. Remove cam timing control valve housing.

a. Remove the gasket.

21. Remove the valve rocker arm.

22. Remove valve rocker shaft.

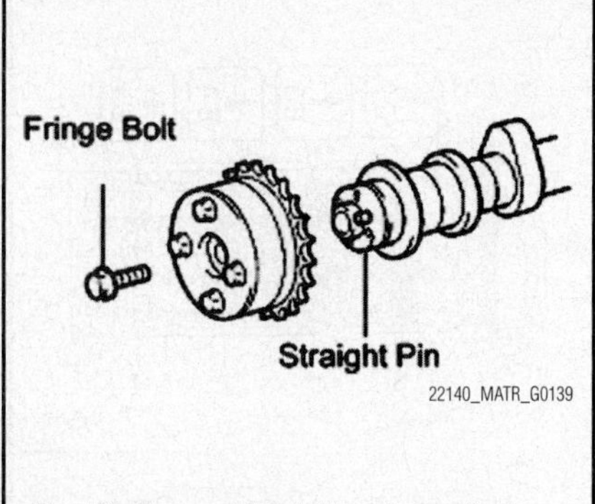

Fig. 72 Remove the bolt and camshaft timing gear assembly—2ZZ-GE engine

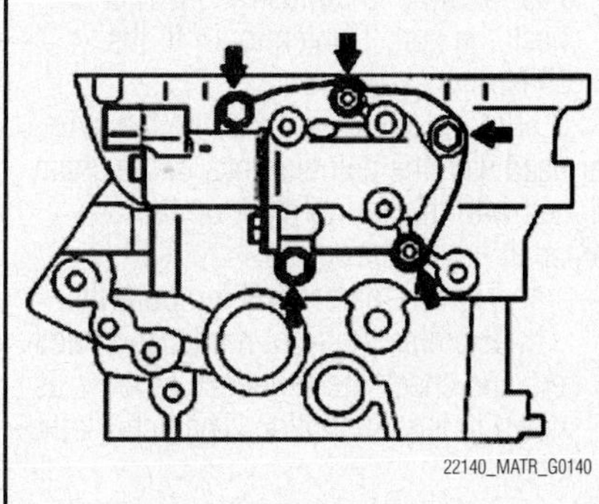

Fig. 73 Remove cam timing control valve housing—2ZZ-GE engine

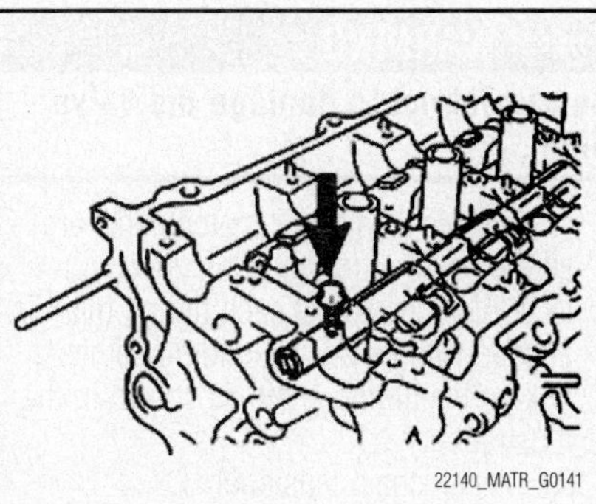

Fig. 74 Remove valve rocker shaft — 2ZZ-GE engine

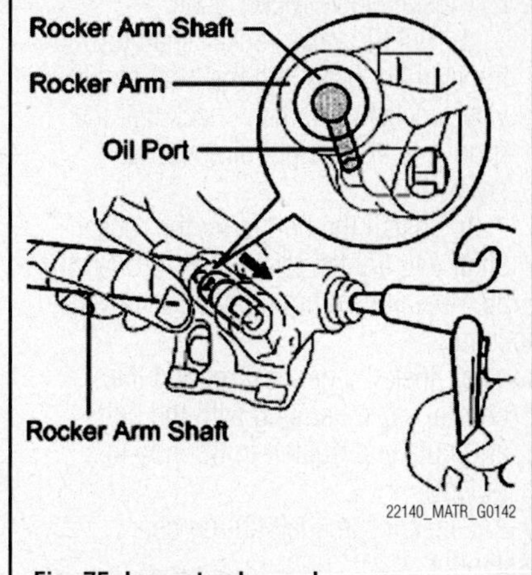

Fig. 75 Inspect valve rocker arm — 2ZZ-GE engine

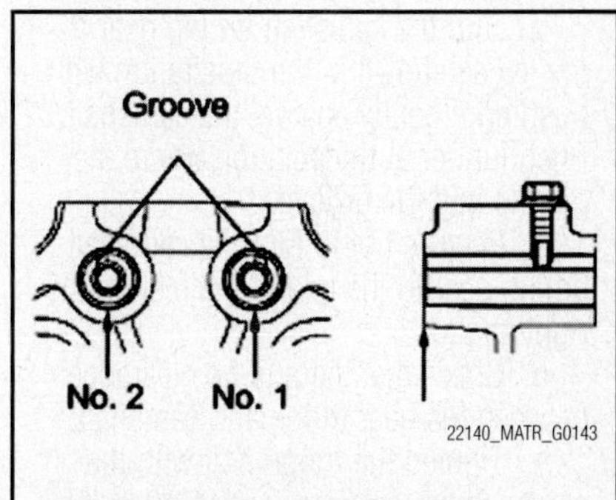

Fig. 76 Check that the rocker shaft groove is at the right direction as shown — 2ZZ-GE engine

To install:

23. Inspect valve rocker arm.

a. Cover the oil paths of the rocker arm shaft except the 2 paths covered with vinyl type.

b. Align the oil path of the rocker arm shaft with the oil path of the rocker arm.

c. Check that the piston inside of the rocker arm moves when 26 psi (180 kpa) of air pressure is applied to the oil paths.

24. Install valve rocker shaft.

a. Put the valve rocker shaft through the cylinder head and the hole of the valve rocker arm, then check that the groove is at the right direction as shown.

b. Install the bolt to fix the rocker shaft and tighten to 7 ft. lbs. (10 Nm).

25. Install valve rocker shaft.
a. Put the valve rocker shaft through the cylinder head and the hole of the valve rocker arm, then check that the groove is at the right direction as shown.
b. Install the bolt to fix the rocker shaft and tighten to 7 ft. lbs. (10 Nm).
26. Install cam timing control valve housing.
a. Install a new gasket and the control valve housing with the bolts and nuts and tighten to 80 inch lbs. (9 Nm).
27. Install camshaft timing gear assembly.
a. Put the camshaft timing gear together with the camshaft with the straight pin off the key groove.
b. Turn the camshaft timing gear to the left as shown in the illustration while pushing it gently against the camshaft. Push further at the position where the pin fits into the groove.
c. Be sure not to turn the camshaft timing gear to the retard angle side (the right angle).
d. Check that there is no clearance between the gear fringe and camshaft.
e. Tighten the fringe bolt with the camshaft timing gear fixed to 40 ft. lbs. (54 Nm).
f. Check that the camshaft timing gear can move to the retard angle side (the right angle), and is locked in the most retarded position.
28. Install the camshaft.
a. Apply engine oil to the camshaft journal parts.
b. Install the timing chain onto the camshaft timing gear with the painted link aligned with the timing mark of the camshaft timing sprocket as shown in the illustration.
c. Examine the front marks and numbers and check that the sequence order is as shown in the illustration. Then check the bolt. Tighten to 14 ft. lbs. (19 Nm).

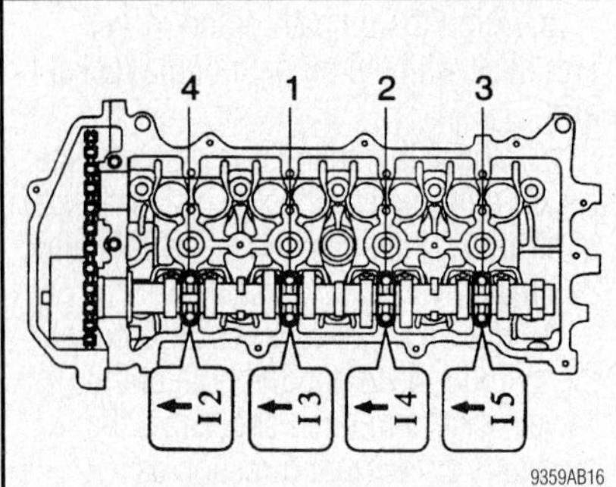

Fig. 77 Camshaft cap bolt tightening sequence—2ZZ-GE engine

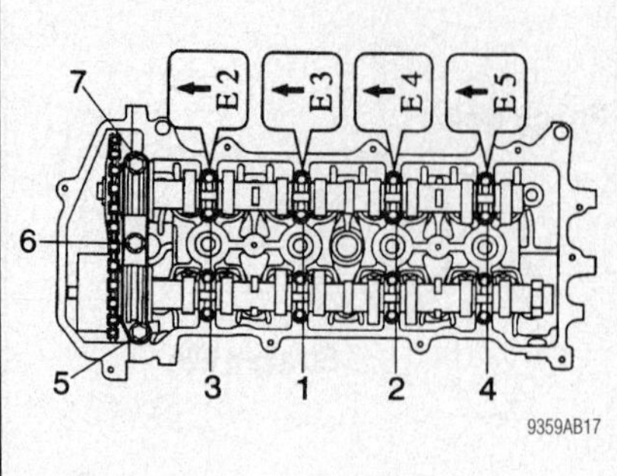

Fig. 78 Camshaft cap bolt tightening sequence—2ZZ-GE engine

29. Put the No. 2 camshaft on the cylinder head with the painted links of the chain aligned with the timing mark on the camshaft timing sprockets.
a. Tighten the set bolt temporarily.
b. Examine the front marks and numbers and check the sequence order is as shown in the illustration. Then check the bolts.
c. Hold the hexagonal lobe of the camshaft, then tighten the camshaft timing gear set bolt to 40 ft. lbs. (54 Nm).

⁂ WARNING

Be careful not to damage the valve lifter.

d. Check that the 2 matchmarks are aligned on the timing chain cover surface. Also check that each timing mark is aligned with its corresponding painted link on the timing chain as shown in the illustration.
30. Install chain tensioner.
a. Check that the o-ring is clean, then set the hook as shown in the illustration.
b. Apply engine oil to the chain tensioner, then install it. Tighten to 80 inch lbs. (9 Nm).
c. When installing the tensioner, set the hook again if the hook releases the plunger.
d. Turn the crankshaft counterclockwise to disconnect the plunger knock pin from the hook.
e. Turn the crankshaft clockwise, then check that the slipper is pushed by the plunger.
f. If the plunger is not expended, push the slipper into the chain tensioner with a screwdriver so that the hook is released from the knock pin and the plunger can be extended.

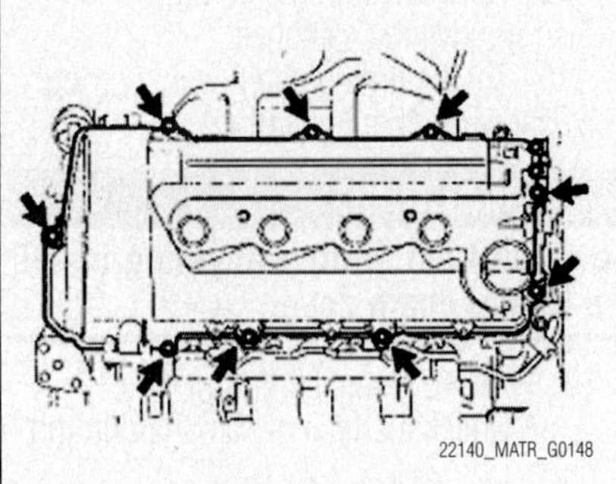

Fig. 79 Uniformly tighten the bolts, in the several passes—2ZZ-GE engine

31. Check and adjust the valve clearance, as necessary.
32. Install cylinder head cover.
a. Install a new o-ring onto the cylinder head cover.
b. Remove any old sealant.
c. Apply sealant to 2 locations as shown in the illustration.

➡Remove any oil from the contact surface. Install the cylinder head cover within 3 minutes of applying sealant. Do not put in engine oil for at least 2 hours after installation.

d. Install the cylinder head cover, wire harness protector and bracket. Uniformly tighten the bolts, in the several passes in the sequence shown to 7 ft. lbs. (10 Nm).
e. Install the union to connector tube hose clamp. Tighten to 62 inch lbs. (7 Nm).
33. Install the fuel tube clamp and tighten the bolt to 66 inch lbs. (8 Nm).
a. Install the wire harness protector and tighten the nuts to 80 inch lbs. (9 Nm).
b. Install a new gasket and ventilation tube. Tighten the nut to 7 ft. lbs. (10 Nm) and the bolt to 18 ft. lbs. (24 Nm).
34. Install ignition coil assembly. Tighten to 80 inch lbs. (9 Nm).
35. Install V-ribbed belt tensioner. Tighten the nut to 21 ft. lbs. (29 Nm) and the bolt to 74 ft. lbs. (100 Nm).
36. Install engine mounting insulator. Tighten to 38 ft. lbs. (52 Nm).
37. Install generator assembly.
38. Install generator bracket.
39. Install air switching valve assembly.
40. Install suction hose. Tighten to 87 inch lbs. (10 Nm).
41. Connect battery negative terminal.
42. Check engine oil leakage.

CRANKSHAFT DAMPER

REMOVAL & INSTALLATION

See Figures 80 through 82.

1. Before servicing the vehicle, refer to the precautions section.
2. Disconnect the negative battery cable.
3. Remove the accessory drive belt.
4. Raise and suitably support the vehicle.
5. Remove the crankshaft pulley, using the special tool.

To install:

6. Lubricate the front seal and the sealing surface of the crankshaft pulley with chassis grease.
7. Install the crankshaft pulley.
   a. Align the pulley set key with the key groove of the pulley and slide on the pulley.
   b. Use the special tools to install the bolt and tighten to 102 ft. lbs. (138 Nm)

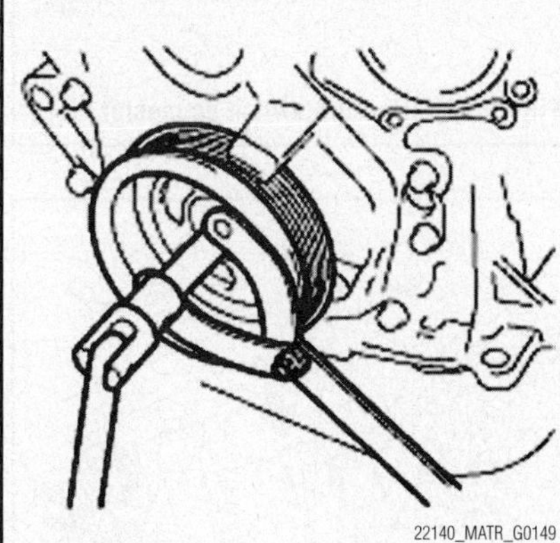

Fig. 80 Removing the crankshaft pulley using the special tool SST: 09960-10010—1ZZ-FE engine

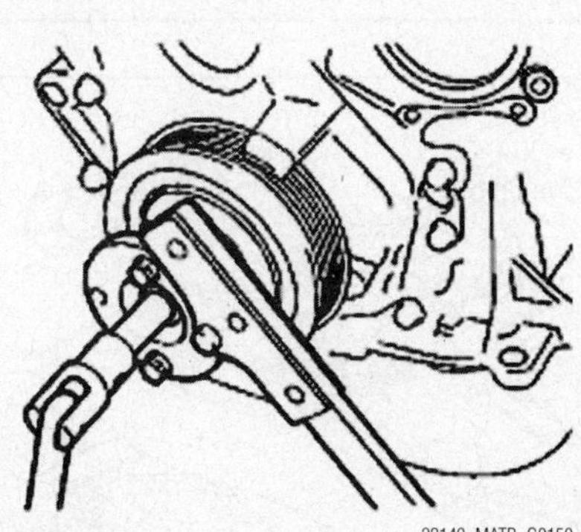

Fig. 81 Removing the crankshaft pulley bolt using the special tools SST: 09213-70011, and SST: 09330-00021—2ZZ-GE engine

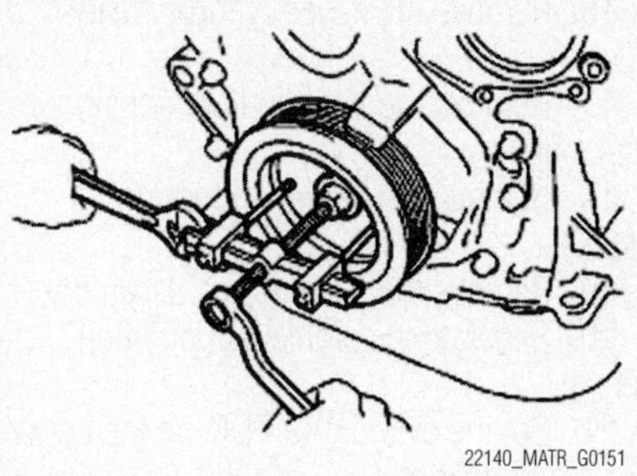

Fig. 82 Removing the crankshaft pulley using the special tool SST: 09950-50013—2ZZ-GE engine

for 1ZZ-FE engine or to 87 ft. lbs. (118 Nm) on 2ZZ-GE engines.
   c. Turn the crankshaft counterclockwise and disconnect the plunger knock pin from the hook.
   d. Turn the crankshaft clockwise and check that the slipper is pushed by the plunger. If the plunger does not spring out, press the slipper into the chain tensioner with a screwdriver so that the hook is released from the knock pin and the plunger springs out.
8. Install the right splash shield.
9. Lower the vehicle.
10. Install the accessory drive belt.
11. Connect the negative battery cable.

CRANKSHAFT FRONT SEAL

REMOVAL & INSTALLATION

1ZZ-FE Engine

See Figure 83.

1. Before servicing the vehicle, refer to the precautions section.
2. Raise and safely support the vehicle securely on jackstands.
3. Remove engine undercover.
4. Remove front wheel.
5. Remove the fan and generator v belt.

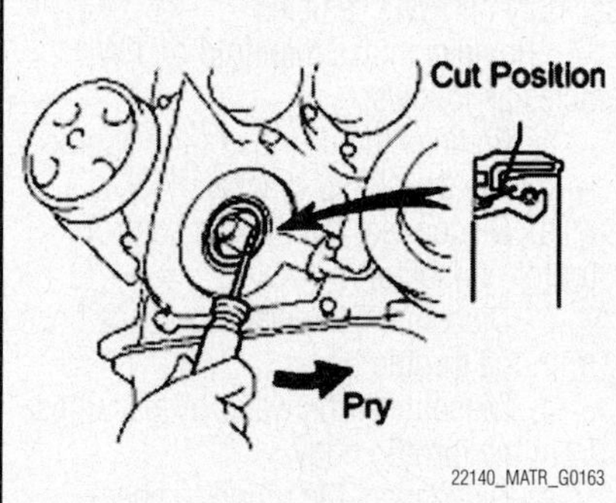

Fig. 83 Using a knife, cut off the oil seal lip and remove the timing chain oil seal

6. Remove the engine mounting insulator.
7. Remove crankshaft pulley.
8. Using a knife, cut off the oil seal lip.
9. Remove timing chain oil seal.

To install:

10. Install timing chain or belt cover oil seal.
   a. Apply multi-purpose grease to a new oil seal lip.
   b. Keep the lip free of forcign objects.
   c. Using a seal installer and a hammer, tap the new oil seal in until its surface is flush with the timing chain cover edge.
   d. Tap the oil seal in a vertical position.
   e. Wipe off extra grease on the crankshaft.
11. Install crankshaft pulley.
12. Install the engine mounting insulator.
13. Install front wheel.
14. Check engine oil leakage.

2ZZ-GE Engine

1. Before servicing the vehicle, refer to the precautions section.
2. Raise and safely support the vehicle securely on jackstands.
3. Remove front wheel.
4. Remove engine undercover.
5. Remove the upper cylinder head cover .
6. Remove wire harness protector.
7. Separate suction hose.
8. Remove air switching valve.
9. Remove fan and generator v belt.
10. Remove engine mounting insulator.
11. Remove crankshaft pulley.
12. Remove timing chain oil seal.
   a. Using a knife, cut off the oil seal lip.
   b. Using a screwdriver with its tip wrapped with tape, pry out the oil seal.
   c. After the removal, check if the crankshaft is not damaged. If damaged, smooth the surface with a 400-grit sandpaper.

To install:

13. Install timing chain oil seal.
   a. Apply a light coat of multi-purpose grease to a new oil seal lip.
   b. Keep the lip free of foreign materials.
   c. Using a seal installer and a hammer, tap in the oil seal until its surface is flush with the rear oil seal retainer edge.
   d. Wipe any extra grease off the crankshaft.
14. Install crankshaft pulley.
15. Install engine mounting insulator.

16. Install air switching valve.
17. Install suction hose.
18. Install front wheel.
19. Check engine oil leakage.
20. Install the upper cylinder head cover.

CYLINDER HEAD

REMOVAL & INSTALLATION

1ZZ-FE Engine

See Figures 84 through 92.

1. Before servicing the vehicle, refer to the precautions section.
2. Discharge the fuel system pressure.
3. Disconnect battery negative terminal.
4. Raise and safely support the vehicle securely on jackstands.
5. Remove front wheel.
6. Remove the engine undercover.
7. Remove the upper cylinder head cover.
8. Drain and recycle the engine coolant.
9. Separate the front exhaust pipe assembly.
10. Remove the manifold stay.
11. Remove the fan and alternator V-belt.
12. Separate the vane pump assembly.
13. Remove the alternator assembly.
14. Remove the engine mounting insulator.
15. Remove the ignition coil assembly.

Timing Mark
Timing Mark
1 2 3 4 5 6 7 8
Point Marks
Timing Chain Cover Surface
Groove
22140_MATR_G0162

Fig. 84 Set the cylinder to TDC on the compression stroke. Check that the point marks of the camshaft timing gears are in a straight line on the timing chain cover surface as shown

16. Remove the lower cylinder head cover.
17. Remove the V-ribbed belt tensioner assembly.
18. Remove the engine mounting bracket.
19. Remove the water pump assembly.
20. Separate the crankshaft position sensor.
21. Set the cylinder to TDC on the compression stroke.
    a. Turn the crankshaft pulley until its groove and timing mark "0" of the timing chain cover are aligned.
    b. Check that the point marks of the camshaft timing gears are in a straight line on the timing chain cover surface as shown in the illustration.
    c. If not, turn the crankshaft 1 revolution (360°) to align the marks as above.
22. Remove the crankshaft pulley.
23. Remove the chain tensioner assembly.
24. Remove the timing chain cover.
25. Remove the timing chain cover oil seal.
26. Remove the crankshaft position plate.
27. Remove the chain tensioner slipper.
28. Remove the timing chain.
29. Remove the chain vibration damper.
30. Remove air cleaner hose on FWD vehicles as follows:
    a. Disconnect the variable vacuum switching valve connector.
    b. Remove the clamp of the engine wire.
    c. Disconnect the fuel vapor feed hose from the vacuum switching valve.
    d. Loosen the hose clamp bolts, then remove the air cleaner hose.
31. Remove air cleaner hose on AWD vehicles as follows:
    a. Loosen the hose clamp bolts, then remove the air cleaner hose.
32. Remove oil level gage.
33. Separate accelerator control cable assembly on AWD vehicles.
34. Remove intake manifold on FWD vehicles as follows:
    a. Remove the clamps of the engine wire from the clamp bracket.
    b. Disconnect the throttle control motor connector.
    c. Disconnect the water by-pass hose from the throttle body.
    d. Disconnect the water by-pass hose from the throttle body.
    e. Disconnect the union to check valve hose from the intake manifold.
    f. Remove the bolts, nuts and clamp brackets.
    g. Remove the intake manifold and throttle body.
    h. Remove the gasket from the intake manifold.
    i. Remove intake manifold on AWD vehicles as follows:
    j. Remove the clamps of the engine wire from the clamp bracket.
    k. Disconnect the throttle position sensor connector.
    l. Disconnect the idle air control valve connector.

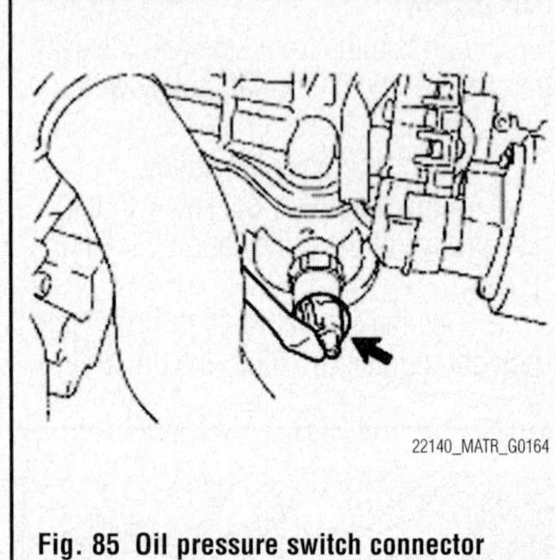

Fig. 85 Oil pressure switch connector

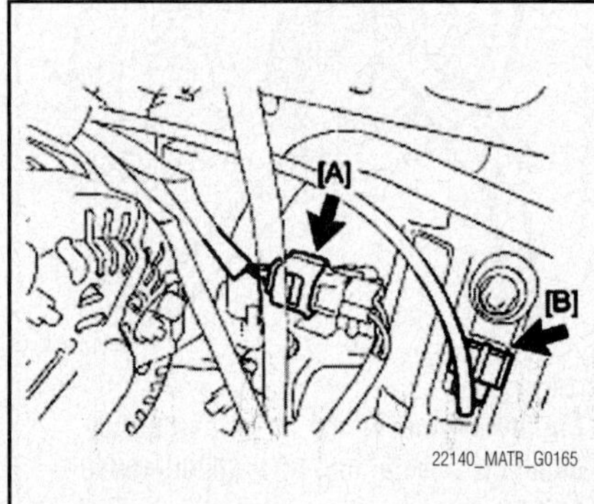

Fig. 86 Crankshaft position sensor connector

Fig. 87 Camshaft position sensor connector

m. Disconnect the water by-pass hose from the throttle body.

n. Disconnect the union to check valve hose from the intake manifold.

o. Remove the bolts, nuts and clamp brackets.

p. Remove the intake manifold and throttle body.

q. Remove the gasket from the intake manifold.

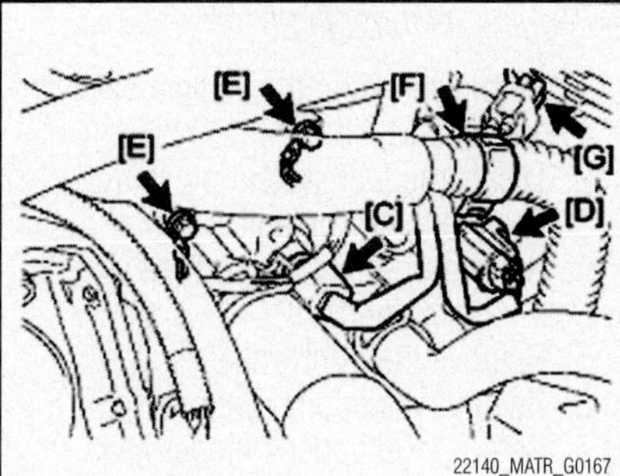

Fig. 88 Coolant temperature sensor connector (C), radio setting condenser connector (D), engine grounds (E), clamp (F) and heated oxygen sensor connector (G).

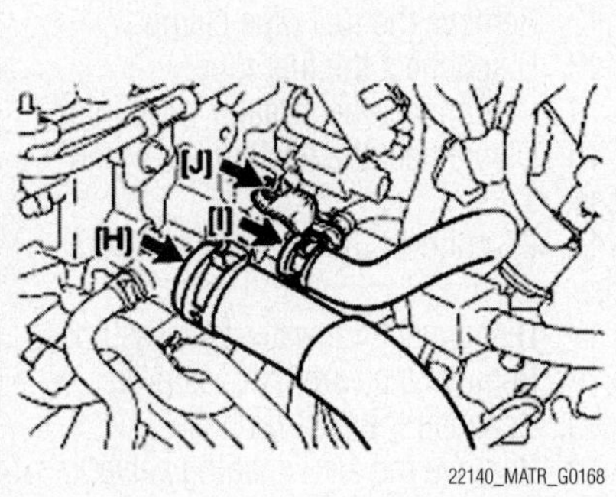

Fig. 89 Radiator hose inlet (H), heater inlet water hose (I) and the bolt (J).

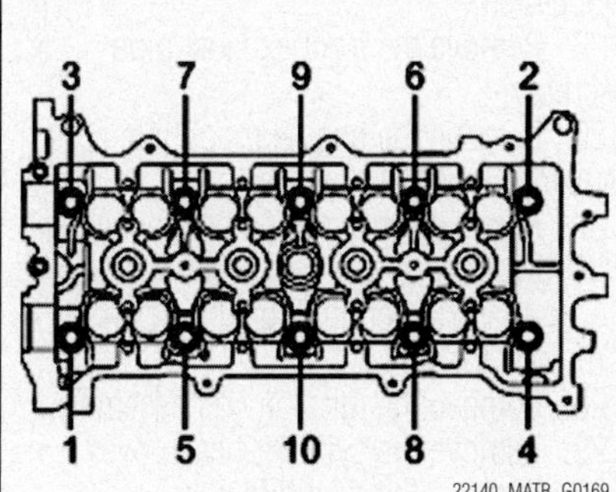

Fig. 90 Using several steps, uniformly loosen and remove the cylinder head bolts in the sequence shown

35. Remove camshaft timing oil control valve assembly.

36. Remove camshaft.

a. Using several steps, uniformly loosen and remove the bearing cap bolts in the sequence shown in the illustration.

b. Remove the bearing caps, then remove the camshafts from the cylinder head.

37. Disconnect the fuel tube using special tool SST: 09268-21010.

38. Remove cylinder head.

a. Disconnect the engine wire.

b. Disconnect the fuel injector connectors.

c. Disconnect the oil pressure switch connector.

d. Disconnect the crankshaft position sensor connector.

e. Disconnect the knock sensor connector.

f. Disconnect the camshaft position sensor connector.

g. Disconnect the engine coolant temperature sensor connector.

h. Disconnect the radio setting condenser connector.

i. Disconnect the engine grounds.

j. Remove the clamp from the clamp bracket.

k. Disconnect the heated oxygen sensor connector.

l. Disconnect the radiator hose inlet.

m. Disconnect the heater inlet water hose.

n. Remove the water by-pass pipe.

o. Using several steps, uniformly loosen and remove the cylinder head bolts in the sequence shown in the illustration. Remove the cylinder head bolts and plate washers.

✲✲ WARNING

Be careful not to drop the plate washers into the cylinder head. Head warpage or cracking could result from removing the bolts in the wrong order.

p. Lift the cylinder head front the engine block.

39. Remove cylinder head gasket.

To install:

40. Install cylinder head gasket.

a. Place a new cylinder head gasket on the cylinder block surface with the lot number stamp upward.

✲✲ WARNING

Make sure that the cylinder head gasket is mounted in the correct direction. Place the cylinder head gently in order to avoid damaging the gasket.

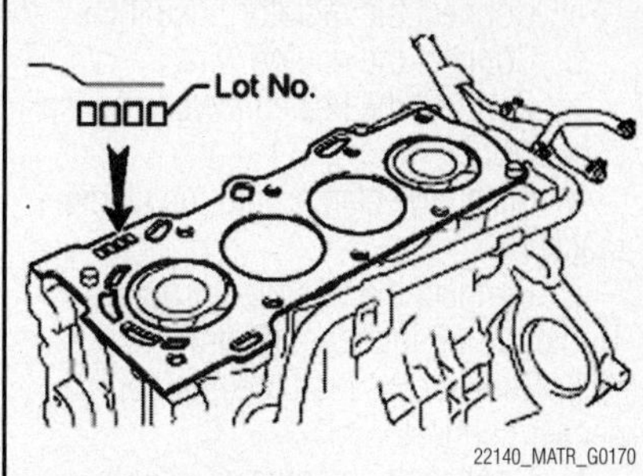

Fig. 91 Place a new cylinder head gasket on the cylinder block surface with the lot number stamp upward

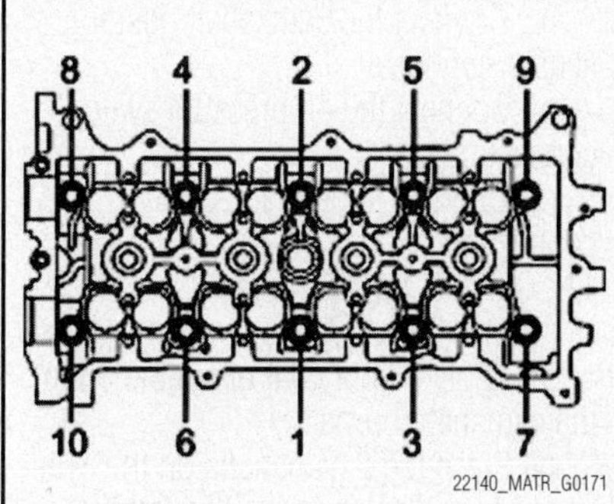

Fig. 92 Using several steps, uniformly install and tighten the cylinder head bolts and plate washers in the sequence shown—1ZZ-FE engine

41. Install cylinder head.

✲✲ WARNING

The cylinder head bolts should be tightened in 2 successive steps.

a. Apply a light coat of engine oil to the threads of the cylinder head bolts.

b. Using several steps, uniformly install and tighten the cylinder head bolts and plate washers in the sequence shown in the illustration. Tighten as follows:

- Tighten to 36 ft. lbs. (49 Nm).
- Mark the front of the cylinder head bolts with paint.
- Tighten the cylinder head bolts and additional 90°in the same sequence.
- Check that the paint mark of each bolt is at a 90°angle to the front.

c. Install the water by-pass pipe. Tighten to 80 inch lbs. (9 Nm).

d. Connect the heater inlet water hose.

e. Connect the radiator hose inlet.

f. Connect the engine wire.

g. Connect the heated oxygen sensor connector.

h. Install the clamp onto the clamp bracket.

i. Connect the 2 engine grounds. Tighten to 80 inch lbs. (9 Nm).

j. Connect the radio setting condenser connector.

k. Connect the engine coolant temperature sensor connector.

l. Connect the camshaft position sensor connector.

m. Connect the knock sensor connector.

n. Connect the crankshaft position sensor connector.

o. Connect the oil pressure switch connector.

p. Connect the fuel injector connectors.

42. Connect the fuel tube.
43. Install camshaft.

a. Apply a light coat of engine oil to the camshaft journals.

b. Place the 2 camshafts on the cylinder head as shown in the illustration.

c. Examine the front marks and numbers and check that the sequence order is as shown in the illustration. Then tighten the bearing caps closest to the timing cover to 17 ft. lbs. bolts (23 Nm) and the remaining bearing caps to 10 ft. lbs. (13 Nm).

44. Install camshaft timing oil control valve assembly.

a. Apply a light coat of engine oil to a new o-ring, then install it onto the camshaft timing oil control valve.

b. Install the camshaft timing oil control valve. Tighten the bolt to 80 inch lbs. (9 Nm).

45. Install intake manifold on FWD vehicles as follows:

a. Install a new gasket onto the intake manifold.

b. Install the intake manifold and throttle body with the brackets, bolts and nuts. Tighten to 22 ft. lbs. (30 Nm).

c. Connect the union to check valve hose to the intake manifold.

d. Connect the water by-pass hose to the throttle body.

e. Connect the throttle control motor connector.

f. Install the clamps of the engine wire onto the clamp bracket.

46. Install intake manifold on AWD vehicles as follows:

a. Install a new gasket onto the intake manifold.

b. Install the intake manifold and throttle body with the brackets, bolts and nuts. Tighten to 22 ft. lbs. (30 Nm).

c. Connect the union to check valve hose to the intake manifold.

d. Connect the water by-pass hose No. 2 to the throttle body.

e. Connect the water by-pass hose to the throttle body.

f. Connect the idle speed control valve connector.

g. Connect the throttle position sensor connector.

h. Install the clamps of the engine wire onto the clamp bracket.

47. Install accelerator control cable assembly on AWD vehicles.
48. Install oil level gage.
49. Install air cleaner hose on FWD vehicles as follows:

a. Install the air cleaner hose with the hose clamp bolts.

b. Connect the fuel vapor feed hose to the vacuum switching valve.

c. Connect the fuel vapor feed hose to the vacuum switching valve.

d. Install the clamp of the engine wire.

e. Connect the vacuum switching valve connector.

50. Install air cleaner hose on AWD vehicles as follows:

a. Install the air cleaner hose with the 2 hose clamp bolts.

b. Install the chain vibration damper.

51. Install the timing chain.
52. Install the chain tensioner slipper.
53. Install the crankshaft position sensor plate.
54. Install the timing chain cover oil seal using a seal installer.
55. Install the timing chain cover.
56. Install the chain tensioner assembly.
57. Install the crankshaft pulley.
58. Install the crankshaft position sensor.
59. Install the water pump assembly.
60. Install the engine mounting bracket.
61. Install the V-ribbed belt tensioner assembly.
62. Install the cylinder head cover.
63. Install the ignition coil assembly.
64. Install the engine mounting insulator.
65. Install the alternator assembly.
66. Install the vane pump assembly.
67. Install fan and alternator V-belt.
68. Install the manifold stay.
69. Install front exhaust pipe assembly.
70. Add engine coolant.
71. Connect the negative battery cable.
72. Check for the engine coolant leakage.
73. Check engine oil leakage.
74. Check fuel leak.
75. Check exhaust gas leak.
76. Inspect the ignition timing.
77. Inspect the engine idling speed.
78. Inspect the compression.
79. Install upper cylinder head cover.
80. Install engine undercover.
81. Install front wheel.
82. Lower the vehicle.

2ZZ-GE Engine

See Figures 93 through 95.

1. Before servicing the vehicle, refer to the precautions section.
2. Discharge fuel system pressure.
3. Drain engine coolant.
4. Raise and safely support the vehicle securely on jackstands.
5. Remove the front wheel.
6. Remove the engine undercover.
7. Remove the upper cylinder head cover.
8. Disconnect the battery negative terminal.
9. Separate the accelerator control cable assembly.
10. Remove the air cleaner cap.
11. Disconnect the union to connector tube hose from the brake booster.
12. Remove the fuel pipe clamp.
13. Disconnect the fuel tube.
14. Disconnect the radiator hose inlet.
15. Disconnect the heater inlet water hose.
16. Disconnect the heater outlet water hose.
17. Disconnect the water by-pass hoses.
18. Remove the wire harness protector.
19. Separate the suction hose.
20. Remove the air switching valve assembly.
21. Remove the fan and alternator V-belt.
22. Remove the alternator bracket.
23. Remove the alternator assembly.
24. Separate the compressor and magnetic clutch.
25. Remove the front exhaust pipe assembly.
26. Remove the engine mounting insulator.
27. Remove the V-ribbed belt tensioner assembly.

a. Jack the engine up and down to remove the bolt.

28. Remove the ignition coil assembly.
29. Remove the cylinder head cover.
30. Remove the manifold stay.
31. Remove the exhaust manifold heat insulator.
32. Remove the exhaust manifold.
33. Remove the water pump pulley.

34. Remove the water pump assembly.
35. Remove the transverse engine mounting bracket.
36. Remove the crankshaft position sensor.
37. Set the No.1 cylinder to TDC on the compression stroke.
a. Turn the crankshaft pulley until its groove and the timing mark "0" of the timing chain cover are aligned.
b. Check that the point marks of the camshaft timing sprocket and the variable valve timing sprocket are as shown in the illustration.
c. If not, turn the crankshaft 1 revolution so as to align the marks as above.
38. Remove the crankshaft pulley.
39. Remove the chain tensioner assembly.

WARNING

Do not turn the crankshaft without the chain tensioner.

40. Remove the timing chain cover.

WARNING

Be careful not to damage the contact surfaces of the timing chain cover, cylinder head and cylinder block.

41. Remove the timing chain cover oil seal.
42. Remove the crankshaft position sensor plate.
43. Remove the chain tensioner slipper.
44. Remove the chain vibration damper.
45. Remove the timing chain.
46. Remove the oil level gauge guide and gauge.
47. Remove the bolts and nut, then remove the surge tank stay.
48. Remove the intake manifold.
a. Remove the bolts and nuts, then remove the intake manifold and gasket.

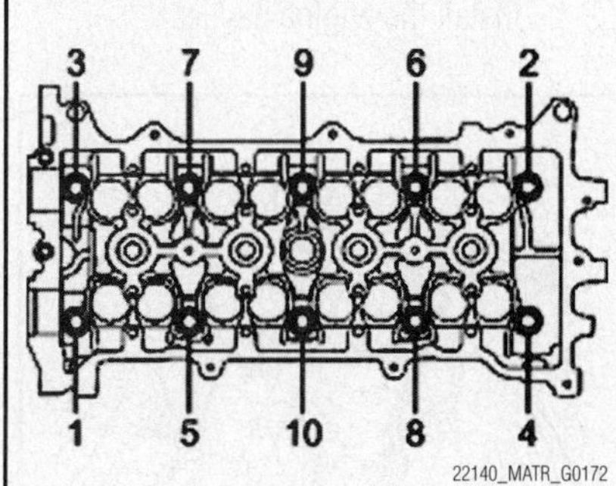

Fig. 93 Using several steps, uniformly install and tighten the cylinder head bolts and plate washers in the sequence shown—2ZZ-GE engine

49. Remove the camshaft.
a. Uniformly loosen and remove the bearing cap bolts, in several passes in the sequence shown, then remove the 9 bearing caps of the intake and exhaust camshafts.
50. Remove the cylinder head.
a. Disconnect the water by-pass pipe from the cylinder head.
b. Uniformly loosen the cylinder head bolts, in several passes in the sequence shown.

WARNING

Be careful not to drop the washers into the cylinder head. Head warpage or cracking could result from removing the bolts in the incorrect order.

c. Remove the cylinder head from the cylinder block.

To install:

51. Install the cylinder head gasket.
a. Place a new cylinder head gasket on the cylinder block surface with the lot No. Stamp upward.

WARNING

Make sure that the cylinder head gasket is mounted in the correct direction when installing. Place the cylinder head gently in order to avoid damaging the gasket.

b. Apply sealant to the cylinder head gasket as shown in the illustration.

WARNING

Avoid applying an excessive amount of sealant to the surface.

➡The parts must be assembled within 3 minutes of applying. Otherwise, the sealant must be removed and reapplied. Immediately remove the nozzle from the tube, then reinstall the cap.

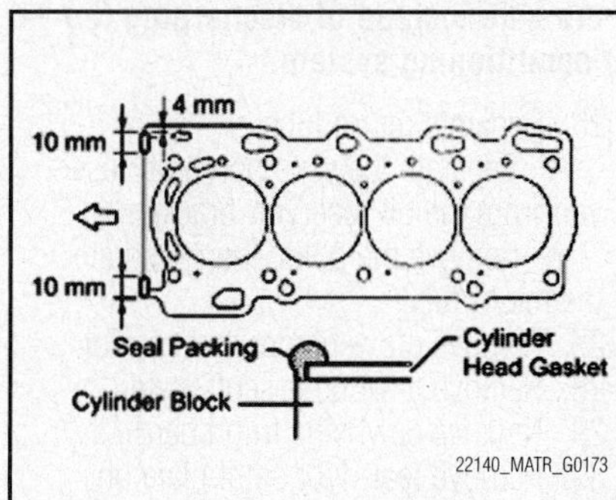

Fig. 94 Apply sealant to the cylinder head gasket as shown —2ZZ-GE engine

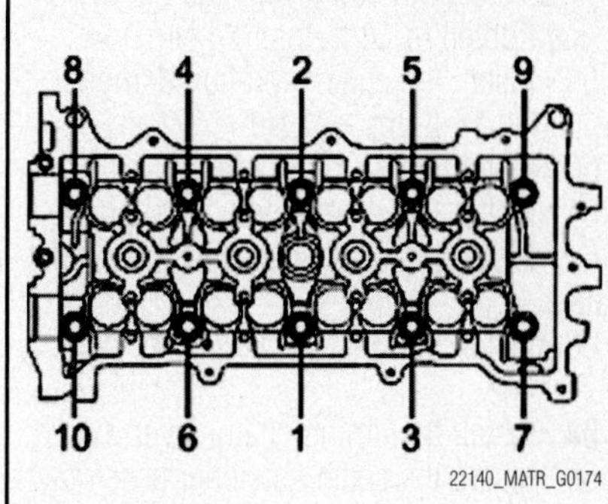

Fig. 95 Uniformly tighten the cylinder head bolts in 2 steps using the sequence shown—2ZZ-GE engine

52. Install the cylinder head.
a. Place the cylinder head on the cylinder block.
b. The cylinder head bolts are tightened in 2 steps.
c. Apply a light coat of engine oil to the threads and under the heads of the cylinder head bolts.
d. Install and uniformly tighten the 10 cylinder head bolts with the plate washers, in 2 steps in the sequence shown.
- Tighten the bolts to 26 ft. lbs. (35 Nm).
- Mark the front of each cylinder head bolt with paint.
- Tighten the cylinder head bolts an additional 180°rotation in sequence.
- Check that the paint marks on each bolts are at a 180°angle from the original position.

53. Install the water by-pass pipe. Tighten to 16 ft. lbs. (21 Nm).
54. Install the camshaft.
a. Apply a light coat of engine oil to the camshaft journals.
b. Place the 2 camshafts on the cylinder head with the cam lobes facing as shown in the illustration.
c. Examine the front marks and numbers and check that the sequence order is as shown in the illustration. Then tighten the bolts to 14 ft. lbs. (19 Nm).
55. Install the intake manifold.
56. Install the a new gasket and the intake manifold. Tighten the fasteners "A" to 25 ft. lbs. (34 Nm) and "B" 34 ft. lbs. (46 Nm).
57. Install the surge tank stay and tighten to 18 ft. lbs. (24 Nm).
58. Install the oil level gage guide.
a. Apply a light coat of engine oil to the new o-ring, then install it onto the oil level gage guide.

b. Install the oil level gage guide bolt and tighten to 18 ft. lbs. (24 Nm).

59. Install the chain vibration damper. Tighten to 15 ft. lbs. (21 Nm)
60. Install the timing chain.
61. Install the chain tensioner slipper.
62. Install the crankshaft position sensor plate.
63. Install the timing chain cover oil seal.
64. Install the timing chain cover.
65. Install the chain tensioner assembly.
66. Install the crankshaft pulley.
67. Install the crankshaft position sensor.
68. Install the transverse engine mounting bracket.
69. Install the water pump assembly.
70. Install the water pump pulley.
71. Install the exhaust manifold.
72. Install the exhaust manifold heat insulator.
73. Install the manifold stay.
74. Install the cylinder head cover.
75. Install the ignition coil assembly.
76. Install the V-ribbed belt tensioner assembly.
77. Install the engine mounting insulator.
78. Install the exhaust pipe assembly front.
79. Install the compressor and magnetic clutch.
80. Install the alternator assembly.
81. Install the alternator bracket.
82. Install the air switching valve assembly.
83. Install the suction hose. Tighten to 87 inch lbs. (10 Nm).
84. Connect the negative battery cable.
85. Install the front wheel.
86. Add engine coolant.
87. Check engine coolant leakage.
88. Check engine oil leakage.
89. Check exhaust gas leakage.
90. Inspect ignition timing.
91. Inspect engine idling speed.
92. Inspect compression.
93. Install the upper cylinder head cover.

ENGINE ASSEMBLY

REMOVAL & INSTALLATION

See Figures 96 through 101.

1. Before servicing the vehicle, refer to the precautions section.
2. Discharge fuel system pressure.
3. Disconnect battery negative terminal.
4. Remove front wheels.
5. Remove engine undercover.
6. Remove cylinder head cover.
7. Drain engine coolant.
8. Drain automatic transaxle fluid on AWD vehicles.
9. Drain transfer oil on AWD vehicles.
10. Disconnect radiator hose inlet and outlet.
11. Disconnect radiator hose outlet.
12. Disconnect oil cooler inlet and outlet hose on automatic transaxle equipped vehicles.
13. Remove radiator (w/ air conditioning system).
14. Remove battery, tray and carrier.
15. Remove air cleaner cap and case.
    a. Disconnect the clamp of the engine wire.
    b. Remove the bolts from the air cleaner case.
    c. Disconnect the air cleaner case from the air cleaner inlet.
16. Separate accelerator control cable on AWD vehicles.
17. Separate cruise control actuator on vehicles equipped with cruise control.
    a. Disconnect the actuator connector.
    b. Remove the bolts, then separate the actuator from the body.
18. Disconnect fuel tube.
19. Disconnect union to check valve hose.
20. Disconnect heater inlet and outlet hose.
21. Separate transmission control cable.
22. Separate clutch release cylinder on manual transaxle equipped vehicles.
23. Remove fan and alternator V-belt.
24. Remove alternator.
25. Separate air conditioning system compressor and magnetic clutch.

➡Secure the compressor and hoses off to the side instead of discharging the air conditioning system.

26. Separate return tube
    a. Separate the vane pump oil reservoir from the oil reservoir bracket.
    b. Remove the bolts, then separate the return tube.
27. Remove glove compartment door.
28. Remove front door scuff plate.
29. Remove cowl side trim board.
30. Remove rear door scuff plate on AWD vehicles.
31. Remove lower center pillar molding on AWD vehicles.
32. Remove front seat on AWD vehicles.
33. Disconnect engine wiring harnesses.
    a. Disconnect the engine wire from the control module and junction block.
    b. Pull out the engine wire.
    c. Remove the engine room relay block cover.
    d. Disconnect the connectors.
    e. Disconnect the engine wire harness by pulling it upward.
    f. Remove the clamp of the engine wire.
    g. Remove the bolt, then disconnect the ground terminal.
34. Remove propeller w/center bearing shaft on AWD vehicles.
35. Remove center exhaust pipe on AWD vehicles.
36. Remove front floor panel brace on FWD vehicles.
37. Remove front exhaust pipe.
38. Remove column hole cover silencer sheet.
39. Separate steering intermediate shaft.
40. Remove front axle shaft nut.
41. Separate front stabilizer link.
42. Separate front speed sensor.
43. Separate tie rod end.
44. Separate lower front suspension arm.
45. Separate front axle.
46. Remove engine with transaxle
    a. Install the engine lifter.
    b. Remove the bolts and nuts, then remove the right side engine mounting insulator.
    c. Remove the through bolt and nut, then separate the left side engine mounting insulator.
    d. Remove the bolts as shown in the illustration.
    e. Carefully remove the engine with transaxle from the engine compartment.
    f. Install the engine hanger.

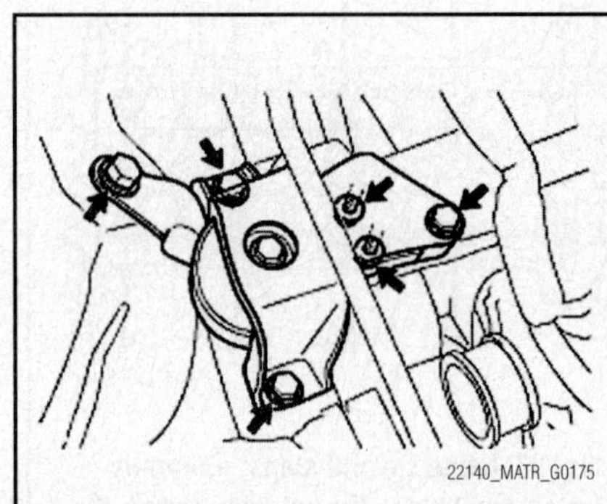

Fig. 96 Right side engine mounting insulator

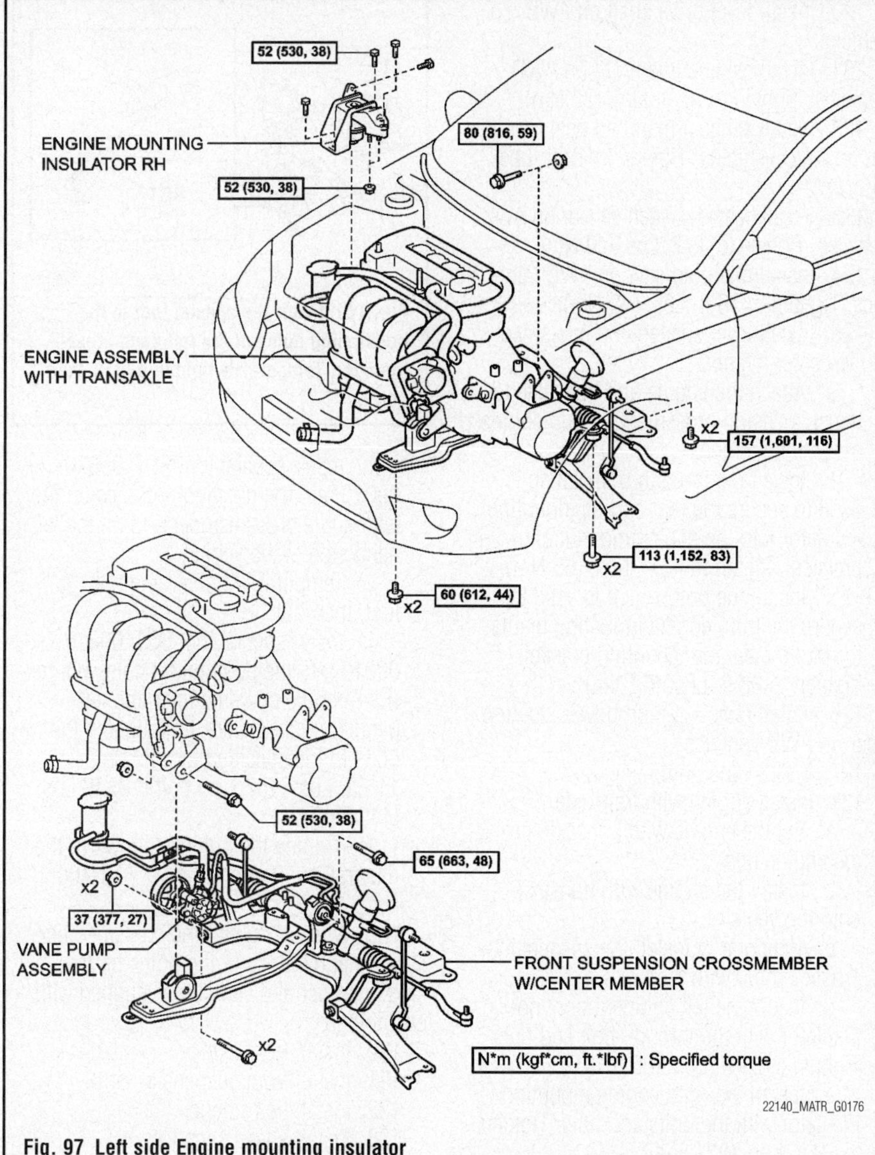

Fig. 97 Left side Engine mounting insulator

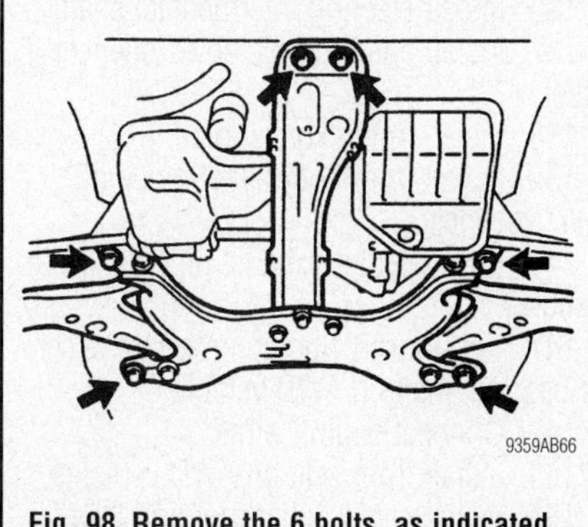

Fig. 98 Remove the 6 bolts, as indicated by arrows

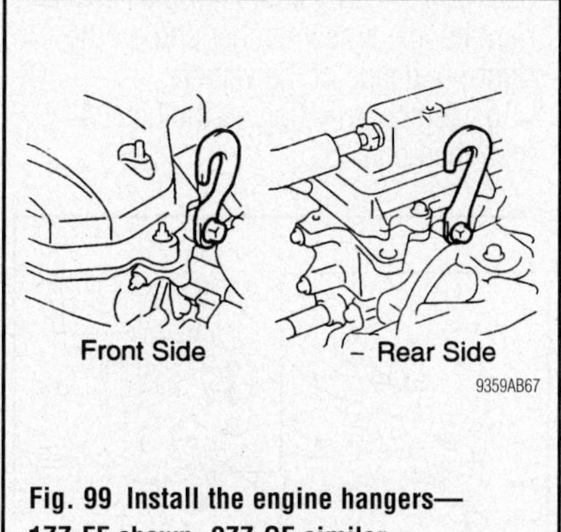

Fig. 99 Install the engine hangers—1ZZ-FE shown, 2ZZ-GE similar

47. Separate vane pump.
48. Remove rack & pinion power steering gear on AWD vehicles.
49. Remove front suspension crossmember w/center member
    a. Remove the through bolt and nut, then separate the engine mounting insulator from the engine mounting bracket.
    b. Separate the engine and transaxle from the suspension crossmember and the engine mounting member.
50. Remove manifold stay on AWD vehicles.
51. Remove heated oxygen sensor on AWD vehicles.
52. Remove exhaust manifold heat insulator on AWD vehicles.
53. Remove exhaust manifold on AWD vehicles.
54. Remove front drive shaft on AWD vehicles.
55. Remove engine mounting bracket on AWD vehicles.
56. Remove transfer stiffener plate on AWD vehicles.
57. Remove transfer case on AWD vehicles.
58. Remove starter.
59. Remove manual transaxle on manual transaxle equipped vehicles.
60. Remove automatic transaxle.
61. Remove clutch cover on manual transaxle equipped vehicles.
62. Remove clutch disc on manual transaxle equipped vehicles.
63. Remove flywheel on manual transaxle equipped vehicles.
64. Remove drive plate & ring gear on automatic transaxle equipped vehicles.
65. Remove ignition coil.
66. Remove fuel delivery pipe.
67. Remove oil level gage.
68. Remove intake manifold.
69. Remove oil level gage guide
    a. Disconnect the crankshaft position sensor clamp.
    b. Remove the bolt, then remove the oil level gage guide.
    c. Disconnect the knock sensor clamp.
    d. Remove the water by-pass pipe.
    e. Remove the gasket from the cylinder block.
70. Remove water inlet.
71. Remove thermostat.
72. Remove engine oil pressure switch.
73. Remove camshaft position sensor.
74. Remove crankshaft position sensor.
75. Remove knock sensor.
76. Remove v-ribbed belt tensioner.
77. Remove manifold stay on FWD vehicles
78. Disconnect the air fuel ratio sensor connector then remove exhaust manifold heat insulator on FWD vehicles.
79. Remove exhaust manifold on FWD vehicles remove the gasket from the cylinder head.
80. Remove engine coolant temperature sensor.
81. Remove radio setting condenser

82. Remove water by-pass hose no.2
83. Remove radiator hose inlet
84. Remove heater inlet water hose

To install:

85. Install heater inlet water hose.
86. Install radiator hose inlet.
87. Install water by-pass hose.
88. Install radio setting condenser.
89. Install engine coolant temperature sensor.
90. Install exhaust manifold on FWD vehicles. Tighten to 27 ft. Lbs. (37 Nm).
91. Install exhaust manifold heat insulator on FWD vehicles. Tighten to 13 ft. Lbs. (18 Nm).
92. Connect the air fuel ratio sensor connector.
93. Install manifold stay on FWD vehicles. Tighten to 36 ft. Lbs. (49 Nm).
94. Install v-ribbed belt tensioner.
95. Install knock sensor. Tighten to 15 ft. Lbs. (20 Nm).
96. Install crankshaft position sensor. Tighten to 80 inch lbs. (9 Nm).
97. Install camshaft position sensor. Tighten to 80 inch lbs. (9 Nm).
98. Install engine oil pressure switch.
99. Install thermostat.
100. Install water inlet.
101. Install water by-pass pipe. Tighten to 80 inch lbs. (9 Nm).
102. connect the knock sensor clamp.
103. Apply a light coat of engine oil to a new o-ring and install it onto the oil level gage guide. Tighten to 10 ft. Lbs. (13 Nm).
104. Connect the crankshaft position sensor clamp.
105. Install intake manifold.
106. Install oil level gage.
107. Install fuel delivery pipe.
108. Install ignition coil.
109. Install flywheel on manual transaxle equipped vehicles.
110. Install drive plate & ring gear on automatic transaxle equipped vehicles.
111. Install clutch disc on manual transaxle equipped vehicles.
112. Install clutch cover on manual transaxle equipped vehicles.
113. Install automatic transaxle.
114. Install manual transaxle on manual transaxle equipped vehicles.
115. Install starter.
116. Install transfer on AWD vehicles.
117. Install transfer stiffener plate on AWD vehicles.
118. Install transfer stiffener plate center on AWD vehicles.
119. Install engine mounting bracket on AWD vehicles.
120. Install front drive shaft on AWD vehicles.
121. Install exhaust manifold on AWD vehicles. Tighten to 27 ft. Lbs. (37 Nm).
122. Install exhaust manifold heat insulator on AWD vehicles. Tighten to 13 ft. Lbs. (18 Nm).
123. Install heated oxygen sensor on AWD vehicles. Tighten to 33 ft. Lbs. (44 Nm).
124. Install manifold stay on AWD vehicles. Tighten to 36 ft. Lbs. (49 Nm).
125. Install front suspension crossmember w/center member
    a. Attach the engine and transaxle to the suspension crossmember and engine mounting member.
    b. Install the through bolt, which is used to secure the rear engine mounting insulator into the engine mounting bracket. Tighten to 48 ft. Lbs. (65 Nm)
    c. Install the bolt, which is used to secure the front engine mounting insulator into the engine mounting bracket. Tighten to 38 ft. Lbs. (52 Nm).
126. Install rack & pinion power steering gear on AWD vehicles.
127. Install vane pump.
128. Install engine with transaxle
    a. Set the engine with transaxle on the engine lifter.
    b. Install the engine with transaxle onto the vehicle.
    c. Temporarily install the suspension crossmember with the bolts.
    d. Install the left engine mounting insulator with the through bolt and nut. Tighten to 59 ft. Lbs. (80 Nm)
    e. Install the right engine mounting insulator with the bolts and nuts. Tighten to 38 ft. Lbs. (52 Nm)
    f. Insert special tool SST: 09670-00010 into the positioning holes on the right handle crossmember and on the right-hand side of the vehicle.
    g. Temporarily tighten the bolt "A" first, then bolt "B".

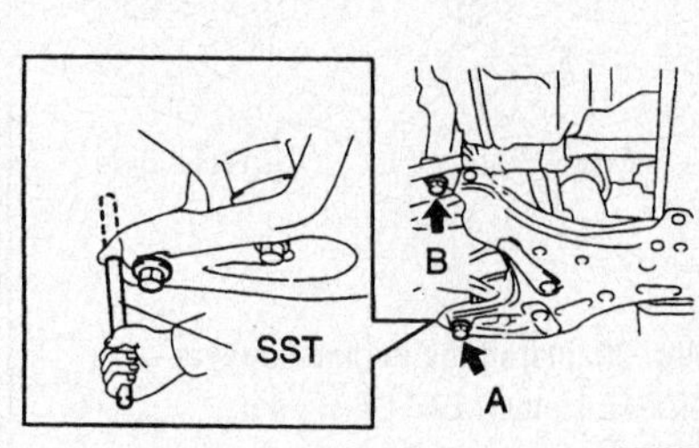

Fig. 100 Insert the special tool to the positioning holes of the right hand cross-member. Temporarily tighten bolt A, then bolt B

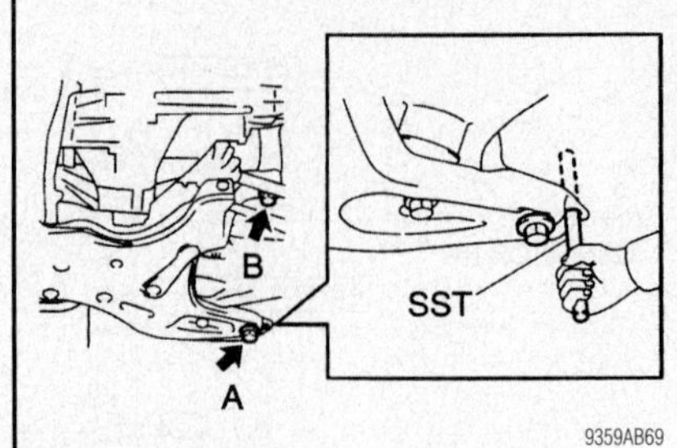

Fig. 101 Insert the special tool to the positioning holes of the left hand cross-member. Temporarily tighten bolt A, then bolt B

    h. Insert special tool SST: 09670-00010 into the positioning holes on the left handle crossmember and on the left-hand side of the vehicle.
    i. Temporarily tighten the bolt "A" first, then bolt "B".
    j. Insert special tool SST: 09670-00010 into the positioning holes on the right-handle crossmember and left-handle of the vehicle, then tighten the bolts to the specified torque. Bolt a 116 ft. Lbs. (157 Nm) bolt b 83 ft. Lbs. (113 Nm).
    k. Tighten the bolts as shown in the illustration. Tighten to 44 ft. Lbs. (60 Nm).
129. When installing the crossmember, check that the positioning holes on the crossmember and vehicle are aligned with each other.
130. Install front axle.
131. Install front suspension arm.
132. Install tie rod end.
133. Install speed sensor front.
134. Install front stabilizer link.
135. Install front axle shaft nut.
136. Install steering intermediate shaft.
137. Install column hole cover silencer sheet.
138. Install front exhaust pipe.
139. Install front floor panel brace on FWD vehicles.
140. Install center exhaust pipe on AWD vehicles.
141. Temporarily tighten propeller w/center bearing shaft on AWD vehicles.
142. Connect engine wires.
143. Install front seat on AWD vehicles.
144. Install center pillar molding lower on AWD vehicles.
145. Install rear door scuff plate on AWD vehicles.
146. Install cowl side trim board.
147. Install front door scuff plate.
148. Install glove compartment door.
149. Install return tube.

150. Install the vane pump oil reservoir onto the oil reservoir bracket.
151. Install air conditioning system compressor and magnetic clutch.
152. Install alternator.
153. Install fan and alternator V- belt.
154. Install clutch release cylinder on manual transaxle equipped vehicles.
155. Install transmission control cable.
156. Connect heater outlet water hose.
157. Connect heater inlet water hose.
158. Connect union to check valve hose.
159. Connect fuel tube.
160. Install cruise control actuator on vehicles equipped with cruise control. Tighten to 53 inch lbs. (6 Nm)
161. Connect the actuator connector.
162. Install accelerator control cable on AWD vehicles.
163. Connect the air cleaner case to the air cleaner inlet.
164. Install the air cleaner case with the 3 bolts. Tighten to 62 inch lbs. (7 Nm).
165. Connect the clamp of the engine wire.
166. Install air cleaner cap.
167. Install battery carrier. Tighten to 9 ft. Lbs. (13 Nm).
168. Install battery tray.
169. Install battery. Tighten bolt to 44 inch lbs. (5 Nm), nut to 31 inch lbs. (4 Nm) and terminal to 48 inch lbs. (5.4 Nm).
170. Install radiator.
171. Connect oil cooler inlet and outlet hose on automatic transaxle equipped vehicles.
172. Connect radiator hose inlet and outlet.
173. Install front wheels.
174. Add transfer oil on AWD vehicles.
175. Add automatic transaxle fluid on AWD vehicles.
176. Add engine coolant.
177. Connect battery negative terminal
178. Check engine coolant leak
179. Check engine oil leakage
180. Check fuel leak
181. Check exhaust gas leak
182. Inspect and adjust front wheel alignment
183. Check abs speed sensor signal.
184. Install cylinder head cover.
185. Install engine undercover.

EXHAUST MANIFOLD

REMOVAL & INSTALLATION

1ZZ-FE Engine

See Figure 102.

1. Before servicing the vehicle, refer to the precautions section.

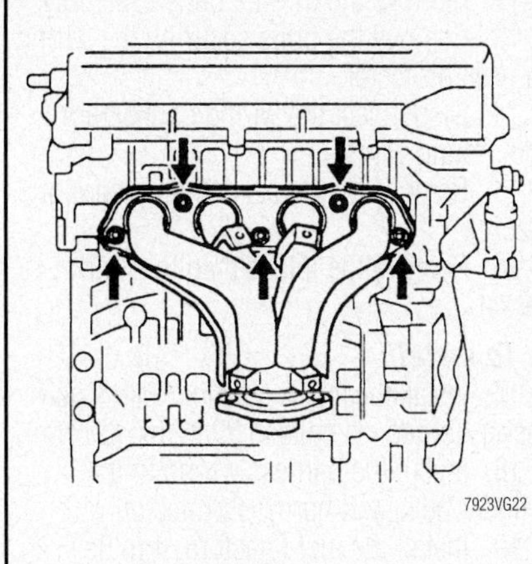

Fig. 102 Exhaust manifold mounting nut locations—1ZZ-FE engine

2. Drain the cooling system.
3. Disconnect the negative battery cable.
4. Remove the drive belt and alternator
5. Remove the air intake duct
6. Remove the accelerator cable
7. Remove the exhaust pipe from the manifold
8. Remove the exhaust manifold support bracket
9. Remove the heat insulator from the dash panel
10. Remove the upper heat insulator
11. Remove the exhaust manifold and gasket
12. If necessary, remove the lower heat insulator from the exhaust manifold.

To install:

13. Install the lower heat insulator on the exhaust manifold. Tighten the bolts to 108 inch lbs. (12 Nm).
14. Install the exhaust manifold using a new gasket. Tighten the nuts, in several passes, to 27 ft. lbs. (37 Nm).
15. Install the upper heat insulator. Tighten the bolts to 108 inch lbs. (12 Nm).
16. Install the heat insulator on the dash panel
17. Install the exhaust manifold support bracket. Tighten the bolts, in an alternating pattern, to 37 ft. lbs. (49 Nm).
18. Install the front exhaust pipe to the manifold. Tighten the bolts to 46 ft. lbs. (62 Nm).
19. Install the oxygen sensor. Tighten the nuts to 14 ft. lbs. (20 Nm).
20. Install the accelerator cable and air intake duct
21. Install the alternator and drive belt
22. Connect the negative battery cable.
23. Fill the cooling system.
24. Start the vehicle, check for leaks and repair if necessary.

2ZZ-GE Engine

See Figures 103 and 104.

1. Before servicing the vehicle, refer to the precautions section.
2. Drain the cooling system.
3. Disconnect the negative battery cable.
4. Remove the drive belt and alternator.
5. Remove the air intake duct.
6. Remove the accelerator cable.
7. Remove the exhaust pipe from the manifold.
8. Remove the exhaust manifold support bracket.
9. Remove the heat insulator from the dash panel.
10. Remove the upper heat insulator.
11. Remove the exhaust manifold and gasket.
12. If necessary, remove the lower heat insulator from the exhaust manifold.

To install:

13. Install the lower heat insulator on the exhaust manifold. Tighten the bolts to 15 ft. lbs. (20 Nm).

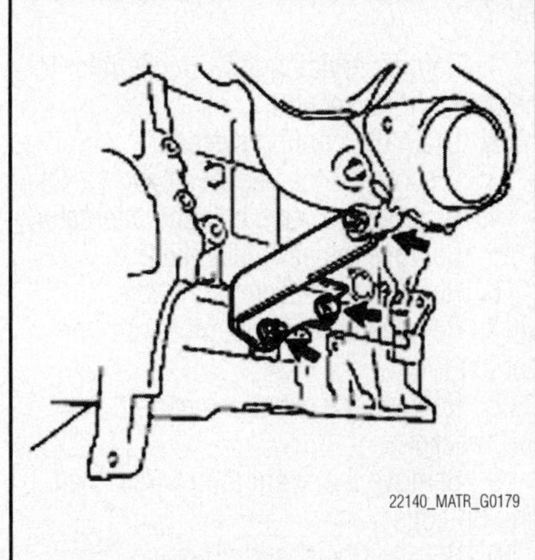

Fig. 103 Exhaust manifold stay mounting bolts—2ZZ-GE engine

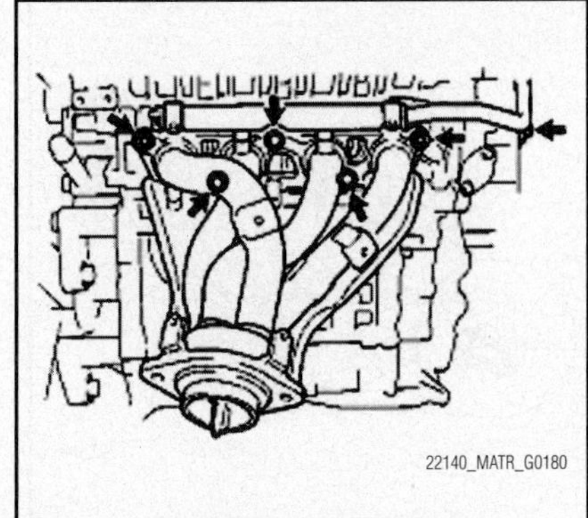

Fig. 104 Exhaust manifold mounting bolts—2ZZ-GE engine

14. Install the exhaust manifold using a new gasket. Tighten the nuts, in several passes to 37 ft. lbs. (50 Nm).
15. Install the upper heat insulator. Tighten the bolts to 15 ft. lbs. (20 Nm).
16. Install the heat insulator on the dash panel.
17. Install the exhaust manifold support bracket. Tighten the bolts to 37 ft. lbs. (49 Nm).
18. Install the front exhaust pipe to the manifold. Tighten the bolts to 46 ft. lbs. (62 Nm).
19. Install the oxygen sensor . Tighten the nuts to 14 ft. lbs. (20 Nm).
20. Install the accelerator cable and air intake duct.
21. Install the alternator and drive belt.
22. Connect the negative battery cable.
23. Fill the cooling system.
24. Start the vehicle, check for leaks and repair if necessary.

INTAKE MANIFOLD

REMOVAL & INSTALLATION

1ZZ-FE Engine

See Figure 105.

1. Before servicing the vehicle, refer to the precautions section.
2. Drain the cooling system.
3. Disconnect the negative battery cable.
4. Remove the drive belt and alternator.
5. Remove the air intake duct.
6. Remove the accelerator cable.
7. Remove the exhaust pipe from the manifold.
8. Remove the exhaust manifold support bracket.
9. Remove the spark plug wires, then ignition coils.
10. Remove the spark plugs.
11. Remove the Positive Crankcase Ventilation (PCV) hoses.
12. Remove the throttle body assembly.
13. Remove the bolts securing the wiring harness protector.
14. Disconnect the wiring connectors and ground wires.
15. Remove the intake manifold support bracket.
16. Remove the intake manifold and gasket.

To install:

17. Install the intake manifold with a new gasket. Torque the bolts to 22 ft. lbs. (30 Nm).
18. Install the harness wiring to the cylinder head and harness protector
19. Install the fuel injectors, throttle body and the PCV hoses
20. Install the spark plugs and ignition coils. Tighten the bolts and nuts to 80 inch lbs. (9 Nm).
21. Install the exhaust manifold and support bracket. Tighten the bolts to 37 ft. lbs. (49 Nm).
22. Install the front exhaust pipe to the manifold. Tighten the bolts to 46 ft. lbs. (62 Nm).
23. Install the oxygen sensor . Tighten the nuts to 14 ft. lbs. (20 Nm).
24. Install the accelerator cable and air intake duct
25. Install the alternator and drive belt
26. Connect the negative battery cable.
27. Fill the cooling system.
28. Start the vehicle, check for leaks and repair if necessary.

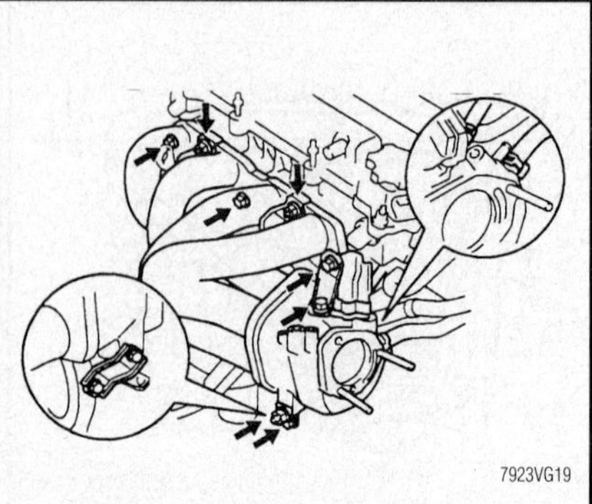

Fig. 105 Intake manifold mounting fastener locations—1ZZ-FE engine

2ZZ-GE Engine

See Figure 106.

1. Before servicing the vehicle, refer to the precautions section.
2. Drain the cooling system.
3. Remove or disconnect the following:
4. Remove the Negative battery cable
5. Remove the Drive belt and alternator
6. Remove the Air intake duct
7. Remove the Accelerator cable
8. Remove the Spark plug wires, then ignition coils
9. Remove the Spark plugs
10. Remove the Positive Crankcase Ventilation (PCV) hoses
11. Remove the Throttle body assembly
12. Remove the Wiring harness
13. Remove the Hoses and tubes connected to the head
14. Remove the Intake manifold support bracket
15. Remove the Intake manifold and gasket

To install:

16. Install the intake manifold with a new gasket. Tighten bolts A to 25 ft. lbs. (34 Nm) and bolt B to 34 ft. lbs. (46 Nm)
17. Install the harness wiring to the cylinder head and harness protector
18. Install the fuel injectors, throttle body and the PCV hoses
19. Install the spark plugs and ignition coils. Tighten the bolts and nuts to 80 inch lbs. (9 Nm).
20. Install the oxygen sensor . Tighten the nuts to 14 ft. lbs. (20 Nm).
21. Install the accelerator cable and air intake duct
22. Install the alternator and drive belt
23. Connect the negative battery cable.
24. Fill the cooling system.
25. Start the vehicle, check for leaks and repair if necessary.

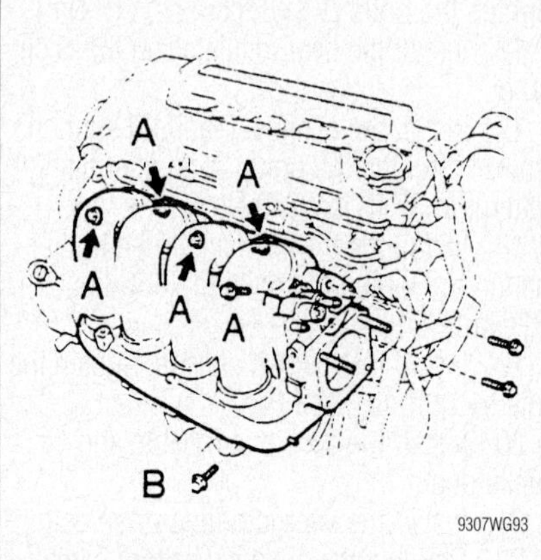

Fig. 106 Intake manifold bolt installation—2ZZ-GE Engine

OIL PAN

REMOVAL & INSTALLATION

See Figures 107 and 108.

1. Before servicing the vehicle, refer to the precautions section.

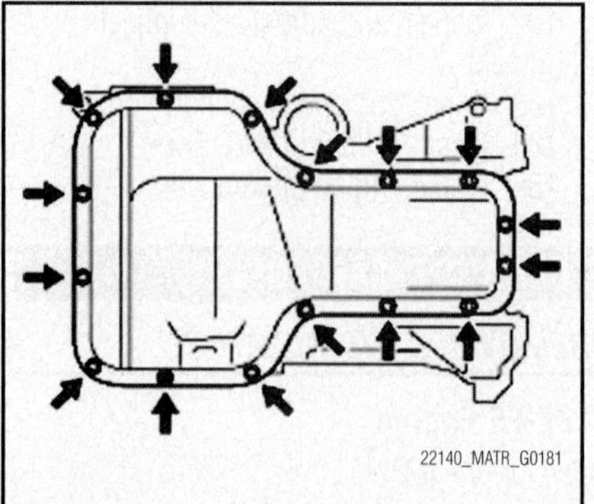

Fig. 107 Oil pan mounting bolt locations

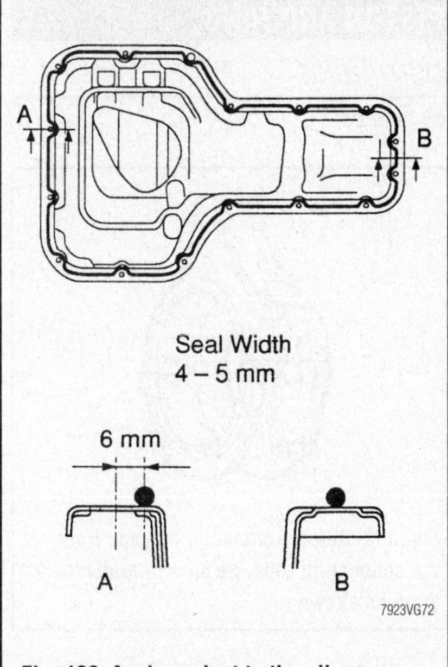

Fig. 108 Apply sealant to the oil pan as shown

2. Drain the engine oil.
3. Disconnect the negative battery cable.
4. Remove the undercovers.
5. Remove the front exhaust pipe.
6. Remove the oil pan mounting bolts and nuts.
7. Remove the oil pan, cutting off the applied sealer.

⁂ WARNING

Be careful not to damage the oil pan contact surface of the bearing cap and oil pan.

To install:

8. Remove any old sealant from the oil pan flange and thoroughly clean the sealing surface.
9. Apply sealant to the oil pan as shown in the illustration.

➡Tighten the bolts and nuts within 3 minutes of applying sealant.

10. Install the oil pan. Tighten the bolts and nuts in several passes to 80 inch lbs. (9 Nm).

➡Tighten the bolts and nuts to the specified torque in two passes.

11. Place a new gasket on the oil pan drain plug, then install it onto the oil pan. Tighten to 27 ft. lbs. (37 Nm).
12. Install the front exhaust pipe.
13. Connect the negative battery cable.
14. Install the undercovers.
15. Fill the engine with clean oil.
16. Start the vehicle, check for leaks and repair if necessary.

OIL PUMP

REMOVAL & INSTALLATION

1ZZ-FE Engine

See Figure 109.

1. Before servicing the vehicle, refer to the precautions section.
2. Disconnect battery negative terminal.
3. Remove the front wheel.
4. Remove the engine undercover.
5. Remove the cylinder head cover.
6. Drain the engine oil.
7. Drain and recycle the engine coolant.
8. Remove the fan and generator v belt.
9. Remove the vane pump assembly.
10. Remove the generator assembly.
11. Remove the engine mounting insulator.
12. Remove the ignition coil.
13. Remove the cylinder head cover.
14. Remove the v-ribbed belt tensioner.
15. Remove the engine mounting bracket.
16. Remove the water pump.
17. Separate crankshaft position sensor.
18. Set the No. 1 cylinder to TDC on the compression stroke.
19. Remove the crankshaft pulley.
20. Remove the chain tensioner.
21. Remove the timing chain cover.
22. Remove the timing chain cover oil seal.
23. Remove the crankshaft position sensor.
24. Remove the chain tensioner slipper.
25. Remove the chain.
26. Remove the chain vibration damper.
27. Remove the bolts, then remove the oil pump.
28. Remove the gasket from the cylinder block.

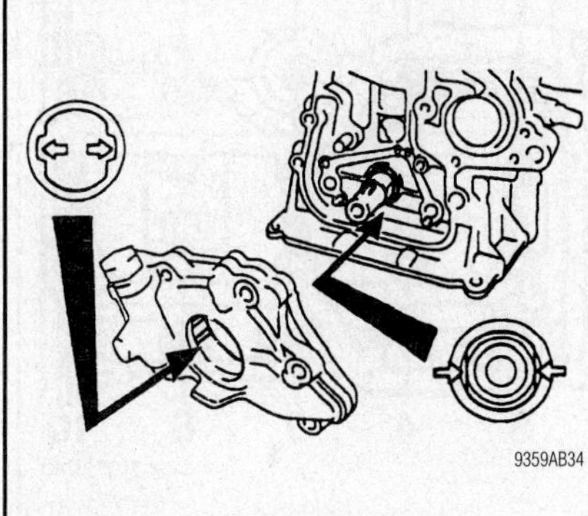
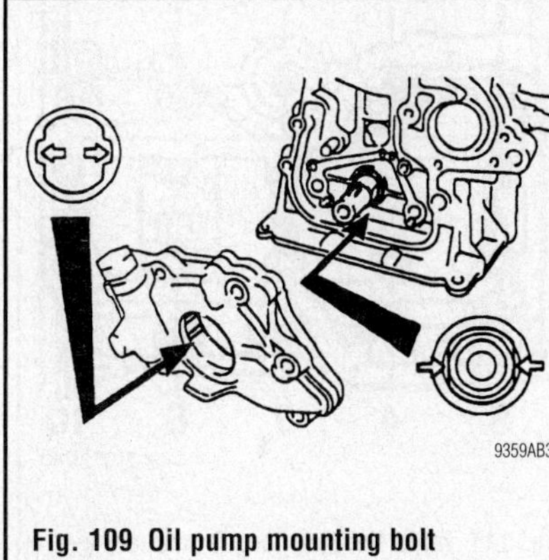

Fig. 109 Oil pump mounting bolt locations

To install:

29. Clean the mounting surface.
30. Install the oil pump, with a new gasket.
31. Engage the spline teeth of the oil pump drive rotor with the larger teeth of the crankshaft, and slide the pump on.
32. Install the oil pump bolts and tighten to 80 inch lbs. (9 Nm).
33. Install the chain vibration damper.
34. Install the chain.
35. Install the chain tensioner slipper.
36. Install the crankshaft position sensor.
37. Install the timing chain cover oil seal.
38. Install the timing chain cover.
39. Install the chain tensioner.
40. Ensure the No. 1 cylinder is set to TDC on the compression stroke.
41. Install the crankshaft pulley.
42. Attach the crankshaft position sensor.
43. Install the water pump.
44. Install the engine mounting bracket.
45. Install the v-ribbed belt tensioner.
46. Install the cylinder head cover.
47. Install the ignition coil.
48. Install the engine mounting insulator.
49. Install the generator assembly.
50. Install the vane pump assembly.
51. Install the fan and generator v belt.
52. Install the cylinder head cover.
53. Install the engine undercover.
54. Install the front wheel.
55. Lower the vehicle.
56. Fill the engine with coolant.
57. Fill the engine with oil.
58. Start the vehicle, check for leaks and repair if necessary.

2ZZ-GE Engine

See Figure 109.

1. Before servicing the vehicle, refer to the precautions section.
2. Disconnect battery negative terminal.
3. Raise and safely support the vehicle securely on jackstands.
4. Drain and recycle the engine coolant.
5. Drain the engine oil
6. Clean and install the oil pan drain plug with a new gasket. Tighten to 27 ft. lbs. (37 Nm).
7. Remove the front wheel.
8. Remove the engine undercover.
9. Remove the cylinder head cover
10. Remove the wire harness protector.
11. Separate suction hose.
12. Remove the air switching valve.
13. Remove the fan and alternator v belt.
14. Remove the alternator bracket.
15. Remove the alternator.

16. Remove the air conditioning system compressor and magnetic clutch.
17. Remove the exhaust pipe front.
18. Remove the engine mounting insulator.
19. Remove the V-ribbed belt tensioner.
20. Remove the ignition coil.
21. Remove the cylinder head cover.
22. Remove the manifold stay.
23. Remove the exhaust manifold heat insulator
24. Remove the exhaust manifold.
25. Remove the water pump pulley.
26. Remove the water pump.
27. Remove the transverse engine mounting bracket.
28. Remove the crankshaft position sensor.
29. Set the No. 1 cylinder to TDC on the compression stroke.
30. Remove the crankshaft pulley.
31. Remove the chain tensioner.
32. Remove the timing chain or belt cover.
33. Remove the timing chain or belt cover oil seal.
34. Remove the crankshaft position sensor plate.
35. Remove the chain tensioner slipper
36. Remove the chain vibration damper
37. Remove the chain.
38. Remove the oil pump.

To install:

39. Clean the mounting surface.
40. Install the oil pump, with a new gasket.
41. Engage the spline teeth of the oil pump drive rotor with the larger teeth of the crankshaft, and slide the pump on.
42. Install the oil pump bolts and tighten to 80 inch lbs. (9 Nm).
43. Install the chain vibration damper
44. Install the chain.
45. Install the chain tensioner slipper.
46. Install the crankshaft position sensor plate.
47. Install the timing chain or belt cover oil seal.
48. Install the timing chain or belt cover.
49. Install the chain tensioner.
50. Install the crankshaft pulley.
51. Install the crankshaft position sensor.
52. Install the transverse engine mounting bracket.
53. Install the water pump.
54. Install the water pump pulley.
55. Install the exhaust manifold.
56. Install the exhaust manifold heat insulator.
57. Install the manifold stay.
58. Install the cylinder head cover.
59. Install the ignition coil.
60. Install the V-ribbed belt tensioner.
61. Install the engine mounting insulator.
62. Install the exhaust pipe front.
63. Install the compressor and magnetic clutch (w/ air conditioning system).
64. Install the alternator.
65. Install the alternator bracket no.1.
66. Install the fan and alternator v belt.
67. Install the suction hose.
68. Install the air switching valve.
69. Install the wire harness protector.
70. Connect battery negative terminal.
71. Install the front wheel.
72. Fill the engine with coolant.
73. Fill the engine with oil.
74. Install the cylinder head cover.
75. Install the engine undercover
76. Install the wheels.
77. Lower the vehicle.

MAIN BEARING TORQUE SEQUENCE

See Figures 110 and 111.

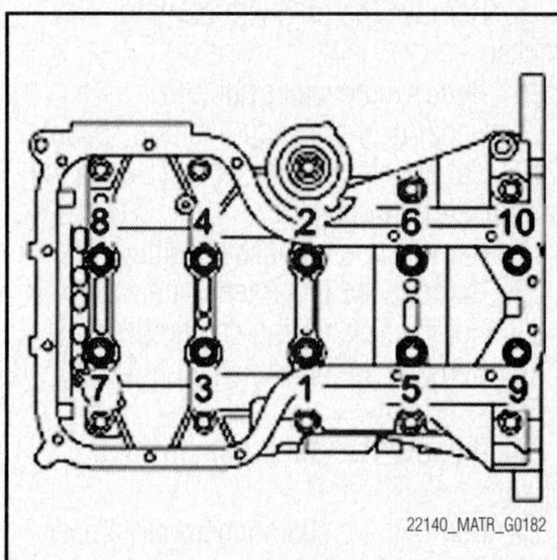

Fig. 110 Main bearing torque sequence—1ZZ-FE engine

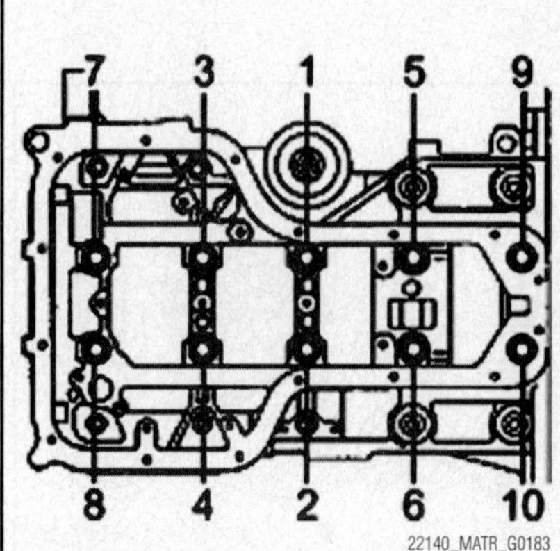

Fig. 111 Main bearing torque sequence—2ZZ-GE engine

PISTON AND RING

POSITIONING

See Figures 112 through 115.

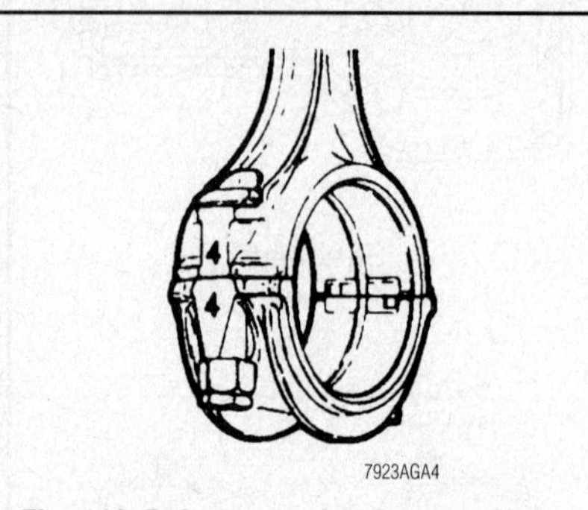

Fig. 112 Before removing the caps from the connecting rods, be sure to matchmark them as shown

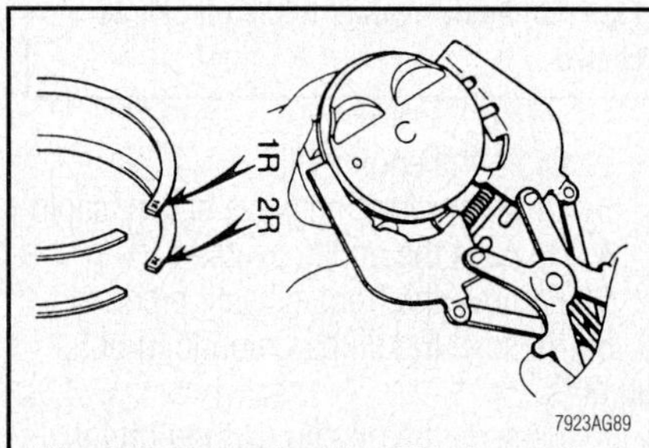

Fig. 113 Piston ring identification mark locations

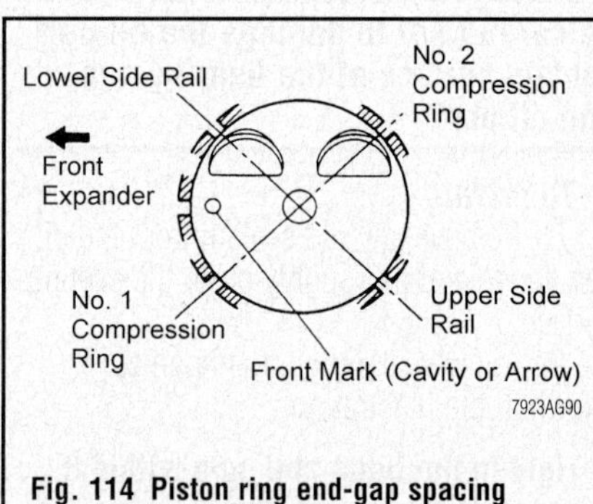

Fig. 114 Piston ring end-gap spacing

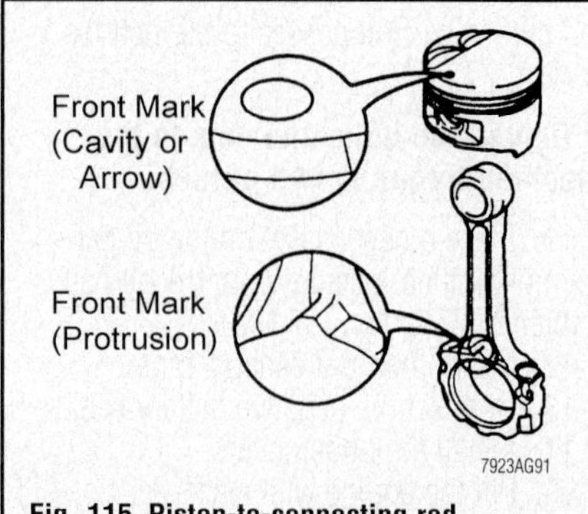

Fig. 115 Piston-to-connecting rod assembly

REAR MAIN SEAL

REMOVAL & INSTALLATION

See Figures 116 and 117.

1. Before servicing the vehicle, refer to the precautions section.
2. Raise and safely support the vehicle securely on jackstands.
3. Remove the transaxle assembly.
4. On manual transaxle equipped vehicles, remove the clutch assembly.
5. Remove the flywheel or flexplate.
6. Use a small sharp knife to cut off the lip of the oil seal. Take great care not to score any metal with the knife.
7. Use a small prytool to pry the old seal from the retaining plate. Be careful not to damage the plate. Protect the tip of the tool with tape and pad the fulcrum point with cloth.
8. Inspect the crankshaft and seal lip contact surfaces for any sign of damage.

➡After removing, check the crankshaft for damage. If damaged, smooth the surface with 400-grit sandpaper.

Fig. 116 Use a small sharp knife to cut off the lip of the oil seal then use a small prytool to pry the old seal from the retaining plate

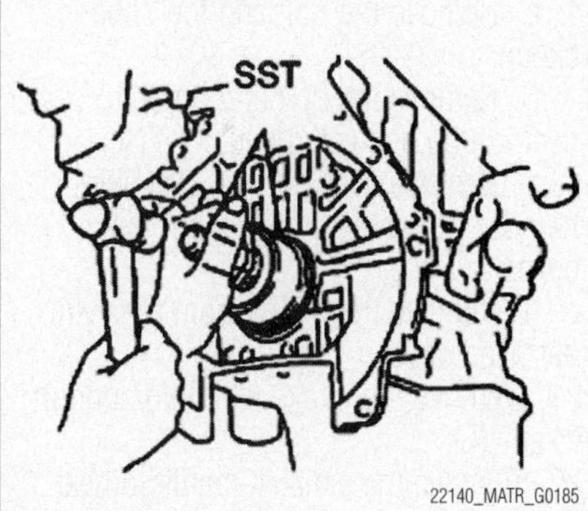

Fig. 117 Use a seal driver of the correct size to install the seal

To install:

9. Apply a light coat of multi-purpose grease to the lip of a new oil seal. Loosely fit the seal into place by hand, making sure it is not crooked.
10. Use a seal driver of the correct size to install the seal. Tap it into place until the surface of the seal is flush with the edge of the housing.
11. Install the flywheel or flexplate.
12. On manual transaxle equipped vehicles, install the clutch assembly.
13. Install the transaxle assembly.
14. Lower the vehicle.

ROCKER ARMS/SHAFTS

REMOVAL & INSTALLATION

2ZZ-GE Engine Only

See Figures 118 and 119.

1. Before servicing the vehicle, refer to the precautions section.
2. Remove the camshafts. Refer to Engine Mechanical, Camshaft, Removal & Installation.
3. Remove the bolt and rocker shaft.
4. Remove the valve rocker arm.

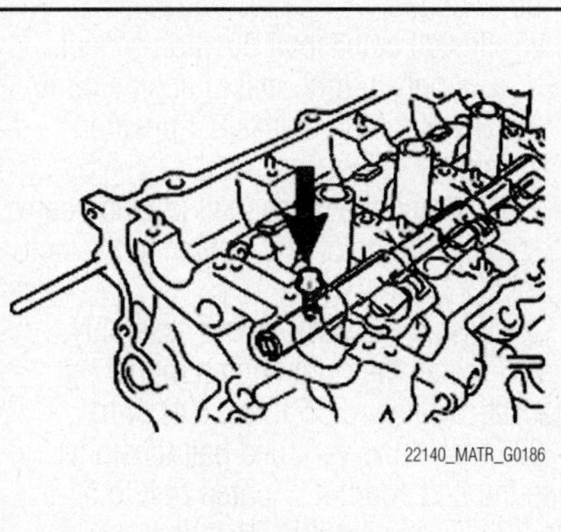

Fig. 118 Remove the bolt and rocker shaft

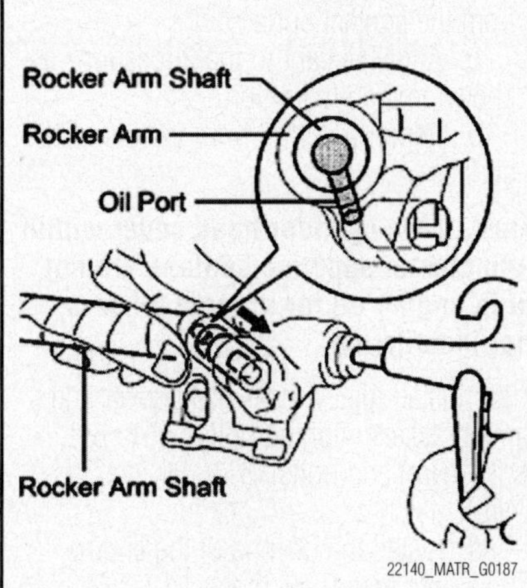

Fig. 119 Check that the piston inside of the rocker arm moves when air pressure is applied to the oil paths

To install:

5. Cover the oil paths of the rocker arm shaft except the 2 paths covered with vinyl type.
6. Align the oil path of the rocker arm shaft with the oil path of the rocker arm.
7. Check that the piston inside of the rocker arm moves when 26 psi (180 kpa) of air pressure is applied to the oil paths.
8. Put the valve rocker shaft through the cylinder head and the hole of the valve rocker arm, then check that the groove is at the right direction as shown.
9. Install the bolt to fix the rocker shaft and tighten to 7 ft. lbs. (10 Nm).

TIMING CHAIN COVER AND SEAL

REMOVAL & INSTALLATION

1ZZ-FE Engine

See Figures 51, 65 through 69, and 121 through 123, 131 through 133.

1. Before servicing the vehicle, refer to the precautions section.
2. Disconnect battery negative terminal.
3. Remove the front wheel.
4. Remove engine undercover.
5. Remove the upper cylinder head cover.
6. Drain and recycle the engine coolant.
7. Remove the fan and alternator V-belt.
8. Separate the vane pump assembly.
9. Remove the alternator assembly.
10. Remove the right engine mounting insulator.
11. Remove the ignition coil assembly.
12. Remove the cylinder head cover.
13. Remove the V-ribbed belt tensioner.

➡Jack the engine up and down to remove the bolt.

14. Remove the right hand engine mounting bracket.
15. Remove the water pump assembly.
16. Separate the crankshaft position sensor.
17. Set the No. 1 cylinder to TDC on the compression stroke.
    a. Turn the crankshaft pulley until its groove and timing mark "0" of the timing chain cover are aligned.
    b. Check that the point marks of the camshaft timing gears are in a straight line on the timing chain cover surface as shown in the illustration.

c. If not, turn the crankshaft 1 revolution (360°) to align the marks as above.
18. Remove the crankshaft pulley.
19. Remove the timing chain tensioner.

⁂ WARNING

Do not turn the crankshaft without the chain tensioner.

20. Remove timing chain cover.
a. Remove the bolts and nuts.
b. Using a Torx® socket, remove the stud bolt.
c. Remove the timing chain cover by prying the portions between the timing chain cover, cylinder head and cylinder block with a screwdriver.

⁂ WARNING

Be careful not to damage the contact surfaces of the timing chain cover, cylinder head and cylinder block.

21. Remove timing chain cover oil seal.
a. Place the timing chain cover on wooden blocks.
b. Using a screwdriver, remove the oil seal.

To install:

22. Install timing chain cover oil seal.
a. Apply a light coat of multi-purpose grease to a new oil seal lip.
b. Place the timing chain cover on wooden blocks.
c. Using a seal installer, tap the oil seal in until its surface is flush with the timing chain cover edge.

➡Keep the lip free of foreign objects.

23. Install timing chain cover.
a. Remove any old packing material from the contact surface.
b. Apply a continuous bead of sealant 0.138 to 0.177 in. (3.5 to 4.5 mm) diameter as shown in the illustration.

➡Remove any oil from the contact surface. Install the chain cover within 3 minutes of applying sealant. Do not start the engine for at least 2 hours after installing the chain cover. Be careful to only apply the specified thickness / volume of sealant in area D.

c. Install the timing chain cover. Tighten bolt and nut "A" to 10 ft. lbs. (13 Nm) and bolt "B" to 14 ft. lbs. (19 Nm).
d. Using a Torx® socket, install the stud bolt and tighten to 84 inch lbs. (10 Nm).

24. Install the chain tensioner.
a. Check that the O-ring is clean, and set the hook as shown in the illustration.
b. Apply a light coat of engine oil to the O-ring.
c. Install the chain tensioner and tighten the nuts to 40 inch lbs. (9 Nm).

⁂ WARNING

Be careful not to twist the O-ring. When installing the chain tensioner, set the hook again if the hook releases the plunger.

d. Turn the crankshaft counterclockwise to disconnect the plunger knock pin from the hook.
e. Turn the crankshaft clockwise, then check that the slipper is pushed by the plunger.
f. If the plunger is not extended, push the slipper into the chain tensioner using a screwdriver so that the hook is released from the knock pin and the plunger can be extended.
25. Install crankshaft position sensor.
a. Apply a light coat of engine oil to the O-ring of the crankshaft position sensor.
b. Install the crankshaft position sensor with the bolts and tighten to 80 inch lbs. (9 Nm).
26. Install the water pump assembly .
27. Install the right engine mounting bracket. Tighten to 35 ft. lbs. (47 Nm).
28. Install the V-ribbed belt tensioner with the bolt and nut. Tighten bolt to 51 ft. lbs. (69 Nm) and nut to 21 ft. lbs. (29 Nm).
29. Install cylinder head cover.
a. Remove any old packing material from the contact surface.
b. Apply sealant to the 2 locations as shown in the illustration.
c. Remove any oil from the contact surface.

➡Install the cylinder head cover within 3 minutes of applying sealant. Do not put in engine oil for at least 2 hours after installation.

30. Install the cylinder head cover and clamp brackets. Tighten bolts "A" to 8 ft. lbs. (11 Nm) and bolts "B" to 80 inch lbs. (9 Nm).
a. Install the clamps of the engine wire onto the clamp bracket.
b. Connect the ventilation hoses.
31. Install ignition coil assembly.
32. Connect battery negative terminal.
33. Check engine oil leakage.
34. Install upper cylinder head cover.
35. Install engine undercover.
36. Lower the vehicle

2ZZ-GE Engine

See Figures 70, 120.

1. Before servicing the vehicle, refer to the precautions section.
2. Drain and recycle the engine coolant.
3. Raise and safely support the vehicle securely on jackstands.
4. Remove the front wheel.
5. Remove the engine undercover.
6. Remove the upper cylinder head cover.
7. Disconnect battery negative terminal.
8. Remove wire harness protector.
9. Separate suction hose.
10. Remove air switching valve assembly.
11. Remove fan and generator V-belt.
12. Remove alternator bracket.
13. Remove alternator assembly.
14. Remove compressor and magnetic clutch. Hang up the hoses instead of detaching.
15. Remove the front exhaust pipe.
16. Remove engine mounting insulator.
a. Place a wooden block between the jack and engine and jack up the engine.
b. Remove the bolts and nuts, then remove the engine mounting insulator .
17. Remove ribbed belt tensioner.

➡Jack the engine up and down to remove the bolt.

18. Disconnect the 4 ignition coil connectors.
19. Remove the ignition coils.
20. Remove lower cylinder head cover.
a. Disconnect the 2 ventilation hoses.
b. Remove the nuts and bolt, then separate the ventilation tube No. 1.
c. Remove the gasket from the cylinder head cover.
d. Remove the wire harness protector.
e. Remove the bolt and fuel tube clamp.
f. Remove the bolt and union to connector tube hose clamp.
g. Remove the bolts, bracket, wire harness protector, and cylinder head cover.
h. Remove the o-ring from the cylinder head cover.
21. Remove the intake manifold support bracket.
22. Remove the exhaust manifold heat shield.
23. Remove the exhaust manifold.
24. Remove the water pump pulley.

25. Remove the water pump.
26. Remove the transverse engine mounting bracket.
27. Remove crankshaft position sensor.
28. Set No. 1 cylinder to TDC on the compression stroke.
a. Turn the crankshaft pulley until its groove and the timing mark "0" of the timing chain cover are aligned.
b. Check that the point marks of the camshaft timing sprocket and the variable valve timing sprocket are as shown in the illustration.
c. If not, turn the crankshaft 1 revolution (360°) so as to align the marks above.
29. Remove chain tensioner.

WARNING

Do not to turn the crankshaft without the chain tensioner.

30. Remove the crankshaft pulley.
31. Remove the timing chain tensioner.

WARNING

Do not turn the crankshaft without the chain tensioner.

32. Remove timing chain cover.
a. Remove the bolts and nuts.
b. Using a Torx® socket, remove the stud bolt.
c. Remove the timing chain cover by prying the portions between the timing chain cover, cylinder head and cylinder block with a screwdriver.

WARNING

Be careful not to damage the contact surfaces of the timing chain cover, cylinder head and cylinder block.

33. Remove timing chain cover oil seal.
a. Place the timing chain cover on wooden blocks.
b. Using a screwdriver, remove the oil seal.

To install:

34. Install timing chain cover oil seal.
a. Apply a light coat of multi-purpose grease to a new oil seal lip.
b. Place the timing chain cover on wooden blocks.
c. Using a seal installer, tap the oil seal in until its surface is flush with the timing chain cover edge.

➡Keep the lip free of foreign objects.

35. Install timing chain cover.
a. Remove any old packing material from the contact surface.
b. Apply a continuous bead of Three Bond 1280 or equivalent sealant 0.118 in. (3 mm) diameter as shown in the illustration.

➡Remove any oil from the contact surface. Install the chain cover within 3 minutes of applying Three Bond 1280 or equivalent sealant. Do not start the engine for at least 2 hours after installing the chain cover. Be careful to only apply the specified thickness / volume of Three Bond 1280 or equivalent sealant in area D.

c. Install the timing chain cover. Tighten M10 bolts and nut "A" to 15 ft. lbs. (21 Nm) and M8 bolts to 8 ft. lbs. (11 Nm).
d. Using a Torx® socket, install the stud bolt and tighten to 84 inch lbs. (10 Nm).
36. Install chain tensioner.
a. Check that the o-ring is clean, then set the hook as shown in the illustration.
b. Apply engine oil to the chain tensioner, then install it. Tighten to 80 inch lbs. (9 Nm).
c. When installing the tensioner, set the hook again if the hook releases the plunger.
d. Turn the crankshaft counterclockwise to disconnect the plunger knock pin from the hook.
e. Turn the crankshaft clockwise, then check that the slipper is pushed by the plunger.
f. If the plunger is not expended, push the slipper into the chain tensioner with a screwdriver so that the hook is released from the knock pin and the plunger can be extended.
37. Install crankshaft position sensor.
a. Apply a light coat of engine oil to the O-ring of the crankshaft position sensor.
b. Install the crankshaft position sensor with the bolts and tighten to 80 inch lbs. (9 Nm).
38. Install the water pump assembly .
39. Install the right engine mounting bracket. Tighten to 35 ft. lbs. (47 Nm).
40. Install the water pump. Tighten to 80 inch lbs. (9 Nm).
41. Install the water pump pulley.
42. Install the exhaust manifold.
43. Install the exhaust manifold heat shield.
44. Install the exhaust manifold support bracket.
45. Install the lower cylinder head cover.
a. Install a new O-ring onto the cover.
b. Remove any old packing material from the contact surface.
c. Apply sealant to the 2 locations as shown in the illustration.
d. Remove any oil from the contact surface.

➡Install the cylinder head cover within 3 minutes of applying sealant. Do not put in engine oil for at least 2 hours after installation.

46. Install the cylinder head cover and clamp brackets. Tighten bolts to 7 ft. lbs. (10 Nm).
a. Install the union to connector tube hose clamp. Tighten to 62 inch lbs. (7 Nm).
47. Install the fuel tube clamp and tighten the bolt to 66 inch lbs. (8 Nm).
a. Install the wire harness protector and tighten the nuts to 80 inch lbs. (9 Nm).
b. Install a new gasket and ventilation tube. Tighten the nut to 7 ft. lbs. (10 Nm) and the bolt to 18 ft. lbs. (24 Nm).
48. Install ignition coil assembly. Tighten to 80 inch lbs. (9 Nm).
49. Install V-ribbed belt tensioner. Tighten the nut to 21 ft. lbs. (29 Nm) and the bolt to 74 ft. lbs. (100 Nm).
50. Install engine mounting insulator. Tighten to 38 ft. lbs. (52 Nm).
51. Install alternator assembly.
52. Install alternator bracket.
53. Install air switching valve assembly.
54. Install suction hose. Tighten to 87 inch lbs. (10 Nm).
55. Connect battery negative terminal.
56. Check engine oil leakage.
57. Install the upper cylinder head cover.

TIMING CHAIN AND SPROCKETS

REMOVAL & INSTALLATION

1ZZ-FE Engine

See Figures 120 through 138.

1. Before servicing the vehicle, refer to the precautions section.
2. Disconnect battery negative terminal.
3. Remove the front wheel.
4. Remove engine undercover.
5. Remove the upper cylinder head cover.
6. Drain and recycle the engine coolant.
7. Remove the fan and alternator V-belt.
8. Separate the vane pump assembly.
9. Remove the alternator assembly.
10. Remove the right engine mounting insulator.

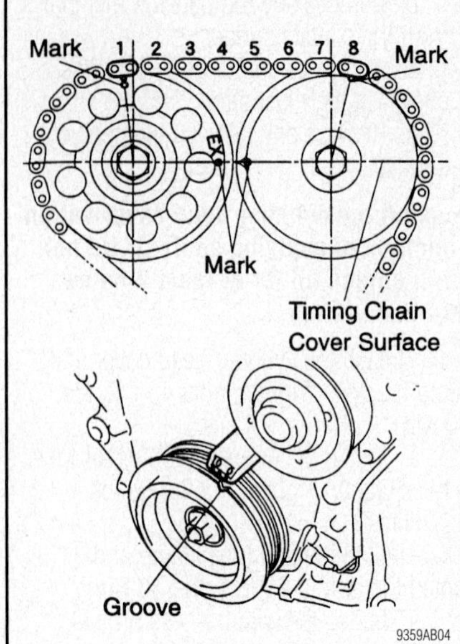

Fig. 120 Proper timing mark alignment for TDC

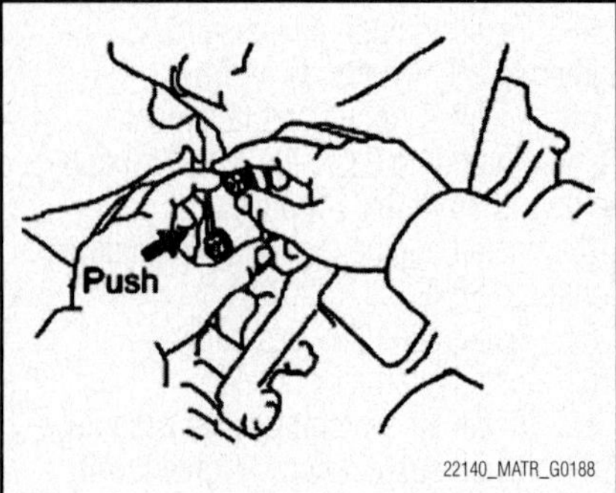

Fig. 121 Remove the timing chain tensioner

11. Remove the ignition coil assembly.
12. Remove the cylinder head cover.
13. Remove the V-ribbed belt tensioner.

➡**Jack the engine up and down to remove the bolt.**

14. Remove the right hand engine mounting bracket.
15. Remove the water pump assembly.
16. Separate the crankshaft position sensor.
17. Set the No. 1 cylinder to TDC on the compression stroke.
    a. Turn the crankshaft pulley until its groove and timing mark "0" of the timing chain cover are aligned.
    b. Check that the point marks of the camshaft timing gears are in a straight line on the timing chain cover surface as shown in the illustration.
    c. If not, turn the crankshaft 1 revolution (360°) to align the marks as above.

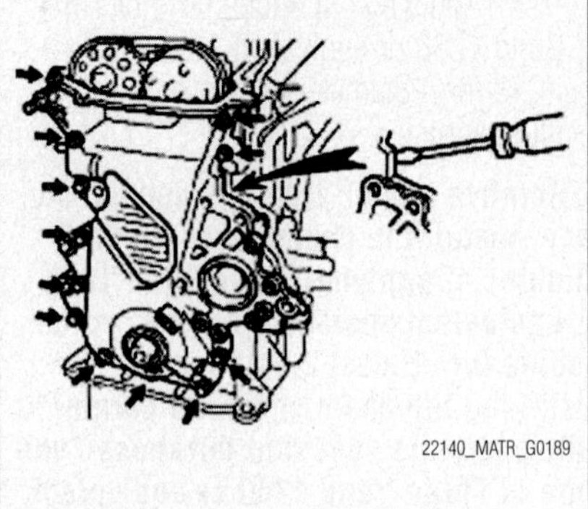

Fig. 122 Timing chain cover fastener locations

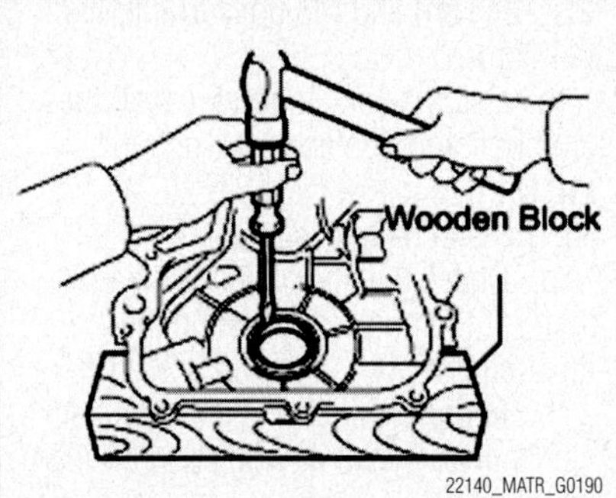

Fig. 123 Place the timing chain cover on wooden blocks. Using a screwdriver, remove the oil seal

18. Remove the crankshaft pulley.
19. Remove the timing chain tensioner.

✲✲ WARNING

Do not turn the crankshaft without the chain tensioner.

20. Remove timing chain cover.
    a. Remove the bolts and nuts.
    b. Using a Torx® socket, remove the stud bolt.
    c. Remove the timing chain cover by prying the portions between the timing chain cover, cylinder head and cylinder block with a screwdriver.

✲✲ WARNING

Be careful not to damage the contact surfaces of the timing chain cover, cylinder head and cylinder block.

21. Remove timing chain cover oil seal.
    a. Place the timing chain cover on wooden blocks.
    b. Using a screwdriver, remove the oil seal.
22. Remove the crankshaft position sensor plate from the crankshaft.
23. Remove the bolt, then remove the chain tensioner slipper.

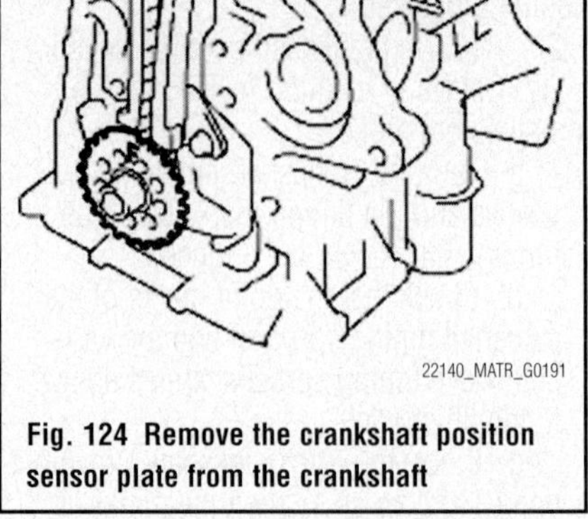

Fig. 124 Remove the crankshaft position sensor plate from the crankshaft

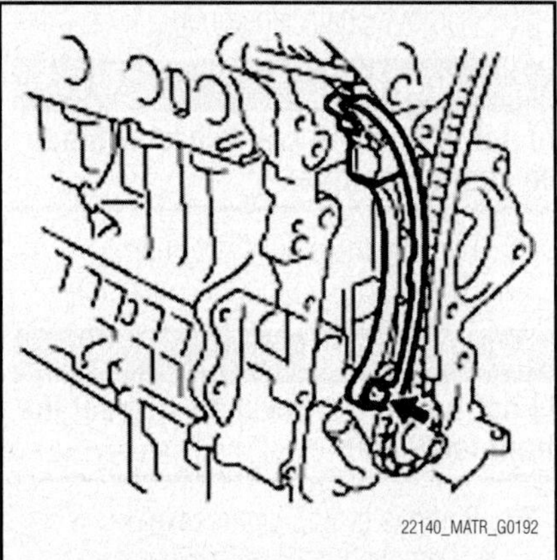

Fig. 125 Timing chain tensioner slipper

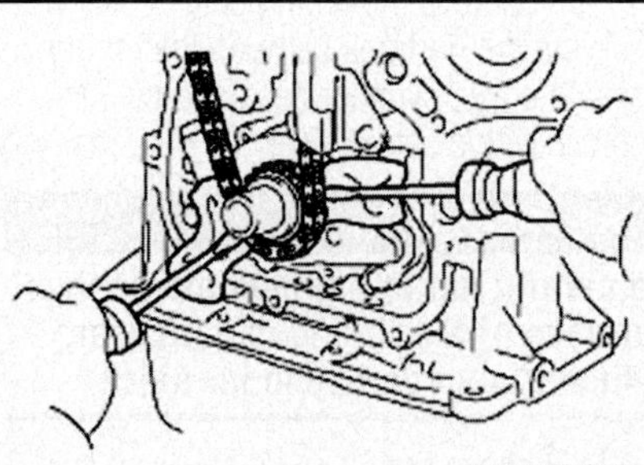

Fig. 126 Using 2 screwdrivers, pry out the chain with the crankshaft timing gear, as shown

24. Remove timing chain.
    a. Using 2 screwdrivers, pry out the chain with the crankshaft timing gear, as shown in the illustration.

➡**Use a shop rag to protect the engine.**

✲✲ WARNING

If revolving the camshafts with the chain off the gears, turn the crankshaft ¼ revolution to prevent the valves from coming into contact with the pistons.

To install:

25. Install timing chain.

a. Turn the hexagonal wrench head portion of the camshafts until the point marks of the 2 camshaft timing gears are aligned.

b. Using a crankshaft pulley bolt, turn the crankshaft to align the timing gear key with the point mark on the oil pump.

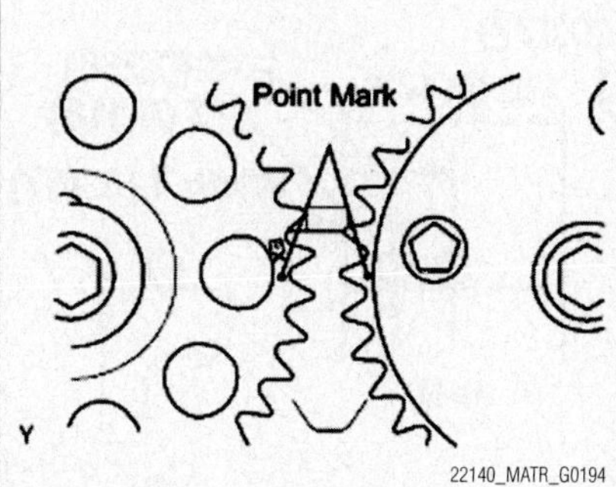

Fig. 127 Turn the hexagonal wrench head portion of the camshafts until the point marks of the 2 camshaft timing gears are aligned

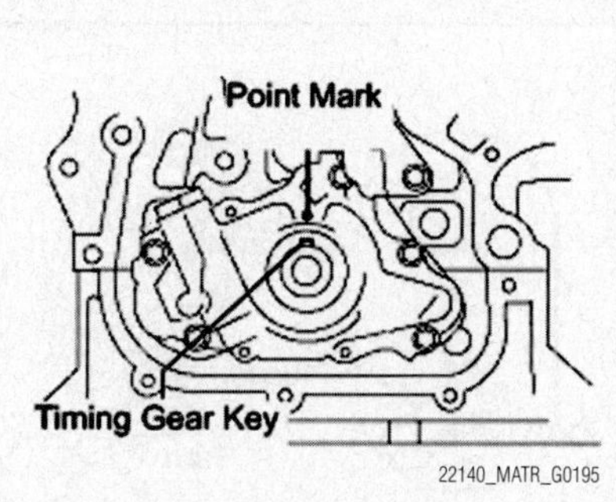

Fig. 128 Using a crankshaft pulley bolt, turn the crankshaft to align the timing gear key with the point mark on the oil pump

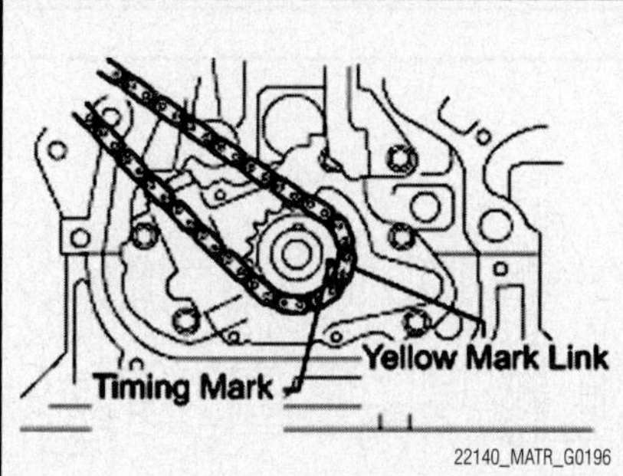

Fig. 129 Install the chain onto the crankshaft timing gear with the yellow mark link aligned with the timing mark on the crankshaft timing gear

c. Install the chain onto the crankshaft timing gear with the yellow mark link aligned with the timing mark on the crankshaft timing gear.

d. Install the crankshaft timing gear.

e. Install the chain onto the camshaft timing gears with the yellow mark links aligned with the timing marks on the camshaft timing gears.

26. Install chain tensioner slipper. Tighten to 14 ft. lbs. (19 Nm).

27. Install the crankshaft position sensor plate with the "F" mark facing forward.

28. Install timing chain cover oil seal.

a. Apply a light coat of multi-purpose grease to a new oil seal lip.

b. Place the timing chain cover on wooden blocks.

c. Using a seal installer, tap the oil seal in until its surface is flush with the timing chain cover edge.

➡Keep the lip free of foreign objects.

29. Install timing chain cover.

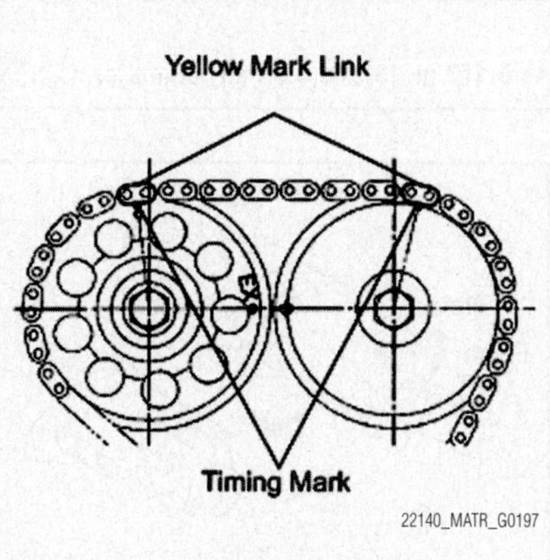

Fig. 130 Install the chain onto the camshaft timing gears with the yellow mark links aligned with the timing marks on the camshaft timing gears

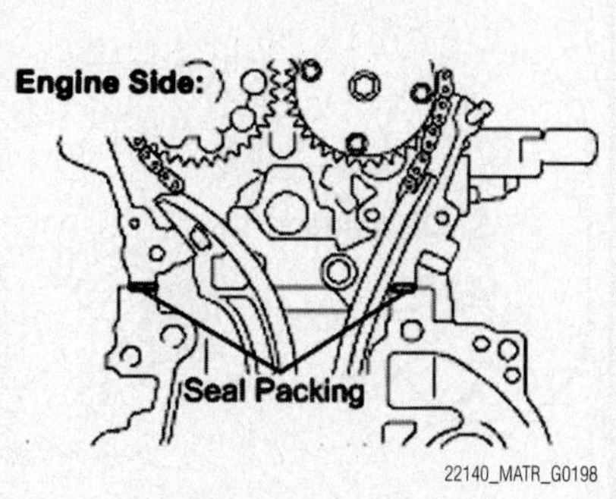

Fig. 131 Apply a continuous bead of sealant 0.138 to 0.177 in. (3.5 to 4.5 mm) diameter as shown—engine side

a. Remove any old packing material from the contact surface.

b. Apply a continuous bead of sealant 0.138 to 0.177 in. (3.5 to 4.5 mm) diameter as shown in the illustration.

➡Remove any oil from the contact surface. Install the chain cover within 3 minutes of applying sealant. Do not start the engine for at least 2 hours after installing the chain cover. Be careful to only apply the specified thickness / volume of sealant in area D.

c. Install the timing chain cover. Tighten bolt and nut "A" to 10 ft. lbs. (13 Nm) and bolt "B" to 14 ft. lbs. (19 Nm).

d. Using a Torx® socket, install the stud bolt and tighten to 84 inch lbs. (10 Nm).

30. Install the chain tensioner.

a. Check that the O-ring is clean, and set the hook as shown in the illustration.

b. Apply a light coat of engine oil to the O-ring.

c. Install the chain tensioner and tighten the nuts to 40 inch lbs. (9 Nm).

✲✲ WARNING

Be careful not to twist the O-ring. When installing the chain tensioner, set the hook again if the hook releases the plunger.

d. Turn the crankshaft counterclockwise to disconnect the plunger knock pin from the hook.

e. Turn the crankshaft clockwise, then check that the slipper is pushed by the plunger.

f. If the plunger is not extended, push the slipper into the chain tensioner using a screwdriver so that the hook is released from the knock pin and the plunger can be extended.

31. Install crankshaft position sensor.

a. Apply a light coat of engine oil to the O-ring of the crankshaft position sensor.

b. Install the crankshaft position sensor with the bolts and tighten to 80 inch lbs. (9 Nm).

32. Install the water pump assembly .

33. Install the right engine mounting bracket. Tighten to 35 ft. lbs. (47 Nm).

34. Install the V-ribbed belt tensioner with the bolt and nut. Tighten bolt to 51 ft. lbs. (69 Nm) and nut to 21 ft. lbs. (29 Nm).

35. Install cylinder head cover.

a. Remove any old packing material from the contact surface.

b. Apply sealant to the 2 locations as shown in the illustration.

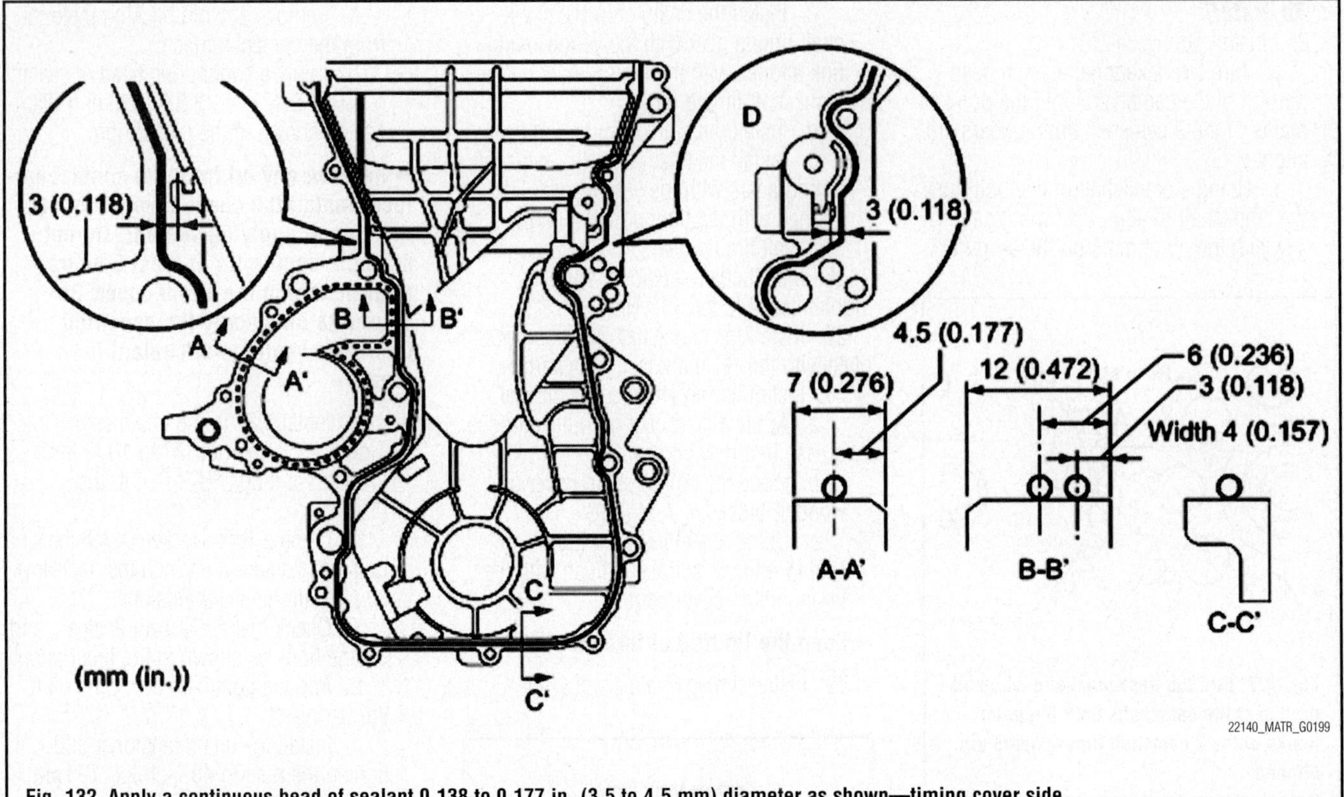

Fig. 132 Apply a continuous bead of sealant 0.138 to 0.177 in. (3.5 to 4.5 mm) diameter as shown—timing cover side

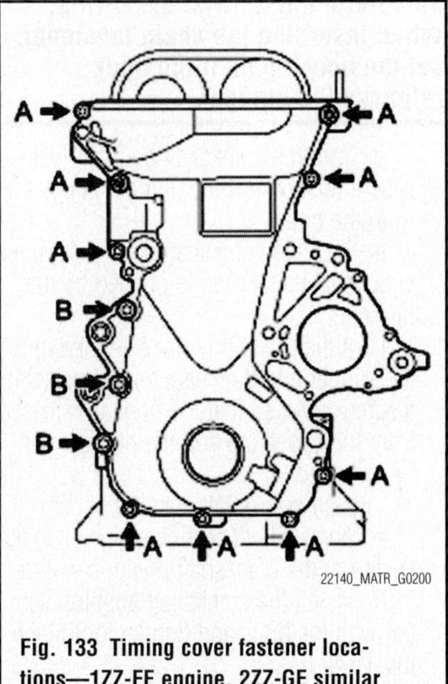

Fig. 133 Timing cover fastener locations—1ZZ-FE engine, 2ZZ-GE similar

c. Remove any oil from the contact surface.

➡Install the cylinder head cover within 3 minutes of applying sealant. Do not put in engine oil for at least 2 hours after installation.

36. Install the cylinder head cover and clamp brackets. Tighten bolts "A" to 8 ft.

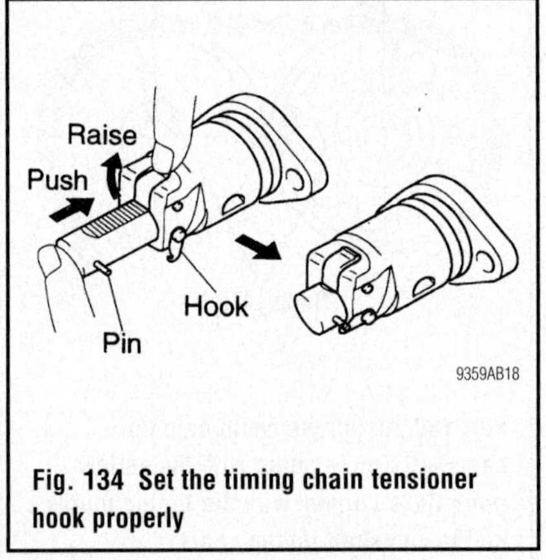

Fig. 134 Set the timing chain tensioner hook properly

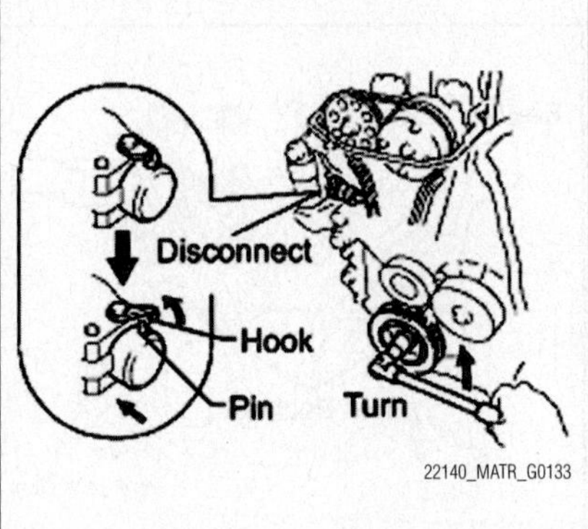

Fig. 135 Turn the crankshaft counterclockwise to disconnect the plunger knock pin from the hook

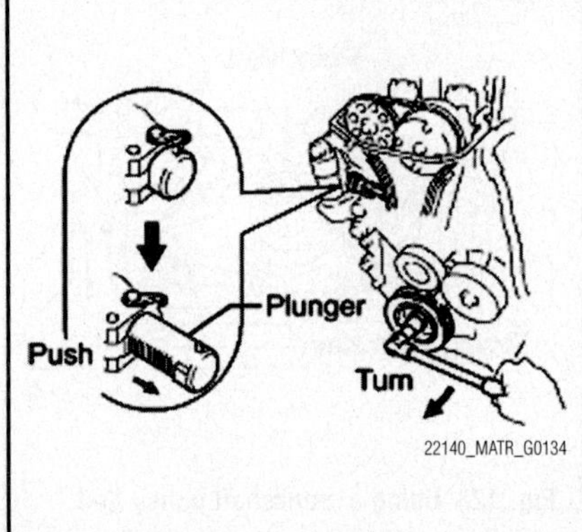

Fig. 136 Turn the crankshaft clockwise, then check that the slipper is pushed by the plunger

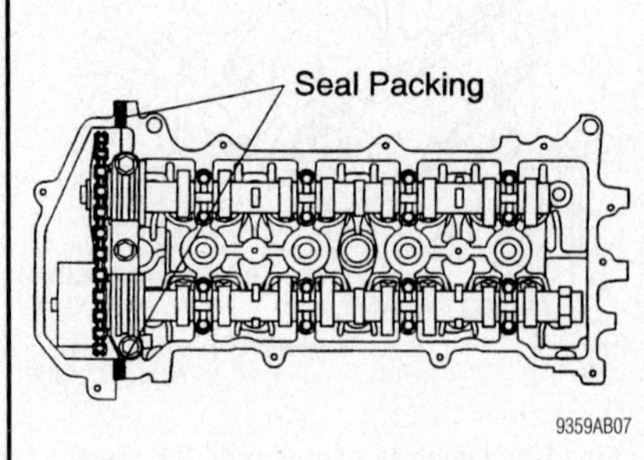

Fig. 137 Apply sealant to the 2 locations as shown

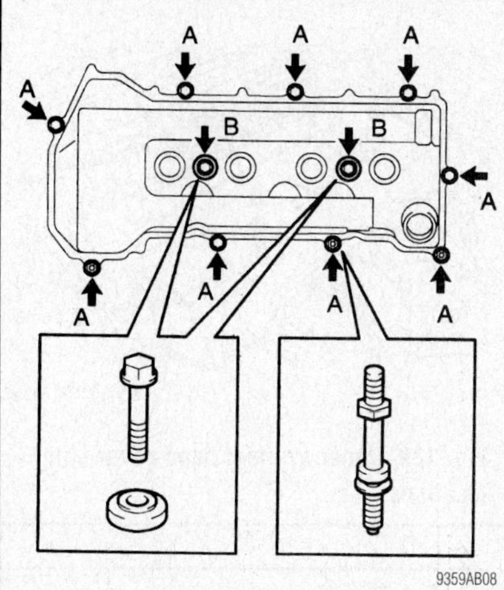

Fig. 138 Cylinder head (valve) cover bolt locations—1ZZ-FE engine

lbs. (11 Nm) and bolts "B" to 80 inch lbs. (9 Nm).

a. Install the clamps of the engine wire onto the clamp bracket.

b. Connect the ventilation hoses.

37. Install ignition coil assembly.
38. Connect battery negative terminal.
39. Check engine oil leakage.
40. Install upper cylinder head cover.
41. Install engine undercover.
42. Lower the vehicle

2ZZ-GE Engine

See Figures 51, 66 through 68, and 121 through 130.

1. Before servicing the vehicle, refer to the precautions section.
2. Drain and recycle the engine coolant.
3. Raise and safely support the vehicle securely on jackstands.
4. Remove the front wheel.
5. Remove the engine undercover.
6. Remove the upper cylinder head cover.
7. Disconnect battery negative terminal.
8. Remove wire harness protector.
9. Separate suction hose.
10. Remove air switching valve assembly.
11. Remove fan and generator V-belt.
12. Remove alternator bracket.
13. Remove alternator assembly.
14. Remove compressor and magnetic clutch. Hang up the hoses instead of detaching.
15. Remove the front exhaust pipe.
16. Remove engine mounting insulator.

a. Place a wooden block between the jack and engine and jack up the engine.

b. Remove the bolts and nuts, then remove the engine mounting insulator .

17. Remove ribbed belt tensioner.

➡Jack the engine up and down to remove the bolt.

18. Disconnect the 4 ignition coil connectors.
19. Remove the ignition coils.
20. Remove lower cylinder head cover.

a. Disconnect the 2 ventilation hoses.

b. Remove the nuts and bolt, then separate the ventilation tube No. 1.

c. Remove the gasket from the cylinder head cover.

d. Remove the wire harness protector.

e. Remove the bolt and fuel tube clamp.

f. Remove the bolt and union to connector tube hose clamp.

g. Remove the bolts, bracket, wire harness protector, and cylinder head cover.

h. Remove the o-ring from the cylinder head cover.

21. Remove the intake manifold support bracket.
22. Remove the exhaust manifold heat shield.
23. Remove the exhaust manifold.
24. Remove the water pump pulley.
25. Remove the water pump.
26. Remove the transverse engine mounting bracket.
27. Remove crankshaft position sensor.
28. Set No. 1 cylinder to TDC on the compression stroke.

a. Turn the crankshaft pulley until its groove and the timing mark "0" of the timing chain cover are aligned.

b. Check that the point marks of the camshaft timing sprocket and the variable valve timing sprocket are as shown in the illustration.

c. If not, turn the crankshaft 1 revolution (360°) so as to align the marks above.

29. Remove chain tensioner.

**** WARNING**

Do not to turn the crankshaft without the chain tensioner.

30. Remove the crankshaft pulley.
31. Remove the timing chain tensioner.

**** WARNING**

Do not turn the crankshaft without the chain tensioner.

32. Remove timing chain cover.

a. Remove the bolts and nuts.

b. Using a Torx® socket, remove the stud bolt.

c. Remove the timing chain cover by prying the portions between the timing chain cover, cylinder head and cylinder block with a screwdriver.

**** WARNING**

Be careful not to damage the contact surfaces of the timing chain cover, cylinder head and cylinder block.

33. Remove timing chain cover oil seal.

a. Place the timing chain cover on wooden blocks.

b. Using a screwdriver, remove the oil seal.

34. Remove the crankshaft position sensor plate from the crankshaft.
35. Remove the bolt, then remove the chain tensioner slipper.
36. Remove the chain vibration damper.
37. Remove timing chain.

a. Using 2 screwdrivers, pry out the chain with the crankshaft timing gear, as shown in the illustration.

➡Use a shop rag to protect the engine.

**** WARNING**

If revolving the camshafts with the chain off the gears, turn the crankshaft ¼ revolution to prevent the valves from coming into contact with the pistons.

To install:

38. Install the chain vibration damper. Tighten to 15 ft. lbs. (21 Nm).
39. Install timing chain.

a. Turn the hexagonal wrench head portion of the camshafts until the point marks of the 2 camshaft timing gears are aligned.

b. Using a crankshaft pulley bolt, turn the crankshaft to align the timing gear key with the point mark on the oil pump.

c. Install the chain onto the crankshaft timing gear with the yellow mark link aligned with the timing mark on the crankshaft timing gear.

d. Install the crankshaft timing gear.

e. Install the chain onto the camshaft timing gears with the yellow mark links aligned with the timing marks on the camshaft timing gears.

40. Install chain tensioner slipper. Tighten to 14 ft. lbs. (19 Nm).
41. Install the crankshaft position sensor plate with the "F" mark facing forward.
42. Install timing chain cover oil seal.

a. Apply a light coat of multi-purpose grease to a new oil seal lip.

b. Place the timing chain cover on wooden blocks.

c. Using a seal installer, tap the oil seal in until its surface is flush with the timing chain cover edge.

➡**Keep the lip free of foreign objects.**

43. Install timing chain cover.
 a. Remove any old packing material from the contact surface.
 b. Apply a continuous bead of Three Bond 1280 or equivalent sealant 0.118 in. (3 mm) diameter as shown in the illustration.

➡**Remove any oil from the contact surface. Install the chain cover within 3 minutes of applying Three Bond 1280 or equivalent sealant. Do not start the engine for at least 2 hours after installing the chain cover. Be careful to only apply the specified thickness / volume of Three Bond 1280 or equivalent sealant in area D.**

 c. Install the timing chain cover. Tighten M10 bolts and nut "A" to 15 ft. lbs. (21 Nm) and M8 bolts to 8 ft. lbs. (11 Nm).
 d. Using a Torx® socket, install the stud bolt and tighten to 84 inch lbs. (10 Nm).

44. Install chain tensioner.
 a. Check that the o-ring is clean, then set the hook as shown in the illustration.
 b. Apply engine oil to the chain tensioner, then install it. Tighten to 80 inch lbs. (9 Nm).
 c. When installing the tensioner, set the hook again if the hook releases the plunger.
 d. Turn the crankshaft counterclockwise to disconnect the plunger knock pin from the hook.
 e. Turn the crankshaft clockwise, then check that the slipper is pushed by the plunger.
 f. If the plunger is not expended, push the slipper into the chain tensioner with a screwdriver so that the hook is released from the knock pin and the plunger can be extended.

45. Install crankshaft position sensor.
 a. Apply a light coat of engine oil to the O-ring of the crankshaft position sensor.
 b. Install the crankshaft position sensor with the bolts and tighten to 80 inch lbs. (9 Nm).

46. Install the water pump assembly .
47. Install the right engine mounting bracket. Tighten to 35 ft. lbs. (47 Nm).
48. Install the water pump. Tighten to 80 inch lbs. (9 Nm).
49. Install the water pump pulley.
50. Install the exhaust manifold.
51. Install the exhaust manifold heat shield.
52. Install the exhaust manifold support bracket.
53. Install the lower cylinder head cover.
 a. Install a new O-ring onto the cover.
 b. Remove any old packing material from the contact surface.
 c. Apply sealant to the 2 locations as shown in the illustration.
 d. Remove any oil from the contact surface.

➡**Install the cylinder head cover within 3 minutes of applying sealant. Do not put in engine oil for at least 2 hours after installation.**

54. Install the cylinder head cover and clamp brackets. Tighten bolts to 7 ft. lbs. (10 Nm).
 a. Install the union to connector tube hose clamp. Tighten to 62 inch lbs. (7 Nm).

55. Install the fuel tube clamp and tighten the bolt to 66 inch lbs. (8 Nm).
 a. Install the wire harness protector and tighten the nuts to 80 inch lbs. (9 Nm).
 b. Install a new gasket and ventilation tube. Tighten the nut to 7 ft. lbs. (10 Nm) and the bolt to 18 ft. lbs. (24 Nm).

56. Install ignition coil assembly. Tighten to 80 inch lbs. (9 Nm).
57. Install V-ribbed belt tensioner. Tighten the nut to 21 ft. lbs. (29 Nm) and the bolt to 74 ft. lbs. (100 Nm).
58. Install engine mounting insulator. Tighten to 38 ft. lbs. (52 Nm).
59. Install alternator assembly.
60. Install alternator bracket.
61. Install air switching valve assembly.
62. Install suction hose. Tighten to 87 inch lbs. (10 Nm).
63. Connect battery negative terminal.
64. Check engine oil leakage.
65. Install the upper cylinder head cover.

VALVE COVERS

REMOVAL & INSTALLATION

1ZZ-FE Engine

See Figures 139 and 140.

1. Before servicing the vehicle, refer to the precautions section.
2. Disconnect the negative battery cable.
3. Remove the upper cylinder head cover.
4. Remove ignition coils.
 a. Disconnect the ignition coil connectors.
 b. Remove the nuts, then separate the engine wire.

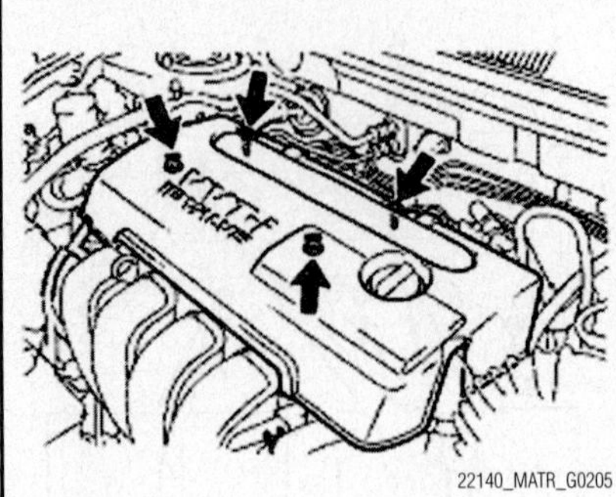

Fig. 139 Upper cylinder head cover bolt locations

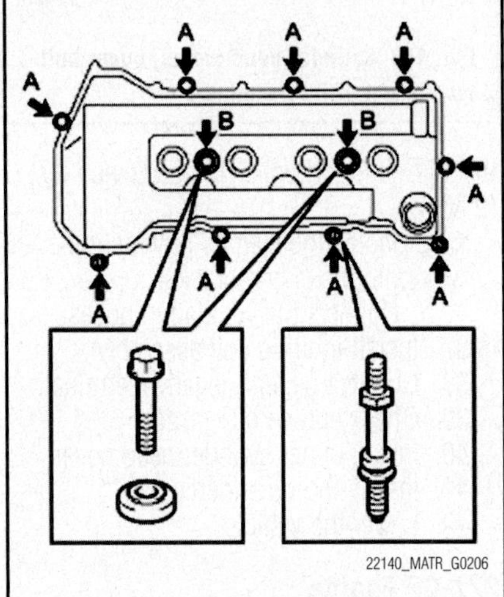

Fig. 140 Lower cylinder head cover bolt locations

 c. Remove the bolts, then remove the ignition coils.

5. Remove lower cylinder head cover
 a. Disconnect the ventilation hose from the cylinder head cover.
 b. Disconnect the ventilation hose from the ventilation valve.
 c. Remove the 3 clamps of the engine wire from the clamp bracket.
 d. Remove the bolts, seal washers, nuts and clamp brackets, then remove the cylinder head cover.

To install:

➡**Use a new cylinder head cover gasket and new seal washers whenever the cylinder head cover replacement procedure is performed.**

6. Ensure that the new gasket and the new seal washers are correctly positioned during the cylinder head cover installation.
7. Install the new cylinder head cover gasket to the cylinder head cover.
8. Install the new seal washers to the cylinder head cover.

9. Install the cylinder head cover to the cylinder head. Secure the cover using nine bolts and two nuts. Tighten cylinder head cover bolts to 8 ft. lbs. (11 Nm) and the stud bolt to 80 inch lbs. (9 Nm).
10. Install the 3 clamps of the engine wire on the clamp bracket
11. Connect the ventilation hose to the ventilation valve.
12. Connect the ventilation hose to the cylinder head cover.
13. Install the ignition coils. Tighten to 80 inch lbs. (9 Nm).
a. Install the engine wire. Tighten to 80 inch lbs. (9 Nm).
b. Connect the ignition coil connectors.

2ZZ-GE Engine

See Figure 141.

1. Before servicing the vehicle, refer to the precautions section.
2. Disconnect the negative battery cable.
3. Remove the upper cylinder head cover.
4. Remove ignition coils.
a. Disconnect the ignition coil connectors.
b. Remove the nuts, then separate the engine wire.
c. Remove the bolts, then remove the ignition coils.
5. Remove lower cylinder head cover
a. Disconnect the ventilation hoses.
b. Remove the nuts and bolt, then separate the ventilation tube.
c. Remove the gasket from the cylinder head cover.
d. Remove the nuts, then separate the wire harness protector.
e. Remove the bolt and fuel tube clamp.
f. Remove the bolt and union to connector tube hose clamp.
g. Remove the bolts, bracket, wire harness protector, and cylinder head cover.
h. Remove the O-ring from the cylinder head cover.

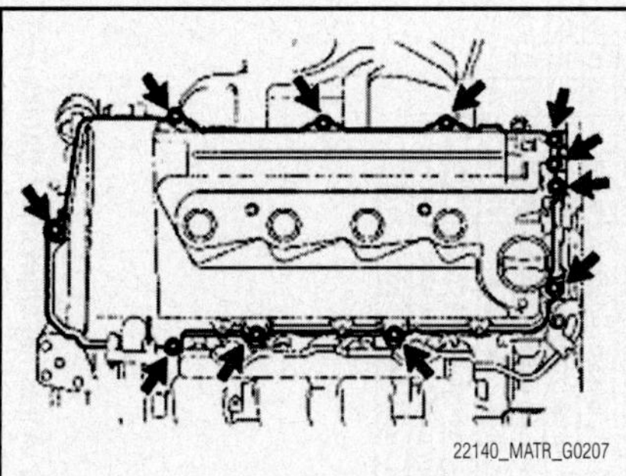

Fig. 141 Lower cylinder head cover bolt locations

To install:

6. Install the lower cylinder head cover.
a. Install a new O-ring onto the cylinder head cover
b. Remove any old sealant from the cylinder head cover.

➡Remove any oil from the contact surface. Install the cylinder head cover within 3 minutes of applying seal packing. Do not put in engine oil for at least 2 hours after installation.

c. Apply seal packing to 2 locations as shown in the illustration.
d. Install the cylinder head cover to the cylinder head. Secure the cover using nine bolts and two nuts. Tighten cylinder head cover bolts to 7 ft. lbs. (12 Nm).
e. Install the union to connector tube hose clamp with the bolt and tighten to 62 inch lbs. (7 Nm).
f. Install the fuel tube clamp with the bolt and tighten to 66 inch lbs. (8 Nm).
g. Install the wire harness protector and tighten to 80 inch lbs. (9 Nm).
h. Install a new gasket and ventilation tube with the nuts and bolt. Tighten nut to 7 ft. lbs. (10 Nm) and bolt to 18 ft. lbs. (24 Nm).
7. Install the ignition coils and tighten to 80 inch lbs. (9 Nm).
8. Connect the ignition coil connectors.

VALVE LASH

ADJUSTMENT

1ZZ-FE Engine

See Figures 142 through 147.

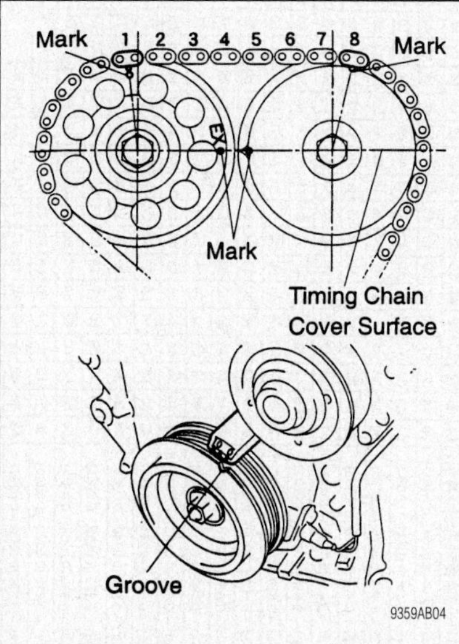

Fig. 142 Proper timing mark alignment for TDC

➡Adjust the valve clearance when the engine is cold.

1. Before servicing the vehicle, refer to the precautions section.
2. Remove the cylinder head cover.
3. Set the No. 1 cylinder to TDC on the compression stroke.
a. Turn the crankshaft pulley until its groove and timing mark "0" of the timing chain cover are aligned.
b. Check that the point marks of the camshaft timing gears are in a straight

No. 1 Cylinder TDC/Compression:

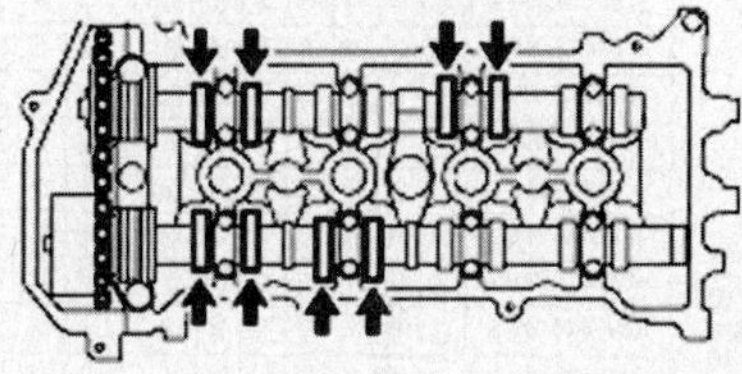

Fig. 143 With the engine at TDC on the compression stroke, check only the valves indicated

No. 4 Cylinder TDC/Compression:

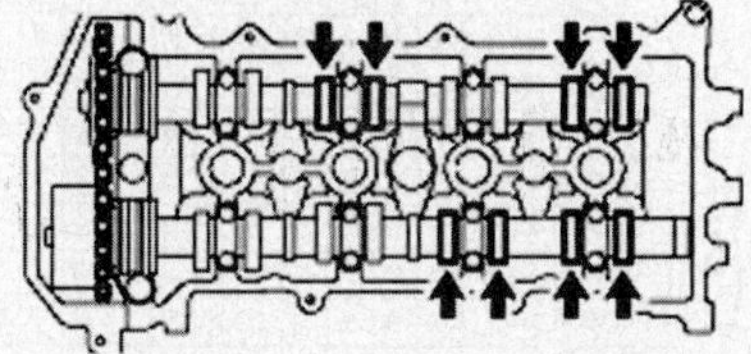

Fig. 144 Turn the crankshaft 1 revolution (360°) to set the No. 4 cylinder to the TDC/compression and check only the valves indicated

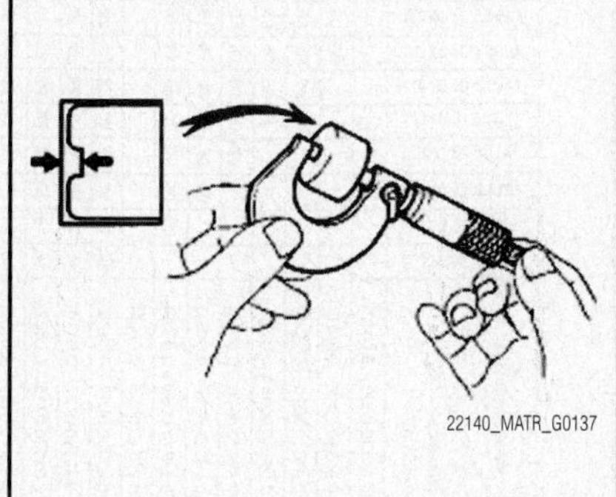

Fig. 145 Using a micrometer, measure the thickness of the removed valve lifters

1ZZ–FE: Valve Lifter Selection Chart (Intake)

Measured clearance mm (in.) \ Installed lifter thickness mm (in.)	5.060 (0.1992)	5.080 (0.2000)	5.100 (0.2008)	5.120 (0.2016)	5.140 (0.2024)	5.160 (0.2031)	5.180 (0.2039)	5.200 (0.2047)	5.210 (0.2051)	5.220 (0.2055)	5.230 (0.2059)	5.240 (0.2063)	5.250 (0.2067)	5.260 (0.2071)	5.270 (0.2075)	5.280 (0.2079)	5.290 (0.2083)	5.300 (0.2087)	5.310 (0.2091)	5.320 (0.2094)	5.330 (0.2098)	5.340 (0.2102)	5.350 (0.2106)	5.360 (0.2110)	5.370 (0.2114)	5.380 (0.2118)	5.390 (0.2122)	5.400 (0.2126)	5.410 (0.2130)	5.420 (0.2134)	5.430 (0.2138)	5.440 (0.2142)	5.450 (0.2146)	5.460 (0.2150)	5.470 (0.2154)	5.480 (0.2157)	5.490 (0.2161)	5.500 (0.2165)	5.510 (0.2169)	5.520 (0.2173)	5.530 (0.2177)	5.540 (0.2181)	5.550 (0.2185)	5.560 (0.2189)	5.570 (0.2193)	5.580 (0.2197)	5.590 (0.2201)	5.600 (0.2205)	5.620 (0.2213)	5.640 (0.2220)	5.660 (0.2228)	5.680 (0.2236)	5.700 (0.2244)	5.720 (0.2252)	5.740 (0.2260)
0.000 – 0.030 (0.0000 – 0.0012)									06	06	06	06	06	08	10	10	12	12	14	14	16	16	18	18	20	20	22	22	24	24	26	26	28	28	30	30	32	32	34	34	36	36	38	38	40	40	42	42	44	46	48	50	52	54	56
0.031 – 0.050 (0.0012 – 0.0020)							06	06	06	06	08	08	10	10	12	12	14	14	16	16	18	18	20	20	22	22	24	24	26	26	28	28	30	30	32	32	34	34	36	36	38	38	40	40	42	42	44	44	46	48	50	52	54	56	58
0.051 – 0.070 (0.0020 – 0.0028)						06	06	06	08	08	10	10	12	12	14	14	16	16	18	18	20	20	22	22	24	24	26	26	28	28	30	30	32	32	34	34	36	36	38	38	40	40	42	42	44	44	46	46	48	50	52	54	56	58	60
0.071 – 0.090 (0.0028 – 0.0035)					06	06	06	08	10	10	12	12	14	14	16	16	18	18	20	20	22	22	24	24	26	26	28	28	30	30	32	32	34	34	36	36	38	38	40	40	42	42	44	44	46	46	48	48	50	52	54	56	58	60	62
0.091 – 0.110 (0.0036 – 0.0043)				06	06	06	08	10	12	12	14	14	16	16	18	18	20	20	22	22	24	24	26	26	28	28	30	30	32	32	34	34	36	36	38	38	40	40	42	42	44	44	46	46	48	48	50	50	52	54	56	58	60	62	64
0.111 – 0.130 (0.0044 – 0.0051)			06	06	06	08	10	12	14	14	16	16	18	18	20	20	22	22	24	24	26	26	28	28	30	30	32	32	34	34	36	36	38	38	40	40	42	42	44	44	46	46	48	48	50	50	52	52	54	56	58	60	62	64	66
0.131 – 0.149 (0.0052 – 0.0059)		06	06	06	08	10	12	14	16	16	18	18	20	20	22	22	24	24	26	26	28	28	30	30	32	32	34	34	36	36	38	38	40	40	42	42	44	44	46	46	48	48	50	50	52	52	54	54	56	58	60	62	64	66	68
0.150 – 0.250 (0.0059 – 0.0098)																																																							
0.251 – 0.270 (0.0099 – 0.0106)	12	14	16	18	20	22	24	26	28	28	30	30	32	32	34	34	36	36	38	38	40	40	42	42	44	44	46	46	48	48	50	50	52	52	54	54	56	56	58	58	60	60	62	62	64	64	66	66	68	70	72	74	74	74	
0.271 – 0.290 (0.0107 – 0.0114)	14	16	18	20	22	24	26	28	30	30	32	32	34	34	36	36	38	38	40	40	42	42	44	44	46	46	48	48	50	50	52	52	54	54	56	56	58	58	60	60	62	62	64	64	66	66	68	68	70	72	74	74	74		
0.291 – 0.310 (0.0115 – 0.0122)	16	18	20	22	24	26	28	30	32	32	34	34	36	36	38	38	40	40	42	42	44	44	46	46	48	48	50	50	52	52	54	54	56	56	58	58	60	60	62	62	64	64	66	66	68	68	70	70	72	74	74	74			
0.311 – 0.330 (0.0122 – 0.0130)	18	20	22	24	26	28	30	32	34	34	36	36	38	38	40	40	42	42	44	44	46	46	48	48	50	50	52	52	54	54	56	56	58	58	60	60	62	62	64	64	66	66	68	68	70	70	72	72	74	74	74				
0.331 – 0.350 (0.0130 – 0.0138)	20	22	24	26	28	30	32	34	36	36	38	38	40	40	42	42	44	44	46	46	48	48	50	50	52	52	54	54	56	56	58	58	60	60	62	62	64	64	66	66	68	68	70	70	72	72	74	74	74	74					
0.351 – 0.370 (0.0138 – 0.0146)	22	24	26	28	30	32	34	36	38	38	40	40	42	42	44	44	46	46	48	48	50	50	52	52	54	54	56	56	58	58	60	60	62	62	64	64	66	66	68	68	70	70	72	72	74	74	74	74	74						
0.371 – 0.390 (0.0146 – 0.0154)	24	26	28	30	32	34	36	38	40	40	42	42	44	44	46	46	48	48	50	50	52	52	54	54	56	56	58	58	60	60	62	62	64	64	66	66	68	68	70	70	72	72	74	74	74	74	74	74							
0.391 – 0.410 (0.0154 – 0.0161)	26	28	30	32	34	36	38	40	42	42	44	44	46	46	48	48	50	50	52	52	54	54	56	56	58	58	60	60	62	62	64	64	66	66	68	68	70	70	72	72	74	74	74	74	74	74	74								
0.411 – 0.430 (0.0162 – 0.0169)	28	30	32	34	36	38	40	42	44	44	46	46	48	48	50	50	52	52	54	54	56	56	58	58	60	60	62	62	64	64	66	66	68	68	70	70	72	72	74	74	74	74	74	74											
0.431 – 0.450 (0.0170 – 0.0177)	30	32	34	36	38	40	42	44	46	46	48	48	50	50	52	52	54	54	56	56	58	58	60	60	62	62	64	64	66	66	68	68	70	70	72	72	74	74	74	74	74	74													
0.451 – 0.470 (0.0178 – 0.0185)	32	34	36	38	40	42	44	46	48	48	50	50	52	52	54	54	56	56	58	58	60	60	62	62	64	64	66	66	68	68	70	70	72	72	74	74	74	74	74	74															
0.471 – 0.490 (0.0185 – 0.0193)	34	36	38	40	42	44	46	48	50	50	52	52	54	54	56	56	58	58	60	60	62	62	64	64	66	66	68	68	70	70	72	72	74	74	74	74	74	74																	
0.491 – 0.510 (0.0193 – 0.0201)	36	38	40	42	44	46	48	50	52	52	54	54	56	56	58	58	60	60	62	62	64	64	66	66	68	68	70	70	72	72	74	74	74	74	74	74																			
0.511 – 0.530 (0.0201 – 0.0209)	38	40	42	44	46	48	50	52	54	54	56	56	58	58	60	60	62	62	64	64	66	66	68	68	70	70	72	72	74	74	74	74	74	74																					
0.531 – 0.550 (0.0209 – 0.0217)	40	42	44	46	48	50	52	54	56	56	58	58	60	60	62	62	64	64	66	66	68	68	70	70	72	72	74	74	74	74	74	74																							
0.551 – 0.570 (0.0217 – 0.0224)	42	44	46	48	50	52	54	56	58	58	60	60	62	62	64	64	66	66	68	68	70	70	72	72	74	74	74	74	74	74																									
0.571 – 0.590 (0.0225 – 0.0232)	44	46	48	50	52	54	56	58	60	60	62	62	64	64	66	66	68	68	70	70	72	72	74	74	74	74	74	74																											
0.591 – 0.610 (0.0233 – 0.0240)	46	48	50	52	54	56	58	60	62	62	64	64	66	66	68	68	70	70	72	72	74	74	74	74	74	74																													
0.611 – 0.630 (0.0241 – 0.0248)	48	50	52	54	56	58	60	62	64	64	66	66	68	68	70	70	72	72	74	74	74	74	74	74																															
0.631 – 0.650 (0.0248 – 0.0256)	50	52	54	56	58	60	62	64	66	66	68	68	70	70	72	72	74	74	74	74	74	74																																	
0.651 – 0.670 (0.0256 – 0.0264)	52	54	56	58	60	62	64	66	68	68	70	70	72	72	74	74	74	74	74	74																																			
0.671 – 0.690 (0.0264 – 0.0272)	54	56	58	60	62	64	66	68	70	70	72	72	74	74	74	74	74	74																																					
0.691 – 0.710 (0.0272 – 0.0280)	56	58	60	62	64	66	68	70	72	72	74	74	74	74	74	74																																							
0.711 – 0.730 (0.0280 – 0.0287)	58	60	62	64	66	68	70	72	74	74	74	74	74	74																																									
0.731 – 0.750 (0.0288 – 0.0295)	60	62	64	66	68	70	72	74	74	74	74	74																																											
0.751 – 0.770 (0.0296 – 0.0303)	62	64	66	68	70	72	74	74	74	74																																													
0.771 – 0.790 (0.0304 – 0.0311)	64	66	68	70	72	74	74	74																																															
0.791 – 0.810 (0.0311 – 0.0319)	66	68	70	72	74	74	74																																																
0.811 – 0.830 (0.0319 – 0.0327)	68	70	72	74	74	74																																																	
0.831 – 0.850 (0.0327 – 0.0335)	70	72	74	74	74																																																		
0.851 – 0.870 (0.0335 – 0.0343)	72	74	74	74																																																			
0.871 – 0.890 (0.0343 – 0.0350)	74	74	74																																																				
0.891 – 0.910 (0.0351 – 0.0358)	74	74																																																					
0.911 – 0.930 (0.0359 – 0.0366)	74																																																						

New lifter thickness mm (in.)

Lifter No.	Thickness	Lifter No.	Thickness	Lifter No.	Thickness
06	5.060 (0.1992)	30	5.300 (0.2087)	54	5.540 (0.2181)
08	5.080 (0.2000)	32	5.320 (0.2094)	56	5.560 (0.2189)
10	5.100 (0.2008)	34	5.340 (0.2102)	58	5.580 (0.2197)
12	5.120 (0.2016)	36	5.360 (0.2110)	60	5.600 (0.2205)
14	5.140 (0.2024)	38	5.380 (0.2118)	62	5.620 (0.2213)
16	5.160 (0.2031)	40	5.400 (0.2126)	64	5.640 (0.2220)
18	5.180 (0.2039)	42	5.420 (0.2134)	66	5.660 (0.2228)
20	5.200 (0.2047)	44	5.440 (0.2142)	68	5.680 (0.2236)
22	5.220 (0.2055)	46	5.460 (0.2150)	70	5.700 (0.2244)
24	5.240 (0.2063)	48	5.480 (0.2157)	72	5.720 (0.2252)
26	5.260 (0.2071)	50	5.500 (0.2165)	74	5.740 (0.2260)
28	5.280 (0.2079)	52	5.520 (0.2173)		

Intake valve clearance (Cold):
0.15 – 0.25 mm (0.006 – 0.010 in.)
EXAMPLE: The 5.250 mm (0.2067 in.) lifter is installed, and the measured clearance is 0.400 mm (0.0157 in.).
Replace the 5.250 mm (0.2067 in.) lifter with a new No. 48 lifter.

9307WG70

Fig. 146 Adjusting shim chart (intake)—1ZZ-FE engine

1ZZ–FE: Valve Lifter Selection Chart (Exhaust)

Measured clearance mm (in.) \ Installed lifter thickness mm (in.)	5.060 (0.1992)	5.080 (0.2000)	5.100 (0.2008)	5.120 (0.2016)	5.140 (0.2024)	5.160 (0.2031)	5.180 (0.2039)	5.200 (0.2047)	5.210 (0.2051)	5.220 (0.2055)	5.230 (0.2059)	5.240 (0.2063)	5.250 (0.2067)	5.260 (0.2071)	5.270 (0.2075)	5.280 (0.2079)	5.290 (0.2083)	5.300 (0.2087)	5.310 (0.2091)	5.320 (0.2094)	5.330 (0.2098)	5.340 (0.2102)	5.350 (0.2106)	5.360 (0.2110)	5.370 (0.2114)	5.380 (0.2118)	5.390 (0.2122)	5.400 (0.2126)	5.410 (0.2130)	5.420 (0.2134)	5.430 (0.2138)	5.440 (0.2142)	5.450 (0.2146)	5.460 (0.2150)	5.470 (0.2154)	5.480 (0.2157)	5.490 (0.2161)	5.500 (0.2165)	5.510 (0.2169)	5.520 (0.2173)	5.530 (0.2177)	5.540 (0.2181)	5.550 (0.2185)	5.560 (0.2189)	5.570 (0.2193)	5.580 (0.2197)	5.590 (0.2201)	5.600 (0.2205)	5.620 (0.2213)	5.640 (0.2220)	5.660 (0.2228)	5.680 (0.2236)	5.700 (0.2244)	5.720 (0.2252)	5.740 (0.2260)
0.000 – 0.030 (0.0000 – 0.0012)																			06	06	06	06	08	08	10	10	12	12	14	14	16	16	18	18	20	20	22	22	24	24	26	26	28	28	30	30	32	32	34	36	38	40	42	44	46
0.031 – 0.050 (0.0012 – 0.0020)																06	06	06	06	06	08	08	10	10	12	12	14	14	16	16	18	18	20	20	22	22	24	24	26	26	28	28	30	30	32	32	34	34	36	38	40	42	44	46	48
0.051 – 0.070 (0.0020 – 0.0028)														06	06	06	06	06	08	08	10	10	12	12	14	14	16	16	18	18	20	20	22	22	24	24	26	26	28	28	30	30	32	32	34	34	36	36	38	40	42	44	46	48	50
0.071 – 0.090 (0.0028 – 0.0035)												06	06	06	06	06	08	08	10	10	12	12	14	14	16	16	18	18	20	20	22	22	24	24	26	26	28	28	30	30	32	32	34	34	36	36	38	38	40	42	44	46	48	50	52
0.091 – 0.110 (0.0036 – 0.0043)											06	06	06	06	08	08	10	10	12	12	14	14	16	16	18	18	20	20	22	22	24	24	26	26	28	28	30	30	32	32	34	34	36	36	38	38	40	40	42	44	46	48	50	52	54
0.111 – 0.130 (0.0044 – 0.0051)									06	06	06	06	08	08	10	10	12	12	14	14	16	16	18	18	20	20	22	22	24	24	26	26	28	28	30	30	32	32	34	34	36	36	38	38	40	40	42	42	44	46	48	50	52	54	56
0.131 – 0.150 (0.0052 – 0.0059)								06	06	06	08	08	10	10	12	12	14	14	16	16	18	18	20	20	22	22	24	24	26	26	28	28	30	30	32	32	34	34	36	36	38	38	40	40	42	42	44	44	46	48	50	52	54	56	58
0.151 – 0.170 (0.0059 – 0.0067)						06	06	06	08	08	10	10	12	12	14	14	16	16	18	18	20	20	22	22	24	24	26	26	28	28	30	30	32	32	34	34	36	36	38	38	40	40	42	42	44	44	46	46	48	50	52	54	56	58	60
0.171 – 0.190 (0.0067 – 0.0075)					06	06	06	08	10	10	12	12	14	14	16	16	18	18	20	20	22	22	24	24	26	26	28	28	30	30	32	32	34	34	36	36	38	38	40	40	42	42	44	44	46	46	48	48	50	52	54	56	58	60	62
0.191 – 0.210 (0.0075 – 0.0083)				06	06	06	08	10	12	12	14	14	16	16	18	18	20	20	22	22	24	24	26	26	28	28	30	30	32	32	34	34	36	36	38	38	40	40	42	42	44	44	46	46	48	48	50	50	52	54	56	58	60	62	64
0.211 – 0.230 (0.0083 – 0.0091)			06	06	06	08	10	12	14	14	16	16	18	18	20	20	22	22	24	24	26	26	28	28	30	30	32	32	34	34	36	36	38	38	40	40	42	42	44	44	46	46	48	48	50	50	52	52	54	56	58	60	62	64	66
0.231 – 0.249 (0.0091 – 0.0098)		06	06	06	08	10	12	14	16	16	18	18	20	20	22	22	24	24	26	26	28	28	30	30	32	32	34	34	36	36	38	38	40	40	42	42	44	44	46	46	48	48	50	50	52	52	54	54	56	58	60	62	64	66	68
0.250 – 0.350 (0.0098 – 0.0138)																																																							
0.351 – 0.370 (0.0138 – 0.0146)	12	14	16	18	20	22	24	26	28	28	30	30	32	32	34	34	36	36	38	38	40	40	42	42	44	44	46	46	48	48	50	50	52	52	54	54	56	56	58	58	60	60	62	62	64	64	66	66	68	70	72	74	74	74	
0.371 – 0.390 (0.0146 – 0.0154)	14	16	18	20	22	24	26	28	30	30	32	32	34	34	36	36	38	38	40	40	42	42	44	44	46	46	48	48	50	50	52	52	54	54	56	56	58	58	60	60	62	62	64	64	66	66	68	68	70	72	74	74	74		
0.391 – 0.410 (0.0154 – 0.0161)	16	18	20	22	24	26	28	30	32	32	34	34	36	36	38	38	40	40	42	42	44	44	46	46	48	48	50	50	52	52	54	54	56	56	58	58	60	60	62	62	64	64	66	66	68	68	70	70	72	74	74	74			
0.411 – 0.430 (0.0162 – 0.0169)	18	20	22	24	26	28	30	32	34	34	36	36	38	38	40	40	42	42	44	44	46	46	48	48	50	50	52	52	54	54	56	56	58	58	60	60	62	62	64	64	66	66	68	68	70	70	72	72	74	74	74				
0.431 – 0.450 (0.0170 – 0.0177)	20	22	24	26	28	30	32	34	36	36	38	38	40	40	42	42	44	44	46	46	48	48	50	50	52	52	54	54	56	56	58	58	60	60	62	62	64	64	66	68	68	70	70	72	72	74	74	74	74	74					
0.451 – 0.470 (0.0178 – 0.0185)	22	24	26	28	30	32	34	36	38	38	40	40	42	42	44	44	46	46	48	48	50	50	52	52	54	54	56	56	58	58	60	60	62	62	64	64	66	66	68	68	70	70	72	72	74	74	74	74	74						
0.471 – 0.490 (0.0185 – 0.0193)	24	26	28	30	32	34	36	38	40	40	42	42	44	44	46	46	48	48	50	50	52	52	54	54	56	56	58	58	60	60	62	62	64	64	66	66	68	68	70	70	72	72	74	74	74	74	74	74							
0.491 – 0.510 (0.0193 – 0.0201)	26	28	30	32	34	36	38	40	42	42	44	44	46	46	48	48	50	50	52	52	54	54	56	56	58	58	60	60	62	62	64	64	66	66	68	68	70	70	72	72	74	74	74	74	74	74									
0.511 – 0.530 (0.0201 – 0.0209)	28	30	32	34	36	38	40	42	44	44	46	46	48	48	50	50	52	52	54	54	56	56	58	58	60	60	62	62	64	64	66	66	68	68	70	70	72	72	74	74	74	74	74	74											
0.531 – 0.550 (0.0209 – 0.0217)	30	32	34	36	38	40	42	44	46	46	48	48	50	50	52	52	54	54	56	56	58	58	60	60	62	62	64	64	66	66	68	68	70	70	72	72	74	74	74	74	74	74													
0.551 – 0.570 (0.0217 – 0.0224)	32	34	36	38	40	42	44	46	48	48	50	50	52	52	54	54	56	56	58	58	60	60	62	62	64	64	66	66	68	68	70	70	72	72	74	74	74	74	74	74															
0.571 – 0.590 (0.0225 – 0.0232)	34	36	38	40	42	44	46	48	50	50	52	52	54	54	56	56	58	58	60	60	62	62	64	64	66	66	68	68	70	70	72	72	74	74	74	74	74	74																	
0.591 – 0.610 (0.0233 – 0.0240)	36	38	40	42	44	46	48	50	52	52	54	54	56	56	58	58	60	60	62	62	64	64	66	66	68	68	70	70	72	72	74	74	74	74	74	74																			
0.611 – 0.630 (0.0241 – 0.0248)	38	40	42	44	46	48	50	52	54	54	56	56	58	58	60	60	62	62	64	64	66	66	68	68	70	70	72	72	74	74	74	74	74	74																					
0.631 – 0.650 (0.0248 – 0.0256)	40	42	44	46	48	50	52	54	56	56	58	58	60	60	62	62	64	64	66	66	68	68	70	70	72	72	74	74	74	74	74	74																							
0.651 – 0.670 (0.0256 – 0.0264)	42	44	46	48	50	52	54	56	58	58	60	60	62	62	64	64	66	66	68	68	70	70	72	72	74	74	74	74	74	74																									
0.671 – 0.690 (0.0264 – 0.0272)	44	46	48	50	52	54	56	58	60	60	62	62	64	64	66	66	68	68	70	70	72	72	74	74	74	74	74	74																											
0.691 – 0.710 (0.0272 – 0.0280)	46	48	50	52	54	56	58	60	62	62	64	64	66	66	68	68	70	70	72	72	74	74	74	74	74	74																													
0.711 – 0.730 (0.0280 – 0.0287)	48	50	52	54	56	58	60	62	64	64	66	66	68	68	70	70	72	72	74	74	74	74	74	74																															
0.731 – 0.750 (0.0288 – 0.0295)	50	52	54	56	58	60	62	64	66	66	68	68	70	70	72	72	74	74	74	74	74	74																																	
0.751 – 0.770 (0.0296 – 0.0303)	52	54	56	58	60	62	64	66	68	68	70	70	72	72	74	74	74	74	74	74																																			
0.771 – 0.790 (0.0304 – 0.0311)	54	56	58	60	62	64	66	68	70	70	72	72	74	74	74	74	74	74																																					
0.791 – 0.810 (0.0311 – 0.0319)	56	58	60	62	64	66	68	70	72	72	74	74	74	74	74	74																																							
0.811 – 0.830 (0.0319 – 0.0327)	58	60	62	64	66	68	70	72	74	74	74	74	74	74																																									
0.831 – 0.850 (0.0327 – 0.0335)	60	62	64	66	68	70	72	74	74	74	74	74																																											
0.851 – 0.870 (0.0335 – 0.0343)	62	64	66	68	70	72	74	74	74	74																																													
0.871 – 0.890 (0.0343 – 0.0350)	64	66	68	70	72	74	74	74																																															
0.891 – 0.910 (0.0351 – 0.0358)	66	68	70	72	74	74	74																																																
0.911 – 0.930 (0.0359 – 0.0366)	68	70	72	74	74	74																																																	
0.931 – 0.950 (0.0367 – 0.0374)	70	72	74	74	74																																																		
0.951 – 0.970 (0.0374 – 0.0382)	72	74	74	74																																																			
0.971 – 0.990 (0.0382 – 0.0390)	74	74	74																																																				
0.991 – 1.010 (0.0390 – 0.0398)	74	74																																																					
1.011 – 1.030 (0.0398 – 0.0406)	74																																																						

New lifter thickness mm (in.)

Lifter No.	Thickness	Lifter No.	Thickness	Lifter No.	Thickness
06	5.060 (0.1992)	30	5.300 (0.2087)	54	5.540 (0.2181)
08	5.080 (0.2000)	32	5.320 (0.2094)	56	5.560 (0.2189)
10	5.100 (0.2008)	34	5.340 (0.2102)	58	5.580 (0.2197)
12	5.120 (0.2016)	36	5.360 (0.2110)	60	5.600 (0.2205)
14	5.140 (0.2024)	38	5.380 (0.2118)	62	5.620 (0.2213)
16	5.160 (0.2031)	40	5.400 (0.2126)	64	5.640 (0.2220)
18	5.180 (0.2039)	42	5.420 (0.2134)	66	5.660 (0.2228)
20	5.200 (0.2047)	44	5.440 (0.2142)	68	5.680 (0.2236)
22	5.220 (0.2055)	46	5.460 (0.2150)	70	5.700 (0.2244)
24	5.240 (0.2063)	48	5.480 (0.2157)	72	5.720 (0.2252)
26	5.260 (0.2071)	50	5.500 (0.2165)	74	5.740 (0.2260)
28	5.280 (0.2079)	52	5.520 (0.2173)		

Exhaust valve clearance (Cold):
0.25 – 0.35 mm (0.010 – 0.014 in.)
EXAMPLE: The 5.340 mm (0.2102 in.) lifter is installed, and the measured clearance is 0.440 mm (0.0173 in.).
Replace the 5.340 mm (0.2102 in.) lifter with a new No. 48 lifter.

9307WG71

Fig. 147 Adjusting shim chart (exhaust)—1ZZ-FE engine

line on the timing chain cover surface as shown in the illustration.

c. If not, turn the crankshaft 1 revolution (360°) to align the marks as above.

4. Check valve clearance.

a. Check only the valves indicated in the illustration.

b. Record any out-of-specification valve clearance measurements. They will be used later to determine the required replacement valve lifters.

c. Turn the crankshaft 1 revolution (360°) to set the No. 4 cylinder to the TDC/compression.

d. Check only the valves indicated in the illustration.

e. Using a feeler gauge, measure the clearance between the valve lifter and camshaft.

f. Record any out-of-specification valve clearance measurements. They will be used later to determine the required replacement valve lifters.

5. If valve clearance needs to be adjusted, proceed as follows:

a. Remove both camshafts.

b. Remove the valve lifters.

c. Using a micrometer, measure the thickness of the removed valve lifters.

d. Calculate the thickness of a new lifter so that the valve clearance comes within the specified values.

e. On the intake side, Thickness of new lifter equals Thickness of used lifter plus (Measured valve clearance minus 0.0079 in. (0.20 mm).

f. On the intake side, Thickness of new lifter equals Thickness of used lifter plus (Measured valve clearance minus 0.0118 in. (0.30 mm).

➡Select a new lifter with a thickness as close to the calculated values as possible.

g. Install the valve lifters and camshafts.

h. Recheck valve clearance.

6. Install cylinder head cover.

2ZZ-GE Engine

See Figures 142, 148 through 153.

➡Adjust the valve clearance when the engine is cold.

1. Before servicing the vehicle, refer to the precautions section.

2. Remove the cylinder head cover.

3. Set the No. 1 cylinder to TDC on the compression stroke.

a. Turn the crankshaft pulley until its groove and timing mark "0" of the timing chain cover are aligned.

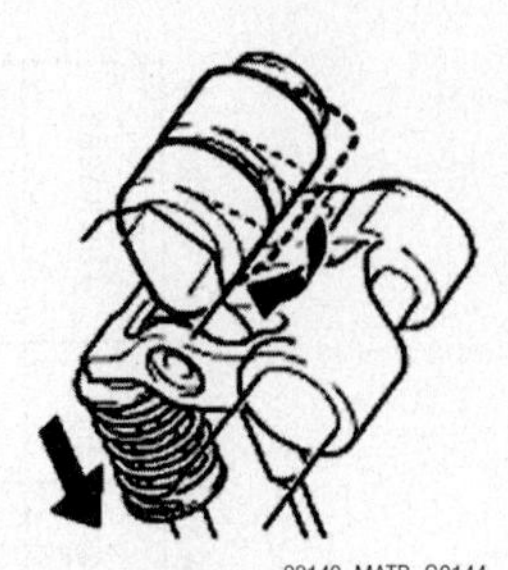

Fig. 148 Turn the crankshaft so that the related rocker arm, where the valve clearance is adjusted, is fully pushed down—2ZZ-GE engine

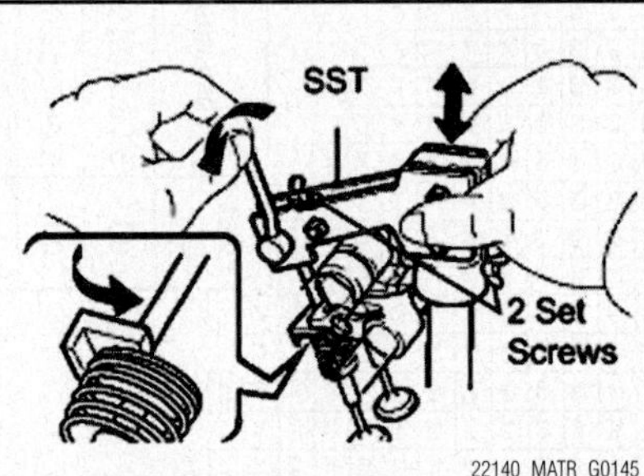

Fig. 149 Install the special tool and operate the lever so that the special tool seat surface comes into contact with the valve retainer, then lock them with the 2 set screws—2ZZ-GE engine

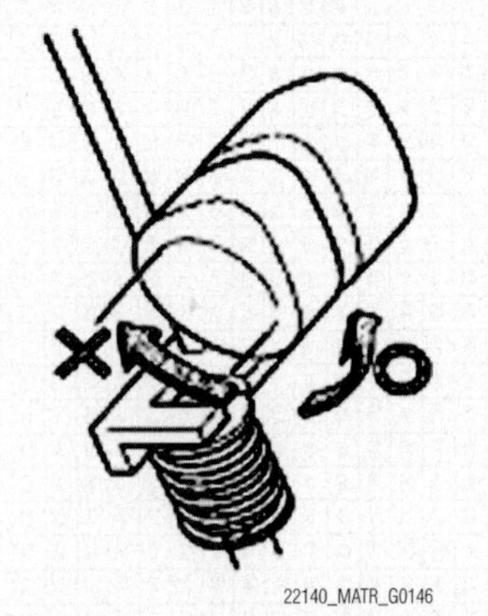

Fig. 150 Turn the crankshaft so that the camshaft is positioned as shown—2ZZ-GE engine

b. Check that the point marks of the camshaft timing gears are in a straight line on the timing chain cover surface as shown in the illustration.

c. If not, turn the crankshaft 1 revolution (360°) to align the marks as above.

4. Inspect the valve clearance.

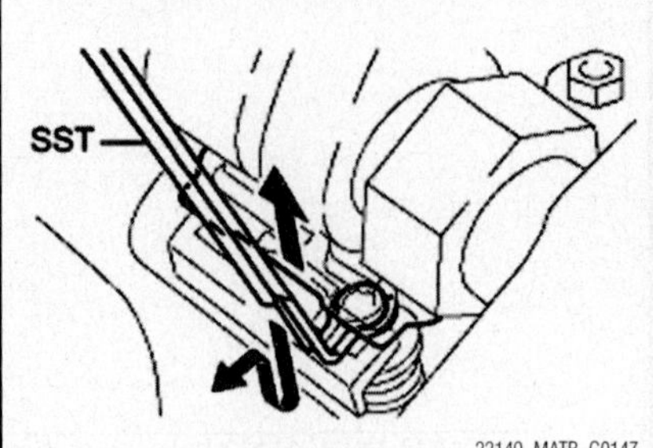

Fig. 151 Lift the rocker arm to make room, then remove the adjusting shim using the special tool—2ZZ-GE engine

a. Using a feeler gauge, measure the clearance between the valve rocker arm and camshaft. Check only the valves indicated in the illustration.

b. Record any out-of-specification valve clearance measurements. They will be used later to determine the required replacement adjusting shim.

c. Turn the crankshaft 1 revolution (360°) to set the No. 4 cylinder to the TDC/compression.

d. Check only the valves indicated in the illustration.

e. Using a feeler gauge, measure the clearance between the valve rocker arm and camshaft.

f. Record any out-of-specification valve clearance measurements. They will be used later to determine the required replacement adjusting shim.

5. If valve clearance needs to be adjusted, proceed as follows:

a. Install special tool SST: 09248-77010

b. Turn the crankshaft so that the related rocker arm, where the valve clearance is adjusted, is fully pushed down.

c. Remove the spark plug, then take off the compression.

d. Insert the special tool into the plug tube.

⁂ WARNING

The special tool cannot be inserted unless the set screw is loosened.

e. Make sure that the camshaft is in the same condition as the previous step.

f. Operate the lever so that the special tool seat surface comes into contact with the valve retainer, then lock them with the 2 set screws.

g. Clearance between the valve retainer and special tool seat surface is not allowed.

Measure clearance mm(in.) \ Installed shim thickness mm(in.)	2.000 (0.0787)	2.020 (0.0795)	2.040 (0.0803)	2.060 (0.0811)	2.080 (0.0819)	2.100 (0.0827)	2.120 (0.0835)	2.140 (0.0843)	2.160 (0.0850)	2.180 (0.0858)	2.190 (0.0862)	2.200 (0.0866)	2.210 (0.0870)	2.220 (0.0874)	2.230 (0.0878)	2.240 (0.0882)	2.250 (0.0886)	2.260 (0.0890)	2.270 (0.0894)	2.280 (0.0898)
0.000 - 0.030 (0.0000 - 0.0012)										00	02	02	04	04	06	06	08	08	10	10
0.031 - 0.050 (0.0012 - 0.0020)								00	02	04	06	06	08	08	10	10	12	12	14	14
0.051 - 0.070 (0.0020 - 0.0028)							00	02	04	06	08	08	10	10	12	12	14	14	16	16
0.071 - 0.090 (0.0028 - 0.0035)					00	02	04	06	08	10	12	12	14	14	16	16	18	18	20	20
0.091 - 0.099 (0.0036 - 0.0039)				00	02	04	06	08	10	12	14	14	16	16	18	18	20	20	22	22
0.100 - 0.160 (0.0039 - 0.0063)																				
0.161 - 0.180 (0.0063 - 0.0071)	06	80	10	12	14	16	18	20	22	24	26	26	28	28	30	30	32	32	34	34
0.181 - 0.200 (0.0071 - 0.0079)	10	12	14	16	18	20	22	24	26	28	30	30	32	32	34	34	36	36	38	38
0.201 - 0.220 (0.0079 - 0.0087)	12	14	16	18	20	22	24	26	28	30	32	32	34	34	36	36	38	38	40	40
0.221 - 0.240 (0.0087 - 0.0094)	16	18	20	22	24	26	28	30	32	34	36	36	38	38	40	40	42	42	44	44
0.241 - 0.260 (0.0095 - 0.0102)	18	20	22	24	26	28	30	32	34	36	38	38	40	40	42	42	44	44	46	46
0.261 - 0.280 (0.0103 - 0.0110)	22	24	26	28	30	32	34	36	38	40	42	42	44	44	46	46	48	48	50	50
0.281 - 0.300 (0.0111 - 0.0118)	24	26	28	30	32	34	36	38	40	42	44	44	46	46	48	48	50	50	52	52
0.301 - 0.320 (0.0119 - 0.0126)	28	30	32	34	36	38	40	42	44	46	48	48	50	50	52	52	54	54	56	56
0.321 - 0.340 (0.0126 - 0.0134)	30	32	34	36	38	40	42	44	46	48	50	50	52	52	54	54	56	56	58	58
0.341 - 0.360 (0.0134 - 0.0142)	34	36	38	40	42	44	46	48	50	52	54	54	56	56	58	58	60	60	62	62
0.361 - 0.380 (0.0142 - 0.0150)	36	38	40	42	44	46	48	50	52	54	56	56	58	58	60	60	62	62	64	64
0.381 - 0.400 (0.0150 - 0.0157)	40	42	44	46	48	50	52	54	56	58	60	60	62	62	64	64	66	66	68	68
0.401 - 0.420 (0.0158 - 0.0165)	42	44	46	48	50	52	54	56	58	60	62	62	64	64	66	66	68	68	70	70
0.421 - 0.440 (0.0166 - 0.0173)	46	48	50	52	54	56	58	60	62	64	66	66	68	68	70	70	72	72	74	74
0.441 - 0.460 (0.0174 - 0.0181)	48	50	52	54	56	58	60	62	64	66	68	68	70	70	72	72	74	74	76	76
0.461 - 0.480 (0.0181 - 0.0189)	52	54	56	58	60	62	64	66	68	70	72	72	74	74	76	76	78	78	80	80
0.481 - 0.500 (0.0189 - 0.0197)	54	56	58	60	62	64	66	68	70	72	74	74	76	76	78	78	80	80		
0.501 - 0.520 (0.0197 - 0.0205)	58	60	62	64	66	68	70	72	74	76	78	78	80	80						
0.521 - 0.540 (0.0205 - 0.0213)	60	62	64	66	68	70	72	74	76	78	80	80								
0.541 - 0.560 (0.0213 - 0.0220)	64	66	68	70	72	74	76	78	80											
0.561 - 0.580 (0.0221 - 0.0228)	66	68	70	72	74	76	78	80												
0.581 - 0.600 (0.0229 - 0.0236)	70	72	74	76	78	80														
0.601 - 0.620 (0.0237 - 0.0244)	72	74	76	78	80															
0.621 - 0.640 (0.0244 - 0.0252)	76	78	80																	
0.641 - 0.660 (0.0252 - 0.0260)	78	80																		
0.661 - 0.680 (0.0260 - 0.0268)	80																			

Measure clearance mm(in.) \ Installed shim thickness mm(in.)	2.290 (0.0902)	2.300 (0.0906)	2.310 (0.0909)	2.320 (0.0913)	2.330 (0.0917)	2.340 (0.0921)	2.340 (0.0921)	2.350 (0.0925)	2.360 (0.0929)	2.370 (0.0933)	2.380 (0.0937)	2.390 (0.0941)	2.400 (0.0945)	2.410 (0.0949)	2.420 (0.0953)	2.430 (0.0957)	2.440 (0.0961)	2.450 (0.0965)	2.460 (0.0969)	2.470 (0.0972)
0.000 - 0.030 (0.0000 - 0.0012)	12	12	14	14	16	16	18	18	20	20	22	22	24	24	26	26	28	28	30	30
0.031 - 0.050 (0.0012 - 0.0020)	16	16	18	18	20	20	22	22	24	24	26	26	28	28	30	30	32	32	34	34
0.051 - 0.070 (0.0020 - 0.0028)	18	18	20	20	22	22	24	24	26	26	28	28	30	30	32	32	34	34	36	36
0.071 - 0.090 (0.0028 - 0.0035)	22	22	24	24	26	26	28	28	30	30	32	32	34	34	36	36	38	38	40	40
0.091 - 0.099 (0.0036 - 0.0039)	24	24	26	26	28	28	30	30	32	32	34	34	36	36	38	38	40	40	42	42
0.100 - 0.160 (0.0039 - 0.0063)																				
0.161 - 0.180 (0.0063 - 0.0071)	36	36	38	38	40	40	42	42	44	44	46	46	48	48	50	50	52	52	54	54
0.181 - 0.200 (0.0071 - 0.0079)	40	40	42	42	44	44	46	46	48	48	50	50	52	52	54	54	56	56	58	58
0.201 - 0.220 (0.0079 - 0.0087)	42	42	44	44	46	46	48	48	50	50	52	52	54	54	56	56	58	58	60	60
0.221 - 0.240 (0.0087 - 0.0094)	46	46	48	48	50	50	52	52	54	54	56	56	58	58	60	60	62	62	64	64
0.241 - 0.260 (0.0095 - 0.0102)	48	48	50	50	52	52	54	54	56	56	58	58	60	60	62	62	64	64	66	66
0.261 - 0.280 (0.0103 - 0.0110)	52	52	54	54	56	56	58	58	60	60	62	62	64	64	66	66	68	68	70	70
0.281 - 0.300 (0.0111 - 0.0118)	54	54	56	56	58	58	60	60	62	62	64	64	66	66	68	68	70	70	72	72
0.301 - 0.320 (0.0119 - 0.0126)	58	58	60	60	62	62	64	64	66	66	68	68	70	70	72	72	74	74	76	76
0.321 - 0.340 (0.0126 - 0.0134)	60	60	62	62	64	64	66	66	68	68	70	70	72	72	74	74	76	76	78	78
0.341 - 0.360 (0.0134 - 0.0142)	64	64	66	66	68	68	70	70	72	72	74	74	76	76	78	78	80	80		
0.361 - 0.380 (0.0142 - 0.0150)	66	66	68	68	70	70	72	72	74	74	76	76	78	78	80	80				
0.381 - 0.400 (0.0150 - 0.0157)	70	70	72	72	74	74	76	76	78	78	80	80								
0.401 - 0.420 (0.0158 - 0.0165)	72	72	74	74	76	76	78	78	80	80										
0.421 - 0.440 (0.0166 - 0.0173)	76	76	78	78	80	80														
0.441 - 0.460 (0.0174 - 0.0181)	78	78	80	80																

Measure clearance mm(in.) \ Installed shim thickness mm(in.)	2.480 (0.0976)	2.490 (0.0980)	2.500 (0.0984)	2.510 (0.0988)	2.520 (0.0992)	2.530 (0.0996)	2.540 (0.1000)	2.550 (0.1004)	2.560 (0.1008)	2.580 (0.1016)	2.600 (0.1024)	2.620 (0.1031)	2.640 (0.1039)	2.660 (0.1047)	2.680 (0.1055)	2.700 (0.1063)	2.720 (0.1071)	2.740 (0.1079)	2.760 (0.1087)	2.780 (0.1094)	2.800 (0.1102)
0.000 - 0.030 (0.0000 - 0.0012)	32	32	34	34	36	36	38	38	40	42	44	46	48	50	52	54	56	58	60	62	64
0.031 - 0.050 (0.0012 - 0.0020)	36	36	38	38	40	40	42	42	44	46	48	50	52	54	56	58	60	62	64	66	68
0.051 - 0.070 (0.0020 - 0.0028)	38	38	40	40	42	42	44	44	46	48	50	52	54	56	58	60	62	64	66	68	70
0.071 - 0.090 (0.0028 - 0.0035)	42	42	44	44	46	46	48	48	50	52	54	56	58	60	62	64	66	68	70	72	74
0.091 - 0.099 (0.0036 - 0.0039)	44	44	46	46	48	48	50	50	52	54	56	58	60	62	64	66	68	70	72	74	76
0.100 - 0.160 (0.0039 - 0.0063)																					
0.161 - 0.180 (0.0063 - 0.0071)	56	56	58	58	60	60	62	62	64	66	68	70	72	74	76	78	80				
0.181 - 0.200 (0.0071 - 0.0079)	60	60	62	62	64	64	66	66	68	70	72	74	76	78	80						
0.201 - 0.220 (0.0079 - 0.0087)	62	62	64	64	66	66	68	68	70	72	74	76	78	80							
0.221 - 0.240 (0.0087 - 0.0094)	66	66	68	68	70	70	72	72	74	76	78	80	82								
0.241 - 0.260 (0.0095 - 0.0102)	68	68	70	70	72	72	74	74	76	78	80										
0.261 - 0.280 (0.0103 - 0.0110)	72	72	74	74	76	76	78	78	80												
0.281 - 0.300 (0.0111 - 0.0118)	74	74	76	76	78	78	80	80													
0.301 - 0.320 (0.0119 - 0.0126)	78	78	80	80																	
0.321 - 0.340 (0.0126 - 0.0134)	80	80																			

Intake valve clearance (Cold):
0.08 – 0.18 mm (0.0031 – 0.0071 in.)

EXAMPLE: The 2.200 mm (0.0826 in.) shim is installed, and the measured clearance is 0.400 mm (0.0157 in.).
Replace the 2.600 mm (0.1024 in.) shim with a new No. 60 shim.

New Shim thickness mm (in.)

Shim No.	Thickness	Shim No.	Thickness	Shim No.	Thickness
00	2.000 (0.0787)	28	2.280 (0.0898)	56	2.560 (0.1008)
02	2.020 (0.0795)	30	2.300 (0.0906)	58	2.580 (0.1016)
04	2.040 (0.0803)	32	2.320 (0.0913)	60	2.600 (0.1024)
06	2.060 (0.0811)	34	2.340 (0.0921)	62	2.620 (0.1031)
08	2.080 (0.0819)	36	2.360 (0.0929)	64	2.640 (0.1039)
10	2.100 (0.0827)	38	2.380 (0.0937)	66	2.660 (0.1047)
12	2.120 (0.0835)	40	2.400 (0.0945)	68	2.680 (0.1055)
14	2.140 (0.0843)	42	2.420 (0.0953)	70	2.700 (0.1063)
16	2.160 (0.0850)	44	2.440 (0.0961)	72	2.720 (0.1071)
18	2.180 (0.0858)	46	2.460 (0.0969)	74	2.740 (0.1079)
20	2.200 (0.0866)	48	2.480 (0.0976)	76	2.760 (0.1087)
22	2.220 (0.0874)	50	2.500 (0.0984)	78	2.780 (0.1094)
24	2.240 (0.0882)	52	2.520 (0.0992)	80	2.800 (0.1102)
26	2.260 (0.0890)	54	2.540 (0.1000)		

9359AB26

Fig. 152 Adjusting shim chart (intake)—2ZZ-GE Engine

Measure clearance mm(in.) \ Installed shim thickness mm(in.)	2.000 (0.0787)	2.020 (0.0795)	2.040 (0.0803)	2.060 (0.0811)	2.080 (0.0819)	2.100 (0.0827)	2.120 (0.0835)	2.140 (0.0843)	2.160 (0.0850)	2.180 (0.0858)	2.190 (0.0862)	2.200 (0.0866)	2.210 (0.0870)	2.220 (0.0874)	2.230 (0.0878)	2.240 (0.0882)	2.250 (0.0886)	2.260 (0.0890)	2.270 (0.0894)	2.280 (0.0898)	2.290 (0.0902)	2.300 (0.0906)	2.310 (0.0909)	2.320 (0.0913)	2.330 (0.0917)	2.340 (0.0921)	2.340 (0.0921)	2.350 (0.0925)	2.360 (0.0929)	2.370 (0.0933)	2.380 (0.0937)	2.390 (0.0941)	2.400 (0.0945)	2.410 (0.0949)	2.420 (0.0953)	2.430 (0.0957)	2.440 (0.0961)	2.450 (0.0965)	2.460 (0.0969)	2.470 (0.0972)	2.480 (0.0976)	2.490 (0.0980)	2.500 (0.0984)	2.510 (0.0988)	2.520 (0.0992)	2.530 (0.0996)	2.540 (0.1000)	2.550 (0.1004)	2.560 (0.1008)	2.580 (0.1016)	2.600 (0.1024)	2.620 (0.1031)	2.640 (0.1039)	2.660 (0.1047)	2.680 (0.1055)	2.700 (0.1063)	2.720 (0.1071)	2.740 (0.1079)	2.760 (0.1087)	2.780 (0.1094)	2.800 (0.1102)
0.000 - 0.030 (0.0000 - 0.0012)																															00	00	02	02	04	04	06	06	08	08	10	10	12	12	14	14	16	16	18	20	22	24	26	28	30	32	34	36	38	40	42
0.031 - 0.050 (0.0012 - 0.0020)																												00	04	04	06	06	08	08	10	10	12	12	14	14	16	16	18	18	20	20	22	22	24	26	28	30	32	34	36	38	40	42	44	46	48
0.051 - 0.070 (0.0020 - 0.0028)																								00	02	02	04	04	06	06	08	08	10	10	12	12	14	14	16	16	18	18	20	20	22	22	24	24	26	28	30	32	34	36	38	40	42	44	46	48	50
0.071 - 0.090 (0.0028 - 0.0035)																					00	00	02	02	04	04	06	06	08	08	10	10	12	12	14	14	16	16	18	18	20	20	22	22	24	24	26	26	28	30	32	34	36	38	40	42	44	46	48	50	52
0.091 - 0.110 (0.0036 - 0.0043)																		00	02	02	04	04	06	06	08	08	10	10	12	12	14	14	16	16	18	18	20	20	22	22	24	24	26	26	28	28	30	30	32	34	36	38	40	42	44	46	48	50	52	54	56
0.111 - 0.130 (0.0044 - 0.0051)															00	00	02	02	04	04	06	06	08	08	10	10	12	12	14	14	16	16	18	18	20	20	22	22	24	24	26	26	28	28	30	30	32	32	34	36	38	40	42	44	46	48	50	52	54	56	58
0.131 - 0.150 (0.0052 - 0.0059)												00	02	02	04	04	06	06	08	08	10	10	12	12	14	14	16	16	18	18	20	20	22	22	24	24	26	26	28	28	30	30	32	32	34	34	36	36	38	40	42	44	46	48	50	52	54	56	58	60	62
0.151 - 0.170 (0.0059 - 0.0067)										00	02	02	04	04	06	06	08	08	10	10	12	12	14	14	16	16	18	18	20	20	22	22	24	24	26	26	28	28	30	30	32	32	34	34	36	36	38	38	40	42	44	46	48	50	52	54	56	58	60	62	64
0.171 - 0.190 (0.0067 - 0.0075)								00	02	04	06	06	08	08	10	10	12	12	14	14	16	16	18	18	20	20	22	22	24	24	26	26	28	28	30	30	32	32	34	34	36	36	38	38	40	40	42	42	44	46	48	50	52	54	56	58	60	62	64	66	68
0.191 - 0.210 (0.0075 - 0.0083)							00	02	04	06	08	08	10	10	12	12	14	14	16	16	18	18	20	20	22	22	24	24	26	26	28	28	30	30	32	32	34	34	36	36	38	38	40	40	42	42	44	44	46	48	50	52	54	56	58	60	62	64	66	68	70
0.211 - 0.230 (0.0083 - 0.0091)					00	02	04	06	08	10	12	12	14	14	16	16	18	18	20	20	22	22	24	24	26	26	28	28	30	30	32	32	34	34	36	36	38	38	40	40	42	42	44	44	46	46	48	48	50	52	54	56	58	60	62	64	66	68	70	72	74
0.231 - 0.239 (0.0091 - 0.0094)				00	02	04	06	08	10	12	14	14	16	16	18	18	20	20	22	22	24	24	26	26	28	28	30	30	32	32	34	34	36	36	38	38	40	40	42	42	44	44	46	46	48	48	50	50	52	54	56	58	60	62	64	66	68	70	72	74	76
0.301 - 0.320 (0.0119 - 0.0126)	06	08	10	12	14	16	18	20	22	24	26	26	28	28	30	30	32	32	34	34	36	36	38	38	40	40	42	42	44	44	46	46	48	48	50	50	52	52	54	54	56	56	58	58	60	60	62	62	64	66	68	70	72	74	76	78	80	80			
0.321 - 0.340 (0.0126 - 0.0134)	08	10	12	14	16	18	20	22	24	26	28	28	30	30	32	32	34	34	36	36	38	38	40	40	42	42	44	44	46	46	48	48	50	50	52	52	54	54	56	56	58	58	60	60	62	62	64	64	66	68	70	72	74	76	78	80					
0.341 - 0.360 (0.0134 - 0.0142)	12	14	16	18	20	22	24	26	28	30	32	32	34	34	36	36	38	38	40	40	42	42	44	44	46	46	48	48	50	50	52	52	54	54	56	56	58	58	60	60	62	62	64	64	66	66	68	68	70	72	74	76	78	80							
0.361 - 0.380 (0.0142 - 0.0150)	16	18	20	22	24	26	28	30	32	34	36	36	38	38	40	40	42	42	44	44	46	46	48	48	50	50	52	52	54	54	56	56	58	58	60	60	62	62	64	64	66	66	68	68	70	70	72	72	74	76	78	80									
0.381 - 0.400 (0.0150 - 0.0157)	18	20	22	24	26	28	30	32	34	36	38	38	40	40	42	42	44	44	46	46	48	48	50	50	52	52	54	54	56	56	58	58	60	60	62	62	64	64	66	66	68	68	70	70	72	72	74	74	76	78	80										
0.401 - 0.420 (0.0158 - 0.0165)	22	24	26	28	30	32	34	36	38	40	42	42	44	44	46	46	48	48	50	50	52	52	54	54	56	56	58	58	60	60	62	62	64	64	66	66	68	68	70	70	72	72	74	74	76	76	78	78	80												
0.421 - 0.440 (0.0166 - 0.0173)	24	26	28	30	32	34	36	38	40	42	44	44	46	46	48	48	50	50	52	52	54	54	56	56	58	58	60	60	62	62	64	64	66	66	68	68	70	70	72	72	74	74	76	76	78	78	80	80													
0.441 - 0.460 (0.0174 - 0.0181)	28	30	32	34	36	38	40	42	44	46	48	48	50	50	52	52	54	54	56	56	58	58	60	60	62	62	64	64	66	66	68	68	70	70	72	72	74	74	76	76	78	78	80	80																	
0.461 - 0.480 (0.0181 - 0.0189)	30	32	34	36	38	40	42	44	46	48	50	50	52	52	54	54	56	56	58	58	60	60	62	62	64	64	66	66	68	68	70	70	72	72	74	74	76	76	78	78	80	80																			
0.481 - 0.500 (0.0189 - 0.0197)	34	36	38	40	42	44	46	48	50	52	54	54	56	56	58	58	60	60	62	62	64	64	66	66	68	68	70	70	72	72	74	74	76	76	78	78	80	80																							
0.501 - 0.520 (0.0197 - 0.0205)	36	38	40	42	44	46	48	50	52	54	56	58	58	58	60	60	62	62	64	64	66	66	68	68	70	70	72	72	74	74	76	76	78	78	80	80																									
0.521 - 0.540 (0.0205 - 0.0213)	40	42	44	46	48	50	52	54	56	58	60	60	62	62	64	64	66	66	68	68	70	70	72	72	74	74	76	76	78	78	80	80																													
0.541 - 0.560 (0.0213 - 0.0220)	42	44	46	48	50	52	54	56	58	60	62	62	64	64	66	66	68	68	70	70	72	72	74	74	76	76	78	78	80	80																															
0.561 - 0.580 (0.0221 - 0.0228)	46	48	50	52	54	56	58	60	62	64	66	66	68	68	70	70	72	72	74	74	76	76	78	78	80	80																																			
0.581 - 0.600 (0.0229 - 0.0236)	48	50	52	54	56	58	60	62	64	66	68	68	70	70	72	72	74	74	76	76	78	78	80	80																																					
0.601 - 0.620 (0.0237 - 0.0244)	52	54	56	58	60	62	64	66	68	70	72	72	74	74	76	76	78	78	80	80																																									
0.621 - 0.640 (0.0244 - 0.0252)	54	56	58	60	62	64	66	68	70	72	74	74	76	76	78	78	80	80																																											
0.641 - 0.660 (0.0252 - 0.0260)	58	60	62	64	66	68	70	72	74	76	78	78	80	80																																															
0.661 - 0.680 (0.0260 - 0.0268)	60	62	64	66	68	70	72	74	76	78	80	80																																																	
0.681 - 0.700 (0.0268 - 0.0276)	64	66	68	70	72	74	76	78	80																																																					
0.701 - 0.720 (0.0276 - 0.0283)	66	68	70	72	74	76	78	80																																																						
0.721 - 0.740 (0.0284 - 0.0291)	70	72	74	76	78	80																																																								
0.741 - 0.760 (0.0292 - 0.0299)	72	74	76	78	80																																																									
0.761 - 0.780 (0.0300 - 0.0307)	76	78	80																																																											
0.781 - 0.800 (0.0307 - 0.0315)	80																																																													
0.801 - 0.820 (0.0315 - 0.0323)	80																																																													

New Shim thickness mm (in.)

Shim No.	Thickness	Shim No.	Thickness	Shim No.	Thickness
00	2.000 (0.0787)	28	2.280 (0.0898)	56	2.560 (0.1008)
02	2.020 (0.0795)	30	2.300 (0.0906)	58	2.580 (0.1016)
04	2.040 (0.0803)	32	2.320 (0.0913)	60	2.600 (0.1024)
06	2.060 (0.0811)	34	2.340 (0.0921)	62	2.620 (0.1031)
08	2.080 (0.0819)	36	2.360 (0.0929)	64	2.640 (0.1039)
10	2.100 (0.0827)	38	2.380 (0.0937)	66	2.660 (0.1047)
12	2.120 (0.0835)	40	2.400 (0.0945)	68	2.680 (0.1055)
14	2.140 (0.0843)	42	2.420 (0.0953)	70	2.700 (0.1063)
16	2.160 (0.0850)	44	2.440 (0.0961)	72	2.720 (0.1071)
18	2.180 (0.0858)	46	2.460 (0.0969)	74	2.740 (0.1079)
20	2.200 (0.0866)	48	2.480 (0.0976)	76	2.760 (0.1087)
22	2.220 (0.0874)	50	2.500 (0.0984)	78	2.780 (0.1094)
24	2.240 (0.0882)	52	2.520 (0.0992)	80	2.800 (0.1102)
26	2.260 (0.0890)	54	2.540 (0.1000)		

Exhaust valve clearance (Cold):
0.22 – 0.32 mm (0.0087 – 0.0126 in.)
EXAMPLE: The 2.200 mm (0.0862 in.) shim is installed, and the measured clearance is 0.500 mm (0.0197 in.).
Replace the 2.540 mm (0.1000 in.) shim with a new No. 54 shim.

9359AB27

Fig. 153 Adjusting shim chart (exhaust)—2ZZ-GE Engine

h. Be careful not to make clearance when inserting SST, since a presence of clearance may unlock the keeper.

i. Lock the set screw on the plug tube side of the special tool.

j. Turn the crankshaft so that the camshaft is positioned as shown in the illustration.

⁂ WARNING

Take care of the rotation direction to prevent the camshaft lobe from interfering with the special tool shaft.

k. Do not turn the crankshaft excessively.

6. Remove the adjusting shim.

a. Lift the rocker arm to make room, then remove the adjusting shim using the special tool.

⁂ WARNING

Do not remove SST with the adjusting shim removed.

➡Setting the special tool from the upper right makes the removal easier. If there is not enough room, reset special tool.

7. Determine the size of the replaced shim.

a. Using a Vernier caliper, measure the thickness of the removed shim.

b. Calculate the thickness of a new shim so that the valve clearance comes within the specified value.

c. For the intake side, the thickness of the new shim equals thickness of the used shim plus (measured valve clearance minus 0.005 in. (0.13 mm) x 1.5.

d. For the intake side, the thickness of the new shim equals thickness of the used shim plus (measured valve clearance minus 0.011 in. (0.27 mm) x 1.5.

8. Install the cylinder head cover.

ENGINE PERFORMANCE & EMISSION CONTROL

See Figures 154 and 155.

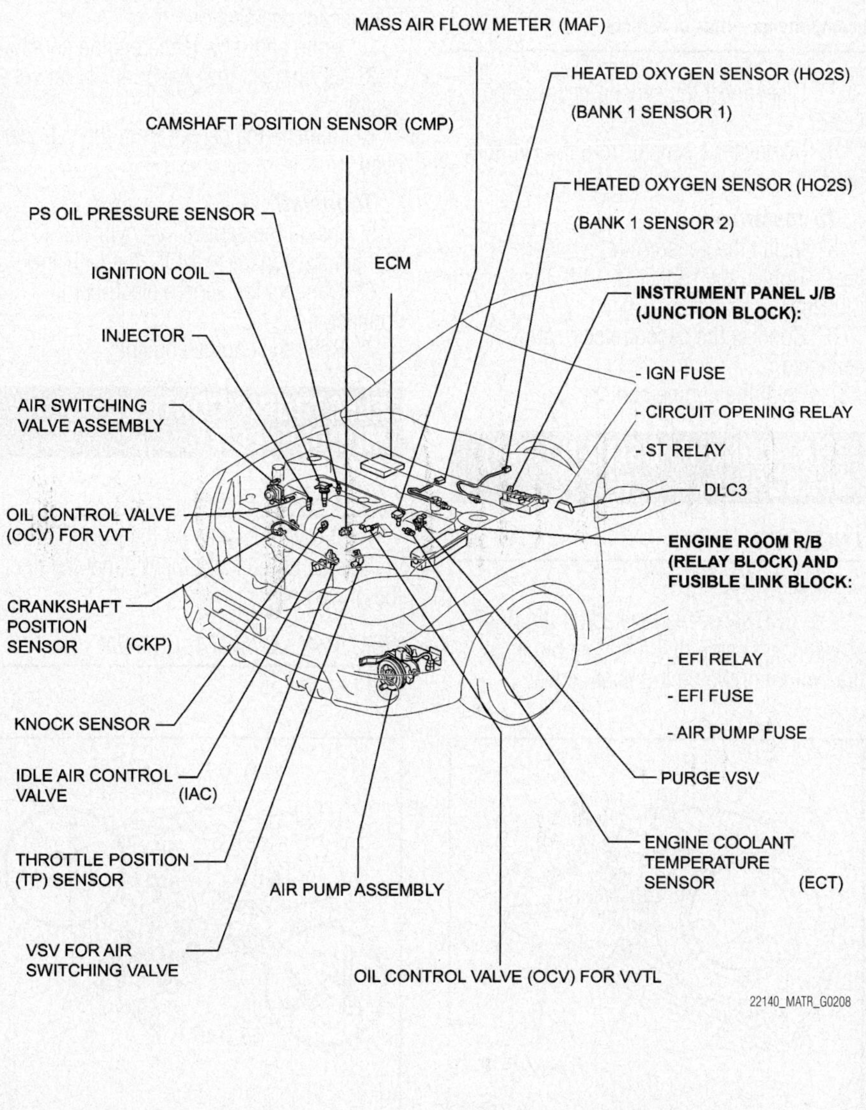

Fig. 154 Engine performance and emission control components—front of vehicle

COMPONENT LOCATIONS

ACCELERATOR PEDAL POSITION (APP) SENSOR

LOCATION

The APP sensor is located on the accelerator pedal bracket.

REMOVAL & INSTALLATION

1. Before servicing the vehicle, refer to the precautions section.
2. Disconnect the negative battery cable.
3. Disconnect the sensor connector.
4. Remove the sensor attaching bolts, then remove the sensor.

To install:

5. Installation is the reverse of removal.
6. Tighten the sensor attaching bolts to 48 inch lbs. (5 Nm).

CAMSHAFT POSITION (CMP) SENSOR

LOCATION

The CMP sensor is located on the front side of the cylinder head.

REMOVAL & INSTALLATION

See Figure 156.

1. Before servicing the vehicle, refer to the precautions section.
2. Disconnect the negative battery cable.
3. Disconnect the sensor connector.
4. Remove the sensor attaching bolt, then remove the sensor.

To install:

5. Installation is the reverse of removal.
6. Tighten the sensor attaching bolts to 48 inch lbs. (5 Nm).

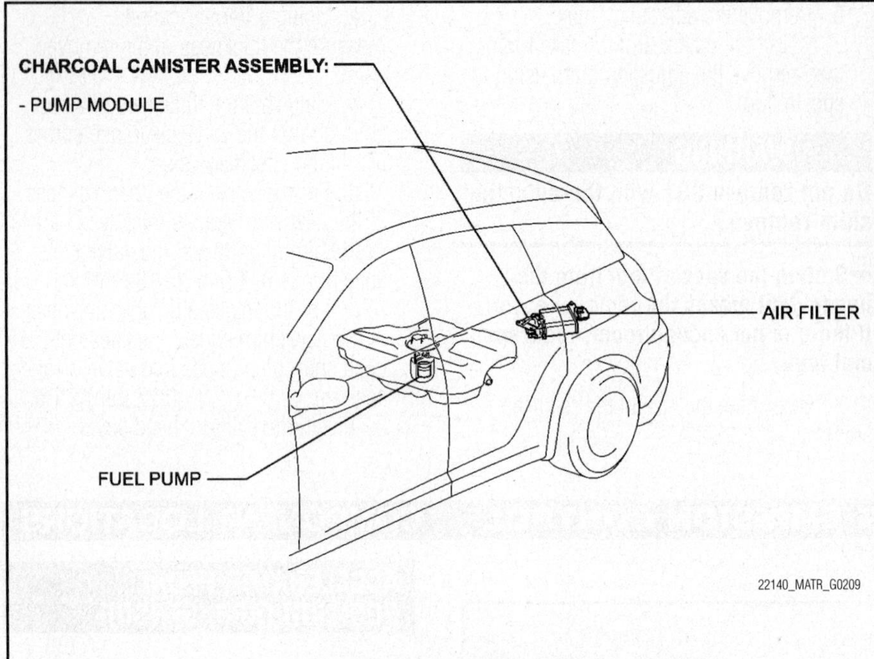

Fig. 155 Engine performance and emission control components—rear of vehicle

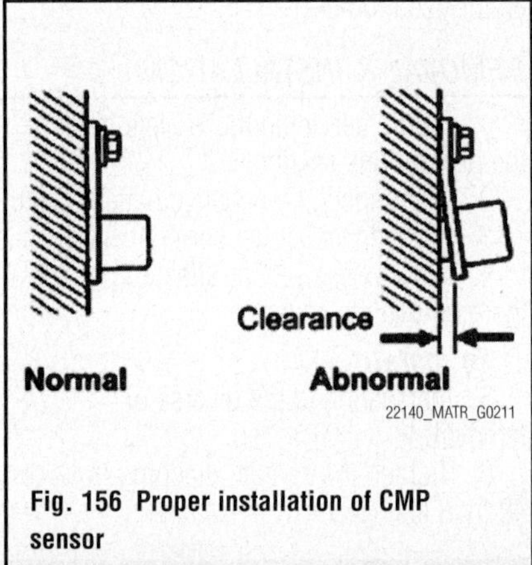

Fig. 156 Proper installation of CMP sensor

COOLANT TEMPERATURE SENSOR

LOCATION

The ECT sensor located next to the upper radiator hose on the cylinder head.

REMOVAL & INSTALLATION

1. Before servicing the vehicle, refer to the precautions section.

✱✱ WARNING

Use care when handling the engine coolant temperature (ECT) sensor. Damage to the ECT sensor will affect proper operation of the fuel injection system.

2. Drain and recycle the engine coolant.
3. Disconnect the sensor electrical connector .
4. Remove the sensor from the cylinder head.

To install:

5. Install the sensor into cylinder head. Tighten the sensor to 14 ft. lbs. (20 Nm).
6. Connect the sensor electrical connector.
7. Refill the engine coolant.

CRANKSHAFT POSITION (CKP) SENSOR

LOCATION

See Figure 157.

The CKP sensor is located on the right (passenger's) side of the engine block, at the bottom of the timing chain cover.

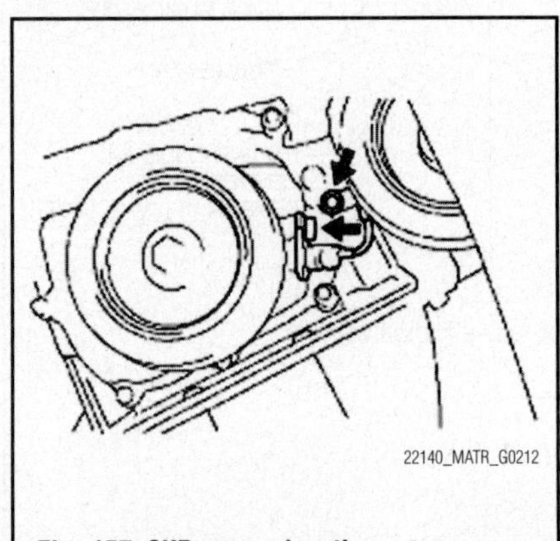

Fig. 157 CKP sensor location

REMOVAL & INSTALLATION

1. Before servicing the vehicle, refer to the precautions section.
2. Disconnect the sensor electrical connector .
3. Remove the attaching bolt and sensor from the timing chain cover.

To install:

4. Install the sensor into cylinder head. Tighten the sensor to 80 inch lbs. (9 Nm).
5. Connect the sensor electrical connector.

ELECTRIC FAN SWITCH

LOCATION

The switch is located next to the upper radiator hose on the cylinder head.

REMOVAL & INSTALLATION

1. Before servicing the vehicle, refer to the precautions section.
2. Drain and recycle the engine coolant.
3. Disconnect the sensor electrical connector .
4. Remove the sensor from the cylinder head.

To install:

5. Install the sensor into cylinder head. Tighten the sensor to 14 ft. lbs. (20 Nm).
6. Connect the sensor electrical connector.
7. Refill the engine coolant.

ELECTRONIC CONTROL MODULE (ECM)

LOCATION

The ECM is located on the right (passenger's) side of the instrument panel, behind the glove box.

REMOVAL & INSTALLATION

See Figure 158.

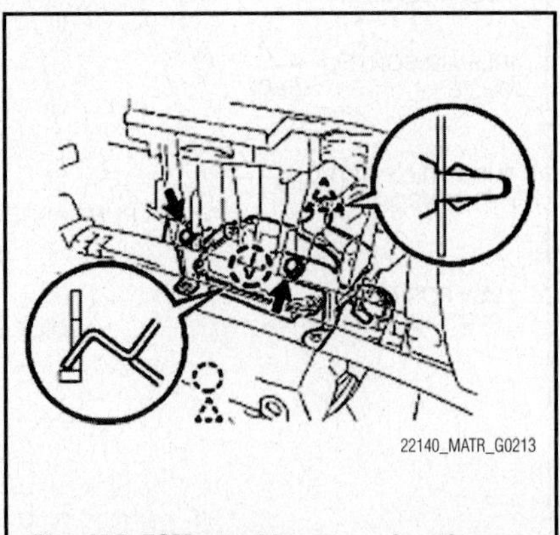

Fig. 158 ECM attaching screw locations

1. Before servicing the vehicle, refer to the precautions section.
2. Disconnect battery negative terminal.
3. Remove glove compartment door assembly.
 a. Remove the screw from the glove compartment door stoppers.
 b. Deform the upper part of the glove compartment door assembly to release the stoppers .
 c. Pull the glove compartment door assembly upward to remove it.
4. Remove ECM.
 a. Remove the 2 clips using a clip remover, then open the cover.
 b. Disconnect the 4 ECM connectors .
 c. Remove the wire harness from the wire harness clamp .
 d. Remove the 2 screws .
 e. Unfasten the claw and clip, and then remove the ECM.
5. Remove ECM cover.
 a. Remove the ECM cover from the ECM.
 b. Remove the screws, then remove the ECM brackets.

To install:

6. Install the ECM brackets and tighten the screws to 28 inch lbs. (3 Nm).
7. Install ECM cover.
 a. Install the ECM with the screws. Tighten to 28 inch lbs. (3 Nm).
 b. Connect the ECM connectors .
 c. Install the wire harness clamp .
 d. Install the cover with the clips.
8. Install glove compartment door assembly.
9. Connect battery negative cable.
10. Using a scan tool, register the VIN in the new ECM.

FUEL LEVEL SENDING UNIT

LOCATION

The fuel level sending unit is located on the fuel sender assembly in the fuel tank.

REMOVAL & INSTALLATION

See Figure 159.

1. Before servicing the vehicle, refer to the precautions section.
2. Disconnect the negative battery cable.
3. Remove the rear left hand seat cushion
4. Remove the rear service hole cover.

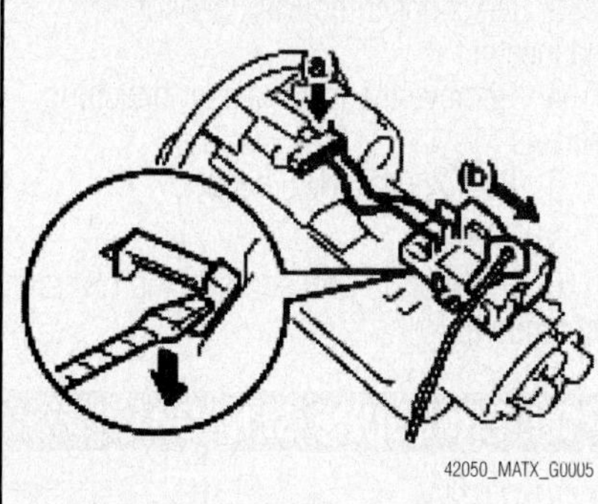

Fig. 159 Removing the fuel sender assembly

5. Disconnect the fuel pump and vapor pressure sensor connector.
6. Properly relieve the fuel system pressure.
7. Remove the bolts and fuel tank protector.
8. Remove the tube joint clip, and pull the fuel tank main tube out.
9. Remove the bolts and fuel tank vent tube set plate.
10. Pull the fuel pump assembly out.

➡Make sure that the arm of the fuel sender gauge does not bend.

11. Remove the gasket from the fuel pump assembly.
12. Using a screwdriver, disconnect the 5 snap clips from the claw holes, and remove the fuel suction support.
13. Remove the fuel pump cushion rubber.
14. Disconnect the fuel sender gage connector.
15. Using a screwdriver with its tip wrapped in tape, unlock the fuel sender gage, and slide it to remove.

To install:

16. Install the components in the reverse order of removal.

HEATED OXYGEN (HO2S) SENSOR

LOCATION

The HO2S sensors are located on each exhaust manifold and on each exhaust pipe after the catalytic converter.

REMOVAL & INSTALLATION

1. Before servicing the vehicle, refer to the precautions section.
2. Raise and safely support the vehicle, if necessary.
3. Disconnect the sensor electrical connector.
4. Remove the sensor.

To install:

5. Coat the threads of the heated oxygen sensor with the anti-seize compound, if necessary. Some new sensors will come pre-coated.
6. Install the heated oxygen sensor tighten to 33 ft. lbs. (44 Nm).
7. Connect the sensor electrical connector.
8. Lower the vehicle, if raised.
9. Connect the negative battery cable.

IDLE AIR CONTROL (IAC) VALVE

LOCATION

The IAC valve is located on the throttle body.

REMOVAL & INSTALLATION

See Figure 160.

1. Before servicing the vehicle, refer to the precautions section.
2. Disconnect the negative battery cable.
3. Disconnect the valve connector.
4. Remove the attaching screws.
5. Remove valve from the throttle body.

To install:

6. Installation is in the reverse order of removal.
7. Always use a new gasket.
8. Tighten the attaching screws to 33 inch lbs. (4 Nm).

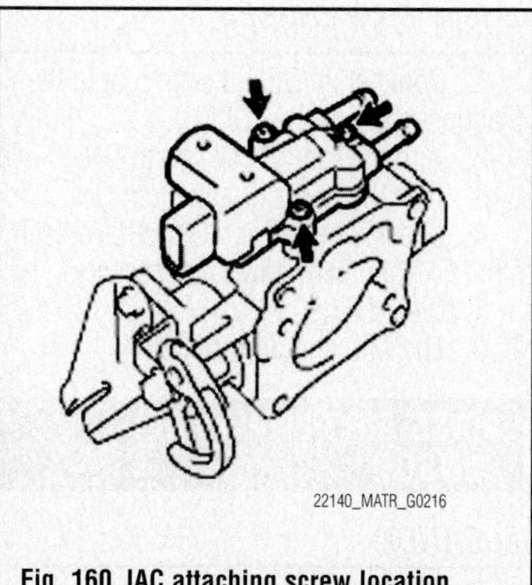

Fig. 160 IAC attaching screw location

INTAKE AIR TEMPERATURE (IAT) SENSOR

The IAT sensor is an integral component of the MAF sensor. Refer to Mass Air Flow (MAF) Sensor.

KNOCK SENSOR (KS)

LOCATION

See Figure 161.

The KS sensor is located at the front of the engine, bolted to the cylinder block.

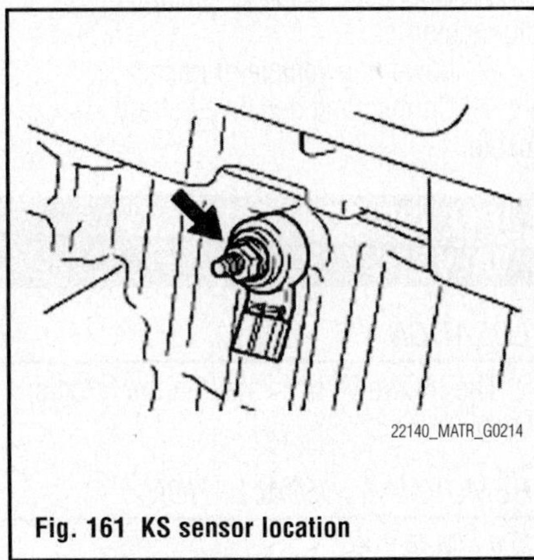

Fig. 161 KS sensor location

REMOVAL & INSTALLATION

1. Before servicing the vehicle, refer to the precautions section.
2. Disconnect the knock sensor connector.
3. Remove the sensor.

To install:

4. Install the sensor and tighten to 29 ft. lbs. (39 Nm).
5. Connect the sensor connector.

MALFUNCTION INDICATOR LIGHT (MIL)

RESET PROCEDURES

1. Connect an OBD II scan tool to the diagnostic connector (DLC3).
2. Turn the ignition switch to ON.
3. Turn the tester or scan tool ON
4. Check whether any DTCs have been stored . Note them down if necessary.
5. Clear DTCs.
6. The MIL should turn off.

MASS AIR FLOW (MAF) SENSOR

LOCATION

The MAF sensor is located on top of the air cleaner assembly.

REMOVAL & INSTALLATION

1. Before servicing the vehicle, refer to the precautions section.
2. Disconnect the negative battery cable.
3. Disconnect the MAF sensor connector.
4. Remove the MAF sensor attaching screws.
5. Remove the MAF sensor.

To install:

6. Installation procedure is the reverse of removal.

OIL PRESSURE SENSOR

LOCATION

See Figure 162.

The oil pressure sensor is located on the oil filter housing.

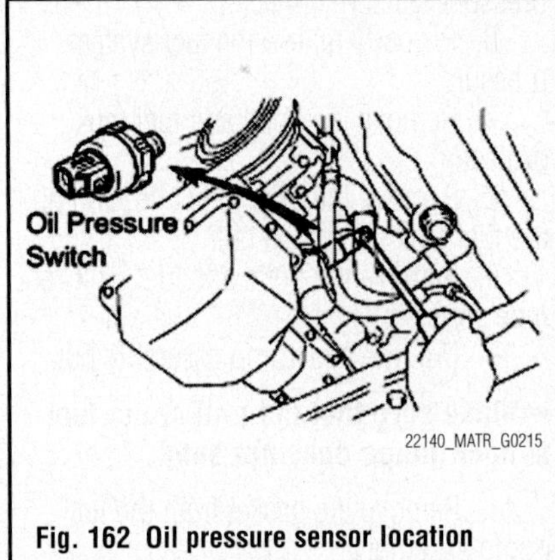

Fig. 162 Oil pressure sensor location

REMOVAL & INSTALLATION

1. Before servicing the vehicle, refer to the precautions section.
2. Disconnect the negative battery cable.
3. Disconnect the sensor connector.
4. Remove the sensor using a deep socket.

To install:

5. Apply a thread locking compound such as Loctite® 242 to 2 or 3 threads of the sensor.
6. Install the sensor and tighten to 11 ft. lbs. (15 Nm).
7. Attach the electrical connector .
8. Connect the negative battery cable.

➡Do not start the engine within 1 hour of installation.

THROTTLE POSITION SENSOR (TPS)

LOCATION

See Figure 163.

The TPS is located at the top of the throttle body.

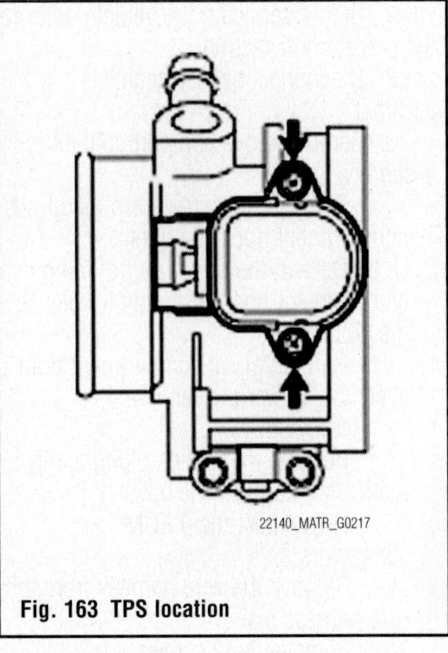

Fig. 163 TPS location

REMOVAL & INSTALLATION

1. Before servicing the vehicle, refer to the precautions section.
2. Disconnect the sensor connector.
3. Remove the sensor attaching screws and remove the sensor.

To install:

4. Install the sensor and tighten to 18 inch lbs. (2 Nm).
5. Attach the electrical connector .

VARIABLE CAMSHAFT TIMING OIL CONTROL SOLENOID

LOCATION

See Figure 164.

The camshaft timing oil control solenoid is mounted to the end of the cylinder head.

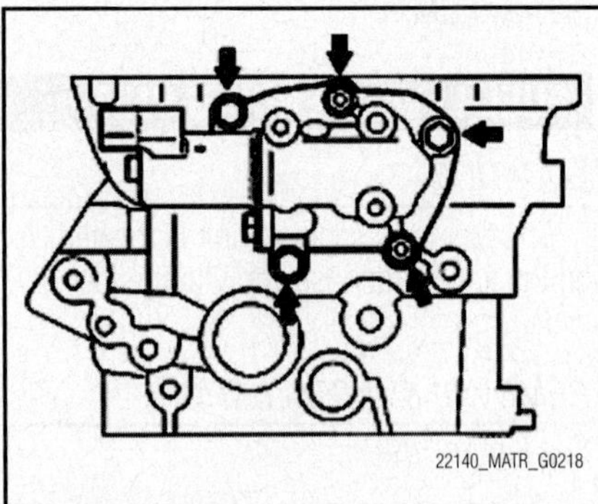

Fig. 164 Camshaft timing oil control solenoid location

REMOVAL & INSTALLATION

1. Before servicing the vehicle, refer to the precautions section.
2. Disconnect the sensor connector.

3. Remove the sensor attaching screws and remove the sensor.

To install:

4. Install the sensor and tighten to 80 inch lbs. (9 Nm).
5. Attach the electrical connector .

VEHICLE SPEED SENSOR (VSS)

LOCATION

See Figure 165.

The VSS is located at the top of the transmission case.

Fig. 165 Vehicle speed sensor location

REMOVAL & INSTALLATION

1. Before servicing the vehicle, refer to the precautions section.
2. Disconnect the sensor connector.
3. Remove the sensor attaching screw and remove the sensor.

To install:

4. Coat the O-ring with transmission fluid.
5. Install the sensor and tighten to 48 inch lbs. (5 Nm).
6. Attach the electrical connector .

FUEL SYSTEM

GASOLINE FUEL INJECTION SYSTEM

FUEL SYSTEM SERVICE PRECAUTIONS

Safety is the most important factor when performing not only fuel system maintenance but any type of maintenance. Failure to conduct maintenance and repairs in a safe manner may result in serious personal injury or death. Maintenance and testing of the vehicle's fuel system components can be accomplished safely and effectively by adhering to the following rules and guidelines.

- To avoid the possibility of fire and personal injury, always disconnect the negative battery cable unless the repair or test procedure requires that battery voltage be applied.
- Always relieve the fuel system pressure prior to disconnecting any fuel system component (injector, fuel rail, pressure regulator, etc.), fitting or fuel line connection. Exercise extreme caution whenever relieving fuel system pressure to avoid exposing skin, face and eyes to fuel spray. Please be advised that fuel under pressure may penetrate the skin or any part of the body that it contacts.
- Always place a shop towel or cloth around the fitting or connection prior to loosening to absorb any excess fuel due to spillage. Ensure that all fuel spillage (should it occur) is quickly removed from engine surfaces. Ensure that all fuel soaked cloths or towels are deposited into a suitable waste container.
- Always keep a dry chemical (Class B) fire extinguisher near the work area.
- Do not allow fuel spray or fuel vapors to come into contact with a spark or open flame.
- Always use a back-up wrench when loosening and tightening fuel line connection fittings. This will prevent unnecessary stress and torsion to fuel line piping.
- Always replace worn fuel fitting O-rings with new. Do not substitute fuel hose or equivalent where fuel pipe is installed.

Before servicing the vehicle, make sure to also refer to the precautions in the beginning of this section as well.

RELIEVING FUEL SYSTEM PRESSURE

⁂ CAUTION

Failure to relieve fuel pressure before repairs or disassembly can cause serious personal injury and/or property damage. Fuel pressure is maintained within the fuel lines, even if the engine is OFFor has not been run in a period of time. This pressure must be safely relieved before any fuel-bearing line or component is loosened or removed. On vehicles equipped with inflatable restraints or air bag systems, wait at least 90 seconds after disconnecting the battery cable before performing any other work. The back-up power will keep the restraint system energized for a period of time after the battery is disconnected.

1. Before servicing the vehicle, refer to the precautions section.
2. Perform the following:
   a. Remove the rear seat cushion.
   b. Remove the rear floor service hole cover.
   c. Disconnect the fuel pump connector.
   d. Start and run the engine, until it stalls.
   e. Turn the ignition key to the **LOCK** position.
   f. Disconnect the negative battery cable.
   g. Connect the fuel pump connector.
   h. Install the service hole cover and rear seat cushion.
   i. Place a catch-pan under the joint to be disconnected. A large quantity of fuel may be released when the joint is opened.
   j. Wear eye or full face protection.
   k. Place a shop towel over the area and slowly release the joint using a wrench of the correct size.
   l. Allow the any fuel left in the line to bleed off slowly before fully disconnecting the joint.
   m. Plug the opened lines.

FUEL FILTER

REMOVAL & INSTALLATION

The fuel filter used on this vehicle is located in the fuel tank and is a integral part of the fuel pump assembly. Normal maintenance of the fuel filter is not required.

FUEL PUMP

REMOVAL & INSTALLATION

FWD

See Figures 166 and 167.

1. Before servicing the vehicle, refer to the precautions section.
2. Remove the rear seat cushion.
3. Remove the rear floor service hole cover.
   a. Remove the screws, then remove the rear floor service hole cover.
   b. Disconnect the fuel pump connector.
4. Discharge the fuel system pressure.
   a. Start the engine.
   b. Turn the ignition switch to **OFF** after the engine stops.

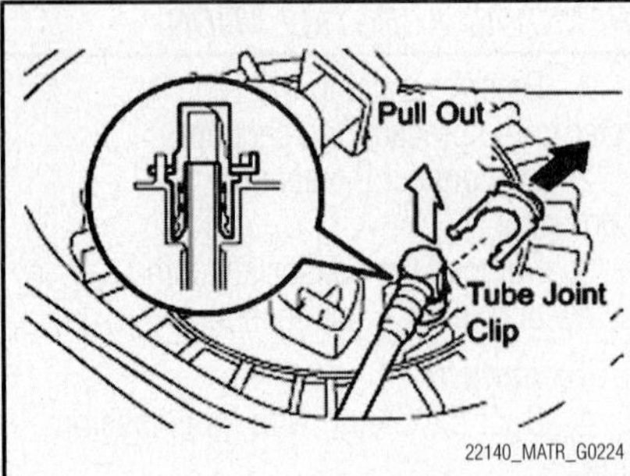

Fig. 166 Disconnect the fuel tank main tube by removing the tube joint clip, then pull the fuel tube joint out of the plug of the fuel suction plate

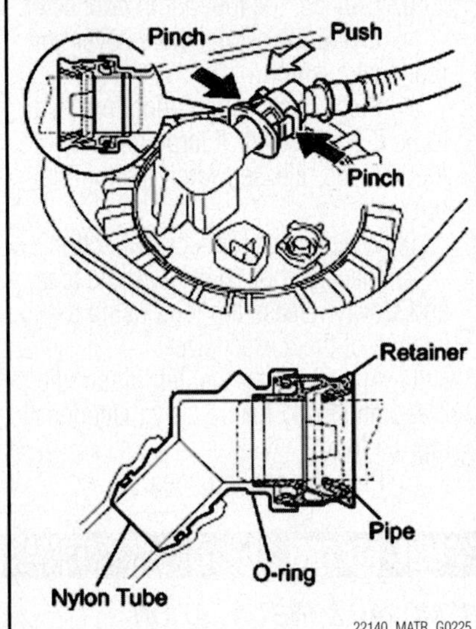

Fig. 167 Disconnect the fuel tank vent hose by pinching the retainer and pull the fuel tank vent connector out of the fuel tank to disconnect the fuel tank vent hose from the fuel suction plate

➡DTC P0171 (system too lean) may be present.

c. Crank the engine again, and check that the engine does not start.

d. Remove the fuel tank cap, and discharge the pressure remaining in the fuel tank completely.

e. Disconnect the battery negative terminal.

5. Remove the fuel tank protector on AWD models.

6. Remove fuel suction pump and gauge assembly

a. Disconnect the fuel tank main tube.

b. Remove the tube joint clip, then pull the fuel tube joint out of the plug of the fuel suction plate.

➡Remove any dirt and foreign objects on the fuel tube joint before performing this work. Do not allow any scratches or foreign objects on the parts when disconnecting, as the fuel tube joint has the o-ring that seals the plug. Perform this work by hand. Do not use any tools. Do not forcibly bend, twist or turn the nylon tube. Protect the disconnected part by covering it with a vinyl bag and gum tape after disconnecting the fuel tank main tube.

c. Disconnect the fuel tank vent hose.

d. Pinch the retainer and pull the fuel tank vent connector out of the fuel tank to disconnect the fuel tank vent hose from the fuel suction plate.

➡If the fuel tank vent connector and pipe are stuck, push and pull them to release.

e. Using special tool SST: 09808-14020, loosen the fuel pump gauge retainer.

f. Remove the fuel pump gauge retainer while holding the fuel suction tube by hand.

g. Remove the fuel suction tube from the fuel tank.

⁂ WARNING

Make sure that the fuel sender gauge arm does not bend.

h. Remove the gasket from the fuel tank.

7. Remove fuel sender gauge assembly.

a. Disconnect the fuel sender gauge connector.

b. Separate the claw fitting. Then slide the fuel sender gauge to remove it from the fuel filter.

8. Remove fuel suction plate.

a. Remove the E-ring.

b. Separate the claw fitting, then remove the fuel pump filter.

c. Separate the 2 claw fittings, then remove the fuel suction plate and spring.

9. Remove fuel pump harness.

a. Disconnect the 2 fuel pump harness connectors.

10. Remove No. 1 fuel suction support.

a. Using a screwdriver with its tip wrapped with tape, separate the 2 claws fittings. Then remove the fuel suction support No. 1.

11. Remove fuel pump filter.

a. Using a screwdriver with its tip wrapped with tape, separate the 5 claw fittings. Then remove the fuel pump filter.

12. Remove fuel pressure regulator assembly.

a. Pull the fuel pressure regulator out of the fuel filter.

b. Remove the 2 o-rings from the fuel pressure regulator.

13. Remove fuel pump.

a. Pull the fuel pump out of the fuel filter.

b. Remove the o-ring from the fuel pump.

To install:

14. Install fuel pump.

a. Apply a light coat of gasoline or spindle oil to a new o-ring, then install it into the fuel pump.

b. Apply a light coat of gasoline or spindle oil to the o-ring of the fuel pump again, then push the fuel pump into the fuel filter.

➡Make sure that the o-ring is not cracked or jammed when installing.

15. Install fuel pressure regulator assembly.

a. Apply a light coat of gasoline or spindle oil to 2 new o-rings, then install them onto the fuel pressure regulator

b. Apply a light coat of gasoline or spindle oil to the o-rings of the fuel pressure regulator again, then push the fuel pressure regulator into the fuel filter.

16. Install fuel pump filter

a. Align the 5 claw positions, then install the fuel pump filter onto the fuel filter.

17. Install fuel suction support.

a. Align the claw positions, then install the fuel suction support No. 1 onto fuel suction plate.

18. Install fuel pump harness.

a. Connect the 2 fuel pump harness connectors.

19. Install fuel suction plate.

a. Align the 2 claw positions, then install the fuel suction plate and spring into the fuel suction case.

b. Align the claw position, then install the fuel pump filter onto the fuel suction case.

c. Install a new e-ring.

20. Install fuel sender gauge assembly.

a. Slide the fuel sender gauge to fit the claw.

b. Connect the fuel sender gauge connector.

21. Install fuel suction pump and gauge tube assembly.

a. Install a new gaskel onto the fuel tank.

b. Set the fuel suction tube to the fuel tank.

c. Align the protrusion of the fuel suction tube with the notch of the fuel tank.

d. While holding the fuel suction tube by hand, align the marks of a new fuel pump gauge retainer and fuel tank as shown, then install the fuel pump gauge retainer.

e. Using the special tool and your hand, tighten it by 2 revolution so that the mark of the fuel pump gauge retainer comes within in the range shown in the illustration.

f. Connect the fuel tank vent hose.

g. Align the fuel tank vent connector with the pipe, then push in the fuel tank vent connector until the retainer makes a "click" sound to connect the fuel tank vent hose to the fuel suction plate.

➡Check that there are no scratches or foreign objects around the connected part of the fuel tank vent connector and pipe before performing this work. After connecting the fuel tank vent hose, check that the fuel tank vent hose is securely connected by pulling the quick connector.

h. Connect the fuel tank main tube.

i. Push the fuel tube joint in the plug of the fuel suction plate, then install the tube joint clip.

➡Check that there are no scratches or foreign objects around the connected part of the fuel tube joint and plug before performing this work. Check that the fuel tube joint is securely inserted to the end.

Check that the tube joint clip is on the collar of the fuel tube joint. After installing the tube joint clip, check that the fuel tank main tube cannot be pulled out.

22. Connect battery negative terminal.
23. Check fuel leakage.
24. Install rear floor service hole cover.
25. Connect the fuel pump connector.
26. Install the rear floor service hole cover with the 4 screws.
27. Install rear seat cushion

AWD

See Figures 168 and 169.

1. Before servicing the vehicle, refer to the precautions section.
2. Remove the rear seat cushion.
3. Remove the rear floor service hole cover.

a. Remove the screws, then remove the rear floor service hole cover.

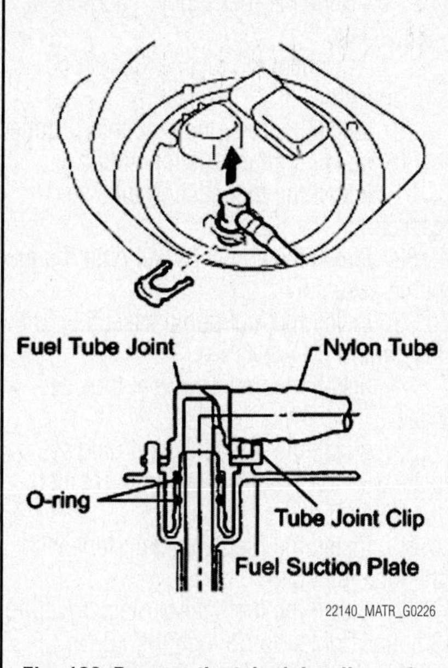

Fig. 168 Remove the tube joint clip, and pull the fuel tank main tube out

b. Disconnect the fuel pump connector.

4. Discharge the fuel system pressure.

a. Start the engine.

b. Turn the ignition switch to **OFF** after the engine stops.

➡DTC P0171 (system too lean) may be present.

c. Crank the engine again, and check that the engine does not start.

d. Remove the fuel tank cap, and discharge the pressure remaining in the fuel tank completely.

e. Disconnect the battery negative terminal.

5. Remove the fuel tank protector on AWD models.
6. Disconnect the fuel tank main tube.

a. Remove the tube joint clip, then pull the fuel tube joint out of the plug of the fuel suction plate.

➡Remove any dirt and foreign objects on the fuel tube joint before performing this work. Do not allow any scratches or foreign objects on the parts when disconnecting, as the fuel tube joint has the o-ring that seals the plug. Perform this work by hand. Do not use any tools. Do not forcibly bend, twist or turn the nylon tube. Protect the disconnected part by covering it with a vinyl bag and gum tape after disconnecting the fuel tank main tube. When the fuel tube joint and plug are stuck, pinch the fuel tank main tube between fingers, and turn it carefully to disconnect the fuel tank main tube.

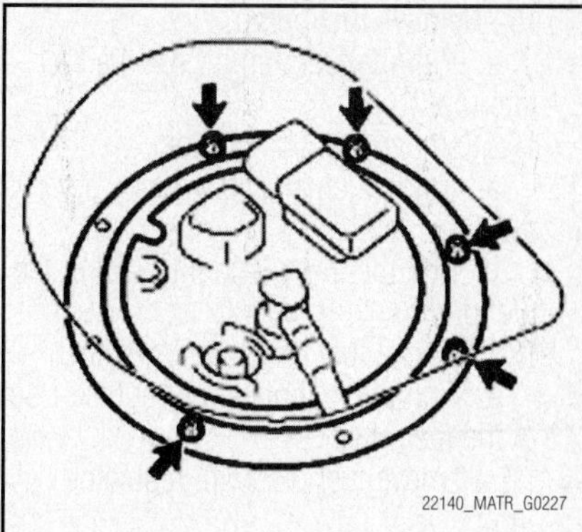

Fig. 169 Remove the bolts and fuel tank vent tube set plate

7. Remove fuel tank vent tube set plate.

a. Remove the bolts and fuel tank vent tube set plate.

8. Remove fuel suction pump and gauge.

a. Pull the fuel suction pump and gauge assembly.

b. Do not damage the fuel pump filter.

➡Make sure that the arm of the fuel sender gauge does not bent.

9. Remove fuel suction tube set gasket.

a. Remove the gasket from the fuel pump assembly.

10. Remove fuel suction support.

a. Using a screwdriver, disconnect the 5 snap claws from the claw holes, and remove the fuel suction support.

b. Do not damage the fuel suction support.

11. Remove fuel pump cushion rubber.

a. Remove the fuel pump cushion rubber.

12. Remove fuel sender gauge.

a. Disconnect the fuel sender gage connector.

b. Using a screwdriver with its tip wrapped in tape, unlock the fuel sender gage, and slide it remove.

13. Remove fuel suction plate.

a. Disconnect the fuel pump connector.

b. Using a screwdriver, disconnect the 4 snap claws from the claw holes, and pull the fuel suction plate out.

⁂ WARNING

Do not damage the fuel suction plate.

14. Remove vapor pressure sensor.

a. Remove the fuel pump harness from the fuel suction plate.

15. Remove fuel pump harness.

a. Remove the fuel pump harness from the fuel suction plate.

16. Remove fuel pump.
   a. Pull the fuel pump from the fuel filter out.
17. Remove fuel pump filter.
   a. Using a small screwdriver, pry out the clip.
   b. Pull the fuel pump filter out of the fuel pump.
18. Remove fuel pressure regulator.
   a. Pull the fuel pressure regulator out of the fuel filter.
19. Remove fuel pressure regulator o-ring.
   a. Remove an O-ring from the fuel pressure regulator.

To install:

20. Install the fuel pressure regulator o-ring.
   a. Apply a light coat of gasoline or spindle oil to a new o-ring, and Install the it onto the fuel pressure regulator.
21. Install the fuel pressure regulator assembly.
   a. Apply a light coat of gasoline or spindle oil to 2 new o-rings, then Install the them onto the fuel pressure regulator.
   b. Apply a light coat of gasoline or spindle oil to the o-rings of the fuel pressure regulator again, then push the fuel pressure regulator into the fuel filter.
   c. Make sure that the o-ring is not cracked or jammed when install.
22. Install the fuel pump filter with a new clip.
23. Install the fuel pump.
   a. Apply a light coat of gasoline or spindle oil to an o-ring, and Install the fuel pump onto the fuel filter.
24. Connect the fuel pump harness.
25. Install the vapor pressure sensor with the tube joint clip.

➡Check that there are no scratches or foreign objects on the connecting parts. Check that the fuel tube joint is securely inserted into the end. Check that the tube joint clip is on the collar of the fuel tube joint. After installing the tube joint clip, check that the fuel tube joint cannot be pulled out. .

26. Install the fuel suction plate.
   a. Align the 2 claw positions, then Install the fuel suction plate and spring into the fuel suction case.
   b. Align the claw position, then Install the fuel pump filter onto the fuel suction case.
27. Install the a new e-ring.
28. Install the fuel sender gauge assembly.
   a. Slide the fuel sender gauge to fit the claw.
   b. Connect the fuel sender gauge connector.
29. Install the fuel pump cushion rubber.
30. Install the fuel suction support.
31. Install the fuel suction tube set gasket.
32. Install the a new gasket onto the fuel pump assembly.
33. Install the fuel pump assembly to the fuel tank.
34. Install the fuel tank vent tube set plate.
35. Install the fuel tank vent tube set plate with the 5 bolts. Tighten to 53 inch lbs. (6 Nm).
36. Install the fuel tank main tube with the tube joint clip.
37. Install the fuel tank protector with the 3 bolts. Tighten to 53 inch lbs. (6 Nm).
38. Connect battery negative terminal.
39. Check fuel leakage.
40. Install rear floor service hole cover.
41. Connect the fuel pump connector.
42. Install the rear floor service hole cover with the 4 screws.
43. Install rear seat cushion.

FUEL RAIL & INJECTORS

REMOVAL & INSTALLATION

See Figures 170 and 171.

1. Before servicing the vehicle, refer to the precautions section.
2. Properly relieve the fuel system pressure.

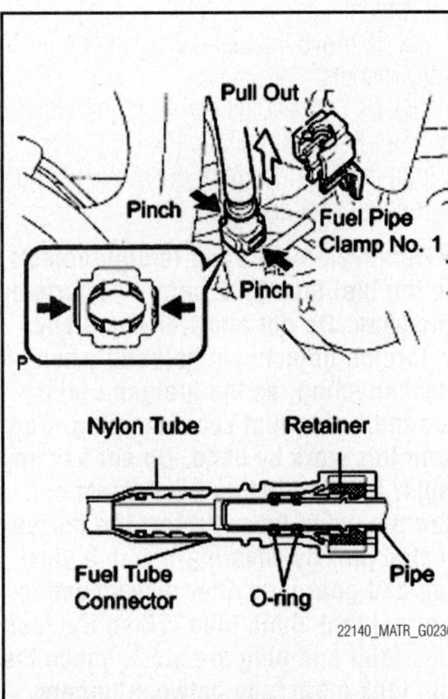

Fig. 170 Disconnecting the fuel pipe clamp—1ZZ-FE engine

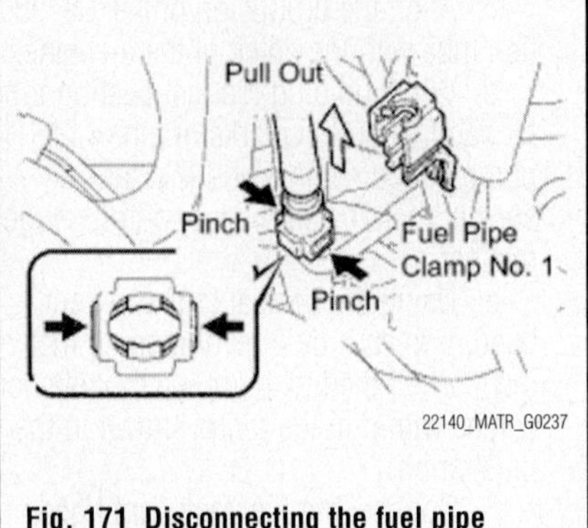

Fig. 171 Disconnecting the fuel pipe clamp—2ZZ-GE engine

3. Disconnect the negative battery cable.
4. Remove the upper cylinder head cover.
5. Remove the ventilation hose.
6. Remove the engine wire, unplugging the injector connectors and clamps.
7. Disconnect the cam position sensor connector.
8. Remove the fuel pipe clamp.
9. Remove the fuel line/tube.

WARNING

Be careful not to drop the fuel injectors when removing the delivery pipe.

10. Remove the fuel delivery pipe with the injectors attached.
11. Remove the delivery pipe and injectors.
12. Remove the spacers from the head.
13. Remove the injectors from the delivery pipe.
14. Remove the O-ring and grommet from each injector.

To install:

15. Install the new grommets.
16. Install the new O-rings coated with light machine oil.
17. Install the injectors on the delivery pipe.

➡Coat the contact point on the pipe with light machine oil and twist the injectors into place. The connector should face outward.

18. Install the spacers.

➡Coat the seats in the head where the injectors contact, with light machine oil.

19. Install the delivery pipe and injectors.
20. Loosely install the hold-down bolts and check that the injectors rotate smoothly. If they don't, the probable cause is incorrect

O-ring installation. Torque the delivery pipe hold-down bolts to 14 ft. lbs. (19 Nm) and the fuel pipe bolt to 80 inch lbs. (9 Nm).

21. Install the engine wire, attaching the injector connectors and clamps.
22. Connect the cam position sensor connector.
23. Install the fuel line/tube.
24. Install the ventilation hose.
25. Install the upper cylinder head cover.

FUEL TANK

REMOVAL & INSTALLATION

FWD

See Figure 172.

1. Remove rear seat cushion.
2. Remove rear floor service hole cover.
3. Remove fuel suction with pump and gauge assembly.
4. Drain fuel.
5. Remove front exhaust pipe assembly.
   a. Using a clip remover, remove the clip .
   b. Pull off the floor carpet front .
   c. Disconnect the heated oxygen sensor connector .
   d. Remove the grommet, then pull the connector to the outside of the vehicle .
   e. Remove nuts and the floor panel brace front .
   f. Remove the bolts and compression springs .
   g. Remove the front exhaust pipe from the exhaust pipe supports.
   h. Remove the gasket from the exhaust manifold .
   i. Remove the gasket from the front exhaust pipe .
6. Remove fuel tank protector.
   a. Remove the bolts, then remove the fuel tank protector.
7. Disconnect No. parking brake cable assembly.
   a. Remove the bolts, then disconnect the parking brake cable.
8. Disconnect No. parking brake cable assembly.
   a. Remove the bolts, then disconnect the parking brake cable.
9. Remove fuel tank assembly.
   a. Disconnect the fuel tank main tube.
   b. Pinch the tab of the retainer of the fuel tube connector to remove the lock claws and push it down as shown in the illustration.
   c. Pull the fuel tank main tube out of the pipe.

➡Remove any dirt or foreign objects on the fuel tube connector before performing this work. Do not allow any scratches or foreign objects on the parts when disconnecting, as the fuel tube connector has the O-ring that seals the pipe. Perform this work by hand. Do not use any tools. Do not forcibly bend, twist or turn the nylon tube. Protect the disconnected part by covering it with a vinyl bag after disconnecting the fuel tank main tube. If the fuel tube connector and pipe are stuck, push and pull them to release.

   d. Disconnect the fuel tank breather tube .
   e. Remove the checker of the fuel tube connector from the pipe.
   f. Pinch the retainer of the fuel tube connector, then pull the fuel tube connector out of the pipe.
   g. Disconnect the fuel tank vent hose.
   h. Pinch the tab of the retainer of the fuel tube connector to remove the lock claws and slide it as shown in the illustration.
   i. Pull the fuel tank vent hose out of the pipe.
   j. Using a screwdriver, unfasten the claw. Then remove the fuel tank filler pipe cover from the fuel tank filler pipe .
   k. Loosen the hose clamp bolt, then disconnect the fuel tank filler pipe hose from the fuel tank .
   l. Hold the fuel tank using a transmission jack .
   m. Remove the 4 bolts, then remove the fuel tank bands.
   n. Operate the transmission jack, then remove the fuel tank .
10. Remove fuel tank main tube.
   a. Remove the fuel tank main tube from the fuel tank.
11. Remove fuel tank vent hose.
   a. Remove the fuel vent hose from the fuel tank clamp.

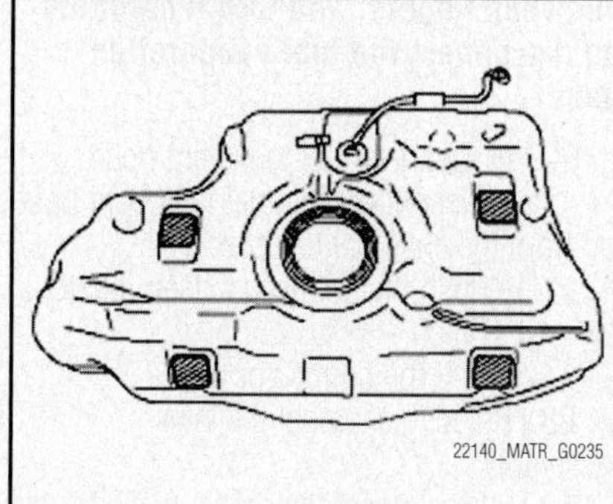

Fig. 172 Fuel tank cushion locations — FWD

12. Remove fuel tank cushion.
   a. Remove the fuel tank cushions from the fuel tank.

To install:

13. Install fuel tank cushion.
   a. Install 4 new fuel tank cushions onto the fuel tank.
14. Install fuel tank vent hose.
   a. Install the fuel onto the fuel tube clamp.
15. Install fuel tank main tube.
   a. Install the fuel tank main tube onto the fuel tank.
16. Install fuel tank assembly.
   a. Set the fuel tank on a transmission jack .
   b. Operate the transmission jack, then install the fuel tank into the vehicle .
   c. Install the fuel tank bands with the bolts. Tighten 29 ft. lbs. (39 Nm).
   d. Connect the fuel tank to filler pipe hose to the fuel tank .

   Make sure that the hose clamp is at the correct direction when installing.

   e. Align the claw position, then install the fuel tank filler pipe cover onto the fuel tank filler pipe .
   f. Connect the fuel tank vent hose .
   g. Align the fuel tube connector with the pipe, then push the fuel tube connector in until it comes into contact with the seat to connect the fuel tank vent hose to the pipe.
   h. Slide the retainer of the fuel tube connector to the claws lock as shown in the illustration.
   i. Connect the fuel tank breather tube.
   j. Align the fuel tube connector with the pipe, then push the fuel tube connector in until the retainer makes a "click" sound to connect the fuel tank breather tube to the pipe.
   k. Install the checker onto the pipe.
   l. Connect the fuel tank main tube .
   m. Align the fuel tube connector with the pipe, then push the fuel tube connector into until it comes into contact with the seat to connect the fuel tank main tube to the pipe, then push the retainer up until the claws lock.
17. Connect No. 3 parking brake cable assembly.
   a. Align the turn prevention positions, then connect the parking brake cable with the bolts. Tighten to 48 inch lbs (5 Nm).
18. Connect No. 2 parking brake cable assembly.
   a. Align the turn prevention positions, then connect the parking brake cable with the bolts. Tighten to 48 inch lbs (5 Nm).

19. Install front exhaust pipe assembly.
    a. Check the free length .
    b. Using Vernier calipers, measure the free length of the compression spring. Minimum for exhaust manifold side is 1.6339 in. (41.5 mm) and exhaust pipe tail side is 1.5158 in. (38.5 mm). If the free length is less than minimum, replace the compression spring.
    c. Using a wooden block and plastic-faced hammer, tap a new gasket into the exhaust manifold .

    Make sure that the gasket is in the correct direction when installing.

➡Do not reuse a removed gasket. Never push the gasket into the exhaust manifold by installing the front exhaust pipe onto the exhaust manifold.

    d. Using a wooden block and plastic-faced hammer, tap a new gasket into the exhaust tail .

➡Make sure that the gasket is in the correct direction when installing.

    e. Install the front exhaust pipe onto the exhaust pipe supports.
    f. Tighten the compression springs and bolts. Tighten to 32 ft. lbs. (43 Nm).
    g. Install the floor panel brace front with nuts. Tighten to 22 ft. lbs. (30 Nm).
    h. Pull the connector back inside the vehicle, then install the grommet .
    i. Connect the heated oxygen sensor connector .
    j. Install the floor carpet front with the clip .
20. Add fuel.
21. Install fuel suction with pump and gauge tube assembly.
22. Connect the negative battery cable.
23. Check for fuel leaks.
24. Check for exhaust gas leak.
25. Install rear floor service hole cover.
26. Install fuel tank protector.
27. Install rear seat cushion.

AWD

See Figure 173.

1. Remove rear seat cushion.
2. Remove rear floor service hole cover.
3. Discharge fuel system pressure.
4. Remove No. 2 fuel tank protector.
5. Remove fuel tank main tube.
6. Remove fuel tank vent tube set plate.
7. Remove fuel pump assembly.
8. Remove fuel suction tube set gasket.
9. Drain fuel.
10. Remove front door scuff plate.
11. Remove cowl side trim board.
12. Remove rear door scuff plate.
13. Remove center pillar lower molding.
14. Remove front seat assembly.
15. Remove center exhaust pipe assembly.
    a. Disconnect the oxygen sensor connector.
    b. Remove the exhaust pipe supports.
    c. Remove the bolt and clamp .
    d. Remove the bolts, compression springs and exhaust pipe .
16. Remove propeller with center bearing shaft assembly.
17. Remove the nuts and front floor heat insulator.
18. Disconnect parking brake cable assembly.
19. Disconnect the charcoal canister outlet hose from the charcoal canister.
20. Disconnect fuel evaporation tube.
    a. Disconnect the fuel tank to canister tube from the charcoal canister .
    b. Remove the fuel tank to canister tube from the evaporation vent tube clamps .
    c. Disconnect the fuel emission tube from the charcoal canister .
    d. Remove the fuel tank to canister tube from the evaporation vent tube clamps .
21. Disconnect fuel evaporation tube.
    a. Pinch the fuel tube connector clip and then pull the fuel evaporation tube out of the pipe.

➡Remove any dirt and foreign objects on the fuel tube joint before performing this work. Do not allow any scratches or foreign objects on the parts when disconnecting, as the fuel tube joint has the O-ring that seals the plug. Do not use any tool. Do not forcibly bend, twist or turn the nylon tube. Protect the disconnected part by covering it with a vinyl bag and gum tape after disconnecting the fuel pump tube. When the fuel tube connector and pipe are stuck, pinch the fuel evaporation tube between fingers, and turn it carefully to disconnect the fuel evaporation tube.

22. Disconnect fuel tank vent hose.
    a. Disconnect the fuel tank bent hose from the charcoal canister.
    b. Push the connector deep inside.
    c. Pinch.
    d. Pull the connector out.
23. Remove charcoal canister assembly.
    a. Disconnect the pressure switching valve connector.
    b. Remove the bolts and charcoal canister.
24. Disconnect fuel tank main tube.
    a. Pinch the fuel tube connector clip, and remove the fuel tube connector clip.
    b. Pull the fuel tank main tube out.

➡Protect the disconnected part by covering it with a vinyl bag and gum tape after disconnecting the fuel pump tube. When the fuel tube connector and pipe are stuck, pinch the fuel tank main tube between fingers, and turn it carefully to disconnect the fuel tank main tube.

25. Disconnect the fuel inlet pipe fuel hose from the fuel tank.
26. Remove fuel tank assembly.
    a. Hold the fuel tank using a transmission jack .
    b. Remove the bolts and fuel tank bands .
    c. Operate the transmission jack and then disconnect the 2 fuel tank breather hoses .
27. Remove the fuel tank main tube from the fuel tube clamp.
28. Remove the bolt and fuel tube clamp.
29. Remove the fuel breather hose from the fuel tank.
30. Remove fuel evaporation tube.
    a. Pinch the fuel connector clip and then pull the fuel emission tube out.
31. Remove the bolt and evaporation vent tube clamp.
32. Remove the bolts and check valve protector.
33. Remove fuel tank breather tube sub assembly.
    a. Pinch the fuel tube connector clip and then pull the breather tube out.
    b. Remove the breather tube from the fuel tube bracket and clamp.
34. Remove the fuel tube clamp from the fuel tank.
35. Remove fuel tank vent hose.
    a. Disconnect the fuel tank vent hose from the charcoal canister.
    b. Push the connector deep inside.
    c. Pinch.
    d. Pull the connector out.
    e. Remove the fuel tank vent hose from the fuel tube clamp.
36. Remove the bolt and fuel tube clamp.
37. Remove the 3 bolt, check the value support and fuel tank inlet value.
38. Remove the gasket from the fuel tank inlet valve.
39. Remove the fuel tank protector.
    a. Using a drill, cut the flange portion of the rivet.
    b. Wind tape around the drill blade so that approximately 0.197 in. (5 mm) of the blade protrude from the tape.

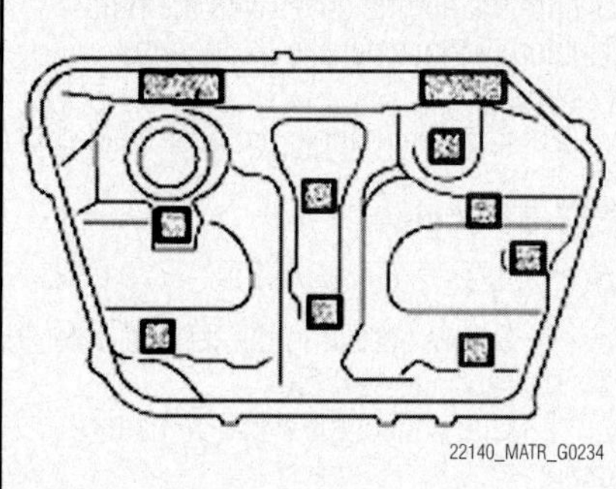

Fig. 173 Fuel tank cushion locations—AWD

⁂ WARNING

Drilling the hole with a drill can lead to damage to the rivet hole or the drill itself.

⁂ CAUTION

Be careful when performing this operation as the cut rivet is hot.

40. Remove the 10 fuel tank cushions from the fuel tank.

To install:

41. Install the fuel tank cushion onto the fuel tank.
42. Install fuel tank protector.
    a. Using a riveter, install the fuel tank protector with the 4 new rivets.
    Install the rivets while holding the materials tightly together to ensure a firm connection.
43. Install a new gasket onto the fuel tank inlet valve.
44. Install the fuel tank inlet valve and check valve support with the bolts. Tighten to 49 inch lbs. (5 Nm).
45. Install the fuel tube clamp with the bolt. Tighten to 53 inch lbs. (6 Nm).
46. Install fuel tank vent hose.
    a. Install the fuel tank vent hose to the fuel pump clamp.
    b. Connect the fuel tank vent hose to the charcoal canister.
47. Install the fuel tube clamp to the fuel tank.
48. Install fuel tank breather tube.
    a. Push the breather tube in the pipe until fuel tube connector makes a "click" sound.

➡Check if there is any damage or foreign objects on the connected part. After connecting, check if the fuel tube connector and pipe are securely connected by pulling them.

49. Install the check valve protector with bolts. Tighten to 31 inch lbs. (4 Nm).
50. Install the evaporation vent tube clamp with the bolt. Tighten to 53 inch lbs. (6 Nm).
51. Install fuel evaporation tube.
    a. Push the fuel emission tube in the pipe until fuel tube connector makes a "click" sound.
52. Install fuel tank breather hose.
    a. Install the fuel tank breather hose to the fuel tank.
53. Install the fuel tube clamp with the bolt. Tighten to 53 inch lbs. (6 Nm).
54. Install fuel tank main tube.
    a. Push the fuel tube connector in the pipe until fuel tube connector, and install the fuel tube connector clip.
55. Install fuel tank assembly.
    a. Set the transmission jack to the fuel tank.
    b. Operate the transmission jack and then connect the fuel tank breather hoses.
    c. Install fuel tank bands with the bolts. Tighten to 29 ft. lbs. (6 Nm).
56. Connect fuel tank inlet pipe fuel hose.
    a. Connect the fuel tank inlet pope fuel hose to the fuel tank.
57. Connect fuel tank main tube.
    a. Connect the fuel tank main tube.
58. Install charcoal canister assembly.
    a. Install the charcoal canister with bolts. Tighten to 14 ft. lbs. (19 Nm).
    b. Connect pressure switching valve connector.
59. Connect fuel tank vent hose.
    a. Connect the fuel tank vent hose to the charcoal canister.
60. Connect fuel evaporation tube.
    a. Push the fuel evaporation tube in the pipe until fuel tube connector makes a "click" sound.
61. Connect the fuel evaporation tube.
62. Connect the charcoal canister outlet hose.
63. Connect parking brake cable assembly.
    a. Install the parking brake cable with bolts. Tighten to 48 inch lbs. (5 Nm).
64. Install front floor heat insulator.
    a. Install the front floor heat insulator with nuts. Tighten to 49 inch lbs. (6 Nm).
65. Install propeller with center bearing shaft assembly.
66. Install center exhaust pipe assembly.
    a. Using Vernier calipers, measure the free length of the compression spring. Free length should be 1.5748 in. (40 mm).
    b. Install a new gasket onto the front exhaust pipe.
    c. Install a new gasket onto the center exhaust pipe.
    d. Install the center exhaust pipe with the compression springs, clamp and bolts. Tighten the front to 32 ft. lbs. (43 Nm) and the rear to 24 ft. lbs. (32 Nm).
    e. Install the exhaust pipe supports.
    f. Connect the oxygen sensor connector.
67. Install front seat assembly.
68. Install center pillar lower molding.
69. Install rear door scuff plate.
70. Install cowl side trim board.
71. Install front door scuff plate.
72. Install fuel suction tube set gasket.
73. Install fuel suction with pump and gauge assembly.
74. Install fuel tank vent tube set plate.
75. connect fuel tank main tube.
76. Install No. 2 fuel tank protector.
77. connect battery negative terminal.
78. inspect fuel leak.
79. inspect exhaust gas leak.
80. Install rear floor service hole cover.
81. Install rear seat cushion.

IDLE SPEED

ADJUSTMENT

Idle speed is maintained by the Powertrain Control Module (PCM). No adjustment is possible.

THROTTLE BODY

REMOVAL & INSTALLATION

1ZZ-FE 2WD Engine

See Figure 174.

1. Before servicing the vehicle, refer to the precautions section.
2. Remove the upper cylinder head cover.

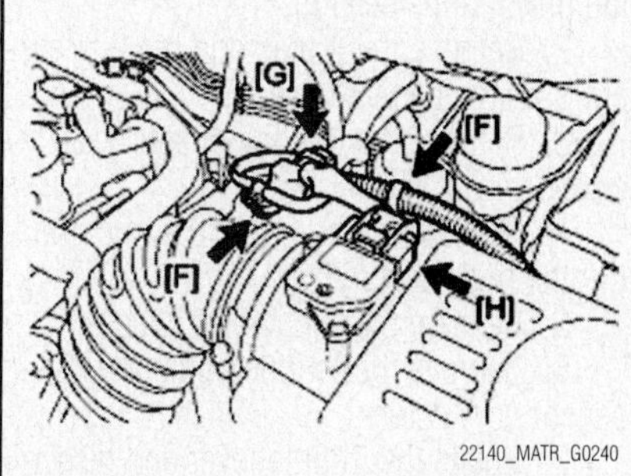

Fig. 174 Wire harness clamps (F), VSV connector (G) and mass air flow meter connector (H)—1ZZ-FE engine

3. Remove the air cleaner and air inlet duct from the throttle body assembly.
   a. Disconnect the mass air flow meter connector.
   b. Disconnect the vacuum solenoid valve connector.
   c. Remove the wire harness clamps.
   d. Disconnect the fuel vapor feed hose.
4. Remove the ventilation hose from the cylinder head.
5. Remove the throttle body assembly.
   a. Disconnect the electrical connector of the throttle control motor.
6. Drain the cooling system.
   a. Remove the coolant hoses from the throttle body.
   b. Remove the bolts and nuts from the throttle body.
7. Remove the throttle body and the gasket from the intake manifold.

To install:

8. Install the throttle body and using a new gasket.
   a. Install the bolts and nuts and tighten to 80 inch lbs. (9 Nm).
   b. Install the coolant hoses from the throttle body.
   c. Connect the electrical connector of the throttle control motor.
9. Install the ventilation hose from the cylinder head.
10. Install the air cleaner and air inlet duct from the throttle body assembly.
   a. Connect the fuel vapor feed hose.
   b. Install the wire harness clamps.
   c. Connect the vacuum solenoid valve connector.
   d. Connect the mass air flow meter connector.
11. Install the upper cylinder head cover.
12. Fill the cooling system.

1ZZ-FE AWD Engine

See Figure 175.

1. Before servicing the vehicle, refer to the precautions section.
2. Remove the 2 nuts and the 2 retainers from the engine cover.
3. Remove the upper cylinder head cover.
4. Remove the air inlet duct from the throttle body assembly.
5. Remove the ventilation hose.
6. Remove the throttle position (TP) sensor connector.
7. Rotate the throttle lever and disconnect the accelerator cable and the throttle valve cable, if equipped.
8. Disconnect the electrical connector of the idle air control (IAC) valve.

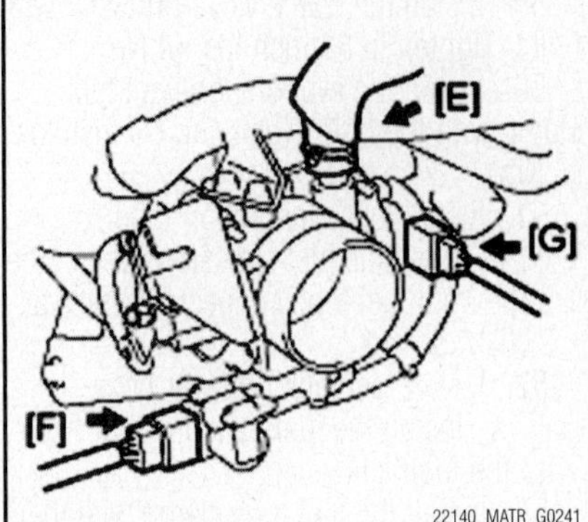

Fig. 175 PCV hose (E), throttle body idle air control valve assembly connector (F) and throttle position sensor connector (G)—1ZZ-FE AWD engine

9. Drain the cooling system.
10. Remove the coolant hoses from the throttle body.
11. Remove the bolts and the accelerator control cable bracket.
12. Remove the fasteners and the throttle body bracket.
13. Remove the throttle body fasteners.
14. Remove the fasteners and the manifold support bracket.
15. Remove the throttle body and the gasket from the intake manifold.
16. Clean any remaining throttle body gasket from the intake manifold.

To install:

17. Install a new throttle body gasket onto the intake manifold.
18. Install the throttle body onto the intake manifold with the 3 fasteners.
19. Install the throttle body bracket and the bolts to the throttle body.
20. Install the throttle body fasteners. Tighten the fasteners to 22 ft. lbs. (30 Nm).
21. Install the manifold support bracket to the throttle body with 2 bolts. Install the 2 fasteners and tighten to 22 ft. lbs. (30 Nm).
22. Install the accelerator control cable bracket to the throttle body with 2 bolts. Install the 2 fasteners and tighten to 10 ft. lbs. (13 Nm).
23. Connect the coolant hoses to the throttle body.
24. Connect the ventilation hose to the throttle body.
25. Connect the IAC valve electrical connector.
26. Connect the TP sensor electrical connector.
27. Install the air inlet duct.
28. Refill the cooling system.
29. Install the engine cover to the engine and secure engine cover with the nuts. Secure the engine cover with the nuts. Tighten the fasteners to 62 inch lbs (7 Nm).
30. Install the upper cylinder head cover.

2ZZ-GE Engine

See Figures 176 and 177.

1. Before servicing the vehicle, refer to the precautions section.
2. Disconnect the negative battery cable.
3. Drain and recycle the engine coolant.
4. Remove the upper head cover.
5. Remove the throttle cable from the cable support, loosen the nut and disconnect the cable end from the throttle lever to separate the control cable.
6. Remove the air cleaner and inlet hose assembly.
   a. Disconnect the mass air flow meter connector.
   b. Disconnect the VSV connectors.
   c. Remove the 2 wire harness clamps.
   d. Disconnect the fuel vapor feed hose from the VSV.
   e. Disconnect the vacuum hoses.
   f. Loosen the air cleaner hose clamp bolt.
   g. Unlock the 2 clamp hooks.
7. Remove the throttle body.
   a. Remove the throttle position (TP) sensor connector.

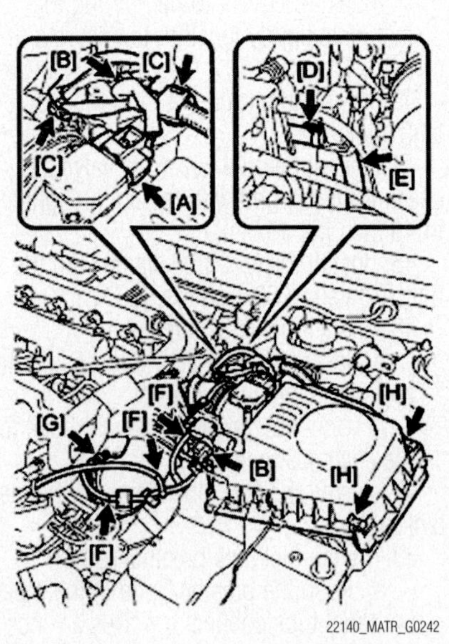

Fig. 176 Mass air flow meter connector (A), VSV connectors (B), wire harness clamps (C), fuel vapor feed hose (D), fuel vapor feed hose (E), vacuum hoses (F), air cleaner hose clamp bolt (G) and clamp hooks (H)—2ZZ-GE engine

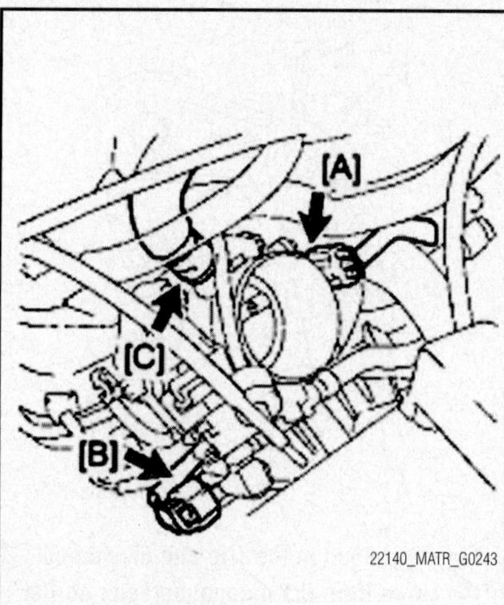

Fig. 177 Throttle position sensor connector (A), ISC valve connector (B) and ventilation hose (C)—2ZZ-GE engine

b. Disconnect the electrical connector of the idle speed control (ISC) valve.

c. Disconnect the ventilation hose from the throttle body.

d. Remove the coolant hoses from the throttle body.

e. Remove the 2 throttle body nuts and the accelerator control cable bracket.

8. Remove the gasket from the intake manifold.

To install:

9. Clean the gasket mating surfaces.

10. Install the throttle body.

a. Install the gasket on the intake manifold.

b. Install the throttle body nuts and the accelerator control cable bracket. Tighten the fasteners to 16 ft. lbs. (22 Nm).

11. Install the air cleaner and inlet hose assembly.

a. Install the clamp hooks.

b. Tighten the air cleaner hose clamp to 18 inch lbs. (2 Nm).

c. Connect the vacuum hoses.

d. Connect the fuel vapor feed hose from the VSV.

e. Install the 2 wire harness clamps.

f. Connect the VSV connectors.

g. Connect the mass air flow meter connector.

12. Install accelerator control cable.

a. Install the accelerator control cable into the throttle lever.

b. Tighten the nut.

c. Connect the accelerator control cable to the accelerator control cable support.

13. Install the upper head cover.

14. Drain and recycle the engine coolant.

15. Connect the negative battery cable.

HEATING & AIR CONDITIONING SYSTEM

BLOWER MOTOR

REMOVAL & INSTALLATION

See Figures 178 through 180.

1. Before servicing the vehicle, refer to the precautions section.

2. Disconnect the negative battery cable.

3. Remove the screw from the glove compartment door stoppers.

4. Press in on the upper part of the glove compartment door to release the stoppers.

5. Remove the glove compartment.

6. Remove the 2 clips retaining the ECM cover, then open the cover.

7. Disconnect the 4 ECM connectors.

8. Remove the wire harness from the wire harness clamp and the 2 screws

9. Unfasten the claw and clip, and then remove the ECM.

10. Unplug the blower motor connector.

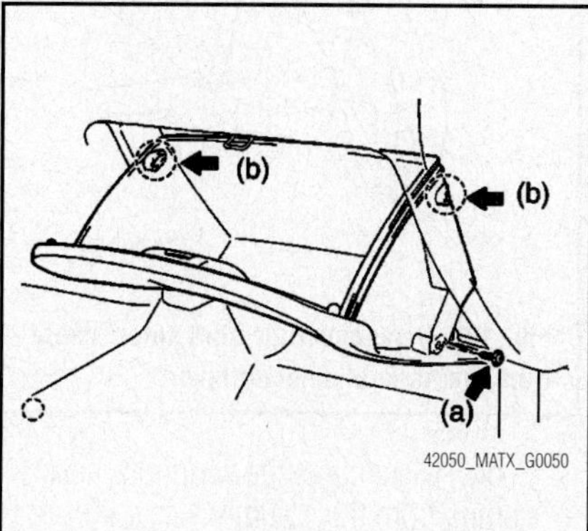

Fig. 178 Location of the glove compartment screws and stoppers

11. Unfasten the blower motor retainers and remove the blower motor.

To install:

12. Install the blower motor, tighten the fasteners and attach the electrical connector.

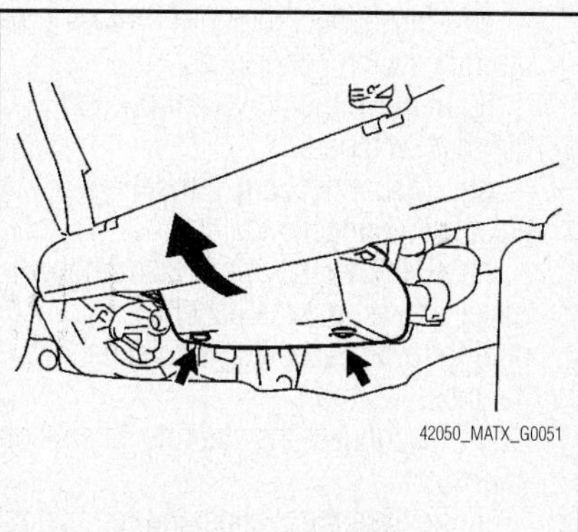

Fig. 179 Remove the ECM cover

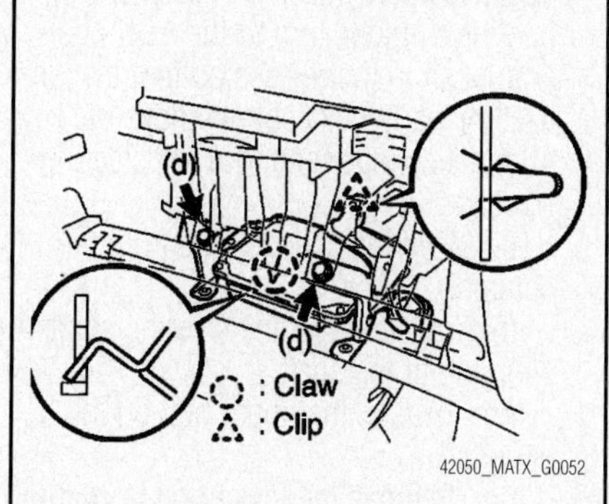

Fig. 180 Unfasten the claw and clip, and then remove the ECM

13. Install the ECM with the 2 screws and tighten to 27 in. lbs. (3 Nm).

14. Attach the ECM connectors and the harness clamp.

15. Install the ECM cover.

16. Install glove compartment.

17. Connect the negative battery cable.

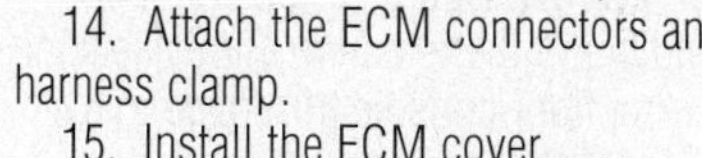

HEATER CORE

REMOVAL & INSTALLATION

See Figures 181 through 190.

1. Before servicing the vehicle, refer to the precautions section.

2. Drain the cooling system.

3. Discharge and recover the A/C system refrigerant using approved equipment.

4. Disconnect negative battery cable

5. Remove the heater hoses from the core

6. Remove the evaporator inlet and outlet tubes from the evaporator and cap the lines to avoid system contamination

7. Remove the instrument panel.

a. Disable the air bag system.

b. Using a taped flat–bladed tool, carefully pry the retaining clips attaching the center trim plate to the instrument panel.

c. Disconnect the A/C switch, hazard switch; rear defogger switch and passenger seat belt indicator switch electrical connections.

d. Remove the radio retaining screws, clamp from the radio bracket, and slide the radio forward to disconnect the power

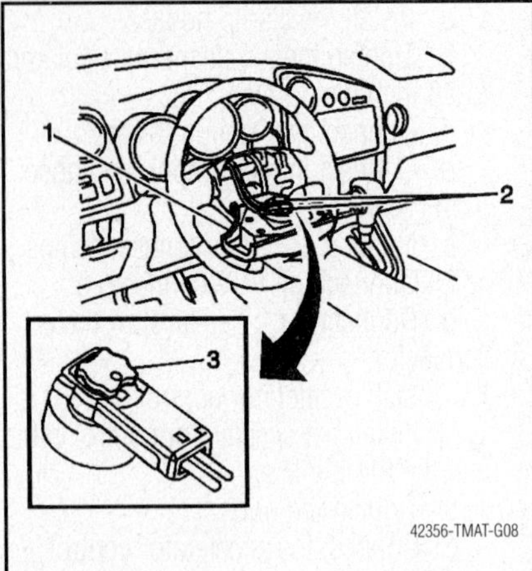

Fig. 181 Exploded view of the CPA assembly

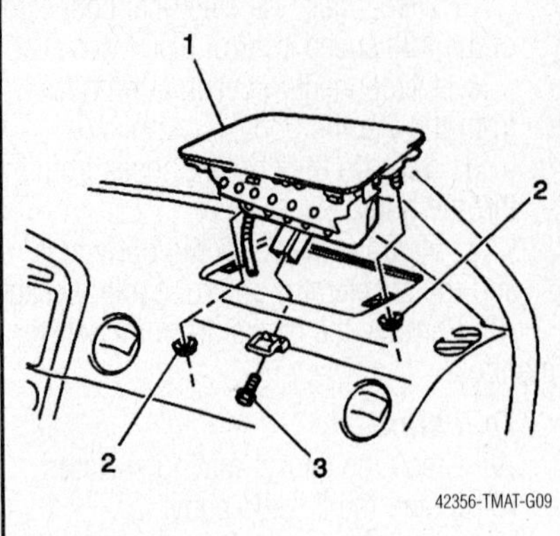

Fig. 182 Remove the instrument panel module connectors and the passenger air bag assembly

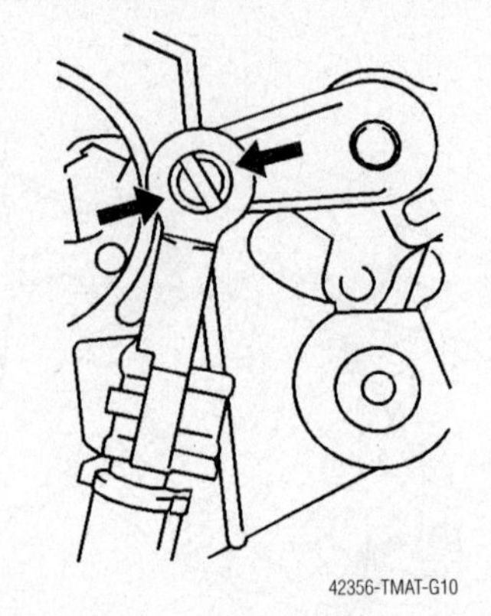

Fig. 183 Push in the clip and disconnect the cable from the manual selector shifter assembly

and antenna connections. Remove the radio.

e. Remove the A/C switch and screw.

f. Remove the hazard switch.

g. Remove the rear defogger switch.

h. Remove the manual transmission shift knob.

i. Using a taped flat–bladed tool, carefully pry the retaining clips attaching the front floor console trim plate to the floor console assembly.

j. Disconnect the 2 cigar lighter connectors.

k. Disconnect the accessory power receptacle connectors.

l. Remove the cigar lighters and power receptacle.

m. Place both wheels in the straight ahead position.

n. Remove the bolts from the steering wheel module.

o. Release the Connector Position Assurance (CPA) from the inflator module.

p. Disconnect the steering wheel module connectors.

q. Remove the steering wheel module.

r. Matchmark the steering wheel nut–to–shaft position, then remove the steering wheel nut and the wheel.

s. Remove the upper and lower steering column cover screws and the covers.

t. Disconnect the turn signal/headlamp assembly connectors.

u. Remove the turn signal/headlamp switch assembly

v. Remove the wiper switch by depressing the tab.

w. Remove the glove box.

x. Disconnect the instrument panel connector.

y. Remove the instrument panel module connectors and the passenger air bag assembly.

z. Remove the cluster trim plate by disengaging the clips.

aa. Remove the cluster screw and disengage the 2 lower clips.

bb. Disconnect the cluster electrical connectors and remove the cluster.

cc. Remove the windshield trim panel moldings.

dd. Using a taped flat–bladed tool, carefully pry the retaining clips attaching the instrument panel left trim plate to the instrument panel.

ee. Disconnect the power mirror and dimmer switch connectors.

ff. Remove the power mirror and dimmer switches.

gg. Disconnect any remaining electrical connections.

hh. Remove the upper instrument panel screws and the panel by pulling towards the rear to disengage the tabs.

ii. Disconnect the steering wheel coil connector.

jj. Release the 3 clips and remove the coil assembly.

kk. If the vehicle is equipped with an automatic transmission, insert the key into the cylinder, turn to the ACC position, push in the release button, disconnect the park lock cable, remove the key from the cylinder and lock the steering wheel.

ll. Move the silencer pad from the column.

mm. Matchmark the steering shaft coupling to the shaft.

nn. Loosen the upper bolt on the coupling.

oo. Remove the lower bolt from the coupling.

pp. Move the coupling onto the column shaft.

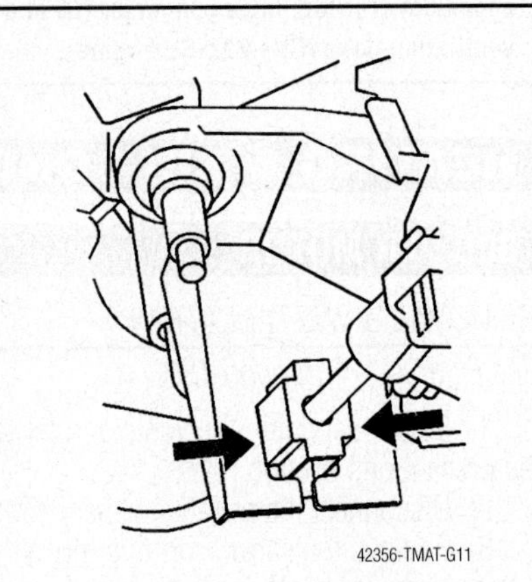

Fig. 184 Using a suitable prytool, disconnect the park lock cable from the bracket

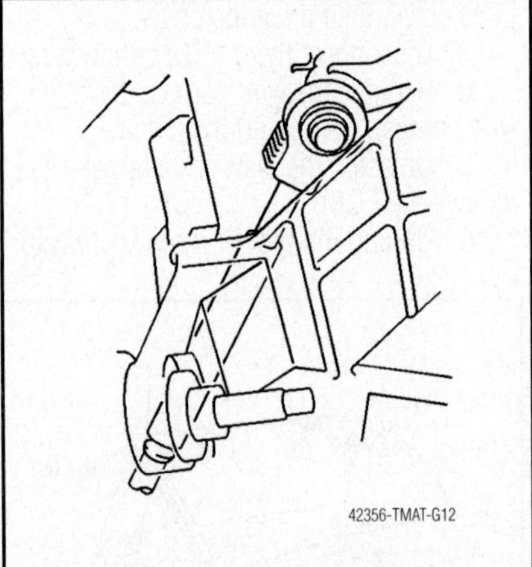

Fig. 185 Disconnect the shift select cable from the manual selector lever

qq. Disconnect the wiring harness clamps from the column.

rr. Remove the bolts and the column.

ss. Remove the body hinge trim panels.

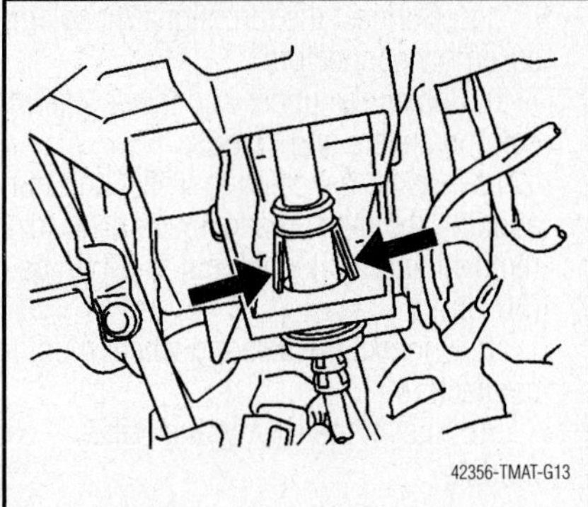

Fig. 186 Using a suitable prytool, disconnect the shift select cable from the shift lever plate

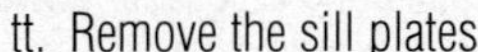

tt. Remove the sill plates.

uu. Remove the front floor console storage door.

vv. Remove the screws attaching the console to the instrument panel, pull the console rewards and up and remove the front floor console.

ww. Remove the HVAC retaining screw; disconnect the electrical connectors and module control, temperature control and A/C cables. Remove the unit.

xx. Push in the clip and disconnect the cable from the manual selector shifter assembly.

yy. Using a suitable prytool, disconnect the park lock cable from the bracket.

zz. Disconnect the shift select cable from the manual selector lever.

aaa. Using a suitable prytool, disconnect the shift select cable from the shift lever plate.

bbb. Disconnect the electrical connectors and the wire harness clip.

ccc. Remove the nuts from the selector and remove the selector.

ddd. Disconnect the hood release cable from the release handle.

eee. Remove the 8 bolts, 4 push retainers and the wire harness clamps from the lower instrument panel and remove the panel.

fff. Disengage the wiring harness clips, remove the bolts retaining the lower instrument panel pad and the pad.

ggg. Remove the ground cable.

hhh. Remove the connector housing bracket from the right instrument panel center support brace.

iii. Remove the left brace nut, right brace nut, left brace bolt, right brace bolt, left center support brace and right center support brace.

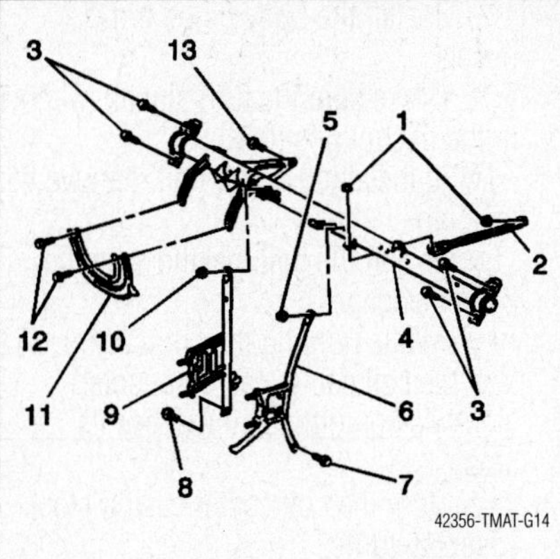

Fig. 187 Exploded view of the instrument panel reinforcement

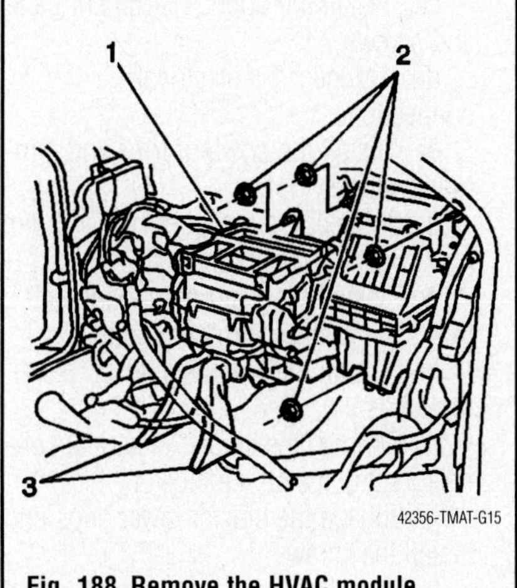

Fig. 188 Remove the HVAC module

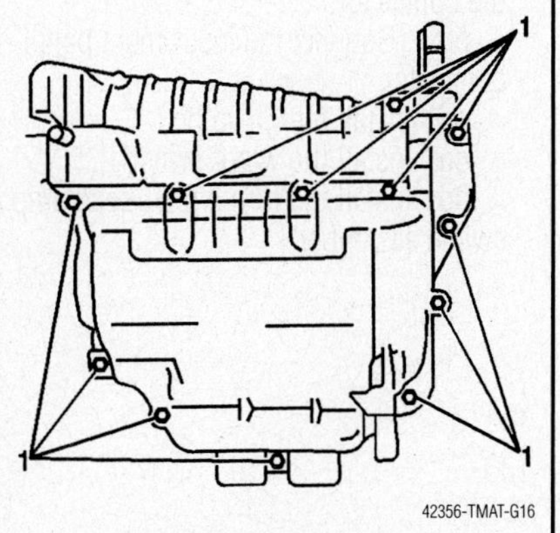

Fig. 189 Remove the 1bolts from the core case to access the heater core

jjj. Remove the windshield defroster nozzle duct from the heater case.

kkk. Remove the bolts and the nuts from the instrument panel reinforcement at the hinge pillars.

lll. Remove the instrument panel reinforcement.

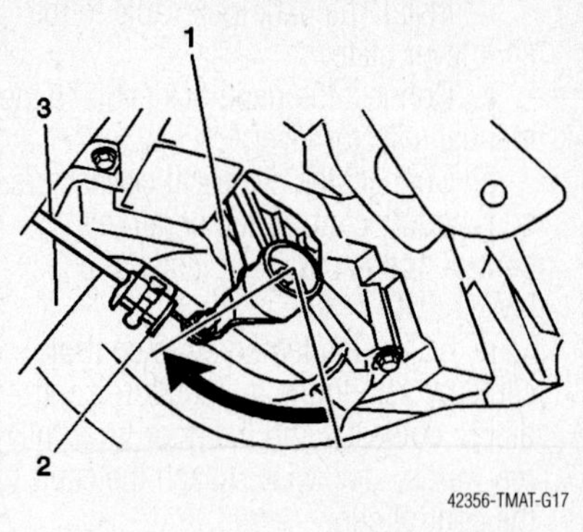

Fig. 190 Location of the components used to adjust the temperature control cable (3) clip (2), door lever (1).

8. Remove the HVAC module.

a. Disconnect the blower motor connector.

b. Disconnect the rear ducts from the HVAC module.

9. Remove the heater core.

a. Remove the bolts from the core case and lift out the core.

To install:

10. Install the heater core.

a. Install the bolts from the core case and tighten to 89 inch lbs. (10 Nm).

11. Install the HVAC module.

a. Connect the rear ducts to the HVAC module.

b. Connect the blower motor connector.

c. Install the instrument panel reinforcement.

d. Install the bolts and the nuts to the instrument panel reinforcement at the hinge pillars. Tighten to 21 ft. lbs. (28 Nm).

e. Install the windshield defroster nozzle duct to the heater case.

f. Install left center support brace and right center support brace. Tighten the nuts and bolt to 15 ft. lbs. (20 Nm).

g. Install the connector housing bracket.

h. Connect the ground cable.

12. Install the lower instrument panel pad and the bolts and attach the wiring harness clips.

a. Connect the hood release cable to the release handle.

b. Install the lever and tighten the nuts to 12 ft. lbs. (18 Nm).

c. Connect the manual selector electrical connections.

d. Attach the shift cable to the shift lever plate.

e. Connect the shift select cable to the selector lever.

f. Install the park lock cable to the shift lever plate.

g. Connect the park lock cable to the manual selector lever.

h. Connect the electrical connectors and module control, temperature control and A/C cables. Install the HVAC unit.

i. Adjust the temperature control cable by setting the temperature control dial to coldest. Hold the door lever fully rearwards, clockwise. Attach the cable to the control clip.

j. Adjust the Mode linkage by setting the dial to defrost. Hold the door lever fully rearwards, clockwise. Attach the cable to the control clip.

k. Install the front floor console and tighten the screws.

l. Install the front floor console storage door.

m. Install the sill plates.

n. Install the column.

o. Install the bolts and tighten the lower bolt to 16 ft. lbs. (21 Nm) and the 2 upper bolts to 16 ft. lbs. (21 Nm).

p. Align the matchmarks made prior to removal.

q. Lower the coupling onto the shaft. Install the bolts and tighten to 26 ft. lbs. (35 Nm).

r. Connect the wiring harness clamps to the column.

s. Move the silencer pad to the column.

t. If the vehicle is equipped with an automatic transmission, insert the key into the cylinder, turn to the ACC position, and insert the park lock cable making sure the release button engages. Make sure the key will not rotate to the lock position unless the shifter is in the park position, remove the key from the cylinder and lock the steering wheel.

u. Install the body hinge trim panels.

v. Make sure the turn signal switch is in the neutral position.

w. If installing a new coil, remove the lock pin.

x. Install the coil making sure the 3 clips engage.

y. While holding the coil casing, turn the coil center casing counterclockwise until the coil reaches its stop.

z. Turn the coil center casing clockwise 2½ turns.

aa. Align the center casing with the arrow on the outer casing.

bb. Connect the coil electrical connector.

cc. Install the upper instrument panel and screws.

dd. Connect the electrical connections.

ee. Install the power mirror and dimmer switches.

ff. Connect the power mirror and dimmer switch connectors.

gg. Install the instrument panel left trim plate to the instrument panel.

hh. Install the windshield trim panel moldings.

ii. Connect the cluster electrical connectors and install the cluster.

jj. Engage the cluster lower clips and install the screw.

kk. Install the cluster trim plate.

ll. Install the passenger air bag assembly and the instrument panel module connectors.

mm. Connect the instrument panel connector.

nn. Install the glove box.

oo. Install the wiper switch.

pp. Install the turn signal/headlamp switch assembly

qq. Connect the turn signal/headlamp assembly connectors.

rr. Install the upper and lower steering column covers and screws.

ss. Install the steering wheel and nut aligning the matchmarks made prior to removal and tighten the nut to 37 ft. lbs. (50 Nm).

tt. Connect the steering wheel module connectors.

uu. Install the CPA to the inflator module.

vv. Install the steering wheel module and tighten the retainers to 78 inch lbs. (9 Nm)

ww. Install cigar lighters and power receptacle.

xx. Connect the accessory power receptacle connectors.

yy. Connect the 2 cigar lighter connectors.

zz. Install the front floor console trim plate to the floor console assembly.

aaa. Install the manual transmission shift knob.

bbb. Install the rear defogger switch.

ccc. Install the hazard switch.

ddd. Install the A/C switch and screw.

eee. Install the radio.

fff. Connect the A/C switch, hazard switch, rear defogger switch and passenger seat belt indicator switch electrical connections.

ggg. Install the center trim plate to the instrument panel.

hhh. Connect the evaporator inlet and outlet tubes.

13. Connect the heater hoses to the core.
14. Connect the negative battery cable.
15. Evacuate and recharge the A/C system.
16. Fill the cooling system.

STEERING

POWER RACK & PINION STEERING GEAR

REMOVAL & INSTALLATION

See Figure 191.

1. Disconnect the negative battery cable.
2. Center the front wheels.
3. Remove air bag module.
4. Remove steering wheel.
5. Remove front wheels.
6. Remove engine undercover.
7. Disconnect tie rod ends.
8. Remove column hole cover silencer sheet.
9. Disconnect steering intermediate shaft.
10. Using special tool SST: 09023-38401, disconnect pressure feed and return tubes.
11. Remove the bolt and disconnect the tube clamp.
12. Remove the nut and disconnect the front stabilizer link assembly from the stabilizer bar.
13. Remove the bolt and nuts and disconnect the front suspension arm from the lower ball joint.
14. Remove hood.
15. Remove upper cylinder head cover.
16. Suspend engine assembly.
    a. Install the engine hangers with the bolts in the correct direction.
    b. Attach the engine chain hoist to the engine hangers.

✲✲ WARNING

Do not attempt to hang the engine by hooking the chain to any other parts.

17. Remove front suspension crossmember.
    a. Remove the 2 bolts and disconnect the center member from the front engine mounting insulator.
    b. Remove the 2 bolts and disconnect the center member from the frame.
    c. Remove the bolt and 3 nuts, disconnect the rear engine mounting insulator from the crossmember.
    d. Using a transmission jack, support the crossmember.
    e. Remove the 4 bolts and front suspension crossmember sub-assembly with the steering gear assembly.
18. Remove steering column hole cover.
19. Remove steering intermediate shaft.
    a. Place matchmarks on the intermediate shaft with control valve.
    b. Remove the bolt and steering intermediate shaft.
20. Remove rack & pinion power steering gear.

To install:

21. Install the power steering gear assembly with the bolts and nuts. Tighten to 43 ft. lbs. (58 Nm) for FWD and 60 ft. lbs. (82 Nm) for AWD.
22. Install the steering intermediate shaft.
    a. Align the matchmarks on the steering intermediate shaft with steering pinion shaft.
    b. Install the bolt and tighten to 26 ft. lbs. (35 Nm).
23. Install steering column hole cover.
24. Install front suspension crossmember.
    a. Using special tool SST: 09670-00010, align the holes of the front suspension member and body, and temporarily tighten the bolt "A" to 116 ft. lbs. (157 Nm) then bolt "B" to 83 ft. lbs. (113 Nm).
    b. Connect the rear engine mounting insulator to the crossmember with the bolt and nuts. Tighten to 38 ft. lbs. (52 Nm).
    c. Install the center member to the frame with the bolts. Tighten to 44 ft. lbs. (60 Nm).
    d. Connect the front engine mounting insulator to the center member with the bolts. Tighten to 38 ft. lbs. (52 Nm).
25. Connect the front suspension lower arm to the lower ball joint with the bolt and nuts. Tighten to 66 ft. lbs. (89 Nm).
26. Connect the front stabilizer link assembly with the nut. Tighten to 55 ft. lbs. (74 Nm).
27. Using the special tool, connect the return and pressure feed tubes. Tighten to 17 ft. lbs. (23 Nm).
    a. Connect the tube clamp with the bolt. Tighten to 69 inch lbs. (8 Nm).
28. Connect the tie rod end with the nut. Tighten to 36 inch lbs. (49 Nm).
    a. Install a new cotter pin.
    b. If the holes for a new cotter pin are not aligned, tighten the nut further up to 60°.
29. Install engine undercover.
30. Install front wheels.
31. Center front wheel.
32. Install column hole cover.
33. Add power steering fluid.
34. Bleed power steering fluid.
35. Inspect for fluid leaks.
36. Install upper cylinder head cover.
37. Install and adjust the hood.
38. Center spiral cable.
39. Install steering wheel.
40. Install air bag module.
41. Inspect and adjust front wheel alignment.
42. Check for air bag DTC codes.

POWER STEERING PUMP

REMOVAL & INSTALLATION

See Figures 192 and 193.

1. Disconnect the negative battery cable.
2. Remove the right side front wheel.
3. Drain power steering fluid
4. Remove the right hand engine undercover.
5. Remove the fan and alternator belt.
6. Remove the clip and disconnect the oil reservoir to pump hose No.1.
7. Using special tool SST:09023-38401, disconnect the pressure feed tube assembly.
8. Remove the bolt and disconnect the pressure feed tube clamp.
9. Disconnect the oil pressure switch connector.
10. Remove the bolts, nuts and pump assembly.

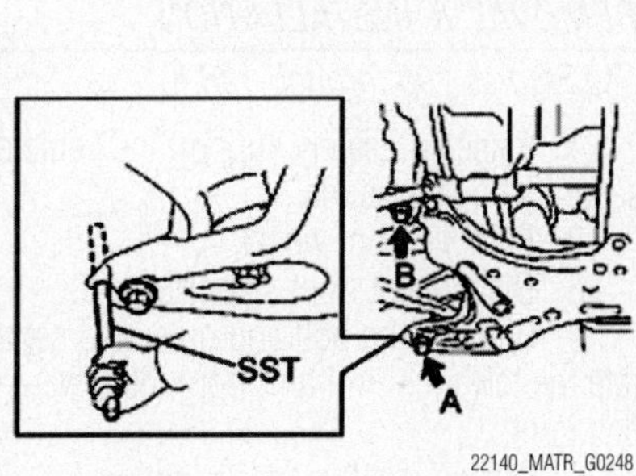

Fig. 191 Using the special tool, align the holes of the front suspension member

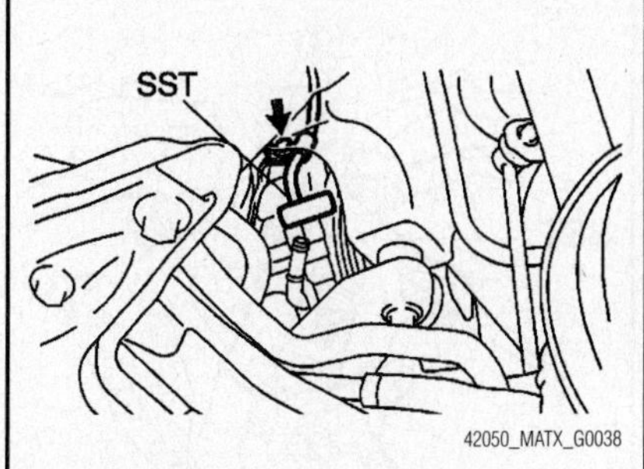

Fig. 192 Using special tool 09023-38401, disconnect the pressure feed tube assembly

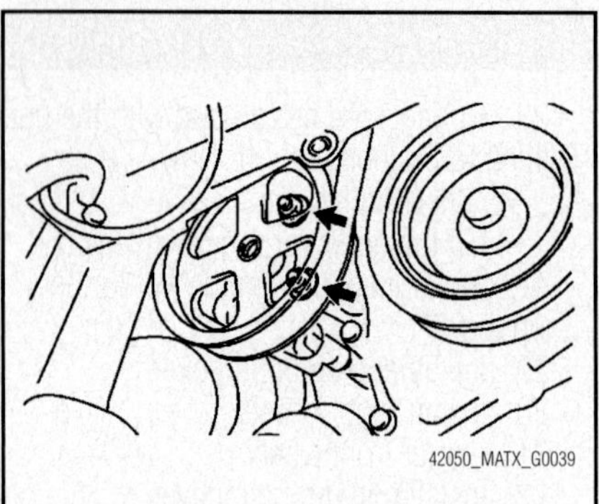

Fig. 193 Location of the power steering pump assembly

To install:

11. Install the pump assembly with the bolts and nuts. Tighten to 27 ft. lbs. (37 Nm).
12. Connect the oil pressure switch connector.

➡Be careful that the oil does not adhere to the connector.

13. Using the special tool, connect the pressure feed tube assembly. Tighten to 30 ft. lbs. (41 Nm) using a torque wrench with a fulcrum length of 13.58 in. (345 mm).
14. Connect the pressure feed tube clamp with the bolt. Tighten to 69 in. lbs. (8 Nm).
15. Connect the oil reservoir to pump hose No.1 with the clip.
16. Install the drive belt.
17. Install the wheel.
18. Fill and bleed the power steering system.
19. Install the engine undercover.

BLEEDING

1. Check the fluid level and top off as needed.
2. Jack up the front of the vehicle and support it with safety stands.
3. With the engine OFF, slowly turn the wheel from lock to lock slowly several times.
4. Lower the vehicle and start the engine and allow to idle for a few minutes.
5. With the engine warm and at idle, turn the steering wheel to the left or right lock position and hold it there for 2 to 3 seconds , then turn the wheel to the opposite side lock and hold for 2 to 3 seconds.
6. Repeat these last two steps several times.
7. Turn the vehicle off and check the fluid level.
8. If there is a lot of foam in the reservoir, check the system for leaks and repair, then bleed the system again.

SUSPENSION

FRONT SUSPENSION

LOWER BALL JOINT

REMOVAL & INSTALLATION

See Figure 194.

1. Before servicing the vehicle, refer to the precautions section.
2. Raise and safely support the vehicle securely on jackstands.
3. Remove the front wheel.
4. Depress the brake pedal and loosen the hub nut.
5. Disconnect the ABS speed sensor, if equipped
6. Remove the cotter pin and nut from the tie rod end. Using a tie rod end removal tool, separate the tie rod end from the steering knuckle.
7. Separate the lower control arm ball joint, using a suitable puller
   a. Disconnect the front halfshaft from the steering knuckle.

SST

9359AB71

Fig. 194 Removing the ball joint from the knuckle

   b. Remove the lower ball joint cotter pin and castle nut.
   c. Using a ball joint tool, separate the lower ball joint from the steering knuckle.

To install:

8. Connect the lower control arm.
   a. Connect the lower ball joint to the lower arm. Tighten the castle nut to 76 ft. lbs. (103 Nm).
   b. Install a new cotter pin.
9. Connect the front halfshaft to the steering knuckle.
10. Connect the tie rod end to the steering knuckle.
11. Connect the ABS speed sensor.
12. Install the hub nut.
13. Install the front wheel.
14. Lower the vehicle.
15. Check and adjust the alignment, if needed.

LOWER CONTROL ARM

REMOVAL & INSTALLATION

See Figures 195 through 198.

1. Raise and safely support the vehicle securely on jackstands.
2. Remove front wheel.
3. Disconnect front stabilizer link.
4. Remove the bolt and nuts, and separate the lower control arm from the lower ball joint.
5. Disconnect the rack & pinion power steering gear and suspend it out of the way.
6. Attach the lifting hooks and suspend engine assembly.

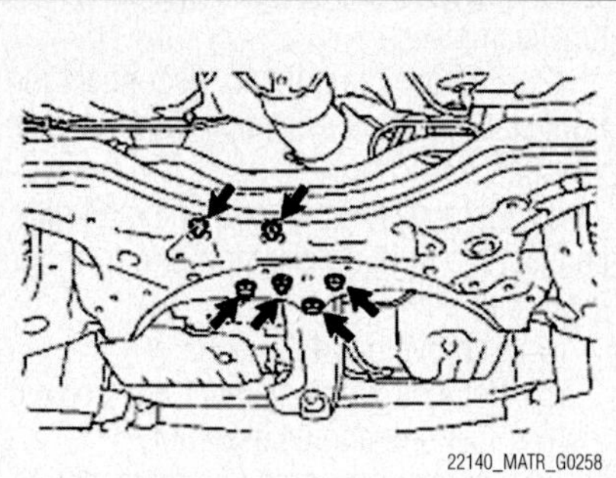

Fig. 195 Front suspension crossmember bolt locations—FWD

Fig. 196 Front suspension crossmember bolt locations—AWD

7. Separate front suspension crossmember.
   a. On AWD models, remove the bolts, disconnect the stabilizer bar front from the front suspension crossmember sub-assembly.

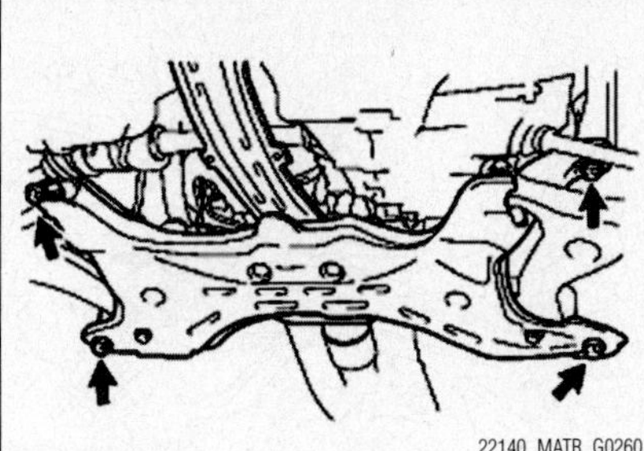

Fig. 197 Remove the bolts as illustrated to free the front suspension crossmember from the vehicle

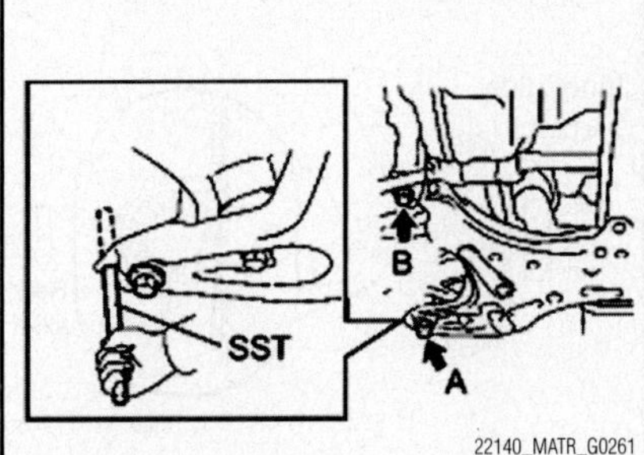

Fig. 198 Tighten the bolt "A" then bolt "B" temporarily as shown

b. Remove the bolts and nuts, disconnect the transverse engine mounting insulator and center engine mounting member from the front suspension crossmember.

c. Remove the bolts.

d. Lower the transmission jack and remove the front suspension crossmember.

8. Remove lower control arm.

a. Remove the bolts, nut and lower control arm from the front suspension crossmember.

To install:

9. Install the lower control arm and temporarily tighten the bolts and nut.

10. Install front suspension crossmember.

a. Lift the front suspension crossmember up with a transmission jack.

b. Insert special tool SST: 09670-00010 into the base hole of the cross-member and side of the vehicle.

11. Tighten the bolt "A" then bolt "B" temporarily as shown.

a. After securing both sides, tighten the bolt "A" to 116 ft. lbs. (157 Nm) then bolt "B" to 83 ft. lbs. (113 Nm) temporarily as illustrated.

b. Connect the transverse engine mounting insulator and center engine mounting member to the front suspension crossmember. Tighten the bolts to 38 ft. lbs. (52 Nm).

c. On AWD models, install the stabilizer bar front and bolts to the front suspension crossmember. Tighten the bolts to 14 ft. lbs. (19 Nm).

12. Install rack & pinion power steering gear. On AWD models, tighten the bolts to 94 ft. lbs. (127 Nm). On FWD models, tighten the bolts to 43 ft. lbs. (58 Nm).

13. Install the lower control arm with the nuts and bolt to the lower ball joint. Tighten the bolts to 66 ft. lbs. (89 Nm).

14. Install front stabilizer link .

15. Stabilize suspension.

a. Install the front wheel

b. Lower the vehicle.

c. Bounce the vehicle up and down several times to stabilize the suspension.

d. Fully tighten lower control arm. Tighten to 101 ft. lbs. (137 Nm).

16. Inspect and adjust front wheel alignment.

MACPHERSON STRUT

REMOVAL & INSTALLATION

See Figures 199 and 200.

1. Before servicing the vehicle, refer to the precautions section.

2. Raise and safely support the vehicle securely on jackstands.

3. Remove the wheel.

4. Remove the stabilizer link from the strut.

➡If the ball joint turns together with the nut, use a hexagon (6 mm) wrench to hold the stud.

5. Remove the bolt, and disconnect the brake hose from the strut.

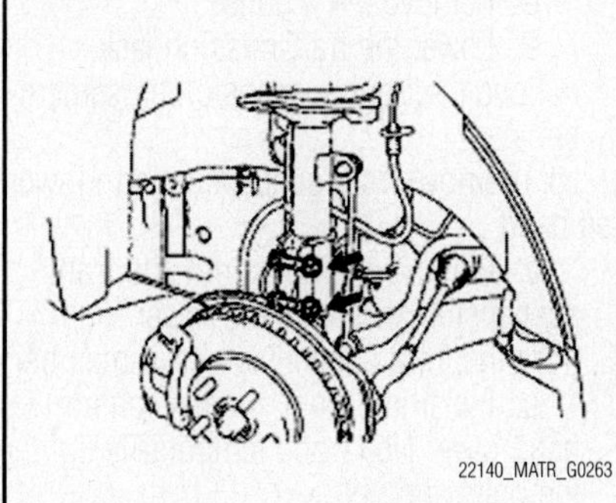

Fig. 199 Lower strut mounting bolts and nuts

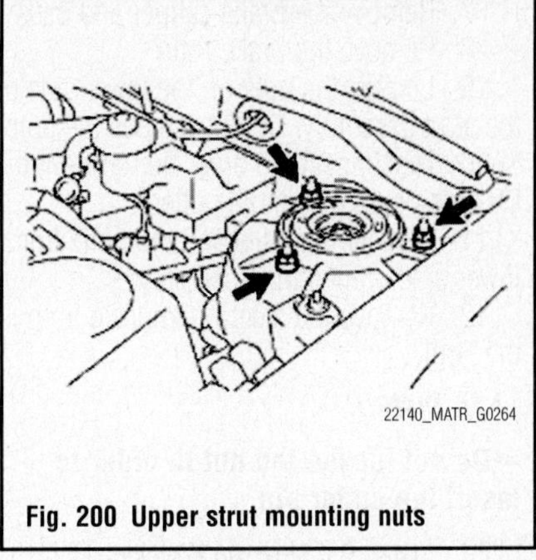

Fig. 200 Upper strut mounting nuts

6. Remove the with abs brakes, speed sensor wiring harness from the strut.

7. Remove the lower strut bolts and nuts.

8. Remove the upper strut nuts.

➡When removing the bolt, stop the bolt from rotating and loosen the nut.

9. Remove the upper strut nuts.

10. Remove the strut from the steering knuckle.

11. Remove the strut.

12. Install the strut.

13. Install the nuts holding the strut to the strut tower. Tighten the nuts to 29 ft. lbs. (39 nm).

14. Install the 2 lower strut bolts and nuts. Tighten to 162 ft. lbs. (220 Nm).

15. Install the brake line to the steering knuckle. Tighten the line bolt to 21 ft. lbs. (29 nm).

16. Install the secure the wiring harness, if equipped with abs.

17. Install the stabilizer link. Tighten the nut to 55 ft. lbs. (74 nm).

18. Install the wheel and lower the vehicle..

19. Check and adjust the alignment, if needed.

STEERING KNUCKLE

REMOVAL & INSTALLATION

1. Raise and support the vehicle.

2. Remove the front tire and wheel assembly.

3. If the vehicle has ABS, remove the ABS wheel speed sensor from the steering knuckle. Position the sensor to the side.

4. Unstake the front wheel drive shaft nut.

5. Remove the drive shaft nut from the drive axle while an assistant presses the brake pedal.

6. Remove the retainers from the lower control arm and separate the control arm from the knuckle.

7. Remove the brake caliper and pads.
8. Remove the brake rotor.
9. Loosen the nuts on the lower side of the strut assembly. Do not remove the bolts.
10. Disconnect the outer tie rod from the knuckle using a suitable puller.
11. Remove the nuts and the bolts from lower side of the strut assembly.
12. Remove the steering knuckle from the strut.

To install:

➡Do not loosen the nut in order to insert the cotter pin.

13. Install the steering knuckle. Tighten the nuts and bolts at the lower side of the strut assembly to162 ft. lbs. (220 Nm).

➡Do not loosen the nut in order to insert the cotter pin.

14. Position the outer tie rod and the nut onto the steering knuckle. Tighten the tie rod nut to 36 ft. lbs. (49 Nm). Tighten the nut up to 1/6 additional turn in order to insert the cotter pin.
15. Install a NEW cotter pin to the tie rod ball joint stud. Bend the cotter pin ends in order to retain the nut.
16. Install the front brake rotor.
17. Install a NEW drive shaft nut while an assistant applies the brakes. Tighten the nut to 159 ft. lbs. (216 Nm).
18. Stake the drive shaft nut into the slot on the wheel drive shaft.
19. Install the lower control arm.
20. If the vehicle has ABS, install the ABS wheel speed sensor to the knuckle.
21. Install the tire and wheel assembly.
22. Lower the vehicle.
23. Inspect and adjust the front wheel alignment as needed.

STABILIZER BAR

REMOVAL & INSTALLATION

See Figures 201 through 205.

1. Before servicing the vehicle, refer to the precautions section.
2. Remove the wheels.
3. Remove the stabilizer bar links on both sides.
4. Disconnect the lower control arms from the steering knuckles.
5. Separate the power steering gear assembly.
6. Support the engine by using a suitable support assembly.
7. Remove the front suspension crossmember.
    a. On FWD models, remove the bolts and nuts, disconnect the transverse engine mounting insulator and center engine mounting member from the front suspension crossmember.
    b. On AWD models, remove the bolts and disconnect the stabilizer bar front from the front suspension crossmember .
    c. On AWD models, remove the bolt and nuts, disconnect the transverse engine mounting insulator and center engine mounting member from the front suspension crossmember .
    d. Remove the 4 bolts.
    e. Lower the transmission jack, remove the front suspension crossmember.
8. Remove front suspension arm lower left hand .
    a. Remove the bolts, nut and front suspension arm left hand lower No.1 from the front suspension crossmember .
    b. Install the front suspension arm left hand lower No. 1 and temporarily tighten the bolts and nut.
9. Remove the stabilizer bar bolts retaining the brackets and remove the brackets; bushings and stabilizer bar.

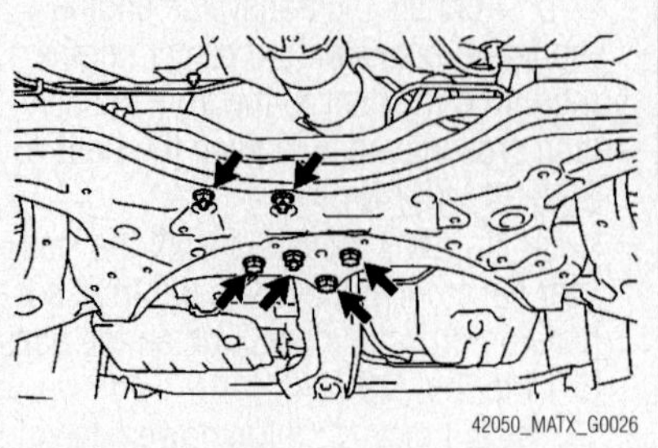

Fig. 201 Remove the bolts and nuts, disconnect the transverse engine mounting insulator and center engine mounting member—FWD models

Fig. 202 Remove the bolt and nuts, disconnect the transverse engine mounting insulator and center engine mounting member from the front suspension crossmember—AWD models

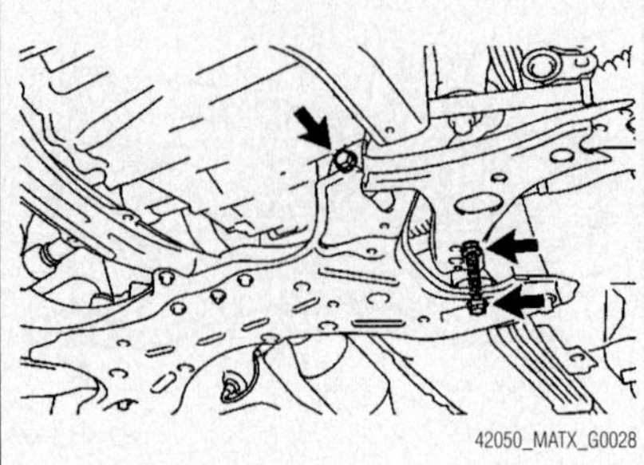

Fig. 203 Install the front suspension arm left hand lower No. 1 and temporarily tighten the bolts and nut

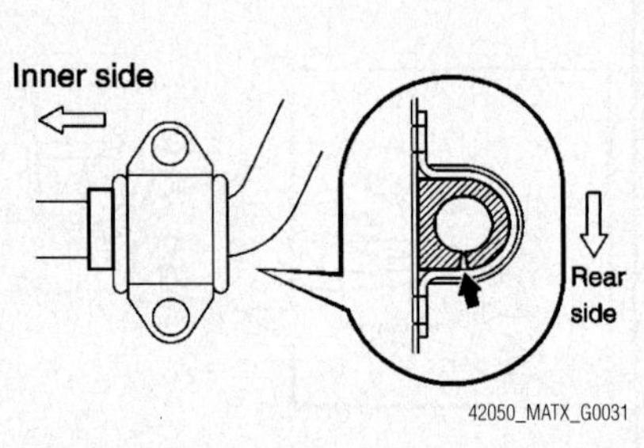

Fig. 204 Install the stabilizer bar with the bushing inner side stopper on the stabilizer bar

To install:

10. Install the stabilizer bar with the bushing inner side stopper on the stabilizer bar as illustrated. Tighten the bracket bolts to 14 ft. lbs. (19 Nm).
11. Install the left hand crossmember with an A/T position.
    a. Raise the front suspension crossmember up with a transmission jack.
    b. Insert special tool 09670-00010 to the base hole of the right hand side crossmember and right side of the vehicle.
    c. Tighten the bolt temporarily in the order A and B as illustrated.
    d. Insert special tool 09670-00010 to the base hole of the left hand side of crossmember and left side of the vehicle.
    e. Tighten the bolt temporarily in the order A and B as illustrated.
    f. Insert special tool 09670-00010 to the base hole of the right hand side of crossmember and right side of the vehicle.
    g. Then tighten the bolt A to 116 ft. lbs. (157 Nm) and bolt B to 83 ft. lbs. (113 Nm).
    h. Insert special tool 09670-00010 to the base hole of the left side of crossmember and left side of the vehicle.

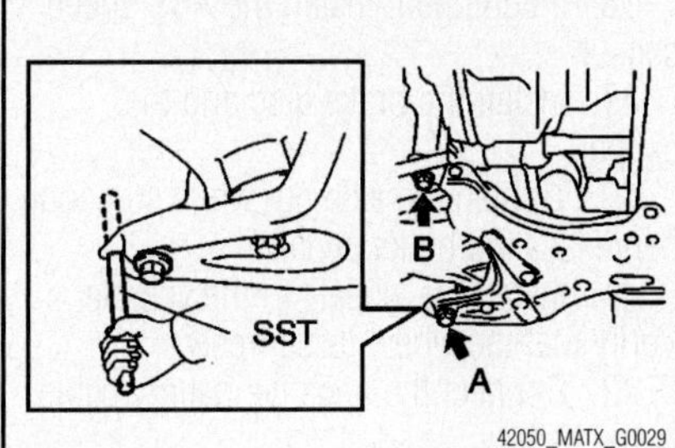

Fig. 205 Insert special tool 09670-00010 to the base hole of the right hand side crossmember and right side of the vehicle. Also not locations of bolts A and B.

i. Connect the transverse engine mounting insulator and center engine mounting member to the front suspension crossmember .

j. On FWD models, tighten the bolts and nuts to 38 ft. lbs. (52 Nm).

k. On AWD models, install the bolt and nuts and tighten to 38 ft. lbs. (52 Nm).

12. Install the power steering gear assembly and tighten the retainers to 43 ft. lbs. (58 Nm) on FWD models or 94 ft. lbs. (127 Nm).

13. Install the brackets and bushings. Prior to installation, place the elastic bushings to the inside of the bushing on the stabilizer bar. Tighten the bolts to 14 ft. lbs. (19 Nm).

14. Connect the lower control arms to the steering knuckles. Tighten the lower arm retainers to 66 ft. lbs. (89 Nm) and the

15. Install the stabilizer bar links and tighten the nuts to 55 ft. lbs. (74 Nm).

16. Install the wheels and tighten to 76 ft. lbs. (103 Nm).

17. Perform a wheel alignment check.

WHEEL HUB AND BEARING

REMOVAL & INSTALLATION

See Figures 206 through 209.

1. Before servicing the vehicle, refer to the precautions section.
2. Remove or disconnect the following:
3. Remove the wheels.
4. Remove the hub nut.
5. Remove the front stabilizer link.
6. Remove the anti-lock brake system (ABS) speed sensor.
7. Remove the brake caliper.
8. Remove the rotor.
9. Remove the tie rod end from the steering knuckle.
10. Remove the lower control arm ball joint.
11. Remove the front halfshaft from the hub, using a mallet to tap it out. Be careful not to damage the boot or speed sensor.
12. Loosen the nuts on the lower side of the strut assembly. Do not remove at this time.
13. Remove the lower ball joint using a puller.
14. Remove the tie rod end from the steering knuckle.
15. Remove the steering knuckle from the lower control arm.
16. Remove the knuckle from the strut assembly.
17. Remove the hub.

➡Cover the halfshaft boot with a shop rag to protect it from any damage.

18. Clamp the steering knuckle in a vise and remove the dust deflector. Remove the nut holding the steering knuckle to the ball joint. Press the ball joint out of the steering knuckle.
19. Remove the inner axle seal.
20. Using a Torx®wrench, remove the bolts securing the dust cover.

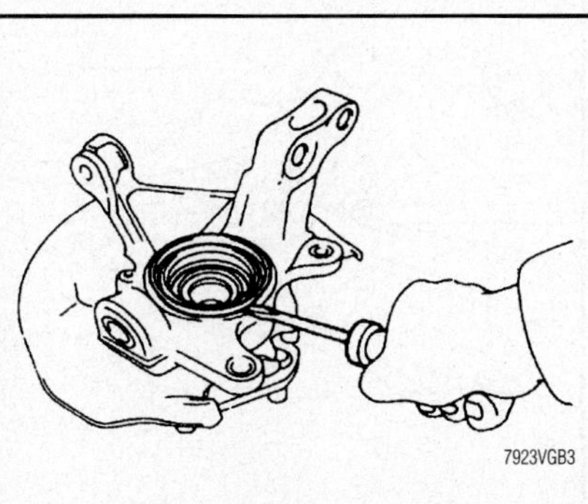

Fig. 206 Removing the inner axle seal from the hub assembly

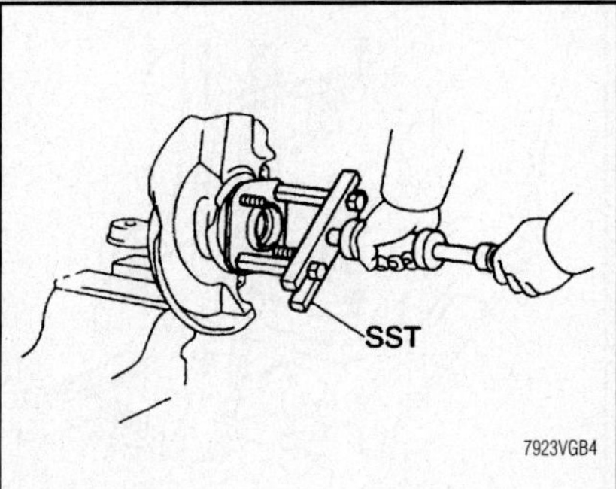

Fig. 207 Removing the axle hub from the knuckle

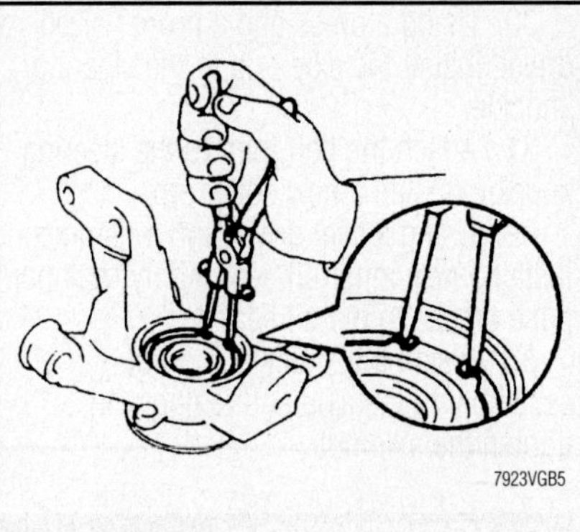

Fig. 208 Removing the snap ring from the knuckle before pressing out the bearing

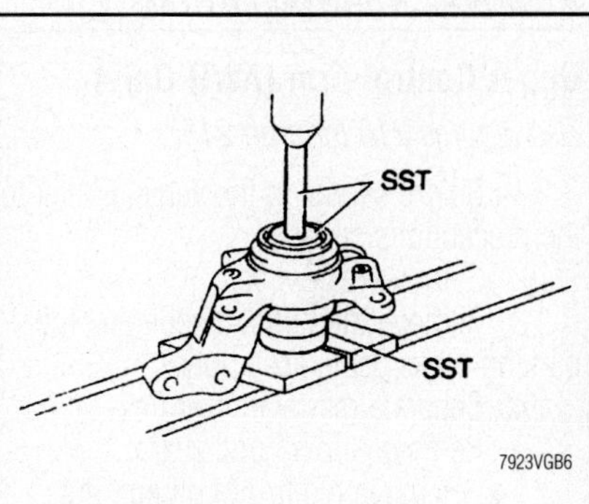

Fig. 209 Removing the bearing from the steering knuckle using a press

21. Using hub puller, remove the hub and backing plate from the steering knuckle.
22. Using a proper sized driver and a press, remove the inner hub race from the axle hub.
23. Using seal removal tool, remove the outer axle seal.
24. Using snap ring pliers, remove the snap ring from the inner side of the steering knuckle.
25. Using a proper sized driver and a press, remove the bearing from the steering knuckle. The bearing is pressed from the front of the steering knuckle and is removed through the back of the steering knuckle.

To install:

26. Using a proper sized driver and a press, install a new bearing to the steering knuckle.
27. Install the snap ring to the steering knuckle using snap ring pliers.
28. Using a seal driver and a hammer, install a new outer oil seal. Apply multipurpose grease to the oil seal lip.
29. Place the dust cover on the steering knuckle. Tighten the bolts: 78 inch lbs. (9 Nm).

30. Using a press and a proper sized driver, install the axle hub to the steering knuckle.

31. Attach the ball joint to the steering knuckle. Install a new cotter pin.

32. Using a seal driver and a hammer, install a new inner oil seal. Apply multipurpose grease to the oil seal lip.

33. Install the knuckle and hub assembly to the axle and temporarily tighten the axle nut.

34. Connect the knuckle assembly to the lower strut bracket. Temporarily insert the mounting bolts from the rear and install the nuts making sure the matchmarks made earlier are in alignment.

35. Connect the lower ball joint to lower arm.

36. Connect the tie rod end to the knuckle.

37. Tighten the bolts on the lower side of the strut assembly.

38. If equipped, install the ABS speed sensor.

39. Install the brake disc and the caliper.

40. Tighten the axle nut while someone depresses the brake pedal.

41. Install the wheels to the vehicle. Verify that the wheel turns freely.

42. Connect the negative battery cable to the battery.

43. Check alignment.

SUSPENSION

REAR SUSPENSION

CONTROL ARMS/LINKS

REMOVAL & INSTALLATION

Upper Control Arm (AWD Only)

See Figures 210 through 213.

1. Before servicing the vehicle, refer to the precautions section.
2. Remove rear wheel.
3. Remove the bolts, disconnect the parking brake cable assembly.
4. Separate rear stabilizer link.
5. Remove upper control arm.
   a. Remove the front bolt and nut.
   b. Support the upper control arm assembly securely.
   c. Make matchmarks on the rear suspension toe adjust plate, rear suspension toe adjust cam and upper control arm.
   d. Remove the nut rear suspension toe adjust plate and rear suspension toe adjust cam sub-assembly.
   e. Remove the bolt and nut, disconnect the rear strut.

➡When removing the bolt, stop the nut from rotating and loosen the bolt.

   f. Remove the bolt and nut, separate the rear side upper control arm.
   g. Remove the front side bolts and the rear lower suspension arm

To install:

6. Install upper control arm.
   a. Install the rear suspension support stopper and upper control arm bracket, temporarily tighten a new nut.
   b. Install the rear side upper control arm, temporarily tighten bolt and nut.

➡When installing the bolt, stop the nut from rotating and tighten the bolt.

   c. Install the bolts and tighten to 48 ft. lbs. (65 Nm).
   a. Install the axle carrier to the upper control arm assembly, temporarily tighten a new bolt and nut.
   b. Insert the rear suspension toe adjust cam sub-assembly into the front side of the vehicle, temporarily tighten with nut through the rear suspension toe adjust plate No. 2.
   c. Install the rear shock absorber with coil spring to the upper control arm , temporarily tighten the bolt and nut.
7. Install rear stabilizer link.
8. Connect the parking brake cable with the bolts and tighten to 48 inch lbs. (5 Nm).
9. Stabilize the suspension by bouncing the vehicle up and down several times.
10. Fully tighten upper control arm. Tighten the nut to 81 ft. lbs. (110 Nm) and the bolt to 55 ft. lbs. (74 Nm).
11. Tighten the strut attaching nut to 103 ft. lbs. (140 Nm).
12. Align the matchmarks on the rear suspension toe adjust and tighten the nut to 55 ft. lbs. (74 Nm).
13. Inspect and adjust rear wheel alignment.

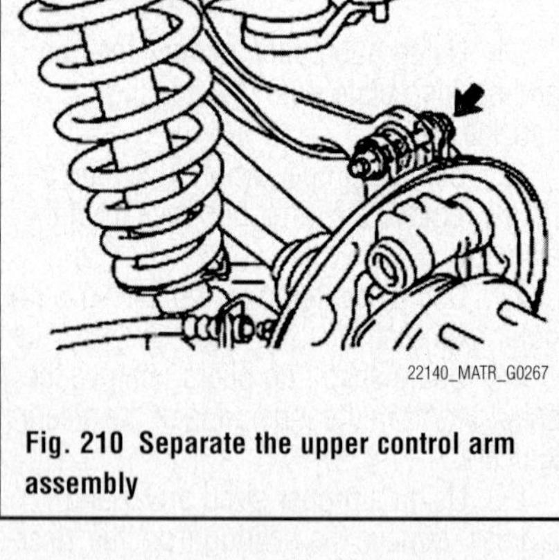

Fig. 210 Separate the upper control arm assembly

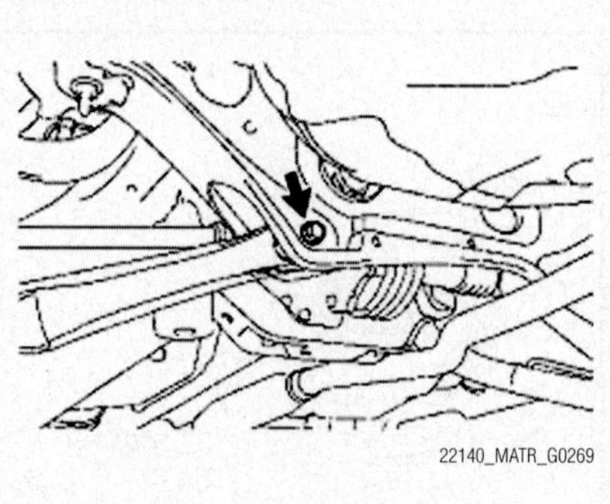
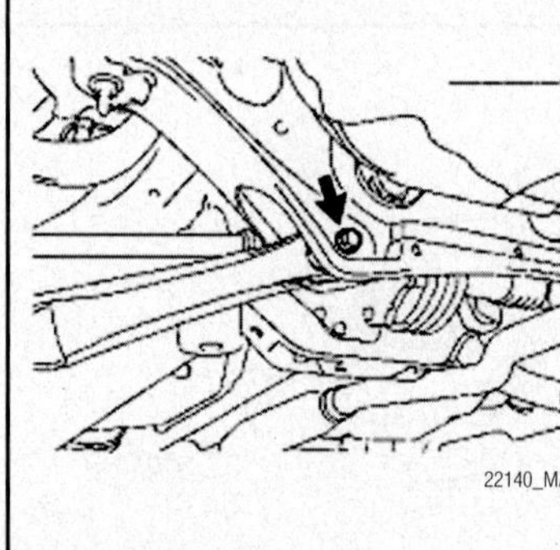

Fig. 212 Disconnect the rear side of the upper control arm

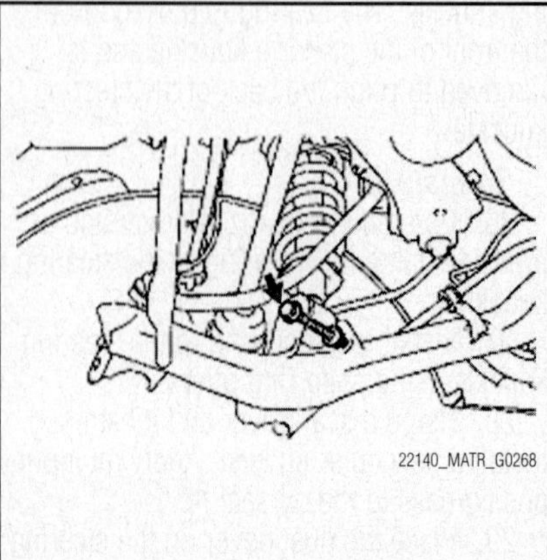

Fig. 211 Disconnect the rear strut

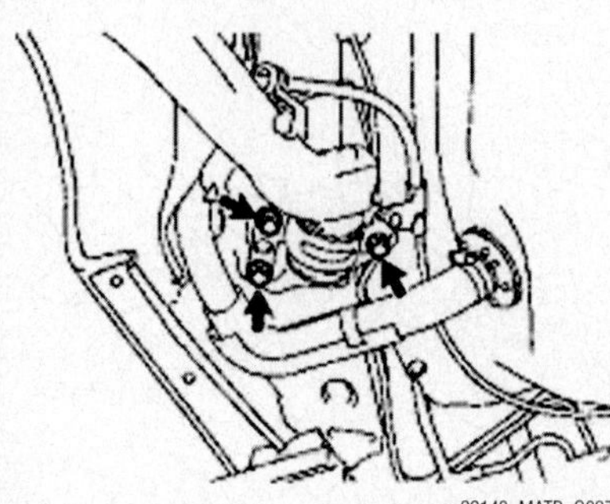

Fig. 213 Remove the front side bolts and separate the upper control arm

MACPHERSON STRUTS

REMOVAL & INSTALLATION

See Figures 214 through 216.

1. Before servicing the vehicle, refer to the precautions section.

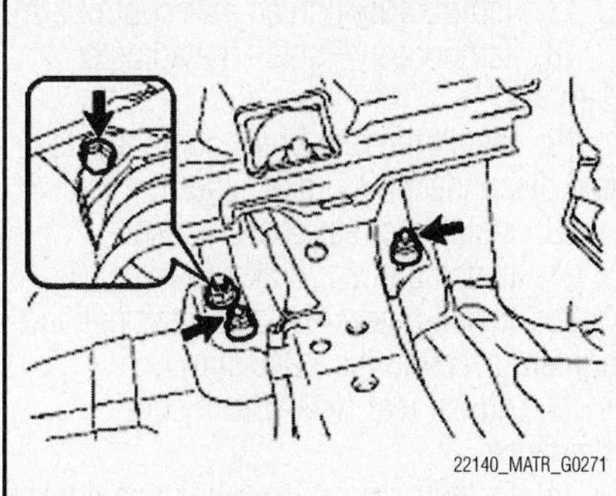

Fig. 214 Upper strut attaching nuts and bolt—FWD shown , AWD similar

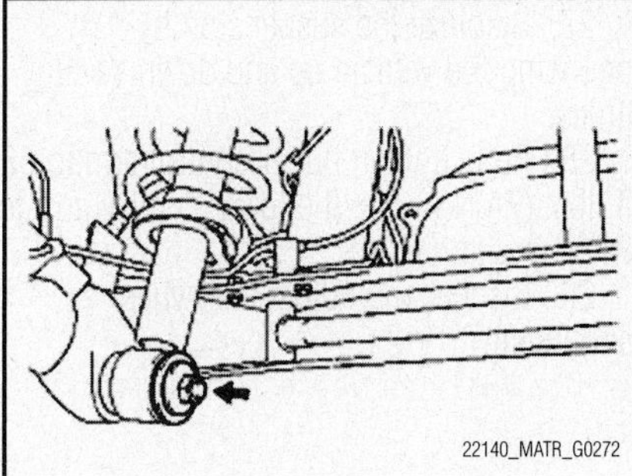

Fig. 215 Lower strut attaching nuts and bolt—FWD shown, AWD similar

2. Raise and safely support the vehicle securely on jackstands.
3. Remove the rear wheel.
4. Rear deck board, luggage compartment tray and any trim necessary to access the strut towers.
5. Shock absorber head cover.
6. On FWD vehicles, support the axle beam with a jack.
   a. Remove the strut tower nuts and bolt.
   b. Remove the lower strut nut, cushion retainer and strut.
7. On AWD vehicles, separate the rear stabilizer link.
   a. Support the rear control arm.
   b. Remove the bolt and nut from the rear control arm.
   c. Remove the strut tower nuts.
   d. Remove the rear control arm bolts.
   e. Press the rear control arm down to the outside of the vehicle, then remove the strut.

To install:

8. Install the strut on the vehicle. Tighten the strut-to-strut tower nuts to 59 ft. lbs. (80 Nm).
9. Connect the lower end of the strut and install the nut and cushion retainer/bolt snug. Do not fully tighten at this time.

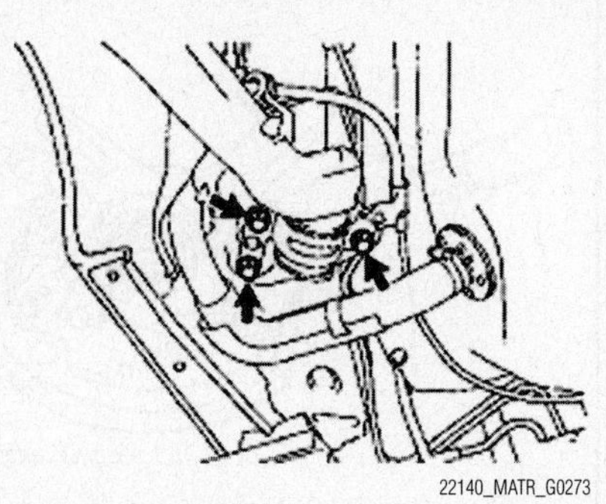

Fig. 216 Remove the 3 bolts from the control arm, press the arm down to the outside of the vehicle and remove the rear strut—AWD

10. On AWD vehicles, connect the rear control arm. Tighten the bolts to 48 ft. lbs. (65 Nm).
11. On AWD vehicles, connect the rear stabilizer link.
12. Install the strut head cover
13. Install the trunk tray, deckboard and any other trim pieces removed.
14. Install the wheel and lower the vehicle.
15. With the vehicle's weight on the suspension, tighten the bolt holding the strut to the axle carrier to 59 ft. lbs. (80 Nm) for FWD vehicles, or 103 ft. lbs. (140 Nm) for AWD vehicles.
16. Check and adjust the rear wheel alignment.

WHEEL HUB AND BEARING

REMOVAL & INSTALLATION

FWD

See Figure 217.

1. Before servicing the vehicle, refer to the precautions section.
2. Raise and safely support the vehicle securely on jackstands.
3. Remove the wheel.
4. Remove the brake drum or rotor.
5. Remove the ABS wheel speed sensor or skid control sensor, as applicable.
6. Remove the hub retaining bolts
7. Separate the hub from the knuckle.

To install:

8. Install the hub on the knuckle and Tighten the bolts to 45 ft. lbs. (61 Nm).
9. Install the ABS wheel speed sensor or skid control sensor, as applicable.
10. Install the brake drum or rotor.
11. Install the wheel.
12. Lower the vehicle.
13. Check and adjust the alignment, if needed.

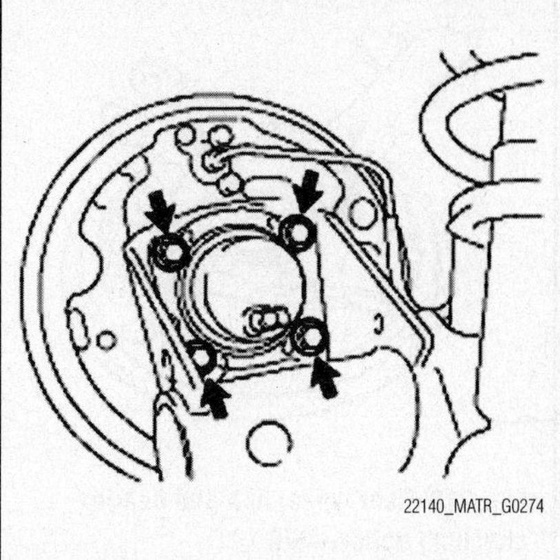

Fig. 217 Rear wheel hub and bearing attaching bolts—FWD

AWD

See Figures 218 and 219.

1. Raise and safely support the vehicle securely on jackstands.
2. Remove rear wheel.
3. Remove the nut and separate the rear stabilizer link.

➡If the ball joint turns together with the nut, use a hexagon wrench (5 mm) to hold the stud.

4. Remove rear axle shaft nut.
5. Remove rear brake drum.
6. Separate rear speed sensor.
7. Remove the rear brake components and disconnect the brake hose.
8. Separate the lower control arm.
   a. Place matchmarks on the camber adjust cam. Remove the nut, bolt and camber adjust cam.
   b. Remove the nut and bolt, separate the lower control arm.
9. Separate upper control arm assembly.
   a. Remove the nut and bolt, separate the upper control arm assembly.
10. Remove rear axle.
   a. Using a plastic hammer, separate the axle from the drive shaft, then remove it.

⁂ WARNING

Be careful not to damage the boot and speed sensor rotor.

11. Remove rear wheel bearing dust deflector.
   a. Using a screw driver and a hammer, remove the dust deflector.

⁂ WARNING

Be careful not to damage the axle carrier.

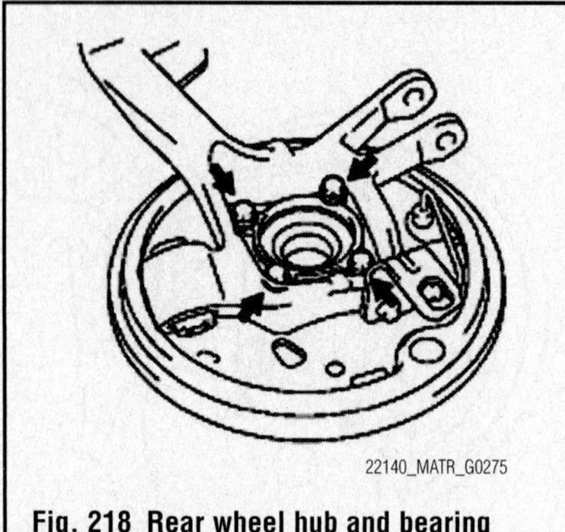

Fig. 218 Rear wheel hub and bearing attaching bolts—AWD

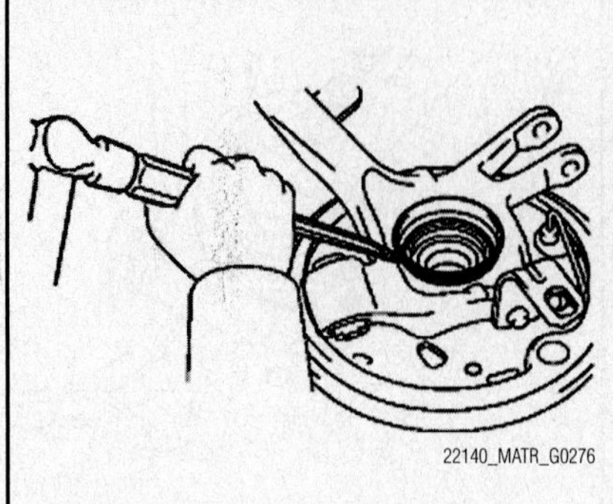

Fig. 219 Removing the rear wheel hub dust boot—AWD

12. Remove the bolts, hub & bearing and carrier.

To install:

13. Install the hub & bearing and carrier. Tighten bolts to 41 ft. lbs. (56 Nm).
14. Install the rear wheel bearing dust deflector using a seal driver
15. Temporarily tighten upper control arm assembly.
16. Install rear axle by fitting the splines, insert the front drive shaft into the axle.

WARNING

Be careful not to damage the boot and speed sensor rotor.

17. Temporarily tighten rear control arm.
18. Temporarily tighten the adjuster cam.
19. Install the rear brake components and disconnect the brake hose.
20. Install the rear speed sensor.
21. Install the rear brake drum.
22. Install a new rear axle shaft nut and tighten to 159 ft. lbs. (216 Nm).
23. Adjust rear brake drum shoe clearance
24. Install rear stabilizer link and tighten to 32 ft. lbs (44 Nm).
25. Bleed the brake system.
26. Install the wheel and lower the vehicle.
27. Stabilize the suspension by bouncing the vehicle up and down several times.
28. Fully tighten upper control arm to 55 ft. lbs. (74 Nm) and the lower control arm to 81 ft. lbs. (110 Nm).
29. Inspect and adjust rear wheel alignment.

TOYOTA

Prius

19

SPECIFICATIONS AND MAINTENANCE CHARTS

VEHICLE AND ENGINE IDENTIFICATION

Engine						
Code	Liters (cc)	Cu. in.	Cyl.	Fuel Sys.	Engine Type	Eng. Mfg.
1NZ-FXE	1.5 (1497)	91.4	4	SFI	DOHC	Toyota

SFI: Sequential Multiport Fuel Injection

DOHC: Dual Overhead Camshaft

Model Year	
Code	Year
6	2006
7	2007
8	2008

22140_PRIU_C0001

GENERAL ENGINE SPECIFICATIONS

Year	Engine ID/VIN	Engine Displacement Liters (cc)	Fuel System Type	Net Horsepower @ rpm	Net Torque @ rpm (ft. lbs.)	Bore x Stroke (in.)	Com-pression Ratio	Oil Pressure @ rpm
2006	1NZ-FXE	1.5 (1497)	SFI	110@5000	82@4200	2.95x3.33	13.0:1	22-80@2500
2007	1NZ-FXE	1.5 (1497)	SFI	110@5000	82@4200	2.95x3.33	13.0:1	22-80@2500
2008	1NZ-FXE	1.5 (1497)	SFI	110@5000	82@4200	2.95x3.33	13.0:1	22-80@2500

SFI: Sequential Multiport Fuel Injection

22140_PRIU_C0002

GASOLINE ENGINE TUNE-UP SPECIFICATIONS

Year	Engine Displacement Liters	Engine ID/VIN	Spark Plugs Gap (in.)	Ignition Timing (deg.)		Fuel Pump (psi)	Idle Speed (rpm)		Valve Clearance	
				MT	AT		MT	AT	In.	Ex.
2006	1.5	1NZ-FXE	0.043	—	8-12B	44-50 ①	—	950-1050	0.007-0.009	0.011-0.013
2007	1.5	1NZ-FXE	0.043	—	8-12B	44-50 ①	—	950-1050	0.007-0.009	0.011-0.013
2008	1.5	1NZ-FXE	0.043	—	8-12B	44-50 ①	—	950-1050	0.007-0.009	0.011-0.013

Note: The Vehicle Emission Control Information label often reflects specification changes made during production.

The label figures must be used if they differ from those in this chart.

B: Before top dead center

① At idle

22140_PRIU_C0003

CAPACITIES

Year	Model	Engine Displacement Liters	Engine ID/VIN	Engine Oil with Filter (qts.)	Transaxle (pts.) 5-Spd	Transaxle (pts.) Auto.	Fuel Tank (gal.)	Cooling System (qts.)
2006	Prius	1.5	1NZ-FXE	3.9	—	7.6 ①	11.9	②
2007	Prius	1.5	1NZ-FXE	3.9	—	7.6 ①	11.9	②
2008	Prius	1.5	1NZ-FXE	3.9	—	7.6 ①	11.9	②

Note: All capacities are approximate. Add fluid gradualy and check to be sure a proper fluid level is obtained.

① Specification for Hybrid transaxle.

② Gasoline engine: 5.2 quarts
Electric motor, inverter and converter: 2.7 quarts

22140_PRIU_C0004

FLUID SPECIFICATIONS

Year	Model	Engine Displacement Liters	Engine ID/VIN	Engine Oil	Auto. Trans.	Manual Trans.	Drive Axle	Transfer Case	Power Steering Fluid	Brake Master Cylinder
2006	Prius	1.5	1NZ-FXE	5W-30	ATF WS	—	—	—	—	DOT 3
2007	Prius	1.5	1NZ-FXE	5W-30	ATF WS	—	—	—	—	DOT 3
2008	Prius	1.5	1NZ-FXE	5W-30	ATF WS	—	—	—	—	DOT 3

DOT: Department Of Transpotation

22140_PRIU_C0013

VALVE SPECIFICATIONS

Year	Engine ID/VIN	Engine Displacement Liters	Seat Angle (deg.)	Face Angle (deg.)	Spring Test Pressure (lbs. @ in.)	Spring Installed Height (in.)	Stem-to-Guide Clearance (in.) Intake	Stem-to-Guide Clearance (in.) Exhaust	Stem Diameter (in.) Intake	Stem Diameter (in.) Exhaust
2006	1NZ-FXE	1.5	45	45	40.5-44.5@ 0.9880	2.353	0.0001-0.0031	0.0012-0.0039	0.1957-0.1963	0.1955-0.1961
2007	1NZ-FXE	1.5	45	45	40.5-44.5@ 0.9880	2.353	0.0001-0.0031	0.0012-0.0039	0.1957-0.1963	0.1955-0.1961
2008	1NZ-FXE	1.5	45	45	40.5-44.5@ 0.9880	2.353	0.0001-0.0031	0.0012-0.0039	0.1957-0.1963	0.1955-0.1961

22140_PRIU_C0005

CAMSHAFT AND BEARING SPECIFICATIONS CHART

All measurements are given in inches.

Year	Engine Displ. Liters	Engine VIN	Journal Dia.	Brg. Oil Clearance	Shaft End-play	Runout	Journal Bore	Lobe Lift Intake	Lobe Lift Exhaust
2006	1.5	1NZ-FXE	①	0.0001-0.0024	0.0016-0.0037	0.0012	NA	1.6657-1.6697	1.7341-1.7380
2007	1.5	1NZ-FXE	①	0.0001-0.0024	0.0016-0.0037	0.0012	NA	1.6657-1.6697	1.7341-1.7380
2008	1.5	1NZ-FXE	①	0.0001-0.0024	0.0016-0.0037	0.0012	NA	1.6657-1.6697	1.7341-1.7380

NA - Not available

① No 1: 1.3563-1.3569
All others: 0.9035-0.9041

22140_PRIU_C0014

CRANKSHAFT AND CONNECTING ROD SPECIFICATIONS

All measurements are given in inches.

Year	Engine ID/VIN	Engine Displacement Liters	Crankshaft Main Brg. Journal Dia.	Crankshaft Main Brg. Oil Clearance	Crankshaft Shaft End-play	Crankshaft Thrust on No.	Connecting Rod Journal Diameter	Connecting Rod Oil Clearance	Connecting Rod Side Clearance
2006	1NZ-FXE	1.5	1.8106-1.8110	0.0004-0.0028	0.0035-0.0120	NA	1.5745-1.5748	0.0006-0.0024	0.0063-0.0142
2007	1NZ-FXE	1.5	1.8106-1.8110	0.0004-0.0028	0.0035-0.0120	NA	1.5745-1.5748	0.0006-0.0024	0.0063-0.0142
2008	1NZ-FXE	1.5	1.8106-1.8110	0.0004-0.0028	0.0035-0.0120	NA	1.5745-1.5748	0.0006-0.0024	0.0063-0.0142

NA: Not Available

22140_PRIU_C0006

PISTON AND RING SPECIFICATIONS

All measurements are given in inches.

Year	Engine ID/VIN	Engine Displacement Liters	Piston Clearance	Ring Gap Top Compression	Ring Gap Bottom Compression	Ring Gap Oil Control	Ring Side Clearance Top Compression	Ring Side Clearance Bottom Compression	Ring Side Clearance Oil Control
2006	1NZ-FXE	1.5	0.0018-0.0032	0.0079-0.0240	0.0118-0.0472	0.0039-0.0453	0.0008-0.0028	0.0008-0.0024	0.0008-0.0024
2007	1NZ-FXE	1.5	0.0018-0.0032	0.0079-0.0240	0.0118-0.0472	0.0039-0.0453	0.0008-0.0028	0.0008-0.0024	0.0008-0.0024
2008	1NZ-FXE	1.5	0.0018-0.0032	0.0079-0.0240	0.0118-0.0472	0.0039-0.0453	0.0008-0.0028	0.0008-0.0024	0.0008-0.0024

22140_PRIU_C0007

TORQUE SPECIFICATIONS

All readings in ft. lbs.

Year	Engine ID/VIN	Engine Displacement Liters	Cylinder Head Bolts	Main Bearing Bolts	Rod Bearing Bolts	Crankshaft Damper Bolts	Flywheel Bolts	Manifold Intake	Manifold Exhaust	Spark Plugs	Lug Nut
2006	1NZ-FXE	1.5	①	②	③	95	36	15	20	13	76
2007	1NZ-FXE	1.5	①	②	③	95	36	15	20	13	76
2008	1NZ-FXE	1.5	①	②	③	95	36	15	20	13	76

① Step 1: 21 ft. lbs. (29 Nm)
Step 2: turn head bolts 90 degrees
Step 3: turn head bolts 90 degrees

② Step 1: 16 ft. lbs. (22 Nm)
Step 2: turn bearing cap bolts 90 degrees

③ Step 1: 11 ft. lbs. (15 Nm)
Step 2: turn bearing cap bolts 90 degrees

④ Step 1: 62 ft. lbs. (84 Nm)
Step 2: turn flywheel bolts 90 degrees

22140_PRIU_C0008

WHEEL ALIGNMENT

Year	Model		Caster Range (+/-Deg.)	Caster Preferred Setting (Deg.)	Camber Range (+/-Deg.)	Camber Preferred Setting (Deg.)	Toe-in (in.)	Steering Axis Inclination (Deg.)
2006	Prius	F	0.75	+3.17	0.75	-0.58	0+/-0.08	12.58+/-0.75
		R	—	—	0.50	-1.50	0.12+/-0.10	—
2007	Prius	F	0.75	+3.17	0.75	-0.58	0+/-0.08	12.58+/-0.75
		R	—	—	0.50	-1.50	0.12+/-0.10	—
2008	Prius	F	0.75	+3.17	0.75	-0.58	0+/-0.08	12.58+/-0.75
		R	—	—	0.50	-1.50	0.12+/-0.10	—

22140_PRIU_C0009

TIRE, WHEEL AND BALL JOINT SPECIFICATIONS

		OEM Tires		Tire Pressures (psi)		Wheel	Ball Joint
Year	Model	Standard	Optional	Front	Rear	Size	Inspection
2006	Prius	P185/65R15	None	35	33	6-JJ	①
2007	Prius	P185/65R15	None	35	33	6-JJ	①
2008	Prius	P185/65R15	None	35	33	6-JJ	①

OEM: Original Equipment Manufacturer

PSI: Pounds Per Square Inch

① Replace if any measurable movement is found.

22140_PRIU_C0010

BRAKE SPECIFICATIONS

All measurements in inches unless noted

		Brake Disc			Brake Drum Diameter			Minimum Lining Thickness		Brake Caliper	
		Original	Minimum	Maximum	Original Inside	Max. Wear	Max. Machine			Bracket bolts	Mounting bolts
Year	Model	Thickness	Thickness	Runout	Diameter	Limit	Diameter	Front	Rear	(ft. lbs.)	(ft. lbs.)
2006	Prius	0.866	0.787	0.002	7.874	7.913	7.913	0.039	0.039	81	25
2007	Prius	0.866	0.787	0.002	7.874	7.913	7.913	0.039	0.039	81	25
2008	Prius	0.866	0.787	0.002	7.874	7.913	7.913	0.039	0.039	81	25

22140_PRIU_C0011

SCHEDULED MAINTENANCE INTERVALS
TOYOTA—PRIUS

		VEHICLE MILEAGE INTERVAL (x1000)												
TO BE SERVICED	TYPE OF SERVICE	7.5	15	22.5	30	37.5	45	52.5	60	67.5	75	82.5	90	97.5
Engine oil & filter	R	✓	✓	✓	✓	✓	✓	✓	✓	✓	✓	✓	✓	✓
Hybrid transaxle fluid	S/I	✓	✓	✓	✓	✓	✓	✓	✓	✓	✓	✓	✓	✓
Drive axle boots	S/I	✓	✓	✓	✓	✓	✓	✓	✓	✓	✓	✓	✓	✓
Gear shift control operation	S/I	✓	✓	✓	✓	✓	✓	✓	✓	✓	✓	✓	✓	✓
Inspect & rotate tires	S/I	✓	✓	✓	✓	✓	✓	✓	✓	✓	✓	✓	✓	✓
Power steering system	S/I	✓	✓	✓	✓	✓	✓	✓	✓	✓	✓	✓	✓	✓
Suspension system	S/I	✓	✓	✓	✓	✓	✓	✓	✓	✓	✓	✓	✓	✓
Brake discs & pads	S/I	✓		✓		✓		✓		✓		✓		✓
Brake shoes & drums	S/I	✓		✓		✓		✓		✓		✓		✓
Brake hoses & pipes	S/I	✓		✓		✓		✓		✓		✓		✓
Brake fluid	S/I		✓		✓		✓		✓		✓		✓	
Brake pedal	S/I		✓		✓		✓		✓		✓		✓	
Cooling system, hoses & connections	S/I		✓		✓		✓		✓		✓		✓	
Fuel tank, cap & lines	S/I		✓		✓		✓		✓		✓		✓	
Air cleaner filter element	R				✓				✓				✓	
Engine coolant	R				✓				✓				✓	
Spark plugs	R				✓				✓				✓	
Drive belt	S/I				✓				✓				✓	
Exhaust system	S/I				✓				✓				✓	

R: Replace S/I: Service or Inspect

① Replace every 60,000 miles.

FREQUENT OPERATION MAINTENANCE (SEVERE SERVICE)

If a vehicle is operated under any of the following conditions it is considered severe service:

- Extremely dusty areas.

- 50% or more of the vehicle operation is in 32°C (90°F) or higher temperatures, or constant operation in temperatures below 0°C (32°F).

- Prolonged idling (vehicle operation in stop and go traffic).

- Frequent short running periods (engine does not warm to normal operating temperatures).

- Police, taxi, delivery usage or trailer towing usage.

Oil & oil filter: change every 3000 miles.

Brake discs & pads: service or inspect initially at 3000 miles, 6000 miles, & every 12,000 miles thereafter.

Brake hoses & pipes: service or inspect initially at 3000 miles, 6000 miles & every 12,000 miles thereafter.

Air cleaner filter element: service or inspect ever 3000 miles & replace every 30,000 miles (if not replaced previously).

Hybrid transaxle fluid: service or inspect every 6000 miles & replace every 15,000 miles (if not replaced previously).

Inspect & rotate tires: service or inspect every 6000 miles.

Power steering system: service or inspect every 6000 miles.

Steering system: service or inspect every 6000 miles.

Suspension system: service or inspect every 6000 miles.

Drive belts: service or inspect every 15,000 miles.

Exhaust system: service or inspect every 15,000 miles.

22140_PRIU_C0012

PRECAUTIONS

Before servicing any vehicle, please be sure to read all of the following precautions, which deal with personal safety, prevention of component damage, and important points to take into consideration when servicing a motor vehicle:

- Never open, service or drain the radiator or cooling system when the engine is hot; serious burns can occur from the steam and hot coolant.
- Observe all applicable safety precautions when working around fuel. Whenever servicing the fuel system, always work in a well-ventilated area. Do not allow fuel spray or vapors to come in contact with a spark, open flame, or excessive heat (a hot drop light, for example). Keep a dry chemical fire extinguisher near the work area. Always keep fuel in a container specifically designed for fuel storage; also, always properly seal fuel containers to avoid the possibility of fire or explosion. Refer to the additional fuel system precautions later in this section.
- Fuel injection systems often remain pressurized, even after the engine has been turned **OFF**. The fuel system pressure must be relieved before disconnecting any fuel lines. Failure to do so may result in fire and/or personal injury.
- Brake fluid often contains polyglycol ethers and polyglycols. Avoid contact with the eyes and wash your hands thoroughly after handling brake fluid. If you do get brake fluid in your eyes, flush your eyes with clean, running water for 15 minutes. If eye irritation persists, or if you have taken brake fluid internally, IMMEDIATELY seek medical assistance.
- The EPA warns that prolonged contact with used engine oil may cause a number of skin disorders, including cancer. You should make every effort to minimize your exposure to used engine oil. Protective gloves should be worn when changing oil. Wash your hands and any other exposed skin areas as soon as possible after exposure to used engine oil. Soap and water, or waterless hand cleaner should be used.
- All new vehicles are now equipped with an air bag system, often referred to as a Supplemental Restraint System (SRS) or Supplemental Inflatable Restraint (SIR) system. The system must be disabled before performing service on or around system components, steering column, instrument panel components, wiring and sensors. Failure to follow safety and disabling procedures could result in accidental air bag deployment, possible personal injury and unnecessary system repairs.
- Always wear safety goggles when working with, or around, the air bag system. When carrying a non-deployed air bag, be sure the bag and trim cover are pointed away from your body. When placing a non-deployed air bag on a work surface, always face the bag and trim cover upward, away from the surface. This will reduce the motion of the module if it is accidentally deployed. Refer to the additional air bag system precautions later in this section.
- Clean, high quality brake fluid from a sealed container is essential to the safe and proper operation of the brake system. You should always buy the correct type of brake fluid for your vehicle. If the brake fluid becomes contaminated, completely flush the system with new fluid. Never reuse any brake fluid. Any brake fluid that is removed from the system should be discarded. Also, do not allow any brake fluid to come in contact with a painted surface; it will damage the paint.
- Never operate the engine without the proper amount and type of engine oil; doing so WILL result in severe engine damage.
- Timing belt maintenance is extremely important. Many models utilize an interference-type, non-freewheeling engine. If the timing belt breaks, the valves in the cylinder head may strike the pistons, causing potentially serious (also time-consuming and expensive) engine damage. Refer to the maintenance interval charts for the recommended replacement interval for the timing belt, and to the timing belt section for belt replacement and inspection.
- Disconnecting the negative battery cable on some vehicles may interfere with the functions of the on-board computer system(s) and may require the computer to undergo a relearning process once the negative battery cable is reconnected.
- When servicing drum brakes, only disassemble and assemble one side at a time, leaving the remaining side intact for reference.
- Only an MVAC-trained, EPA-certified automotive technician should service the air conditioning system or its components.

BRAKES

ELECTRICALLY CONTROLLED BRAKE (ECB) SYSTEM

** CAUTION

Dust and dirt accumulating on brake parts during normal use may contain asbestos fibers from production or aftermarket brake linings. Breathing excessive concentrations of asbestos fibers can cause serious bodily harm. Exercise care when servicing brake parts. Do not sand or grind brake lining unless equipment used is designed to contain the dust residue. Do not clean brake parts with compressed air or by dry brushing. Cleaning should be done by dampening the brake components with a fine mist of water, then wiping the brake components clean with a dampened cloth. Dispose of cloth and all residue containing asbestos fibers in an impermeable container with the appropriate label. Follow practices prescribed by the Occupational Safety and Health Administration (OSHA) and the Environmental Protection Agency (EPA) for the handling, processing, and disposing of dust or debris that may contain asbestos fibers.

GENERAL INFORMATION

Toyota Hybrid System II (THS-II) is used on these vehicles. THS-II operates on the same basic concepts as the original Toyota THS system, but incorporates improvements to the motor generators, the battery, and the engine.

These vehicles also use:

- Antilock Brake System (ABS): Pressure to each of the four brake cylinders can be individually modulated. Pressure sensors within the brake actuator monitor the individual wheel circuits.
- Electrically Controlled Brake (ECB): Master cylinder pressure (from the driver's foot on the brake pedal) is used as a signal and does not directly apply the brakes. Instead, all braking pressure (normal and ABS) is developed by a pump and accumulator. Pressure sensors monitor brake master cylinder output and the accumulator circuit.
- Electronic Brake Force Distribution (EBD): Modifies braking force at individual wheels to compensate both for changes in vehicle load (occupants and cargo) and to

increase braking efficiency during cornering.

- Brake Assist (BA): Monitors the speed with which the driver depresses the brake pedal and provides additional brake fluid pressure during emergency stops.
- Traction Control (TRAC): Helps prevent drive wheel spin during acceleration using ABS and control of engine power output.
- Vehicle Stability Control (VSC): Helps prevent skidding or spinning while cornering. VSC controls ABS and engine power output to help keep the vehicle traveling in the intended direction, even if the driver enters a turn too fast or steers the vehicle too sharply for road conditions.
- Vehicle Dynamics Integration management (VDIM): Combines a range of vehicle stability control functions into a single, seamless process for overall vehicle dynamic driveability. In addition to the braking control offered by ABS, VDIM controls elements of transmission and engine operation to completely manage the manner in which a vehicle is maneuvered.

SERVICE PRECAUTIONS

- Do not remove or install the Enhanced VSC or Electronically Controlled Brake parts such as the steering sensor, yaw rate sensor or brake pedal stroke sensor except when required, as they need to be adjusted correctly after removal or installation.
- Be sure to perform preparation before work and confirmation after work is completed by following the directions in the repair manual when working on the Enhanced VSC or Electronically Controlled Brake system.
- Be sure to remove and install the ECU, actuator, each sensor, etc. with the power switch OFF unless otherwise specified in the inspection procedure.
- Be sure to remove the 2 main relays before removal and installation, or replacement of the Enhanced VSC or Electronically Controlled Brake parts.
- The removal or installation of the actuator, master cylinder or stroke simulator as well as some other procedures can cause the fluid level to drop below the fluid reservoir port. If this happens when performing such work, be sure to remove the 2 motor relays until the bleeding of the air in the pipeline is completed.
- When the pump motor is operated with the air in the brake actuator hose, bleeding the air becomes difficult due to air in the actuator.
- The skid control ECU may operate the stroke simulator and drive the pump motor even when the power switch is OFF.
- The Electronically Controlled Brake system has its own auxiliary power source. This system can be operated after disconnecting the negative terminal from the auxiliary battery (12 V) until the discharge is completed.
- With the power switch OFF, the skid control ECU can be operated for 2 minutes after the brake operation is finished.
- Wait for 2 minutes after turning the power switch OFF, stopping the brake pedal operation and closing the driver door before removing the main relay and motor relay.
- The above situation occurs when the pump motor operates to prepare for the next operation just before brake control system turns off.
- When removing and installing the ECU, actuator and each sensor, be sure to check that the normal display is output in Test Mode inspection and in DTC output inspection after installing all the parts.

BRAKES — BLEEDING THE BRAKE SYSTEM

BLEEDING PROCEDURE

Bleeding the brakes requires the use of a Toyota Techstream, or equivalent scan tool.

Follow tool manufacturer's procedures when bleeding the hydraulic brake system.

BRAKES — FRONT DISC BRAKES

BRAKE CALIPER

REMOVAL & INSTALLATION

See Figures 1 and 2.

1. Before servicing the vehicle, refer to the precautions.
2. With the power switch **OFF** to prohibit brake control, remove the No. 1 and No. 2 motor relays.

➡If the pump motor operates while there is air remaining inside the brake actuator hose, the air will enter the actuator, resulting in difficulty in air bleeding.

3. Raise and safely support the front of the vehicle.
4. Remove the front wheel.
5. Remove the bolt and the gasket attaching the brake hose to the caliper.
6. Disconnect the brake hose, and plug the openings in the caliper and the brake hose to prevent fluid loss and contamination.
7. Hold the front disc brake cylinder slide pin using a wrench.
8. Remove the 2 caliper mounting bolts and remove the caliper from the vehicle.
9. Remove the 2 disc brake pads from the front disc brake cylinder mounting.
10. Remove the anti-squeal shims from the disc brake pads.
11. Remove the front disc brake pad support plates.
12. Remove the slide pins and bushing dust boots from the front disc brake cylinder mounting, if necessary.
13. Remove the front disc brake cylinder mounting from the steering knuckle, if necessary.

To install:

14. Install the front disc brake cylinder mounting to the steering knuckle, if

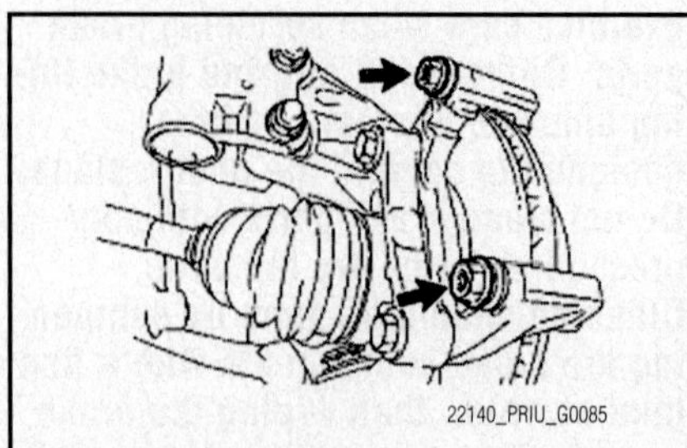

Fig. 1 Remove the front disc brake cylinder slide pins from the mounting

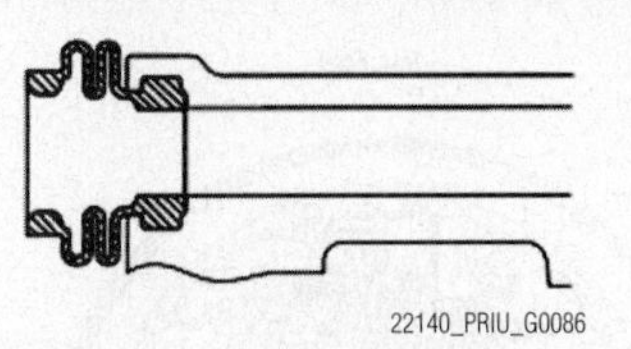

Fig. 2 Apply lithium soap base glycol grease to seal surface of 2 new bushing dust boots and install on to the front disc brake cylinder mounting

removed. Torque the 2 mounting bolts to 81 ft. lbs. (109 Nm).

15. Install new bushing dust boots and slide pins into the front disc brake cylinder mounting, if necessary. Apply lithium soap base glycol grease to the sealing, sliding and fitting areas before installation.
16. Install the front disc brake pad support plates.
17. Install the anti-squeal shims onto the disc brake pads.
18. Install the disc brake pads onto the front disc brake cylinder mounting.
19. Install the brake caliper onto the mounting along with the 2 caliper mounting bolts. Torque the 2 mounting bolts to 25 ft. lbs (34 Nm).
20. Connect the brake hose to the caliper with bolt and a new gasket. Tighten the caliper brake hose bolt to 24 ft. lbs. (33 Nm).
21. Fill the master cylinder to the proper level with clean brake fluid.
22. Bleed air from front and rear brake systems.
23. Recheck the fluid level.
24. Install the wheel and lower the vehicle.
25. Repeatedly press the brake pedal to bring the pads in contact with the rotor.
26. Check and clear the DTCs.

DISC BRAKE PADS

REMOVAL & INSTALLATION

1. Before servicing the vehicle, refer to the precautions.
2. Raise and safely support the front of the vehicle.
3. Remove the front wheel.
4. Hold the front disc brake cylinder slide pin using a wrench.

➡Caliper removal is not necessary to service the brake pads.

5. Remove the 2 caliper mounting bolts and separate the caliper from the mounting and support from the vehicle with wire. Do not allow the caliper to hang low enough so to put tension on the brake hose.
6. Remove the brake pads.

To install:

7. Install the anti-squeal shims onto the disc brake pads.
8. Install the disc brake pads.
9. Install the brake caliper onto the mounting along with the 2 caliper mounting bolts. Torque the 2 mounting bolts to 25 ft. lbs (34 Nm).
10. Install the wheel.
11. Lower the vehicle.
12. Repeatedly press the brake pedal to bring the pads in contact with the rotor.

BRAKES

REAR DRUM BRAKES

BRAKE DRUM

REMOVAL & INSTALLATION

See Figure 3.

1. Before servicing the vehicle, refer to the precautions.
2. Raise and safely support the rear of the vehicle.
3. Remove the rear wheels.
4. Put match marks on the rear brake drum and the axle hub.
5. Release the parking brake and remove the brake drum.
6. If the brake drum cannot be removed easily, do the following steps.
   a. Remove the plug and insert a screwdriver through the hole in the backing plate.
   b. Using another screwdriver, reduce the brake shoe adjuster by turning the adjusting wheel.

To install:

7. Aligning the match marks, install the brake drum.
8. Adjust the drum brake shoe clearance.
   a. Temporarily install 2 hub nuts.
   b. Remove the hole plug, and turn the adjuster to expand the shoe until the drum locks up.
   c. Rotate the adjuster back by 8 notches.
   d. Install the hole plug.
   e. Remove the 2 hub nuts.
9. Install the wheel. Torque the wheel lugs to 76 ft. lbs. (103 Nm).
10. Lower the vehicle.
11. Check brake fluid level in the reservoir and add to proper level with clean brake fluid, if necessary.

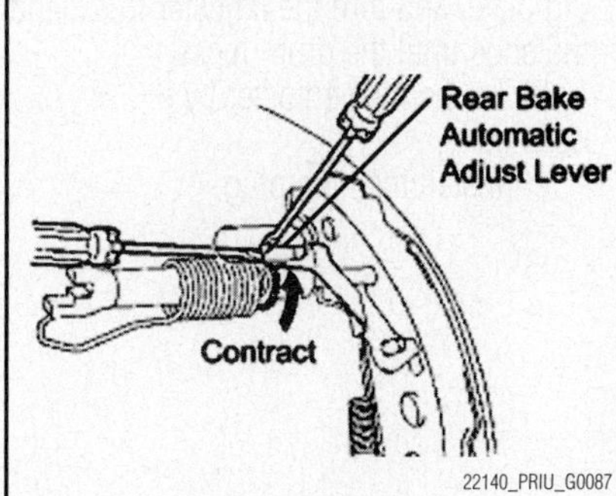

Fig. 3 Release the parking brake and remove the rear brake drum

BRAKE SHOES

REMOVAL & INSTALLATION

See Figures 4 through 7.

1. Before servicing the vehicle, refer to the precautions.
2. Remove the rear wheels.
3. Remove the rear brake drums.
4. Disconnect the spring and parking brake shoe adjuster set from the brake shoes.
5. Remove the cup, shoe hold-down spring and pin.
6. Remove front brake shoe.
7. Remove the parking brake shoe adjuster set from the rear brake shoe.
8. Remove the automatic adjusting lever spring and automatic adjusting lever.
9. Remove anchor spring from the rear brake shoe.
10. Remove the cup, shoe hold-down spring and pin.
11. Using needle-nose pliers, disconnect the parking brake cable, and remove the rear brake shoe.

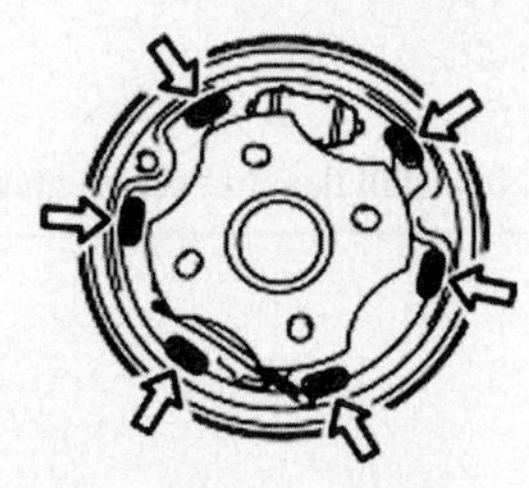

Fig. 4 Apply high temperature grease to the shoe attached surface of the backing plate

12. Using a screwdriver, remove the C-washer and parking brake lever.

To install:

13. Apply high-temperature grease to the shoe attached surface of the backing plate.

➡There should be no oil or grease adhering to the friction surfaces of the shoe lining and the drum.

14. Using needle-nose pliers, install the parking brake lever to the rear brake shoe with a new C-washer.
15. Using needle-nose pliers, connect the parking brake cable to the parking brake lever.
16. Install the rear brake shoe, pin, shoe hold-down spring and cup.
17. Install the automatic adjusting lever and automatic adjusting lever spring to the front brake shoe.

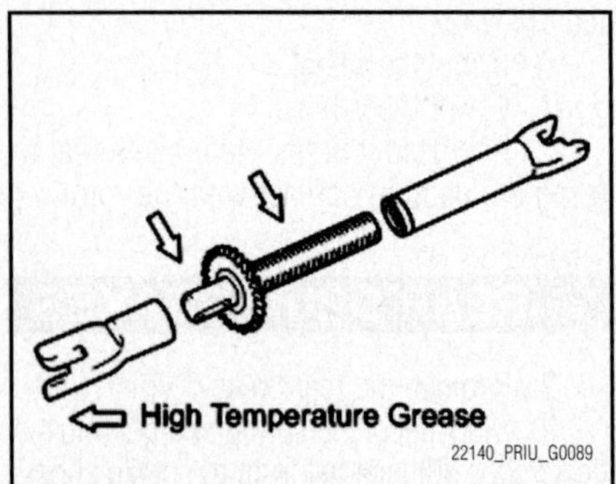

Fig. 5 Apply high temperature grease to the adjusting bolt and assemble the strut set

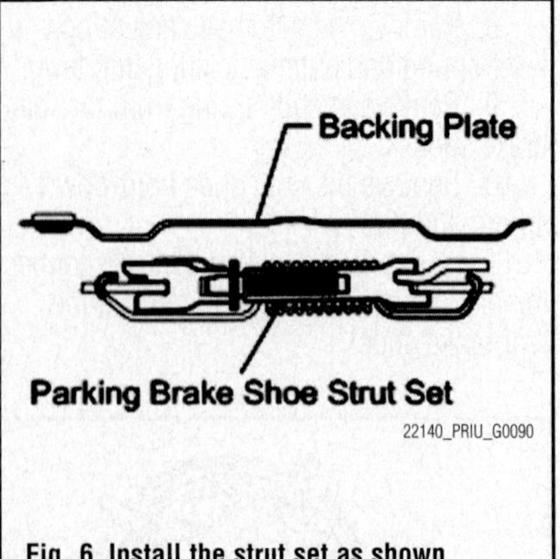

Fig. 6 Install the strut set as shown

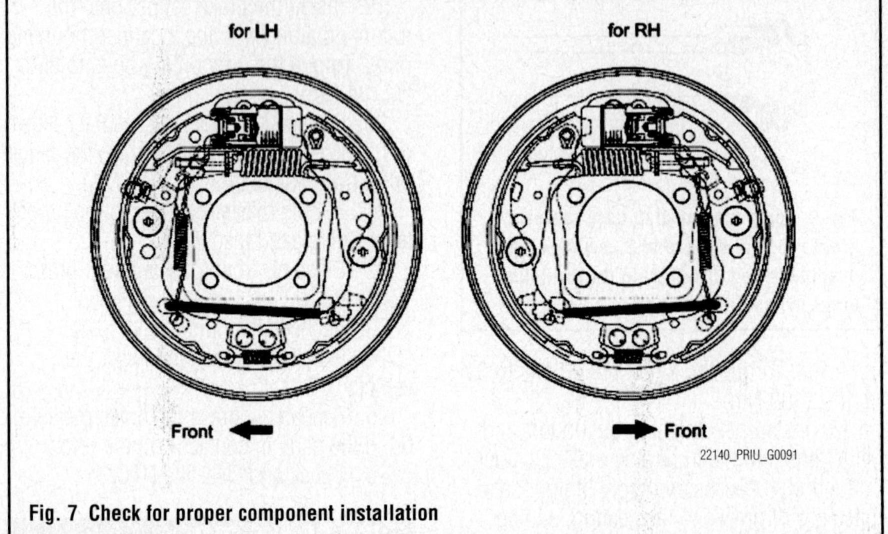

Fig. 7 Check for proper component installation

18. Apply high-temperature grease to the adjusting bolt and assemble the parking brake shoe adjuster set.
19. Install the rear brake shoe return spring to the adjuster set and install set.
20. Connect the anchor spring to the front and rear brake shoes.
21. Install the front brake shoe, pin, shoe hold-down spring and cup.
22. Connect the shoe return spring to the front and rear brake shoes.
23. Check to make sure that all the components of the brake shoe assembly are installed properly.
24. Install the brake drums.
25. Install the wheels.
26. Lower the vehicle.
27. Press the brake pedal 3–5 times to adjust the brake shoe clearance.

ADJUSTMENT

See Figure 8.

1. With the brake drum off the vehicle:
   a. Measure the brake drum inside diameter and the diameter of the brake shoes using a brake shoe caliper.
   b. Check that the difference between the diameter is 0.024 in. (0.6 mm).
   c. Adjust the brake shoes as necessary.

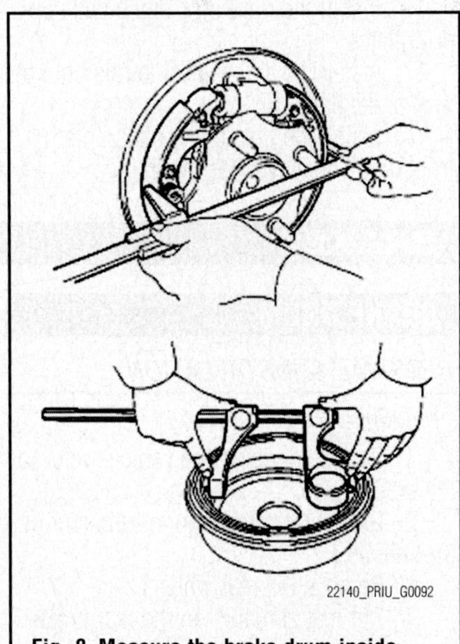

Fig. 8 Measure the brake drum inside diameter and the diameter of the brake shoes using a brake shoe caliper

2. With the brake drum installed on the vehicle:
   a. Remove the hole plug in the backing plate, and turn the adjuster to expand the shoe until the drum locks.
   b. Loosen the adjuster by 8 notches.
   c. Install the hole plug.

BRAKES

PARKING BRAKE

PARKING BRAKE SHOES

REMOVAL & INSTALLATION

The rear drum brake shoes serve as the parking brakes. Refer to Rear Drum Brakes, Brake Shoes, Removal & Installation.

ADJUSTMENT

The rear drum brake shoes serve as the parking brakes. Refer to Rear Drum Brakes, Brake Shoes, Adjustment.

CHASSIS ELECTRICAL

AIR BAG (SUPPLEMENTAL RESTRAINT SYSTEM)

GENERAL INFORMATION

⁂ CAUTION

Some vehicles are equipped with an air bag system. The system must be disarmed before performing service on, or around, system components, the steering column, instrument panel components, wiring and sensors. Failure to follow the safety precautions and the disarming procedure could result in accidental air bag deployment, possible injury and unnecessary system repairs.

SERVICE PRECAUTIONS

Disconnect and isolate the battery negative cable before beginning any airbag system component diagnosis, testing, removal, or installation procedures. Allow system capacitor to discharge for two minutes before beginning any component service. This will disable the airbag system. Failure to disable the airbag system may result in accidental airbag deployment, personal injury, or death.

Do not place an intact undeployed airbag face down on a solid surface. The airbag will propel into the air if accidentally deployed and may result in personal injury or death.

When carrying or handling an undeployed airbag, the trim side (face) of the airbag should be pointing towards the body to minimize possibility of injury if accidental deployment occurs. Failure to do this may result in personal injury or death.

Replace airbag system components with OEM replacement parts. Substitute parts may appear interchangeable, but internal differences may result in inferior occupant protection. Failure to do so may result in occupant personal injury or death.

Wear safety glasses, rubber gloves, and long sleeved clothing when cleaning powder residue from vehicle after an airbag deployment. Powder residue emitted from a deployed airbag can cause skin irritation. Flush affected area with cool water if irritation is experienced. If nasal or throat irritation is experienced, exit the vehicle for fresh air until the irritation ceases. If irritation continues, see a physician.

Do not use a replacement airbag that is not in the original packaging. This may result in improper deployment, personal injury, or death.

The factory installed fasteners, screws and bolts used to fasten airbag components have a special coating and are specifically designed for the airbag system. Do not use substitute fasteners. Use only original equipment fasteners listed in the parts catalog when fastener replacement is required.

During, and following, any child restraint anchor service, due to impact event or vehicle repair, carefully inspect all mounting hardware, tether straps, and anchors for proper installation, operation, or damage. If a child restraint anchor is found damaged in any way, the anchor must be replaced. Failure to do this may result in personal injury or death.

Deployed and non-deployed airbags may or may not have live pyrotechnic material within the airbag inflator.

Do not dispose of driver/passenger/curtain airbags or seat belt tensioners unless you are sure of complete deployment. Refer to the Hazardous Substance Control System for proper disposal.

Dispose of deployed airbags and tensioners consistent with state, provincial, local, and federal regulations.

After any airbag component testing or service, do not connect the battery negative cable. Personal injury or death may result if the system test is not performed first.

If the vehicle is equipped with the Occupant Classification System (OCS), do not connect the battery negative cable before performing the OCS Verification Test using the scan tool and the appropriate diagnostic information. Personal injury or death may result if the system test is not performed properly.

Never replace both the Occupant Restraint Controller (ORC) and the Occupant Classification module (OCM) at the same time. If both require replacement, replace one, then perform the Airbag System test before replacing the other.

Both the ORC and the OCM store Occupant Classification System (OCS) calibration data, which they transfer to one another when one of them is replaced. If both are replaced at the same time, an irreversible fault will be set in both modules and the OCS may malfunction and cause personal injury or death.

If equipped with OCS, the Seat Weight Sensor is a sensitive, calibrated unit and must be handled carefully. Do not drop or handle roughly. If dropped or damaged, replace with another sensor. Failure to do so may result in occupant injury or death.

If equipped with OCS, the front passenger seat must be handled carefully as well. When removing the seat, be careful when setting on floor not to drop. If dropped, the sensor may be inoperative, could result in occupant injury, or possibly death.

If equipped with OCS, when the passenger front seat is on the floor, no one should sit in the front passenger seat. This uneven force may damage the sensing ability of the seat weight sensors. If sat on and damaged, the sensor may be inoperative, could result in occupant injury, or possibly death.

DISARMING THE SYSTEM

To avoid personal injury when working on vehicles equipped with an air bag, the negative battery cable must be disconnected and at least 90 seconds must elapse before working on the system. Failure to do so may result in deployment of the air bag.

ARMING THE SYSTEM

To rearm the air bag system, reconnect the battery cable (s).

CLOCKSPRING CENTERING

See Figure 9.

1. Before servicing the vehicle, refer to the precautions.
2. Install the spiral cable.
   a. Check that the front wheels are facing straight ahead.
   b. Set the turn signal switch to the neutral position.

➡Make sure that the turn signal switch is in the neutral position, as the pin of the turn signal switch may be snapped.

   c. Install the spiral cable.

➡When replacing the spiral cable with a new one, remove the lock pin before installing the steering wheel.

d. Connect the connector to the spiral cable.

⁂ WARNING

When handling the airbag connector, do not damage the airbag wire harness.

3. Slowly rotate the spiral cable counterclockwise by hand until it feels firm.
4. Rotate the spiral cable clockwise approximately 2.5 turns to align the marks.
5. The spiral cable should rotate approximately 2.5 turns to the left and right from the center.

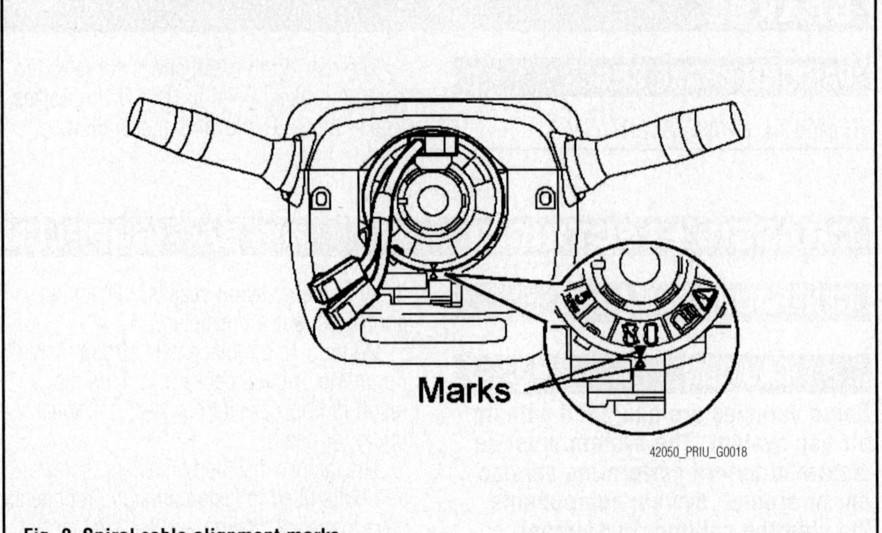

Fig. 9 Spiral cable alignment marks

DRIVETRAIN

AUTOMATIC TRANSAXLE ASSEMBLY

REMOVAL & INSTALLATION

See Figures 10 through 12.

1. Before servicing the vehicle, refer to the precautions.
2. Disconnect the negative battery cable.

➡Wait at least 90 seconds after disconnecting the cable from the negative (-) battery terminal to prevent airbag and seat belt pretensioner activation.

3. While wearing insulating gloves, slide up the lever of the service plug grip and remove while turning the lever to the left. Be sure to insulate the service plug with insulating tape.

⁂ CAUTION

Do not touch the high voltage connectors and terminals for 5 minutes after removing the service plug grip.

4. Drain engine coolant and transaxle fluid.
5. Remove the front wheels.
6. Remove the engine under covers.
7. Remove the hood.
8. Remove the left and right front wiper arms.
9. Remove the cowl top front panel.
10. While wearing insulating gloves, remove the inverter with converter assembly.
    a. Remove the left and right engine under covers.

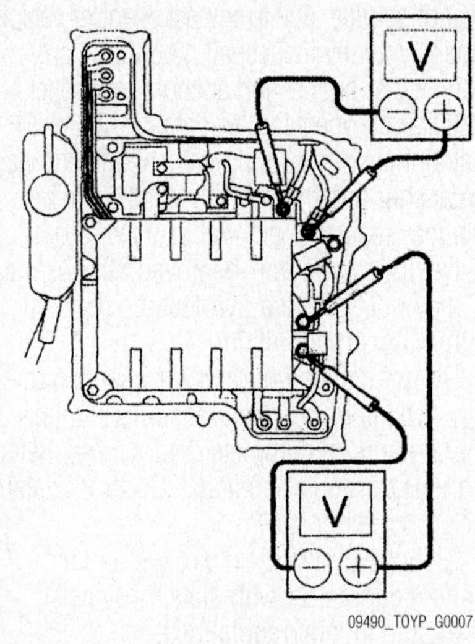

Fig. 10 Verifying that there is 0 volt (s) at the inverter with converter

b. Drain the high voltage coolant.
c. Remove the radiator support opening cover.
d. Remove the inverter cover.
e. Verify that there is 0 volt (s) at the inverter with converter, using a voltmeter utilizing a measuring range of DC 400 volt (s) or more.
f. Again, verify that there is 0 volt (s) by measuring the voltage between the terminals of the three phase connector (U-V, V-W, U-W), using a voltmeter utilizing a measuring range of DC 400 volt (s) or more.
g. Disconnect the No. 1, No. 2 and No. 6 inverter cooling hoses.
h. Disconnect the No. 1 circuit breaker sensor by first moving the outer section away toward the wire side.
i. Disconnect the 2 frame wire connectors from the inverter with converter assembly and protect the electrode and connector parts with insulating tape.
j. Use a small screwdriver to lift up the green lock pin, then disconnect the connector for the air conditioning inverter.
k. Disconnect any remaining wiring connectors including the engine main wiring harness.
l. Remove the mounting bolts and disconnect the MG1 and MG2 power cables. Protect the connector parts with insulating tape.

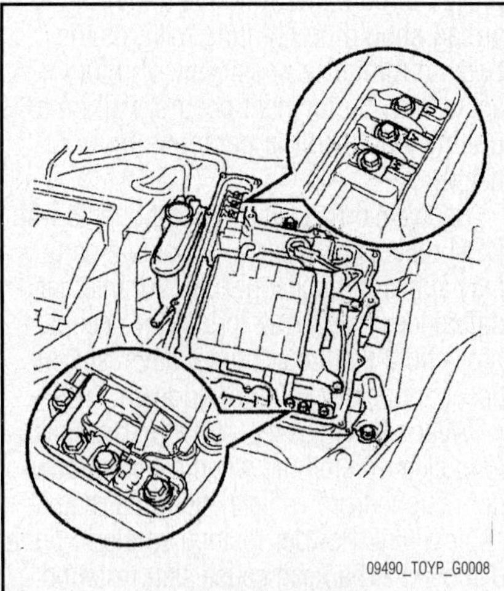

Fig. 11 Measuring voltage between the terminals of the three phase connector

m. Remove the 3 mounting bolts and the inverter with converter assembly.

11. Remove the air cleaner assembly.

12. Remove the associated wiring harnesses and clamps.

13. Remove the 3 bolts and transmission case cover from the transaxle.

14. Remove the bolt and ground wire.

15. Remove the 4 clamps and 3 cooling hoses from inverter.

16. Remove the front exhaust pipe.

17. Remove the left and right front axle hub nuts.

18. Remove the left front stabilizer link.

19. Remove the left and right tie rod ends.

20. Remove the left and right front lower control arms.

21. Remove the left and right front axle hub/bearing assemblies.

22. Remove the left and right drive shafts.

23. Remove the engine torque rod.

24. Remove the front suspension crossmember.

25. Remove the 2 PCV hoses.

26. Install the No. 1 (12281-22021) and No. 2 (12281-15040) engine hangers in the correct direction. Torque the hanger bolt (91512-B1016) to 28 ft. lbs. (38 Nm).

27. Attach an engine chain hoist to the engine hangers to keep the engine suspended.

WARNING

Do not attempt to hang the engine by hooking the chain to any other parts.

28. Using a transmission jack, support the hybrid vehicle transaxle.

29. Remove the nut and disconnect the No. 3 engine mounting bracket from the engine mounting insulator.

30. Remove the 2 bolts, starter cover and housing side cover.

31. Remove the 6 bolts and transaxle unit.

➡To avoid damage to the knock pin, do not pry the connecting portion of the high voltage transaxle and the engine.

32. Remove the No. 6 clamp and inverter cooling hose.

33. Remove the 3 bolts and No. 3 engine mounting bracket.

34. Remove the 6 bolts and clamp.

To install:

35. Install the clamp in position with the 6 bolts and tighten to 80 inch lbs. (9 Nm).

36. Install the No. 3 engine mounting bracket with the 3 bolts and tighten to 38 ft. lbs. (52 Nm).

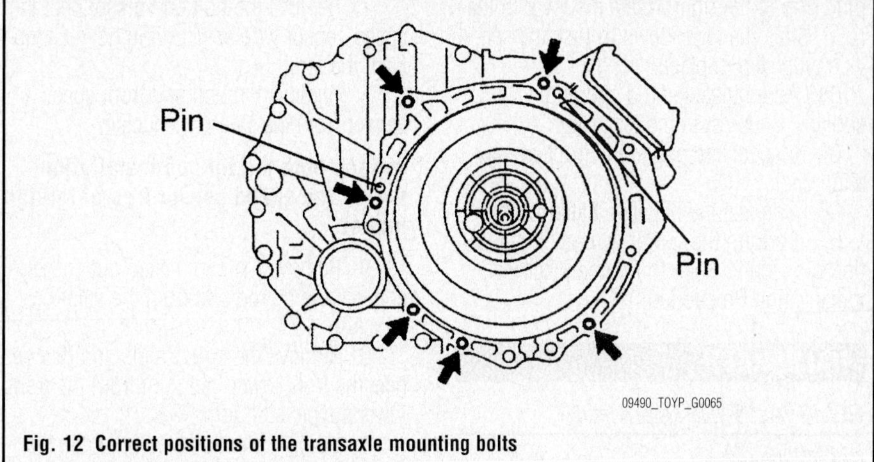

Fig. 12 Correct positions of the transaxle mounting bolts

37. Connect the No. 6 inverter cooling hose and install the clamp.

38. Install the transaxle to the engine in the vehicle. Tighten the 6 bolts to 24 ft. lbs. (33 Nm).

➡Ensure that the knock pin is installed on the engine side. Place the transaxle in a horizontal position and align the knock pin to its hole. Then tighten the 6 bolts in the correct positions.

39. Install the housing side cover and starter cover and tighten the 2 bolts to 23 ft. lbs. (32 Nm).

40. Install the No. 3 engine mounting bracket to the engine mounting insulator with the nut. Torque the nut to 59 ft. lbs. (80 Nm).

41. Remove the engine hangers.

42. Install the 2 PCV hoses.

43. Install the front suspension crossmember. Torque the left and right front corner bolts to 83 ft. lbs. (113 Nm). Torque the left and right rear corner bolts to 116 ft. lbs. (157 Nm).

44. Install the engine torque rod. Torque the 2 nuts and 2 bolts, at the engine, to 74 ft. lbs. (100 Nm). Torque the 2 nuts and 2 bolts, at the body side, to 41 ft. lbs. (56 Nm).

45. Install the left and right drive shafts.

46. Install the left and right front axle hub/bearing assemblies.

47. Install the left and right front lower control arms.

48. Install the left and right tie rod ends.

49. Install the left front stabilizer link.

50. Install the left and right front axle hub nuts.

51. Install the front exhaust pipe.

52. Install the 3 cooling hoses to the inverter with the 4 clamps.

53. Install the ground wire and bolt tightened to 80 inch lbs. (9 Nm).

54. Install the transmission case cover to the transaxle with 3 bolts tightened to 62 inch lbs. (7 Nm).

55. Install the associated wiring harnesses and clamps.

56. Install the air cleaner assembly.

57. While wearing insulating gloves, install the inverter with converter assembly.

a. Install the inverter with converter assembly and tighten the three mounting bolts to 16 ft. lbs. (21 Nm).

b. Connect the MG1 and MG2 power cables and tighten the mounting bolts to 71 inch lbs. (8 Nm).

c. Connect any remaining wiring connectors including the engine main wiring harness. Be sure to insert the grommet of the engine main wiring harness into the U-shaped groove of the inverter case.

d. Engage the connector for the air conditioning inverter and secure by pushing in the lock pin.

e. Connect the 2 frame wire connectors to the inverter with converter assembly.

f. Connect the No. 1 circuit breaker sensor.

g. Connect the No. 6, No. 2 and No. 1 inverter cooling hoses.

h. Install the inverter cover and tighten the mounting fasteners to 8 ft. lbs. (11 Nm).

i. Install the radiator support opening cover.

j. Install the left and right engine under covers.

58. Install the cowl top front panel.

59. Install the left and right front wiper arms.

60. Install the hood.

61. Install the engine under covers.

62. Install the front wheels.

63. While wearing insulating gloves, joint the service plug grip with the high voltage battery. While pushing the service

plug grip to the right, rotate the lever to the right. Slide the lever down to lock the service plug grip in place.

64. Add transaxle fluid and engine coolant. Check systems for leaks.

65. Inspect and adjust the front wheel alignment.

66. Connect the negative battery cable.

67. Perform the power window initialization procedure. Refer to Engine Electrical, Initialization Procedure.

HALFSHAFT

REMOVAL & INSTALLATION

See Figure 13.

1. Before servicing the vehicle, refer to the precautions.
2. Drain transaxle fluid.
3. After draining the fluid, tighten the drain plug, along with a new gasket to 29 ft. lbs. (39 Nm).
4. Raise and safely support the vehicle.
5. Remove the front wheel.
6. Unstake the axle hub nut and using a 30mm socket, remove the axle hub nut.

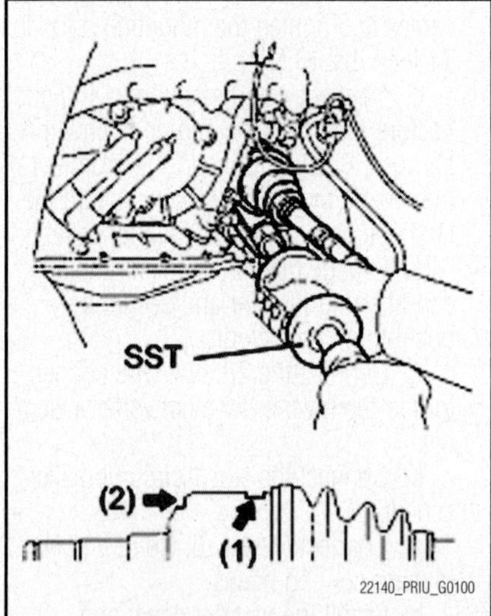

Fig. 13 Hook the special tool SST: 09520-01010 and 09520-24010 claw in position to remove the front drive shaft

7. Remove the bolt and disconnect the speed sensor wire and flexible hose clamp from the strut.
8. Remove the bolt and front speed sensor from the steering knuckle.

➡Keep both the tip and installation part of the speed sensor free of foreign matter.

9. Remove the clip, castle nut and disconnect the tie rod end from the steering knuckle.
10. Remove the bolt, 2 nuts and disconnect the front lower ball joint from the front lower suspension arm.
11. Using a plastic-faced hammer, tap the end of the front drive shaft and disengage the fitting between the front drive shaft and front axle. If it is difficult to disengage, tap the end of the front drive shaft with a brass bar and hammer.
12. Push the front axle outward from the vehicle to remove the front drive shaft from the front axle. Be careful not to push the front axle outward from the vehicle more than necessary to remove it.

➡Be careful not to damage the rubber boots. Hang the drive shaft down with a string or equivalent.

13. Remove the front fender apron seal.
14. Hook the SST 09520-01010 and 09520-24010 claw in position to remove the front drive shaft.

⁂ CAUTION

Be careful not to damage the oil seal. Be careful not to damage the front drive shaft boot. Be careful not to drop the front drive shaft.

15. Check for noticeable looseness when turning the joint up and down, left and right, and in the thrust direction.
16. Check for cracks, damage or grease leaks on the joint boot.

➡Carry the drive shaft levelly.

To install:

17. Apply ATF to the spline of the inboard joint.
18. Align the spline of the front drive shaft and insert the front drive shaft using a brass bar and hammer.

➡Face the snap ring cut area downward. Be careful not to damage the oil seal. Be careful not to damage the front drive shaft boot.

19. Install the front fender apron seal.
20. Install the front drive shaft dust cover (right drive shaft).
21. Push the front axle outward from the vehicle to align the spline of the front drive shaft with the front axle and insert.
22. Connect the front suspension lower arm to the front lower ball joint and tighten the bolt and 2 nuts to 66 ft. lbs. (89 Nm).
23. Connect the tie rod end to the steering knuckle and install it with the castle nut. Torque the castle nut to 36 ft. lbs. (49 Nm) and install a new cotter pin.

➡The cotter pin hole alignment should be done after tightening the castle nut up to 60 degrees beyond the torque specification.

24. Connect the front speed sensor wire and flexible hose clamp to the strut with the bolt.
25. Install the front speed sensor to the steering knuckle and tighten the bolt to 71 inch lbs. (8 Nm).
26. Using a 30mm socket wrench, install a new hub nut and tighten to 159 ft. lbs. (216 Nm).
27. Using a chisel and hammer, stake the hub nut.
28. Install the front wheel.
29. Refill the transaxle fluid to the proper level.
30. Connect the negative battery cable.
31. Perform the power window initialization procedure. Refer to Engine Electrical, Initialization Procedure.
32. Check the front wheel alignment.
33. Check the ABS speed sensor signal.

ENGINE COOLING

THERMOSTAT

REMOVAL & INSTALLATION

See Figure 14.

1. Before servicing the vehicle, refer to the precautions.
2. Remove radiator support opening cover.
3. Remove the left engine under cover.
4. Remove the right engine under cover.
5. Drain engine coolant.
6. Remove the 2 nuts and water inlet with radiator outlet hose.
7. Remove the thermostat.
8. Remove the gasket from the thermostat.

To install:

9. Install a new gasket to the thermostat.
10. Install the thermostat so the jiggle valve faces upward.
11. Install the water inlet with radiator outlet hose with the 2 nuts and tighten to 9.0 Nm (80 in. lbs)
12. Add engine coolant.
13. Check for engine coolant leaks.
14. Install the right engine under cover.
15. Install the left engine under cover.
16. Install radiator support opening cover.

Fig. 14 Positioning the thermostat

WATER PUMP

REMOVAL & INSTALLATION

See Figure 15.

1. Before servicing the vehicle, refer to the precautions.
2. Disconnect the negative battery cable.
3. Drain the cooling system into a suitable container and tighten the drain plug.
4. Remove the radiator support opening cover, if necessary.
5. Remove the right engine under cover.
6. Remove the left engine under cover, if necessary.
7. Remove the accessory drive belt.
8. Remove the right engine mounting insulator, if necessary.
9. Remove the water pump pulley.
10. Remove the water pump mounting bolts and nuts.
11. Remove the water pump and gasket from engine block.

To install:

12. Install the water pump and new gasket to engine block.
13. Install the water pump mounting bolts and nuts. Tighten to 8 ft. lbs. (11 Nm).
14. Install the water pump pulley and mounting bolts. Tighten to 11 ft. lbs. (15 Nm).
15. Install the right engine mounting insulator , if necessary. Torque to 38 ft. lbs. (52 Nm).
16. Install the accessory drive belt.
17. Refill the engine cooling system.
18. Connect the negative battery cable.
19. Perform the power window initialization procedure. Refer to Engine Electrical, Initialization Procedure.
20. Start the engine and top off the coolant as necessary.
21. Check the cooling system for leaks.
22. Install the right engine under cover
23. Install the left engine under cover, if necessary
24. Install the radiator support opening cover, if necessary

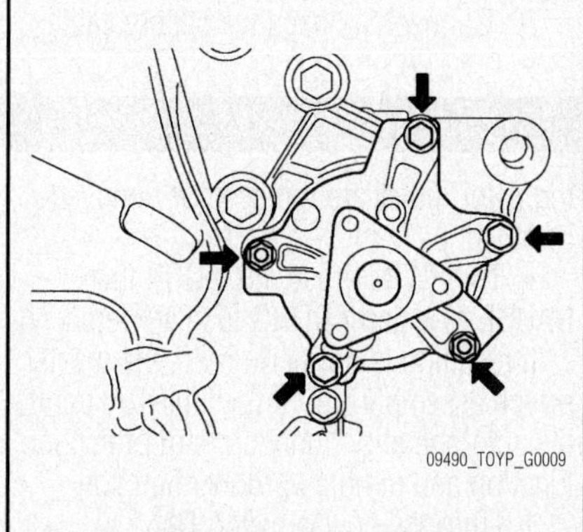

Fig. 15 Water pump and mounting bolts

ENGINE ELECTRICAL

CHARGING SYSTEM

ALTERNATOR

The Toyota Prius, being a hybrid vehicle that utilizes both electric and gasoline (internal combustion) engine power for mobility, does not require (or come equipped with) an alternator as a part of its charging system. The Toyota Hybrid system replaces the alternator with a pair of electrical motor-generators, a computerized shunt system to control them, a mechanical power splitter that acts as a second differential, and a battery pack that serves as an energy reservoir.

ENGINE ELECTRICAL — DISTRIBUTORLESS IGNITION SYSTEM

ADJUSTMENT

The ignition system is controlled by the Powertrain Control Module (PCM). No adjustment is possible.

FIRING ORDER

See Figure 16.

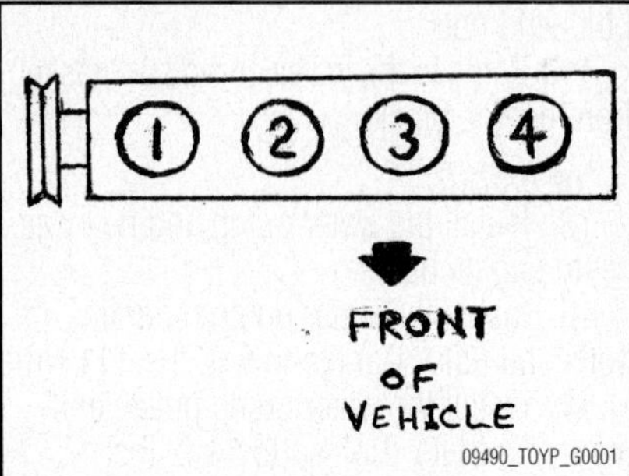

Fig. 16 1.5L I4 Hybrid engine Firing order: 1–3–4–2 Distributorless ignition system

IGNITION COIL

REMOVAL & INSTALLATION

1. Disconnect the negative battery cable.

➡Wait at least 90 seconds after disconnecting the cable from the negative (-) battery terminal to prevent airbag and seat belt pretensioner activation.

2. Disconnect the engine room relay block.
3. Disconnect the 4 ignition coil connectors.
4. Remove the 4 bolts and pull out the 4 ignition coils.

To install:

5. Install the 4 ignition coils with the 4 bolts. Tighten to 80 inch lbs. (9 Nm).
6. Connect the 4 ignition coil connectors.
7. Install the relay block with the 2 bolts. Tighten to 74 inch lbs. (8 Nm).
8. Connect the negative battery cable.
9. Perform the power window initialization procedure. Refer to Engine Electrical, Initialization Procedure.

IGNITION TIMING

ADJUSTMENT

The ignition timing is controlled by the Powertrain Control Module (PCM). No adjustment is possible.

SPARK PLUGS

REMOVAL & INSTALLATION

1. Remove the ignition coils.
2. Using a 16 mm plug wrench, remove the spark plugs.

To install:

3. Using a 16 mm plug wrench, install the spark plugs and tighten to 17.5 Nm (13 ft lbs)
4. Reinstall the ignition coils. Tighten to 80 inch lbs. (9 Nm).

ENGINE ELECTRICAL — INITIALIZATION

INITIALIZATION PROCEDURE

Resetting the power window regulator motor (initializing the pulse sensor) is necessary if:

- The battery terminal cable is disconnected
- The power window regulator master switch assembly, wire harness, power window regulator switch, power window regulator assembly and power window regulator motor are replaced or removed / installed
- The PWR H-fuse, FR DOOR fuse, GAUGE fuse and ECU-IG fuse are replaced.

If resetting is not performed, the master switch assembly will not be able to operate the auto operation function, jam protection function and remote up/down function.

1. Turn the power switch **ON**.
2. Open the power window halfway by pressing the power window switch.
3. Fully pull up on the switch until the power window is fully closed and continue to hold the switch for at least 1 second.
4. Check that the auto up/down function operates normally. If the function operates normally, reset operations are complete.
5. If the function operates abnormally, disconnect the negative battery cable for 10 seconds.
6. Connect the battery cable.
7. Perform steps 1–4 again.

ENGINE ELECTRICAL — STARTING SYSTEM

STARTER

The Toyota Prius, being a hybrid vehicle that utilizes both electric and gasoline (internal combustion) engine power for mobility, does not require (or come equipped with) a starter motor as a part of its starting system. The function of the starter motor is performed by a pair of electrical motor-generators, a computerized shunt system to control them, a mechanical power splitter that acts as a second differential, and a battery pack that serves as an energy reservoir.

HYBRID CONTROL SYSTEM

PRECAUTIONS

- Before inspecting the high-voltage system, take safety precautions to prevent electrical shocks, such as wearing insulated gloves and removing the service plug grip. After removing the service plug grip, put it in your pocket to prevent other technicians from reconnecting it while you are servicing the high-voltage system.
- Turning the power switch **ON (READY)** with the service plug grip removed could cause a malfunction. Therefore, do not turn the power switch **ON (READY)** unless instructed by the repair manual.
- After disconnecting the service plug grip, wait for at least 5 minutes before touching any of high-voltage connectors or terminals No. .
- At least 5 minutes are required to discharge the high-voltage condenser inside the inverter.
- Since liquid leakage may occur, wear protective goggles when checking inside the high voltage battery.
- Wear insulated gloves, turn the power switch **OFF** , and disconnect the negative terminal of the auxiliary battery before touching any of the orange-colored wires of the high-voltage system.
- Turn the power switch **OFF** before performing a resistance check.
- Turn the power switch **OFF** before disconnecting or reconnecting any connector.
- To install the service plug grip, the lever must be flipped and locked downward. Once it is locked in place, it turns the interlock switch **ON**. Make sure to lock it securely because if you leave it unlocked, the system will output a DTC pertaining to the interlock switch system.
- When the warning light is illuminated or the battery has been disconnected and reconnected, pressing the power switch may not start the system on the first attempt. If so, press the power switch again.

With the power switch's power mode changed to **ON (IG)** , disconnect the battery. If the key is not in the key slot during reconnection, DTC B2799 may be output.

SERVICE PLUG GRIP

REMOVAL & INSTALLATION

See Figures 17 through 22.

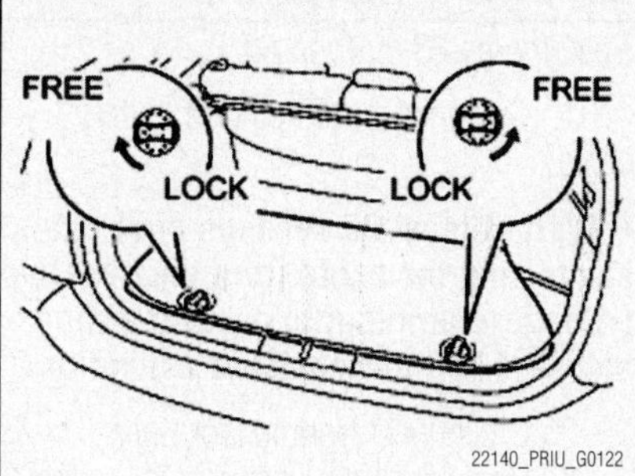

Fig. 17 Turn the knob to release the rear No. 2 floor board lock

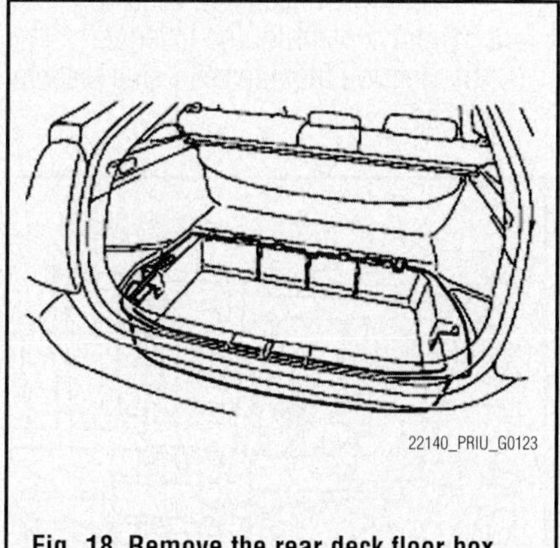

Fig. 18 Remove the rear deck floor box

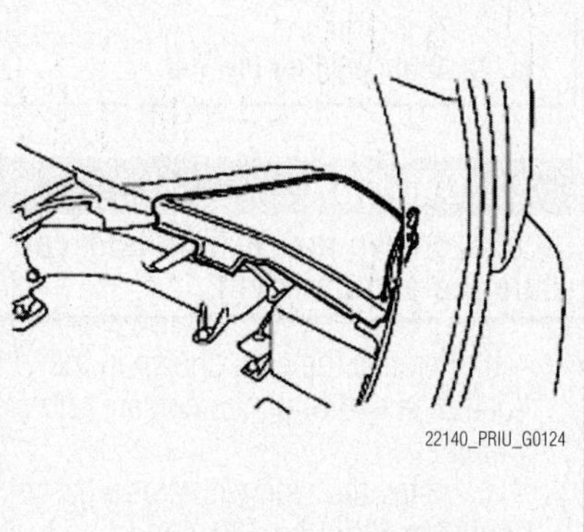

Fig. 19 Remove the rear No. 3 floor board

⁂ WARNING

After removing the service plug grip, do not operate the power switch as it may damage the hybrid vehicle control ECU.

1. Remove rear No. 2 floor board.
   a. As shown in the illustration, turn the knob to release the lock.
   b. Remove the rear floor board No. 2.
2. Remove rear deck floor box.
   a. remove the deck floor box rear shown in the illustration.
3. Remove rear No. 3 floor board.
   a. Remove the rear floor board shown in the illustration.
4. Disconnect the 12 volt auxiliary (not hybrid high voltage) battery negative cable.
5. Remove service plug grip.

⁂ CAUTION

Wear insulating gloves.

   a. Slide up the lever of the service plug grip. Remove the service plug grip while turning the lever to the left.
   b. Insulate the service plug with insulating tape.

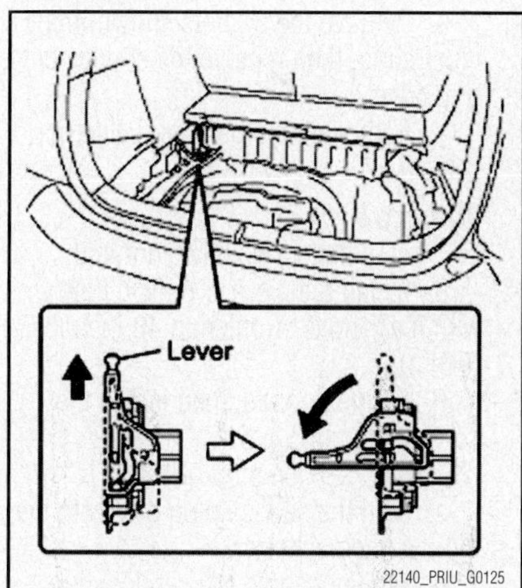

Fig. 20 Slide up the lever of the service plug grip. Remove the service plug grip while turning the lever to the left

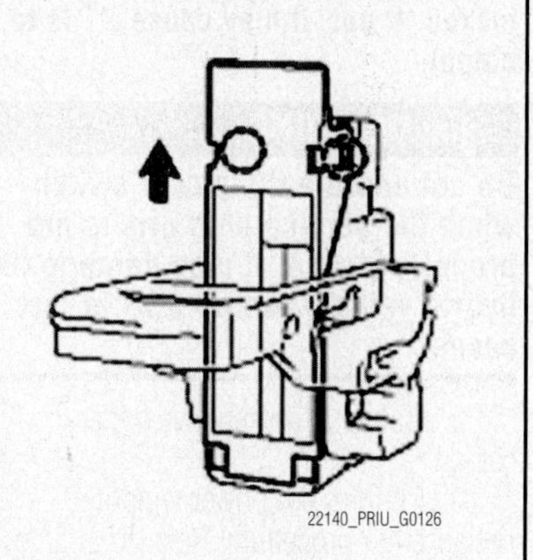

Fig. 21 Remove the 2 joints shown in the illustration, then remove the service plug grip cover

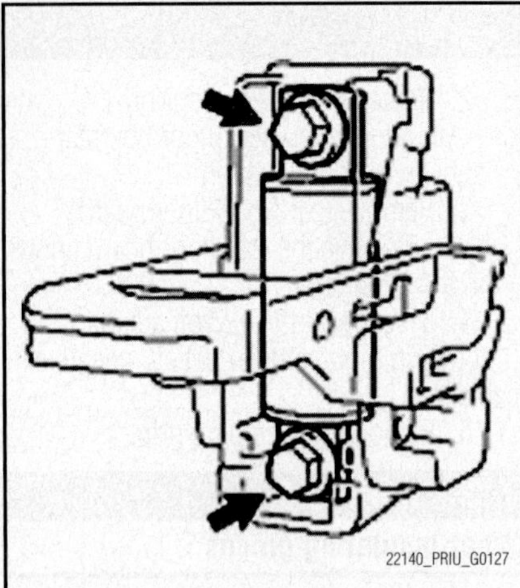

Fig. 22 Remove the 2 bolts and electric vehicle fuse

6. Remove the hybrid battery positive terminal.
   a. Remove the 2 joints shown in the illustration, then remove the service plug grip cover.
   b. Remove the 2 bolts and electric vehicle fuse.

To install:

7. Install battery positive terminal.
   a. Install the electric vehicle fuse with the 2 bolts. Tighten to 48 inch lbs. (5 Nm).
   b. Fit the 2 joints, then install the service plug grip cover.
8. Install service plug grip
   a. Join the service plug grip with the high voltage battery.
   b. While pushing the service plug grip to the right, rotate the lever to the right.
   c. Slide the lever down to lock the service plug grip in place.

➡The service plug grip must be locked. If not, it may cause DTCs to output.

**** WARNING**

Do not operate the power switch when the service plug grip is not properly locked. It may damage the hybrid vehicle control ECU in rare cases.

9. Connect the negative battery cable.
10. Perform the power window initialization procedure. Refer to Engine Electrical, Initialization Procedure.
11. Install rear No. 3 floor board.
12. Install rear deck floor box.
13. Install rear No. 2 floor board.

WATER PUMP WITH MOTOR

REMOVAL & INSTALLATION

See Figures 23 through 25.

1. Disconnect the negative battery cable.

➡Wait at least 90 seconds after disconnecting the cable from the negative (-) battery terminal to prevent airbag and seat belt pretensioner activation.

2. Disconnect front fender liner.
3. Remove center engine under cover.
4. Remove front spoiler cover.
5. Remove front bumper cover.
6. Remove front bumper energy absorber.
7. Remove left headlight assembly.
8. Drain coolant for the inverter.
   a. Remove the transaxle-side reserve tank cap.

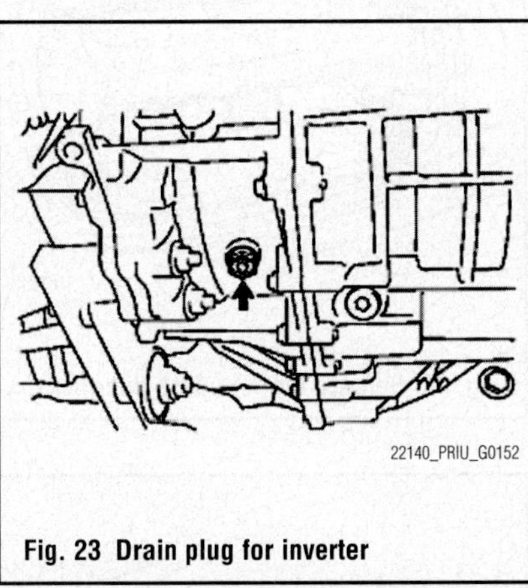

Fig. 23 Drain plug for inverter

**** CAUTION**

Do not remove the reserve tank cap while the engine is hot.

   b. Remove the plug shown in the illustration and drain the coolant into a container.
   c. Install the plug with a new gasket. Tighten to 29 ft. lbs. (39 Nm).
9. Remove the bolt and nut, then disconnect the water valve.
10. Disconnect the 2 inverter cooling hoses.
11. Remove the bolt, then disconnect the water pump with motor assembly from the vehicle.
12. Disconnect the connector and remove the water pump with motor assembly.

To install:

13. Install the water pump with motor assembly. Tighten to 62 inch lbs. (7 Nm).

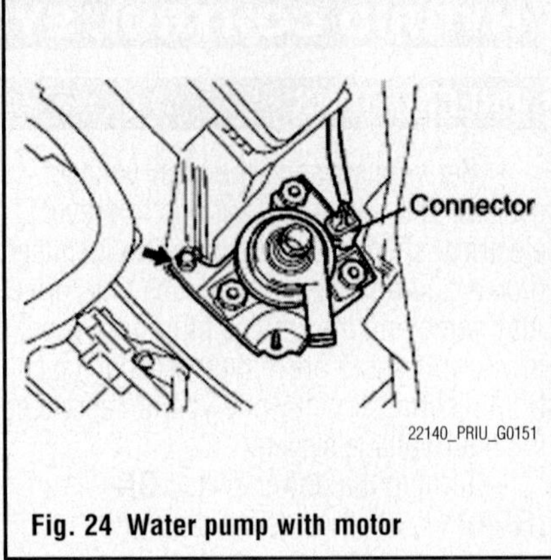

Fig. 24 Water pump with motor

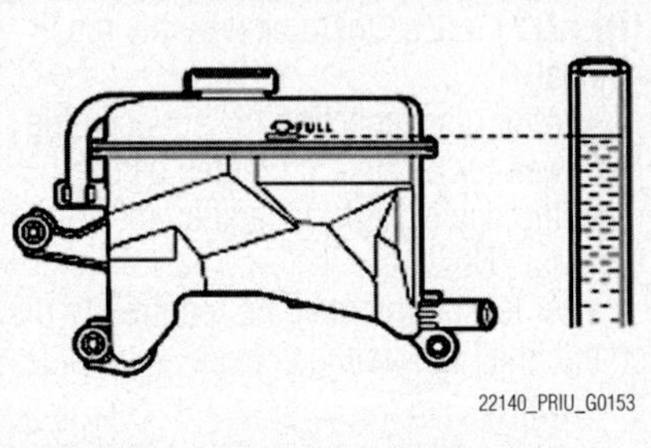

Fig. 25 Add coolant until the level of coolant in the hose attached to the bleeder tank reaches the same level as the FULL line of the reserve tank

   a. Connect the connector.
   b. Connect the 2 inverter cooling hoses.
14. Install the water valve with the bolt and nut. Tighten to 62 inch lbs. (7 Nm).
15. Install left headlight assembly.
16. Install front bumper energy absorber.
17. Install front bumper cover.
18. Install front spoiler cover.
19. Install center engine under cover.
20. Install front fender liner.
21. Add coolant for the inverter.
   a. Loosen the bleeder plug and connect a hose.
   b. Insert one end of the hose into the bleeder tank.
   c. Add coolant until the level of coolant in the hose attached to the bleeder tank reaches the same level as the **FULL** line of the reserve tank.
   d. Close the bleeder plug.
   e. Turn the power switch **ON** and run the water pump for approximately 20 seconds.
   f. Turn the power switch **OFF.**
   g. Loosen the bleeder plug and bleed the air from the transaxle.
   h. Add coolant into the bleeder tank.
   i. Repeat the previous three steps again.

j. Coolant system air bleeding is complete when water pump noise becomes softer and coolant circulation in reserve tank improves.

k. If air remains in the coolant system, the water pump noise becomes louder and the coolant circulation in the reserve tank becomes worse.

l. Turn the power switch **ON** and run the water pump for approximately 5 minutes after completing air bleeding of the coolant system.

➡Ensure that the bleeder plug is closed.

m. Add coolant until the reserve tank is filled up to the **FULL** mark.

22. Connect the negative battery cable.
23. Perform the power window initialization procedure. Refer to Engine Electrical, Initialization Procedure.
24. Inspect the engine cooling system for leakage.

ENGINE MECHANICAL

ACCESSORY DRIVE BELTS

ACCESSORY BELT ROUTING

See Figures 26 and 27.

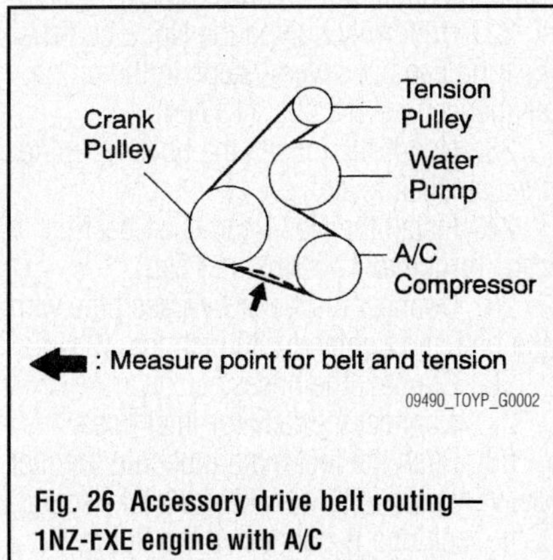

Fig. 26 Accessory drive belt routing—1NZ-FXE engine with A/C

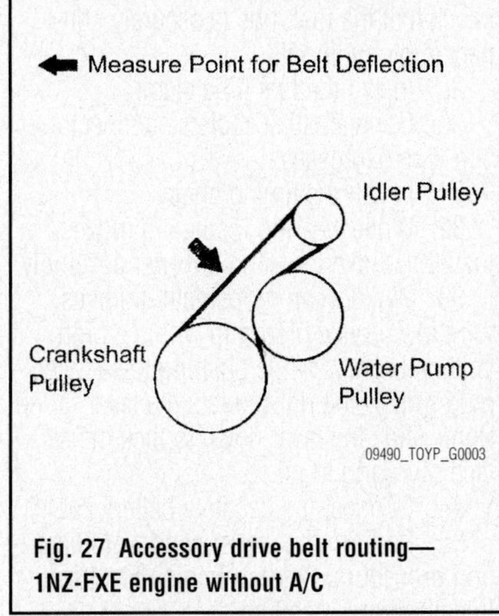

Fig. 27 Accessory drive belt routing—1NZ-FXE engine without A/C

INSPECTION

Inspect the drive belt for signs of glazing or cracking. A glazed belt will be perfectly smooth from slippage, while a good belt will have a slight texture of fabric visible. Cracks will usually start at the inner edge of the belt and run outward. All worn or damaged drive belts should be replaced immediately.

ADJUSTMENT

See Figure 28.

1. Using a belt tension gauge, measure the belt tension.

a. New belt tension should be 99 to 121 ft. lbs. used belt tension should be 55 to 77 ft. lbs.

➡A "new belt" is a belt which has been used less than 5 minutes on a running engine. A "used belt" is a belt which has been used on a running engine for 5 minutes or more.

b. If tension is not as specified, adjust the belt by loosening nut "A", then turning adjust bolt "B" to provide the proper belt tension.

2. When not using a belt tension gauge, measure the belt deflection.

a. Press the belt at the center of the longest span using 22 ft. lbs. (10 kg) of force.

b. New belt tension should be 0.35 to 0.47 in. (9.0 to 12.0 mm), used belt tension should be 0.43 to 0.59 in. (11 to 15 mm)

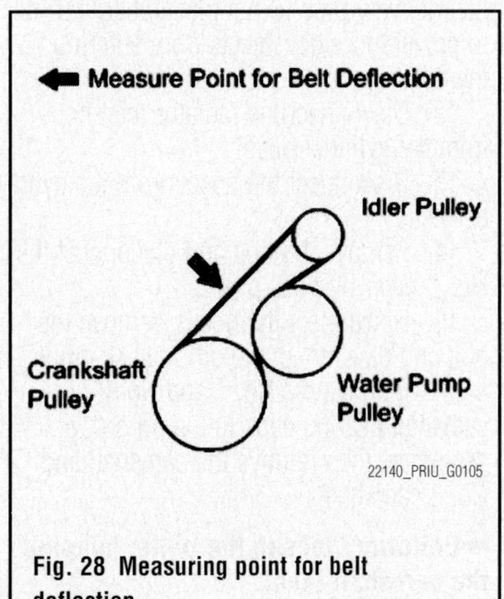

Fig. 28 Measuring point for belt deflection

REMOVAL & INSTALLATION

See Figure 29.

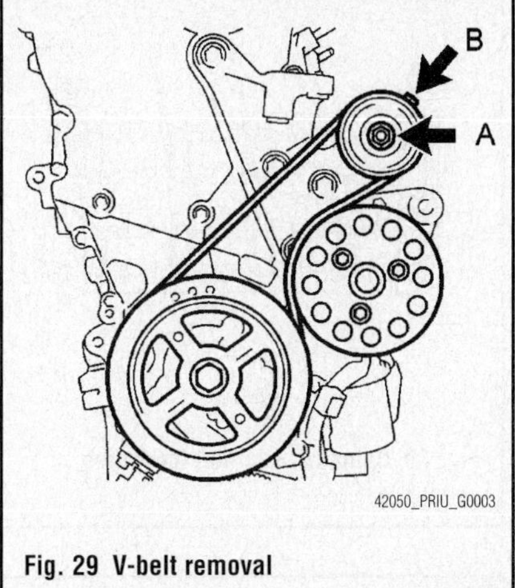

Fig. 29 V-belt removal

1. Before servicing the vehicle, refer to the precautions.
2. Remove the engine under cover.
3. Loosen the clamp, then disconnect the air cleaner inlet from the air cleaner case.
4. Referring to the illustration, loosen nut "A", then turn adjust bolt "B" to relieve the V-ribbed belt tension.
5. Remove the belt.

To install:

6. Install the V-ribbed belt to each pulley.

➡After installing the drive belt, check that it fits properly in the ribbed grooves. Check with your hands to confirm that the belt has not slipped out of the groove on the bottom of the crankshaft pulley. After installing a new belt, run the engine for approximately 5 minutes and then recheck the tension.

7. Turn adjust bolt "B" to adjust the V-ribbed belt tension.
8. Tighten nut "A" to 40 Nm (30 ft. lbs).

CAMSHAFT AND VALVE LIFTERS

REMOVAL & INSTALLATION

See Figures 30 through 33.

1. Before servicing the vehicle, refer to the precautions.

2. Properly relieve the fuel system pressure.
3. Drain engine oil.
4. Disconnect the negative battery cable.

➡Wait at least 90 seconds after disconnecting the cable from the negative (-) battery terminal to prevent airbag and seat belt pretensioner activation.

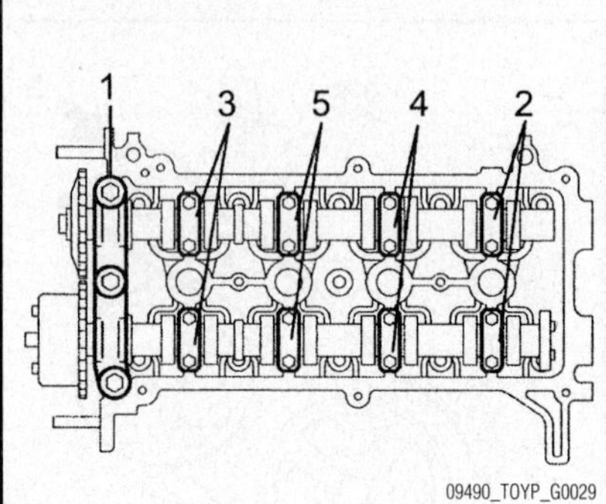

Fig. 30 Camshaft cap bolt loosening sequence

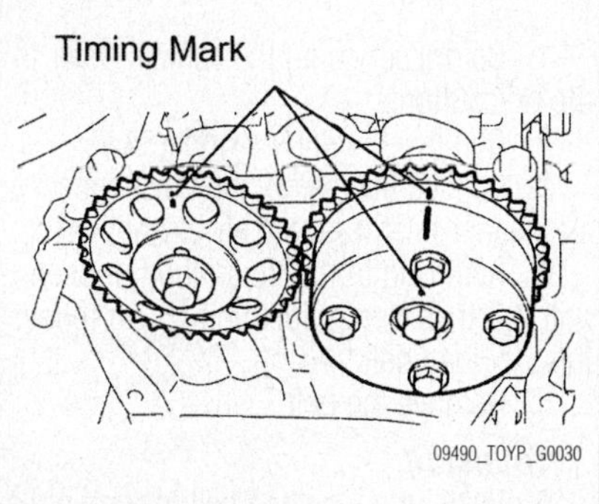

Fig. 31 Timing mark on the camshaft timing gear facing upward

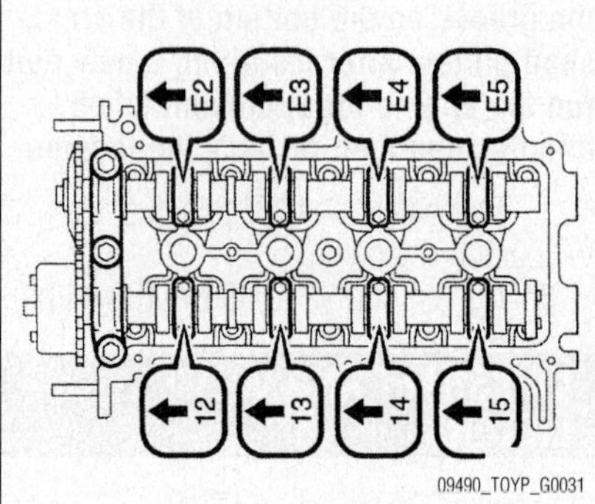

Fig. 32 Check the front marks and Numbers on the No. 1 and No. 2 camshaft bearing caps

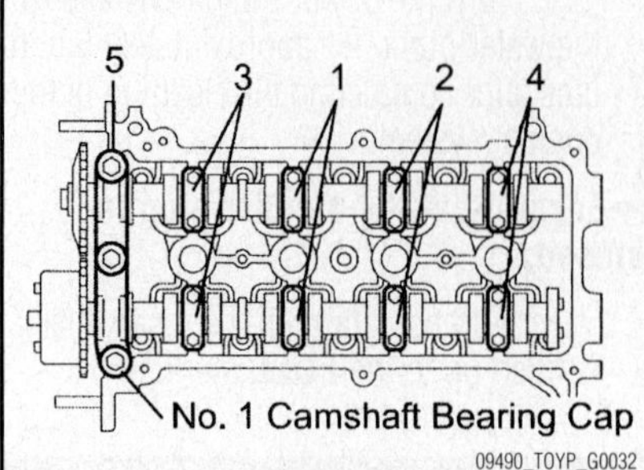

Fig. 33 Camshaft cap bolt tightening sequence

5. While wearing insulating gloves, slide up the lever of the service plug grip and remove while turning the lever to the left. Be sure to insulate the service plug with insulating tape.

CAUTION

Do not touch the high voltage connectors and terminals for 5 minutes after removing the service plug grip.

6. While wearing insulating gloves, remove the inverter with converter assembly.
7. Remove timing chain.
8. Disconnect associated connectors and wire harnesses.
9. Remove the fuel pipe clamp.
10. Disconnect the fuel tube from the fuel delivery pipe. Even if the fuel tube is stuck and cannot be disconnected, do not use any tools. Push and pull the parts with the quick connector pinched to disconnect the tube.
11. Cover the disconnected fuel tube and fuel delivery pipe with a plastic bag in order to prevent foreign objects from entering them.
12. Disconnect the radiator inlet hose from the cylinder head.
13. Disconnect the hoses from the cylinder head.
14. Remove the bolt and disconnect the No. 1 water by-pass pipe.
15. Disconnect the hoses, remove the bolt and disconnect the oil dipstick guide.
16. Remove the No. 1 and No. 2 camshaft bearing caps in the proper sequence, then remove the camshaft and No. 2 camshaft.

➡Uniformly loosen the bolts, keeping the camshaft level.

17. Remove the valve lifters. Keep the valve lifters in the correct order so that they can be returned to their original locations when reassembling.

To install:

18. Apply a light coat of engine oil to the valve lifter and install the valve lifter. Check that the valve lifter rotates smoothly by hand.

➡If turning the camshaft with the chain removed, turn the crankshaft counterclockwise by 40 degrees from TDC/compression.

19. Apply engine oil to the cam and cylinder head journals.
20. Place the camshaft and No. 2 camshaft on the cylinder head with the timing mark on the camshaft timing gear facing upward.
21. Check the front marks and Numbers on the No. 1 and No. 2 camshaft bearing caps, then temporarily install them.
22. Uniformly tighten the No. 2 camshaft bearing caps in several steps in the proper sequence to 9.6 ft. lbs. (13 Nm).
23. Uniformly loosen the bolts, keeping the camshaft level.
24. Install the No. 1 camshaft bearing cap. Torque to 17 ft. lbs. (23 Nm).
25. Connect the water by-pass pipe with the bolt and tighten to 80 inch lbs. (9 Nm).
26. Connect the hoses.
27. Connect the radiator inlet hose.
28. Push the fuel main tube into the fuel delivery pipe until it makes a "click" sound. If the fuel tube is connected too tightly, apply a light coat of engine oil to the tip of the fuel delivery pipe. After connecting, check that the fuel tube is securely connected by pulling it.
29. Install the fuel pipe clamp.
30. Connect all associated connectors and wire harnesses.
31. Install the timing chain.
32. While wearing insulating gloves, install the inverter with converter assembly.
33. While wearing insulating gloves, joint the service plug grip with the high voltage battery. While pushing the service plug grip to the right, rotate the lever to the right. Slide the lever down to lock the service plug grip in place.
34. Connect the negative battery cable.
35. Perform the power window initialization procedure. Refer to Engine Electrical, Initialization Procedure.
36. Refill engine oil.
37. Start the engine and check for leaks, check for abnormal noises, shock slippage, correct shift points and smooth operation.

INSPECTION

1. Inspect the camshaft for runout.
    a. Place the camshaft on V-blocks.
    b. Using a dial indicator, measure the circle runout at the center journal.

c. If the circle runout is greater than the maximum, replace the camshaft.
2. Inspect the cam lobes.
a. Using a micrometer, measure the cam lobe height.
b. If the cam lobe height is less than the minimum, replace the camshaft.
3. Inspect the camshaft journals.
a. Using a micrometer, measure the journal diameter.
b. If the journal diameter is not as specified, check the oil clearance.
4. Inspect camshaft thrust clearance.
a. Install the camshafts.
b. Using a dial indicator, measure the thrust clearance while moving the camshaft back and forth.
c. If the thrust clearance is greater than the maximum, replace the camshaft. If necessary, replace the bearing caps and the cylinder head.
5. Inspect camshaft oil clearance.
a. Clean the bearing caps and the camshaft journals.
b. Place the camshafts on the cylinder head.
c. Lay a strip of Plastigage across each of the camshaft journals.
d. Install and tighten the bearing caps.

➡Do not turn the camshaft.

e. Remove the bearing caps.
f. Measure the Plastigage at its widest point.
g. If the oil clearance is greater than the maximum, replace the camshaft. If necessary, replace the bearing caps and the cylinder head.
h. Completely remove the Plastigage.

CRANKSHAFT DAMPER

REMOVAL & INSTALLATION

See Figures 34 and 35.

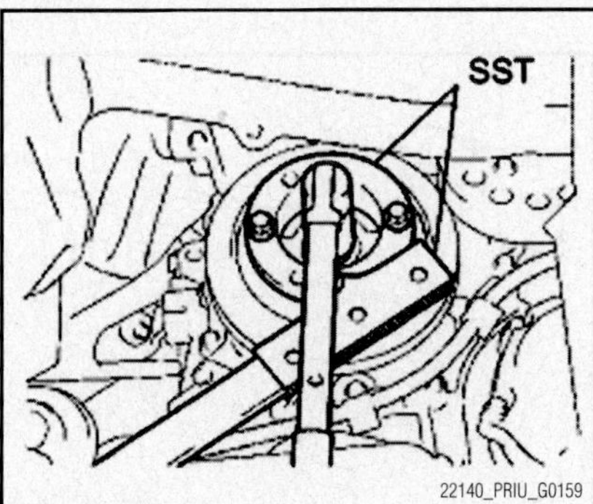

Fig. 34 Using the special tool, hold the crankshaft pulley and loosen the crankshaft bolt

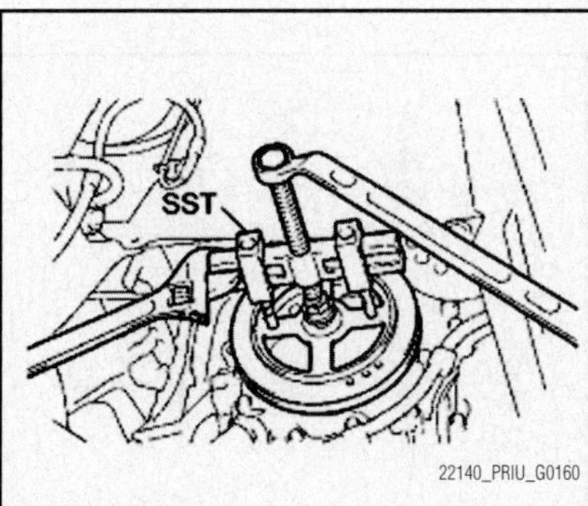

Fig. 35 Using the special tool, remove the crankshaft damper

1. Before servicing the vehicle, refer to the precautions.
2. Remove right engine under cover.
3. Loosen the clamp, and disconnect the air cleaner inlet from the air cleaner case.
4. Remove drive belt.
a. Loosen the locknut.
b. Turn adjust bolt, and then release the belt tension.
c. Remove the belt.
5. Remove crankshaft pulley.
6. Using special tool SST: 09213-58013, hold the crankshaft pulley and loosen the crankshaft bolt.

➡When installing the special tool, be careful that the bolt which holds it does not interfere with the chain cover.

a. Loosen the crankshaft bolt until 2 to 3 threads of the bolt are tightened to the crankshaft.
b. Using special tool SST:09950-50013, remove the crankshaft damper.
c. Remove the crankshaft bolt.

To install:

7. Install crankshaft pulley.
a. Align the hole of the crankshaft damper with the straight pin, then install the crankshaft damper.
b. Using special tool SST: 09213-58013, hold the crankshaft damper and tighten the crankshaft bolt to 95 ft. lbs. (128 Nm).
8. Install drive belt.
a. Temporarily install the belt on each pulley.
b. Adjust drive belt tension.
c. Turn adjust bolt to adjust the belt tension.
d. Tighten the locknut to 30 ft. lbs. 40 Nm).
e. Check drive belt tension.
9. Connect the air cleaner inlet to the air cleaner case and tighten the clamp to 27 inch lbs. (3 Nm).
10. Install right engine under cover.

CRANKSHAFT FRONT SEAL

REMOVAL & INSTALLATION

See Figure 36.

1. Before servicing the vehicle, refer to the precautions.
2. Remove the crankshaft damper.
3. Using a knife, cut off the lip of the oil seal.
4. Using a screwdriver with the tip wrapped in tape, pry out the oil seal.

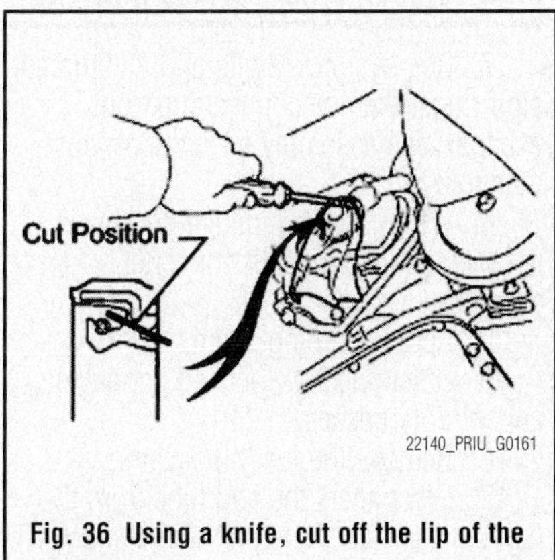

Fig. 36 Using a knife, cut off the lip of the oil seal

To install:

➡After removal, check if the crankshaft is damaged. If it is damaged, smooth the surface with 400-grit sandpaper.

5. Apply multi-purpose grease to the lip of a new oil seal. Keep the lip free of foreign objects.
6. Using seal installer and a hammer, tap in the oil seal until its surface is flush with the timing chain cover edge.

➡Be careful not to tap the oil seal at an angle.

7. Wipe any extra grease off the crankshaft.
8. Install the crankshaft damper.

CYLINDER HEAD

REMOVAL & INSTALLATION

See Figures 37 through 40.

1. Before servicing the vehicle, refer to the precautions.
2. Properly relieve the fuel system pressure.
3. Drain engine oil.

4. Drain transaxle and engine coolant.

5. Disconnect the negative battery cable.

➡Wait at least 90 seconds after disconnecting the cable from the negative (-) battery terminal to prevent airbag and seat belt pretensioner activation.

6. While wearing insulating gloves, slide up the lever of the service plug grip and remove while turning the lever to the left. Be sure to insulate the service plug with insulating tape.

CAUTION

Do not touch the high voltage connectors and terminals for 5 minutes after removing the service plug grip.

7. Remove the 2 bolts and 2 compression rings, then disconnect the front exhaust pipe assembly from the exhaust manifold.

8. While wearing insulating gloves, remove the inverter with converter assembly.

9. Remove timing chain.

10. Disconnect associated connectors and wire harnesses.

11. Remove the fuel pipe clamp.

12. Disconnect the fuel tube from the fuel delivery pipe. Even if the fuel tube is stuck and cannot be disconnected, do not use any tools. Push and pull the parts with the quick connector pinched to disconnect the tube.

13. Cover the disconnected fuel tube and fuel delivery pipe with a plastic bag in order to prevent foreign objects from entering them.

14. Disconnect the radiator inlet hose from the cylinder head.

15. Disconnect the hoses from the cylinder head as shown in the illustration.

16. Remove the bolt and disconnect the No. 1 water by-pass pipe.

17. Disconnect the hoses, remove the bolt and disconnect the oil dipstick guide.

18. Remove the camshafts.

19. Using an 8mm bi-hexagon wrench, loosen the cylinder head bolts in several steps in the proper sequence. Then remove the cylinder head bolts and washer.

CAUTION

When removing the bolt, do not drop the washer into the engine. Removing the cylinder head bolts in the wrong order may cause damage to the cylinder head.

20. Remove the cylinder head and gasket.

To install:

21. Install the cylinder head along with a new gasket.

22. Apply sealant with a 0.177–0.217 inch (4.5–5.5mm) thickness. Install the cylinder head within 3 minutes of applying sealant.

23. Apply a light coat of engine oil to the threads of the cylinder head bolts.

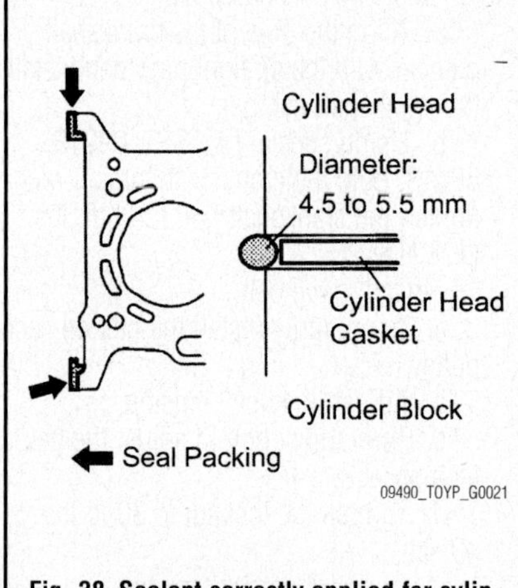

Fig. 38 Sealant correctly applied for cylinder head installation

24. Using several steps, install and tighten the 10 cylinder head bolts and plate washers uniformly with an 8mm bi-hexagon wrench in the proper sequence to 21 ft. lbs. (29 Nm).

25. Mark the front of the cylinder head bolt with paint.

26. Retighten the cylinder head bolts by an additional 90 degrees and then another 90 degrees. Check that the paint mark is now 180 degrees opposite to the front.

27. Install the camshafts.

28. Install the oil dipstick guide with the bolt and tighten to 80 inch lbs. (9 Nm). Connect the hose.

29. Connect the water by-pass pipe with the bolt and tighten to 80 inch lbs. (9 Nm).

30. Connect the hoses.

31. Connect the radiator inlet hose.

32. Push the fuel main tube into the fuel delivery pipe until it makes a "click" sound. If the fuel tube is connected too tightly, apply a light coat of engine oil to the tip of the fuel delivery pipe. After connecting, check that the fuel tube is securely connected by pulling it.

33. Install the fuel pipe clamp.

34. Connect all associated connectors and wire harnesses.

35. Install the timing chain.

36. While wearing insulating gloves, install the inverter with converter assembly.

37. Install the front exhaust assembly.

38. While wearing insulating gloves, joint the service plug grip with the high voltage battery. While pushing the service plug grip to the right, rotate the lever to the right. Slide the lever down to lock the service plug grip in place.

39. Connect the negative battery cable.

40. Perform the power window initialization procedure. Refer to Engine Electrical, Initialization Procedure.

41. Refill engine oil.

42. Refill transaxle and engine coolant.

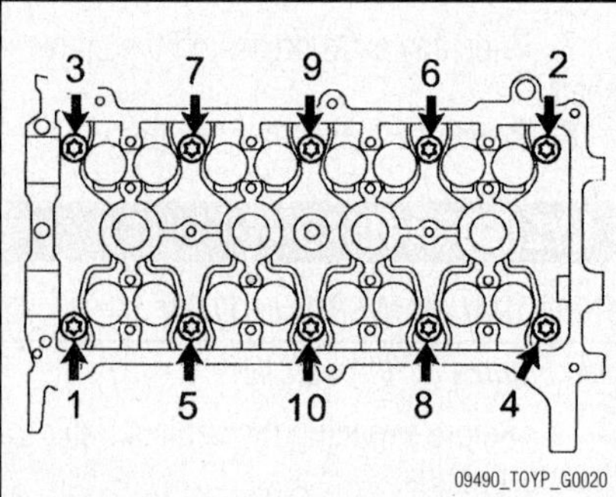

Fig. 37 Cylinder head bolt loosening sequence

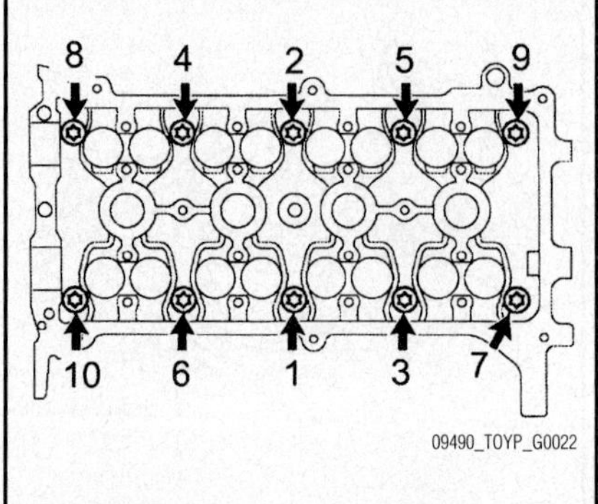

Fig. 39 Cylinder head bolt tightening sequence

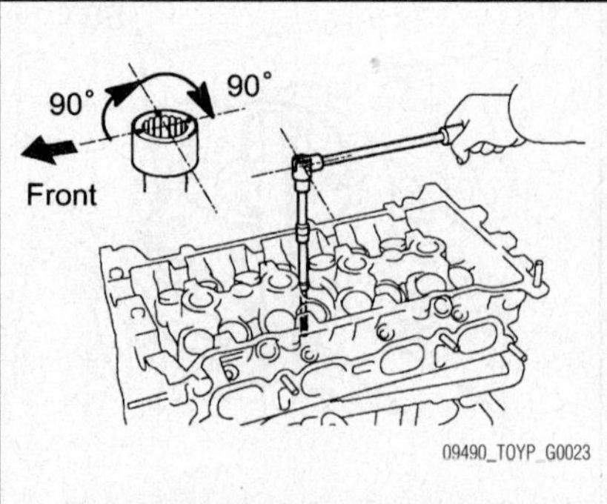

Fig. 40 Retightening the cylinder head bolts by an additional 90 degrees

43. Start the engine and check for leaks, check for abnormal noises, shock slippage, correct shift points and smooth operation.
44. Recheck transaxle and engine coolant.

ENGINE ASSEMBLY

REMOVAL & INSTALLATION

CAUTION

The hybrid system uses high voltage circuits, so improper handling could cause electric shock or leakage. During service, be sure to follow procedures.

1. Before servicing the vehicle, refer to the precautions.
2. Drain engine coolant, engine oil and transaxle fluid.
3. Remove the rear No. 2 floor board.
4. Remove the rear deck floor box.
5. Remove the rear No. 3 floor board.
6. Properly relieve the fuel system pressure.
7. Disconnect the negative battery cable.

➡Wait at least 90 seconds after disconnecting the cable from the negative (-) battery terminal to prevent airbag and seat belt pretensioner activation.

8. While wearing insulating gloves, slide up the lever of the service plug grip and remove while turning the lever to the left. Be sure to insulate the service plug with insulating tape.

CAUTION

Do not touch the high voltage connectors and terminals For 5 minutes after removing the service plug grip.

9. Remove front wheels.
10. Install the engine under covers.
11. Install the left and right front wiper arms.
12. Install the hood-to-cowl top seal.
13. Install the left and right cowl top ventilator louvers.
14. Install the windshield wiper motor and link assembly.
15. Install the front outer cowl top panel.
16. While wearing insulating gloves, remove the inverter with converter assembly.
17. Remove the radiator assembly.
18. Remove the No. 3 inverter cooling hose.
19. Remove the No. 1 heat storage water by-pass hose.
20. Remove the engine ground cable.
21. Remove the air cleaner inlet hose from air cleaner case.
22. Remove the air cleaner assembly.
23. Remove the vacuum switching valve assembly.
24. Remove the fuel pipe clamp.
25. Remove the fuel tube from fuel delivery pipe by pinching the quick connector with your hand and pulling.
26. Remove the heater water hose.
27. Remove the No. 3 heat storage water by-pass hose.
28. Remove the connector from ECM, then pull engine wire harness toward the engine compartment side.
29. Remove the harness and harness clamp from the engine room main relay block.
30. Remove the ground cable.
31. Remove the wiring harness from compressor.
32. Remove the compressor unit from engine. secure the compressor, still connected to the high- and low-pressure hoses, to the side of the vehicle with rope.
33. Disconnect the steering sliding yoke.
    a. Hold the steering wheel in the straight ahead position, using the seat belt looped through the steering wheel and buckled securely so the steering wheel will not turn. This will prevent an open circuit of the spiral cable.
    b. Remove the 2 clips, then remove the column hole cover silencer.
    c. Loosen the bolt on the steering column side of the sliding yoke.
    d. Remove the bolt on the steering gear side of the sliding yoke.
    e. Put paint marks on the sliding yoke and intermediate shaft, then disconnect the sliding yoke.
34. Remove or disconnect the following:
35. Remove the front exhaust pipe.
36. Remove the left and right front axle hub nuts.
37. Remove the left front stabilizer link.
38. Remove the left and right tie rod ends.
39. Remove the left and right front lower control arms.
40. Remove the left and right front axle hub/bearing assemblies.
41. Remove the left and right drive shafts.
42. Remove the engine torque rod.
43. Remove the front suspension crossmember.
44. Secure the engine/transaxle assembly onto a lifting device.
45. Remove the right engine mounting bracket and right engine mounting insulator.
46. Remove the left engine mounting bracket and left engine mounting insulator.
47. While operating the engine lifter, lower the engine and transaxle assembly out of the vehicle slowly and carefully. Make sure the engine is clear of all wiring, hoses and cables.
48. Separate the engine from the transaxle unit.

To install:

49. Connect the transaxle assembly to the engine assembly.
50. Install the engine and transaxle assembly into the engine compartment using the engine lifting device.

right engine mounting bracket and right engine mounting insulator. torque the nuts and bolts to 38 ft. lbs. (52 Nm).

51. Install the left engine mounting bracket and left engine mounting insulator. torque the nut to 59 ft. lbs. (80 Nm).
52. Install the front suspension crossmember. torque the left and right front corner bolts to 83 ft. lbs. (113 Nm). torque the left and right rear corner bolts to 116 ft. lbs. (157 Nm).
53. Install the engine torque rod. torque the 2 nuts and 2 bolts, at the engine, to 74 ft. lbs. (100 Nm), torque the 2 nuts and 2 bolts, at the body side, to 41 ft. lbs. (56 Nm).
54. Install the left and right drive shafts.
55. Install the left and right front axle hub/bearing assemblies.
56. Install the left and right front lower control arms.
57. Install the left and right tie rod ends.
58. Install the left front stabilizer link.
59. Install the new left and right front axle hub nuts and torque to 159 ft. lbs. (216 Nm). using a chisel and hammer, stake the hub nut.
60. Install the front exhaust pipe.
61. Connect the steering sliding yoke.
    a. Align the paint marks on the sliding yoke and intermediate shaft, then connect the sliding yoke.
    b. Torque the sliding yoke bolts to 26 ft. lbs. (35 Nm).
    c. Install the column hole cover silencer with the 2 clips.
    d. Remove the seat belt from the steering wheel.
62. Install the compressor unit to the engine. torque the three mounting bolts to 18 ft. lbs. (25 Nm).
63. Install the wiring harness to the compressor.
64. Install the ground cable.
65. Install the harness and harness clamp to the engine room main relay block.

66. Install the wire harness connector to the ECM.
67. Install the No. 3 heat storage water by-pass hose.
68. Install the heater water hose.
69. Install the fuel tube to the fuel delivery pipe by pushing the quick connector until it makes a "click" sound. Check to make sure the connection is secure by gently, but firmly pulling on it.
70. Install the fuel pipe clamp.
71. Install the VSV assembly and torque bolt to 5.5 ft. lbs. (7.5 Nm).
72. Install the air cleaner assembly.
73. Install the air cleaner inlet hose to the air cleaner case.
74. Install the engine ground cable.
75. Install the No. 1 heat storage water by-pass hose.
76. Install the No. 3 inverter cooling hose.
77. Install the radiator assembly.
78. While wearing insulating gloves, install the inverter with converter assembly.
   a. Install the inverter with converter assembly and tighten the three mounting bolts to 16 ft. lbs. (21 Nm).
   b. Connect the MG1 and MG2 power cables and tighten the mounting bolts to 71 inch lbs. (8 Nm).
   c. Connect any remaining wiring connectors including the engine main wiring harness. Be sure to insert the grommet of the engine main wiring harness into the U-shaped groove of the inverter case.
   d. Engage the connector for the air conditioning inverter and secure by pushing in the lock pin.
   e. Connect the 2 frame wire connectors to the inverter with converter assembly.
   f. Connect the No. 1 circuit breaker sensor.
   g. Connect the No. 6, No. 2 and No. 1 inverter cooling hoses.
   h. Install the inverter cover and tighten the mounting fasteners to 8 ft. lbs. (11 Nm).
   i. Install the radiator support opening cover.
   j. Install the left and right engine under covers.
79. Install the front outer cowl top panel.
80. Install the windshield wiper motor and link assembly.
81. Install the left and right cowl top ventilator louvers.
82. Install the hood-to-cowl top seal.
83. Install the left and right front wiper arms.
84. While wearing insulating gloves, joint the service plug grip with the high voltage battery. While pushing the service plug grip to the right, rotate the lever to the right. Slide the lever down to lock the service plug grip in place.
85. Connect the negative battery cable.
86. Perform the power window initialization procedure. Refer to Engine Electrical, Initialization Procedure.
87. Install the rear No. 3 floor board.
88. Install the rear deck floor box.
89. Install the rear No. 2 floor board.
90. Refill engine coolant, engine oil and transaxle fluid.
91. Install or connect the following:
92. Install the front wheels.
93. Install the engine under covers.
94. Check the front wheel alignment.
95. Run the engine and verify that there are no fuel, coolant, transaxle or exhaust leaks.

EXHAUST MANIFOLD

REMOVAL & INSTALLATION

See Figures 41 and 42.

Toyota recommends this procedure be performed with the cylinder head removed. It may be possible to remove the exhaust manifold without removing the cylinder head.

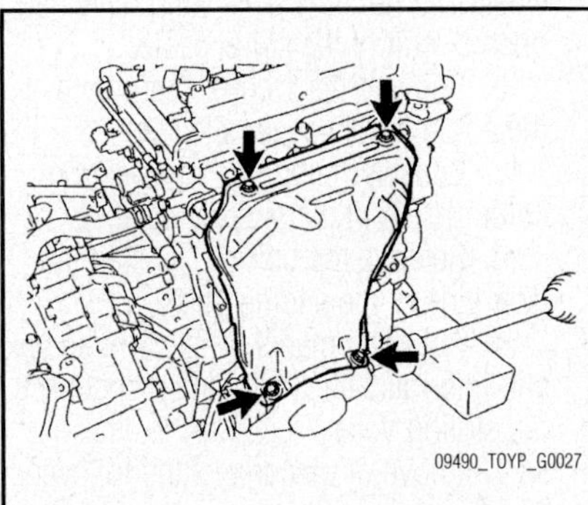

Fig. 41 Exhaust manifold insulator bolt locations

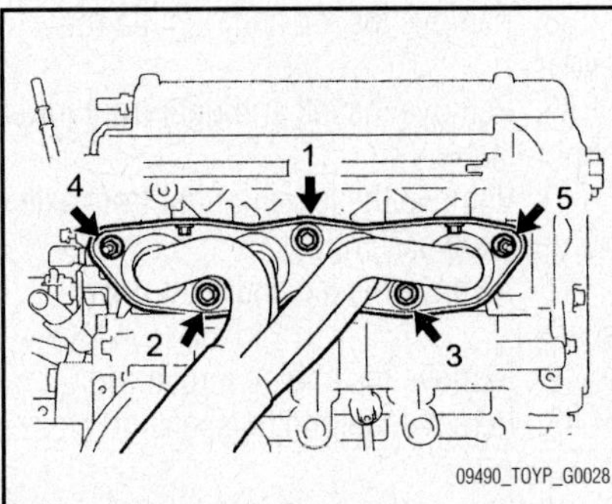

Fig. 42 Exhaust manifold nut and bolt tightening sequence

✲✲ CAUTION

To avoid the danger of being burned, do not service the exhaust system while it is hot.

Service should be performed only after the system cools down.

1. Before servicing the vehicle, refer to the precautions.
2. Properly relieve the fuel system pressure.
3. Disconnect the negative battery cable.

➡Wait at least 90 seconds after disconnecting the cable from the negative (-) battery terminal to prevent airbag and seat belt pretensioner activation.

4. While wearing insulating gloves, slide up the lever of the service plug grip and remove while turning the lever to the left. Be sure to insulate the service plug with insulating tape.

✲✲ CAUTION

Do not touch the high voltage connectors and terminals For 5 minutes after removing the service plug grip.

5. Remove the cylinder head.
6. Remove the 4 bolts and exhaust manifold insulator.
7. Remove the 3 bolts and 2 nuts, then remove the exhaust manifold.
8. Clean the sealing surfaces of the exhaust manifold and the cylinder head.

To install:

9. Install a new gasket, then install the exhaust manifold.
10. Tighten the 3 bolts and 2 nuts in the proper sequence to 20 ft. lbs. (27 Nm).
11. Install the exhaust manifold insulator with the 4 bolts and tighten to 71 inch lbs. (8 Nm).
12. Install the cylinder head.
13. Install the service plug.
14. Connect the negative battery cable.
15. Perform the power window initialization procedure. Refer to Engine Electrical, Initialization Procedure.
16. Run the engine and check for exhaust leaks.

FLYWHEEL

REMOVAL & INSTALLATION

See Figures 43 and 44.

1. Before servicing the vehicle, refer to the precautions.

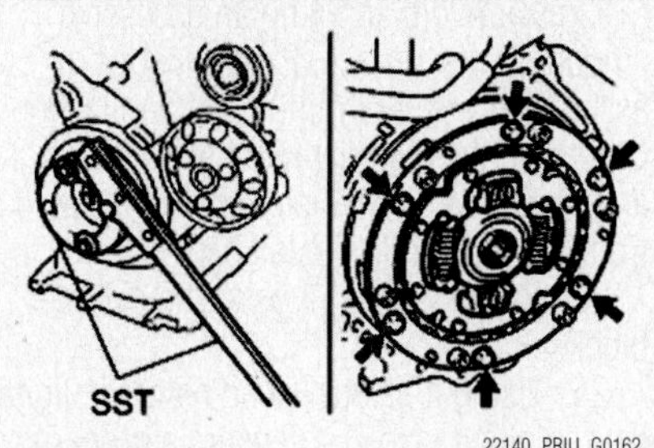

22140_PRIU_G0162

Fig. 43 Using the special tool, hold the crankshaft, remove the 6 bolts, then remove the input damper and input damper cover

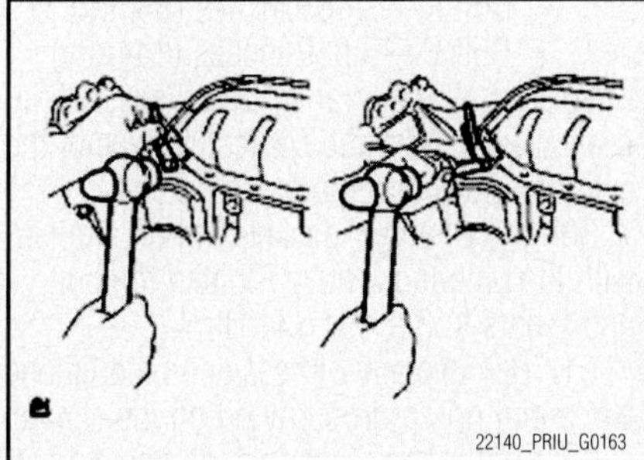
22140_PRIU_G0163

Fig. 44 Using the special tool, hold the crankshaft, remove the 6 bolts and flywheel

2. Remove hybrid vehicle transaxle assembly.
3. Remove transmission input damper assembly.
4. Remove the 6 bolts, then remove the input damper and input damper cover.
5. Remove the 6 bolts and flywheel.

To install:

6. Apply adhesive Part No. 08833-00070, THREE BOND 1324, or equivalent to the 2 or 3 threads of the bolt end.
7. Install the flywheel with the 6 bolts.
    a. 49 Nm (36 ft. lbs)
    b. Plus 90 degrees
8. Install the input damper and tighten the bolts to 20 Nm (15 ft. lbs)
9. Install hybrid vehicle transaxle assembly.

INTAKE MANIFOLD

REMOVAL & INSTALLATION

See Figures 45 through 47.

Toyota recommends this procedure be performed with the cylinder head removed. It may be possible to remove the intake manifold without removing the cylinder head.

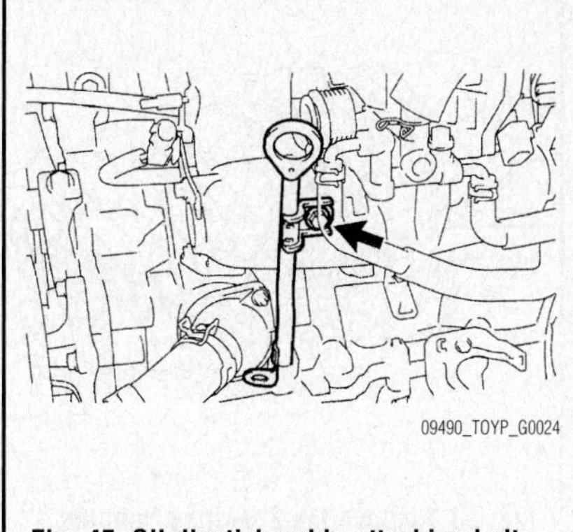
09490_TOYP_G0024

Fig. 45 Oil dipstick guide attaching bolt

09490_TOYP_G0025

Fig. 46 Intake manifold hoses/ harnesses

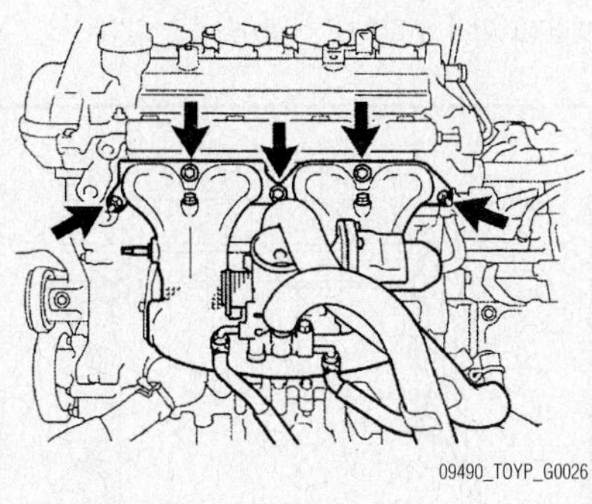
09490_TOYP_G0026

Fig. 47 Intake manifold mounting bolts and nuts

1. Before servicing the vehicle, refer to the precautions.
2. Disconnect the negative battery cable.

➡Wait at least 90 seconds after disconnecting the cable from the negative (-) battery terminal to prevent airbag and seat belt pretensioner activation.

3. Drain the engine coolant.
4. Remove the air cleaner assembly.
5. Remove the cylinder head.
6. Remove the oil dipstick guide.
7. Remove the bolt and knock control sensor with bracket.
8. Disconnect the wiring harness from the bracket.
9. Disconnect the ventilation hose.
10. Disconnect the water by-pass hose.
11. Remove the 3 bolts and 2 nuts, then remove the intake manifold and gasket.

To install:

12. Install a new gasket, then install the intake manifold with the bolts, nuts and brackets. Uniformly tighten the bolts and nuts in several passes to 15 ft. lbs. (20 Nm).
13. Connect the water by-pass hose.
14. Connect the ventilation hose.
15. Install the knock control sensor with bracket with the bolt and tighten to 80 inch lbs. (9 Nm).
16. Apply engine oil to a new O-ring, then install it to the dipstick guide.
17. Install the dipstick guide with the bolt and tighten to 80 inch lbs. (9 Nm).
18. Install the cylinder head.
19. Install the dipstick.
20. Install the air cleaner assembly.
21. Refill the engine coolant.
22. Connect the negative battery cable.
23. Perform the power window initialization procedure. Refer to Engine Electrical, Initialization Procedure.
24. Start the engine and check for coolant leaks.

OIL PAN

REMOVAL & INSTALLATION

See Figures 48 through 53.

Toyota recommends this procedure be performed with the engine assembly removed. It may be possible to remove the

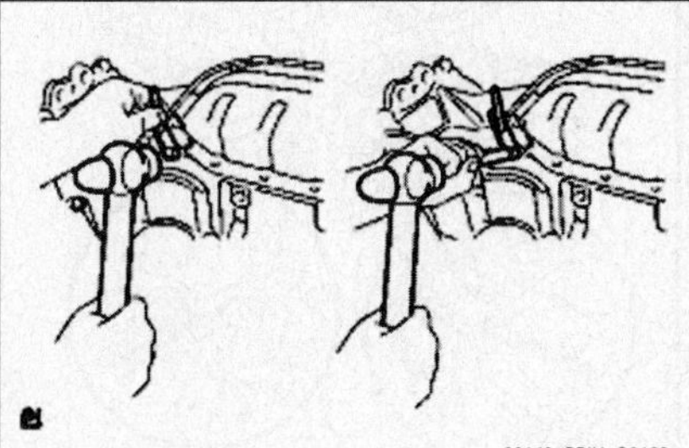
22140_PRIU_G0163

Fig. 48 Insert the blade of oil pan seal cutter between the oil pan and No. 2 oil pan, cut off the applied sealer and remove the No. 2 oil pan

oil pan without removing the engine assembly.

1. Before servicing the vehicle, refer to the precautions.
2. Disconnect the negative battery cable.

➡Wait at least 90 seconds after disconnecting the cable from the negative (-) battery terminal to prevent airbag and seat belt pretensioner activation.

3. Raise and safely support the vehicle securely on jackstands.
4. Drain engine oil.
5. Remove engine assembly.
6. Install engine to engine stand.
7. Remove the timing chain.
8. Remove the cylinder head.
9. Remove engine wire.
10. Remove the 2 nuts, bolt and water bypass pipe.
11. Remove the thermostat.
12. Remove the knock sensor.
13. Remove the oil pressure switch.
14. Remove the engine coolant drain union.
15. Remove the oil filter.
16. Using a 12mm hexagon wrench, remove the oil filter union.
17. Remove the 9 bolts and 2 nuts of the No. 2 oil pan.
18. Insert the blade of SST 09032-00100 between the oil pan No. 1 and oil pan No. 2, and cut off applied sealer and remove the oil pan. Be careful not to the damage the oil pan contact surface of the oil pan No. 1. or the oil pan No. 2 flange.

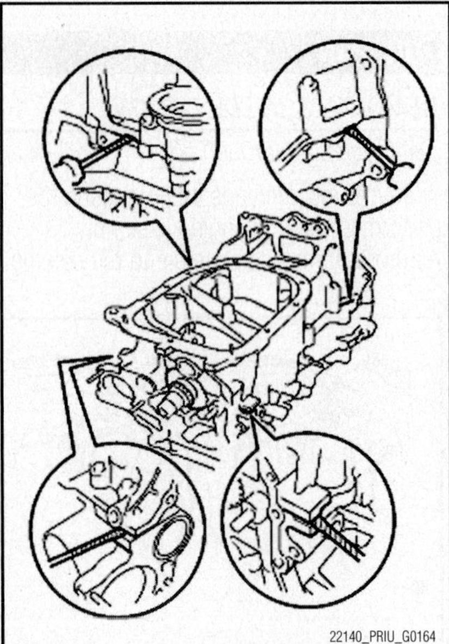

Fig. 49 Using a screwdriver, remove the oil pan by prying between the cylinder block and oil pan

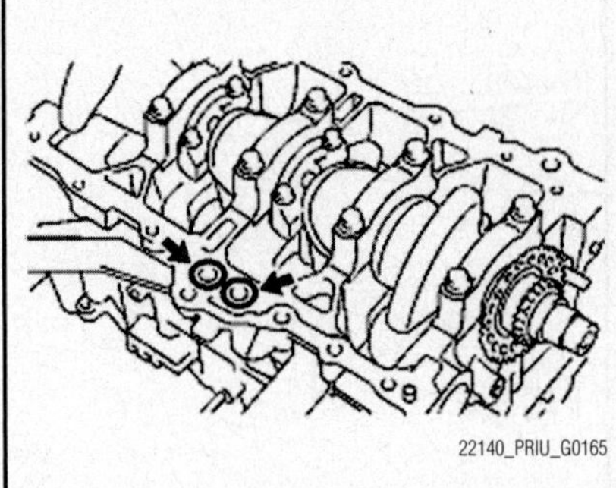

Fig. 50 Remove the 2 O-rings from the cylinder block

19. Remove the bolt and 2 nuts, oil strainer and gasket.
20. Uniformly loosen and remove the 13 bolts, in several passes.
21. Using screwdriver remove the oil pan No. 1 by prying the portions between the cylinder block and oil pan No. 1.
22. Remove the 2 O-rings from the cylinder block.

To install:

23. Remove any old sealant material and be careful not to drop any oil on the contact surface of the oil pan No. 1 and cylinder block. Using a razor blade and gasket scraper, remove all the old sealant material from the gasket surfaces and sealing grooves. Thoroughly clean all components to remove all the loose material. Using a non-residue solvent, clean both sealing surfaces.
24. Apply sealant to the oil pan No. 1 with a seal width of 0.08–0.12 inch (2–3mm). Avoid applying an excessive amount to the surface. Parts must be assembled within 3 minutes of application. Otherwise the material must be removed and reapplied. Immediately remove nozzle from the tube and reinstall cap.

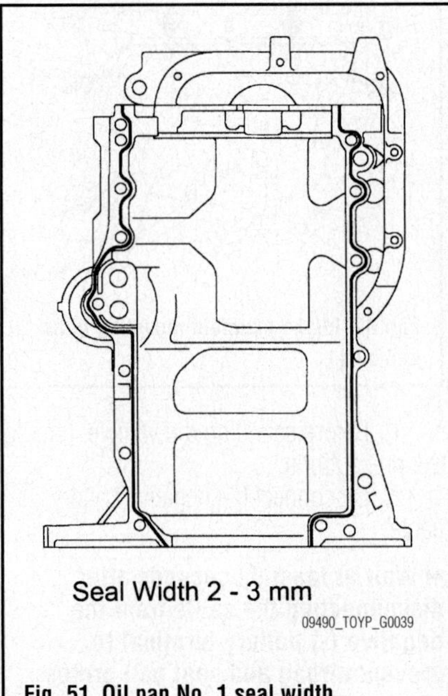

Fig. 51 Oil pan No. 1 seal width

25. Install new O-rings to the cylinder block.
26. Using a plastic-faced hammer, lightly tap the oil pan No. 1 to ensure a proper fit.
27. Install and uniformly tighten the 13 bolts, in several passes, in sequence to 18 ft. lbs. (24 Nm).
28. Each bolt indicated in the illustration shown is the following length:
    - Bolt A—1.929 inches (49mm)
    - Bolt B—3.465 inches (88mm)
    - Bolt C—5.669 inches (144mm)
29. Install rear crankshaft oil seal. Wipe sealant away from the contact surface of the cylinder block assembly and oil seal.
30. Install a new gasket, and oil strainer with the bolt and 2 nuts. Tighten the bolt and 2 nuts to 8 ft. lbs. (11 Nm).
31. Remove any old sealant material and be careful not to drop any oil on the contact surface of the main bearing cap and oil pan. Using a razor blade and gasket scraper, remove all the old sealant material from the gasket surfaces and sealing grooves. Thoroughly clean all components to remove

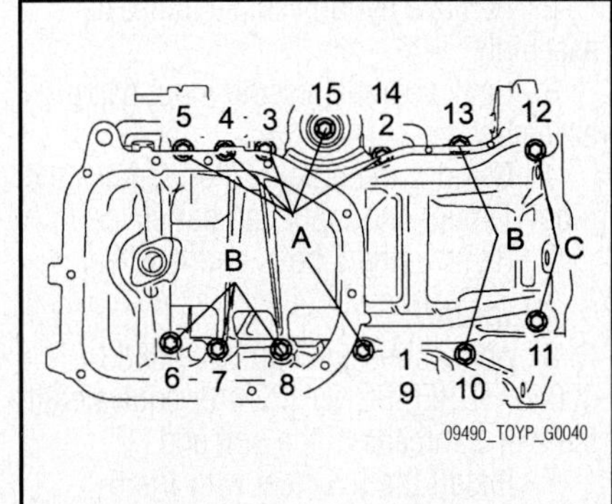

Fig. 52 Oil pan No. 1 bolt identification and tightening sequence

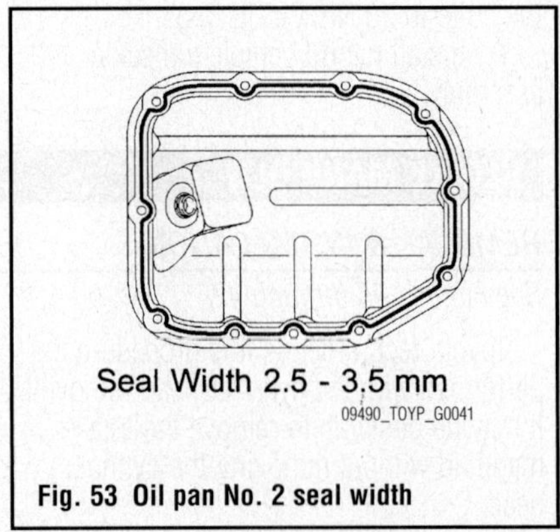

Fig. 53 Oil pan No. 2 seal width

all the loose material. Using a non-residue solvent, clean both sealing surfaces.

32. Apply sealant to the oil pan with a seal width of 0.098–0.138 inch (2.5–3.5mm). Avoid applying an excessive amount to the surface. Parts must be assembled within 3 minutes of application. Otherwise the material must be removed and reapplied. Immediately remove nozzle from the tube and reinstall cap.

33. Install the oil pan with the 9 bolts and 2 nuts. Uniformly tighten the bolts and nuts in several passes to 80 inch lbs. (9 Nm).

34. Install the oil filter union and tighten to 21 ft. lbs. (30 Nm).

35. Install the oil filter.

36. Apply adhesive to 2 or 3 threads and install the engine coolant drain union. Torque the union to 25 ft. lbs. (35 Nm) and after applying the specified torque, rotate the drain union clockwise until its drain port is facing downward.

37. Install the knock sensor and tighten to 29 ft. lbs. (39 Nm).

38. Install the oil pressure switch.

39. Install the thermostat.

40. Install the water bypass pipe and tighten bolts to 80 inch lbs. (9 Nm)

41. Install the engine wire.

42. Install the cylinder head.

43. Install the timing chain.

44. Remove the engine assembly from the engine stand.

45. Install the engine assembly into the vehicle.

46. Refill the engine with oil.

47. Connect the negative battery cable.

48. Perform the power window initialization procedure. Refer to Engine Electrical, Initialization Procedure.

49. Perform the power window initialization procedure. Refer to Engine Electrical, Initialization Procedure.

50. Start the engine and check for leaks.

OIL PUMP

REMOVAL & INSTALLATION

See Figures 54 through 56.

1. Before servicing the vehicle, refer to the precautions.

2. Disconnect the negative battery cable.

➡Wait at least 90 seconds after disconnecting the cable from the negative (-) battery terminal to prevent airbag and seat belt pretensioner activation.

3. Drain the engine oil.

4. Remove the timing chain cover.

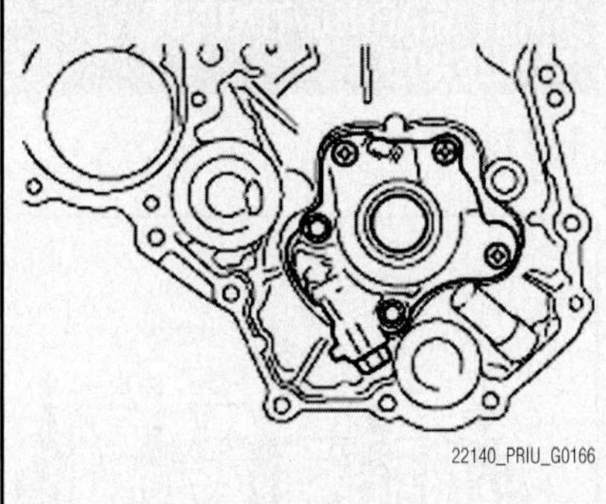

Fig. 54 Remove the 2 bolts and 3 screws, then remove the oil pump cover

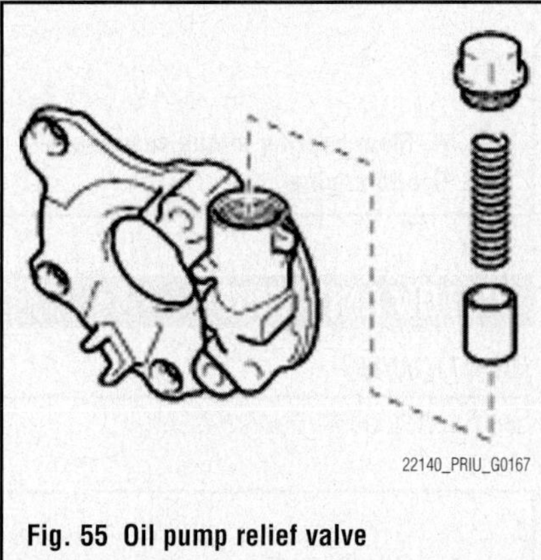

Fig. 55 Oil pump relief valve

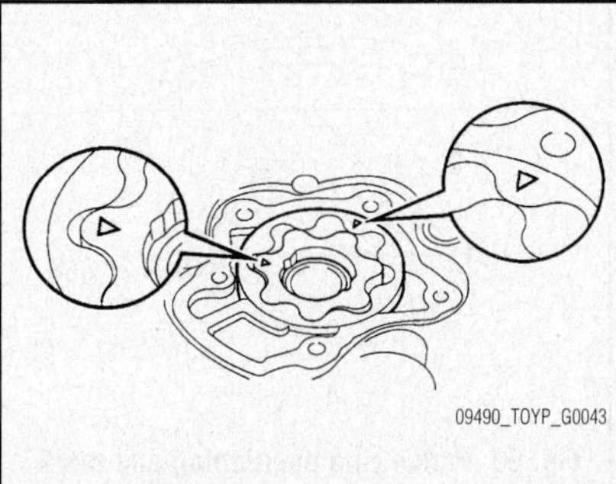

Fig. 56 Install the rotors into timing chain cover with the marks facing the oil pump cover side

5. Remove the 2 bolts, 3 screws and oil pump cover from the timing chain cover.

6. Remove the drive and driven rotors.

7. Remove the plug, spring and relief valve.

To install:

8. Insert the relief valve and spring into the pump body hole, and install the plug. Torque the plug to 18 ft. lbs. (25 Nm).

9. Place the drive and driven rotors into timing chain cover with the marks facing the oil pump cover side.

10. Install the oil pump cover to the timing chain cover with the 2 bolts and 3 screws. Torque the bolts to 78 inch lbs. (8.8 Nm), and the screws to 8 ft. lbs. (10.3 Nm).

11. Install the timing chain cover (refer to the timing chain procedure).

12. Refill engine with engine oil.

13. Connect the negative battery cable.

14. Perform the power window initialization procedure. Refer to Engine Electrical, Initialization Procedure.

15. Perform the power window initialization procedure. Refer to Engine Electrical, Initialization Procedure.

16. Start the engine and check the engine oil pressure.

17. Check that no leaks are present.

INSPECTION

See Figures 57 through 60.

1. Before servicing the vehicle, refer to the precautions.

2. Remove the oil pump from the engine and disassemble it.

3. Coat the relief valve with engine oil and check that it falls smoothly into the valve hole by its own weight.

4. If it doesn't, replace the relief valve. If necessary, replace the oil pump assembly.

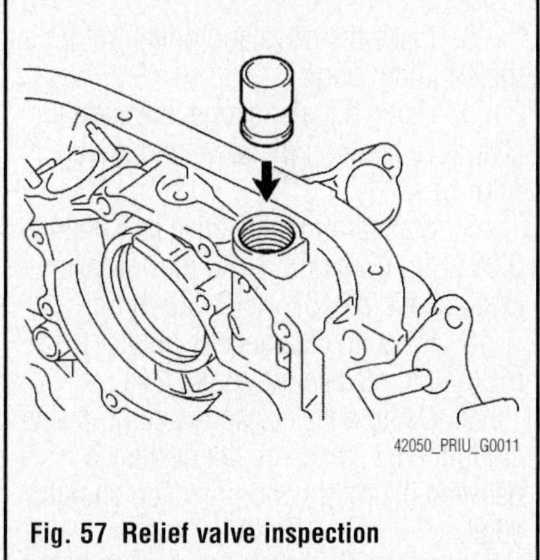

Fig. 57 Relief valve inspection

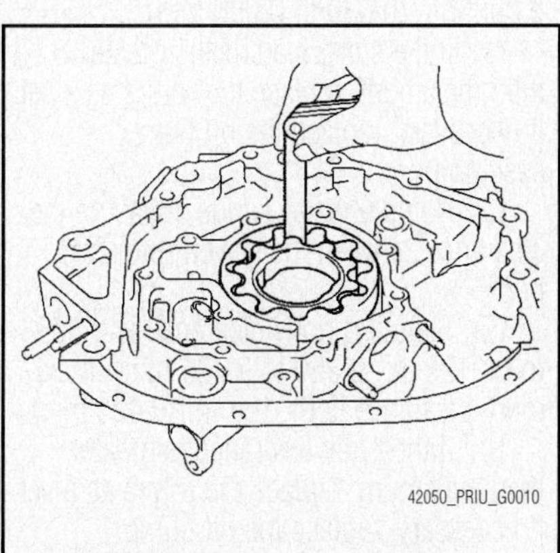

Fig. 58 Measuring rotor tip clearance

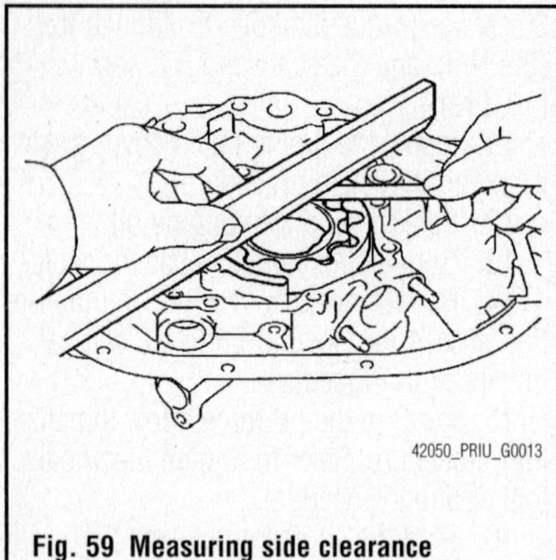

Fig. 59 Measuring side clearance

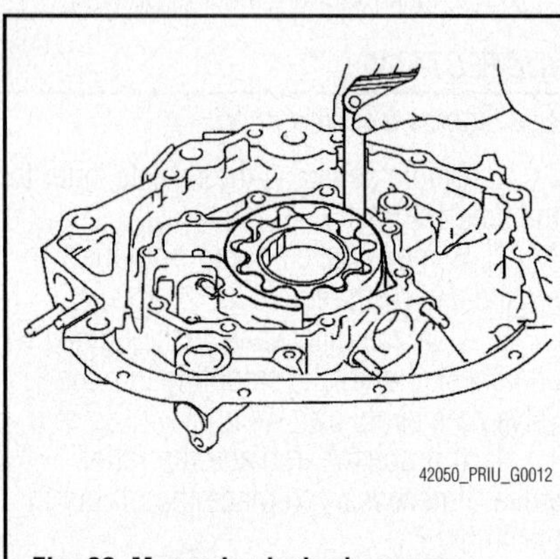

Fig. 60 Measuring body clearance

5. Place the drive and driven rotors into the oil pump body.
6. Using a feeler gauge, measure the clearance between the drive and driven rotor tips.
7. Standard tip clearance is 0.0024 to 0.0071 in. (0.060 to 0.180 mm). Maximum clearance is 0.0167 in. (0.425 mm).
8. If the tip clearance is greater than maximum, replace the rotors as a set.
9. Using a feeler gauge and precision straight edge, measure the clearance between the rotors and precision straight edge.
10. Standard side clearance is 0.0012 to 0.0035 in. (0.030 to 0.090 mm).
11. If the side clearance is greater than maximum, replace the rotors as a set. If necessary, replace the oil pump assembly.
12. Using a feeler gauge, measure the clearance between the driven rotor and body.
13. Standard body clearance is 0.0098 to 0.0128 in. (0.250 to 0.325 mm). Maximum clearance is 0.0167 in. (0.425 mm).
14. If the body clearance is greater than maximum, replace the rotors as a set. If necessary, replace the oil pump assembly.

MAIN BEARING TORQUE SEQUENCE

See Figure 61.

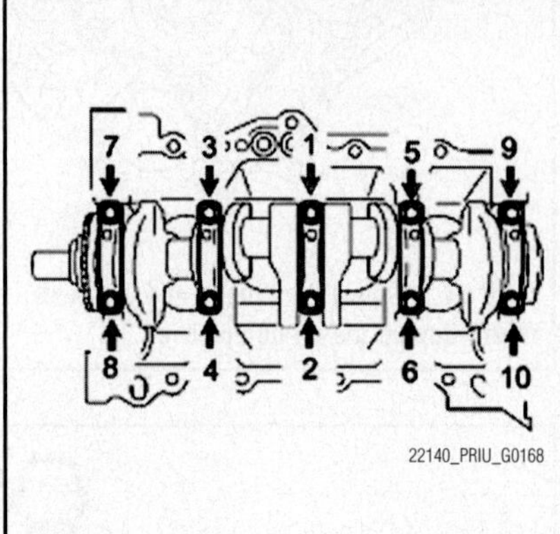

Fig. 61 Main bearing torque sequence—1.5L Hybrid engine

PISTON AND RING

POSITIONING

See Figures 62 through 64.

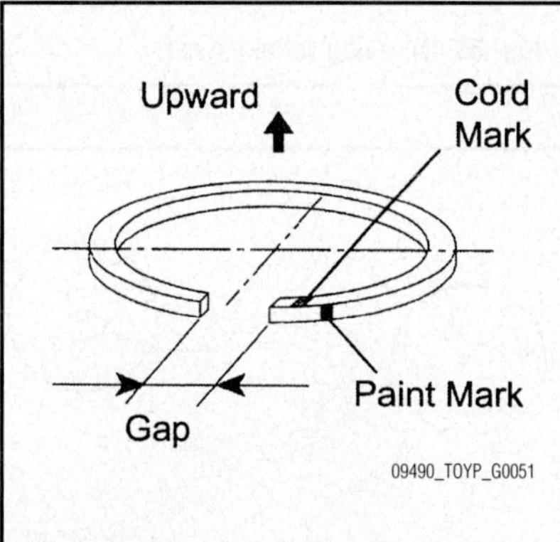

Fig. 62 Piston ring positioning and mark locations—1.5L Hybrid engine

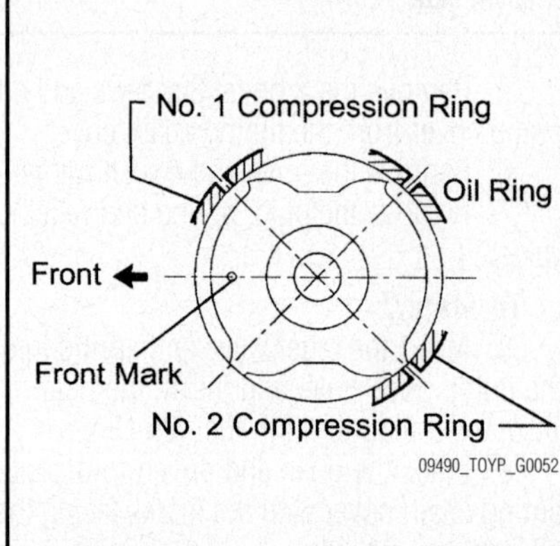

Fig. 63 Piston ring positioning—1.5L Hybrid engine

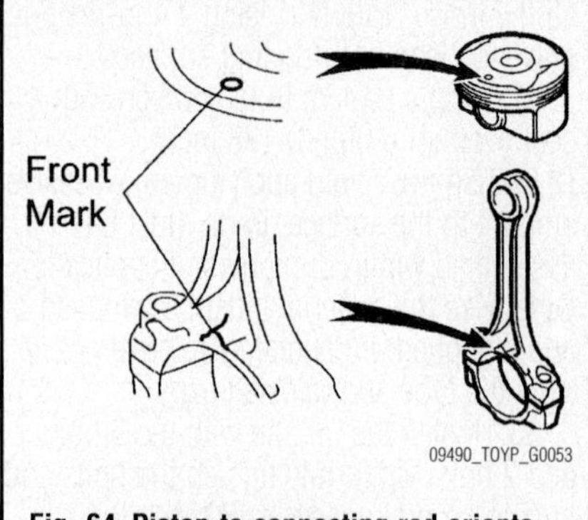

Fig. 64 Piston-to-connecting rod orientation—1.5L Hybrid engine

REAR MAIN SEAL

REMOVAL & INSTALLATION

See Figures 65 and 66.

1. Before servicing the vehicle, refer to the precautions.
2. Remove flywheel from the crankshaft.
3. Using a knife, cut off the lip of the oil seal.
4. Using a screwdriver with the tip wrapped in tape, carefully pry out the oil seal without scratching the sealing surface of the crankshaft.

➡Check if the crankshaft is damaged. If it is damaged, smooth the surface with 400-grit sandpaper.

To install:

5. Apply multi-purpose grease to the lip of the new seal.

➡Keep the lip free of foreign materials.

6. Install the seal in the retainer using a suitable seal driver. Wipe any extra grease off the crankshaft.
7. Install the flywheel.

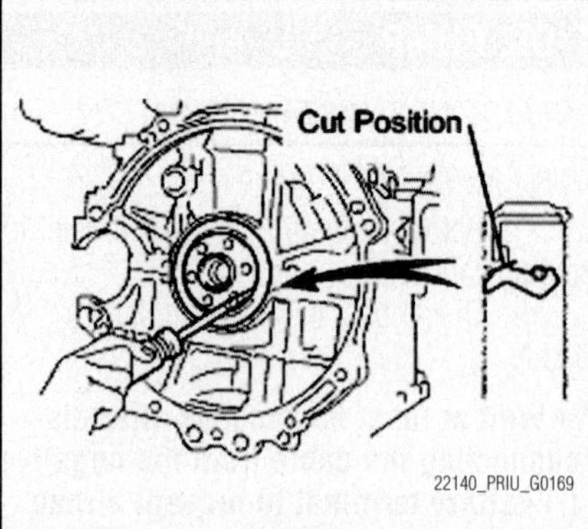

Fig. 65 Using a knife, cut off the lip of the oil seal

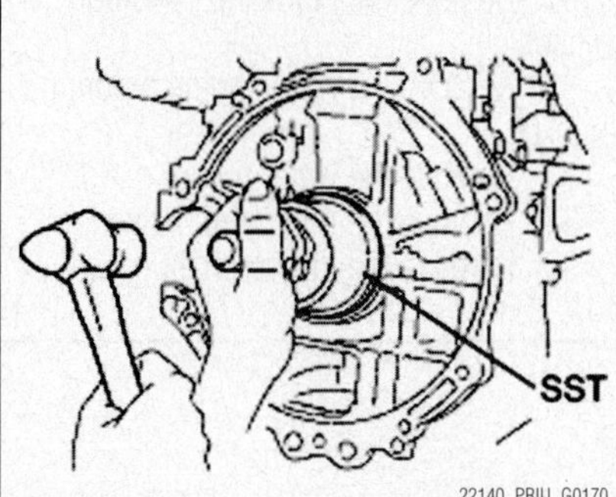

Fig. 66 Using a seal installer and a hammer, tap in the oil seal until its surface is flush with the rear oil seal retainer edge

ROCKER ARMS/SHAFTS

REMOVAL & INSTALLATION

The 1.5L hybrid engine does not utilize rocker arms/shafts, the camshaft directly actuates the valves.

TIMING CHAIN COVER AND SEAL

REMOVAL & INSTALLATION

See Figures 67 through 73.

On this engine, the timing chain cover also functions as a housing for the oil pump and may be referred to as the oil pump or oil pump cover.

1. Before servicing the vehicle, refer to the precautions.
2. Properly relieve the fuel system pressure.

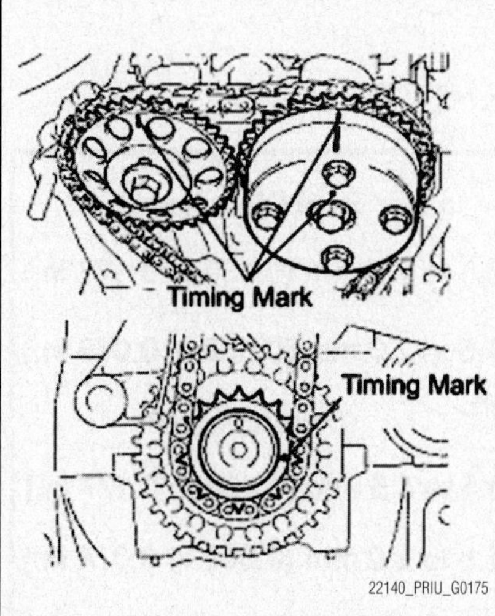

Fig. 67 Aligning both timing marks on the camshaft timing sprocket and valve timing controller assembly

3. Disconnect the negative battery cable.

➡Wait at least 90 seconds after disconnecting the cable from the negative (-) battery terminal to prevent airbag and seat belt pretensioner activation.

4. While wearing insulating gloves, slide up the lever of the service plug grip and remove while turning the lever to the left. Be sure to insulate the service plug with insulating tape.

⁂ CAUTION

Do not touch the high voltage connectors and terminals For 5 minutes after removing the service plug grip.

5. Remove the outer front cowl top panel.
6. Remove right engine under cover.
7. Drain the engine coolant.
8. Remove the air cleaner assembly.
9. Remove the brake fluid level sensor connector.
10. Remove the brake fluid reservoir tank mounting bolts and suspend it with rope.
11. Remove the brake fluid reservoir tank bracket.
12. Remove the ignition connectors.
13. Remove the fuel injector connectors.
14. Remove the vacuum switching valve connectors.
15. Remove the camshaft position sensor connector.
16. Remove the water temperature connector.
17. Remove the camshaft timing oil control valve connector.
18. Remove the air cleaner inlet hose.
19. Remove the engine coolant reservoir tank.

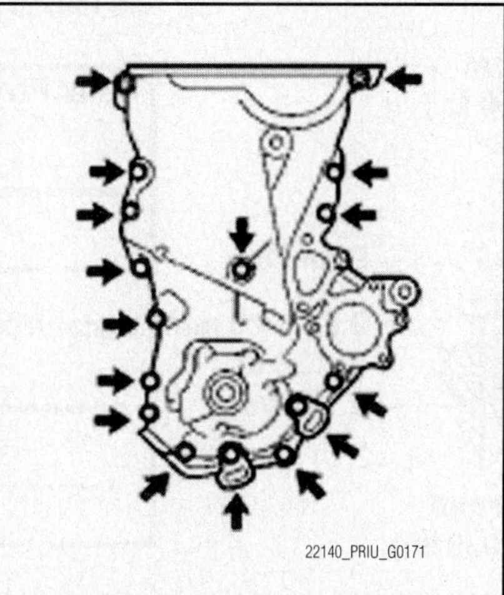

Fig. 68 Timing chain attaching bolt locations

20. Remove the vacuum suction valve from engine mounting insulator.
21. Remove the accessory drive belt.
22. Place a floor jack under the engine to support it, with a block of wood between the engine and the jack.
23. Remove the right engine mounting insulator.
24. Remove the engine wiring from cylinder head cover.
25. Remove the ignition coils.
26. Remove the PCV hoses.

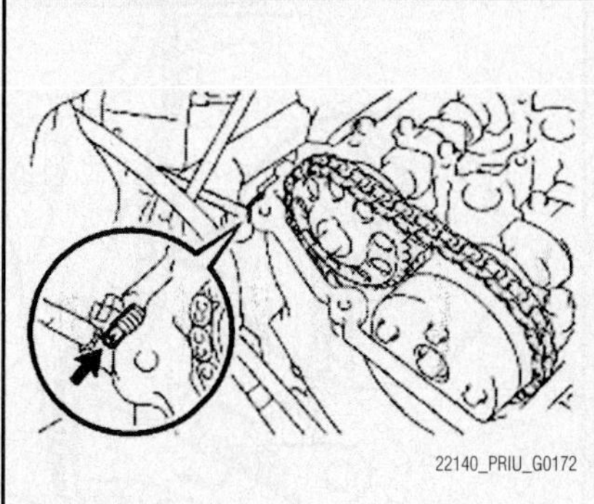

Fig. 69 Using an E8 Torx® socket wrench, remove the stud bolt

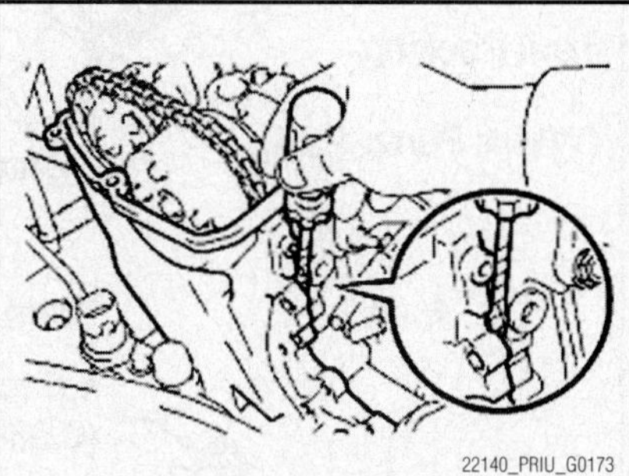

Fig. 70 Using a screwdriver with its tip wrapped in tape, remove the cover by prying out between the cylinder head and cylinder block

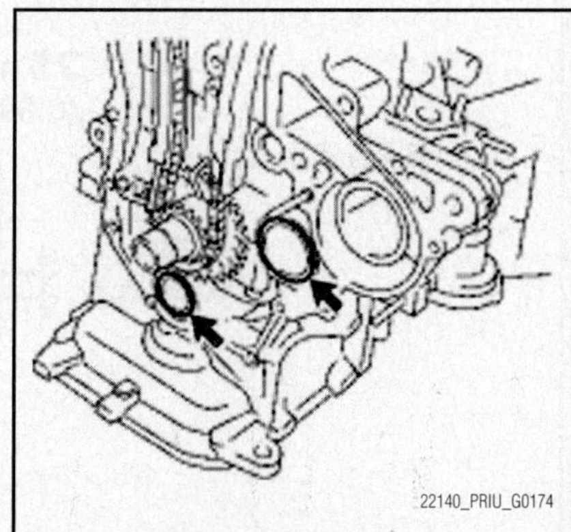

Fig. 71 Remove the 2 O-rings

27. Remove the 7 bolts, 2 seal washers, 2 nuts, cylinder head cover and gasket.

28. Set the No. 1 cylinder to Top Dead Center/compression by turning the crankshaft pulley and aligning its groove with timing mark **"0"** of the timing chain cover.

29. Check that both timing marks on the camshaft timing sprocket and valve timing controller assembly are facing right up. If not, turn the crankshaft 1 revolution (360 degrees) and align the marks.

30. Remove the crankshaft pulley bolt and the crankshaft pulley.

31. Remove the crankshaft position sensor.

32. Remove the right engine mounting bracket.

33. Remove the water pump.

34. Remove the oil control valve.

35. Remove the timing chain cover.

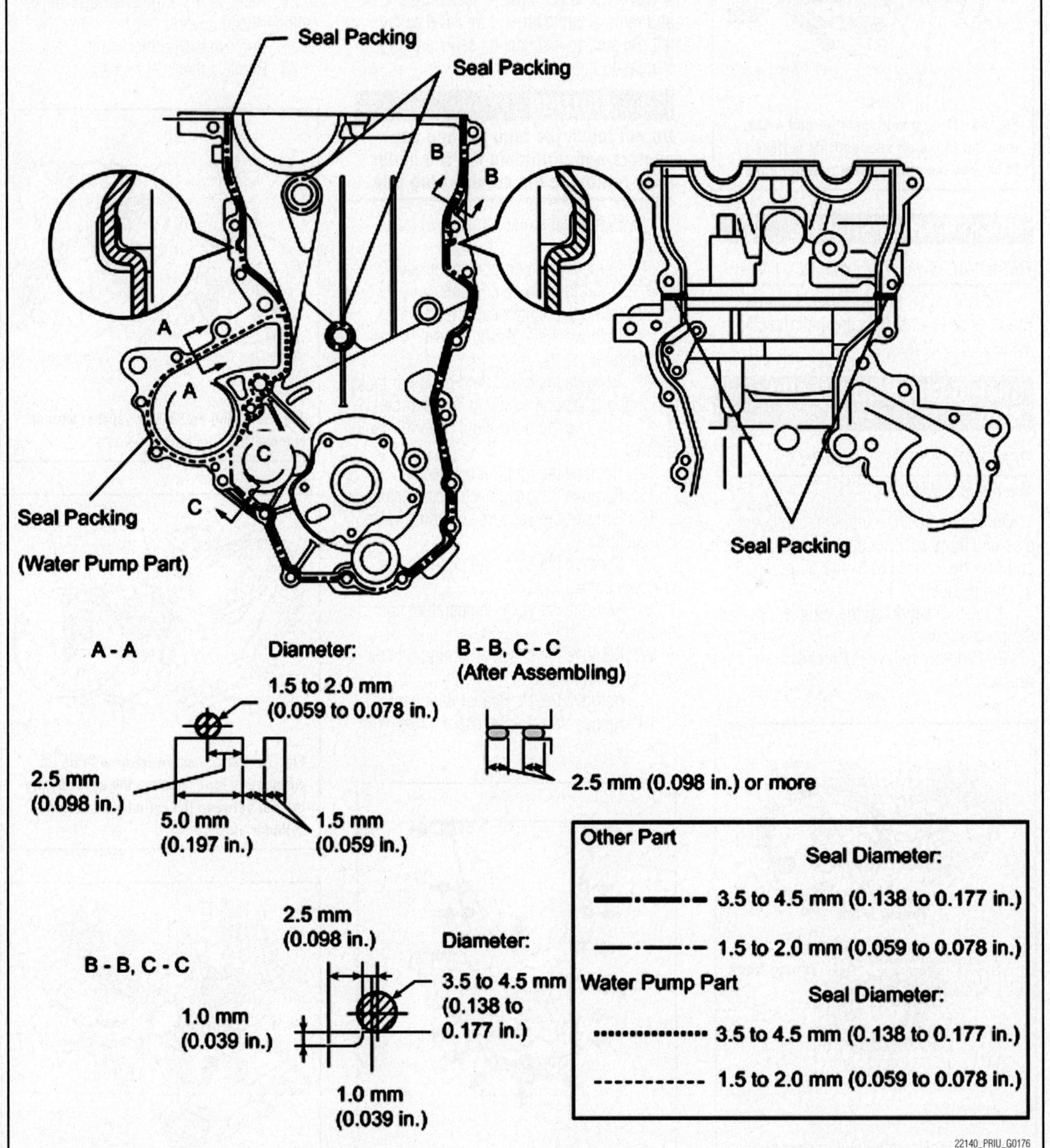

Fig. 72 Apply sealant to the timing chain cover as shown

36. Using a screwdriver, remove the oil seal. Tape the screwdriver tip before use.

To install:

37. Using SST 09950-60010, 09950-70010 and a hammer, tap in a new oil seal until its surface is flush with the timing chain cover edge. Be careful not to tap the oil seal at an angle. Keep the lip free of foreign objects.

38. Apply multi-purpose grease to the lip of the oil seal.

39. Remove any old sealant material and be careful not to drop any oil on the contact surfaces of the timing chain cover, cylinder head and cylinder block. Using a razor blade and a gasket scraper, remove all the old sealant material from the gasket surfaces and sealing grooves. Thoroughly clean all components to remove all the loose material. Using a non-residue solvent, clean both sealing surfaces.

40. Apply sealant to the timing chain cover. Install a nozzle that has been cut to a 0.16–0.20 inch (4–5mm) opening. Sealant shall be accumulated in the groove to a depth of 0.10 inch (2.5mm) or more. Avoid applying an excessive amount to the surface. Parts must be assembled within 3 minutes of application. Otherwise the material must be removed and reapplied. Immediately remove nozzle from the tube and reinstall cap.

41. Install 2 new O-rings to the cylinder block and oil pan No. 1.

42. Install the timing chain cover, new O-ring and water pump with the bolts and nuts. Uniformly tighten the bolts and nut in several passes to the following torque values:

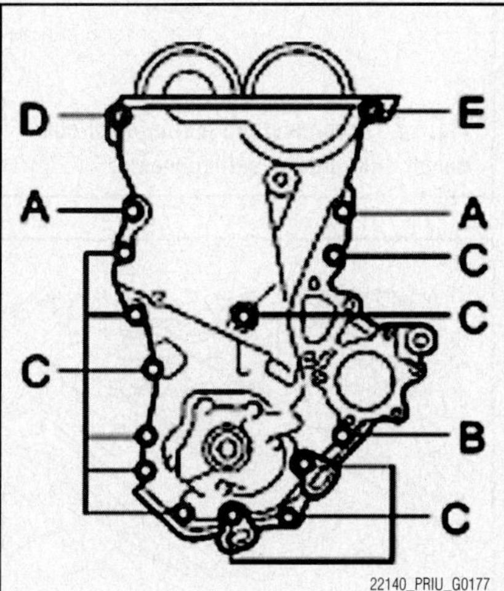

Fig. 73 Correct installation of the timing chain cover and water pump bolts and nuts

- Bolt A, 0.787 inch (20mm)—18 ft. lbs. (24 Nm)
- Bolt B, 1.181 inch (30mm)—8 ft. lbs. (11 Nm)
- Bolt C, 1.378 inch (35mm)—8 ft. lbs. (11 Nm)
- Bolt D, 0.787 inch (20mm)—18 ft. lbs. (24 Nm)
- Bolt E, 1.378 inch (35mm)—18 ft. lbs. (24 Nm)

➡Pay attention not to wrap the chain and slipper over the chain cover seal line. After installing the chain cover, must install the mounting bracket and water pump within 15 minutes.

43. Apply sealant to threads of the engine mounting bracket mounting bolt, but do not apply sealant to 2 or 3 threads of the bolt end.

44. Install the right mounting bracket with the 4 bolts and tighten to 41 ft. lbs. (55 Nm).

45. Install the crankshaft position sensor and tighten the bolt at the sensor to 66 inch lbs. (7.5 Nm), and the other bolts to 8 ft. lbs. (11 Nm).

46. Install the oil control valve and tighten to 66 inch lbs. (7.5 Nm).

47. Install the crankshaft pulley.

a. Clean the crankshaft pulley inside.

b. Install the pin to the crankshaft.

c. Align the hole in the crank pulley with the pin position and install the crank pulley.

d. Using SST 09213-70011 and 09330-00021, install the pulley bolt and tighten to 94 ft. lbs. (128 Nm).

48. Remove any old sealant material and apply sealant to 2 locations as shown in the illustration.

49. Install the gasket to the cylinder head cover.

➡Part must be assembled within 3 minutes of sealant application. Otherwise the material must be remove and reapplied.

50. Install the cylinder head cover and cable bracket with the 7 bolts, 2 seal washers and 2 nuts. Uniformly tighten the bolts and nuts, in the several passes, in the sequence to 7 ft. lbs. (10 Nm).

51. Connect the 2 PCV hoses to the cylinder head cover.

52. Connect the engine wire to cylinder head cover.

53. Install the ignition coils.

54. Install the RH engine mounting insulator with the 5 bolts and 2 nuts.

55. Install VSV to right engine mounting insulator

56. Install the drive belt.

57. Install the engine coolant reservoir tank.

58. Install the air inlet.

59. Connect the Camshaft timing oil control valve connector.

60. Connect the water temperature sensor connector.

61. Connect the camshaft position sensor connector.

62. Connect the 2 VSV connectors.

63. Connect the 4 injector connectors.

64. Connect the 4 ignition connectors.

65. Install the air cleaner assembly.

66. Install brake fluid reservoir tank.

67. Install the outer front cowl top panel assembly.

68. Fill the engine with coolant.

69. Install the engine under covers.

70. Connect the negative battery cable and high voltage battery service plug.

71. Road test the vehicle and check for abnormal noises, shock slippage, correct shift points and smooth operation.

72. Recheck the engine and transaxle fluids.

TIMING CHAIN AND SPROCKETS

REMOVAL & INSTALLATION

See Figures 74 through 77.

1. Before servicing the vehicle, refer to the precautions.

2. Properly relieve the fuel system pressure.

3. Disconnect the negative battery cable.

➡Wait at least 90 seconds after disconnecting the cable from the negative (-) battery terminal to prevent airbag and seat belt pretensioner activation.

4. While wearing insulating gloves, slide up the lever of the service plug grip

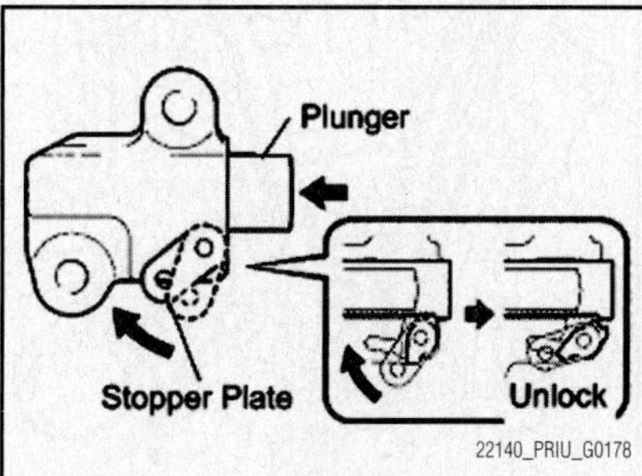

Fig. 74 Lift up the stopper plate, then unlock the plunger then push in the plunger to the end with the plunger unlocked

and remove while turning the lever to the left. Be sure to insulate the service plug with insulating tape.

CAUTION

Do not touch the high voltage connectors and terminals for 5 minutes after removing the service plug grip.

5. Remove the timing chain cover.
6. Remove the chain tensioner.
   a. Lift up the stopper plate, then unlock the plunger.
   b. Push in the plunger to the end with the plunger unlocked.
   c. Lower the stopper plate with the plunger pushed to the end, then lock the plunger.
   d. Insert a 3.0 mm (0.118 in.) diameter bar into the hole of the stopper plate with the plunger locked.

➡If the stopper plate is not completely lowered and a 0.118 in. (3.0 mm) diameter bar cannot be inserted, unlock and pull out the plunger slightly. Then the stopper plate will be completely lowered and a 0.118 in. (3.0 mm) diameter bar can be inserted easily.

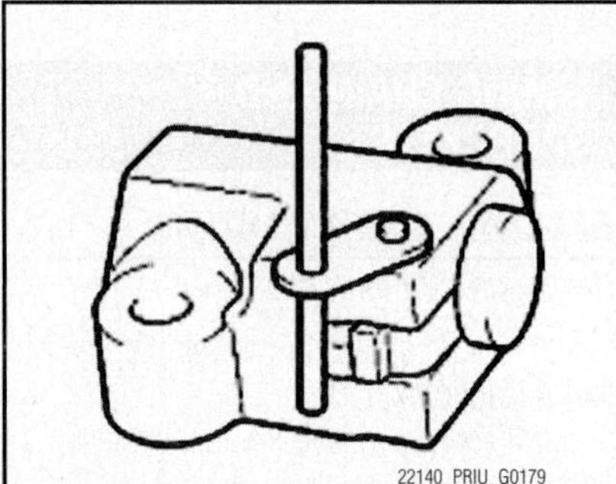

Fig. 75 Insert a 0.118 in. (3.0 mm) diameter bar into the hole of the stopper plate with the plunger locked

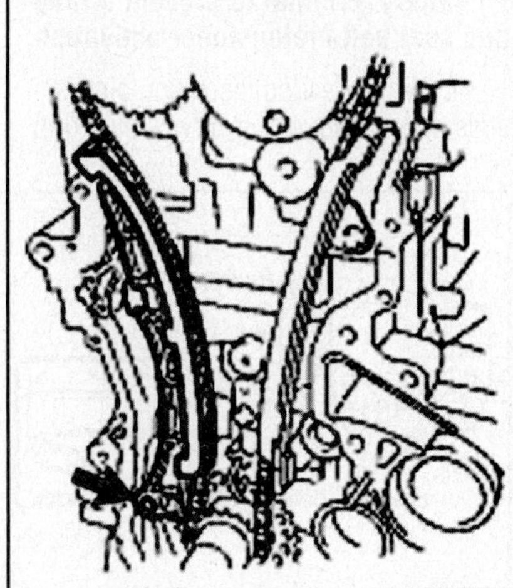

Fig. 76 Remove the bolt and tensioner slipper

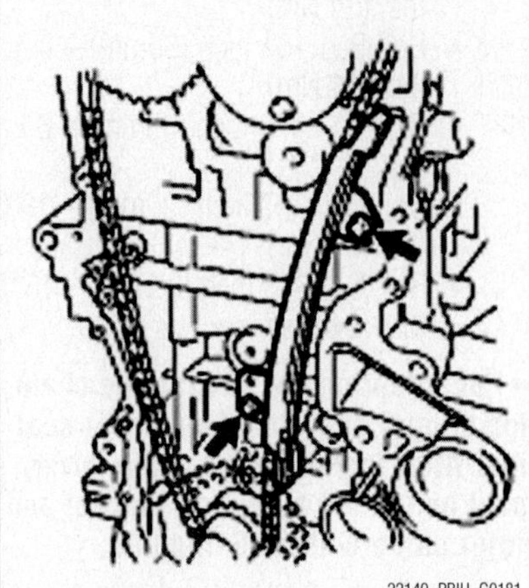

Fig. 77 Remove the bolts and chain vibration damper

   e. Remove the 2 bolts and chain tensioner.
7. Remove the chain tensioner slipper.
8. Remove the chain vibration damper.
9. Remove the timing chain.

To install:

10. Install the timing chain.
11. After setting the crankshaft at ATDC 40–140 degrees, set cams of intake and exhaust timing sprockets at ATDC 20 degrees and then the reset the crankshaft at ATDC 20 degrees.
12. Install the chain vibration damper with the 2 bolts and tighten to 80 inch lbs. (9 Nm).
13. Align the match marks of timing chain mark plate (Yellow), camshaft timing sprocket, camshaft timing gear and crankshaft timing sprocket to install the timing chain. To prevent the exhaust camshaft from spring back, turn it using a wrench and set it at the mark on a chain.
14. Install the chain tensioner slipper.
15. While rotating the lock plate of the tensioner up-ward, push in the plunger of the tensioner.
16. While rotating the lock plate of the tensioner down-ward, insert a bar of 0.098 inch (2.5mm) into the holes in the lock plate.
17. Install the chain tensioner with the 2 bolts and tighten to 80 inch lbs. (9 Nm).
18. Remove the bar from the chain tensioner.
19. Check the tension between the intake and exhaust camshaft timing sprocket.
20. Install the timing chain cover.
21. While wearing insulating gloves, install the service plug grip.
22. Connect the negative battery cable.
23. Perform the power window initialization procedure. Refer to Engine Electrical, Initialization Procedure.

VALVE COVER

REMOVAL & INSTALLATION

See Figures 78 through 82.

1. Before servicing the vehicle, refer to the precautions.
2. Disconnect the fuel injector connector and ignition coil connector.
3. Remove the 3 bolts and wire harness clamp, then disconnect the wire harness.
4. Remove the brake master cylinder reservoir cover.
5. Disconnect the ventilation hose and ventilation hose No. 2.
6. Remove the bolt and ignition coil.

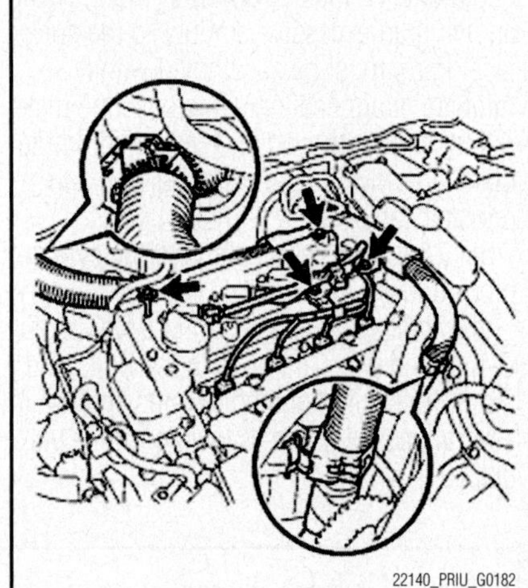

Fig. 78 Disconnect the fuel injector connector and ignition coil connector

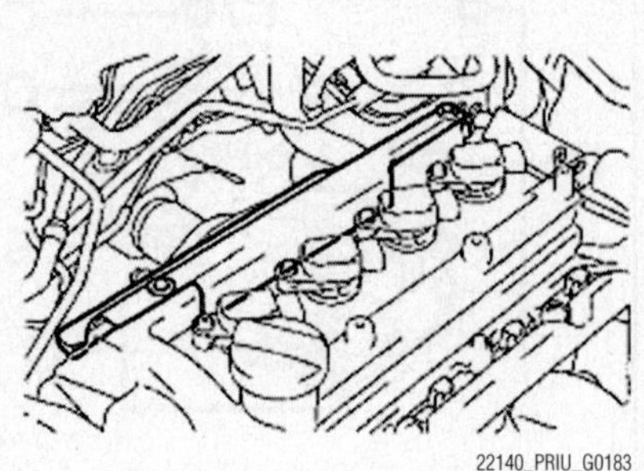

Fig. 79 Remove the 3 bolts and wire harness clamp, then disconnect the wire harness

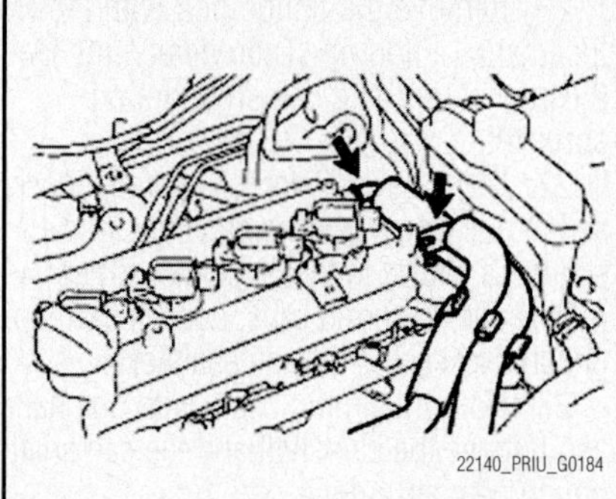
Fig. 80 Disconnect the ventilation hose and No. 2 ventilation hose

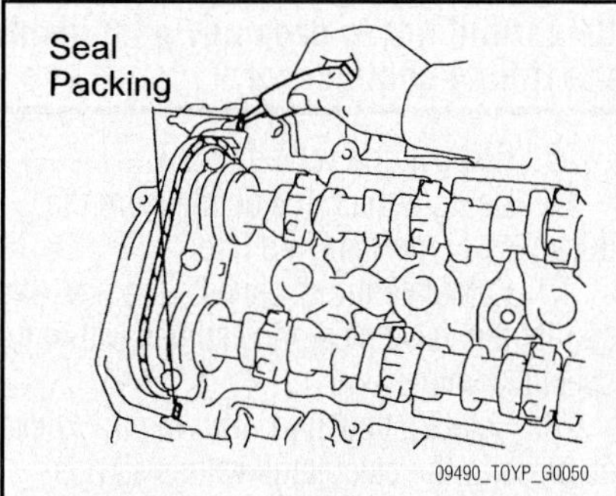

Fig. 81 Prior to installing the cylinder head cover, apply sealant to 2 locations as shown

7. Remove the 9 bolts and 2 nuts, then remove the cylinder head cover.

To install:

8. Apply Three Bond 1207B or equivalent sealant to the 2 locations shown in the illustration, then install the cylinder head cover.

➡Remove any oil from the contact surface. Install the cylinder head cover within 3 minutes after applying seal packing. Do not start the engine within 2 hours of installing.

9. Install the cylinder head cover with the 9 bolts and 2 nuts.

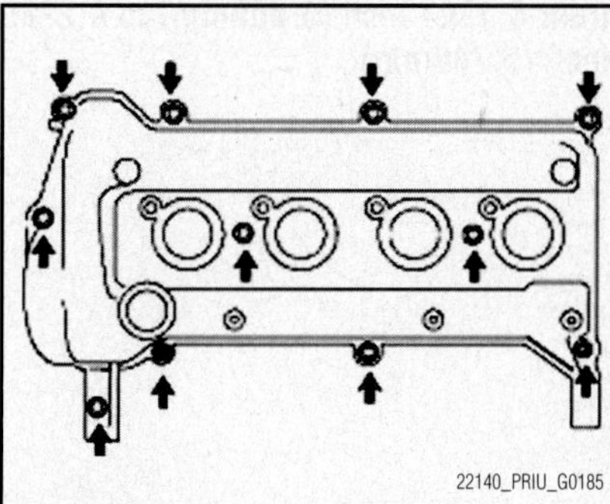
Fig. 82 Cylinder head cover bolt torque sequence

10. Using several steps, tighten the bolts and nuts to 7 ft. lbs. (10 Nm) in the sequence shown in the illustration.

11. Install the ignition coil with the bolt. Tighten to 80 inch lbs. (9 Nm).

12. Install the brake master cylinder reservoir cover to the cylinder head cover.

13. Install the wire harness and brake master cylinder reservoir cover with the 3 bolts. Tighten to 80 inch lbs. (9 Nm).

14. Connect the fuel injector connector and ignition coil connector.

VALVE LASH

ADJUSTMENT

See Figures 83 through 88.

➡Inspect and adjust the valve clearance when the engine is cold.

1. Before servicing the vehicle, refer to the precautions.
2. Set the No. 1 cylinder to TDC/compression.
   - Turn the crankshaft pulley until its timing notch and timing mark 0 of the chain cover are aligned.
   - Check that both timing marks on the camshaft timing sprocket and camshaft timing gear are facing upward as shown in the illustration. If not, turn the crankshaft 1 complete revolution (360 degrees) and align the marks as above.
3. Check the valves indicated in the illustration. Using a feeler gauge, measure the clearance between the valve lifter and camshaft.
   - Standard intake valve clearance (Cold): 0.007–0.009 inch (0.17–0.23mm)
   - Standard exhaust valve clearance (Cold): 0.011–0.013 inch (0.27–0.33mm)
4. Record any out-of-specification valve clearance measurements. They will be used later to determine the required replacement lifter.
5. Turn the crankshaft 1 complete revolution until its timing notch and timing mark 0 of the chain cover are aligned.
6. Check the valves indicated in the illustration. Using a feeler gauge, measure the clearance between the valve lifter and camshaft.
   - Standard intake valve clearance (Cold): 0.007–0.009 inch (0.17–0.23mm)
   - Standard exhaust valve clearance (Cold): 0.011–0.013 inch (0.27–0.33mm)
7. Record any out-of-specification valve clearance measurements. They will be used later to determine the required replacement lifter.
8. Set the No. 1 cylinder to TDC/compression.
9. Turn the crankshaft pulley until its timing notch and timing mark 0 of the chain cover are aligned.
10. Check that both timing marks on the camshaft timing sprocket and valve timing

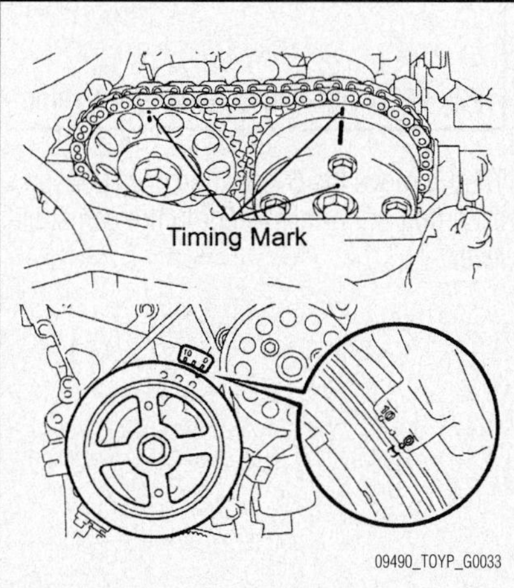

Fig. 83 Check that both timing marks on the camshaft timing sprocket and camshaft timing gear are facing upward

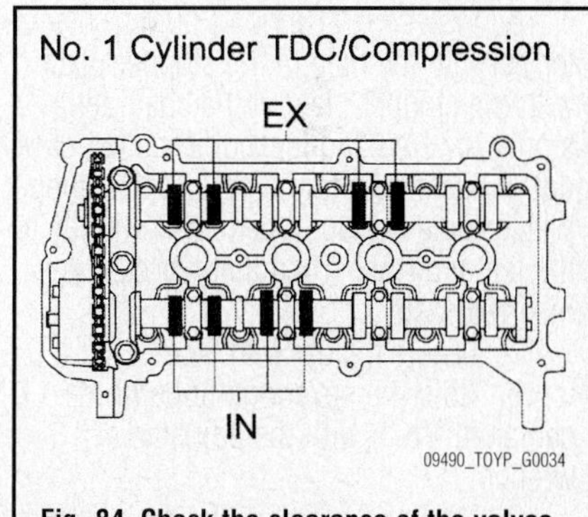

Fig. 84 Check the clearance of the valves indicated (No. 1 Cylinder)

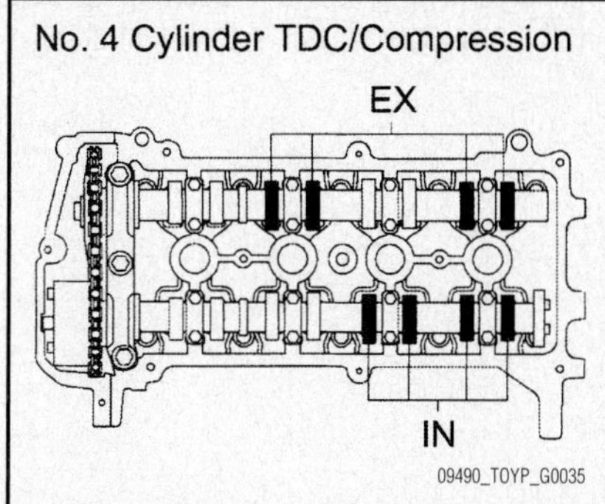

Fig. 85 Check the clearance of the following valves (No. 4 cylinder)

controller assembly are facing upward. If not, turn the crankshaft 1 complete revolution (360 degrees) and align the marks as above.

11. Put paint marks on the timing chain where the timing marks of the camshaft timing sprocket and the camshaft timing gear are located.

12. Using an 8mm hexagon wrench, remove the screw plug.

13. Insert a screwdriver into the service hole of the chain tensioner to hold the stopper plate of the chain tensioner at an upward position. Lifting up the stopper plate of the chain tensioner unlocks the plunger.

14. Keeping the stopper plate of the chain tensioner lifted, slightly rotate the hexagonal lobe of the No. 2 camshaft to the right with an adjustable wrench so the plunger of the chain tensioner is pushed. When the camshaft No. 2 is slightly rotated to the right, the plunger is pushed.

15. Keeping the adjustable wrench installed, remove the screwdriver with the plunger pushed. Do not move the adjustable wrench.

➡Removing the screwdriver lowers the stopper plate and locks the plunger.

16. Insert a 0.118 inch (3.0mm) diameter bar into the hole of the stopper plate with the stopper plate of the chain tensioner lowered and locked. If the bar cannot be inserted into the hole of the stopper plate, rotate the No. 2 camshaft slightly to the left and right. Then that bar can be inserted easily.

17. Secure the bar with tape.

18. Hold the hexagonal lobe of the camshaft No. 2 with the adjustable wrench.

19. Using SST 09023-38400, loosen the bolt.

20. Using several steps, uniformly loosen and remove the 11 bearing cap bolts in the sequence shown in the illustration.

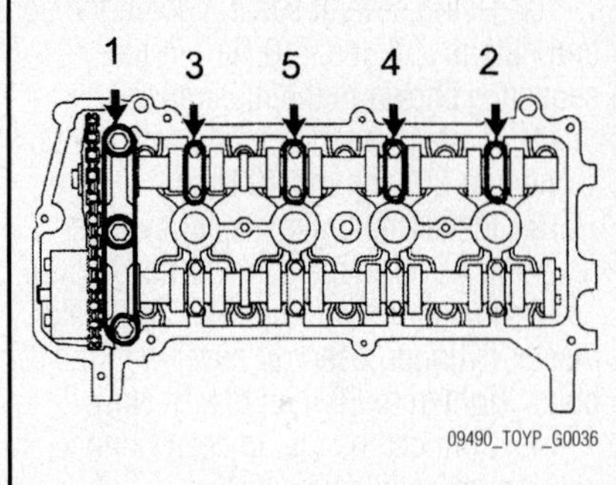

Fig. 86 Remove the 11 bearing cap bolts in the sequence

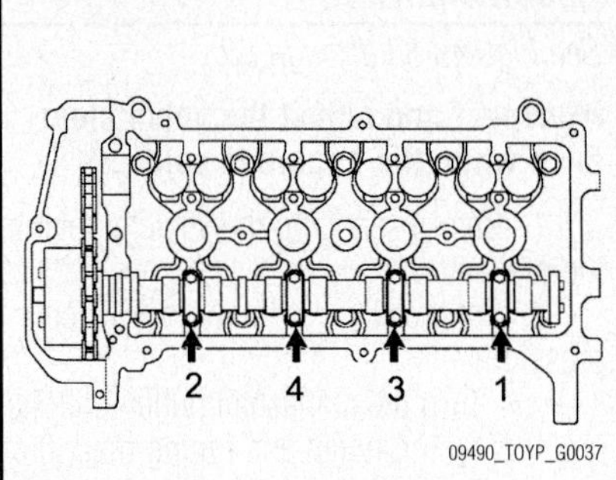

Fig. 87 Remove the 8 bearing cap bolts in the sequence

Fig. 88 Tie the timing chain with a string

Then remove the 5 bearing caps. Loosen each bolt uniformly, keeping the camshaft level.

21. Remove the flange bolt with the No. 2 camshaft lifted up. Then detach the No. 2 camshaft and the camshaft timing sprocket.

22. Using several steps, uniformly loosen and remove the 8 bearing cap bolts in the sequence shown in the illustration. Then remove the 4 bearing caps. Loosen each bolt uniformly, keeping the camshaft level.

23. Hold the timing chain with one hand, and remove the camshaft and the camshaft timing gear assembly.

24. Tie the timing chain with a string as shown in the illustration.

**** CAUTION**

Be careful not to drop anything inside the timing chain cover.

25. Remove the valve lifters.

26. Using a micrometer, measure the thickness of the removed lifter.

27. Calculate the thickness of a new lifter so that the valve clearance comes within the specified value.

28. Select a new lifter with the thickness as close to the calculated values as possible.

- EXAMPLE: (Intake) Measured valve clearance equals 0.0158 inch (0.40mm)
- 0.0158 inch (0.40mm) minus 0.0079 inch (0.20mm) equals 0.0079 inch (0.20mm) (Measured minus Specification equals Excess clearance)
- Used lifter measurement equals 0.2067 inch (5.25mm)
- 0.0079 inch (0.20mm) plus 0.2067 inch (5.25mm) equals 0.2146 inch (5.45mm) (Excess clearance plus Used lifter equals Ideal new lifter)
- Closest new lifter equals 5.45 mm (0.2146 in.); select lifter (0.2150 inch (5.46mm))

➡Lifters are available in 35 sizes in increments of 0.0008 inch (0.020mm), from 0.1992 inch (5.060mm) to 0.2260 inch (5.740mm).

ENGINE PERFORMANCE & EMISSION CONTROLS

COMPONENT LOCATIONS

See Figures 89 through 97.

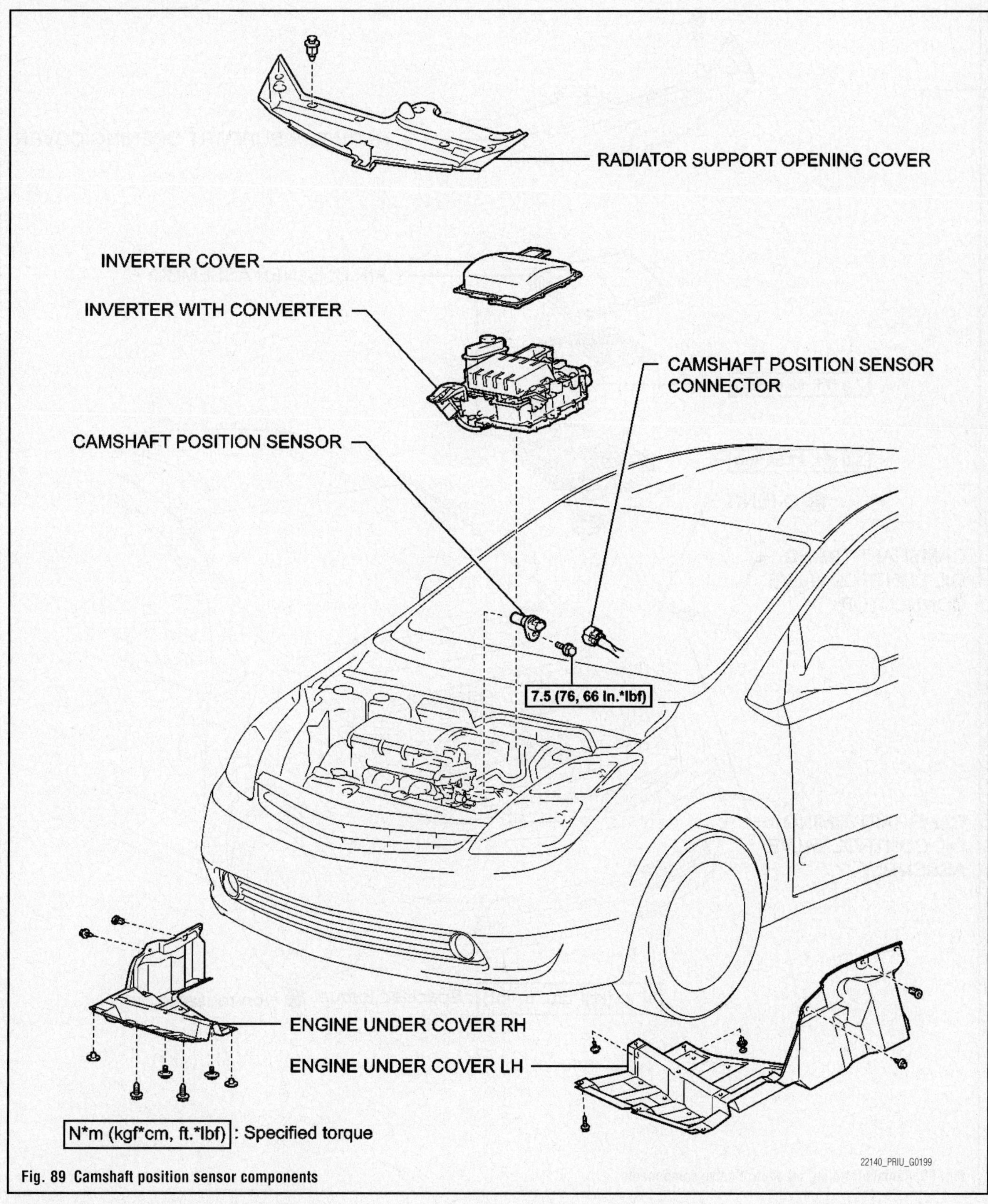

Fig. 89 Camshaft position sensor components

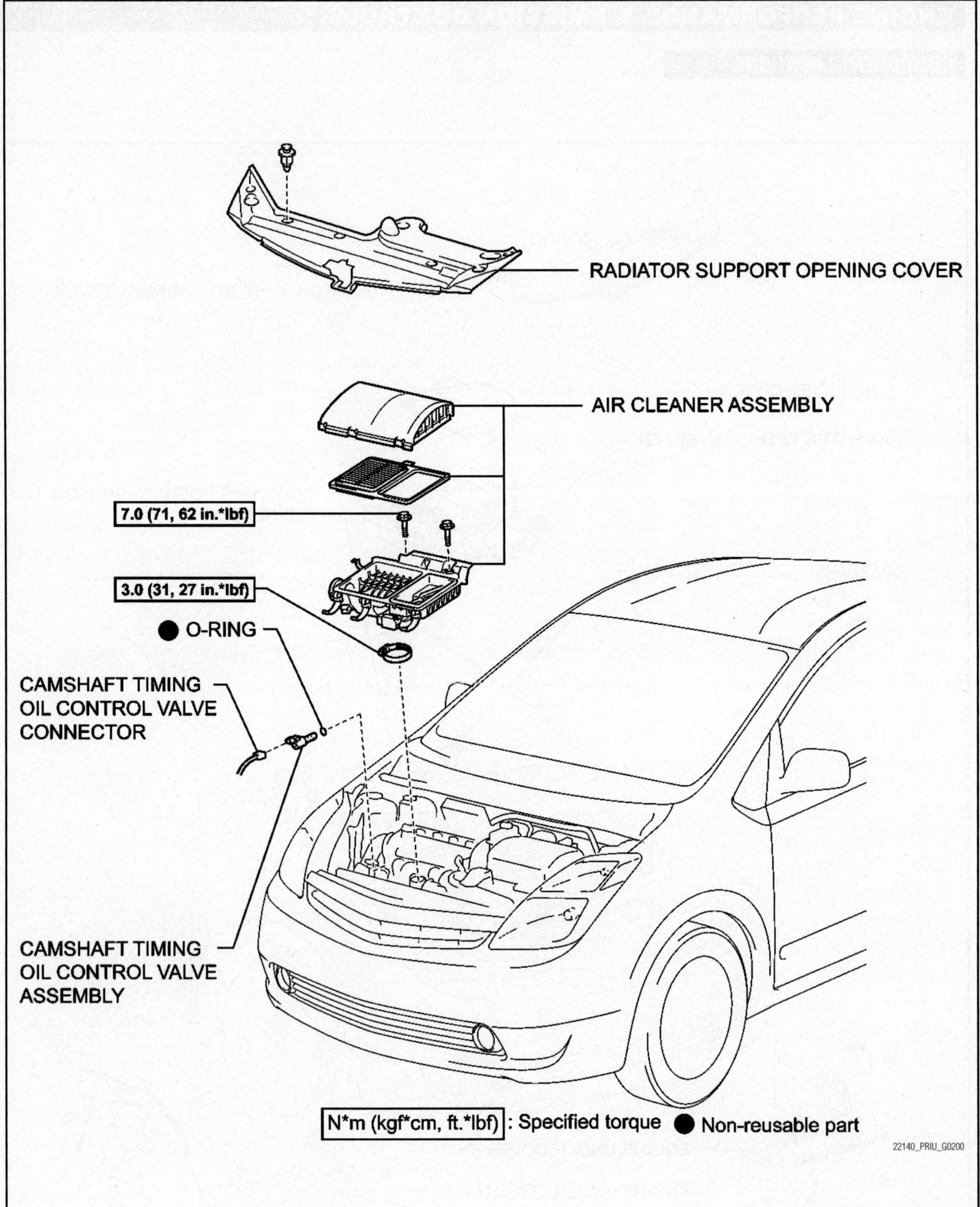

Fig. 90 Camshaft timing oil control valve components

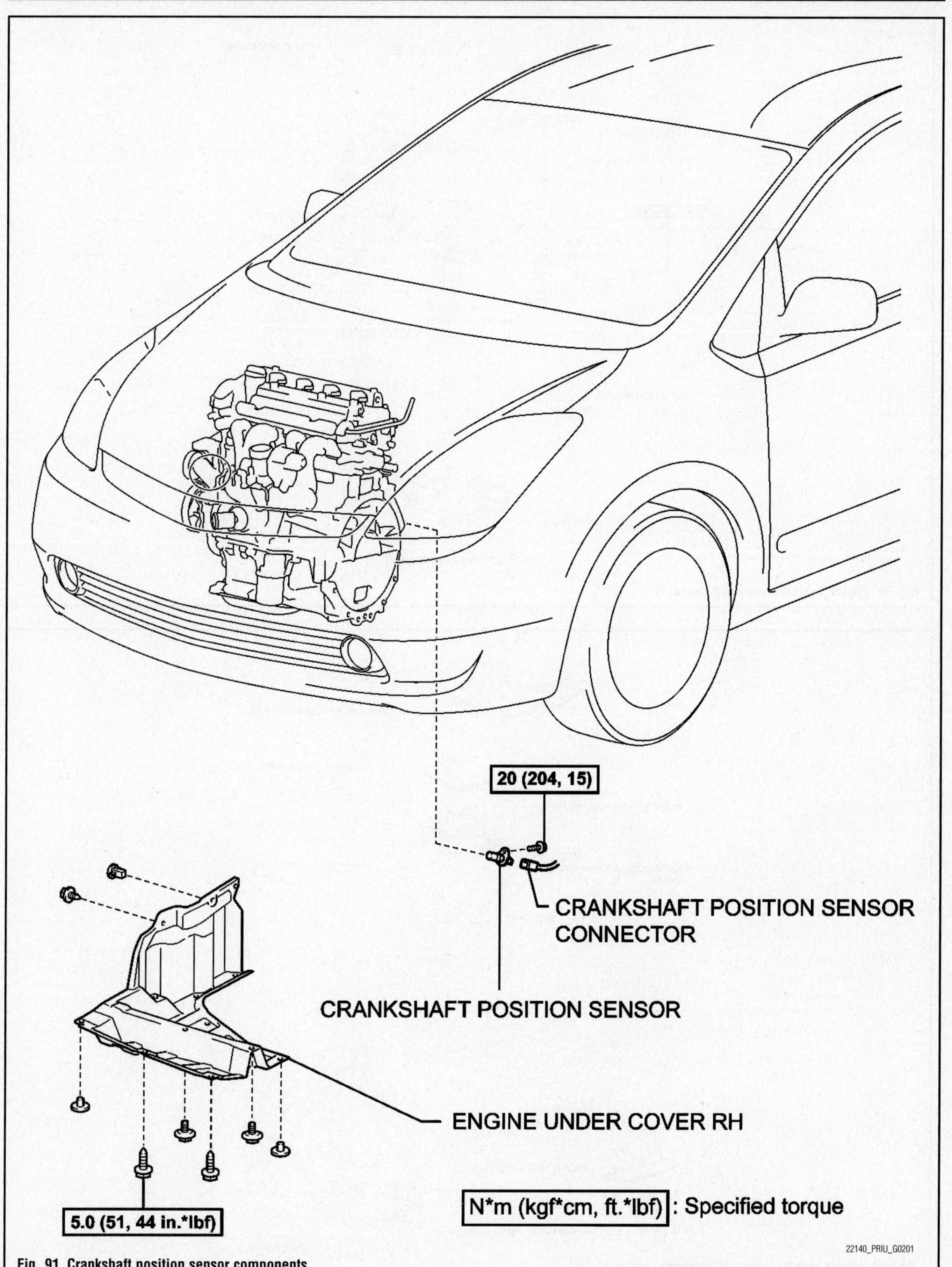

Fig. 91 Crankshaft position sensor components

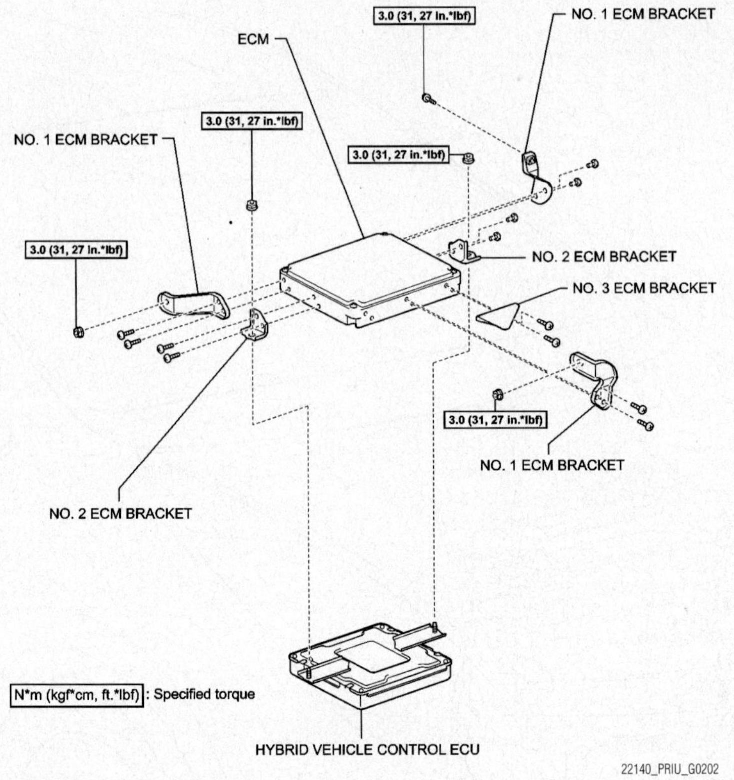

Fig. 92 Engine control module components

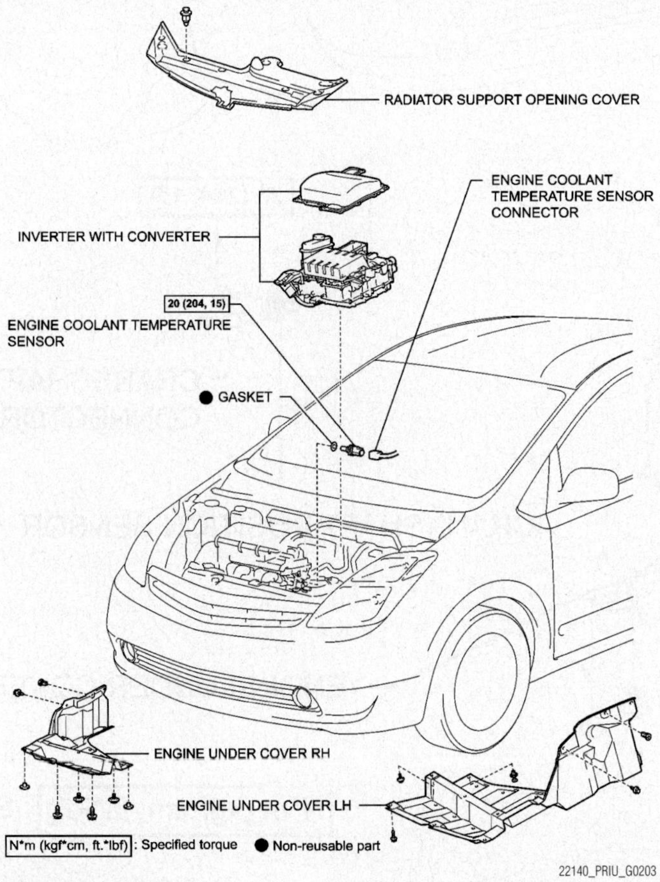

Fig. 93 Engine coolant temperature sensor components

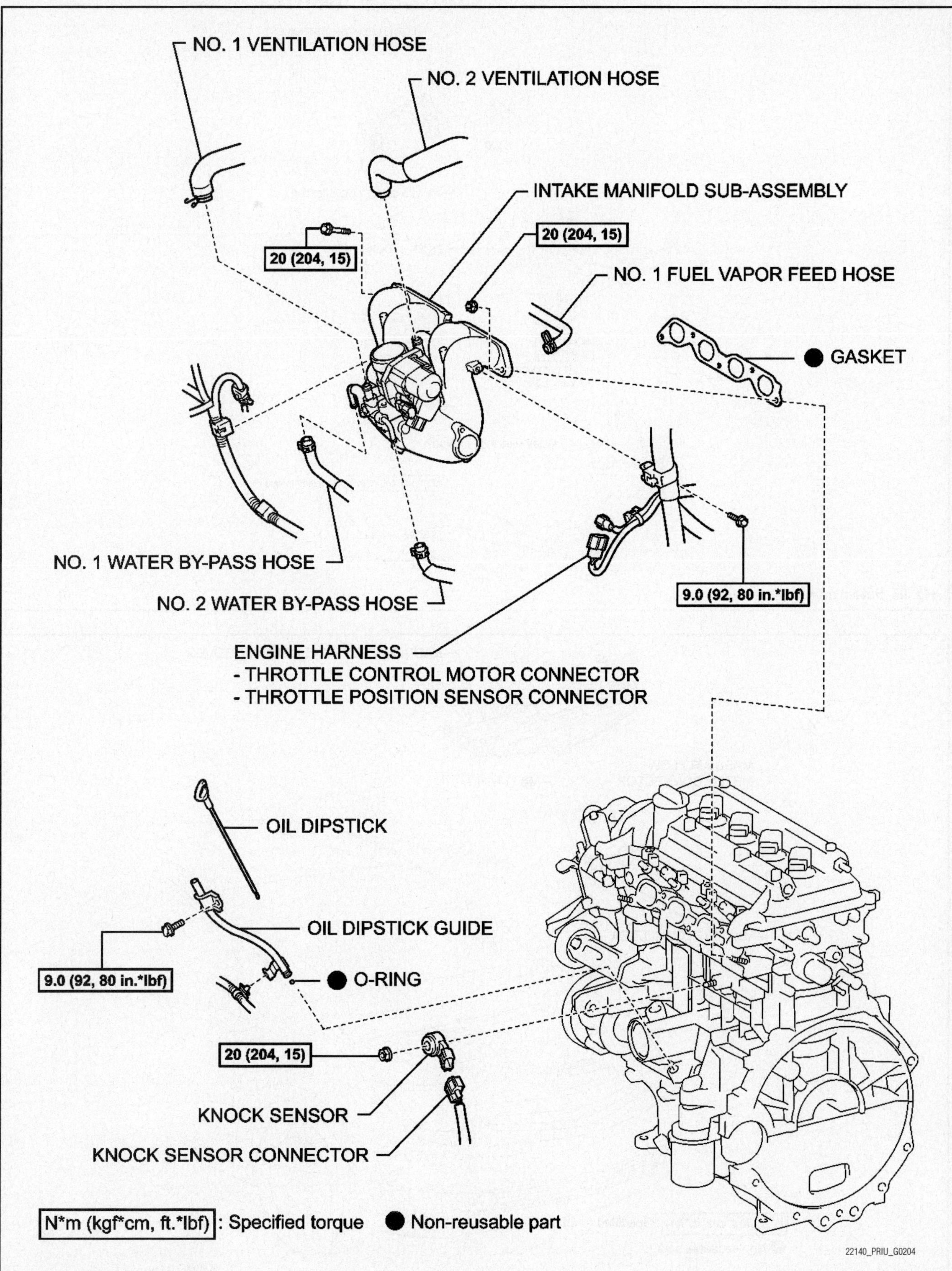

Fig. 94 Knock sensor components

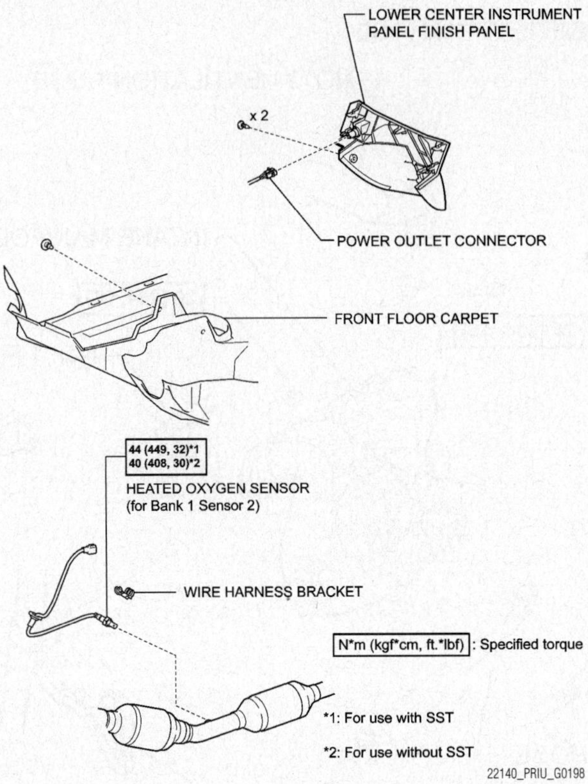

LOWER CENTER INSTRUMENT PANEL FINISH PANEL

x 2

POWER OUTLET CONNECTOR

FRONT FLOOR CARPET

44 (449, 32)*1
40 (408, 30)*2

HEATED OXYGEN SENSOR (for Bank 1 Sensor 2)

WIRE HARNESS BRACKET

N*m (kgf*cm, ft.*lbf) : Specified torque

*1: For use with SST

*2: For use without SST

22140_PRIU_G0198

Fig. 95 Heated oxygen sensor components

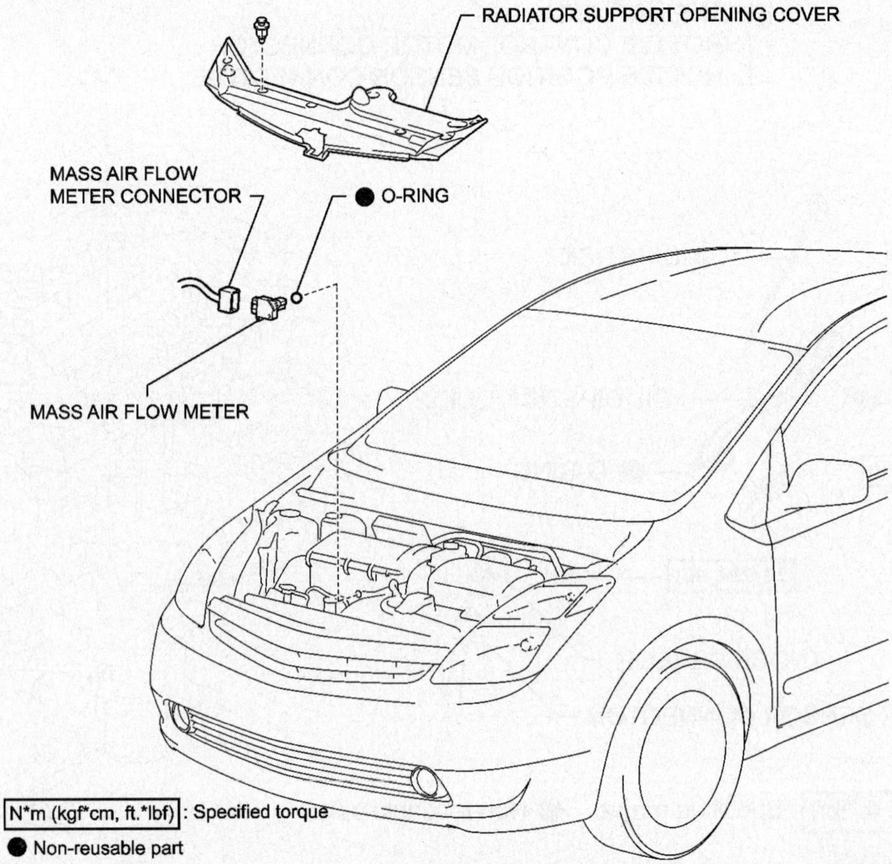

RADIATOR SUPPORT OPENING COVER

MASS AIR FLOW METER CONNECTOR

● O-RING

MASS AIR FLOW METER

N*m (kgf*cm, ft.*lbf) : Specified torque

● Non-reusable part

22140_PRIU_G0205

Fig. 96 Mass air flow sensor components

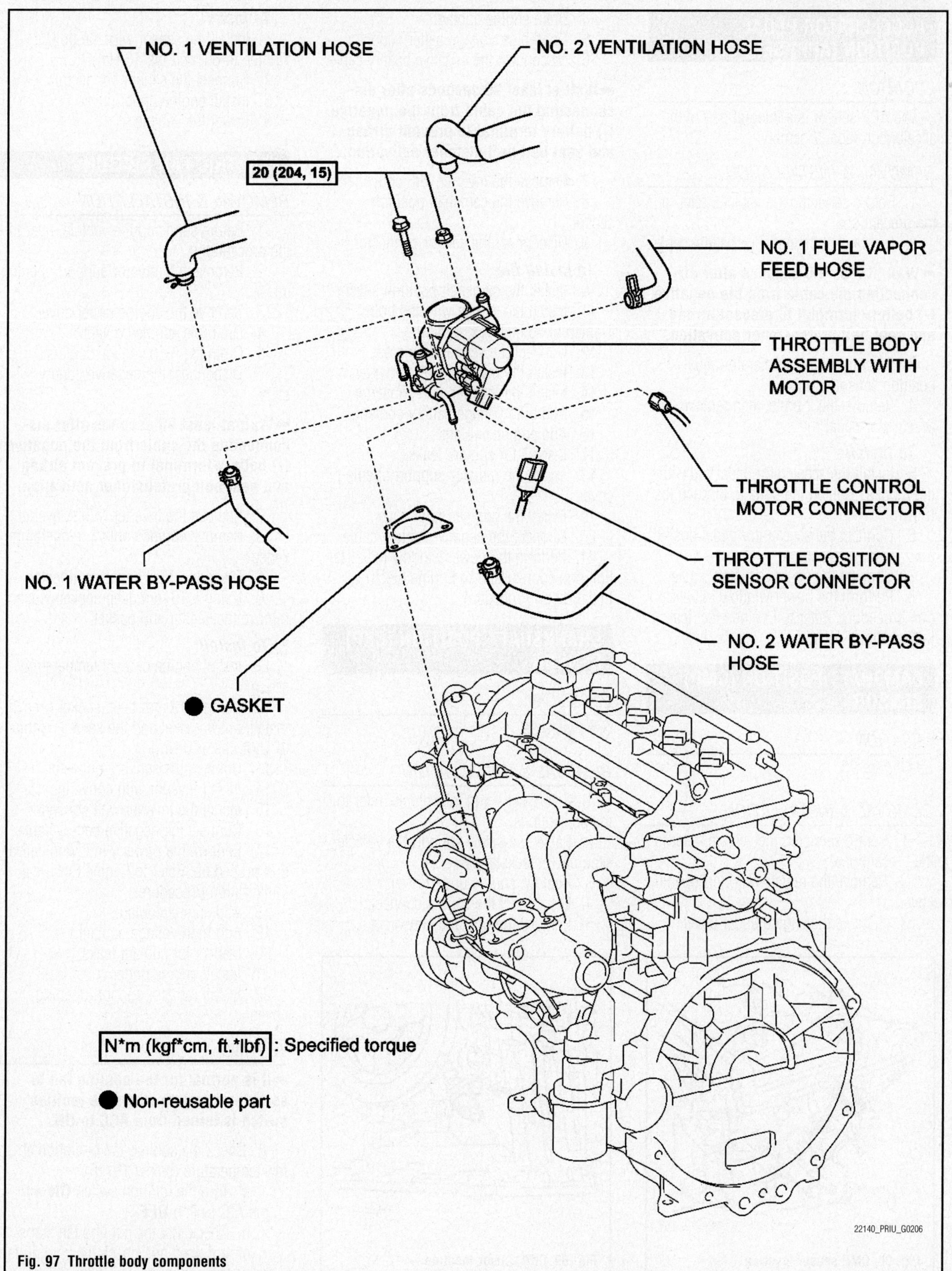

Fig. 97 Throttle body components

ACCELERATOR PEDAL POSITION (APP) SENSOR

LOCATION

The APP sensor is a integral part of the accelerator pedal assembly.

REMOVAL & INSTALLATION

1. Before servicing the vehicle, refer to the precautions.
2. Disconnect the negative battery cable.

➡Wait at least 90 seconds after disconnecting the cable from the negative (-) battery terminal to prevent airbag and seat belt pretensioner activation.

3. Disconnect the accelerator pedal position sensor connector.
4. Remove the 2 bolts, then remove the accelerator pedal rod.

To install:

5. Install the accelerator pedal rod with the 2 bolts. Tighten the bolts to 66 inch lbs. (8 Nm).
6. Connect the accelerator pedal position sensor connector.
7. Connect the negative battery cable.
8. Perform the power window initialization procedure. Refer to Engine Electrical, Initialization Procedure.

CAMSHAFT POSITION (CMP) SENSOR

LOCATION

See Figure 98.

REMOVAL & INSTALLATION

1. Before servicing the vehicle, refer to the precautions.
2. Remove the radiator support opening cover.
3. Remove the engine under cover.

Fig. 98 CMP sensor location

4. Drain engine coolant.
5. Drain high voltage battery coolant.
6. Disconnect the negative battery cable.

➡Wait at least 90 seconds after disconnecting the cable from the negative (-) battery terminal to prevent airbag and seat belt pretensioner activation.

7. Remove the inverter with converter.
8. Remove the camshaft position sensor.
9. Disconnect the sensor connector.

To Install the:

10. Install the camshaft position sensor.
11. Install the sensor with the bolt. Tighten to 66 inch lbs. (8 Nm).
12. Connect the sensor connector.
13. Install the inverter with converter.
14. Install the inverter with converter.
15. Add high voltage battery coolant.
16. Add engine coolant.
17. Inspect for coolant leaks.
18. Install the radiator support opening cover.
19. Install the engine under cover.
20. Connect the negative battery cable.
21. Perform the power window initialization procedure. Refer to Engine Electrical, Initialization Procedure.

CRANKSHAFT POSITION (CKP) SENSOR

LOCATION

See Figure 99.

REMOVAL & INSTALLATION

1. Before servicing the vehicle, refer to the precautions.
2. Raise and safely support the vehicle securely on jackstands.
3. Remove engine under cover.
4. Disconnect the sensor connector.
5. Remove the bolt and sensor.

Fig. 99 CKP sensor location

To install:

6. Install the sensor with the bolt. Tighten to 66 inch lbs. (8 Nm).
7. Connect the sensor connector.
8. Install engine under cover.
9. Lower the vehicle.

ELECTRIC FAN SWITCH

REMOVAL & INSTALLATION

1. Before servicing the vehicle, refer to the precautions.
2. Remove the radiator support opening cover.
3. Remove the engine under cover.
4. Drain the engine coolant.
5. Drain the high voltage coolant.
6. Disconnect the negative battery cable.

➡Wait at least 90 seconds after disconnecting the cable from the negative (-) battery terminal to prevent airbag and seat belt pretensioner activation.

7. Remove the inverter with converter.
8. Remove engine coolant temperature sensor.
9. Disconnect the sensor connector.
10. Using a 19 mm deep socket wrench, remove the sensor and gasket.

To install:

11. Install engine coolant temperature sensor.
12. Using a 19 mm deep socket wrench, install a new gasket and the sensor. Tighten to 15 ft. lbs. (20 Nm).
13. Connect the sensor connector.
14. Install inverter with converter.
15. Install the inverter with converter.
16. Connect the negative battery cable.
17. Perform the power window initialization procedure. Refer to Engine Electrical, Initialization Procedure.
18. Add engine coolant.
19. Add high voltage coolant.
20. Inspect for coolant leaks.
21. Install engine under cover.
22. Install radiator support opening cover.

TESTING

➡It is normal for the cooling fan to sometimes rotate when the ignition switch is turned from ACC to ON.

1. Check the cooling fan operation at low temperature (below 181°F).
   a. Turn the ignition switch **ON** with the A/C switch **OFF** .
   b. Check that the cooling fan stops.
   c. If it does not, check the cooling fan relay and engine coolant temperature

sensor, and check if there is a disconnection or circuit open between them.

d. Disconnect the engine coolant temperature sensor connector.

e. Check that the cooling fan rotates.

f. If it does not, check the fuses, cooling fan relay, ECM and cooling fan, and check for a short in the circuit between the cooling fan relay and engine coolant temperature sensor.

g. Reconnect the engine coolant temperature sensor connector.

2. Check cooling fan operation at high temperature (above 199°F).

a. Start the engine, and raise the coolant temperature to above 199°F.

b. Check that the A/C switch is OFF.

c. Check that the cooling fan rotates.

d. If it does not, check the fuses, cooling fan relay, ECM, cooling fan and engine coolant temperature sensor.

3. Check cooling fan.

a. Disconnect the cooling fan connector.

b. Connect the battery and an ammeter to the cooling fan.

c. Check that the cooling fan rotates smoothly, and check the reading on the ammeter. Amperage should be 9.2 to 11.0 amps.

d. Reconnect the cooling fan connector.

ELECTRONIC CONTROL MODULE (ECM)

LOCATION

See Figure 100.

REMOVAL & INSTALLATION

1. Before servicing the vehicle, refer to the precautions.

2. Disconnect the negative battery cable.

➡Wait at least 90 seconds after disconnecting the cable from the negative (-) battery terminal to prevent airbag and seat belt pretensioner activation.

3. Remove the instrument panel.

4. Remove heater to register duct.

5. Disconnect the 4 ECM connectors.

6. Disconnect the 4 hybrid vehicle control ECU connectors.

7. Remove the 2 nuts and bolt, and ECM with bracket.

To install:

8. Install the ECM with bracket with the 2 nuts and bolt. Tighten 27 inch lbs. (3 Nm).

9. Connect the 4 hybrid vehicle control ECU connectors.

10. Connect the 4 ECM connectors.

11. Install No. 3 heater to register duct.

12. Install instrument panel.

13. Connect the negative battery cable.

14. Perform the power window initialization procedure. Refer to Engine Electrical, Initialization Procedure.

ENGINE COOLANT TEMPERATURE (ECT) SENSOR

LOCATION

See Figure 101.

REMOVAL & INSTALLATION

1. Before servicing the vehicle, refer to the precautions.

2. Remove the radiator support opening cover.

3. Raise and safely support the vehicle securely on jackstands.

4. Remove the engine under cover.

5. Drain the engine coolant.

6. Drain the high voltage coolant.

7. Disconnect the negative battery cable.

➡Wait at least 90 seconds after disconnecting the cable from the negative (-) battery terminal to prevent airbag and seat belt pretensioner activation.

8. Remove the inverter with converter.

9. Remove engine coolant temperature sensor.

10. Disconnect the sensor connector.

11. Using a 19 mm deep socket wrench, remove the sensor and gasket.

To install:

12. Install engine coolant temperature sensor.

13. Using a 19 mm deep socket wrench, install a new gasket and the sensor. Tighten to 15 ft. lbs. (20 Nm).

14. Connect the sensor connector.

15. Install inverter with converter.

16. Install the inverter with converter.

17. Connect the negative battery cable.

18. Perform the power window initialization procedure. Refer to Engine Electrical, Initialization Procedure.

19. Add engine coolant.

20. Add high voltage coolant.

21. Inspect for coolant leaks.

22. Install engine under cover.

23. Lower the vehicle.

24. Install radiator support opening cover.

FUEL LEVEL SENDING UNIT

LOCATION

See Figure 102.

REMOVAL & INSTALLATION

1. Before servicing the vehicle, refer to the precautions.

2. Remove the rear seat cushion.

a. Detach the seat cushions 2 front hooks from the vehicle body by choosing a hook to detach first. Place your hands near one of the hooks, then lift the seat cushion to detach the hook.

b. Repeat for the other hook.

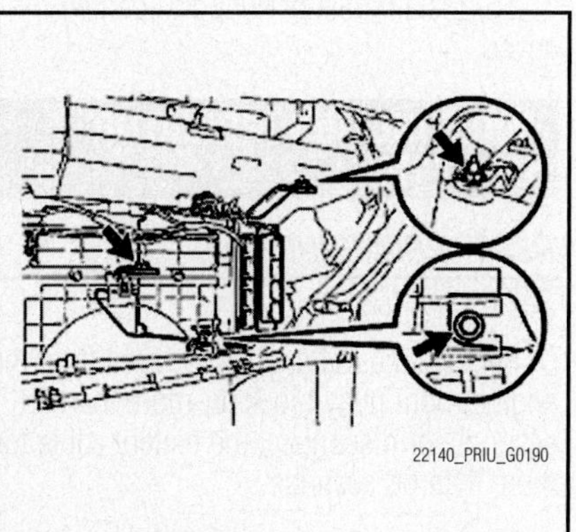

Fig. 100 ECM location

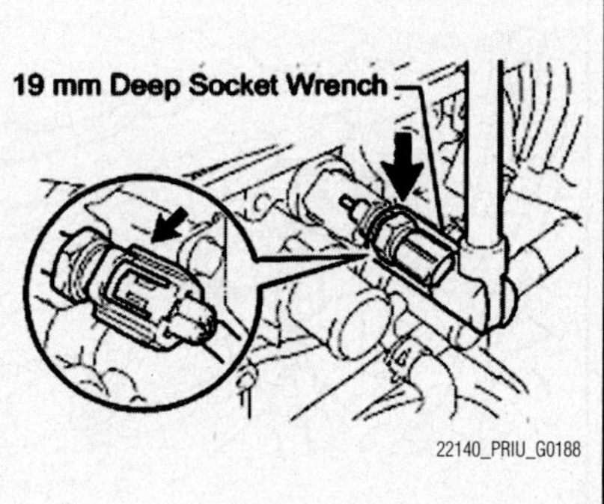

Fig. 101 ECT sensor location

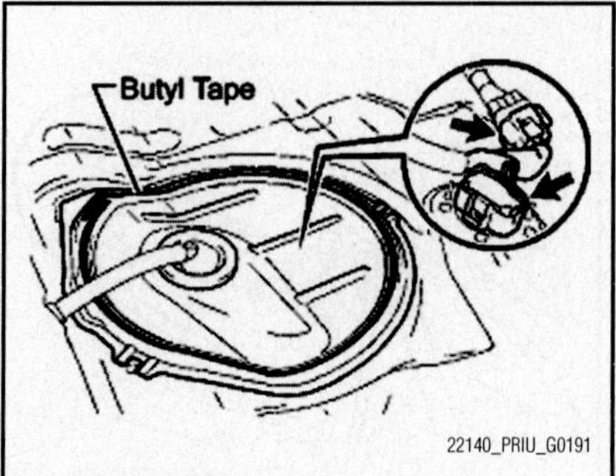

Fig. 102 Fuel level sending unit location

c. Detach the seat cushions rear hook.

d. Remove the seat cushion.

3. Remove the butyl tape and rear floor service hole cover.

4. Disconnect the fuel pump connector.

5. Disconnect the wire to wire connector.

6. Remove the sending unit attaching bolts.

7. Carefully remove the sending unit from the fuel tank.

To install:

8. Using a new gasket as necessary, install the sending unit.

9. Carefully tighten the sending unit attaching bolts.

10. Attach new butyl tape to the rear floor service hole cover.

11. Connect the wire to wire connector.

12. Connect the fuel pump connector.

13. Install the rear floor service hole cover while adjusting it to the 3 convex parts of the floor panel.

➡Be careful that the rear floor service hole cover does not overlap the convex parts of the floor panel when installing.

14. Install the rear seat cushion by engaging the 3 seat hooks.

HEATED OXYGEN (HO2S) SENSOR

LOCATION

See Figure 103.

REMOVAL & INSTALLATION

1. Disconnect the negative battery cable.

➡Wait at least 90 seconds after disconnecting the cable from the negative (-) battery terminal to prevent airbag and seat belt pretensioner activation.

2. Remove lower center instrument panel finish panel.

3. Remove heated oxygen sensor.

a. Using a clip remover, remove the clip.

b. Fold back the floor carpet front.

c. Disconnect the sensor connector.

d. Remove the grommet of the sensor from the vehicle.

e. Remove the wire harness clamp bracket from the sensor.

f. Using special tool SST: 09224-00010, remove the sensor.

To install:

4. Install heated oxygen sensor.

a. Using special tool SST: 09224-00010, install the sensor. Tighten to 30ft. lbs. (40 Nm).

b. Install the wire harness clamp bracket to the sensor.

c. Install the grommet of the sensor to the vehicle.

d. Connect the sensor connector.

5. Install the floor carpet front with the clip.

6. Connect cable to negative battery terminal.

7. Inspect for exhaust gas leak.

8. Install lower center instrument panel finish panel.

9. Connect the negative battery cable.

10. Perform the power window initialization procedure. Refer to Engine Electrical, Initialization Procedure.

Fig. 103 HO2S location

KNOCK SENSOR (KS)

LOCATION

See Figure 104.

REMOVAL & INSTALLATION

1. Disconnect the negative battery cable.

➡Wait at least 90 seconds after disconnecting the cable from the negative (-) battery terminal to prevent airbag and seat belt pretensioner activation.

2. Remove radiator support opening cover.

3. Remove engine under cover.

4. Drain engine coolant.

5. Remove air cleaner assembly.

6. Remove oil dipstick guide.

a. Remove the dipstick.

b. Disconnect the wire harness clamp.

c. Remove the bolt and dipstick guide.

7. Remove intake manifold.

8. Disconnect the knock sensor connector.

9. Remove the nut and sensor.

To install:

10. Install knock sensor.

11. Install the knock sensor with the nut. Tighten to 15 ft. lbs. (20 Nm).

➡Be careful to install the knock sensor in the correct direction.

12. Connect the knock sensor connector.

13. Install intake manifold.

14. Install oil dipstick guide.

a. Apply a light coat of engine oil to a new o-ring and install it to the dipstick guide.

b. Install the dipstick guide with the bolt. Tighten to 80 inch lbs. (9 Nm).

➡Be careful that the o-ring is not cracked or jammed when installing it.

c. Connect the wire harness clamp.

d. Install the dipstick.

15. Install air cleaner assembly.

16. Connect the negative battery cable.

17. Perform the power window initialization procedure. Refer to Engine Electrical, Initialization Procedure.

18. Add engine coolant.

19. Inspect for coolant leak.

20. Install engine under cover.

21. Install radiator support opening cover.

Fig. 104 KS sensor location

MALFUNCTION INDICATOR LIGHT (MIL)

RESET PROCEDURES

See Figure 105.

Remove the EFI and ETCS fuses from the engine room relay block for more than 60 seconds, or disconnect the battery cable for more than 60 seconds.

➡After reconnecting the battery cable, perform the power window initialization

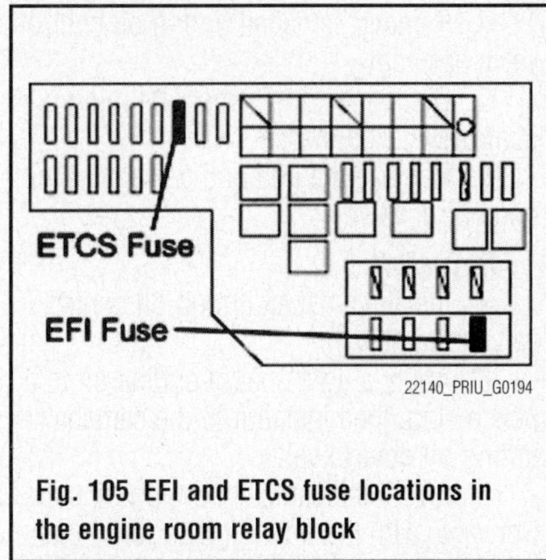

Fig. 105 EFI and ETCS fuse locations in the engine room relay block

procedure. Refer to Interior, Power Windows, Initialization Procedure.

MASS AIR FLOW (MAF) SENSOR

LOCATION

See Figure 106.

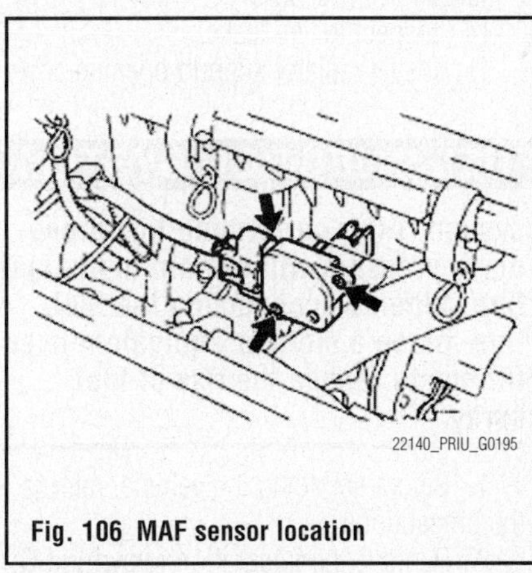

Fig. 106 MAF sensor location

REMOVAL & INSTALLATION

1. Disconnect the negative battery cable.

➡**Wait at least 90 seconds after disconnecting the cable from the negative (-) battery terminal to prevent airbag and seat belt pretensioner activation.**

2. Remove radiator support opening cover.
3. Disconnect the MAF meter connector.
4. Remove the 2 screws and MAF meter.

To install:

5. Install a new O-ring to the MAF meter.
6. Install the MAF meter with the 2 screws.
7. Connect the MAF meter connector.
8. Connect the negative battery cable.
9. Perform the power window initialization procedure. Refer to Engine Electrical, Initialization Procedure.

OIL PRESSURE SENSOR

LOCATION

See Figure 107.

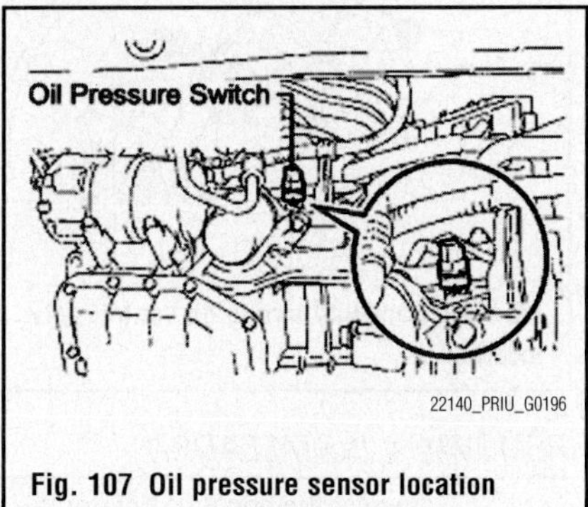

Fig. 107 Oil pressure sensor location

REMOVAL & INSTALLATION

1. Raise and safely support the vehicle securely on jackstands.
2. Disconnect the oil pressure switch connector.
3. Using a 24 mm deep socket wrench, remove the oil pressure switch.

To install:

4. Apply Three Bond 1324 or equivalent adhesive to the threads.
5. Using a 24 mm deep socket wrench, install the oil pressure switch. Tighten to 11 ft. lbs. (15 Nm).

➡**Do not start the engine within 1 hour of installation.**

6. Connect the oil pressure switch connector.
7. Lower the vehicle.

THROTTLE POSITION SENSOR (TPS)

LOCATION

See Figure 108.

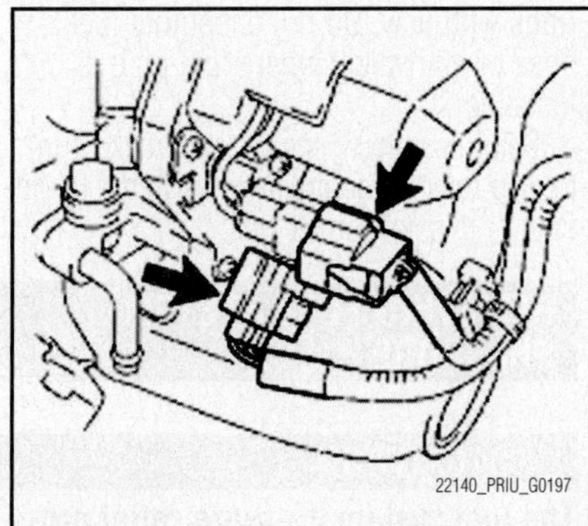

Fig. 108 Throttle position sensor location

REMOVAL & INSTALLATION

1. Disconnect the negative battery cable.

➡**Wait at least 90 seconds after disconnecting the cable from the negative (-) battery terminal to prevent airbag and seat belt pretensioner activation.**

2. Remove engine under cover.
3. Drain engine coolant.
4. Remove the 6 clips and radiator support opening cover.
5. Remove air cleaner assembly
   a. Disconnect the MAF sensor connector.
   b. Disconnect the wire harness from the wire harness clamp.
   c. Loosen the hose clamp bolt, and then disconnect the No. 1 air cleaner inlet.
   d. Remove the 2 bolts.
   e. Loosen the hose clamp bolt, and then remove the air cleaner.
6. Remove throttle body assembly with motor.
   a. Disconnect the ventilation hose.
   b. Disconnect the No. 2 ventilation hose.
   c. Disconnect the No. 1 fuel vapor feed hose.
   d. Disconnect the water by-pass hose.
   e. Disconnect the No. 2 water by-pass hose.
   f. Disconnect the throttle control motor connector.
   g. Disconnect the throttle position sensor connector.
   h. Remove the bolt, 2 nuts and throttle with motor body.
   i. Remove the gasket from the intake manifold.

To install:

7. Install throttle body assembly with motor
   a. Install a new gasket to the intake manifold.
   b. Install the throttle with motor body with the bolt and 2 nuts. Tighten to 15 ft. lbs. (20 Nm).
   c. Connect the throttle position sensor connector.
   d. Connect the throttle control motor connector.
   e. Connect the No. 2 water by-pass hose.
   f. Connect the water by-pass hose.
   g. Connect the No. 1 fuel vapor feed hose.
   h. Connect the No. 2 ventilation hose.
   i. Connect the ventilation hose.

8. Install air cleaner assembly
   a. Install the air cleaner with the 2 bolts. Tighten to 62 inch lbs. (7 Nm).
   b. Tighten the hose clamp bolt to 27 inch lbs. (3 Nm).
   c. Connect the No. 1 air cleaner inlet, and tighten the hose clamp bolt to 27 inch lbs. (3 Nm).
   d. Connect the MAF sensor meter connector.
9. Connect the negative battery cable.
10. Perform the power window initialization procedure. Refer to Engine Electrical, Initialization Procedure.
11. Add engine coolant
12. Inspect for coolant leaks.
13. Install engine under cover.
14. Install radiator support opening cover.

VARIABLE CAMSHAFT TIMING OIL CONTROL SOLENOID

LOCATION

See Figure 109.

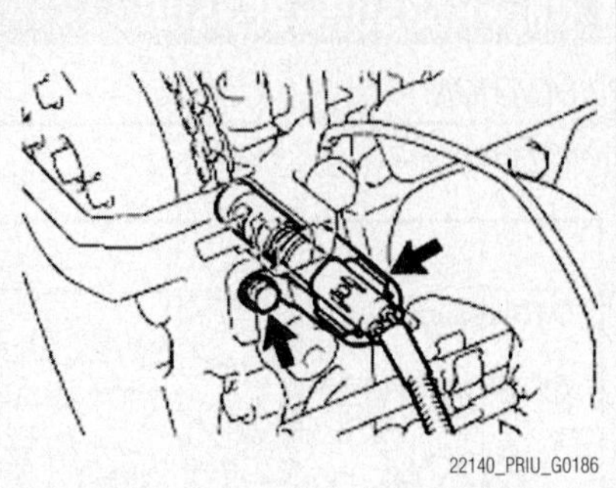

Fig. 109 Camshaft timing oil control valve location

REMOVAL & INSTALLATION

1. Disconnect the negative battery cable.

➡Wait at least 90 seconds after disconnecting the cable from the negative (-) battery terminal to prevent airbag and seat belt pretensioner activation.

2. Remove radiator support opening cover.
3. Remove air cleaner assembly.
4. Remove camshaft timing oil control valve assembly.
5. Disconnect the camshaft timing oil control valve connector.
6. Remove the bolt and camshaft timing oil control valve.

To install:

7. Install camshaft timing oil control valve assembly.
8. Apply a light coat of engine oil to a new o-ring, then install it to the camshaft timing oil control valve.
9. Install the camshaft timing oil control valve with the bolt. Tighten to 66 inch lbs. (8 Nm).

➡Be careful that the o-ring is not cracked or jammed when installing it.

10. Connect the camshaft timing oil control valve connector.
11. Install air cleaner assembly.
12. Connect the negative battery cable.
13. Perform the power window initialization procedure. Refer to Engine Electrical, Initialization Procedure.
14. Inspect for oil leaks.
15. Install radiator support opening cover.

FUEL SYSTEM — GASOLINE FUEL INJECTION SYSTEM

FUEL SYSTEM SERVICE PRECAUTIONS

Safety is the most important factor when performing not only fuel system maintenance but any type of maintenance. Failure to conduct maintenance and repairs in a safe manner may result in serious personal injury or death. Maintenance and testing of the vehicle's fuel system components can be accomplished safely and effectively by adhering to the following rules and guidelines.

- To avoid the possibility of fire and personal injury, always disconnect the negative battery cable unless the repair or test procedure requires that battery voltage be applied.
- Always relieve the fuel system pressure prior to disconnecting any fuel system component (injector, fuel rail, pressure regulator, etc.), fitting or fuel line connection. Exercise extreme caution whenever relieving fuel system pressure to avoid exposing skin, face and eyes to fuel spray. Please be advised that fuel under pressure may penetrate the skin or any part of the body that it contacts.
- Always place a shop towel or cloth around the fitting or connection prior to loosening to absorb any excess fuel due to spillage. Ensure that all fuel spillage (should it occur) is quickly removed from engine surfaces. Ensure that all fuel soaked cloths or towels are deposited into a suitable waste container.
- Always keep a dry chemical (Class B) fire extinguisher near the work area.
- Do not allow fuel spray or fuel vapors to come into contact with a spark or open flame.
- Always use a back-up wrench when loosening and tightening fuel line connection fittings. This will prevent unnecessary stress and torsion to fuel line piping.
- Always replace worn fuel fitting O-rings with new. Do not substitute fuel hose or equivalent where fuel pipe is installed.

Before servicing the vehicle, make sure to also refer to the precautions in the beginning of this section as well.

RELIEVING FUEL SYSTEM PRESSURE

⁂ CAUTION

The fuel system pressure relief procedure must be performed before disconnecting any part of the fuel system. After performing this procedure, pressure will remain in the fuel line. When disconnecting the fuel line, place a cloth or equivalent over fittings to reduce the risk of fuel spray.

1. Before servicing the vehicle, refer to the precautions.
2. Remove the integration relay (unit C: C/OPN relay) from the engine room junction block.
3. Start the vehicle and allow the engine to run until it stops, then turn the power switch **OFF**. This may set off a trouble code (DTC P0171: system too lean).
4. Check that the engine does not start.
5. Remove the fuel filler cap from the filler neck to release the fuel vapor pressure in the fuel tank.
6. Disconnect the negative battery cable.

➡Wait at least 90 seconds after disconnecting the cable from the negative (-) battery terminal to prevent airbag and seat belt pretensioner activation.

7. Install the integration relay (unit C: C/OPN relay) to the engine room junction block.

8. After servicing the fuel system, connect the negative battery cable.
9. Start the engine and check for leaks in the system.

FUEL FILTER

REMOVAL & INSTALLATION

The fuel filter for Prius models is an integral component of the tank-mounted fuel pump assembly and is not regularly serviced.

FUEL PUMP

REMOVAL & INSTALLATION

1. Before servicing the vehicle, refer to the precautions.
2. Relieve the pressure from the fuel system.
3. Disconnect the negative battery cable.

➡Wait at least 90 seconds after disconnecting the cable from the negative (-) battery terminal to prevent airbag and seat belt pretensioner activation.

4. Remove the fuel tank.
5. Remove the trap canister with pump module.
   a. Disconnect the VSV connector.
   b. Remove the clamp from the fuel tank vent hose and canister hose.
   c. Remove the fuel tank vent hose from the 2 fuel tube clamps.
   d. Remove the 2 bolts and trap canister with pump module and disconnect the ground terminal of the fuel tank wire.
   e. Remove the gasket from the fuel tank.
   f. Remove the 2 clamps from the trap canister with pump module.

To install:

6. Install the trap canister with pump module.
   a. Install a new gasket to the fuel tank.
   b. Insert the trap canister with pump module to the fuel tank. Be careful that the gasket does not drop in the fuel tank.
   c. Install the 2 clamps to the trap canister with pump module.
   d. Install the trap canister with pump module and connect the ground terminal of the fuel tank wire with the 2 bolts. Tighten the 2 bolts to 53 inch lbs. (6 Nm).
   e. Install the fuel tank vent hose to the 2 fuel tube clamps.
   f. Install the clamp to the fuel tank vent hose and canister hose.
   g. Connect the VSV connector.
7. Install the fuel tank.
8. Connect the negative battery cable.
9. Perform the power window initialization procedure. Refer to Engine Electrical, Initialization Procedure.
10. Check for fuel leaks and proper fuel pressure.

FUEL RAIL & INJECTORS

REMOVAL & INSTALLATION

See Figures 110 through 112.

1. Before servicing the vehicle, refer to the precautions.
2. Relieve the pressure from the fuel system.
3. Disconnect the negative battery cable.
4. Remove the windshield wipers and windshield wiper motor and link assembly, if necessary.
5. Remove the front outer cowl top panel, if necessary.
6. Remove the air cleaner assembly.
7. Remove the brake master cylinder reservoir.
   a. Disconnect the brake fluid level switch connector.
   b. Remove the 2 bolts.
   c. Disconnect the claw fitting, then remove the brake master cylinder reservoir.
8. Remove the reservoir bracket.
   a. Remove the No. 2 fuel vapor feed hose from the hose clamp.
   b. Remove the connector clamp.
   c. Remove the wire harness clamp, the 3 bolts and reservoir bracket.
9. Remove the cylinder head cover.
10. Remove fuel delivery pipe.
   a. Disconnect the fuel tube from the fuel delivery pipe.
   b. Remove the No. 1 fuel pipe clamp.
   c. Pinch the retainer of the fuel tube connector, and then pull out the fuel tube connector to disconnect the fuel tube from the fuel delivery pipe.

⁂ CAUTION

Be careful not to drop the injectors when removing the delivery pipe.

   d. Disconnect the 4 injector connectors from injector.
   e. Remove the 3 bolts and delivery pipe together with the 4 injectors and fuel pipe.
   f. Remove the 2 spacers from the cylinder head.
   g. Pull out the 4 injectors from the delivery pipe.
   h. Remove the O-ring and grommet from each injector.

To install:

11. Install the grommet to each injector.
12. Apply a light coat of spindle oil or gasoline to new O-ring and install them to each injector.
13. Apply a light coat of spindle oil or gasoline to the surface of the fuel delivery pipe which contacts the O-ring of the fuel injector.
14. Apply a light coat of spindle oil or gasoline to the O-ring again, and install the fuel injector by turning it right and left while pushing it onto the fuel delivery pipe.

➡Be careful that the O-ring is not cracked or jammed when installing it.

15. Check that the fuel injector rotates smoothly. If the fuel injector does not rotate, replace the O-ring.
16. Install the 4 injectors.
17. Position the injector connector outward.
18. Install 4 new insulators to the cylinder head.
19. Install the 2 delivery pipe spacers to the cylinder head.
20. Place the delivery pipe and fuel pipe together with the 4 injectors in position on the cylinder head and then temporarily tighten the 3 bolts.
21. Check that the injectors rotate smoothly. If the fuel injectors do not rotate, replace the O-ring.
22. Tighten the 2 bolts holding the delivery pipe to the cylinder head and tighten to 14 ft. lbs. (19 Nm). Tighten the bolt holding the fuel pipe to the cylinder head to 80 inch lbs. (9 Nm).
23. Align the fuel tube connector with the pipe, then push in the fuel tube connector until the retainer makes a "click" sound to connect the fuel tube to the fuel delivery pipe.
24. Install the No. 1 fuel pipe clamp.
25. Install the cylinder head cover.
26. Install the reservoir bracket and tighten the 3 bolts to 75 inch lbs. (8.5 Nm).
   a. Install the wire harness clamp and the connector clamp.
   b. Install the No. 2 fuel vapor feed hose to the hose clamp.
   c. Connect the claw fitting.
27. Install the brake master cylinder reservoir and tighten the 2 mounting bolts to 75 inch lbs. (8.5 Nm).
   a. Connect the brake fluid level switch connector.

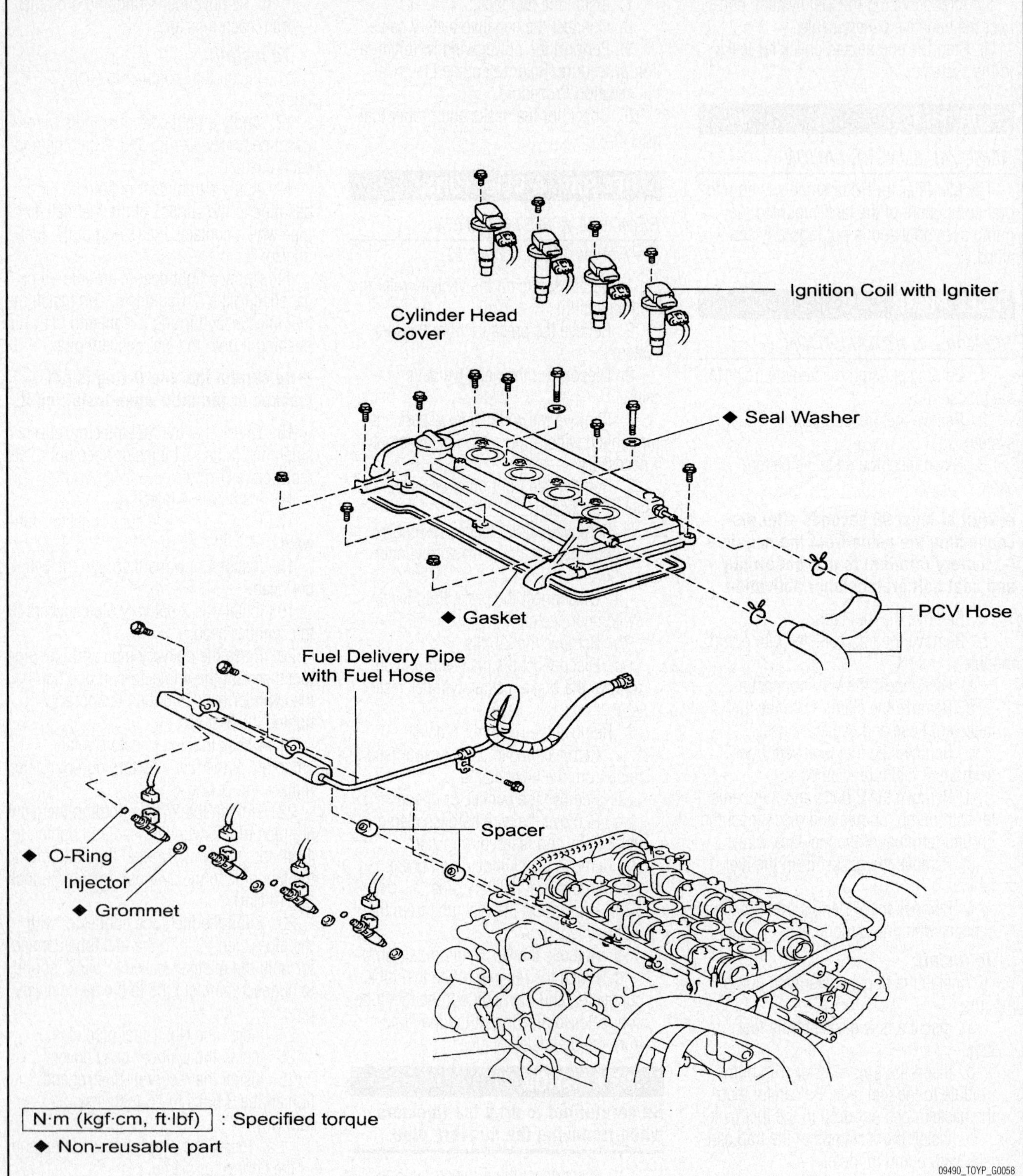

Fig. 110 Fuel injectors and cylinder head cover.

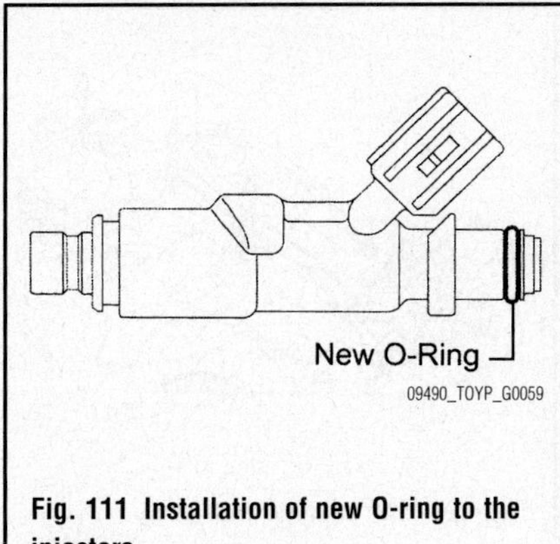

Fig. 111 Installation of new O-ring to the injectors.

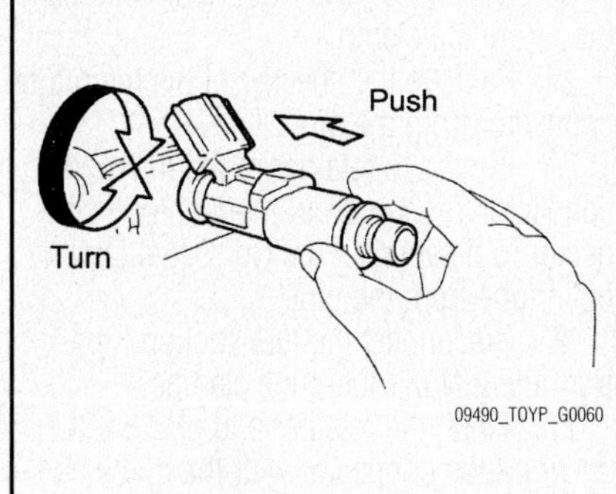

Fig. 112 Installation of the fuel injectors to the fuel delivery pipe.

28. Install the air cleaner assembly.
29. Install the front outer cowl top panel, if necessary. Tighten the panel bolts to 57 inch lbs. (6.4 Nm), and the No. 2 engine room relay block bolts to 74 inch lbs. (8.4 Nm).
30. Install the windshield wiper motor and link assembly and windshield wipers, if necessary.
31. Connect the negative battery cable.
32. Perform the power window initialization procedure. Refer to Engine Electrical, Initialization Procedure.
33. With the ignition **ON** and the engine **OFF** check for leaks.

FUEL TANK

REMOVAL & INSTALLATION

See Figures 113 through 125.

1. Before servicing the vehicle, refer to the precautions.
2. Relieve the pressure from the fuel system.
3. Disconnect the negative battery cable.

➡Wait at least 90 seconds after disconnecting the cable from the negative (-) battery terminal to prevent airbag and seat belt pretensioner activation.

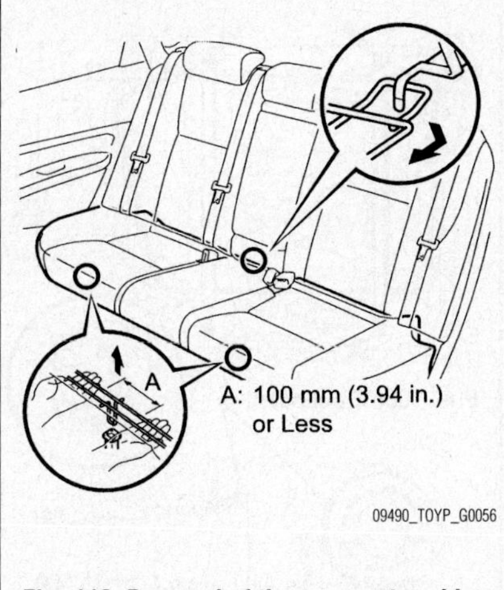

Fig. 113 Removal of the rear seat cushion assembly

4. Remove instrument panel finish panel lower center.
5. Remove the front floor panel brace and front exhaust pipe assembly.
6. Remove the rear seat cushion.
   a. Detach the seat cushions 2 front hooks from the vehicle body by choosing a hook to detach first. Place your hands near one of the hooks, then lift the seat cushion to detach the hook.
   b. Repeat for the other hook.
   c. Detach the seat cushions rear hook.
   d. Remove the seat cushion.
7. Remove the butyl tape and rear floor service hole cover.

⁂ CAUTION

Remove dirt or foreign objects on the fuel line connectors before any disconnecting procedures. Do not allow any scratches or foreign objects on the parts when disconnecting them as the fuel connectors have O-rings that seal the pipes. Perform such work by hand. Do not use any tools. Do not forcibly bend, twist or turn the nylon tube. Protect the connecting part by covering it with a plastic bag after disconnecting the tube. If the connector and pipe are stuck, push and pull them to release them.

8. Disconnect the fuel pump connector.
9. Disconnect the wire-to-wire connector.
10. Pinch the retainer of the fuel tube connector, then pull out the fuel tube connector to disconnect the fuel tank to canister tube from the pipe.
11. Remove the checker of the fuel tube connector from the pipe.

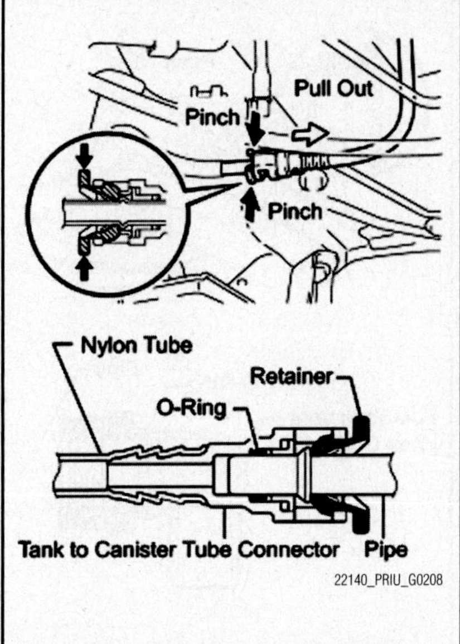

Fig. 114 Disconnect the fuel tank to canister tube from the pipe

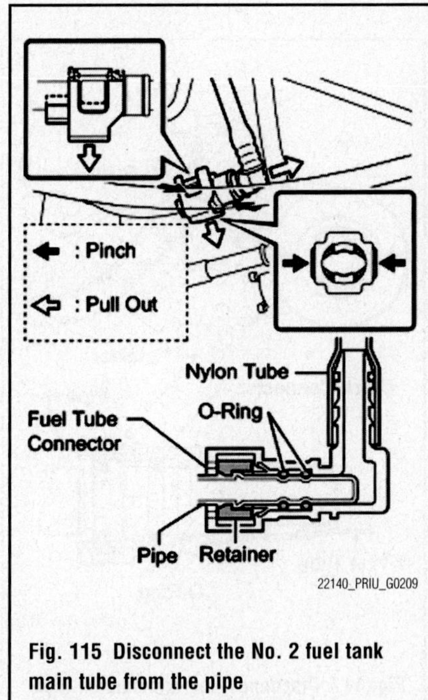

Fig. 115 Disconnect the No. 2 fuel tank main tube from the pipe

12. Pinch the retainer of the fuel tube connector, and then pull out the fuel tube connector to disconnect the No. 2 fuel tank main tube from the pipe.
13. Pinch the retainer and pull out the fuel tank vent hose connector with the fuel tank vent hose connector pushed to the pipe side to disconnect the fuel tank vent hose from the canister filter.
14. Pinch the retainer and pull out the suction tube connector with the suction tube connector pushed to the pipe side to disconnect the fuel suction tube from the fuel tank to filler pipe.

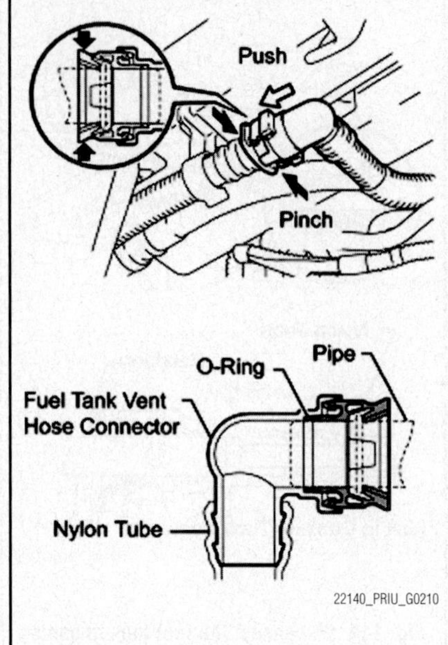

Fig. 116 Disconnect the fuel tank vent hose from the canister filter

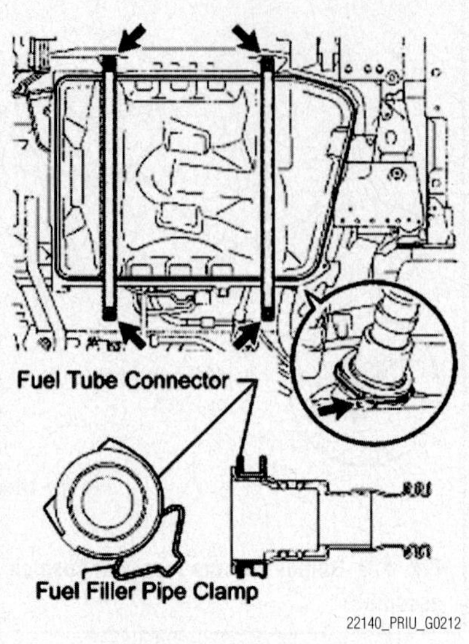

Fig. 118 Remove the fuel filler pipe clamp and fuel tube connector from the fuel tank inlet pipe

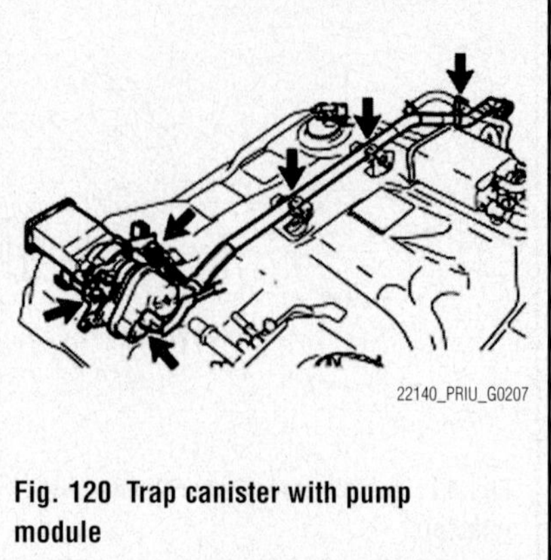

Fig. 120 Trap canister with pump module

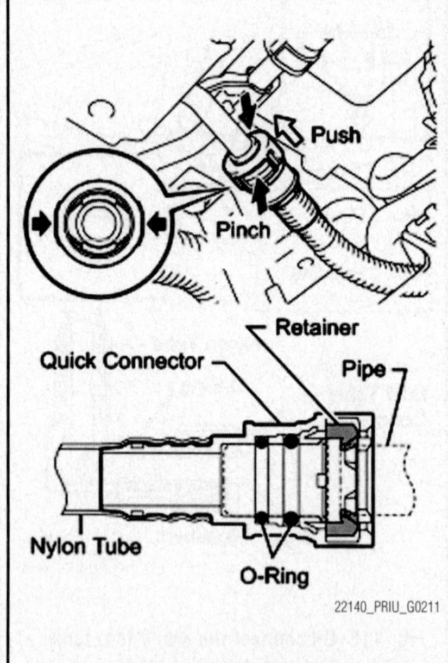

Fig. 117 Disconnect the fuel suction tube from the fuel tank to filler pipe

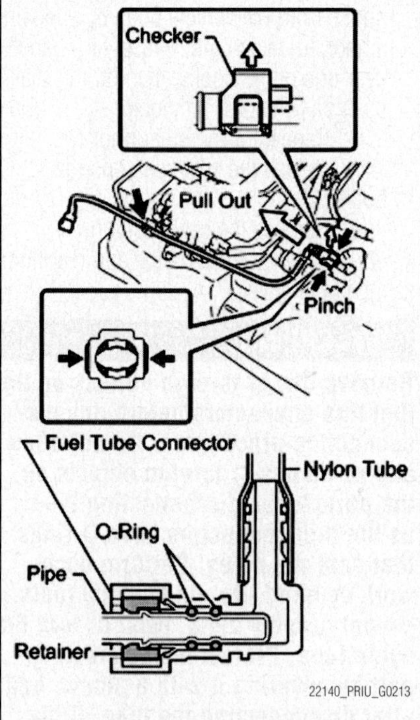

Fig. 119 Remove the checker of the main tube connector from the pipe

15. Pinch the retainer and pull out the No. 1 canister tube connector with the No. 1 canister tube connector pushed to the pipe side to disconnect the No. 1 canister tube from the fuel tank to filler pipe.
16. Set a transmission jack to the fuel tank.
17. Remove the fuel filler pipe clamp and fuel tube connector from the fuel tank inlet pipe.
18. Remove the 4 bolts and No. 1 fuel tank band right and left.
19. Operate the transmission jack, and then disconnect the fuel tank inlet pipe.
20. Operate the transmission jack, and then remove the fuel tank.
21. Remove the 3 nuts and rear fuel tank bracket.
22. Disconnect the wire to wire connector from the rear fuel tank bracket.
23. Disconnect the No. 2 fuel tank main tube from the clamp.
24. Remove the checker of the main tube connector from the pipe.
25. Pinch the retainer of the main tube connector, then pull out the fuel tube connector to disconnect the No. 2 fuel tank main tube from the pipe.
26. Disconnect the fuel suction tube from the 2 No. 1 fuel tube clamps.
27. Pinch the retainer and pull out the suction tube connector with the quick connector pushed to the pipe side to disconnect the fuel suction tube from the pipe.
28. Disconnect the fuel tank to canister tube from the clamp.
29. Disconnect the fuel tank to canister tube from the 2 No. 1 fuel tube clamps.
30. Remove the fuel tank to canister tube from the fuel tank.
31. Remove the trap canister with pump module.
    a. Disconnect the VSV connector.
    b. Remove the clamp from the fuel tank vent hose and canister hose.
    c. Remove the fuel tank vent hose from the 2 fuel tube clamps.
    d. Remove the 2 bolts and trap canister with pump module and disconnect the ground terminal of the fuel tank wire.
    e. Remove the gasket from the fuel tank.
    f. Remove the 2 clamps from the trap canister with pump module.
32. Pinch the retainer and pull out the fuel tank vent hose connector with the fuel tank vent hose connector pushed to the fuel tank vent hose side to disconnect the fuel tank vent hose from the trap with outlet valve canister.
33. Remove fuel tank wire.
    a. Remove the clamp as shown in the illustration "A".
    b. Disconnect the VSV connector as shown in the illustration "B".

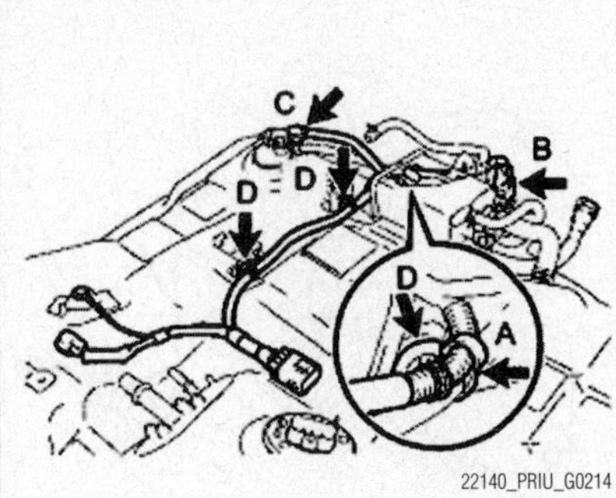

Fig. 121 Remove the clamp (A), disconnect the VSV connector (B), disconnect the vapor pressure sensor connector (C) and remove the 3 wire harness clamps (D).

Fig. 122 Canister attaching points

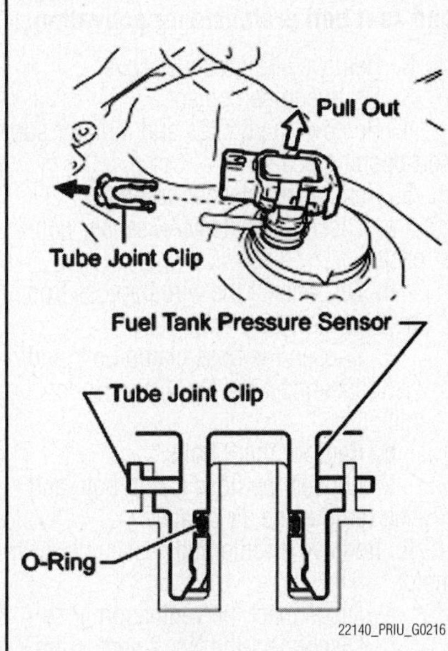

Fig. 123 Remove the tube joint clip, then pull out the fuel tank pressure sensor from the fuel tank retainer

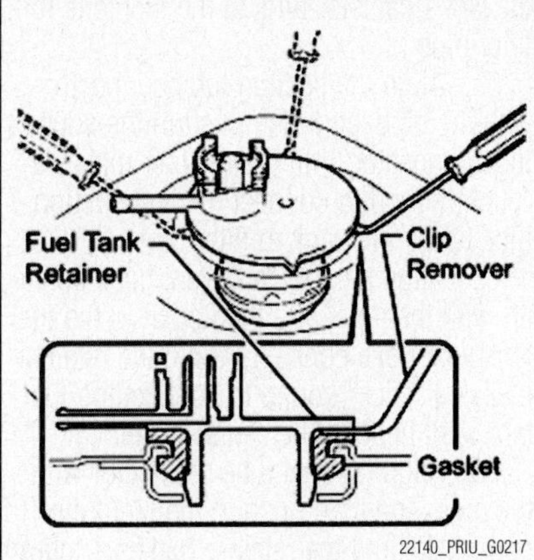

Fig. 124 Insert a clip remover between the fuel tank retainer and gasket, then remove the fuel tank retainer by lifting it little by little

c. Disconnect the vapor pressure sensor connector as shown in the illustration "C".

d. Remove the 3 wire harness clamps as shown in the illustration "D".

34. Remove canister.

a. Disconnect the canister hose from the fuel tank retainer.

b. Disconnect the No. 1 canister outlet hose from the fuel tank.

c. Remove the bolt, 2 nuts and canister.

d. Remove the nut from the fuel tank.

35. Remove fuel tank pressure sensor.

a. Remove the tube joint clip, then pull out the fuel tank pressure sensor from the fuel tank retainer.

36. Remove fuel tank retainer.

a. Insert a clip remover between the fuel tank retainer and gasket, then remove the fuel tank retainer by lifting it little by little.

✲✲ WARNING

The fuel tank retainer is made of resin and easily damaged if removed or installed forcibly. Handle the part correctly to ensure proper sealing. After removing the fuel tank retainer, check that the contact surface of the fuel tank retainer on the fuel tank is not damaged.

b. Remove the gasket from the fuel tank.

37. Remove the 2 clamps from the fuel tank.

38. Remove the 9 cushions from the fuel tank.

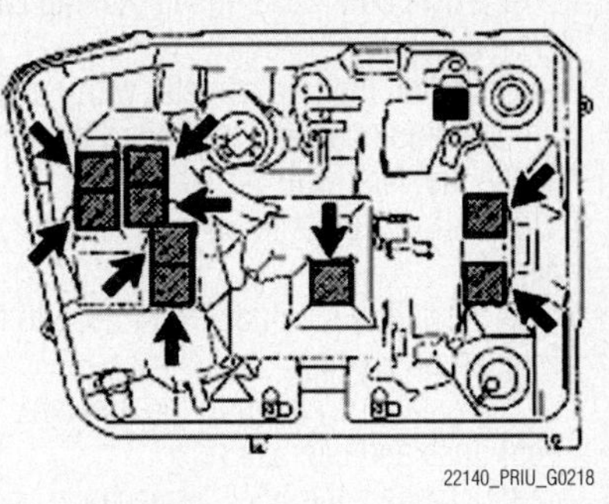

Fig. 125 Remove the 9 cushions from the fuel tank

To install:

39. Install 9 new cushions to the fuel tank.

40. Install the 2 clamps to the fuel tank.

41. Install fuel tank retainer.

a. Install a new gasket to the fuel tank.

b. While being careful that the gasket does not drop in the fuel tank, insert the fuel tank retainer into the fuel tank so the protrusion of the fuel tank retainer is in the middle of the 2 convex pats of the fuel tank.

42. Push the fuel tank pressure sensor to the plug of the fuel tank retainer, and then install the tube joint clip.

43. Install canister.

a. Install the nut to the fuel tank.

b. Install the canister with the bolt and 2 nuts. Tighten to 53 inch lbs. (6 Nm).

c. Connect the No. 1 canister outlet hose to the fuel tank.

d. Connect the canister hose to the fuel tank retainer.

44. Install fuel tank wire.

a. Install the 3 wire harness clamps.

b. Connect the vapor pressure sensor connector.

c. Connect the VSV connector.

d. Install the clamp.

45. Install fuel tank vent hose.

a. Align the fuel tank vent hose connector with the pipe, then push in the fuel tank vent hose connector until the retainer makes a "click" sound to install the fuel tank vent hose to the trap canister with pump module.

46. Install the trap canister with pump module.

a. Install a new gasket to the fuel tank.

b. Insert the trap canister with pump module to the fuel tank. Be careful that the gasket does not drop in the fuel tank.

c. Install the 2 clamps to the trap canister with pump module.

d. Install the trap canister with pump module and connect the ground terminal of the fuel tank wire with the 2 bolts. Tighten the 2 bolts to 53 inch lbs. (6 Nm).

e. Install the fuel tank vent hose to the 2 fuel tube clamps.

f. Install the clamp to the fuel tank vent hose and canister hose.

g. Connect the VSV connector.

47. Install the fuel tank to canister tube to the canister's hose.
48. Connect the fuel tank to canister tube to the 2 No. 1 fuel tube clamps.

CAUTION

Check that there are no scratches or foreign objects around any connected part of the fuel line connectors and pipes before these procedures. After connecting any fuel line tubes, check that the tube is securely connected by pulling on the connector.

49. Align the suction tube connector with the pipe, and then push in the suction tube connector until the retainer makes a "click" sound to install the fuel suction tube to the pipe.
50. Connect the fuel suction tube to the 2 No. 1 fuel tube clamps.
51. Align the main tube connector with the pipe, and then push in the main tube connector until the retainer makes a "click" sound to install the No. 2 fuel tank main tube to the pipe. Install the checker to the pipe.
52. Connect the No. 2 fuel tank main tube to the clamp.
53. Connect the connector clamp to the rear fuel tank bracket.
54. Install the rear fuel tank bracket with the 3 nuts and tighten to 53 inch lbs. (6 Nm).
55. Set the fuel tank to a transmission jack.
56. Operate the transmission jack, and then install the fuel tank to the vehicle.
57. Operate the transmission jack, and then connect the fuel tank inlet pipe.
58. Install the No. 1 fuel tank band right and left with the 4 bolts. Torque the 4 bolts to 29 ft. lbs. (39 Nm).
59. Install the fuel tube connector and fuel filler pipe clamp to the fuel tank inlet pipe.
60. Align the No. 1 canister tube connector with the pipe, and then push in the No. 1 canister tube connector until the retainer makes a "click" sound to connect the No. 1 canister tube to the fuel tank to filler pipe.
61. Align the suction tube connector with the pipe, and then push in the suction tube connector until the retainer makes a "click" sound to connect the fuel suction tube to the fuel tank to filler pipe.
62. Align the fuel tank vent hose connector with the pipe, and then push in the fuel tank vent hose connector until the retainer makes a "click" sound to connect the fuel tank vent hose to the canister filter.
63. Align the fuel tube connector with the pipe, and then push in the fuel tube connector until the retainer makes a "click" sound to connect the No. 2 fuel tank main tube to the pipe. Install the checker to the pipe.
64. Align the fuel tank to canister tube connector with the pipe, and then push in the fuel tank to canister tube connector until the retainer makes a "click" sound to connect the fuel tank to canister tube to the pipe.
65. Install the front exhaust pipe
66. Connect the negative battery cable.
67. Perform the power window initialization procedure. Refer to Engine Electrical, Initialization Procedure.
68. Check for fuel and exhaust leaks.
69. Install the front floor panel brace.
70. Install the instrument panel finish panel lower center.
71. Attach new butyl tape to the rear floor service hole cover.
72. Connect the wire-to-wire connector.
73. Connect the fuel pump connector.
74. Install the rear floor service hole cover while adjusting it to the 3 convex parts of the floor panel.

WARNING

Be careful that the rear floor service hole cover does not overlap the convex parts of the floor panel when installing.

75. Install the rear seat cushion by engaging the 3 seat hooks.

IDLE SPEED

ADJUSTMENT

Idle speed is maintained by the Electronic Control Module (ECM). No adjustment is possible.

THROTTLE BODY

REMOVAL & INSTALLATION

See Figures 126 through 129.

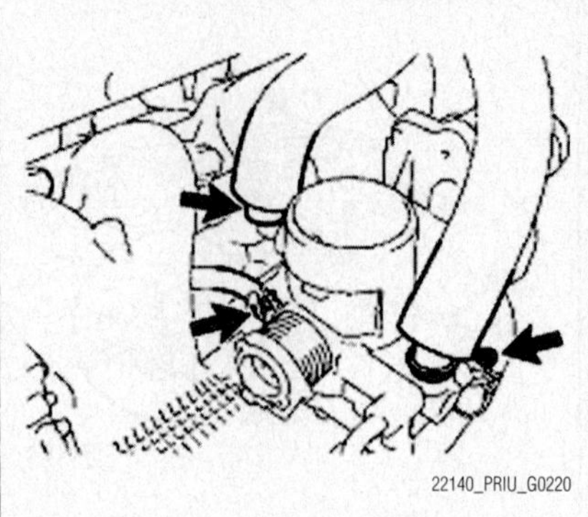

Fig. 126 Disconnect the ventilation hose

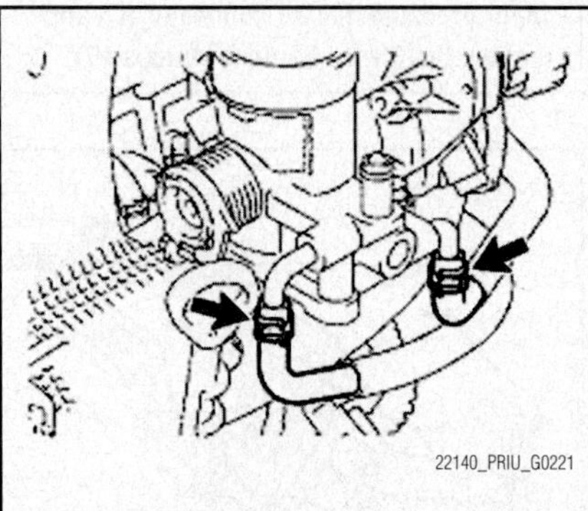

Fig. 127 Disconnect the water by-pass hose

1. Disconnect the negative battery cable.

➡Wait at least 90 seconds after disconnecting the cable from the negative (-) battery terminal to prevent airbag and seat belt pretensioner activation.

2. Remove engine under cover.
3. Drain engine coolant.
4. Remove the 6 clips and radiator support opening cover.
5. Remove air cleaner assembly

a. Disconnect the MAF sensor connector.

b. Disconnect the wire harness from the wire harness clamp.

c. Loosen the hose clamp bolt, and then disconnect the No. 1 air cleaner inlet.

d. Remove the 2 bolts.

e. Loosen the hose clamp bolt, and then remove the air cleaner.

6. Remove throttle body assembly with motor.

a. Disconnect the ventilation hose.

b. Disconnect the No. 2 ventilation hose.

c. Disconnect the No. 1 fuel vapor feed hose.

d. Disconnect the water by-pass hose.

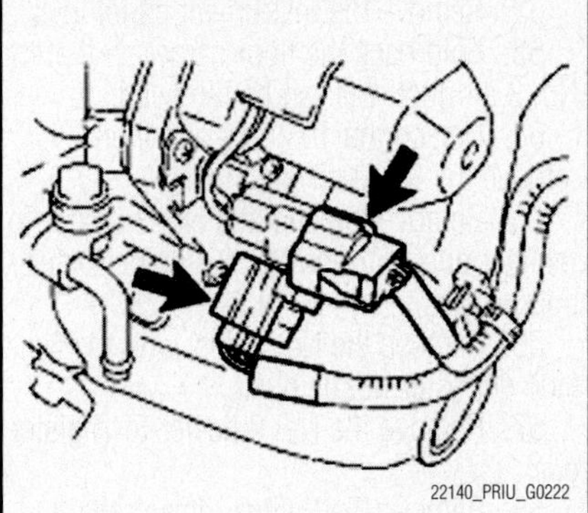
Fig. 128 Disconnect the throttle control motor and throttle position sensor connector

e. Disconnect the No. 2 water by-pass hose.

f. Disconnect the throttle control motor connector.

g. Disconnect the throttle position sensor connector.

h. Remove the bolt, 2 nuts and throttle with motor body.

i. Remove the gasket from the intake manifold.

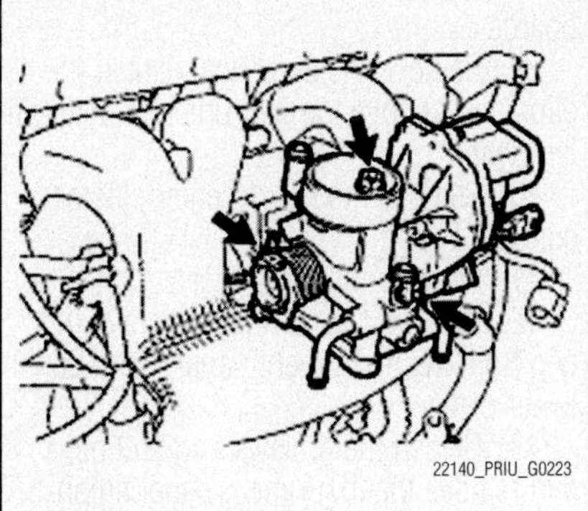
Fig. 129 Throttle body fastener locations

To install:

7. Install throttle body assembly with motor

a. Install a new gasket to the intake manifold.

b. Install the throttle with motor body with the bolt and 2 nuts. Tighten to 15 ft. lbs. (20 Nm).

c. Connect the throttle position sensor connector.

d. Connect the throttle control motor connector.

e. Connect the No. 2 water by-pass hose.

f. Connect the water by-pass hose.

g. Connect the No. 1 fuel vapor feed hose.

h. Connect the No. 2 ventilation hose.

i. Connect the ventilation hose.

8. Install air cleaner assembly

a. Install the air cleaner with the 2 bolts. Tighten to 62 inch lbs. (7 Nm).

b. Tighten the hose clamp bolt to 27 inch lbs. (3 Nm).

c. Connect the No. 1 air cleaner inlet, and tighten the hose clamp bolt to 27 inch lbs. (3 Nm).

d. Connect the MAF sensor meter connector.

9. Connect the negative battery cable.

10. Perform the power window initialization procedure. Refer to Engine Electrical, Initialization Procedure.

11. Add engine coolant

12. Inspect for coolant leaks.

13. Install engine under cover.

14. Install radiator support opening cover.

HEATING & AIR CONDITIONING SYSTEM

AIR CONDITIONING UNIT

REMOVAL & INSTALLATION

See Figures 130 through 133.

1. Before servicing the vehicle, refer to the precautions.
2. Disconnect the negative battery cable.

➡Wait at least 90 seconds after disconnecting the cable from the negative (-) battery terminal to prevent airbag and seat belt pretensioner activation.

3. Place front wheels in the straight ahead position.

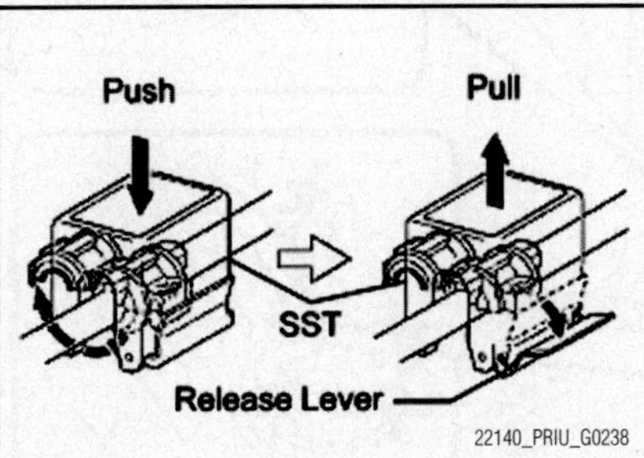

Fig. 130 Push down on special tool with your thumb while holding the pipe with both hands. Pull special tool until the stopper touches the pipe

4. Drain the cooling system.
5. Recover refrigerant from A/C system.
6. Disconnect the suction hose.

a. Check SST 09870-00015 installation direction. Set SST so that the stopper side is on the piping clamp lock side.

b. Install SST on the piping clamp.

c. Push down on special tool with your thumb while holding the pipe with both hands. Be careful not to bend the pipe.

d. Pull special tool until the stopper touches the pipe.

e. Raise SST stopper and remove the piping clamp with SST from the pipe.

f. Remove the piping clamp from SST.

g. Disconnect the suction hose by hand or using a screwdriver.

h. Remove the 2 O-rings from the suction hose. Do not apply excessive force to the suction hose. Seal the opening of the disconnected part using vinyl tape to prevent moisture and foreign matter from entering it.

7. Disconnect the cooler refrigerant liquid pipe to cooler unit by using the same procedures described for the suction hose utilizing SST 09870-00025.
8. Slide the clip and disconnect the heater water hoses. Do not apply excessive force to the water hoses. Prepare a drain pan or cloth for when the cooling water leaks.
9. Detach the 2 claws and 4 clips, and remove the instrument panel register.
10. Remove the 2 screws and disconnect the hood lock control cable.
11. Detach the 4 claws, 5 clips and disconnect all connectors and remove the instrument panel finish panel.
12. Detach the 3 claws, 4 clips and disconnect the connector and remove the instrument panel finish panel.
13. Detach the claw, 5 clips and disconnect the connector and remove the No. 3 instrument panel register.
14. Detach the 6 clips and remove the No. 4 instrument panel register.
15. Detach the 2 claws and 4 clips, and remove the No. 2 instrument panel register.
16. Remove the multi-display assembly.
17. Remove the glove compartment door stopper from the glove compartment door.
18. While pushing in the sides of the glove compartment door, open the door to release it from the 2 stoppers.
19. Open the door until it is horizontal.
20. Pull the glove compartment door toward the rear of the vehicle to detach the 2 hinges and remove the glove compartment door.

21. Remove the instrument panel cushion.
22. Remove the instrument cluster finish panel end.
23. Using a screwdriver, remove the 2 instrument panel hole covers.
24. Remove the 2 screws, detach the claw and 2 clips, and remove the glove compartment door.
25. Remove the No. 1 instrument panel speaker.
26. Remove the left and right front pillar garnish corner pieces.
27. Remove the left and right front pillar garnishes.
28. Disconnect the passenger airbag connector.
29. Remove the 3 screws, 2 bolts, pull up the instrument panel to detach the 6 claws
and 3 clips.
30. Remove the instrument panel safety pad.
31. Remove the NO. 2 and No. 3 steering wheel lower covers.
32. Remove the air bag module.
33. Remove the steering wheel.
34. Remove the tilt lever bracket.
35. Remove the steering column cover.
36. Remove the spiral cable.
37. Remove left and right front door scuff plates.
38. Remove left and right cowl side trim boards.
39. Remove the 2 screws, detach the 4 clips and remove the instrument cluster finish panel.
40. Remove the No. 1 center cluster cushion.
41. Remove the radio receiver.
42. Using a clip remover, remove the clip from the instrument panel finish panel lower center.
43. Detach the 4 claws and 2 clips, and remove the instrument panel finish panel.
44. Detach the 6 claws and remove the glove compartment door lock cover.
45. Remove the 4 screws and glove compartment door lock.
46. Detach the 4 claws and remove the instrument panel finish panel lower.
47. Detach the 2 claws and remove the No. 1 instrument panel under cover.
48. Remove power steering ECU.
49. Remove the shift lever assembly lower.
50. Using a clip remover, remove the 8 clips from the duct and lower instrument panel.
51. Remove the 6 bolts, 2 screws, disconnect all connectors, and detach all clamps.
52. Remove the instrument panel lower.
53. Fold back the floor carpet so that the No. 3 air duct rear can be removed.
54. Detach the 10 claws and then remove the No. 3 air duct rear.
55. Remove the clip and No. 3 heater to register duct with the No. 2 side defroster nozzle duct.
56. Remove the No. 1 duct with the No. 1 side defroster nozzle duct.
57. Remove the No. 2 heater-to-register duct.
58. Remove the 2 clips, detach the 3 claws and then remove the defroster nozzle.
59. Remove the transaxle control ECU assembly.
60. Remove the ECM.
61. Remove the network gateway ECU.
62. Detach the clamp and disconnect the harness.
63. Remove the 2 bolts, nut and instrument panel brace.
64. Remove the air conditioning amplifier assembly by disconnecting the connector and removing the 2 screws.
65. Separate steering column from under the lower instrument panel.
66. Disconnect each connector and remove each clamp. Disconnect the wire harness.

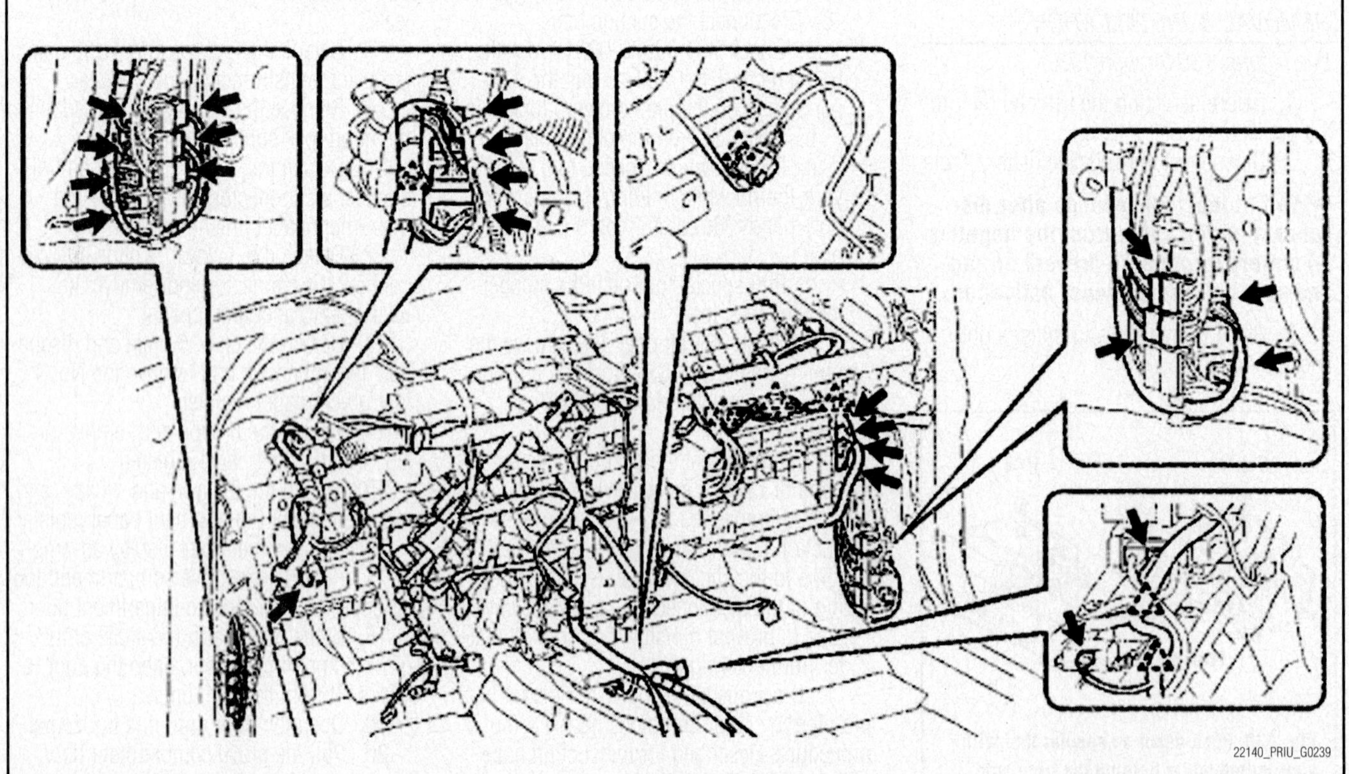

Fig. 131 Disconnect each connector and remove each clamp. Disconnect the wire harness

67. Remove the 7 bolts, 2 nuts and then remove the instrument panel reinforcement with the air conditioner unit.

68. Remove the air conditioning unit assembly.

a. Disconnect the 5 connectors.

b. Detach the 9 clamps and disconnect the wire harness.

c. Remove the bolt.

d. Detach the clamp and disconnect the junction connector.

e. Remove the 2 screws and air conditioning unit from the instrument panel reinforcement.

To install:

69. Install air conditioning tube and accessory assembly.

a. Sufficiently apply compressor oil ND-OIL 11 to 2 new O-rings and the fitting surface. Install the 2 O-rings to the air conditioning tube assembly.

CAUTION

Do not use any compressor oil other than ND-OIL 11. If any compressor oil other than ND-OIL 11 is used, compressor motor insulation performance may decrease, resulting in a leakage of electric power.

b. Install the air conditioning tube to the No. 1 cooler evaporator, placing the cooler expansion valve between them. Using a 0.16 inch (4mm) hexagon wrench, install the 2 hexagon bolts and tighten to 30 inch lbs. (3.5 Nm).

c. Install the expansion cover with the 2 screws.

70. Install the air outlet control servo motor and attach the claw.

71. Attach the claw and install the 2 screws.

72. Install the air mix control servo motor with the 2 screws.

73. Install the defroster nozzle lower and attach the 2 claws.

74. Install the blower and then attach the fittings with the air conditioner and install the 2 screws.

75. Install the air conditioning unit assembly.

a. Install the air conditioning unit to the instrument panel reinforcement with the 2 screws. Use repair screws (part No. 90159-70003) if the screws removed before cannot be tightened.

b. Attach the clamp to connect the junction connector.

c. Install the bolt.

d. Attach the 9 clamps to connect the wire harness.

e. Connect the 5 connectors.

76. Install the instrument panel reinforcement with the 7 bolts.

77. Install the 2 nuts to the air conditioning unit and temporarily tighten them.

78. Connect each connector and install each clamp. Then connect the wire harness.

79. Install the air conditioning amplifier assembly with the 2 screws and engage the connector.

80. Place steering column into position underneath the lower instrument panel and tighten the mounting bolts.

81. Install the instrument panel brace with the 2 bolts and nut.

82. Attach the clamp to connect the harness.

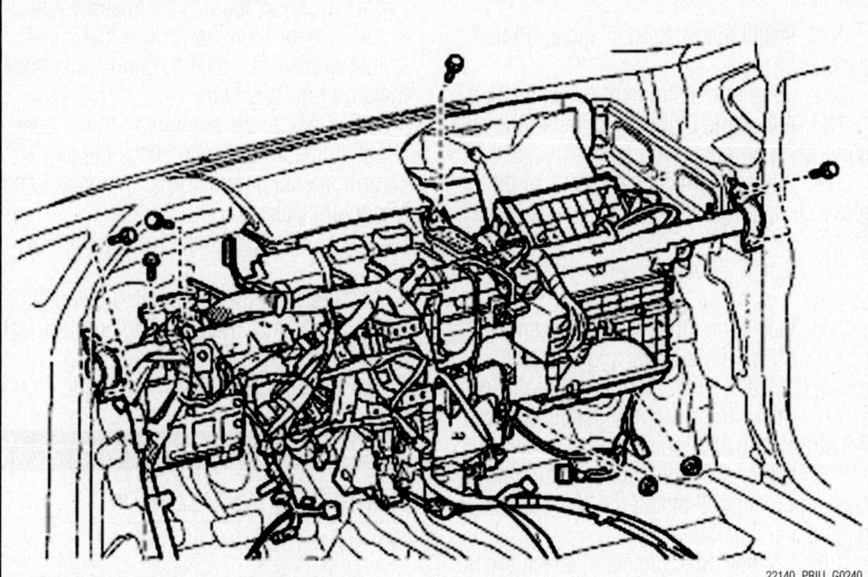

Fig. 132 Remove the 7 bolts, 2 nuts and then remove the instrument panel reinforcement with the air conditioner unit

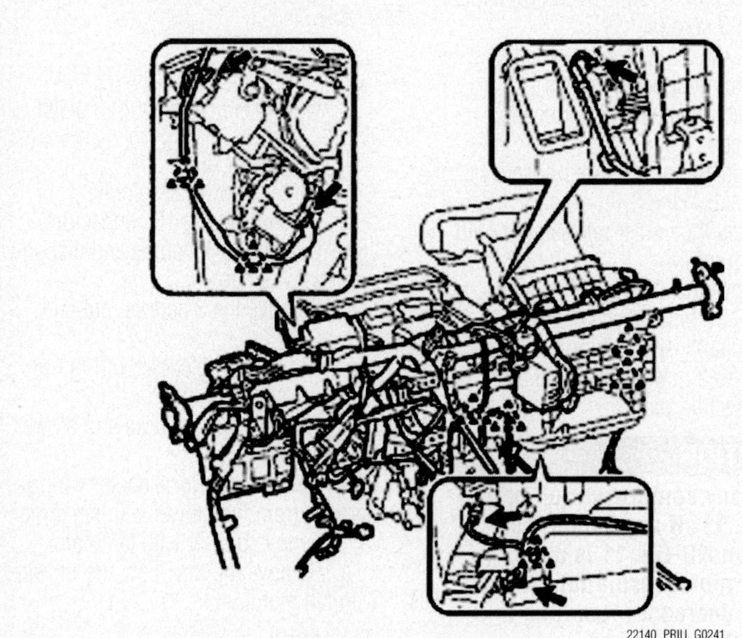

Fig. 133 Disconnect the 5 connectors. Detach the 9 clamps and disconnect the wire harness to remove the air conditioning unit.

83. Fully tighten the air conditioning unit with the 2 nuts. Tighten the left nut first, then the right nut second.
84. Install the network gateway ECU.
85. Install the ECM.
86. Install the transaxle control ECU assembly.
87. Install the defroster nozzle and attach the 3 claws.
88. Install the 2 clips.
89. Install the No. 2 heater-to-register duct.
90. Install the duct with the No. 1 side defroster nozzle duct.
91. Install the clip and then the duct with the No. 2 side defroster nozzle duct.
92. Attach the 10 claws to install the No. 3 air duct rear, then return the carpet to its original position.
93. Install the lower instrument panel.
94. Connect all the connectors and attach all the clamps.
95. Install the 2 screws, 6 bolts, 8 clips to the duct and lower instrument panel.
96. Install the lower shift lever assembly.
97. Install the power steering ECU.
98. Attach the 2 claws to install the No. 1 instrument panel under cover.
99. Attach the 4 claws to install the instrument panel finish panel.
100. Install the glove compartment door lock with the 4 screws.
101. Attach the 6 claws to install the glove compartment door lock cover.
102. Attach the 4 claws and 2 clips to install the instrument panel finish panel lower center. Install the clip.
103. Install the radio receiver.
104. Install the No. 1 center cluster cushion.
105. Attach the 4 claws and 2 clips to install the instrument cluster finish panel assembly center. Install the 2 screws.
106. Install left and right cowl side trim boards.
107. Install left and right front door scuff plates.
108. Install the spiral cable.
109. Install the steering column cover.
110. Install the tilt lever bracket.
111. Install the steering wheel and tighten the center nut to 37 ft. lbs. (50 Nm).
112. Install the air bag module.
113. Install the NO. 2 and No. 3 steering wheel lower covers.
114. Attach the 6 claws and 3 clips to install the instrument panel safety pad. Install the 2 bolts and 3 screws.
115. Connect the passenger airbag connector.
116. Install left and right front pillar garnishes.
117. Install left and right front pillar garnish corner pieces.
118. Install the No. 1 instrument panel speaker (w/ JBL Sound System).
119. Attach the claw and 2 clips to install the glove compartment door. Install the 2 screws.
120. Install the 2 instrument panel hole covers.
121. Install the instrument cluster finish panel end.
122. Install the instrument panel cushion.
123. Attach the 2 hinges to install the glove compartment door.
124. While pushing in the sides of the glove compartment door, engage it to the 2 stoppers.
125. Install glove compartment door stopper.
126. Install the multi-display assembly.
127. Attach the 2 claws and 4 clips to install the No. 2 instrument panel register.
128. Attach the 6 clips to install the No. 4 instrument panel register.
129. Connect the connector, attach the claw and 5 clips to install the No. 3 instrument panel register.
130. Connect the connector, attach the 3 claws and 4 clips to install the upper instrument panel finish panel.
131. Connect the connectors, attach the 4 claws and 5 clips to install the lower instrument panel finish panel.
132. Connect the hood lock control cable and install the 2 screws.
133. Attach the 2 claws and 4 clips to install the instrument panel register.
134. Connect the heater water hose A and slide the clip.
135. Connect the heater water hose B and slide the clip.
136. Connect the cooler refrigerant liquid pipe to cooler unit.
   a. Remove the attached vinyl tape from the pipe's disconnected part.
   b. Sufficiently apply compressor oil ND-OIL 11 to 2 new O-rings and the pipe's connecting part.

CAUTION

Do not use any compressor oil other than ND-OIL 11. If any compressor oil other than ND-OIL 11 is used, compressor motor insulation performance may decrease, resulting in a leakage of electric power.

   c. Install the O-rings to the pipe.
   d. Insert the pipe joint into the cooler unit fitting hole securely.
   e. Using the piping clamp, connect the cooler unit refrigerant liquid pipe. Ensure that the piping clamp is securely engaged.
137. Connect the suction hose by using the same procedures described for the cooler unit refrigerant liquid pipe.
138. Refill the cooling system.
139. Evacuate and recharge the refrigerant.
140. Connect the negative battery cable.
141. Perform the power window initialization procedure. Refer to Engine Electrical, Initialization Procedure.
142. Operate the engine to normal operating temperatures; then, check the climate control operation and check for coolant and refrigerant leaks.
143. Connect the negative battery cable.
144. Perform the power window initialization procedure. Refer to Engine Electrical, Initialization Procedure.
145. Check the SRS warning light.

BLOWER MOTOR

REMOVAL & INSTALLATION

See Figures 134 through 136.

1. Before servicing the vehicle, refer to the precautions.
2. Remove instrument panel with passenger airbag assembly.
3. Remove lower instrument panel.
4. Remove transmission control ECU assembly.
5. Remove ECM.
6. Remove network gateway ECU.
7. Remove No. 1 heater to register duct.
8. Remove air duct.
9. Remove blower assembly.
   a. Disconnect the 10 connectors.
   b. Detach the 4 clamps and disconnect the wire harness.
   c. Remove the 3 screws, nut and blower.
10. Remove No. 2 cooler wiring harness.
   a. Remove the 5 screws and blower motor cover.
   b. Remove the No. 2 cooler wiring harness from the blower with fan motor.
11. Remove blower with fan motor.
   a. Remove the 3 screws and blower with fan motor.

To install:

12. Install blower with fan motor.
   a. Install the blower with fan motor with the 3 screws.
13. Install No. 2 cooler wiring harness.

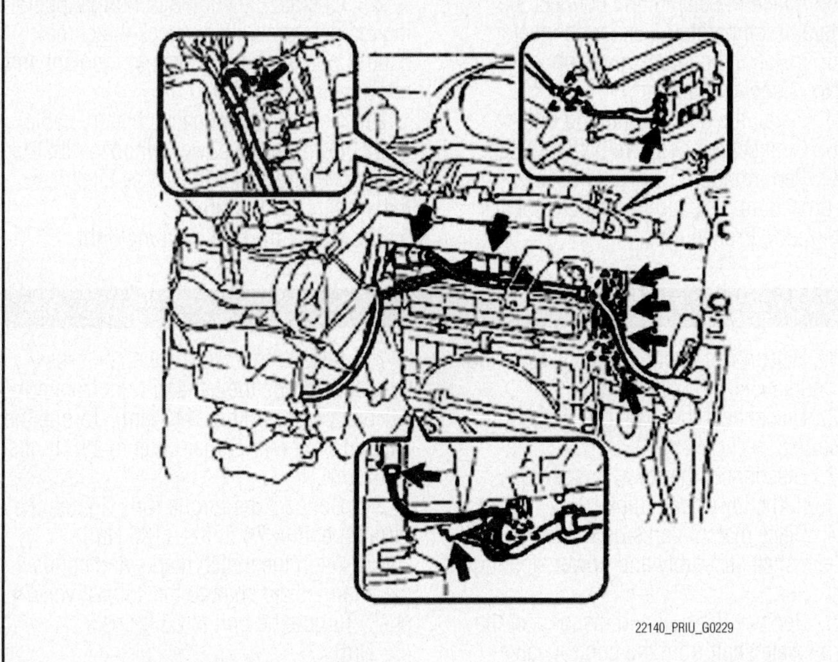

Fig. 134 Blower assembly connectors

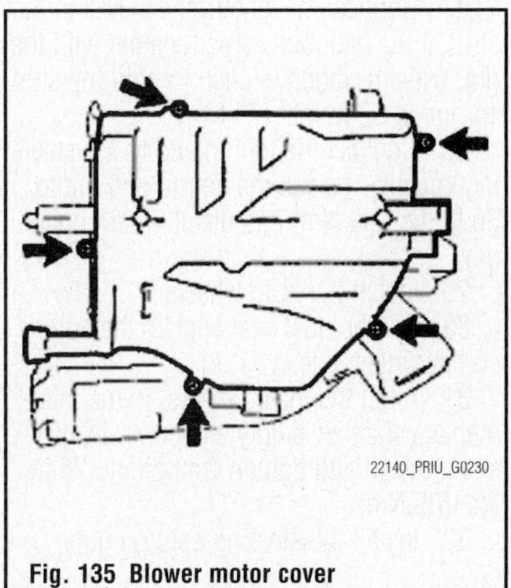

Fig. 135 Blower motor cover

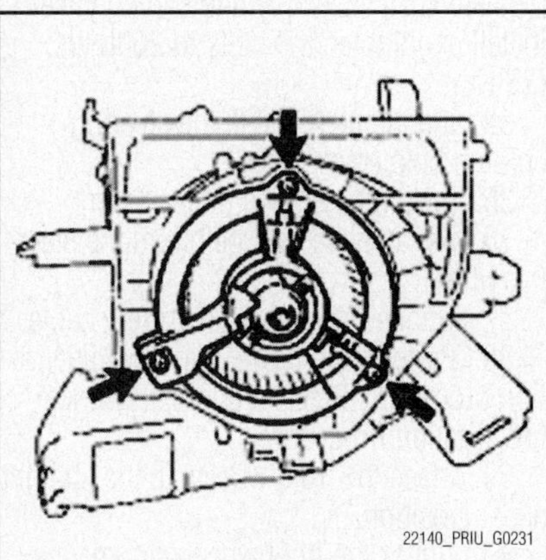

Fig. 136 Blower motor attaching screws

a. Install the No. 2 cooler wiring harness to the blower with fan motor.

b. Install the blower motor cover with the 5 screws.

14. Install blower assembly.

a. Install the blower with the 3 screws.

15. Use a repair screw (part No. 90159-70003) if the screw removed before cannot be tightened.

a. Attach the 4 clamps to connect the wire harness.

b. Connect the connectors.

c. Install the 10 connectors of the transponder key ECU.

16. Install air duct.

a. Install the air duct with the 2 screws.

17. Install No. 1 heater to register duct.

18. Install network gateway ECU.

19. Install ECU.

20. Install transmission control ECU assembly.

21. Install lower instrument panel.

22. Install instrument panel with passenger airbag assembly.

HEATER CORE

REMOVAL & INSTALLATION

See Figures 137 and 138.

1. Before servicing the vehicle, refer to the precautions.

2. Disconnect the negative battery cable.

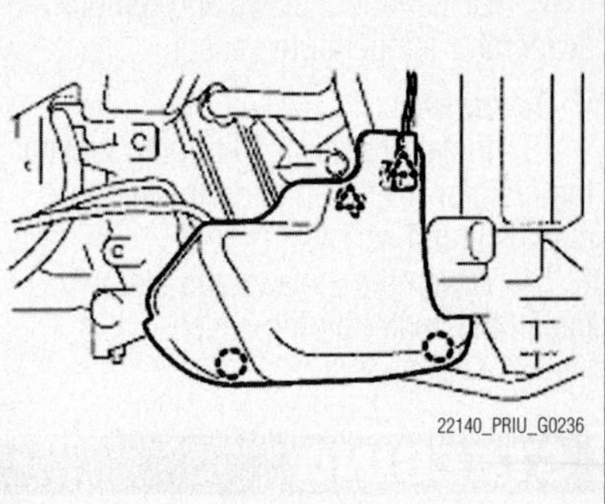

Fig. 137 Detach the clamp and 2 claws, and remove the heater piping cover

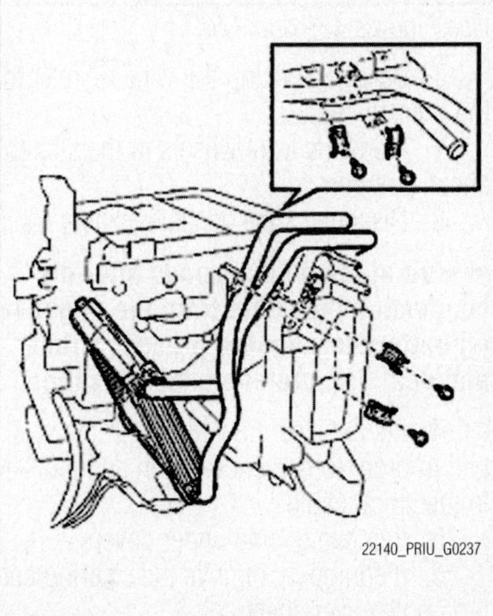

Fig. 138 Remove the radiator heater unit from the air conditioner radiator

➡Wait at least 90 seconds after disconnecting the cable from the negative (-) battery terminal to prevent airbag and seat belt pretensioner activation.

3. Remove the air conditioning unit.

4. Detach the clamp and disconnect the evaporator temperature sensor connector.

5. Detach the clamp and 2 claws, and remove the heater piping cover.

6. Remove the 4 screws and 4 clamps.

7. Remove the radiator heater unit from the air conditioner radiator.

➡Prepare a drain pan or cloth for when the cooling water leaks.

8. Remove the 2 screws and No. 1 air duct.

9. Remove the 4 screws and quick heater assembly.

10. Remove the 7 screws and No. 1 cooler evaporator from the heater case.

11. Remove the 2 O-rings from the No. 1 cooler evaporator.

12. Detach the 2 claws and remove the evaporator temperature sensor.

To install:

13. Install the air conditioning radiator to the radiator heater unit and install the 4 clamps and 4 screws.
14. Attach the 2 claws and clamp to install the heater piping cover.
15. Attach the clamp and connect the evaporator temperature sensor connector.
16. Install the cooler expansion valve to the No. 1 cooler evaporator.
17. Install the air conditioning unit.
18. Connect the negative battery cable.
19. Perform the power window initialization procedure. Refer to Engine Electrical, Initialization Procedure.
20. Operate the engine to normal operating temperatures; then, check the climate control operation and check for coolant and refrigerant leaks.
21. Connect the negative battery cable.
22. Perform the power window initialization procedure. Refer to Engine Electrical, Initialization Procedure.
23. Check the SRS warning light.

STEERING

POWER RACK & PINION STEERING GEAR

REMOVAL & INSTALLATION

See Figures 139 and 140.

1. Before servicing the vehicle, refer to the precautions.
2. Place the front wheels in the straight ahead position.
3. Disconnect the negative battery cable.

➡Wait at least 90 seconds after disconnecting the cable from the negative (-) battery terminal to prevent airbag and seat belt pretensioner activation.

4. Fix the steering wheel with the seat belt in order to prevent rotation and damage to the spiral cable.
5. Remove engine under covers.
6. If equipped, remove the steering column hole cover sheet.
7. Put match marks in the sliding yoke and intermediate shaft.
8. Loosen the top bolt and remove the lower bolt to separate the sliding yoke.
9. If necessary, remove the steering column hole cover from the body. Be careful not to damage the clips.
10. Remove the front wheels.
11. Disconnect the tie rod ends from the steering knuckles.
12. Disconnect the stabilizer links from the struts.
13. Disconnect the lower suspension arm from the lower ball joint.
14. Place match marks on the intermediate shaft assembly and power steering gear.
15. Remove the bolt and disconnect the intermediate shaft from the control valve shaft.
16. Disconnect the torque rod.
17. Remove the front suspension crossmember assembly.
18. Remove the stabilizer bar mounting brackets and the stabilizer bar.
19. Remove the 4 bolts and power steering gear assembly from the front suspension crossmember.

To install:

20. Install the steering column hole cover to the steering gear.
21. Install the power steering gear assembly with the 4 new bolts to the front suspension crossmember. Torque the bolts to 43 ft. lbs. (58 Nm).
22. Install the stabilizer bar and stabilizer bar mounting brackets. Torque the mounting bracket bolts to 14 ft. lbs. (19 Nm).
23. Install the front suspension crossmember. Torque the left and right front corner bolts to 83 ft. lbs. (113 Nm). Torque the left and right rear corner bolts to 116 ft. lbs. (157 Nm).
24. Connect the torque rod. Tighten the through bolt to 74 ft. lbs. (100 Nm).
25. Align the match marks and connect the intermediate shaft to the control valve shaft. Torque the bolt to 26 ft. lbs. (35 Nm).
26. Connect the left and right ball joints to the front lower control arms. Torque to 66 ft. lbs. (89 Nm).
27. Connect the stabilizer bar link to the strut. If the ball joint turns together with the nut, use a hexagon wrench to hold the stud. Torque to 55 ft. lbs. (74 Nm).
28. Connect the tie rod end to the steering knuckle. Torque the tie rod end nut to 36 ft. lbs. (49 Nm) and install a new cotter pin.
29. Install the front wheels.
30. Put the dust seal back to the engine compartment side.
31. Align the match marks on the intermediate shaft assembly and control valve shaft. Install and tighten the bolts to 26 ft. lbs. (35 Nm)
32. Install the steering column hole cover.
33. Align the match marks on the intermediate shaft assembly and sliding yoke. Install and tighten the bolts to 26 ft. lbs. (35 Nm)
34. If equipped, install the steering column hole cover sheet.
35. Install the engine under covers.
36. Remove the seat belt from the steering wheel.
37. Connect the negative battery cable.
38. Perform the power window initialization procedure. Refer to Engine Electrical, Initialization Procedure.
39. Place the front wheels in the straight ahead position.
40. Check the front wheel alignment.

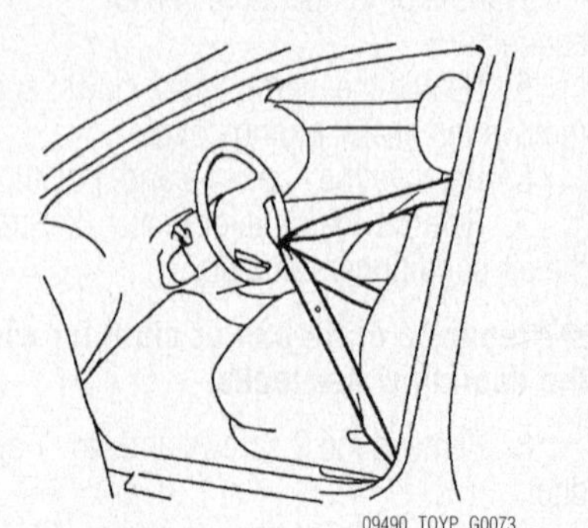

Fig. 139 Steering wheel in fixed position using seat belt

Fig. 140 Remove the clip labeled "A" and disconnect the hole cover from the body. Do not damage the clip labeled "B"

COIL SPRING

REMOVAL & INSTALLATION

This vehicle uses a MacPherson strut front suspension. Refer to Front Suspension, MacPherson Strut, Removal & Installation.

LOWER BALL JOINT

REMOVAL & INSTALLATION

See Figures 141 and 142.

1. Before servicing the vehicle, refer to the precautions.
2. Remove the steering knuckle with the hub/bearing assembly. Refer to Front Suspension, Steering Knuckle, Removal & Installation.
3. Mount the steering knuckle in a vise.
4. Remove the cotter pin and nut.
5. Using SST 09628-62011, remove the lower ball joint.

To install:

6. Install the lower ball joint and tighten the nut to 52 ft. lbs. (71 Nm).
7. Install a new cotter pin. If the holes for the cotter pin are not aligned, tighten the nut further up to 60 degrees.
8. Install the steering knuckle with the hub/bearing assembly.
9. Check the ABS speed sensor signal.

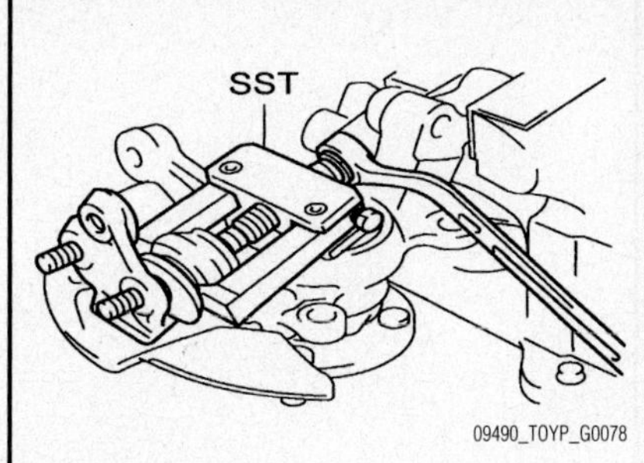

Fig. 142 Separate the ball joint from the steering knuckle.

10. Check and adjust the front wheel alignment.
11. Check for proper speed sensor signal.

LOWER CONTROL ARM

REMOVAL & INSTALLATION

See Figure 143.

1. Before servicing the vehicle, refer to the precautions.
2. Place front wheels facing straight ahead.
3. Disconnect the negative battery cable.
4. Raise and safely support the front of the vehicle. Let the control arms hang free.
5. Remove the front wheels

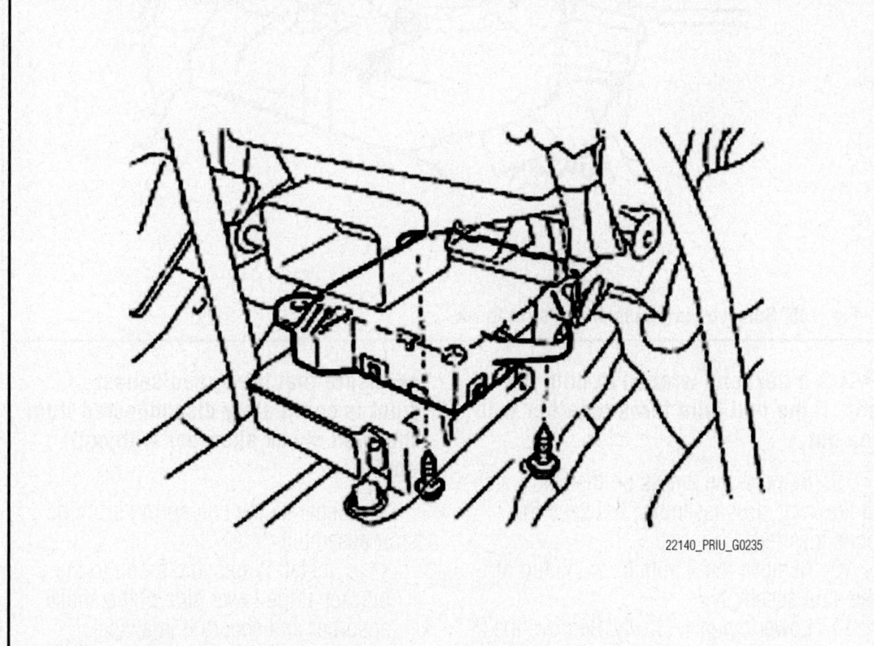

Fig. 141 Lower ball joint attaching bolt/nuts

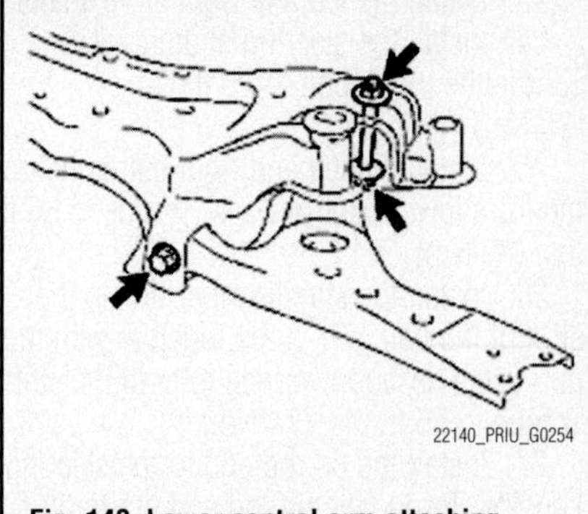

Fig. 143 Lower control arm attaching points

6. Remove the steering column hole cover silencer, if necessary.
7. Remove the engine under cover.
8. Remove the front exhaust pipe assembly, if necessary.
9. Remove the tie rod end cotter pin and nut.
10. Remove the tie rod end from the steering knuckle.
11. Remove the stabilizer bar link from the strut. if the ball joint turns together with the nut, use a hexagon wrench to hold the stud.
12. Remove the bolt and 2 nuts, and disconnect the lower suspension arm from the lower ball joint.
13. Remove the loosen the 2 lower suspension arm set bolts.
14. Remove the steering sliding yoke.
15. Remove the drive shafts, if necessary.
16. Support the suspension member with a transmission jack.
17. Remove the bolt and nut, disconnect the torque rod from the suspension member.
18. Remove the 4 bolts and disconnect the suspension member from the body.
19. Remove the 2 lower suspension arm set bolts and disconnect the lower suspension arm from suspension crossmember.

To install:

20. Install the front lower control arm to the suspension crossmember and temporarily tighten the front suspension lower No. 1 arm with the 2 bolts and nut.
21. Install the front suspension crossmember. Torque the left and right front corner bolts to 83 ft. lbs. (113 Nm). Torque the left and right rear corner bolts to 116 ft. lbs. (157 Nm).
22. Install the torque rod. Tighten the 2 nuts and 2 bolts, at the engine, to 74 ft. lbs.

(100 Nm). Torque the 2 bolts, at the body side, to 41 ft. lbs. (56 Nm).

23. Install the left and right drive shafts.

24. Install the steering sliding yoke. Torque the pinch bolt to 26 ft. lbs. (35 Nm).

25. Install the left and right ball joints to the front lower control arms. Torque to 66 ft. lbs. (89 Nm).

26. Install the stabilizer bar link to the strut. If the ball joint turns together with the nut, use a hexagon wrench to hold the stud. Torque to 55 ft. lbs. (74 Nm).

27. Install the tie rod end to the steering knuckle. Torque the tie rod end nut to 36 ft. lbs. (49 Nm) and install a new clip.

28. Install the front wheels

29. Install the front exhaust pipe assembly, if necessary

30. Install the steering column hole cover silencer, if necessary

31. Lower the vehicle and bounce it up and down several times to stabilize the front suspension.

32. Fully tighten the 2 front lower control arm bolts. Torque to 101 ft. lbs. (137 Nm). Keep the nut from rotating while tightening the rear-side bolt. Lower the tires to the ground using a 4-post lift.

33. Connect the negative battery cable.

34. Perform the power window initialization procedure. Refer to Engine Electrical, Initialization Procedure.

35. Check the front wheel alignment.

MACPHERSON STRUT

REMOVAL & INSTALLATION

See Figures 144 through 148.

1. Before servicing the vehicle, refer to the precautions.

2. Disconnect the negative battery cable.

➡Wait at least 90 seconds after disconnecting the cable from the negative (-) battery terminal to prevent airbag and seat belt pretensioner activation.

3. Remove front wheel.

4. Remove front wipers and front wiper motor link assembly.

5. Remove the bolt and disconnect the ABS speed sensor wire harness clamp.

6. Remove the bolt and disconnect the flexible hose from the shock absorber bracket.

7. Place a wooden block on a jack, and support the front suspension lower No. 1 arm with the jack.

8. Remove the nut and separate the front stabilizer link from the strut assembly.

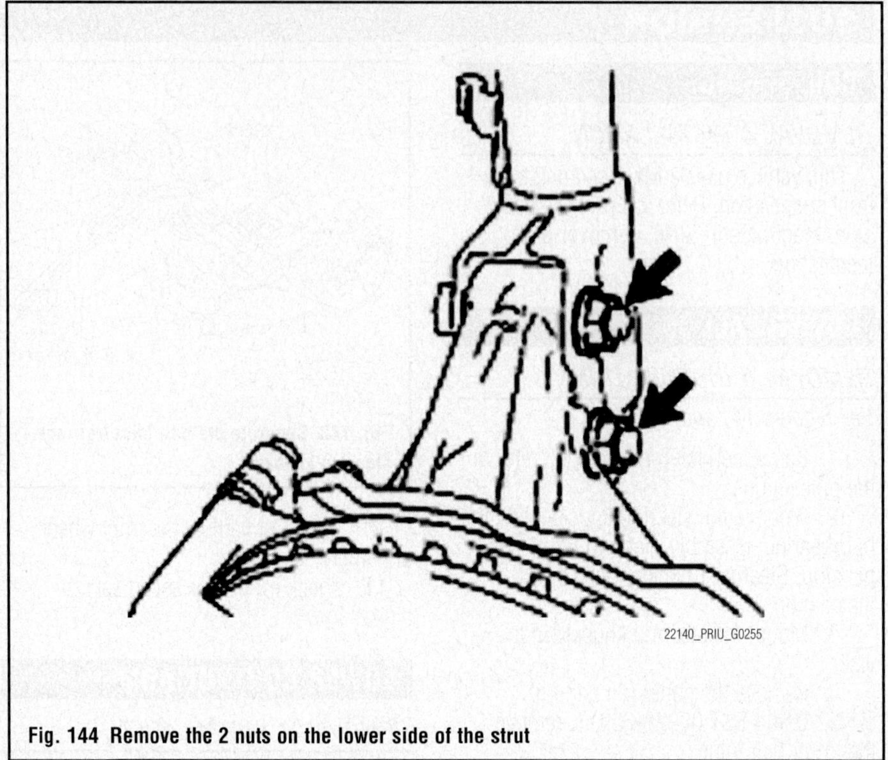

Fig. 144 Remove the 2 nuts on the lower side of the strut

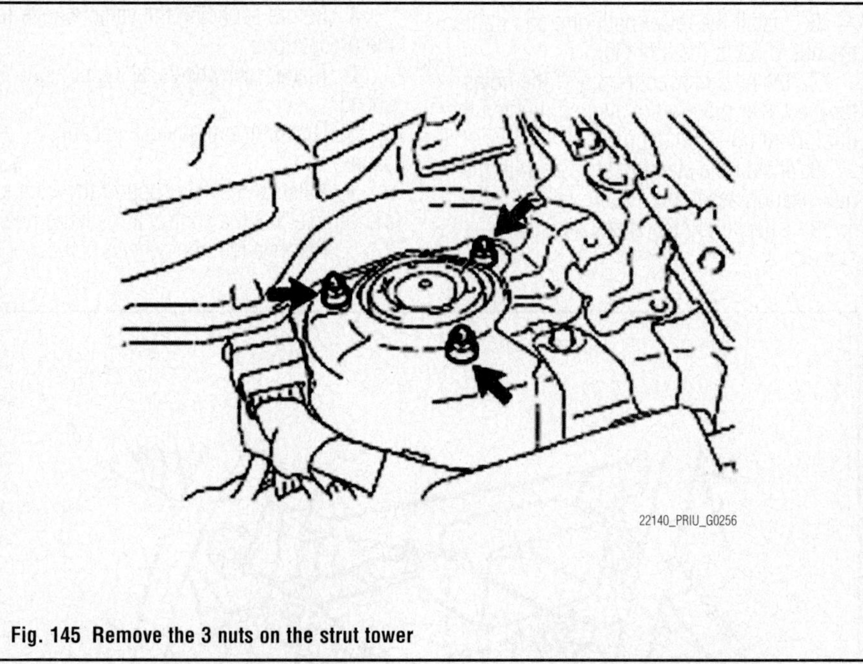

Fig. 145 Remove the 3 nuts on the strut tower

➡Use a hexagon wrench to hold the stud if the ball joint turns together with the nut.

9. Remove the 2 nuts on the lower side of the front strut assembly, but keep the bolts inserted.

10. Remove the 3 nuts from the top of the strut assembly.

11. Lower the jack slowly. Remove the 2 bolts on the lower side and the front strut assembly.

➡Ensure that the speed sensor front is completely disconnected from the front shock absorber with coil spring.

12. Separate the coil spring from the strut assembly.

a. Install 2 nuts and a bolt to the bracket at the lower side of the shock absorber and secure it in a vise.

b. Using SST 09727-30021, compress the coil spring.

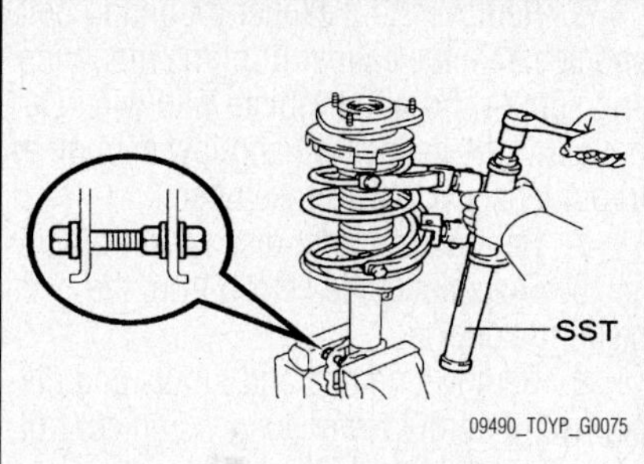

Fig. 146 Securing the strut assembly in a vise while installing the spring compressor tool.

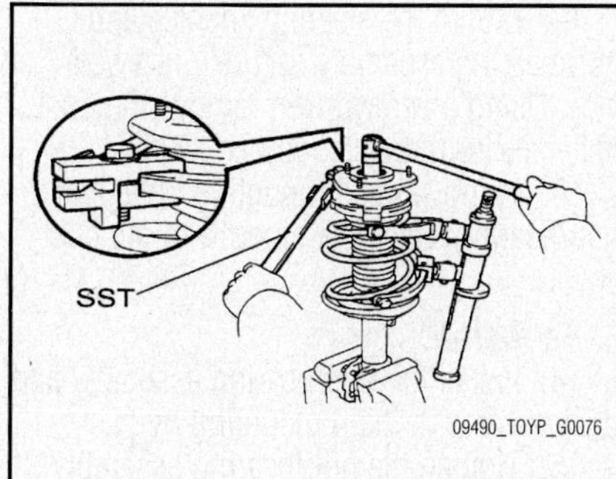

Fig. 147 Removing the nut while holding the spring seat.

c. Remove the cap from the suspension support.

d. Using SST 09729-22031 to hold the spring seat, remove the nut.

e. Remove the suspension support, dust seal, spring seat, upper insulator, coil spring, spring bumper and lower insulator.

To install:

13. Install the coil spring by assembling the strut assembly.

a. Install the lower insulator onto the strut.

b. Install the spring bumper to piston rod.

c. Using SST 09727-30021, compress the coil spring. Do not use an impact wrench. It will damage the SST.

d. Install the coil spring to the shock absorber, fitting the lower end of the coil spring into the gap of the spring lower seat.

e. Install the upper insulator.

f. Install the spring seat to the shock absorber with the arrow mark facing to the outside of the vehicle.

g. Install the dust seal and suspension support.

h. Using SST 09729-22031 to hold

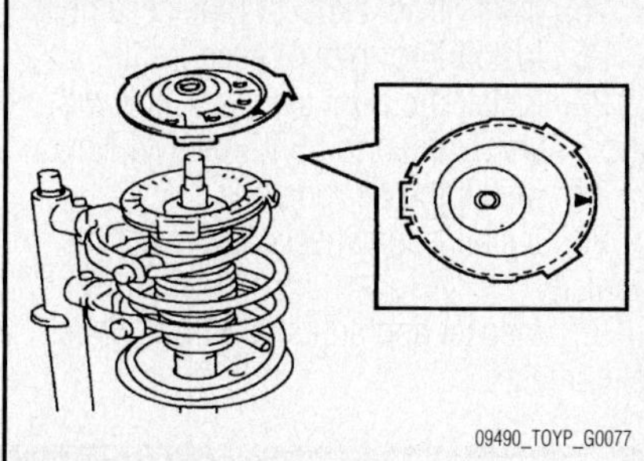

Fig. 148 Install the spring seat to the shock absorber with the arrow mark facing to the outside of the vehicle.

the suspension support, install a new nut and torque to 34 ft. lbs. (47 Nm).

i. Remove the tool.

j. Apply MP grease No.2 into the suspension support. Do not touch grease on rubber surface of upper support.

k. Install the cap.

14. Install the strut assembly to the steering knuckle and tighten the steering knuckle-to-strut assembly nuts and the bolts to 113 ft. lbs. (153 Nm)

15. Connect the stabilizer shaft link to strut assembly and tighten the stabilizer link-to-strut assembly nut to 55 ft. lbs. (74 Nm)

16. Install the flexible brake hose to the shock absorber bracket.

17. Connect the ABS speed sensor wire harness clamp.

18. Install the front wiper motor link assembly and the front wipers.

19. Install the front wheel.

20. Lower the vehicle.

21. Install the nuts securing the strut assembly to the body of the vehicle and tighten to 29 ft. lbs. (39 Nm).

22. Connect the negative battery cable.

23. Perform the power window initialization procedure. Refer to Engine Electrical, Initialization Procedure.

STEERING KNUCKLE

REMOVAL & INSTALLATION

1. Before servicing the vehicle, refer to the precautions.
2. Raise and safely support the vehicle.
3. Remove the front wheel.
4. Remove the front axle hub nut.
5. Disconnect the front ABS speed sensor connector.
6. Remove the 2 caliper mounting bolts and separate the caliper from the mounting and support from the vehicle with wire. Do not allow the caliper to hang low enough as to put tension on the brake hose.
7. Remove the front disc brake rotor.
8. Disconnect the tie rod from the steering knuckle.
9. Remove the bolt and 2 nuts and disconnect the front lower suspension control arm from the lower ball joint.
10. Separate the front drive shaft from the hub/bearing and steering knuckle assembly.
11. Separate the steering knuckle from the strut assembly.

To install:

12. Install the steering knuckle to the strut assembly. Torque the nuts and bolts to 113 ft. lbs. (153 Nm)
13. Install the front drive shaft.
14. Install the lower ball joint to the front lower suspension control arm. Torque the bolt and 2 nuts to 66 ft. lbs. (89 Nm).
15. Connect the tie rod end to the steering knuckle. Torque the tie rod end nut to 36 ft. lbs. (49 Nm) and install a new clip.
16. Install the front disc brake rotor.
17. Install the front disc brake caliper.
18. Connect the front ABS speed sensor connector.
19. Install the front axle hub nut. Torque the nut to 159 ft. lbs. (216 Nm), then stake the hub nut, using a chisel and hammer.
20. Install the front wheel.
21. Inspect and adjust the front wheel alignment.
22. Check ABS speed sensor signal.

STABILIZER BAR

REMOVAL & INSTALLATION

See Figures 149 and 150.

1. Place front wheels facing straight ahead.
2. Remove column hole cover silencer sheet.
3. Disconnect steering sliding yoke.
4. Disconnect No. 1 steering column hole cover.
5. Raise and safely support the vehicle securely on jackstands.
6. Remove front wheel.
7. Remove the 2 nuts and front stabilizer link.

➡Use a 6 mm hexagon wrench to hold the stud if the ball joint turns together with the nut.

8. Disconnect tie rod ends.
9. Remove floor panel brace front.
10. Remove front exhaust pipe assembly.
11. Remove the 2 bolts and the No. 1 stabilizer bracket from the front suspension crossmember.
12. Disconnect front stabilizer bracket.

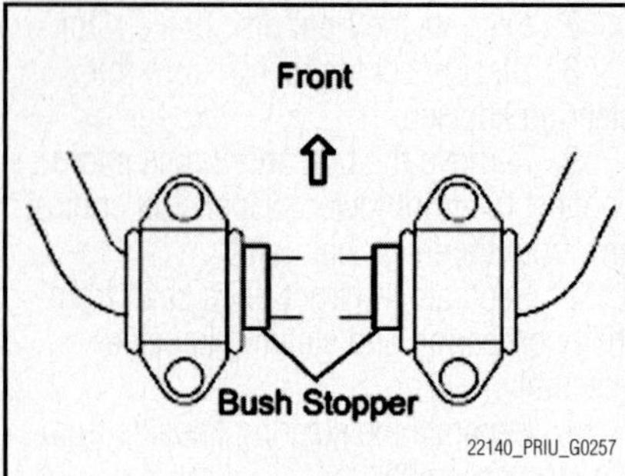

Fig. 149 Install the bushing to the outer side of the bushing stopper on the stabilizer bar

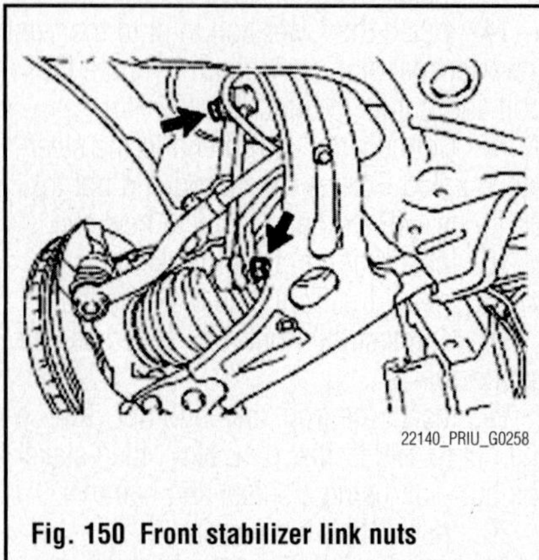

Fig. 150 Front stabilizer link nuts

13. Remove steering intermediate shaft.
14. Remove steering column hole cover.
15. Remove steering gear assembly.
16. Remove the 2 front stabilizer bar bushes from the stabilizer bar.
17. Remove the front stabilizer bar from the right side of the vehicle.

To install:

18. Insert the front stabilizer bar from the right side of the vehicle.
19. Install the bushing to the outer side of the bushing stopper on the stabilizer bar.

➡Place the cutout of the stabilizer bushing facing the rear side. Ensure the right and left deviation of the stabilizer bar is 0.20 in. (5 mm) or less.

20. Install steering gear assembly.
21. Install steering column hole cover.
22. Install steering intermediate shaft.
23. Install the stabilizer bracket the front suspension crossmember with the 2 bolts. Tighten to 14 ft. lbs. (19 Nm).
24. Connect front stabilizer bracket.
25. Install front exhaust pipe assembly.
26. Install floor panel brace front.
27. Install the front stabilizer link with the 2 nuts. Tighten to 55 ft. lbs. (74 Nm).
28. Install the tie rod ends.
29. Install front wheels and lower the vehicle.
30. Inspect and adjust the front wheel alignment

WHEEL HUB AND BEARING

REMOVAL & INSTALLATION

See Figures 151 and 152.

1. Before servicing the vehicle, refer to the precautions.
2. Raise and safely support the vehicle.
3. Remove the front wheel.
4. Remove the front axle hub nut.
5. Disconnect the front ABS speed sensor connector.

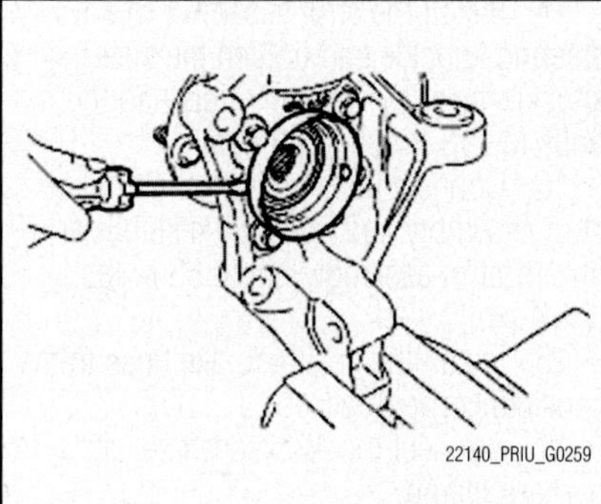

Fig. 151 Using a screwdriver, remove the deflector from the steering knuckle

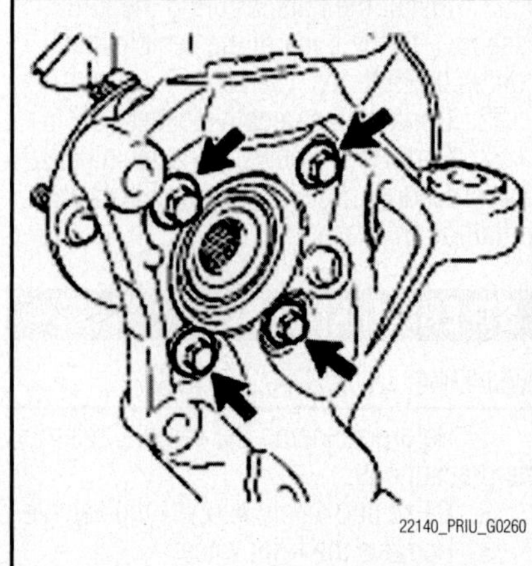

Fig. 152 Front wheel hub attaching bolts

6. Remove the 2 caliper mounting bolts and separate the caliper from the mounting and support from the vehicle with wire. Do not allow the caliper to hang low enough as to put tension on the brake hose.
7. Remove the front disc brake rotor.
8. Disconnect the tie rod from the steering knuckle.
9. Remove the bolt and 2 nuts and disconnect the front lower suspension control arm from the lower ball joint.
10. Separate the front drive shaft from the hub/bearing and steering knuckle assembly.
11. Separate the steering knuckle from the strut assembly.
12. Mount the steering knuckle and hub/bearing assembly securely in a soft vice. Using a screwdriver, remove the dust deflector from the steering knuckle.
13. Remove the 4 mounting bolts and hub/bearing assembly along with the dust cover.

To install:

14. Install the hub/bearing assembly and dust cover with the 4 mounting bolts.
15. Tighten the hub/bearing assembly mounting bolts to 41 ft. lbs. (56 Nm).
16. Using SST 09950-70010, 09608-320-10 and 09950-60020, press in a new dust deflector.
17. Install the steering knuckle to the strut assembly. Torque the nuts and bolts to 113 ft. lbs. (153 Nm)
18. Install the front drive shaft.
19. Install the lower ball joint to the front lower suspension control arm. Torque the bolt and 2 nuts to 66 ft. lbs. (89 Nm).
20. Connect the tie rod end to the steering knuckle. Torque the tie rod end nut to 36 ft. lbs. (49 Nm) and install a new clip.
21. Install the front disc brake rotor.
22. Install the front disc brake caliper.
23. Connect the front ABS speed sensor connector.
24. Install the front axle hub nut. Torque the nut to 159 ft. lbs. (216 Nm), then stake the hub nut, using a chisel and hammer.
25. Install the front wheel.
26. Inspect and adjust the front wheel alignment.
27. Check ABS speed sensor signal.

ADJUSTMENT

The front wheel bearings are a cartridge type design and cannot be adjusted.

SUSPENSION — REAR SUSPENSION

AXLE HOUSING

REMOVAL & INSTALLATION

See Figures 153 and 154.

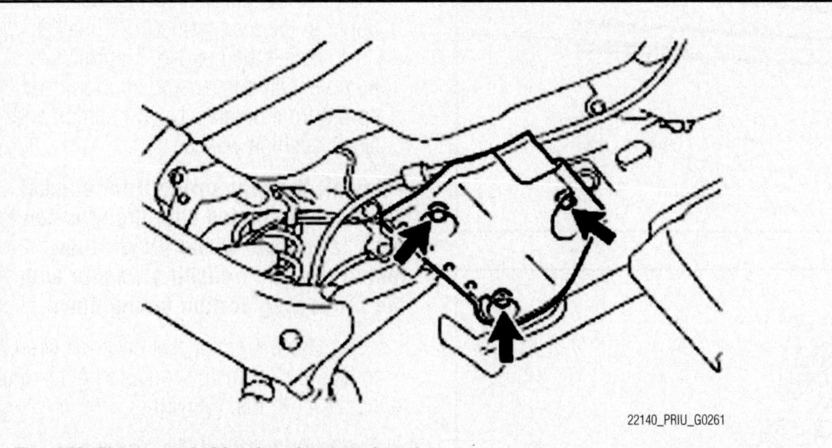

Fig. 153 Floor panel side plate fastener location

1. Before servicing the vehicle, refer to the precautions.
2. Raise and safely support the rear of the vehicle.
3. Remove the rear wheels.
4. Remove the left and right floor panel side plates, if equipped.
5. Remove the left rear height control sensor and rear sensor connecting bracket, if equipped.
6. Disconnect brake lines from the flexible hose and remove clip. Use a container to catch the brake fluid as it drains out.
7. Disconnect the ABS speed sensor wire harness and parking brake cable clamps from the axle beam.
8. Remove rear wheel hub/bearing assemblies.
9. Disconnect the rear struts from the rear axle beam.
10. Remove the 2 bolts, nuts and rear axle beam with the 2 brackets from the vehicle.

To install:

11. Install the rear axle beam into the vehicle along with the 2 nuts and bolts.

After adjusting the vehicle height by pushing down or lifting up the body, torque the rear axle beam nuts to 66 ft. lbs. (90 Nm).

12. Connect the rear struts to the rear axle beam.
13. Install the rear wheel hub/bearing assemblies.
14. Connect the ABS speed sensor wire harness and parking brake cable clamps to the rear axle beam.
15. Connect the brake lines to the flexible hose and install clip. Tighten the line to 11 ft. lbs. (15 Nm).
16. Install the left rear height control sensor and rear sensor connecting bracket, if equipped.
17. Install the left and right floor panel side plates, if equipped.
18. Bleed the brake system.
19. Install the rear wheels and lower the vehicle.
20. Check the ABS sensor signal.

COIL SPRING

REMOVAL & INSTALLATION

This vehicle uses a MacPherson strut rear suspension. Refer to Rear Suspension, MacPherson Strut, Removal & Installation.

MACPHERSON STRUTS

REMOVAL & INSTALLATION

See Figures 155 through 158.

1. Before servicing the vehicle, refer to the precautions.
2. Remove the rear No. 2 floor board.
3. Remove the rear deck floor box.
4. Remove the rear deck trim cover.
5. Remove the tonneau cover assembly.
6. Remove the left rear seatback assembly.
7. Remove the rear No. 1 floor board.
8. Remove the left rear side seatback frame.
9. Remove the rear No. 4 floor board.
10. Remove the left deck floor box.
11. Remove the left deck trim side panel assembly
12. Remove the battery carrier bracket.
13. Remove the rear wheel (s).
14. Support the rear axle beam with a jack. Insert a wooden block between the jack and the rear axle beam to prevent damage.
15. Remove the 2 nuts from the rear strut assembly (upper side).
16. Remove the rear strut assembly (upper side) bolt from the under-side of the vehicle.
17. Remove the nut and spacer from the rear strut assembly (lower side).

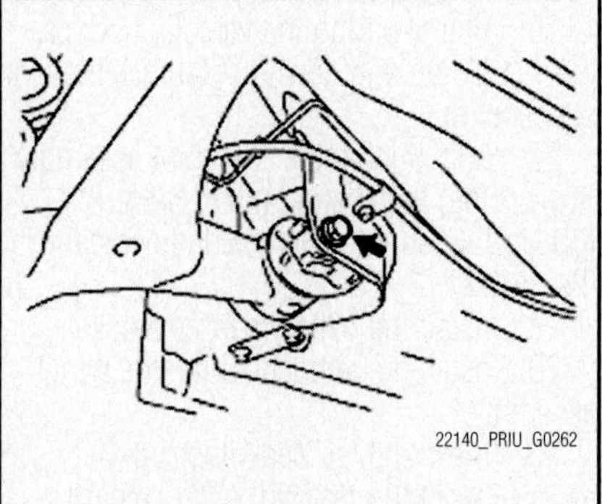

Fig. 154 Remove the rear axle beam attaching bolts

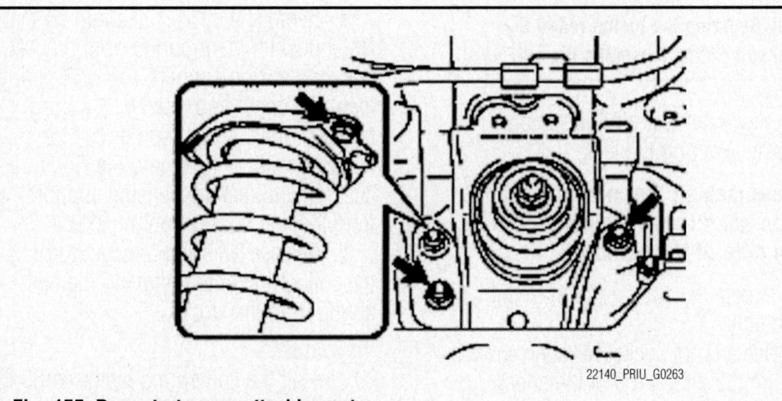

Fig. 155 Rear strut upper attaching nuts

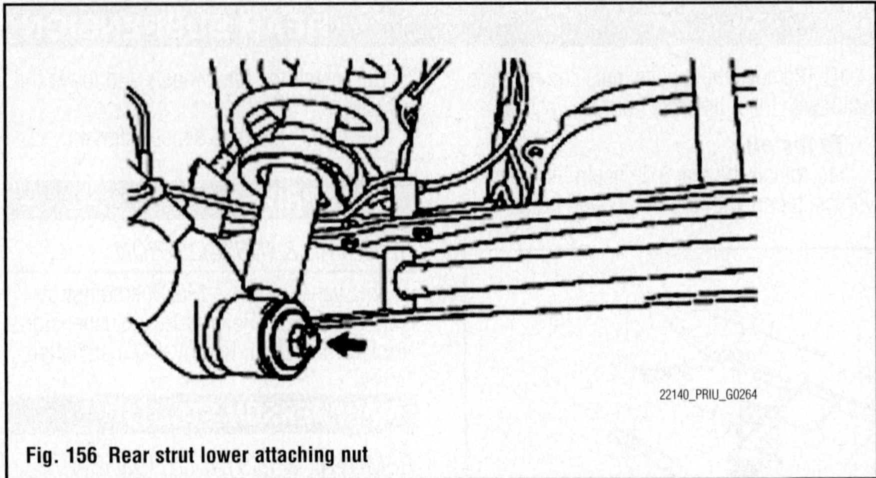

Fig. 156 Rear strut lower attaching nut

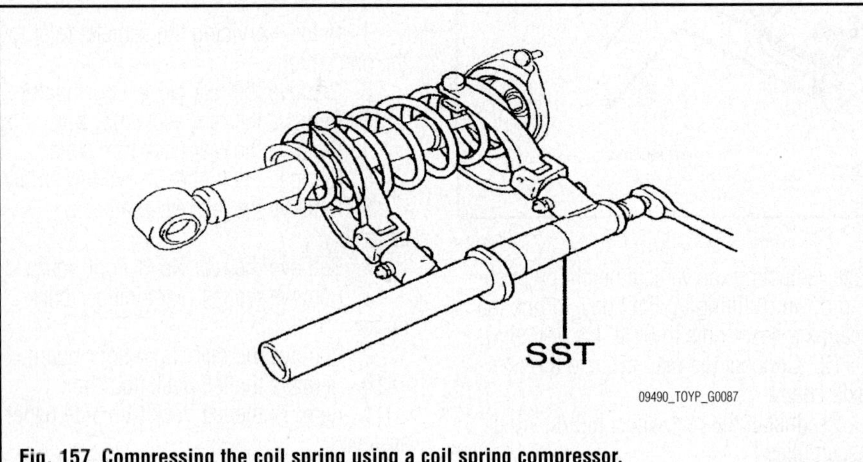

Fig. 157 Compressing the coil spring using a coil spring compressor.

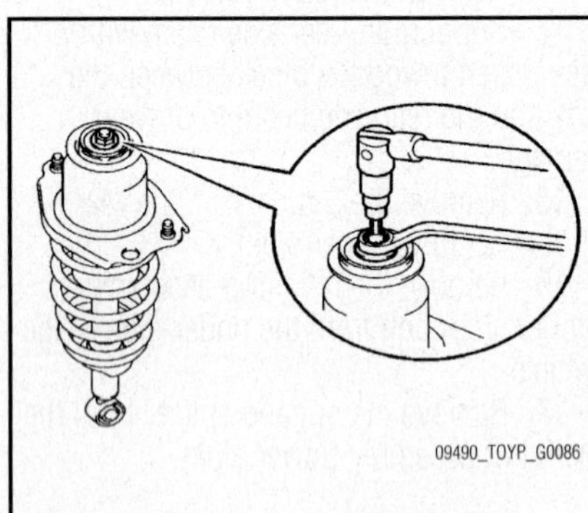

Fig. 158 Securing the piston rod of the shock absorber and loosening the nut.

18. Remove the rear strut assembly while slowly lowering the jack.

➡Seat the jack so that no extra load is placed on the strut assembly on the opposite side of the vehicle.

19. Separate the coil spring from the strut assembly.

a. Use a 6mm socket hexagon wrench to secure the piston rod of the shock absorber and loosen the nut.

➡Do not remove the nut. Sufficiently insert the hexagon wrench.

b. Attach SST 09727-30021 to the coil spring so that the upper and lower hooks of the installed area are as wide as possible.

c. Compress the coil spring until it moves freely.

➡Do not use an impact wrench. It will damage SST.

d. Remove the nut.

e. Remove the No. 1 cushion washer, No. 1 cushion, rear spring front bracket, rear suspension support, rear coil spring insulator upper and rear No. 1 spring bumper. The shock absorber can be replaced without removing the No. 1 cushion and rear suspension support from the rear spring front bracket.

f. Release SST and remove it from the coil spring after removing the coil spring from the strut.

To install:

20. Install the coil spring by assembling the strut assembly.

a. Using SST 09727-30021, compress the coil spring.

➡Do not use an impact wrench. It will damage SST.

b. Fit the coil spring end into the recessed part of the strut assembly lower seat.

c. Fit the rear coil spring insulator upper to the rear spring front bracket.

d. Install the rear No. 1 spring bumper, rear suspension support, rear spring front bracket, No. 1 cushion and No. 1 cushion washer.

➡Install the rear spring front bracket so that it is aligned with the strut lower bush, as shown in the illustration. Install the No. 1 cushion washer with the protruding portion facing down.

e. Use a 6mm socket hexagon wrench to fix the strut piston rod and tighten the nut to 41 ft. lbs. (56 Nm).

➡Sufficiently insert the hexagon wrench.

f. Release SST and remove it from the coil spring.

➡Do not use an impact wrench. It will damage SST. Remove SST while confirming the direction of the rear spring front bracket.

21. Install the rear strut assembly to the rear axle beam. Place the spacer and temporarily tighten the nut.

22. Install the rear strut assembly (upper side) to the vehicle by slowly raising the rear axle beam on a jack. Insert a wooden block between the jack and the rear axle beam to prevent damage.

23. Do not raise the rear axle beam more than necessary. Securely insert the rear spring front bracket stud bolt into the vehicle.

24. Tighten the bolt and 2 nuts of the rear strut assembly (upper side) to 59 ft. lbs. (80 Nm).

25. Install the rear wheel.

26. After lowering the vehicle, bounce the vehicle up and down to stabilize the rear suspension.

27. Fully tighten the rear strut assembly (lower side) installation nut to 59 ft. lbs. (80 Nm). Ensure the vehicle is lowered to the ground.

28. Install the battery carrier bracket.

29. Install the left deck trim side panel assembly.

30. Install the left deck floor box.

31. Install the rear No. 4 floor board.

32. Install the left rear side seatback frame.

33. Install the rear No. 1 floor board.
34. Install the left rear seatback assembly.
35. Install the tonneau cover assembly.
36. Install the rear deck trim cover.
37. Install the rear deck floor box
38. Install the rear No. 2 floor board.
39. Inspect the rear wheel alignment.

WHEEL HUB AND BEARING

REMOVAL & INSTALLATION

See Figure 159.

1. Before servicing the vehicle, refer to the precautions.
2. Raise and safely support the vehicle.
3. Remove the rear wheel.
4. Remove the rear brake drum.
5. Disconnect the rear ABS speed sensor connector.
6. Remove automatic adjusting lever, shoe adjuster and adjusting lever spring from the brake shoes assembly, if necessary.
7. Remove the hub bolts and hub/bearing assembly.

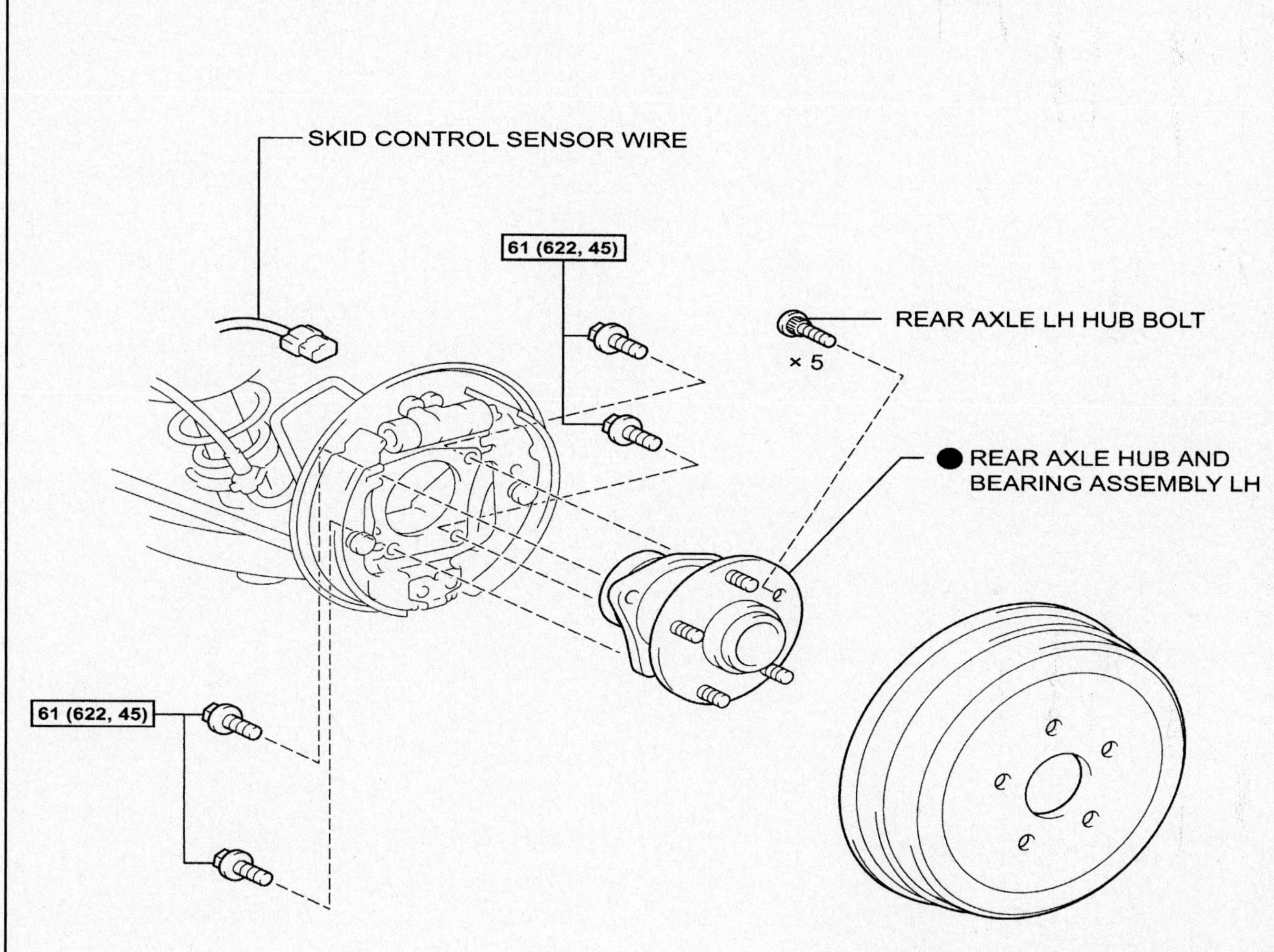

Fig. 159 Rear hub mounting bolts

To install:

8. Install the hub/bearing assembly.
9. Tighten the hub assembly bolts to 45 ft. lbs. (61 Nm).
10. Install adjusting lever spring, shoe adjuster and automatic adjusting lever from the brake shoes assembly, if removed earlier.
11. Connect the rear ABS speed sensor connector.
12. Install the rear brake drum.
13. Install the rear wheel.
14. Lower the vehicle.
15. Inspect the rear wheel alignment.
16. Check to ensure that the brakes are free from dragging and that proper braking is obtained.

TOYOTA

RAV4

20

SPECIFICATIONS AND MAINTENANCE CHARTS

ENGINE AND VEHICLE IDENTIFICATION

Engine						
Code/VIN ①	Liters (cc)	Cu. In.	Cyl.	Fuel Sys.	Engine Type	Eng. Mfg.
③ ④	2.4 (2362)	144	4	SFI	DOHC	Toyota
2GR-FE/K	3.5 (3498)	213	6	SFI	DOHC	Toyota

Model Year	
Code ②	Year
6	2006
7	2007
8	2008

SFI: Sequential Fuel Injection

DOHC: Double Overhead Camshaft

① 5th digit of the vehicle identification number (VIN)

② 10th digit of the Vehicle Identification Number (VIN)

③ 2AZ-FE/D and/or 2AZ-FE/E (2006 engines)

④ 2AZ-FE/E (2007-2008 engines)

22140_RAV4_C0001

GENERAL ENGINE SPECIFICATIONS

Year	Model	Engine Displacement Liters	Engine Series Code/VIN	Net Horsepower @ rpm	Net Torque @ rpm (ft. lbs.)	Bore x Stroke (in.)	Com-pression Ratio	Oil Pressure @ rpm
2006	RAV4	2.4	①	166@6000	165@4000	3.48x3.78	NA	②
	RAV4	3.5	2GR-FE/K	269@6200	246@4700	NA	NA	③
2007	RAV4	2.4	2AZ-FE/E	166@6000	165@4000	3.48x3.78	NA	②
	RAV4	3.5	2GR-FE/K	269@6200	246@4700	NA	NA	③
2008	RAV4	2.4	2AZ-FE/E	166@6000	165@4000	3.48x3.78	NA	②
	RAV4	3.5	2GR-FE/K	269@6200	246@4700	NA	NA	③

NA: Not Available

① 2AZ-FE/D and/or 2AZ-FE/E

② 4.3 psi or more at idle

③ 11.6 psi or more at idle

22140_RAV4_C0002

ENGINE TUNE-UP SPECIFICATIONS

Year	Engine Displacement Liters	Engine Code/VIN	Spark Plug Gap (in.)	Ignition Timing (deg.)	Fuel Pump (psi)	Idle Speed (rpm) MT	Idle Speed (rpm) AT	Valve Clearance Intake	Valve Clearance Exhaust
2006	2.4	①	0.043	②	44-50	—	600-700	0.0075-0.0114	0.0150-0.0189
	3.5	2GR-FE/K	0.043	③	44-50	—	600-700	NA	NA
2007	2.4	2AZ-FE/E	0.043	②	44-50	—	600-700	0.0075-0.0114	0.0150-0.0189
	3.5	2GR-FE/K	0.043	③	44-50	—	600-700	NA	NA
2008	2.4	2AZ-FE/E	0.043	②	44-50	—	600-700	0.0075-0.0114	0.0150-0.0189
	3.5	2GR-FE/K	0.043	③	44-50	—	600-700	NA	NA

NOTE: The Vehicle Emission Control Information label often reflects specification changes made during production.

The label figures must be used if they differ from those in this chart.

NA: Not Available

① 2AZ-FE/D and/or 2AZ-FE/E

② 5-15 degrees BTDC when using intelligent tester tool.
8-12 degrees BTDC at idle. Connect SST when not using intelligent tester tool.
5-15 degrees BTDC at idle. Disconnect SST when not using intelligent tester tool.

③ 8-12 degrees BTDC when using intelligent tester tool.
8-12 degrees BTDC at idle. Connect SST when not using intelligent tester tool.
5-15 degrees BTDC at idle. Disconnect SST when not using intelligent tester tool.

22140_RAV4_C0003

CAPACITIES

Year	Model	Engine Displacement Liters	Engine Code/VIN	Engine Oil with Filter (qts.)	Transmission (pts.) 5-Spd	Transmission (pts.) Auto.*	Transfer Case (pts.)	Drive Axle Front (pts.)	Drive Axle Rear (pts.)	Fuel Tank (gal.)	Cooling System (qts.)
2006	RAV4	2.4	①	4.5	—	7.4	1.0	—	1.0	15.9	7.2
	RAV4	3.5	2GR-FE/K	6.4	—	NA	1.0	—	1.0	15.9	②
2007	RAV4	2.4	2AZ-FE/E	4.5	—	7.4	1.0	—	1.0	15.9	7.2
	RAV4	3.5	2GR-FE/K	6.4	—	NA	1.0	—	1.0	15.9	②
2008	RAV4	2.4	2AZ-FE/E	4.5	—	7.4	③	—	1.0	15.9	7.2
	RAV4	3.5	2GR-FE/K	6.4	—	NA	1.0	—	1.0	15.9	②

*After draining, add the following amounts, then fill to the cold full line.

NA: Not Available

① 2AZ-FE/D and/or 2AZ-FE/E

② STD: 9.4
TWG: 9.8

③ 0.48 quart

22140_RAV4_C0004

FLUID SPECIFICATIONS

Year	Model	Engine Displacement Liters (VIN)	Engine Oil	Auto. Trans.	Drive Axle	Power Steering Fluid	Brake Master Cylinder
2006	RAV4	2.4 (D or E)	①	ATF World Standard	②	NA	DOT 3
	RAV4	3.5 (K)	5W-30	ATF World Standard	③	NA	DOT 3
2007	RAV4	2.4 (E)	①	ATF World Standard	②	NA	DOT 3
	RAV4	3.5 (K)	5W-30	ATF World Standard	③	NA	DOT 3
2008	RAV4	2.4 (E)	①	ATF World Standard	②	NA	DOT 3
	RAV4	3.5 (K)	5W-30	ATF World Standard	③	NA	DOT 3

DOT: Department Of Transportation

NA: Not Available

Note: If specification disagrees with specification in owners manual, use specification in owners manaual

① 0W-20 above 40 degrees F. 5W-20 below 40 degrees F

② API GL-5 SAE 90 above 0 degrees F. API GL-5 SAE 80W-90 below 0 degrees

③ API GL-5 SAE 85W-90 above 0 degrees F. API GL-5 SAE 80W-90 below 0 degrees

22140_RAV4_C0014

VALVE SPECIFICATIONS

Year	Engine Displacement Liters	Engine Code/VIN	Seat Angle (deg.)	Face Angle (deg.)	Spring Test Pressure (lbs. @ in.)	Spring Installed Height (in.)	Stem-to-Guide Clearance (in.)		Stem Diameter (in.)	
							Intake	Exhaust	Intake	Exhaust
2006	2.4	①	NA	NA	NA	1.8670	0.0010-0.0024	0.0012-0.0026	0.2154-0.2159	0.2152-0.2158
	3.5	2GR-FE/K	NA	NA	NA	1.7898	0.0010-0.0024	0.0012-0.0026	0.2154-0.2159	0.2151-0.2157
2007	2.4	2AZ-FE/E	NA	NA	NA	1.8670	0.0010-0.0024	0.0012-0.0026	0.2154-0.2159	0.2152-0.2158
	3.5	2GR-FE/K	NA	NA	NA	1.7898	0.0010-0.0024	0.0012-0.0026	0.2154-0.2159	0.2151-0.2157
2008	2.4	2AZ-FE/E	NA	NA	NA	1.8670	0.0010-0.0024	0.0012-0.0026	0.2154-0.2159	0.2152-0.2158
	3.5	2GR-FE/K	NA	NA	NA	1.7898	0.0010-0.0024	0.0012-0.0026	0.2154-0.2159	0.2151-0.2157

NA: Not Available

① 2AZ-FE/D and/or 2AZ-FE/E

22140_RAV4_C0005

CAMSHAFT SPECIFICATIONS

All measurements in inches unless noted

Year	Engine Displacement Liters	Engine Code/VIN	Journal Dia.	Brg. Oil Clearance	Shaft End-play ①	Circle Runout	Lobe Height Intake	Lobe Height Exhaust
2006	2.4	②	③	④	⑤	0.0012	1.8624-1.8664	1.8104-1.8143
	3.5	2GR-FE/K	⑥	⑦	0.0031-0.0051	0.0016	1.7447-1.7487	1.7426-1.7465
2007	2.4	2AZ-FE/E	③	④	⑤	0.0012	1.8624-1.8664	1.8104-1.8143
	3.5	2GR-FE/K	⑥	⑦	0.0031-0.0051	0.0016	1.7447-1.7487	1.7426-1.7465
2008	2.4	2AZ-FE/E	③	④	⑤	0.0012	1.8624-1.8664	1.8104-1.8143
	3.5	2GR-FE/K	⑥	⑦	0.0031-0.0051	0.0016	1.7447-1.7487	1.7426-1.7465

① Thrust clearance

② 2AZ-FE/D and/or 2AZ-FE/E

③ No. 1: 1.4162-1.4167
All others: 0.9039-0.9045

④ Intake No. 1: 0.0003-0.0015
All others: 0.0010-0.0024
Exhaust No1: 0.0016-0.0031
All others: 0.0010-0.0024

⑤ Intake: 0.0016-0.0037
Exhaust: 0.0032-0.0053

⑥ No1: 1.4152-1.4157
All others: 1.0220-1.0226

⑦ No. 1: 0.0016-0.0031
All others: 0.0010-0.0024

22140_RAV4_C0006

CRANKSHAFT AND CONNECTING ROD SPECIFICATIONS

All measurements are given in inches.

Year	Engine Displ. Liters	Engine Code/VIN	Crankshaft Main Brg. Journal Dia.	Crankshaft Main Brg. Oil Clearance	Crankshaft Shaft End-play	Crankshaft Thrust on No.	Connecting Rod Journal Diameter	Connecting Rod Oil Clearance	Connecting Rod Side Clearance
2006	2.4	①	2.1649-2.1654	0.0003-0.0009	NA	NA	1.8894-1.8898	0.0009-0.0019	0.0063-0.0143
	3.5	2GR-FE/K	2.4011-2.4016	NA	0.0016-0.0094	NA	NA	0.0018-0.0026	NA
2007	2.4	2AZ-FE/E	2.1649-2.1654	0.0003-0.0009	NA	NA	1.8894-1.8898	0.0009-0.0019	0.0063-0.0143
	3.5	2GR-FE/K	2.4011-2.4016	NA	0.0016-0.0094	NA	NA	0.0018-0.0026	NA
2008	2.4	2AZ-FE/E	2.1649-2.1654	0.0003-0.0009	NA	NA	1.8894-1.8898	0.0009-0.0019	0.0063-0.0143
	3.5	2GR-FE/K	2.4011-2.4016	NA	0.0016-0.0094	NA	NA	0.0018-0.0026	NA

NA: Not Available

① 2AZ-FE/D and/or 2AZ-FE/E

22140_RAV4_C0007

PISTON AND RING SPECIFICATIONS

All measurements are given in inches.

Year	Engine Displ. Liters	Engine Code/VIN	Piston Clearance	Ring Gap Top Comp.	Ring Gap Bottom Comp.	Ring Gap Oil Control	Ring Side Clearance Top Comp.	Ring Side Clearance Bottom Comp.	Ring Side Clearance Oil Control
2006	2.4	①	NA	0.0094-0.0122 ②	0.0130-0.0169 ②	0.0040-0.0118 ②	0.0008-0.0028	0.0008-0.0024	0.0008-0.0028
	3.5	2GR-FE/K	NA	0.0098-0.0138	0.0197-0.0236	0.0039-0.0157	0.0008-0.0028	0.0008-0.0024	0.0028-0.0059
2007	2.4	2AZ-FE/E	NA	0.0094-0.0122 ②	0.0130-0.0169 ②	0.0040-0.0118 ②	0.0008-0.0028	0.0008-0.0024	0.0008-0.0028
	3.5	2GR-FE/K	NA	0.0098-0.0138	0.0197-0.0236	0.0039-0.0157	0.0008-0.0028	0.0008-0.0024	0.0028-0.0059
2008	2.4	2AZ-FE/E	NA	0.0094-0.0122 ②	0.0130-0.0169 ②	0.0040-0.0118 ②	0.0008-0.0028	0.0008-0.0024	0.0008-0.0028
	3.5	2GR-FE/K	NA	0.0098-0.0138	0.0197-0.0236	0.0039-0.0157	0.0008-0.0028	0.0008-0.0024	0.0028-0.0059

NA: Not Available

① 2AZ-FE/D and/or 2AZ-FE/E

② Measure the piston ring a little beyond the bottom of the ring travel, 4.33 inch from the top of the cylinder block

22140_RAV4_C0008

TORQUE SPECIFICATIONS

All readings in ft. lbs.

Year	Engine Displacement Liters	Engine Code/VIN	Cylinder Head Bolts	Main Bearing Bolts	Rod Bearing Bolts	Crankshaft Damper Bolts	Flywheel Bolts	Manifold Intake	Manifold Exhaust	Spark Plugs	Oil Pan Drain Plug
2006	2.4	①	②	③	④	133	72	22	27	18	30
	3.5	2GR-FE/K	⑤	⑥	④	184	132	15	15	13	30
2007	2.4	2AZ-FE/E	②	③	④	133	72	22	27	18	30
	3.5	2GR-FE/K	⑤	⑥	④	184	132	15	15	13	30
2008	2.4	2AZ-FE/E	②	③	④	133	72	22	27	18	30
	3.5	2GR-FE/K	⑤	⑥	④	184	132	15	15	13	30

① 2AZ-FE/D and/or 2AZ-FE/E

② Step 1: 52 ft. lbs.
Step 2: plus 90 degrees

③ Step 1: 30 ft. lbs.
Step 2: plus 90 degrees

④ Step 1: 18 ft. lbs.
Step 2: plus 90 degrees

⑤ Step 1: 27 ft. lbs.
Step 2: plus 90 degrees
Step 3: plus 90 degrees
Bolt should be 22 ft. lbs. on Bank 2

⑥ Step 1: 45 ft. lbs.
Step 2: plus 90 degrees
Main bearing cap bolt: 38 ft. lbs.

22140_RAV4_C0009

WHEEL ALIGNMENT

Year	Model		Caster Range (+/-Deg.)	Caster Preferred Setting (Deg.)	Camber Range (+/-Deg.)	Camber Preferred Setting (Deg.)	Toe-in (in.)	Steering Axis Inclination (Deg.)
2006	RAV4	2WD	0.75	+2.00	0.75	-0.42	0+/-0.08	11+/-0.75
		4WD	0.75	+1.92	0.75	-0.33	0.04+/-0.08	10.75+/-0.75
2007	RAV4	2WD	0.75	+2.00	0.75	-0.42	0+/-0.08	11+/-0.75
		4WD	0.75	+1.92	0.75	-0.33	0.04+/-0.08	10.75+/-0.75
2008	RAV4	2WD	0.75	+2.00	0.75	-0.42	0+/-0.08	11+/-0.75
		4WD	0.75	+1.92	0.75	-0.33	0.04+/-0.08	10.75+/-0.75

22140_RAV4_C0010

TIRE, WHEEL AND BALL JOINT SPECIFICATIONS

Year	Model	OEM Tires Standard	OEM Tires Optional	Tire Pressures (psi) Front	Tire Pressures (psi) Rear	Wheel Size	Ball Joint Inspection	:Lug Nut Torque (ft. lbs.)
2006	RAV4	①	①	②	②	NA	NA	76
2007	RAV4	①	①	②	②	NA	NA	76
2008	RAV4	①	①	②	②	NA	NA	76

NA: Not Available

OEM: Original Equipment Manufacturer

PSI: Pounds Per Square Inch

① Base model: P215/70R16, P225/65R17. Optional P235/55R18
Sport model: P235/55R18
Limited model: P225/65R17

② 32 PSI. However if placard on vehicle disagrees with this specification, use the specification on vehicle placard.

22140_RAV4_C0011

BRAKE SPECIFICATIONS

All measurements in inches unless noted

Year	Model		Brake Disc Original Thickness	Brake Disc Minimum Thickness	Brake Disc Maximum Runout	Brake Drum Diameter Original Inside Diameter	Brake Drum Diameter Max. Wear Limit	Brake Drum Diameter Maximum Machine Diameter	Minimum Lining Thickness	Brake Caliper Bracket Bolts (ft. lbs.)	Brake Caliper Mounting Bolts (ft. lbs.)
2006	RAV4	F	①	②	0.0020	—	—	—	0.039	72	25
		R	0.472	0.413	0.0059	—	—	—	0.039	65	20
2007	RAV4	F	①	②	0.0020	—	—	—	0.039	72	25
		R	0.472	0.413	0.0059	—	—	—	0.039	65	20
2008	RAV4	F	①	②	0.0020	—	—	—	0.039	72	25
		R	0.472	0.413	0.0059	—	—	—	0.039	65	20

F: Front

R: Rear

① 275 disc (15" disc): 0.984
296 disc (16" disc) : 1.102

② 275 disc (15" disc): 0.886
296 disc (16" disc): 0.984

22140_RAV4_C0012

SCHEDULED MAINTENANCE INTERVALS

TOYOTA—RAV4

TO BE SERVICED	TYPE OF SERVICE	VEHICLE MILEAGE INTERVAL (x1000)																		
		5	10	15	20	25	30	35	40	45	50	55	60	65	70	75	80	85	90	95
Automatic transmission and differential fluid	S/I			✓			✓			✓			✓			✓			✓	
Ball joints and boots	S/I			✓			✓			✓			✓			✓			✓	
Brake system	S/I			✓			✓			✓			✓			✓			✓	
Charcoal canister	S/I												✓							
Drive belts	S/I						✓						✓						✓	
Driveshaft bushing	L						✓						✓						✓	
Engine coolant	R						✓						✓						✓	
Engine oil & filter	R	✓	✓	✓	✓	✓	✓	✓	✓	✓	✓	✓	✓	✓	✓	✓	✓	✓	✓	✓
Exhaust pipes & mounts	S/I			✓			✓			✓			✓			✓			✓	
Fuel tank cap gasket	S/I						✓						✓						✓	
Halfshaft boots & flange bolts	S/I			✓			✓			✓			✓			✓			✓	
Limited slip differential fluid	R						✓						✓						✓	
Manual transmission and differential fluid	S/I						✓						✓						✓	
Platinum spark plugs	R												✓							
Propeller shaft bolts	S/I			✓			✓			✓			✓			✓			✓	
Steering linkage	S/I			✓			✓			✓			✓			✓			✓	
Tires (rotate)	S/I	✓	✓	✓	✓	✓	✓	✓	✓	✓	✓	✓	✓	✓	✓	✓	✓	✓	✓	✓
Transfer case and differential fluid	S/I			✓			✓			✓			✓			✓			✓	
Valves	S/I												✓							

R: Replace S/I: Service or Inspect L: Lubricate

FREQUENT OPERATION MAINTENANCE (SEVERE SERVICE)

If a vehicle is operated under any of the following conditions it is considered severe service:

- Towing a trailer or using a camper or car-top carrier.

- Repeated short trips of less than 5 miles in temperatures below freezing.

- Excessive idling or low-speed driving for long distances as in heavy commercial use, such as delivery, taxi or police cars.

- Operating on rough, muddy or salt-covered roads.

- Operating on unpaved or dusty roads.

Oil filter: service or inspect every 5000 miles or 4 months, whichever occurs first.

Brake linings and discs or drums: service or inspect every 5000 miles or 4 months, whichever occurs first.

Steering linkage: service or inspect every 5000 miles or 4 months, whichever occurs first.

Ball joints and boots: service or inspect every 5000 miles or 4 months, whichever occurs first.

Brake discs & pads (front): service or inspect every 6000 miles.

Halfshaft boots: service or inspect every 5000 miles or 4 months. Retighten the flange bolts, whichever occurs first.

Body chassis bolts and nuts: service or inspect every 5000 miles or 4 months, whichever occurs first.

Transmission and differential fluid: replace every 15,000 miles or 12 months, whichever occurs first.

Transfer case and differential fluid: replace every 15,000 miles or 12 months, whichever occurs first.

Timing belt: replace every 60,000 miles or 48 months, whichever occurs first.

22140_RAV4_C0013

PRECAUTIONS

Before servicing any vehicle, please be sure to read all of the following precautions, which deal with personal safety, prevention of component damage, and important points to take into consideration when servicing a motor vehicle:

- Never open, service or drain the radiator or cooling system when the engine is hot; serious burns can occur from the steam and hot coolant.
- Observe all applicable safety precautions when working around fuel. Whenever servicing the fuel system, always work in a well-ventilated area. Do not allow fuel spray or vapors to come in contact with a spark, open flame, or excessive heat (a hot drop light, for example). Keep a dry chemical fire extinguisher near the work area. Always keep fuel in a container specifically designed for fuel storage; also, always properly seal fuel containers to avoid the possibility of fire or explosion. Refer to the additional fuel system precautions later in this section.
- Fuel injection systems often remain pressurized, even after the engine has been turned **OFF**. The fuel system pressure must be relieved before disconnecting any fuel lines. Failure to do so may result in fire and/or personal injury.
- Brake fluid often contains polyglycol ethers and polyglycols. Avoid contact with the eyes and wash your hands thoroughly after handling brake fluid. If you do get brake fluid in your eyes, flush your eyes with clean, running water for 15 minutes. If eye irritation persists, or if you have taken brake fluid internally, IMMEDIATELY seek medical assistance.
- The EPA warns that prolonged contact with used engine oil may cause a number of skin disorders, including cancer. You should make every effort to minimize your exposure to used engine oil. Protective gloves should be worn when changing oil. Wash your hands and any other exposed skin areas as soon as possible after exposure to used engine oil. Soap and water, or waterless hand cleaner should be used.
- All new vehicles are now equipped with an air bag system, often referred to as a Supplemental Restraint System (SRS) or Supplemental Inflatable Restraint (SIR) system. The system must be disabled before performing service on or around system components, steering column, instrument panel components, wiring and sensors. Failure to follow safety and disabling procedures could result in accidental air bag deployment, possible personal injury and unnecessary system repairs.
- Always wear safety goggles when working with, or around, the air bag system. When carrying a non-deployed air bag, be sure the bag and trim cover are pointed away from your body. When placing a non-deployed air bag on a work surface, always face thc bag and trim cover upward, away from the surface. This will reduce the motion of the module if it is accidentally deployed. Refer to the additional air bag system precautions later in this section.
- Clean, high quality brake fluid from a sealed container is essential to the safe and proper operation of the brake system. You should always buy the correct type of brake fluid for your vehicle. If the brake fluid becomes contaminated, completely flush the system with new fluid. Never reuse any brake fluid. Any brake fluid that is removed from the system should be discarded. Also, do not allow any brake fluid to come in contact with a painted surface; it will damage the paint.
- Never operate the engine without the proper amount and type of engine oil; doing so WILL result in severe engine damage.
- Timing belt maintenance is extremely important. Many models utilize an interference-type, non-freewheeling engine. If the timing belt breaks, the valves in the cylinder head may strike the pistons, causing potentially serious (also time-consuming and expensive) engine damage. Refer to the maintenance interval charts for the recommended replacement interval for the timing belt, and to the timing belt section for belt replacement and inspection.
- Disconnecting the negative battery cable on some vehicles may interfere with the functions of the on-board computer system(s) and may require the computer to undergo a relearning process once the negative battery cable is reconnected.
- When servicing drum brakes, only disassemble and assemble one side at a time, leaving the remaining side intact for reference.
- Only an MVAC-trained, EPA-certified automotive technician should service the air conditioning system or its components.

BRAKES

ANTI-LOCK BRAKE SYSTEM (ABS)

GENERAL INFORMATION

PRECAUTIONS

- Certain components within the ABS system are not intended to be serviced or repaired individually.
- Do not use rubber hoses or other parts not specifically specified for and ABS system. When using repair kits, replace all parts included in the kit. Partial or incorrect repair may lead to functional problems and require the replacement of components.
- Lubricate rubber parts with clean, fresh brake fluid to ease assembly. Do not use shop air to clean parts; damage to rubber components may result.
- Use only DOT 3 brake fluid from an unopened container.
- If any hydraulic component or line is removed or replaced, it may be necessary to bleed the entire system.
- A clean repair area is essential. Always clean the reservoir and cap thoroughly before removing the cap. The slightest amount of dirt in the fluid may plug an orifice and impair the system function. Perform repairs after components have been thoroughly cleaned; use only denatured alcohol to clean components. Do not allow ABS components to come into contact with any substance containing mineral oil; this includes used shop rags.
- The Anti-Lock control unit is a microprocessor similar to other computer units in the vehicle. Ensure that the ignition switch is **OFF** before removing or installing controller harnesses. Avoid static electricity discharge at or near the controller.
- If any arc welding is to be done on the vehicle, the control unit should be unplugged before welding operations begin.

BRAKES — BLEEDING THE BRAKE SYSTEM

BLEEDING PROCEDURE

BLEEDING PROCEDURE

Master Cylinder

1. Before servicing the vehicle, refer to the Precautions Section.
2. Clamp the master cylinder in a vise.

➡ Only clamp the master cylinder by its mounting flange. Do not clamp the piston rod, reservoir, seal or body.

3. Thread the bleeder tube tool or equivalent into each master cylinder outlet port.
4. Tighten each tube to 120 inch lbs.
5. Flex the bleeder tubes and put the open ends into the fluid reservoir, as far down as possible to keep them below the fluid level.

➡Be sure the tubes stay below the surface of the brake fluid once the reservoir is filled to the proper level.

6. Fill the reservoir with the proper grade and type brake fluid.
7. Use a wooden dowel as a push rod and slowly depress the master cylinder pistons. Release the pressure and allow the pistons to return to their normal detent.
8. Repeat several times. Be sure the brake fluid level remains constant.
9. Remove the bleeder tool from the master cylinder. Plug the outlet ports. Install the fill cap.
10. Remove the master cylinder from the vise. Install the master cylinder.

Brake Line

1. Before servicing the vehicle, refer to the Precautions Section.
2. Remove the bleeder plug cap. Connect the vinyl tube to the bleeder plugs.
3. Depress the brake pedal several times. Loosen the bleeder plug with the pedal depressed.
4. When the fluid stops coming out, immediately tighten the bleeder plug and release the brake pedal.
5. Repeat the above steps until all air is removed from the brake lines.
6. Tighten the bleeder plug to 73 inch lbs.
7. Install the cap.
8. Repeat the above procedure for each wheel.

BRAKES — FRONT DISC BRAKES

✲✲ CAUTION

Dust and dirt accumulating on brake parts during normal use may contain asbestos fibers from production or aftermarket brake linings. Breathing excessive concentrations of asbestos fibers can cause serious bodily harm. Exercise care when servicing brake parts. Do not sand or grind brake lining unless equipment used is designed to contain the dust residue. Do not clean brake parts with compressed air or by dry brushing. Cleaning should be done by dampening the brake components with a fine mist of water, then wiping the brake components clean with a dampened cloth. Dispose of cloth and all residue containing asbestos fibers in an impermeable container with the appropriate label. Follow practices prescribed by the Occupational Safety and Health Administration (OSHA) and the Environmental Protection Agency (EPA) for the handling, processing, and disposing of dust or debris that may contain asbestos fibers.

BRAKE CALIPER

REMOVAL & INSTALLATION

See Figures 1 and 2.

1. Before servicing the vehicle, refer to the Precautions Section.
2. Raise and safely support the vehicle.
3. Remove the wheel(s).
4. If the caliper is being replaced, remove the union bolt and 2 washers and remove the flexible brake hose from the caliper. Use a suitable container to catch the brake fluid as it drains out. Discard the washers.
5. Hold the sliding pin and loosen the 2 caliper mounting bolts. Remove the bolts and remove the caliper from the torque plate.
6. Remove the brake pads and brake hardware.

To install:

7. Install the brake pads and brake hardware.
8. Install the caliper to the torque plate with the 2 mounting bolts.
9. Reconnect the flexible brake hose to the caliper with 2 new washers and the union bolt. Torque the union bolt to 22 ft. lbs. (30 Nm).
10. Refill the master cylinder with brake fluid and bleed the brake system.
11. Check for proper operation and make sure there are no leaks.

DISC BRAKE PADS

REMOVAL INSTALLATION

See Figures 1 and 2.

1. Before servicing the vehicle, refer to the Precautions Section.
2. Raise and safely support the vehicle.
3. Remove the wheel(s).
4. Temporarily install 2 wheel stud nuts to hold the brake rotor in place.
5. If necessary, siphon a sufficient quantity of brake fluid from the master cylinder reservoir to prevent any brake fluid from overflowing the master cylinder when removing or installing new pads. This may be necessary, as the piston must be forced into the caliper bore to provide sufficient clearance when installing the pads.
6. Grasp the caliper from behind and carefully pull it towards you. This will start to seat the piston(s) in its bore. Using a C-clamp or other suitable tool, press the piston the remaining way into the caliper. Be careful not to cock the piston in the bore. Also, do not force the piston or the caliper and piston may be damaged.
7. Hold the sliding pin and loosen the 2 caliper mounting bolts. Remove the bolts and remove the caliper from the torque plate.
8. Secure the caliper assembly out of the way with a wire; so as not to stress the flexible hose.
9. Slide out the old brake pads along with any anti-squeal shims, springs, pad wear indicators and pad support plates. Make sure to note the position of all assorted pad hardware.

To install:

10. Check the brake disc (rotor) for thickness and run-out. Inspect the caliper and piston assembly for breaks, cracks, fluid seepage or other damage. Overhaul or replace as necessary.

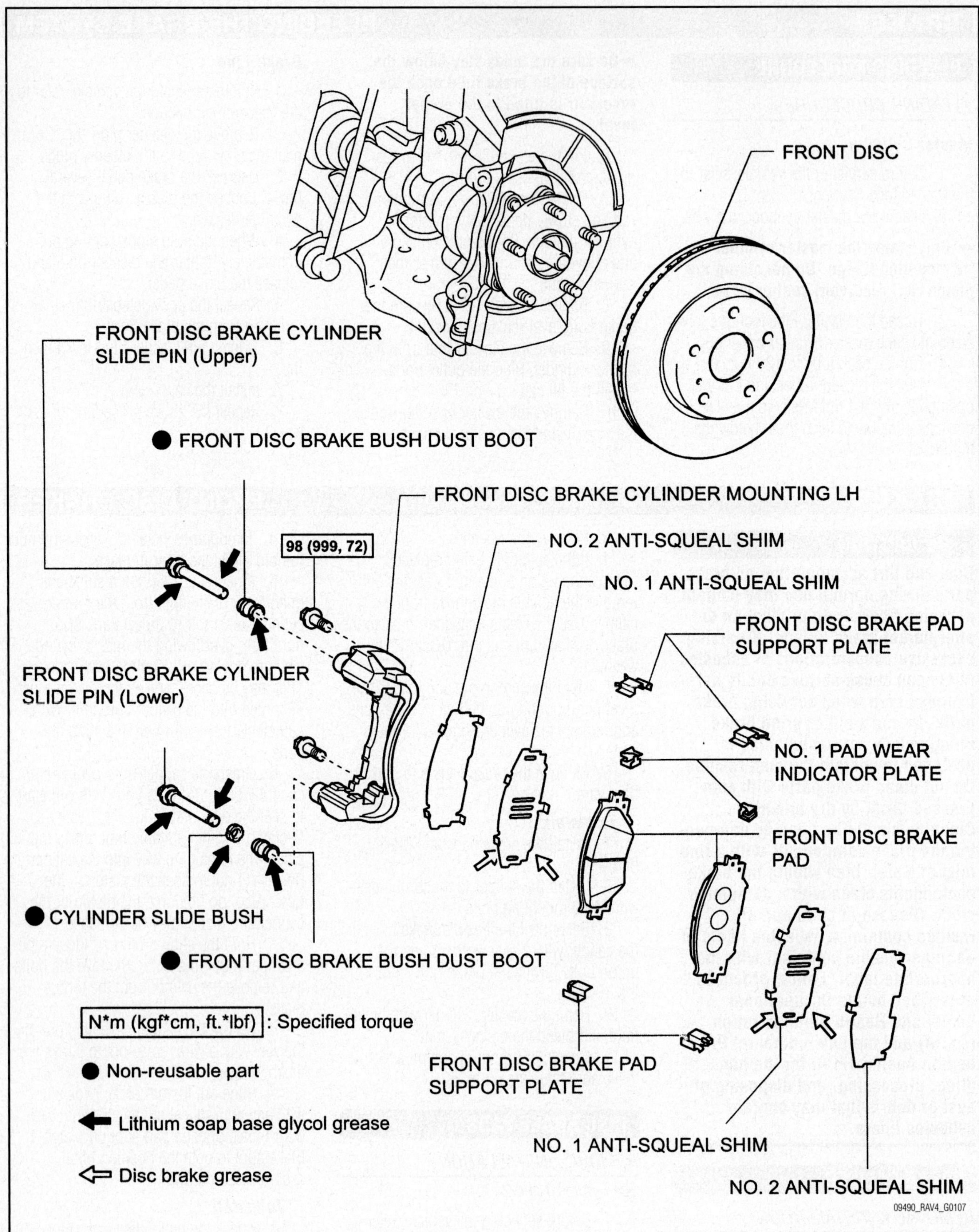

Fig. 1 Front brake assembly and related components (275 disc)

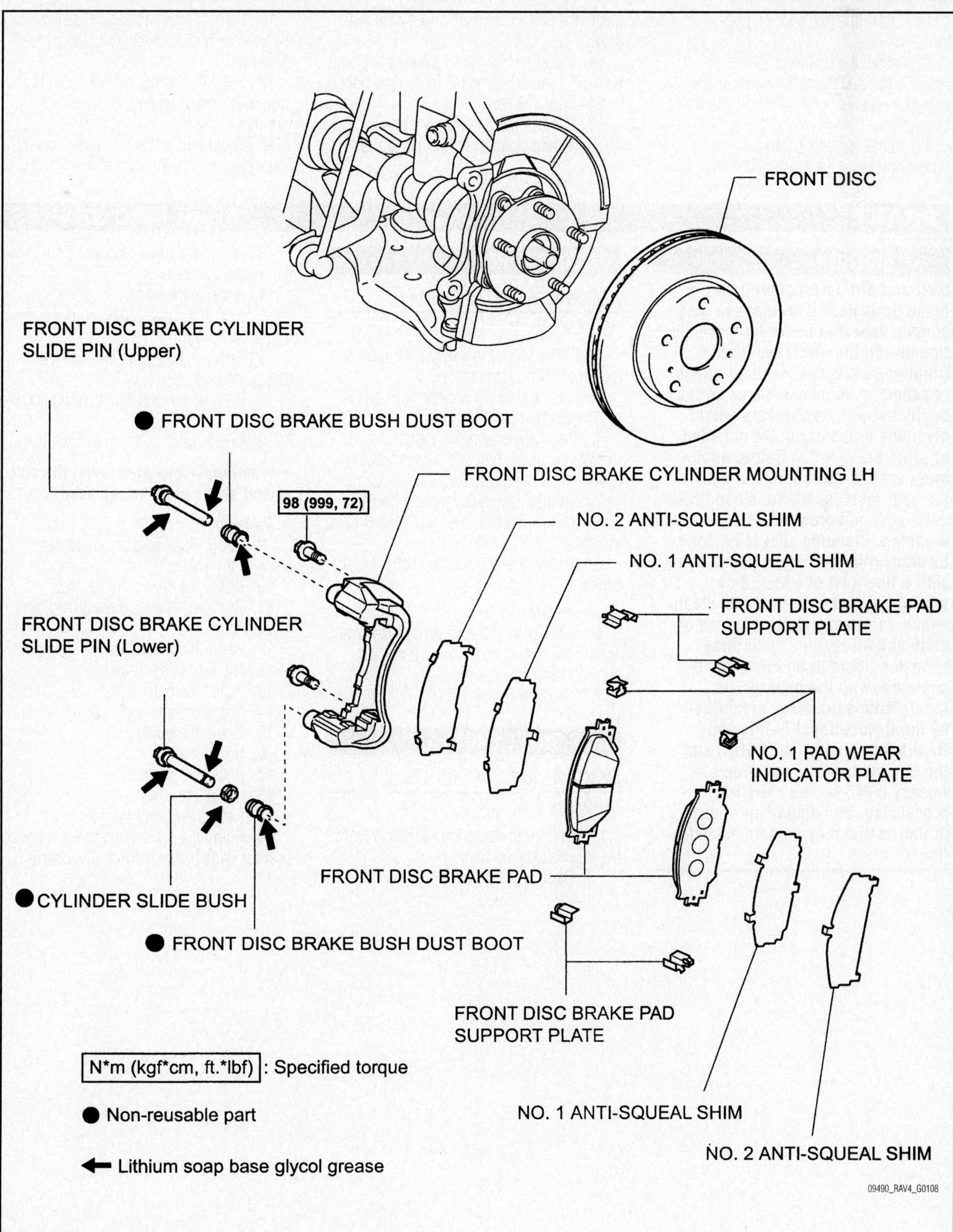

Fig. 2 Front brake assembly and related components (296 disc)

11. Install the pad support plates into the torque plate.

12. Install the pad wear indicators onto the pads. Be sure the arrow on the indicator plate is pointing in the direction of rotation.

13. Install the anti-squeal shims on the outside of each pad and then install the pad assemblies into the torque plate.

14. Install the caliper to the torque plate with the 2 mounting bolts. Torque the bolts to 20 ft. lbs. (26 Nm).

15. Remove the 2 temporary wheel stud nuts and check that the rotor turns freely.

16. Reinstall the wheel(s). Safely lower the vehicle, and road-test for proper brake operation.

17. Refill the master cylinder with the proper grade and type brake fluid, as required.

18. Be sure to pump the brakes several times prior to moving the vehicle.

BRAKES

REAR DISC BRAKES

✲✲ CAUTION

Dust and dirt accumulating on brake parts during normal use may contain asbestos fibers from production or aftermarket brake linings. Breathing excessive concentrations of asbestos fibers can cause serious bodily harm. Exercise care when servicing brake parts. Do not sand or grind brake lining unless equipment used is designed to contain the dust residue. Do not clean brake parts with compressed air or by dry brushing. Cleaning should be done by dampening the brake components with a fine mist of water, then wiping the brake components clean with a dampened cloth. Dispose of cloth and all residue containing asbestos fibers in an impermeable container with the appropriate label. Follow practices prescribed by the Occupational Safety and Health Administration (OSHA) and the Environmental Protection Agency (EPA) for the handling, processing, and disposing of dust or debris that may contain asbestos fibers.

BRAKE CALIPER

REMOVAL & INSTALLATION

See Figure 3.

1. Before servicing the vehicle, refer to the Precautions Section.
2. Raise and safely support the vehicle.
3. Remove the wheel.
4. If the caliper is being replaced, remove the union bolt and disconnect the brake line. If possible, plug the line to prevent fluid loss. Otherwise place a container under the line to catch the fluid. Discard the washers.
5. Remove the 2 mounting bolts and the caliper.

To install:

6. Installation is the reverse of removal. Torque the brake line union bolt to 22 ft. lbs. (30 Nm). Use new washers.
7. Bleed the brakes.

DISC BRAKE PADS

REMOVAL AND INSTALLATION

See Figure 3.

1. Before servicing the vehicle, refer to the Precautions Section.
2. Raise and safely support the vehicle.
3. Remove a small amount of fluid from the master cylinder.
4. Remove the wheel.
5. Pry off the pad protector, being careful not to bend it.
6. Remove the spring clip and the 2 pad guide pins.
7. Remove the anti-rattle spring and the pads.
8. Remove the shims from each pad.

➡When replacing worn pads, the anti-squeal shims must be replaced.

To install:

9. Apply disc brake grease to both sides of the shims.
10. Install the shims on the pads.
11. Compress the piston with a piston forcing tool.
12. Install the inner pad with the wear indicator plate facing down.
13. Install the outer pad.
14. Install the anti-rattle spring.
15. Install the guide pins.
16. Install the clip.
17. Install the protector.
18. Install the wheel.
19. Refill the master cylinder.
20. Pump the brake pedal a few times to seat the pads before moving the vehicle.

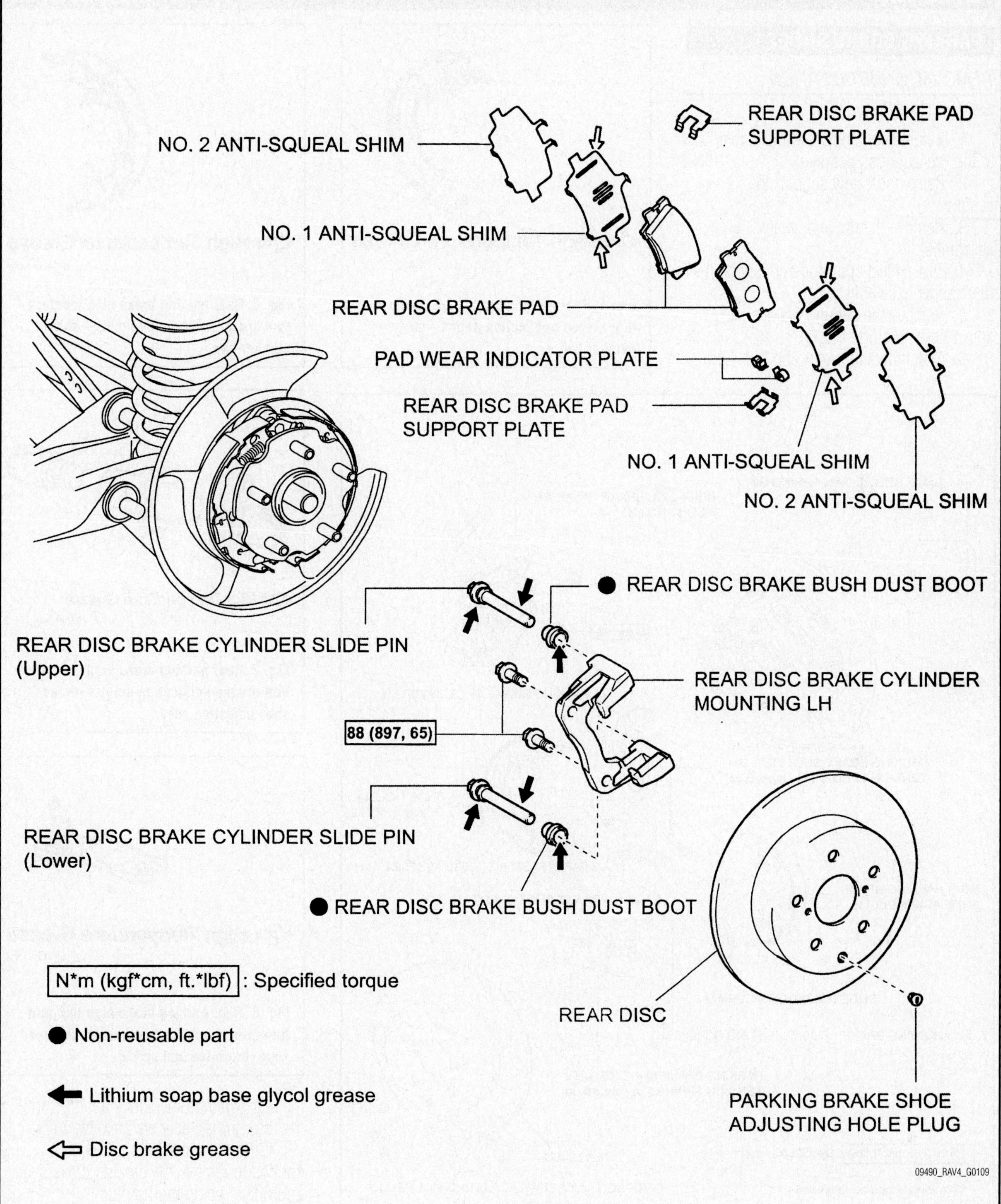

Fig. 3 Rear brake assembly and related components

PARKING BRAKE SHOES

REMOVAL & INSTALLATION

See Figures 4 through 9.

1. Before servicing the vehicle, refer to the Precautions Section.
2. Raise and safely support the vehicle.
3. Remove the tire and wheel assemblies.
4. Disconnect and properly support the rear caliper assembly.
5. Remove the parking brake shoe adjusting hole plug.
6. Matchmark the rotor and hub. Turn the rear parking brake shoe adjuster until the rotor turns freely. Remove the rotor.
7. Remove the two shoe tension springs to remove the parking brake shoe return tension spring (upper).
8. Remove the parking brake shoe strut.
9. Remove the two shoe tension springs to remove the parking brake shoe return tension spring (lower).
10. Remove the set screw and remove the parking brake shoe adjusting screw set.

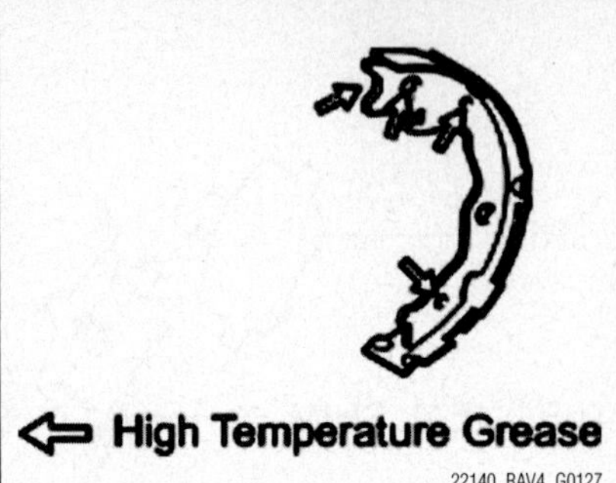

Fig. 5 Rear parking brake high temperature grease application points—shoe number two

Fig. 6 Rear parking brake high temperature grease application points—shoe number one

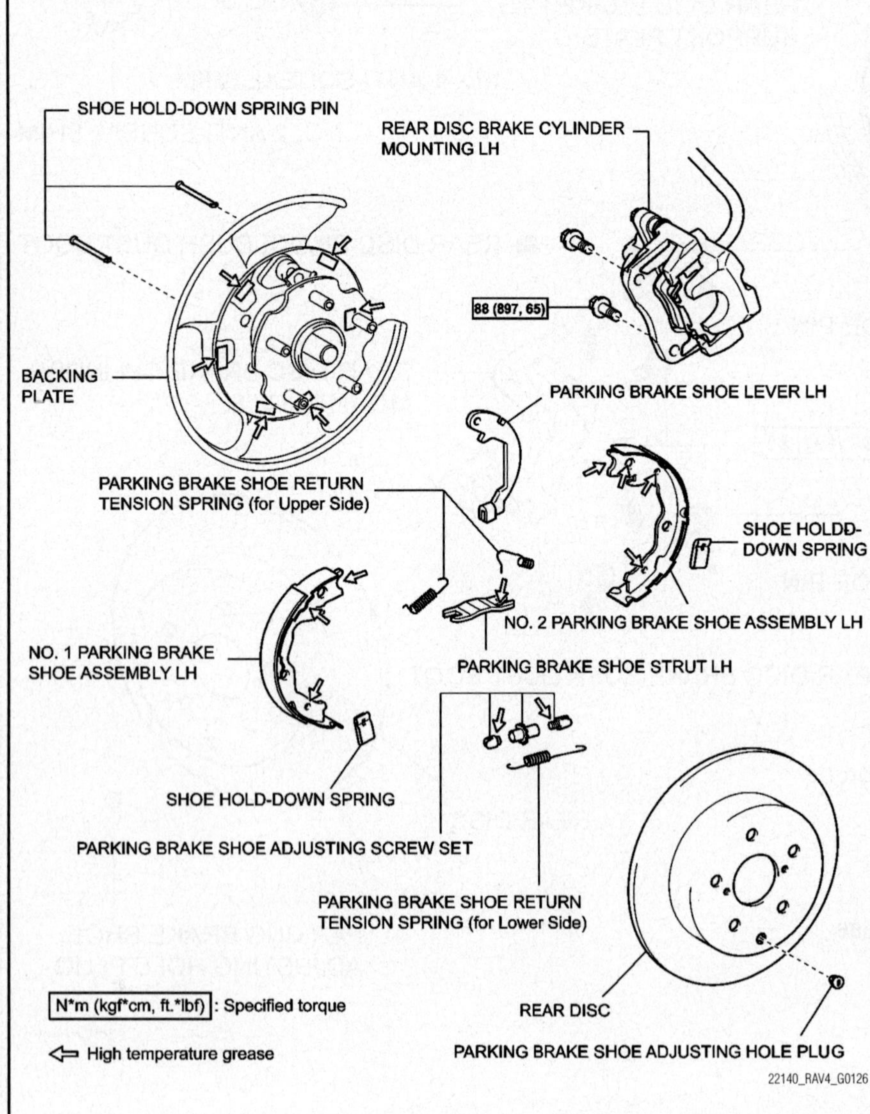

Fig. 4 Rear parking brake shoes and related components

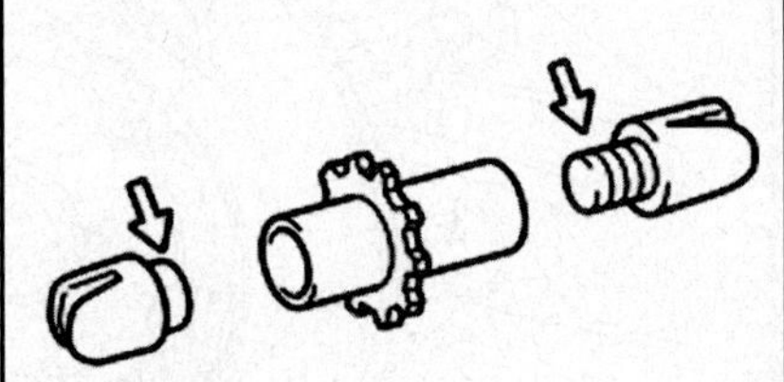

Fig. 7 Rear parking brake high temperature grease application points—brake shoe adjusting set

Fig. 8 Rear parking brake high temperature grease application points—strut and shoe return tension spring

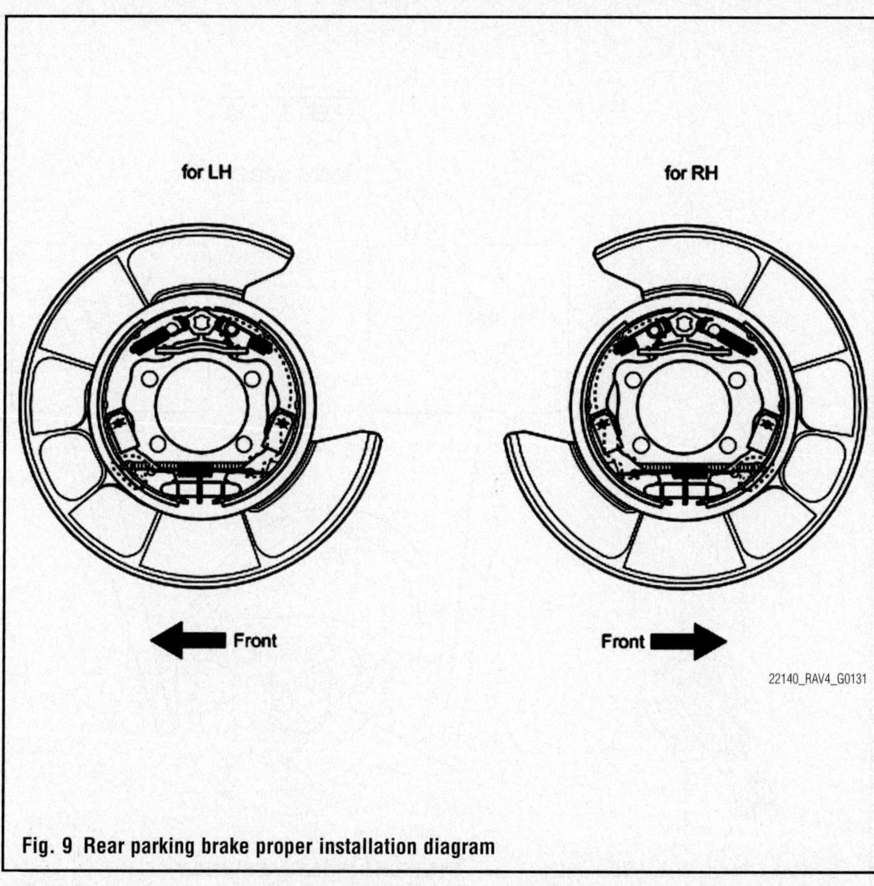

Fig. 9 Rear parking brake proper installation diagram

11. Press the shoe hold-down spring to remove the pin, shoe hold-down spring and shoe in order to remove the number one shoe assembly.
12. Press the shoe hold-down spring to remove the pin, shoe hold-down spring and shoe in order to remove the number two shoe assembly.
13. Remove the parking brake shoe lever.

To install:

14. Installation is the reverse of the removal procedure.
15. Be sure to apply high temperature grease to the shoe contact surfaces.
16. Be sure to apply high temperature grease to the brake shoe adjusting set.
17. Be sure to apply high temperature grease to the strut and shoe return tension spring.
18. Be sure that the rear parking brake shoes are correctly installed. There should be no oil or grease adhering to the friction surfaces of the shoe lining or disc. Correct as required.
19. Adjust the rear parking brake shoe clearance.
20. Tighten the wheel lug nuts to 76 ft. lbs.
21. Check the parking brake lever travel, adjust as required.

CHASSIS ELECTRICAL

AIR BAG (SUPPLEMENTAL RESTRAINT SYSTEM)

GENERAL INFORMATION

⁂ CAUTION

These vehicles are equipped with an air bag system. The system must be disarmed before performing service on, or around, system components, the steering column, instrument panel components, wiring and sensors. Failure to follow the safety precautions and the disarming procedure could result in accidental air bag deployment, possible injury and unnecessary system repairs.

SERVICE PRECAUTIONS

Disconnect and isolate the battery negative cable before beginning any airbag system component diagnosis, testing, removal, or installation procedures. Allow system capacitor to discharge for two minutes before beginning any component service. This will disable the airbag system. Failure to disable the airbag system may result in accidental airbag deployment, personal injury, or death.

Do not place an intact undeployed airbag face down on a solid surface. The airbag will propel into the air if accidentally deployed and may result in personal injury or death.

When carrying or handling an undeployed airbag, the trim side (face) of the airbag should be pointing towards the body to minimize possibility of injury if accidental deployment occurs. Failure to do this may result in personal injury or death.

Replace airbag system components with OEM replacement parts. Substitute parts may appear interchangeable, but internal differences may result in inferior occupant protection. Failure to do so may result in occupant personal injury or death.

Wear safety glasses, rubber gloves, and long sleeved clothing when cleaning powder residue from vehicle after an airbag deployment. Powder residue emitted from a deployed airbag can cause skin irritation. Flush affected area with cool water if irritation is experienced. If nasal or throat irritation is experienced, exit the vehicle for fresh air until the irritation ceases. If irritation continues, see a physician.

Do not use a replacement airbag that is not in the original packaging. This may result in improper deployment, personal injury, or death.

The factory installed fasteners, screws and bolts used to fasten airbag components have a special coating and are specifically designed for the airbag system. Do not use substitute fasteners. Use only original equipment fasteners listed in the parts catalog when fastener replacement is required.

During, and following, any child restraint anchor service, due to impact event or vehicle repair, carefully inspect all mounting hardware, tether straps, and anchors for proper installation, operation, or damage. If a child restraint anchor is found damaged in any way, the anchor must be replaced. Failure to do this may result in personal injury or death.

Deployed and non-deployed airbags may or may not have live pyrotechnic material within the airbag inflator.

Do not dispose of driver/passenger/curtain airbags or seat belt tensioners unless you are sure of complete deployment. Refer to the Hazardous Substance Control System for proper disposal.

Dispose of deployed airbags and tensioners consistent with state, provincial, local, and federal regulations.

After any airbag component testing or service, do not connect the battery negative cable. Personal injury or death may result if the system test is not performed first.

If the vehicle is equipped with the Occupant Classification System (OCS), do not connect the battery negative cable before performing the OCS Verification Test using the scan tool and the appropriate diagnostic information. Personal injury or death may result if the system test is not performed properly.

Never replace both the Occupant Restraint Controller (ORC) and the Occupant Classification Module (OCM) at the same time. If both require replacement, replace one, then perform the Airbag System test before replacing the other.

Both the ORC and the OCM store Occupant Classification System (OCS) calibration data, which they transfer to one another when one of them is replaced. If both are replaced at the same time, an irreversible fault will be set in both modules and the OCS may malfunction and cause personal injury or death.

If equipped with OCS, the Seat Weight Sensor is a sensitive, calibrated unit and must be handled carefully. Do not drop or handle roughly. If dropped or damaged, replace with another sensor. Failure to do so may result in occupant injury or death.

If equipped with OCS, the front passenger seat must be handled carefully as well. When removing the seat, be careful when setting on floor not to drop. If dropped, the sensor may be inoperative, could result in occupant injury, or possibly death.

If equipped with OCS, when the passenger front seat is on the floor, no one should sit in the front passenger seat. This uneven force may damage the sensing ability of the seat weight sensors. If sat on and damaged, the sensor may be inoperative, could result in occupant injury, or possibly death.

DISARMING THE SYSTEM

To avoid personal injury when working on vehicles equipped with an air bag, the negative battery cable must be disconnected and at least 90 seconds must elapse before working on the system. Failure to do so may result in deployment of the air bag and seat belt pretensioner activation.

ARMING THE SYSTEM

Connect the negative battery. Wait 2 minutes before performing and service operations.

CLOCKSPRING CENTERING

See Figures 10 through 13.

1. Before servicing the vehicle, refer to the Precautions Section.

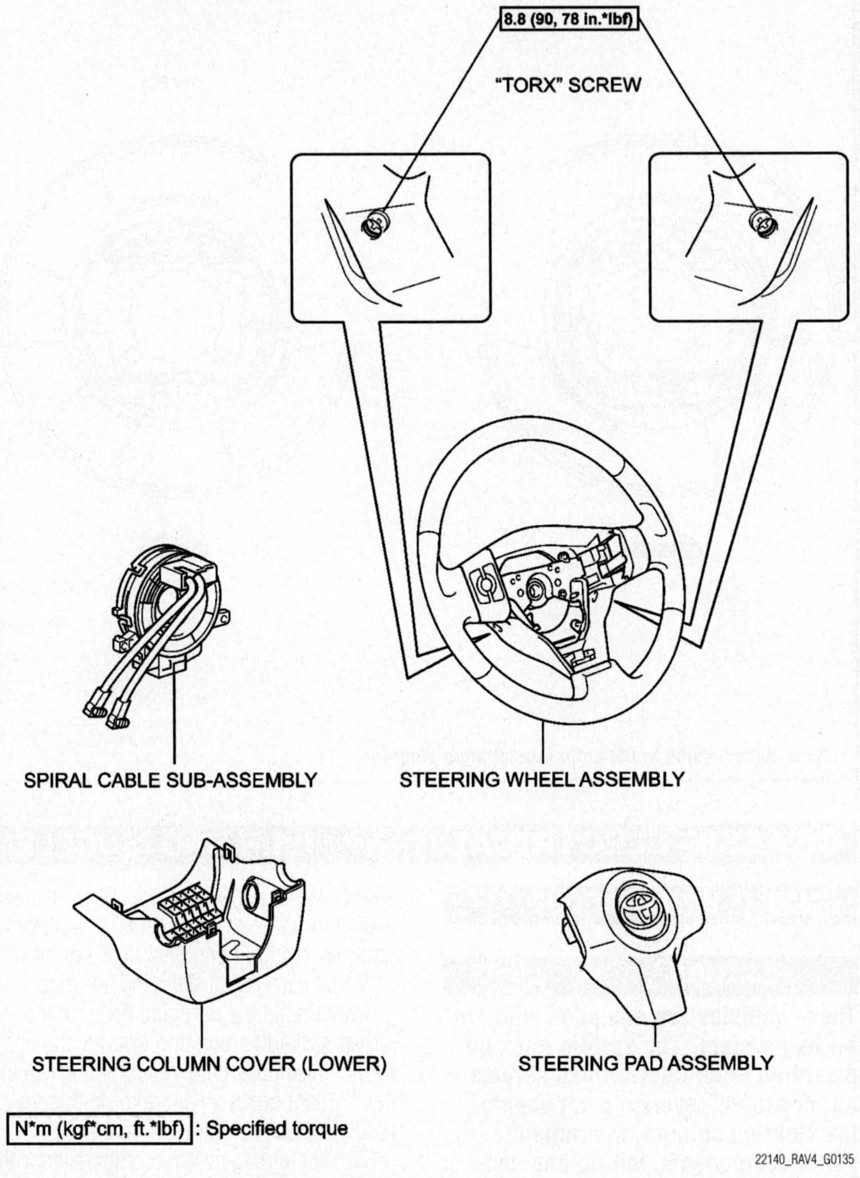

8.8 (90, 78 in.*lbf)
"TORX" SCREW
SPIRAL CABLE SUB-ASSEMBLY
STEERING WHEEL ASSEMBLY
STEERING COLUMN COVER (LOWER)
STEERING PAD ASSEMBLY
N*m (kgf*cm, ft.*lbf) : Specified torque
22140_RAV4_G0135

Fig. 10 Spiral cable and related components

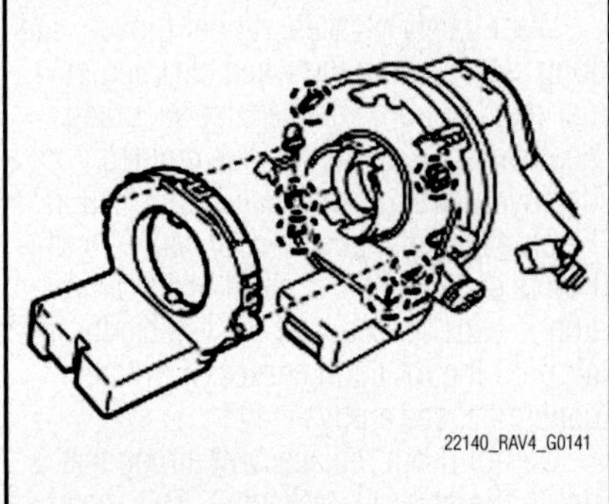

Fig. 11 Spiral cable and sensor disconnection points

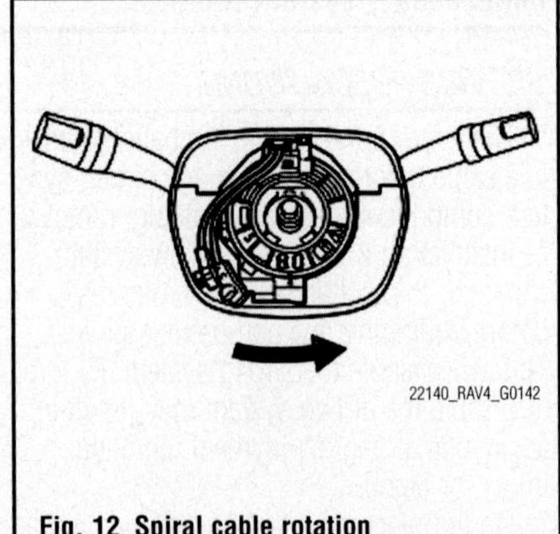

Fig. 12 Spiral cable rotation

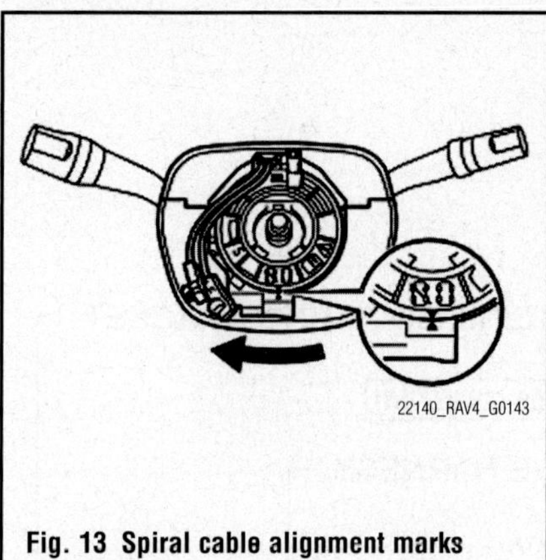

Fig. 13 Spiral cable alignment marks

2. Disconnect the negative battery cable.

➡**Wait at least 90 seconds after disconnecting the negative battery cable to prevent air bag and seat belt pretensioner activation.**

➡**When prying on plastic parts, use a blunt prytool with a taped end to prevent marring of the plastic.**

➡**Be sure that the front wheels are in the straight ahead position.**

3. To remove the driver's side air bag loosen the two screws until the groove along each screw circumference catches on the screw case, Pull the air bag (steering pad) from the steering wheel.

➡**Properly support the assembly. Do not pull on the wire harness.**

4. Disconnect the air bag electrical connectors. Disconnect the horn connector and remove the air bag (steering pad).

5. Matchmark the steering wheel retaining nut. Remove the nut. Using the proper removal tool, remove the steering wheel.

6. Release the tilt lever. Press both sides of the lower steering column cover to disengage the two claws.

7. Insert your finger thru the tilt lever opening, in the lower steering column cover, and disengage the claw. Disengage the two claws and remove the cover.

➡**The claws can be disengaged by pushing the lower steering column cover toward the front of the vehicle while pulling the upper steering column cover.**

8. Disengage the three claws and two pins and separate the upper steering column cover.

9. To remove the spiral cable assembly, disconnect all connectors from the turn signal switch with the spiral cable.

To install:

➡**Do not use the sensor, if it has been dropped. Do not use the sensor if it is cracked, chipped or dented. Do not use the speed sensor if the wire harness connector is scratched, cracked or damaged**

10. Align the locating pins, attach the six claws and install the sensor to the spiral cable.

11. Install the spiral cable.

➡**If replacing the spiral cable with a new one be sure to remove the lock pin before installing the steering wheel.**

12. Connect the connectors to the spiral cable.

➡**When handling the air bag connector, do not damage the air bag wire harness.**

13. Install the upper and lower steering column covers.

14. To position the spiral cable, slowly rotate the spiral cable counterclockwise by hand until it feels firm.

➡**Do not use the air bag wire harness to turn the spiral cable.**

15. Rotate the spiral cable clockwise 2.5 turns to align the marks.

➡**Do not use the air bag wire harness to turn the spiral cable.**

16. The spiral cable will rotate approximately 2.5 turns to the left and right from the center.

17. Install the steering wheel. Be sure to use the matchmarks made during the removal procedure. Torque the retaining nut to 37 ft. lbs.

18. Install the driver's side air bag (steering pad). Tighten the retaining screws to 78 inch lbs.

19. Connect the negative battery cable.

➡**When connecting the battery cable after installing the steering angle sensor, be sure that the front tires and the steering wheel are positioned straight ahead before and after connecting the cable to the battery.**

20. Check the SRS system for proper operation. Correct as required.

DRIVETRAIN

AUTOMATIC TRANSAXLE ASSEMBLY

REMOVAL & INSTALLATION

2WD Models

See Figures 14 and 15.

1. Before servicing the vehicle, refer to the Precautions Section.

2. Properly relieve the fuel system pressure.

3. Disconnect the negative battery cable.

➡**Wait at least 90 seconds after disconnecting the negative battery cable before starting any repair work to prevent air bag and seat belt pretensioner activation.**

4. Drain the transaxle fluid.

5. Remove the engine and transaxle assembly from the vehicle and position it in a suitable holding fixture.

6. Remove the starter. Disconnect the wiring harnesses.

7. Remove the transaxle oil cooler.

8. Remove the rear engine mounting bracket.

9. Remove the front engine mounting bracket.

10. Remove the engine mounting bracket, left side.

11. Remove the flywheel housing inspection cover.

12. Remove the six torque converter to flywheel retaining bolts.

13. Remove the lower side transaxle to engine retaining bolts. Remove the upper side transaxle to engine retaining bolts.

14. Separate the transaxle from the engine.

15. Remove the transaxle dipstick tube assembly. Remove the number one transaxle control cable bracket.

16. Remove the torque converter from the transaxle.

To install:

17. When installing the torque converter use a caliper to measure dimension "A", between the transaxle and the end surface of the drive plate.

18. Using the caliper and a straight edge, measure the dimension "B" and check that "B" is greater than "A". Measurement should be 0.039 inch or more.

19. On U140F and U241E transaxles,

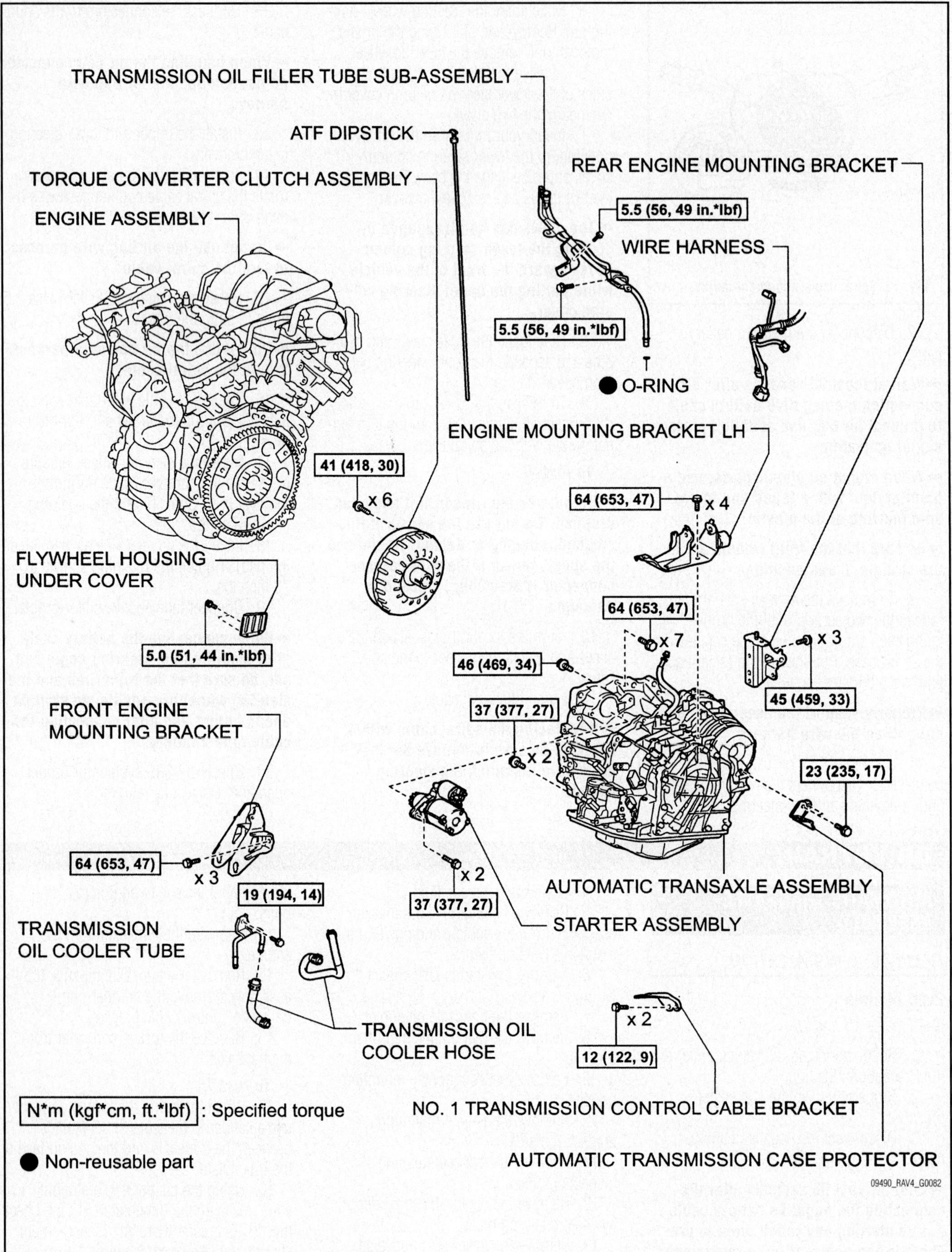

Fig. 14 Automatic transaxle and related components—2WD

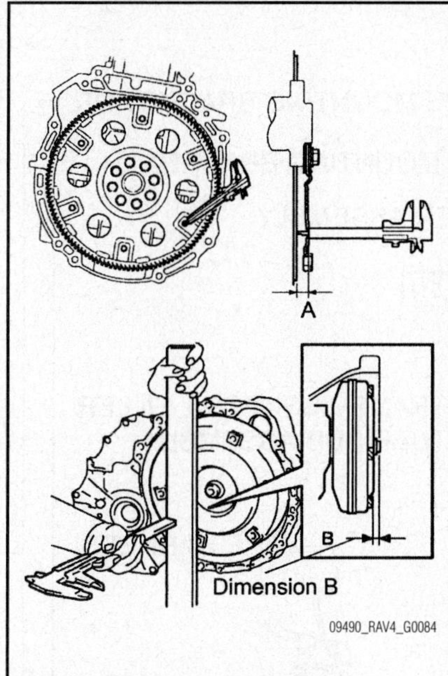

Fig. 15 Torque converter installation measurement points

tighten the lower side transaxle to engine retaining bolts to 27 ft. lbs. Tighten the upper side transaxle to engine retaining bolts to 47 ft. lbs. (for the top three bolts) and 34 ft. lbs. (for the bottom lower bolts, on each side).

20. On U151E and U151F transaxles, tighten the lower side transaxle to engine retaining bolts to 27 ft. lbs. (for the bottom two bolts). And 34 ft. lbs. (for the upper bolt). Tighten the upper side transaxle to engine retaining bolts to 47 ft. lbs.

21. Tighten the torque converter to flex-plate retaining bolts to 30 ft. lbs.

22. Continue the installation in the reverse order of the removal procedure.

➡Perform the automatic transmission initialization procedure when replacing the automatic transmission, engine or ECM.

23. Turn the ignition switch OFF.

24. Connect the intelligent tester tool together with the controller area network vehicle interface module (CAN VIN) to the DLC3.

25. Turn the ignition switch to the ON position.

26. Push the intelligent tester tool main switch to the ON position.

27. Select the items, DIAGNOSIS/ENHANCED OBD II.

28. Perform the reset memory procedure from the ENGINE menu.

➡After performing the reset memory, be sure to perform the road test procedure. For road test procedure information, refer to the intelligent tester instruction manual.

4WD Models

See Figure 16.

1. Before servicing the vehicle, refer to the Precautions Section.
2. Properly relieve the fuel system pressure.
3. Disconnect the negative battery cable.

➡Wait at least 90 seconds after disconnecting the negative battery cable before starting any repair work to prevent air bag and seat belt pretensioner activation.

4. Drain the transaxle fluid. Drain the transfer case fluid.
5. Remove the engine and transaxle assembly from the vehicle and position it in a suitable holding fixture.
6. Remove the starter. Disconnect the wiring harnesses.
7. Remove the transaxle oil cooler.
8. Remove the right rear engine mounting bracket.
9. Remove the front rear engine mounting bracket.
10. Remove the left engine mounting bracket.
11. Remove the transfer case retaining bolts. Separate the transfer case for the transaxle.
12. Remove the flywheel housing inspection cover.
13. Remove the six torque converter to flywheel retaining bolts.
14. Remove the lower side transaxle to engine retaining bolts. Remove the upper side transaxle to engine retaining bolts.
15. Separate the transaxle from the engine.
16. Remove the transaxle dipstick tube assembly. Remove the number one transaxle control cable bracket.
17. Remove the torque converter from the transaxle.

To install:

18. When installing the torque converter use a caliper to measure dimension "A", between the transaxle and the end surface of the drive plate.
19. Using the caliper and a straight edge, measure the dimension "B" and check that "B" is greater than "A". Measurement should be 0.039 inch or more.
20. On U140F and U241E transaxles, tighten the lower side transaxle to engine retaining bolts to 27 ft. lbs. Tighten the upper side transaxle to engine retaining bolts to 47 ft. lbs. (for the top three bolts) and 34 ft. lbs. (for the bottom lower bolts, on each side).
21. On U151E and U151F transaxles, tighten the lower side transaxle to engine retaining bolts to 27 ft. lbs. (for the bottom two bolts). And 34 ft. lbs. (for the upper bolt). Tighten the upper side transaxle to engine retaining bolts to 47 ft. lbs.
22. Tighten the torque converter to flex-plate retaining bolts to 30 ft. lbs.
23. Tighten the transfer case to transaxle retaining bolts to 51 ft. lbs.
24. Continue the installation in the reverse order of the removal procedure.

➡Perform the automatic transmission initialization procedure when replacing the automatic transmission, engine or ECM.

25. Turn the ignition switch OFF.
26. Connect the intelligent tester tool together with the controller area network vehicle interface module (CAN VIN) to the DLC3.
27. Turn the ignition switch to the ON position.
28. Push the intelligent tester tool main switch to the ON position.
29. Select the items, DIAGNOSIS/ENHANCED OBD II.
30. Perform the reset memory procedure from the ENGINE menu.

➡After performing the reset memory, be sure to perform the road test procedure. For road test procedure information, refer to the intelligent tester instruction manual.

FRONT HALFSHAFT

REMOVAL & INSTALLATION

See Figures 17 and 18.

1. Before servicing the vehicle, refer to the Precautions Section.
2. Disconnect the negative battery cable.

➡Wait at least 90 seconds after disconnecting the negative battery cable before starting any repair work to prevent air bag and seat belt pretensioner activation.

3. Raise and support the vehicle safely.
4. Remove the tire and wheel assembly. Remove the front axle hub nut.
5. Drain the transaxle fluid.
6. Disconnect the left and right speed sensors.
7. Remove the left and right brake calipers.

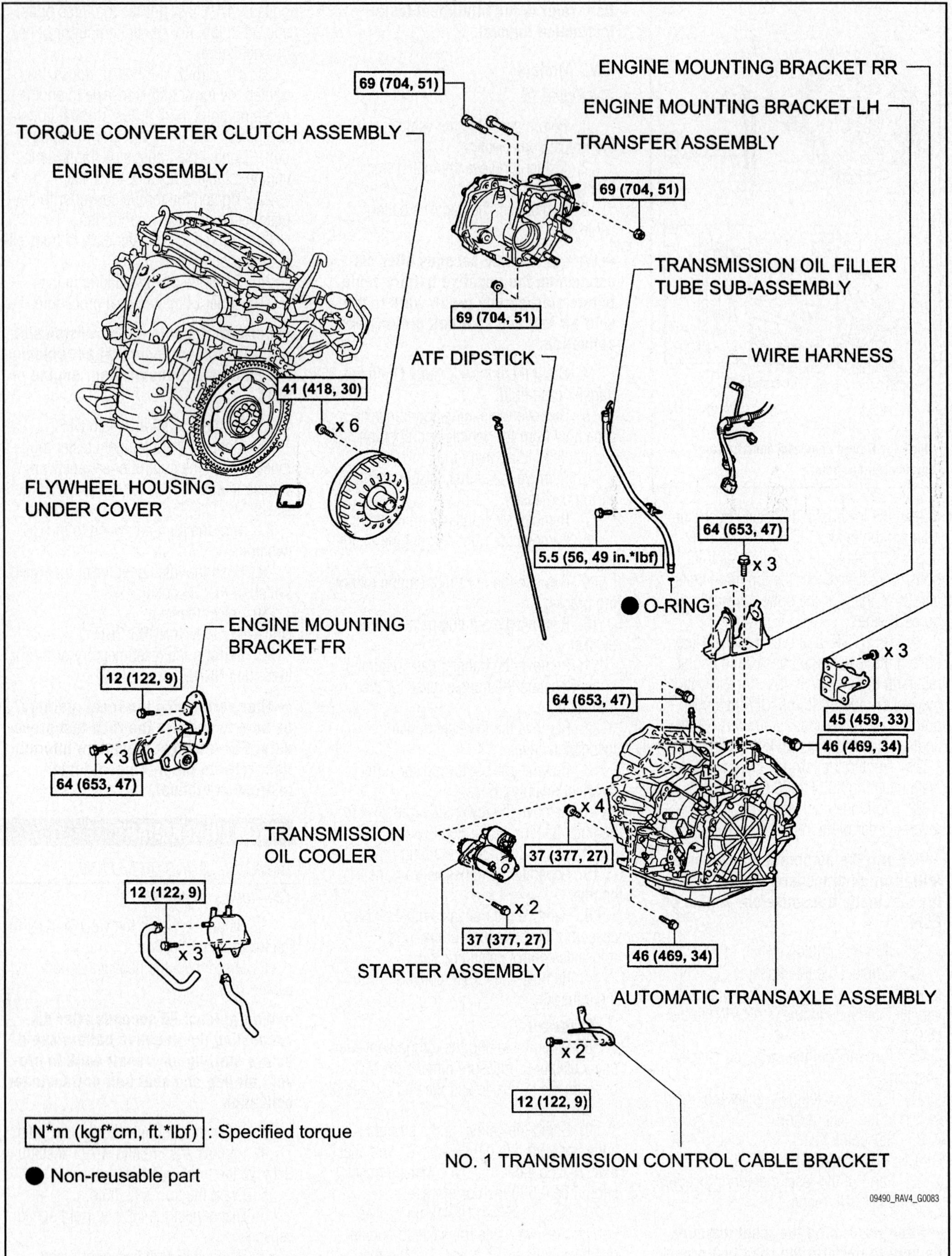

Fig. 16 Automatic transaxle and related components—4WD

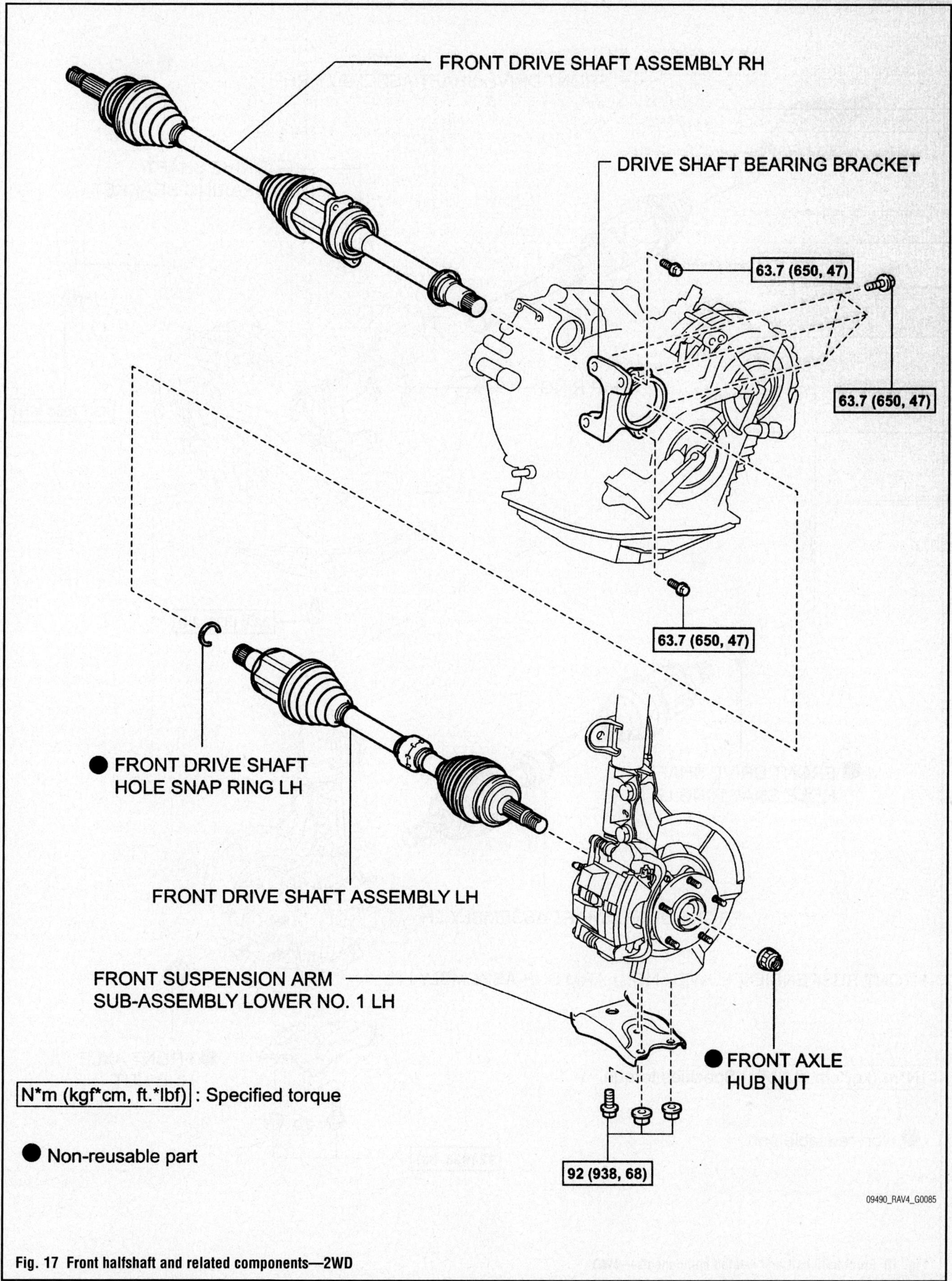

Fig. 17 Front halfshaft and related components—2WD

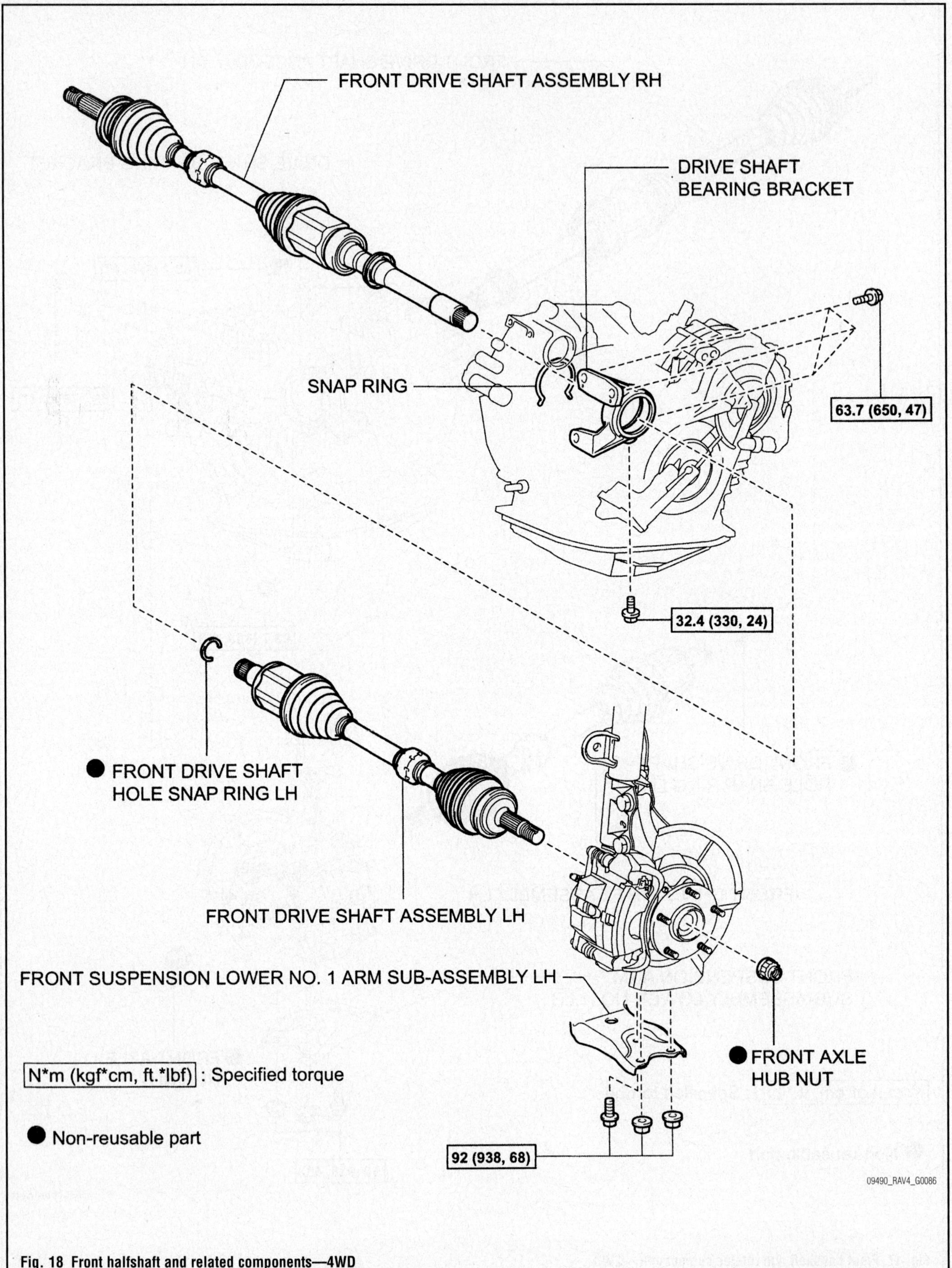

Fig. 18 Front halfshaft and related components—4WD

8. Disconnect the left and right front stabilizer link assemblies.

9. Disconnect the left and right front lower number one arm subassemblies.

10. Matchmark the halfshaft and the axle hub, left and right side.

➡Do not punch the marks.

11. Using a plastic hammer, disconnect the steering knuckle with the axle hub, left and right side.

➡Be careful not to damage the boot and speed sensor rotor. Do not push out excessively the halfshaft from the axle assembly.

12. Disconnect the left and right tie rod subassemblies.

13. On the left side, using tool SST09520-01010, or equivalent remove the front halfshaft.

14. On 2WD, to remove the right halfshaft remove the two bolts and pull out the halfshaft together with the halfshaft bearing case. Remove the halfshaft from the transaxle.

➡Be careful not to damage the boot and speed sensor rotor. Do not push out excessively the halfshaft from the axle assembly.

15. On 4WD, to remove the right halfshaft use a brass bar and hammer to remove the halfshaft.

➡Do not damage the oil seal, boot or allow the halfshaft to fall out.

16. Support the front axle assembly.

➡The hub bearing could be damaged if it is subjected to the vehicle weight. If it is necessary to place weight on the hub bearing, such as moving it when the halfshaft is removed, support it using too SST09608-16042, or equivalent.

To install:

17. Installation is the reverse of the removal procedure.

18. On vehicles manufactured thru 01/2006 tighten the front axle hub nut to 159 ft. lbs. On vehicles manufactured after 01/2006 and equipped with the 2.4L engine, tighten the axle hub nut to 159 ft. lbs. On vehicles manufactured after 01/2006 and equipped with the 3.5L engine, tighten the axle hub nut to 215 ft. lbs.

19. On 2WD, tighten the right side bearing bracket bolts to 47 ft. lbs.

20. On 4WD, tighten the right side bearing bracket bolts to 24 ft. lbs.

21. Check and adjust the wheel alignment, as required.

22. Be sure to fill the transaxle with the proper grade and type transaxle fluid.

23. Start the engine and check for leaks.

TRANSFER CASE ASSEMBLY

REMOVAL & INSTALLATION

The transfer case is part of the engine/transaxle assembly and is removed with those units.

REAR HALFSHAFT

REMOVAL & INSTALLATION

See Figures 19 and 20.

1. Before servicing the vehicle, refer to the Precautions Section.

2. Disconnect the negative battery cable.

➡Wait at least 90 seconds after disconnecting the negative battery cable before starting any repair work to prevent air bag and seat belt pretensioner activation.

3. Raise and support the vehicle safely.

4. Remove the tire and wheel assembly.

5. Drain the differential oil.

6. Remove the tailpipe assembly.

7. Remove the driveshaft with the center bearing.

8. Remove the rear axle shaft nut.

9. Support the rear differential using a suitable jack.

10. Fix the nuts in place and remove bolt "A" "B" and "C". Do not loosen the nuts, loosen the bolts. Slowly lower the jack and tilt the rear differential carrier, as shown in the illustration.

11. Using a suitable tool disconnect the left and right rear halfshafts from the differential carrier.

12. On 2WD, disconnect the skid control sensor wire.

13. On 4WD, disconnect the rear speed sensor, left side.

14. Put matchmarks on the halfshaft and the axle hub. Do not punch the marks.

15. Using a plastic faced hammer, separate the halfshaft from the axle hub.

➡Be careful not to damage the boot and speed sensor rotor. Do not excessively push out the halfshaft from the axle.

16. Support the rear halfshaft assembly.

➡The hub bearing could be damaged if it is subjected to the vehicle weight. If it is necessary to place weight on the hub bearing, such as moving it when the halfshaft is removed, support it using too SST09608-16042, or equivalent.

To install:

17. Installation is the reverse of the removal procedure.

18. Tighten the differential carrier bolts to 63 ft. lbs. for bolt "A", 103 ft. lbs. for bolt "B".

19. Check and adjust the wheel alignment, as required.

20. Be sure to fill the differential with the proper grade and type fluid.

21. Start the engine and check for leaks.

CV-BOOTS INSPECTION

Inspect the halfshaft for damaged boots. If the boots are damaged or leaking grease they must be replaced. If the halfshaft is bent, replace as required.

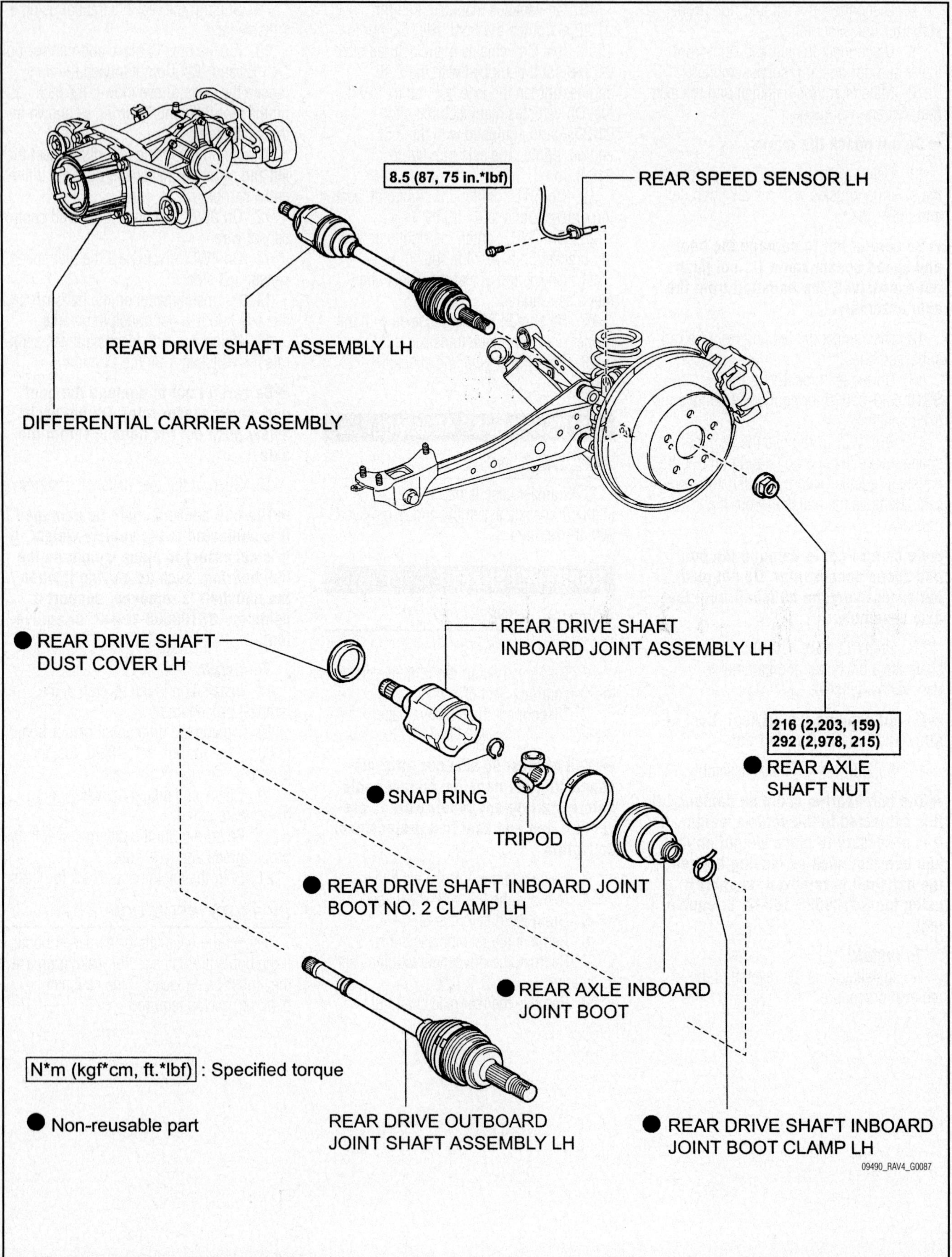

Fig. 19 Rear halfshaft and related components

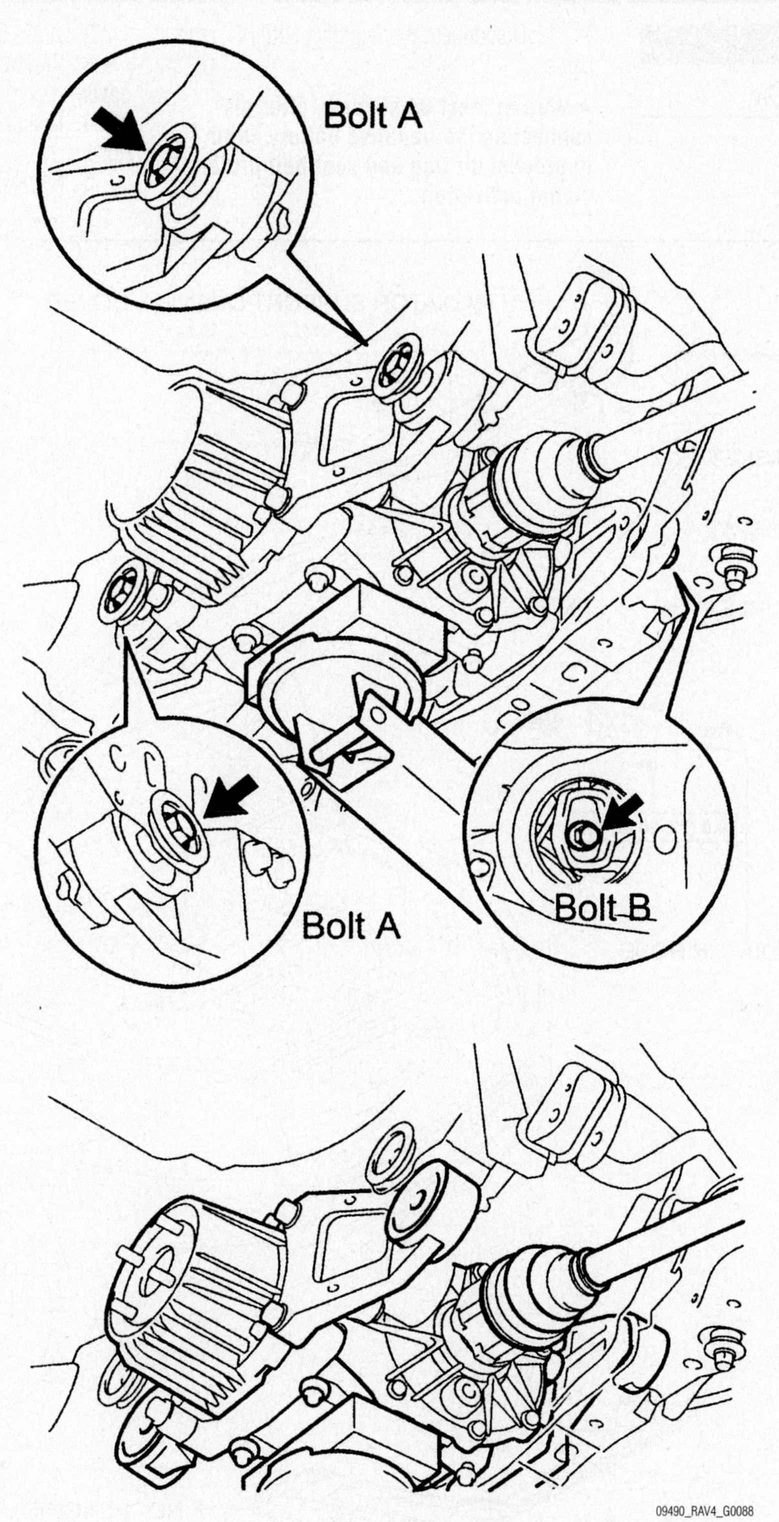

Fig. 20 Rear differential bolt locations

ENGINE COOLING

THERMOSTAT

REMOVAL & INSTALLATION

2.4L Engine

See Figures 21 and 22.

1. Disconnect the negative battery cable.

➡**Wait at least 90 seconds after disconnecting the negative battery cable to prevent air bag and seat belt pretensioner activation.**

2. Drain the engine coolant.
3. Remove the radiator support opening cover.
4. Disconnect the number two radiator hose.
5. Remove the water inlet retaining

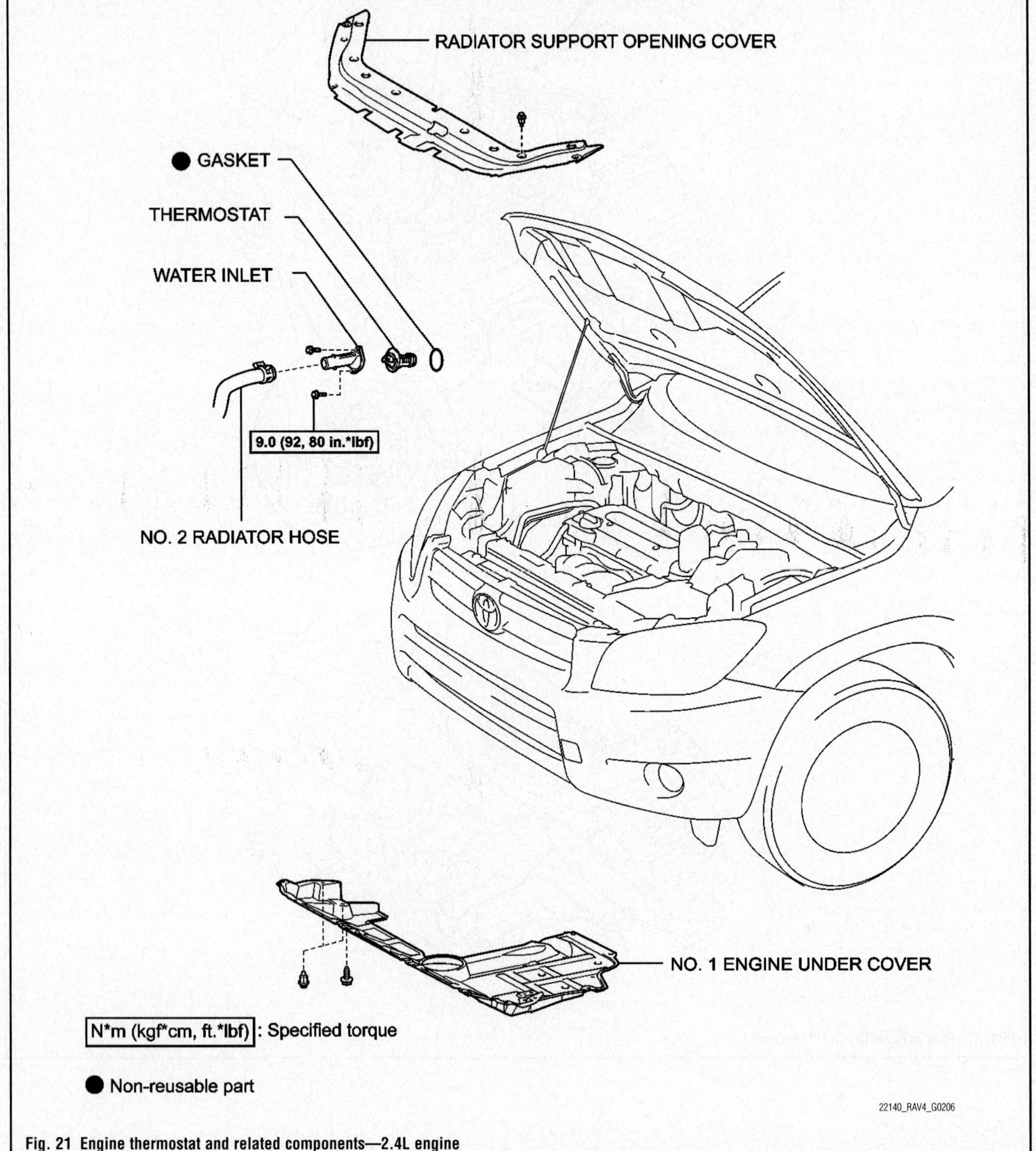

Fig. 21 Engine thermostat and related components—2.4L engine

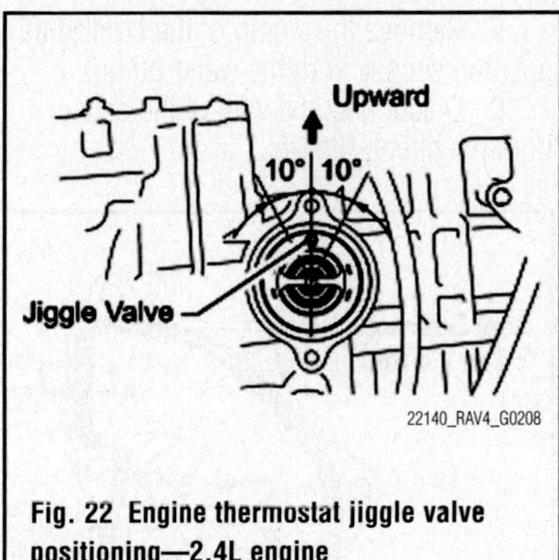

Fig. 22 Engine thermostat jiggle valve positioning—2.4L engine

nuts. Remove the water inlet from the cylinder block.

6. Remove the thermostat from its mounting. Discard the gasket.

To install:

7. Installation is the reverse of the removal procedure.

8. Be sure to install a new gasket onto the thermostat.

9. Install the thermostat with the jiggle valve upward. The jiggle valve may be set within 10 degrees on either side of the assembly. See illustration.

10. Tighten the retaining nuts to 80 inch lbs.

11. Fill the engine with the proper grade and type coolant.

12. Start the engine and check for leaks, correct as required.

3.5L Engine

See Figures 22 through 24.

1. Disconnect the negative battery cable.

➡Wait at least 90 seconds after disconnecting the negative battery cable to prevent air bag and seat belt pretensioner activation.

2. Drain the engine coolant.

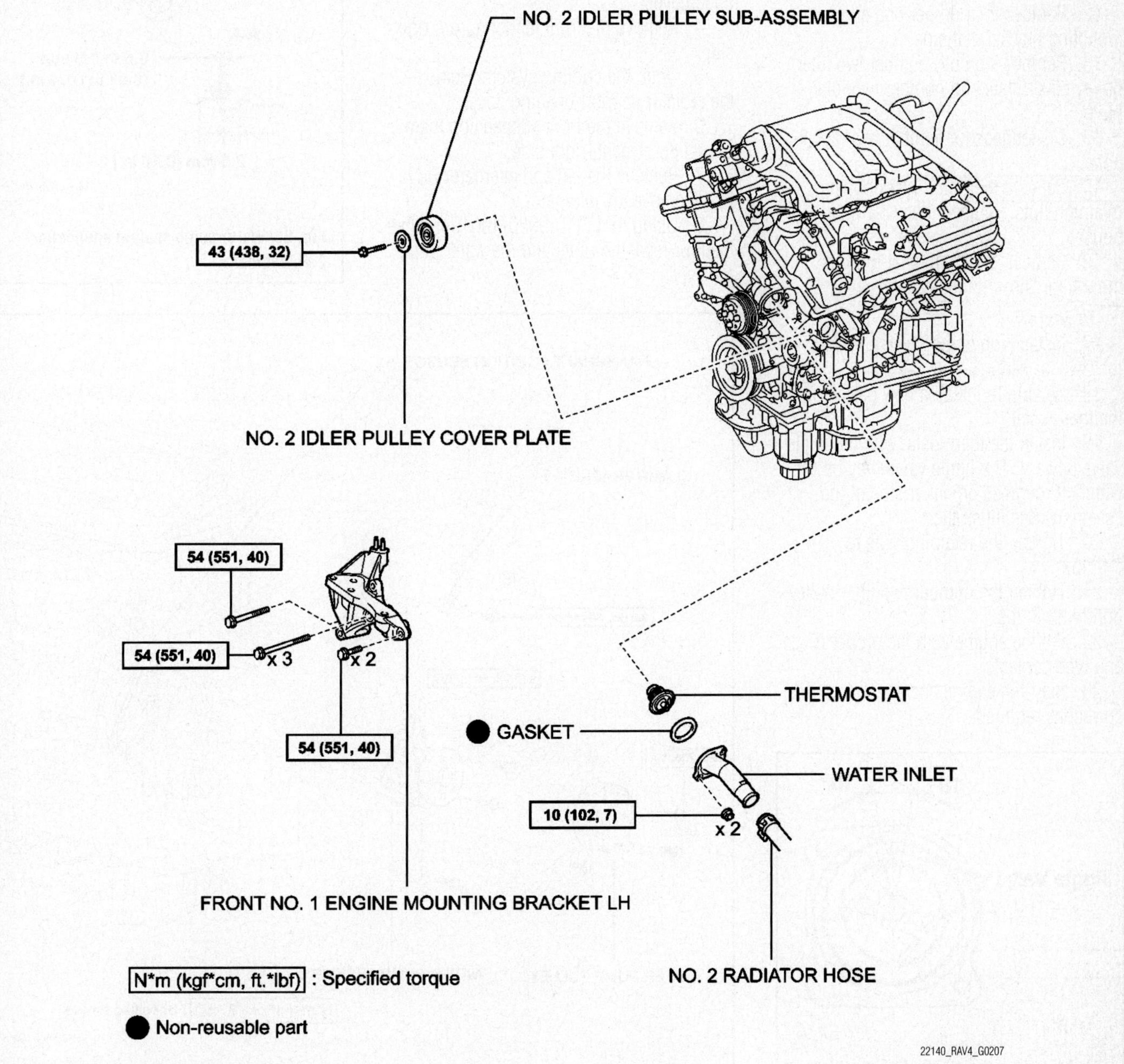

NO. 2 IDLER PULLEY SUB-ASSEMBLY
43 (438, 32)
NO. 2 IDLER PULLEY COVER PLATE
54 (551, 40)
54 (551, 40)
x 3
x 2
54 (551, 40)
THERMOSTAT
● GASKET
WATER INLET
10 (102, 7)
x 2
FRONT NO. 1 ENGINE MOUNTING BRACKET LH
N*m (kgf*cm, ft.*lbf) : Specified torque
NO. 2 RADIATOR HOSE
● Non-reusable part
22140_RAV4_G0207

Fig. 23 Engine thermostat and related components—3.5L engine

3. Remove the right tire and wheel assembly.
4. Remove the number one engine cover.
5. Remove the engine under cover rear, right side.
6. Remove the front suspension member reinforcement, right side.
7. Remove the V-bank cover assembly.
8. Remove the radiator support opening cover.
9. Remove the radiator reservoir tank assembly.
10. Remove the fan and alternator belt.
11. Remove the engine mounting insulator, right side five retaining bolts. Remove the component.
12. Remove the number one engine mounting bracket, left side.
13. Remove the bolt, number two idler pulley cover plate and number two idler pulley.
14. Disconnect the number two radiator hose.
15. Remove the water inlet housing retaining nuts. Remove the water inlet housing.
16. Remove the thermostat from its mounting. Discard the gasket.

To install:

17. Installation is the reverse of the removal procedure.
18. Be sure to install a new gasket onto the thermostat.
19. Install the thermostat with the jiggle valve upward. The jiggle valve may be set within 10 degrees on either side of the assembly. See illustration.
20. Tighten the retaining nuts to 7 ft. lbs.
21. Tighten the number two idler pulley bolt to 32 ft. lbs.
22. Fill the engine with the proper grade and type coolant.
23. Start the engine and check for leaks, correct as required.

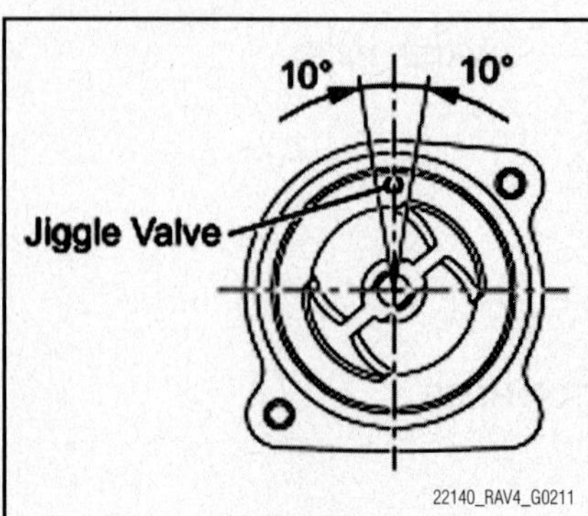

Fig. 24 Engine thermostat jiggle valve positioning—3.5L engine

WATER PUMP

REMOVAL & INSTALLATION

2.4L Engine

See Figures 25 and 26.

1. Before servicing the vehicle, refer to the Precautions Section.
2. Disconnect the negative battery cable.

➡Wait at least 90 seconds after disconnecting the negative battery cable before starting any repair work to prevent air bag and seat belt pretensioner activation.

3. Remove the number one engine undercover.
4. Remove the front fender apron, right side.
5. Drain the cooling system. Remove the radiator support opening cover.
6. Remove the front suspension member reinforcement, right side.
7. Remove the fan and alternator drive belt. Remove the alternator.
8. Using tool SST09960-10010 remove the four retaining bolts and the water pump pulley.
9. Remove the clamp of the crankshaft position sensor from the water pump.
10. Disconnect the wire of the sensor from the clamp bracket.

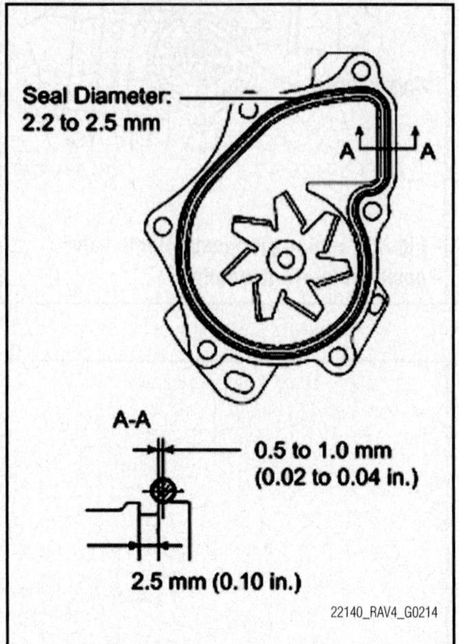

Fig. 26 Water pump sealant application—2.4L engine

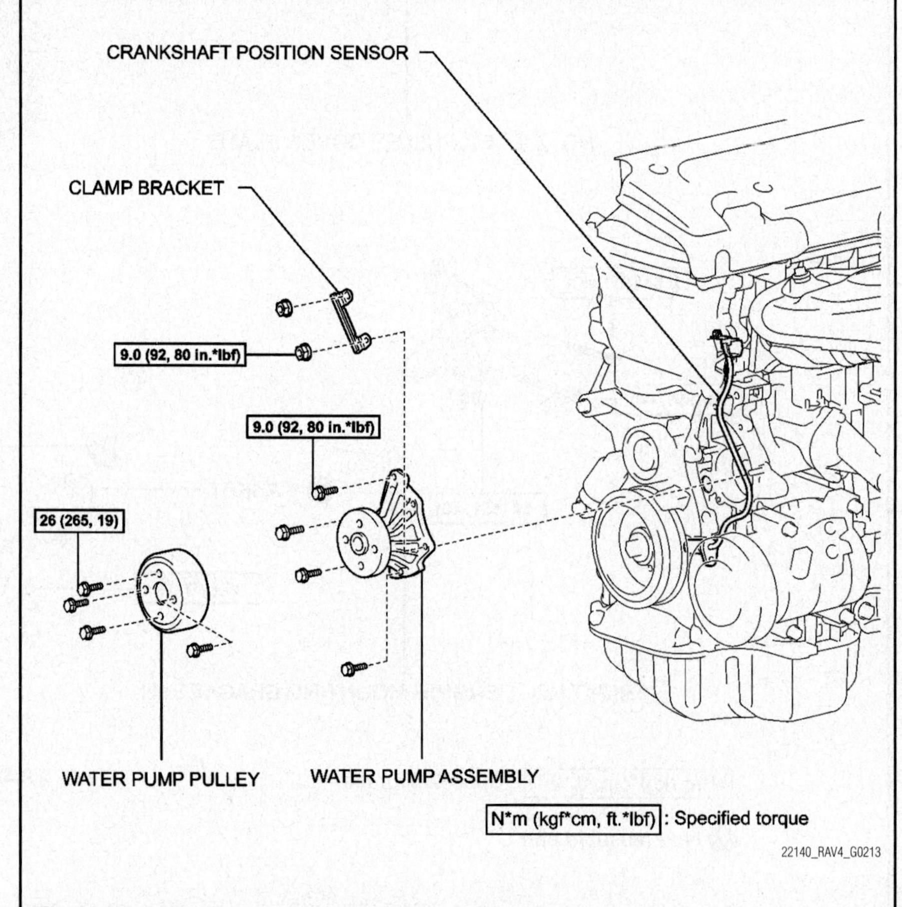

Fig. 25 Water pump and related components—2.4L engine

11. Remove the four water pump retaining bolts, two nuts and clamp bracket. Remove the water pump from the engine.

To install:

12. Apply a 2.5mm wide bead of RTV gasket material to the pump sealing surface as shown.

➡Install the pump with 5 minutes of applying the sealer or the sealer will have to be removed and new sealer applied.

13. Install the pump and torque the nuts and bolts to 80 inch lbs. (9 Nm).

14. The remainder of installation is the reverse of removal. Refill the cooling system.

15. Start the vehicle and check for leaks, correct ass required.

3.5L Engine

See Figures 27 through 29.

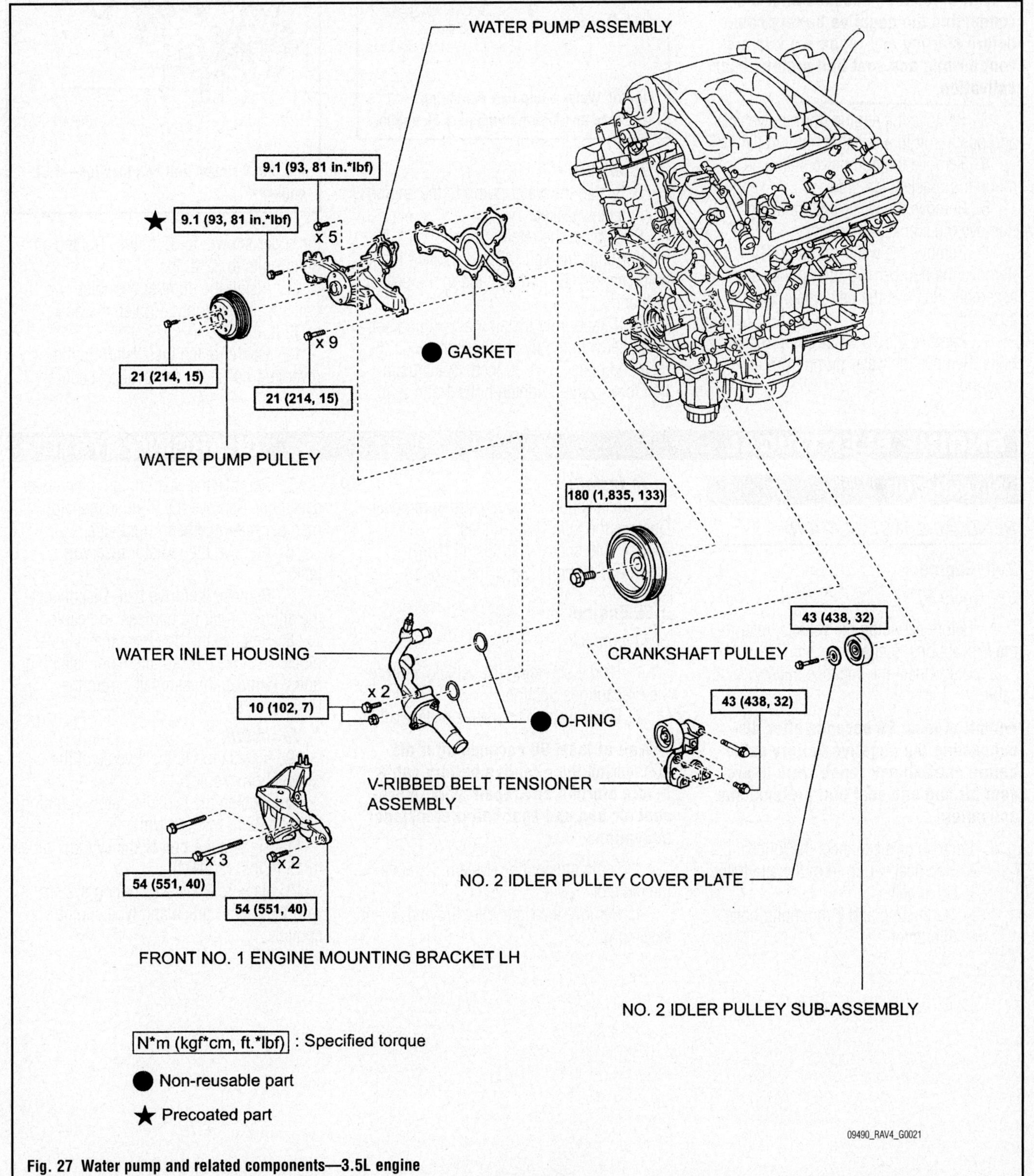

Fig. 27 Water pump and related components—3.5L engine

➡In order to replace the water pump, the manufacturer recommends removing the engine from the vehicle.

1. Before servicing the vehicle, refer to the Precautions Section.
2. Disconnect the negative battery cable.

➡Wait at least 90 seconds after disconnecting the negative battery cable before starting any repair work to prevent air bag and seat belt pretensioner activation.

3. Remove the engine from the vehicle and position it in a suitable holding fixture.
4. Remove the number one engine mounting bracket, left side.
5. Remove the water inlet housing. Remove the crankshaft pulley.
6. Remove the water pump pulley. Remove the number two idler pulley sub-assembly. Remove the belt tensioner assembly.
7. Remove the 16 water pump retaining bolts. Remove the water pump from its mounting.

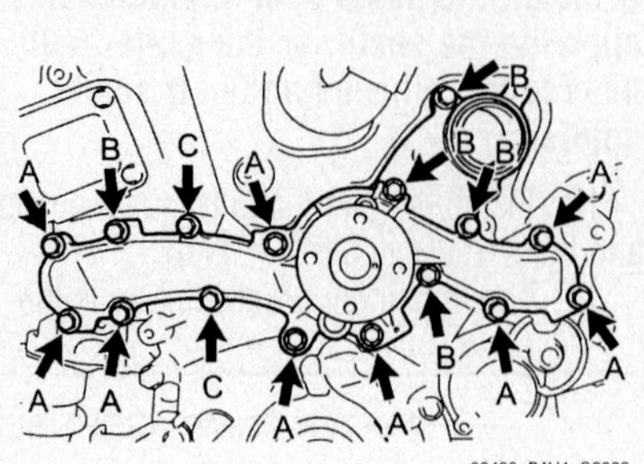

Fig. 28 Water pump bolt tightening sequence and bolt markings—3.5L engine

To install:

8. Install the water pump to the engine using a new gasket.
9. Tighten the retaining bolts to 15 ft. lbs. for bolts marked "A" and 81 inch lbs for bolts marked "B" and "C" and in the proper sequence.
10. Temporarily install the V-ribbed belt tensioner with the five bolts. Bolt "A" 2.76 inch and bolts "B" 1.30 inch. See illustration for location. Tighten bolts 1 and 2, in

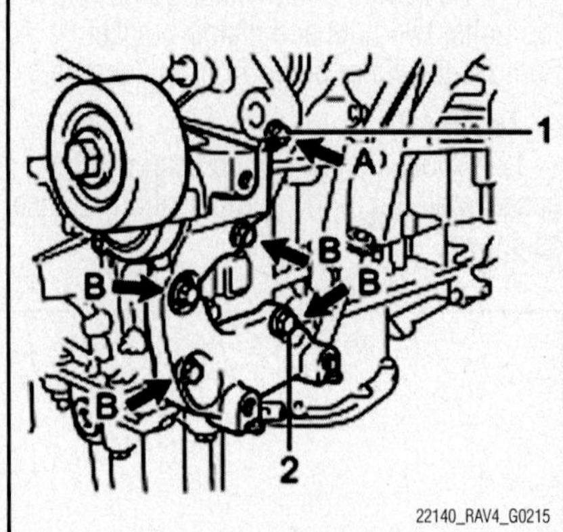

Fig. 29 V-ribbed belt bolt location—3.5L engine

the order shown, to 32 ft. lbs. Tighten all other bolts to 32 ft. lbs.

11. Install the number two idler pulley sub-assembly. Tighten the bolt to 32 ft. lbs.
12. Continue the installation in the reverse order of the removal procedure.

ENGINE ELECTRICAL

CHARGING SYSTEM

ALTERNATOR

REMOVAL & INSTALLATION

2.4L Engine

See Figure 30.

1. Before servicing the vehicle, refer to the Precautions Section.
2. Disconnect the negative battery cable.

➡Wait at least 90 seconds after disconnecting the negative battery cable before starting any repair work to prevent air bag and seat belt pretensioner activation.

3. Remove or disconnect the following:
   - Electrical wiring from the alternator
   - Drive belt
   - 1 adjusting and 2 mounting bolts
   - Alternator

To install:

4. Installation is the reverse of removal. Observe the following torques:
   - M8 bolts: 15 ft. lbs. (21Nm)
   - M10 bolts: 38 ft. lbs. (52Nm)

3.5L Engine

See Figure 31.

1. Before servicing the vehicle, refer to the Precautions Section.
2. Disconnect the negative battery cable.

➡Wait at least 90 seconds after disconnecting the negative battery cable before starting any repair work to prevent air bag and seat belt pretensioner activation.

3. Drain the cooling system. Remove the radiator.
4. Remove the right front tire and wheel assembly.
5. Remove the rear engine undercover assembly. Remove the front suspension member reinforcement, right side.
6. Remove the radiator reservoir tank.
7. Remove the drive belt. Disconnect the alternator wiring harness and cable.
8. Remove the bolt from the cylinder block. Remove the two alternator retaining bolts. Remove the alternator from the vehicle.

To install:

9. Installation is the reverse of the removal procedure.
10. Tighten the two alternator mounting bolts to 32 ft. lbs. (43 Nm).
11. Tighten the bolt to the cylinder block to 15 ft. lbs. (20 NM).
12. Be sure to fill the cooling system with the proper grade and type engine coolant.

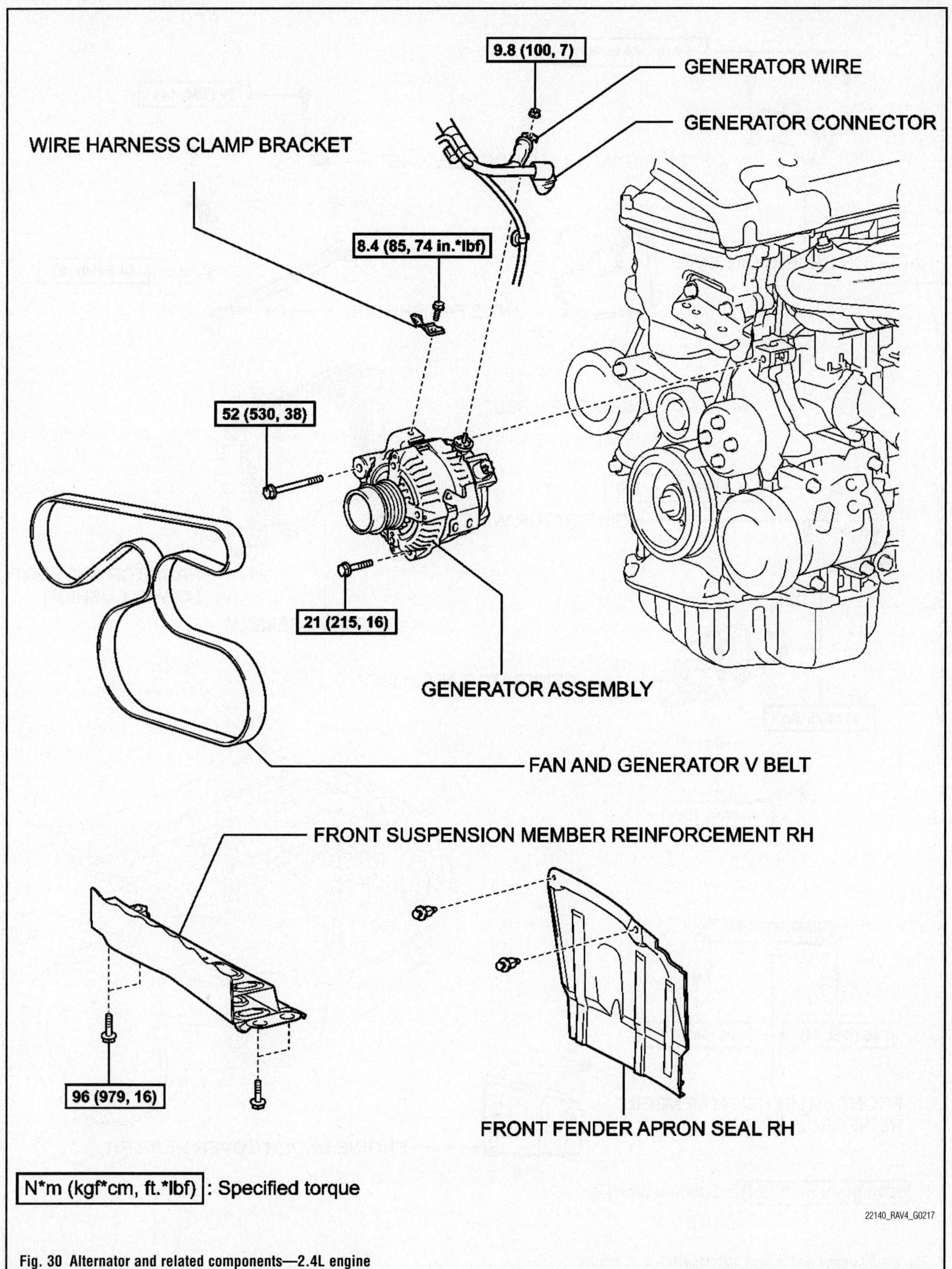

Fig. 30 Alternator and related components—2.4L engine

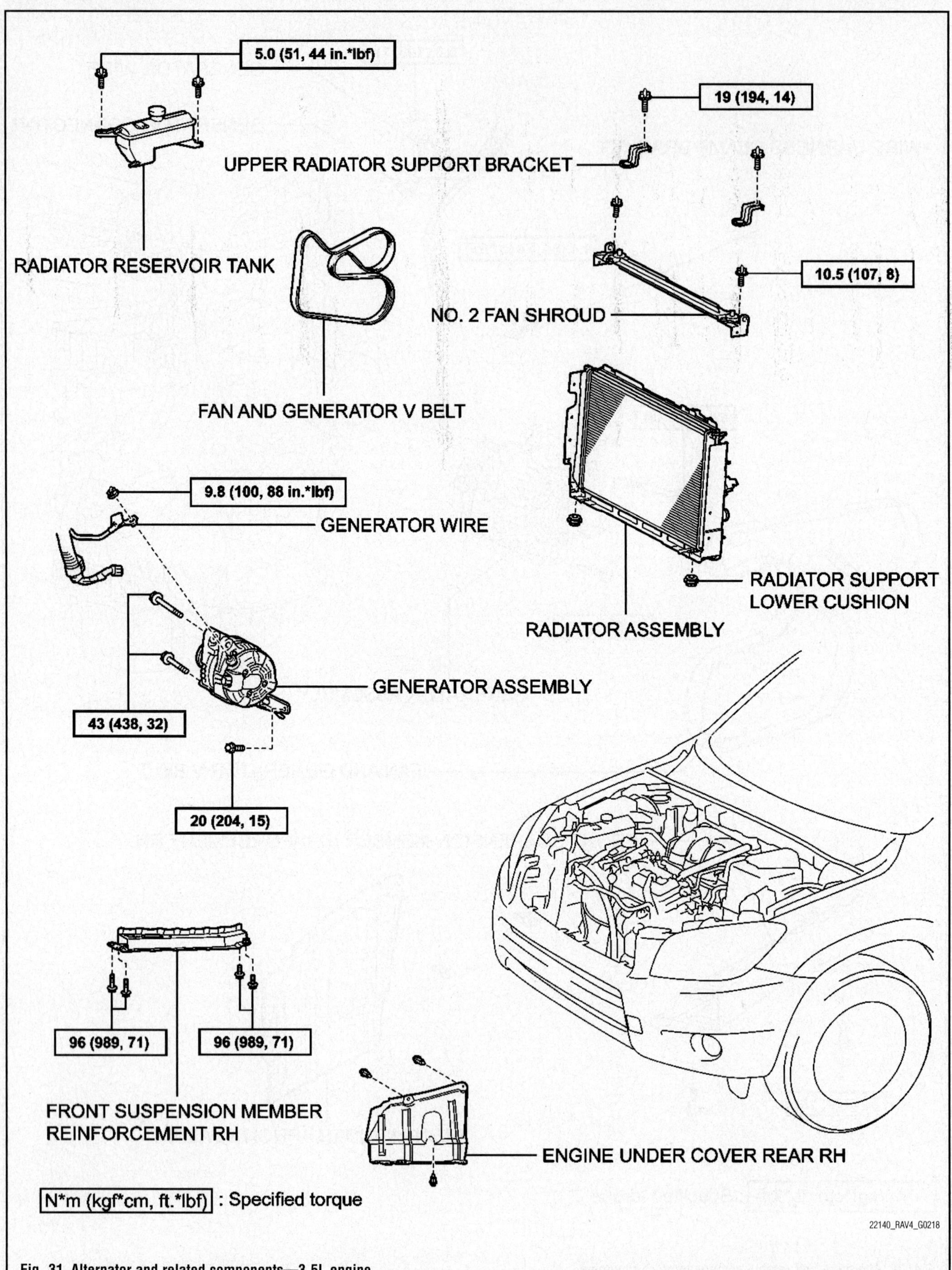

Fig. 31 Alternator and related components—3.5L engine

FIRING ORDER

See Figure 32.

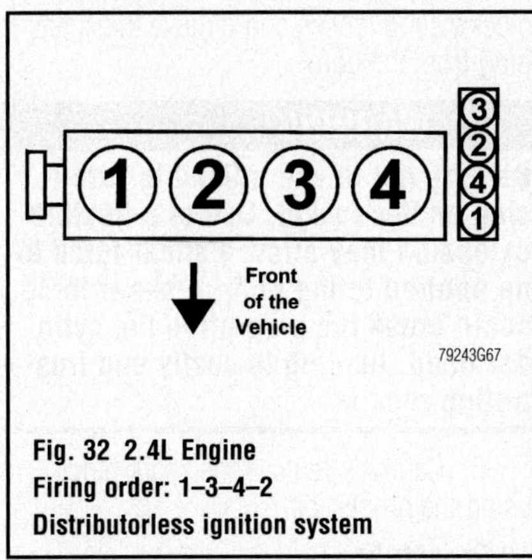

Fig. 32 2.4L Engine Firing order: 1–3–4–2 Distributorless ignition system

IGNITION COIL

REMOVAL & INSTALLATION

2.4L Engine

See Figure 33.

1. Before servicing the vehicle, refer to the Precautions Section.
2. Disconnect the negative battery cable.

➡Wait at least 90 seconds after disconnecting the negative battery cable before starting any repair work to prevent air bag and seat belt pretensioner activation.

3. Disconnect the four ignition coil connectors.
4. Remove the four retaining bolts.
5. Remove the four ignition coils.

To install:

6. Installation is the reverse of the removal procedure.
7. Tighten the four ignition coil bolts to 80 inch lbs.

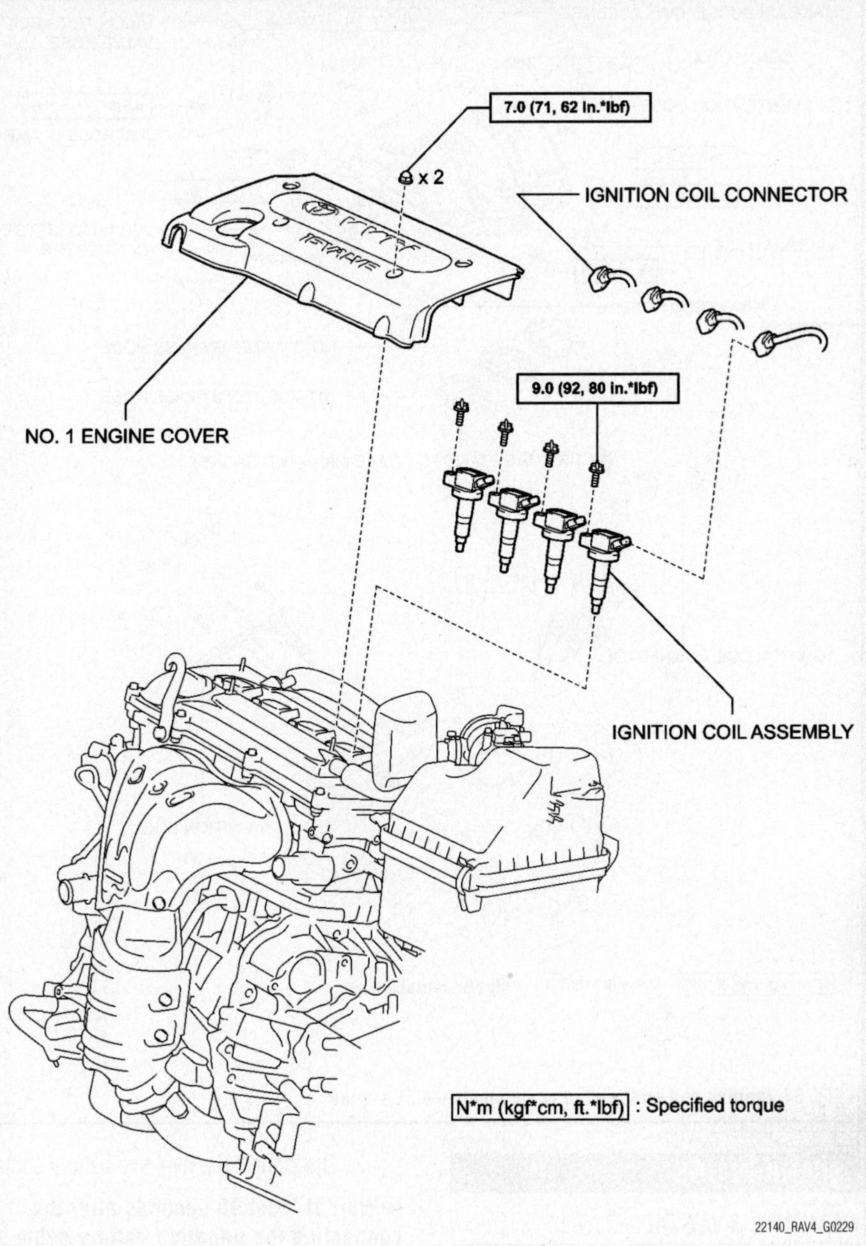

Fig. 33 Ignition coil and related components—2.4L engine

3.5L Engine

See Figure 34.

➡In order to replace the ignition coil pack, the manufacturer recommends removing the engine from the vehicle.

1. Before servicing the vehicle, refer to the Precautions Section.
2. Disconnect the negative battery cable.

➡Wait at least 90 seconds after disconnecting the negative battery cable before starting any repair work to prevent air bag and seat belt pretensioner activation.

3. Remove the engine from the vehicle. Position the assembly in a suitable holding fixture.
4. Remove the intake air surge tank assembly.
5. Disconnect the six ignition coil connectors.
6. Remove the six retaining bolts.
7. Remove the six ignition coils.

To install:

8. Installation is the reverse of the removal procedure.
9. Tighten the six ignition coil bolts to 7 ft. lbs.

IGNITION TIMING

All engines are equipped with a Distributorless Ignition System (DIS). Ignition timing can be checked, but no timing adjustment is possible.

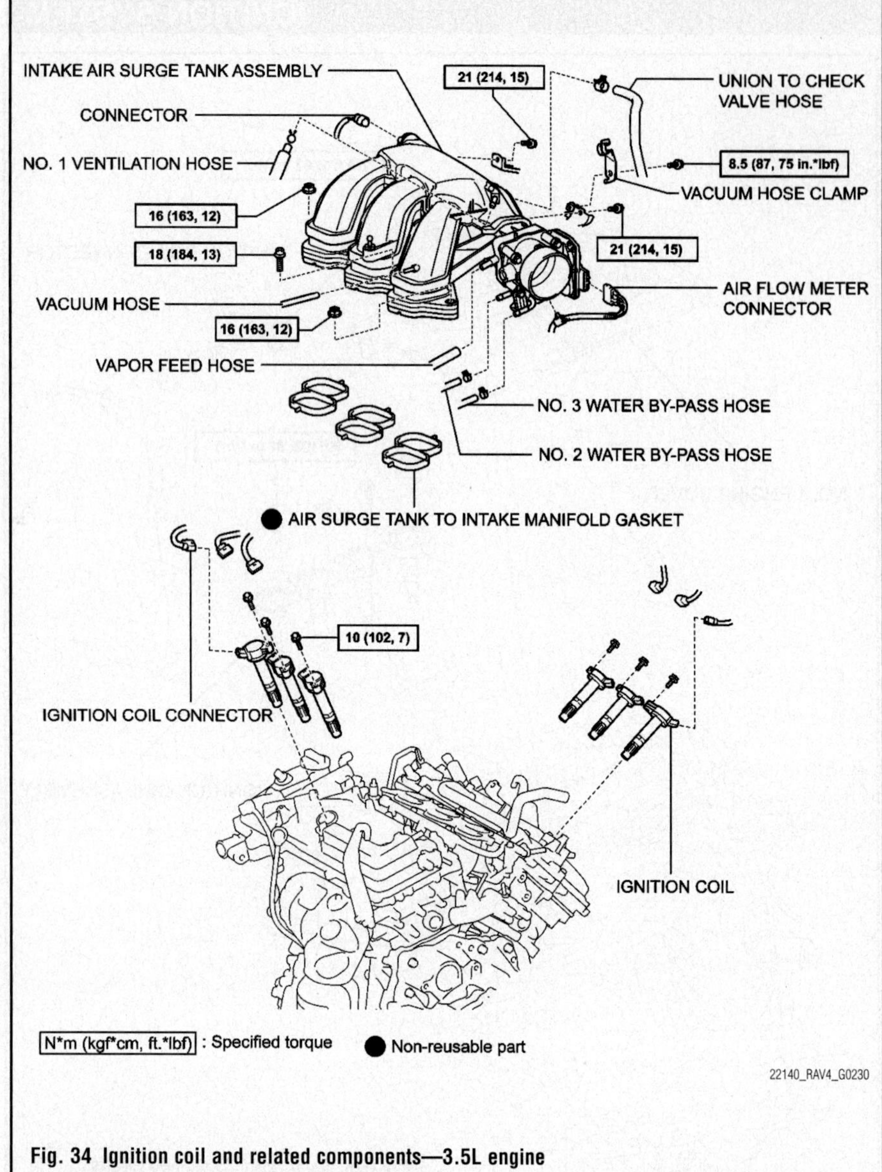

Fig. 34 Ignition coil and related components—3.5L engine

SPARK PLUGS

REMOVAL & INSTALLATION

➡Remove the spark plugs when the engine is cold, if possible, to prevent damage to the threads. If removal of the plugs is difficult, apply a few drops of penetrating oil or silicone spray to the area around the base of the plug, and allow it a few minutes to work.

1. Before servicing the vehicle, refer to the Precautions Section.
2. Disconnect the negative battery cable.

➡Wait at least 90 seconds after disconnecting the negative battery cable before starting any repair work to prevent air bag and seat belt pretensioner activation.

3. Remove the necessary components in order to gain access to the spark plugs.
4. Use compressed air, blow any water or debris from the spark plug well to assure that no harmful contaminants are allowed to enter the combustion chamber when the spark plug is removed. If compressed air is not available, use a rag or a brush to clean the area.
5. Using a spark plug socket that is equipped with a rubber insert to properly hold the plug, turn the spark plug counterclockwise to loosen and remove the spark plug from the bore.

※※ WARNING

Be sure not to use a flexible extension on the socket. Use of a flexible extension may allow a shear force to be applied to the plug. A shear force could break the plug off in the cylinder head, leading to costly and frustrating repairs.

6. Carefully remove the spark plugs, using the proper tool.

To install:

7. Installation is the reverse of the removal procedure.
8. Using a wire feeler gauge, check and adjust the spark plug gap. When using a gauge, the proper size should pass between the electrodes with a slight drag. The next larger size should not be able to pass while the next smaller size should pass freely.
9. Carefully thread the plug into the bore by hand. If resistance is felt before the plug is almost completely threaded, back the plug out and begin threading again. In small, hard to reach areas, an old spark plug wire and boot could be used as a threading tool. The boot will hold the plug while you twist the end of the wire and the wire is supple enough to twist before it would allow the plug to crossthread.

※※ WARNING

Do not use the spark plug socket to thread the plugs. Always carefully thread the plug by hand or using an old plug wire to prevent the possibility of crossthreading and damaging the cylinder head bore.

10. Carefully tighten the spark plug. If the plug you are installing is equipped with a crush washer, seat the plug, then tighten about ¼turn to crush the washer. If you are installing a tapered seat plug, tighten the plug to specifications provided by the vehicle or plug manufacturer.

STARTER

REMOVAL & INSTALLATION

2.4L Engine

See Figure 35.

1. Before servicing the vehicle, refer to the Precautions Section.
2. Disconnect the negative battery cable.

➡Wait at least 90 seconds after disconnecting the negative battery cable before starting any repair work to prevent air bag and seat belt pretensioner activation.

3. Disconnect the positive battery cable. Remove the battery clamp. Remove the battery insulator. Remove the battery.

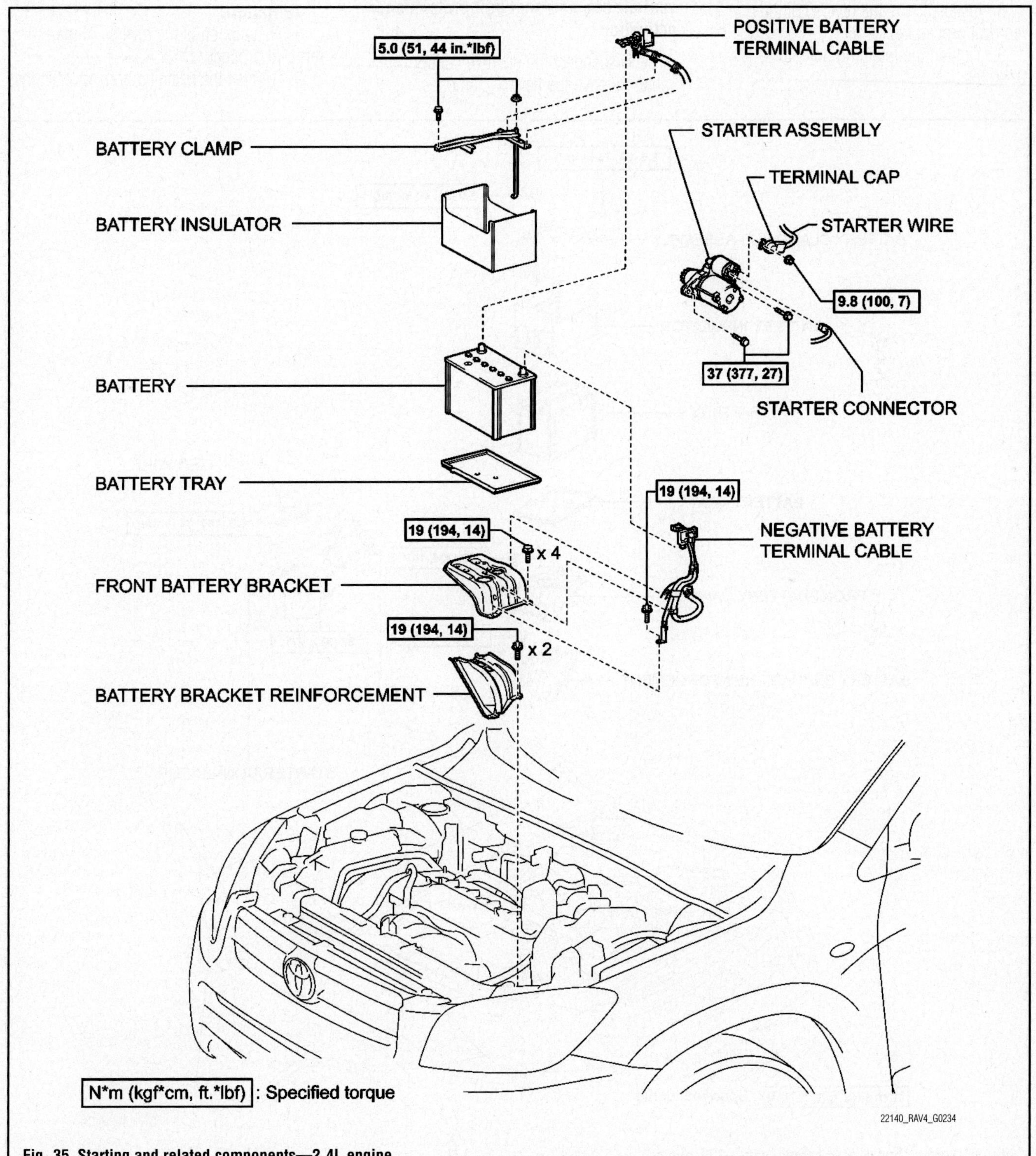

Fig. 35 Starting and related components—2.4L engine

4. Remove the front battery bracket. Remove the battery bracket reinforcement.
5. Disconnect the starter connector.
6. Open the terminal cap. Remove the nut and disconnect the starter wire.
7. Remove the starter retaining bolts. Remove the starter from the engine.

To install:

8. Installation is the reverse of the removal procedure.
9. Tighten the retaining bolts to 27 ft. lbs.

3.5L Engine

See Figure 36.

1. Before servicing the vehicle, refer to the Precautions Section.
2. Disconnect the negative battery cable.

➡Wait at least 90 seconds after disconnecting the negative battery cable before starting any repair work to prevent air bag and seat belt pretensioner activation.

3. Disconnect the positive battery cable. Remove the battery clamp. Remove the battery insulator. Remove the battery.
4. Remove the front battery bracket. Remove the battery bracket reinforcement.
5. Disconnect the starter connector.
6. Open the terminal cap. Remove the nut and disconnect the starter wire.
7. Remove the starter retaining bolts. Remove the starter from the engine.

To install:

8. Installation is the reverse of the removal procedure.
9. Tighten the retaining bolts to 27 ft. lbs.

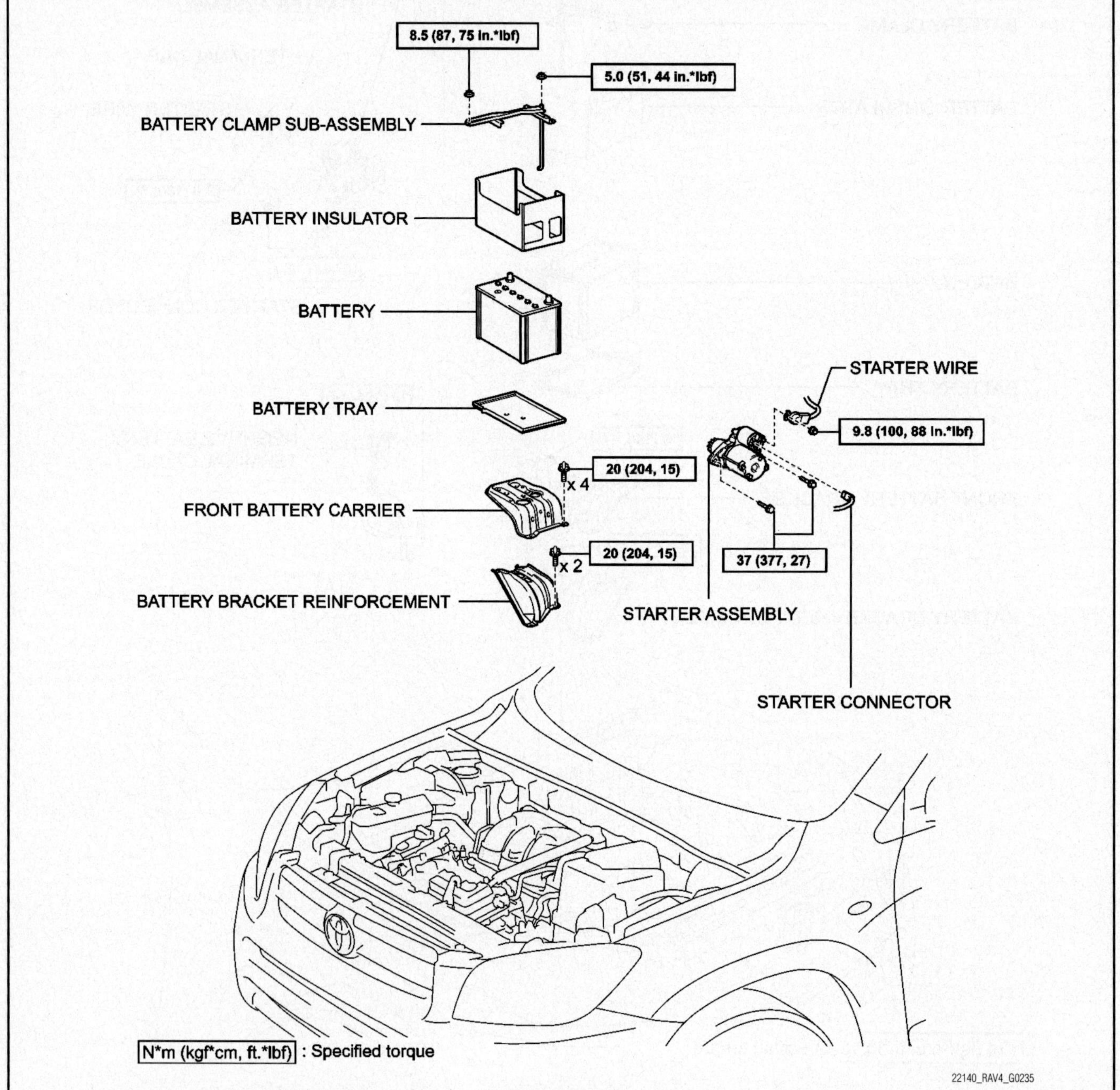

Fig. 36 Starting and related components—3.5L engine

ENGINE MECHANICAL

➡Disconnecting the negative battery cable may interfere with the functions of the on board computer systems and may require the computer to undergo a relearning process, once the negative battery cable is reconnected.

ACCESSORY DRIVE BELTS

ACCESSORY BELT ROUTING

See Figures 37 and 38.

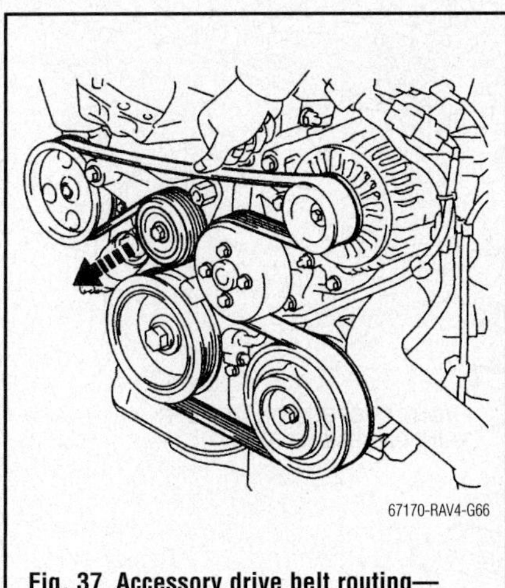

Fig. 37 Accessory drive belt routing—2.4L engine

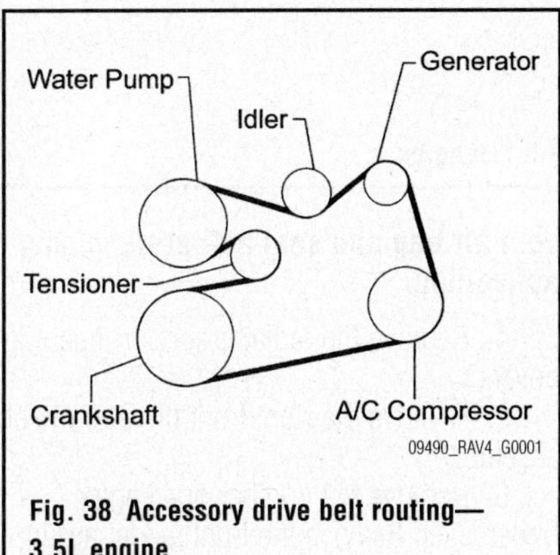

Fig. 38 Accessory drive belt routing—3.5L engine

INSPECTION

Visually check the drive belt for excessive wear, frayed cords, missing chunks of belt, etc. If defects exist, replace the belt.

➡Cracks on the rid side of the drive belt are considered normal.

REMOVAL & INSTALLATION

2.4L Engine

See Figures 39 and 40.

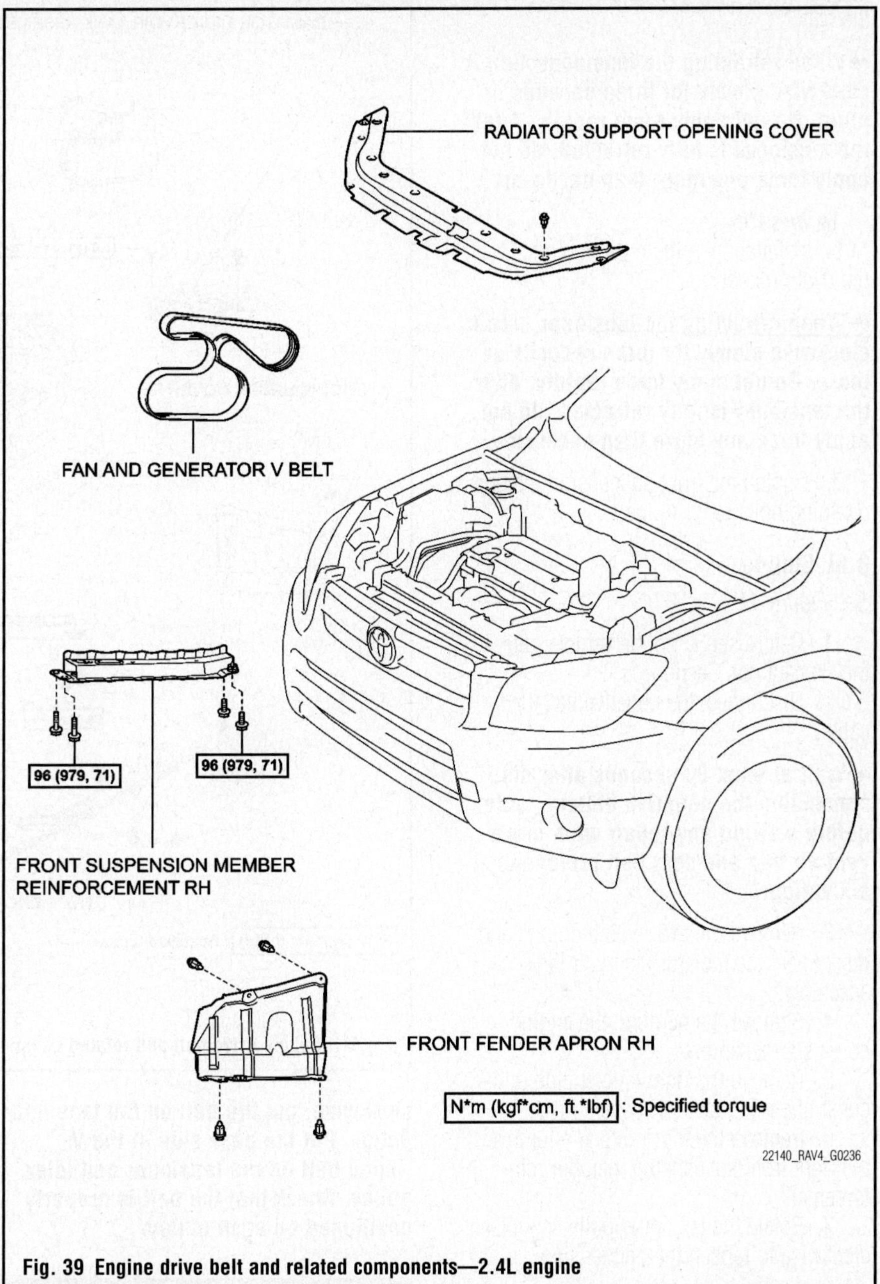

Fig. 39 Engine drive belt and related components—2.4L engine

1. Before servicing the vehicle, refer to the Precautions Section.
2. Disconnect the negative battery cable.

➡Wait at least 90 seconds after disconnecting the negative battery cable before starting any repair work to prevent air bag and seat belt pretensioner activation.

3. Raise and support the vehicle safely. Remove the right side tire and wheel assembly.
4. Remove the four bolts and remove the right front suspension member reinforcement.

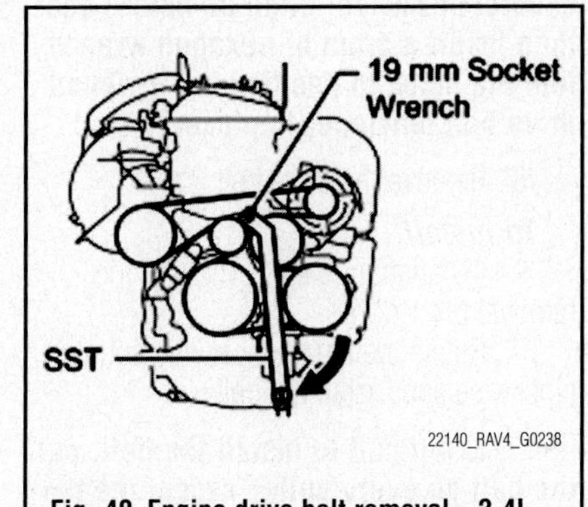

Fig. 40 Engine drive belt removal—2.4L engine

5. Using the proper tool, loosen the V-belt tensioner arm clockwise. Remove the belt.

➡When retracting the tensioner, turn it clockwise slowly for three seconds or more. Do not apply force rapidly. After the tensioner is fully retracted, do not apply force any more than necessary.

To install:

6. Installation is the reverse of the removal procedure.

➡When retracting the tensioner, turn it clockwise slowly for three seconds or more. Do not apply force rapidly. After the tensioner is fully retracted, do not apply force any more than necessary.

7. Tighten the front suspension member retaining bolts to 71 ft. lbs.

3.5L Engine

See Figure 41.

1. Before servicing the vehicle, refer to the Precautions Section.
2. Disconnect the negative battery cable.

➡Wait at least 90 seconds after disconnecting the negative battery cable before starting any repair work to prevent air bag and seat belt pretensioner activation.

3. Remove the two retaining bolts and disconnect the radiator reservoir tank assembly.
4. Remove the number one engine cover sub-assembly.
5. Remove the right side engine under cover clips. Remove the cover.
6. Remove the four bolts and remove the right front suspension member reinforcement.
7. Rotate the tensioner pulley counterclockwise to loosen the belt tension.

➡While turning the belt tensioner counterclockwise, align its holes, and then insert a 5mm bi-hexagon wrench into the holes to position the V-ribbed drive belt tensioner in place.

8. Remove the drive belt.

To install:

9. Installation is the reverse of the removal procedure.
10. Rotate the tensioner pulley counterclockwise and install the belt.

➡If it is difficult to install the belt, put the belt on every pulley except the tensioner pulley. While releasing the belt tension, by turning the belt counterclockwise, put the belt on the tensioner pulley. Put the back side of the V-ribbed belt on the tensioner and idler pulley. Check that the belt is properly positioned on each pulley.

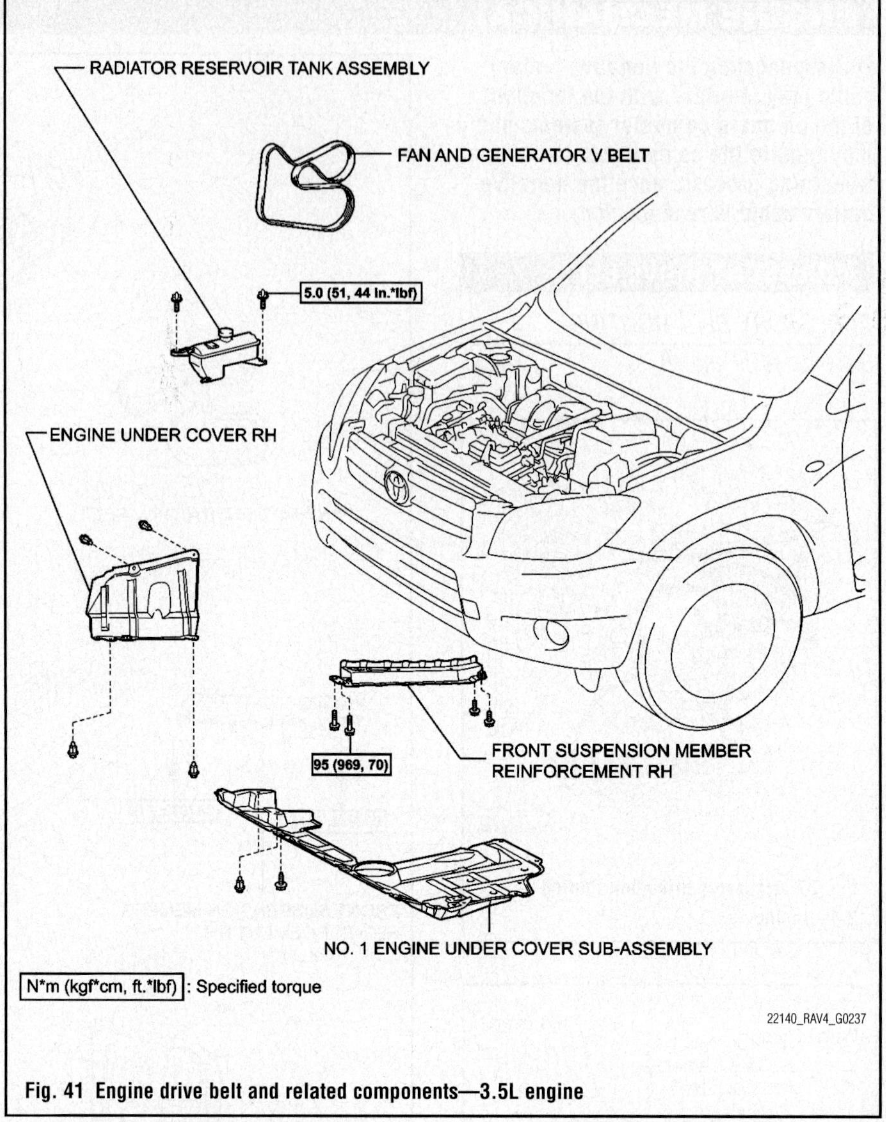

Fig. 41 Engine drive belt and related components—3.5L engine

CAMSHAFT AND VALVE LIFTERS

REMOVAL & INSTALLATION

2.4L Engine

See Figures 42 through 47.

1. Before servicing the vehicle, refer to the Precautions Section.
2. Properly relieve the fuel system pressure.
3. Disconnect the negative battery cable.

➡Wait at least 90 seconds after disconnecting the negative battery cable before starting any repair work to prevent air bag and seat belt pretensioner activation.

4. Remove the radiator support opening cover.
5. Remove the right front tire and wheel assembly.
6. Remove the number one engine undercover. Remove the front fender apron, right side. Remove the number one engine cover.
7. Drain the engine coolant. Drain the engine oil.
8. Remove the air cleaner assembly.
9. Disconnect the purge line hose from the throttle body. Disconnect the water bypass hoses from the throttle body. Disconnect the number one throttle body hose from the throttle body.
10. Disconnect the throttle position sensor and control motor connector. Disconnect the wire harness clamp. Disconnect the fuel tube from the clamp.

11. Remove the four throttle body retaining bolts and remove the throttle body assembly. Remove the gasket.

12. Remove the fuel delivery pipe sub-assembly.

13. Remove the intake manifold. Remove the intake manifold insulator.

14. Remove the front exhaust pipe. Remove the oil dipstick. Remove the oil dipstick guide.

15. Remove the manifold stay. Remove the number two manifold stay.

16. Remove the number one exhaust manifold heat insulator. Disconnect the air/fuel ratio sensor connector. Remove the five nuts and remove the exhaust manifold converter sub-assembly.

17. Disconnect the radiator hose.

18. Disconnect the radio setting condenser connector. Disconnect the engine oil pressure switch connector. Disconnect the engine coolant temperature sensor connector.

19. Disconnect the camshaft position sensor connector. Remove the bolt and ground cable.

20. Remove the front suspension member reinforcement, right side.

21. Remove the alternator. Remove the right side engine mounting insulator. Remove the idler pulley.

22. Remove the ignition coil assembly. Remove the spark plugs.

23. Remove the cylinder head cover retaining bolts. Remove the cylinder head cover from the engine.

24. To remove the belt tensioner assembly, lift the engine upward, using a transmission jack. Remove the bolt, nut and the belt tensioner assembly.

➡Do not lift the engine more than necessary.

25. Remove the crankshaft position sensor.

26. Remove the oil pan.

27. Turn the crankshaft pulley until its groove and the timing mark "0" of the timing chain cover are aligned. Check that each timing mark of the camshaft timing gear and sprocket is aligned with each timing mark located on the NO. 1 and No. 2 bearing caps, as shown in the illustration.

➡If not rotate the engine 360 degrees to align the timing marks.

28. To remove the number two camshaft, paint matchmarks on the chain in alignment with the timing marks on the camshaft timing gear and camshaft timing sprocket.

29. Remove the two nuts, tensioner and gasket. While holding the camshaft with a wrench, loosen the camshaft timing set bolt.

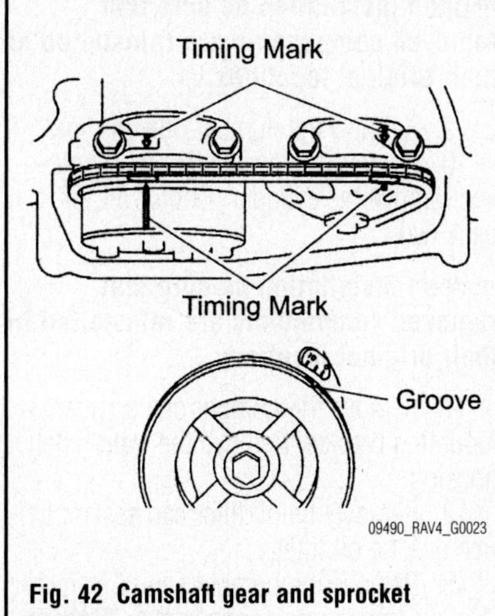

Fig. 42 Camshaft gear and sprocket alignment—2.4L engine

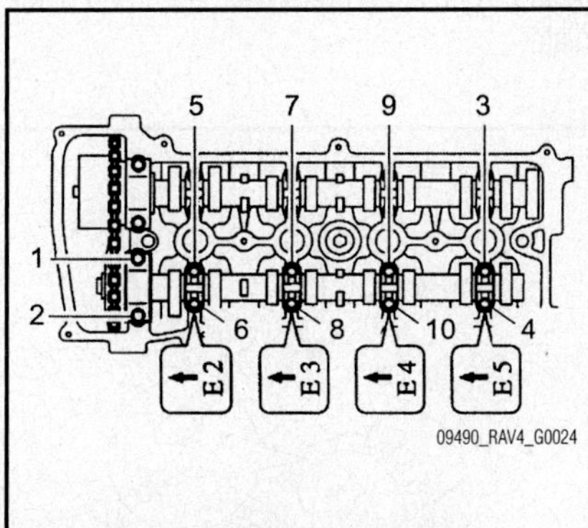

Fig. 43 Camshaft number two bolt removal sequence—2.4L engine

30. Using several steps, uniformly loosen and remove the ten bearing cap bolts in the proper sequence. Remove the five bearing caps.

31. While holding the number two camshaft, remove the camshaft timing sprocket set bolt.

➡Remove the camshaft timing sprocket from the camshaft with the timing chain wrapped on the sprocket.

32. Remove the camshaft timing sprocket from the timing chain.

33. To remove the number one camshaft, using several steps uniformly loosen and remove the ten bearing cap bolts in the proper sequence. Remove the five bearing caps.

34. Remove the camshaft and camshaft timing gear while holding the timing chain with your hand.

35. Tie the timing chain with a string to the side of the engine block.

➡Be careful not to drop anything inside the timing chain cover.

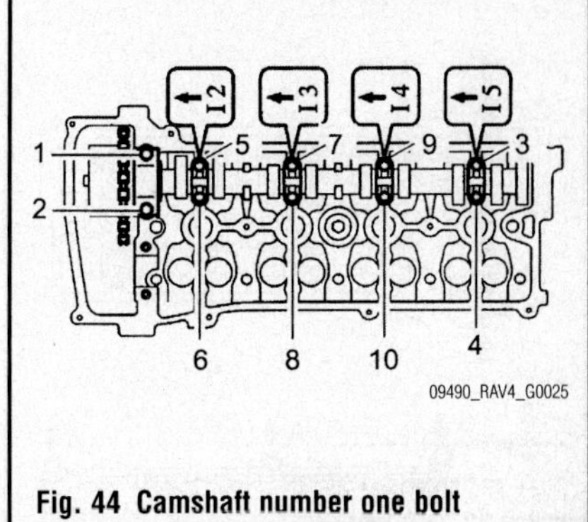

Fig. 44 Camshaft number one bolt removal sequence—2.4L engine

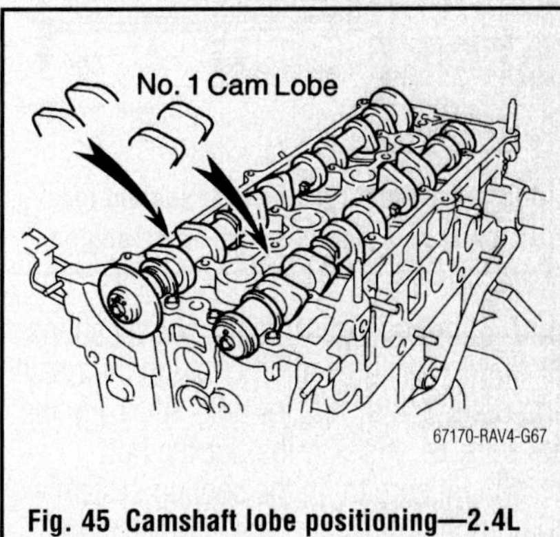

Fig. 45 Camshaft lobe positioning—2.4L engine

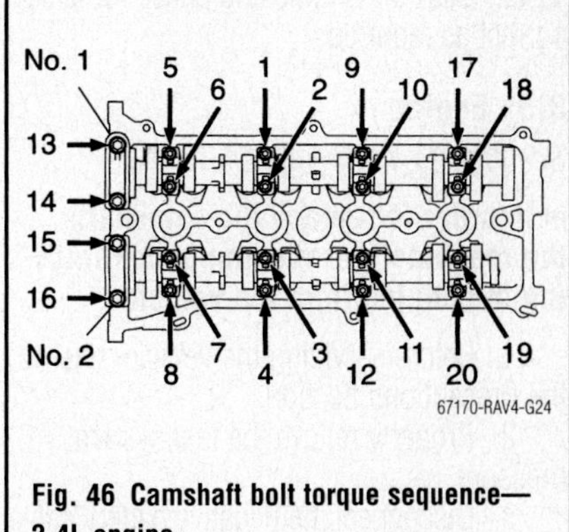

Fig. 46 Camshaft bolt torque sequence—2.4L engine

To install:

36. Installation is the reverse of the removal procedure.

37. Tighten the camshaft retaining bolts to 22 ft. lbs. for No. 1 and No. 2 bearing cap, and to 80 inch lbs. for the No. 3 bearing cap.

➡Be sure to install the camshafts with the timing mark of the camshaft timing gear on top.

38. When installing the throttle body be sure to use a new gasket. Tighten the retaining bolts to 22 ft. lbs.

39. When installing the cylinder head cover, apply seal packing (three bond

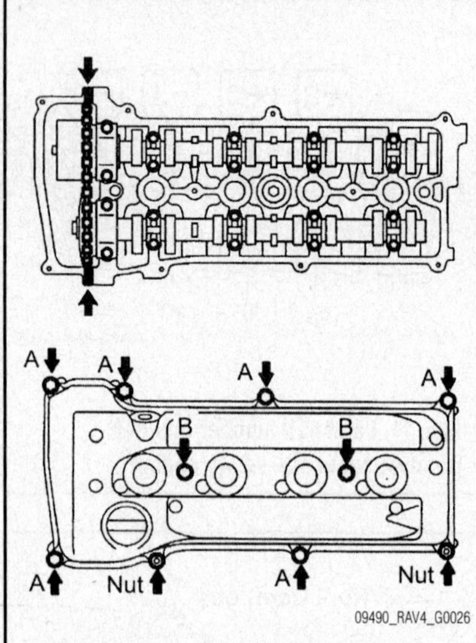

Fig. 47 Cylinder head cover sealant location and bolt identification—2.4L engine

1207B, or equivalent) to the two locations, shown in the illustration. Tighten the retaining bolts to 8 ft. lbs. for bolt "A", 10 ft. lbs. for bolt "B" and 8 ft. lbs. for the nut.

40. Be sure to fill the cooling system with the proper grade and type engine coolant.

41. Be sure to fill the engine with the proper grade and type engine oil.

42. Start the engine and check for leaks, correct as required.

3.5L Engine

See Figures 48 through 51.

➡In order to replace the camshafts, the manufacturer recommends removing the engine from the vehicle.

1. Before servicing the vehicle, refer to the Precautions Section.
2. Properly relieve the fuel system pressure.
3. Disconnect the negative battery cable.

➡Wait at least 90 seconds after disconnecting the negative battery cable before starting any repair work to prevent air bag and seat belt pretensioner activation.

4. Remove the engine and position it in a suitable holding fixture.
5. Remove the oil filler cap sub-assembly. Remove the spark plugs.
6. Remove the camshaft timing control valve assembly.
7. Remove the VVT sensor. Remove the water inlet. Remove the oil pipe.
8. Remove the twelve retaining bolts and remove the cylinder head cover for bank one.

➡Upon installation be sure that removed components are reinstalled in their original locations.

9. Remove the number one oil pipe.
10. Remove the twelve retaining bolts and remove the cylinder head cover for bank two.

➡Upon installation be sure that removed components are reinstalled in their original locations.

11. Remove the number one engine mounting bracket. Remove the water inlet housing.
12. Remove the oil filter cap assembly. Remove the oil filter.
13. Remove the number two oil pan. Remove the oil strainer assembly. Remove the number one oil pan. Remove the seven retaining bolts and remove the oil pan baffle plate.

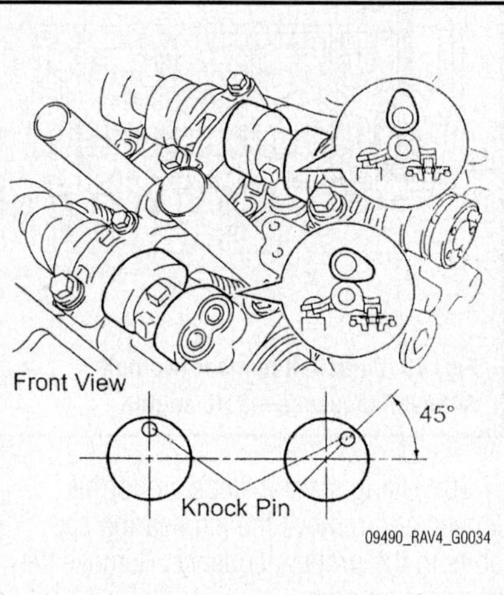

Fig. 48 Knock pin alignment (bank one)—3.5L engine

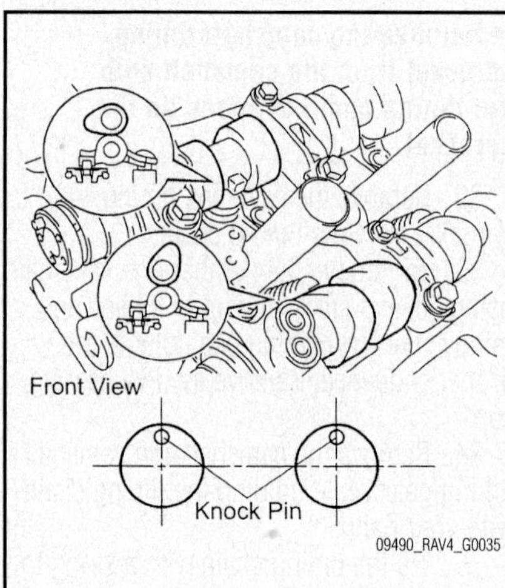

Fig. 49 Knock pin alignment (bank two)—3.5L engine

14. Remove the crankshaft pulley. Remove the water pump.
15. Remove the timing chain.
16. Remove the three camshaft gaskets (two in front, one in rear).
17. Make sure that the knock pin of the camshaft is positioned, as shown in the illustration.
18. Uniformly loosen and remove the eight bearing cap bolts in the proper sequence. Uniformly loosen and remove the twelve bearing cap bolts in the proper sequence. Remove the five bearing caps. Remove the camshafts.

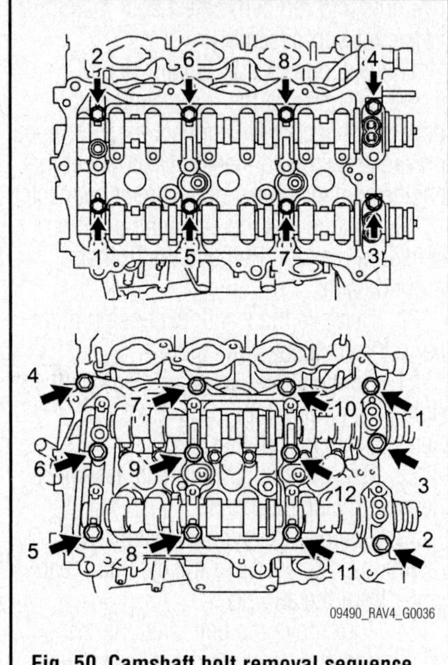

Fig. 50 Camshaft bolt removal sequence (bank one)—3.5L engine

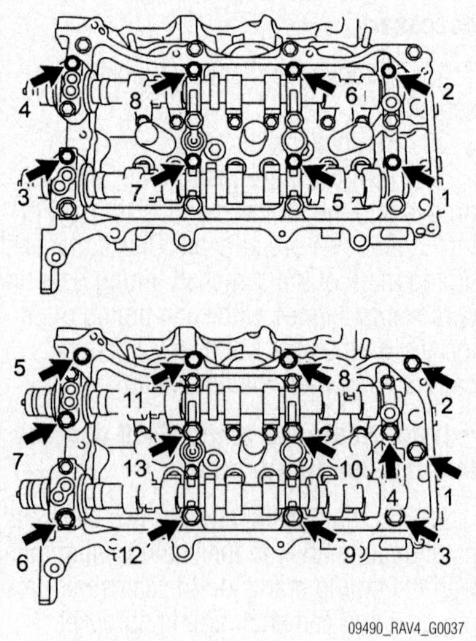

Fig. 51 Camshaft bolt removal sequence (bank two)—3.5L engine

19. If removing the lifters, remove the camshaft housing sub-assembly by prying between the cylinder head and camshaft housing with a suitable pry tool.

➡Be careful not to damage the contact surfaces of the cylinder head and the camshaft housing.

20. Remove the valve rocker arms. Remove the valve lash adjusters.

➡Be sure to keep parts arranged in a logical order for reinstallation in their exact locations.

To install:

21. Installation is the reverse of the removal procedure.
22. When installing the cylinder head cover, apply seal packing, three bond 1207B or equivalent, as shown in the illustration.

➡Remove any oil from the contact surfaces. Install the component within three minutes after applying the sealant. Do not start the engine for two hours after the installation.

23. Install the cover retaining bolts and torque to 15 ft. lbs for bolt "A" and to 7 ft. lbs. for all other bolts and in the proper sequence.

CRANKSHAFT FRONT SEAL

REMOVAL & INSTALLATION

2.4L Engine

1. Before servicing the vehicle, refer to the Precautions Section.
2. Properly relieve the fuel system pressure.
3. Disconnect the negative battery cable.

➡Wait at least 90 seconds after disconnecting the negative battery cable before starting any repair work to prevent air bag and seat belt pretensioner activation.

4. Remove the timing chain cover.
5. Remove the timing cover oil seal, from the cover using the proper tool.

To install:

6. Installation is the reverse of the removal procedure.

3.5L Engine

➡In order to replace the timing cover oil seal, the manufacturer recommends removing the engine from the vehicle.

1. Before servicing the vehicle, refer to the Precautions Section.
2. Properly relieve the fuel system pressure.
3. Disconnect the negative battery cable.

➡Wait at least 90 seconds after disconnecting the negative battery cable before starting any repair work to prevent air bag and seat belt pretensioner activation.

4. Remove the engine and position it in a suitable holding fixture.
5. Remove the timing chain cover.
6. Remove the timing cover oil seal, from the cover using the proper tool.

To install:

7. Installation is the reverse of the removal procedure.

CYLINDER HEAD

REMOVAL & INSTALLATION

2.4L Engine

See Figures 52 through 58.

1. Before servicing the vehicle, refer to the Precautions Section.
2. Properly relieve the fuel system pressure.
3. Disconnect the negative battery cable.

➡Wait at least 90 seconds after disconnecting the negative battery cable before starting any repair work to prevent air bag and seat belt pretensioner activation.

4. Remove the radiator support opening cover.
5. Remove the right front tire and wheel assembly.
6. Remove the number one engine undercover. Remove the front fender apron, right side. Remove the number one engine cover.
7. Drain the engine coolant. Drain the engine oil.
8. Remove the air cleaner assembly.
9. Disconnect the purge line hose from the throttle body. Disconnect the water bypass hoses from the throttle body. Disconnect the number one throttle body hose from the throttle body.
10. Disconnect the throttle position sensor and control motor connector. Disconnect the wire harness clamp. Disconnect the fuel tube from the clamp.
11. Remove the four throttle body retaining bolts and remove the throttle body assembly. Remove the gasket.
12. Remove the fuel delivery pipe sub-assembly.
13. Remove the intake manifold. Remove the intake manifold insulator.
14. Remove the front exhaust pipe. Remove the oil dipstick. Remove the oil dipstick guide.
15. Remove the manifold stay. Remove the number two manifold stay.
16. Remove the number one exhaust manifold heat insulator. Disconnect the air/fuel ratio sensor connector. Remove the five nuts and remove the exhaust manifold converter sub-assembly.
17. Disconnect the radiator hose.
18. Disconnect the radio setting condenser connector. Disconnect the engine oil pressure switch connector. Disconnect the engine coolant temperature sensor connector.
19. Disconnect the camshaft position sensor connector. Remove the bolt and ground cable.
20. Remove the front suspension member reinforcement, right side.
21. Remove the alternator. Remove the right side engine mounting insulator. Remove the idler pulley.
22. Remove the ignition coil assembly. Remove the spark plugs.
23. Remove the cylinder head cover retaining bolts. Remove the cylinder head cover from the engine.
24. To remove the belt tensioner assembly, lift the engine upward, using a transmission jack. Remove the bolt, nut and the belt tensioner assembly.

➡Do not lift the engine more than necessary.

25. Remove the crankshaft position sensor.
26. Remove the oil pan.
27. Turn the crankshaft pulley until its groove and the timing mark "0" of the timing chain cover are aligned. Check that each timing mark of the camshaft timing gear and sprocket is aligned with each timing mark located on the No. 1 and No. 2 bearing caps, as shown in the illustration.

➡If not rotate the engine 360 degrees to align the timing marks.

28. To remove the number two camshaft, paint matchmarks on the chain in alignment with the timing marks on the camshaft timing gear and camshaft timing sprocket.
29. Remove the two nuts, tensioner and gasket. While holding the camshaft with a wrench, loosen the camshaft timing set bolt.
30. Using several steps, uniformly loosen and remove the ten bearing cap bolts in the proper sequence. Remove the five bearing caps.
31. While holding the number two camshaft, remove the camshaft timing sprocket set bolt.

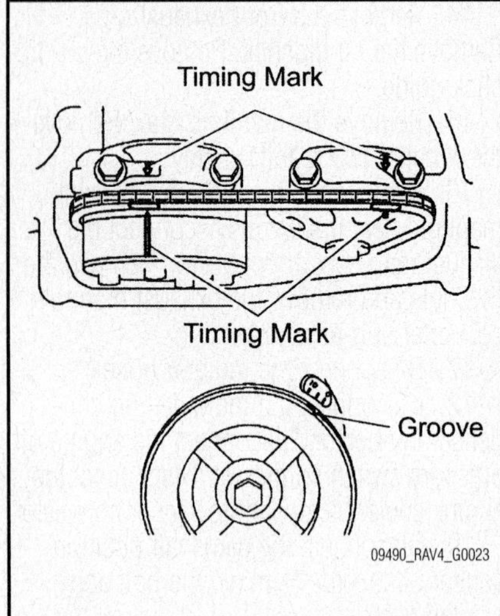

Fig. 52 Camshaft gear and sprocket alignment—2.4L engine

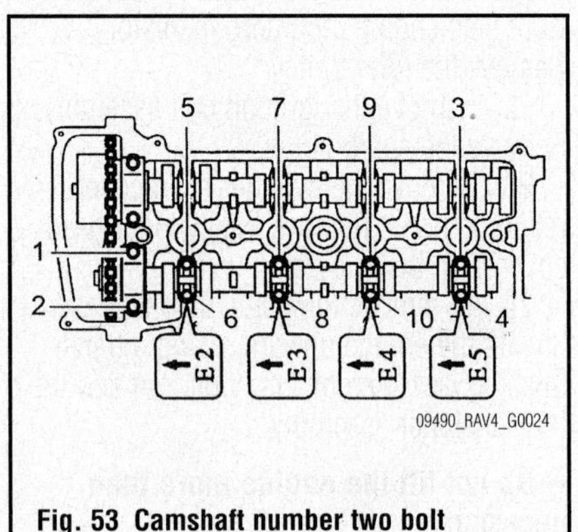

Fig. 53 Camshaft number two bolt removal sequence—2.4L engine

➡**Remove the camshaft timing sprocket from the camshaft with the timing chain wrapped on the sprocket.**

32. Remove the camshaft timing sprocket from the timing chain.
33. To remove the number one camshaft, using several steps uniformly loosen and remove the ten bearing cap bolts in the proper sequence. Remove the five bearing caps.
34. Remove the camshaft and camshaft timing gear while holding the timing chain with your hand.
35. Tie the timing chain with a string to the side of the engine block.

➡**Be careful not to drop anything inside the timing chain cover.**

36. Remove the crankshaft pulley. Remove the engine mounting bracket, right side.
37. Remove the timing chain cover. Remove the number one crankshaft position sensor plate. Remove the timing chain guide.

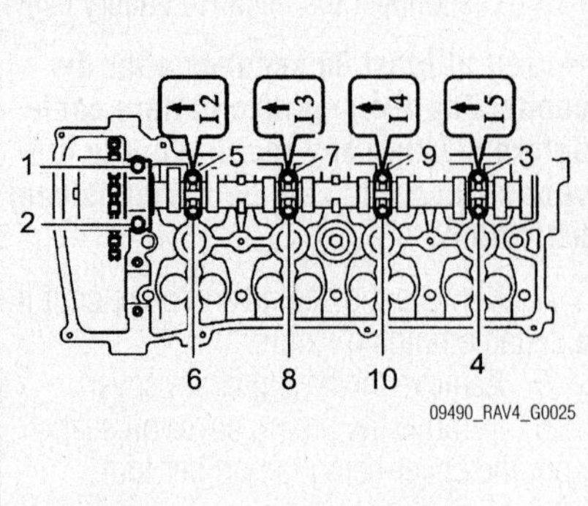

Fig. 54 Camshaft number one bolt removal sequence—2.4L engine

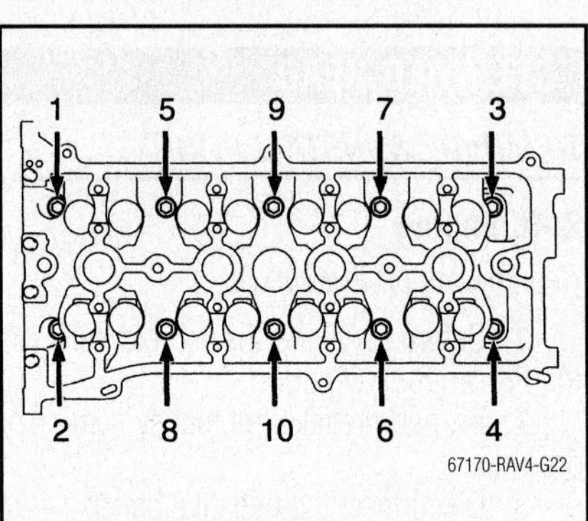

Fig. 55 Cylinder head bolt removal sequence—2.4L engine

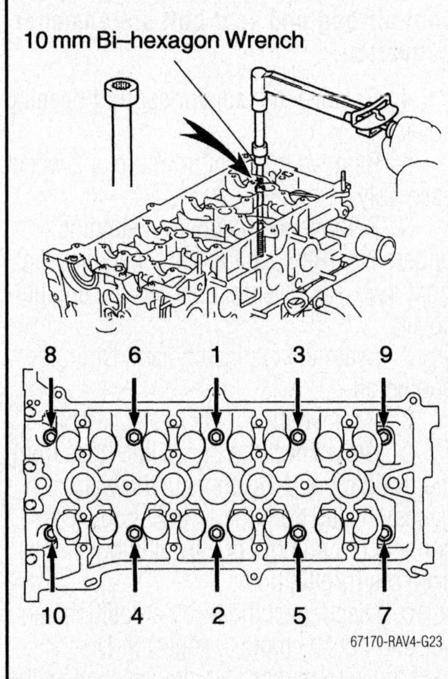

Fig. 56 Cylinder head bolt torque sequence—2.4L engine

38. Remove the chain tensioner slipper. Remove the number one chain vibration damper. Remove the chain sub-assembly.
39. Remove the camshaft timing oil control valve assembly.

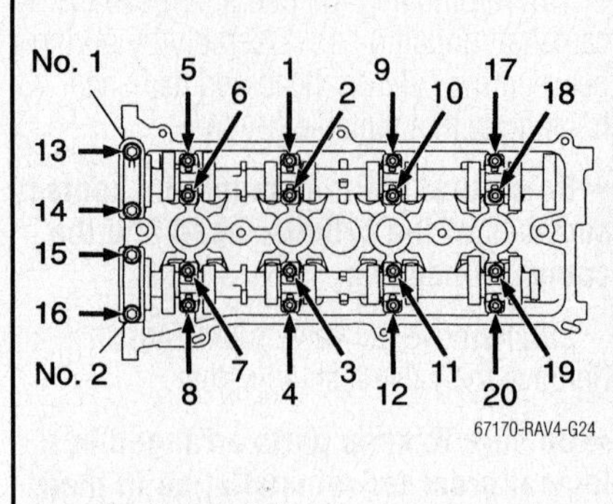

Fig. 57 Camshaft bolt torque sequence—2.4L engine

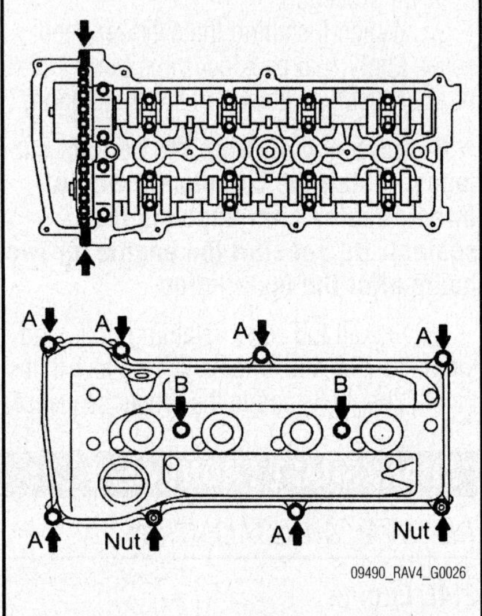

Fig. 58 Cylinder head cover sealant location and bolt identification—2.4L engine

40. Using several steps, uniformly loosen and remove the ten cylinder head bolts in the proper sequence.

➡**Head warpage or cracking could result from removing the bolts in the wrong order.**

41. Carefully remove the cylinder head from the engine. Remove and discard the head gasket.

To install:

42. Thoroughly clean all gasket surfaces.
43. Install a new gasket with the identification lot number upwards.
44. Carefully install the head.
45. Apply a light coating of engine oil to the threads and under the head of each bolt. Install the bolts and tighten them evenly and in several passes, in the sequence shown, to 52 ft. lbs. (70 Nm).
46. Matchmark the head of each bolt and the cylinder head. Tighten each bolt

in sequence an additional 90 degrees. Check that the paint mark is now at a 90 degree angle to the front.

47. Install the camshafts. Tighten the retaining bolts to 22 ft. lbs. for No.1 and No.2 bearing cap, and to 80 inch lbs. for the No. 3 bearing cap.

➡Be sure to install the camshafts with the timing mark of the camshaft timing gear on top.

48. When installing the throttle body be sure to use a new gasket. Tighten the retaining bolts to 22 ft. lbs.

49. When installing the cylinder head cover, apply seal packing (three bond 1207B, or equivalent) to the two locations, shown in the illustration. Tighten the retaining bolts to 8 ft. lbs. for bolt "A", 10 ft. lbs. for bolt "B" and 8 ft. lbs. for the nut.

50. Continue the installation in the reverse order of the removal procedure.

51. Be sure to fill the cooling system with the proper grade and type engine coolant.

52. Be sure to fill the engine with the proper grade and type engine oil.

53. Start the engine and check for leaks, correct as required.

3.5L Engine

See Figures 59 through 79.

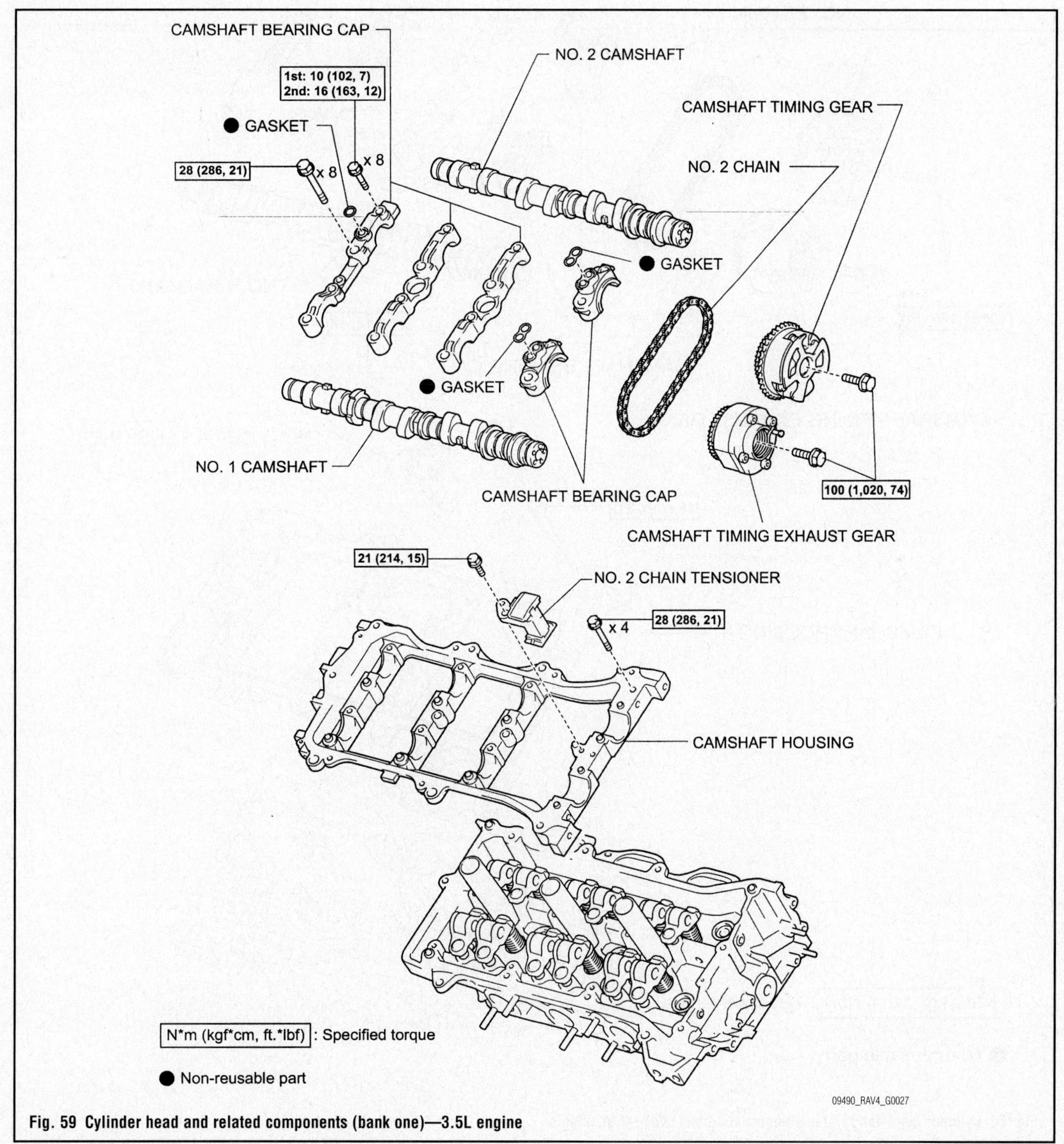

Fig. 59 Cylinder head and related components (bank one)—3.5L engine

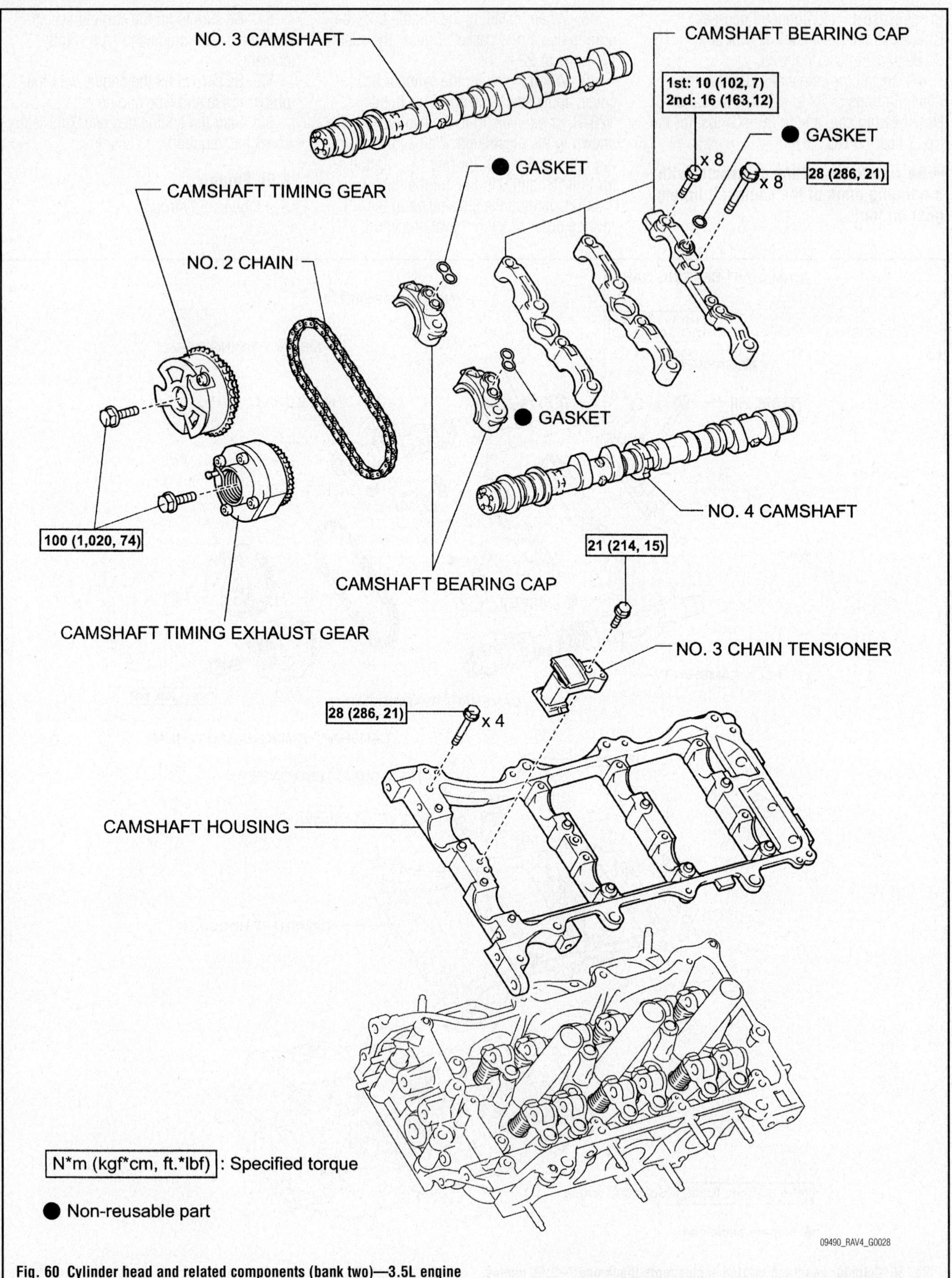

Fig. 60 Cylinder head and related components (bank two)—3.5L engine

➡In order to replace the cylinder head gasket, the manufacturer recommends removing the engine from the vehicle.

1. Before servicing the vehicle, refer to the Precautions Section.
2. Properly relieve the fuel system pressure.
3. Disconnect the negative battery cable.

➡Wait at least 90 seconds after disconnecting the negative battery cable before starting any repair work to prevent air bag and seat belt pretensioner activation.

4. Remove the engine and position it in a suitable holding fixture.
5. Remove the oil filler cap sub-assembly. Remove the spark plugs.
6. Remove the camshaft timing control valve assembly.
7. Remove the VVT sensor. Remove the water inlet. Remove the oil pipe.
8. Remove the twelve retaining bolts and remove the cylinder head cover for bank one.

➡Upon installation be sure that removed components are reinstalled in their original locations.

9. Remove the number one oil pipe.
10. Remove the twelve retaining bolts and remove the cylinder head cover for bank two.

➡Upon installation be sure that removed components are reinstalled in their original locations.

11. Remove the number one engine mounting bracket. Remove the water inlet housing.
12. Remove the oil filter cap assembly. Remove the oil filter.
13. Remove the number two oil pan. Remove the oil strainer assembly. Remove the number one oil pan. Remove the seven retaining bolts and remove the oil pan baffle plate.
14. Remove the crankshaft pulley. Remove the water pump.
15. Remove the timing chain.
16. Remove the camshafts.
17. Remove the camshaft housing sub-assembly by prying between the cylinder head and camshaft housing with a suitable pry tool.

➡Be careful not to damage the contact surfaces of the cylinder head and the camshaft housing.

18. Remove the valve rocker arms. Remove the valve lash adjusters.

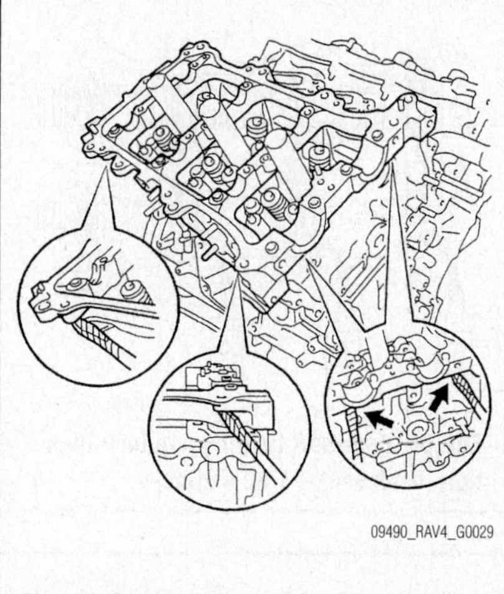

Fig. 61 Camshaft housing sub-assembly removal points—3.5L engine

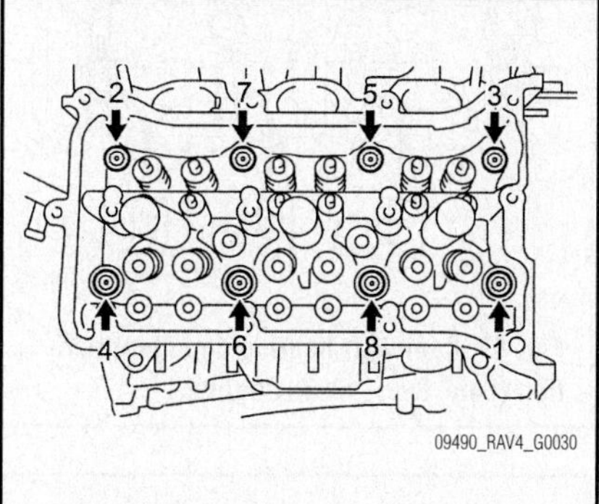

Fig. 62 Cylinder head bolt removal sequence (bank one)—3.5L engine

➡Be sure to keep parts arranged in a logical order for reinstallation in their exact locations.

19. Uniformly loosen the eight cylinder head retaining bolts for bank one in the proper sequence. Remove the bolts and washers. Remove the cylinder head from the engine and discard the gasket.

➡Be careful not to drop the washers into the cylinder head. Head warpage or cracking could occur from removing the bolts in an incorrect order. Be sure to keep all components labeled and in the correct order for reinstallation

20. To remove the cylinder head for bank two, first, uniformly loosen and remove the two bolts, as shown in the illustration. Then, uniformly loosen the eight cylinder head retaining bolts in the proper sequence. Remove the bolts and washers. Remove the cylinder head from the engine and discard the gasket.

➡Be careful not to drop the washers into the cylinder head. Head warpage or cracking could occur from removing the bolts in an incorrect order. Be sure to keep all components labeled and in the correct order for reinstallation

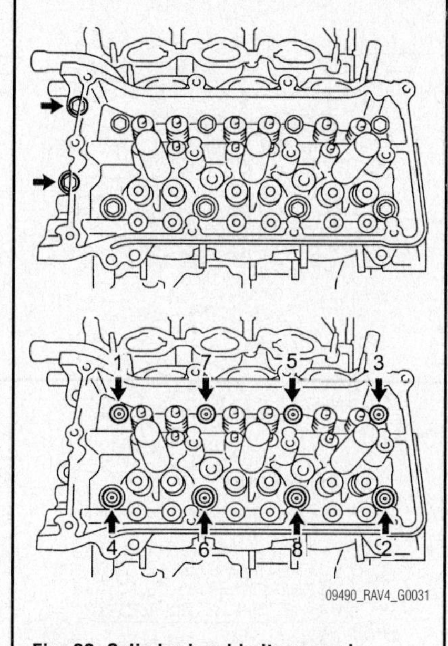

Fig. 63 Cylinder head bolt removal sequence (bank two)—3.5L engine

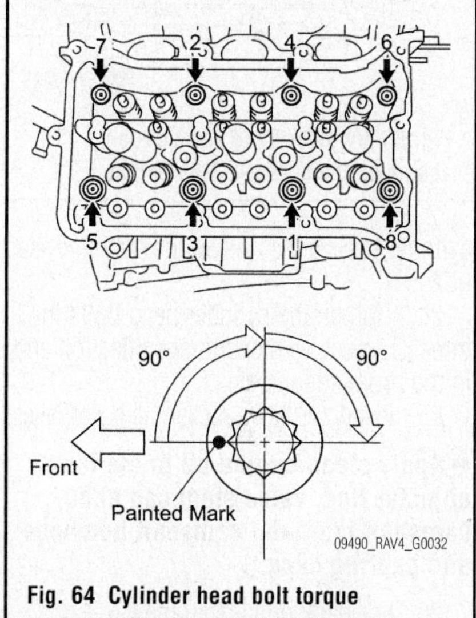

Fig. 64 Cylinder head bolt torque sequence (bank one)—3.5L engine

To install:

21. Position a new cylinder head gasket on the cylinder block. Be sure that the front face of the lot number, stamped on the gasket, is facing upward.
22. Position the cylinder head on the cylinder block.

➡Ensure that no oil is on the mounting surface of the cylinder head.

23. Apply a thin coat of clean engine oil to the threads and under the heads of the

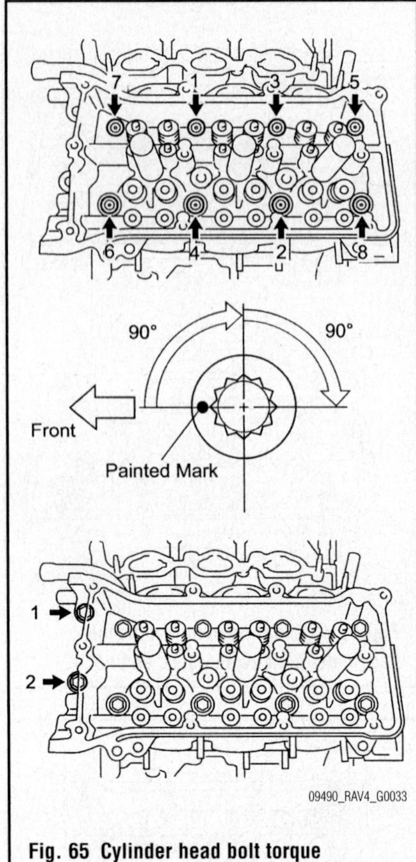

Fig. 65 Cylinder head bolt torque sequence (bank two)—3.5L engine

cylinder head bolts. Install the cylinder head bolts.

24. Tighten the cylinder head bolts in three progressive steps to specification and in the proper sequence.

25. Install the valve rocker subassemblies.

➡Apply clean engine oil to the lash adjuster tips, valve stem cap ends, camshaft journals, camshaft housings and bearing caps.

26. For bank one install the No. 1 camshaft and No 2 camshaft to the camshaft housing. Be sure the identification numbers on the camshaft bearing caps are in the proper position and direction. Temporarily install the eight bolts in the proper order. Torque to 7 ft. lbs.

27. For bank two install the No. 3 camshaft and No 4 camshaft to the camshaft housing. Be sure the identification numbers on the camshaft bearing caps are in the proper position and direction. Temporarily install the eight bolts in the proper order. Torque to 7 ft. lbs.

28. To install the camshaft housing subassembly for bank one, apply a continuous bead (0.138–0.177 inch) of three bond 1207B or equivalent as shown in the illustration.

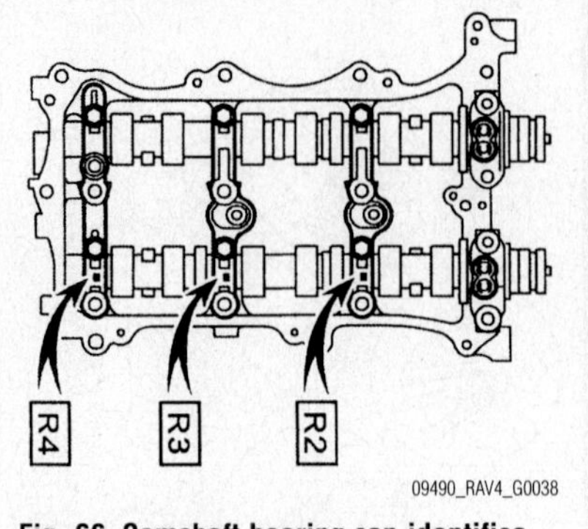

Fig. 66 Camshaft bearing cap identification (bank one)—3.5L engine

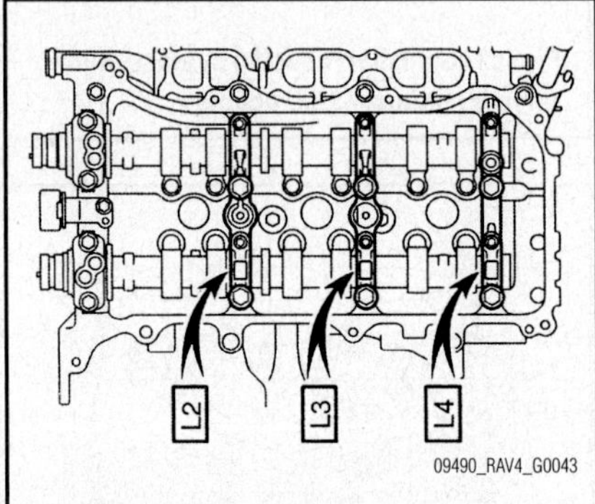

Fig. 67 Camshaft bearing cap identification (bank two)—3.5L engine

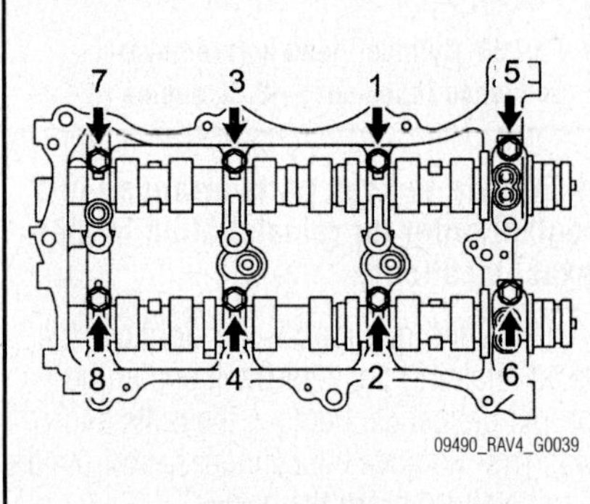

Fig. 68 Camshaft bearing cap bolt location (bank one)—3.5L engine

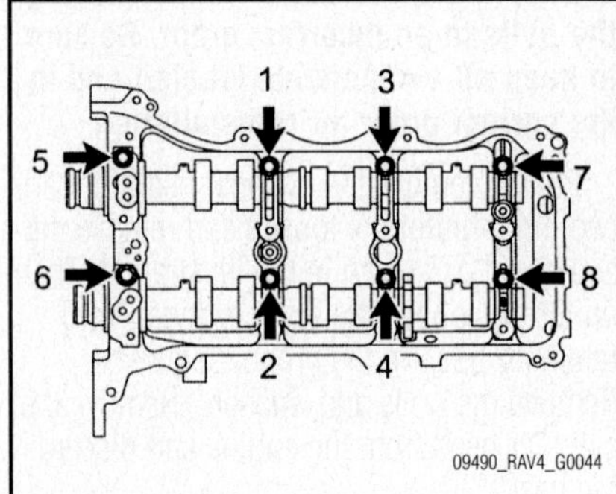

Fig. 69 Camshaft bearing cap bolt location (bank two)—3.5L engine

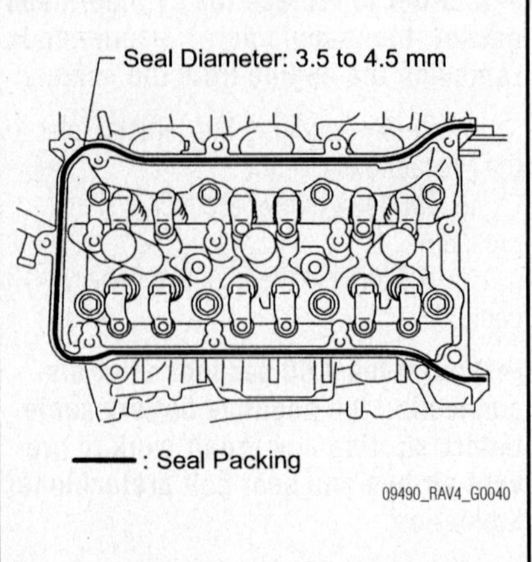

Fig. 70 Sealant application (bank one)—3.5L engine

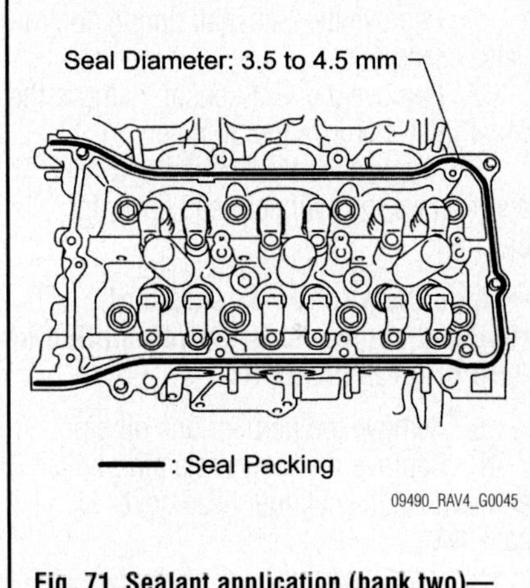

Fig. 71 Sealant application (bank two)—3.5L engine

➡Remove any oil from the contact surfaces. Install the component within three minutes after applying the sealant. Do not start the engine for two hours after the installation.

29. To install the camshaft housing subassembly for bank two, apply a continuous bead (0.138–0.177 inch) of three bond 1207B or equivalent as shown in the illustration.

➡Remove any oil from the contact surfaces. Install the component within three minutes after applying the sealant. Do not start the engine for two hours after the installation.

30. Install the camshaft housing and tighten the bolts to 21 ft. lbs and in the proper sequence.

➡Make sure that the knock pin of the camshaft is positioned as shown in the illustration, before installing the housing.

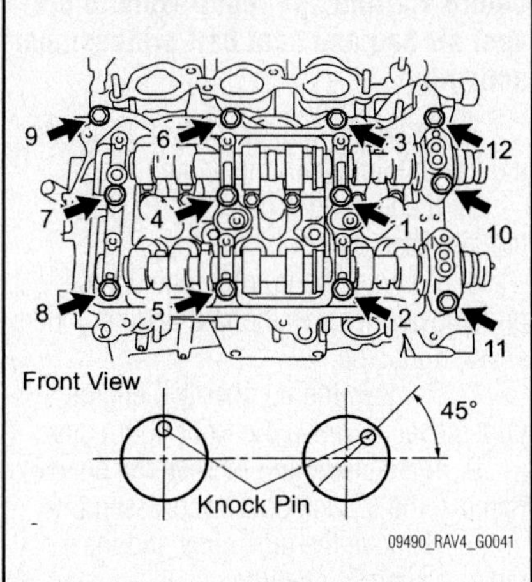

Fig. 72 Camshaft housing bolt torque sequence (bank one)—3.5L engine

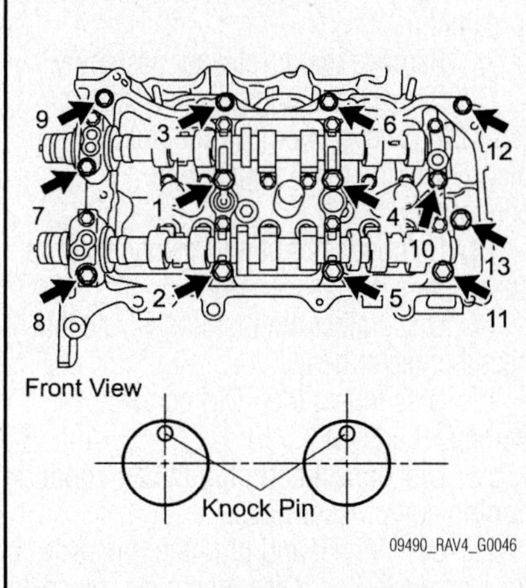

Fig. 73 Camshaft housing bolt torque sequence (bank two)—3.5L engine

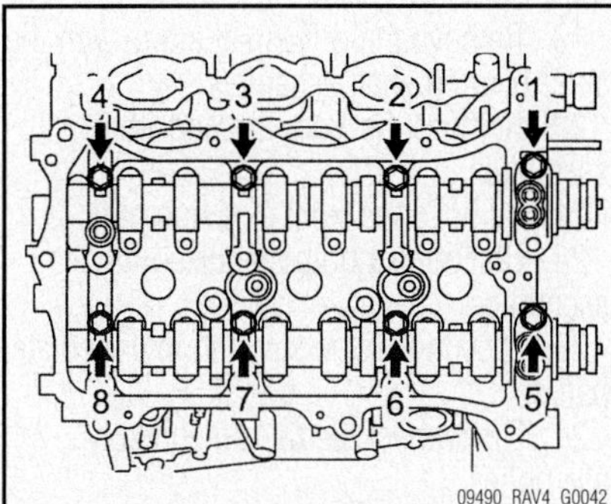

Fig. 74 Camshaft bearing cap bolt location (bank one)—3.5L engine

31. Tighten the eight bolts to 12 ft. lbs. and in the proper sequence. Be sure to clean any seal packing. Install three new gaskets.

32. Continue the installation in the reverse order of the removal procedure.

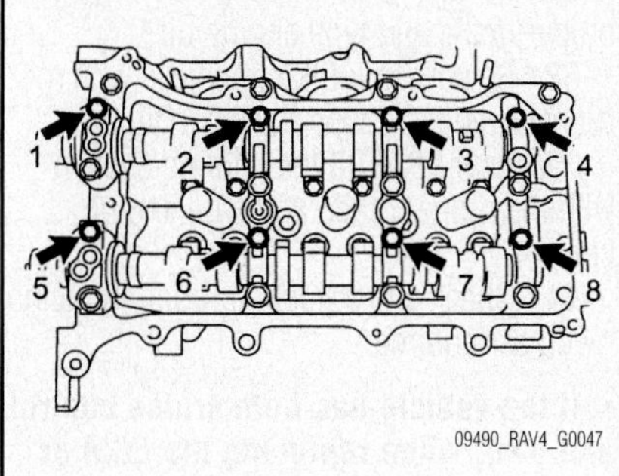

Fig. 75 Camshaft bearing cap bolt location (bank two)—3.5L engine

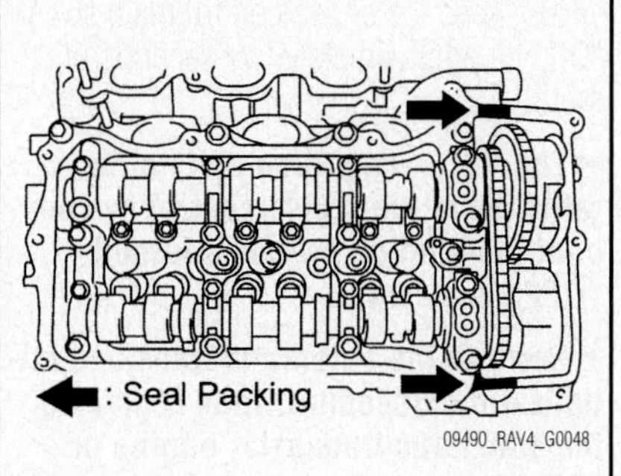

Fig. 76 Cylinder head cover seal packing application (bank one)—3.5L engine

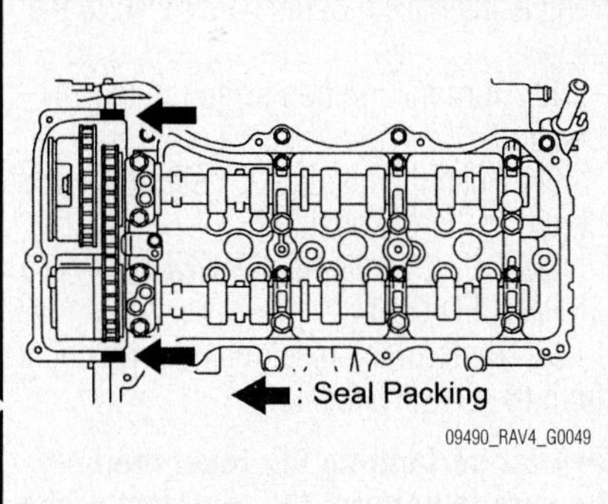

Fig. 77 Cylinder head cover seal packing application (bank two)—3.5L engine

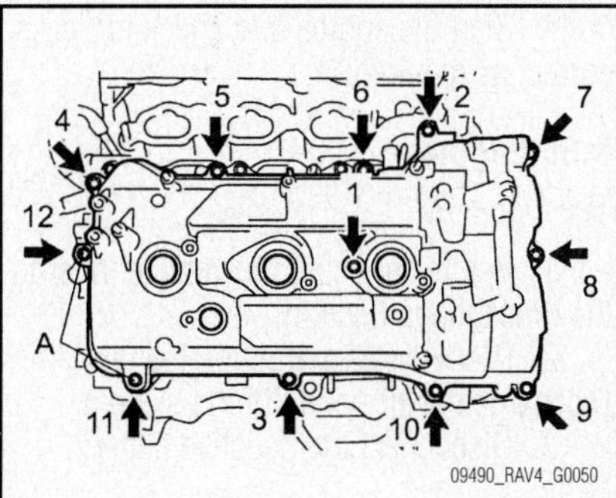

Fig. 78 Cylinder head cover bolt torque sequence (bank one)—3.5L engine

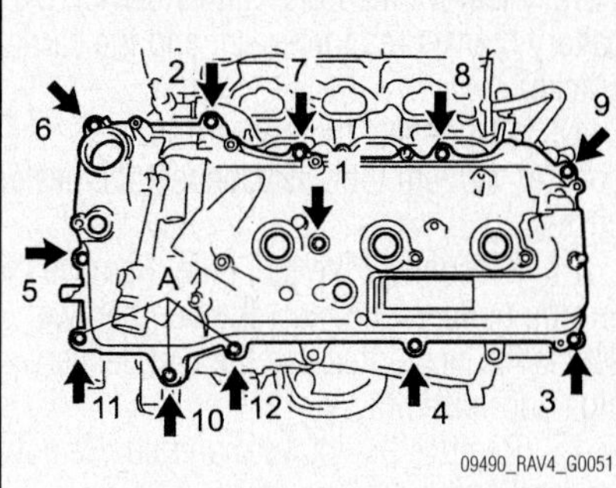

Fig. 79 Cylinder head cover bolt torque sequence (bank two)—3.5L engine

33. When installing the cylinder head cover, apply seal packing, three bond 1207B or equivalent, as shown in the illustration.

➡Remove any oil from the contact surfaces. Install the component within three minutes after applying the sealant. Do not start the engine for two hours after the installation.

34. Install the cover retaining bolts and torque to 15 ft. lbs for bolt "A" and to 7 ft. lbs. for all other bolts and in the proper sequence.

ENGINE ASSEMBLY

REMOVAL & INSTALLATION

2.4L Engine

1. Before servicing the vehicle, refer to the Precautions Section.
2. Properly discharge the fuel system.
3. Disconnect the negative battery cable.

➡Wait at least 90 seconds after disconnecting the negative battery cable before starting any repair work to prevent air bag and seat belt pretensioner activation.

4. Drain the engine coolant. Drain the transaxle fluid. Drain the engine oil.
5. Matchmark and remove the hood. Remove the radiator support opening cover.
6. Disconnect the positive battery cable. Remove the battery. Remove the battery tray.
7. Remove the front tire and wheel assemblies.
8. Remove the number one and number two engine undercovers.
9. Remove the front fender apron, right side.
10. Remove the radiator. Remove the radiator reservoir.
11. Remove the purge VSV assembly. Remove the air cleaner assembly.

12. Remove the front battery carrier, the battery bracket reinforcement and the carrier bracket.

13. Disconnect the hoses from the booster vacuum tube. Disconnect the heater hoses.

14. Disconnect the fuel tube. Remove the throttle body. Disconnect the number two ventilation hose. Remove the fuel delivery pipe sub-assembly.

15. Remove the intake manifold. Remove the intake manifold insulator.

16. Remove the transaxle control cable assembly.

17. Remove the front suspension member reinforcement, right side.

18. Remove the drive belt. Remove the fan assembly.

19. Remove the air conditioning compressor from its mounting and position it to the side. Do not disconnect the refrigerant lines.

20. Remove the ECM.

21. Remove the engine room relay block cover. Remove the nut and disconnect the two engine wire connectors. Remove the bolt and engine wire cover.

22. Disconnect the engine wire from the engine wire cover. Disconnect the ground cable. Remove the nut from the positive battery terminal to disconnect the engine wire.

23. Disconnect the ground cable from the clamp located near the starter. Remove the bolt and ground cable.

24. Remove the front exhaust pipe assembly. Remove the front halfshaft assembly.

25. Remove the driveshaft, if equipped with 4WD.

26. Remove the front and rear engine mounting insulators.

27. Attach a suitable engine removal fixture to the engine using engine hanger tools 12281-28010 and 12282-28010 or equivalent.

➡**Do not attempt to hang the engine by hooking the chain to any other part.**

28. Carefully remove the engine assembly out of the vehicle.

➡**Be sure the assembly is clear of all wiring, hoses and cables before removing it.**

29. Separate the engine and transaxle, as required.

To install:

30. Installation is the reverse of the removal procedure.

31. Be sure to fill the engine with the proper grade and type engine oil.

32. Be sure to fill the transaxle with the proper grade and type transaxle fluid.

33. Be sure to fill the cooling system with the proper grade and type engine coolant.

34. Check and adjust the front end alignment, as required.

➡**If the vehicle has both cruise control and VSC, when replacing the ECM or the engine perform VSC recognition.**

35. Turn the ignition switch "ON".

36. After waiting five seconds, turn the cruise control main switch "ON".

37. Keep the cruise control main switch "ON" for approximately five seconds or more.

➡**The VSC recognition will not be reflected in the cruise control system until the ignition switch is turned "OFF" and "ON".**

➡**Perform the automatic transaxle initialization procedure when replacing the automatic transaxle, engine or ECM.**

38. Turn the ignition switch OFF.

39. Connect the intelligent tester tool together with the controller area network vehicle interface module (CAN VIN) to the DLC3.

40. Turn the ignition switch to the ON position.

41. Push the intelligent tester tool main switch to the ON position.

42. Select the items, DIAGNOSIS/ ENHANCED OBD II.

43. Perform the reset memory procedure from the ENGINE menu.

➡**After performing the reset memory, be sure to perform the road test procedure. For road test procedure information, refer to the intelligent tester instruction manual.**

44. Start the engine and check for leaks, correct as required.

3.5L Engine

See Figure 80.

1. Before servicing the vehicle, refer to the Precautions Section.

2. Properly discharge the fuel system. Discharge the air conditioning system.

3. Disconnect the negative battery cable.

➡**Wait at least 90 seconds after disconnecting the negative battery cable before starting any repair work to prevent air bag and seat belt pretensioner activation.**

4. Drain the engine coolant. Drain the transaxle fluid. Drain the engine oil.

5. If equipped with 4WD, drain the transfer case oil.

6. Remove the number one engine undercover. Remove the rear engine undercover, right side.

7. Remove the number two engine undercover. Remove the front floor cover.

8. Matchmark and remove the hood. Remove the V bank cover sub-assembly.

9. Remove the nine clips and the radiator support opening cover.

10. Disconnect the positive battery cable. Remove the battery. Remove the battery tray. Remove the bolts and the battery bracket. Remove the battery bracket reinforcement.

11. Remove the air cleaner assembly.

12. Remove the engine room upper relay block cover. Remove the nut and disconnect the three engine wire connectors.

13. Remove the nut and disconnect the starter wire.

14. Disconnect the fuel hoses. Disconnect the neater hoses.

15. Disconnect the ECM connectors. Remove the ECM.

16. Disconnect and plug the air conditioning compressor lines.

17. Disconnect and plug the transaxle fluid cooler hoses. Disconnect the transaxle control cable assembly.

18. Disconnect the front exhaust pipe assembly. Disconnect the center exhaust pipe assembly.

19. Remove the driveshaft along with the center bearing shaft assembly.

20. Remove the front tire and wheel assemblies.

21. Remove the front axle hub nuts.

22. Disconnect both stabilizer link assemblies.

23. Disconnect the steering intermediate shaft.

24. Disconnect the tie rod end sub-assemblies.

25. Remove the halfshafts.

26. Attach a suitable engine removal fixture to the engine using engine hanger tools 12281-31120 and 12282-31100 or equivalent.

➡**Do not raise the engine more than necessary. If the engine is raised excessively, the vehicle may also be lifted up.**

27. Position an engine lifter underneath the engine.

28. Remove the column hole cover. Loosen the bolt holding the intermediate shaft and slide the intermediate shaft. Matchmark the intermediate shaft and pinion.

29. Remove the two bolts and nuts, and disconnect the engine mounting insulator, right side.

30. Remove the two bolts and nuts, and disconnect the engine mounting insulator, left side.

31. Remove the six bolts and the front suspension member brace, right and left sides.

32. Remove the six bolts, crossmember and suspension member.

33. Using a chain block, slowly remove the engine assembly from the vehicle and the intermediate shaft from the pinion. Position the assembly in a suitable holding fixture.

➡Make sure that the engine is clear of all wiring, hoses and cables before removing it. While lowering the assembly do not allow it to come in contact with the vehicle.

34. Separate the engine, transaxle, front suspension crossmember assembly (including steering gear) and transfer case, as required.

To install:

35. Installation is the reverse of the removal procedure.

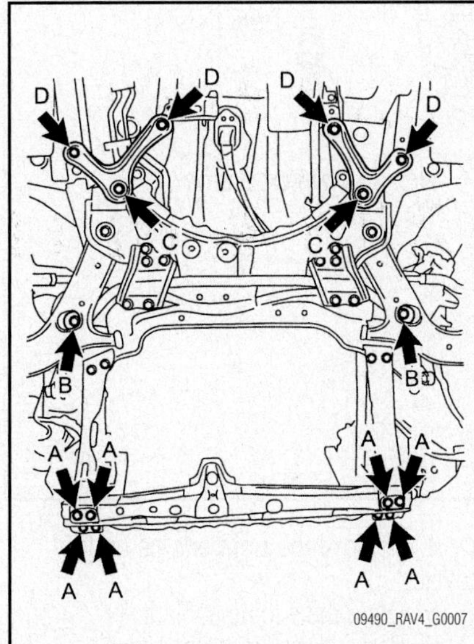

Fig. 80 Suspension crossmember bolt locations—3.5L engine

36. Tighten the suspension member and crossmember bolts to 71 ft. lbs. for bolt "A", 107 ft. lbs. for bolt "B".

37. Tighten the member brace rear bolts to 107 ft. lbs. for bolt "C" and 69 ft. lbs. for bolt "D".

38. Be sure to fill the engine with the proper grade and type engine oil.

39. Be sure to fill the transaxle with the proper grade and type transaxle fluid.

40. Be sure to fill the cooling system with the proper grade and type engine coolant.

41. Recharge the air conditioning system.

42. Check and adjust the front end alignment, as required.

➡If the vehicle has both cruise control and VSC, when replacing the ECM or the engine perform VSC recognition.

43. Turn the ignition switch "ON".

44. After waiting five seconds, turn the cruise control main switch "ON".

45. Keep the cruise control main switch "ON" for approximately five seconds or more.

➡The VSC recognition will not be reflected in the cruise control system until the ignition switch is turned "OFF" and "ON".

➡Perform the automatic transaxle initialization procedure when replacing the automatic transaxle, engine or ECM.

46. Turn the ignition switch OFF.

47. Connect the intelligent tester tool together with the controller area network vehicle interface module (CAN VIN) to the DLC3.

48. Turn the ignition switch to the ON position.

49. Push the intelligent tester tool main switch to the ON position.

50. Select the items, DIAGNOSIS/ ENHANCED OBD II.

51. Perform the reset memory procedure from the ENGINE menu.

➡After performing the reset memory, be sure to perform the road test procedure. For road test procedure information, refer to the intelligent tester instruction manual.

52. Start the engine and check for leaks, correct as required.

EXHAUST MANIFOLD

REMOVAL & INSTALLATION

2.4L Engine

See Figure 81.

1. Before servicing the vehicle, refer to the Precautions Section.

2. Disconnect the negative battery cable.

➡Wait at least 90 seconds after disconnecting the negative battery cable before starting any repair work to prevent air bag and seat belt pretensioner activation.

3. Remove the necessary components in order to gain access to the exhaust manifold retaining bolts.

4. Disconnect the front exhaust pipe retaining bolts.

5. Remove the exhaust manifold retaining bolts. Remove the exhaust manifold from the engine.

To install:

6. Thoroughly clean all gasket surfaces.

7. Install the exhaust manifold using new gaskets. Install the exhaust manifold nuts and torque them to specification.

3.5L Engine

➡In order to replace the exhaust manifold, the manufacturer recommends removing the engine from the vehicle.

1. Before servicing the vehicle, refer to the Precautions Section.

2. Disconnect the negative battery cable.

➡Wait at least 90 seconds after disconnecting the negative battery cable before starting any repair work to prevent air bag and seat belt pretensioner activation.

3. Remove the necessary components in order to gain access to the exhaust manifold retaining bolts.

4. Disconnect the front exhaust pipe retaining bolts.

5. Remove the exhaust manifold retaining bolts. Remove the exhaust manifold from the engine.

To install:

6. Thoroughly clean all gasket surfaces.

7. Install the exhaust manifold using new gaskets. Install the exhaust manifold nuts and torque them to specification.

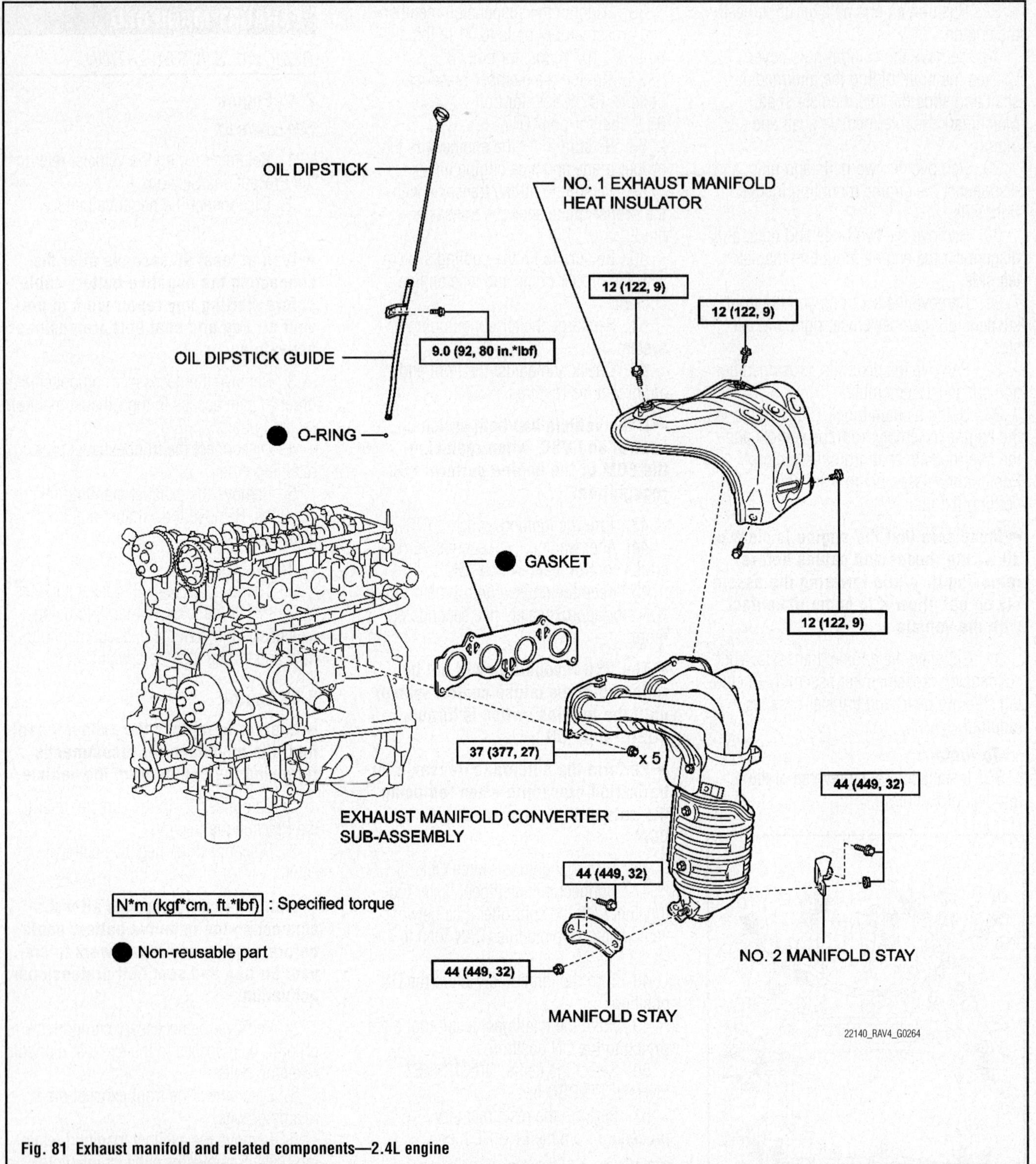

Fig. 81 Exhaust manifold and related components—2.4L engine

INTAKE MANIFOLD

REMOVAL & INSTALLATION

2.4L Engine

See Figures 82 and 83.

1. Before servicing the vehicle, refer to the Precautions Section.
2. Properly discharge the fuel system pressure.
3. Disconnect the negative battery cable.

➡Wait at least 90 seconds after disconnecting the negative battery cable before starting any repair work to prevent air bag and seat belt pretensioner activation.

4. Remove the right engine under-cover.
5. Drain the engine coolant.
6. Remove the air cleaner cap.
7. Remove the throttle body assembly.
8. Remove the fuel delivery pipe. Remove the fuel tube.
9. Remove the heater water inlet hose. Remove the heater water outlet hose.

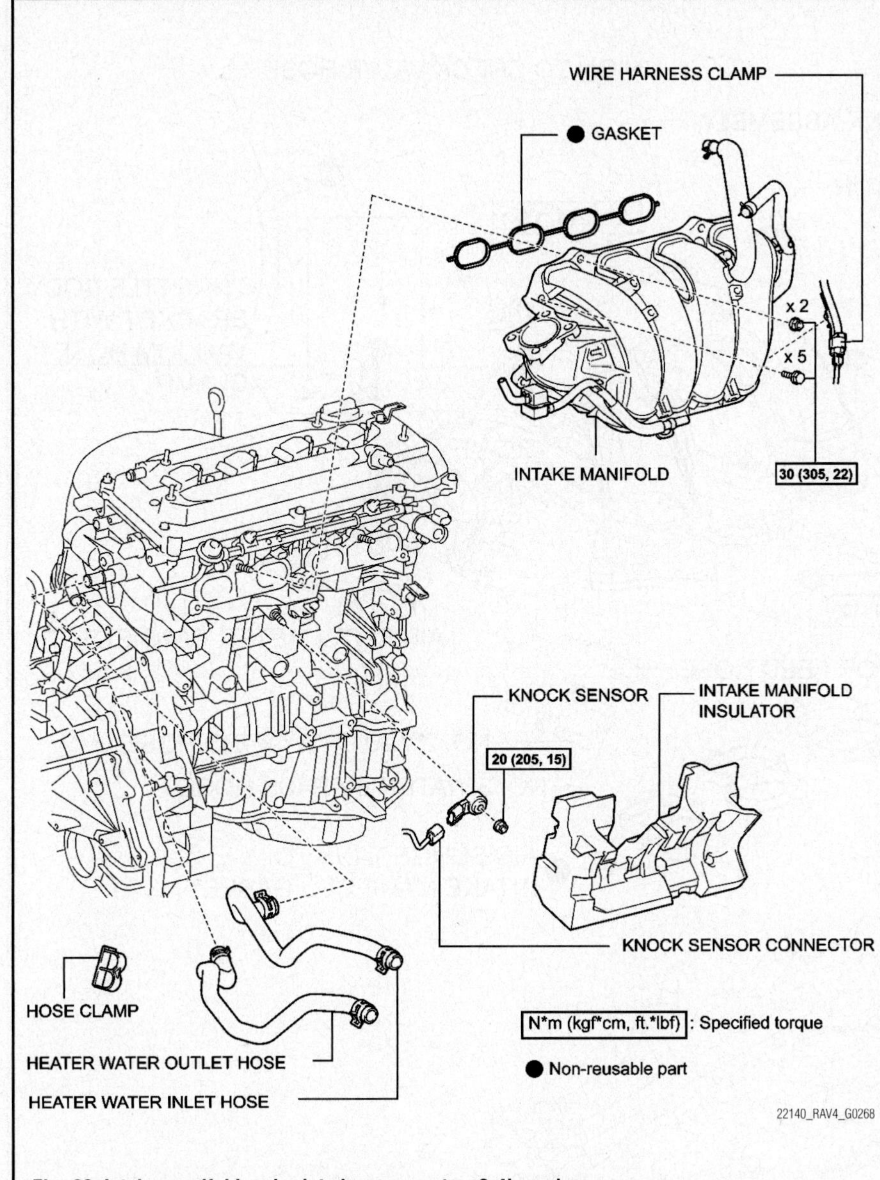

Fig. 82 Intake manifold and related components—2.4L engine

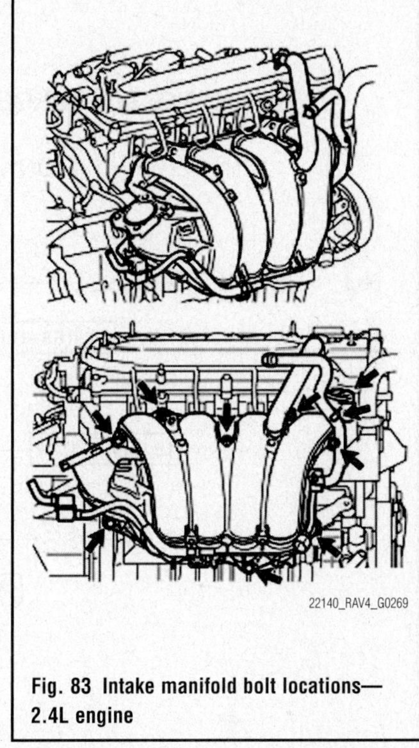

Fig. 83 Intake manifold bolt locations—2.4L engine

10. Disconnect the union to check valve hose from the power brake booster.
11. Disconnect the camshaft timing oil control valve connector. Remove the wire harness clamp.
12. Remove the union to check valve hose from the vacuum hose clamp.
13. Remove the five intake manifold retaining bolts. Remove the intake manifold from the engine. Discard the gasket.
14. As required, remove the intake manifold insulator assembly from the cylinder block.

To install:

15. Thoroughly clean all gasket surfaces.
16. Install the intake manifold, using a new gasket.
17. Tighten the nuts and bolts evenly and in several passes, to 22 ft. lbs. (30 Nm).
18. Continue the installation in the reverse order of the removal procedure.
19. Be sure to fill the engine with the proper grade and type engine coolant.
20. Start the engine and check for leaks, correct as required.

3.5L Engine

See Figures 84 and 85.

➡In order to replace the intake manifold, the manufacturer recommends removing the engine from the vehicle.

1. Before servicing the vehicle, refer to the Precautions Section.
2. Disconnect the negative battery cable.

➡Wait at least 90 seconds after disconnecting the negative battery cable before starting any repair work to prevent air bag and seat belt pretensioner activation.

3. Disconnect the two water by pass hoses from the throttle body.
4. Disconnect the vapor feed hose. Disconnect the throttle body connector and clamp.
5. Disconnect the number two ventilation hose. Disconnect the union to check valve hose.
6. Remove the bolt and the vacuum hose clamp. Disconnect the connector.
7. Remove the intake air surge tank retaining bolts. Remove the two nuts, two bolts and the assembly from the engine. Discard the gaskets.
8. Remove the fuel rail assembly.
9. Remove the intake manifold retaining bolts. Remove the intake manifold from the engine.

To install:

10. Thoroughly clean all gasket surfaces. Be sure to align the port holes of the gasket and cylinder head. Be careful of the installation direction.
11. Install the intake manifold, using a new gasket.
12. Tighten the nuts and bolts evenly and in several passes, to specification.
13. Install the intake air surge tank assembly, using new gaskets. Tighten the retaining bolts to 15 ft. lbs. and the nuts to 12 ft. lbs.

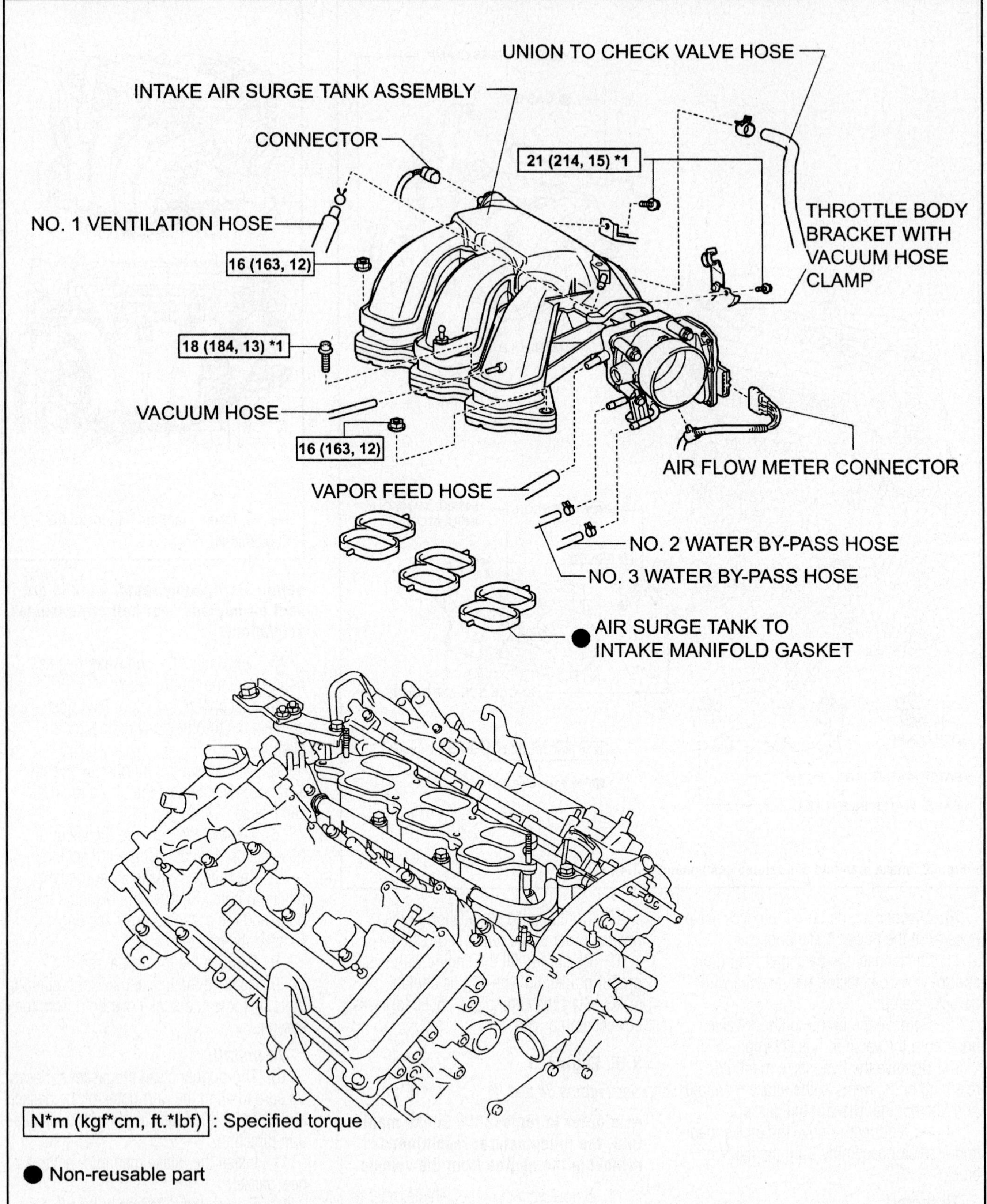

Fig. 84 Intake manifold and related components—3.5L engine

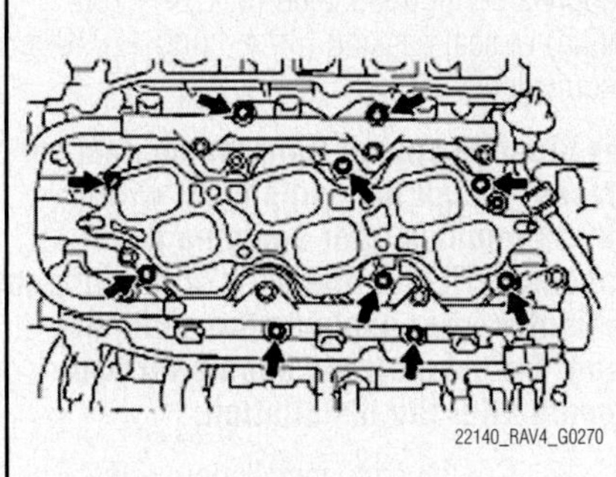

Fig. 85 Intake manifold bolt locations—3.5L engine

➡Do not allow oil to contact the retaining bolts and nuts.

14. Continue the installation in the reverse order of the removal procedure.

OIL PAN

REMOVAL & INSTALLATION

2.4L Engine

See Figures 86 and 87.

1. Before servicing the vehicle, refer to the Precautions Section.
2. Disconnect the negative battery cable.

➡Wait at least 90 seconds after disconnecting the negative battery cable before starting any repair work to prevent air bag and seat belt pretensioner activation.

3. Remove the engine undercover. Drain the engine oil. Remove the oil filter.
4. Install engine hanger tool 12281-28010 and 12282-28010. Install the engine lifting fixture to the engine hanger tools.
5. Remove the twelve retaining bolts and two nuts.
6. Insert the blade of tool SST09032-00100 between the crankcase, chain cover and the oil pan, cut off the applied sealer and remove the oil pan from the engine.

To install:

7. Clean all gasket surfaces completely.
8. Apply a continuous bead (0.118–0.157 inch) of seal packing (three bond 1207B or equivalent).

To install:

9. Clean all gasket surfaces completely.
10. Install the retaining bolts and nuts. Tighten the retaining bolts to 80 inch lbs. and in the proper sequence.
11. Continue the installation in the reverse order of the removal procedure.

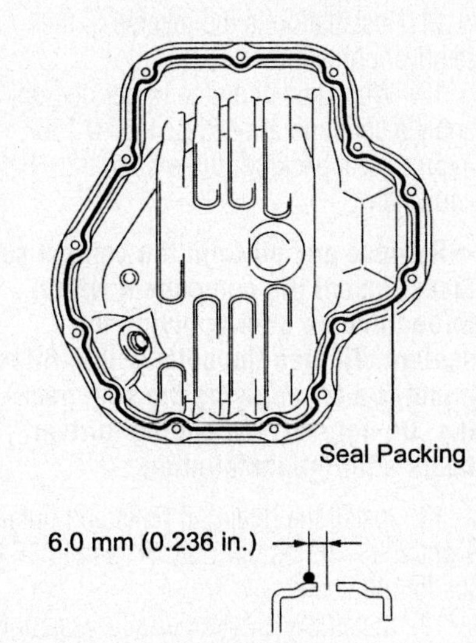

Fig. 86 Oil pan sealant application—2.4L engine

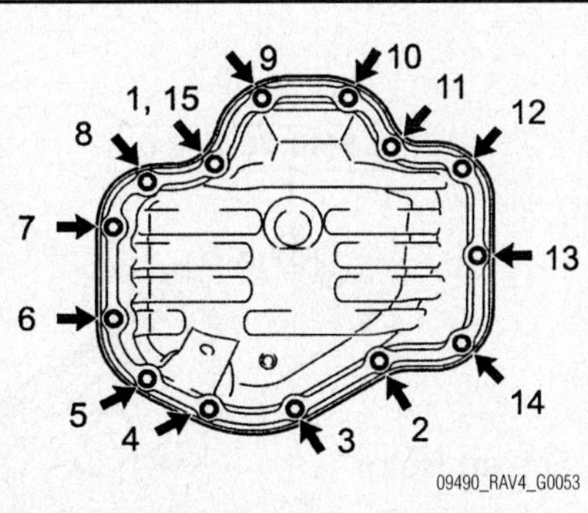

Fig. 87 Oil pan bolt torque sequence—2.4L engine

3.5L Engine

See Figures 88 through 91.

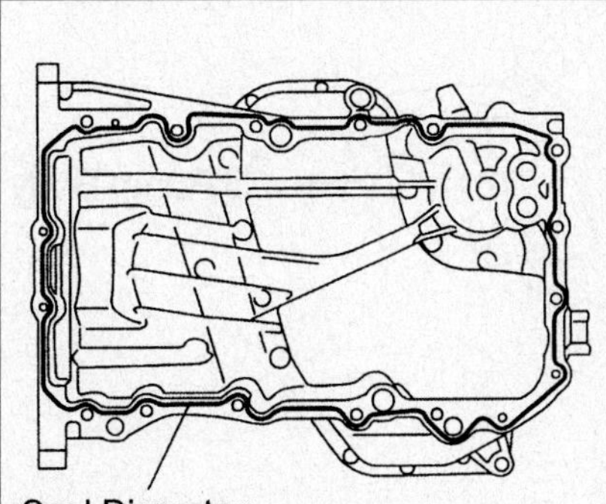

Fig. 88 Upper oil pan sealant application—3.5L engine

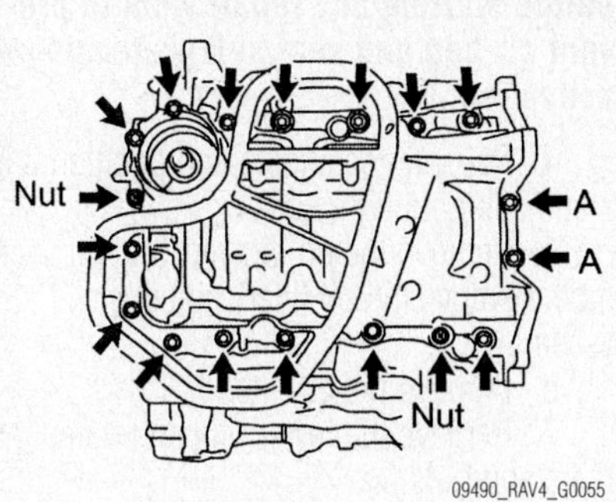

Fig. 89 Upper oil pan bolt torque sequence—3.5L engine

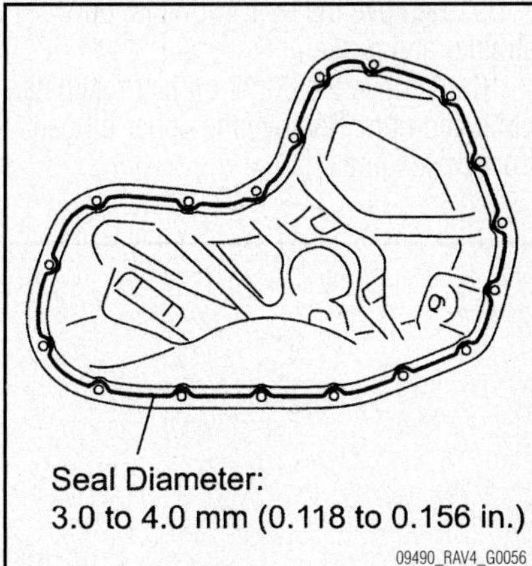

Fig. 90 Lower oil pan sealant application—3.5L engine

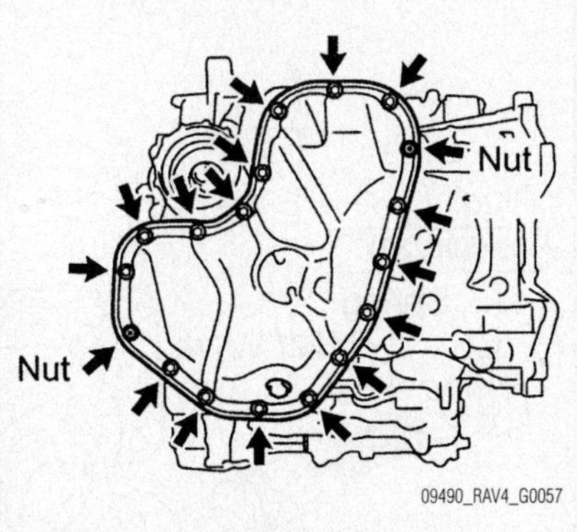

Fig. 91 Lower oil pan bolt torque sequence—3.5L engine

➡In order to replace the engine oil pan, the manufacturer recommends removing the engine from the vehicle.

1. Before servicing the vehicle, refer to the Precautions Section.
2. Properly relieve the fuel system pressure.
3. Disconnect the negative battery cable.

➡Wait at least 90 seconds after disconnecting the negative battery cable

before starting any repair work to prevent air bag and seat belt pretensioner activation.

4. Remove the engine and position it in a suitable holding fixture.
5. Remove the oil dipstick and dipstick tube. Remove the oil filler cap sub-assembly.
6. Remove the oil filter.
7. Remove the lower oil pan retaining bolts and nuts.
8. Insert the blade of tool SST09032-00100 at the oil pan, cut off the applied sealer and remove the lower oil pan from the engine.
9. Remove the bolt, two nuts, oil strainer and gasket.
10. Remove the upper oil pan retaining bolts and nuts. Remove the upper oil pan from the engine.

To install:

11. Installation is the reverse of the removal procedure.
12. When installing the upper oil pan, apply a continuous bead (0.118–0.156 inch) of seal packing (three bond 1207B or equivalent).

➡Remove any oil from the contact surfaces. Install the component within three minutes after applying the sealant. Tighten the bolts within fifteen minutes after applying the seal packing. Do not start the engine for two hours after the installation.

13. Install the retaining bolts and nuts. Tighten to 7 ft. lbs. for bolt "A" and to 15 ft. lbs for all others.
14. Install the oil strainer sub-assembly. Tighten the retaining bolt to 7 ft. lbs.
15. When installing the lower oil pan, apply a continuous bead (0.118–0.156 inch) of seal packing (three bond 1207B or equivalent).

➡Remove any oil from the contact surfaces. Install the component within three minutes after applying the sealant. Tighten the bolts within fifteen minutes after applying the seal packing. Do not start the engine for two hours after the installation.

16. Continue the installation in the reverse order of the removal procedure.

OIL PUMP

REMOVAL & INSTALLATION

2.4L Engine

See Figure 92.

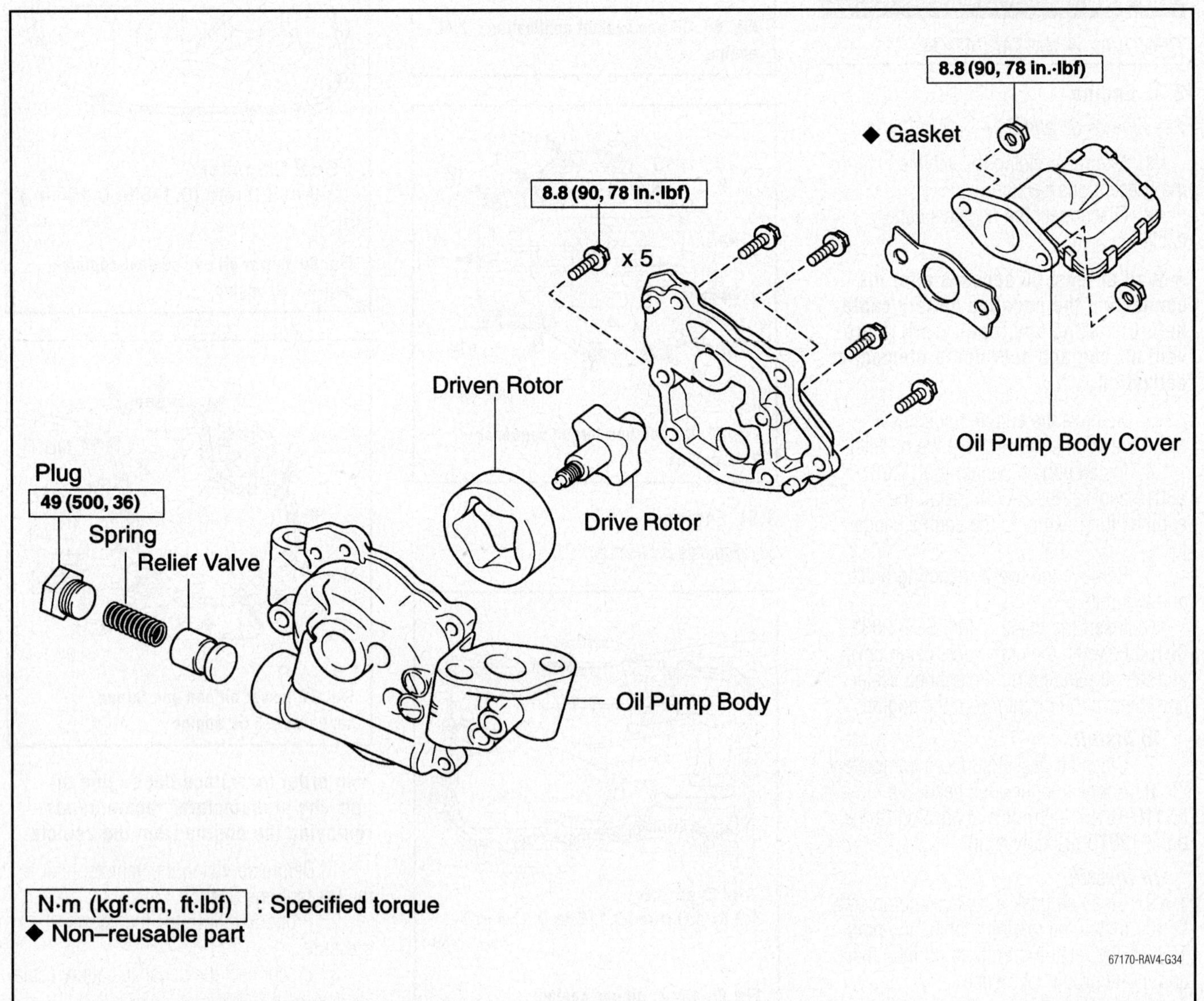

Fig. 92 Oil Pump and related components—2.4L engine

1. Before servicing the vehicle, refer to the Precautions Section.
2. Disconnect the negative battery cable.

➡**Wait at least 90 seconds after disconnecting the negative battery cable before starting any repair work to prevent air bag and seat belt pretensioner activation.**

3. Remove the timing chain sub-assembly.
4. Remove the three oil pump retaining bolts.
5. Remove the oil pump from its mounting. Discard the gasket.

To install:

6. Install the pump with a new gasket. Torque to 14 ft. lbs. (19 Nm).
7. Continue the installation in the reverse order of the removal procedure.

3.5L Engine

See Figures 93 through 98.

➡**In order to replace the engine oil pump, the manufacturer recommends removing the engine from the vehicle.**

1. Before servicing the vehicle, refer to the Precautions Section.
2. Properly relieve the fuel system pressure.
3. Disconnect the negative battery cable.

➡**Wait at least 90 seconds after disconnecting the negative battery cable before starting any repair work to prevent air bag and seat belt pretensioner activation.**

4. Remove the engine and position it in a suitable holding fixture.
5. Remove the lower oil pan. Remove the oil strainer sub-assembly. Remove the upper oil pan.
6. Remove the two O-rings from the oil pump. Remove the water inlet housing. Remove the oil pipe.
7. Remove the cylinder head cover retaining bolts. Remove the cylinder head cover.
8. Remove the crankshaft pulley.
9. Remove the twenty three bolts and two nuts. Remove the timing chain cover with the oil pump.

➡**When removing the timing chain cover be careful when prying between the timing chain cover and cylinder head or cylinder block. Do not damage the contact surfaces.**

10. Remove the oil pump relief valve.

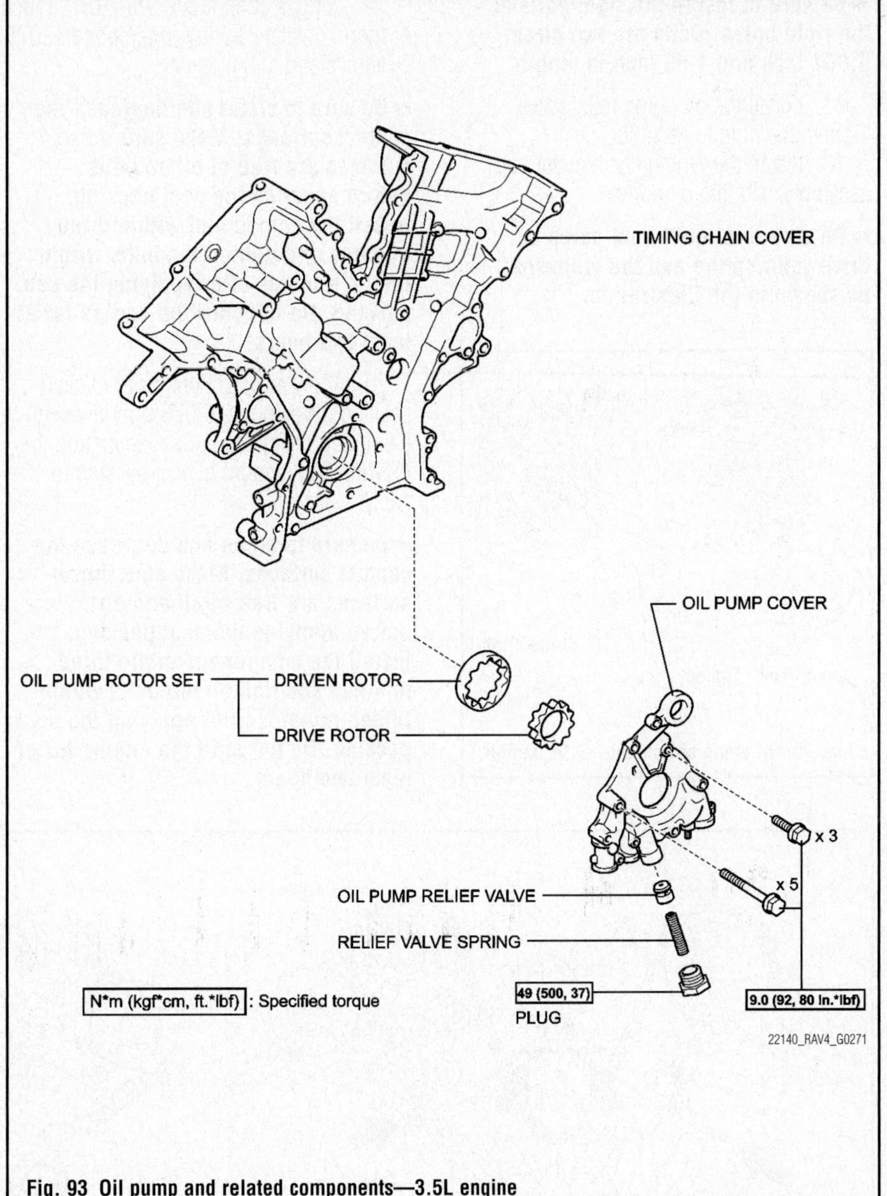

Fig. 93 Oil pump and related components—3.5L engine

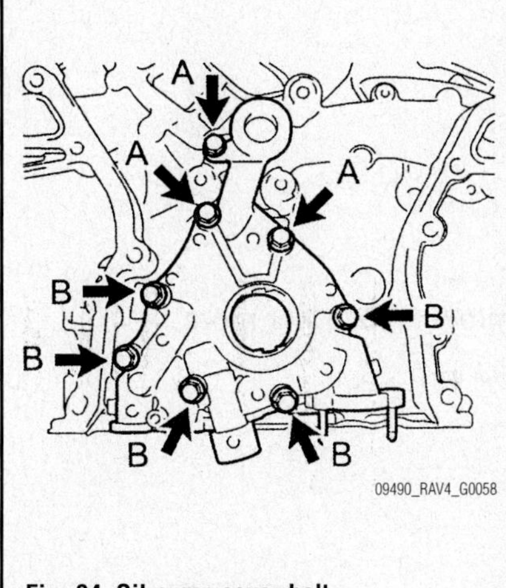

Fig. 94 Oil pump cover bolt identification—3.5L engine

Remove the eight bolts and the oil pump cover, drive rotor and driven rotor.

11. Repair or replace defective parts as required.
12. The clearance between the drive rotor and driven rotor should be 0.0024–0.0063 inch.
13. The clearance between the timing chain cover and driven rotor should be 0.0098–0.0128 inch.
14. The clearance between the rotors and the precision measuring straightedge should be 0.0012–0.0035 inch.

To install:

15. Replace the timing chain cover oil seal.
16. Install the oil pump cover. Tighten the retaining bolts to 80 inch lbs. in an alternating sequence.

➡Be sure to install the right bolts in the right holes. Bolts are two sizes, 0.087 inch and 1.58 inch in length.

17. Install the oil pump relief valve. Tighten the plug to 36 ft. lbs.

18. Install the timing chain cover sub-assembly, with the oil pump.

➡Be sure to align the oil pump's drive rotor spline and the crankshaft, as shown in the illustration.

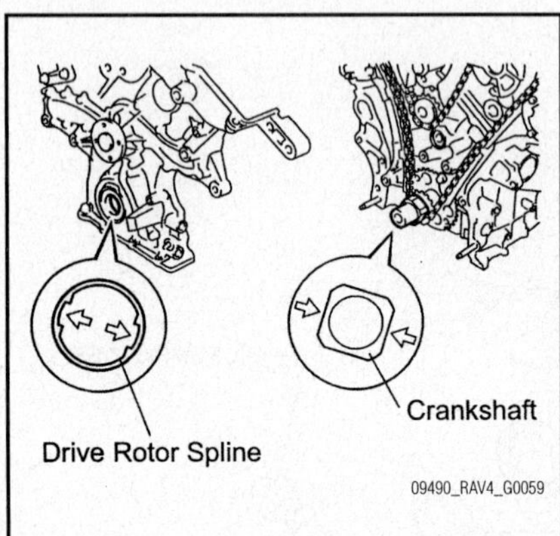

Fig. 95 Oil pump alignment—3.5L engine

19. Apply a continuous bead (0.118 inch or more) of seal packing (three bond 1207B or equivalent) to the engine.

➡Be sure to clean and degrease the contact surfaces. Make sure these surfaces are free of oil and dirt before applying the seal packing. Install the component within three minutes and tighten the bolts within fifteen minutes after applying the seal packing. Do not start the engine for at least two hours.

20. Apply a continuous bead of seal packing (three bond 1207B or equivalent) to the timing chain cover. See illustration for location and amount of seal packing to apply.

➡Be sure to clean and degrease the contact surfaces. Make sure these surfaces are free of oil and dirt before applying the seal packing. Install the component within three minutes and tighten the bolts within fifteen minutes after applying the seal packing. Do not start the engine for at least two hours.

21. Install the timing chain cover. Tighten the retaining bolts and nuts in the following order, area "1", area "2", area "3" and area "4". Tighten the bolts and nuts to 15 ft. lbs for area "1" area "2" and area "3". Tighten bolt "A" in area "4" to 32 ft. lbs. Tighten all other bolts in area "4" to 15 ft. lbs.

➡Be sure that there is no oil on the bolts. Bolt "A" is 1.57 inch long, bolt "B" is 2.17 inch long and bolt "C" is 0.98 inch long.

22. When installing the cylinder head cover, apply seal packing, three bond 1207B or equivalent.

➡Remove any oil from the contact surfaces. Install the component within three minutes after applying the sealant. Do not start the engine for two hours after the installation.

23. Install the cover retaining bolts and torque to 15 ft. lbs for bolt "A" and to 7 ft. lbs. for all other bolts and in the proper sequence.

24. Continue the installation is the reverse order of the removal procedure.

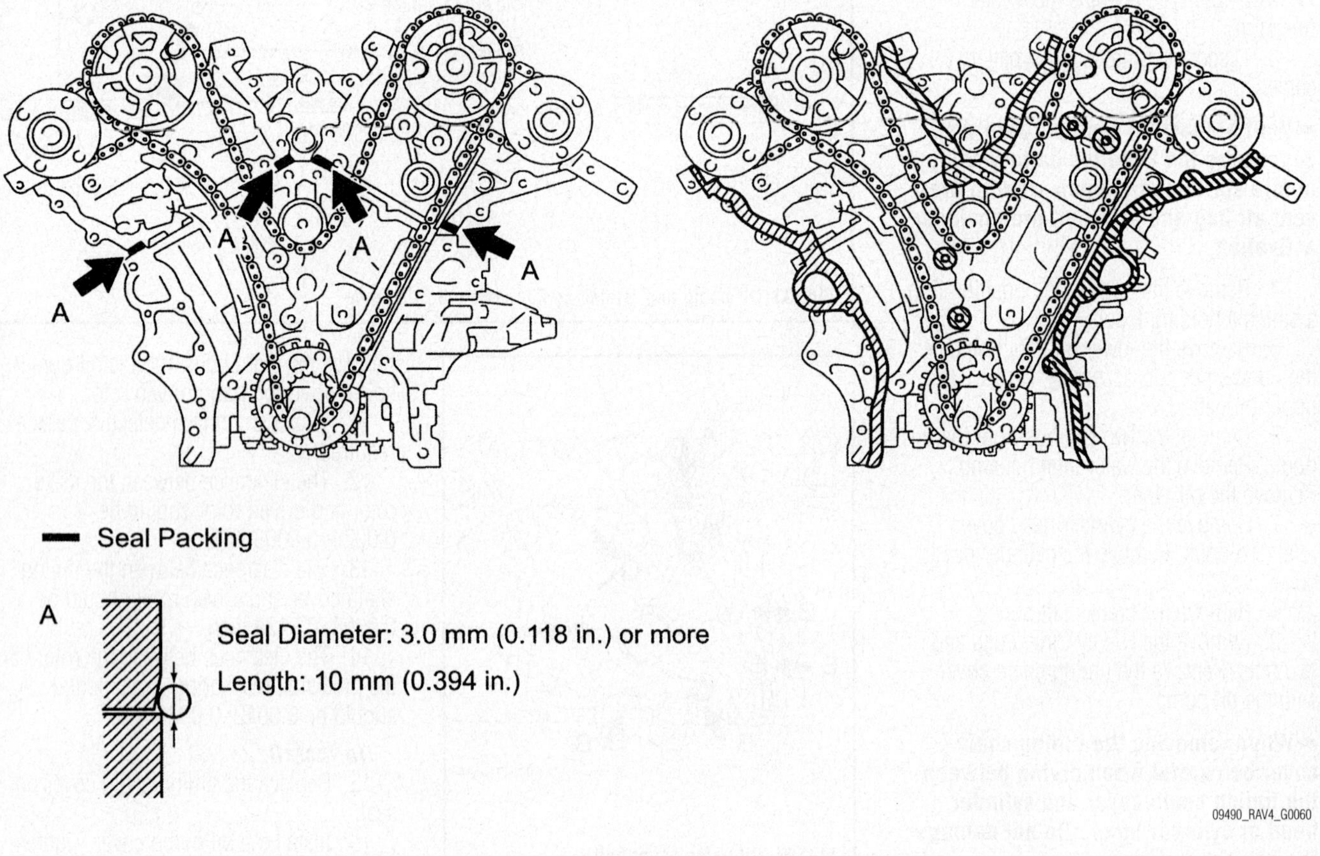

Fig. 96 Timing chain cover sealant application (to engine block)—3.5L engine

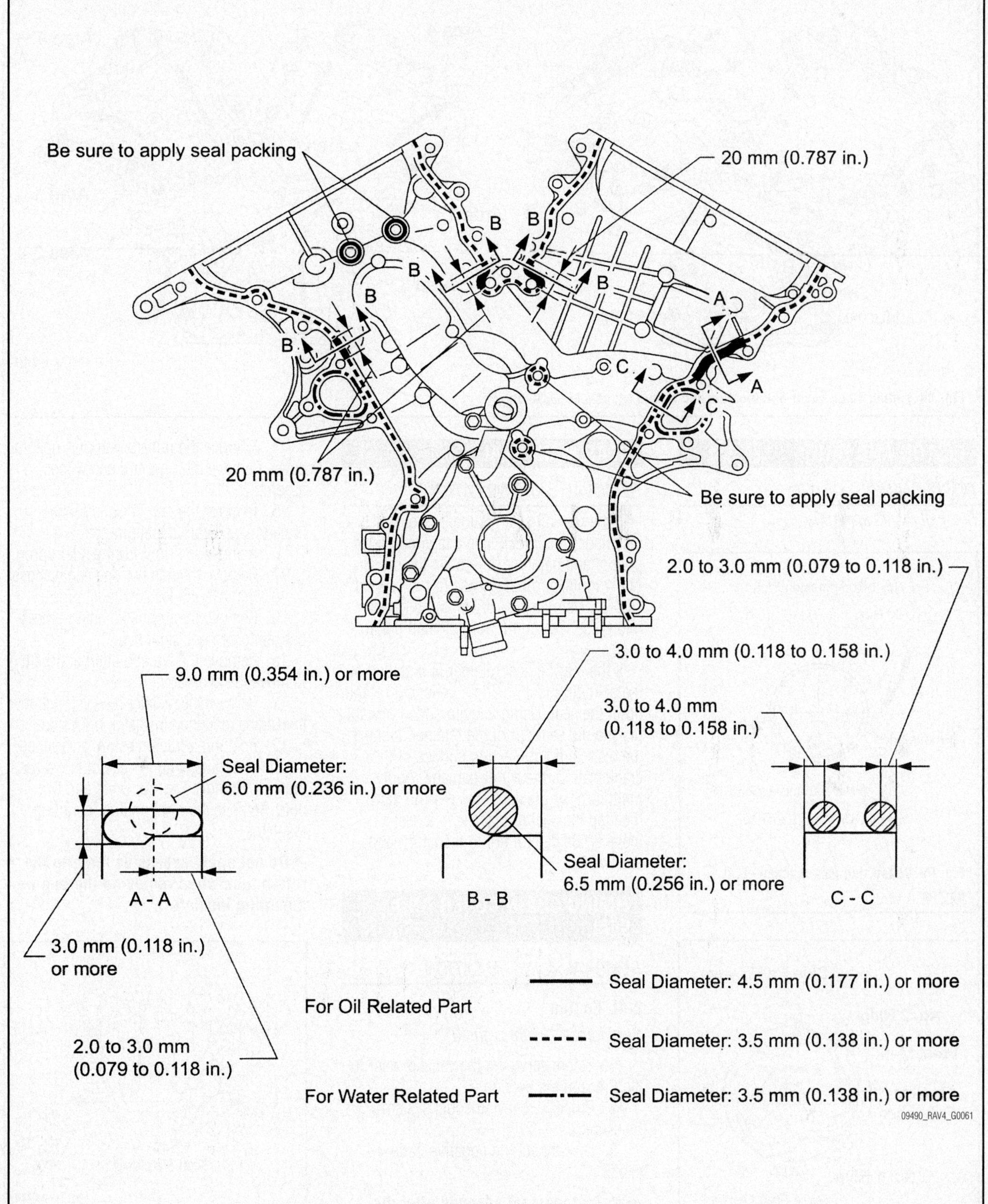

Fig. 97 Timing chain cover sealant application (to timing cover)—3.5L engine

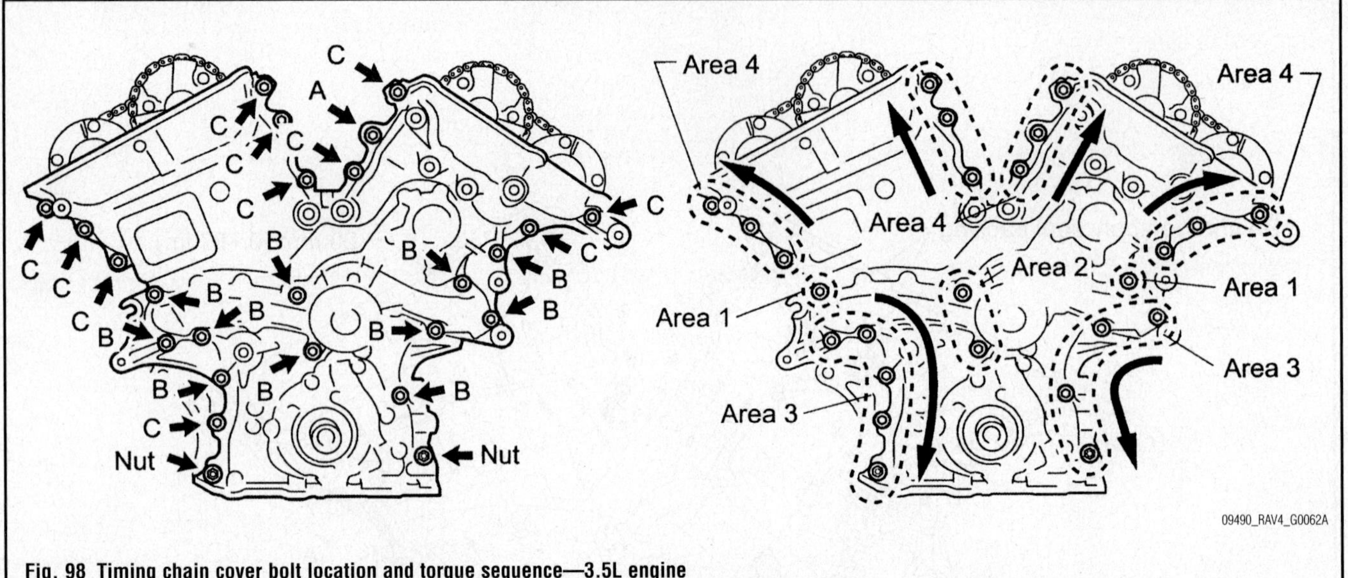

Fig. 98 Timing chain cover bolt location and torque sequence—3.5L engine

PISTON AND RING

POSITIONING

See Figures 99 and 100.

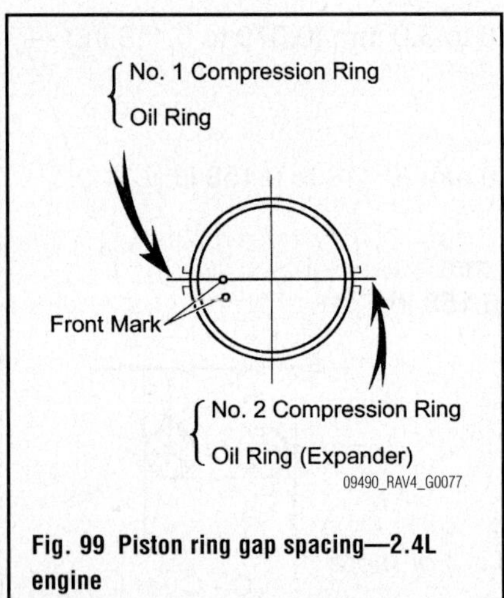

Fig. 99 Piston ring gap spacing—2.4L engine

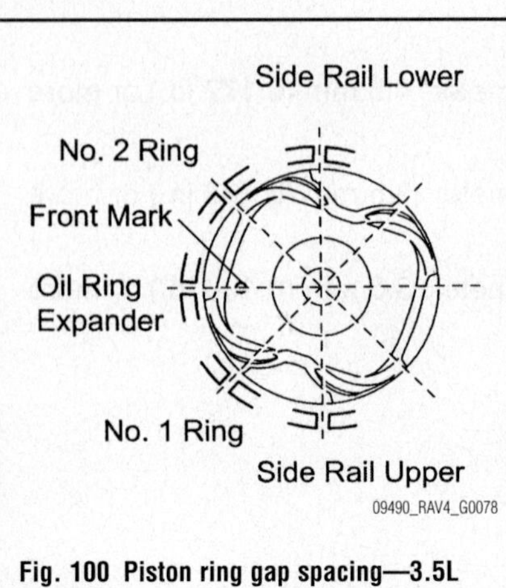

Fig. 100 Piston ring gap spacing—3.5L engine

REAR MAIN SEAL

REMOVAL & INSTALLATION

If the rear oil seal retainer is not installed to the block, use a tapered ended prytool and hammer to remove the oil seal. Apply multi-purpose grease to the new oil seal lip. Using a seal driver, tap the seal into place. Be careful not to install it slantwise.

If the rear oil seal retainer is installed on the cylinder block, using a knife, cut off the lip of the seal. Using a taped ended prytool, pry the old seal out of the retainer. Inspect the oil seal lip contacting surface of the crankshaft for cracks or damage. Apply multipurpose grease to the new oil seal, then tap the seal in place with a seal installer. Be careful not to install the seal slantwise.

TIMING CHAIN COVER AND SEAL

REMOVAL & INSTALLATION

2.4L Engine

See Figures 101 through 105.

1. Before servicing the vehicle, refer to the Precautions Section.
2. Properly relieve the fuel system pressure.
3. Disconnect the negative battery cable.

➡Wait at least 90 seconds after disconnecting the negative battery cable before starting any repair work to prevent air bag and seat belt pretensioner activation.

4. Remove the radiator support opening cover. Remove the right tire and wheel assembly.
5. Remove the number one engine under cover. Drain the engine oil.
6. Remove the right front fender apron.
7. Remove the number one engine cover.
8. Remove the front exhaust pipe.
9. Remove the front suspension member reinforcement, right side.
10. Remove the fan and alternator belt. Remove the alternator.
11. Remove the radiator reservoir. Remove the engine mounting insulator, right side.
12. Position a transmission jack underneath the engine, with a piece of wood on the jack. Remove the four bolts and two nuts. Remove the right engine mounting insulator.

➡Do not apply excessive force to the return tube when removing the engine mounting insulator.

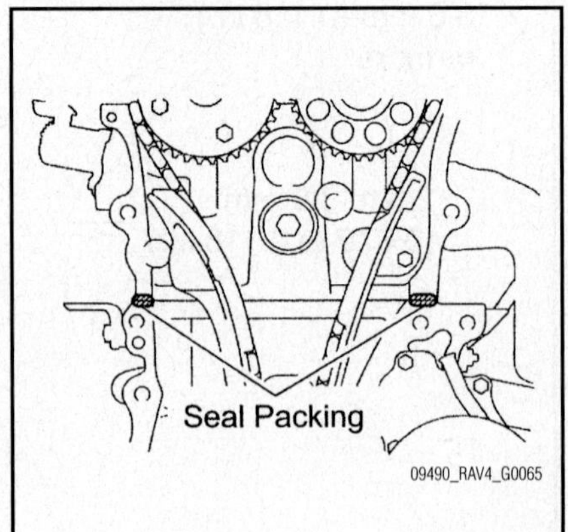

Fig. 101 Seal packing application locating points—2.4L engine

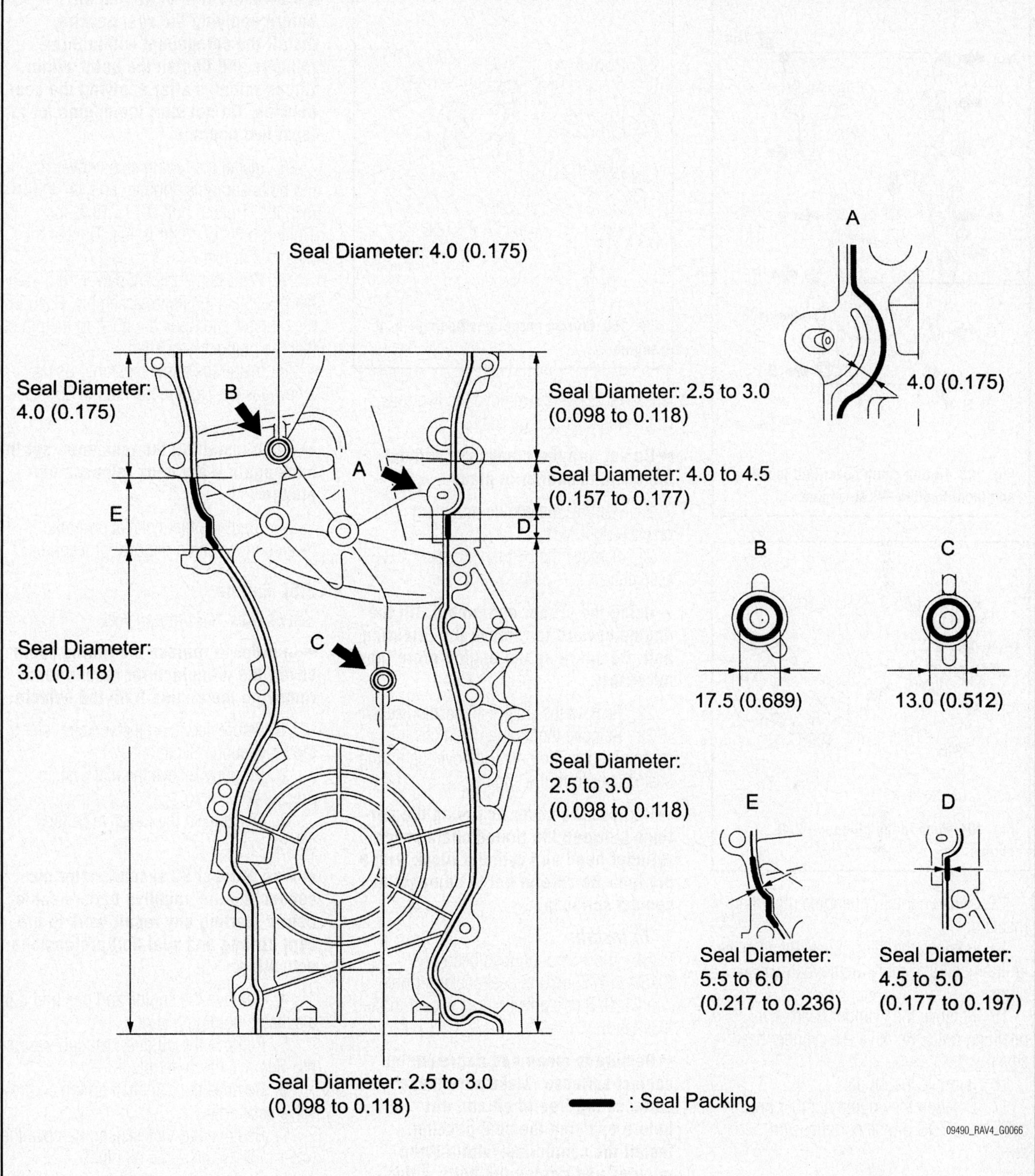

Fig. 102 Timing chain cover seal packing application locating points—2.4L engine

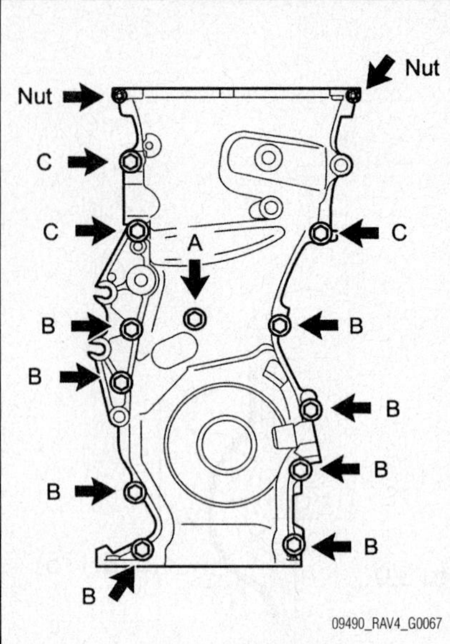

Fig. 103 Timing chain cover bolt location and identification—2.4L engine

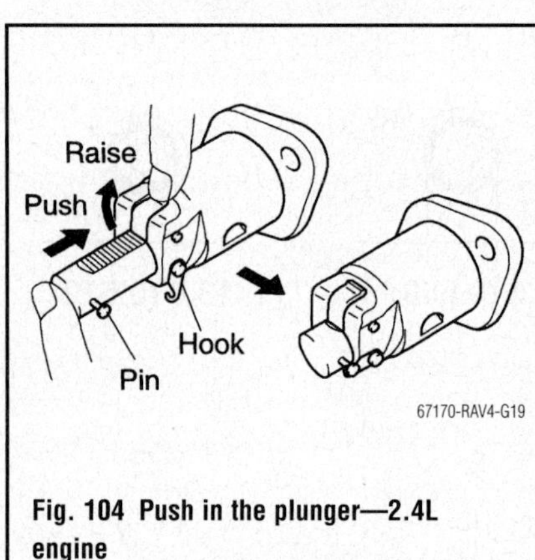

Fig. 104 Push in the plunger—2.4L engine

13. Remove the front engine mounting insulator.
14. Remove the idler pulley. Remove the ignition coil assembly. Remove the spark plugs.
15. Remove the cylinder head cover retaining bolts. Remove the cylinder head cover.
16. Remove the oil pan.
17. Position the engine number one cylinder at TDC on the compression stroke.

➡Check that the timing mark of the camshaft timing gear and sprocket is aligned with the timing mark located on the number one and number two bearing caps. If not rotate the engine 360 degrees.

18. Remove the crankshaft pulley.
19. To remove the number one chain

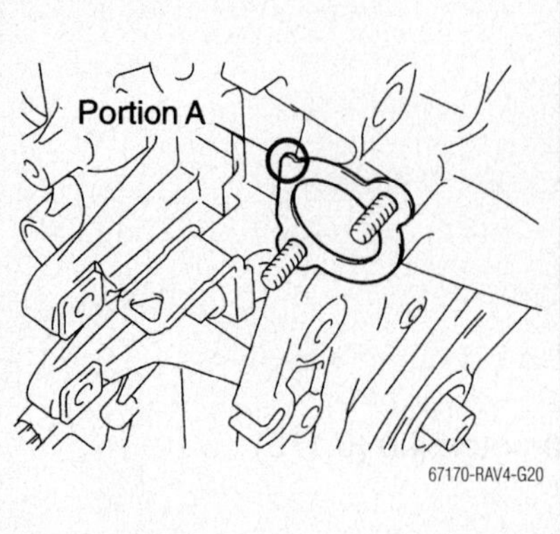

Fig. 105 Correct gasket positioning—2.4L engine

tensioner assembly, remove the two nuts, chain tensioner and gasket.

➡Do not turn the crankshaft without the chain tensioner in place.

20. Remove the engine mounting bracket, right side.
21. Remove the belt tensioner assembly.

➡Using the proper equipment, lift the engine upward to remove the retaining bolt. Do not raise the engine more than necessary.

22. Remove the crankshaft position sensor.
23. Remove the timing chain cover retaining bolts and nuts. Remove the timing chain cover from the engine.

➡Remove the cover by prying the portions between the timing chain cover, cylinder head and cylinder block with a pry tool. Be careful not to damage the contact surfaces.

To install:

24. Apply a continuous bead (0157–0.177 inch) of seal packing (three bond 1207B or equivalent) as shown in the illustration.

➡Be sure to clean and degrease the contact surfaces. Make sure these surfaces are free of oil and dirt before applying the seal packing. Install the component within three minutes and tighten the bolts within fifteen minutes after applying the seal packing. Do not start the engine for at least two hours.

25. Apply a continuous bead of seal packing (three bond 1207B or equivalent) as shown in the illustration.

➡Be sure to clean and degrease the contact surfaces. Make sure these surfaces are free of oil and dirt before applying the seal packing. Install the component within three minutes and tighten the bolts within fifteen minutes after applying the seal packing. Do not start the engine for at least two hours.

26. Install the timing chain cover retaining bolts and nuts. Tighten bolt "A" to 80 inch lbs. Tighten bolt "B" to 18 ft. lbs. Tighten bolt "C" to 41 ft. lbs. Tighten the nuts to 8 ft. lbs.
27. Release the ratchet pawl of the number one chain tensioner assembly. Push in the plunger and hook the hook to the pin so that the plunger is pushed in.
28. Install the chain tensioner using a new gasket. Tighten the nuts to 80 inch lbs.

➡When installing the tensioner, set the hook again if the hook releases the plunger.

29. Continue the installation in the reverse order of the removal procedure.

3.5L Engine

See Figures 106 through 108.

➡In order to replace the timing chain cover, the manufacturer recommends removing the engine from the vehicle.

1. Before servicing the vehicle, refer to the Precautions Section.
2. Properly relieve the fuel system pressure.
3. Disconnect the negative battery cable.

➡Wait at least 90 seconds after disconnecting the negative battery cable before starting any repair work to prevent air bag and seat belt pretensioner activation.

4. Remove the engine and position it in a suitable holding fixture.
5. Remove the oil filler cap sub-assembly. Remove the spark plugs.
6. Remove the camshaft timing control valve assembly.
7. Remove the VVT sensor. Remove the water inlet. Remove the oil pipe.
8. Remove the twelve retaining bolts and remove the cylinder head cover for bank one.

➡Upon installation be sure that removed components are reinstalled in their original locations.

9. Remove the number one oil pipe.
10. Remove the twelve retaining bolts

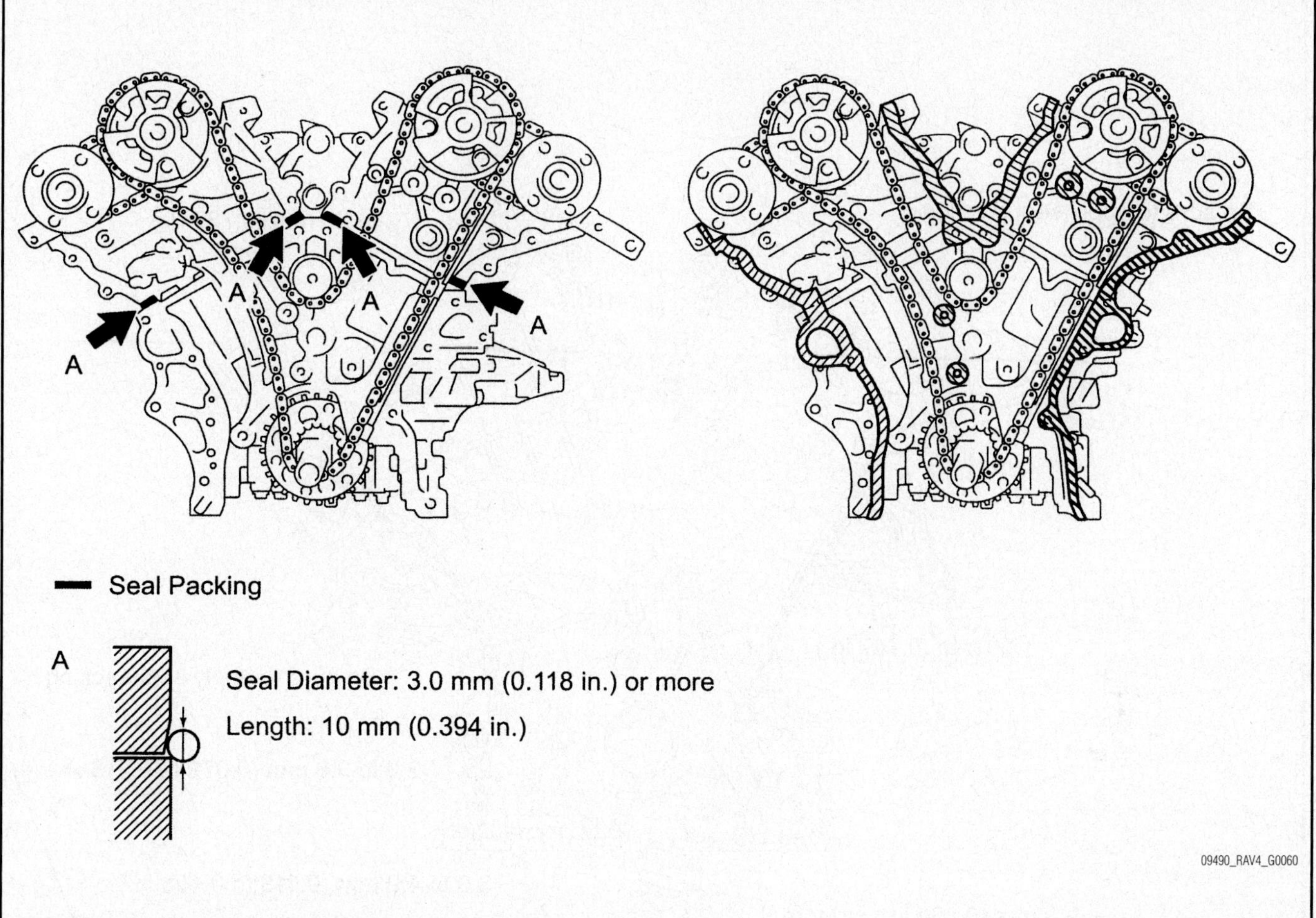

Fig. 106 Timing chain cover sealant application (to engine block)—3.5L engine

and remove the cylinder head cover for bank two.

➡**Upon installation be sure that removed components are reinstalled in their original locations.**

11. Remove the number one engine mounting bracket. Remove the water inlet housing.
12. Remove the oil filter cap assembly. Remove the oil filter.
13. Remove the number two oil pan. Remove the oil strainer assembly. Remove the number one oil pan. Remove the seven retaining bolts and remove the oil pan baffle plate.
14. Remove the crankshaft pulley. Remove the water pump.
15. Remove the twenty three bolts and two nuts, and remove the timing chain front cover.

➡**Carefully remove the cover by prying between the timing chain cover and cylinder head or cylinder block, using a pry tool. Do not damage the contact surfaces.**

To install:

16. Replace the timing chain cover oil seal.
17. Install the timing chain cover sub-assembly, with the oil pump.

➡**Be sure to align the oil pump's drive rotor spline and the crankshaft, as shown in the illustration.**

18. Apply a continuous bead (0.118 inch or more) of seal packing (three bond 1207B or equivalent) to the engine.

➡**Be sure to clean and degrease the contact surfaces. Make sure these surfaces are free of oil and dirt before applying the seal packing. Install the component within three minutes and tighten the bolts within fifteen minutes after applying the seal packing. Do not start the engine for at least two hours.**

19. Apply a continuous bead of seal packing (three bond 1207B or equivalent) to the timing chain cover. See illustration for location and amount of seal packing to apply.

➡**Be sure to clean and degrease the contact surfaces. Make sure these surfaces are free of oil and dirt before applying the seal packing. Install the component within three minutes and tighten the bolts within fifteen minutes after applying the seal packing. Do not start the engine for at least two hours.**

20. Install the timing chain cover. Tighten the retaining bolts and nuts in the following order, area "1", area "2", area "3" and area "4". Tighten the bolts and nuts to 15 ft. lbs for area "1" area "2" and area "3". Tighten bolt "A" in area "4" to 32 ft. lbs. Tighten all other bolts in area "4" to 15 ft. lbs.

➡**Be sure that there is no oil on the bolts. Bolt "A" is 1.57 inch long, bolt "B" is 2.17 inch long and bolt "C" is 0.98 inch long.**

21. When installing the cylinder head cover, apply seal packing, three bond 1207B or equivalent.

➡**Remove any oil from the contact surfaces. Install the component within**

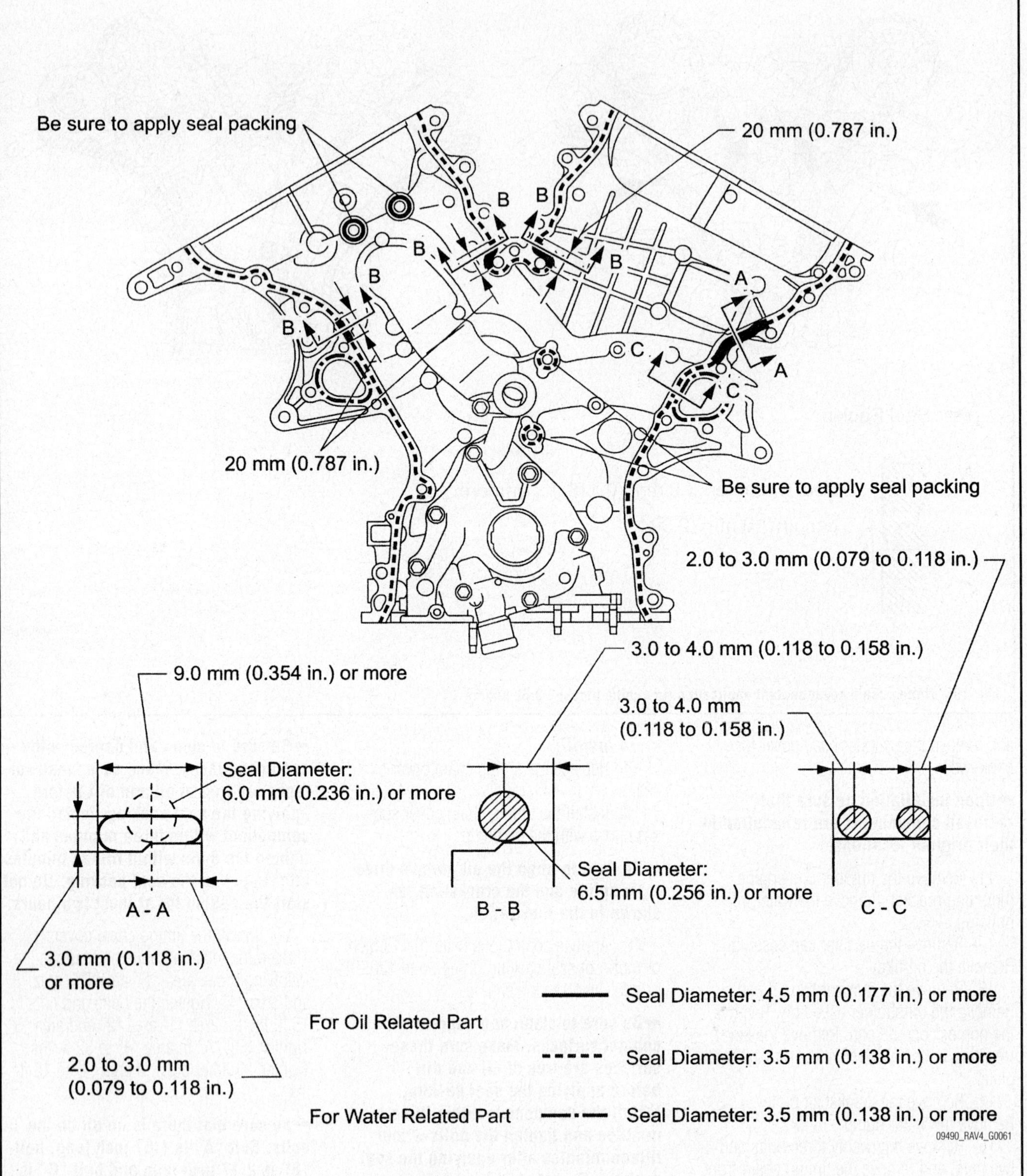

Fig. 107 Timing chain cover sealant application (to timing cover)—3.5L engine

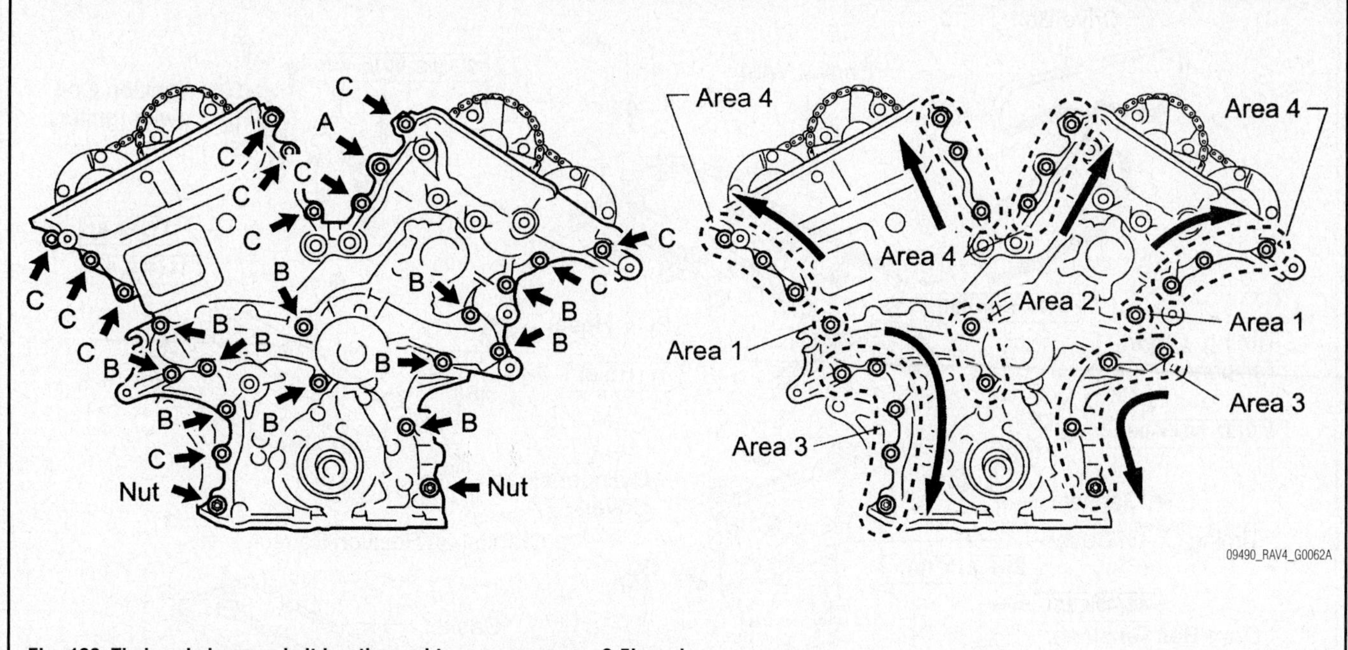

Fig. 108 Timing chain cover bolt location and torque sequence—3.5L engine

three minutes after applying the sealant. Do not start the engine for two hours after the installation.

22. Install the cover retaining bolts and torque to 15 ft. lbs for bolt "A" and to 7 ft. lbs. for all other bolts and in the proper sequence.
23. Continue the installation is the reverse order of the removal procedure.

TIMING CHAIN AND SPROCKETS

REMOVAL & INSTALLATION

2.4L Engine

See Figures 109 through 117.

1. Before servicing the vehicle, refer to the Precautions Section.
2. Properly relieve the fuel system pressure.
3. Disconnect the negative battery cable.

➡Wait at least 90 seconds after disconnecting the negative battery cable before starting any repair work to prevent air bag and seat belt pretensioner activation.

4. Remove the radiator support opening cover. Remove the right tire and wheel assembly.
5. Remove the number one engine under cover. Drain the engine oil.
6. Remove the right front fender apron.
7. Remove the number one engine cover.
8. Remove the front exhaust pipe.
9. Remove the front suspension member reinforcement, right side.
10. Remove the fan and alternator belt. Remove the alternator.
11. Remove the radiator reservoir. Remove the engine mounting insulator, right side.
12. Position a transmission jack underneath the engine, with a piece of wood on the jack. Remove the four bolts and two nuts. Remove the right engine mounting insulator.

➡Do not apply excessive force to the return tube when removing the engine mounting insulator.

13. Remove the front engine mounting insulator.
14. Remove the idler pulley. Remove the ignition coil assembly. Remove the spark plugs.
15. Remove the cylinder head cover retaining bolts. Remove the cylinder head cover.
16. Remove the oil pan.
17. Position the engine number one cylinder at TDC on the compression stroke.

➡Check that the timing mark of the camshaft timing gear and sprocket is aligned with the timing mark located on the number one and number two bearing caps. If not rotate the engine 360 degrees.

18. Remove the crankshaft pulley.
19. To remove the number one chain tensioner assembly, remove the two nuts, chain tensioner and gasket.

➡Do not turn the crankshaft without the chain tensioner in place.

20. Remove the engine mounting bracket, right side.
21. Remove the belt tensioner assembly.

➡Using the proper equipment, lift the engine upward to remove the retaining bolt. Do not raise the engine more than necessary.

22. Remove the crankshaft position sensor.
23. Remove the timing chain cover retaining bolts and nuts. Remove the timing chain cover from the engine.

➡Remove the cover by prying the portions between the timing chain cover, cylinder head and cylinder block with a pry tool. Be careful not to damage the contact surfaces.

24. Remove the number one crankshaft position sensor plate. Remove the bolt and the timing chain guide.
25. Remove the bolt and the chain tensioner slipper.
26. Remove the two bolts and the number one chain vibration damper. Remove the timing chain.
27. Remove the crankshaft timing sprocket.
28. To remove the number two timing chain, turn the crankshaft ninety degrees counterclockwise to align the adjusting hole

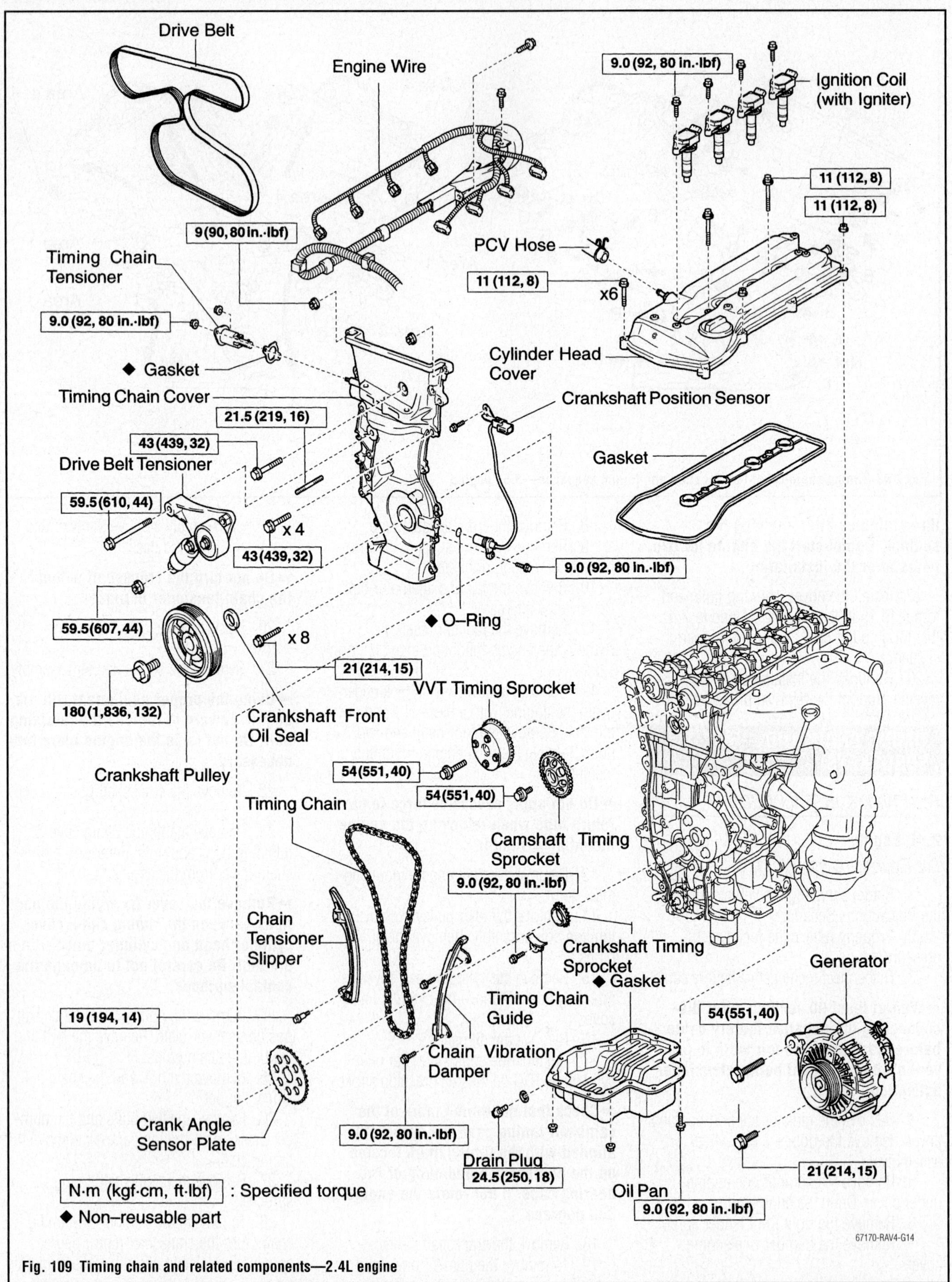

Fig. 109 Timing chain and related components—2.4L engine

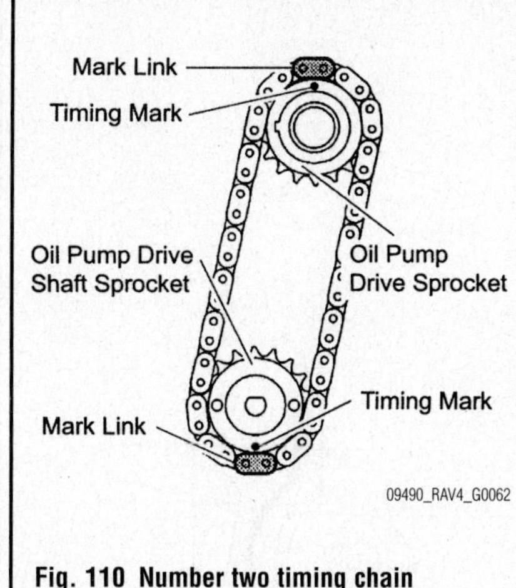

Fig. 110 Number two timing chain alignment—2.4L engine

of the oil pump driveshaft sprocket with the groove of the oil pump.

29. Insert a 4mm diameter bar into the adjusting hole of the oil pump driveshaft sprocket to lock the gear in position. Remove the nut.

30. Remove the bolt, chain tensioner plate and spring. Remove the oil pump drive sprocket, oil pump driveshaft and number two timing chain.

To install:

31. Drive a new seal into place with a seal installer until it is flush with the cover. Coat the seal lip with multi-purpose grease.

32. Check the timing chain with the chain full stretched. The length of 16 consecutive links should be 122.6mm (4.827 inches) max. If it's longer than that, replace it.

33. To install the number two timing chain, position the crankshaft key in the left horizontal position. Turn the cutout of the driveshaft so that it faces upward.

34. Align the yellow links with the timing marks of each gear. Install the sprockets onto the crankshaft and oil pump shaft with the chain wrapped on the gears.

35. Temporarily tighten the oil pump driveshaft sprocket with the nut.

36. Insert the damper spring into the adjusting hole, and then install the chain tensioner plate with the bolt. Tighten the bolt to 9 ft. lbs.

37. Align the adjusting hole of the oil pump driveshaft sprocket with the groove of the oil pump.

38. Insert a 4mm diameter bar into the adjusting hole of the oil pump driveshaft gear to lock the gear in position. Tighten the nut to 22 ft. lbs.

39. Install the crankshaft timing sprocket. Install the number one chain vibration damper. Tighten the bolts to 80 inch lbs.

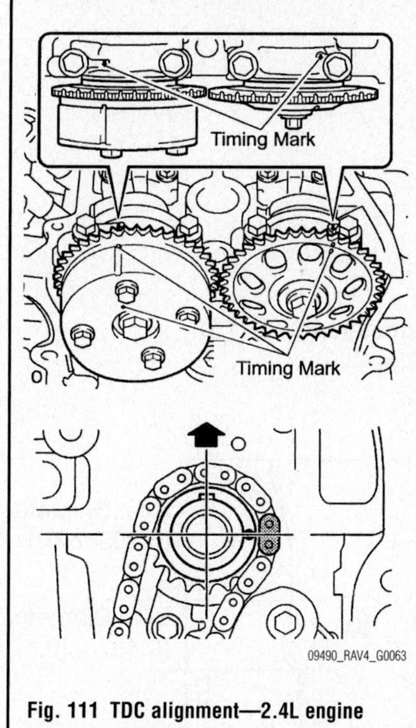

Fig. 111 TDC alignment—2.4L engine

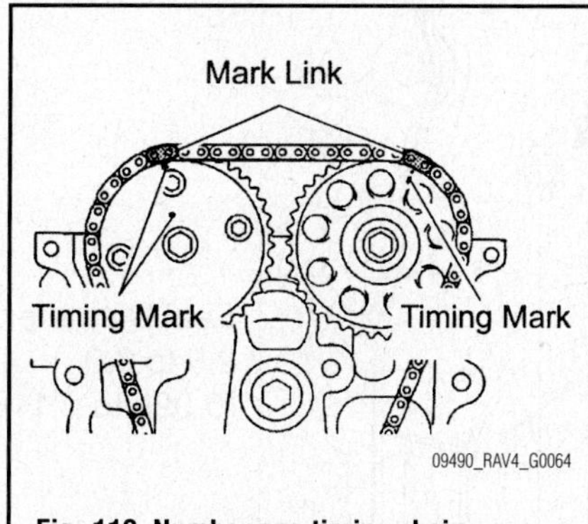

Fig. 112 Number one timing chain alignment—2.4L engine

40. To install the number one timing chain, be sure the number one cylinder is at TDC on the compression stroke. Position the key on the crankshaft upward.

41. Install the chain on the crankshaft sprocket with the gold or orange mark aligned with the timing mark on the crankshaft.

42. Align the gold or yellow links with each timing mark located on the camshaft timing gear and sprocket. Install the chain.

43. Install the chain tensioner slipper. Tighten the bolt to 14 ft. lbs.

44. Install the timing chain guide and bolt. Tighten to 80 inch lbs.

45. Install the number one crankshaft position sensor plate. Be sure the "F" mark is facing forward.

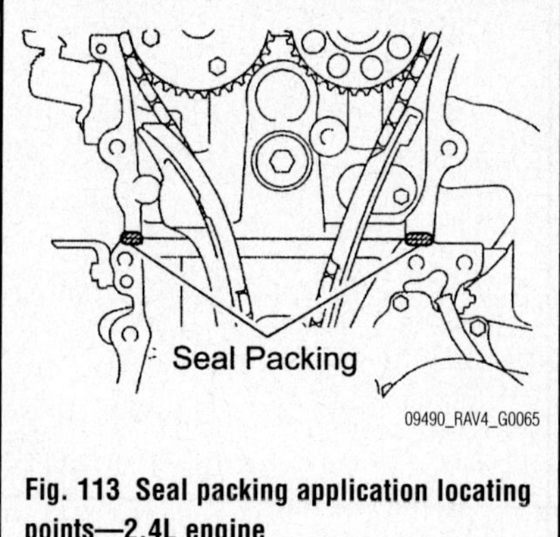

Fig. 113 Seal packing application locating points—2.4L engine

46. To install the timing chain cover, remove any old packing material. Be careful not to drop oil on the contact surfaces.

47. Apply a continuous bead (0157–0.177 inch) of seal packing (three bond 1207B or equivalent) as shown in the illustration.

➡Be sure to clean and degrease the contact surfaces. Make sure these surfaces are free of oil and dirt before applying the seal packing. Install the component within three minutes and tighten the bolts within fifteen minutes after applying the seal packing. Do not start the engine for at least two hours.

48. Apply a continuous bead of seal packing (three bond 1207B or equivalent) as shown in the illustration.

➡Be sure to clean and degrease the contact surfaces. Make sure these surfaces are free of oil and dirt before applying the seal packing. Install the component within three minutes and tighten the bolts within fifteen minutes after applying the seal packing. Do not start the engine for at least two hours.

49. Install the timing chain cover retaining bolts and nuts. Tighten bolt "A" to 80 inch lbs. Tighten bolt "B" to 18 ft. lbs. Tighten bolt "C" to 41 ft. lbs. Tighten the nuts to 8 ft. lbs.

50. Release the ratchet pawl of the number one chain tensioner assembly. Push in the plunger and hook the hook to the pin so that the plunger is pushed in.

51. Install the chain tensioner using a new gasket. Tighten the nuts to 80 inch lbs.

➡When installing the tensioner, set the hook again if the hook releases the plunger.

52. Continue the installation in the reverse order of the removal procedure.

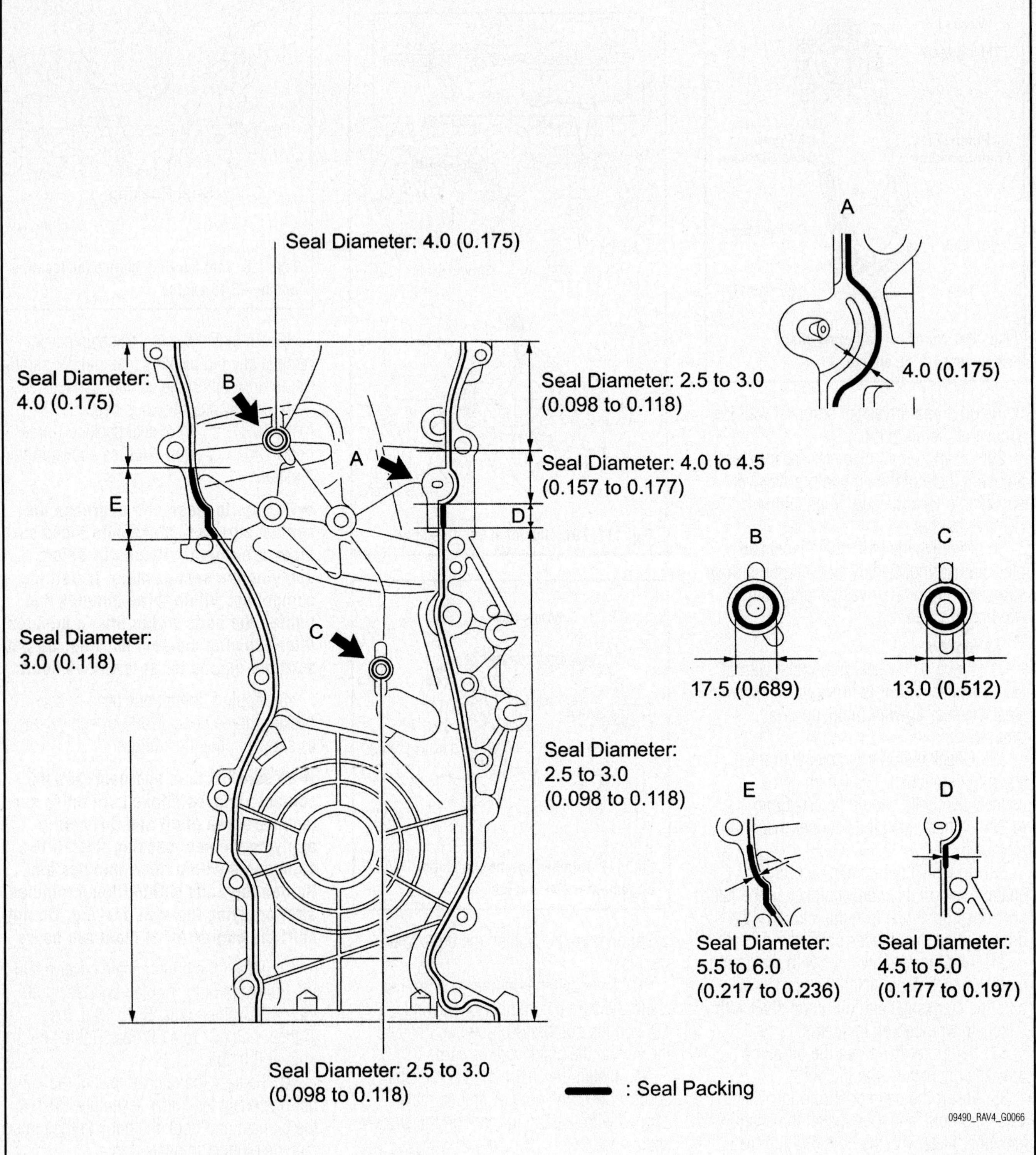

Fig. 114 Timing chain cover seal packing application locating points—2.4L engine

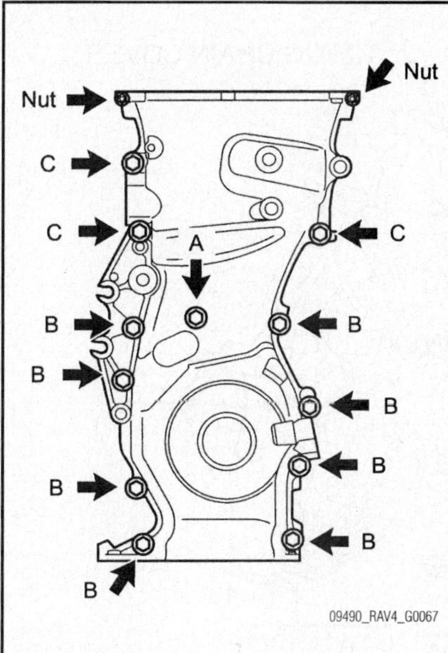

Fig. 115 Timing chain cover bolt location and identification—2.4L engine

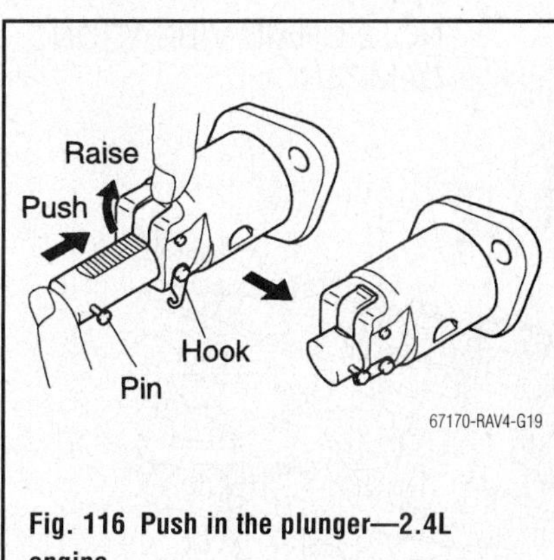

Fig. 116 Push in the plunger—2.4L engine

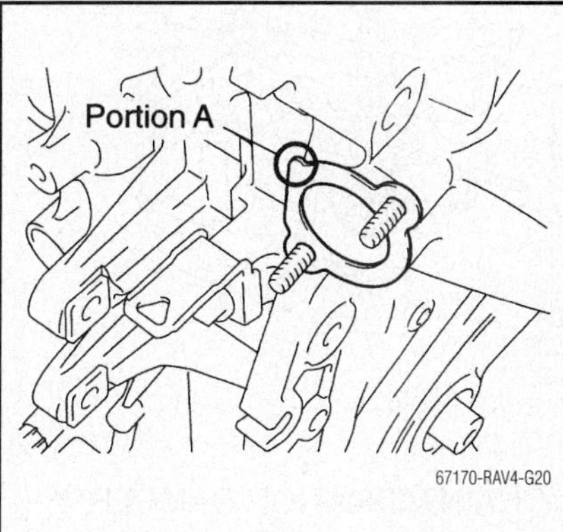

Fig. 117 Correct gasket positioning—2.4L engine

3.5L Engine

See Figures 118 through 129.

➡In order to replace the timing chains and gears, the manufacturer recommends removing the engine from the vehicle.

1. Before servicing the vehicle, refer to the Precautions Section.
2. Properly relieve the fuel system pressure.
3. Disconnect the negative battery cable.

➡Wait at least 90 seconds after disconnecting the negative battery cable before starting any repair work to prevent air bag and seat belt pretensioner activation.

4. Remove the engine and position it in a suitable holding fixture.
5. Remove the oil filler cap sub-assembly. Remove the spark plugs.
6. Remove the camshaft timing control valve assembly.
7. Remove the VVT sensor. Remove the water inlet. Remove the oil pipe.
8. Remove the twelve retaining bolts and remove the cylinder head cover for bank one.

➡Upon installation be sure that removed components are reinstalled in their original locations.

9. Remove the number one oil pipe.
10. Remove the twelve retaining bolts and remove the cylinder head cover for bank two.

➡Upon installation be sure that removed components are reinstalled in their original locations.

11. Remove the number one engine mounting bracket. Remove the water inlet housing.
12. Remove the oil filter cap assembly. Remove the oil filter.
13. Remove the number two oil pan. Remove the oil strainer assembly. Remove the number one oil pan. Remove the seven retaining bolts and remove the oil pan baffle plate.
14. Remove the crankshaft pulley. Remove the water pump.
15. Remove the twenty three bolts and two nuts, and remove the timing chain front cover.

➡Carefully remove the cover by prying between the timing chain cover and cylinder head or cylinder block, using a pry tool. Do not damage the contact surfaces.

16. Position the number one piston at TDC on the compression stroke. Temporarily tighten the pulley set bolt. Set the timing mark on the crank angle sensor plate to the right side block bore center line.
17. Check that the timing marks of the camshaft timing gears are aligned with the timing marks of the bearing cap, as shown in the illustration. If not rotate the crankshaft 360 degrees, to align the marks.
18. Move the stopper plate upward to release the lock and push the plunger deep into the tensioner. Move the stopper plate downward to set the lock, and insert a hexagon wrench into the stopper plate hole. Remove the two bolts and the chain tensioner.
19. Remove the chain tensioner slipper.
20. Turn the crankshaft counterclockwise 10 degrees to loosen the chain of the crankshaft timing gear.
21. Remove the chain from the crankshaft timing gear and place it on the crankshaft.
22. Turn the camshaft gear on the bank one (right bank) clockwise (about 60 degrees) and position as shown in the illustration. Remove the chain.
23. Remove the number two idle gear shaft, idle sprocket and the number one idle gear shaft.
24. Remove the two bolts and the number one chain vibration damper. Remove the two bolts and the number two vibration damper.
25. Remove the pulley set bolt. Remove the crankshaft timing gear from the crankshaft. Remove the two pulley set keys from the crankshaft.
26. On bank one, while raising up the number two chain tensioner, insert a pin (0.039 inch) into the hole to hold it in place. Hold the hexagonal portion of the camshaft with a wrench, and remove the two bolts and two camshaft gears. Remove the number two chain. Remove the bolt and the number two chain tensioner.

➡Be careful not to damage the cylinder head with the wrench. Do not disassemble the camshaft timing gear.

27. On bank two, while raising up the number two chain tensioner, insert a pin (0.039 inch) into the hole to hold it in place. Hold the hexagonal portion of the camshaft with a wrench, and remove the two bolts and two camshaft gears. Remove the number two chain. Remove the bolt and the number three chain tensioner.

➡Be careful not to damage the cylinder head with the wrench. Do not disassemble the camshaft timing gear.

To install:

28. Install the number two chain tensioner with the bolt. Tighten the bolt to 15 ft. lbs. While pushing in the tensioner, insert a 0.039 inch pin into the hole to hold it in place.

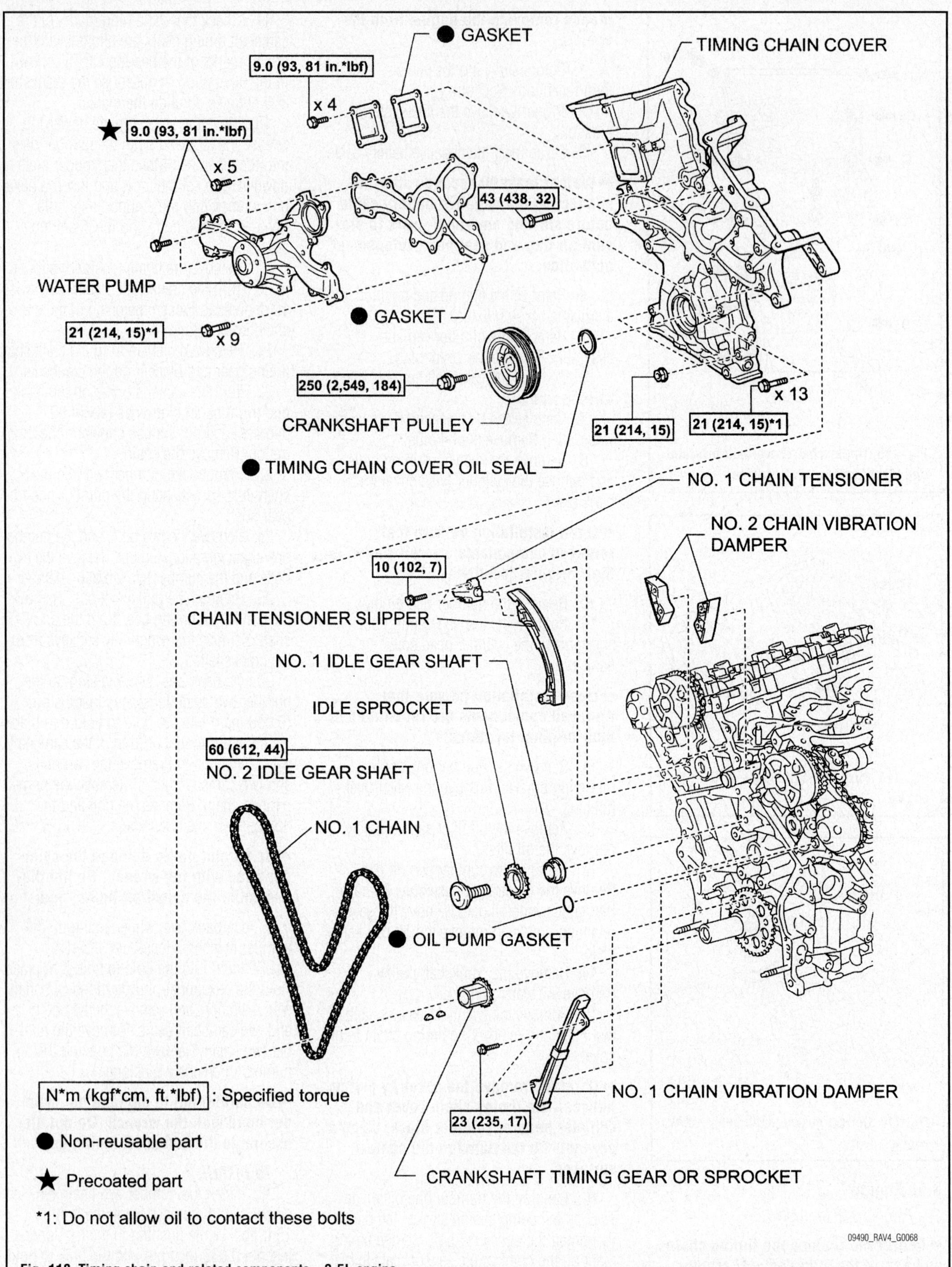

Fig. 118 Timing chain and related components—3.5L engine

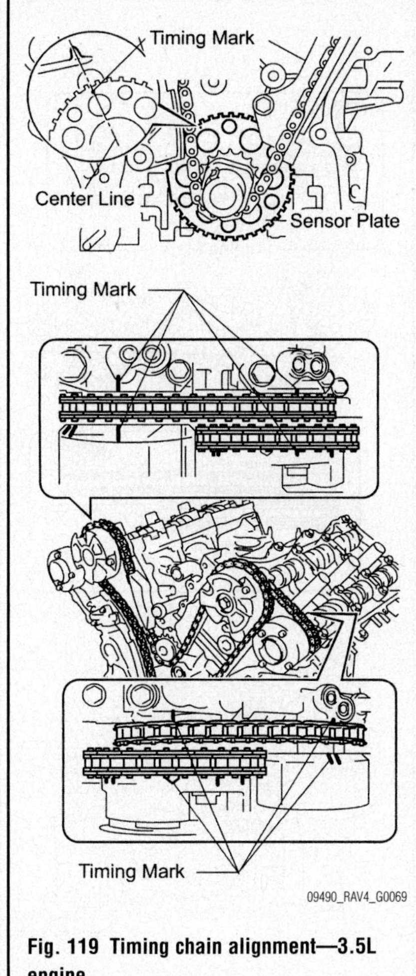

Fig. 119 Timing chain alignment—3.5L engine

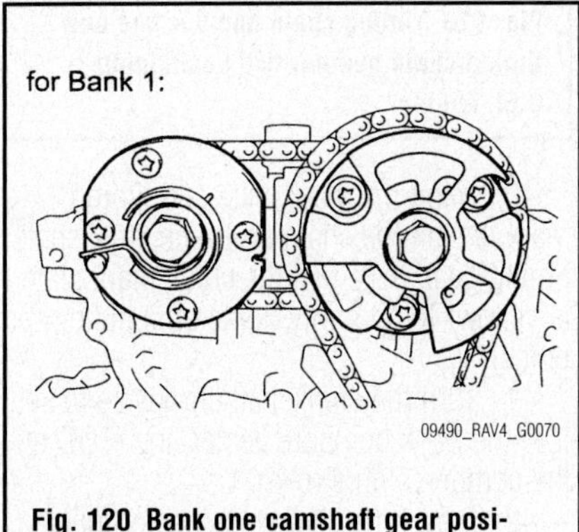

Fig. 120 Bank one camshaft gear positioning—3.5L engine

29. Install the number three chain tensioner assembly. Tighten the bolt to 15 ft. lbs. While pushing in the tensioner, insert a 0.039 inch pin into the hole to hold it in place.

30. Align the mark plate (yellow) with the timing marks (1 dot mark for bank one and two dot mark for bank two) of the camshaft timing gears, as shown in the illustration.

31. Apply a small amount of engine oil to the bolt threads and the bolt sealing surface.

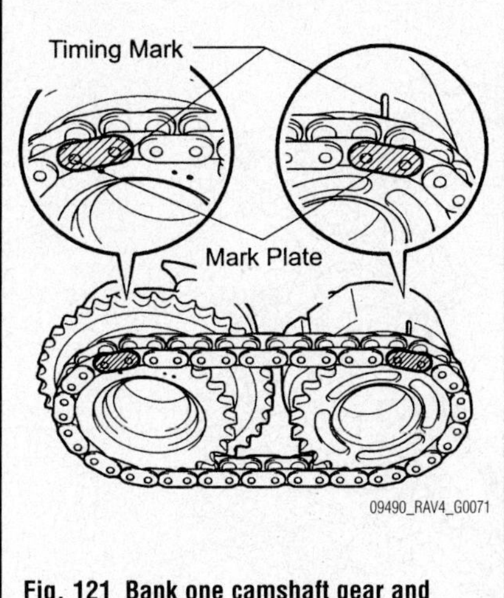

Fig. 121 Bank one camshaft gear and chain alignment—3.5L engine

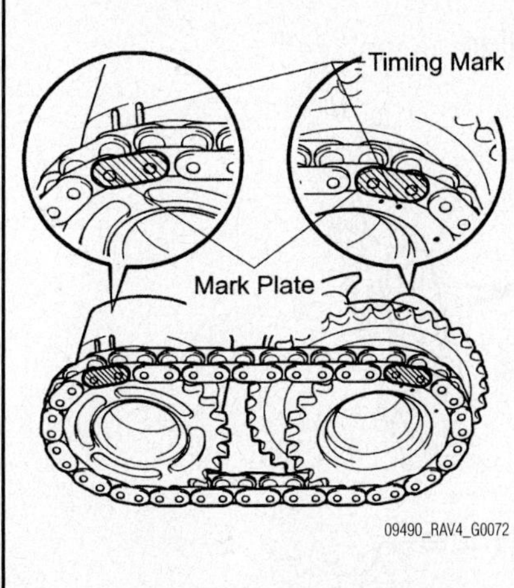

Fig. 122 Bank two camshaft gear and chain alignment—3.5L engine

Align the knock pin of the camshaft with the pin hole on the camshaft timing gear.

32. Install the camshaft timing gear and the camshaft timing exhaust gear with the number two chain installed.

33. Hold the hexagonal portion of the camshaft with a wrench, and tighten the bolts to 74 ft. lbs. Remove the pin.

34. Install the crankshaft gear or sprocket. Install the vibration damper. Tighten the bolts to 17 ft. lbs.

35. Install the number two chain vibration damper.

36. Apply a light coat of clean engine oil to the rotating surface of the number one idle gear shaft.

37. Temporarily install the number one idle gear shaft and idle sprocket with the number two idle gear shaft while aligning the knock pin of the number one idle gear shaft with the knock pin groove of the cylinder block. Tighten the bolt to 44 ft. lbs.

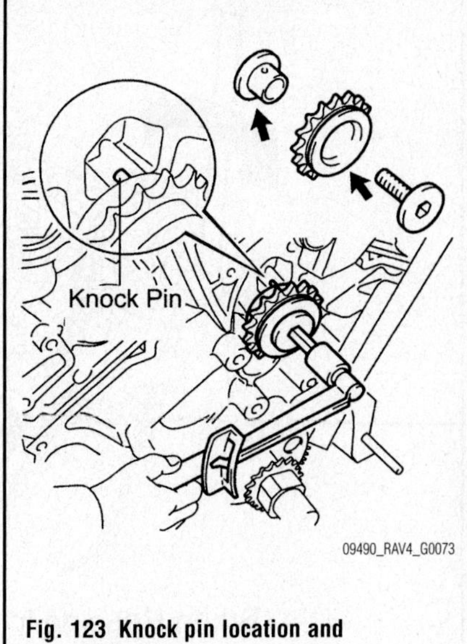

Fig. 123 Knock pin location and installation—3.5L engine

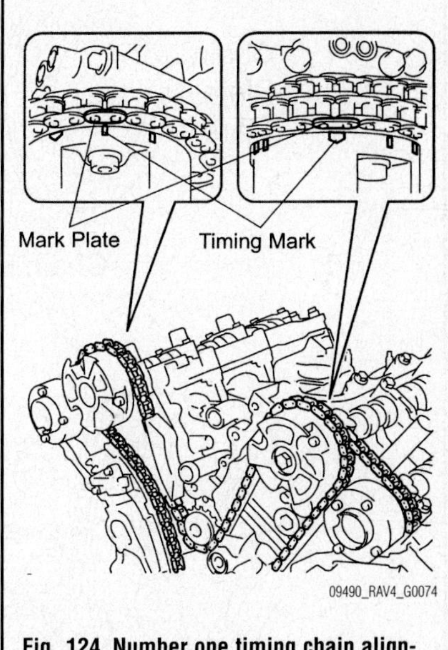

Fig. 124 Number one timing chain alignment—3.5L engine

➡Be careful of the idle gear direction.

38. Align the mark plate and the timing mark of the number one timing chain.

➡Do not pass the chain over the crankshaft, just put it on. The chain mark plate is orange.

39. Turn the camshaft timing gear on bank one counterclockwise to tighten the chain between both banks.

➡If reusing the idle sprocket, align one of the idle sprocket's chain plate marks with one of the chain's chain plates when installing the idle sprocket.

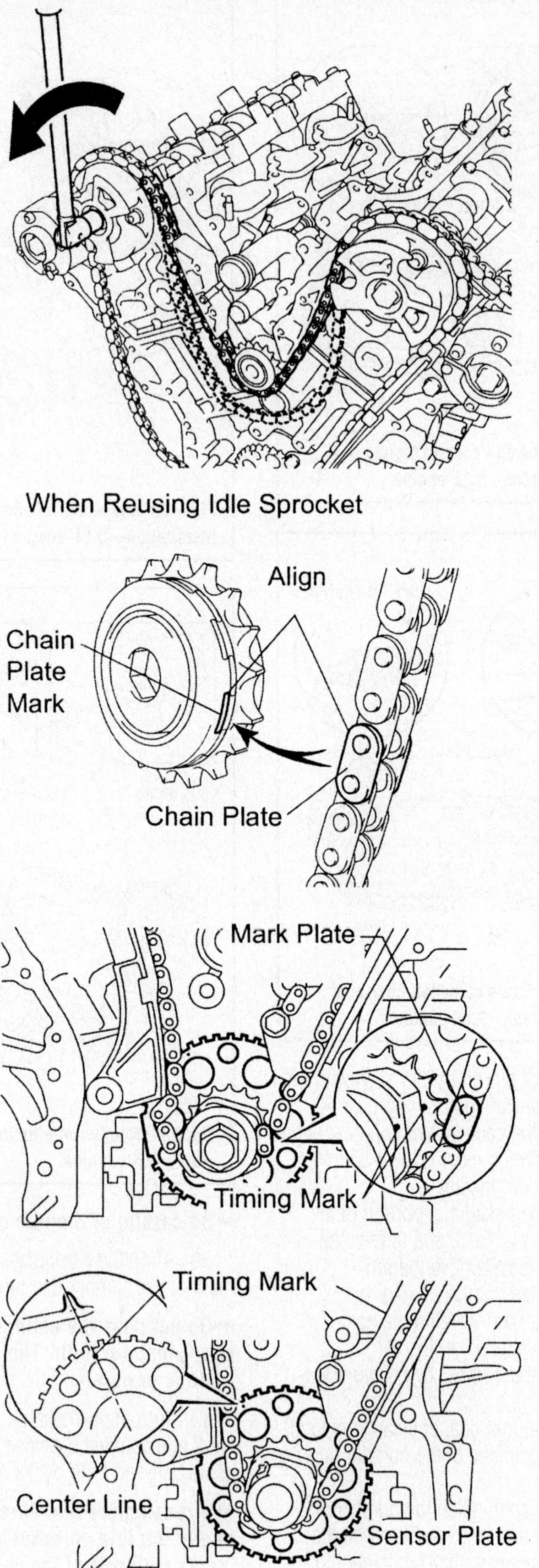

Fig. 125 Idle sprocket and mark plate alignment—3.5L engine

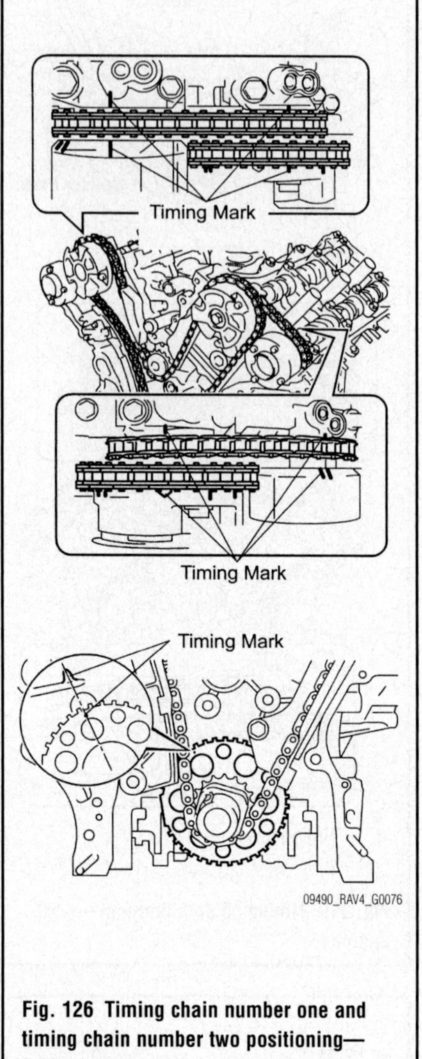

Fig. 126 Timing chain number one and timing chain number two positioning—3.5L engine

40. Align the mark plate and timing mark. Install the chain on to the crankshaft timing gear. Note that the chain mark plate is yellow. Temporarily tighten the pulley set bolt.
41. Turn the crankshaft clockwise to set it to the bank one bore center line (TDC on the compression stroke).
42. Install the chain tensioner slipper.
43. To install the number one chain tensioner assembly, move the stopper plate upward to release the lock. Push the plunger deep into the tensioner.
44. Move the stopper plate downward to set the lock. Insert a hexagon wrench into the hole of the stopper plate. Install the chain tensioner bolts and tighten to 7 ft. lbs.
45. Remove the lock pin of the chain tensioner. Check that each timing mark is aligned with the crankshaft at TDC of the compression stroke.

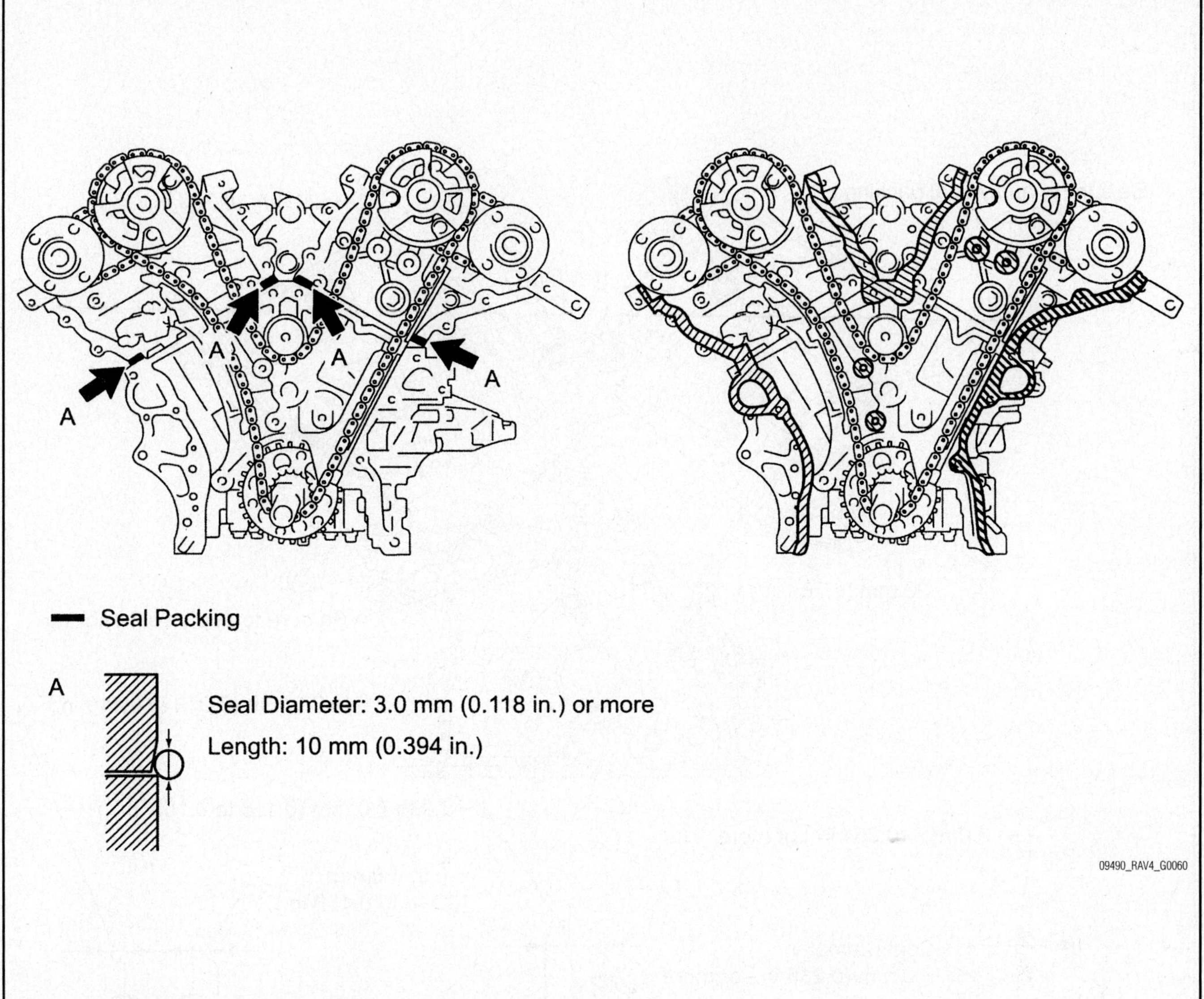

Fig. 127 Timing chain cover sealant application (to engine block)—3.5L engine

46. Replace the timing chain cover oil seal.

47. Install the timing chain cover sub-assembly, with the oil pump.

➡Be sure to align the oil pump's drive rotor spline and the crankshaft, as shown in the illustration.

48. Apply a continuous bead (0.118 inch or more) of seal packing (three bond 1207B or equivalent) to the engine.

➡Be sure to clean and degrease the contact surfaces. Make sure these surfaces are free of oil and dirt before applying the seal packing. Install the component within three minutes and tighten the bolts within fifteen minutes after applying the seal packing. Do not start the engine for at least two hours.

49. Apply a continuous bead of seal packing (three bond 1207B or equivalent) to the timing chain cover. See illustration for location and amount of seal packing to apply.

➡Be sure to clean and degrease the contact surfaces. Make sure these surfaces are free of oil and dirt before applying the seal packing. Install the component within three minutes and tighten the bolts within fifteen minutes after applying the seal packing. Do not start the engine for at least two hours.

50. Install the timing chain cover. Tighten the retaining bolts and nuts in the following order, area "1", area "2", area "3" and area "4". Tighten the bolts and nuts to 15 ft. lbs for area "1" area "2" and area "3". Tighten bolt "A" in area "4" to 32 ft. lbs. Tighten all other bolts in area "4" to 15 ft. lbs.

➡Be sure that there is no oil on the bolts. Bolt "A" is 1.57 inch long, bolt "B" is 2.17 inch long and bolt "C" is 0.98 inch long.

51. When installing the cylinder head cover, apply seal packing, three bond 1207B or equivalent.

➡Remove any oil from the contact surfaces. Install the component within three minutes after applying the sealant. Do not start the engine for two hours after the installation.

52. Install the cover retaining bolts and torque to 15 ft. lbs for bolt "A" and to 7 ft. lbs. for all other bolts and in the proper sequence.

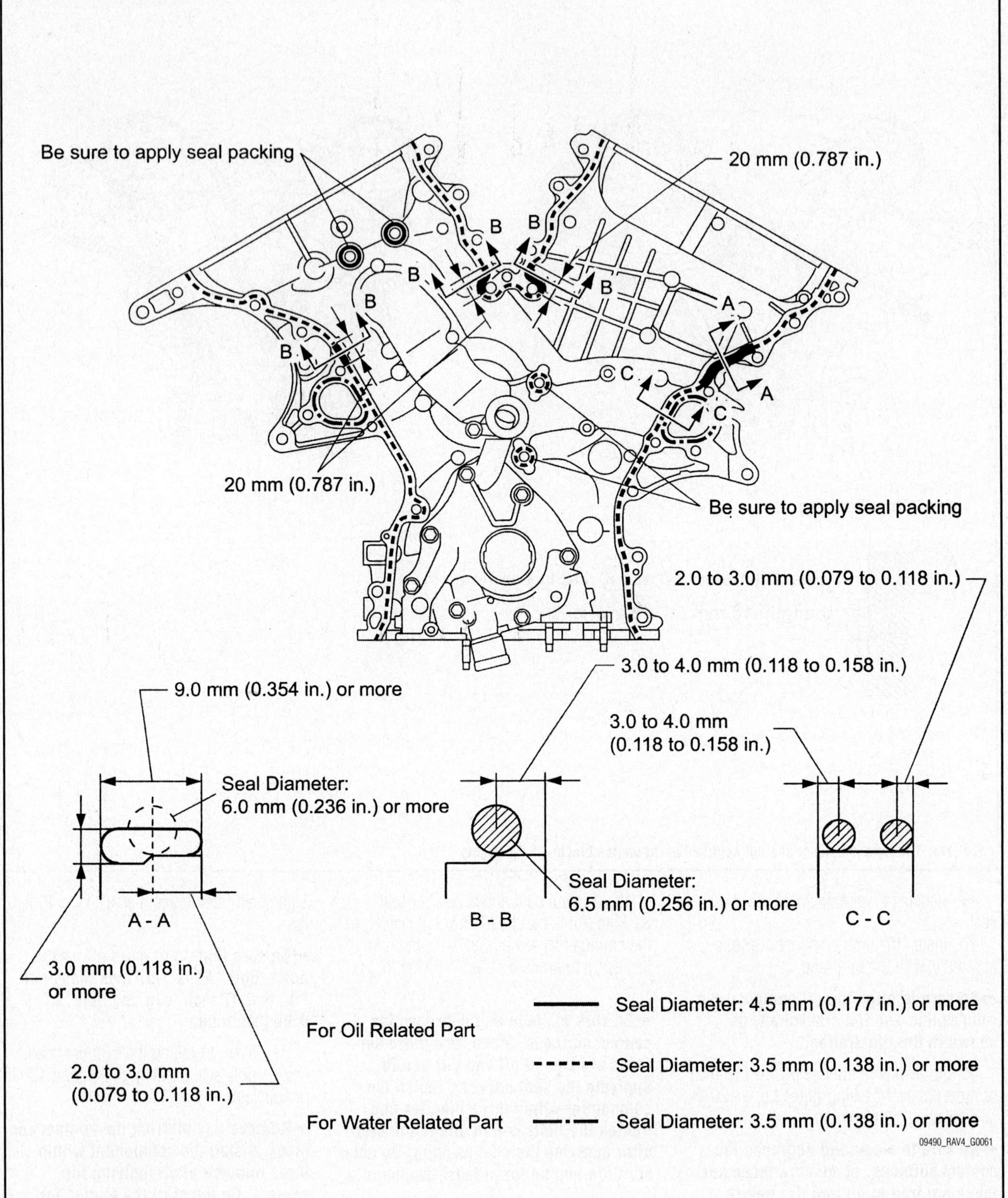

Fig. 128 Timing chain cover sealant application (to timing cover)—3.5L engine

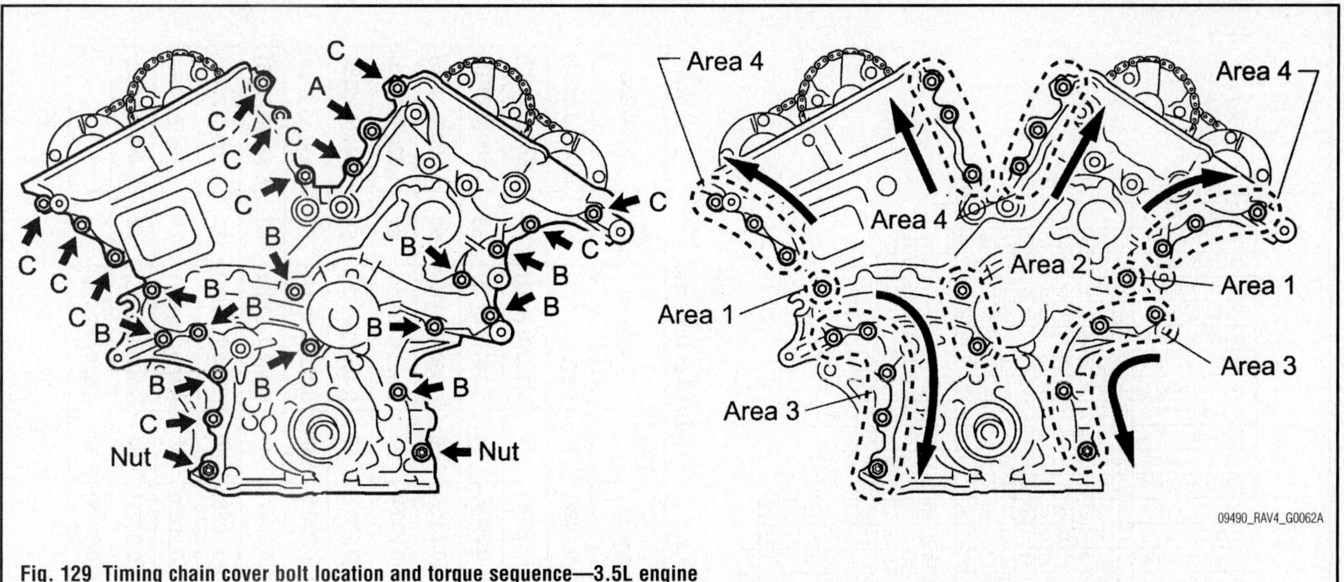

Fig. 129 Timing chain cover bolt location and torque sequence—3.5L engine

53. Continue the installation is the reverse order of the removal procedure.

VALVE LASH

ADJUSTMENT

2.4L Engine

See Figures 130 through 135.

1. Before servicing the vehicle, refer to the Precautions Section.
2. Disconnect the negative battery cable.

➡Wait at least 90 seconds after disconnecting the negative battery cable before starting any repair work to prevent air bag and seat belt pretensioner activation.

3. Remove the right front wheel and tire assembly. Remove the engine undercover. Remove the front fender apron, right side.
4. Remove the number one engine cover.
5. Remove the ignition coil. Remove the spark plugs.
6. Remove the cylinder head cover.
7. Position the number one cylinder to TDC on the compression stroke. Check and record the clearance on the number one and two intake valves and on the number one and three exhaust valves.
8. Rotate the engine 360 degrees and set the number four cylinder to TDC on the compression stroke. Check and record the clearance on the number three and four intake valves and on the number two and four exhaust valves.

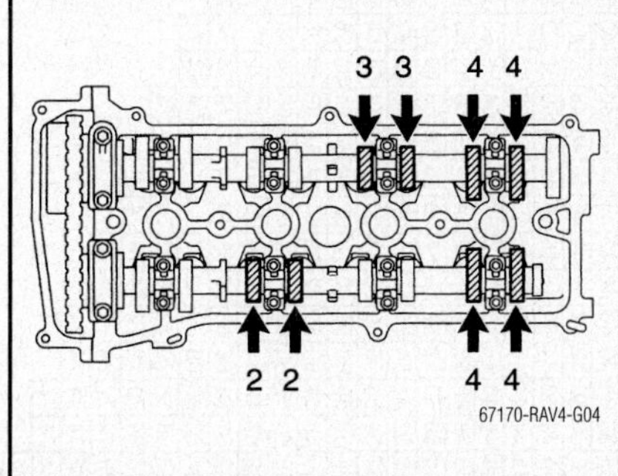

Fig. 131 Check the clearance on the Nos. 3 and 4 intake valves and the Nos. 2 and 4 exhaust valves—2.4L engine

9. If adjustment is necessary, remove the camshafts. Remove the valve lifters.
10. Using a micrometer, measure the lifter thickness.
11. Calculate the thickness of a new liter to bring the valve clearance into the proper range.
12. Select a lifter with a thickness as close as possible to correct the specified value, from the accompanying charts. Lifters are available in 35 sizes in 0.0008 inch (0.020mm) increments from 5.060mm to 5.740mm.
13. Reinstall removed components in the reverse order of the removal procedure.
14. When installing the cylinder head cover, apply seal packing (three bond 1207B, or equivalent) to the two locations, shown in the illustration. Tighten the retaining bolts to 8 ft. lbs. for bolt "A", 10 ft. lbs. for bolt "B" and 8 ft. lbs. for the nut.

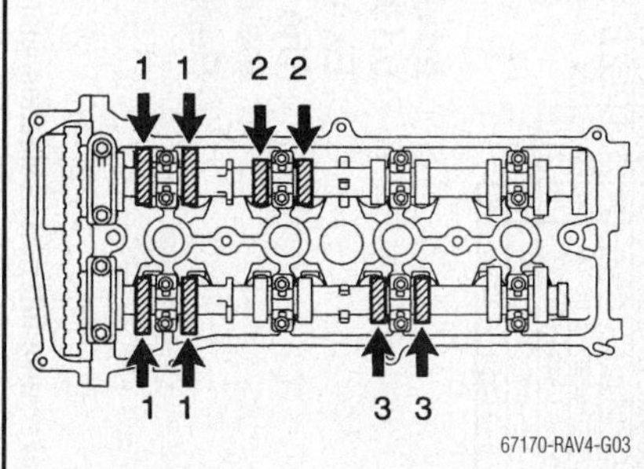

Fig. 130 Check the clearance on the Nos. 1 and 2 intake valves and the Nos. 1 and 3 exhaust valves—2.4L engine

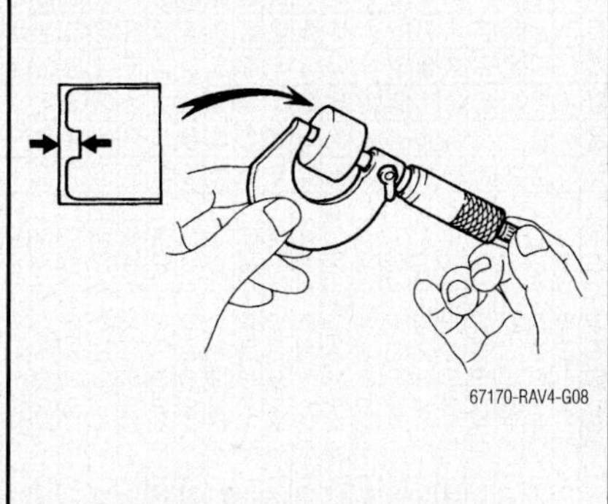

Fig. 132 Using a micrometer, measure the lifter thickness—2.4L engine

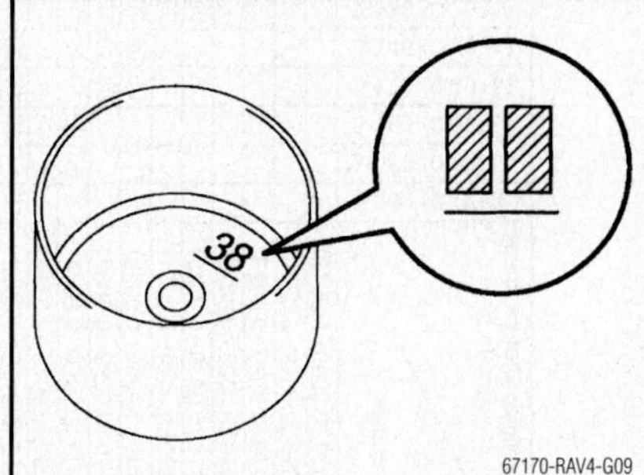

Fig. 133 An ID number inside the lifter shows the two decimal place size—2.4L engine

Valve Lifter Selection Chart (Intake)

Measure clearance mm (in.) \ Installed lifter thickness mm (in.)	5.060 (0.1992)	5.080 (0.2000)	5.100 (0.2008)	5.120 (0.2016)	5.140 (0.2024)	5.160 (0.2031)	5.180 (0.2039)	5.200 (0.2047)	5.210 (0.2051)	5.220 (0.2055)	5.230 (0.2059)	5.240 (0.2063)	5.250 (0.2067)	5.260 (0.2071)	5.270 (0.2075)	5.280 (0.2079)	5.290 (0.2083)	5.300 (0.2087)	5.310 (0.2091)	5.320 (0.2094)	5.330 (0.2098)	5.340 (0.2102)	5.350 (0.2106)	5.360 (0.2110)	5.370 (0.2114)	5.380 (0.2118)	5.390 (0.2122)	5.400 (0.2126)	5.410 (0.2130)	5.420 (0.2134)	5.430 (0.2138)	5.440 (0.2142)	5.450 (0.2146)	5.460 (0.2150)	5.470 (0.2154)	5.480 (0.2157)	5.490 (0.2161)	5.500 (0.2165)	5.510 (0.2169)	5.520 (0.2173)	5.530 (0.2177)	5.540 (0.2181)	5.550 (0.2185)	5.560 (0.2189)	5.570 (0.2193)	5.580 (0.2197)	5.590 (0.2201)	5.600 (0.2205)	5.620 (0.2213)	5.640 (0.2220)	5.660 (0.2228)	5.680 (0.2236)	5.700 (0.2244)	5.720 (0.2252)	5.740 (0.2260)
0.000-0.030 (0.0000-0.0012)													06	06	08	08	10	10	12	12	14	14	16	16	18	18	20	20	22	22	24	24	26	26	28	28	30	30	32	32	34	34	36	36	38	38	40	40	42	44	46	48	50	52	54
0.031-0.050 (0.0012-0.0020)									06	06	06	08	08	10	10	12	12	14	14	16	16	18	18	20	20	22	22	24	24	26	26	28	28	30	30	32	32	34	34	36	36	38	38	40	40	42	42	44	46	48	50	52	54	56	58
0.051-0.070 (0.0020-0.0028)								06	08	08	08	10	10	12	12	14	14	16	16	18	18	20	20	22	22	24	24	26	26	28	28	30	30	32	32	34	34	36	36	38	38	40	40	42	42	44	44	46	48	50	52	54	56	58	60
0.071-0.090 (0.0028-0.0035)							06	08	10	10	10	12	12	14	14	16	16	18	18	20	20	22	22	24	24	26	26	28	28	30	30	32	32	34	34	36	36	38	38	40	40	42	42	44	44	46	46	48	50	52	54	56	58	60	62
0.091-0.110 (0.0036-0.0043)						06	08	10	12	12	12	14	14	16	16	18	18	20	20	22	22	24	24	26	26	28	28	30	30	32	32	34	34	36	36	38	38	40	40	42	42	44	44	46	46	48	48	50	52	54	56	58	60	62	64
0.111-0.130 (0.0044-0.0051)					06	08	10	12	14	14	14	16	16	18	18	20	20	22	22	24	24	26	26	28	28	30	30	32	32	34	34	36	36	38	38	40	40	42	42	44	44	46	46	48	48	50	50	52	54	56	58	60	62	64	66
0.131-0.150 (0.0052-0.0059)				06	08	10	12	14	16	16	16	18	18	20	20	22	22	24	24	26	26	28	28	30	30	32	32	34	34	36	36	38	38	40	40	42	42	44	44	46	46	48	48	50	50	52	52	54	56	58	60	62	64	66	68
0.151-0.170 (0.0059-0.0067)			06	08	10	12	14	16	18	18	18	20	20	22	22	24	24	26	26	28	28	30	30	32	32	34	34	36	36	38	38	40	40	42	42	44	44	46	46	48	48	50	50	52	52	54	54	56	58	60	62	64	66	68	70
0.171-0.189 (0.0067-0.0074)		06	08	10	12	14	16	18	20	20	20	22	22	24	24	26	26	28	28	30	30	32	32	34	34	36	36	38	38	40	40	42	42	44	44	46	46	48	48	50	50	52	52	54	54	56	56	58	60	62	64	66	68	70	72
0.190-0.290 (0.0075-0.0114)																																																							
0.291-0.310 (0.0115-0.0122)	12	14	16	18	20	22	24	26	28	28	30	30	32	32	34	34	36	36	38	38	40	40	42	42	44	44	46	46	48	48	50	50	52	52	54	54	56	56	58	58	60	60	62	62	64	64	66	66	68	70	72	74	74	74	
0.311-0.330 (0.0122-0.0130)	14	16	18	20	22	24	26	28	30	30	32	32	34	34	36	36	38	38	40	40	42	42	44	44	46	46	48	48	50	50	52	52	54	54	56	56	58	58	60	60	62	62	64	64	66	66	68	68	70	72	74	74	74		
0.331-0.350 (0.0130-0.0138)	16	18	20	22	24	26	28	30	32	32	34	34	36	36	38	38	40	40	42	42	44	44	46	46	48	48	50	50	52	52	54	54	56	56	58	58	60	60	62	62	64	64	66	66	68	68	70	70	72	74	74	74			
0.351-0.370 (0.0138-0.0146)	18	20	22	24	26	28	30	32	34	34	36	36	38	38	40	40	42	42	44	44	46	46	48	48	50	50	52	52	54	54	56	56	58	58	60	60	62	62	64	64	66	66	68	68	70	70	72	72	74	74	74				
0.371-0.390 (0.0146-0.0154)	20	22	24	26	28	30	32	34	36	36	38	38	40	40	42	42	44	44	46	46	48	48	50	50	52	52	54	54	56	56	58	58	60	60	62	62	64	64	66	66	68	68	70	70	72	72	74	74	74	74					
0.391-0.410 (0.0154-0.0161)	22	24	26	28	30	32	34	36	38	38	40	40	42	42	44	44	46	46	48	48	50	50	52	52	54	54	56	56	58	58	60	60	62	62	64	64	66	66	68	68	70	70	72	72	74	74	74	74							
0.411-0.430 (0.0162-0.0169)	24	26	28	30	32	34	36	38	40	40	42	42	44	44	46	46	48	48	50	50	52	52	54	54	56	56	58	58	60	60	62	62	64	64	66	66	68	68	70	70	72	72	74	74	74	74									
0.431-0.450 (0.0170-0.0177)	26	28	30	32	34	36	38	40	42	42	44	44	46	46	48	48	50	50	52	52	54	54	56	56	58	58	60	60	62	62	64	64	66	66	68	68	70	70	72	72	74	74	74	74											
0.451-0.470 (0.0178-0.0185)	28	30	32	34	36	38	40	42	44	44	46	46	48	48	50	50	52	52	54	54	56	56	58	58	60	60	62	62	64	64	66	66	68	68	70	70	72	72	74	74	74	74													
0.471-0.490 (0.0185-0.0193)	30	32	34	36	38	40	42	44	46	46	48	48	50	50	52	52	54	54	56	56	58	58	60	60	62	62	64	64	66	66	68	68	70	70	72	72	74	74	74	74															
0.491-0.510 (0.0193-0.0201)	32	34	36	38	40	42	44	46	48	48	50	50	52	52	54	54	56	56	58	58	60	60	62	62	64	64	66	66	68	68	70	70	72	72	74	74	74	74																	
0.511-0.530 (0.0201-0.0209)	34	36	38	40	42	44	46	48	50	50	52	52	54	54	56	56	58	58	60	60	62	62	64	64	66	66	68	68	70	70	72	72	74	74	74	74																			
0.531-0.550 (0.0209-0.0217)	36	38	40	42	44	46	48	50	52	52	54	54	56	56	58	58	60	60	62	62	64	64	66	66	68	68	70	70	72	72	74	74	74	74																					
0.551-0.570 (0.0217-0.0224)	38	40	42	44	46	48	50	52	54	54	56	56	58	58	60	60	62	62	64	64	66	66	68	68	70	70	72	72	74	74	74	74																							
0.571-0.590 (0.0225-0.0232)	40	42	44	46	48	50	52	54	56	56	58	58	60	60	62	62	64	64	66	66	68	68	70	70	72	72	74	74	74	74																									
0.591-0.610 (0.0233-0.0240)	42	44	46	48	50	52	54	56	58	58	60	60	62	62	64	64	66	66	68	68	70	70	72	72	74	74	74	74																											
0.611-0.630 (0.0241-0.0248)	44	46	48	50	52	54	56	58	60	60	62	62	64	64	66	66	68	68	70	70	72	72	74	74	74	74																													
0.631-0.650 (0.0248-0.0256)	46	48	50	52	54	56	58	60	62	62	64	64	66	66	68	68	70	70	72	72	74	74	74	74																															
0.651-0.670 (0.0256-0.0264)	48	50	52	54	56	58	60	62	64	64	66	66	68	68	70	70	72	72	74	74	74	74																																	
0.671-0.690 (0.0264-0.0272)	50	52	54	56	58	60	62	64	66	66	68	68	70	70	72	72	74	74	74	74																																			
0.691-0.710 (0.0272-0.0280)	52	54	56	58	60	62	64	66	68	68	70	70	72	72	74	74	74	74																																					
0.711-0.730 (0.0280-0.0287)	54	56	58	60	62	64	66	68	70	70	72	72	74	74	74	74																																							
0.731-0.750 (0.0288-0.0295)	56	58	60	62	64	66	68	70	72	72	74	74	74	74																																									
0.751-0.770 (0.0296-0.0303)	58	60	62	64	66	68	70	72	74	74	74	74																																											
0.771-0.790 (0.0304-0.0311)	60	62	64	66	68	70	72	74	74	74																																													
0.791-0.810 (0.0311-0.0319)	62	64	66	68	70	72	74																																																
0.811-0.830 (0.0319-0.0327)	64	66	68	70	72	74																																																	
0.831-0.850 (0.0327-0.0335)	66	68	70	72	74																																																		
0.851-0.870 (0.0335-0.0343)	68	70	72	74																																																			
0.871-0.890 (0.0343-0.0350)	70	72	74																																																				
0.891-0.910 (0.0351-0.0358)	72	74																																																					
0.911-0.930 (0.0359-0.0366)	74																																																						

Intake valve clearance (Cold):
0.19 to 0.29 mm (0.008 to 0.011 in.)

EXAMPLE:
The 5.250 mm (0.2067 in.) lifter is installed, and the measured clearance is 0.400 mm (0.0157 in.).
Replace the 5.250 mm (0.2067 in.) lifter with a new No. 42 lifter.

New Lifter Thickness mm (in.)

Lifter No.	Thickness	Lifter No.	Thickness	Lifter No.	Thickness
06	5.060 (0.1992)	30	5.300 (0.2087)	54	5.540 (0.2181)
08	5.080 (0.2000)	32	5.320 (0.2094)	56	5.560 (0.2189)
10	5.100 (0.2008)	34	5.340 (0.2102)	58	5.580 (0.2197)
12	5.120 (0.2016)	36	5.360 (0.2110)	60	5.600 (0.2205)
14	5.140 (0.2024)	38	5.380 (0.2118)	62	5.620 (0.2213)
16	5.160 (0.2031)	40	5.400 (0.2126)	64	5.640 (0.2220)
18	5.180 (0.2039)	42	5.420 (0.2134)	66	5.660 (0.2228)
20	5.200 (0.2047)	44	5.440 (0.2142)	68	5.680 (0.2236)
22	5.220 (0.2055)	46	5.460 (0.2150)	70	5.700 (0.2244)
24	5.240 (0.2063)	48	5.480 (0.2157)	72	5.720 (0.2252)
26	5.260 (0.2071)	50	5.500 (0.2165)	74	5.740 (0.2260)
28	5.280 (0.2079)	52	5.520 (0.2173)		

67170-RAV4-G10

Fig. 134 Intake valve lifter size selection chart—2.4L engine

Valve Lifter Selection Chart (Exhaust)

Measure clearance mm (in.) \ Installed lifter thickness mm (in.)	5.060 (0.1992)	5.080 (0.2000)	5.100 (0.2008)	5.120 (0.2016)	5.140 (0.2024)	5.160 (0.2031)	5.180 (0.2039)	5.200 (0.2047)	5.210 (0.2051)	5.220 (0.2055)	5.230 (0.2059)	5.240 (0.2063)	5.250 (0.2067)	5.260 (0.2071)	5.270 (0.2075)	5.280 (0.2079)	5.290 (0.2083)	5.300 (0.2087)	5.310 (0.2091)	5.320 (0.2094)	5.330 (0.2098)	5.340 (0.2102)	5.350 (0.2106)	5.360 (0.2110)	5.370 (0.2114)	5.380 (0.2118)	5.390 (0.2122)	5.400 (0.2126)	5.410 (0.2130)	5.420 (0.2134)	5.430 (0.2138)	5.440 (0.2142)	5.450 (0.2146)	5.460 (0.2150)	5.470 (0.2154)	5.480 (0.2157)	5.490 (0.2161)	5.500 (0.2165)	5.510 (0.2169)	5.520 (0.2173)	5.530 (0.2177)	5.540 (0.2181)	5.550 (0.2185)	5.560 (0.2189)	5.570 (0.2193)	5.580 (0.2197)	5.590 (0.2201)	5.600 (0.2205)	5.620 (0.2213)	5.640 (0.2220)	5.660 (0.2228)	5.680 (0.2236)	5.700 (0.2244)	5.720 (0.2252)	5.740 (0.2260)
0.000-0.030 (0.0000-0.0012)																							06	06	08	08	10	10	12	12	14	14	16	16	18	18	20	20	22	22	24	24	26	26	28	28	30	30	32	34	36	38	40	42	44
0.031-0.050 (0.0012-0.0020)																					06	06	08	08	10	10	12	12	14	14	16	16	18	18	20	20	22	22	24	24	26	26	28	28	30	30	32	32	34	36	38	40	42	44	46
0.051-0.070 (0.0020-0.0028)																			06	06	08	08	10	10	12	12	14	14	16	16	18	18	20	20	22	22	24	24	26	26	28	28	30	30	32	32	34	34	36	38	40	42	44	46	48
0.071-0.090 (0.0028-0.0035)																	06	06	08	08	10	10	12	12	14	14	16	16	18	18	20	20	22	22	24	24	26	26	28	28	30	30	32	32	34	34	36	36	38	40	42	44	46	48	50
0.091-0.110 (0.0036-0.0043)															06	06	08	08	10	10	12	12	14	14	16	16	18	18	20	20	22	22	24	24	26	26	28	28	30	30	32	32	34	34	36	36	38	38	40	42	44	46	48	50	52
0.111-0.130 (0.0044-0.0051)													06	06	08	08	10	10	12	12	14	14	16	16	18	18	20	20	22	22	24	24	26	26	28	28	30	30	32	32	34	34	36	36	38	38	40	40	42	44	46	48	50	52	54
0.131-0.150 (0.0052-0.0059)											06	06	08	08	10	10	12	12	14	14	16	16	18	18	20	20	22	22	24	24	26	26	28	28	30	30	32	32	34	34	36	36	38	38	40	40	42	42	44	46	48	50	52	54	56
0.151-0.170 (0.0059-0.0067)									06	06	08	08	10	10	12	12	14	14	16	16	18	18	20	20	22	22	24	24	26	26	28	28	30	30	32	32	34	34	36	36	38	38	40	40	42	42	44	44	46	48	50	52	54	56	58
0.171-0.190 (0.0067-0.0075)								06	08	08	10	10	12	12	14	14	16	16	18	18	20	20	22	22	24	24	26	26	28	28	30	30	32	32	34	34	36	36	38	38	40	40	42	42	44	44	46	46	48	50	52	54	56	58	60
0.191-0.210 (0.0075-0.0083)							06	08	10	10	12	12	14	14	16	16	18	18	20	20	22	22	24	24	26	26	28	28	30	30	32	32	34	34	36	36	38	38	40	40	42	42	44	44	46	46	48	48	50	52	54	56	58	60	62
0.211-0.230 (0.0083-0.0091)						06	08	10	12	12	14	14	16	16	18	18	20	20	22	22	24	24	26	26	28	28	30	30	32	32	34	34	36	36	38	38	40	40	42	42	44	44	46	46	48	48	50	50	52	54	56	58	60	62	64
0.231-0.250 (0.0091-0.0098)					06	08	10	12	14	14	16	16	18	18	20	20	22	22	24	24	26	26	28	28	30	30	32	32	34	34	36	36	38	38	40	40	42	42	44	44	46	46	48	48	50	50	52	52	54	56	58	60	62	64	66
0.251-0.270 (0.0099-0.0106)				06	08	10	12	14	16	16	18	18	20	20	22	22	24	24	26	26	28	28	30	30	32	32	34	34	36	36	38	38	40	40	42	42	44	44	46	46	48	48	50	50	52	52	54	54	56	58	60	62	64	66	68
0.271-0.290 (0.0107-0.0114)			06	08	10	12	14	16	18	18	20	20	22	22	24	24	26	26	28	28	30	30	32	32	34	34	36	36	38	38	40	40	42	42	44	44	46	46	48	48	50	50	52	52	54	54	56	56	58	60	62	64	66	68	70
0.291-0.299 (0.0115-0.0118)		06	08	10	12	14	16	18	20	20	22	22	24	24	26	26	28	28	30	30	32	32	34	34	36	36	38	38	40	40	42	42	44	44	46	46	48	48	50	50	52	52	54	54	56	56	58	58	60	62	64	66	68	70	72
0.300-0.400 (0.0118-0.0157)																																																							
0.401-0.420 (0.0158-0.0165)	12	14	16	18	20	22	24	26	26	28	28	30	30	32	32	34	34	36	36	38	38	40	40	42	42	44	44	46	46	48	48	50	50	52	52	54	54	56	56	58	58	60	60	62	62	64	64	66	68	70	72	74	74	74	
0.421-0.440 (0.0166-0.0173)	14	16	18	20	22	24	26	28	28	30	30	32	32	34	34	36	36	38	38	40	40	42	42	44	44	46	46	48	48	50	50	52	52	54	54	56	56	58	58	60	60	62	62	64	64	66	66	68	70	72	74	74	74		
0.441-0.460 (0.0174-0.0181)	16	18	20	22	24	26	28	30	30	32	32	34	34	36	36	38	38	40	40	42	42	44	44	46	46	48	48	50	50	52	52	54	54	56	56	58	58	60	60	62	62	64	64	66	66	68	68	70	72	74	74	74			
0.461-0.480 (0.0181-0.0189)	18	20	22	24	26	28	30	32	32	34	34	36	36	38	38	40	40	42	42	44	44	46	46	48	48	50	50	52	52	54	54	56	56	58	58	60	60	62	62	64	64	66	66	68	68	70	70	72	74	74	74				
0.481-0.500 (0.0189-0.0197)	20	22	24	26	28	30	32	34	34	36	36	38	38	40	40	42	42	44	44	46	46	48	48	50	50	52	52	54	54	56	56	58	58	60	60	62	62	64	64	66	66	68	68	70	70	72	72	74	74	74					
0.501-0.520 (0.0197-0.0205)	22	24	26	28	30	32	34	36	36	38	38	40	40	42	42	44	44	46	46	48	48	50	50	52	52	54	54	56	56	58	58	60	60	62	62	64	64	66	66	68	68	70	70	72	72	74	74	74	74						
0.521-0.540 (0.0205-0.0213)	24	26	28	30	32	34	36	38	38	40	40	42	42	44	44	46	46	48	48	50	50	52	52	54	54	56	56	58	58	60	60	62	62	64	64	66	66	68	68	70	70	72	72	74	74	76	76	74							
0.541-0.560 (0.0213-0.0220)	26	28	30	32	34	36	38	40	40	42	42	44	44	46	46	48	48	50	50	52	52	54	54	56	56	58	58	60	60	62	62	64	64	66	66	68	68	70	70	72	72	74	74	74	74	74	74								
0.561-0.580 (0.0221-0.0228)	28	30	32	34	36	38	40	42	42	44	44	46	46	48	48	50	50	52	52	54	54	56	56	58	58	60	60	62	62	64	64	66	66	68	68	70	70	72	72	74	74	74	74	74	74										
0.581-0.600 (0.0229-0.0236)	30	32	34	36	38	40	42	44	44	46	46	48	48	50	50	52	52	54	54	56	56	58	58	60	60	62	62	64	64	66	66	68	68	70	70	72	72	74	74	74	74	74	74												
0.601-0.620 (0.0237-0.0244)	32	34	36	38	40	42	44	46	46	48	48	50	50	52	52	54	54	56	56	58	58	60	60	62	62	64	64	66	66	68	68	70	70	72	72	74	74	74	74	74	74														
0.621-0.640 (0.0244-0.0252)	34	36	38	40	42	44	46	48	48	50	50	52	52	54	54	56	56	58	58	60	60	62	62	64	64	66	66	68	68	70	70	72	72	74	74	74	74	74	74																
0.641-0.660 (0.0252-0.0260)	36	38	40	42	44	46	48	50	50	52	52	54	54	56	56	58	58	60	60	62	62	64	64	66	66	68	68	70	70	72	72	74	74	74	74	74	74																		
0.661-0.680 (0.0260-0.0268)	38	40	42	44	46	48	50	52	52	54	54	56	56	58	58	60	60	62	62	64	64	66	66	68	68	70	70	72	72	74	74	74	74	74	74																				
0.681-0.700 (0.0268-0.0276)	40	42	44	46	48	50	52	54	54	56	56	58	58	60	60	62	62	64	64	66	66	68	68	70	70	72	72	74	74	74	74	74	74																						
0.701-0.720 (0.0276-0.0283)	42	44	46	48	50	52	54	56	56	58	58	60	60	62	62	64	64	66	66	68	68	70	70	72	72	74	74	74	74	74	74																								
0.721-0.740 (0.0284-0.0291)	44	46	48	50	52	54	56	58	58	60	60	62	62	64	64	66	66	68	68	70	70	72	72	74	74	74	74	74	74																										
0.741-0.760 (0.0292-0.0299)	46	48	50	52	54	56	58	60	60	62	62	64	64	66	66	68	68	70	70	72	72	74	74	74	74	74	74																												
0.761-0.780 (0.0300-0.0307)	48	50	52	54	56	58	60	62	62	64	64	66	66	68	68	70	70	72	72	74	74	74	74	74	74																														
0.781-0.800 (0.0307-0.0315)	50	52	54	56	58	60	62	64	64	66	66	68	68	70	70	72	72	74	74	74	74	74	74																																
0.801-0.820 (0.0315-0.0323)	52	54	56	58	60	62	64	66	66	68	68	70	70	72	72	74	74	74	74	74	74																																		
0.821-0.840 (0.0323-0.0331)	54	56	58	60	62	64	66	68	68	70	70	72	72	74	74	74	74	74	74																																				
0.841-0.860 (0.0331-0.0339)	56	58	60	62	64	66	68	70	70	72	72	74	74	74	74	74	74																																						
0.861-0.880 (0.0339-0.0346)	58	60	62	64	66	68	70	72	72	74	74	74	74	74	74																																								
0.881-0.900 (0.0347-0.0354)	60	62	64	66	68	70	72	74	74	74	74	74	74																																										
0.901-0.920 (0.0355-0.0362)	62	64	66	68	70	72	74	74	74	74	74																																												
0.921-0.940 (0.0363-0.0370)	64	66	68	70	72	74	74	74	74																																														
0.941-0.960 (0.0370-0.0378)	66	68	70	72	74	74	74																																																
0.961-0.980 (0.0378-0.0386)	68	70	72	74	74	74																																																	
0.981-1.000 (0.0386-0.0394)	70	72	74	74	74																																																		
1.001-1.020 (0.0394-0.0402)	72	74	74	74																																																			
1.021-1.040 (0.0402-0.0409)	74	74	74																																																				
1.041-1.060 (0.0410-0.0417)	74	74																																																					
1.061-1.080 (0.0418-0.0425)	74																																																						

Exhaust valve clearance (Cold):
0.30 to 0.40 mm (0.012 to 0.016 in.)

EXAMPLE:
The 5.340 mm (0.2102 in.) lifter is installed, and the measured clearance is 0.430 mm (0.0169 in.).
Replace the 5.340 mm (0.2102 in.) lifter with a new No. 42 lifter.

New Lifter Thickness mm (in.)

Lifter No.	Thickness	Lifter No.	Thickness	Lifter No.	Thickness
06	5.060 (0.1992)	30	5.300 (0.2087)	54	5.540 (0.2181)
08	5.080 (0.2000)	32	5.320 (0.2094)	56	5.560 (0.2189)
10	5.100 (0.2008)	34	5.340 (0.2102)	58	5.580 (0.2197)
12	5.120 (0.2016)	36	5.360 (0.2110)	60	5.600 (0.2205)
14	5.140 (0.2024)	38	5.380 (0.2118)	62	5.620 (0.2213)
16	5.160 (0.2031)	40	5.400 (0.2126)	64	5.640 (0.2220)
18	5.180 (0.2039)	42	5.420 (0.2134)	66	5.660 (0.2228)
20	5.200 (0.2047)	44	5.440 (0.2142)	68	5.680 (0.2236)
22	5.220 (0.2055)	46	5.460 (0.2150)	70	5.700 (0.2244)
24	5.240 (0.2063)	48	5.480 (0.2157)	72	5.720 (0.2252)
26	5.260 (0.2071)	50	5.500 (0.2165)	74	5.740 (0.2260)
28	5.280 (0.2079)	52	5.520 (0.2173)		

67170-RAV4-G11

Fig. 135 Exhaust valve lifter size selection chart—2.4L engine

ENGINE PERFORMANCE & EMISSION CONTROL

ACCELERATOR PEDAL POSITION (APP) SENSOR

LOCATION

See Figure 136.

The accelerator pedal rod is located on the accelerator pedal.

REMOVAL & INSTALLATION

1. Before servicing the vehicle, refer to the Precautions Section.
2. Disconnect the negative battery cable.

➡Wait at least 90 seconds after disconnecting the negative battery cable before starting any repair work to prevent air bag and seat belt pretensioner activation.

3. Disconnect the electrical connector.
4. Remove the retaining bolts.
5. Remove the component from its mounting.

To install:

6. Installation is the reverse of the removal procedure.
7. Tighten the retaining bolts to 48 inch lbs.

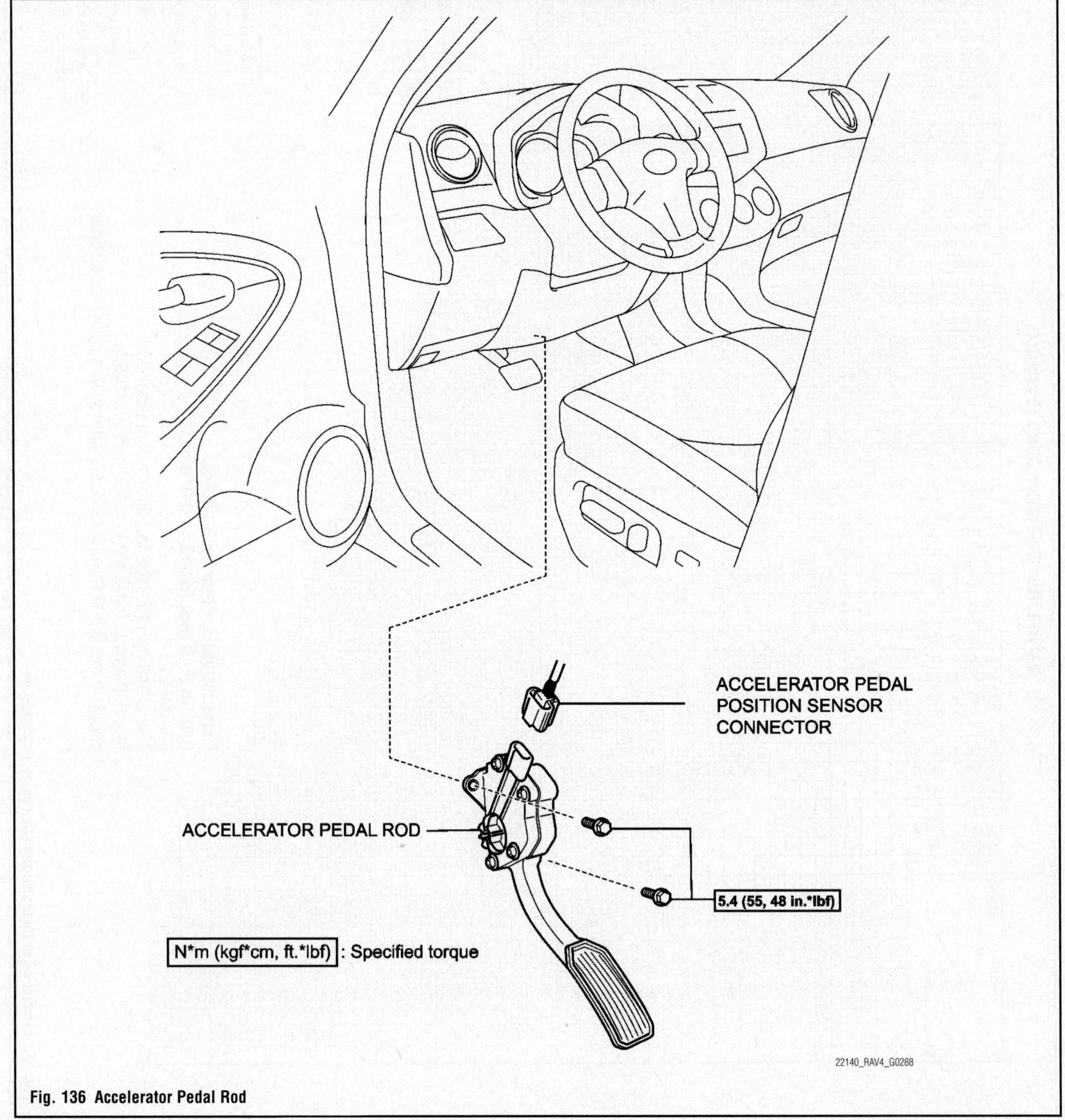

Fig. 136 Accelerator Pedal Rod

AIR FUEL (AF) SENSOR

LOCATION

See Figures 137 and 138.

REMOVAL & INSTALLATION

See Figures 139 and 140.

1. Before servicing the vehicle, refer to the Precautions Section.
2. Disconnect the negative battery cable.

➡Wait at least 90 seconds after disconnecting the negative battery cable before starting any repair work to prevent air bag and seat belt pretensioner activation.

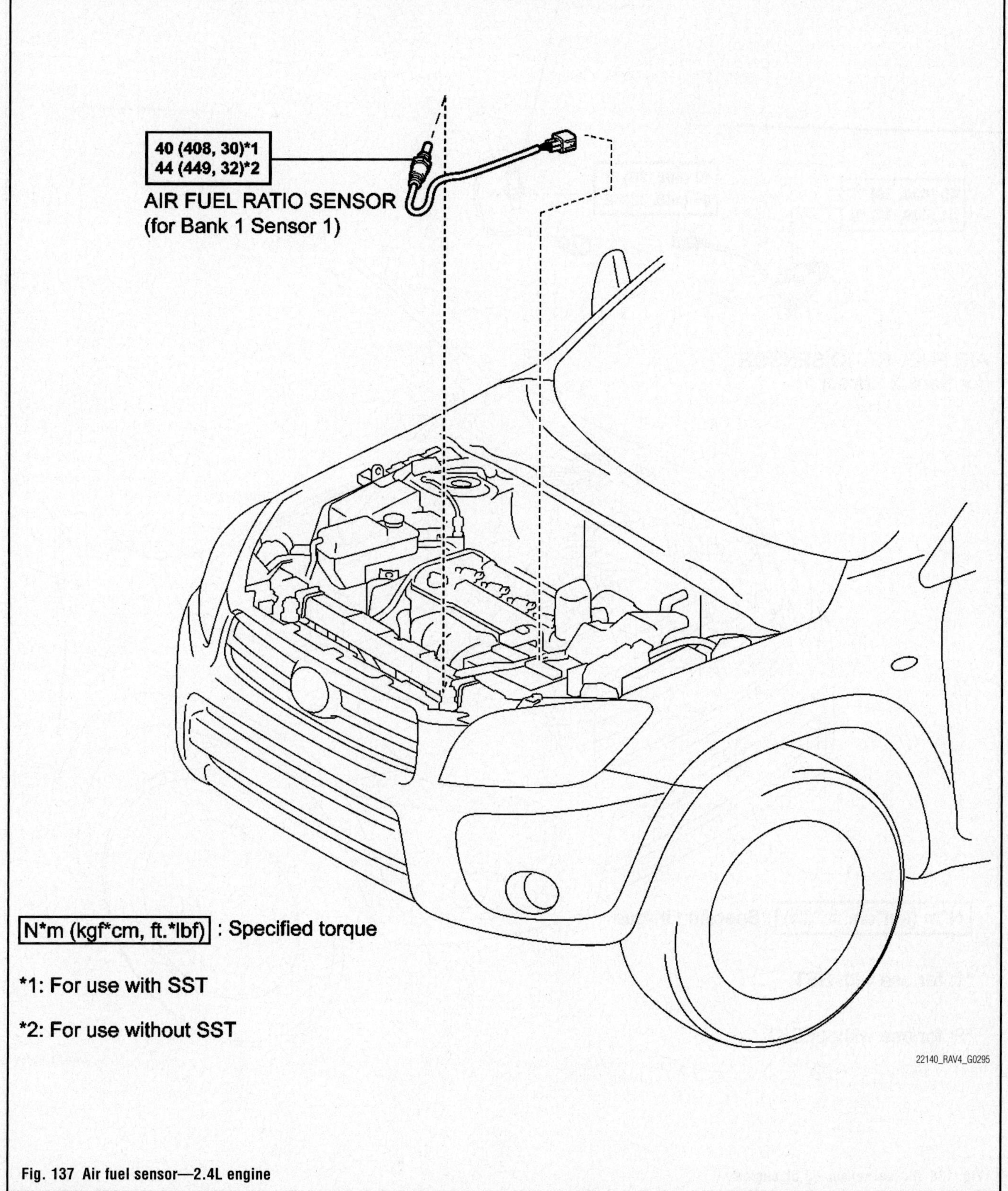

Fig. 137 Air fuel sensor—2.4L engine

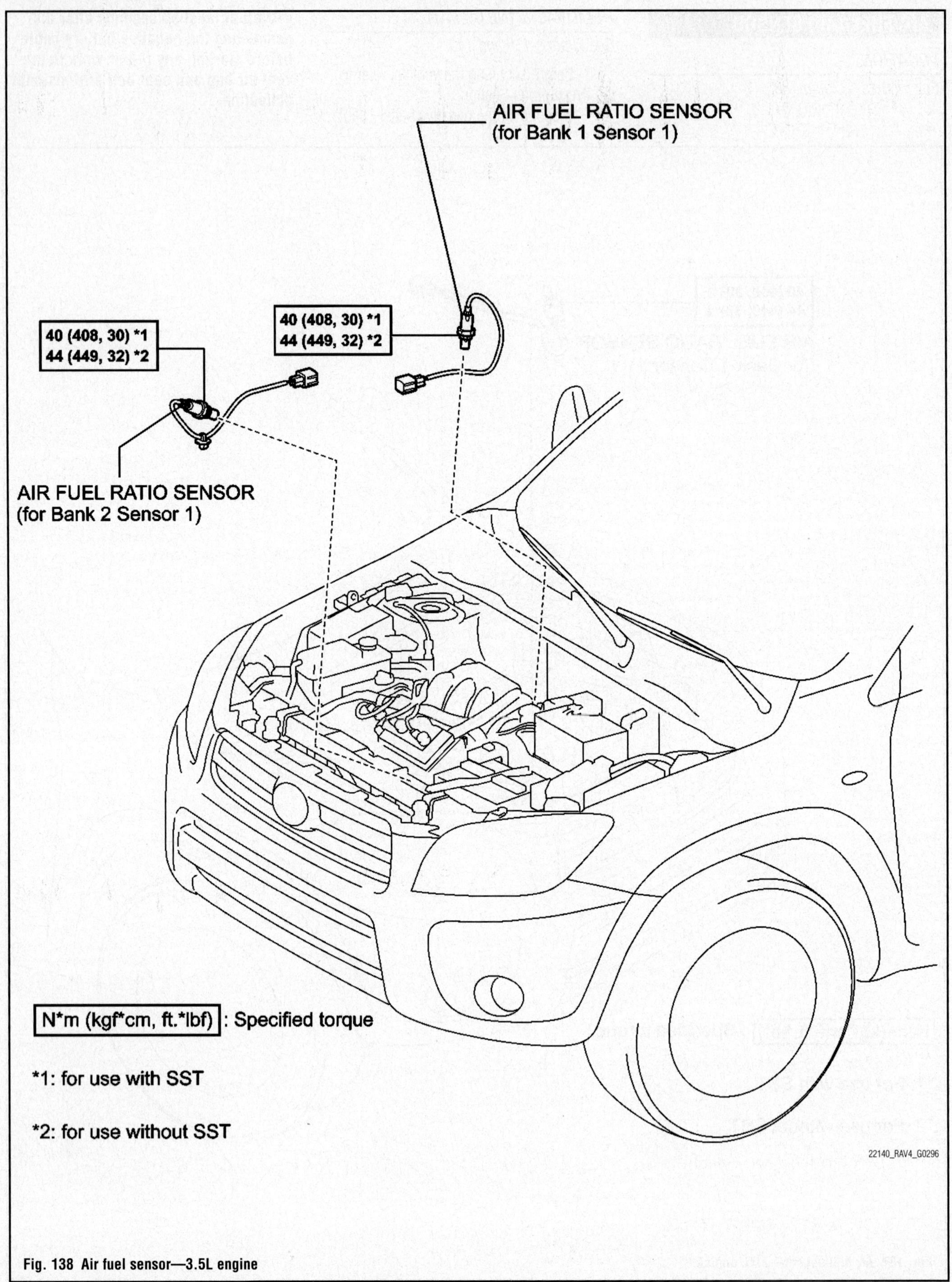

Fig. 138 Air fuel sensor—3.5L engine

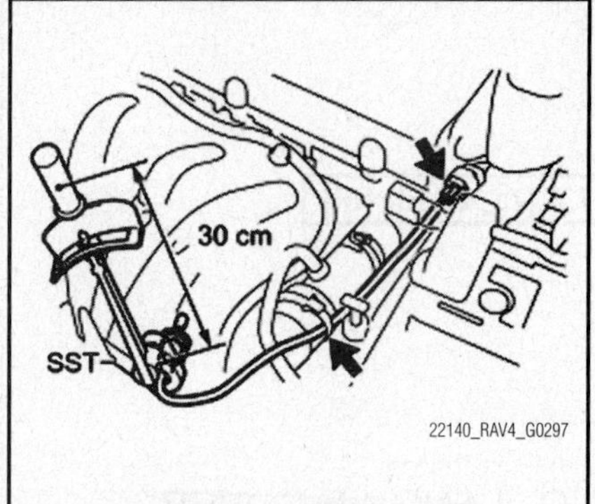

Fig. 139 Air fuel sensor torque wrench positioning—2.4L engine

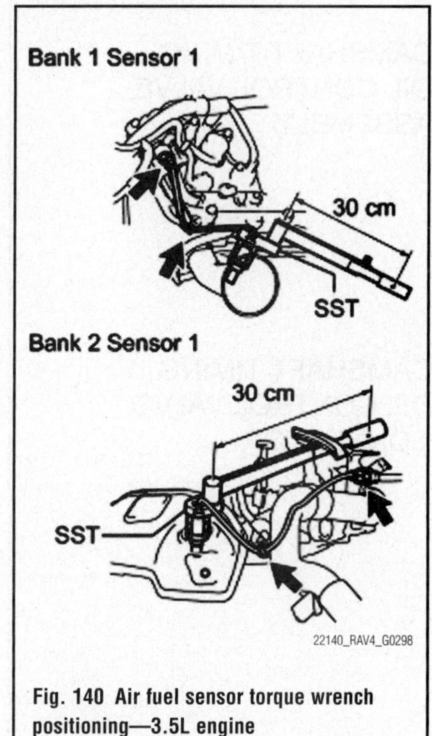

Fig. 140 Air fuel sensor torque wrench positioning—3.5L engine

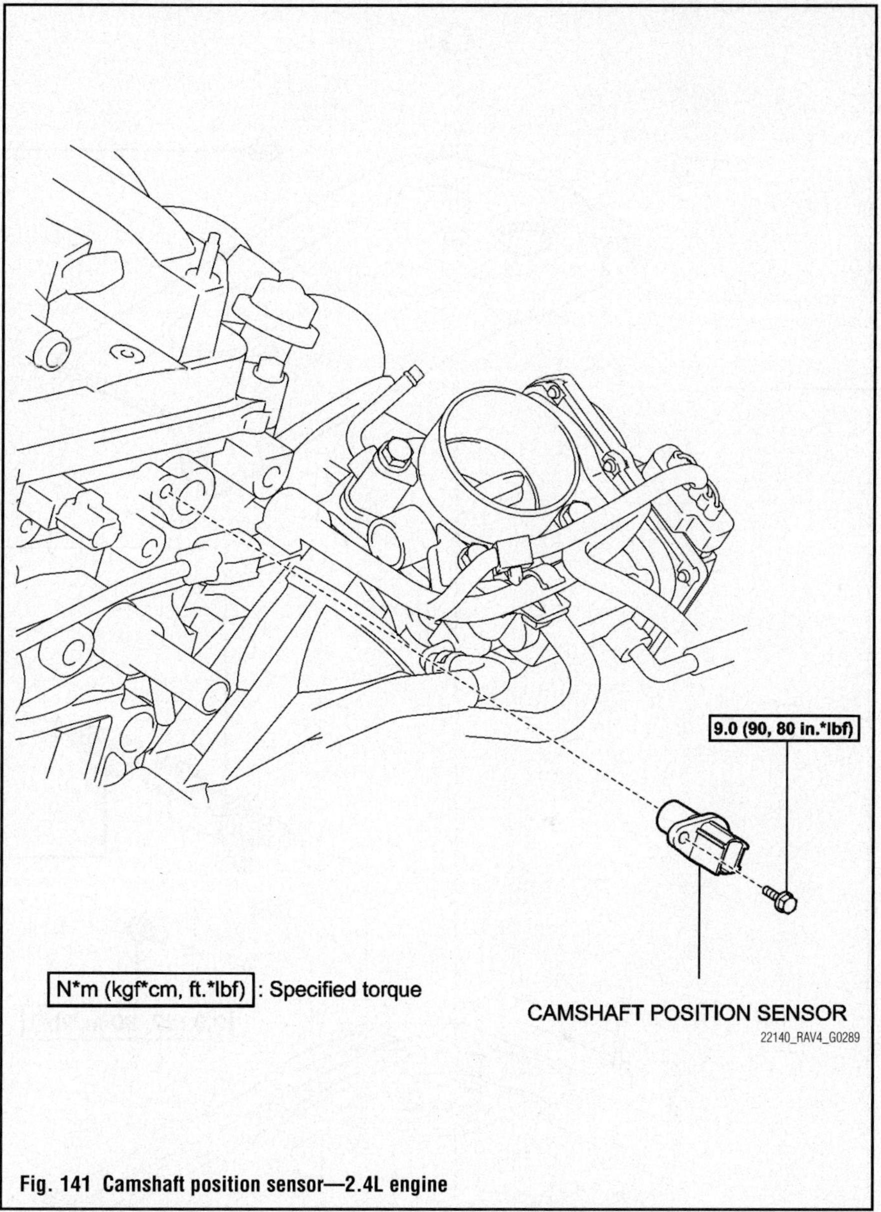

Fig. 141 Camshaft position sensor—2.4L engine

3. Disconnect the electrical connector.
4. Remove the component from its mounting, using the sensor removal tool SST:09224-00010 or equivalent.

To install:

5. Installation is the reverse of the removal procedure.
6. Tighten the sensor to 32 ft. lbs. Be sure to use a torque wrench with a fulcrum length of 11.81 inch.
7. Check for exhaust leaks.

CAMSHAFT POSITION (CMP) SENSOR

LOCATION

2.4L Engine

See Figure 141.

REMOVAL & INSTALLATION

2.4L Engine

1. Before servicing the vehicle, refer to the Precautions Section.
2. Disconnect the negative battery cable.

➡Wait at least 90 seconds after disconnecting the negative battery cable before starting any repair work to prevent air bag and seat belt pretensioner activation.

3. Remove the number one engine cover.
4. Remove the air cleaner assembly.
5. Disconnect the electrical connector.
6. Remove the retaining bolt.
7. Remove the component from its mounting.

To install:

8. Installation is the reverse of the removal procedure.
9. Be sure that the sensor O-ring is not cracked or jammed when installing the sensor.
10. Apply a light coat of clean engine oil to the O-ring.
11. Tighten the retaining bolt to 80 inch lbs.

CAMSHAFT TIMING OIL CONTROL VALVE (OCV)

LOCATION

See Figures 142 and 143.

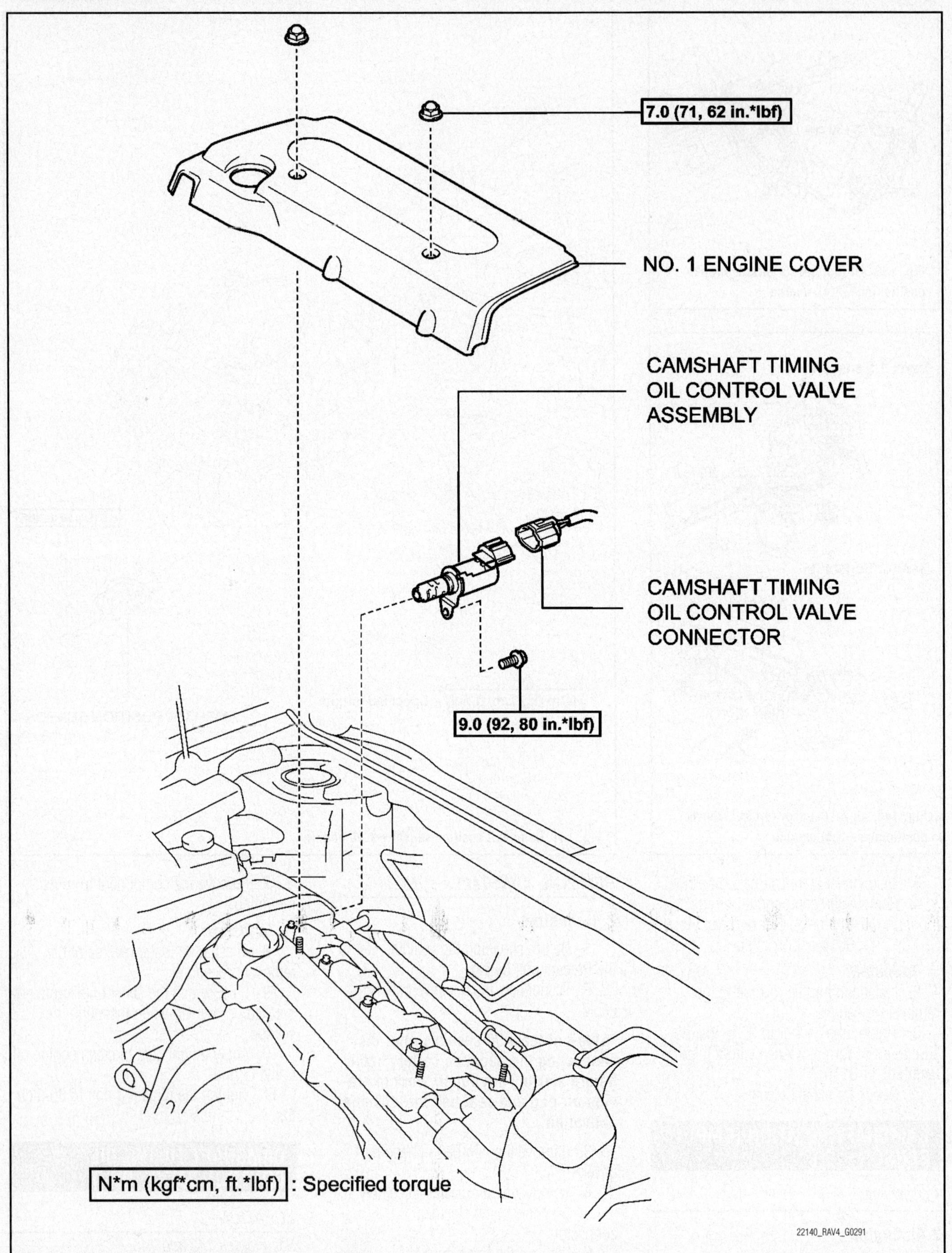

Fig. 142 Camshaft timing oil control valve—2.4L engine

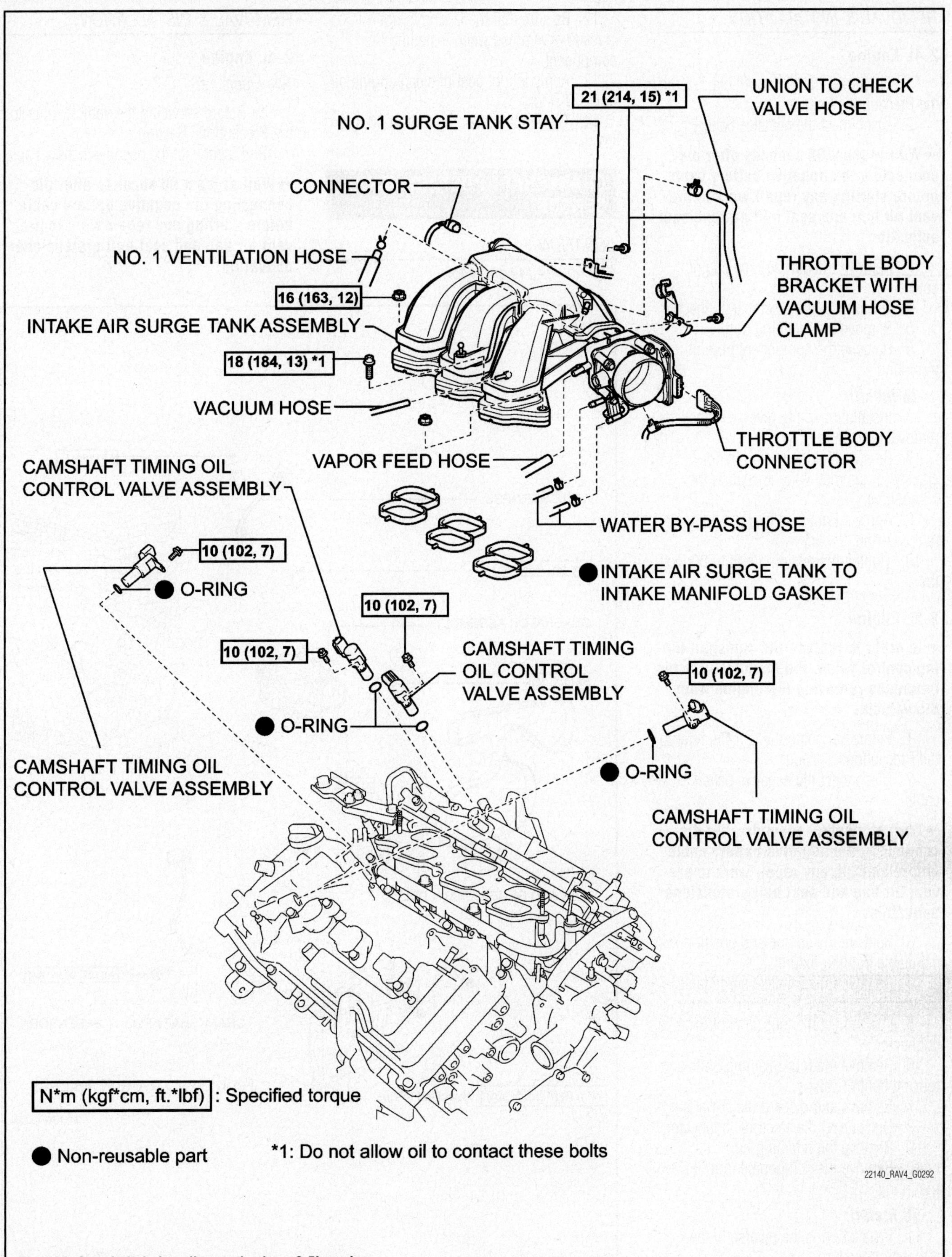

Fig. 143 Camshaft timing oil control valve—3.5L engine

REMOVAL & INSTALLATION

2.4L Engine

1. Before servicing the vehicle, refer to the Precautions Section.
2. Disconnect the negative battery cable.

➡Wait at least 90 seconds after disconnecting the negative battery cable before starting any repair work to prevent air bag and seat belt pretensioner activation.

3. Remove the number one engine cover.
4. Disconnect the electrical connector.
5. Remove the retaining bolt.
6. Remove the component from its mounting.

To install:

7. Installation is the reverse of the removal procedure.
8. Be sure that the O-ring is not cracked or jammed when installing the component.
9. Apply a light coat of clean engine oil to the O-ring.
10. Tighten the retaining bolt to 80 inch lbs.

3.5L Engine

➡In order to replace the camshaft timing control valve, the manufacturer recommends removing the engine from the vehicle.

1. Before servicing the vehicle, refer to the Precautions Section.
2. Disconnect the negative battery cable.

➡Wait at least 90 seconds after disconnecting the negative battery cable before starting any repair work to prevent air bag and seat belt pretensioner activation.

3. Remove the engine and position it in a suitable holding fixture.
4. Remove the air intake surge tank assembly.
5. Disconnect the four oil control valve connectors.
6. Remove the four retaining bolts and four oil control valves.
7. Remove and discard the O-rings.
8. Disconnect the electrical connector.
9. Remove the retaining bolt.
10. Remove the component from its mounting.

To install:

11. Installation is the reverse of the removal procedure.
12. Be sure that the O-ring is not cracked or jammed when installing the component.
13. Apply a light coat of clean engine oil to the O-ring.
14. Tighten the retaining bolt to 80 inch lbs.

CRANKSHAFT POSITION (CKP) SENSOR

LOCATION

See Figures 144 and 145.

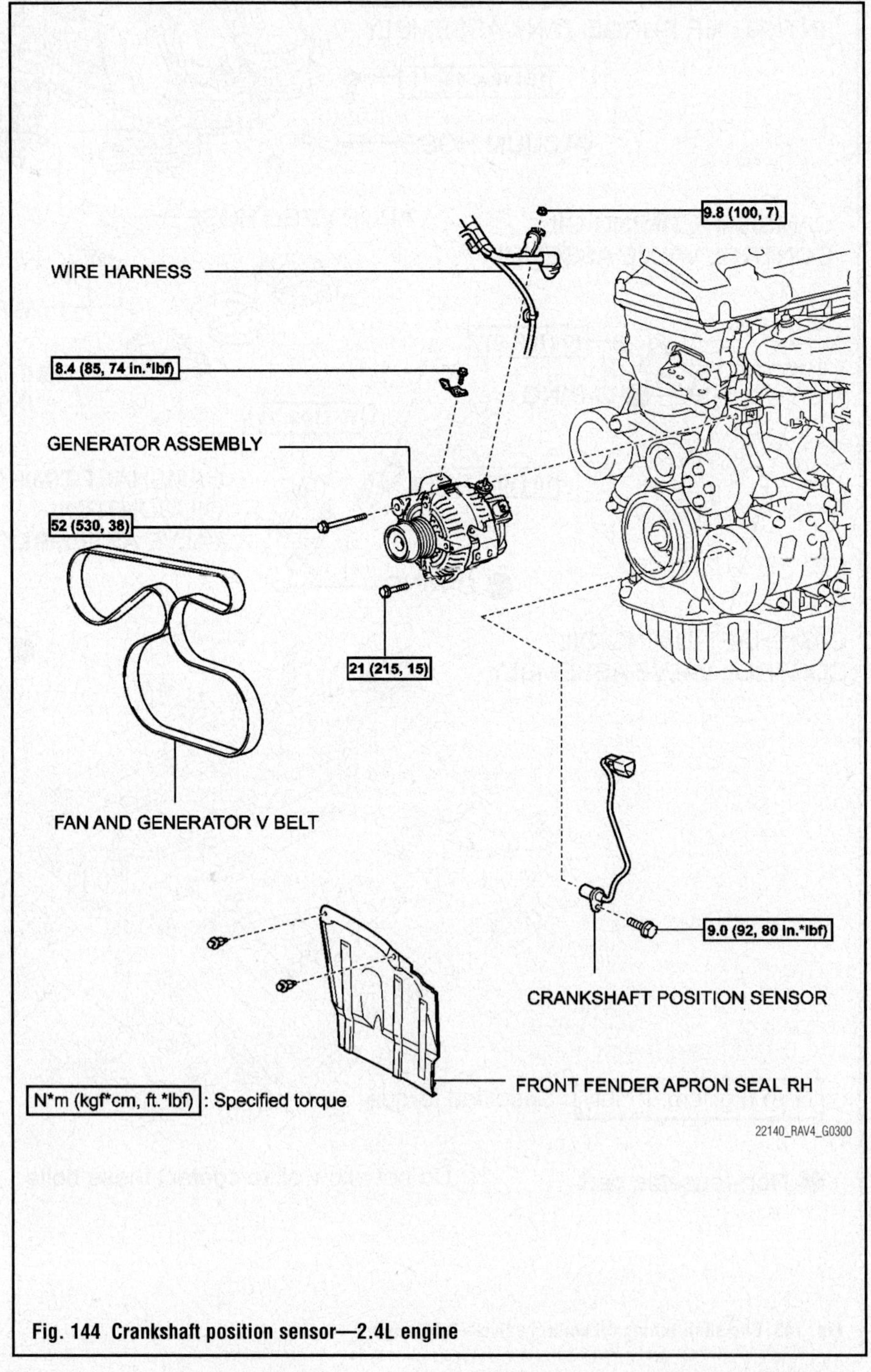

Fig. 144 Crankshaft position sensor—2.4L engine

REMOVAL & INSTALLATION

2.4L Engine

See Figure 146.

1. Before servicing the vehicle, refer to the Precautions Section.
2. Disconnect the negative battery cable.

➡Wait at least 90 seconds after disconnecting the negative battery cable before starting any repair work to prevent air bag and seat belt pretensioner activation.

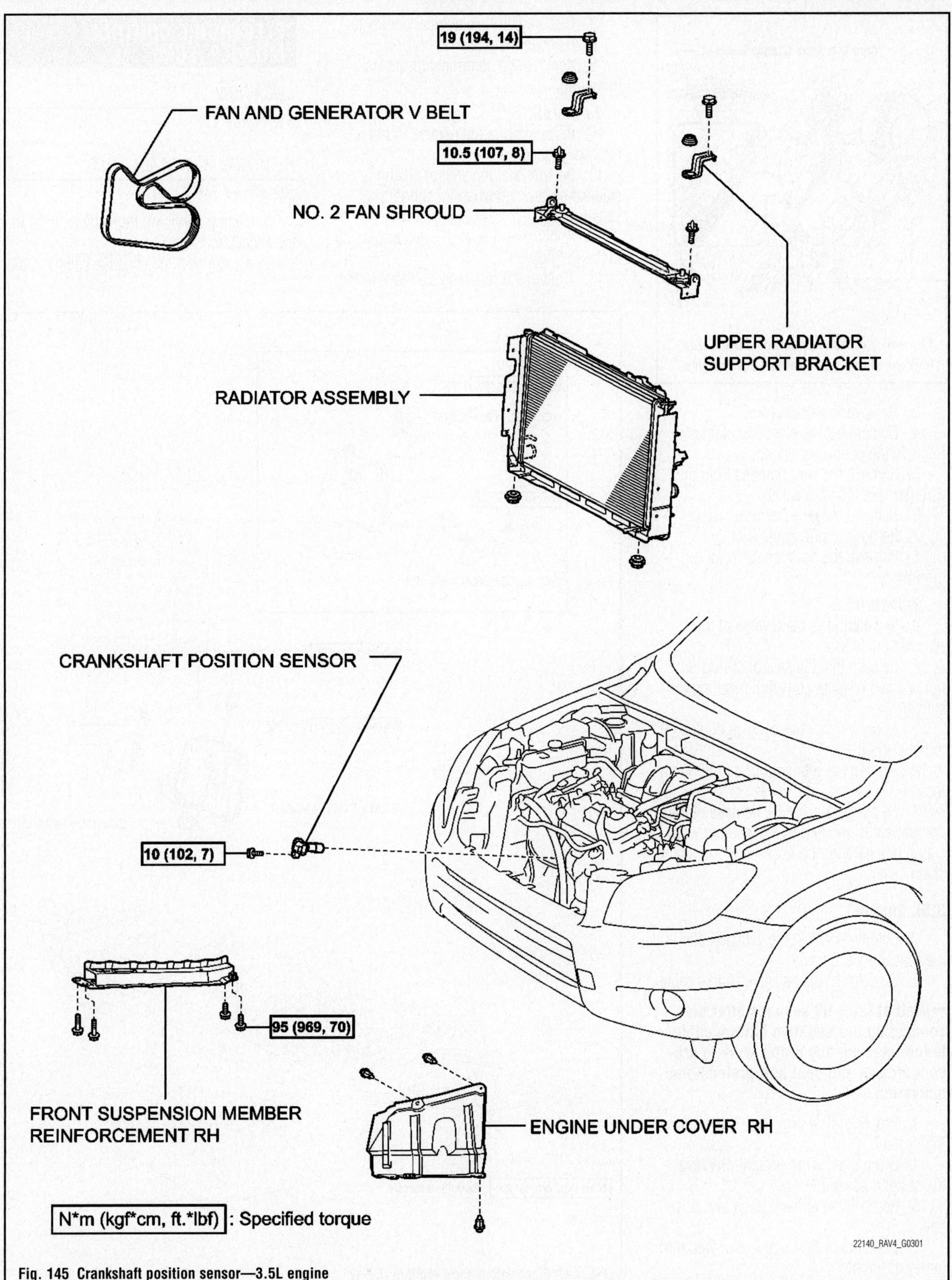

Fig. 145 Crankshaft position sensor—3.5L engine

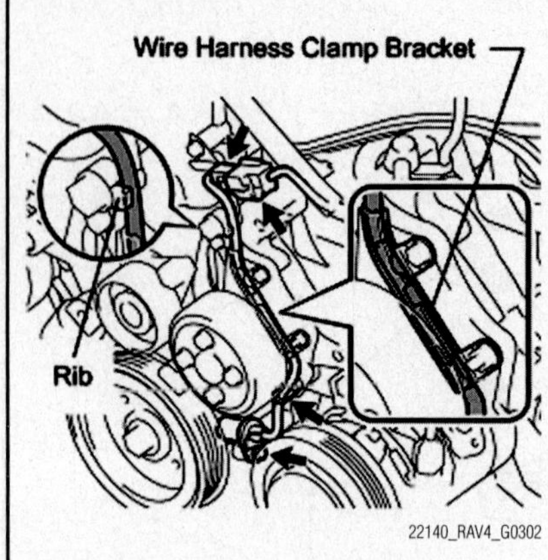

Fig. 146 Crankshaft position sensor wire harness clamp positioning—2.4L engine

3. Remove the alternator.
4. Disconnect the electrical connector. Remove the connector clamp.
5. Remove the wire harness from the wire harness clamp bracket.
6. Remove the wire harness clamp.
7. Remove the retaining bolt.
8. Remove the component from its mounting.

To install:

9. Installation is the reverse of the removal procedure.
10. Be sure that the sensor O-ring is not cracked or jammed when installing the sensor.
11. Apply a light coat of clean engine oil to the O-ring.
12. Tighten the retaining bolt to 80 inch lbs.
13. Be sure that the wire harness of the component is secured to the wire harness clamp bracket thru the back rib of the timing chain cover.

3.5L Engine

1. Before servicing the vehicle, refer to the Precautions Section.
2. Disconnect the negative battery cable.

➡Wait at least 90 seconds after disconnecting the negative battery cable before starting any repair work to prevent air bag and seat belt pretensioner activation.

3. Properly discharge the air conditioning system.
4. Drain the engine coolant. Remove the radiator assembly.
5. Remove the fan and alternator drive belt.
6. Disconnect the compressor with the pulley assembly.
7. Disconnect the electrical connector.
8. Remove the retaining bolt.
9. Remove the component from its mounting.

To install:

10. Installation is the reverse of the removal procedure.
11. Be sure that the sensor O-ring is not cracked or jammed when installing the sensor.
12. Apply a light coat of clean engine oil to the O-ring.
13. Tighten the retaining bolt to 7 ft. lbs.

ELECTRONIC CONTROL MODULE (ECM)

LOCATION

See Figure 147.

REMOVAL & INSTALLATION

See Figure 148.

1. Before servicing the vehicle, refer to the Precautions Section.
2. Disconnect the negative battery cable.

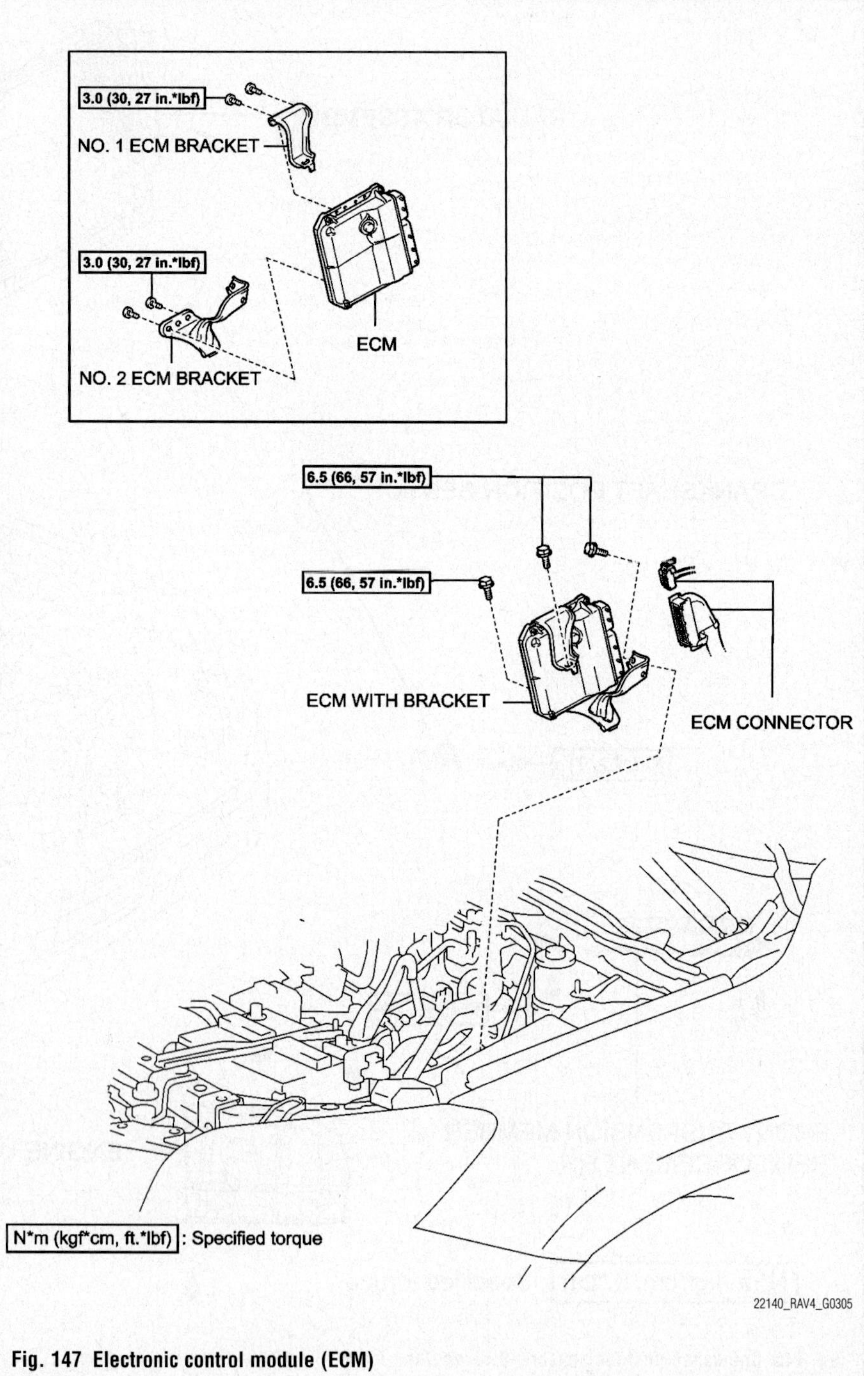

Fig. 147 Electronic control module (ECM)

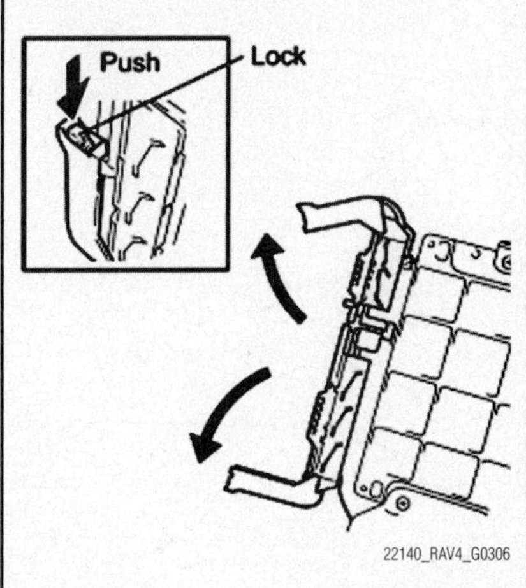

Fig. 148 Electronic control module (ECM) removal

➡Wait at least 90 seconds after disconnecting the negative battery cable before starting any repair work to prevent air bag and seat belt pretensioner activation.

3. Remove the air cleaner cap.
4. Remove the air cleaner case.
5. Remove the ECM bolts and bracket.
6. Disconnect the electrical connectors.

➡After disconnecting the connectors be sure that no foreign matter or water comes in contact with the connecting part of the connector.

7. Raise the two levers while pushing the locks on the two levers, and disconnect the two ECM connectors.
8. Remove the component with the bracket.
9. Remove the four screws and the two brackets.

To install:

10. Installation is the reverse of the removal procedure.

ENGINE COOLANT TEMPERATURE (ECT) SENSOR

LOCATION

See Figures 149 and 150.

REMOVAL & INSTALLATION

1. Before servicing the vehicle, refer to the Precautions Section.
2. Disconnect the negative battery cable.

➡Wait at least 90 seconds after disconnecting the negative battery cable before starting any repair work to prevent air bag and seat belt pretensioner activation.

3. Drain and properly dispose of the engine coolant.
4. On 2.4L engine, remove the air cleaner cap and air cleaner case.
5. Disconnect the electrical connectors.
6. Remove the component using tool SST: 09817-33190, or equivalent.

To install:

7. Installation is the reverse of the removal procedure.
8. be sure to use a new gasket.
9. Tighten the component to 14 ft. lbs.

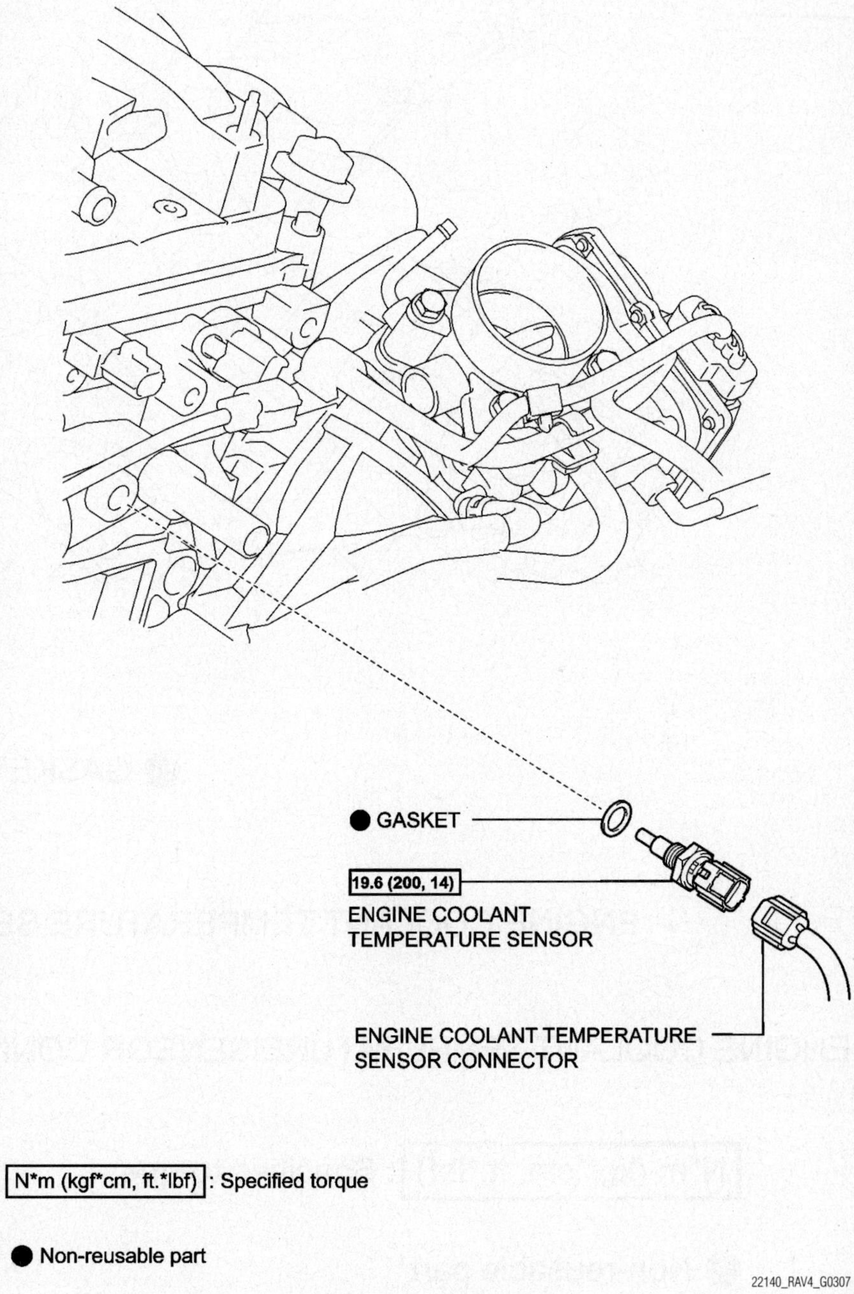

● GASKET
19.6 (200, 14)
ENGINE COOLANT TEMPERATURE SENSOR
ENGINE COOLANT TEMPERATURE SENSOR CONNECTOR
N*m (kgf*cm, ft.*lbf) : Specified torque
● Non-reusable part
22140_RAV4_G0307

Fig. 149 Engine coolant temperature sensor—2.4L engine

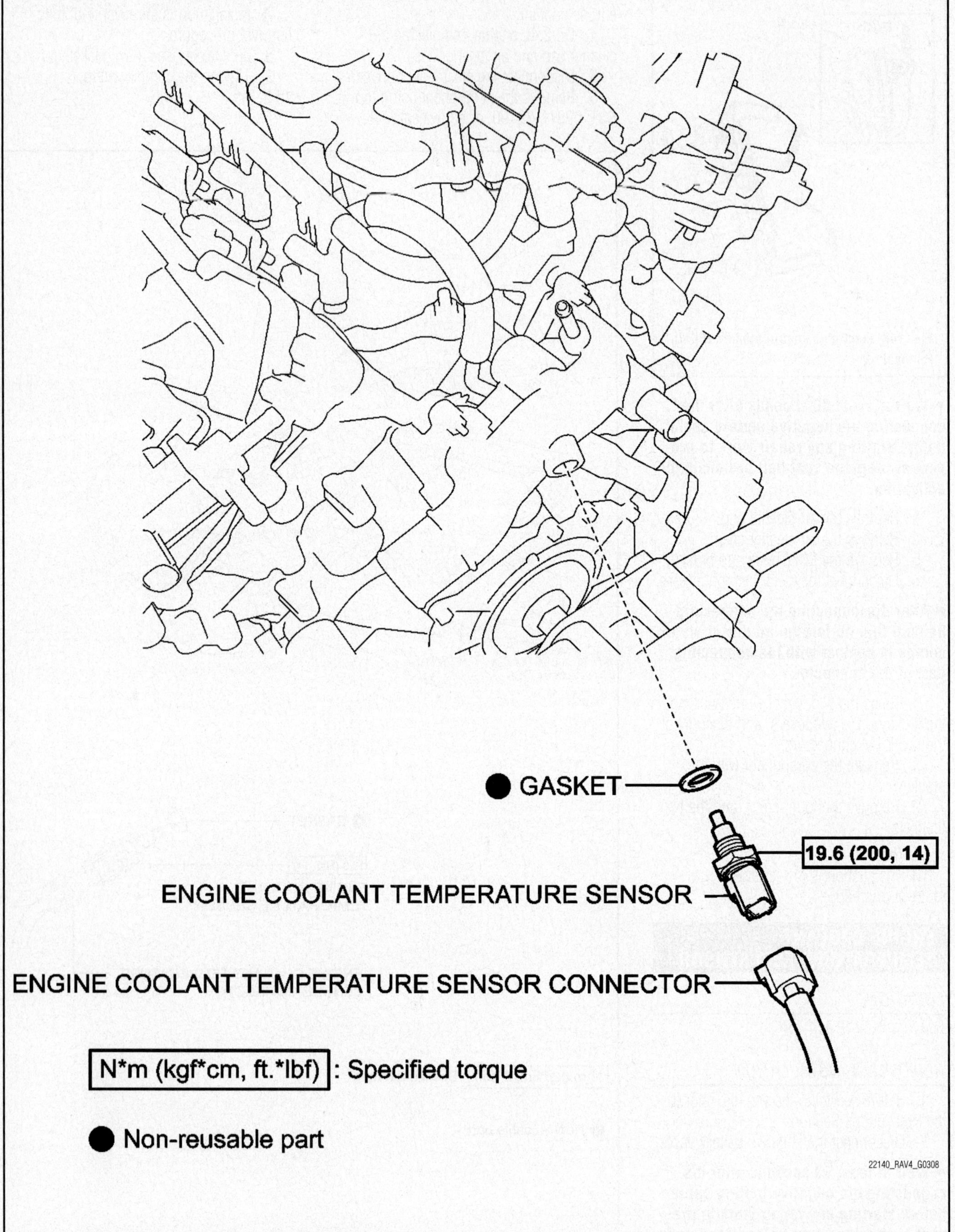

Fig. 150 Engine coolant temperature sensor—3.5L engine

HEATED OXYGEN (HO2S) SENSOR

LOCATION

See Figures 151 and 152.

REMOVAL & INSTALLATION

See Figures 153 and 154.

1. Before servicing the vehicle, refer to the Precautions Section.
2. Disconnect the negative battery cable.

➡Wait at least 90 seconds after disconnecting the negative battery cable before starting any repair work to prevent air bag and seat belt pretensioner activation.

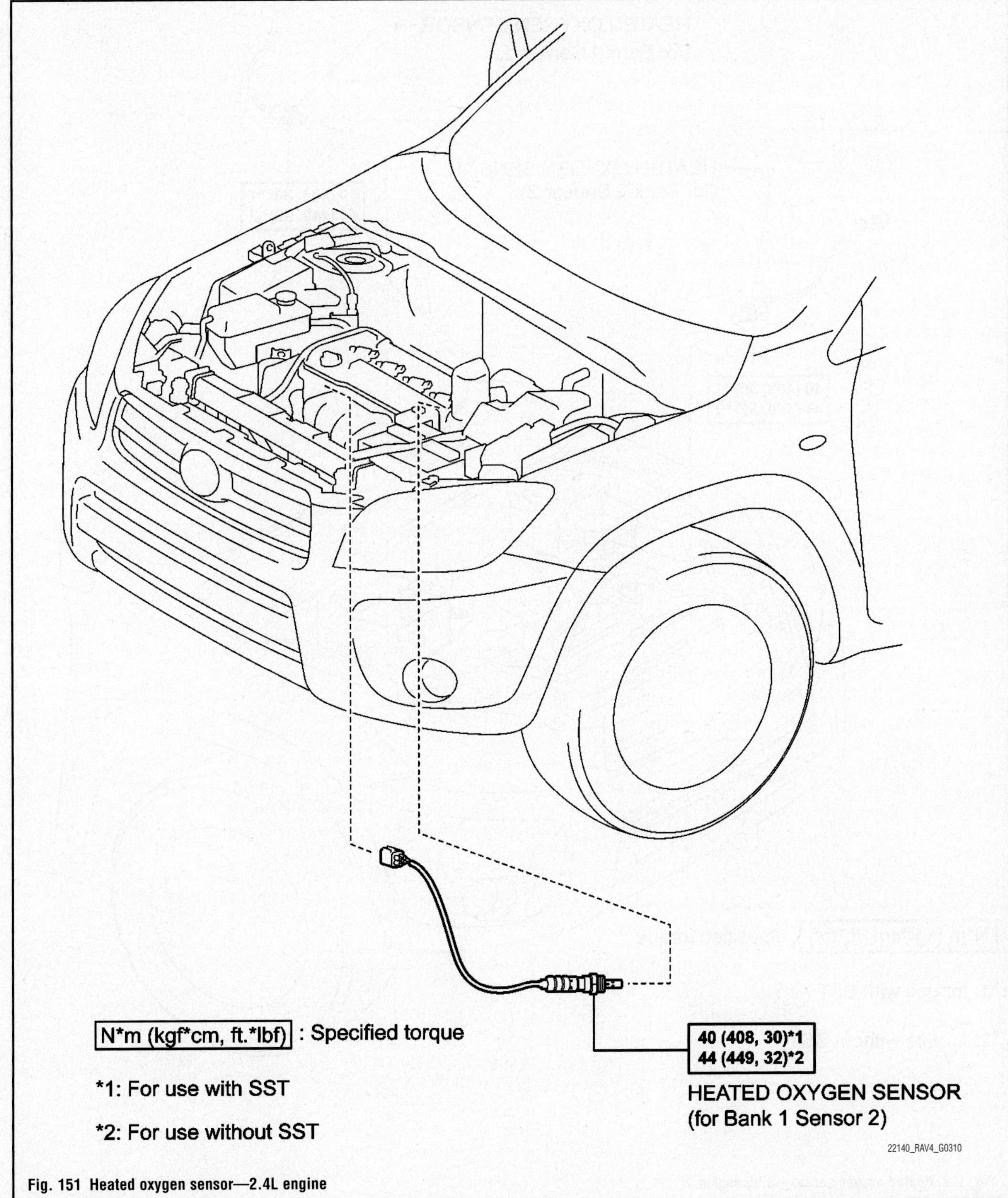

Fig. 151 Heated oxygen sensor—2.4L engine

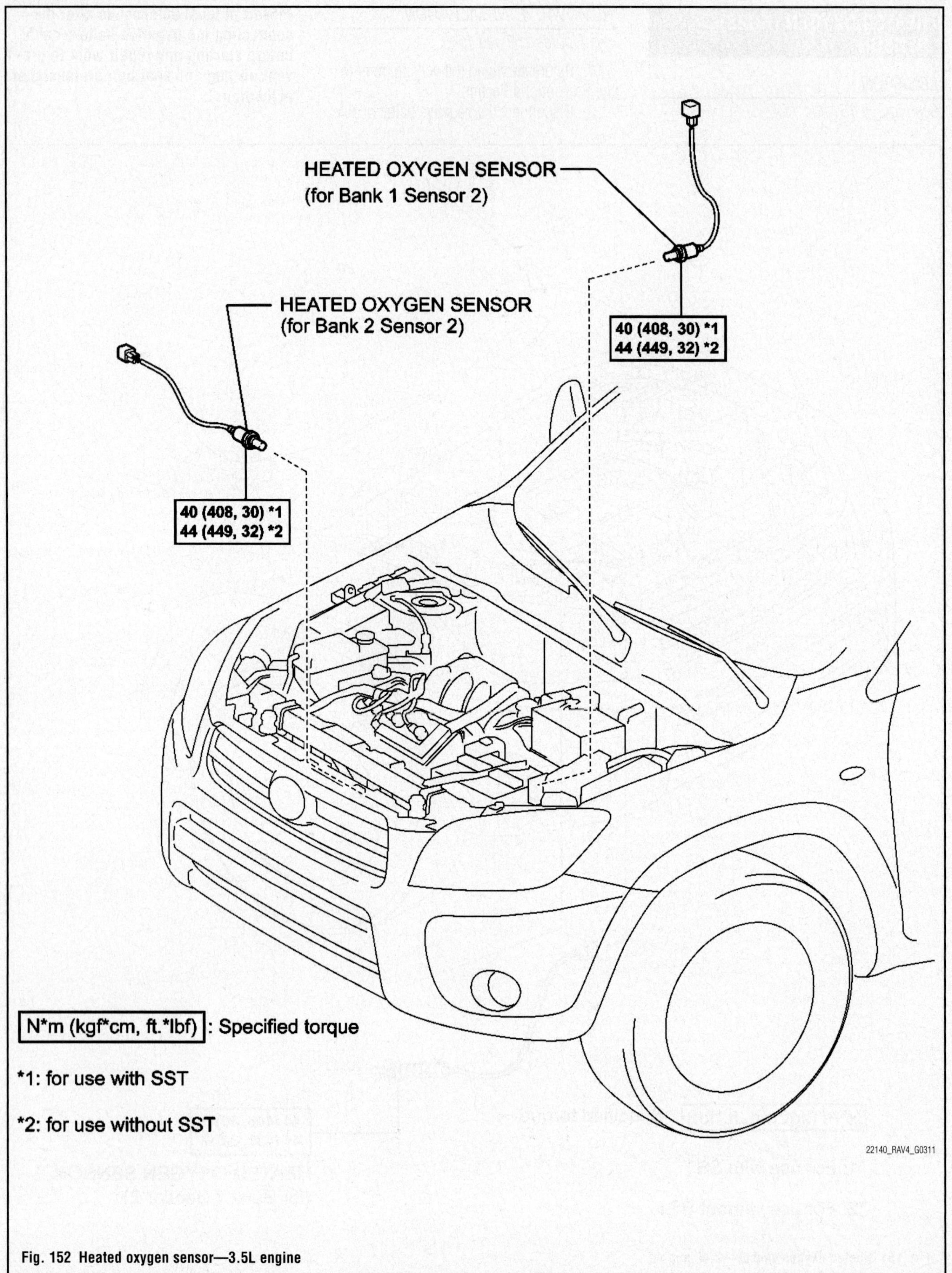

Fig. 152 Heated oxygen sensor—3.5L engine

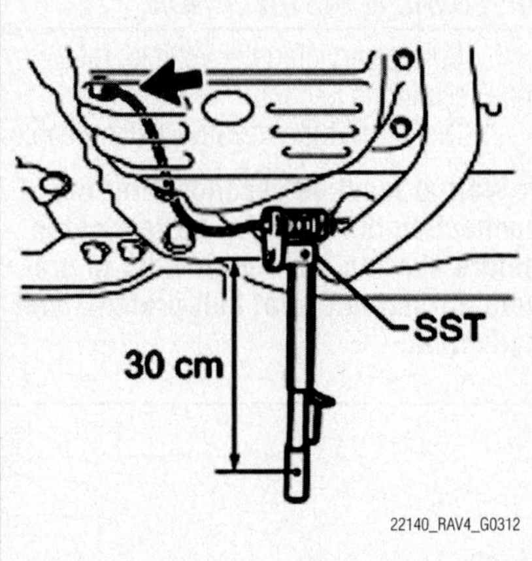

Fig. 153 Heated oxygen sensor torque wrench positioning—2.4L engine

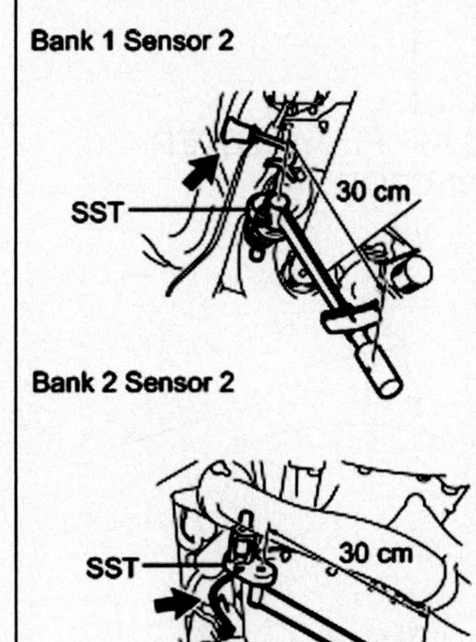

Fig. 154 Heated oxygen sensor torque wrench positioning—3.5L engine

3. Disconnect the electrical connector.
4. Remove the component from its mounting, using the sensor removal tool SST:09224-00010 or equivalent.

To install:

5. Installation is the reverse of the removal procedure.
6. Tighten the sensor to 32 ft. lbs. Be sure to use a torque wrench with a fulcrum length of 11.81 inch.
7. Check for exhaust leaks.

KNOCK SENSOR (KS)

LOCATION

See Figures 155 and 156.

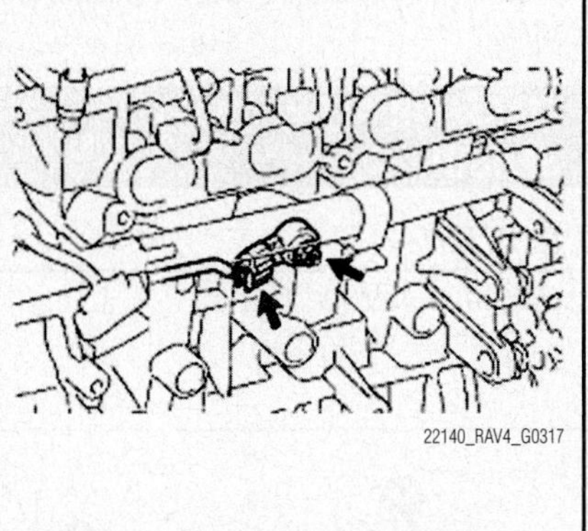

Fig. 155 Knock sensor—2.4L engine

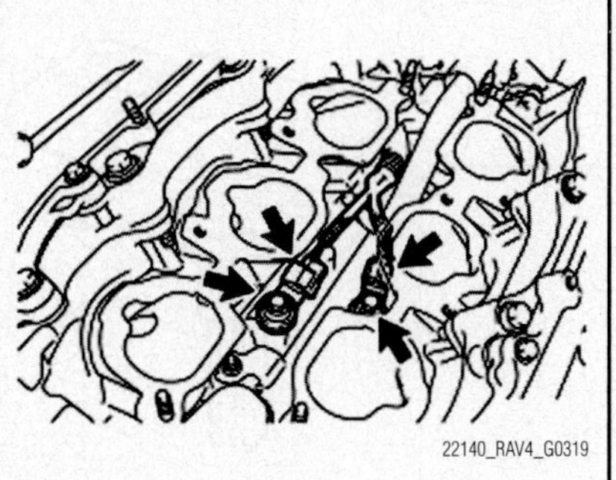

Fig. 156 Knock oxygen sensor—3.5L engine

REMOVAL & INSTALLATION

4.2L Engine

See Figure 157.

1. Before servicing the vehicle, refer to the Precautions Section.
2. Disconnect the negative battery cable.

➡Wait at least 90 seconds after disconnecting the negative battery cable before starting any repair work to prevent air bag and seat belt pretensioner activation.

3. Properly discharge the fuel system pressure.
4. Drain the engine coolant and properly dispose of it.
5. Remove the number one engine cover.

Remove the air cleaner cap.

6. Remove the throttle body assembly.
7. Remove the fuel delivery pipe. Disconnect the fuel tube.
8. Remove the water inlet hose. Remove the water outlet hose.
9. Remove the intake manifold. Discard the gasket.
10. Remove the intake manifold insulator.

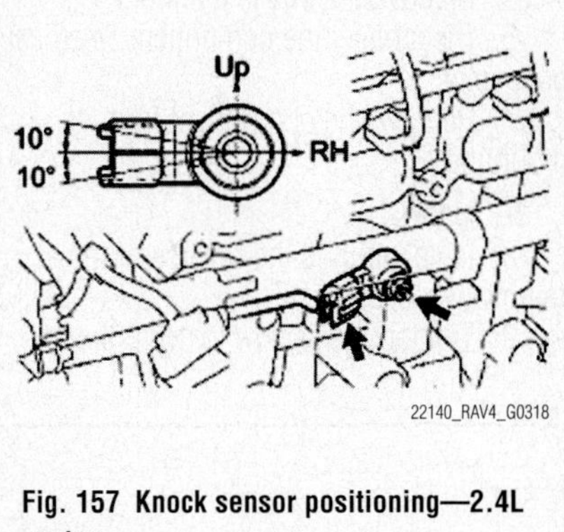

Fig. 157 Knock sensor positioning—2.4L engine

11. Disconnect the component electrical connector.
12. Remove the component from its mounting.

To install:

13. Installation is the reverse of the removal procedure.
14. Tighten the sensor to 15 ft. lbs.
15. Be sure that the sensor is positioned correctly

3.5L Engine

See Figure 158.

➡In order to replace the knock sensor, the manufacturer recommends removing the engine from the vehicle.

1. Before servicing the vehicle, refer to the Precautions Section.
2. Disconnect the negative battery cable.

➡Wait at least 90 seconds after disconnecting the negative battery cable before starting any repair work to prevent air bag and seat belt pretensioner activation.

3. Remove the engine and position it in a suitable holding fixture.
4. Remove the air intake surge tank assembly.

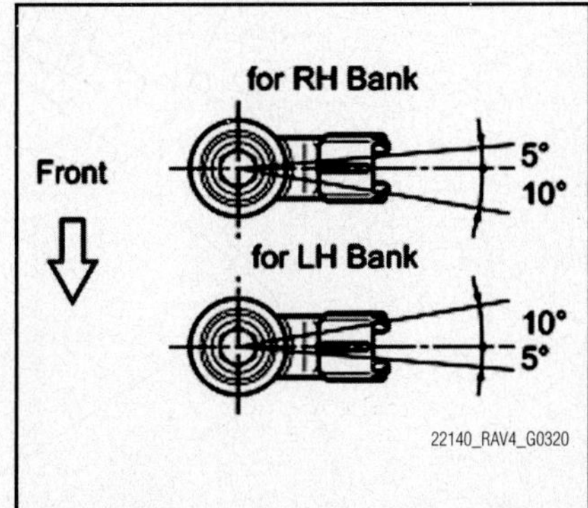

Fig. 158 Knock sensor positioning—3.5L engine

5. Remove the intake manifold.
6. Disconnect the component electrical connector.
7. Remove the component from its mounting.

To install:

8. Installation is the reverse of the removal procedure.
9. Tighten the sensor to 15 ft. lbs.
10. Be sure that the sensor is positioned correctly

MASS AIR FLOW (MAF) SENSOR

LOCATION

See Figures 159 and 160.

REMOVAL & INSTALLATION

1. Before servicing the vehicle, refer to the Precautions Section.
2. Disconnect the negative battery cable.

➡Wait at least 90 seconds after disconnecting the negative battery cable before starting any repair work to prevent air bag and seat belt pretensioner activation.

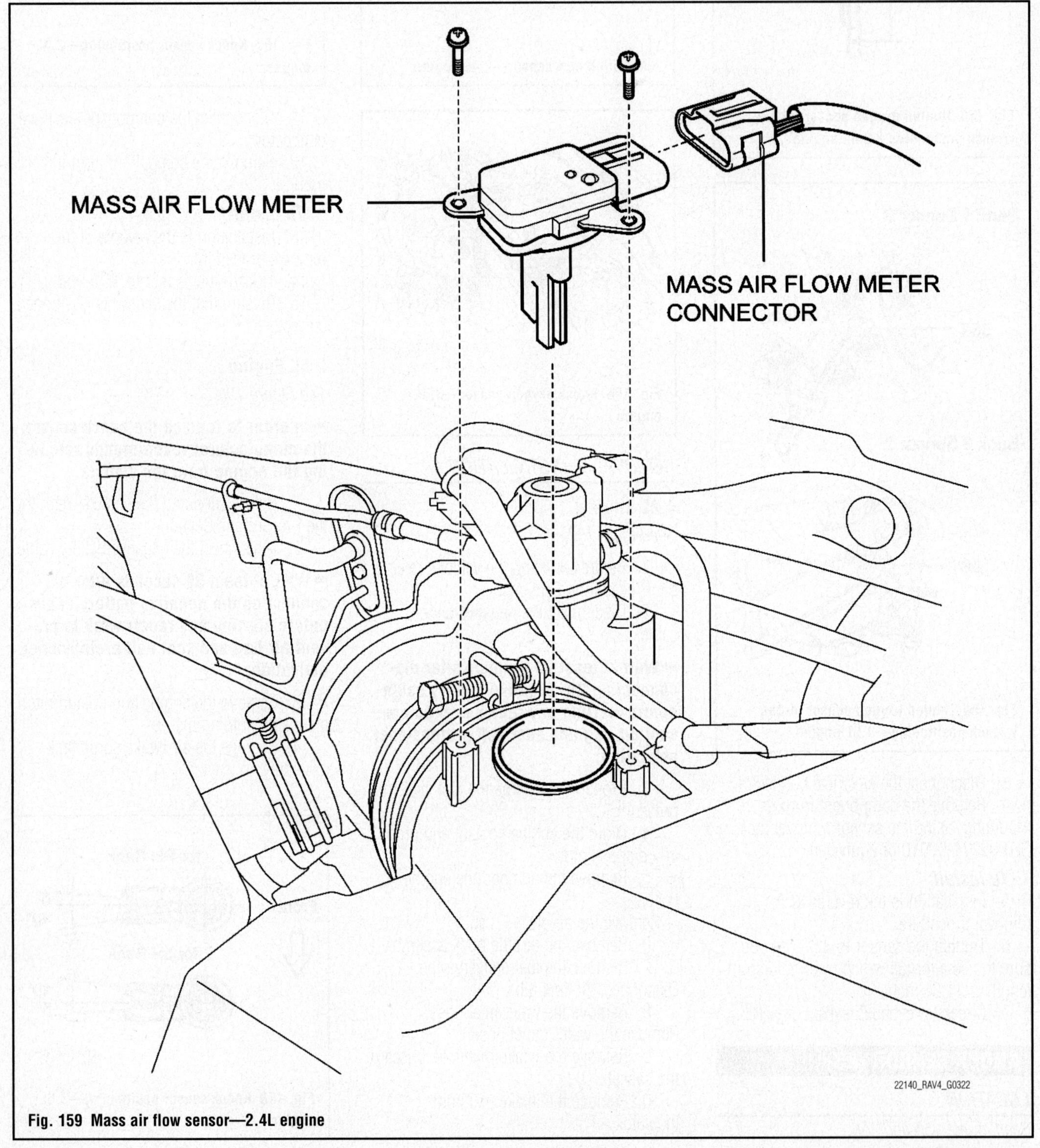

Fig. 159 Mass air flow sensor—2.4L engine

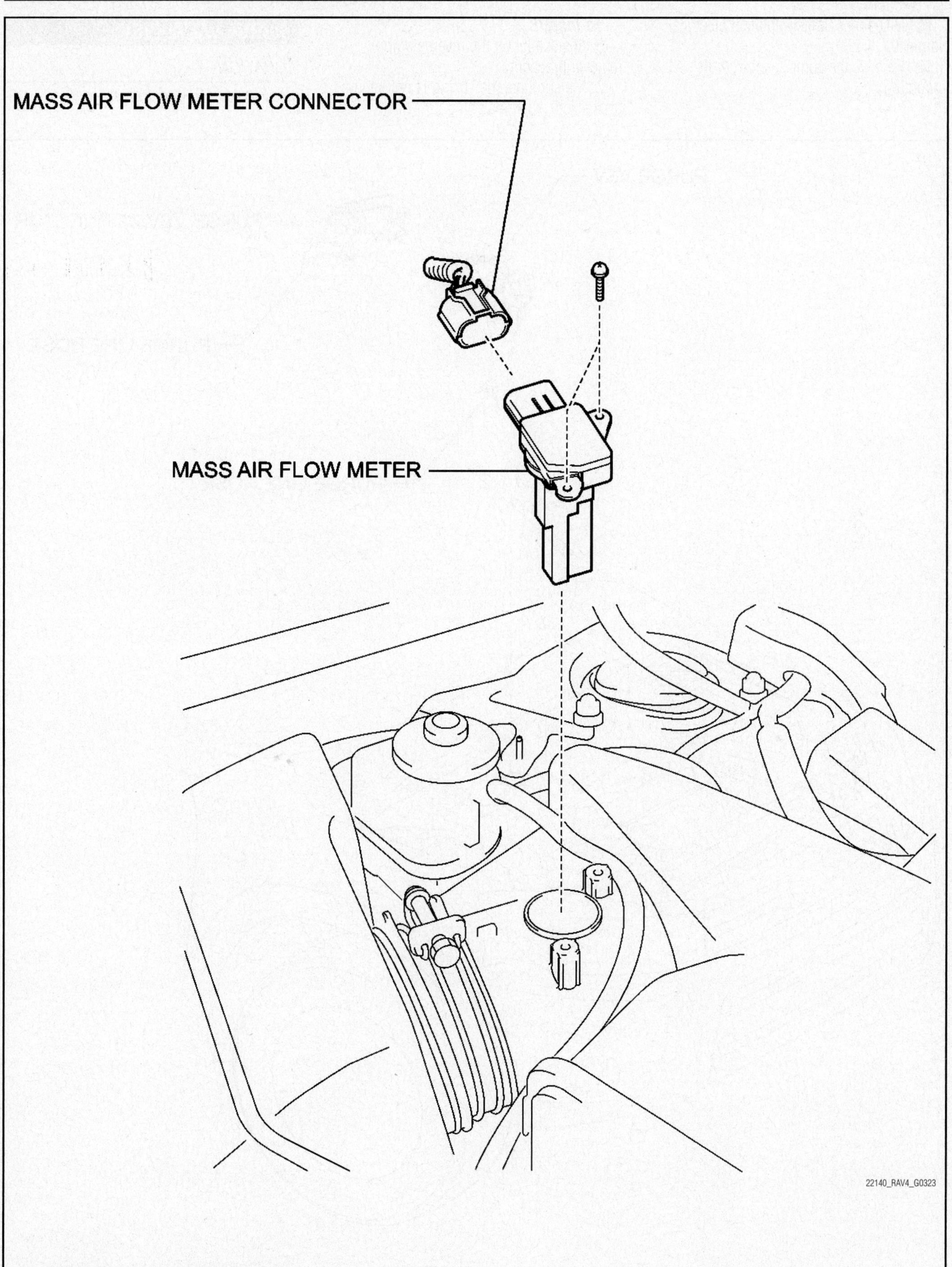

Fig. 160 Mass air flow sensor—3.5L engine

3. Disconnect the component electrical connector.
4. Remove the component from its mounting.

To install:

5. Installation is the reverse of the removal procedure.
6. Be sure that the O-ring is not jammed or cracked, when installing it.

PURGE VALVE (VSV)

LOCATION

See Figures 161 and 162.

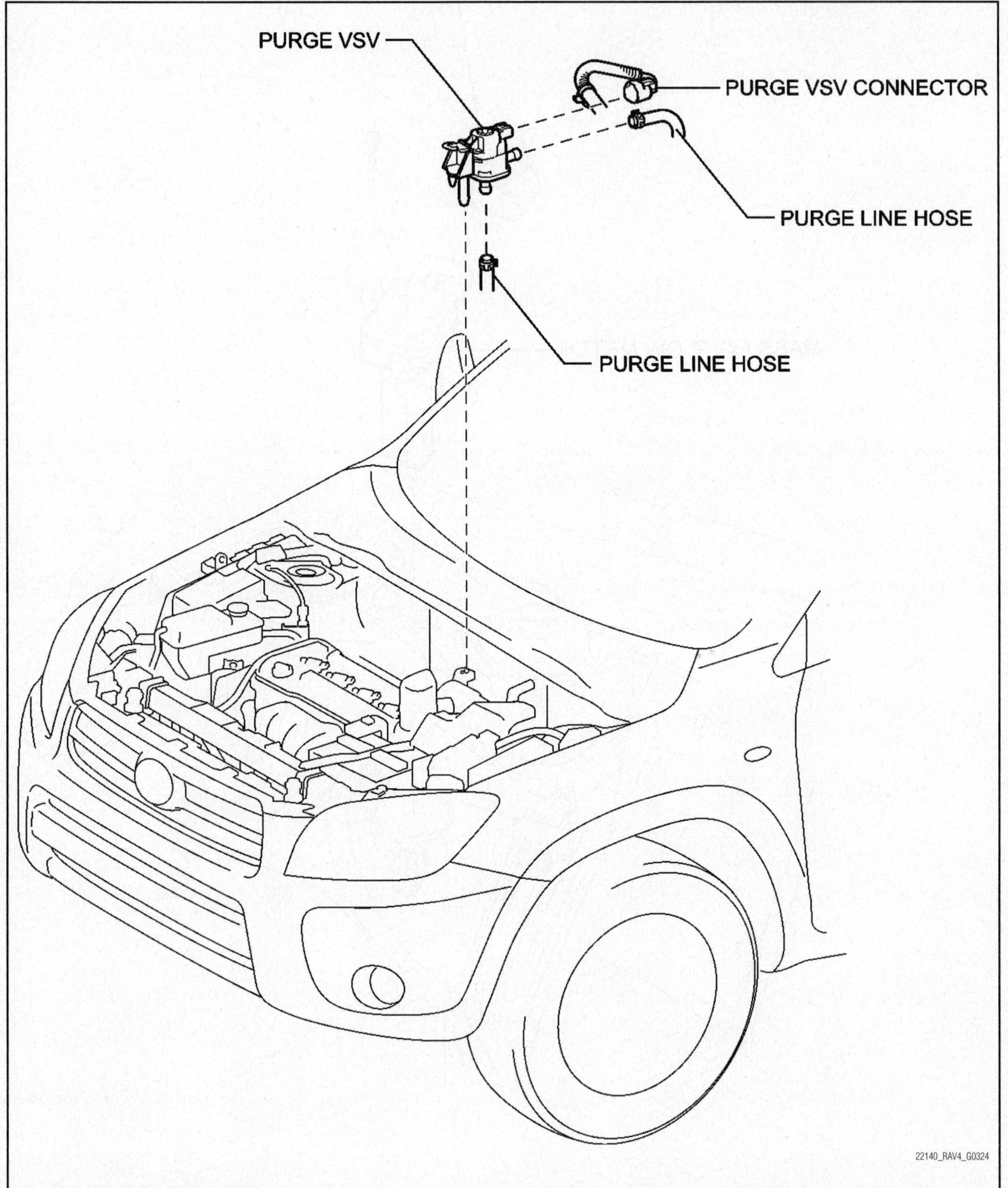

Fig. 161 Purge valve—2.4L engine

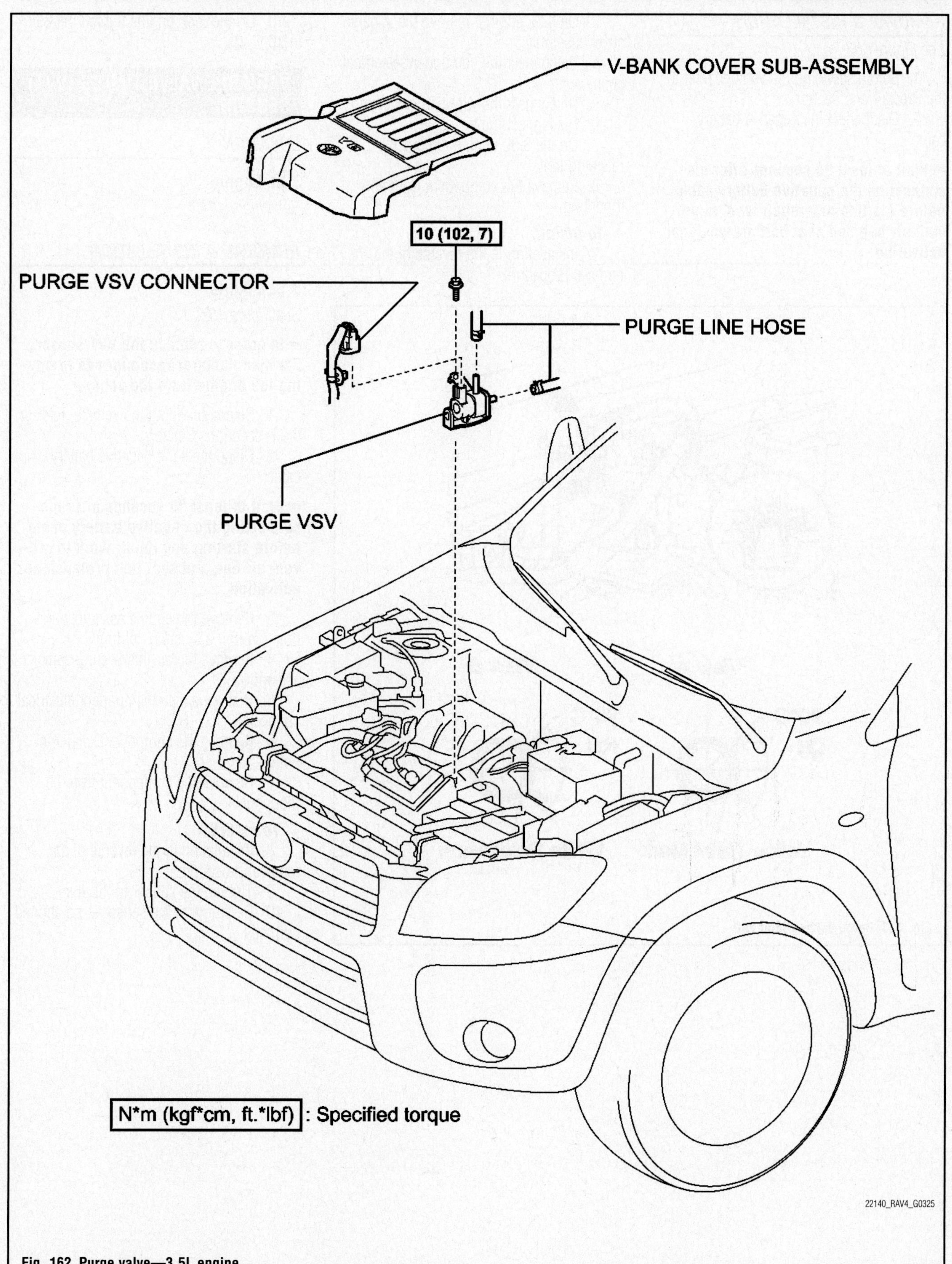

Fig. 162 Purge valve—3.5L engine

REMOVAL & INSTALLATION

See Figure 163.

1. Before servicing the vehicle, refer to the Precautions Section.
2. Disconnect the negative battery cable.

➡Wait at least 90 seconds after disconnecting the negative battery cable before starting any repair work to prevent air bag and seat belt pretensioner activation.

3. On 3.5L engine, remove the V-bank cover assembly.
4. Disconnect the component electrical connector.
5. Disconnect the wire harness clamp.
6. Disconnect the hoses.
7. On the 3.5L engine remove the retaining bolt.
8. Remove the component from its mounting.

To install:

9. Installation is the reverse of the removal procedure.
10. On the 3.5L engine tighten the bolt to 10 ft. lbs.

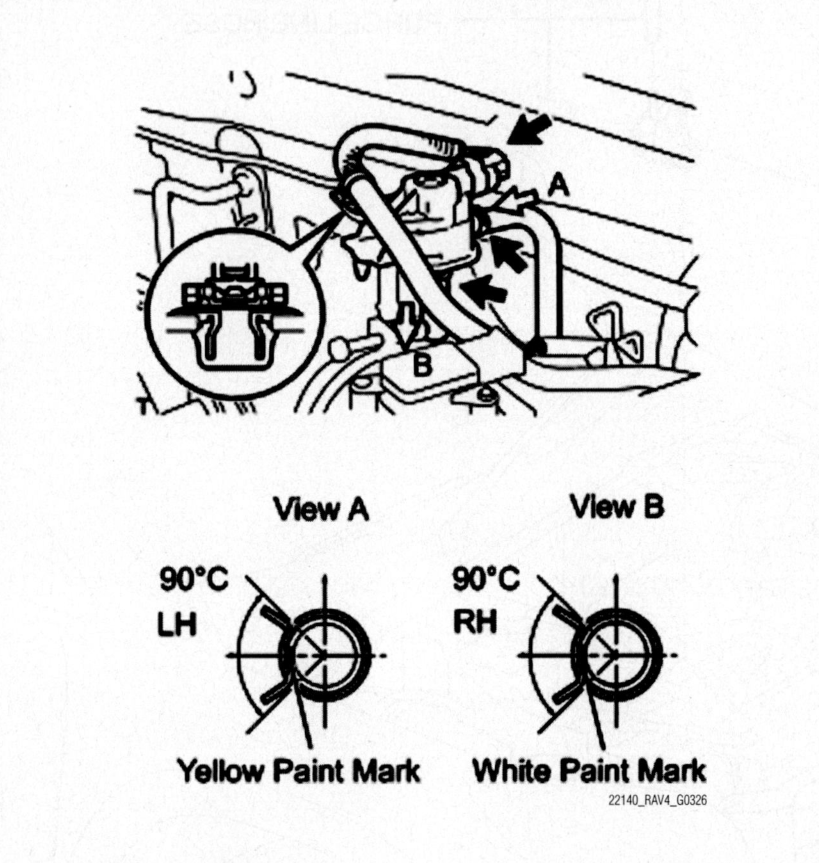

Fig. 163 Purge valve alignment

VARIABLE VALVE TIMING (VVT) SENSOR

LOCATION

3.5L Engine

See Figure 164.

REMOVAL & INSTALLATION

3.5L Engine

See Figure 165.

➡In order to replace the VVT sensor, the manufacturer recommends removing the engine from the vehicle.

1. Before servicing the vehicle, refer to the Precautions Section.
2. Disconnect the negative battery cable.

➡Wait at least 90 seconds after disconnecting the negative battery cable before starting any repair work to prevent air bag and seat belt pretensioner activation.

3. Remove the engine assembly and position it in a suitable holding fixture.
4. Remove the air intake surge tank assembly.
5. Disconnect the component electrical connector.
6. Remove the component retaining bolts.
7. Remove the component from its mounting.

To install:

8. Installation is the reverse of the removal procedure.
9. Tighten the sensor to 7ft. lbs.
10. Be sure that the sensor is positioned correctly

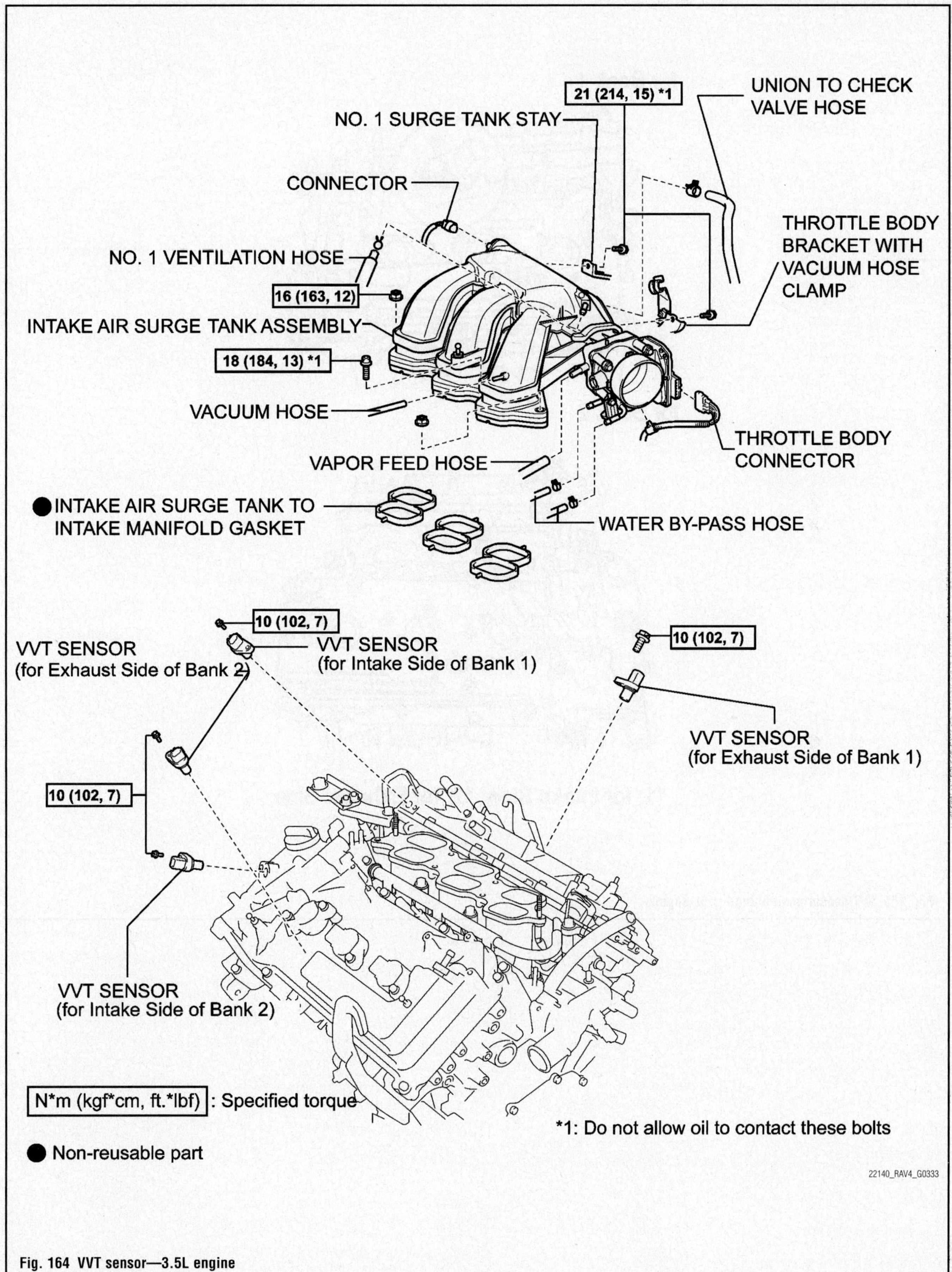

Fig. 164 VVT sensor—3.5L engine

for Bank 1

for Bank 2

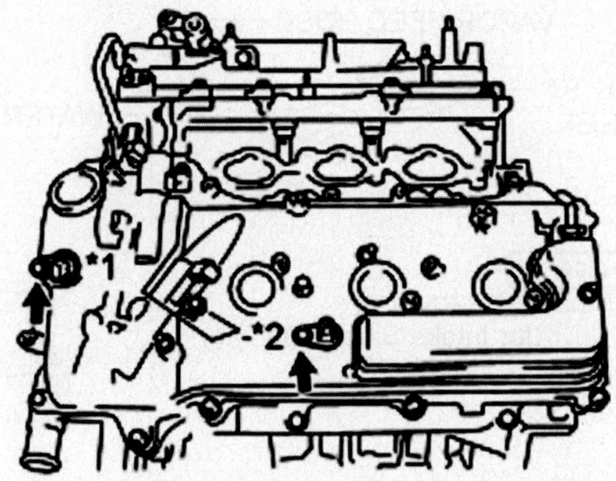

*1: for Intake Side *2: for Exhaust Side

22140_RAV4_G0334

Fig. 165 VVT sensor positioning—3.5L engine

FUEL SYSTEM SERVICE PRECAUTIONS

Safety is the most important factor when performing not only fuel system maintenance but any type of maintenance. Failure to conduct maintenance and repairs in a safe manner may result in serious personal injury or death. Maintenance and testing of the vehicle's fuel system components can be accomplished safely and effectively by adhering to the following rules and guidelines.

- To avoid the possibility of fire and personal injury, always disconnect the negative battery cable unless the repair or test procedure requires that battery voltage be applied.
- Always relieve the fuel system pressure prior to disconnecting any fuel system component (injector, fuel rail, pressure regulator, etc.), fitting or fuel line connection. Exercise extreme caution whenever relieving fuel system pressure to avoid exposing skin, face and eyes to fuel spray. Please be advised that fuel under pressure may penetrate the skin or any part of the body that it contacts.
- Always place a shop towel or cloth around the fitting or connection prior to loosening to absorb any excess fuel due to spillage. Ensure that all fuel spillage (should it occur) is quickly removed from engine surfaces. Ensure that all fuel soaked cloths or towels are deposited into a suitable waste container.
- Always keep a dry chemical (Class B) fire extinguisher near the work area.
- Do not allow fuel spray or fuel vapors to come into contact with a spark or open flame.
- Always use a back-up wrench when loosening and tightening fuel line connection fittings. This will prevent unnecessary stress and torsion to fuel line piping.
- Always replace worn fuel fitting O-rings with new Do not substitute fuel hose or equivalent where fuel pipe is installed.

Before servicing the vehicle, make sure to also refer to the precautions in the beginning of this section as well.

RELIEVING FUEL SYSTEM PRESSURE

2.4L Engine

➡The following procedure must be performed before disconnecting any part of the fuel system. After performing this procedure, pressure will remain in the fuel line. Use care when disconnecting any fuel lines.

1. Before servicing the vehicle, refer to the Precautions Section.
2. Remove the console box.
3. Disconnect the connector.
4. Start the engine. After the engine has stopped, turn the ignition switch to the OFF position.

➡DTC P0171 (system lean) may set.

5. Check that the engine does not start.
6. Disconnect the negative battery cable.
7. Remove the fuel tank cap.
8. Connect the electrical connector. Install the console box.

3.5L Engine

➡The following procedure must be performed before disconnecting any part of the fuel system. After performing this procedure, pressure will remain in the fuel line. Use care when disconnecting any fuel lines.

1. Before servicing the vehicle, refer to the Precautions Section.
2. Remove the console box.
3. Disconnect the connector.
4. Start the engine. After the engine has stopped, turn the ignition switch to the OFF position.

➡DTC P0171/P0172 (system lean) may set.

5. Check that the engine does not start.
6. Disconnect the negative battery cable.
7. Remove the fuel tank cap.
8. Connect the electrical connector. Install the console box.

FUEL FILTER

REMOVAL & INSTALLATION

2.4L and 3.5L Engines

The filter is part of the fuel pump module and is not normally serviced.

FUEL RAIL & INJECTORS

REMOVAL & INSTALLATION

2.4L Engine

See Figures 166 through 169.

1. Before servicing the vehicle, refer to the Precautions Section.
2. Properly relieve the fuel system pressure.
3. Disconnect the negative battery cable.

➡Wait at least 90 seconds after disconnecting the negative battery cable before starting any repair work to prevent air bag and seat belt pretensioner activation.

4. Remove the air cleaner cap sub-assembly.
5. To disconnect the fuel main tube, remove the fuel tube from the fuel hose clamp. Remove the fuel pipe clamp.

➡Do not forcibly bend, kink or twist the tube.

6. Hold the fuel tube connector and install tool SST 09268-21010, or equivalent. Turn the tool to align the retainer inside the fuel tube connector with the chamfered part of the tool.
7. Insert the tool into the fuel tube and hold it. Push the fuel tube connector toward the tool.
8. Mount the retainer of the fuel tube connector onto the chamfered part of the tool.
9. Slide the tool and the fuel tube connector together toward the fuel tube until they make a "click", then disconnect the fuel tube.
10. Properly drain the remaining fuel inside the fuel tube. Cover the fuel tube with a clean plastic bag, to protect the disconnected part.
11. Disconnect the number two ventilation hose from the ventilation valve.
12. Remove the two wiring harness clamps. Disconnect the four fuel injector electrical connectors.
13. Remove the two retaining bolts. Remove the fuel rail along with the fuel injectors.

➡Be careful not to drop the fuel injectors when removing the injector rail.

14. Remove the two delivery pipe spacers from the cylinder head.
15. Remove the four insulators from the cylinder head.
16. As required, remove the injectors from the fuel rail.

To install:

17. Installation is the reverse of the removal procedure.
18. Installation is the reverse of removal.
19. Apply a light coat gasoline to the new fuel injector O-rings and install them on to the injectors.
20. Apply a light coat of gasoline to the part of the injector rail that contacts the O-ring of the injector.
21. Install the injectors on the fuel rail.

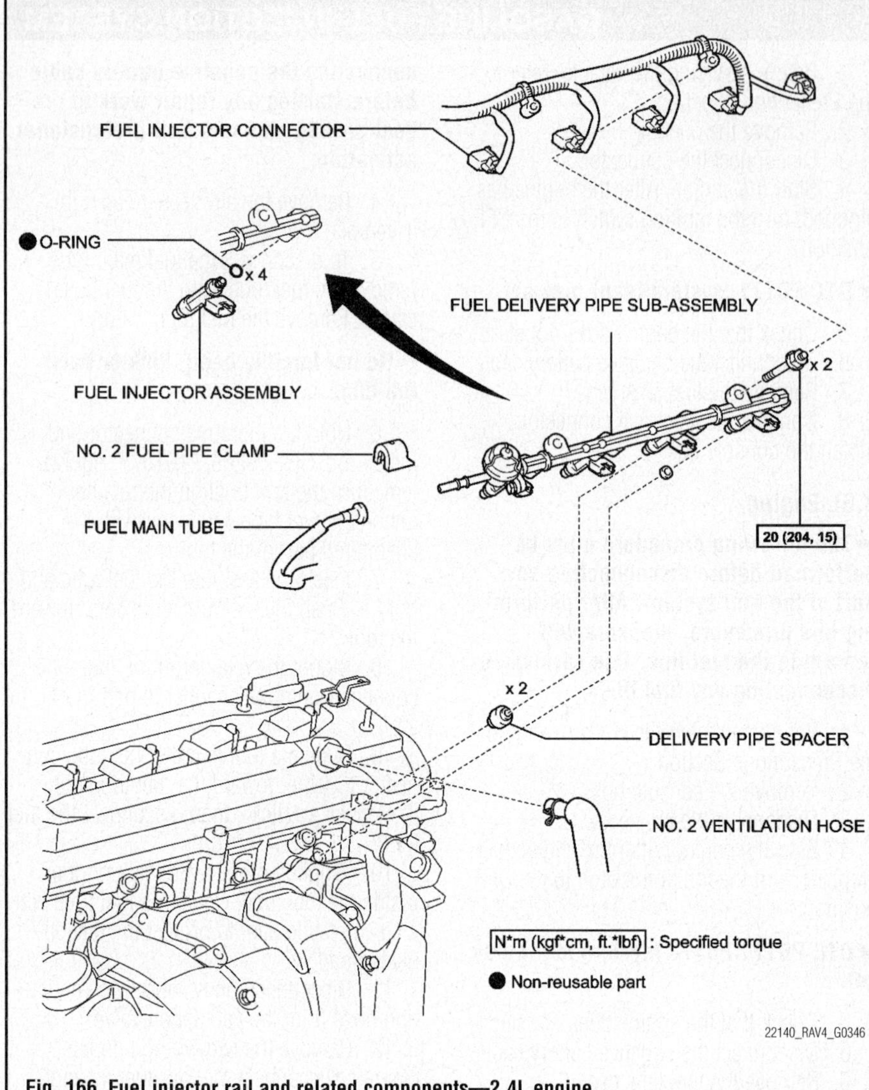

Fig. 166 Fuel injector rail and related components—2.4L engine

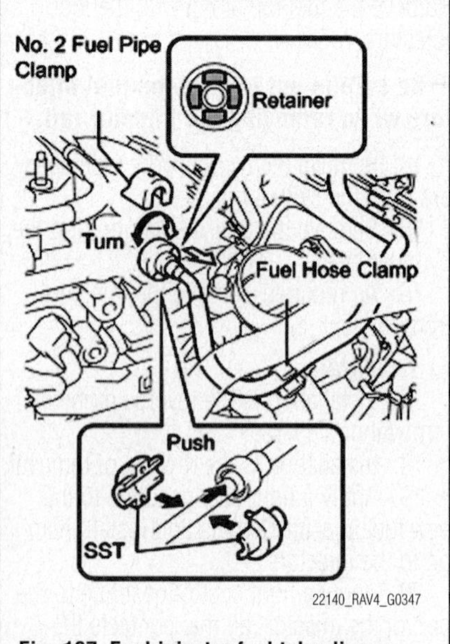

Fig. 167 Fuel injector fuel tube disconnection—2.4L engine

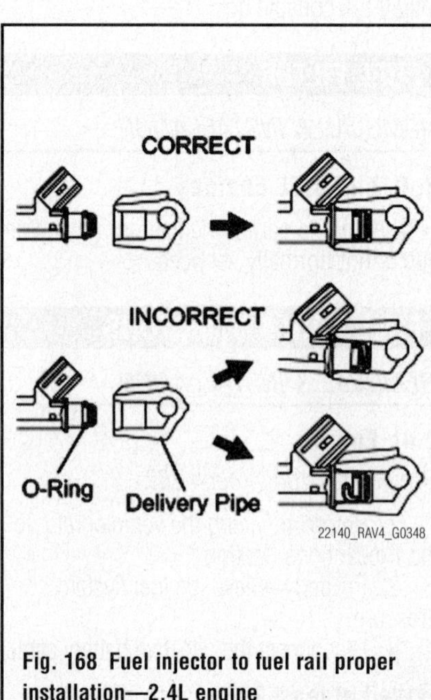

Fig. 168 Fuel injector to fuel rail proper installation—2.4L engine

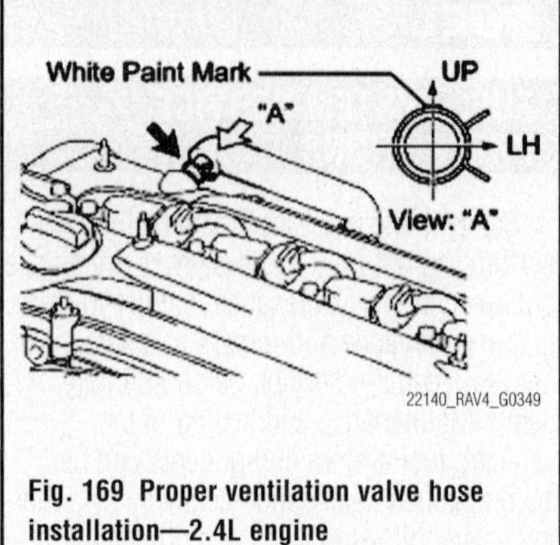

Fig. 169 Proper ventilation valve hose installation—2.4L engine

➡Be sure that the O-rings are not damaged or jammed when you are installing them.

22. Check that the injector rotates freely. If it does not, replace the O-ring.
23. Install the four new insulators to the cylinder head. Install the two delivery pipe spacers to the cylinder head.
24. Install the fuel rail together with the fuel injectors, then temporarily tighten the two retaining bolts.
25. Check that the fuel injectors rotate smoothly. If it does not replace the O-ring.
26. Tighten the two retaining bolts to 15 ft. lbs.
27. When installing the number two ventilation, see the illustration for proper positioning.

➡Make sure that the paint mark and hose clamp are at the correct angle when installing the hose.

28. Once the installation is complete be sure to inspect for fuel leaks, using the Techstream tool.
29. Connect the tool to the DLC3.
30. Turn the ignition switch to the ON position. Push the Techstream main switch on.

➡Do not start the engine.

31. Select the active test mode on the tool. Refer to the tools instruction book for additional details.
32. Check that there are no fuel system leaks.
33. Turn the ignition switch OFF.
34. Disconnect the tool.

3.5L Engine

See Figure 170.

➡In order to replace the fuel rail and injectors, the manufacturer recommends removing the engine from the vehicle.

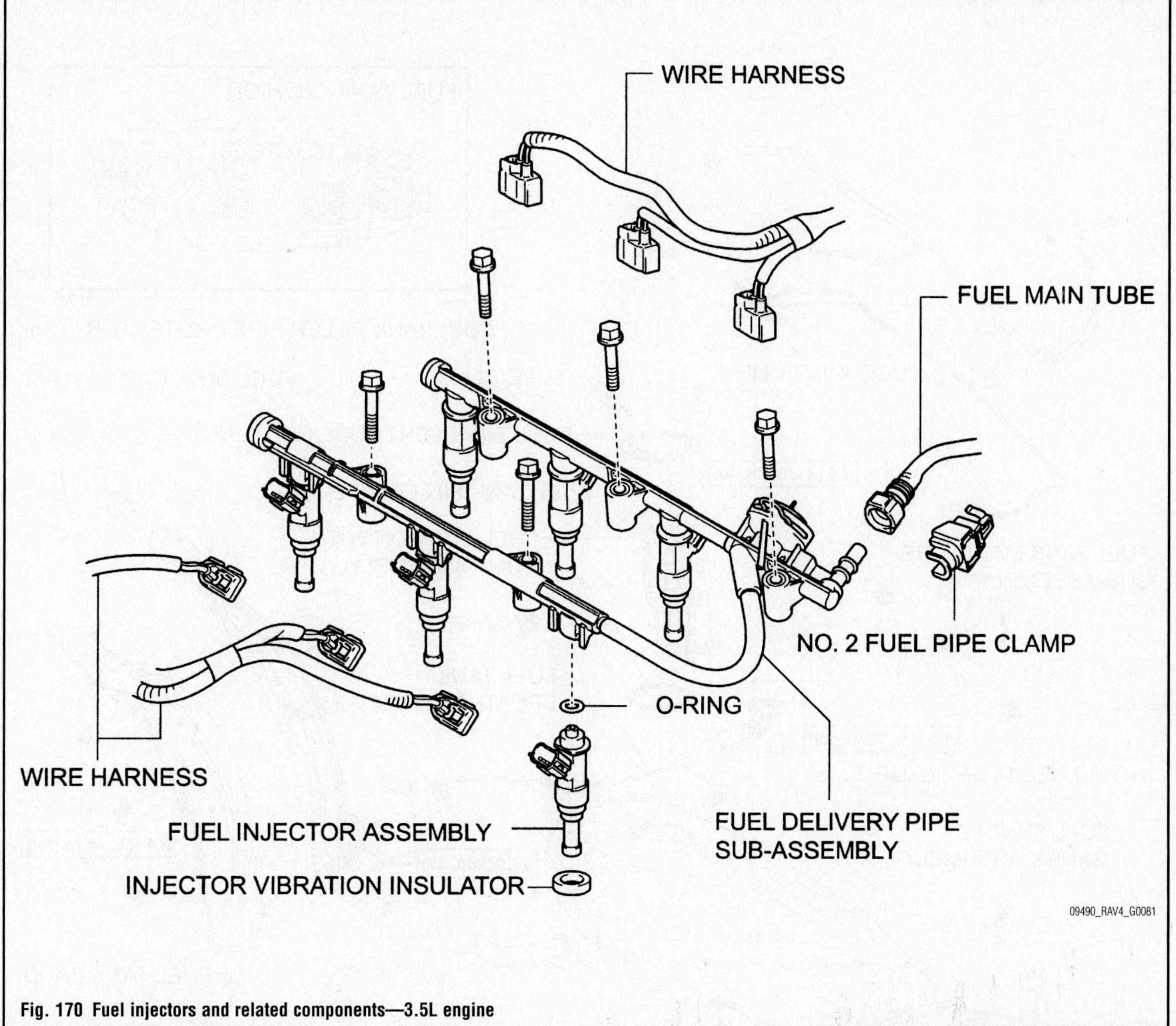

Fig. 170 Fuel injectors and related components—3.5L engine

1. Before servicing the vehicle, refer to the Precautions Section.
2. Properly relieve the fuel system pressure.
3. Disconnect the negative battery cable.

➡Wait at least 90 seconds after disconnecting the negative battery cable before starting any repair work to prevent air bag and seat belt pretensioner activation.

4. Remove the engine from the vehicle and position it in a suitable holding fixture.
5. Remove the intake air surge tank assembly.
6. Disconnect the number two main tube. Pinch the tube connector and then pull out the pipe.
7. Disconnect the six injector connectors. Remove the five bolts and the fuel delivery rail together with the injectors.
8. Remove the six insulators from the intake manifold.
9. Pull out the injectors from the rail. Remove and discard the O-rings.

To install:

10. Installation is the reverse of removal.
11. Coat the new O-rings with gasoline. Push the injectors onto the pipes and make sure they rotate freely.
12. Position the assembly onto the head and install the bolts finger tight. Make sure that the injectors still rotate freely. If not, replace the O-rings. Torque the bolts to 15 ft. lbs. (20 Nm).
13. Start the engine and check for leaks, correct as required.

FUEL PUMP

REMOVAL & INSTALLATION

2.4L Engine

See Figure 171.

➡The fuel tank must first be removed from the vehicle. Be sure to check and adjust the fuel level as required, before removing the fuel tank. Take all the necessary precautions to avoid safety and fuel disposal problems.

1. Before servicing the vehicle, refer to the Precautions Section.
2. Properly relieve the fuel system pressure.
3. Disconnect the negative battery cable.

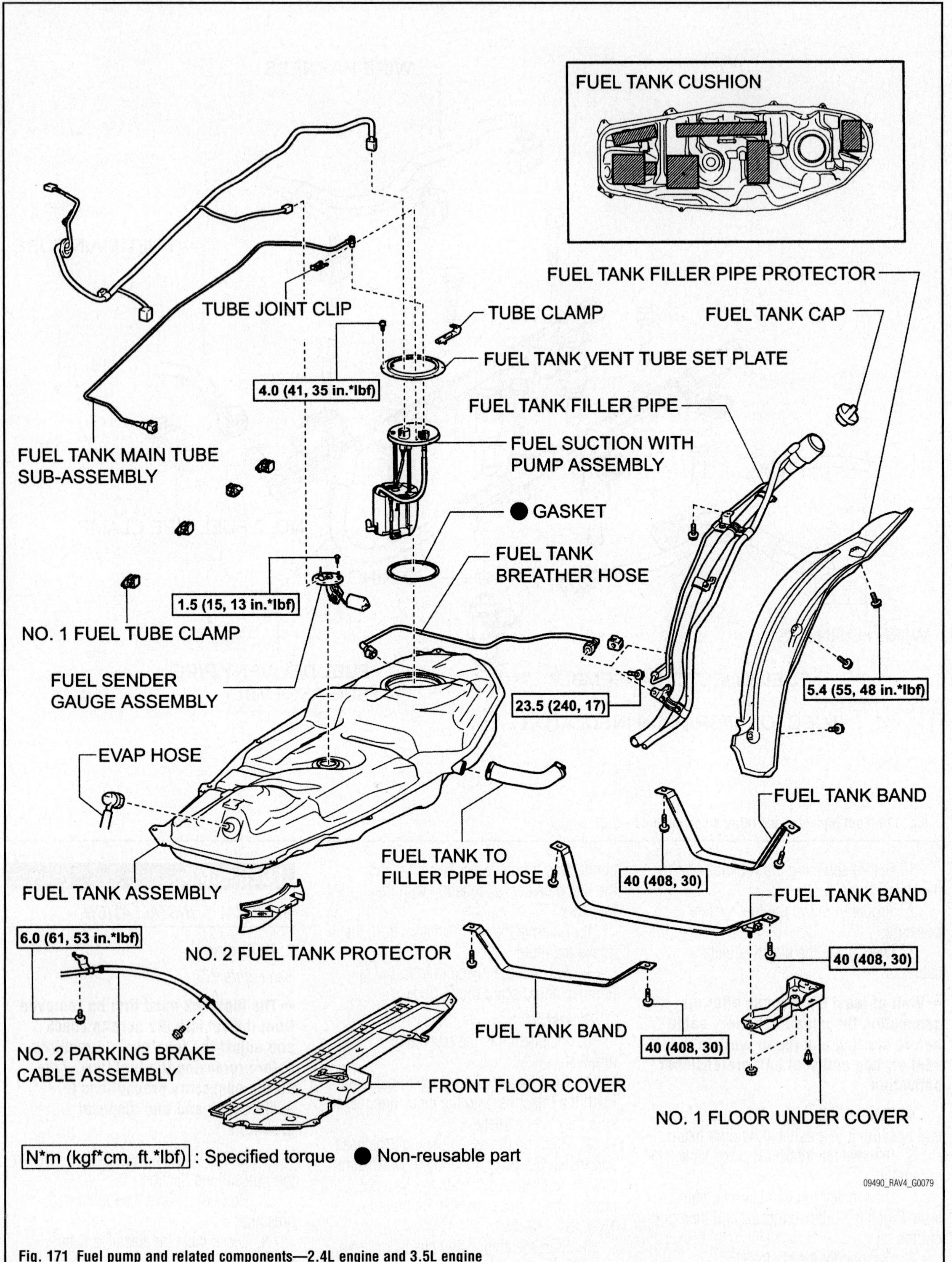

Fig. 171 Fuel pump and related components—2.4L engine and 3.5L engine

➡Wait at least 90 seconds after disconnecting the negative battery cable before starting any repair work to prevent air bag and seat belt pretensioner activation.

4. Remove the front floor carpet.
5. Disconnect the number two parking brake cable assembly.
6. Disconnect the fuel tank main tube assembly.
7. Disconnect the fuel tank to filler pipe hose.
8. Disconnect the fuel tank breather hose.
9. Remove the fuel tank filler pipe.
10. Position a suitable jack under the fuel tank. Remove the six bolts and three fuel tank bands.
11. Slightly lower the suitable jack.

➡Be careful not to cut the wing nuts.

12. Fold back about half of each cushion rubber so that the wire harness can be detached.
13. Disconnect the fuel pump connector and sender gauge connector.
14. Detach the wire harness from the four clamps. Carefully remove the fuel tank from the vehicle.
15. Remove the joint clip and fuel tank main tube.
16. Remove the eight bolts and the fuel tank vent tube set plate.
17. Disconnect the fuel hose and remove the fuel pump assembly from the fuel tank. Discard the gasket.

To install:

18. Installation is the reverse of the removal procedure.
19. Tighten the eight retaining bolts to 35 inch lbs, and in an alternating sequence.
20. Tighten the fuel tank retaining bolts to 30 ft. lbs.
21. Tighten the fuel tank filler pipe bolts to 17 ft. lbs.
22. Start the engine and check for leaks. Correct as required.

3.5L Engine

See Figure 171.

➡The fuel tank must first be removed from the vehicle. Be sure to check and adjust the fuel level as required, before removing the fuel tank. Take all the necessary precautions to avoid safety and fuel disposal problems.

1. Before servicing the vehicle, refer to the Precautions Section.
2. Properly relieve the fuel system pressure.
3. Disconnect the negative battery cable.

➡Wait at least 90 seconds after disconnecting the negative battery cable before starting any repair work to prevent air bag and seat belt pretensioner activation.

4. Remove the front floor carpet.
5. Disconnect the number two parking brake cable assembly.
6. Disconnect the fuel tank main tube assembly.
7. Disconnect the fuel tank to filler pipe hose.
8. Disconnect the fuel tank breather hose.
9. Remove the fuel tank filler pipe.
10. Position a suitable jack under the fuel tank. Remove the six bolts and three fuel tank bands.
11. Slightly lower the suitable jack.

➡Be careful not to cut the wing nuts.

12. Fold back about half of each cushion rubber so that the wire harness can be detached.
13. Disconnect the fuel pump connector and sender gauge connector.
14. Detach the wire harness from the four clamps. Carefully remove the fuel tank from the vehicle.
15. Remove the joint clip and fuel tank main tube.
16. Remove the eight bolts and the fuel tank vent tube set plate.
17. Disconnect the fuel hose and remove the fuel pump assembly from the fuel tank. Discard the gasket.

To install:

18. Installation is the reverse of the removal procedure.
19. Tighten the eight retaining bolts to 35 inch lbs, and in an alternating sequence.
20. Tighten the fuel tank retaining bolts to 30 ft. lbs.
21. Tighten the fuel tank filler pipe bolts to 17 ft. lbs.
22. Start the engine and check for leaks. Correct as required.

FUEL TANK

REMOVAL & INSTALLATION

See Figures 172 through 175.

1. Before servicing the vehicle, refer to the Precautions Section.
2. Properly relieve the fuel system pressure.
3. Disconnect the negative battery cable.

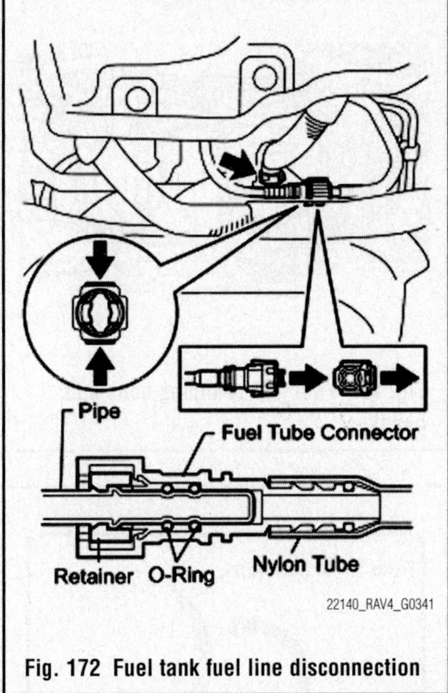

Fig. 172 Fuel tank fuel line disconnection

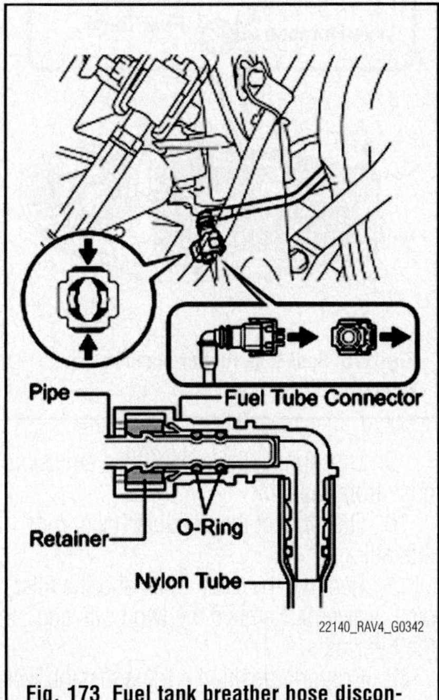

Fig. 173 Fuel tank breather hose disconnection

➡Wait at least 90 seconds after disconnecting the negative battery cable before starting any repair work to prevent air bag and seat belt pretensioner activation.

4. Remove the fuel cap.
5. Raise and support the vehicle safely.
6. Remove the nut, bolt and three clips in order to remove the front floor cover.
7. Remove the two bolts and disconnect the parking brake cover.
8. Disconnect the fuel tank main tube sub-assembly.

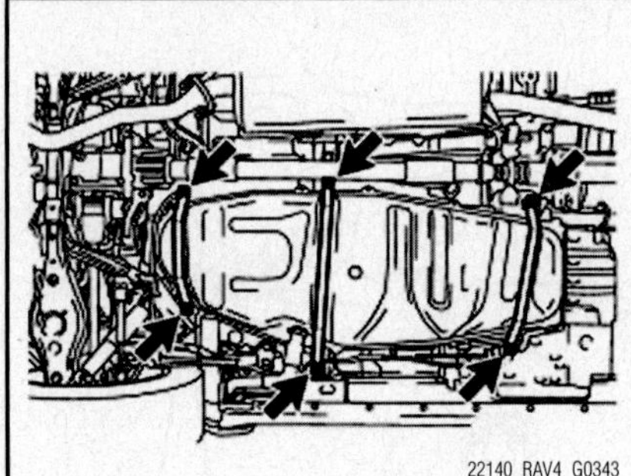

Fig. 174 Fuel tank retaining bolts and bands

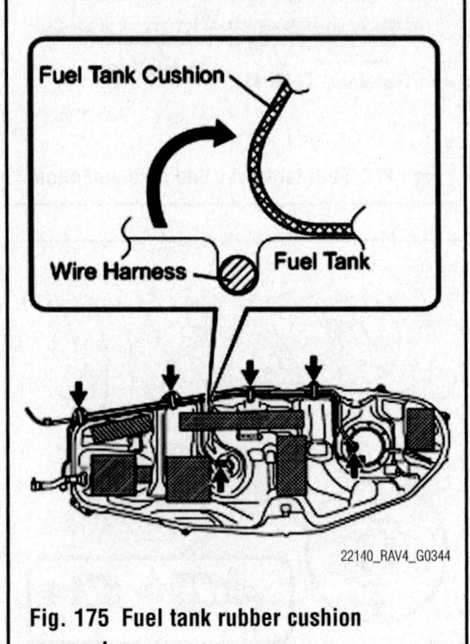

Fig. 175 Fuel tank rubber cushion removal

9. Disconnect the evaporative emission hose from the tank.
10. Disconnect the breather hose from the tank.
11. Remove the three bolts and the filler pipe protector. Remove the two bolts and the filler pipe.
12. Properly position a transmission jack underneath the fuel tank assembly.
13. Remove the six bolts and three fuel tank retaining bands.
14. Carefully lower the transmission jack.

➡Be careful not to cut any electrical wiring or hoses.

15. Fold back about half of each cushion rubber so that the wire harness can be detached from the fuel tank.
16. Disconnect the fuel pump electrical connector and sender gauge connector.

➡Check the connector for dirt, mud and other contamination, prior to disconnection. Do not use any tools to disconnect the connector.

17. Detach the wire harness from the four clamps.
18. Remove the fuel tank from the vehicle.

To install:

19. Installation is the reverse of the removal procedure.
20. Tighten the six band retaining bolts to 30 ft. lbs.
21. When connecting tube connectors push firmly until a "click" sound is heard.

➡Check that the connector and pipe are securely connected by trying to pull them apart.

22. Start the engine and check for leaks. Correct as required.

IDLE SPEED

ADJUSTMENT

Idle speed is maintained by the Powertrain Control Module (PCM). No adjustment is necessary or possible.

THROTTLE BODY

REMOVAL & INSTALLATION

2.4L Engine

See Figures 176 through 178.

1. Before servicing the vehicle, refer to the Precautions Section.
2. Properly relieve the fuel system pressure.
3. Disconnect the negative battery cable.

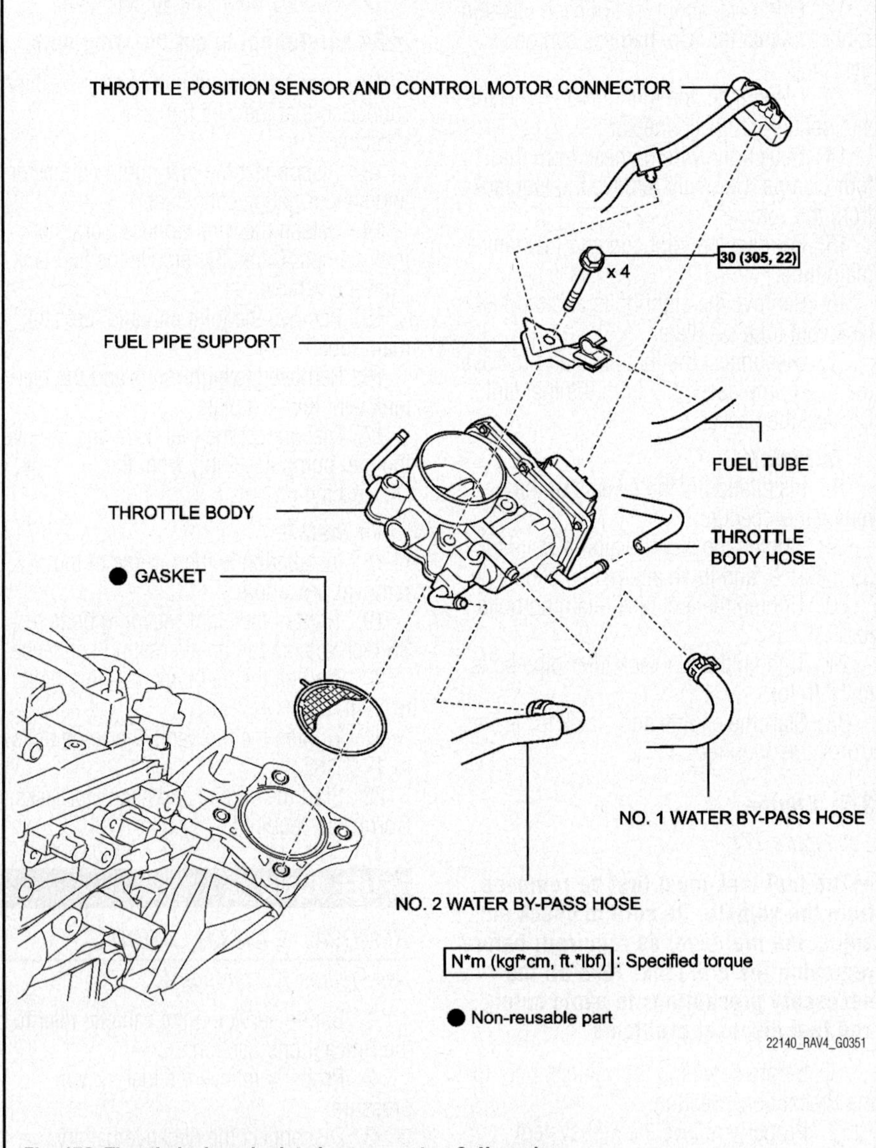

Fig. 176 Throttle body and related components—2.4L engine

➡Wait at least 90 seconds after disconnecting the negative battery cable before starting any repair work to prevent air bag and seat belt pretensioner activation.

4. Drain and properly dispose of the engine coolant.
5. Remove the number one engine cover retaining nuts. Remove the cover.
6. Disconnect the mass air flow meter connector. Disconnect the purge connector. Disconnect the four wire harness clamps.
7. Disconnect the number two ventilation hose from the air cleaner hose. Disconnect the purge line hose from the clamp.
8. Lock the number one air cleaner hose clamp and then disconnect the number one air cleaner hose from the throttle body.
9. Unfasten the two hook clamps and remove the air cleaner cap.
10. Remove the filter element from the case.

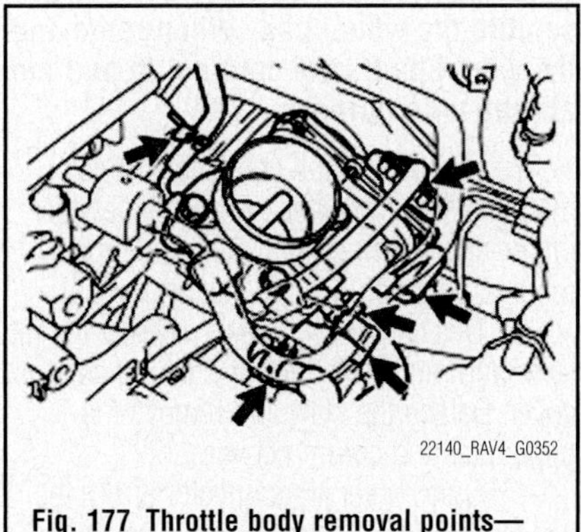

Fig. 177 Throttle body removal points—2.4L engine

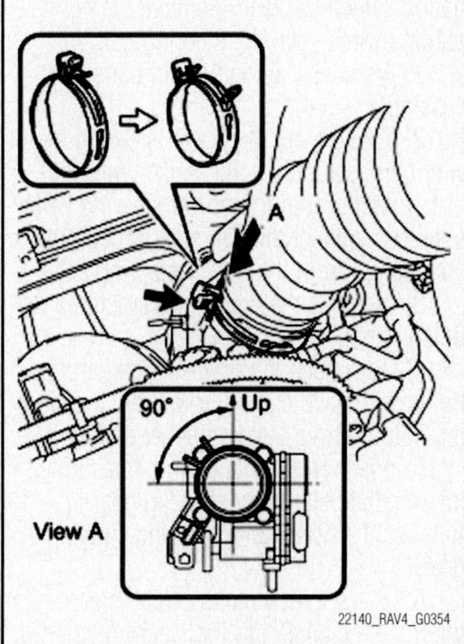

Fig. 178 Proper air cleaner hose installation—2.4L engine

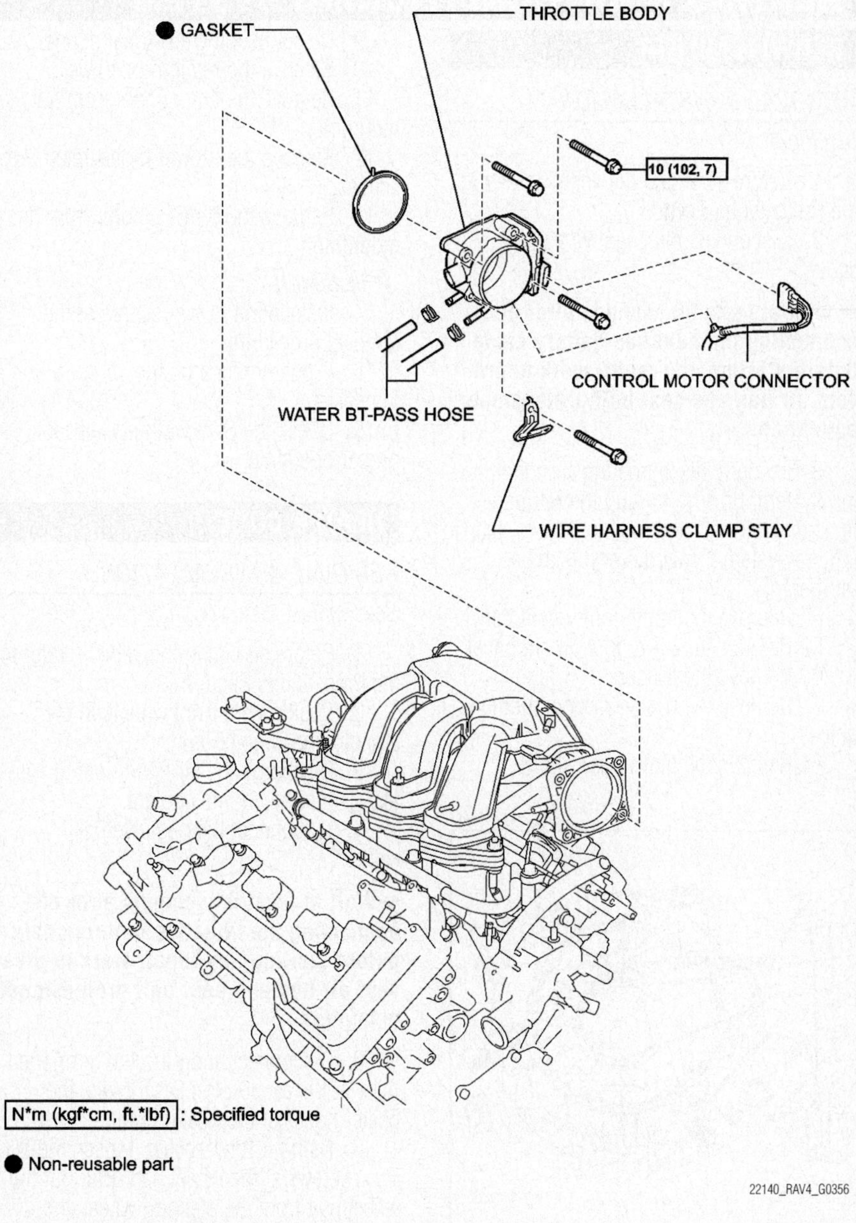

11. Disconnect the purge line hose from the throttle body.
12. Disconnect the water bypass hose from the throttle body. Disconnect the number two water bypass hose from the throttle body.
13. Disconnect the number one water bypass hose from the throttle body.
14. Disconnect the throttle position sensor and the control motor connector. Disconnect the wire harness clamp.
15. Disconnect the fuel tube from the clamp.
16. Remove the throttle body retaining bolts.
17. Remove the component from its mounting. Remove and discard the gasket from the intake manifold.

To install:

18. Installation is the reverse of removal.
19. Be sure to use a new gasket.
20. Tighten the retaining bolts to 22 ft. lbs.
21. Be sure that the air clean hose is properly installed.
22. Start the engine and check for leaks. Correct as required.

3.5L Engine

See Figures 179 and 180.

Fig. 179 Throttle body and related components—3.5L engine

1. Before servicing the vehicle, refer to the Precautions Section.
2. Properly relieve the fuel system pressure.
3. Disconnect the negative battery cable.

➡Wait at least 90 seconds after disconnecting the negative battery cable before starting any repair work to prevent air bag and seat belt pretensioner activation.

4. Drain and properly dispose of the engine coolant.
5. Remove the air cleaner sub-assembly.
6. Disconnect the two water bypass hoses from the throttle body.

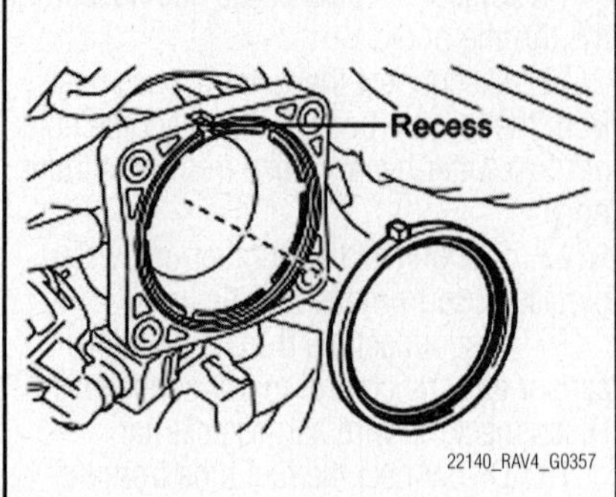

Fig. 180 Proper gasket positioning—3.5L engine

7. Disconnect the control motor connector.
8. Remove the four throttle body retaining bolts. Remove the wire harness clamp stay.
9. Remove the component from its mounting.
10. Discard the gasket.

To install:

11. Installation is the reverse of removal.
12. Be sure to use a new gasket.
13. Tighten the retaining bolts to 7 ft. lbs.
14. Start the engine and check for leaks. Correct as required.

HEATING & AIR CONDITIONING SYSTEM

BLOWER MOTOR

REMOVAL & INSTALLATION

See Figure 181.

1. Before servicing the vehicle, refer to the Precautions Section.
2. Disconnect the negative battery cable.

➡Wait at least 90 seconds after disconnecting the negative battery cable before starting any repair work to prevent air bag and seat belt pretensioner activation.

3. Properly discharge the air conditioning system. Be sure to plug the refrigerant lines to avoid dirt and moisture from entering the system. Properly recycle the ac refrigerant.
4. Remove the upper instrument panel.
5. Remove the lower instrument panel.
6. Remove the air duct.
7. Remove the heater/air conditioning unit.
8. Remove the blower assembly.

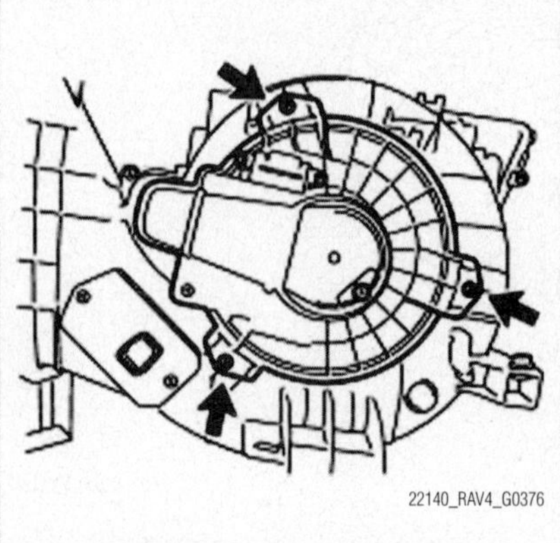

Fig. 181 Blower motor retaining screws

9. Disconnect the electrical connector.
10. Remove the retaining screws.
11. Remove the component from its mounting.
12. Remove the blower motor retaining screws.
13. Remove the blower motor from its mounting.

To install:

14. Installation is the reverse of the removal procedure.
15. Properly recharge the air conditioning system.
16. Check the SRS warning light for proper operation.

HEATER CORE

REMOVAL & INSTALLATION

See Figures 182 through 187.

1. Before servicing the vehicle, refer to the Precautions Section.
2. Position the front wheels in the straight ahead position.
3. Discharge the air conditioning system. Drain the engine coolant.
4. Disconnect the negative battery cable.

➡Wait at least 90 seconds after disconnecting the negative battery cable before starting any repair work to prevent air bag and seat belt pretensioner activation.

5. Disconnect and plug the refrigerant lines at the evaporator. Disconnect the heater hoses at the heater core.
6. Using a Torx®socket, loosen the two Torx®screws at the wheel pad. Pull out the wheel pad from the steering wheel and disconnect the air bag connector. Remove the wheel pad.

➡If the air bag connector is disconnected with the ignition switch "ON", DTC's will be recorded. When storing the wheel pad, keep the upper surface of the pad facing upward. Never disassemble the wheel pad. When removing the wheel pad, take care not to pull the air bag wire harness.

7. Disconnect the connector. Matchmark the steering wheel. Remove the steering wheel retaining nut. Using the proper removal tool, remove the steering wheel.
8. Detach the four claws, release the tilt lever and remove the lower steering column cover. Detach the claw and remove the upper steering column cover.
9. Disconnect all connectors from the turn signal switch and the spiral cable. Remove the spiral cable.
10. Detach the clamp holding the combination switch in place. Remove the combination switch from the steering column.
11. Remove the instrument panel sub-assembly.
12. Disconnect the power steering motor wire harness and torque sensor wire harness clamps from the power steering ECU side. Disconnect the two steering column connectors from the power steering ECU.
13. Remove the lower instrument panel finish panel.
14. Turn back the driver's side carpet. Remove the two clips and remove the steering column hole cover silencer sheet.
15. Place matchmarks on the steering intermediate shaft and steering gear. Remove the bolt and detach the steering gear.
16. Place matchmarks on the steering intermediate shaft and the steering column. Remove the bolt and detach the steering intermediate shaft from the steering column.

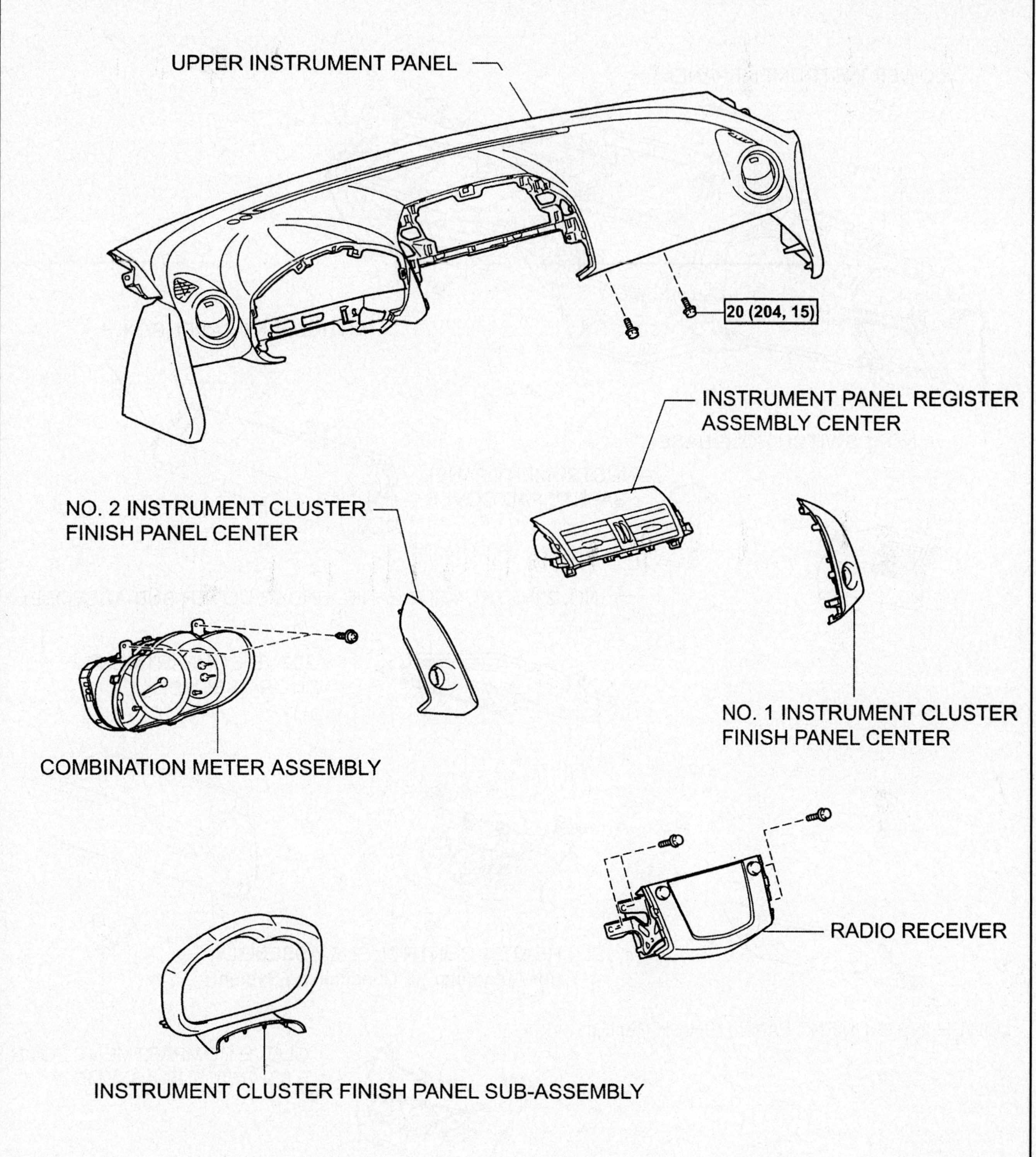

Fig. 182 Upper instrument panel and related components

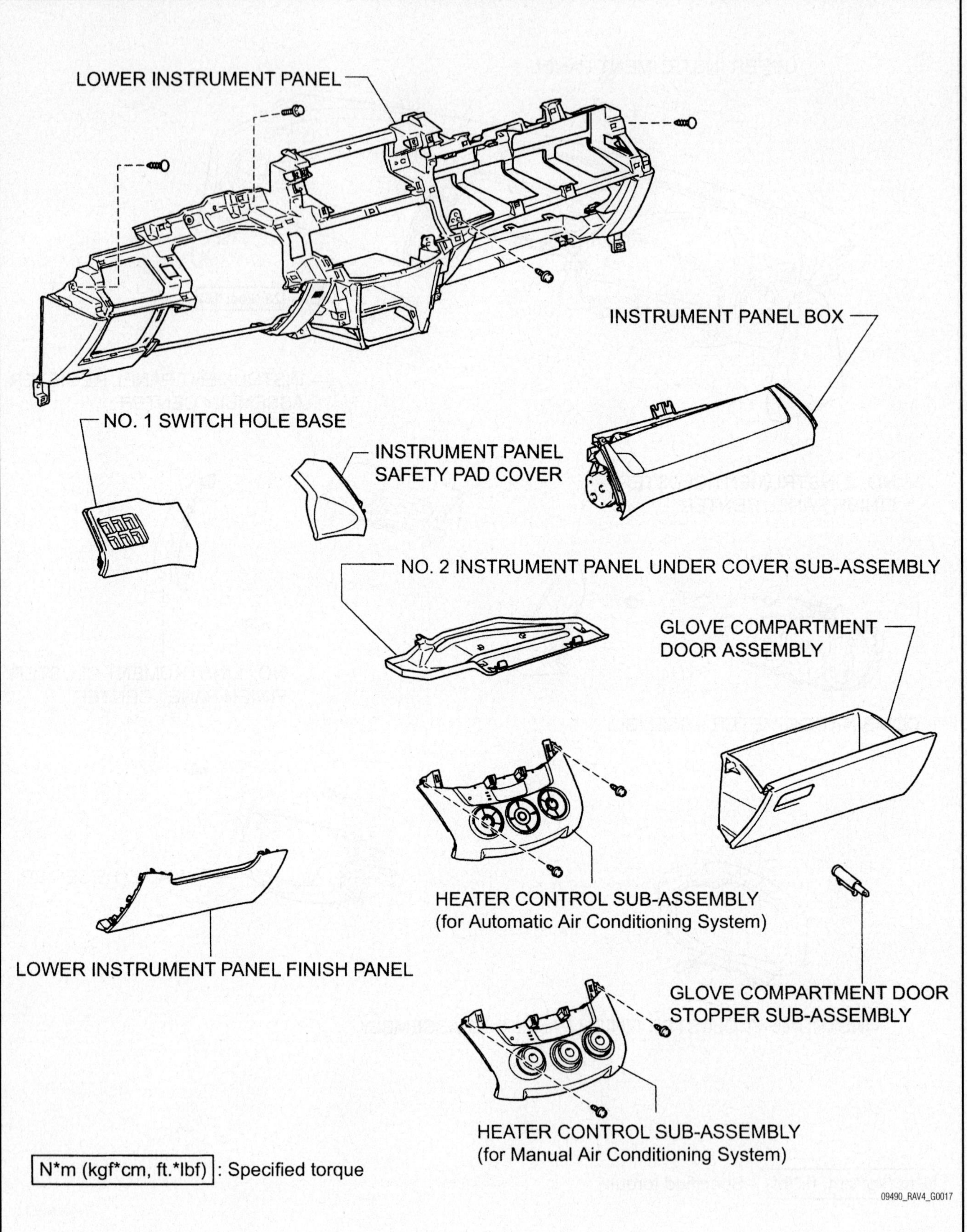

Fig. 183 Lower instrument panel and related components

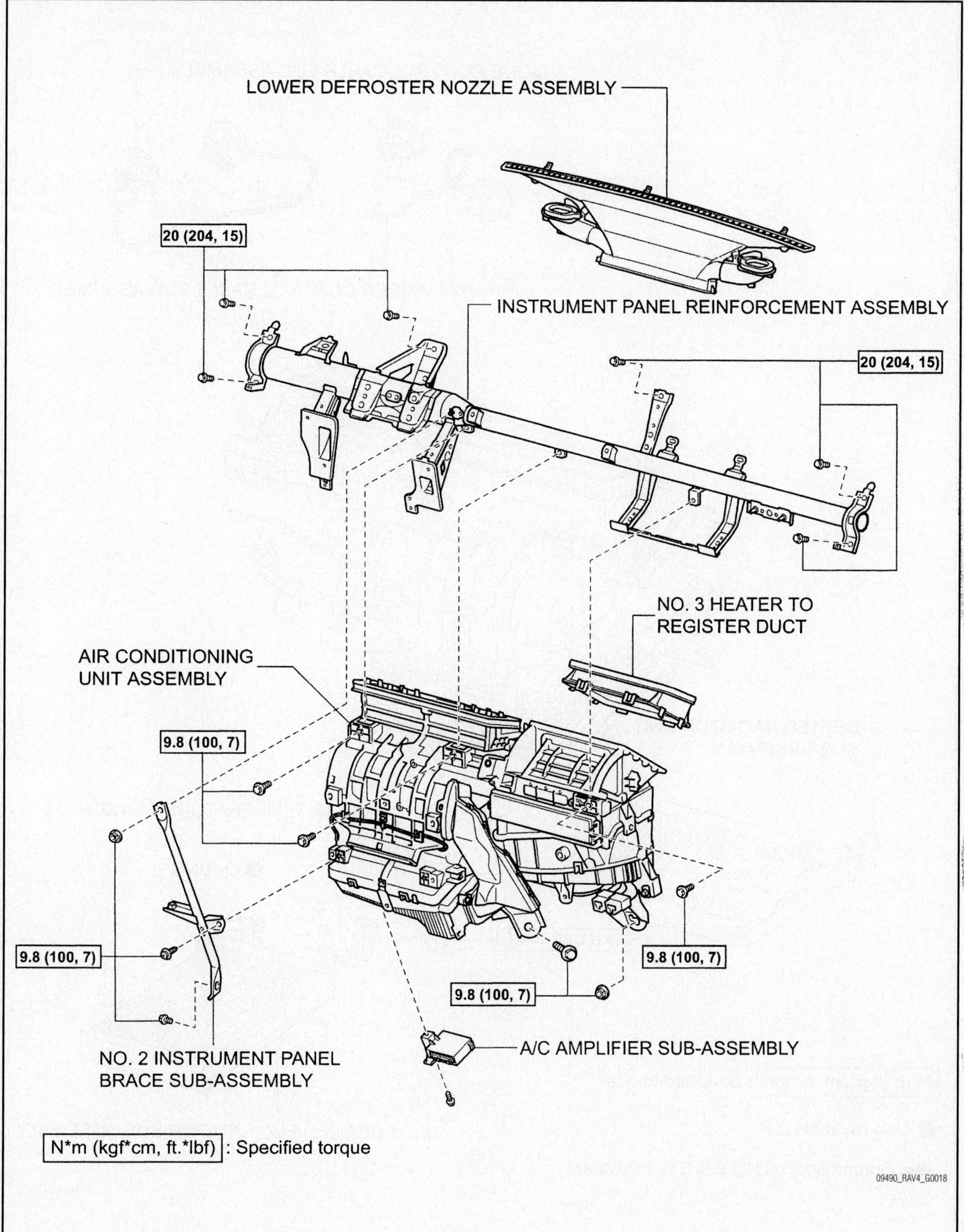

Fig. 184 Reinforcement panel and related components

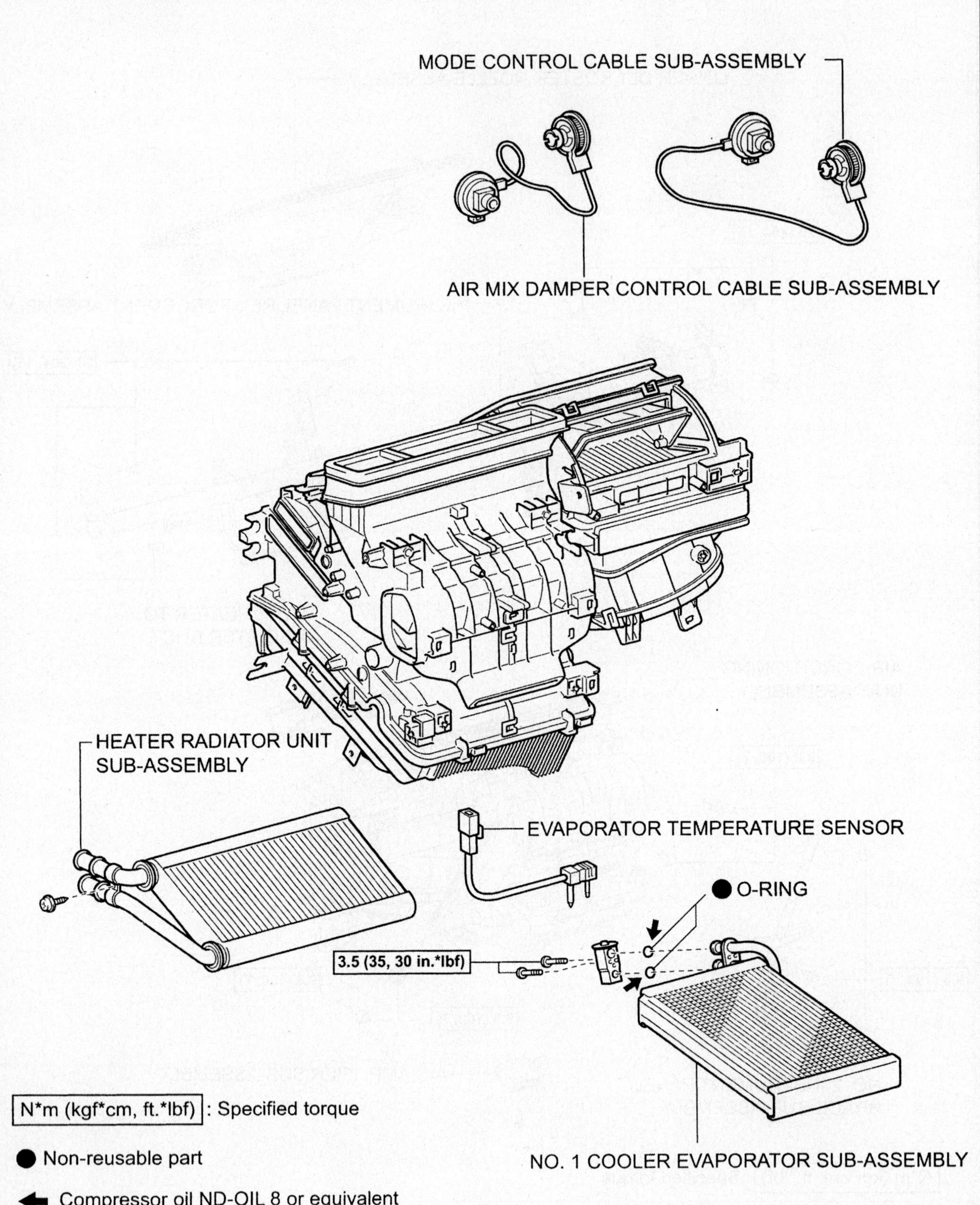

Fig. 185 Manual air conditioning heater core and related components

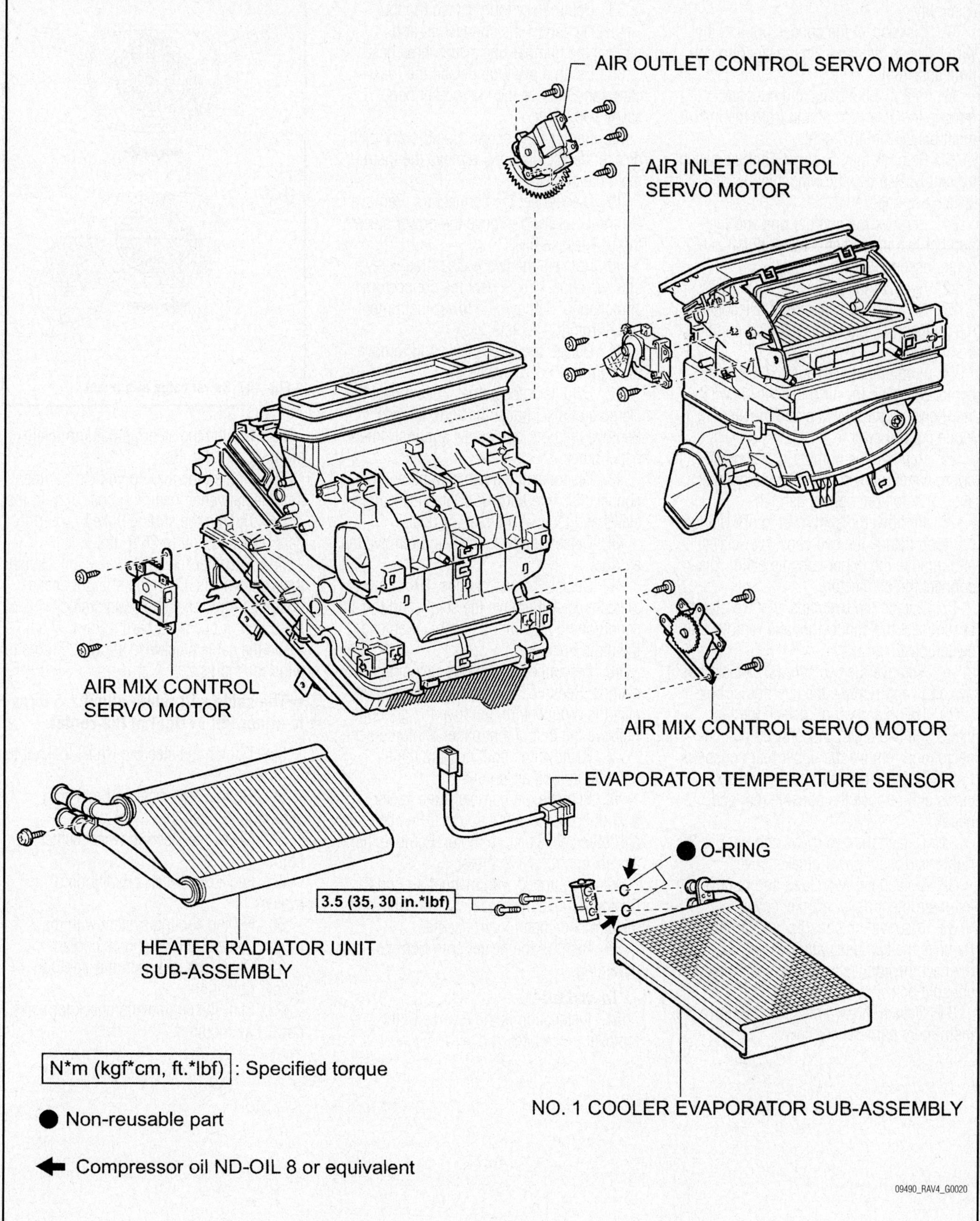

Fig. 186 Automatic air conditioning heater core and related components

17. Remove the brake pedal support assembly.

18. Disconnect the connectors and the wire harness clamps from the steering column assembly.

19. Remove the bolts and nuts and remove the steering column from the instrument panel reinforcement.

20. Remove the headlight dimmer switch assembly. Remove the windshield wiper switch assembly.

21. Remove the number one and the number two instrument cluster finish panel center assemblies.

22. Remove the radio.

23. Remove the two screws. Disconnect the connectors and remove the air conditioning control.

24. If equipped with automatic climate control, remove the air mix control switch, the blower control switch and the air vent mode control switch.

25. Remove the number one console upper panel garnish. Remove the number two console upper panel garnish.

26. Remove the shift lever knob. Using a pry tool, detach the two clips, four claws and remove the upper console panel. Disconnect the connector.

27. Detach the two clips and two claws. Disconnect the connectors and remove the switch base.

28. Remove the two screws. Detach the two clips and remove the cup holder box.

29. Using a pry tool, detach the four claws and two clips. Disconnect the connector and remove the upper rear console.

30. Using a pry tool, detach the six claws and remove the console rear end panel.

31. Detach the two claws and remove the right instrument panel bracket cover.

32. Detach the two claws and remove the left instrument panel bracket cover.

33. Remove the console box carpet. Remove the two bolts and two screws. Disconnect the connector and remove the rear console box.

34. Detach the two claws and remove the instrument panel under cover.

35. Remove the glove box door assembly.

36. Using a pry tool, detach the four clips. Disconnect the connectors and remove the number one switch hole base.

37. Using a pry tool, detach the four claws and remove the instrument panel safety pad cover.

38. Remove the screw. Using a pry tool, detach the six clips and remove the instrument panel box.

39. Disconnect the connectors. Remove the three nuts and remove the power steering ECU assembly.

40. Remove the two bolts, three screws and two clips. Disconnect the connectors and clamps. Remove the lower instrument panel from the vehicle.

41. Detach the three claws and remove the defroster nozzle assembly.

42. Fold back the carpet. Disconnect the clamp and disconnect the wire harness. Remove the bolt, nut, screw and instrument panel brace.

43. Disconnect the two clamps and disconnect the wire harness. Detach the three claws and remove the rear air duct.

44. Detach the two claws and remove the air duct.

45. Detach the two clamps. Disconnect the connector. Remove the screw and the air conditioning amplifier assembly. Remove the drain hose.

46. Disconnect the twelve retaining clamps. Remove the four bolts and disconnect the ground wire. Remove the six bolts. Remove the instrument panel reinforcement.

47. Remove the bolt, nut and the air conditioner unit assembly.

48. Remove the number three heater register duct. Remove the air duct. Remove the air outlet control servo motor. Remove the air mix control servo motor.

49. If equipped with manual air conditioning remove the mode control cable and the airmix damper control cable.

50. Remove the heater core from its mounting.

To install:

51. Installation is the reverse of the removal procedure.

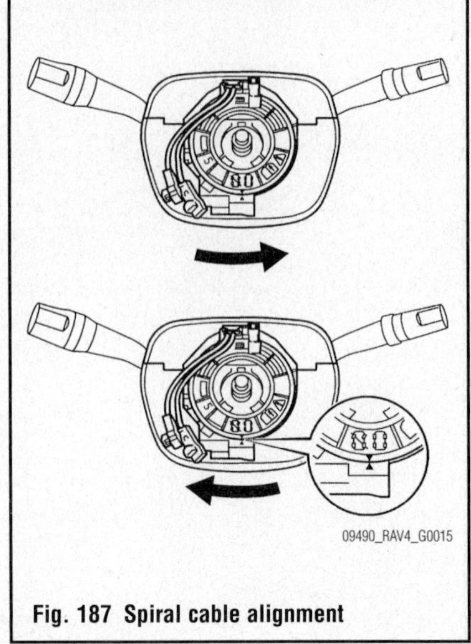

Fig. 187 Spiral cable alignment

52. Tighten the steering column retaining bolts to 18 ft. lbs.

53. Tighten the sliding yoke to number intermediate shaft retaining bolt to 26 ft. lbs.

54. Tighten the sliding yoke to steering gear retaining bolt to 26 ft. lbs.

55. To center the spiral cable, check that the front wheels are in the straight ahead position. Turn the cable counterclockwise by hand until it becomes hard to turn. Then rotate the cable clockwise about 2.5 turns to align the marks.

➡The cable will rotate about 2.5 turns to either left or right of the center.

56. Tighten the steering wheel locknut to 37 ft. lbs.

57. Tighten the wheel pad retaining screws to 78 inch lbs.

58. Check the steering wheel center point.

59. Recharge the air conditioning system.

60. Fill the cooling system with the proper grade and type engine coolant.

61. Check the SRS warning light for proper operation.

62. Start the engine and check for leaks, correct as required.

STEERING

POWER RACK & PINION STEERING GEAR

REMOVAL & INSTALLATION

See Figure 188.

1. Before servicing the vehicle, refer to the Precautions Section.
2. Disconnect the negative battery cable.

➡Wait at least 90 seconds after disconnecting the negative battery cable before starting any repair work to prevent air bag and seat belt pretensioner activation.

3. Place the wheels in a straight-ahead position.
4. Raise and support the vehicle safely.
5. Remove the tire and wheel assemblies.
6. Disconnect the right and left tie rod ends, using tool SST09628-62011 or equivalent.
7. Remove the floor carpet. Remove the two clips and the column hole silencer cover.
8. Use the seat belt to position the steering wheel in order to avoid breakage of the spiral cable.

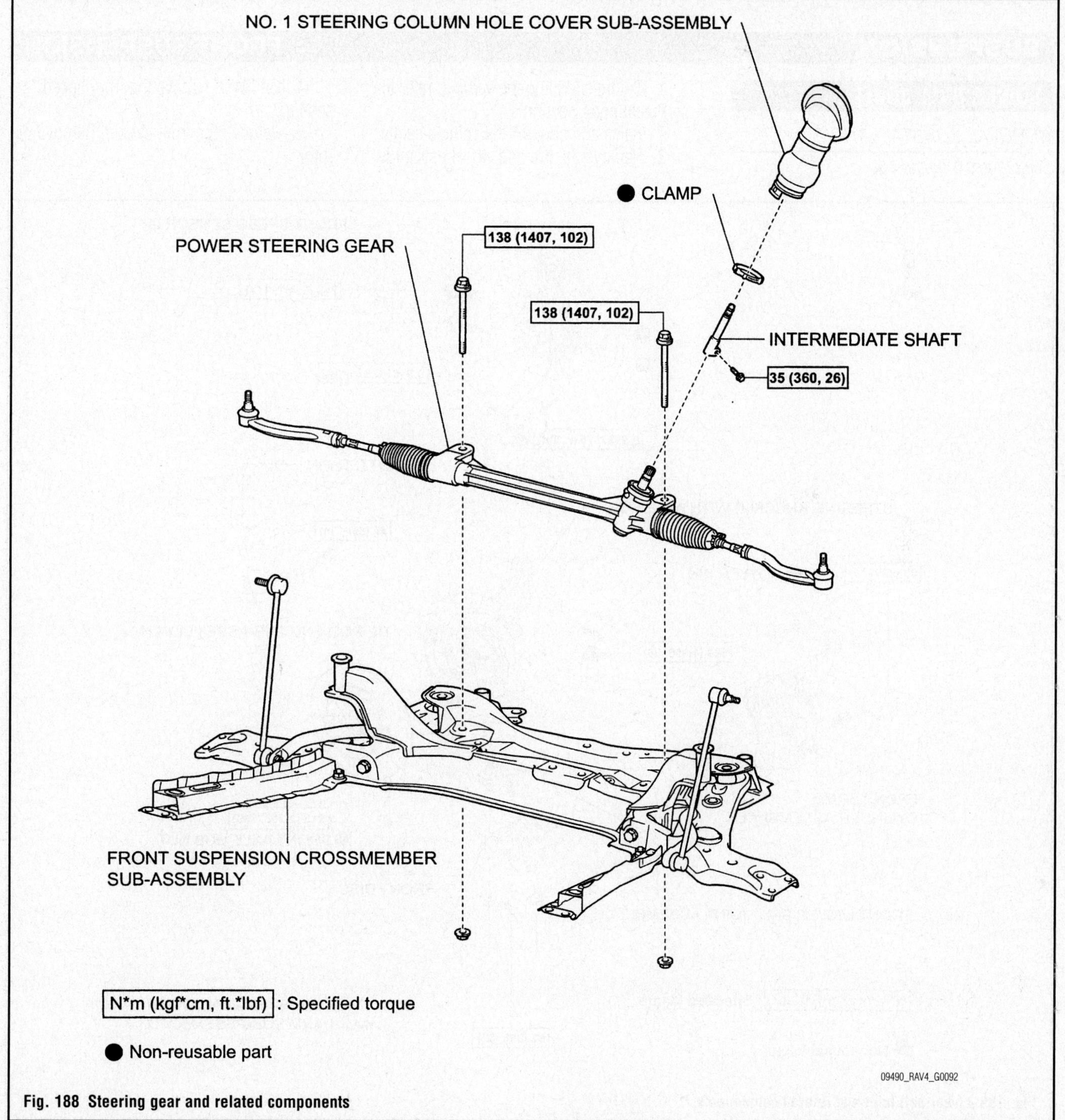

Fig. 188 Steering gear and related components

9. Matchmark the sliding yoke of the steering intermediate shaft. Remove the bolt and disconnect the sliding yoke.

10. Remove the bottom clip and detach the upper clip from the body and disconnect the number one steering column hole cover. Be careful not to damage the clips.

11. Remove the engine and transaxle assembly from the vehicle and position it in a suitable holding fixture.

12. Remove the clamp and disconnect the number one column hole cover from the steering gear.

13. Matchmark the intermediate shaft of the steering gear. Remove the bolt and disconnect the steering intermediate shaft from the steering gear.

14. Remove the two bolts, two nuts and the steering gear from the crossmember. Be sure to keep the nut from rotating while turning the bolt.

To install:

15. Installation is the reverse of the removal procedure.

16. Tighten the steering gear retaining bolts to 102 ft. lbs. Be sure to keep the nut from rotating while turning the bolt.

17. Tighten the intermediate shaft retaining bolt to 26 ft. lbs.

18. Tighten the sliding yoke retaining bolt to 26 ft. lbs.

19. Tighten the tie rod end castle nut to 36 ft. lbs. If the holes for the clip are not aligned, tighten the nut an additional 60 degrees. Be sure to use a new cotter pin.

20. Check and adjust the alignment, as required.

SUSPENSION

FRONT SUSPENSION

LOWER BALL JOINT

REMOVAL & INSTALLATION

See Figures 189 and 190.

1. Before servicing the vehicle, refer to the Precautions Section.
2. Raise and support the vehicle safely.
3. Remove the tire and wheel assembly.
4. On 2WD, remove the front speed sensor.
5. Remove the front caliper. Remove the rotor.

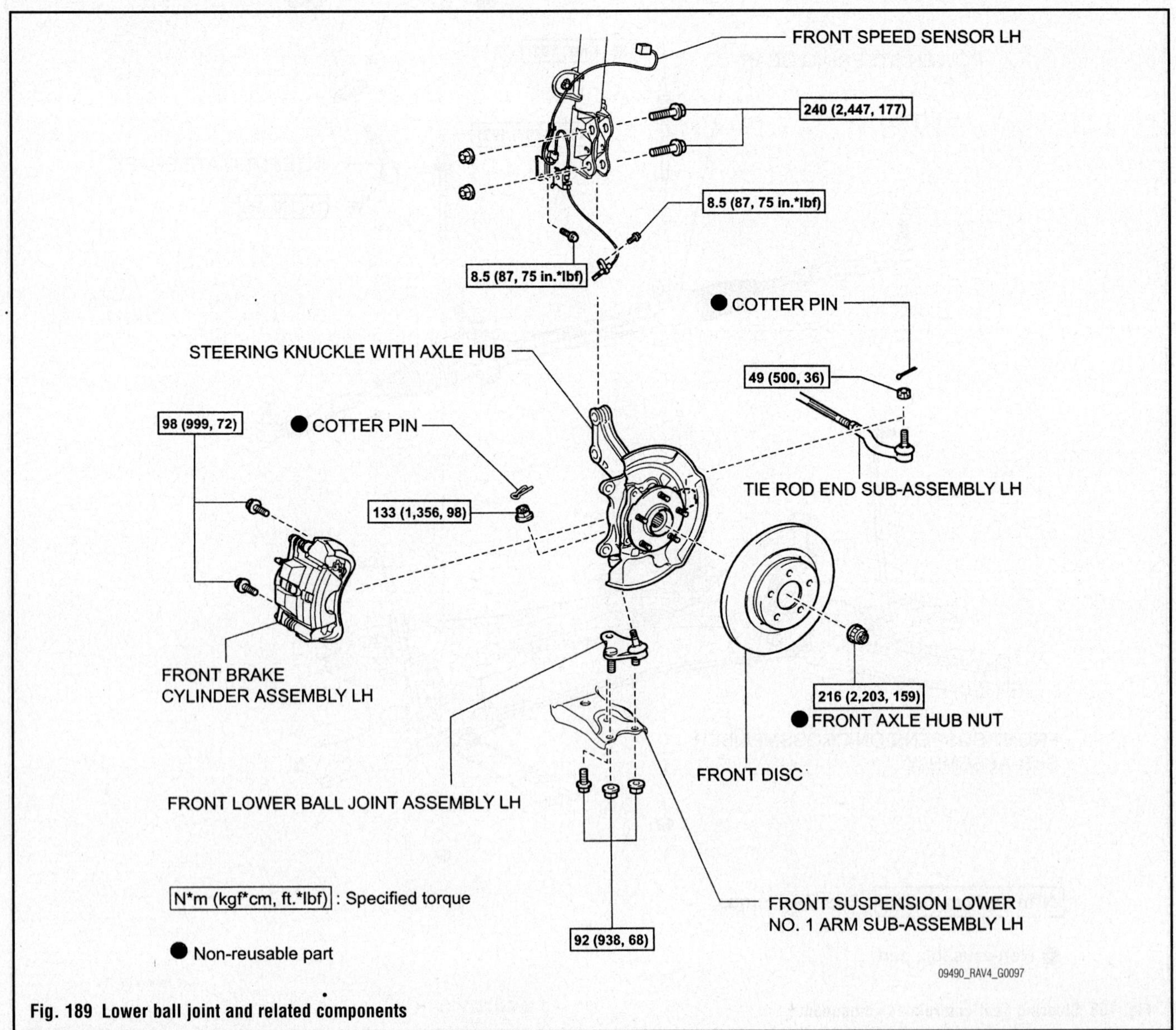

Fig. 189 Lower ball joint and related components

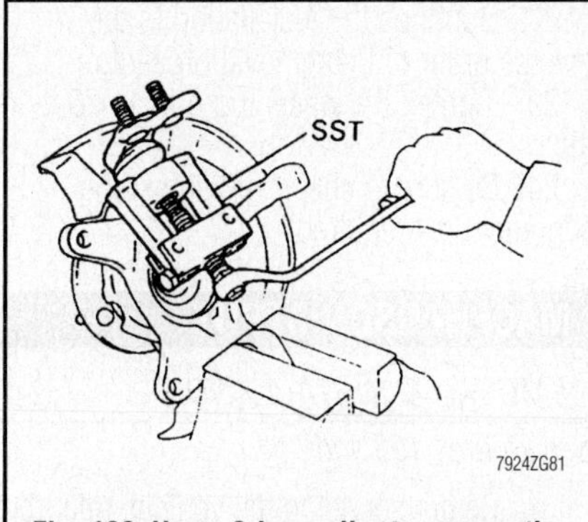

Fig. 190 Use a 2-jaw puller to remove the lower ball joint

6. Remove the front axle hub nut.

7. To disconnect the front suspension lower number one arm, remove the bolt and two nuts. Disconnect the lower arm from the ball joint.

8. Using the proper tools, disconnect the tie rod end.

9. Remove the steering knuckle with the axle hub, using the proper removal tools.

10. Remove the cotter pin and nut. Using tool SST09628-62011, remove the lower ball joint.

To install:

11. Installation is the reverse of the removal procedure.

12. Tighten the two ball joint to the lower arm nuts to 68 ft. lbs.

13. Tighten the lower ball joint to the steering knuckle to 98 ft. lbs.

14. Tighten the wheel lug nuts to 76 ft. lbs.

15. Be sure to check and adjust the alignment, as required.

LOWER CONTROL ARM

REMOVAL AND & INSTALLATION

See Figures 191 and 192.

1. Before servicing the vehicle, refer to the Precautions Section.

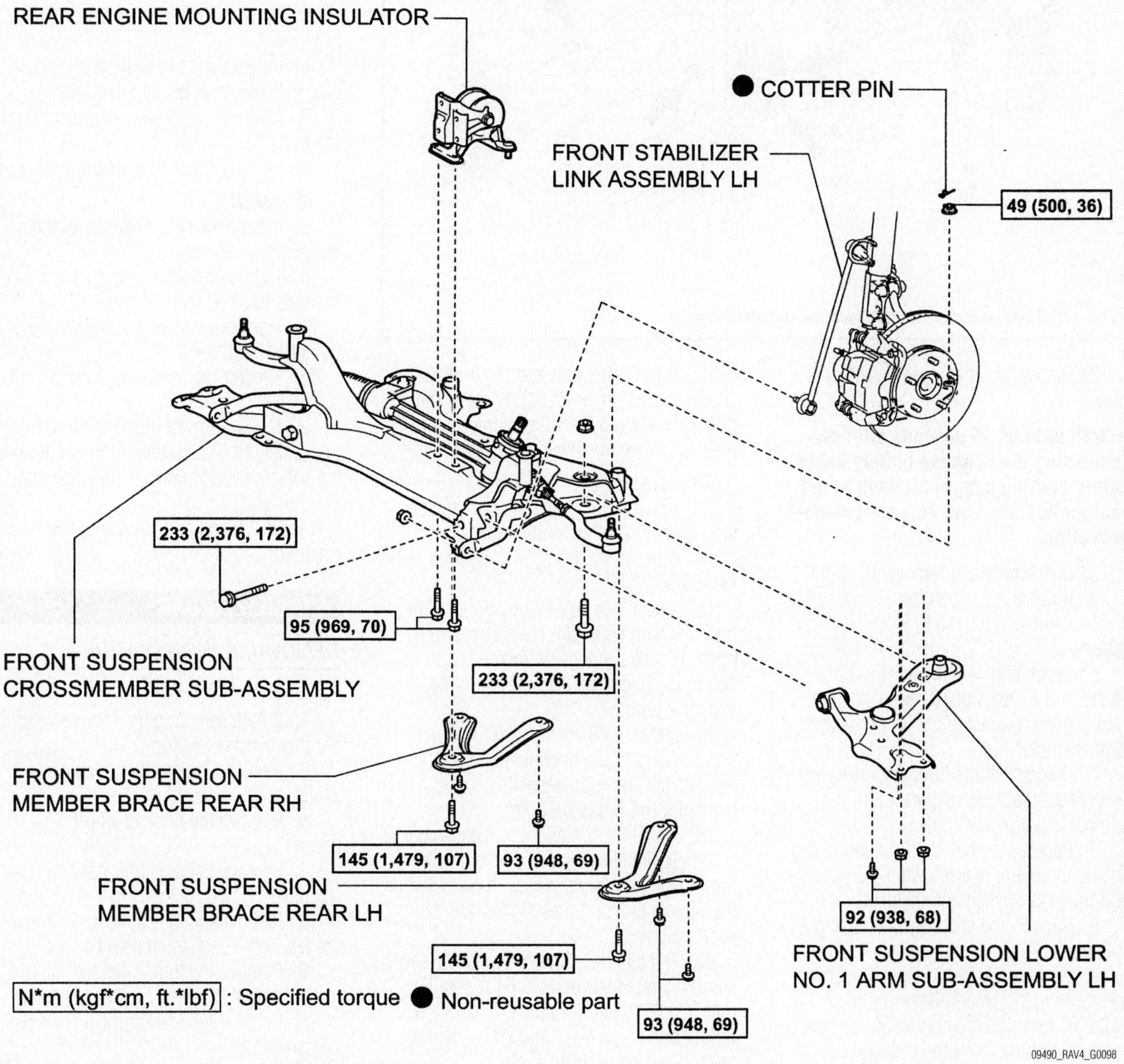

Fig. 191 Lower control arm and related components

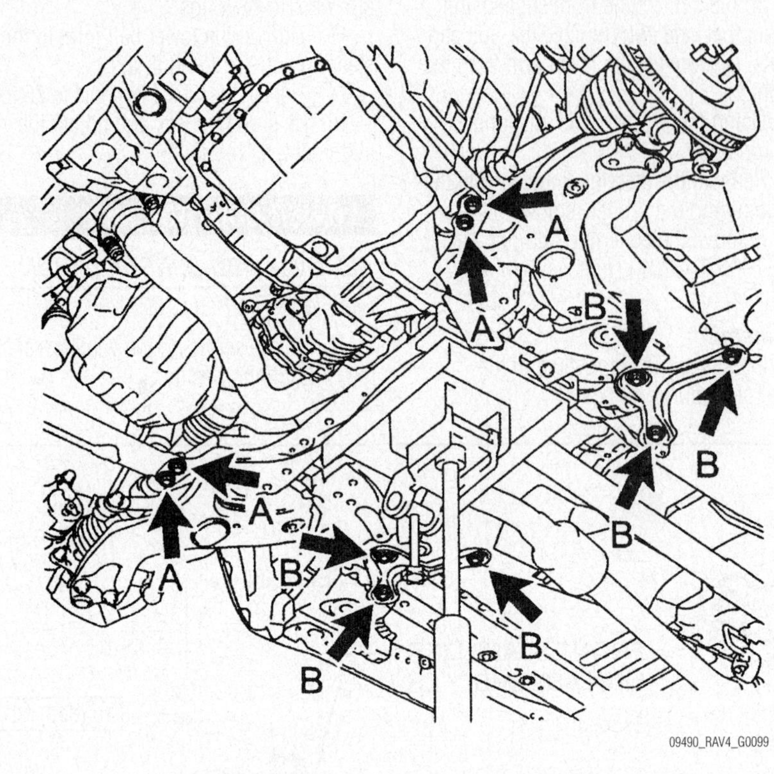

Fig. 192 Front crossmember bolt location and identification

2. Disconnect the negative battery cable.

➡Wait at least 90 seconds after disconnecting the negative battery cable before starting any repair work to prevent air bag and seat belt pretensioner activation.

3. Matchmark and remove the hood.
4. Raise and support the vehicle safely.
5. Remove the tire and wheel assembly.
6. Install engine hanger tools 12281-28010 and 12282-28010. Suspend the engine assembly using the proper engine removal tools.
7. Disconnect the front stabilizer links. Disconnect the front suspension lower number one arm assembly.
8. Remove the two nuts, two bolts and engine mounting rear insulator. Remove the bolt from the suspension member.
9. Support the crossmember, using a suitable jack.
10. Remove the four bolts "A" from the member reinforcement. Remove the six bolts "B" from the member reinforcement.
11. Carefully lower the jack and disconnect the crossmember from the vehicle.
12. Remove the bolt and nut from the suspension member (front). Remove the bolt and nut from the suspension member (rear). Remove the lower control arm.

To install:

13. Temporarily install the front suspension lower arm in its mounting.
14. Connect the front crossmember sub-assembly.
15. Install, but do not fully tighten, the four retaining bolts "A", the six retaining bolts "B", the suspension member with the bolt to the body and the rear mounting insulator bolts.
16. Install the front stabilizer link.
17. Connect the front suspension lower number one arm sub-assembly. Tighten the two bolts and nut to 68 ft. lbs.
18. Install the tire and wheel assembly.
19. To stabilize the suspension, lower the vehicle to ground height. Press down on the vehicle several times to stabilize the suspension.
20. Tighten the front crossmember sub-assembly retaining bolts to 64 ft. lbs. for bolt "A", 69 ft. lbs. for bolt "B", 107 ft. lbs. for the bolt to the body and 70 ft. lbs. for the rear engine insulator bolts.
21. Tighten the front suspension lower number one arm sub-assembly.
22. Continue the installation in the reverse order of the removal procedure.
23. Tighten the wheel lug nuts to 76 ft. lbs.
24. Be sure to check and adjust the alignment, as required.

MACPHERSON STRUT

REMOVAL & INSTALLATION

See Figures 193 and 194.

1. Before servicing the vehicle, refer to the Precautions Section.
2. Raise and support the vehicle safely.
3. Remove the tire and wheel assembly.
4. Remove the front speed sensor.
5. Remove the stabilizer link assembly.
6. Remove the two bolts and disconnect the strut from the steering knuckle.
7. Remove the three strut upper retaining bolts.
8. Remove the strut from the vehicle.

To install:

9. Installation is the reverse of the removal procedure.
10. Tighten the three upper strut retaining nuts to 37 ft. lbs.
11. Tighten the lower strut retaining bolts to 177 ft. lbs.
12. Tighten the wheel lug nuts to 76 ft. lbs.
13. To stabilize the suspension, lower the vehicle to ground height. Press down on the vehicle several times to stabilize the suspension.
14. Check and adjust the alignment, as required.

STABILIZER BAR

REMOVAL & INSTALLATION

See Figures 195 and 196.

1. Before servicing the vehicle, refer to the Precautions Section.
2. Raise and support the vehicle safely.
3. Remove the tire and wheel assembly.
4. Remove the stabilizer bar link retaining nuts.
5. Remove the four bolts and remove the left front suspension member brace.
6. Remove the four bolts and remove the right front suspension member brace.
7. Remove the stabilizer bar from the crossmember.
8. Remove the bushings from the stabilizer bar.

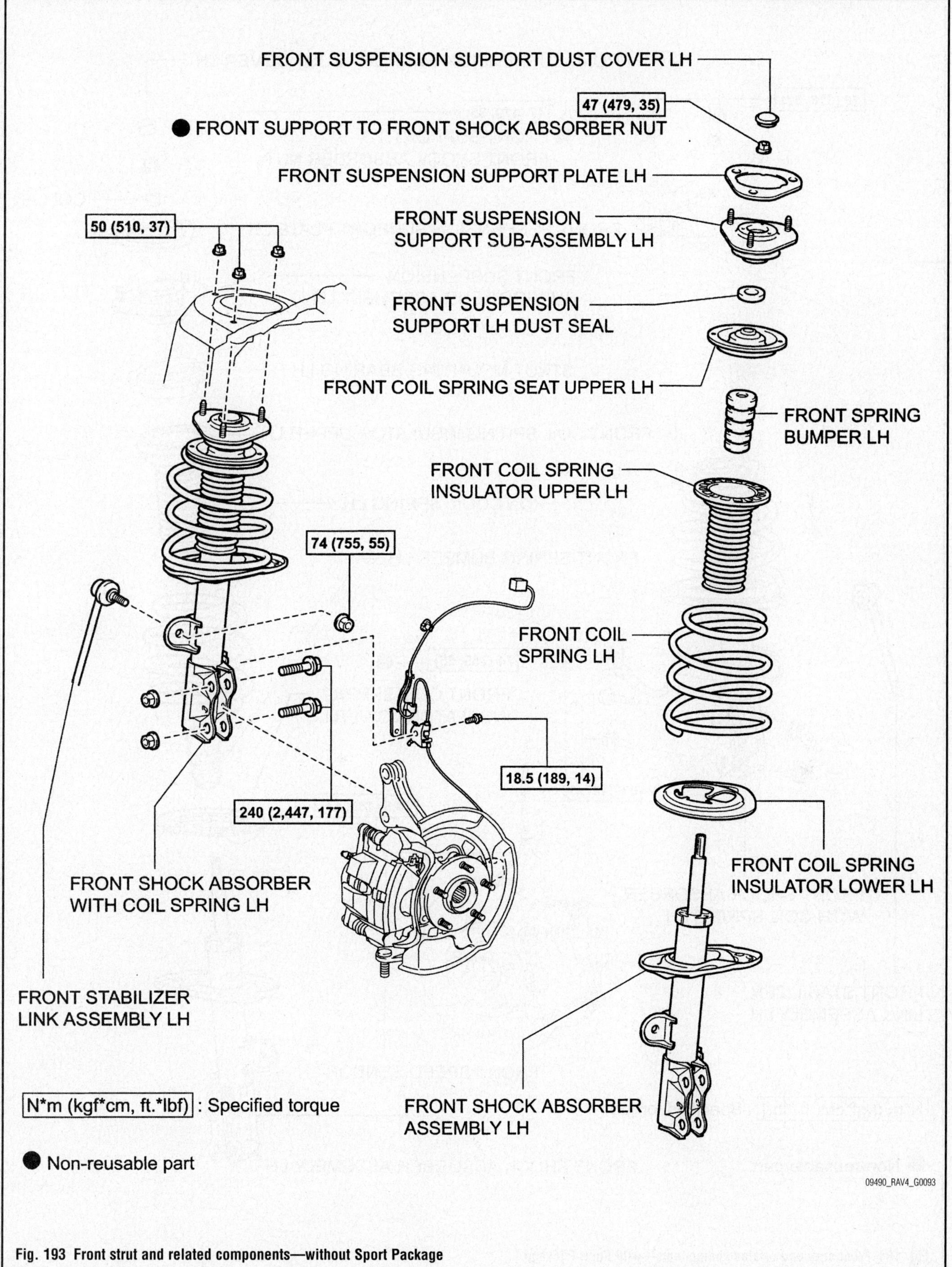

Fig. 193 Front strut and related components—without Sport Package

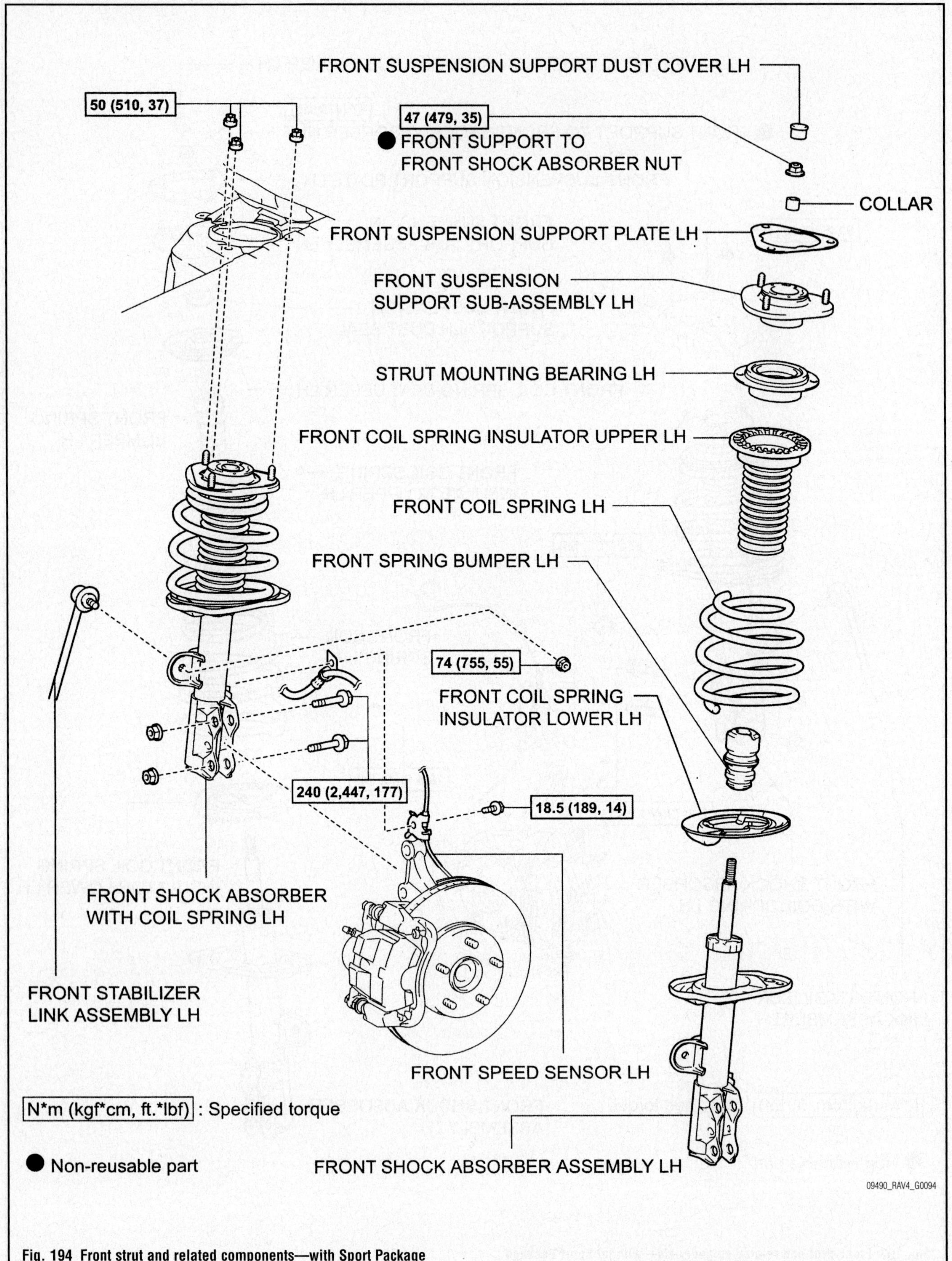

Fig. 194 Front strut and related components—with Sport Package

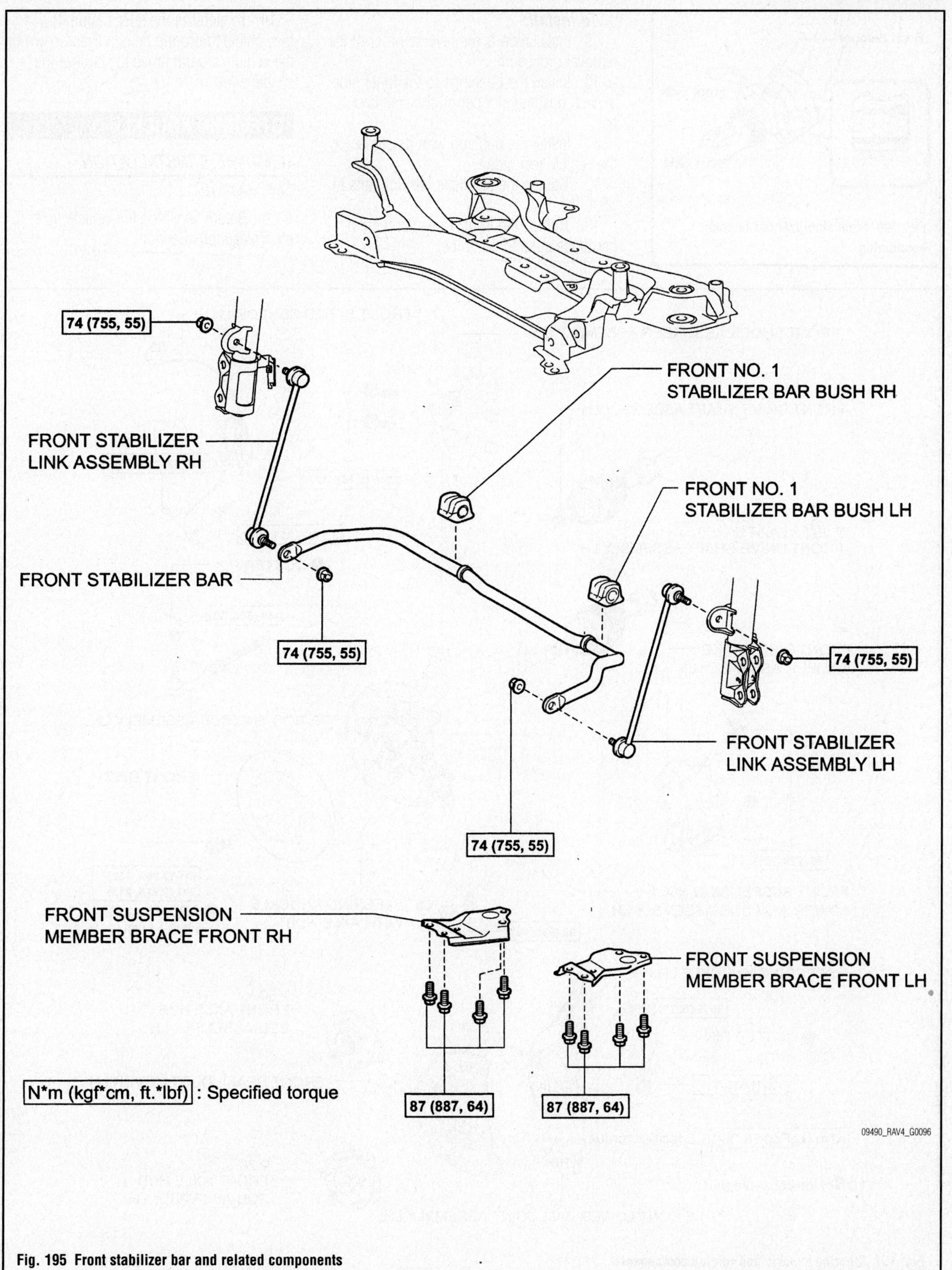

Fig. 195 Front stabilizer bar and related components

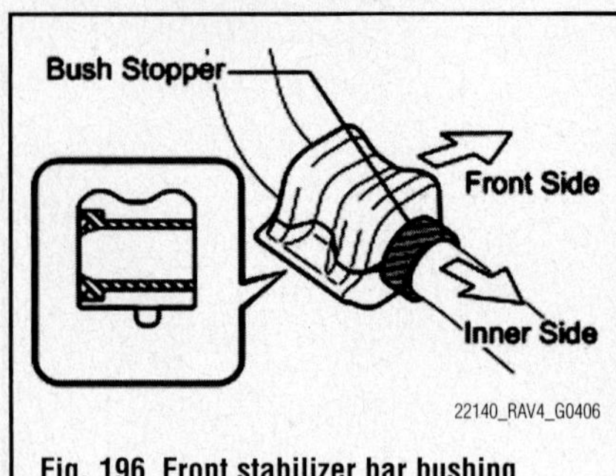

Fig. 196 Front stabilizer bar bushing positioning

To install:

9. Installation is the reverse order of the removal procedure.
10. Install the bushings to the inner side of each bushing stopper on the stabilizer bar.
11. Install the bushing with its slit facing the vehicle rear side.
12. Tighten the stabilizer bar link nuts to 55 ft. lbs.
13. Tighten the front suspension member brace bolts to 64 ft. lbs.
14. To stabilize the suspension, lower the vehicle to ground height. Press down on the vehicle several times to stabilize the suspension.

STEERING KNUCKLE

REMOVAL & INSTALLATION

See Figure 197.

1. Before servicing the vehicle, refer to the Precautions Section.

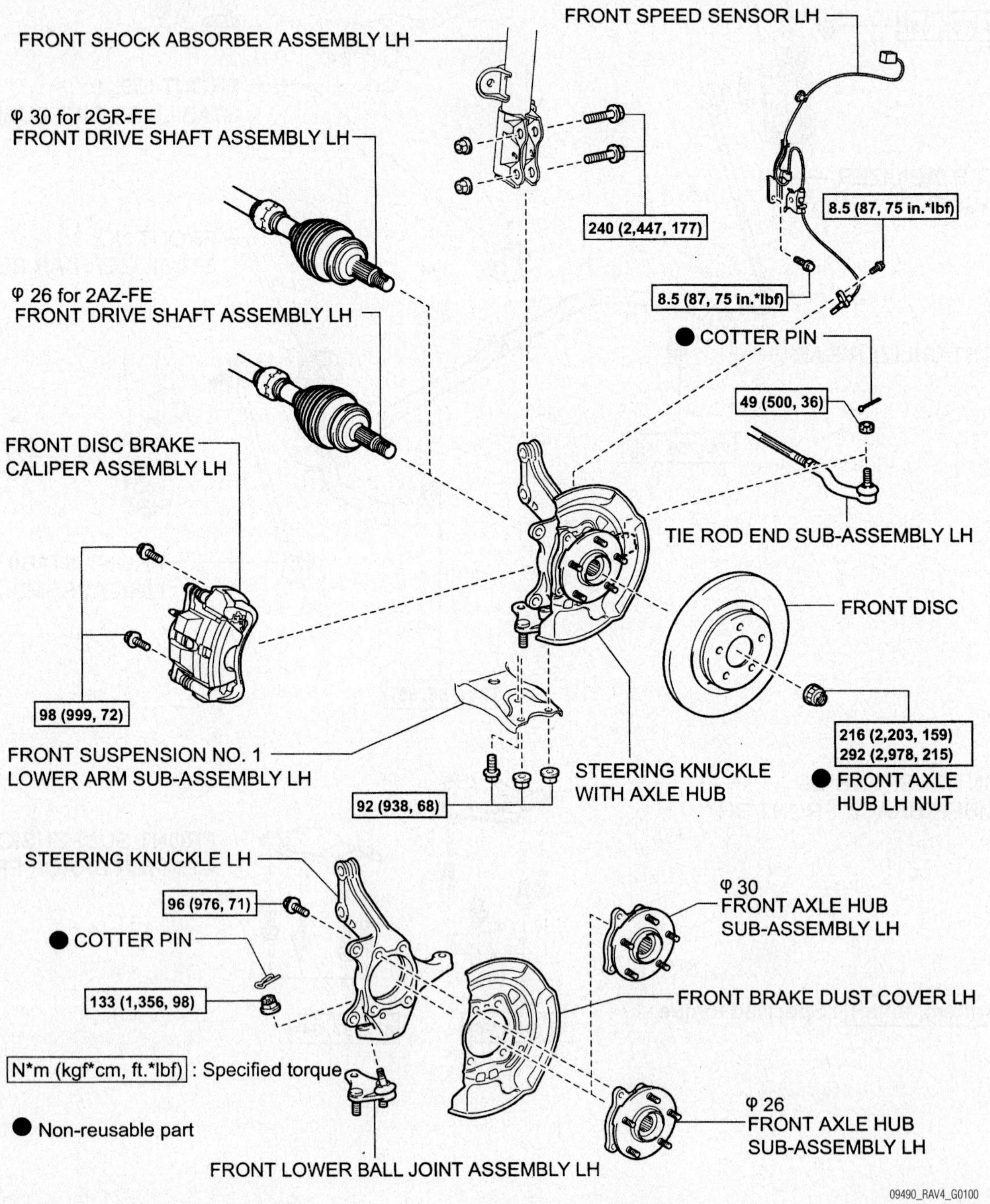

Fig. 197 Steering knuckle and related components

2. Disconnect the negative battery cable.

➡Wait at least 90 seconds after disconnecting the negative battery cable before starting any repair work to prevent air bag and seat belt pretensioner activation.

3. Raise and support the vehicle safely.
4. Drain the transaxle fluid.
5. Remove the tire and wheel assembly. Remove the front axle hub nut.
6. Remove the front speed sensor. Remove the brake caliper. Remove the rotor.
7. Disconnect the tie rod end, using the proper tools.
8. Disconnect the front suspension number one lower arm sub-assembly.
9. Remove the two bolts and two nuts. Disconnect the strut from the steering knuckle.
10. Matchmark the halfshaft and the axle hub.
11. Remove the steering knuckle with the axle hub.

➡Be careful not to damage the boot and the speed sensor rotor. Do not excessively push out the halfshaft from the axle assembly.

12. Remove the four bolts and the axle hub from the steering knuckle. Remove the dust cover from the steering knuckle.

➡Do not place the hub and bearing's magnet rotor side so that it is facing downward. Do not allow the magnet rotor side to become damaged or contact foreign matter.

To install:

13. Installation is the reverse of the removal procedure.
14. Tighten the four axle hub bolts to 71 ft. lbs.
15. Tighten the two steering knuckle to axle hub bolts to 177 ft. lbs.
16. Tighten the front axle hub nut to 159 ft. lbs. on vehicles manufactured from 11/2005 to 01/2006.
17. Tighten the front axle hub nut to 159 ft. lbs. for vehicles manufactured after 01/2006 and equipped with the 2.4L engine, for vehicles equipped with the 3.5L engine tighten the nut to 215 ft. lbs.
18. Be sure to refill the transaxle with the proper grade and type transaxle fluid.
19. Start the engine and check for leaks, correct as required.
20. Using the Techstream tool, or equivalent check the speed sensor signal.
21. Check and adjust the alignment, as required.

WHEEL HUB AND BEARING

REMOVAL & INSTALLATION

See Figure 198.

1. Before servicing the vehicle, refer to the Precautions Section.
2. Disconnect the negative battery cable.

➡Wait at least 90 seconds after disconnecting the negative battery cable before starting any repair work to prevent air bag and seat belt pretensioner activation.

3. Raise and support the vehicle safely.
4. Drain the transaxle fluid.
5. Remove the tire and wheel assembly. Remove the front axle hub nut.
6. Remove the front speed sensor. Remove the brake caliper. Remove the rotor.
7. Disconnect the tie rod end, using the proper tools.
8. Disconnect the front suspension number one lower arm sub-assembly.
9. Remove the two bolts and two nuts. Disconnect the strut from the steering knuckle.
10. Matchmark the halfshaft and the axle hub.
11. Remove the steering knuckle with the axle hub.

➡Be careful not to damage the boot and the speed sensor rotor. Do not excessively push out the halfshaft from the axle assembly.

12. Remove the four bolts and the axle hub from the steering knuckle. Remove the dust cover from the steering knuckle.

➡Do not place the hub and bearing's magnet rotor side so that it is facing downward. Do not allow the magnet rotor side to become damaged or contact foreign matter.

To install:

13. Installation is the reverse of the removal procedure.
14. Tighten the four axle hub bolts to 71 ft. lbs.
15. Tighten the two steering knuckle to axle hub bolts to 177 ft. lbs.
16. Tighten the front axle hub nut to 159 ft. lbs. on vehicles manufactured from 11/2005 to 01/2006.
17. Tighten the front axle hub nut to 159 ft. lbs. for vehicles manufactured after 01/2006 and equipped with the 2.4L engine, for vehicles equipped with the 3.5L engine tighten the nut to 215 ft. lbs.
18. Be sure to refill the transaxle with the proper grade and type transaxle fluid.
19. Start the engine and check for leaks, correct as required.
20. Using the Techstream tool, or equivalent check the speed sensor signal.
21. Check and adjust the alignment, as required.

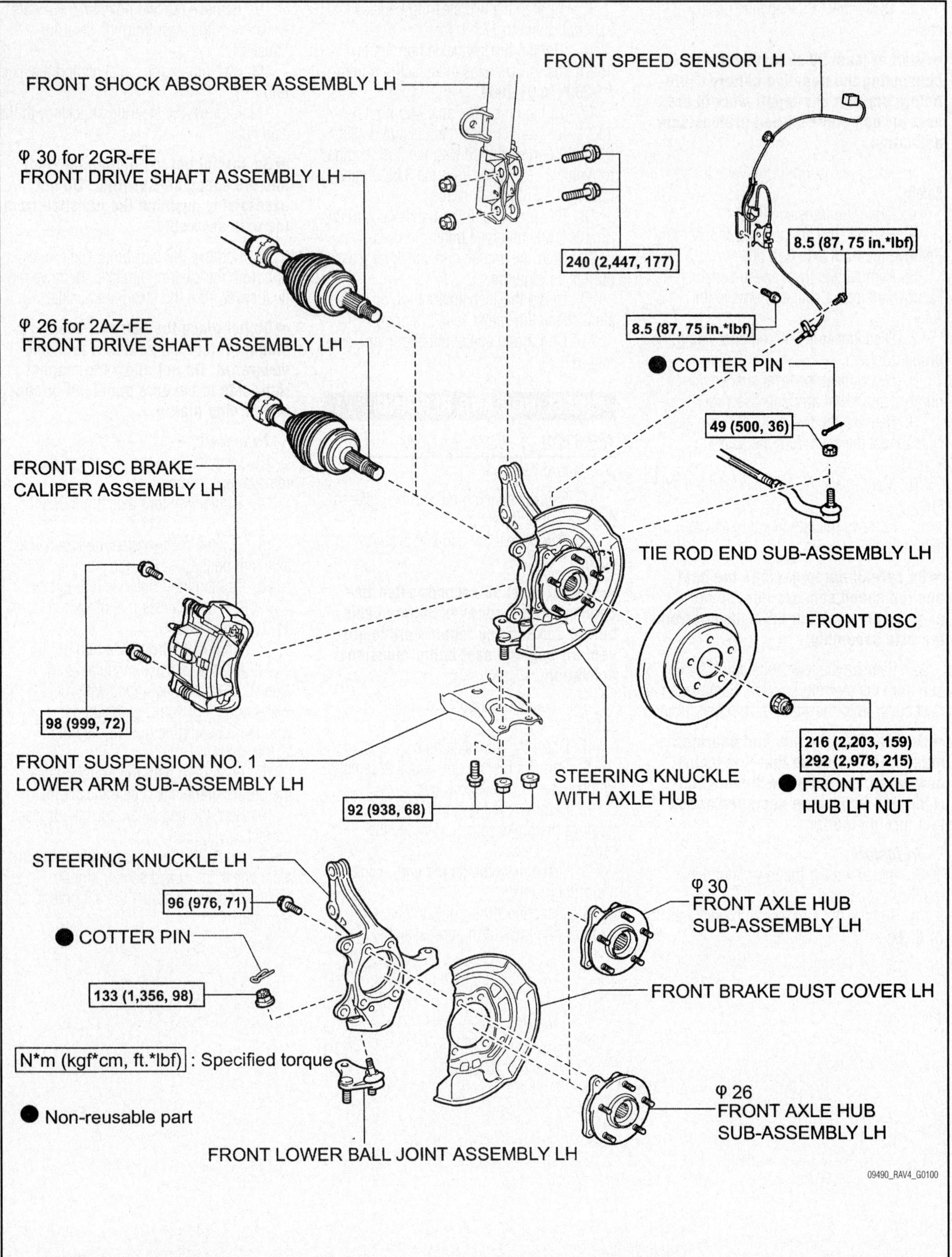

Fig. 198 Front hub and related components

COIL SPRING

REMOVAL & INSTALLATION

See Figure 199.

1. Before servicing the vehicle, refer to the Precautions Section.
2. Raise and support the vehicle safely.
3. Remove the tire and wheel assembly.
4. On 2WD, remove the skid control sensor wire.
5. On 4WD, remove the rear speed sensor wire.
6. Disconnect the number two parking brake cable assembly.
7. Disconnect the rear stabilizer link assembly.
8. To disconnect the rear suspension number two arm assembly, loosen the bolt from the suspension member side. Support the number two suspension arm with a suitable jack.

➡Do not remove the bolt, only loosen it.

9. Remove the bolt and nut from the axle carrier side. Slowly lower the jack and disconnect the number two suspension arm from the axle carrier.
10. Remove the upper spring insulator. Remove the spring. Remove the lower insulator.

To install:

11. Installation is the reverse of the removal procedure.
12. Do not apply final tightening torque to the rear suspension number two arm assembly until the suspension is stabilized.
13. To stabilize the suspension, lower the vehicle to ground height. Press down on the vehicle several times to stabilize the suspension.
14. Check and adjust the alignment, as required.

CONTROL ARMS/LINKS

REMOVAL & INSTALLATION

See Figures 200 and 201.

1. Before servicing the vehicle, refer to the Precautions Section.
2. Raise and support the vehicle safely.
3. Remove the tire and wheel assembly.
4. Support the lower control arm assembly.
5. Remove the bolt and two nuts from the suspension member and axle carrier.
6. Disconnect the suspension arm from the axle carrier. Be careful not to damage the dust cover.

To install:

7. Installation is the reverse of the removal procedure.
8. Do not apply final tightening torque to the rear suspension number two arm

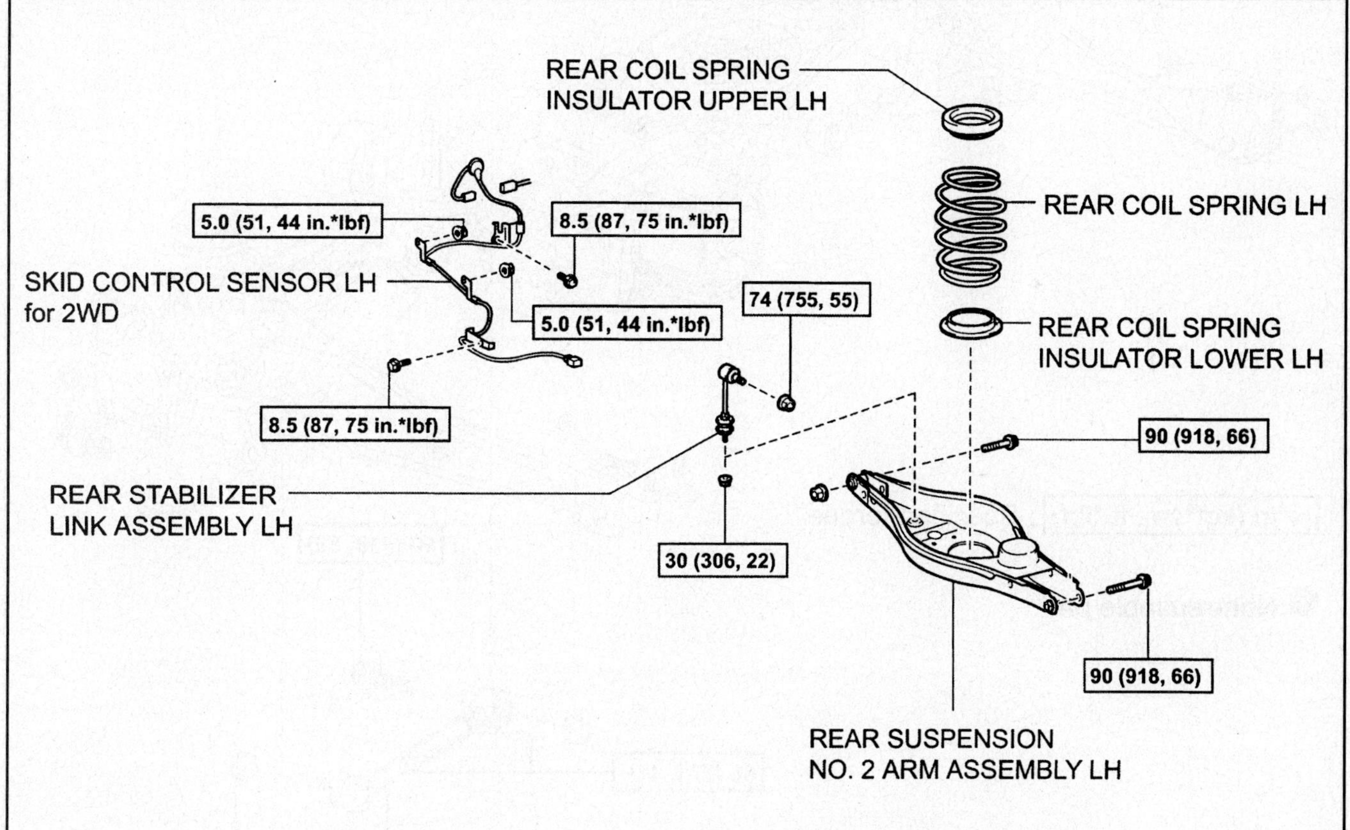

Fig. 199 Rear coil spring and related components

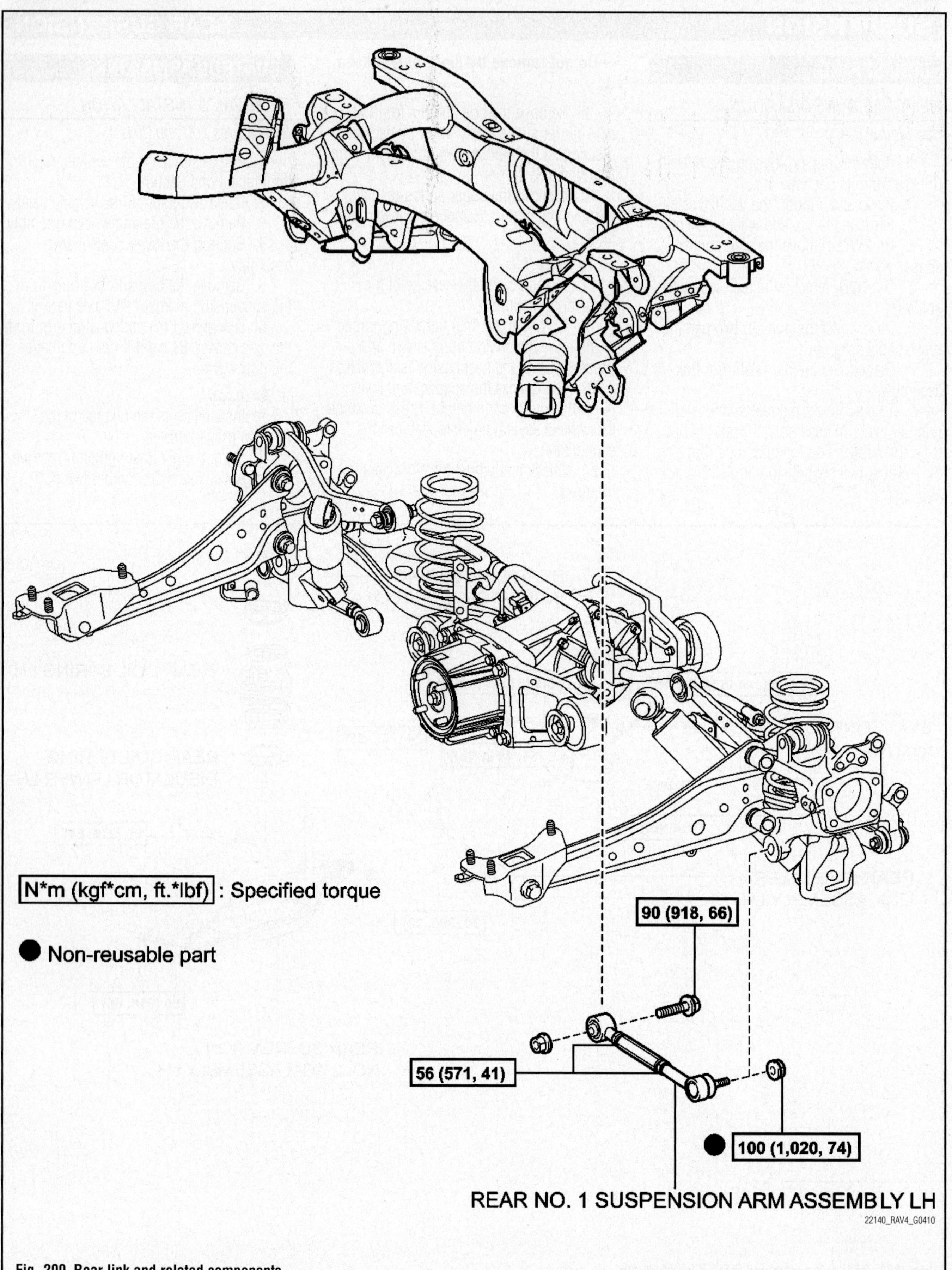

Fig. 200 Rear link and related components

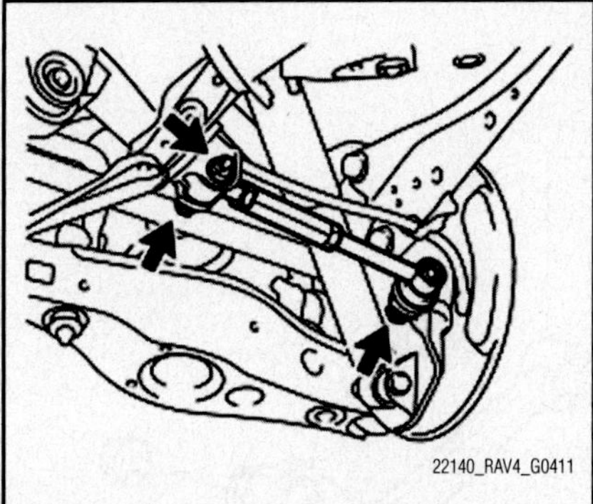

Fig. 201 Rear link retaining bolt location points

assembly until the suspension is stabilized.

9. Tighten the bolts to 66 ft. lbs. Tighten the nut to 74 ft. lbs.

10. Tighten the wheel lug nuts to 76 ft. lbs.

LOWER CONTROL ARM

REMOVAL & INSTALLATION

See Figure 202.

1. Before servicing the vehicle, refer to the Precautions Section.
2. Raise and support the vehicle safely.
3. Remove the tire and wheel assembly.
4. On 2WD, remove the skid control sensor wire.
5. On 4WD, remove the rear speed sensor wire.
6. Disconnect the number two parking brake cable assembly.
7. Disconnect the rear stabilizer link assembly.
8. To disconnect the rear suspension number two arm assembly, loosen the bolt from the suspension member side. Support the number two suspension arm with a suitable jack.

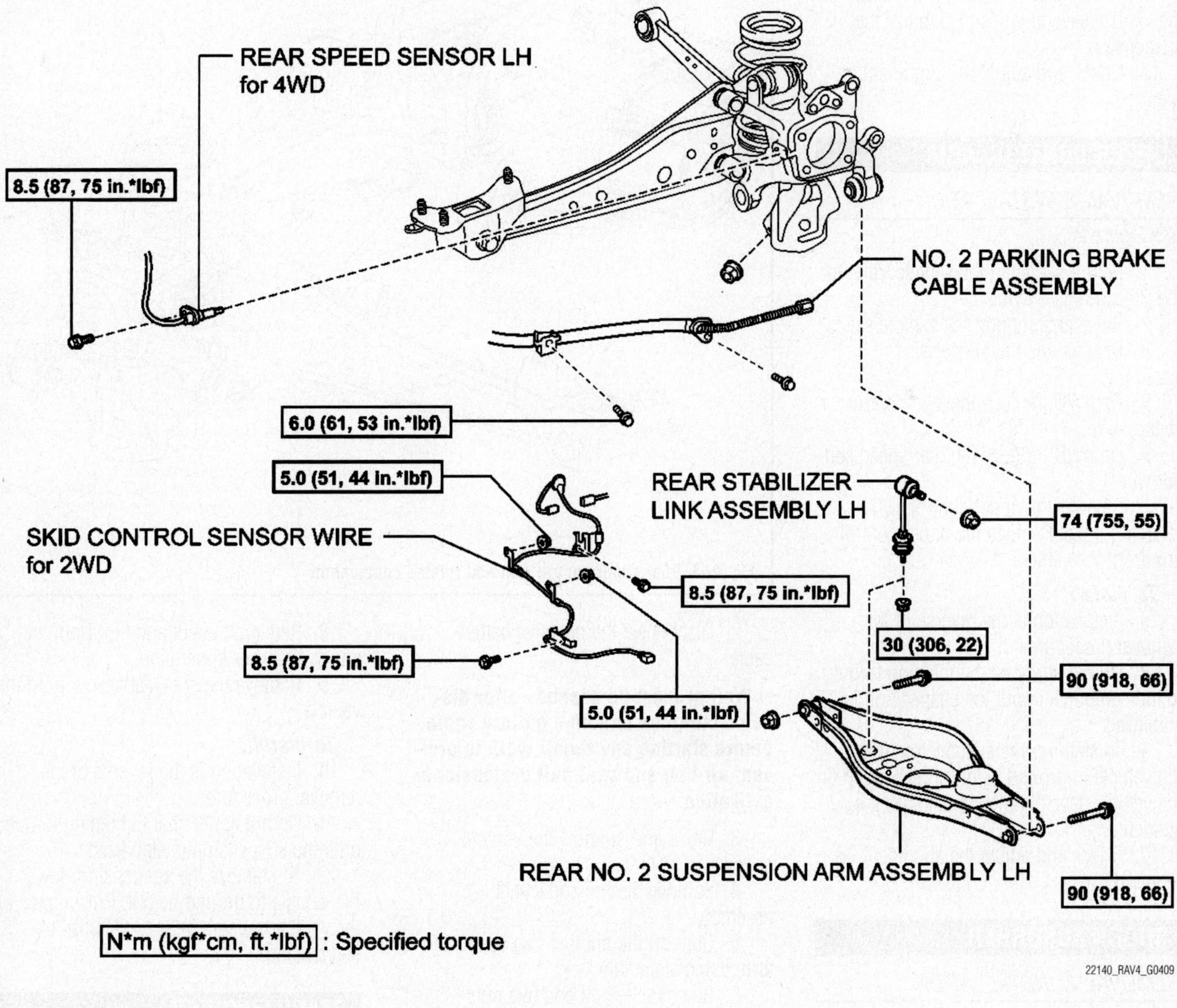

Fig. 202 Rear lower control arm and related components

➡Do not remove the bolt, only loosen it.

9. Remove the bolt and nut from the axle carrier side. Slowly lower the jack and disconnect the number two suspension arm from the axle carrier.
10. Remove the upper spring insulator. Remove the spring. Remove the lower insulator.
11. Remove the bolt, nut and suspension arm from the suspension member.

To install:

12. Installation is the reverse of the removal procedure.
13. Do not apply final tightening torque to the rear suspension number two arm assembly until the suspension is stabilized.
14. To stabilize the suspension, lower the vehicle to ground height. Press down on the vehicle several times to stabilize the suspension.
15. Check and adjust the alignment, as required.

UPPER CONTROL ARM

REMOVAL & INSTALLATION

See Figure 203.

1. Before servicing the vehicle, refer to the Precautions Section.
2. Raise and support the vehicle safely.
3. Remove the tire and wheel assembly.
4. On 2WD, remove the skid control sensor wire.
5. On 4WD, remove the rear speed sensor wire.
6. Remove the upper control arm retaining bolts. Remove the upper control arm from the vehicle.

To install:

7. Installation is the reverse of the removal procedure.
8. Do not apply final tightening torque to the component until the suspension is stabilized.
9. To stabilize the suspension, lower the vehicle to ground height. Press down on the vehicle several times to stabilize the suspension.
10. Check and adjust the alignment, as required.

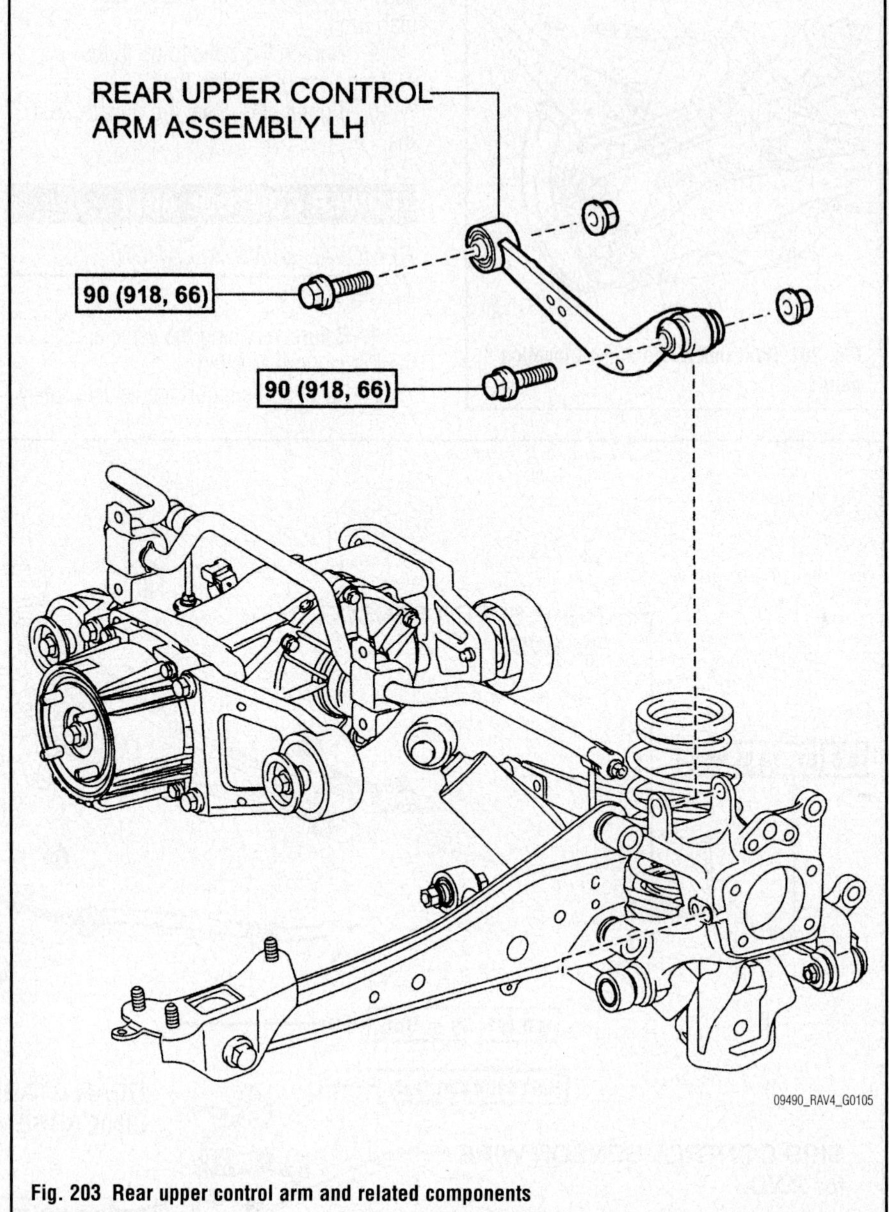

Fig. 203 Rear upper control arm and related components

SHOCK ABSORBER

REMOVAL & INSTALLATION

See Figure 204.

1. Before servicing the vehicle, refer to the Precautions Section.
2. Disconnect the negative battery cable.

➡Wait at least 90 seconds after disconnecting the negative battery cable before starting any repair work to prevent air bag and seat belt pretensioner activation.

3. Raise and support the vehicle safely.
4. Remove the tire and wheel assembly.
5. Support the number two suspension arm, using a suitable jack.
6. Remove the bolt and two nuts from the suspension member and axle carrier.
7. Remove the two bolts and disconnect the shock absorber with the bracket.
8. Remove the nut and bolt from the shock absorber upper side.
9. Remove the shock absorber from the vehicle.

To install:

10. Installation is the reverse of the removal procedure.
11. Do not apply final tightening torque until the suspension is stabilized.
12. To stabilize the suspension, lower the vehicle to ground height. Press down on the vehicle several times to stabilize the suspension.

STABILIZER BAR

REMOVAL & INSTALLATION

See Figure 205.

1. Before servicing the vehicle, refer to the Precautions Section.
2. Raise and support the vehicle safely.
3. Remove the tire and wheel assembly.
4. Remove the nut and disconnect the link from the suspension number two arm. Remove the nut and the link from the stabilizer bar.
5. Remove the rear number two suspension arms (lower control arm).
6. Remove the coil springs.
7. Remove the stabilizer bracket retaining bolts. Remove the stabilizer bar from the vehicle. Remove the bushings from the bar.

➡When removing the bar be sure not to damage the sensor wire, brake hose etc.

To install:

8. Installation is the reverse of the removal procedure.
9. Tighten the stabilizer bar links to 55 ft. lbs.
10. Tighten the stabilizer bracket and bushing bolts to 44 ft. lbs.
11. Install each bushing to the outer side of the bushing stopper on each stabilizer bar. Install each bushing with its slit facing the vehicle front side.
12. Do not apply final tightening torque until the suspension is stabilized.
13. To stabilize the suspension, lower the vehicle to ground height. Press down on the vehicle several times to stabilize the suspension.
14. Check and adjust the alignment, as required.

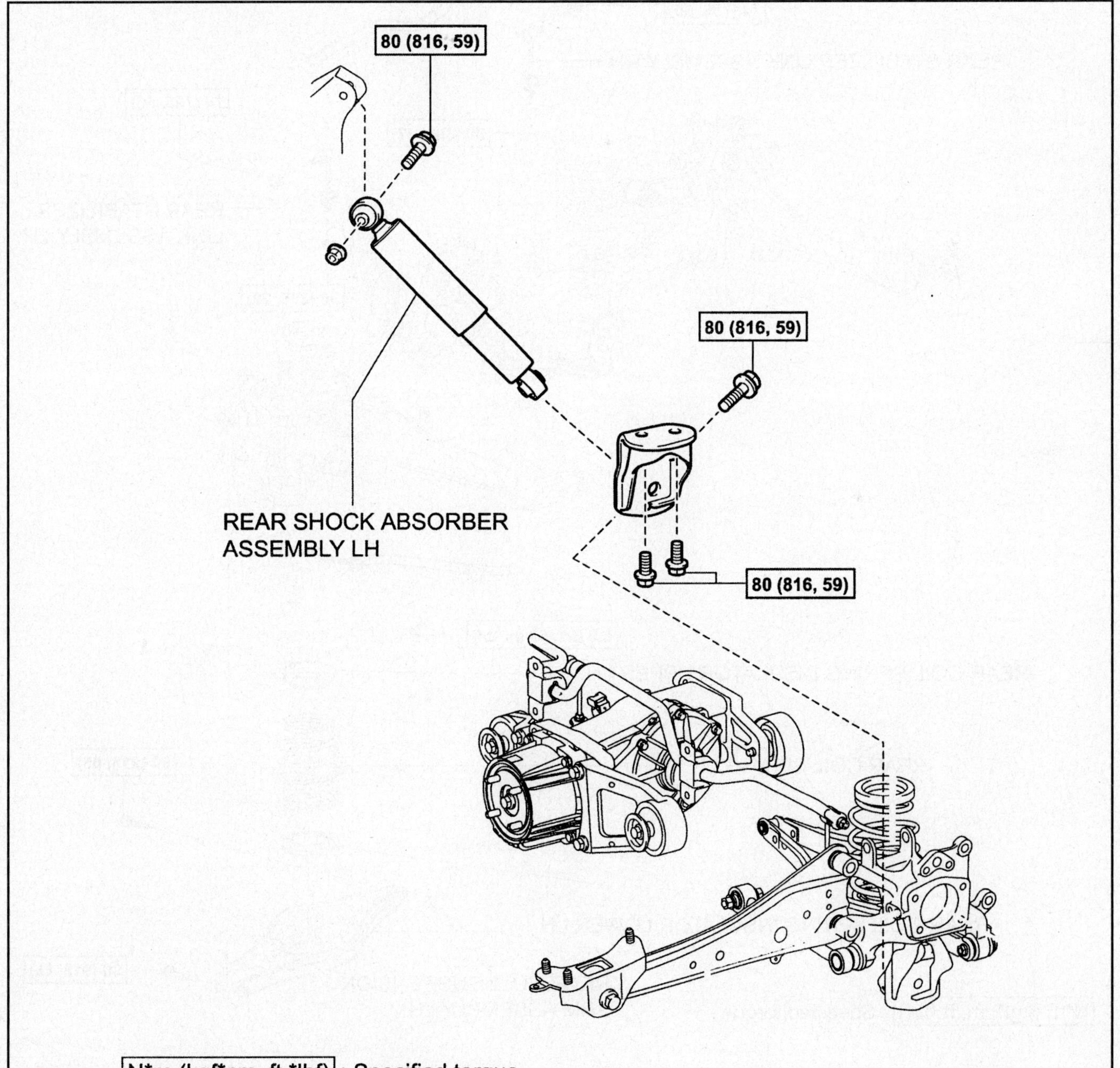

Fig. 204 Rear shock absorber and related components

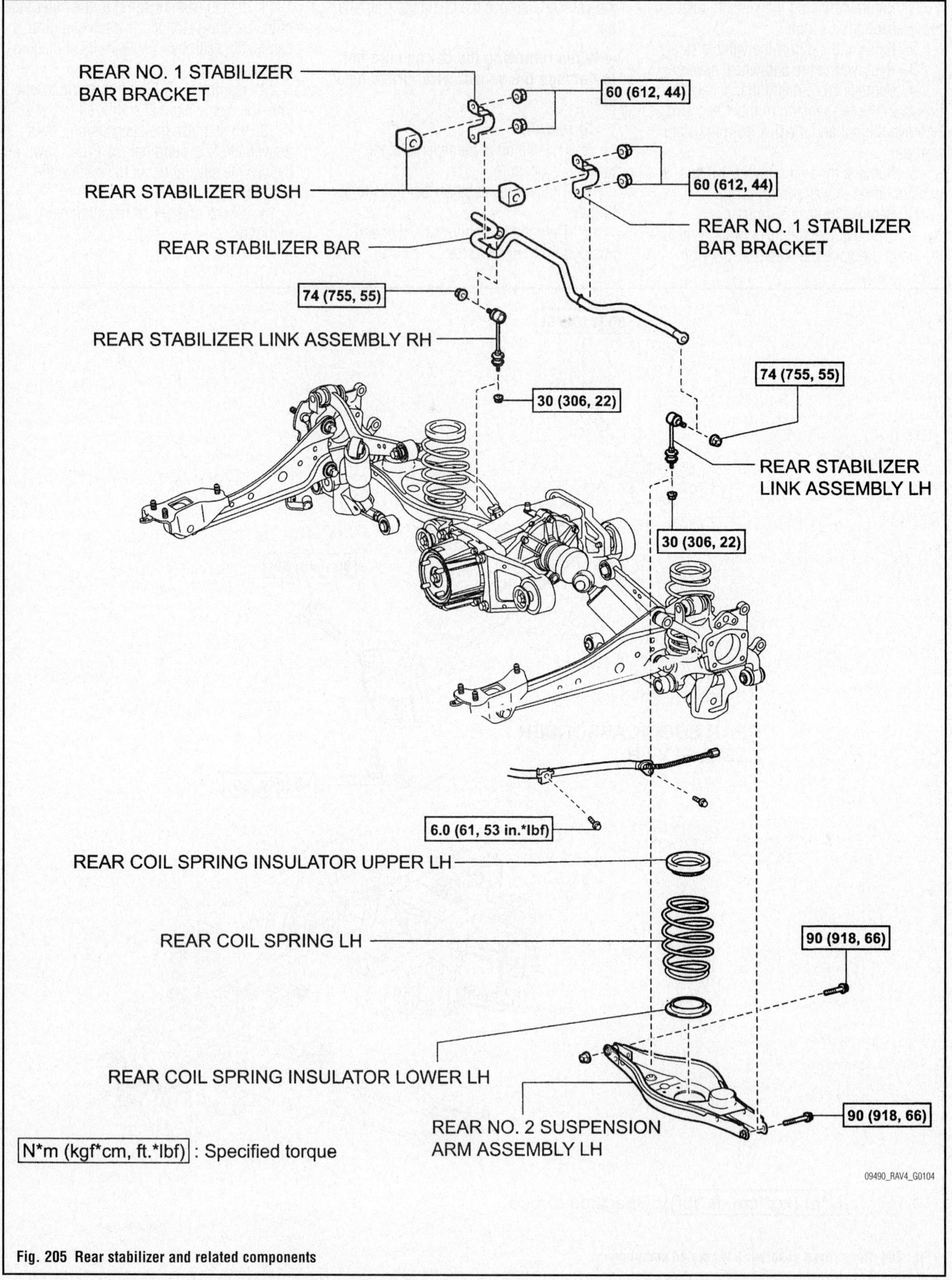

Fig. 205 Rear stabilizer and related components

WHEEL HUB AND BEARING

REMOVAL & INSTALLATION

See Figures 206 through 208.

1. Before servicing the vehicle, refer to the Precautions Section.
2. Properly relieve the fuel system pressure.
3. Disconnect the negative battery cable.

➡Wait at least 90 seconds after disconnecting the negative battery cable

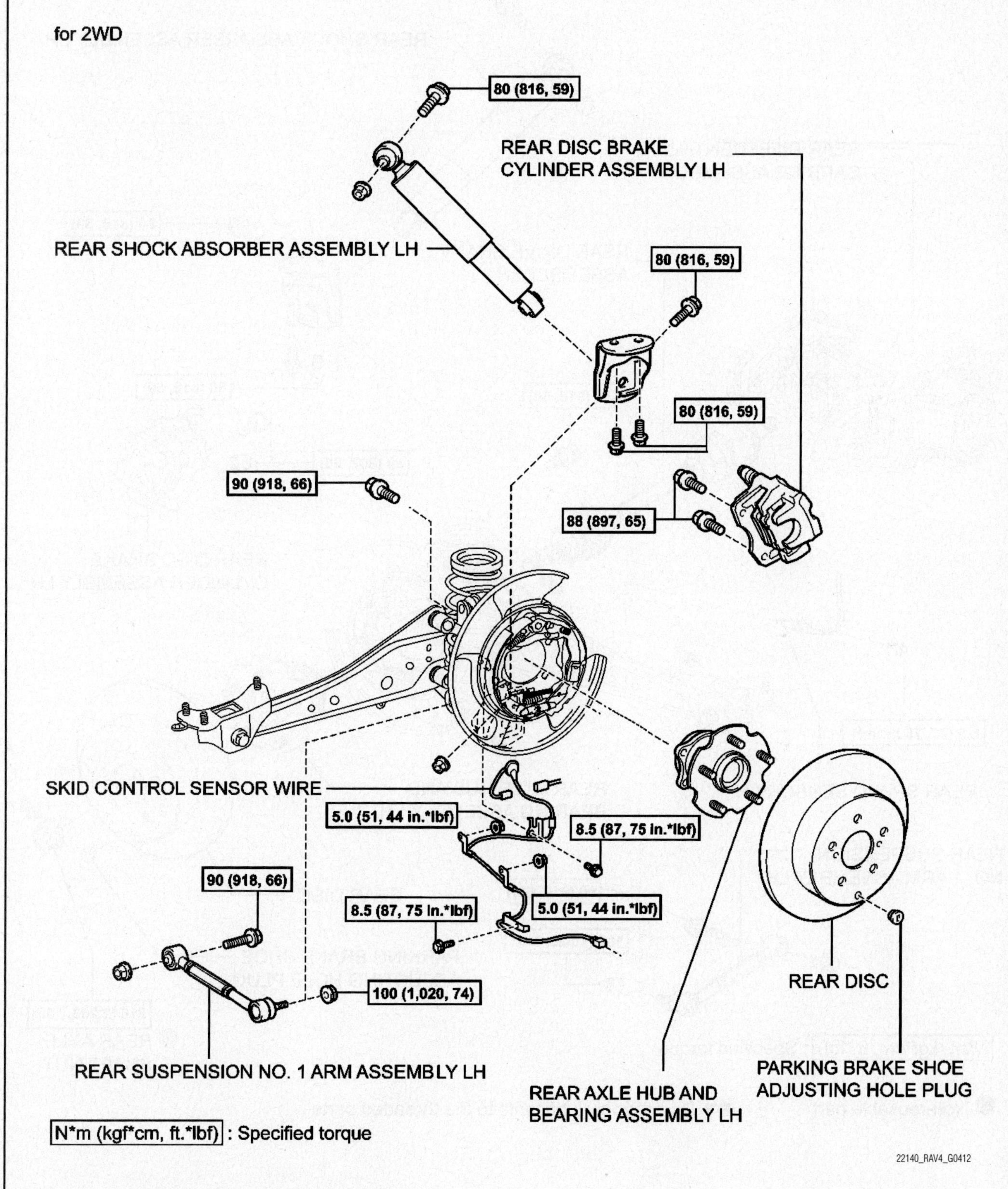

Fig. 206 Rear hub and related components—2WD vehicles

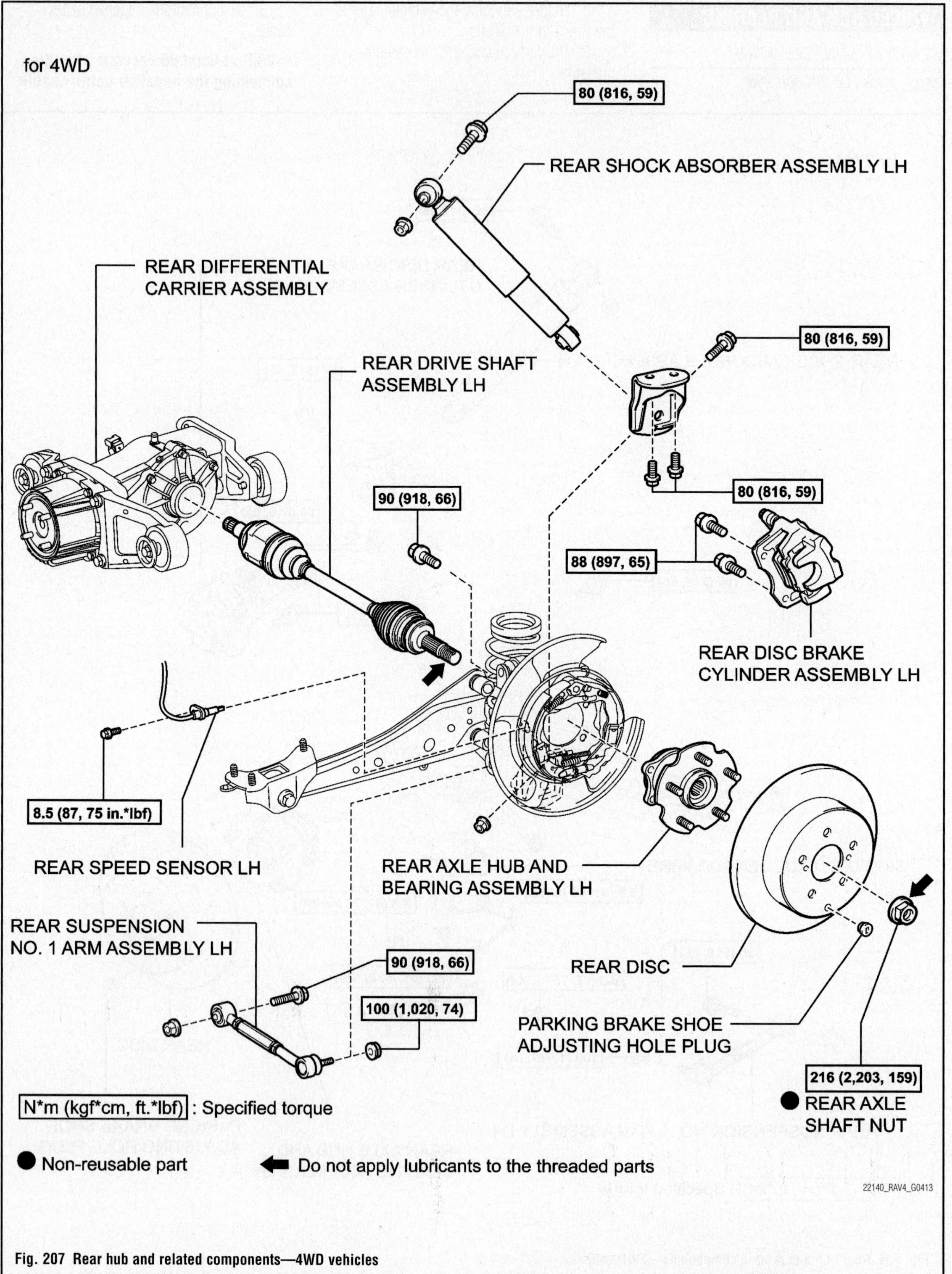

Fig. 207 Rear hub and related components—4WD vehicles

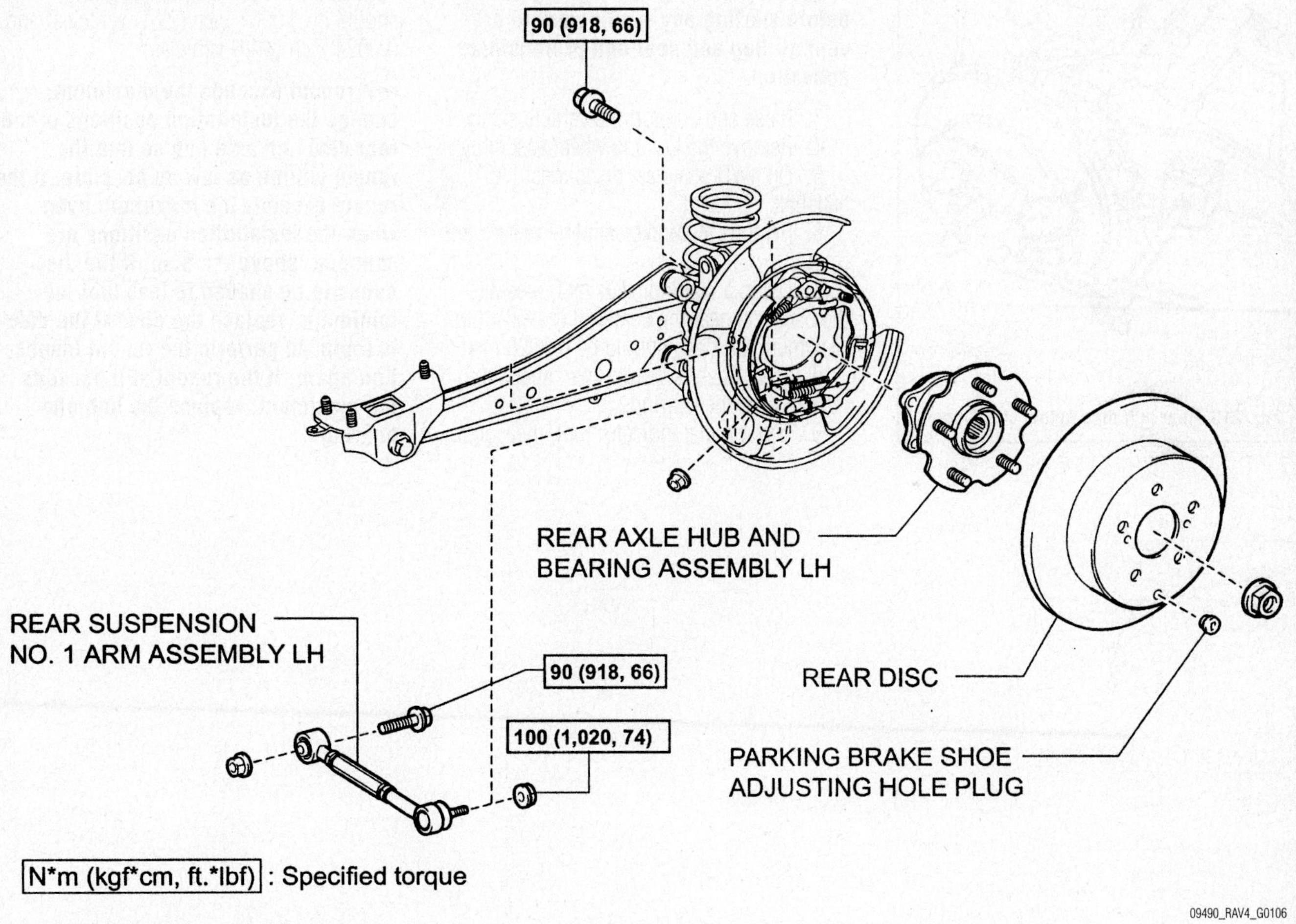

Fig. 208 Rear hub and bearing assembly

before starting any repair work to prevent air bag and seat belt pretensioner activation.

4. Raise and support the vehicle safely.
5. Remove the tire and wheel assembly.
6. On 4WD, remove the rear axle shaft nut.
7. Remove the caliper. Remove the rotor.
8. On 2WD remove the skid control sensor wire.
9. On 4WD, remove the rear speed sensor.
10. Remove the rear suspension number one arm.
11. Disconnect the shock absorber from the axle carrier.
12. On 2WD, remove the four bolts, and the axle hub and bearing from the axle carrier.
13. On 4WD, matchmark the halfshaft and the axle hub and bearing. Remove the four bolts, and the axle hub and bearing from the axle carrier.

➡Do not place the hub and bearing's magnet rotor side so that it is facing downward. Do not allow the magnet rotor side to become damaged or contact foreign matter.

To install:

14. Installation is the reverse of the removal procedure.
15. Tighten the four hub and bearing bolts to 68 ft. lbs.
16. Do not apply final tightening torque to the component until the suspension is stabilized.
17. To stabilize the suspension, lower the vehicle to ground height. Press down on the vehicle several times to stabilize the suspension.
18. Check and adjust the alignment, as required.

ADJUSTMENT

See Figures 209 and 210.

Fig. 209 Rear hub backlash check point

1. Before servicing the vehicle, refer to the Precautions Section.
2. Disconnect the negative battery cable.

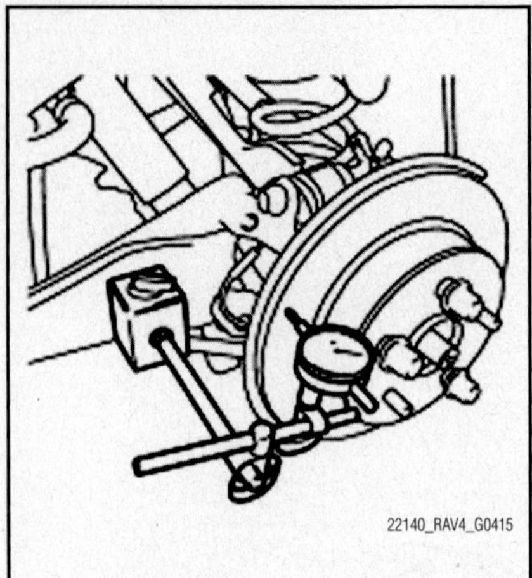

Fig. 210 Rear hub disc runout check point

➡Wait at least 90 seconds after disconnecting the negative battery cable before starting any repair work to prevent air bag and seat belt pretensioner activation.

3. Raise and support the vehicle safely.
4. Remove the tire and wheel assembly.
5. On 4WD vehicles, disconnect the halfshaft.
6. Remove the brake caliper. Remove the rotor.
7. Using a dial indicator tool, measure the backlash near the center of the axle hub. Maximum backlash should be 0.0020 inch.
8. If backlash is greater than specification, replace the bearing.
9. Using a dial indicator tool, measure the disc runout (0.39 inch) inside the outer edge of the disc. Maximum deviation should be 0.0031 inch (2WD vehicles) and 0.0024 inch (4WD vehicles).

➡If runout exceeds the maximum, change the installation positions of the rear disc and axle hub so that the runout will be as low as possible. If the runout exceeds the maximum even when the installation positions are changed, shave the disc. If the disc needs to be shaved to less that the minimum, replace the disc. If the disc is replaced perform the runout inspection again. If the runout still exceeds the maximum, replace the hub and bearing.

TOYOTA

Sienna

21

SPECIFICATIONS AND MAINTENANCE CHARTS

ENGINE AND VEHICLE IDENTIFICATION

Engine							Model Year	
Code ①	Liters (cc)	Cu. In.	Cyl.	Fuel Sys.	Engine Type	Eng. Mfg.	Code ②	Year
3MZ-FE	3.3 (3310)	NA	6	SFI	DOHC	Toyota	7	2007
2GR-FE	3.5 (3456)	NA	6	SFI	DOHC	Toyota	8	2008

MFI: Multi-port Fuel Injection

DOHC: Double Overhead Camshaft

① Stamped on the left side of the engine block

② 10th digit of the Vehicle Identification Number (VIN)

22140_SIEN_C0001

GENERAL ENGINE SPECIFICATIONS

Year	Model	Engine Displacement Liters	Engine Series ID	Net Horsepower @ rpm	Net Torque @ rpm (ft. lbs.)	Bore x Stroke (in.)	Com-pression Ratio	Oil Pressure @ rpm
2007	Sienna	3.5	2GR-FE	268@6200	248@4700	3.70x3.27	NA	43-78@3000
2008	Sienna	3.5	2GR-FE	268@6200	248@4700	3.70x3.27	NA	43-78@3000

22140_SIEN_C0003

ENGINE TUNE-UP SPECIFICATIONS

Year	Engine Displacement Liters	Engine ID	Spark Plug Gap (in.)	Ignition Timing (deg.)	Fuel Pump (psi)	Idle Speed (rpm)	Valve Clearance	
							Intake	Exhaust
2007	3.5	2GR-FE	0.039-0.043	①	44-50	550-650	0.006-0.010	0.010-0.014
2008	3.5	2GR-FE	0.039-0.043	①	44-50	550-650	0.006-0.010	0.010-0.014

NOTE: The Vehicle Emission Control Information label often reflects specification changes made during production.

The label figures must be used if they differ from those in this chart.

B: Before top dead center

① With terminal TC and CG of DLC3 connected: 8-12 degrees BTDC
With terminal TC and CG of DLC3 disconnected: 7-24 degrees BTDC

22140_SIEN_C0002

CAPACITIES

Year	Model	Engine Displacement Liters	Engine ID	Engine Oil with Filter (qts.)	Transmission (pts.) 5-Spd	Transmission (pts.) Auto.	Transfer Case (pts.)	Drive Axle Front (pts.)	Drive Axle Rear (pts.)	Fuel Tank (gal.)	Cooling System (qts.)
2007	Sienna	3.5	2GR-FE	5.0	—	①	2.0	—	2.0	20.0	12.4
2008	Sienna	3.5	2GR-FE	5.0	—	①	2.0	—	2.0	20.0	12.4

① 2WD: 7.4 pts.
4WD: 7.6 pts.

22140_SIEN_C0004

VALVE SPECIFICATIONS

Year	Engine Displacemen Liters	Engine ID	Seat Angle (deg.)	Face Angle (deg.)	Spring Test Pressure (lbs. @ in.)	Spring Installed Height (in.)	Stem-to-Guide Clearance (in.) Intake	Stem-to-Guide Clearance (in.) Exhaust	Stem Diameter (in.) Intake	Stem Diameter (in.) Exhaust
2007	3.5	2GR-FE	45	40.5	41.9-46.3@ 1.437	1.331	0.0010-0.0024	0.0012-0.0026	0.2154-0.2159	0.2152 0.2157
2008	3.5	2GR-FE	45	40.5	41.9-46.3@ 1.437	1.331	0.0010-0.0024	0.0012-0.0026	0.2154-0.2159	0.2152 0.2157

22140_SIEN_C0005

CRANKSHAFT AND CONNECTING ROD SPECIFICATIONS

All measurements are given in inches.

Year	Engine Displacement Liters	Engine ID	Crankshaft Main Brg. Journal Dia.	Crankshaft Main Brg. Oil Clearance	Crankshaft Shaft End-play	Crankshaft Thrust on No.	Connecting Rod Journal Diameter	Connecting Rod Oil Clearance	Connecting Rod Side Clearance
2007	3.5	2GR-FE	2.4011-2.4016	①	0.0016-0.0095	2	2.0863-2.0866	0.0015-0.0026	0.0059-0.0118
2008	3.5	2GR-FE	2.4011-2.4016	①	0.0016-0.0095	2	2.0863-2.0866	0.0015-0.0026	0.0059-0.0118

① Journals 1 and 4: 0.0006 - 0.0013 in.
Journals 2 and 3: 0.0010 - 0.0018 in.

22140_SIEN_C0006

PISTON AND RING SPECIFICATIONS

All measurements are given in inches.

Year	Engine Displ. Liters	Engine ID	Piston Clearance	Ring Gap Top Comp.	Ring Gap Bottom Comp.	Ring Gap Oil Control	Ring Side Clearance Top Comp.	Ring Side Clearance Bottom Comp.	Ring Side Clearance Oil Control
2007	3.5	2GR-FE	0.0013-0.0023	0.0118-0.0157	0.0197-0.0236	0.0059-0.0157	0.0012-0.0031	0.0008-0.0024	0.0012-0.0043
2008	3.5	2GR-FE	0.0013-0.0023	0.0118-0.0157	0.0197-0.0236	0.0059-0.0157	0.0012-0.0031	0.0008-0.0024	0.0012-0.0043

22140_SIEN_C0007

TORQUE SPECIFICATIONS

All readings in ft. lbs.

Year	Engine Displacement Liters	Engine ID	Cylinder Head Bolts	Main Bearing Bolts	Rod Bearing Bolts	Crankshaft Damper Bolts	Flywheel Bolts	Manifold Intake	Manifold Exhaust	Spark Plugs	Oil Pan Drain Plug
2007	3.5	2GR-FE	①	②	③	162	61	11	36	18	33
2008	3.5	2GR-FE	①	②	③	162	61	11	36	18	33

① Step 1: 40 ft. lbs.
Step 2: Plus 90 degrees
Recessed bolt: 13 ft. lbs.

② 6-point bolts: 20 ft. lbs.
12-point bolts:
Step 1: 16 ft. lbs.
Step 2: Plus 90 degrees

③ Step 1: 18 ft. lbs.
Step 2: Plus 90 degrees

22140_SIEN_C0008

WHEEL ALIGNMENT

Year	Model		Caster Range (+/-Deg.)	Caster Preferred Setting (Deg.)	Camber Range (+/-Deg.)	Camber Preferred Setting (Deg.)	Toe-in (in.)	Inside Wheel Angle (Deg.)
2007	Sienna							
	FWD	F	0.75	2.80	0.75	-0.28	0+/-0.08	42.75
		R	—	—	0.50	-1.37	0.11+/-0.12	—
	4WD	F	0.75	2.67	0.75	-0.28	0+/-0.08	42.7
		R	—	—	0.50	-1.42	0.05+/-0.12	—
2008	Sienna							
	FWD	F	0.75	2.80	0.75	-0.28	0+/-0.08	42.75
		R	—	—	0.50	-1.37	0.11+/-0.12	—
	4WD	F	0.75	2.67	0.75	-0.28	0+/-0.08	42.7
		R	—	—	0.50	-1.42	0.05+/-0.12	—

22140_SIEN_C0010

TIRE, WHEEL AND BALL JOINT SPECIFICATIONS

Year	Model	OEM Tires Standard	OEM Tires Optional	Tire Pressures (psi) Front	Tire Pressures (psi) Rear	Wheel Size	Ball Joint Inspection	Lugnut Torque (ft. lbs.)
2007	Sienna	P215/65R16 96T	P225/60R17 98T	35	35	6.5	①	76
2008	Sienna	P215/65R16 96T	P225/60R17 98T	35	35	6.5	①	76

OEM: Original Equipment Manufacturer

NA: Information not available

PSI: Pounds Per Square Inch

① Ball joint turning torque should be 8.7-30 inch lbs.

22140_SIEN_C0011

BRAKE SPECIFICATIONS

All measurements in inches unless noted

Year	Model		Brake Disc Original Thickness	Brake Disc Minimum Thickness	Brake Disc Maximum Runout	Brake Drum Diameter Original Inside Diameter	Brake Drum Diameter Maximum Machine Diameter	Minimum Lining Thickness	Brake Caliper Bracket Bolts (ft. lbs.)	Brake Caliper Mounting Bolts (ft. lbs.)
2007	Sienna	F	1.102	1.024	0.0020	—	—	0.039	79	25
		R	0.472	0.413	0.0039	10.00	10.08	0.039	65	25
2008	Sienna	F	1.102	1.024	0.0020	—	—	0.039	79	25
		R	0.472	0.413	0.0039	10.00	10.08	0.039	65	25

F: Front

R: Rear

22140_SIEN_C0012

SCHEDULED MAINTENANCE INTERVALS

TOYOTA—SIENNA

TO BE SERVICED	TYPE OF SERVICE	VEHICLE MILEAGE INTERVAL (x1000)																		
		5	10	15	20	25	30	35	40	45	50	55	60	65	70	75	80	85	90	95
Automatic transmission and differential fluid	S/I			✓			✓			✓			✓			✓			✓	
Ball joints and boots	S/I			✓			✓			✓			✓			✓			✓	
Brake linings, discs/drums, lines & hoses	S/I			✓			✓			✓			✓			✓			✓	
Charcoal canister	S/I												✓							
Drive belts	S/I						✓						✓						✓	
Engine coolant	R						✓						✓						✓	
Engine oil & filter	R	✓	✓	✓	✓	✓	✓	✓	✓	✓	✓	✓	✓	✓	✓	✓	✓	✓	✓	✓
Exhaust pipes & mounts	S/I			✓			✓			✓			✓			✓			✓	
Fuel lines & connections, fuel tank vapor vent system hoses, fuel tank band	S/I						✓						✓						✓	
Fuel tank cap gasket	S/I						✓						✓						✓	
Halfshaft boots & flange bolts	S/I			✓			✓			✓			✓			✓			✓	
Non-platinum spark plugs	R						✓						✓						✓	
Platinum spark plugs	R												✓							
Rack and pinion assembly	S/I			✓			✓			✓			✓			✓			✓	
Steering linkage	S/I			✓			✓			✓			✓			✓			✓	
Valves	S/I												✓							

R: Replace S/I: Service or Inspect L: Lubricate

FREQUENT OPERATION MAINTENANCE (SEVERE SERVICE)

If a vehicle is operated under any of the following conditions it is considered severe service:

- Towing a trailer or using a camper or car-top carrier.

- Repeated short trips of less than 5 miles in temperatures below freezing.

- Excessive idling or low-speed driving for long distances as in heavy commercial use, such as delivery, taxi or police cars.

- Operating on rough, muddy or salt-covered roads.

- Operating on unpaved or dusty roads.

Oil filter: service or inspect every 5000 miles or 4 months, whichever occurs first.

Brake linings and discs or drums: service or inspect every 5000 miles or 4 months, whichever occurs first.

Steering linkage: service or inspect every 5000 miles or 4 months, whichever occurs first.

Ball joints and boots: service or inspect every 5000 miles or 4 months, whichever occurs first.

Brake discs & pads (front): service or inspect every 6000 miles.

Halfshaft boots: service or inspect every 5000 miles or 4 months. Retighten the flange bolts, whichever occurs first.

Body chassis bolts and nuts: service or inspect every 5000 miles or 4 months, whichever occurs first.

Transmission and differential fluid: replace every 15,000 miles or 12 months, whichever occurs first.

Timing belt: replace every 60,000 miles or 48 months, whichever occurs first.

22140_SIEN_C0009

PRECAUTIONS

Before servicing any vehicle, please be sure to read all of the following precautions, which deal with personal safety, prevention of component damage, and important points to take into consideration when servicing a motor vehicle:

- Never open, service or drain the radiator or cooling system when the engine is hot; serious burns can occur from the steam and hot coolant.
- Observe all applicable safety precautions when working around fuel. Whenever servicing the fuel system, always work in a well-ventilated area. Do not allow fuel spray or vapors to come in contact with a spark, open flame, or excessive heat (a hot drop light, for example). Keep a dry chemical fire extinguisher near the work area. Always keep fuel in a container specifically designed for fuel storage; also, always properly seal fuel containers to avoid the possibility of fire or explosion. Refer to the additional fuel system precautions later in this section.
- Fuel injection systems often remain pressurized, even after the engine has been turned **OFF**. The fuel system pressure must be relieved before disconnecting any fuel lines. Failure to do so may result in fire and/or personal injury.
- Brake fluid often contains polyglycol ethers and polyglycols. Avoid contact with the eyes and wash your hands thoroughly after handling brake fluid. If you do get brake fluid in your eyes, flush your eyes with clean, running water for 15 minutes. If eye irritation persists, or if you have taken brake fluid internally, IMMEDIATELY seek medical assistance.
- The EPA warns that prolonged contact with used engine oil may cause a number of skin disorders, including cancer. You should make every effort to minimize your exposure to used engine oil. Protective gloves should be worn when changing oil. Wash your hands and any other exposed skin areas as soon as possible after exposure to used engine oil. Soap and water, or waterless hand cleaner should be used.
- All new vehicles are now equipped with an air bag system, often referred to as a Supplemental Restraint System (SRS) or Supplemental Inflatable Restraint (SIR) system. The system must be disabled before performing service on or around system components, steering column, instrument panel components, wiring and sensors. Failure to follow safety and disabling procedures could result in accidental air bag deployment, possible personal injury and unnecessary system repairs.
- Always wear safety goggles when working with, or around, the air bag system. When carrying a non-deployed air bag, be sure the bag and trim cover are pointed away from your body. When placing a non-deployed air bag on a work surface, always face the bag and trim cover upward, away from the surface. This will reduce the motion of the module if it is accidentally deployed. Refer to the additional air bag system precautions later in this section.
- Clean, high quality brake fluid from a sealed container is essential to the safe and proper operation of the brake system. You should always buy the correct type of brake fluid for your vehicle. If the brake fluid becomes contaminated, completely flush the system with new fluid. Never reuse any brake fluid. Any brake fluid that is removed from the system should be discarded. Also, do not allow any brake fluid to come in contact with a painted surface; it will damage the paint.
- Never operate the engine without the proper amount and type of engine oil; doing so WILL result in severe engine damage.
- Timing belt maintenance is extremely important. Many models utilize an interference-type, non-freewheeling engine. If the timing belt breaks, the valves in the cylinder head may strike the pistons, causing potentially serious (also time-consuming and expensive) engine damage. Refer to the maintenance interval charts for the recommended replacement interval for the timing belt, and to the timing belt section for belt replacement and inspection.
- Disconnecting the negative battery cable on some vehicles may interfere with the functions of the on-board computer system(s) and may require the computer to undergo a relearning process once the negative battery cable is reconnected.
- When servicing drum brakes, only disassemble and assemble one side at a time, leaving the remaining side intact for reference.
- Only an MVAC-trained, EPA-certified automotive technician should service the air conditioning system or its components.

BRAKES — ANTI-LOCK BRAKE SYSTEM (ABS)

GENERAL INFORMATION

When conventional brakes are applied in an emergency stop or on ice, one or more wheels may lock. This may result in loss of steering control and vehicle stability. The purpose of the Anti-lock Brake System (ABS) is to prevent lock up when traction is marginal or under heavy braking conditions. This system offers many benefits allowing the driver increased safety and control during braking. Anti-lock braking operates only at speeds above 6 mph (10 km/h).

Under normal braking conditions, the ABS functions the same as a standard brake system with a diagonally split master cylinder and conventional vacuum assist.

If wheel lock is detected during the brake application, the system will enter anti-lock mode. During anti-lock mode, hydraulic pressure in the four wheel circuits is modulated to prevent any one wheel from locking. Each wheel circuit is designed with a set of electrical valves and hydraulic line to provide modulation, although for vehicle stability, both rear wheel valves receive the same electrical signal. The system can build or reduce pressure at each wheel, depending on signals generated by the Wheel Speed Sensors (WSS) at each wheel and received at the control module.

Anti-lock braking systems (ABS) are available on all Honda models. When this system engages, some audible noise as well as pulses in the brake pedal may occur. Do not be alarmed; this is normal system operation.

BRAKES — BLEEDING THE BRAKE SYSTEM

BLEEDING PROCEDURE

See Figure 1.

➡**This bleeding procedure is for the bleeding ABS System bleeding the base brake system.**

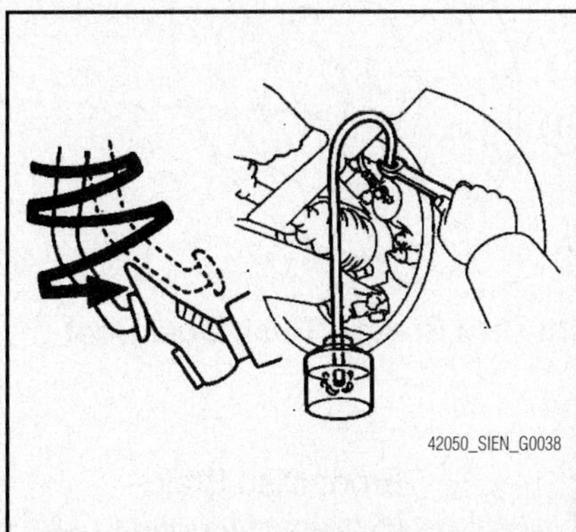

42050_SIEN_G0038

Fig. 1 Bleeding the brake system—Sienna

➡**Do not reuse the drained fluid. Use only clean DOT 3 Brake Fluid from an unopened container.**

⁂ WARNING

Make sure no dirt or other foreign matter is allowed to contaminate the brake fluid.

⁂ WARNING

Do not spill brake fluid on the vehicle, it may damage the paint; if brake fluid does contact the paint, wash it off immediately with water.

1. The reservoir on the master cylinder must be at the MAX level mark at the start of the bleeding procedure and checked after bleeding each brake caliper. Add fluid as required.
2. Make sure the brake fluid level in the reservoir is at the MAX level line.
3. Slide a piece of clear plastic hose over the first bleed screw, and submerge the other end in a container of new brake fluid.
4. Have someone slowly pump the brake pedal several times, then apply steady pressure.
5. Starting at the left-front, loosen the brake bleed screw to allow air to escape from the system. Then tighten the bleed screw securely. Bleed the right rear, right front and left rear using the same procedure.
6. Repeat the procedure for each wheel until air bubbles no longer appear in the fluid.
7. Refill the master cylinder reservoir to the MAX level line.

BRAKES — FRONT DISC BRAKES

⁂ CAUTION

Dust and dirt accumulating on brake parts during normal use may contain asbestos fibers from production or aftermarket brake linings. Breathing excessive concentrations of asbestos fibers can cause serious bodily harm. Exercise care when servicing brake parts. Do not sand or grind brake lining unless equipment used is designed to contain the dust residue. Do not clean brake parts with compressed air or by dry brushing. Cleaning should be done by dampening the brake components with a fine mist of water, then wiping the brake components clean with a dampened cloth. Dispose of cloth and all residue containing asbestos fibers in an impermeable container with the appropriate label. Follow practices prescribed by the Occupational Safety and Health Administration (OSHA) and the Environmental Protection Agency (EPA) for the handling, processing, and disposing of dust or debris that may contain asbestos fibers.

BRAKE CALIPER

REMOVAL & INSTALLATION

See Figure 2.

1. Before servicing the vehicle, refer to the precautions section.
2. Disconnect the negative battery cable from the battery.
3. Raise and support the vehicle safely.
4. Remove the wheels.
5. Disconnect the brake hose from the caliper by removing the union bolt and 2 gaskets. Plug the end of the hose to prevent loss of fluid.
6. Remove the bolts that attach the caliper to the torque plate.
7. Lift the bottom of the caliper up and remove the caliper assembly.

To install:

8. Grease the caliper slides and bolts with lithium grease or equivalent. Install the caliper and secure with the bolts. Torque the bolts to 25 ft. lbs. (34 Nm).
9. Reconnect the brake hose to the caliper, using 2 new washers. Make sure the flexible hose lock is securely in the lock hole of the caliper. Torque the union bolt to 21 ft. lbs. (29 Nm). Also, verify that the brake hose is not twisted.
10. Fill the brake system to the proper level and bleed the brake system.
11. Install the tire and wheel assembly.
12. Top off the brake fluid level in the master cylinder. Check for leaks and proper brake operation.
13. Connect the negative battery cable to the battery.

DISC BRAKE PADS

REMOVAL & INSTALLATION

1. Before servicing the vehicle, refer to the precautions section.
2. Raise and safely support the front of the vehicle.
3. Remove the front wheels and temporarily fasten the rotor disc with the hub nuts.
4. Hold the sliding pin on the bottom of the caliper and loosen the installation bolt.
5. Remove the lower installation bolt.
6. Lift up the caliper and suspend it securely. Do not remove the upper installation bolt.
7. Remove the following parts:
   - The 2 anti-squeal springs.
   - The 2 brake pads.
   - The 4 anti-squeal shims.
   - The 4 pad support plates.

To install:

8. Install the pad support plates.
9. Install a pad wear indicator plate to the pad. Install the anti-squeal shims and support plates to each pad.

➡**It recommended that a suitable anti-squeal compound be applied to both sides of the inner anti-squeal shim.**

10. Draw out a small amount of brake fluid from the brake reservoir. Press in the caliper piston with a suitable tool.

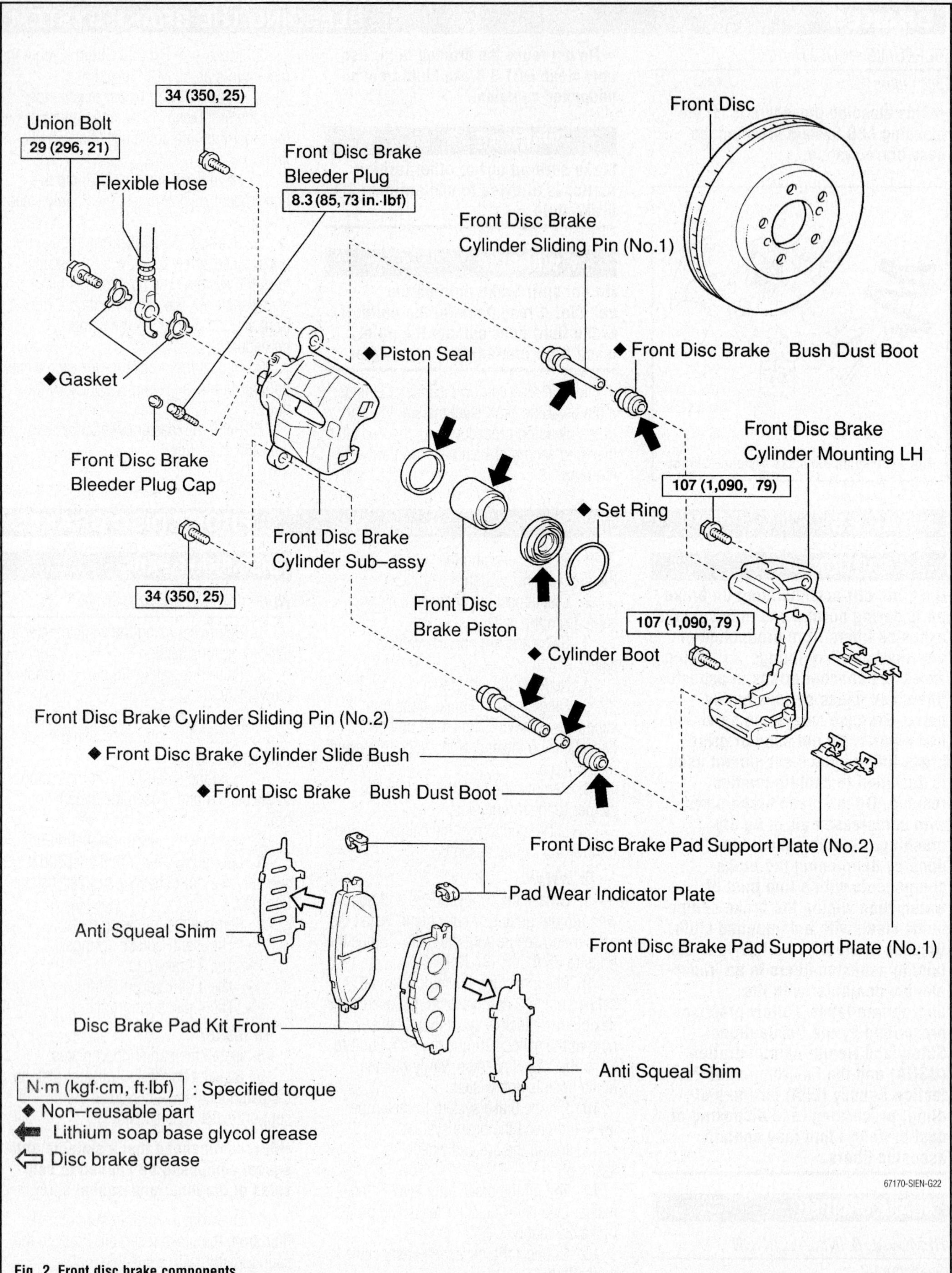

Fig. 2 Front disc brake components

11. Press the brake piston in carefully so the boot will not become wedged.

12. Install the 2 pads so that the wear indicator plate is facing upward. Do not allow oil or grease to get in the rubbing face of the pads.

13. Lower and install the caliper. Torque the sliding main pin to 25 ft. lbs. (34 Nm).

➡When installing the sliding main pin, be careful that the plug installed in the torque plate does not come loose.

14. Install the front wheels and lower the vehicle.

15. Check the fluid level in the master cylinder and add as necessary. Be sure to pump the brake pedal a few times before road-testing the vehicle.

BRAKES

REAR DISC BRAKES

⁂ CAUTION

Dust and dirt accumulating on brake parts during normal use may contain asbestos fibers from production or aftermarket brake linings. Breathing excessive concentrations of asbestos fibers can cause serious bodily harm. Exercise care when servicing brake parts. Do not sand or grind brake lining unless equipment used is designed to contain the dust residue. Do not clean brake parts with compressed air or by dry brushing. Cleaning should be done by dampening the brake components with a fine mist of water, then wiping the brake components clean with a dampened cloth. Dispose of cloth and all residue containing asbestos fibers in an impermeable container with the appropriate label. Follow practices prescribed by the Occupational Safety and Health Administration (OSHA) and the Environmental Protection Agency (EPA) for the handling, processing, and disposing of dust or debris that may contain asbestos fibers.

BRAKE CALIPER

REMOVAL & INSTALLATION

See Figure 3.

1. Before servicing the vehicle, refer to the precautions section.
2. Remove the wheel.
3. Disconnect and plug the brake line.
4. Remove the brake hose.
5. Hold the slide pin and remove the 2 caliper mounting bolts. Lift off the caliper.
6. Installation is the reverse of removal. Refill the system and bleed the brakes. Torque the mounting bolts to 25 ft. lbs. (34 Nm). Torque the brake hose-to-caliper to 17 ft. lbs. (23 Nm). Torque the steel brake line to 11 ft. lbs. (15 Nm).

DISC BRAKE PADS

REMOVAL & INSTALLATION

1. Before servicing the vehicle, refer to the precautions section.
2. Raise and safely support the rear of the vehicle.
3. Remove the rear wheels and temporarily fasten the rotor disc with the hub nuts.
4. Hold the sliding pin on the bottom of the caliper and loosen the installation bolt.
5. Remove the lower installation bolt.
6. Lift up the caliper and suspend it securely. Do not remove the upper installation bolt.
7. Remove the following parts:
   - The 2 anti-squeal springs.
   - The 2 brake pads.
   - The 4 anti-squeal shims.
   - The 4 pad support plates.

To install:

8. Install the pad support plates.
9. Install a pad wear indicator plate to the pad. Install the anti-squeal shims and support plates to each pad.

➡It recommended that a suitable anti-squeal compound (available at your local parts house) be applied to both sides of the inner anti-squeal shim.

10. Draw out a small amount of brake fluid from the brake reservoir. Press in the caliper piston with a suitable tool.
11. Press the brake piston in carefully so the boot will not become wedged.
12. Install the 2 pads so that the wear indicator plate is facing upward. Do not allow oil or grease to get in the rubbing face of the pads.
13. Lower and install the caliper. Torque the sliding main pin to 25 ft. lbs. (34 Nm).

➡When installing the sliding main pin, be careful that the plug installed in the torque plate does not come loose.

14. Install the rear wheels and lower the vehicle.
15. Check the fluid level in the master cylinder and add as necessary. Be sure to pump the brake pedal a few times before road-testing the vehicle.

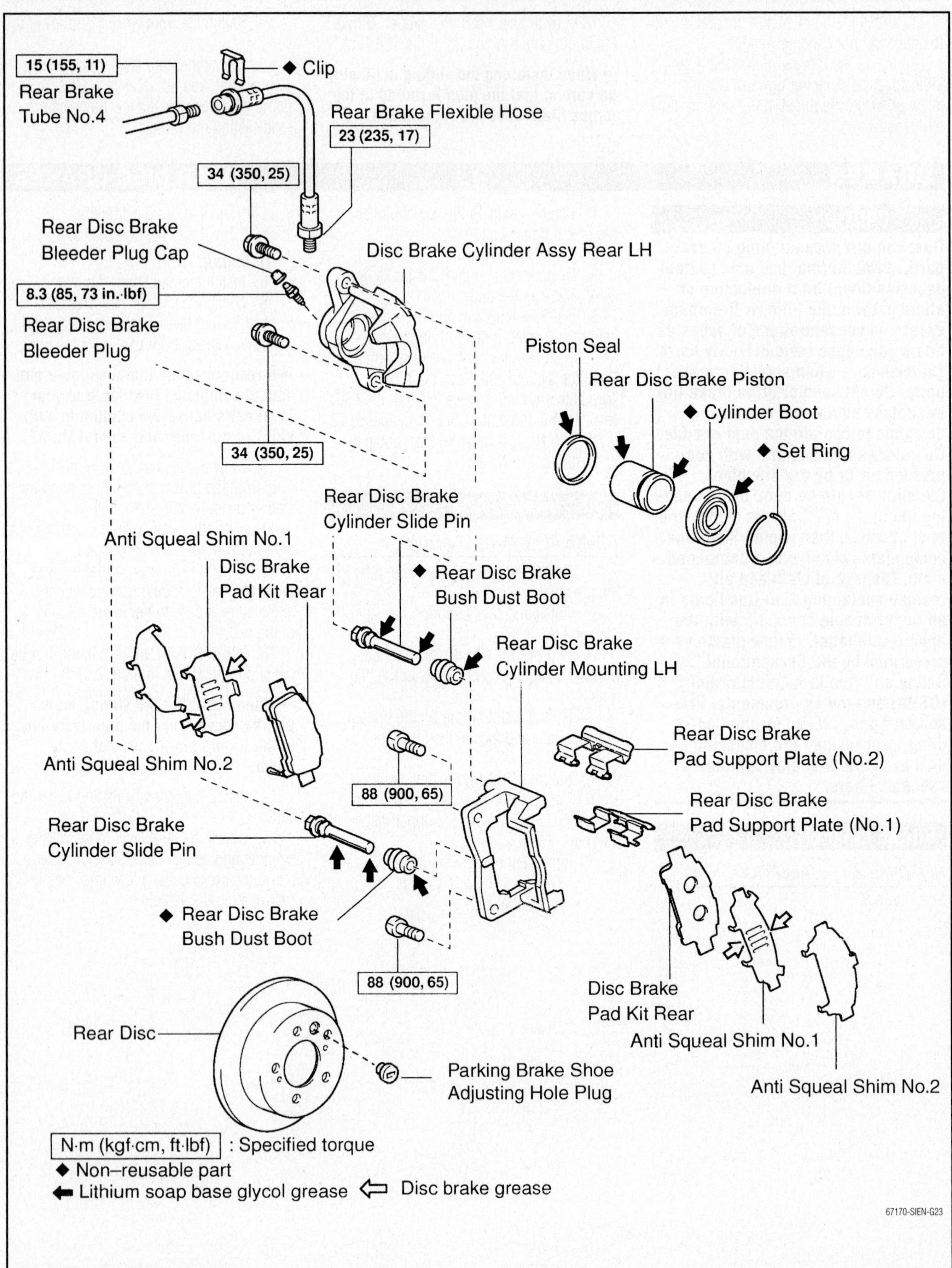

Fig. 3 Rear disc brake components

⁂ CAUTION

Dust and dirt accumulating on brake parts during normal use may contain asbestos fibers from production or aftermarket brake linings. Breathing excessive concentrations of asbestos fibers can cause serious bodily harm. Exercise care when servicing brake parts. Do not sand or grind brake lining unless equipment used is designed to contain the dust residue. Do not clean brake parts with compressed air or by dry brushing. Cleaning should be done by dampening the brake components with a fine mist of water, then wiping the brake components clean with a dampened cloth. Dispose of cloth and all residue containing asbestos fibers in an impermeable container with the appropriate label. Follow practices prescribed by the Occupational Safety and Health Administration (OSHA) and the Environmental Protection Agency (EPA) for the handling, processing, and disposing of dust or debris that may contain asbestos fibers.

BRAKE DRUM

REMOVAL & INSTALLATION

1. Before servicing the vehicle, refer to the precautions section.
2. Release the parking brake.
3. Remove the rear wheel.
4. Remove the inspection plug from the brake drum.
5. Insert a suitable regular screwdriver through the plug hole in the drum and hold the adjusting lever away from the adjuster.
6. Insert a second regular screwdriver and turn the brake shoe adjuster to reduce the size of the adjuster.
7. Remove the brake drum.

To install:

8. Measure the brake drum inside diameter and the diameter of the installed brake shoe assembly.
9. Turn the brake shoe adjuster to either retract or expand the brake shoe assembly until the diameter of the brake she assembly is slightly less than that of the drum.
10. Install the drum, and turn the wheel/drum assembly. There should be a hardly noticeable drag as the drum is turned.
11. Install the wheel and pump the brake pedal several times to fully seat the brakes.
12. Start the vehicle and very slowly drive back and forwards. Driving backwards and applying the brake pedal causes the rear brake adjuster to self adjust.

➡Do not drive backwards at high speeds, for long distances or apply the brakes too hard. A slow speed under 1 mile per hour for a short distance and plying the brake will help the adjuster to self adjust.

BRAKE SHOES

REMOVAL & INSTALLATION

See Figure 4.

1. Before servicing the vehicle, refer to the precautions section.
2. Disconnect the negative battery cable from the battery.
3. Loosen the rear wheel lug nuts slightly. Release the parking brake.
4. Block the front wheels, raise the rear of the vehicle, and safely support it with jackstands.
5. Remove the wheel lug nuts and the wheel.
6. Remove the brake drum retaining screws, if equipped. Remove the brake drum.
7. If the drum is difficult to remove, perform the following:
    a. Insert the end of a bent wire (a coat hanger will do nicely) through the hole in the brake drum and hold the automatic adjusting lever away from the adjuster.
    b. Reduce the brake shoe adjustment by turning the adjuster bolt with a brake tool.
    c. The drum should now be loose enough to remove without much effort.
8. Carefully unhook the return spring from the leading (front) brake shoe.
9. Press the hold down spring retainer in and turn the pin on the front brake shoe.
10. Remove the hold down spring, retainers and the pin for the front brake shoe.
11. Pull out the brake shoe and unhook the anchor spring from the lower edge.
12. Remove the hold down spring from the trailing (rear) shoe. Pull the shoe out with the adjuster, automatic adjuster assembly and springs attached. Disconnect the parking brake cable. Remove the tension/return and anchor springs from the rear shoe.
13. Unhook the adjusting lever spring from the rear shoe and then remove the automatic adjuster assembly.

To install:

14. Inspect the shoes for signs of unusual wear or scoring.
15. Check the wheel cylinder for any sign of fluid seepage or frozen pistons.
16. Clean and inspect the brake backing plate and all other components. Check that the brake drum inner diameter is within specified limits. Lubricate the backing plate at the positions the brakes come in contact with the backing plate. Also lubricate the anchor plate.
17. Mount the automatic adjuster assembly onto a new rear brake shoe.
18. Connect the parking brake cable to the rear shoe and then install the automatic adjusting lever, spring and E-ring. Position the rear shoe so the lower end rides in the anchor plate and the upper end is against the boot of the wheel cylinder.
19. Install the pin and the hold down spring. Press the retainer down over the pin and rotate the pin so the crimped edge is held by the retainer.
20. Place the front brake into position and install the anchor spring between the front and rear shoes. Stretch the spring enough so the front shoe will fit as the rear did. Install the hold down spring, pin and retainer to the front brake shoe.
21. Connect the return spring to the front brake shoe.
22. Check the operation of the automatic adjuster mechanism:
    a. Apply the parking brake lever and verifying the adjusting bolt turns.
    b. Adjust the strut to where it is the shortest possible length.
    c. Install the brake drum.
    d. Apply the parking brake lever until the clicking sound can no longer be heard.
23. Check the clearance between the brake shoes and drum:
    a. Remove the brake drum.
    b. Measure the brake drum inside diameter and diameter of the brake shoes. The difference is "Shoe-to-drum clearance" and should be approximately 0.024 inch (0.6mm). If incorrect, check the parking brake system.

➡A special brake caliper tool is required to gauge the brake drum inside diameter and shoe-to-drum clearance. However it is not required to perform brake shoe adjustment.

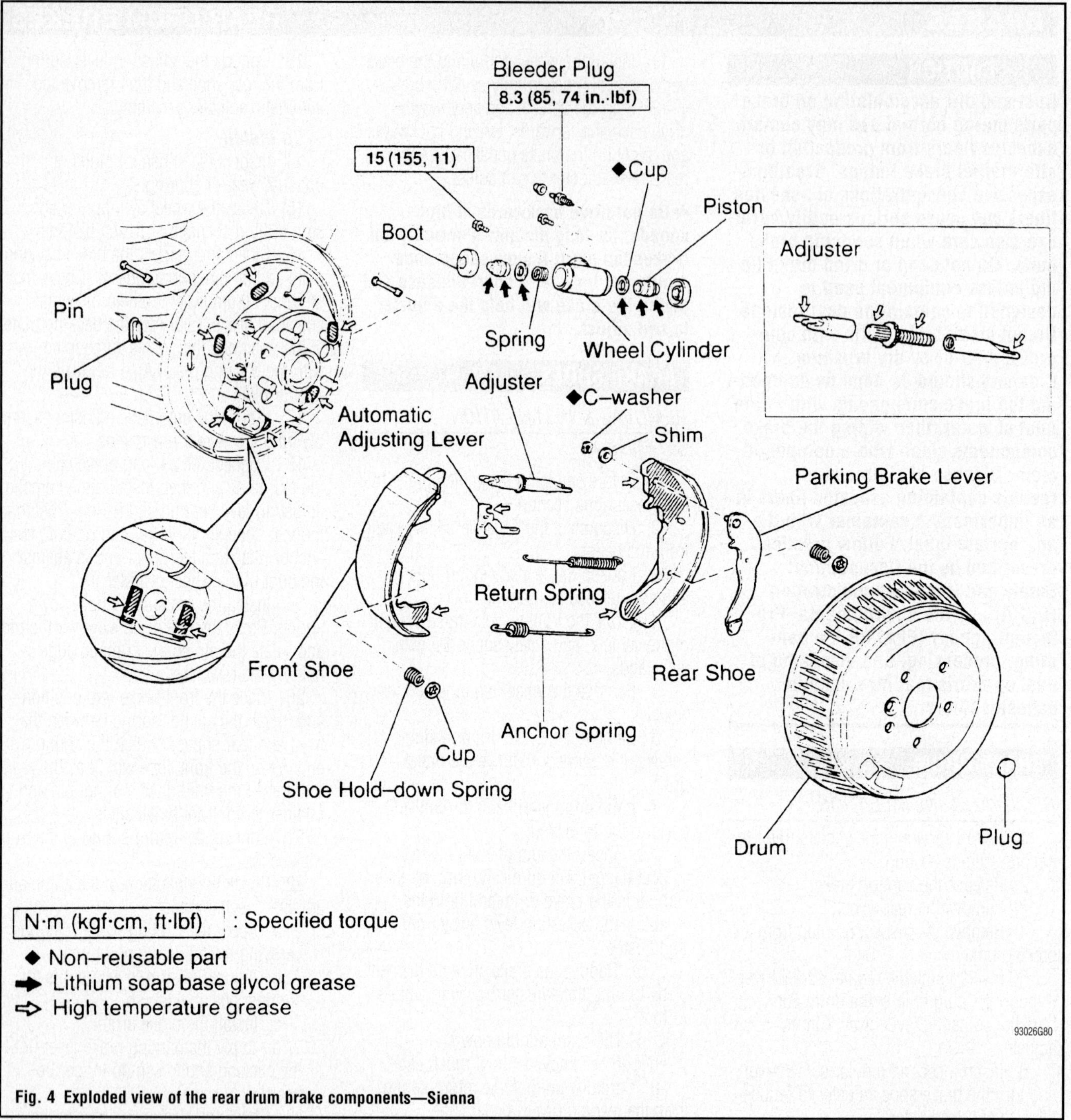

Fig. 4 Exploded view of the rear drum brake components—Sienna

24. Install the brake drum.
25. Adjust the brake pedal until a slight drag is felt when the drum is spun by hand.
26. Pull the parking lever all the way up until a clicking sound can no longer be heard. Check the clearance between brake shoes and brake drum.
27. Install the rear wheels, tighten the wheel lug nuts and lower the vehicle.
28. Retighten the wheel lug nuts and pump the brake pedal a few times before moving the vehicle. Adjust the rear brakes again if necessary.
29. Check the level of brake fluid in the master cylinder, then perform a test drive.
30. Connect the negative battery cable to the battery.

ADJUSTMENT

1. Before servicing the vehicle, refer to the precautions section.
2. Remove the rear wheels.
3. Measure the brake drum inside diameter and the diameter of the installed brake shoe assembly.
4. Turn the brake shoe adjuster to either retract or expand the brake shoe assembly until the diameter of the brake she assembly is slightly less than that of the drum.
5. Install the drum, and turn the adjuster to expand the shoes until the drum locks.
6. Back off the adjuster 15 notches and install the hole plug.

PARKING BRAKE SHOES

REMOVAL & INSTALLATION

See Figure 5.

1. Before servicing the vehicle, refer to the precautions section.
2. Raise and safely support the vehicle.
3. Remove the wheels.
4. Remove the brake caliper.
5. Use a suitable piece of wire to suspend the caliper assembly from the upper control arm. This will prevent the weight of the caliper from being supported by the brake flex hose which will damage the hose.
6. Remove the caliper bracket bolts and remove the bracket.
7. Place match marks on the rotor and hub.
8. Remove the rotor. If the disc cannot be removed easily, turn the parking brake adjuster until the rotor turns freely and remove the rotor.
9. Use needle nosed pliers to remove the left hand No.1 parking brake shoe return springs.
10. Disconnect the tension spring from the left hand No.1 parking brake shoe.
11. Slide out the left hand No.1 parking brake shoe assembly and remove it.
12. Remove the left hand No.1 hold down spring and cups, then remove the hold down spring pin.
13. Remove the left hand No.1 parking brake shoe adjuster assembly.
14. Remove the left hand No.1 parking brake shoe strut.
15. Disconnect the return spring from the left hand No.2 parking brake shoe and remove the tension spring.
16. Slide out the left hand No.2 parking brake shoe assembly and remove it.
17. Remove the left hand No.2 hold down spring and cups, then remove the hold down spring pin.
18. Use needle nosed pliers to disconnect the parking brake cable from the parking brake shoe lever and the left hand No.2 parking brake shoe.

To install:

19. Apply high temperature grease to the surfaces where the parking brake shoe and backing plate contact.
20. Install the shoe lever and shim to the rear shoe using a new C-washer.

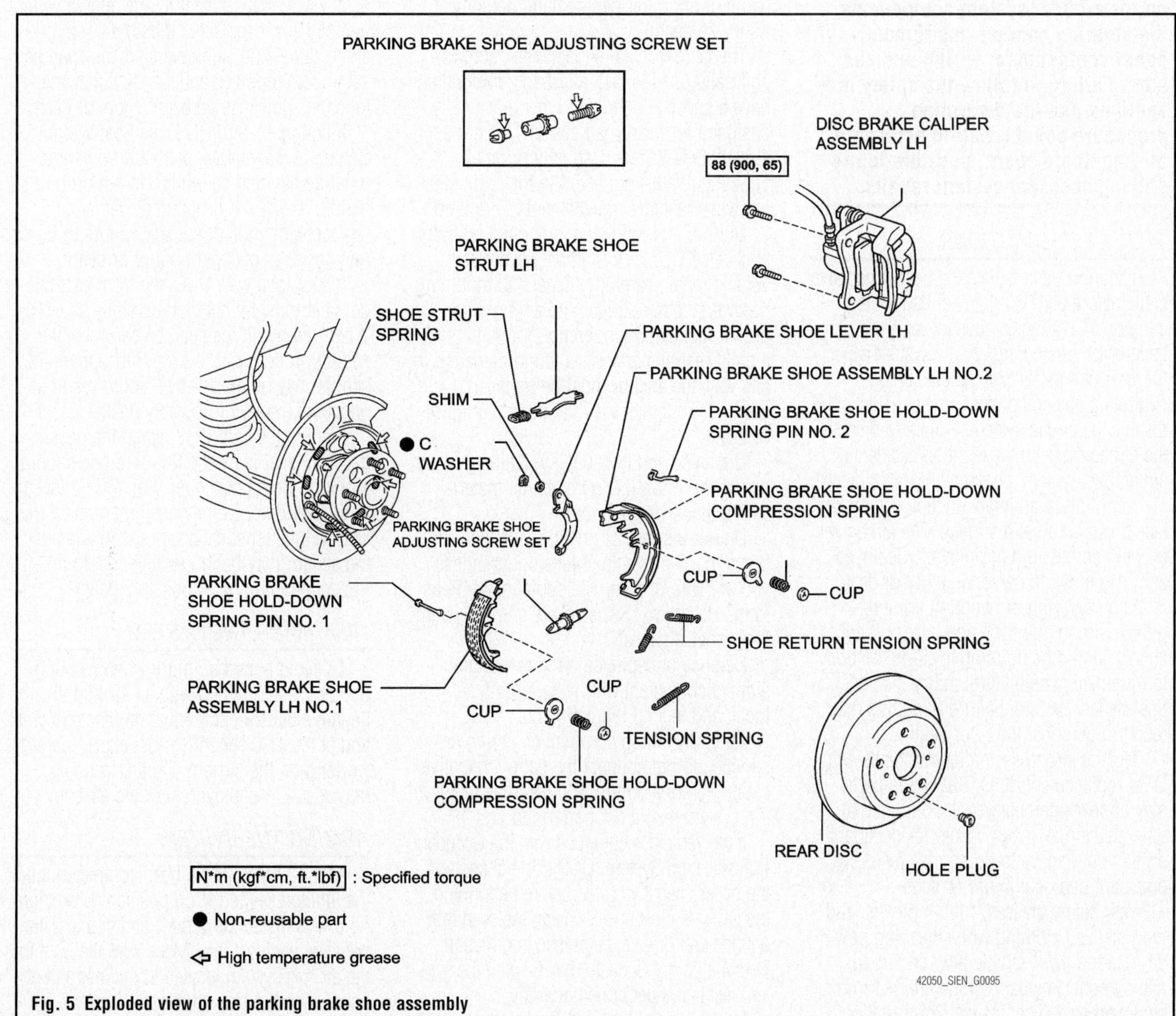

Fig. 5 Exploded view of the parking brake shoe assembly

21. Measure the distance between the C-washer and shoe, the clearance should be 0.0138 inch (0.35mm). If the clearance is more than specified, use a 3, 6 Or 9mm shim to reduce the clearance to the proper specification.

22. Using needle nosed plies, connect the parking brake cable to the parking brake shoe lever.

23. Install the left hand No.2 parking brake shoe assembly with the hold down compression spring pin, compression spring and cups.

24. Apply high temperature grease to the adjuster assembly moving components and install the adjuster assembly.

25. Install the parking brake shoe strut.

26. Install the left hand No.1 parking brake shoe assembly with the hold down compression spring pin, compression spring and cups.

27. Install the tension spring and 2 return springs.

28. Adjust the parking brake.

29. Install the rotor and align the match marks.

30. Install the caliper bracket and tighten the bracket to steering knuckle bolts to 65 ft. lbs. (88 Nm).

31. Install the caliper and wheels.

CHASSIS ELECTRICAL

AIR BAG (SUPPLEMENTAL RESTRAINT SYSTEM)

GENERAL INFORMATION

⁂ CAUTION

Some vehicles are equipped with an air bag system. The system must be disarmed before performing service on, or around, system components, the steering column, instrument panel components, wiring and sensors. Failure to follow the safety precautions and the disarming procedure could result in accidental air bag deployment, possible injury and unnecessary system repairs.

SERVICE PRECAUTIONS

Disconnect and isolate the battery negative cable before beginning any airbag system component diagnosis, testing, removal, or installation procedures. Allow system capacitor to discharge for two minutes before beginning any component service. This will disable the airbag system. Failure to disable the airbag system may result in accidental airbag deployment, personal injury, or death.

Do not place an intact undeployed airbag face down on a solid surface. The airbag will propel into the air if accidentally deployed and may result in personal injury or death.

When carrying or handling an undeployed airbag, the trim side (face) of the airbag should be pointing towards the body to minimize possibility of injury if accidental deployment occurs. Failure to do this may result in personal injury or death.

Replace airbag system components with OEM replacement parts. Substitute parts may appear interchangeable, but internal differences may result in inferior occupant protection. Failure to do so may result in occupant personal injury or death.

Wear safety glasses, rubber gloves, and long sleeved clothing when cleaning powder residue from vehicle after an airbag deployment. Powder residue emitted from a deployed airbag can cause skin irritation. Flush affected area with cool water if irritation is experienced. If nasal or throat irritation is experienced, exit the vehicle for fresh air until the irritation ceases. If irritation continues, see a physician.

Do not use a replacement airbag that is not in the original packaging. This may result in improper deployment, personal injury, or death.

The factory installed fasteners, screws and bolts used to fasten airbag components have a special coating and are specifically designed for the airbag system. Do not use substitute fasteners. Use only original equipment fasteners listed in the parts catalog when fastener replacement is required.

During, and following, any child restraint anchor service, due to impact event or vehicle repair, carefully inspect all mounting hardware, tether straps, and anchors for proper installation, operation, or damage. If a child restraint anchor is found damaged in any way, the anchor must be replaced. Failure to do this may result in personal injury or death.

Deployed and non-deployed airbags may or may not have live pyrotechnic material within the airbag inflator.

Do not dispose of driver/passenger/curtain airbags or seat belt tensioners unless you are sure of complete deployment. Refer to the Hazardous Substance Control System for proper disposal.

Dispose of deployed airbags and tensioners consistent with state, provincial, local, and federal regulations.

After any airbag component testing or service, do not connect the battery negative cable. Personal injury or death may result if the system test is not performed first.

If the vehicle is equipped with the Occupant Classification System (OCS), do not connect the battery negative cable before performing the OCS Verification Test using the scan tool and the appropriate diagnostic information. Personal injury or death may result if the system test is not performed properly.

Never replace both the Occupant Restraint Controller (ORC) and the Occupant Classification Module (OCM) at the same time. If both require replacement, replace one, then perform the Airbag System test before replacing the other.

Both the ORC and the OCM store Occupant Classification System (OCS) calibration data, which they transfer to one another when one of them is replaced. If both are replaced at the same time, an irreversible fault will be set in both modules and the OCS may malfunction and cause personal injury or death.

If equipped with OCS, the Seat Weight Sensor is a sensitive, calibrated unit and must be handled carefully. Do not drop or handle roughly. If dropped or damaged, replace with another sensor. Failure to do so may result in occupant injury or death.

If equipped with OCS, the front passenger seat must be handled carefully as well. When removing the seat, be careful when setting on floor not to drop. If dropped, the sensor may be inoperative, could result in occupant injury, or possibly death.

If equipped with OCS, when the passenger front seat is on the floor, no one should sit in the front passenger seat. This uneven force may damage the sensing ability of the seat weight sensors. If sat on and damaged, the sensor may be inoperative, could result in occupant injury, or possibly death.

DISARMING THE SYSTEM

To avoid personal injury when working on vehicles equipped with an air bag, the negative battery cable must be disconnected and at least 90 seconds must elapse before working on the system. Failure to do so may result in deployment of the air bag.

ARMING THE SYSTEM

After all repairs have been completed, turn the ignition key to the on position. Make sure no one is inside the vehicle and connect the negative battery cable. Make sure the light for the air bag system located in the instrument panel does not stay illuminated.

DRIVETRAIN

AUTOMATIC TRANSAXLE ASSEMBLY

REMOVAL & INSTALLATION

See Figures 6 and 7.

1. Remove the engine/transaxle assembly. See Engine Removal and Installation".
2. Remove the left and right halfshafts.
3. Disconnect all wiring between the engine and transaxle.
4. Remove the starter.
5. Disconnect all cables between the engine and transaxle.
6. Remove the oil filler tube.
7. Remove the oil cooler tubes.

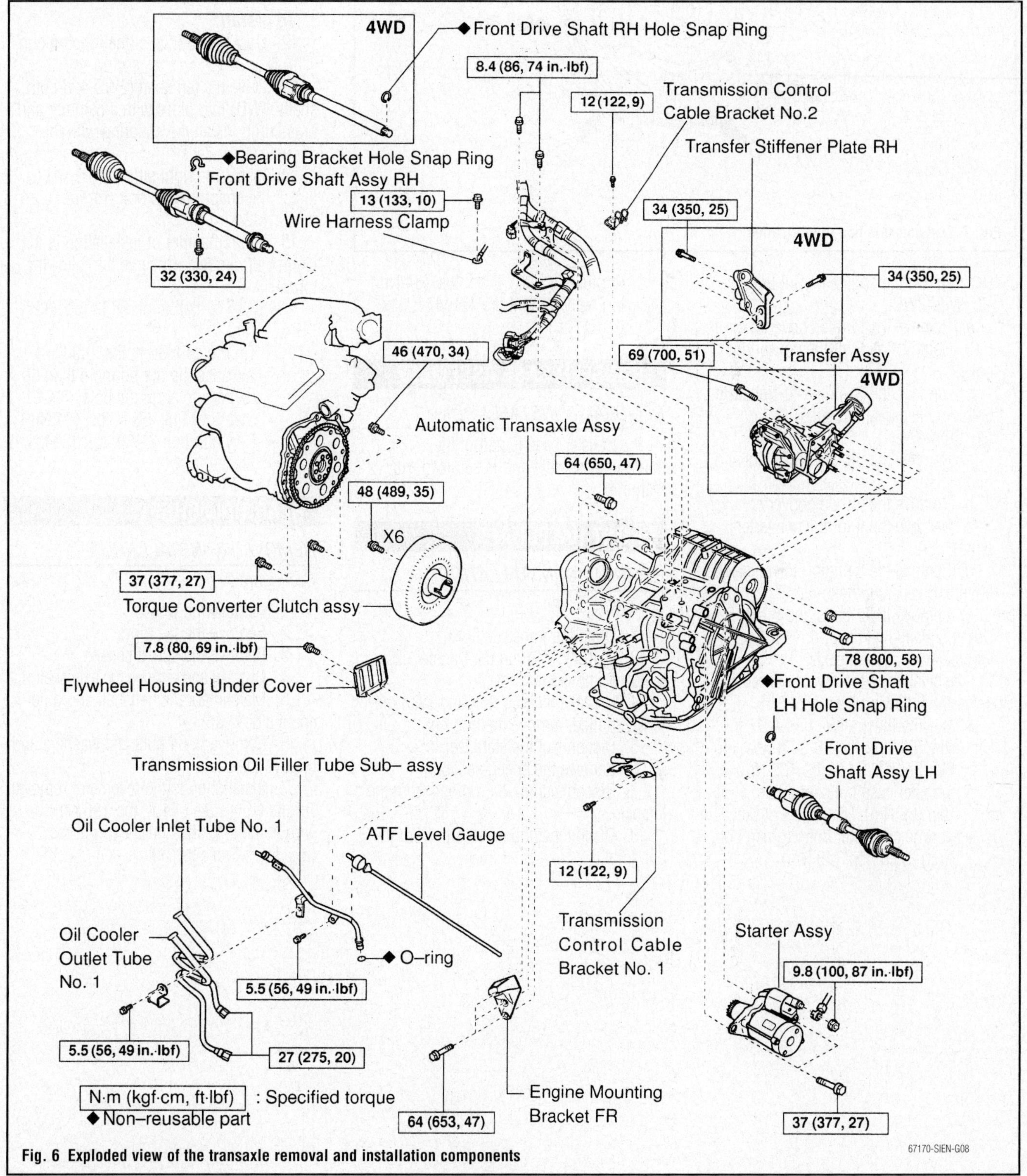

Fig. 6 Exploded view of the transaxle removal and installation components

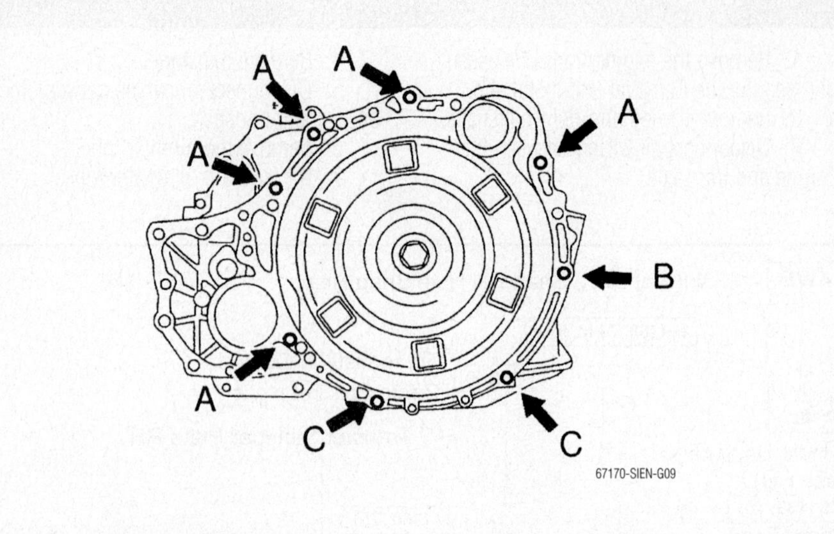

Fig. 7 Transmission bolt identification

8. Remove the engine mount bracket.
9. With 2WD:
   a. Remove the torque converter cover.
   b. Remove the 6 torque converter bolts.
   c. Remove the 8 engine-to-transaxle bolts and separate the units.
10. With 4WD:
   a. Remove the 5 bolts and 1 nut and remove the transfer case stiffener plate.
   b. Remove the torque converter cover.
   c. Remove the 6 torque converter bolts.
   d. Remove the 8 engine-to-transaxle bolts and separate the units.
   e. Remove the 2 bolts and 6 nuts. Using a plastic hammer, drive the transfer case from the transaxle.
11. Installation is the reverse of removal. Observe the following torques:
   - See the illustration: bolt A 47 ft. lbs. (64 Nm); bolt B 34 ft. lbs. (46 Nm); bolt C 27 ft. lbs. (37 Nm)
   - Transfer case bolts and nuts: 51 ft. lbs. (69 Nm). Use a new gasket.
   - Torque converter bolts (green bolt first): 35 ft. lbs. (48 Nm)
   - Stiffener plate: 25 ft. lbs. (34 Nm)
   - Engine mount bracket: 47 ft. lbs. (64 Nm)

TRANSFER CASE ASSEMBLY

REMOVAL & INSTALLATION

The transfer case is part of the transaxle assembly and is serviced with that unit.

FRONT HALFSHAFT

REMOVAL & INSTALLATION

See Figure 8.

1. Drain the transaxle fluid.
2. With 4WD, drain the transfer case.
3. Remove the wheel.
4. Unstake the hub nut, and, with the brake applied, remove the hub nut.
5. Disconnect the stabilizer link.
6. Remove the speed sensor.
7. Disconnect the tie rod end from the knuckle.
8. Disconnect the lower arm from the ball joint.
9. Using a plastic hammer, drive the halfshaft from the hub.
10. On the left side with 2WD and both sides with 4WD, using a slide hammer with adapter, pull the halfshaft from the transaxle.
11. On the right side, with 2WD, remove the halfshaft bearing bracket snapring. Remove the bolt and the halfshaft from the bearing bracket.

To install:

12. Coat the splines of the inboard end with clean ATF.
13. Drive the left shaft (2WD and both shafts 4WD) into place with a hammer and brass drift. Install the snapring with the opening downward.
14. Install the right side (2WD) shaft. Install the snapring and bolt. Torque to 24 ft. lbs. (32 Nm).
15. The remainder of installation is the reverse of removal. Observe the following torques:
   - Arm-to-ball joint: 94 ft. lbs. (127 Nm)
   - Tie rod end: 36 ft. lbs. (49 Nm). Advance the nut no more than 60 degrees to align the hole.
   - Stabilizer link: 55 ft. lbs. (74 Nm)
   - New hub nut: 217 ft. lbs. (294 Nm). Stake the nut.

REAR HALFSHAFT

REMOVAL & INSTALLATION

See Figure 9.

1. Remove the wheel.
2. Remove the tail pipe.
3. Remove the speed sensor.
4. Unstake and remove the axle shaft nut.
5. Matchmark the halfshaft and differential side gear.
6. Remove the 4 nuts and washers and remove the shaft.
7. Installation is the reverse of removal. Torque the 4 nuts t 41 ft. lbs. (56 Nm). Torque the axle shaft nut to 159 ft. lbs. (216 Nm). Stake the nut.

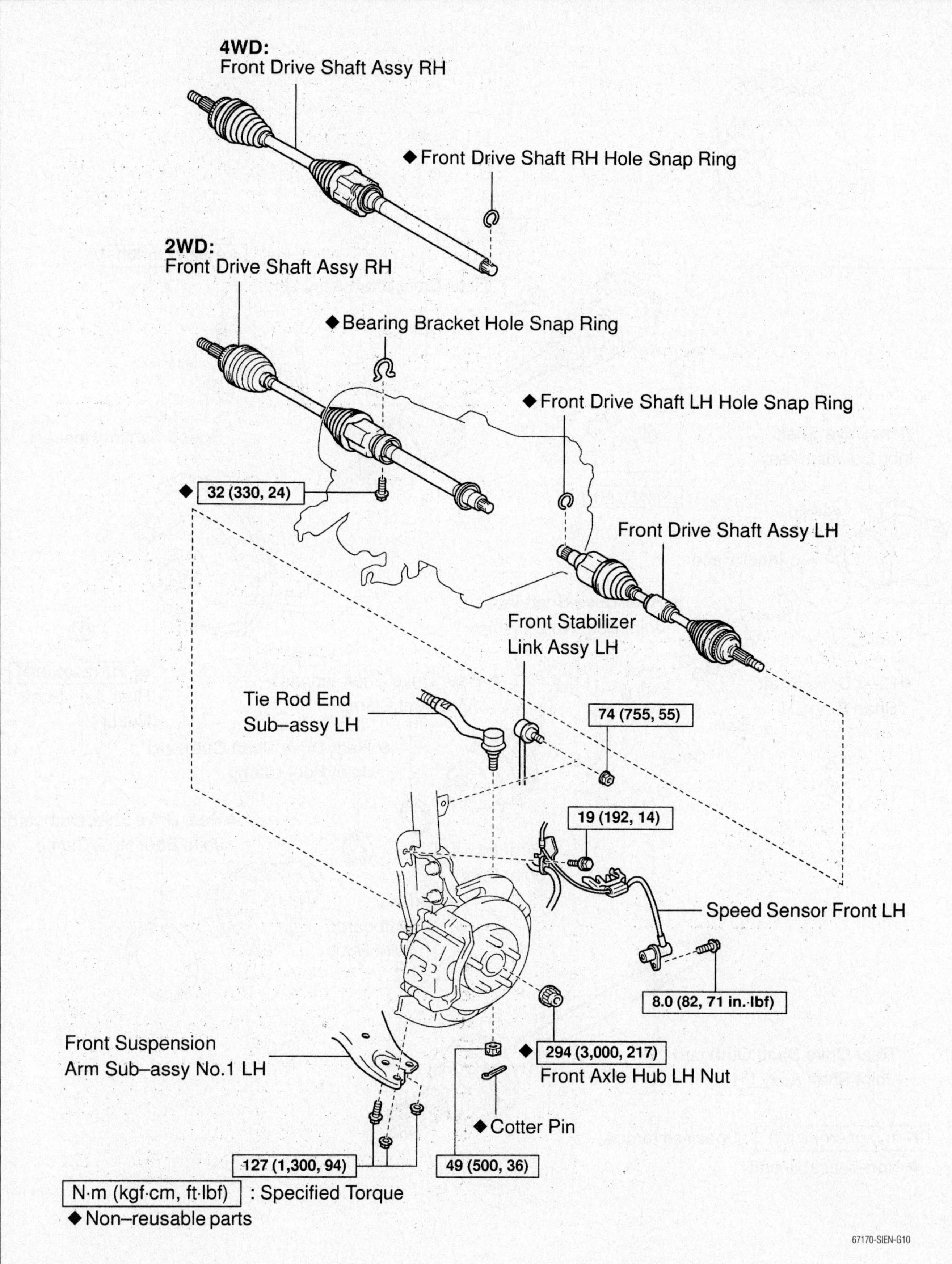

Fig. 8 Front halfshaft exploded view

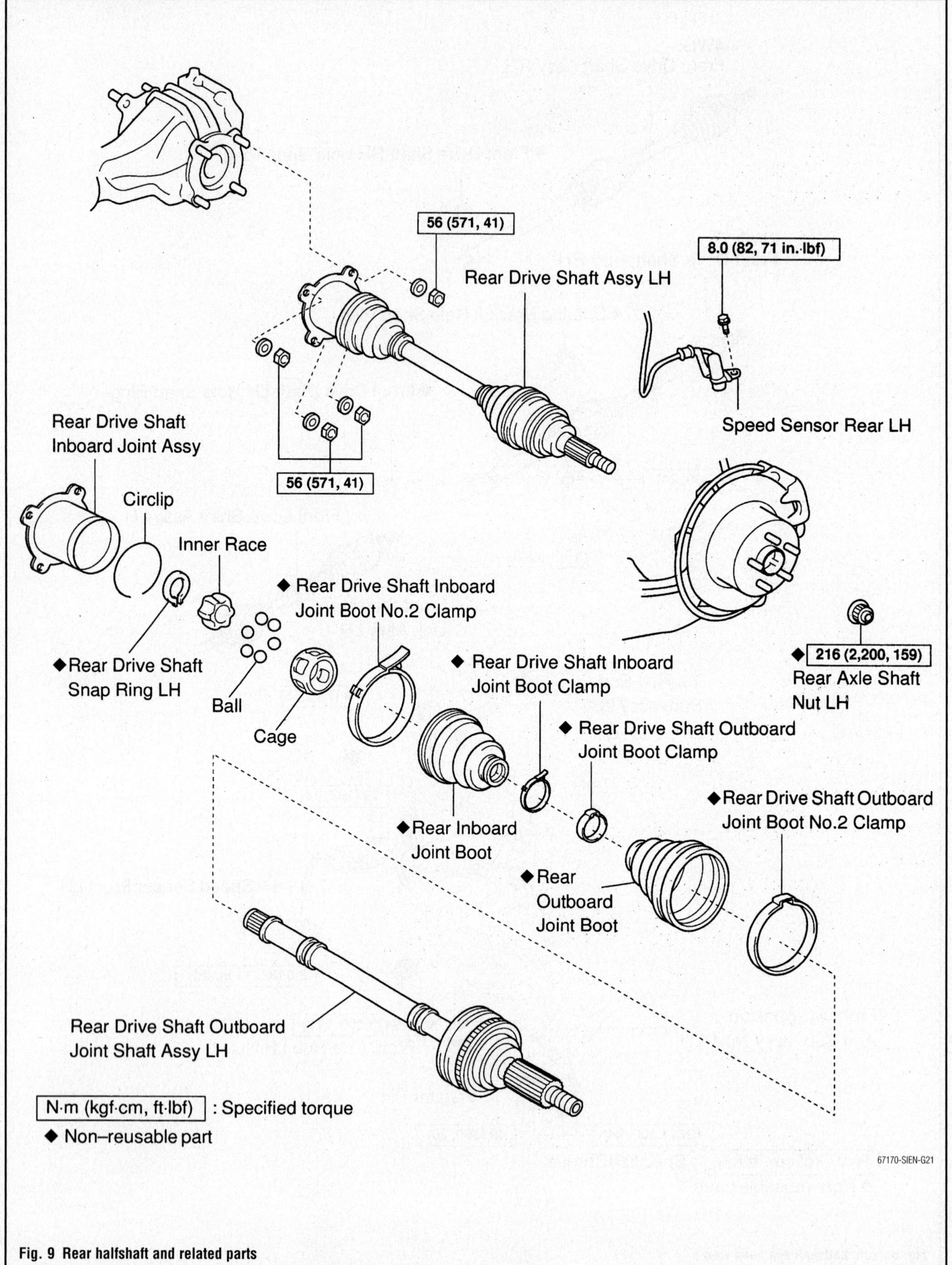

Fig. 9 Rear halfshaft and related parts

ENGINE COOLING

THERMOSTAT

REMOVAL & INSTALLATION

1. Disconnect the negative battery cable.
2. Drain and recycle the engine coolant.
3. Remove the battery.
4. Remove the air cleaner assembly and bracket.
5. Remove the starter motor.
6. Disconnect the wire harness clamp from the water inlet.
7. Remove the 3 nuts, water inlet and the thermostat.
8. Remove the thermostat gasket and clean the gasket mating surfaces.

To install:

9. Install a new thermostat gasket then place the thermostat in position. The thermostat jiggle valve must be aligned with the upper stud bolt.

➡The jiggle valve must be within 15 degrees of either side of the stud bolt.

10. Install the water inlet and tighten the nuts to 71 inch lbs. (8 Nm) for the nut and 14 ft. lbs. (20 Nm) for the bolt.
11. Install the starter.
12. Connect the wire harness clamp to the water inlet.
13. Install the air cleaner bracket, air cleaner assembly and hose.
14. Fill the cooling system
15. Start the vehicle, check for leaks and recheck coolant level.

WATER PUMP

REMOVAL & INSTALLATION

See Figure 10.

1. Before servicing the vehicle, refer to the precautions section.
2. Properly relieve the fuel system pressure.
3. Disconnect the negative battery cable.
4. Drain and recycle the engine coolant.
5. Remove the right hand front wheel.
6. Remove the wiper arms.
7. Remove the cowl panel.
8. Remove the wiper link assembly.
9. Remove the cowl top to inner brace.
10. Remove the cowl top cover outer sub assembly.
11. Remove the right hand apron seal.
12. Remove the A/C compressor to crankshaft pulley belt.
13. Remove the vane pump belt.
14. Remove the engine moving control rod bolts and rod.
15. Remove the number 2 engine mounting stay as follows:
    a. Remove the bolt and wire harness bracket.
    b. Remove the bolt, stay and bracket.
16. Remove the timing belt covers 1 and 2. Refer to the timing belt removal and installation procedure.
17. Remove the right hand engine mount bracket.
18. Remove the timing belt guide and timing belt. Refer to the timing belt removal and installation procedure.
19. Remove the timing belt guide number 2. Refer to the timing belt removal and installation procedure.
20. Remove the camshaft pulley and lower timing cover. Refer to the timing belt removal and installation procedure.
21. Remove the timing belt idler sub assembly 1. Refer to the timing belt removal and installation procedure.
22. Loosen the bolts and remove the water pump and gasket.

To install:

23. Clean all gasket mating surfaces and install a new gasket.
24. Install the water pump and tighten the bolts to 71 inch lbs. (8 Nm).
25. Install the remaining components in the reverse order of removal. Refer to the timing belt removal and installation procedure in this manual for all timing belt related component installation.
26. Fill the cooling system and check for leaks.

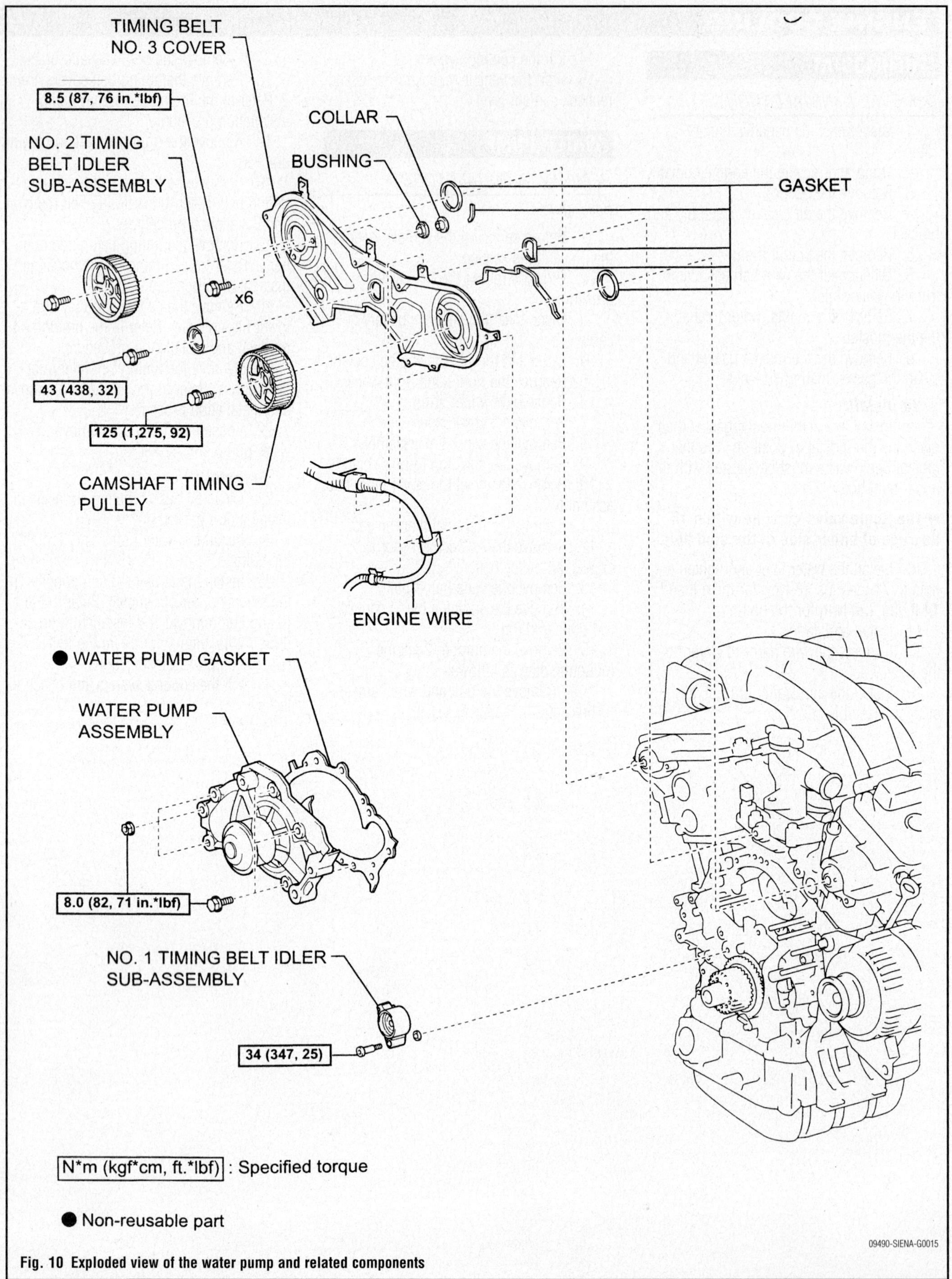

Fig. 10 Exploded view of the water pump and related components

ENGINE ELECTRICAL — CHARGING SYSTEM

ALTERNATOR

REMOVAL & INSTALLATION

1. Before servicing the vehicle, refer to the precautions section.
2. Remove or disconnect the following:
   - Alternator electrical connectors
   - Wiring harness from the clip
   - Pivot bolt
   - Plate washer
   - Adjusting lockbolt
   - Drive belt
   - Alternator

To install:

3. Install or connect the following:
   - Alternator
   - Drive belt. Tension the belt to 170–180 lbs. for a new belt or 95–135 lbs. for a used belt.
   - Adjusting lockbolt. Tighten the bolt to 13 ft. lbs. (18 Nm).
   - Plate washer
   - Pivot bolt. Tighten the bolt to 43 ft. lbs. (58 Nm).
   - Wiring harness from the clip
   - Alternator electrical connectors

VOLTAGE REGULATOR

REMOVAL & INSTALLATION

The voltage regulator is serviced with the alternator as an assembly.

ENGINE ELECTRICAL — DISTRIBUTORLESS IGNITION SYSTEM

FIRING ORDER

See Figure 11.

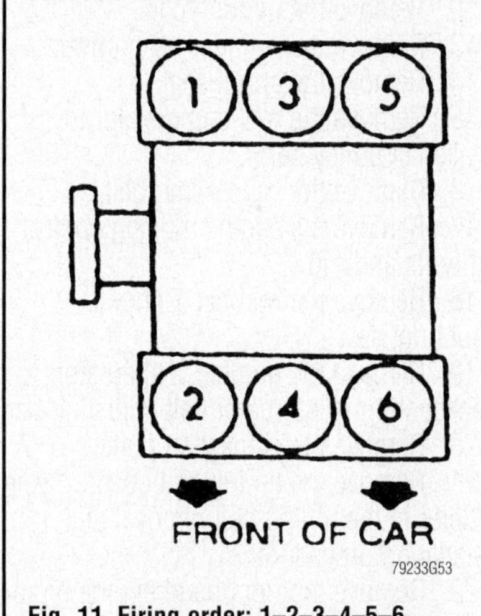

Fig. 11 Firing order: 1–2–3–4–5–6 Distributorless ignition system

IGNITION COIL

REMOVAL & INSTALLATION

➡It is a good idea to remove and reinstall the coils one at a time to prevent the coils being installed out of order.

1. Disconnect the negative battery cable.
2. Disconnect the spark plug wire from the coil.
3. Disconnect the electrical connectors from the coil.
4. Remove the coil bolts and the coils.

To install:

5. Install the coil and tighten the bolts to 69 inch lbs. (8 Nm).
6. Connect the wiring and spark plug wires
7. Connect the negative battery cable.

IGNITION TIMING

ADJUSTMENT

The ignition timing is controlled by the Powertrain Control Module (PCM). No adjustment is necessary or possible.

SPARK PLUGS

REMOVAL & INSTALLATION

When removing spark plugs, work on them one at a time to avoid confusion. Do not remove the plug wires at the same time, unless you number them, they may become mixed up. Take a minute before you begin and number the wires with tape.

1. Disconnect the negative battery cable, and if the vehicle has been run recently, allow the engine to thoroughly cool.
2. Carefully twist the spark plug wire boot to loosen it, then pull upward and remove the boot from the plug. Be sure to pull on the boot and not on the wire, otherwise the connector located inside the boot may become separated.
3. Using compressed air, blow any water or debris from the spark plug well to assure that no harmful contaminants are allowed to enter the combustion chamber when the spark plug is removed. If compressed air is not available, use a rag or a brush to clean the area.

➡Remove the spark plugs when the engine is cold to prevent damage to the threads. If the plugs is hard to remove, apply a few drops of a penetrating lubricant to the area around the base of the plug, and allow it sit for a few minutes to work.

4. Using a spark plug socket that is equipped with a rubber insert to hold the plug, turn the spark plug counterclockwise to loosen and remove the spark plug from the bore.

**** WARNING**

It is advisable not to use a flexible extension on the socket when removing or installing spark plugs. This may allow a shear force to be applied to the plug. A shear force could break the plug off in the cylinder head, leading to costly and frustrating repairs.

To install:

5. Inspect the spark plug boot for damage and replace if necessary
6. Check and adjust the spark plug gap
7. Carefully thread the plug into the bore by hand. If resistance is felt before the plug is almost completely threaded, back the plug out and begin threading again.
8. Carefully tighten the spark plug to 18 ft. lbs.
9. Apply a small amount of silicone dielectric compound to the end of the spark plug lead or inside the spark plug boot to prevent sticking, then install the boot to the spark plug and push until it clicks into place.

STARTER

REMOVAL & INSTALLATION

1. Before servicing the vehicle, refer to the precautions section.
2. Remove the battery and tray.
3. Remove the air cleaner assembly and inlet tubes.
4. Remove the air cleaner bracket.
5. Remove the wiring from the starter.
6. Remove the 2 bolts and lower the starter.
7. Installation is the reverse of removal. Torque the starter bolts to 26 ft. lbs. (37 Nm).

ENGINE MECHANICAL

ACCESSORY DRIVE BELTS

ACCESSORY BELT ROUTING

See Figure 12.

Refer to the accompanying illustration for accessory belt routing.

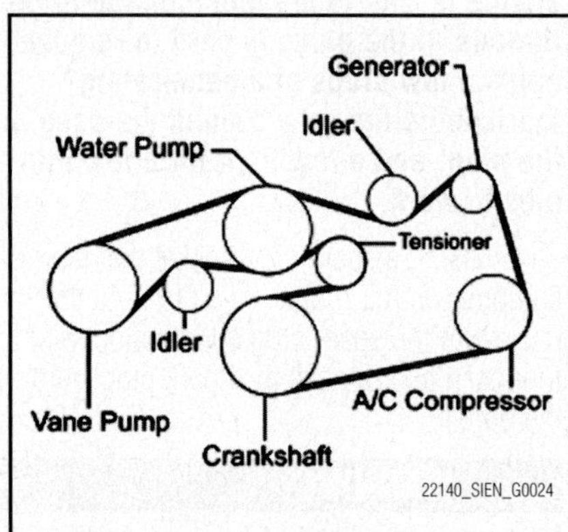

Fig. 12 Accessory drive belt routing–3.5L engine

INSPECTION

Inspect the drive belt for signs of glazing or cracking. A glazed belt will be perfectly smooth from slippage, while a good belt will have a slight texture of fabric visible. Cracks will usually start at the inner edge of the belt and run outward. All worn or damaged drive belts should be replaced immediately.

REMOVAL & INSTALLATION

See Figures 13 and 14.

1. Before servicing the vehicle, refer to the precautions section.
2. Remove the right hand front wheel and apron seal.
3. Remove the A/C compressor to crankshaft pulley belt by loosening bolts A and B, then loosen adjuster bolt C and remove the belt.
4. Remove the vane pump belt by loosening bolts A and B, then remove the belt.

To install:

5. Install the vane pump belt and tighten the bolts to 32 ft. lbs. (44 Nm).

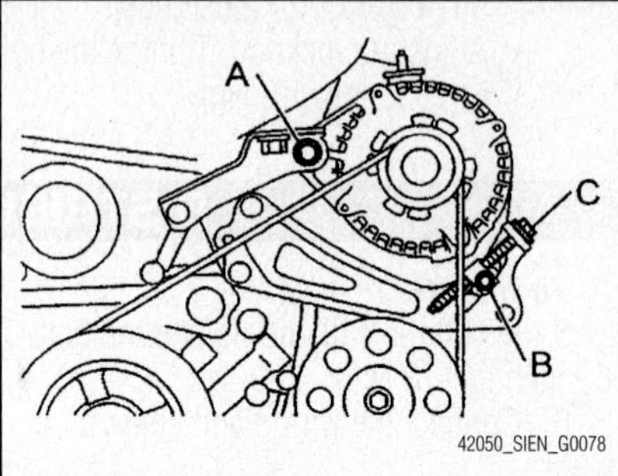

Fig. 13 A/C compressor to crankshaft pulley belt bolt locations

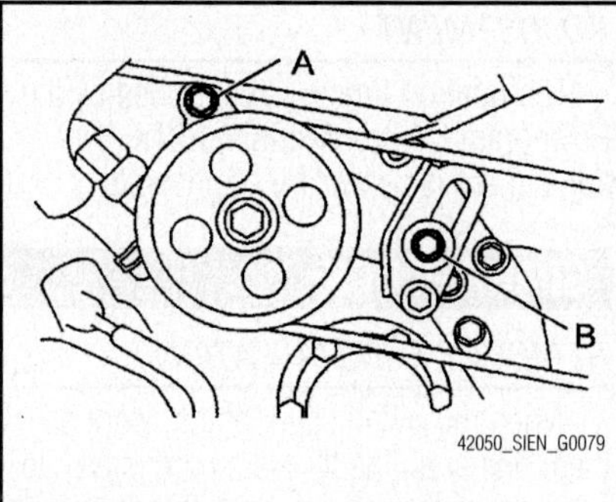

Fig. 14 Vane pump pulley belt bolt locations

6. Install the A/C compressor to crankshaft pulley belt over the pulleys. Tighten adjuster bolt C to adjust the belt tension.
7. Tighten bolt A to 43 ft. lbs. (58 Nm) and bolt B to 13 ft. lbs. (18 Nm).
8. Install the apron seal and front wheel. Tighten the wheel nuts to 76 ft. lbs. (103 Nm)

CAMSHAFT AND VALVE LIFTERS

REMOVAL & INSTALLATION

Left Side

See Figures 15 through 28.

1. Before servicing the vehicle, refer to the precautions section.
2. Properly relieve the fuel system pressure.
3. Drain and recycle the engine coolant.
4. Disconnect the negative battery cable.
5. Remove the right hand front wheel.
6. Remove the wiper arms.
7. Remove the cowl panel.
8. Remove the wiper link assembly.
9. Remove the cowl top to inner brace.
10. Remove the cowl top cover outer sub assembly.
11. Remove the radiator inlet hose.
12. Remove the ignition coil.
13. Remove the cylinder head cover.
14. Remove the apron seal.
15. Remove the A/C compressor to crankshaft pulley belt.
16. Remove the vane pump belt.
17. Remove the engine moving control rod bolts and rod.
18. Remove the number 2 engine mounting stay.
19. Remove the number 2 alternator bracket and the alternator belt adjusting bar.
20. Remove the crankshaft pulley.
21. Remove the timing belt covers 1 and 2. Refer to the timing belt removal and installation procedure.
22. Remove the right hand engine mount bracket.
23. Remove the timing belt guide number 2. Refer to the timing belt removal and installation procedure.
24. Remove the timing belt.
25. Remove the timing belt guide number 2 idler sub assembly. Refer to the timing belt removal and installation procedure.
26. Remove the camshaft pulley and lower timing cover. Refer to the timing belt removal and installation procedure.
27. Remove the timing belt number 3 cover. Refer to the timing belt removal and installation procedure.
28. Align the timing marks of the camshaft drive and driven gears by using a wrench to turn the camshaft.
29. Attach the exhaust camshaft sub gear to the main gear with a 16–20mm long x 6mm thread diameter bolt and tighten to 48 inch lbs. (5 Nm).

➡When removing the camshaft, be sure the torsional spring force of the

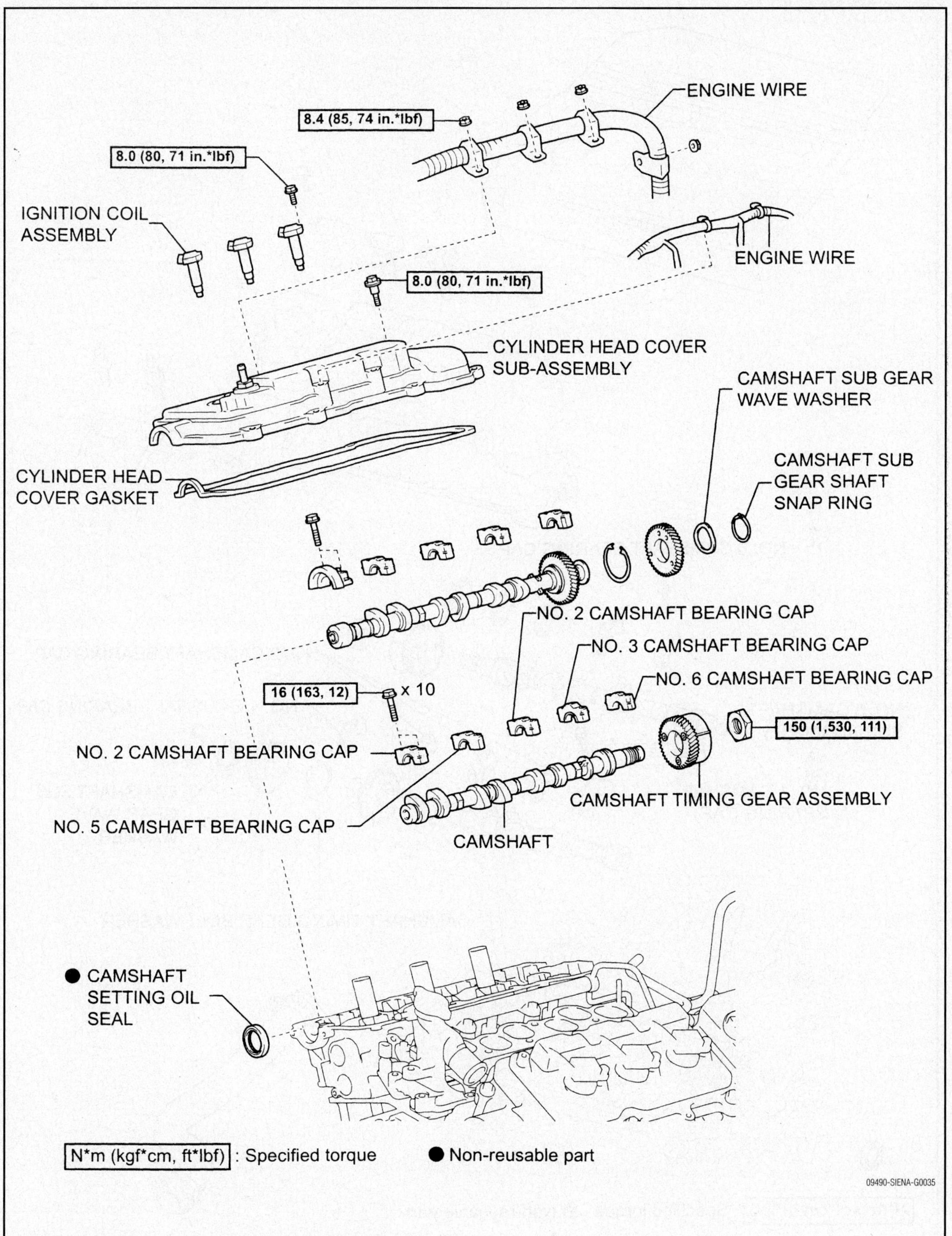

Fig. 15 Exploded view of the right side camshaft assemblies

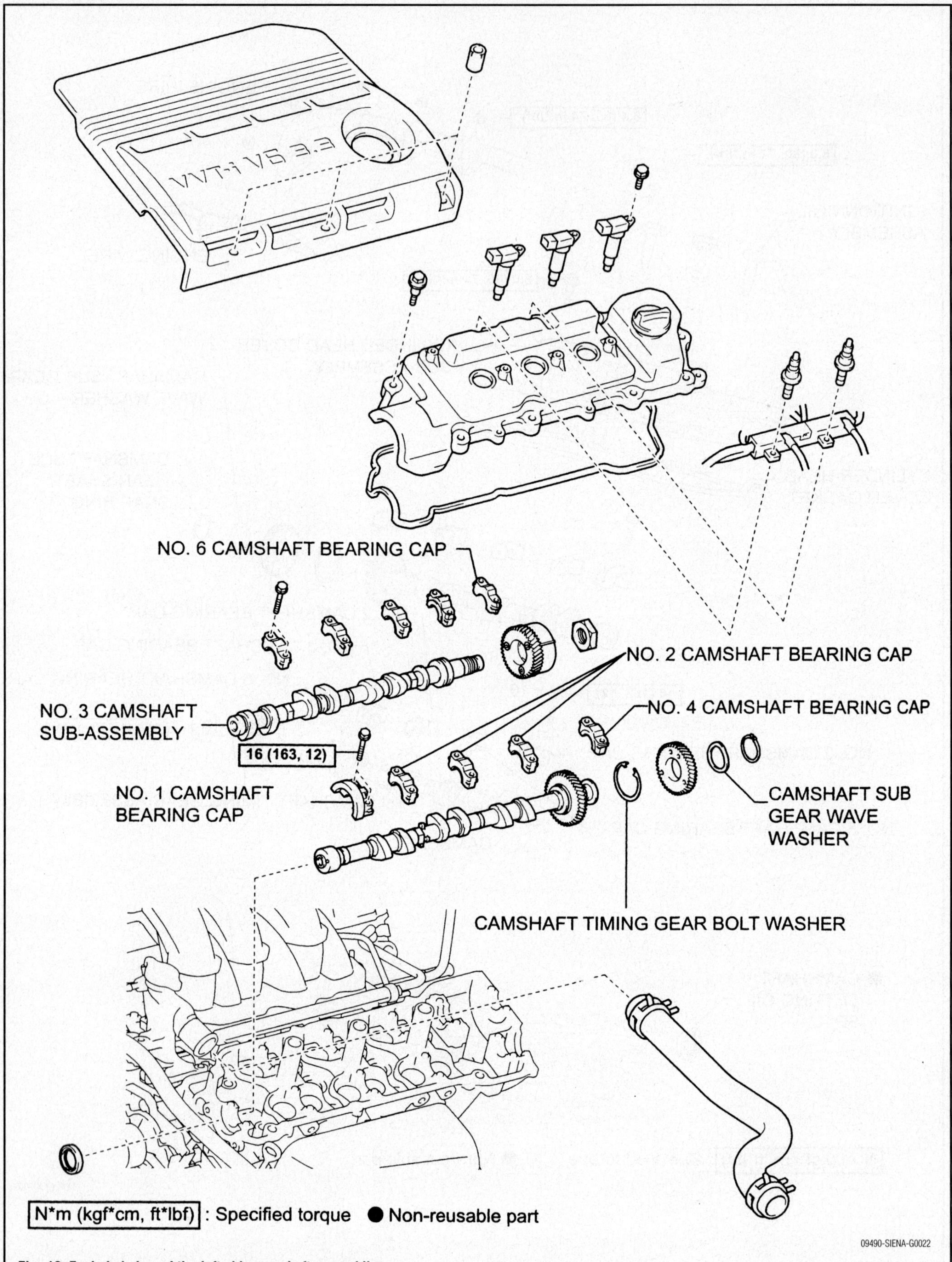

Fig. 16 Exploded view of the left side camshaft assemblies

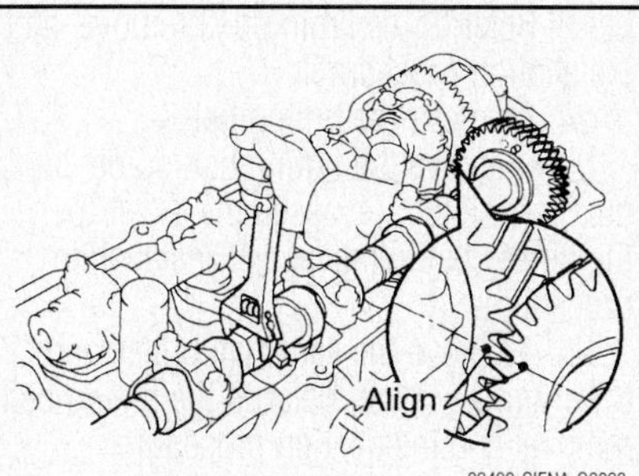

Fig. 17 Align the timing marks of the left side camshaft drive and driven gears by using a wrench to turn the camshaft

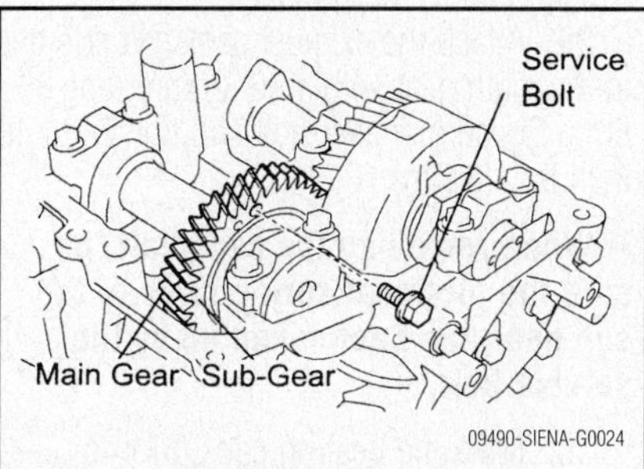

Fig. 18 Attach the left side exhaust camshaft sub gear to the main gear with a 16–20mm long x 6mm thread diameter bolt

sub gear has been taken up by the service bolt.

30. Mark the bearing caps prior to removal so they can be installed in their original positions.
31. Loosen the number 3 camshaft cap bolts in sequence using several steps.
32. Loosen the number 4 camshaft cap bolts in sequence using several steps.

To install:

⁂ WARNING

The camshaft has a small thrust clearance, make sure to keep the camshaft level during installation to avoid damage to the camshaft and cylinder head.

33. Apply clean engine oil to the camshaft thrust and journal locations.
34. Place the number 4 camshaft at a 90 degree angle of timing on the cylinder head as illustrated.
35. Apply a multipurpose grease to a new oil seal lip and install the seal. Make sure to not turn the seal lip over and insert the seal until it stops. Remove any packing material from the seal surface.
36. Apply seal packing to the number 1 bearing cap as illustrated. Install the cap within 5 minutes of applying the packing. Do not let the seal come into contact with engine oil until at least two hours after it has been installed.

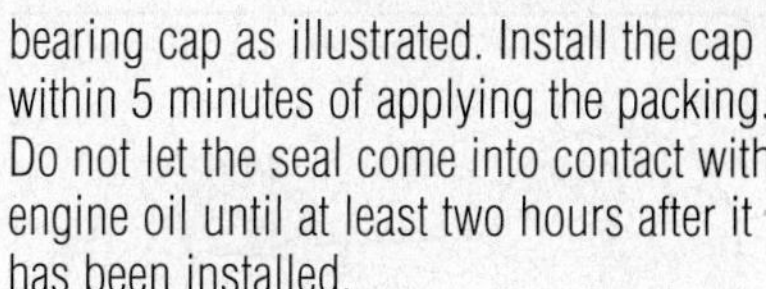

37. Install the number 4 camshaft bearing caps in their original locations, apply a light coat of oil to the cap bolt thread and using several passes; tighten the bolts to 12 ft. lbs. (16 Nm) in the sequence shown.
38. Align the timing marks on the number 3 camshaft as shown.
39. Install the number 3 camshaft bearing caps in their original locations, apply

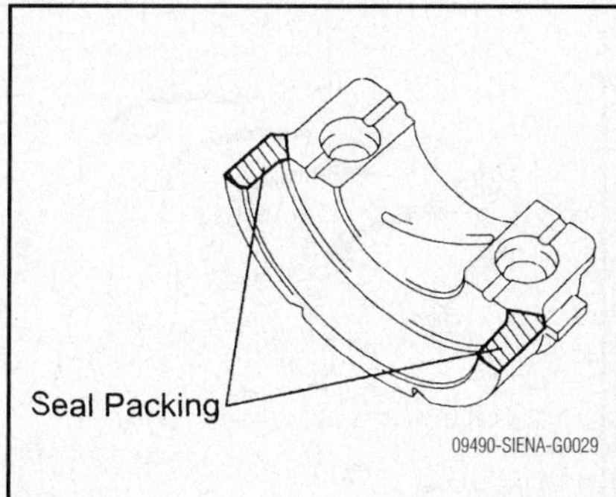

Fig. 23 Apply seal packing to the number 1 bearing cap at the locations shown

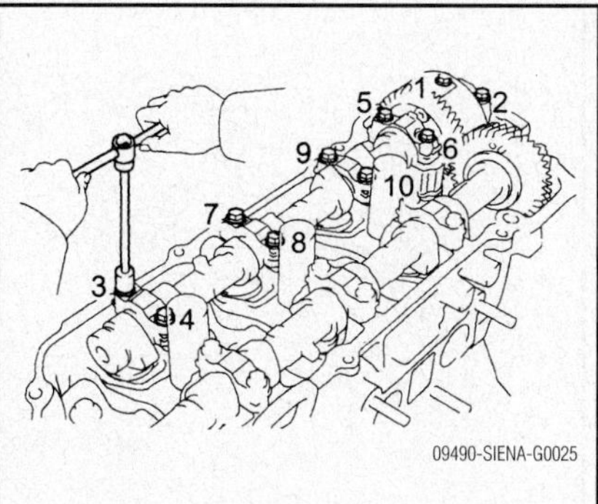

Fig. 19 Loosen the number 3 camshaft cap bolts in sequence

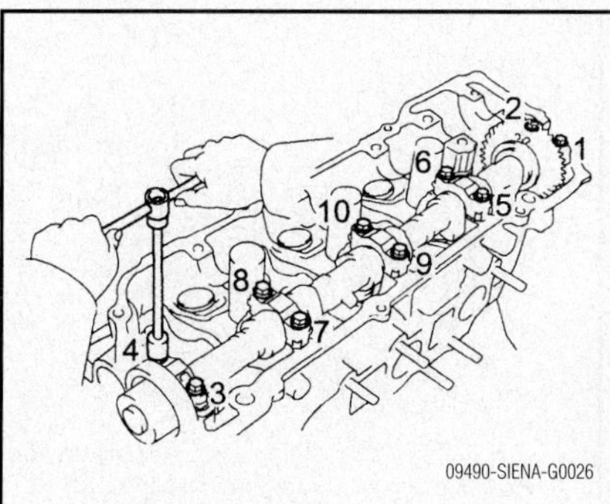

Fig. 20 Loosen the number 4 camshaft cap bolts in sequence

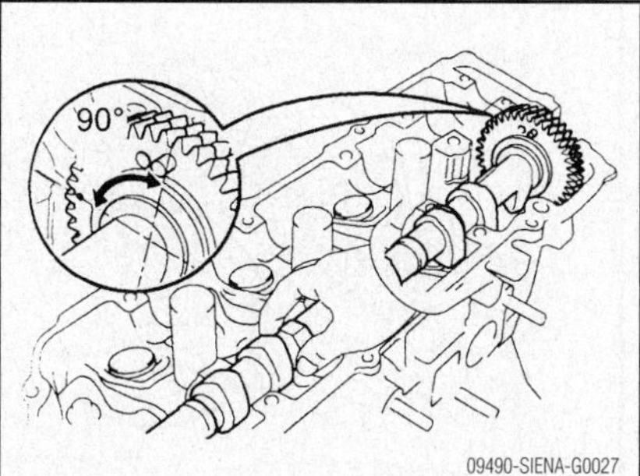

Fig. 21 Place the number 4 camshaft at a 90 degree angle of timing on the cylinder head

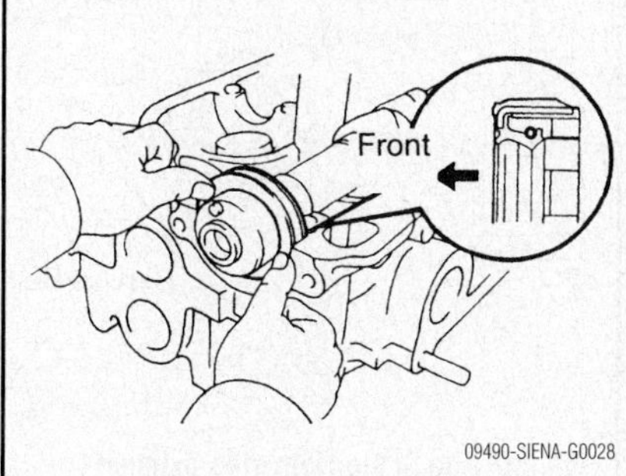

Fig. 22 Make sure to not turn the seal lip over and insert the seal until it stops

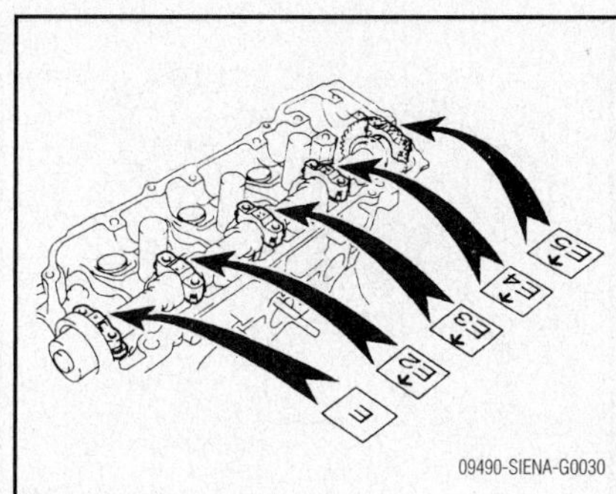

Fig. 24 Install the number 4 camshaft bearing caps in their original locations

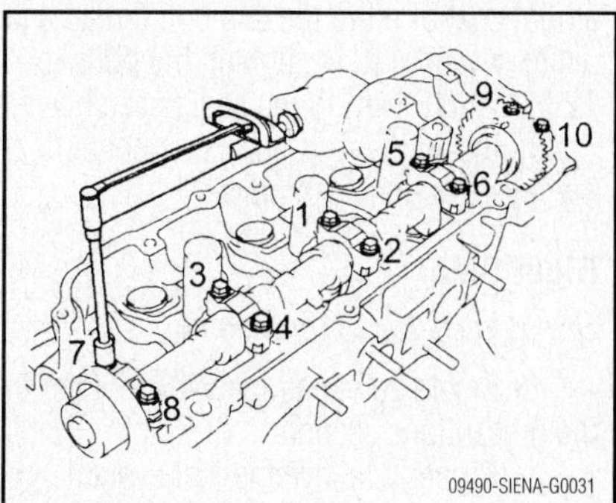

Fig. 25 Tighten the number 4 camshaft bearing caps in this sequence

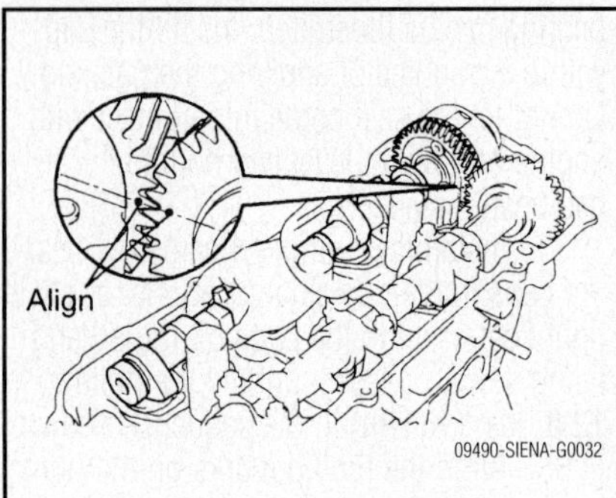

Fig. 26 Align the timing marks on the number 3 camshaft as shown

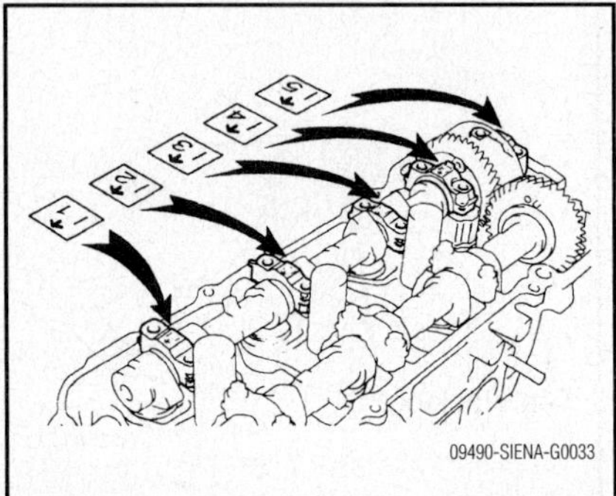

Fig. 27 Install the number 3 camshaft bearing caps in their original locations

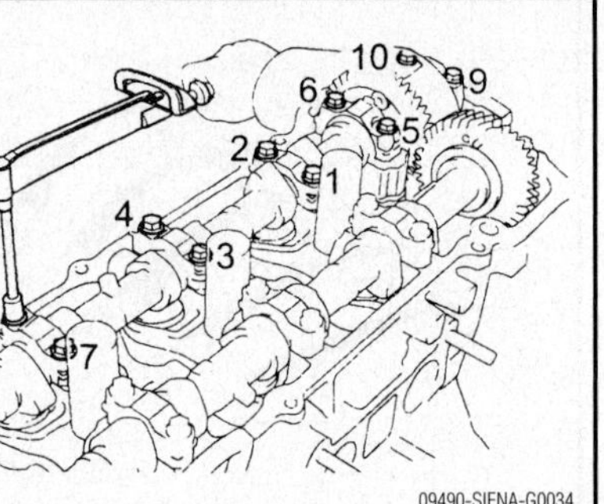

Fig. 28 Tighten the number 3 camshaft bearing caps in this sequence

a light coat of oil to the cap bolt thread and using several passes; tighten the bolts to 12 ft. lbs. (16 Nm) in the sequence shown.

40. Install the remaining components in the reverse order of removal.

Right Side

See Figures 15, 29 through 40.

1. Before servicing the vehicle, refer to the precautions section.
2. Properly relieve the fuel system pressure.
3. Drain and recycle the engine coolant.
4. Disconnect the negative battery cable.
5. Remove the left hand front wheel.
6. Remove the wiper arms.
7. Remove the cowl panel.
8. Remove the wiper link assembly.
9. Remove the cowl top to inner brace.
10. Remove the cowl top cover outer sub assembly.
11. Remove the radiator inlet hose.
12. Remove the ignition coil.
13. Remove the cylinder head cover.
14. Remove the apron seal.
15. Remove the A/C compressor to crankshaft pulley belt.
16. Remove the vane pump belt.
17. Remove the engine moving control rod bolts and rod.
18. Remove the number 2 engine mounting stay.
19. Remove the number 2 alternator bracket and the alternator belt adjusting bar.
20. Remove the crankshaft pulley.
21. Remove the timing belt covers 1 and 2. Refer to the timing belt removal and installation procedure.
22. Remove the right hand engine mount bracket.
23. Remove the timing belt guide number 2. Refer to the timing belt removal and installation procedure.
24. Remove the timing belt.
25. Remove the timing belt guide number 2 idler sub assembly. Refer to the timing belt removal and installation procedure.
26. Remove the camshaft pulley and lower timing cover. Refer to the timing belt removal and installation procedure.
27. Remove the timing belt number 3 cover. Refer to the timing belt removal and installation procedure.
28. Align the timing marks of the camshaft drive and driven gears by using a wrench to turn the camshaft.
29. Attach the exhaust camshaft sub gear to the main gear with a 16–20mm long x 6mm thread diameter bolt and tighten to 48 inch lbs. (5 Nm).

➡When removing the camshaft, be sure the torsional spring force of the sub gear has been taken up by the service bolt.

30. Mark the bearing caps prior to removal so they can be installed in their original positions.
31. Loosen the number 1 camshaft cap bolts in sequence using several steps.

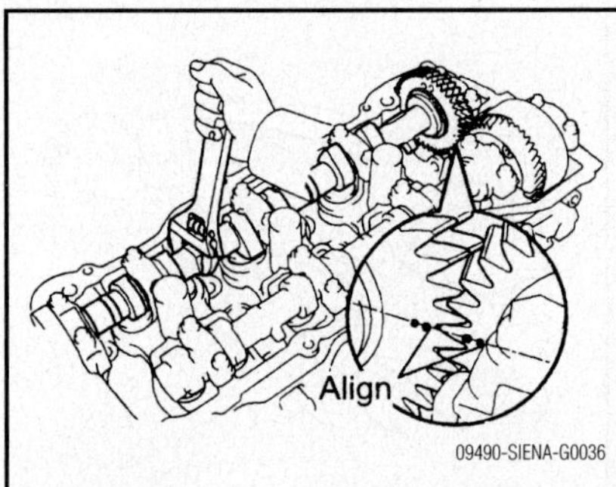

Fig. 29 Align the timing marks of the right side camshaft drive and driven gears by using a wrench to turn the camshaft

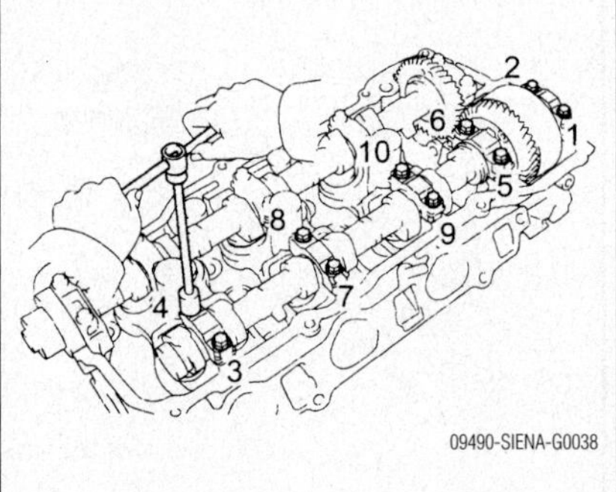

Fig. 31 Loosen the number 1 camshaft cap bolts in sequence

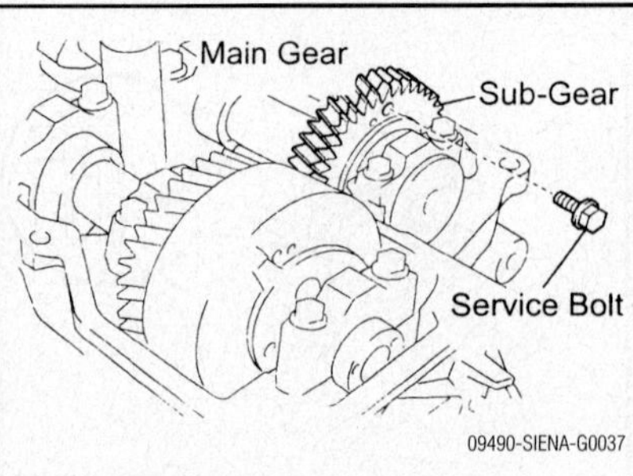

Fig. 30 Attach the right side exhaust camshaft sub gear to the main gear with a 16–20mm long x 6mm thread diameter bolt

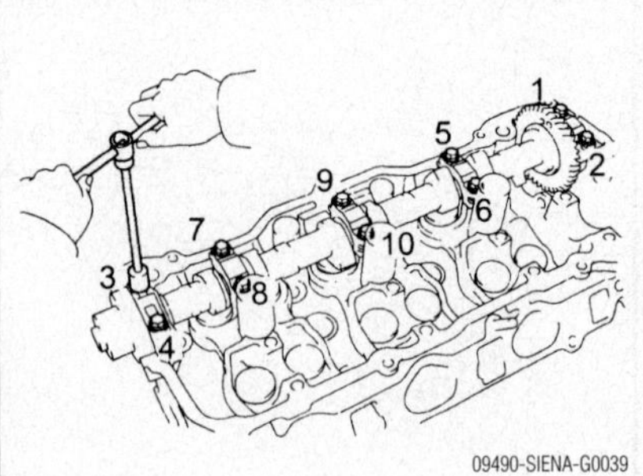

Fig. 32 Loosen the number 2 camshaft cap bolts in sequence

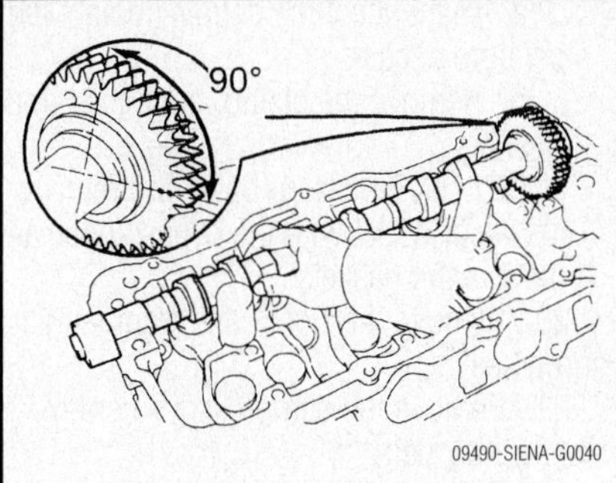

Fig. 33 Place the number 2 camshaft at a 90 degree angle of timing on the cylinder head

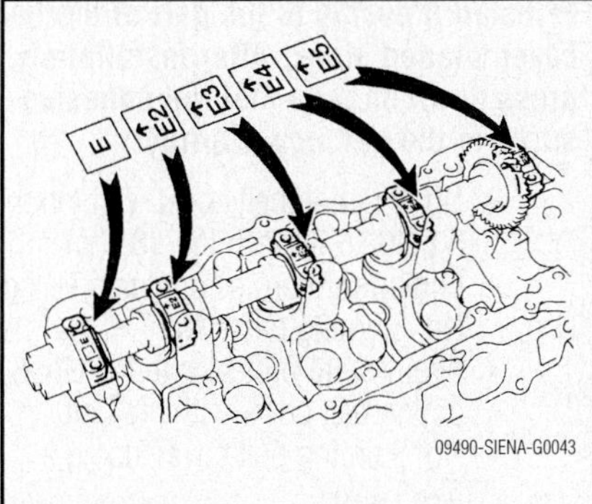

Fig. 36 Install the number 2 camshaft bearing caps in their original locations

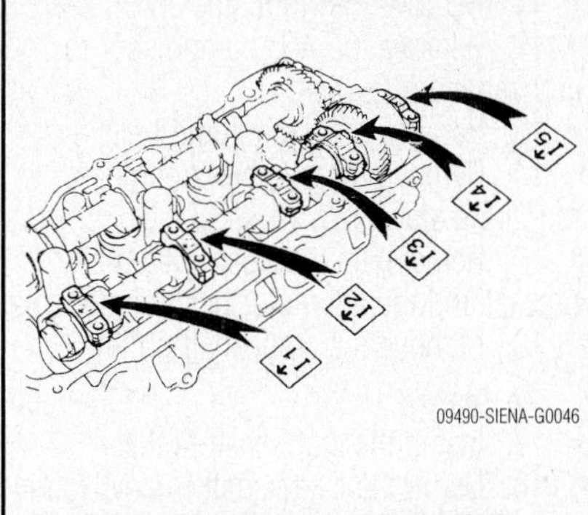

Fig. 39 Install the number 1 camshaft bearing caps in their original locations

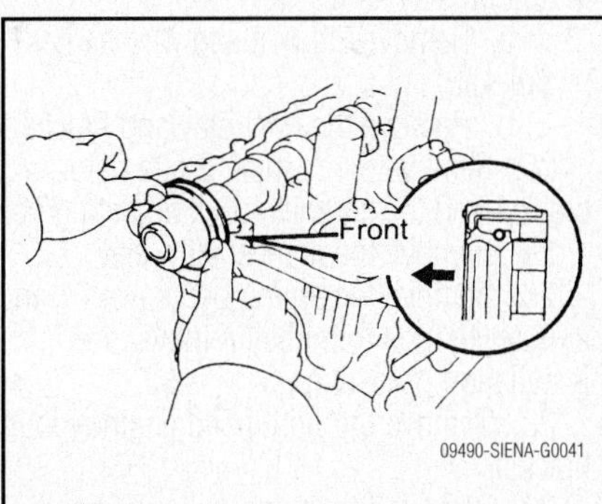

Fig. 34 Make sure to not turn the seal lip over and insert the seal until it stops

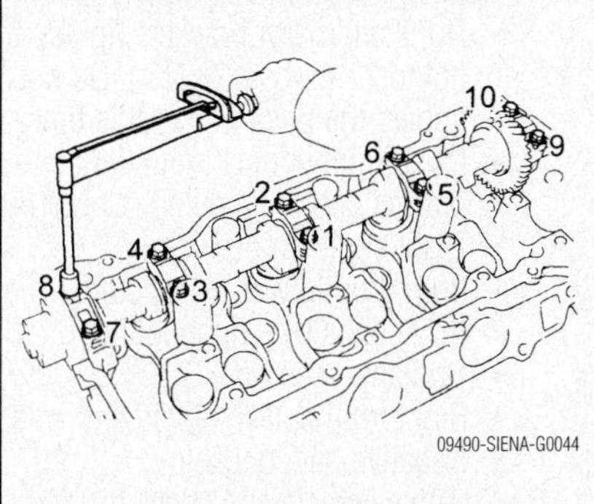

Fig. 37 Tighten the number 2 camshaft bearing caps in this sequence

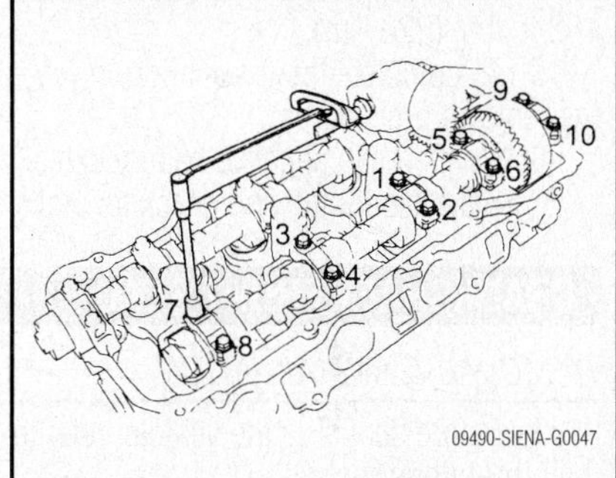

Fig. 40 Tighten the number 1 camshaft bearing caps in this sequence

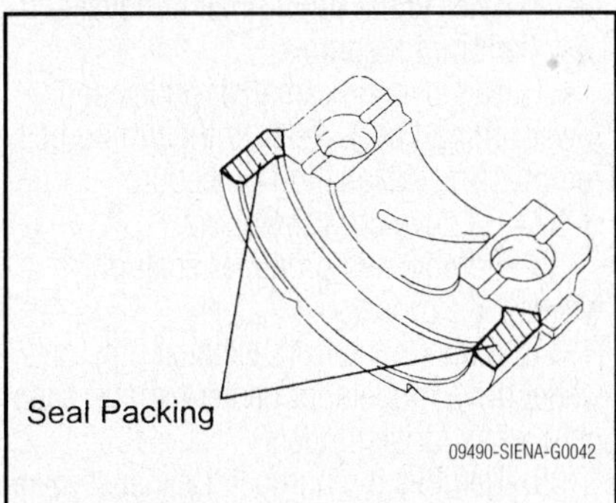

Fig. 35 Apply seal packing to the number 1 bearing cap as illustrated

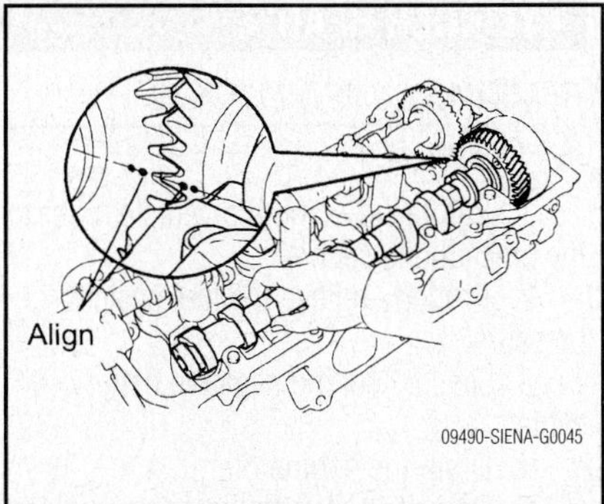

Fig. 38 Align the timing marks on the number 1 camshaft as shown

32. Loosen the number 2 camshaft cap bolts in sequence using several steps.

To install:

⁂ WARNING

The camshaft has a small thrust clearance, make sure to keep the camshaft level during installation to avoid damage to the camshaft and cylinder head.

33. Apply clean engine oil to the camshaft thrust and journal locations.

34. Place the number 2 camshaft at a 90 degree angle of timing on the cylinder head as illustrated.

35. Apply a multipurpose grease to a new oil seal lip and install the seal. Make sure to not turn the seal lip over and insert the seal until it stops. Remove any packing material from the seal surface.

36. Apply seal packing to the number 1 bearing cap as illustrated. Install the cap within 5 minutes of applying the packing. Do not let the seal come into contact with engine oil until at least two hours after it has been installed.

37. Install the number 2 camshaft bearing caps in their original locations, apply a light coat of oil to the cap bolt thread and using several passes; tighten the bolts to 12 ft. lbs. (16 Nm) in the sequence shown.

38. Align the timing marks on the number 1 camshaft as shown.

39. Install the number 1 camshaft bearing caps in their original locations, apply a light coat of oil to the cap bolt thread and using several passes; tighten the bolts to 12 ft. lbs. (16 Nm) in the sequence shown.

40. Install the remaining components in the reverse order of removal.

CRANKSHAFT DAMPER

REMOVAL & INSTALLATION

1. Before servicing the vehicle, refer to the precautions section.
2. Properly relieve the fuel system pressure.
3. Drain and recycle the engine coolant.
4. Disconnect the negative battery cable.
5. Remove the right hand front wheel.
6. Remove the wiper arms.
7. Remove the cowl panel.
8. Remove the wiper link assembly.
9. Remove the cowl top to inner brace.
10. Remove the cowl top cover outer sub assembly.

11. Remove the right side apron seal.
12. Remove the A/C compressor to crankshaft pulley belt.
13. Remove the vane pump belt.
14. Remove the number 2 engine mounting stay.
15. Remove the number 2 alternator bracket and the alternator belt adjusting bar.
16. Remove the crankshaft pulley.

To install:

17. Install the crankshaft pulley.
18. Using the Crankshaft Pulley Holding tool 09213-54015, Bolt tool 91651-60855 and Companion Flange Holding tool 09330-00021, or equivalent, install the crankshaft pulley bolt and torque the bolt to 162 ft. lbs. (220 Nm).
19. To complete the installation, reverse the removal procedures.
20. Connect the negative battery cable.
21. Start the engine and check for leaks.

CRANKSHAFT FRONT SEAL

REMOVAL & INSTALLATION

1. Before servicing the vehicle, refer to the precautions section.
2. Remove or disconnect the following:
   - Engine coolant reservoir tank and the alternator belt
   - Right front wheel and the splash shield
   - Power steering pump drive belt, by loosening both bolts
   - Both ground wire connectors
   - Right engine mounting stay
   - Engine moving control rod and the No. 2 right engine mount bracket

➡To extract the engine bracket and control rod, raise the engine slightly.

   - No. 2 alternator bracket
   - Crankshaft pulley bolt, using a pry-bar and wrench or Crankshaft Pulley Holding tool 09213-54015 and Flange Holding tool 09330-00021
   - Crankshaft pulley, using a puller
   - No. 1 timing belt cover
3. Remove the No. 2 timing belt cover, as follows:
   - Engine wire protector from the No. 3 (rear) timing belt cover
   - Engine wire protector clamp from the No. 3 timing belt cover
   - 5 bolts from the No. 2 timing belt cover
   - No. 2 cover

To install:

4. Install or connect the following:
   - No. 2 timing belt cover, using a new gasket

➡Install it evenly to the part of the belt cover shaded black. After installation, press down on it so that the adhesive sticks to the belt cover firmly.

   - No. 2 timing belt cover. Tighten the 5 bolts to 74 inch lbs. (8 Nm).
   - Engine wire protector clamp to the No. 3 timing belt cover
   - Engine wire protector to the No. 3 timing belt cover with the bolt
   - No. 3 timing belt cover, using a new gasket
   - Tighten the 4 No. 1 timing belt cover bolts to 74 inch lbs. (8 Nm).
   - Crankshaft pulley. Tighten the bolt to 159 ft. lbs. (215 Nm).
   - No. 2 alternator bracket. Tighten the nut to 21 ft. lbs. (28 Nm). Do not tighten the pivot bolt at this time.
   - No. 2 right engine mounting bracket and the moving control rod
   - Right engine mount stay
   - Both ground wire connectors
   - Drive belts by adjusting them
   - Coolant reservoir
   - Right front splash shield and wheel
   - Negative battery cable
5. Start the vehicle and check for any leaks.
6. Recheck the ignition timing.

CYLINDER HEAD

REMOVAL & INSTALLATION

See Figures 41 and 42.

1. Before servicing the vehicle, refer to the precautions section.
2. Properly relieve the fuel system pressure.
3. Disconnect the negative battery cable.
4. Drain the engine oil.
5. Drain and recycle the engine coolant.
6. Remove the right hand front wheel.
7. Remove the wiper arms.
8. Remove the cowl panel.
9. Remove the wiper link assembly.
10. Remove the cowl top to inner brace.
11. Remove the cowl top cover outer sub assembly.
12. Remove the engine cover.
13. Remove the air cleaner assembly.
14. Remove the air surge tank.
15. Disconnect the fuel pipe sub assembly.
16. Disconnect the heater hose inlet pipe.
17. Remove the intake manifold.
18. Disconnect the radiator inlet hose.
19. Disconnect the water outlet as follows:
    a. Disconnect the engine coolant temperature sensor.
    b. Remove the clamp, bolts nuts and washers.
    c. Lock the clamp open and remove the water outlet with the bypass hose and remove the gaskets.
20. Remove the right hand front fender apron seal.
21. Remove the compressor to crankshaft pulley belt.
22. Remove the vane pump belt.
23. Remove the engine moving control rod.
24. Remove the number 2 engine mounting stay as follows:
    a. Remove the bolt and wire harness bracket.
    b. Remove the bolt, stay and bracket.
25. Remove the number 2 alternator bracket and the alternator belt adjusting bar.
26. Remove the crankshaft pulley.
27. Remove the timing belt covers 1 and 2. Refer to the timing belt removal and installation procedure.
28. Remove the right hand engine mount bracket.
29. Remove the timing belt guide and timing belt. Refer to the timing belt removal and installation procedure.
30. Remove the timing belt guide number 2. Refer to the timing belt removal and installation procedure.
31. Remove the camshaft pulley and lower timing cover. Refer to the timing belt removal and installation procedure.
32. Remove the vane pump.
33. Remove the propeller shaft on 4 wheel drive models.
34. Remove the front exhaust pipe on 2 wheel drive models, or center exhaust pipe on 4 wheel drive models.
35. Remove the number 1 exhaust manifold heat insulator.
36. Remove the manifold stay.
37. Disconnect the oxygen sensor and remove the right hand exhaust manifold.
38. Disconnect the front exhaust pipe assembly.
39. Remove the number 3 manifold converter and number e exhaust manifold hear insulator.
40. Disconnect the oxygen sensor and remove the left hand exhaust manifold.
41. Remove the dipstick tube.
42. Remove the ignition coil.
43. Remove the cylinder head covers.
44. Remove the camshafts and their sub assemblies.
45. Disconnect the VVT sensor connector.
46. Disconnect the camshaft timing oil

control valve connector and wire harness clamp.

47. Remove the hexagon bolt on the right hand side head.

48. On the left hand side, make sure to disconnect the ground cable, wire harness clamp bracket and water inlet. Remove the hexagon bolt.

49. Be sure the engine is at/or near ambient temperature and remove the 2 (1 on each head) 8mm recessed hex bolts. Loosen and remove the 8 head bolts evenly, in 3 passes, in the reverse order of the installation sequence. Carefully lift the head from the engine; if necessary to pry the head loose, take great care not to damage the mating surfaces. Place the head on wood blocks in a clean work area.

➡If the cylinder head bolts are loosened out of sequence, warpage or cracking could result.

50. Remove the cylinder head gasket. With a gasket scraper, carefully remove all the old gasket material from the cylinder head and engine block surfaces.

To install:

51. Place the new cylinder head gasket onto the cylinder block.

52. Install the cylinder head, in

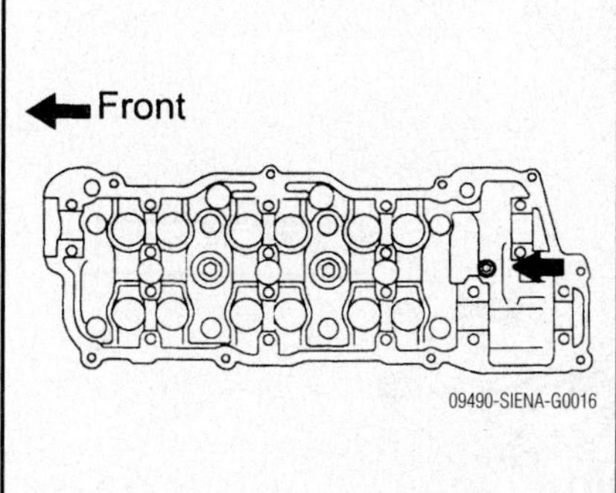

Fig. 41 Remove the hexagon bolt on the right hand side head

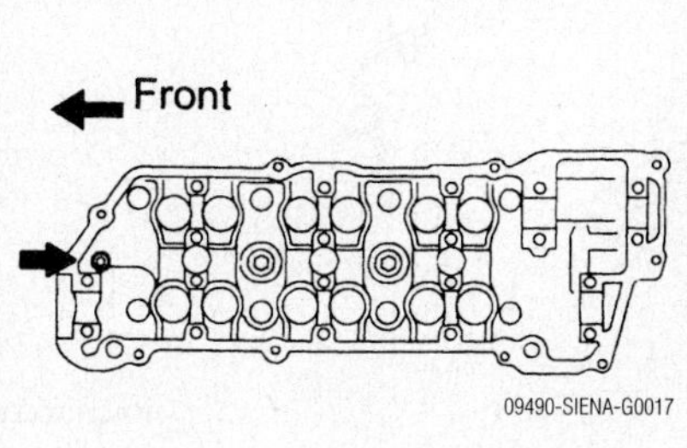

Fig. 42 Remove the hexagon bolt on the left hand side head

sequence, using several steps, as follows:

- Cylinder head onto the gasket
- Cylinder head bolts lubricated with clean engine oil
- Tighten the bolts in sequence in 3 steps to 40 ft. lbs. (54 Nm).

➡If any bolt does not meet the torque, replace it.

- Mark the forward edge of each bolt with paint, then tighten each bolt, in proper sequence, an additional 90 degrees.
- Check that each painted mark is now at a 90 degrees angle to the front

➡The paint mark applied to the bolt in the 9 o'clock position and should now be in the 12 o'clock position.

- Tighten the hexagon bolts, lubricated with engine oil. Tighten to 14 ft. lbs. (18 Nm).

53. Tighten the water outlet bolt to 14 ft. lbs. (18 Nm).

54. Install the camshafts.

55. Check and adjust the valves.

56. Apply sealant to the cylinder heads where the camshaft supports meet the cylinder heads. Install the cylinder head covers, using new gaskets.

57. Install the remaining components in the reverse order of removal.

58. Fill the engine with correct grade and amount of oil.

59. Fill the cooling system.

60. Start the engine and check for leaks.

61. Bleed the air from the cooling system.

62. Road test the vehicle and check for unusual noise, shock, slippage, correct shift points and smooth operation.

63. Recheck the coolant and engine oil levels.

ENGINE ASSEMBLY

REMOVAL & INSTALLATION

See Figures 43 through 56.

1. Before servicing the vehicle, refer to the precautions section.
2. Properly relieve the fuel system pressure.
3. Disconnect the negative battery cable.
4. Remove the right hand front wheel.
5. Remove the engine under cover.
6. Remove the fender liners.
7. Remove the apron seals.
8. Drain and recycle the engine coolant.
9. Drain the engine oil.
10. Drain the transaxle fluid.
11. If equipped, drain the transfer oil.
12. Remove the wiper arms.
13. Remove the cowl panel.
14. Remove the wiper link assembly.
15. Remove the cowl top to inner brace.
16. Remove the cowl top cover outer sub assembly.
17. Remove the engine cover.
18. Remove the battery.
19. Remove the air cleaner assembly and bracket.
20. Remove the A/C compressor to crankshaft pulley belt.
21. Remove the alternator.
22. Remove the engine moving control rod bolts and rod.
23. Remove the number 2 engine mounting stay as follows:
    a. Remove the bolt and wire harness bracket.
    b. Remove the bolt, stay and bracket.
24. Remove the number 2 alternator bracket and the alternator belt adjusting bar.
25. Separate the compressor and magnetic clutch.
26. Separate the transmission control assembly as follows:
    a. Remove the nut from the control shaft lever and disconnect the control cable assembly from the lever.
    b. Remove the clip and disconnect the cable from the bracket.
    c. Disconnect the cable assembly from the transmission bracket.
    d. Remove the nuts and remove the cable control assembly.
27. Disconnect the following hoses:
    - Union to check valve hose
    - Number 1 fuel vapor feed hose
    - Heater hoses
    - Radiator hoses
    - Oil cooler hoses
    - Oil reservoir to pump hose
28. Disconnect the steering gear outlet return tube.
29. Remove the glove box door.
30. Disconnect the harness from the PCM.
31. Disconnect the harness from the junction block by removing the nut, separating the harness, then using a screwdriver release the junction block and pull the wire upwards. Pull out the harness and remove the body ground.
32. On 4 wheel drive models, remove the propeller shaft.
33. On 4 wheel drive models, remove the center exhaust pipe assembly.
34. On 2 wheel drive models, remove the front exhaust pipe assembly.

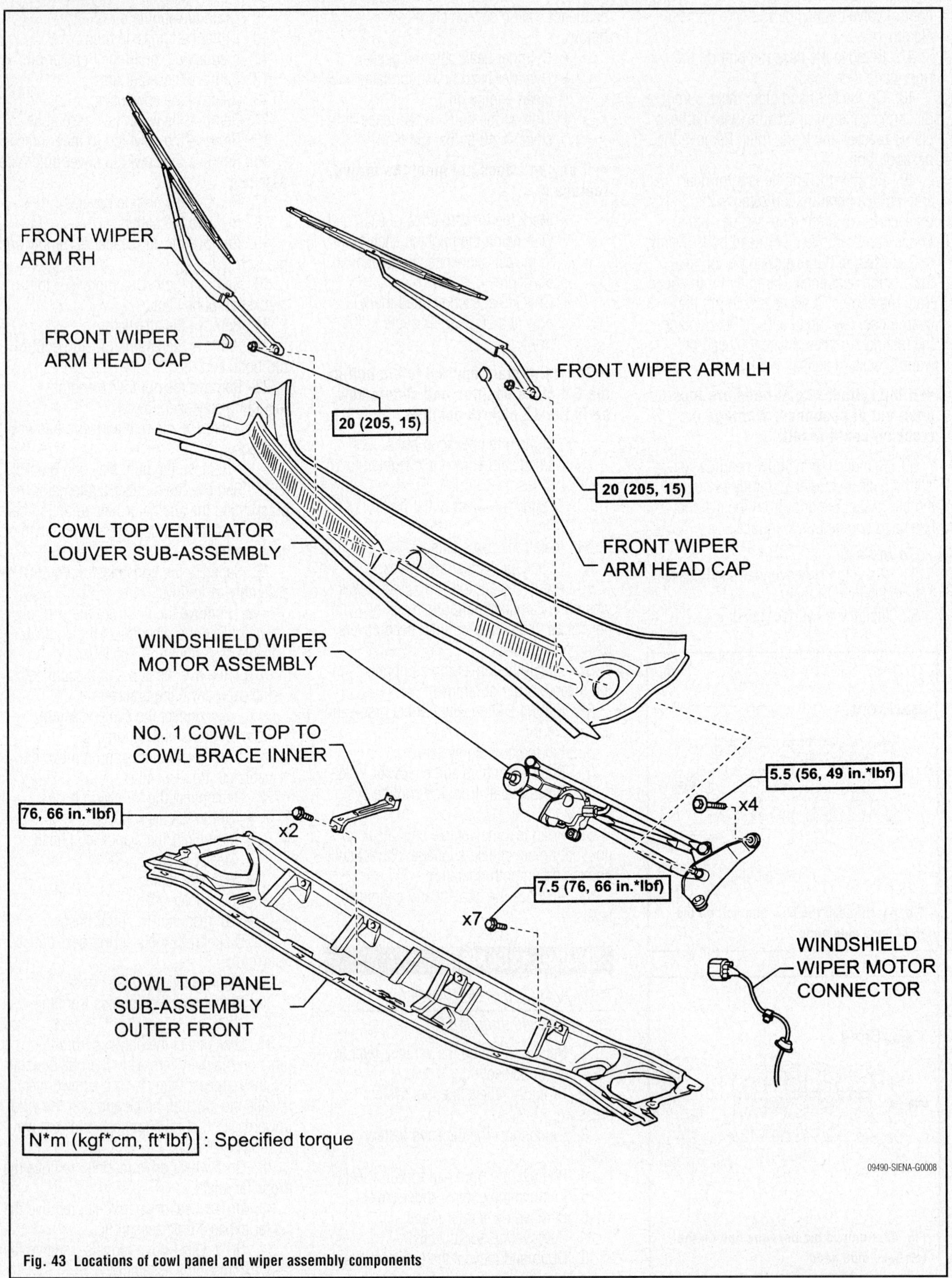

Fig. 43 Locations of cowl panel and wiper assembly components

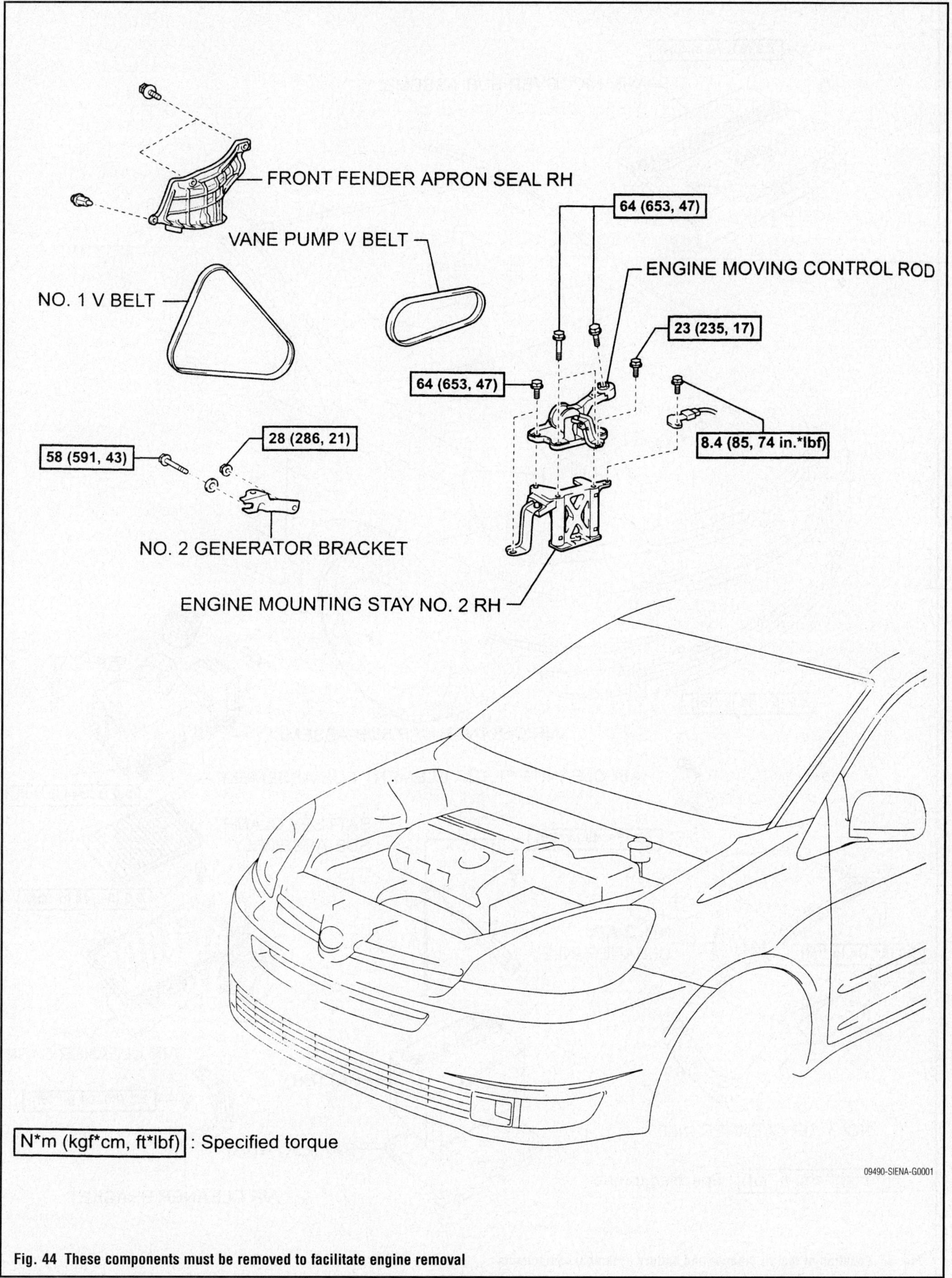

Fig. 44 These components must be removed to facilitate engine removal

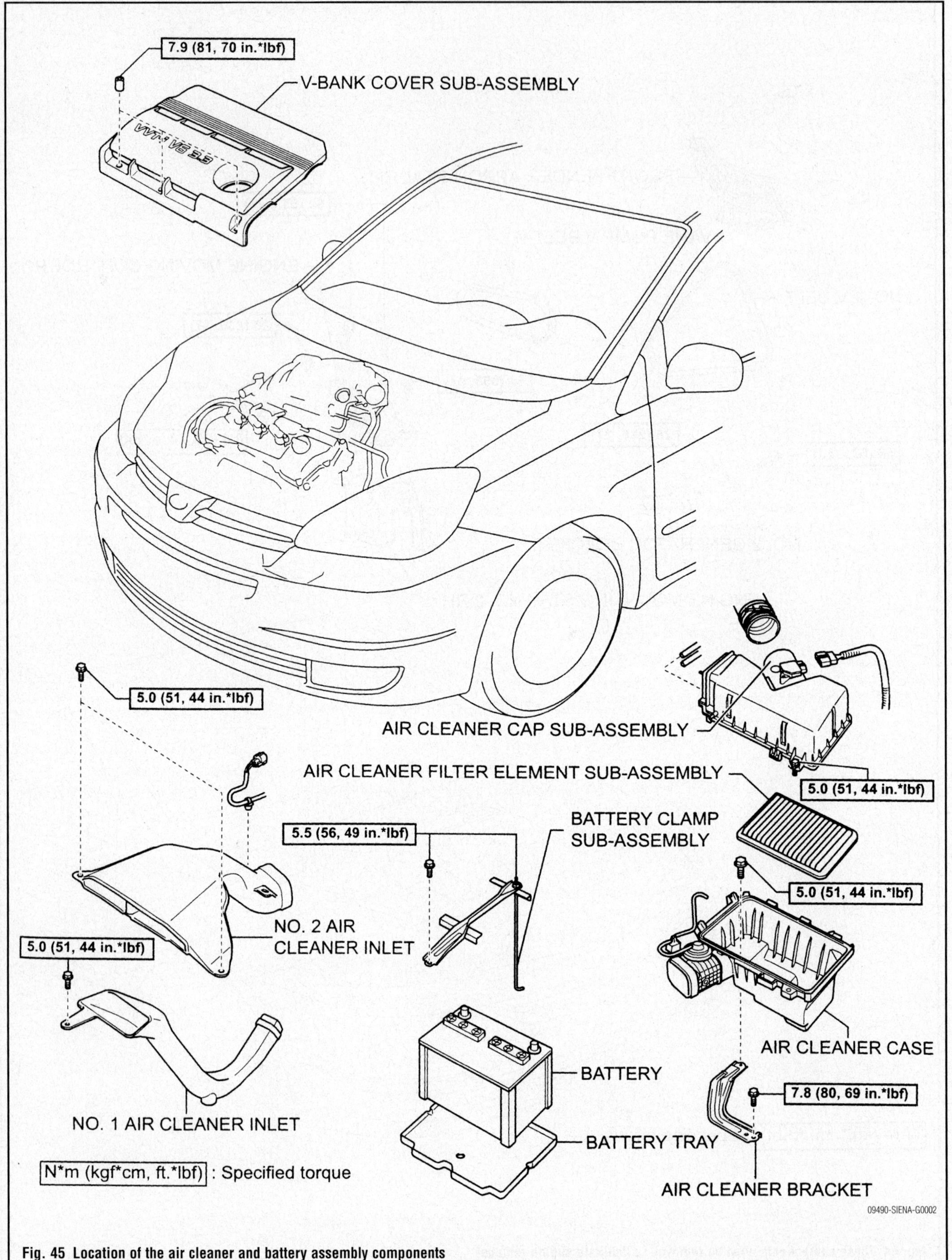

Fig. 45 Location of the air cleaner and battery assembly components

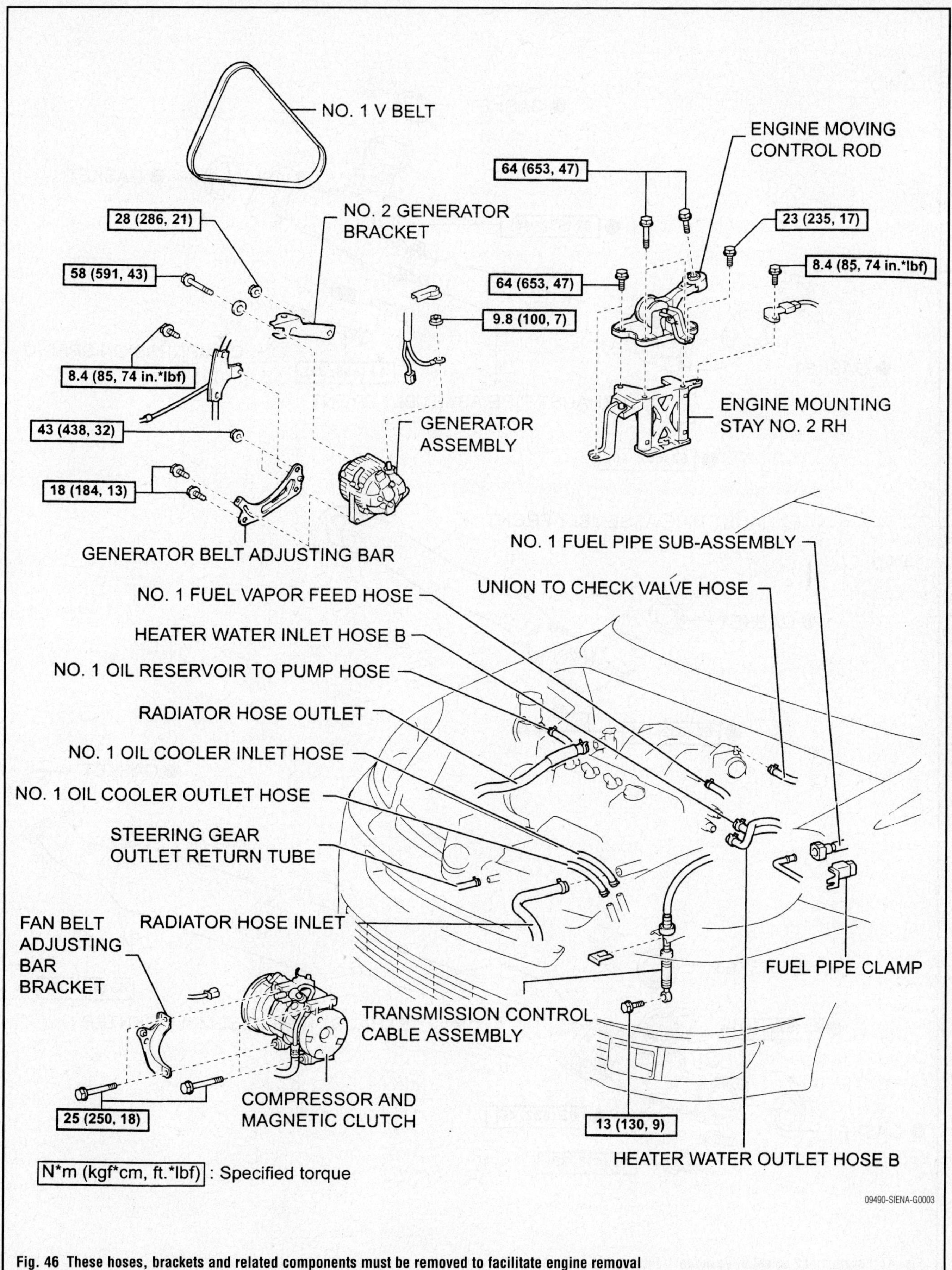

Fig. 46 These hoses, brackets and related components must be removed to facilitate engine removal

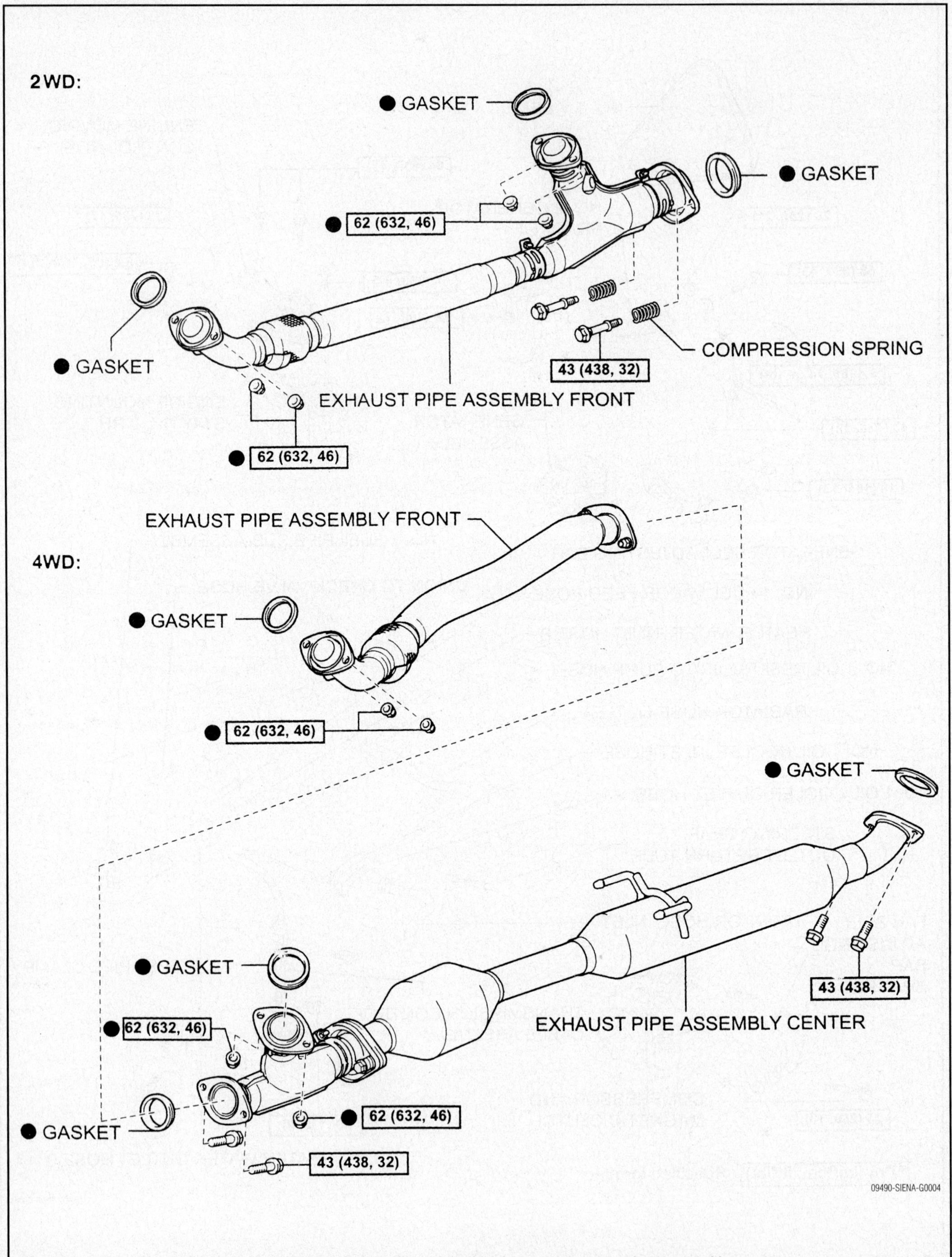

Fig. 47 Location of 2 wheel drive model front exhaust assembly and 4 wheel drive model center exhaust assembly

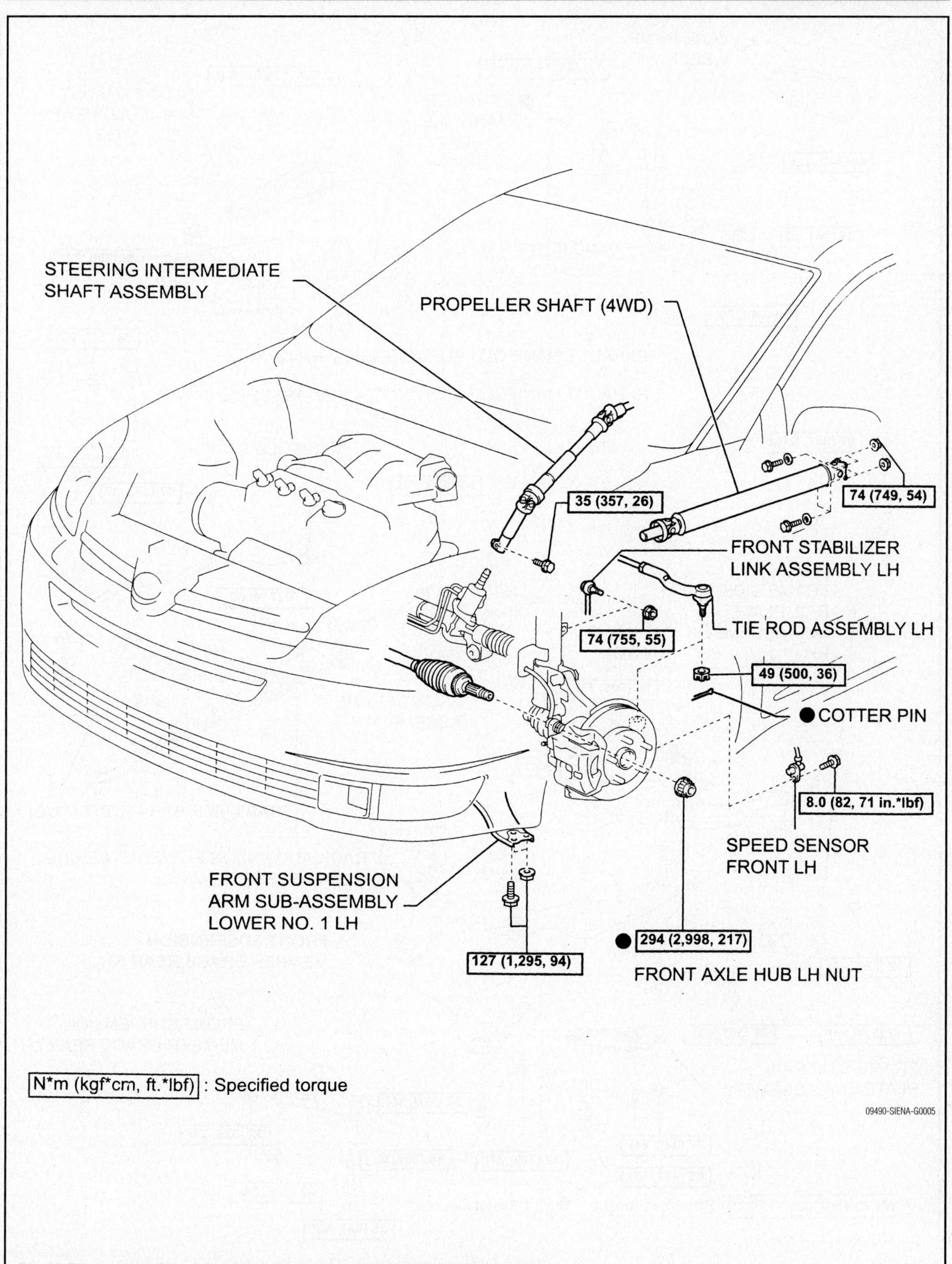

Fig. 48 These suspension components must be removed to facilitate engine removal

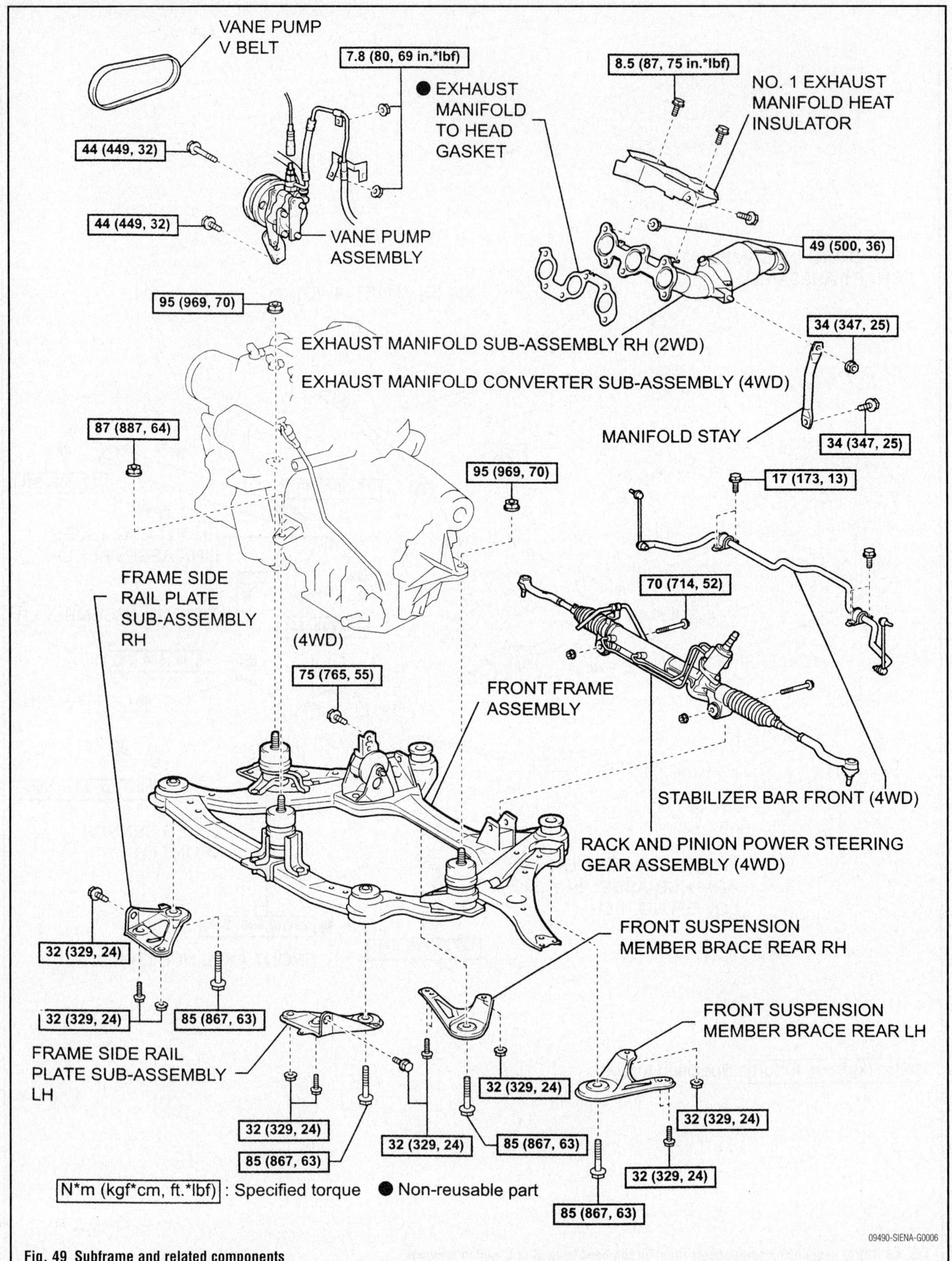

Fig. 49 Subframe and related components

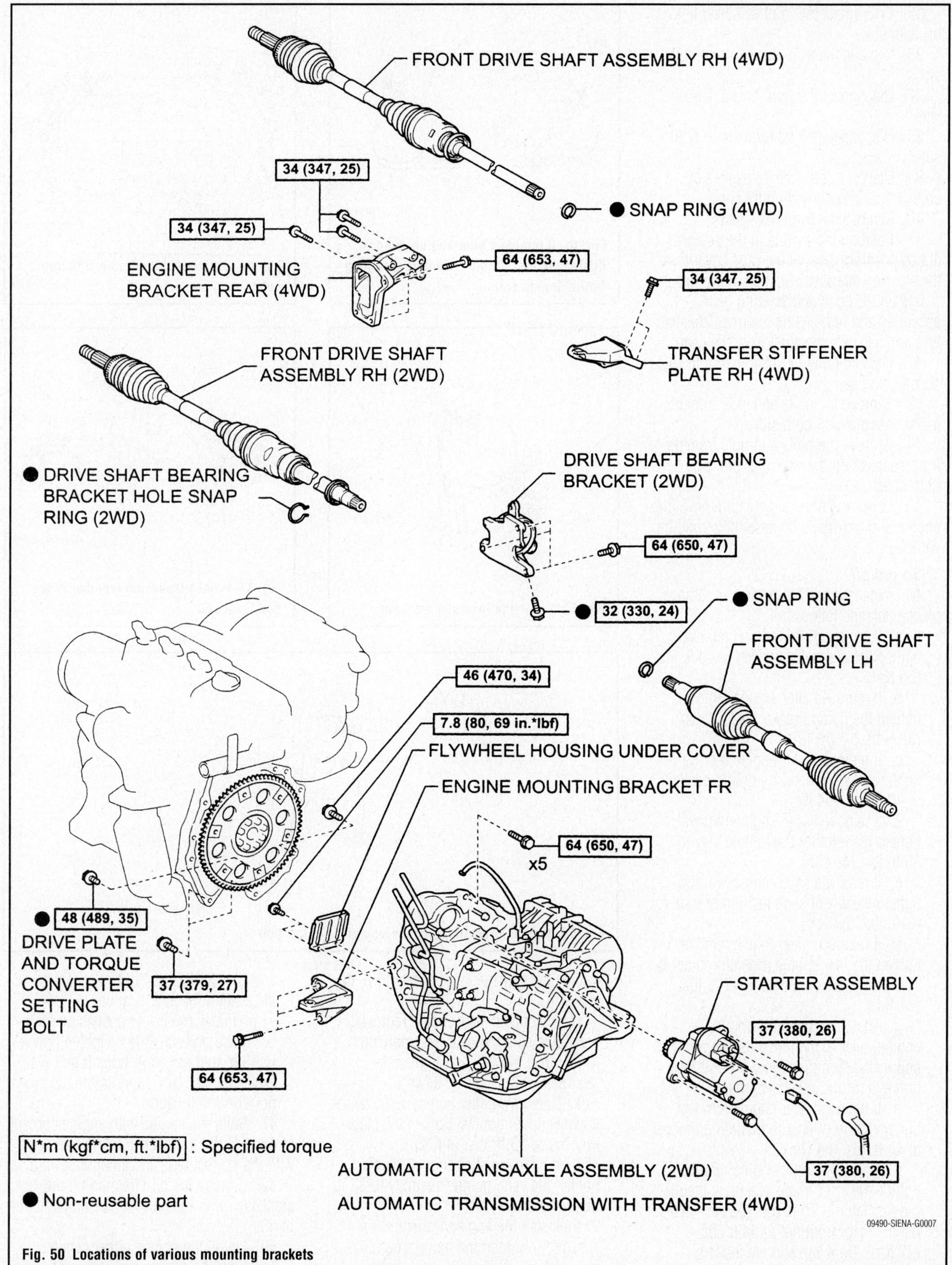

Fig. 50 Locations of various mounting brackets

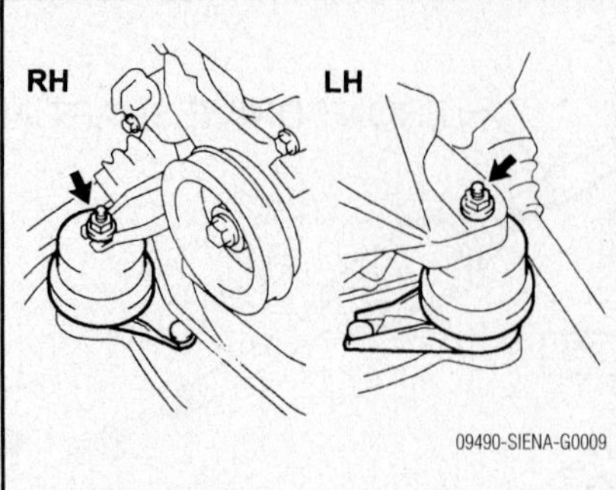

Fig. 51 If removed after engine removal, tighten the front frame assembly mounting insulator nuts and . . . .

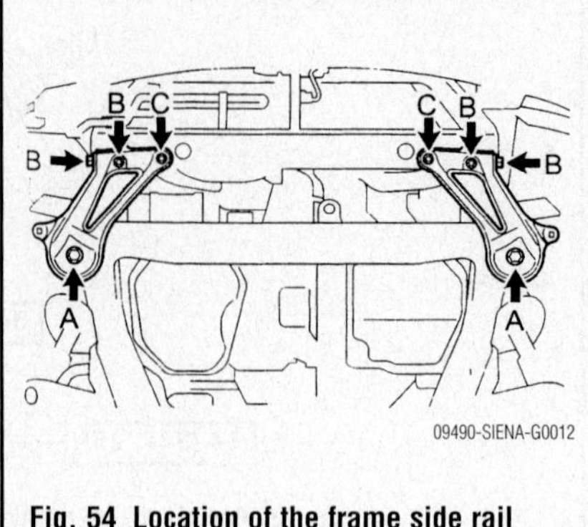

Fig. 54 Location of the frame side rail bolts

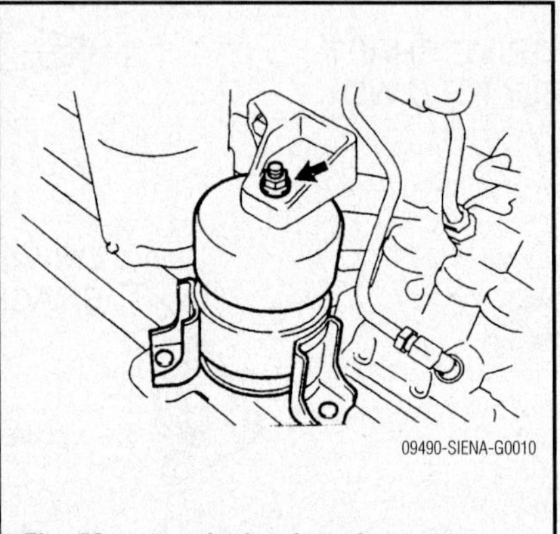

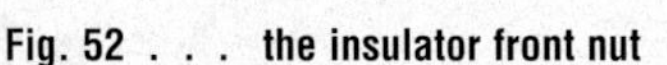
Fig. 52 . . . the insulator front nut

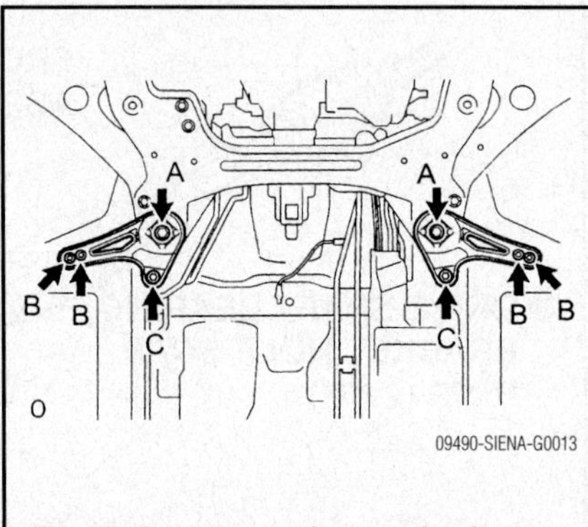

Fig. 55 Front suspension member brace bolt locations

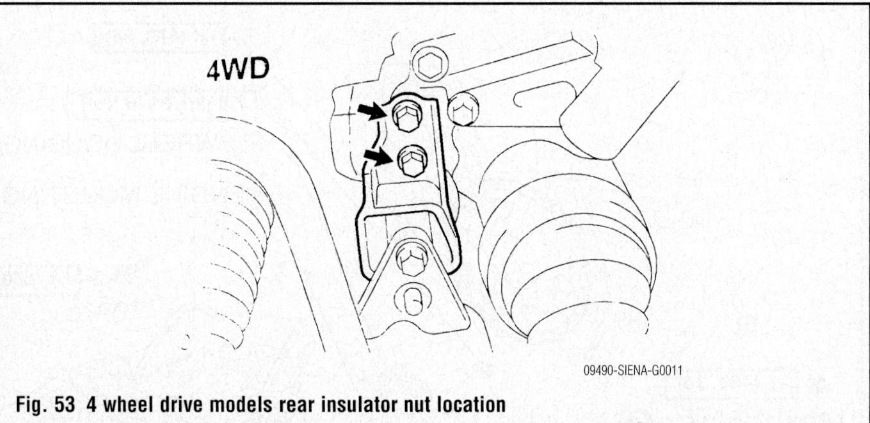

Fig. 53 4 wheel drive models rear insulator nut location

35. Disconnect the front stabilizer link on both sides.
36. Remove the axle hub nut from both sides.
37. Disconnect the front speed sensor wire.
38. Disconnect the tie rod end on both sides.
39. Disconnect the front suspension control arm assembly on both sides.
40. Remove the front half shafts.
41. Position the wheels in the straight ahead position, remove the dust cover from the steering intermediate shaft. make match marks on the shaft and steering gear assembly and remove the intermediate shaft bolt and separate the shaft from the gear.
42. Install a dolly or suitable tool to support the engine.
43. Remove the bolts and nuts from the frame side plate on both sides.
44. Remove the bolts and nuts from the front suspension member rear brace on both sides.
45. Lower the dolly slowly to remove the engine and transmission assembly from the vehicle.

To install:

46. Installation is the reverse of removal, please note the following:
    a. If removed after engine removal, tighten the alternator bracket to 43 ft. lbs. (58 Nm).
    b. If removed after engine removal, tighten the pump bracket to 24 ft. lbs. (32 Nm).
    c. If removed after engine removal, tighten the right hand mounting bracket to 40 ft. lbs. (54 Nm).
    d. If removed after engine removal, tighten the number 2 manifold stay to 36 ft. lbs. (49 Nm).
    e. If removed after engine removal, tighten the wheel shaft bearing bracket to 47 ft. lbs. (64 Nm).
    f. If removed after engine removal, tighten the rear engine mounting bracket on 4 wheel drive models to 47 ft. lbs. (64 Nm).
    g. Attach the transmission to the engine, refer to the transmission removal and installation procedure for bolt location and torque specifications.
    h. If removed after engine removal, tighten the front engine mounting bracket to 47 ft. lbs. (64 Nm).
    i. If removed after engine removal, tighten the front frame assembly mounting insulator to 70 ft. lbs. (95 Nm) and the insulator front nut to 64 ft. lbs. (87 Nm). On 4 wheel drive models tighten the rear insulator nut to 55 ft. lbs. (75 Nm).
    j. If removed after engine removal, install the steering gear assembly on 4 wheel drive models and tighten the retainers to 52 ft. lbs. (70 Nm).
    k. If removed after engine removal, on 4 wheel drive models, tighten the stabilizer bar to 13 ft. lbs. (44 Nm).
    l. If removed after engine removal, tighten the vane pump assembly to 32 ft. lbs. (44 Nm).
    m. Install the engine/transmission assembly. Tighten the frame side rail bolts A to 63 ft. lbs. (85 Nm), bolts B and C to 24 ft. lbs. (32 Nm). refer to the illustration for bolt location.
    n. Install the front suspension member brace on both sides. Tighten bolt A to 63 ft. lbs. (85 Nm), bolts B and C to 24 ft. lbs. (32 Nm). refer to the illustration for bolt location.
47. Refill the engine with engine coolant.
48. Fill the transmission and transfer with the correct type and amount of fluid.
49. Change the oil filter and fill the crankcase with the correct type and amount of oil.
50. Install the engine under covers.
51. Start the engine and check for leaks.

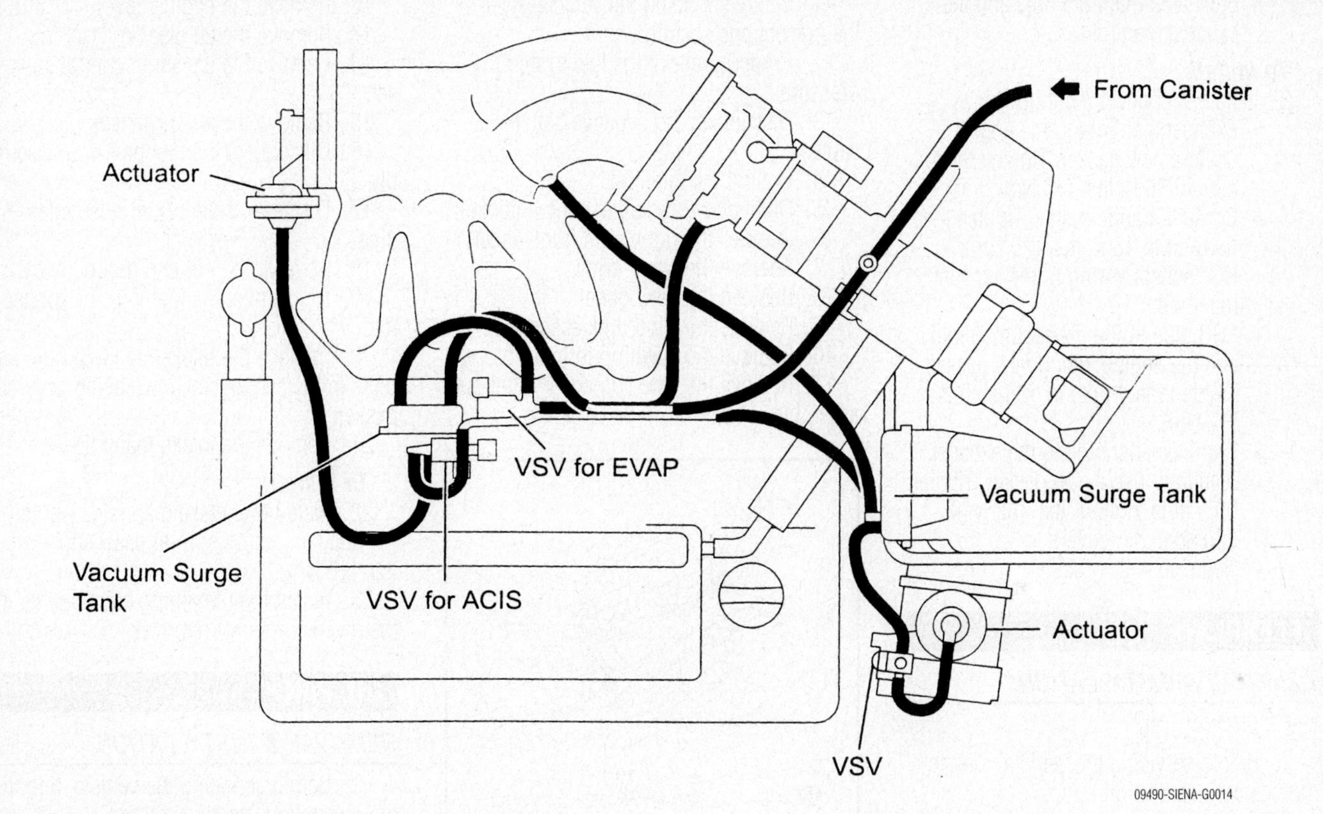

Fig. 56 Vacuum hose routing

EXHAUST MANIFOLD

REMOVAL & INSTALLATION

Front Manifold

See Figures 57 and 58.

➡Removing the oil filter helps gain access to a lower bolt in the front exhaust manifold.

1. Before servicing the vehicle, refer to the precautions section.
2. Remove or disconnect the following:
   - Negative battery cable
   - Engine under covers

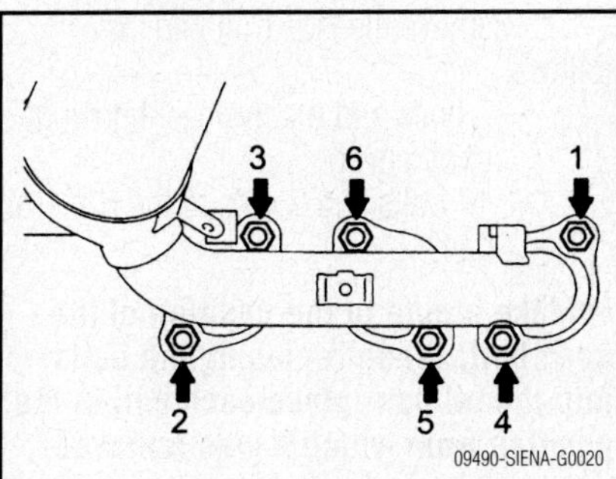

Fig. 57 Remove the exhaust manifold nuts and bolts in the sequence shown

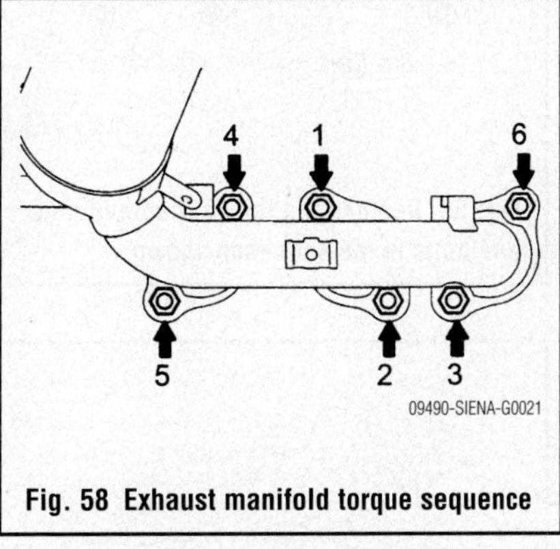

Fig. 58 Exhaust manifold torque sequence

- Front exhaust pipe from the exhaust manifolds, by removing the nuts

➡Check for access to some of the manifold lower bolts, if so remove any possible.

- Heated Oxygen (HO_2) sensor
- Exhaust manifold stay, by removing the bolt and nut
- Remaining exhaust manifold nuts; then, separate the exhaust manifold from the engine

To install:

3. Install or connect the following:
   - Exhaust manifold, using a new gasket. Uniformly, tighten the bolts to 36 ft. lbs. (49 Nm).
   - Exhaust manifold stay. Tighten the nut/bolt to 15 ft. lbs. (20 Nm).
   - Heated Oxygen (HO_2) sensor to the exhaust manifold
   - Front exhaust pipe to the exhaust manifold, using a new gasket. Tighten both nuts to 46 ft. lbs. (62 Nm).
   - Engine under covers
   - Negative battery cable

Rear Manifold

1. Before servicing the vehicle, refer to the precautions section.
2. Remove or disconnect the following:
   - Negative battery cable
   - Engine under covers
   - Front exhaust pipe from both exhaust manifolds, from below the engine
   - Exhaust Gas Recirculation (EGR) pipe from the rear exhaust manifold, by removing the 4 nuts
   - Heated Oxygen (HO_2) sensor wiring, from the right exhaust manifold
   - Exhaust manifold stay

- 6 exhaust manifold nuts and the exhaust manifold

To install:

3. Install or connect the following:
   - Exhaust manifold to the engine, using a new gasket. Tighten the 6 nuts to 36 ft. lbs. (49 Nm).
   - Exhaust manifold stay. Tighten the nut/bolt to 15 ft. lbs. (20 Nm).
   - HO_2 sensor wiring to the exhaust manifold
   - EGR pipe to the exhaust manifold and the engine, using new gaskets. Tighten the 4 nuts to 108 inch lbs. (12 Nm).
   - Front exhaust pipe to the exhaust manifold, use a new gasket. Tighten both nuts to 46 ft. lbs. (62 Nm).
   - Engine under covers
   - Negative battery cable

FLEXPLATE

REMOVAL & INSTALLATION

See Figure 59.

1. Before servicing the vehicle, refer to the precautions section.
2. Remove the transmission assembly.
3. Hold the crankshaft and unfasten the flywheel bolts.
4. Remove the flywheel.

To install:

5. Apply anti-lock compound to 2 or 3 threads of the flywheel bolts.
6. Install the front spacer, flywheel and rear plate.
7. Hold the crankshaft and tighten the bolts in a star pattern to 61 ft. lbs. (83 Nm).
8. Install the transaxle assembly.

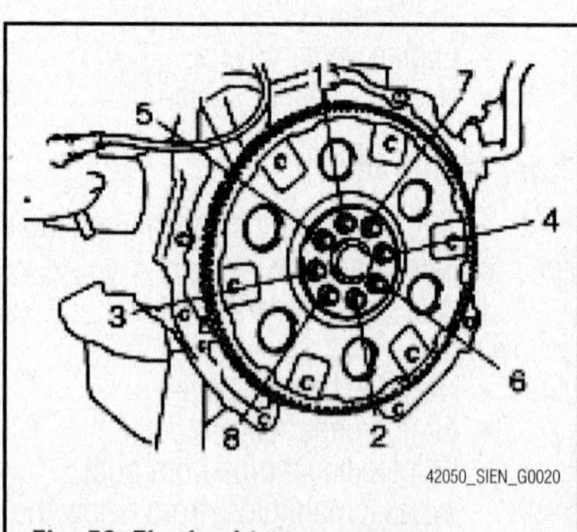

Fig. 59 Flywheel torque sequence—Sienna

INTAKE MANIFOLD

REMOVAL & INSTALLATION

See Figures 60 and 61.

1. Before servicing the vehicle, refer to the precautions section.
2. Properly relieve the fuel system pressure.
3. Disconnect the negative battery cable.
4. Drain the engine oil.
5. Drain and recycle the engine coolant.
6. Remove the right hand front wheel.
7. Remove the wiper arms.
8. Remove the cowl panel.
9. Remove the wiper link assembly.
10. Remove the cowl top to inner brace.
11. Remove the cowl top cover outer sub assembly.

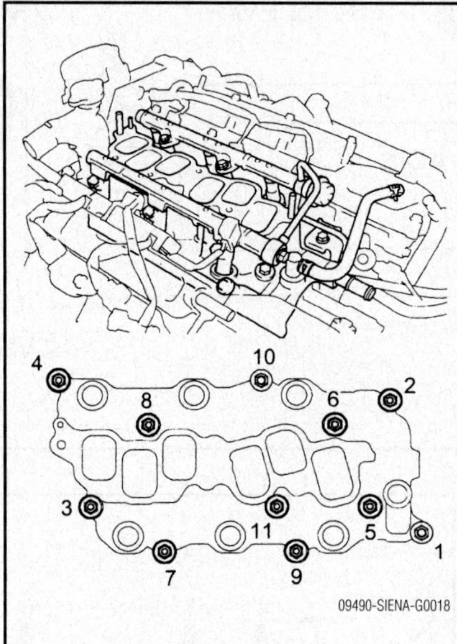

Fig. 60 Remove the intake manifold nuts and bolts in the sequence shown

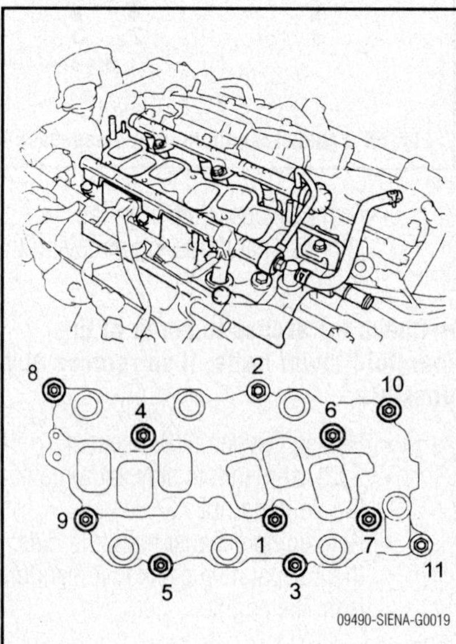

Fig. 61 Tighten the intake manifold nuts and bolts in the sequence shown

12. Remove the engine cover.
13. Remove the air cleaner assembly.
14. Remove the emission control valve set.
15. Remove the air surge tank.
16. Disconnect the fuel pipe sub assembly.
17. Disconnect the heater hose inlet pipe.
18. Remove the nut and ground cable.
19. Disconnect the fuel injector connectors.
20. Remove the intake manifold nuts and bolts in the sequence shown using several passes.
21. Remove the intake manifold.

To install:

22. Install the intake manifold. Tighten the retainers using several passes to 11 ft. lbs. (15 Nm).
23. Install the remaining components in the reverse order of removal.

OIL PAN

REMOVAL & INSTALLATION

1. Before servicing the vehicle, refer to the precautions section.
2. Remove or disconnect the following:
   - Right front wheel
   - Fender apron seal
   - Engine undercover
3. Drain the engine oil from the engine.
4. Remove or disconnect the following:
   - Front exhaust pipe
   - Front exhaust pipe bracket from the No. 1 oil pan
   - Flywheel housing undercover
   - 10 bolts and 2 nuts to the No. 2 oil pan
5. Insert the blade of the Oil Pan Seal Cutting tool 09032-00100 between the No. 1 and No. 2 oil pans. Clean the surfaces of the oil pans.
6. Remove or disconnect the following:
   - 3 oil strainer nuts and gasket
7. Remove the No. 1 oil pan, as follows:
   - 2 bolts and the flywheel housing undercover
   - 17 bolts and 2 nuts to the No. 1 oil pan

➡Make a note of the position of the each bolt. When replacing the bolts into the oil pan, place each bolt in the position from which it was removed.

   - Oil pan, by prying the portions between the cylinder block and the oil pan

➡**Be careful not to damage the contact surfaces.**

- Baffle plate from the No. 1 oil pan

To install:

8. Clean all mating surfaces of the oil pans.
9. Install the baffle plate to the No. 1 oil pan and tighten to 69 inch lbs. (8 Nm).
10. Install the No. 1 oil pan, as follows:
   a. Using a non residue solvent, clean both sealing surfaces to the oil pan.
   b. Apply liquid sealant to the oil pan and engine block.
   c. Install the oil pan with the 17 bolts and 2 nuts. Uniformly tighten the bolts and nuts in several passes.
   d. Tighten the No. 1 oil pan bolts, as follows:
      - 10mm head bolt: 69 inch lbs. (8 Nm)
      - 12mm head bolt: 14 ft. lbs. (20 Nm)
      - 14mm head bolt: 27 ft. lbs. (37 Nm)
   e. Install the flywheel housing undercover with the 2 bolts. Tighten the bolts to 69 inch lbs. (8 Nm).
11. Install the oil strainer with the 3 nuts. Tighten the nuts to 69 inch lbs. (8 Nm).
12. Install the No. 2 oil pan, as follows:
   a. Using a non residue solvent, clean both sealing surfaces to the oil pan.
   b. Apply liquid sealant to the oil pan and engine block.
   c. Install the No. 2 oil pan with the 10 bolts and 2 nuts. Uniformly tighten the bolts and nuts in several passes. Tighten the bolts to 69 inch lbs. (8 Nm).
13. Install or connect the following:
   - Flywheel housing undercover
   - Front exhaust pipe bracket to the No. 1 oil pan. Tighten the bolts to 15 ft. lbs. (21 Nm).
14. Install the front exhaust pipe, as follows:
   - Temporarily install the 3 new gaskets and the front exhaust pipe with the 2 bolts and 6 nuts
   - Tighten the 4 exhaust manifolds-to-front exhaust pipe nuts to 46 ft. lbs. (62 Nm).
   - Tighten the both front exhaust pipe-to-center exhaust pipe nuts/bolts to 41 ft. lbs. (56 Nm).
   - Bracket. Tighten both bolts to 14 ft. lbs. (19 Nm).
   - Support stay. Tighten both bolts to 22 ft. lbs. (29 Nm).
15. Install or connect the following:
   - Engine undercover
   - Right fender apron seal
   - Right front wheel
16. Fill the engine with oil.
17. Start the engine and check for leaks.

OIL PUMP

REMOVAL & INSTALLATION

See Figure 62.

1. Before servicing the vehicle, refer to the precautions section.
2. Remove or disconnect the following:
   - Oil pan
   - Crankshaft Position (CKP) sensor
   - 9 oil pump bolts

➡**Make a note of the position of the each bolt. When replacing the bolts into the oil pump body, place each bolt in the position from which it was removed.**

   - Oil pump body, by prying between the oil pump and main bearing cap
   - O-ring from the cylinder block
   - Plug, gasket, spring and relief valve from the oil pump body
   - 9 screws, pump body cover, drive and driven rotors

To install:

3. Install or connect the following:
   - Driven rotors, drive, pump body cover, using the 9 screws
   - Oil pump relief valve, spring, gasket and the plug to the oil pump body
   - New O-ring on the cylinder block
4. Using a non residue solvent, clean both sealing surfaces to the oil pump.
5. Apply liquid sealant to the oil pump and engine block.
6. Install or connect the following:
   - Oil pump

➡**Be sure to engage the spline teeth of the oil pump drive gear with the large teeth of the crankshaft.**

   - 9 oil pump bolts. Tighten the bolts in several passes. Tighten bolt A to 71 inch lbs. (8 Nm), bolt B to 14 ft. lbs. (20 Nm) and bolt C to 32 ft. lbs. (43 Nm). refer to the illustration for bolt locations.
   - CKP sensor. Tighten the bolt to 69 inch lbs. (8 Nm).
   - Baffle plate to the No. oil pan. Tighten to 69 inch lbs. (8 Nm).
   - No. 1 oil pan, oil strainer and No. 2 oil pan
7. Refill the engine with oil.
8. Start the engine and inspect for leaks.
9. Recheck the engine oil level.

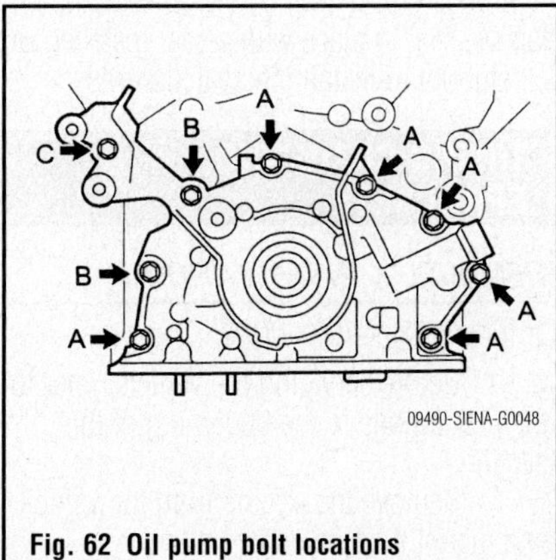

Fig. 62 Oil pump bolt locations

PISTON AND RING

POSITIONING

See Figures 63 through 65.

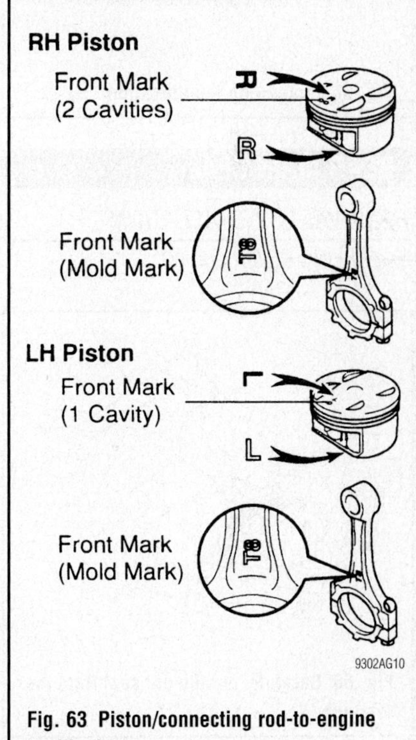

Fig. 63 Piston/connecting rod-to-engine positioning

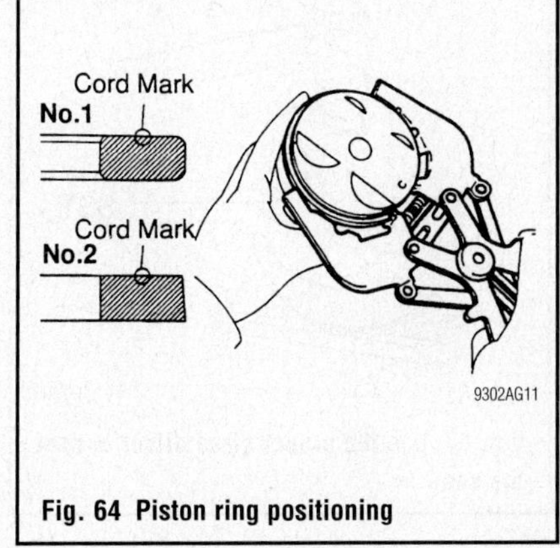

Fig. 64 Piston ring positioning

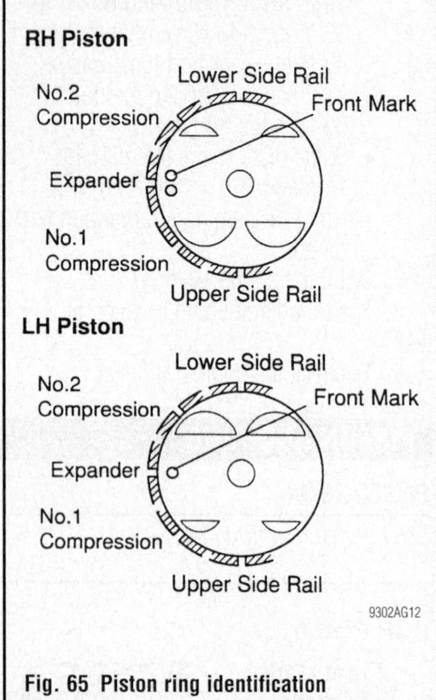

Fig. 65 Piston ring identification

REAR MAIN SEAL

REMOVAL & INSTALLATION

See Figures 66 through 69.

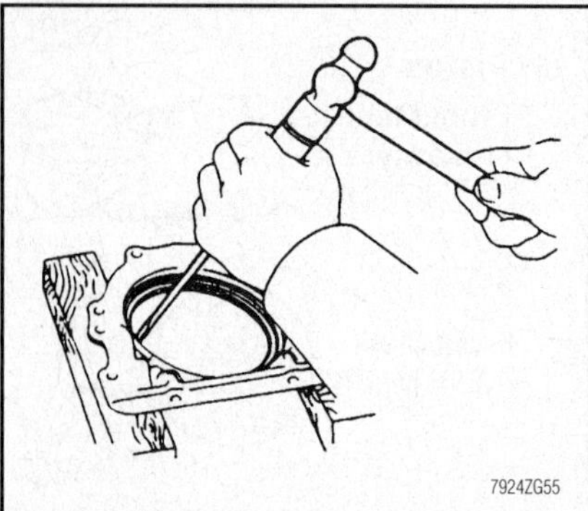

Fig. 66 Carefully tap the old seal from the retainer

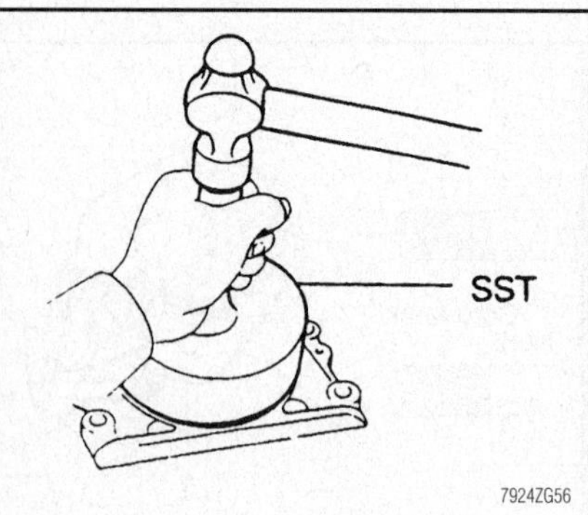

Fig. 67 Use the proper sized driver to seat the seal

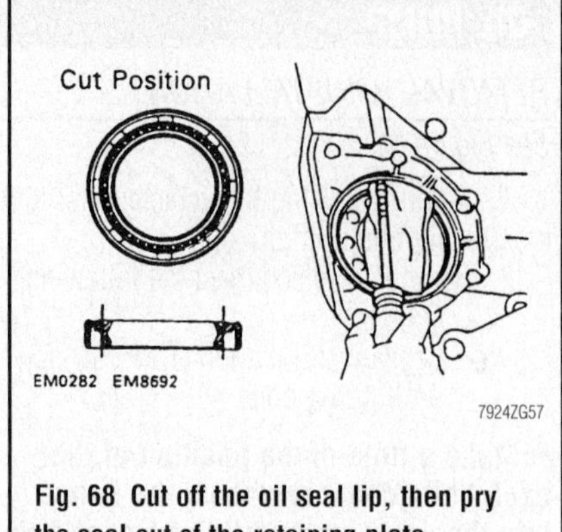

Fig. 68 Cut off the oil seal lip, then pry the seal out of the retaining plate

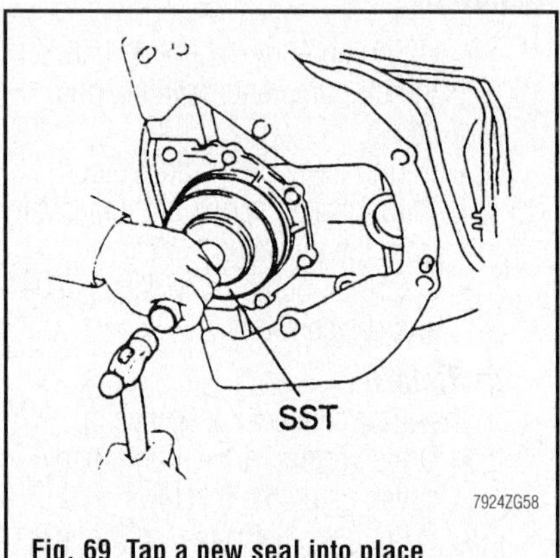

Fig. 69 Tap a new seal into place

If the rear oil seal retainer is not installed to the block, use a tapered ended screw-driver and hammer to remove the oil seal. Apply multi-purpose grease to the new oil seal lip. Using a seal driver, tap the seal into place. Be careful not to install it slantwise.

1. Before servicing the vehicle, refer to the precautions section.

If the rear oil seal retainer is installed on the cylinder block, using a knife, cut off the lip of the seal. Using a taped ended prytool, pry the old seal out of the retainer. Inspect the oil seal lip contacting surface of the crankshaft for cracks or damage. Apply multipurpose grease to the new oil seal, then tap the seal in place with a seal installer. Be careful not to install the seal slantwise.

TIMING CHAIN AND SPROCKETS

REMOVAL & INSTALLATION

See Figures 70 through 92.

1. Before servicing the vehicle, refer to the precautions in the beginning of this section.
2. Remove the engine from the vehicle and mount it on an engine stand.

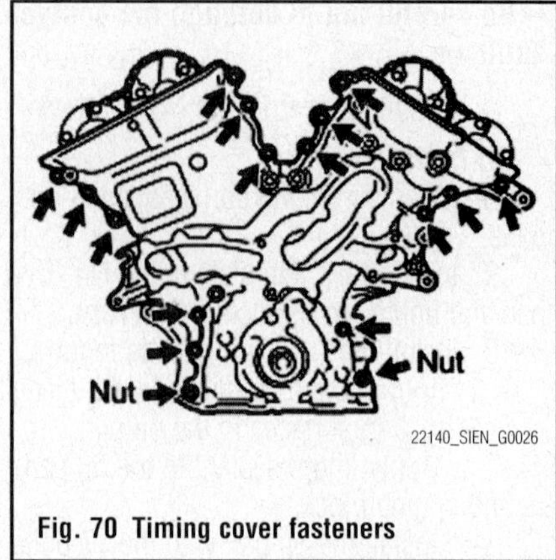

Fig. 70 Timing cover fasteners

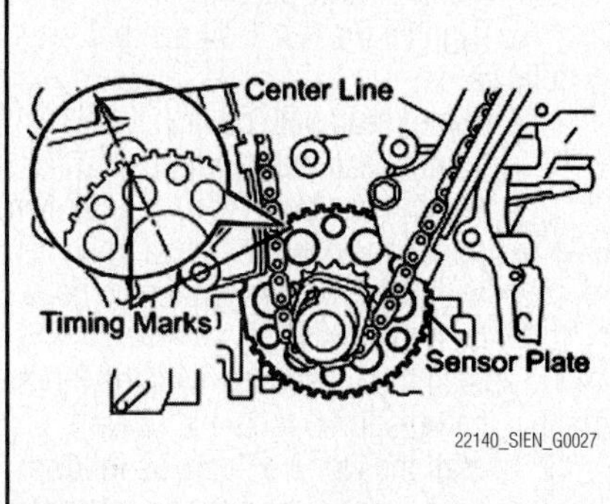

Fig. 71 Crank Angle Sensor Plate timing marks

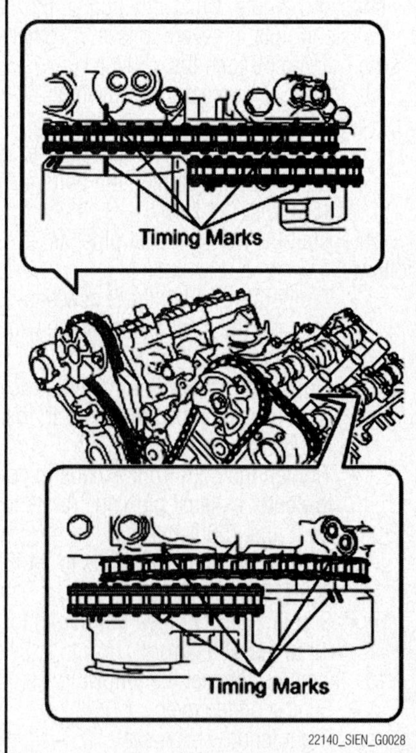

Fig. 72 Camshaft timing marks

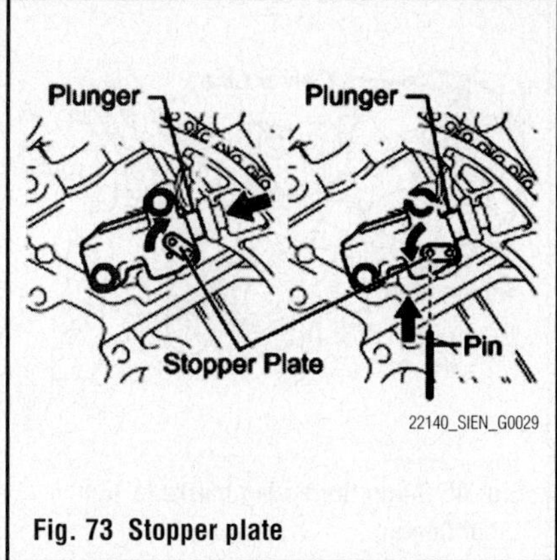

Fig. 73 Stopper plate

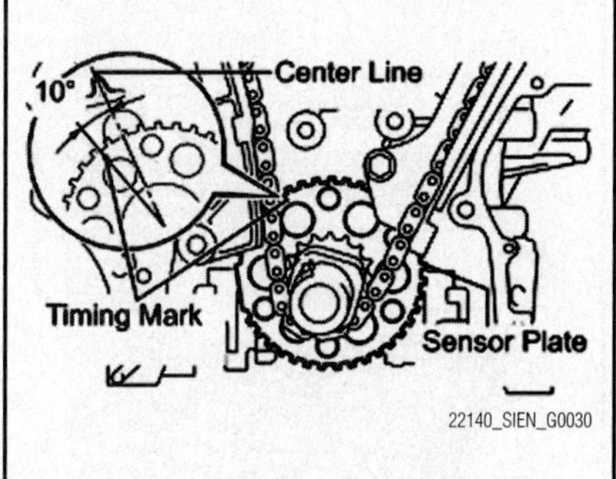

Fig. 74 Rotate the crankshaft 10° counter-clockwise

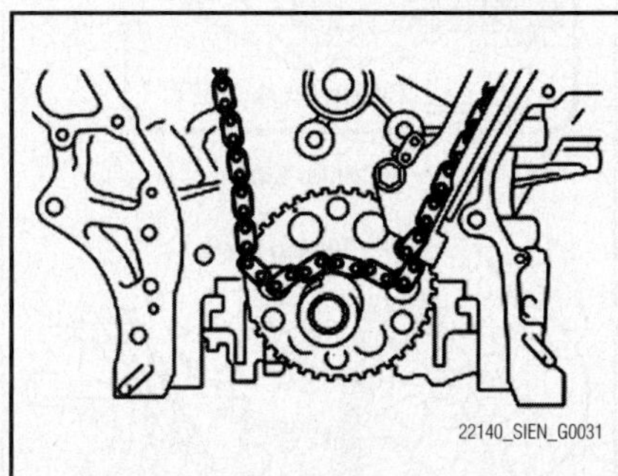

Fig. 75 Remove the chain sub-assembly from the crankshaft timing sprocket and place it on the crankshaft

3. Remove the 15 bolts and 2 nuts as shown in the illustration.

4. Remove the timing chain cover sub-assembly by prying between the timing chain cover and cylinder head sub-assembly or cylinder block sub-assembly with a prytool.

➡Be careful not to damage the contact surfaces of the cylinder head, cylinder block and chain cover.

5. Remove the 4 bolts, chain cover plate and chain cover plate gasket.

6. Temporarily tighten the pulley set bolt.

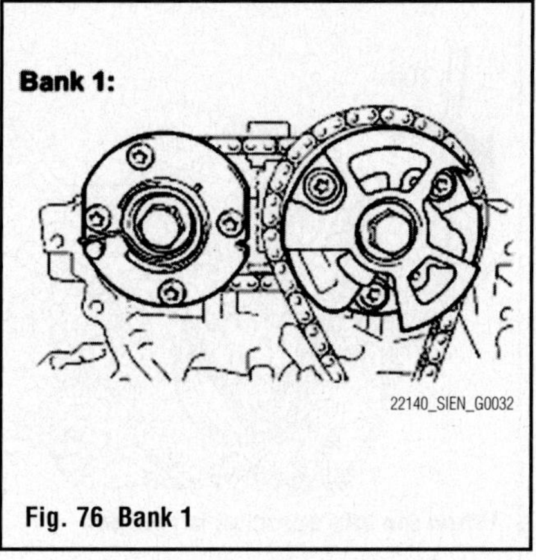

Fig. 76 Bank 1

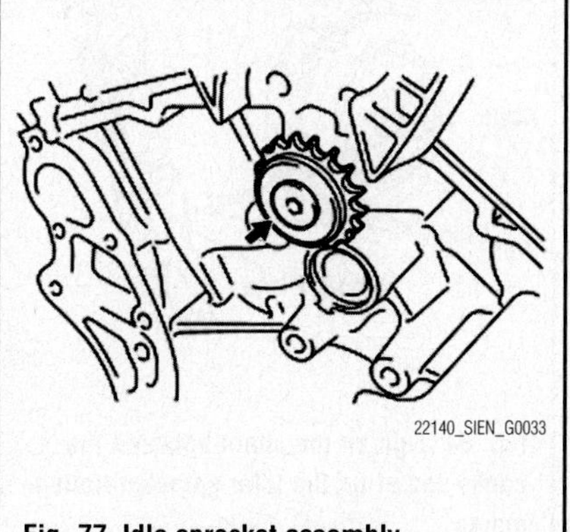
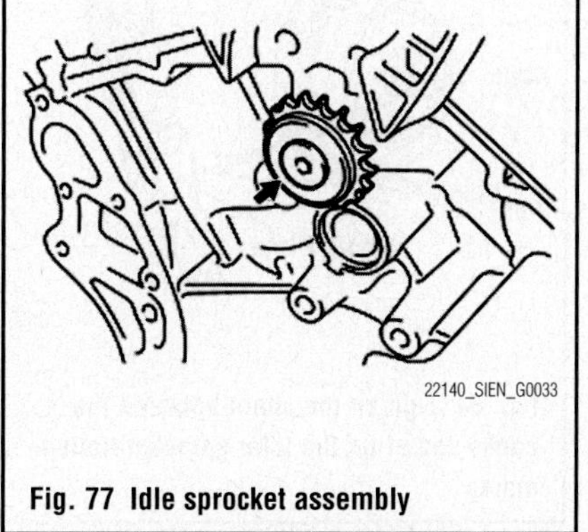

Fig. 77 Idle sprocket assembly

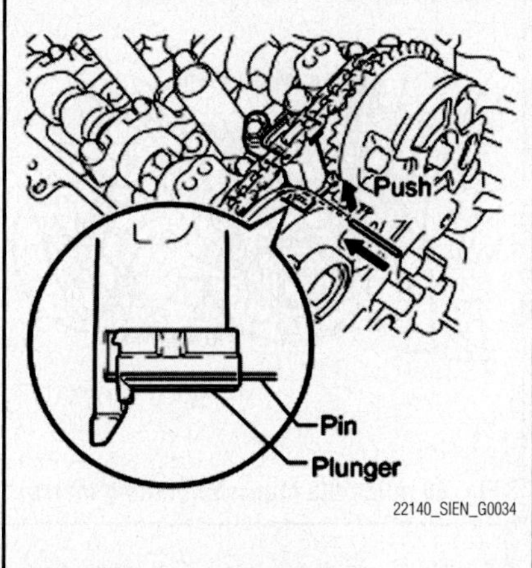

Fig. 78 No. 2 chain tensioner

7. Set the timing mark on the crank angle sensor plate to the RH block bore center line (TDC / compression).

8. Check that the timing marks of the camshaft timing gears are aligned with those of the bearing cap as shown in the illustration.

9. If not, turn the crankshaft 1 revolution (360°) and align the timing marks as above.

10. Move the stopper plate upward to release the lock, and push the plunger deep into the tensioner.

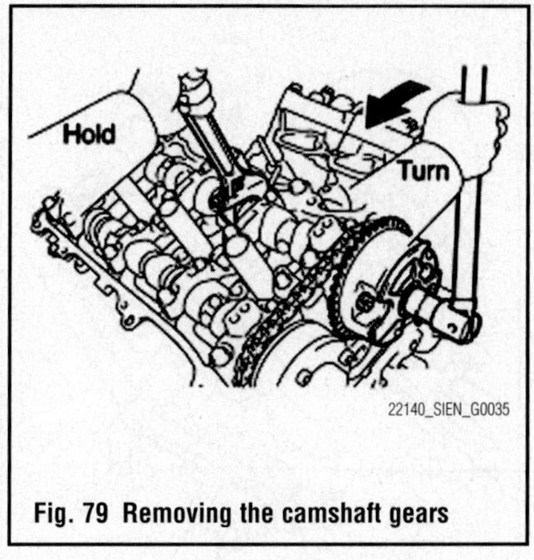

Fig. 79 Removing the camshaft gears

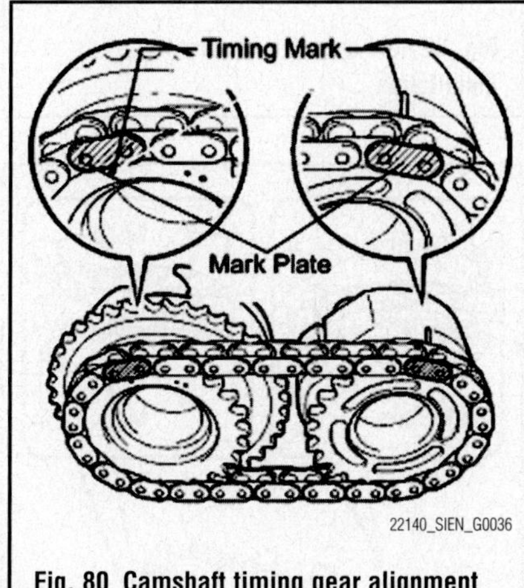

Fig. 80 Camshaft timing gear alignment

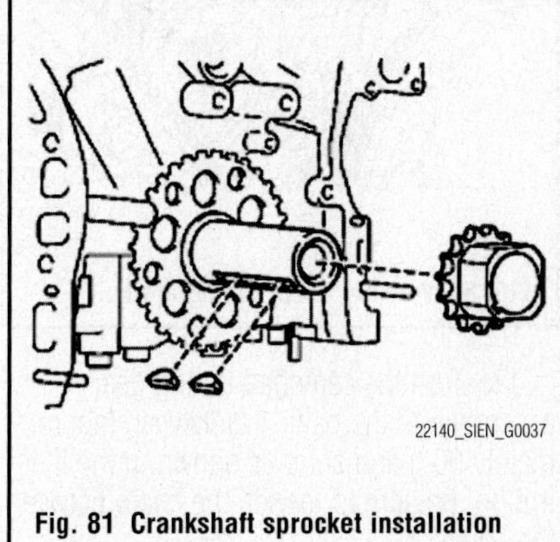

Fig. 81 Crankshaft sprocket installation

11. Move the stopper plate downward to set the lock, and insert a pin of 1.27 mm (0.05 in.) into the stopper plate's hole.

12. Remove the 2 bolts and No. 1 chain tensioner assembly.

13. Remove the chain tensioner slipper.

14. Turn the crankshaft counterclockwise 10° to loosen the chain sub-assembly of the crankshaft timing sprocket.

15. Remove the pulley set bolt.

16. Remove the chain sub-assembly from the crankshaft timing sprocket and place it on the crankshaft.

Fig. 82 Idle sprocket assembly installation

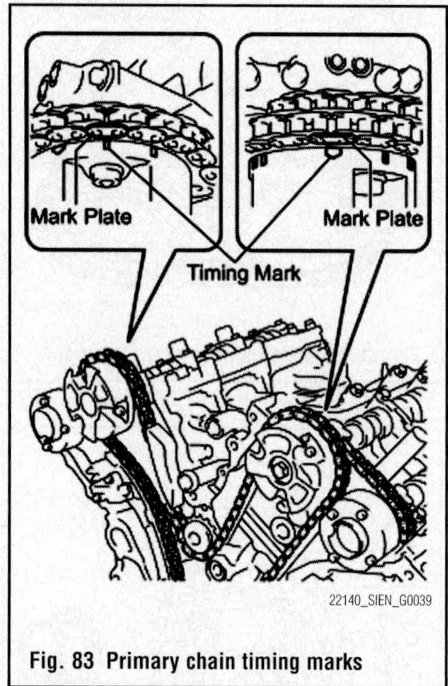

Fig. 83 Primary chain timing marks

17. Turn the camshaft timing gear assembly on the bank 1 clockwise (approximately 60°) and set it as shown in the illustration. Be sure to loosen the chain between the banks.

18. Remove the chain.

19. Using a 10 mm hexagon wrench, remove the No. 2 idle gear shaft, idle sprocket assembly and No. 1 idle gear shaft.

20. Remove the 2 bolts and No. 1 chain vibration damper.

21. Remove the 2 No. 2 chain vibration dampers.

22. Remove the crankshaft timing sprocket from the crankshaft.

23. Remove the 2 keys from the crankshaft.

24. While raising the No. 2 chain tensioner assembly, insert a pin of 1.0 mm

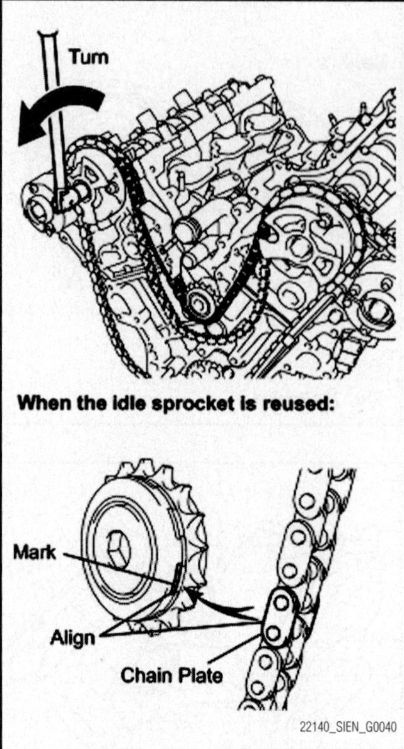

Fig. 84 Tighten the chain between the banks and align the idler sprocket timing marks

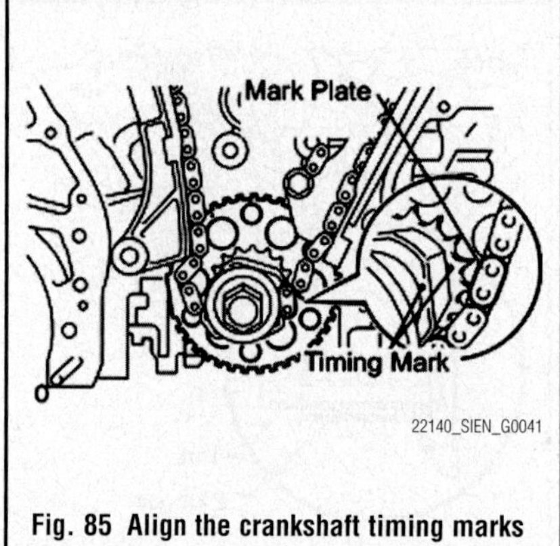

Fig. 85 Align the crankshaft timing marks

(0.039 in.) into the hole to fix the No. 2 chain tensioner.

25. Hold the hexagonal portion of the camshaft with a wrench, and remove the 2 bolts and 2 camshaft timing gear assemblies.

➡**Be careful not to damage the cylinder head with the wrench.**

➡**Do not disassemble the camshaft timing gear assemblies.**

26. Remove the No. 2 chain assembly.

27. Remove the bolt and No. 2 chain tensioner assembly.

28. While pushing down the No. 3 chain tensioner assembly (Bank 2), insert a pin of

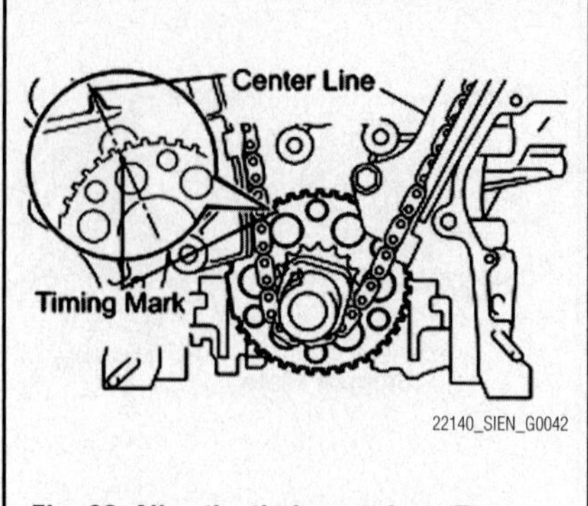

Fig. 86 Align the timing marks to Top Dead Center

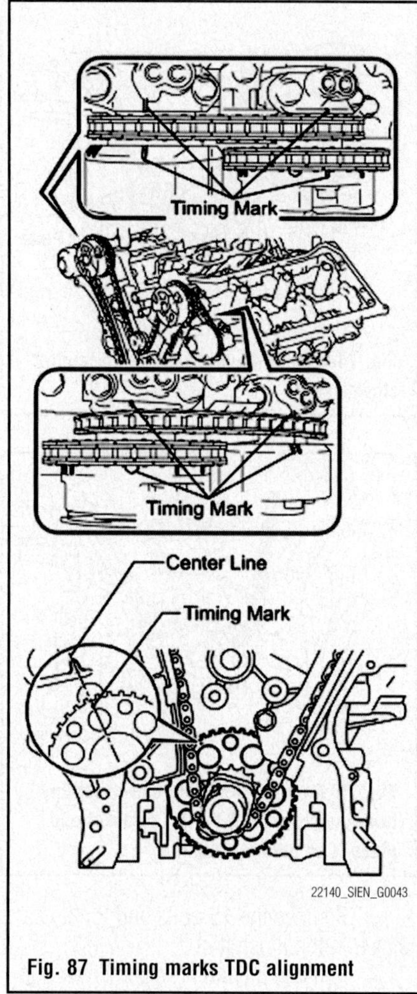

Fig. 87 Timing marks TDC alignment

1.0 mm (0.039 in.) into the hole to fix the No. 3 chain tensioner assembly.

29. Hold the hexagonal portion of the camshaft with a wrench, and remove the 2 bolts and 2 camshaft timing gear assemblies.

➡**Be careful not to damage the cylinder head with the wrench.**

➡**Do not disassemble the camshaft timing gear assemblies.**

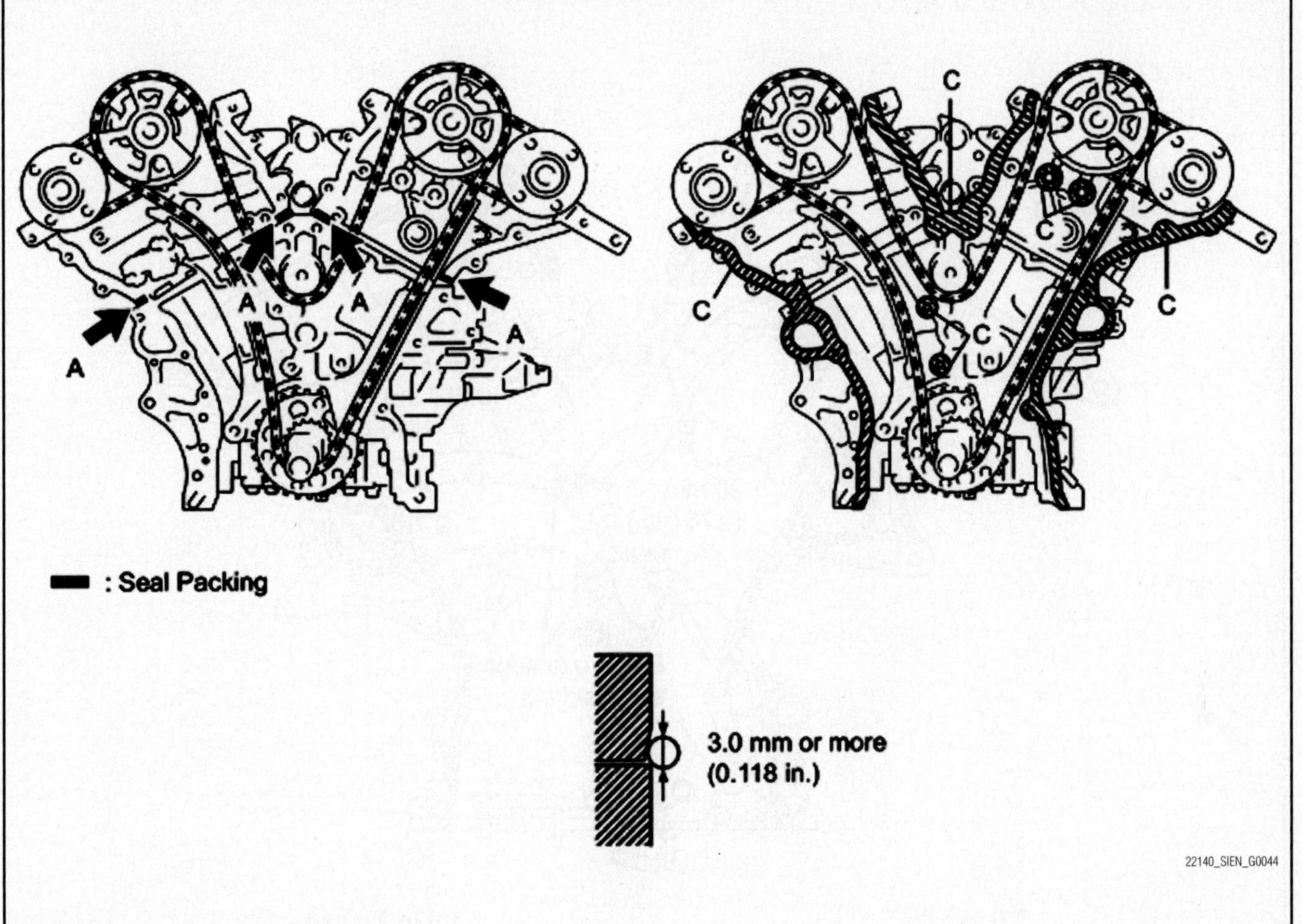

Fig. 88 Apply sealant to the engine as shown

30. Remove the No. 2 chain sub-assembly.

31. Remove the bolt and No. 3 chain tensioner assembly.

To install:

32. For Bank 1, install the No. 2 chain tensioner assembly and tighten the bolt to 15 ft. lbs. (21 Nm).

33. While pushing in the tensioner, insert a pin of 1.0 mm (0.039 in.) into the hole to fix it.

34. Align the mark plate (yellow) with the timing marks (1-dot mark) of the camshaft timing gears as shown in the illustration.

➡Apply a light coat of engine oil to the bolt threads and bolt-seating surface.

35. Align the knock pin of the camshaft with the pin hole of the camshaft timing gear. Install the camshaft timing gear and camshaft timing exhaust gear RH with the No. 2 chain sub-assembly installed.

36. Hold the hexagonal portion of the camshaft with a wrench, and tighten the 2 bolts and camshaft timing gear assemblies to 74 ft. lbs. (100 Nm).

37. Remove the pin from the No. 2 chain tensioner assembly.

38. For Bank 2, install the No. 2 chain tensioner assembly and tighten the bolt to 15 ft. lbs. (21 Nm).

39. While pushing in the tensioner, insert a pin of 1.0 mm (0.039 in.) into the hole to fix it.

40. Align the mark plate (yellow) with the timing marks (1-dot mark) of the camshaft timing gears as shown in the illustration.

➡Apply a light coat of engine oil to the bolt threads and bolt-seating surface.

41. Align the knock pin of the camshaft with the pin hole of the camshaft timing gear. Install the camshaft timing gear and camshaft timing exhaust gear LH with the No. 2 chain sub-assembly installed.

42. Hold the hexagonal portion of the camshaft with a wrench, and tighten the 2 bolts and camshaft timing gear assemblies to 74 ft. lbs. (100 Nm).

43. Remove the pin from the No. 2 chain tensioner assembly.

44. Install the No. 1 chain vibration damper and tighten the bolts to 17 ft. lbs. (23 Nm).

45. Install the 2 No. 2 chain vibration dampers.

46. Install the 2 keys and crankshaft timing sprocket as shown in the illustration.

47. Apply a light coat of engine oil to the rotating surface of the No. 1 idle gear shaft.

48. Temporarily install the No. 1 idle gear shaft and idle sprocket with the No. 2 idle gear shaft while aligning the knock pin of the No. 1 idle gear with the knock pin groove of the cylinder block.

➡Be careful of the idle gear direction.

49. Check that no foreign objects are on the No. 1 and No. 2 idle gear shafts.

50. Using a 10 mm hexagon wrench, tighten the No. 2 idle gear shaft to 44 ft. lbs. (60 Nm).

51. After installing the idle sprocket assembly, check that the idle sprocket turns smoothly.

52. Align the mark plate and timing

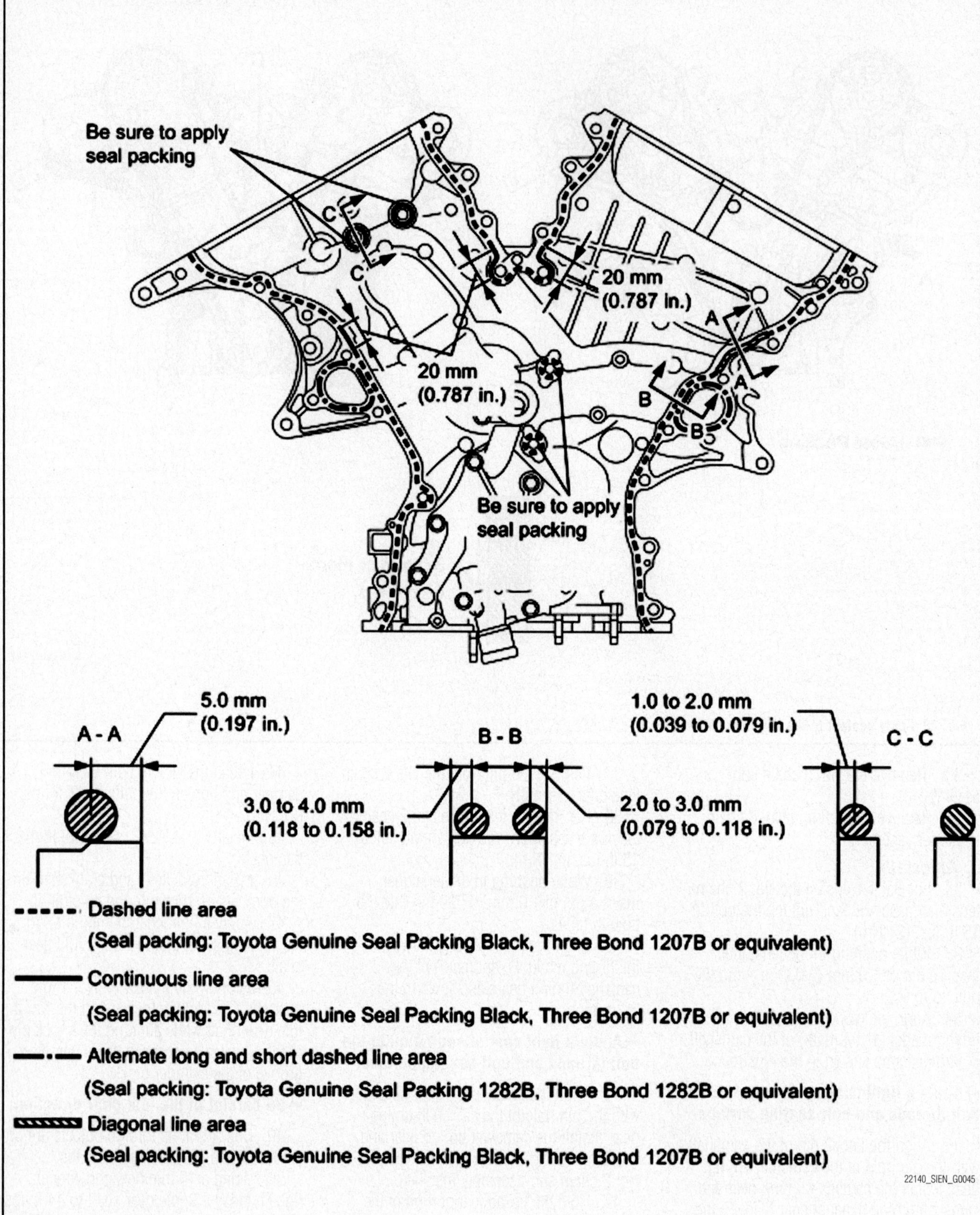

Fig. 89 Apply sealant to the timing cover as shown

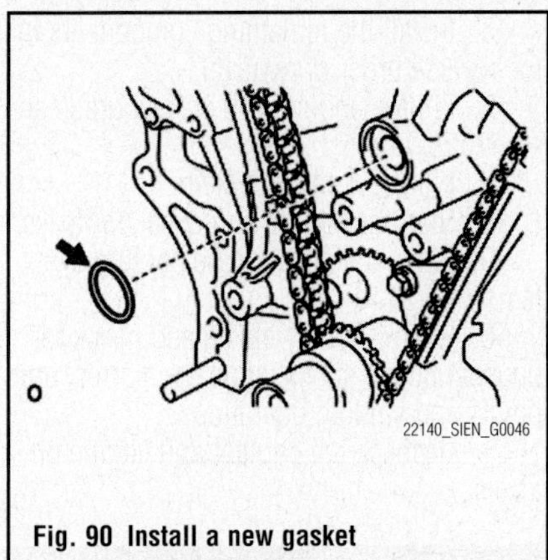

Fig. 90 Install a new gasket

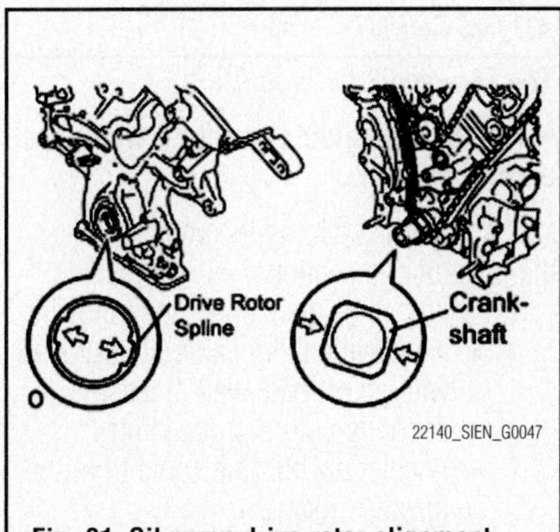

Fig. 91 Oil pump drive rotor alignment

marks as shown in the illustration and install the chain.

➡**The camshaft mark plate is orange.**

53. Do not pass the chain over the crankshaft, just put it on it.
54. Turn the camshaft timing gear assembly on the bank 1 counterclockwise to tighten the chain between the banks.
55. When the idle sprocket is reused, align the chain plate with the mark where the plate had been in order to tighten the chain between the banks.
56. Align the mark plate and timing marks as shown in the illustration and install the chain onto the crankshaft timing sprocket.

➡**The crankshaft mark plate is yellow.**

57. Temporarily tighten the pulley set bolt.
58. Turn the crankshaft clockwise to set it to the RH block bore center line (TDC / compression).
59. Install the chain tensioner slipper.
60. Move the stopper plate upward to release the lock, and push the plunger deep into the tensioner.
61. Move the stopper plate downward to set the lock, and insert a hexagon wrench into the hole of the stopper plate.
62. Install the chain tensioner and tighten the bolts to 84 inch lbs. (10 Nm).
63. Remove the hexagon wrench of the chain tensioner. Check that each timing mark is aligned with the crankshaft at the TDC / compression.
64. Remove the pulley set bolt.
65. Apply seal packing in a continuous line to the engine unit as shown in the following illustration. Use Toyota Genuine Seal Packing Black, Three Bond 1207B or equivalent
66. Be sure to clean and degrease the contact surfaces, especially the surfaces indicated by C in the illustration.
67. When the contact surfaces are wet, wipe them with an oil-free cloth before applying seal packing.
68. Install the chain cover within 3 minutes.

➡**Do not start the engine for at least 2 hours after installing.**

69. Apply seal packing in a continuous line to the timing chain cover as shown in the following illustration.

➡**When the contact surfaces are wet, wipe them with an oil-free cloth before applying seal packing.**

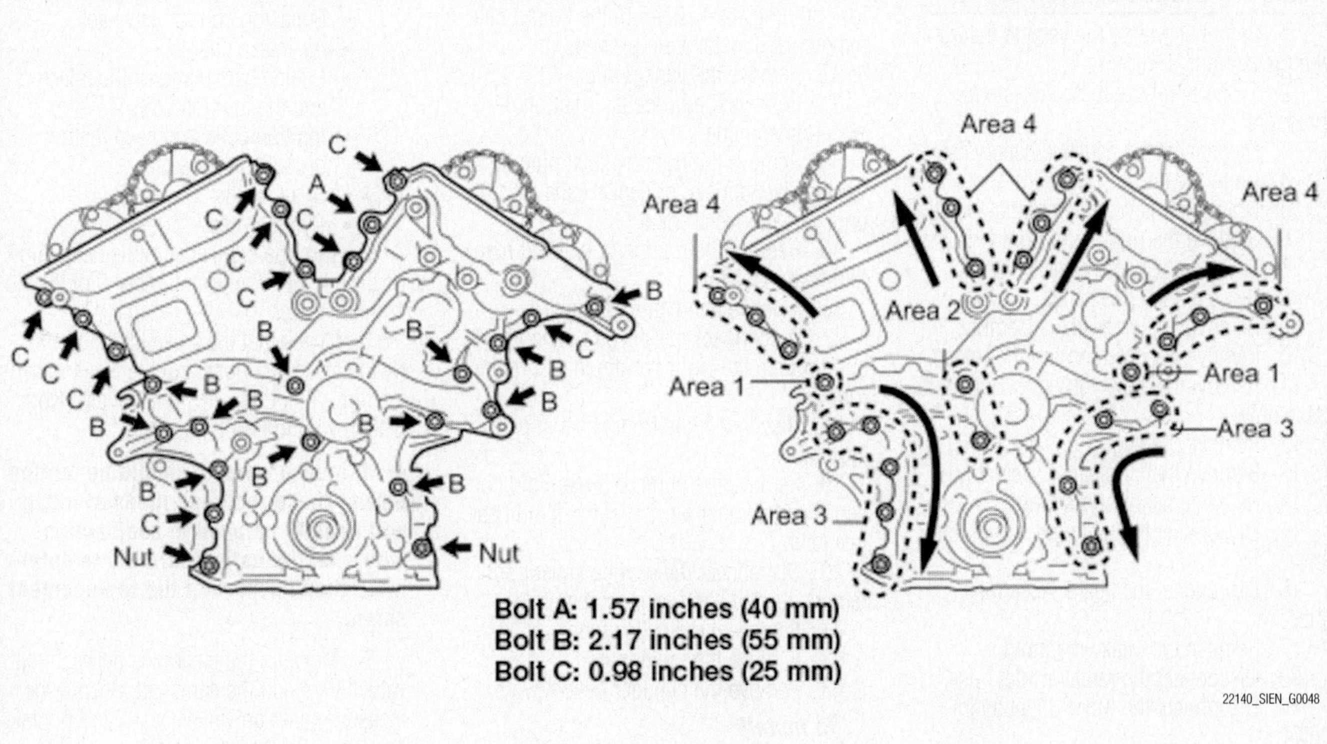

Fig. 92 Timing cover fastener locations and torque sequence

Install the chain cover within 3 minutes and tighten the bolts within 15 minutes after applying seal packing. Do not start the engine for at least 2 hours after installing.

70. Install a new gasket.
71. Align the oil pump's drive rotor spline and the crankshaft as shown in the illustration. Install the spline and chain cover to the crankshaft.
72. Install the cover bolts and nuts as shown.
73. Temporarily tighten the timing chain cover with the 23 bolts and 2 nuts.

➡Make sure that there is no oil on the bolt and nut threads.

74. Fully tighten the bolts as follows:
 a. Step 1: Area 1 and Area 2 to 15 ft. lbs. (21 Nm)
 b. Step 2: Fully tighten the bolts in Area 3 to 15 ft. lbs. (21 Nm)
 c. Step 3: Tighten bolt A to 32 ft. lbs. (43 Nm)
 d. Step 4: Tighten bolts in Area 4 to 15 ft. lbs. (21 Nm)
75. Install the engine to the vehicle. Check for leaks and proper operation.

VALVE COVERS

REMOVAL & INSTALLATION

1. Before servicing the vehicle, refer to the precautions section.
2. Properly relieve the fuel system pressure.
3. Disconnect the negative battery cable.
4. Drain the engine oil.
5. Drain and recycle the engine coolant.
6. Remove the right hand front wheel.
7. Remove the wiper arms.
8. Remove the cowl panel.
9. Remove the wiper link assembly.
10. Remove the cowl top to inner brace.
11. Remove the cowl top cover outer sub assembly.
12. Remove the engine cover.
13. Remove the air cleaner assembly.
14. Remove the air surge tank.
15. Disconnect the fuel pipe sub assembly.
16. Disconnect the heater hose inlet pipe.
17. Remove the intake manifold.
18. Disconnect the radiator inlet hose.
19. Disconnect the water outlet as follows:
 a. Disconnect the engine coolant temperature sensor.
 b. Remove the clamp, bolts nuts and washers.
 c. Lock the clamp open and remove the water outlet with the bypass hose and remove the gaskets.
20. Remove the right hand front fender apron seal.
21. Remove the compressor to crankshaft pulley belt.
22. Remove the vane pump belt.
23. Remove the engine moving control rod.
24. Remove the number 2 engine mounting stay as follows:
 a. Remove the bolt and wire harness bracket.
 b. Remove the bolt, stay and bracket.
25. Remove the number 2 alternator bracket and the alternator belt adjusting bar.
26. Remove the crankshaft pulley.
27. Remove the timing belt covers 1 and 2. Refer to the timing belt removal and installation procedure.
28. Remove the right hand engine mount bracket.
29. Remove the timing belt guide and timing belt. Refer to the timing belt removal and installation procedure.
30. Remove the timing belt guide number 2. Refer to the timing belt removal and installation procedure.
31. Remove the camshaft pulley and lower timing cover. Refer to the timing belt removal and installation procedure.
32. Remove the vane pump.
33. Remove the propeller shaft on 4 wheel drive models.
34. Remove the front exhaust pipe on 2 wheel drive models, or center exhaust pipe on 4 wheel drive models.
35. Remove the number 1 exhaust manifold heat insulator.
36. Remove the manifold stay.
37. Disconnect the oxygen sensor and remove the right hand exhaust manifold.
38. Disconnect the front exhaust pipe assembly.
39. Remove the number 3 manifold converter and number e exhaust manifold hear insulator.
40. Disconnect the oxygen sensor and remove the left hand exhaust manifold.
41. Remove the dipstick tube.
42. Remove the ignition coil.
43. Remove the cylinder head covers.

To install:

44. Apply sealant to the cylinder heads where the camshaft supports meet the cylinder heads. Install the cylinder head covers, using new gaskets.
45. Install the remaining components in the reverse order of removal.
46. Fill the engine with correct grade and amount of oil.
47. Fill the cooling system.
48. Start the engine and check for leaks.
49. Bleed the air from the cooling system.
50. Road test the vehicle and check for unusual noise, shock, slippage, correct shift points and smooth operation.
51. Recheck the coolant and engine oil levels.

VALVE LASH

ADJUSTMENT

See Figures 93 through 95.

➡Adjust the valve clearance when the engine is cold.

1. Before servicing the vehicle, refer to the precautions section.
2. Remove or disconnect the following:
 - Negative battery cable. If equipped with an air bag, wait at least 90 seconds before proceeding.
 - Accelerator/throttle cable from the throttle linkage
 - Air cleaner cover, air flow meter and air duct assembly
 - V-bank cover
 - Emission control valve set
 - Air intake chamber
 - Engine harness from the injectors and the ignition coils
 - Ignition coils and keep them in order for reassembly
 - Spark plugs
 - Cylinder head covers
3. Turn the crankshaft pulley and align its groove with the timing mark **0** of the No. 1 timing cover.
4. Check that the valve lifters on the No. 1 intake are loose and the No. 1 exhaust are tight. If not, turn the crankshaft 1 complete revolution (360 degrees).

➡All measurements should be written down. These recorded measurements will need to be used in conjunction with a mathematical formula to determine the thickness of the replacement shims.

5. Measure the clearance between the valve lifters and the camshaft. Record the measurements on valves No. 1 and 6 intake; No. 2 and 3 exhaust.
 c. The intake valve clearance cold is 0.006–0.010 in. (0.15–0.25mm).
 d. The exhaust valve clearance cold is 0.010–0.014 in. (0.25–0.35mm).

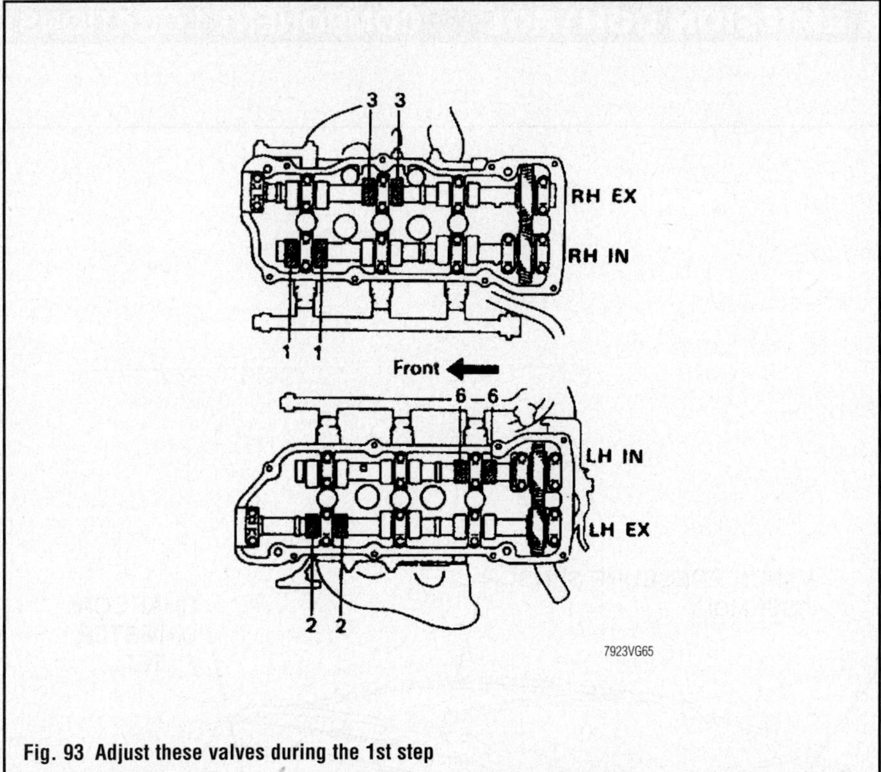

Fig. 93 Adjust these valves during the 1st step

6. Turn the crankshaft ⅔ of a revolution (240 degrees). Record the measurements on valves No. 2 and 3 intake; No. 4 and 5 exhaust.

7. Turn the crankshaft another ⅔ of a revolution. Record the measurements on valves No. 4 and 5 intake; No. 1 and 6 exhaust.

8. Remove the adjusting shim by turning the crankshaft to position the cam lobe of the camshaft in the up position on the valve to be adjusted. Using a small thin flat bladed tool, turn the valve lifter so that the notches are perpendicular to the camshaft. Press down the valve lifter with tool 09248-55010 part A. Place too 09248-55010 part B between the camshaft and the valve lifter; remove part A.

9. Remove the adjusting shim with a magnet and a small screwdriver.

10. Determine the replacement adjusting shim size by either using the charts or the following formulas:

- Intake: N = T + (A - 0.008 in./0.020mm)
- Exhaust: N = T + (A - 0.012 in./0.30mm)
- T = Thickness of removed shim
- A = Measured valve clearance
- N = Thickness of new shim

11. Select a new shim with a thickness as close as possible to the calculated value. Install the new replacement shim.

➡Shims are available in 17 sizes in increments of 0.0020 in. (0.050mm), from 0.0984 in. (2.500mm) to 0.1299 in. (3.300mm).

12. Recheck the valve clearance.

13. Install or connect the following:
- Cylinder head covers
- Spark plugs and the ignition coils
- Engine wiring harness to the injectors and the coils
- Intake chamber
- Emission control valve set
- V-bank cover
- Air flow meter, air duct and air cleaner cover
- Negative battery cable

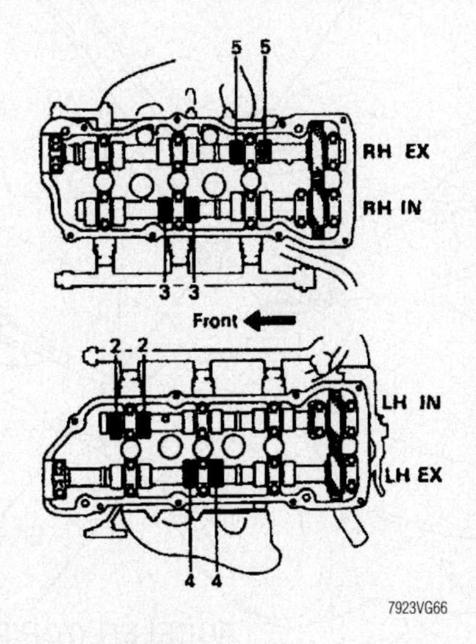

Fig. 94 Adjust these valves during the 2nd step

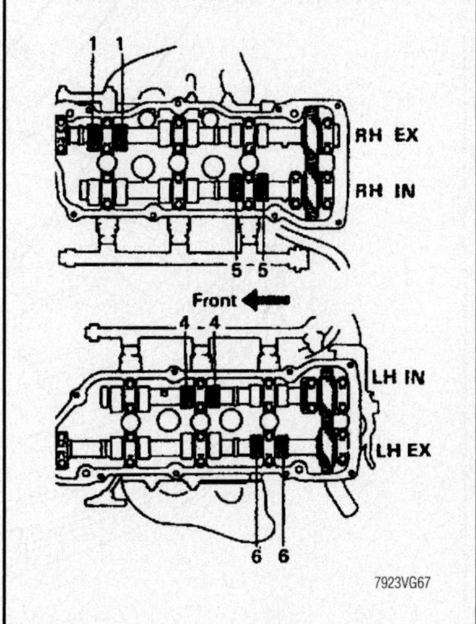

Fig. 95 Adjust these valves during the 3rd step

See Figures 96 through 100.

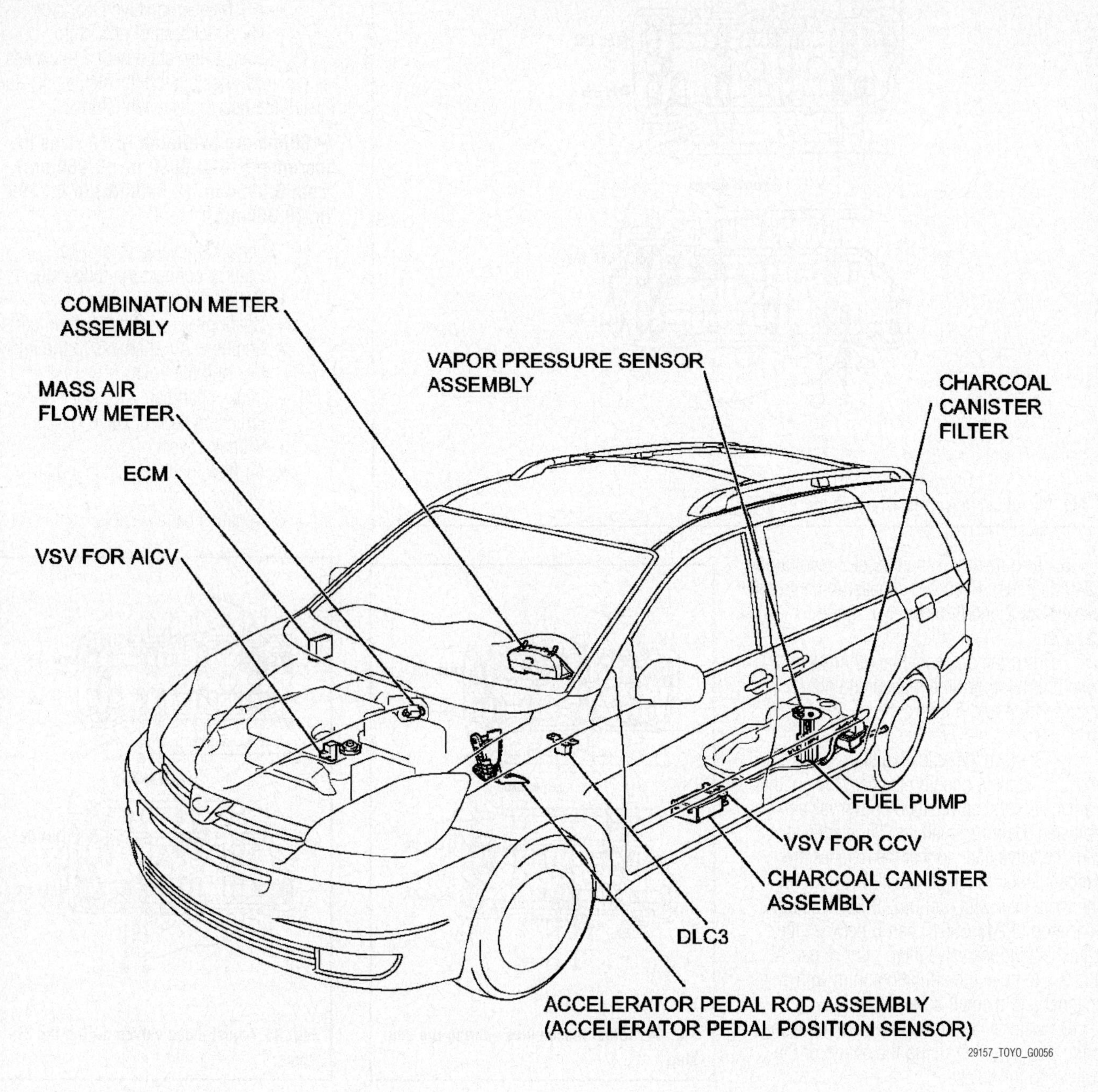

Fig. 96 Sienna vehicle component locations

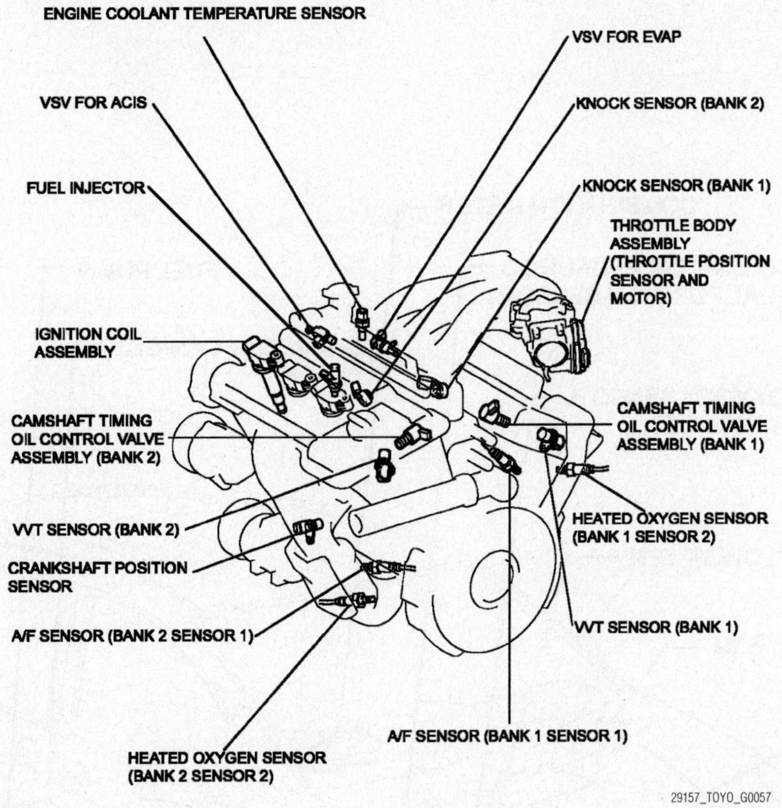

ENGINE COOLANT TEMPERATURE SENSOR
VSV FOR EVAP
VSV FOR ACIS
KNOCK SENSOR (BANK 2)
FUEL INJECTOR
KNOCK SENSOR (BANK 1)
THROTTLE BODY ASSEMBLY (THROTTLE POSITION SENSOR AND MOTOR)
IGNITION COIL ASSEMBLY
CAMSHAFT TIMING OIL CONTROL VALVE ASSEMBLY (BANK 2)
CAMSHAFT TIMING OIL CONTROL VALVE ASSEMBLY (BANK 1)
HEATED OXYGEN SENSOR (BANK 1 SENSOR 2)
VVT SENSOR (BANK 2)
CRANKSHAFT POSITION SENSOR
VVT SENSOR (BANK 1)
A/F SENSOR (BANK 2 SENSOR 1)
A/F SENSOR (BANK 1 SENSOR 1)
HEATED OXYGEN SENSOR (BANK 2 SENSOR 2)
29157_TOYO_G0057

Fig. 97 Sienna engine component locations

Location of EVAP (Evaporative Emission) System:
Canister
Fuel Tank
Recirculation Pipe
Pump Module
Fuel Tank Cap
Air Filter
EVAP Hose (To Throttle Body)
Purge VSV
To Purge VSV EVAP Hose (from Canister)
22140_SIEN_G0001

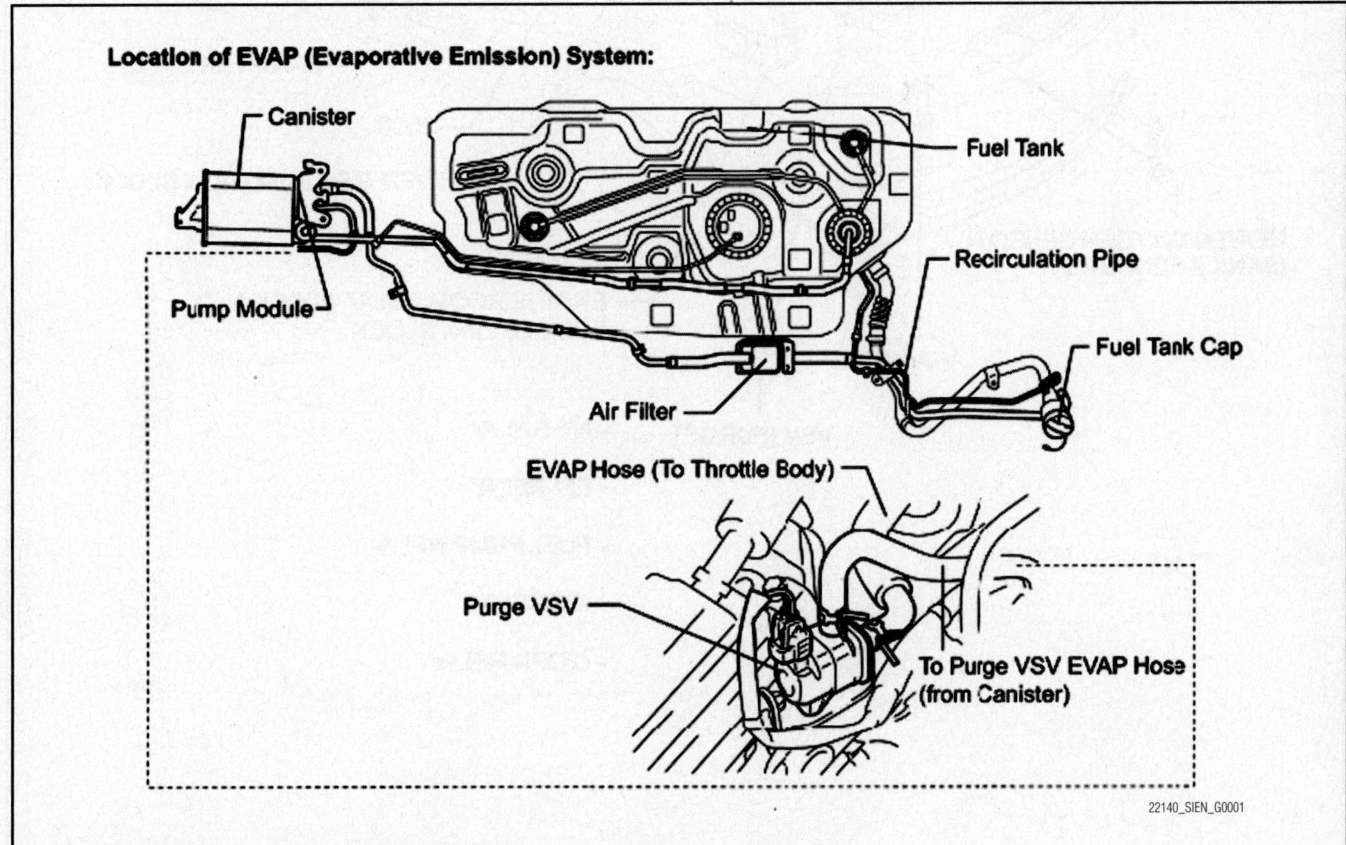

Fig. 98 Sienna Evaporative Emissions component locations—3.5L engine

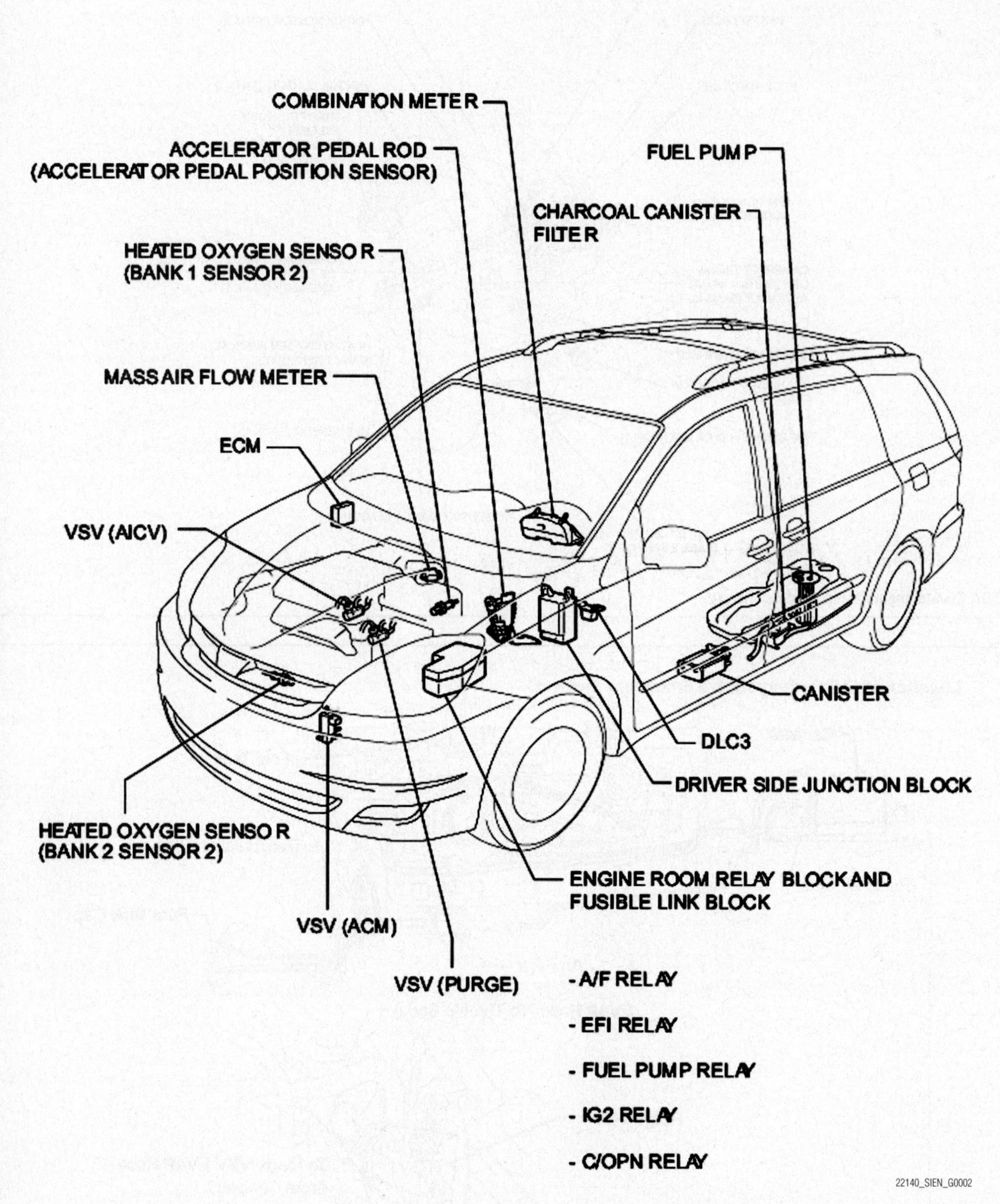

Fig. 99 Sienna vehicle component locations—3.5L engine

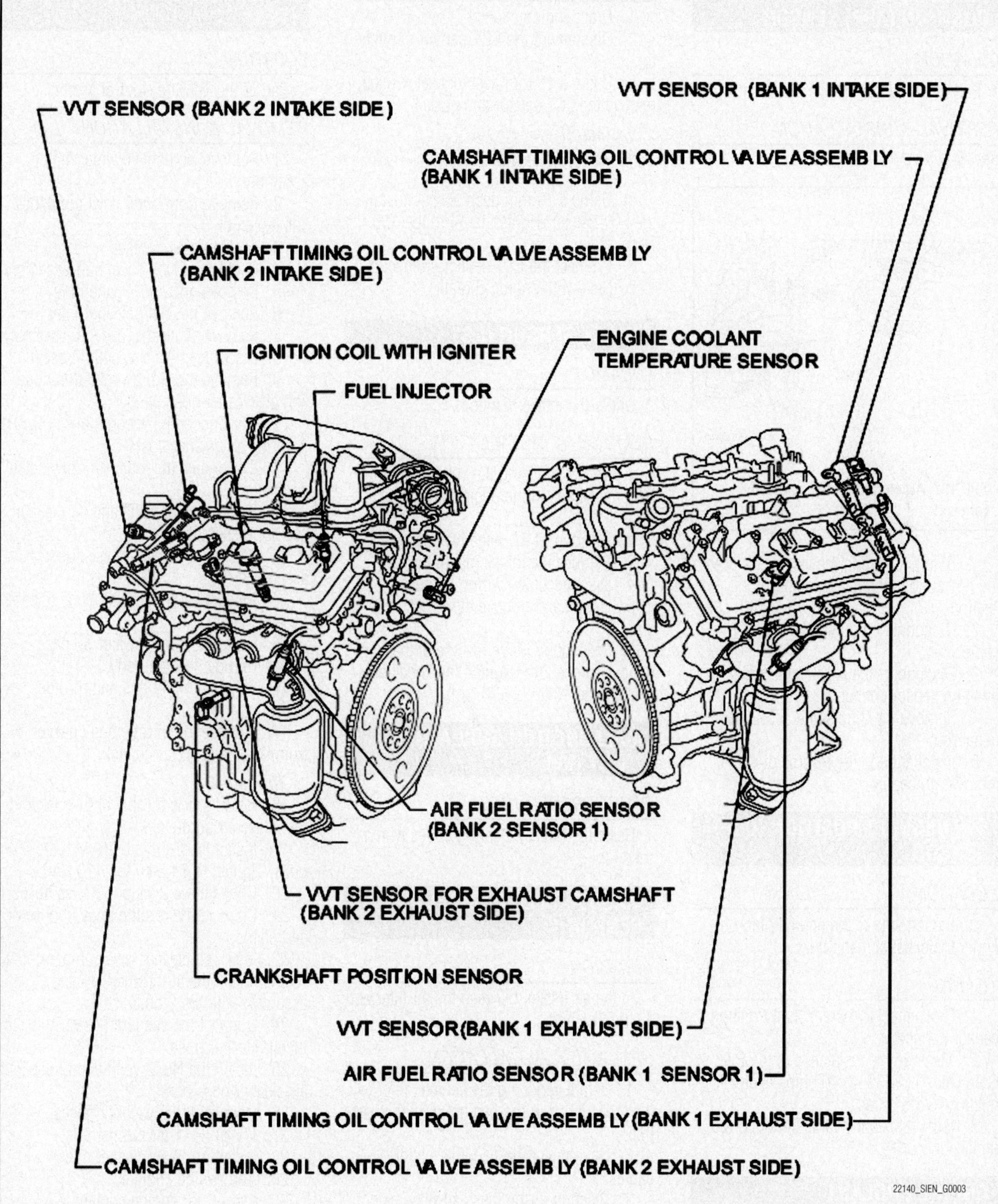

Fig. 100 Sienna engine component locations—3.5L engine

ACCELERATOR PEDAL POSITION (APP) SENSOR

LOCATION

Part of the accelerator pedal assembly.

REMOVAL & INSTALLATION

See Figure 101.

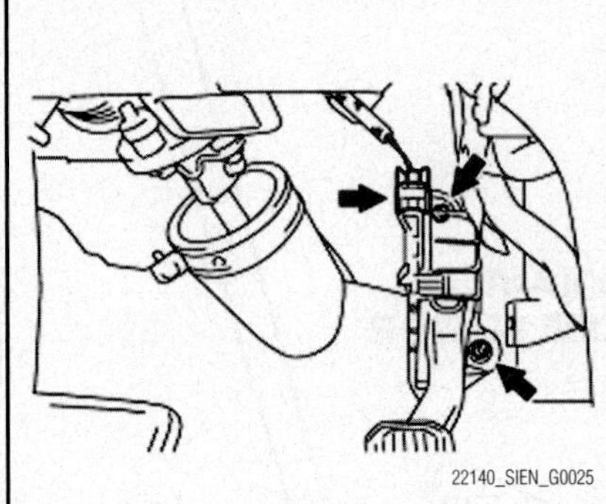

Fig. 101 Accelerator pedal assembly removal

1. Before servicing the vehicle, refer to the precautions in the beginning of this section.
2. Disconnect the negative battery cable.
3. Disconnect the accelerator pedal position sensor connector.
4. Remove the 2 nuts and accelerator pedal rod.
5. Installation is the reverse of the removal procedure.

CRANKSHAFT POSITION (CKP) SENSOR

LOCATION

On the left side of the engine block, under the exhaust manifold.

TESTING

1. Disconnect the Crankshaft Position sensor connector.
2. Measure the resistance of the CKP. Cold value is 1630–2740 ohms. Hot value is 2065–3225 ohms.
3. If the resistance value is not correct, replace the CKP.

ENGINE COOLANT TEMPERATURE (ECT) SENSOR

LOCATION

Mounted in the crossover pipe at the rear of the engine.

REMOVAL & INSTALLATION

1. Drain engine coolant.
2. Disconnect the ECT sensor connector.
3. Using a 19 mm deep socket wrench, remove the ECT sensor and gasket.

To install:

4. Install a new gasket to the ECT sensor.
5. Using a 19 mm deep socket, install the ECT sensor. Tighten to 14 ft. lbs. (20 Nm).
6. Connect the ECT sensor connector.
7. Refill with engine coolant.

KNOCK SENSOR (KS)

LOCATION

Under the intake manifold.

REMOVAL & INSTALLATION

1. Before servicing the vehicle, refer to the precautions in the beginning of this section.
2. Disconnect the negative battery cable.
3. Remove the intake manifold.
4. Disconnect the wiring harness.
5. Unbolt and remove the knock sensors.
6. Installation is the reverse of the removal procedure. Tighten the knock sensor bolts to 15 ft. lbs.(20 Nm).

MALFUNCTION INDICATOR LIGHT (MIL)

RESET PROCEDURES

Reset the MIL by clearing codes with a scan tool.

MASS AIR FLOW (MAF) SENSOR

LOCATION

In the air intake, between the air filter and the throttle body.

REMOVAL & INSTALLATION

1. Before servicing the vehicle, refer to the precautions in the beginning of this section.
2. Disconnect the negative battery cable.
3. Disconnect the mass air flow meter connector.
4. Remove the 2 screws and mass air flow meter.
5. Installation is the reverse of the removal procedure.

THROTTLE POSITION SENSOR (TPS)

LOCATION

Part of the Throttle Control Motor.

REMOVAL & INSTALLATION

1. Remove windshield wiper motor assembly
2. Remove front outer cowl top panel sub-assembly
3. Drain engine coolant
4. Remove v-bank cover sub-assembly
5. Remove no. 2 air cleaner inlet
6. Remove no. 1 air cleaner inlet
7. Remove air cleaner cap sub-assembly
8. Disconnect the 3 vacuum hoses.
9. Remove the No. 2 ventilation hose and air cleaner hose band.
10. Disconnect the vacuum hose (EVAP) from the air cleaner hose.
11. Disconnect the mass air flow meter connector.
12. Remove the 2 bolts and air cleaner cap sub-assembly.
13. Remove air cleaner case sub-assembly
14. Disconnect the throttle body connector and clamp.
15. Disconnect the 2 water by-pass hoses from the throttle body.
16. Remove the 4 bolts and throttle body.
17. Remove the throttle body gasket from the intake air surge tank.

To install:

18. Install a new throttle body gasket to the intake air surge tank.
19. Install the throttle body with the 4 bolts. Tighten to 84 inch lbs. (10 Nm).
20. Connect the 2 water by-pass hoses.
21. Connect the throttle body connector and clamp.
22. Install air cleaner case sub-assembly.
23. Install the air cleaner cap sub-assembly with the 2 bolts.
24. Connect the vacuum hose (EVAP) to the air cleaner hose.
25. Install the No. 2 ventilation hose and air cleaner hose band.
26. Connect the 3 vacuum hoses..
27. Install no. 1 air cleaner inlet.
28. Install no. 2 air cleaner inlet
29. Add engine coolant.
30. Inspect for engine coolant leak.
31. Install v-bank cover sub-assembly.
32. Install front outer cowl top panel sub-assembly.
33. Install windshield wiper motor assembly.

FUEL SYSTEM SERVICE PRECAUTIONS

Safety is the most important factor when performing not only fuel system maintenance but any type of maintenance. Failure to conduct maintenance and repairs in a safe manner may result in serious personal injury or death. Maintenance and testing of the vehicle's fuel system components can be accomplished safely and effectively by adhering to the following rules and guidelines.

- To avoid the possibility of fire and personal injury, always disconnect the negative battery cable unless the repair or test procedure requires that battery voltage be applied.
- Always relieve the fuel system pressure prior to disconnecting any fuel system component (injector, fuel rail, pressure regulator, etc.), fitting or fuel line connection. Exercise extreme caution whenever relieving fuel system pressure to avoid exposing skin, face and eyes to fuel spray. Please be advised that fuel under pressure may penetrate the skin or any part of the body that it contacts.
- Always place a shop towel or cloth around the fitting or connection prior to loosening to absorb any excess fuel due to spillage. Ensure that all fuel spillage (should it occur) is quickly removed from engine surfaces. Ensure that all fuel soaked cloths or towels are deposited into a suitable waste container.
- Always keep a dry chemical (Class B) fire extinguisher near the work area.
- Do not allow fuel spray or fuel vapors to come into contact with a spark or open flame.
- Always use a back-up wrench when loosening and tightening fuel line connection fittings. This will prevent unnecessary stress and torsion to fuel line piping.
- Always replace worn fuel fitting O-rings with new. Do not substitute fuel hose or equivalent where fuel pipe is installed.

Before servicing the vehicle, make sure to also refer to the precautions in the beginning of this section as well.

RELIEVING FUEL SYSTEM PRESSURE

1. Remove the fuel circuit opening relay from the engine compartment relay block.

Start the engine. After the engine stops, turn the ignition to OFF.

2. Check that the engine won't start.
3. Remove the fuel tank cap.
4. Disconnect the battery ground cable.
5. Install the relay.

FUEL FILTER

REMOVAL & INSTALLATION

See Figure 102.

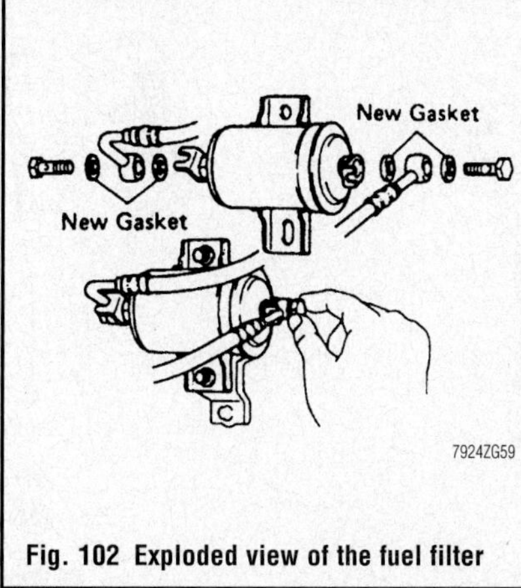

Fig. 102 Exploded view of the fuel filter

1. Before servicing the vehicle, refer to the precautions section.
2. Disconnect the negative battery cable.
3. Relieve the fuel system pressure.

➡The fuel filter is located in the engine compartment, at the inlet line to the fuel rail.

4. Remove or disconnect the following:
   - Inlet and outlet lines from the filter
   - Fuel filter

To install:

5. Install or connect the following:
   - Fuel filter, using new O-rings. Tighten the lines to 22 ft. lbs. (29 Nm).
   - Negative battery cable
6. Start the engine and check for leaks.

FUEL PUMP

REMOVAL & INSTALLATION

See Figures 103 through 106.

1. Before servicing the vehicle, refer to the precautions section.
2. Discharge the fuel system pressure.
3. Remove the charcoal canister cover.
4. Remove the fuel tank filler hose cover.
5. Disconnect the fuel tank vent hose.
6. Disconnect the fuel tank main tube.
7. Disconnect the filler hose.
8. Remove the wire harness clamps.
9. Place a jack under the tank, remove the bolts and the support bands.
10. Disconnect all remaining wiring and lower the tank.
11. Remove the tube joint clip and pull out the fuel main tube from the fuel pump module.
12. Remove the main tube from the tank.
13. Using tool 09808-14020, or equivalent lock ring tool, remove the lockring from the fuel pump module.
14. Remove the module from the tank.
15. Remove the joint clip and remove the pressure sensor.
16. Remove the sender assembly.
17. Wrap the tip of a small screwdriver with tape and disconnect the 4 snap retainers, and remove the fuel suction plate.
18. Disconnect the snap retainers, disconnect the connector, and remove the fuel pump.
19. Remove the O-ring and spacer from the pump.
20. Installation is the reverse of removal. Use a new O-ring coated with clean gasoline.

➡Prior to assembly, all new parts must be stored at room temperature for a minimum of 12 hours.

21. Make sure that the gasket groove is clean. Use a new gasket. Align the arrow on the fuel suction tube and the tank suction support.
22. Align the marks on the fuel pump module retainer and the fuel tank.
23. Position the retainer on the module and push down. Hold the module and turn the retainer by hand, one complete turn.

➡Make sure that the anti-rotation tab is in the groove during tightening. The "S" arrow on the fuel tank indicates "0" degrees position. Make sure that the retainer isn't cross-threaded.

24. Using the special tool, torque the retainer to 59–67 ft. lbs. (80–90 Nm). The triangle mark on the retainer should be about 1½ turn from the start.

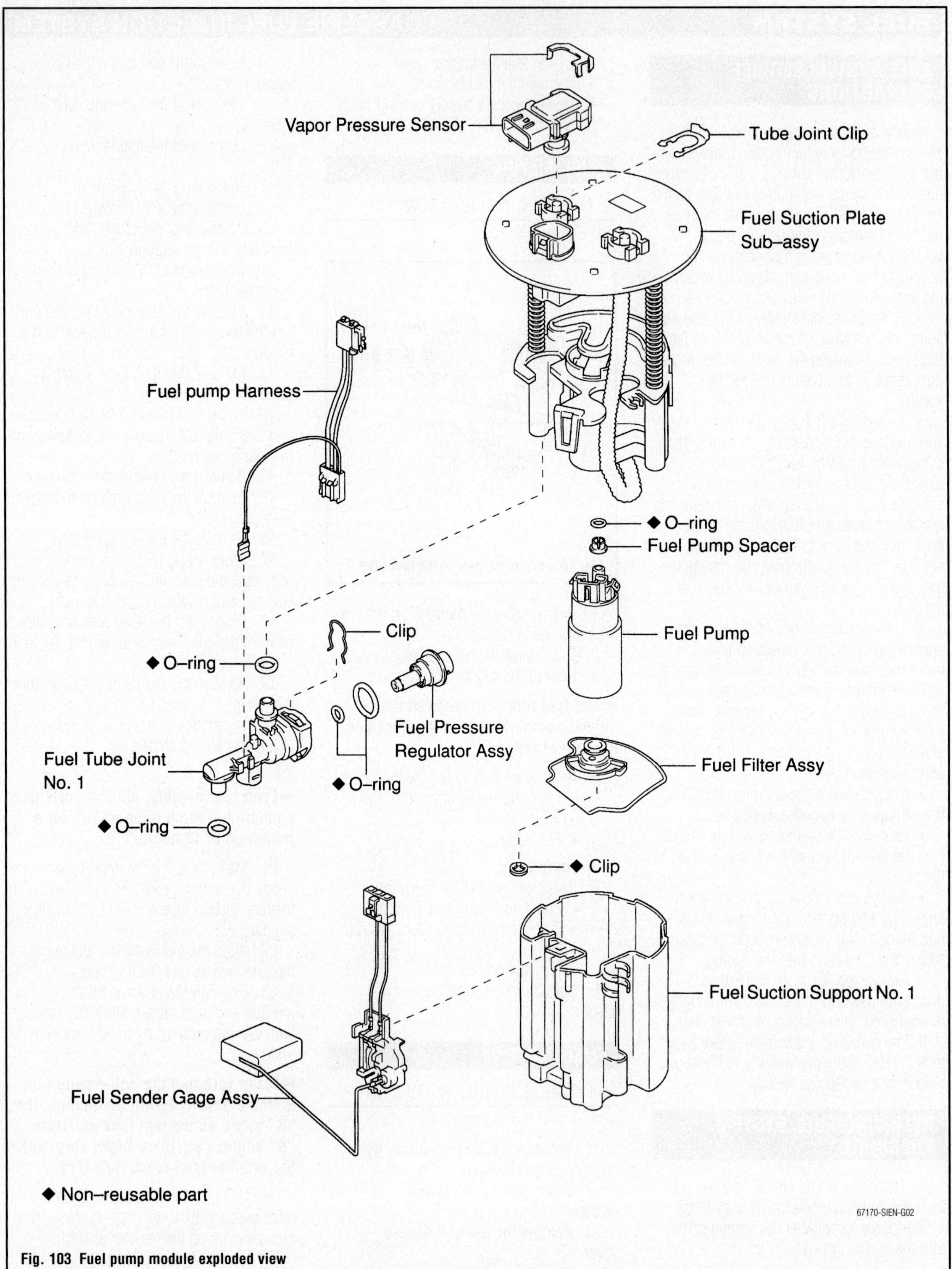

Fig. 103 Fuel pump module exploded view

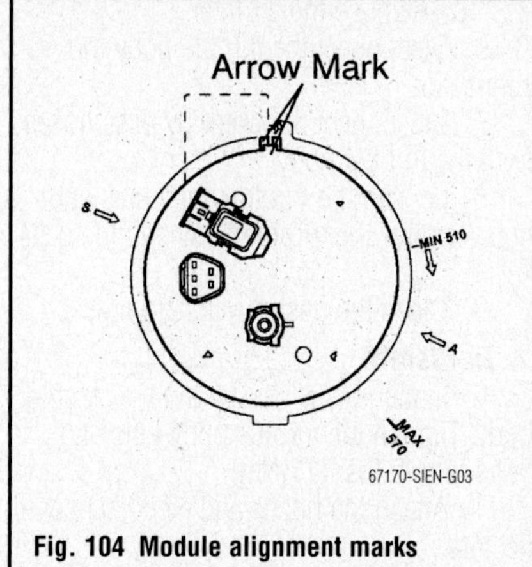

Fig. 104 Module alignment marks

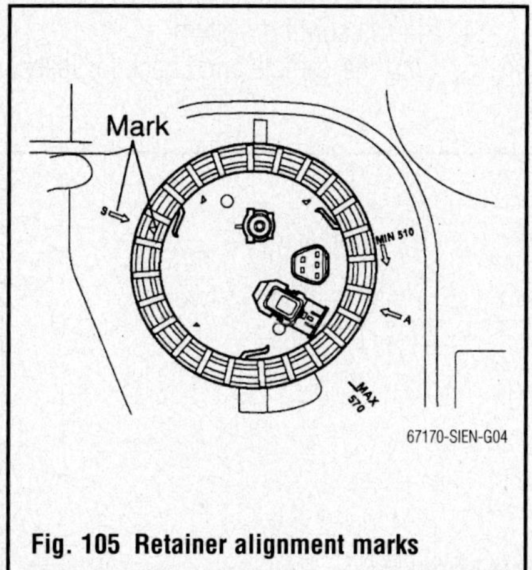

Fig. 105 Retainer alignment marks

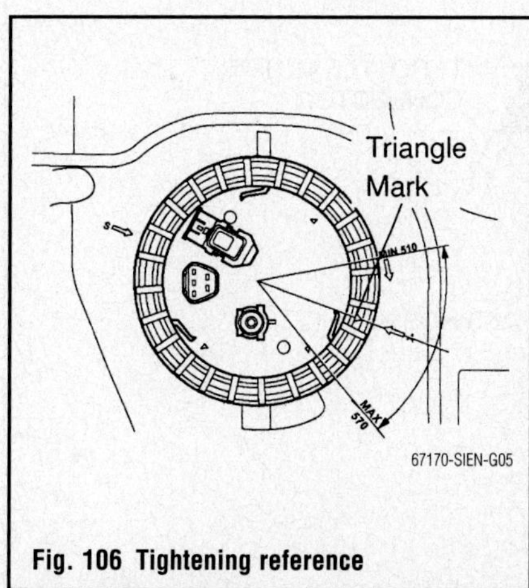

Fig. 106 Tightening reference

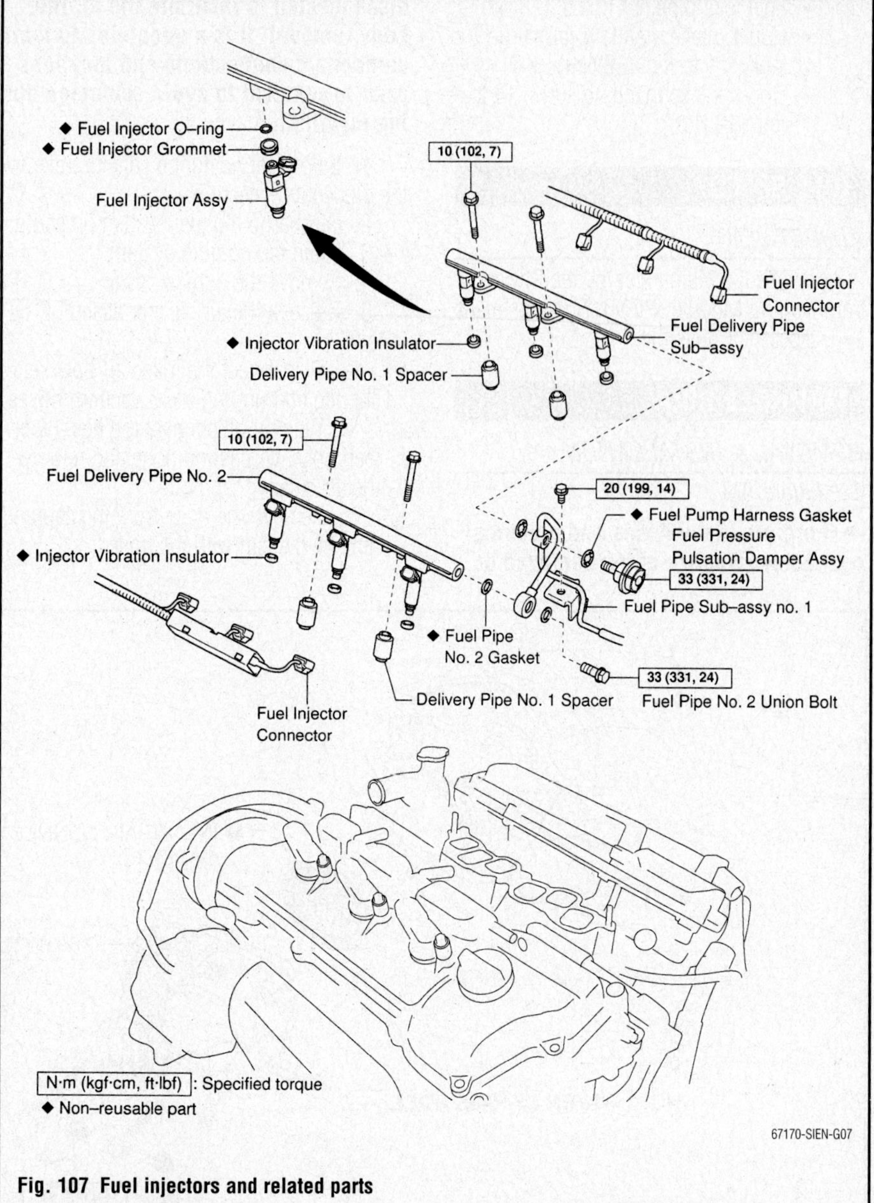

Fig. 107 Fuel injectors and related parts

FUEL RAIL & INJECTORS

REMOVAL & INSTALLATION

See Figure 107.

1. Before servicing the vehicle, refer to the precautions section.
2. Relieve the fuel system pressure.
3. Drain the coolant.
4. Remove the wiper arms.
5. Remove the wiper motor.
6. Remove the cowl tops.
7. Remove the V-bank cover.
8. Remove the air cleaner assembly.
9. Remove the emission control valve set.
10. Remove the upper intake manifold (intake air surge tank). Discard the gasket.
11. Remove the fuel pipe sub-assembly.
12. Disconnect the wiring at the injectors.
13. Remove the 4 bolts and 2 delivery pipe along with the injectors.
14. Remove the delivery pipe spacers and insulators from the manifold.
15. Pull each injector from the pipe.

To install:

16. Install new O-rings on each injector. Apply a light coating of gasoline to the O-rings and mating points on the pipes.
17. Using a twisting motion, install the injectors on the pipes.

➡Be careful to avoid twisting the O-rings. After installation, check that the injectors turn smoothly. If not, use new O-rings.

18. Install the pipes and injectors.
19. Loosely install the bolts and make sure that the injectors still turn freely. If not, replace the O-rings.
20. Torque the bolts to 84 inch lbs. (10 Nm).
21. The remainder of installation is the reverse of removal. Observe the following torques:
    - Fuel line union bolt: 24 ft. lbs. (33 Nm)
    - Pulsation damper: 24 ft. lbs. (33 Nm)

- Fuel feed pipe: 14 ft. lbs. (20 Nm)
- Upper intake manifold (air surge tank): 21 ft. lbs. (28 Nm)
- Upper intake manifold stays: 14 ft. lbs. (20 Nm)

IDLE SPEED

ADJUSTMENT

Idle speed is maintained by the Powertrain Control Module (PCM). No adjustment is necessary or possible.

THROTTLE BODY

REMOVAL & INSTALLATION

See Figure 108.

➡There are many hoses and electrical connectors that are either removed or disconnected to facilitate the throttle body removal. It is a good idea to mark component connections and locations prior to removal to avoid confusion during installation.

1. Before servicing the vehicle, refer to the precautions section.
2. Discharge the fuel system pressure.
3. Drain the cooling system.
4. Remove the engine cover.
5. Remove the air cleaner assembly as follows:
   a. Disconnect the mass air flow sensor, tag and unplug the 3 vacuum hoses.
   b. Tag and disconnect the fuel vapor feed hose, ventilation hose and remove the air cleaner cap bolts.
   c. Remove the hose from the cap and remove the cap with the hose.
   d. Remove the air filter.
6. Disconnect the throttle body motor connector.
7. Disconnect the water by pass hoses from the throttle body.
8. Loosen the throttle body nuts and nuts, remove the throttle body and the gasket.
9. Clean the gasket mating surfaces.

To install:

10. Install a new gasket and the throttle body. Tighten the throttle body nuts and bolts to 8 ft. lbs. (11 Nm).
11. Attach the hoses and electrical connectors.
12. Install the air cleaner assembly.
13. Install the remaining components.
14. Fill the cooling system.
15. Start the vehicle and check for leaks.

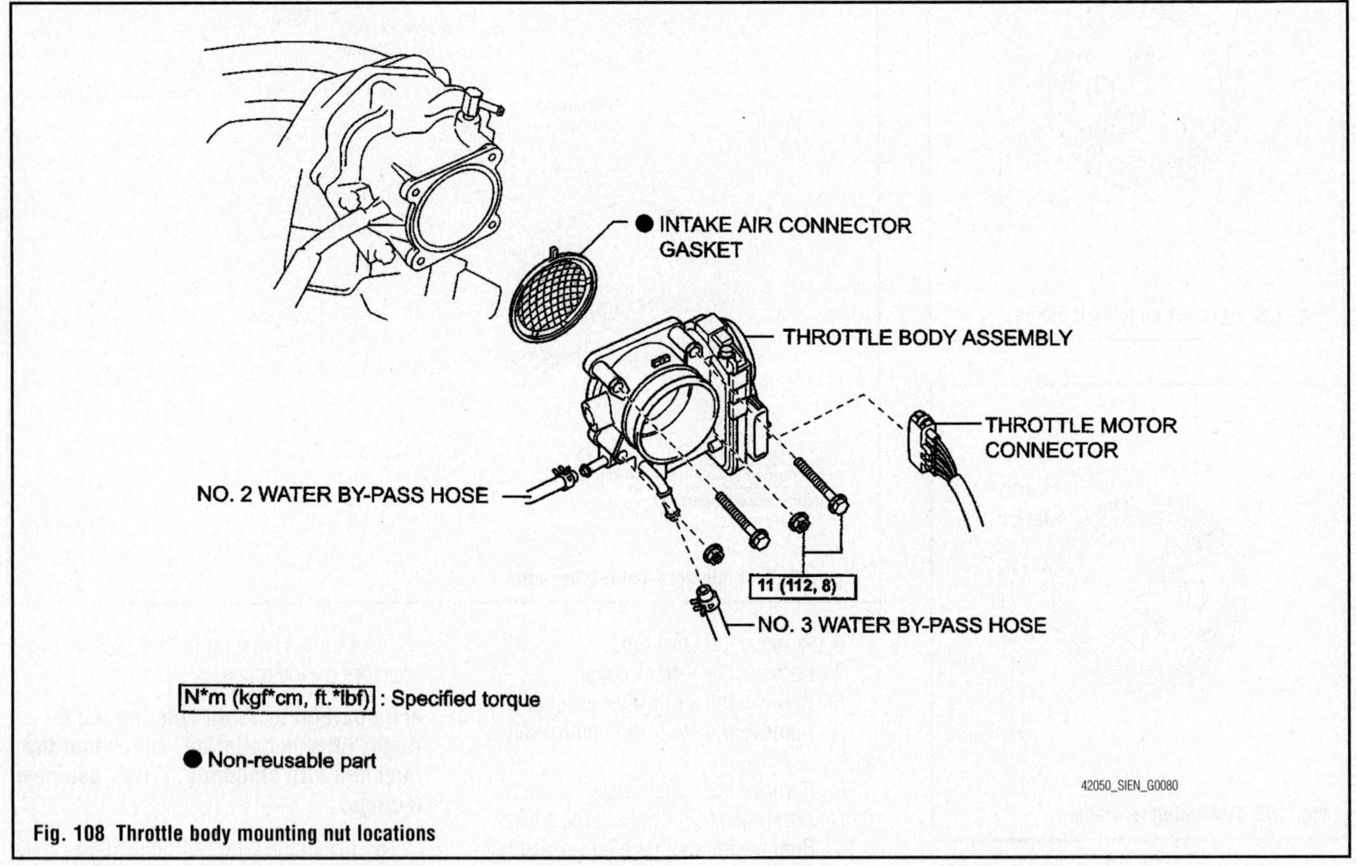

Fig. 108 Throttle body mounting nut locations

HEATING & AIR CONDITIONING SYSTEM

BLOWER MOTOR

REMOVAL & INSTALLATION

See Figure 109.

1. Before servicing the vehicle, refer to the precautions section.
2. Disconnect the negative battery cable.
3. Remove the instrument panel assembly.
4. Remove the clip and the heater to foot No. 1 duct.
5. Remove the ECM.
6. Remove the stereo amplifier assembly.
7. Release the clamps, unfasten the nuts and the wiring harness.
8. Remove the bolt, screws, nut and the blower assembly.
9. Remove the 3 screws and the blower motor.

To install:

10. Install the blower motor and tighten the screws.
11. Install the blower assembly, tighten the fasteners and attach the wiring connector.
12. Install the stereo amplifier assembly.
13. Install the ECM.
14. Install the heater to foot No. 1 duct and clip.

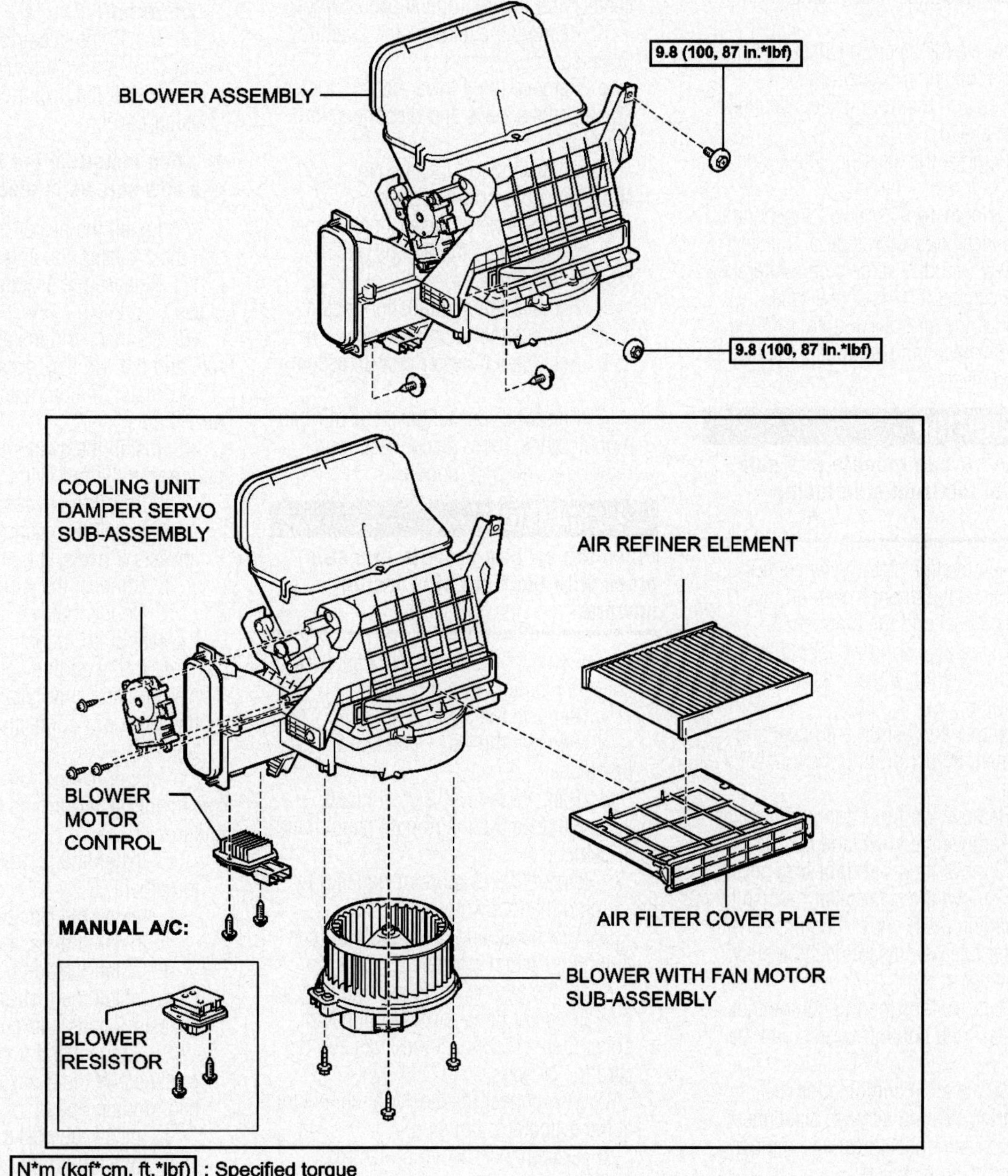

Fig. 109 Exploded view of the front blower motor assembly

15. Install the instrument panel assembly.
16. Connect the negative battery cable.

HEATER CORE

REMOVAL & INSTALLATION

See Figures 110 through 115.

1. Before servicing the vehicle, refer to the precautions section.
2. Disconnect the negative battery cable.
3. Drain the cooling system into a clean container for reuse.
4. Disconnect the heater hoses from the heater core.
5. Remove the steering wheel by performing the following procedure:
   a. Position the front wheels facing straight-ahead.
   b. Remove the steering wheel side covers.
   c. Using a Torx® wrench, loosen the 2 screws located at each side of the steering wheel until the screw's circumference groove catches on the screw case.
   d. Pull the air bag module from the steering wheel and disconnect the electrical connector.

✱✱ CAUTION

Place the air bag module in a safe place with the front side facing upward.

   e. Remove the steering wheel nut.
   f. Place alignment marks on the steering wheel and the main shaft.
   g. Using a steering wheel puller, press the steering wheel from the steering column.
6. Remove the instrument panel and reinforcement by performing the following procedure:
   a. Remove the front door scuff plates.
   b. Remove the cowl side boards.
   c. Remove the front door trim covers.
   d. Remove the front pillar garnish by disengaging the 5 clips. If equipped with a tweeter speaker, disconnect the electrical connector.
   e. Remove the steering column covers-to-steering column screws and the covers.
   f. Remove the combination switch-to-steering column screws, disconnect the electrical connector(s) and remove the combination switch.
   g. Remove the 2 hood open lever screws and the hood open lever.
   h. Remove the 2 lower finish panel bolts and disengage the panel from the 3 clips.
   i. Remove the 2 No. 1 safety pad insert bolts and the insert.
   j. Remove the 2 No. 2 finish panel bolts and disengage the panel from the 4 clips.
   k. In the left side of the glove compartment, pry out the glove box door finish plate and disconnect the air bag module connector.
   l. Remove the glove box 3 nuts and 2 screws and the glove box.
   m. Remove the center cluster finish panel by disengaging the claw (bottom center) and 4 clips (one at each corner).
   n. Remove the ashtray, the 2 ashtray receptacle box screws.
   o. Remove the 4 lower center cluster finish panel screws and disconnect the connector.
   p. Remove the clock, the No. 1 and No. 2 registers from the panel.
   q. Remove the 3 cluster finish panel screws, disengage the 8 clips and remove the panel.
   r. Remove the combination meter.
   s. Remove the radio assembly.
   t. Remove the heater control assembly.
   u. Remove 2 passenger's side air bag module bolts; then, disconnect and remove the air bag module.

✱✱ CAUTION

Place the air bag module in a safe place with the front side facing upward.

   v. Remove the instrument panel-to-chassis 5 bolts and nut.
   w. Remove the audio amplifier.
   x. Remove the No. 1 and No. 2 braces.
   y. Remove the No. 2 cowl brace.
   z. Remove the instrument panel reinforcement.
7. Remove the evaporator housing by performing the following procedure:
   a. Discharge and recover the air conditioning system refrigerant.
   b. In the engine compartment, remove the refrigerant lines-to-cowl connector bolts; then, disconnect the lines and discard the O-rings.
   c. Disconnect the electrical connector at the evaporator housing.
   d. Disconnect the wiring harness clamp.
   e. Remove the evaporator housing-to-chassis 2 rivets, 3 bolts and nut.
   f. Remove the evaporator housing.
8. Remove the 4 defroster nozzle nuts and the nozzle.
9. Disconnect and remove the theft deterrent and the wireless door lock ECUs.
10. Release the 2 air duct claws and the air duct.
11. Remove the 2 heater housing-to-chassis rivets and the heater housing.

➡When installing the heater housing, use new screws in place of the rivets.

12. Remove the heater core-to-heater housing cover.
13. Remove both heater core screws and clamps; then, remove the heater core.

To install:

14. Install the heater core and both heater core screws and clamps.
15. Install the heater core-to-heater housing cover.

➡When installing the heater housing, use new screws in place of the rivets.

16. Install the heater housing-to-chassis and the 2 heater housing screws.
17. Release the air duct and the air duct claws.
18. Connect and install the theft deterrent and the wireless door lock ECUs.
19. Install the defroster nozzle and the 4 nozzle nuts.
20. Install the evaporator housing by performing the following procedure:
   a. Install the evaporator housing.
   b. Install the evaporator housing-to-chassis 2 rivets, 3 bolts and nut.
   c. Connect the wiring harness clamp.
   d. Connect the electrical connector at the evaporator housing.
   e. In the engine compartment, use new O-rings and install the refrigerant lines-to-cowl connector and install the bolts.
21. Install the instrument panel and reinforcement by performing the following procedure:
   a. Install the instrument panel reinforcement.
   b. Install the No. 2 cowl brace.
   c. Install the No. 1 and No. 2 braces.
   d. Install the audio amplifier.
   e. Install the instrument panel-to-chassis 5 bolts and nut.
   f. Connect and install the air bag module and the 2 passenger's side air bag module bolts.
   g. Install the heater control assembly.
   h. Install the radio assembly.
   i. Install the combination meter.
   j. Install the cluster finish panel, engage the 8 clips and install the panel screws.

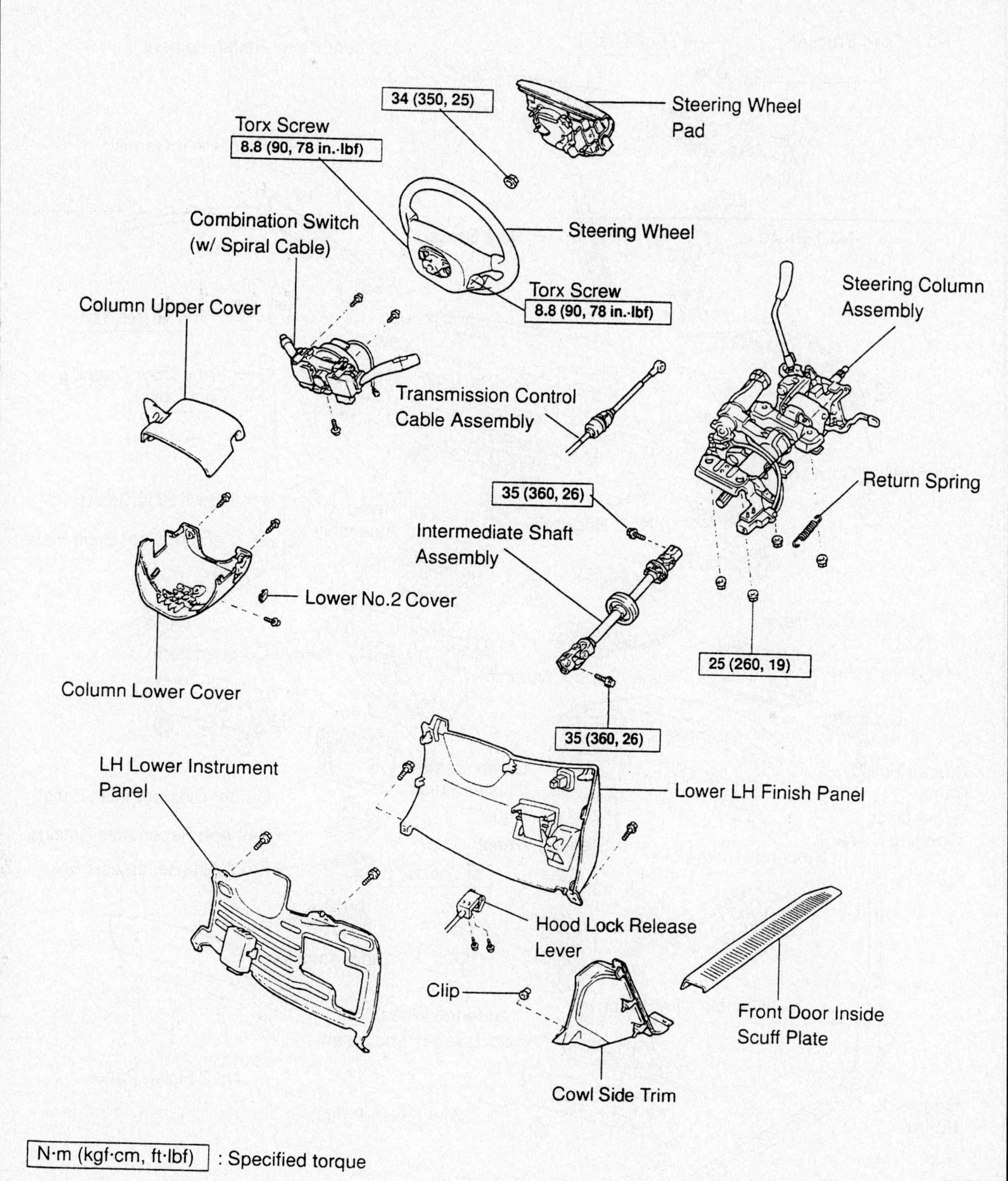

Fig. 110 Exploded view of the steering wheel, steering column and related components

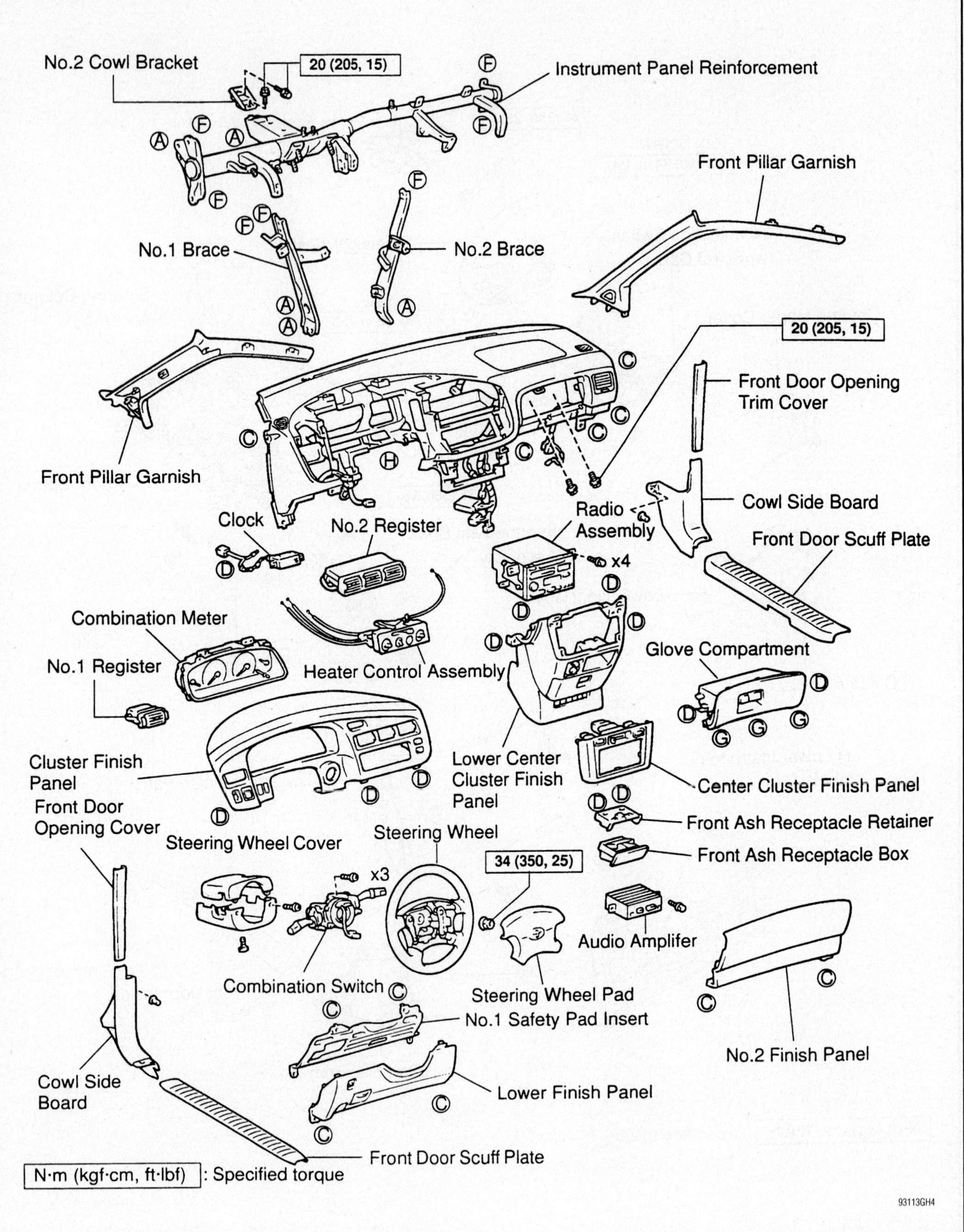

Fig. 111 Exploded view of the instrument panel and related components

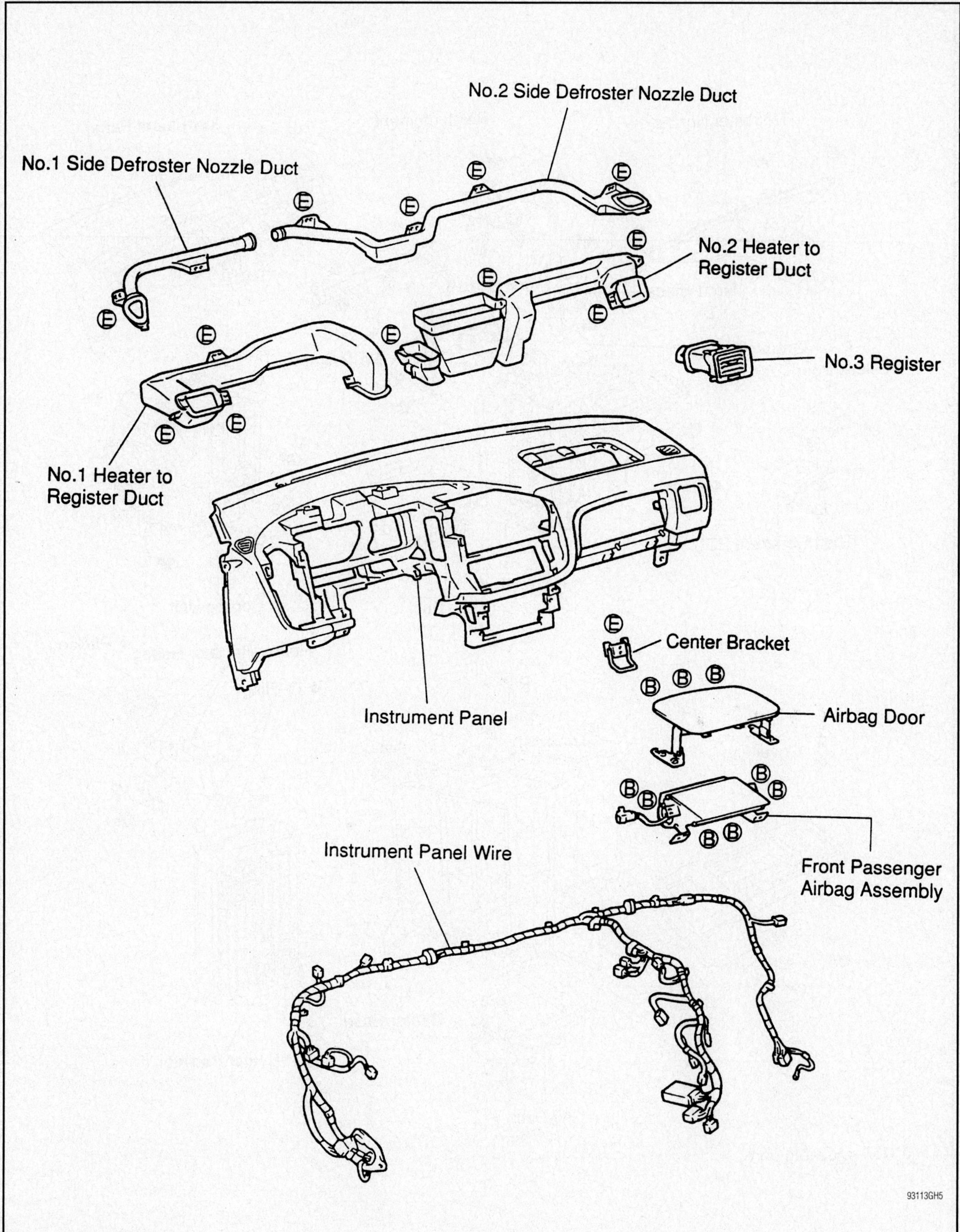

Fig. 112 Exploded view of the ventilation system and related components

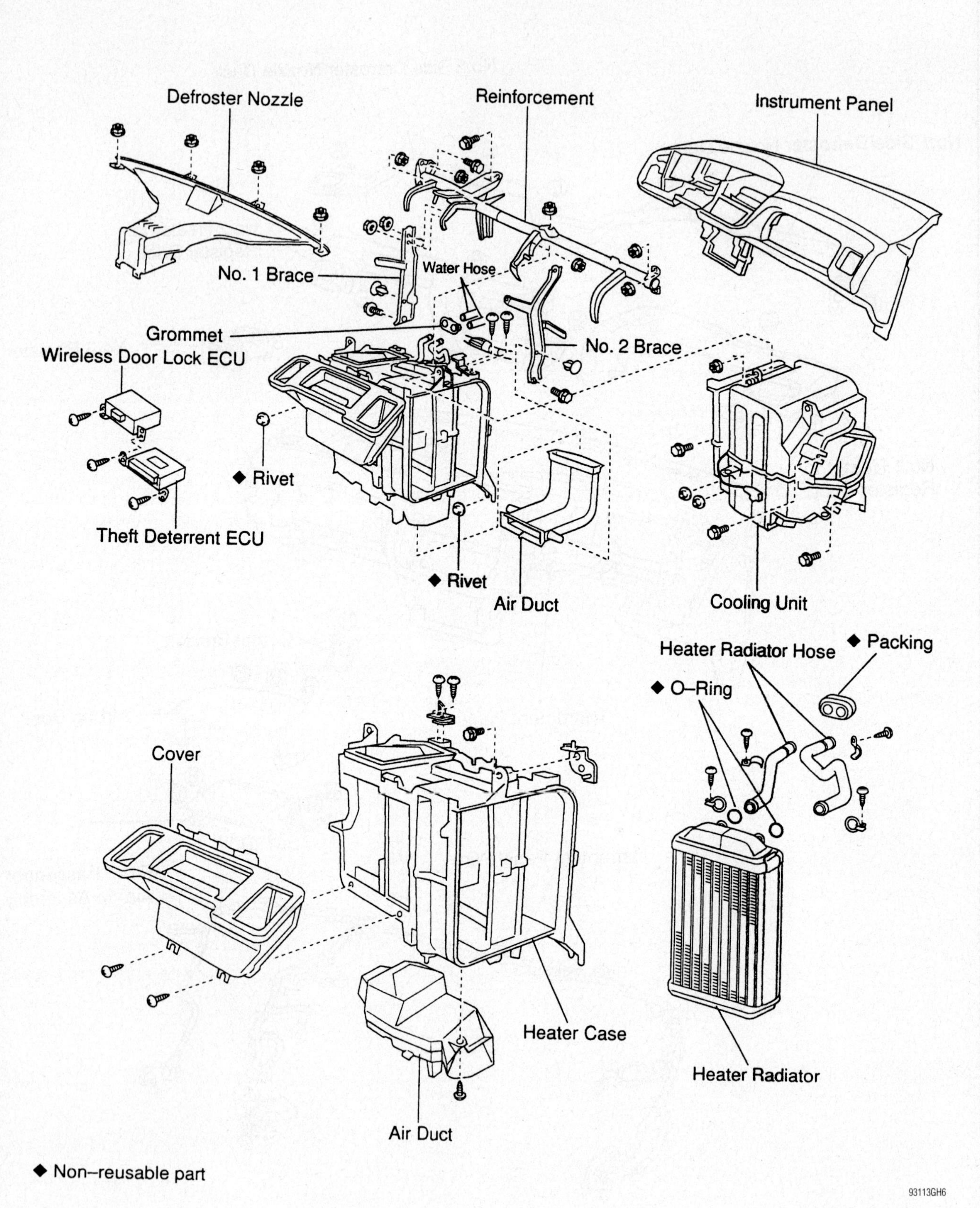

Fig. 113 Exploded view of the heater core, heater housing, evaporator housing and related components

k. Install the No. 1 and No. 2 registers and the clock to the panel.

l. Connect the lower center cluster finish panel connector and install the 4 lower center cluster finish panel screws.

m. Install the 2 ashtray receptacle box screws and the ashtray.

n. Install the center cluster finish panel by engaging the 4 clips (1 at each corner) and the claw (bottom center).

o. Install the glove box and the glove box 3 nuts and 2 screws.

p. In the left side of the glove compartment, connect the air bag module connector and install the glove box door finish plate.

q. Install the No. 2 finish panel, engage the 4 panel clips and install the 3 panel bolts.

r. Install the No. 1 safety pad insert and the 2 insert bolts.

s. Install the finish panel, engage the 3 finish panel clips and install 2 lower finish panel bolts.

t. Install the hood open lever and the 2 hood open lever screws.

u. Install the combination switch, connect the electrical connector(s) and install the combination switch-to-steering column screws.

v. Install the steering column covers and the covers-to-steering column screws.

w. Install the front pillar garnish by engaging the 5 clips. If equipped with a tweeter speaker, connect the electrical connector.

x. Install the front door trim covers.

y. Install the cowl side boards.

z. Install the front door scuff plates.

22. Install the steering wheel by performing the following procedure:

a. Install the steering wheel to the steering column.

b. Align the steering wheel-to-main shaft marks.

c. Install the steering wheel nut and torque the nut to 25 ft. lbs. (34 Nm).

d. Install the air bag module to the steering wheel and connect the electrical connector.

e. Using a Torx® wrench, tighten the steering wheel screws to 78 inch lbs. (8.8 Nm).

f. Install the steering wheel side covers.

23. Connect the heater hoses to the heater core.

24. Refill the cooling system.

25. Connect the negative battery cable.

26. Evacuate and charge the air conditioning system.

27. Run the engine to normal operating temperatures; then, check the climate control operation and check for leaks.

AUXILIARY HEATING & AIR CONDITIONING SYSTEM

BLOWER MOTOR

REMOVAL & INSTALLATION

See Figure 114.

1. Before servicing the vehicle, refer to the precautions section.
2. Disconnect the negative battery cable.
3. Disconnect the negative battery cable.
4. Evacuate the A/C system using approved equipment.
5. Remove the scuff plates from the back door and right hand rear door.
6. Remove the quarter trim panel garnish.
7. Remove the rear headliner garnish.
8. Remove the rear right window side garnish assembly.
9. Unfasten the clips and remove the rear air duct.
10. Unfasten the claw retainers and remove the cover sub assembly.
11. Remove the air duct assembly.
12. Unfasten the heater hose clamp and disconnect the heater hose.
13. Unfasten the 2 bolts and disconnect the A/C tube and accessory assembly.
14. Unfasten the connectors and clamps, unplug the wiring harness and remove the blower assembly.
15. Unfasten the bolts and remove the blower assembly.

To install:

16. Installation is the reverse of removal.

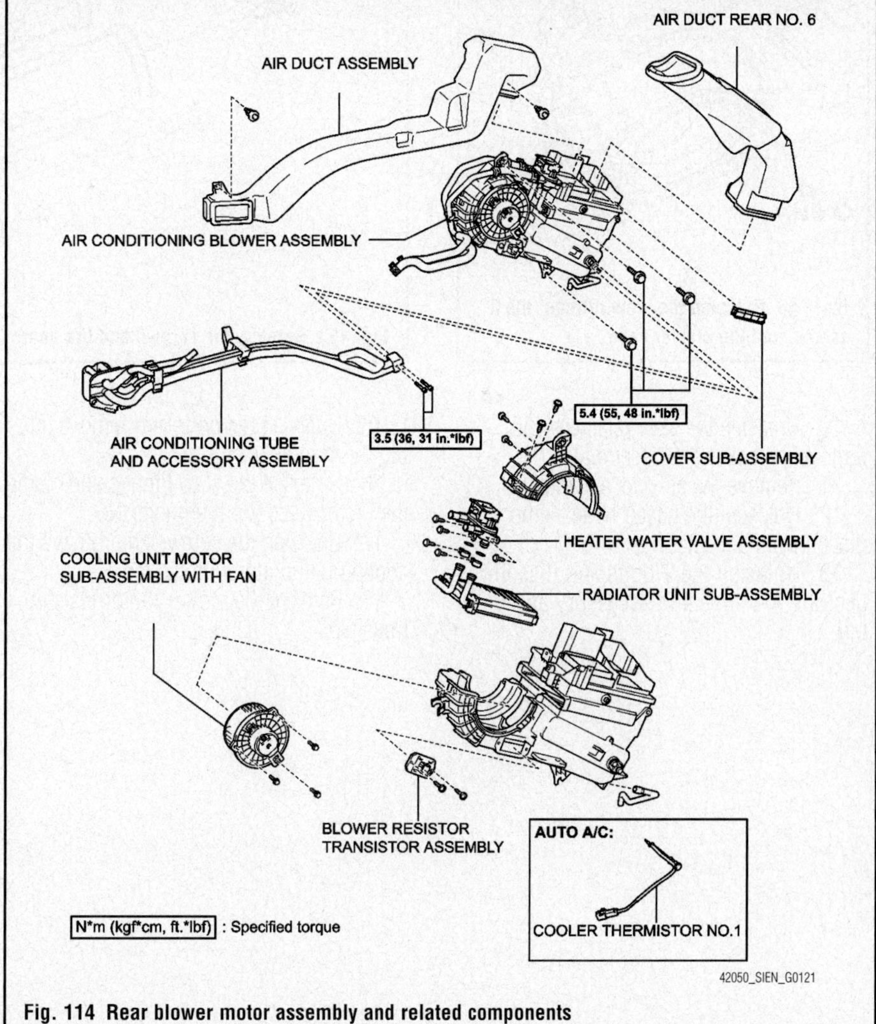

Fig. 114 Rear blower motor assembly and related components

HEATER CORE

REMOVAL & INSTALLATION

See Figures 115 through 117.

1. Before servicing the vehicle, refer to the precautions section.
2. Disconnect the negative battery cable.
3. Drain and recycle the engine coolant.
4. Evacuate the A/C system using approved equipment.
5. Remove the scuff plates from the back door and right hand rear door.
6. Remove the quarter trim panel garnish.
7. Remove the rear headliner garnish.
8. Remove the rear right window side garnish assembly.
9. Unfasten the clips and remove the rear air duct.

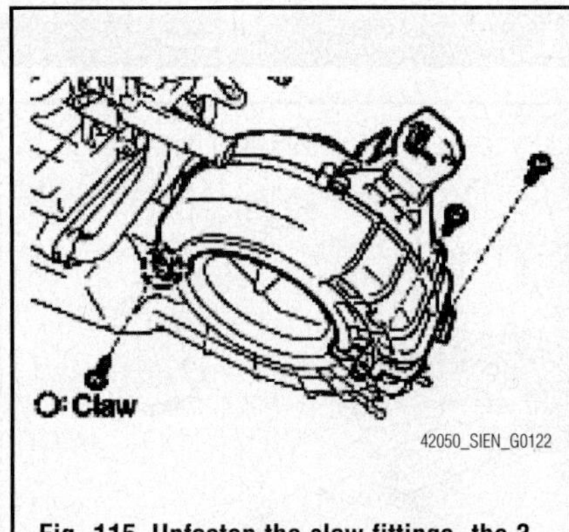

Fig. 115 Unfasten the claw fittings, the 3 screws and the blower case

10. Unfasten the claw retainers and remove the cover sub assembly.
11. Remove the air duct assembly.
12. Unfasten the heater hose clamp and disconnect the heater hose.
13. Unfasten the 2 bolts and disconnect the A/C tube and accessory assembly.

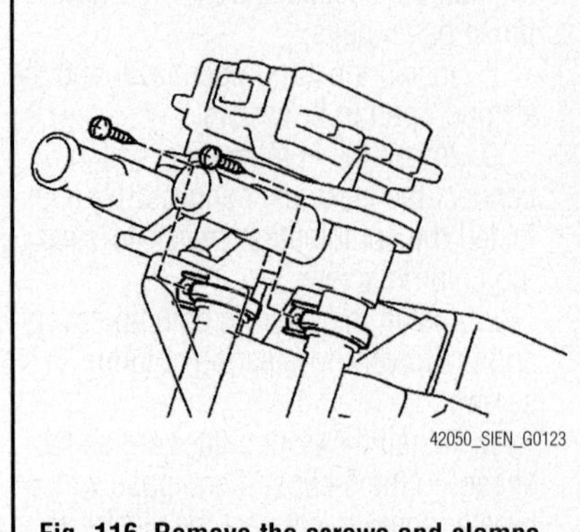

Fig. 116 Remove the screws and clamps

14. Unfasten the connectors and clamps, unplug the wiring harness and remove the blower assembly.
15. Unfasten the bolts and remove the blower assembly.
16. Unfasten the claw fittings and clamp, then remove cooler thermistor No. 1.
17. Unfasten the screws and remove the cooling unit motor with the fan.
18. Remove the blower motor resistor transistor.
19. Remove the rear heater core unit as follows:
   a. Unfasten the claw fittings, the 3 screws and the blower case.
   b. Remove the screws and clamps.
   c. Remove the screws and the heater water valve. Remove the O-rings from the valve.
   d. Remove the heater core.

To install:

20. Install the heater core.
21. Install new O-rings on the water valve, install the valve and the screws.
22. Install the valve screws and clamps.
23. Install the blower case.
24. Install the blower motor resistor.
25. Install the cooling fan motor assembly with the fan.

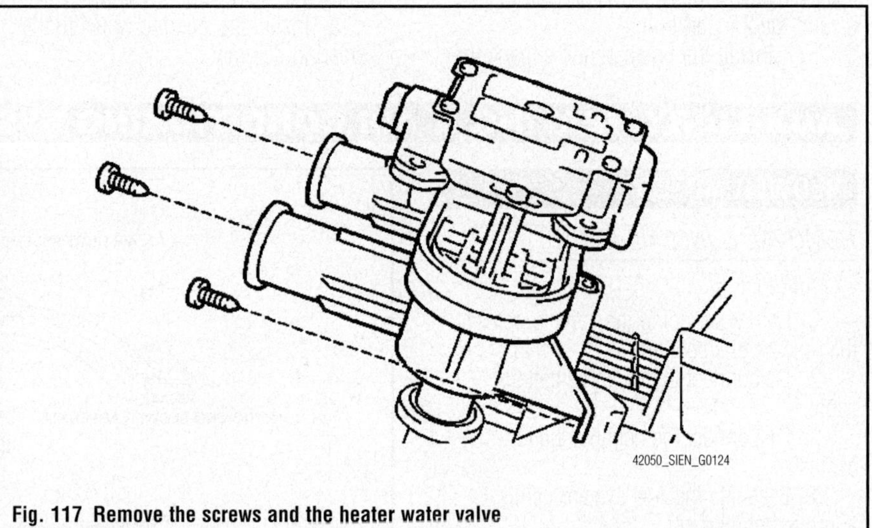

Fig. 117 Remove the screws and the heater water valve

26. Install the No. 1 thermistor.
27. Install the remaining components in the reverse order of removal.
28. Recharge the A/C system.
29. Fill the cooling system, start the vehicle, allow to reach operating system; turn on the rear heat and check for leaks.

STEERING

POWER RACK & PINION STEERING GEAR

REMOVAL & INSTALLATION

See Figure 118.

1. Place the wheels in a straight-ahead position.
2. Remove the wheels.
3. Matchmark and remove the intermediate shaft from the gear.
4. Remove the steering column hole cover from the dash panel.
5. Disconnect the tie rod ends.
6. Disconnect the stabilizer bar.
7. Disconnect the pressure and return lines.

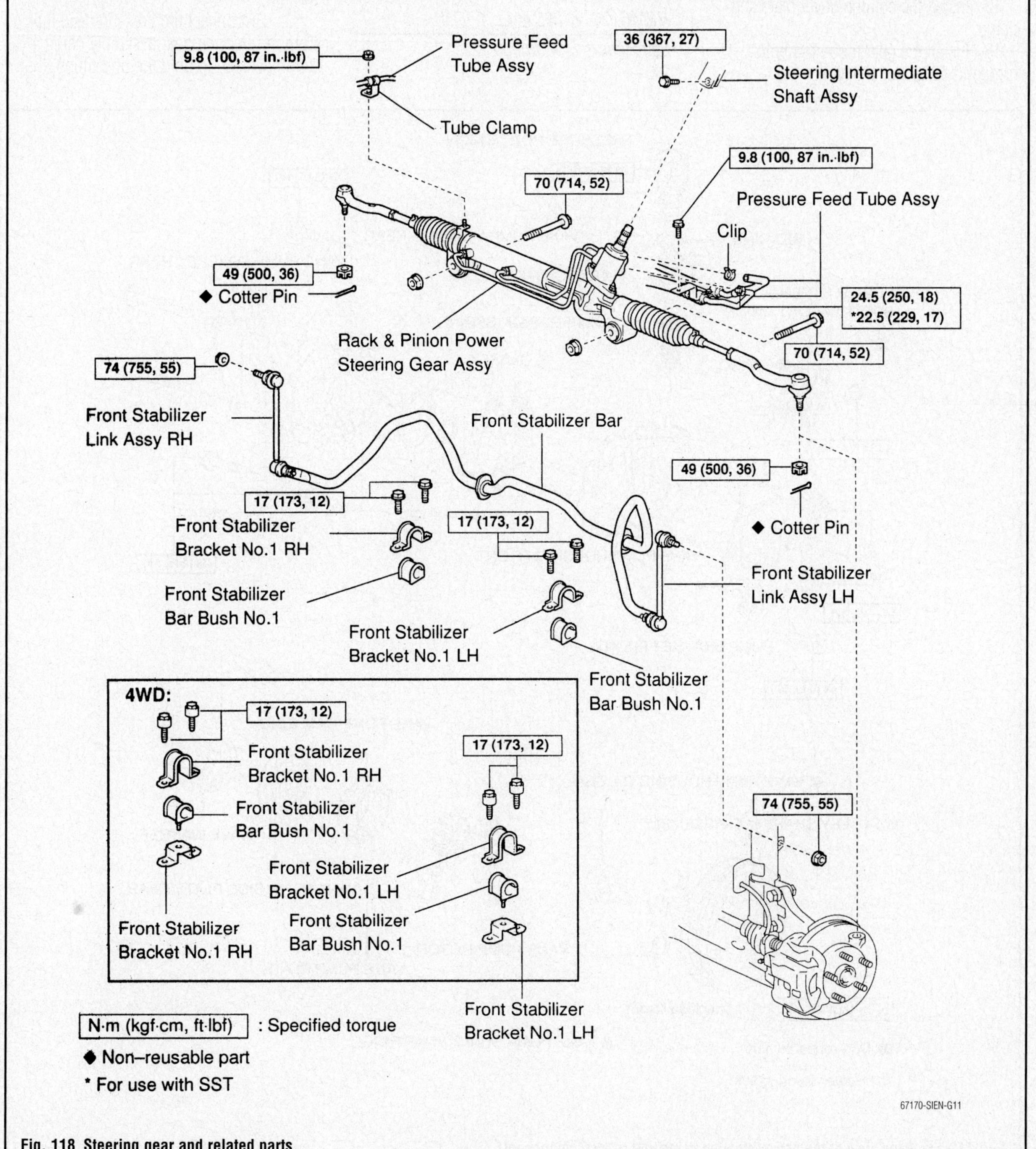

Fig. 118 Steering gear and related parts

8. Remove the 2 gear mounting bolts.
9. Matchmark the intermediate shaft extension and control valve shaft.
10. Remove the intermediate shaft extension bolt and remove the gear.

To install:

11. Place the gear in position. Align the matchmarks and connect the intermediate extension shaft. Torque to 27 ft. lbs. (36 Nm).
12. Install the column cover plate and clamp.
13. Install the gear mounting bolts. Torque to 52 ft. lbs. (70 Nm).
14. Connect the stabilizer bar. Torque to 12 ft. lbs. (17 Nm).
15. Connect the tie rods ends. Torque to 36 ft. lbs. (49 Nm).
16. Connect the intermediate shaft. Torque to 26 ft. lbs. (36 Nm).
17. Check the alignment.

POWER STEERING PUMP

REMOVAL & INSTALLATION

See Figures 119 through 121.

1. Before servicing the vehicle, refer to the precautions section.
2. Remove the right hand front wheel.
3. Drain the power steering fluid.
4. Remove the right side fender apron.
5. Disconnect the clip and disconnect oil reservoir to No. 1 pump hose.
6. Disconnect the connector from the oil pressure switch and remove the switch.
7. Disconnect the pressure feed tube from the steering gear assembly using a 24mm wrench to hold the port union, then

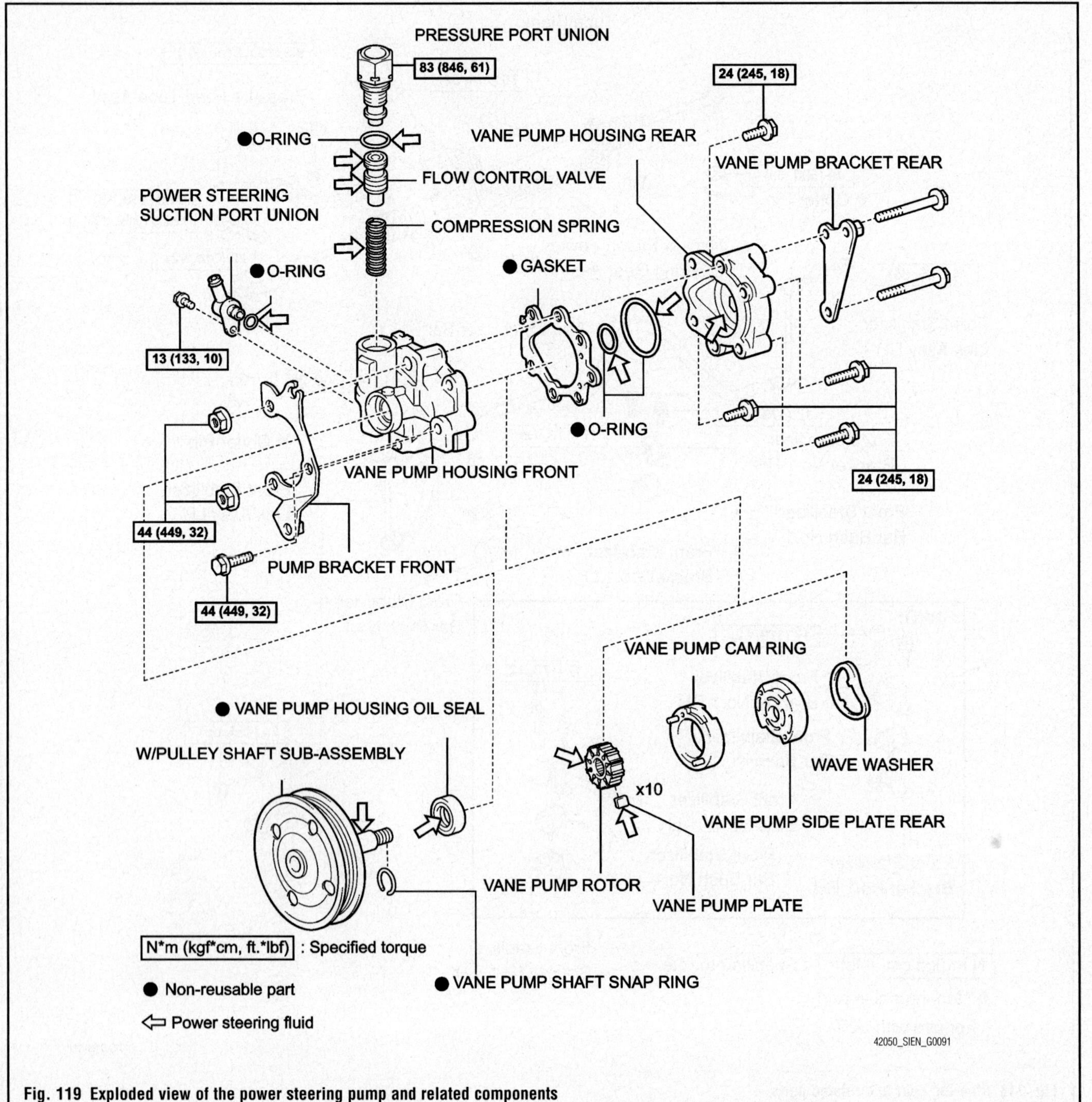

Fig. 119 Exploded view of the power steering pump and related components

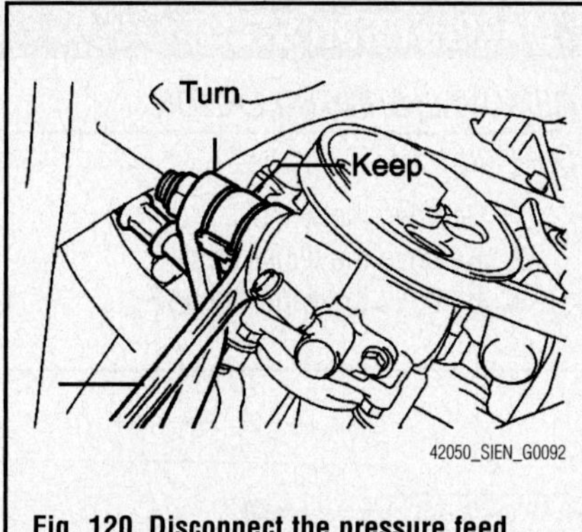

Fig. 120 Disconnect the pressure feed tube from the steering gear assembly

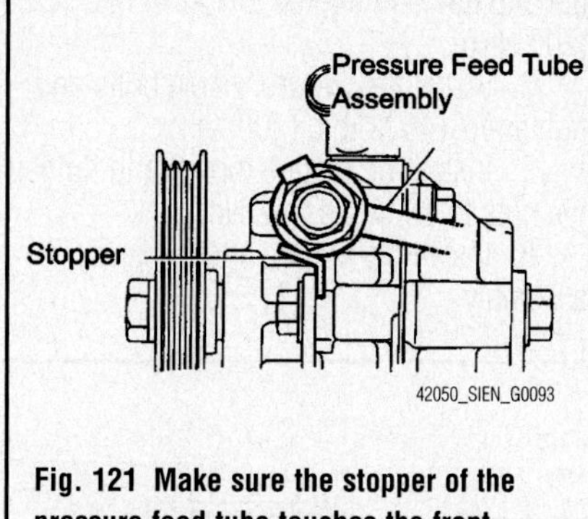

Fig. 121 Make sure the stopper of the pressure feed tube touches the front bracket

remove the union bolt and gasket. Disconnect the feed tube assembly.

8. Remove the drive belt.
9. Remove the pump bolts and the pump.

To install:

10. Install the pump assembly and hand tighten the 2 bolts.
11. Install the drive belt and adjust the belt tension.
12. Tighten the bolts to 32 ft. lbs. (44 Nm).
13. Install a new gasket and the pressure feed tube using the union bolt. Tighten the union bolt to 38 ft. lbs. (52 Nm)

➡Make sure the stopper of the tube touches the front bracket as illustrated.

14. Install the oil pressure switch and tighten to 15 ft. lbs. (21 Nm).
15. Attach the oil pressure switch connector.
16. Connect the return hose and clip.
17. Install the right side fender apron.
18. Install the right hand front wheel.
19. Fill the power steering fluid reservoir.
20. Bleed the power steering system.

BLEEDING

1. Before servicing the vehicle, refer to the precautions section.
2. Check the fluid level and top off as needed.
3. Jack up the front of the vehicle and support it with safety stands.
4. With the engine OFF, turn the wheel from lock to lock slowly several times.
5. Lower the vehicle and start the engine and allow to idle for a few minutes.
6. With the engine warm and at idle, turn the steering wheel to the left or right lock position and hold it there for 2 to 3 seconds , then turn the wheel to the opposite side lock and hold for 2 to 3 seconds.
7. Repeat these last two steps several times.
8. Turn the vehicle off and check the fluid level.
9. if there is a lot of foam in the reservoir, check the system for leaks and repair, then bleed the system again.

SUSPENSION

FRONT SUSPENSION

LOWER BALL JOINT

REMOVAL & INSTALLATION

1. Remove the wheel.
2. Remove the hub nut.
3. Remove the speed sensor.
4. Remove the caliper, and hang it out of the way.
5. Remove the rotor.
6. Remove the lower arm from the ball joint.
7. Remove the lower ball joint nut and cotter pin.
8. Using a puller, remove the ball joint from the knuckle.
9. Installation is the reverse of removal. Torque the ball joint stud nut to 91 ft. lbs. (123 Nm). Torque the arm-to-ball joint to 94 ft. lbs. (127 Nm).Torque the hub nut to 217 ft. lbs. (294 Nm) and stake it.

LOWER CONTROL ARM

REMOVAL & INSTALLATION

See Figure 122.

1. Before servicing the vehicle, refer to the precautions section.
2. Remove the engine and transaxle assembly.

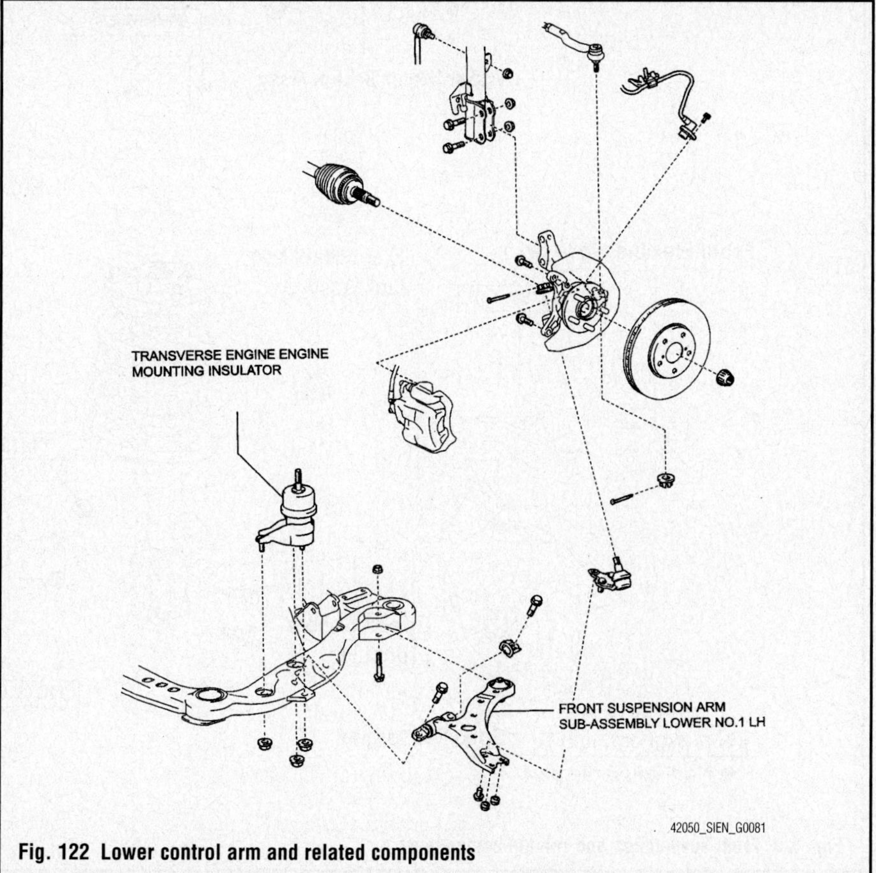

Fig. 122 Lower control arm and related components

3. Remove the nuts and the transverse engine mount.
4. Remove the two bolts on the front of the lower arm. Remove the nut and bolt, then remove the arm.

To install:

5. Install the arm assembly, install the nut and bolt and tighten to 152 ft. lbs. (206 Nm).
6. Install the two front side bolts and tighten to 148 ft. lbs. (200 Nm).
7. Install the engine mount and tighten the nuts to 64 ft. lbs. (87 Nm).
8. Install the engine and transaxle assembly.

MACPHERSON STRUT

REMOVAL & INSTALLATION

See Figure 123.

1. Remove the wheel.
2. Remove the wiper arms.
3. Remove the wiper motor.

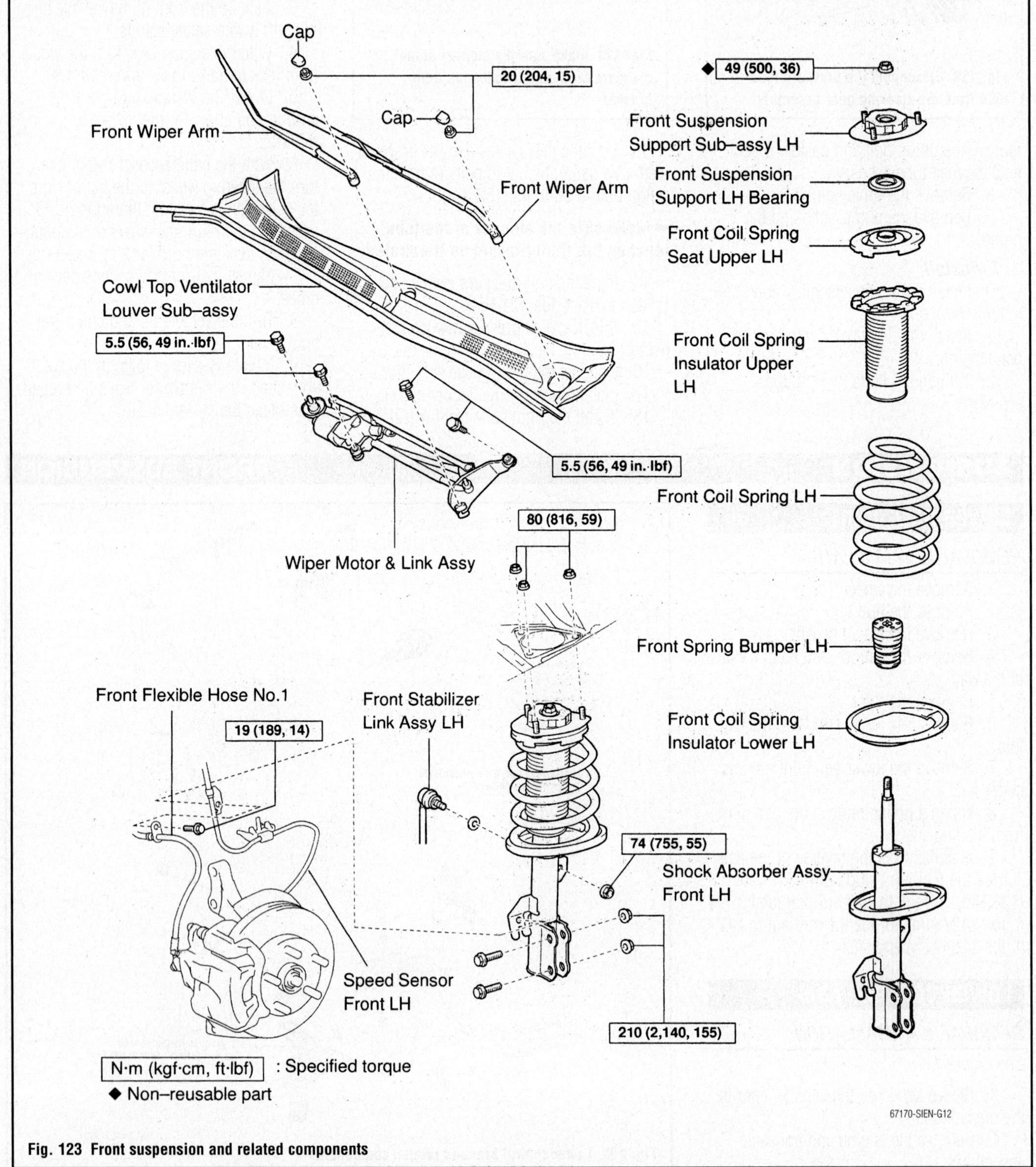

Fig. 123 Front suspension and related components

4. Remove the top cowl.
5. Remove the stabilizer link from the strut.
6. Loosen the strut rod locknut. Don't remove it.
7. Remove the brake hose bracket from the strut.
8. Remove the lower strut bolts.
9. Remove the 3 upper strut nuts.
10. Remove the strut.

To install:

11. Installation is the reverse of removal. Observe the following torques:
    - Upper nuts: 59 ft. lbs. (80 Nm)
    - Lower nuts/bolts: 155 ft. lbs. (210 Nm)
    - Strut rod locknut: 36 ft. lbs. (49 Nm)
    - Stabilizer link: 55 ft. lbs. (74 Nm)

STEERING KNUCKLE

REMOVAL & INSTALLATION

1. Before servicing the vehicle, refer to the precautions section.
2. Remove or disconnect the following:
    - Front wheels
    - Fender apron seal
3. Check the bearing backlash and axle hub deviation, as follows:
    a. Remove the 2 brake caliper set bolts.
    b. Hang the caliper using stiff wire on the shock absorber assembly.
    c. Remove the rotor.
    d. Place a dial indicator near the center of the axle hub and check the backlash in the bearing shaft direction.
    e. Backlash maximum should read 0.0020 inch (0.05mm). If greater than specified, replace the bearing.
    f. Using the dial indicator, check the deviation at the surface of the axle hub outside and hub bolt. Maximum is 0.0020 inch (0.05mm). If greater than specified, replace the axle hub.
4. Install the rotor and caliper assembly.
5. Remove or disconnect the following:
    - Cotter pin (discard it) and lock cap off the center hub nut
    - Driveshaft locknut, by applying the front brakes
    - Tie rod end, from the steering knuckle
    - Left and right stabilizer end brackets, from the lower arms
    - Both nuts and the lower arm from the ball joint
    - Driveshaft from the axle hub. Secure the shaft aside using wire.

**** WARNING**

Be careful not to damage the shaft boot or Anti-lock Brake System (ABS) sensor rotor.

- Both brake caliper mounting bolts and the caliper.

➡Support caliper from the vehicle using wire.

- Brake rotor
- Sensor from the steering knuckle, if equipped with ABS
- Both nuts from the lower end of the shock
- Steering knuckle and hub assembly

To install:

6. Install or connect the following:
    - Steering knuckle/hub assembly and temporarily install the lower shock bolts
    - Lower ball joint to the lower arm. Tighten the bolt and nuts to 94 ft. lbs. (127 Nm).
    - Tie rod to the knuckle. Tighten the nut to 36 ft. lbs. (49 Nm).
    - New cotter pin
    - Tighten the lower shock nuts to 156 ft. lbs. (211 Nm).
    - Both side stabilizer end brackets to the lower arm.
    - Front ABS sensor. Tighten it to 69 inch lbs. (8 Nm).
    - Front brake rotor and caliper.
    - Driveshaft locknut, by applying the brakes. Tighten it to 217 ft. lbs. (294 Nm).
    - Lock cap and new cotter pin
    - Front fender apron seal
    - Front wheel. Tighten the lug nuts to 76 ft. lbs. (103 Nm).

STABILIZER BAR

REMOVAL & INSTALLATION

2WD Models

See Figures 124 and 125.

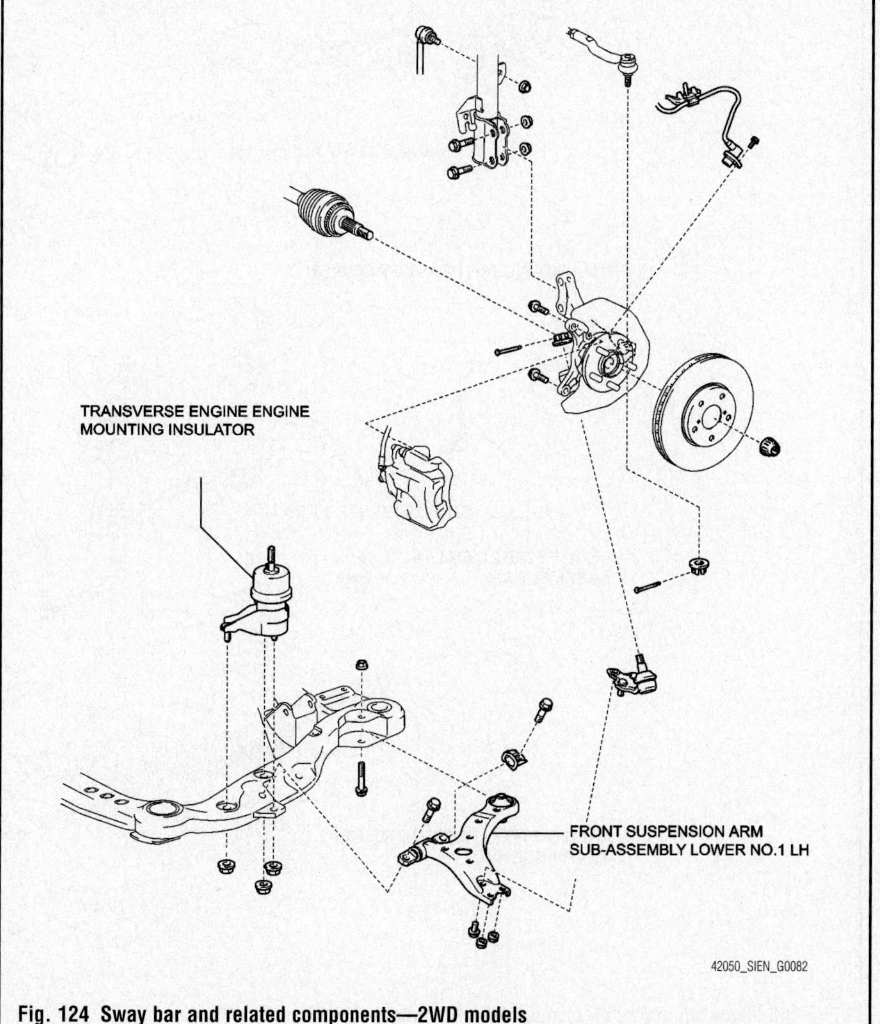

Fig. 124 Sway bar and related components—2WD models

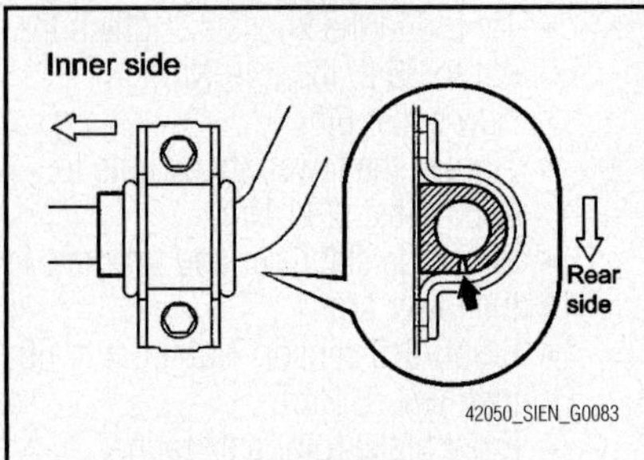

Fig. 125 Install the stabilizer bar No.1 bushing with the outer side of the bushing stopper on the stabilizer bar—2WD

1. Before servicing the vehicle, refer to the precautions section.
2. Remove the wheels.
3. Remove the stabilizer bar links on both sides.
4. Remove the bolts for the stabilizer bar brackets. Remove the brackets from both sides.
5. Disconnect the tie rod ends from the steering knuckle.
6. Disconnect the power steering pressure tube.
7. Separate the steering intermediate shaft.
8. Remove the power steering gear assembly.
9. Remove the stabilizer bar No.1 bushing and the stabilizer bar.
10. Remove the power steering gear set bolts and nuts, then remove the stabilizer bar from the left side.

To install:

11. Install the stabilizer bar.
12. Install the stabilizer bar No.1 bushing with the outer side of the bushing stopper on the stabilizer bar as illustrated.
13. Install the steering intermediate shaft and the power steering gear, then connect the pressure feed tube.
14. Attach the tie rod end, tighten the nut to 36 ft. lbs. (49 Nm) and install a new cotter pin.
15. Install the brackets and bushings. Tighten the bolts to 12 ft. lbs. (17 Nm).
16. Install the stabilizer bar links and tighten the nuts to 55 ft. lbs. (74 Nm).
17. Install the wheels and tighten to 76 ft. lbs. (103 Nm).
18. Perform a wheel alignment check.

4WD Models

See Figures 126 and 127.

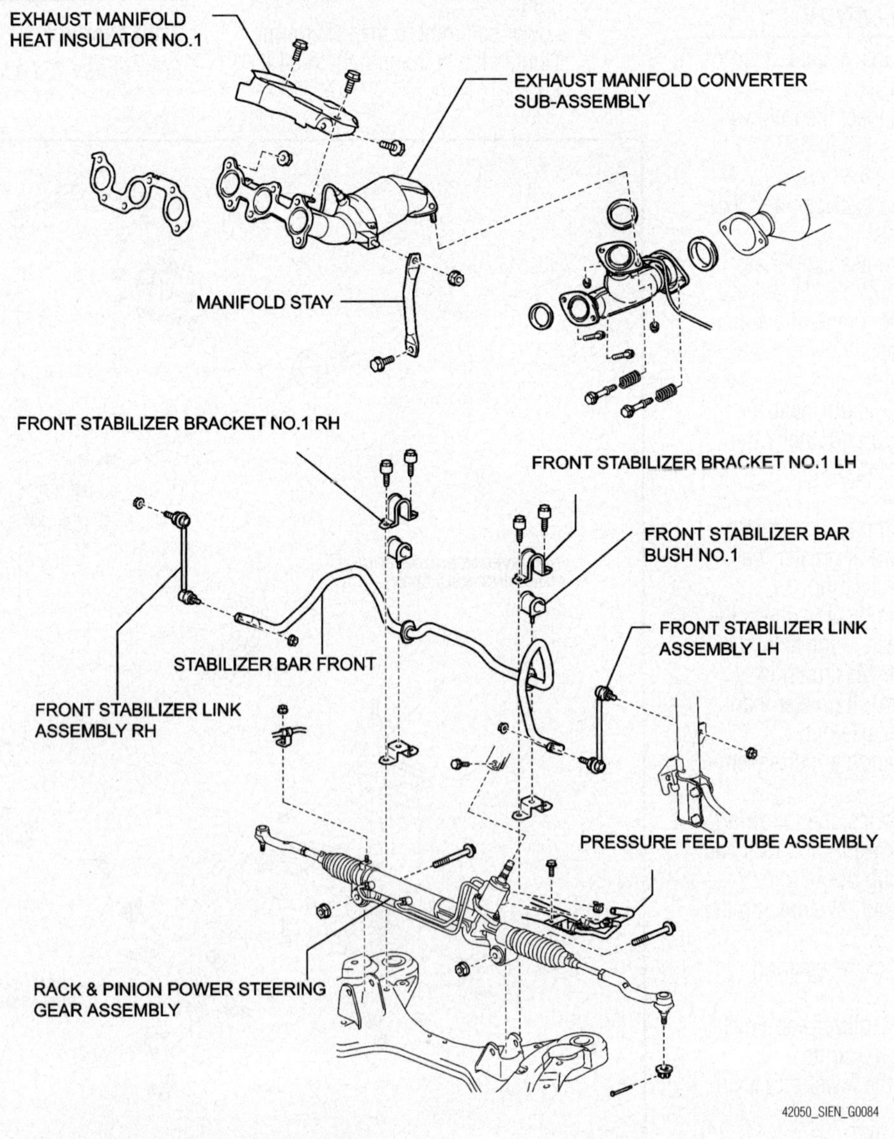

Fig. 126 Sway bar and related components—4 wheel drive Sienna

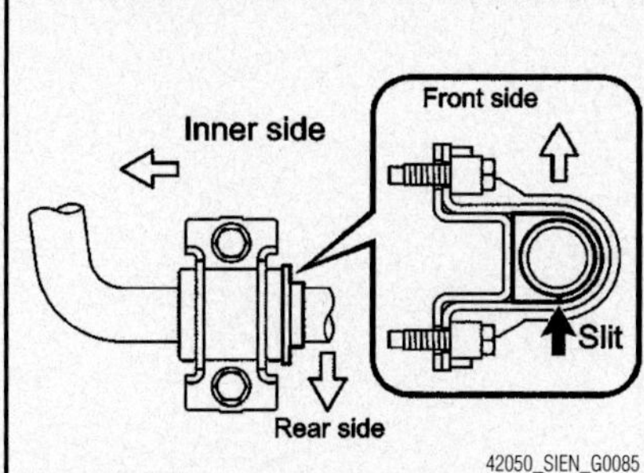

Fig. 127 Install the stabilizer bar No.1 bushing with the outer side of the bushing stopper on the stabilizer bar—4WD

1. Before servicing the vehicle, refer to the precautions section.
2. Remove the wheels.
3. Remove the stabilizer bar links on both sides.
4. Remove the exhaust pipe assembly.
5. Remove the bolts for the stabilizer bar brackets. Remove the brackets from both sides.
6. Disconnect the tie rod ends from the steering knuckle.
7. Disconnect the power steering pressure tube.
8. Separate the steering intermediate shaft.
9. Remove the power steering gear assembly.
10. Remove the exhaust manifold heat shield, manifold stay and manifold converter sub-assembly.
11. Remove the stabilizer bar No.1 bushing and the stabilizer bar.
12. Remove the power steering gear set bolts and nuts, then remove the stabilizer bar from the left side.

To install:

13. Install the stabilizer bar.

➡Install the bushings with the slit facing the rear of the vehicle.

14. Install the stabilizer bar No.1 bushing with the outer side of the bushing stopper on the stabilizer bar as illustrated.
15. Remove the exhaust manifold converter sub-assembly, manifold stay and heat shield.
16. Install the steering intermediate shaft and the power steering gear, then connect the pressure feed tube.
17. Attach the tie rod end, tighten the nut to 36 ft. lbs. (49 Nm) and install a new cotter pin.
18. Install the brackets and bushings. Tighten the bolts to 12 ft. lbs. (17 Nm).
19. Install the stabilizer bar links and tighten the nuts to 55 ft. lbs. (74 Nm).
20. Install the exhaust pipe assembly.
21. Install the wheels and tighten to 76 ft. lbs. (103 Nm).
22. Perform a wheel alignment check.

WHEEL HUB AND BEARING

REMOVAL & INSTALLATION

See Figures 128 and 129.

1. Before servicing the vehicle, refer to the precautions section.
2. Remove or disconnect the following:
   - Front wheels
   - Fender apron seal
3. Check the bearing backlash and axle hub deviation, as follows:
   a. Remove the 2 brake caliper set bolts.
   b. Hang the caliper using stiff wire on the shock absorber assembly.
   c. Remove the rotor.
   d. Place a dial indicator near the center of the axle hub and check the backlash in the bearing shaft direction.
   e. Backlash maximum should read 0.0020 inch (0.05mm). If greater than specified, replace the bearing.
   f. Using the dial indicator, check the deviation at the surface of the axle hub outside and hub bolt. Maximum is 0.0020 inch (0.05mm). If greater than specified, replace the axle hub.
4. Install the rotor and caliper assembly.
5. Remove or disconnect the following:
   - Cotter pin (discard it) and lock cap off the center hub nut
   - Driveshaft locknut, by applying the front brakes
   - Tie rod end, from the steering knuckle
   - Left and right stabilizer end brackets, from the lower arms
   - Both nuts and the lower arm from the ball joint
   - Driveshaft from the axle hub. Secure the shaft aside using wire.

⁂ WARNING

Be careful not to damage the shaft boot or Anti-lock Brake System (ABS) sensor rotor.

   - Both brake caliper mounting bolts and the caliper.

➡Support caliper from the vehicle using wire.

   - Brake rotor
   - Sensor from the steering knuckle, if equipped with ABS
   - Both nuts from the lower end of the shock
   - Steering knuckle and hub assembly
6. Clamp the steering knuckle in a vise with soft jaws to protect the knuckle.
7. Remove or disconnect the following:
   - Dust deflector from the hub, using a screwdriver
   - Bearing inner oil seal, by prying it from the knuckle
   - Snapring from the knuckle bore
   - Dust deflector from the steering knuckle
   - Axle hub, by pulling it from the dust deflector, using a 2-armed mechanical puller
   - Inner (inside) bearing race from the bearing, using the puller
   - Sensor control rotor from the axle hub, using Torx® wrench
   - Outer bearing race, using the puller
   - Outer bearing seal, using the puller
8. Position the inner (outside) race inside the bearing.
9. Using a brass rod, tap the bearing from the steering knuckle.

To install:

10. Clean all the oil seal and bearing seating surfaces with a clean, dry rag.
11. Install or connect the following:
   - Bearing into the bore, using a Bearing Driver tool 09608-32010 and a press
   - New outer oil seal, driving it into the steering knuckle, by inserting the seal side lip into the factory tool
   - Brake disc cover to the steering knuckle with the bolts
12. Apply multi-purpose grease between the oil seal lip, oil seal and bearing.
13. Install or connect the following:
   - Hub, by pressing it into the knuckle
   - New snapring into the knuckle
   - New oil seal, by pressing it into the knuckle once lubricated with multi-purpose grease
   - Dust deflector, by pressing it into the knuckle.

⁂ WARNING

Align the speed sensor holes in the dust deflector and steering knuckle, if equipped with ABS.

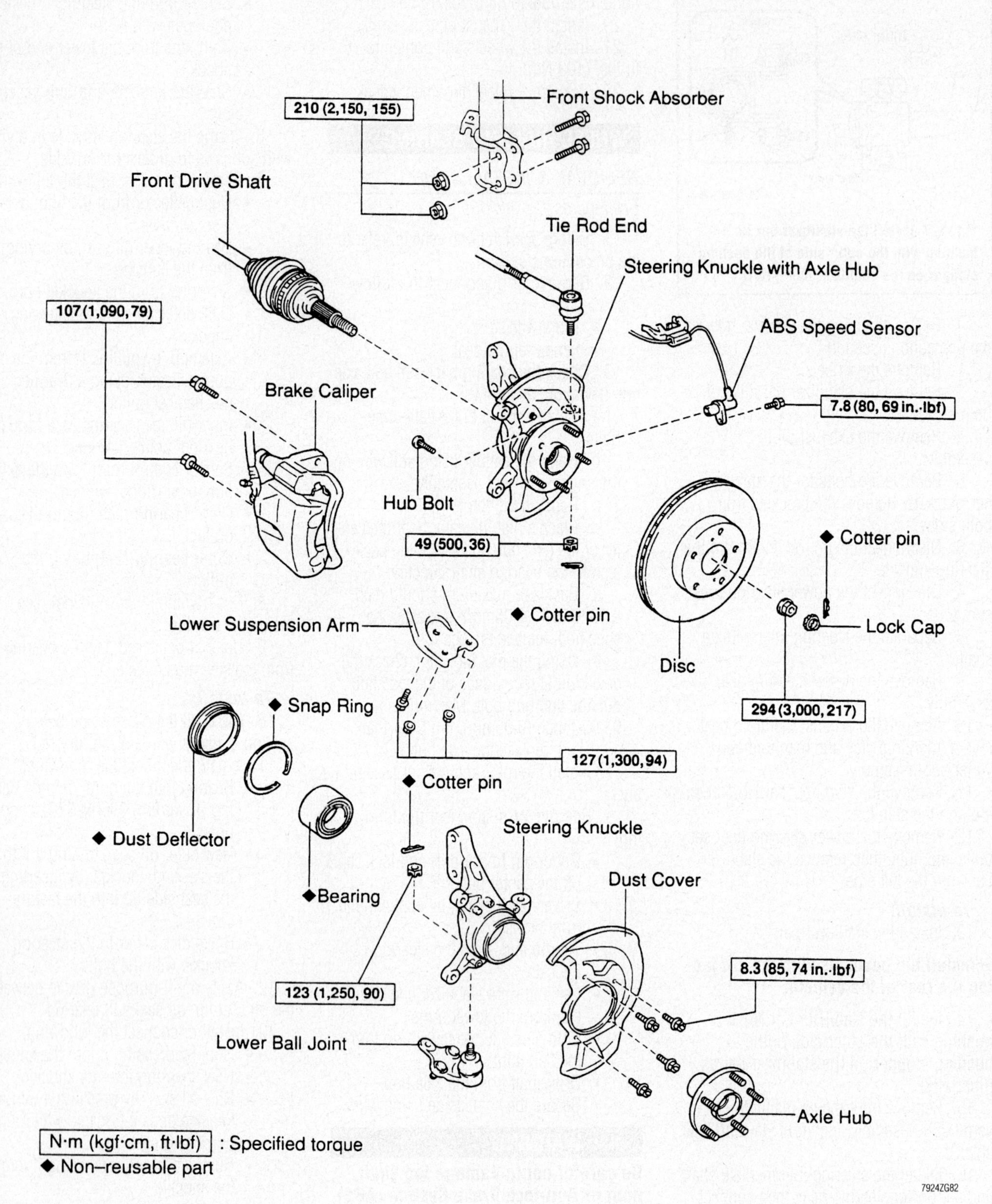

Fig. 128 Exploded view of the front hub, bearing and steering knuckle assembly

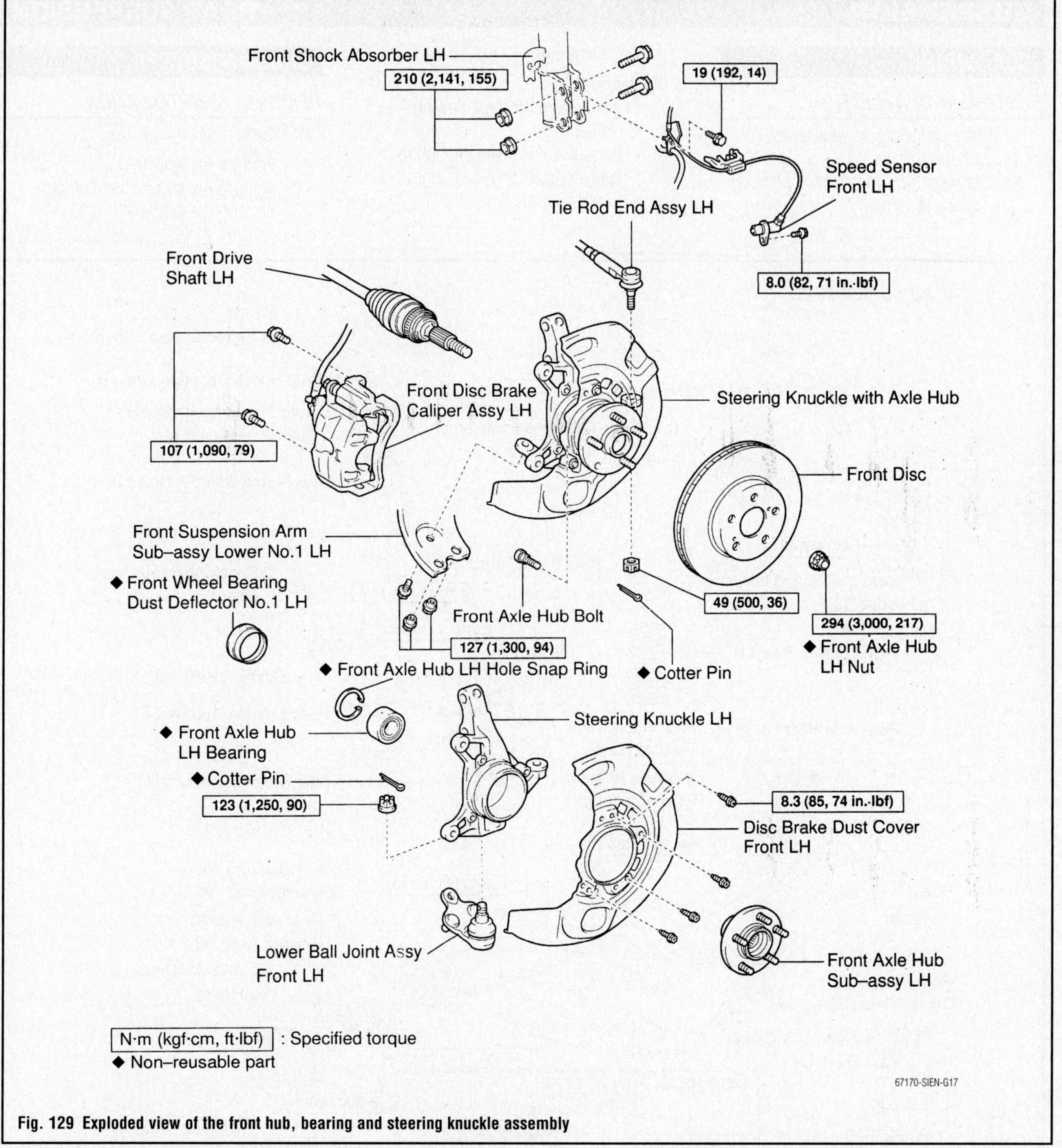

Fig. 129 Exploded view of the front hub, bearing and steering knuckle assembly

- Ball joint to the steering knuckle. Tighten the bolts to 94 ft. lbs. (127 Nm).
- Steering knuckle/hub assembly and temporarily install the lower shock bolts
- Lower ball joint to the lower arm. Tighten the bolt and nuts to 94 ft. lbs. (127 Nm).
- Tie rod to the knuckle. Tighten the nut to 36 ft. lbs. (49 Nm).
- New cotter pin
- Tighten the lower shock nuts to 156 ft. lbs. (211 Nm).
- Both side stabilizer end brackets to the lower arm. Tighten the fasteners to 55 ft. lbs. (74 Nm).
- Front ABS sensor. Tighten it to 69 inch lbs. (8 Nm).
- Front brake rotor and caliper.
- Driveshaft locknut, by applying the brakes. Tighten it to 217 ft. lbs. (294 Nm).
- Lock cap and new cotter pin
- Front fender apron seal
- Front wheel. Tighten the lug nuts to 76 ft. lbs. (103 Nm).

ADJUSTMENT

The wheel bearings are sealed units and are not adjustable.

COIL SPRING

REMOVAL & INSTALLATION

1. Before servicing the vehicle, refer to the precautions section.
2. Remove or disconnect the following:
   - Shock absorbers
   - Coil springs

To install:

3. Install or connect the following:
   - Coil springs
   - Raise the axle beam enough to apply tension on the springs
   - Shock absorbers

SHOCK ABSORBER

REMOVAL & INSTALLATION

See Figures 130 through 132.

1. Remove the wheel.
2. Remove the shock absorber cap.
3. Support the axle with a jack.

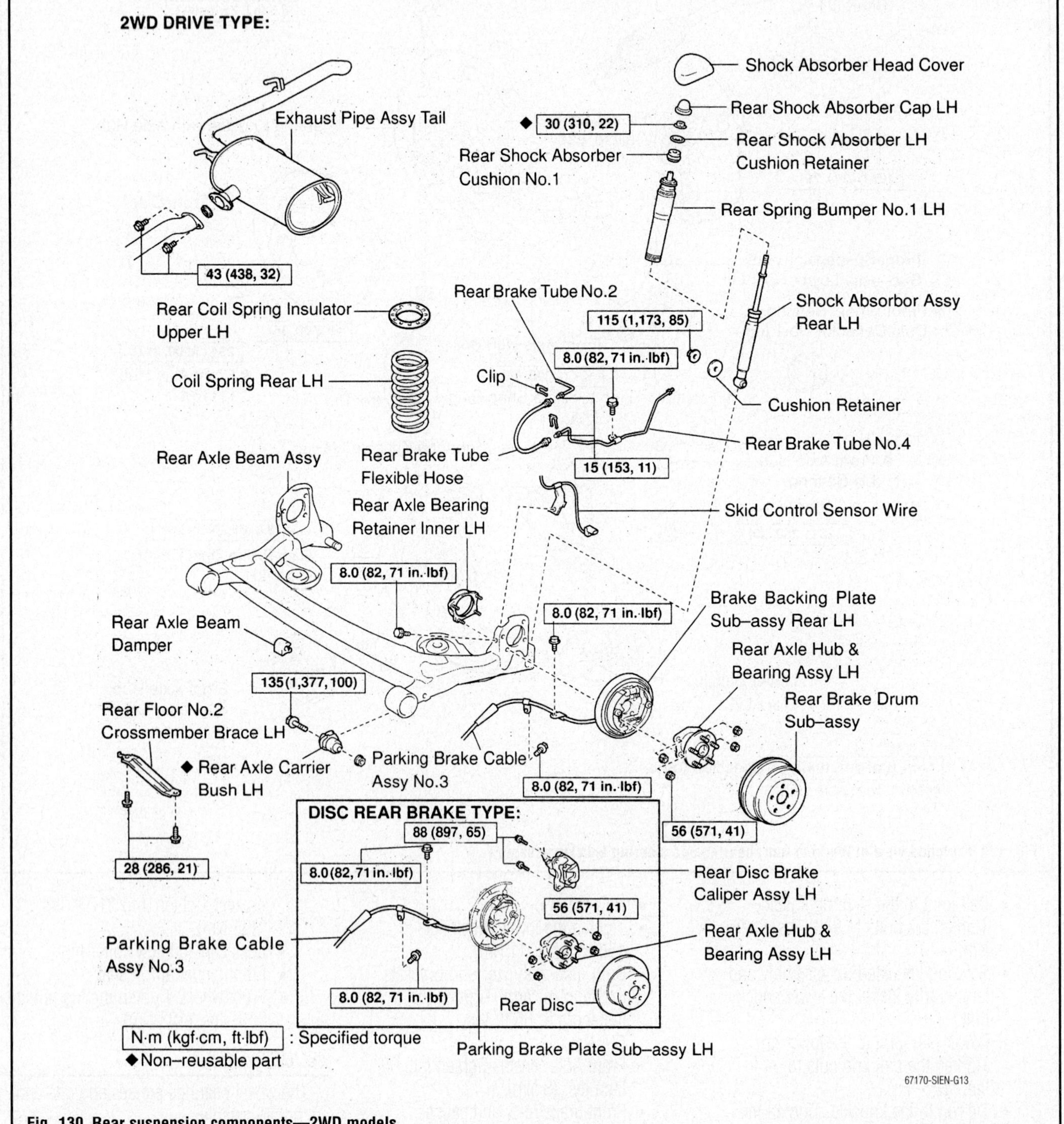

Fig. 130 Rear suspension components—2WD models

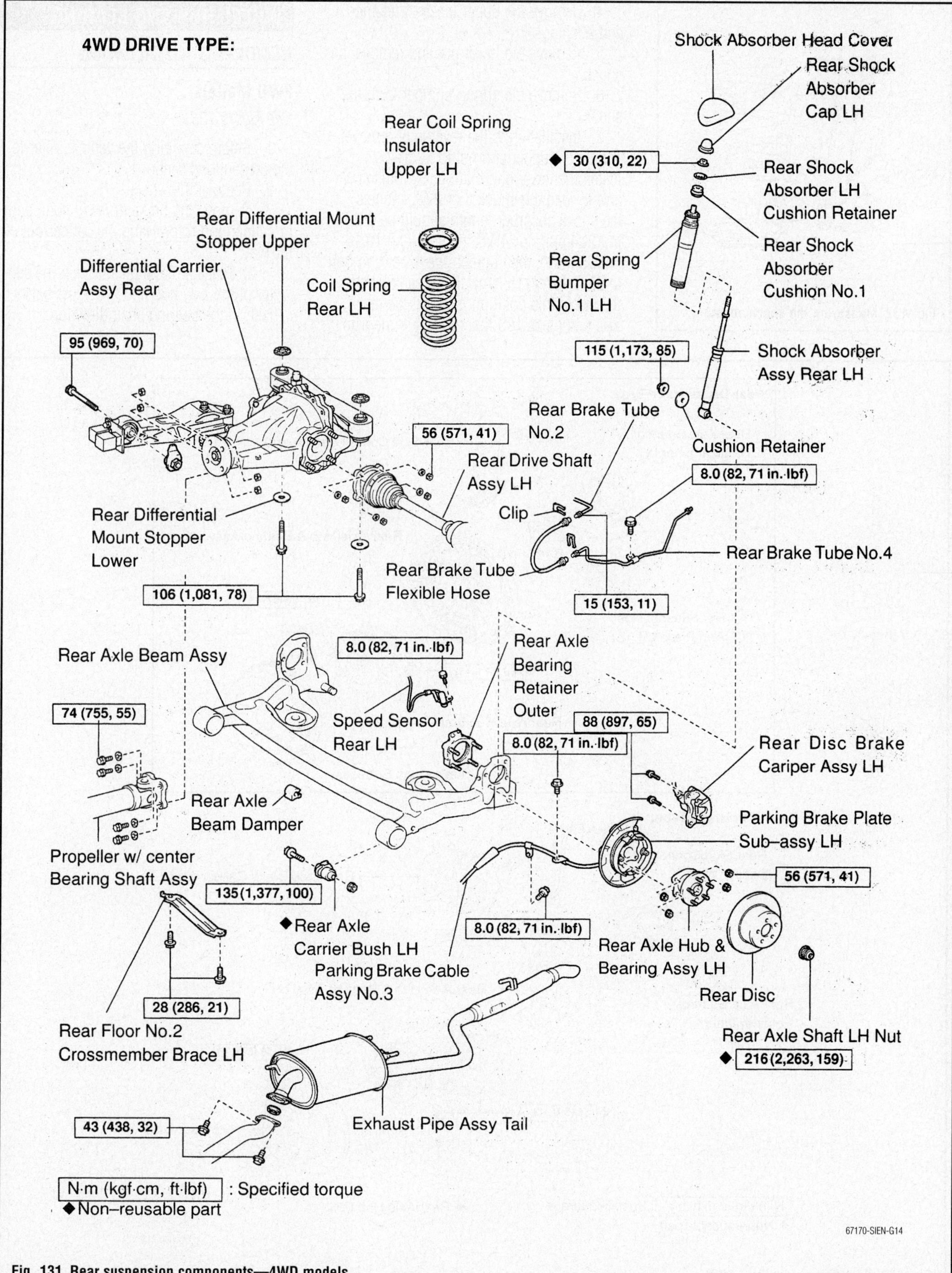

Fig. 131 Rear suspension components—4WD models

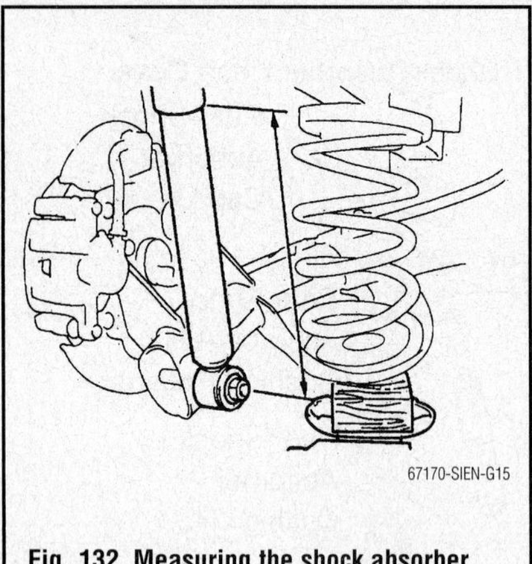

Fig. 132 Measuring the shock absorber

4. Remove the upper locknut, retainer and bushing.
5. Remove the lower nut and remove the shock.
6. Remove the spring bumper from the shock.
7. Installation is the reverse of removal. Torque the upper end nut to 22 ft. lbs. Install the lower end nut loosely. Raise the axle to load the shock. For 2WD models, the shock absorber length should be 9.22 in. (234mm); for 4WD, it should be 10.16 in. (258mm), then tighten the lower end nut to 85 ft. lbs. (115 Nm). If you can't reach the nut in this position, support the rear axle and place 198 lbs. (90 kg) in the trunk.

WHEEL HUB AND BEARING

REMOVAL & INSTALLATION

2WD Models

See Figure 133.

1. Before servicing the vehicle, refer to the precautions section.
2. Remove the wheel.
3. Remove the brake drum (if equipped) or caliper and rotor. Hang the caliper out of the way.
   a. Place a dial indicator near the center of the axle hub and check the backlash in the bearing shaft direction.

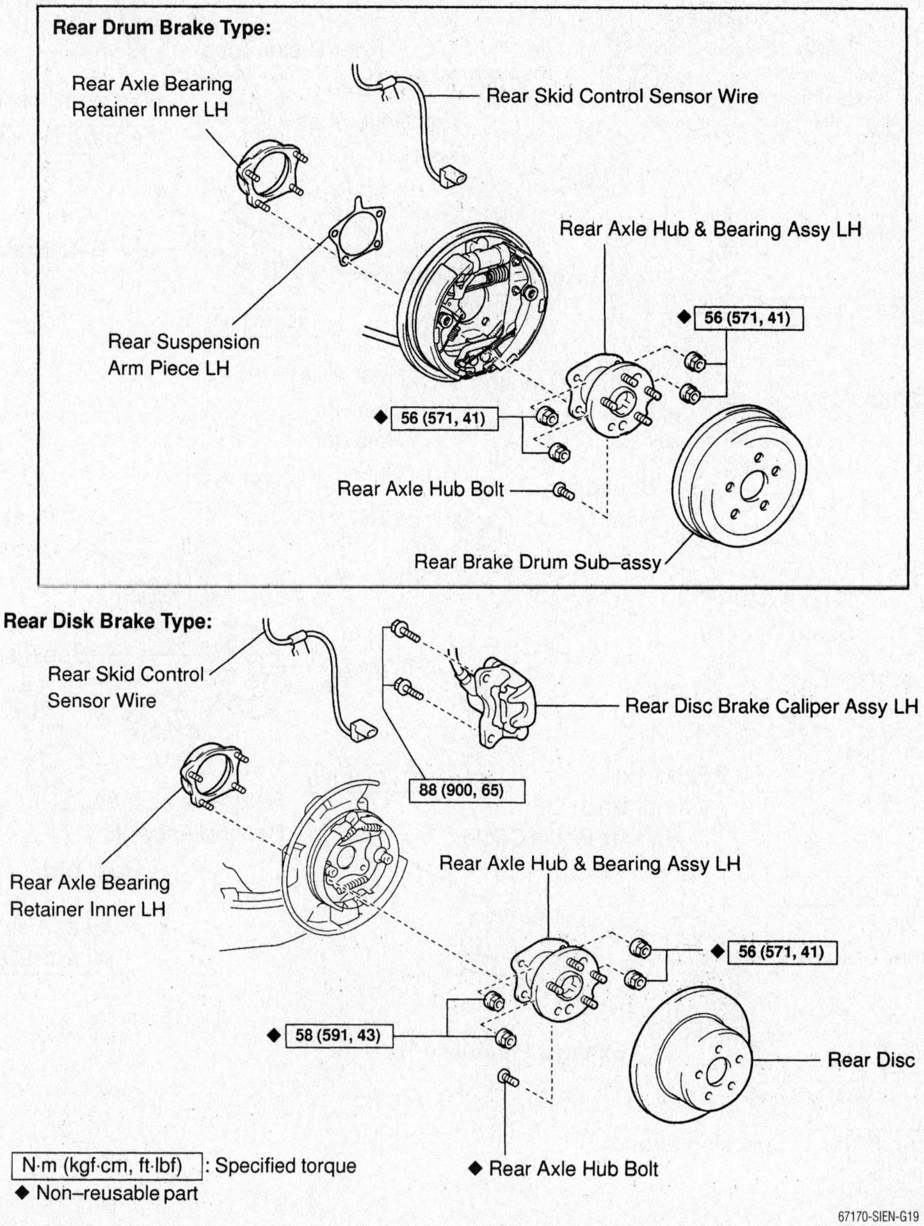

Fig. 133 Rear hub assembly— 2WD

b. Backlash maximum should read 0.0020 inch (0.05mm). If greater than specified, replace the bearing.

c. Using the dial indicator, check the deviation at the surface of the axle hub outside and hub bolt. Maximum is 0.0020 inch (0.05mm). If greater than specified, replace the axle hub.

4. Remove the ABS sensor wire.

5. Remove the 4 bolts and the hub/bearing assembly.

6. Installation is the reverse of removal. Torque the hub bolts to 41 ft. lbs. (56 Nm).

4WD Models

See Figure 134.

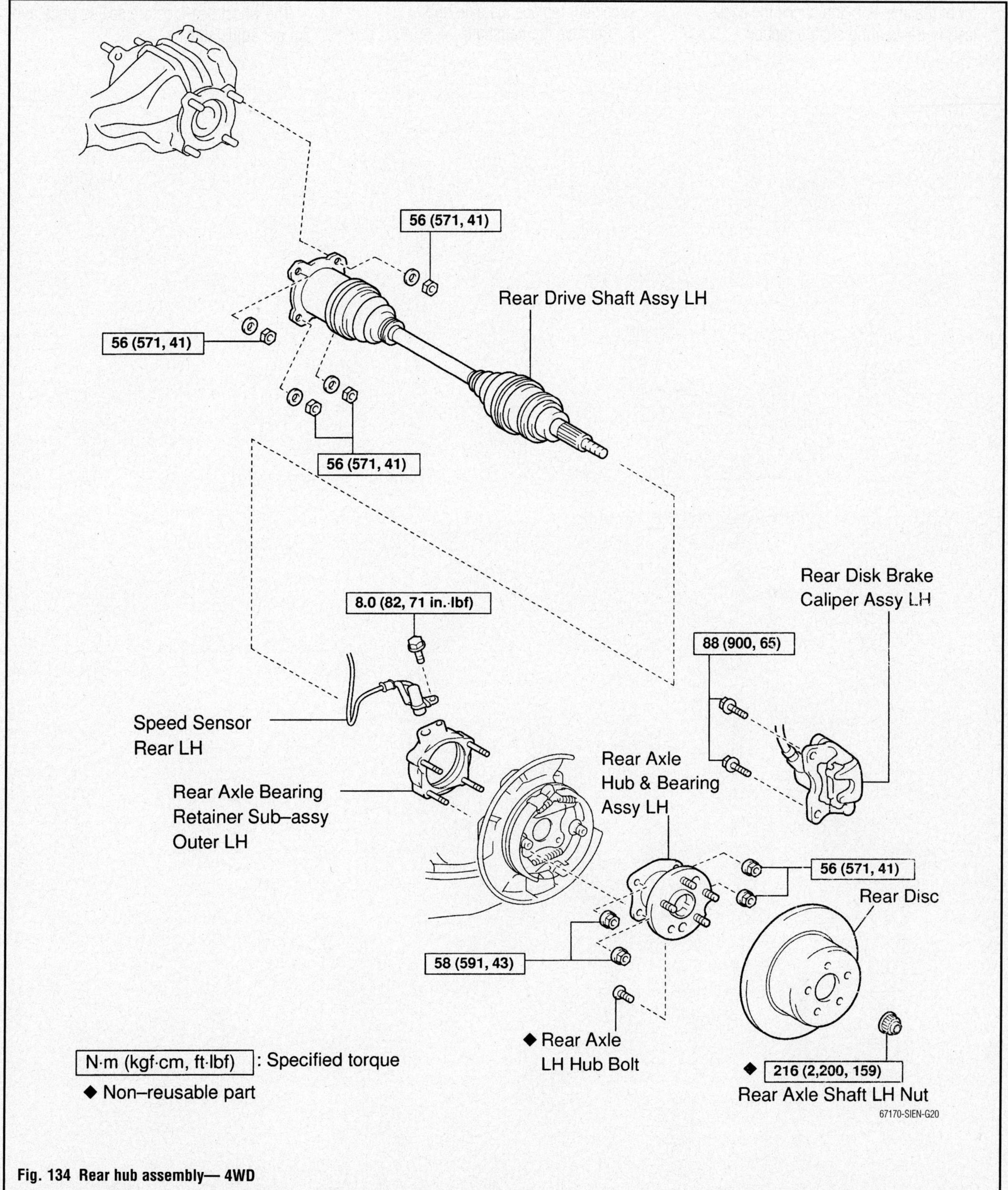

Fig. 134 Rear hub assembly— 4WD

1. Before servicing the vehicle, refer to the precautions section.
2. Remove the wheel.
3. Unstake and remove the axle shaft nut.
4. Remove the caliper and rotor. Hang the caliper out of the way.
   a. Place a dial indicator near the center of the axle hub and check the backlash in the bearing shaft direction.
   b. Backlash maximum should read 0.0020 inch (0.05mm). If greater than specified, replace the bearing.
   c. Using the dial indicator, check the deviation at the surface of the axle hub outside and hub bolt. Maximum is 0.0020 inch (0.05mm). If greater than specified, replace the axle hub.
5. Remove the halfshaft.
6. Remove the ABS sensor wire.
7. Remove the 4 bolts and the hub/bearing assembly.
8. Installation is the reverse of removal. Torque the hub bolts to 41 ft. lbs. (56 Nm).

ADJUSTMENT

The wheel bearings are sealed units and are not adjustable.

TOYOTA

Sequoia • Tundra

22

SPECIFICATIONS AND MAINTENANCE CHARTS

ENGINE AND VEHICLE IDENTIFICATION

Engine							Model Year	
Code ① ②	Liters (cc)	Cu. In.	Cyl.	Fuel Sys.	Engine Type	Eng. Mfg.	Code ③	Year
1GR-FE/U	4.0 (3956)	241	6	MFI	DOHC	Toyota	7	2007
2UZ-FE/T	4.7 (4664)	285	8	SFI	DOHC	Toyota		
3UR-FE	5.7 (5663)	346	8	SFI	DOHC	Toyota		

SFI: Sequential Fuel Injection

MFI: Multi-port Fuel Injection

DOHC: Double Overhead Camshaft

① Except 1GR-FE/U engine: stamped on the left side of the engine block. Engine ID is fifth character of the VIN number.

② 1GR-FE/U engine: stamped on the right side of the engine block. Ingine ID is the fifth character of the VIN number.

③ 10th digit of the VIN number

22140_SEQU_C0001

GENERAL ENGINE SPECIFICATIONS

Year	Model	Engine Displacement Liters	Engine Series Code/ID	Net Horsepower @ rpm	Net Torque @ rpm (ft. lbs.)	Bore x Stroke (in.)	Com-pression Ratio	Oil Pressure @ rpm
2007	Tundra	4.0	1GR-FE/U	236@5200	266@4000	3.70x3.74	NA	43-85@3000
		4.7	2UZ-FE/T	271@5400	313@3400	3.70x3.70	9.6:1	43-85@3000
		5.7	3UR-FE	383@5600	403@3600	3.70x4.02	10.2:1	45-65@3000
	Sequoia	4.7	2UZ-FE/T	271@5400	313@3400	3.70x3.70	9.6:1	43-85@3000

22140_SEQU_C0002

ENGINE TUNE-UP SPECIFICATIONS

Year	Engine Displacement Liters	Engine Code/ID	Spark Plug Gap (in.)	Ignition Timing (deg.)	Fuel Pump (psi)	Idle Speed (rpm)		Valve Clearance	
						MT	AT	Intake	Exhaust
2007	4.0	1GR-FE/U	0.039-0.043	7-24B ①	40.8-41.7	650-750	650-750	0.006-0.010	0.011-0.015
	4.7	2UZ-FE/T	0.043	8-12 ①	38-44	650-750	650-750	0.006-0.009	0.011-0.014
	5.7	3UR-FE	0.043	8-12 ①	41-42	650-750	650-750	N/A	N/A

NOTE: The Vehicle Emission Control Information label often reflects specification changes made during production.

The label figures must be used if they differ from those in this chart.

B: Before top dead center

① With terminals TE1 and E1 connected to DLC1. Adjustments are not possible or necessary.

22140_SEQU_C0003

CAPACITIES

Year	Model	Engine Displacement Liters	Engine Code/ID	Engine Oil with Filter (qts.)	Transmission (qts.) 5-Spd	Transmission (qts.) Auto.	Transfer Case (qts.)	Drive Axle Front (qts.)	Drive Axle Rear (qts.)	Fuel Tank (gal.)	Cooling System (qts.)
2007	Tundra	4.0	1GR-FE/U	4.8	①	3.2	1.2	2.2	②	26.4	10.1
		4.7	2UZ-FE/T	6.5	①	3.2	1.2	2.2	②	26.4	10.3
		5.7	3UR-FE	7.0	①	3.2	1.2	2.2	③	26.4	13.7
	Sequoia	4.7	2UZ-FE/T	6.5	N/A	3.2	1.5	1.2	4.1	26.4	④

NA: Not Available

① Six speed: 4.0

② 4.0L & 4.7L standard bed: 4.3
4.0L & 4.7L long bed: 4.9

③ 5.7L regular cab standard bed: 3.6
5.7 regular cab long bed: 3.8
5.7 double cab standard bed: 3.8
5.7 double cab long bed: 4.0
5.7 crew cab 2wd: 4.0
5.7 crew cab 4wd: 3.8

④ With rear heater 15.3
without rear heater: 13.5

22140_SEQU_C0004

VALVE SPECIFICATIONS

Year	Engine Displ. Liters	Engine Code/ID	Seat Angle (deg.)	Face Angle (deg.)	Spring Test Pressure (lbs. @ in.)	Spring Installed Height (in.)	Stem-to-Guide Clearance (in.) Intake	Stem-to-Guide Clearance (in.) Exhaust	Stem Diameter (in.) Intake	Stem Diameter (in.) Exhaust
2007	4.0	1GR-FE/U	NA	44.5	41.9-46.3@ 1.311	1.882	0.0010-0.0024	0.0012-0.0026	0.2154 0.2159	0.2152 0.2157
	4.7	2UZ-FE/T	45	44.5	45.9-50.7@ 1.378	1.380	0.0010-0.0024	0.0012-0.0026	0.2154-0.2159	0.2152-0.2157
	5.7	3UR-FE	45	44.5	N/A	N/A	0.0009-0.0023	0.0011-0.0025	0.215-0.2160	0.215-0.2160

NA: Not Available

22140_SEQU_C0005

CRANKSHAFT AND CONNECTING ROD SPECIFICATIONS

All measurements are given in inches.

	Engine		Crankshaft				Connecting Rod		
Year	Displ. Liters	Engine Code/ID	Main Brg. Journal Dia.	Main Brg. Oil Clearance	Shaft End-play	Thrust on No.	Journal Diameter	Oil Clearance	Side Clearance
2007	4.0	1GR-FE/U	2.8342-2.8346	0.0007-0.0012	NA	2	2.2044-2.2047	0.0010-0.0018	0.0059-0.0118
	4.7	2UZ-FE/T	2.6373-2.6378	0.0016-0.0023	0.0008-0.0087	3	2.0465-2.0472	0.0011-0.0021	0.0063-0.0138
	5.7	3UR-FE	2.6373-2.6378	①	0.0008-0.0087	3	2.3230-2.3235	0.0010-0.0019	0.0059-0.0217

NA: Not Available

① Nos. 1 and 2: 0.0011-0.0018

All others: 0.0016-0.0023

22140_SEQU_C0007

PISTON AND RING SPECIFICATIONS

All measurements are given in inches.

	Engine			Ring Gap			Ring Side Clearance		
Year	Displ. Liters	Engine Code/ID	Piston Clearance	Top Compression	Bottom Compression	Oil Control	Top Compression	Bottom Compression	Oil Control
2007	4.0	1GR-FE/U	0.0031-0.0040	0.0118-0.0157	0.0157-0.0197	0.0039-0.0157	0.0008-0.0028	0.0008-0.0024	0.0028-0.0060
	4.7	2UZ-FE/T	0.0035-0.0044	0.0118-0.0197	0.0157-0.0256	0.0051-0.0189	0.0012-0.0031	0.0012-0.0028	SNUG
	5.7	3UR-FE	0.0016-0.0024	0.0098-0.0138	0.0157-0.0197	0.0039-0.0157	0.0007-0.0028	0.0007-0.0027	0.0028-0.0057

22140_SEQU_C0008

TORQUE SPECIFICATIONS

All readings in ft. lbs.

Year	Engine Displacement Liters	Engine Code/ID	Cylinder Head Bolts	Main Bearing Bolts	Rod Bearing Bolts	Crankshaft Damper Bolts	Flywheel Bolts	Manifold Intake	Manifold Exhaust	Spark Plugs	Oil Pan Drain Plug
2007	4.0	1GR-FE/U	①	②	③	185	61	19	16	13	30
	4.7	2UZ-FE/T	④	⑤	③	181	⑥	13	32	13	29
	5.7	3UR-FE	⑦	⑧	⑨	221	⑩	15	15	13	29

① Right side: 27 ft. lbs., then + 90 degrees, then +90 degrees
Left side (recessed head): 27 ft. lbs. Then plus 180 degrees
Left side (0.55 inch head): 22 ft. lbs.

② 12 point head: 45 ft' lbs., then +90 degrees
12mm head: 18 ft. lbs.

③ Step 1: 18 ft. lbs.
Step 2: Plus 90 degrees

④ Step 1: 30 ft. lbs.
Step 2: Plus 90 degrees
Step 3: Plus 90 degrees

⑤ Step 1: 20 ft. lbs.
Step 2: Plus 90 degrees

⑥ Step 1: 35 ft. lbs.
Step 2: Plus 90 degrees

⑦ Step 1: 27 ft. lbs.
Step 2: Plus 90 degrees
Step 3: Plus 90 degrees
12 mm head 15 ft. lbs.

⑧ Inside position 45 ft. lbs.
Step 1: Outside position 20 ft. lbs.
Step 2: Plus 90 degrees
Step 3: Cylinder block side position 33 ft. lbs.

⑨ Step 1: 30 ft. lbs.
Step 2: Plus 90 degrees

⑩ Step 1: 22 ft. lbs.
Step 2: Plus 90 degrees

22140_SEQU_C0009

WHEEL ALIGNMENT

Year	Model	Caster Range (+/-Deg.)	Caster Preferred Setting (Deg.)	Camber Range (+/-Deg.)	Camber Preferred Setting (Deg.)	Toe-in (in.)	Steering Axis Inclination (Deg.)
2007	Tundra	NA	NA	NA	NA	NA	NA
	Sequoia	NA	NA	NA	NA	NA	NA

NOTE: All alignment figures based on nominal ride height and standard tires

NA: Not Available

22140_SEQU_C0011

TIRE, WHEEL AND BALL JOINT SPECIFICATIONS

Year	Model	OEM Tires		Tire Pressures (psi)		Wheel	Ball Joint	Lug Nut Torque
		Standard	Optional	Front	Rear	Size	Inspection	(ft. lbs.)
2007	Tundra	P255/70R18	P275/65R18	①	①	8J	②	③
	Sequoia	P245/70R16	P265/65R17	①	①	8J	②	③

OEM: Original Equipment Manufacturer

PSI: Pounds Per Square Inch

STD: Standard

OPT: Optional

NS: Not specified by manufacturer

NA: Not available

① See placard on vehicle

Lower: turning torque within 1-22 inch lbs.

② Turning torque 89 inch lbs.

③ Steel wheel 154 ft. lbs.; aluminum wheels 97 ft. lbs.

22140_SEQU_C0012

BRAKE SPECIFICATIONS

All measurements in inches unless noted

Year	Model		Brake Disc Original Thickness	Brake Disc Minimum Thickness	Brake Disc Maximum Runout	Minimum Lining Thickness	Brake Caliper Bracket Bolts (ft. lbs.)	Brake Caliper Mounting Bolts (ft. lbs.)
2007	Tundra	F	1.260	1.140	0.0020	0.039	—	90
		R	0.709	0.630	0.0080	0.039	70	65
	Sequoia	F	1.102	1.024	0.0020	0.039	—	91
		R	0.709	0.630	0.0039	0.039	77	65

F: Front

R: Rear

22140_SEQU_C0010

SCHEDULED MAINTENANCE INTERVALS

TOYOTA—SEQUOIA, TUNDRA

TO BE SERVICED	TYPE OF SERVIC E	VEHICLE MILEAGE INTERVAL (x1000)																		
		5	10	15	20	25	30	35	40	45	50	55	60	65	70	75	80	85	90	95
Automatic transmission and differential fluid	S/I			✓			✓			✓			✓			✓			✓	
Ball joints and boots	S/I			✓			✓			✓			✓			✓			✓	
Brake system	S/I			✓			✓			✓			✓			✓			✓	
Charcoal canister	S/I												✓							
Drive belts	S/I						✓						✓						✓	
Driveshaft bushing	L						✓						✓						✓	
Engine coolant	R						✓						✓						✓	
Engine oil & filter	R	✓	✓	✓	✓	✓	✓	✓	✓	✓	✓	✓	✓	✓	✓	✓	✓	✓	✓	✓
Exhaust system	S/I			✓			✓			✓			✓			✓			✓	
Fuel lines	S/I						✓						✓						✓	
Fuel tank cap gasket	S/I						✓						✓						✓	
Halfshaft boots & flange bolts	S/I			✓			✓			✓			✓			✓			✓	
Limited slip differential fluid	R						✓						✓						✓	
Manual transmission and differential fluid	S/I						✓						✓						✓	
Non-platinum spark plugs	R						✓						✓						✓	
Platinum spark plugs	R												✓							
Propeller shaft (4WD)	L			✓			✓			✓			✓			✓			✓	
Propeller shaft bolts	S/I			✓			✓			✓			✓			✓			✓	
Steering gear	S/I			✓			✓			✓			✓			✓			✓	
Steering linkage	S/I			✓			✓			✓			✓			✓			✓	
Tires (rotate)	S/I	✓	✓	✓	✓	✓	✓	✓	✓	✓	✓	✓	✓	✓	✓	✓	✓	✓	✓	✓
Valves	S/I												✓							

R: Replace S/I: Service or Inspect L: Lubricate

FREQUENT OPERATION MAINTENANCE (SEVERE SERVICE)

If a vehicle is operated under any of the following conditions it is considered severe service:

- Towing a trailer or using a camper or car-top carrier.

- Repeated short trips of less than 5 miles in temperatures below freezing.

- Excessive idling or low-speed driving for long distances as in heavy commercial use, such as delivery, taxi or police cars.

- Operating on rough, muddy or salt-covered roads.

- Operating on unpaved or dusty roads.

Oil filter: service or inspect every 5000 miles or 4 months, whichever occurs first.

Brake linings and discs or drums: service or inspect every 5000 miles or 4 months, whichever occurs first.

Steering linkage: service or inspect every 5000 miles or 4 months, whichever occurs first.

Ball joints and boots: service or inspect every 5000 miles or 4 months, whichever occurs first.

Brake discs & pads (front): service or inspect every 6000 miles.

Halfshaft boots: service or inspect every 5000 miles or 4 months. Retighten the flange bolts, whichever occurs first.

Body chassis bolts and nuts: service or inspect every 5000 miles or 4 months, whichever occurs first.

Transmission and differential fluid: replace every 15,000 miles or 12 months, whichever occurs first.

Transfer case and differential fluid: replace every 15,000 miles or 12 months, whichever occurs first.

22140_SEQU_C0013

PRECAUTIONS

Before servicing any vehicle, please be sure to read all of the following precautions, which deal with personal safety, prevention of component damage, and important points to take into consideration when servicing a motor vehicle:

- Never open, service or drain the radiator or cooling system when the engine is hot; serious burns can occur from the steam and hot coolant.
- Observe all applicable safety precautions when working around fuel. Whenever servicing the fuel system, always work in a well-ventilated area. Do not allow fuel spray or vapors to come in contact with a spark, open flame, or excessive heat (a hot drop light, for example). Keep a dry chemical fire extinguisher near the work area. Always keep fuel in a container specifically designed for fuel storage; also, always properly seal fuel containers to avoid the possibility of fire or explosion. Refer to the additional fuel system precautions later in this section.
- Fuel injection systems often remain pressurized, even after the engine has been turned **OFF**. The fuel system pressure must be relieved before disconnecting any fuel lines. Failure to do so may result in fire and/or personal injury.
- Brake fluid often contains polyglycol ethers and polyglycols. Avoid contact with the eyes and wash your hands thoroughly after handling brake fluid. If you do get brake fluid in your eyes, flush your eyes with clean, running water for 15 minutes. If eye irritation persists, or if you have taken brake fluid internally, IMMEDIATELY seek medical assistance.
- The EPA warns that prolonged contact with used engine oil may cause a number of skin disorders, including cancer. You should make every effort to minimize your exposure to used engine oil. Protective gloves should be worn when changing oil. Wash your hands and any other exposed skin areas as soon as possible after exposure to used engine oil. Soap and water, or waterless hand cleaner should be used.
- All new vehicles are now equipped with an air bag system, often referred to as a Supplemental Restraint System (SRS) or Supplemental Inflatable Restraint (SIR) system. The system must be disabled before performing service on or around system components, steering column, instrument panel components, wiring and sensors. Failure to follow safety and disabling procedures could result in accidental air bag deployment, possible personal injury and unnecessary system repairs.
- Always wear safety goggles when working with, or around, the air bag system. When carrying a non-deployed air bag, be sure the bag and trim cover are pointed away from your body. When placing a non-deployed air bag on a work surface, always face the bag and trim cover upward, away from the surface. This will reduce the motion of the module if it is accidentally deployed. Refer to the additional air bag system precautions later in this section.
- Clean, high quality brake fluid from a sealed container is essential to the safe and proper operation of the brake system. You should always buy the correct type of brake fluid for your vehicle. If the brake fluid becomes contaminated, completely flush the system with new fluid. Never reuse any brake fluid. Any brake fluid that is removed from the system should be discarded. Also, do not allow any brake fluid to come in contact with a painted surface; it will damage the paint.
- Never operate the engine without the proper amount and type of engine oil; doing so WILL result in severe engine damage.
- Timing belt maintenance is extremely important. Many models utilize an interference-type, non-freewheeling engine. If the timing belt breaks, the valves in the cylinder head may strike the pistons, causing potentially serious (also time-consuming and expensive) engine damage. Refer to the maintenance interval charts for the recommended replacement interval for the timing belt, and to the timing belt section for belt replacement and inspection.
- Disconnecting the negative battery cable on some vehicles may interfere with the functions of the on-board computer system(s) and may require the computer to undergo a relearning process once the negative battery cable is reconnected.
- When servicing drum brakes, only disassemble and assemble one side at a time, leaving the remaining side intact for reference.
- Only an MVAC-trained, EPA-certified automotive technician should service the air conditioning system or its components.

BRAKES — ANTI-LOCK BRAKE SYSTEM (ABS)

GENERAL INFORMATION

1. The skid control ECU calculates vehicle stability tendency based on signals from the speed sensor, yaw rate (deceleration) sensor and steering angle sensor. In addition, it evaluates the results of the calculations to determine whether any control actions (control of the engine output torque by electronic throttle control and of the wheel brake pressure by the brake actuator) should be implemented.

BRAKES — BLEEDING THE BRAKE SYSTEM

BLEEDING PROCEDURE

Without VSC

Master Cylinder

➡Immediately wash off any brake fluid that comes into contact with any painted surfaces.

➡Depressing the brake pedal with the reservoir cap removed will cause the fluid to spray.

➡When bleeding, maintain the amount of fluid in the reservoir between the Min. and Max. lines.

1. Before servicing the vehicle, refer to the precautions in the beginning of this section.
2. Fill reservoir with dot3 brake fluid.
3. Using SST 09023-00101, disconnect the brake lines from the master cylinder.
4. Slowly depress the brake pedal and hold it there.
5. Block the outer holes with your fingers, and release the brake pedal.
6. Repeat 3 or 4 times.

Brake Lines

1. Before servicing the vehicle, refer to the precautions in the beginning of this section.
2. Connect the vinyl tube to the bleeder plug.
3. Depress the brake pedal several times, then loosen the bleeder plug with the pedal held down.

4. At the point where the fluid stops coming out, tighten the bleeder plug, then release the brake pedal.
5. Repeat until all the air in the fluid has been bled out.
6. Repeat the above procedure to bleed the air out of the brake line for each wheel.
7. Check the fluid level and add fluid if necessary.

With VSC

Master Cylinder

➡Immediately wash off any brake fluid that comes into contact with any painted surfaces.

➡Depressing the brake pedal with the reservoir cap removed will cause the fluid to spray.

➡When bleeding, maintain the amount of fluid in the reservoir between the Min. and Max. lines.

1. Before servicing the vehicle, refer to the precautions in the beginning of this section.
2. Fill reservoir with dot3 brake fluid.
3. Turn the ignition switch to ON, and wait until the pump motor has stopped.
4. Turn the ignition switch to OFF, and depress the brake pedal more than 20 times.
5. Repeat 5 times.
6. Turn the ignition switch to ON, and check that the pump stops after approximately 8 to 14 seconds.

Front Brake Lines

➡Immediately wash off any brake fluid that comes into contact with any painted surfaces.

➡Depressing the brake pedal with the reservoir cap removed will cause the fluid to spray.

➡When bleeding, maintain the amount of fluid in the reservoir between the Min. and Max. lines.

1. Before servicing the vehicle, refer to the precautions in the beginning of this section.
2. Turn the ignition switch to ON, and wait until the pump motor has stopped.
3. (1) Connect the vinyl tube to the brake caliper.
4. Depress the brake pedal several times, then loosen the bleeder plug with the pedal held down.
5. At the point when the fluid stops coming out, tighten the bleeder plug, then release the brake pedal.
6. Repeat until all the air in the fluid has been bled out.
7. Repeat the above procedures to bleed the other brake line.

Rear Brake Lines

➡Immediately wash off any brake fluid that comes into contact with any painted surfaces.

➡Depressing the brake pedal with the reservoir cap removed will cause the fluid to spray.

➡When bleeding, maintain the amount of fluid in the reservoir between the Min. and Max. lines.

1. Before servicing the vehicle, refer to the precautions in the beginning of this section.
2. Connect the vinyl tube to the wheel cylinder.
3. Depress the brake pedal, hold it, and then loosen the bleeder plug.
4. Loosen the bleeder plug and release air.
5. When the air is completely bled out of the brake fluid through the bleeder plug, tighten the bleeder plug.
6. Repeat the above procedures to bleed the other brake line.

BRAKES — FRONT DISC BRAKES

⁂ CAUTION

Dust and dirt accumulating on brake parts during normal use may contain asbestos fibers from production or aftermarket brake linings. Breathing excessive concentrations of asbestos fibers can cause serious bodily harm. Exercise care when servicing brake parts. Do not sand or grind brake lining unless equipment used is designed to contain the dust residue. Do not clean brake parts with compressed air or by dry brushing. Cleaning should be done by dampening the brake components with a fine mist of water, then wiping the brake components clean with a dampened cloth. Dispose of cloth and all residue containing asbestos fibers in an impermeable container with the appropriate label. Follow practices prescribed by the Occupational Safety and Health Administration (OSHA) and the Environmental Protection Agency (EPA) for the handling, processing, and disposing of dust or debris that may contain asbestos fibers.

BRAKE CALIPER

REMOVAL & INSTALLATION

See Figure 1.

1. Disconnect the negative battery cable from the battery.
2. Raise and support the vehicle safely.
3. Remove the wheels.
4. Disconnect the brake hose from the caliper. Plug the end of the hose to prevent loss of fluid.
5. Remove the bolts that attach the caliper to the torque plate.
6. Lift the bottom of the caliper up and remove the caliper assembly.

To install:

7. Grease the caliper slides and bolts with lithium grease or equivalent. Install the caliper and secure with the bolts. Torque the bolts to 90 ft. lbs. (123 Nm).
8. Connect the brake hose to the caliper. Torque 11 ft. lbs. (15 Nm).
9. Fill the brake system to the proper level and bleed the brake system.
10. Install the tire and wheel assembly.
11. Top off the brake fluid level in the master cylinder. Check for leaks and proper brake operation.
12. Connect the negative battery cable to the battery.

DISC BRAKE PADS

REMOVAL & INSTALLATION

See Figure 1.

1. Raise the vehicle and support it safely.
2. Remove the wheels.
3. Remove the clip, pins and anti-rattle spring.
4. Withdraw the pads and remove the anti-squeal shims.

To install:

5. Before installing the new pads, check the disc thickness and disc runout.

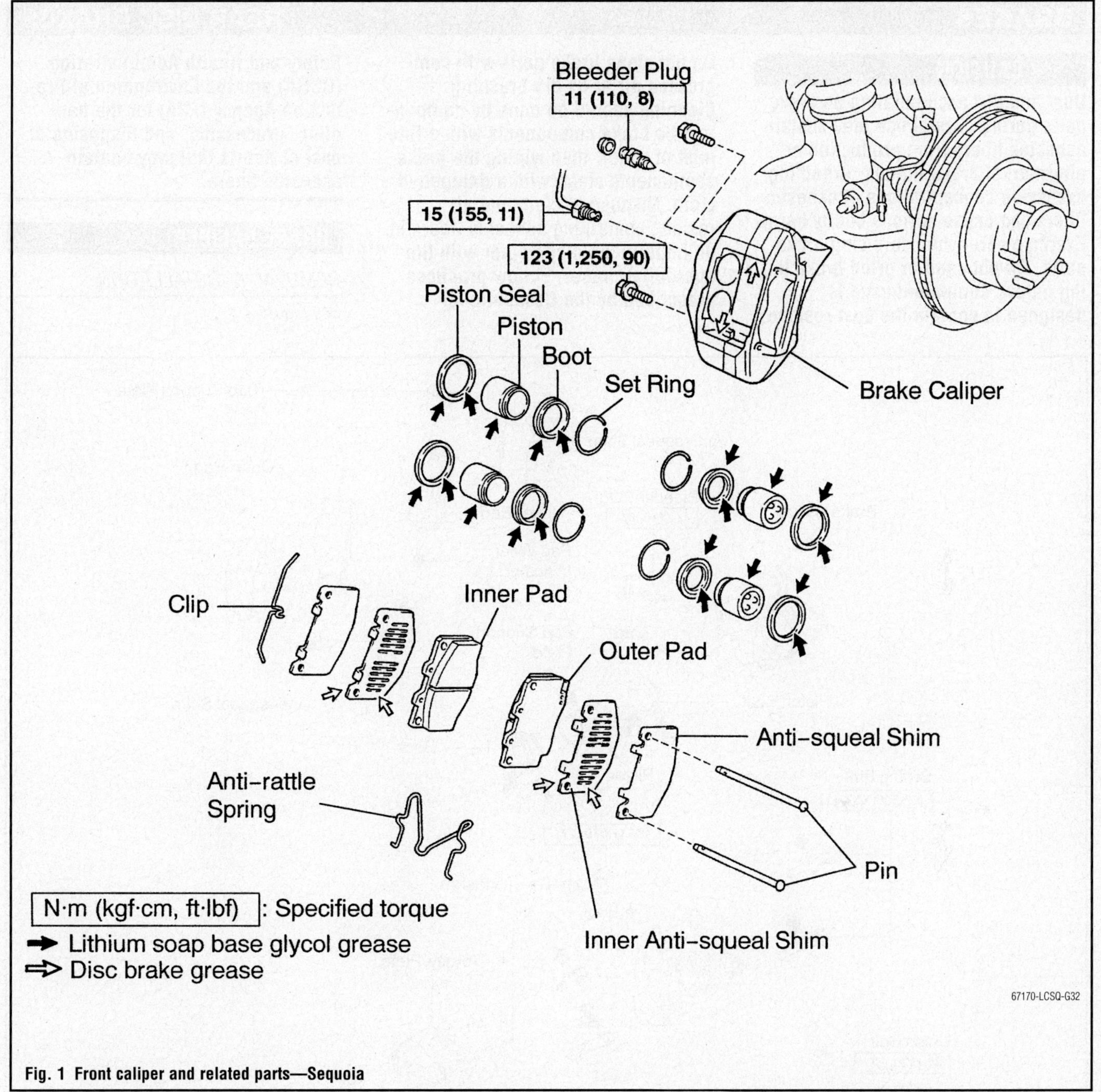

Fig. 1 Front caliper and related parts—Sequoia

6. Siphon out a small amount of brake fluid from the reservoir.
7. Press in the pistons with a hammer handle or equivalent.
8. Apply disc brake grease to both sides of the inner anti-squeal shim. Install the anti-squeal shims to the new pads.
9. Install the pads.
10. Install the anti-rattle springs and pins. Install the clip.
11. Install the wheels.
12. Check and adjust the fluid level. Apply the brake pedal several times.
13. Road-test the vehicle for proper operation.

CAUTION

Dust and dirt accumulating on brake parts during normal use may contain asbestos fibers from production or aftermarket brake linings. Breathing excessive concentrations of asbestos fibers can cause serious bodily harm. Exercise care when servicing brake parts. Do not sand or grind brake lining unless equipment used is designed to contain the dust residue. Do not clean brake parts with compressed air or by dry brushing. Cleaning should be done by dampening the brake components with a fine mist of water, then wiping the brake components clean with a dampened cloth. Dispose of cloth and all residue containing asbestos fibers in an impermeable container with the appropriate label. Follow practices prescribed by the Occupational Safety and Health Administration (OSHA) and the Environmental Protection Agency (EPA) for the handling, processing, and disposing of dust or debris that may contain asbestos fibers.

BRAKE CALIPER

REMOVAL & INSTALLATION

See Figure 2.

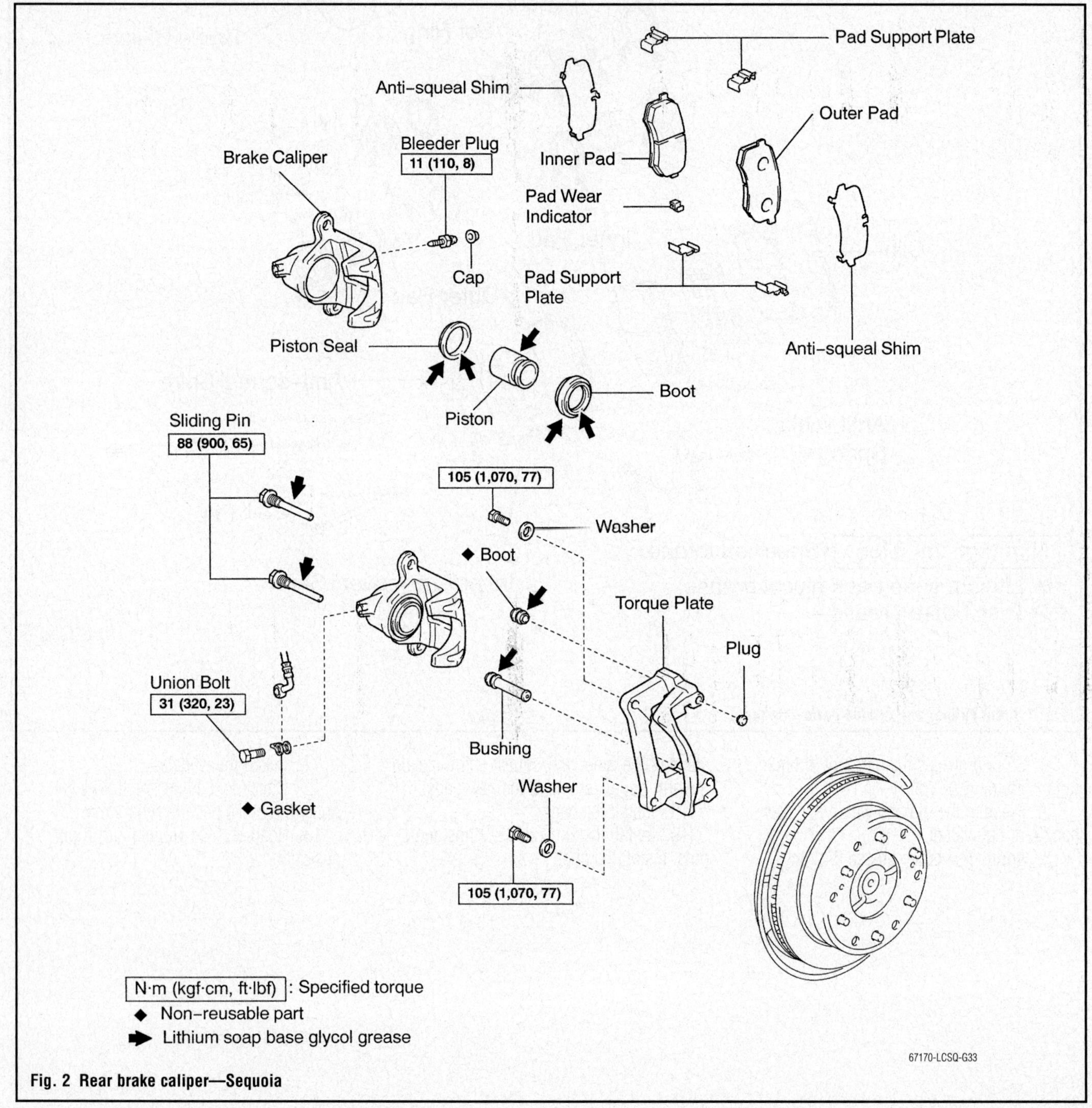

Fig. 2 Rear brake caliper—Sequoia

1. Disconnect the negative battery cable from the battery.
2. Raise and support the vehicle safely.
3. Remove the wheels.
4. Disconnect the brake hose from the caliper by removing the union bolt and 2 gaskets. Plug the end of the hose to prevent loss of fluid.
5. Remove the 2 sliding pins.
6. Lift the bottom of the caliper up and remove the caliper assembly.

To install:

7. Grease the caliper slides and pins with silicone grease or equivalent. Install the caliper and secure with the bolts. Torque the pins to 65 ft. lbs. (883 Nm).
8. Connect the brake hose to the caliper, using 2 new washers. Torque the union bolt to 22 ft. lbs. (30 Nm).
9. Fill the brake system to the proper level and bleed the brake system.
10. Install the tire and wheel assembly.
11. Top off the brake fluid level in the master cylinder. Check for leaks and proper brake operation.
12. Connect the negative battery cable to the battery.

DISC BRAKE PADS

REMOVAL & INSTALLATION

See Figure 2.

1. Raise the vehicle and support it safely.
2. Remove the wheels.
3. Remove the brake caliper and suspend it so the hose is not stretched.
4. Remove the brake pads, anti-squeal shim, pad support plates and wear indicators.

To install:

5. Before installing the new pads, check the disc thickness and disc runout.
6. Install the pad support plates.
7. Install the pad wear indicator plates on each pad.
8. Install the anti-squeal shim to the outer pad. Install the pads.
9. Install the brake caliper.
10. Install the wheels.
11. Apply the brake pedal several times.
12. Road-test the vehicle for proper operation.

BRAKES — PARKING BRAKE

PARKING BRAKE SHOES

REMOVAL & INSTALLATION

See Figure 3.

1. Before servicing the vehicle, refer to the precautions in the beginning of this section.
2. Raise and safely support the vehicle.
3. Remove the rear wheel.
4. Remove the 2 mounting bolts and remove the disc brake assembly.
5. Suspend the disc brake securely and so the hose is not stretched.
6. Release the parking brake lever.
7. Place matchmarks on the disc and rear axle hub.
8. Remove the disc.

➡If the disc cannot be removed easily, turn the shoe adjuster until the wheel turns freely.

9. Using needle-nose pliers, remove the 2 shoe return springs.

➡At the time of reassembly, install the strut with the spring facing forward.

10. Slide the front shoe toward outside and remove the shoe adjuster.
11. Using a needle-nose pliers, disconnect the anchor spring and tension spring from the front shoe.
12. Using a needle-nose pliers, disconnect the anchor spring and tension spring from the rear shoe.
13. Installation is the reverse of removal.

ADJUSTMENT

1. Before servicing the vehicle, refer to the precautions in the beginning of this section.
2. Turn the adjuster and expand the shoes until the disc locks.
3. Return the adjuster 8 notches.
4. Depress the parking brake pedal with 147 N (33 lbs).
5. Drive the vehicle at about 50 km/h (31 mph) on a safe, level and dry road for about 400 meters (0.25 mile) in this condition.
6. Repeat this procedure 2 or 3 times.

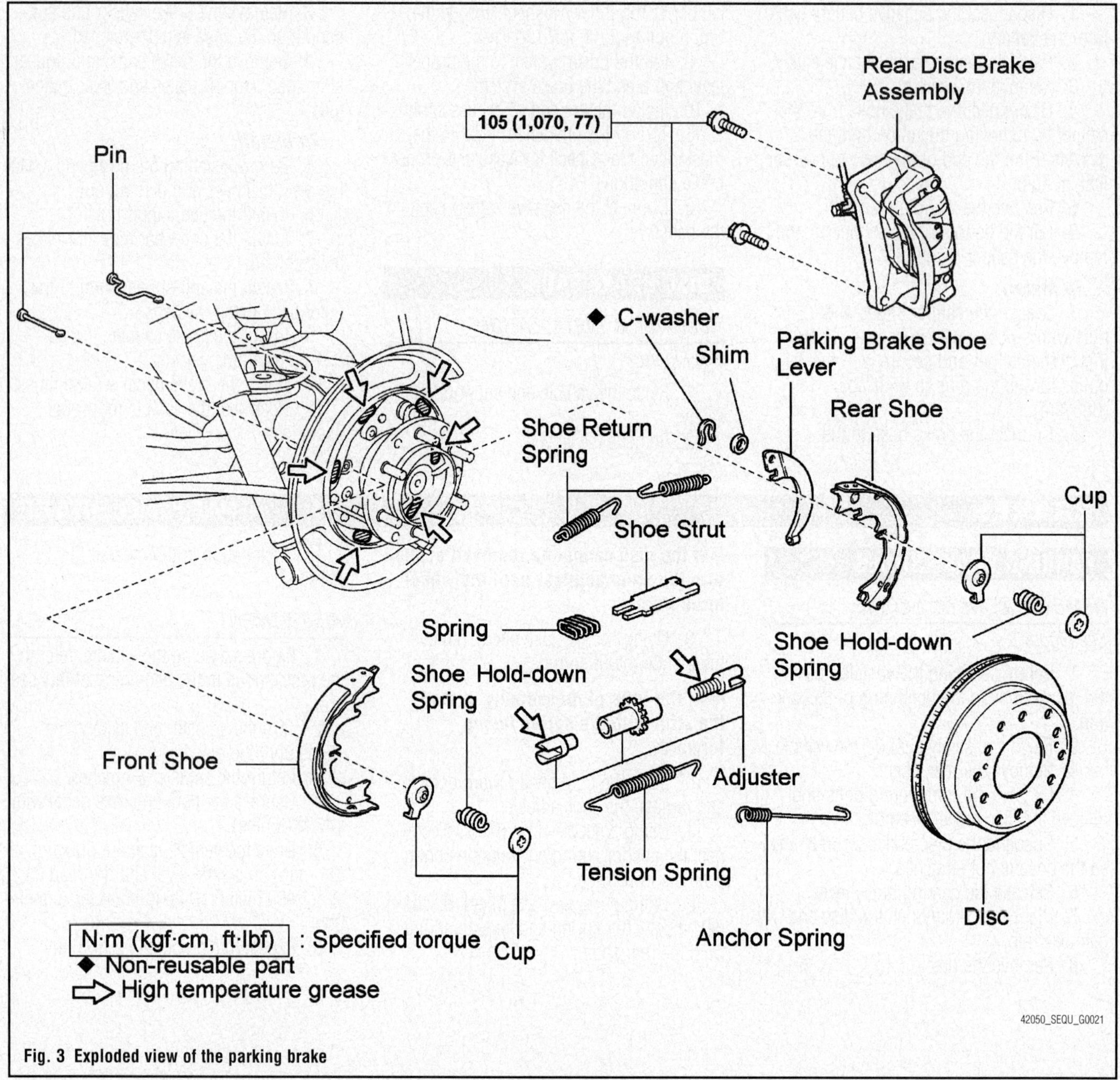

Fig. 3 Exploded view of the parking brake

CHASSIS ELECTRICAL

AIR BAG (SUPPLEMENTAL RESTRAINT SYSTEM)

GENERAL INFORMATION

** CAUTION

Some vehicles are equipped with an air bag system. The system must be disarmed before performing service on, or around, system components, the steering column, instrument panel components, wiring and sensors. Failure to follow the safety precautions and the disarming procedure could result in accidental air bag deployment, possible injury and unnecessary system repairs.

SERVICE PRECAUTIONS

Disconnect and isolate the battery negative cable before beginning any airbag system component diagnosis, testing, removal, or installation procedures. Allow system capacitor to discharge for two minutes before beginning any component service. This will disable the airbag system. Failure to disable the airbag system may result in accidental airbag deployment, personal injury, or death.

Do not place an intact undeployed airbag face down on a solid surface. The airbag will propel into the air if accidentally deployed and may result in personal injury or death.

When carrying or handling an undeployed airbag, the trim side (face) of the airbag should be pointing towards the body to minimize possibility of injury if accidental deployment occurs. Failure to do this may result in personal injury or death.

Replace airbag system components with OEM replacement parts. Substitute parts may appear interchangeable, but internal differences may result in inferior occupant protection. Failure to do so may result in occupant personal injury or death.

Wear safety glasses, rubber gloves, and long sleeved clothing when cleaning powder residue from vehicle after an airbag deployment. Powder residue emitted from a deployed airbag can cause skin irritation. Flush affected area with cool water if irritation is experienced. If nasal or throat irritation is experienced, exit the vehicle for fresh air until the irritation ceases. If irritation continues, see a physician.

Do not use a replacement airbag that is not in the original packaging. This may result in improper deployment, personal injury, or death.

The factory installed fasteners, screws and bolts used to fasten airbag components have a special coating and are specifically designed for the airbag system. Do not use substitute fasteners. Use only original equipment fasteners listed in the parts catalog when fastener replacement is required.

During, and following, any child restraint anchor service, due to impact event or vehicle repair, carefully inspect all mounting hardware, tether straps, and anchors for proper installation, operation, or damage. If a child restraint anchor is found damaged in any way, the anchor must be replaced. Failure to do this may result in personal injury or death.

Deployed and non-deployed airbags may or may not have live pyrotechnic material within the airbag inflator.

Do not dispose of driver/passenger/curtain airbags or seat belt tensioners unless you are sure of complete deployment. Refer to the Hazardous Substance Control System for proper disposal.

Dispose of deployed airbags and tensioners consistent with state, provincial, local, and federal regulations.

After any airbag component testing or service, do not connect the battery negative cable. Personal injury or death may result if the system test is not performed first.

If the vehicle is equipped with the Occupant Classification System (OCS), do not connect the battery negative cable before performing the OCS Verification Test using the scan tool and the appropriate diagnostic information. Personal injury or death may result if the system test is not performed properly.

Never replace both the Occupant Restraint Controller (ORC) and the Occupant Classification Module (OCM) at the same time. If both require replacement, replace one, then perform the Airbag System test before replacing the other.

Both the ORC and the OCM store Occupant Classification System (OCS) calibration data, which they transfer to one another when one of them is replaced. If both are replaced at the same time, an irreversible fault will be set in both modules and the OCS may malfunction and cause personal injury or death.

If equipped with OCS, the Seat Weight Sensor is a sensitive, calibrated unit and must be handled carefully. Do not drop or handle roughly. If dropped or damaged, replace with another sensor. Failure to do so may result in occupant injury or death.

If equipped with OCS, the front passenger seat must be handled carefully as well. When removing the seat, be careful when setting on floor not to drop. If dropped, the sensor may be inoperative, could result in occupant injury, or possibly death.

If equipped with OCS, when the passenger front seat is on the floor, no one should sit in the front passenger seat. This uneven force may damage the sensing ability of the seat weight sensors. If sat on and damaged, the sensor may be inoperative, could result in occupant injury, or possibly death.

DISARMING THE SYSTEM

To avoid personal injury when working on vehicles equipped with an air bag, the negative battery cable must be disconnected and at least 90 seconds must elapse before working on the system. Failure to do so may result in deployment of the air bag.

ARMING THE SYSTEM

Connect the negative battery and wait 2 minutes before performing and work on the vehicle.

DRIVETRAIN

AUTOMATIC TRANSMISSION ASSEMBLY

REMOVAL & INSTALLATION

See Figures 4 through 7.

1. Disconnect cable from negative battery terminal.

➡After the ignition switch is turned OFF, the navigation system requires approximately 90 seconds to record various types of memory and settings. As a result, after turning the ignition switch OFF, wait 90 seconds or more before disconnecting the cable from the negative (-) battery terminal.

2. Remove no. 1 engine under cover.
3. Drain automatic transmission fluid.
4. Drain engine coolant.
5. Remove the propeller shaft.
6. Remove the front exhaust pipe.
7. Remove the nut and clip, and disconnect the transmission control cable.
8. On V6 engine, remove starter.
9. Remove the bolt and flywheel housing cover.
10. Turn the crankshaft to gain access to the 6 bolts and remove each bolt while holding the crankshaft pulley setting bolt with a wrench.
11. Support the transmission with a transmission jack. Lift the transmission slightly from the crossmember.
12. Remove the 4 bolts of the rear engine mounting insulator.
13. Remove the 4 nuts, 4 washers, 4 bolts and crossmember.
14. Remove the 4 bolts and rear engine mounting insulator from the transmission.
15. Remove the bolt and disconnect the ground cable.
16. Tilt the transmission downward.

➡Make sure the cooling fan does not contact the fan shroud.

17. Disconnect the park/neutral position switch connector, transmission wire connector and 2 speed sensor connectors.
18. Disconnect the 4 connector clamps and 4 harness clamps.
19. Remove the bolt and disconnect the harness clamp and wire harness.
20. Remove the 2 bolts and disconnect the 2 water by-pass hoses from the transmission oil cooler, and separate the by-pass pipe from the automatic transmission.
21. For V6 engine, remove the 9 bolts and transmission.
22. For V8 engine, remove the 10 bolts and transmission.

To install:

23. For V6 engine, install the transmission with the 9 bolts.
24. For V8 engine, install the transmission with the 10 bolts.

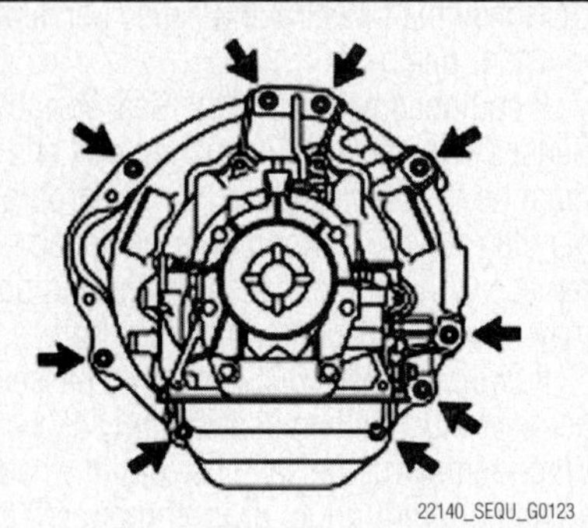

Fig. 4 Transmission bolt locations V6 engine

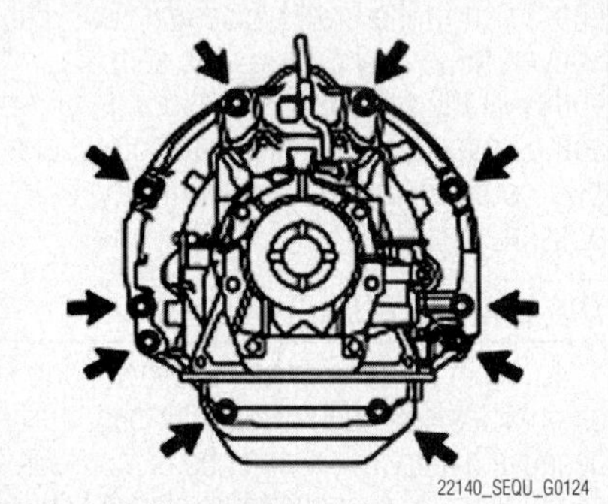

Fig. 5 Transmission bolt locations V8 engine

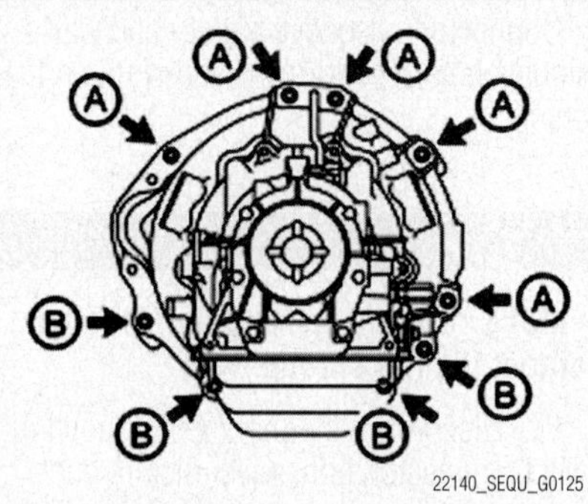

Fig. 6 Transmission tightening sequence V6 engine

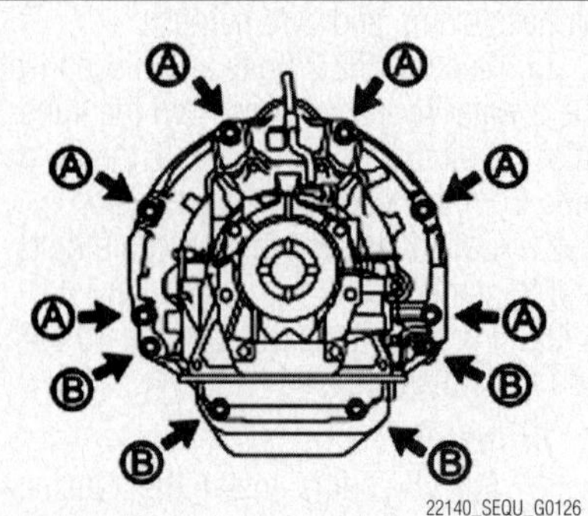

Fig. 7 Transmission tightening sequence V8 engine

25. For both engines, tighten 17 mm bolts to 52 ft. lbs. (72 Nm) and 14 mm bolts to 27 ft. lbs. (37 Nm).
26. Connect the 2 water by-pass hoses with the 2 clips and install the water by-pass pipe with the bolt and tighten to 13 ft. lbs.(18 Nm).
27. Connect the park/neutral position switch connector, transmission wire connector and 2 speed sensor connectors.
28. Connect the harness clamp with the bolt.
29. Connect the ground cable with the bolt and tighten to 49 inch lbs. (5.5 Nm).
30. Install the rear engine mounting insulator to the transmission with the 4 bolts and tighten to 30 ft. lbs. (40 Nm).
31. Install the frame crossmember to the rear engine mounting insulator with the 4 bolts and tighten to 12 ft. lbs. (16 Nm).
32. Install the frame crossmember with the 4 bolts, 4 washers and 4 nuts and tighten to 81 ft. lbs. (110 Nm).
33. Turn the crankshaft to gain access to the 6 bolts and install each bolt while holding the crankshaft pulley setting bolt with a wrench and tighten to 35 ft. lbs. (48 Nm).
34. Install the flywheel housing under cover with the bolt.
35. For V6 engines, install starter.
36. Connect the control cable with a new clip and nut and tighten bolt to 10 ft. lbs. (14 Nm).
37. Install the front exhaust pipe.
38. Install the propeller shaft.
39. Connect cable to negative battery terminal and perform initialization, if necessary.
40. Add transmission fluid and coolant.
41. Adjust shift lever position, if necessary.
42. Check for fluid and exhaust leaks.

TRANSFER CASE ASSEMBLY

REMOVAL & INSTALLATION

1. Drain transfer oil.
2. Remove the automatic transmission with transfer assembly.
3. Remove the 8 transfer adaptor rear mounting bolts.
4. Pull the transfer straight up and remove it from the transmission.

➡Take care not to damage the rear adaptor oil seal with the transfer input gear spline.

To install:

5. Install the transfer with the 8 bolts and tighten to 30 ft. lbs. (40 Nm).
6. Install the automatic transmission with transfer assembly.
7. Add transfer oil.
8. Check transfer oil.
9. Checks for leaks.

FRONT AXLE SHAFT, BEARING & SEAL

REMOVAL & INSTALLATION

1. Before servicing the vehicle, refer to the precautions in the beginning of this section.
2. Remove or disconnect the following:
   - Front wheel
   - Under cover
3. Drain the differential oil.
4. Remove or disconnect the following:
   - Remove the 2 bolts and disconnect the disc brake caliper from the steering knuckle
   - Remove front disc
   - Grease cap (4WD)
   - Cotter pin and lock cap
   - Halfshaft locknut by applying the brakes
   - Remove the 4 bolts
   - For 4WD, using a plastic-faced hammer, tap out the front drive shaft from the front axle hub.
   - Remove the axle hub and dust cover from the steering knuckle
   - Remove the O-ring from the axle hub.

To install:

5. Install or connect the following:
   - After applying grease to the O-ring, install the O-ring to the axle hub
   - For 4WD, connect the front drive shaft to the front axle hub
   - Install the dust cover and axle hub to the steering knuckle with the 4 bolts and tighten to 73 ft. lbs. (99 Nm)
   - Clean the threaded parts on the drive shaft and axle hub nut using a non-residue solvent.

➡Be sure to perform this work for a new drive shaft. Keep the threaded parts free of oil and foreign objects.

6. Using a 39 mm socket wrench, install the hub nut and tighten to 249 ft. lbs. (338 Nm).
   - Lock cap and a new cotter pin
   - Grease cap
   - Front disc
   - Connect the front disc brake caliper and install 2 new bolts and tighten to 73 ft. lbs. (99 Nm)
7. Refill the differential with oil.

8. Install or connect the following:
   - Under cover
   - Front wheel

REAR AXLE SHAFT, BEARING & SEAL

REMOVAL & INSTALLATION

1. Before servicing the vehicle, refer to the precautions in the beginning of this section.
2. Remove or disconnect the following:
   - Rear wheel
   - Rear caliper
   - Brake disc
3. Remove or disconnect the following:
   - Anti-lock Brake System (ABS) speed sensor from the rear axle housing, if equipped
   - Parking brake cable
   - Parking brake shoe assemblies
   - Remove the 4 nuts and rear axle shaft together with the parking brake plate

⁂ WARNING

Be careful not to damage the oil seal.

   - O-ring from the rear axle housing
   - Inner side oil seal using tool 09308-00010

To install:

4. Install or connect the following:
   - New O-ring to the rear axle
   - Install the rear axle shaft and parking brake plate with the 4 nuts and tighten to 44 ft. lbs. (60 Nm)
   - Anti-lock Brake System (ABS) speed sensor to the rear axle housing, if equipped
   - Parking brake cable assembly
   - Parking brake shoe assembly
   - Install rear disc
   - Adjust parking brake shoe clearance
   - Connect the rear disc brake cylinder and install 2 new bolts and tighten to 70 ft. lbs. (95 Nm)

➡Do not twist the flexible hose. Make sure that the bolts are free from damage and foreign matter. Do not over tighten the bolts.

   - Rear wheel

ENGINE COOLING

THERMOSTAT

REMOVAL & INSTALLATION

V6 Engine

1. Remove No. 1 engine under cover.
2. Drain engine coolant.
3. Remove V-bank cover.
4. Disconnect radiator hose outlet.
5. Remove the 3 nuts, water inlet with thermostat and gasket.

To install:

6. To install., reverse removal procedure and tighten nuts to 80 inch lbs. (9.0 Nm).

4.7L V8 engine

See Figure 8.

1. Before servicing the vehicle, refer to the precautions in the beginning of this section.
2. Drain engine coolant.
3. Remove the 3 nuts and disconnect the water inlet from the water inlet housing.
4. Remove the thermostat.
5. Remove the gasket from the thermostat.

To install:

6. Install a new gasket to the thermostat.
7. Insert the thermostat into the water inlet housing with the jiggle valve facing straight upward.

➡The jiggle valve may be set within 30° of either side of the prescribed position.

8. Install the water inlet with the 3 nuts and tighten to 19 Nm (14 ft. lbs.).
9. Fill with engine coolant.
10. Start engine and check for coolant leaks.
11. Recheck engine coolant level.

5.7L V8 Engine

See Figures 9 through 11.

1. Remove No. 1 engine under cover.
2. Drain engine coolant.
3. Remove V-bank cover sub-assembly.
4. Disconnect outlet radiator hose.
5. Disconnect the No. 2 and No. 3 air hoses.
6. Disconnect the air pump's connector.
7. Disconnect the air pump connector clamp's holder.

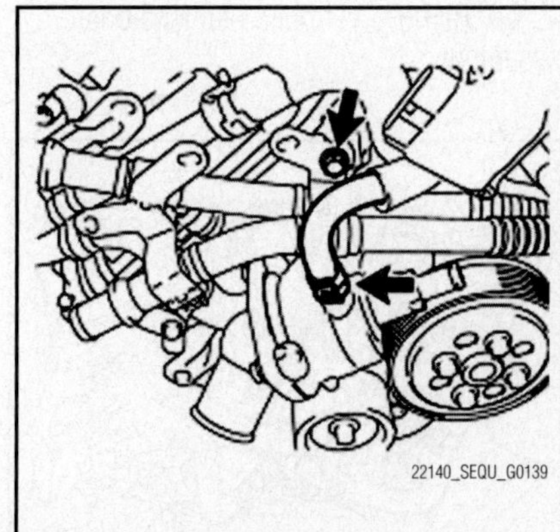

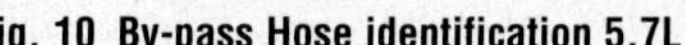
Fig. 10 By-pass Hose identification 5.7L

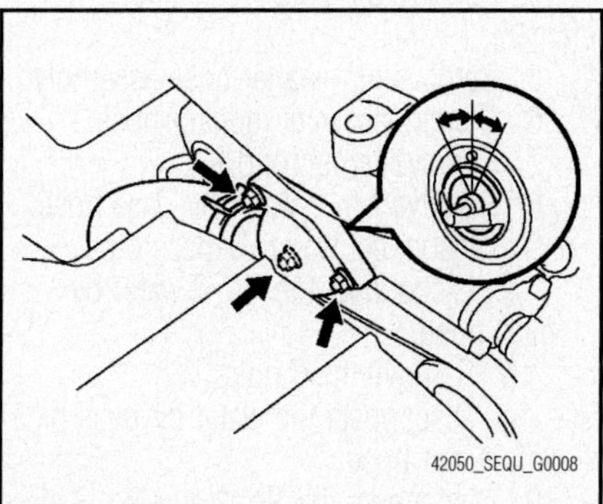

Fig. 8 Thermostat positioning and installation

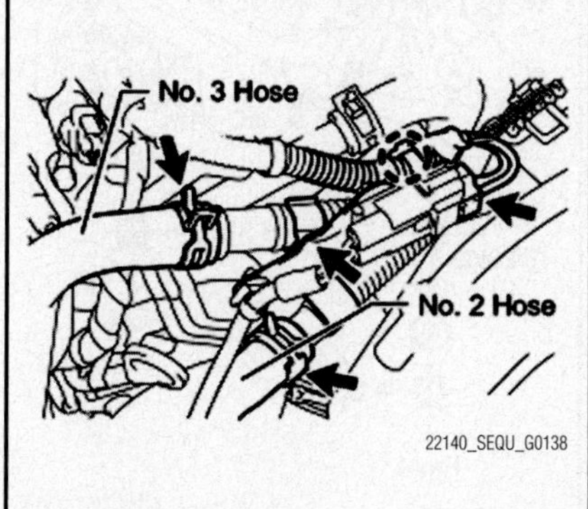

Fig. 9 Hose identification 5.7L

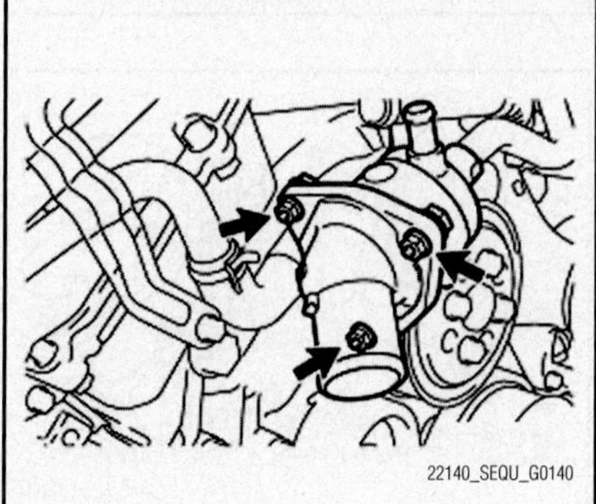

Fig. 11 Thermostat nut location 5.7L

8. Using a clip remover, detach the wire harness clamp.
9. Disconnect the No. 5 water by-pass hose.
10. Remove the air tube bracket's bolt.
11. Remove the 3 nuts, water inlet with thermostat and gasket.

WATER PUMP

REMOVAL & INSTALLATION

V6 Engine

See Figures 12 through 15.

1. Remove No. 1 engine under cover.
2. Drain engine coolant.
3. Remove V-bank cover.
4. Disconnect radiator hose inlet and outlet.
5. Remove fan shroud.
6. Remove No. 2 air cleaner hose.
7. Remove air cleaner assembly with element.
8. Disconnect the 2 oil cooler hoses.
9. Disconnect the 5 water by-pass hoses.
10. Remove the 5 bolts and water inlet.
11. Remove the gasket from the water outlet pipe.
12. Remove the 2 bolts, 2 cover plates and 2 idler pulleys.
13. Remove V-ribbed belt tensioner assembly.

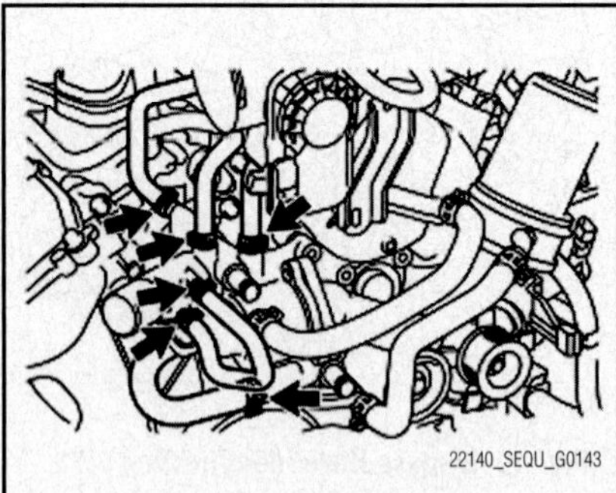

Fig. 12 Water bypass hose location

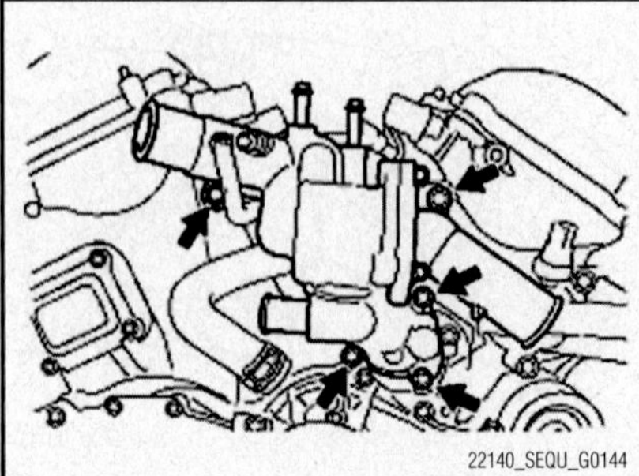

Fig. 13 Water inlet bolt location

14. Remove generator assembly.
15. Remove the 17 bolts, water pump and gasket.

To install:

16. Install a new gasket and the water pump with the 17 bolts. Tighten 10 mm head bolts to 80 inch lbs. (9.0 Nm) and 12 mm head bolts to 17 ft. lbs. (23 Nm).
17. Install generator assembly.
18. Install v-ribbed belt tensioner assembly.
19. Install the 2 idler pulleys and 2 cover plates with the 2 bolts an tighten to 29 ft. lbs. (39 Nm).
20. Install a new gasket to the water outlet pipe.
21. Install a new gasket to the water pump.
22. Apply soapy water to the gasket.
23. Install the water inlet with the 5 bolts and tighten to 80 inch lbs. (9.0 Nm).
24. Connect the 5 water by-pass hoses.
25. Connect the 2 oil cooler hoses.
26. Install fan shroud.
27. Connect radiator hose inlet and outlet.

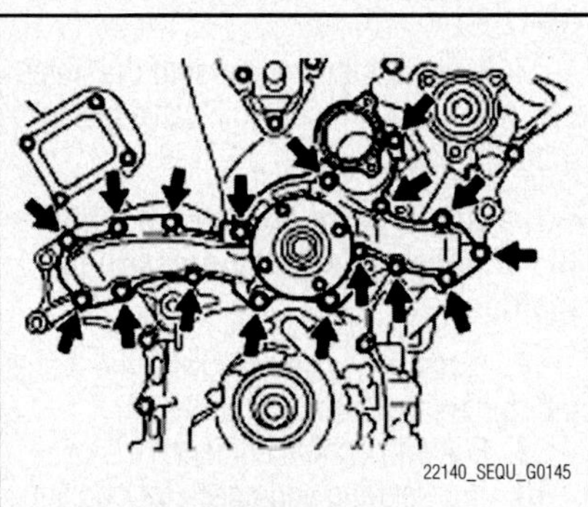

Fig. 14 Water pump bolt location

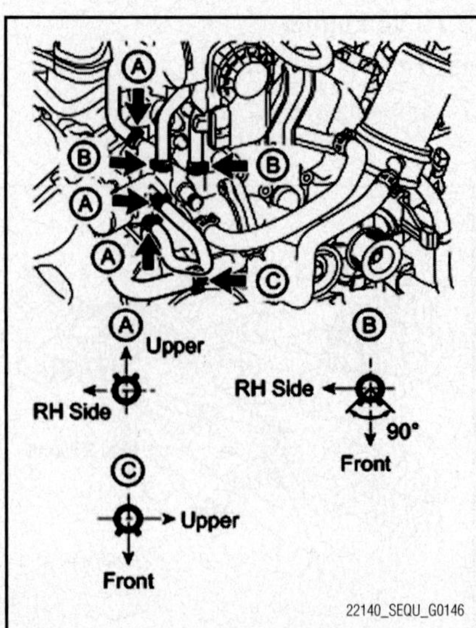

Fig. 15 Bypass hose clamp location

28. Install air cleaner assembly with element.
29. Install No. 2 air cleaner hose.
30. Add engine coolant.
31. Inspect for coolant leak.
32. Install V-bank cover.
33. Install No. 1 engine under cover.

4.7L V8 Engine

See Figures 16 and 17.

1. Before servicing the vehicle, refer to the precautions in the beginning of this section.
2. Drain the cooling system.
3. Remove or disconnect the following:
   - Negative battery cable
   - Timing belt
   - No. 2 idler pulley
   - Radiator hose
   - Bypass hose
   - Water inlet housing assembly
   - Water pump

To install:

4. Install or connect the following:
   - Water pump. Use a new gasket and tighten the bolts to 15 ft. lbs. (21 Nm). Tighten the stud bolt and nut to 13 ft. lbs. (18 Nm).
   - Water inlet housing assembly. Use a new O-ring and apply sealant as shown. Tighten the bolts to 13 ft. lbs. (18 Nm).
   - Bypass hose
   - Radiator hose
   - No. 2 idler pulley
   - Timing belt
   - Negative battery cable
5. Fill the cooling system.
6. Start the engine and check for leaks.

5.7L V8 Engine

See Figure 18.

1. Remove No. 1 engine under cover.
2. Drain engine coolant.
3. Remove V-bank cover sub-assembly.
4. Remove air cleaner assembly with element.
5. Remove air cleaner hose assembly.
6. Disconnect inlet radiator hose.
7. Remove fan shroud.
8. Remove No. 1 water by-pass hose.
9. Disconnect water by-pass pipe.
   a. Disconnect the No. 6 water by-pass hose.
   b. Remove the 3 bolts.
   c. Disconnect the water by-pass pipe with water hose.
10. Disconnect the No. 2 and No. 3 air hoses.
11. Disconnect the air pump's connector.

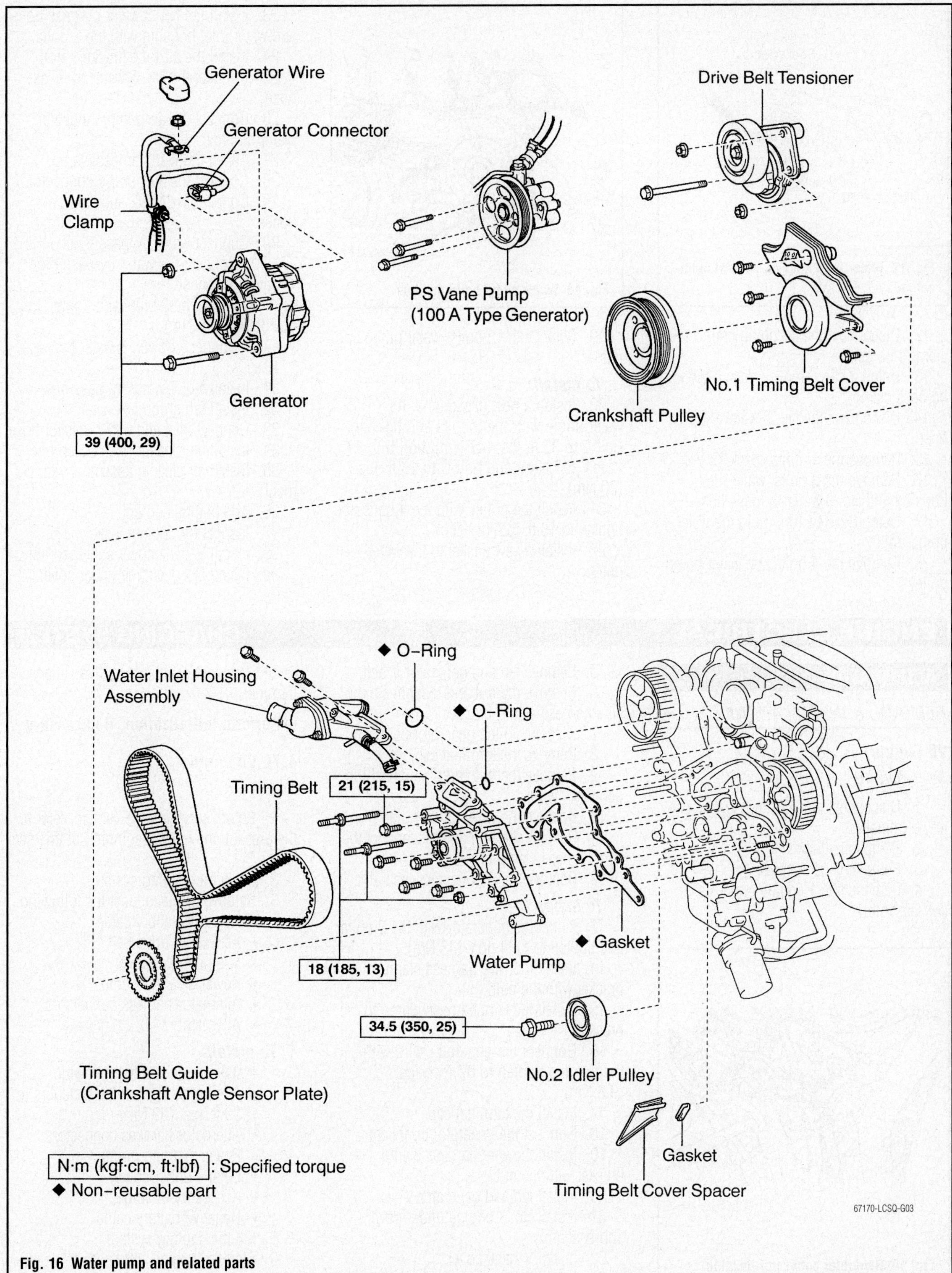

Fig. 16 Water pump and related parts

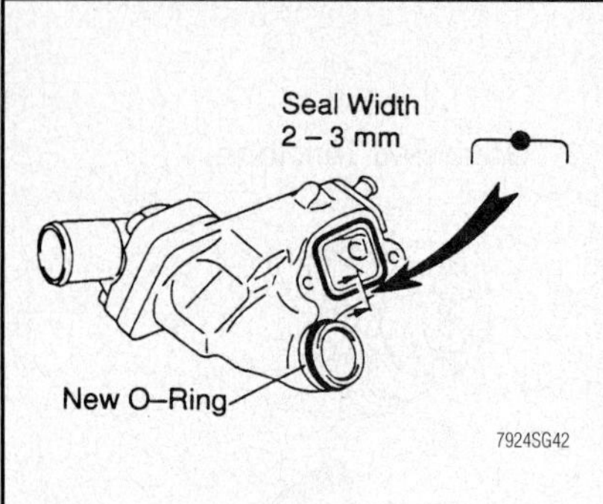

Fig. 17 Water inlet housing sealant application

12. Disconnect the air pump connector clamp's holder.
13. Using a clip remover, detach the wire harness clamp.
14. Disconnect the No. 5 water by-pass hose.
15. Remove the air tube bracket's bolt.
16. Remove the 3 bolts, water inlet housing and No. 3 water by-pass hose.
17. Using Service Tool, hold the water pump pulley.
18. Remove the 4 bolts and water pump pulley.

Fig. 18 Water pump bolt locations

19. Remove the 8 bolts, water pump and gasket.

To install:

20. Install a new gasket and the water pump with the 8 bolts and tighten: Bolt A to 35 ft. lbs. (47 Nm), bolt B to 17 ft. lbs. (23 Nm), and bolt C to 15 ft. lbs. (20 Nm).
21. Install the pulley with the 4 bolts and tighten to 15 ft. lbs. (21 Nm).
22. Install a new gasket to the water pump.
23. Install the No. 3 water by-pass hose and water inlet housing with the 3 bolts.
24. Install the air tube bracket's bolt.
25. Connect the No. 5 water by-pass hose.
26. Connect the air pump connector clamp's holder.
27. Attach the wire harness clamp.
28. Connect the air pump's connector.
29. Connect the No. 2 and No. 3 air hoses.
30. Connect water by-pass pipe.
    a. Connect the water by-pass pipe with water hose.
    b. Install the 3 bolts and tighten to 7 ft. lbs. (10 Nm).
    c. Connect the No. 6 water by-pass hose.
31. Install No. 1 water by-pass hose.
32. Install fan shroud.
33. Connect inlet and outlet radiator hose.
34. Install air cleaner hose assembly.
35. Install air cleaner assembly with element.
36. Add engine coolant.
37. Inspect for coolant leak.
38. Install V-bank cover sub-assembly.
39. Install No. 1 engine under cover.

ENGINE ELECTRICAL

CHARGING SYSTEM

ALTERNATOR

REMOVAL & INSTALLATION

V6 Engine

See Figure 19.

1. Disconnect cable from negative battery terminal.
2. Remove battery.
3. Remove V-bank cover.
4. Remove No. 1 engine under cover sub-assembly.

Fig. 19 Generator connector location

5. Remove fan and generator V belt.
6. Remove the nut and disconnect the wire harness clamp bracket.
7. Disconnect the generator connector.
8. Remove the terminal cap.
9. Remove the nut and disconnect the generator wire.
10. Remove the wire harness clamp.
11. Remove the bolt and disconnect the wire harness clamp bracket.
12. Remove the 2 bolts and generator.

To install:

13. Install the generator with the 2 bolts and tighten to 32 ft. lbs. (43 Nm).
14. Install the wire harness clamp bracket with the bolt.
15. Install the wire harness clamp to the bracket.
16. Connect the generator wire with the nut and tighten to 87 inch lbs. (9.8 Nm).
17. Install the terminal cap.
18. Connect the generator connector.
19. Install the wire harness clamp bracket with the nut.
20. Install fan and generator V belt.
21. Install No. 1 engine under cover sub-assembly.
22. Install V-bank cover.
23. Install battery.
24. Connect cable to negative battery terminal.

➡Perform initialization, if necessary.

4.7L V8 Engine

See Figure 20.

1. Before servicing the vehicle, refer to the precautions in the beginning of this section.
2. Drain the cooling system.
3. Remove or disconnect the following:
   - Negative battery cable
   - Accessory drive belt
   - Engine under cover
   - Power steering pump
   - Alternator harness connectors
   - Alternator

To install:

4. Install or connect the following:
   - Alternator. Tighten the fasteners to 29 ft. lbs. (39 Nm).
   - Alternator harness connectors
   - Power steering pump
   - Engine under cover
   - Accessory drive belt
   - Negative battery cable
5. Fill the cooling system.
6. Start the engine and check for leaks.

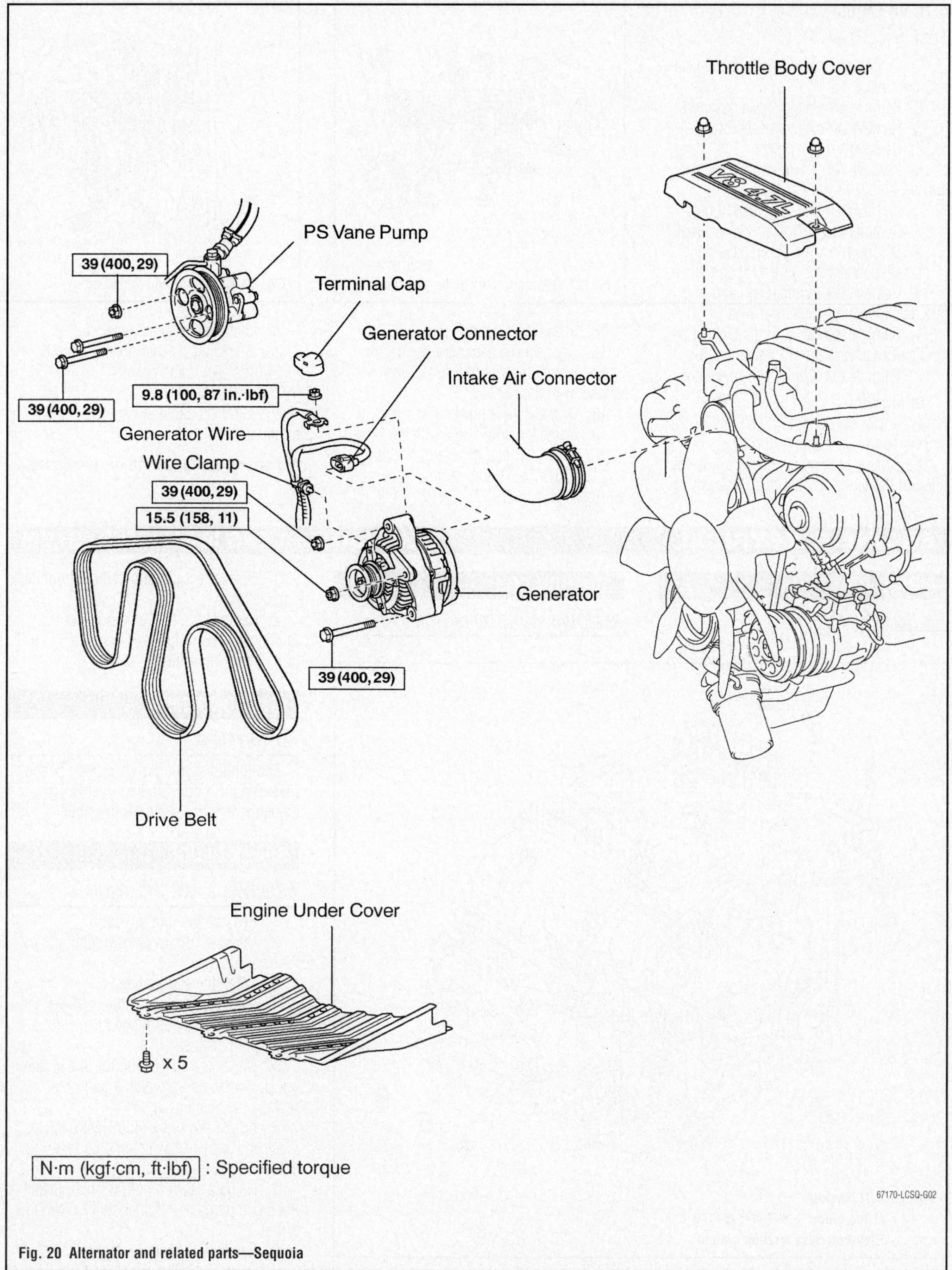

Fig. 20 Alternator and related parts—Sequoia

5.7L V8 Engine

See Figures 21 and 22.

1. Disconnect cable from negative battery terminal.
2. Remove air cleaner hose assembly.
3. Remove air cleaner assembly.
4. Remove V-bank cover.
5. Remove No. 1 engine under cover sub-assembly.
6. Remove fan and generator V belt.
7. Remove the 5 clips and apron seal.
8. Disconnect vane pump assembly.
9. Disconnect the generator connector.
10. Remove the terminal cap and nut, and disconnect the generator wire.
11. Remove the bolt and disconnect the wire harness bracket from the generator.
12. Remove the 3 bolts and generator.

To install:

13. Install the generator with the 3 bolts and tighten to 32 ft. lbs. (43 Nm).
14. Connect the generator wire with the nut and tighten to 87 inch lbs. (9.8 Nm).

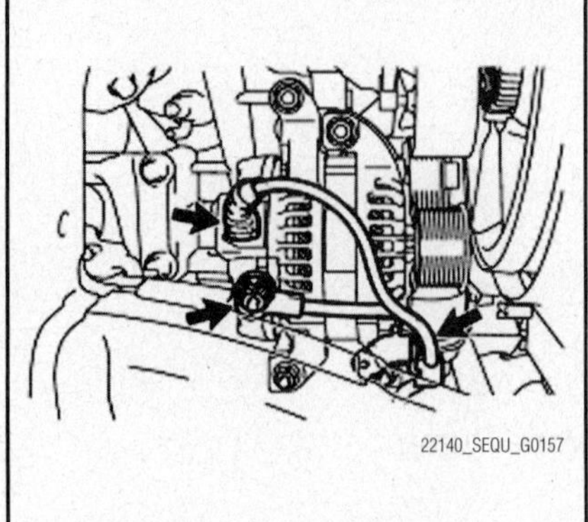

Fig. 21 Generator connector location

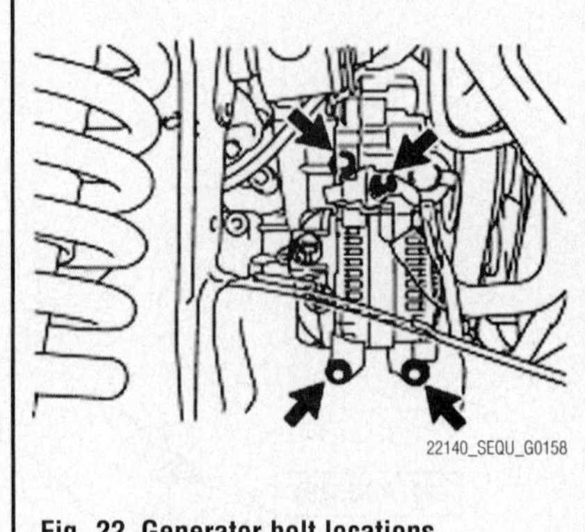

Fig. 22 Generator bolt locations

15. Install the terminal cap.
16. Connect the generator connector.
17. Install the wire harness clamp bracket with the nut.
18. Connect vane pump assembly.
19. Install the apron seal with the 5 clips.
20. Install fan and generator V belt.
21. Install No. 1 engine under cover sub-assembly.
22. Install air cleaner assembly.
23. Install air cleaner hose assembly.
24. Install V-bank cover.
25. Install battery.
26. Connect cable to negative battery terminal.

➡**Perform initialization, if necessary.**

ENGINE ELECTRICAL — DISTRIBUTORLESS IGNITION SYSTEM

FIRING ORDER

See Figure 23.

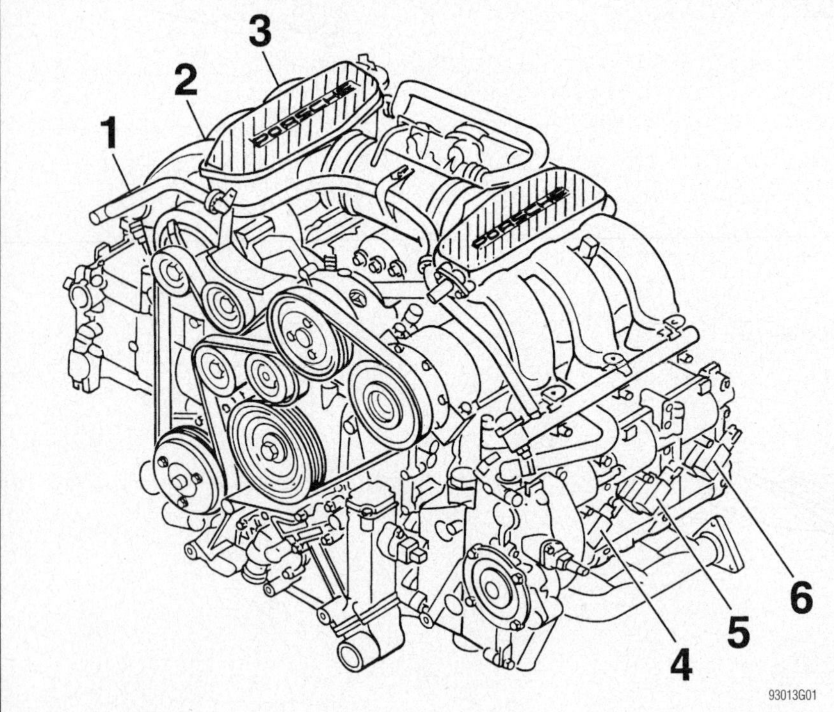

Fig. 23 4.7L Engine
Firing order: 1–8–4–3–6–5–7–2
Distributorless ignition system

IGNITION COIL

REMOVAL & INSTALLATION

1. Disconnect ignition coil (with igniter) connectors.
2. Remove the bolt, and pull out the ignition coil (with igniter). Remove the ignition coils (with igniter).

IGNITION TIMING

ADJUSTMENT

The ignition timing is controlled by the Powertrain Control Module (PCM). No adjustment is necessary or possible.

SPARK PLUGS

REMOVAL & INSTALLATION

1. Remove the ignition coils.
2. Using a 16 mm plug wrench, remove the spark plugs.
3. Clean the spark plugs.
4. If the electrode has traces of wet carbon, allow it to dry and then clean with a spark plug cleaner.
5. Check the spark plug for thread damage and insulator damage. If abnormal, replace the spark plug.
6. Adjust the spark plug electrode gap. Electrode gap for new spark plug is 1.0 to 1.1 mm (0.039 to 0.043 in.).
7. Using a 16 mm plug wrench, install the spark plugs and tighten to 17.5 Nm (13 ft lbs)
8. Reinstall the ignition coils.

STARTER

REMOVAL & INSTALLATION

V6 Engine

See Figure 24.

1. Disconnect cable from negative battery terminal.
2. Remove the terminal cap.
3. Remove the nut and disconnect the starter wire.
4. Disconnect the starter connector.
5. Remove the 2 bolts and starter assembly.

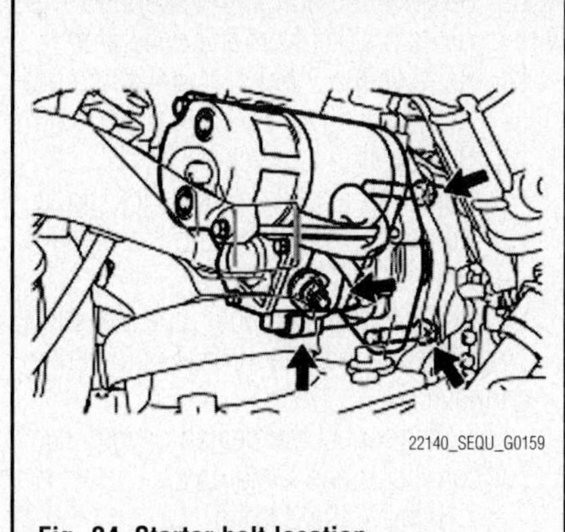

Fig. 24 Starter bolt location

To install:

6. Install the starter with the 2 bolts and tighten to 43 ft. lbs. (58 Nm).
7. Connect the starter connector.
8. Connect the starter wire with the nut and tighten nut to 87 inch lbs. (9.8 Nm).
9. Install the terminal cap.
10. Connect cable to negative battery terminal.

➡Some systems need to be initialized after the cable is reconnected.

4.7L V8 Engine

See Figure 25.

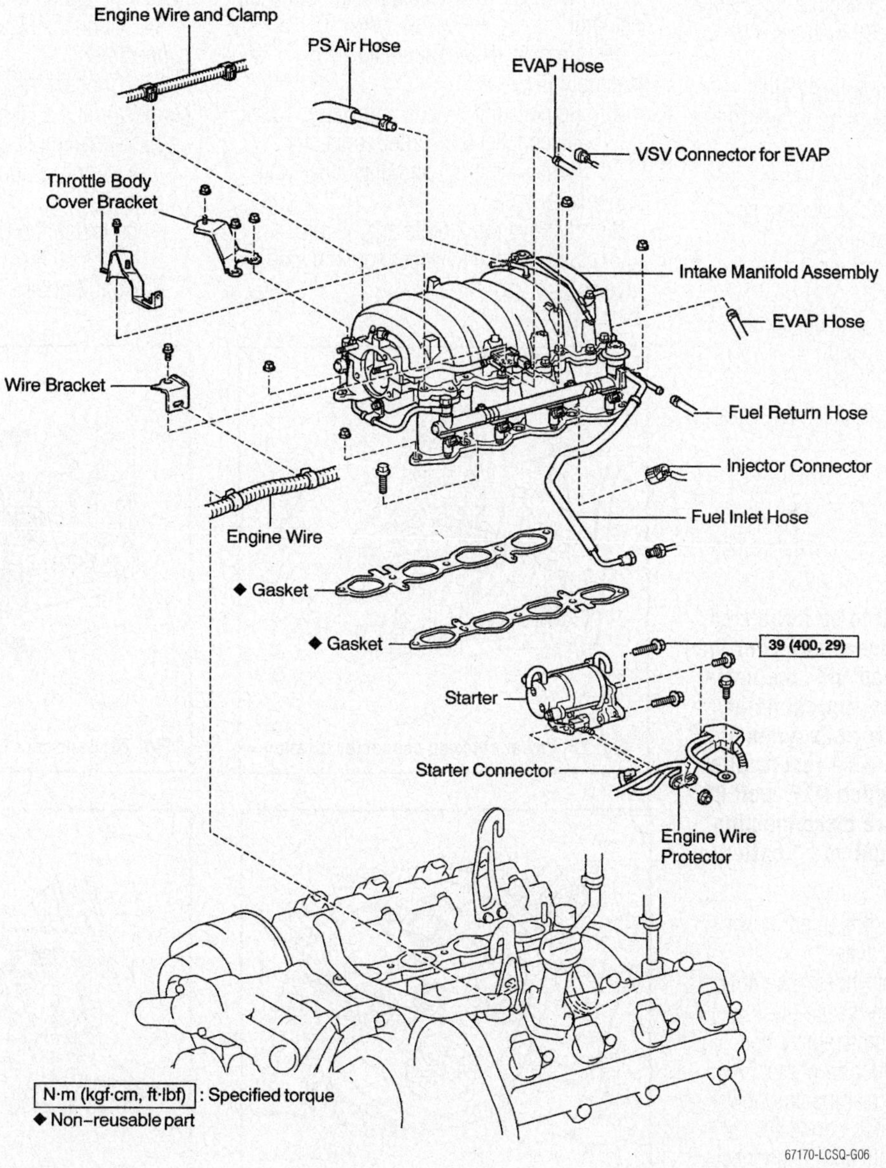

Fig. 25 Starter removal—Sequoia

1. Before servicing the vehicle, refer to the precautions in the beginning of this section.
2. Drain the cooling system.
3. Relieve the fuel system pressure.
4. Remove or disconnect the following:
   - Negative battery cable
   - Engine appearance cover
   - Air intake tube
   - Intake manifold
   - Water bypass assembly
   - No. 4 air injection system hose
   - Air pump assembly
   - Starter motor mounting bolts
   - Starter wiring connectors
   - Starter motor

To install:

5. Install or connect the following:
   - Starter motor
   - Starter wiring connectors. Tighten the cable nut to 86 inch lbs. (10 Nm).
   - Starter motor mounting bolts. Tighten the bolts to 18 ft. lbs. (25 Nm).
   - Air pump assembly
   - No. 4 air injection system hose
   - Water bypass assembly
   - Intake manifold
   - Air intake tube
   - Engine appearance cover
   - Negative battery cable
6. Fill the cooling system.
7. Start the engine and check for leaks.

5.7L V8 Engine

See Figures 26 through 31.

1. Disconnect cable from negative battery terminal.

➡Some systems need to be initialized after the cable is reconnected. After the ignition switch is turned OFF, the navigation system requires approximately 90 seconds to record various types of memory and settings. As a result, after turning the ignition switch OFF, wait 90 seconds or more before disconnecting the cable from the negative (-) battery terminal.

2. Remove No. 1 engine under cover.
3. Remove V-bank cover.
4. Remove air cleaner hose assembly.
5. Remove air cleaner assembly.
6. Remove fan and generator V belt.
7. Remove front fender apron seal RH.
8. Disconnect the 2 clamps and power steering oil pressure switch connector.
9. Remove the 2 bolts and disconnect the vane pump
10. Disconnect the generator connector.
11. Remove the terminal cap and nut, and disconnect the generator wire.
12. Remove the bolt and disconnect the wire harness bracket from the generator.
13. Remove the 2 bolts, 2 nuts and generator.
14. Remove the dipstick.
15. Remove the bolt and dipstick guide.
16. Remove the O-ring from the dipstick guide.
17. Remove front exhaust pipe assembly.
    a. Disconnect the air fuel ratio sensor connector.
    b. Disconnect the heated oxygen sensor connector and 2 clamps.
    c. Remove the 2 bolts.
    d. Remove the 3 nuts and front exhaust pipe.
    e. Remove the 2 gaskets.
18. Remove No. 1 exhaust manifold heat insulator.
19. Remove exhaust manifold sub-assembly RH.
20. Remove the 3 bolts and starter cover.
21. Disconnect the starter connector.
22. Remove the nut and disconnect the starter wire.
23. Remove the 2 bolts and starter.
24. Remove the flywheel housing side cover.

To install:

25. Install the flywheel housing side cove.
26. Install the starter with the 2 bolts and tighten to 27 ft. lbs. (37 Nm).
27. Install the starter wire with the nut and tighten to 87 inch lbs. (9.8 Nm).
28. Connect the starter connector.
29. Install the starter cover with the 3 bolts and tighten to 9 ft. lbs. (12 Nm).
30. Install exhaust manifold sub-assembly RH.
31. Install No. 1 exhaust manifold heat insulator.
32. Install a new gasket and the exhaust pipe to the exhaust manifold RH with 3 new nuts and tighten to 40 ft. lbs. (54 Nm).
33. Install a new gasket and the exhaust pipe to the center exhaust pipe with the 2 bolts and tighten to 35 ft. lbs. (48 Nm).
34. Connect the air fuel ratio sensor connector.
35. Connect the heated oxygen sensor connector and 2 clamps.
36. Install dipstick.
    a. Apply a light coat of engine oil to a new O-ring.
    b. Install the O-ring to the guide.
    c. Install the dipstick guide with the bolt and tighten to 7 ft. lbs. (10 Nm).

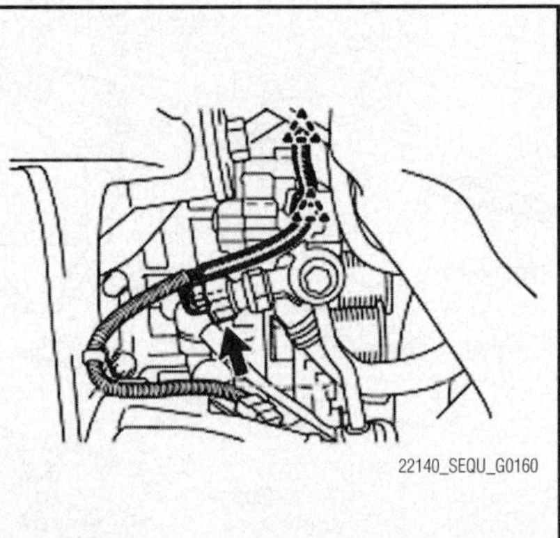

Fig. 26 Power steering connector location

Fig. 28 Generator bolt location

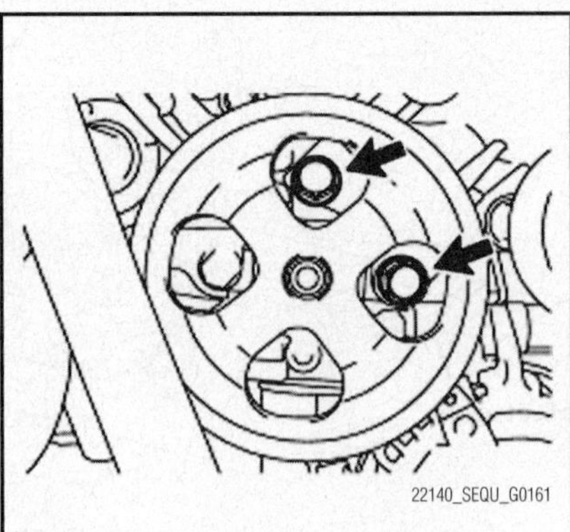

Fig. 27 Vane pump bolt location

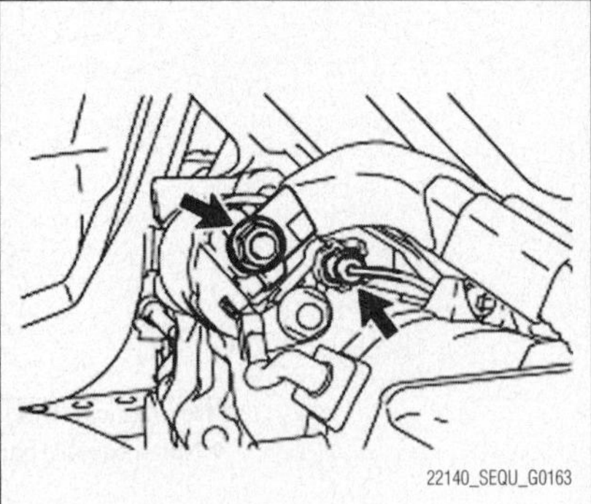

Fig. 29 Starter connection location

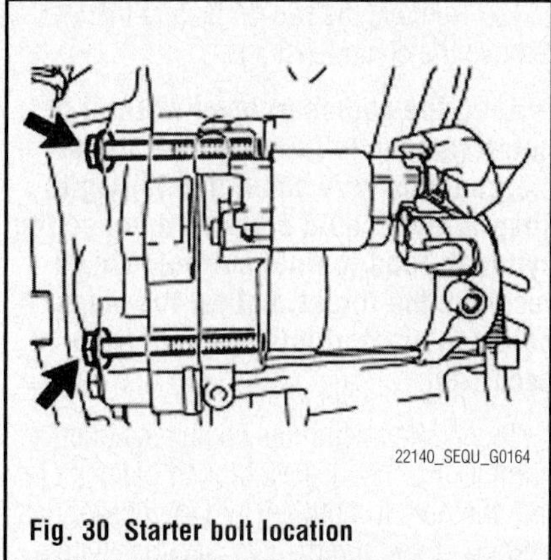

Fig. 30 Starter bolt location

37. Install the generator with the 2 bolts and 2 nuts and tighten to 32 ft. lbs. (43 Nm).
38. Connect the generator connector.

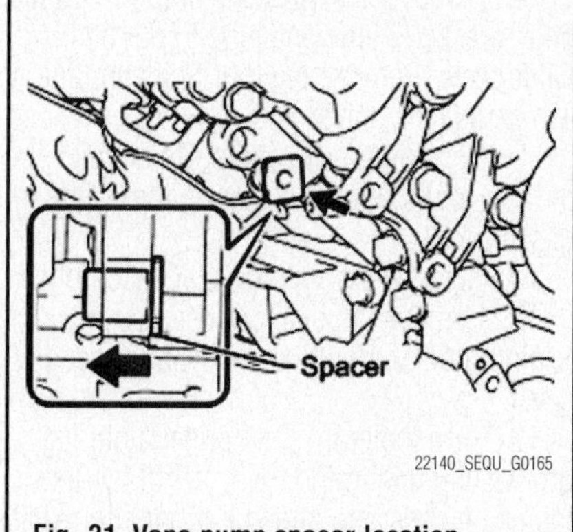

Fig. 31 Vane pump spacer location

39. Connect the generator wire with the nut and tighten to 87 inch lbs. (9.8 Nm).
40. Install the terminal cap.
41. Install the harness bracket to the generator with the bolt.
42. Connect vane pump assembly.
    a. Connect the vane pump to the timing chain cover with the 2 bolts and tighten to 21 ft. lbs. (28 Nm).
    b. Connect the 2 clamps and power steering oil pressure switch connector.
43. Install front fender apron seal RH.
44. Install fan and generator V belt.
45. Install air cleaner assembly.
46. Install air cleaner hose assembly.
47. Install V-bank cover.
48. Install No. 1 engine under cover.
49. Connect cable to negative battery terminal.

➡**Some systems need to be initialized after the cable is reconnected.**

ENGINE MECHANICAL

ACCESSORY DRIVE BELTS

ACCESSORY BELT ROUTING

See Figures 32 through 34.

Refer to the accompanying illustrations for belt routing.

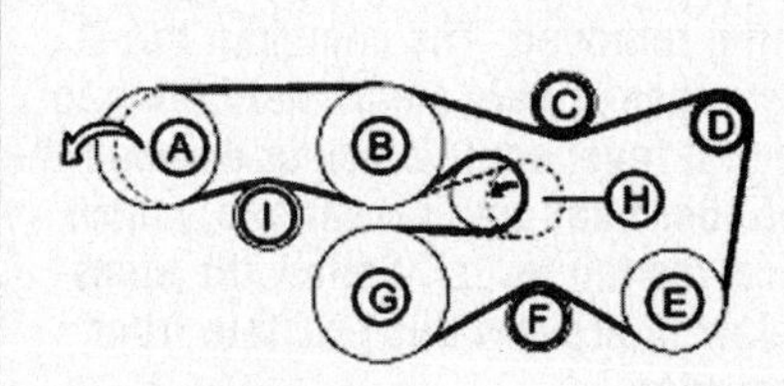

A: Vane Pump
B: Water Pump
C: No. 2 Idler
D: Generator
E: Cooler Compressor
F: No. 1 Idler
G: Crankshaft
H: V-Ribbed Belt Tensioner
I: No. 2 Idler

22140_SEQU_G0166

Fig. 32 Drive belt routing V6 engine

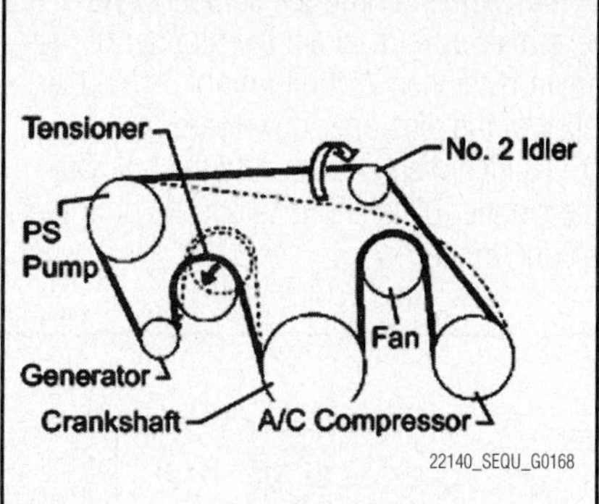

Fig. 33 Drive belt routing 4.7L V8 engine

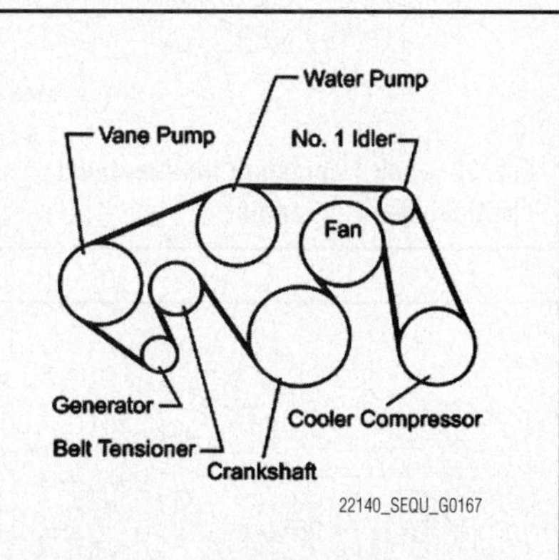

Fig. 34 Drive belt routing 5.7L V8 engine

INSPECTION

Inspect the drive belt for signs of glazing or cracking. A glazed belt will be perfectly smooth from slippage, while a good belt will have a slight texture of fabric visible. Cracks will usually start at the inner edge of the belt and run outward. All worn or damaged drive belts should be replaced immediately.

ADJUSTMENT

The belt does not require adjustment.

REMOVAL & INSTALLATION

See Figure 35.

1. Before servicing the vehicle, refer to the precautions in the beginning of this section.
2. Loosen the drive belt tension by turning the drive belt tensioner counterclockwise, and remove the drive belt.
3. Installation is the reverse of removal.

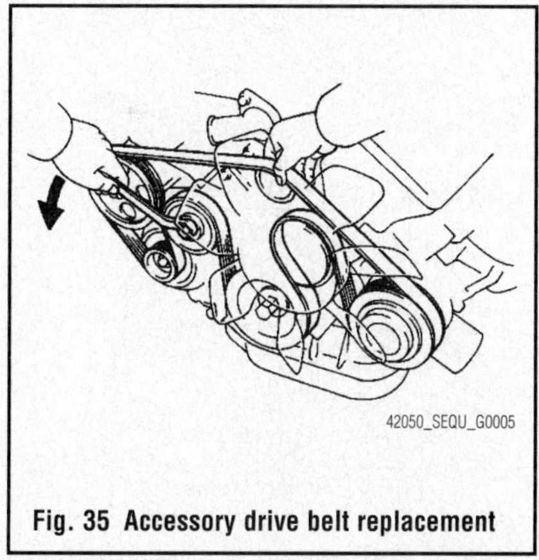

Fig. 35 Accessory drive belt replacement

CAMSHAFT AND VALVE LIFTERS

REMOVAL & INSTALLATION

V6 Engine

Bank 1

See Figures 36 through 50.

1. Before servicing the vehicle, refer to the Precautions Section.

2. Properly disarm the SRS system.

3. Disconnect the negative battery cable.

4. Drain the engine coolant. Remove the V bank cover.

5. Remove the air cleaner assembly.

6. Disconnect the two water bypass hoses. Disconnect the fuel vapor feed hose. Disconnect the ventilation hose. Disconnect the VSV connectors.

7. Disconnect the throttle body motor connector. Separate the three wire harness clamps and hose clamp.

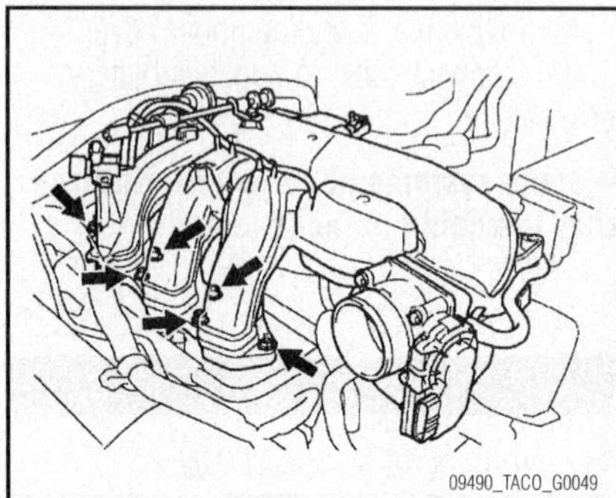

Fig. 36 Intake surge tank bolt locations—4.0L engine

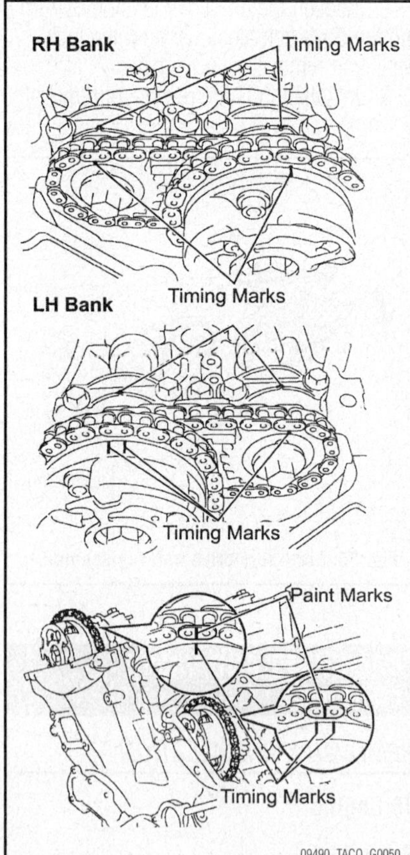

Fig. 37 Bank 1 camshaft timing mark alignment—4.0L engine

8. Remove the two bolts and the throttle body bracket. Remove the bolt and the oil baffle plate. Remove the four bolts and the two serge tank stays.

9. Remove the two nuts. Remove the four bolts, intake air surge tank and gasket.

10. Remove the ignition coil assembly.

11. Remove the cylinder head cover retaining bolts. Remove the cylinder head cover.

12. Turn the crankshaft pulley until its groove and the timing mark "0" of the timing chain cover are aligned. If not aligned at TDC of the compression stroke, turn the crankshaft one complete revolution, in the direction of rotation. Paint alignment marks on the number one chain links corresponding to the timing marks of the camshaft timing gears.

13. Remove the four bolts, and then remove the timing chain cover plate and gasket.

14. While turning the stopper plate of the tensioner upward, push the plunger of the chain tensioner. While turning the stopper plate of the tensioner downward, insert a 0.118 inch diameter bar into the holes in the stopper plate and tensioner to hold the stopper plate.

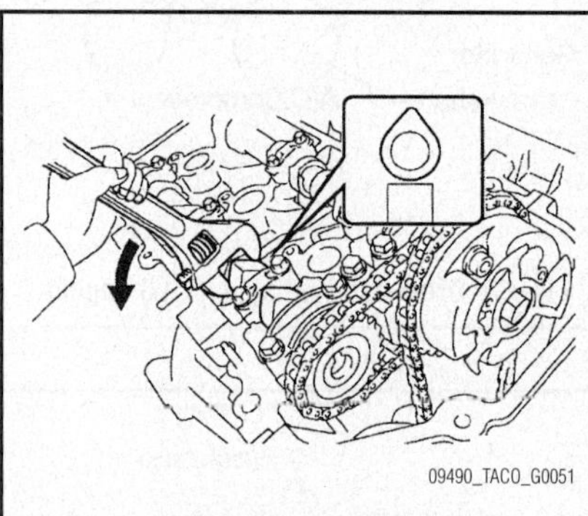

Fig. 38 Bank 1 camshaft lobe removal positioning—4.0L engine

15. Remove the two bolts, and then remove the chain tensioner.

➡Keep the camshaft level while it is being removed. The camshaft thrust clearance is very small and failing to keep it level could crack or damage the cylinder head journal surface, which receives the thrust. Follow the steps below to prevent this problem from occurring.

16. While raising the chain tensioner number two insert a 0.039 inch diameter pin into the hole to hold it. Hold the hexagonal portion of the number two camshaft with a wrench. Remove the camshaft timing gear set bolt.

17. Separate the camshaft timing gear from the number two camshaft. Rotate the camshafts counterclockwise, using a wrench, so that the cam lobes of the number one cylinder face in the direction shown.

18. Using several steps, loosen and remove the eight bearing cap bolts uniformly and in the proper removal sequence. Remove the four bearing caps and the number two camshaft.

19. Remove the number two chain tensioner bolt. Remove the number chain tensioner and camshaft timing gear.

➡Keep the camshaft level while it is being removed. The camshaft thrust clearance is very small and failing to keep it level could crack or damage the cylinder head journal surface, which receives the thrust. Follow the steps below to prevent this problem from occurring.

20. Hold the hexagonal portion of the number one camshaft, with a wrench. Loosen the camshaft timing gear set bolt.

➡Do not disassemble the camshaft timing gear assembly.

21. Slide the camshaft timing gear and

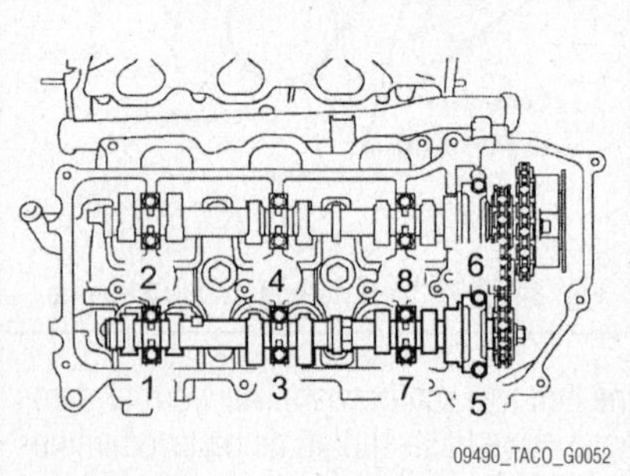

Fig. 39 Bank 1 camshaft number two bearing cap bolt removal sequence—4.0L engine

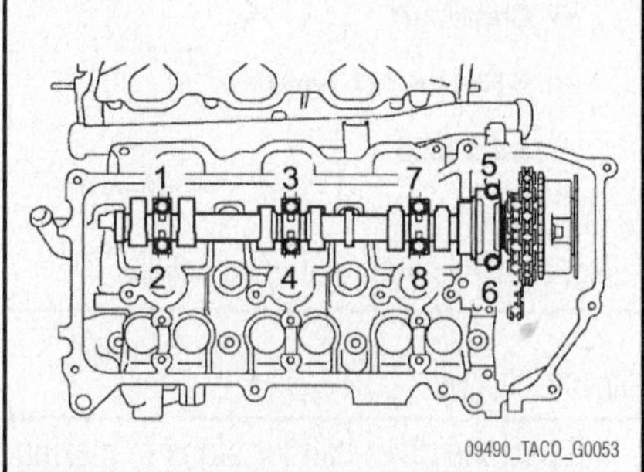

Fig. 40 Bank 1 camshaft number one bearing cap bolt removal sequence—4.0L engine

separate the number one chain from the camshaft timing gear.

22. Rotate the number one camshaft counterclockwise, using the wrench so that the cam lobes on the number one cylinder face downward.

23. Using several steps, loosen and remove the eight bearing cap bolts uniformly and in the proper removal sequence. Remove the four bearing caps.

24. Remove the camshaft timing gear set bolt with the number one camshaft lifted up. Remove the number one camshaft and camshaft timing gear with the number two chain.

25. Tie the number one chain to the side. Be careful not to drop anything inside the timing chain cover.

To install:

➡Keep the camshaft level while it is being installed. The camshaft thrust clearance is very small and failing to keep it level could crack or damage the cylinder head journal surface, which receives the thrust.

26. Align the yellow mark link with the timing mark (1 dot mark) of the camshaft timing gear. Apply new engine oil to the thrust portion and journal of the camshafts.

27. Temporarily install the number one chain onto the number two chain of the camshaft timing gear.

28. Align the knock pin hole of the camshaft timing gear with the knock pin of the number one camshaft. Insert the number one camshaft into the camshaft timing gear.

29. Temporarily install the camshaft timing gear set bolt. Install the number one camshaft onto the right cylinder head with the cam lobes of the number one cylinder facing downward, as indicated in the illustration.

30. Install the four bearing caps, in the proper location. Apply a light coat of engine oil to the threads and under the heads of the cap bolts.

31. Using several steps, uniformly install and tighten the bearing cap bolts in the proper sequence to 80 inch lbs. for the 10mm bolts and 18 ft. lbs. for the 12mm bolts.

32. Rotate the number one camshaft clockwise, using a wrench so that the timing mark of the camshaft timing gear is aligned with the timing mark of the camshaft bearing cap.

33. Align the paint mark of the number one chain with the timing mark of the camshaft timing gear.

34. Hold the hexagonal portion of the number one camshaft with a wrench, and tighten the camshaft timing gear set bolt to 74 ft. lbs.

35. While pushing in on the number two chain tensioner, insert a 0.039 inch pin into the hole to hold it.

36. Temporarily install the camshaft timing gear and chain tensioner number two and align the yellow mark links with the timing marks (1 dot mark) on the camshaft timing gears. Tighten the bolt to 14 ft. lbs.

➡Keep the camshaft level while it is being installed. The camshaft thrust clearance is very small and failing to keep it level could crack or damage the cylinder head journal surface, which receives the thrust.

37. Install the number two camshaft onto the right cylinder head with the cam lobes of the number one cylinder facing upward, as indicated in the illustration.

38. Install the four bearing caps in the proper location. Apply a light coat of clean engine oil to the threads and under the heads of the bolts.

39. Using several steps, uniformly install and tighten the eight bearing cap bolts in sequence to 80 inch lbs, for the 10mm head

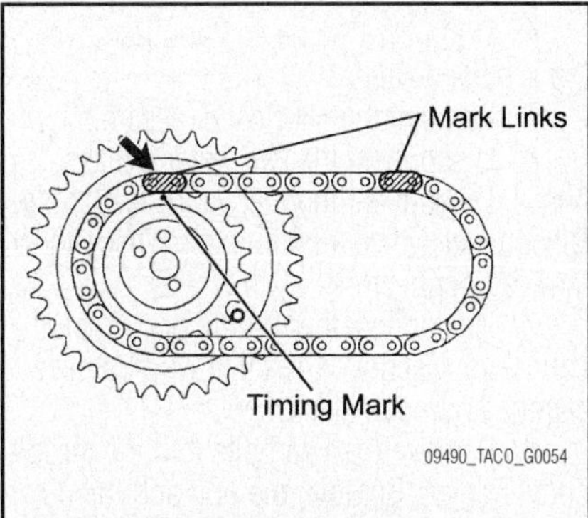

Fig. 41 Bank 1 camshaft number one alignment marks—4.0L engine

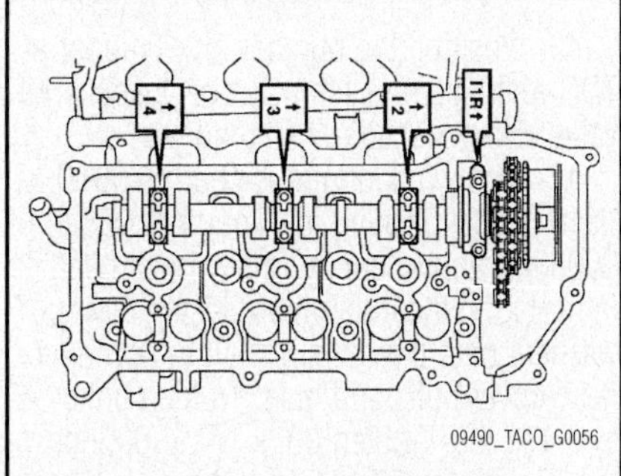

Fig. 43 Bank 1 camshaft number one bearing cap bolt installation sequence—4.0L engine

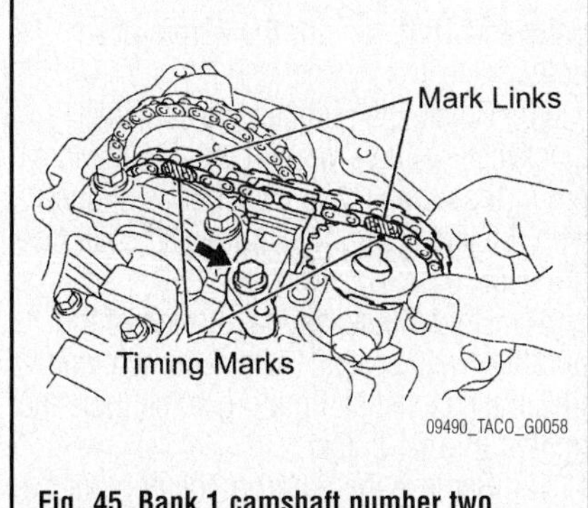

Fig. 45 Bank 1 camshaft number two alignment marks—4.0L engine

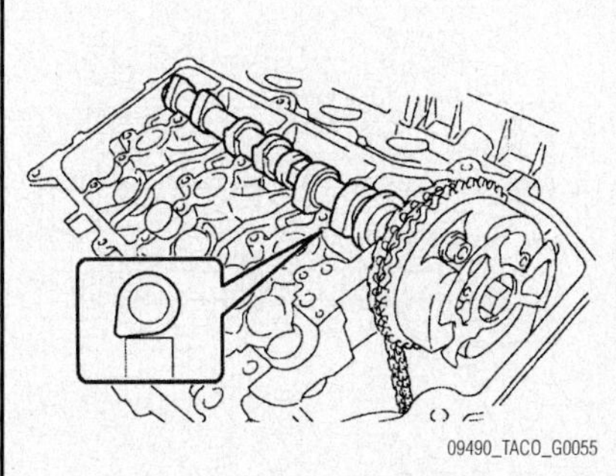

Fig. 42 Bank 1 camshaft number one lobe installation positioning—4.0L engine

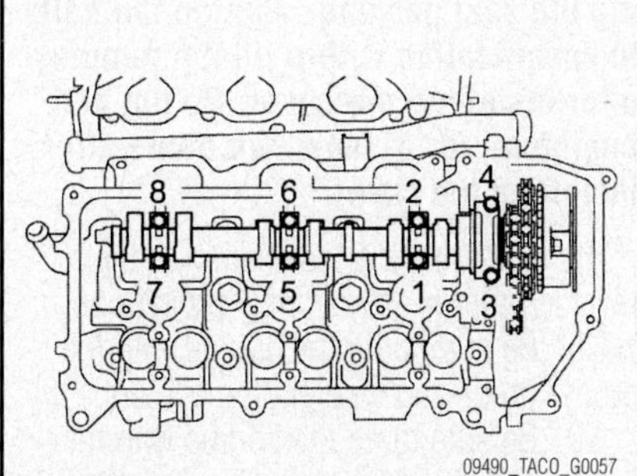

Fig. 44 Bank 1 camshaft number one bearing cap bolt torque sequence—4.0L engine

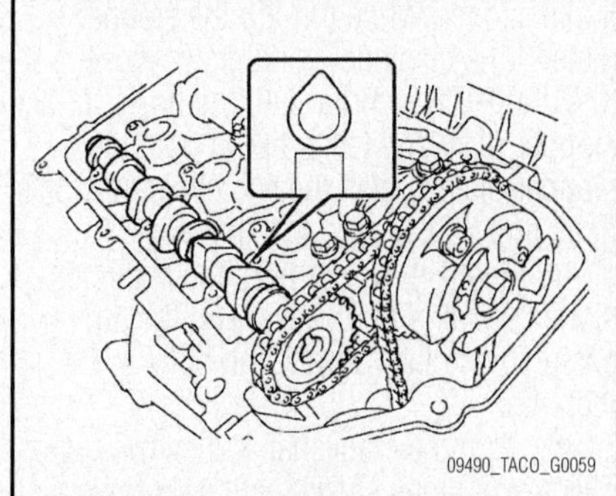

Fig. 46 Bank 1 camshaft number two lobe installation positioning—4.0L engine

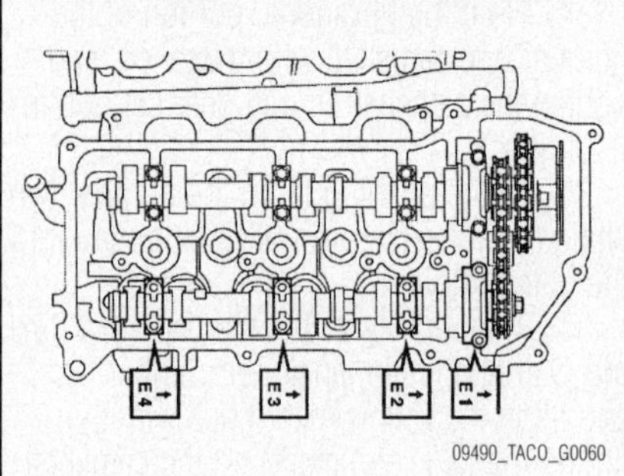

Fig. 47 Bank 1 camshaft number two bearing cap bolt installation sequence—4.0L engine

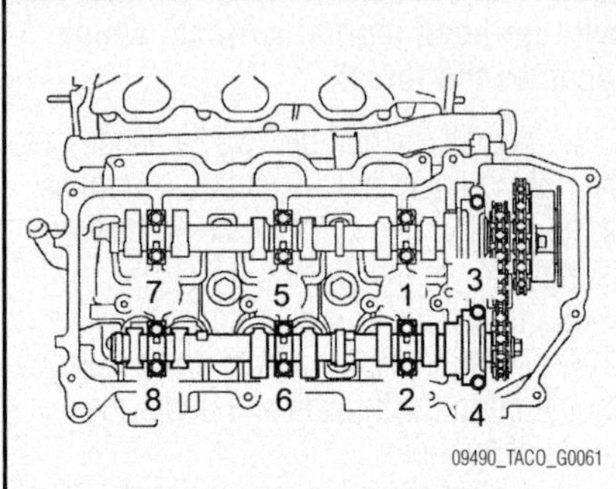

Fig. 48 Bank 1 camshaft number two bearing cap bolt torque sequence—4.0L engine

bolts and 18 ft. lbs. for the 12mm head bolts.

40. Rotate the number two camshaft clockwise, using a wrench, so that the lock pin of the number two camshaft is aligned with the knock pin hole of the camshaft timing gear.

41. Hold the hexagonal portion of the number two camshaft, with a wrench, and install the camshaft timing gear set bolt and tighten it to 74 ft. lbs.

42. Remove the pin from the number two chain tensioner.

43. While turning the stopper plate of the tensioner clockwise, push in the plunger of the tensioner. While turning the stopper plate of the tensioner counterclockwise, insert a 0.138 inch bar into the holes in the stopper plate and tensioner to hold the stopper plate. Install the two chain tensioner bolts and tighten to 7.4 ft. lbs.

44. Remove the bar from the chain tensioner. Install a new gasket and the timing chain cover plate. Torque the bolts to 80 inch lbs.

45. Turn the crankshaft pulley two complete revolutions slowly until its groove and the timing mark "0" of the timing chain cover are aligned.

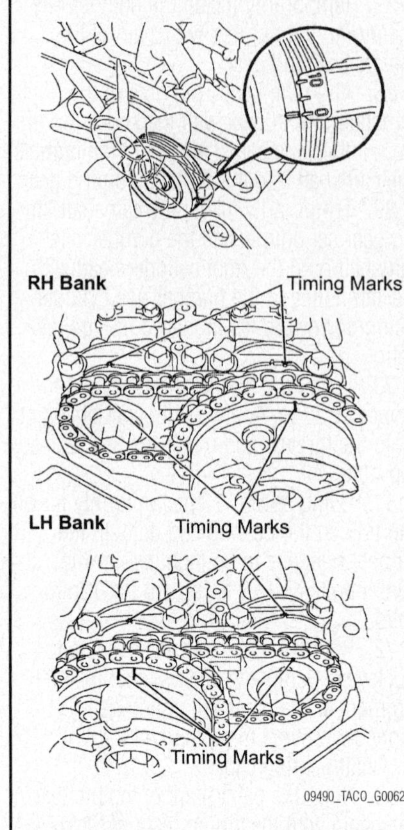

Fig. 49 Bank 1 and Bank 2 timing mark alignment—4.0L engine

46. Position the number one cylinder at TDC of the compression stroke. Inspect the valve clearance, adjust as required.

47. Apply a continuous bead (0.08–0.12 inch) of seal packing, part number 08826-00080 or equivalent, to the cylinder head as indicated in the illustration. Install the seal washers onto the bolts. Install the cylinder head cover bolts and nuts. Tighten bolts "A" to 7.4 ft. lbs. Tighten bolts "B" to 80 inch lbs. Tighten nuts to 80 inch lbs.

➡Be sure to remove any oil from the contact surfaces of the cylinder head cover and the cylinder head. Install the cover within three minutes after applying the seal packing. Tighten the bolts to specification within fifteen minutes after installing the cover. Do not add engine oil for at least two hours after installing the cover.

48. Continue the installation in the reverse order of the removal procedure.

49. Be sure to fill the engine with the proper grade and type engine coolant.

50. Be sure to fill the engine with the proper grade and type engine oil.

51. Start the engine and check for leaks. Correct as required.

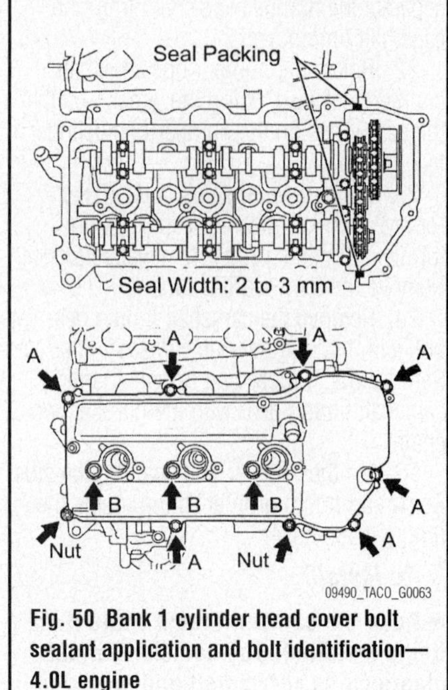

Fig. 50 Bank 1 cylinder head cover bolt sealant application and bolt identification—4.0L engine

Bank 2

See Figures 51 through 59.

1. Before servicing the vehicle, refer to the Precautions Section.
2. Properly disarm the SRS system.
3. Disconnect the negative battery cable.
4. Drain the engine coolant. Remove the V bank cover.
5. Remove the air cleaner assembly.
6. Disconnect the two water bypass hoses. Disconnect the fuel vapor feed hose. Disconnect the ventilation hose. Disconnect the VSV connectors.
7. Disconnect the throttle body motor connector. Separate the three wirc harness clamps and hose clamp.
8. Remove the two bolts and the throttle body bracket. Remove the bolt and the oil baffle plate. Remove the four bolts and the two serge tank stays.

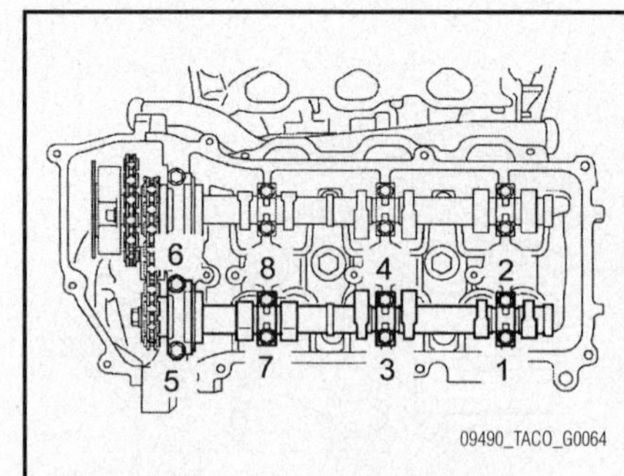

Fig. 51 Bank 2 camshaft number four bearing cap bolt removal sequence—4.0L engine

9. Remove the two nuts. Remove the four bolts, intake air surge tank and gasket.

10. Remove the ignition coil assembly.

11. Remove the cylinder head cover retaining bolts. Remove the cylinder head cover.

12. Turn the crankshaft pulley until its groove and the timing mark "0" of the timing chain cover are aligned. If not aligned at TDC of the compression stroke, turn the crankshaft one complete revolution, in the direction of rotation. Paint alignment marks on the number one chain links corresponding to the timing marks of the camshaft timing gears.

13. While turning the stopper plate of the tensioner upward, push in the plunger of the chain tensioner. While turning the stopper plate of the tensioner downward, insert a 0.138 inch bar into the holes in the stopper plate and tensioner to hold the stopper plate. Remove the two bolts and then remove the number one chain tensioner assembly.

➡Never rotate the crankshaft with the chain tensioner removed. When rotating the camshaft with the tensioner removed, rotate the crankshaft counterclockwise forty degrees from TDC, first.

➡Keep the camshaft level while it is being removed. The camshaft thrust clearance is very small and failing to keep it level could crack or damage the cylinder head journal surface, which receives the thrust. Follow the steps below to prevent this problem from occurring.

14. While pushing down on chain tensioner number three insert a 0.039 inch diameter pin into the hole to hold it. Hold the hexagonal portion of the number four camshaft with a wrench. Remove the camshaft timing gear set bolt.

15. Separate the camshaft timing gear from the number four camshaft.

16. Using several steps, loosen and remove the eight bearing cap bolts uniformly and in the proper removal sequence. Remove the four bearing caps and the number four camshaft.

17. Remove the number three chain tensioner bolt. Remove the number chain tensioner and camshaft timing gear.

➡Keep the camshaft level while it is being removed. The camshaft thrust clearance is very small and failing to keep it level could crack or damage the cylinder head journal surface, which receives the thrust. Follow the steps below to prevent this problem from occurring.

18. Release the chain tension between the camshaft gear (LH bank) and the crankshaft timing gear by turning the crankshaft pulley counterclockwise slightly.

19. Hold the hexagonal portion of the number three camshaft, with a wrench. Loosen the camshaft timing gear set bolt.

➡Do not disassemble the camshaft timing gear assembly.

20. Slide the camshaft timing gear and separate the number one chain from the camshaft timing gear.

21. Using several steps, loosen and remove the eight bearing cap bolts uniformly and in the proper removal sequence. Remove the four bearing caps.

22. Remove the camshaft timing gear set bolt with the number three camshaft lifted up. Remove the number three camshaft and camshaft timing gear with the number two chain.

23. Tie the number one chain to the side. Be careful not to drop anything inside the timing chain cover.

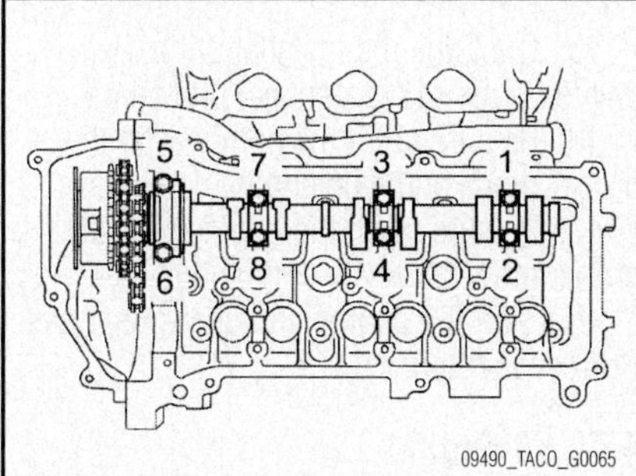

Fig. 52 Bank 2 camshaft number three bearing cap bolt removal sequence—4.0L engine

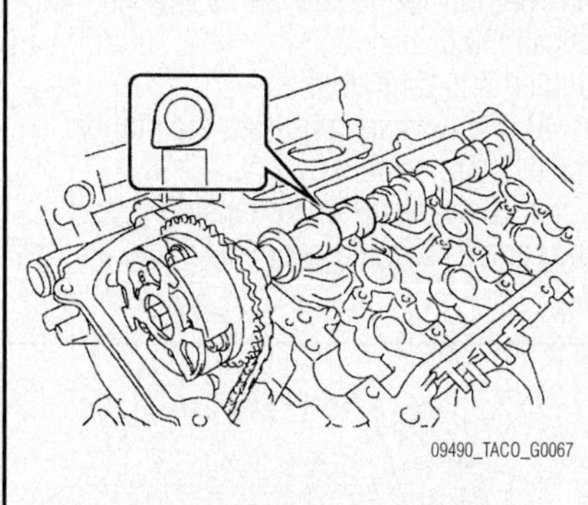

Fig. 54 Bank 2 camshaft number three lobe installation positioning—4.0L engine

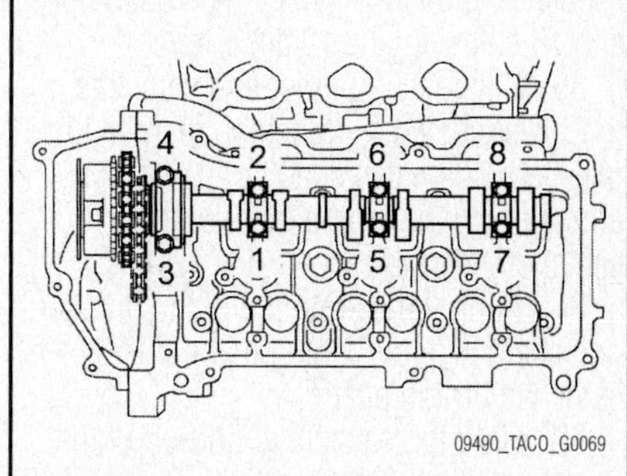

Fig. 56 Bank 2 camshaft number three bearing cap bolt torque sequence—4.0L engine

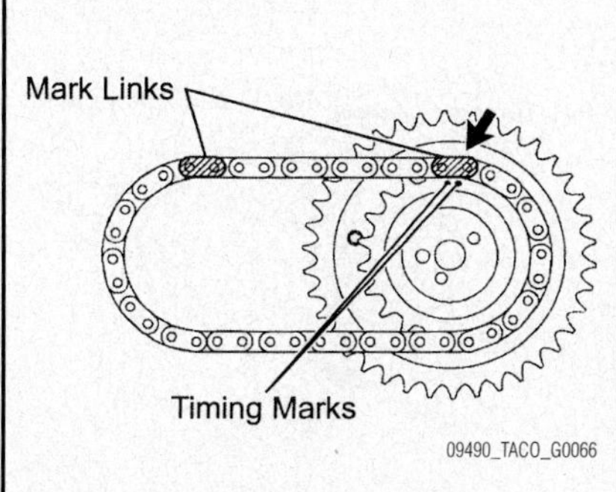

Fig. 53 Bank 2 camshaft number three alignment marks—4.0L engine

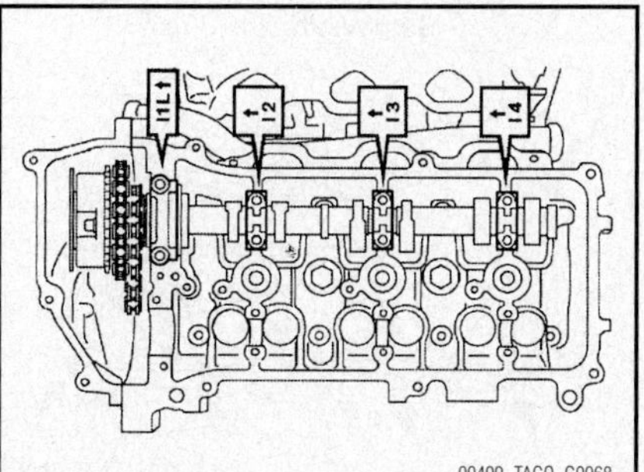

Fig. 55 Bank 2 camshaft number three bearing cap bolt installation sequence—4.0L engine

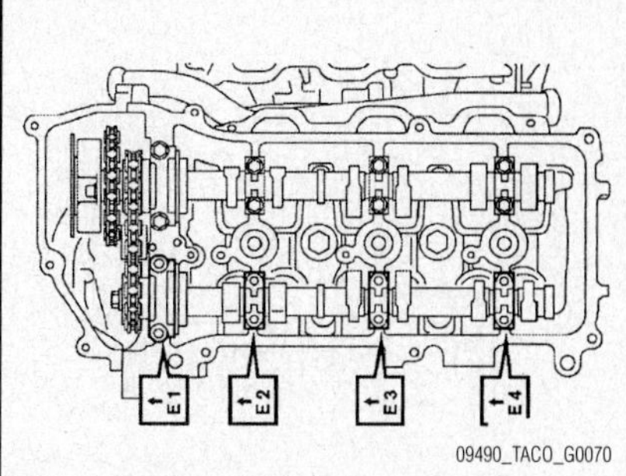

Fig. 57 Bank 2 camshaft number four bearing cap bolt installation sequence—4.0L engine

To install:

➡Keep the camshaft level while it is being installed. The camshaft thrust clearance is very small and failing to keep it level could crack or damage the cylinder head journal surface, which receives the thrust.

24. Align the yellow mark link with the timing mark (2 dot mark) of the camshaft timing gear. Apply new engine oil to the thrust portion and journal of the camshafts.
25. Temporarily install the number one chain onto the number two chain of the camshaft timing gear.
26. Align the knock pin hole of the camshaft timing gear with the knock pin of the number three camshaft. Insert the number three camshaft into the camshaft timing gear.
27. Temporarily install the camshaft timing gear set bolt. Install the number three camshaft onto the left cylinder head with the cam lobes of the number two cylinder facing downward, as indicated in the illustration.
28. Install the four bearing caps, in the proper location. Apply a light coat of engine oil to the threads and under the heads of the cap bolts.
29. Using several steps, uniformly install and tighten the bearing cap bolts in the proper sequence to 80 inch lbs. for the 10mm bolts and 18 ft. lbs. for the 12mm bolts.
30. Rotate the number one camshaft clockwise, using a wrench so that the timing mark of the camshaft timing gear is aligned with the timing mark of the camshaft bearing cap.
31. Align the paint mark of the number one chain with the timing mark of the camshaft timing gear.
32. Hold the hexagonal portion of the number three camshaft with a wrench, and tighten the camshaft timing gear set bolt to 74 ft. lbs.
33. While pushing in on the number three chain tensioner, insert a 0.039 inch pin into the hole to hold it.
34. Temporarily install the camshaft timing gear and chain tensioner number three and align the yellow mark links with the timing marks (1 dot mark and 2 dot marks) on the camshaft timing gears. Tighten the bolt to 14 ft. lbs.

➡Keep the camshaft level while it is being installed. The camshaft thrust clearance is very small and failing to keep it level could crack or damage the cylinder head journal surface, which receives the thrust.

35. Align the knock pin hole in the camshaft timing gear with the knock pin of the number four camshaft, and insert the number four camshaft into the camshaft timing gear.
36. Temporarily install the camshaft timing gear set bolt.
37. Install the four bearing caps in the proper location. Apply a light coat of clean engine oil to the threads and under the heads of the bolts.
38. Using several steps, uniformly install and tighten the eight bearing cap bolts in sequence to 80 inch lbs, for the 10mm head bolts and 18 ft. lbs. for the 12mm head bolts.
39. Hold the hexagonal portion of the number four camshaft, with a wrench, and install the camshaft timing gear set bolt and tighten it to 74 ft. lbs.
40. Remove the pin from the number three chain tensioner.
41. Release the chain tension between the camshaft timing gear (RH bank) and the crankshaft timing gear by turning the crankshaft pulley clockwise slightly.
42. While turning the stopper plate of the tensioner clockwise, push in the plunger of the tensioner. While turning the stopper plate of the tensioner counterclockwise, insert a 0.138 inch bar into the holes in the stopper plate and tensioner to hold the stopper plate. Install the two chain tensioner bolts and tighten to 7.4 ft. lbs. Remove the bar from the chain tensioner.
43. Position the number one cylinder at TDC of the compression stroke. Inspect the valve clearance, adjust as required.
44. Apply a continuous bead (0.08–0.12 inch) of seal packing, part number 08826-00080 or equivalent, to the cylinder head as indicated in the illustration. Install the seal washers onto the bolts. Install the cylinder head cover bolts and nuts. Tighten bolts "A" to 7.4 ft. lbs. Tighten bolts "B" to 80 inch lbs. Tighten nuts to 80 inch lbs.

➡Be sure to remove any oil from the contact surfaces of the cylinder head cover and the cylinder head. Install the cover within three minutes after applying the seal packing. Tighten the bolts to specification within fifteen minutes after installing the cover. Do not add engine oil for at least two hours after installing the cover.

45. Continue the installation in the reverse order of the removal procedure.
46. Be sure to fill the engine with the proper grade and type engine coolant.
47. Be sure to fill the engine with the proper grade and type engine oil.
48. Start the engine and check for leaks. Correct as required.

4.7L V8 Engine

See Figures 60 through 70.

1. Before servicing the vehicle, refer to the precautions in the beginning of this section.

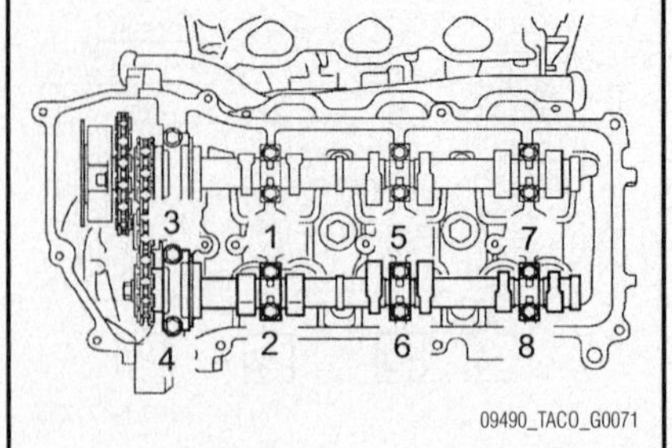

09490_TACO_G0071

Fig. 58 Bank 2 camshaft number four bearing cap bolt torque sequence—4.0L engine

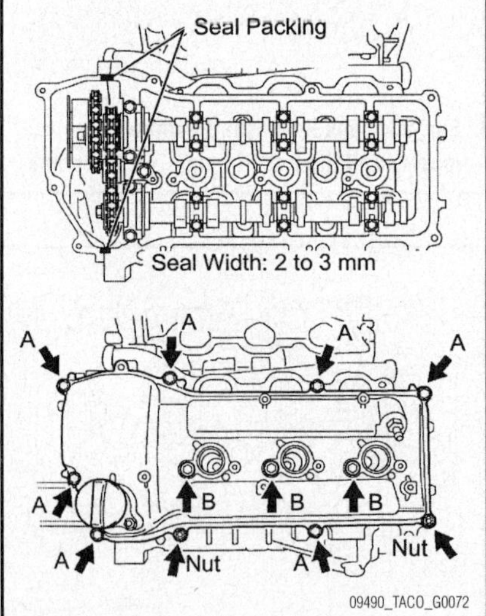

Fig. 59 Bank 2 cylinder head cover bolt sealant application and bolt identification—4.0L engine

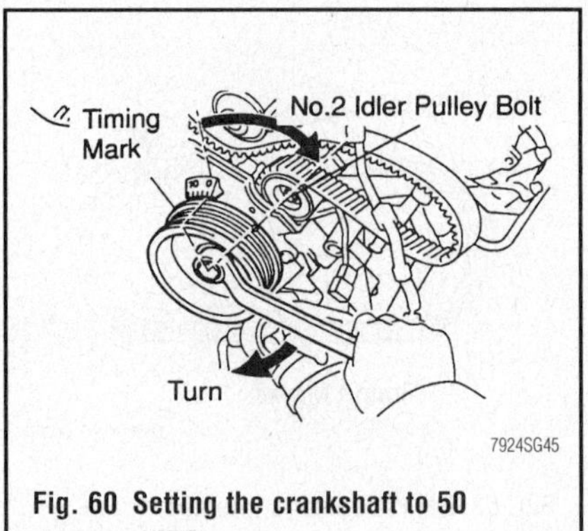

Fig. 60 Setting the crankshaft to 50 degrees ATDC

2. Drain the cooling system.
3. Relieve the fuel system pressure.
4. Remove or disconnect the following:
   - Negative battery cable
   - Engine under covers
   - Engine appearance cover
   - Air intake hose
   - Accessory drive belt
   - Cooling fan
   - Radiator
   - Idler pulley
   - Upper and middle timing belt covers
   - A/C compressor
   - Cooling fan bracket
   - Alternator
   - Accessory drive belt tensioner
5. Set the engine to Top Dead Center (TDC) with the camshaft sprocket timing marks aligned with the rear cover timing marks.
6. Rotate the crankshaft to 50 degrees After TDC as shown. The crankshaft pulley timing mark should align with the center of the No. 2 idler pulley bolt.
7. Remove or disconnect the following:
   - Crankshaft pulley
   - Lower timing cover
   - Timing belt.
   - Camshaft timing sprockets
   - Camshaft Position (CMP) sensor
   - Ignition coils
   - Valve cover
   - Timing belt rear covers
8. Rotate the right bank camshafts as necessary to access the exhaust camshaft sub-gear service bolt hole and install a 6mm x 1.0mm bolt.

➡Keep all valvetrain components in order for assembly.

9. Align the right bank camshaft 1 dot timing marks to a **10** degree angle as shown.
10. Loosen the bearing cap bolts in sequence and in several passes.
11. Remove the right bank camshafts.
12. Rotate the left bank camshafts as necessary to access the exhaust camshaft sub-gear service bolt hole and install a 6mm x 1.0mm bolt.
13. Align the left bank camshaft 2 dot timing marks as shown.
14. Loosen the bearing cap bolts in sequence and in several passes.
15. Remove the left bank camshafts.
16. Remove the valve lifters and shims.

To install:

17. Ensure that the crankshaft is at 50 degrees After TDC.
18. Install or connect the following:
   - Valve lifters and shims in their original positions
   - Right bank camshafts with the 1 dot timing marks at 10 degrees
   - Left bank camshafts with the 2 dot timing marks aligned
   - Left and right bank camshaft bearing caps in their original positions. Apply sealant to the front bearing caps as shown.
   - Camshaft oil seals
19. The bearing cap bolts vary in length and are identified as follows:
   - A: 3.70 inches (94mm)
   - B: 2.83 inches (72mm)
   - C: 0.98 inches (25mm)
   - D: 2.05 inches (52mm)
   - E: 1.50 inches (38mm)
20. Bolts in positions **A**, **B** and **C** are installed dry.
21. Lubricate the threads and under the contact flange for bolts in positions **D** and **E**.
22. Install oil feed pipes and the bearing cap bolts according to position in the illustrations.
23. Tighten the camshaft bearing bolts in sequence and in several passes to the following specifications:
   - Bolt C: 66 inch lbs. (7.5 Nm)
   - All others: 12 ft. lbs. (16 Nm)

Fig. 61 Camshaft service bolt installation

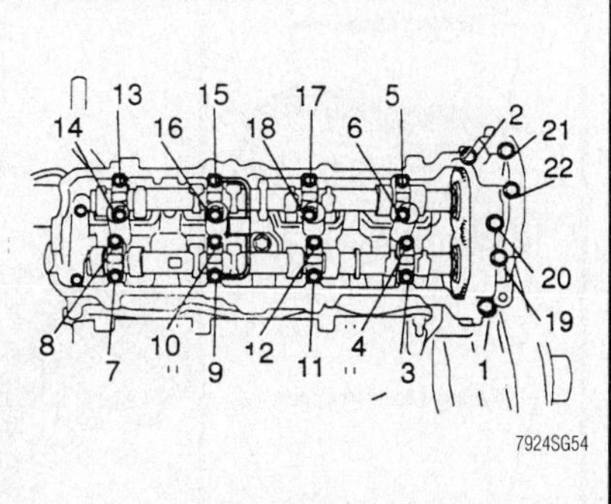

Fig. 63 Right bank camshaft bearing cap loosening sequence

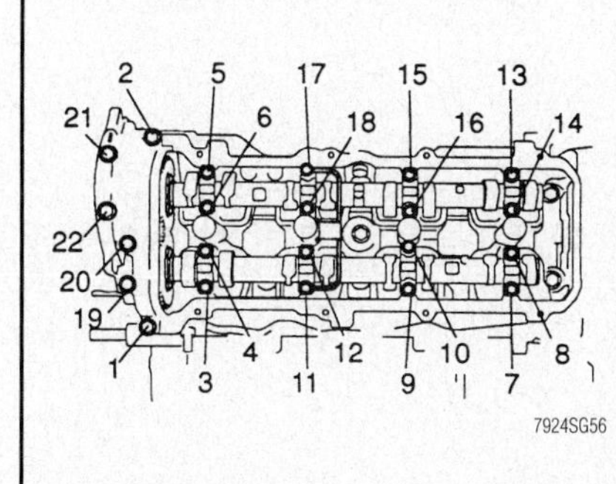

Fig. 65 Left bank camshaft bearing cap loosening sequence

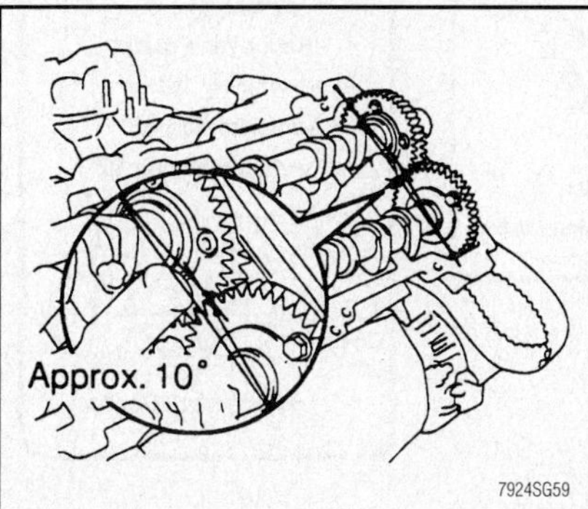

Fig. 62 Right bank camshaft timing mark (1 dot marks) alignment

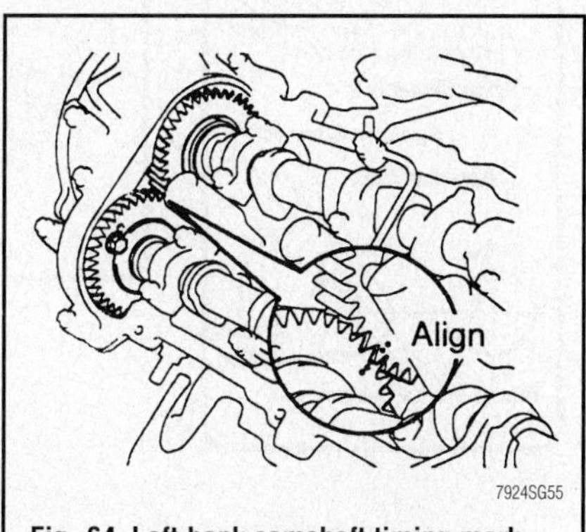

Fig. 64 Left bank camshaft timing mark (2 dot marks) alignment

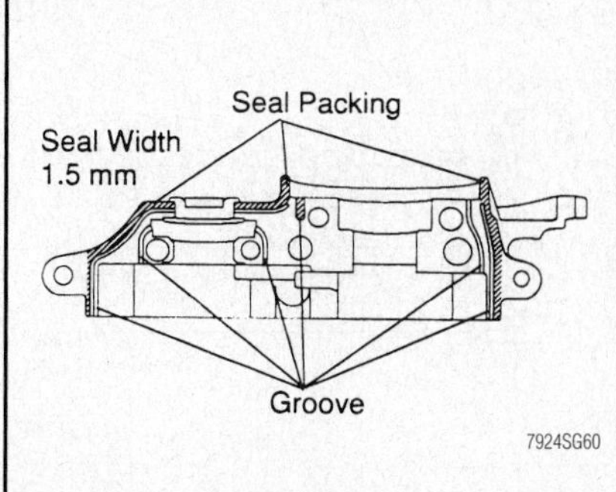

Fig. 66 Apply a 1.5mm bead of sealant to the front bearing caps

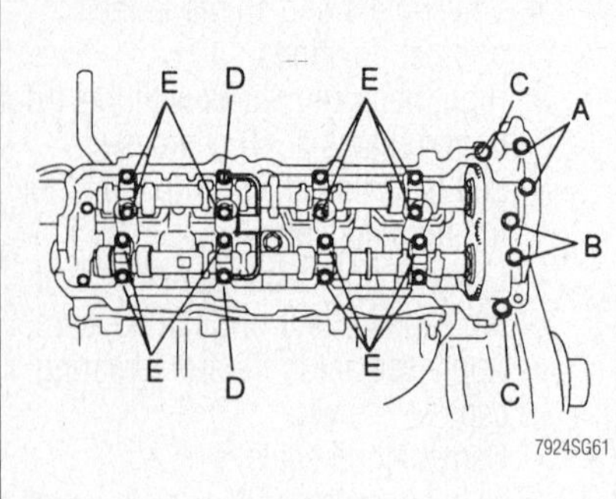

Fig. 67 Right bank bearing cap bolt location

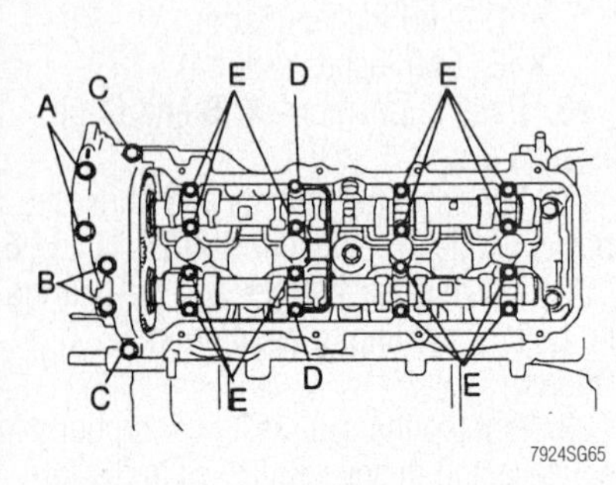

Fig. 68 Left camshaft bearing cap bolt locations

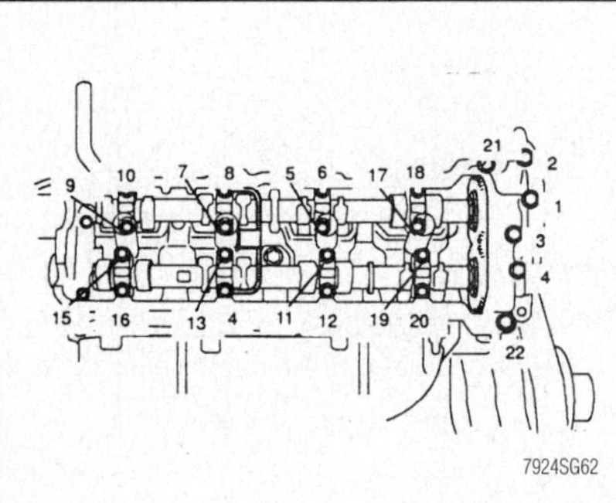

Fig. 69 Right bank camshaft bearing cap bolt torque sequence

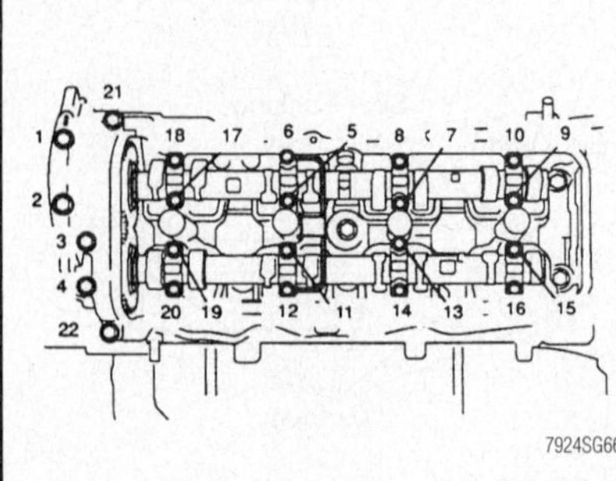

Fig. 70 Left bank camshaft bearing cap bolt torque sequence

24. Remove the service bolts from the exhaust camshaft gears.
25. Install or connect the following:
- Timing belt rear covers
- Valve cover
- Ignition coils
- CMP sensor
- Camshaft timing sprockets. Tighten the bolts to 80 ft. lbs. (108 Nm).
- Timing belt
- Lower timing cover
- Crankshaft pulley. Tighten the bolt to 181 ft. lbs. (245 Nm).
- Accessory drive belt tensioner
- Alternator
- Cooling fan bracket
- A/C compressor
- Upper and middle timing belt covers
- Idler pulley. Tighten the bolt to 27 ft. lbs. (37 Nm).
- Radiator
- Cooling fan
- Accessory drive belt
- Air intake hose
- Engine appearance cover
- Engine under covers
- Negative battery cable

26. Fill the cooling system.
27. Start the engine and check for leaks.

5.7L V8 Engine

See Figures 71 through 89.

1. Discharge fuel system pressure.
2. Disconnect cable from negative battery terminal.
3. Remove front exhaust pipe assembly.
4. Remove front fender apron seal.
5. Remove front fender apron seal rear.
6. Remove No. 2 steering intermediate shaft sub-assembly.
7. Remove exhaust manifold heat insulator.
8. Remove exhaust manifold sub-assembly.
9. Remove the timing chain cover.
10. Set No. 1 cylinder to TDC / compression.
11. Remove chain tensioner assembly.
12. Remove chain tensioner slipper.
13. Remove chain vibration damper.
14. Remove chain sub-assembly.
15. Remove chain tensioner assembly.
16. Make sure that the knock pin of the camshaft is positioned as shown.

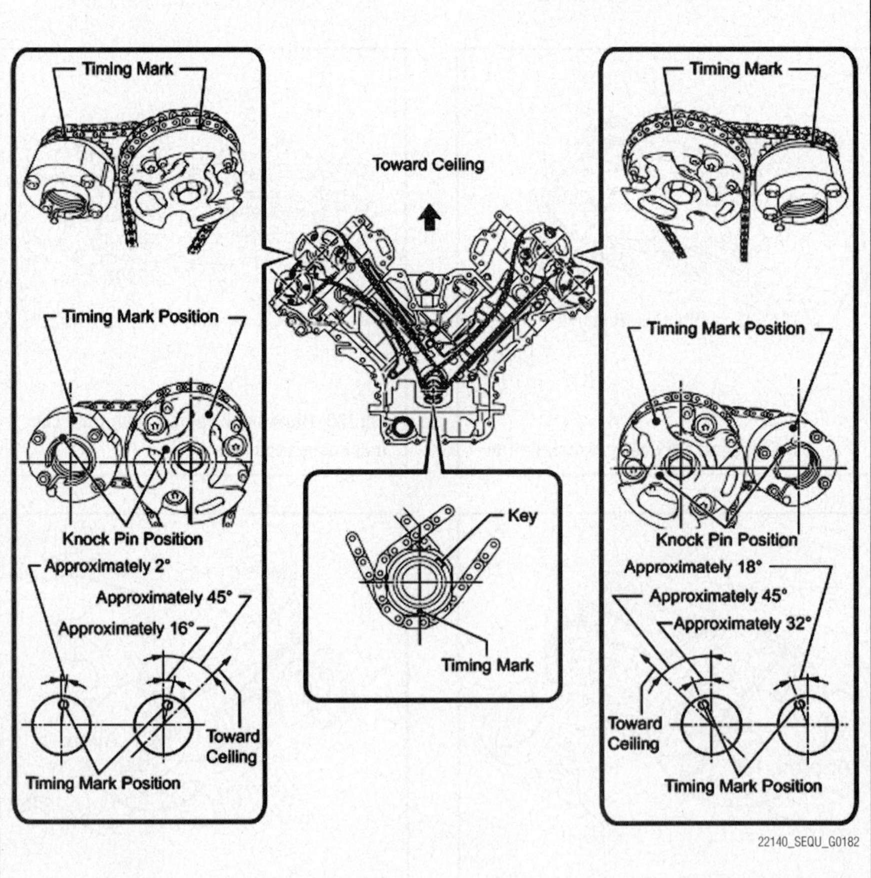

Fig. 71 Positioning engine to TDC

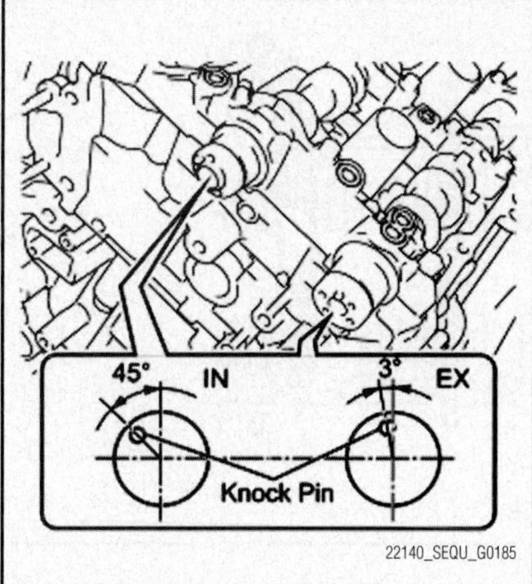

Fig. 72 Camshaft knock pin positioning LH

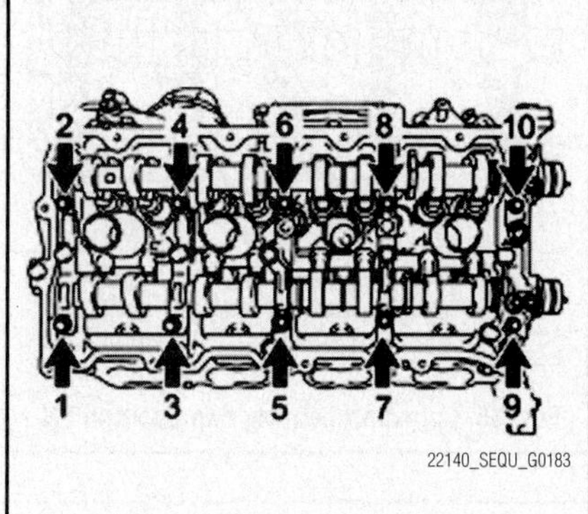

Fig. 73 Camshaft cap removal LH

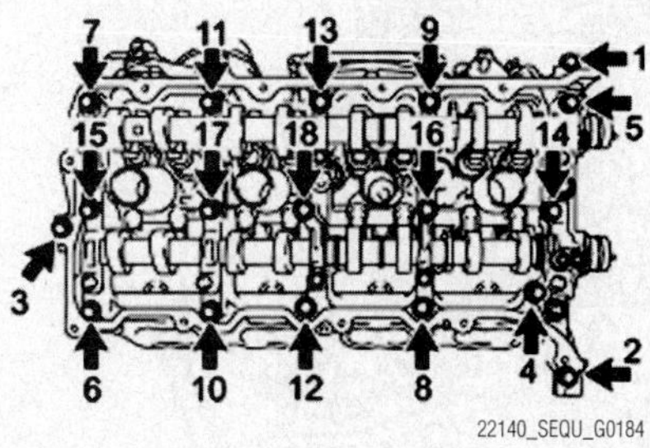

Fig. 74 Camshaft cap removal LH

17. Uniformly loosen and remove the 10 bearing cap bolts in the sequence LH.

18. Uniformly loosen and remove the 18 bearing cap bolts in the sequence LH.

➡Uniformly loosen the bolts while keeping the camshaft level.

19. Remove the 6 bearing caps.

20. Remove the No. 3 and No. 4 camshafts.

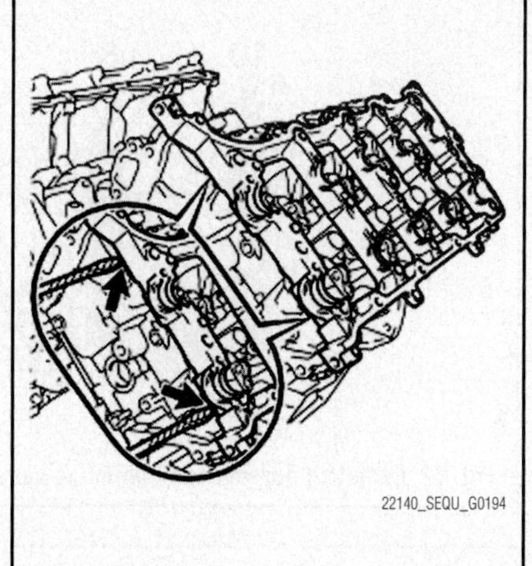

Fig. 75 Camshaft housing removal LH

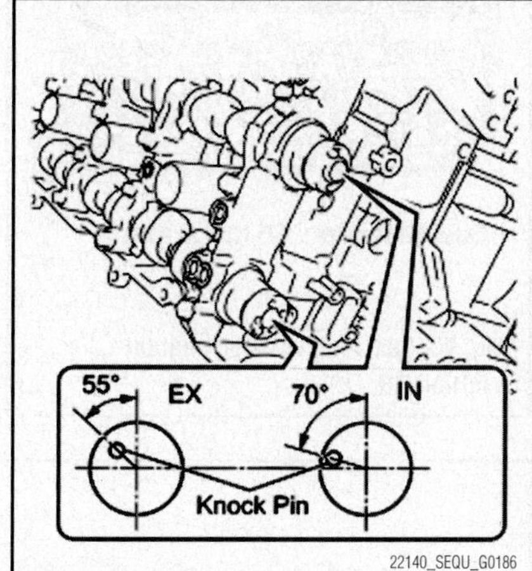

Fig. 76 Camshaft knock pin positioning RH

21. Remove the camshaft housing by prying between the cylinder head and camshaft housing with a screwdriver.

➡Be careful not to damage the contact surfaces of the cylinder head and camshaft housing.

22. Remove the 16 valve rocker arms from the cylinder head.

23. Remove the 16 valve lash adjusters from the cylinder head.

24. Remove the 16 valve stem caps from the cylinder head.

➡Arrange the removed parts in the correct order.

25. Make sure that the knock pin of the camshaft is positioned as shown in the illustration.

26. Uniformly loosen and remove the 10 bearing cap bolts in the sequence RH.

27. Uniformly loosen and remove the 18 bearing cap bolts in the sequence RH.

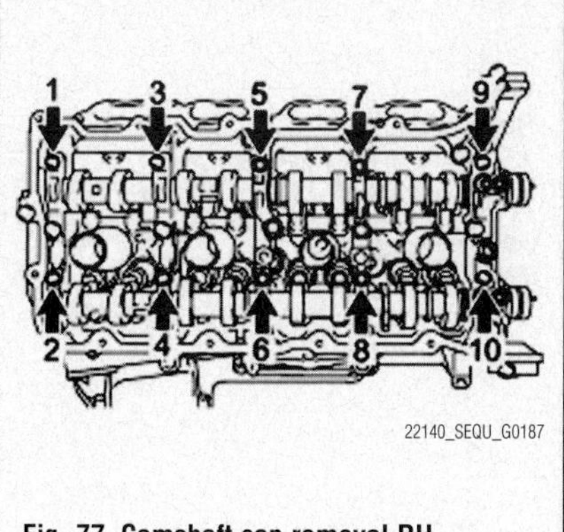

Fig. 77 Camshaft cap removal RH

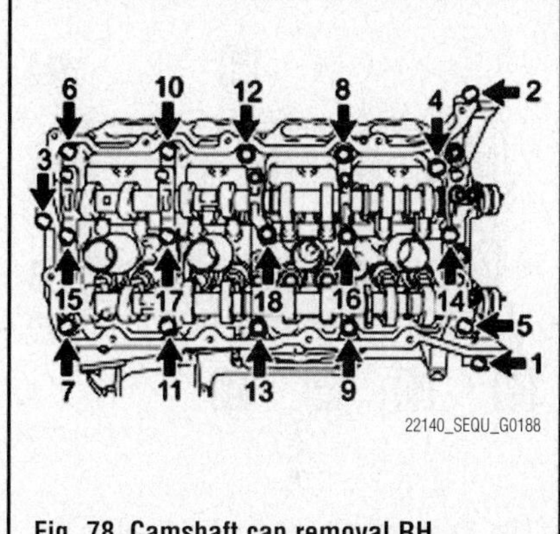

Fig. 78 Camshaft cap removal RH

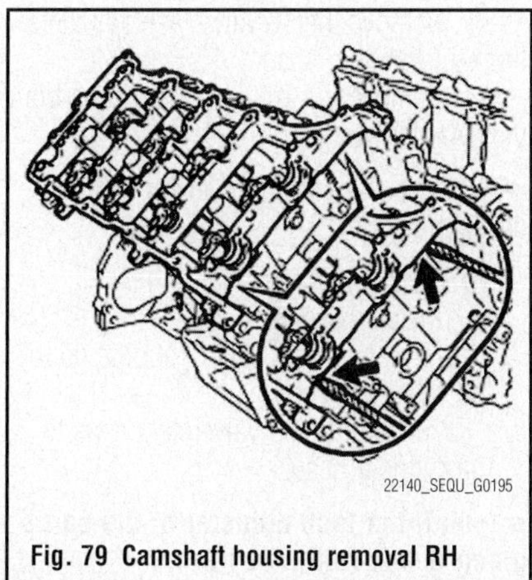

Fig. 79 Camshaft housing removal RH

➡Uniformly loosen the bolts while keeping the camshaft level.

28. Remove the 6 bearing caps.

29. Remove the No. 1 and No. 2 camshafts.

30. Remove the camshaft housing by prying between the cylinder head and camshaft housing with a screwdriver.

➡Be careful not to damage the contact surfaces of the cylinder head and camshaft housing.

31. Remove the 16 valve rocker arms from the cylinder head.

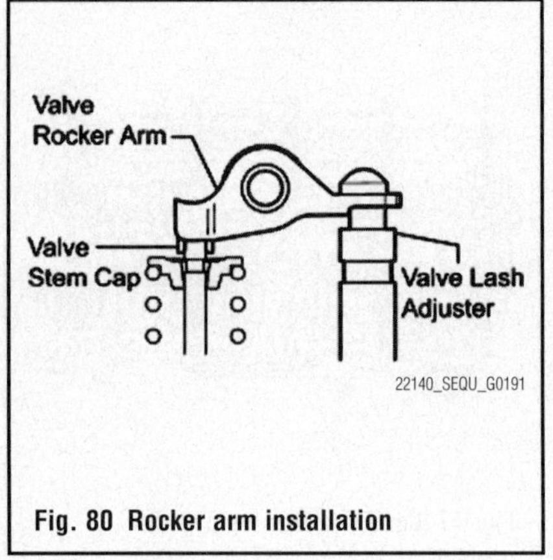

Fig. 80 Rocker arm installation

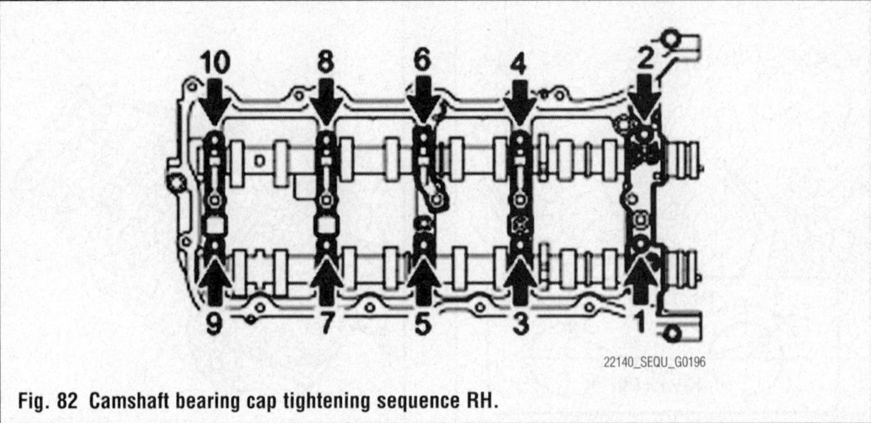

Fig. 82 Camshaft bearing cap tightening sequence RH.

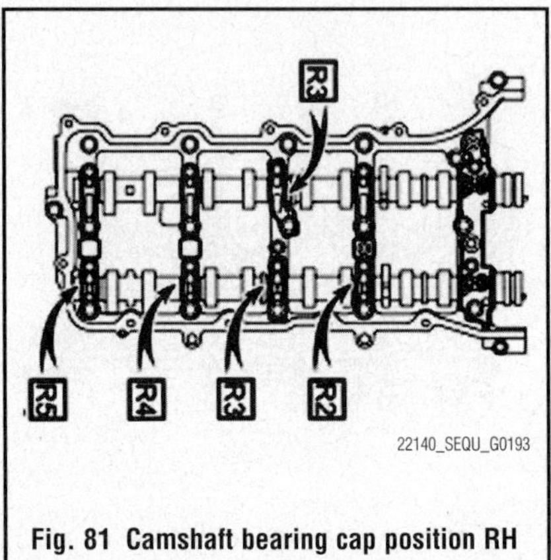

Fig. 81 Camshaft bearing cap position RH

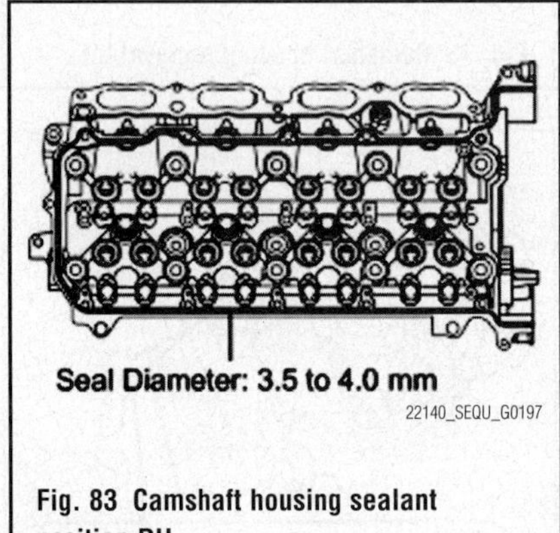

Fig. 83 Camshaft housing sealant position RH.

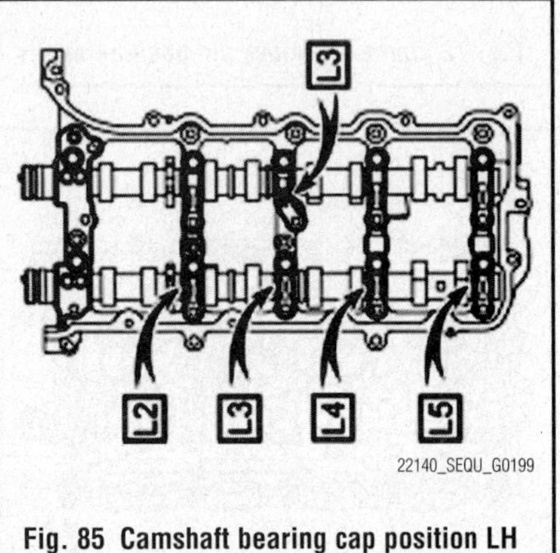

Fig. 85 Camshaft bearing cap position LH

32. Remove the 16 valve lash adjusters from the cylinder head.

33. Remove the 16 valve stem caps from the cylinder head.

➡**Arrange the removed parts in the correct order.**

To install:

34. Install valve stem cap.
   a. Apply a light coat of engine oil to the valve stem caps.
   b. Install the 32 valve stem caps to the cylinder head.

➡**Install the lash adjuster at the same place it was removed from.**

35. Install the 32 valve rocker arms.

36. Apply a light coat of engine oil to the camshaft journals, camshaft housings and bearing caps.

37. Install the No. 1 and No. 2 camshafts to the camshaft housing.

38. Confirm the marks and numbers on the camshaft bearing caps and place them in their proper positions and directions.

39. Temporarily install the 10 bolts in the order shown in the illustration.

40. Make sure that the valve rocker arms are installed properly.

41. Apply seal packing in a continuous line.

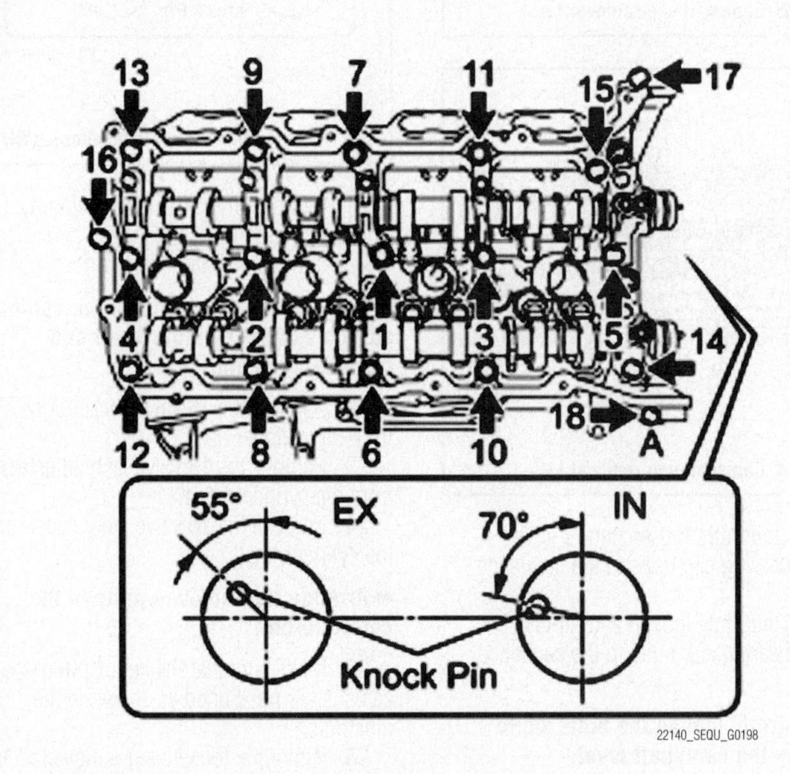

Fig. 84 Camshaft housing tightening sequence RH

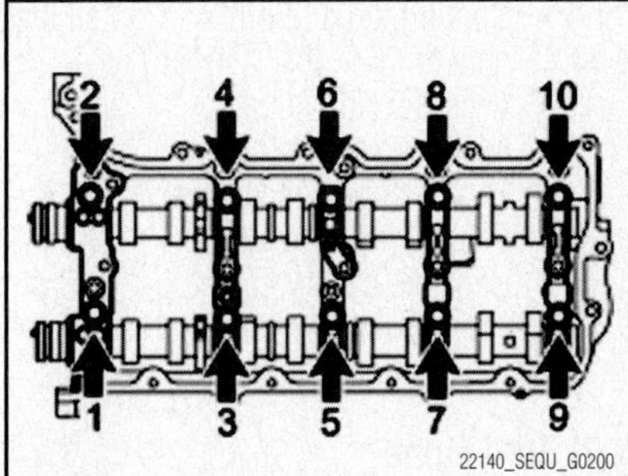

Fig. 86 Camshaft bearing cap tightening sequence LH

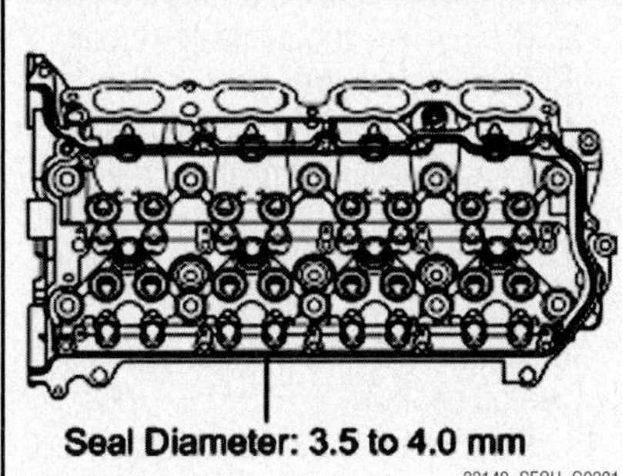

Fig. 87 Camshaft housing sealant position LH

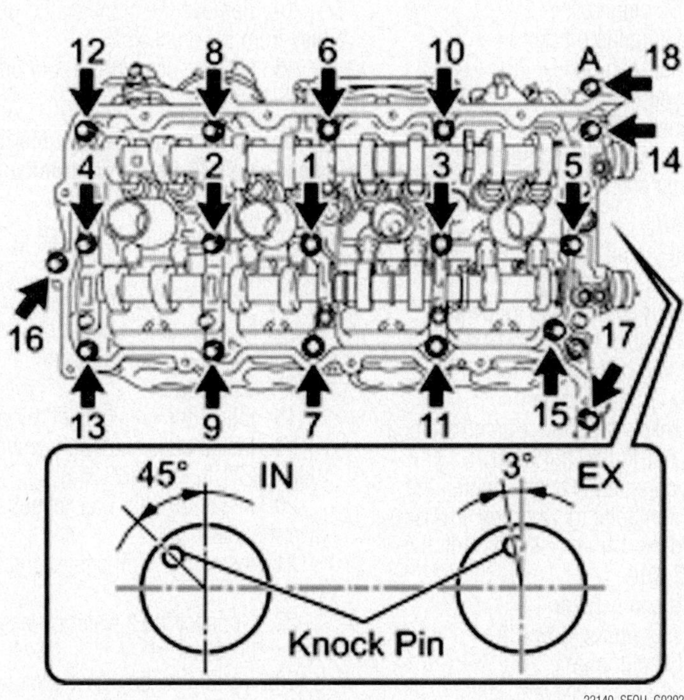

Fig. 88 Camshaft housing tightening sequence LH

➡Remove any oil from the contact surface. Install the camshaft housing within 3 minutes and tighten the bolts within 15 minutes after applying seal packing. Do not start the engine for at least 2 hours after the installation.

42. Install the camshaft housing, and install the 18 bolts in the order and tighten bolt "A" to 7 ft. lbs. (10 Nm), and all others to 22 ft. lbs. (30 Nm).

➡Make sure that the knock pin of the camshaft is positioned as shown in the illustration before installing the camshaft housing.

43. Tighten the 10 bolts in the order to 12 ft. lbs. (16 Nm).
44. Apply a light coat of engine oil to the camshaft journals, camshaft housings and bearing caps.
45. Install the No. 1 and No. 2 camshafts to the camshaft housing.
46. Confirm the marks and numbers on the camshaft bearing caps and place them in their proper positions and directions.
47. Temporarily install the 10 bolts in the order shown in the illustration.
48. Make sure that the valve rocker arms are installed properly.

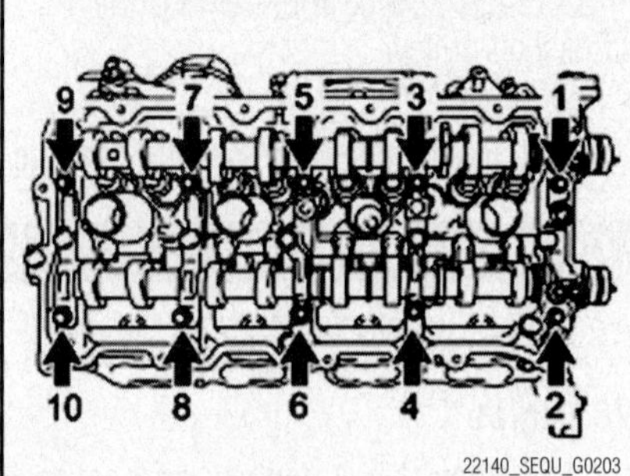

Fig. 89 Camshaft bearing cap tightening sequence LH

49. Apply seal packing in a continuous line.

➡Remove any oil from the contact surface. Install the camshaft housing within 3 minutes and tighten the bolts within 15 minutes after applying seal packing. Do not start the engine for at least 2 hours after the installation.

50. Install the camshaft housing, and install the 18 bolts in the order and tighten bolt "A" to 7 ft. lbs. (10 Nm), and all others to 22 ft. lbs. (30 Nm).

➡Make sure that the knock pin of the camshaft is positioned as shown in the illustration before installing the camshaft housing.

51. Tighten the 10 bolts in the order to 12 ft. lbs. (16 Nm).
52. Install chain tensioner assembly.
53. Install chain sub-assembly.
54. Install chain tensioner slipper.
55. Install chain tensioner assembly.
56. Install chain vibration damper.
57. Tighten camshaft timing gear.
58. Check No. 1 cylinder to TDC.
59. Install the timing chain cover.
60. Install exhaust manifold sub-assembly.
61. Install exhaust manifold heat insulator.
62. Install No.2 steering intermediate shaft sub-assembly.
63. Install front fender apron seal rear.
64. Install front fender apron seal.
65. Install front exhaust pipe assembly.
66. Connect cable to negative battery terminal.

CRANKSHAFT DAMPER

REMOVAL & INSTALLATION

1. See Crankshaft Front Seal for access to damper.
2. Loosen crankshaft pulley bolt.

3. Using Service Tool, remove the crankshaft pulley.

To install:

4. To install, reverse removal procedure.

CRANKSHAFT FRONT SEAL

REMOVAL & INSTALLATION

V6 Engine

1. Remove No. 1 engine under cover.
2. Drain engine coolant.
3. Remove V-bank cover.
4. Disconnect radiator hose inlet and outlet.
5. Loosen the 4 nuts holding the fluid coupling fan.
6. Remove the fan and generator V-belt.
7. Disconnect the reservoir hose from the radiator tank upper.
8. Remove the 2 bolts holding the fan shroud.
9. Remove the 4 nuts of the fluid coupling fan, and then remove the shroud together with the coupling fan.
10. Remove the fan pulley from the water pump.
11. Using a clip remover, remove the 6 clips and side deflectors.
12. Remove the 4 bolts and radiator.
13. Remove crankshaft pulley.
14. Using a screwdriver, pry out the oil seal.

➡Be careful not to damage the crankshaft. Wrap the tip of the screwdriver with tape.

To install:

15. Apply MP grease to a new oil seal lip.
16. Using Service Tool and a hammer, tap in the oil seal until its surface is flush with the timing chain cover edge.
17. Install crankshaft pulley and tighten bolt to 184 ft. lbs. (250 Nm).
18. Set the radiator bracket hooks to the radiator support holes.
19. Install the radiator with the 4 bolts and tighten to 13 ft. lbs. (18 Nm).
20. Install the deflectors with the 6 clips.
21. Install the fan pulley to the water pump.
22. Place the shroud together with the coupling fan between the radiator and engine.
23. Install the fluid coupling fan to the water pump with the 4 nuts. Tighten the nuts as much as possible by hand.
24. Attach the shroud's claws to the radiator as shown in the illustration.
25. Install the shroud with the 2 bolts and tighten to 58 inch lbs. (6.5 Nm)
26. Connect the reservoir hose to the radiator tank upper
27. Install the fan and generator V-belt
28. Tighten the 4 nuts of the fluid coupling fan to 15 ft. lbs. (21 Nm)
29. Connect radiator hose inlet and outlet
30. Add engine coolant
31. Inspect for coolant leak
32. Install V-bank cover
33. Install No. 1 engine under cover

4.7L V8 Engine

1. Before servicing the vehicle, refer to the precautions in the beginning of this section.
2. Drain the cooling system.
3. Remove or disconnect the following:
   - Negative battery cable
   - Engine under cover
   - Engine appearance cover
   - Air intake assembly
   - Accessory drive belt
   - Cooling fan and pulley
   - Radiator
   - Drive belt idler pulley
   - Camshaft Position (CMP) sensor connector
   - Upper timing covers
   - Oil cooler pipe
   - Center timing cover
   - A/C compressor
   - Cooling fan bracket
   - Crankshaft pulley
   - Lower timing cover
   - Timing belt.
   - Crankshaft timing sprocket
   - Front crankshaft seal

To install:

4. Install the oil seal so that it is flush with the oil pump housing.
5. Install or connect the following:
   - Crankshaft timing sprocket
   - Timing belt
   - Lower timing cover
   - Crankshaft pulley. Tighten the bolt to 181 ft. lbs. (245 Nm).
   - Cooling fan bracket. Tighten the 12mm bolts to 12 ft. lbs. (16 Nm) and the 14mm bolts to 24 ft. lbs. (32 Nm).
   - A/C compressor
   - Center timing cover
   - Oil cooler pipe
   - Upper timing covers
   - CMP sensor connector
   - Drive belt idler pulley. Tighten the bolt to 27 ft. lbs. (37 Nm).
   - Radiator
   - Cooling fan and pulley. Tighten the nuts to 16 ft. lbs. (21 Nm).
   - Accessory drive belt
   - Air intake assembly
   - Engine appearance cover
   - Engine under cover
   - Negative battery cable
6. Fill the cooling system.
7. Start the engine and check for leaks.

5.7L V8 Engine

1. Remove V-bank cover sub-assembly.
2. Remove No. 1 engine under cover.
3. Drain engine coolant.
4. Remove inlet radiator hose.
5. Remove fan and generator V belt.
6. Remove fan shroud.
7. Remove the 4 bolts, 2 stabilizer brackets and 2 stabilizer bushes. Then disconnect the stabilizer bar.
8. Remove oil pressure sender gauge assembly.
9. Remove oil filter bracket sub-assembly (w/o oil cooler).
10. Remove the 3 bolts and disconnect the 2 water by-pass hoses from the oil cooler.
11. Remove No. 1 oil cooler bracket (w/ oil cooler).
12. Remove oil filter bracket sub-assembly (w/ oil cooler).
13. Remove crankshaft pulley.
14. Remove the crankshaft timing gear key from the crankshaft.
15. Using a screwdriver, pry out the oil seal.

➡Do not damage the surface of the oil seal press fit hole and crankshaft.

To install:

16. Apply MP grease to the lip of a new oil seal.
17. Using Service Tool and a hammer, tap in the oil seal to a depth between 0 to 1.0 mm (0 to 0.0394 in.) from the timing chain cover edge.
18. Install the crankshaft timing gear key.
19. Install crankshaft pulley and tighten bolt to 221 ft. lbs. (300 Nm).
20. Install oil filter bracket sub-assembly (w/oil cooler).
21. Install no. 1 oil cooler bracket (w/oil cooler).
22. Connect the 2 water by-pass hoses to the oil cooler.
23. Install the 3 by-pass pipe bolts and tighten to 7 ft. lbs. (10 Nm).
24. Install oil filter bracket sub-assembly (w/o oil cooler).
25. Install oil pressure sender gauge assembly.

26. Connect the stabilizer bar and install the 2 stabilizer bushes and 2 stabilizer brackets with the 4 bolts and tighten to 51 ft. lbs. (69 Nm).
27. Install fan shroud.
28. Install fan and generator v belt.
29. Install inlet radiator hose.
30. Add engine coolant.
31. Inspect for leaks.
32. Install No. 1 engine under cover.
33. Install V-bank cover sub-assembly.

CYLINDER HEAD

REMOVAL & INSTALLATION

V6 Engine

Bank 1

See Figures 36 through 50 and 90 through 92.

1. Before servicing the vehicle, refer to the Precautions Section.
2. Properly disarm the SRS system.
3. Disconnect the negative battery cable.
4. Drain the engine coolant. Remove the V bank cover.
5. Remove the air cleaner assembly.
6. Disconnect the two water bypass hoses. Disconnect the fuel vapor feed hose. Disconnect the ventilation hose. Disconnect the VSV connectors.
7. Disconnect the throttle body motor connector. Separate the three wire harness clamps and hose clamp.
8. Remove the two bolts and the throttle body bracket. Remove the bolt and the oil baffle plate. Remove the four bolts and the two serge tank stays.
9. Remove the two nuts. Remove the four bolts, intake air surge tank and gasket.
10. Remove the ignition coil assembly.
11. Remove the cylinder head cover retaining bolts. Remove the cylinder head cover.
12. Turn the crankshaft pulley until its groove and the timing mark "0" of the timing chain cover are aligned. If not aligned at TDC of the compression stroke, turn the crankshaft one complete revolution, in the direction of rotation. Paint alignment marks on the number one chain links corresponding to the timing marks of the camshaft timing gears.
13. Remove the four bolts, and then remove the timing chain cover plate and gasket.
14. While turning the stopper plate of the tensioner upward, push the plunger of the chain tensioner. While turning the stopper plate of the tensioner downward, insert a 0.118 inch diameter bar into the holes in the stopper plate and tensioner to hold the stopper plate.
15. Remove the two bolts, and then remove the chain tensioner.

➡Keep the camshaft level while it is being removed. The camshaft thrust clearance is very small and failing to keep it level could crack or damage the cylinder head journal surface, which receives the thrust. Follow the steps below to prevent this problem from occurring.

16. While raising the chain tensioner number two insert a 0.039 inch diameter pin into the hole to hold it. Hold the hexagonal portion of the number two camshaft with a wrench. Remove the camshaft timing gear set bolt.
17. Separate the camshaft timing gear from the number two camshaft. Rotate the camshafts counterclockwise, using a wrench, so that the cam lobes of the number one cylinder face in the direction shown.
18. Using several steps, loosen and remove the eight bearing cap bolts uniformly and in the proper removal sequence. Remove the four bearing caps and the number two camshaft.
19. Remove the number two chain tensioner bolt. Remove the number chain tensioner and camshaft timing gear.

➡Keep the camshaft level while it is being removed. The camshaft thrust clearance is very small and failing to keep it level could crack or damage the cylinder head journal surface, which receives the thrust. Follow the steps below to prevent this problem from occurring.

20. Hold the hexagonal portion of the number one camshaft, with a wrench. Loosen the camshaft timing gear set bolt.

➡Do not disassemble the camshaft timing gear assembly.

21. Slide the camshaft timing gear and separate the number one chain from the camshaft timing gear.
22. Rotate the number one camshaft counterclockwise, using the wrench so that the cam lobes on the number one cylinder face downward.
23. Using several steps, loosen and remove the eight bearing cap bolts uniformly and in the proper removal sequence. Remove the four bearing caps.
24. Remove the camshaft timing gear set bolt with the number one camshaft lifted up. Remove the number one camshaft and camshaft timing gear with the number two chain.

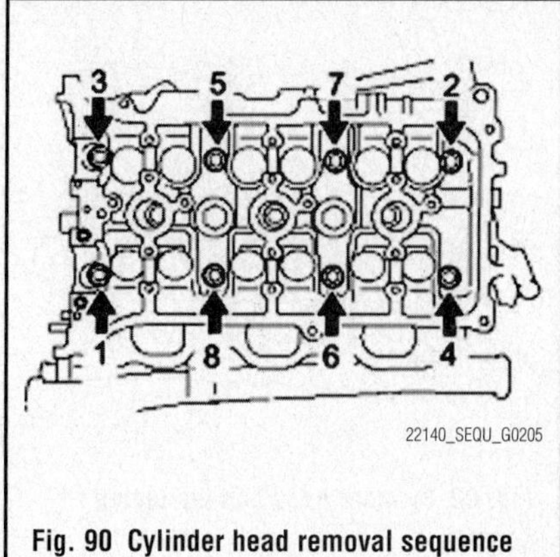

Fig. 90 Cylinder head removal sequence

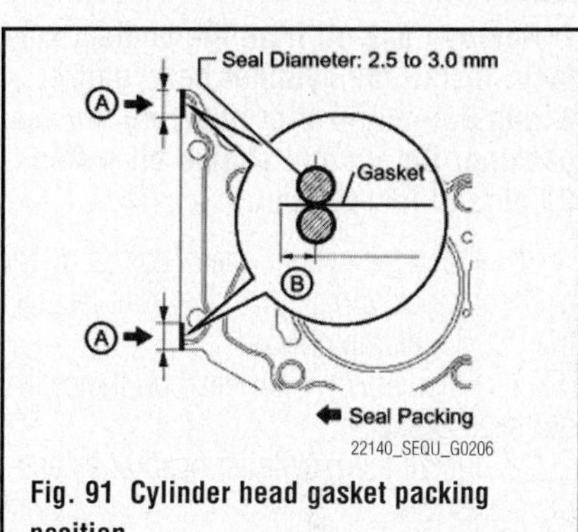

Fig. 91 Cylinder head gasket packing position

25. Tie the number one chain to the side. Be careful not to drop anything inside the timing chain cover.
26. Using a 10 mm bi-hexagon wrench, uniformly loosen the 8 cylinder head bolts in the sequence shown in the illustration. Remove the 8 cylinder head bolts and plate washers.

➡Be careful not to drop the plate washers into the cylinder head. Cylinder head warpage or cracking could result from removing bolts in the incorrect order.

27. Lift the cylinder head from the dowels on the cylinder block, and place the cylinder head on wooden blocks on a bench.

➡Be careful not to damage the contact surfaces of the cylinder head and cylinder block.

28. Remove cylinder head gasket.

To install:

29. Install cylinder head gasket.
   a. Remove any old packing (FIPG) material and be careful not to drop any oil on the contact surfaces of the cylinder head and cylinder block.
   b. Apply seal packing to a new cylinder head gasket.

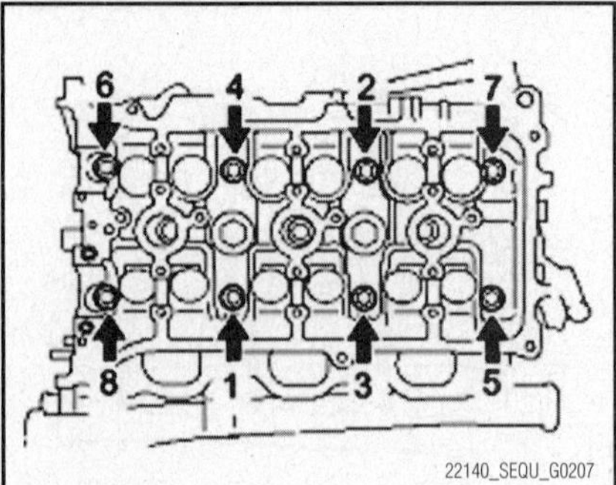

Fig. 92 Cylinder head bolt tightening sequence

➡Remove any oil from the contact surface. Install the cylinder head gasket within 3 minutes after applying the seal packing. Do not add engine oil within 2 hours of installation.

30. Place the cylinder head gasket on the cylinder block surface with the front face of the Lot No. stamp upward.
31. Place the cylinder head on the cylinder head gasket.
32. Install the 8 cylinder head bolts and plate washers.
    a. Apply a light coat of engine oil to the threads and under the heads of the cylinder head bolts.
    b. Install the plate washer to the cylinder head bolt.
33. Using a 10 mm bi-hexagon wrench, install and uniformly tighten the 8 cylinder head bolts to 27 ft. lbs. (36 Nm) with the plate washers in several steps in sequence.

➡If any one of the cylinder head bolts does not meet the torque specification, replace the cylinder head bolt.

34. Mark the cylinder head bolt heads with paint.
35. Tighten the cylinder head bolts another 90° in the sequence.
36. Tighten the cylinder head bolts another 90° in the sequence for a total of 180° from torque.
37. Check that the painted marks are now facing correctly.
38. Seal packing will seep out from the engine's front side. Thoroughly wipe off seeped out seal packing.

➡Keep the camshaft level while it is being installed. The camshaft thrust clearance is very small and failing to keep it level could crack or damage the cylinder head journal surface, which receives the thrust.

39. Align the yellow mark link with the timing mark (1 dot mark) of the camshaft timing gear. Apply new engine oil to the thrust portion and journal of the camshafts.
40. Temporarily install the number one chain onto the number two chain of the camshaft timing gear.
41. Align the knock pin hole of the camshaft timing gear with the knock pin of the number one camshaft. Insert the number one camshaft into the camshaft timing gear.
42. Temporarily install the camshaft timing gear set bolt. Install the number one camshaft onto the right cylinder head with the cam lobes of the number one cylinder facing downward, as indicated in the illustration.
43. Install the four bearing caps, in the proper location. Apply a light coat of engine oil to the threads and under the heads of the cap bolts.
44. Using several steps, uniformly install and tighten the bearing cap bolts in the proper sequence to 80 inch lbs. for the 10mm bolts and 18 ft. lbs. for the 12mm bolts.
45. Rotate the number one camshaft clockwise, using a wrench so that the timing mark of the camshaft timing gear is aligned with the timing mark of the camshaft bearing cap.
46. Align the paint mark of the number one chain with the timing mark of the camshaft timing gear.
47. Hold the hexagonal portion of the number one camshaft with a wrench, and tighten the camshaft timing gear set bolt to 74 ft. lbs.
48. While pushing in on the number two chain tensioner, insert a 0.039 inch pin into the hole to hold it.
49. Temporarily install the camshaft timing gear and chain tensioner number two and align the yellow mark links with the timing marks (1 dot mark) on the camshaft timing gears. Tighten the bolt to 14 ft. lbs.

➡Keep the camshaft level while it is being installed. The camshaft thrust clearance is very small and failing to keep it level could crack or damage the cylinder head journal surface, which receives the thrust.

50. Install the number two camshaft onto the right cylinder head with the cam lobes of the number one cylinder facing upward, as indicated in the illustration.
51. Install the four bearing caps in the proper location. Apply a light coat of clean engine oil to the threads and under the heads of the bolts.
52. Using several steps, uniformly install and tighten the eight bearing cap bolts in sequence to 80 inch lbs, for the 10mm head bolts and 18 ft. lbs. for the 12mm head bolts.
53. Rotate the number two camshaft clockwise, using a wrench, so that the lock pin of the number two camshaft is aligned with the knock pin hole of the camshaft timing gear.
54. Hold the hexagonal portion of the number two camshaft, with a wrench, and install the camshaft timing gear set bolt and tighten it to 74 ft. lbs.
55. Remove the pin from the number two chain tensioner.
56. While turning the stopper plate of the tensioner clockwise, push in the plunger of the tensioner. While turning the stopper plate of the tensioner counterclockwise, insert a 0.138 inch bar into the holes in the stopper plate and tensioner to hold the stopper plate. Install the two chain tensioner bolts and tighten to 7.4 ft. lbs.
57. Remove the bar from the chain tensioner. Install a new gasket and the timing chain cover plate. Torque the bolts to 80 inch lbs.
58. Turn the crankshaft pulley two complete revolutions slowly until its groove and the timing mark "0" of the timing chain cover are aligned.
59. Position the number one cylinder at TDC of the compression stroke. Inspect the valve clearance, adjust as required.
60. Apply a continuous bead (0.08–0.12 inch) of seal packing, part number 08826-00080 or equivalent, to the cylinder head as indicated in the illustration. Install the seal washers onto the bolts. Install the cylinder head cover bolts and nuts. Tighten bolts "A" to 7.4 ft. lbs. Tighten bolts "B" to 80 inch lbs. Tighten nuts to 80 inch lbs.

➡Be sure to remove any oil from the contact surfaces of the cylinder head cover and the cylinder head. Install the cover within three minutes after applying the seal packing. Tighten the bolts to specification within fifteen minutes after installing the cover. Do not add engine oil for at least two hours after installing the cover.

61. Continue the installation in the reverse order of the removal procedure.
62. Be sure to fill the engine with the proper grade and type engine coolant.
63. Be sure to fill the engine with the proper grade and type engine oil.
64. Start the engine and check for leaks. Correct as required.

Bank 2

See Figures 51 through 59 and 93 through 97.

1. Before servicing the vehicle, refer to the Precautions Section.

2. Properly disarm the SRS system.

3. Disconnect the negative battery cable.

4. Drain the engine coolant. Remove the V bank cover.

5. Remove the air cleaner assembly.

6. Disconnect the two water bypass hoses. Disconnect the fuel vapor feed hose. Disconnect the ventilation hose. Disconnect the VSV connectors.

7. Disconnect the throttle body motor connector. Separate the three wire harness clamps and hose clamp.

8. Remove the two bolts and the throttle body bracket. Remove the bolt and the oil baffle plate. Remove the four bolts and the two serge tank stays.

9. Remove the two nuts. Remove the four bolts, intake air surge tank and gasket.

10. Remove the ignition coil assembly.

11. Remove the cylinder head cover retaining bolts. Remove the cylinder head cover.

12. Turn the crankshaft pulley until its groove and the timing mark "0" of the timing chain cover are aligned. If not aligned at TDC of the compression stroke, turn the crankshaft one complete revolution, in the direction of rotation. Paint alignment marks on the number one chain links corresponding to the timing marks of the camshaft timing gears.

13. While turning the stopper plate of the tensioner upward, push in the plunger of the chain tensioner. While turning the stopper plate of the tensioner downward, insert a 0.138 inch bar into the holes in the stopper plate and tensioner to hold the stopper plate. Remove the two bolts and then remove the number one chain tensioner assembly.

➡Never rotate the crankshaft with the chain tensioner removed. When rotating the camshaft with the tensioner removed, rotate the crankshaft counterclockwise forty degrees from TDC, first.

➡Keep the camshaft level while it is being removed. The camshaft thrust clearance is very small and failing to keep it level could crack or damage the cylinder head journal surface, which receives the thrust. Follow the steps below to prevent this problem from occurring.

14. While pushing down on chain tensioner number three insert a 0.039 inch diameter pin into the hole to hold it. Hold the hexagonal portion of the number four camshaft with a wrench. Remove the camshaft timing gear set bolt.

15. Separate the camshaft timing gear from the number four camshaft.

16. Using several steps, loosen and remove the eight bearing cap bolts uniformly and in the proper removal sequence. Remove the four bearing caps and the number four camshaft.

17. Remove the number three chain tensioner bolt. Remove the number chain tensioner and camshaft timing gear.

➡Keep the camshaft level while it is being removed. The camshaft thrust clearance is very small and failing to keep it level could crack or damage the cylinder head journal surface, which receives the thrust. Follow the steps below to prevent this problem from occurring.

18. Release the chain tension between the camshaft gear (LH bank) and the crankshaft timing gear by turning the crankshaft pulley counterclockwise slightly.

19. Hold the hexagonal portion of the number three camshaft, with a wrench. Loosen the camshaft timing gear set bolt.

➡Do not disassemble the camshaft timing gear assembly.

20. Slide the camshaft timing gear and separate the number one chain from the camshaft timing gear.

21. Using several steps, loosen and remove the eight bearing cap bolts uniformly and in the proper removal sequence. Remove the four bearing caps.

22. Remove the camshaft timing gear set bolt with the number three camshaft lifted up. Remove the number three camshaft and camshaft timing gear with the number two chain.

23. Tie the number one chain to the side. Be careful not to drop anything inside the timing chain cover.

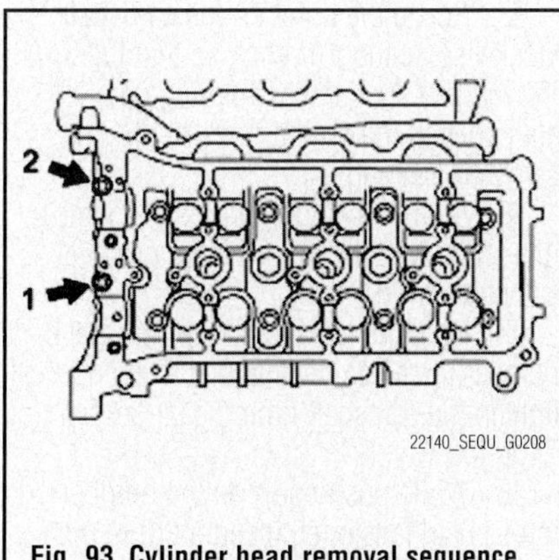

Fig. 93 Cylinder head removal sequence

24. Uniformly loosen and remove the 2 cylinder head bolts in the sequence.

25. Using a 10 mm bi-hexagon wrench, uniformly loosen the 8 cylinder head bolts in the sequence shown in the illustration. Remove the 8 cylinder head bolts and plate washers.

➡Be careful not to drop the plate washers into the cylinder head. Cylinder head warpage or cracking could result from removing bolts in the incorrect order.

26. Lift the cylinder head from the dowels on the cylinder block, and place the cylinder head on wooden blocks on a bench.

➡Be careful not to damage the contact surfaces of the cylinder head and cylinder block.

27. Remove cylinder head gasket.

To install:

28. Install cylinder head gasket.

a. Remove any old packing (FIPG) material and be careful not to drop any oil on the contact surfaces of the cylinder head and cylinder block.

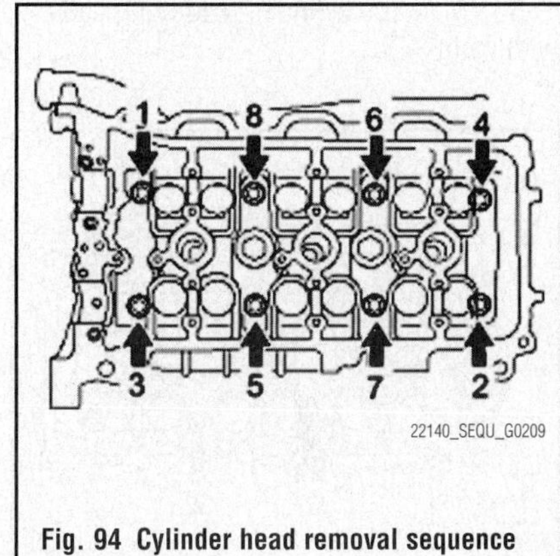

Fig. 94 Cylinder head removal sequence

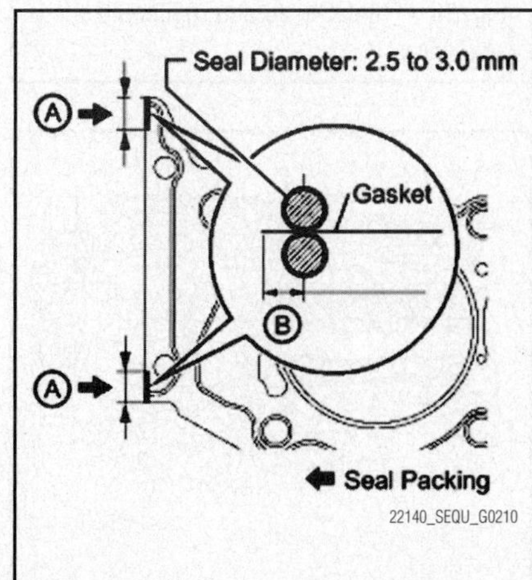

Fig. 95 Cylinder head gasket packing position

b. Apply seal packing to a new cylinder head gasket.

➡Remove any oil from the contact surface. Install the cylinder head gasket within 3 minutes after applying the seal packing. Do not add engine oil within 2 hours of installation.

29. Place the cylinder head gasket on the cylinder block surface with the front face of the Lot No. stamp upward.
30. Place the cylinder head on the cylinder head gasket.
31. Install the 8 cylinder head bolts and plate washers.
 a. Apply a light coat of engine oil to the threads and under the heads of the cylinder head bolts
 b. Install the plate washer to the cylinder head bolt.
32. Using a 10 mm bi-hexagon wrench, install and uniformly tighten the 8 cylinder head bolts to 27 ft. lbs. (36 Nm) with the plate washers in several steps in sequence.

➡If any one of the cylinder head bolts does not meet the torque specification, replace the cylinder head bolt.

33. Mark the cylinder head bolt heads with paint.

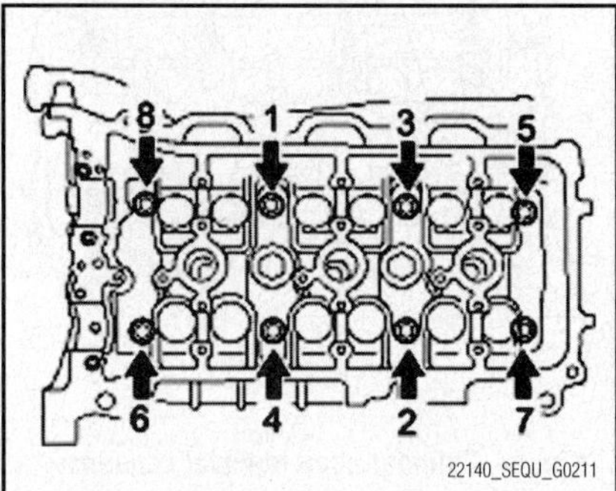

Fig. 96 Cylinder head bolt tightening sequence

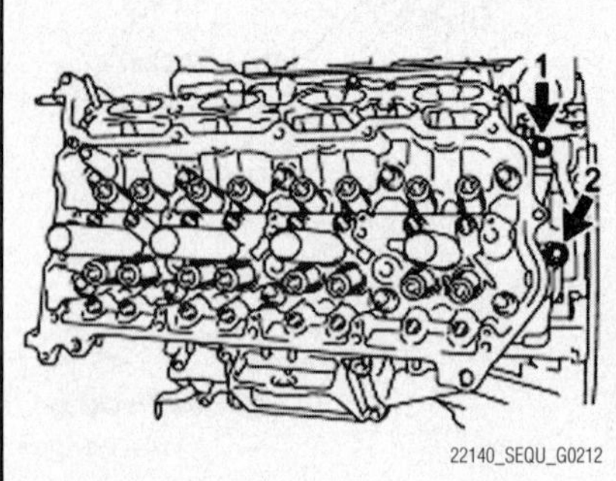

Fig. 97 Cylinder head bolt tightening sequence

34. Tighten the cylinder head bolts another 90° in the sequence.
35. Tighten the cylinder head bolts another 90° in the sequence for a total of 180° from torque.
36. Check that the painted marks are now facing correctly.
37. Install the 2 bolts in the order and tighten to 22 ft. lbs. (30 Nm).
38. Seal packing will seep out from the engine's front side. Thoroughly wipe off seeped out seal packing.

➡Keep the camshaft level while it is being installed. The camshaft thrust clearance is very small and failing to keep it level could crack or damage the cylinder head journal surface, which receives the thrust.

39. Align the yellow mark link with the timing mark (2 dot mark) of the camshaft timing gear. Apply new engine oil to the thrust portion and journal of the camshafts.
40. Temporarily install the number one chain onto the number two chain of the camshaft timing gear.
41. Align the knock pin hole of the camshaft timing gear with the knock pin of the number three camshaft. Insert the number three camshaft into the camshaft timing gear.
42. Temporarily install the camshaft timing gear set bolt. Install the number three camshaft onto the left cylinder head with the cam lobes of the number two cylinder facing downward, as indicated in the illustration.
43. Install the four bearing caps, in the proper location. Apply a light coat of engine oil to the threads and under the heads of the cap bolts.
44. Using several steps, uniformly install and tighten the bearing cap bolts in the proper sequence to 80 inch lbs. for the 10mm bolts and 18 ft. lbs. for the 12mm bolts.
45. Rotate the number one camshaft clockwise, using a wrench so that the timing mark of the camshaft timing gear is aligned with the timing mark of the camshaft bearing cap.
46. Align the paint mark of the number one chain with the timing mark of the camshaft timing gear.
47. Hold the hexagonal portion of the number three camshaft with a wrench, and tighten the camshaft timing gear set bolt to 74 ft. lbs.
48. While pushing in on the number three chain tensioner, insert a 0.039 inch pin into the hole to hold it.
49. Temporarily install the camshaft timing gear and chain tensioner number three and align the yellow mark links with the timing marks (1 dot mark and 2 dot marks) on the camshaft timing gears. Tighten the bolt to 14 ft. lbs.

➡Keep the camshaft level while it is being installed. The camshaft thrust clearance is very small and failing to keep it level could crack or damage the cylinder head journal surface, which receives the thrust.

50. Align the knock pin hole in the camshaft timing gear with the knock pin of the number four camshaft, and insert the number four camshaft into the camshaft timing gear.
51. Temporarily install the camshaft timing gear set bolt.
52. Install the four bearing caps in the proper location. Apply a light coat of clean engine oil to the threads and under the heads of the bolts.
53. Using several steps, uniformly install and tighten the eight bearing cap bolts in sequence to 80 inch lbs, for the 10mm head bolts and 18 ft. lbs. for the 12mm head bolts.
54. Hold the hexagonal portion of the number four camshaft, with a wrench, and install the camshaft timing gear set bolt and tighten it to 74 ft. lbs.
55. Remove the pin from the number three chain tensioner.
56. Release the chain tension between the camshaft timing gear (RH bank) and the crankshaft timing gear by turning the crankshaft pulley clockwise slightly.
57. While turning the stopper plate of the tensioner clockwise, push in the plunger of the tensioner. While turning the stopper plate of the tensioner counterclockwise, insert a 0.138 inch bar into the holes in the stopper plate and tensioner to hold the stopper plate. Install the two chain tensioner bolts and tighten to 7.4 ft. lbs. Remove the bar from the chain tensioner.
58. Position the number one cylinder at TDC of the compression stroke. Inspect the valve clearance, adjust as required.
59. Apply a continuous bead (0.08–0.12 inch) of seal packing, part number 08826-00080 or equivalent, to the cylinder head as indicated in the illustration. Install the seal washers onto the bolts. Install the cylinder head cover bolts and nuts. Tighten bolts "A" to 7.4 ft. lbs. Tighten bolts "B" to 80 inch lbs. Tighten nuts to 80 inch lbs.

➡Be sure to remove any oil from the contact surfaces of the cylinder head

cover and the cylinder head. Install the cover within three minutes after applying the seal packing. Tighten the bolts to specification within fifteen minutes after installing the cover. Do not add engine oil for at least two hours after installing the cover.

60. Continue the installation in the reverse order of the removal procedure.
61. Be sure to fill the engine with the proper grade and type engine coolant.
62. Be sure to fill the engine with the proper grade and type engine oil.
63. Start the engine and check for leaks. Correct as required.

4.7L V8 Engine

See Figures 98 through 101.

1. Before servicing the vehicle, refer to the precautions in the beginning of this section.
2. Drain the cooling system.
3. Relieve the fuel system pressure.
4. Remove or disconnect the following:
   - Battery and tray
   - Engine appearance cover
   - Engine under covers
   - Air intake assembly
   - Accessory drive belt
   - A/C compressor and bracket
   - Cooling fan and bracket
   - Radiator
   - Idler pulley
   - Front covers
   - Timing belt.
   - Camshaft sprockets
   - Camshaft Position (CMP) sensor
   - Power steering pump
   - Exhaust front pipes
   - Transmission dipstick tube
   - Ignition coils
   - Rear timing belt covers
   - Fuel lines
   - Intake manifold
   - Water inlet housing assembly
   - Front and rear water bypass joints
   - Engine lifting eyes
   - Oil dipstick tube
   - Valve covers
   - Camshafts
   - Cylinder heads with the exhaust manifolds attached. Loosen the bolts in the sequence shown.

To install:

5. Install the cylinder heads with new gaskets. Tighten the bolts in sequence as follows:
   a. Step 1: 24 ft. lbs. (32 Nm).
   b. Step 2: Plus 180 degrees.
6. Install or connect the following:
   - Camshafts
   - Valve covers
   - Oil dipstick tube
   - Engine lifting eyes
   - Front and rear water bypass joints
   - Water inlet housing assembly
   - Intake manifold
   - Fuel lines
   - Rear timing belt covers
   - Ignition coils
   - Transmission dipstick tube
   - Exhaust front pipes
   - Power steering pump
   - CMP sensor
   - Camshaft sprockets
   - Timing belt
   - Front covers
   - Idler pulley
   - Radiator
   - Cooling fan and bracket
   - A/C compressor and bracket
   - Accessory drive belt
   - Air intake assembly
   - Engine under covers
   - Engine appearance cover
   - Battery and tray
7. Fill the cooling system.
8. Start the engine and check for leaks.

RH Cylinder Head
LH Cylinder Head
Front
7924SG46

Fig. 100 Cylinder head torque sequence

5.7L Engine

See Figure 71.

1. Discharge fuel system pressure.
2. Drain engine coolant.
3. Disconnect cable from negative battery terminal.
4. Remove front exhaust pipe assembly.
5. Remove front fender apron seal.
6. Remove front fender apron seal rear.
7. Remove No. 2 steering intermediate shaft sub-assembly.
8. Remove exhaust manifold heat insulator.
9. Remove exhaust manifold sub-assembly.
10. Remove the timing chain cover.
11. Set No. 1 cylinder to TDC / compression.
12. Remove chain tensioner assembly.
13. Remove chain tensioner slipper.
14. Remove chain vibration damper.
15. Remove chain sub-assembly.
16. Remove chain tensioner assembly.

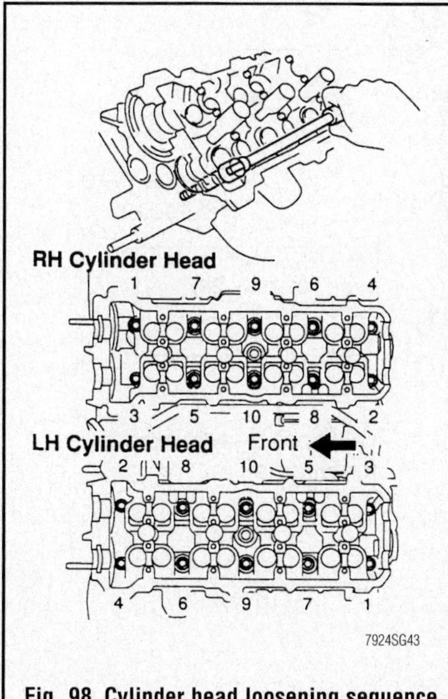

Fig. 98 Cylinder head loosening sequence

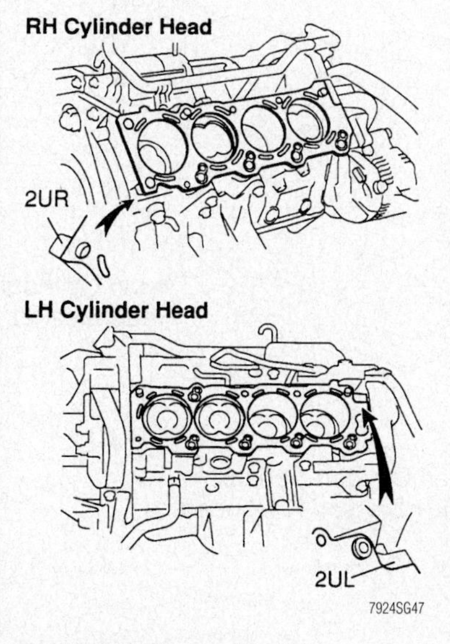

Fig. 99 Cylinder head gasket identification

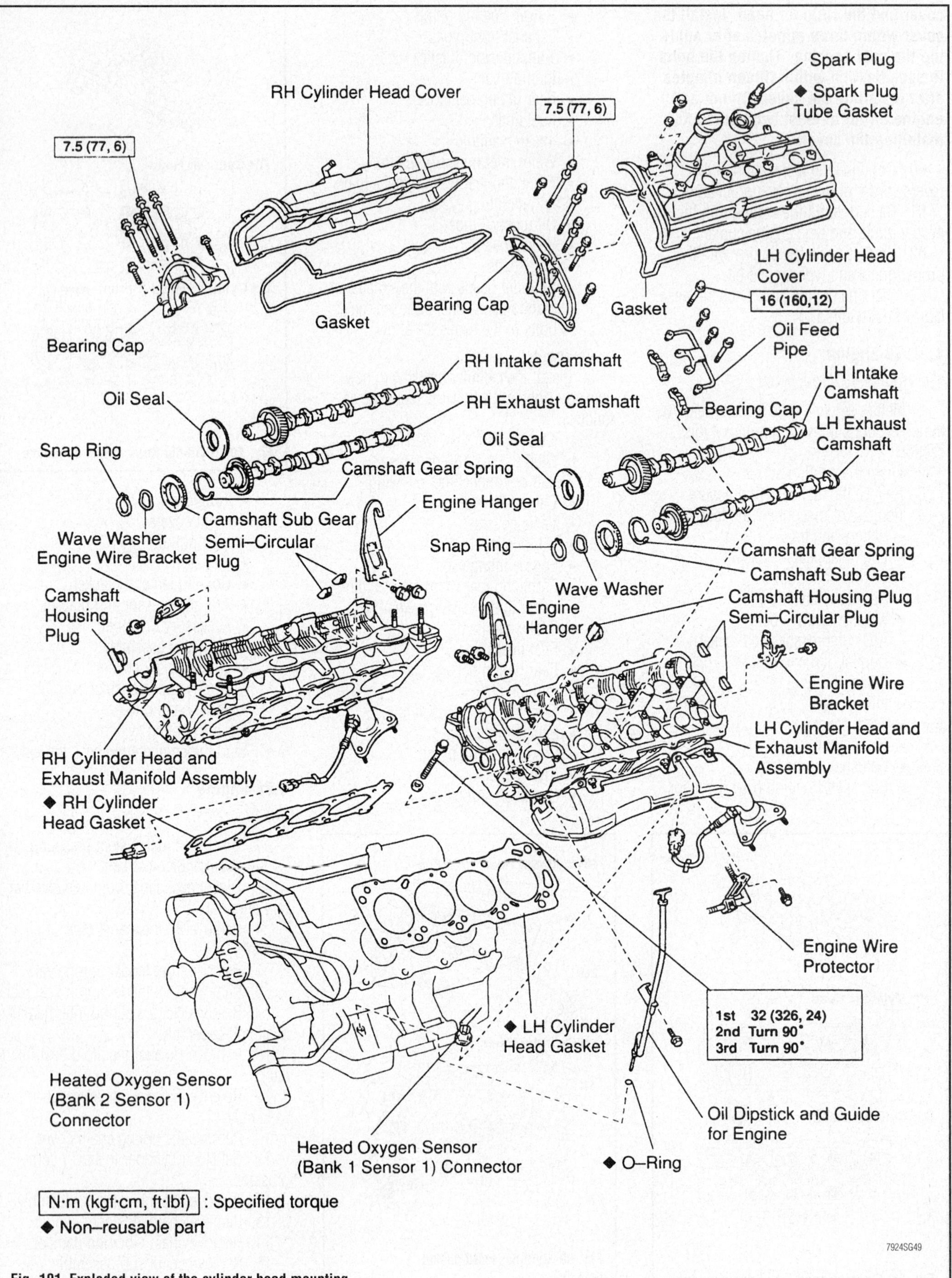

Fig. 101 Exploded view of the cylinder head mounting

Bank 1 Cylinder Head Removal

See Figures 72 through 74 and 102 through 104.

1. Make sure that the knock pin of the camshaft is positioned as shown.
2. Uniformly loosen and remove the 10 bearing cap bolts in the sequence LH.
3. Uniformly loosen and remove the 18 bearing cap bolts in the sequence LH.

➡Uniformly loosen the bolts while keeping the camshaft level.

4. Remove the 6 bearing caps.
5. Remove the No. 3 and No. 4 camshafts.
6. Remove the camshaft housing by prying between the cylinder head and camshaft housing with a screwdriver.

➡Be careful not to damage the contact surfaces of the cylinder head and camshaft housing.

7. Remove the 16 valve rocker arms from the cylinder head.
8. Remove the 16 valve lash adjusters from the cylinder head.
9. Remove the 16 valve stem caps from the cylinder head.

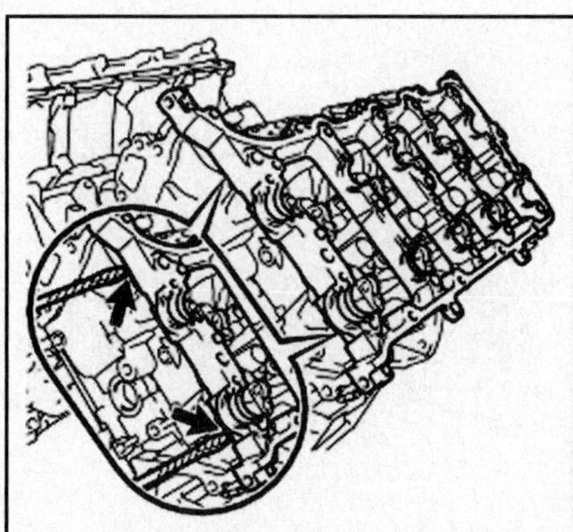

Fig. 102 Camshaft housing removal LH

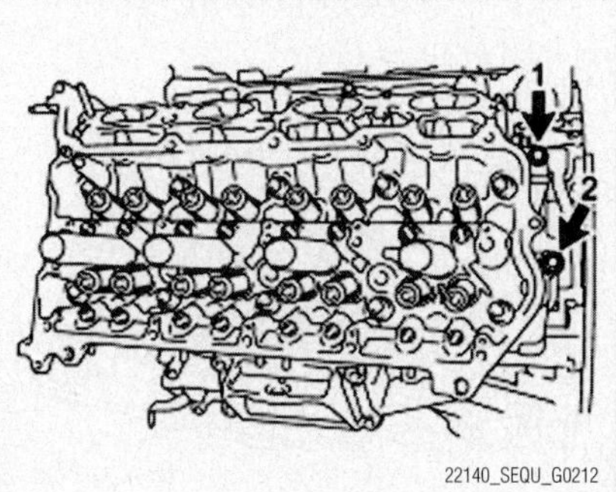

Fig. 103 Cylinder head removal sequence LH

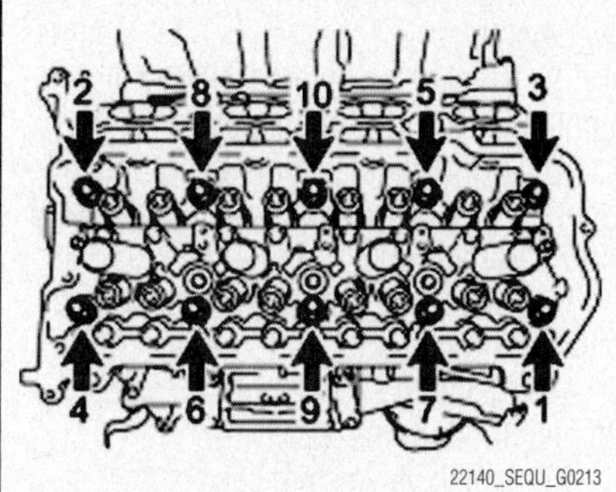

Fig. 104 Cylinder head removal sequence LH

➡Arrange the removed parts in the correct order.

10. Uniformly loosen and remove the 2 bolts in the sequence.
11. Using a 10 mm bi-hexagon wrench, uniformly loosen the 10 cylinder head bolts in the sequence shown in the illustration. Remove the 10 cylinder head bolts and plate washers.

➡Be careful not to drop washers into the cylinder head. Head warpage or cracking could result from removing bolts in an incorrect order.

12. Remove cylinder head gasket LH.

Bank 2 Cylinder Head Removal

See Figures 76 through 78 and 105 through 107.

1. Make sure that the knock pin of the camshaft is positioned as shown in the illustration.
2. Uniformly loosen and remove the 10 bearing cap bolts in the sequence RH.
3. Uniformly loosen and remove the 18 bearing cap bolts in the sequence RH.

➡Uniformly loosen the bolts while keeping the camshaft level.

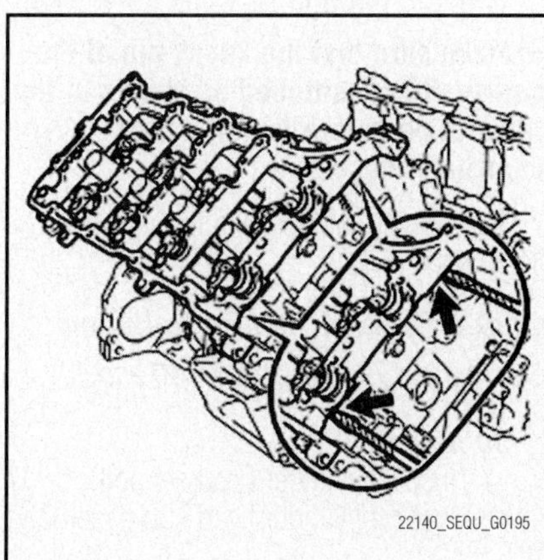

Fig. 105 Camshaft housing removal RH

4. Remove the 6 bearing caps.
5. Remove the No. 1 and No. 2 camshafts.
6. Remove the camshaft housing by prying between the cylinder head and camshaft housing with a screwdriver.

➡Be careful not to damage the contact surfaces of the cylinder head and camshaft housing.

7. Remove the 16 valve rocker arms from the cylinder head.
8. Remove the 16 valve lash adjusters from the cylinder head.
9. Remove the 16 valve stem caps from the cylinder head.

➡Arrange the removed parts in the correct order.

10. Uniformly loosen and remove the 2 bolts in the sequence.
11. Using a 10 mm bi-hexagon wrench, uniformly loosen the 10 cylinder head bolts in the sequence.

➡Be careful not to drop washers into the cylinder head. Head warpage or cracking could result from removing bolts in an incorrect order.

12. Remove cylinder head gasket RH.

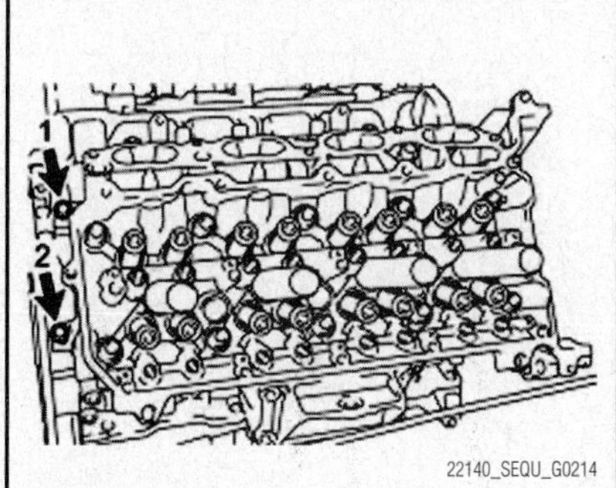

Fig. 106 Cylinder head removal sequence RH

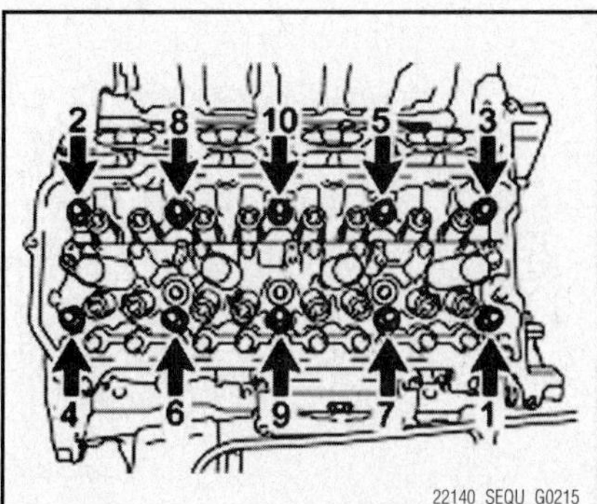

Fig. 107 Cylinder head removal sequence RH

Bank 1 Cylinder Head Installation

See Figures 80, 85 through 89, 108 and 109.

To install:

1. Inspect cylinder head set bolt.
2. Inspect cylinder head sub-assembly LH.
3. Clean the cylinder block with solvent.
4. Place the cylinder head gasket on the cylinder block surface with the front face of the Lot No. stamp upward.
5. Place the cylinder head on the cylinder block.

➡Ensure that no oil is on the mounting surface of the cylinder head. Gently place the cylinder head in order not to damage the gasket with the bottom part of the head.

6. Apply a light coat of engine oil to the threads and under the heads of the cylinder head bolts.
7. Using a 10 mm bi-hexagon wrench, install and uniformly tighten the 10 cylinder head bolts to 27 ft. lbs. (36 Nm) with the plate washers in several steps, in the sequence.

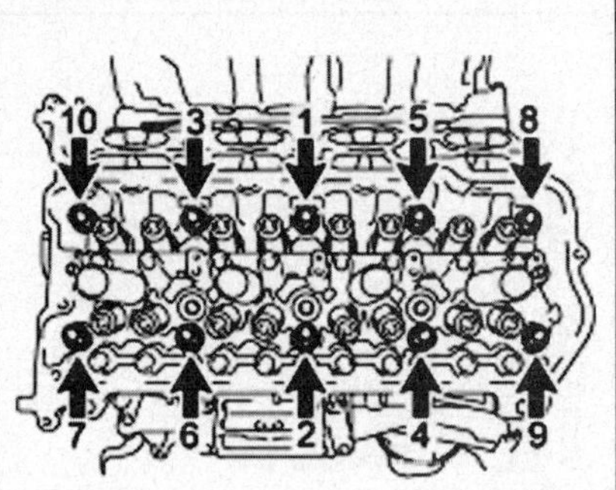

Fig. 108 Cylinder head installation sequence LH

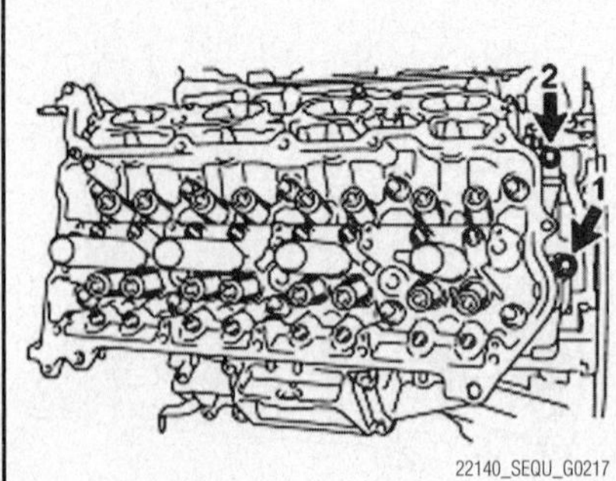

Fig. 109 Cylinder head installation sequence LH

8. Mark each cylinder head bolt head with paint.
9. Tighten the cylinder head bolts another 90 ° in the sequence.
10. Tighten the cylinder head bolts another 90 ° in the sequence for a total of 180 ° from torque.
11. Check that the painted marks are now facing correctly.
12. Install the 2 bolts in the order and tighten to 15 ft. lbs. (21 Nm).
13. Install valve stem cap.
    a. Apply a light coat of engine oil to the valve stem caps.
    b. Install the 16 valve stem caps to the cylinder head.

➡Install the lash adjuster at the same place it was removed from.

14. Install the 16 valve rocker arms.
15. Apply a light coat of engine oil to the camshaft journals, camshaft housings and bearing caps.
16. Install the No. 1 and No. 2 camshafts to the camshaft housing.
17. Confirm the marks and numbers on the camshaft bearing caps and place them in their proper positions and directions.
18. Temporarily install the 10 bolts in the order shown in the illustration.
19. Make sure that the valve rocker arms are installed properly.
20. Apply seal packing in a continuous line.

➡Remove any oil from the contact surface. Install the camshaft housing within 3 minutes and tighten the bolts within 15 minutes after applying seal packing. Do not start the engine for at least 2 hours after the installation.

21. Install the camshaft housing, and install the 18 bolts in the order and tighten bolt "A" to 7 ft. lbs. (10 Nm), and all others to 22 ft. lbs. (30 Nm).

➡Make sure that the knock pin of the camshaft is positioned as shown in the illustration before installing the camshaft housing.

22. Tighten the 10 bolts in the order to 12 ft. lbs. (16 Nm).

Bank 2 Cylinder Head Installation

See Figures 79 through 83, 110 and 111.

To install:

1. Inspect cylinder head set bolt.
2. Inspect cylinder head sub-assembly RH.
3. Clean the cylinder block with solvent.
4. Place the cylinder head gasket on the cylinder block surface with the front face of the Lot No. stamp upward.
5. Place the cylinder head on the cylinder block.

➡Ensure that no oil is on the mounting surface of the cylinder head. Gently place the cylinder head in order not to damage the gasket with the bottom part of the head.

6. Apply a light coat of engine oil to the threads and under the heads of the cylinder head bolts.
7. Using a 10 mm bi-hexagon wrench, install and uniformly tighten the 10 cylinder head bolts to 27 ft. lbs. (36 Nm) with the plate washers in several steps, in the sequence .
8. Mark each cylinder head bolt head with paint.
9. Tighten the cylinder head bolts another 90 ° in the sequence.
10. Tighten the cylinder head bolts another 90 ° in the sequence for a total of 180 ° from torque.
11. Check that the painted marks are now facing correctly.
12. Install the 2 bolts in the order and tighten to 15 ft. lbs. (21 Nm).

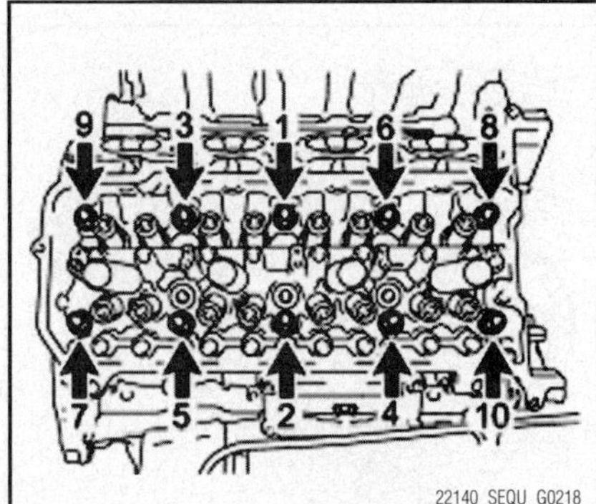

Fig. 110 Cylinder head installation sequence RH

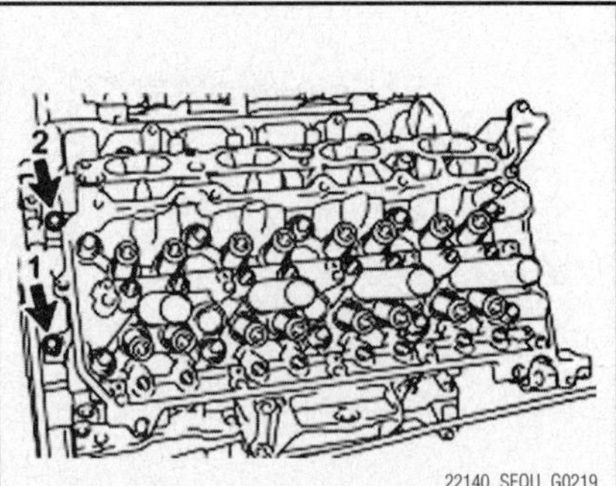

Fig. 111 Cylinder head installation sequence RH

13. Install valve stem cap.
   a. Apply a light coat of engine oil to the valve stem caps.
   b. Install the 16 valve stem caps to the cylinder head.

➡**Install the lash adjuster at the same place it was removed from.**

14. Install the 16 valve rocker arms.
15. Apply a light coat of engine oil to the camshaft journals, camshaft housings and bearing caps.
16. Install the No. 1 and No. 2 camshafts to the camshaft housing.
17. Confirm the marks and numbers on the camshaft bearing caps and place them in their proper positions and directions.
18. Temporarily install the 10 bolts in the order shown in the illustration.
19. Make sure that the valve rocker arms are installed properly.
20. Apply seal packing in a continuous line.

➡**Remove any oil from the contact surface. Install the camshaft housing within 3 minutes and tighten the bolts within 15 minutes after applying seal packing. Do not start the engine for at least 2 hours after the installation.**

21. Install the camshaft housing, and install the 18 bolts in the order and tighten bolt "A" to 7 ft. lbs. (10 Nm), and all others to 22 ft. lbs. (30 Nm).

➡**Make sure that the knock pin of the camshaft is positioned as shown in the illustration before installing the camshaft housing.**

22. Tighten the 10 bolts in the order to 12 ft. lbs. (16 Nm).
23. Install No. 2 chain tensioner assembly RH.
24. Install No. 1 chain sub-assembly RH.
25. Install No. 1 chain vibration damper RH.
26. Install No 1 chain tensioner slipper RH.
27. Install No. 1 chain tensioner assembly RH.
28. Install No. 3 chain tensioner assembly.
29. Install No. 1 chain sub-assembly LH.
30. Install No 1 chain tensioner slipper LH.
31. Install No. 1 chain tensioner assembly LH.
32. Install No. 1 chain vibration damper LH.
33. Tighten camshaft timing gears.
34. Check No. 1 cylinder to TDC.
35. Install the timing chain cover.
36. Install exhaust manifold sub-assembly.
37. Install exhaust manifold heat insulator.
38. Install No.2 steering intermediate shaft sub-assembly.
39. Install front fender apron seal rear.
40. Install front fender apron seal.
41. Install front exhaust pipe assembly.
42. Add engine coolant.
43. Connect cable to negative battery terminal.
44. Perform initialization, if necessary.
45. Inspect for leaks.
46. Inspect idle speed.

ENGINE ASSEMBLY

REMOVAL & INSTALLATION

V6 Engine

See Figures 112 through 122.

1. Discharge fuel system pressure.
2. Disconnect cable from negative battery terminal.
3. Drain engine coolant.
4. Drain automatic transmission fluid.
5. Drain engine oil.
6. Disconnect hood support assembly.
7. Remove the 4 bolts and hood.
8. Remove the front wiper motor and link.
9. Remove front cowl top outer panel sub-assembly.
10. Remove V-bank cover.
11. Remove No. 2 air cleaner hose.
12. Remove air cleaner assembly with element.
13. Remove No. 1 engine under cover.
14. Remove intake air surge tank.
15. Disconnect engine room RH side wire as follows :
   - Disconnect the ECM connector labeled A as shown in the illustration
   - Remove the junction block cover
   - Disconnect the 7 connectors from junction block labeled B shown in the illustration
   - Using a clip remover, detach the 2 wire harness clamps labeled B shown in the illustration
   - Remove the 2 bolts and ground cable labeled C shown in the illustration
   - Disconnect the 2 air pump connectors labeled D shown in the illustration
   - Using a clip remover, detach the 2 wire harness clamps labeled E shown in the illustration

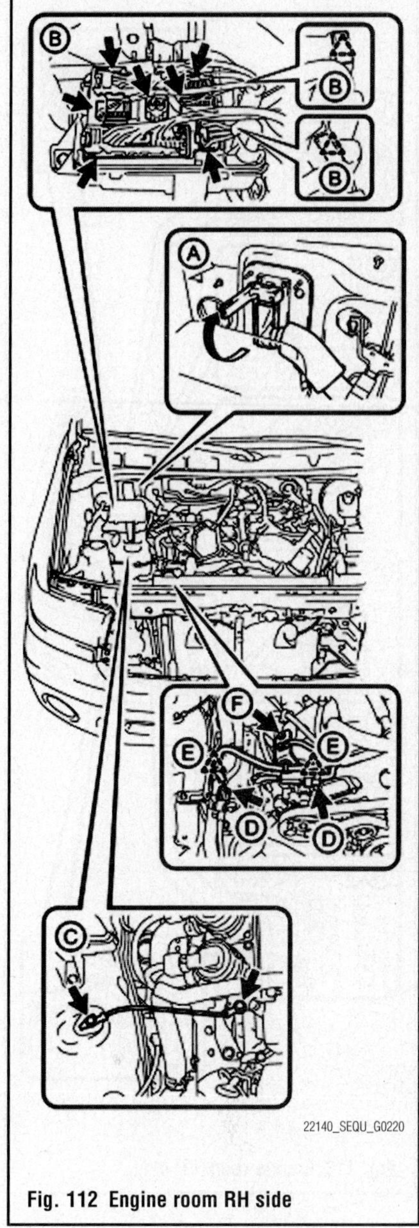

Fig. 112 Engine room RH side

   - Remove the bolt and bracket labeled F shown in the illustration
16. Disconnect engine room LH side wire as follows :
   - Disconnect the 4 connectors from the air injection control driver labeled A shown in the illustration
   - Remove the engine room relay block cover
   - Disconnect the 2 connectors from the engine room junction block labeled B shown in the illustration
   - Remove the nut from the engine room junction block, and disconnect the No. 2 engine wire's cover labeled C shown in the illustration
   - Remove the nut and disconnect the wire harness bracket from the inner

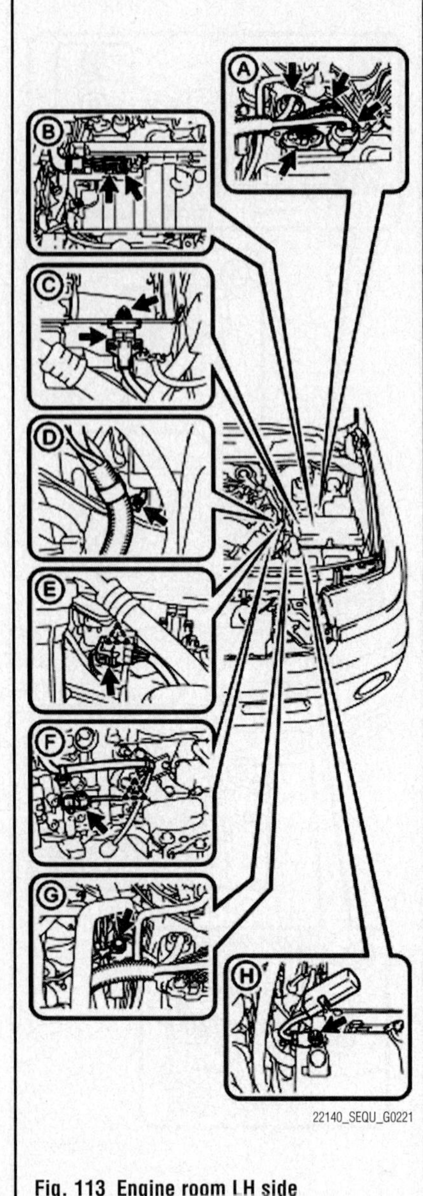

Fig. 113 Engine room LH side

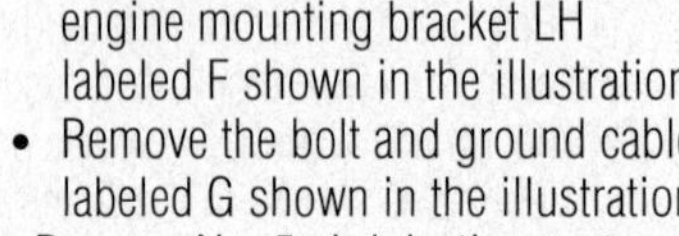

body panel side labeled D shown in the illustration

- Disconnect the connector, and using a clip remover, detach the front ABS junction connector's clamp from the engine room relay block side labeled E shown in the illustration
- Disconnect the connector from the crankshaft position sensor labeled F shown in the illustration
- Using a clip remover, detach the wire harness clamp from the front engine mounting bracket LH labeled F shown in the illustration
- Remove the bolt and ground cable labeled G shown in the illustration

17. Remove No. 5 air injection system hose.

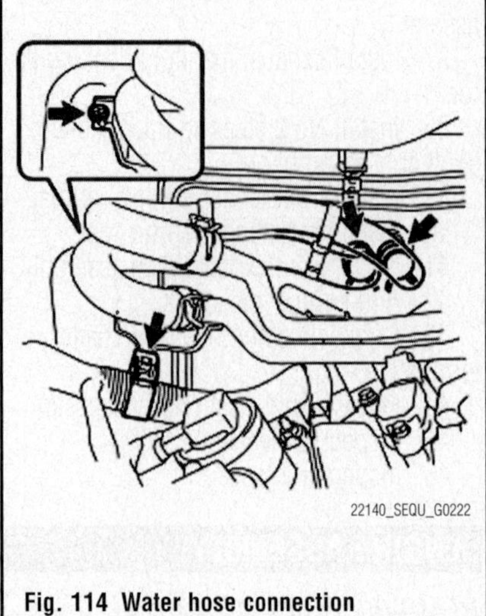

Fig. 114 Water hose connection

18. Remove No. 4 air injection system hose.
19. Remove No. 1 air tube.
20. Disconnect No. 2 air tube.
21. Remove air switching valve assembly.
22. Remove front No. 3 exhaust pipe sub-assembly (for regular cab standard deck).
23. Remove front No. 2 exhaust pipe assembly (for regular cab standard deck).
24. Remove front No. 2 exhaust pipe assembly (except regular cab standard deck).
25. Remove front exhaust pipe assembly.
26. Remove exhaust manifold sub-assembly RH.
27. Remove No. 2 exhaust manifold heat insulator.
28. Remove exhaust manifold sub-assembly LH.
29. Disconnect radiator hose inlet.
30. Disconnect radiator hose outlet.
31. Remove radiator side deflector RH.
32. Remove radiator side deflector LH.
33. Remove radiator assembly.
34. Remove the wire harness clamp.
35. Disconnect the heater water inlet and outlet hose.
36. Remove the nut and heater water pipe bracket.
37. Disconnect the power steering oil pressure switch connector.
38. Remove the 2 bolts, and disconnect the vane pump.

➡Make sure to suspend the vane pump securely.

39. Remove generator assembly.
40. Remove the 2 nuts, 2 bolts and disconnect the cooler compressor.
41. Disconnect the No. 1 and No. 2 fuel pipe.

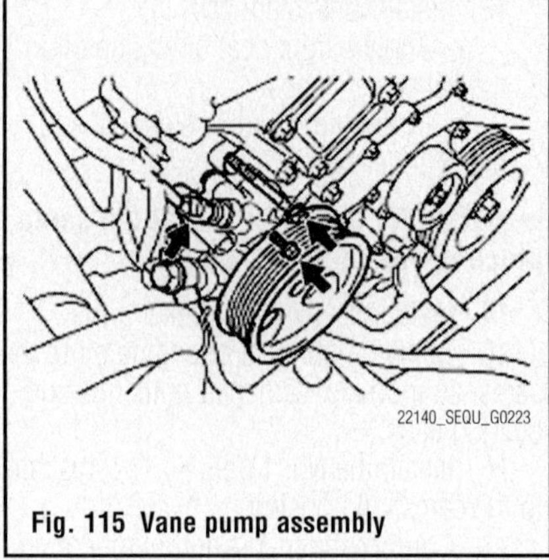

Fig. 115 Vane pump assembly

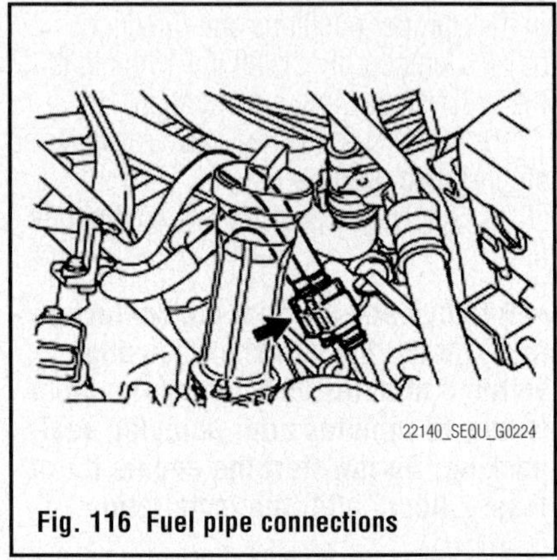

Fig. 116 Fuel pipe connections

42. Remove the fuel pipe clamp from the fuel tube connector.
43. Pinch the connector and disconnect the connector and pipe.
44. Check for dirt or mud on the seal surface of the disconnected pipe. Clean if necessary.
45. To protect the disconnected pipe and connector from damage and contamination, cover it with a plastic bag.
46. Remove battery.
47. Remove front fender apron seal LH.
48. Remove front fender apron seal RH.
49. Remove starter assembly.
50. Remove the flywheel housing side cover.
51. Turn the crankshaft to gain access to the 6 bolts and remove the 6 bolts while holding the crankshaft pulley set bolt with a wrench.
52. Remove the flywheel housing under cover.
53. Remove propeller shaft assembly.
54. Disconnect transmission control cable assembly.
55. Support automatic transmission assembly.
56. Remove the 4 bolts of the rear No. 1 engine mounting insulator.
57. Remove the 4 nuts, 4 washers, 4 bolts and No. 3 frame crossmember.

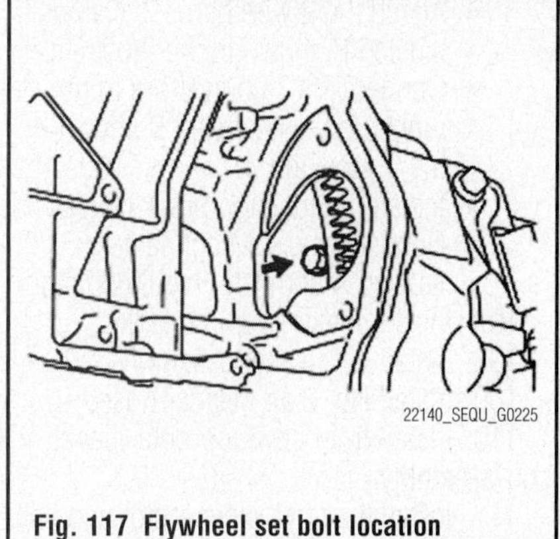

Fig. 117 Flywheel set bolt location

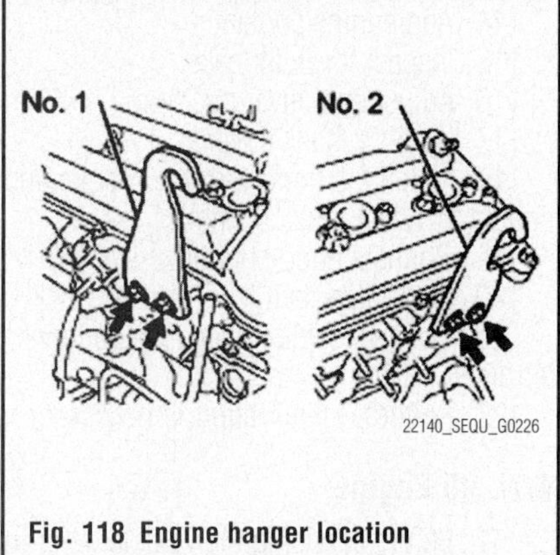

Fig. 118 Engine hanger location

58. Remove the 4 bolts and engine mounting insulator.
59. Disconnect wire harness and connector.
60. Disconnect no. 1 water by-pass pipe.
61. Remove automatic transmission assembly.
62. Remove engine assembly.
   a. Install the 2 engine hangers with the 4 bolts and tighten to 24 ft. lbs. (33 Nm) as shown in the illustration.
   b. Attach the engine sling device and hang the engine with the chain block.
   c. Remove the 2 nuts and 4 bolts from the front engine mounting insulator LH and RH.
   d. Lift the engine out of the vehicle carefully.

➡**Make sure the engine is clear of all wiring and hoses.**

To install:

63. Attach the engine sling device and hang the engine with the chain block.
64. Slowly lower the engine into the engine compartment.
65. Install the front engine mounting insulator LH and tighten bolt to 58 ft. lbs. (79 Nm) and nut to 53 ft. lbs. (72 Nm).
66. Install the front engine mounting insulator RH and tighten bolt to 58 ft. lbs. (79 Nm) and nut to 53 ft. lbs. (72 Nm).

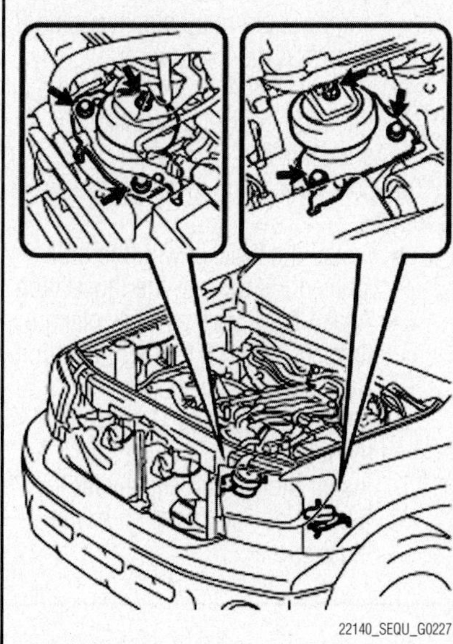

Fig. 119 Engine insulators

67. Remove the 2 engine hangers and 4 bolts.
68. Install automatic transmission assembly.
69. Connect No. 1 water by-pass pipe.
70. Connect wire harness and connector.
71. Install rear No. 1 engine mounting insulator.
72. Install No. 3 frame crossmember sub-assembly.
73. Install the flywheel housing under cover.
74. Install drive plate and torque converter clutch setting bolt and tighten to 35 ft. Lbs. (48 nm).
75. Install starter assembly.
76. Install front fender apron seal RH and LH.
77. Install battery.
78. Check that there is no damage or contamination in the connected part of pipe.
79. Align the axis of the connector with the axis of the pipe. Push the pipe into the connector until the connector makes a "click" sound. If the connection is tight, apply a small amount of fresh engine oil on the tip of the pipe.
80. After having finished the connection, try to pull apart the pipe and the connector and confirm that they are securely connected.
81. Install the fuel pipe clamp to the connector.
82. Install the cooler compressor with the 2 nuts and 2 bolts and tighten to 18 ft. lbs. (25 Nm) in several steps in the sequence shown in the illustration.

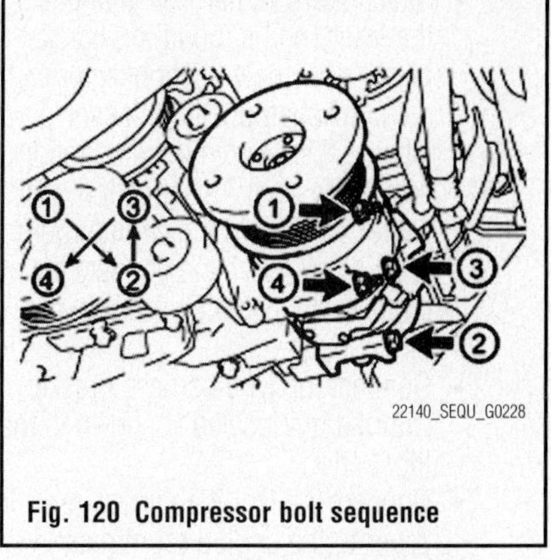

Fig. 120 Compressor bolt sequence

83. Connect the cooler compressor connector.
84. Install generator assembly.
85. Connect vane pump assembly.
86. Install the heater water pipe bracket with the nut.
87. Connect the heater water inlet and outlet hose.
88. Connect the wire harness clamp.
89. Install radiator assembly.
90. Install radiator side deflector RH and LH.
91. Install fan shroud.
92. Connect radiator hose outlet and inlet.
93. Install exhaust manifold sub-assembly lh.
94. Install No. 2 exhaust manifold heat insulator.
95. Install exhaust manifold sub-assembly RH.
96. Install front exhaust pipe assembly.
97. Install front No. 2 exhaust pipe assembly (except regular cab standard deck).
98. Install front No. 2 exhaust pipe assembly (for regular cab standard deck).
99. Install front No. 3 exhaust pipe sub-assembly (for regular cab standard deck).
100. Install air switching valve assembly.
101. Install No. 2 air tube.
102. Install No. 1 air tube.
103. Connect No. 4 air injection system hose.
104. Connect No. 5 air injection system hose.
105. Connect engine room LH side wire as follows:
   - Install the battery positive cable with the nut labeled H shown in the illustration
   - Connect the battery positive cable cover
   - Install the ground cable with the bolt labeled G shown in the illustration

- Attach the wire harness clamp to the front engine mounting bracket LH, and connect the connector to the crankshaft position sensor labeled F shown in the illustration
- Attach the front ABS junction connector's clamp to the engine room relay block side, and connect the connector labeled E shown in the illustration.
- Connect the wire harness bracket with the nut labeled D shown in the illustration
- Connect the No. 2 engine wire's cover to the engine room relay block, and install the No. 2 engine wire with the nut labeled C shown in the illustration

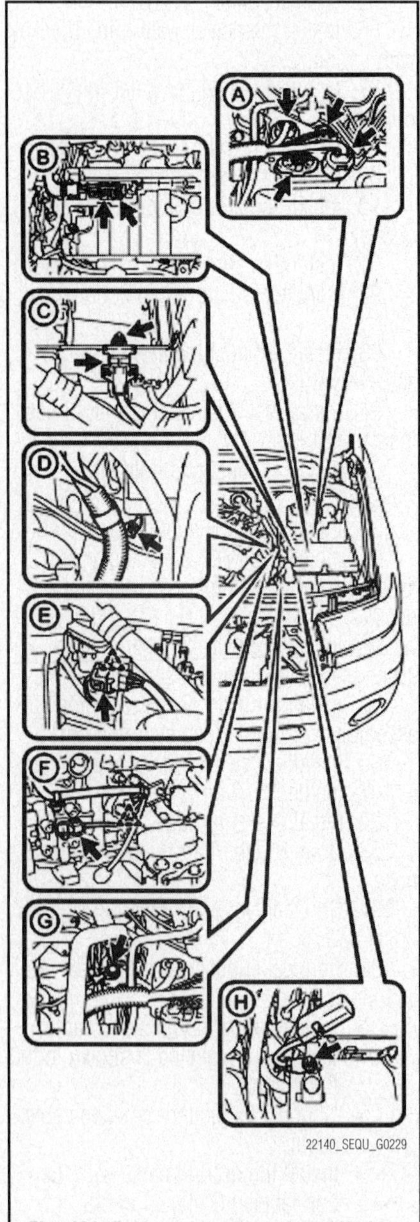

Fig. 121 Engine wire connections LH side

- Connect the 2 connectors to the engine room relay block labeled B shown in the illustration
- Connect the 4 connectors to the air injection control driver labeled A shown in the illustration

106. Connect engine room RH side wire as follows :

- Install the bracket with the bolt labeled F shown in the illustration
- Attach the 2 wire harness clamps labeled E shown in the illustration
- Connect the 2 air pump connectors labeled D shown in the illustration
- Install the ground cable with the 2 bolts labeled C shown in the illustration

Fig. 122 Engine wire connections RH side

- Attach the 2 wire harness clamps labeled B shown in the illustration
- Connect the 7 connectors to the junction block labeled B shown in the illustration
- Install the junction block cover
- Connect the ECM connector labeled A shown in the illustration

107. Install intake air surge tank.
108. Install air cleaner assembly.
109. Install No. 2 air cleaner hose.
110. Install front cowl top outer panel sub-assembly.
111. Install the front wiper motor and link.
112. Add engine oil.
113. Add automatic transmission fluid.
114. Add engine coolant.
115. Inspect for fluid leaks.
116. Adjust shift lever position.
117. Install V-bank cover.
118. Install the hood with the 4 bolts and tighten to 120 ft. Lbs. (13 nm).
119. Connect hood support assembly.
120. Inspect coolant level at reservoir.
121. Connect cable to negative battery terminal.
122. Perform initialization, if necessary.

4.7L V8 Engine

1. Before servicing the vehicle, refer to the precautions in the beginning of this section.
2. Relieve the fuel system pressure.
3. Drain the cooling system.
4. Drain the engine oil.
5. Remove or disconnect the following:

- Battery and tray
- Hood
- Engine appearance cover
- Air intake pipe
- Engine under covers
- Coolant recovery tank
- Radiator hoses
- Radiator and fan shroud
- Accessory drive belt
- Cooling fan and pulley
- Powertrain Control Module (PCM) harness connectors and pass the wiring harness through the firewall
- Accelerator cable
- Power steering vacuum hoses
- Alternator harness connectors
- Heater hoses
- Engine control wiring harness and grommet at the firewall
- Ground cable connector
- Fuel lines
- Evaporative Emissions (EVAP) canister hoses
- Wire clamp at right inner fender
- Negative battery cable at the relay box and right inner fender

- Positive battery cable
- Center console
- Transmission shift lever assembly
- Transfer case shift lever and rod
- Exhaust front pipes
- Stabilizer bar
- Front and rear driveshafts
- A/C compressor
- Power steering pump

6. Attach a hoist to the engine lifting eyes.
7. Remove or disconnect the following:
   - Transfer case skid plate
   - Left and right motor mounts
   - Transmission mount crossmember
8. Attach a hoist to the engine lifting eyes and raise the powertrain out of the vehicle.

To install:

9. Lower the powertrain into the vehicle.
10. Install or connect the following:
    - Transmission mount crossmember. Tighten the bolts to 37 ft. lbs. (50 Nm) and the nuts to 55 ft. lbs. (74 Nm).
    - Transfer case skid plate
    - Left and right motor mounts. Tighten the fasteners to 22 ft. lbs. (30 Nm).
    - Power steering pump. Tighten the bolts to 13 ft. lbs. (17 Nm).
    - A/C compressor. Tighten the bolts to 36 ft. lbs. (49 Nm).
    - Front driveshaft. Tighten the fasteners to 59 ft. lbs. (80 Nm).
    - Rear driveshaft. Tighten the fasteners to 78 ft. lbs. (106 Nm).
    - Stabilizer bar. Tighten the bracket bolts to 13 ft. lbs. (18 Nm) and the link nuts to 18 ft. lbs. (25 Nm).
    - Exhaust front pipes
    - Transfer case shift lever and rod
    - Transmission shift lever assembly
    - Center console
    - Positive battery cable
    - Negative battery cable at the relay box and right inner fender
    - Wire clamp at right inner fender
    - EVAP canister hoses
    - Fuel lines
    - Ground cable connector
    - Engine control wiring harness and grommet at the firewall
    - Heater hoses
    - Alternator harness connectors
    - Power steering vacuum hoses
    - Accelerator cable
    - PCM harness connectors
    - Cooling fan and pulley
    - Accessory drive belt
    - Radiator and fan shroud
    - Radiator hoses
    - Coolant recovery tank
    - Engine under covers
    - Air intake pipe
    - Engine appearance cover
    - Hood
    - Battery and tray
11. Fill the crankcase to the correct level.
12. Fill the cooling system.
13. Start the engine and check for leaks.

5.7L Engine

See Figures 123 through 133.

1. Disconnect engine room LH side wire as follows :
   - Disconnect the 2 camshaft timing control valve connectors
   - Disconnect the 4 ignition coil connectors
   - Disconnect the 2 VVT sensor connectors
   - Disconnect the camshaft position sensor connector
   - Disconnect the vacuum switching valve connector (for ACIS).
   - Disconnect the purge VSV connector
   - Disconnect the engine coolant temperature sensor connector
   - Disconnect the fuel injector connector
   - Disconnect the noise filter connector
   - Remove the bolt and disconnect the 7 clamps
2. Disconnect engine room RH side wire as follows :
   - Disconnect the 2 camshaft timing control valve connectors
   - Disconnect the 4 ignition coil connectors
   - Disconnect the 2 VVT sensor connectors
   - Disconnect the fuel injector connector
   - Disconnect the noise filter connector
   - Disconnect the throttle sensor connector
   - Disconnect the oil pressure sender gauge connector
   - Disconnect the 7 clamps
3. Disconnect engine room rear side wire as follows :
   - Disconnect the 2 air switching valve connectors
   - Disconnect the 5 clamps
   - Remove the 3 bolts
4. Remove the engine wire.
5. Remove the fuel tube.
6. Remove the 2 water drain cock plugs from the water drain cocks.
7. Remove the 2 water drain cocks from the cylinder block.
8. Remove front engine mounting insulators.
9. Remove the 4 bolts and mounting brackets RH an LH.

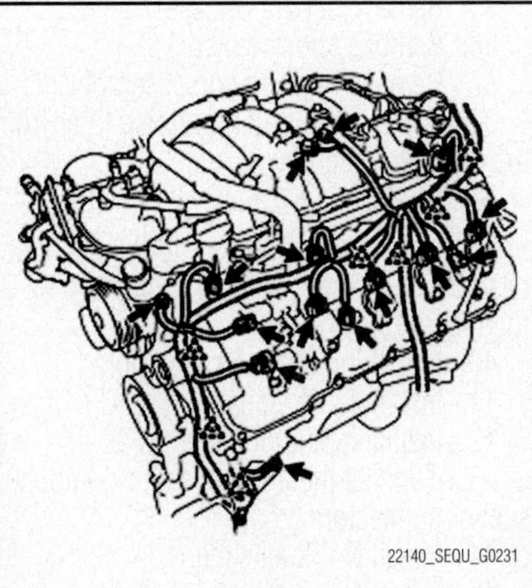

Fig. 123 Engine wire connections LH side

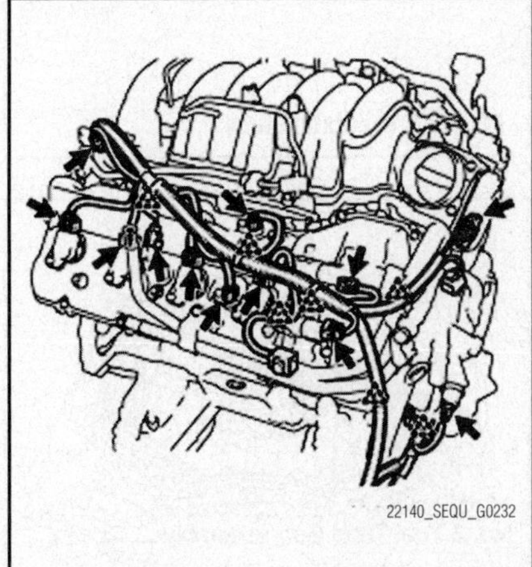

Fig. 124 Engine wire connections RH side

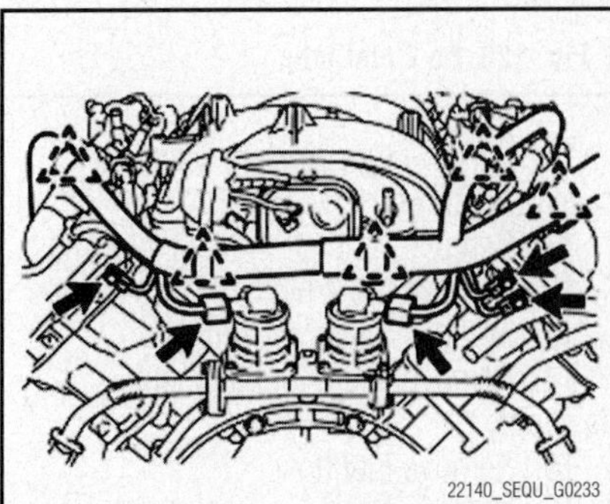

Fig. 125 Engine wire connections rear side

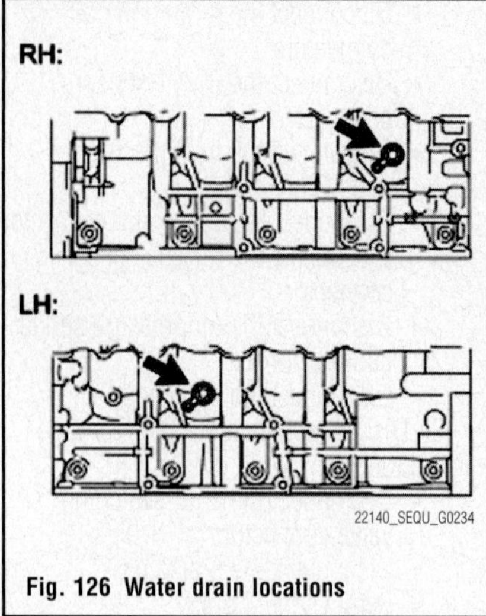

Fig. 126 Water drain locations

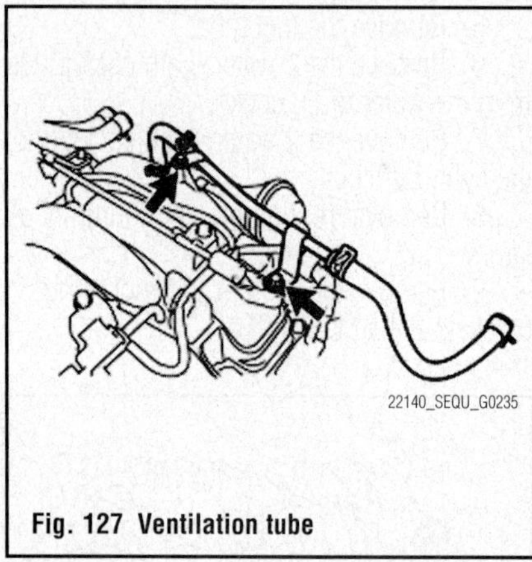

Fig. 127 Ventilation tube

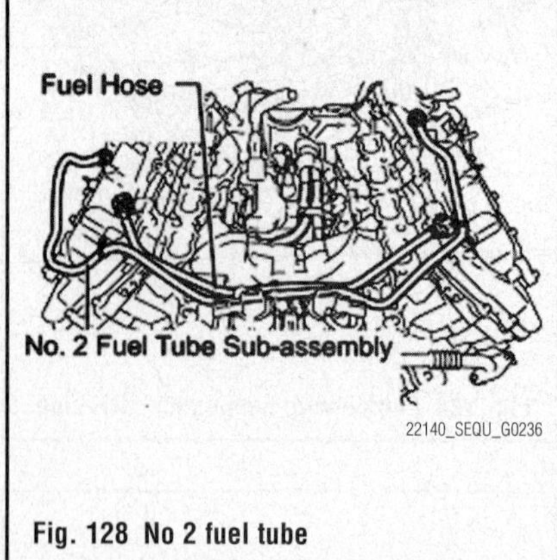

Fig. 128 No 2 fuel tube

10. Remove fuel delivery pipe sub-assembly RH and LH .
11. Remove ventilation hose assembly.
12. Remove the 2 bolts and No. 1 tube.
13. Remove intake manifold
14. Remove No. 2 fuel tube sub-assembly
15. Remove fuel hose.
16. Remove No. 2 water by-pass pipe.
17. Remove No. 1 water by-pass hose.
18. Remove front water by-pass joint.
19. Remove No. 2 engine cover.

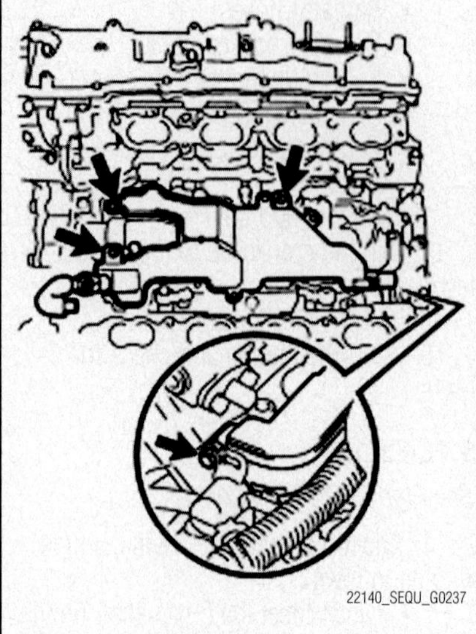

Fig. 129 Separator case

20. Remove No. 1 engine cover.
21. Remove water pump pulley.
22. Remove No. 1 idler pulley sub-assembly.
23. Remove fan bracket sub-assembly.
24. Remove V-ribbed belt tensioner assembly.
25. Remove oil pressure sender gauge assembly.
26. Remove air tube.
27. Remove water inlet housing.
28. Remove water by-pass pipe sub-assembly.
29. Remove air switching valve assembly.
30. Remove No. 2 air tube.
31. Remove No. 3 air tube.
32. Disconnect the 4 knock sensor connectors.
33. Disconnect the 3 clamps. Then remove the engine wire.
34. Remove No. 2 cylinder head cover.
35. Remove separator case.
36. Remove knock sensor.
37. Remove ignition coil assembly.
38. Remove the bolt and noise filter from the cylinder head cover.

To install:

39. Install the noise filter to the cylinder head cover with the bolt and tighten to 62 inch lbs. (7.0 Nm).
40. Install ignition coil assembly.
41. Install knock sensor.
42. Install separator case.
43. Connect the 3 clamps and 4 knock sensor connectors.
44. Install No. 3 air tube.
45. Install No. 2 air tube.
46. Install air switching valve assembly.
47. Install water inlet housing.
48. Install air tube.
49. Install oil pressure sender gauge assembly.
50. Install No. 1 idler pulley sub-assembly.
51. Install water pump pulley.
52. Install No. 1 engine cover.
53. Install No. 2 engine cover.
54. Install front water by-pass joint.
55. Install No. 1 water by-pass hose.
56. Install No. 2 water by-pass pipe.
57. Install fuel hose.
58. Install No. 2 fuel tube sub-assembly.
59. Install intake manifold.
60. Install the No. 1 tube with the 2 bolts.
61. Install ventilation hose assembly.
62. Install fuel delivery pipe sub-assembly RH and LH.
63. Install the mounting bracket with the 4 bolts and tighten to 26 ft. Lbs. (35 nm) RH and LH.
64. Install front engine mounting insulator RH.
65. Apply adhesive to 2 or 3 threads of the drain cocks.
66. Install the water drain cocks.
67. Further tighten the drain cocks within 360° so that the drain cock pipes are within the range shown in the illustration.

➡Do not rotate the drain cocks more than 1 revolution (360°) after tightening the drain cocks to the specified torque. Do not loosen the drain cocks to adjust them. If an adjustment is necessary, remove the drain cocks and reinstall them.

68. Install the water drain cock plugs to the water drain cocks and tighten to 10 ft. lbs. (13 Nm).
69. Install front engine mounting insulator.
70. Install the fuel tube.
71. Connect engine room rear side wire as follows :
    - Install the 3 bolts as shown in the illustration
    - Connect the 5 clamps
    - Connect the 2 air switching valve connectors
72. Connect engine room RH side wire as follows :
    - Connect the 7 clamps
    - Connect the oil pressure sender gauge connector
    - Connect the throttle sensor connector
    - Connect the noise filter connector
    - Connect the fuel injector connector
    - Connect the 2 VVT sensor connectors

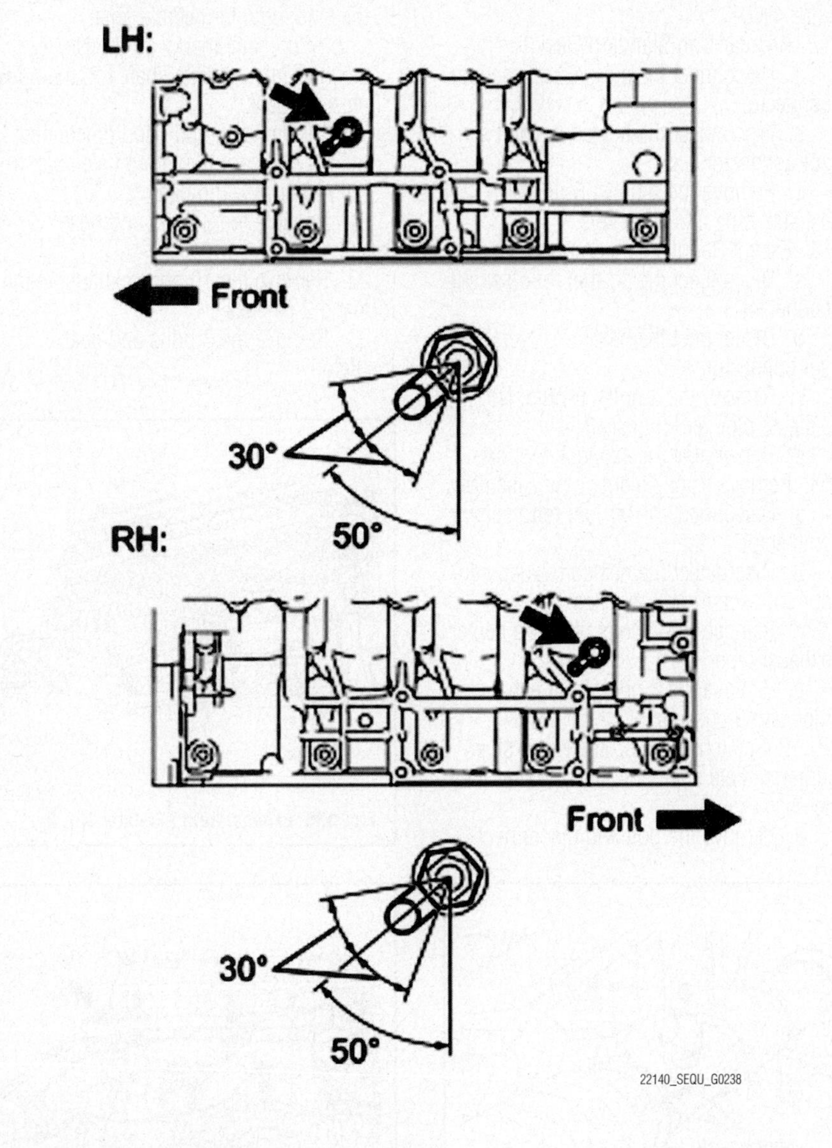

Fig. 130 Water cock positions

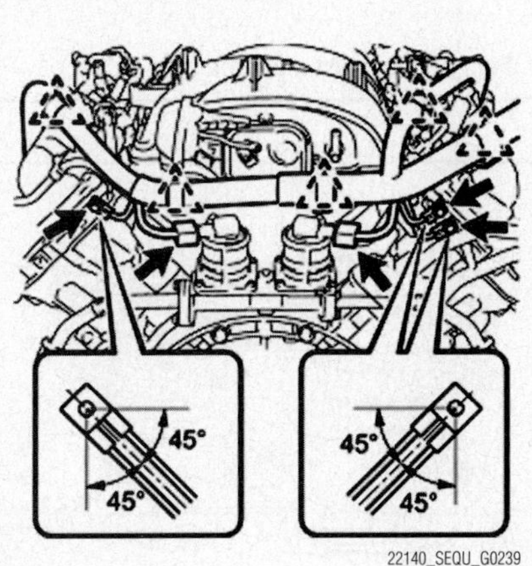

Fig. 131 Engine wire connections rear side

- Connect the 4 ignition coil connectors
- Connect the 2 camshaft timing control valve connectors

73. Connect engine room LH side wire as follows :

- Connect the 7 clamps and install the engine wire with the bolt
- Connect the noise filter connector
- Connect the fuel injector connector
- Connect the engine coolant temperature sensor connector
- Connect the purge VSV connector
- Connect the 2 VVT sensor connectors
- Connect the vacuum switching valve connector (for ACIS).
- Connect the camshaft position sensor connector

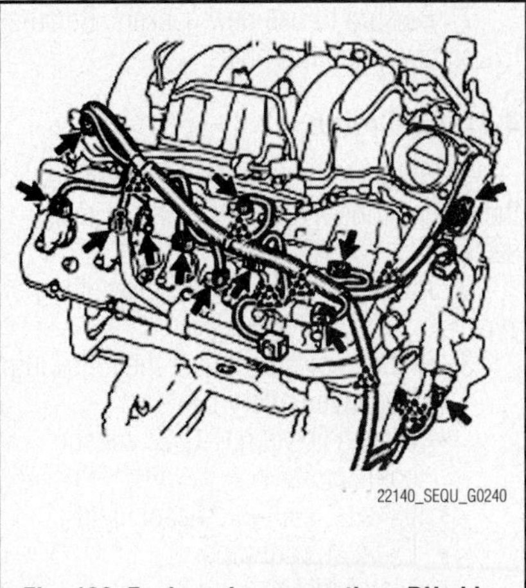

Fig. 132 Engine wire connections RH side

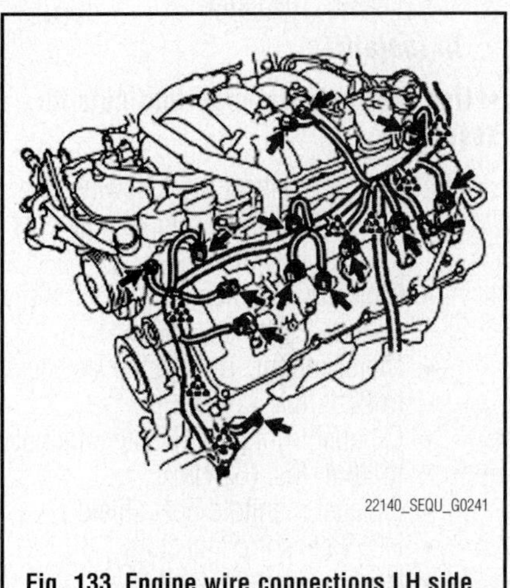

Fig. 133 Engine wire connections LH side

- Connect the 2 VVT sensor connectors
- Connect the 4 ignition coil connectors
- Connect the 2 camshaft timing control valve connector

EXHAUST MANIFOLD

REMOVAL & INSTALLATION

V6 Engine

1. Before servicing the vehicle, refer to the Precautions Section.
2. Disconnect the negative battery cable.
3. Disconnect the exhaust manifold to exhaust flange nuts.
4. Remove the necessary components to gain access to the exhaust manifold retaining bolts.
5. Remove the exhaust manifold retaining nuts. Discard the nuts. Remove the exhaust manifold from the engine.

To install:

6. Installation is the reverse of the removal procedure.

7. Be sure to use new gaskets. Be sure to use new retaining nuts.

4.7L V8 Engine

1. Before servicing the vehicle, refer to the precautions in the beginning of this section.
2. Attach a hoist to the engine lifting eyes.
3. Remove or disconnect the following:
   - Negative battery cable
   - Heated Oxygen (HO_2S) sensor connectors
   - Exhaust manifold heat shield
   - Exhaust front pipe
   - Motor mount
   - Motor mount bracket
   - Exhaust manifold

To install:

➡Use new exhaust manifold nuts for assembly.

4. Install or connect the following:
   - Exhaust manifold. Tighten the nuts to 32 ft. lbs. (44 Nm).
   - Motor mount bracket. Tighten the bolts to 27 ft. lbs. (36 Nm).
   - Motor mount. Tighten the fasteners to 22 ft. lbs. (30 Nm).
   - Exhaust front pipe. Tighten the nuts to 46 ft. lbs. (62 Nm).
   - Exhaust manifold heat shield
   - HO_2S sensor connectors
   - Negative battery cable
5. Start the engine and check for leaks.

5.7L V8 Engine

See Figures 134 through 141.

1. Disconnect cable from negative battery terminal.
2. Remove No. 1 engine under cover.
3. Drain engine coolant.
4. Remove V-bank cover sub-assembly.
5. Remove air cleaner hose assembly.
6. Disconnect inlet and outlet radiator hose.
7. Remove fan and generator v-belt.
8. Remove fan shroud.
9. Remove the 6 clips and fender apron seal.
10. Remove the 5 clips and fender apron seal.
11. Remove engine oil level dipstick guide.
12. Disconnect vane pump assembly.
13. Remove generator assembly.
14. Disconnect cooler compressor assembly.
15. Remove the 2 bolts and heat insulator 4WD.
16. Remove front propeller shaft assembly (for 4WD).
17. Regular Cab Standard Deck:
    a. Disconnect the air fuel ratio sensor connector.
    b. Disconnect the heated oxygen sensor connector.
    c. Remove the 2 bolts, 5 nuts, No. 2 exhaust pipe and 2 gaskets.
18. Except Regular Cab Standard Deck:
    a. Disconnect the air fuel ratio sensor connector.
    b. Disconnect the heated oxygen sensor connector.
    c. Remove the 2 bolts, 3 nuts, No. 2 exhaust pipe and 2 gaskets.
    d. Disconnect the exhaust support.
19. Remove front exhaust pipe assembly.
    a. Disconnect the air fuel ratio sensor connector.
    b. Disconnect the heated oxygen sensor connector and 2 clamps.
    c. Remove the 2 bolts from the center exhaust pipe.
    d. Remove the 3 nuts, front exhaust pipe and 2 gaskets.
20. Put matchmarks on the No. 2 steering intermediate shaft and steering intermediate shaft.
    a. Remove the bolt and disconnect the No. 2 steering intermediate shaft from the steering intermediate shaft.
    b. Put matchmarks on the No. 2 steering intermediate shaft and the power steering gear.
    c. Remove the bolt and disconnect the No. 2 steering intermediate shaft from the power steering gear.
21. Remove the 3 bolts and heat insulator.
22. Remove the 10 nuts, exhaust manifold and 2 gaskets.
23. Remove the 3 bolts and heat insulator.

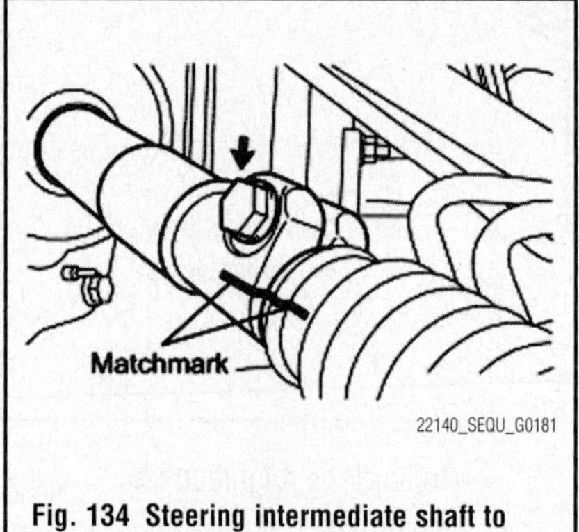

Fig. 134 Steering intermediate shaft to steering intermediate shaft

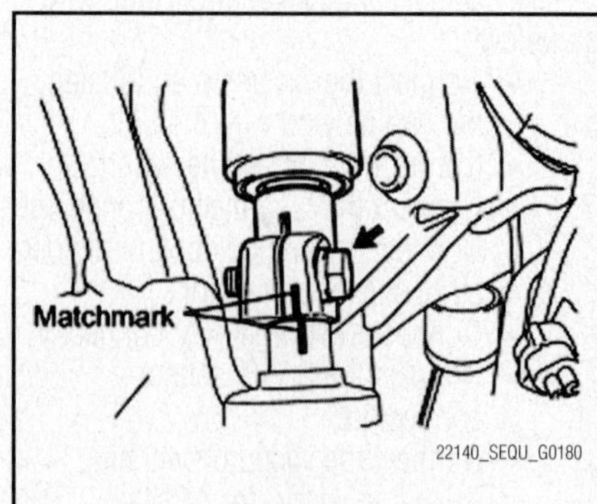

Fig. 135 Steering intermediate shaft to the power steering gear

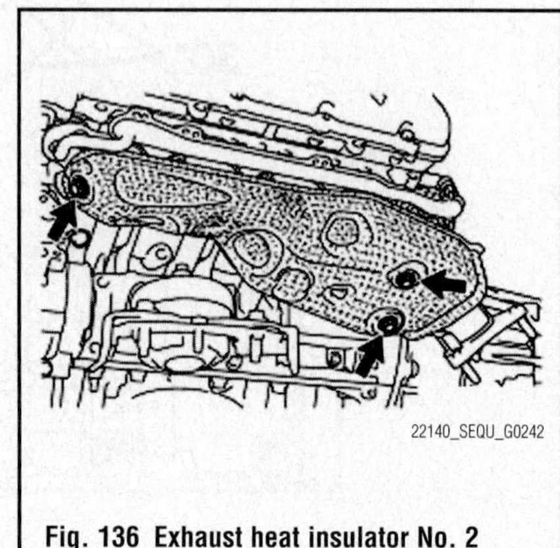

Fig. 136 Exhaust heat insulator No. 2

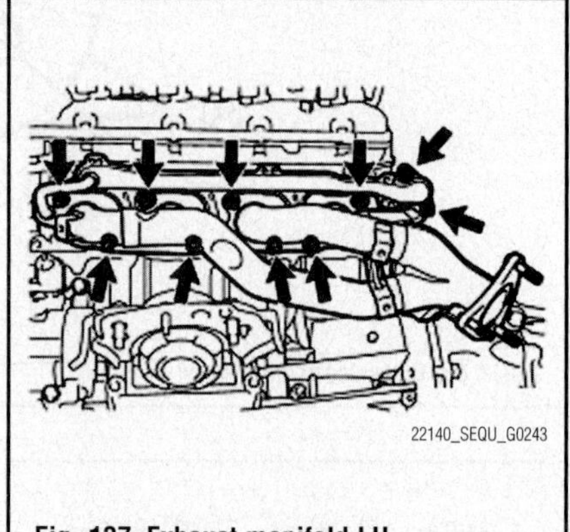

Fig. 137 Exhaust manifold LH

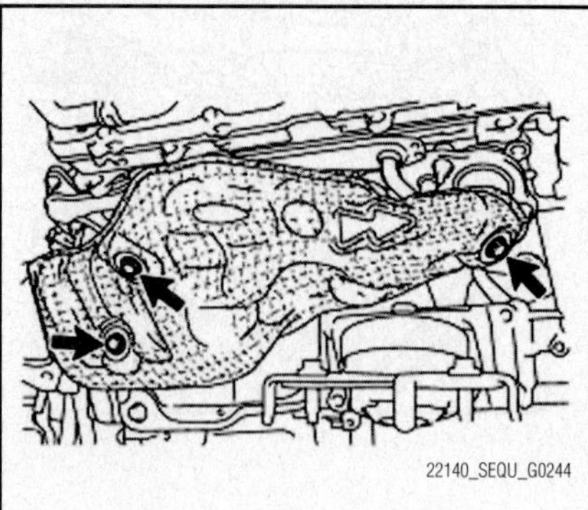

Fig. 138 Exhaust heat insulator No. 1

24. Remove the 10 nuts, exhaust manifold and 2 gaskets.

To install:

25. Install a new gasket to the cylinder head and a new gasket to the No. 2 air tube.

➡Install the exhaust manifold gasket with the gasket tab facing toward the front of the engine. Install the air tube gasket with the gasket's claws facing the tube side.

26. Temporarily install the exhaust manifold and then uniformly tighten 8 new nuts that are not labeled A.
27. Tighten the new nuts labeled A in the illustration to 7 ft. lbs. (10 Nm) and remaining nuts to 15 ft. lbs. (21 Nm).
28. Install a new gasket to the cylinder head and a new gasket to the No. 3 air tube.

➡Install the exhaust manifold gasket with the gasket tab facing toward the front of the engine. Install the air tube gasket with the gasket's claws facing the tube side.

29. Temporarily install the exhaust manifold and then uniformly tighten 8 new nuts that are not labeled A.

Fig. 139 Exhaust manifold tightening sequence RH

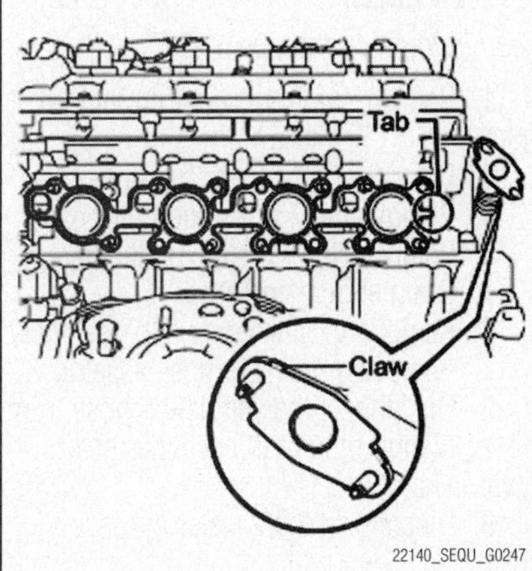

Fig. 140 Exhaust manifold gasket position LH

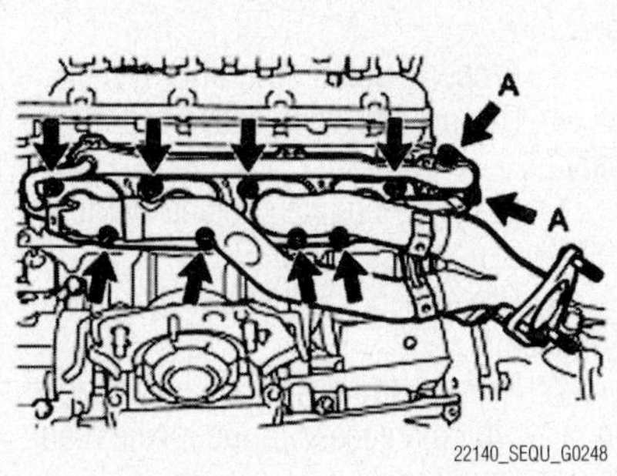

Fig. 141 Exhaust manifold tightening sequence LH

30. Tighten the new nuts labeled A in the illustration to 7 ft. lbs. (10 Nm) and remaining nuts to 15 ft. lbs. (21 Nm).
31. Install the heat insulator with the 3 bolts and tighten to 7 ft. lbs. (10 Nm).
32. Align the matchmarks and insert the No. 2 intermediate shaft into the intermediate shaft.
33. Align the matchmarks and insert the No. 2 intermediate shaft into the power steering gear.
34. Install the 2 bolt and tighten to 26 ft. lbs. (35 Nm).
35. Install front exhaust pipe assembly.
    a. Install a new gasket and the front exhaust pipe to the exhaust manifold RH with 3 new nuts and tighten to 40 ft. lbs. (54 Nm).
    b. Install a new gasket and the front exhaust pipe to the center exhaust pipe with the 2 bolts and tighten to 35 ft. lbs. (48 Nm).
    c. Connect the air fuel ratio sensor connector.
    d. Connect the heated oxygen sensor connector and 2 clamps.
36. Except Regular Cab Standard Deck:
    a. Connect the front No. 2 exhaust pipe to the exhaust support.
    b. Install a new gasket and the front No. 2 exhaust pipe to the exhaust manifold LH with 3 new nuts and tighten to 40 ft. lbs. (54 Nm).
    c. Install a new gasket and the front No. 2 exhaust pipe to the center exhaust pipe with the 2 bolts and tighten to 35 ft. lbs. (48 Nm).
    d. Connect the air fuel ratio sensor connector.
    e. Connect the heated oxygen sensor connector.
37. Regular Cab Standard Deck :
    a. Connect the front No. 2 exhaust pipe to the exhaust support.
    b. Install a new gasket and the front No. 2 exhaust pipe to the exhaust manifold LH with 3 new nuts and tighten to 40 ft. lbs. (54 Nm).
    c. Install a new gasket and the front No. 2 exhaust pipe to the center exhaust pipe with the 2 bolts and tighten to 35 ft. lbs. (48 Nm).
    d. Connect the air fuel ratio sensor connector.
    e. Connect the heated oxygen sensor connector.
38. Install front propeller shaft assembly (4WD).
39. Install the heat insulator with the 2 bolts (4WD).
40. Connect cooler compressor assembly.
41. Install generator assembly.
42. Connect vane pump assembly.
43. Install engine oil level dipstick guide.
44. Install fan shroud.
45. Install fan and generator v belt.
46. Connect outlet and inlet radiator hose.
47. Install air cleaner assembly.
48. Install air cleaner hose assembly.
49. Install front fender apron seal rear LH.
50. Install front fender apron seal LH.
51. Install front fender apron seal rear RH.
52. Install front fender apron seal RH.
53. Install No. 1 engine under cover.
54. Install V-bank cover sub-assembly.
55. Add engine coolant.
56. Inspect for exhaust gas leak.

➡If gas is leaking, tighten the areas necessary to stop the leak. Replace damaged parts as necessary.

FLEXPLATE

REMOVAL & INSTALLATION

See Figures 142 through 144.

1. Before servicing the vehicle, refer to the precautions in the beginning of this section.
2. Remove the transmission.
3. Remove the bolts and the Flexplate.

To install:

4. Install the flexplate.
5. Coat two or three threads of each mounting bolt with Part No. 008833-00070 Three Bond 1324 or equivalent.

➡Do not reuse the flywheel installation bolts

6. Tighten the bolts in sequence as follows:
    a. For V6 engine, 61 ft. lbs. (83 Nm)
    b. For 4.7L V8 engine, 36 ft. lbs. (49 Nm) Plus 90°.

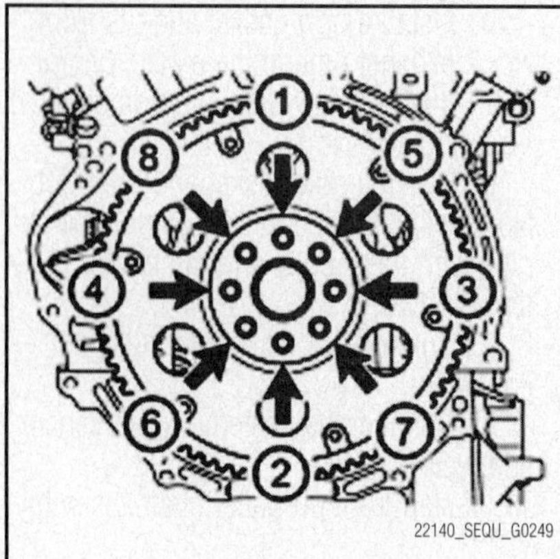

Fig. 142 Flexplate torque sequence V6 engine

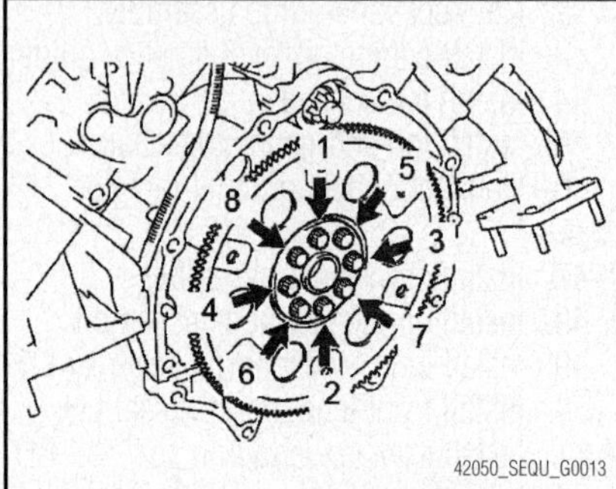

Fig. 143 Flexplate torque sequence 4.7L V8 engine

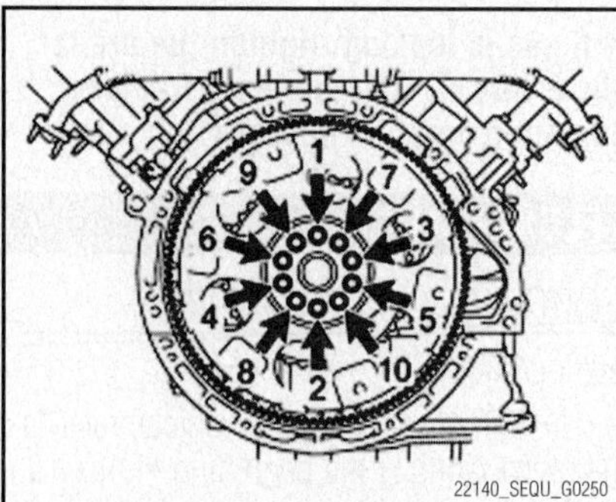

Fig. 144 Flexplate torque sequence 5.7L V8 engine

c. For 5.7L V8 engine, 22 ft. lbs. (30 Nm) plus 90°.

7. Install the transmission.

INTAKE MANIFOLD

REMOVAL & INSTALLATION

V6 Engine

See Figure 145.

1. Before servicing the vehicle, refer to the Precautions Section.
2. Properly relieve the fuel system pressure.
3. Properly disarm the SRS system.
4. Disconnect the negative battery cable.
5. Drain the engine coolant. Remove the air cleaner assembly.
6. Disconnect the fuel injector wiring connectors.
7. Remove the necessary components in order to gain access to the intake manifold retaining bolts.
8. Remove the intake manifold retaining bolts. Remove the intake manifold from the engine.

To install:

9. Clean all surfaces.
10. Install a new gasket on each cylinder head.

➡Align the ports of the gasket and the cylinder head. Be careful of the installation direction. Position the intake manifold to the engine.

11. Install the retaining bolts and tighten to specification, in an alternating sequence.
12. Continue the installation in the reverse order of the removal procedure.
13. Be sure to fill the cooling system with the proper grade and type engine coolant.
14. Start the engine and check for leaks. Correct as required.

4.7L V8 Engine

See Figures 147 and 148.

1. Before servicing the vehicle, refer to the precautions in the beginning of this section.
2. Drain the cooling system.
3. Relieve the fuel system pressure.
4. Remove or disconnect the following:
   - Negative battery cable
   - Engine appearance cover
   - Accelerator cable
   - Throttle Position (TP) sensor connector
   - Accelerator pedal position sensor
   - Throttle motor connector
   - Evaporative Emissions (EVAP) vacuum switching valve connector
   - Fuel injector connectors
   - Engine Coolant Temperature (ECT) sensor connector
   - ETC gauge sender connector
   - Heated Oxygen (HO_2S) sensor connectors
   - Fuel pressure regulator vacuum hose
   - Positive Crankcase Ventilation (PCV) valve and hose
   - EVAP hoses
   - Power steering vacuum hoses
   - Water bypass hose
   - Engine control wiring harness clamps
   - Cylinder head ground cables
   - Intake manifold wire harness protector
   - EVAP pipe
   - Engine appearance cover brackets
   - Intake manifold

To install:

5. Install or connect the following:
   - Intake manifold. Tighten the fasteners to 13 ft. lbs. (18 Nm).
   - Engine appearance cover brackets
   - EVAP pipe
   - Intake manifold wire harness protector
   - Cylinder head ground cables
   - Engine control wiring harness clamps
   - Water bypass hose
   - Power steering vacuum hoses
   - EVAP hoses
   - PCV valve and hose
   - Fuel pressure regulator vacuum hose
   - HO_2S sensor connectors
   - ETC gauge sender connector
   - ECT sensor connector
   - Fuel injector connectors
   - EVAP vacuum switching valve connector
   - Throttle motor connector
   - Accelerator pedal position sensor
   - TP sensor connector
   - Accelerator cable
   - Engine appearance cover
   - Negative battery cable
6. Fill the cooling system.
7. Start the engine and check for leaks.

5.7L V8 Engine

See Figures 148 through 152.

1. Remove the front wiper motor and link.
2. Disconnect the 2 washer hoses.
3. Remove the 7 bolts and outer panel.
4. Remove No. 1 engine under cover.
5. Drain engine coolant.
6. Remove V-bank cover sub-assembly.
7. Remove air cleaner hose assembly.
8. Disconnect the ventilation hose from the ventilation pipe of the cylinder head cover LH and RH.
9. Disconnect the 2 water by-pass hoses.
10. Disconnect the throttle body connector.
11. Disconnect the No. 1 ventilation hose.

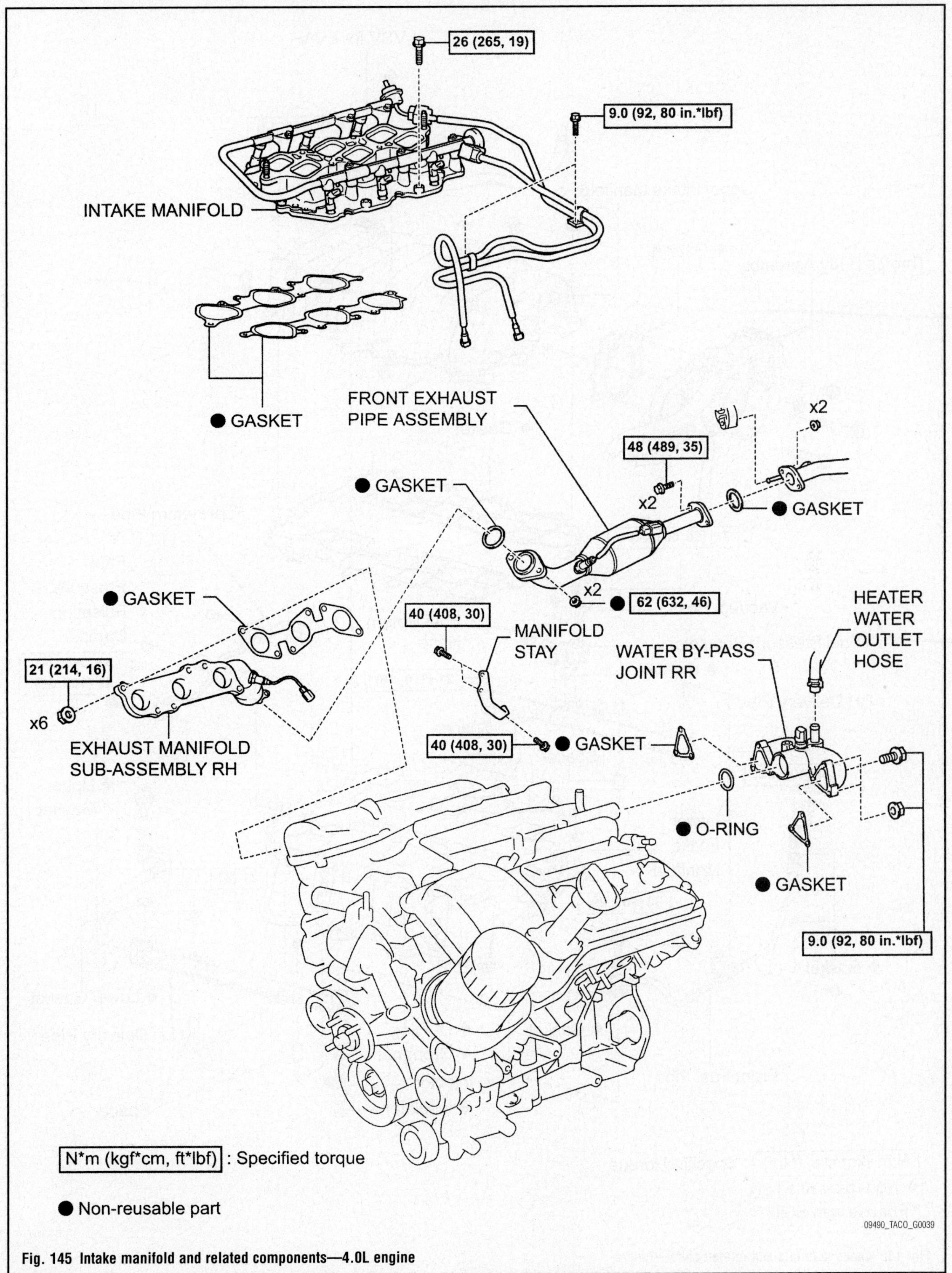

Fig. 145 Intake manifold and related components—4.0L engine

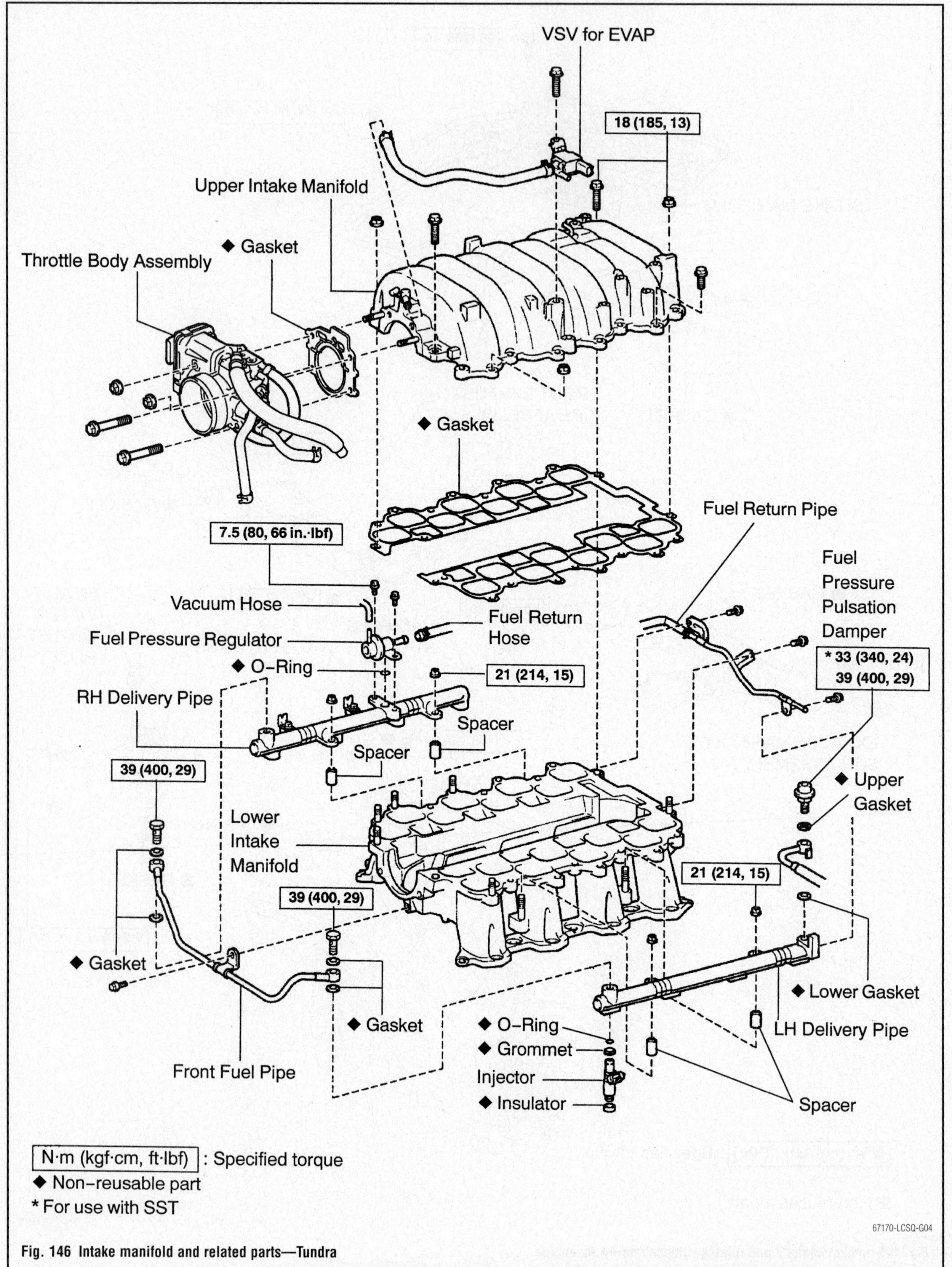

Fig. 146 Intake manifold and related parts—Tundra

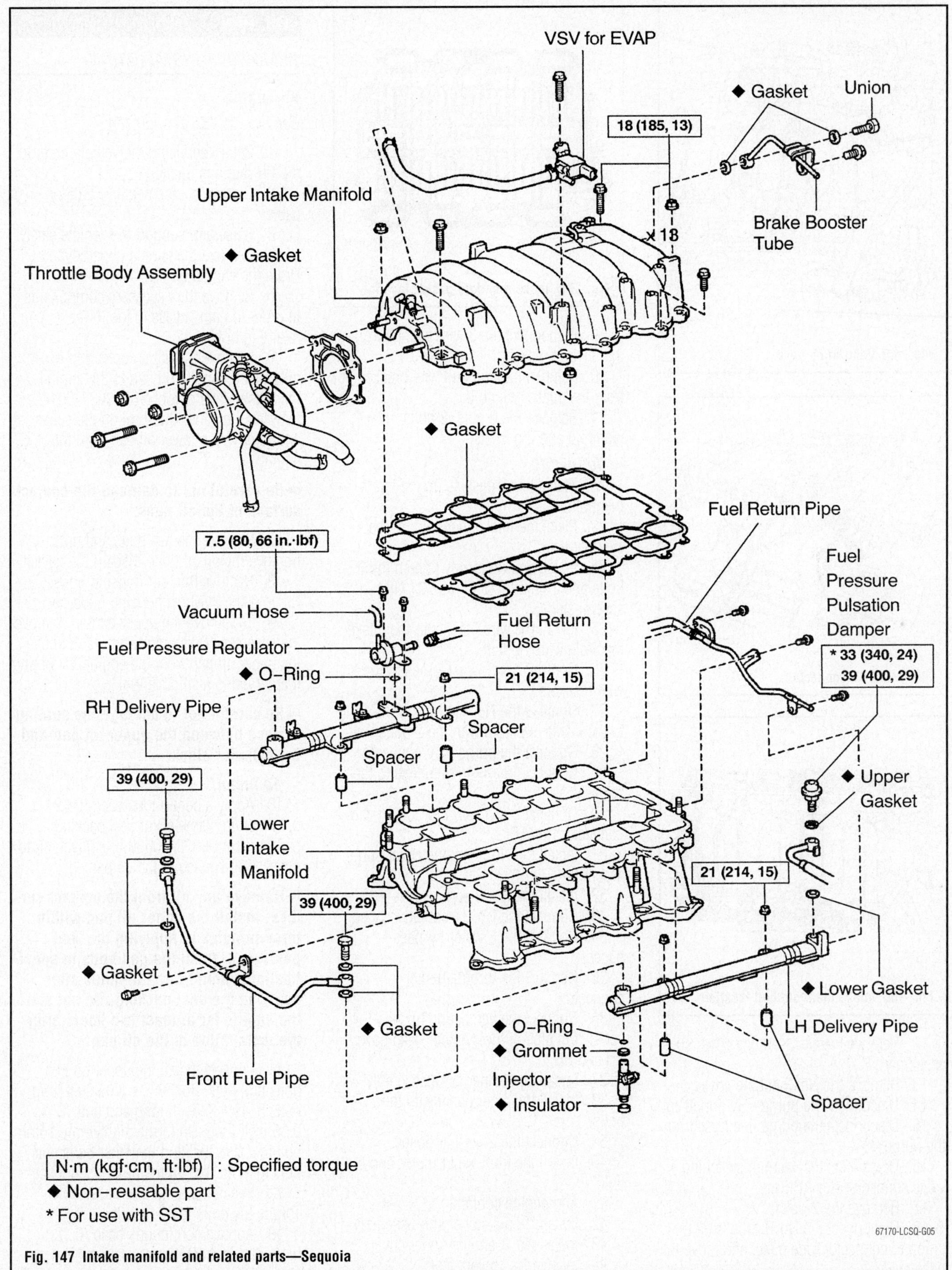

Fig. 147 Intake manifold and related parts—Sequoia

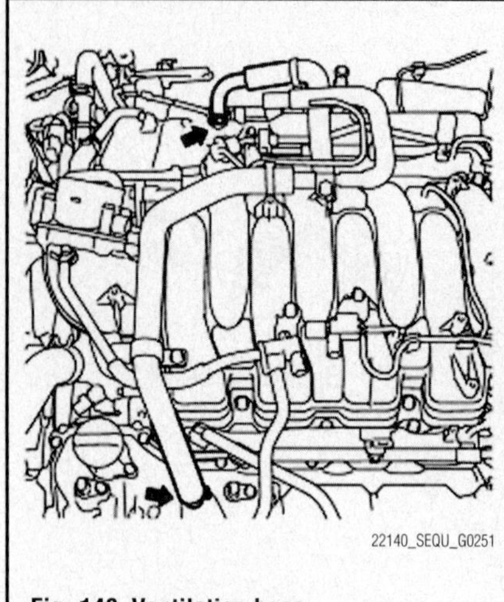

Fig. 148 Ventilation hose

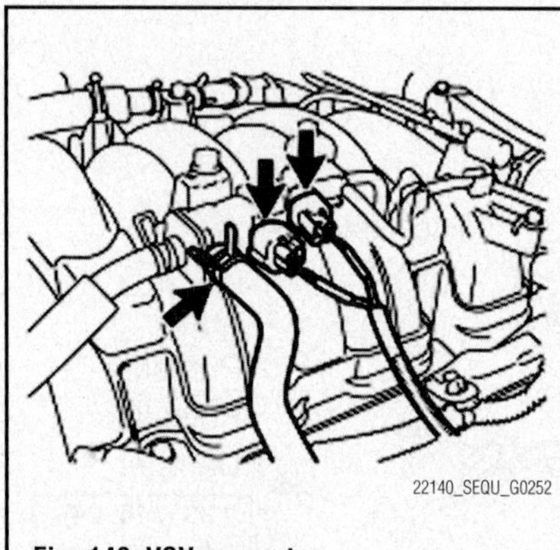

Fig. 149 VSV connector

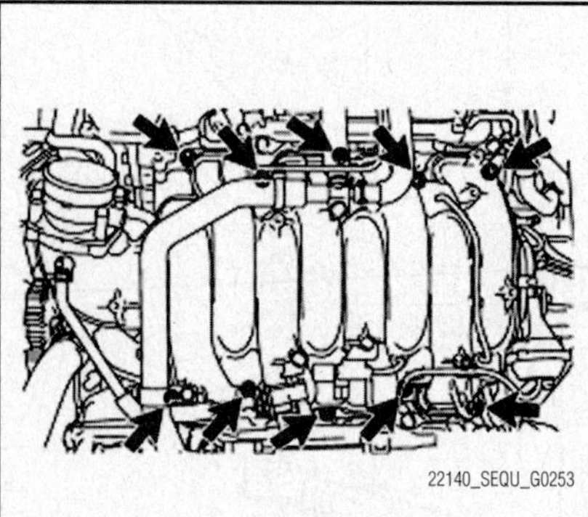

Fig. 150 Intake manifold bolt location

12. Remove the No. 1 engine cover sub-assembly.
13. Remove the No. 3 engine cover.
14. Disconnect the purge VSV connector.
15. Disconnect the purge line hose from the purge VSV.
16. Disconnect the vacuum switching valve connector (for ACIS).
17. Remove the 2 bolts.
18. Disconnect the No. 1 tube from the union to connector tube hose, and move the hose aside.

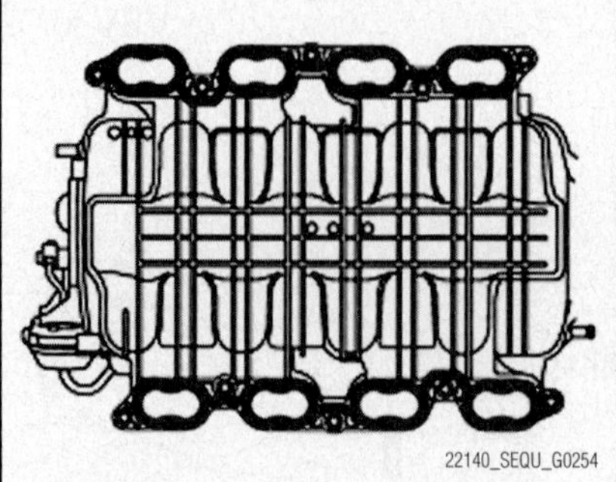

Fig. 151 Intake manifold gasket location

19. Disconnect the 3 wire clamps from the 3 wire brackets.
20. Remove the bolt and wire bracket from the intake manifold.
21. Remove the 2 nuts, 8 bolts, intake manifold and 2 gaskets.

To install:

22. Place 2 new gaskets on the intake manifold.
23. Place the intake manifold on the cylinder head.
24. Install and uniformly tighten the 8 bolts and 2 nuts to 15 ft. lbs. (21 Nm) in several steps.
25. Install the wire bracket to the intake manifold with the bolt.
26. Connect the 3 wire clamps to the 3 wire brackets.
27. Connect the No. 1 tube and install it to the intake manifold with the 2 bolts.
28. Connect the purge VSV connector.
29. Connect the purge line hose to the purge VSV.
30. Connect the vacuum switching valve connector (for ACIS).
31. Install the No. 1 engine cover sub-assembly.
32. Install the No. 3 engine cover.
33. Connect the No. 1 ventilation hose.
34. Connect the 2 water by-pass hoses.
35. Connect the throttle body connector.
36. Connect the ventilation hose to the ventilation pipe of the cylinder head cover LH and RH.
37. Install air cleaner hose assembly.
38. Install the outer panel with the 7 bolts.
39. Connect the 2 washer hoses.
40. Install the front wiper motor and link.
41. Add engine coolant.
42. Install V-bank cover sub-assembly.
43. Install No. 1 engine under cover.
44. Inspect for coolant leak.

OIL PAN

REMOVAL & INSTALLATION

V6 Engine

See Figures 152 through 154.

1. Before servicing the vehicle, refer to the Precautions Section.
2. Disconnect the negative battery cable.
3. Raise and support the vehicle safely.
4. Remove the engine undercover. Drain the engine oil.
5. Remove the necessary components in order to gain access to the lower oil pan retaining bolts.
6. Remove the fifteen bolts and two nuts that retain the oil pan to the engine. Insert the blade of tool SST09032-00100 between the pans. Cut through the sealer and remove the lower oil pan from the engine.

➡Be careful not to damage the contact surface of the oil pans.

7. Remove the two bolts and nuts. Remove the oil strainer. Discard the gasket.
8. Remove the four housing bolts. Remove the flywheel housing undercover.
9. To remove the upper oil pan, remove the seventeen bolts and two nuts. Remove the upper oil pan from the engine, by prying it apart using a suitable tool.

➡Be careful not to damage the sealing surface between the upper oil pan and the cylinder block.

To install:

10. Apply a continuous bead (0.12–0.16 inch in diameter) of seal packing, part number 08826-00080 or equivalent, to the sealing surface of the oil pan.

➡Remove any oil from the contact surface. Install the upper oil pan within three minutes of applying the seal packing. Tighten the pan bolts to specification within fifteen minutes after applying the seal packing. Do not start the engine for at least two hours after the installation of the oil pan.

11. Loosely install the upper oil pan bolts and nuts. Bolt "A" is 0.98 inch long, bolt "B" is 1.77 inch long and bolt "C" is 0.55 inch long. Uniformly tighten the 14mm bolt to 7.0 ft. lbs, and the other bolts and nuts to 17 ft. lbs., in the proper sequence.
12. Install the oil strainer assembly. Torque the bolts to 80 inch lbs.
13. Apply a continuous bead (0.12–0.16 inch in diameter) of seal packing,

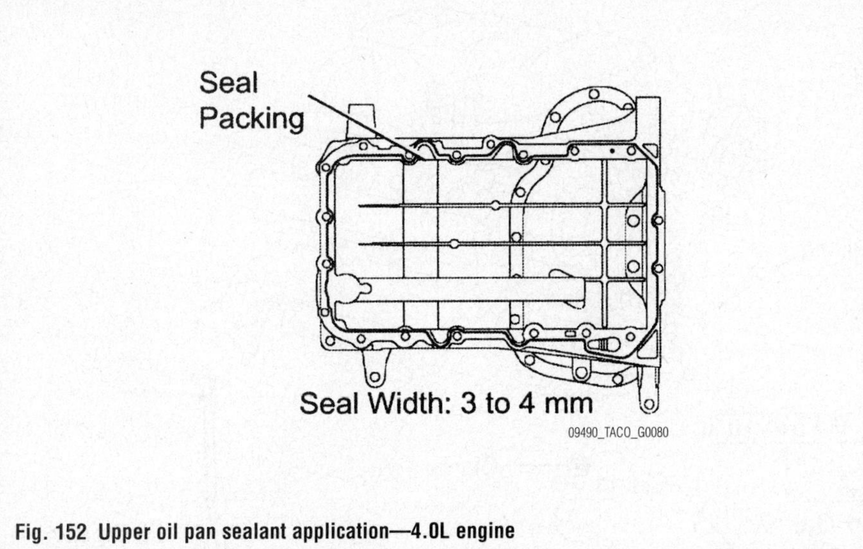

Fig. 152 Upper oil pan sealant application—4.0L engine

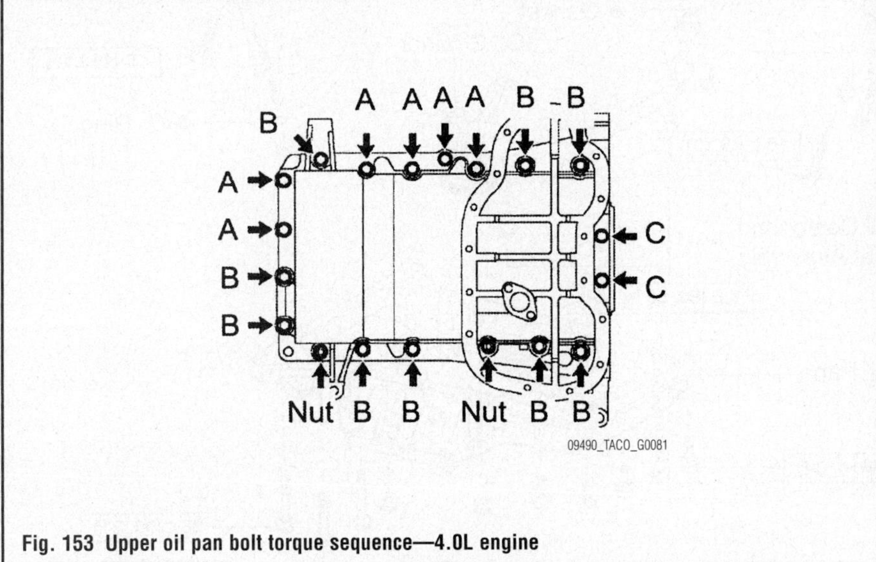

Fig. 153 Upper oil pan bolt torque sequence—4.0L engine

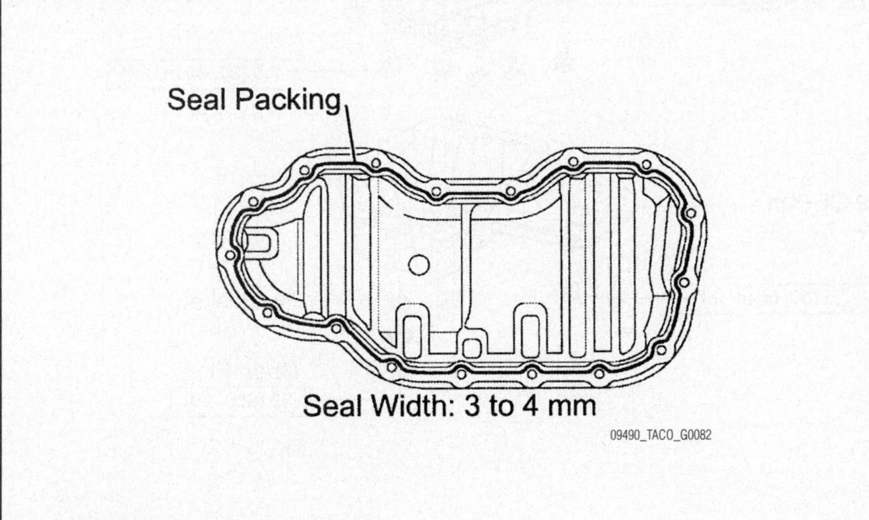

Fig. 154 Lower oil pan sealant application—4.0L engine

part number 08826-00080 or equivalent, to the sealing surface of the oil pan.

➡Remove any oil from the contact surface. Install the lower oil pan within three minutes of applying the seal packing. Tighten the pan bolts to specification within fifteen minutes after applying the seal packing. Do not start the engine for at least two hours after the installation of the oil pan.

14. Loosely install the lower oil pan bolts and nuts. Uniformly tighten the bolts to 80 inch lbs, and the nuts to 7.0 ft. lbs., in several steps.
15. Continue the installation in the reverse order of the removal procedure.
16. Be sure to fill the engine with the proper grade and type engine coolant.
17. Be sure to fill the engine with the proper grade and type engine oil.
18. Start the engine and check for leaks. Correct as required.

4.7L V8 Engine

See Figures 155 through 159.

1. Before servicing the vehicle, refer to the precautions in the beginning of this section.
2. Remove the engine from the vehicle and mount it on a stand.
3. Remove or disconnect the following:
   - Oil dipstick tube
   - Lower oil pan
   - Oil pan baffle
   - Upper oil pan

To install:

4. The upper oil pan bolts are different lengths and are identified as follows:
   - A: 0.79 inch (20mm) w/10mm head
   - B: 0.98 inch (25mm) w/12mm head
   - C: 2.36 inch (60mm) w/12mm head
   - D: 1.38 inch (35mm) w/10mm head
5. Apply silicone sealant to the upper oil pan as shown.
6. Install the upper oil pan and tighten the fasteners in several passes to the following specifications:
   - 10mm: 66 inch lbs. (7.5 Nm)
   - 12mm: 21 ft. lbs. (28 Nm)
7. Install or connect the following:
   - Oil pan baffle. Tighten the fasteners to 66 inch lbs. (7.5 Nm).
   - Lower oil pan. Tighten the fasteners in several passes to 66 inch lbs. (7.5 Nm).
   - Oil dipstick tube
8. Install the engine.

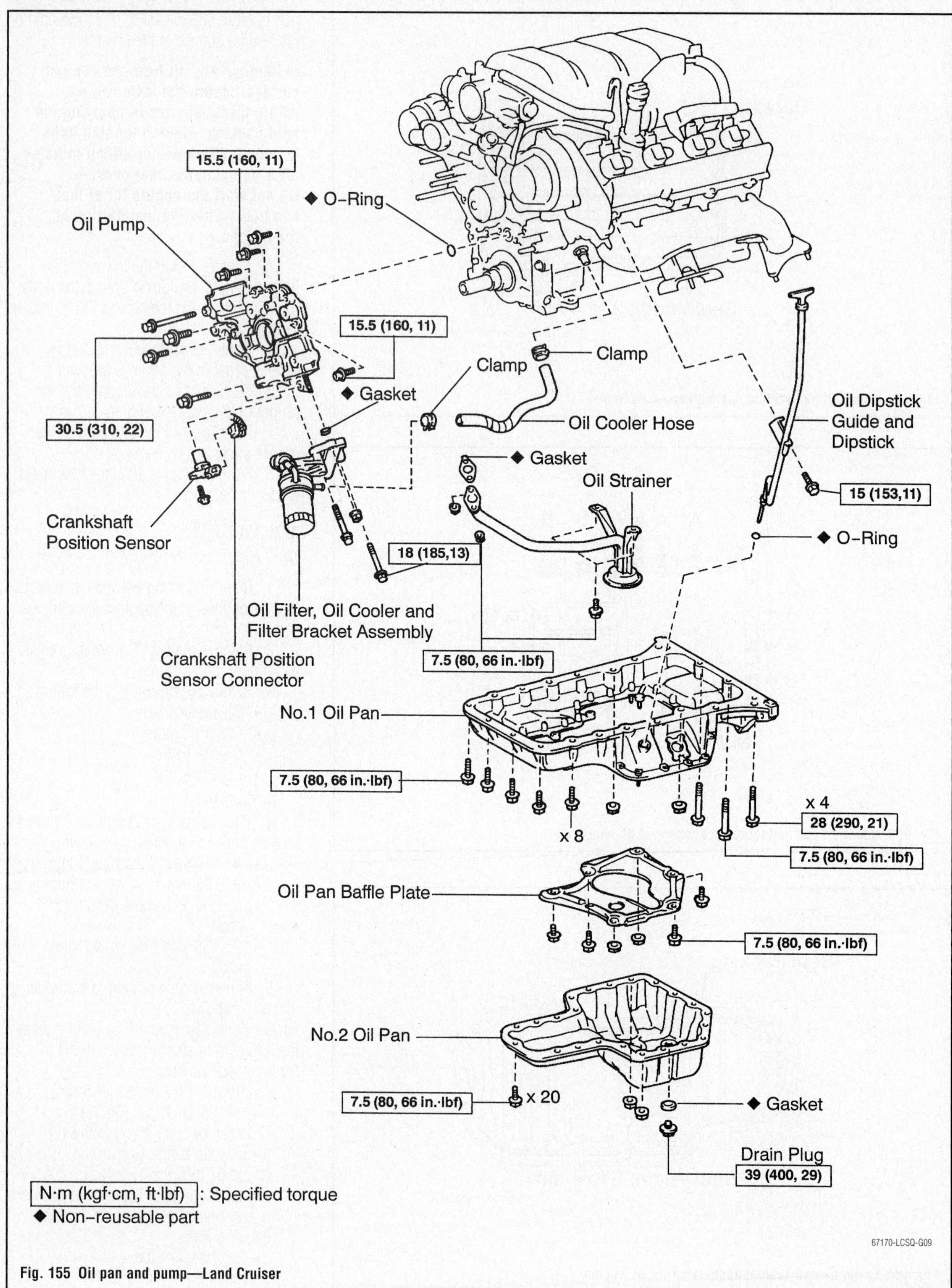

Fig. 155 Oil pan and pump—Land Cruiser

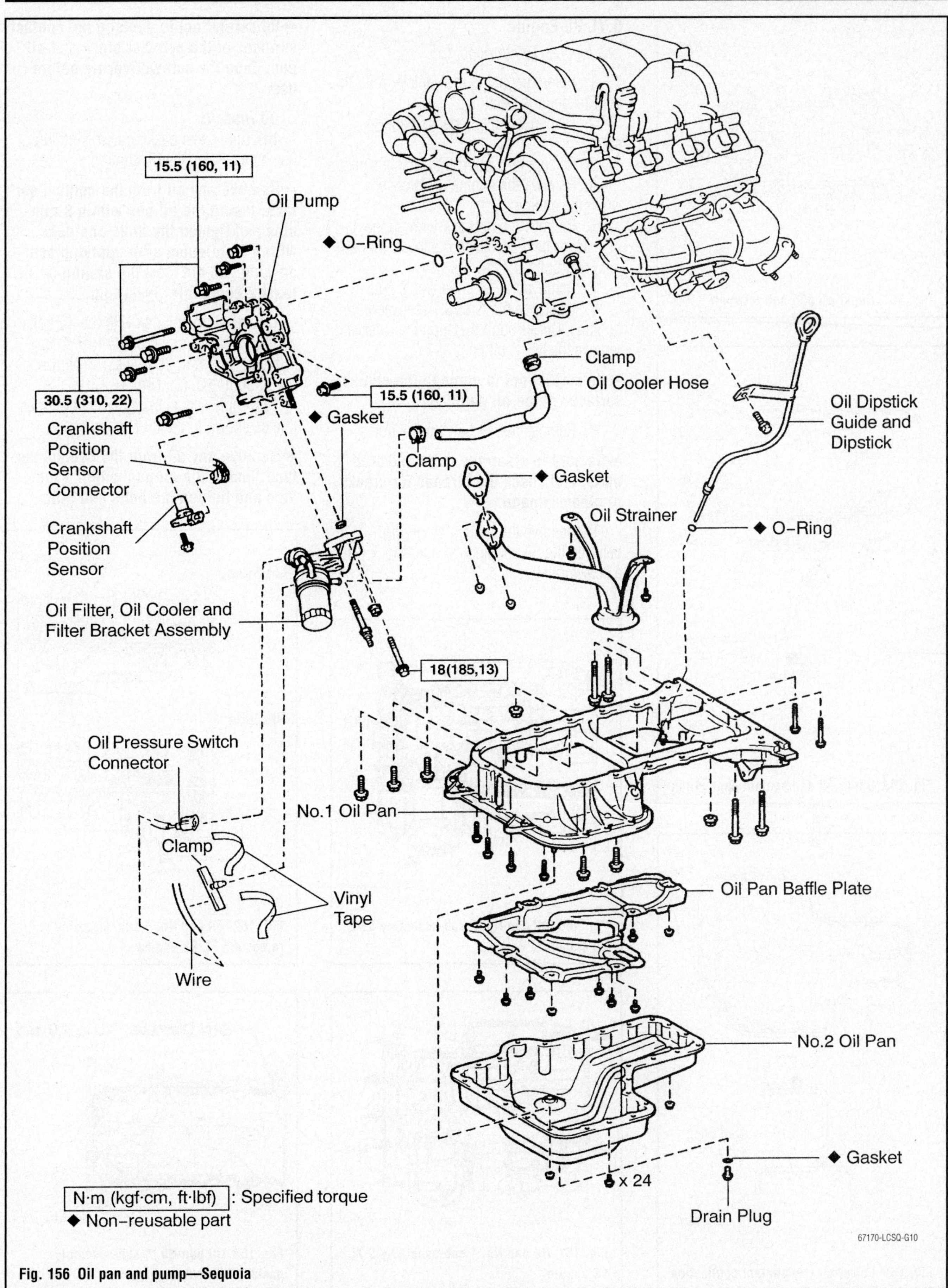

Fig. 156 Oil pan and pump—Sequoia

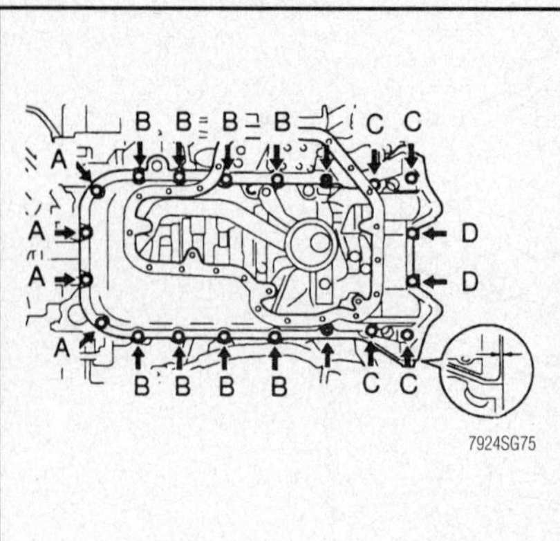

Fig. 157 Upper oil pan bolt location

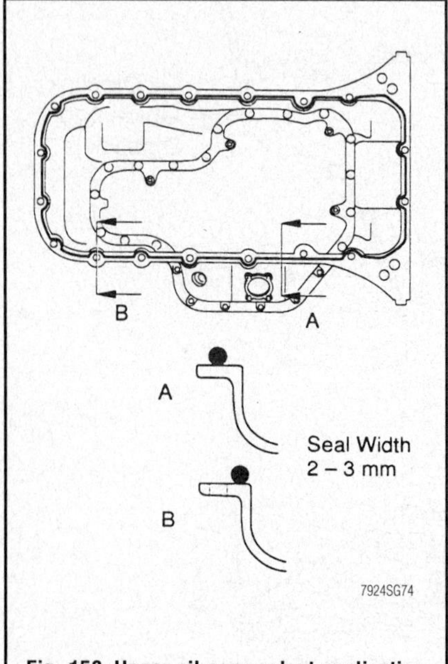

Fig. 158 Upper oil pan sealant application

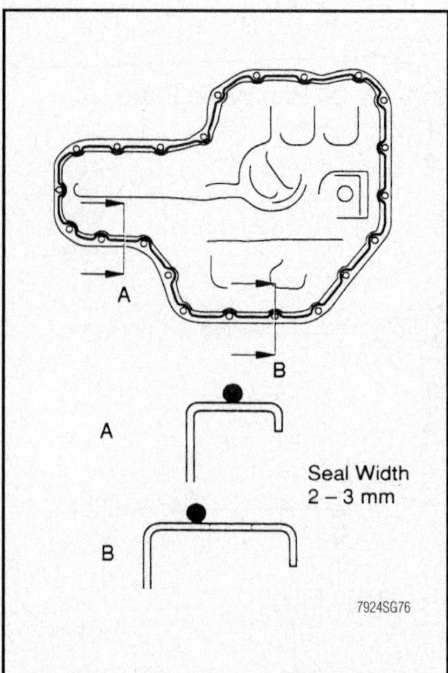

Fig. 159 Lower oil pan sealant application

5.7L V8 Engine

See Figures 160 through 166.

1. Before servicing the vehicle, refer to the Precautions Section.
2. Disconnect the negative battery cable.
3. Raise and support the vehicle safely.
4. Remove the engine undercover. Drain the engine oil.
5. Remove the necessary components in order to gain access to the lower oil pan retaining bolts.
6. Remove the 14 bolts and 2 nuts.
7. Insert the blade of SST between the oil pans. Cut through the applied sealer and remove the No. 2 oil pan.

➡Be careful not to damage the contact surfaces of the oil pans.

8. Remove the 14 bolts and 2 nuts.

➡Be sure to clean the bolts and stud bolts, and check the threads for cracks or other damage.

9. Remove the oil pan by prying between the oil pan and cylinder block with a screwdriver.

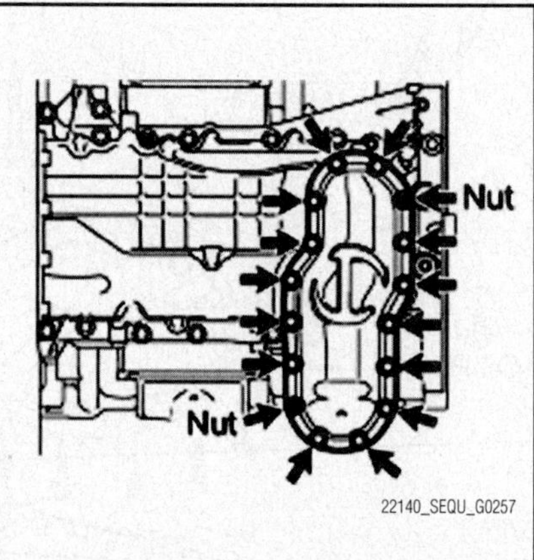

Fig. 160 Oil pan No. 2 sub-assembly 5.7L V8 engine

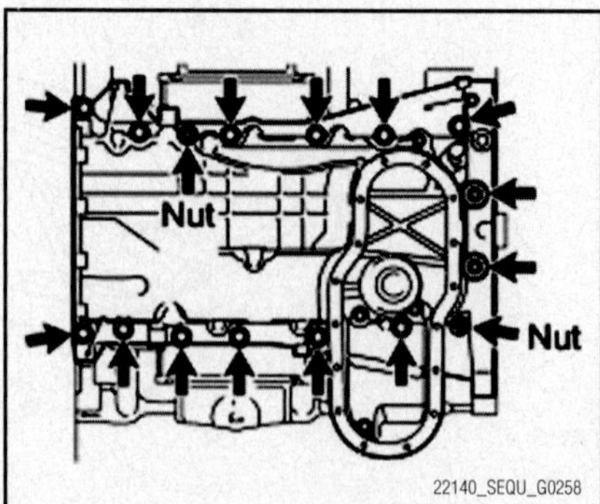

Fig. 161 Oil pan No. 1 sub-assembly 5.7L V8 engine

➡Be careful not to damage the contact surfaces of the cylinder block and oil pan. Tape the screwdriver tip before use.

To install:

10. Apply seal packing in a continuous line as shown in the illustration.

➡Remove any oil from the contact surface. Install the oil pan within 3 minutes and tighten the bolts and nuts within 15 minutes after applying seal packing. Do not start the engine for at least 2 hours after installing.

11. Install the oil pan with the 14 bolts and 2 nuts and tighten bolt A to 7 ft. lbs. (10 Nm), bolt B to 26 ft. lbs. (35 Nm) and nut to 26 ft. lbs. (35 Nm).
12. Apply seal packing in a continuous line as shown in the illustration.

➡Remove any oil from the contact surface. Install the oil pan within 3 minutes and tighten the bolts and nuts

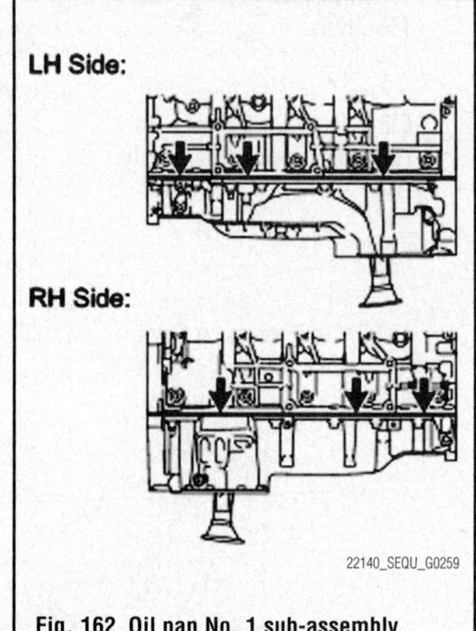

Fig. 162 Oil pan No. 1 sub-assembly removal 5.7L V8 engine

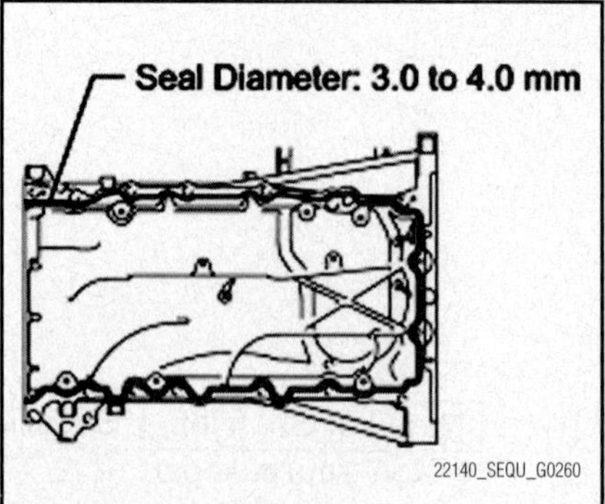

Fig. 163 Oil pan No. 1 sub-assembly packing 5.7L V8 engine

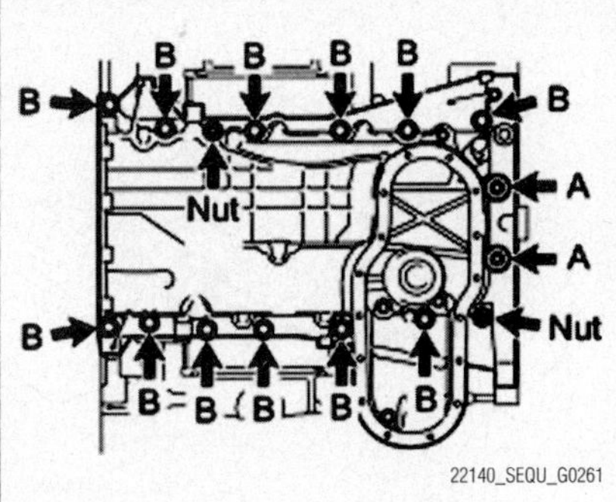

Fig. 164 Oil pan No. 1 sub-assembly tightening sequence 5.7L V8 engine

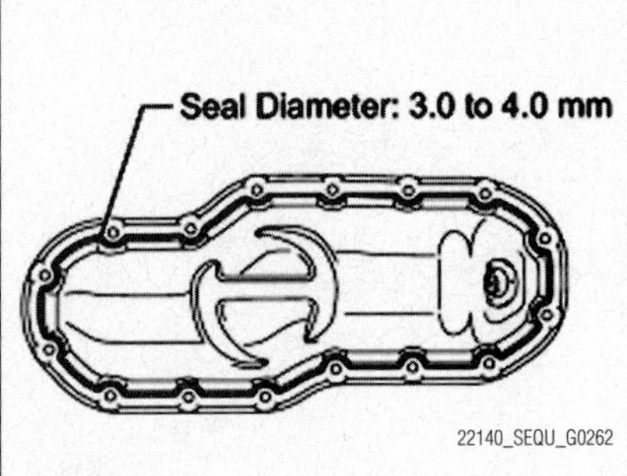

Fig. 165 Oil pan No. 2 sub-assembly packing 5.7L V8 engine

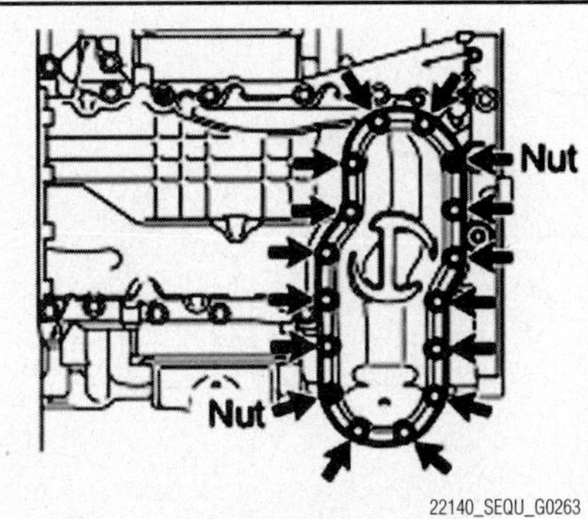

Fig. 166 Oil pan No. 1 sub-assembly tightening sequence 5.7L V8 engine

within 15 minutes after applying seal packing. Do not start the engine for at least 2 hours after installing.

13. Install the oil pan with the 14 bolts and 2 nuts and tighten bolts to 7 ft. lbs. (10 Nm).
14. Continue the installation in the reverse order of the removal procedure.
15. Be sure to fill the engine with the proper grade and type engine coolant.
16. Be sure to fill the engine with the proper grade and type engine oil.
17. Start the engine and check for leaks. Correct as required.

OIL PUMP

REMOVAL & INSTALLATION

V6 Engine

See Figures 167 through 171.

1. Before servicing the vehicle, refer to the Precautions Section.
2. Properly relieve the fuel system pressure.
3. Disconnect the negative battery cable. Disconnect the positive battery cable. Remove the battery.
4. Drain the engine coolant. Remove the radiator.
5. Remove the air cleaner assembly.
6. Separate the vane pump assembly. Remove the alternator. Remove the air conditioning compressor and position it to the side.
7. Remove the belt tensioner assembly. Remove the idler pulley number two sub-assembly. Remove the idler pulley number one subassembly. Remove the crankshaft pulley.
8. Remove the lower oil pan. Remove the oil strainer and pickup tube assembly. Remove the upper oil pan.
9. Remove the intake manifold. Remove the cylinder head cover assembly.
10. Remove the camshaft timing oil control valve assembly. Remove the VVT sensor. Remove the oil filter bracket sub-assembly.
11. Remove the timing chain case cover retaining bolts. Remove the cover from the engine. Remove the O-ring from the left cylinder head.

➡Carefully remove the cover by prying between the cover and the cylinder head or block with a suitable tool. Be sure to cover the tip of the suitable tool prior to usage. Be careful not to damage the contact surfaces of the cylinder block, cylinder head and timing chain cover.

➡The oil pump gears are located inside the timing case cover.

12. Remove the three bolts and remove the oil pipe. Remove the two O-rings. Remove the seven oil pump cover bolts. Remove the gears from the timing chain case cover. Remove the oil pump relief valve.

To install:

13. Coat the oil pump relief valve with clean engine oil. Install the plug, using a new gasket and tighten to 36 ft. lbs.
14. Coat the oil pump gears with clean engine oil. Position the gears in the timing chain case cover with the identification marks facing oil pump cover side. Install the cover. Alternately tighten the bolts to 80 inch lbs. Install the oil pipe, tighten the bolts to 80 inch lbs.
15. Install a new front case oil seal. Install a new O-ring onto the left cylinder head.
16. Apply continuous beads (0.12–0.16 inch in diameter) of seal packing, part number 08826-00080 or equivalent to the four locations shown in the illustration.
17. Apply continuous beads (0.12–0.16 inch in diameter) of seal packing, part number 08826-00080 or equivalent to all parts except the water pump part: for the water pump part use, part number 08826-00080 or equivalent, to the timing chain cover. Do not apply seal packing to portion "A" in the illustration.

➡Remove any oil from the contact surfaces. Install the timing chain case cover within three minutes and tighten the bolts within fifteen minutes of applying the seal packing.

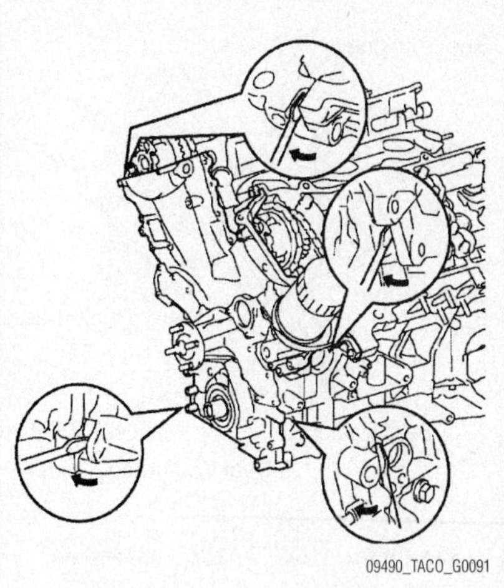

Fig. 167 Timing chain case cover removal points—4.0L engine

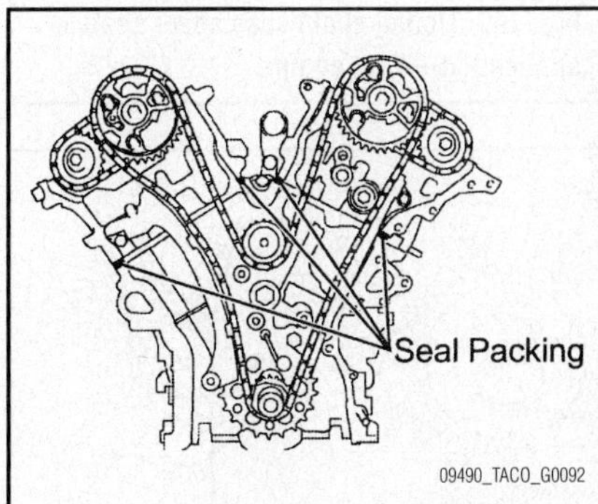

Fig. 168 Timing chain case cover seal packing locating points—4.0L engine

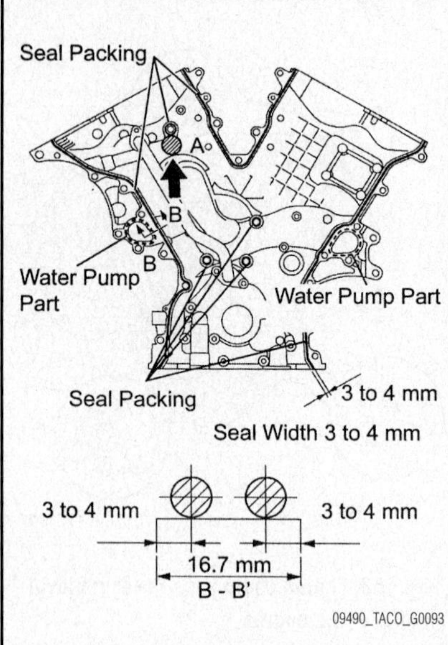

Fig. 169 Timing chain case cover sealant application—4.0L engine

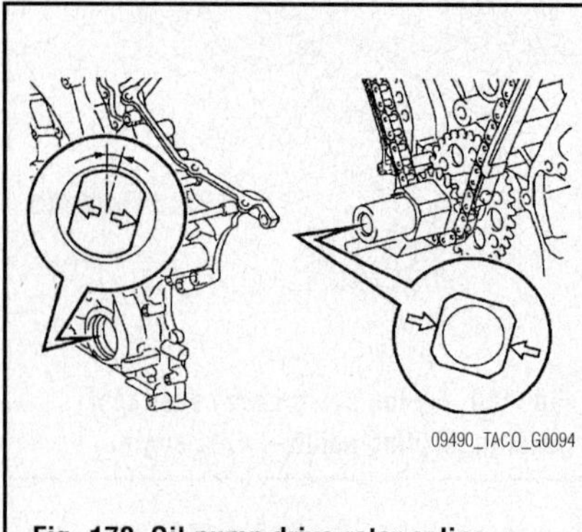

Fig. 170 Oil pump drive rotor spline alignment—4.0L engine

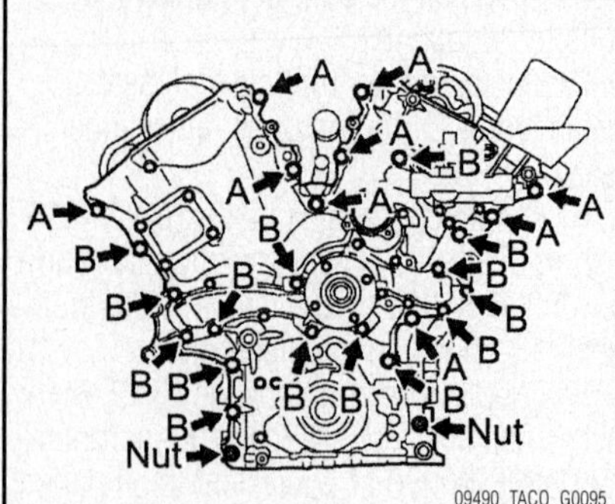

Fig. 171 Timing chain case cover bolt location and torque sequence—4.0L engine

18. Align the key way of the oil pump drive motor with the rectangular portion of the crankshaft timing gear and slide the timing chain case cover into place.

19. Install the timing chain case cover bolts. Tighten the bolts and nuts uniformly in several steps to 17 ft. lbs.

➡Do not wrap the chain and slipper over the timing chain case cover seal line.

20. Continue the installation in the reverse order of the removal procedure.

21. When installing the cylinder head cover apply a continuous bead (0.08–0.12 inch) of seal packing, part number 08826-00080 or equivalent, to the cylinder head. Install the seal washers onto the bolts. Install the cylinder head cover bolts and nuts. Tighten bolts "A" to 7.4 ft. lbs. Tighten bolts "B" to 80 inch lbs. Tighten nuts to 80 inch lbs.

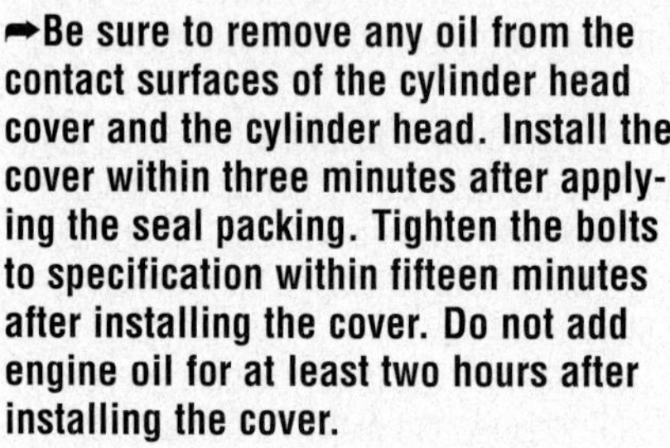

➡Be sure to remove any oil from the contact surfaces of the cylinder head cover and the cylinder head. Install the cover within three minutes after applying the seal packing. Tighten the bolts to specification within fifteen minutes after installing the cover. Do not add engine oil for at least two hours after installing the cover.

22. Be sure to fill the engine with the proper grade and type engine coolant.

23. Be sure to fill the engine with the proper grade and type engine oil.

24. Start the engine and check for leaks, correct as required.

4.7L V8 Engine

See Figures 172 through 174.

1. Before servicing the vehicle, refer to the precautions in the beginning of this section.
2. Remove the engine from the vehicle and mount it on a stand.
3. Remove or disconnect the following:
   - Front cover
   - Timing belt.
   - Timing belt idler pulleys
   - Crankshaft timing sprocket
   - Oil dipstick tube
   - Oil filter and bracket
   - Crankshaft Position (CKP) sensor
   - Oil pan and baffle
   - Oil pump pickup tube
   - Oil pump

To install:

4. The upper oil pan bolts are different lengths and are identified as follows:
   - A: 1.38 inch (35mm) w/12mm head
   - B: 1.97 inch (50mm) w/12mm head
   - C: 4.17 inch (106mm) w/12mm head

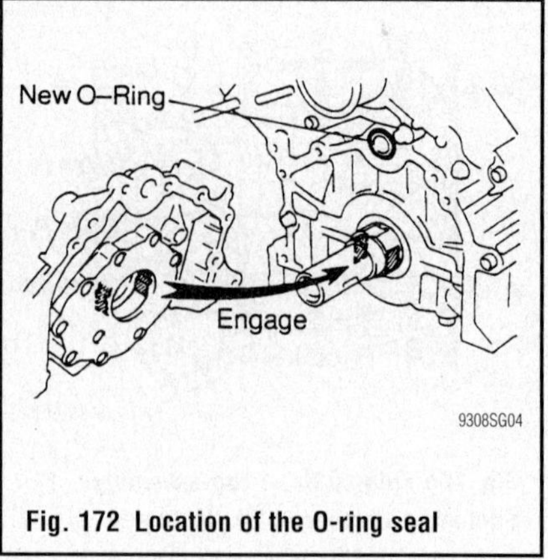

Fig. 172 Location of the O-ring seal

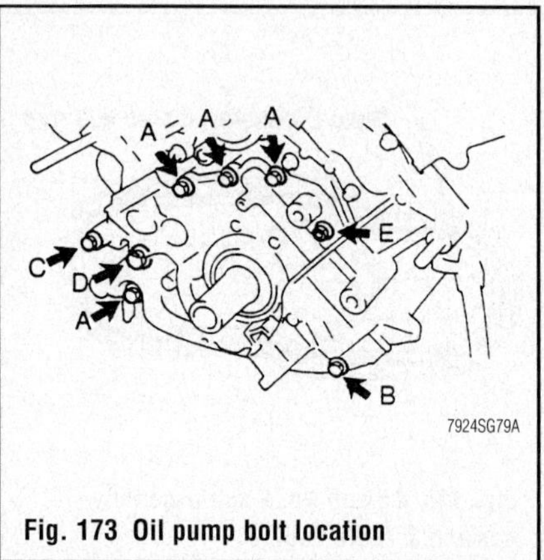

Fig. 173 Oil pump bolt location

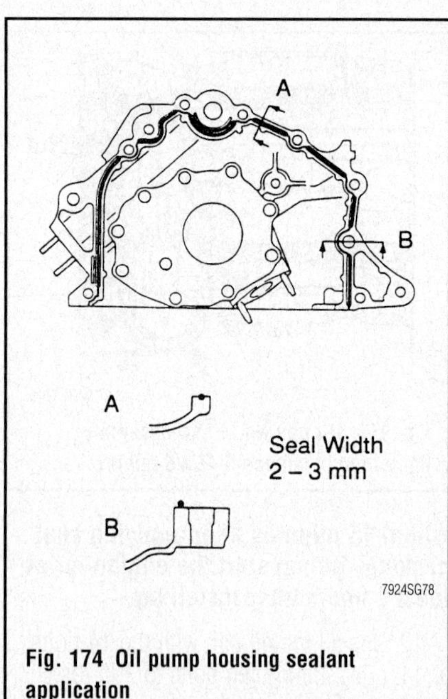

Fig. 174 Oil pump housing sealant application

   - D: 1.57 inch (40mm) w/14mm head
   - E: 1.18 inch (30mm) w/6mm hex head

5. Install a new O-ring on the engine block.
6. Apply silicone sealant to the oil pump housing as shown.

7. Install the oil pump. Tighten the bolts in several passes to the following specifications:
   - 12mm: 11 ft. lbs. (15.5 Nm)
   - 14mm: 22 ft. lbs. (30.5 Nm)
   - 6mm Hex: 11 ft. lbs. (15.5 Nm)
8. Install or connect the following:
   - Oil pump pickup tube. Tighten the bolts to 66 inch lbs. (7.5 Nm).
   - Oil pan and baffle
   - CKP sensor
   - Oil filter and bracket. Tighten the bolts to 13 ft. lbs. (18 Nm).
   - Oil dipstick tube
   - Crankshaft timing sprocket
   - Timing belt idler pulleys
   - Timing belt
   - Front cover
9. Install the engine

5.7L V8 Engine

See Figures 175 through 184.

1. Discharge fuel system pressure.
2. Disconnect cable from negative battery terminal.
3. Drain engine oil.
4. Drain engine coolant.
5. Remove the radiator.
6. Remove air cleaner hose assembly.
7. Remove air cleaner assembly.
8. Engine Room LH Side:
   a. Remove the engine room relay block cover.
   b. Disconnect the 2 connectors and detach the 2 clamps from the engine room junction block.
   c. Disconnect the 4 air injection control driver connectors.
   d. Disconnect the 2 wire harness clamps.
   e. Disconnect the injector connector.
   f. Disconnect the 4 ignition coil connectors.
   g. Disconnect the 2 VVT sensor connectors.
   h. Disconnect the 4 clamps.
   i. Remove the 2 bolts and ground wire.
   j. Disconnect the noise filter connector.
   k. Disconnect the engine coolant temperature sensor connector.
   l. Disconnect the 2 camshaft timing oil control valve connectors.
   m. Disconnect the camshaft position sensor connector.
   n. Disconnect the 3 clamps.
   o. Disconnect the cooler compressor connector.
9. Engine Room RH Side.
   a. Disconnect the 2 camshaft timing oil control valve connectors.
   b. Disconnect the 4 ignition coil connectors.
   c. Disconnect the injector connector.
   d. Disconnect the 2 VVT sensor connectors.
   e. Disconnect the noise filter connector.
   f. Remove the 2 bolts and ground wire.
   g. Disconnect the 2 air pump connectors.
   h. Disconnect the throttle position sensor and throttle control motor connector.
   i. Disconnect the 5 clamps.
   j. Disconnect the 2 clamps and power steering oil pressure switch connector.
   k. For 4WD: Disconnect the 2 clamps and power steering oil pressure switch connector.
10. Remove the intake manifold.
11. Disconnect the No. 2 and No. 3 air hoses and the 2 clamps.
12. Remove the 2 bolts and disconnect the heater 3 hoses.
13. Remove the 2 bolts, 2 nuts and 2 stud bolts, and disconnect the cooler compressor.

➡It is not necessary to completely remove the compressor. With the hoses connected to the compressor, hang the compressor on the vehicle body with a rope.

14. Remove the 2 bolts and disconnect the fuel tube.
15. Remove oil filter element.
16. Remove oil pressure sender gauge assembly.
17. Disconnect the 4 hoses and remove the water by-pass pipe.
18. Remove No. 1 oil cooler bracket (w/ oil cooler).
19. Remove the 2 bolts, 2 nuts and oil filter bracket.
20. Remove the 2 O-rings.
21. Remove engine oil level dipstick guide.
22. Remove the 2 bolts and disconnect the vane pump.
23. Disconnect the generator connector.
24. Remove the terminal cap and nut, and disconnect the generator wire.
25. Remove the bolt and disconnect the wire harness bracket from the generator.
26. Remove the 2 bolts, 2 nuts and generator.
27. Remove the 2 stud bolts.
28. Remove the No. 1 water by-pass hose by disconnect the hose from the water inlet housing and front water by-pass joint.
29. Remove the bolt and disconnect the air tube.
30. Remove the 2 bolts and hoses and water by-pass pipe.
31. Disconnect the No. 2 water by-pass hose from the water by-pass joint.
32. Remove the 4 nuts, water by-pass joint and 2 gaskets.
33. Remove No.1 and No. 2 engine covers.
34. Remove the bolt, disconnect the 2 hoses, and remove the air tube.
35. Remove the 3 bolts, water inlet housing and gasket.
36. Remove the 4 bolts and water pump pulley.
37. Remove the bolt and idler pulley.
38. Remove the 4 bolts and fan bracket.
39. Remove the standard bolt, 6 mm hexagon wrench bolt and belt tensioner.
40. Remove the 8 bolts and 8 ignition coils.
41. Remove cylinder head cover subassembly LH.
    a. Remove the 14 bolts, seal washer, cylinder head cover and gasket.
    b. Remove the 5 gaskets from the camshaft bearing caps (No. 2, No. 3).
42. Remove cylinder head cover subassembly RH.
    a. Remove the bolt and noise filter.

Fig. 175 Air injection control connectors

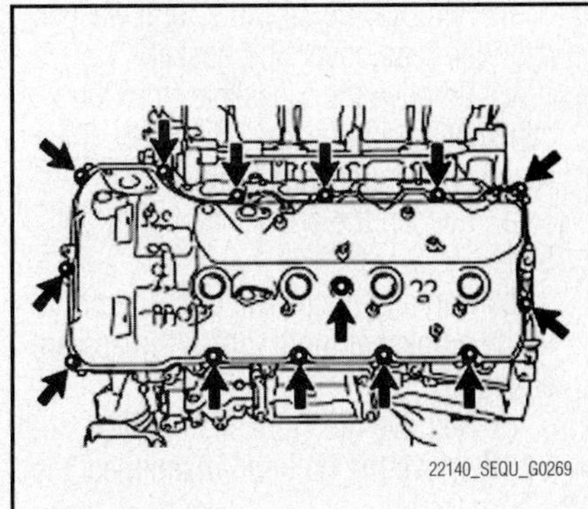

Fig. 176 Cylinder head cover LH

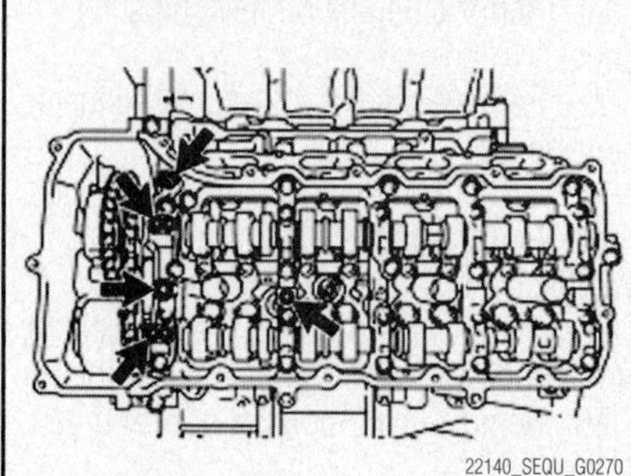

Fig. 177 Cylinder head cover gasket locations

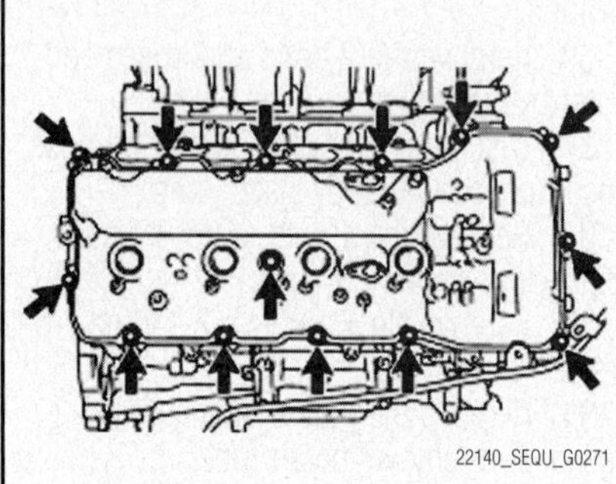

Fig. 178 Cylinder head cover RH

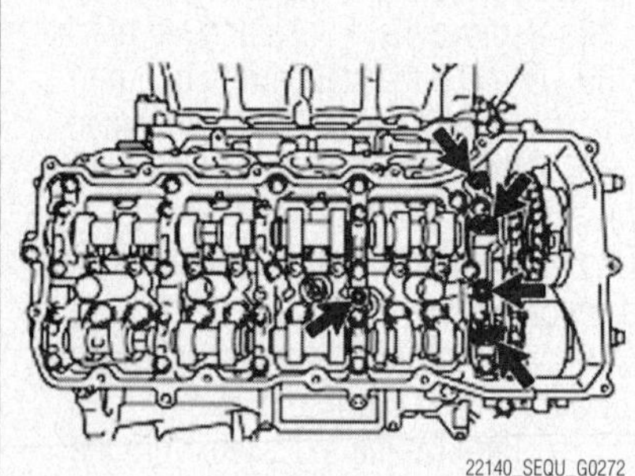

Fig. 179 Cylinder head cover gasket locations

b. Remove the 14 bolts, seal washer, cylinder head cover and gasket.

c. Remove the 5 gaskets from the camshaft bearing caps (No. 1, No. 3).

43. Remove crankshaft pulley.

a. Loosen the crankshaft pulley set bolt.

b. Partially install the pulley set bolt to the crankshaft until 2 or 3 threads are engaged.

c. Remove the crankshaft pulley.

44. Remove the bolt and disconnect the wire harness bracket.

45. Remove timing chain cover sub-assembly.

a. Remove the 28 bolts and nut shown in the Timing Chain and Sprockets procedure in this section.

b. Remove the timing chain cover, as outlined in the Timing Chain and Sprockets procedure in this section.

46. Remove the oil pump gasket from the cylinder block.

47. Remove the O-ring from the oil pan.

48. Remove the water inlet pipe.

49. Remove the 2 O-rings from the water inlet pipe.

To install:

50. Apply soapy water to 2 new O-rings and install them to the inlet pipe.

51. Install the inlet pipe to the No. 1 heat exchanger cover.

52. Install timing chain cover sub-assembly.

a. Apply a light coat of engine oil to a new oil pump gasket.

b. Install the oil pump gasket.

c. Apply a light coat of engine oil to a new O-ring.

d. Install the O-ring.

e. Apply seal packing in a continuous line to the timing chain cover as shown in the Timing Chain and Sprockets procedure in this section.

f. Align the oil pump's drive rotor spline and the crankshaft as shown in the illustration. Install the spline and chain cover to the crankshaft.

g. Temporarily install the timing chain cover with the 28 bolts and nut.

53. Tighten the 5 bolts in several steps in the sequence shown to 35 ft. lbs. (47 Nm).

a. Temporarily install the fan bracket with the 4 bolts.

b. Temporarily install the belt tensioner with the standard bolt and 6 mm hexagon wrench bolt.

c. Tighten the 23 bolts labeled 12 to 35 and nut in several steps in the sequence to 17 ft. lbs. (23 Nm).

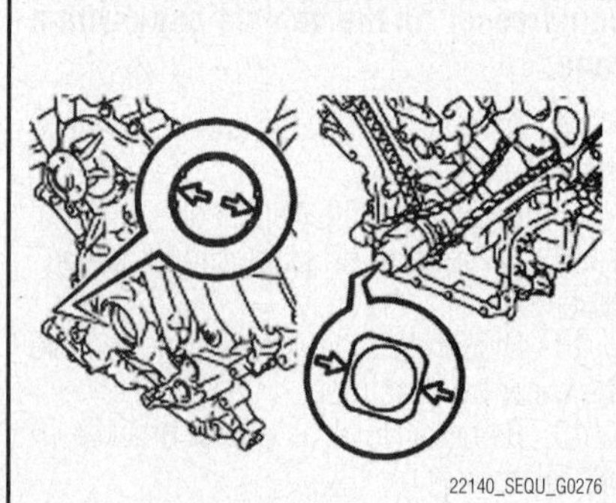

Fig. 180 Oil pump alignment

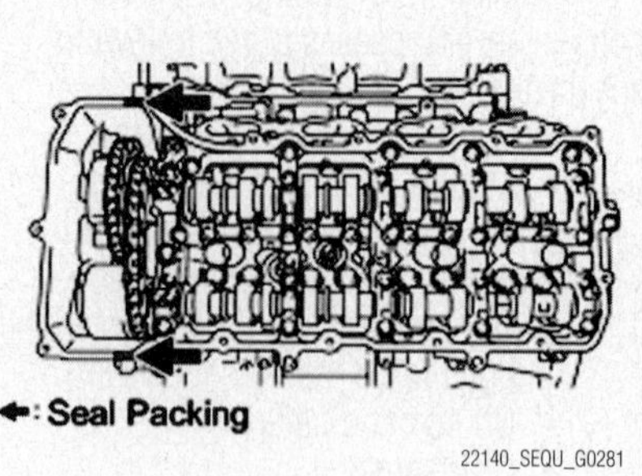

Fig. 181 Cylinder head cover packing location LH

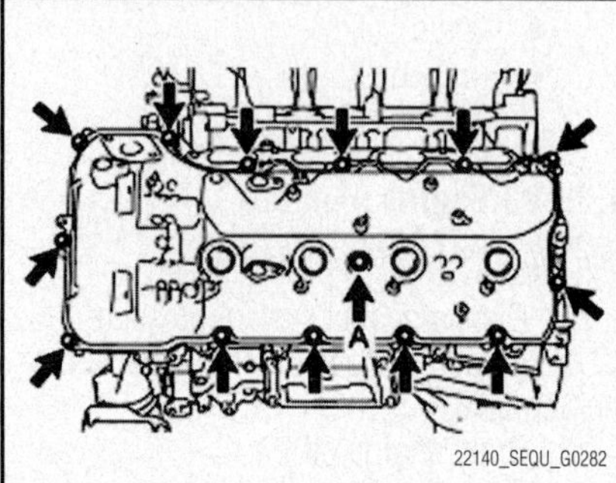

Fig. 182 Cylinder head cover bolt tightening sequence LH

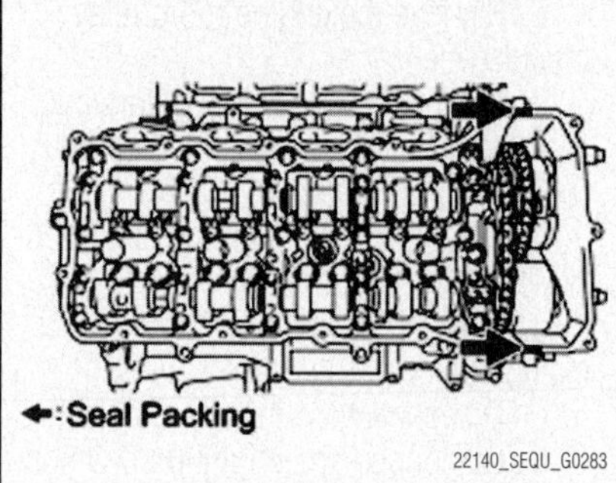

Fig. 183 Cylinder head cover packing location RH

54. Install cylinder head cover sub-assembly LH.

a. Install 5 new gaskets to the camshaft bearing caps (No. 2, No. 3).

b. Install the gasket to the cylinder head cover.

c. Apply seal packing as shown.

d. Install the cylinder head cover washer with a new seal and the 14 bolts and tighten bolt A to 15 ft. lbs. (21 Nm) and remaining bolts to 9 ft. lbs. (12 Nm).

55. Install cylinder head cover sub-assembly RH.

a. Install 5 new gaskets to the camshaft bearing caps (No. 1, No. 3).

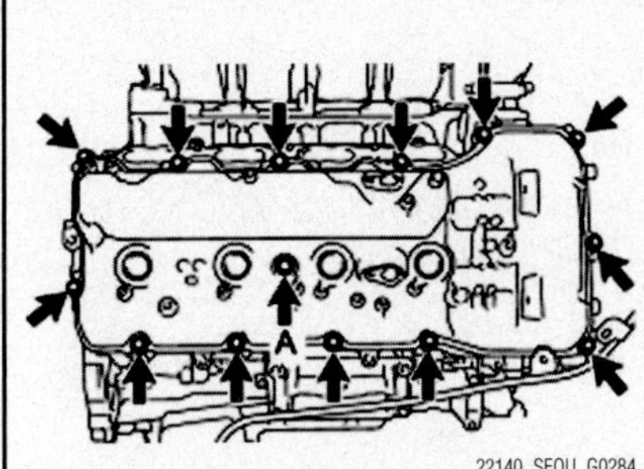

Fig. 184 Cylinder head cover bolt tightening sequence RH

b. Install the gasket to the cylinder head cover.

c. Apply seal packing as shown.

d. Install the cylinder head cover washer with a new seal and the 14 bolts and tighten bolt A to 15 ft. lbs. (21 Nm) and remaining bolts to 9 ft. lbs. (12 Nm).

e. Install the noise filter to the cylinder head cover with the bolt and tighten to 62 inch lbs. (7.0 Nm).

56. Install the 8 ignition coils with the 8 bolts and tighten to 7 ft. lbs. (10 Nm).

57. Align the pulley set key with the key groove of the pulley, and slide on the pulley and tighten to 221 ft. Lbs. (300 Nm).

58. Connect the bracket to the timing chain cover with the bolt.

59. Install the idler pulley with the bolt and tighten to 32 ft. lbs. (43 Nm).

60. Temporarily install the water pump pulley with the 4 bolts and tighten to 15 ft. lbs. (21 Nm).

61. Install a new water cover gasket to the timing chain cover.

62. Install the water inlet with the 3 bolts and tighten to 15 ft. lbs. (21 Nm).

63. Connect the 2 air tube hoses.

64. Connect the 2 hoses.

65. Install No. 1 and No 2 engine covers.

66. Install 2 new gaskets and the water by-pass joint with the 4 nuts and tighten to 15 ft. lbs. (21 Nm).

67. Connect the No. 2 water by-pass hose to the water by-pass joint.

68. Install the water by-pass pipe.

69. Connect the air tube.

70. Install the No. 1 water by-pass hose by connecting the hose to the water inlet housing and front water by-pass joint.

71. Install the 2 stud bolts.

72. Install the generator with the 2 bolts and 2 nuts and tighten to 32 ft. lbs. (43 Nm).

73. Connect the generator wire with the nut and tighten to 87 inch lbs. (9.8 Nm).

74. Connect the wire harness bracket to the generator.

75. Connect the vane pump to the timing chain cover with the 2 bolts and tighten to 21 ft. lbs. (28 Nm).

76. Install engine oil level dipstick guide.

77. Install oil filter bracket (w/o Oil Cooler).

a. Apply a light coat of engine oil to 2 new O-rings.

b. Install the 2 O-rings to the timing chain cover.

c. Install the oil filter bracket with the 2 bolts and 2 nuts and tighten to 26 ft. lbs. (35 Nm).

78. Install No. 1 oil cooler bracket.

79. Install No. 2 water by-pass pipe.

80. Install oil pressure sender gauge assembly.

81. Install oil filter element.

82. Connect No. 2 fuel tube sub-assembly.

83. Install the cooler compressor with the 2 stud bolts and tighten to 7 ft. lbs. (10 Nm).

84. Install the 2 bolts and 2 nuts and tighten to 18 ft. lbs. (235 Nm).

85. Connect the 3 heater hoses.

86. Connect the No. 2 and No. 3 air hoses.

87. Connect the 2 clamps and air pump wire harness.

88. Install the intake manifold.

89. Engine Room RH Side:

a. Connect the 5 clamps.

b. Connect the throttle position sensor and throttle control motor connector.

c. Connect the 2 air pump connectors.

d. Install the ground wire with the 2 bolts.

e. Connect the noise filter connector.

f. Connect the 2 VVT sensor connectors.

g. Connect the injector connector.

h. Connect the 4 ignition coil connectors.

i. Connect the 2 camshaft timing oil control valve connectors.

j. Connect the 2 clamps and power steering oil pressure switch connector.

k. For 4WD: Connect the clamp and junction connector.

90. Engine Room LH Side:

a. Connect the cooler compressor connector.

b. Connect the 3 clamps.

c. Connect the camshaft position sensor connector.

d. Connect the 2 camshaft timing oil control valve connectors.

e. Connect the engine coolant temperature sensor connector.

f. Connect the 4 clamps.

g. Install the ground wire with the 2 bolts.

h. Connect the noise filter connector.

i. Connect the 2 VVT sensor connectors.

j. Connect the 4 ignition coil connectors.

k. Connect the injector connector.

l. Connect the 2 wire harness clamps.

m. Connect the 4 air injection control driver connectors.

n. Connect the 2 connectors and attach the 2 clamps to the engine room junction block.

o. Install the engine room relay block cover.

91. Install air cleaner assembly.

92. Install air cleaner hose assembly.

93. Install the radiator.

94. Add engine oil.

95. Add engine coolant.

96. Connect cable to negative battery terminal.

97. Inspect for leaks.

98. Check engine oil level.

99. Perform initializations, if necessary.

MAIN BEARING TORQUE SEQUENCE

See Figures 185 through 189.

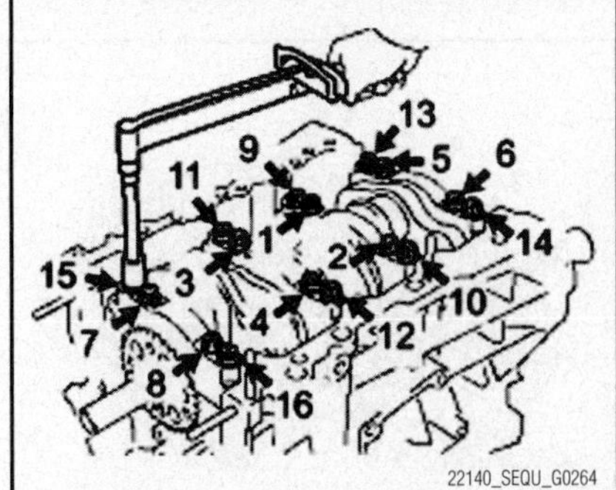

Fig. 185 Main bearing torque sequence—V6 Engine

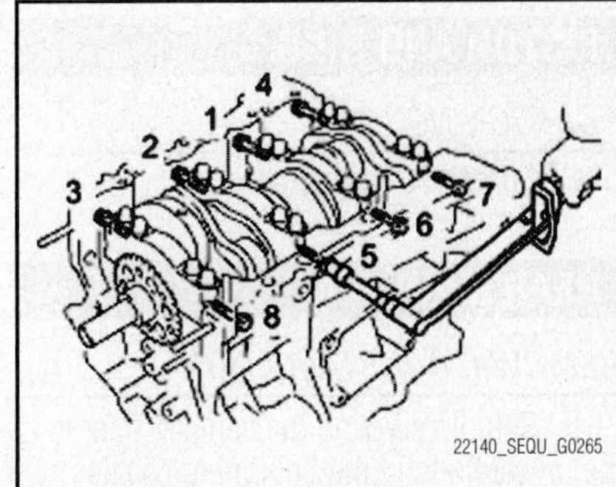

Fig. 186 Main bearing torque sequence—V6 Engine

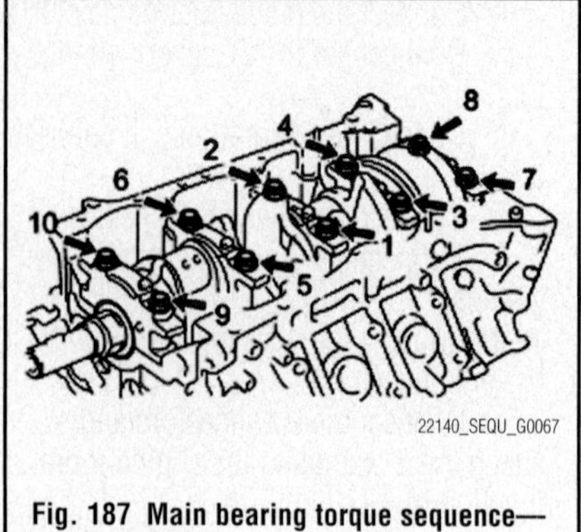

Fig. 187 Main bearing torque sequence—4.7L V8 Engine

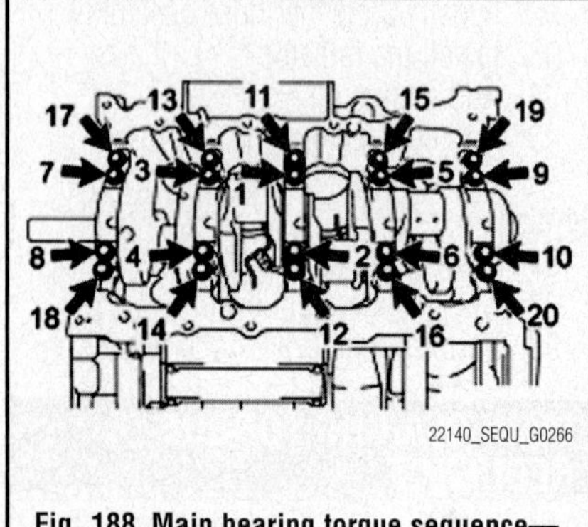

Fig. 188 Main bearing torque sequence—5.7L V8 Engine

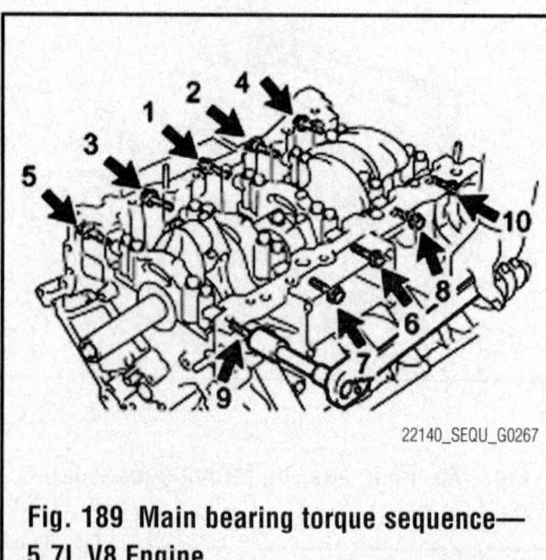

Fig. 189 Main bearing torque sequence—5.7L V8 Engine

PISTON AND RING

POSITIONING

See Figures 190 through 192.

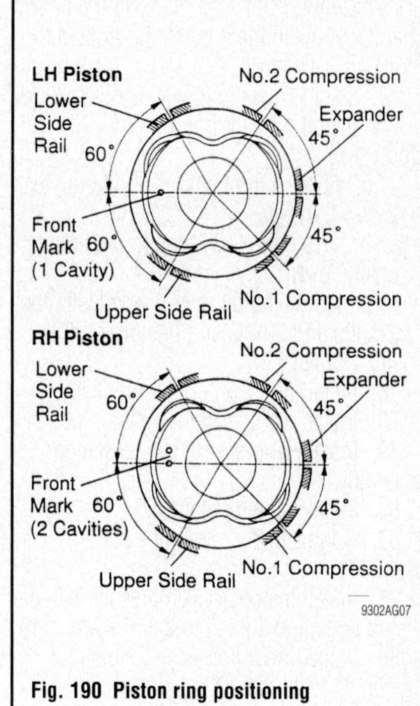

Fig. 190 Piston ring positioning

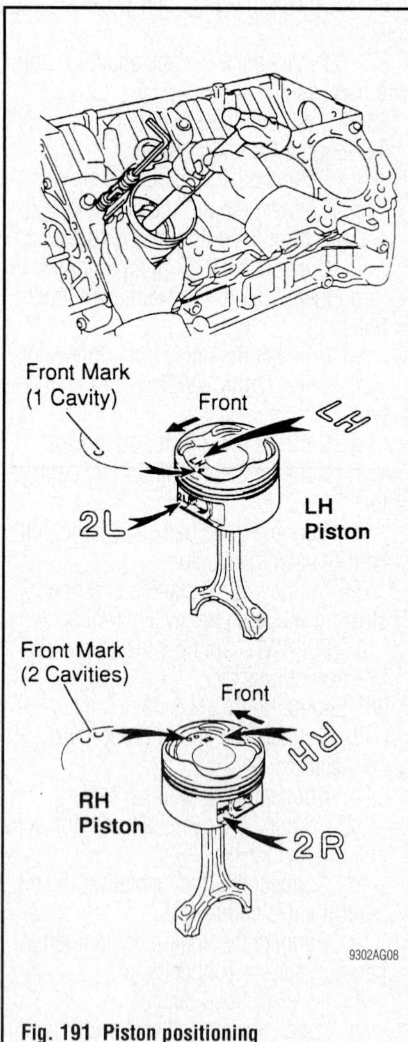
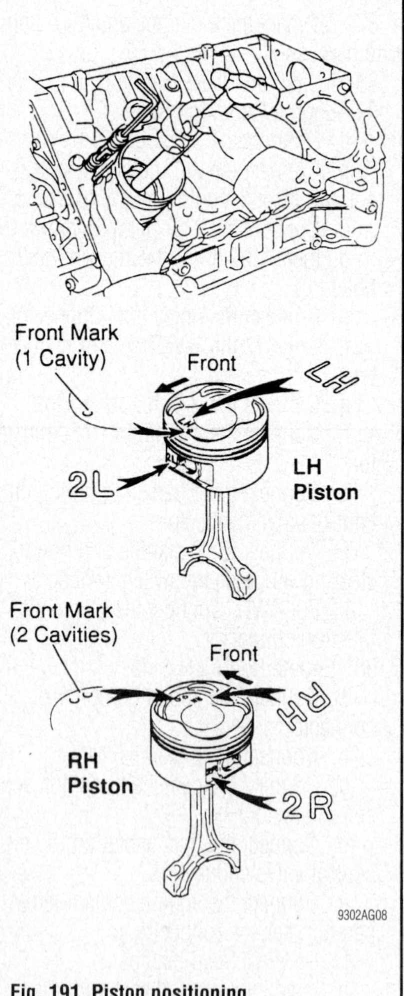

Fig. 191 Piston positioning

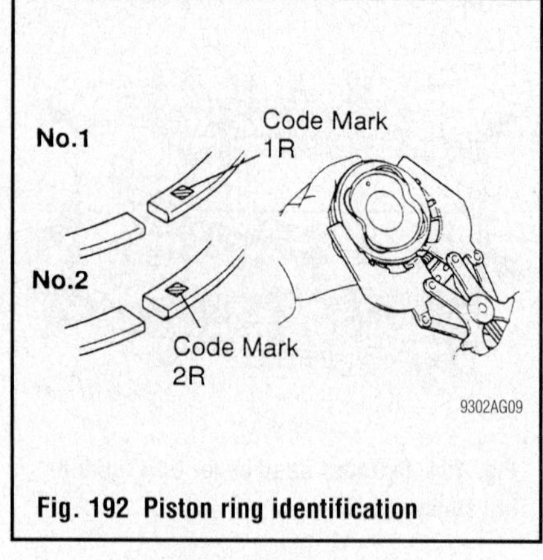

Fig. 192 Piston ring identification

REAR MAIN SEAL

REMOVAL & INSTALLATION

1. Before servicing the vehicle, refer to the precautions in the beginning of this section.
2. Remove the transmission and flywheel from the vehicle.
3. Cut off the rubber lip portion of the seal with a sharp knife.
4. Pry out the oil seal.

To install:

5. Install the rear main seal so that it is flush with the seal retainer housing.
6. Install flywheel/driveplate and tighten the bolts:
   a. For V6 engine, 61 ft. lbs. (83 Nm).
   b. For 4.7L V8 engine to 35 ft. lbs. (48 Nm) plus a 90 degree turn.
   c. For 5.7L V8 engine to 22 ft. lbs. (30 Nm) plus a 90 degree turn.
7. Transmission.

TIMING BELT FRONT COVER

REMOVAL & INSTALLATION

See Timing Belt and Sprockets.

TIMING BELT AND SPROCKETS

REMOVAL & INSTALLATION

4.7L V8 Engine

See Figures 193 through 203.

1. Disconnect the negative battery cable.
2. Raise and safely support the vehicle.
3. Remove the oil pan protector and the engine under cover.
4. Drain the cooling system and store the coolant for refilling purposes.
5. Lower the vehicle and remove the battery clamp cover.
6. From the top of the engine, remove the fuel return hose, the engine cover nuts/bolts and the cover.
7. Remove the air cleaner and the intake air connector assembly.
8. Remove the cooling fan pulley by performing the following procedures:
   a. Loosen the 4 fan clutch-to-fan pulley nuts.

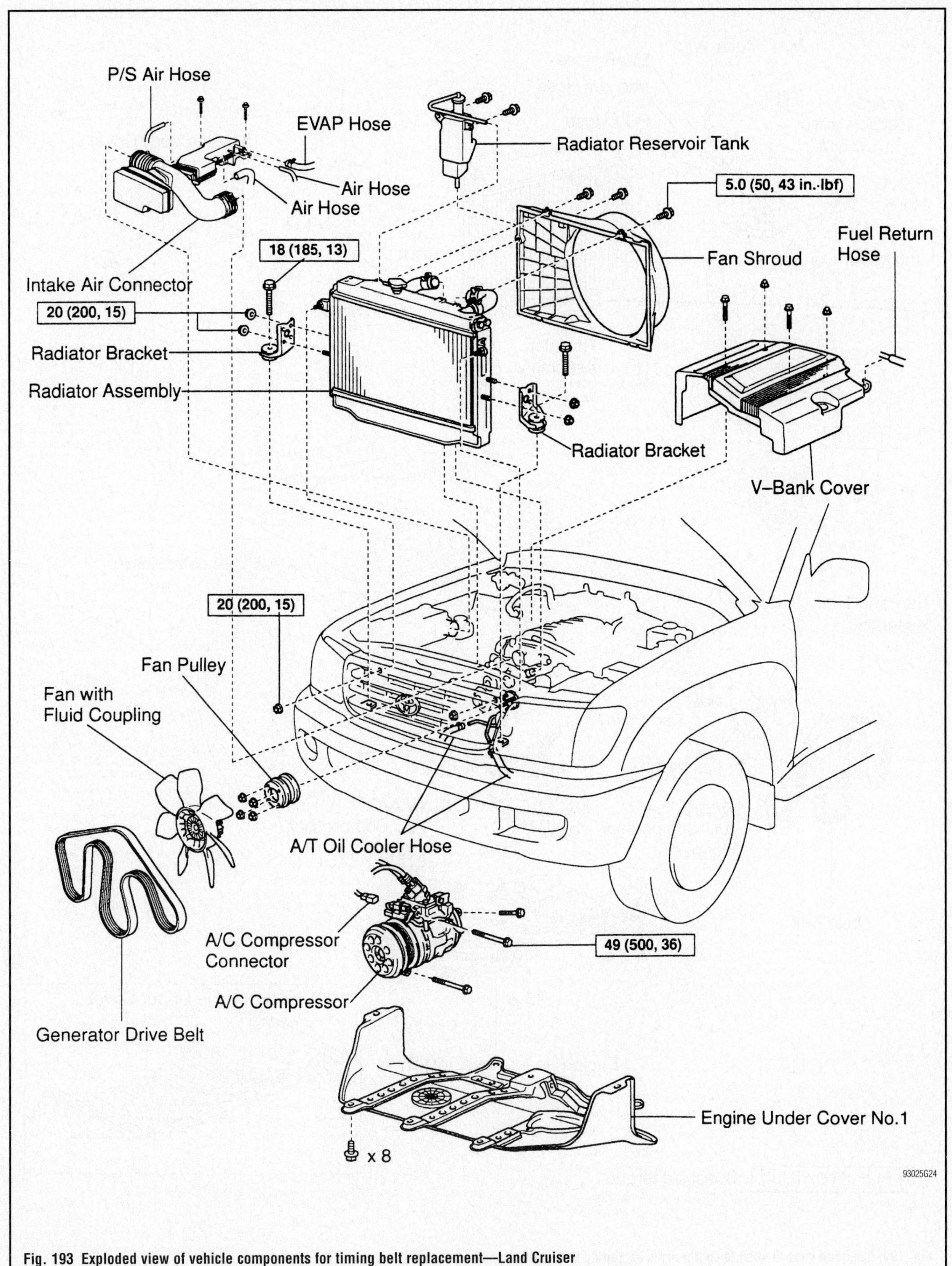

Fig. 193 Exploded view of vehicle components for timing belt replacement—Land Cruiser

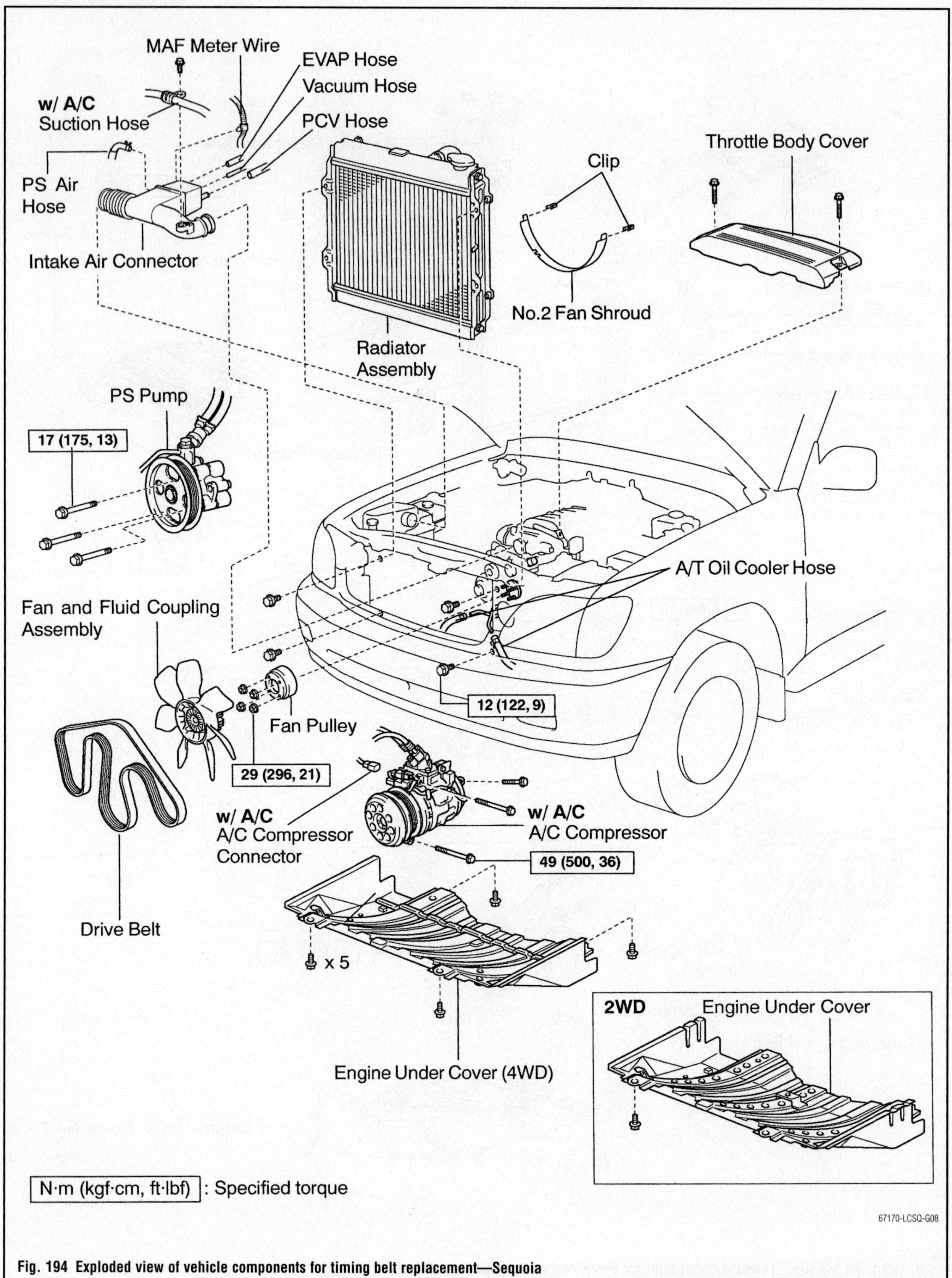

Fig. 194 Exploded view of vehicle components for timing belt replacement—Sequoia

b. Using a box-end wrench on the serpentine drive belt tensioner bolt, rotate the tensioner counterclockwise and remove the drive belt.

➡The serpentine drive belt tensioner bolt is a left-hand thread.

c. Remove the fan clutch-to-fan pulley nuts, the fan, the clutch assembly and the fan pulley.

9. Remove the radiator by performing the following procedures:

a. Disconnect the upper, lower and reservoir hoses from the radiator.

b. Disconnect and plug the automatic transmission oil cooler at the radiator. Disconnect the automatic transmission oil cooler hoses from the fan shroud clamp.

c. Remove the radiator reservoir tank.

d. Remove the fan shroud-to-radiator bolts and the shroud.

e. Remove the 2 upper radiator-to-chassis nuts.

f. Remove the middle radiator-to-chassis nut/bolts and brackets.

g. Carefully, lift the radiator from the vehicle.

10. Remove the serpentine drive belt idler pulley bolt, cover plate and pulley.

11. Remove the right side (No. 3) timing belt cover.

12. Remove the left side (No. 3) timing belt cover by performing the following procedures:

a. Disconnect the engine wire from both wire clamps.

b. Disconnect the camshaft position sensor wire from the wire clamp on the left-side (No.3) timing belt over.

c. Disconnect the sensor connector from the connector bracket.

d. Disconnect the sensor connector.

e. Remove the wire grommet from the left-side (No. 3) timing belt cover.

f. Remove the oil cooler tube bolts and tube.

13. Remove the middle (No. 2) timing belt cover bolts and cover.

14. Remove the cooling fan bracket nuts/bolts and bracket.

➡If reusing the timing belt, make sure that there are 3 installation marks on the belt; if there are none, install them.

15. Using the Crankshaft Pulley Holding tool 09213-70010, Bolt tool 90105-08076 and Companion Flange Holding tool 09330-00021, or equivalent, loosen the crankshaft pulley bolt.

16. Position the No. 1 cylinder to approximately 50 degrees After Top Dead Center (ATDC) of the compression

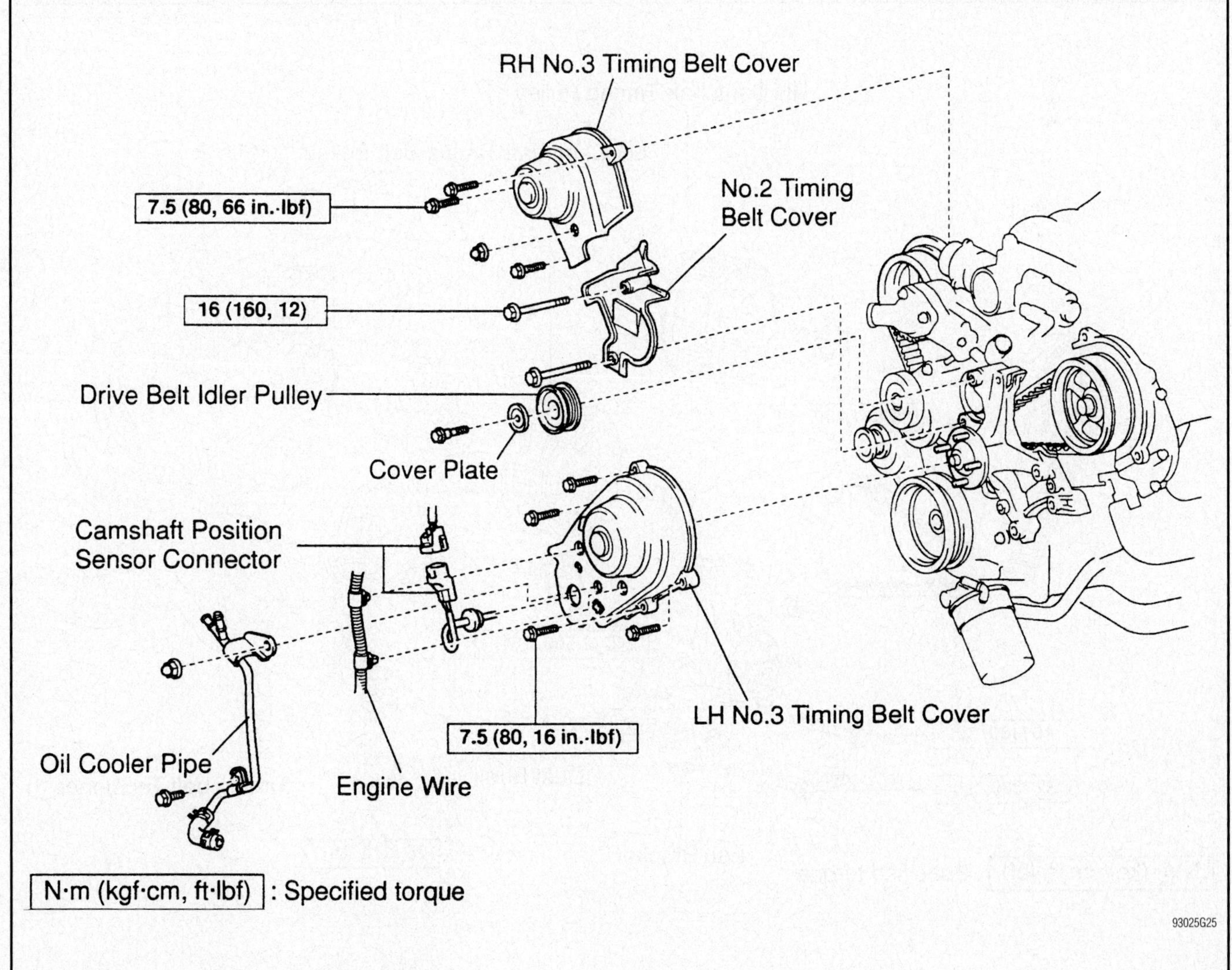

Fig. 195 Exploded view of upper timing belt covers

stroke by performing the following procedures:

a. Rotate the crankshaft pulley (CLOCKWISE) to align its groove with the timing mark "0" on the lower (No. 1) timing belt cover.

b. Check that the camshaft sprocket timing marks are aligned with the rear timing belt plate marks; if not, rotate the crankshaft 1 revolution (360 degrees).

c. Rotate the crankshaft pulley approximately 50 degrees (CLOCKWISE) and align the crankshaft pulley timing mark between the centers of the crankshaft pulley bolt and the idler pulley bolt.

WARNING

If the timing belt is disengaged, having the crankshaft pulley in the wrong angle can cause the valve to come into contact with the piston when removing the camshaft pulley.

17. Remove the crankshaft pulley bolt.

➡If reusing the timing belt and the installation marks have disappeared, place new installation marks on the timing belt to match the camshaft timing sprocket marks.

➡To avoid meshing the timing sprocket and the timing belt, secure one with a string; then, place matchmarks on the timing belt and the right-side camshaft timing sprocket.

18. Remove the timing belt tensioner bolts and the tensioner.

19. Using the Camshaft Holding tool 09960-10010, or equivalent, slightly turn the left-side camshaft sprocket clockwise to loosen the tension spring. Then, disconnect the timing belt from the camshaft sprockets.

20. Remove the alternator by performing the following procedures:

a. Disconnect the electrical connector from the alternator.

b. Remove the rubber cap/nut and disconnect the battery wire from the alternator.

c. Disconnect the wire clamp from the alternator cord clip.

d. Remove the alternator-to-engine nuts/bolts and the alternator.

21. Remove the serpentine drive belt tensioner nuts/bolts and the tensioner.

22. Using the Crankshaft Puller Assembly tool 09950-50012, or equivalent, press the crankshaft pulley from the crankshaft.

WARNING

DO NOT rotate the crankshaft pulley.

23. Remove the lower (No. 1) timing belt cover bolts and the cover.

24. Remove the timing belt guide, spacer and the timing belt.

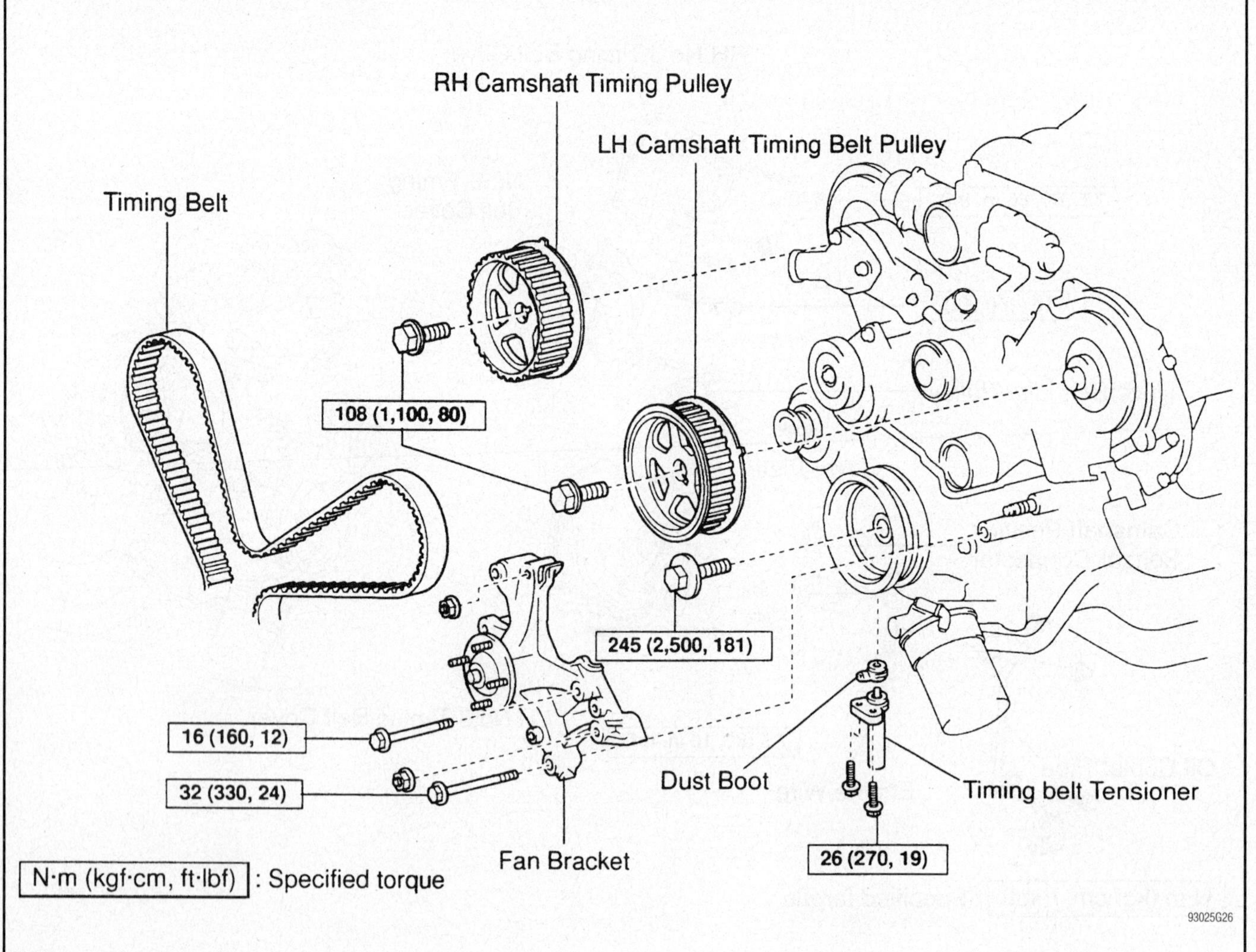

Fig. 196 Exploded view of upper timing sprockets and components

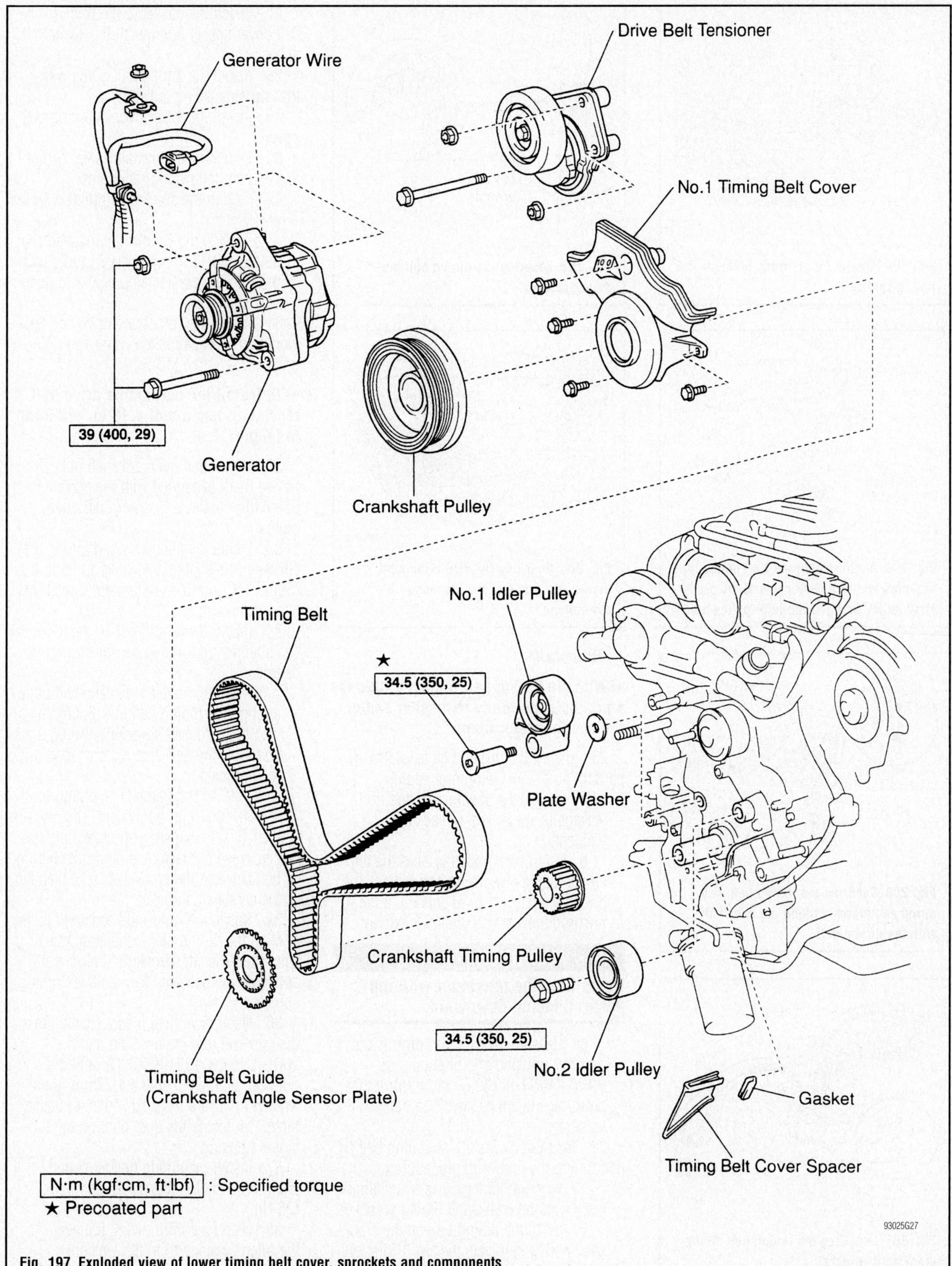

Fig. 197 Exploded view of lower timing belt cover, sprockets and components

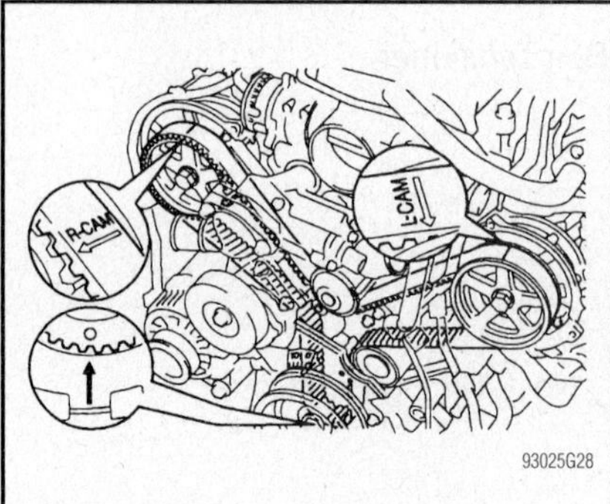

Fig. 198 Alignment of timing belt with the timing sprockets

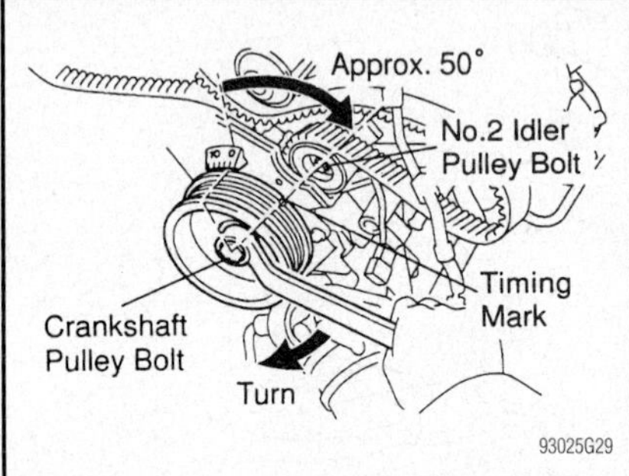

Fig. 199 Aligning of crankshaft pulley timing mark with the center line of the crankshaft pulley bolt and the idler pulley bolt

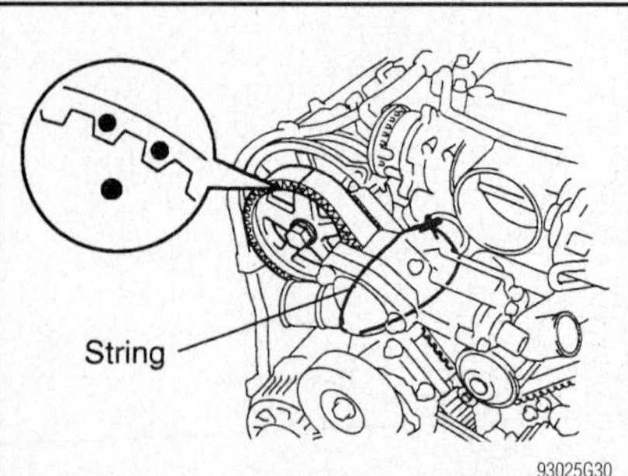

Fig. 200 Securing the timing belt with string and matchmarking the camshaft with the timing belt

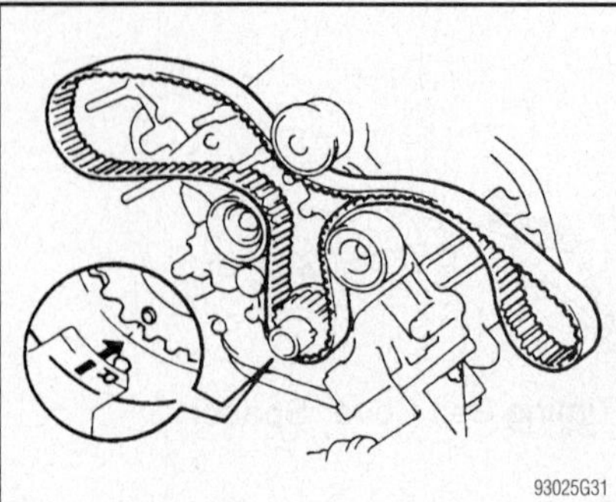

Fig. 201 Installing the timing belt on the crankshaft sprocket

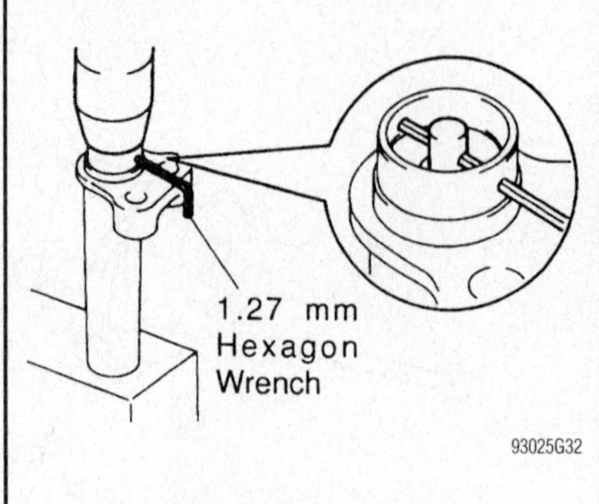

Fig. 202 Securing the timing belt tensioner pushrod

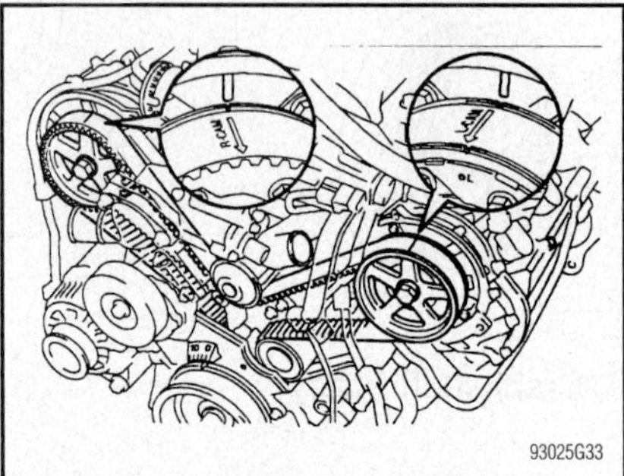

Fig. 203 Checking the TDC alignment marks after rotating the crankshaft 2 revolutions

To install:

➡With the timing belt removed, this is a perfect opportunity to inspect and/or replace the water pump.

25. Inspect the timing belt tensioner by performing the following procedures:
   a. Inspect the seal for leakage; if leakage is suspected, replace the tensioner.
   b. Using both hands to hold the tensioner facing upward, strongly press the pushrod against a solid surface. If the pushrod moves, replace the tensioner.

⁂ WARNING

Never hold the tensioner with the pushrod facing downward.

   c. Measure the pushrod protrusion from the housing end, it should be 0.413–0.453 in. (10.5–11.5mm). If the protrusion is not as specified, replace the tensioner.
26. Temporarily install the timing belt by performing the following procedures:
   a. Align the timing belt's installation mark with the crankshaft timing sprocket.
   b. Install the timing belt on the crankshaft timing sprocket, the No. 1 idler pulley and the No. 2 idler pulley.
27. Install the gasket to the timing belt cover spacer and install the cover spacer.
28. Install the timing belt guide with the cup side facing outward.
29. Install the lower (No. 1) timing belt cover.
30. Install the crankshaft pulley by performing the following procedures:
   a. Align the crankshaft pulley with the crankshaft key.
   b. Using the Crankshaft Installer tool 09223-46011, or equivalent, and a hammer, tap the crankshaft pulley into position.
31. Install the serpentine drive belt tensioner and torque the tensioner-to-engine bolts to 12 ft. lbs. (16 Nm).

➡To install the serpentine drive belt tensioner, use a bolt 4.18 in. (106mm) in length.

32. Check that the crankshaft pulley's timing mark is aligned with the centers of the idler pulley and crankshaft pulley bolts.
33. Install the alternator and torque the alternator-to-engine nuts/bolts to 29 ft. lbs. (39 Nm). Connect the alternator's electrical connectors and clip.
34. Install the timing belt to the left-side camshaft by performing the following procedures:
   a. Rotate the left-side camshaft pulley to align the timing belt installation mark with the camshaft sprocket's timing mark and slide the belt onto the camshaft timing sprocket.
   b. Using the Camshaft Holding tool 09960-10010, or equivalent, slightly turn the left-side camshaft sprocket counterclockwise to place tension on the timing belt between the crankshaft sprocket and the camshaft sprocket.
35. Rotate the right-side camshaft pulley to align the timing belt installation mark with the camshaft sprocket's timing mark and slide the belt onto the camshaft timing sprocket.
36. Using a vertical press, slowly press the pushrod into the housing using 200–2205 lbs. (981–9807 N) until the holes align, then, install a 1.27mm Allen® wrench to secure the pushrod and release the press. Install the dust boot on the tensioner housing.
37. Install the timing belt tensioner and torque the bolts to 19 ft. lbs. (26 Nm).
38. Using a pair of pliers, remove the Allen® wrench from the tensioner housing.

39. Check the valve timing by performing the following procedure:
   a. Temporarily install the crankshaft pulley bolt.
   b. Slowly, rotate the crankshaft pulley 2 revolutions (CLOCKWISE) and realign the TDC marks.

➡If the pulley/sprocket timing marks do not realign, remove the timing belt and reinstall it.

40. Using the Crankshaft Pulley Holding tool 09213-70010, Bolt tool 90105-08076 and Companion Flange Holding tool 09330-00021, or equivalent, torque the crankshaft pulley bolt to 181 ft. lbs. (245 Nm).
41. Install the cooling fan bracket and torque the 12mm (head size) bolt to 12 ft. lbs. (16 Nm) and the 14mm (head size) bolt to 24 ft. lbs. (32 Nm).
42. Install the air conditioning compressor.
43. Install the middle (No. 2) timing belt cover and torque the bolts to 12 ft. lbs. (16 Nm).
44. Install the upper right-side (No. 3) timing belt cover and torque the bolts to 66 inch lbs. (7.5 Nm).
45. Install the upper left-side (No. 3) timing belt cover by performing the following procedures:
   a. Install the oil cooler tube and bolt.
   b. Feed the Camshaft Position Sensor (CPS) through the left-side (No. 3) timing belt cover hole.
   c. Install the left-side (No. 3) timing belt cover and torque the bolts to 66 inch lbs. (7.5 Nm).
   d. Install the wire grommet to the left-side (No. 3) timing belt cover.
   e. Install the sensor connector to the connector bracket and connect the sensor connector.
   f. Install the sensor wire and the engine wire to the clamps on the left-side (No. 3) timing belt cover.
46. Install the drive belt idler pulley and cover plate; then, torque the pulley bolt to 27 ft. lbs. (37 Nm).
47. To complete the installation, reverse the removal procedures.
48. Refill the cooling system and connect the negative battery cable.

TIMING CHAIN COVER AND SEAL

REMOVAL & INSTALLATION

See Timing Chain and Sprockets.

TIMING CHAIN AND SPROCKETS

REMOVAL & INSTALLATION

V6 Engine

See Figures 204 and 205.

1. Before servicing the vehicle, refer to the Precautions Section.
2. Properly relieve the fuel system pressure.
3. Disconnect the negative battery cable. Disconnect the positive battery cable. Remove the battery.
4. Remove the V bank cover.
5. Drain the engine coolant. Remove the radiator.
6. Loosen the fluid coupling assembly (fan). Remove the drive belt. Remove the fan assembly.
7. Remove the air cleaner assembly. Remove the oil level gauge. Remove the water inlet.
8. Separate the vane pump assembly (power steering pump). Remove the alternator. Remove the air conditioning compressor and position it to the side.
9. Remove the belt tensioner assembly. Remove the idler pulley number two subassembly. Remove the idler pulley number one subassembly. Remove the crankshaft pulley.
10. Remove the lower oil pan. Remove the oil strainer and pickup tube assembly. Remove the upper oil pan.
11. Remove the intake manifold. Remove the ignition coil assembly. Remove the cylinder head cover assembly.
12. Remove the camshaft timing oil control valve assembly. Remove the VVT sensor. Remove the oil filter bracket subassembly.

➡Carefully remove the cover by prying between the cover and the cylinder head or block with a suitable tool. Be sure to cover the tip of the suitable tool prior to usage. Be careful not to damage the contact surfaces of the cylinder block, cylinder head and timing chain cover.

13. Using the crankshaft pulley set bolt, turn the crankshaft to align the crankshaft set key with the timing line of the cylinder block. If not aligned at TDC of the compression stroke, turn the crankshaft one complete revolution, in the direction of rotation
14. While turning the stopper plate of the tensioner upward, push the plunger of the chain tensioner. While turning the stopper plate of the tensioner downward, insert a 0.138 inch diameter bar into the holes in the stopper plate and tensioner to hold the stopper plate.
15. Remove the two bolts, and then remove the chain tensioner.
16. Remove the chain tensioner slipper. Remove the idle gear shaft number two, idle gear number one and idle gear shaft number one.
17. Remove the number two chain vibration damper. Remove the timing chain subassembly.

To install:

18. Install the chain tensioner slipper.
19. While turning the stopper plate of the tensioner clockwise, push in the plunger of the chain tensioner. While turning the stopper plate of the tensioner counterclockwise, insert a 0.138 inch diameter bar into the holes in the stopper plate and tensioner to hold the stopper plate.
20. Install the chain tensioner. Tighten the bolts to 7.4 ft. lbs.
21. Position the engine at TDC on the compression stroke. Align the camshaft timing gears and bearing caps. Using the crankshaft pulley set bolt, align the crankshaft set key with the timing line of the cylinder.
22. Align the yellow mark line with the timing mark of the crankshaft timing link. Align the orange mark links with the timing marks of the camshaft timing gears, and install the chain.
23. Install the number two chain vibration damper.
24. Apply a light coat of clean engine oil to the rotating surface of the idle gear shaft number one.
25. Temporarily install the idle gear shaft number one together with idle gear shaft number two, while aligning the knock pin of idle gear shaft number one with the knock pin groove of the cylinder block.

➡Be care of the idle gear direction.

26. Tighten the idle gear shaft number two to 44 ft. lbs. Remove the bar from the chain tensioner.
27. Install a new front case oil seal. Install a new O-ring onto the left cylinder head.
28. Apply continuous beads (0.12–0.16 inch in diameter) of seal packing, part number 08826-00080 or equivalent to the four locations shown in the illustration.
29. Apply continuous beads (0.12–0.16 inch in diameter) of seal packing, part number 08826-00080 or equivalent to all parts except the water pump part: for the water pump part use, part number

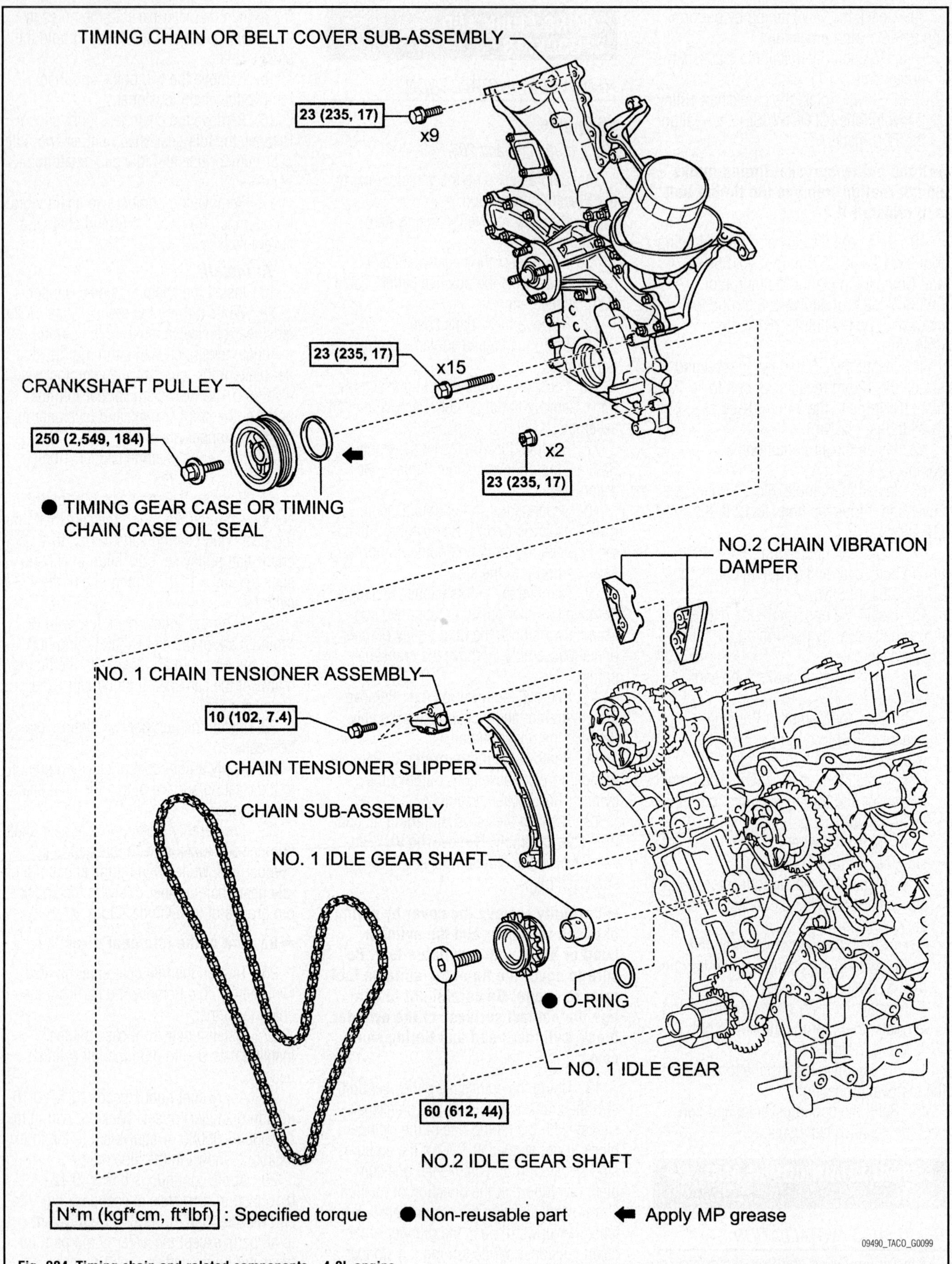

Fig. 204 Timing chain and related components—4.0L engine

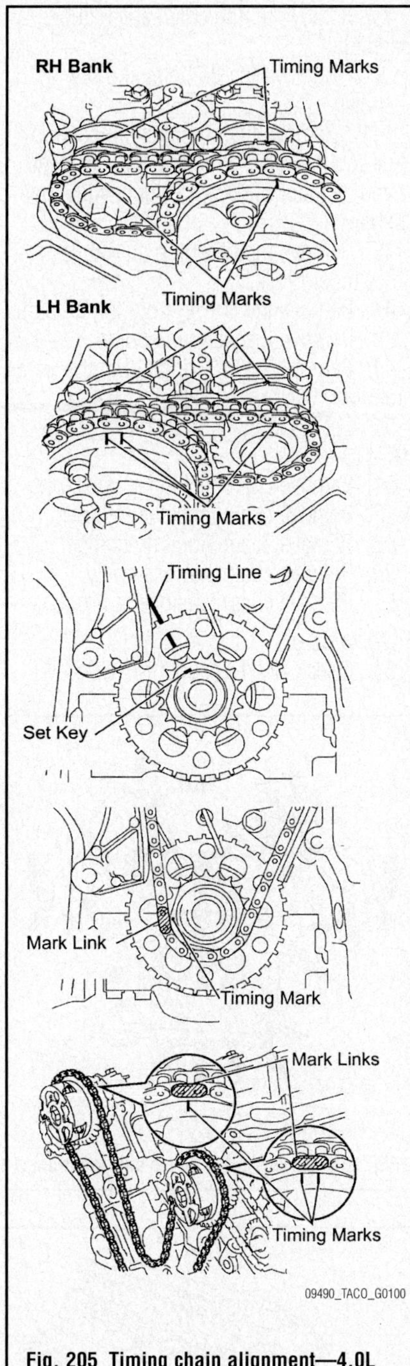

Fig. 205 Timing chain alignment—4.0L engine

08826-00080 or equivalent, to the timing chain cover. Do not apply seal packing to portion "A".

➡Remove any oil from the contact surfaces. Install the timing chain case cover within three minutes and tighten the bolts within fifteen minutes of applying the seal packing.

30. Align the key way of the oil pump drive motor with the rectangular portion of the crankshaft timing gear and slide the timing chain case cover into place.

31. Install the timing chain case cover bolts. Tighten the bolts and nuts uniformly in several steps to 17 ft. lbs.

➡Do not wrap the chain and slipper over the timing chain case cover seal line.

32. Continue the installation in the reverse order of the removal procedure.

33. When installing the cylinder head cover apply a continuous bead (0.08–0.12 inch) of seal packing, part number 08826-00080 or equivalent, to the cylinder head. Install the seal washers onto the bolts. Install the cylinder head cover bolts and nuts. Tighten bolts "A" to 7.4 ft. lbs. Tighten bolts "B" to 80 inch lbs. Tighten nuts to 80 inch lbs.

➡Be sure to remove any oil from the contact surfaces of the cylinder head cover and the cylinder head. Install the cover within three minutes after applying the seal packing. Tighten the bolts to specification within fifteen minutes after installing the cover. Do not add engine oil for at least two hours after installing the cover.

34. Be sure to fill the engine with the proper grade and type engine coolant.

35. Be sure to fill the engine with the proper grade and type engine oil.

36. Start the engine and check for leaks, correct as required

5.7L V8 Engine

See Figures 176, 206 through 239.

1. Discharge fuel system pressure.
2. Disconnect cable from negative battery terminal.
3. Drain engine oil.
4. Drain engine coolant.
5. Remove the radiator.
6. Remove air cleaner hose assembly.
7. Remove air cleaner assembly.
8. Engine Room LH Side:
   a. Remove the engine room relay block cover.
   b. Disconnect the 2 connectors and detach the 2 clamps from the engine room junction block.
   c. Disconnect the 4 air injection control driver connectors.
   d. Disconnect the 2 wire harness clamps.
   e. Disconnect the injector connector.
   f. Disconnect the 4 ignition coil connectors.
   g. Disconnect the 2 VVT sensor connectors.
   h. Disconnect the 4 clamps.
   i. Remove the 2 bolts and ground wire.
   j. Disconnect the noise filter connector.
   k. Disconnect the engine coolant temperature sensor connector.
   l. Disconnect the 2 camshaft timing oil control valve connectors.
   m. Disconnect the camshaft position sensor connector.
   n. Disconnect the 3 clamps.
   o. Disconnect the cooler compressor connector.
9. Engine Room RH Side.
   a. Disconnect the 2 camshaft timing oil control valve connectors.
   b. Disconnect the 4 ignition coil connectors.
   c. Disconnect the injector connector.
   d. Disconnect the 2 VVT sensor connectors.
   e. Disconnect the noise filter connector.
   f. Remove the 2 bolts and ground wire.
   g. Disconnect the 2 air pump connectors.
   h. Disconnect the throttle position sensor and throttle control motor connector.
   i. Disconnect the 5 clamps.
   j. Disconnect the 2 clamps and power steering oil pressure switch connector.
   k. For 4WD: Disconnect the 2 clamps and power steering oil pressure switch connector.
10. Remove the intake manifold.
11. Disconnect the No. 2 and No. 3 air hoses and the 2 clamps.
12. Remove the 2 bolts and disconnect the heater 3 hoses.
13. Remove the 2 bolts, 2 nuts and 2 stud bolts, and disconnect the cooler compressor.

➡It is not necessary to completely remove the compressor. With the hoses connected to the compressor, hang the compressor on the vehicle body with a rope.

14. Remove the 2 bolts and disconnect the fuel tube.
15. Remove oil filter element.
16. Remove oil pressure sender gauge assembly.
17. Disconnect the 4 hoses and remove the water by-pass pipe.
18. Remove No. 1 oil cooler bracket (w/ oil cooler).
19. Remove the 2 bolts, 2 nuts and oil filter bracket.

20. Remove the 2 O-rings.
21. Remove engine oil level dipstick guide.
22. Remove the 2 bolts and disconnect the vane pump.
23. Disconnect the generator connector.
24. Remove the terminal cap and nut, and disconnect the generator wire.
25. Remove the bolt and disconnect the wire harness bracket from the generator.
26. Remove the 2 bolts, 2 nuts and generator.
27. Remove the 2 stud bolts.
28. Remove the No. 1 water by-pass hose by disconnect the hose from the water inlet housing and front water by-pass joint.
29. Remove the bolt and disconnect the air tube.
30. Remove the 2 bolts and hoses and water by-pass pipe.
31. Disconnect the No. 2 water by-pass hose from the water by-pass joint.
32. Remove the 4 nuts, water by-pass joint and 2 gaskets.
33. Remove No.1 and No. 2 engine covers .
34. Remove the bolt, disconnect the 2 hoses, and remove the air tube.
35. Remove the 3 bolts, water inlet housing and gasket.
36. Remove the 4 bolts and water pump pulley.
37. Remove the bolt and idler pulley.
38. Remove the 4 bolts and fan bracket.
39. Remove the standard bolt, 6 mm hexagon wrench bolt and belt tensioner.
40. Remove the 8 bolts and 8 ignition coils.
41. Remove cylinder head cover sub-assembly LH.
    a. Remove the 14 bolts, seal washer, cylinder head cover and gasket.
    b. Remove the 5 gaskets from the camshaft bearing caps (No. 2, No. 3).
42. Remove cylinder head cover sub-assembly RH.
    a. Remove the bolt and noise filter.
    b. Remove the 14 bolts, seal washer, cylinder head cover and gasket.
    c. Remove the 5 gaskets from the camshaft bearing caps (No. 1, No. 3).
43. Remove crankshaft pulley.
    a. Loosen the crankshaft pulley set bolt.
    b. Partially install the pulley set bolt to the crankshaft until 2 or 3 threads are engaged.
    c. Remove the crankshaft pulley.
44. Remove the bolt and disconnect the wire harness bracket.

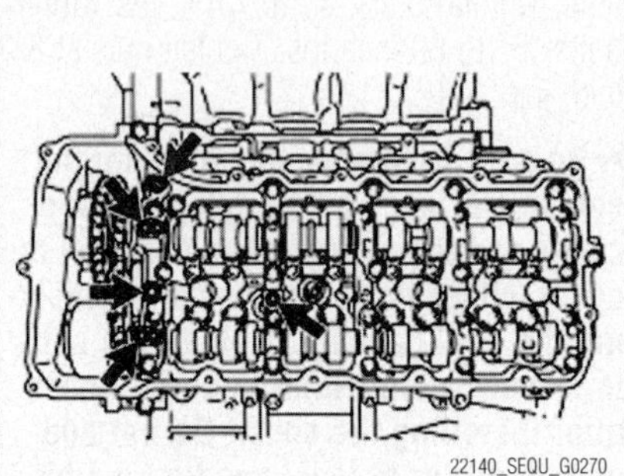

Fig. 207 Cylinder head cover gasket locations

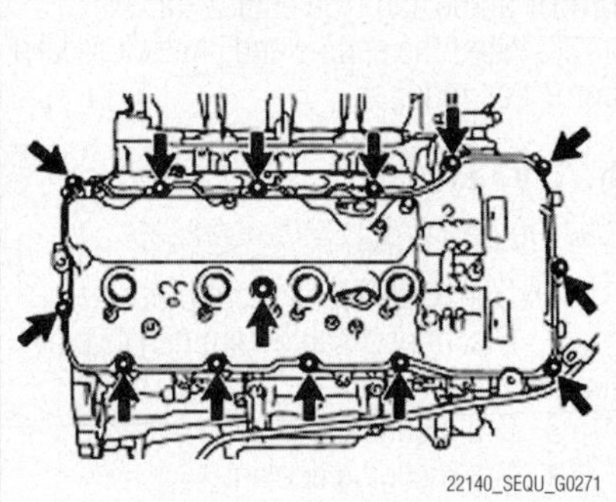

Fig. 208 Cylinder head cover RH

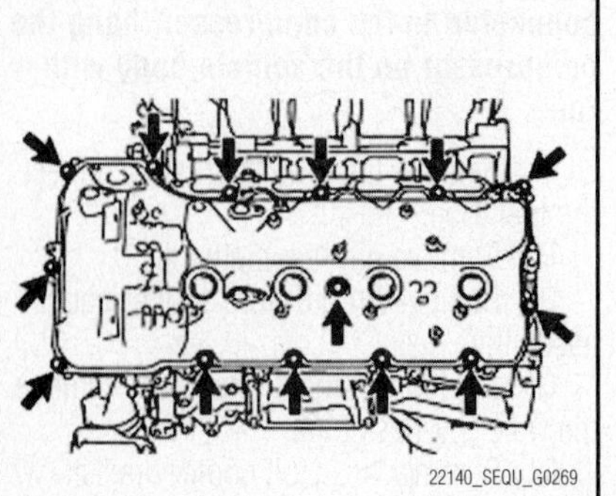

Fig. 206 Cylinder head cover LH

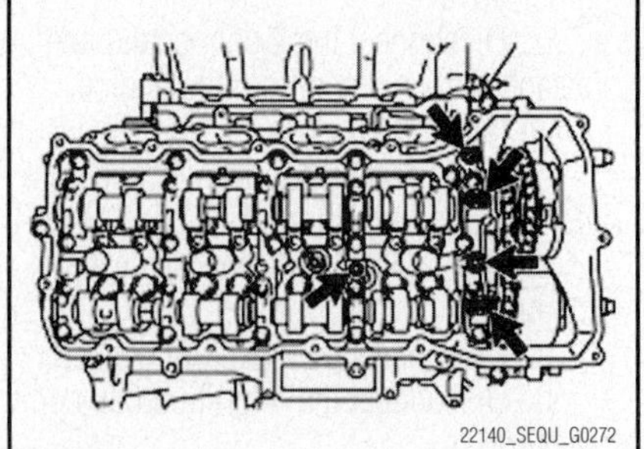

Fig. 209 Cylinder head cover gasket locations

45. Remove timing chain cover sub-assembly.
    a. Remove the 28 bolts and nut shown.
    b. Remove the timing chain cover by prying between it and the cylinder head and cylinder block with a screwdriver as shown
46. Remove the oil pump gasket from the cylinder block.
47. Remove the O-ring from the oil pan.
48. Remove the water inlet pipe.
49. Remove the 2 O-rings from the water inlet pipe.
50. Set No. 1 cylinder to TDC / compression.
51. Remove chain tensioner assembly.
52. Remove chain tensioner slipper.
53. Remove chain vibration damper.
54. Remove chain sub-assembly.
55. Remove chain tensioner assembly.

To install:

56. Check No. 1 cylinder to TDC.

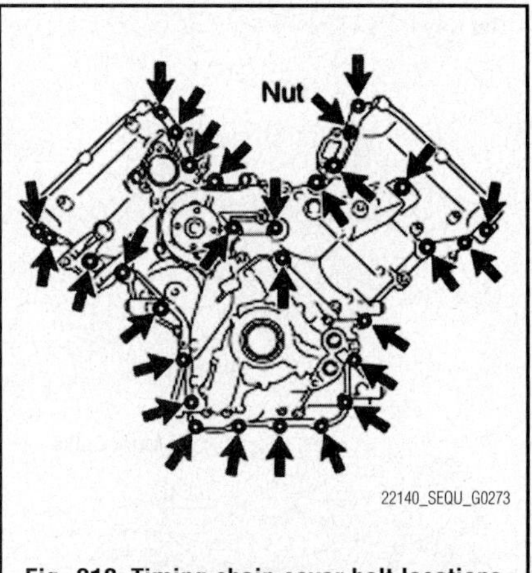

Fig. 210 Timing chain cover bolt locations

22140_SEQU_G0274

Fig. 211 Timing chain cover pry locations

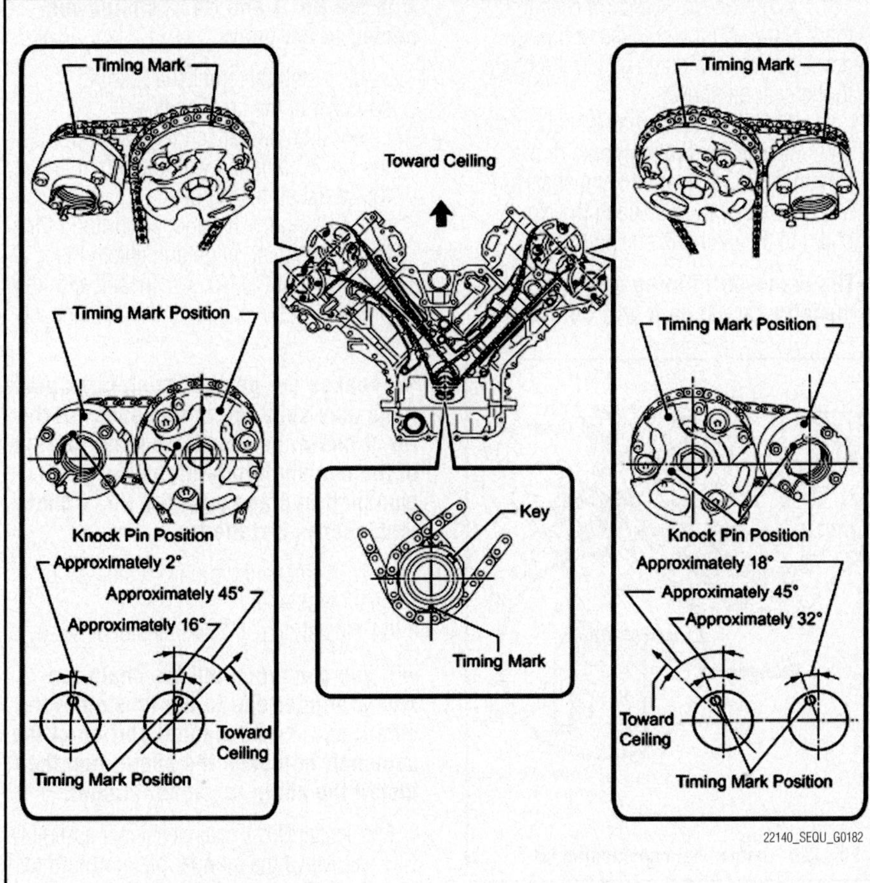

Fig. 212 Positioning engine to TDC

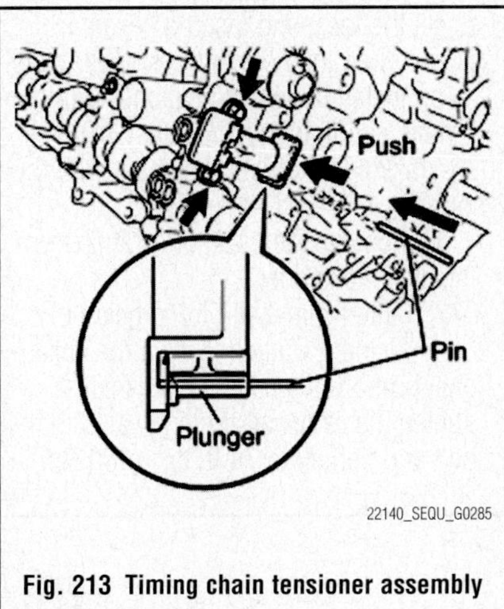

Fig. 213 Timing chain tensioner assembly No. 2

57. Install No.2 chain tensioner assembly. While raising up the No. 2 chain tensioner, insert a pin of &phiv$1.0 mm (0.0394 in.) into the hole to fix it in place.

58. Install No. 1 chain sub-assembly RH.

a. Align the No. 1 chain's orange mark plates with the camshaft timing gear's timing mark, and attach the chain to the gear as shown.

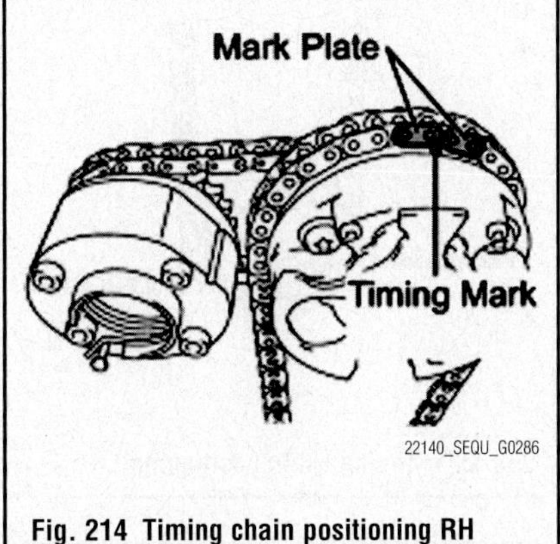

Fig. 214 Timing chain positioning RH

b. Align the No. 1 chain's orange mark plate with the crankshaft timing gear's timing mark, and attach the chain to the gear as shown.

c. Align the No. 2 chain's mark plates (yellow) with the timing marks of the camshaft timing gear and camshaft timing exhaust gear, and attach the No. 2 chain to the gears as shown.

➡The crankshaft timing gear and camshaft exhaust gear will be installed with the No. 1 and No. 2 chains connected to the gears.

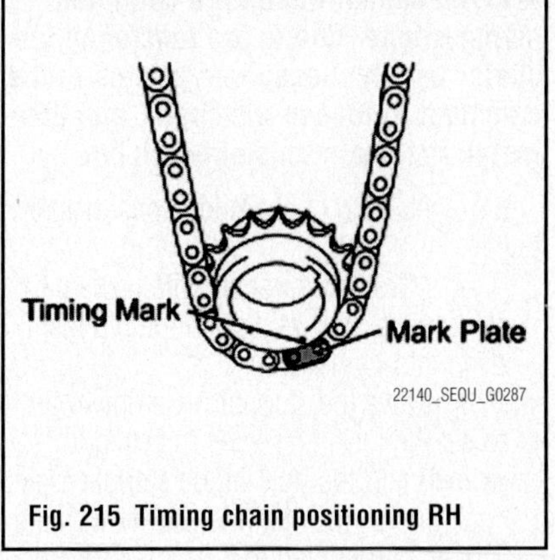

Fig. 215 Timing chain positioning RH

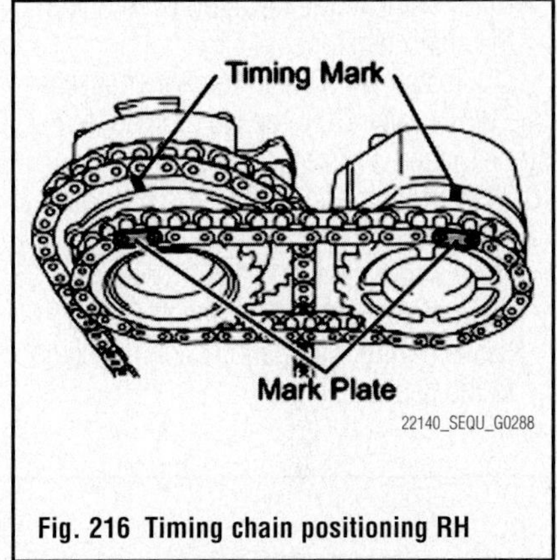

Fig. 216 Timing chain positioning RH

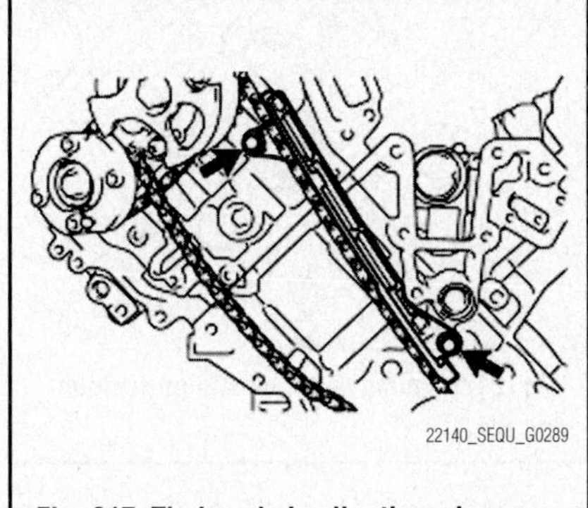

Fig. 217 Timing chain vibrations damper positioning RH

d. Install the crankshaft timing sprocket to the crankshaft.

e. Align and attach the knock pin of the No. 1 camshaft with the pin hole of the camshaft timing gear.

f. Using the hexagonal portion of the No. 2 camshaft, align and attach the knock pin of the No. 2 camshaft with the pin hole of the camshaft timing exhaust gear.

g. Remove the pin from the No. 2 chain tensioner.

59. Install the vibration damper with the 2 bolts.

60. Install No. 1 chain tensioner slipper.

➡If you cannot install the chain tensioner slipper due to the tension of the chain, use the hexagonal portion of the camshaft to loosen the chain, and then install the chain tensioner slipper

61. Install No. 1 chain tensioner assembly.
    a. Move the stopper plate upward to release the lock, and push the plunger deep into the tensioner.
    b. Move the stopper plate downward to set the lock, and insert a hexagon wrench into the hole of the stopper plate.
    c. Install the chain tensioner with the 2 bolts and tighten to 7 ft. lbs. (10 Nm).
    d. Remove the hexagon wrench from the chain tensioner.
62. Install No.3 chain tensioner assembly. While raising up the No. 3 chain tensioner, insert a pin of &phiv$1.0 mm (0.0394 in.) into the hole to fix it in place.
63. Install No. 1 chain sub-assembly LH.
    a. Align the No. 1 chain's orange mark plates with the camshaft timing gear's timing mark, and attach the chain to the gear as shown.

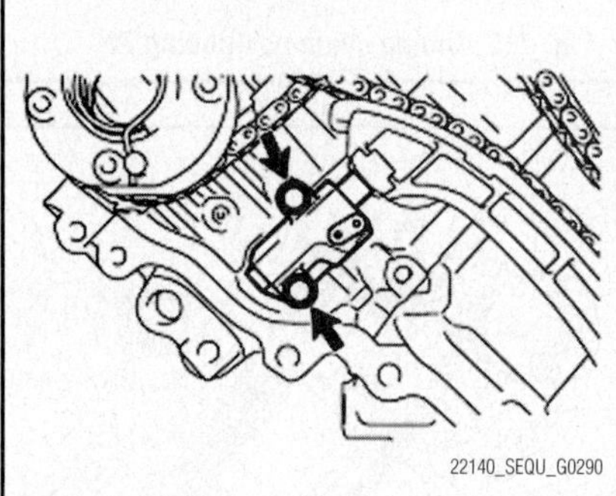

Fig. 218 Timing chain tensioner positioning RH

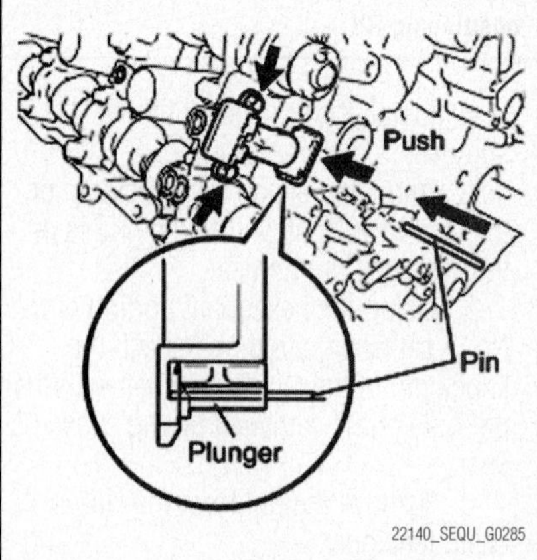

Fig. 219 Timing chain tensioner assembly No. 2

    b. Align the No. 1 chain's orange mark plate with the crankshaft timing gear's timing mark, and attach the chain to the gear as shown
    c. Align the No. 2 chain's mark plates (yellow) with the timing marks of the camshaft timing gear and camshaft timing exhaust gear, and attach the No. 2 chain to the gears as shown

➡The crankshaft timing gear and camshaft exhaust gear will be installed with the No. 1 and No. 2 chains connected to the gears.

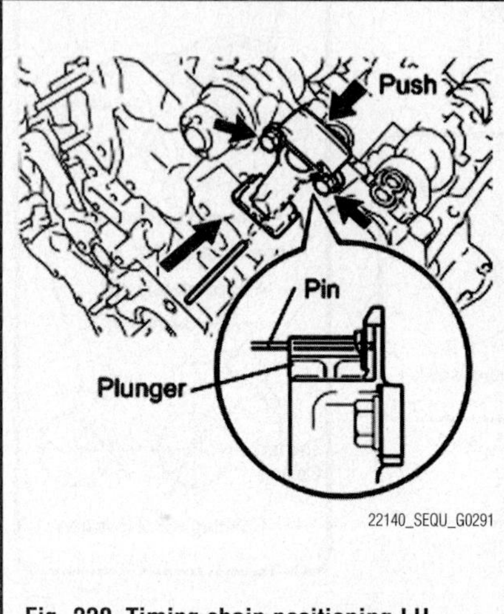

Fig. 220 Timing chain positioning LH

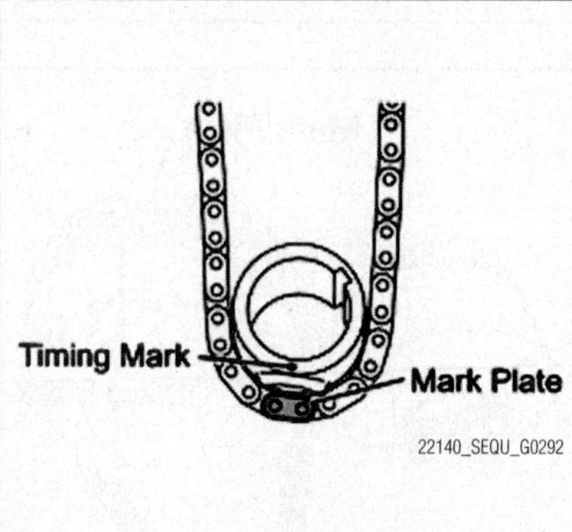

Fig. 221 Timing chain positioning LH

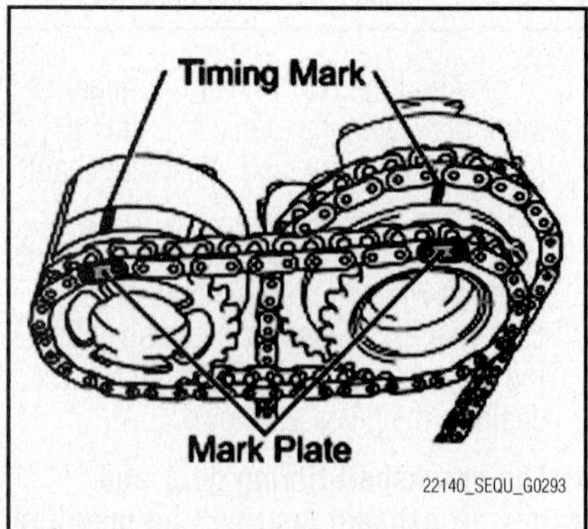

Fig. 222 Timing chain positioning LH

    d. Install the crankshaft timing sprocket to the crankshaft.
    e. Align and attach the knock pin of the No. 3 camshaft with the pin hole of the camshaft timing gear.
    f. Using the hexagonal portion of the No. 4 camshaft, align and attach the knock pin of the No. 4 camshaft with the pin hole of the camshaft timing exhaust gear.

➡Because the gears' timing mark positions may shift due to looseness of the No. 1 chain, use the hexagonal portion of the camshaft to hold the No. 3 camshaft in place until the No. 1 chain tensioner is installed.

    g. Remove the pin from the No. 2 chain tensioner.
64. Install No. 1 chain tensioner slipper.

➡If you cannot install the chain tensioner slipper due to the tension of the chain, use the hexagonal portion of the camshaft to loosen the chain, and then install the chain tensioner slipper.

65. Install No. 1 chain tensioner assembly.
    a. Move the stopper plate upward to release the lock, and push the plunger deep into the tensioner.
    b. Move the stopper plate downward to set the lock, and insert a hexagon wrench into the hole of the stopper plate.
    c. Install the chain tensioner with the 2 bolts and tighten to 7 ft. lbs. (10 Nm).
66. Install the vibration damper with the 2 bolts
    a. Remove the hexagon wrench from the chain tensioner.
67. Tighten camshaft timing gears LH.
    a. Using a wrench to hold the hexagonal portion of the No. 3 camshaft, tighten the camshaft timing gear with the bolt and tighten to 74 ft. lbs. (100 Nm).

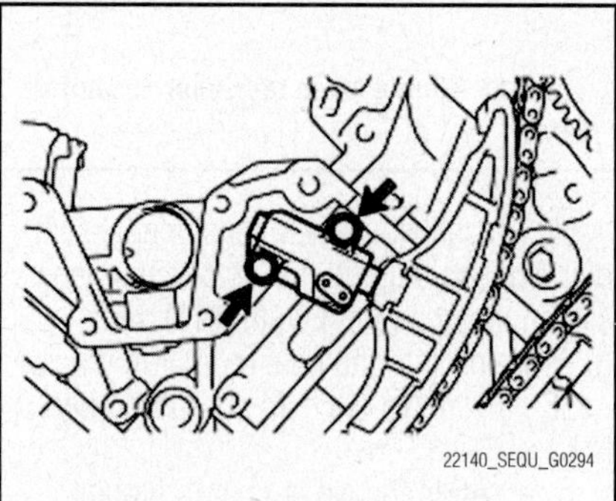

Fig. 223 Timing chain tensioner positioning LH

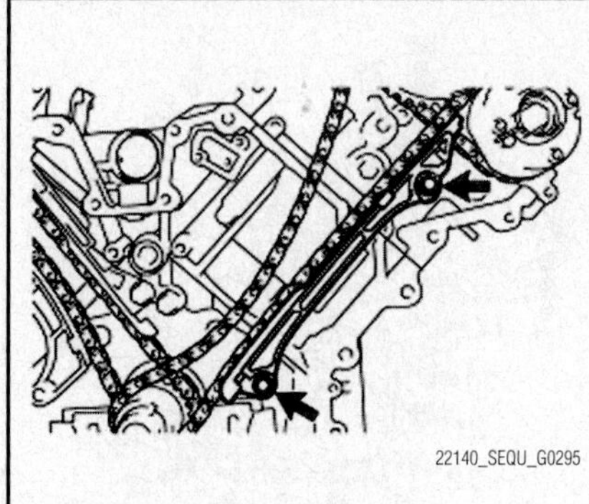

Fig. 224 Timing chain vibrations damper positioning RH

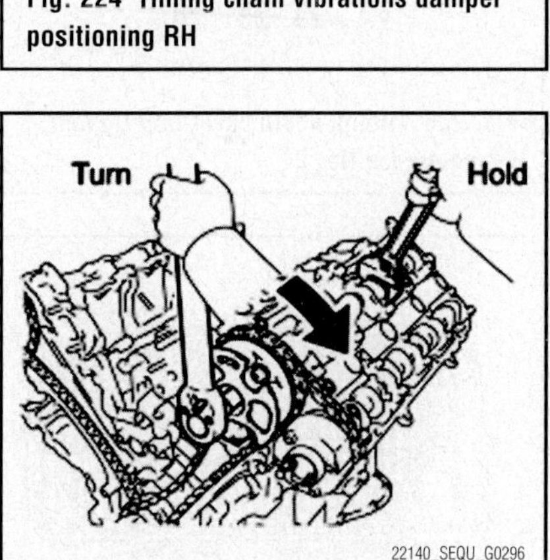
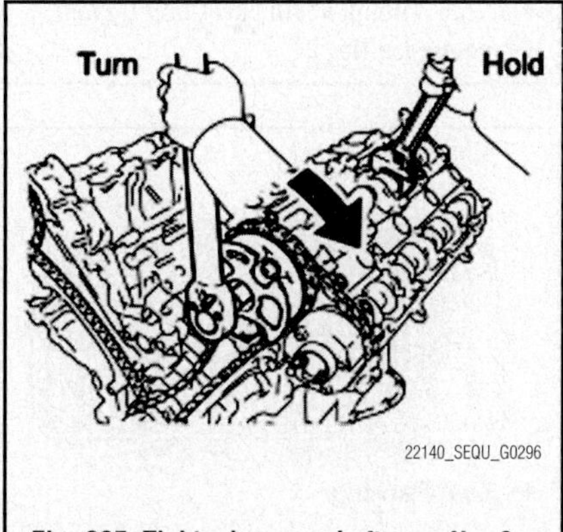

Fig. 225 Tightening camshaft gear No. 3

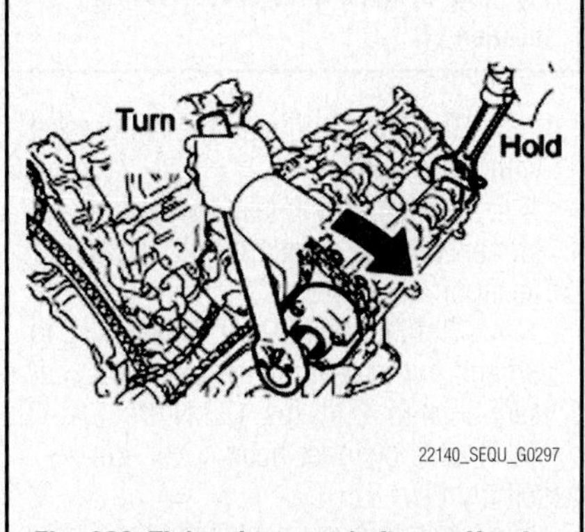

Fig. 226 Tightening camshaft gear No. 4

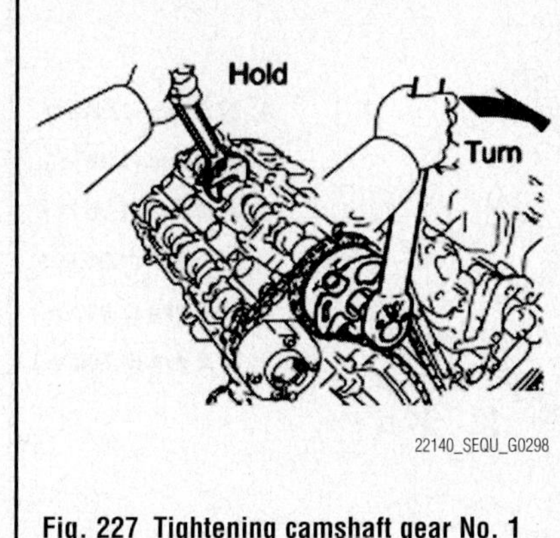

Fig. 227 Tightening camshaft gear No. 1

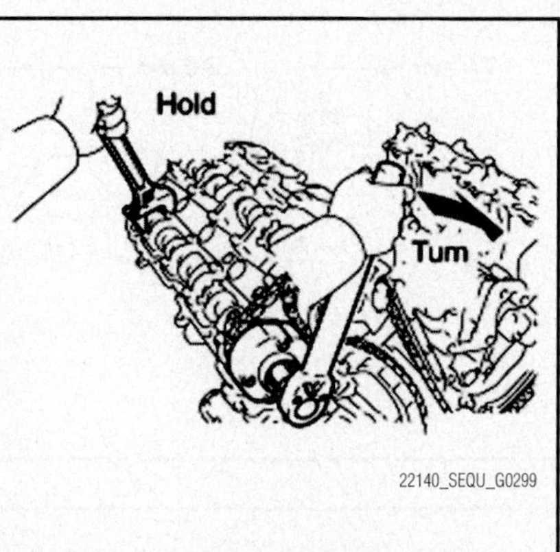

Fig. 228 Tightening camshaft gear No. 2

b. Using a wrench to hold the hexagonal portion of the No. 4 camshaft, tighten the camshaft timing gear with the bolt and tighten to 74 ft. lbs. (100 Nm).

68. Tighten camshaft timing gears RH.

a. Using a wrench to hold the hexagonal portion of the No. 1 camshaft, tighten the camshaft timing gear with the bolt and tighten to 74 ft. lbs. (100 Nm).

b. Using a wrench to hold the hexagonal portion of the No. 2 camshaft, tighten the camshaft timing gear with the bolt and tighten to 74 ft. lbs. (100 Nm).

69. Check No. 1 cylinder to TDC / compression.

a. Temporarily install the pulley set bolt.

b. Rotate the crankshaft clockwise, and check that the timing marks on the crankshaft timing gear and camshaft timing gears are as shown.

c. Remove the crankshaft pulley bolt.

70. Apply soapy water to 2 new O-rings and install them to the inlet pipe.

71. Install the inlet pipe to the No. 1 heat exchanger cover.

72. Install timing chain cover sub-assembly.

a. Apply a light coat of engine oil to a new oil pump gasket.

b. Install the oil pump gasket.

c. Apply a light coat of engine oil to a new O-ring.

d. Install the O-ring.

e. Apply seal packing in a continuous line to the timing chain cover as shown.

f. Align the oil pump's drive rotor spline and the crankshaft as shown in the illustration. Install the spline and chain cover to the crankshaft.

g. Temporarily install the timing chain cover with the 28 bolts and nut.

h. Tighten the 5 bolts in several steps in the sequence shown to 35 ft. lbs. (47 Nm).

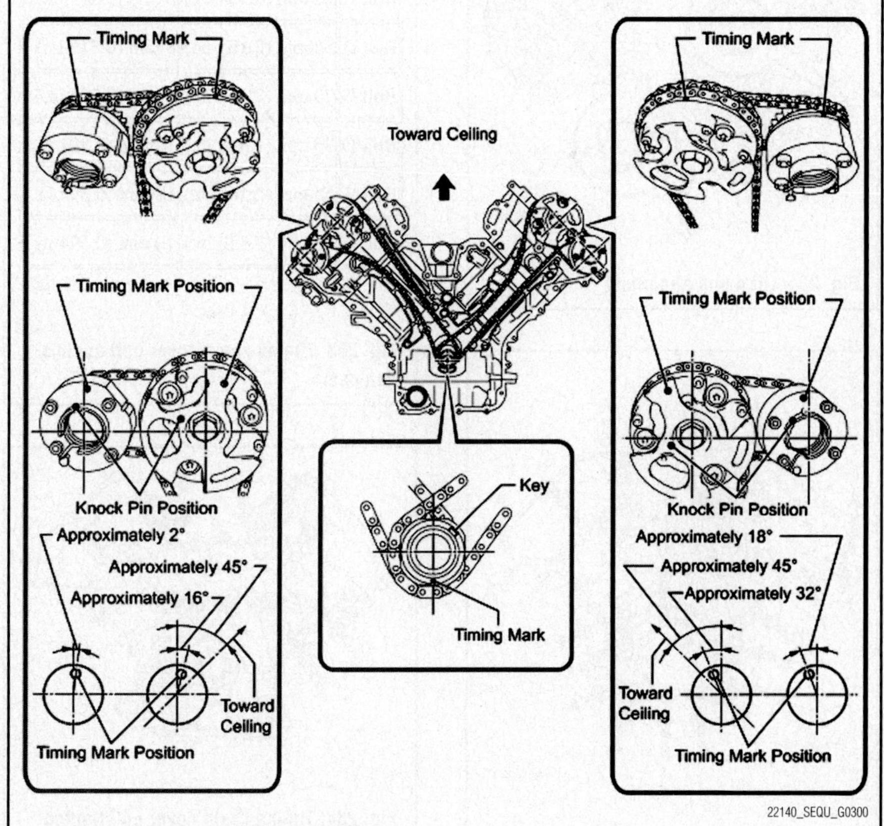

Fig. 229 Proper timing chain positioning

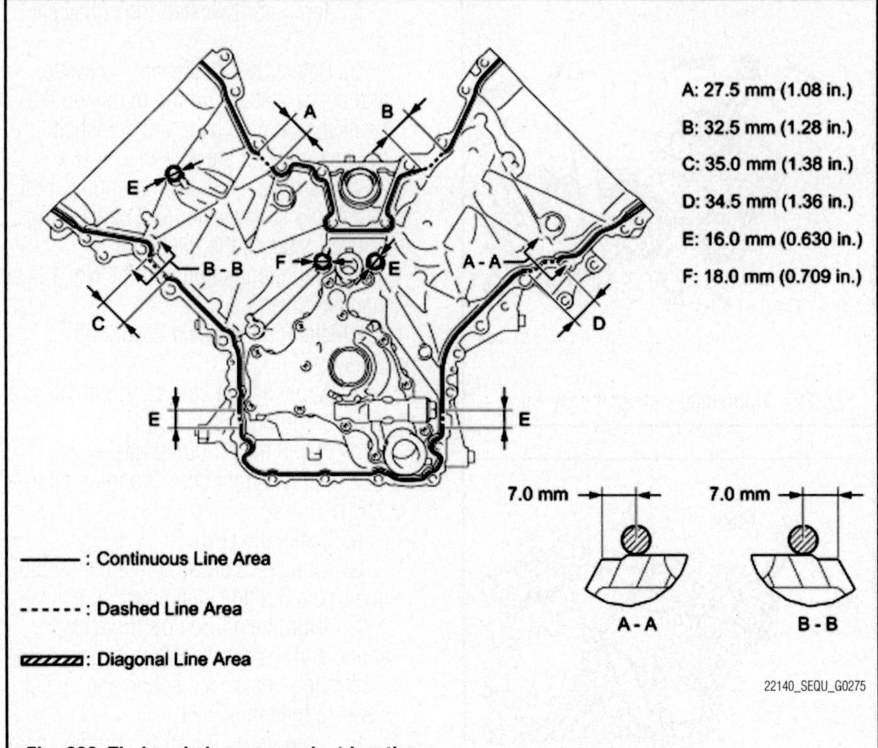

Fig. 230 Timing chain cover sealant locations

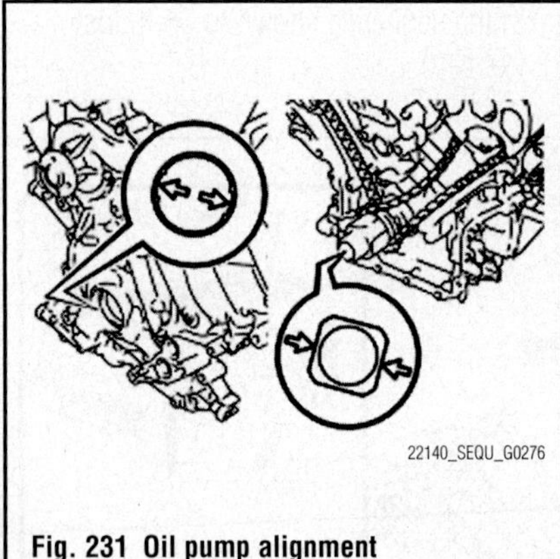

Fig. 231 Oil pump alignment

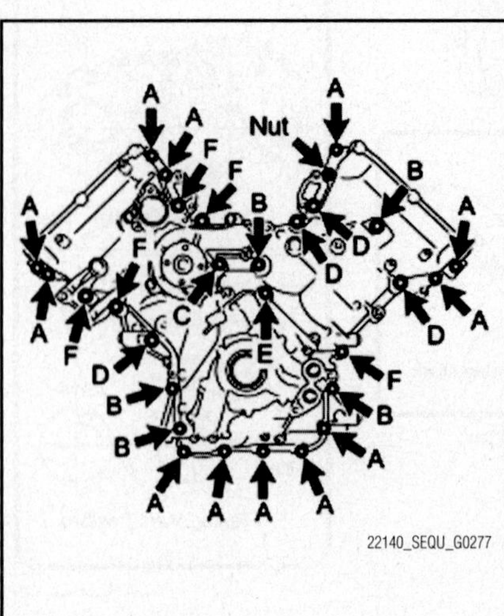

Fig. 232 Timing chain cover bolt locations

Item	Length	Thread diameter
Bolt A	25 mm (0.984 in.)	8 mm (0.315 in.)
Bolt B	55 mm (2.165 in.)	8 mm (0.315 in.)
Bolt C	70 mm (2.756 in.)	8 mm (0.315 in.)
Bolt D	35 mm (1.378 in.)	10 mm (0.394 in.)
Bolt E	55 mm (2.165 in.)	10 mm (0.394 in.)
Bolt F	80 mm (3.150 in.)	10 mm (0.394 in.)

22140_SEQU_G0278

Fig. 233 Timing chain cover bolt application chart

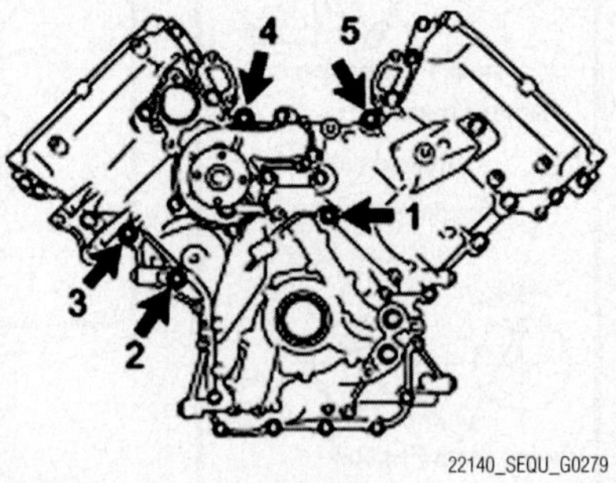

22140_SEQU_G0279

Fig. 234 Timing chain cover bolt tightening sequence No. 1

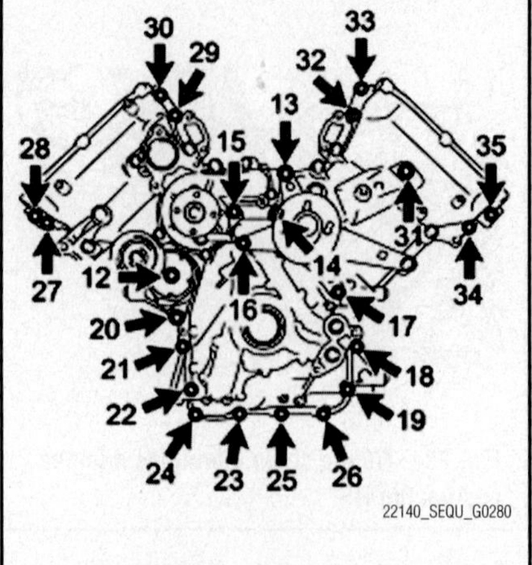

Fig. 235 Timing chain cover bolt tightening sequence No. 2

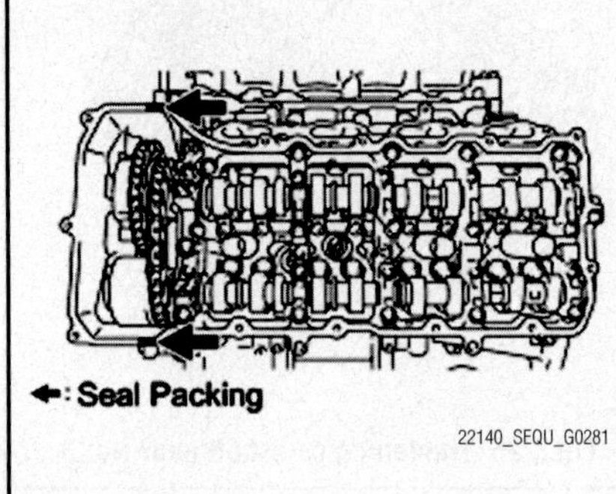

Fig. 236 Cylinder head cover packing location LH

i. Temporarily install the fan bracket with the 4 bolts.

j. Temporarily install the belt tensioner with the standard bolt and 6 mm hexagon wrench bolt.

k. Tighten the 23 bolts labeled 12 to 35 and nut in several steps in the sequence to 17 ft. lbs. (23 Nm).

73. Install cylinder head cover sub-assembly LH.

a. Install 5 new gaskets to the camshaft bearing caps (No. 2, No. 3).

b. Install the gasket to the cylinder head cover.

c. Apply seal packing as shown.

d. Install the cylinder head cover washer with a new seal and the 14 bolts and tighten bolt A to 15 ft. lbs. (21 Nm) and remaining bolts to 9 ft. lbs. (12 Nm).

74. Install cylinder head cover sub-assembly RH.

a. Install 5 new gaskets to the camshaft bearing caps (No. 1, No. 3).

b. Install the gasket to the cylinder head cover.

c. Apply seal packing as shown.

d. Install the cylinder head cover

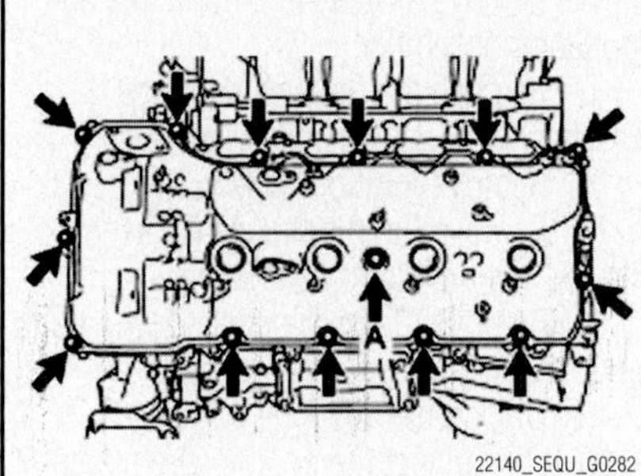

Fig. 237 Cylinder head cover bolt tightening sequence LH

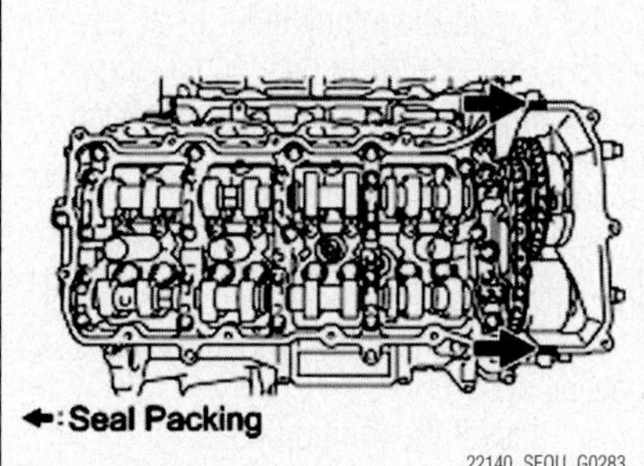

Fig. 238 Cylinder head cover packing location RH

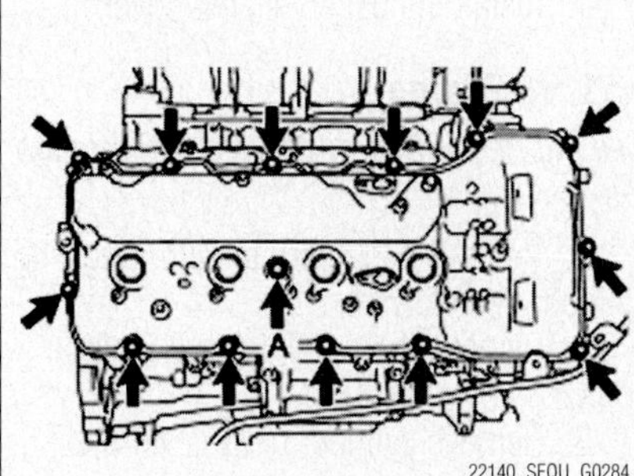

Fig. 239 Cylinder head cover bolt tightening sequence RH

washer with a new seal and the 14 bolts and tighten bolt A to 15 ft. lbs. (21 Nm) and remaining bolts to 9 ft. lbs. (12 Nm).

e. Install the noise filter to the cylinder head cover with the bolt and tighten to 62 inch lbs. (7.0 Nm).

75. Install the 8 ignition coils with the 8 bolts and tighten to 7 ft. lbs. (10 Nm).

76. Align the pulley set key with the key groove of the pulley, and slide on the pulley and tighten to 221 ft. Lbs. (300 Nm).

77. Connect the bracket to the timing chain cover with the bolt.

78. Install the idler pulley with the bolt and tighten to 32 ft. lbs. (43 Nm).

79. Temporarily install the water pump pulley with the 4 bolts and tighten to 15 ft. lbs. (21 Nm).

80. Install a new water cover gasket to the timing chain cover.

81. Install the water inlet with the 3 bolts and tighten to 15 ft. lbs. (21 Nm).

82. Connect the 2 air tube hoses.

83. Connect the 2 hoses.

84. Install No. 1 and No 2 engine covers.

85. Install 2 new gaskets and the water by-pass joint with the 4 nuts and tighten to 15 ft. lbs. (21 Nm).

86. Connect the No. 2 water by-pass hose to the water by-pass joint.

87. Install the water by-pass pipe.

88. Connect the air tube.

89. Install the No. 1 water by-pass hose by connecting the hose to the water inlet housing and front water by-pass joint.

90. Install the 2 stud bolts.

91. Install the generator with the 2 bolts and 2 nuts and tighten to 32 ft. lbs. (43 Nm).

92. Connect the generator wire with the nut and tighten to 87 inch lbs. (9.8 Nm).

93. Connect the wire harness bracket to the generator.

94. Connect the vane pump to the timing chain cover with the 2 bolts and tighten to 21 ft. lbs. (28 Nm).

95. Install engine oil level dipstick guide.

96. Install oil filter bracket (w/o Oil Cooler).

a. Apply a light coat of engine oil to 2 new O-rings.

b. Install the 2 O-rings to the timing chain cover.

c. Install the oil filter bracket with the 2 bolts and 2 nuts and tighten to 26 ft. lbs. (35 Nm).

97. Install No. 1 oil cooler bracket.

98. Install No. 2 water by-pass pipe.

99. Install oil pressure sender gauge assembly.

100. Install oil filter element.

101. Connect No. 2 fuel tube sub-assembly.

102. Install the cooler compressor with the 2 stud bolts and tighten to 7 ft. lbs. (10 Nm).

103. Install the 2 bolts and 2 nuts and tighten to 18 ft. lbs. (235 Nm).

104. Connect the 3 heater hoses.

105. Connect the No. 2 and No. 3 air hoses.

106. Connect the 2 clamps and air pump wire harness.

107. Install the intake manifold.

108. Engine Room RH Side:

a. Connect the 5 clamps.

b. Connect the throttle position sensor and throttle control motor connector.

c. Connect the 2 air pump connectors.

d. Install the ground wire with the 2 bolts.

e. Connect the noise filter connector.

f. Connect the 2 VVT sensor connectors.

g. Connect the injector connector.

h. Connect the 4 ignition coil connectors.

i. Connect the 2 camshaft timing oil control valve connectors.

j. Connect the 2 clamps and power steering oil pressure switch connector.

k. For 4WD: Connect the clamp and junction connector.

109. Engine Room LH Side:

a. Connect the cooler compressor connector.

b. Connect the 3 clamps.

c. Connect the camshaft position sensor connector.

d. Connect the 2 camshaft timing oil control valve connectors.

e. Connect the engine coolant temperature sensor connector.

f. Connect the 4 clamps.

g. Install the ground wire with the 2 bolts.

h. Connect the noise filter connector.

i. Connect the 2 VVT sensor connectors.

j. Connect the 4 ignition coil connectors.

k. Connect the injector connector.

l. Connect the 2 wire harness clamps.

m. Connect the 4 air injection control driver connectors.

n. Connect the 2 connectors and attach the 2 clamps to the engine room junction block.

o. Install the engine room relay block cover.

110. Install air cleaner assembly.

111. Install air cleaner hose assembly.

112. Install the radiator.

113. Add engine oil.

114. Add engine coolant.

115. Connect cable to negative battery terminal.

116. Inspect for leaks.

117. Check engine oil level.

118. Perform initializations, if necessary.

VALVE COVERS

REMOVAL & INSTALLATION

V6 Engine

See Figures 240 and 241.

1. Before servicing the vehicle, refer to the precautions in the beginning of this section.

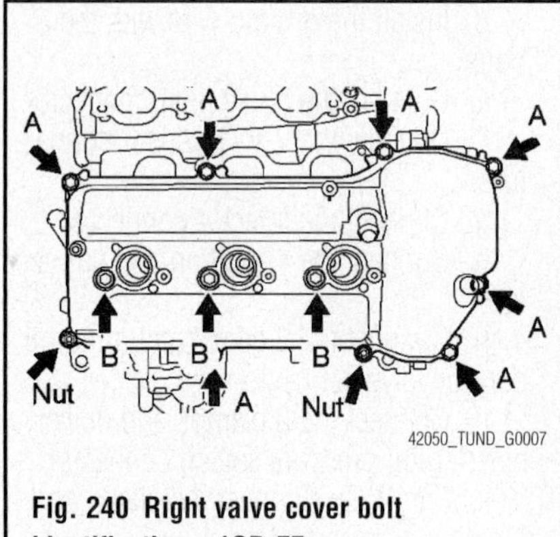

Fig. 240 Right valve cover bolt identification—1GR-FE

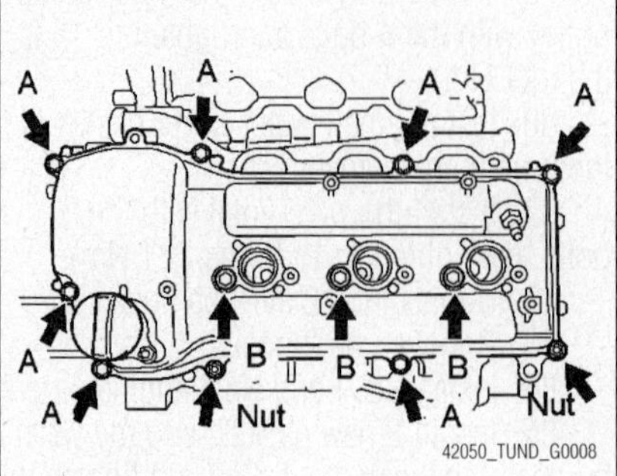

Fig. 241 Left valve cover bolt identification—1GR-FE

2. Disconnect cable from negative battery terminal.
3. Drain engine coolant.
4. Remove v-bank cover.
5. Remove air cleaner assembly.
6. Disconnect the 2 water by-pass hoses.
7. Disconnect the fuel vapor feed hose.
8. Disconnect the ventilation hose.
9. Disconnect the 2 VSV connectors.
10. Disconnect the throttle body w/ motor connector.
11. Separate the 3 wire harness clamps and hose clamp.
12. Remove the nut, then separate the clutch flexible hose bracket from the surge tank stay (w/ Manual Transmission).
13. Remove the 2 bolts and throttle body bracket.
14. Remove the bolt and oil baffle plate.
15. Remove the 4 bolts and 2 surge tank stays.
16. Remove the 2 nuts.
17. Remove the 4 bolts, intake air surge tank and gasket.
18. Disconnect the 3 connectors.
19. Remove the 3 bolts, then remove the 3 ignition coils.
20. Remove the 10 bolts, 3 seal washers, 2 nuts, right cylinder head cover and gasket.
21. Remove the 10 bolts, 3 seal washers, 2 nuts, left cylinder head cover and gasket.

To install:

22. Remove any old packing material and be careful not to drop any oil on the contact surfaces of the cylinder head, timing chain cover and cylinder head covers.
23. Install the cylinder head covers within 3 minutes of applying seal packing Part No. 08826-00080 or equivalent. Tighten the cylinder head cover bolts and nuts within 15 minutes of installing the cylinder head covers. Otherwise, the seal packing mush be removed and reapplied.
24. Install the seal washers onto the bolts.
25. Install the cylinder head cover with the 10 bolts and 2 nuts. Tighten the bolts and nuts uniformly in several steps as follows:
    a. Bolts labeled A: 10 Nm (7.4 ft. lbs.).
    b. Bolts labeled B: 9.0 Nm (80 in. lbs.).
    c. Nuts: 9.0 Nm (80 in. lbs.).
26. Install the ignition coils and tighten the bolts to 10 Nm (7.4 ft. lbs.).
27. Install a new gasket onto the intake air surge tank.
28. Install the intake air surge tank with the 4 bolts and tighten to 28 Nm (21 ft. lbs).
29. Install the 2 intake air surge tank nuts and tighten to 28 Nm (21 ft. lbs).
30. Install the 2 surge tank stays with the 4 bolts and tighten to 21 Nm (15 ft. lbs).
31. Install the oil baffle plate with the bolt and tighten to 9.0 Nm (80 in. lbs.).
32. Install the throttle body bracket with the 2 bolts and tighten to 21 Nm (15 ft. lbs).
33. Install the clutch flexible hose bracket with the nut (w/ Manual Transmission) and tighten to 24 Nm (18 ft. lbs).
34. Install the 3 wire harness clamps and hose clamp.
35. Connect the throttle body w/ motor connector.
36. Connect the 2 VSV connectors.
37. Connect the ventilation hose.
38. Connect the fuel vapor feed hose.
39. Connect the 2 water by-pass hoses.
40. Install air cleaner assembly.
41. Connect cable to negative battery terminal.
42. Add engine coolant.
43. Check for engine coolant leakage.
44. Install the V-bank cover with the 2 nuts and tighten to 7.5 Nm (66 in. lbs.)

4.7L V8 Engine

1. Before servicing the vehicle, refer to the precautions in the beginning of this section.
2. Remove battery clamp cover.
3. Remove throttle body cover.
4. Remove air cleaner and intake air connector assembly.
5. Remove the upper timing belt covers.
6. Remove ignition coils.
7. Remove the 9 bolts, seal washers and the right cylinder head cover.
8. Remove the oil dipstick for the transmission.
9. Disconnect the PCV hose.
10. Disconnect the engine wire clamp from the wire bracket on the cylinder head cover.
11. Remove the 9 bolts, 9 seal washers and the left cylinder head cover.

To install:

12. Install the left cylinder head cover.
13. Connect the engine wire clamp to the wire bracket on the cylinder head cover.
14. Connect the PCV hose.
15. Replace the oil dipstick for the transmission.
16. Install the right cylinder head cover.
17. Install the ignition coils.
18. Install the upper timing belt covers.
19. Install the air cleaner and intake air connector assembly.
20. Install the throttle body cover.
21. Install the battery clamp cover.

5.7L V8 Engine

See Figures 176 through 179, 242 through 245.

1. Remove necessary to gain access to cylinder head covers.
2. Remove the 8 bolts and 8 ignition coils.
3. Remove cylinder head cover sub-assembly LH.
    a. Remove the 14 bolts, seal washer, cylinder head cover and gasket.
    b. Remove the 5 gaskets from the camshaft bearing caps (No. 2, No. 3).
4. Remove cylinder head cover sub-assembly RH.
    a. Remove the bolt and noise filter.
    b. Remove the 14 bolts, seal washer, cylinder head cover and gasket.
    c. Remove the 5 gaskets from the camshaft bearing caps (No. 1, No. 3).

To install:

5. Install cylinder head cover sub-assembly LH.
    a. Install 5 new gaskets to the camshaft bearing caps (No. 2, No. 3).
    b. Install the gasket to the cylinder head cover.
    c. Apply seal packing as shown.

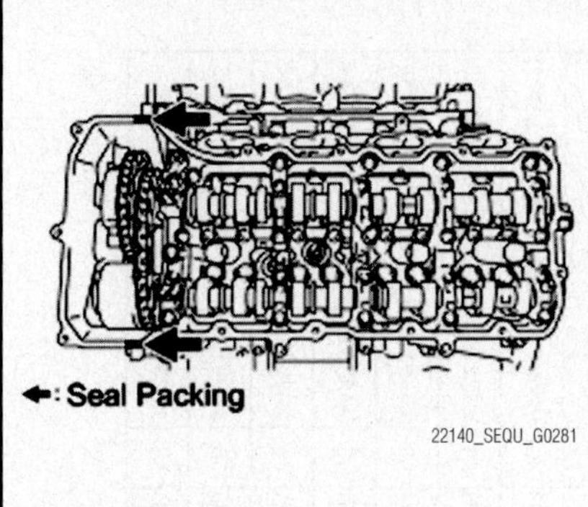

Fig. 242 Cylinder head cover packing location LH

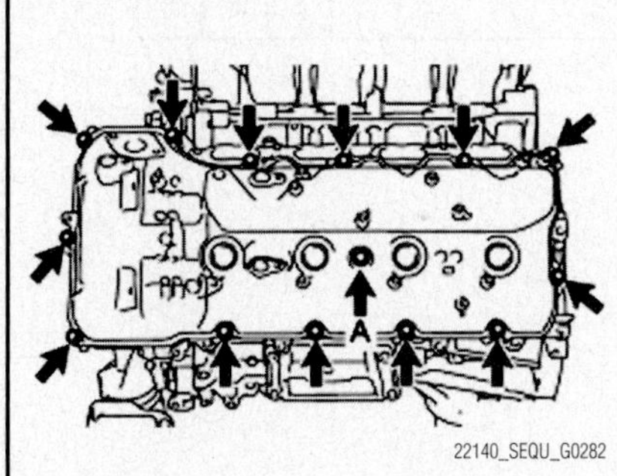

Fig. 243 Cylinder head cover bolt tightening sequence LH

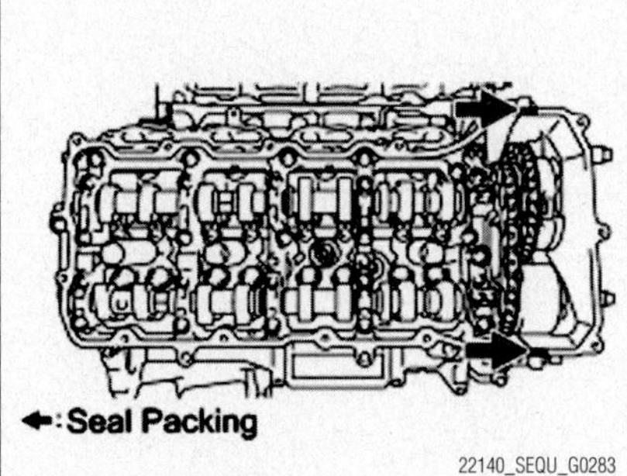

Fig. 244 Cylinder head cover packing location RH

d. Install the cylinder head cover washer with a new seal and the 14 bolts and tighten bolt A to 15 ft. lbs. (21 Nm) and remaining bolts to 9 ft. lbs. (12 Nm).

6. Install cylinder head cover sub-assembly RH.

a. Install 5 new gaskets to the camshaft bearing caps (No. 1, No. 3).

b. Install the gasket to the cylinder head cover.

c. Apply seal packing as shown.

d. Install the cylinder head cover washer with a new seal and the 14 bolts and tighten bolt A to 15 ft. lbs. (21 Nm)

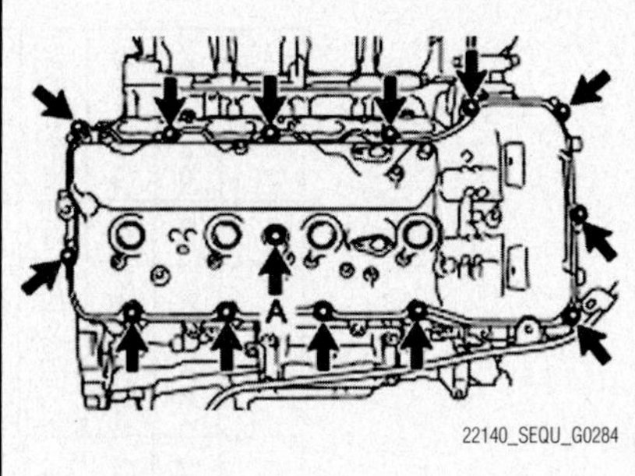

Fig. 245 Cylinder head cover bolt tightening sequence RH

and remaining bolts to 9 ft. lbs. (12 Nm).

e. Install the noise filter to the cylinder head cover with the bolt and tighten to 62 inch lbs. (7.0 Nm).

7. Install the 8 ignition coils with the 8 bolts and tighten to 7 ft. lbs. (10 Nm).

8. Reinstall items accessed to remove cylinder head covers.

VALVE LASH

ADJUSTMENT

V6 Engine

See Figures 246 through 248.

1. Before servicing the vehicle, refer to the Precautions Section.
2. Disconnect the negative battery cable.
3. Drain the engine coolant. Remove the V bank cover.
4. Remove the air cleaner assembly.
5. Disconnect the two water bypass hoses. Disconnect the fuel vapor feed hose. Disconnect the ventilation hose. Disconnect the VSV connectors.
6. Disconnect the throttle body motor connector. Separate the three wire harness clamps and hose clamp.
7. Remove the two bolts and the throttle body bracket. Remove the bolt and the oil baffle plate. Remove the four bolts and the two surge tank stays.
8. Remove the two nuts. Remove the four bolts, intake air surge tank and gasket.
9. Remove the ignition coil assembly.
10. Remove the cylinder head cover retaining bolts. Remove the cylinder head cover.
11. Turn the crankshaft pulley until its groove and the timing mark "0" of the timing chain cover are aligned. If not aligned at TDC of the compression stroke, turn the crankshaft one complete revolution, in the direction of rotation.

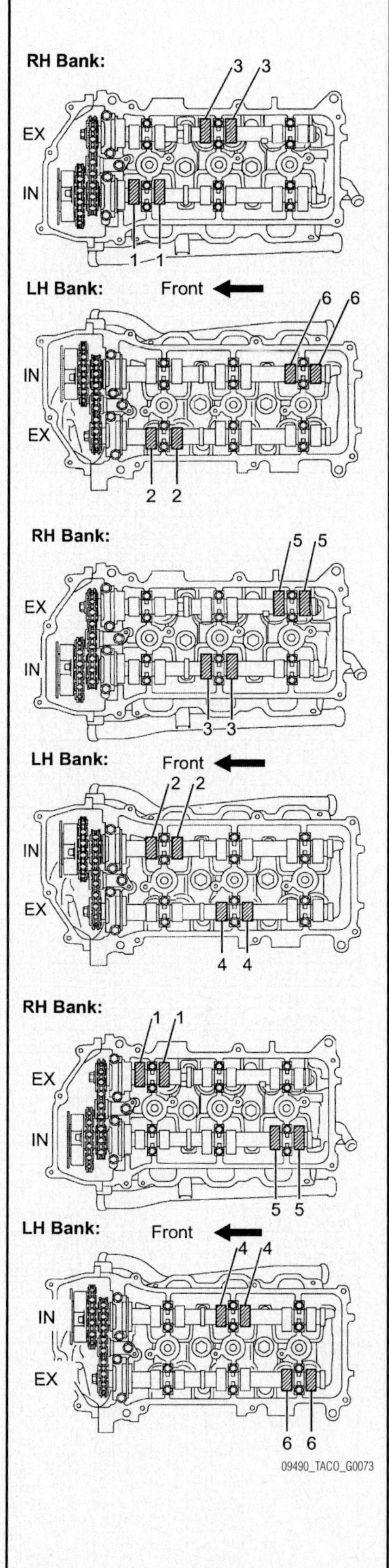

Fig. 246 Valve clearance location—4.0L engine

	5.060 (0.1992)	5.080 (0.2000)	5.100 (0.2008)	5.120 (0.2016)	5.140 (0.2024)	5.160 (0.2031)	5.180 (0.2039)	5.200 (0.2047)	5.210 (0.2051)	5.220 (0.2055)	5.230 (0.2059)	5.240 (0.2063)	5.250 (0.2067)	5.260 (0.2071)	5.270 (0.2075)	5.280 (0.2079)	5.290 (0.2083)	5.300 (0.2087)	5.310 (0.2091)	5.320 (0.2094)	5.330 (0.2098)	5.340 (0.2102)	5.350 (0.2106)	5.360 (0.2110)	5.370 (0.2114)	5.380 (0.2118)	5.390 (0.2122)	5.400 (0.2126)	5.410 (0.2130)	5.420 (0.2134)	5.430 (0.2138)	5.440 (0.2142)	5.450 (0.2146)	5.460 (0.2150)	5.470 (0.2154)	5.480 (0.2157)	5.490 (0.2161)	5.500 (0.2165)	5.510 (0.2169)	5.520 (0.2173)	5.530 (0.2177)	5.540 (0.2181)	5.550 (0.2185)	5.560 (0.2189)	5.570 (0.2193)	5.580 (0.2197)	5.590 (0.2201)	5.600 (0.2205)	5.620 (0.2213)	5.640 (0.2220)	5.660 (0.2228)	5.680 (0.2236)	5.700 (0.2244)	5.720 (0.2252)	5.740 (0.2260)
0.000 - 0.020 (0.0000 - 0.0008)													06	08	08	10	10	12	12	14	14	16	16	18	18	20	20	22	22	24	24	26	26	28	28	30	30	32	32	34	34	36	36	38	38	40	40	42	44	46	48	50	52	54	56
0.021 - 0.040 (0.0008 - 0.0016)											06	08	08	10	10	12	12	14	14	16	16	18	18	20	20	22	22	24	24	26	26	28	28	30	30	32	32	34	34	36	36	38	38	40	40	42	42	44	46	48	50	52	54	56	58
0.041 - 0.060 (0.0016 - 0.0024)									06	08	08	10	10	12	12	14	14	16	16	18	18	20	20	22	22	24	24	26	26	28	28	30	30	32	32	34	34	36	36	38	38	40	40	42	42	44	44	46	48	50	52	54	56	58	60
0.061 - 0.080 (0.0024 - 0.0031)								08	08	10	10	12	12	14	14	16	16	18	18	20	20	22	22	24	24	26	26	28	28	30	30	32	32	34	34	36	36	38	38	40	40	42	42	44	44	46	46	48	50	52	54	56	58	60	62
0.081 - 0.100 (0.0032 - 0.0039)							08	10	10	12	12	14	14	16	16	18	18	20	20	22	22	24	24	26	26	28	28	30	30	32	32	34	34	36	36	38	38	40	40	42	42	44	44	46	46	48	48	50	52	54	56	58	60	62	64
0.101 - 0.120 (0.0040 - 0.0047)						08	10	12	12	14	14	16	16	18	18	20	20	22	22	24	24	26	26	28	28	30	30	32	32	34	34	36	36	38	38	40	40	42	42	44	44	46	46	48	48	50	50	52	54	56	58	60	62	64	66
0.121 - 0.140 (0.0048 - 0.0055)					08	10	12	14	14	16	16	18	18	20	20	22	22	24	24	26	26	28	28	30	30	32	32	34	34	36	36	38	38	40	40	42	42	44	44	46	46	48	48	50	50	52	52	54	56	58	60	62	64	66	68
0.141 - 0.149 (0.0056 - 0.0059)				06	08	10	12	14	16	16	18	18	20	20	22	22	24	24	26	26	28	28	30	30	32	32	34	34	36	36	38	38	40	40	42	42	44	44	46	46	48	48	50	50	52	52	54	54	56	58	60	62	64	66	68
0.251 - 0.270 (0.0099 - 0.0106)	12	14	16	18	20	22	24	26	28	28	30	30	32	32	34	34	36	36	38	38	40	40	42	42	44	44	46	46	48	48	50	50	52	52	54	54	56	56	58	58	60	60	62	62	64	64	66	66	68	70	72				
0.271 - 0.290 (0.0107 - 0.0114)	14	16	18	20	22	24	26	28	30	30	32	32	34	34	36	36	38	38	40	40	42	42	44	44	46	46	48	48	50	50	52	52	54	54	56	56	58	58	60	60	62	62	64	64	66	66	68	68	70	72					
0.291 - 0.310 (0.0115 - 0.0122)	16	18	20	22	24	26	28	30	32	32	34	34	36	36	38	38	40	40	42	42	44	44	46	46	48	48	50	50	52	52	54	54	56	56	58	58	60	60	62	62	64	64	66	66	68	68	70	70	72						
0.311 - 0.330 (0.0122 - 0.0130)	18	20	22	24	26	28	30	32	34	34	36	36	38	38	40	40	42	42	44	44	46	46	48	48	50	50	52	52	54	54	56	56	58	58	60	60	62	62	64	64	66	66	68	68	70	70	72	72							
0.331 - 0.350 (0.0130 - 0.0138)	20	22	24	26	28	30	32	34	36	36	38	38	40	40	42	42	44	44	46	46	48	48	50	50	52	52	54	54	56	56	58	58	60	60	62	62	64	64	66	66	68	68	70	70	72	72	74								
0.351 - 0.370 (0.0138 - 0.0146)	22	24	26	28	30	32	34	36	38	38	40	40	42	42	44	44	46	46	48	48	50	50	52	52	54	54	56	56	58	58	60	60	62	62	64	64	66	66	68	68	70	70	72	72	74										
0.371 - 0.390 (0.0146 - 0.0154)	24	26	28	30	32	34	36	38	40	40	42	42	44	44	46	46	48	48	50	50	52	52	54	54	56	56	58	58	60	60	62	62	64	64	66	66	68	68	70	70	72	72	74												
0.391 - 0.410 (0.0154 - 0.0161)	26	28	30	32	34	36	38	40	42	42	44	44	46	46	48	48	50	50	52	52	54	54	56	56	58	58	60	60	62	62	64	64	66	66	68	68	70	70	72	72	74														
0.411 - 0.430 (0.0162 - 0.0169)	28	30	32	34	36	38	40	42	44	44	46	46	48	48	50	50	52	52	54	54	56	56	58	58	60	60	62	62	64	64	66	66	68	68	70	70	72	72	74																
0.431 - 0.450 (0.0170 - 0.0177)	30	32	34	36	38	40	42	44	46	46	48	48	50	50	52	52	54	54	56	56	58	58	60	60	62	62	64	64	66	66	68	68	70	70	72	72	74																		
0.451 - 0.470 (0.0178 - 0.0185)	32	34	36	38	40	42	44	46	48	48	50	50	52	52	54	54	56	56	58	58	60	60	62	62	64	64	66	66	68	68	70	70	72	72	74																				
0.471 - 0.490 (0.0185 - 0.0193)	34	36	38	40	42	44	46	48	50	50	52	52	54	54	56	56	58	58	60	60	62	62	64	64	66	66	68	68	70	70	72	72	74																						
0.491 - 0.510 (0.0193 - 0.0201)	36	38	40	42	44	46	48	50	52	52	54	54	56	56	58	58	60	60	62	62	64	64	66	66	68	68	70	70	72	72	74																								
0.511 - 0.530 (0.0201 - 0.0209)	38	40	42	44	46	48	50	52	54	54	56	56	58	58	60	60	62	62	64	64	66	66	68	68	70	70	72	72	74																										
0.531 - 0.550 (0.0209 - 0.0217)	40	42	44	46	48	50	52	54	56	56	58	58	60	60	62	62	64	64	66	66	68	68	70	70	72	72	74																												
0.551 - 0.570 (0.0217 - 0.0224)	42	44	46	48	50	52	54	56	58	58	60	60	62	62	64	64	66	66	68	68	70	70	72	72	74																														
0.571 - 0.590 (0.0225 - 0.0232)	44	46	48	50	52	54	56	58	60	60	62	62	64	64	66	66	68	68	70	70	72	72	74																																
0.591 - 0.610 (0.0233 - 0.0240)	46	48	50	52	54	56	58	60	62	62	64	64	66	66	68	68	70	70	72	72	74																																		
0.611 - 0.630 (0.0241 - 0.0248)	48	50	52	54	56	58	60	62	64	64	66	66	68	68	70	70	72	72	74																																				
0.631 - 0.650 (0.0248 - 0.0256)	50	52	54	56	58	60	62	64	66	66	68	68	70	70	72	72	74																																						
0.651 - 0.670 (0.0256 - 0.0264)	52	54	56	58	60	62	64	66	68	68	70	70	72	72	74																																								
0.671 - 0.690 (0.0264 - 0.0272)	54	56	58	60	62	64	66	68	70	70	72	72	74																																										
0.691 - 0.710 (0.0272 - 0.0280)	56	58	60	62	64	66	68	70	72	72	74																																												
0.711 - 0.730 (0.0280 - 0.0287)	58	60	62	64	66	68	70	72	74																																														
0.731 - 0.750 (0.0288 - 0.0295)	60	62	64	66	68	70	72																																																
0.751 - 0.770 (0.0296 - 0.0303)	62	64	66	68	70	72																																																	
0.771 - 0.790 (0.0304 - 0.0311)	64	66	68	70	72																																																		
0.791 - 0.810 (0.0311 - 0.0319)	66	68	70	72																																																			
0.811 - 0.830 (0.0319 - 0.0327)	68	70	72																																																				
0.831 - 0.850 (0.0327 - 0.0335)	70	72																																																					
0.851 - 0.870 (0.0335 - 0.0343)	72																																																							

06	5.060 (0.1992)	30	5.300 (0.2087)	54	5.540 (0.2181)
08	5.080 (0.2000)	32	5.320 (0.2094)	56	5.560 (0.2189)
10	5.100 (0.2008)	34	5.340 (0.2102)	58	5.580 (0.2197)
12	5.120 (0.2016)	36	5.360 (0.2110)	60	5.600 (0.2205)
14	5.140 (0.2024)	38	5.380 (0.2118)	62	5.620 (0.2213)
16	5.160 (0.2031)	40	5.400 (0.2126)	64	5.640 (0.2220)
18	5.180 (0.2039)	42	5.420 (0.2134)	66	5.660 (0.2228)
20	5.200 (0.2047)	44	5.440 (0.2142)	68	5.680 (0.2236)
22	5.220 (0.2055)	46	5.460 (0.2150)	70	5.700 (0.2244)
24	5.240 (0.2063)	48	5.480 (0.2157)	72	5.720 (0.2252)
26	5.260 (0.2071)	50	5.500 (0.2165)	74	5.740 (0.2260)
28	5.280 (0.2079)	52	5.520 (0.2173)		

09490_TACO_G0074

Fig. 247 Shim selection chart, part one—4.0L engine

	5.060 (0.1992)	5.080 (0.2000)	5.100 (0.2008)	5.120 (0.2016)	5.140 (0.2024)	5.160 (0.2031)	5.180 (0.2039)	5.200 (0.2047)	5.210 (0.2051)	5.220 (0.2055)	5.230 (0.2059)	5.240 (0.2063)	5.250 (0.2067)	5.260 (0.2071)	5.270 (0.2075)	5.280 (0.2079)	5.290 (0.2083)	5.300 (0.2087)	5.310 (0.2091)	5.320 (0.2094)	5.330 (0.2098)	5.340 (0.2102)	5.350 (0.2106)	5.360 (0.2110)	5.370 (0.2114)	5.380 (0.2118)	5.390 (0.2122)	5.400 (0.2126)	5.410 (0.2130)	5.420 (0.2134)	5.430 (0.2138)	5.440 (0.2142)	5.450 (0.2146)	5.460 (0.2150)	5.470 (0.2154)	5.480 (0.2157)	5.490 (0.2161)	5.500 (0.2165)	5.510 (0.2169)	5.520 (0.2173)	5.530 (0.2177)	5.540 (0.2181)	5.550 (0.2185)	5.560 (0.2189)	5.570 (0.2193)	5.580 (0.2197)	5.590 (0.2201)	5.600 (0.2205)	5.620 (0.2213)	5.640 (0.2220)	5.660 (0.2228)	5.680 (0.2236)	5.700 (0.2244)	5.720 (0.2252)	5.740 (0.2260)
0.000 - 0.020 (0.0000 - 0.0008)																											06	08	08	10	10	12	12	14	14	16	16	18	18	20	20	22	22	24	24	26	26	28	30	32	34	36	38	40	42
0.021 - 0.040 (0.0008 - 0.0016)																									06	08	08	10	10	12	12	14	14	16	16	18	18	20	20	22	22	24	24	26	26	28	28	30	32	34	36	38	40	42	44
0.041 - 0.060 (0.0016 - 0.0024)																							06	08	08	10	10	12	12	14	14	16	16	18	18	20	20	22	22	24	24	26	26	28	28	30	30	32	34	36	38	40	42	44	46
0.061 - 0.080 (0.0024 - 0.0031)																					06	08	08	10	10	12	12	14	14	16	16	18	18	20	20	22	22	24	24	26	26	28	28	30	30	32	32	34	36	38	40	42	44	46	48
0.081 - 0.100 (0.0032 - 0.0039)																			06	08	08	10	10	12	12	14	14	16	16	18	18	20	20	22	22	24	24	26	26	28	28	30	30	32	32	34	34	36	38	40	42	44	46	48	50
0.101 - 0.120 (0.0040 - 0.0047)																	06	08	08	10	10	12	12	14	14	16	16	18	18	20	20	22	22	24	24	26	26	28	28	30	30	32	32	34	34	36	36	38	40	42	44	46	48	50	52
0.121 - 0.140 (0.0048 - 0.0055)															06	08	08	10	10	12	12	14	14	16	16	18	18	20	20	22	22	24	24	26	26	28	28	30	30	32	32	34	34	36	36	38	38	40	42	44	46	48	50	52	54
0.141 - 0.160 (0.0056 - 0.0063)													06	08	08	10	10	12	12	14	14	16	16	18	18	20	20	22	22	24	24	26	26	28	28	30	30	32	32	34	34	36	36	38	38	40	40	42	44	46	48	50	52	54	56
0.161 - 0.180 (0.0063 - 0.0071)											06	08	08	10	10	12	12	14	14	16	16	18	18	20	20	22	22	24	24	26	26	28	28	30	30	32	32	34	34	36	36	38	38	40	40	42	42	44	46	48	50	52	54	56	58
0.181 - 0.200 (0.0071 - 0.0079)									06	08	08	10	10	12	12	14	14	16	16	18	18	20	20	22	22	24	24	26	26	28	28	30	30	32	32	34	34	36	36	38	38	40	40	42	42	44	44	46	48	50	52	54	56	58	60
0.201 - 0.220 (0.0079 - 0.0087)								08	08	10	10	12	12	14	14	16	16	18	18	20	20	22	22	24	24	26	26	28	28	30	30	32	32	34	34	36	36	38	38	40	40	42	42	44	44	46	46	48	50	52	54	56	58	60	62
0.221 - 0.240 (0.0087 - 0.0094)							08	10	10	12	12	14	14	16	16	18	18	20	20	22	22	24	24	26	26	28	28	30	30	32	32	34	34	36	36	38	38	40	40	42	42	44	44	46	46	48	48	50	52	54	56	58	60	62	64
0.241 - 0.260 (0.0095 - 0.0102)						08	10	12	12	14	14	16	16	18	18	20	20	22	22	24	24	26	26	28	28	30	30	32	32	34	34	36	36	38	38	40	40	42	42	44	44	46	46	48	48	50	50	52	54	56	58	60	62	64	66
0.261 - 0.280 (0.0103 - 0.0110)					08	10	12	14	14	16	16	18	18	20	20	22	22	24	24	26	26	28	28	30	30	32	32	34	34	36	36	38	38	40	40	42	42	44	44	46	46	48	48	50	50	52	52	54	56	58	60	62	64	66	68
0.281 - 0.289 (0.0111 - 0.0114)				06	08	10	12	14	16	16	18	18	20	20	22	22	24	24	26	26	28	28	30	30	32	32	34	34	36	36	38	38	40	40	42	42	44	44	46	46	48	48	50	50	52	52	54	54	56	58	60	62	64	66	68
0.391 - 0.410 (0.0154 - 0.0161)	12	14	16	18	20	22	24	26	28	28	30	30	32	32	34	34	36	36	38	38	40	40	42	42	44	44	46	46	48	48	50	50	52	52	54	54	56	56	58	58	60	60	62	62	64	64	66	66	68	70	72				
0.411 - 0.430 (0.0162 - 0.0169)	14	16	18	20	22	24	26	28	30	30	32	32	34	34	36	36	38	38	40	40	42	42	44	44	46	46	48	48	50	50	52	52	54	54	56	56	58	58	60	60	62	62	64	64	66	66	68	68	70	72					
0.431 - 0.450 (0.0170 - 0.0177)	16	18	20	22	24	26	28	30	32	32	34	34	36	36	38	38	40	40	42	42	44	44	46	46	48	48	50	50	52	52	54	54	56	56	58	58	60	60	62	62	64	64	66	66	68	68	70	70	72						
0.451 - 0.470 (0.0178 - 0.0185)	18	20	22	24	26	28	30	32	34	34	36	36	38	38	40	40	42	42	44	44	46	46	48	48	50	50	52	52	54	54	56	56	58	58	60	60	62	62	64	64	66	66	68	68	70	70	72	72							
0.471 - 0.490 (0.0185 - 0.0193)	20	22	24	26	28	30	32	34	36	36	38	38	40	40	42	42	44	44	46	46	48	48	50	50	52	52	54	54	56	56	58	58	60	60	62	62	64	64	66	66	68	68	70	70	72	72	74								
0.491 - 0.510 (0.0193 - 0.0201)	22	24	26	28	30	32	34	36	38	38	40	40	42	42	44	44	46	46	48	48	50	50	52	52	54	54	56	56	58	58	60	60	62	62	64	64	66	66	68	68	70	70	72	72	74										
0.511 - 0.530 (0.0201 - 0.0209)	24	26	28	30	32	34	36	38	40	40	42	42	44	44	46	46	48	48	50	50	52	52	54	54	56	56	58	58	60	60	62	62	64	64	66	66	68	68	70	70	72	72	74												
0.531 - 0.550 (0.0209 - 0.0217)	26	28	30	32	34	36	38	40	42	42	44	44	46	46	48	48	50	50	52	52	54	54	56	56	58	58	60	60	62	62	64	64	66	66	68	68	70	70	72	72	74														
0.551 - 0.570 (0.0217 - 0.0224)	28	30	32	34	36	38	40	42	44	44	46	46	48	48	50	50	52	52	54	54	56	56	58	58	60	60	62	62	64	64	66	66	68	68	70	70	72	72	74																
0.571 - 0.590 (0.0225 - 0.0232)	30	32	34	36	38	40	42	44	46	46	48	48	50	50	52	52	54	54	56	56	58	58	60	60	62	62	64	64	66	66	68	68	70	70	72	72	74																		
0.591 - 0.610 (0.0233 - 0.0240)	32	34	36	38	40	42	44	46	48	48	50	50	52	52	54	54	56	56	58	58	60	60	62	62	64	64	66	66	68	68	70	70	72	72	74																				
0.611 - 0.630 (0.0241 - 0.0248)	34	36	38	40	42	44	46	48	50	50	52	52	54	54	56	56	58	58	60	60	62	62	64	64	66	66	68	68	70	70	72	72	74																						
0.631 - 0.650 (0.0248 - 0.0256)	36	38	40	42	44	46	48	50	52	52	54	54	56	56	58	58	60	60	62	62	64	64	66	66	68	68	70	70	72	72	74																								
0.651 - 0.670 (0.0256 - 0.0264)	38	40	42	44	46	48	50	52	54	54	56	56	58	58	60	60	62	62	64	64	66	66	68	68	70	70	72	72	74																										
0.671 - 0.690 (0.0264 - 0.0272)	40	42	44	46	48	50	52	54	56	56	58	58	60	60	62	62	64	64	66	66	68	68	70	70	72	72	74																												
0.691 - 0.710 (0.0272 - 0.0280)	42	44	46	48	50	52	54	56	58	58	60	60	62	62	64	64	66	66	68	68	70	70	72	72	74																														
0.711 - 0.730 (0.0280 - 0.0287)	44	46	48	50	52	54	56	58	60	60	62	62	64	64	66	66	68	68	70	70	72	72	74																																
0.731 - 0.750 (0.0288 - 0.0295)	46	48	50	52	54	56	58	60	62	62	64	64	66	66	68	68	70	70	72	72	74																																		
0.751 - 0.770 (0.0296 - 0.0303)	48	50	52	54	56	58	60	62	64	64	66	66	68	68	70	70	72	72	74																																				
0.771 - 0.790 (0.0304 - 0.0311)	50	52	54	56	58	60	62	64	66	66	68	68	70	70	72	72	74																																						
0.791 - 0.810 (0.0311 - 0.0319)	52	54	56	58	60	62	64	66	68	68	70	70	72	72	74																																								
0.811 - 0.830 (0.0319 - 0.0327)	54	56	58	60	62	64	66	68	70	70	72	72	74																																										
0.831 - 0.850 (0.0327 - 0.0335)	56	58	60	62	64	66	68	70	72	72	74																																												
0.851 - 0.870 (0.0335 - 0.0343)	58	60	62	64	66	68	70	72	74																																														
0.871 - 0.890 (0.0343 - 0.0350)	60	62	64	66	68	70	72																																																
0.891 - 0.910 (0.0351 - 0.0358)	62	64	66	68	70	72																																																		
0.911 - 0.930 (0.0359 - 0.0366)	64	66	68	70	72																																																			
0.931 - 0.950 (0.0367 - 0.0374)	66	68	70	72																																																				
0.951 - 0.970 (0.0374 - 0.0382)	68	70	72																																																					
0.971 - 0.990 (0.0382 - 0.0390)	70	72																																																						
0.991 - 1.010 (0.0390 - 0.0398)	72																																																							

06	5.060 (0.1992)	30	5.300 (0.2087)	54	5.540 (0.2181)
08	5.080 (0.2000)	32	5.320 (0.2094)	56	5.560 (0.2189)
10	5.100 (0.2008)	34	5.340 (0.2102)	58	5.580 (0.2197)
12	5.120 (0.2016)	36	5.360 (0.2110)	60	5.600 (0.2205)
14	5.140 (0.2024)	38	5.380 (0.2118)	62	5.620 (0.2213)
16	5.160 (0.2031)	40	5.400 (0.2126)	64	5.640 (0.2220)
18	5.180 (0.2039)	42	5.420 (0.2134)	66	5.660 (0.2228)
20	5.200 (0.2047)	44	5.440 (0.2142)	68	5.680 (0.2236)
22	5.220 (0.2055)	46	5.460 (0.2150)	70	5.700 (0.2244)
24	5.240 (0.2063)	48	5.480 (0.2157)	72	5.720 (0.2252)
26	5.260 (0.2071)	50	5.500 (0.2165)	74	5.740 (0.2260)
28	5.280 (0.2079)	52	5.520 (0.2173)		

09490_TACO_G0075

Fig. 248 Shim selection chart, part two—4.0L engine

12. Using a feeler gauge, check and record the valve clearance on the following valves: right bank, exhaust number three and intake number one, left bank exhaust number two and intake number six.

13. Rotate the crankshaft 240 degrees clockwise and using a feeler gauge, check and record the following valves: right bank, exhaust number five and intake number three, left bank exhaust number four and intake number two.

14. Rotate the crankshaft 240 degrees clockwise and using a feeler gauge, check and record the following valves: right bank, exhaust number one and intake number five, left bank exhaust number six and intake number four.

15. If adjustment is required, position the engine at TDC on the compression stroke.

16. Place paint marks on the number one chain links corresponding to the timing marks of the camshaft timing gears.

17. Remove the chain tensioner assembly number one. Remove the number two camshaft.

18. Remove the chain tensioner assembly number two. Remove the camshaft.

19. Remove the number four camshaft subassembly. Remove the chain tensioner assembly number three.

20. Remove the number three camshaft subassembly.

21. Remove the valve lifters.

22. Determine the replacement adjusting shim size according to the following formula or use the adjusting shim charts.

23. Using a micrometer, measure the thickness of the removed shim. Calculate the thickness of a new shim so that the valve clearance comes within the specified value.

- T: Thickness of the removed shim
- A: Measured valve clearance
- N: Thickness of the new shim

a. Intake: N = T + A
b. Exhaust: N = T + A

24. Select a new lifter with a thickness as close as possible to the calculated value.

25. Install removed components in the reverse order of the removal procedure.

26. When installing the cylinder head cover, apply a continuous bead (0.08–0.12 inch) of seal packing, part number 08826-00080 or equivalent, to the cylinder head as indicated in the illustration. Install the seal washers onto the bolts. Install the cylinder head cover bolts and nuts. Tighten bolts "A" to 7.4 ft. lbs. Tighten bolts "B" to 80 inch lbs. Tighten nuts to 80 inch lbs.

➡Be sure to remove any oil from the contact surfaces of the cylinder head cover and the cylinder head. Install the cover within three minutes after applying the seal packing. Tighten the bolts to specification within fifteen minutes after installing the cover. Do not add engine oil for at least two hours after installing the cover.

27. Continue the installation in the reverse order of the removal procedure.

28. Be sure to fill the engine with the proper grade and type engine coolant.

29. Be sure to fill the engine with the proper grade and type engine oil.

30. Start the engine and check for leaks. Correct as required.

4.7L V8 Engine

See Figures 249 and 250.

➡Measure valve clearance with the engine cold.

1. Before servicing the vehicle, refer to the precautions in the beginning of this section.
2. Drain the cooling system.
3. Remove or disconnect the following:
   - Negative battery cable
   - Ignition coils
   - Valve covers
4. Set the engine to the top of the compression stroke with the valves closed for the cylinder to be measured.
5. Check the valve clearance. The valve clearance specifications are as follows:
   - Intake: 0.006–0.010 in. (0.15–0.25mm)
   - Exhaust: 0.010–0.014 in. (0.25–0.35mm)
6. Record the measurements for each valve.
7. When all valve clearances have been measured, remove the camshafts.
8. Remove the valve shims and measure them. Note this measurement along with the clearance measurement recorded earlier.
9. Using the valve clearance and shim thickness measurements, find replacement shims in the Adjusting Shim Selection charts.
10. Install or connect the following:
    - Replacement valve shims
    - Camshafts
    - Valve covers
    - Ignition coils
    - Negative battery cable
11. Fill the cooling system.
12. Start the engine and check for leaks.

5.7L V8 Engine

Valve lash is not adjustable for this engine.

Measured clearance mm (in.) \ Installed shim thickness mm (in.)	2.000 (0.0787)	2.020 (0.0795)	2.040 (0.0803)	2.060 (0.0811)	2.080 (0.0819)	2.100 (0.0827)	2.120 (0.0835)	2.140 (0.0843)	2.160 (0.0850)	2.180 (0.0858)	2.200 (0.0866)	2.210 (0.0870)	2.220 (0.0874)	2.230 (0.0878)	2.240 (0.0882)	2.250 (0.0886)	2.260 (0.0890)	2.270 (0.0894)	2.280 (0.0898)	2.290 (0.0902)	2.300 (0.0906)	2.310 (0.0909)	2.320 (0.0913)	2.330 (0.0917)	2.340 (0.0921)	2.350 (0.0925)	2.360 (0.0929)	2.370 (0.0933)	2.380 (0.0937)	2.390 (0.0941)	2.400 (0.0945)	2.410 (0.0949)	2.420 (0.0953)	2.430 (0.0957)	2.440 (0.0961)	2.450 (0.0965)	2.460 (0.0969)	2.470 (0.0972)	2.480 (0.0976)	2.490 (0.0980)	2.500 (0.0984)	2.510 (0.0988)	2.520 (0.0992)	2.530 (0.0996)	2.540 (0.1000)	2.550 (0.1004)	2.560 (0.1008)	2.570 (0.1012)	2.580 (0.1016)	2.590 (0.1020)	2.600 (0.1024)	2.620 (0.1031)	2.640 (0.1039)	2.660 (0.1047)	2.680 (0.1055)	2.700 (0.1063)	2.720 (0.1071)	2.740 (0.1079)	2.760 (0.1087)	2.780 (0.1094)	2.800 (0.1102)
0.000-0.030 (0.0000-0.0012)							00	00	00	00	02	02	04	04	06	06	08	08	10	10	12	12	14	14	16	16	18	18	20	20	22	22	24	24	26	26	28	28	30	30	32	32	34	34	36	36	38	38	40	40	42	44	46	48	50	52	54	56	58	60	62
0.031-0.050 (0.0012-0.0020)						00	00	00	00	02	04	06	06	08	08	10	10	12	12	14	14	16	16	18	18	20	20	22	22	24	24	26	26	28	28	30	30	32	32	34	34	36	36	38	38	40	40	42	42	44	44	46	48	50	52	54	56	58	60	62	64
0.051-0.070 (0.0020-0.0028)					00	00	00	00	02	04	06	08	08	10	10	12	12	14	14	16	16	18	18	20	20	22	22	24	24	26	26	28	28	30	30	32	32	34	34	36	36	38	38	40	40	42	42	44	44	46	46	48	50	52	54	56	58	60	62	64	66
0.071-0.090 (0.0028-0.0035)				00	00	00	00	02	04	06	08	10	10	12	12	14	14	16	16	18	18	20	20	22	22	24	24	26	26	28	28	30	30	32	32	34	34	36	36	38	38	40	40	42	42	44	44	46	46	48	48	50	52	54	56	58	60	62	64	66	68
0.091-0.110 (0.0036-0.0043)			00	00	00	00	02	04	06	08	10	12	12	14	14	16	16	18	18	20	20	22	22	24	24	26	26	28	28	30	30	32	32	34	34	36	36	38	38	40	40	42	42	44	44	46	46	48	48	50	50	52	54	56	58	60	62	64	66	68	70
0.111-0.130 (0.0044-0.0051)		00	00	00	00	02	04	06	08	10	12	14	14	16	16	18	18	20	20	22	22	24	24	26	26	28	28	30	30	32	32	34	34	36	36	38	38	40	40	42	42	44	44	46	46	48	48	50	50	52	52	54	56	58	60	62	64	66	68	70	72
0.131-0.149 (0.0052-0.0059)		00	00	00	02	04	06	08	10	12	14	16	16	18	18	20	20	22	22	24	24	26	26	28	28	30	30	32	32	34	34	36	36	38	38	40	40	42	42	44	44	46	46	48	48	50	50	52	52	54	54	56	58	60	62	64	66	68	70	72	74
0.150-0.250 (0.0059-0.0098)																																																													
0.251-0.270 (0.0099-0.0106)	06	08	10	12	14	16	18	20	22	24	26	28	28	30	30	32	32	34	34	36	36	38	38	40	40	42	42	44	44	46	46	48	48	50	50	52	52	54	54	56	56	58	58	60	60	62	62	64	64	66	66	68	70	72	74	76	78	80	80	80	
0.271-0.290 (0.0107-0.0114)	08	10	12	14	16	18	20	22	24	26	28	30	30	32	32	34	34	36	36	38	38	40	40	42	42	44	44	46	46	48	48	50	50	52	52	54	54	56	56	58	58	60	60	62	62	64	64	66	66	68	68	70	72	74	76	78	80	80	80		
0.291-0.310 (0.0115-0.0122)	10	12	14	16	18	20	22	24	26	28	30	32	32	34	34	36	36	38	38	40	40	42	42	44	44	46	46	48	48	50	50	52	52	54	54	56	56	58	58	60	60	62	62	64	64	66	66	68	68	70	70	72	74	76	78	80	80	80			
0.311-0.330 (0.0122-0.0130)	12	14	16	18	20	22	24	26	28	30	32	34	34	36	36	38	38	40	40	42	42	44	44	46	46	48	48	50	50	52	52	54	54	56	56	58	58	60	60	62	62	64	64	66	66	68	68	70	70	72	72	74	76	78	80	80	80				
0.331-0.350 (0.0130-0.0138)	14	16	18	20	22	24	26	28	30	32	34	36	36	38	38	40	40	42	42	44	44	46	46	48	48	50	50	52	52	54	54	56	56	58	58	60	60	62	62	64	64	66	66	68	68	70	70	72	72	74	74	76	78	80	80	80					
0.351-0.370 (0.0138-0.0146)	16	18	20	22	24	26	28	30	32	34	36	38	38	40	40	42	42	44	44	46	46	48	48	50	50	52	52	54	54	56	56	58	58	60	60	62	62	64	64	66	66	68	68	70	70	72	72	74	74	76	76	78	80	80	80						
0.371-0.390 (0.0146-0.0154)	18	20	22	24	26	28	30	32	34	36	38	40	40	42	42	44	44	46	46	48	48	50	50	52	52	54	54	56	56	58	58	60	60	62	62	64	64	66	66	68	68	70	70	72	72	74	74	76	76	78	78	80	80	80							
0.391-0.410 (0.0154-0.0161)	20	22	24	26	28	30	32	34	36	38	40	42	42	44	44	46	46	48	48	50	50	52	52	54	54	56	56	58	58	60	60	62	62	64	64	66	66	68	68	70	70	72	72	74	74	76	76	78	78	80	80	80	80								
0.411-0.430 (0.0162-0.0169)	22	24	26	28	30	32	34	36	38	40	42	44	44	46	46	48	48	50	50	52	52	54	54	56	56	58	58	60	60	62	62	64	64	66	66	68	68	70	70	72	72	74	74	76	76	78	78	80	80	80	80	80									
0.431-0.450 (0.0170-0.0177)	24	26	28	30	32	34	36	38	40	42	44	46	46	48	48	50	50	52	52	54	54	56	56	58	58	60	60	62	62	64	64	66	66	68	68	70	70	72	72	74	74	76	76	78	78	80	80	80	80	80	80										
0.451-0.470 (0.0178-0.0185)	26	28	30	32	34	36	38	40	42	44	46	48	48	50	50	52	52	54	54	56	56	58	58	60	60	62	62	64	64	66	66	68	68	70	70	72	72	74	74	76	76	78	78	80	80	80	80	80	80	80											
0.471-0.490 (0.0185-0.0193)	28	30	32	34	36	38	40	42	44	46	48	50	50	52	52	54	54	56	56	58	58	60	60	62	62	64	64	66	66	68	68	70	70	72	72	74	74	76	76	78	78	80	80	80	80	80	80	80													
0.491-0.510 (0.0193-0.0201)	30	32	34	36	38	40	42	44	46	48	50	52	52	54	54	56	56	58	58	60	60	62	62	64	64	66	66	68	68	70	70	72	72	74	74	76	76	78	78	80	80	80	80	80	80	80															
0.511-0.530 (0.0201-0.0209)	32	34	36	38	40	42	44	46	48	50	52	54	54	56	56	58	58	60	60	62	62	64	64	66	66	68	68	70	70	72	72	74	74	76	76	78	78	80	80	80	80	80	80	80																	
0.531-0.550 (0.0209-0.0217)	34	36	38	40	42	44	46	48	50	52	54	56	56	58	58	60	60	62	62	64	64	66	66	68	68	70	70	72	72	74	74	76	76	78	78	80	80	80	80	80	80	80																			
0.551-0.570 (0.0217-0.0224)	36	38	40	42	44	46	48	50	52	54	56	58	58	60	60	62	62	64	64	66	66	68	68	70	70	72	72	74	74	76	76	78	78	80	80	80	80	80	80	80																					
0.571-0.590 (0.0225-0.0232)	38	40	42	44	46	48	50	52	54	56	58	60	60	62	62	64	64	66	66	68	68	70	70	72	72	74	74	76	76	78	78	80	80	80	80	80	80	80																							
0.591-0.610 (0.0233-0.0240)	40	42	44	46	48	50	52	54	56	58	60	62	62	64	64	66	66	68	68	70	70	72	72	74	74	76	76	78	78	80	80	80	80	80	80	80																									
0.611-0.630 (0.0241-0.0248)	42	44	46	48	50	52	54	56	58	60	62	64	64	66	66	68	68	70	70	72	72	74	74	76	76	78	78	80	80	80	80	80	80	80																											
0.631-0.650 (0.0248-0.0256)	44	46	48	50	52	54	56	58	60	62	64	66	66	68	68	70	70	72	72	74	74	76	76	78	78	80	80	80	80	80	80	80																													
0.651-0.670 (0.0256-0.0264)	46	48	50	52	54	56	58	60	62	64	66	68	68	70	70	72	72	74	74	76	76	78	78	80	80	80	80	80	80	80																															
0.671-0.690 (0.0264-0.0272)	48	50	52	54	56	58	60	62	64	66	68	70	70	72	72	74	74	76	76	78	78	80	80	80	80	80	80	80																																	
0.691-0.710 (0.0272-0.0280)	50	52	54	56	58	60	62	64	66	68	70	72	72	74	74	76	76	78	78	80	80	80	80	80	80	80																																			
0.711-0.730 (0.0280-0.0287)	52	54	56	58	60	62	64	66	68	70	72	74	74	76	76	78	78	80	80	80	80	80	80	80																																					
0.731-0.750 (0.0288-0.0295)	54	56	58	60	62	64	66	68	70	72	74	76	76	78	78	80	80	80	80	80	80	80																																							
0.751-0.770 (0.0296-0.0303)	56	58	60	62	64	66	68	70	72	74	76	78	78	80	80	80	80	80	80	80																																									
0.771-0.790 (0.0304-0.0311)	58	60	62	64	66	68	70	72	74	76	78	80	80	80	80	80	80	80																																											
0.791-0.810 (0.0311-0.0319)	60	62	64	66	68	70	72	74	76	78	80	80	80	80	80	80																																													
0.811-0.830 (0.0319-0.0327)	62	64	66	68	70	72	74	76	78	80	80	80	80	80																																															
0.831-0.850 (0.0327-0.0335)	64	66	68	70	72	74	76	78	80	80	80	80																																																	
0.851-0.870 (0.0335-0.0343)	66	68	70	72	74	76	78	80	80	80																																																			
0.871-0.890 (0.0343-0.0350)	68	70	72	74	76	78	80	80	80																																																				
0.891-0.910 (0.0351-0.0358)	70	72	74	76	78	80	80	80																																																					
0.911-0.930 (0.0359-0.0366)	72	74	76	78	80	80	80																																																						
0.931-0.950 (0.0367-0.0374)	74	76	78	80	80	80																																																							
0.951-0.970 (0.0374-0.0382)	76	78	80	80	80																																																								
0.971-0.990 (0.0382-0.0390)	78	80	80	80																																																									
0.991-1.010 (0.0390-0.0398)	80	80	80																																																										
1.011-1.030 (0.0398-0.0406)	80	80																																																											
1.031-1.050 (0.0406-0.0413)	80																																																												

Intake valve clearance (Cold):
0.15 – 0.25 mm (0.006 – 0.010 in.)

EXAMPLE:
The 2.300 mm (0.0906 in.) shim is installed, and the measured clearance is 0.440 mm (0.0173 in.). Replace the 2.300 mm (0.0906 in.) shim with a No. 54 shim.

New shim thickness mm (in.)

Shim No.	Thickness	Shim No.	Thickness	Shim No.	Thickness
00	2.000 (0.0787)	28	2.280 (0.0898)	56	2.560 (0.1008)
02	2.020 (0.0795)	30	2.300 (0.0906)	58	2.580 (0.1016)
04	2.040 (0.0803)	32	2.320 (0.0913)	60	2.600 (0.1024)
06	2.060 (0.0811)	34	2.340 (0.0921)	62	2.620 (0.1031)
08	2.080 (0.0819)	36	2.360 (0.0929)	64	2.640 (0.1039)
10	2.100 (0.0827)	38	2.380 (0.0937)	66	2.660 (0.1047)
12	2.120 (0.0835)	40	2.400 (0.0945)	68	2.680 (0.1055)
14	2.140 (0.0843)	42	2.420 (0.0953)	70	2.700 (0.1063)
16	2.160 (0.0850)	44	2.440 (0.0961)	72	2.720 (0.1071)
18	2.180 (0.0858)	46	2.460 (0.0969)	74	2.740 (0.1079)
20	2.200 (0.0866)	48	2.480 (0.0976)	76	2.760 (0.1087)
22	2.220 (0.0874)	50	2.500 (0.0984)	78	2.780 (0.1094)
24	2.240 (0.0882)	52	2.520 (0.0992)	80	2.800 (0.1102)
26	2.260 (0.0890)	54	2.540 (0.1000)		

7924SG71

Fig. 249 Intake valve clearance shim selection chart

Measured clearance mm (in.) \ Installed shim thickness mm (in.)	2.000 (0.0787)	2.020 (0.0795)	2.040 (0.0803)	2.060 (0.0811)	2.080 (0.0819)	2.100 (0.0827)	2.120 (0.0835)	2.140 (0.0843)	2.160 (0.0850)	2.180 (0.0858)	2.200 (0.0866)	2.210 (0.0870)	2.220 (0.0874)	2.230 (0.0878)	2.240 (0.0882)	2.250 (0.0886)	2.260 (0.0890)	2.270 (0.0894)	2.280 (0.0898)	2.290 (0.0902)	2.300 (0.0906)	2.310 (0.0909)	2.320 (0.0913)	2.330 (0.0917)	2.340 (0.0921)	2.350 (0.0925)	2.360 (0.0929)	2.370 (0.0933)	2.380 (0.0937)	2.390 (0.0941)	2.400 (0.0945)	2.410 (0.0949)	2.420 (0.0953)	2.430 (0.0957)	2.440 (0.0961)	2.450 (0.0965)	2.460 (0.0969)	2.470 (0.0972)	2.480 (0.0976)	2.490 (0.0980)	2.500 (0.0984)	2.510 (0.0988)	2.520 (0.0992)	2.530 (0.0996)	2.540 (0.1000)	2.550 (0.1004)	2.560 (0.1008)	2.570 (0.1012)	2.580 (0.1016)	2.590 (0.1020)	2.600 (0.1024)	2.620 (0.1031)	2.640 (0.1039)	2.660 (0.1047)	2.680 (0.1055)	2.700 (0.1063)	2.720 (0.1071)	2.740 (0.1079)	2.760 (0.1087)	2.780 (0.1094)	2.800 (0.1102)
0.000–0.030 (0.0000–0.0012)												00	00	00	00	00	00	00	00	00	02	02	04	04	06	06	08	08	10	10	12	12	14	14	16	16	18	18	20	20	22	22	24	24	26	26	28	28	30	30	32	34	36	38	40	42	44	46	48	50	52
0.031–0.050 (0.0012–0.0020)											00	00	00	00	00	00	00	02	02	04	04	06	06	08	08	10	10	12	12	14	14	16	16	18	18	20	20	22	22	24	24	26	26	28	28	30	30	32	32	34	34	36	38	40	42	44	46	48	50	52	54
0.051–0.070 (0.0020–0.0028)										00	00	00	00	00	00	02	02	04	04	06	06	08	08	10	10	12	12	14	14	16	16	18	18	20	20	22	22	24	24	26	26	28	28	30	30	32	32	34	34	36	36	38	40	42	44	46	48	50	52	54	56
0.071–0.090 (0.0028–0.0035)									00	00	00	00	00	02	02	04	04	06	06	08	08	10	10	12	12	14	14	16	16	18	18	20	20	22	22	24	24	26	26	28	28	30	30	32	32	34	34	36	36	38	38	40	42	44	46	48	50	52	54	56	58
0.091–0.110 (0.0036–0.0043)								00	00	00	00	02	02	04	04	06	06	08	08	10	10	12	12	14	14	16	16	18	18	20	20	22	22	24	24	26	26	28	28	30	30	32	32	34	34	36	36	38	38	40	40	42	44	46	48	50	52	54	56	58	60
0.111–0.130 (0.0044–0.0051)							00	00	00	00	02	04	04	06	06	08	08	10	10	12	12	14	14	16	16	18	18	20	20	22	22	24	24	26	26	28	28	30	30	32	32	34	34	36	36	38	38	40	40	42	42	44	46	48	50	52	54	56	58	60	62
0.131–0.150 (0.0052–0.0059)						00	00	00	00	02	04	06	06	08	08	10	10	12	12	14	14	16	16	18	18	20	20	22	22	24	24	26	26	28	28	30	30	32	32	34	34	36	36	38	38	40	40	42	42	44	44	46	48	50	52	54	56	58	60	62	64
0.151–0.170 (0.0059–0.0067)					00	00	00	00	02	04	06	08	08	10	10	12	12	14	14	16	16	18	18	20	20	22	22	24	24	26	26	28	28	30	30	32	32	34	34	36	36	38	38	40	40	42	42	44	44	46	46	48	50	52	54	56	58	60	62	64	66
0.171–0.190 (0.0067–0.0075)				00	00	00	00	02	04	06	08	10	10	12	12	14	14	16	16	18	18	20	20	22	22	24	24	26	26	28	28	30	30	32	32	34	34	36	36	38	38	40	40	42	42	44	44	46	46	48	48	50	52	54	56	58	60	62	64	66	68
0.191–0.210 (0.0075–0.0083)			00	00	00	00	02	04	06	08	10	12	12	14	14	16	16	18	18	20	20	22	22	24	24	26	26	28	28	30	30	32	32	34	34	36	36	38	38	40	40	42	42	44	44	46	46	48	48	50	50	52	54	56	58	60	62	64	66	68	70
0.211–0.230 (0.0083–0.0091)		00	00	00	00	02	04	06	08	10	12	14	14	16	16	18	18	20	20	22	22	24	24	26	26	28	28	30	30	32	32	34	34	36	36	38	38	40	40	42	42	44	44	46	46	48	48	50	50	52	52	54	56	58	60	62	64	66	68	70	72
0.231–0.249 (0.0091–0.0098)		00	00	00	02	04	06	08	10	12	14	16	16	18	18	20	20	22	22	24	24	26	26	28	28	30	30	32	32	34	34	36	36	38	38	40	40	42	42	44	44	46	46	48	48	50	50	52	52	54	54	56	58	60	62	64	66	68	70	72	74
0.250–0.350 (0.0098–0.0138)																																																													
0.351–0.370 (0.0138–0.0146)	06	08	10	12	14	16	18	20	22	24	26	28	28	30	30	32	32	34	34	36	36	38	38	40	40	42	42	44	44	46	46	48	48	50	50	52	52	54	54	56	56	58	58	60	60	62	62	64	64	66	66	68	70	72	74	76	78	80	80	80	
0.371–0.390 (0.0146–0.0154)	08	10	12	14	16	18	20	22	24	26	28	30	30	32	32	34	34	36	36	38	38	40	40	42	42	44	44	46	46	48	48	50	50	52	52	54	54	56	56	58	58	60	60	62	62	64	64	66	66	68	68	70	72	74	76	78	80	80	80		
0.391–0.410 (0.0154–0.0161)	10	12	14	16	18	20	22	24	26	28	30	32	32	34	34	36	36	38	38	40	40	42	42	44	44	46	46	48	48	50	50	52	52	54	54	56	56	58	58	60	60	62	62	64	64	66	66	68	68	70	70	72	74	76	78	80	80	80			
0.411–0.430 (0.0162–0.0169)	12	14	16	18	20	22	24	26	28	30	32	34	34	36	36	38	38	40	40	42	42	44	44	46	46	48	48	50	50	52	52	54	54	56	56	58	58	60	60	62	62	64	64	66	66	68	68	70	70	72	72	74	76	78	80	80	80				
0.431–0.450 (0.0170–0.0177)	14	16	18	20	22	24	26	28	30	32	34	36	36	38	38	40	40	42	42	44	44	46	46	48	48	50	50	52	52	54	54	56	56	58	58	60	60	62	62	64	64	66	66	68	68	70	70	72	72	74	74	76	78	80	80	80					
0.451–0.470 (0.0178–0.0185)	16	18	20	22	24	26	28	30	32	34	36	38	38	40	40	42	42	44	44	46	46	48	48	50	50	52	52	54	54	56	56	58	58	60	60	62	62	64	64	66	66	68	68	70	70	72	72	74	74	76	76	78	80	80	80						
0.471–0.490 (0.0185–0.0193)	18	20	22	24	26	28	30	32	34	36	38	40	40	42	42	44	44	46	46	48	48	50	50	52	52	54	54	56	56	58	58	60	60	62	62	64	64	66	66	68	68	70	70	72	72	74	74	76	76	78	78	80	80	80							
0.491–0.510 (0.0193–0.0201)	20	22	24	26	28	30	32	34	36	38	40	42	42	44	44	46	46	48	48	50	50	52	52	54	54	56	56	58	58	60	60	62	62	64	64	66	66	68	68	70	70	72	72	74	74	76	76	78	78	80	80	80	80								
0.511–0.530 (0.0201–0.0209)	22	24	26	28	30	32	34	36	38	40	42	44	44	46	46	48	48	50	50	52	52	54	54	56	56	58	58	60	60	62	62	64	64	66	66	68	68	70	70	72	72	74	74	76	76	78	78	80	80	80	80	80									
0.531–0.550 (0.0209–0.0217)	24	26	28	30	32	34	36	38	40	42	44	46	46	48	48	50	50	52	52	54	54	56	56	58	58	60	60	62	62	64	64	66	66	68	68	70	70	72	72	74	74	76	76	78	78	80	80	80	80	80	80										
0.551–0.570 (0.0217–0.0224)	26	28	30	32	34	36	38	40	42	44	46	48	48	50	50	52	52	54	54	56	56	58	58	60	60	62	62	64	64	66	66	68	68	70	70	72	72	74	74	76	76	78	78	80	80	80	80	80	80	80											
0.571–0.590 (0.0225–0.0232)	28	30	32	34	36	38	40	42	44	46	48	50	50	52	52	54	54	56	56	58	58	60	60	62	62	64	64	66	66	68	68	70	70	72	72	74	74	76	76	78	78	80	80	80	80	80	80	80													
0.591–0.610 (0.0233–0.0240)	30	32	34	36	38	40	42	44	46	48	50	52	52	54	54	56	56	58	58	60	60	62	62	64	64	66	66	68	68	70	70	72	72	74	74	76	76	78	78	80	80	80	80	80	80	80															
0.611–0.630 (0.0241–0.0248)	32	34	36	38	40	42	44	46	48	50	52	54	54	56	56	58	58	60	60	62	62	64	64	66	66	68	68	70	70	72	72	74	74	76	76	78	78	80	80	80	80	80	80	80																	
0.631–0.650 (0.0248–0.0256)	34	36	38	40	42	44	46	48	50	52	54	56	56	58	58	60	60	62	62	64	64	66	66	68	68	70	70	72	72	74	74	76	76	78	78	80	80	80	80	80	80	80																			
0.651–0.670 (0.0256–0.0264)	36	38	40	42	44	46	48	50	52	54	56	58	58	60	60	62	62	64	64	66	66	68	68	70	70	72	72	74	74	76	76	78	78	80	80	80	80	80	80	80																					
0.671–0.690 (0.0264–0.0272)	38	40	42	44	46	48	50	52	54	56	58	60	60	62	62	64	64	66	66	68	68	70	70	72	72	74	74	76	76	78	78	80	80	80	80	80	80	80																							
0.691–0.710 (0.0272–0.0280)	40	42	44	46	48	50	52	54	56	58	60	62	62	64	64	66	66	68	68	70	70	72	72	74	74	76	76	78	78	80	80	80	80	80	80	80																									
0.711–0.730 (0.0280–0.0287)	42	44	46	48	50	52	54	56	58	60	62	64	64	66	66	68	68	70	70	72	72	74	74	76	76	78	78	80	80	80	80	80	80	80																											
0.731–0.750 (0.0288–0.0295)	44	46	48	50	52	54	56	58	60	62	64	66	66	68	68	70	70	72	72	74	74	76	76	78	78	80	80	80	80	80	80	80																													
0.751–0.770 (0.0296–0.0303)	46	48	50	52	54	56	58	60	62	64	66	68	68	70	70	72	72	74	74	76	76	78	78	80	80	80	80	80	80	80																															
0.771–0.790 (0.0304–0.0311)	48	50	52	54	56	58	60	62	64	66	68	70	70	72	72	74	74	76	76	78	78	80	80	80	80	80	80	80																																	
0.791–0.810 (0.0311–0.0319)	50	52	54	56	58	60	62	64	66	68	70	72	72	74	74	76	76	78	78	80	80	80	80	80	80	80																																			
0.811–0.830 (0.0319–0.0327)	52	54	56	58	60	62	64	66	68	70	72	74	74	76	76	78	78	80	80	80	80	80	80	80																																					
0.831–0.850 (0.0327–0.0335)	54	56	58	60	62	64	66	68	70	72	74	76	76	78	78	80	80	80	80	80	80	80																																							
0.851–0.870 (0.0335–0.0343)	56	58	60	62	64	66	68	70	72	74	76	78	78	80	80	80	80	80	80	80																																									
0.871–0.890 (0.0343–0.0350)	58	60	62	64	66	68	70	72	74	76	78	80	80	80	80	80	80	80																																											
0.891–0.910 (0.0351–0.0358)	60	62	64	66	68	70	72	74	76	78	80	80	80	80	80	80																																													
0.911–0.930 (0.0359–0.0366)	62	64	66	68	70	72	74	76	78	80	80	80	80	80																																															
0.931–0.950 (0.0367–0.0374)	64	66	68	70	72	74	76	78	80	80	80	80																																																	
0.951–0.970 (0.0374–0.0382)	66	68	70	72	74	76	78	80	80	80																																																			
0.971–0.990 (0.0382–0.0390)	68	70	72	74	76	78	80	80	80																																																				
0.991–1.010 (0.0390–0.0398)	70	72	74	76	78	80	80	80																																																					
1.011–1.030 (0.0398–0.0406)	72	74	76	78	80	80	80																																																						
1.031–1.050 (0.0406–0.0413)	74	76	78	80	80	80																																																							
1.051–1.070 (0.0414–0.0421)	76	78	80	80	80																																																								
1.071–1.090 (0.0422–0.0429)	78	80	80	80																																																									
1.091–1.110 (0.0430–0.0437)	80	80	80																																																										
1.111–1.130 (0.0437–0.0445)	80	80																																																											
1.131–1.150 (0.0445–0.0453)	80																																																												

Exhaust valve clearance (Cold):
0.25 – 0.35 mm (0.010 – 0.014 in.)

EXAMPLE:
The 2.300 mm (0.0906 in.) shim is installed, and the measured clearance is 0.440 mm (0.0173 in.). Replace the 2.300 mm (0.0906 in.) shim with a No. 44 shim.

New shim thickness — mm (in.)

Shim No.	Thickness	Shim No.	Thickness	Shim No.	Thickness
00	2.000 (0.0787)	28	2.280 (0.0898)	56	2.560 (0.1008)
02	2.020 (0.0795)	30	2.300 (0.0906)	58	2.580 (0.1016)
04	2.040 (0.0803)	32	2.320 (0.0913)	60	2.600 (0.1024)
06	2.060 (0.0811)	34	2.340 (0.0921)	62	2.620 (0.1031)
08	2.080 (0.0819)	36	2.360 (0.0929)	64	2.640 (0.1039)
10	2.100 (0.0827)	38	2.380 (0.0937)	66	2.660 (0.1047)
12	2.120 (0.0835)	40	2.400 (0.0945)	68	2.680 (0.1055)
14	2.140 (0.0843)	42	2.420 (0.0953)	70	2.700 (0.1063)
16	2.160 (0.0850)	44	2.440 (0.0961)	72	2.720 (0.1071)
18	2.180 (0.0858)	46	2.460 (0.0969)	74	2.740 (0.1079)
20	2.200 (0.0866)	48	2.480 (0.0976)	76	2.760 (0.1087)
22	2.220 (0.0874)	50	2.500 (0.0984)	78	2.780 (0.1094)
24	2.240 (0.0882)	52	2.520 (0.0992)	80	2.800 (0.1102)
26	2.260 (0.0890)	54	2.540 (0.1000)		

7924SG72

Fig. 250 Exhaust valve clearance shim selection chart

ENGINE PERFORMANCE & EMISSION CONTROL

ACCELERATOR PEDAL POSITION (APP) SENSOR

LOCATION

See Figure 251.

REMOVAL & INSTALLATION

1. Disconnect a accelerator pedal connector.
2. Remove the 2 nuts and accelerator pedal assembly.

To install:

➡Be care not to give a shock to the accelerator pedal assembly. Be care not to disassemble the accelerator pedal assembly.

3. Install the accelerator pedal assembly with the 2 nuts. Tighten nuts to 44 inch lbs. (5.0 Nm)
4. Connect a accelerator pedal connector.

AIR BYPASS VALVE (ABV)

LOCATION

See Figure 252.

REMOVAL & INSTALLATION

See Figure 253.

1. Drain engine coolant.
2. Remove v-bank cover sub-assembly.
3. Remove air cleaner hose assembly.
4. Disconnect fuel hose.

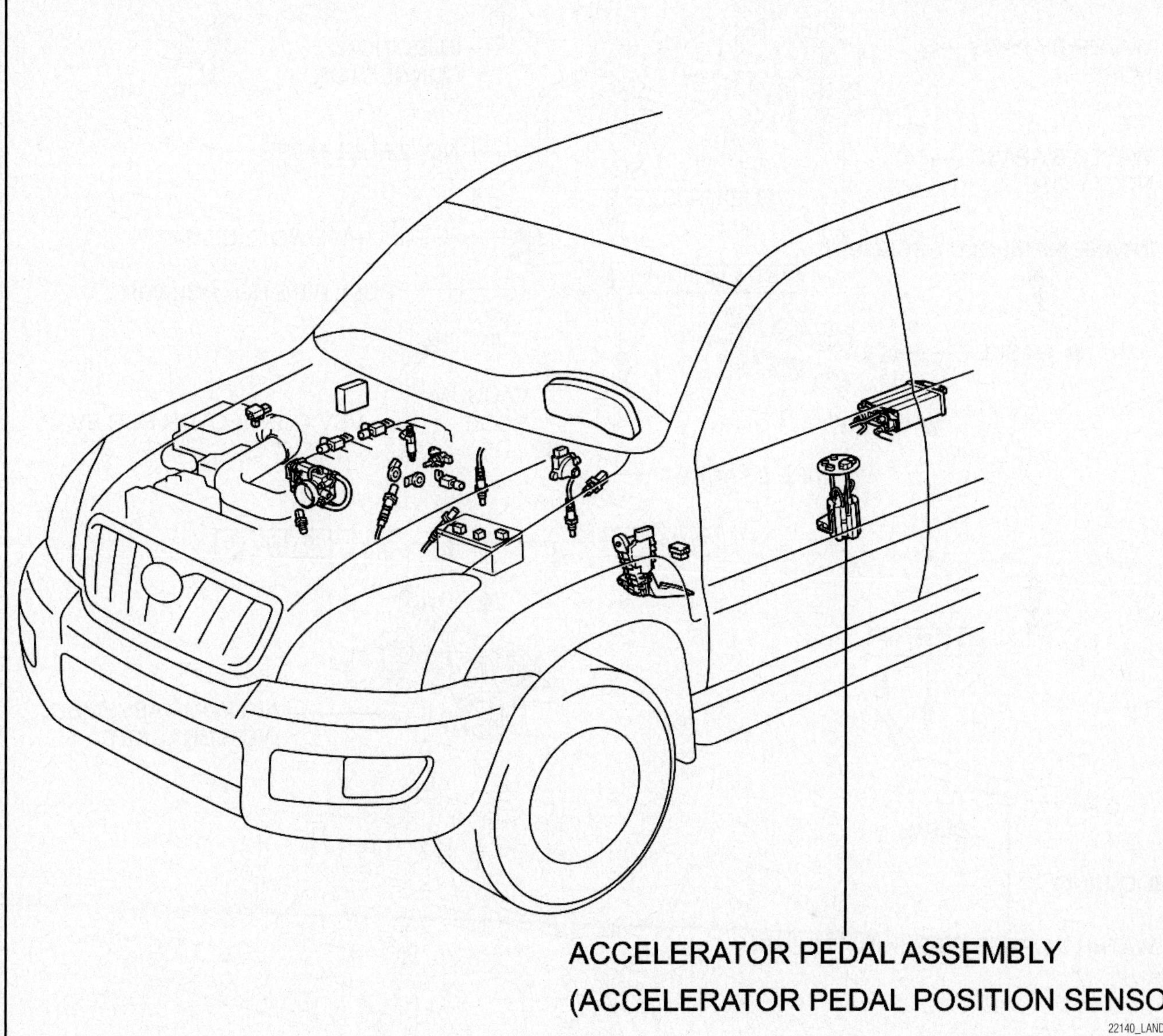

Fig. 251 Accelerator Pedal Position Location

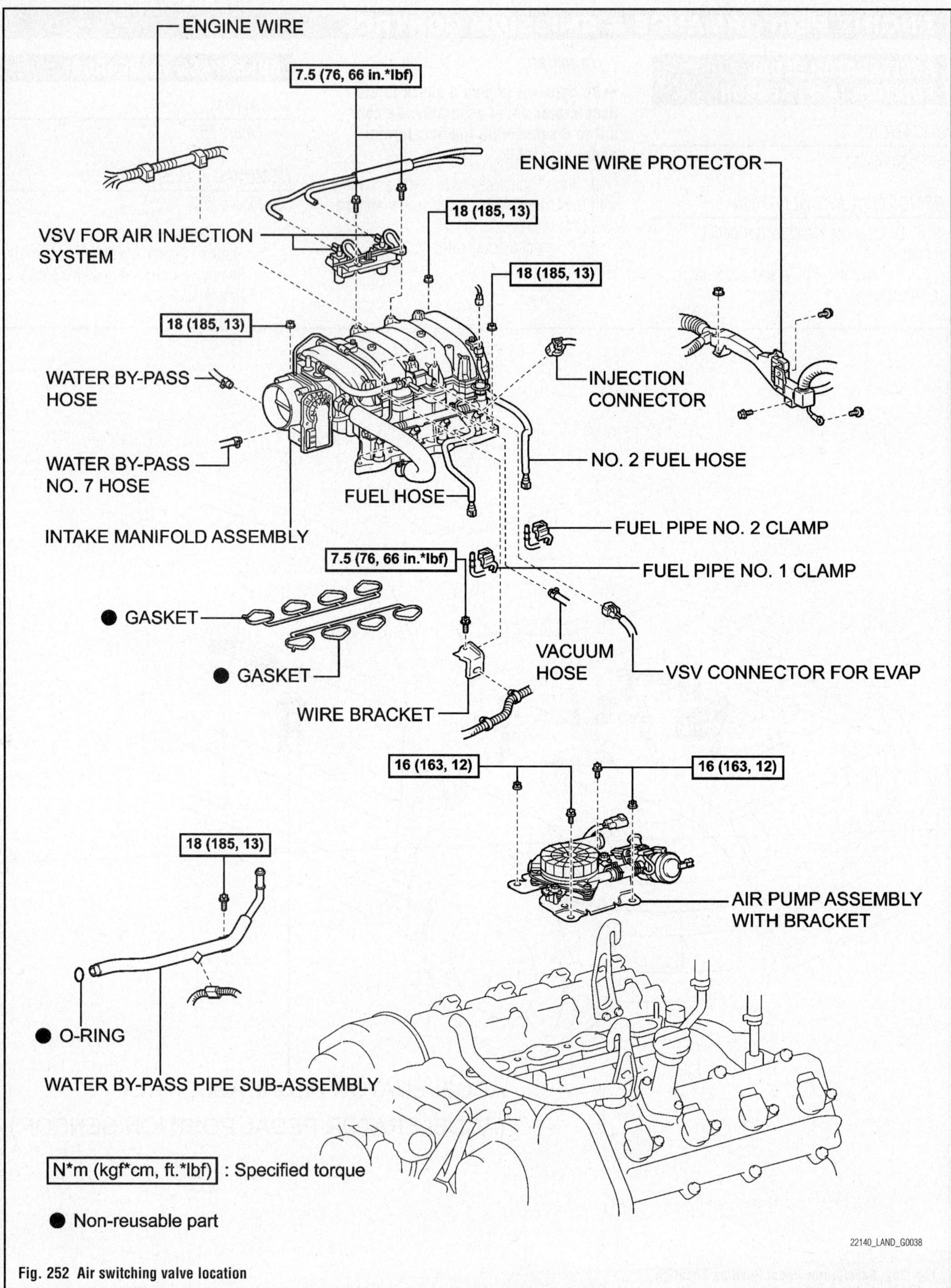

Fig. 252 Air switching valve location

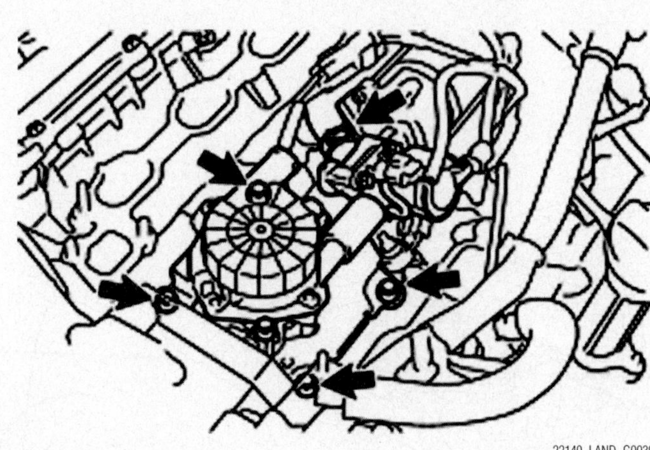

Fig. 253 Removing air switching valve

5. Disconnect no. 2 fuel hose.
6. Disconnect connectors from intake manifold.
7. Disconnect hoses from intake manifold.
8. Remove intake manifold assembly.
9. Remove water by-pass pipe sub-assembly.
10. Disconnect the no.2 air hose from the air switching valve.
11. Disconnect the air switching valve connector.
12. Disconnect the pressure sensor connector for the air injection system.
13. Remove the 4 bolts and air pump assembly with bracket.

To install:

14. Install the air pump assembly with the 4 bolts. Tighten bolts to 7 ft. lbs. (10 Nm).
15. Connect the pressure sensor connector for the air injection system.
16. Connect the air switching valve connector.
17. Connect the No. 2 air hose to the air switching valve.
18. Install water by-pass pipe sub-assembly.
19. Install intake manifold assembly.
20. Connect hoses to intake manifold.
21. Connect connectors to intake manifold.
22. Connect no. 2 fuel hose.
23. Connect fuel hose.
24. Install air cleaner hose assembly.
25. Install v-bank cover sub-assembly.
26. Add engine coolant.
27. Check for engine coolant leak.
28. Check for fuel leak.

CAMSHAFT POSITION (CMP) SENSOR

LOCATION

See Figure 255.

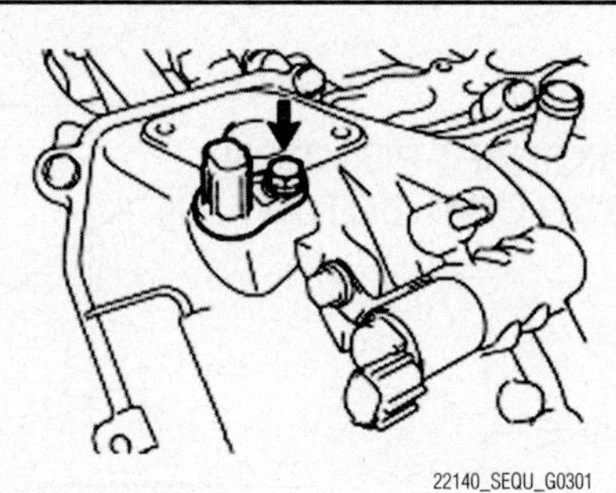

Fig. 255 Camshaft Position Sensor location 5.7L V8

REMOVAL & INSTALLATION

4.7L V8 Engine

1. Drain engine coolant.
2. Remove V-bank cover sub-assembly.
3. Remove fan and generator v belt.
4. Remove oil cooler pipe.
5. Remove timing belt cover sub-assembly No. 3 LH.
6. Disconnect the camshaft position sensor connector.
7. Remove the bolt, stud bolt and camshaft position sensor.

To install:

8. Install the camshaft position sensor with the bolt and stud bolt. Tighten bolt to 66 inch lbs. (7.5 Nm).
9. Reconnect the camshaft position sensor connector.
10. Install timing belt cover sub-assembly No. 3 LH.
11. Install oil cooler pipe.
12. Install fan and generator v belt.
13. Add engine coolant.
14. Check for engine coolant leaks.
15. Install V-bank cover sub-assembly.

5.7L V8 Engine

1. Remove V-bank cover sub-assembly.
2. Disconnect the sensor connector.
3. Remove the bolt and sensor.

To install:

4. Install the sensor with the bolt and tighten to 7 ft. lbs. (10 Nm).
5. Connect the sensor connector.
6. Install V-bank cover sub-assembly.

CRANKSHAFT POSITION (CKP) SENSOR

LOCATION

See Figures 256 through 258.

REMOVAL & INSTALLATION

V6 Engine

1. Disconnect cable from negative battery terminal.

➡Wait at least 90 seconds after disconnecting the cable from the negative (-) battery terminal to prevent airbag and seat belt pretensioner activation.

2. Remove no. 1 engine under cover.
3. Remove fan and generator v belt.
4. Remove generator assembly.
5. Disconnect cooler compressor assembly.
6. Disconnect the sensor connector.
7. Remove the bolt and sensor.

To install:

8. Install crankshaft position sensor.
9. Install the sensor with the bolt. Tighten bolt to 57 inch lbs. (6.5 nm).
10. Connect the sensor connector.
11. Reverse removal procedure.
12. Connect cable to negative battery terminal.
13. Perform initialization, if necessary.

4.7L V8 Engine

1. Disconnect cable from negative battery terminal.

➡Wait at least 90 seconds after disconnecting the cable from the negative (-) battery terminal to prevent airbag and seat belt pretensioner activation.

2. Remove no. 1 engine under cover.

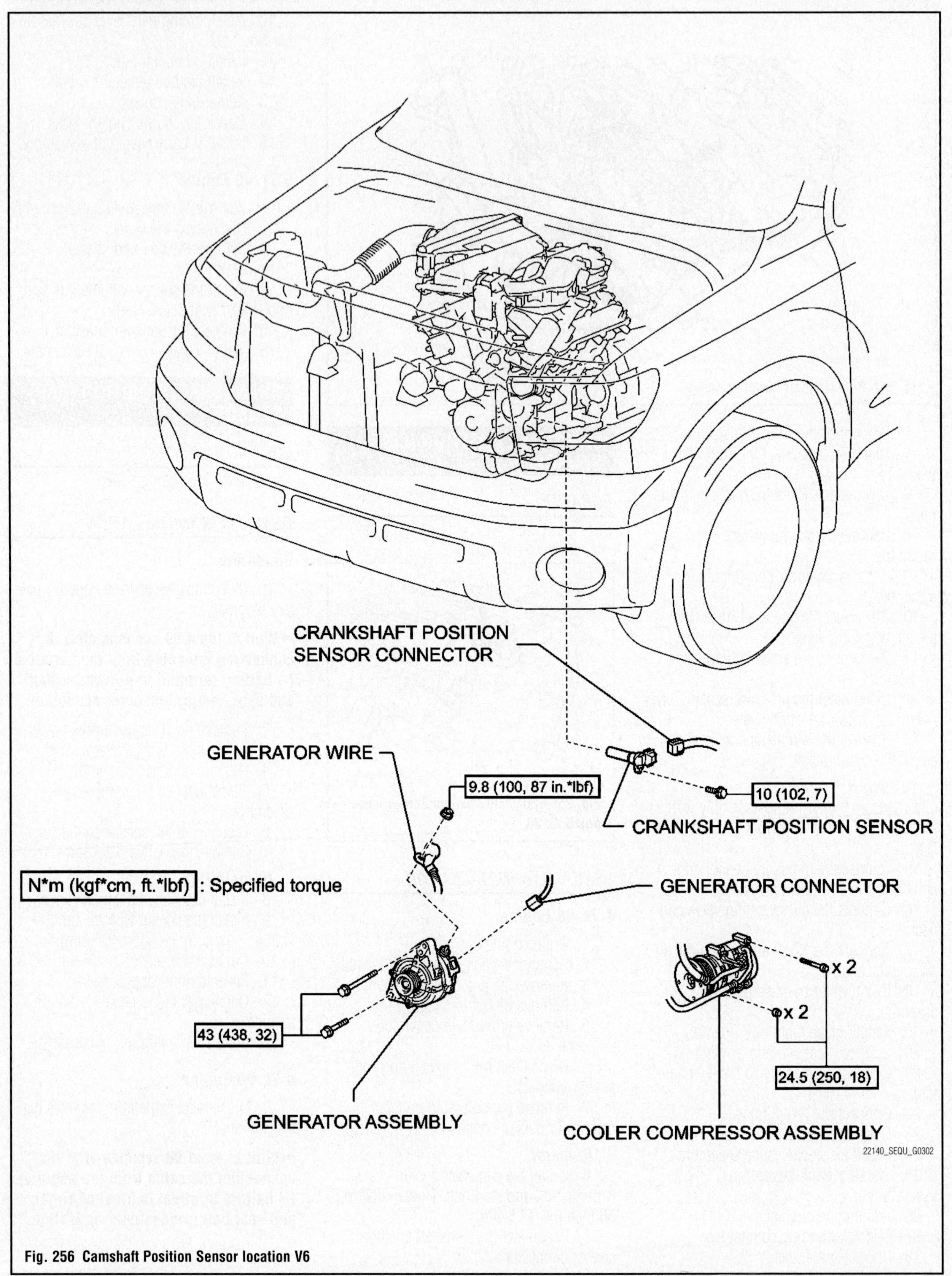

Fig. 256 Camshaft Position Sensor location V6

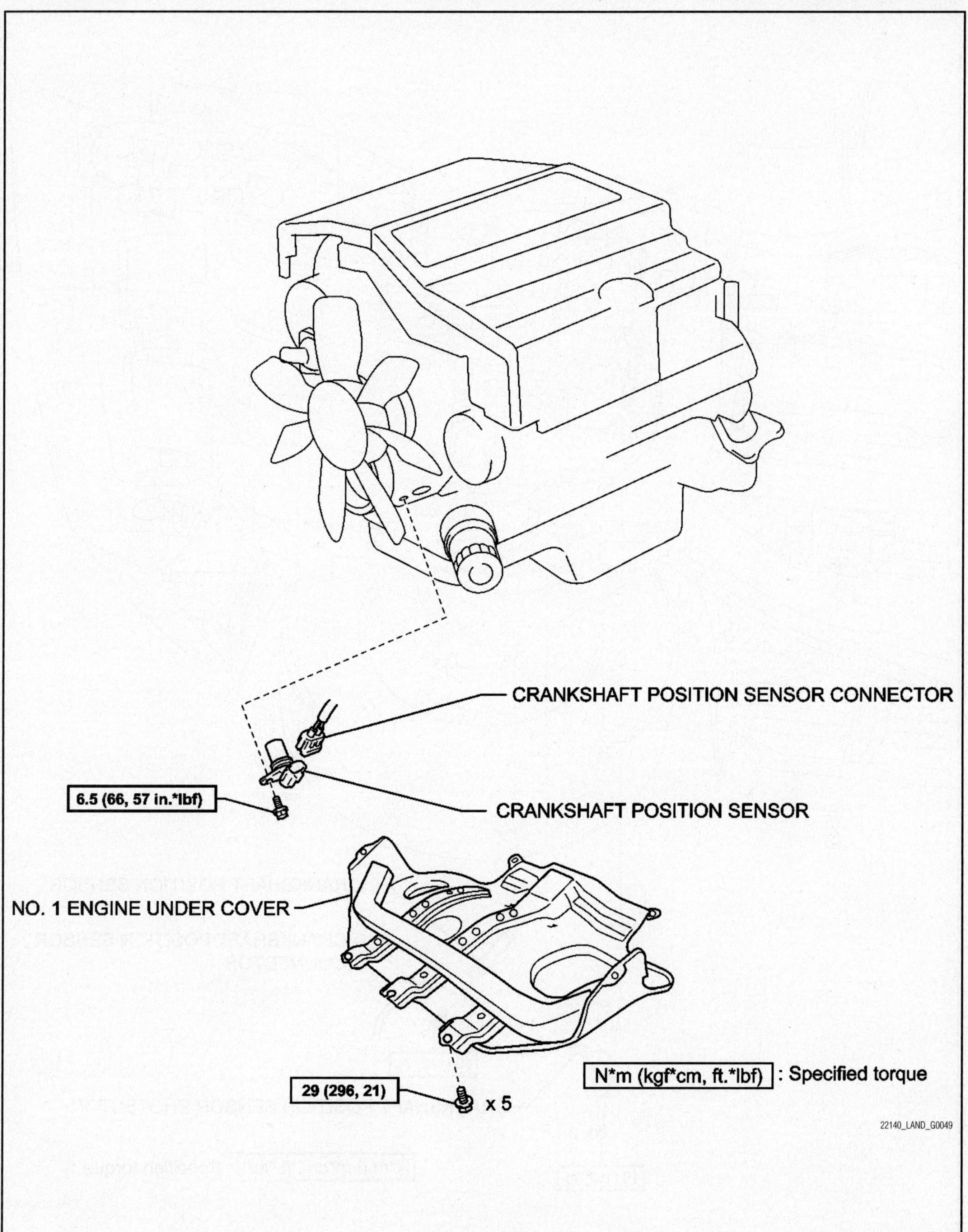

Fig. 257 Crankshaft Position Sensor location 4.7L V8

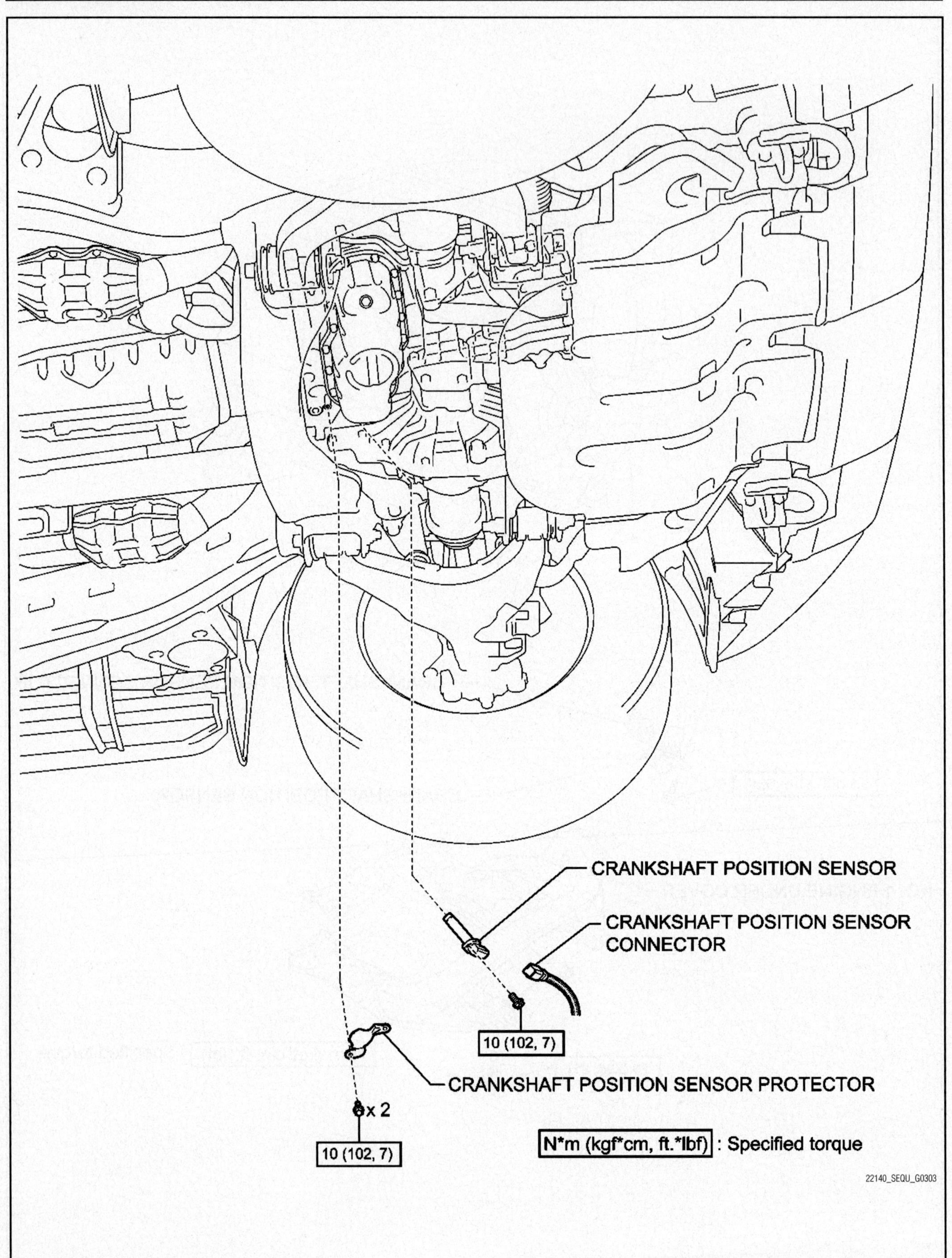

Fig. 258 Camshaft Position Sensor location 5.7L V8

3. Disconnect the sensor connector.
4. Remove the bolt and sensor.

To install:

5. Install crankshaft position sensor.
6. Install the sensor with the bolt. Tighten bolt to 57 inch lbs. (6.5 nm).
7. Connect the sensor connector.
8. Install no. 1 engine under cover.
9. Connect cable to negative battery terminal.
10. Perform initialization, if necessary.

5.7L V8 Engine

1. Remove the 2 bolts and sensor protector
2. Disconnect the sensor connector.
3. Remove the bolt and sensor

To install:

4. Install the sensor with the bolt and tighten to 7 ft. lbs. (10 Nm)
5. Connect the sensor connector
6. Install the sensor protector with the 2 bolts and tighten to 7 ft. lbs. (10 Nm)

ELECTRONIC CONTROL MODULE (ECM)

LOCATION

See Figure 259.

REMOVAL & INSTALLATION

1. Disconnect cable from negative battery terminal.

➡Wait at least 90 seconds after disconnecting the cable from the negative (-) battery terminal to prevent airbag and seat belt pretensioner activation.

2. Remove glove box compartment door.
3. Remove the 2 screws and glove compartment door.
4. Remove no. 2 finish panel lower.
5. Remove the 3 screws and no. 2 finish panel lower.

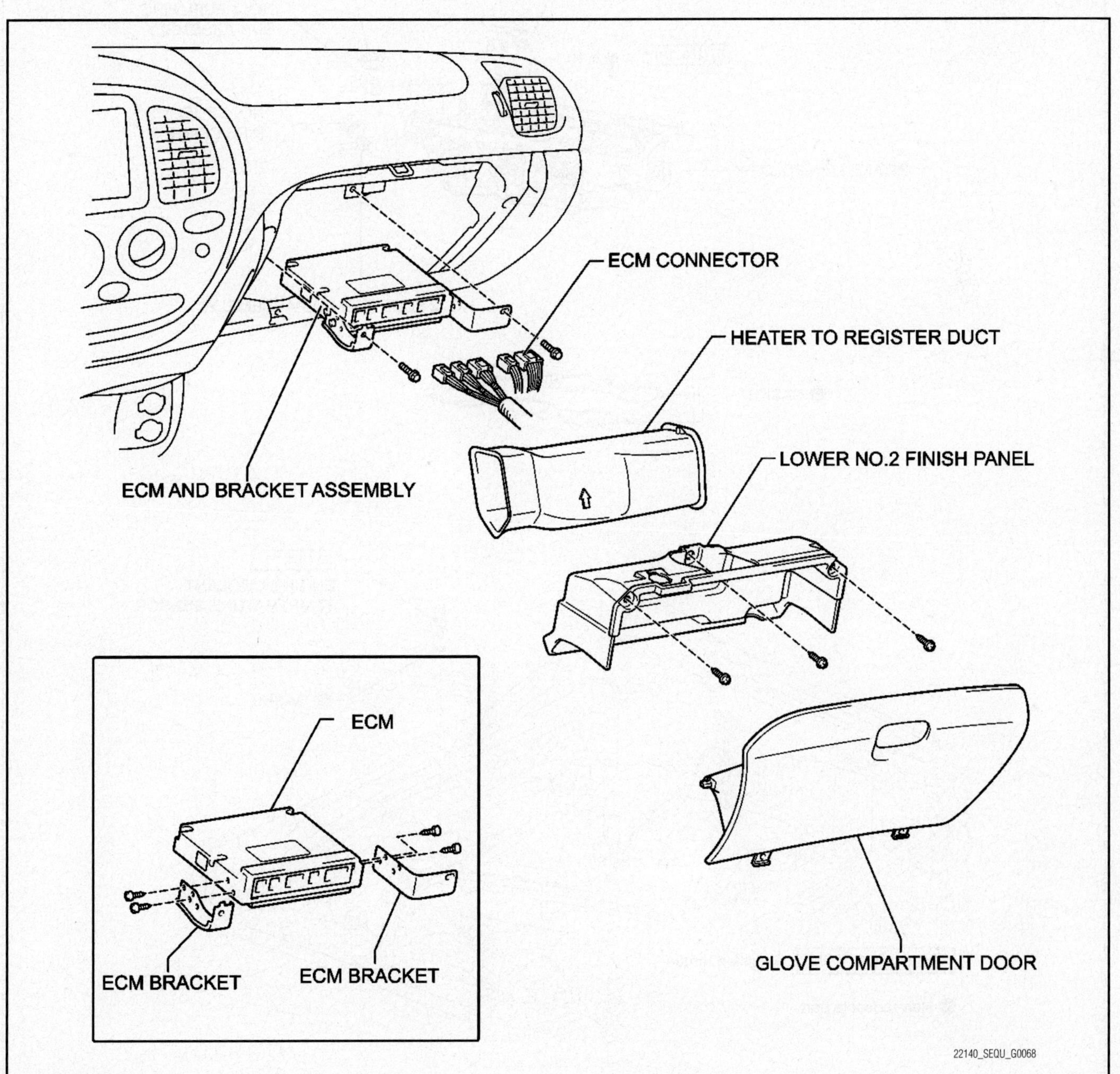

Fig. 259 ECM location

6. Disconnect the 5 ECM connectors.
7. Remove the 3 screws and ECM.

To install:

8. Install the ECM with the 3 screws, tighten to 49 inch lbs. (5.5 Nm).
9. Connect the 5 ECM connectors.

➡Be sure to securely connect the connectors.

10. Install the No. 2 finish panel lower with the 3 screws.
11. Install the glove compartment door with the 2 screws.
12. Connect cable to negative battery terminal.
13. Perform initialization, if necessary.

➡Certain systems need to be initialized after disconnecting and reconnecting the cable from the negative (-) battery terminal.

ENGINE COOLANT TEMPERATURE (ECT) SENSOR

LOCATION

See Figures 260 through 262.

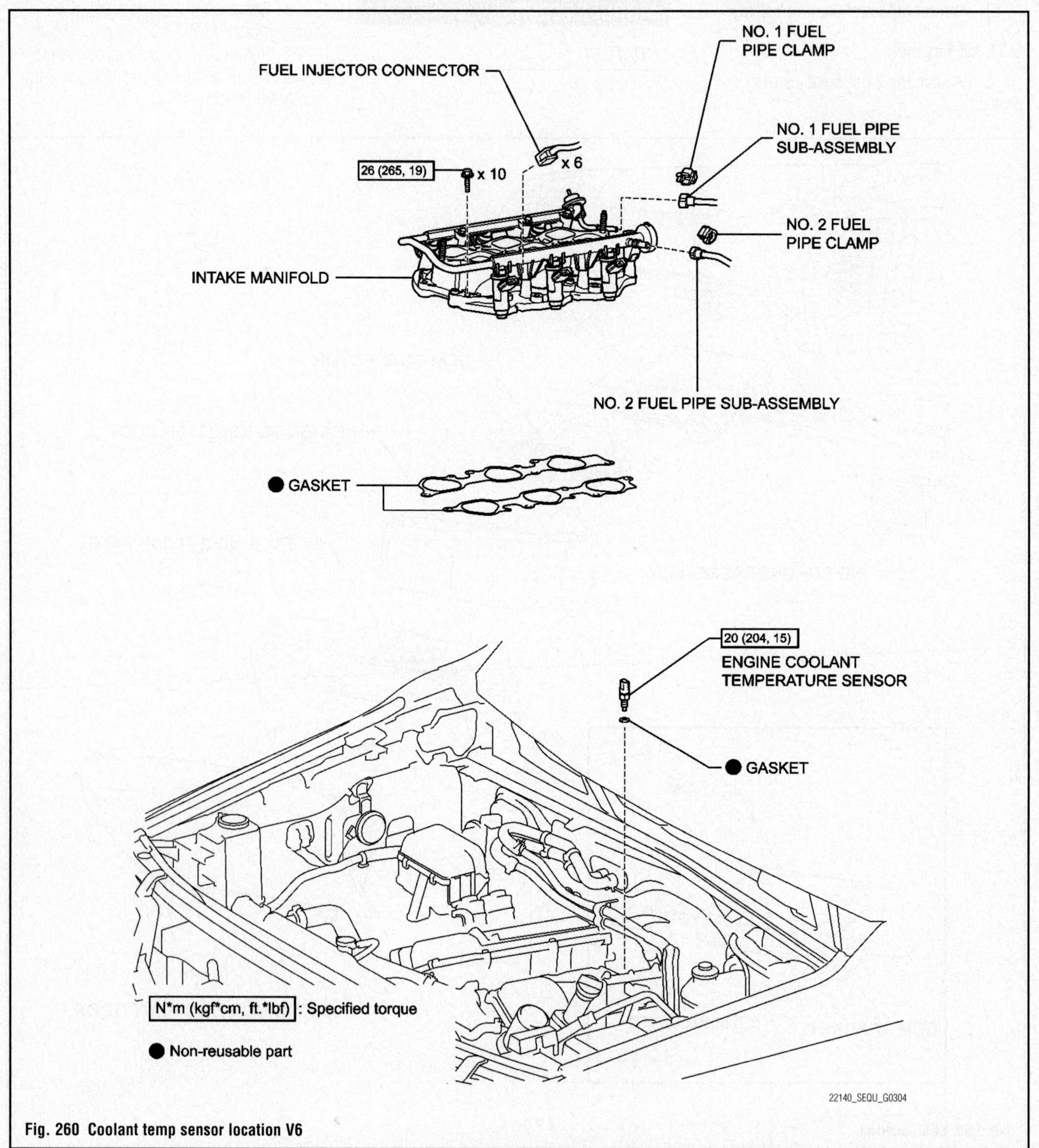

Fig. 260 Coolant temp sensor location V6

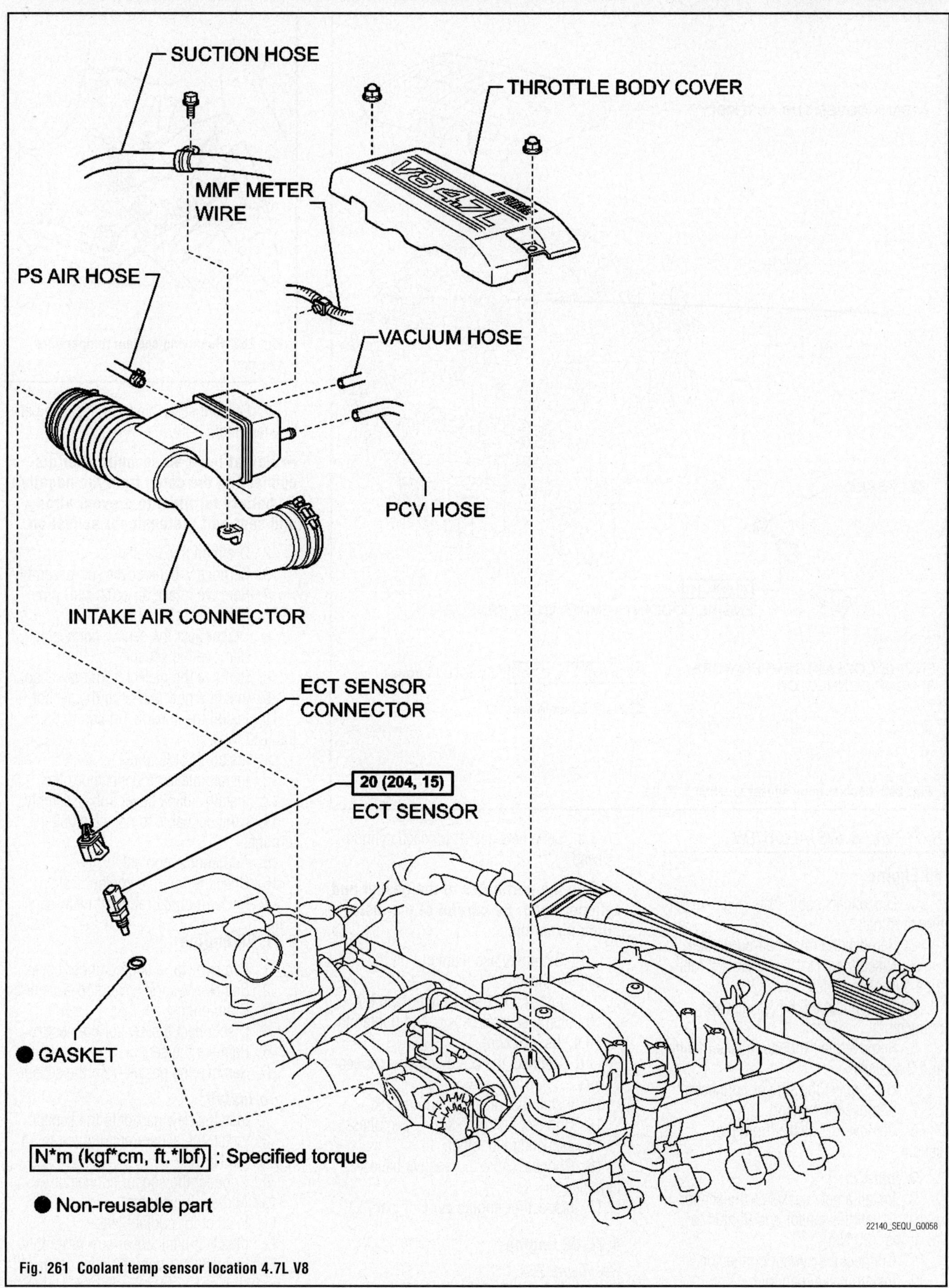

Fig. 261 Coolant temp sensor location 4.7L V8

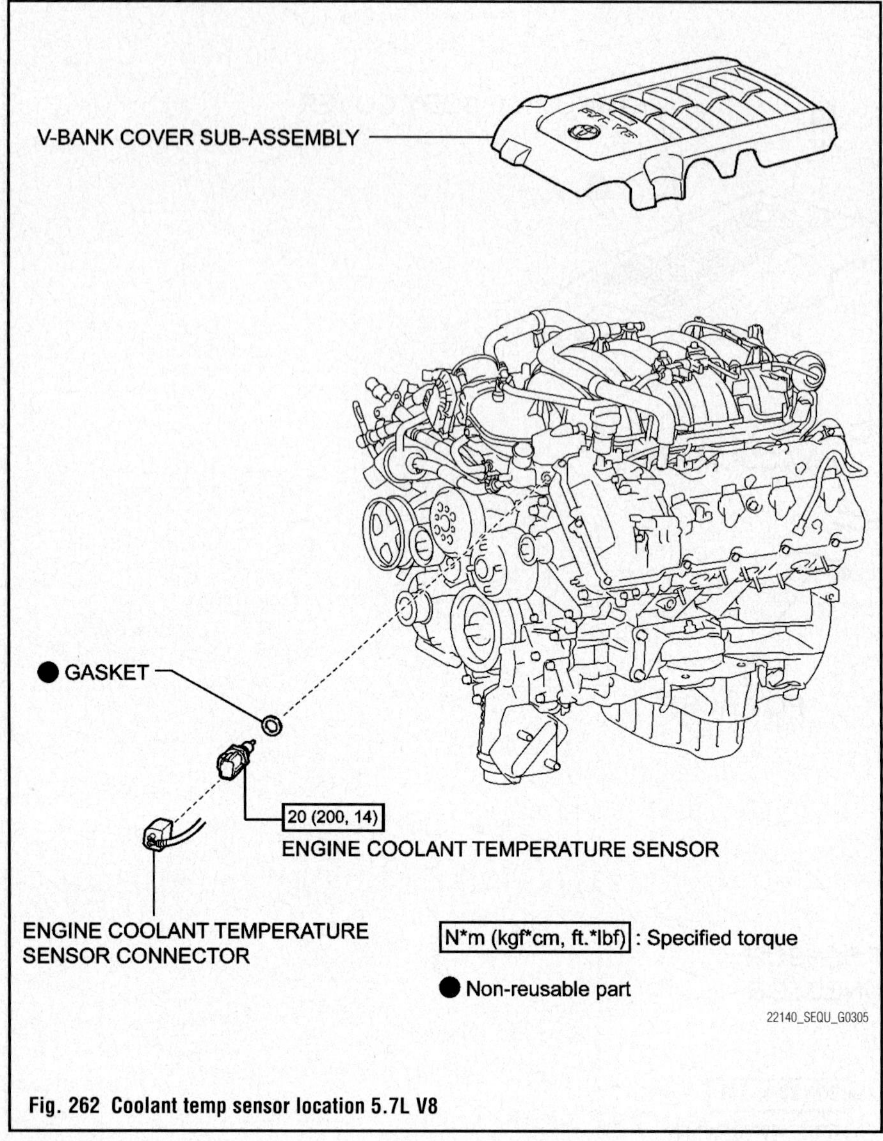

Fig. 262 Coolant temp sensor location 5.7L V8

REMOVAL & INSTALLATION

V6 Engine

1. Disconnect cable from negative battery terminal.
2. Remove the intake air surge tank.
3. Disconnect No. 1 and No.1 2 fuel pipe sub-assemblies.
4. Disconnect the 6 fuel injector connectors.
5. Remove the 10 bolts, intake manifold and 2 gaskets.
6. Disconnect the sensor connector.
7. Remove the sensor.
8. Remove the gasket from the sensor.

To install:

9. Install a new gasket to the sensor.
10. Install the sensor and tighten to 15 ft. lbs. (20 Nm).
11. Connect the sensor connector.
12. Install intake manifold.
    a. Set a new gasket on each cylinder head.

➡Align the port holes of the gasket and cylinder head. Be careful of the installation direction.

    b. Set the intake manifold on the cylinder heads.
    c. Install and uniformly tighten the 10 bolts in several passes and tighten to 19 ft. lbs. (26 Nm).
    d. Connect the 6 fuel injector connectors.
13. Install the intake air surge tank.
14. Connect fuel pipe sub-assemblies.
15. Add engine coolant.
16. Connect cable to negative battery terminal.
17. Inspect for engine coolant leak.

4.7L V8 Engine

See Figure 263.

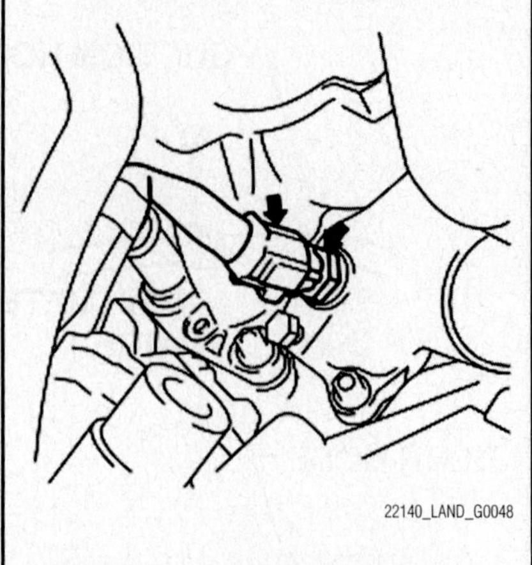

Fig. 263 Removing coolant temperature sensor

1. Disconnect cable from negative battery terminal.

➡Wait at least 90 seconds after disconnecting the cable from the negative (-) battery terminal to prevent airbag and seat belt pretensioner activation.

2. Drain engine coolant.
3. Remove v-bank cover sub-assembly.
4. Remove intake air connector pipe.
5. Remove throttle body.
6. Disconnect the sensor connector.
7. Remove the sensor.
8. Remove the gasket from the sensor.
9. Install a new gasket to the sensor.
10. Install the sensor. Tighten to 15 ft. lbs. (20 Nm).
11. Install throttle body.
12. Install intake air connector pipe.
13. Install v-bank cover sub-assembly.
14. Connect cable to negative battery terminal.
15. Add engine coolant.
16. Check for engine coolant leaks.
17. Perform initialization, if necessary.

5.7L V8 Engine

1. Remove No. 1 engine under cover.
2. Remove V-bank cover sub-assembly.
3. Drain engine coolant.
4. Disconnect the sensor connector.
5. Remove the sensor.
6. Remove the gasket from the sensor.

To install:

7. Install a new gasket to the sensor.
8. Install the sensor and tighten to 14 ft. lbs. (20 Nm).
9. Connect the sensor connector.
10. Add engine coolant.
11. Inspect for coolant leak.
12. Install V-bank cover sub-assembly.
13. Install No. 1 engine under cover.

HEATED OXYGEN (HO2S) SENSOR

LOCATION

See Figures 264 through 266.

REMOVAL & INSTALLATION

CAUTION

Wear protective gloves when removing the sensor. The exhaust pipe assembly is extremely hot immediately after the engine has stopped. Confirm that the exhaust pipe assembly has cooled down before removing it.

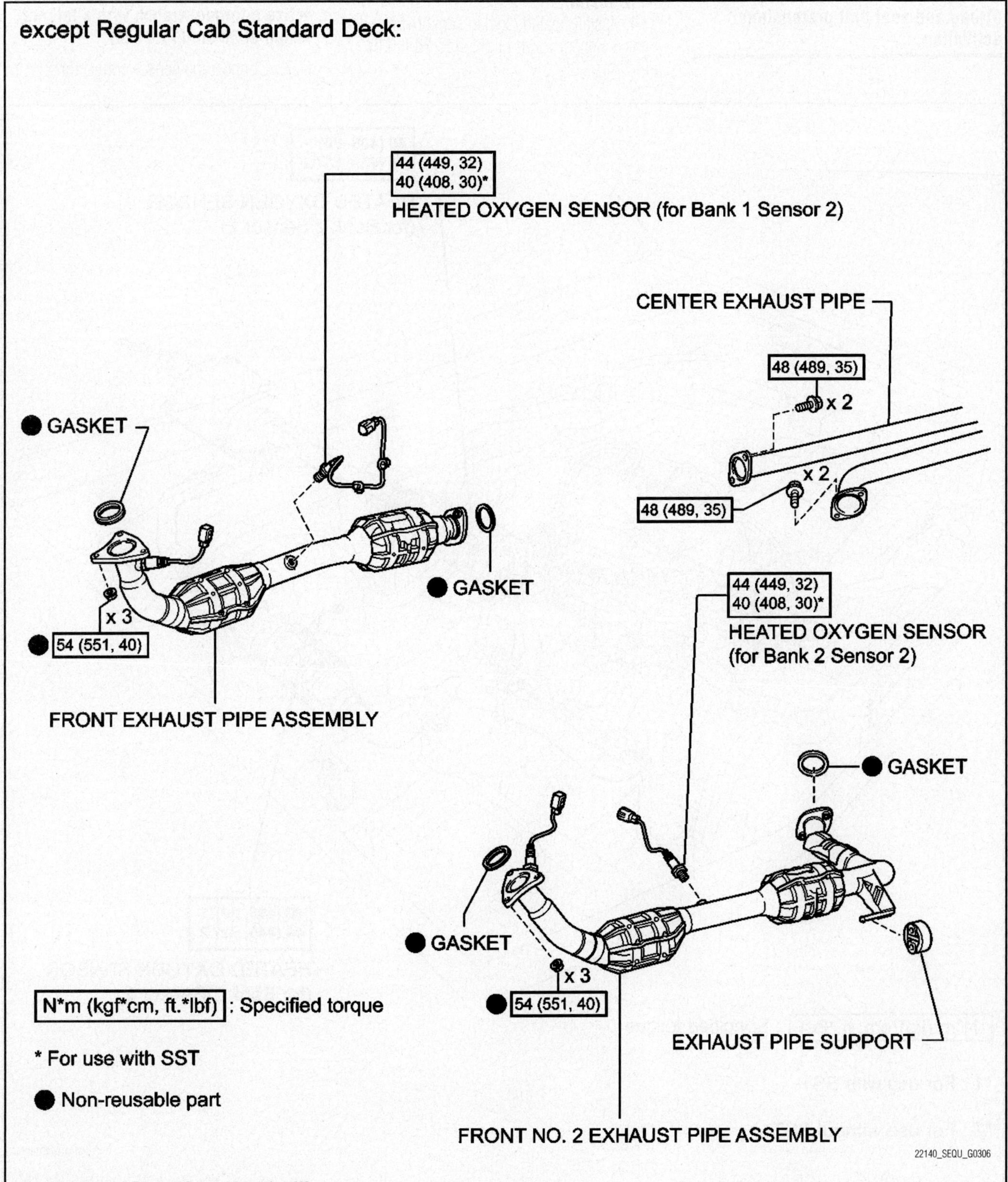

Fig. 264 Heated Oxygen Sensor Location V6

1. Disconnect cable from negative battery terminal.

⁂ CAUTION

Wait at least 90 seconds after disconnecting the cable from the negative (-) battery terminal to prevent airbag and seat belt pretensioner activation.

2. Disconnect the sensor connector and remove heated oxygen sensor (for Bank 1 Sensor 2).
3. Disconnect the sensor connector and remove heated oxygen sensor (for Bank 2 Sensor 2).

To install:

4. Install heated oxygen sensor (for Bank 1 Sensor 2) and tighten to 32 ft. Lbs. (44 Nm.).

➡Use a torque wrench with a fulcrum length of 30 cm (11.81 in).

5. Connect the sensor connector.
6. Install heated oxygen sensor (for Bank 2 Sensor 2) and tighten to 32 ft. Lbs. (44 Nm.).

➡Use a torque wrench with a fulcrum length of 30 cm (11.81 in).

7. Connect the sensor connector.

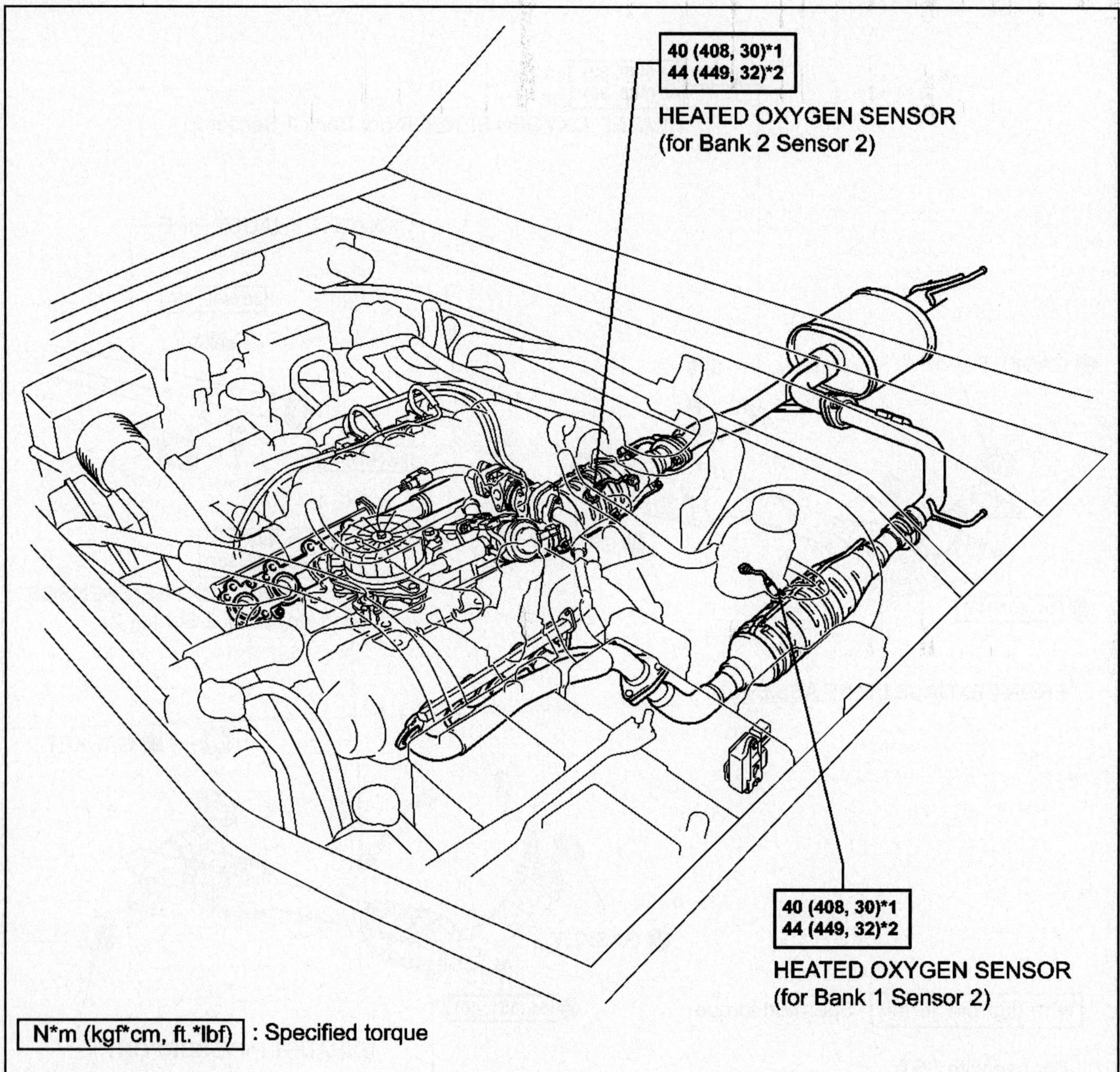

Fig. 265 Heated Oxygen Sensor Location 4.7L V8

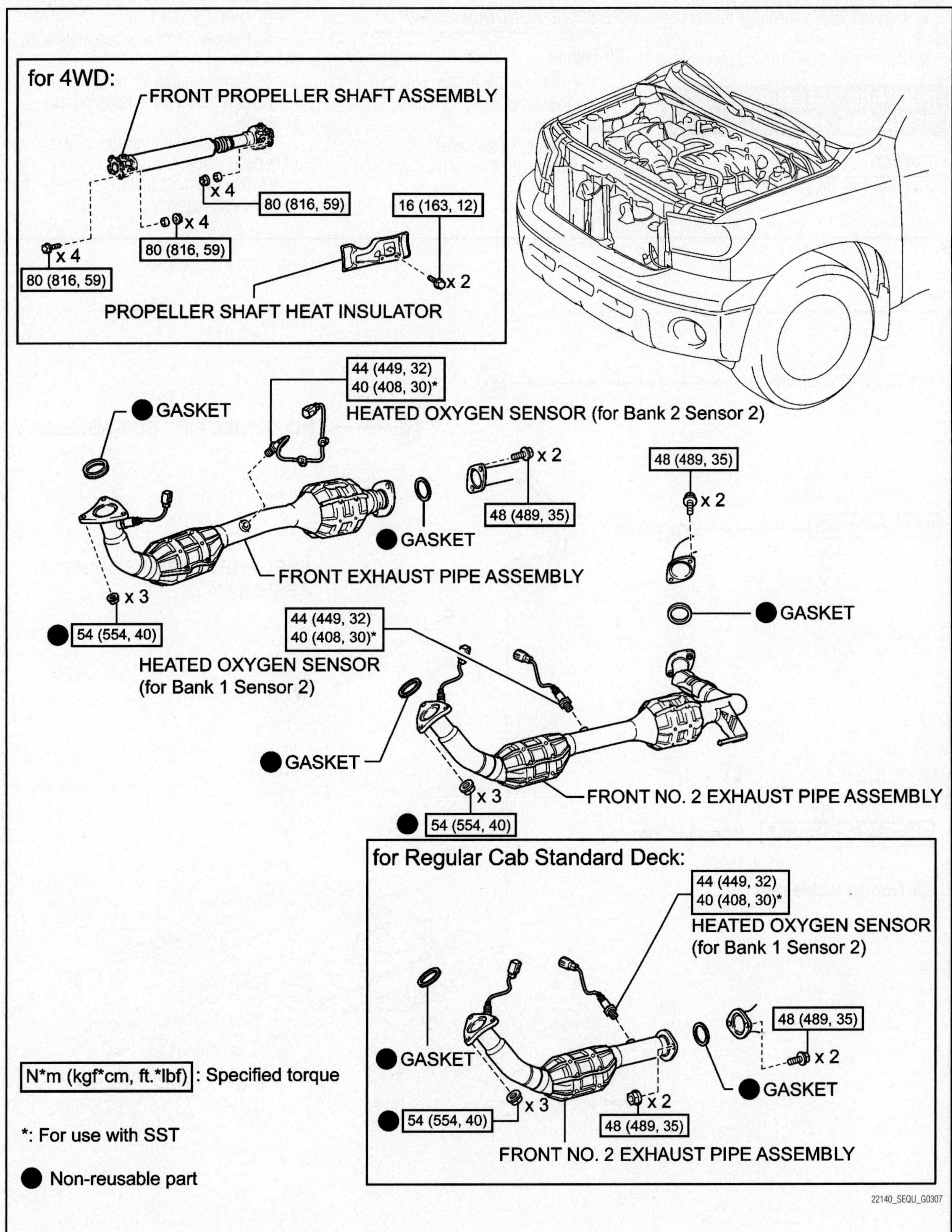

Fig. 266 Heated Oxygen Sensor Location 5.7L V8

8. Connect cable to negative battery terminal.
9. Perform initialization, if necessary.

INJECTION PRESSURE REGULATOR (IPR)

LOCATION

See Figures 267 through 269.

REMOVAL & INSTALLATION

V6 Engine

1. Discharge fuel system pressure.
2. Disconnect cable from negative battery terminal.
3. Remove V-bank cover.
4. Remove the front wiper motor and link.
5. Drain coolant.
6. Remove air cleaner assembly with element.
7. Remove intake air surge tank.
8. Disconnect No. 2 fuel pipe sub-assembly.
9. Disconnect the vacuum sensing hose from the pressure regulator.
10. Remove the 2 bolts, then remove the fuel pressure regulator.

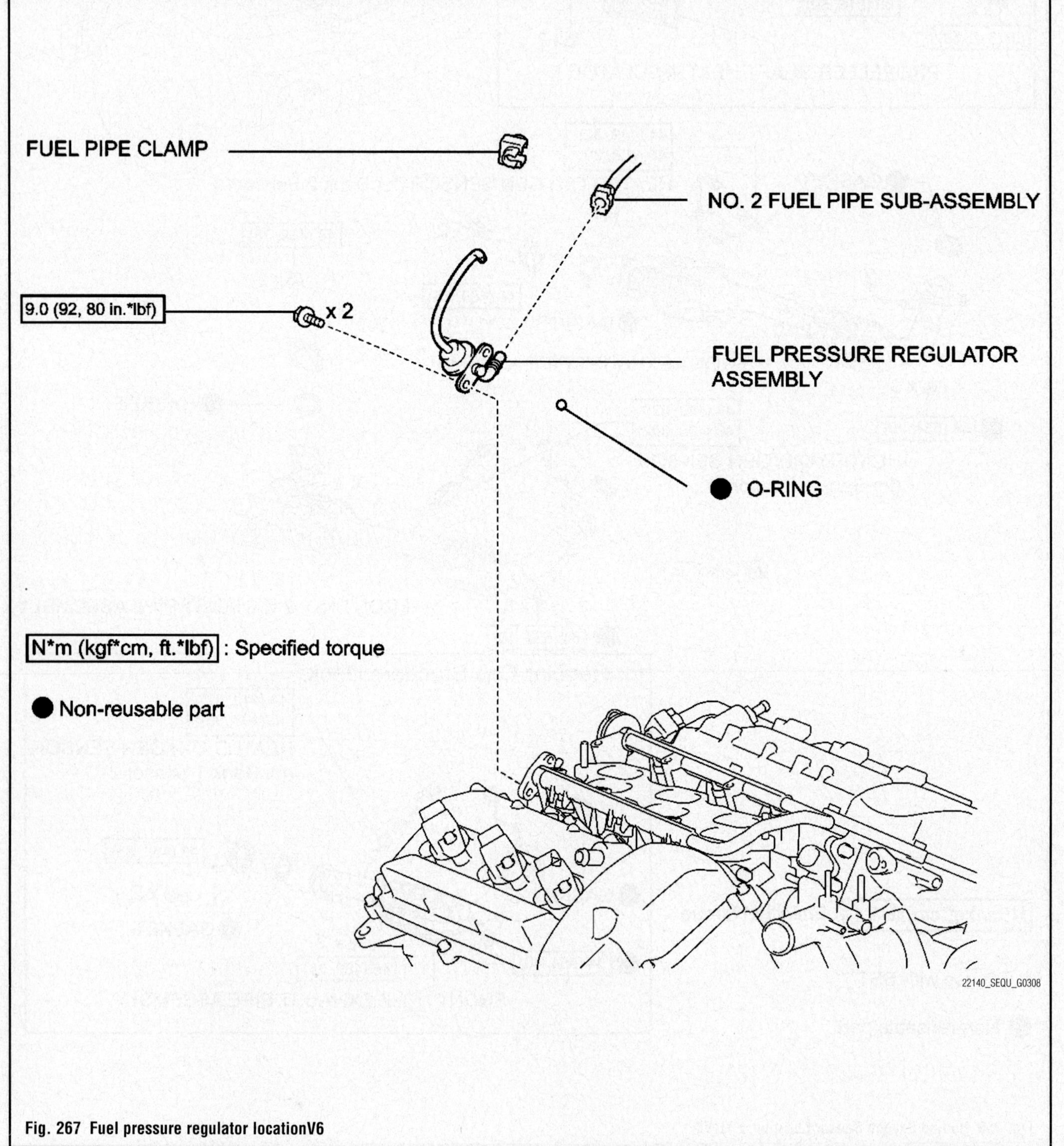

Fig. 267 Fuel pressure regulator locationV6

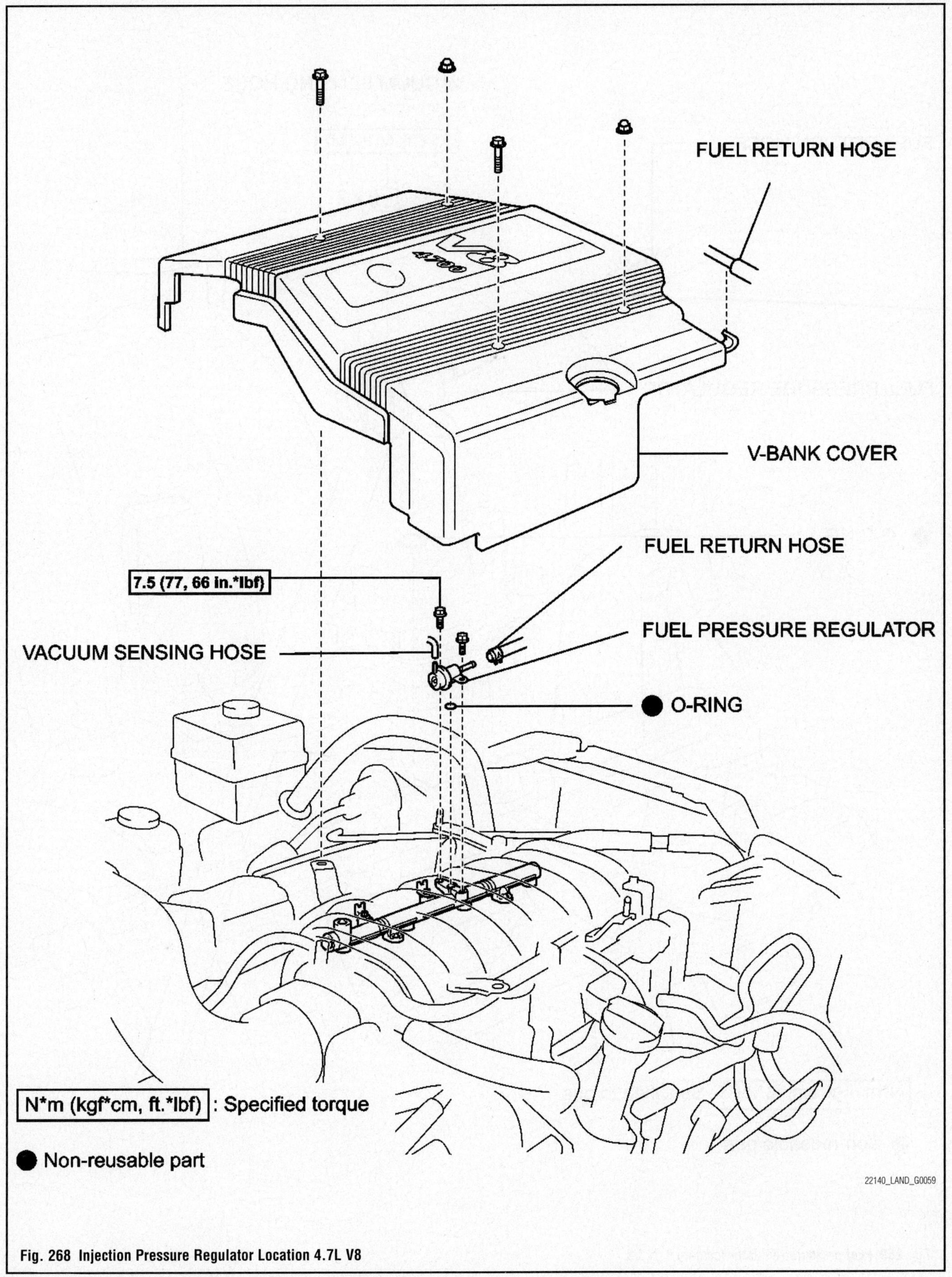

Fig. 268 Injection Pressure Regulator Location 4.7L V8

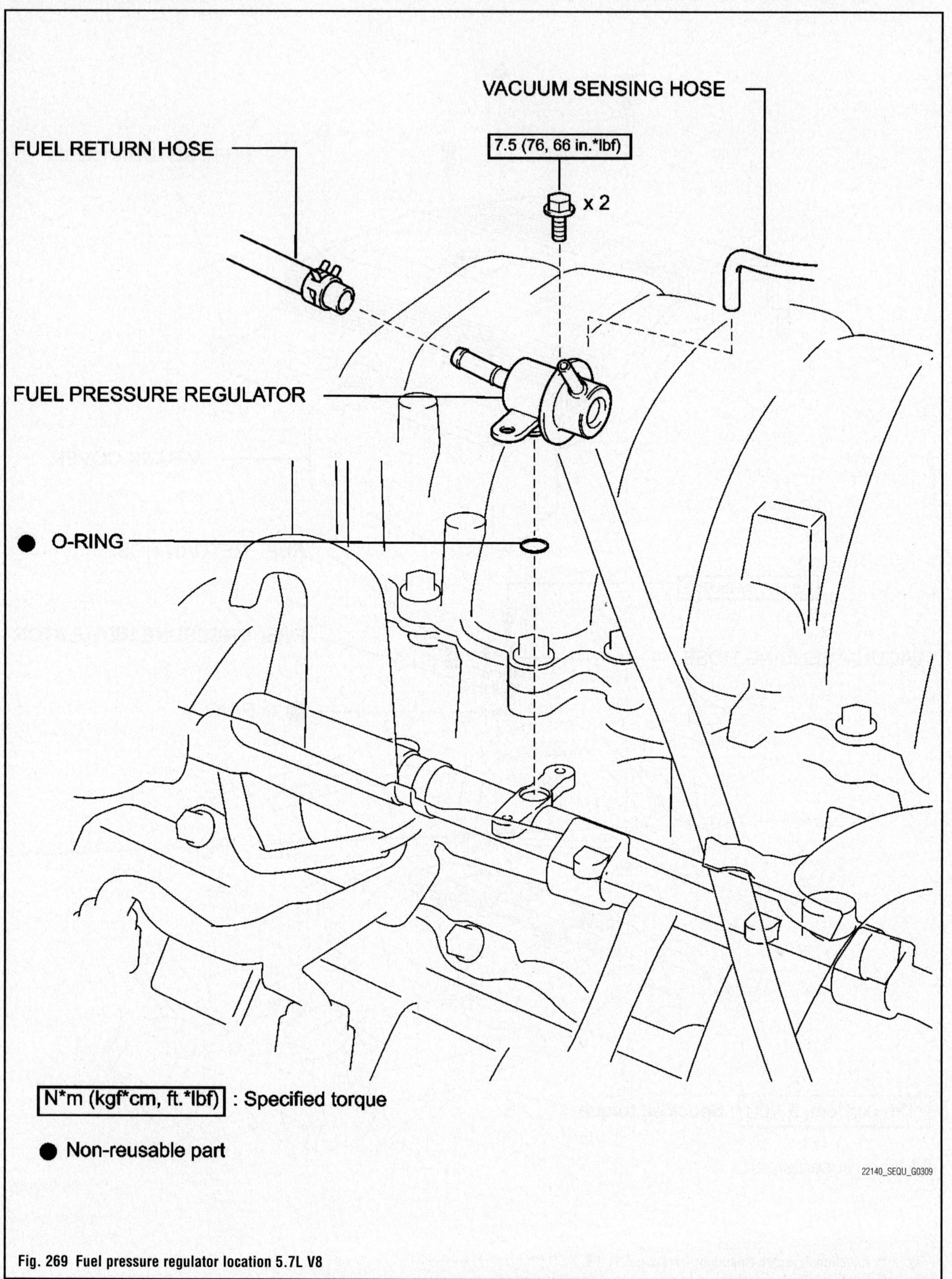

Fig. 269 Fuel pressure regulator location 5.7L V8

11. Remove the O-ring from the fuel pressure regulator.
12. Remove the vacuum hose.

To install:

13. Apply a light coat of gasoline to a new O-ring, and install it to the pressure regulator.
14. While turning the pressure regulator left and right, install it to the delivery pipe.
15. Install the pressure regulator with the 2 bolts and tighten bolts to 66 inch lbs. (7.5 Nm).
16. Connect No. 2 fuel pipe sub-assembly.
17. Install intake air surge tank.
18. Install air cleaner assembly with element.
19. Add engine coolant.
20. Install the front wiper motor and link.
21. Install No. 1 engine under cover.
22. Check for leaks.
23. Perform initialization, if necessary.

4.7L V8 Engine

See Figure 270.

1. Discharge fuel system pressure.
2. Disconnect cable from negative battery terminal.
3. Remove V-bank cover.
4. Disconnect the vacuum sensing hose from the pressure regulator.
5. Disconnect the fuel return hose from the pressure regulator
6. Remove the 2 bolts, and pull out the pressure regulator.
7. Remove the O-ring from the pressure regulator.

To install:

8. Apply a light coat of gasoline to a new O-ring, and install it to the pressure regulator.
9. While turning the pressure regulator left and right, install it to the delivery pipe.
10. Install the pressure regulator with the 2 bolts and tighten bolts to 66 inch lbs. (7.5 Nm).
11. Install V-bank cover.
12. Install battery cable.
13. Check for leaks.

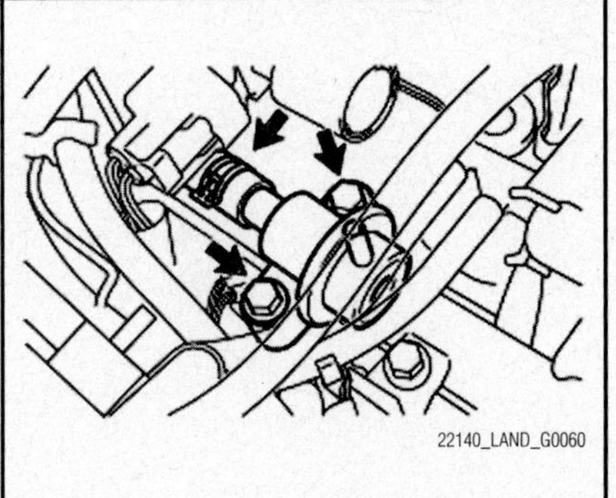

Fig. 270 Removing Injection Pressure Regulator

5.7L V8 Engine

See Figure 271.

1. Discharge fuel system pressure.
2. Disconnect cable from negative battery terminal.
3. Remove V-bank cover.
4. Disconnect the No. 2 fuel tube from the pressure regulator.
5. Disconnect the vacuum sensing hose from the pressure regulator.

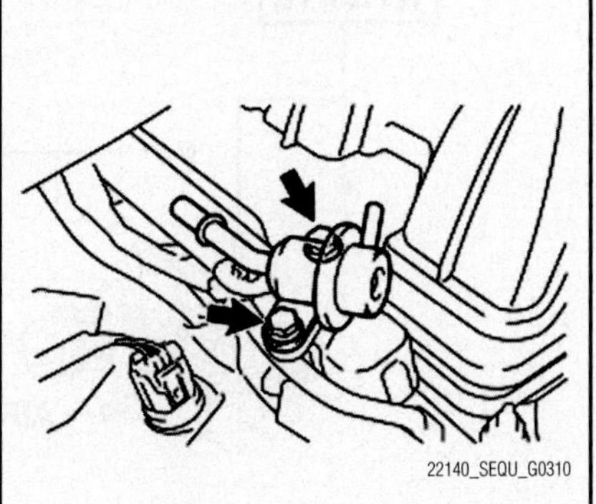

Fig. 271 Removing Injection Pressure Regulator

6. Remove the 2 bolts, and pull out the pressure regulator.
7. Remove the O-ring from the pressure regulator.
8. Apply a light coat of gasoline or spindle oil to a new O-ring, and install it to the pressure regulator.
9. While turning the pressure regulator left and right, install it to the delivery pipe.
10. Install the 2 bolts and tighten to 7 ft. lbs. (10 Nm).
11. Connect the vacuum sensing hose to the pressure regulator.
12. Connect the No. 2 fuel tube to the pressure regulator.
13. Install V-bank cover.
14. Install battery cable.
15. Check for leaks.

INTAKE AIR TEMPERATURE (IAT) SENSOR

LOCATION

The Intake Air Temperature (IAT) sensor, built into the Mass Air Flow (MAF) meter.

REMOVAL & INSTALLATION

See Mass Air Flow Meter.

KNOCK SENSOR (KS)

LOCATION

See Figures 272 through 274.

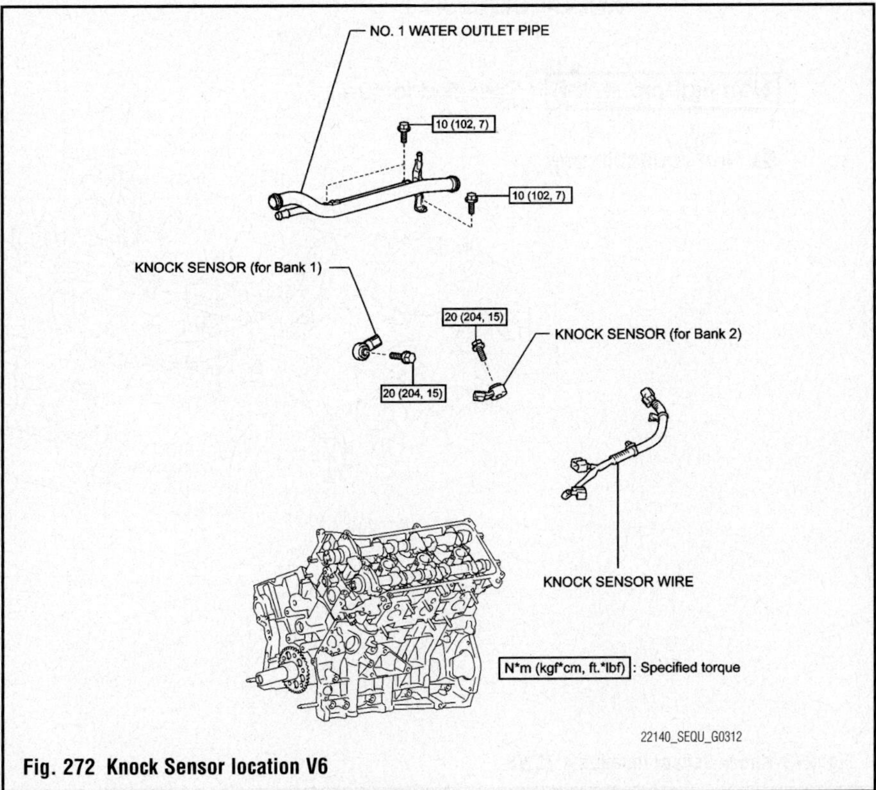

Fig. 272 Knock Sensor location V6

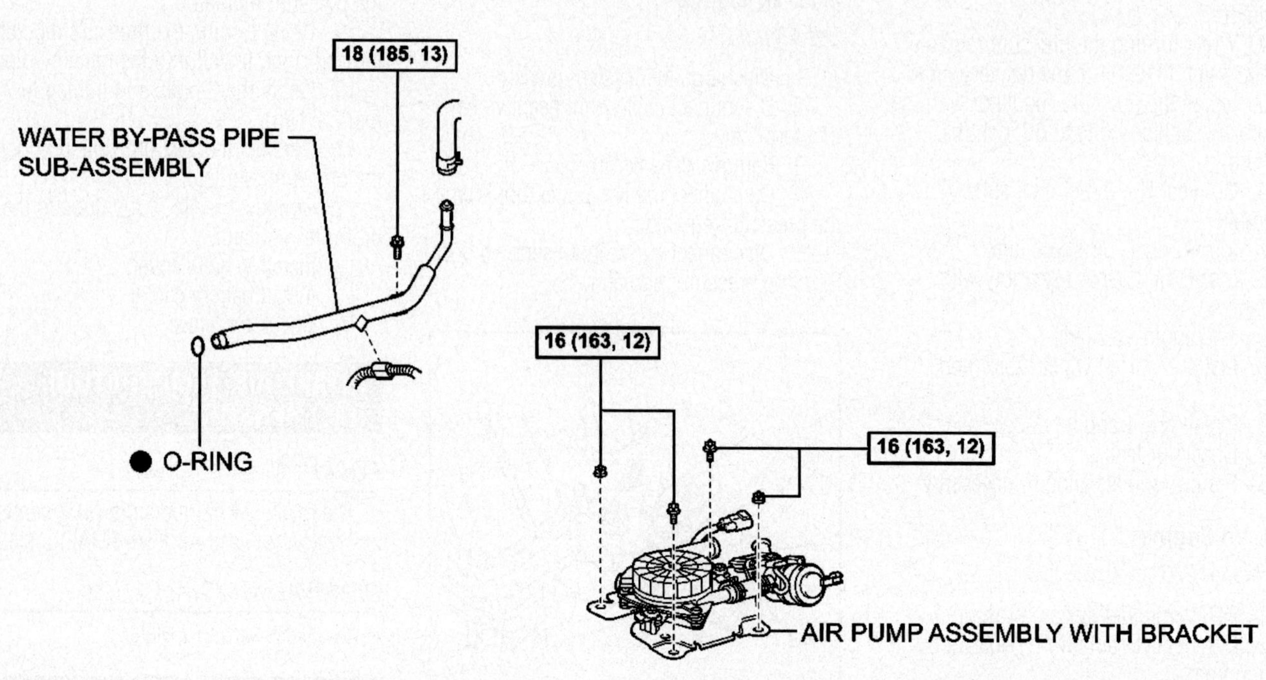

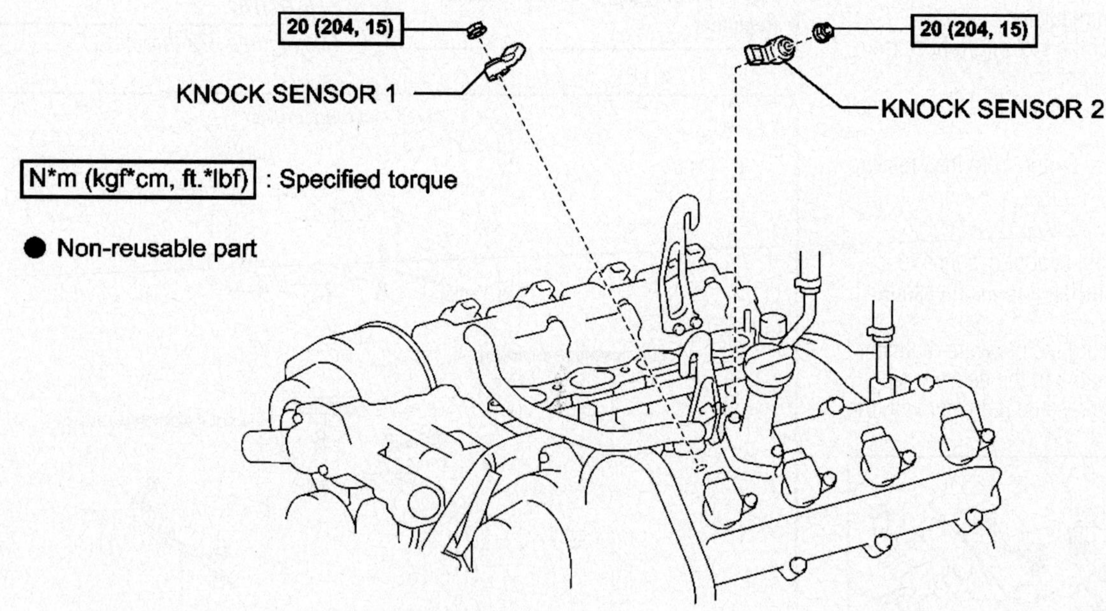

Fig. 273 Knock Sensor location 4.7L V8

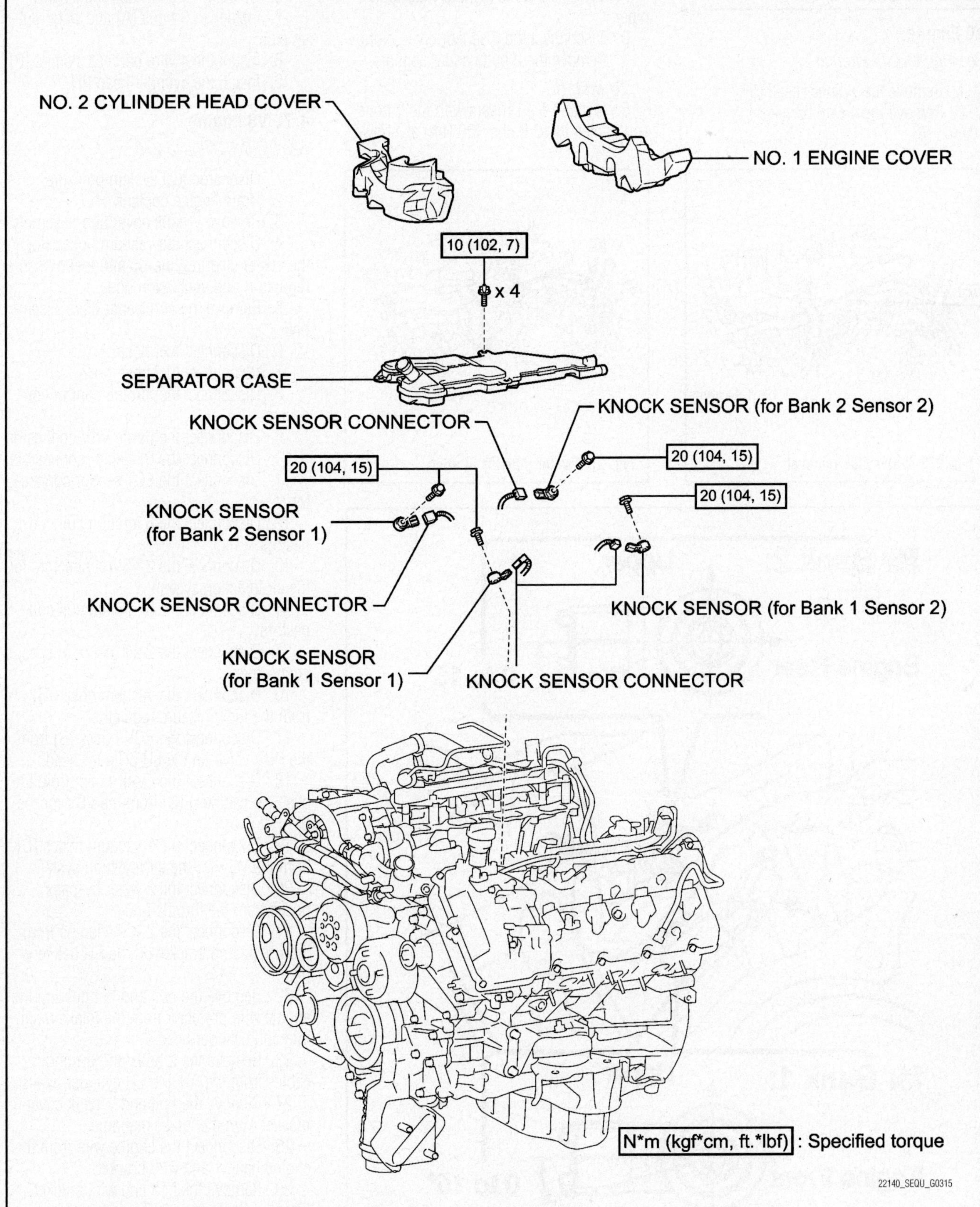

Fig. 274 Knock Sensor location 5.7L V8

REMOVAL & INSTALLATION

V6 Engine

See Figures 275 through 277.

1. Remove the cylinder head RH.
2. Remove the 4 wire harness clamps [A].

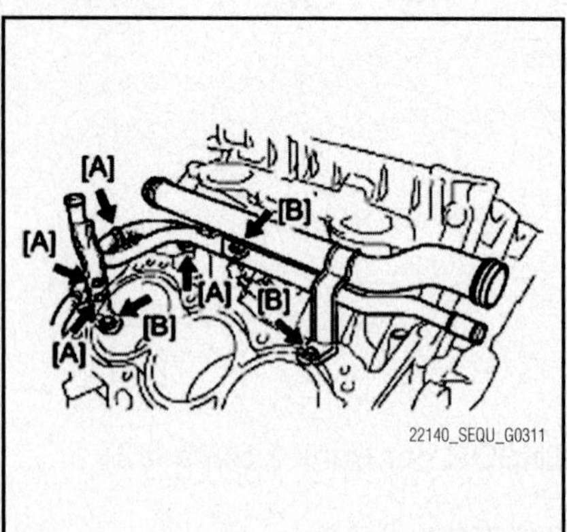

Fig. 275 Water pipe removal

Remove the 3 bolts [B] and water outlet pipe.

3. Disconnect the 2 sensor connectors.
4. Remove the 2 bolts and 2 sensors.

To install:

5. Install the 2 sensors with the 2 bolts and tighten to 15 ft. lbs. (20 Nm) as shown.

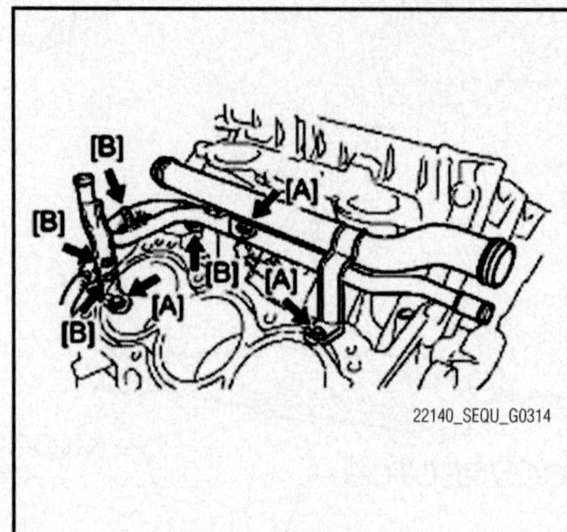

Fig. 277 Water pipe installation

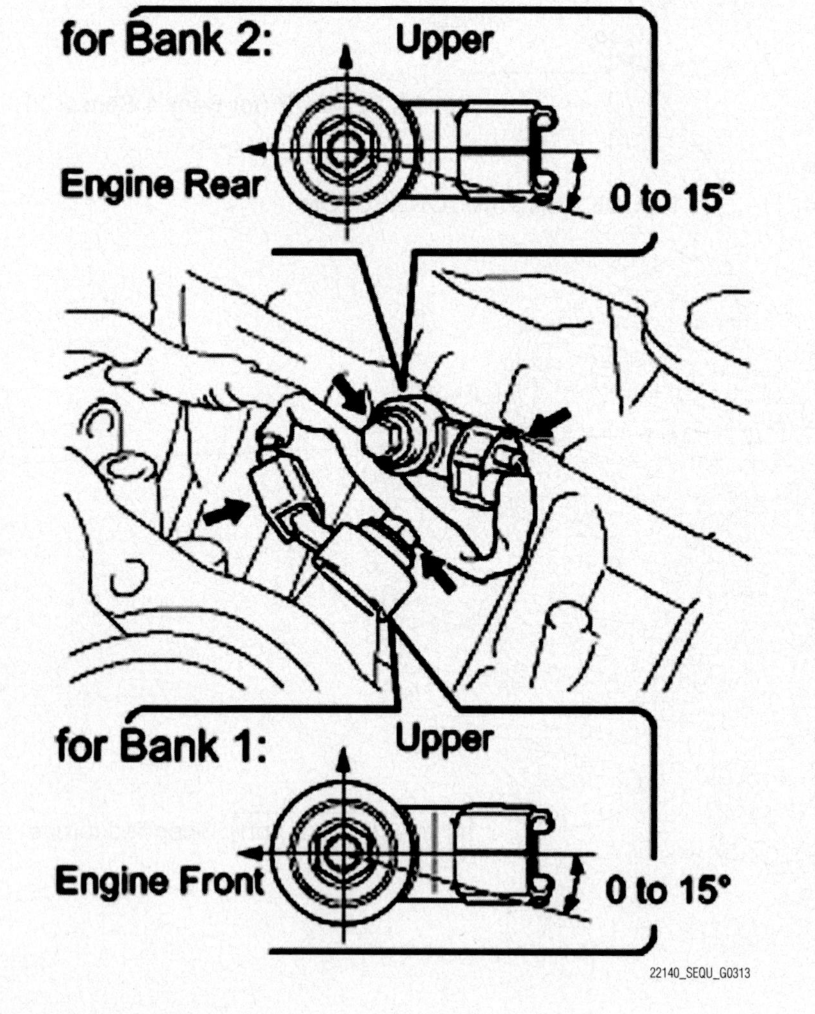

Fig. 276 Knock sensor installation

6. Connect the 2 sensor connectors.
7. Install the 3 bolts [A] and water outlet pipe.
8. Install the 4 wire harness clamps [B].
9. Install the cylinder head RH.

4.7L V8 Engine

See Figures 278 and 279.

1. Discharge fuel system pressure.
2. Drain engine coolant.
3. Remove v-bank cover sub-assembly.
4. Disconnect the vacuum hoses (for the power steering idle-up and fuel pressure regulator) and ventilation hose.
5. Remove the air cleaner hose assembly.
6. Disconnect fuel hose.
7. Disconnect fuel hose no.2.
8. Disconnect the throttle control connector.
9. Disconnect the purge VSV connector.
10. Disconnect the 8 injector connectors.
11. Disconnect the ECT sensor connector.
12. Disconnect the 8 ignition coil connectors.
13. Disconnect the 2 VSV connectors for the air injection system.
14. Disconnect the 8 ignition coil connectors.
15. Disconnect the 2 air fuel ratio sensor connectors.
16. Disconnect the vacuum hose [A] from the fuel pressure regulator.
17. Disconnect the PCV hoses [B] from the PCV valve on the LH cylinder head.
18. Disconnect the EVAP hose (from the charcoal canister) [C] from the VSV for the EVAP.
19. Disconnect the 2 vacuum hoses [D] from the VSV for the air injection system.
20. Disconnect the 2 water by-pass hoses from the throttle body.
21. Disconnect the 2 wire clamps from the wire clamp bracket on the RH delivery pipe.
22. Remove the bolt and nut holding the engine wire protector from the intake manifold and cylinder head.
23. Remove the 2 bolts and ground cables from the RH and LH cylinder heads.
24. Remove the bolt and V-bank cover bracket from the intake manifold.
25. Disconnect the engine wire from the engine hanger and wire bracket.
26. Remove the bolt and wire bracket from the intake manifold.
27. Remove the 6 bolts, 4 nuts, intake manifold assembly and 2 gaskets.
28. Remove air pump assembly with bracket.

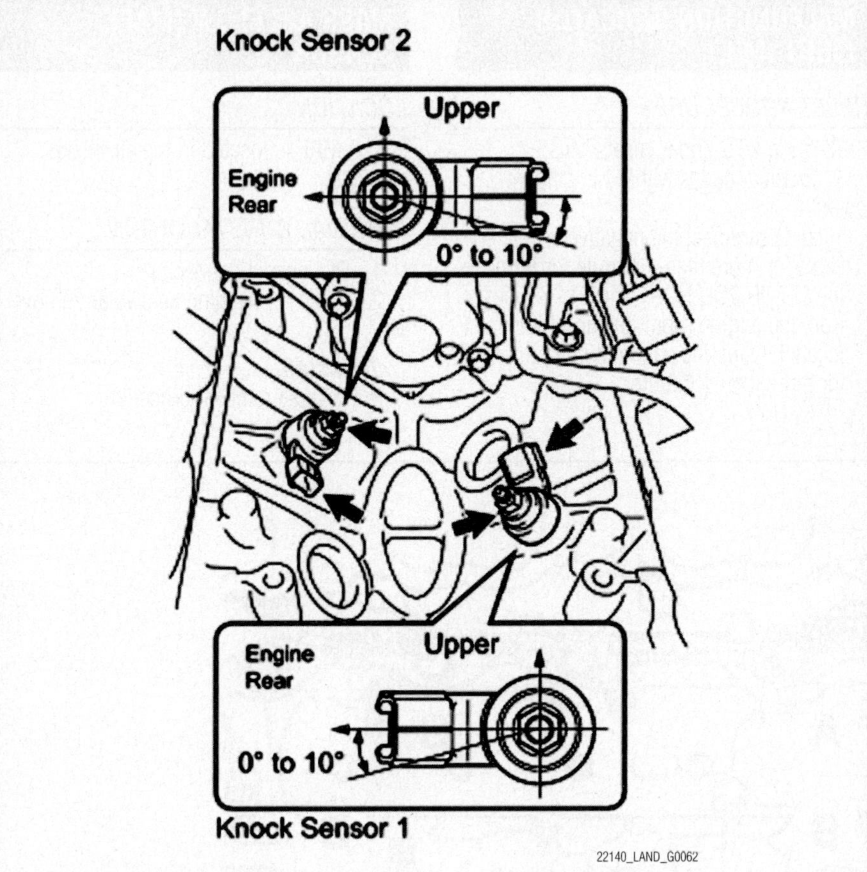

Fig. 278 Installing Knock Sensor

29. Remove knock sensor.
30. Disconnect the 2 knock sensor connectors.

To install:

31. Install the 2 knock sensors with the 2 nuts as shown in the illustration and tighten nuts to 15 ft. lbs. (20 Nm).
32. Connect the 2 knock sensor connectors.
33. Place 2 new gaskets on the intake manifold.
34. Place the intake manifold on the cylinder heads.

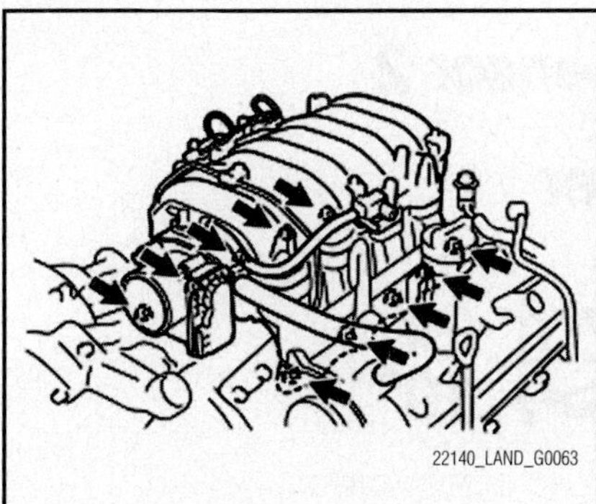

Fig. 279 Intake manifold tightening bolts

35. Install and uniformly tighten the 6 bolts and 4 nuts in several steps to 13 ft. lbs. (18 Nm).
36. Install the V-bank cover bracket to the intake manifold.
37. Install the wire bracket to the intake manifold with the bolt.
38. Connect the engine wire to the engine hanger and wire bracket.
39. Connect the wire protector to the intake manifold and cylinder heads with the bolt and nut.
40. Install the 2 ground cables with the 2 bolts to the RH and LH cylinder heads.
41. Connect the 2 water by-pass hoses to the throttle body.
42. Connect the 2 wire clamps to the wire clamp bracket on the RH delivery pipe.
43. Connect the vacuum hose to the fuel pressure regulator.
44. Connect the PCV hose to the PCV valve on the LH cylinder head.
45. Connect the EVAP hose (from the charcoal canister) to the purge VSV.
46. Connect the 2 vacuum hoses to the VSV for the air injection system.
47. Connect the throttle control connector.
48. Connect the 2 VSV connectors for the air injection system.
49. Connect the purge VSV connector.
50. Connect the 8 injector connectors.
51. Connect the ECT sensor connector.
52. Connect the 8 ignition coil connectors.
53. Connect the 2 air fuel ratio sensor connectors.
54. Install fuel hose no.2.
55. Install fuel hose.
56. Install v-bank cover sub-assembly.
57. Add engine coolant.
58. Check for engine coolant leaks.
59. Check for fuel leaks.

5.7 V8 Engine

See Figures 280 through 282.

1. Remove the intake manifold.
2. Remove No. 2 cylinder head cover.
3. Remove No. 1 engine cover.
4. Remove the 4 bolts and separator case.
5. Disconnect the 4 knock sensor connectors.
6. Remove the 4 bolts and 4 knock sensors.

To install:

7. Install the 4 sensors with the 4 bolts so that the sensors are angled as shown in the illustration.
8. Connect the 4 sensor connectors.
9. Install the separator case.
10. Install No. 1 engine cover.
11. Install No. 2 cylinder head cover.
12. Install the intake manifold.

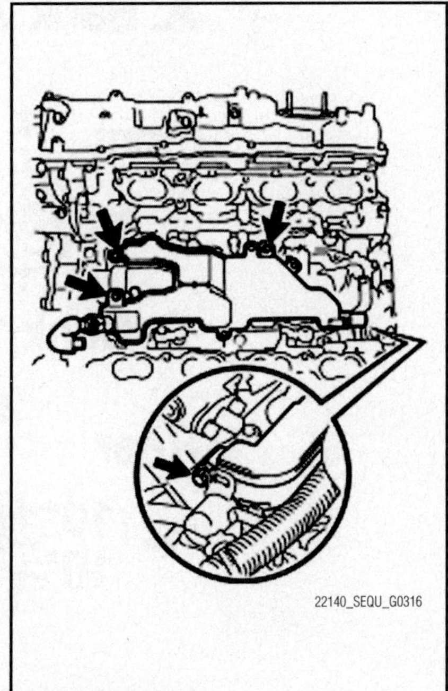

Fig. 280 Separator case

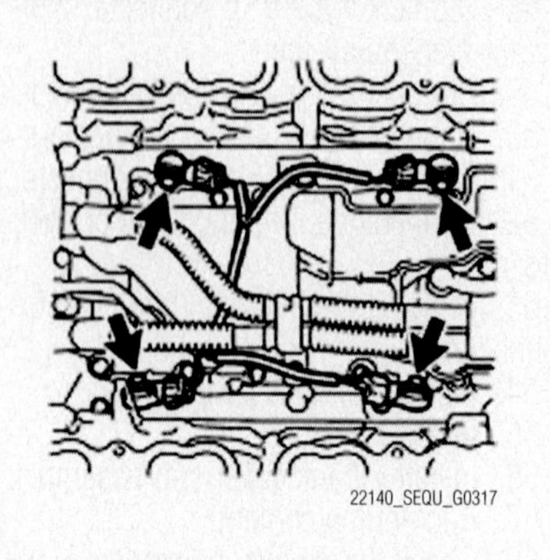

Fig. 281 Removing knock sensors

MALFUNCTION INDICATOR LIGHT (MIL)

RESET PROCEDURES

Clearing DTC codes resets MIL.

1. To clear codes without intelligent tester:
   a. Disconnect the negative (-) battery cable for more than 1 minute, or Remove the EFI OR ECD NO. 1 and ETCS fuses from the engine room junction block located inside the engine compartment for more than 1 minute.

MASS AIR FLOW (MAF) SENSOR

LOCATION

The MAF is located in the air intake snorkel.

REMOVAL & INSTALLATION

1. Disconnect connector.
2. Remove attaching screws an remove MAF.

To install:

3. Reverse removal procedure.

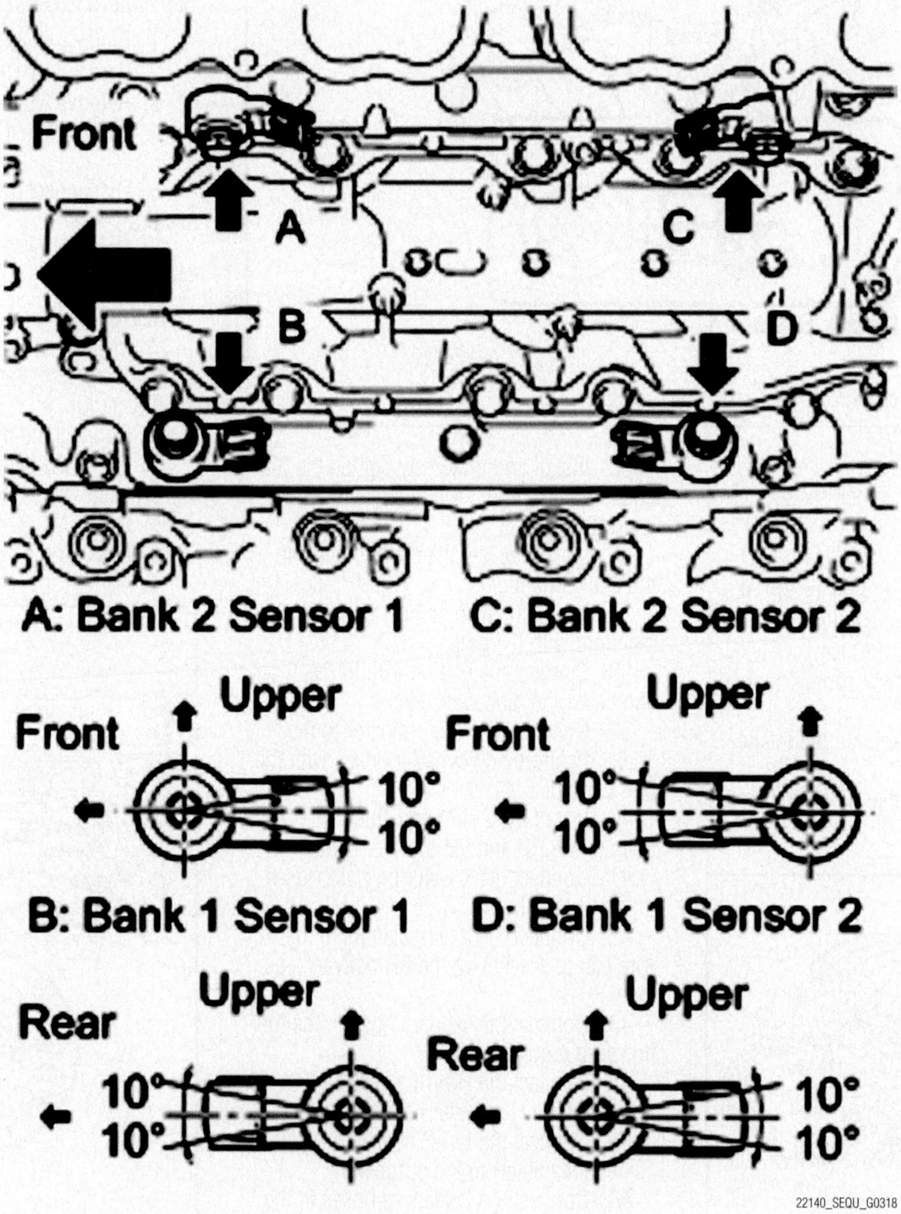

Fig. 282 Installing knock sensors

VARIABLE CAMSHAFT TIMING OIL CONTROL SOLENOID

LOCATION

See Figures 283 and 284.

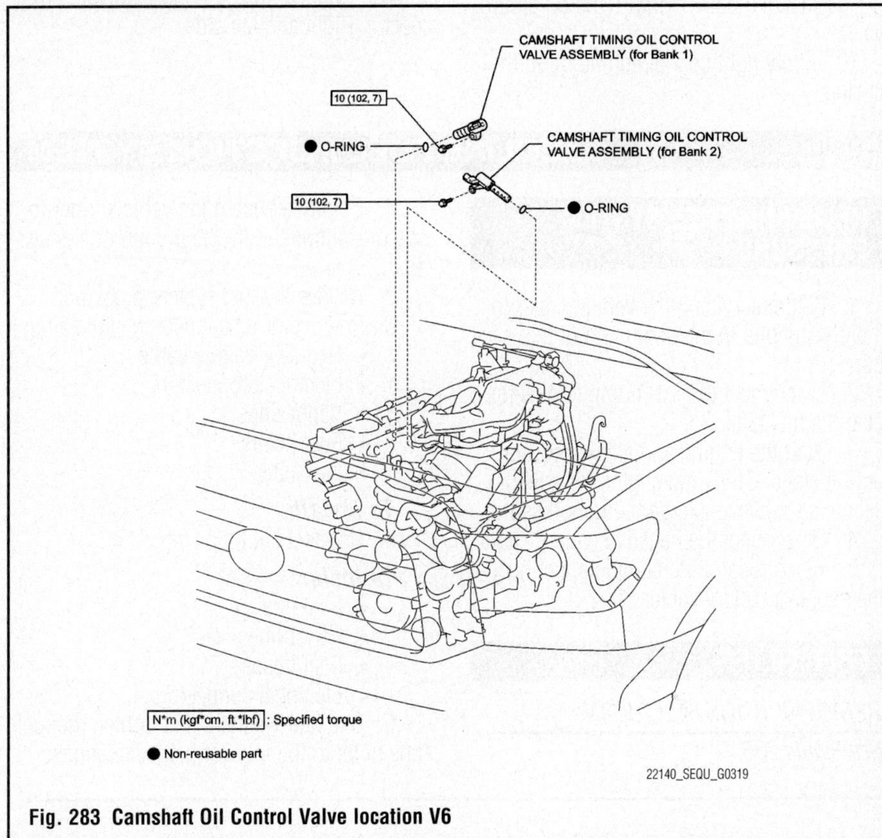

Fig. 283 Camshaft Oil Control Valve location V6

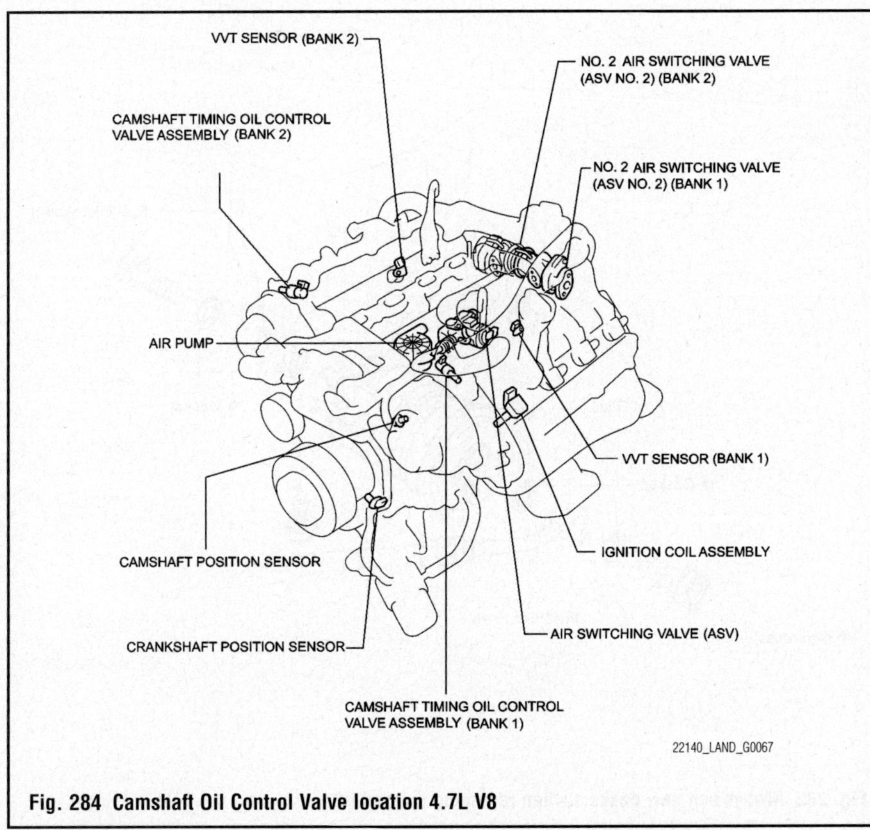

Fig. 284 Camshaft Oil Control Valve location 4.7L V8

REMOVAL & INSTALLATION

V6 Engine

1. Disconnect cable from negative battery terminal.
2. Remove No. 1 engine under cover.
3. Drain engine coolant.
4. Remove the front wiper motor and link assembly.
5. Remove front cowl top outer panel sub-assembly.
6. Remove V-bank cover.
7. Remove No. 2 air cleaner hose.
8. Remove air cleaner assembly with element.
9. Remove throttle body with motor assembly.
10. Disconnect the oil control valve connector.
11. Remove the bolt and oil control valve for bank 1.
12. Remove the O-ring from the oil control valve.
13. Disconnect the oil control valve connector.
14. Remove the bolt and oil control valve for bank 2.
15. Remove the O-ring from the oil control valve.

To install:

16. Apply a light coat of engine oil to a new O-ring of the oil control valve.
17. Install the oil control valve on bank 1 with the bolt and tighten to 7 ft. lbs. (10 Nm).
18. Connect the camshaft timing oil control valve connector.
19. Repeat the same procedure for bank 2.
20. Install throttle body with motor assembly.
21. Install air cleaner assembly with element.
22. Install No. 2 air cleaner hose.
23. Connect cable to negative battery terminal.
24. Add engine coolant.
25. Install V-bank cover.
26. Install front cowl top outer panel sub-assembly.
27. Install the front wiper motor and link assembly.
28. Install No. 1 engine under cover.
29. Check for leaks.
30. Perform initialization, if necessary.

4.7L V8 Engine

1. Disconnect cable from negative battery terminal.
2. Remove the 2 nuts, 2 bolts and V-bank cover.
3. Disconnect the 3 hoses.

4. Remove the 2 bolts.
5. Loosen the 2 clamp bolts and remove the intake air connector.
6. Disconnect the oil control valve connector. Right and left side.
7. Remove the bolt and oil control valve. Right and left side.
8. Remove the O-ring from the oil control valve. Right and left side.

To install:

9. Install a new O-ring to the oil control valve.
10. Apply light coat of engine oil to the O-ring.
11. Install the oil control valve with the bolt and tighten bolt to 66 inch lbs. (7.5 Nm). Right and left side.
12. Connect the oil control valve connector. Right and left side.

FUEL SYSTEM — GASOLINE FUEL INJECTION SYSTEM

FUEL SYSTEM SERVICE PRECAUTIONS

Safety is the most important factor when performing not only fuel system maintenance but any type of maintenance. Failure to conduct maintenance and repairs in a safe manner may result in serious personal injury or death. Maintenance and testing of the vehicle's fuel system components can be accomplished safely and effectively by adhering to the following rules and guidelines.

- To avoid the possibility of fire and personal injury, always disconnect the negative battery cable unless the repair or test procedure requires that battery voltage be applied.
- Always relieve the fuel system pressure prior to disconnecting any fuel system component (injector, fuel rail, pressure regulator, etc.), fitting or fuel line connection. Exercise extreme caution whenever relieving fuel system pressure to avoid exposing skin, face and eyes to fuel spray. Please be advised that fuel under pressure may penetrate the skin or any part of the body that it contacts.
- Always place a shop towel or cloth around the fitting or connection prior to loosening to absorb any excess fuel due to spillage. Ensure that all fuel spillage (should it occur) is quickly removed from engine surfaces. Ensure that all fuel soaked cloths or towels are deposited into a suitable waste container.
- Always keep a dry chemical (Class B) fire extinguisher near the work area.
- Do not allow fuel spray or fuel vapors to come into contact with a spark or open flame.
- Always use a back-up wrench when loosening and tightening fuel line connection fittings. This will prevent unnecessary stress and torsion to fuel line piping.
- Always replace worn fuel fitting O-rings with new. Do not substitute fuel hose or equivalent where fuel pipe is installed.

Before servicing the vehicle, make sure to also refer to the precautions in the beginning of this section as well.

RELIEVING FUEL SYSTEM PRESSURE

1. Before servicing the vehicle, refer to the precautions in the beginning of this section.
2. Disconnect the fuel pump connector near the fuel tank.
3. Start the engine and allow it to run until it stalls. Crank the engine for a few seconds to relieve additional fuel pressure.
4. Disconnect the negative battery cable.
5. When repairs are complete, connect the negative battery cable.

FUEL FILTER

REMOVAL & INSTALLATION

See Figure 285.

1. Before servicing the vehicle, refer to the precautions in the beginning of this section.
2. Relieve the fuel system pressure.
3. Remove or disconnect the following:
   - Negative battery cable
   - Electrical connectors
   - Vapor lines
   - Fuel hoses
   - Fuel filter

To install:

4. Connect the following:

To install:

   - Fuel filter.
   - Vapor lines
   - Fuel hoses
   - Electrical connectors

5. Use new washers and tighten the fuel line bolts to the following specifications:

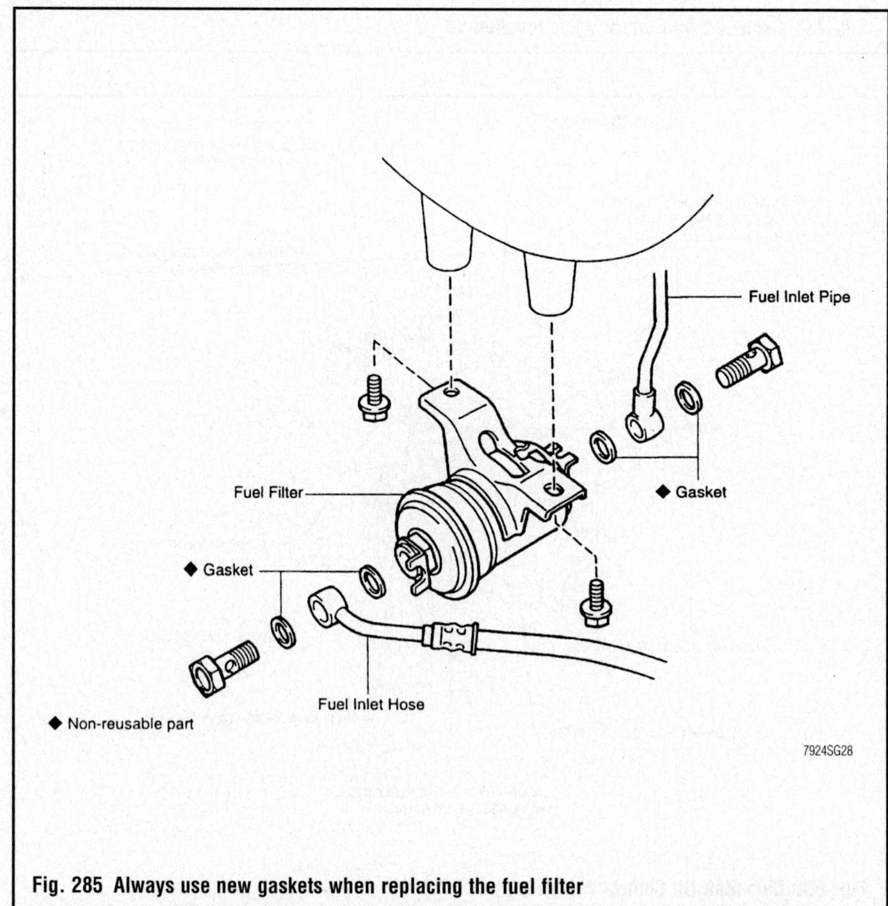

Fig. 285 Always use new gaskets when replacing the fuel filter

- Banjo bolt fittings: 21 ft. lbs. (29 Nm)
- Flare nut fitting: 28 ft. lbs. (38 Nm)

6. Connect the negative battery cable.
7. Start the engine and check for leaks.

FUEL PUMP

REMOVAL & INSTALLATION

See Figure 286.

1. Before servicing the vehicle, refer to the precautions in the beginning of this section.
2. Relieve the fuel system pressure.
3. Remove or disconnect the following:
   - Negative battery cable
   - Fuel tank
   - Fuel pump harness connector
   - Fuel lines
   - Fuel pump module

To install:

4. Install or connect the following:
   - Fuel pump module. Tighten the bolts to 35 inch lbs. (4 Nm).
   - Fuel lines
   - Fuel pump harness connector
   - Fuel tank
   - Negative battery cable
5. Start the engine and check for leaks.

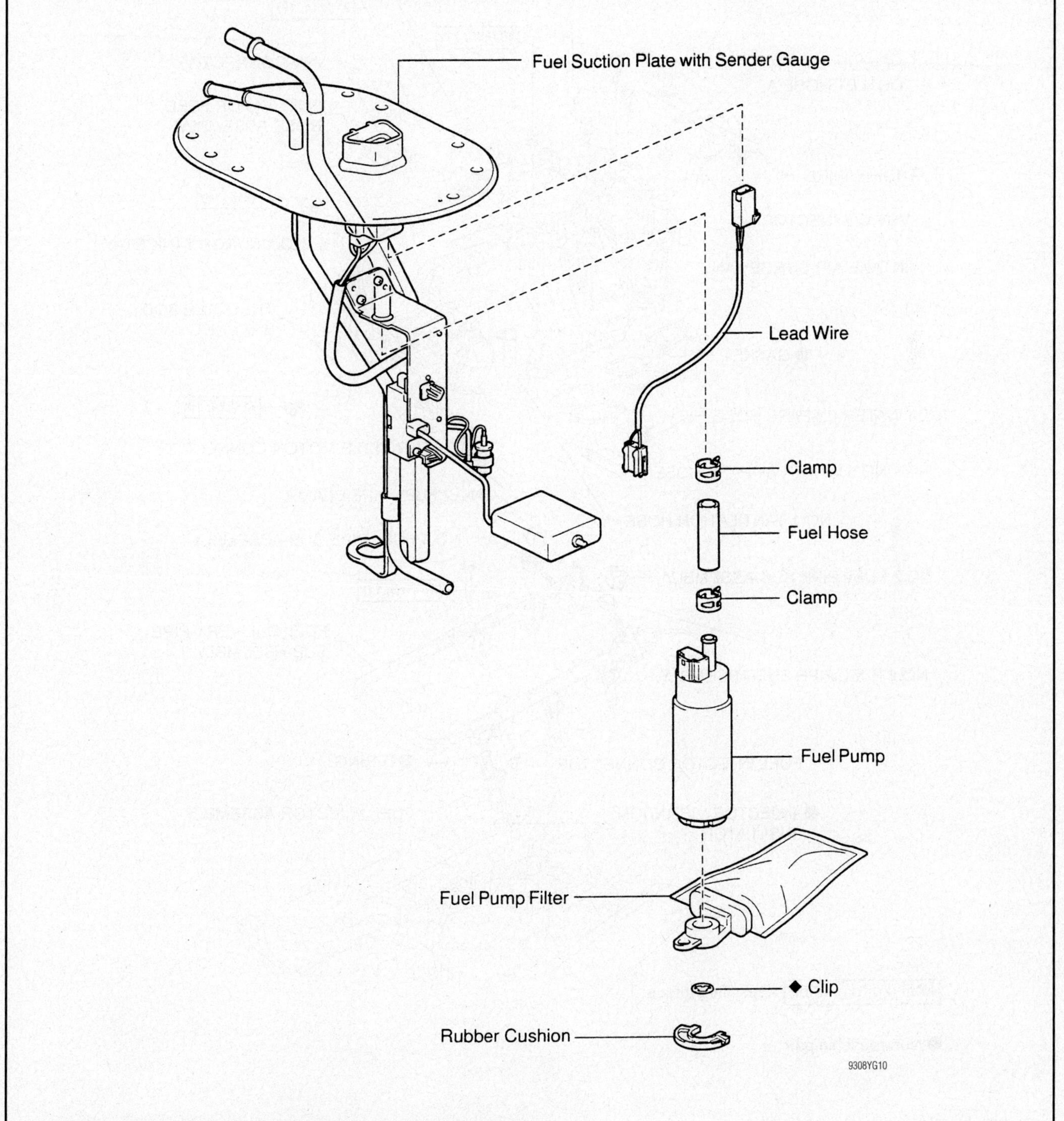

Fig. 286 Exploded view of the fuel pump and related components—Sequoia

FUEL RAIL & INJECTORS

REMOVAL & INSTALLATION

V6 Engine

See Figure 287.

1. Before servicing the vehicle, refer to the Precautions Section.
2. Relieve the fuel system pressure.
3. Disconnect the negative battery cable.
4. Drain the engine coolant. Remove the V bank cover.
5. Remove the air cleaner assembly.
6. Remove the intake manifold.
7. Disconnect and plug the fuel line hoses.
8. Disconnect the fuel injector connectors. Remove the six bolts and the fuel rail together with the injectors.
9. Pull the injectors out of the fuel rail.

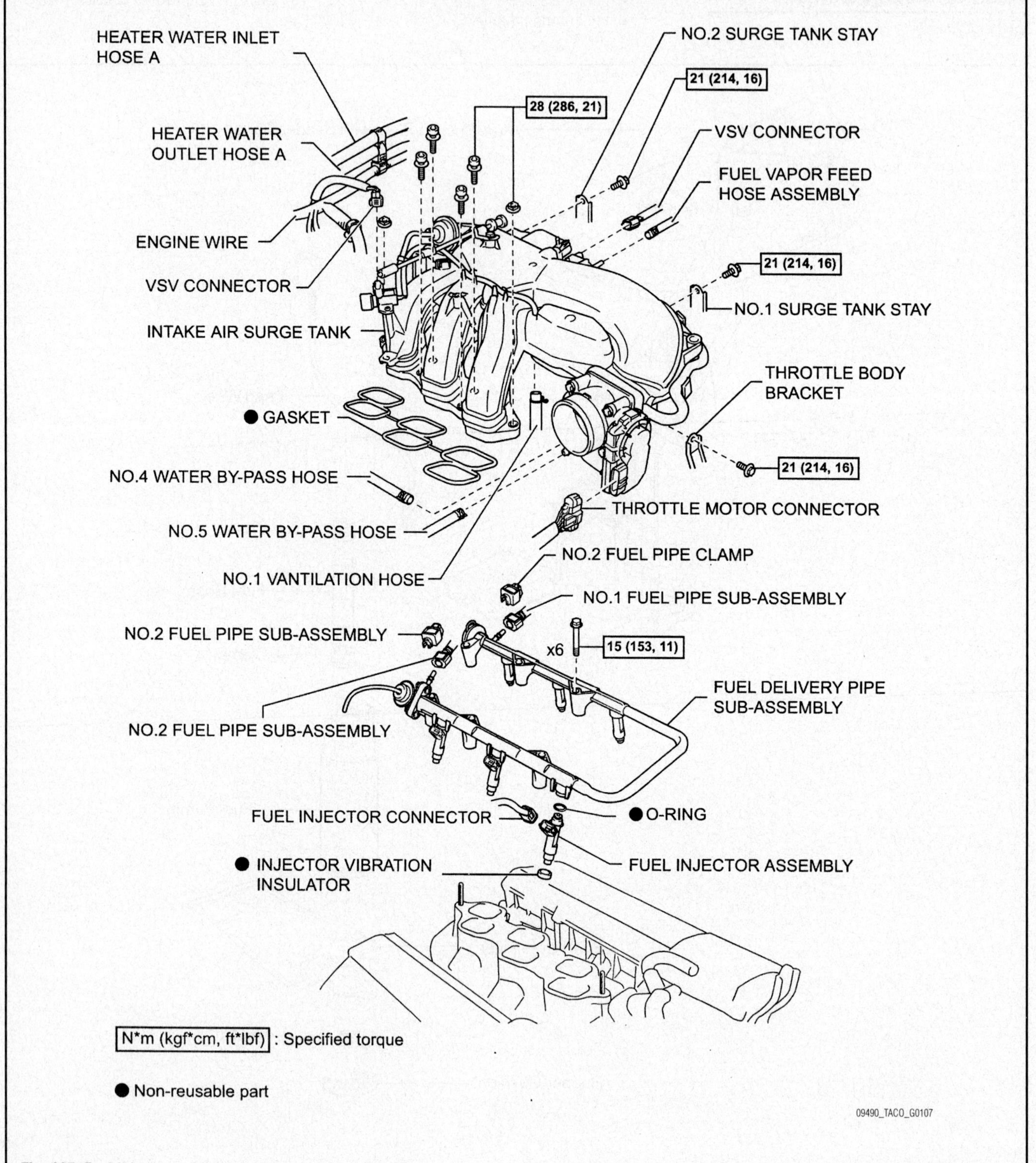

Fig. 287 Fuel injector and related components—4.0L engine

To install:

10. Installation is the reverse of the removal procedure.

11. Be sure to use new gaskets and O-rings, as required.

12. Be sure to fill the cooling system with the proper grade and type engine coolant.

13. Start the engine and check for leaks, correct as required.

4.7L V8 Engine

See Figures 288 and 289.

1. Before servicing the vehicle, refer to the precautions in the beginning of this section.
2. Relieve the fuel system pressure.
3. Remove or disconnect the following:
   - Negative battery cable
   - Engine appearance cover
   - Air intake tube
   - Fuel lines
   - Fuel pulsation damper
   - Fuel pressure regulator vacuum line
   - Accelerator cable and bracket
   - Positive Crankcase Ventilation (PCV) valve and hose
   - Evaporative Emissions (EVAP) vacuum switching valve
   - Engine appearance cover brackets
   - Fuel injector harness connectors
   - Engine harness protector
   - Fuel supply manifold crossover pipe
   - Fuel supply manifolds with injectors attached
   - Fuel injectors

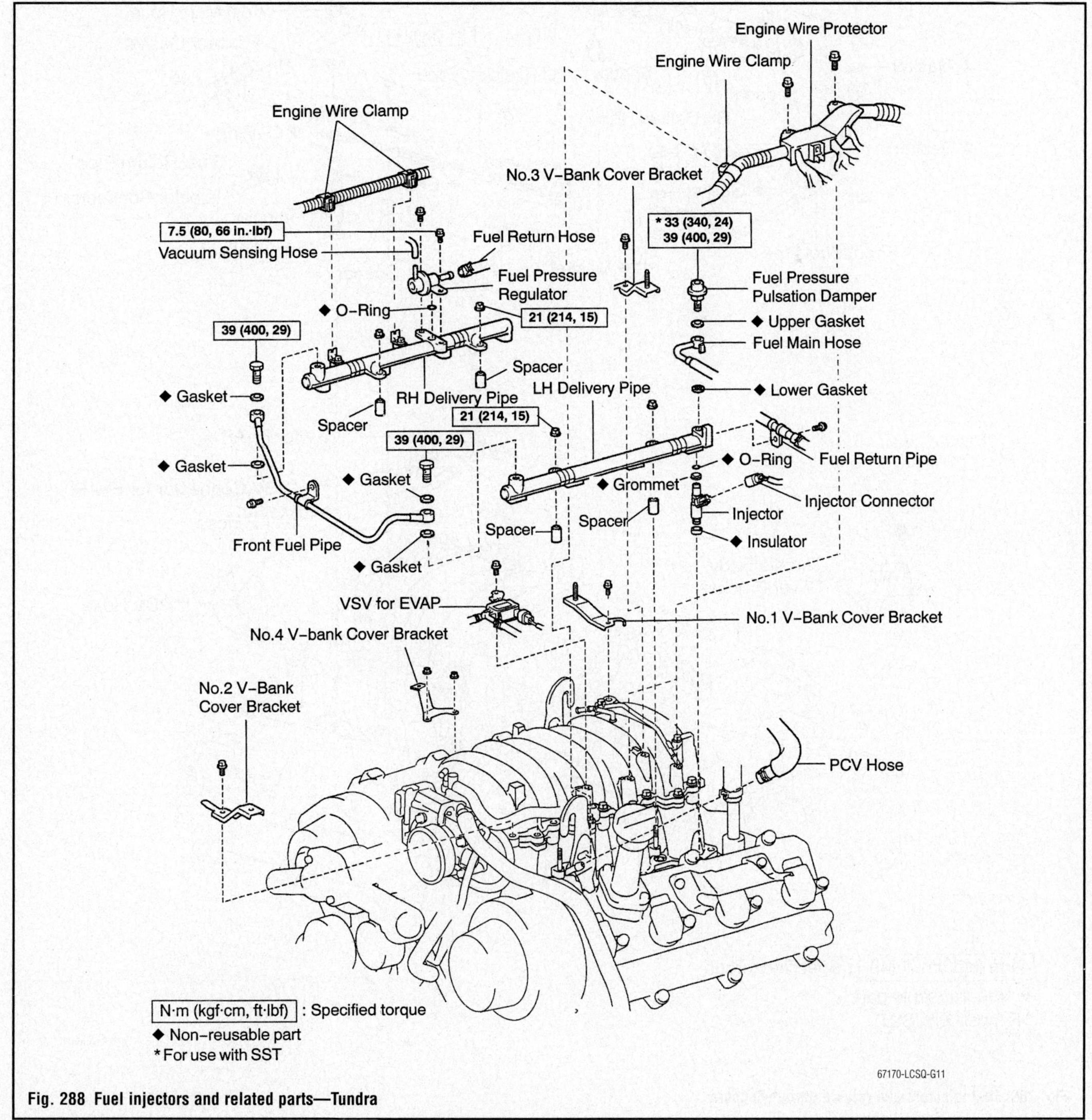

Fig. 288 Fuel injectors and related parts—Tundra

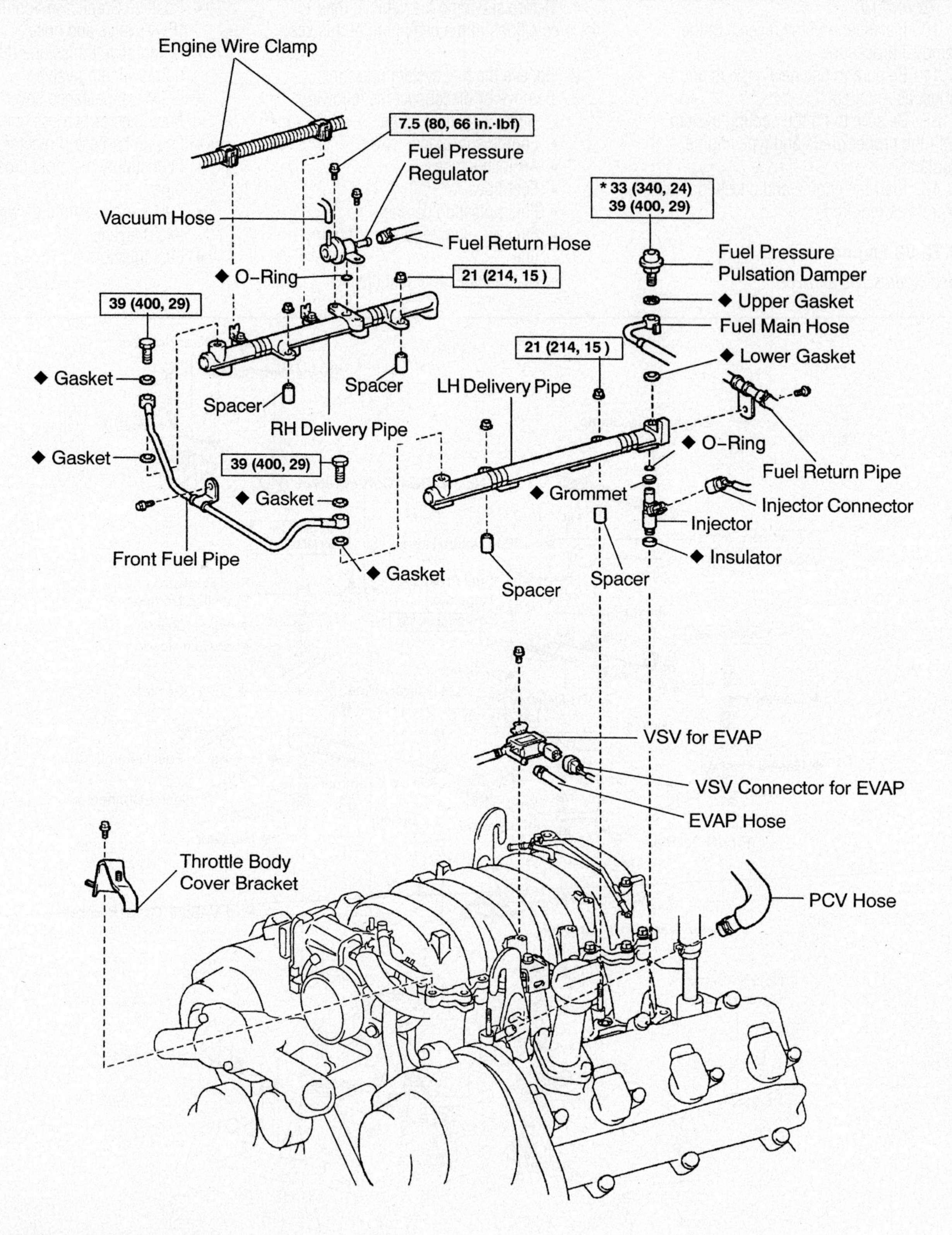

Fig. 289 Fuel injectors and related parts—Sequoia

To install:

4. Install the fuel injectors to the supply manifold with new O-ring seals and new grommets.
5. Install new injector insulators to the intake manifold.
6. Install or connect the following:
   - Fuel supply manifolds with injectors attached. Tighten the bolts to 66 inch lbs. (7.5 Nm).
   - Fuel supply manifold crossover pipe. Tighten the bolts to 29 ft. lbs. (39 Nm).
   - Engine harness protector
   - Fuel injector harness connectors
   - Engine appearance cover brackets
   - EVAP vacuum switching valve
   - PCV valve and hose
   - Accelerator cable and bracket
   - Fuel pressure regulator vacuum line
   - Fuel pulsation damper
   - Fuel lines
   - Air intake tube
   - Engine appearance cover
   - Negative battery cable
7. Start the engine and check for leaks.

5.7L V8 Engine

See Figures 290 and 291.

1. Properly discharge the fuel system.
2. Disconnect cable from negative battery terminal.
3. Remove V-bank cover.
4. Remove air cleaner hose assembly.
5. Disconnect the No. 2 fuel tube from the fuel pressure regulator.
6. Disconnect the No. 1 fuel tube from the fuel delivery pipes.
7. Disconnect the fuel tube from the fuel delivery pipe LH.
8. Disconnect the ventilation hose and No. 6 wire harness connector.
9. Remove the 2 bolts and fuel delivery pipe RH.
10. Remove the 2 delivery pipe spacers and 4 insulators from the intake manifold.
11. Disconnect the No. 7 wire harness connector.
12. Remove the No. 3 engine cover.
13. Remove the 2 bolts and fuel delivery pipe LH.
14. Remove the 2 delivery pipe spacers and 4 insulators from the intake manifold.
15. Remove the fuel injector from the fuel delivery pipe, and then disconnect the injector connector.

To install:

16. Attach the 2 clamps to install the No. 6 wire harness to the delivery pipe.
17. Attach the 3 clamps to install the No. 7 wire harness to the delivery pipe LH.
18. Apply gasoline or spindle oil to a new O-ring and install it to the injector.
19. Connect the injector connector.
20. Check that each injector is installed to the delivery pipe facing the direction shown in the illustration.
21. Install the 2 delivery pipe spacers and 4 insulators to the cylinder head LH.
22. Install the delivery pipe (with injectors) to the cylinder head LH.
23. Install the 2 bolts and tighten to 15 ft. lbs. (21 Nm).

➡Make sure that the part of the injector labeled B is between the parts of the delivery pipe labeled A.

24. Connect the No. 6 wire harness connector.
25. Install the No. 3 engine cover.
26. Install the 2 delivery pipe spacers and 4 insulators to the cylinder head RH.
27. Install the delivery pipe (with injectors) to the cylinder head RH.
28. Install the 2 bolts and tighten to 15 ft. lbs. (21 Nm).
29. Install the No. 1 engine cover.
30. Connect the No. 7 wire harness connector.
31. Connect the ventilation hose.
32. Connect the fuel tube to the fuel delivery pipe LH.
33. Connect the No. 1 fuel tube to the delivery pipes.
34. Connect the No. 2 fuel tube to the fuel pressure regulator.
35. Install air cleaner hose assembly.
36. Install V-bank cover.
37. Connect cable to negative battery terminal.
38. Check for fuel leak.
39. Perform initialization, if necessary.

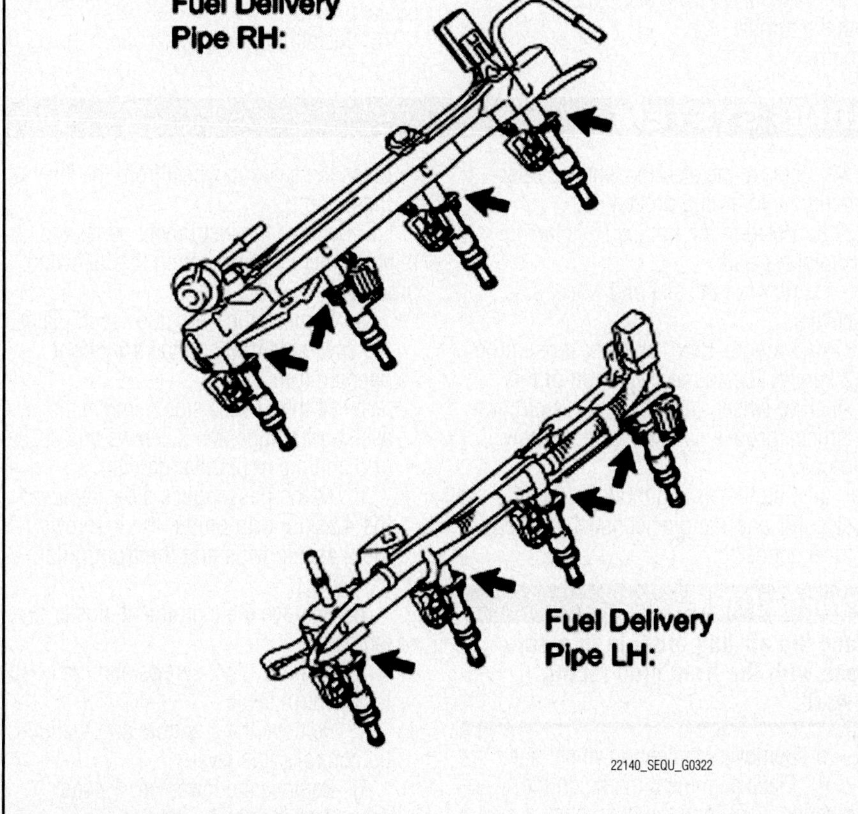

Fig. 290 Fuel injector positioning

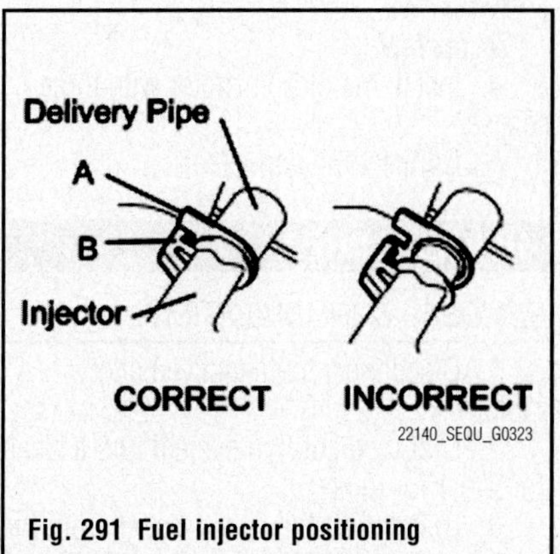

Fig. 291 Fuel injector positioning

FUEL TANK

REMOVAL & INSTALLATION

1. Discharge fuel system pressure.
2. Remove spare tire.
3. Disconnect fuel pump connector.
4. Remove the 2 bolts, 2 nuts and fuel tank protector.
5. Disconnect fuel main tube and return tube.
6. Push the connector deep into the charcoal canister to release the locking tab.
7. Pinch and pull out the connector.
8. Loosen the hose clamp bolt and disconnect the fuel inlet hose from the fuel filler pipe.
9. Disconnect the breather tube.
10. Set up a transmission jack under the fuel tank.
11. Remove the 2 bolts and disconnect the 2 fuel tank bands from the fuel tank.
12. Operate the transmission jack and remove the fuel tank.

To install:

13. Set up the fuel tank to the transmission jack.
14. Operate the transmission jack and install the fuel tank.
15. Install the 2 fuel tank bands with the 2 bolts and tighten to 45 ft. lbs. (62 Nm).
16. Connect the fuel inlet hose to the filler pipe and install the clamp.
17. Connect the breather tube.
18. Connect the fuel tank vent hose to the charcoal canister.
19. Connect fuel tank main tube and return tube.
20. Install the fuel tank protector with the 2 bolts and 2 nuts.
21. Connect fuel pump connector.
22. Check for fuel leaks.
23. Install spare tire.

IDLE SPEED

ADJUSTMENT

Idle speed is maintained by the Powertrain Control Module (PCM). No adjustment is necessary or possible.

THROTTLE BODY

REMOVAL & INSTALLATION

See Figure 292.

1. Before servicing the vehicle, refer to the precautions in the beginning of this section.
2. Remove throttle body cover.
3. Drain engine coolant.
4. For V6 engine, remove the front wiper motor and link assembly.
5. Remove intake air connector.
6. Disconnect the throttle control connector.
7. Disconnect the water bypass hoses from the throttle body.

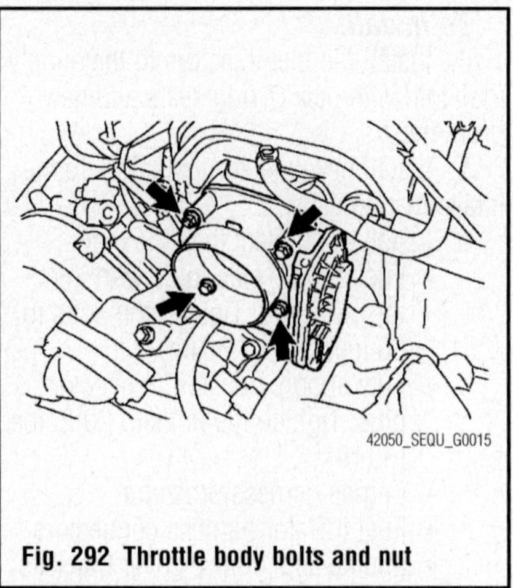

Fig. 292 Throttle body bolts and nut

8. Remove the nut and 3 bolts, and remove the throttle body from the intake manifold.

To install:

9. Install the throttle body and tighten to 14 Nm (10 ft. lbs.).
10. Connect the water bypass hoses to the throttle body.
11. Connect the throttle control connector.
12. Install intake air connector.
13. For V6 engine, install the front wiper motor and link assembly
14. Fill with engine coolant.
15. Start engine and check for engine coolant leaks.
16. Install throttle body cover.

HEATING & AIR CONDITIONING SYSTEM

BLOWER MOTOR

REMOVAL & INSTALLATION

1. Before servicing the vehicle, refer to the precautions in the beginning of this section.
2. Disconnect the connector.
3. Remove the three screws and the blower motor.

To install:

4. Install the blower motor with three screws.
5. Connect the connector.

HEATER CORE

REMOVAL & INSTALLATION

1. Disconnect the negative battery cable.
2. Drain the cooling system into a clean container for reuse.
3. Disconnect the heater hoses from the heater core.
4. Remove the steering wheel by performing the following procedure:
   a. Position the front wheels facing straight-ahead.
   b. Remove the steering wheel side covers.
   c. Using a Torx® wrench, loosen the 2 screws located at each side of the steering wheel until the screw's circumference groove catches on the screw case.
   d. Pull the air bag module from the steering wheel and disconnect the electrical connector.

CAUTION

Place the air bag module in a safe place with the front side facing upward.

   e. Remove the steering wheel nut.
   f. Place alignment marks on the steering wheel and the main shaft.
   g. Using a steering wheel puller, press the steering wheel from the steering column.
5. Remove the instrument panel and reinforcement by performing the following procedure:
   a. Remove the front door scuff plates, the cowl side trim and the front door opening trim.
   b. At the driver's side, remove the 2 assist grip plugs, the 2 screws and assist grip and the front pillar garnish.
   c. At the passenger's side, remove the 4 assist grip plugs, the 4 screws, the 2 assist grips and the front pillar garnish.
   d. Remove the instrument cluster finish panel.
   e. Remove the 2 screws and the hood lock control cable.
   f. Remove the 2 screws and the fuel lid control cable lever.
   g. Remove the lower No. 1 panel screw and the panel.
   h. Remove the lower left side panel.

i. Remove the 3 steering column cover screws and the covers.

j. At the steering column, disconnect the electrical connectors; then, remove the clamp, the 3 screws and the combination switch.

k. Remove the No. 2 heater-to-register duct screw and the duct.

l. Remove the steering column-to-instrument panel bolts and the steering column.

m. At the combination meter, disconnect the electrical connectors; then, remove the 4 screws and the combination meter.

n. Remove the glove compartment door stoppers, the 2 screws and the glove box door.

o. At the passenger's side air bag module, remove the No. 1 undercover, pull the air bag connector up from the undercover and disconnect it; then, remove the air bag.

⁂ CAUTION

Place the air bag module in a safe place with the front side facing upward.

p. Remove the 3 lower No. 2 panel screws and the panel.

q. Remove the center cluster; then, pry the center cluster from the dash by prying the 8 clips in the following order:

- Left side
- Right side
- Top left side
- Top right side

r. Remove the 4 radio screws, pull the radio outward, disconnect the electrical connectors and remove the radio.

s. At the rear console panel, remove the transfer shift lever knob; then, pry the panel upward disengaging the 4 clips (2 on each side) and remove the panel.

t. At the rear of the console, remove the 2 rear end panel-to-console screws; then, pry the end panel rearward disengaging the 2 clips and remove the panel.

u. If not equipped with a rear air conditioning system, disconnect the connector and control cable; then, remove the 3 rear heater control panel screws and the panel.

v. Remove the 4 rear console box-to-chassis screws/bolts and the console box.

w. Remove the center lower cluster finish panel by prying panel rearward disengaging the 5 clips; then, disconnect the electrical connector.

x. Remove the 2 front console-to-chassis bolts/screws, disengage the 2 clips and remove the console.

y. At the instrument panel, disconnect the junction connectors (the connectors can be disconnected by loosening the bolts), the instrument panel-to-chassis 8 bolts and 2 nuts. Using an assistant, remove the instrument panel.

z. Disconnect the electrical connector and remove the ECM.

aa. Remove the No. 3 and No. 4 heater-to-register ducts.

bb. Remove the floor brace, the No. 1 brace and the reinforcement.

6. Remove the evaporator housing by performing the following procedure:

a. Discharge and recover the air conditioning system refrigerant.

b. Remove the air conditioning liquid line clamp.

c. Remove the air conditioning suction line clamp.

d. Disconnect both air conditioning lines and plug the openings to prevent contamination. Discard the 4 O-rings.

e. Remove the antenna relay electrical connector, the 2 screws and the relay.

f. Remove the evaporator housing-to-chassis 4 screws/2 nuts and the housing.

7. Remove the heater housing by performing the following procedure:

a. Remove the defroster nozzle.

b. Disconnect the electrical connector.

c. Remove the 4 nuts and the heater housing.

8. Remove the heater core-to-heater housing packing, the screw, the bracket, the clamp and the heater core.

To install:

9. Install the heater core, the clamp, the bracket, the screw and the heater core-to-heater housing packing.

10. Install the heater housing by performing the following procedure:

a. Install the heater housing and the 4 nuts.

b. Connect the electrical connector.

c. Install the defroster nozzle.

11. Install the evaporator housing by performing the following procedure:

a. Install the evaporator housing and the housing-to-chassis 4 screws and 2 nuts.

b. Install the antenna relay, the 2 screws and the electrical connector.

c. Using new O-rings, connect both air conditioning lines.

d. Install the air conditioning liquid line and suction line clamp.

12. Install the instrument panel and reinforcement by performing the following procedure:

a. Install the reinforcement, the No. 1 brace and the floor brace.

b. Install the No. 3 and No. 4 heater-to-register ducts.

c. Install the ECM and connect the electrical connector.

d. Using an assistant, install the instrument panel, connect the junction connectors, the instrument panel-to-chassis 8 bolts and 2 nuts.

e. Install the front the console, engage the 2 clips and install the 2 console-to-chassis bolts/screws.

f. Connect the electrical connector; then, install the center lower cluster finish panel by engaging the 5 clips.

g. Install the console box and the 4 rear console box-to-chassis screws/bolts.

h. If not equipped with a rear air conditioning system, install rear heater control panel, the 3 panel screws; then, connect the connector and control cable.

i. Install the rear of the console and engage the 2 clips; then, install the 2 rear end panel-to-console screws.

j. Install the rear console panel and engage the 4 clips (2 on each side); then, install the transfer shift lever knob.

k. Install the radio, connect the electrical connectors and the 4 radio screws.

l. Install the center cluster and engage the 8 center cluster clips.

m. Install the lower No. 2 panel and the 3 panel screws.

n. Install the passenger's side air bag module, connect it and install the No. 1 undercover.

o. Install the glove box door, the 2 screws and the glove compartment door stoppers.

p. Install the combination meter and the 4 screws; then, connect the electrical connectors.

q. Install the steering column and the steering column-to-instrument panel bolts.

r. Install the No. 2 heater-to-register duct and the duct screw.

s. At the steering column, install the combination switch, the 3 screws and the clamp; then, connect the electrical connectors.

t. Install the steering column covers and the 3 covers screws.

u. Install the lower left side panel.

v. Install the lower No. 1 panel and the panel screw.

w. Install the fuel lid control cable lever and the 2 screws.

x. Install the hood lock control cable and the 2 screws.

y. Install the instrument cluster finish panel.

z. At the passenger's side, install the front pillar garnish, the 2 assist grips, the 4 screws and the 4 assist grip plugs.

aa. At the driver's side, install the front pillar garnish, assist grip, the 2 screws and the 2 assist grip plugs.

bb. Install the front door scuff plates, the cowl side trim and the front door opening trim.

13. Install the steering wheel by performing the following procedure:

a. Install the steering wheel to the steering column.

b. Align the steering wheel-to-main shaft marks.

c. Install the steering wheel nut and torque to 25 ft. lbs. (34 Nm).

d. Install the air bag module to the steering wheel and connect the electrical connector.

e. Using a Torx® wrench, tighten the 2 screws located at each side of the steering wheel to 78 inch lbs. (8.8 Nm).

f. Install the steering wheel side covers.

14. Connect the heater hoses to the heater core.

15. Refill the cooling system.

16. Connect the negative battery cable.

a. Evacuate and charge the air conditioning system refrigerant.

17. Run the engine to normal operating temperatures; then, check the climate control operation and check for leaks.

AUXILIARY HEATING & AIR CONDITIONING SYSTEM

BLOWER MOTOR

REMOVAL & INSTALLATION

1. Before servicing the vehicle, refer to the precautions in the beginning of this section.
2. Remove the rear cooling unit from the vehicle.
3. Remove the 2 screws and the A/C amplifier.
4. Disconnect the connector.
5. Remove the 2 screws and the power transistor.
6. Remove the wire harness.
7. Remove the 11 screws, 3 holding springs and separate the cooling unit cases.
8. Disconnect the connector.
9. Remove the blower case.
10. Remove the 3 screws and the blower motor.

To install:

11. Installation is the reverse of the removal procedure.

HEATER CORE

REMOVAL & INSTALLATION

1. Disconnect the negative battery cable.
2. Drain the cooling system into a clean container for reuse.
3. Disconnect the heater hoses from the rear heater core.
4. Remove the front seats.
5. Remove the rear heater control assembly.
6. Remove the rear console box.
7. Remove the front console box cover.
8. Remove the lower center cluster finish panel.
9. Remove the front door scuff plates.
10. Remove the cowl side trim.
11. Remove the rear door scuff plates.
12. Remove the center pillar garnishes.
13. Slide the carpet rearward.
14. Remove the cooler bracket bolts and the bracket.
15. Remove the rear heater duct bolt/screw and the duct.
16. Disconnect the rear heater housing electrical connector.
17. Remove the 3 rear heater housing-to-chassis bolts and the heater housing.
18. Remove the heater core-to-heater housing 3 screws and 2 clamps.
19. Remove the heater core from the heater housing.

To install:

20. Install the heater core to the heater housing.
21. Install the heater core-to-heater housing 3 screws and 2 clamps.
22. Install the heater housing and the 3 rear heater housing-to-chassis bolts.
23. Connect the rear heater housing electrical connector.
24. Install the rear heater duct and the duct bolt/screw.
25. Install the cooler bracket and the bracket bolts.
26. Slide the carpet rearward.
27. Install the center pillar garnishes.
28. Install the rear door scuff plates.
29. Install the cowl side trim.
30. Install the front door scuff plates.
31. Install the lower center cluster finish panel.
32. Install the front console box cover.
33. Install the rear console box.
34. Install the rear heater control assembly.
35. Install the front seats.
36. Connect the heater hoses to the rear heater core.
37. Refill the cooling system.
38. Connect the negative battery cable.

STEERING

POWER RACK & PINION STEERING GEAR

REMOVAL & INSTALLATION

See Figures 293 through 295.

1. Before servicing the vehicle, refer to the precautions in the beginning of this section.
2. Position the front wheels in the straight-ahead position.
3. Remove or disconnect the following:
   - Engine under cover
   - Steering wheel pad
   - Steering wheel
   - Left and right outer tie-rod ends from the steering knuckles
4. Put matchmarks on the No. 2 steering intermediate shaft and steering intermediate shaft.
   a. Remove the bolt and disconnect the No. 2 steering intermediate shaft from the steering intermediate shaft.
   b. Put matchmarks on the No. 2 steering intermediate shaft and the power steering gear
   c. Remove the bolt and disconnect the No. 2 steering intermediate shaft from the power steering gear
5. Remove or disconnect the following:
   - Clamp plate
   - Pressure feed and return tubes from the power steering gear, using tool 09631-22020
   - Power steering gear assembly

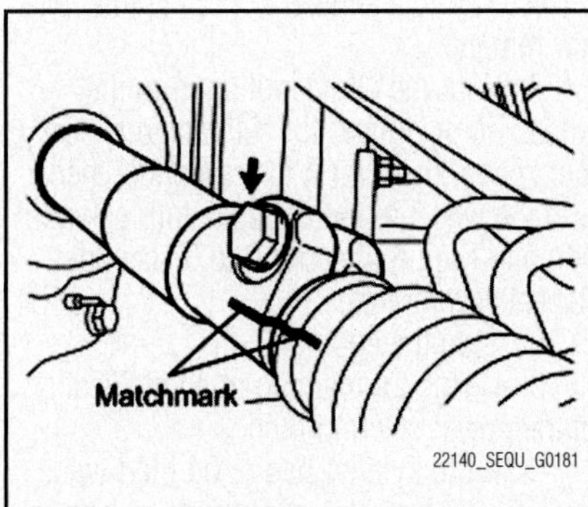

Fig. 293 Steering intermediate shaft and steering intermediate shaft

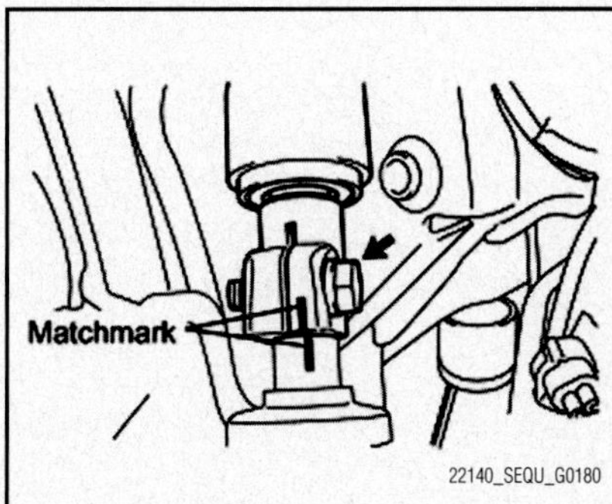

Fig. 294 Steering intermediate shaft and the power steering gear

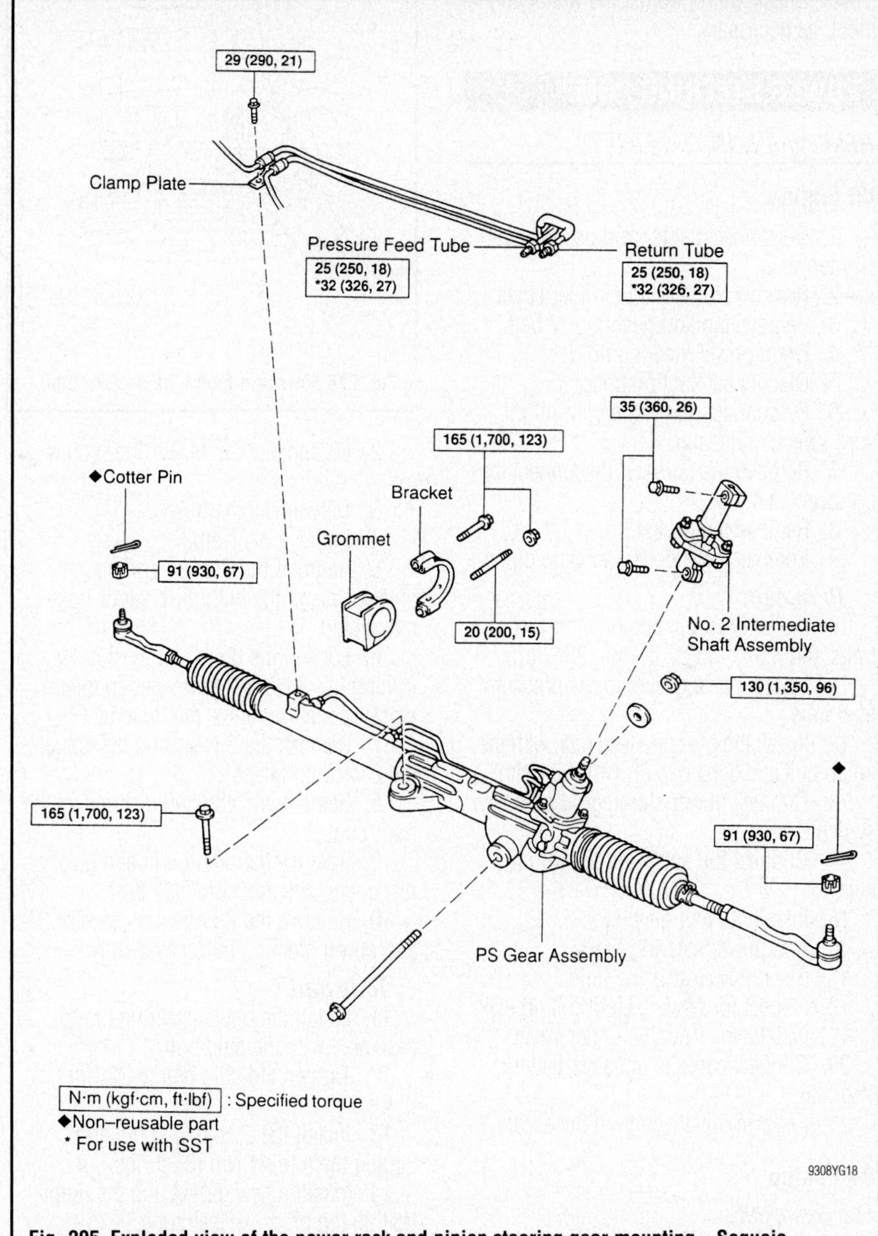

Fig. 295 Exploded view of the power rack and pinion steering gear mounting—Sequoia

To install:

6. Install or connect the following:
   - Power steering gear assembly. Torque the set bolt to 123 ft. lbs. (165 Nm) and the set nut/bolt to 96 ft. lbs. (91 Nm).
   - Pressure feed and return tubes to the power steering gear. Torque them to 27 ft. lbs. (32 Nm), using tool 09631-22020.
   - Clamp plate. Torque the bolt to 21 ft. lbs. (29 Nm).
   - No. 2 intermediate shaft to the steering gear input shaft
   - Left and right outer tie-rod ends to the steering knuckles. Torque the nuts to 67 ft. lbs. (91 Nm).
   - Steering wheel. Torque the nut to 26 ft. lbs. (35 Nm).
   - Steering wheel pad
   - Engine under cover

7. Fill and bleed the power steering system.
8. Check and/or adjust the wheel alignment, as necessary.

POWER STEERING PUMP

REMOVAL & INSTALLATION

V6 Engine

1. Disconnect cable from negative battery terminal.
2. Remove No. 1 engine under cover.
3. Remove fan and generator V belt.
4. Drain power steering fluid.
5. Disconnect suction hose.
6. Disconnect power steering oil pressure switch connector.
7. Remove the bolt and disconnect the pressure feed tube.
8. Remove the gasket.
9. Remove the 2 bolts and vane pump.

To install:

10. Install the vane pump with the 2 bolts and tighten to 21 ft. lbs. (28 Nm).
11. Install a new gasket to the pressure feed tube.
12. Install the pressure feed tube with the union bolt and tighten to 38 ft lbs. (51 Nm).
13. Connect power steering oil pressure switch connector.
14. Connect the suction hose with the clip.
15. Install fan and generator V belt.
16. Add power steering fluid.
17. Bleed power steering fluid.
18. Inspect for power steering fluid leak.
19. Install No. 1 engine under cover.
20. Connect cable to negative battery terminal.
21. Perform initialization, if necessary.

V8 Engine

See Figure 296.

1. Before servicing the vehicle, refer to the precautions in the beginning of this section.

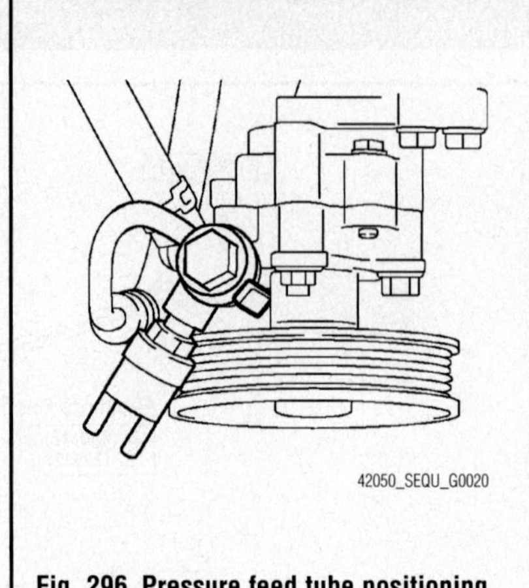

Fig. 296 Pressure feed tube positioning

2. Disconnect the MAF meter connector.
3. Disconnect the hoses.
4. Remove the clamp.
5. Remove the 3 bolts and air cleaner assembly with air cleaner hose connected.
6. Loosen the drive belt tension by turning the drive belt tensioner counterclockwise, and remove the drive belt.
7. Remove the 2 clips and disconnect the 2 vacuum hoses.
8. Remove the clip and disconnect the return hose.
9. Remove the union bolt and gasket, disconnect the pressure feed tube.
10. Remove the 2 bolts, nut, stud bolt and power steering pump assembly.

To install:

11. Install the power steering pump assembly with the stud bolt.
12. Tighten the stud bolt to 22 Nm (16 ft. lbs.).
13. Install the 2 bolts and nut and tighten them to 44 Nm (33 ft. lbs.).
14. Install a new gasket and the union bolt on the pressure feed tube.

➡Make sure that the stopper of the pressure feed tube contacts the power steering pump body as shown in the illustration.

15. Tighten the union bolt to 46.5 Nm (34 ft. lbs.)
16. Connect the return hose with the clip.
17. Connect the 2 vacuum hoses and install the 2 clips.
18. Loosen the drive belt tension by turning the drive belt tensioner counterclockwise, and install the belt.
19. Install the air cleaner assembly with air cleaner hose and the 3 bolts.
20. Install the clamp.
21. Connect the MAF meter connector.
22. Fill with power steering fluid and bleed the system.

BLEEDING

1. Before servicing the vehicle, refer to the precautions in the beginning of this section.
2. Check fluid level.
3. Jack up front of vehicle and support it with stands.
4. With the engine stopped, turn the wheel slowly from lock to lock several times.
5. Lower the vehicle.
6. Start the engine and run at idle for a few minutes.
7. With the engine idling, turn the wheel left or right to the full lock position and keep it there for 2 to 3 seconds, then turn the wheel to the opposite full lock position and keep it there for 2 to 3 seconds. Repeat several times.
8. Stop the engine.
9. Check for foaming or emulsification of the power steering fluid.
10. If the system has to be bled twice specifically because of foaming or emulsification, check for fluid leaks in the system.
11. Check fluid level.

COIL SPRING

REMOVAL & INSTALLATION

See Figures 297 and 298.

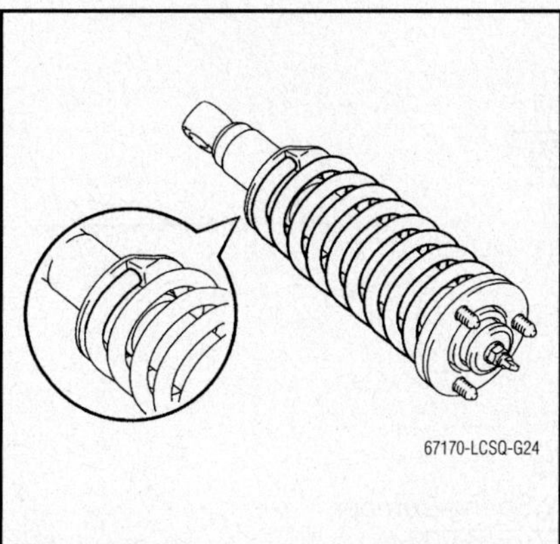

Fig. 297 Correct spring positioning—Sequoia

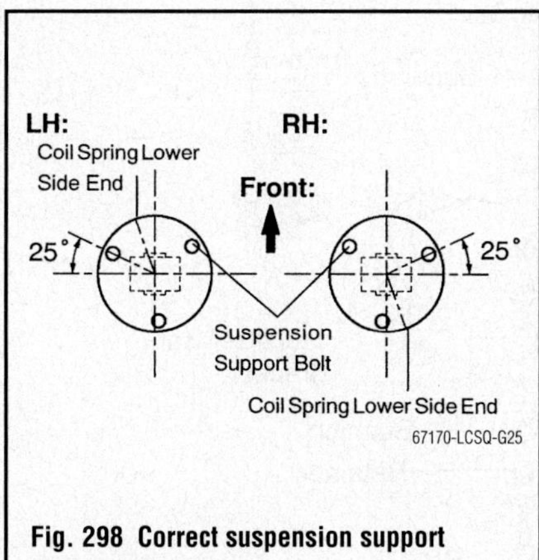

Fig. 298 Correct suspension support positioning—Sequoia

1. Remove the strut.
2. Using a compressor, compress the coil spring.

➡A compressor with a force of 2,860 lbs. (12,740 N) or more must be used. Make sure that the suspension support is free from the spring. Do not compress the spring more than necessary. Do not position the spring with the upper end towards you.

3. Remove the center nut.
4. Remove the retainers, bushing support and spring.
5. Assembly is the reverse of disassembly. See the accompanying illustration for correct positioning of the suspension support and spring. Torque the nut to 18 ft. lbs. (25 Nm).

LOWER BALL JOINT

REMOVAL & INSTALLATION

Sequoia

See Figure 299.

1. Before servicing the vehicle, refer to the precautions section.
2. Remove or disconnect the following:
   - Front wheel
   - 4 lower ball joint set bolts
   - Tie-rod end from the lower ball joint, using tool 09610-20012
   - Lower ball joint nut.
   - Lower ball joint from the lower control arm, using tool 09628-62011

To install:

3. Install or connect the following:
   - New lower ball joint to the lower control arm stud. Torque the bolts to 117 ft. lbs. (159 Nm).
   - New cotter pin
   - Tie-rod end to the lower ball joint. Torque the nut to 67 ft. lbs. (91 Nm).
   - Lower ball joint set bolts. Torque the 4 bolts to 48 ft. lbs. (65 Nm).
   - Front wheel.
4. Check and/or adjust the front wheel alignment.

Tundra

For Ball Joint replacement, see Lower Control Arm

LOWER CONTROL ARM

REMOVAL & INSTALLATION

Sequoia

See Figures 300 and 301.

1. Before servicing the vehicle, refer to the precautions in the beginning of this section.
2. Remove front wheel.
3. Disconnect the tie-rod end, as follows:
   - Cotter pin and nut
   - Tie-rod end from the lower ball joint, using tool 09610-20012
4. Remove or disconnect the following:
   - Power steering gear set bolts and nuts
   - Stabilizer bar link from the lower control arm
   - Strut from the lower control arm
5. Disconnect the lower ball joint, as follows:
   - Cotter pin and nut
   - Lower ball joint from the lower control arm
6. Matchmark both front and rear cam plates and chassis frame.

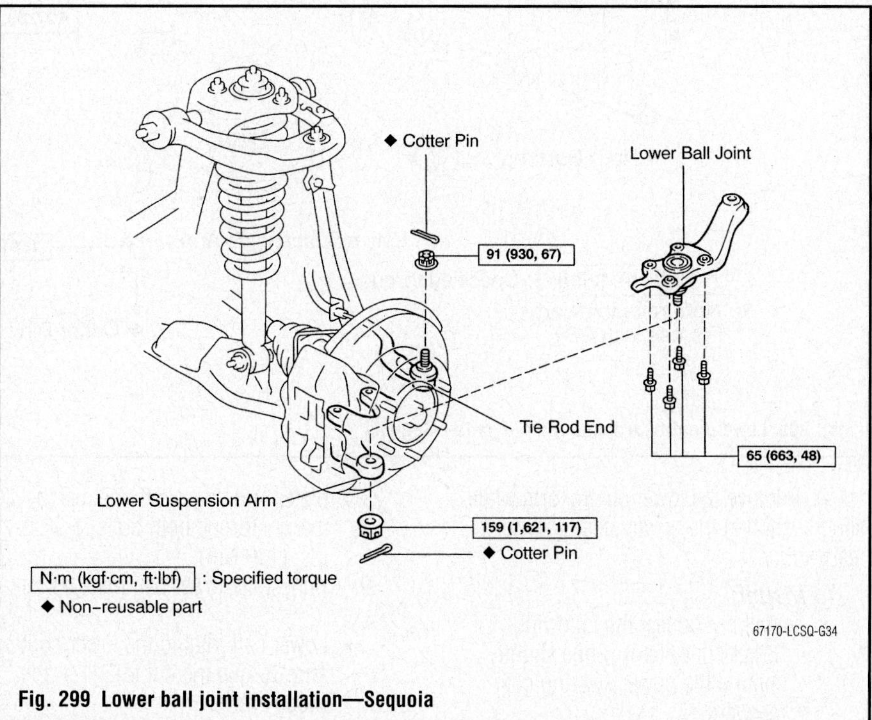

Fig. 299 Lower ball joint installation—Sequoia

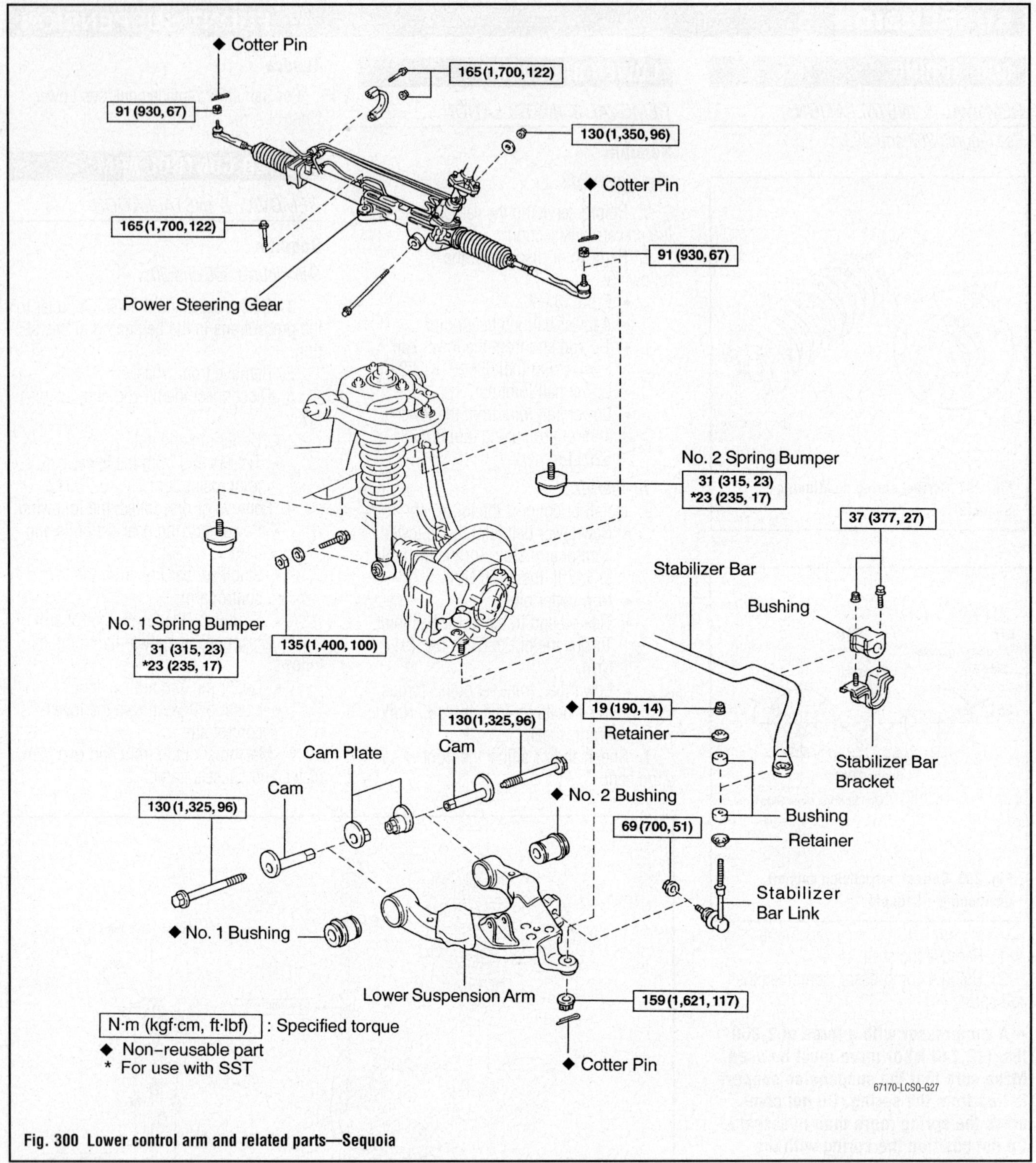

Fig. 300 Lower control arm and related parts—Sequoia

7. Remove the lower control arm while slightly shifting the power steering gear rearward.

To install:

8. Install or connect the following:
   - Lower control arm while slightly shifting the power steering gear rearward
   - Align both front and rear cam plates and chassis frame matchmarks. Torque both bolts to 96 ft. lbs. (130 Nm).
9. Connect the lower ball joint, as follows:
   - Lower ball joint to the lower control arm. Torque the nut to 221 ft. lbs. (300 Nm).
   - New cotter pin
10. Install or connect the following:
   - Strut to the lower control arm. Torque the nut/bolt to 100 ft. lbs. (135 Nm).
   - Stabilizer bar link to the lower control arm. Torque the nut to 51 ft. lbs. (69 Nm).
   - Power steering gear set bolts and nuts. Torque the set bolt and clamp

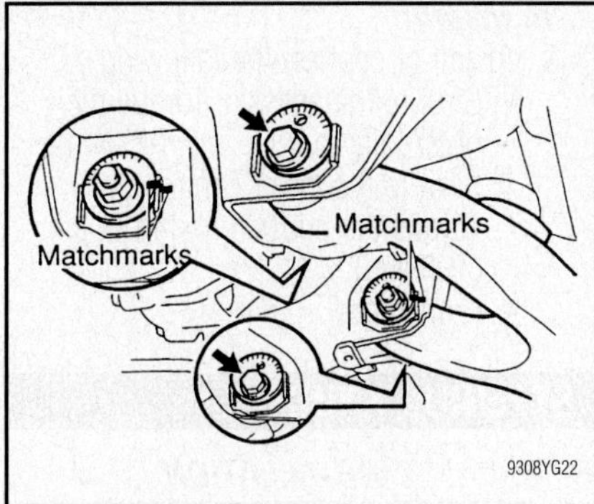

Fig. 301 View of the lower control arm's cam plate alignment—Sequoia

nut/bolt to 122 ft. lbs. and the set nut/bolt to 96 ft. lbs. (130 Nm)

- Tie-rod end to the lower ball joint. Torque the nut to 67 ft. lbs. (91 Nm).
- New cotter pin
- Front wheel.

11. Check and/or adjust the wheel alignment.

Tundra

1. Remove front wheel.
2. Remove front stabilizer link assembly LH.
3. Remove the nut and bolt, and disconnect the shock absorber from the lower side.
4. Remove the 2 bolts and disconnect the attachment from the steering knuckle.
5. Place matchmarks on the No. 2 camber adjusting cam and No. 2 toe adjusting plate.
6. Remove the nut, washer, No. 2 camber adjusting cam, camber adjusting cam assembly, bolt, toe adjusting cam, No. 2 toe adjusting plate and front No. 1 suspension arm lower LH.
7. Remove the cotter pin and nut.
8. Using Service Tool, remove the lower ball joint attachment.

To install:

9. Install the lower suspension arm with the camber adjusting cam, No. 2 camber adjusting cam, No. 2 toe adjusting plate, toe adjusting cam, bolt, nut and washer.
10. Align the matchmarks on the No. 2 camber adjusting cam and No. 2 toe adjusting plate. Temporarily install the bolt and nut.
11. Connect the attachment to the steering knuckle with the 2 bolts and tighten to 221 ft. lbs. (300 Nm).
12. Connect the shock absorber with the bolt and nut.
13. Install front stabilizer link assembly LH.
14. Stabilize suspension.
15. Tighten front no. 1 suspension arm lower sub-assembly LH to 207 ft. lbs. (280 Nm).
16. Tighten front shock absorber with coil spring to 144 ft. lbs. (195 Nm).
17. Install front wheel.
18. Inspect and adjust the front wheel alignment.

MACPHERSON STRUT

REMOVAL & INSTALLATION

Sequoia

See Figure 302.

1. Before servicing the vehicle, refer to the precautions in the beginning of this section.
2. Remove or disconnect the following:
   - Front wheel
   - Strut-to-lower control arm nut/bolt and the strut
   - Strut-to-chassis nuts/bolts and the strut

To install:

3. Install or connect the following:
   - Strut to the chassis. Torque the nuts/bolts to 47 ft. lbs. (64 Nm).
   - Strut to the lower control arm. Torque the nut/bolt to 100 ft. lbs. (135 Nm).
   - Front wheel

Tundra

See Figure 303.

1. Before servicing the vehicle, refer to the precautions in the beginning of this section.
2. Remove or disconnect the following:
   - Front wheel
   - Strut-to-lower control arm nut/bolt and the strut
   - Strut-to-chassis nuts/bolts and the strut

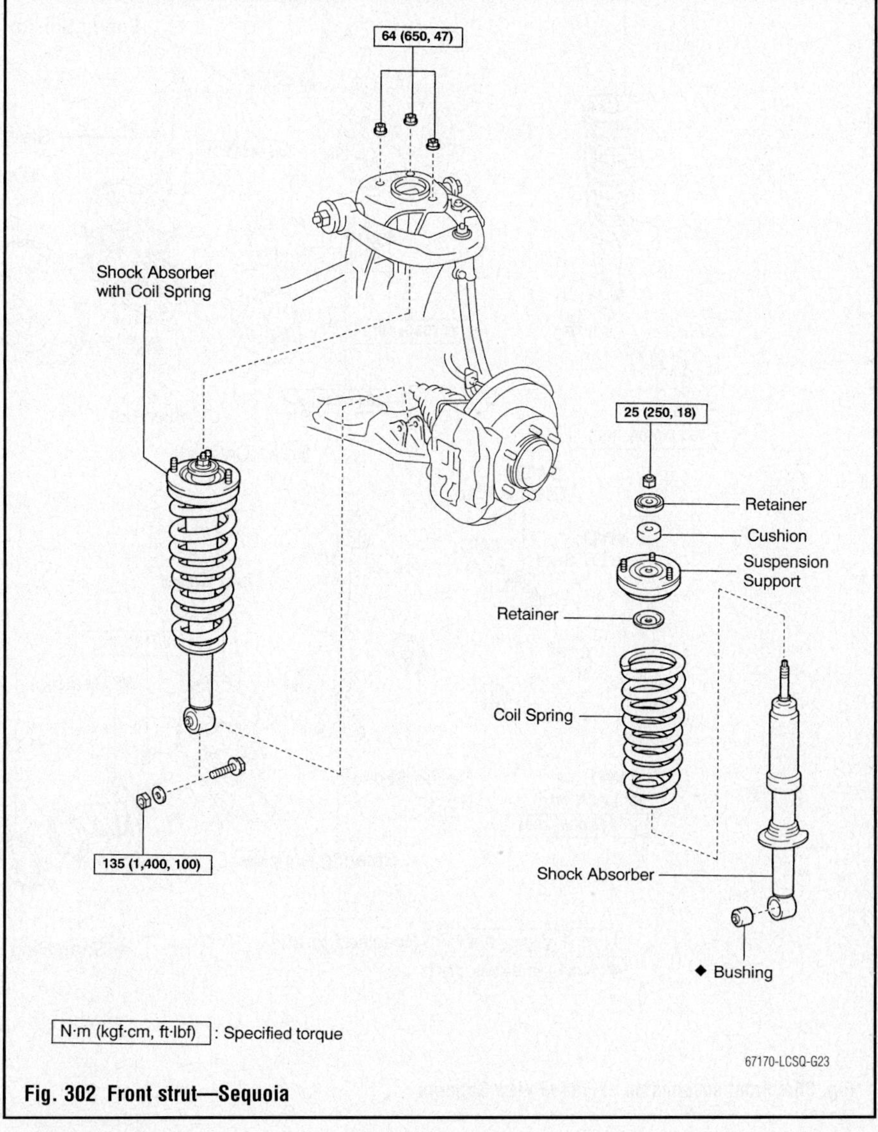

Fig. 302 Front strut—Sequoia

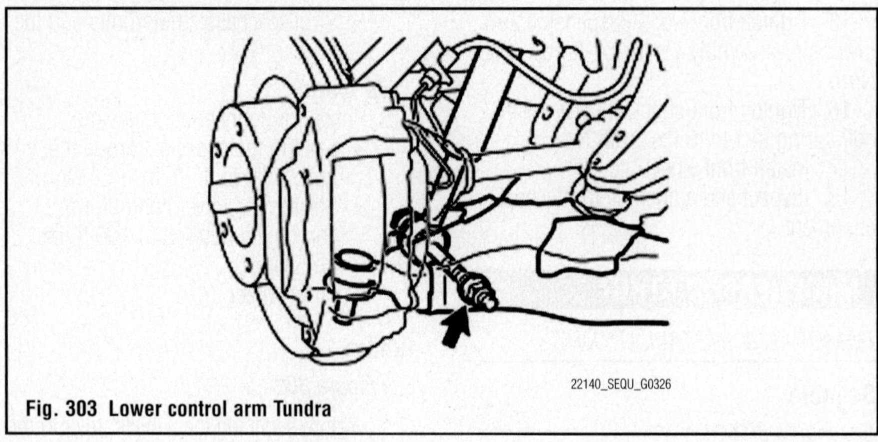
22140_SEQU_G0326

Fig. 303 Lower control arm Tundra

To install:

3. Install or connect the following:
   - Strut to the chassis. Torque the nuts/bolts to 33 ft. lbs. (45 Nm).
   - Strut to the lower control arm. Torque the nut/bolt to 144 ft. lbs. (195 Nm).
   - Front wheel

STEERING KNUCKLE

REMOVAL & INSTALLATION

Sequoia

See Figure 304.

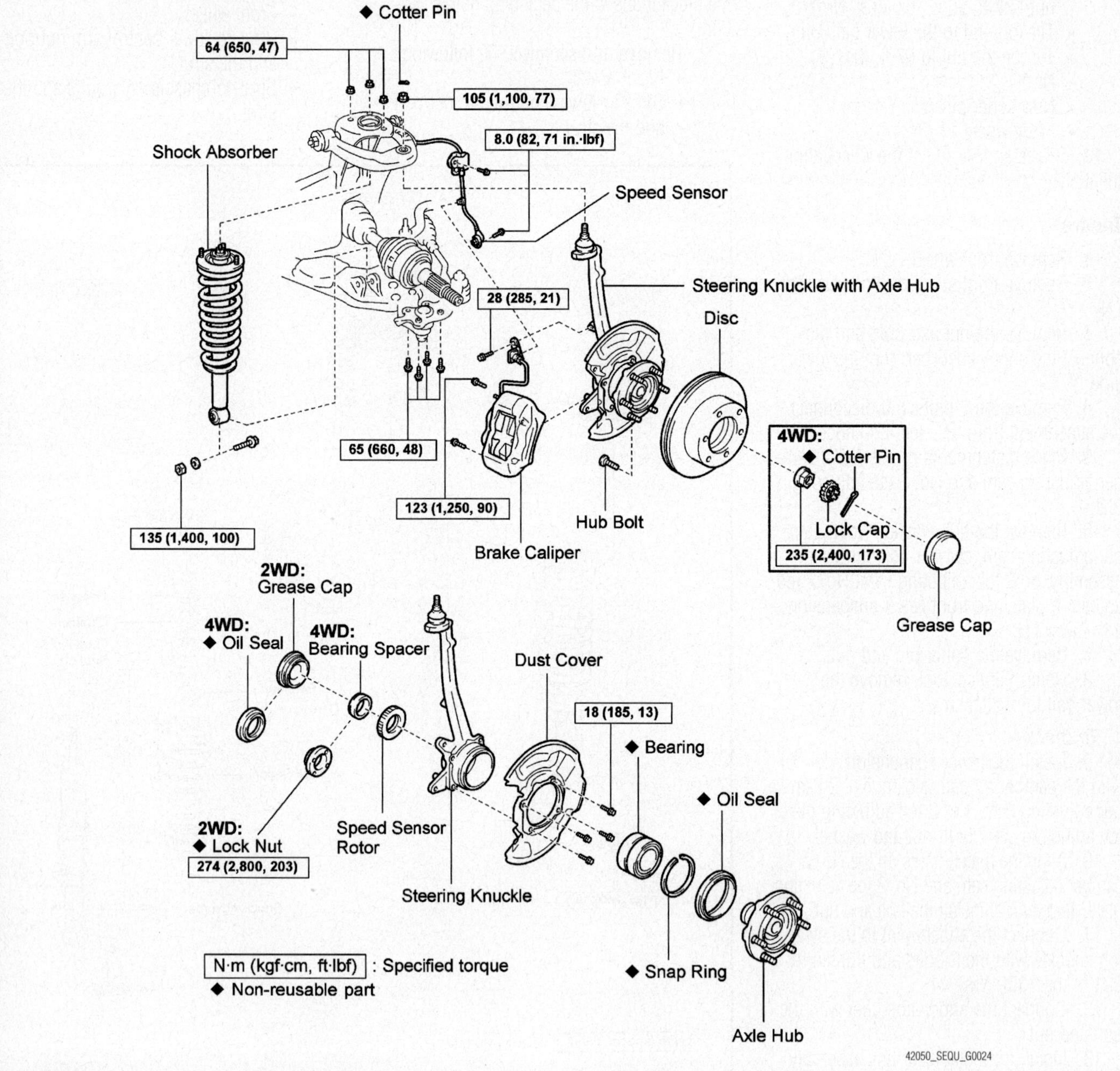

Fig. 304 Front suspension exploded view Sequoia

Refer to the accompanying illustration for steering knuckle mounting.

Tundra

1. Disconnect cable from negative battery terminal.
2. Remove the front axle hub.
3. Disconnect the speed sensor from the steering knuckle.
4. Disconnect tie rod end sub-assembly.
   a. Remove the cotter pin and nut.
   b. Disconnect the tie rod end LH from the steering knuckle.
5. Disconnect front lower ball joint attachment.
6. Support the front suspension lower arm with a jack.
7. Remove the clip and the nut.
8. Disconnect the upper ball joint from the steering knuckle.
9. Remove the steering knuckle.

To install:

10. Install the front suspension upper arm to the steering knuckle with the nut and tighten to 81 ft. lbs. (110 Nm).
11. Install a new clip.

➡If the holes for the clip are not aligned, tighten the nut up to another 60°.

12. Connect front lower ball joint attachment.
13. Connect tie rod end sub-assembly.
14. Install the front axle hub and tighten nut to 249 ft. lbs. (338 Nm).
15. Connect the front speed sensor to the steering knuckle.
16. Attach the clip to the steering knuckle.
17. Install front wheel.
18. Inspect and adjust the front wheel alignment.

STABILIZER BAR

REMOVAL & INSTALLATION

Sequoia

1. Remove the nut and disconnect the stabilizer bar links from the lower suspension arm.

➡If the ball joint turns together with the nut, use a hexagon (6 mm) wrench to hold the stud.

2. Remove the 2 bolts, nuts and stabilizer bar with the cushions and brackets.
3. Remove the 2 brackets and cushions from the stabilizer bar.
4. Hold the stabilizer bar link, and remove the nut.
5. Remove the stabilizer bar link, 2 retainers and bushings from each end of the stabilizer bar.

To install:

6. Install the 2 bushings, retainers and stabilizer bar link to the stabilizer bar.
7. Hold the stabilizer bar link, and install a new nut and tighten to 14 ft. lbs. (19 Nm).
8. Install the 2 bushings with their cutout facing to the rearward of the stabilizer bar.
9. Install the stabilizer bar and 2 brackets with the nuts and bolts and tighten to 27 ft. lbs. (37 Nm).
10. Connect front stabilizer links to suspension lower arm and tighten to 51 ft. lbs. (69 Nm).

➡If the ball joint turns together with the nut, use a hexagon (6 mm) wrench to hold the nut.

Tundra

1. Remove the nut and disconnect the stabilizer bar links from the lower suspension arm.

➡If the ball joint turns together with the nut, use a hexagon (6 mm) wrench to hold the stud.

2. Remove the 2 bolts, nuts and stabilizer bar with the cushions and brackets.
3. Remove the 2 brackets and cushions from the stabilizer bar.
4. Hold the stabilizer bar link, and remove the nut.
5. Remove the stabilizer bar link, 2 retainers and bushings from each end of the stabilizer bar.

To install:

6. Install the 2 bushings, retainers and stabilizer bar link to the stabilizer bar.

➡Temporarily tighten all suspension components until stabilized. After stabilization, tighten bolts and nuts to specifications.

7. Hold the stabilizer bar link, and install a new nut and tighten to 111 ft. lbs. (150 Nm).
8. Install the 2 bushings with their cutout facing to the rearward of the stabilizer bar.
9. Install the stabilizer bar and 2 brackets with the nuts and bolts and tighten to 51 ft. lbs. (69 Nm).
10. Connect front stabilizer links to suspension lower arm and tighten to 89 ft. lbs. (120 Nm).

➡If the ball joint turns together with the nut, use a hexagon (6 mm) wrench to hold the nut.

UPPER BALL JOINT

REMOVAL & INSTALLATION

Sequoia

See Figure 305.

1. Before servicing the vehicle, refer to the precautions in the beginning of this section.
2. Remove or disconnect the following:
   - Front wheel
   - Steering knuckle with the axle hub
   - Wire and boot
   - Snapring
   - Upper ball joint from the steering knuckle, using a deep socket wrench and tool 09050-40011, or equivalent press.

To install:

3. Install or connect the following:
   - New upper ball joint to the steering knuckle, using a deep socket and tool 09309-37010, or equivalent press.
   - New snapring
4. Using a torque wrench, inspect the upper ball joint rotation, as follows:
   a. Flip the ball joint back-and-forth 5 times.
   b. Using a torque wrench, continuously turn the nut 1 turn in 2–4 seconds.
   c. Take the reading on the 5th turn; it should be 6–39 inch lbs. (0.7–4.4 Nm). If not, replace the upper ball joint.
5. Install or connect the following:
   - New boot secured with a wire
   - Front wheel.
6. Check and/or adjust the front wheel alignment.

Tundra

For Ball Joint replacement, see Upper Control Arm

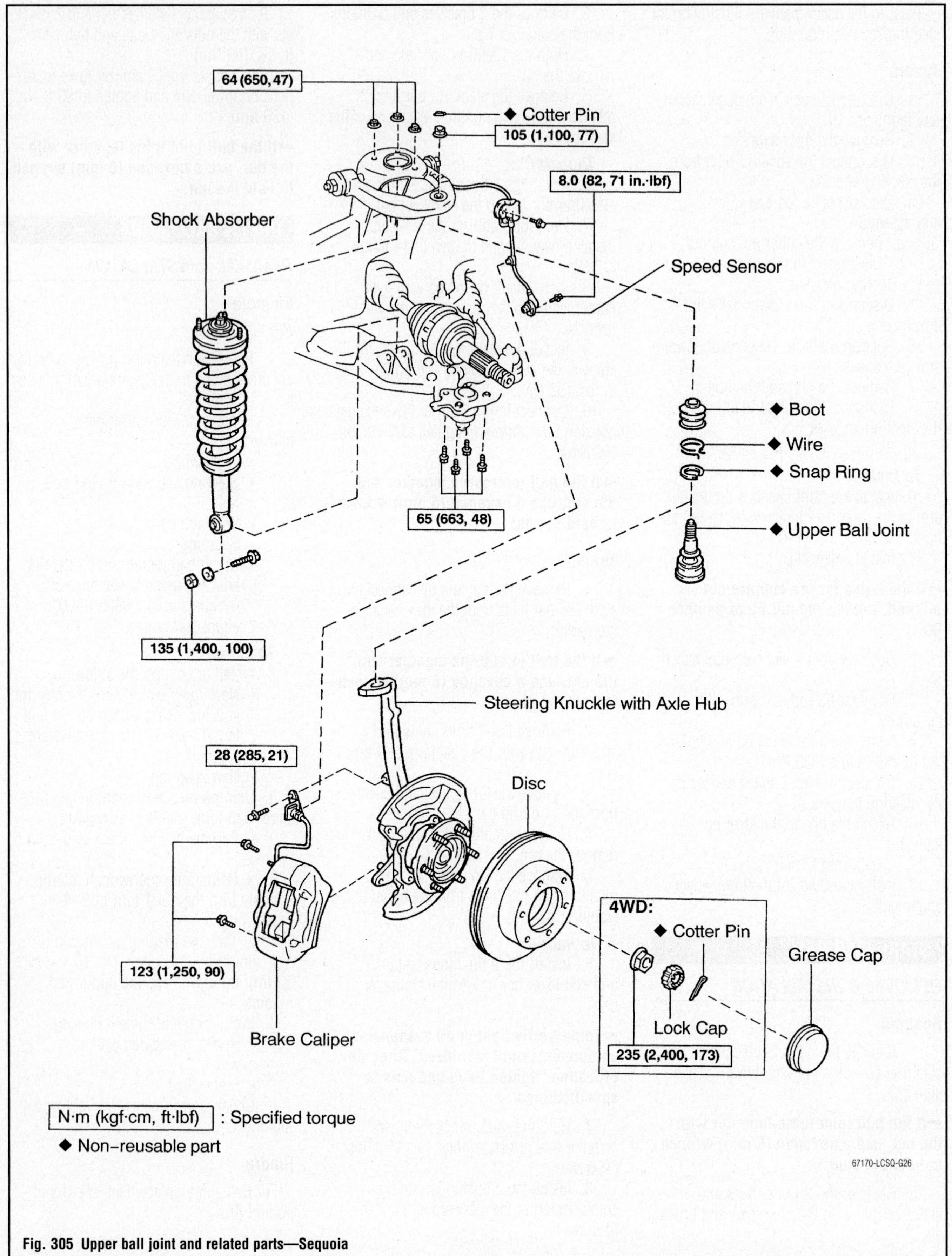

Fig. 305 Upper ball joint and related parts—Sequoia

UPPER CONTROL ARM

REMOVAL & INSTALLATION

Sequoia

See Figure 306.

1. Before servicing the vehicle, refer to the precautions in the beginning of this section.
2. Remove or disconnect the following:
   - Front wheel
   - Strut
   - Wheel speed sensor harness, if equipped with Anti-lock Brake System (ABS)
3. Upper ball joint, as follows:
   - Cotter pin and loosen the nut
   - Upper ball joint from the upper control arm, using tool 09950-40011

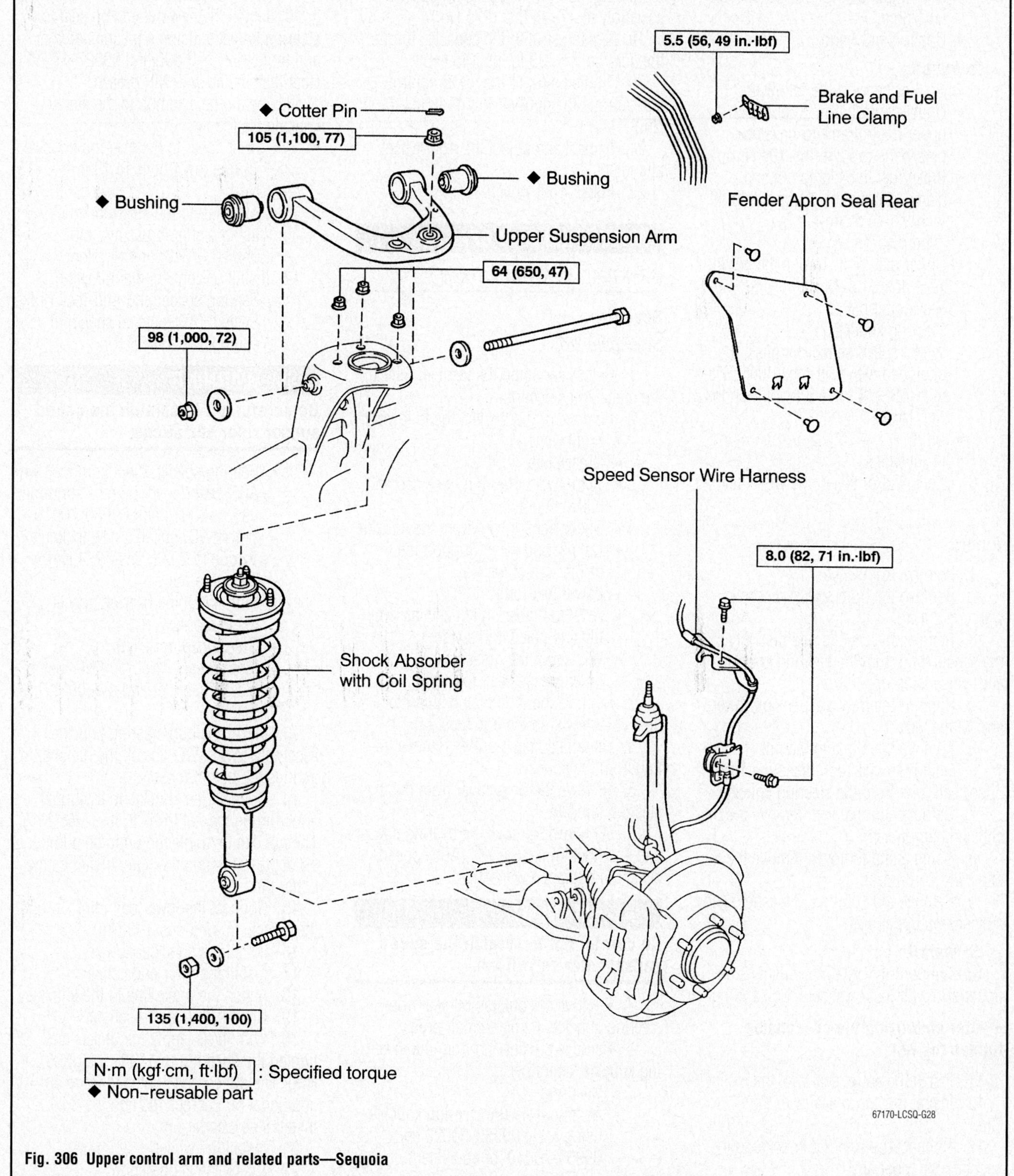

Fig. 306 Upper control arm and related parts—Sequoia

- Steering knuckle, support it securely
- Upper ball joint nut

4. Remove or disconnect the following:
   - 4 clips and the fender apron seal
   - Brake/fuel line clamp nut and clamp
   - Both upper control arm-to-chassis nuts/bolts
   - Upper control arm

To install:

5. Install or connect the following:
   - Upper control arm. Torque both upper control arm-to-chassis nuts/bolts to 72 ft. lbs. (98 Nm).
   - Brake/fuel line clamp nut and clamp. Torque the clamp nut to 49 inch lbs. (5.5 Nm).
   - Fender apron seal
   - Upper ball joint. Torque the nut to 77 ft. lbs. (105 Nm).
   - New cotter pin
   - Steering knuckle
   - Wheel speed sensor harness, if equipped with Anti-lock Brake System (ABS). Torque it to 71 inch lbs. (8.0 Nm).
   - Strut
   - Front wheel
6. Check and/or adjust the wheel alignment.

Tundra

1. Remove front wheel.
2. Remove the front shock absorber with coil spring.
3. Remove the 2 bolts and disconnect the sensor wire from the steering knuckle and suspension upper arm.
4. Support the front suspension lower arm with a jack.
5. Remove the clip and the nut.
6. Using Service Tool, disconnect the upper ball joint from the steering knuckle.
7. Using a clip remover, remove the 6 clips and apron seal.
8. Using a clip remover, remove the 5 clips and apron seal.
9. Remove the nut, bolt, 2 washers and suspension upper arm.

To install:

10. Temporarily install the suspension upper arm with the 2 washers, bolt and nut.

➡After stabilizing the suspension, tighten the nut.

11. Install the apron seal with the 5 clips.
12. Install the apron seal with the 6 clips.
13. Connect the steering knuckle to the suspension upper arm.
14. Install the nut and a new cotter pin and tighten to 81 ft. lbs. (110 Nm).
15. Connect the sensor wire to the steering knuckle and upper arm.
16. Temporarily install front shock absorber with coil spring.
17. Stabilize suspension.
18. Tighten front suspension upper arm assembly to 173 ft. lbs. (235 Nm).
19. Tighten strut to the chassis. Torque the nuts/bolts to 33 ft. lbs. (45 Nm).
20. Tighten strut to the lower control arm. Torque the nut/bolt to 144 ft. lbs. (195 Nm).
21. Inspect and adjust the front wheel alignment.
22. Install front wheel.

WHEEL HUB AND BEARING

REMOVAL & INSTALLATION

Sequoia

See Figure 307.

1. Before servicing the vehicle, refer to the precautions section.
2. Remove or disconnect the following:
   - Front wheel
   - Grease cap
   - With 4wd, cotter pin, lock cap and halfshaft nut
   - Speed sensor wire from the knuckle
   - Caliper and rotor. Support the caliper out of the way.
   - Lower ball joint
   - Axle hub/steering knuckle assembly and place it in a vise
   - With 2wd, the grease cap
   - Inner grease seal, for 4WD
3. Remove the 4 bolts and shift the dust cover towards the hub side.
4. Using a suitable puller, remove the hub from the knuckle.
5. Remove the dust cover from the steering knuckle.
6. With 4wd, remove the bearing spacer and Anti-lock Brake System (ABS) speed sensor, if equipped with ABS

⁂ WARNING

Be careful not to scratch the speed sensor rotor serrations.

7. Remove the outside oil seal from steering knuckle, using a small prybar.
8. Remove the bearing from the steering knuckle, as follows:
   - Snapring
   - Bearing from the steering knuckle, using tools 09950-60020 and 09950-70010, or equivalent.

To install:

9. Install the bearing to the steering knuckle, as follows:
   - Bearing to the steering knuckle, using a press and tools 09950-60020 and 09527-17011, or equivalent
   - New snapring
10. Install the new outside oil seal to steering knuckle, using a plastic hammer and tools 09223-15030 and 09527-17011. Coat the seal lip with MP grease.
11. Install the axle hub to the steering knuckle, as follows:
    - Dust cover to the steering knuckle. Torque the 4 bolts to 13 ft. lbs. (18 Nm).
    - Axle hub to the steering knuckle, using a shop press and tool 09649-17010, or equivalent.
12. Install or connect the following:
    - Bearing spacer and Anti-lock Brake System (ABS) speed sensor, if equipped with ABS

⁂ WARNING

Be careful not to scratch the speed sensor rotor serrations.

- Bearing spacer, if not equipped with ABS, using a shop press and tools 09950-60010 and 09950-70010
- With 2WD, install a new locknut. Torque to 203 ft. lbs. (274 Nm). Stake the nut.
- With 4WD, the bearing spacer using a press
- Grease cap, for 2WD
- Inner grease seal, for 4WD, using a plastic hammer and tool 09527-17011

13. Install the hub/steering knuckle assembly. With 4WD, install the halfshaft and install the nut loosely.
14. Connect the knuckle to the upper arm. Torque the nut to 77 ft. lbs. (105 Nm). The nut can be tightened up to an additional 60 degrees to align the hole. Install a new cotter pin.
15. Connect the lower ball joint. Torque the 4 bolts to 48 ft. lbs. (65 Nm).
16. Install the shock absorber.
17. Install the rotor and caliper.
18. Attach the brake line to the knuckle.
19. Connect the speed sensor.
20. With 4WD, apply the brakes and tighten the shaft nut to 173 ft. lbs. (235 Nm). The nut can be tightened up to an additional 60 degrees to align the hole. Install a new cotter pin.
21. Install the wheel.
22. Check the alignment.

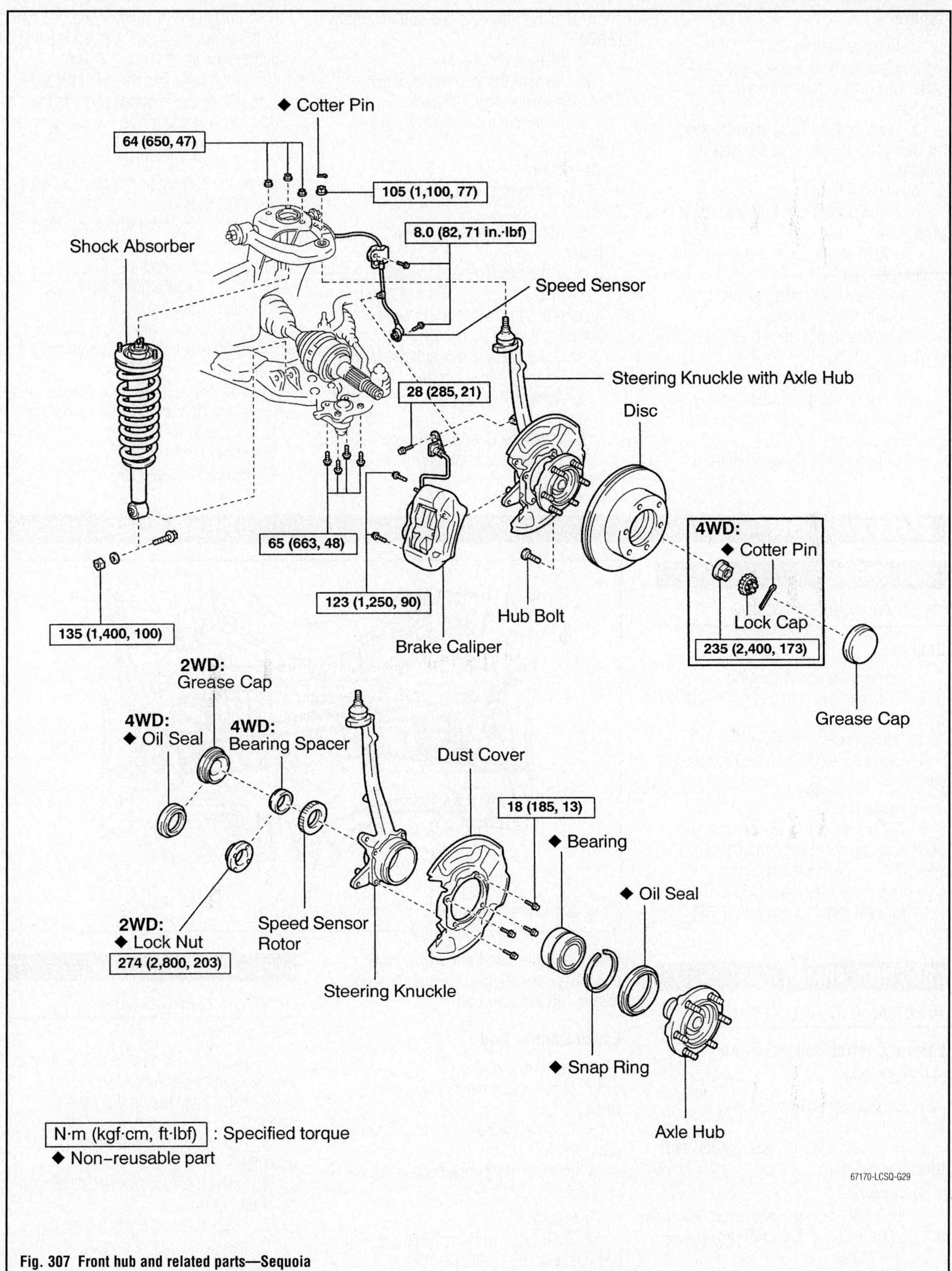

Fig. 307 Front hub and related parts—Sequoia

Tundra

1. Remove front wheel.
2. Remove the bolt and disconnect the brake tube bracket from the steering knuckle.
3. Remove the 2 bolts and disconnect the disc brake caliper from the steering knuckle.
4. Remove front disc.
5. Remove front axle hub grease cap (4WD).
   a. Remove the cotter pin and adjusting lock cap.
   b. Using a 39 mm socket wrench, remove the axle hub nut.
6. Remove front axle hub sub-assembly (4WD).
   a. Remove the 4 bolts.
   b. Using a plastic-faced hammer, tap out the front drive shaft from the front axle hub.
   c. Remove the O-ring from the axle hub.
7. Remove front axle hub sub-assembly (2WD).
   a. Remove the 4 bolts.
   b. Remove the axle hub and dust cover from the steering knuckle.
   c. Remove the O-ring from the axle hub.

To install:

8. Install front axle hub sub-assembly (2WD).
   a. Apply MP grease to a new O-ring.
   b. Install the O-ring to the axle hub.
   c. Install the dust cover and axle hub to the steering knuckle with the 4 bolts and tighten to 73 ft. lbs. (99 Nm).
9. Install front axle hub sub-assembly (4WD).
   a. Apply MP grease to a new O-ring.
   b. Install the O-ring to the axle hub.
   c. Connect the front drive shaft to the front axle hub.
   d. Install the dust cover and axle hub to the steering knuckle with the 4 bolts and tighten to 73 ft. lbs. (99 Nm).
10. Install front axle hub nut (4WD).
   a. Clean the threaded parts on the drive shaft and axle hub nut using a non-residue solvent.
   b. Using a 39 mm socket wrench, install the hub nut and tighten to 249 ft. lbs. (338 Nm).
   c. Install the front wheel adjusting lock cap and a new cotter pin.
11. Inspect the front axle hub.
12. Install the axle hub grease cap.
13. Install front disc.
14. Connect the front disc brake caliper and install 2 new bolts and tighten to 73 ft. lbs. (99 Nm).
15. Connect the brake tube bracket to the steering knuckle with the bolt and tighten to 21 ft. lbs. (29 Nm).
16. Install front wheel.
17. Check the speed sensor signal.

SUSPENSION — REAR SUSPENSION

COIL SPRING

REMOVAL & INSTALLATION

Sequoia

1. Remove the shock absorber.
2. Disconnect the left and right stabilizer bar links.
3. Disconnect the lateral rod.
4. Lower the axle slowly and remove the spring.

To install:

5. Installation is the reverse of removal. Make sure that the lower end of the coil spring is correctly installed, against the stop. Torque the later rod to 103 ft. lbs. (140 Nm); the stabilizer bar links to 51 ft. lbs. (69 Nm).

CONTROL ARMS/LINKS

REMOVAL & INSTALLATION

Lateral Control Rod Assembly

See Figure 308.

1. Support rear axle housing with jack.
2. Remove the 2 bolts, nut, washer and lateral control rod.

To install:

3. To install, reverse removal procedure and tighten to 103 ft. lbs. (140 Nm) and 96 ft. lbs. (130 Nm).

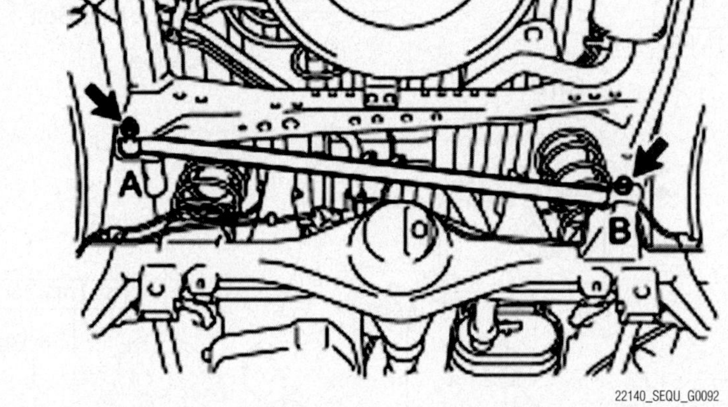

Fig. 308 Lateral control rod

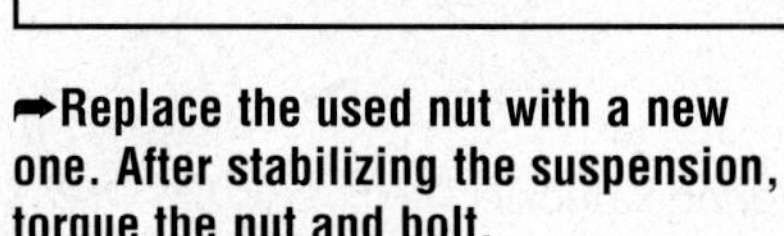

➡Replace the used nut with a new one. After stabilizing the suspension, torque the nut and bolt.

Lower Control Rod

1. Remove rear wheel.
2. Support rear axle housing with jack.
3. Remove the bolt and parking brake cable bracket.
4. Remove the 2 nuts, bolts and lower control arm.

To install:

5. To install, reverse removal procedure and tighten to 96 ft. lbs. (130 Nm).

LEAF SPRING

REMOVAL & INSTALLATION

See Figure 309.

1. Before servicing the vehicle, refer to the Precautions Section.
2. Raise and support the vehicle safely.
3. Remove the tire and wheel assembly.
4. Lower the floor jack to take tension off of the spring.
5. Support the axle with a floor jack.
6. Remove or disconnect the following:

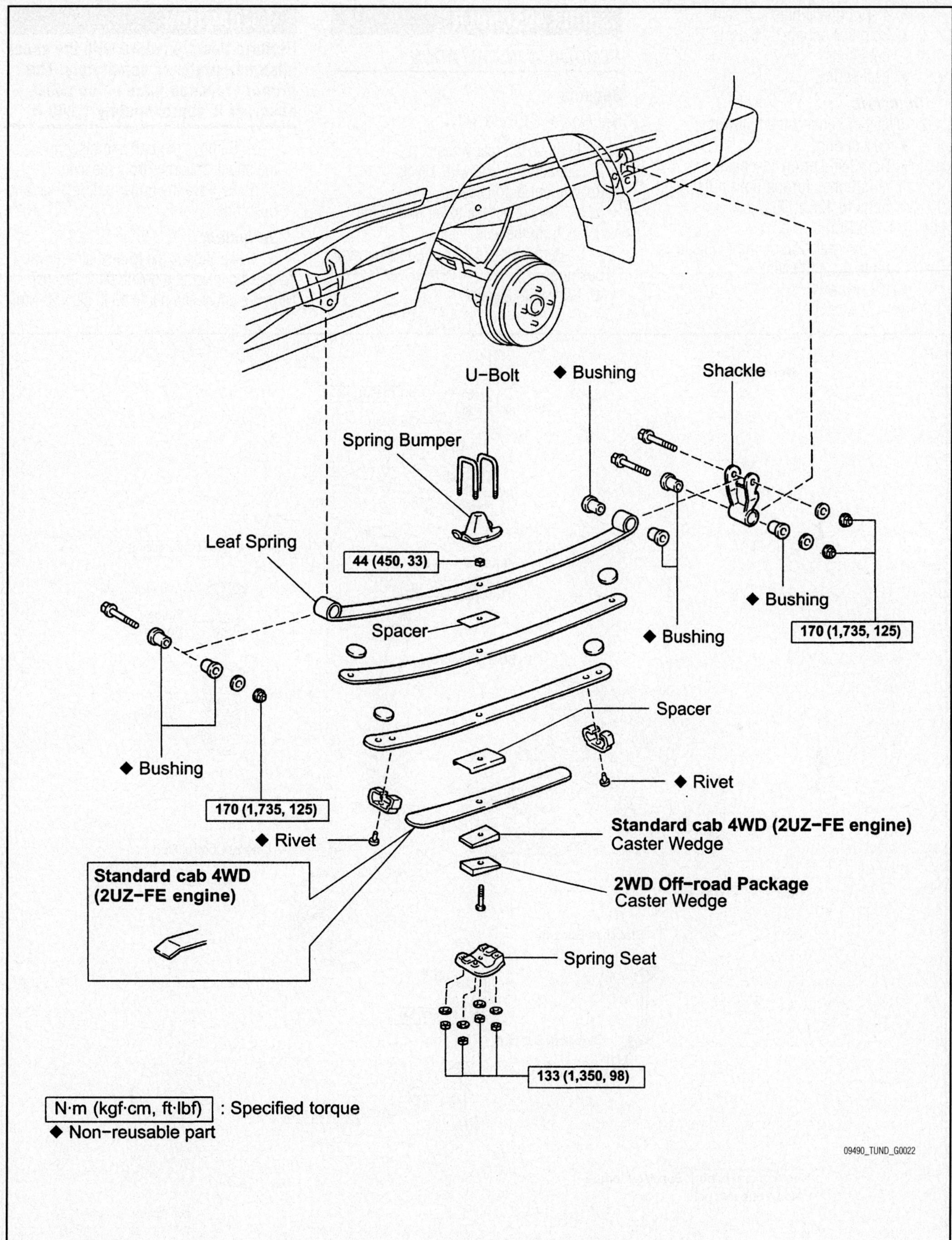

Fig. 309 Rear spring and related components

- 4 spring seat nuts and seat
- Both leaf spring-to-chassis nuts/bolts
- Leaf spring

To install:

7. Install or connect the following:
   - Leaf spring
   - Both leaf spring-to-chassis nuts/bolts. Torque both nuts/bolts to 125 ft. lbs. (170 Nm).
   - Spring seat. Torque the 4 nuts to 98 ft. lbs. (133 Nm).
   - Rear wheel

SHOCK ABSORBER

REMOVAL & INSTALLATION

Sequoia

See Figures 310 and 311.

1. Remove the rear wheels.
2. Support the axle with a jack.
3. Without Auto Leveler:
   a. Disconnect the lower end of the shock from the axle.
   b. Remove the upper nut, retainer and bushings.
4. With Auto Leveler:

CAUTION

Perform this procedure with the shock absorber stretched completely. The primary reaction force of the shock absorber is approximately 1,000 N.

   a. Remove the bolt and disconnect the shock absorber from the axle.
   b. Remove the upper nut, retainer and bushings.

To install:

5. Installation is the reverse of removal. Torque the lower end bolt to 64 ft. lbs. (87 Nm); the upper end nut to 43 ft. lbs. (58 Nm).

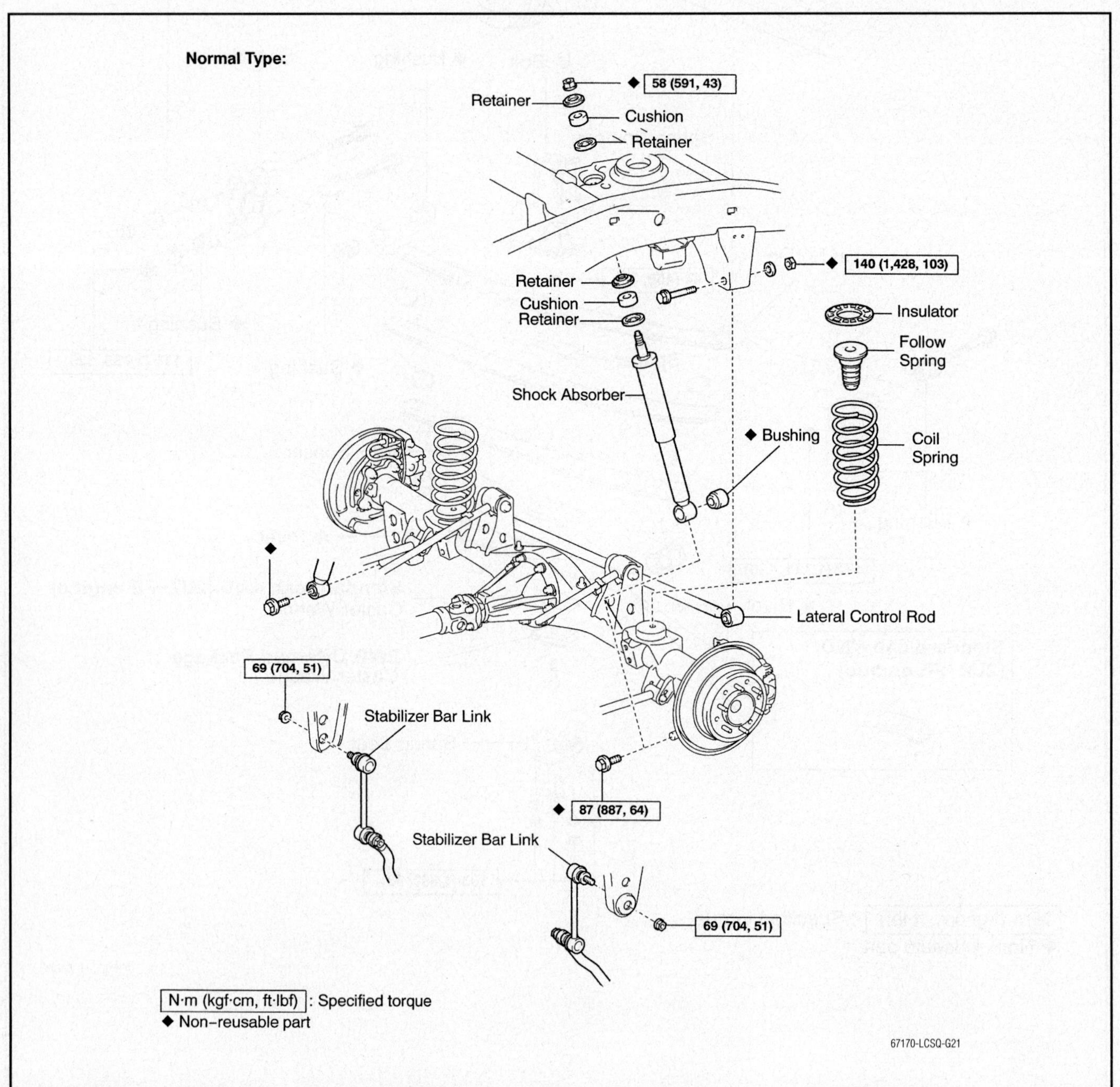

Fig. 310 Sequoia rear suspension, without Auto Leveler

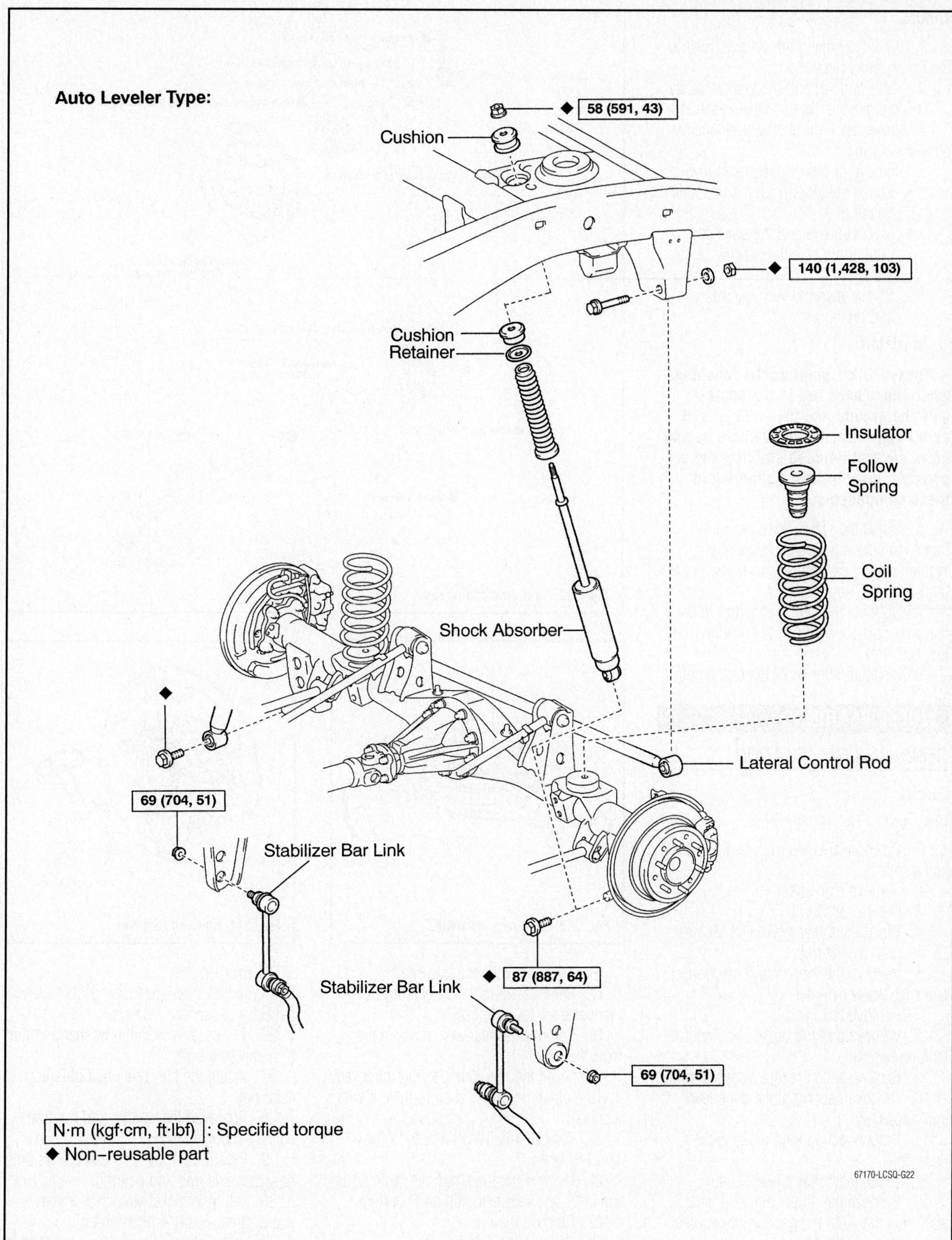

Fig. 311 Sequoia rear suspension, with Auto Leveler

Tundra

1. Before servicing the vehicle, refer to the Precautions Section.
2. Raise and support the vehicle safely.
3. Remove the tire and wheel assembly.
4. Lower the floor jack to take tension off of the spring.
5. Remove or disconnect the following:
   - Shock absorber from the rear axle housing
   - Nut, retainers and the cushions holding the shock absorber to the frame
   - Shock absorber with the washers and bushings

To install:

➡Always fully tighten rubber bushings when the wheels are in full contact with the ground and the vehicle is at curb height. Bounce the vehicle up and down several times to stabilize the suspension, prior to final tightening of these components.

6. Install the shock absorber to the frame with the washers and bushings. Tighten the shock absorber-to-frame nut to 64 ft. lbs. (87 Nm).
7. Connect the shock absorber to the rear axle housing. Tighten the bolt to 15 ft. lbs. (20 Nm).
8. Install the tire and wheel assembly.

WHEEL BEARINGS

REMOVAL & INSTALLATION

Tundra

See Figures 312 through 317.

1. Disconnect cable from negative battery terminal.
2. Remove rear wheel.
3. Drain brake fluid.
4. Disconnect rear brake flexible hose.
5. Remove the clip.
6. Remove the 2 bolts and disconnect rear disc brake cylinder.
7. Remove rear disc.
8. Remove parking brake shoe return tension spring.
9. Remove parking brake shoe assembly.
10. Remove parking brake shoe lever sub-assembly.
11. Disconnect parking brake cable assembly.
12. Disconnect rear speed sensor.
13. Remove the 4 nuts and rear axle shaft together with the parking brake plate.
14. Remove the O-ring.
15. Using a snap ring expander, remove the snap ring.

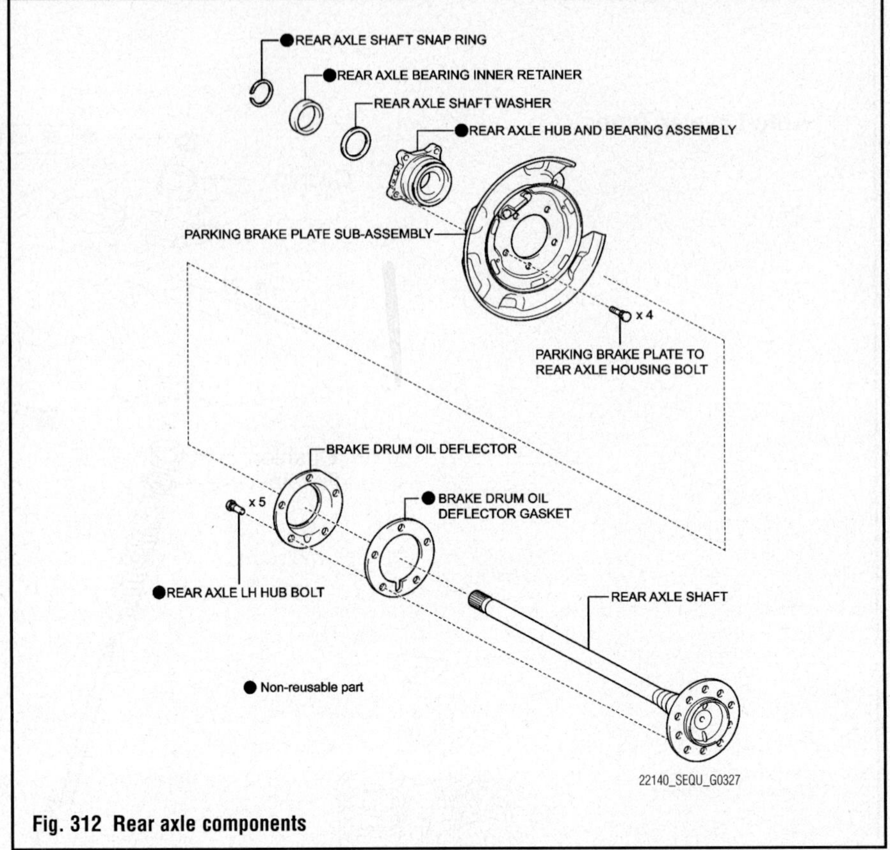

Fig. 312 Rear axle components

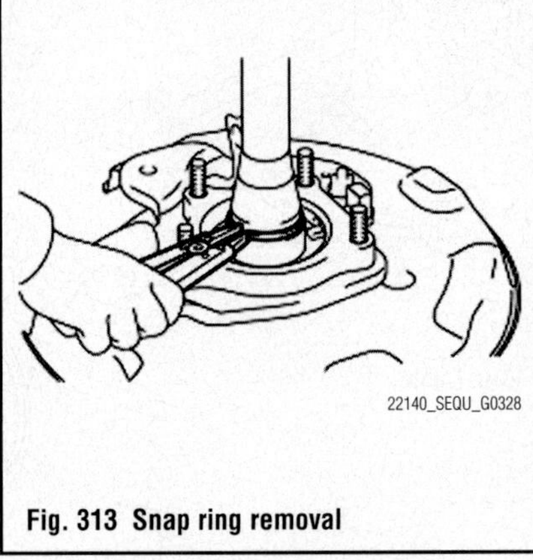

Fig. 313 Snap ring removal

Fig. 314 Housing removal

16. Press out the rear axle shaft.
17. Remove the rear axle bearing inner retainer from the axle hub.
18. Remove the rear axle shaft washer from the axle hub.
19. Grind the rear axle bearing inner race surface using a grinder, then remove it with a chisel.
20. Temporarily install 4 nuts to the housing bolts.
21. Using a hammer, remove the 4 housing bolts and rear axle hub and bearing.
22. Remove the nuts.
23. Remove the 5 hub bolts.
24. Remove the deflector and deflector gasket from the rear axle shaft.

To install:

25. Install a new deflector gasket and deflector to the rear axle shaft.
26. Temporarily install a washer and nut to 5 new hub bolts.
27. Install the hub bolts by tightening each nut.
28. Removed the washer and nut from each hub bolt.
29. Position the parking brake plate on a new rear axle hub and bearing.
30. Using 2 socket wrenches and a press, press in the 4 housing bolts.
31. Install the washer and a new retainer to the axle hub.

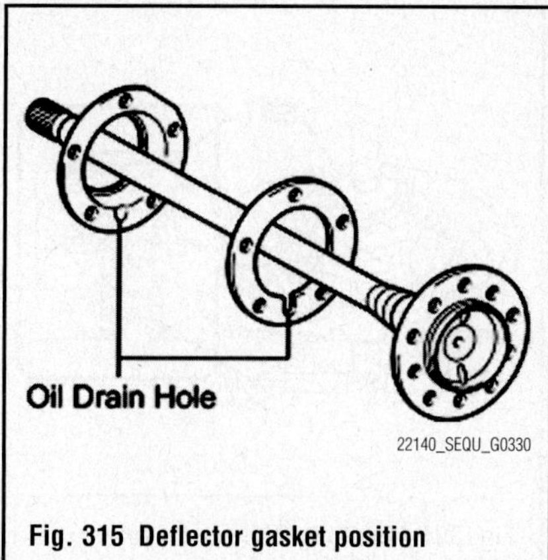

Fig. 315 Deflector gasket position

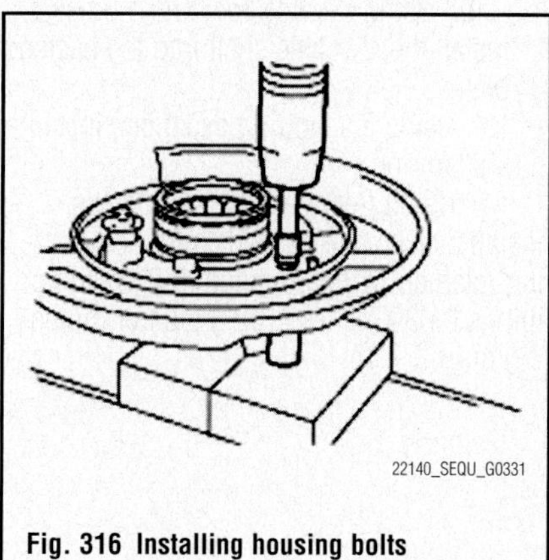

Fig. 316 Installing housing bolts

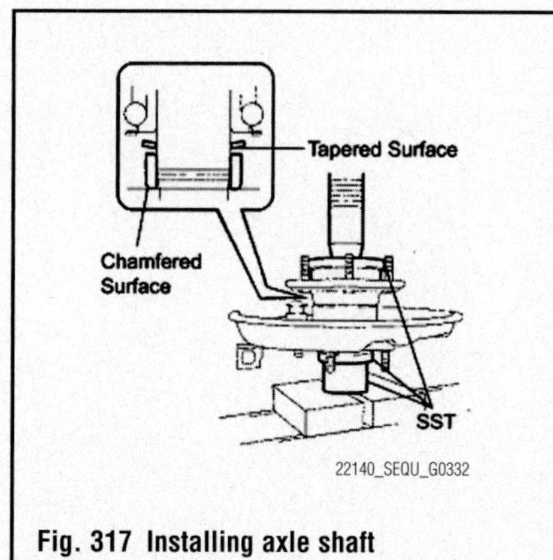

Fig. 317 Installing axle shaft

➡**Install the washer with its tapered surface facing downward. Install the retainer with its chamfered surface facing downward.**

32. Press in the rear axle shaft.
33. Using a snap ring expander, install a new snap ring.
34. Apply MP grease to the lip of the oil seal.
35. Install the O-ring to the axle shaft.
36. Install the rear axle shaft and parking brake plate with the 4 nuts and tighten to 44 ft. lbs. (60 Nm).
37. Inspect the rear axle shaft.
38. Connect rear speed sensor.
39. Connect parking brake cable assembly.
40. Install parking brake shoe lever sub-assembly.
41. Install parking brake shoe assemblies.
42. Install parking brake shoe return tension spring.
43. Check parking brake installation.
44. Install rear disc.
45. Adjust parking brake shoe clearance.
46. Connect the rear disc brake cylinder and install 2 new bolts and tighten to 70 ft. lbs. (95 Nm).
47. Set the flexible hose to the connecting point with the brake tube, and then install a new clip.
48. Connect the flexible hose to the brake tube while holding the flexible hose with a wrench.
49. Connect cable to negative battery terminal.
50. Fill reservoir with brake fluid.
51. Bleed air from brakes.
52. Install rear wheel.
53. Inspect differential oil.
54. Inspect for differential oil leak.
55. Check parking brake pedal travel.

WHEEL HUB AND BEARING

REMOVAL & INSTALLATION

See Figures 318 through 324.

1. Remove rear axle bearing retainer inner.
2. Remove speed sensor rotor.
3. Attach the 4 nuts to the serration bolts and remove the serration bolts from the backing plate using a hammer.

➡**Do not reuse the nuts previously removed from the vehicle.**

4. Grind the retainer and sensor rotor surfaces using a grinder, then pry them out with a chisel and hammer.

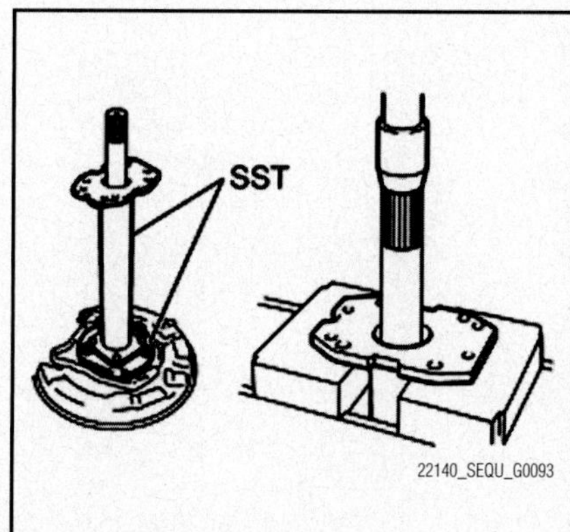

Fig. 318 Service tool installation

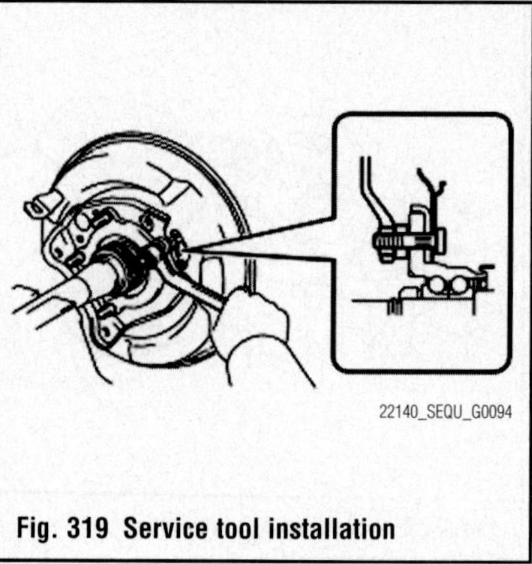

Fig. 319 Service tool installation

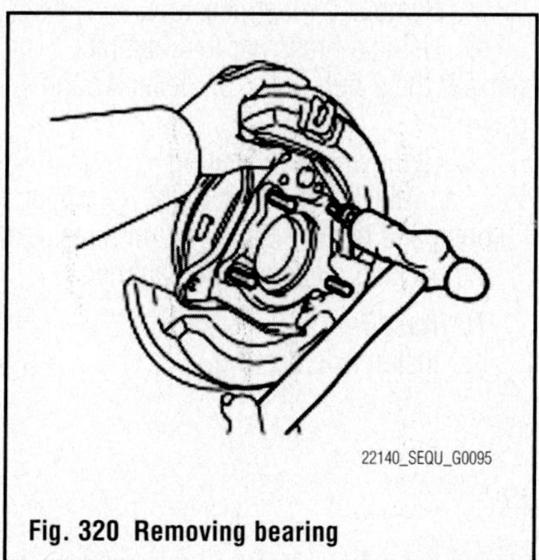

Fig. 320 Removing bearing

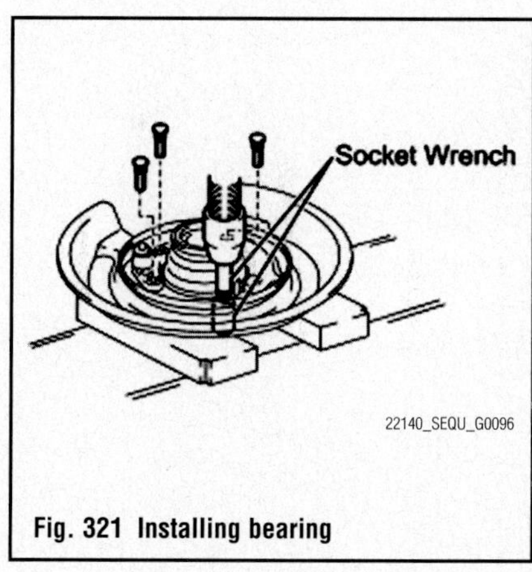

Fig. 321 Installing bearing

5. Using a snap ring expander, remove the snap ring.
6. Remove rear axle shaft.
   a. Attach the washers and nut to the serration bolts, then torque the nuts to install the serration bolts to the backing plate.
   b. Remove the 4 nuts from the serration bolts.
   c. Position Service Tool on the backing plate with the 4 nuts.
   d. Using a press, remove the rear axle shaft, spring washer and bearing retainer from the backing plate.

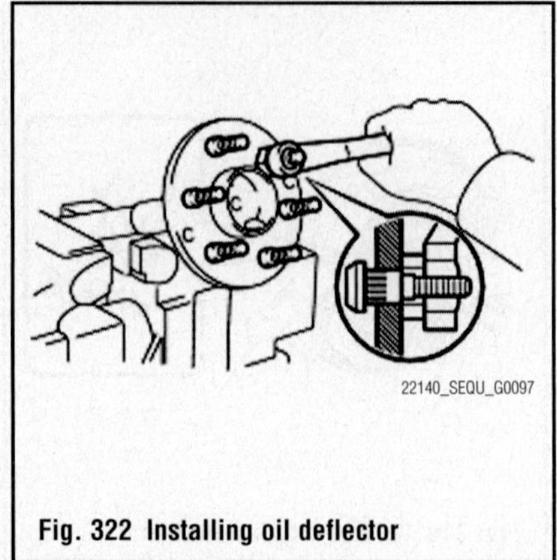

Fig. 322 Installing oil deflector

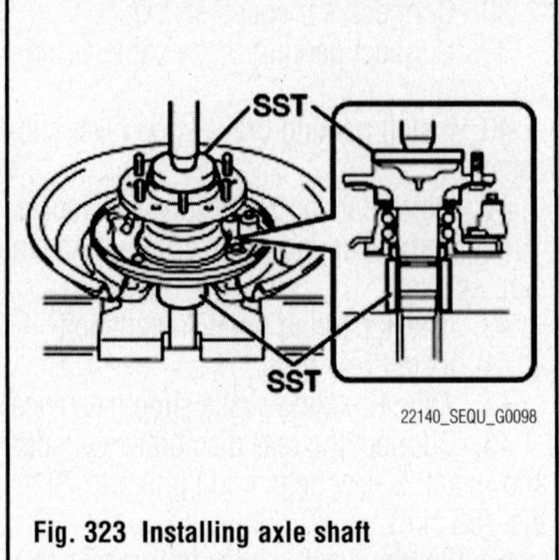

Fig. 323 Installing axle shaft

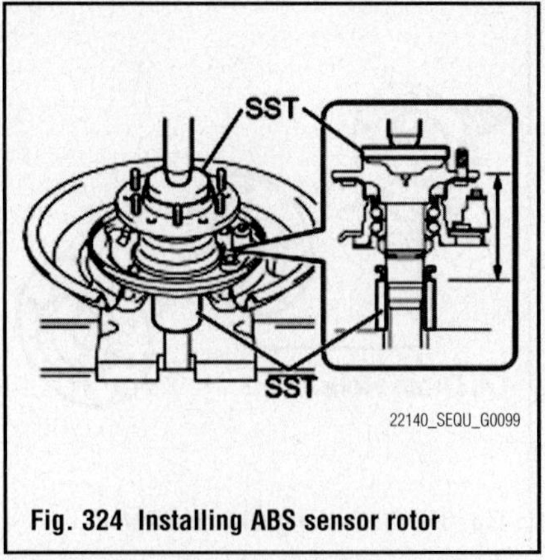

Fig. 324 Installing ABS sensor rotor

7. Remove the Service Tool.
8. Using a brass bar and hammer, remove the 6 hub bolts, oil deflector and gasket.
9. Removing axle bearing.
   a. Attach the 4 nuts to the serration bolts and remove the serration bolts from the backing plate using a hammer.

To install:

10. Install axle bearing.
   a. Position the backing plate on the bearing assembly and using a press and 2 socket wrenches, install the serration bolts.
11. Position a new gasket and oil deflector on the axle shaft and install a washer and nut to a new hub bolt, and install the hub bolt by torquing the nut.
12. Install rear axle shaft.
   a. Install the backing plate, spring washer and bearing retainer on the rear axle shaft.
   b. Using Service Tool and a press, install the rear axle shaft into the backing plate.
13. Using a snap ring expander, install a new snap ring.
14. Using Service Tool and a press, install a new speed sensor rotor and bearing retainer. Standard length 5.228 plus or minus 0.039 inches (132.8 plus or minus 1.0 mm).

TOYOTA

Tacoma

23

SPECIFICATIONS AND MAINTENANCE CHARTS

ENGINE AND VEHICLE IDENTIFICATION

Engine							Model Year	
Code/ID	Liters (cc)	Cu. In.	Cyl.	Fuel Sys.	Engine Type	Eng. Mfg.	Code ①	Year
2TR-FE	2.7 (2693)	164	4	MFI	DOHC	Toyota	7	2007
1GR-FE	4.0 (3956)	241	6	MFI	DOHC	Toyota	8	2008

MFI: Multi-port Fuel Injection

DOHC: Double Overhead Camshaft

① 10th digit of the VIN number

22140_TACO_C0001

GENERAL ENGINE SPECIFICATIONS

Year	Model	Engine Displacement Liters	Engine Series Code/ID	Net Horsepower @ rpm	Net Torque @ rpm (ft. lbs.)	Bore x Stroke (in.)	Com-pression Ratio	Oil Pressure @ rpm
2007	Tacoma	2.7	2TR-FE	159@5200	236@5200	3.74x3.74	NA	43-85@3000
		4.0	1GR-FE	180@3800	266@4000	3.70x3.74	NA	23-75@3000
2008	Tacoma	2.7	2TR-FE	159@5200	236@5200	3.74x3.74	NA	43-85@3000
		4.0	1GR-FE	180@3800	266@4000	3.70x3.74	NA	23-75@3000

NA: Not Available

22140_TACO_C0002

ENGINE TUNE-UP SPECIFICATIONS

Year	Engine Displacement Liters	Engine Code/ID	Spark Plug Gap (in.)	Ignition Timing (deg.)	Fuel Pump (psi)	Idle Speed (rpm) MT	Idle Speed (rpm) AT	Valve Clearance Intake	Valve Clearance Exhaust
2007	2.7	2TR-FE	0.039-0.043	3-7B ①	40.8-41.7	600-700	600-700	②	②
	4.0	1GR-FE	0.039-0.043	8-12B ①	40.8-41.7	650-750	650-750	0.006-0.010	0.011-0.015
2008	2.7	2TR-FE	0.039-0.043	3-7B ①	40.8-41.7	600-700	600-700	②	②
	4.0	1GR-FE	0.039-0.043	8-12B ①	40.8-41.7	650-750	650-750	0.006-0.010	0.011-0.015

NOTE: The Vehicle Emission Control Information label often reflects specification changes made during production.

The label figures must be used if they differ from those in this chart.

B: Before top dead center

① With terminals TC and CG of the DLC3 connected

② Automatic adjustment

22140_TACO_C0003

CAPACITIES

Year	Model	Engine Displacement Liters	Engine Code/ID	Engine Oil with Filter (qts.)	Transmission (pts.) 5-Spd	Transmission (pts.) Auto.	Transfer Case (pts.)	Drive Axle Front (pts.)	Drive Axle Rear (pts.)	Fuel Tank (gal.)	Cooling System (qts.)
2007	Tacoma	2.7	2TR-FE	5.8	①	②	2.2	3.2	③	21.1	9.1
		4.0	1GR-FE	④	3.6	②	2.2	3.2	③	21.1	⑤
2008	Tacoma	2.7	2TR-FE	5.8	①	②	2.2	3.2	③	21.1	9.1
		4.0	1GR-FE	④	3.6	②	2.2	3.2	③	21.1	⑤

① 2WD: 5.4
4WD: 4.6

② 4 speed: 4.0
5 speed: 6.0

③ 2WD: 7.0
4WD: 6.0

④ 2WD, except PreRunner: 4.5
4WD and PreRunner: 5.2

⑤ MT: 10.3
AT: 10.1

22140_TACO_C0004

VALVE SPECIFICATIONS

Year	Engine Displacement Liters	Engine Code/ID	Seat Angle (deg.)	Face Angle (deg.)	Spring Test Pressure (lbs. @ in.)	Spring Installed Height (in.)	Stem-to-Guide Clearance (in.) Intake	Stem-to-Guide Clearance (in.) Exhaust	Stem Diameter (in.) Intake	Stem Diameter (in.) Exhaust
2007	2.7	2TR-FE	45	NA	NA	1.9106	0.0010-0.0024	0.0012-0.0026	0.2154-0.2159	0.2151-0.2157
	4.0	1GR-FE	NA	44.5	41.9-46.3@1.311	1.882	0.0010-0.0024	0.0012-0.0026	0.2154-0.2159	0.2152-0.2158
2008	2.7	2TR-FE	45	NA	NA	1.9106	0.0010-0.0024	0.0012-0.0026	0.2154-0.2159	0.2151-0.2157
	4.0	1GR-FE	NA	44.5	41.9-46.3@1.311	1.882	0.0010-0.0024	0.0012-0.0026	0.2154-0.2159	0.2152-0.2158

NA: Not Available

22140_TACO_C0005

CAMSHAFT SPECIFICATIONS

All measurements in inches unless noted

Year	Engine Displacement Liters	Engine Code/ID	Journal Dia.	Brg. Oil Clearance	Shaft End-play ①	Circle Runout	Lobe Height Intake	Lobe Height Exhaust
2007	2.7	2TR-FE	②	③	0.0039-0.0090	0.0012	1.6872-1.6911	1.6872 1.6911
	4.0	1GR-FE	④	⑤	0.0160-0.0350	0.0024	⑥	⑥
2008	2.7	2TR-FE	②	③	0.0039-0.0090	0.0012	1.6872-1.6911	1.6872 1.6911
	4.0	1GR-FE	④	⑤	0.0160-0.0350	0.0024	⑥	⑥

NA: Not Available

① Thrust clearance

② No. 1: 1.4153-1.4159
All others: 1.0614-1.0620

③ No. 1: 0.0014-0.0029
All others: 0.0010-0.0024

④ No. 1: 1.4162-1.4167
All others: 0.9039-0.9045

⑤ No. 1: 0.0016-0.0031
All others: 0.0010-0.0024

⑥ No. 1 camshaft: 1.7389-1.7428
No. 2 camshaft: 1.7551-1.7591
No. 3 camshaft (sub assembly): 1.7389-1.7428
No. 2 camshaft (sub assembly): 1.7551-1.7591
No. 2 camshaft: 1.7551-1.7591

22140_TACO_C0006

CRANKSHAFT AND CONNECTING ROD SPECIFICATIONS

All measurements are given in inches.

Year	Engine Displacement Liters	Engine Code/ID	Crankshaft Main Brg. Journal Dia.	Crankshaft Main Brg. Oil Clearance	Crankshaft Shaft End-play	Crankshaft Thrust on No.	Connecting Rod Journal Diameter	Connecting Rod Oil Clearance	Connecting Rod Side Clearance
2007	2.7	2TR-FE	①	②	0.0008-0.0087	NA	NA	0.0009-0.0019	0.0059-0.0138
	4.0	1GR-FE	2.8342-2.8346	0.0007-0.0012	NA	NA	NA	0.0010-0.0018	0.0059-0.0118
2008	2.7	2TR-FE	①	②	0.0008-0.0087	NA	NA	0.0009-0.0019	0.0059-0.0138
	4.0	1GR-FE	2.8342-2.8346	0.0007-0.0012	NA	NA	NA	0.0010-0.0018	0.0059-0.0118

NA: Not Available

① No. 3: 2.3615-2.3620
All others: 2.3619-2.3622

② No. 3: 0.0012-0.0022
All others: 0.0009-0.0019

22140_TACO_C0007

PISTON AND RING SPECIFICATIONS

All measurements are given in inches.

	Engine			Ring Gap			Ring Side Clearance		
Year	Displacement Liters	Engine Code/ID	Piston Clearance	Top Compression	Bottom Compression	Oil Control	Top Compression	Bottom Compression	Oil Control
2007	2.7	2TR-FE	0.0007-0.0020	0.0087-0.0134	0.0177-0.0224	0.0039-0.0157	0.0008-0.0030	0.0008-0.0026	0.0008-0.0028
	4.0	1GR-FE	0.0031-0.0040	0.0118-0.0157	0.0157-0.0197	0.0039-0.0157	0.0008-0.0028	0.0008-0.0024	0.0028-0.0060
2008	2.7	2TR-FE	0.0007-0.0020	0.0087-0.0134	0.0177-0.0224	0.0039-0.0157	0.0008-0.0030	0.0008-0.0026	0.0008-0.0028
	4.0	1GR-FE	0.0031-0.0040	0.0118-0.0157	0.0157-0.0197	0.0039-0.0157	0.0008-0.0028	0.0008-0.0024	0.0028-0.0060

22140_TACO_C0008

TORQUE SPECIFICATIONS

All readings in ft. lbs.

Year	Engine Displacement Liters	Engine Code/ID	Cylinder Head Bolts	Main Bearing Bolts	Rod Bearing Bolts	Crankshaft Damper Bolts	Flywheel Bolts	Manifold Intake	Manifold Exhaust	Spark Plugs	Oil Pan Drain Plug
2007	2.7	2TR-FE	①	②	③	192	④	18	27	13	28
	4.0	1GR-FE	⑤	⑥	③	185	61	19	16	13	30
2008	2.7	2TR-FE	①	②	③	192	④	18	27	13	28
	4.0	1GR-FE	⑤	⑥	③	185	61	19	16	13	30

NA: Information not available

① Step 1: 29 ft. lbs.
Step 2: Plus 90 degrees
Step 3: Plus 90 degrees

② Step 1: 29 ft. lbs.
Step 2: Plus 90 degrees

③ Step 1: 18 ft. lbs.
Step 2: Plus 90 degrees

④ AT: 55 ft. lbs.
MT: 20 ft. lbs. plus 90 degrees

⑤ Step 1: 27 ft. lbs.
Step 2: Plus 180 degrees
Left side 14mm bolt: 22 ft. lbs.

⑥ 12 pointed head: 45 ft. lbs. Then plus 90 degrees
12mm head: 18 ft. lbs.

22140_TACO_C0009

WHEEL ALIGNMENT

Year	Model	Caster Range (+/-Deg.)	Caster Preferred Setting (Deg.)	Camber Range (+/-Deg.)	Camber Preferred Setting (Deg.)	Toe-in (in.)	Steering Axis Inclination (Deg.)
2007	2WD exc. PreRunner	0.75	+0.67	0.75	0	0.06+/-0.08	10.00
	4WD & PreRunner	0.75	0.30	0.75	+1.62	0.06+/-0.08	10.40
2008	2WD exc. PreRunner	0.75	+0.67	0.75	0	0.06+/-0.08	10.00
	4WD & PreRunner	0.75	0.30	0.75	+1.62	0.06+/-0.08	10.40

All alignment figures based on nominal ride height and standard tires

22140_TACO_C0010

TIRE, WHEEL AND BALL JOINT SPECIFICATIONS

Year	Model	OEM Tires Standard	OEM Tires Optional	Tire Pressures (psi) Front	Tire Pressures (psi) Rear	Wheel Size	Ball Joint Inspection ①	Lug Nut Torque (ft. lbs.)
2007	Reg. Cab 2wd	P215/70R15	None	29	32	NA	②	85
	Reg. Cab 4wd	P245/75R16	None	29	29	NA	②	85
	Reg. Cab PreRunner	P245/75R16	None	29	29	NA	②	85
	Extended Cab 2wd	P215/70R15	None	③	③	NA	②	85
	Extended Cab 4wd	P245/75R16	None	③	③	NA	②	85
	Extended Cab PreRunner	P245/75R16	None	③	③	NA	②	85
	X-Runner	P255/45R18	None	29	29	NA	②	85
	Double Cab	P245/75R16	None	29	29	NA	②	85
2008	Reg. Cab 2wd	P215/70R15	None	29	32	NA	②	85
	Reg. Cab 4wd	P245/75R16	None	29	29	NA	②	85
	Reg. Cab PreRunner	P245/75R16	None	29	29	NA	②	85
	Extended Cab 2wd	P215/70R15	None	③	③	NA	②	85
	Extended Cab 4wd	P245/75R16	None	③	③	NA	②	85
	Extended Cab PreRunner	P245/75R16	None	③	③	NA	②	85
	X-Runner	P255/45R18	None	29	29	NA	②	85
	Double Cab	P245/75R16	None	29	29	NA	②	85

OEM: Original Equipment Manufacturer
PSI: Pounds Per Square Inch
NA: Information not available
STD: Standard
OPT: Optional
① Torque required in inch lbs. to rotate ball joint when removed from the knuckle
② 2wd exc. PreRunner: Upper 4-30 inch lbs.; lower 0.8-30 inch lbs.
4wd and PreRunner: Upper 6-39 inch lbs.; lower 1-22 inch lbs.
Lower ball joint excessive play, all models: 0.020 inch
③ See driver's side door placard

22140_TACO_C0011

BRAKE SPECIFICATIONS

All measurements in inches unless noted

Year	Model	Brake Disc Original Thickness	Brake Disc Minimum Thickness	Brake Disc Maximum Runout	Brake Drum Diameter Original Inside Diameter	Brake Drum Diameter Maximum Machine Diameter	Minimum Lining Thickness Front	Minimum Lining Thickness Rear	Brake Caliper Bracket Bolts (ft. lbs.)	Brake Caliper Mounting Bolts (ft. lbs.)
2007	2WD	0.984	0.906	0.0020	10.00	10.08	0.039	—	80	27
	①	1.102	1.024	0.0020	10.00	10.08	—	0.039	91	—
2008	2WD	0.984	0.906	0.0020	10.00	10.08	0.039	—	80	27
	①	1.102	1.024	0.0020	10.00	10.08	—	0.039	91	—

① 4WD and PreRunner

22140_TACO_C0012

SCHEDULED MAINTENANCE INTERVALS

TOYOTA—TACOMA

TO BE SERVICED	TYPE OF	VEHICLE MILEAGE INTERVAL (x1000)																		
		5	10	15	20	25	30	35	40	45	50	55	60	65	70	75	80	85	90	95
Automatic transmission and differential fluid	S/I			✓			✓			✓			✓			✓			✓	
Ball joints and boots	S/I			✓			✓			✓			✓			✓			✓	
Brake linings, discs/drums, lines & hoses	S/I			✓			✓			✓			✓			✓			✓	
Charcoal canister	S/I												✓							
Drive belts	S/I						✓						✓						✓	
Driveshaft bushing (4WD)	L						✓						✓						✓	
Engine coolant	R						✓						✓						✓	
Engine oil & filter	R	✓	✓	✓	✓	✓	✓	✓	✓	✓	✓	✓	✓	✓	✓	✓	✓	✓	✓	✓
Exhaust pipes & mounts	S/I			✓			✓			✓			✓			✓			✓	
Fuel lines & connections, fuel tank vapor vent system hoses, fuel tank band	S/I						✓						✓						✓	
Fuel tank cap gasket	S/I						✓						✓						✓	
Halfshaft boots & flange bolts	S/I			✓			✓			✓			✓			✓			✓	
Limited slip differential fluid	R						✓						✓						✓	
Manual transmission and differential fluid	S/I						✓						✓						✓	
Non-platinum spark plugs	R						✓						✓						✓	
Platinum spark plugs	R												✓							
Propeller shaft (4WD)	L			✓			✓			✓			✓			✓			✓	
Propeller shaft bolts	S/I			✓			✓			✓			✓			✓			✓	
Rack and pinion assembly	S/I			✓			✓			✓			✓			✓			✓	
Rear wheel bearing	L						✓						✓						✓	
Rotate tires	S/I	✓	✓	✓	✓	✓	✓	✓	✓	✓	✓	✓	✓	✓	✓	✓	✓	✓	✓	✓
Steering linkage	S/I			✓			✓			✓			✓			✓			✓	
Valves	S/I												✓							

R: Replace S/I: Service or Inspect L: Lubricate

FREQUENT OPERATION MAINTENANCE (SEVERE SERVICE)

If a vehicle is operated under any of the following conditions it is considered severe service:

- Towing a trailer or using a camper or car-top carrier.

- Repeated short trips of less than 5 miles in temperatures below freezing.

- Excessive idling or low-speed driving for long distances as in heavy commercial use, such as delivery, taxi or police cars.

- Operating on rough, muddy or salt-covered roads.

- Operating on unpaved or dusty roads.

Oil filter: service or inspect every 5000 miles or 4 months, whichever occurs first.

Brake linings and discs or drums: service or inspect every 5000 miles or 4 months, whichever occurs first.

Steering linkage: service or inspect every 5000 miles or 4 months, whichever occurs first.

Ball joints and boots: service or inspect every 5000 miles or 4 months, whichever occurs first.

Brake discs & pads (front): service or inspect every 6000 miles.

Halfshaft boots: service or inspect every 5000 miles or 4 months. Retighten the flange bolts, whichever occurs first.

Body chassis bolts and nuts: service or inspect every 5000 miles or 4 months, whichever occurs first.

Transmission and differential fluid: replace every 15,000 miles or 12 months, whichever occurs first.

Transfer case and differential fluid: replace every 15,000 miles or 12 months, whichever occurs first.

Timing belt: replace every 60,000 miles or 48 months, whichever occurs first.

22140_TACO_C0013

PRECAUTIONS

Before servicing any vehicle, please be sure to read all of the following precautions, which deal with personal safety, prevention of component damage, and important points to take into consideration when servicing a motor vehicle:

- Never open, service or drain the radiator or cooling system when the engine is hot; serious burns can occur from the steam and hot coolant.
- Observe all applicable safety precautions when working around fuel. Whenever servicing the fuel system, always work in a well-ventilated area. Do not allow fuel spray or vapors to come in contact with a spark, open flame, or excessive heat (a hot drop light, for example). Keep a dry chemical fire extinguisher near the work area. Always keep fuel in a container specifically designed for fuel storage; also, always properly seal fuel containers to avoid the possibility of fire or explosion. Refer to the additional fuel system precautions later in this section.
- Fuel injection systems often remain pressurized, even after the engine has been turned **OFF**. The fuel system pressure must be relieved before disconnecting any fuel lines. Failure to do so may result in fire and/or personal injury.
- Brake fluid often contains polyglycol ethers and polyglycols. Avoid contact with the eyes and wash your hands thoroughly after handling brake fluid. If you do get brake fluid in your eyes, flush your eyes with clean, running water for 15 minutes. If eye irritation persists, or if you have taken brake fluid internally, IMMEDIATELY seek medical assistance.
- The EPA warns that prolonged contact with used engine oil may cause a number of skin disorders, including cancer. You should make every effort to minimize your exposure to used engine oil. Protective gloves should be worn when changing oil. Wash your hands and any other exposed skin areas as soon as possible after exposure to used engine oil. Soap and water, or waterless hand cleaner should be used.
- All new vehicles are now equipped with an air bag system, often referred to as a Supplemental Restraint System (SRS) or Supplemental Inflatable Restraint (SIR) system. The system must be disabled before performing service on or around system components, steering column, instrument panel components, wiring and sensors. Failure to follow safety and disabling procedures could result in accidental air bag deployment, possible personal injury and unnecessary system repairs.
- Always wear safety goggles when working with, or around, the air bag system. When carrying a non-deployed air bag, be sure the bag and trim cover are pointed away from your body. When placing a non-deployed air bag on a work surface, always face the bag and trim cover upward, away from the surface. This will reduce the motion of the module if it is accidentally deployed. Refer to the additional air bag system precautions later in this section.
- Clean, high quality brake fluid from a sealed container is essential to the safe and proper operation of the brake system. You should always buy the correct type of brake fluid for your vehicle. If the brake fluid becomes contaminated, completely flush the system with new fluid. Never reuse any brake fluid. Any brake fluid that is removed from the system should be discarded. Also, do not allow any brake fluid to come in contact with a painted surface; it will damage the paint.
- Never operate the engine without the proper amount and type of engine oil; doing so WILL result in severe engine damage.
- Timing belt maintenance is extremely important. Many models utilize an interference-type, non-freewheeling engine. If the timing belt breaks, the valves in the cylinder head may strike the pistons, causing potentially serious (also time-consuming and expensive) engine damage. Refer to the maintenance interval charts for the recommended replacement interval for the timing belt, and to the timing belt section for belt replacement and inspection.
- Disconnecting the negative battery cable on some vehicles may interfere with the functions of the on-board computer system(s) and may require the computer to undergo a relearning process once the negative battery cable is reconnected.
- When servicing drum brakes, only disassemble and assemble one side at a time, leaving the remaining side intact for reference.
- Only an MVAC-trained, EPA-certified automotive technician should service the air conditioning system or its components.

BRAKES

ANTI-LOCK BRAKE SYSTEM (ABS)

When there is a malfunction in the contact point of the terminals or installation problems with any parts, removal and reinstallation of the suspected problem parts may return the system to the normal condition either completely or temporarily.

In order to determine the location of the malfunction, be sure to check the conditions at the time the malfunction occurred, through data such as DTC and freeze frame data outputs. Record this information before disconnecting any connectors and removing or installing any parts.

Be sure to remove and install the brake actuator and each sensor with the ignition switch off, unless specified in the inspection procedures.

When removing and installing the brake actuator assembly and each sensor, be sure to check that the normal display is output during a test mode inspection and a DTC output inspection after installing all the parts.

GENERAL INFORMATION

See Figure 1.

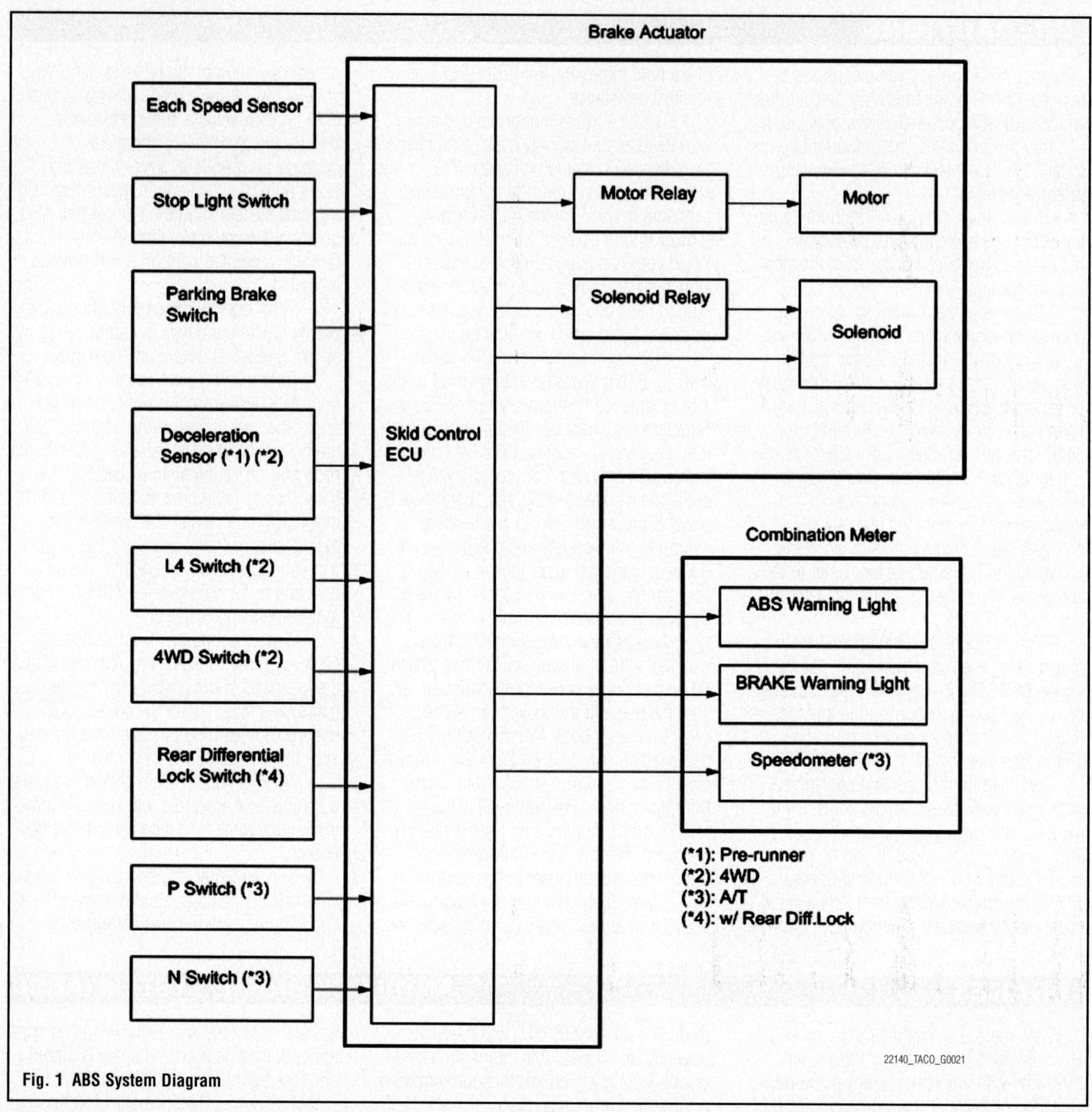

Fig. 1 ABS System Diagram

BLEEDING PROCEDURE

With VSC

Master Cylinder

➡Immediately wash off any brake fluid that comes into contact with any painted surfaces.

➡Depressing the brake pedal with the reservoir cap removed will cause the fluid to spray.

➡When bleeding, maintain the amount of fluid in the reservoir between the Min. and Max. lines.

1. Before servicing the vehicle, refer to the precautions in the beginning of this section.

Fill reservoir with dot3 brake fluid.

2. Turn the ignition switch to ON, and wait until the pump motor has stopped.
3. Turn the ignition switch to OFF, and depress the brake pedal more than 20 times.
4. Repeat 5 times.
5. Turn the ignition switch to ON, and check that the pump stops after approximately 8 to 14 seconds.

Front Brake Lines

➡Immediately wash off any brake fluid that comes into contact with any painted surfaces.

➡Depressing the brake pedal with the reservoir cap removed will cause the fluid to spray.

➡**When bleeding, maintain the amount of fluid in the reservoir between the Min. and Max. lines.**

1. Before servicing the vehicle, refer to the precautions in the beginning of this section.
2. Turn the ignition switch to ON, and wait until the pump motor has stopped.
3. (1) Connect the vinyl tube to the brake caliper.
4. Depress the brake pedal several times, then loosen the bleeder plug with the pedal held down.
5. At the point when the fluid stops coming out, tighten the bleeder plug, then release the brake pedal.
6. Repeat until all the air in the fluid has been bled out.
7. Repeat the above procedures to bleed the other brake line.

Rear Brakes

➡**Immediately wash off any brake fluid that comes into contact with any painted surfaces.**

➡**Depressing the brake pedal with the reservoir cap removed will cause the fluid to spray.**

➡**When bleeding, maintain the amount of fluid in the reservoir between the Min. and Max. lines.**

1. Before servicing the vehicle, refer to the precautions in the beginning of this section.
2. Connect the vinyl tube to the wheel cylinder.
3. Depress the brake pedal, hold it, and then loosen the bleeder plug.
4. Loosen the bleeder plug and release air.
5. When the air is completely bled out of the brake fluid through the bleeder plug, tighten the bleeder plug.
6. Repeat the above procedures to bleed the other brake line.

Without VSC

Master Cylinder

➡**Immediately wash off any brake fluid that comes into contact with any painted surfaces.**

➡**Depressing the brake pedal with the reservoir cap removed will cause the fluid to spray.**

➡**When bleeding, maintain the amount of fluid in the reservoir between the Min. and Max. lines.**

1. Before servicing the vehicle, refer to the precautions in the beginning of this section.
2. Fill reservoir with dot3 brake fluid.
3. Using SST 09023-00101, disconnect the brake lines from the master cylinder.
4. Slowly depress the brake pedal and hold it there.
5. Block the outer holes with your fingers, and release the brake pedal.
6. Repeat 3 or 4 times.

Brake Lines

1. Before servicing the vehicle, refer to the precautions in the beginning of this section.
2. Connect the vinyl tube to the bleeder plug.
3. Depress the brake pedal several times, then loosen the bleeder plug with the pedal held down.
4. At the point where the fluid stops coming out, tighten the bleeder plug, then release the brake pedal.
5. Repeat until all the air in the fluid has been bled out.
6. Repeat the above procedure to bleed the air out of the brake line for each wheel.
7. Check the fluid level and add fluid if necessary.

BRAKES

FRONT DISC BRAKES

⁂ CAUTION

Dust and dirt accumulating on brake parts during normal use may contain asbestos fibers from production or aftermarket brake linings. Breathing excessive concentrations of asbestos fibers can cause serious bodily harm. Exercise care when servicing brake parts. Do not sand or grind brake lining unless equipment used is designed to contain the dust residue. Do not clean brake parts with compressed air or by dry brushing. Cleaning should be done by dampening the brake components with a fine mist of water, then wiping the brake components clean with a dampened cloth. Dispose of cloth and all residue containing asbestos fibers in an impermeable container with the appropriate label. Follow practices prescribed by the Occupational Safety and Health Administration (OSHA) and the Environmental Protection Agency (EPA) for the handling, processing, and disposing of dust or debris that may contain asbestos fibers.

BRAKE CALIPER

REMOVAL & INSTALLATION

See Figures 2 and 3.

1. Before servicing the vehicle, refer to the Precautions Section.
2. Disconnect the negative battery cable from the battery.
3. Raise and support the vehicle safely.
4. Remove the wheels.
5. Disconnect the brake hose from the caliper by removing the union bolt and 2 gaskets. Plug the end of the hose to prevent loss of fluid.
6. Remove the bolts that attach the caliper to its mounting.
7. Lift the bottom of the caliper up and remove the caliper assembly.

To install:

8. Grease the caliper slides and bolts with lithium grease or equivalent. Install the caliper and secure with the bolts.
9. Connect the brake hose to the caliper, using 2 new washers. Make sure the flexible hose lock is securely in the lock hole of the caliper. Torque the union bolt to 22 ft. lbs. (30 Nm).
10. Fill the brake system to the proper level and bleed the brake system.
11. Install the tire and wheel assembly.
12. Top off the brake fluid level in the master cylinder. Check for leaks and proper brake operation.
13. Connect the negative battery cable to the battery.

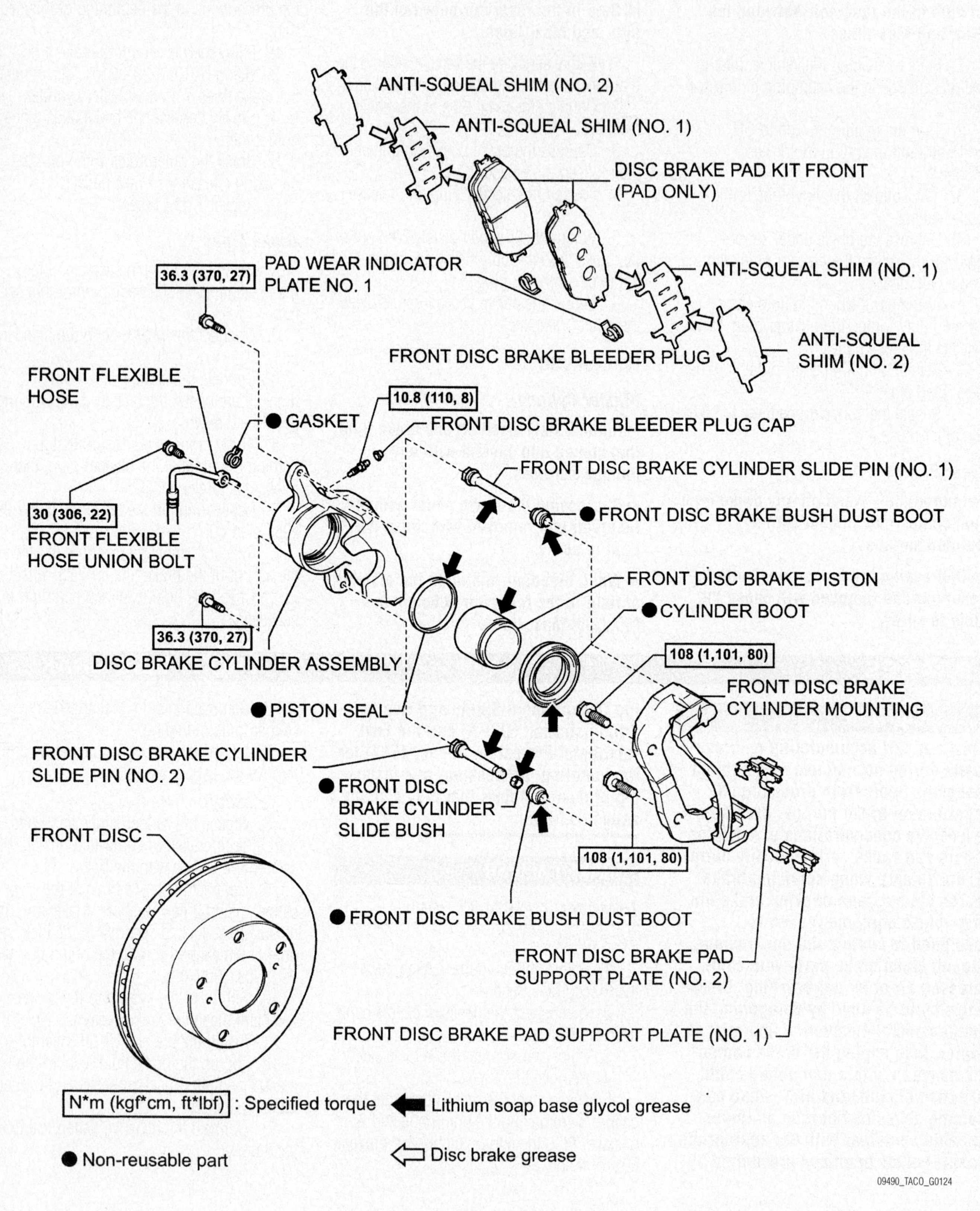

Fig. 2 Brake caliper and related components—2WD

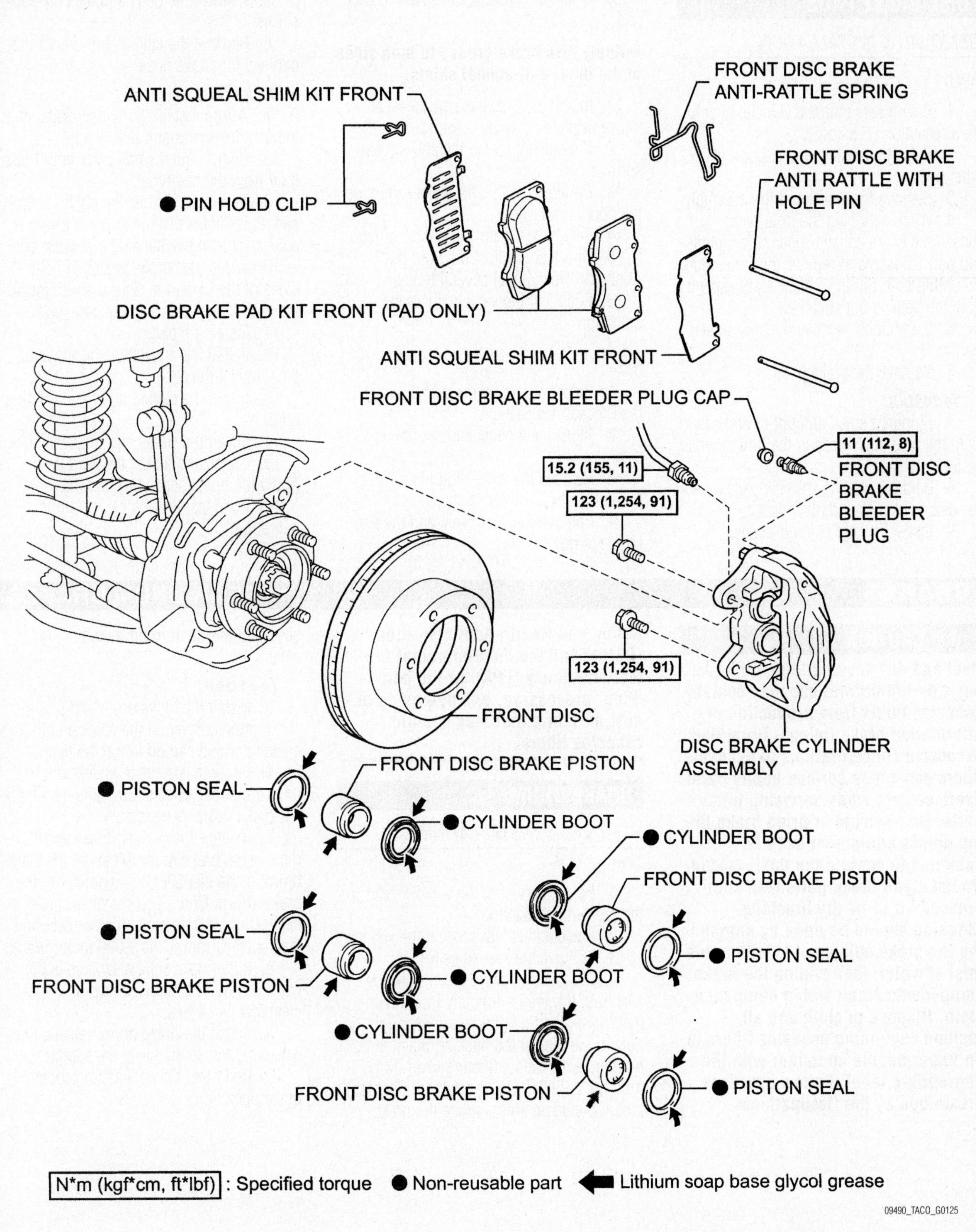

Fig. 3 Brake caliper and related components—4WD and PreRunner

DISC BRAKE PADS

REMOVAL & INSTALLATION

2WD

1. Before servicing the vehicle, refer to the Precautions Section.
2. Raise the vehicle and support it safely.
3. Remove the wheel and tire assembly.
4. When servicing the front pads, loosen the brake caliper upper side mounting bolt. Loosen and remove the lower side mounting bolt. Lift the caliper and suspend it so the hose is not stretched.
5. If equipped, remove the anti-squeal spring.
6. Remove the brake pads.

To install:

7. Siphon a small amount of brake fluid from the reservoir. Press in the brake caliper piston with the proper tool.
8. Before installing the new pads, check the disc thickness and disc runout.
9. Install the pad support plates.
10. Install the anti-squeal shims to each pad.

➡Apply disc brake grease to both sides of the inner anti-squeal shims.

11. Install the disc pads so the wear indicator plate is facing downward.
12. If removed, install the anti-squeal springs.
13. Carefully install the brake caliper so the boot is not wedged.
14. Install the wheel and tire assembly.
15. Check and adjust the fluid level. Apply the brake pedal several times.
16. Road test the vehicle for proper operation.

4WD

1. Before servicing the vehicle, refer to the Precautions Section.
2. Raise the vehicle and support it safely.
3. Remove the wheel and tire assembly.
4. Remove the clip, pins, and the anti-rattle spring.
5. Remove the pads and the anti-squeal shims.
6. Remove the caliper, but do not disconnect the brake hose.

To install:

7. Before installing the new pads, check the disc thickness and disc runout.
8. Siphon out a small amount of brake fluid from the reservoir.
9. Temporarily install the old inner brake pad. Press in the pistons with a C-clamp or equivalent. Remove the old inner brake pad.
10. Apply disc brake grease to both sides of the inner anti-squeal shim. Install the anti-squeal shims to the new pads.
11. Install the pads.
12. Install the anti-rattle springs and pins. Install the clip.
13. Install the caliper and the mounting bolts.
14. Install the wheel and tire assembly.
15. Check and adjust the fluid level. Apply the brake pedal several times.
16. Road test the vehicle for proper operation.

BRAKES

REAR DRUM BRAKES

⁂ CAUTION

Dust and dirt accumulating on brake parts during normal use may contain asbestos fibers from production or aftermarket brake linings. Breathing excessive concentrations of asbestos fibers can cause serious bodily harm. Exercise care when servicing brake parts. Do not sand or grind brake lining unless equipment used is designed to contain the dust residue. Do not clean brake parts with compressed air or by dry brushing. Cleaning should be done by dampening the brake components with a fine mist of water, then wiping the brake components clean with a dampened cloth. Dispose of cloth and all residue containing asbestos fibers in an impermeable container with the appropriate label. Follow practices prescribed by the Occupational Safety and Health Administration (OSHA) and the Environmental Protection Agency (EPA) for the handling, processing, and disposing of dust or debris that may contain asbestos fibers.

BRAKE DRUM

REMOVAL & INSTALLATION

See Figure 4.

1. Before servicing the vehicle, refer to the Precautions Section.
2. Raise and safely support the vehicle.
3. Remove the rear wheel(s).
4. Remove the brake drum from the axle hub. If there is difficulty in removing the drum, insert a suitable tool through the hole in the rear of the backing plate, and hold the automatic adjusting lever away from the adjuster. Using another suitable tool at the same time, reduce the brake shoe adjuster by turning the adjusting wheel.

To install:

5. Install the brake drum and pull the parking brake lever all the way up until a clicking sound can no longer be heard.
6. Verify that the rear wheels will not turn. If the rear wheels turn, adjust the parking brake cable as necessary.
7. Release the parking brake and remove the brake drum. Measure the brake drum inside diameter and diameter of the brake shoes. Check that the difference between the diameters is the correct shoe clearance. Clearance is 0.020 inch (5mm).
8. If the brake shoe clearance is not correct, adjust the brake shoes until the clearance is correct.
9. Install the brake drum, replace the wheel(s), and safely lower the vehicle.
10. Road-test the vehicle for proper brake operation.

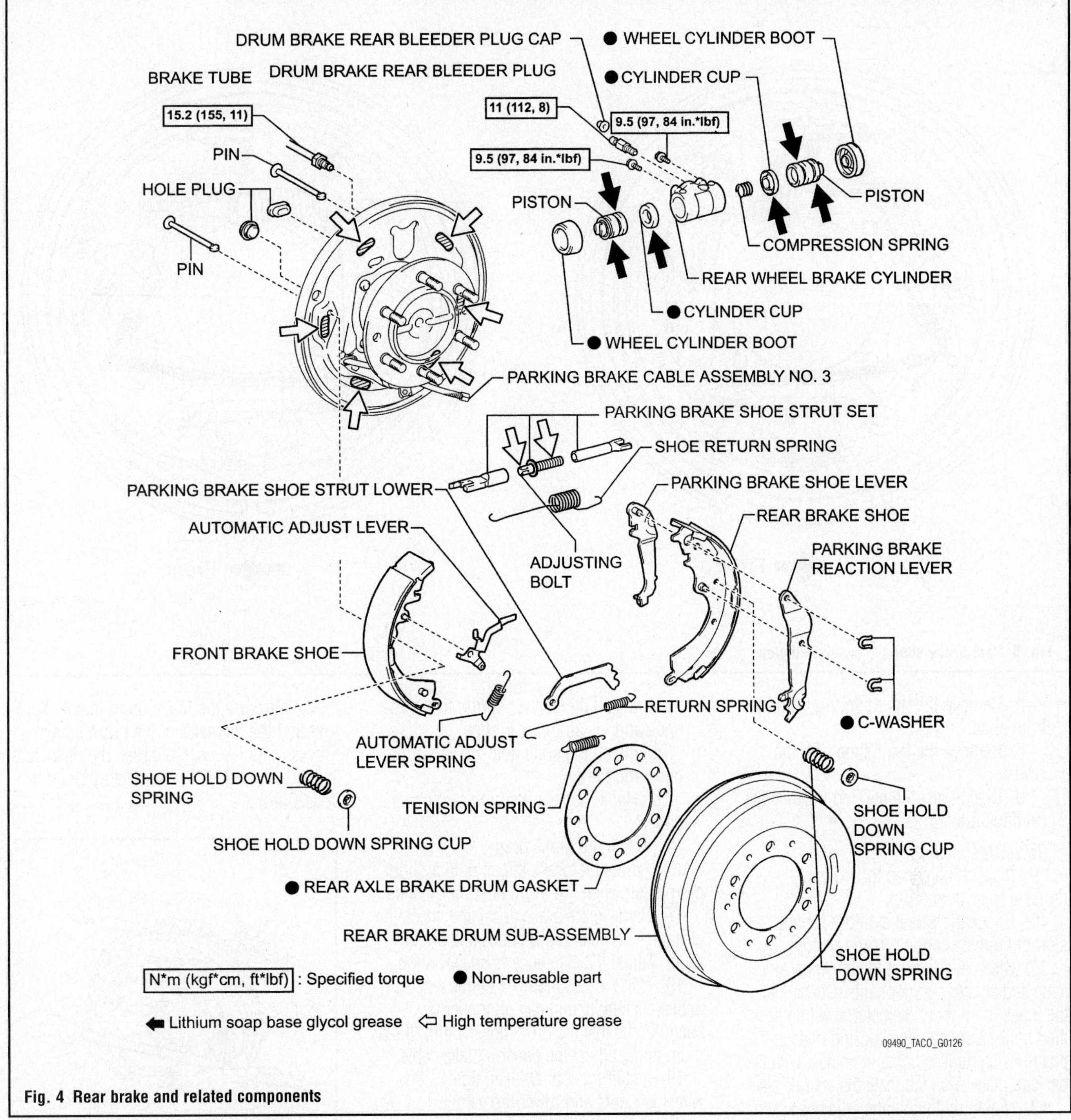

Fig. 4 Rear brake and related components

BRAKE SHOES

REMOVAL & INSTALLATION

See Figure 5.

1. Before servicing the vehicle, refer to the Precautions Section.
2. Loosen the rear wheel lug nuts slightly.
3. Raise and support the vehicle safely.
4. Remove the wheel lug nuts and the wheel.
5. Remove the brake drum.
6. If the drum is difficult to remove, perform the following:
    a. Insert a flat prying tool through the hole in the brake drum and hold the automatic adjusting lever away from the adjuster.
    b. Reduce the brake shoe adjustment by turning the adjuster bolt with a brake tool.
    c. The drum should now be loose enough to remove without much effort.
7. Remove the rear shoe.
    a. Carefully unhook the return spring from the brake shoe.
    b. Remove the shoe hold-down spring, cups and the pin.
    c. Disconnect the anchor spring from the rear shoe and remove the rear shoe.
    d. Disconnect the anchor spring from the front shoe.
8. Remove the front shoe.
    a. Remove the shoe hold-down spring, cups and pin.

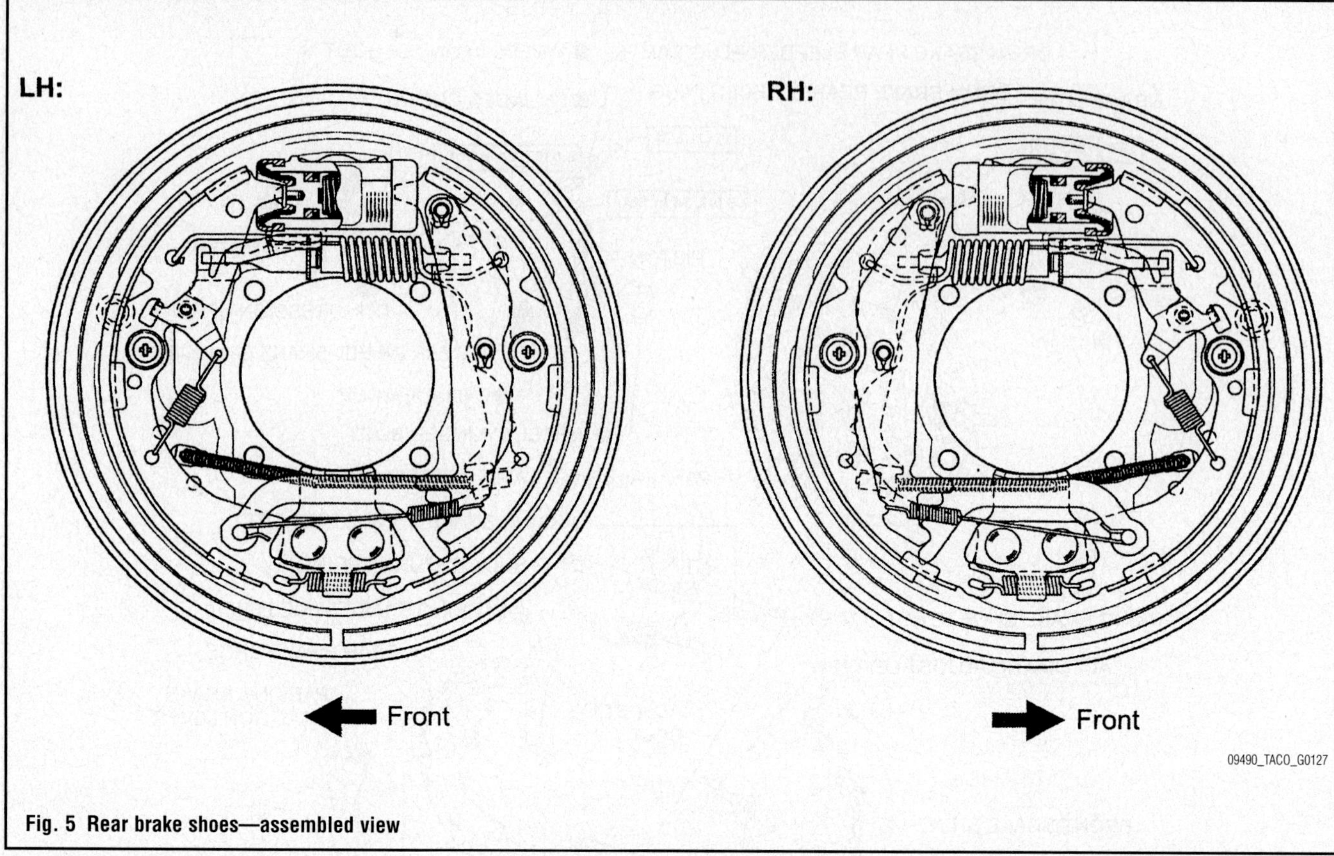

Fig. 5 Rear brake shoes—assembled view

b. Remove the return spring from the front shoe.

c. Remove the front shoe with the adjuster.

d. Disconnect the parking brake cable from the front shoe.

To install:

9. Inspect the shoes for signs of unusual wear or scoring.

10. Check the wheel cylinder for any sign of fluid seepage or frozen pistons.

11. Clean and inspect the brake backing plate and all other components. Check that the brake drum inner diameter is within specified limits. Lubricate the backing plate at the positions the brakes come in contact with the backing plate. Also lubricate the anchor plate.

12. Mount the automatic adjuster assembly onto a new rear brake shoe.

13. Install the front shoe.

a. Install the parking brake cable to the front shoe.

b. Install the front shoe with the adjuster.

c. Install the return spring to the front shoe.

d. Install the shoe hold-down spring, cups and pin.

14. Install the rear shoe.

a. Install the anchor spring to the front shoe.

b. Install the anchor spring to the rear shoe and install the rear shoe.

c. Install the shoe hold-down spring, cups and the pin.

d. Hook the return spring to the brake shoe.

15. Install the brake drum.

16. Adjust the brake shoes until a slight drag is felt when the drum is spun by hand.

17. Remove the brake drum and check the clearance between brake shoes and brake drum. Adjust the clearance to specification.

18. Pull the parking lever all the way up until a clicking sound can no longer be heard. Verify that the drum doesn't turn. If the drum turns, adjust the parking brake cable.

19. Install the rear wheels, tighten the wheel lug nuts and lower the vehicle.

20. Retighten the wheel lug nuts and pump the brake pedal a few times before moving the vehicle. Adjust the rear brakes again if necessary.

21. Check the level of brake fluid in the master cylinder, and then perform a test drive.

22. Connect the negative battery cable to the battery.

ADJUSTMENT

See Figure 6.

1. Before servicing the vehicle, refer to the precautions in the beginning of this section.

2. Measure the brake drum inside diameter and the diameter of the brake shoes. Check that difference between the diameters is the specified shoe clearance of 0.5 mm (0.020 in.).

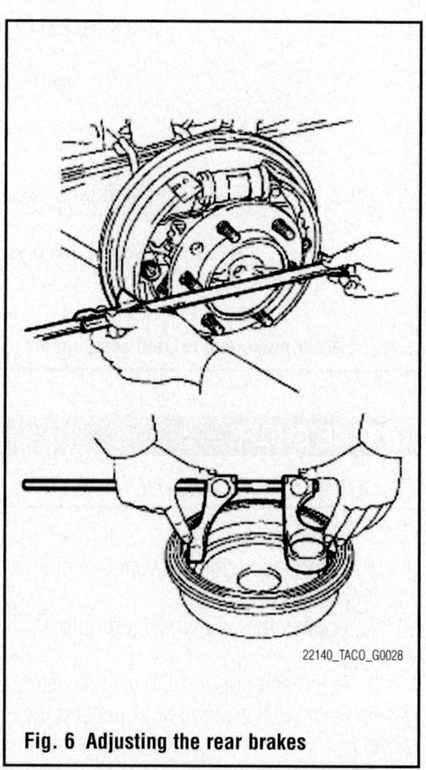

Fig. 6 Adjusting the rear brakes

BRAKES

PARKING BRAKE

PARKING BRAKE SHOES

REMOVAL & INSTALLATION

The rear drum brake shoes serve as the parking brakes. Refer to the procedures under Rear Drum Brakes.

CHASSIS ELECTRICAL

AIR BAG (SUPPLEMENTAL RESTRAINT SYSTEM)

GENERAL INFORMATION

CAUTION

Some vehicles are equipped with an air bag system. The system must be disarmed before performing service on, or around, system components, the steering column, instrument panel components, wiring and sensors. Failure to follow the safety precautions and the disarming procedure could result in accidental air bag deployment, possible injury and unnecessary system repairs.

SERVICE PRECAUTIONS

Disconnect and isolate the battery negative cable before beginning any airbag system component diagnosis, testing, removal, or installation procedures. Allow system capacitor to discharge for two minutes before beginning any component service. This will disable the airbag system. Failure to disable the airbag system may result in accidental airbag deployment, personal injury, or death.

Do not place an intact undeployed airbag face down on a solid surface. The airbag will propel into the air if accidentally deployed and may result in personal injury or death.

When carrying or handling an undeployed airbag, the trim side (face) of the airbag should be pointing towards the body to minimize possibility of injury if accidental deployment occurs. Failure to do this may result in personal injury or death.

Replace airbag system components with OEM replacement parts. Substitute parts may appear interchangeable, but internal differences may result in inferior occupant protection. Failure to do so may result in occupant personal injury or death.

Wear safety glasses, rubber gloves, and long sleeved clothing when cleaning powder residue from vehicle after an airbag deployment. Powder residue emitted from a deployed airbag can cause skin irritation. Flush affected area with cool water if irritation is experienced. If nasal or throat irritation is experienced, exit the vehicle for fresh air until the irritation ceases. If irritation continues, see a physician.

Do not use a replacement airbag that is not in the original packaging. This may result in improper deployment, personal injury, or death.

The factory installed fasteners, screws and bolts used to fasten airbag components have a special coating and are specifically designed for the airbag system. Do not use substitute fasteners. Use only original equipment fasteners listed in the parts catalog when fastener replacement is required.

During, and following, any child restraint anchor service, due to impact event or vehicle repair, carefully inspect all mounting hardware, tether straps, and anchors for proper installation, operation, or damage. If a child restraint anchor is found damaged in any way, the anchor must be replaced. Failure to do this may result in personal injury or death.

Deployed and non-deployed airbags may or may not have live pyrotechnic material within the airbag inflator.

Do not dispose of driver/passenger/curtain airbags or seat belt tensioners unless you are sure of complete deployment. Refer to the Hazardous Substance Control System for proper disposal.

Dispose of deployed airbags and tensioners consistent with state, provincial, local, and federal regulations.

After any airbag component testing or service, do not connect the battery negative cable. Personal injury or death may result if the system test is not performed first.

If the vehicle is equipped with the Occupant Classification System (OCS), do not connect the battery negative cable before performing the OCS Verification Test using the scan tool and the appropriate diagnostic information. Personal injury or death may result if the system test is not performed properly.

Never replace both the Occupant Restraint Controller (ORC) and the Occupant Classification Module (OCM) at the same time. If both require replacement, replace one, then perform the Airbag System test before replacing the other.

Both the ORC and the OCM store Occupant Classification System (OCS) calibration data, which they transfer to one another when one of them is replaced. If both are replaced at the same time, an irreversible fault will be set in both modules and the OCS may malfunction and cause personal injury or death.

If equipped with OCS, the Seat Weight Sensor is a sensitive, calibrated unit and must be handled carefully. Do not drop or handle roughly. If dropped or damaged, replace with another sensor. Failure to do so may result in occupant injury or death.

If equipped with OCS, the front passenger seat must be handled carefully as well. When removing the seat, be careful when setting on floor not to drop. If dropped, the sensor may be inoperative, could result in occupant injury, or possibly death.

If equipped with OCS, when the passenger front seat is on the floor, no one should sit in the front passenger seat. This uneven force may damage the sensing ability of the seat weight sensors. If sat on and damaged, the sensor may be inoperative, could result in occupant injury, or possibly death.

DISARMING THE SYSTEM

To avoid personal injury when working on vehicles equipped with an air bag, the negative battery cable must be disconnected and at least 90 seconds must elapse before working on the system. Failure to do so may result in deployment of the air bag.

ARMING THE SYSTEM

Reconnect the negative battery cable. Wait 2 minutes before performing any service.

DRIVETRAIN

AUTOMATIC TRANSMISSION ASSEMBLY

REMOVAL & INSTALLATION

A340E

See Figure 7.

1. Before servicing the vehicle, refer to the Precautions Section.
2. Disconnect the negative battery cable.
3. Remove the driveshaft.

➡Some vehicles are also equipped with a center bearing assembly, remove if equipped.

4. Drain the transmission fluid.
5. Remove the transmission dipstick. Remove the two bolts and remove the transmission oil filler tube assembly.
6. Remove the three bolts and clamps. Using tool SST09023-12701, disconnect the oil cooler outlet tube number one.
7. Remove the three bolts and clamps. Using tool SST09023-12701, disconnect the oil cooler inlet tube number one.
8. Remove the manifold stay.
9. On cable type, disconnect the transmission control cable assembly.
10. On floor shift (rod type), remove the clip and separate the floor shift gear shifting rod from the transmission.
11. Support the transmission assembly, with a transmission jack.
12. Remove the eight bolts and two front suspension member brackets.
13. Disconnect the heated oxygen sensor connector. Remove the four bolts from the frame crossmember subassembly number three.
14. Remove the four nuts, four bolts and frame crossmember number three.
15. Remove the starter.
16. Tilt the transmission assembly downward. Disconnect all electrical wires and hoses that will interfere with the removal of the transmission.

➡When tilting the transmission assembly downward, make sure that the cooling fan does not hit the radiator fan shroud.

17. Remove the flywheel housing dust shield. Remove the torque converter to flexplate retaining bolts.
18. Remove the transmission assembly retaining bolts. Remove the transmission from the vehicle.

To install:

19. Installation is the reverse of the removal procedure.
20. When installing the torque converter on to the transmission mainshaft, use a calipers and straight edge and measure between the transmission and the end of the surface of the driveplate. Specification should be 0.8772 inch.
21. Use a calipers and straight edge and measure the dimension "B" shown in the illustration and check that "B" is greater than "A". Specification should be 0.9165 inch or more.
22. Tighten the three upper transmission to engine retaining bolts to 52 ft. lbs. and the four lower bolts to 27 ft. lbs.
23. Tighten the torque converter to flexplate retaining bolts to 35 ft. lbs.
24. Be sure to fill the transmission with the proper grade and type transmission fluid.
25. Start the engine and check for leaks, correct as required.

➡Perform the automatic transmission initialization procedure when replacing the automatic transmission, engine or ECM.

26. On the A340E automatic transmission, initialization is completed by connecting and disconnecting the negative battery cable.

A750E

1. Before servicing the vehicle, refer to the Precautions Section.

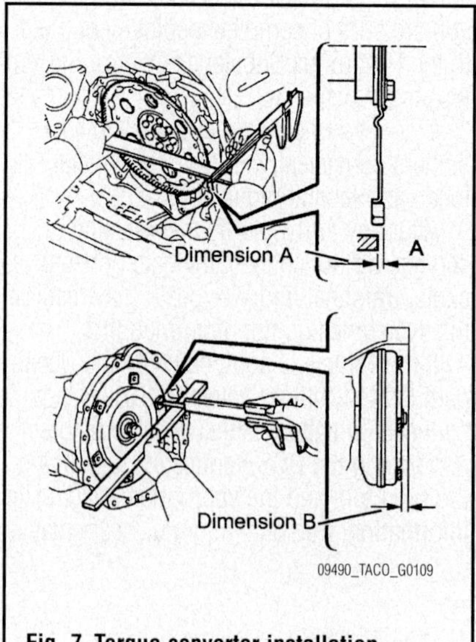

Fig. 7 Torque converter installation

2. Disconnect the negative battery cable.
3. Remove the number one engine undercover assembly. Remove the number two engine undercover assembly.
4. Disconnect the oxygen sensor. Remove the exhaust pipe assembly.
5. Remove the driveshaft and center bearing assembly.
6. Drain the transmission fluid. Remove the number one manifold stay. Remove the number two manifold stay.
7. Remove the three bolts and clamps. Using tool SST09023-12701, disconnect the oil cooler outlet tube number one.
8. Remove the three bolts and clamps. Using tool SST09023-12701, disconnect the oil cooler inlet tube number one.
9. Disconnect the transmission control cable assembly. Disconnect all electrical wires and hoses that will interfere with the removal of the transmission.
10. Remove the starter.
11. Remove the flywheel housing dust shield cover.
12. Support the transmission assembly, with a transmission jack.
13. Remove the four bolts from the frame crossmember subassembly number three.
14. Remove the four nuts, four bolts and frame crossmember number three.
15. Remove the torque converter to flexplate retaining bolts.
16. Remove the transmission assembly retaining bolts. Remove the transmission from the vehicle.

To install:

17. Installation is the reverse of the removal procedure.
18. When installing the torque converter on to the transmission mainshaft, use a calipers and straight edge and measure between the transmission and the end of the surface of the driveplate, dimension "A".
19. Use a calipers and straight edge and measure the dimension "B" shown in the illustration and check that "B" is greater than "A". Specification should be plus 1mm or more.
20. Tighten the five upper transmission to engine retaining bolts to 53 ft. lbs. and the four lower bolts to 27 ft. lbs.
21. Tighten the torque converter to flexplate retaining bolts to 35 ft. lbs.
22. Be sure to fill the transmission with the proper grade and type transmission fluid.

23. Start the engine and check for leaks, correct as required.

➡**Perform the automatic transmission initialization procedure when replacing the automatic transmission, engine or ECM.**

24. On the A750E automatic transmission, initialization is performed using the intelligent tester tool.
25. Turn the ignition switch OFF.
26. Connect the intelligent tester tool together with the controller area network vehicle interface module (CAN VIN) to the DLC3.
27. Turn the ignition switch to the ON position.
28. Push the intelligent tester tool main switch to the ON position.
29. Select the items, DIAGNOSIS/ ENHANCED OBD II.
30. Perform the reset memory procedure from the ENGINE menu.

➡**After performing the reset memory, be sure to perform the road test procedure. For road test procedure information, refer to the intelligent tester instruction manual.**

MANUAL TRANSMISSION ASSEMBLY

REMOVAL & INSTALLATION

1. Before servicing the vehicle, refer to the Precautions Section.
2. Disconnect the negative battery cable.
3. Remove the upper console rear panel subassembly.
4. Remove the rear console box assembly. Remove the front console box. Remove the shift lever boot assembly. Remove the floor shift lever assembly.
5. Remove the number two engine undercover subassembly.
6. Drain the transmission oil.
7. Remove the front exhaust pipe assembly.
8. Remove the front driveshaft, if equipped.
9. Remove the driveshaft.

➡**Some vehicles are also equipped with a center bearing assembly, remove if equipped.**

10. Remove the starter. Remove the manifold stay.
11. On five speed, separate the clutch release cylinder assembly by removing the two bolts.
12. On six speed, remove the number one clutch housing cover. Remove the clutch release cylinder assembly. Remove the clutch accumulator assembly.
13. Using the proper support equipment, support the transmission (2WD) or transmission/transfer case (4WD).
14. Remove the bolts from the engine rear mounting. Remove the bolts, nuts and frame crossmember number three sub assembly.
15. Remove the bolts and the engine mounting insulator rear number one.

➡**Be sure that the transmission (2WD) or transmission/transfer case (4WD) is properly supported.**

16. Tilt the transmission assembly downward. Disconnect all electrical wires and hoses that will interfere with the removal of the transmission (2WD) or transmission/transfer case (4WD).

➡**When tilting the transmission assembly downward, make sure that the cooling fan does not hit the radiator fan shroud.**

17. Remove the transmission assembly retaining bolts. Remove the transmission (2WD) or transmission/transfer case (4WD) from the vehicle.

➡**If equipped with 4WD, remove the transmission and the transfer case as one unit.**

18. On 4WD separate the transmission from the transfer case as required.

To install:

19. Installation is the reverse of the removal procedure.
20. On five speed, tighten the three upper transmission to engine retaining bolts to 53 ft. lbs. and the four lower bolts to 27 ft. lbs.
21. On five speed, tighten the four upper transmission to engine retaining bolts to 53 ft. lbs. and the five lower bolts to 28 ft. lbs.
22. Be sure to fill the transmission with the proper grade and type oil.
23. Start the engine and check for leaks, correct as required.

CLUTCH DRIVEN DISC & PRESSURE PLATE

REMOVAL & INSTALLATION

See Figure 8.

1. Before servicing the vehicle, refer to the Precautions Section.
2. Remove or disconnect the following:
    - Negative battery cable
    - Transmission assembly

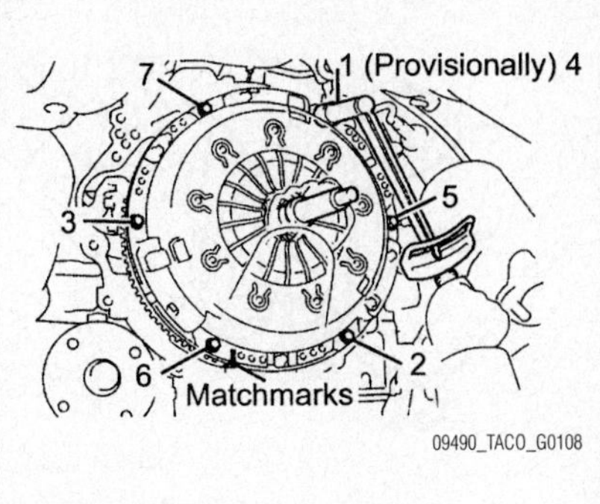

Fig. 8 Clutch cover bolt tightening sequence

3. Matchmark the clutch cover to the flywheel.
4. At the clutch cover, loosen each bolt 1 turn until spring tension is released.
5. Remove or disconnect the following:
    - Clutch cover set bolts and the clutch cover with the clutch disc.
    - Release bearing retaining clip and withdraw it
    - Release fork and boot assembly

To install:

6. Install or connect the following:
    - Clutch disc onto the flywheel, using a clutch disc alignment tool
    - Clutch cover, position it onto the flywheel and if reusing the old pressure plate, align the matchmarks.
    - Clutch cover. Tighten the bolts in a crisscross pattern to 14 ft. lbs. (19 Nm).
7. Lubricate the release fork pivot and contact points, the release bearing, bearing hub and input shaft spline surfaces with a suitable molybdenum disulfide lithium based or multi-purpose grease.
8. Install or connect the following:
    - Boot, release fork, hub and the bearing assemblies
    - Transmission
    - Negative battery cable

CLUTCH MASTER CYLINDER

REMOVAL & INSTALLATION

See Figures 9 and 10.

1. Before servicing the vehicle, refer to the precautions in the beginning of this section.
2. Drain clutch fluid.
3. Disconnect clutch reservoir tube.
4. Using SST 09023-00101, disconnect clutch master cylinder to flexible hose tube.
5. Remove the left front door scuff plate.

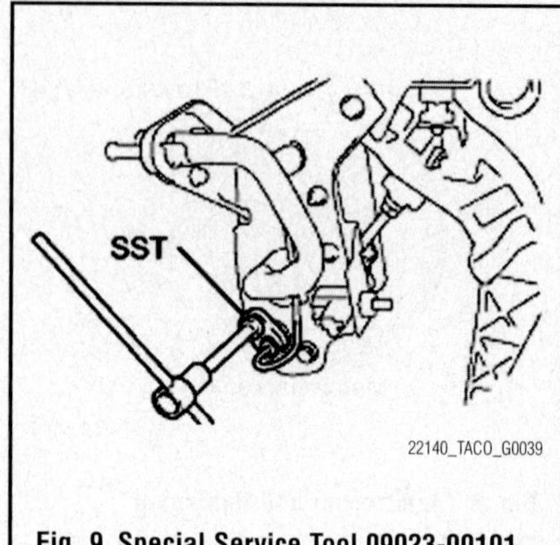

Fig. 9 Special Service Tool 09023-00101

6. Remove the left cowl side trim board.
7. Remove lower no. 1 instrument panel finish panel.
8. Separate driver side junction block.
9. Remove clutch pedal with clutch master cylinder.
10. Disconnect the clutch start switch connector.
11. Remove the 2 nuts and bolt, then remove the clutch pedal with clutch master cylinder.
12. Remove clutch master cylinder assembly.
13. Disconnect the clutch reservoir tube.
14. Using SST 09023-00101, disconnect the 2 way tube from the clutch master cylinder.
15. Remove the nut and clutch start switch from the clutch pedal support.
16. Remove the 2 bolts, then remove the clutch master cylinder.

To install:

17. Install the clutch master cylinder with the 2 bolts and tighten to 11 ft. lbs. (16 Nm).
18. Install the clutch start switch with the nut and tighten to 12 ft. lbs. (16 Nm).
19. Using SST 09023-00101, connect the clutch master cylinder to 2 way tube and tighten to 11 ft. lbs. (15 Nm).
20. Connect the clutch reservoir tube.
21. Install the clutch pedal with clutch master cylinder onto the vehicle with the 2 nuts and bolt. Tighten the nuts to 10 ft. lbs. (14 Nm) and the bolt to 18 ft. lbs. (24 Nm).
22. Connect the clutch start switch connector.
23. Connect driver side junction block.
24. Install lower no. 1 instrument panel finish panel.
25. Install cowl side trim board.
26. Install front door scuff plate.
27. Using SST 09023-00101, connect the flexible hose tube and tighten to 11 ft. lbs. (15 Nm).
28. Connect clutch reservoir tube.

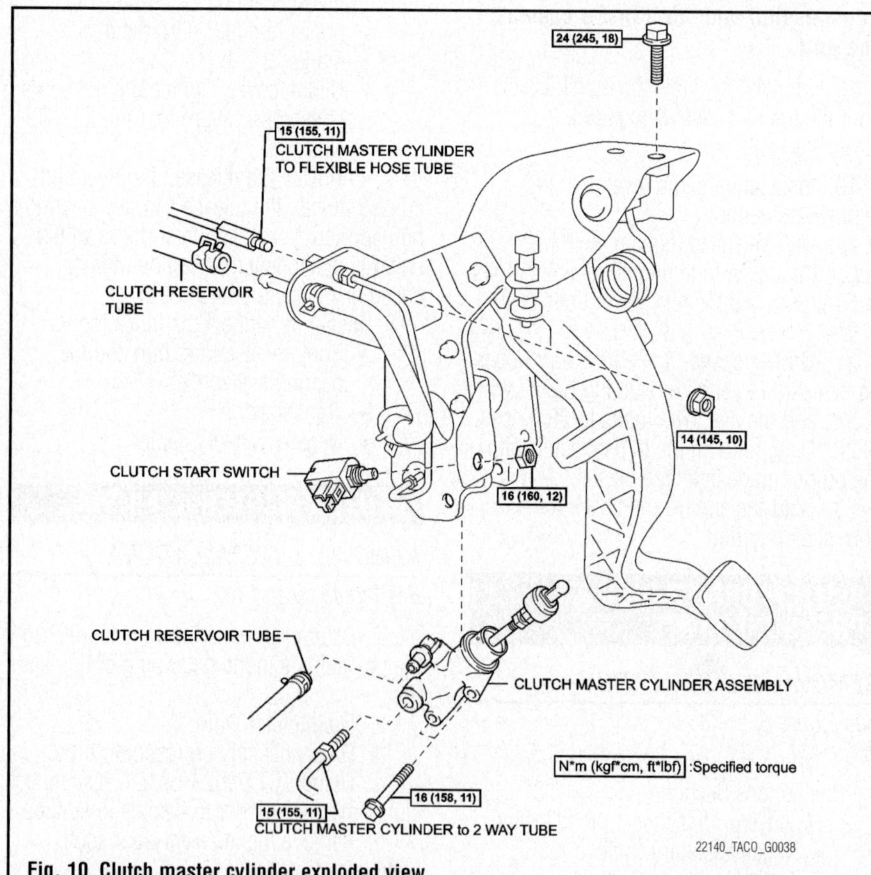

Fig. 10 Clutch master cylinder exploded view

29. Bleed clutch pipe line.
30. Inspect and adjust clutch pedal sub-assembly.
31. Check for clutch fluid leakage.
32. Check fluid level in reservoir.

CLUTCH SLAVE CYLINDER

REMOVAL & INSTALLATION

1. Before servicing the vehicle, refer to the precautions in the beginning of this section.
2. Drain clutch fluid
3. Disconnect clutch release cylinder to flexible hose tube
4. Using SST 09023-00101, disconnect the flexible hose tube.
5. Remove the 2 bolts and pull out the clutch release cylinder.

To install:

6. Install the clutch release cylinder with the 2 bolts and tighten to 104 inch lbs. (12 Nm).
7. Using SST 09023-00101, connect the flexible hose tube and tighten to 11ft. lbs. (15 Nm).
8. Fill the brake reservoir tank with brake fluid and bleed the clutch system.
9. Check for clutch fluid leakage.

CLUTCH HYDRAULIC SYSTEM BLEEDING

1. Before servicing the vehicle, refer to the Precautions Section.
2. Fill the clutch reservoir with brake fluid. Check the reservoir level frequently and add fluid as needed.
3. Connect one end of a vinyl tube to the bleeder plug on the slave cylinder and submerge the other end into a clear container half-filled with brake fluid.
4. Slowly pump the clutch pedal several times.
5. Have an assistant hold the clutch pedal down and loosen the bleeder plug until fluid and/or air starts to run out of the bleeder plug. Close the bleeder plug while the pedal is held to the floor.
6. Repeat Steps 2 and 3 until all the air bubbles are removed from the system.
7. Tighten the bleeder plug when all the air is gone.
8. Refill the master cylinder to the proper level as required.
9. Check the system for leaks.

TRANSFER CASE ASSEMBLY

REMOVAL & INSTALLATION

1. Before servicing the vehicle, refer to the Precautions Section.

2. Disconnect the negative battery cable.
3. Drain the transfer case oil.
4. Remove the four bolts and remove the transfer case lower protector.
5. Remove the transmission assembly.
6. Remove the eight transfer adaptor rear mounting bolts. Pull the transfer case straight up and remove it from the transmission.

➡Take care not to damage the adaptor rear oil seal with the transfer case input gear spline.

To install:

7. Installation is the reverse of the removal procedure.
8. Tighten the eight transfer case mounting bolts to 17 ft. lbs.
9. Be sure to fill the transfer case with the proper grade and type transmission fluid.
10. Start the engine and check for leaks, correct as required.

FRONT HALFSHAFT

REMOVAL & INSTALLATION

See Figure 11.

1. Before servicing the vehicle, refer to the Precautions Section.
2. Disconnect the negative battery cable.

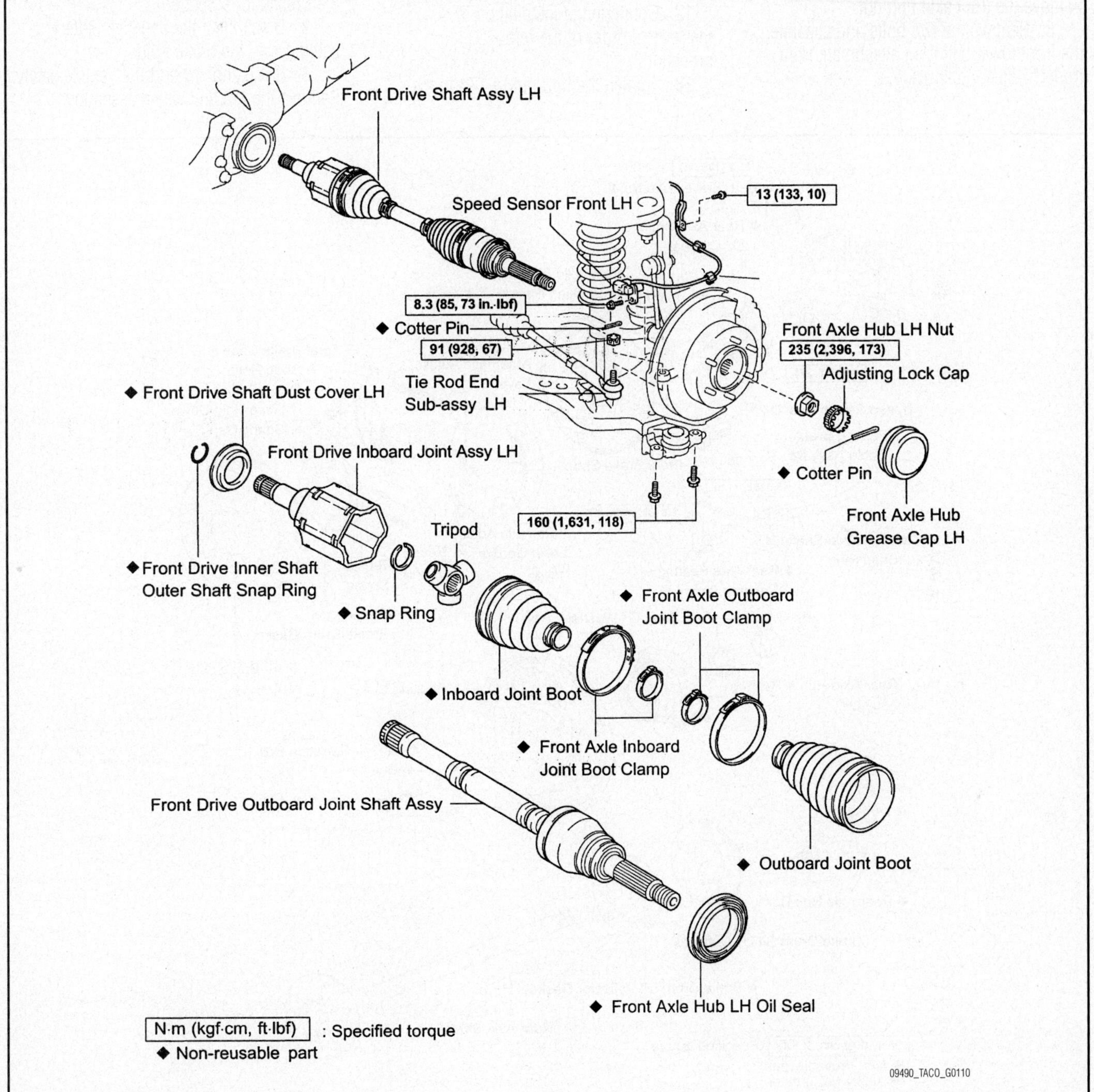

Fig. 11 Halfshaft and related components

3. Raise and support the vehicle safely. Remove the tire and wheel assembly.
4. Drain the differential.
5. Remove the bolt and separate the front speed sensor. Disengage the two clamps. Remove the bolt and separate the speed sensor wire harness from the steering knuckle.
6. Remove the cotter pin and nut. Using tool SST09628-62011 or equivalent, separate the tie rod end from the steering knuckle.
7. Using a suitable tool and hammer, remove the front axle hub grease cap. Remove the cotter pin and adjusting cap. Remove the front axle hub nut.
8. Remove the two bolts and separate the front lower ball joint attachment front from the steering knuckle.
9. Using tool SST09520-01010, and SST09520-24010 remove the halfshaft. Be careful not to damage the oil seal.

To install:

10. Coat the spline of the inboard joint shaft with clean ATF.
11. Align the shaft splines and install the halfshaft.

➡Set the snapring with the opening side facing downward. Be careful not to damage the oil seal.

12. Continue the installation in the reverse order of the removal procedure.
13. Tighten the hub nut to 173 ft. lbs.
14. Be sure to fill the differential with the proper type and grade lubricant.
15. Check and correct leaks, as required.
16. Check and adjust the alignment, as required.

REAR AXLE SHAFT, BEARING & SEAL

REMOVAL & INSTALLATION

See Figures 12 and 13.

1. Before servicing the vehicle, refer to the Precautions Section.
2. Disconnect the negative battery cable. Drain the brake fluid.
3. Raise and support the vehicle safely. Remove the tire and wheel assembly.

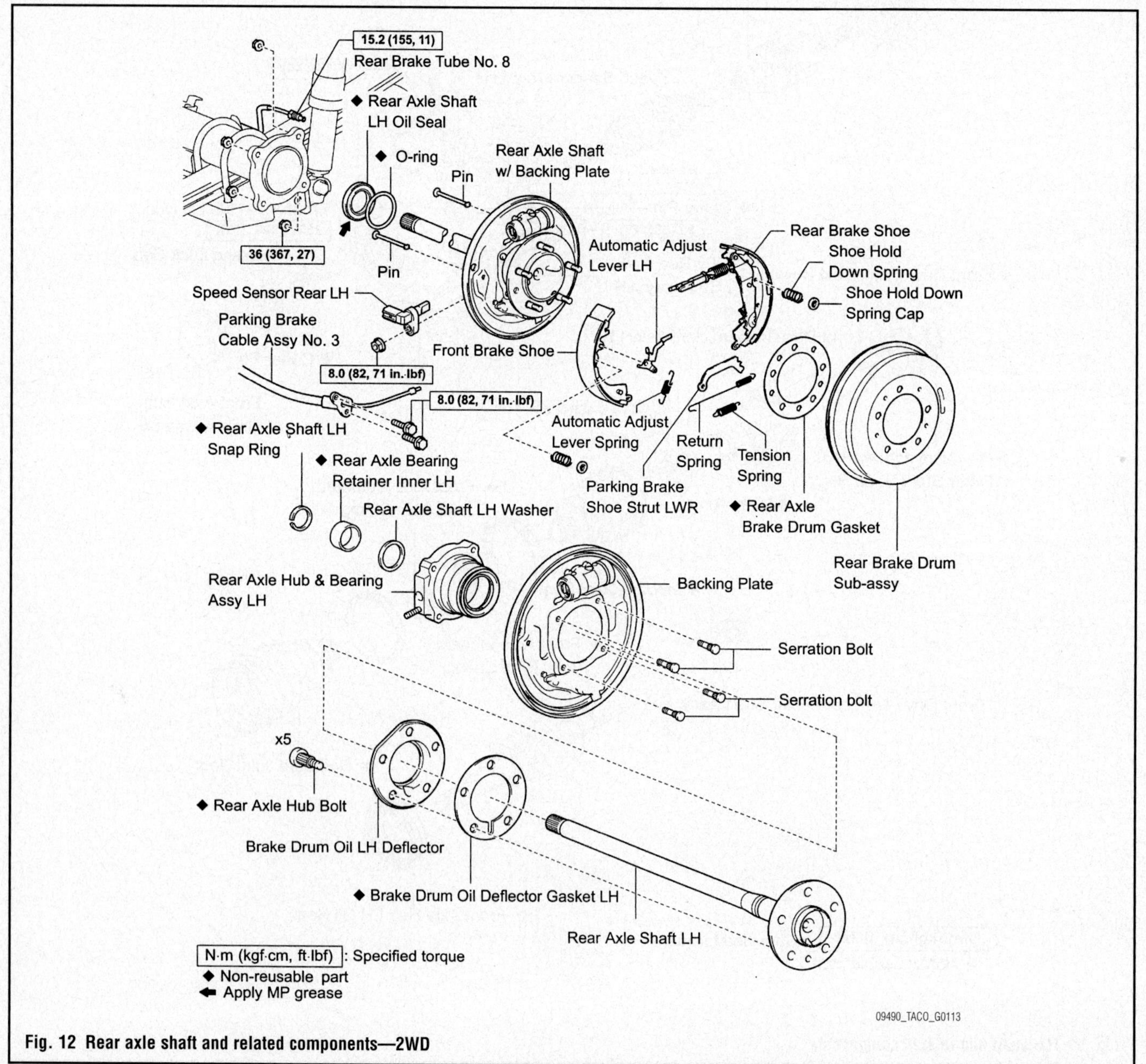

Fig. 12 Rear axle shaft and related components—2WD

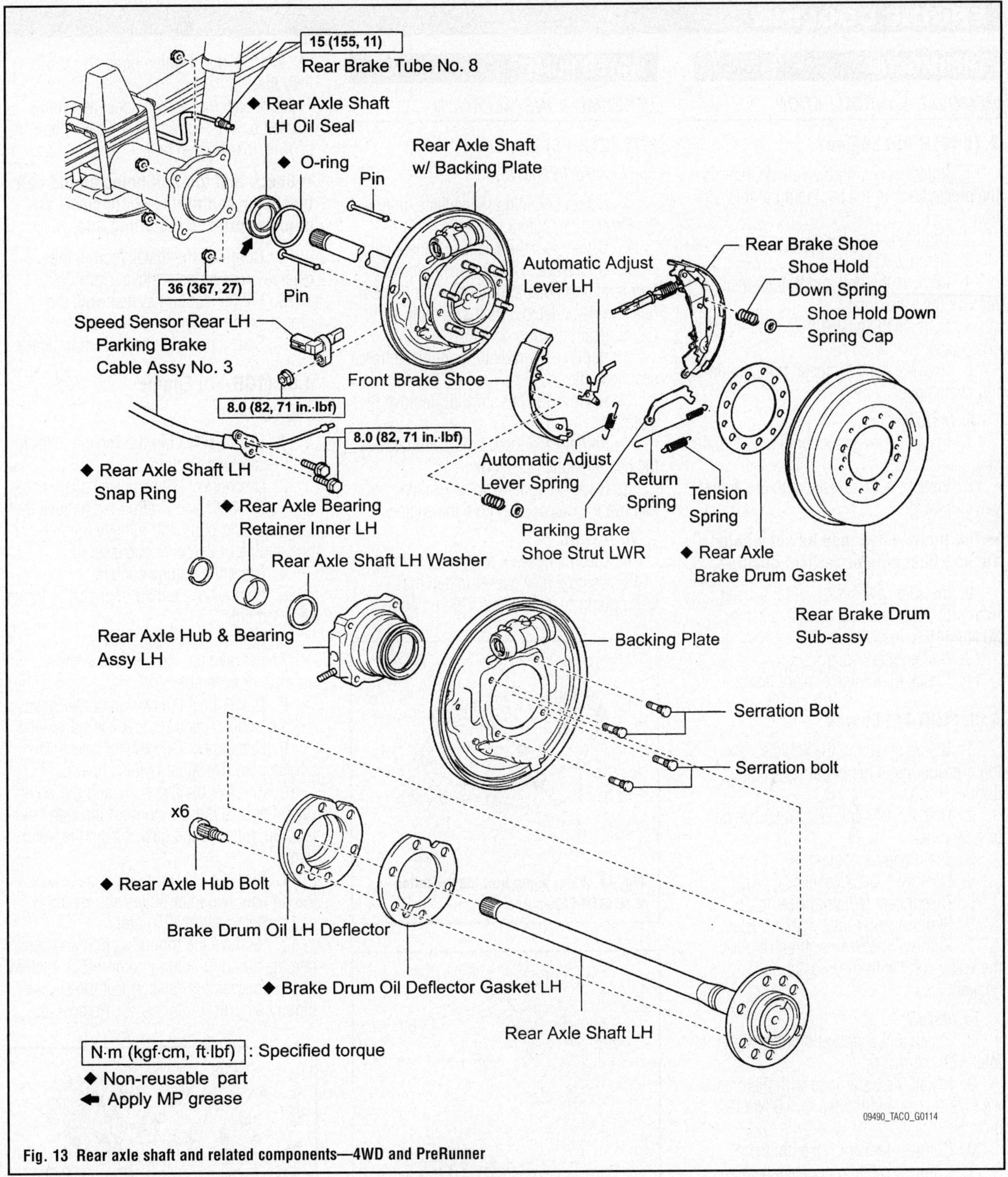

Fig. 13 Rear axle shaft and related components—4WD and PreRunner

4. Remove the brake drum. Remove the brake shoes.

5. Remove the rear speed sensor.

6. Remove the two bolts and disconnect the parking brake cable from the backing plate. Disconnect the brake line at the brake backing plate.

7. Remove the four nuts and the rear axle shaft and backing plate. Remove the O-ring.

8. Remove the rear axle shaft oil seal using tool SST09308-00010.

To install:

9. Installation is the reverse of the removal procedure.

10. Be sure to fill the master cylinder with the proper grade and type brake fluid.

11. Bleed the brakes, as required.

ENGINE COOLING

THERMOSTAT

REMOVAL & INSTALLATION

2.7L (2TR-FE) Engine

1. Before servicing the vehicle, refer to the precautions in the beginning of this section.
2. Drain engine coolant.
3. Remove water inlet.
4. Remove the bolt and 2 nuts, then remove the water inlet.
5. Remove the gasket from the timing chain cover.
6. Remove the thermostat from the timing chain cover.

To install:

7. Install a new gasket onto the thermostat.
8. Install the thermostat with the jiggle valve upward.

➡The jiggle valve may be set within 10° to either side of vertical position.

9. Install a new gasket and the water inlet with the bolt and 2 nuts. Tighten to 20 Nm (15 ft. lbs.).
10. Add engine coolant.
11. Check for engine coolant leakage.

4.0L (1GR-FE) Engine

1. Before servicing the vehicle, refer to the precautions in the beginning of this section.
2. Disconnect cable from negative battery terminal.
3. Drain engine coolant.
4. Remove v-bank cover.
5. Disconnect radiator hose no. 2
6. Remove water inlet w/thermostat
7. Remove the 3 nuts, then remove the water inlet with thermostat and gasket.

To install:

8. Install a new gasket onto the water inlet with thermostat.
9. Install the water inlet with thermostat with the 3 nuts and tighten to 9.0 Nm (80 inch lbs.)
10. Connect radiator hose no. 2.
11. Connect cable to negative battery terminal.
12. Add engine coolant.
13. Check for engine coolant leakage.
14. Install v-bank cover.
15. Install the V-bank cover with the 2 nuts.

WATER PUMP

REMOVAL & INSTALLATION

2.7L (2TR-FE) Engine

See Figures 14 and 15.

1. Before servicing the vehicle, refer to the Precautions Section.
2. Disconnect the negative battery cable.
3. On 4WD and PreRunner, remove the four retaining bolts and remove the number one engine undercover subassembly.
4. Drain the engine coolant.
5. Remove the radiator support to frame seal, left side.
6. Remove the fan shroud. Remove the alternator.
7. Remove the three bolts, and remove the belt tensioner assembly.
8. Remove the water pump retaining bolts. Remove the water pump from the engine.

To install:

9. Clean all gasket mounting surfaces.
10. Using a new gasket install the water pump to the engine. Tighten bolts "A" to 15 ft. lbs. Tighten bolts "B" to 80 inch lbs.
11. Install the belt tensioner assembly. Tighten bolt "B" to 30 ft. lbs. Tighten bolt "A" to 16 ft. lbs. Tighten bolt "C" to 32 ft. lbs.

➡Check that the bolt holes on the belt tensioner and timing chain cover are aligned, prior to installing bolt "C".

12. Continue the installation in the reverse order of the removal procedure.
13. Fill the cooling system with the proper grade and type engine coolant.
14. Start the engine and check for leaks.

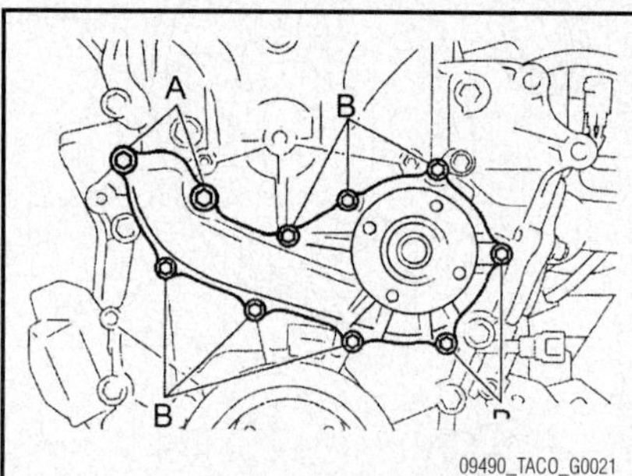

Fig. 14 Water pump bolt identification—2.7L (2TR-FE) engine

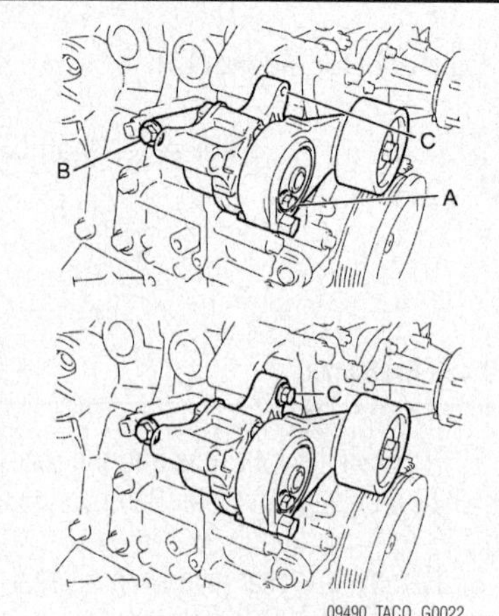

Fig. 15 Belt tensioner bolt identification and location—2.7L (2TR-FE) engine

4.0L (1GR-FE) Engine

See Figure 16.

1. Before servicing the vehicle, refer to the Precautions Section.
2. Disconnect the negative battery cable.
3. On 4WD and PreRunner, remove the four retaining bolts and remove the number one engine undercover subassembly.
4. Drain the engine coolant.
5. Remove the radiator support to frame seal, left side.
6. Remove the V bank cover.
7. Remove the fan shroud. Remove the air cleaner assembly.
8. Disconnect the two oil cooler hoses (with oil cooler) and remove the water inlet.
9. Disconnect the radiator hoses. Disconnect the five water bypass hoses.
10. Remove the five bolts and the water inlet. Remove the O-ring from the water outlet pipe. Remove the gasket from the water pump.
11. Remove the two bolts and remove the number two idler pulley subassembly.
12. Remove the alternator.
13. Remove the mounting bolt and separate the air conditioning compressor suction hose subassembly. Disconnect the air condition compressor connector. Remove the

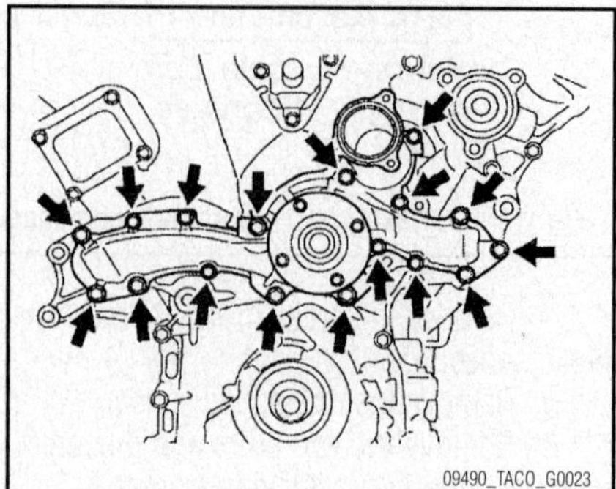

Fig. 16 Water pump bolt identification—4.0L (1GR-FE) engine

four bolts and separate the compressor from the belt tensioner assembly.

14. Remove belt tensioner assembly.

15. Remove the water pump retaining bolts. Remove the water pump from the engine.

To install:

16. Clean all gasket mounting surfaces.

17. Using a new gasket install the water pump to the engine. Tighten bolts "A" to 80 inch lbs. Tighten bolts "B" to 17 ft. lbs.

18. Install the belt tensioner assembly.

19. Continue the installation in the reverse order of the removal procedure.

20. Fill the cooling system with the proper grade and type engine coolant.

21. Start the engine and check for leaks.

ENGINE ELECTRICAL — CHARGING SYSTEM

ALTERNATOR

REMOVAL & INSTALLATION

On some vehicles, the alternator is mounted very low on the engine. It may be necessary to remove the gravel shield and work from beneath the vehicle in order to gain access to the alternator. Replacing the alternator while the engine is cold is recommended.

1. Before servicing the vehicle, refer to the Precautions Section.
2. Disconnect the negative battery cable.
3. Remove the V bank cover.
4. Remove the radiator support to frame seal, left side.
5. Remove the radiator fan shroud.
6. Disconnect the alternator harness wiring.
7. Remove the drive belt.
8. Remove the alternator retaining bolts.
9. Remove the alternator from the vehicle.

To install:

10. Installation is the reverse of the removal procedure.
11. Tighten the alternator retaining bolts to 32 ft. lbs. (43 Nm.).

ENGINE ELECTRICAL — DISTRIBUTORLESS IGNITION SYSTEM

IGNITION COIL

REMOVAL & INSTALLATION

➡It is a good idea to remove and reinstall the coils one at a time to prevent the coils being installed out of order.

1. Disconnect the negative battery cable.
2. Disconnect the spark plug wire from the coil.
3. Disconnect the electrical connectors from the coil.
4. Remove the coil bolts and the coils.

To install:

5. Install the coil and tighten the bolts to 69 inch lbs. (8 Nm).
6. Connect the wiring and spark plug wires
7. Connect the negative battery cable.

FIRING ORDERS

See Figures 17 and 18.

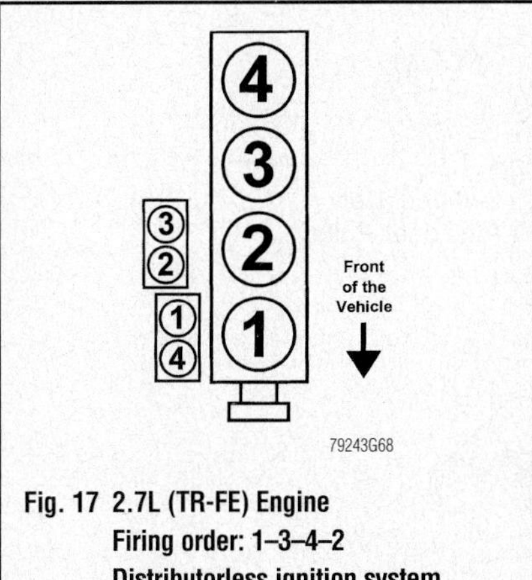

Fig. 17 2.7L (TR-FE) Engine Firing order: 1–3–4–2 Distributorless ignition system

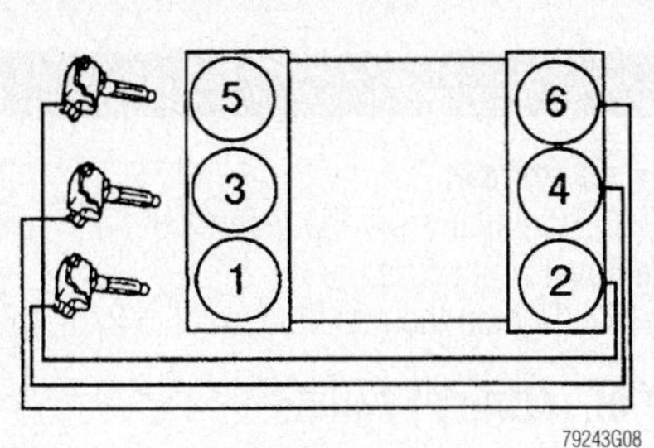

Fig. 18 4.0L (1GR-FE) Engine Firing order: 1–2–3–4–5–6 Distributorless ignition system

IGNITION TIMING

INSPECTION

2.7L (2TR-FE) Engine

See Figure 19.

With Intelligent Tester

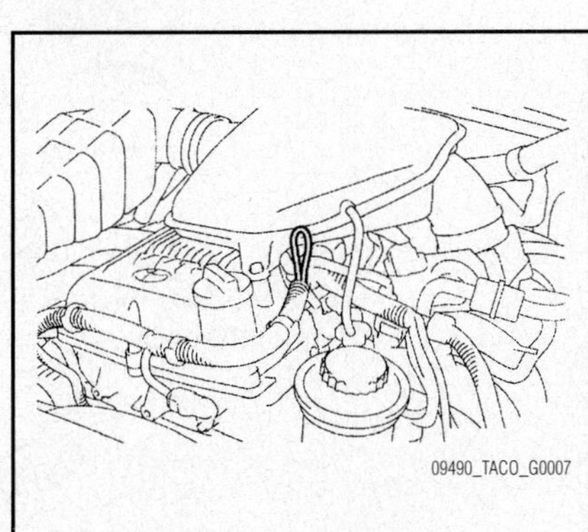

Fig. 19 Wiring harness location—2.7L (2TR-FE) engine

1. Connect the tool to the DLC3.
2. Turn the ignition switch ON.
3. Turn the tester ON.
4. Start the engine and allow it to warm up.
5. Select the following menu items: DIAGNOSIS/ENHANCED OBDII/DATA LIST/PRIMARY/IGN ADVANCE. Check that the ignition timing advances immediately when the engine speed is increased.

Without Intelligent Tester

1. Turn the ignition switch ON.
2. Start the engine and allow it to warm up.
3. Install the timing light.

➡Be sure to use a timing light that detects the first signal. After checking be sure to wrap the wire harness with tape.

4. Using too SST09843-18040, connect terminals 13 (TC) and 4 (CG) of the DLC3.

➡Be careful not to connect the wrong terminals, as engine damage may occur.

5. Check that the ignition timing is within specification. Be sure all electrical systems are OFF.

4.0L (1GR-FE) Engine

See Figure 20.

With Intelligent Tester

1. Warm up the engine.
2. Connect the tool to the DLC3.
3. Select the following menu items:

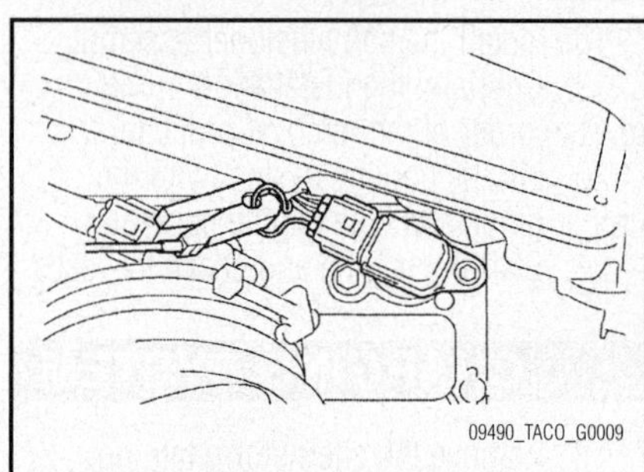
Fig. 20 Wiring harness location—4.0L (1GR-FE) engine

DIAGNOSIS/ENHANCED OBDII/DATA LIST/PRIMARY/IGN ADVANCE.

4. Inspect the ignition timing during idling.
5. Check that the ignition timing advances immediately when the engine speed is increased.

Without Intelligent Tester

1. Using too SST09843-18040, connect terminals 13 (TC) and 4 (CG) of the DLC3.

➡Be careful not to connect the wrong terminals, as engine damage may occur.

2. Remove the air cleaner. Pull out the wire harness.
3. Connect the test probe of the timing light to the wire of the ignition coil connector for the No. 1 cylinder.

➡Be sure to use a timing light that detects the first signal. After checking be sure to wrap the wire harness with tape.

4. Check the ignition timing during idling. Check that the ignition timing is within specification. Be sure all electrical systems are OFF.

ADJUSTMENT

All engines use a distributorless ignition system referred to as Direct Ignition System (DIS). All spark advance is permanently set by the PCM.

SPARK PLUGS

REMOVAL & INSTALLATION

1. Remove the ignition coils.
2. Using a 16 mm plug wrench, remove the spark plugs.
3. Clean the spark plugs.
4. If the electrode has traces of wet carbon, allow it to dry and then clean with a spark plug cleaner.
5. Check the spark plug for thread damage and insulator damage. If abnormal, replace the spark plug.
6. Adjust the spark plug electrode gap. Electrode gap for new spark plug is 1.0 to 1.1 mm (0.039 to 0.043 in.).
7. Using a 16 mm plug wrench, install the spark plugs and tighten to 17.5 Nm (13 ft. lbs.)
8. Reinstall the ignition coils.

ENGINE ELECTRICAL

STARTING SYSTEM

STARTER

REMOVAL & INSTALLATION

2.7L (TR-FE) Engine

1. Before servicing the vehicle, refer to the Precautions Section.
2. Disconnect the negative battery cable.
3. Remove the terminal cap. Disconnect the electrical connections.
4. Disconnect the positive battery cable.
5. Remove the starter retaining bolts. Remove the starter from the vehicle.

To install:

6. Installation is the reverse of the removal procedure.
7. Tighten the retaining bolts to 27 ft. lbs.

4.0L (1GR-FE) Engine

1. Before servicing the vehicle, refer to the Precautions Section.
2. Disconnect the negative battery cable.
3. Remove the engine undercover assembly.
4. On 2WD and PreRunner, remove the number two manifold stay.
5. On 4WD vehicles, remove the number two exhaust front pipe assembly. Remove the five clips and then remove the front fender splash shield, left side. Remove the number two steering intermediate shaft.
6. Disconnect the starter electrical connectors.
7. Disconnect the positive battery cable.
8. Remove the starter retaining bolts. Remove the starter from the vehicle.

To install:

9. Installation is the reverse of the removal procedure.
10. Tighten the retaining bolts to 27 ft. lbs.

ENGINE MECHANICAL

ACCESSORY DRIVE BELTS

ACCESSORY BELT ROUTING

See Figures 21 and 22.

INSPECTION

Inspect the drive belt for signs of glazing or cracking. A glazed belt will be perfectly smooth from slippage, while a good belt will have a slight texture of fabric visible. Cracks will usually start at the inner edge of the belt and run outward. All worn or damaged drive belts should be replaced immediately.

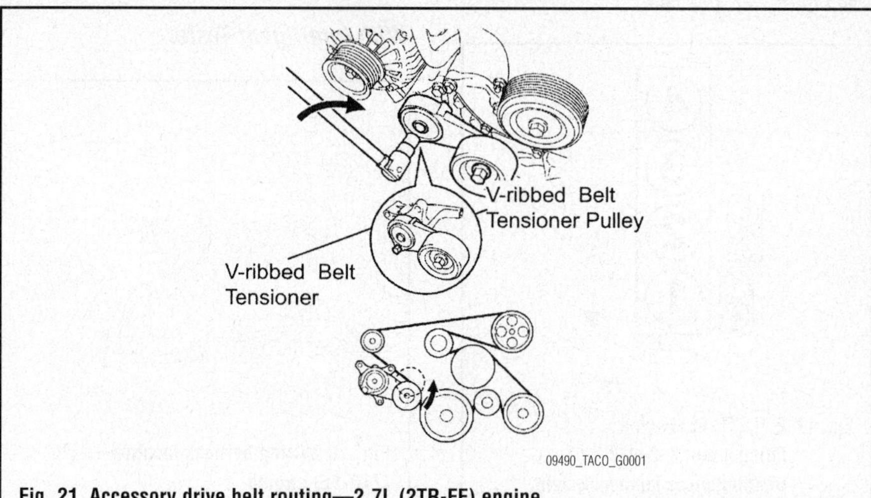

Fig. 21 Accessory drive belt routing—2.7L (2TR-FE) engine

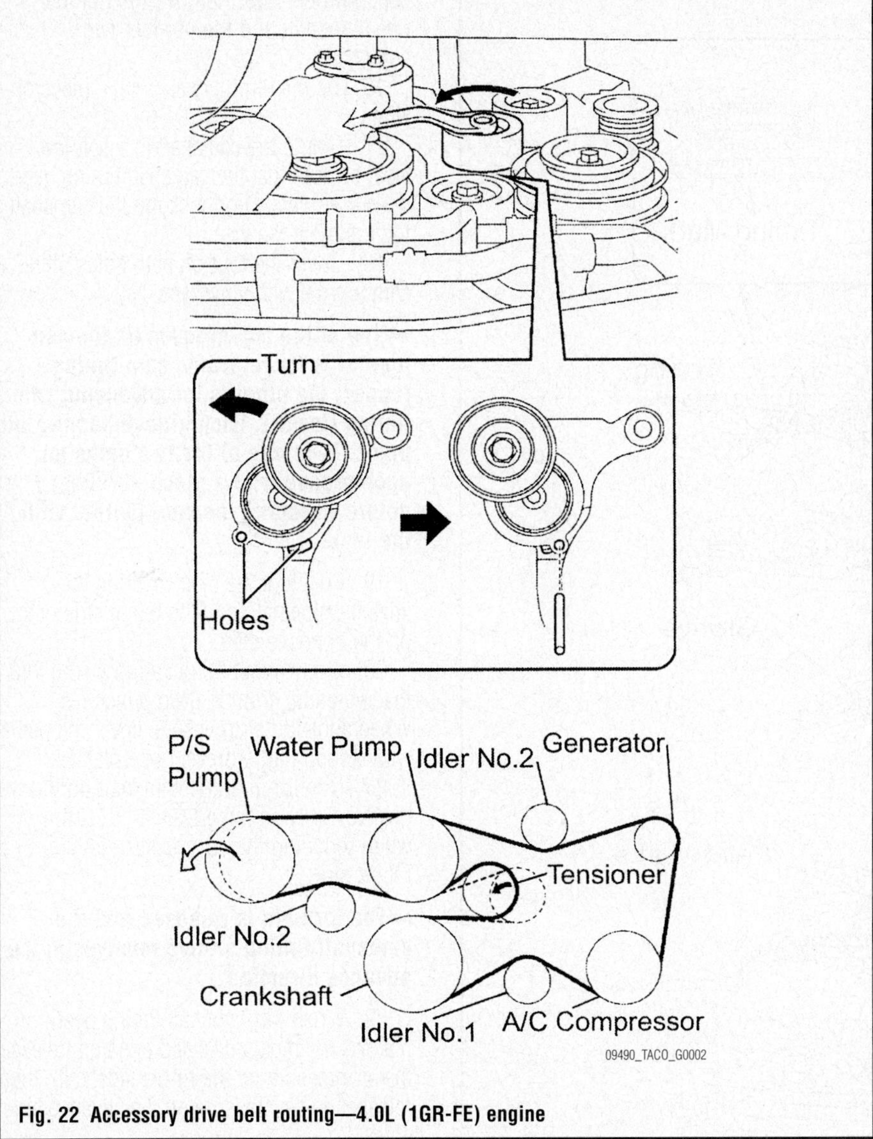

Fig. 22 Accessory drive belt routing—4.0L (1GR-FE) engine

REMOVAL & INSTALLATION

See Figure 23.

1. Before servicing the vehicle, refer to the precautions in the beginning of this section.
2. Loosen the drive belt tension by turning the drive belt tensioner counterclockwise, and remove the drive belt.
3. Installation is the reverse of removal.

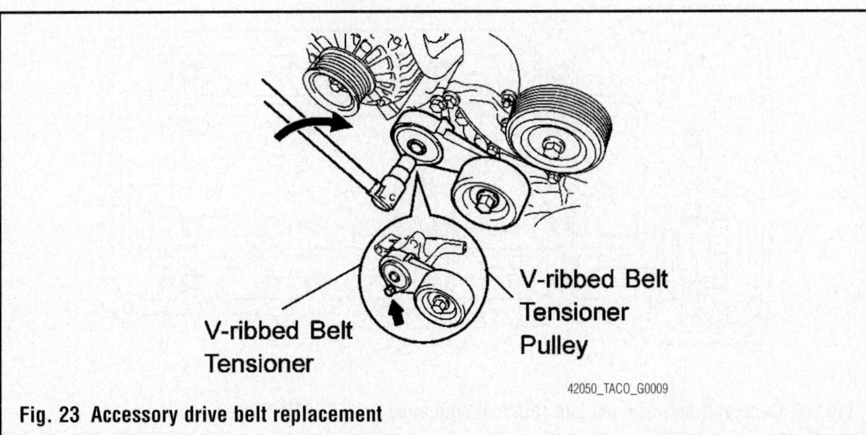

Fig. 23 Accessory drive belt replacement

CAMSHAFT AND VALVE LIFTERS

REMOVAL & INSTALLATION

2.7L (2TR-FE) Engine

See Figures 24 through 31.

1. Before servicing the vehicle, refer to the Precautions Section.
2. Disconnect the negative battery cable.
3. On vehicles equipped with 4WD, remove the engine undercover subassembly.
4. Drain the engine coolant. Remove the radiator support to frame seal, left side. Remove the fan shroud.
5. Remove the air cleaner cap subassembly. Remove the intake air connector.
6. Disconnect the ignition coil connectors. Disconnect the throttle body motor connector. Disconnect the VSV connector.
7. Disconnect the camshaft position sensor connector. Disconnect the engine wire harness clamps. Remove the ignition coils. Disconnect the PCV hose.
8. Remove the cylinder head cover retaining bolts. Remove the cylinder head cover.
9. Remove the two timing chain guide bolts. Remove the timing chain guide. Remove the O-ring.
10. Position the number one cylinder at TDC on the compression stroke.

➡Turn the crankshaft pulley clockwise to align the timing mark notch with the timing mark "0". Paint marks on the timing chain plates that align with the timing marks on the camshaft timing gear.

11. Hold the hexagonal lobe of the number two camshaft, with a suitable tool. Loosen the bolt. Remove the head straight screw plug. Insert a suitable tool into the service hole of the chain tensioner to hold the stopper plate of the chain tensioner lifted up.

➡Lifting up the stopper plate of the chain tensioner unlocks the plunger.

12. While keeping the stopper plate of the chain tensioner lifted up, slightly rotate the hexagonal lobe of the number two camshaft clockwise so that the plunger of the chain tensioner is pushed. Be careful not to damage the camshaft oil delivery pipe.

➡With the wrench still installed, remove the suitable tool with the plunger still pushed in. Do not remove the wrench. Removing the suitable tool lifts down the stopper plate and locks the plunger.

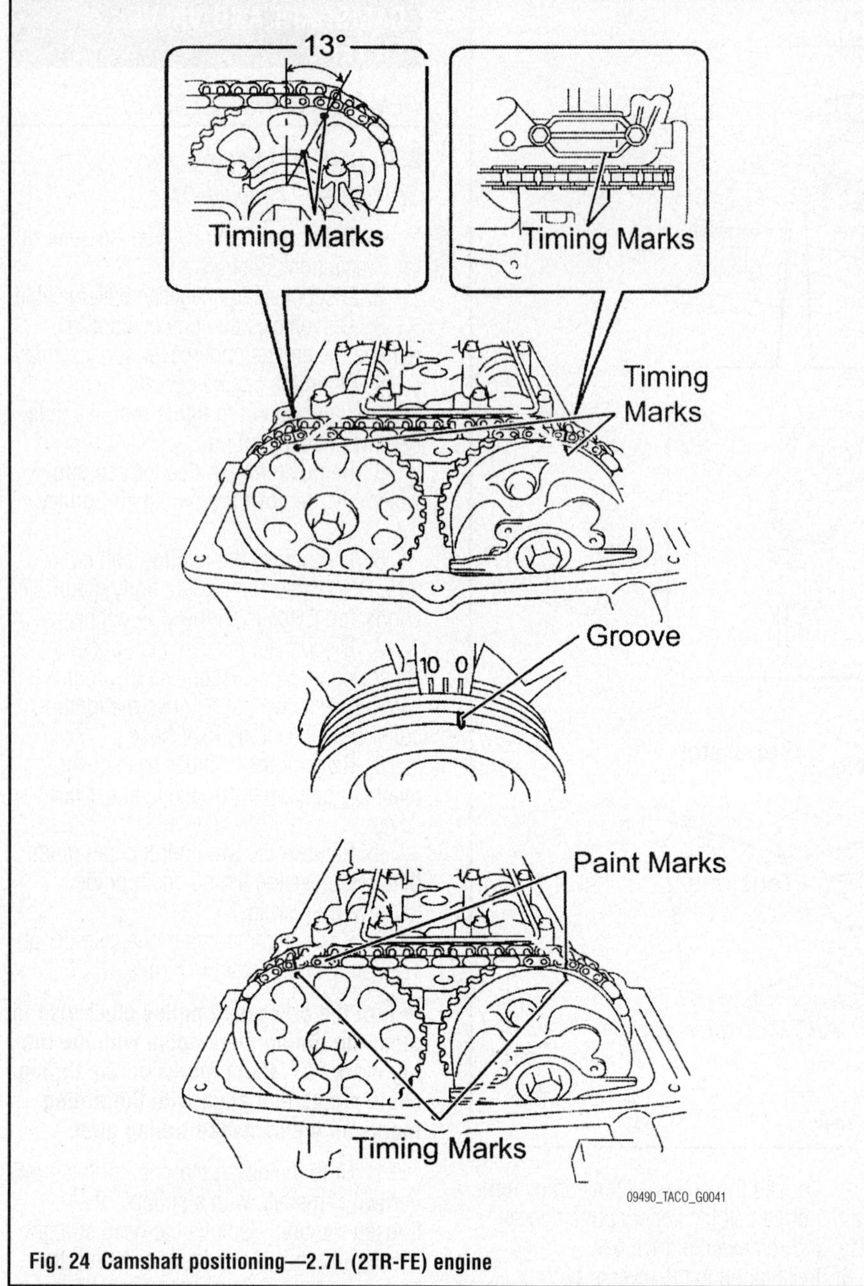

Fig. 24 Camshaft positioning—2.7L (2TR-FE) engine

13. Insert a 0.118 inch diameter bar into the hole of the stopper plate with the stopper plate of the chain tensioner lifter down and locked. Secure the bar with tape.

➡**If the bar cannot be installed, rotate the number two camshaft slightly to the left and right. Then insert the bar.**

14. Remove the camshaft timing gear bolt. Remove the gear.

15. Using several steps, uniformly loosen and remove the camshaft bearing cap bolts in the proper sequence. Remove the camshaft oil delivery pipe and O-ring.

Remove the camshaft bearing cap number one and eight camshaft bearing caps number two. Remove the number one camshaft and the number two camshaft.

16. Tie the timing chain with a piece of wire.

17. Clamp the camshaft in a soft jaw vise, be sure that the camshaft timing gear does not rotate. Do not clamp the camshaft too tightly in the vise.

18. Cover the four oil path holes of the cam journal with vinyl tape.

➡**One of the two grooves on the cam journal is for retarding cam timing (upper) the other is for advancing cam timing (lower). Each groove has two oil paths. Plug one of the two paths for each groove with a piece of rubber before wrapping the cam journal with the tape.**

19. Puncture the tape covering the advance side path and the retard side path on the opposite side.

20. Apply about 29 psi air pressure into the two paths, from the two punctures. When applying air pressure, cover the paths with a shop rag to prevent oil splashes.

21. Confirm that the camshaft timing gear revolves in the advance direction, when reducing the air pressure on the retard side.

➡**The lock pin is released and the camshaft timing groove revolves in the advance direction.**

22. When the camshaft timing gear reaches the most advanced position, release the air pressure on the retard side path, then release the air pressure on the advance side path.

➡**If the air pressure on the advance path is released first, the camshaft timing gear assembly occasionally shifts in the retard direction abruptly. This may**

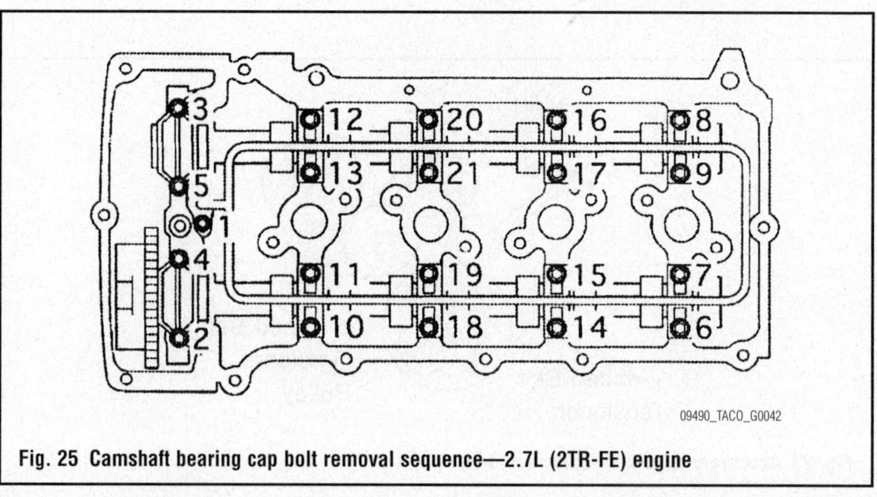

Fig. 25 Camshaft bearing cap bolt removal sequence—2.7L (2TR-FE) engine

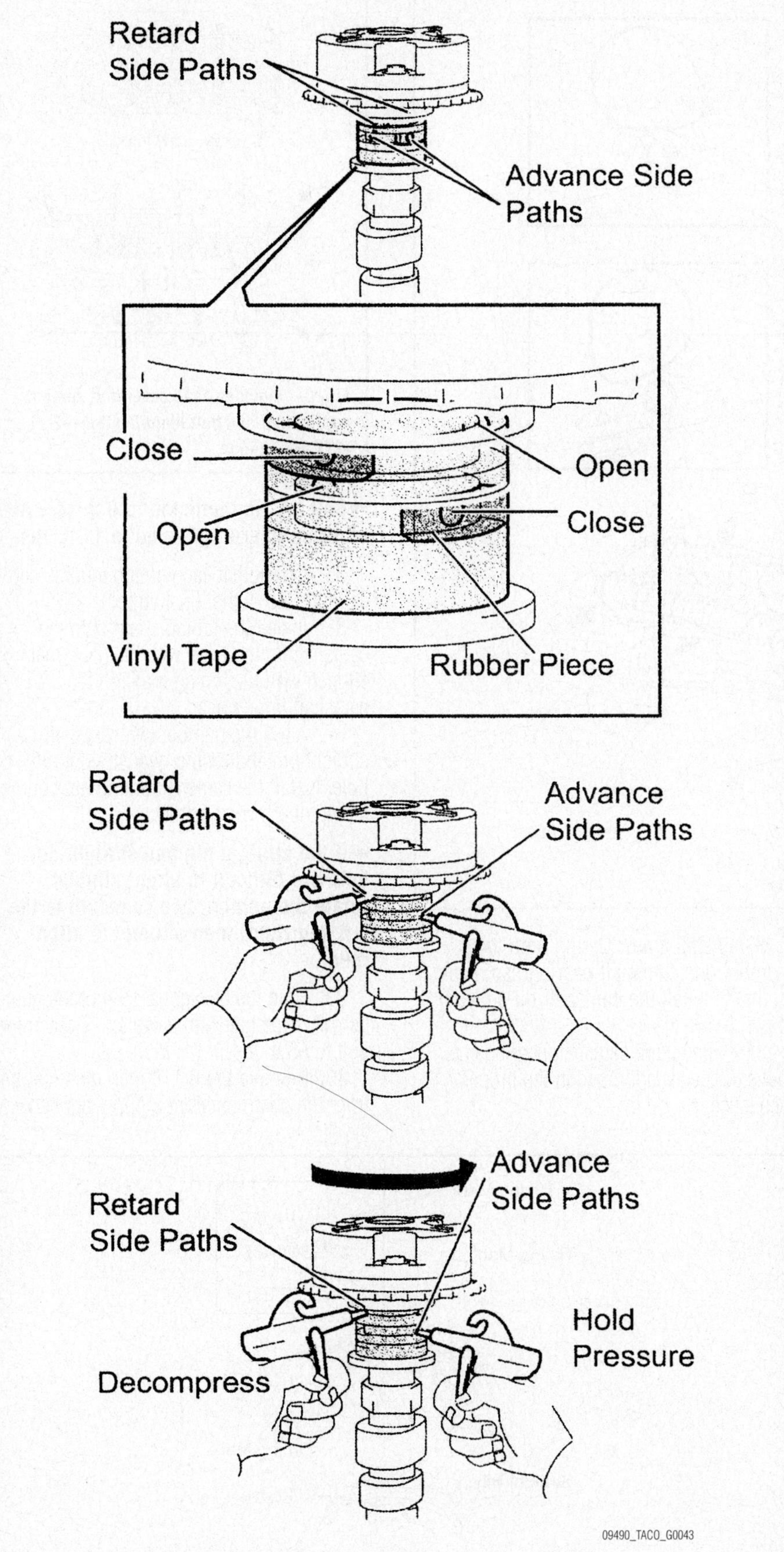

Fig. 26 Camshaft timing gear removal—2.7L (2TR-FE) engine

damage the lock pin. Be sure to release the air pressure on the retard side first.

23. Remove the fringe bolt of the camshaft timing gear.

➡**Do not remove the other three bolts.**

To install:

24. Clean all surfaces.
25. Put the camshaft timing gear and the camshaft together by aligning the key groove and the straight pin.
26. Gently press the gear against the camshaft and turn the gear. Push further at the position where the pin fits the groove.

➡**Be sure not to turn the camshaft timing gear to the retard angle side (to the right angle).**

27. Check that there is no clearance between the gear's fringe and the camshaft. Tighten the fringe bolt to 58 ft. lbs.
28. Check that the camshaft timing gear can move to the retard side (the right angle) and is locked in the extreme retard position.
29. Check that the valve rocker arm is correctly installed. Apply clean engine oil to the camshaft's cam portion and the cylinder head journals.
30. Install the chain onto the camshaft timing gear, with the painted mark of the link aligned with the timing mark of the camshaft timing gear.

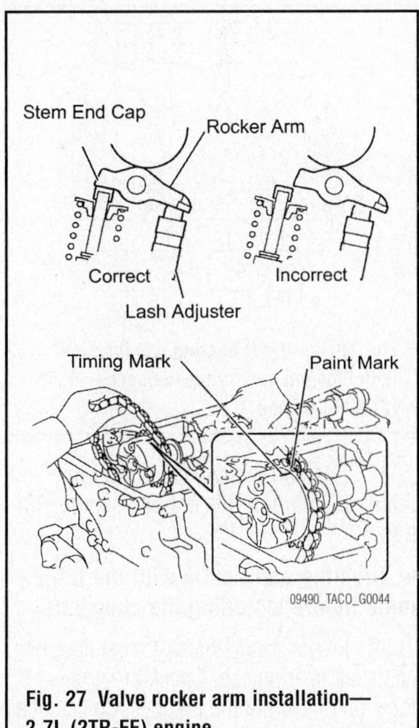

Fig. 27 Valve rocker arm installation—2.7L (2TR-FE) engine

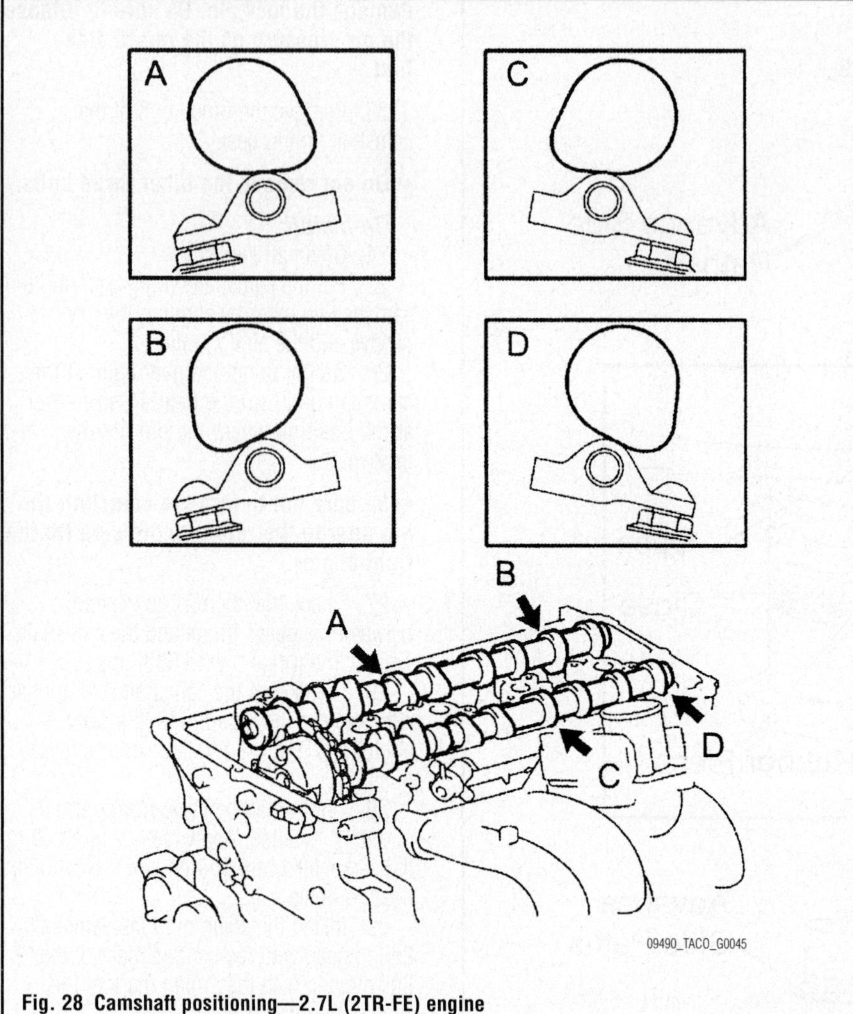

Fig. 28 Camshaft positioning—2.7L (2TR-FE) engine

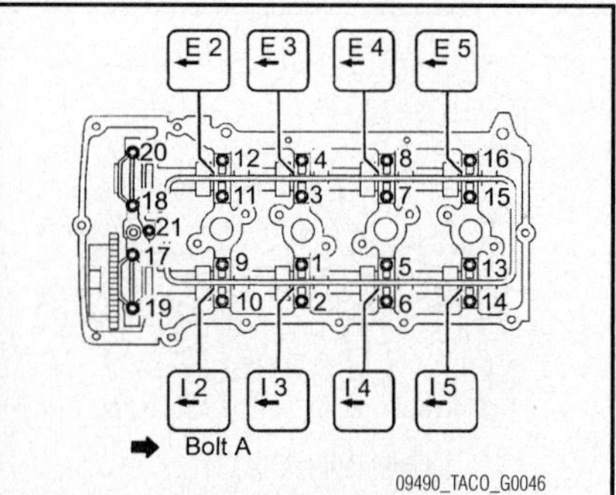

Fig. 29 Camshaft bearing cap location, identification and torque sequence—2.7L (2TR-FE) engine

31. Position the two camshafts in there mounting on the engine, see illustration.

➡Align the paint mark with the timing mark before installing the camshaft.

32. Provisionally install the number one camshaft bearing cap. Check the proper location of each of the number two camshaft bearing caps and install them.

33. Install a new O-ring onto the number one camshaft bearing cap. Provisionally install the camshaft oil delivery pipe.

34. Tighten the camshaft bearing cap bolts to specification and in the proper sequence.

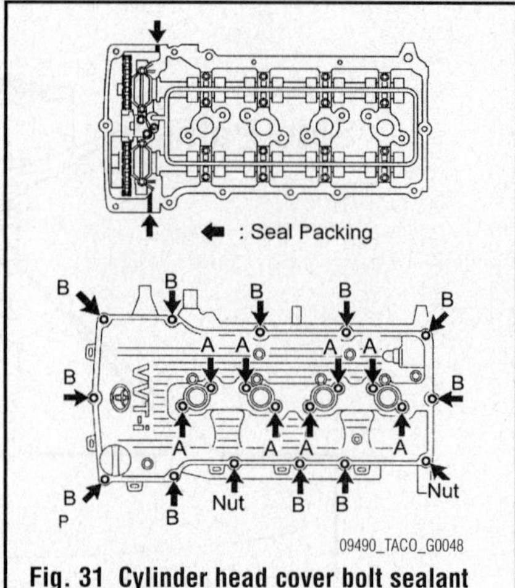

Fig. 31 Cylinder head cover bolt sealant application and bolt identification—2.7L (2TR-FE) engine

➡Bolt "A" is tightened to 9 ft. lbs. All other bolts are tightened to 11 ft. lbs.

35. Check that each timing mark is set as indicated in the illustration.

36. Install the timing chain onto the camshaft timing gear, with the paint mark aligned with the timing mark on the camshaft timing gear.

37. Align the number two camshaft straight pin and timing gear straight pin hole. Install the camshaft timing gear onto the number two camshaft.

➡If the straight pin and straight pin hole are difficult to align, slightly rotate the number two camshaft to the left and right, then attempt to align them.

38. Hold the hexagonal lobe of the number two camshaft with a wrench. Tighten the bolt to 58 ft. lbs.

39. Remove the 0.118 inch diameter bar from the chain tensioner. Apply adhesive,

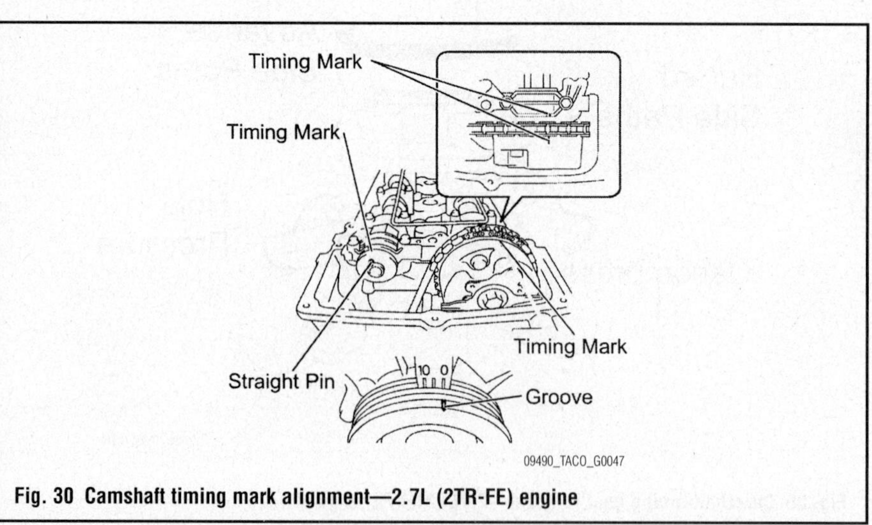

Fig. 30 Camshaft timing mark alignment—2.7L (2TR-FE) engine

part number 08833-00070 or equivalent, to two or three threads of the timing gear case with head straight screw plug. Install the timing gear case with head straight screw plug and tighten to 12 ft. lbs.

40. Install a new O-ring onto the camshaft bearing cap. Install the two timing chain guide bolts. Tighten to 7 ft. lbs.

41. Apply seal packing, part number 08826-00080 or equivalent, to the cylinder head as indicated in the illustration. Provisionally install the cylinder head cover bolts and nuts. Tighten bolts "A" to 80 inch lbs. Tighten bolts "B" to 80 inch lbs. Retighten bolts "A" to 80 inch lbs.

➡Be sure to remove any oil from the contact surfaces of the cylinder head cover and the cylinder head. Install the cover within three minutes after applying the seal packing. Do not add engine oil for at least two hours after installing the cover.

42. Continue the installation in the reverse order of the removal procedure.

43. Be sure to fill the engine with the proper grade and type engine coolant.

44. Be sure to fill the engine with the proper grade and type engine oil.

45. Start the engine and check for leaks. Correct as required.

4.0L (1GR-FE) Engine

Bank 1

See Figures 32 through 46.

1. Before servicing the vehicle, refer to the Precautions Section.
2. Disconnect the negative battery cable.
3. Drain the engine coolant. Remove the V bank cover.
4. Remove the air cleaner assembly.
5. Disconnect the two water bypass hoses. Disconnect the fuel vapor feed hose. Disconnect the ventilation hose. Disconnect the VSV connectors.

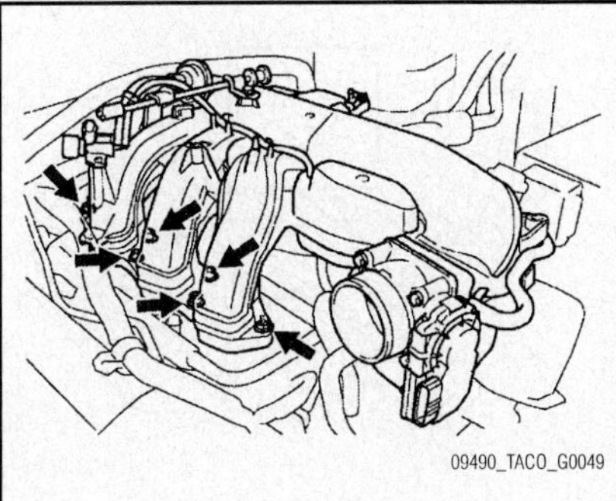

Fig. 32 Intake surge tank bolt locations—4.0L (1GR-FE) engine

6. Disconnect the throttle body motor connector. Separate the three wire harness clamps and hose clamp.
7. If equipped with manual transmission, remove the nut, then separate the clutch flexible hose bracket from the surge tank stay.
8. Remove the two bolts and the throttle body bracket. Remove the bolt and the oil baffle plate. Remove the four bolts and the two serge tank stays.
9. Remove the two nuts. Remove the four bolts, intake air surge tank and gasket.
10. Remove the ignition coil assembly.
11. Remove the cylinder head cover retaining bolts. Remove the cylinder head cover.
12. Turn the crankshaft pulley until its groove and the timing mark "0" of the timing chain cover are aligned. If not aligned at TDC of the compression stroke, turn the crankshaft one complete revolution, in the direction of rotation. Paint alignment marks on the number one chain links corresponding to the timing marks of the camshaft timing gears.
13. Remove the four bolts, and then remove the timing chain cover plate and gasket.
14. While turning the stopper plate of the tensioner upward, push the plunger of the chain tensioner. While turning the stopper plate of the tensioner downward, insert a 0.118 inch diameter bar into the holes in the stopper plate and tensioner to hold the stopper plate.
15. Remove the two bolts, and then remove the chain tensioner.

➡Keep the camshaft level while it is being removed. The camshaft thrust clearance is very small and failing to keep it level could crack or damage the cylinder head journal surface, which receives the thrust. Follow the steps below to prevent this problem from occurring.

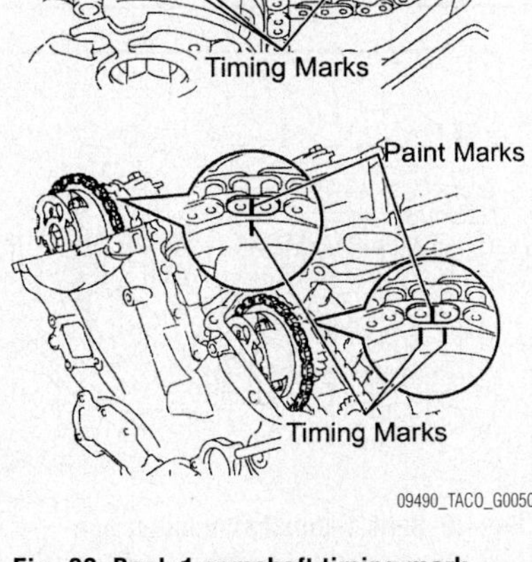

Fig. 33 Bank 1 camshaft timing mark alignment—4.0L (1GR-FE) engine

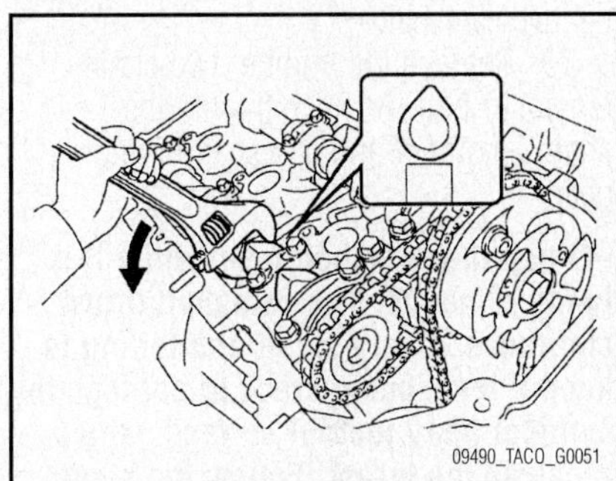

Fig. 34 Bank 1 camshaft lobe removal positioning—4.0L (1GR-FE) engine

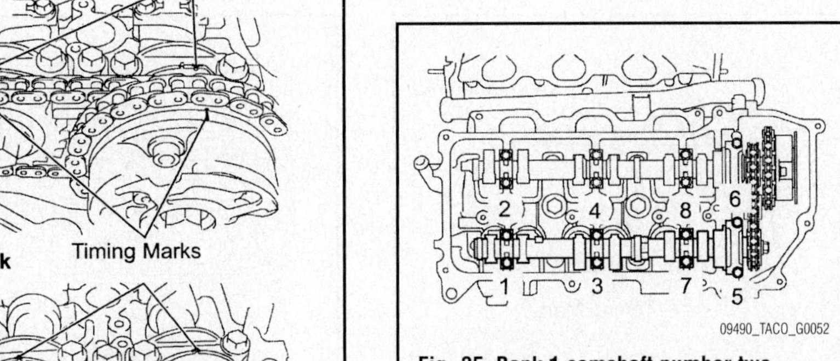

Fig. 35 Bank 1 camshaft number two bearing cap bolt removal sequence—4.0L (1GR-FE) engine

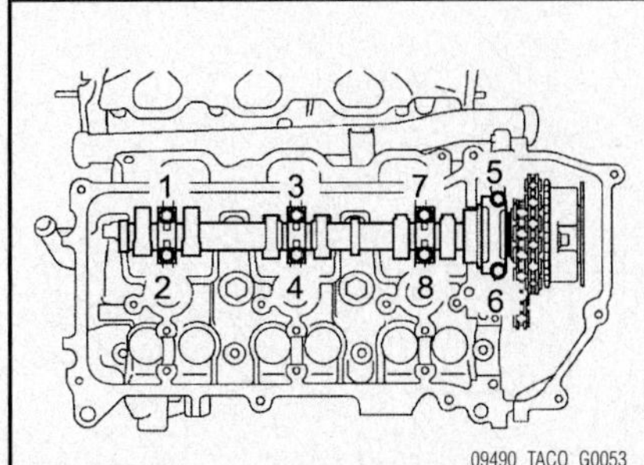

Fig. 36 Bank 1 camshaft number one bearing cap bolt removal sequence—4.0L (1GR-FE) engine

16. While raising the chain tensioner number two insert a 0.039 inch diameter pin into the hole to hold it. Hold the hexagonal portion of the number two camshaft with a wrench. Remove the camshaft timing gear set bolt.

17. Separate the camshaft timing gear from the number two camshaft. Rotate the camshafts counterclockwise, using a wrench, so that the cam lobes of the number one cylinder face in the direction shown.

18. Using several steps, loosen and remove the eight bearing cap bolts uniformly and in the proper removal sequence. Remove the four bearing caps and the number two camshaft.

19. Remove the number two chain tensioner bolt. Remove the number chain tensioner and camshaft timing gear.

➡Keep the camshaft level while it is being removed. The camshaft thrust clearance is very small and failing to keep it level could crack or damage the cylinder head journal surface, which receives the thrust. Follow the steps below to prevent this problem from occurring.

20. Hold the hexagonal portion of the number one camshaft, with a wrench. Loosen the camshaft timing gear set bolt.

➡Do not disassemble the camshaft timing gear assembly.

21. Slide the camshaft timing gear and separate the number one chain from the camshaft timing gear.

22. Rotate the number one camshaft counterclockwise, using the wrench so that the cam lobes on the number one cylinder face downward.

23. Using several steps, loosen and remove the eight bearing cap bolts uniformly and in the proper removal sequence. Remove the four bearing caps.

24. Remove the camshaft timing gear set bolt with the number one camshaft lifted up. Remove the number one camshaft and camshaft timing gear with the number two chain.

25. Tie the number one chain to the side. Be careful not to drop anything inside the timing chain cover.

To install:

➡Keep the camshaft level while it is being installed. The camshaft thrust clearance is very small and failing to keep it level could crack or damage the cylinder head journal surface, which receives the thrust.

26. Align the yellow mark link with the timing mark (1 dot mark) of the camshaft timing gear. Apply new engine oil to the thrust portion and journal of the camshafts.

27. Temporarily install the number one chain onto the number two chain of the camshaft timing gear.

28. Align the knock pin hole of the camshaft timing gear with the knock pin of the number one camshaft. Insert the number one camshaft into the camshaft timing gear.

29. Temporarily install the camshaft timing gear set bolt. Install the number one camshaft onto the right cylinder head with the cam lobes of the number one cylinder facing downward, as indicated in the illustration.

30. Install the four bearing caps, in the proper location. Apply a light coat of engine oil to the threads and under the heads of the cap bolts.

31. Using several steps, uniformly install and tighten the bearing cap bolts in the proper sequence to 80 inch lbs. for the 10mm bolts and 18 ft. lbs. for the 12mm bolts.

32. Rotate the number one camshaft clockwise, using a wrench so that the timing mark of the camshaft timing gear is

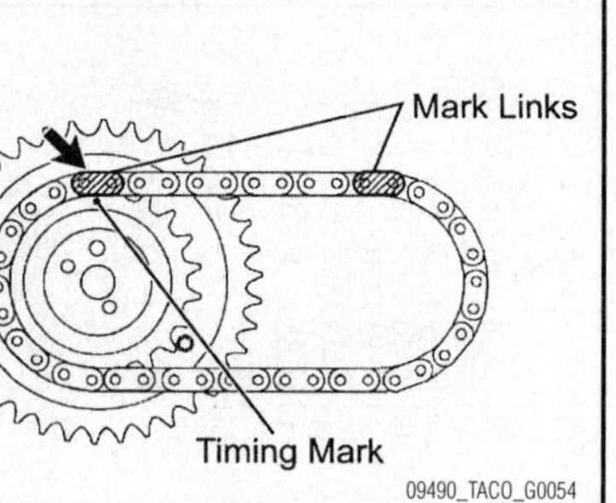

Fig. 37 Bank 1 camshaft number one alignment marks—4.0L (1GR-FE) engine

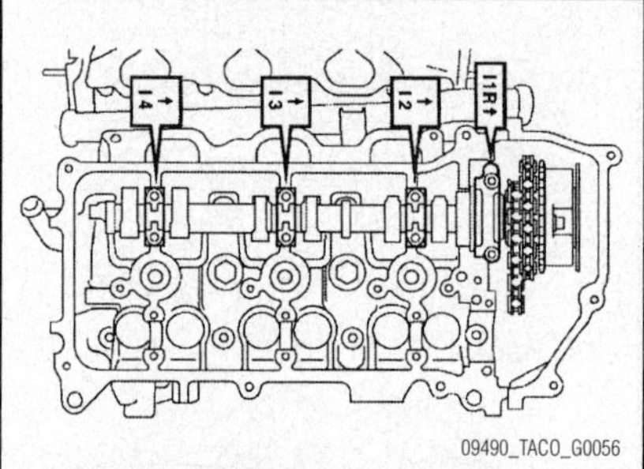

Fig. 39 Bank 1 camshaft number one bearing cap bolt installation sequence—4.0L (1GR-FE) engine

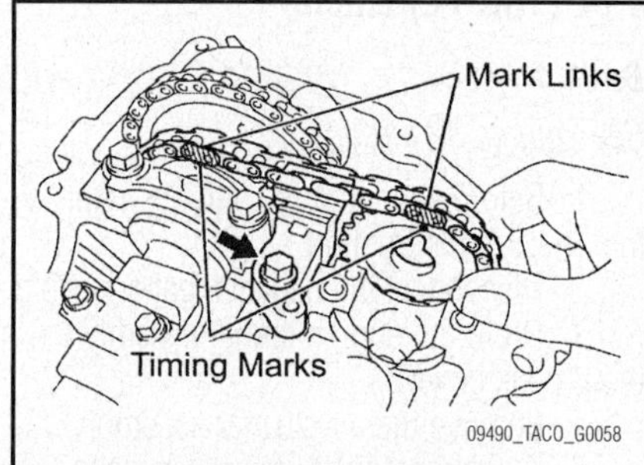

Fig. 41 Bank 1 camshaft number two alignment marks—4.0L (1GR-FE) engine

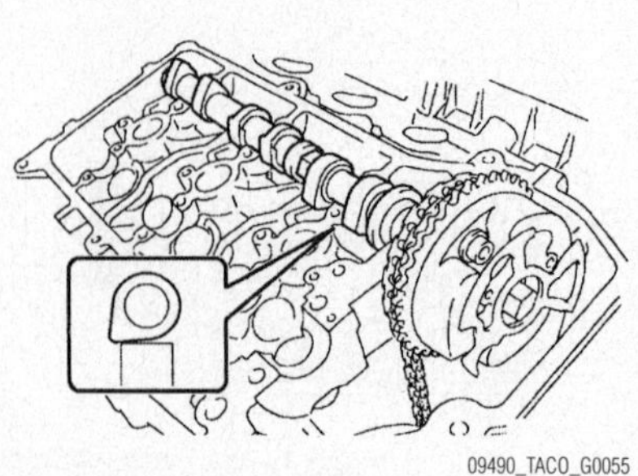

Fig. 38 Bank 1 camshaft number one lobe installation positioning—4.0L (1GR-FE) engine

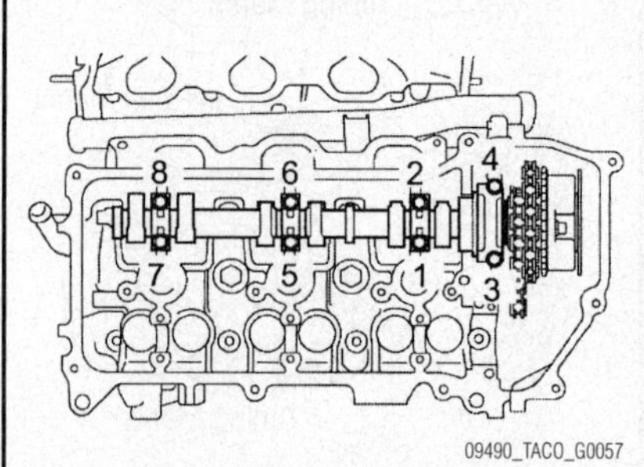

Fig. 40 Bank 1 camshaft number one bearing cap bolt torque sequence—4.0L (1GR-FE) engine

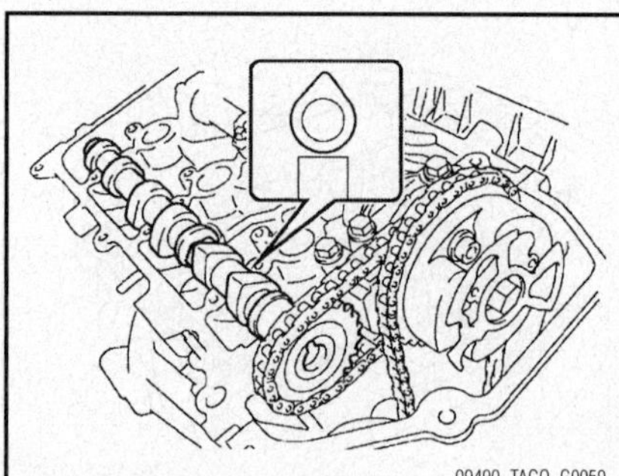

Fig. 42 Bank 1 camshaft number two lobe installation positioning—4.0L (1GR-FE) engine

aligned with the timing mark of the camshaft bearing cap.

33. Align the paint mark of the number one chain with the timing mark of the camshaft timing gear.

34. Hold the hexagonal portion of the number one camshaft with a wrench, and tighten the camshaft timing gear set bolt to 74 ft. lbs.

35. While pushing in on the number two chain tensioner, insert a 0.039 inch pin into the hole to hold it.

36. Temporarily install the camshaft timing gear and chain tensioner number two and align the yellow mark links with the timing marks (1 dot mark) on the camshaft timing gears. Tighten the bolt to 14 ft. lbs.

➡Keep the camshaft level while it is being installed. The camshaft thrust clearance is very small and failing to keep it level could crack or damage the cylinder head journal surface, which receives the thrust.

37. Install the number two camshaft onto the right cylinder head with the cam lobes of the number one cylinder facing upward, as indicated in the illustration.

38. Install the four bearing caps in the proper location. Apply a light coat of clean engine oil to the threads and under the heads of the bolts.

39. Using several steps, uniformly install and tighten the eight bearing cap bolts in sequence to 80 inch lbs, for the 10mm head bolts and 18 ft. lbs. for the 12mm head bolts.

40. Rotate the number two camshaft clockwise, using a wrench, so that the lock pin of the number two camshaft is aligned with the knock pin hole of the camshaft timing gear.

41. Hold the hexagonal portion of the number two camshaft, with a wrench, and install the camshaft timing gear set bolt and tighten it to 74 ft. lbs.

42. Remove the pin from the number two chain tensioner.

43. While turning the stopper plate of the tensioner clockwise, push in the plunger of the tensioner. While turning the stopper plate of the tensioner counterclockwise, insert a 0.138 inch bar into the holes in the stopper plate and tensioner to hold the stopper plate. Install the two chain tensioner bolts and tighten to 7.4 ft. lbs.

44. Remove the bar from the chain tensioner. Install a new gasket and the timing chain cover plate. Torque the bolts to 80 inch lbs.

45. Turn the crankshaft pulley two complete revolutions slowly until its groove and the timing mark "0" of the timing chain cover are aligned.

46. Position the number one cylinder at TDC of the compression stroke. Inspect the valve clearance, adjust as required.

47. Apply a continuous bead (0.08–0.12 inch) of seal packing, part number 08826-00080 or equivalent, to the cylinder head as indicated in the illustration. Install the seal washers onto the bolts. Install the cylinder head cover bolts and nuts. Tighten bolts "A" to 7.4 ft. lbs. Tighten bolts "B" to 80 inch lbs. Tighten nuts to 80 inch lbs.

➡Be sure to remove any oil from the contact surfaces of the cylinder head cover and the cylinder head. Install the cover within three minutes after applying the seal packing. Tighten the bolts to specification within fifteen minutes after installing the cover. Do not add engine oil for at least two hours after installing the cover.

48. Continue the installation in the reverse order of the removal procedure.

49. Be sure to fill the engine with the proper grade and type engine coolant.

50. Be sure to fill the engine with the proper grade and type engine oil.

51. Start the engine and check for leaks. Correct as required.

Bank 2

See Figures 47 through 55.

1. Before servicing the vehicle, refer to the Precautions Section.

2. Disconnect the negative battery cable.

3. Drain the engine coolant. Remove the V bank cover.

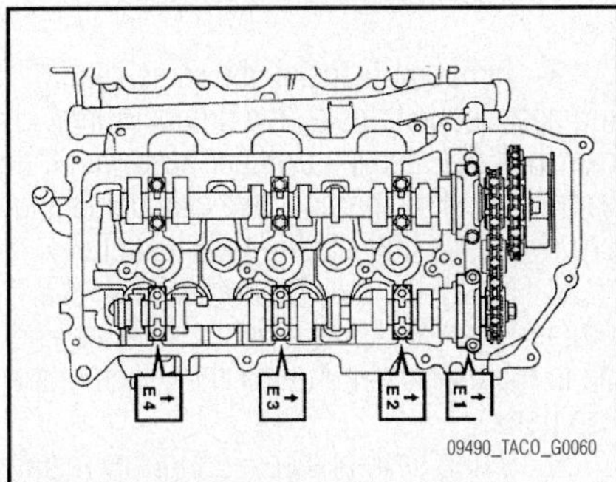

Fig. 43 Bank 1 camshaft number two bearing cap bolt installation sequence—4.0L (1GR-FE) engine

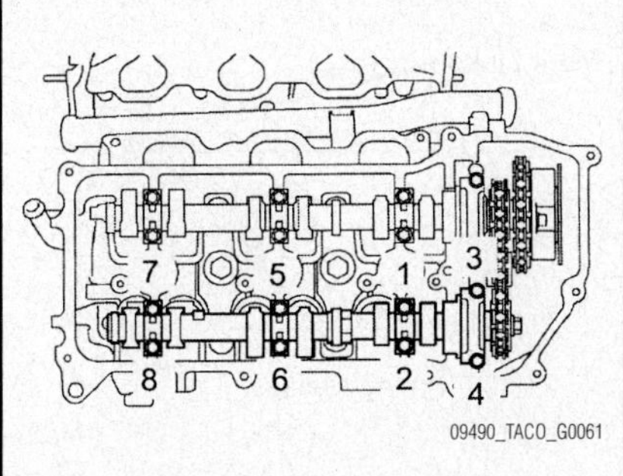

Fig. 44 Bank 1 camshaft number two bearing cap bolt torque sequence—4.0L (1GR-FE) engine

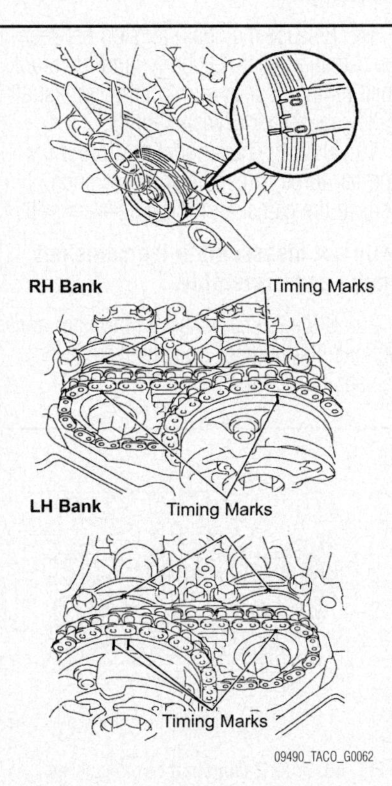

Fig. 45 Bank 1 and Bank 2 timing mark alignment—4.0L (1GR-FE) engine

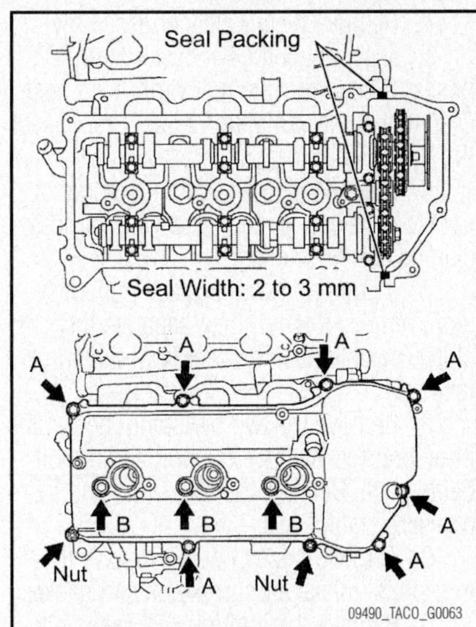

Fig. 46 Bank 1 cylinder head cover bolt sealant application and bolt identification—4.0L (1GR-FE) engine

4. Remove the air cleaner assembly.

5. Disconnect the two water bypass hoses. Disconnect the fuel vapor feed hose. Disconnect the ventilation hose. Disconnect the VSV connectors.

6. Disconnect the throttle body motor connector. Separate the three wire harness clamps and hose clamp.

7. If equipped with manual transmission, remove the nut, then separate the clutch flexible hose bracket from the surge tank stay.

8. Remove the two bolts and the throttle body bracket. Remove the bolt and the oil baffle plate. Remove the four bolts and the two serge tank stays.

9. Remove the two nuts. Remove the four bolts, intake air surge tank and gasket.

10. Remove the ignition coil assembly.

11. Remove the cylinder head cover retaining bolts. Remove the cylinder head cover.

12. Turn the crankshaft pulley until its groove and the timing mark "0" of the timing chain cover are aligned. If not aligned at TDC of the compression stroke, turn the crankshaft one complete revolution, in the direction of rotation. Paint alignment marks on the number one chain links corresponding to the timing marks of the camshaft timing gears.

13. While turning the stopper plate of the tensioner upward, push in the plunger of the chain tensioner. While turning the stopper plate of the tensioner downward, insert a 0.138 inch bar into the holes in the stopper plate and tensioner to hold the stopper plate. Remove the two bolts and then remove the number one chain tensioner assembly.

➡Never rotate the crankshaft with the chain tensioner removed. When rotating the camshaft with the tensioner removed, rotate the crankshaft counterclockwise forty degrees from TDC, first.

➡Keep the camshaft level while it is being removed. The camshaft thrust clearance is very small and failing to keep it level could crack or damage the cylinder head journal surface, which receives the thrust. Follow the steps below to prevent this problem from occurring.

14. While pushing down on chain tensioner number three insert a 0.039 inch diameter pin into the hole to hold it. Hold the hexagonal portion of the number four camshaft with a wrench. Remove the camshaft timing gear set bolt.

15. Separate the camshaft timing gear from the number four camshaft.

16. Using several steps, loosen and remove the eight bearing cap bolts uniformly and in the proper removal sequence. Remove the four bearing caps and the number four camshaft.

17. Remove the number three chain tensioner bolt. Remove the number chain tensioner and camshaft timing gear.

➡Keep the camshaft level while it is being removed. The camshaft thrust clearance is very small and failing to keep it level could crack or damage the cylinder head journal surface, which receives the thrust. Follow the steps below to prevent this problem from occurring.

18. Release the chain tension between the camshaft gear (LH bank) and the crankshaft timing gear by turning the crankshaft pulley counterclockwise slightly.

19. Hold the hexagonal portion of the number three camshaft, with a wrench. Loosen the camshaft timing gear set bolt.

➡Do not disassemble the camshaft timing gear assembly.

20. Slide the camshaft timing gear and separate the number one chain from the camshaft timing gear.

21. Using several steps, loosen and remove the eight bearing cap bolts uniformly and in the proper removal sequence. Remove the four bearing caps.

22. Remove the camshaft timing gear set bolt with the number three camshaft lifted up. Remove the number three camshaft and camshaft timing gear with the number two chain.

23. Tie the number one chain to the side. Be careful not to drop anything inside the timing chain cover.

To install:

➡Keep the camshaft level while it is being installed. The camshaft thrust clearance is very small and failing to keep it level could crack or damage the cylinder head journal surface, which receives the thrust.

24. Align the yellow mark link with the timing mark (2 dot mark) of the camshaft timing gear. Apply new engine oil to the thrust portion and journal of the camshafts.

25. Temporarily install the number one chain onto the number two chain of the camshaft timing gear.

26. Align the knock pin hole of the camshaft timing gear with the knock pin of the number three camshaft. Insert the number three camshaft into the camshaft timing gear.

27. Temporarily install the camshaft timing gear set bolt. Install the number three camshaft onto the left cylinder head with the cam lobes of the number two cylinder facing downward, as indicated in the illustration.

28. Install the four bearing caps, in the proper location. Apply a light coat of engine oil to the threads and under the heads of the cap bolts.

29. Using several steps, uniformly install and tighten the bearing cap bolts in the proper sequence to 80 inch lbs. for the 10mm bolts and 18 ft. lbs. for the 12mm bolts.

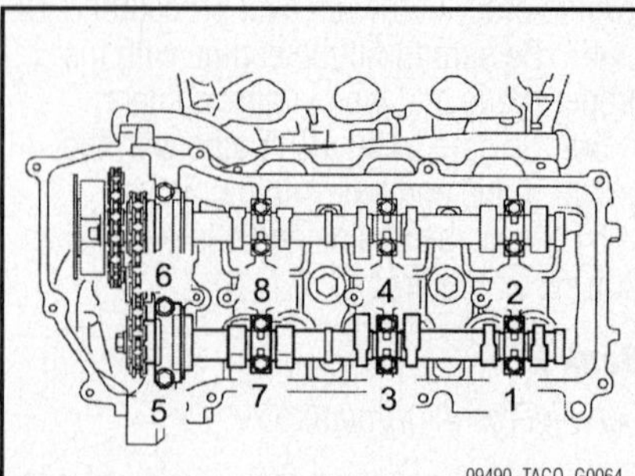

Fig. 47 Bank 2 camshaft number four bearing cap bolt removal sequence—4.0L (1GR-FE) engine

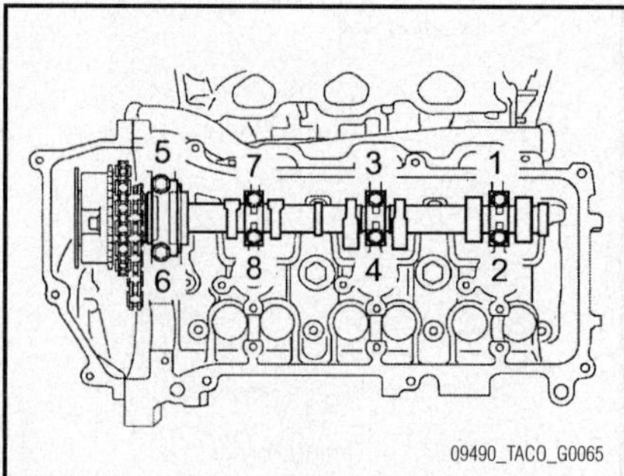

Fig. 48 Bank 2 camshaft number three bearing cap bolt removal sequence—4.0L (1GR-FE) engine

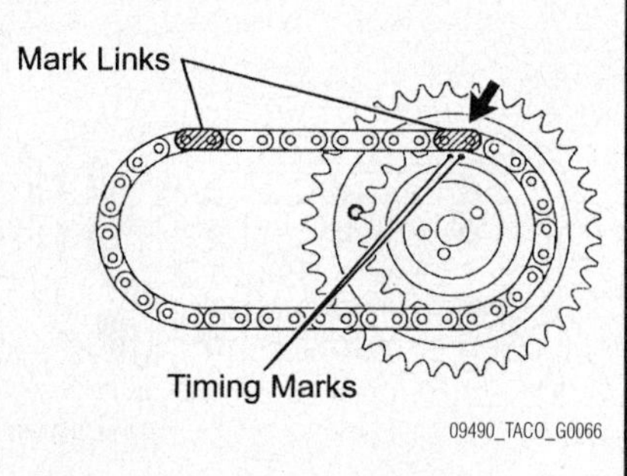

Fig. 49 Bank 2 camshaft number three alignment marks—4.0L (1GR-FE) engine

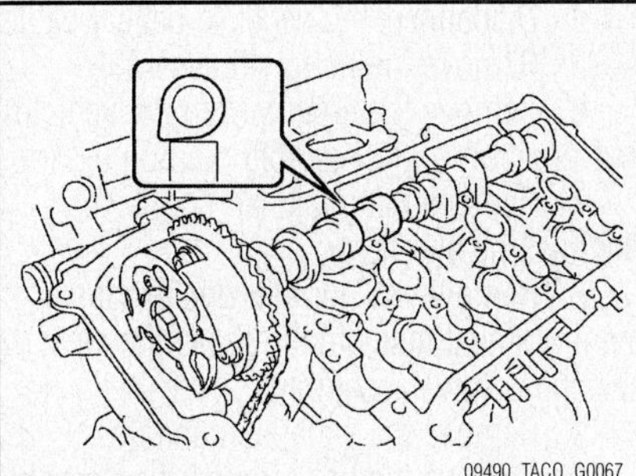

Fig. 50 Bank 2 camshaft number three lobe installation positioning—4.0L (1GR-FE) engine

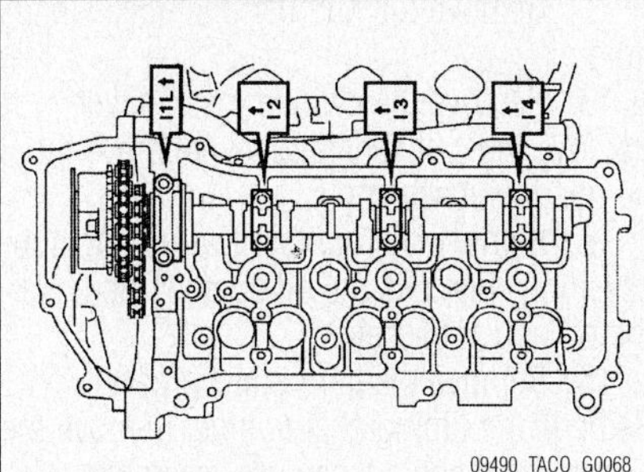

Fig. 51 Bank 2 camshaft number three bearing cap bolt installation sequence—4.0L (1GR-FE) engine

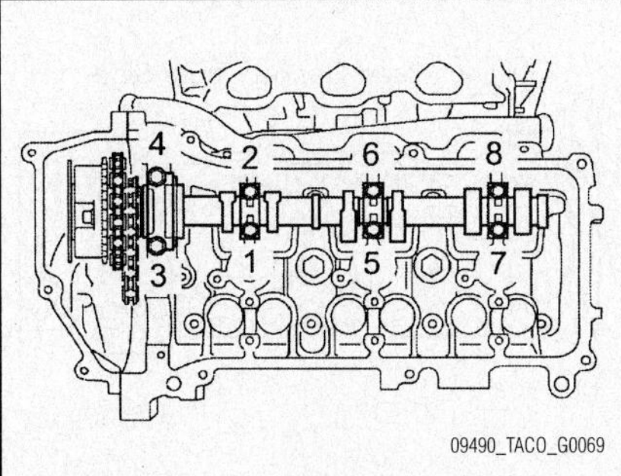

Fig. 52 Bank 2 camshaft number three bearing cap bolt torque sequence—4.0L (1GR-FE) engine

30. Rotate the number one camshaft clockwise, using a wrench so that the timing mark of the camshaft timing gear is aligned with the timing mark of the camshaft bearing cap.

31. Align the paint mark of the number one chain with the timing mark of the camshaft timing gear.

32. Hold the hexagonal portion of the number three camshaft with a wrench, and tighten the camshaft timing gear set bolt to 74 ft. lbs.

33. While pushing in on the number three chain tensioner, insert a 0.039 inch pin into the hole to hold it.

34. Temporarily install the camshaft timing gear and chain tensioner number three and align the yellow mark links with the timing marks (1 dot mark and 2 dot marks) on the camshaft timing gears. Tighten the bolt to 14 ft. lbs.

➡Keep the camshaft level while it is being installed. The camshaft thrust clearance is very small and failing to keep it level could crack or damage the cylinder head journal surface, which receives the thrust.

35. Align the knock pin hole in the camshaft timing gear with the knock pin of the number four camshaft, and insert the number four camshaft into the camshaft timing gear.

36. Temporarily install the camshaft timing gear set bolt.

37. Install the four bearing caps in the proper location. Apply a light coat of clean engine oil to the threads and under the heads of the bolts.

38. Using several steps, uniformly install and tighten the eight bearing cap bolts in sequence to 80 inch lbs, for the 10mm head bolts and 18 ft. lbs. for the 12mm head bolts.

39. Hold the hexagonal portion of the number four camshaft, with a wrench, and install the camshaft timing gear set bolt and tighten it to 74 ft. lbs.

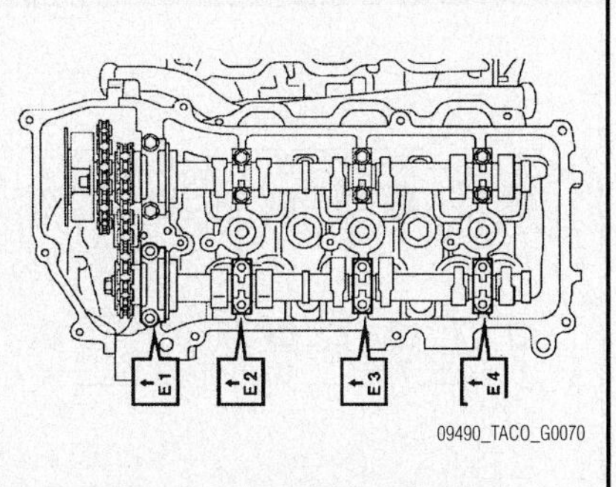

Fig. 53 Bank 2 camshaft number four bearing cap bolt installation sequence—4.0L (1GR-FE) engine

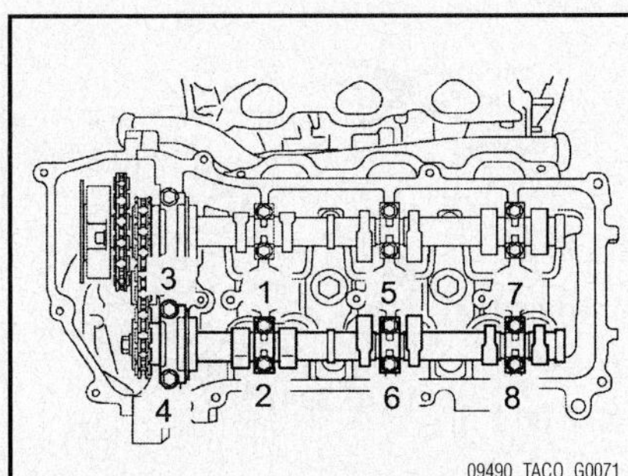

Fig. 54 Bank 2 camshaft number four bearing cap bolt torque sequence—4.0L (1GR-FE) engine

40. Remove the pin from the number three chain tensioner.

41. Release the chain tension between the camshaft timing gear (RH bank) and the crankshaft timing gear by turning the crankshaft pulley clockwise slightly.

42. While turning the stopper plate of the tensioner clockwise, push in the plunger of the tensioner. While turning the stopper plate of the tensioner counterclockwise, insert a 0.138 inch bar into the holes in the stopper plate and tensioner to hold the stopper plate. Install the two chain tensioner bolts and tighten to 7.4 ft. lbs. Remove the bar from the chain tensioner.

43. Position the number one cylinder at TDC of the compression stroke. Inspect the valve clearance, adjust as required.

44. Apply a continuous bead (0.08–0.12 inch) of seal packing, part number 08826-00080 or equivalent, to the cylinder head as indicated in the illustration. Install the seal washers onto the bolts. Install the cylinder head cover bolts and nuts. Tighten bolts "A" to 7.4 ft. lbs. Tighten bolts "B" to 80 inch lbs. Tighten nuts to 80 inch lbs.

➡Be sure to remove any oil from the contact surfaces of the cylinder head cover and the cylinder head. Install the cover within three minutes after applying the seal packing. Tighten the bolts to specification within fifteen minutes after installing the cover. Do not add engine oil for at least two hours after installing the cover.

45. Continue the installation in the reverse order of the removal procedure.

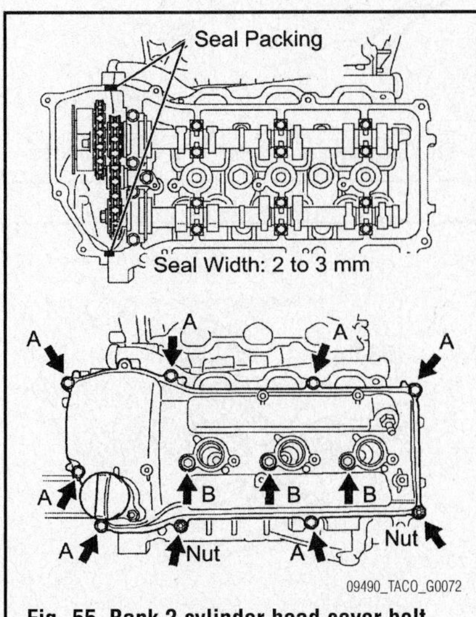

Fig. 55 Bank 2 cylinder head cover bolt sealant application and bolt identification—4.0L (1GR-FE) engine

46. Be sure to fill the engine with the proper grade and type engine coolant.

47. Be sure to fill the engine with the proper grade and type engine oil.

48. Start the engine and check for leaks. Correct as required.

CRANKSHAFT DAMPER

REMOVAL & INSTALLATION

2.7L (2TR-FE)

See Figures 56 and 57.

1. Before servicing the vehicle, refer to the precautions in the beginning of this section.
2. Remove the accessory drive belt.
3. Using SST 09213-54015, fix the pulley and loosen the pulley bolt.
4. Using the pulley set bolt and SST 09950-50013, remove the crankshaft pulley.

To install:

5. Installation is the reverse of the removal procedure. Tighten the crankshaft bolt to 192 ft. lbs (260 Nm).

4.0L (1GR-FE) Engine

See Figures 56 and 57.

1. Before servicing the vehicle, refer to the precautions in the beginning of this section.
2. Remove the accessory drive belt.
3. Using SST 09213-54015, fix the pulley and loosen the pulley bolt.
4. Using the pulley set bolt and SST 09950-50013, remove the crankshaft pulley.

To install:

5. Installation is the reverse of the removal procedure. Tighten the crankshaft bolt to 184 ft. lbs (250 Nm).

CRANKSHAFT FRONT SEAL

REMOVAL & INSTALLATION

See Timing Chain Cover and Seal.

CYLINDER HEAD

REMOVAL & INSTALLATION

2.7L (2TR-FE) Engine

See Figures 58 through 61.

➡The engine must first be removed from the vehicle.

1. Before servicing the vehicle, refer to the Precautions Section.
2. Properly discharge the fuel system pressure.
3. Properly disarm the SRS system.
4. Disconnect the negative battery cable.
5. Remove the hood subassembly.
6. Remove the engine from the vehicle and position it in a suitable holding fixture.
7. Remove the intake air connector. Remove the alternator.
8. Remove the number one exhaust manifold heat insulator. Remove the air switching valve assembly.
9. Remove the exhaust manifold.
10. Remove the belt tensioner assembly. Remove the number one idler pulley subassembly. Remove the idle pulley assembly and bracket.
11. Remove the crankshaft position sensor. Remove the camshaft position sensor.
12. Remove the number one intake manifold.
13. Remove the cylinder head cover. Remove the crankshaft pulley. Remove the oil gauge subassembly.
14. Remove the number two oil pan subassembly. Remove the oil strainer. Remove the oil pap subassembly.
15. Remove the timing chain cover. Remove the timing chain guide. Remove the number one chain tensioner assembly.
16. Remove the chain tensioner slipper. Remove the number one chain vibration damper. Remove the chain subassembly.

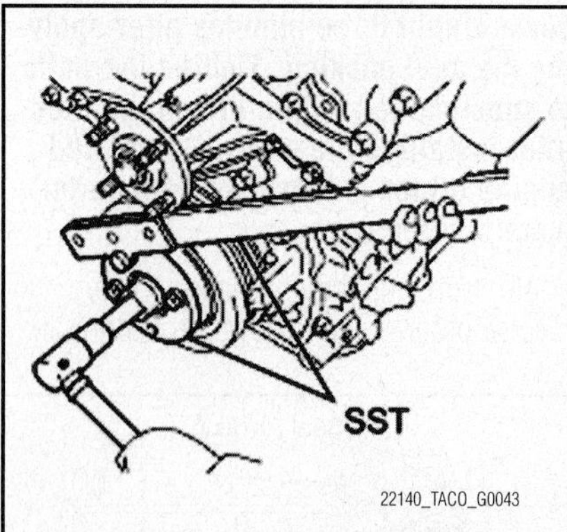

Fig. 56 Special Service Tool 09213-54015

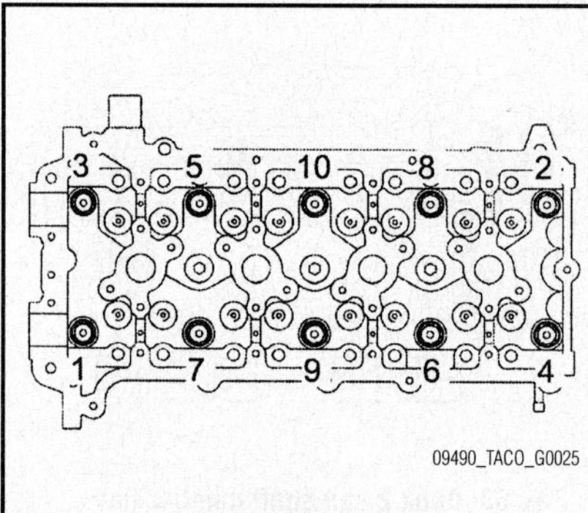

Fig. 58 Cylinder head bolt loosening sequence—2.7L (2TR-FE) engine

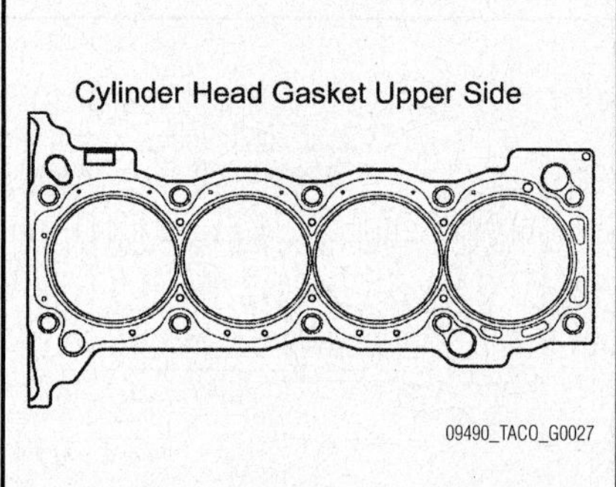

Fig. 60 Cylinder head gasket positioning—2.7L (2TR-FE) engine

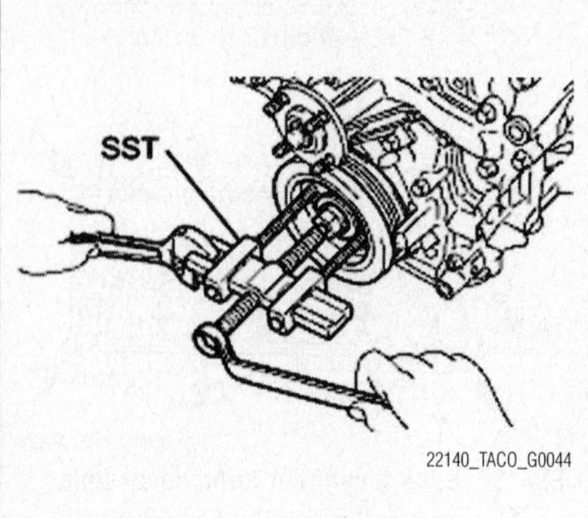

Fig. 57 Special Service Tool 09950-50013

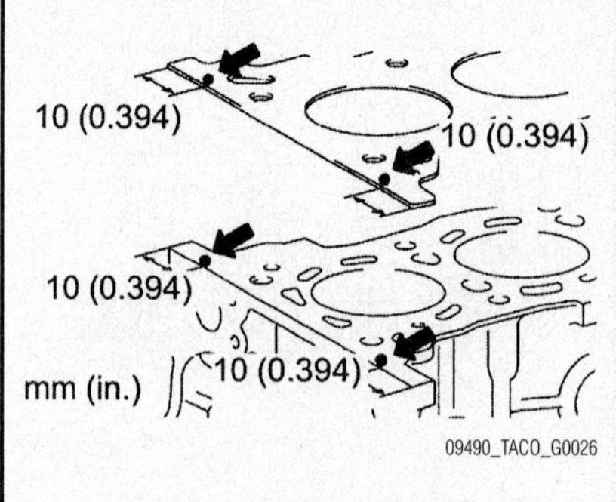

Fig. 59 Cylinder head gasket sealant application—2.7L (2TR-FE) engine

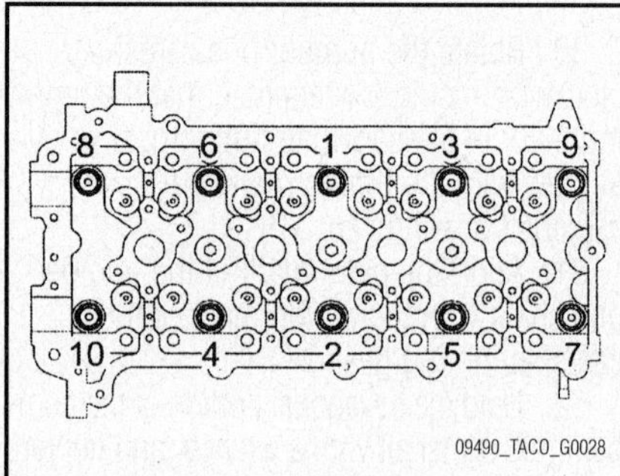

Fig. 61 Cylinder head bolt torque sequence—2.7L (2TR-FE) engine

17. Remove the camshaft. Remove the valve rocker arms.

18. Remove the valve lash adjuster assembly.

19. Disconnect the water hoses. Disconnect the engine coolant temperature sensor wire. Remove the bolts and separate the wire from the harnesses. Remove the bolt, and then separate the wire harness bracket.

20. Loosen the cylinder head retaining bolts in several steps and in the proper removal sequence. Remove the bolts.

➡Be careful not to drop the washers into the cylinder head. Head warpage and cracking could result in removing the bolts in the wrong order.

To install:

21. Before installing, thoroughly clean the gasket mating surfaces and check for warpage.

22. Apply continuous beads of seal packing, part number 08826-00080 or equivalent, to the cylinder block upper side and cylinder head gasket upper side. Bead width should be 0.15–0.28 inch.

➡Remove any oil from the contact surface. Install the cylinder head gasket within three minutes after applying the seal packing. Install the cylinder head bolts within fifteen minutes after applying the seal packing. Do not fill the engine with oil for at least four hours.

➡Be sure to position the cylinder head gasket on the engine block with the lot number stamp facing upward.

➡Be sure that the head gasket is installed in the correct direction. Position the gasket gently in order to avoid damaging the gasket with the bottom part of the cylinder head.

23. Apply a light coat of clean engine oil to the cylinder head bolts. Tighten the cylinder head bolts, in two successive steps, to specification and in the proper sequence.

24. Continue the installation in the reverse of the removal procedure.

25. Be sure to fill the cooling system with the proper grade and type engine coolant.

26. Be sure to fill the engine oil with the proper grade and type oil.

➡ Do not fill the engine with oil for at least four hours after the cylinder head has been installed.

27. Start the engine and check for leaks. Correct as required.

4.0L (1GR-FE) Engine

See Figures 62 through 70.

2WD–Bank 1

➡The engine must first be removed from the vehicle.

1. Before servicing the vehicle, refer to the Precautions Section.

2. Properly discharge the fuel system pressure.

3. Properly disarm the SRS system.

4. Disconnect the negative battery cable. Disconnect the positive battery cable. Remove the battery.

5. Remove the hood subassembly.

6. Remove the engine from the vehicle and position it in a suitable holding fixture.

7. Remove the timing chain or belt cover subassembly.

8. Remove the chain subassembly.

9. Remove the number one cool inlet. Remove the front exhaust pipe assembly. Remove the three bolts and remove the exhaust manifold stay.

10. Disconnect the air fuel ratio sensor connector. Remove the six nuts; remove the exhaust manifold and gasket.

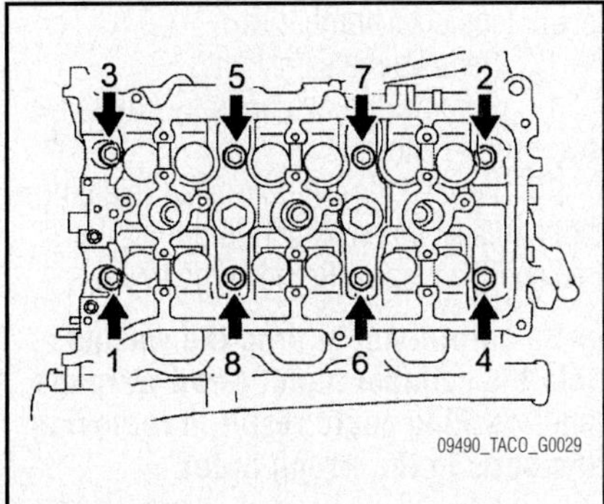

Fig. 62 Bank 1 cylinder head bolt loosening sequence—4.0L (1GR-FE) engine

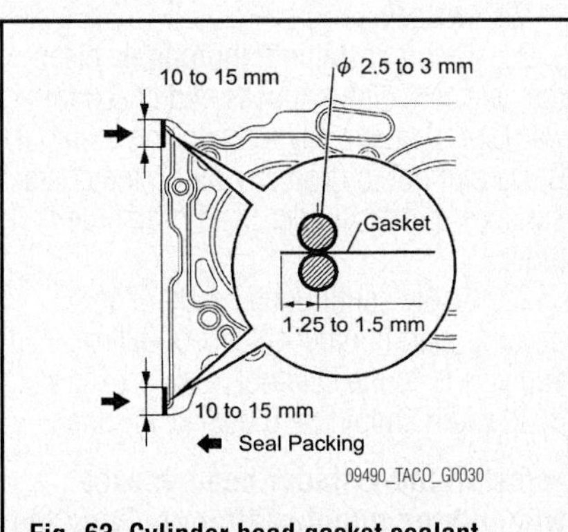

Fig. 63 Cylinder head gasket sealant application—4.0L (1GR-FE) engine

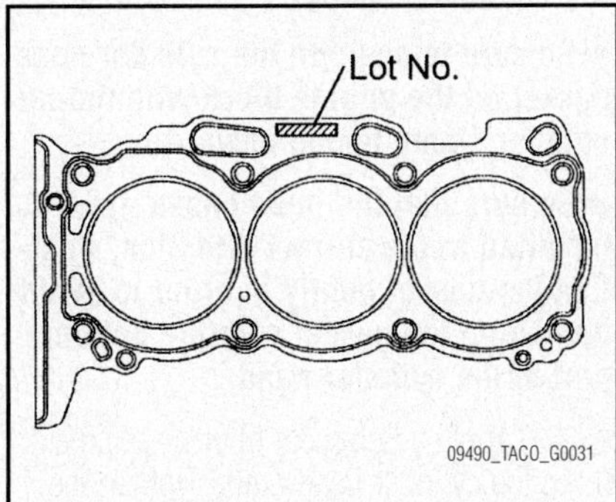

Fig. 64 Bank 1 cylinder head gasket positioning—4.0L (1GR-FE) engine

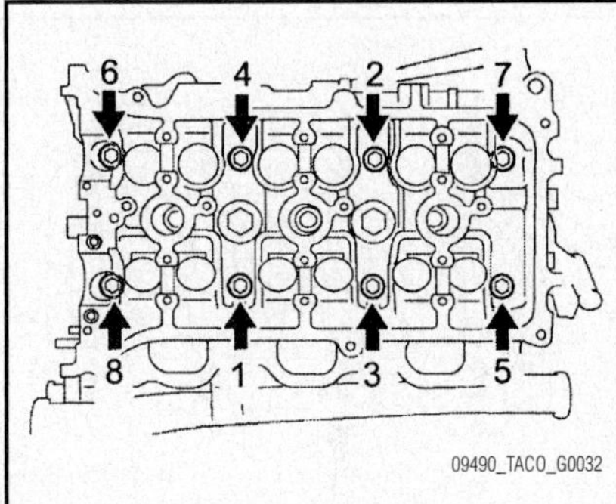

Fig. 65 Bank 1 cylinder head bolt torque sequence—4.0L (1GR-FE) engine

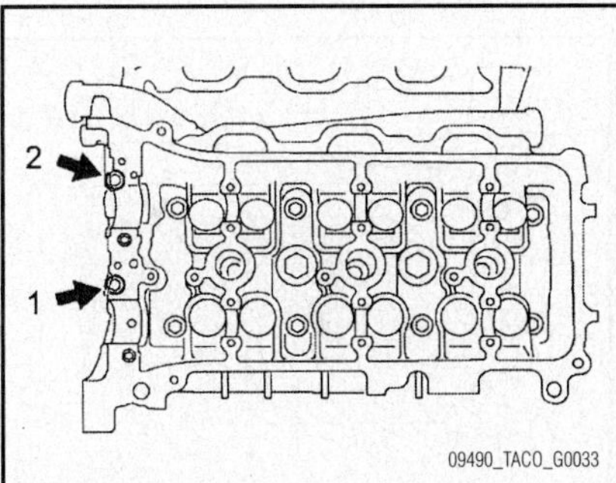

Fig. 66 Bank 2 recessed cylinder head bolt loosening sequence—4.0L (1GR-FE) engine

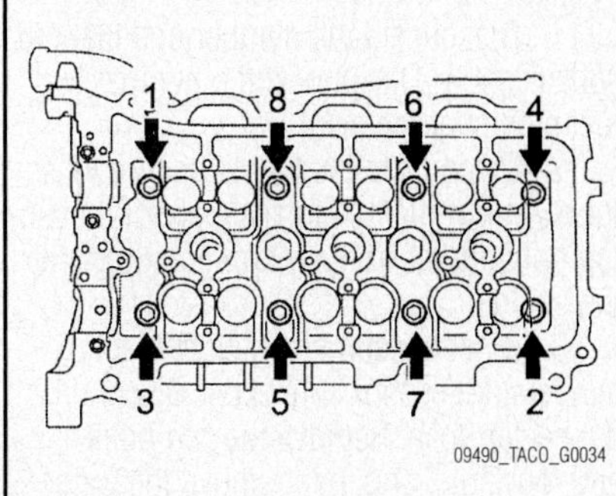

Fig. 67 Bank 2 cylinder head bolt loosening sequence—4.0L (1GR-FE) engine

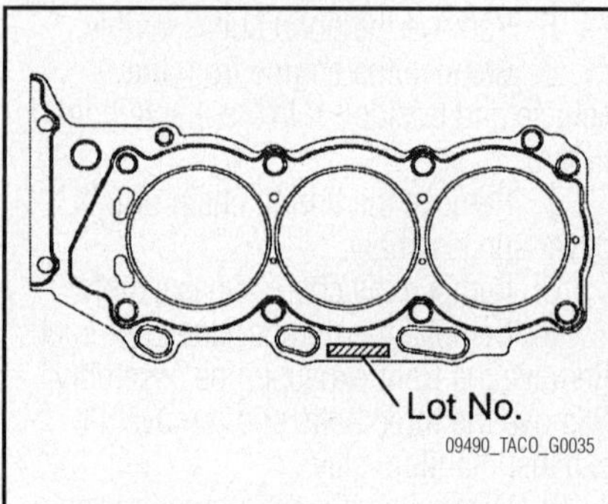

Fig. 68 Bank 2 cylinder head gasket positioning—4.0L (1GR-FE) engine

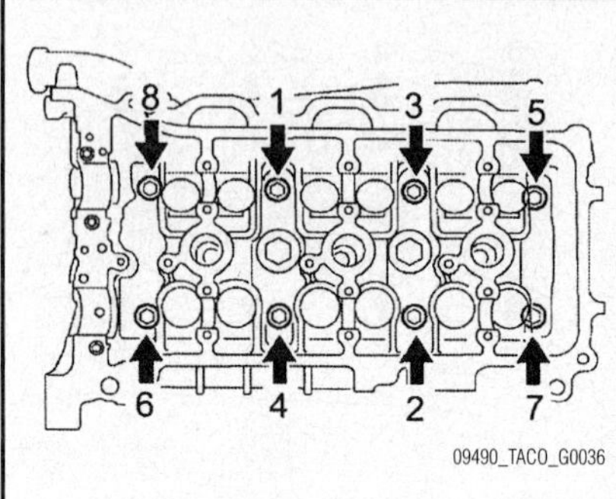

Fig. 69 Bank 2 Cylinder head bolt torque sequence—4.0L (1GR-FE) engine

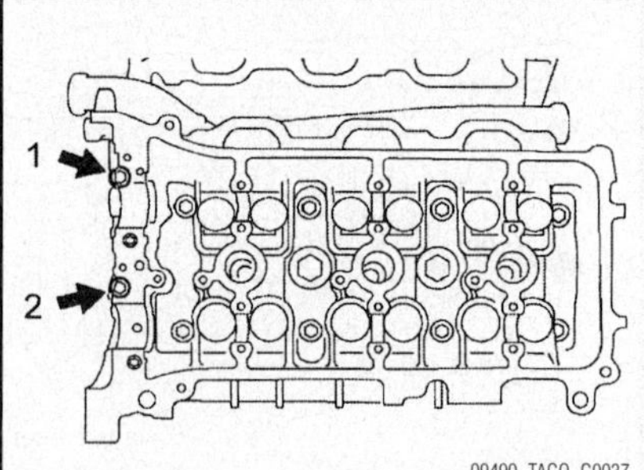

Fig. 70 Bank 2 recessed cylinder head bolt torque sequence—4.0L (1GR-FE) engine

11. Disconnect the number one fuel pipe subassembly. Disconnect the number two fuel pipe subassembly.
12. Disconnect the fuel injector connectors. Remove the ten intake manifold retaining bolts. Remove the intake manifold from the engine.
13. Disconnect the engine coolant temperature sensor connector. Disconnect the heater hose. Remove the two bolts and four nuts, and then remove the water bypass joint RR and two gaskets. Remove the O-ring from the water outlet hose.
14. While raising the chain tensioner number two, insert a 0.039 inch diameter pin, into the hole. Hold the hexagonal portion of the camshaft in place, using a wrench.

➡**Be careful not to damage the cylinder head and valve lifter with the wrench.**

15. Remove the two bolts. Remove the camshaft timing gear, camshaft timing gear assembly and timing chain number two.

➡**Do not disassemble the camshaft timing gear assembly.**

16. Remove the bolt and then remove the chain tensioner number two.
17. Remove the camshafts.
18. Remove the two bolts. Separate the two ground cables.
19. Loosen the cylinder head retaining bolts in several steps and in the proper removal sequence. Remove the bolts.

➡**Be careful not to drop the washers into the cylinder head. Head warpage and cracking could result in removing the bolts in the wrong order.**

20. Lift the cylinder head from the dowels on the cylinder block and remove it from the engine.

To install:

21. Before installing, thoroughly clean the gasket mating surfaces and check for warpage. Remove any old packing material. Be careful not to drop any oil on the contact surfaces of the cylinder head or engine block.
22. Apply continuous beads of seal packing, part number 08826-00080 or equivalent, to the cylinder head gasket. Bead width should be 0.098–0.118 inch.

➡**Install the cylinder head gasket within three minutes after applying the seal packing. Install the cylinder head bolts within fifteen minutes after applying the seal packing.**

➡**Be sure to position the cylinder head gasket on the engine block with the lot number stamp facing upward.**

➡**Be sure that the head gasket is installed in the correct direction. Position the gasket gently in order to avoid damaging the gasket with the bottom part of the cylinder head.**

23. Apply a light coat of clean engine oil to the cylinder head bolts. Tighten the cylinder head bolts, in two successive steps, to specification and in the proper sequence.

➡**Replace defective bolts, as required.**

24. Continue the installation in the reverse of the removal procedure.
25. Be sure to fill the cooling system with the proper grade and type engine coolant.
26. Be sure to fill the engine oil with the proper grade and type oil.

➡ **Do not fill the engine with oil for at least four hours after the cylinder head has been installed.**

27. Start the engine and check for leaks. Correct as required.

2WD–Bank 2

➡**The engine must first be removed from the vehicle.**

1. Before servicing the vehicle, refer to the Precautions Section.
2. Properly discharge the fuel system pressure.
3. Properly disarm the SRS system.
4. Disconnect the negative battery cable. Disconnect the positive battery cable. Remove the battery.
5. Remove the hood subassembly.
6. Remove the engine from the vehicle and position it in a suitable holding fixture.
7. Remove the timing chain or belt cover subassembly.
8. Remove the chain subassembly.
9. Remove the number one cool inlet. Remove the number two front exhaust pipe assembly. Remove the three bolts and remove the exhaust manifold stay.
10. Disconnect the air fuel ratio sensor connector. Remove the six nuts; remove the exhaust manifold and gasket.
11. Disconnect the fuel injector connectors. Remove the ten intake manifold retaining bolts. Remove the intake manifold from the engine.
12. Disconnect the engine coolant temperature sensor connector. Disconnect the heater hose. Remove the two bolts and four nuts, and then remove the water bypass joint RR and two gaskets. Remove the O-ring from the water outlet hose.
13. Remove the two bolts and then remove the chain vibration damper number one.
14. While pushing down on chain tensioner number two, insert a 0.039 inch diameter pin, into the hole. Hold the hexagonal portion of the camshaft in place, using a wrench.

➡**Be careful not to damage the cylinder head and valve lifter with the wrench.**

15. Remove the two bolts. Remove the camshaft timing gear, camshaft timing gear assembly and timing chain number two.

➡**Do not disassemble the camshaft timing gear assembly.**

16. Remove the bolt and then remove the chain tensioner number three.
17. Remove the camshafts.
18. Remove the two bolts. Separate the two ground cables.
19. Loosen the cylinder head retaining bolts in several steps and in the proper removal sequence. Remove the bolts.

➡**Be careful not to drop the washers into the cylinder head. Head warpage and cracking could result in removing the bolts in the wrong order.**

20. Lift the cylinder head from the dowels on the cylinder block and remove it from the engine.

To install:

21. Before installing, thoroughly clean the gasket mating surfaces and check for warpage. Remove any old packing material. Be careful not to drop any oil on the contact surfaces of the cylinder head or engine block.
22. Apply continuous beads of seal packing, part number 08826-00080 or equivalent, to the cylinder head gasket. Bead width should be 0.098–0.118 inch.

➡**Install the cylinder head gasket within three minutes after applying the seal packing. Install the cylinder head bolts within fifteen minutes after applying the seal packing.**

➡**Be sure to position the cylinder head gasket on the engine block with the lot number stamp facing upward.**

➡**Be sure that the head gasket is installed in the correct direction. Position the gasket gently in order to avoid damaging the gasket with the bottom part of the cylinder head.**

23. Apply a light coat of clean engine oil to the cylinder head bolts. Tighten the cylinder head bolts, in two successive steps, to specification and in the proper sequence.

➡**Replace defective bolts, as required.**

24. Continue the installation in the reverse of the removal procedure.
25. Be sure to fill the cooling system with the proper grade and type engine coolant.
26. Be sure to fill the engine oil with the proper grade and type oil.

➡ **Do not fill the engine with oil for at least four hours after the cylinder head has been installed.**

27. Start the engine and check for leaks. Correct as required.

4WD and PreRunner–Bank 1

1. Before servicing the vehicle, refer to the Precautions Section.
2. Properly discharge the fuel system pressure.
3. Properly disarm the SRS system.
4. Disconnect the negative battery cable. Disconnect the positive battery cable. Remove the battery.
5. Remove the power steering gear assembly.
6. Remove the front differential carrier assembly.
7. Remove the timing chain or belt cover subassembly.
8. Remove the chain subassembly.
9. Remove the number one cool inlet. Remove the front exhaust pipe assembly. Remove the three bolts and remove the exhaust manifold stay.
10. Disconnect the air fuel ratio sensor connector. Remove the six nuts; remove the exhaust manifold and gasket.
11. Disconnect the number one fuel pipe subassembly. Disconnect the number two fuel pipe subassembly.
12. Disconnect the fuel injector connectors. Remove the ten intake manifold retaining bolts. Remove the intake manifold from the engine.
13. Disconnect the engine coolant temperature sensor connector. Disconnect the heater hose. Remove the two bolts and four nuts, and then remove the water bypass joint RR and two gaskets. Remove the O-ring from the water outlet hose.
14. While raising the chain tensioner number two, insert a 0.039 inch diameter pin, into the hole. Hold the hexagonal portion of the camshaft in place, using a wrench.

➡**Be careful not to damage the cylinder head and valve lifter with the wrench.**

15. Remove the two bolts. Remove the camshaft timing gear, camshaft timing gear assembly and timing chain number two.

➡**Do not disassemble the camshaft timing gear assembly.**

16. Remove the bolt and then remove the chain tensioner number two.
17. Remove the camshafts.
18. Remove the two bolts. Separate the two ground cables.
19. Loosen the cylinder head retaining bolts in several steps and in the proper removal sequence. Remove the bolts.

➡**Be careful not to drop the washers into the cylinder head. Head warpage and cracking could result in removing the bolts in the wrong order.**

20. Lift the cylinder head from the dowels on the cylinder block and remove it from the engine.

To install:

21. Before installing, thoroughly clean the gasket mating surfaces and check for warpage. Remove any old packing material. Be careful not to drop any oil on the contact surfaces of the cylinder head or engine block.
22. Apply continuous beads of seal packing, part number 08826-00080 or equivalent, to the cylinder head gasket. Bead width should be 0.098–0.118 inch.

➡**Install the cylinder head gasket within three minutes after applying the seal packing. Install the cylinder head bolts within fifteen minutes after applying the seal packing.**

➡**Be sure to position the cylinder head gasket on the engine block with the lot number stamp facing upward.**

➡**Be sure that the head gasket is installed in the correct direction. Position the gasket gently in order to avoid damaging the gasket with the bottom part of the cylinder head.**

23. Apply a light coat of clean engine oil to the cylinder head bolts. Tighten the cylinder head bolts, in two successive steps, to specification and in the proper sequence.

➡**Replace defective bolts, as required.**

24. Continue the installation in the reverse of the removal procedure.
25. Be sure to fill the cooling system with the proper grade and type engine coolant.
26. Be sure to fill the engine oil with the proper grade and type oil.

➡ **Do not fill the engine with oil for at least four hours after the cylinder head has been installed.**

27. Start the engine and check for leaks. Correct as required.

4WD and PreRunner–Bank 2

1. Before servicing the vehicle, refer to the Precautions Section.
2. Properly discharge the fuel system pressure.
3. Properly disarm the SRS system.

4. Disconnect the negative battery cable. Disconnect the positive battery cable. Remove the battery.
5. Remove the power steering gear assembly.
6. Remove the front differential carrier assembly.
7. Remove the timing chain or belt cover subassembly.
8. Remove the chain subassembly.
9. Remove the number one cool inlet. Remove the number two front exhaust pipe assembly. Remove the three bolts and remove the exhaust manifold stay.
10. Disconnect the air fuel ratio sensor connector. Remove the six nuts; remove the exhaust manifold and gasket.
11. Disconnect the fuel injector connectors. Remove the ten intake manifold retaining bolts. Remove the intake manifold from the engine.
12. Disconnect the engine coolant temperature sensor connector. Disconnect the heater hose. Remove the two bolts and four nuts, and then remove the water bypass joint RR and two gaskets. Remove the O-ring from the water outlet hose.
13. Remove the two bolts and then remove the chain vibration damper number one.
14. While pushing down on chain tensioner number two, insert a 0.039 inch diameter pin, into the hole. Hold the hexagonal portion of the camshaft in place, using a wrench.

➡Be careful not to damage the cylinder head and valve lifter with the wrench.

15. Remove the two bolts. Remove the camshaft timing gear, camshaft timing gear assembly and timing chain number two.

➡Do not disassemble the camshaft timing gear assembly.

16. Remove the bolt and then remove the chain tensioner number three.
17. Remove the camshafts.
18. Remove the two bolts. Separate the two ground cables.
19. Loosen the cylinder head retaining bolts in several steps and in the proper removal sequence. Remove the bolts.

➡Be careful not to drop the washers into the cylinder head. Head warpage and cracking could result in removing the bolts in the wrong order.

20. Lift the cylinder head from the dowels on the cylinder block and remove it from the engine.

To install:

21. Before installing, thoroughly clean the gasket mating surfaces and check for warpage. Remove any old packing material. Be careful not to drop any oil on the contact surfaces of the cylinder head or engine block.
22. Apply continuous beads of seal packing, part number 08826-00080 or equivalent, to the cylinder head gasket. Bead width should be 0.098–0.118 inch.

➡Install the cylinder head gasket within three minutes after applying the seal packing. Install the cylinder head bolts within fifteen minutes after applying the seal packing.

➡Be sure to position the cylinder head gasket on the engine block with the lot number stamp facing upward.

➡Be sure that the head gasket is installed in the correct direction. Position the gasket gently in order to avoid damaging the gasket with the bottom part of the cylinder head.

23. Apply a light coat of clean engine oil to the cylinder head bolts. Tighten the cylinder head bolts, in two successive steps, to specification and in the proper sequence.

➡Replace defective bolts, as required.

24. Continue the installation in the reverse of the removal procedure.
25. Be sure to fill the cooling system with the proper grade and type engine coolant.
26. Be sure to fill the engine oil with the proper grade and type oil.

➡ Do not fill the engine with oil for at least four hours after the cylinder head has been installed.

27. Start the engine and check for leaks. Correct as required.

ENGINE ASSEMBLY

REMOVAL & INSTALLATION

2.7L (2TR-FE) Engine

See Figure 72.

1. Before servicing the vehicle, refer to the Precautions Section.
2. Properly disable the SRS system.
3. Properly relieve the fuel system pressure.
4. Disconnect the negative battery cable. Disconnect the positive battery cable.
5. Matchmark and remove the hood.
6. On 4WD and PreRunner, remove the number one engine cover.
7. On 4WD and PreRunner (regular cab) remove the number two engine cover.
8. Drain the engine coolant. Remove the battery. Remove the battery tray. Drain the engine oil.
9. Remove the radiator support to frame seal, left side.
10. Remove the fan shroud. Remove the air cleaner cap sub assembly. Remove the air cleaner filter element subassembly. Remove the air cleaner case.
11. Separate the vane pump. Remove the radiator hoses. Separate the air conditioning compressor from the engine and position it to the side.
12. Disconnect the heater hoses from the heater core.
13. Disconnect the fuel line hoses. Separate the fuel vapor feed hose assembly. Disconnect the number one air injection hose.
14. Disconnect all engine wiring harnesses and connectors in the engine compartment.

➡Disconnect the connector from the ECM, then pull the engine wire harness to the compartment side. Do not forcibly pull the wire harness to the engine compartment side.

15. Remove the exhaust pipe assembly. Remove the front exhaust pipe assembly.
16. Remove the transmission.
17. Install engine hanger, tool 12281-75040 and retainer bolts 91552-A1020. Attach a suitable engine lifting fixture.
18. Remove the engine retaining bolts and mounts.

➡Check to be sure nothing will interfere with the removal of the engine.

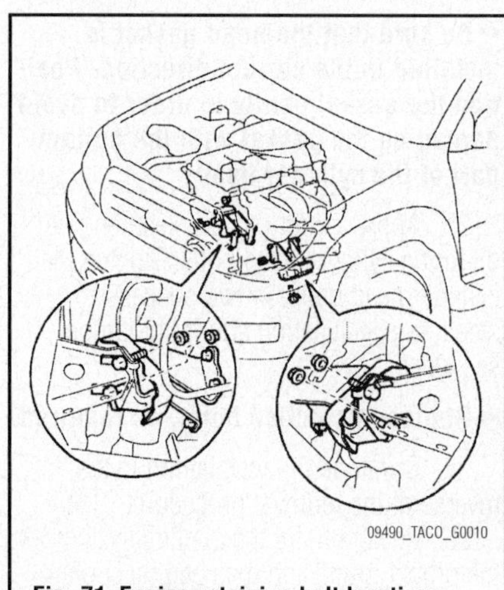

Fig. 71 Engine retaining bolt locations—2.7L (2TR-FE) engine

Check to be sure all required wires, hoses and connectors are disconnected.

19. Remove the engine from the vehicle.

To install:

20. Installation is the reverse of the removal procedure.
21. Tighten the four mounting insulator bolts to 28 ft. lbs.
22. Be sure to fill the cooling system with the proper grade and type engine coolant.
23. Be sure to fill the engine with the proper grade and type engine oil.
24. Check ignition timing. Check idle speed.
25. Start the engine and check for leaks, correct as required.

4.0L (1GR-FE) Engine

See Figure 72.

1. Before servicing the vehicle, refer to the Precautions Section.
2. Properly disable the SRS system.
3. Properly relieve the fuel system pressure.
4. Disconnect the negative battery cable. Disconnect the positive battery cable.
5. Disconnect the windshield washer hose. Matchmark and remove the hood.
6. Drain the engine coolant. Remove the battery. Remove the battery tray. Drain the engine oil.
7. Remove the radiator support to frame seal, left side.
8. Remove the fan shroud. Remove the radiator hoses. Remove the radiator.
9. Remove the V bank cover.
10. Remove the transmission. If equipped with manual transmission, remove the clutch cover and clutch.
11. Remove the fan pulley. Remove the vane pump assembly. Remove the alternator.

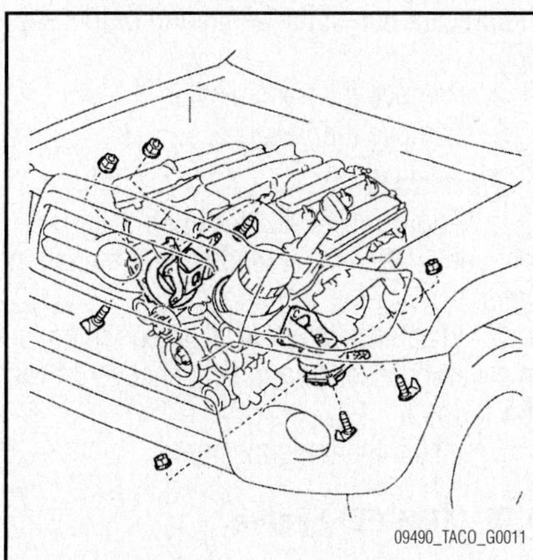

Fig. 72 Engine retaining bolt locations—4.0L (1GR-FE) engine

12. Separate the air conditioning compressor from the engine and position it to the side.
13. Disconnect the heater hoses from the heater core. Disconnect the fuel line hoses. Remove the intake air surge tank.
14. To separate the main engine wire harness, remove the glove box door. Remove the instrument panel finish panel subassembly, lower right side. Disconnect the ECM connectors. Disconnect the connectors from the ECU, if equipped with 4WD. Pull the harness into the engine compartment. Disconnect the front differential connector, if equipped with 4WD. Disconnect the three connectors from the engine compartment relay block. Separate the engine wire from the engine compartment relay block.
15. Disconnect the ground cable from the cylinder head.
16. Remove the front exhaust pipe assembly.
17. Install engine hanger, tool 12281-31070 and 12282-31050 and retainer bolts 90119-08177. Attach a suitable engine lifting fixture.
18. Remove the engine retaining bolts and mounts.

➡Check to be sure nothing will interfere with the removal of the engine. Check to be sure all required wires, hoses and connectors are disconnected.

19. Remove the engine from the vehicle.

To install:

20. Installation is the reverse of the removal procedure.
21. Tighten the four mounting insulator bolts to 28 ft. lbs.
22. If equipped with automatic transmission, perform the automatic transmission initialization procedure listed below.

➡Perform the automatic transmission initialization procedure when replacing the automatic transmission, engine or ECM.

23. On the A340E automatic transmission, initialization is completed by connecting and disconnecting the negative battery cable.
24. On the A750E automatic transmission, initialization is performed using the intelligent tester tool.
25. Turn the ignition switch OFF.
26. Connect the intelligent tester tool together with the controller area network vehicle interface module (CAN VIN) to the DLC3.
27. Turn the ignition switch to the ON position.
28. Push the intelligent tester tool main switch to the ON position.
29. Select the items, DIAGNOSIS/ ENHANCED OBD II.
30. Perform the reset memory procedure from the ENGINE menu.

➡After performing the reset memory, be sure to perform the road test procedure. For road test procedure information, refer to the intelligent tester instruction manual.

31. Be sure to fill the cooling system with the proper grade and type engine coolant.
32. Be sure to fill the engine with the proper grade and type engine oil.
33. Check ignition timing. Check idle speed.
34. Start the engine and check for leaks, correct as required.

EXHAUST MANIFOLD

REMOVAL & INSTALLATION

2.7L (2TR-FE) Engine

See Figure 73.

1.Before servicing the vehicle, refer to the Precautions Section.
2. Disconnect the negative battery cable.
3. Disconnect the exhaust manifold to exhaust flange nuts.
4. Remove the necessary components to gain access to the exhaust manifold retaining bolts.
5. Remove the exhaust manifold retaining nuts. Discard the nuts. Remove the exhaust manifold from the engine.

To install:

6. Installation is the reverse of the removal procedure.
7. Be sure to use new gaskets. Be sure to use new retaining nuts.

4.0L (1GR-FE) Engine

1. Before servicing the vehicle, refer to the Precautions Section.
2. Disconnect the negative battery cable.
3. Disconnect the exhaust manifold to exhaust flange nuts.
4. Remove the necessary components to gain access to the exhaust manifold retaining bolts.
5. Remove the exhaust manifold retaining nuts. Discard the nuts. Remove the exhaust manifold from the engine.

To install:

6. Installation is the reverse of the removal procedure.

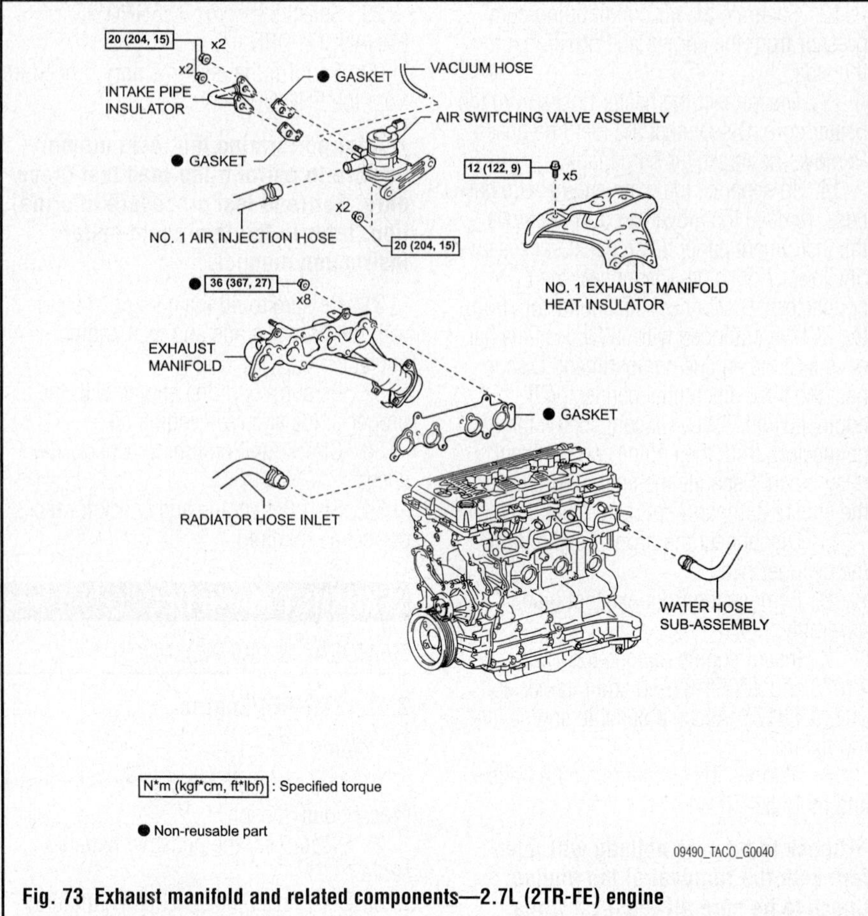

Fig. 73 Exhaust manifold and related components—2.7L (2TR-FE) engine

7. Be sure to use new gaskets. Be sure to use new retaining nuts.

FLYWHEEL

REMOVAL & INSTALLATION

2.7L (2TR-FE) Engines

See Figure 74.

1. Before servicing the vehicle, refer to the precautions in the beginning of this section.
2. Remove the transmission.

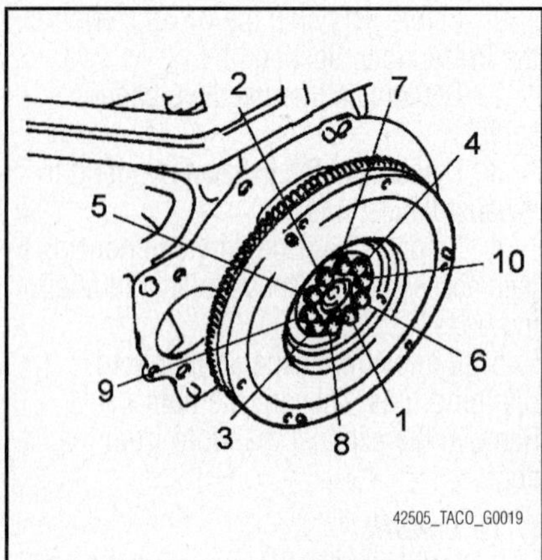

Fig. 74 Flywheel torque sequence

3. Remove the clutch assembly.
4. Remove the flywheel.

To install:

5. Apply THREE BOND 1324 or equivalent adhesive to the threads of the mounting bolts.
6. Install the flywheel. Tighten the bolts in sequence as follows:
   a. Step 1: 27 Nm (20 ft. lbs.)
   b. Step 2: Plus 90°
7. Install the clutch assembly.
8. Install the transmission.

4.0L (1GR-FE) Engines

See Figure 75.

1. Before servicing the vehicle, refer to the precautions in the beginning of this section.
2. Remove the transmission.
3. Remove the clutch assembly.
4. Remove the flywheel.

To install:

5. Apply THREE BOND 1324 or equivalent adhesive to the threads of the mounting bolts.
6. Install the flywheel. Tighten the bolts in several passes in sequence to 85 Nm (63 ft. lbs.).

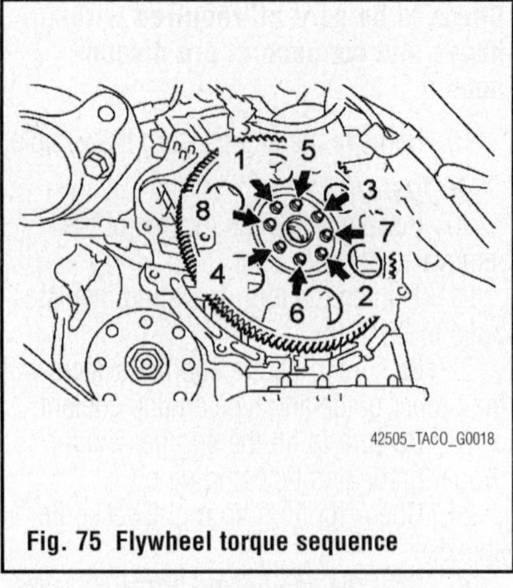

Fig. 75 Flywheel torque sequence

7. Install the clutch assembly.
8. Install the transmission.

FLEXPLATE

REMOVAL & INSTALLATION

2.7L (2TR-FE) Engines

See Figure 76.

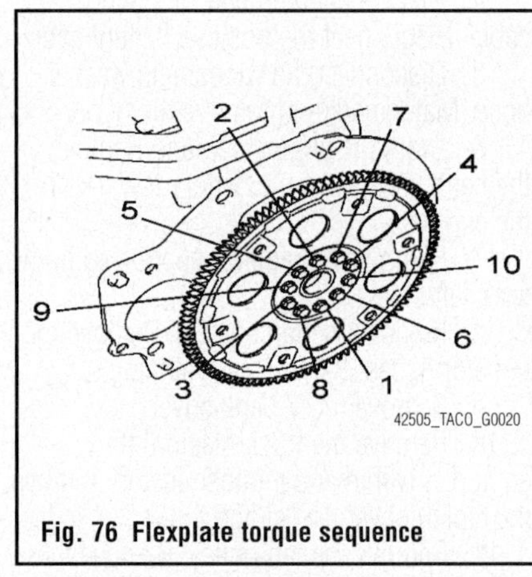

Fig. 76 Flexplate torque sequence

1. Before servicing the vehicle, refer to the precautions in the beginning of this section.
2. Remove the transmission.
3. Remove the flexplate.

To install:

4. Apply THREE BOND 1324 or equivalent adhesive to the threads of the mounting bolts.
5. Install the flexplate. Tighten the bolts in several passes and in sequence to 74 Nm (55 ft. lbs.).
6. Install the transmission.

4.0L (1GR-FE) Engine

See Figure 77.

1. Before servicing the vehicle, refer to the precautions in the beginning of this section.

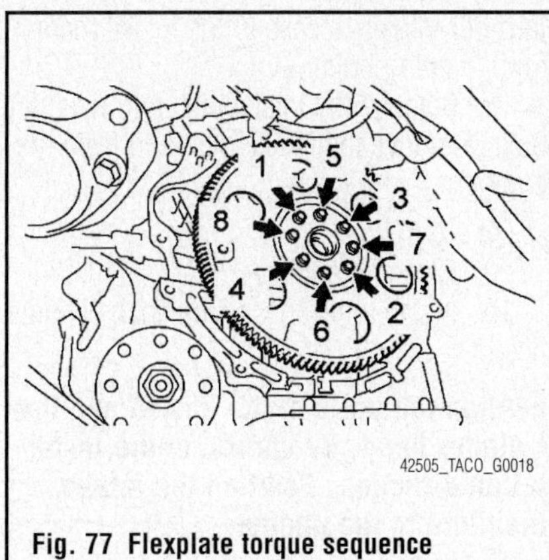

Fig. 77 Flexplate torque sequence

2. Remove the transmission.
3. Remove the flexplate.

To install:

4. Apply THREE BOND 1324 or equivalent adhesive to the threads of the mounting bolts.
5. Install the flexplate. Tighten the bolts in several passes in sequence to 83 Nm (61 ft. lbs.).
6. Install the clutch assembly.
7. Install the transmission.

INTAKE MANIFOLD

REMOVAL & INSTALLATION

2.7L (2TR-FE) Engine

See Figure 78.

1. Before servicing the vehicle, refer to the Precautions Section.
2. Properly relieve the fuel system pressure.
3. Properly disarm the SRS system
4. Disconnect the negative battery cable.
5. On 4WD and PreRunner, remove the engine undercover subassembly.
6. Drain the engine coolant. Remove the air intake connector.
7. Remove the throttle body and motor assembly. Disconnect the fuel hoses.
8. Disconnect the fuel vapor feed hose from the VSR. Disconnect the vacuum hose. Remove the bolt, and then remove the clamp bracket.
9. Disconnect the number two water bypass hose. Disconnect the number three ventilation hose. Disconnect the VSV connector.
10. Disengage the engine wire harness clamp. Disconnect the air conditioning compressor magnetic clutch connector.
11. Disengage the wire harness clamp.

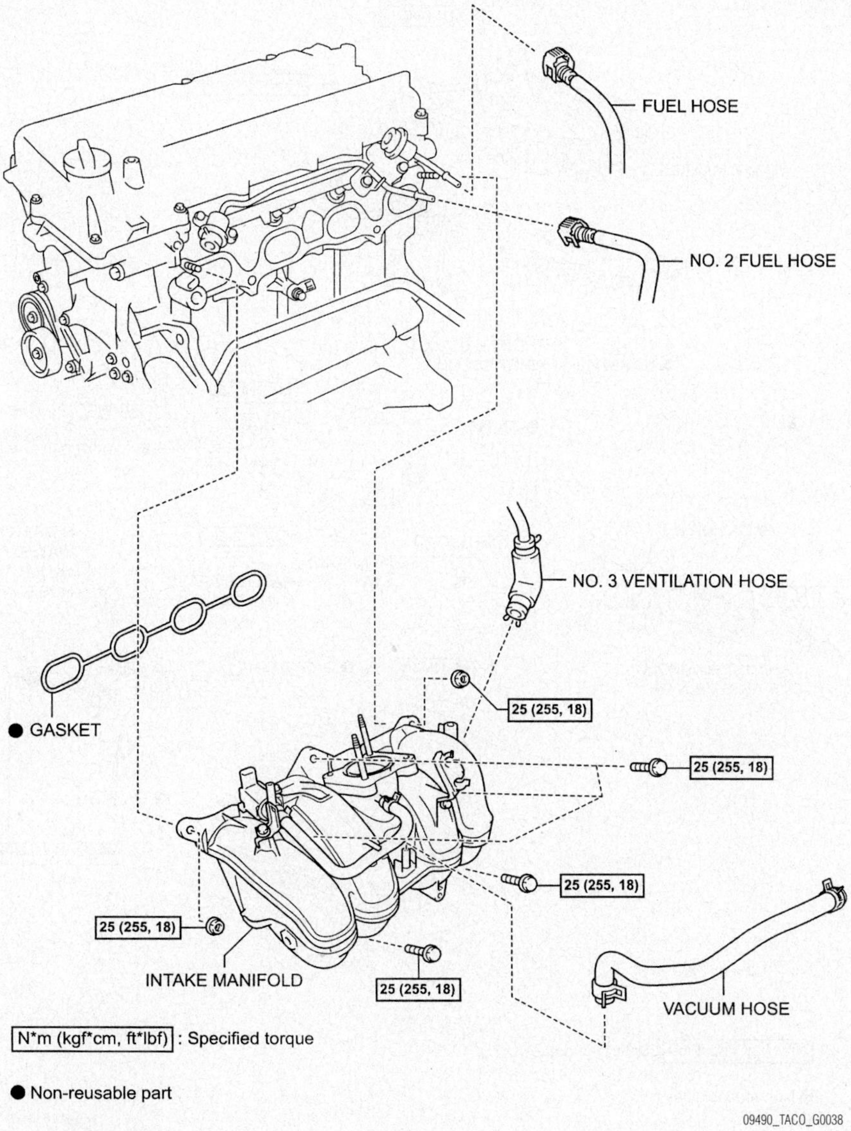

Fig. 78 Intake manifold and related components—2.7L (2TR-FE) engine

Remove the bolt and harness clamp bracket. Disconnect the three connectors.

12. Remove the retaining nut inside the relay block. Disconnect the engine wire harness from the relay block.

13. Remove the five bolts and two nuts retaining the intake manifold in place. Remove the intake manifold from the engine.

To install:

14. Clean all surfaces.

15. Install a new gasket onto the intake manifold. Position the intake manifold to the engine.

16. Install the retaining bolts and tighten to specification, in an alternating sequence.

17. Continue the installation in the reverse order of the removal procedure.

18. Be sure to fill the cooling system with the proper grade and type engine coolant.

19. Start the engine and check for leaks. Correct as required.

4.0L (1GR-FE) Engine

See Figure 79.

1. Before servicing the vehicle, refer to the Precautions Section.

2. Properly relieve the fuel system pressure.

3. Properly disarm the SRS system

4. Disconnect the negative battery cable.

5. Drain the engine coolant. Remove the air cleaner assembly.

6. Disconnect the fuel injector wiring connectors.

7. Remove the necessary components in order to gain access to the intake manifold retaining bolts.

8. Remove the intake manifold retaining bolts. Remove the intake manifold from the engine.

To install:

9. Clean all surfaces.

10. Install a new gasket on each cylinder head.

➡Align the ports of the gasket and the cylinder head. Be careful of the installation direction. Position the intake manifold to the engine.

11. Install the retaining bolts and tighten to specification, in an alternating sequence.

12. Continue the installation in the reverse order of the removal procedure.

13. Be sure to fill the cooling system

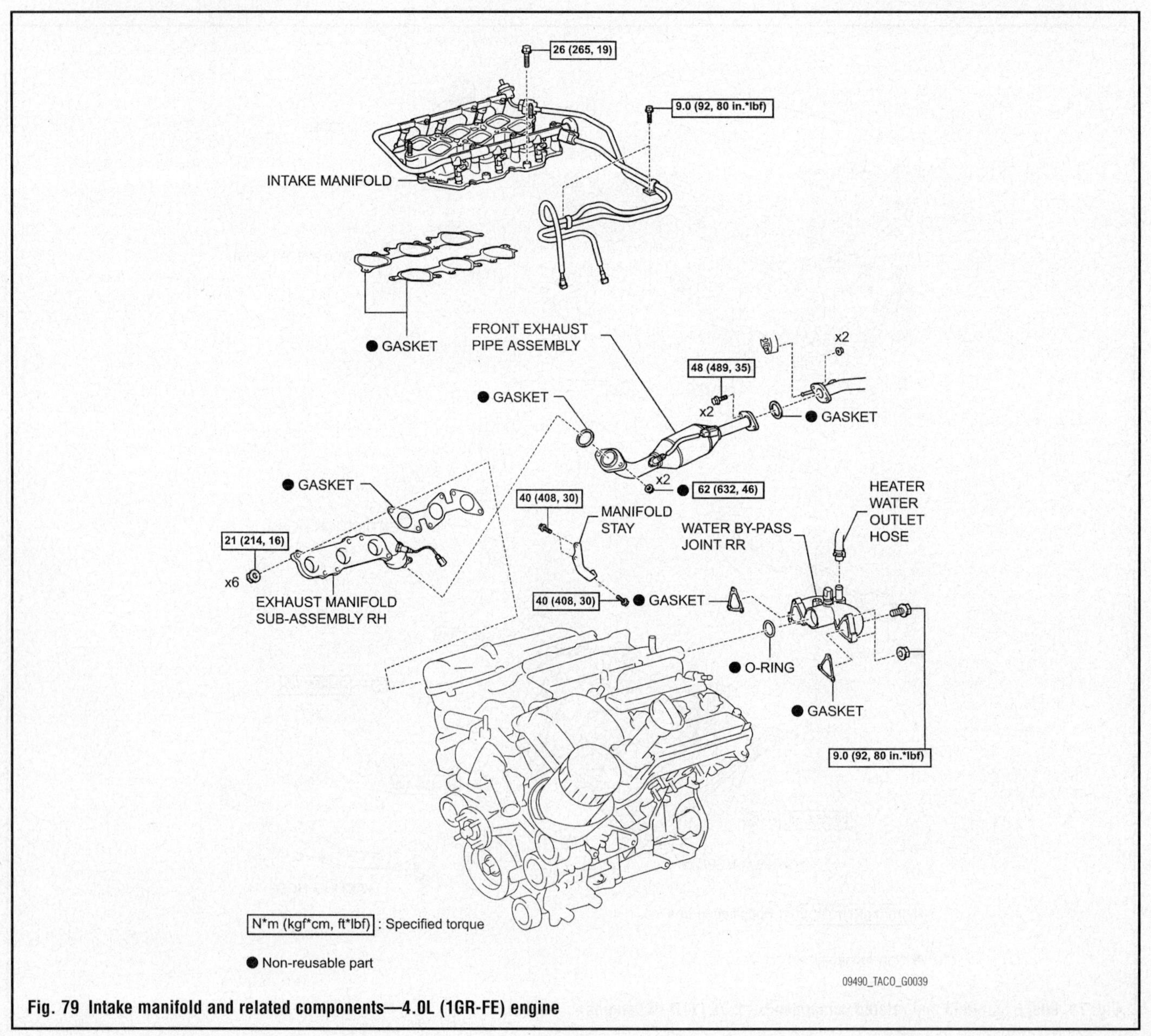

Fig. 79 Intake manifold and related components—4.0L (1GR-FE) engine

with the proper grade and type engine coolant.

14. Start the engine and check for leaks. Correct as required.

OIL PAN

REMOVAL & INSTALLATION

2.7L (2TR-FE) Engine

See Figures 80 through 83.

1. Before servicing the vehicle, refer to the Precautions Section.
2. Disconnect the negative battery cable.
3. Raise and support the vehicle safely.
4. Remove the engine undercover. Drain the engine oil.
5. Remove the necessary components in order to gain access to the lower oil pan retaining bolts.
6. Remove the eighteen bolts and two nuts. Insert the blade of tool SST09032-00100 between the pans. Cut through the sealer and remove the lower oil pan from the engine.

➡Be careful not to damage the contact surface of the oil pans.

7. Remove the two bolts and nuts. Remove the oil strainer. Discard the gasket.
8. To remove the upper oil pan, remove the sixteen bolts and two nuts. Remove the upper oil pan from the engine, by prying it apart using a suitable tool.

➡Be careful not to damage the sealing surface between the upper oil pan and the cylinder block.

To install:

9. Apply a continuous bead (0.079–0.118 inch in diameter) of seal packing, part number 08826-00080 or equivalent, to the sealing surface of the oil pan.

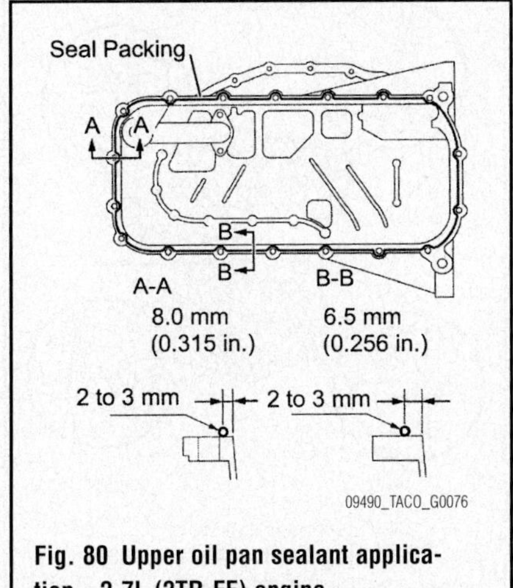

Fig. 80 Upper oil pan sealant application—2.7L (2TR-FE) engine

➡Remove any oil from the contact surface. Install the upper oil pan within three minutes of applying the seal packing. Do not start the engine for at least two hours after the installation of the oil pan.

10. Loosely install the upper oil pan bolts and nuts. Bolt "A" is 0.79 inch long and bolt "B" is 1.57 inch long. Uniformly tighten the bolts to 19 ft. lbs, in the proper sequence.
11. Install the oil strainer assembly. Torque the bolts to 19 ft. lbs.
12. Apply a continuous bead (0.118–0.157 inch in diameter) of seal packing, part number 08826-00080 or equivalent, to the sealing surface of the oil pan.

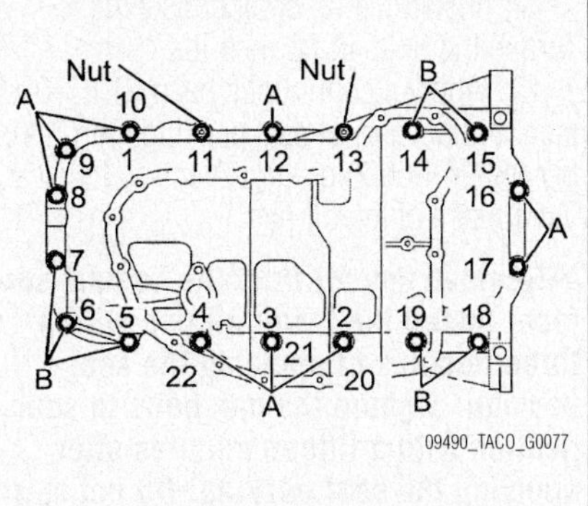

Fig. 81 Upper oil pan bolt torque sequence—2.7L (2TR-FE) engine

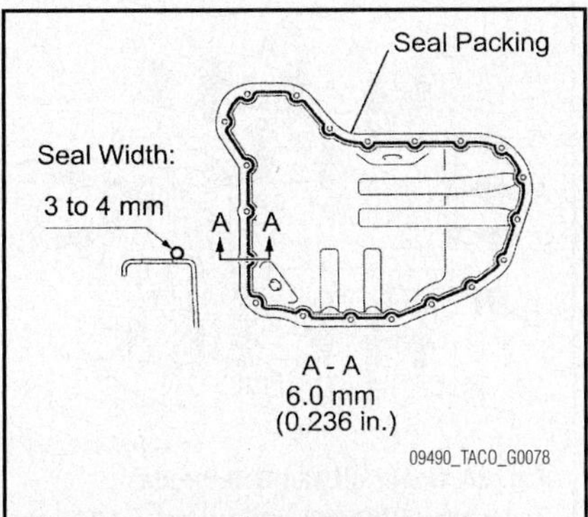

Fig. 82 Lower oil pan sealant application—2.7L (2TR-FE) engine

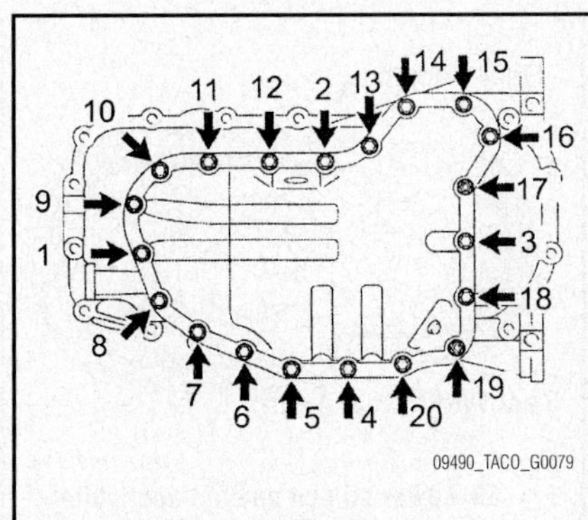

Fig. 83 Lower oil pan bolt torque sequence—2.7L (2TR-FE) engine

➡Remove any oil from the contact surface. Install the lower oil pan within three minutes of applying the seal packing. Do not start the engine for at least two hours after the installation of the oil pan.

13. Loosely install the lower oil pan bolts and nuts. Uniformly tighten the bolts to 80 inch lbs, in the proper sequence.
14. Continue the installation in the reverse order of the removal procedure.
15. Be sure to fill the engine with the proper grade and type engine coolant.
16. Be sure to fill the engine with the proper grade and type engine oil.
17. Start the engine and check for leaks. Correct as required.

4.0L (1GR-FE) Engine

See Figures 84 through 89.

1. Before servicing the vehicle, refer to the Precautions Section.
2. Disconnect the negative battery cable.
3. Raise and support the vehicle safely.
4. Remove the engine undercover. Drain the engine oil.

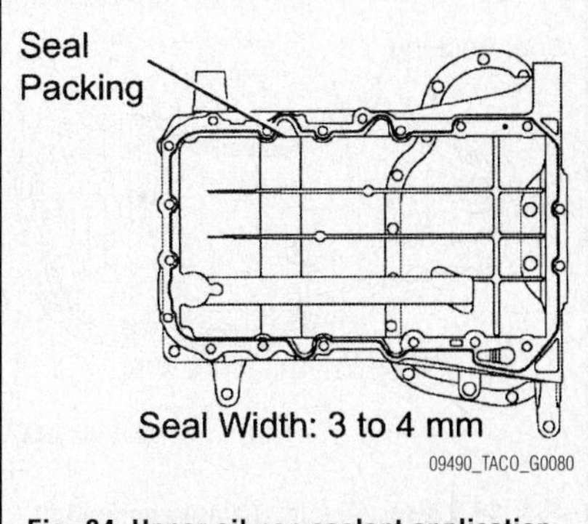

Fig. 84 Upper oil pan sealant application (2WD)—4.0L (1GR-FE) engine

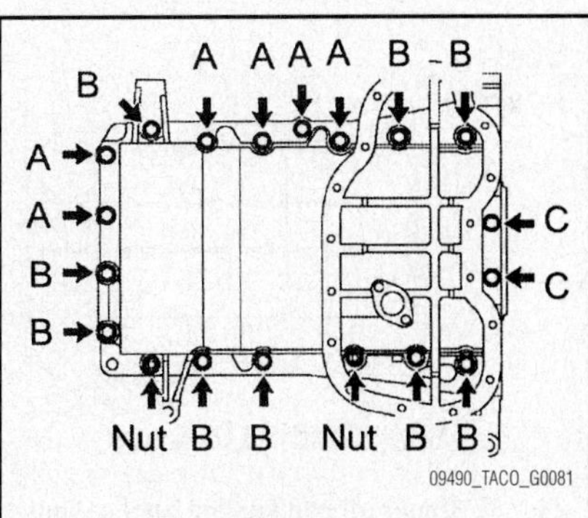

Fig. 85 Upper oil pan bolt torque sequence (2WD)—4.0L (1GR-FE) engine

5. Remove the necessary components in order to gain access to the lower oil pan retaining bolts.

6. Remove the fifteen bolts and two nuts (2WD vehicles) and ten bolts and two nuts (4WD and PreRunner). Insert the blade of tool SST09032-00100 between the pans. Cut through the sealer and remove the lower oil pan from the engine.

➡Be careful not to damage the contact surface of the oil pans.

7. Remove the two bolts and nuts. Remove the oil strainer. Discard the gasket.

8. On 4WD and PreRunner, remove the four housing bolts. Remove the flywheel housing undercover.

9. To remove the upper oil pan, remove the seventeen bolts and two nuts. Remove the upper oil pan from the engine, by prying it apart using a suitable tool.

➡Be careful not to damage the sealing surface between the upper oil pan and the cylinder block.

To install:

10. Apply a continuous bead (0.12–0.16 inch in diameter) of seal packing, part number 08826-00080 or equivalent, to the sealing surface of the oil pan.

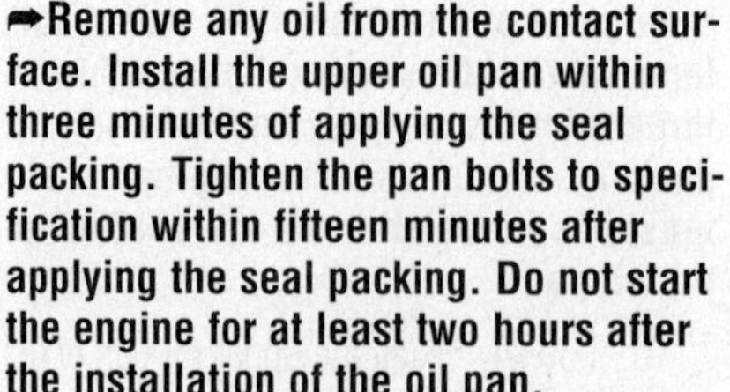

➡Remove any oil from the contact surface. Install the upper oil pan within three minutes of applying the seal packing. Tighten the pan bolts to specification within fifteen minutes after applying the seal packing. Do not start the engine for at least two hours after the installation of the oil pan.

11. Loosely install the upper oil pan bolts and nuts. Bolt "A" is 0.98 inch long, bolt "B" is 1.77 inch long and bolt "C" is 0.55 inch long. Uniformly tighten the 10mm bolt head to 7.4 ft. lbs, and the 12mm bolt head to 16 ft. lbs., in the proper sequence.

12. Install the oil strainer assembly. Torque the bolts to 80 inch lbs.

13. Apply a continuous bead (0.12–0.16 inch in diameter) of seal packing, part number 08826-00080 or equivalent, to the sealing surface of the oil pan.

➡Remove any oil from the contact surface. Install the lower oil pan within three minutes of applying the seal packing. Tighten the pan bolts to specification within fifteen minutes after applying the seal packing. Do not start the engine for at least two hours after the installation of the oil pan.

14. Loosely install the lower oil pan bolts and nuts. Uniformly tighten the bolts to 80 inch lbs, and the nuts to 7.4 ft. lbs., in several steps.

15. Continue the installation in the reverse order of the removal procedure.

16. Be sure to fill the engine with the proper grade and type engine coolant.

17. Be sure to fill the engine with the proper grade and type engine oil.

18. Start the engine and check for leaks. Correct as required.

OIL PUMP

REMOVAL & INSTALLATION

2.7L (2TR-FE) Engine

See Figures 90 through 94.

1. Before servicing the vehicle, refer to the Precautions Section.

2. Disconnect the negative battery cable.

3. Remove the engine and position it in a suitable holding fixture.

4. Remove the air intake connector. Remove the alternator.

5. Remove the belt tensioner assembly. Remove the idler pulley subassembly. Remove the air conditioning idler pulley assembly and bracket.

6. Remove the crankshaft position sensor. Remove the camshaft position sensor.

7. Remove the intake manifold. Remove the cylinder head cover.

8. Remove the crankshaft pulley. Remove the oil level gauge subassembly.

9. Remove the lower oil pan. Remove the oil strainer assembly. Remove the upper oil pan.

10. Remove the two nuts and separate the water bypass pipe number one.

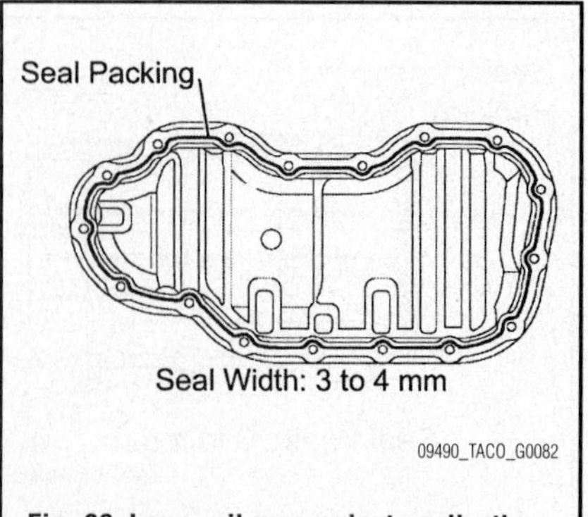

Fig. 86 Lower oil pan sealant application (2WD)—4.0L (1GR-FE) engine

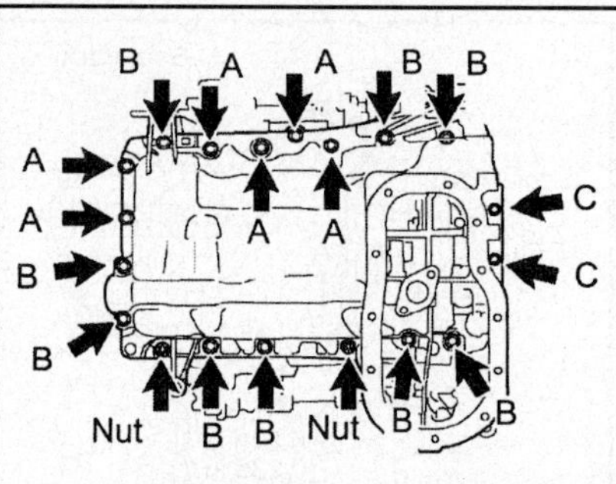

Fig. 88 Upper oil pan bolt torque sequence (4WD and PreRunner)—4.0L (1GR-FE) engine

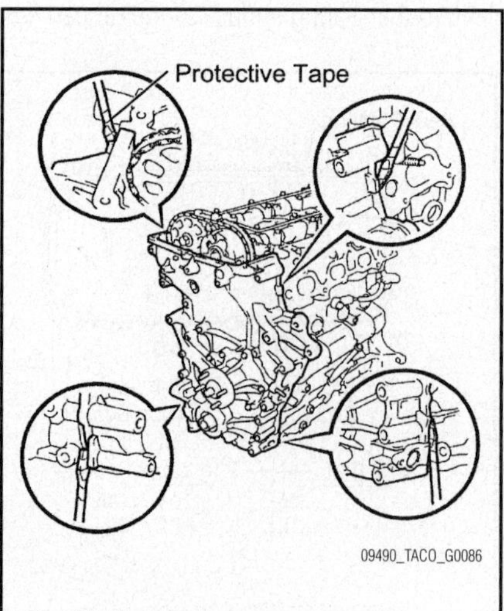

Fig. 90 Timing chain cover removal points—2.7L (2TR-FE) engine

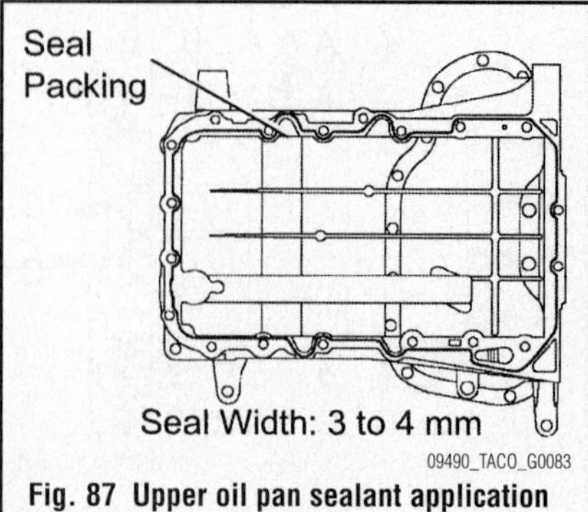

Fig. 87 Upper oil pan sealant application (4WD and PreRunner)—4.0L (1GR-FE) engine

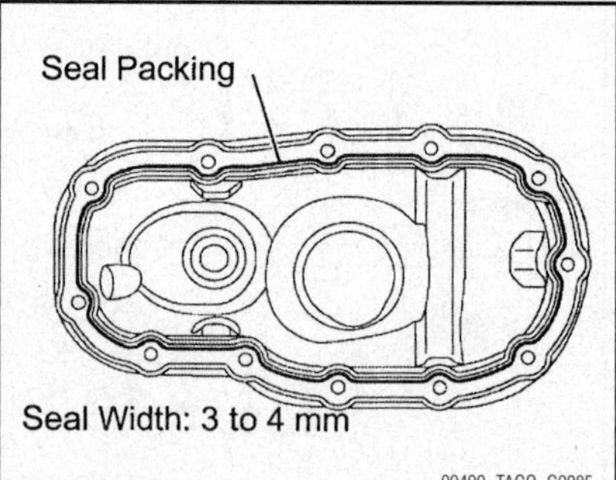

Fig. 89 Lower oil pan sealant application (4WD and PreRunner)—4.0L (1GR-FE) engine

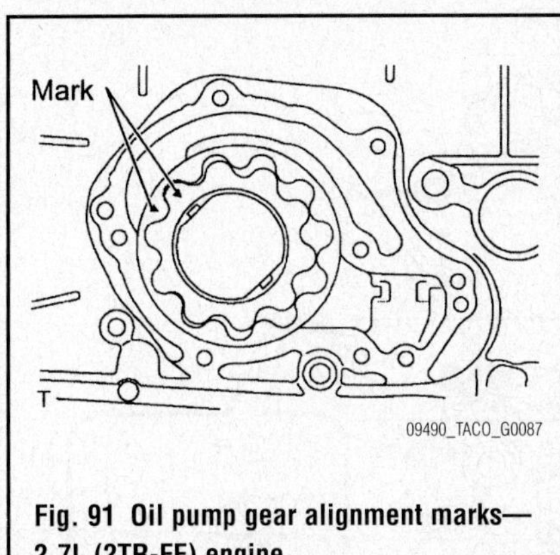

Fig. 91 Oil pump gear alignment marks—2.7L (2TR-FE) engine

11. Remove the nineteen bolts and two nuts retaining the timing chain case cover to its mounting. Remove the timing chain case cover from the engine.

➡Carefully remove the cover by prying between the cover and the cylinder head or block with a suitable tool. Be sure to cover the tip of the suitable tool prior to usage. Be careful not to damage the contact surfaces of the cylinder block, cylinder head and timing chain cover.

➡The oil pump gears are located inside the timing case cover.

12. Remove and discard the O-rings. Remove the head straight screw plug. Remove the water inlet. Remove the thermostat. Remove the oil seal.

13. Remove the oil pump relief valve. Remove the seven oil pump cover bolts. Remove the gears from the timing chain case cover.

To install:

14. Coat the oil pump gears with clean engine oil. Position the gears in the timing chain case cover with the identification marks facing outward. Check that the rotors revolve smoothly. Install the cover. Alternately tighten the bolts to 80 inch lbs.

15. Coat the oil pump relief valve with clean engine oil. Install the plug, using a new gasket and tighten to 36 ft. lbs.

16. Install a new front case oil seal. Install the thermostat. Install the water inlet.

17. Apply adhesive, part number 08833-00070 or equivalent to the head straight screw plug. Install the plug and tighten to 12 ft. lbs. Install four new O-rings onto the timing chain case cover.

18. Apply continuous beads of seal packing, part number 08826-00080 or equivalent as shown in the illustration.

➡Remove any oil from the contact surfaces. Install the timing chain case cover within three minutes and tighten the bolts within fifteen minutes of applying the seal packing. Do not start the engine for at least four hours after installation of the cover.

19. Align the oil pump drive rotor spline and the crankshaft, as indicated in the illustration. Install the spline and timing chain case cover onto the crankshaft.

20. Loosely install the timing chain case cover retaining bolts and nuts.

➡If the vehicle is equipped with air conditioning install the bolts that hold the idle pulley bracket in place when installing the idle pulley, as they are for this purpose.

21. Fully tighten the bolts and nuts, except bolts "A" in the following order: Area 1, Area 3 and then Area 2 to 15 ft. lbs.

22. Fully tighten the bolts "A" in the following order: Area 2 and then Area 3 to 34 ft. lbs.

23. Fully tighten the bolts "E" in Area 4 to 15 ft. lbs.

24. Continue the installation in the reverse order of the removal procedure.

25. When installing the cylinder head cover, apply seal packing, part number 08826-00080 or equivalent, to the cylinder head. Provisionally install the cylinder head cover bolts and nuts. Tighten bolts "A" to 80 inch lbs. Tighten bolts "B" to 80 inch lbs. Retighten bolts "A" to 80 inch lbs.

➡Be sure to remove any oil from the contact surfaces of the cylinder head cover and the cylinder head. Install the cover within three minutes after applying the seal packing. Do not add engine oil for at least two hours after installing the cover.

26. Be sure to fill the engine with the proper grade and type engine coolant.

27. Be sure to fill the engine with the proper grade and type engine oil.

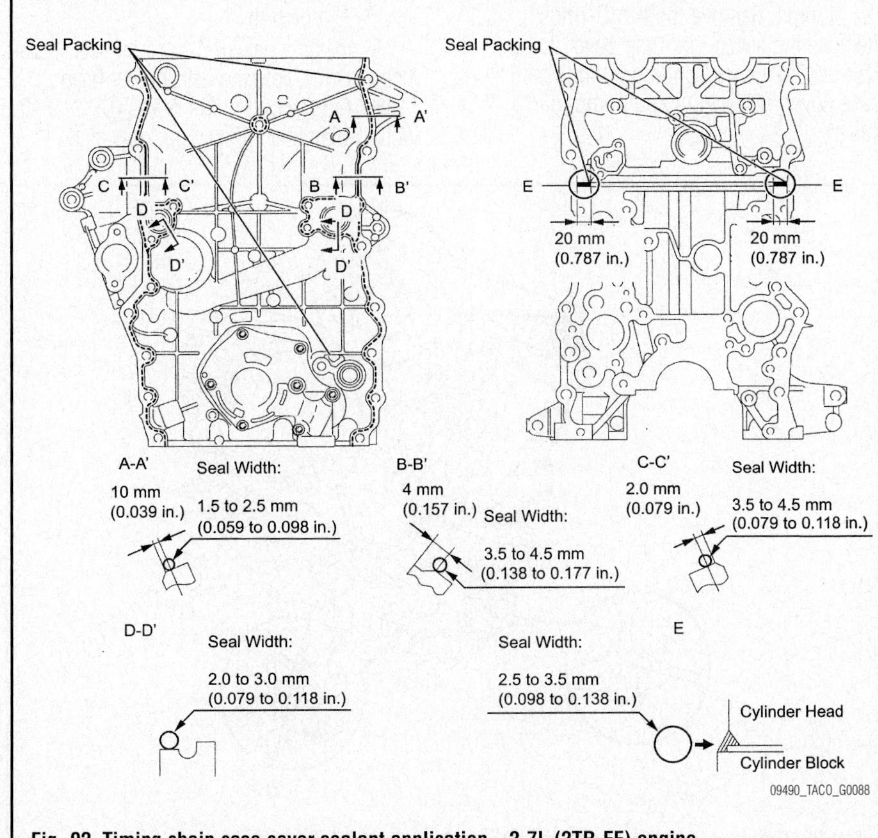

Fig. 92 Timing chain case cover sealant application—2.7L (2TR-FE) engine

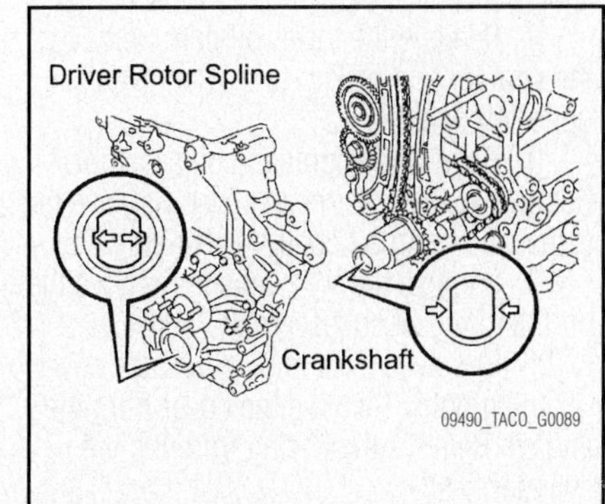

Fig. 93 Oil pump drive rotor spline alignment—2.7L (2TR-FE) engine

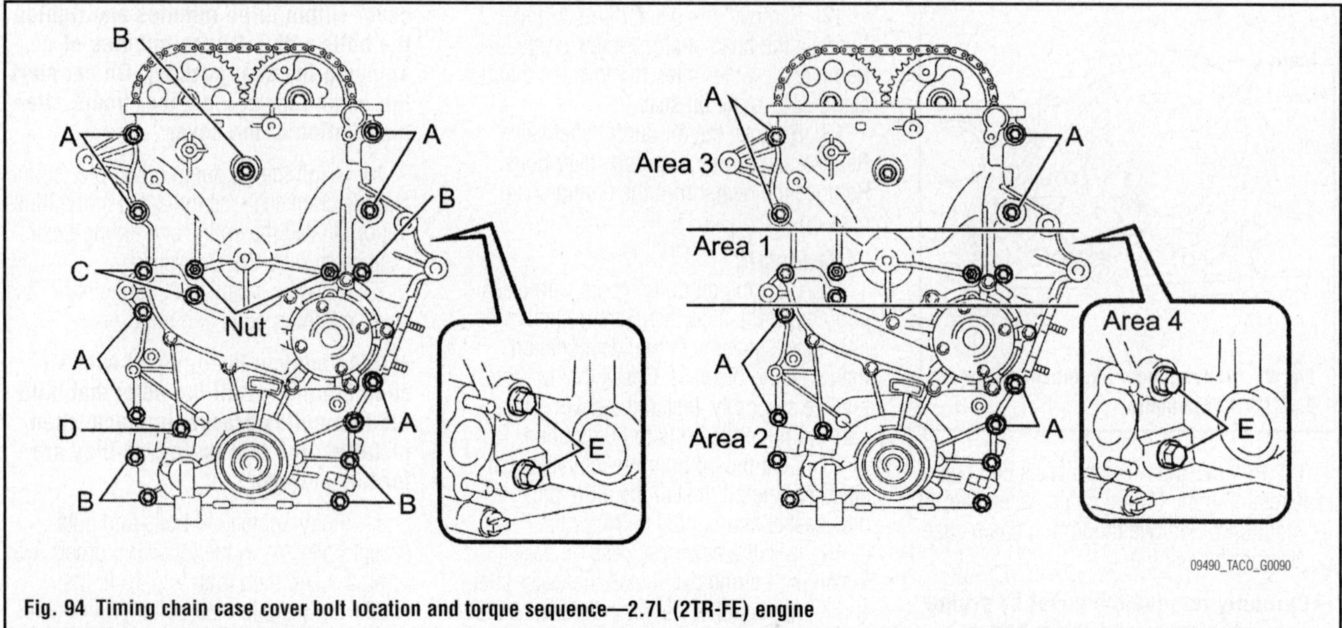

Fig. 94 Timing chain case cover bolt location and torque sequence—2.7L (2TR-FE) engine

28. Start the engine and check for leaks, correct as required.

4.0L (1GR-FE) Engine

See Figures 95 through 99.

2WD

1. Before servicing the vehicle, refer to the Precautions Section.
2. Properly relieve the fuel system pressure.
3. Disconnect the negative battery cable. Disconnect the positive battery cable. Remove the battery.
4. Drain the engine coolant. Drain the engine oil.
5. Remove the engine from the vehicle and position it in a suitable holding fixture.
6. Remove the oil level gauge guide. Remove the water inlet. Remove the belt tensioner.
7. Remove the idler pulley number two subassembly. Remove the idler pulley number one sub assembly. Remove the crankshaft pulley.
8. Remove the lower oil pan. Remove the strainer and pickup tube. Remove the upper oil pan.
9. Remove the ignition coil assembly. Remove the cylinder head cover. Remove the camshaft timing oil control valve assembly.
10. Remove the VVT sensor. Remove the oil filter bracket subassembly.
11. Remove the timing chain case cover retaining bolts. Remove the cover from the engine. Remove the O-ring from the left cylinder head.

➡Carefully remove the cover by prying between the cover and the cylinder head or block with a suitable tool. Be sure to cover the tip of the suitable tool prior to usage. Be careful not to damage the contact surfaces of the cylinder block, cylinder head and timing chain cover.

➡The oil pump gears are located inside the timing case cover.

12. Remove the three bolts and remove the oil pipe. Remove the two O-rings. Remove the seven oil pump cover bolts. Remove the gears from the timing chain case cover. Remove the oil pump relief valve.

To install:

13. Coat the oil pump relief valve with clean engine oil. Install the plug, using a new gasket and tighten to 36 ft. lbs.
14. Coat the oil pump gears with clean engine oil. Position the gears in the timing chain case cover with the identification marks facing oil pump cover side. Install the cover. Alternately tighten the bolts to 80 inch lbs. Install the oil pipe, tighten the bolts to 80 inch lbs.
15. Install a new front case oil seal. Install a new O-ring onto the left cylinder head.
16. Apply continuous beads (0.12–0.16 inch in diameter) of seal packing, part

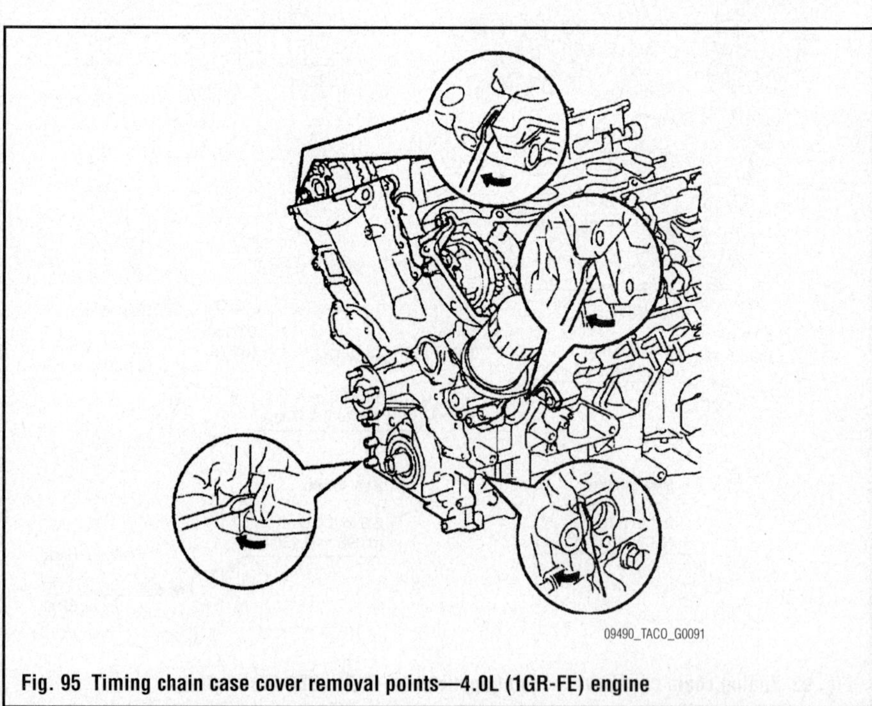

Fig. 95 Timing chain case cover removal points—4.0L (1GR-FE) engine

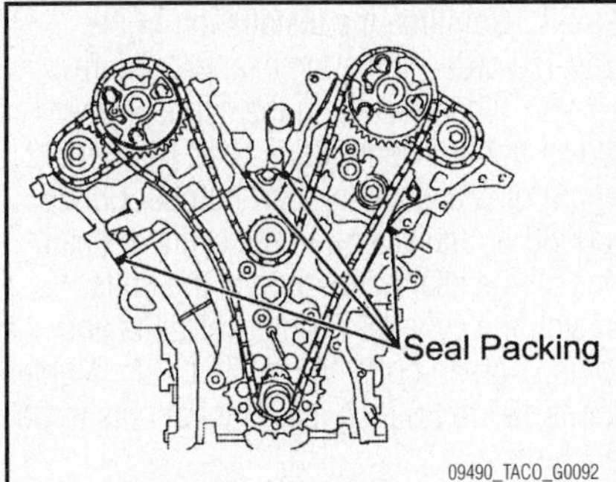

Fig. 96 Timing chain case cover seal packing locating points—4.0L (1GR-FE) engine

number 08826-00080 or equivalent to the four locations shown in the illustration.

17. Apply continuous beads (0.12–0.16 inch in diameter) of seal packing, part number 08826-00080 or equivalent to all parts except the water pump part: for the water pump part use, part number 08826-00080 or equivalent, to the timing chain cover. Do not apply seal packing to portion "A" in the illustration.

➡Remove any oil from the contact surfaces. Install the timing chain case cover within three minutes and tighten the bolts within fifteen minutes of applying the seal packing.

18. Align the key way of the oil pump drive motor with the rectangular portion of the crankshaft timing gear and slide the timing chain case cover into place.

19. Install the timing chain case cover bolts. Tighten the bolts and nuts uniformly in several steps to 17 ft. lbs.

➡Do not wrap the chain and slipper over the timing chain case cover seal line.

20. Continue the installation in the reverse order of the removal procedure.

21. When installing the cylinder head cover apply a continuous bead (0.08–0.12 inch) of seal packing, part number 08826-00080 or equivalent, to the cylinder head. Install the seal washers onto the bolts. Install the cylinder head cover bolts and nuts. Tighten bolts "A" to 7.4 ft. lbs. Tighten bolts "B" to 80 inch lbs. Tighten nuts to 80 inch lbs.

➡Be sure to remove any oil from the contact surfaces of the cylinder head cover and the cylinder head. Install the cover within three minutes after applying the seal packing. Tighten the bolts to specification within fifteen minutes after installing the cover. Do not add engine oil for at least two hours after installing the cover.

22. Be sure to fill the engine with the proper grade and type engine coolant.

23. Be sure to fill the engine with the proper grade and type engine oil.

24. Start the engine and check for leaks, correct as required.

4WD and PreRunner

1. Before servicing the vehicle, refer to the Precautions Section.

2. Properly relieve the fuel system pressure.

3. Disconnect the negative battery cable. Disconnect the positive battery cable. Remove the battery.

4. Drain the engine coolant. Drain the engine oil.

5. Remove the power steering gear assembly.

6. If equipped with 4WD, remove the front differential carrier assembly.

7. Remove the V bank cover. Remove the radiator support to frame seal, left side. Remove the fan shroud.

8. Remove the air cleaner assembly. Remove the oil level gauge. Remove the water inlet.

9. Separate the vane pump assembly. Remove the alternator. Remove the air conditioning compressor and position it to the side.

10. Remove the belt tensioner assembly. Remove the idler pulley number two subassembly. Remove the idler pulley number one subassembly. Remove the crankshaft pulley.

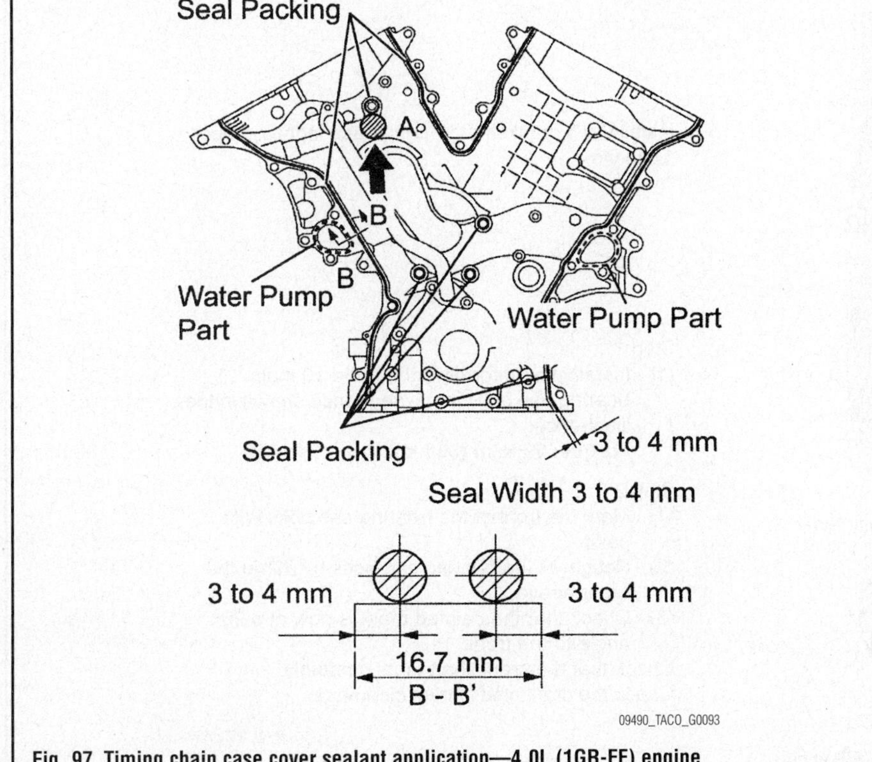

Fig. 97 Timing chain case cover sealant application—4.0L (1GR-FE) engine

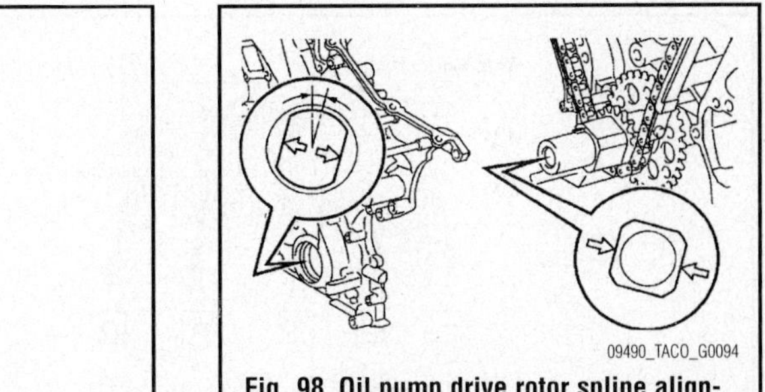

Fig. 98 Oil pump drive rotor spline alignment—4.0L (1GR-FE) engine

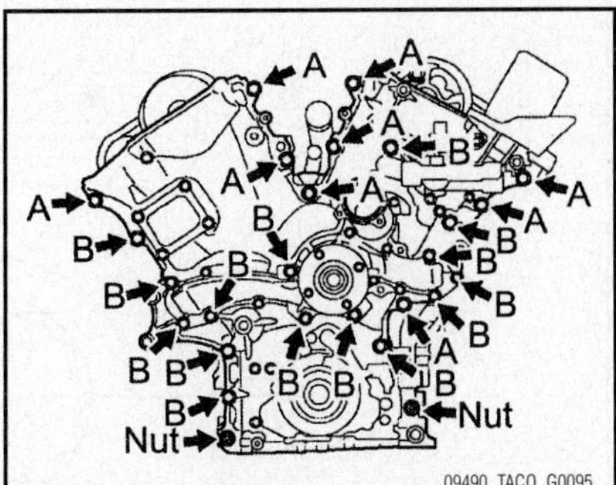

Fig. 99 Timing chain case cover bolt location and torque sequence—4.0L (1GR-FE) engine

11. Remove the lower oil pan. Remove the oil strainer and pickup tube assembly. Remove the upper oil pan.

12. Remove the intake manifold. Remove the cylinder head cover assembly.

13. Remove the camshaft timing oil control valve assembly. Remove the VVT sensor. Remove the oil filter bracket subassembly.

➡Carefully remove the cover by prying between the cover and the cylinder head or block with a suitable tool. Be sure to cover the tip of the suitable tool prior to usage. Be careful not to damage the contact surfaces of the cylinder block, cylinder head and timing chain cover.

➡The oil pump gears are located inside the timing case cover.

14. Remove the three bolts and remove the oil pipe. Remove the two O-rings. Remove the seven oil pump cover bolts. Remove the gears from the timing chain case cover. Remove the oil pump relief valve.

To install:

15. Coat the oil pump relief valve with clean engine oil. Install the plug, using a new gasket and tighten to 36 ft. lbs.

16. Coat the oil pump gears with clean engine oil. Position the gears in the timing chain case cover with the identification marks facing oil pump cover side. Install the cover. Alternately tighten the bolts to 80 inch lbs. Install the oil pipe, tighten the bolts to 80 inch lbs.

17. Install a new front case oil seal. Install a new O-ring onto the left cylinder head.

18. Apply continuous beads (0.12–0.16 inch in diameter) of seal packing, part number 08826-00080 or equivalent to the four locations shown in the illustration.

19. Apply continuous beads (0.12–0.16 inch in diameter) of seal packing, part number 08826-00080 or equivalent to all parts except the water pump part: for the water pump part use, part number 08826-00080 or equivalent, to the timing chain cover. Do not apply seal packing to portion "A" in the illustration.

➡Remove any oil from the contact surfaces. Install the timing chain case cover within three minutes and tighten the bolts within fifteen minutes of applying the seal packing.

20. Align the key way of the oil pump drive motor with the rectangular portion of the crankshaft timing gear and slide the timing chain case cover into place.

21. Install the timing chain case cover bolts. Tighten the bolts and nuts uniformly in several steps to 17 ft. lbs.

➡Do not wrap the chain and slipper over the timing chain case cover seal line.

22. Continue the installation in the reverse order of the removal procedure.

23. When installing the cylinder head cover apply a continuous bead (0.08–0.12 inch) of seal packing, part number 08826-00080 or equivalent, to the cylinder head. Install the seal washers onto the bolts. Install the cylinder head cover bolts and nuts. Tighten bolts "A" to 7.4 ft. lbs. Tighten bolts "B" to 80 inch lbs. Tighten nuts to 80 inch lbs.

➡Be sure to remove any oil from the contact surfaces of the cylinder head cover and the cylinder head. Install the cover within three minutes after applying the seal packing. Tighten the bolts to specification within fifteen minutes after installing the cover. Do not add engine oil for at least two hours after installing the cover.

24. Be sure to fill the engine with the proper grade and type engine coolant.

25. Be sure to fill the engine with the proper grade and type engine oil.

26. Start the engine and check for leaks, correct as required.

MAIN BEARING TORQUE SEQUENCE

See Figures 100 and 101.

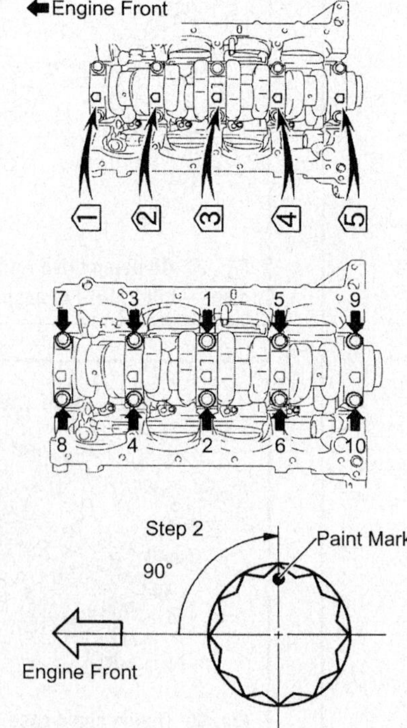

Install the 5 crankshaft bearing caps in their proper locations.

Step 1

(1) Install and uniformly tighten the 10 main bearing cap bolts in the sequence shown in the illustration.
Torque: 39 N m (398 kgf cm, 29 ft. lbf)

Step 2

(1) Mark the front of the bearing cap bolts with paint.
(2) Retighten the bearing cap bolts by 90° in the order above.
(3) Check that the painted mark is now at a 90° angle to the front.

Check that the crankshaft turns smoothly.
Check the crankshaft thrust clearance.

09490_TACO_G0005

Fig. 100 Main bearing torque sequence—2.7L (2TR-FE) engine

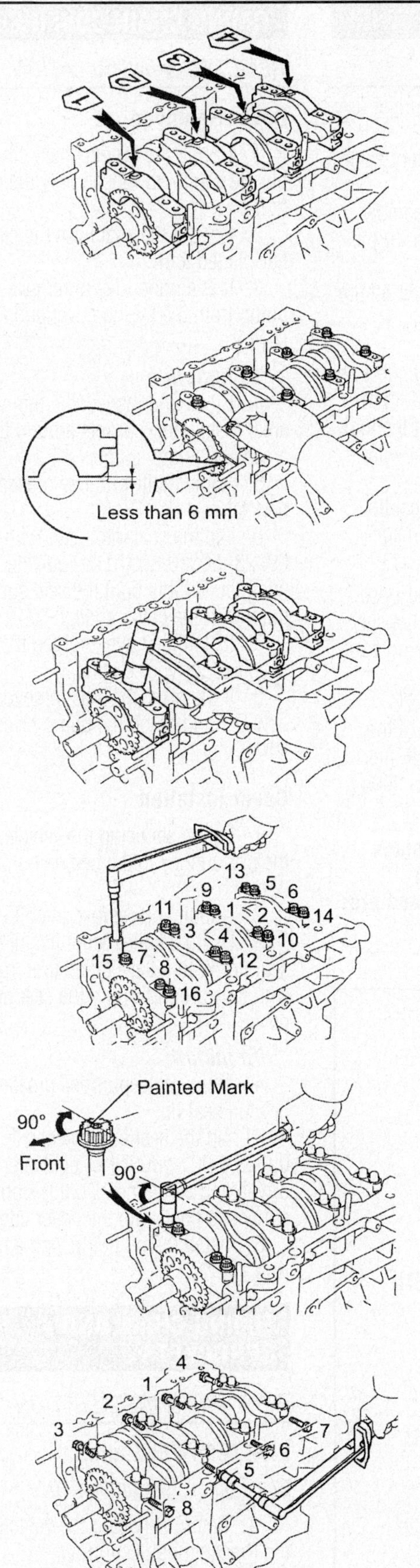

Examine the front marks and numbers, check the sequence number is as shown in the illustration and install the bearing caps on the cylinder block.
Apply a light coat of engine oil to the threads of bearing cap bolts.
Temporarily install the 8 main bearing cap bolts in the inside positions.

Install the main bearing caps. Tighten the 2 bolts for each bearing cap until the clearance between the bearing cap and the cylinder block is under 6 mm (0.23 in.).

Using a plastic-faced hammer, lightly tap the bearing cap to ensure a proper fit.
Apply a light coat of engine oil to the threads of main bearing cap bolts.

Install the 16 main bearing cap bolts. Using several steps, tighten the bolts uniformly in the sequence shown in the illustration.
Torque: 61 N m (622 kgf cm, 45 ft. lbf)

Mark the front side of the bearing cap bolts with paint.
Retighten the bearing cap bolts 90° in the sequence as shown.
Check that the painted mark is now at a 90° angle from the front.
Check that the crankshaft turns smoothly.

Using several steps, tighten the 8 main bearing cap bolts uniformly in the sequence shown in the illustration.
Torque: 25 N m (255 kgf cm, 18 ft. lbf)

09490_TACO_G0006

Fig. 101 Main bearing torque sequence—4.0L (1GR-FE) engine

PISTON AND RING

POSITIONING

See Figures 102 through 104.

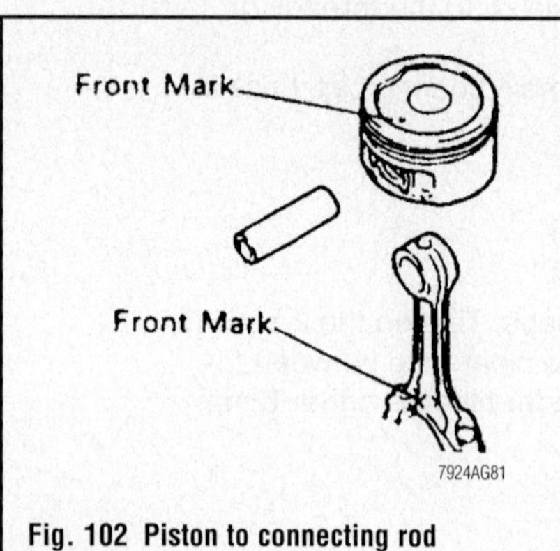

Fig. 102 Piston to connecting rod assembly

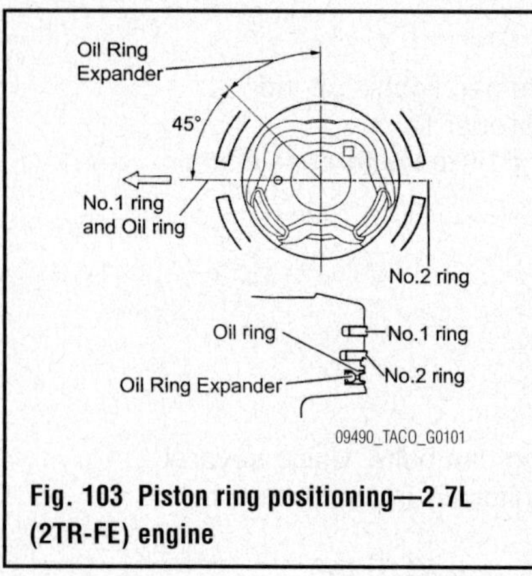

Fig. 103 Piston ring positioning—2.7L (2TR-FE) engine

Fig. 104 Piston ring positioning—4.0L (1GR-FE) engine

REAR MAIN SEAL

REMOVAL & INSTALLATION

1. Before servicing the vehicle, refer to the Precautions Section.
2. Disconnect the negative battery cable.
3. Remove the transmission.
4. Remove the clutch cover assembly and flywheel (manual transmission) or the flexplate (automatic transmission).
5. Use a small, sharp knife to cut off the lip of the oil seal. Take great care not to score any metal with the knife.
6. Use a small prybar to pry the old seal from the retaining plate. Be careful not to damage the plate. Protect the tip of the tool with tape and pad the fulcrum point with cloth.
7. Inspect the crankshaft and seal lip contact surfaces for any sign of damage.

To install:

8. Apply a light coat of multi-purpose grease to the lip of a new oil seal. Loosely fit the seal into place by hand, making sure it is not crooked.
9. Use a seal driver such as (SST 09223–15030 and 09950–70010) of the correct size to install the seal. Tap it into place until the surface of the seal is flush with the edge of the housing.

➡Use the correct tools. Homemade substitutes may install the seal crooked, resulting in oil leaks and premature seal failure.

TIMING CHAIN COVER SEAL

REMOVAL & INSTALLATION

Cover Removed

1. Unbolt the timing chain cover assembly. Be careful to loosen only the correct bolts.
2. Pry out the seal from the cover with a flat-bladed tool.
3. It is a good idea to remove the oil pump from the timing cover and replace the O-ring.

To install:

4. Clean and inspect the timing cover area. Install new gaskets around the dowel areas and pump spline.
5. Apply multi-purpose grease to the new oil seal lip.
6. Tap the seal into place with SST 09223–50010/60010 or equivalent, and a hammer. Do this until the seal surface is flush with the cover edge.
7. Install the cover, tighten the bolts as specified for your engine.
8. If the oil pump was removed, install a new O-ring behind the pump prior to installation.

Cover Installed

1. Before servicing the vehicle, refer to the precautions in the beginning of this section.
2. Unbolt and remove the oil pump.
3. Using a knife, carefully cut off the oil seal lip. With a flat-bladed tool, (preferably with tape around it) pry the seal from the cover.

To install:

4. Apply multi-purpose grease to the new oil seal lip.
5. Tap the seal into place with SST 09223–50010/60011 or equivalent seal driver, and a hammer. Do this until the seal surface is flush with the cover edge.
6. Install the oil pump with a new O-ring.

TIMING CHAIN AND SPROCKETS

REMOVAL & INSTALLATION

2.7L (2TR-FE) Engine

See Figures 105 through 107.

1. Before servicing the vehicle, refer to the Precautions Section.
2. Disconnect the negative battery cable.

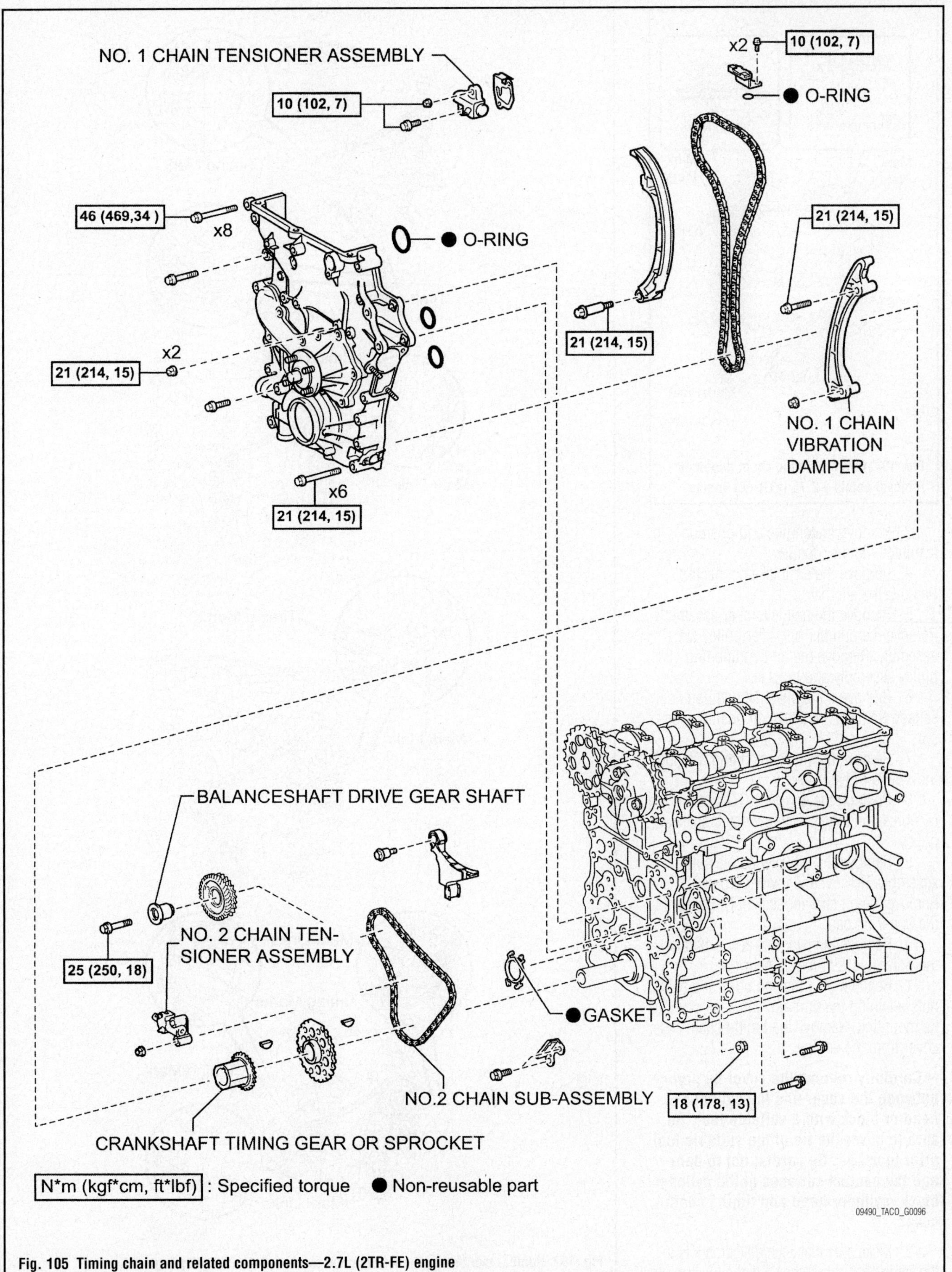

Fig. 105 Timing chain and related components—2.7L (2TR-FE) engine

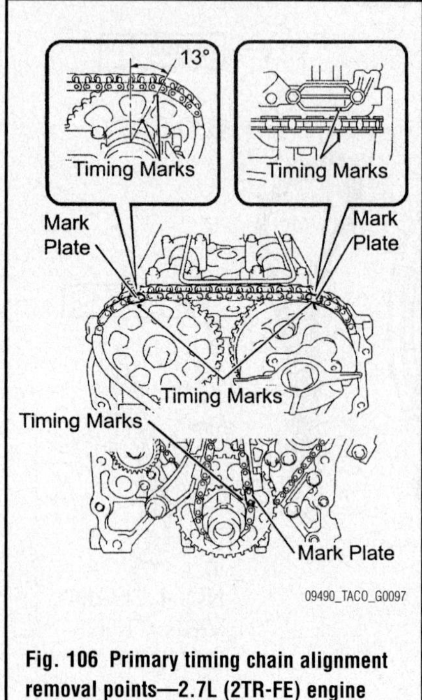

Fig. 106 Primary timing chain alignment removal points—2.7L (2TR-FE) engine

3. Remove the engine and position it in a suitable holding fixture.

4. Remove the air intake connector. Remove the alternator.

5. Remove the belt tensioner assembly. Remove the number one idler pulley sub-assembly. Remove the air conditioning idler pulley assembly and bracket.

6. Remove the crankshaft position sensor. Remove the camshaft position sensor.

7. Remove the intake manifold. Remove the cylinder head cover.

8. Position the engine at TDC of the compression stroke. Remove the crankshaft pulley.

9. Remove the oil level gauge sub-assembly. Remove the lower oil pan. Remove the oil strainer assembly. Remove the upper oil pan.

10. Remove the two nuts and separate the water bypass pipe number one.

11. Remove the nineteen bolts and two nuts retaining the timing chain case cover to its mounting. Remove the timing chain case cover from the engine.

➡Carefully remove the cover by prying between the cover and the cylinder head or block with a suitable tool. Be sure to cover the tip of the suitable tool prior to usage. Be careful not to damage the contact surfaces of the cylinder block, cylinder head and timing chain cover.

12. Make sure that each matchmark is in the same position as shown in the illustration.

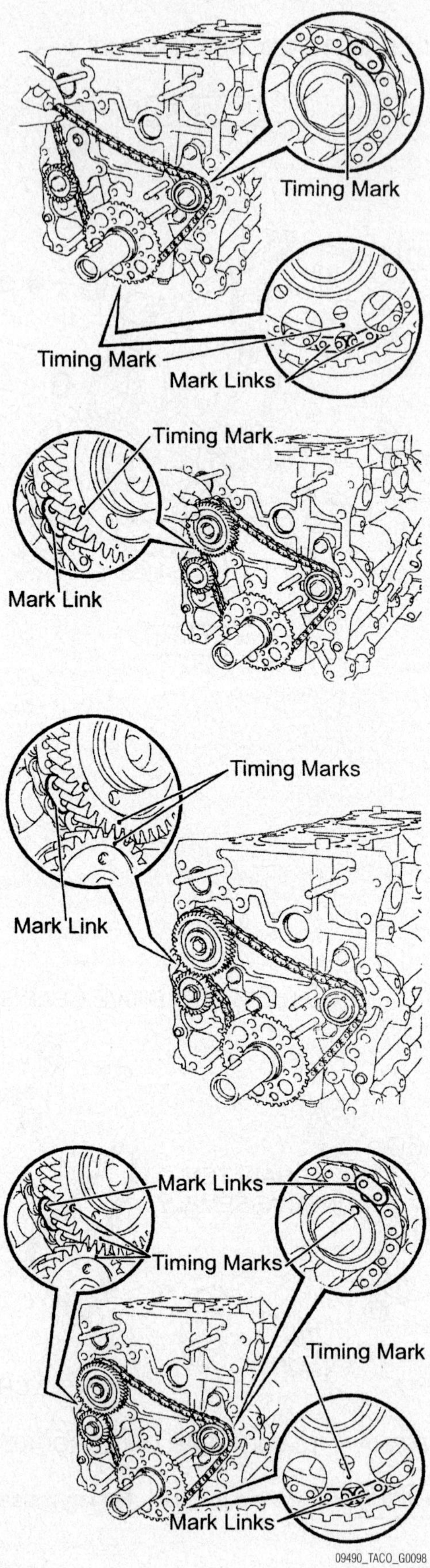

Fig. 107 Number two timing chain alignment marks—2.7L (2TR-FE) engine

Remove the two bolts, timing chain guide and O-ring.

13. Move the stopper plate upward to release the lock, and push the plunger deep into the tensioner.

14. Move the stopper plate downward to set the lock. Insert a 0.118 inch diameter bar into the stopper plate hole. Remove the bolt, nut, number one chain tensioner and gasket.

➡**When the number one chain tensioner is removed do not rotate the crankshaft. When the chain is removed and the camshaft needs to be rotated, rotate the crankshaft 90 degrees to the right.**

15. Remove the bolt and chain tensioner slipper. Remove the two bolts and remove the number one chain vibration damper. Remove the primary timing chain sub-assembly.

16. Remove the crankshaft timing gear or sprocket. Remove the bolt and remove the number two chain vibration damper.

17. Remove the two bolts and remove the number three chain vibration damper.

18. Remove the nut and the number two chain tensioner assembly. Remove the bolt, balance shaft drive gear shaft and balance shaft drive gear. Remove the crankshaft timing sprocket number two and chain.

To install:

19. Install the chain with its marks aligned with the timing marks on the crankshaft timing sprocket and balance shaft timing sprocket.

20. Bring the other mark link of the crankshaft timing sprocket behind the large timing mark of the balance shaft drive gear.

21. Insert the balance shaft drive gear shaft through the balance shaft drive gear so that it fits into the thrust plate hole.

22. Align the small timing mark of the balance shaft drive gear with the timing mark of the balance shaft timing gear.

23. Install the bolt onto the balance shaft drive gear and tighten it to 18 ft. lbs.

24. Check that the timing mark is aligned with the corresponding mark link.

25. Install the number two chain tensioner assembly. Tighten the nut to 13 ft. lbs.

➡**Assemble the chain tensioner with the 0.118 inch diameter bar installed, then remove the bar after assembly. When doing this avoid pushing the vibration damper against the chain.**

26. Install the number three chain vibration damper with the two bolts. Tighten the bolts to 13 ft. lbs.

27. Install the chain vibration damper number two bolt and tighten it to 20 ft. lbs. Remove the pin from the chain tensioner and release the plunger.

28. Install the crankshaft timing gear or sprocket.

29. Install the number one chain vibration damper bolt and nut. Tighten to 15 ft. lbs.

30. Install the primary timing chain onto the sprocket and gear with the painted marks aligned with the timing marks on the sprocket and gear.

➡**The camshaft mark plate is orange. The crankshaft mark plate is yellow.**

31. Use a rope to tie the chain of the crankshaft timing sprocket. Tie the rope near the sprocket.

➡**After the chain tensioner has been installed, remove the rope. The rope is used to prevent gear jumping.**

32. Install the tensioner slipper and tighten the bolt to 15 ft. lbs.

33. Install the number one chain tensioner assembly, using a new gasket. Tighten the bolts to 7 ft. lbs.

34. Install a new front case oil seal. Install the thermostat. Install the water inlet.

35. Apply adhesive, part number 08833-00070 or equivalent to the head straight screw plug. Install the plug and tighten to 12 ft. lbs. Install four new O-rings onto the timing chain case cover.

36. Apply continuous beads of seal packing, part number 08826-00080 or equivalent.

➡**Remove any oil from the contact surfaces. Install the timing chain case cover within three minutes and tighten the bolts within fifteen minutes of applying the seal packing. Do not start the engine for at least four hours after installation of the cover.**

37. Align the oil pump drive rotor spline and the crankshaft. Install the spline and timing chain case cover onto the crankshaft.

38. Loosely install the timing chain case cover retaining bolts and nuts.

➡**If the vehicle is equipped with air conditioning install the bolts that hold the idle pulley bracket in place when installing the idle pulley, as they are for this purpose.**

39. Fully tighten the bolts and nuts, except bolts "A" in the following order: Area 1, Area 3 and then Area 2 to 15 ft. lbs.

40. Fully tighten the bolts "A" in the following order: Area 2 and then Area 3 to 34 ft. lbs.

41. Fully tighten the bolts "E" in Area 4 to 15 ft. lbs.

42. Continue the installation in the reverse order of the removal procedure.

43. When installing the cylinder head cover, apply seal packing, part number 08826-00080 or equivalent, to the cylinder head. Provisionally install the cylinder head cover bolts and nuts. Tighten bolts "A" to 80 inch lbs. Tighten bolts "B" to 80 inch lbs. Retighten bolts "A" to 80 inch lbs.

➡**Be sure to remove any oil from the contact surfaces of the cylinder head cover and the cylinder head. Install the cover within three minutes after applying the seal packing. Do not add engine oil for at least two hours after installing the cover.**

44. Be sure to fill the engine with the proper grade and type engine coolant.

45. Be sure to fill the engine with the proper grade and type engine oil.

46. Start the engine and check for leaks, correct as required.

4.0L (1GR-FE) Engine

See Figures 108 and 109.

2WD

1. Before servicing the vehicle, refer to the Precautions Section.

2. Properly relieve the fuel system pressure.

3. Disconnect the negative battery cable. Disconnect the positive battery cable. Remove the battery.

4. Drain the engine coolant. Drain the engine oil.

5. Remove the engine from the vehicle and position it in a suitable holding fixture.

6. Remove the oil level gauge guide. Remove the water inlet. Remove the belt tensioner.

7. Remove the idler pulley number two subassembly. Remove the idler pulley number one sub assembly. Remove the crankshaft pulley.

8. Remove the lower oil pan. Remove the strainer and pickup tube. Remove the upper oil pan.

9. Remove the intake manifold. Remove the ignition coil assembly. Remove the cylinder head cover. Remove the camshaft timing oil control valve assembly.

10. Remove the VVT sensor. Remove the oil filter bracket subassembly.

11. Remove the timing chain case cover retaining bolts. Remove the cover from the engine. Remove the O-ring from the left cylinder head.

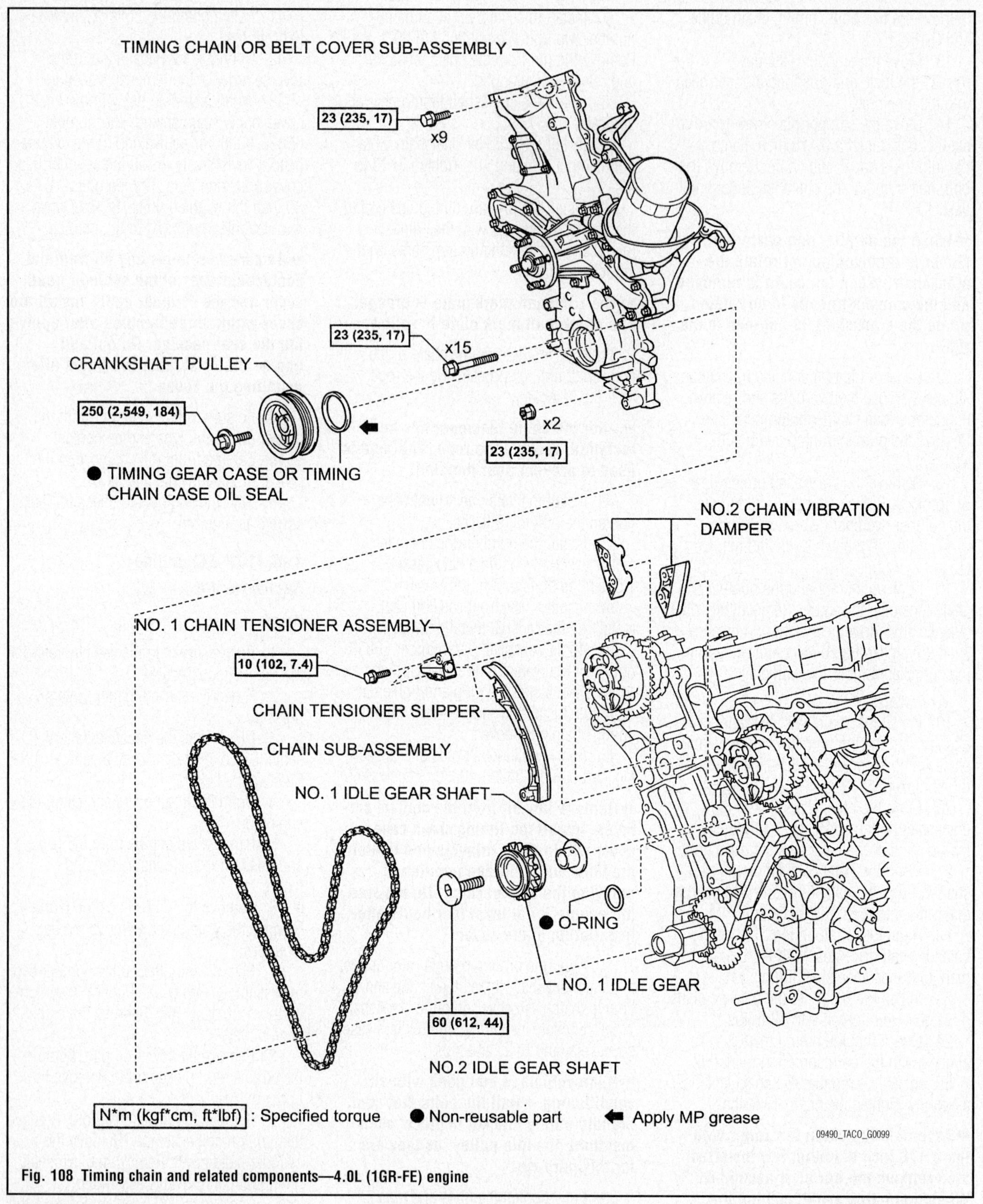

Fig. 108 Timing chain and related components—4.0L (1GR-FE) engine

➡Carefully remove the cover by prying between the cover and the cylinder head or block with a suitable tool. Be sure to cover the tip of the suitable tool prior to usage. Be careful not to damage the contact surfaces of the cylinder block, cylinder head and timing chain cover.

12. Using the crankshaft pulley set bolt, turn the crankshaft to align the crankshaft set key with the timing line of the cylinder block. If not aligned at TDC of the compression stroke, turn the crankshaft one complete revolution, in the direction of rotation

13. While turning the stopper plate of the tensioner upward, push the plunger of the chain tensioner. While turning the stopper plate of the tensioner downward, insert a 0.138 inch diameter bar into the holes in the stopper plate and tensioner to hold the stopper plate.

14. Remove the two bolts, and then remove the chain tensioner.

15. Remove the chain tensioner slipper. Remove the idle gear shaft number two, idle gear number one and idle gear shaft number one.

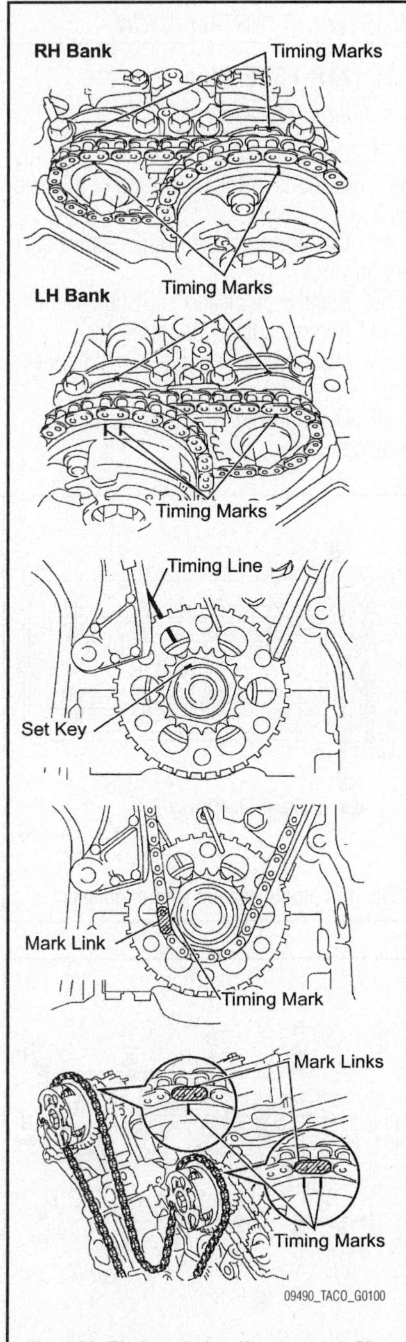

Fig. 109 Timing chain alignment—4.0L (1GR-FE) engine

16. Remove the number two chain vibration damper. Remove the timing chain subassembly.

To install:

17. Install the chain tensioner slipper.

18. While turning the stopper plate of the tensioner clockwise, push in the plunger of the chain tensioner. While turning the stopper plate of the tensioner counterclockwise, insert a 0.138 inch diameter bar into the holes in the stopper plate and tensioner to hold the stopper plate.

19. Install the chain tensioner. Tighten the bolts to 7.4 ft. lbs.

20. Position the engine at TDC on the compression stroke. Align the camshaft timing gears and bearing caps. Using the crankshaft pulley set bolt, align the crankshaft set key with the timing line of the cylinder.

21. Align the yellow mark line with the timing mark of the crankshaft timing link. Align the orange mark links with the timing marks of the camshaft timing gears, and install the chain.

22. Install the number two chain vibration damper.

23. Apply a light coat of clean engine oil to the rotating surface of the idle gear shaft number one.

24. Temporarily install the idle gear shaft number one together with idle gear shaft number two, while aligning the knock pin of idle gear shaft number one with the knock pin groove of the cylinder block.

➡Be care of the idle gear direction.

25. Tighten the idle gear shaft number two to 44 ft. lbs. Remove the bar from the chain tensioner.

26. Install a new front case oil seal. Install a new O-ring onto the left cylinder head.

27. Apply continuous beads (0.12–0.16 inch in diameter) of seal packing, part number 08826-00080 or equivalent to the four locations shown in the illustration.

28. Apply continuous beads (0.12–0.16 inch in diameter) of seal packing, part number 08826-00080 or equivalent to all parts except the water pump part: for the water pump part use, part number 08826-00080 or equivalent, to the timing chain cover. Do not apply seal packing to portion "A".

➡Remove any oil from the contact surfaces. Install the timing chain case cover within three minutes and tighten the bolts within fifteen minutes of applying the seal packing.

29. Align the key way of the oil pump drive motor with the rectangular portion of the crankshaft timing gear and slide the timing chain case cover into place.

30. Install the timing chain case cover bolts. Tighten the bolts and nuts uniformly in several steps to 17 ft. lbs.

➡Do not wrap the chain and slipper over the timing chain case cover seal line.

31. Continue the installation in the reverse order of the removal procedure.

32. When installing the cylinder head cover apply a continuous bead (0.08–0.12 inch) of seal packing, part number 08826-00080 or equivalent, to the cylinder head. Install the seal washers onto the bolts. Install the cylinder head cover bolts and nuts. Tighten bolts "A" to 7.4 ft. lbs. Tighten bolts "B" to 80 inch lbs. Tighten nuts to 80 inch lbs.

➡Be sure to remove any oil from the contact surfaces of the cylinder head cover and the cylinder head. Install the cover within three minutes after applying the seal packing. Tighten the bolts to specification within fifteen minutes after installing the cover. Do not add engine oil for at least two hours after installing the cover.

33. Be sure to fill the engine with the proper grade and type engine coolant.

34. Be sure to fill the engine with the proper grade and type engine oil.

35. Start the engine and check for leaks, correct as required.

4WD and PreRunner

1. Before servicing the vehicle, refer to the Precautions Section.

2. Properly relieve the fuel system pressure.

3. Disconnect the negative battery cable. Disconnect the positive battery cable. Remove the battery.

4. Drain the engine coolant. Drain the engine oil.

5. Remove the power steering gear assembly.

6. If equipped with 4WD, remove the front differential carrier assembly.

7. Remove the V bank cover. Remove the radiator support to frame seal, left side. Remove the fan shroud.

8. Remove the air cleaner assembly. Remove the oil level gauge. Remove the water inlet.

9. Separate the vane pump assembly. Remove the alternator. Remove the air conditioning compressor and position it to the side.

10. Remove the belt tensioner assembly. Remove the idler pulley number two subassembly. Remove the idler pulley number one subassembly. Remove the crankshaft pulley.

11. Remove the lower oil pan. Remove the oil strainer and pickup tube assembly. Remove the upper oil pan.

12. Remove the intake manifold. Remove the ignition coil assembly. Remove the cylinder head cover assembly.

13. Remove the camshaft timing oil control valve assembly. Remove the VVT sensor. Remove the oil filter bracket subassembly.

➡Carefully remove the cover by prying between the cover and the cylinder head or block with a suitable tool. Be sure to cover the tip of the suitable tool prior to usage. Be careful not to damage the contact surfaces of the cylinder block, cylinder head and timing chain cover.

14. Using the crankshaft pulley set bolt, turn the crankshaft to align the crankshaft set key with the timing line of the cylinder block. If not aligned at TDC of the compression stroke, turn the crankshaft one complete revolution, in the direction of rotation

15. While turning the stopper plate of the tensioner upward, push the plunger of the chain tensioner. While turning the stopper plate of the tensioner downward, insert a 0.138 inch diameter bar into the holes in the stopper plate and tensioner to hold the stopper plate.

16. Remove the two bolts, and then remove the chain tensioner.

17. Remove the chain tensioner slipper. Remove the idle gear shaft number two, idle gear number one and idle gear shaft number one.

18. Remove the number two chain vibration damper. Remove the timing chain subassembly.

To install:

19. Install the chain tensioner slipper.

20. While turning the stopper plate of the tensioner clockwise, push in the plunger of the chain tensioner. While turning the stopper plate of the tensioner counterclockwise, insert a 0.138 inch diameter bar into the holes in the stopper plate and tensioner to hold the stopper plate.

21. Install the chain tensioner. Tighten the bolts to 7.4 ft. lbs.

22. Position the engine at TDC on the compression stroke. Align the camshaft timing gears and bearing caps. Using the crankshaft pulley set bolt, align the crankshaft set key with the timing line of the cylinder.

23. Align the yellow mark line with the timing mark of the crankshaft timing link. Align the orange mark links with the timing marks of the camshaft timing gears, and install the chain.

24. Install the number two chain vibration damper.

25. Apply a light coat of clean engine oil to the rotating surface of the idle gear shaft number one.

26. Temporarily install the idle gear shaft number one together with idle gear shaft number two, while aligning the knock pin of idle gear shaft number one with the knock pin groove of the cylinder block.

➡Be care of the idle gear direction.

27. Tighten the idle gear shaft number two to 44 ft. lbs. Remove the bar from the chain tensioner.

28. Install a new front case oil seal. Install a new O-ring onto the left cylinder head.

29. Apply continuous beads (0.12–0.16 inch in diameter) of seal packing, part number 08826-00080 or equivalent to the four locations shown in the illustration.

30. Apply continuous beads (0.12–0.16 inch in diameter) of seal packing, part number 08826-00080 or equivalent to all parts except the water pump part: for the water pump part use, part number 08826-00080 or equivalent, to the timing chain cover. Do not apply seal packing to portion "A".

➡Remove any oil from the contact surfaces. Install the timing chain case cover within three minutes and tighten the bolts within fifteen minutes of applying the seal packing.

31. Align the key way of the oil pump drive motor with the rectangular portion of the crankshaft timing gear and slide the timing chain case cover into place.

32. Install the timing chain case cover bolts. Tighten the bolts and nuts uniformly in several steps to 17 ft. lbs.

➡Do not wrap the chain and slipper over the timing chain case cover seal line.

33. Continue the installation in the reverse order of the removal procedure.

34. When installing the cylinder head cover apply a continuous bead (0.08–0.12 inch) of seal packing, part number 08826-00080 or equivalent, to the cylinder head. Install the seal washers onto the bolts. Install the cylinder head cover bolts and nuts. Tighten bolts "A" to 7.4 ft. lbs. Tighten bolts "B" to 80 inch lbs. Tighten nuts to 80 inch lbs.

➡Be sure to remove any oil from the contact surfaces of the cylinder head cover and the cylinder head. Install the cover within three minutes after applying the seal packing. Tighten the bolts to specification within fifteen minutes after installing the cover. Do not add engine oil for at least two hours after installing the cover.

35. Be sure to fill the engine with the proper grade and type engine coolant.

36. Be sure to fill the engine with the proper grade and type engine oil.

37. Start the engine and check for leaks, correct as required.

VALVE COVERS

REMOVAL & INSTALLATION

2.7L (2TR-FE) Engine

See Figures 110 and 111.

1. Before servicing the vehicle, refer to the precautions in the beginning of this section.

2. Disconnect cable from negative battery terminal

3. Remove air cleaner cap sub-assembly.

4. Remove intake air connector.

5. Disconnect the ignition coil connectors.

6. Disconnect the throttle with motor body connector.

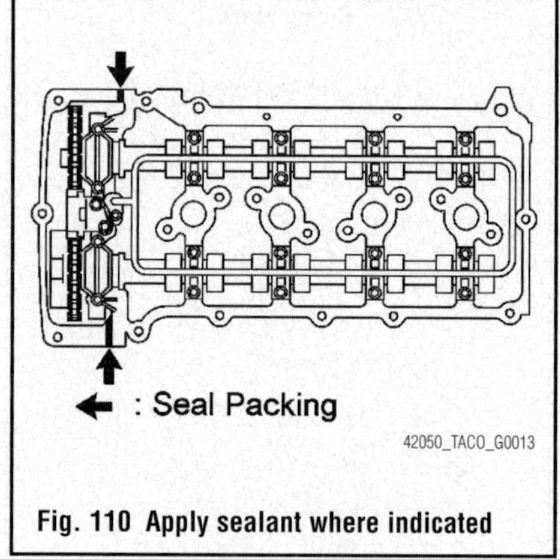

Fig. 110 Apply sealant where indicated

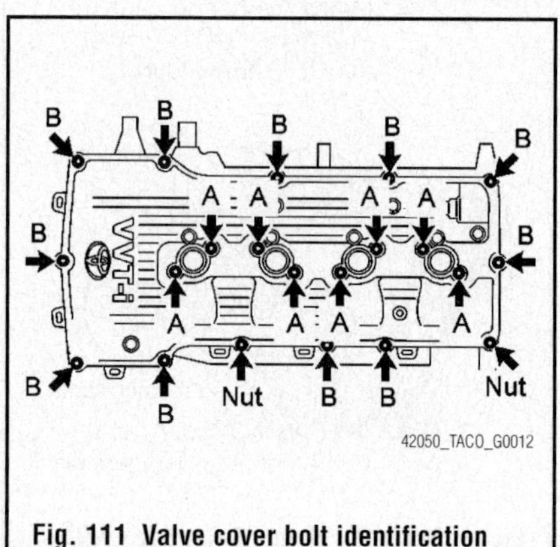

Fig. 111 Valve cover bolt identification

7. Disconnect the VSV connector.
8. Disconnect the camshaft position sensor connector.
9. Disconnect the engine wire harness clamps.
10. Remove the bolts, then remove the ignition coils.
11. Disconnect the ventilation hose.
12. Remove the 19 bolts and 2 nuts, then remove the cylinder head cover.

To install:

13. Apply seal packing Part No. 08826-00080 or equivalent to the 2 locations shown in the illustration.

➡Install the cylinder head cover within 3 minutes of applying the seal packing.

➡Do not apply engine oil for at least 2 hours after installation.

14. Install the cylinder head cover with the 19 bolts and 2 nuts as follows:
    a. Step 1: Tighten bolts marked A to 9.0 Nm (80 inch lbs.)
    b. Step 2: Tighten bolts marked B and nuts to 9.0 Nm (80 inch lbs.)
    c. Step 3: Retighten bolts marked A to 9.0 Nm (80 inch lbs.)
15. Connect the ventilation hose.
16. Install the ignition coils with the bolts and tighten to 9.0 Nm (80 inch lbs.)
17. Install the engine wire harness clamps.
18. Connect the camshaft position sensor connector.
19. Connect the VSV connector.
20. Connect the throttle with motor body connector.
21. Connect the ignition coil connectors.
22. Install intake air connector.
23. Install air cleaner cap sub-assembly.
24. Connect cable to negative battery terminal.

4.0L (1GR-FE)

See Figures 112 and 113.

1. Before servicing the vehicle, refer to the precautions in the beginning of this section.
2. Disconnect cable from negative battery terminal.
3. Drain engine coolant.
4. Remove v-bank cover.
5. Remove air cleaner assembly.
6. Disconnect the 2 water by-pass hoses.
7. Disconnect the fuel vapor feed hose.
8. Disconnect the ventilation hose.
9. Disconnect the 2 VSV connectors.
10. Disconnect the throttle body w/ motor connector.
11. Separate the 3 wire harness clamps and hose clamp.
12. Remove the nut, then separate the clutch flexible hose bracket from the surge tank stay (w/ Manual Transmission).
13. Remove the 2 bolts and throttle body bracket.
14. Remove the bolt and oil baffle plate.
15. Remove the 4 bolts and 2 surge tank stays.
16. Remove the 2 nuts.
17. Remove the 4 bolts, intake air surge tank and gasket.
18. Disconnect the 3 connectors.
19. Remove the 3 bolts, then remove the 3 ignition coils.
20. Remove the 10 bolts, 3 seal washers, 2 nuts, right cylinder head cover and gasket.
21. Remove the 10 bolts, 3 seal washers, 2 nuts, left cylinder head cover and gasket.

To install:

22. Remove any old packing material and be careful not to drop any oil on the contact surfaces of the cylinder head, timing chain cover and cylinder head covers.
23. Install the cylinder head covers within 3 minutes of applying seal packing Part No. 08826-00080 or equivalent. Tighten the cylinder head cover bolts and nuts within 15 minutes of installing the cylinder head covers. Otherwise, the seal packing mush be removed and reapplied.

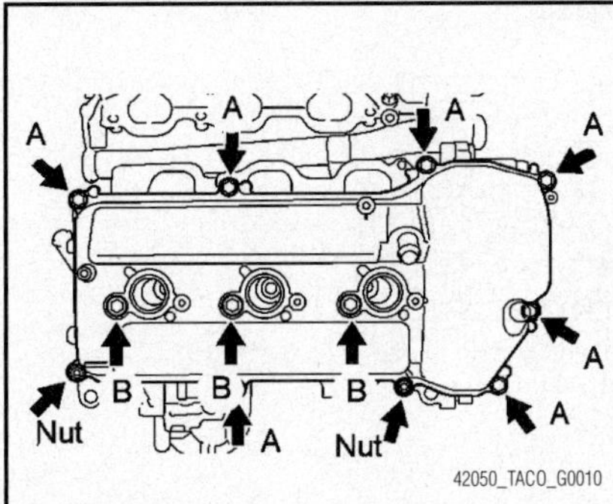

Fig. 112 Right valve cover bolt identification

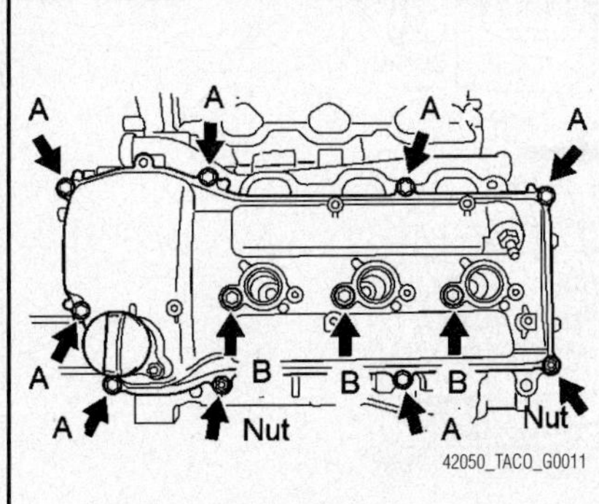

Fig. 113 Left valve cover bolt identification

24. Install the seal washers onto the bolts.
25. Install the cylinder head cover with the 10 bolts and 2 nuts. Tighten the bolts and nuts uniformly in several steps as follows:
    a. Bolts labeled A: 10 Nm (7.4 ft. lbs.)
    b. Bolts labeled B: 9.0 Nm (80 inch lbs.)
    c. Nuts: 9.0 Nm (80 inch lbs.)
26. Install the ignition coils and tighten the bolts to 10 Nm (7.4 ft. lbs.)
27. Install a new gasket onto the intake air surge tank.
28. Install the intake air surge tank with the 4 bolts and tighten to 28 Nm (21 ft. lbs)
29. Install the 2 intake air surge tank nuts and tighten to 28 Nm (21 ft. lbs)
30. Install the 2 surge tank stays with the 4 bolts and tighten to 21 Nm (15 ft. lbs)
31. Install the oil baffle plate with the bolt and tighten to 9.0 Nm (80 inch lbs.)
32. Install the throttle body bracket with the 2 bolts and tighten to 21 Nm (15 ft. lbs)
33. Install the clutch flexible hose bracket with the nut (w/ Manual Transmission) and tighten to 24 Nm (18 ft. lbs)
34. Install the 3 wire harness clamps and hose clamp.
35. Connect the throttle body w/ motor connector.
36. Connect the 2 VSV connectors.
37. Connect the ventilation hose.
38. Connect the fuel vapor feed hose.
39. Connect the 2 water by-pass hoses.
40. Install air cleaner assembly.
41. Connect cable to negative battery terminal.
42. Add engine coolant.
43. Check for engine coolant leakage.
44. Install the V-bank cover with the 2 nuts and tighten to 7.5 Nm (66 inch lbs.)

VALVE LASH

ADJUSTMENT

2.7L (2TR-FE) Engine

This engine is equipped with hydraulic lash adjusters.

4.0L (1GR-FE) Engine

See Figures 114 through 116.

1. Before servicing the vehicle, refer to the Precautions Section.
2. Disconnect the negative battery cable.
3. Drain the engine coolant. Remove the V bank cover.

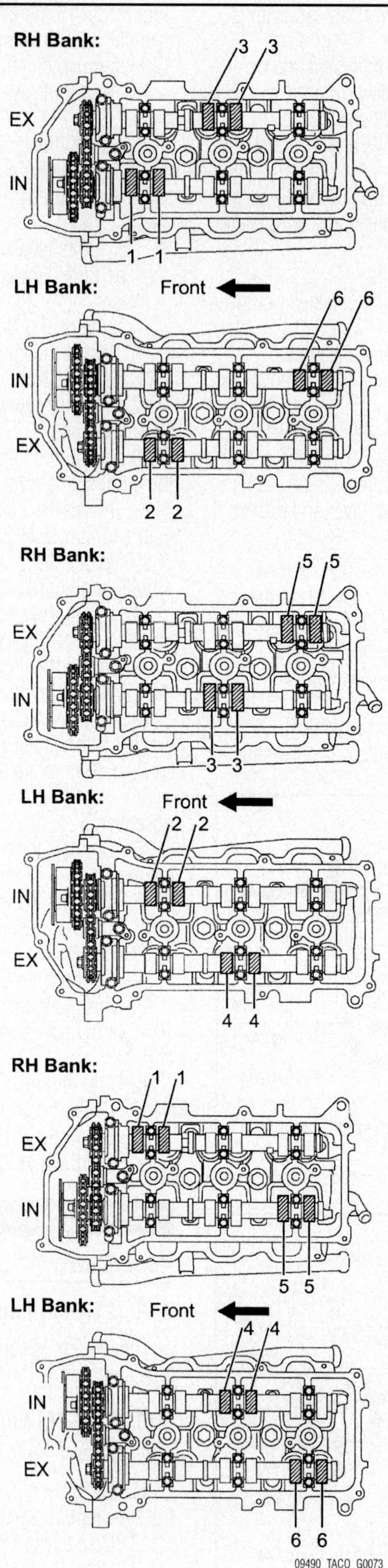

Fig. 114 Valve clearance location—4.0L (1GR-FE) engine

4. Remove the air cleaner assembly.
5. Disconnect the two water bypass hoses. Disconnect the fuel vapor feed hose. Disconnect the ventilation hose. Disconnect the VSV connectors.
6. Disconnect the throttle body motor connector. Separate the three wire harness clamps and hose clamp.
7. If equipped with manual transmission, remove the nut, then separate the clutch flexible hose bracket from the surge tank stay.
8. Remove the two bolts and the throttle body bracket. Remove the bolt and the oil baffle plate. Remove the four bolts and the two serge tank stays.
9. Remove the two nuts. Remove the four bolts, intake air surge tank and gasket.
10. Remove the ignition coil assembly.
11. Remove the cylinder head cover retaining bolts. Remove the cylinder head cover.
12. Turn the crankshaft pulley until its groove and the timing mark "0" of the timing chain cover are aligned. If not aligned at TDC of the compression stroke, turn the crankshaft one complete revolution, in the direction of rotation.
13. Using a feeler gauge, check and record the valve clearance on the following valves: right bank, exhaust number three and intake number one, left bank exhaust number two and intake number six.
14. Rotate the crankshaft 240 degrees clockwise and using a feeler gauge, check and record the following valves: right bank, exhaust number five and intake number three, left bank exhaust number four and intake number two.
15. Rotate the crankshaft 240 degrees clockwise and using a feeler gauge, check and record the following valves: right bank, exhaust number one and intake number five, left bank exhaust number six and intake number four.
16. If adjustment is required, position the engine at TDC on the compression stroke.
17. Place paint marks on the number one chain links corresponding to the timing marks of the camshaft timing gears.
18. Remove the chain tensioner assembly number one. Remove the number two camshaft.
19. Remove the chain tensioner assembly number two. Remove the camshaft.
20. Remove the number four camshaft subassembly. Remove the chain tensioner assembly number three.
21. Remove the number three camshaft subassembly.
22. Remove the valve lifters.

	5.060 (0.1992)	5.080 (0.2000)	5.100 (0.2008)	5.120 (0.2016)	5.140 (0.2024)	5.160 (0.2031)	5.180 (0.2039)	5.200 (0.2047)	5.210 (0.2051)	5.220 (0.2055)	5.230 (0.2059)	5.240 (0.2063)	5.250 (0.2067)	5.260 (0.2071)	5.270 (0.2075)	5.280 (0.2079)	5.290 (0.2083)	5.300 (0.2087)	5.310 (0.2091)	5.320 (0.2094)	5.330 (0.2098)	5.340 (0.2102)	5.350 (0.2106)	5.360 (0.2110)	5.370 (0.2114)	5.380 (0.2118)	5.390 (0.2122)	5.400 (0.2126)	5.410 (0.2130)
0.000 - 0.020 (0.0000 - 0.0008)													06	08	08	10	10	12	12	14	14	16	16	18	18	20	20	22	22
0.021 - 0.040 (0.0008 - 0.0016)											06	08	08	10	10	12	12	14	14	16	16	18	18	20	20	22	22	24	24
0.041 - 0.060 (0.0016 - 0.0024)									06	08	08	10	10	12	12	14	14	16	16	18	18	20	20	22	22	24	24	26	26
0.061 - 0.080 (0.0024 - 0.0031)								08	08	10	10	12	12	14	14	16	16	18	18	20	20	22	22	24	24	26	26	28	28
0.081 - 0.100 (0.0032 - 0.0039)							08	10	10	12	12	14	14	16	16	18	18	20	20	22	22	24	24	26	26	28	28	30	30
0.101 - 0.120 (0.0040 - 0.0047)						08	10	12	12	14	14	16	16	18	18	20	20	22	22	24	24	26	26	28	28	30	30	32	32
0.121 - 0.140 (0.0048 - 0.0055)					08	10	12	14	14	16	16	18	18	20	20	22	22	24	24	26	26	28	28	30	30	32	32	34	34
0.141 - 0.149 (0.0056 - 0.0059)				06	08	10	12	14	16	16	18	18	20	20	22	22	24	24	26	26	28	28	30	30	32	32	34	34	36
0.251 - 0.270 (0.0099 - 0.0106)	12	14	16	18	20	22	24	26	28	28	30	30	32	32	34	34	36	36	38	38	40	40	42	42	44	44	46	46	48
0.271 - 0.290 (0.0107 - 0.0114)	14	16	18	20	22	24	26	28	30	30	32	32	34	34	36	36	38	38	40	40	42	42	44	44	46	46	48	48	50
0.291 - 0.310 (0.0115 - 0.0122)	16	18	20	22	24	26	28	30	32	32	34	34	36	36	38	38	40	40	42	42	44	44	46	46	48	48	50	50	52
0.311 - 0.330 (0.0122 - 0.0130)	18	20	22	24	26	28	30	32	34	34	36	36	38	38	40	40	42	42	44	44	46	46	48	48	50	50	52	52	54
0.331 - 0.350 (0.0130 - 0.0138)	20	22	24	26	28	30	32	34	36	36	38	38	40	40	42	42	44	44	46	46	48	48	50	50	52	52	54	54	56
0.351 - 0.370 (0.0138 - 0.0146)	22	24	26	28	30	32	34	36	38	38	40	40	42	42	44	44	46	46	48	48	50	50	52	52	54	54	56	56	58
0.371 - 0.390 (0.0146 - 0.0154)	24	26	28	30	32	34	36	38	40	40	42	42	44	44	46	46	48	48	50	50	52	52	54	54	56	56	58	58	60
0.391 - 0.410 (0.0154 - 0.0161)	26	28	30	32	34	36	38	40	42	42	44	44	46	46	48	48	50	50	52	52	54	54	56	56	58	58	60	60	62
0.411 - 0.430 (0.0162 - 0.0169)	28	30	32	34	36	38	40	42	44	44	46	46	48	48	50	50	52	52	54	54	56	56	58	58	60	60	62	62	64
0.431 - 0.450 (0.0170 - 0.0177)	30	32	34	36	38	40	42	44	46	46	48	48	50	50	52	52	54	54	56	56	58	58	60	60	62	62	64	64	66
0.451 - 0.470 (0.0178 - 0.0185)	32	34	36	38	40	42	44	46	48	48	50	50	52	52	54	54	56	56	58	58	60	60	62	62	64	64	66	66	68
0.471 - 0.490 (0.0185 - 0.0193)	34	36	38	40	42	44	46	48	50	50	52	52	54	54	56	56	58	58	60	60	62	62	64	64	66	66	68	68	70
0.491 - 0.510 (0.0193 - 0.0201)	36	38	40	42	44	46	48	50	52	52	54	54	56	56	58	58	60	60	62	62	64	64	66	66	68	68	70	70	72
0.511 - 0.530 (0.0201 - 0.0209)	38	40	42	44	46	48	50	52	54	54	56	56	58	58	60	60	62	62	64	64	66	66	68	68	70	70	72	72	74
0.531 - 0.550 (0.0209 - 0.0217)	40	42	44	46	48	50	52	54	56	56	58	58	60	60	62	62	64	64	66	66	68	68	70	70	72	72	74		
0.551 - 0.570 (0.0217 - 0.0224)	42	44	46	48	50	52	54	56	58	58	60	60	62	62	64	64	66	66	68	68	70	70	72	72	74				
0.571 - 0.590 (0.0225 - 0.0232)	44	46	48	50	52	54	56	58	60	60	62	62	64	64	66	66	68	68	70	70	72	72	74						
0.591 - 0.610 (0.0233 - 0.0240)	46	48	50	52	54	56	58	60	62	62	64	64	66	66	68	68	70	70	72	72	74								
0.611 - 0.630 (0.0241 - 0.0248)	48	50	52	54	56	58	60	62	64	64	66	66	68	68	70	70	72	72	74										
0.631 - 0.650 (0.0248 - 0.0256)	50	52	54	56	58	60	62	64	66	66	68	68	70	70	72	72	74												
0.651 - 0.670 (0.0256 - 0.0264)	52	54	56	58	60	62	64	66	68	68	70	70	72	72	74														
0.671 - 0.690 (0.0264 - 0.0272)	54	56	58	60	62	64	66	68	70	70	72	72	74																
0.691 - 0.710 (0.0272 - 0.0280)	56	58	60	62	64	66	68	70	72	72	74																		
0.711 - 0.730 (0.0280 - 0.0287)	58	60	62	64	66	68	70	72	74																				
0.731 - 0.750 (0.0288 - 0.0295)	60	62	64	66	68	70	72																						
0.751 - 0.770 (0.0296 - 0.0303)	62	64	66	68	70	72																							
0.771 - 0.790 (0.0304 - 0.0311)	64	66	68	70	72																								
0.791 - 0.810 (0.0311 - 0.0319)	66	68	70	72																									
0.811 - 0.830 (0.0319 - 0.0327)	68	70	72																										
0.831 - 0.850 (0.0327 - 0.0335)	70	72																											
0.851 - 0.870 (0.0335 - 0.0343)	72																												

	5.420 (0.2134)	5.430 (0.2138)	5.440 (0.2142)	5.450 (0.2146)	5.460 (0.2150)	5.470 (0.2154)	5.480 (0.2157)	5.490 (0.2161)	5.500 (0.2165)	5.510 (0.2169)	5.520 (0.2173)	5.530 (0.2177)	5.540 (0.2181)	5.550 (0.2185)	5.560 (0.2189)	5.570 (0.2193)	5.580 (0.2197)	5.590 (0.2201)	5.600 (0.2205)	5.620 (0.2213)	5.640 (0.2220)	5.660 (0.2228)	5.680 (0.2236)	5.700 (0.2244)	5.720 (0.2252)	5.740 (0.2260)
0.000 - 0.020 (0.0000 - 0.0008)	24	24	26	26	28	28	30	30	32	32	34	34	36	36	38	38	40	40	42	44	46	48	50	52	54	56
0.021 - 0.040 (0.0008 - 0.0016)	26	26	28	28	30	30	32	32	34	34	36	36	38	38	40	40	42	42	44	46	48	50	52	54	56	58
0.041 - 0.060 (0.0016 - 0.0024)	28	28	30	30	32	32	34	34	36	36	38	38	40	40	42	42	44	44	46	48	50	52	54	56	58	60
0.061 - 0.080 (0.0024 - 0.0031)	30	30	32	32	34	34	36	36	38	38	40	40	42	42	44	44	46	46	48	50	52	54	56	58	60	62
0.081 - 0.100 (0.0032 - 0.0039)	32	32	34	34	36	36	38	38	40	40	42	42	44	44	46	46	48	48	50	52	54	56	58	60	62	64
0.101 - 0.120 (0.0040 - 0.0047)	34	34	36	36	38	38	40	40	42	42	44	44	46	46	48	48	50	50	52	54	56	58	60	62	64	66
0.121 - 0.140 (0.0048 - 0.0055)	36	36	38	38	40	40	42	42	44	44	46	46	48	48	50	50	52	52	54	56	58	60	62	64	66	68
0.141 - 0.149 (0.0056 - 0.0059)	36	38	38	40	40	42	42	44	44	46	46	48	48	50	50	52	52	54	54	56	58	60	62	64	66	68
0.251 - 0.270 (0.0099 - 0.0106)	48	50	50	52	52	54	54	56	56	58	58	60	60	62	62	64	64	66	66	68	70	72				
0.271 - 0.290 (0.0107 - 0.0114)	50	52	52	54	54	56	56	58	58	60	60	62	62	64	64	66	66	68	68	70	72					
0.291 - 0.310 (0.0115 - 0.0122)	52	54	54	56	56	58	58	60	60	62	62	64	64	66	66	68	68	70	70	72						
0.311 - 0.330 (0.0122 - 0.0130)	54	56	56	58	58	60	60	62	62	64	64	66	66	68	68	70	70	72	72							
0.331 - 0.350 (0.0130 - 0.0138)	56	58	58	60	60	62	62	64	64	66	66	68	68	70	70	72	72	74								
0.351 - 0.370 (0.0138 - 0.0146)	58	60	60	62	62	64	64	66	66	68	68	70	70	72	72	74										
0.371 - 0.390 (0.0146 - 0.0154)	60	62	62	64	64	66	66	68	68	70	70	72	72	74												
0.391 - 0.410 (0.0154 - 0.0161)	62	64	64	66	66	68	68	70	70	72	72	74														
0.411 - 0.430 (0.0162 - 0.0169)	64	66	66	68	68	70	70	72	72	74																
0.431 - 0.450 (0.0170 - 0.0177)	66	68	68	70	70	72	72	74																		
0.451 - 0.470 (0.0178 - 0.0185)	68	70	70	72	72	74																				
0.471 - 0.490 (0.0185 - 0.0193)	70	72	72	74																						
0.491 - 0.510 (0.0193 - 0.0201)	72	74																								

06	5.060 (0.1992)	30	5.300 (0.2087)	54	5.540 (0.2181)
08	5.080 (0.2000)	32	5.320 (0.2094)	56	5.560 (0.2189)
10	5.100 (0.2008)	34	5.340 (0.2102)	58	5.580 (0.2197)
12	5.120 (0.2016)	36	5.360 (0.2110)	60	5.600 (0.2205)
14	5.140 (0.2024)	38	5.380 (0.2118)	62	5.620 (0.2213)
16	5.160 (0.2031)	40	5.400 (0.2126)	64	5.640 (0.2220)
18	5.180 (0.2039)	42	5.420 (0.2134)	66	5.660 (0.2228)
20	5.200 (0.2047)	44	5.440 (0.2142)	68	5.680 (0.2236)
22	5.220 (0.2055)	46	5.460 (0.2150)	70	5.700 (0.2244)
24	5.240 (0.2063)	48	5.480 (0.2157)	72	5.720 (0.2252)
26	5.260 (0.2071)	50	5.500 (0.2165)	74	5.740 (0.2260)
28	5.280 (0.2079)	52	5.520 (0.2173)		

09490_TACO_G0074

Fig. 115 Shim selection chart, part one—4.0L (1GR-FE) engine

	5.060 (0.1992)	5.080 (0.2000)	5.100 (0.2008)	5.120 (0.2016)	5.140 (0.2024)	5.160 (0.2031)	5.180 (0.2039)	5.200 (0.2047)	5.210 (0.2051)	5.220 (0.2055)	5.230 (0.2059)	5.240 (0.2063)	5.250 (0.2067)	5.260 (0.2071)	5.270 (0.2075)	5.280 (0.2079)	5.290 (0.2083)	5.300 (0.2087)	5.310 (0.2091)	5.320 (0.2094)	5.330 (0.2098)	5.340 (0.2102)	5.350 (0.2106)	5.360 (0.2110)	5.370 (0.2114)	5.380 (0.2118)	5.390 (0.2122)	5.400 (0.2126)	5.410 (0.2130)	5.420 (0.2134)	5.430 (0.2138)	5.440 (0.2142)	5.450 (0.2146)	5.460 (0.2150)	5.470 (0.2154)	5.480 (0.2157)	5.490 (0.2161)	5.500 (0.2165)	5.510 (0.2169)	5.520 (0.2173)	5.530 (0.2177)	5.540 (0.2181)	5.550 (0.2185)	5.560 (0.2189)	5.570 (0.2193)	5.580 (0.2197)	5.590 (0.2201)	5.600 (0.2205)	5.620 (0.2213)	5.640 (0.2220)	5.660 (0.2228)	5.680 (0.2236)	5.700 (0.2244)	5.720 (0.2252)	5.740 (0.2260)
0.000 - 0.020 (0.0000 - 0.0008)																											06	08	08	10	10	12	12	14	14	16	16	18	18	20	20	22	22	24	24	26	26	28	30	32	34	36	38	40	42
0.021 - 0.040 (0.0008 - 0.0016)																										06	08	08	10	10	12	12	14	14	16	16	18	18	20	20	22	22	24	24	26	26	28	28	30	32	34	36	38	40	42	44
0.041 - 0.060 (0.0016 - 0.0024)																							06	08	08	10	10	12	12	14	14	16	16	18	18	20	20	22	22	24	24	26	26	28	28	30	30	32	34	36	38	40	42	44	46
0.061 - 0.080 (0.0024 - 0.0031)																					06	08	08	10	10	12	12	14	14	16	16	18	18	20	20	22	22	24	24	26	26	28	28	30	30	32	32	34	36	38	40	42	44	46	48
0.081 - 0.100 (0.0032 - 0.0039)																			06	08	08	10	10	12	12	14	14	16	16	18	18	20	20	22	22	24	24	26	26	28	28	30	30	32	32	34	34	36	38	40	42	44	46	48	50
0.101 - 0.120 (0.0040 - 0.0047)																	06	08	08	10	10	12	12	14	14	16	16	18	18	20	20	22	22	24	24	26	26	28	28	30	30	32	32	34	34	36	36	38	40	42	44	46	48	50	52
0.121 - 0.140 (0.0048 - 0.0055)															06	08	08	10	10	12	12	14	14	16	16	18	18	20	20	22	22	24	24	26	26	28	28	30	30	32	32	34	34	36	36	38	38	40	42	44	46	48	50	52	54
0.141 - 0.160 (0.0056 - 0.0063)													06	08	08	10	10	12	12	14	14	16	16	18	18	20	20	22	22	24	24	26	26	28	28	30	30	32	32	34	34	36	36	38	38	40	40	42	44	46	48	50	52	54	56
0.161 - 0.180 (0.0063 - 0.0071)											06	08	08	10	10	12	12	14	14	16	16	18	18	20	20	22	22	24	24	26	26	28	28	30	30	32	32	34	34	36	36	38	38	40	40	42	42	44	46	48	50	52	54	56	58
0.181 - 0.200 (0.0071 - 0.0079)									06	08	08	10	10	12	12	14	14	16	16	18	18	20	20	22	22	24	24	26	26	28	28	30	30	32	32	34	34	36	36	38	38	40	40	42	42	44	44	46	48	50	52	54	56	58	60
0.201 - 0.220 (0.0079 - 0.0087)								08	08	10	10	12	12	14	14	16	16	18	18	20	20	22	22	24	24	26	26	28	28	30	30	32	32	34	34	36	36	38	38	40	40	42	42	44	44	46	46	48	50	52	54	56	58	60	62
0.221 - 0.240 (0.0087 - 0.0094)							08	10	10	12	12	14	14	16	16	18	18	20	20	22	22	24	24	26	26	28	28	30	30	32	32	34	34	36	36	38	38	40	40	42	42	44	44	46	46	48	48	50	52	54	56	58	60	62	64
0.241 - 0.260 (0.0095 - 0.0102)						08	10	12	12	14	14	16	16	18	18	20	20	22	22	24	24	26	26	28	28	30	30	32	32	34	34	36	36	38	38	40	40	42	42	44	44	46	46	48	48	50	50	52	54	56	58	60	62	64	66
0.261 - 0.280 (0.0103 - 0.0110)					08	10	12	14	14	16	16	18	18	20	20	22	22	24	24	26	26	28	28	30	30	32	32	34	34	36	36	38	38	40	40	42	42	44	44	46	46	48	48	50	50	52	52	54	56	58	60	62	64	66	68
0.281 - 0.289 (0.0111 - 0.0114)				06	08	10	12	14	16	16	18	18	20	20	22	22	24	24	26	26	28	28	30	30	32	32	34	34	36	36	38	38	40	40	42	42	44	44	46	46	48	48	50	50	52	52	54	54	56	58	60	62	64	66	68
0.391 - 0.410 (0.0154 - 0.0161)	12	14	16	18	20	22	24	26	28	28	30	30	32	32	34	34	36	36	38	38	40	40	42	42	44	44	46	46	48	48	50	50	52	52	54	54	56	56	58	58	60	60	62	62	64	64	66	66	68	70	72
0.411 - 0.430 (0.0162 - 0.0169)	14	16	18	20	22	24	26	28	30	30	32	32	34	34	36	36	38	38	40	40	42	42	44	44	46	46	48	48	50	50	52	52	54	54	56	56	58	58	60	60	62	62	64	64	66	66	68	68	70	72
0.431 - 0.450 (0.0170 - 0.0177)	16	18	20	22	24	26	28	30	32	32	34	34	36	36	38	38	40	40	42	42	44	44	46	46	48	48	50	50	52	52	54	54	56	56	58	58	60	60	62	62	64	64	66	66	68	68	70	70	72
0.451 - 0.470 (0.0178 - 0.0185)	18	20	22	24	26	28	30	32	34	34	36	36	38	38	40	40	42	42	44	44	46	46	48	48	50	50	52	52	54	54	56	56	58	58	60	60	62	62	64	64	66	66	68	68	70	70	72	72
0.471 - 0.490 (0.0185 - 0.0193)	20	22	24	26	28	30	32	34	36	36	38	38	40	40	42	42	44	44	46	46	48	48	50	50	52	52	54	54	56	56	58	58	60	60	62	62	64	64	66	66	68	68	70	70	72	72	74
0.491 - 0.510 (0.0193 - 0.0201)	22	24	26	28	30	32	34	36	38	38	40	40	42	42	44	44	46	46	48	48	50	50	52	52	54	54	56	56	58	58	60	60	62	62	64	64	66	66	68	68	70	70	72	72	74
0.511 - 0.530 (0.0201 - 0.0209)	24	26	28	30	32	34	36	38	40	40	42	42	44	44	46	46	48	48	50	50	52	52	54	54	56	56	58	58	60	60	62	62	64	64	66	66	68	68	70	70	72	72	74
0.531 - 0.550 (0.0209 - 0.0217)	26	28	30	32	34	36	38	40	42	42	44	44	46	46	48	48	50	50	52	52	54	54	56	56	58	58	60	60	62	62	64	64	66	66	68	68	70	70	72	72	74
0.551 - 0.570 (0.0217 - 0.0224)	28	30	32	34	36	38	40	42	44	44	46	46	48	48	50	50	52	52	54	54	56	56	58	58	60	60	62	62	64	64	66	66	68	68	70	70	72	72	74
0.571 - 0.590 (0.0225 - 0.0232)	30	32	34	36	38	40	42	44	46	46	48	48	50	50	52	52	54	54	56	56	58	58	60	60	62	62	64	64	66	66	68	68	70	70	72	72	74
0.591 - 0.610 (0.0233 - 0.0240)	32	34	36	38	40	42	44	46	48	48	50	50	52	52	54	54	56	56	58	58	60	60	62	62	64	64	66	66	68	68	70	70	72	72	74
0.611 - 0.630 (0.0241 - 0.0248)	34	36	38	40	42	44	46	48	50	50	52	52	54	54	56	56	58	58	60	60	62	62	64	64	66	66	68	68	70	70	72	72	74
0.631 - 0.650 (0.0248 - 0.0256)	36	38	40	42	44	46	48	50	52	52	54	54	56	56	58	58	60	60	62	62	64	64	66	66	68	68	70	70	72	72	74
0.651 - 0.670 (0.0256 - 0.0264)	38	40	42	44	46	48	50	52	54	54	56	56	58	58	60	60	62	62	64	64	66	66	68	68	70	70	72	72	74
0.671 - 0.690 (0.0264 - 0.0272)	40	42	44	46	48	50	52	54	56	56	58	58	60	60	62	62	64	64	66	66	68	68	70	70	72	72	74
0.691 - 0.710 (0.0272 - 0.0280)	42	44	46	48	50	52	54	56	58	58	60	60	62	62	64	64	66	66	68	68	70	70	72	72	74
0.711 - 0.730 (0.0280 - 0.0287)	44	46	48	50	52	54	56	58	60	60	62	62	64	64	66	66	68	68	70	70	72	72	74
0.731 - 0.750 (0.0288 - 0.0295)	46	48	50	52	54	56	58	60	62	62	64	64	66	66	68	68	70	70	72	72	74
0.751 - 0.770 (0.0296 - 0.0303)	48	50	52	54	56	58	60	62	64	64	66	66	68	68	70	70	72	72	74
0.771 - 0.790 (0.0304 - 0.0311)	50	52	54	56	58	60	62	64	66	66	68	68	70	70	72	72	74
0.791 - 0.810 (0.0311 - 0.0319)	52	54	56	58	60	62	64	66	68	68	70	70	72	72	74
0.811 - 0.830 (0.0319 - 0.0327)	54	56	58	60	62	64	66	68	70	70	72	72	74
0.831 - 0.850 (0.0327 - 0.0335)	56	58	60	62	64	66	68	70	72	72	74
0.851 - 0.870 (0.0335 - 0.0343)	58	60	62	64	66	68	70	72	74
0.871 - 0.890 (0.0343 - 0.0350)	60	62	64	66	68	70	72
0.891 - 0.910 (0.0351 - 0.0358)	62	64	66	68	70	72
0.911 - 0.930 (0.0359 - 0.0366)	64	66	68	70	72
0.931 - 0.950 (0.0367 - 0.0374)	66	68	70	72
0.951 - 0.970 (0.0374 - 0.0382)	68	70	72
0.971 - 0.990 (0.0382 - 0.0390)	70	72
0.991 - 1.010 (0.0390 - 0.0398)	72

06	5.060 (0.1992)	30	5.300 (0.2087)	54	5.540 (0.2181)
08	5.080 (0.2000)	32	5.320 (0.2094)	56	5.560 (0.2189)
10	5.100 (0.2008)	34	5.340 (0.2102)	58	5.580 (0.2197)
12	5.120 (0.2016)	36	5.360 (0.2110)	60	5.600 (0.2205)
14	5.140 (0.2024)	38	5.380 (0.2118)	62	5.620 (0.2213)
16	5.160 (0.2031)	40	5.400 (0.2126)	64	5.640 (0.2220)
18	5.180 (0.2039)	42	5.420 (0.2134)	66	5.660 (0.2228)
20	5.200 (0.2047)	44	5.440 (0.2142)	68	5.680 (0.2236)
22	5.220 (0.2055)	46	5.460 (0.2150)	70	5.700 (0.2244)
24	5.240 (0.2063)	48	5.480 (0.2157)	72	5.720 (0.2252)
26	5.260 (0.2071)	50	5.500 (0.2165)	74	5.740 (0.2260)
28	5.280 (0.2079)	52	5.520 (0.2173)		

09490_TACO_G0075

Fig. 116 Shim selection chart, part two—4.0L (1GR-FE) engine

23. Determine the replacement adjusting shim size according to the following formula or use the adjusting shim charts.

24. Using a micrometer, measure the thickness of the removed shim. Calculate the thickness of a new shim so that the valve clearance comes within the specified value.

- T: Thickness of the removed shim
- A: Measured valve clearance
- N: Thickness of the new shim

a. Intake: N = T + A
b. Exhaust: N = T + A

25. Select a new lifter with a thickness as close as possible to the calculated value.

26. Install removed components in the reverse order of the removal procedure.

27. When installing the cylinder head cover, apply a continuous bead (0.08–0.12 inch) of seal packing, part number 08826-00080 or equivalent, to the cylinder head as indicated in the illustration. Install the seal washers onto the bolts. Install the cylinder head cover bolts and nuts. Tighten bolts "A" to 7.4 ft. lbs. Tighten bolts "B" to 80 inch lbs. Tighten nuts to 80 inch lbs.

➡Be sure to remove any oil from the contact surfaces of the cylinder head cover and the cylinder head. Install the cover within three minutes after applying the seal packing. Tighten the bolts to specification within fifteen minutes after installing the cover. Do not add engine oil for at least two hours after installing the cover.

28. Continue the installation in the reverse order of the removal procedure.

29. Be sure to fill the engine with the proper grade and type engine coolant.

30. Be sure to fill the engine with the proper grade and type engine oil.

31. Start the engine and check for leaks. Correct as required.

ENGINE PERFORMANCE & EMISSION CONTROL

See Figures 117 and 118.

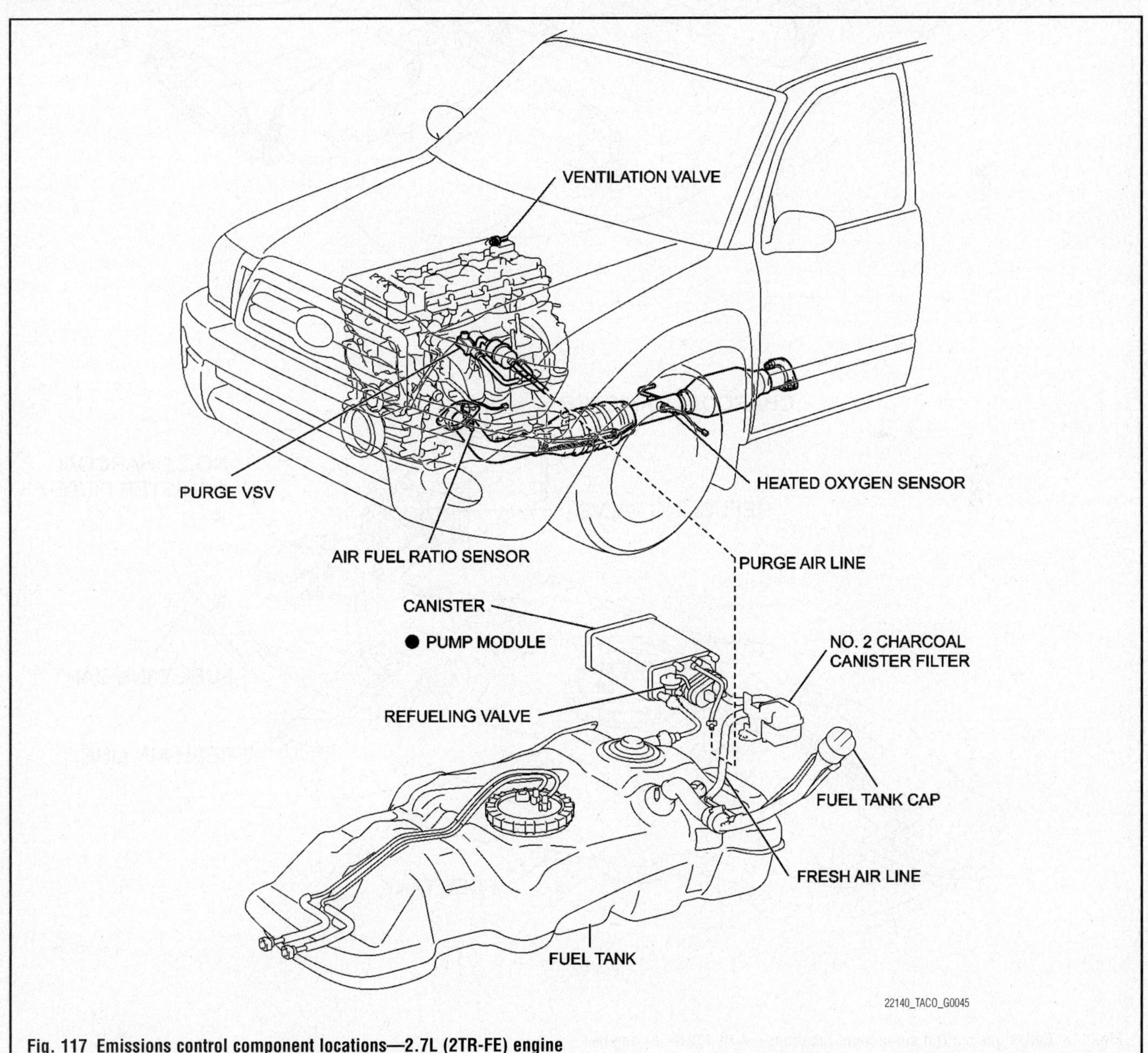

Fig. 117 Emissions control component locations—2.7L (2TR-FE) engine

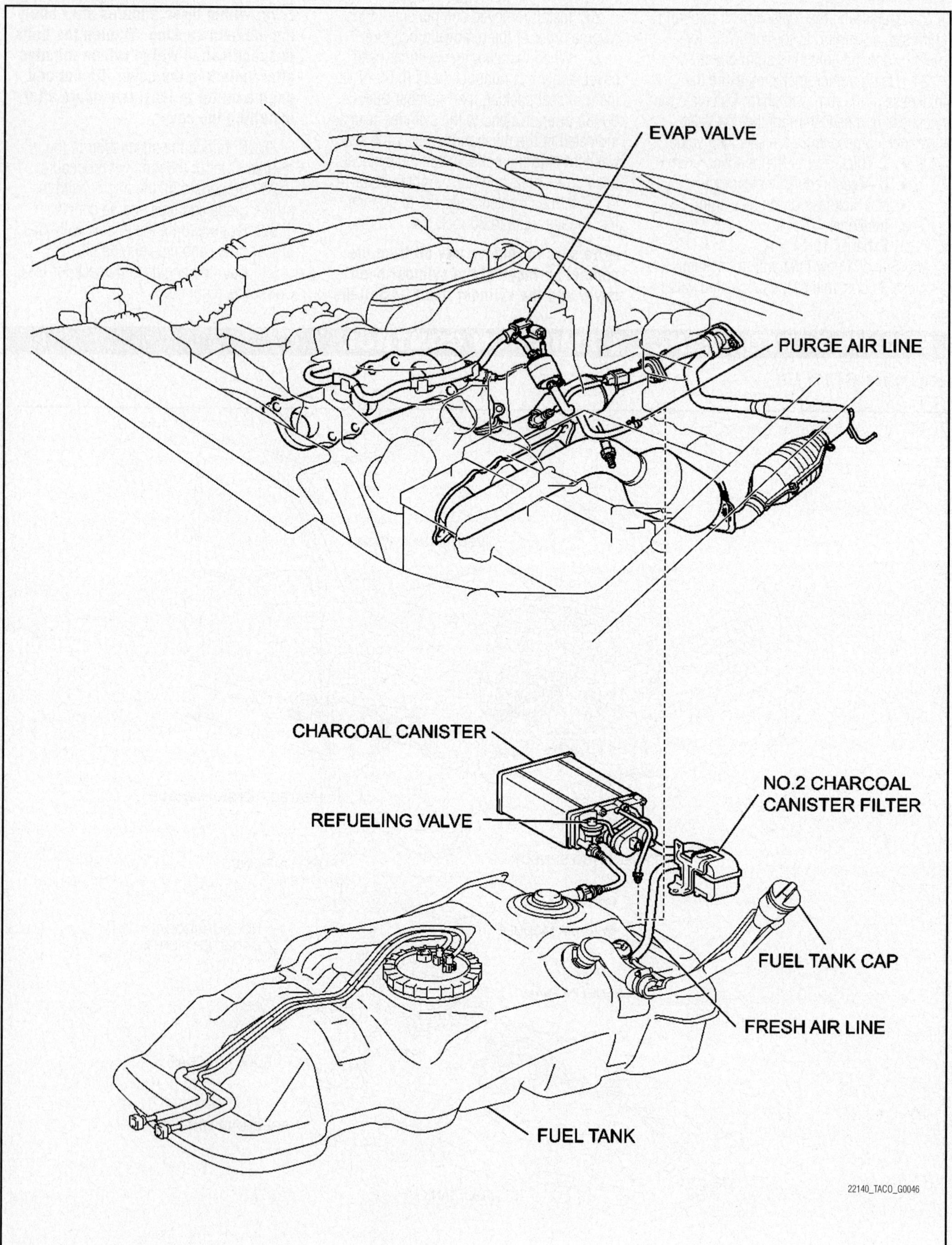

Fig. 118 Emissions control component locations—4.0L (1GR-FE) engine

ACCELERATOR PEDAL POSITION (APP) SENSOR

LOCATION

Part of the Accelerator Pedal assembly

REMOVAL & INSTALLATION

See Figure 119.

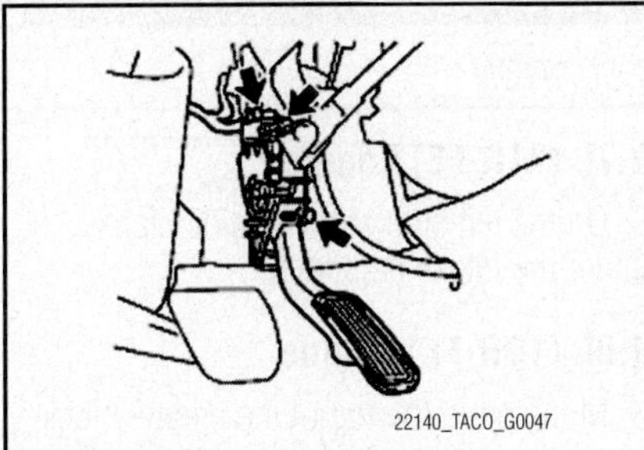

Fig. 119 Accelerator Pedal Position Sensor

1. Before servicing the vehicle, refer to the precautions in the beginning of this section.
2. Disconnect the accelerator position sensor connector.
3. Remove the 2 bolts, then remove the accelerator pedal.

To install:

4. Install the accelerator pedal with the 2 bolts and tighten to 44 inch lbs. (5 Nm).
5. Connect the accelerator position sensor connector.
6. Connect cable to negative battery terminal.

CAMSHAFT POSITION (CMP) SENSOR

LOCATION

At the front of the cylinder head.

REMOVAL & INSTALLATION

See Figure 120.

1. Before servicing the vehicle, refer to the precautions in the beginning of this section.

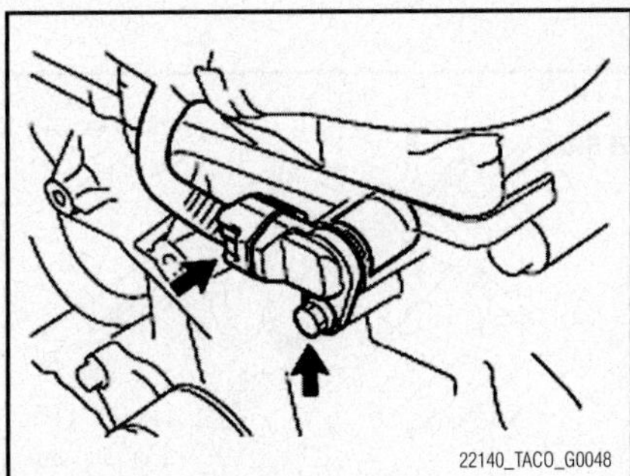

Fig. 120 Camshaft position sensor

2. Disconnect the negative battery cable.
3. Disconnect the camshaft position sensor connector.
4. Remove the bolt, then remove the camshaft position sensor.
5. Installation is the reverse of the removal procedure.

CRANKSHAFT POSITION (CKP) SENSOR

LOCATION

2.7L (2TR-FE) Engine

Lower front left of the engine block, under the A/C compressor bracket.

4.0L (1GR-FE) Engine

Front left of the engine block, under the A/C compressor.

REMOVAL & INSTALLATION

2.7L (2TR-FE) Engine

See Figure 121.

1. Before servicing the vehicle, refer to the precautions in the beginning of this section.
2. Disconnect cable from negative battery terminal.
3. Remove engine under cover sub-assembly no. 1 (for 4wd and pre-runner).
4. Remove fan and generator v belt.
5. Separate compressor and magnetic clutch.
6. Remove idle pulley assembly with bracket.
7. Remove the 5 bolts, then remove the idle pulley with bracket.
8. Disconnect the crankshaft position sensor connector and the 2 wire harness clamps.
9. Remove the bolt, then remove the crankshaft position sensor.

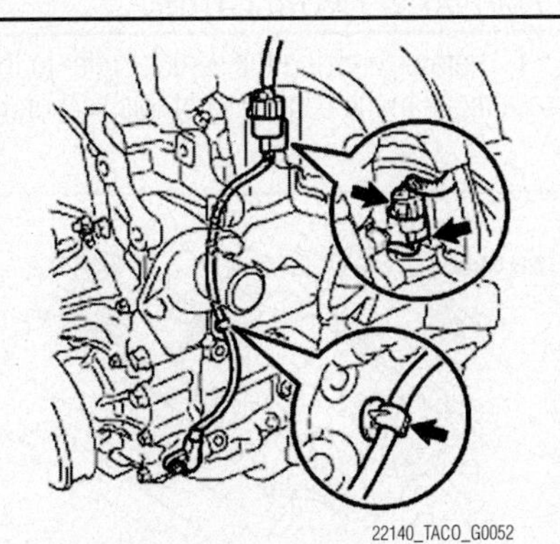

Fig. 121 Crankshaft position sensor and harness—2.7L (2TR-FE) engine

10. Installation is the reverse of the removal procedure.

4.0L (1GR-FE) Engine

See Figure 122.

1. Before servicing the vehicle, refer to the precautions in the beginning of this section.
2. Remove v-bank cover.
3. Remove fan shroud.
4. Remove generator assembly.
5. Separate cooler compressor assembly.
6. Remove the bolt, then separate the suction hose sub-assembly.
7. Disconnect the cooler compressor assembly connector.
8. Remove the 4 bolts, then separate the cooler compressor assembly from the V-ribbed belt tensioner assembly.
9. Disconnect the crankshaft position sensor connector.
10. Remove the bolt, then remove the crankshaft position sensor.

Fig. 122 Crankshaft position sensor — 4.0L (1GR-FE) engine

11. Installation is the reverse of the removal procedure.

ELECTRONIC CONTROL MODULE (ECM)

LOCATION

Behind the glovebox.

REMOVAL & INSTALLATION

See Figure 123.

1. Before servicing the vehicle, refer to the precautions in the beginning of this section.
2. Perform VIN registration when replacing the ECM.
3. Disconnect cable from negative battery terminal
4. Remove glove compartment door assembly
5. Remove instrument panel finish panel sub-assembly lower rh

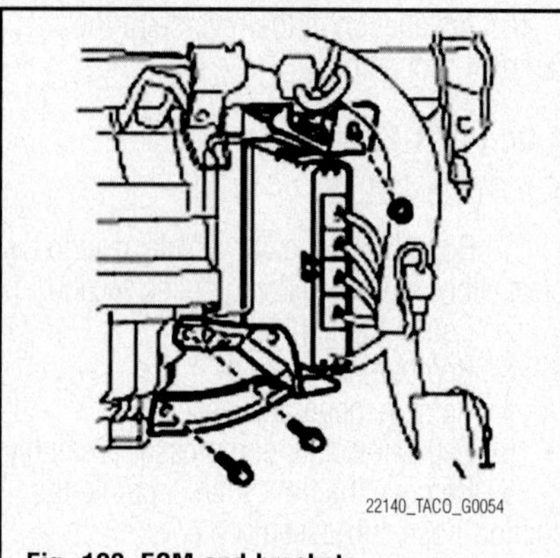

Fig. 123 ECM and brackets

6. Disconnect the 4 connectors.
7. Remove the 2 bolts and nut, then remove the ECM.
8. Remove the 3 screws, then remove the ECM bracket.
9. Remove the 2 screws, then remove the ECM bracket No. 2.

ENGINE COOLANT TEMPERATURE (ECT) SENSOR

LOCATION

Mounted near the thermostat housing.

REMOVAL & INSTALLATION

4.0L (1GR-FE) Engine

See Figure 124.

1. Before servicing the vehicle, refer to the precautions in the beginning of this section.
2. Discharge fuel system pressure.
3. Drain engine coolant.
4. Disconnect cable from negative battery terminal.
5. Remove v-bank cover.
6. Remove air cleaner assembly.
7. Remove intake air surge tank.
8. Disconnect fuel pipe sub-assembly no.2.

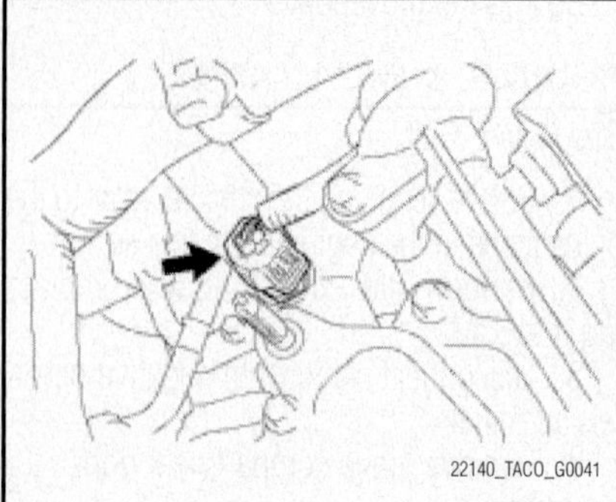

Fig. 124 Coolant Temperature Sensor—4.0L (1GR-FE) engine

9. Disconnect the connector.
10. Using a 19 mm deep socket wrench, remove the water temperature sensor and gasket.

To install:

11. Using a 19 mm deep socket wrench, install the water temperature sensor with a new gasket and tighten to 15 ft. lbs. (20 Nm).
12. Connect the connector.
13. Connect fuel pipe sub-assembly no.2
14. Install intake air surge tank
15. Install air cleaner assembly
16. Connect cable to negative battery terminal
17. Add engine coolant
18. Check for fuel leakage
19. Check for engine coolant leakage
20. Install v-bank cover

HEATED OXYGEN (HO2S) SENSOR

LOCATION

See Figures 125 and 126.

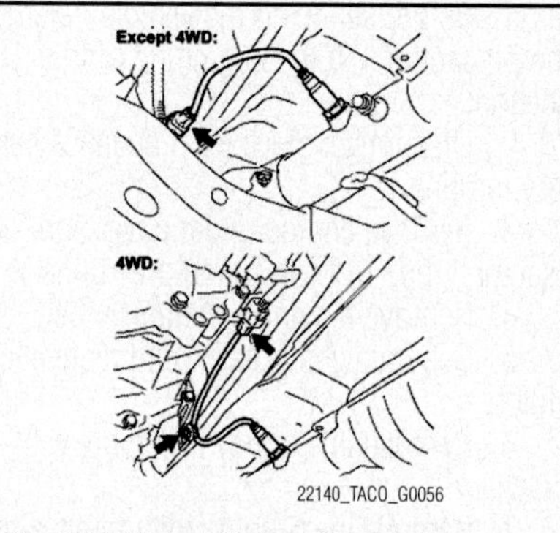

Fig. 125 Heated Oxygen Sensor location—2.7L (2TR-FE)

Mounted in the exhaust, between the engine and the catalytic converter(s).

REMOVAL & INSTALLATION

1. Before servicing the vehicle, refer to the precautions in the beginning of this section.
2. Disconnect the negative battery cable.
3. Disconnect the heated oxygen sensor connector.
4. Remove the heated oxygen sensor from the exhaust pipe.
5. Installation is the reverse of the removal procedure. Tighten the heated oxygen sensor to 33 ft. lbs. (44Nm).

KNOCK SENSOR (KS)

LOCATION

2.7L (2TR-FE) Engine

On the left side of the engine block, under the intake manifold.

4.0L (1GR-FE) Engine

Mounted to the top of the engine block under the intake manifold.

REMOVAL & INSTALLATION

2.7L (2TR-FE) Engine

1. Before servicing the vehicle, refer to the precautions in the beginning of this section.
2. Disconnect cable from negative battery terminal.
3. Remove engine under cover sub-assembly.
4. Drain engine coolant.
5. Remove intake air connector.
6. Remove throttle with motor body assembly.
7. Disconnect fuel hose.
8. Disconnect fuel hose no. 2.
9. Remove intake manifold.
10. Disconnect the fuel vapor feed hose from the VSV.
11. Disconnect the vacuum hose.
12. Remove the bolt, then remove the clamp bracket.
13. Disconnect the water by-pass hose No. 2.
14. Disconnect the ventilation hose No. 3.
15. Disconnect the VSV connector.
16. Disengage the engine wire harness clamp.

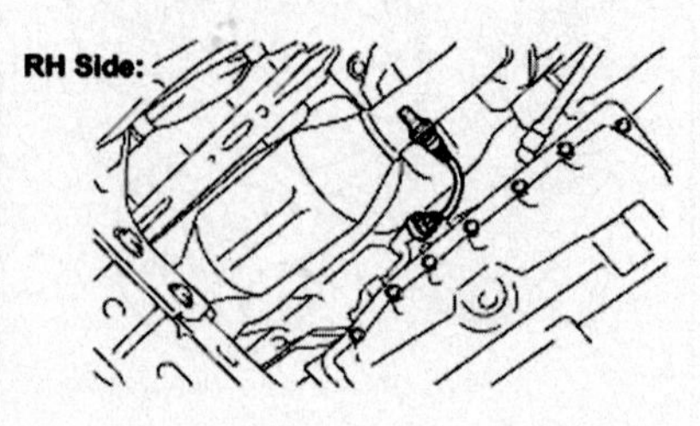

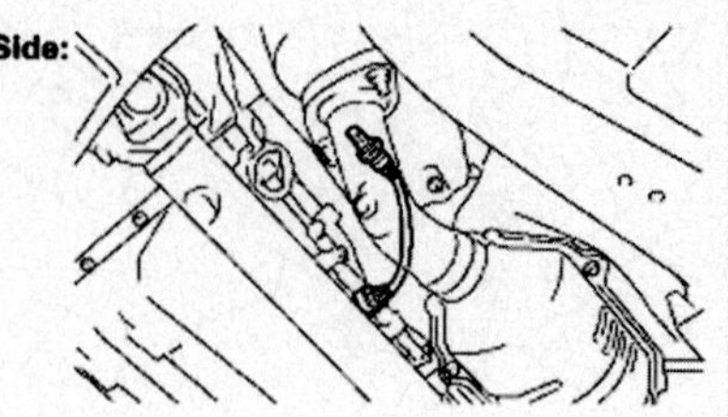

Fig. 126 Heated Oxygen Sensor location—4.0L (1GR-FE)

17. Disconnect the compressor magnetic clutch connector.
18. Disengage the wire harness clamp.
19. Remove the bolt and harness clamp bracket.
20. Disconnect the 3 connectors.
21. Remove the nut shown in the illustration.
22. Disconnect the engine wire harness from the relay block.
23. Remove the 5 bolts and 2 nuts, then remove the intake manifold.
24. Disconnect the knock control sensor connector.
25. Remove the bolt, then remove the knock control sensor.

To install:

26. Installation is the reverse of the removal procedure. Tighten the knock sensor bolt to 15 ft. lbs. (20 Nm).

4.0L (1GR-FE) Engine

2WD

See Figure 127.

1. Before servicing the vehicle, refer to the precautions in the beginning of this section.
2. Discharge fuel system pressure.
3. Remove battery.
4. Drain engine coolant.
5. Drain engine oil.
6. Remove engine assembly.
7. Remove timing chain or belt cover sub-assembly.
8. Remove cool air inlet no. 1.
9. Remove the 2 bolts, then remove the cool air inlet.
10. Remove exhaust front pipe assembly no. 2.

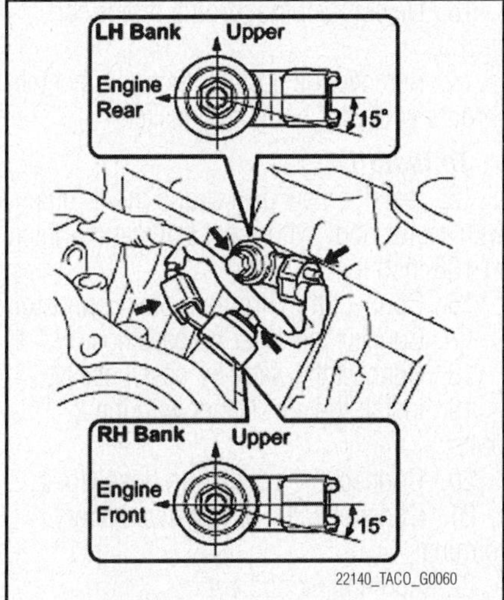

Fig. 127 Knock sensor installation—4.0L (1GR-FE) engine

11. Remove exhaust pipe assembly front.
12. Remove manifold stay.
13. Remove the 3 bolts, then remove the exhaust manifold stay.
14. Remove exhaust manifold sub-assembly rh
15. Disconnect the air fuel ratio sensor connector.
16. Remove the 6 nuts, then remove the exhaust manifold and gasket.
17. Remove intake manifold.
18. Remove water by-pass joint.
19. Remove chain vibration damper no. 1.
20. Remove the 2 bolts, then remove the chain vibration damper No. 1.
21. Remove camshaft timing gears and no.2 chain (rh bank).
22. Remove chain tensioner assembly no.2.
23. Remove camshafts.
24. Remove camshaft bearing no. 2.
25. Remove cylinder head sub-assembly.
26. Disconnect heater water inlet hose.
27. Remove water outlet pipe no. 1.
28. Remove the 4 wire harness clamps.
29. Remove the 3 bolts and water outlet pipe.
30. Disconnect the 2 knock sensor connectors.
31. Remove the 2 bolts and 2 knock sensors.

To install:

32. Installation is the reverse of the removal procedure. Install the knock sensors as shown and tighten to 15 ft. lbs. (20 Nm).

4WD and PreRunner

See Figure 127.

1. Before servicing the vehicle, refer to the precautions in the beginning of this section.
2. Discharge fuel system pressure.
3. Remove battery.
4. Drain engine coolant.
5. Drain engine oil.
6. Remove power steering link assembly.
7. Remove differential carrier assembly front (4wd drive type).
8. Remove timing chain or belt cover sub-assembly.
9. Remove chain sub-assembly.
10. Remove the 2 bolts, then remove the cool air inlet.
11. Remove exhaust front pipe assembly no.2.
12. Remove exhaust pipe assembly front.
13. Remove manifold stay.
14. Remove the 6 bolts, then remove the 2 exhaust manifold stays.
15. Remove exhaust manifold sub-assembly rh.
16. Disconnect fuel pipe sub-assembly no. 1.
17. Disconnect fuel pipe sub-assembly no. 2.
18. Remove intake manifold.
19. Remove water by-pass joint rr.
20. Remove chain vibration damper no. 1.
21. Remove camshaft timing gears and no. 2 chain (rh bank).
22. Remove chain tensioner assembly no. 2.
23. Remove camshafts.
24. Remove camshaft bearing no. 2.
25. Remove cylinder head sub-assembly.
26. Disconnect heater water inlet hose.
27. Remove the 4 wire harness clamps.
28. Remove the 3 bolts and water outlet pipe.
29. Disconnect the 2 knock sensor connectors.
30. Remove the 2 bolts and 2 knock sensors.

To install:

31. Installation is the reverse of the removal procedure. Install the knock sensors as shown and tighten to 15 ft. lbs. (20 Nm).

MALFUNCTION INDICATOR LIGHT (MIL)

RESET PROCEDURES

Reset the MIL by clearing codes with a scan tool.

MASS AIR FLOW (MAF) SENSOR

LOCATION

In the air intake, between the air filter and the throttle body.

REMOVAL & INSTALLATION

See Figure 128.

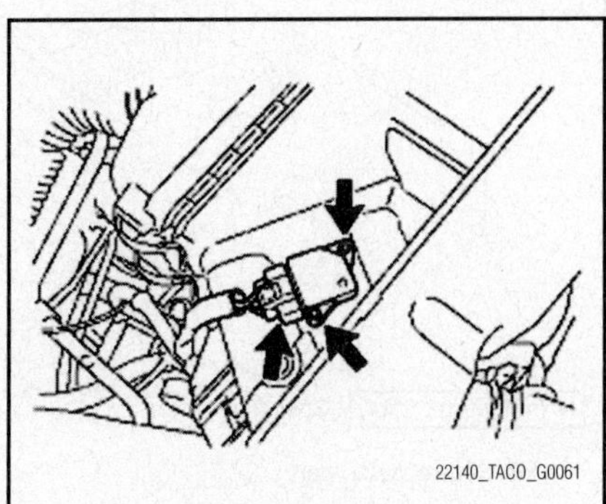

Fig. 128 Mass Air Flow Sensor

1. Before servicing the vehicle, refer to the precautions in the beginning of this section.
2. Disconnect cable from negative battery terminal.
3. Remove v-bank cover.
4. Disconnect the connector.
5. Remove the 2 screws, then remove the mass air flow meter.

THROTTLE POSITION SENSOR (TPS)

LOCATION

Part of the throttle body.

REMOVAL & INSTALLATION

2.7L (2TR-FE) Engine

See Figure 129.

1. Before servicing the vehicle, refer to the precautions in the beginning of this section.
2. Disconnect cable from negative battery terminal
3. Remove no. 1 engine under cover sub-assembly (for 4wd and pre-runner)
4. Drain engine coolant
5. Remove intake air connector
6. Disconnect the pressure sensor connector.
7. Disengage the wire harness clamp.
8. Disconnect the vacuum hose.
9. Disconnect the No. 2 ventilation hose.
10. Disconnect the vacuum hose.
11. Loosen the 2 hose clamp bolts.
12. Remove the 3 bolts, then remove the intake air connector.
13. Disconnect the water by-pass hose.
14. Disconnect the No. 2 water by-pass hose.
15. Disconnect the throttle motor connector.
16. Remove the 2 bolts and 2 nuts, then remove the throttle with motor body.
17. Remove the gasket from the intake manifold.

To install:

18. Install a new gasket onto the intake manifold.
19. Install the throttle with motor body with the 2 bolts and 2 nuts. Tighten to 80 inch lbs. (9 Nm).
20. Connect the water by-pass hose.
21. Connect the water by-pass hose No. 2.
22. Connect the throttle motor connector.
23. Install the intake air connector with the 3 bolts.
24. Tighten the 2 hose clamp bolts.
25. Connect the vacuum hose.
26. Connect the ventilation hose No. 2.
27. Connect the vacuum hose.
28. Engage the wire harness clamp.
29. Connect the pressure sensor connector.
30. Connect cable to negative battery terminal
31. Add engine coolant
32. Check for engine coolant leakage
33. Install engine under cover sub-assembly No.1 (for 4WD and Pre-Runner).

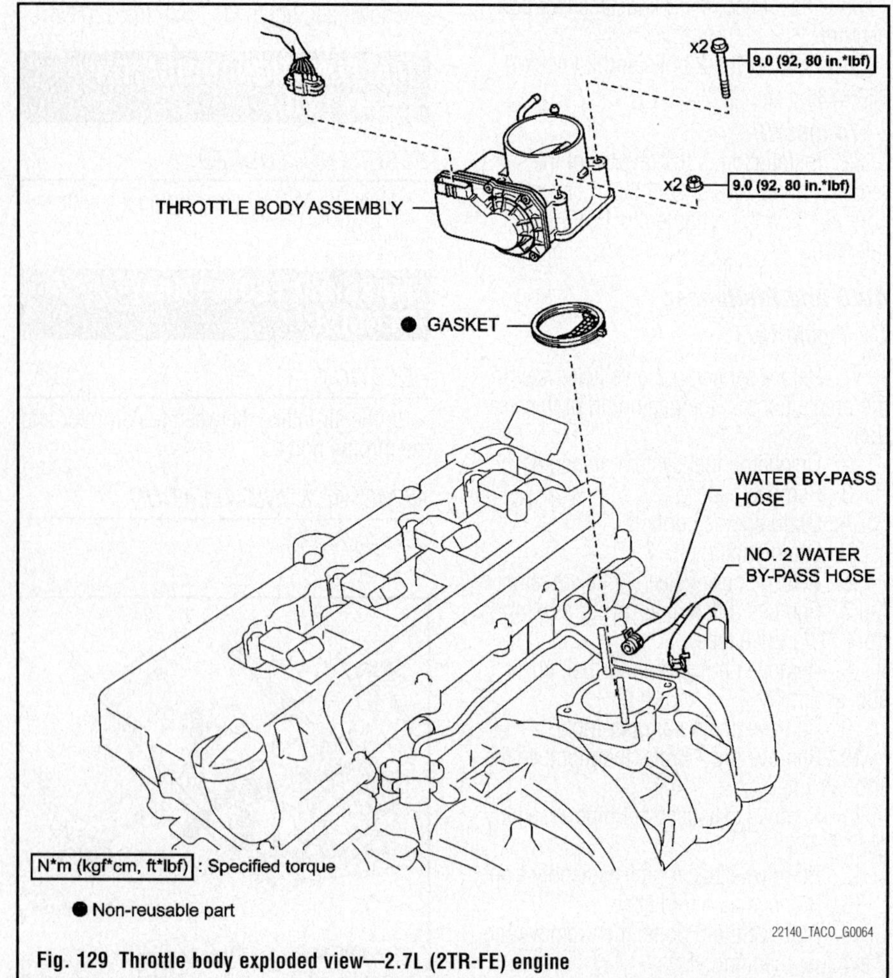

Fig. 129 Throttle body exploded view—2.7L (2TR-FE) engine

4.0L (1GR-FE) Engine

See Figure 130.

1. Before servicing the vehicle, refer to the precautions in the beginning of this section.
2. Disconnect cable from negative battery terminal
3. Drain engine coolant
4. Remove the 2 nuts, then remove the V-bank cover.
5. Disconnect the ventilation hose No. 2.
6. Disconnect the vacuum hose.
7. Disconnect the mass air flow meter connector.
8. Remove the 2 wire harness clamps.
9. Loosen the 2 hose clamps.
10. Remove the 2 bolts, then remove the air cleaner.
11. Disconnect the water by-pass hose No. 5.
12. Disconnect the water by-pass hose No. 4.
13. Disconnect the throttle motor connector.
14. Remove the 4 bolts, then remove the throttle w/ motor body and gasket.

To install:

15. Install a new gasket and the throttle with motor body with the 4 bolts and tighten to 108 inch lbs. (11 Nm).
16. Connect the throttle motor connector.
17. Connect the water by-pass hose No. 4.
18. Connect the water by-pass hose No. 5.
19. Install the air cleaner with the 2 bolts.
20. Connect the ventilation hose No. 2.
21. Connect cable to negative battery terminal
22. Add engine coolant
23. Check for engine coolant leakage
24. Install the V-bank cover with the 2 nuts.

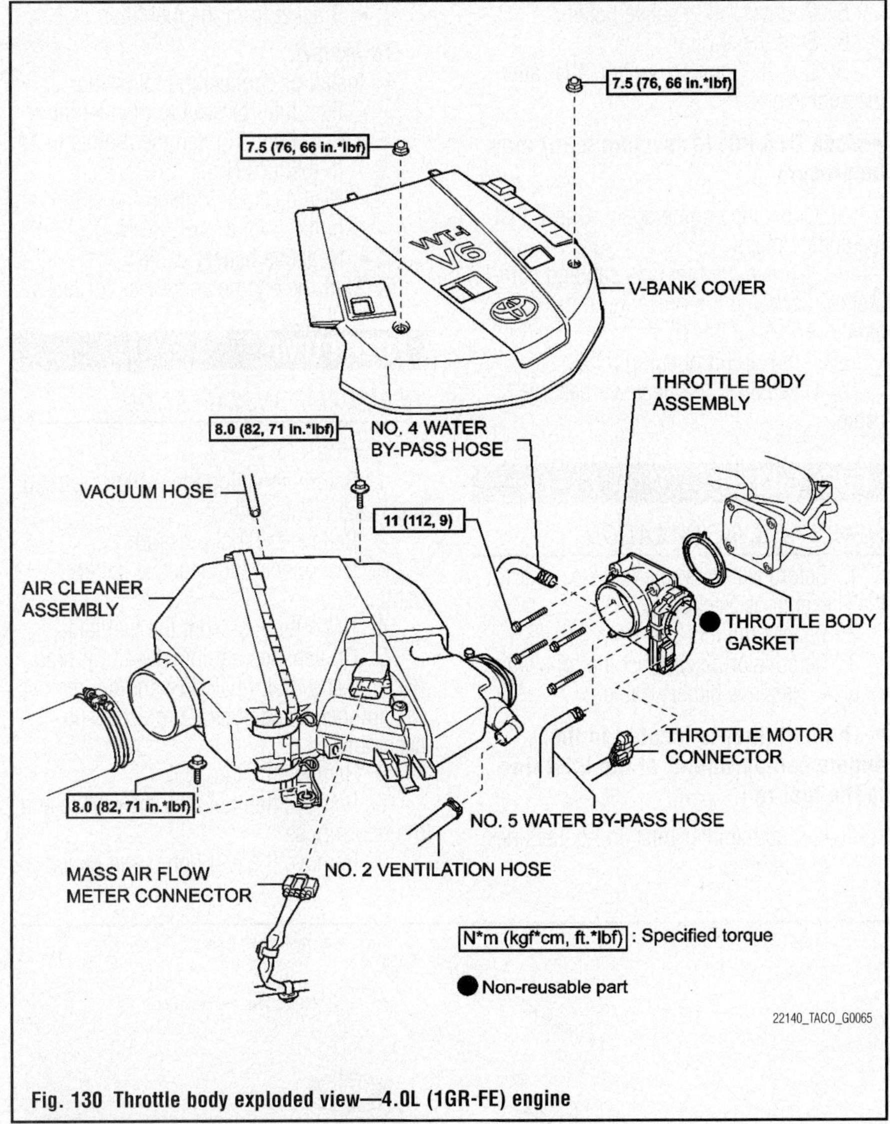

Fig. 130 Throttle body exploded view—4.0L (1GR-FE) engine

VARIABLE CAMSHAFT TIMING OIL CONTROL SOLENOID

REMOVAL & INSTALLATION

1. Before servicing the vehicle, refer to the precautions in the beginning of this section.
2. Drain engine coolant.
3. Disconnect cable from negative battery terminal.
4. Remove v-bank cover.
5. Remove air cleaner assembly.
6. Remove intake air surge tank.
7. Remove camshaft timing oil control valve assembly.
8. Disconnect the 2 connectors.
9. Remove the 2 bolts, then remove the 2 camshaft timing oil control valves.
10. Installation is the reverse of the removal procedure.

VEHICLE SPEED SENSOR (VSS)

REMOVAL & INSTALLATION

See Figure 131.

1. Before servicing the vehicle, refer to the precautions in the beginning of this section.
2. Disconnect the speed sensor connector.
3. Remove the bolt and speed sensor.
4. Remove the O-ring from the speed sensor.
5. Installation is the reverse of the removal procedure.

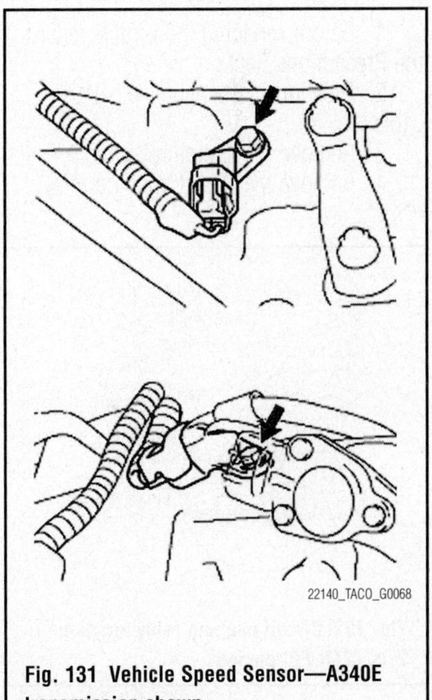

Fig. 131 Vehicle Speed Sensor—A340E transmission shown

FUEL SYSTEM — GASOLINE FUEL INJECTION SYSTEM

FUEL SYSTEM SERVICE PRECAUTIONS

Safety is the most important factor when performing not only fuel system maintenance but any type of maintenance. Failure to conduct maintenance and repairs in a safe manner may result in serious personal injury or death. Maintenance and testing of the vehicle's fuel system components can be accomplished safely and effectively by adhering to the following rules and guidelines.

- To avoid the possibility of fire and personal injury, always disconnect the negative battery cable unless the repair or test procedure requires that battery voltage be applied.
- Always relieve the fuel system pressure prior to disconnecting any fuel system component (injector, fuel rail, pressure regulator, etc.), fitting or fuel line connection. Exercise extreme caution whenever relieving fuel system pressure to avoid exposing skin, face and eyes to fuel spray. Please be advised that fuel under pressure may penetrate the skin or any part of the body that it contacts.
- Always place a shop towel or cloth around the fitting or connection prior to loosening to absorb any excess fuel due to spillage. Ensure that all fuel spillage (should it occur) is quickly removed from

engine surfaces. Ensure that all fuel soaked cloths or towels are deposited into a suitable waste container.

- Always keep a dry chemical (Class B) fire extinguisher near the work area.
- Do not allow fuel spray or fuel vapors to come into contact with a spark or open flame.
- Always use a back-up wrench when loosening and tightening fuel line connection fittings. This will prevent unnecessary stress and torsion to fuel line piping.
- Always replace worn fuel fitting O-rings with new. Do not substitute fuel hose or equivalent where fuel pipe is installed.

Before servicing the vehicle, make sure to also refer to the precautions in the beginning of this section as well.

RELIEVING FUEL SYSTEM PRESSURE

See Figures 132 and 133.

1. Before servicing the vehicle, refer to the Precautions Section.
2. Disconnect the negative battery cable.
3. Remove the engine relay block cover.
4. Remove the circuit opening relay.
5. Connect the negative battery.
6. Start the engine.
7. Turn the ignition switch "ON" after the engine stops.

➡Code DTC P0171 (system lean) may be present.

8. Crank the engine again. Check that the engine stops.
9. Remove the fuel tank cap and completely discharge the pressure in the fuel tank.

Install the circuit opening relay.

10. Disconnect the negative battery cable.

FUEL FILTER

REMOVAL & INSTALLATION

1. Before servicing the vehicle, refer to the Precautions Section.
2. Relieve the fuel system pressure.
3. Remove or disconnect the following:
   - Negative battery cable

➡The fuel filter is located in the engine compartment, at the inlet line to the fuel rail.

- Plug the filter inlet and outlet lines
- Fuel filter
- Bracket from the fuel filter

To install:

4. Install or connect the following:
   - Fuel filter bracket to the fuel filter
   - Fuel filter. Tighten the 2 bolts to 14 ft. lbs. (20 Nm).
   - New gaskets. Tighten the union bolts to 22 ft. lbs. (30 Nm).
   - Negative battery cable
5. Start the engine and check for leaks.

FUEL PUMP

REMOVAL & INSTALLATION

See Figure 134.

1. Before servicing the vehicle, refer to the Precautions Section.
2. Relieve the fuel pressure.
3. Disconnect the negative battery cable.
4. Drain the fuel from the fuel tank.
5. On vehicles equipped with off road package and 4.0L (1GR-FE) engine, remove the number one fuel tank protector sub-assembly.
6. Remove the fuel tank.
7. Remove the fuel tank main hose and the return hose.
8. Remove the fuel pump and gauge

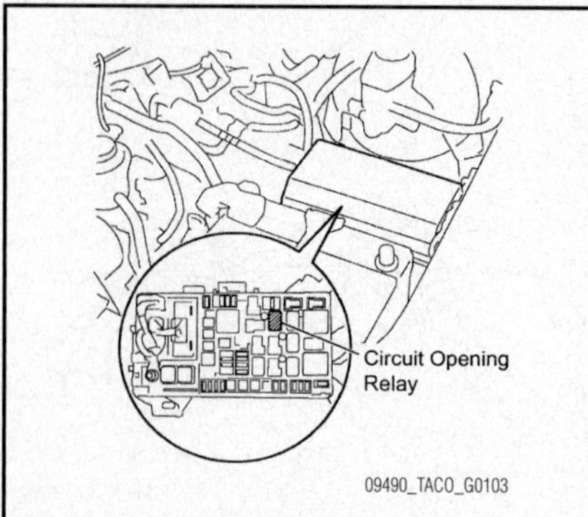

Fig. 132 Circuit opening relay location—2.7L (2TR-FE) engine

Fig. 133 Circuit opening relay location—4.0L (1GR-FE) engine

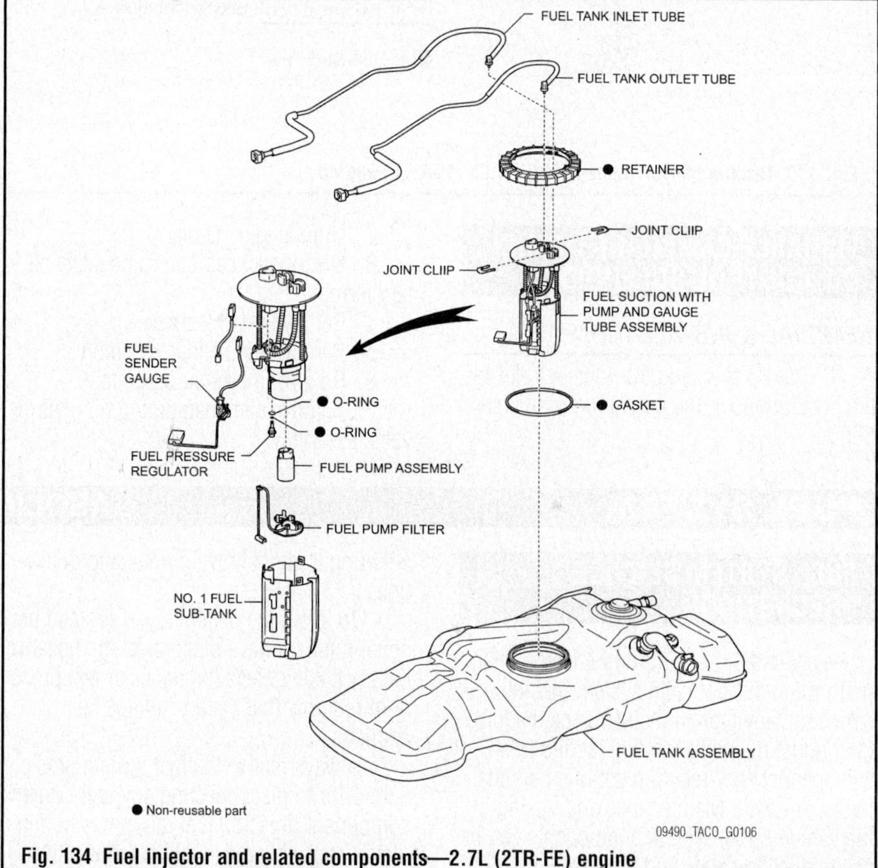

Fig. 134 Fuel injector and related components—2.7L (2TR-FE) engine

assembly retainer from its mounting, using tool SST09808-14020 or equivalent.

9. Pull the fuel pump and gauge assembly out of the fuel tank. Be careful not to bend the arm of the sender. Remove and discard the gasket.

10. Disconnect the connector from the assembly.

11. Disengage the claw fitting and remove the sender gauge by sliding it forward.

12. Disengage the five claw fittings and remove the fuel pump tank. Separate the connector and disengage the clamp.

13. Disengage the clamp and then disconnect the connector.

14. Disengage the five claw fittings and separate the fuel pump from the fuel pump case. Disconnect the connector from the fuel pump.

15. Remove the fuel filter from the fuel pump.

To install:

16. Installation is the reverse of the removal procedure.

17. Start the engine and check for leaks, correct as required.

FUEL RAIL & INJECTORS

REMOVAL & INSTALLATION

2.7L (2TR-FE) Engine

See Figure 135.

1. Before servicing the vehicle, refer to the Precautions Section.
2. Relieve the fuel system pressure.
3. Disconnect the negative battery cable.
4. Remove the engine undercover sub-assembly.
5. Drain the engine coolant. Remove the intake air connector.

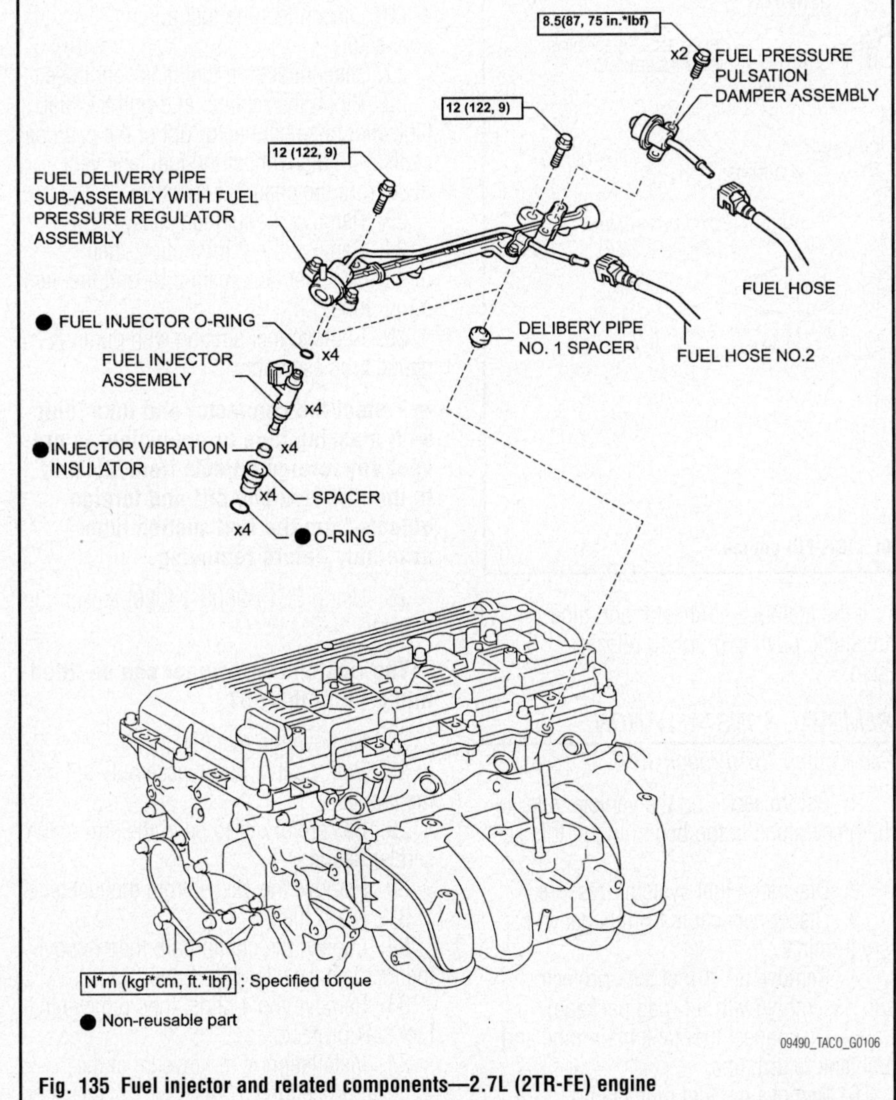

Fig. 135 Fuel injector and related components—2.7L (2TR-FE) engine

6. Remove the throttle body motor assembly.
7. Disconnect and plug the fuel line hoses.
8. Remove the fuel pressure pulsation damper assembly.
9. Disconnect the fuel injector electrical connectors. Disconnect the VSV connector. Disconnect the engine wiring harness clamp.
10. Disconnect the air conditioning compressor clutch connector. Disconnect the wire harness clamp. Remove the bolt and then remove the harness clamp bracket.
11. Remove the two bolts and remove the fuel rail together with the fuel injectors.
12. Remove the fuel rail number one spacers. Remove the four injector vibration insulators. Remove the four spacers.
13. Remove the fuel injectors. Discard all gaskets.

To install:

14. Installation is the reverse of the removal procedure.
15. Be sure to use new gaskets and O-rings, as required.
16. Be sure to fill the cooling system with the proper grade and type engine coolant.
17. Start the engine and check for leaks, correct as required.

4.0L (1GR-FE) Engine

See Figure 136.

1. Before servicing the vehicle, refer to the Precautions Section.
2. Relieve the fuel system pressure.
3. Disconnect the negative battery cable.
4. Drain the engine coolant. Remove the V bank cover.
5. Remove the air cleaner assembly.
6. Remove the intake manifold.
7. Disconnect and plug the fuel line hoses.
8. Disconnect the fuel injector connectors. Remove the six bolts and the fuel rail together with the injectors.
9. Pull the injectors out of the fuel rail.

To install:

10. Installation is the reverse of the removal procedure.
11. Be sure to use new gaskets and O-rings, as required.
12. Be sure to fill the cooling system with the proper grade and type engine coolant.
13. Start the engine and check for leaks, correct as required.

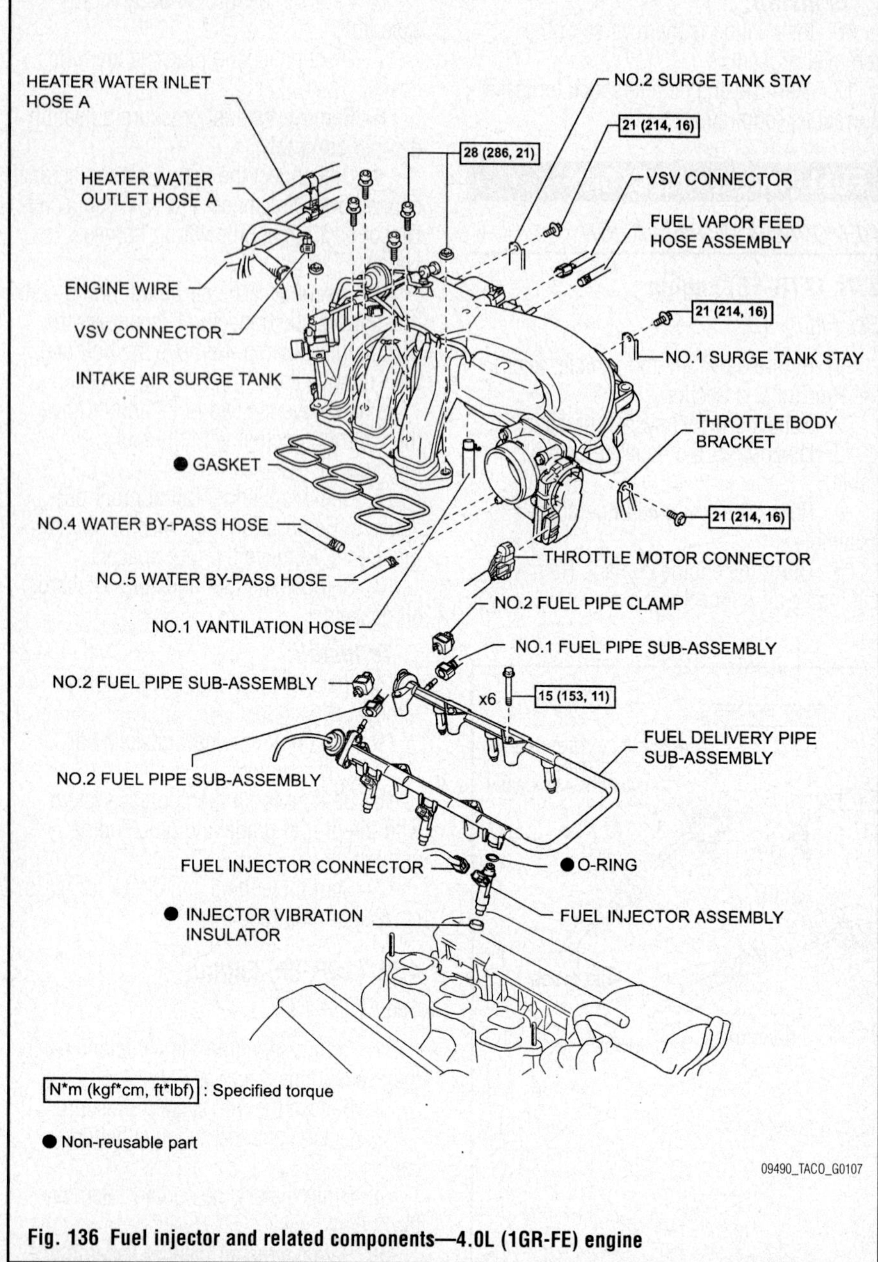

Fig. 136 Fuel injector and related components—4.0L (1GR-FE) engine

FUEL TANK

PRECAUTIONS

Remove any dirt and foreign objects from the fuel tube connector before performing this work.

Do not allow any scratches or foreign objects on the parts when disconnecting, as the fuel tube connector has the O-ring that seals the pipe.

Perform this work by hand. Do not use any tools.

Do not forcibly bend, twist or turn the nylon tube.

Protect the disconnected part by covering it with a vinyl bag after disconnecting the fuel tube.

If the fuel tube connector and pipe are stuck, push and pull to release them.

REMOVAL & INSTALLATION

See Figures 137 through 139.

1. Before servicing the vehicle, refer to the precautions in the beginning of this section.
2. Discharge fuel system pressure.
3. Disconnect cable from negative battery terminal.
4. Remove no. 1 fuel tank protector sub-assembly (with off road package).
5. Disconnect fuel tank main tube and fuel tank return tube.
6. Remove the fuel pipe clamp.
7. Disconnect the fuel tank main tube and fuel tank return tube.
8. Pinch the retainer as illustrated, then pull the fuel tube connectors out of the pipes.
9. Loosen the clamp bolt.
10. Disconnect the fuel breather tube.
11. Remove the checker of the fuel tube connector from the pipe.
12. Pinch the retainer of the fuel tube connector, then pull the tube connector out of the pipe.
13. Hold the fuel tank using the mission jack.
14. Remove the 2 fuel tank bands.
15. Remove the 2 bolts.
16. Remove the 2 clips and 2 pins, then remove the 2 fuel tank bands.
17. Slightly jack the fuel tank down and separate the fuel tank to filler pipe hose from the inlet pipe.
18. Slowly jack down the fuel tank so as not to tear the wire harness and hose.
19. Remove the fuel pump cover.
20. Disconnect the fuel pump connector.
21. Disconnect the fuel tank vent hose.
22. Pinch the retainer and pull the fuel tank vent hose connector out of the charcoal canister to disconnect the fuel tank vent hose from the charcoal canister.
23. Remove the fuel tank assy.
24. Remove the 2 joint clips, then remove the fuel tank main tube and fuel tank return tube.
25. Remove fuel suction with pump & gauge tube assembly

➡Protect the connector and tube joint with masking tape or equivalent to prevent any foreign objects from sticking to them. Clean any dirt and foreign objects from the fuel suction tube assembly before removing.

26. Using SST 09808-14020, loosen the retainer.

➡The ribs on the retainer can be fitted into a tip of the SST.

27. Remove the retainer.
28. Pull the fuel pump assembly out of the fuel tank.
29. Be careful not to bend the arm of the sender gauge.
30. Remove the gasket from the fuel tank.
31. Drain fuel.
32. Loosen the clamp bolt, then remove the fuel tank to filler pipe hose.
33. Remove the 4 clips, then remove the fuel tank protector.
34. Installation is the reverse of the removal procedure.

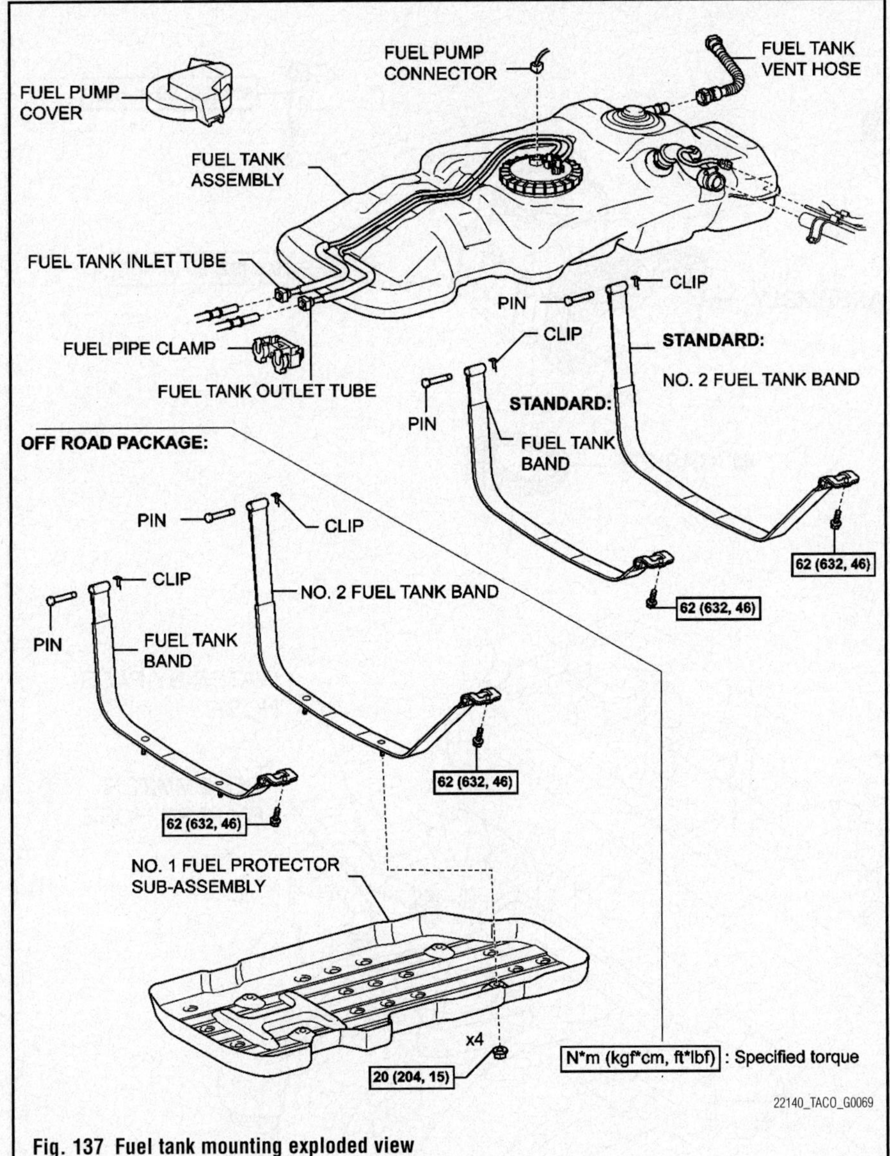

Fig. 137 Fuel tank mounting exploded view

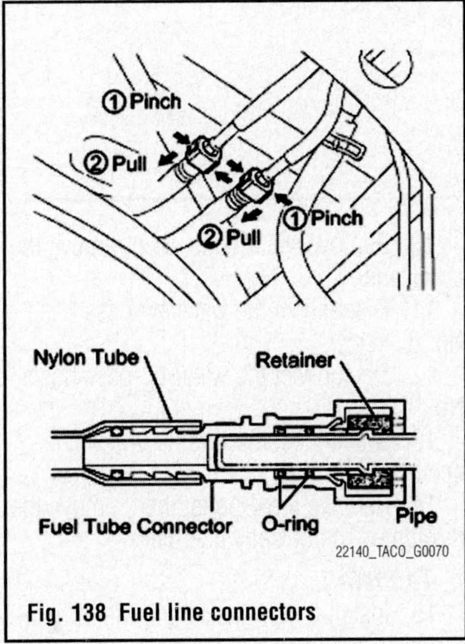

Fig. 138 Fuel line connectors

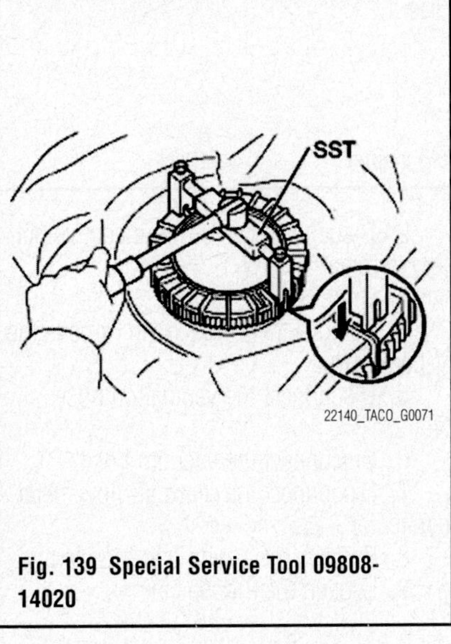

Fig. 139 Special Service Tool 09808-14020

IDLE SPEED

ADJUSTMENT

Idle speed is maintained by the Powertrain Control Module (PCM). No adjustment is necessary or possible.

THROTTLE BODY

REMOVAL & INSTALLATION

2.7L (2TR-FE) Engine

See Figure 140.

1. Before servicing the vehicle, refer to the precautions in the beginning of this section.
2. Disconnect cable from negative battery terminal
3. Remove no. 1 engine under cover sub-assembly (for 4wd and pre-runner)
4. Drain engine coolant
5. Remove intake air connector
6. Disconnect the pressure sensor connector.
7. Disengage the wire harness clamp.
8. Disconnect the vacuum hose.
9. Disconnect the No. 2 ventilation hose.
10. Disconnect the vacuum hose.
11. Loosen the 2 hose clamp bolts.
12. Remove the 3 bolts, then remove the intake air connector.
13. Disconnect the water by-pass hose.
14. Disconnect the No. 2 water by-pass hose.
15. Disconnect the throttle motor connector.
16. Remove the 2 bolts and 2 nuts, then remove the throttle with motor body.
17. Remove the gasket from the intake manifold.

To install:

18. Install a new gasket onto the intake manifold.
19. Install the throttle with motor body with the 2 bolts and 2 nuts. Tighten to 80 inch lbs. (9 Nm).
20. Connect the water by-pass hose.
21. Connect the water by-pass hose No. 2.
22. Connect the throttle motor connector.
23. Install the intake air connector with the 3 bolts.
24. Tighten the 2 hose clamp bolts.
25. Connect the vacuum hose.
26. Connect the ventilation hose No. 2.
27. Connect the vacuum hose.
28. Engage the wire harness clamp.
29. Connect the pressure sensor connector.

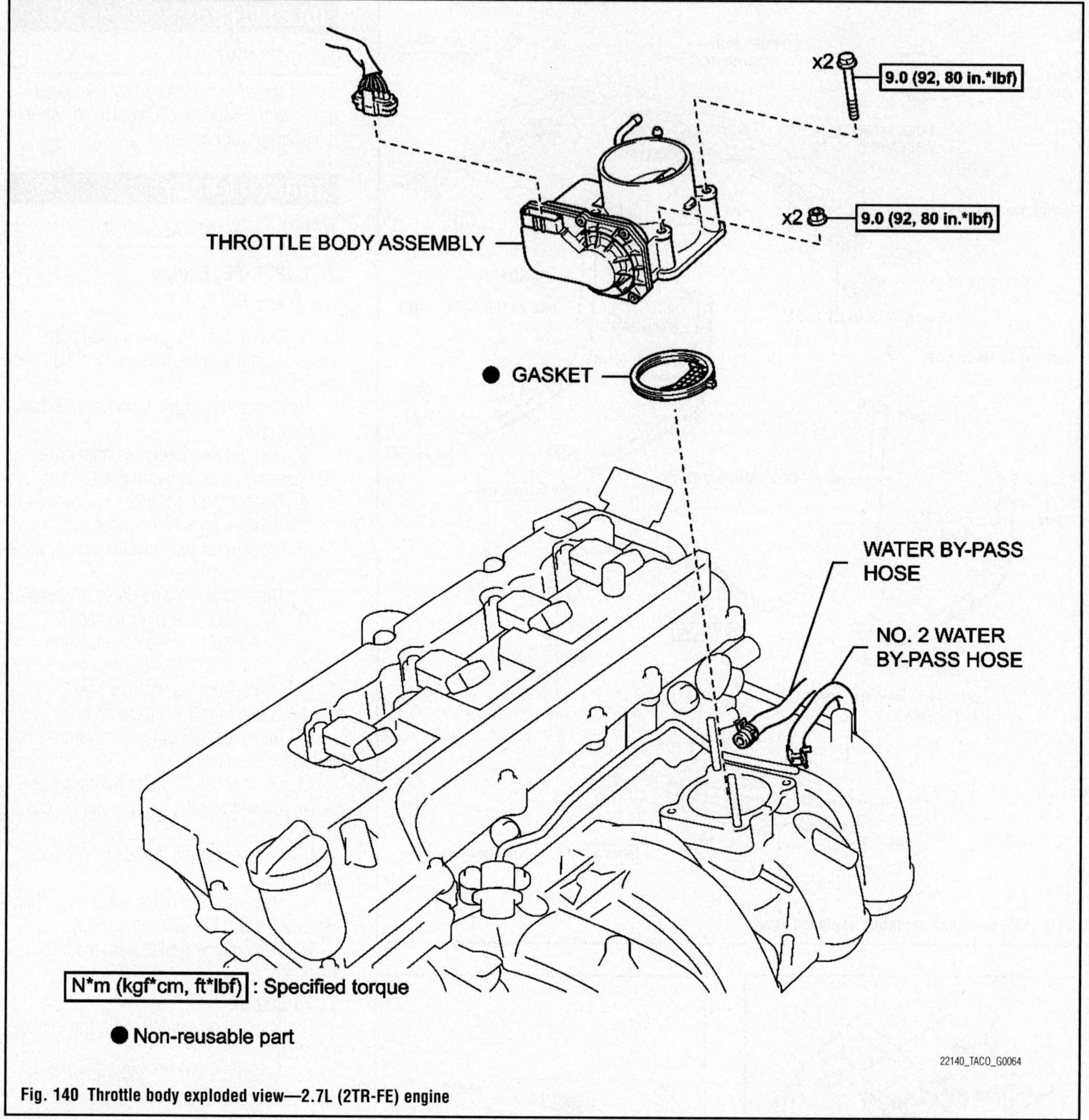

Fig. 140 Throttle body exploded view—2.7L (2TR-FE) engine

30. Connect cable to negative battery terminal
31. Add engine coolant
32. Check for engine coolant leakage
33. Install engine under cover sub-assembly NO.1 (for 4WD and Pre-Runner).

4.0L (1GR-FE) Engine

See Figure 141.

1. Before servicing the vehicle, refer to the precautions in the beginning of this section.
2. Disconnect cable from negative battery terminal
3. Drain engine coolant
4. Remove the 2 nuts, then remove the V-bank cover.
5. Disconnect the ventilation hose No. 2.
6. Disconnect the vacuum hose.
7. Disconnect the mass air flow meter connector.
8. Remove the 2 wire harness clamps.
9. Loosen the 2 hose clamps.
10. Remove the 2 bolts, then remove the air cleaner.
11. Disconnect the water by-pass hose No. 5.
12. Disconnect the water by-pass hose No. 4.
13. Disconnect the throttle motor connector.
14. Remove the 4 bolts, then remove the throttle w/ motor body and gasket.

To install:

15. Install a new gasket and the throttle

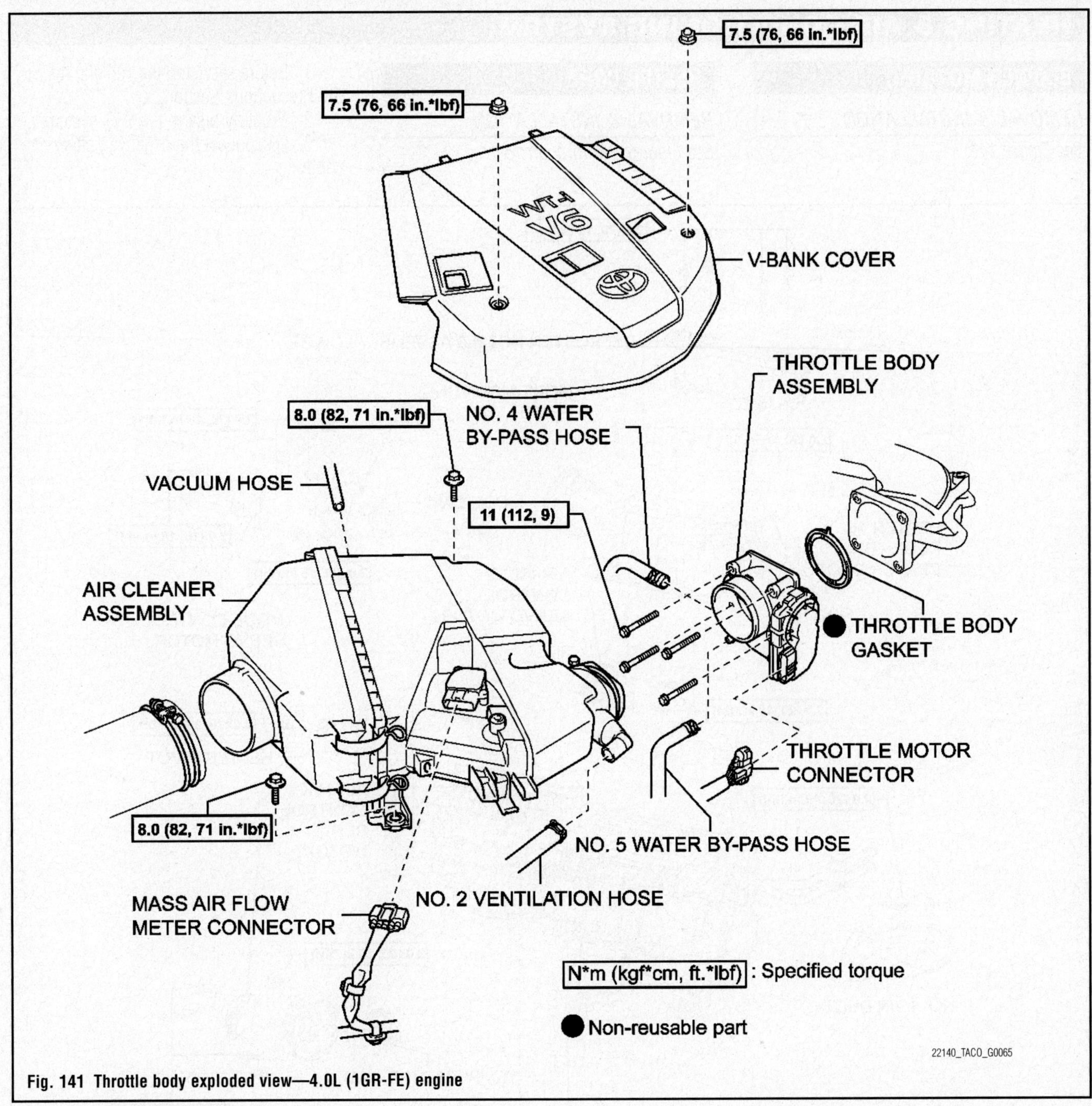

Fig. 141 Throttle body exploded view—4.0L (1GR-FE) engine

with motor body with the 4 bolts and tighten to 108 inch lbs. (11 Nm).

16. Connect the throttle motor connector.
17. Connect the water by-pass hose No. 4.
18. Connect the water by-pass hose No. 5.
19. Install the air cleaner with the 2 bolts.
20. Connect the ventilation hose No. 2.
21. Connect cable to negative battery terminal
22. Add engine coolant
23. Check for engine coolant leakage
24. Install the V-bank cover with the 2 nuts.

HEATING & AIR CONDITIONING SYSTEM

BLOWER MOTOR

REMOVAL & INSTALLATION

See Figure 142.

HEATER CORE

REMOVAL & INSTALLATION

See Figures 143 through 151.

1. Before servicing the vehicle, refer to the Precautions Section.
2. Properly disarm the SRS system.
3. Disconnect the negative battery cable.

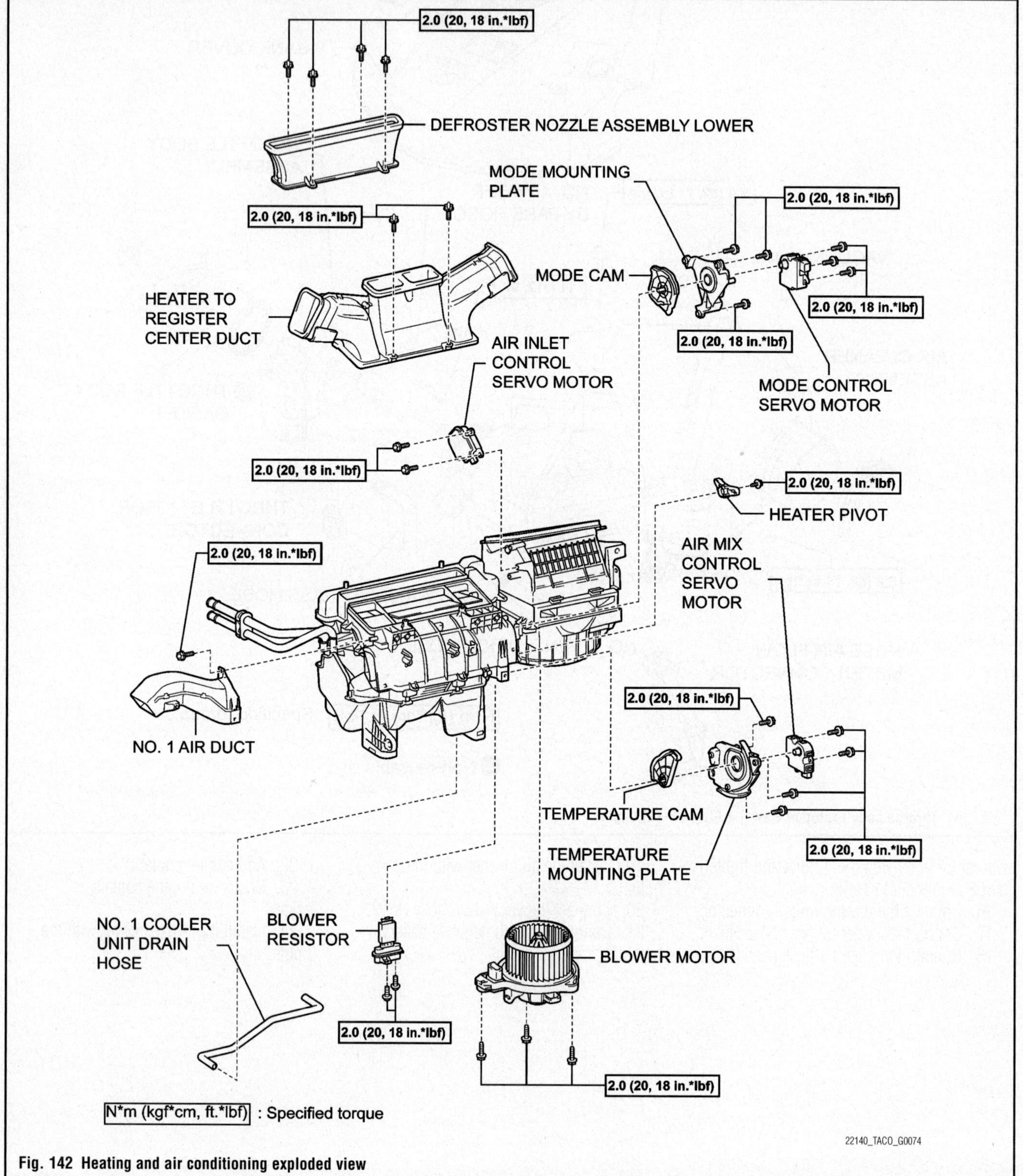

Fig. 142 Heating and air conditioning exploded view

CAUTION

After the negative battery cable has been disconnected, wait at least 1½ minutes for the air bag module to deplete its energy.

4. Properly discharge the air conditioning system. Drain the engine coolant.
5. Disconnect and plug the air condition compressor refrigerant lines.
6. Disconnect the heater hoses at the heater core.
7. Remove the front windshield wiper arm head cap. Remove the wiper arms.
8. Remove the right and left front fender to cowl side seals. Remove the cowl top ventilator louver subassembly.
9. Position the front wheels in the straight ahead position.
10. Remove the lower number two steering wheel cover after disengaging the two claws using the proper tool.
11. Remove the lower number three steering wheel cover after disengaging the two claws using the proper tool.
12. Using a Torx®socket wrench loosen the two retaining screws until the groove along the screw circumference fits in the screw case.
13. Remove the lower number two steer-

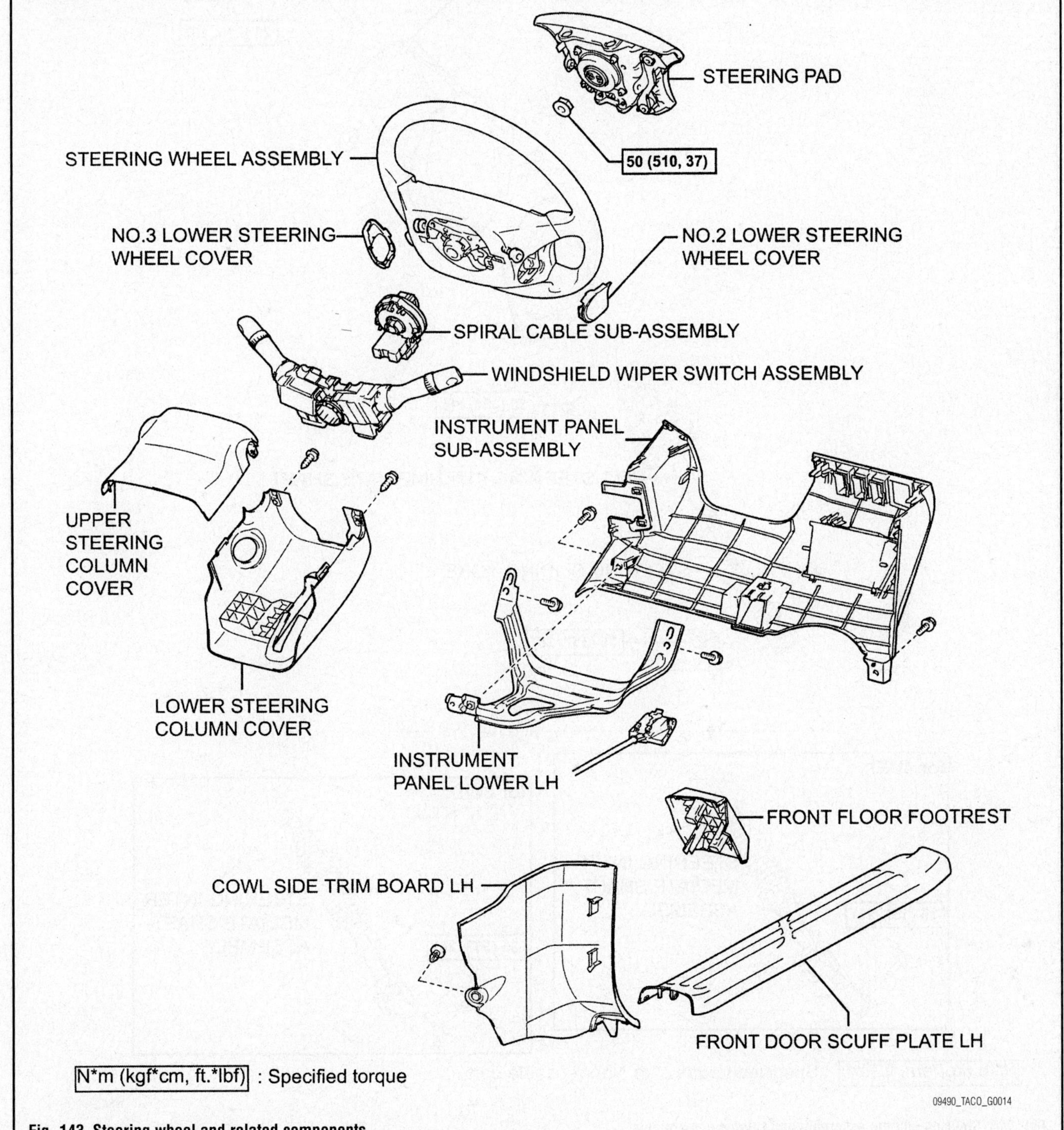

Fig. 143 Steering wheel and related components

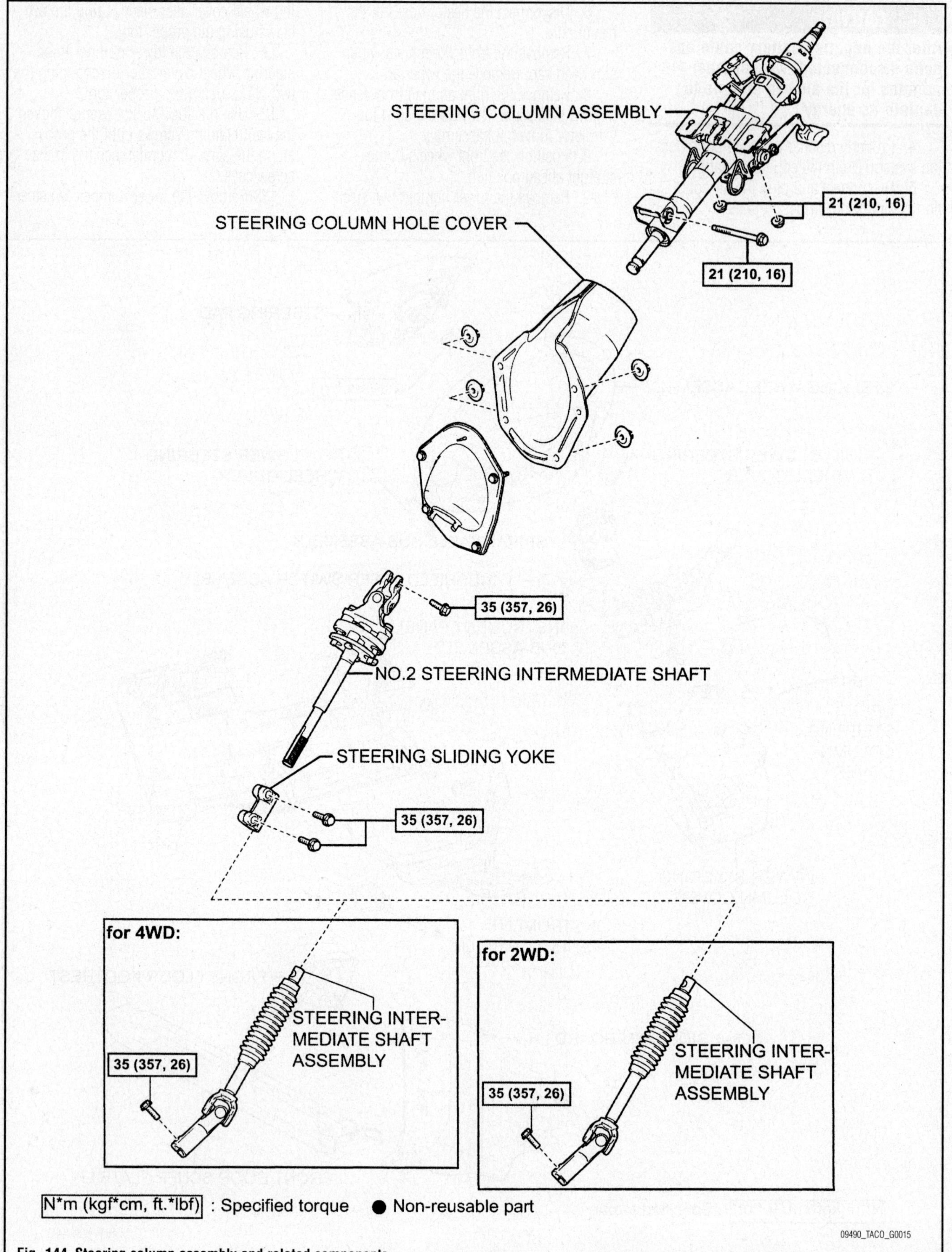

Fig. 144 Steering column assembly and related components

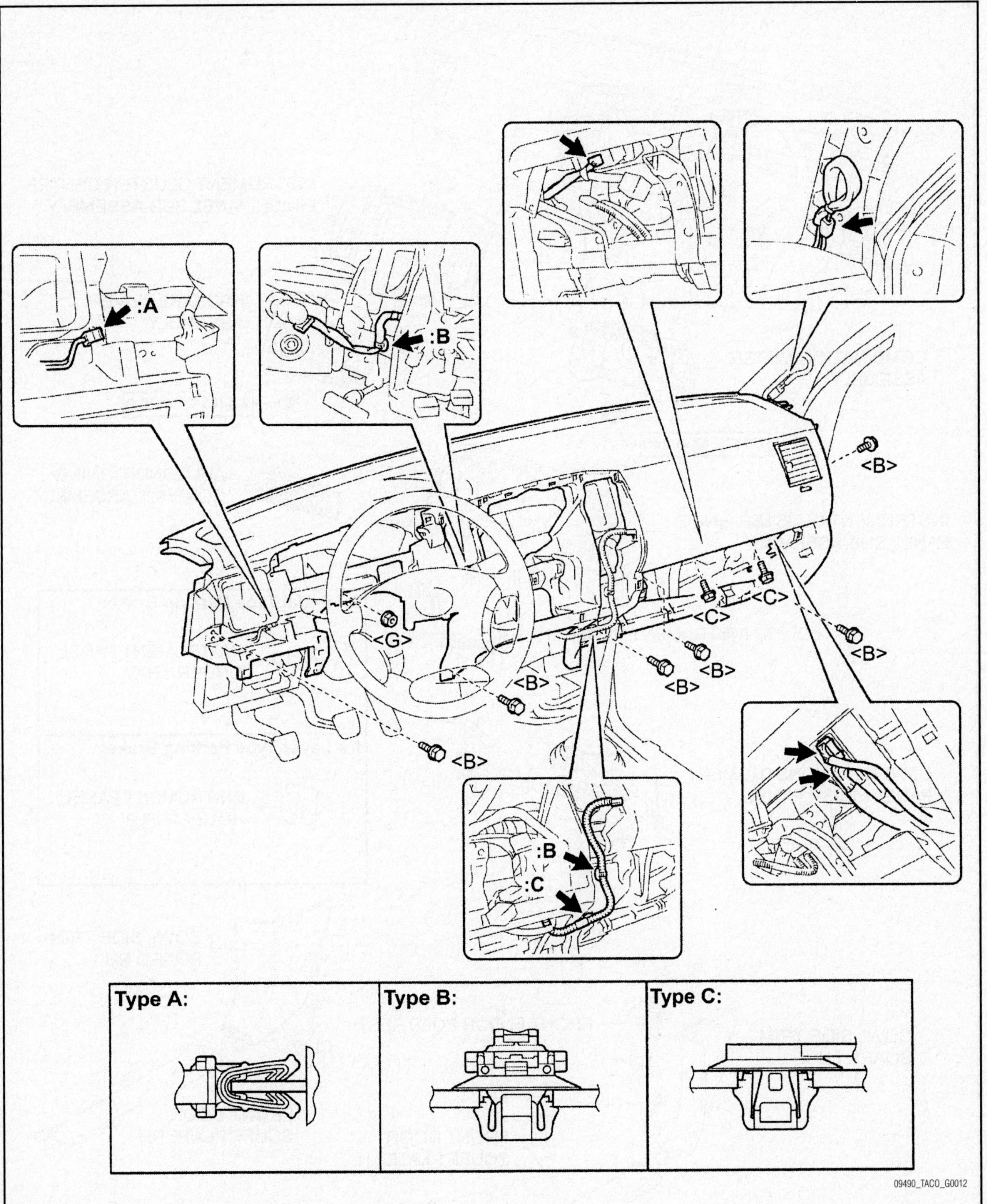

Fig. 145 Instrument panel retaining clip locations

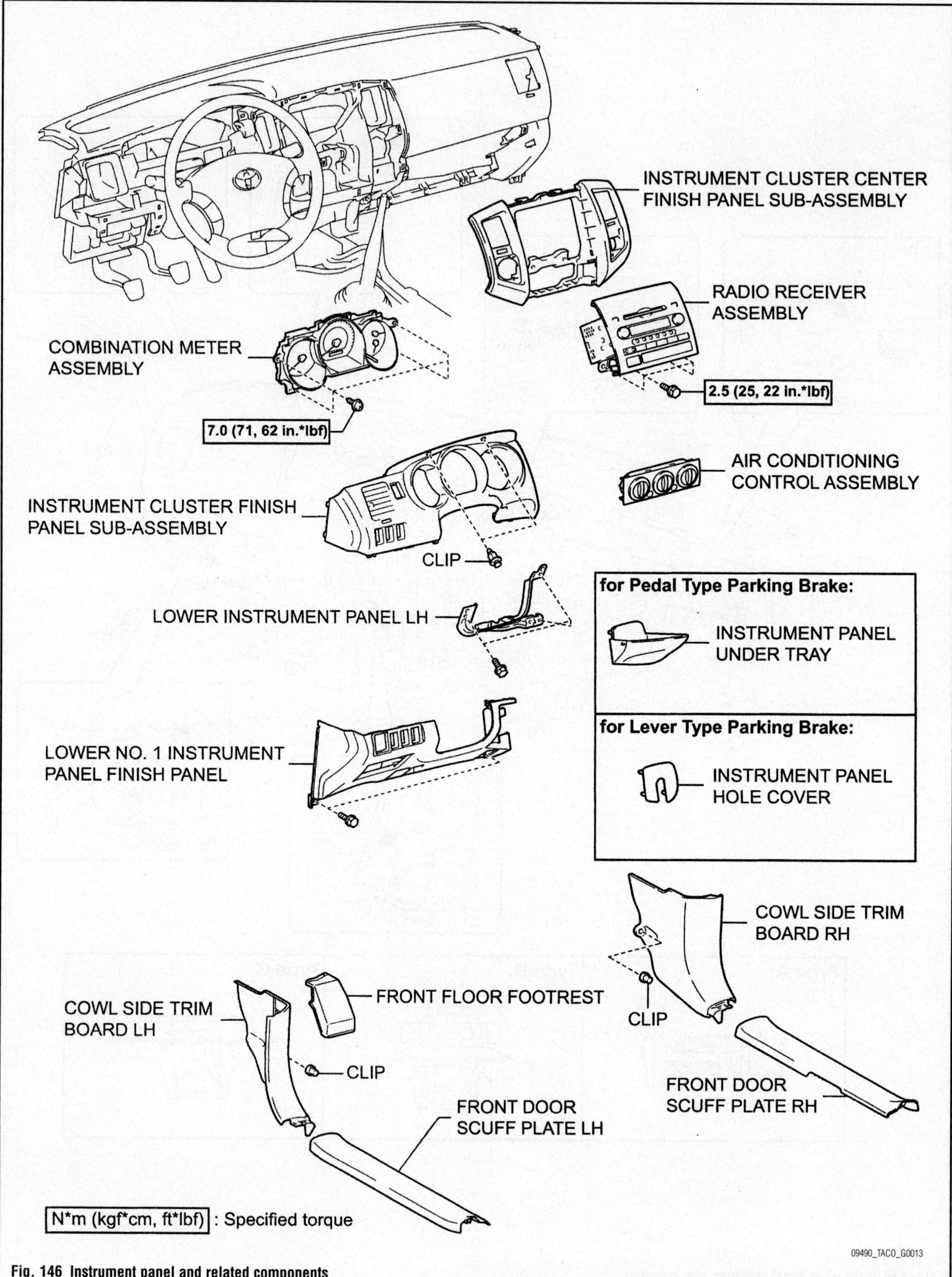

Fig. 146 Instrument panel and related components

A
B
Matchmarks
C
Matchmarks
09490_TACO_G0016

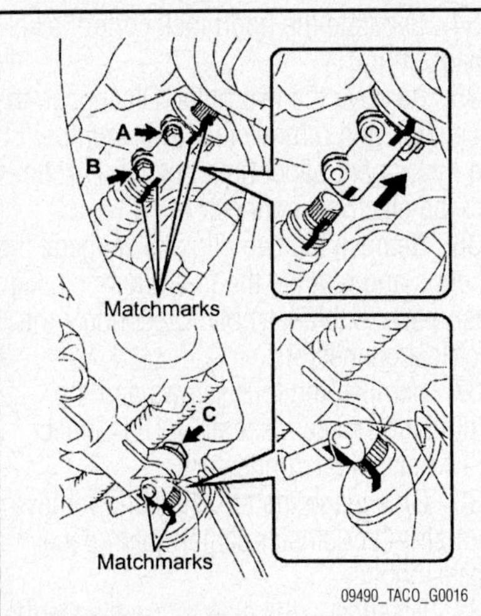

Fig. 147 Matchmarking the steering gear assembly

ing wheel cover after disengaging the two claws using the proper tool.

14. Pull the steering pad out of the steering wheel assembly and support the steering pad using your hand.

➡Take care as not to pull the air bag wire harness.

15. Disconnect the horn ground harness from the steering pad. Disconnect the air bag connectors. Remove the steering pad.

16. Remove the steering wheel retainer nut. Using the proper removal tool, remove the steering wheel from the steering column.

17. Disengage the two claws, using the proper tool, and remove the lower steering column cover.

18. Disengage the claw, using the proper tool, and remove the upper steering column cover.

19. To remove the spiral cable assembly, disconnect the air bag connector and the connector from the spiral cable subassembly. Take care as not to damage the air bag with the harness.

20. Disengage the three claws and remove the spiral cable subassembly.

21. Remove the windshield wiper switch assembly.

22. Remove the front floor footrest. Remove the right and left front door scuff plates. Remove the right and left cowl side trim boards.

23. Separate the right and left front door opening trim weatherstrips. Remove the right and left front pillar garnish.

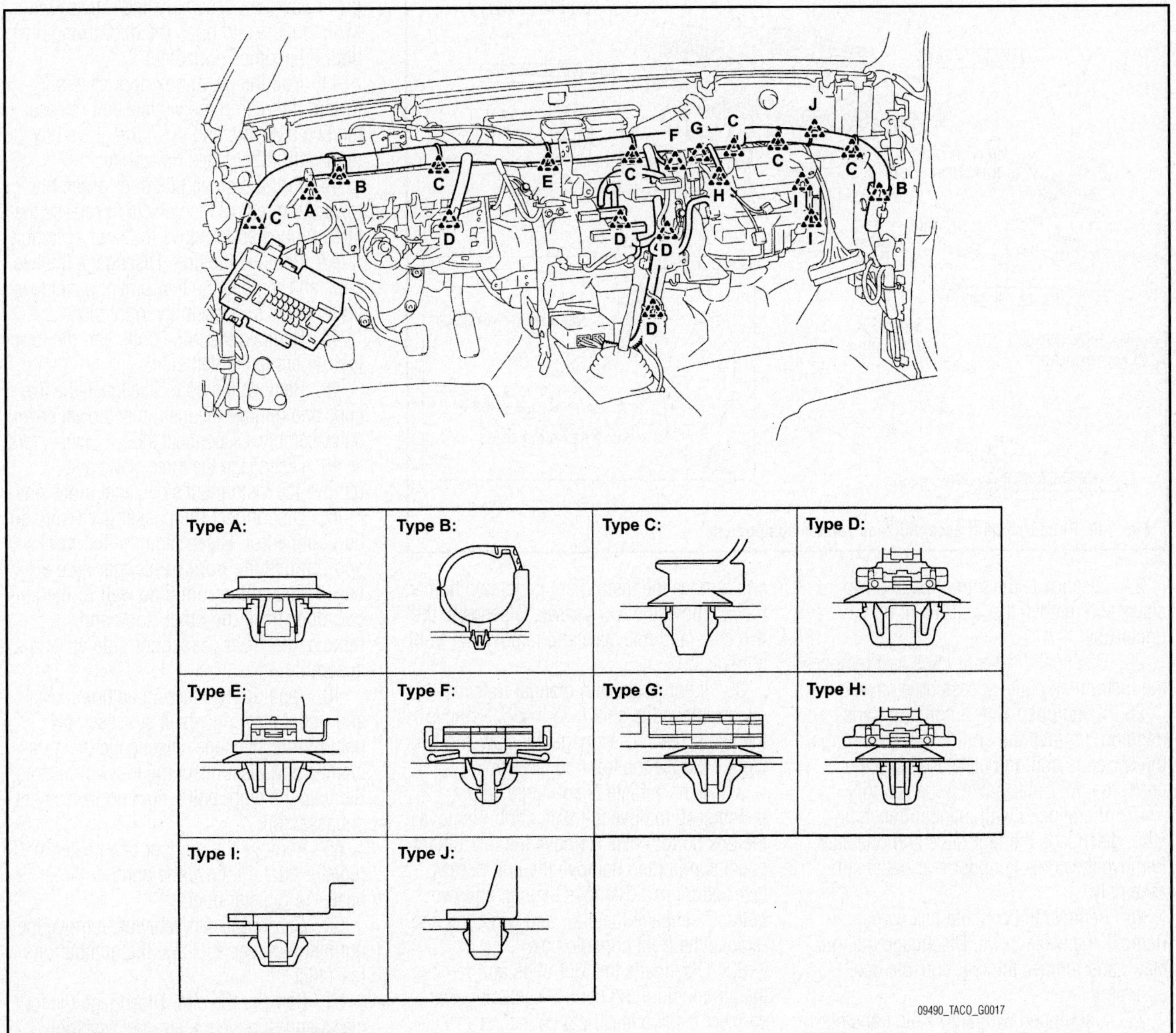

J
F G C
C
B
C
E
C
H
C
B
C
A
B
I
D
D
D
I
D
Type A:
Type B:
Type C:
Type D:
Type E:
Type F:
Type G:
Type H:
Type I:
Type J:
09490_TACO_G0017

Fig. 148 Air conditioning unit clip location

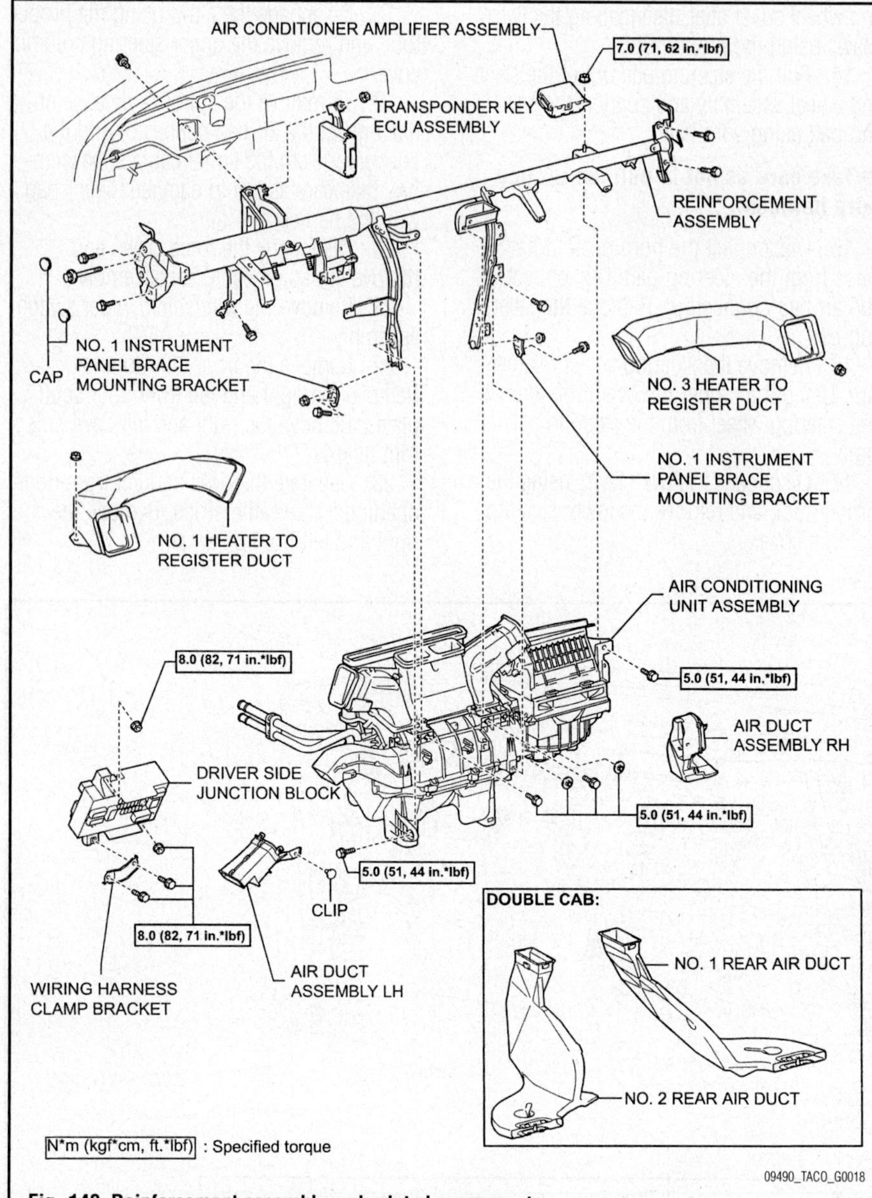

Fig. 149 Reinforcement assembly and related components

24. Disengage the four clips and two claws and remove the instrument panel under tray.
25. Disengage the four clips and remove the instrument panel access hole cover
26. If equipped with automatic transmission, remove the shift knob. Disengage the six clips and one claw. Remove the console upper rear panel subassembly.
27. If equipped with manual transmission, disengage the four clips and one claw. Remove the console upper rear panel subassembly.
28. Remove the console box carpet. Remove the two screws. Disengage the four claws and remove the rear console box assembly.
29. If equipped with automatic transmission, disengage the four claws and two clips and remove the instrument panel cup holder tray. Remove the two screws. Disengage the two clips and one claw and remove the front console box.
30. If equipped with manual transmission, remove the shift lever knob. Remove the two screws. Disengage the claw and two clips. Remove the front console box.
31. If the vehicle is equipped with a bench seat, remove the shift knob. Remove the box bottom mat. Remove the instrument panel cup holder. Remove the number two box bottom mat. Remove the clip and two bolts. Disengage the claw and two clips and remove the front console box.
32. Disengage the four clips and remove the air conditioning control assembly. Disconnect the two electrical connectors.
33. Remove the radio.
34. Separate the hood lock control lever subassembly.
35. Remove the two bolts. Disengage the three clips and remove the lower number one instrument panel finish panel. Disconnect the electrical connectors.
36. Remove the two clips. Disengage the six clips and remove the instrument cluster finish panel sub assembly. Disconnect the electrical connectors.
37. Remove the four screws and the combination meter assembly. Disconnect the two electrical connectors.
38. Disengage the ten clips and remove the instrument cluster center finish panel subassembly.
39. Disengage the claw and separate the glove box door stopper from the glove box door assembly.
40. Slightly deform the upper part of the glove box door assembly and release the two stoppers and open the door assembly until it becomes horizontal.
41. Pull the glove box door assembly toward the rear of the vehicle and release the two stoppers and open the glove box door until it becomes horizontal.
42. Pull the glove box door assembly toward the rear of the vehicle to release the three hinges and remove the door assembly.
43. Remove the bolt. Disengage the two clips and remove the instrument panel lower finish panel subassembly, right side.
44. Remove the three bolts and the lower instrument panel, left side.
45. Remove the bolt. Disengage the five clips and remove the instrument lower cover subassembly. Disconnect the two connectors.
46. Disengage the three claws and remove the instrument side panel, right side.
47. Disconnect the passenger's side air bag connector. Disengage the ten hooks and remove the front passenger side air bag. Release the front side wall of the air bag door from the other hook and remove the front passenger side air bag assembly.
48. Tape up the steering column cover and upper steering wheel subassembly using protective tape. Disengage the four clamps and disconnect the four connectors. Remove the eight bolts, nut and instrument subassembly.
49. Remove the number one heater to register duct. Remove the number three heater to register duct.
50. On double cab vehicles, remove the number two rear duct and the number one rear duct.
51. Remove the clip. Disengage the four claws and remove the air duct assembly heater, left side.

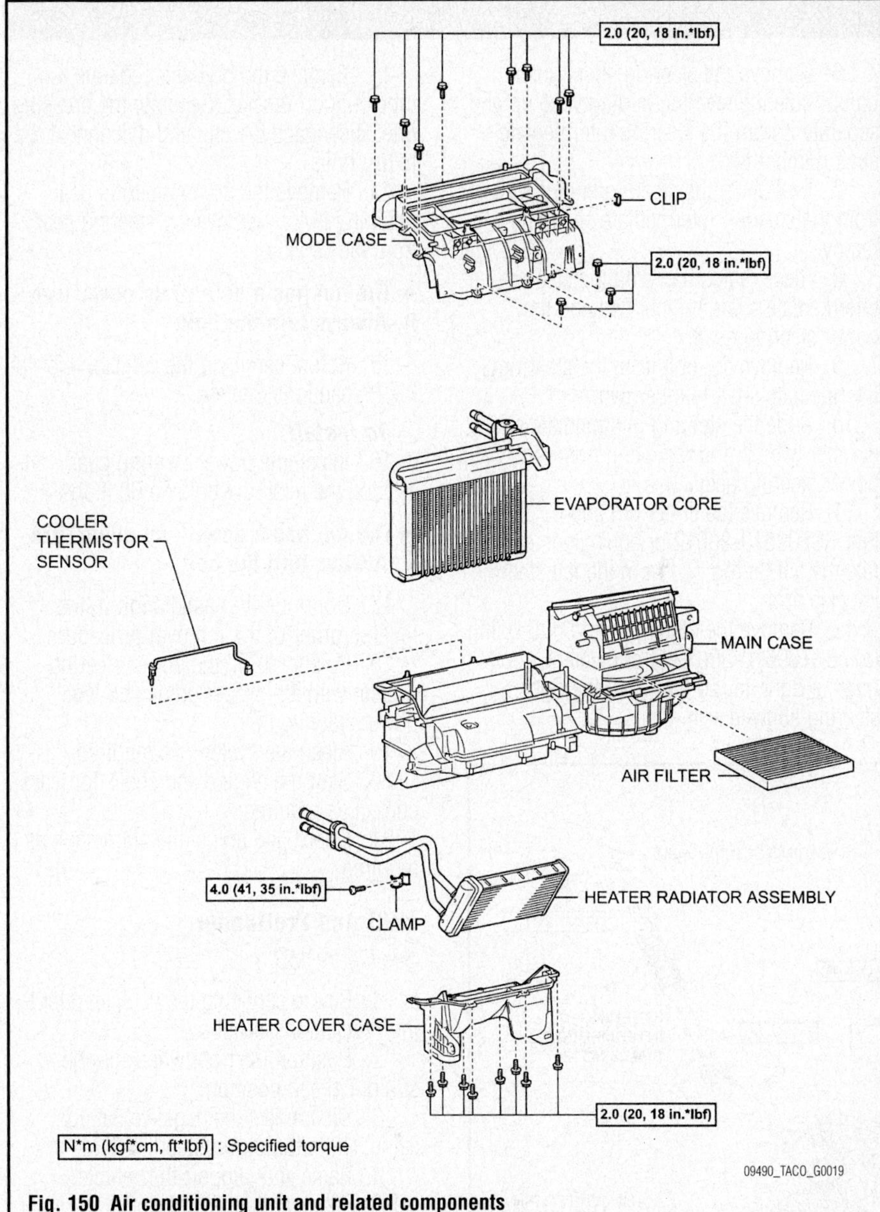

Fig. 150 Air conditioning unit and related components

52. Remove the clip. Disengage the three claws and remove the air duct assembly heater, right side.

53. Remove the bolt, nut and instrument panel brace mounting bracket, left side.

54. Remove the bolt, nut and instrument panel brace mounting bracket, right side.

55. Disengage the four clips and remove the steering column hole cover.

56. Matchmark the steering gear sliding yoke, steering gear intermediate shaft sub-assembly number two and the steering gear intermediate shaft assembly.

57. Remove bolts "A" and "B" from the steering sliding yoke.

58. Slide the steering yoke up and separate it from the steering intermediate shaft subassembly number two.

59. Pull down the steering sliding yoke from the steering intermediate shaft assembly to remove it.

60. Disconnect the steering column electrical connectors. Remove the steering column retaining bolts. Remove the steering column from the vehicle.

61. Disconnect the connector. Remove the nut and the transponder key ECU assembly.

62. Disconnect the connector. Remove the nut and the air conditioner amplifier assembly.

63. Disengage the four clamps and disconnect the six connectors from the air conditioning unit assembly. Remove the three bolts. Disconnect the connector. Disconnect the two air bag connectors. Disengage the twenty clamps.

64. Remove the five bolts and two nuts.

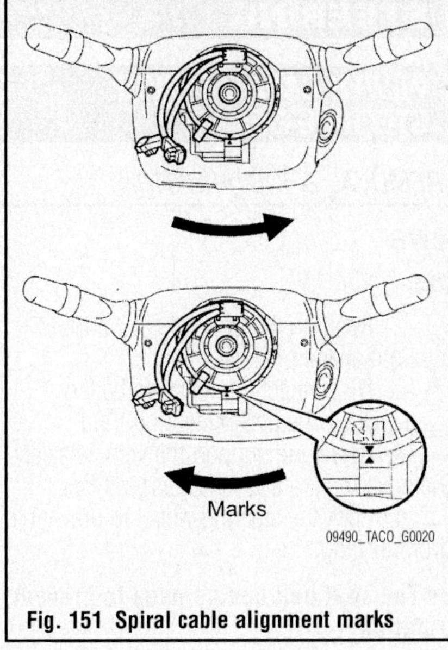

Fig. 151 Spiral cable alignment marks

Remove the two caps and seven bolts. Disengage the reinforcement hook of the air conditioning unit. Remove the reinforcement.

65. Remove the air conditioning unit assembly from the vehicle.

66. Remove the heater to register center duct. Remove the number one air duct. Remove the air filter. Remove the mode control servo motor.

67. Remove the air mix control servo motor. Remove the air inlet control servo motor. Remove the three heater core assembly retaining bolts and the temperature mounting plate. Remove the temperature cam. Remove the eight bolts and the heater cover case. Remove the bolt and clamp. Remove the heater core from its mounting.

To install:

68. Installation is the reverse of the removal procedure.

69. When installing the steering column be sure to align the matchmarks made during the removal procedure. Tighten bolts "A", "B" and "C" to 26 ft. lbs.

70. When installing the spiral cable, check that the ignition switch is in the "OFF" position. Turn the cable counterclockwise by hand until it becomes hard to turn. Turn the cable clockwise about 2½ turns to align the marks.

➡The cable will rotate about 2½ turns both left and right from the center.

71. Refill the cooling system.

72. Evacuate and charge the air conditioning system refrigerant.

73. Run the engine to normal operating temperatures; then, check the climate control operation and check for leaks.

STEERING

POWER RACK & PINION STEERING GEAR

REMOVAL & INSTALLATION

2WD

See Figure 152.

1. Position the front wheels in the straight ahead position.
2. Disconnect the negative battery cable. Drain the power steering fluid.
3. Raise and support the vehicle safely. Remove the tire and wheel assemblies.
4. Lock the steering wheel to prevent it from turning.

➡The seat belt can be used to prevent rotation.

5. Place matchmarks on the steering slider yoke, the steering intermediate shaft number two and the steering intermediate shaft.
6. Remove the steering slider yoke bolts. Slide the steering sliding yoke up and separate it from the steering intermediate shaft number two.
7. Pull down the steering sliding yoke from the steering intermediate shaft and remove it.
8. Place matchmarks on the steering intermediate shaft number two and the power steering gear.
9. Remove the bolt from the steering intermediate shaft number two.
10. Slide the steering intermediate shaft number two up and remove it from the power steering gear.
11. Remove the cotter pin and nut. Using tool SST09610-20012 or equivalent, separate the left tie rod end from the left steering knuckle arm.
12. Remove the cotter pin and nut. Using tool SST09610-20012 or equivalent, separate the right tie rod end from the right steering knuckle arm.

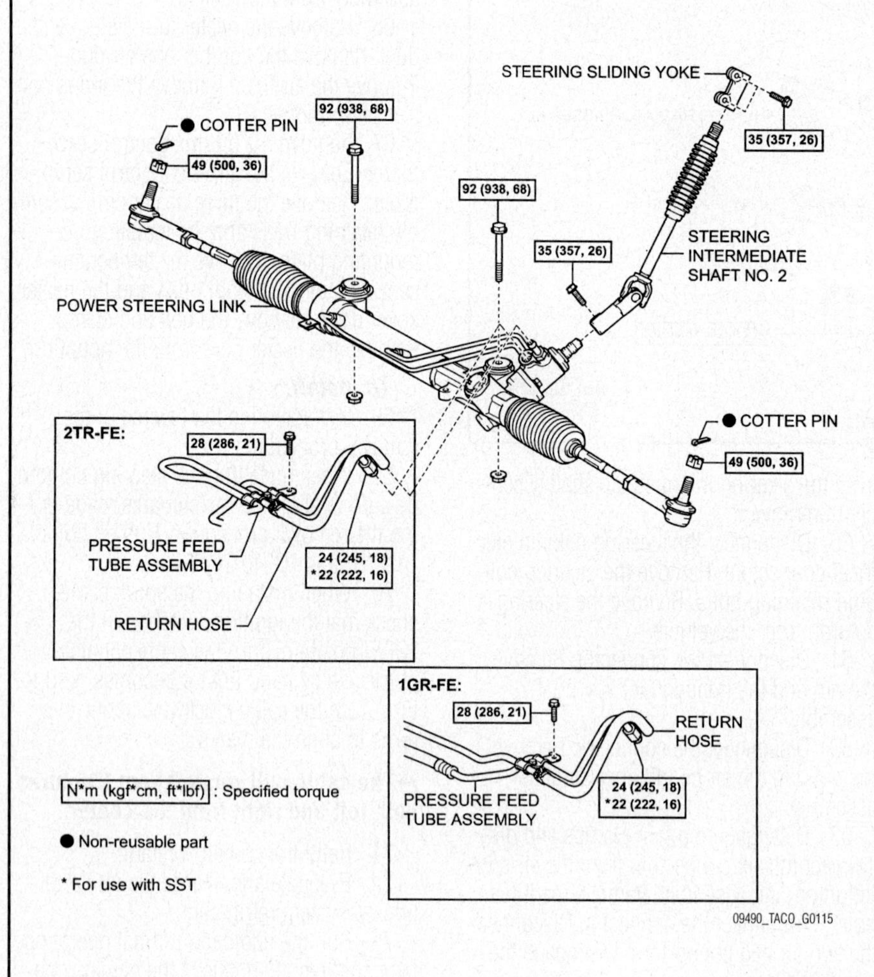

Fig. 152 Power steering gear and related components—2WD

13. Remove the bolt and separate the tube support bracket. Separate the pressure line. Disengage the clip and disconnect the return hose.
14. Remove the power steering gear retaining bolts. Remove the steering gear from the vehicle.

➡The nut has a detent, so never turn it. Always turn the bolt.

15. Before servicing the vehicle, refer to the Precautions Section.

To install:

16. Install the power steering gear. Tighten the retaining bolts to 68 ft. lbs.

➡The nut has a detent, so never turn it. Always turn the bolt.

17. Continue the installation in the reverse order of the removal procedure.
18. Be sure to fill the power steering system with the proper grade and type power steering fluid.
19. Bleed the system, as required.
20. Start the engine and check for leaks, correct as required.
21. Check and adjust the alignment, as required.

4WD And PreRunner

See Figure 153.

1. Before servicing the vehicle, refer to the Precautions Section.
2. Position the front wheels in the straight ahead position.
3. Disconnect the negative battery cable. Drain the power steering fluid.
4. Raise and support the vehicle safely. Remove the tire and wheel assemblies.
5. Remove the number one engine undercover subassembly.
6. Remove the front exhaust pipe assembly. On the 4.0L (1GR-FE) engine, remove the number two exhaust pipe assembly.
7. Remove the driveshaft. Some vehicles also use a center bearing assembly, remove that too.
8. Remove the frame crossmember subassembly.
9. Remove the stabilizer bar.
10. Lock the steering wheel to prevent it from turning.

➡The seat belt can be used to prevent rotation.

11. Place matchmarks on the steering slider yoke, the steering intermediate shaft

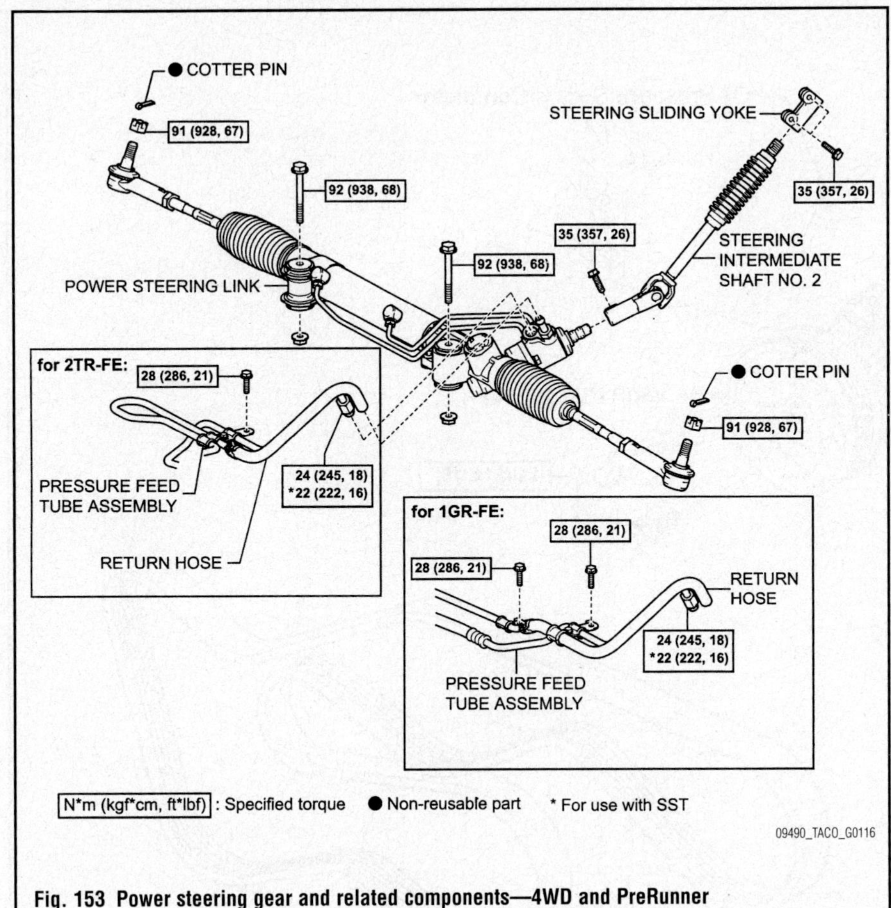

Fig. 153 Power steering gear and related components—4WD and PreRunner

number two and the steering intermediate shaft.

12. Remove the steering slider yoke bolts. Slide the steering sliding yoke up and separate it from the steering intermediate shaft number two.
13. Pull down the steering sliding yoke from the steering intermediate shaft and remove it.
14. Place matchmarks on the steering intermediate shaft number two and the power steering gear.
15. Remove the bolt from the steering intermediate shaft number two.
16. Slide the steering intermediate shaft number two up and remove it from the power steering gear.
17. Remove the cotter pin and nut. Using tool SST09610-20011 or equivalent, separate the left tie rod end from the left steering knuckle arm.
18. Remove the cotter pin and nut. Using tool SST09610-20011 or equivalent, separate the right tie rod end from the right steering knuckle arm.
19. Remove the bolt and separate the tube support bracket. Separate the pressure line. Disengage the clip and disconnect the return hose.
20. Remove the power steering gear retaining bolts. Tilt the transmission and remove the steering gear from the vehicle.

➡The nut has a detent, so never turn it. Always turn the bolt.

To install:

21. Install the power steering gear. Tighten the retaining bolts to 68 ft. lbs.

➡The nut has a detent, so never turn it. Always turn the bolt.

22. Continue the installation in the reverse order of the removal procedure.
23. Be sure to fill the power steering system with the proper grade and type power steering fluid.
24. Bleed the system, as required.
25. Start the engine and check for leaks, correct as required.
26. Check and adjust the alignment, as required.

POWER STEERING PUMP

REMOVAL & INSTALLATION

See Figures 154 and 155.

Refer to the accompanying illustrations for Power Steering Pump mounting.

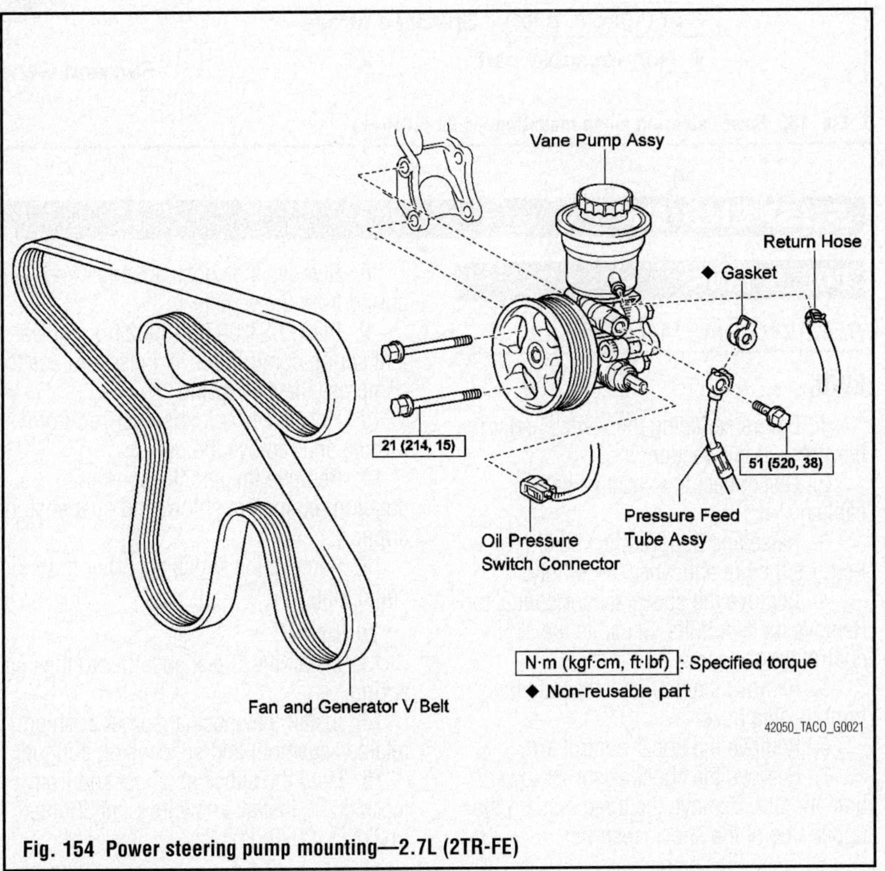

Fig. 154 Power steering pump mounting—2.7L (2TR-FE)

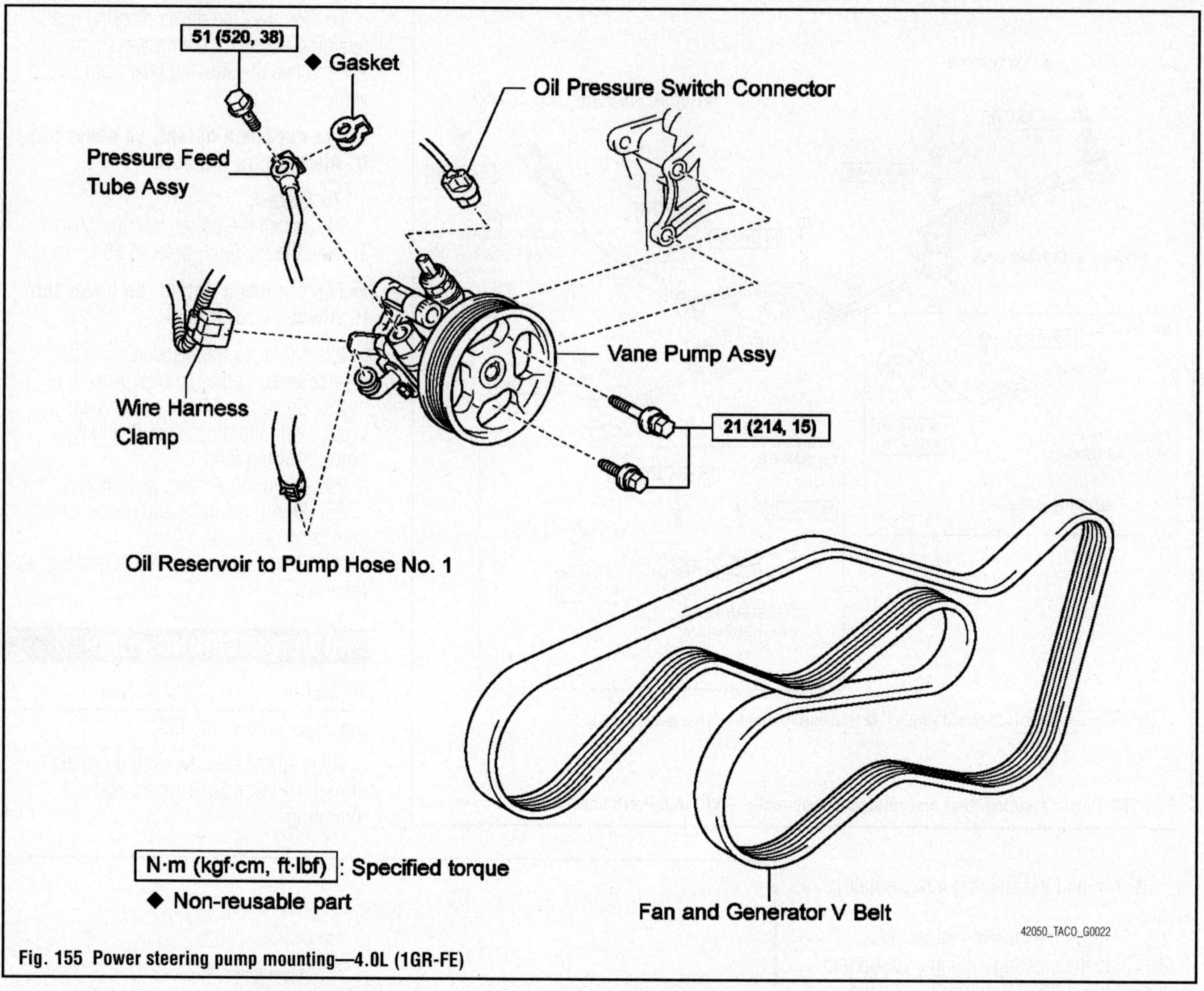

Fig. 155 Power steering pump mounting—4.0L (1GR-FE)

SUSPENSION

FRONT SUSPENSION

COIL SPRING

REMOVAL & INSTALLATION

2WD

1. Before servicing the vehicle, refer to the Precautions Section.
2. Disconnect the negative battery cable.
3. Raise and support the vehicle safely. Remove the tire and wheel assembly.
4. Remove the speed sensor connector. Remove the two bolts. Separate the skid control sensor wire.
5. Remove the bolt and separate the front flexible hose.
6. Remove the upper control arm.
7. Remove the shock absorber lower bolt and nut. Remove the three nuts on the upper side of the shock absorber.
8. Remove the shock absorber and coil spring from the vehicle.
9. Place SST 09727-30021 or equivalent spring compressor in a vise and use to compress the coil spring.
10. Hold the shock absorber rod from turning and remove the nut.
11. Remove the shock absorber cushion retainer, cushion, and suspension support.
12. Remove the shock absorber from the coil spring.

To install:

13. Install the shock absorber to the coil spring.
14. Install the shock absorber cushion retainer, cushion, and suspension support.
15. Hold the shock absorber rod from turning and install a new lock nut. Tighten to 27Nm (20 Ft. Lbs.).
16. Loosen and remove the spring compressor.

➡Always fully tighten rubber bushings when the wheels are in full contact with the ground and the vehicle is at curb height. Bounce the vehicle up and down several times to stabilize the suspension, prior to final tightening of these components.

17. Position the shock absorber to its mounting on the vehicle.
18. On the left side, install the coil spring onto the body with the lower end of the coil spring facing the outer side of the vehicle.
19. On the right side, install the coil spring onto the body with the lower end of the coil spring facing the inner side of the vehicle.

20. Install the upper retaining nuts. Torque to 47 ft. lbs.

21. Temporarily tighten the lower shock retaining bolt.

22. Continue the installation in the reverse order of the removal procedure.

23. Final tightening torque for the lower shock absorber mounting bolt is 61 ft. lbs.

24. Check and adjust the alignment, as required.

4WD And PreRunner

1. Before servicing the vehicle, refer to the Precautions Section.

2. Disconnect the negative battery cable.

3. Raise and support the vehicle safely. Remove the tire and wheel assembly.

4. Remove the engine undercover assembly.

5. Remove the stabilizer bar.

6. Remove the cotter pin and nut. Using the proper tool, separate the tie rod end from the steering knuckle arm.

7. Remove the shock absorber lower bolt and nut. Remove the three nuts on the upper side of the shock absorber.

8. Remove the shock absorber and coil spring from the vehicle.

9. Place SST 09727-30021 or equivalent spring compressor in a vise and use to compress the coil spring.

10. Hold the shock absorber rod from turning and remove the nut.

11. Remove the shock absorber cushion retainer, cushion, and suspension support.

12. Remove the shock absorber from the coil spring.

To install:

13. Install the shock absorber to the coil spring.

14. Install the shock absorber cushion retainer, cushion, and suspension support.

15. Hold the shock absorber rod from turning and install a new lock nut. Tighten to 27Nm (20 Ft. Lbs.).

16. Loosen and remove the spring compressor.

➡Always fully tighten rubber bushings when the wheels are in full contact with the ground and the vehicle is at curb height. Bounce the vehicle up and down several times to stabilize the suspension, prior to final tightening of these components.

17. Position the shock absorber to its mounting on the vehicle.

18. On the left side, install the coil spring onto the body with the lower end of the coil spring facing the outer side of the vehicle.

19. On the right side, install the coil spring onto the body with the lower end of the coil spring facing the inner side of the vehicle.

20. Install the upper retaining nuts. Torque to 47 ft. lbs.

21. Temporarily tighten the lower shock retaining bolt.

22. Continue the installation in the reverse order of the removal procedure.

23. Final tightening torque for the lower shock absorber mounting bolt is 61 ft. lbs.

24. Check and adjust the alignment, as required.

LOWER BALL JOINT

REMOVAL & INSTALLATION

➡The lower ball joint is removed with the upper control arm.

LOWER CONTROL ARM

REMOVAL & INSTALLATION

See Figure 156.

1. Before servicing the vehicle, refer to the Precautions Section.

2. Raise and support the vehicle safely.

3. Remove the tire and wheel assemblies.

4. To check the lower ball joint, install the hub nuts. Using a dial indicator gauge push the hub nut up and down with a force of 66 ft. lbs. Specification should be 0.020 inch.

➡If not within specification, replace the lower control arm.

5. On 2WD vehicles, remove the stabilizer link assembly.

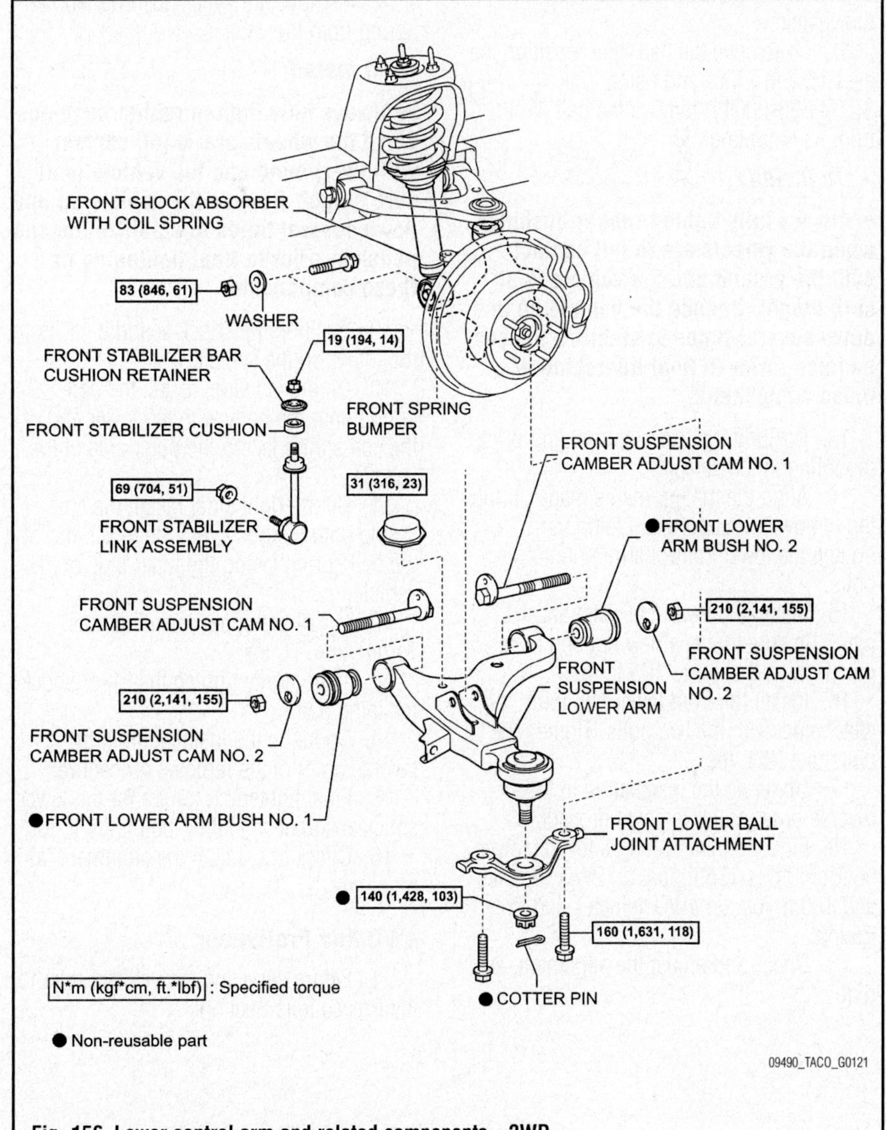

Fig. 156 Lower control arm and related components—2WD

6. Properly support the lower control arm assembly, as required.

7. Remove the bolt, nut and washer. Separate the front shock absorber with the coil spring from the lower control arm.

8. Remove the two bolts and separate the front lower ball joint attachment from the front axle.

9. On 2WD vehicles, place matchmarks on the camber adjusting cam number two. Remove the two nuts, the two number two camber adjusting cams, the two number one camber adjusting cams.

10. On 4WD and PreRunner, place matchmarks on the camber adjusting cam number two. Remove the nut, the camber adjusting cam number two, the camber adjusting cam number one, the bolt, the toe adjust cam and the toe adjust plate number two.

11. Remove the lower control arm from the vehicle.

12. To remove the ball joint, position the assembly in a vise and using tool SST09628-00011, remove the ball joint from its mounting.

To install:

➡Always fully tighten rubber bushings when the wheels are in full contact with the ground and the vehicle is at curb height. Bounce the vehicle up and down several times to stabilize the suspension, prior to final tightening of these components.

13. Position the lower control arm on its mounting in the vehicle.

14. Align the mating marks made during the removal procedure and temporarily tighten the lower control arm mounting bolts.

15. Install the lower ball joint attachment. Be sure to use a new nut and cotter pin. Tighten the nut to 103 ft. lbs.

16. Install the front lower ball joint attachment with the two bolts. Tighten the bolts to 118 ft. lbs.

17. Continue the installation in the reverse order of the removal procedure.

18. Final tightening torque for the lower control arm is 155 ft. lbs. on 2WD vehicles and 100 ft. lbs. on 4WD vehicles and PreRunner.

19. Check and adjust the alignment, as required.

SHOCK ABSORBERS

REMOVAL & INSTALLATION

2WD

1. Before servicing the vehicle, refer to the Precautions Section.

2. Disconnect the negative battery cable.

3. Raise and support the vehicle safely. Remove the tire and wheel assembly.

4. Remove the speed sensor connector. Remove the two bolts. Separate the skid control sensor wire.

5. Remove the bolt and separate the front flexible hose.

6. Remove the upper control arm.

7. Remove the shock absorber lower bolt and nut. Remove the three nuts on the upper side of the shock absorber.

8. Remove the shock absorber and coil spring from the vehicle.

To install:

➡Always fully tighten rubber bushings when the wheels are in full contact with the ground and the vehicle is at curb height. Bounce the vehicle up and down several times to stabilize the suspension, prior to final tightening of these components.

9. Position the shock absorber to its mounting on the vehicle.

10. On the left side, install the coil spring onto the body with the lower end of the coil spring facing the outer side of the vehicle.

11. On the right side, install the coil spring onto the body with the lower end of the coil spring facing the inner side of the vehicle.

12. Install the upper retaining nuts. Torque to 47 ft. lbs.

13. Temporarily tighten the lower shock retaining bolt.

14. Continue the installation in the reverse order of the removal procedure.

15. Final tightening torque for the lower shock absorber mounting bolt is 61 ft. lbs.

16. Check and adjust the alignment, as required.

4WD And PreRunner

1. Before servicing the vehicle, refer to the Precautions Section.

2. Disconnect the negative battery cable.

3. Raise and support the vehicle safely. Remove the tire and wheel assembly.

4. Remove the engine undercover assembly.

5. Remove the stabilizer bar.

6. Remove the cotter pin and nut. Using the proper tool, separate the tie rod end from the steering knuckle arm.

7. Remove the shock absorber lower bolt and nut. Remove the three nuts on the upper side of the shock absorber.

8. Remove the shock absorber and coil spring from the vehicle.

To install:

➡Always fully tighten rubber bushings when the wheels are in full contact with the ground and the vehicle is at curb height. Bounce the vehicle up and down several times to stabilize the suspension, prior to final tightening of these components.

9. Position the shock absorber to its mounting on the vehicle.

10. On the left side, install the coil spring onto the body with the lower end of the coil spring facing the outer side of the vehicle.

11. On the right side, install the coil spring onto the body with the lower end of the coil spring facing the inner side of the vehicle.

12. Install the upper retaining nuts. Torque to 47 ft. lbs.

13. Temporarily tighten the lower shock retaining bolt.

14. Continue the installation in the reverse order of the removal procedure.

15. Final tightening torque for the lower shock absorber mounting bolt is 61 ft. lbs.

16. Check and adjust the alignment, as required.

STEERING KNUCKLE

REMOVAL & INSTALLATION

See Figures 157 and 158.

Refer to the accompanying illustrations for Steering Knuckle mounting.

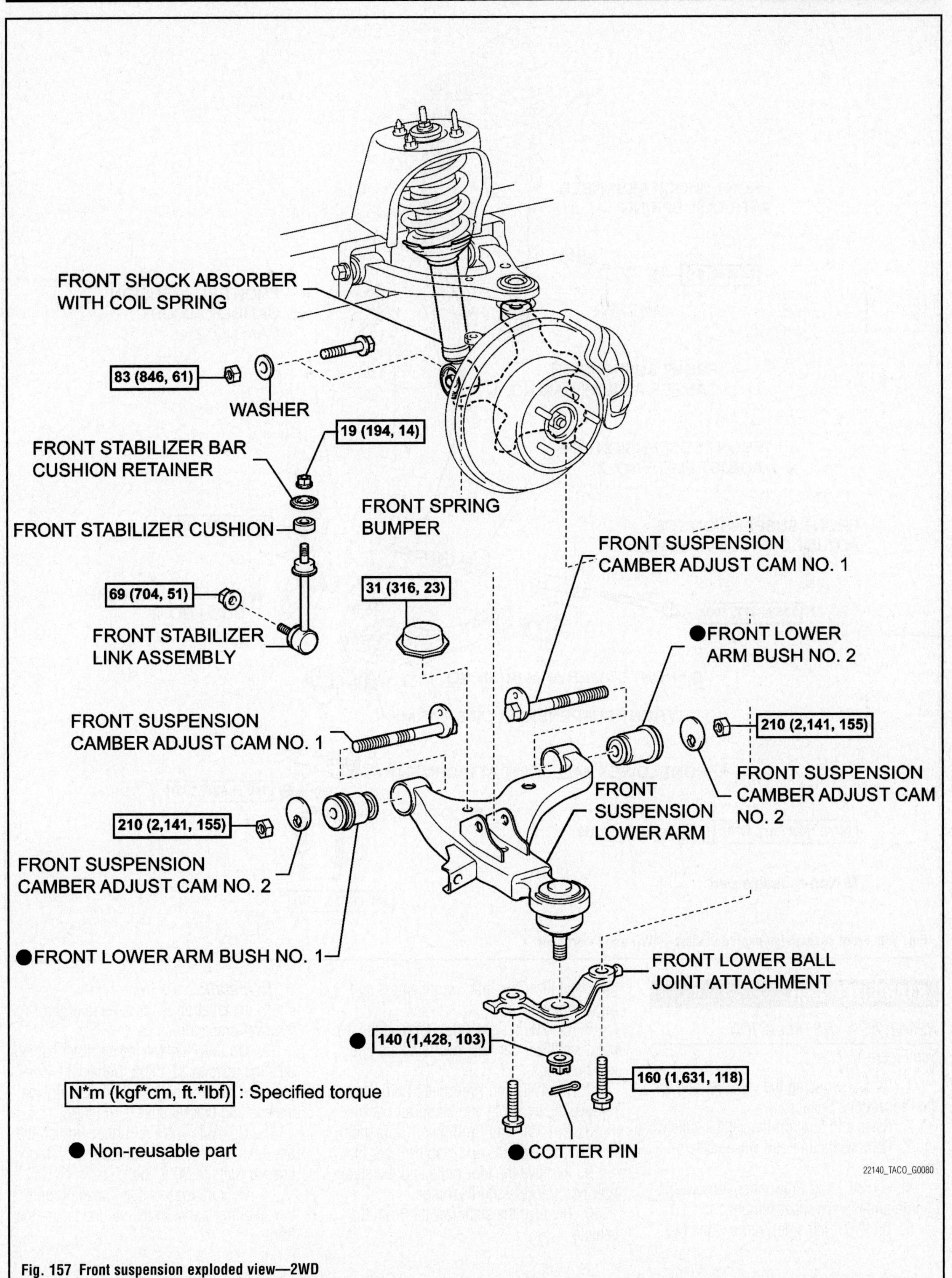

Fig. 157 Front suspension exploded view—2WD

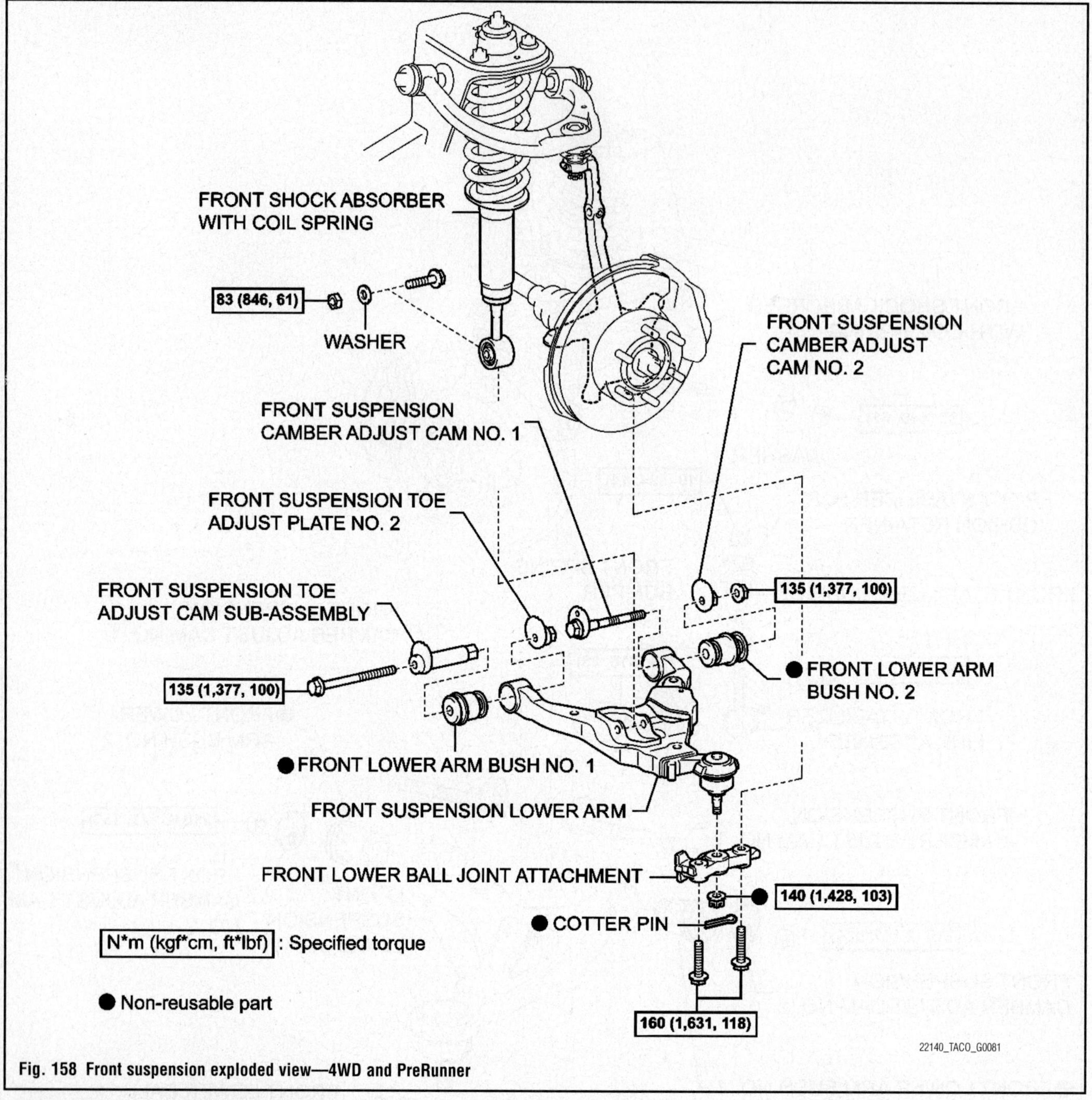

Fig. 158 Front suspension exploded view—4WD and PreRunner

STABILIZER BAR

REMOVAL & INSTALLATION

See Figure 159.

1. Before servicing the vehicle, refer to the Precautions Section.
2. Raise and support the vehicle safely.
3. Remove the tire and wheel assemblies.
4. On 4WD and PreRunner, remove the engine undercover subassembly.
5. On 2WD (left side), remove the two nuts, stabilizer bar link, two retainers and two bushings.
6. On 2WD (right side), remove the two nuts, stabilizer bar link, two retainers and two bushings.
7. On 4WD and PreRunner (left side), remove the two nuts and stabilizer bar link.
8. On 4WD and PreRunner (right side), remove the two nuts and stabilizer bar link.
9. Remove the four bolts and the stabilizer bar brackets and bushings.
10. Remove the stabilizer bar from the vehicle.

To install:

11. Installation is the reverse of the removal procedure.
12. On 2WD, tighten the stabilizer bar side bushing bolts to 14 ft. lbs. Tighten the lower arm side bushing bolts to 51 ft. lbs. Tighten the bushing bracket bolts to 16 ft. lbs.
13. On 4WD and PreRunner, tighten the bar links to 52 ft. lbs. Tighten the bushing bracket bolts to 30 ft. lbs.
14. Be sure to install the cushion onto the stabilizer bar with its cut line facing the front.

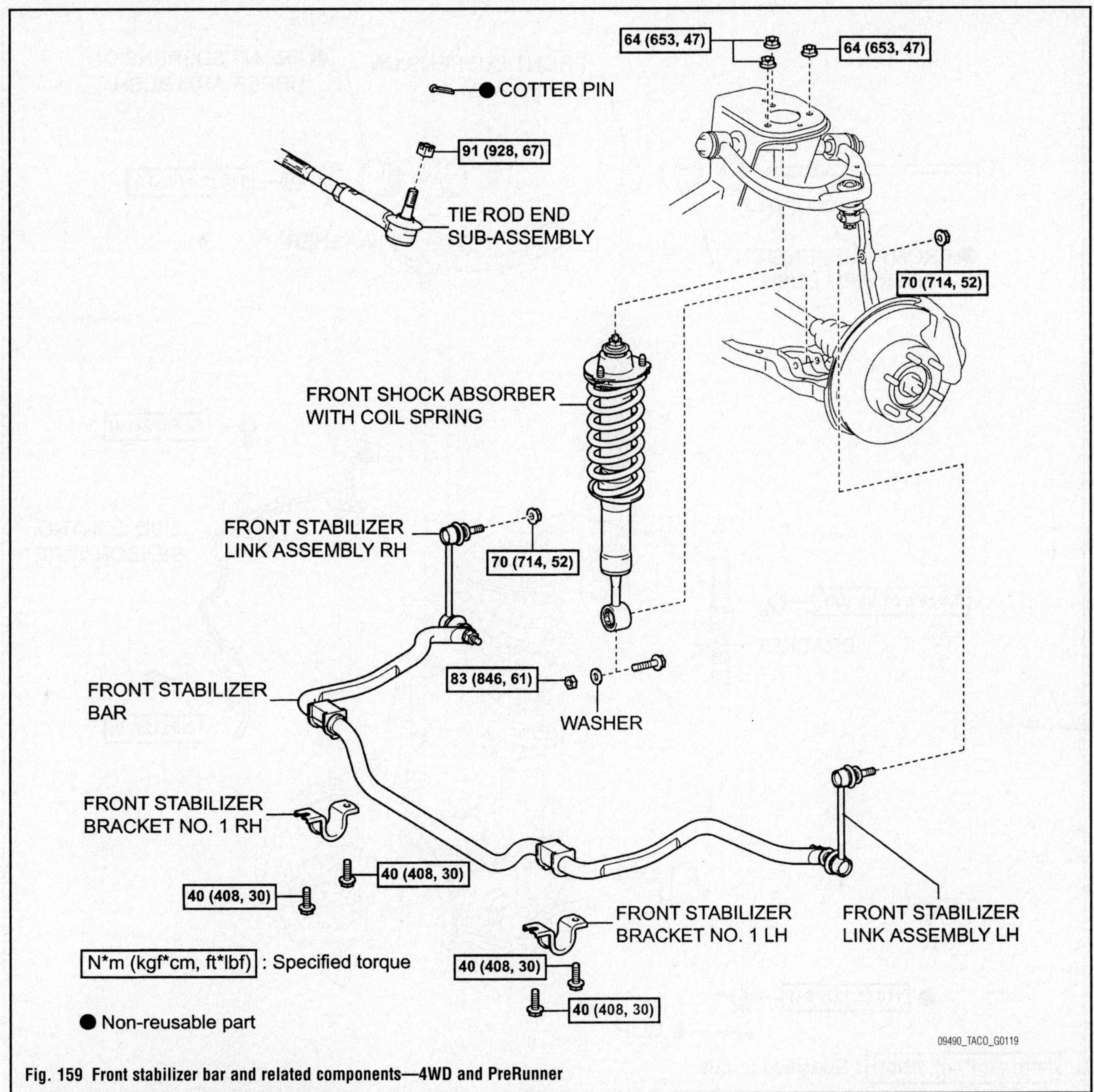

Fig. 159 Front stabilizer bar and related components—4WD and PreRunner

UPPER BALL JOINT

REMOVAL & INSTALLATION

➡The upper ball joint is removed with the upper control arm.

UPPER CONTROL ARM

REMOVAL & INSTALLATION

See Figure 162.

1. Before servicing the vehicle, refer to the Precautions Section.
2. Disconnect the negative battery cable.
3. Raise and support the vehicle safely. Remove the tire and wheel assemblies.
4. Remove the speed sensor connector. Remove the two bolts and separate the skid control sensor wire.
5. On 2WD vehicles, remove the bolt and separate the front flexible hose.
6. Support the front suspension lower control arm using a jack.
7. Remove the clip and nut. Using the tool SST09628-62011, or equivalent separate the upper ball joint from the steering knuckle.
8. On 4WD and PreRunner, remove the bolt and bracket.
9. Remove the two upper control arm retaining bolts.
10. Remove the upper control arm from the vehicle.

To install:

➡Always fully tighten rubber bushings when the wheels are in full contact with the ground and the vehicle is at curb height. Bounce the vehicle up and down several times to stabilize the suspension, prior to final tightening of these components.

11. Position the upper control arm on its mounting in the vehicle.

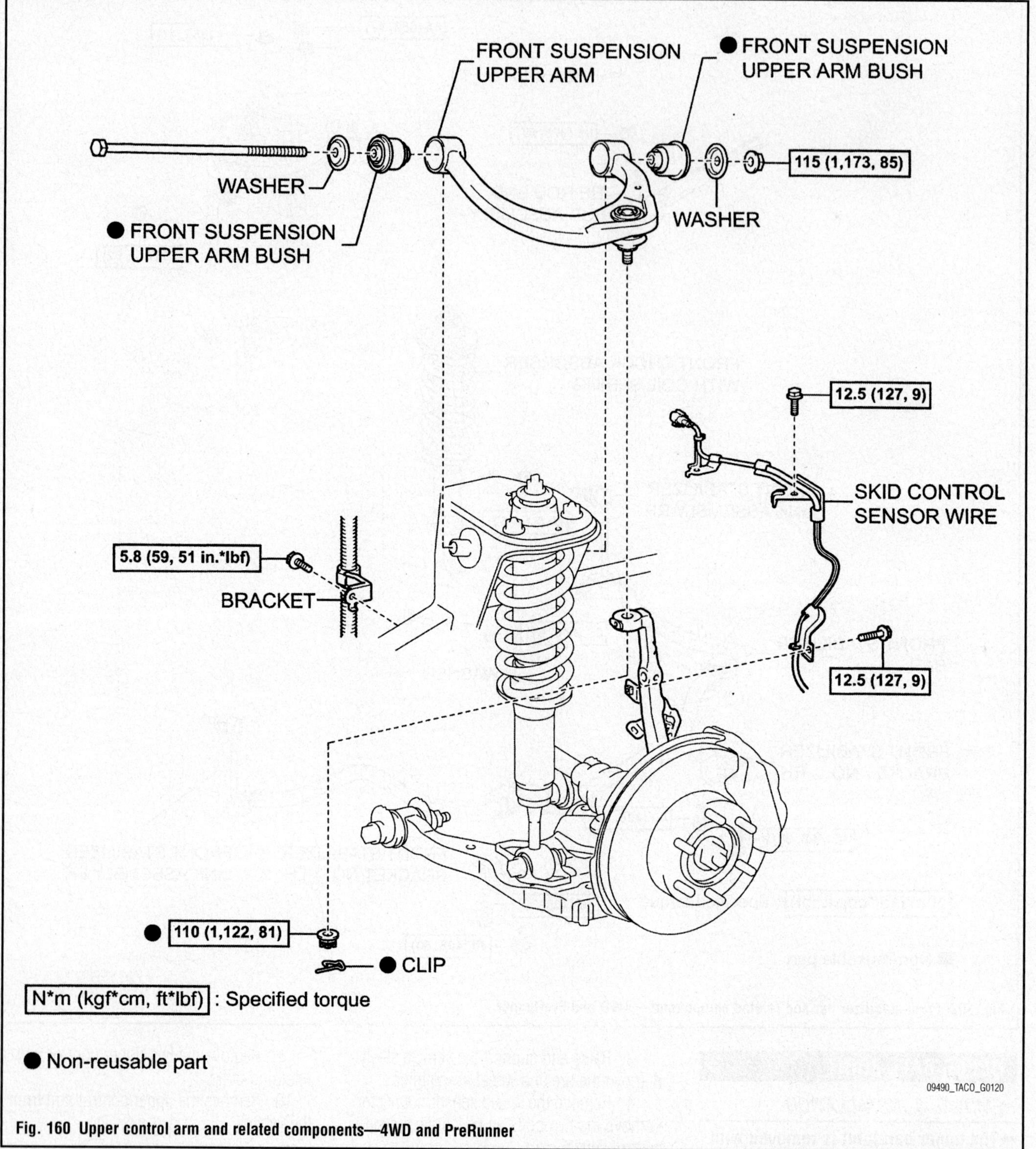

Fig. 160 Upper control arm and related components—4WD and PreRunner

12. Temporarily tighten the upper control arm mounting bolts.

13. Continue the installation in the reverse order of the removal procedure.

14. Final tightening torque for the upper control arm bushings is 60 ft. lbs. on 2WD vehicles and 85 ft. lbs. on 4WD vehicles and PreRunner.

15. Check and adjust the alignment, as required.

WHEEL HUB AND BEARING

REMOVAL & INSTALLATION

2WD

See Figure 161.

1. Before servicing the vehicle, refer to the Precautions Section.

2. Disconnect the negative battery cable. Drain the brake fluid.

3. Raise and support the vehicle safely. Remove the tire and wheel assembly.

4. Remove the speed sensor connector. Remove the two bolts and separate the skid control sensor wire.

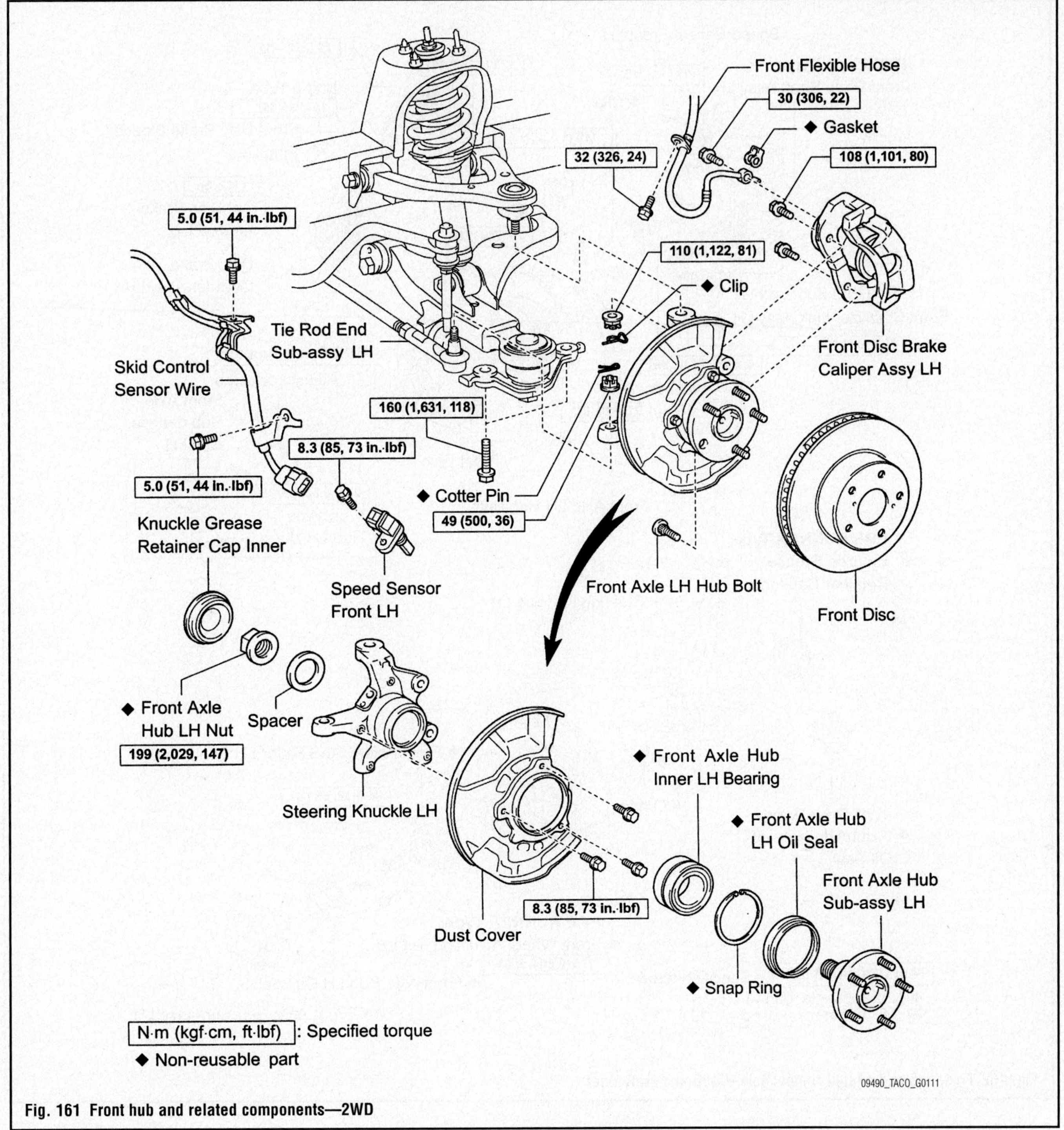

Fig. 161 Front hub and related components—2WD

5. Remove the bolt and separate the front flexible hose.
6. Remove the front brake caliper. Remove the rotor.
7. Remove the cotter pin and nut. Using the proper tool separate the tie rod end from the steering knuckle.
8. Remove the two bolts and separate the front lower ball joint attachment from the front axle.
9. Remove the clip and nut. Using the proper tool, separate the upper ball joint from the steering knuckle.
10. Remove the front axle hub from the vehicle.

To install:

11. Installation is the reverse of the removal procedure.
12. Be sure to fill the master cylinder with the proper grade and type brake fluid.
13. Bleed the brakes, as required.
14. Be sure to check and adjust the alignment, as required.

4WD And PreRunner

See Figure 162.

1. Before servicing the vehicle, refer to the Precautions Section.
2. Disconnect the negative battery cable. Drain the brake fluid.
3. Raise and support the vehicle safely. Remove the tire and wheel assembly.

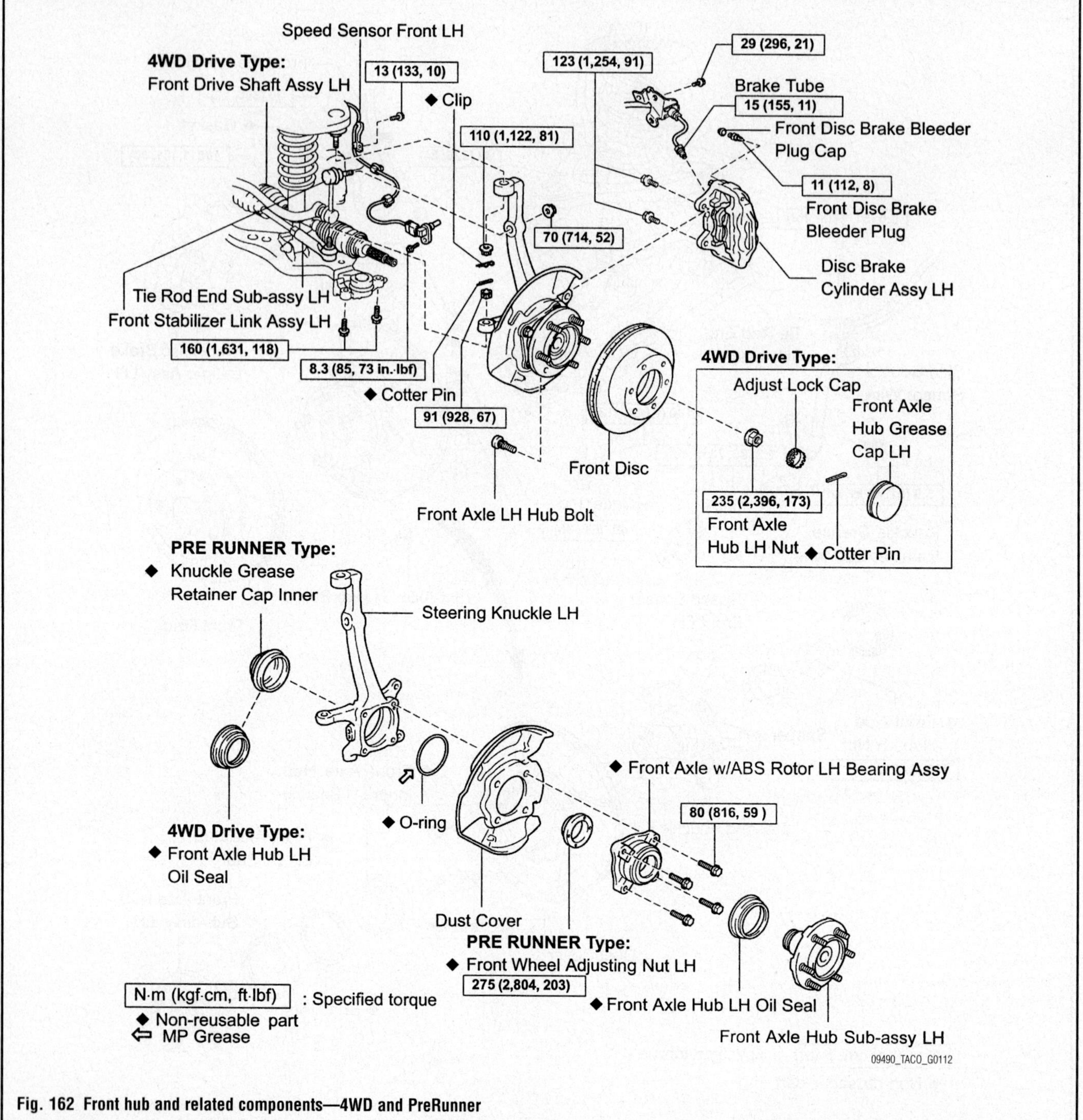

Fig. 162 Front hub and related components—4WD and PreRunner

4. Remove the bolt and separate the speed sensor. Disengage the clamps. Remove the bolt and separate the speed sensor wire harness from the steering knuckle.

5. Remove the front brake caliper. Remove the rotor.

6. Remove the axle hub grease cap. Remove the cotter pin and lock cap. Remove the front axle hub nut.

7. Remove the nut and separate the stabilizer link from the steering knuckle.

8. Remove the cotter pin and nut. Using the proper tool separate the tie rod end from the steering knuckle.

9. Remove the two bolts and separate the front suspension lower arm from the front axle.

10. Properly support the front suspension lower arm with a jack. Remove the clip and nut.

11. Using the proper tool separate the steering knuckle from the front suspension upper control arm.

12. On 4WD, use a plastic hammer and separate the front axle hub from the front driveshaft.

13. Remove the front axle hub from the vehicle.

To install:

14. Installation is the reverse of the removal procedure.

15. Be sure to fill the master cylinder with the proper grade and type brake fluid.

16. Bleed the brakes, as required.

17. Be sure to check and adjust the alignment, as required.

LEAF SPRING

REMOVAL & INSTALLATION

See Figures 163 and 164.

1. Before servicing the vehicle, refer to the Precautions Section.
2. Raise and support the vehicle safely.
3. Remove the tire and wheel assemblies.
4. On 4WD and PreRunner, remove the spare tire.
5. Support the rear axle housing. Remove the bolt nut and washer.
6. On 2WD separate the shock absorber from the rear spring seat.
7. On 4WD and PreRunner, separate the shock absorber from the rear axle housing.
8. Remove the bolt and then separate the parking brake cable.
9. On 2WD remove the two nuts and spring bumper. Remove the four nuts and four washers. Remove the spring seat and two U-bolts.

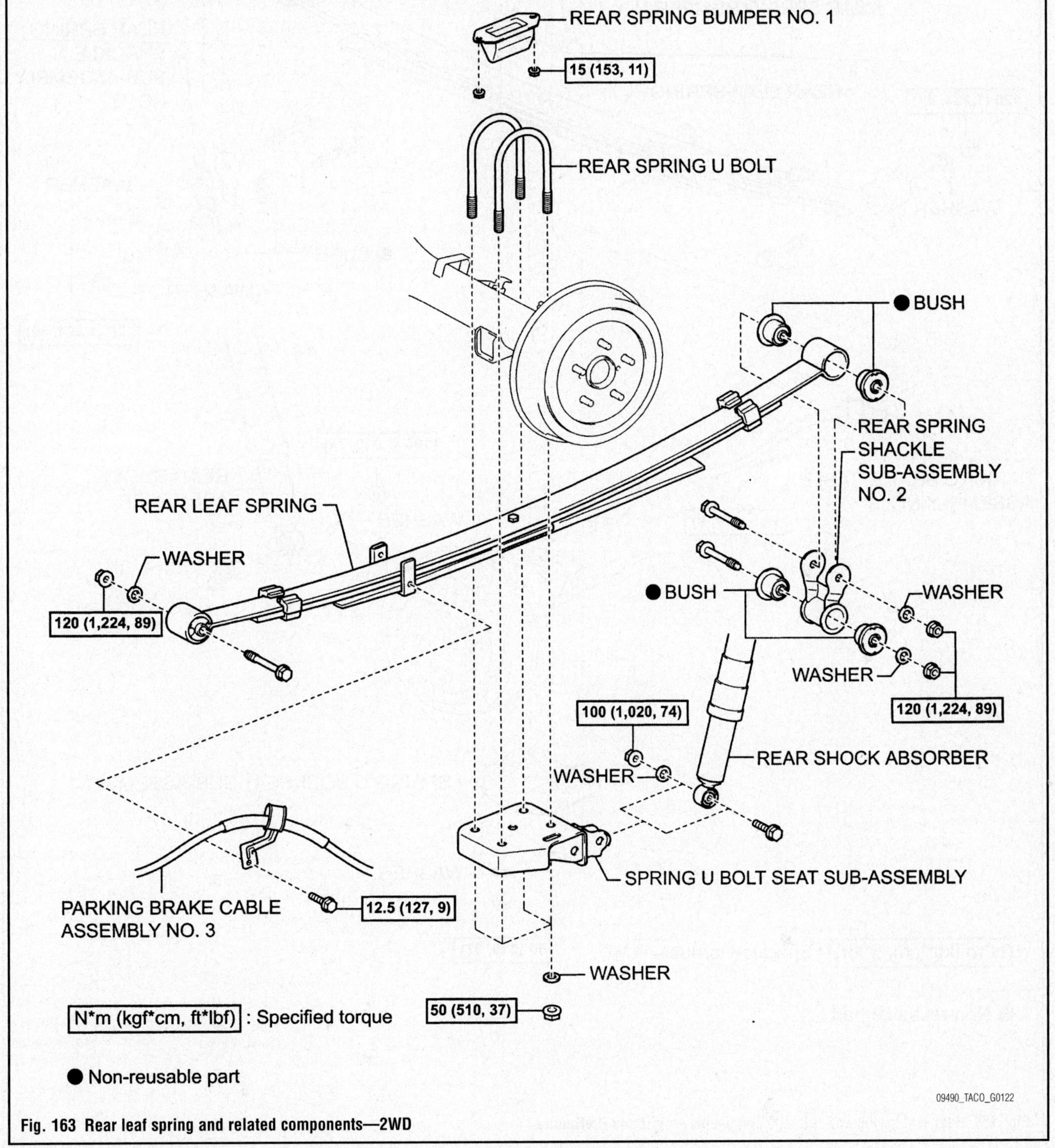

Fig. 163 Rear leaf spring and related components—2WD

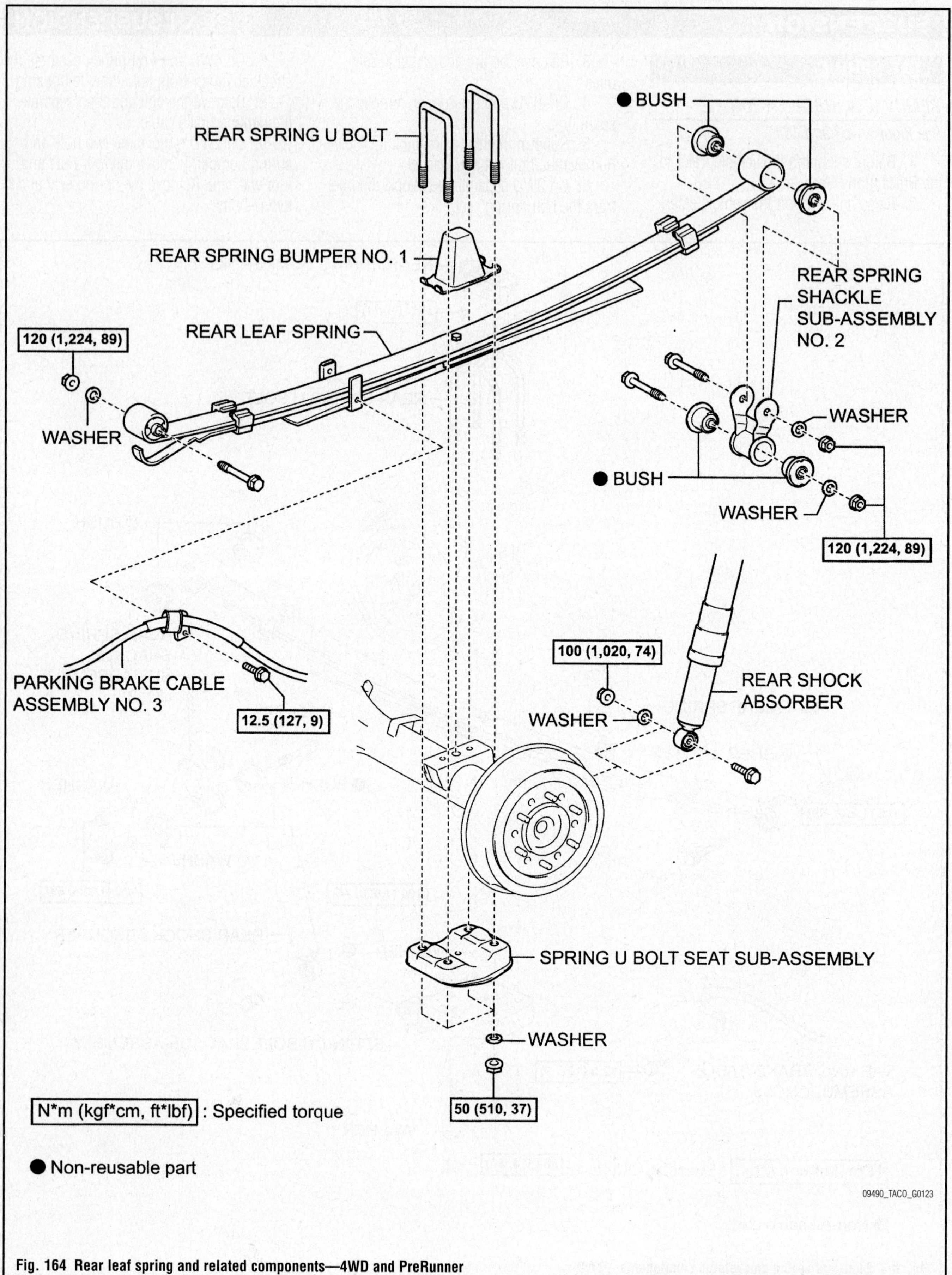

Fig. 164 Rear leaf spring and related components—4WD and PreRunner

10. On 4WD and PreRunner, remove the four nuts and four washers. Remove the spring seat and two U-bolts. Remove the spring bumper.

11. Remove the nut, washer and through bolt. Remove the spring from the vehicle.

➡Be careful not to drop the spring when removing the through bolt.

To install:

➡Always fully tighten rubber bushings when the wheels are in full contact with the ground and the vehicle is at curb height. Bounce the vehicle up and down several times to stabilize the suspension, prior to final tightening of these components.

12. Installation is the reverse of the removal procedure.

13. Tighten the rear spring U-bolts to 37 ft. lbs. Be sure that the lengths of all of the U-bolts under the spring seat are the same.

STABILIZER BAR

REMOVAL & INSTALLATION

1. Before servicing the vehicle, refer to the Precautions Section.
2. Raise and support the vehicle safely.
3. Remove the tire and wheel assemblies.
4. On the left side, remove the two nuts and stabilizer bar link.
5. On the right side, remove the two nuts and stabilizer bar link.
6. Remove the four bolts and the stabilizer bar brackets and bushings.
7. Remove the stabilizer bar from the vehicle.

To install:

8. Installation is the reverse of the removal procedure.
9. Tighten the stabilizer bar and bushing bracket bolts to 20 ft. lbs.
10. Tighten the stabilizer link assembly bolts to 51 ft. lbs
11. Be sure to install the bushing onto the outer side of the mark on the stabilizer bar.

SHOCK ABSORBER

REMOVAL & INSTALLATION

See Figure 165.

1. Before servicing the vehicle, refer to the Precautions Section.

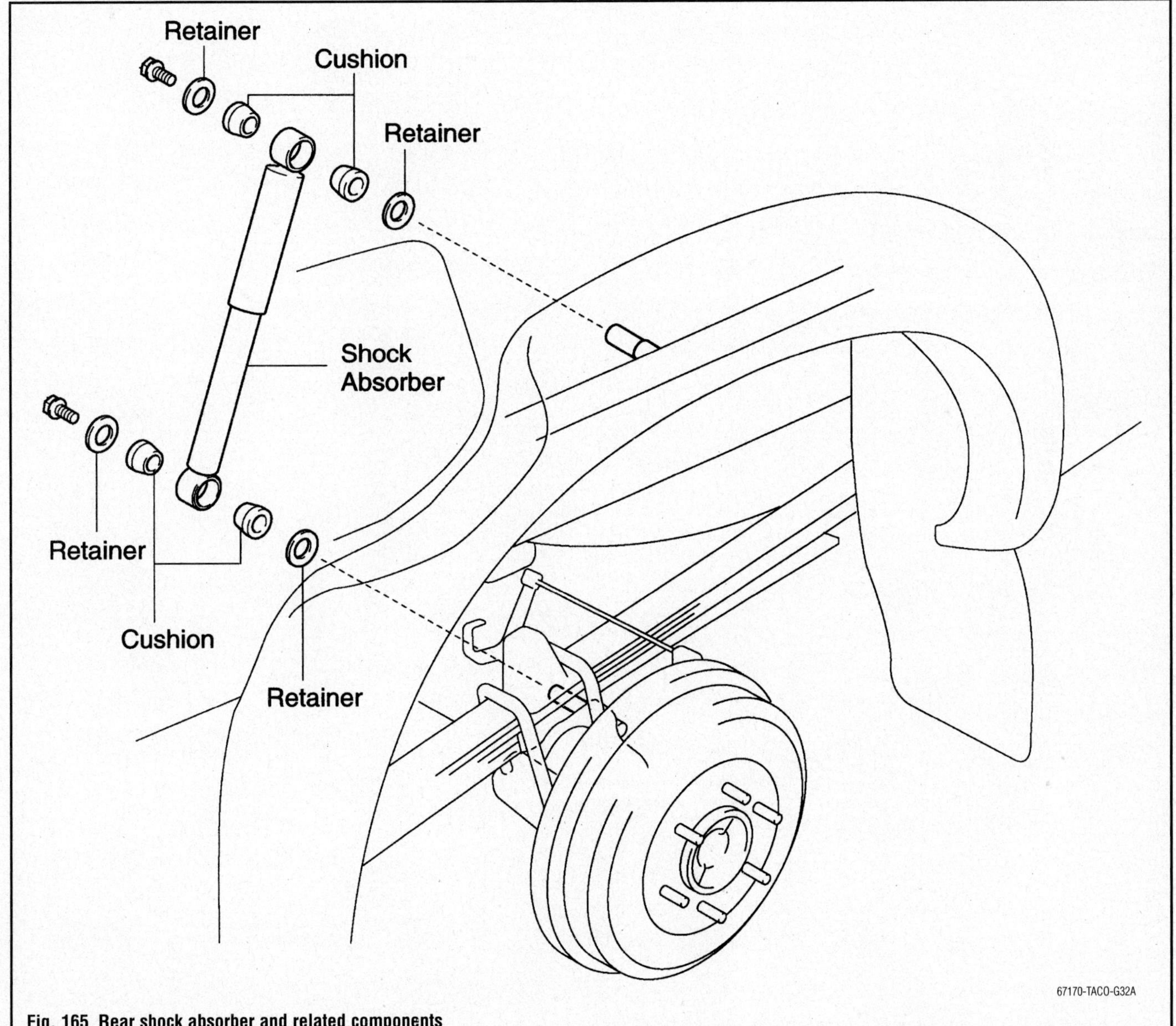

Fig. 165 Rear shock absorber and related components

2. Raise and support the vehicle safely.
3. Remove the tire and wheel assembly.
4. Lower the floor jack to take tension off of the spring.
5. Remove or disconnect the following:
   - Shock absorber from the rear axle housing
   - Nut, retainers and the cushions holding the shock absorber to the frame
   - Shock absorber with the washers and bushings

To install:

➡Always fully tighten rubber bushings when the wheels are in full contact with the ground and the vehicle is at curb height. Bounce the vehicle up and down several times to stabilize the suspension, prior to final tightening of these components.

6. Install the shock absorber to the frame with the washers and bushings.
7. Tighten the shock absorber-to-upper frame nut on to 15 ft. lbs. (20 Nm).
8. Connect the shock absorber to the rear axle housing.
9. Tighten the bolt to 74 ft. lbs. (100 Nm).
10. Install the wheels.

LEXUS AND TOYOTA

Diagnostic Trouble Codes

24

DIAGNOSTIC TROUBLE CODES

OBD II VEHICLE APPLICATIONS

LEXUS

ES350
2007–2008
2007-2008
• 3.5L V6. . . . . 2GR-FE

GS350
2007–2008
• 3.5L V6. . . . . 2GR-FE

GS430
2007
• 4.3L V8. . . . . 3UZ-FE

GS460
2008
• 4.6L V8. . . . . 1UR-FSE

GX470
2007–2008
• 4.7L V8. . . . . 2UZ-FE

IS250
2007–2008
• 2.5L V6. . . . . 4GR-FSE

IS350
2007–2008
3.5L V6. . . . . 2GR-FE

LS460
2007–2008
• 4.6L V8. . . . . 3UZ-FE

LX470
2007
• 4.7L V8. . . . . 2UZ-FE

LX570
• 5.7L V8. . . . . 3UR-FE

RX350
2007–2008
• 3.3L V6. . . . . 2GR-FE

SC430
2007–2008
• 4.3L V8. . . . . 3UZ-FE

TOYOTA

4Runner
2007–2008
• 4.0L V6. . . . . 1GR-FE
• 4.7L V8. . . . . 2UZ-FE

Avalon
2007–2008
• 3.5L V6. . . . . 2GR-FE

Camry
2007–2008
• 2.4L I4 . . . . 2AZ-FE
3.5L V6. . . . . 2GR-FE

Solara
2007–2008
• 2.4L I4 . . . . 2AZ-FE
• 3.5L V6. . . . . 3MZ-FE

Corolla
2007–2008
• 1.8L I4 . . . . 1ZZ-FE

Highlander
2007–2008
• 2.4L I4 . . . . 2AZ-FE
• 3.3L V6. . . . . 3MZ-FE
• 3.5L V6. . . . . 2GR-FE

Land Cruiser
2007–2008
• 4.7L V8. . . . . 2UZ-FE
• 5.7L V8. . . . . 3UR-FE

Matrix
2007–2008
• 1.8L I4 . . . . 1ZZ-FE

RAV 4
2007–2008
• 2.4L I4 . . . . 2AZ-FE
• 3.5L V6. . . . . 2GR-FE

Sequoia
2007–2008
• 4.7L V8. . . . . 2UZ-FE
• 5.7L V8. . . . . 3UR-FE

Sienna
2007–2008
• 3.3L V6. . . . . 3MZ-FE

Tacoma
2007–2008
• 2.7L I4 . . . . 2TR-FE
• 4.0L V6. . . . . 1GR-FE

Tundra
2007–2008
• 4.0L V6. . . . . 1GR-FE
• 4.7L V8. . . . . 2UZ-FE
• 5.7L V8. . . . . 3UR-FE

Yaris
2007–2008
• 1.5L I4 . . . . 1NZ-FE
• 1.5L I4 1NZ-FE. . VID: NCP91, NCP93

OBD II Trouble Code List (P0xxx Codes)

DTC	Trouble Code Title, Conditions & Possible Causes
DTC: P0010 **2T CCM, MIL: Yes** **Years:** 2007, 2008 **Models:** All **Engines:** All **Transmissions:** All	**Camshaft Position 'A' Actuator Circuit (Bank 1) Conditions:** Key on or engine running; and the PCM detected an unexpected voltage condition on the VVT Oil Control Valve Bank 1 circuit. The VVT system controls the intake camshaft in order to provide optimal valve timing during all conditions based signals from the ECT, IAT and TP Sensor. The VVT regulates the intake camshaft angle using oil pressure through the Oil Control Valve. This results in the relative position between the camshaft and crankshaft to become optimal. The result is higher torque, better fuel economy and low emissions. **Possible Causes:** • OCV assembly connector is damaged or loose • OCV assembly control circuit open or shorted to ground • OCV assembly is damaged or has failed • PCM failure
DTC: P0011 **2T CCM, MIL: Yes** **Years:** 2007, 2008 **Models:** All **Engines:** All **Transmissions:** All	**Camshaft Position 'A' Over-Advanced Or System Performance (Bank 1) Conditions:** Engine started, ECT Sensor more than 158°F, vehicle driven at an engine speed of 400-4000 RPM, and the PCM detected the valve timing did not change from the "current" valve timing, or the valve timing remain fixed during testing. The VVT system controls the intake camshaft in order to provide optimal valve timing during all conditions based signals from the ECT, IAT and TP Sensor. The VVT regulates the intake camshaft angle using oil pressure through the Oil Control Valve. This results in the relative position between the camshaft and crankshaft to become optimal. The result is better engine torque, fuel economy and lower emissions. **Possible Causes:** • Engine valve timing malfunction • Camshaft timing oil control valve unit is damaged or has failed • PCM or VVT ECM has failed
DTC: P0012 **2T CCM, MIL: Yes** **Years:** 2007, 2008 **Models:** All **Engines:** All **Transmissions:** All	**Camshaft Position 'A' Over-Retarded (Bank 1) Conditions:** Engine started, ECT Sensor more than 158°F, vehicle driven at an engine speed of 400-4000 RPM, and the PCM detected the valve timing did not change from the "current" valve timing, or that the valve timing remain fixed during the test period. The VVT system controls the intake camshaft in order to provide optimal valve timing during all conditions. The VVT regulates the intake camshaft angle using oil pressure through the OCV. This causes the relative position between the camshaft and crankshaft to become optimal. The result is improved engine torque, better fuel economy and low emissions. **Possible Causes:** • Engine valve timing malfunction • Camshaft timing oil control valve unit is damaged or has failed • PCM or VVT ECM has failed
DTC: P0013 **2T CCM, MIL: Yes** **Years:** 2007, 2008 **Models:** All **Engines:** All **Transmissions:** All	**Camshaft Position 'B' Actuator Circuit / Open (Bank 1) Conditions:** Engine started, ECT Sensor more than 158°F, vehicle driven at an engine speed of 400-4000 RPM, and the PCM detected the valve timing did not change from the "current" valve timing, or that the valve timing remain fixed during the test period. The VVT system controls the intake camshaft in order to provide optimal valve timing during all conditions. The VVT regulates the intake camshaft angle using oil pressure through the OCV. This causes the relative position between the camshaft and crankshaft to become optimal. The result is improved engine torque, better fuel economy and low emissions. **Possible Causes:** • Open or short in OCV for exhaust camshaft (Bank 1) • OCV for exhaust camshaft (for Bank 1) • ECM
DTC: P0014 **2T CCM, MIL: Yes** **Years:** 2007, 2008 **Models:** All **Engines:** All **Transmissions:** All	**Camshaft Position 'B' Timing Over-Advanced or System Performance (Bank 1) Conditions:** Engine started, ECT Sensor more than 158°F, vehicle driven at an engine speed of 400-4000 RPM, and the PCM detected the valve timing did not change from the "current" valve timing, or that the valve timing remain fixed during the test period. The VVT system controls the intake camshaft in order to provide optimal valve timing during all conditions. The VVT regulates the intake camshaft angle using oil pressure through the OCV. This causes the relative position between the camshaft and crankshaft to become optimal. The result is improved engine torque, better fuel economy and low emissions. **Possible Causes:** • Valve timing • Oil control valve (OCV) for exhaust camshaft • OCV filter • Camshaft timing exhaust gear • ECM

DTC	Trouble Code Title, Conditions & Possible Causes
DTC: P0015 **2T CCM, MIL: Yes** **Years:** 2007, 2008 **Models:** All **Engines:** All **Transmissions:** All	**Camshaft Position 'B' Timing Over-Retarded (Bank 1) Conditions:** Engine started, ECT Sensor more than 158°F, vehicle driven at an engine speed of 400-4000 RPM, and the PCM detected the valve timing did not change from the "current" valve timing, or that the valve timing remain fixed during the test period. The VVT system controls the intake camshaft in order to provide optimal valve timing during all conditions. The VVT regulates the intake camshaft angle using oil pressure through the OCV. This causes the relative position between the camshaft and crankshaft to become optimal. The result is improved engine torque, better fuel economy and low emissions. **Possible Causes:** • Valve timing • (OCV) for exhaust camshaft • OCV filter • Camshaft timing exhaust gear • ECM
DTC: P0016 **2T CCM, MIL: Yes** **Years:** 2007, 2008 **Models:** All **Engines:** All **Transmissions:** All	**Camshaft Position-Crankshaft Position (Bank 1 Sensor A) Conditions:** Engine started, engine running, and the PCM detected a deviation between the crankshaft position Sensor signal and the VVT Sensor 1 signal during the test period. The crankshaft position (NE) Sensor consists of a magnet, iron core and pickup coil. The NE Sensor signal plate, installed on the crankshaft-timing pulley, has 34 teeth. This Sensor generates 34 signals for each engine revolution. The PCM detects the crankshaft angle and engine speed based on the NE signal. It detects the correct cylinder based on signals from the VVT 1 Sensor along signals from the crankshaft position Sensor. **Possible Causes:** • Engine valve timing problem • Engine timing chain mechanical problem • OCV • OCV filter • Camshaft timing gear assembly • ECM
DTC: P0017 **2T CCM, MIL: Yes** **Years:** 2006, 2007 **Models:** All **Engines:** All **Transmissions:** All	**Crankshaft Position – Camshaft Position Correlation (Bank 1 Sensor B) Conditions:** Engine started, ECT Sensor more than 158°F, vehicle driven at an engine speed of 400-4000 RPM, and the PCM detected the valve timing did not change from the "current" valve timing, or that the valve timing remain fixed during the test period. The VVT system controls the intake camshaft in order to provide optimal valve timing during all conditions. The VVT regulates the intake camshaft angle using oil pressure through the OCV. This causes the relative position between the camshaft and crankshaft to become optimal. The result is improved engine torque, better fuel economy and low emissions. **Possible Causes:** • Engine timing chain mechanical problem • ECM
DTC: P0018 **2T CCM, MIL: Yes** **Years:** 2007, 2008 **Models:** All **Engines:** All **Transmissions:** All	**Camshaft Position-Crankshaft Position (Bank 2 Sensor A) Conditions:** Engine started, engine running, and the PCM detected a deviation between the crankshaft position Sensor signal and the VVT Sensor 2 signal during the test period. The crankshaft position (NE) Sensor consists of a magnet, iron core and pickup coil. The NE Sensor signal plate, installed on the crankshaft-timing pulley, has 34 teeth. This Sensor generates 34 signals for each engine revolution. The PCM detects the crankshaft angle and engine speed based on the NE signal. It detects the correct cylinder based on signals from the VVT 2 Sensor along signals from the crankshaft position Sensor. **Possible Causes:** • Engine valve timing problem • Engine timing chain mechanical problem • PCM failure
DTC: P0019 **2T CCM, MIL: Yes** **Years:** 2006, 2007 **Models:** All **Engines:** All **Transmissions:** All	**Crankshaft Position – Camshaft Position Correlation (Bank 2 Sensor B) Conditions:** Engine started, ECT Sensor more than 158°F, vehicle driven at an engine speed of 400-4000 RPM, and the PCM detected the valve timing did not change from the "current" valve timing, or that the valve timing remain fixed during the test period. The VVT system controls the intake camshaft in order to provide optimal valve timing during all conditions. The VVT regulates the intake camshaft angle using oil pressure through the OCV. This causes the relative position between the camshaft and crankshaft to become optimal. The result is improved engine torque, better fuel economy and low emissions. **Possible Causes:** • Engine timing chain mechanical problem • ECM

DTC	Trouble Code Title, Conditions & Possible Causes
DTC: P0020 **2T CCM, MIL: Yes** **Years:** 2007, 2008 **Models:** All **Engines:** All **Transmissions:** All	**Camshaft Position Sensor Actuator 'A' Circuit (Bank 2) Conditions:** Key on or engine running; and the PCM detected an unexpected voltage condition on the Camshaft Position Sensor 'A' Bank 2 circuit. The VVT system controls the intake camshaft in order to provide optimal valve timing during all conditions based signals from the ECT, IAT and TP Sensor. The VVT regulates the intake camshaft angle using oil pressure through the OCV. The result is that relative position between the camshaft and crankshaft becomes optimal. The engine has better torque, fuel economy and lower emissions. **Possible Causes:** • OCV assembly connector is damaged or loose • OCV assembly control circuit open or shorted to ground • OCV assembly is damaged or has failed • PCM failure
DTC: P0021 **2T CCM, MIL: Yes** **Years:** 2007, 2008 **Models:** All **Engines:** All **Transmissions:** All	**Camshaft Position 'A' Timing Over-Advanced Or System Performance (Bank 2) Conditions:** Engine started, ECT Sensor more than 158°F, vehicle driven at an engine speed of 400-4000 RPM, and the PCM detected the valve timing did not change from the "current" valve timing, or the valve timing remain fixed during testing. The VVT system controls the intake camshaft in order to provide optimal valve timing during all conditions. The VVT regulates the intake camshaft angle using oil pressure through the Oil Control Valve. The result is that relative position between the camshaft and crankshaft becomes optimal. The engine has better torque, fuel economy and lower emissions. **Possible Causes:** • Engine valve timing malfunction • Camshaft timing oil control valve unit is damaged or has failed • PCM or VVT ECM has failed
DTC: P0022 **2T CCM, MIL: Yes** **Years:** 2007, 2008 **Models:** All **Engines:** All **Transmissions:** All	**Camshaft Position 'A' Timing Over-Retarded (Bank 2) Conditions:** Engine started, ECT Sensor over 158°F, engine speed of 400-4000 RPM, and the PCM detected the valve timing did not change from the "current" valve timing, or the valve timing remain fixed. The VVT system controls the intake camshaft in order to provide optimal valve timing during all conditions. The VVT regulates the intake camshaft angle using oil pressure through the OCV. The result is that relative position between the camshaft and crankshaft becomes optimal. The engine has better torque, fuel economy and lower emissions. **Possible Causes:** • Engine valve timing malfunction • Camshaft timing oil control valve unit is damaged or has failed • PCM or VVT ECM has failed
DTC: P0023 **2T CCM, MIL: Yes** **Years:** 2006, 2007 **Models:** All **Engines:** All **Transmissions:** All	**Camshaft Position 'B' Actuator Circuit/Open (Bank 2) Conditions:** Engine started, ECT Sensor more than 158°F, vehicle driven at an engine speed of 400-4000 RPM, and the PCM detected the valve timing did not change from the "current" valve timing, or that the valve timing remain fixed during the test period. The VVT system controls the intake camshaft in order to provide optimal valve timing during all conditions. The VVT regulates the intake camshaft angle using oil pressure through the OCV. This causes the relative position between the camshaft and crankshaft to become optimal. The result is improved engine torque, better fuel economy and low emissions. **Possible Causes:** • Open or short in OCV for exhaust camshaft (for Bank 2) circuit • OCV for exhaust camshaft (Bank 2) • ECM
DTC: P0024 **2T CCM, MIL: Yes** **Years:** 2006, 2007 **Models:** All **Engines:** All **Transmissions:** All	**Camshaft Position 'B' Timing Over-Advanced or System Performance (Bank 2) Conditions:** Engine started, ECT Sensor more than 158°F, vehicle driven at an engine speed of 400-4000 RPM, and the PCM detected the valve timing did not change from the "current" valve timing, or that the valve timing remain fixed during the test period. The VVT system controls the intake camshaft in order to provide optimal valve timing during all conditions. The VVT regulates the intake camshaft angle using oil pressure through the OCV. This causes the relative position between the camshaft and crankshaft to become optimal. The result is improved engine torque, better fuel economy and low emissions. **Possible Causes:** • Valve timing • Oil control valve (OCV) for exhaust camshaft • OCV filter • Camshaft timing exhaust gear • ECM

DTC	Trouble Code Title, Conditions & Possible Causes
DTC: P0025 **2T CCM, MIL: Yes** **Years:** 2006, 2007 **Models:** All **Engines:** All **Transmissions:** All	**Camshaft Position 'B' Timing Over-Retarded (Bank 2) Conditions:** Engine started, ECT Sensor more than 158°F, vehicle driven at an engine speed of 400-4000 RPM, and the PCM detected the valve timing did not change from the "current" valve timing, or that the valve timing remain fixed during the test period. The VVT system controls the intake camshaft in order to provide optimal valve timing during all conditions. The VVT regulates the intake camshaft angle using oil pressure through the OCV. This causes the relative position between the camshaft and crankshaft to become optimal. The result is improved engine torque, better fuel economy and low emissions. **Possible Causes:** • Valve timing • (OCV) for exhaust camshaft • OCV filter • Camshaft timing exhaust gear • ECM
DTC: P0031 **1T CCM, MIL: Yes** **Years:** 2007, 2008 **Models:** All **Engines:** All **Transmissions:** All	**Oxygen Sensor (Bank 1 Sensor 1) Heater Circuit Low Conditions:** Engine started, and the PCM detected the Oxygen Sensor heater control circuit indicated less than 0.20 amps during the CCM test period. **Possible Causes:** • Open in Air-Fuel Ratio (A/F) Sensor heater circuit • A/F Sensor heater circuit (Sensor 1) • Engine room J/B (EFI relay) • ECM
DTC: P0032 **1T CCM, MIL: Yes** **Years:** 2007, 2008 **Models:** All **Engines:** All **Transmissions:** All	**Oxygen Sensor (Bank 1 Sensor 1) Heater Circuit High Input Conditions:** Engine started, and the PCM detected the Oxygen Sensor heater control circuit indicated more than 2.0 amps during the CCM test period. **Possible Causes:** • Short in A/F Sensor heater circuit • A/F Sensor heater (Sensor 1) • Engine room J/B (EFI realy) • ECM
DTC: P0037 **2T CCM, MIL: Yes** **Years:** 2007, 2008 **Models:** All **Engines:** All **Transmissions:** All	**Oxygen Sensor (Bank 1 Sensor 2) Heater Circuit Low Input Conditions:** Engine started, and the PCM detected the Oxygen Sensor heater control circuit indicated less than 0.20 amps during the CCM test period. **Possible Causes:** • Heated Oxygen Sensor heater control circuit open • Heated Oxygen Sensor heater assembly is damaged or has failed • Engine room J/B (EFI realy) • PCM failure
DTC: P0038 **2T CCM, MIL: Yes** **Years:** 2007, 2008 **Models:** All **Engines:** All **Transmissions:** All	**Oxygen Sensor (Bank 1 Sensor 2) Heater Circuit High Input Conditions:** Engine started, and the PCM detected the Oxygen Sensor heater control circuit indicated more than 2.0 amps during the CCM test period. **Possible Causes:** • Heated Oxygen Sensor heater control circuit is shorted to ground • Heated Oxygen Sensor heater assembly is damaged or has failed • PCM failure
DTC: P0051 **2T CCM, MIL: Yes** **Years:** 2007, 2008 **Models:** All **Engines:** All **Transmissions:** All	**Oxygen (A/F) Sensor (Bank 2 Sensor 1) Heater Circuit Low Input Conditions:** Engine started, and the PCM detected the Oxygen Sensor heater control circuit indicated less than 0.20 amps during the CCM test period. **Possible Causes:** • Heated Oxygen Sensor heater control circuit open • Heated Oxygen Sensor heater assembly is damaged or has failed • PCM failure
DTC: P0052 **2T CCM, MIL: Yes** **Years:** 2007, 2008 **Models:** All **Engines:** All **Transmissions:** All	**Oxygen (A/F) Sensor (Bank 2 Sensor 1) Heater Circuit High Input Conditions:** Engine started, and the PCM detected the Oxygen Sensor heater control circuit indicated more than 2.0 amps during the CCM test period. **Possible Causes:** • Heated Oxygen Sensor heater control circuit is shorted to ground • Heated Oxygen Sensor heater assembly is damaged or has failed • PCM failure

DTC	Trouble Code Title, Conditions & Possible Causes
DTC: P0057 **2T CCM, MIL: Yes** **Years:** 2007, 2008 **Models:** All **Engines:** All **Transmissions:** All	**Oxygen Sensor (Bank 2 Sensor 2) Heater Circuit Low Input Conditions:** Engine started, and the PCM detected the Oxygen Sensor heater control circuit indicated less than 0.20 amps during the CCM test period. **Possible Causes:** • Heated Oxygen Sensor heater control circuit open • Heated Oxygen Sensor heater assembly is damaged or has failed • PCM failure
DTC: P0058 **2T CCM, MIL: Yes** **Years:** 2007, 2008 **Models:** All **Engines:** All **Transmissions:** All	**Oxygen Sensor (Bank 2 Sensor 2) Heater Circuit High Input Conditions:** Engine started, and the PCM detected the Oxygen Sensor heater control circuit indicated more than 2.0 amps during the CCM test period. **Possible Causes:** • Heated Oxygen Sensor heater control circuit is shorted to ground • Heated Oxygen Sensor heater assembly is damaged or has failed • PCM failure
DTC: P0100 **1T CCM, MIL: Yes** **Years:** 2007, 2008 **Models:** All **Engines:** All **Transmissions:** All	**Mass Or Volume Airflow Sensor Circuit Malfunction Conditions:** Engine started, engine running at under 4000 RPM, and the PCM detected an unexpected low or high voltage condition on the Mass Airflow (MAF) Sensor circuit for over 3 seconds during the CCM test. The MAF Sensor on this engine includes a hot wire assembly with an air temperature Sensor, platinum hot wire and control unit mounted in a plastic housing. This airflow meter works on the principle that the hot wire and temperature Sensor located in the intake air bypass of the housing detect any changes in the (incoming) air temperature. **Possible Causes:** • MAF Sensor signal circuit open, shorted to ground or power • MAF Sensor ground circuit open between Sensor and ground • MAF Sensor power circuit open (check the power to the relay) • MAF Sensor has failed, or the PCM failure
DTC: P0101 **2T CCM, MIL: Yes** **Years:** 2007, 2008 **Engines:** All **Transmissions:** All	**Mass Airflow Sensor Signal Range/Performance Conditions:** DTC P0100 not set, engine speed under 900 RPM, throttle valve closed, ECT Sensor over 158°F, and the PCM detected the MAF was above 2.20v; or with the engine speed over 1500 RPM, the throttle valve closed and the TP Sensor over 0.63v, the MAF Sensor was less than 1.06v. This airflow meter works on the principle that the hot wire and temperature Sensor in the intake air bypass of the housing detect any changes in the (incoming) air temperature **Possible Causes:** • MAF Sensor signal circuit open, shorted to ground or power • MAF Sensor is contaminated, damaged or it has failed • PCM failure
DTC: P0102 **2T CCM, MIL: Yes** **Years:** 2007, 2008 **Models:** All **Engines:** All **Transmissions:** All	**Mass Airflow Sensor Circuit Low Input Conditions:** DTC P0100 not set, engine started, and the PCM detected an unexpected low voltage condition on the MAF Sensor circuit during the CCM test period. This airflow meter works on the principle that the hot wire and temperature Sensor located in the intake air bypass of the housing detect any changes in the (incoming) air temperature **Possible Causes:** • MAF Sensor signal circuit open or shorted to ground • MAF Sensor is contaminated, damaged or it has failed • PCM failure
DTC: P0103 **2T CCM, MIL: Yes** **Years:** 2007, 2008 **Models:** All **Engines:** All **Transmissions:** All	**Mass Airflow Sensor Circuit High Input Conditions:** DTC P0100 not set, engine started, and the PCM detected an unexpected high voltage condition on the MAF Sensor circuit during the CCM test period. This airflow meter works on the principle that the hot wire and temperature Sensor located in the intake air bypass of the housing detect any changes in the (incoming) air temperature **Possible Causes:** • MAF Sensor signal circuit is shorted to power • MAF Sensor is contaminated, damaged or it has failed • PCM failure
DTC: P0110 **1T CCM, MIL: Yes** **Years:** 2007, 2008 **Models:** All **Engines:** All **Transmissions:** All	**Intake Air Temperature Sensor Circuit Malfunction Conditions:** Engine started, and the PCM detected an unexpected low or high voltage on the IAT Sensor circuit (Scan Tool reads −40°F or 284°F). This Sensor is located inside the MAF Sensor. **Possible Causes:** • IAT Sensor signal circuit open, shorted to ground or VREF • IAT Sensor ground circuit open between Sensor and ground • IAT Sensor is contaminated, damaged or has failed • PCM failure

DTC	Trouble Code Title, Conditions & Possible Causes
DTC: P0111 **1T CCM, MIL: Yes** **Years:** 2007, 2008 **Models:** All **Engines:** All **Transmissions:** All	**Intake Air Temperature Sensor Gradient Too High Conditions:** Engine started, and the PCM detected an unexpected low or high voltage on the IAT Sensor circuit (Scan Tool reads −40°F or 284°F). This Sensor is located inside the MAF Sensor. **Possible Causes:** • Mass air flow meter assembly
DTC: P0112 **1T CCM, MIL: Yes** **Years:** 2007, 2008 **Models:** All **Engines:** All **Transmissions:** All	**Intake Air Temperature Sensor Circuit Low Input Conditions:** Key on or engine running; and the PCM detected an unexpected low voltage condition on the IAT Sensor circuit for over 500 ms. **Possible Causes:** • IAT Sensor connector is damaged (it may be shorted internally) • IAT Sensor ground circuit is shorted to ground • IAT Sensor is damaged or has failed • PCM failure
DTC: P0113 **1T CCM, MIL: Yes** **Years:** 2007, 2008 **Models:** All **Engines:** All **Transmissions:** All	**Intake Air Temperature Sensor Circuit High Input Conditions:** Key on or engine running; and the PCM detected an unexpected high voltage condition on the IAT Sensor circuit for over 500 ms. **Possible Causes:** • IAT Sensor connector is damaged (it may be open internally) • IAT Sensor ground circuit open • IAT Sensor is damaged or has failed • PCM failure
DTC: P0115 **2T CCM, MIL: Yes** **Years:** 2007, 2008 **Models:** All **Engines:** All **Transmissions:** All	**Engine Coolant Temperature Sensor Circuit Malfunction Conditions:** Key on or engine running; and the PCM detected an unexpected "low" or "high" voltage on the ECT Sensor circuit (Scan Tool reads less than −40°F or low voltage or more than 284°F for high voltage). **Possible Causes:** • ECT Sensor signal circuit open, shorted to ground or VREF • ECT Sensor ground circuit open between Sensor and ground • ECT Sensor is contaminated, damaged or has failed • PCM failure
DTC: P0116 **2T CCM, MIL: Yes** **Years:** 2007, 2008 **Models:** All **Engines:** All **Transmissions:** All	**Engine Coolant Temperature Sensor Range/Performance Conditions:** Engine started, ECT Sensor from 95°F to 140°F, IAT Sensor more than 19.9°F, vehicle driven with several changes in the VSS signals, and the PCM detected the ECT Sensor signal did not increase more than 37.4°F after engine was started and the test period completed. **Possible Causes:** • Check for problems in the cooling system (i.e., coolant, the fan) • ECT Sensor signal circuit or ground circuit has high resistance • ECT Sensor is contaminated, damaged or has failed • PCM failure
DTC: P0117 **2T CCM, MIL: Yes** **Years:** 2007, 2008 **Models:** All **Engines:** All **Transmissions:** All	**Engine Coolant Temperature Sensor Circuit Low Input Conditions:** Key on or engine running; and the PCM detected an unexpected low voltage condition on the ECT Sensor (Scan Tool reads below −40°F). **Possible Causes:** • ECT Sensor signal circuit is shorted to ground • ECT Sensor is damaged or has failed • PCM failure
DTC: P0118 **2T CCM, MIL: Yes** **Years:** 2007, 2008 **Models:** All **Engines:** All **Transmissions:** All	**Engine Coolant Temperature Circuit High Input Conditions:** Key on or engine running; and the PCM detected an unexpected high voltage condition on the ECT Sensor (Scan Tool reads over 284°F). **Possible Causes:** • Open in ECT Sensor circuit • ECT Sensor • ECM

DTC	Trouble Code Title, Conditions & Possible Causes
DTC: P011B **2T CCM, MIL: Yes** **Years:** 2007, 2008 **Models:** All **Engines:** All **Transmissions:** All	**Engine Coolant Temperature/Intake Air Temperature Correlation Conditions:** The engine has two temperature Sensors, an Engine Coolant Temperature (ECT) Sensor and an Intake Air Temperature (IAT) Sensor, to detect the temperature while the engine is in operation. A thermistor, whose resistance value varies according to the temperature, is built into each Sensor. When the temperature is low, the resistance of the thermistor increases. When the temperature is high, the resistance drops. These variations in resistance are transmitted to the ECM as voltage changes. Based on these temperature signals output from the Sensors, the ECM determines the fuel injection time and the ignition timing to control the engine. **Possible Causes:** • IAT Sensor • ECT Sensor • ECM
DTC: P0120 **2T CCM, MIL: Yes** **Years:** 2007, 2008 **Models:** All **Engines:** All **Transmissions:** All	**Throttle Pedal Position Sensor/Switch 'A' Circuit Malfunction Conditions** Key on or engine running; and the PCM detected the TP Sensor 'A' Signal indicated less than 0.10v with the throttle position closed (the switch open), or the TP Sensor signal indicated more than 4.9v at any time. The Electric TP Sensor is mounted on the throttle body. It has two Sensors (the electrical throttle system does not use a cable). **Possible Causes:** • TP Sensor signal circuit open is or shorted to ground • TP Sensor ground circuit open • TP Sensor power circuit open (test VREF circuit at the PCM) • TP Sensor is damaged or has failed • PCM failure
DTC: P0121 **2T CCM, MIL: Yes** **Years:** 2007, 2008 **Models:** All **Engines:** All **Transmissions:** All	**Throttle Pedal Position Sensor Switch 'A' Signal Range/Performance Conditions:** Engine started; and the PCM detected the difference between the TP Sensor VTA1 and VTA2 signal was out-of-range. The TP Sensor, mounted on the throttle body, detects the Throttle Valve opening angle (about 0.70v with the throttle closed). The PCM uses the VTA signal for air/fuel ratio and power increase correction. The Electric TP Sensor is mounted on the throttle body. It has two Sensors (the electrical throttle system does not use a cable). **Possible Causes:** • TP Sensor connector is damaged or open • TP Sensor is damaged or has failed • PCM failure
DTC: P0122 **2T CCM, MIL: Yes** **Years:** 2007, 2008 **Models:** All **Engines:** All **Transmissions:** All	**Throttle Pedal Position Sensor/Switch 'A' Circuit Low Input Conditions:** Engine started, and the PCM detected an unexpected low voltage (below 0.20v) on the VTA1 signal circuit. The TP Sensor detects the Throttle Valve opening angle (0.70v with the throttle closed). The Electric TP Assembly has 2 Sensors (this system does not use a throttle cable). **Possible Causes:** • Throttle control motor and Sensor unit is damaged or has failed • TP Sensor VC (VREF) circuit open, or the VTA1 signal circuit is shorted to ground • PCM failure
DTC: P0123 **2T CCM, MIL: Yes** **Years:** 2007, 2008 **Models:** All **Engines:** All **Transmissions:** All	**Throttle Pedal Position Sensor/Switch 'A' Circuit High Input Conditions:** Engine started, and the PCM detected an unexpected high voltage (over 4.80v) on the VTA1 signal circuit. The TP Sensor detects the Throttle Valve opening angle (0.70v with the throttle closed). The Electric TP Assembly has 2 Sensors (this system does not use a throttle cable). **Possible Causes:** • Throttle control motor and Sensor unit is damaged or has failed • TP Sensor VC (VREF) circuit is shorted to the VTA1 circuit • VTA1 ground circuit open or the VTA1 signal circuit open • PCM failure
DTC: P0125 **1T CCM, MIL: Yes** **Years:** 2007, 2008 **Models:** All **Engines:** All **Transmissions:** All	**Insufficient Coolant Temperature For Closed Loop Conditions:** ECT Sensor less than 19.4°F, engine started, engine runtime over 20 minutes, and the PCM detected the ECT Sensor indicated 68°F or less; or with the ECT Sensor from 19.4°F to 50.0°F at startup, engine runtime over 5 minutes, the PCM detected the ECT Sensor indicated 68°F or less; or with the ECT Sensor more than 19.4°F at startup, and the engine runtime over 5 minutes, the PCM detected the ECT Sensor signal was 68°F or less; or with the ECT Sensor over 50°F at startup, engine runtime over 2 minutes, the PCM detected the ECT Sensor signal did not reach 86°F during the CCM test period. **Possible Causes:** • Check the operation of the thermostat (it may be stuck open) • ECT Sensor signal circuit has high resistance • ECT Sensor has failed • Inspect for low coolant level for an incorrect coolant mixture

DTC	Trouble Code Title, Conditions & Possible Causes
DTC: P0128 **2T CCM, MIL: Yes** **Years:** 2007, 2008 **Models:** All **Engines:** All **Transmissions:** All	**Coolant Thermostat (Coolant Temperature Below Thermostat Regulating Temperature) Conditions:** Engine started, ECT Sensor signal below 140°F at startup, and the PCM detected the ECT Sensor did not reach 167°F after the warmup period expired (engine runtime 5-10 minutes). **Possible Causes:** • Check the operation of the thermostat (it may be stuck open) • ECT Sensor is out-of-calibration or skewed • Inspect for low coolant level or for an incorrect coolant mixture
DTC: P0136 **2T CCM, MIL: Yes** **Years:** 2007, 2008 **Models:** All **Engines:** All **Transmissions:** All	**Oxygen Sensor (Bank 1 Sensor 2) Circuit Malfunction Conditions:** Vehicle driven to a speed of over 25 MPH while in closed loop, and the PCM detected the Heated Oxygen Sensor signal was fixed at more than 400 mv, or that it was fixed at less than 600 mv. **Possible Causes:** • Open or short in HO2 Sensor (for bank 1) circuit • HO2 Sensor (for Bank 1) • HO2 Sensor heater (for Bank 1) • Air Fuel Ratio (A/F) Sensor (for Bank 1) • Integration relay • Gas leak from exhaust system
DTC: P0137 **2T CCM, MIL: Yes** **Years:** 2007, 2008 **Models:** All **Engines:** All **Transmissions:** All	**Oxygen Sensor Circuit Low Voltage (Bank 1 Sensor 2) Conditions:** Vehicle driven to a speed of over 25 MPH while in closed loop, and the PCM detected the Oxygen Sensor signal was fixed at more than 400 mv, or that it was fixed at less than 600 mv. **Possible Causes:** • Open or short in HO2 Sensor (for bank 1) circuit • HO2 Sensor (for Bank 1) • HO2 Sensor heater (for Bank 1) • Air Fuel Ratio (A/F) Sensor (for Bank 1) • Integration relay • Gas leak from exhaust system
DTC: P0138 **2T CCM, MIL: Yes** **Years:** 2007, 2008 **Models:** All **Engines:** All **Transmissions:** All	**Oxygen Sensor Circuit High Voltage (Bank 1 Sensor 2) Conditions:** Vehicle driven to a speed of over 25 MPH while in closed loop, and the PCM detected the Oxygen Sensor signal was fixed at more than 400 mv, or that it was fixed at less than 600 mv. **Possible Causes:** • Open or short in HO2 Sensor (for bank 1) circuit • HO2 Sensor (for Bank 1) • HO2 Sensor heater (for Bank 1) • Air Fuel Ratio (A/F) Sensor (for Bank 1) • Integration relay • Gas leak from exhaust system
DTC: P0139 **2T CCM, MIL: Yes** **Years:** 2007, 2008 **Models:** All **Engines:** All **Transmissions:** All	**Oxygen Sensor Circuit Slow Response (Bank 1 Sensor 2) Conditions:** Vehicle driven to a speed of over 25 MPH while in closed loop, and the PCM detected the Oxygen Sensor signal was fixed at more than 400 mv, or that it was fixed at less than 600 mv. **Possible Causes:** • Open or short in HO2 Sensor (for bank 1) circuit • HO2 Sensor (for Bank 1) • HO2 Sensor heater (for Bank 1) • Air Fuel Ratio (A/F) Sensor (for Bank 1) • Integration relay • Gas leak from exhaust system
DTC: P0141 **2T CCM, MIL: Yes** **Years:** 2007, 2008 **Models:** All **Engines:** All **Transmissions:** All	**Oxygen Sensor (Bank 1 Sensor 2) Heater Circuit Malfunction Conditions:** Engine started, engine running, and the PCM detected the Heated Oxygen Sensor heater current exceeded 2 amps, or that it was 0.25 amps or less. **Possible Causes:** • Open or short in HO2 Sensor heater circuit • HO2 Sensor heater (for Sensor 2) • Integration relay • ECM

DTC	Trouble Code Title, Conditions & Possible Causes
DTC: P0156 **2T CCM, MIL: Yes** **Years:** 2007, 2008 **Models:** All **Engines:** All **Transmissions:** All	**Oxygen Sensor (Bank 2 Sensor 2) Circuit Malfunction Conditions:** Engine started, engine warmup completed, engine idling in closed loop, and the PCM detected the Oxygen Sensor signal was fixed at 400 mv or higher, or that it was fixed from 400-550 mv during the test. **Possible Causes:** • Open or short in HO2 Sensor (for Bank 2) circuit • HO2 Sensor (for Bank 2) • HO2 Sensor heater (for Bank 2) • Air-Fuel (A/F) Sensor (for Bank 2) • Integration relay • Gas leak from exhaust system
DTC: P0157 **2T CCM, MIL: Yes** **Years:** 2007, 2008 **Models:** All **Engines:** All **Transmissions:** All	**Oxygen Sensor Circuit Low Voltage (Bank 2 Sensor 2) Conditions:** Engine started, engine warmup completed, engine idling in closed loop, and the PCM detected the Oxygen Sensor signal was fixed at 400 mv or higher, or that it was fixed from 400-550 mv during the test. **Possible Causes:** • Open or short in HO2 Sensor (for Bank 2) circuit • HO2 Sensor (for Bank 2) • HO2 Sensor heater (for Bank 2) • Air-Fuel (A/F) Sensor (for Bank 2) • Integration relay • Gas leak from exhaust system
DTC: P0158 **2T CCM, MIL: Yes** **Years:** 2007, 2008 **Models:** All **Engines:** All **Transmissions:** All	**Oxygen Sensor Circuit High Voltage (Bank 2 Sensor 2) Conditions:** Engine started, engine warmup completed, engine idling in closed loop, and the PCM detected the Oxygen Sensor signal was fixed at 400 mv or higher, or that it was fixed from 400-550 mv during the test. **Possible Causes:** • Open or short in HO2 Sensor (for Bank 2) circuit • HO2 Sensor (for Bank 2) • HO2 Sensor heater (for Bank 2) • Air-Fuel (A/F) Sensor (for Bank 2) • Integration relay • Gas leak from exhaust system
DTC: P0159 **2T CCM, MIL: Yes** **Years:** 2007, 2008 **Models:** All **Engines:** All **Transmissions:** All	**Oxygen Sensor Circuit Slow Response (Bank 2 Sensor 2) Conditions:** Engine started, engine warmup completed, engine idling in closed loop, and the PCM detected the Oxygen Sensor signal was fixed at 400 mv or higher, or that it was fixed from 400-550 mv during the test. **Possible Causes:** • Open or short in HO2 Sensor (for Bank 2) circuit • HO2 Sensor (for Bank 2) • HO2 Sensor heater (for Bank 2) • Air-Fuel (A/F) Sensor (for Bank 2) • Integration relay • Gas leak from exhaust system
DTC: P0161 **2T CCM, MIL: Yes** **Years:** 2007, 2008 **Models:** All **Engines:** All **Transmissions:** All	**Oxygen Sensor Heater Circuit Malfunction (Bank 2 Sensor 2) Conditions:** Engine started, engine warmup completed, engine idling in closed loop, and the PCM detected the Heated Oxygen Sensor signal was fixed at 400 mv or higher, or that it was fixed from 400-550 mv during the test. **Possible Causes:** • Open or short in HO2 Sensor heater circuit • HO2 Sensor heater (for Sensor 2) • Integration relay • ECM
DTC: P0171 **2T FUEL, MIL: Yes** **Years:** 2007, 2008 **Models:** All **Engines:** All **Transmissions:** All	**Fuel System Too Lean (Bank 1) Conditions:** DTC P0100, P0101, P0105, P0110, P0115, P0120, P0121, P0130, P0133, P0136, P0135, P0136, P0141, P0151, P0156, P0161, P0300, P0301-P0306, P0440, P0500 and P0505 not set, vehicle speed less than 62 MPH with the engine speed over 1500 RPM, ECT Sensor more than 158°F, and the PCM detected the lean fuel trim correction value was over the limit. **Possible Causes:** • A/FS or Heated Oxygen Sensor is contaminated, deteriorated or it has failed • Air leaks after the MAF Sensor, or in the EGR or PCV system • Base engine mechanical fault affecting one or more cylinders • Exhaust leaks located in front of the A/FS or Heated Oxygen Sensor location • Fuel system supplying too little fuel during cruise or idle (faulty fuel pump or fuel filter) • Fuel injector (one or more) dirty or pressure regulator has failed • Vehicle driven low on fuel or until it ran out of fuel

DTC	Trouble Code Title, Conditions & Possible Causes
DTC: P0172 **2T FUEL, MIL: Yes** **Years:** 2007, 2008 **Models:** All **Engines:** All **Transmissions:** All	**Fuel System Too Rich (Bank 1) Conditions:** DTC P0100, P0101, P0105, P0110, P0115, P0120, P0121, P0130, P0133, P0136, P0135, P0136, P0141, P0151, P0156, P0161, P0300, P0301-P0306, P0440, P0500 and P0505 not set, ECT Sensor more than 158°F, vehicle driven at a constant speed of less than 62 MPH with the engine speed over 1500 RPM, and the PCM detected the rich fuel trim correction value was over the limit. **Possible Causes:** • A/FS or Heated Oxygen Sensor is contaminated, deteriorated or it has failed • Base engine mechanical fault affecting one or more cylinders • EVAP system component has failed or canister fuel saturated • Exhaust leaks located in front of the A/FS or Heated Oxygen Sensor location • Fuel control Sensor is out of calibration (i.e., ECT, IAT or MAF) • Fuel delivery system supplying too much fuel during cruise or idle periods (e.g., faulty fuel pump, or faulty pressure regulator) • Fuel injector(s) is leaking or stuck partially open (one or more)
DTC: P0174 **2T FUEL, MIL: Yes** **Years:** 2007, 2008 **Models:** All **Engines:** All **Transmissions:** All	**Fuel System Too Lean (Bank 2) Conditions:** DTC P0100, P0101, P0105, P0110, P0115, P0120, P0121, P0136, P0141, P0151, P0156, P0161, P0300, P0301-P0306, P0440, P0500, P0505, P1130, P1133, P1135, 1150, 53 and P1155 not set, ECT Sensor more than 158°F, vehicle driven at a constant speed of less than 62 MPH with the engine speed over 1500 RPM, and the PCM detected the lean fuel trim correction value was over the limit. **Possible Causes:** • A/FS or Heated Oxygen Sensor is contaminated, deteriorated or it has failed • Air leaks after the MAF Sensor, or in the EGR or PCV system • Base engine mechanical fault affecting one or more cylinders • Exhaust leaks located in front of the A/FS or Heated Oxygen Sensor location • Fuel control Sensor is out of calibration (i.e., ECT, IAT or MAF) • Fuel delivery system supplying too little fuel during cruise or idle periods (e.g., faulty fuel pump or dirty, restricted fuel filter) • Fuel injector (one or more) dirty or pressure regulator has failed • MAF Sensor is contaminated, out-of-calibration or damaged • Vehicle driven low on fuel or until it ran out of fuel
DTC: P0175 **2T FUEL, MIL: Yes** **Years:** 2007, 2008 **Models:** All **Engines:** All **Transmissions:** All	**Fuel System Too Rich (Bank 2) Conditions:** DTC P0100, P0101, P0105, P0110, P0115, P0120, P0121, P0136, P0141, P0151, P0156, P0161, P0300, P0301-P0306, P0440, P0500, P0505, P1130, P1133, P1135, 1150, 53 and P1155 not set, ECT Sensor more than 158°F, vehicle driven at a constant speed of less than 62 MPH with the engine speed over 1500 RPM, and the PCM detected the rich fuel trim correction value was over the limit. **Possible Causes:** • A/FS or Heated Oxygen Sensor is contaminated, deteriorated or it has failed • Base engine mechanical fault affecting one or more cylinders • EVAP system component has failed or canister fuel saturated • Exhaust leaks located in front of the A/FS or Heated Oxygen Sensor location • Fuel control Sensor is out of calibration (i.e., ECT, IAT or MAF) • Fuel delivery system supplying too much fuel during cruise or idle periods (e.g., faulty fuel pump, or faulty pressure regulator) • Fuel injector(s) is leaking or stuck partially open (one or more)
DTC: P0201 **MIL: Yes** **Years:** 2007, 2008 **Models:** All **Engines:** All **Transmissions:** All	**Injector Circuit/Open - (Cylinder 1) Conditions:** Each EDU receives fuel injection request signals from the ECM and converts the signals to high voltage / high current injector operation signals to operate the fuel injectors (for direction injection). The fuel injection sequence occurs in the following order: No.1, No. 8, No. 7, No. 3, No. 6, No. 5, No. 4, No. 2. The ECM monitors each EDU at all times. If drivers or injectors are malfunctioning, the EDU sends fuel injector operation condition fail signals (INJ1 to INJ4) to the ECM. **Possible Causes:** Either of following are detected (1 trip detection logic): • INJ1 signal (IJ1) is not input for 20 consecutive revolutions • INJ1 signal (IJ6) is not input for 20 consecutive revolutions, and INJ1 signal (IJ1) is not input for 15 consecutive revolutions

DTC	Trouble Code Title, Conditions & Possible Causes
DTC: P0202 **MIL: Yes** **Years:** 2007, 2008 **Models:** All **Engines:** All **Transmissions:** All	**Injector Circuit/Open - (Cylinder 2) Conditions:** Each EDU receives fuel injection request signals from the ECM and converts the signals to high voltage / high current injector operation signals to operate the fuel injectors (for direction injection). The fuel injection sequence occurs in the following order: No.1, No. 8, No. 7, No. 3, No. 6, No. 5, No. 4, No. 2. The ECM monitors each EDU at all times. If drivers or injectors are malfunctioning, the EDU sends fuel injector operation condition fail signals (INJ1 to INJ4) to the ECM. **Possible Causes:** Either of following are detected (1 trip detection logic): • INJ4 signal (IJ2) is not input for 20 consecutive revolutions • INJ4 signal (IJ3) is not input for 20 consecutive revolutions, and INJ4 signal (IJ2) is not input for 15 consecutive revolutions
DTC: P0203 **MIL: Yes** **Years:** 2007, 2008 **Models:** All **Engines:** All **Transmissions:** All	**Injector Circuit/Open - (Cylinder 3) Conditions:** Each EDU receives fuel injection request signals from the ECM and converts the signals to high voltage / high current injector operation signals to operate the fuel injectors (for direction injection). The fuel injection sequence occurs in the following order: No.1, No. 8, No. 7, No. 3, No. 6, No. 5, No. 4, No. 2. The ECM monitors each EDU at all times. If drivers or injectors are malfunctioning, the EDU sends fuel injector operation condition fail signals (INJ1 to INJ4) to the ECM. **Possible Causes:** Either of following are detected (1 trip detection logic): • INJ4 signal (IJ3) is not input for 20 consecutive revolutions • INJ4 signal (IJ2) is not input for 20 consecutive revolutions, and INJ4 signal (IJ2) is not input for 15 consecutive revolutions
DTC: P0204 **MIL: Yes** **Years:** 2007, 2008 **Models:** All **Engines:** All **Transmissions:** All	**Injector Circuit/Open - (Cylinder 4) Conditions:** Each EDU receives fuel injection request signals from the ECM and converts the signals to high voltage / high current injector operation signals to operate the fuel injectors (for direction injection). The fuel injection sequence occurs in the following order: No.1, No. 8, No. 7, No. 3, No. 6, No. 5, No. 4, No. 2. The ECM monitors each EDU at all times. If drivers or injectors are malfunctioning, the EDU sends fuel injector operation condition fail signals (INJ1 to INJ4) to the ECM. **Possible Causes:** Either of following are detected (1 trip detection logic): • INJ2 signal (IJ4) is not input for 20 consecutive revolutions • INJ2 signal (IJ7) is not input for 20 consecutive revolutions, and INJ2 signal (IJ4) is not input for 15 consecutive revolutions
DTC: P0205 **MIL: Yes** **Years:** 2007, 2008 **Models:** All **Engines:** All V6, V8 **Transmissions:** All	**Injector Circuit/Open - (Cylinder 5) Conditions:** Each EDU receives fuel injection request signals from the ECM and converts the signals to high voltage / high current injector operation signals to operate the fuel injectors (for direction injection). The fuel injection sequence occurs in the following order: No.1, No. 8, No. 7, No. 3, No. 6, No. 5, No. 4, No. 2. The ECM monitors each EDU at all times. If drivers or injectors are malfunctioning, the EDU sends fuel injector operation condition fail signals (INJ1 to INJ4) to the ECM. **Possible Causes:** Either of following are detected (1 trip detection logic): • INJ3 signal (IJ5) is not input for 20 consecutive revolutions • INJ3 signal (IJ8) is not input for 20 consecutive revolutions, and INJ3 signal (IJ5) is not input for 15 consecutive revolutions
DTC: P0206 **MIL: Yes** **Years:** 2007, 2008 **Models:** All **Engines:** All V6, V8 **Transmissions:** All	**Injector Circuit/Open - (Cylinder 6) Conditions:** Each EDU receives fuel injection request signals from the ECM and converts the signals to high voltage / high current injector operation signals to operate the fuel injectors (for direction injection). The fuel injection sequence occurs in the following order: No.1, No. 8, No. 7, No. 3, No. 6, No. 5, No. 4, No. 2. The ECM monitors each EDU at all times. If drivers or injectors are malfunctioning, the EDU sends fuel injector operation condition fail signals (INJ1 to INJ4) to the ECM. **Possible Causes:** Either of following are detected (1 trip detection logic): • INJ1 signal (IJ6) is not input for 20 consecutive revolutions • INJ1 signal (IJ1) is not input for 20 consecutive revolutions, and INJ1 signal (IJ6) is not input for 15 consecutive revolutions
DTC: P0207 **MIL: Yes** **Years:** 2007, 2008 **Models:** All **Engines:** All V8 **Transmissions:** All	**Injector Circuit/Open - (Cylinder 7) Conditions:** Each EDU receives fuel injection request signals from the ECM and converts the signals to high voltage / high current injector operation signals to operate the fuel injectors (for direction injection). The fuel injection sequence occurs in the following order: No.1, No. 8, No. 7, No. 3, No. 6, No. 5, No. 4, No. 2. The ECM monitors each EDU at all times. If drivers or injectors are malfunctioning, the EDU sends fuel injector operation condition fail signals (INJ1 to INJ4) to the ECM. **Possible Causes:** Either of following are detected (1 trip detection logic): • INJ2 signal (IJ7) is not input for 20 consecutive revolutions • INJ2 signal (IJ4) is not input for 20 consecutive revolutions, and INJ2 signal (IJ7) is not input for 15 consecutive revolutions

DTC	Trouble Code Title, Conditions & Possible Causes
DTC: P0208 **MIL: Yes** **Years:** 2007, 2008 **Models:** All **Engines:** All V8 **Transmissions:** All	**Injector Circuit/Open - (Cylinder 8) Conditions:** Each EDU receives fuel injection request signals from the ECM and converts the signals to high voltage / high current injector operation signals to operate the fuel injectors (for direction injection). The fuel injection sequence occurs in the following order: No.1, No. 8, No. 7, No. 3, No. 6, No. 5, No. 4, No. 2. The ECM monitors each EDU at all times. If drivers or injectors are malfunctioning, the EDU sends fuel injector operation condition fail signals (INJ1 to INJ4) to the ECM. **Possible Causes:** Either of following are detected (1 trip detection logic): • INJ3 signal (IJ8) is not input for 20 consecutive revolutions • INJ3 signal (IJ5) is not input for 20 consecutive revolutions, and INJ3 signal (IJ8) is not input for 15 consecutive revolutions
DTC: P0220 **2T CCM, MIL: Yes** **Years:** 2007, 2008 **Models:** All **Engines:** All **Transmissions:** All	**Throttle Pedal Position Sensor/Switch 'B' Circuit Malfunction Conditions:** Key on or engine running; and the PCM detected the TP Sensor 'B' Signal indicated less than 0.50v with the throttle position closed (the switch open), or the TP Sensor signal indicated more than 4.9v at any time. The Electric TP Sensor is mounted on the throttle body. It has two Sensors (the electrical throttle system does not use a cable). **Possible Causes:** • TP Sensor signal circuit open or shorted to ground • TP Sensor ground circuit open • TP Sensor power circuit open (check VREF circuit at PCM) • TP Sensor is damaged or has failed • PCM failure
DTC: P0222 **2T CCM, MIL: Yes** **Years:** 2007, 2008 **Models:** All **Engines:** All **Transmissions:** All	**Throttle Pedal Position Sensor Switch B Circuit Low Input Conditions:** Key on or engine running; and the PCM detected an unexpected high voltage condition (more than 4.97v) on the VTA2 circuit. The Electric TP Sensor is mounted on the throttle body. It has two Sensors (the electrical throttle system does not use a cable). **Possible Causes:** • Electric TP Sensor connector is damaged or open • Electric TP Sensor circuit open or shorted to VREF • Electric TP Sensor is damaged or has failed • ECM
DTC: P0223 **2T CCM, MIL: Yes** **Years:** 2007, 2008 **Models:** All **Engines:** All **Transmissions:** All	**Throttle Pedal Position Sensor Switch B Circuit High Input Conditions:** Key on or engine running; and the PCM detected an unexpected high voltage condition (more than 4.97v) on the VTA2 circuit. The Electric TP Sensor is mounted on the throttle body. It has two Sensors (the electrical throttle system does not use a cable). **Possible Causes:** • Electric TP Sensor connector is damaged or open • Electric TP Sensor circuit open or shorted to VREF • Electric TP Sensor is damaged or has failed • ECM
DTC: P0230 **2T CCM, MIL: Yes** **Years:** 2007, 2008 **Models:** All **Engines:** All **Transmissions:** All	**Fuel Pump Primary Circuit Malfunction Conditions:** Engine started; and the PCM detected an unexpected voltage on the Fuel Pump Primary control circuit (from ST terminal to Starter Relay coil and to the STA terminal of the PCM). **Possible Causes:** • Circuit opening relay is damaged or has failed • Fuel pump relay control circuit open or shorted to ground • Fuel pump relay is damaged or has failed • Fuel pump is damaged or has failed • ECM
DTC: P0300 **2T MISFIRE, MIL: Yes** **Years:** 2007, 2008 **Models:** All **Engines:** All **Transmissions:** All	**Multiple Cylinder Misfire Detected Conditions:** **Trouble Code Conditions** DTC P0100, P0101, P0102, P0103, P0105, P0110, P0112, P0113, P0115, P0117, P0118, P0120, P0121, P0122, P0123, P0125, P0335, P0340, P0500, P0505 and P0510 not set, engine started, vehicle driven to a speed of over 3 MPH for 1 minute, and the PCM detected a misfire rate of 1-2% (High Emissions 2T), or a misfire rate of 6-30% (Catalyst Damaging 1T) in two or more cylinders. **Note: If the misfire is severe, the MIL will flash on/off on the 1st trip. Look at the misfire ratio for all of the cylinders on the Scan Tool. The cylinder with the highest misfire ratio should be checked first.** **Possible Causes:** • Air leak in the intake manifold, or in the EGR or PCV system • Base engine mechanical fault that affects two or more cylinders • EGR valve is stuck open or the PCV system has a vacuum leak • Fuel delivery component fault that affects two or more cylinders (e.g., contaminated, dirty or sticking fuel injectors) • Ignition system fault (coil or plug) that affects several cylinders • Mass airflow meter is contaminated, or its signal is out of range

DTC	Trouble Code Title, Conditions & Possible Causes
DTC: P0301 **2T MISFIRE, MIL: Yes** **Years:** 2007, 2008 **Models:** All **Engines:** All **Transmissions:** All	**Cylinder 1 Misfire Detected Conditions:** DTC P0100, P0101, P0115, P0116, P0120, P0121, P0335, P0340 and P0500 not set, engine running, vehicle speed over 3 MPH, and the PCM detected a misfire condition in Cylinder 1 in the 200 (Catalyst) or 1000-RPM (High Emissions) revolution range. **Note: If the misfire is severe, the MIL will flash on/off on the 1st trip.** **Possible Causes:** • Air leak in the intake manifold, or in the EGR or PCV system • Base engine mechanical fault that affects only one cylinder • Fuel delivery component fault that affects only one cylinder (e.g., a contaminated, dirty or sticking fuel injector) • Ignition system problem (coil or plug) that affects one cylinder • TSB EG006-02 (1/02) contains information related to this code
DTC: P0302 **2T MISFIRE, MIL: Yes** **Years:** 2007, 2008 **Models:** All **Engines:** All **Transmissions:** All	**Cylinder 2 Misfire Detected Conditions:** DTC P0100, P0101, P0115, P0116, P0120, P0121, P0335, P0340 and P0500 not set, engine running, vehicle speed over 3 MPH, and the PCM detected a misfire condition in Cylinder 2 in the 200 (Catalyst) or 1000-RPM (High Emissions) revolution range. **Note: If the misfire is severe, the MIL will flash on/off on the 1st trip.** **Possible Causes:** • Air leak in the intake manifold, or in the EGR or PCV system • Base engine mechanical fault that affects only one cylinder • Fuel delivery component fault that affects only one cylinder (e.g., a contaminated, dirty or sticking fuel injector) • Ignition system problem (coil or plug) that affects one cylinder • TSB EG006-02 (1/02) contains information related to this code
DTC: P0303 **2T MISFIRE, MIL: Yes** **Years:** 2007, 2008 **Models:** All **Engines:** All **Transmissions:** All	**Cylinder 3 Misfire Detected Conditions:** DTC P0100, P0101, P0115, P0116, P0120, P0121, P0335, P0340 and P0500 not set, engine running, vehicle speed over 3 MPH, and the PCM detected a misfire condition in Cylinder 3 in the 200 (Catalyst) or 1000-RPM (High Emissions) revolution range. **Note: If the misfire is severe, the MIL will flash on/off on the 1st trip.** **Possible Causes:** • Air leak in the intake manifold, or in the EGR or PCV system • Base engine mechanical fault that affects only one cylinder • Fuel delivery component fault that affects only one cylinder (e.g., a contaminated, dirty or sticking fuel injector) • Ignition system problem (coil or plug) that affects one cylinder • TSB EG006-02 (1/02) contains information related to this code
DTC: P0304 **2T MISFIRE, MIL: Yes** **Years:** 2007, 2008 **Models:** All **Engines:** All **Transmissions:** All	**Cylinder 4 Misfire Detected Conditions:** DTC P0100, P0101, P0115, P0116, P0120, P0121, P0335, P0340 and P0500 not set, engine running, vehicle speed over 3 MPH, and the PCM detected a misfire condition in Cylinder 4 in the 200 (Catalyst) or 1000-RPM (High Emissions) revolution range. **Note: If the misfire is severe, the MIL will flash on/off on the 1st trip.** **Possible Causes:** • Air leak in the intake manifold, or in the EGR or PCV system • Base engine mechanical fault that affects only one cylinder • Fuel delivery component fault that affects only one cylinder (e.g., a contaminated, dirty or sticking fuel injector) • Ignition system problem (coil or plug) that affects one cylinder • TSB EG006-02 (1/02) contains information related to this code
DTC: P0305 **2T MISFIRE, MIL: Yes** **Years:** 2007, 2008 **Models:** All **Engines:** All V6, V8 **Transmissions:** All	**Cylinder 5 Misfire Detected Conditions:** DTC P0100, P0101, P0105, P0110, P0115, P0120, P0121, P0335, P0340 and P0500 not set, engine running, vehicle speed over 3 MPH, and the PCM detected a misfire condition in Cylinder 5 in the 200 (Catalyst) or 1000-RPM (High Emissions) revolution range. **Note: If the misfire is severe, the MIL will flash on/off on the 1st trip.** **Possible Causes:** • Air leak in the intake manifold, or in the EGR or PCV system • Base engine mechanical fault that affects only one cylinder • Fuel delivery component fault that affects only one cylinder (e.g., a contaminated, dirty or sticking fuel injector) • Ignition system problem (coil or plug) that affects one cylinder
DTC: P0306 **2T MISFIRE, MIL: Yes** **Years:** 2007, 2008 **Models:** All **Engines:** All V6, V8 **Transmissions:** All	**Cylinder 6 Misfire Detected Conditions:** DTC P0100, P0101, P0102, P0103, P0105, P0110, P0112, P0113, P0115, P0117, P0118, P0120, P0121, P0122, P0123, P0125, P0335, P0340, P0500, P0505 and P0510 not set, engine started, vehicle driven to a speed of over 3 MPH for 1 minute, and the PCM detected a misfire rate of 1-2% (High Emissions 2T), or a misfire rate of 6-30% (Catalyst Damaging 1T) in Cylinder 6. **Note: If the misfire is severe, the MIL will flash on/off on the 1st trip.** **Possible Causes:** • Base engine mechanical fault that affects only Cylinder 6 • EGR valve is stuck open or the PCV system has a vacuum leak • Fuel delivery component fault that affects only Cylinder 6 (e.g., a contaminated, dirty or sticking fuel injector) • Ignition system problem (coil or plug) that affects Cylinder 6 • Mass airflow meter is contaminated, or its signal is out of range

DTC	Trouble Code Title, Conditions & Possible Causes
DTC: P0307 **2T MISFIRE, MIL: Yes** **Years:** 2007, 2008 **Models:** All **Engines:** All V8 **Transmissions:** All	**Cylinder 7 Misfire Detected Conditions:** **Trouble Code Conditions** DTC P0100, P0101, P0102, P0103, P0105, P0110, P0112, P0113, P0115, P0117, P0118, P0120, P0121, P0122, P0123, P0125, P0335, P0340, P0500, P0505 and P0510 not set, engine started, vehicle driven to a speed of over 3 MPH for 1 minute, and the PCM detected a misfire rate of 1-2% (High Emissions 2T), or a misfire rate of 6-30% (Catalyst Damaging 1T) in Cylinder 7. **Note: If the misfire is severe, the MIL will flash on/off on the 1st trip.** **Possible Causes:** • Base engine mechanical fault that affects only Cylinder 7 • EGR valve is stuck open or the PCV system has a vacuum leak • Fuel delivery component fault that affects only Cylinder 7 (e.g., a contaminated, dirty or sticking fuel injector) • Ignition system problem (coil or plug) that affects Cylinder 7 • Mass airflow meter is contaminated, or its signal is out of range
DTC: P0308 **2T MISFIRE, MIL: Yes** **Years:** 2007, 2008 **Models:** All **Engines:** All V8 **Transmissions:** All	**Cylinder 8 Misfire Detected Conditions:** **Trouble Code Conditions** DTC P0100, P0101, P0102, P0103, P0105, P0110, P0112, P0113, P0115, P0117, P0118, P0120, P0121, P0122, P0123, P0125, P0335, P0340, P0500, P0505 and P0510 not set, engine started, vehicle driven to a speed of over 3 MPH for 1 minute, and the PCM detected a misfire rate of 1-2% (High Emissions 2T), or a misfire rate of 6-30% (Catalyst Damaging 1T) in Cylinder 8. **Note: If the misfire is severe, the MIL will flash on/off on the 1st trip.** **Possible Causes:** • Base engine mechanical fault that affects only Cylinder 8 • EGR valve is stuck open or the PCV system has a vacuum leak • Fuel delivery component fault that affects only Cylinder 8 (e.g., a contaminated, dirty or sticking fuel injector) • Ignition system problem (coil or plug) that affects Cylinder 8 • Mass airflow meter is contaminated, or its signal is out of range
DTC: P0327 **1T CCM, MIL: Yes** **Years:** 2007, 2008 **Models:** All **Engines:** All **Transmissions:** All	**Knock Sensor 1 Circuit Low Input (Bank 1) Conditions:** Engine started, engine speed from 1500-5500 RPM, and the PCM detected an unexpected low voltage on the Knock Sensor 1 circuit. **Possible Causes:** • Verify that the Knock Sensor (KS) is tightened to specification • Knock Sensor signal circuit is shorted to ground • Knock Sensor is damaged or has failed • PCM failure
DTC: P0328 **1T CCM, MIL: Yes** **Years:** 2007, 2008 **Models:** All **Engines:** All **Transmissions:** All	**Knock Sensor 1 Circuit High Input (Bank 1) Conditions:** Engine started, engine speed from 1500-5500 RPM, and the PCM detected an unexpected high voltage on the Knock Sensor 1 circuit. **Possible Causes:** • Verify that the Knock Sensor (KS) is tightened to specification • Knock Sensor signal circuit open or shorted to power • Knock Sensor is damaged or has failed • PCM failure
DTC: P032C **1T CCM, MIL: Yes** **Years:** 2007, 2008 **Models:** All **Engines:** All **Transmissions:** All	**Knock Sensor 3 Circuit Low Conditions:** Engine started, engine speed from 1500-5500 RPM, and the PCM detected an unexpected low voltage on the Knock Sensor 1 circuit. **Possible Causes:** • Short in knock Sensor (for Bank 2 Sensor 2) circuit • Knock Sensor (for Bank 1 Sensor 2) • ECM
DTC: P032D **1T CCM, MIL: Yes** **Years:** 2007, 2008 **Models:** All **Engines:** All **Transmissions:** All	**Knock Sensor 3 Circuit High Conditions:** Engine started, engine speed from 1500-5500 RPM, and the PCM detected an unexpected high voltage on the Knock Sensor 1 circuit. **Possible Causes:** • Open in knock Sensor (for Bank 2 Sensor 2) circuit • Knock Sensor (for Bank 1 Sensor 2) • ECM
DTC: P0332 **1T CCM, MIL: Yes** **Years:** 2007, 2008 **Models:** All **Engines:** All **Transmissions:** All	**Knock Sensor 2 Circuit Low Input (Bank 2) Conditions:** Engine started, engine speed from 1500-5500 RPM, and the PCM detected an unexpected low voltage on the Knock Sensor 2 circuit. **Possible Causes:** • Verify that the Knock Sensor (KS) is tightened to specification • Knock Sensor signal circuit is shorted to ground • Knock Sensor had failed, or the PCM failure

DTC	Trouble Code Title, Conditions & Possible Causes
DTC: P0333 **1T CCM, MIL: Yes** **Years:** 2007, 2008 **Models:** All **Engines:** All **Transmissions:** All	**Knock Sensor 2 Circuit High Input (Bank 2) Conditions:** Engine speed from 1500-5500 RPM, and the PCM detected an unexpected high voltage on the Knock Sensor 2 circuit. **Possible Causes:** • Verify that the Knock Sensor (KS) is tightened to specification • Knock Sensor signal circuit open or shorted to power • Knock Sensor has failed, or the PCM failure
DTC: P0335 **1T CCM, MIL: Yes** **Years:** 2007, 2008 **Models:** All **Engines:** All **Transmissions:** All	**Crankshaft Position Sensor 'A' Circuit Malfunction Conditions:** Engine cranking; and the PCM did not detect any CKP Sensor 'A' signals, or with the engine speed over 600 RPM, it did not receive any CKP Sensor signals, or the CKP signal was lost. **Possible Causes:** • CKP Sensor 'A' signal circuit open, shorted to ground or shorted to system power • CKP Sensor 'A' signal ground circuit open • CKP Sensor 'A' is damaged or has failed
DTC: P0337 **2T CCM, MIL: Yes** **Years:** 2007, 2008 **Models:** All **Engines:** All **Transmissions:** All	**Crankshaft Position Sensor 'A' Circuit Low Input Conditions:** Engine running at idle or cruise speed for one minute, and the PCM detected a variation between the CKP Sensor and the CMP Sensor signals. **Possible Causes:** • Open or short in CKP Sensor circuit • CKP Sensor • CKP Sensor plate • ECM
DTC: P0338 **2T CCM, MIL: Yes** **Years:** 2007, 2008 **Models:** All **Engines:** All **Transmissions:** All	**Crankshaft Position Sensor 'A' Circuit High Input Conditions:** Engine running at idle or cruise speed for one minute, and the PCM detected a variation between the CKP Sensor and the CMP Sensor signals. **Possible Causes:** • Open or short in CKP Sensor circuit • CKP Sensor • CKP Sensor plate • ECM
DTC: P0339 **2T CCM, MIL: Yes** **Years:** 2007, 2008 **Models:** All **Engines:** All **Transmissions:** All	**Crankshaft Position Sensor 'A' Circuit Intermittent Conditions:** Engine started, STA signal indicating "off", engine runtime over 3 seconds since STA switched from "on" to "off", engine speed over 1000 RPM, and the PCM did not detect any CKP Sensor 'A' signals for 500 ms. The crankshaft position (NE) Sensor consists of a magnet, iron core and pickup coil. The NE Sensor signal plate, which has 34 teeth, installed on the crankshaft-timing pulley. This Sensor generates 34 signals for each engine revolution. The PCM detects the crankshaft angle and engine speed based on the NE signal. **Possible Causes:** • CKP Sensor signal circuit open, shorted to ground or power • CKP Sensor signal ground circuit open • Crankshaft timing pulley is damaged or out of alignment • CKP Sensor has failed, or the PCM failure
DTC: P033C **2T CCM, MIL: Yes** **Years:** 2007, 2008 **Models:** All **Engines:** All **Transmissions:** All	**Knock Sensor 4 Circuit Low Input Conditions:** Engine started, STA signal indicating "off", engine runtime over 3 seconds since STA switched from "on" to "off", engine speed over 1000 RPM, and the PCM did not detect any CKP Sensor 'A' signals for 500 ms. The crankshaft position (NE) Sensor consists of a magnet, iron core and pickup coil. The NE Sensor signal plate, which has 34 teeth, installed on the crankshaft-timing pulley. This Sensor generates 34 signals for each engine revolution. The PCM detects the crankshaft angle and engine speed based on the NE signal. **Possible Causes:** • Short in knock Sensor (for Bank 2 Sensor 2) circuit • Know Sensor (for Bank 2 Sensor 2) • ECM
DTC: P033D **2T CCM, MIL: Yes** **Years:** 2007, 2008 **Models:** All **Engines:** All **Transmissions:** All	**Knock Sensor 4 Circuit High Input Conditions:** Engine started, STA signal indicating "off", engine runtime over 3 seconds since STA switched from "on" to "off", engine speed over 1000 RPM, and the PCM did not detect any CKP Sensor 'A' signals for 500 ms. The crankshaft position (NE) Sensor consists of a magnet, iron core and pickup coil. The NE Sensor signal plate, which has 34 teeth, installed on the crankshaft-timing pulley. This Sensor generates 34 signals for each engine revolution. The PCM detects the crankshaft angle and engine speed based on the NE signal. **Possible Causes:** • Open in knock Sensor (for Bank 2 Sensor 2) circuit • Know Sensor (for Bank 2 Sensor 2) • ECM

DTC	Trouble Code Title, Conditions & Possible Causes
DTC: P0340 **2T CCM, MIL: Yes** **Years:** 2007, 2008 **Models:** All **Engines:** All **Transmissions:** All	**Camshaft Position Sensor Circuit Malfunction Conditions:** Engine cranking; and the PCM did not detect any CMP Sensor signals, or with the engine speed over 600 RPM, it did not detect any CMP signals, or the CMP signal was interrupted. **Possible Causes:** • CMP Sensor signal circuit open, shorted to ground or power • CMP Sensor signal ground circuit open • CMP Sensor has failed, or the PCM failure
DTC: P0340 **2T CCM, MIL: Yes** **Years:** 2007, 2008 **Models:** All **Engines:** All **Transmissions:** All	**Camshaft Position Sensor 'A' Circuit (Bank 1 or Single Sensor) Conditions:** Engine cranking; and the PCM did not detect any CMP Sensor signals, or with the engine speed over 600 RPM, it did not detect any CMP signals, or the CMP signal was interrupted. **Possible Causes:** • Open or short in CMP Sensor circuit • CMP Sensor • Camshaft • Jumped tooth of timing chain • ECM
DTC: P0342 **2T CCM, MIL: Yes** **Years:** 2007, 2008 **Models:** All **Engines:** All **Transmissions:** All	**Camshaft Position Sensor 'A' Circuit Low Input Conditions:** Engine cranking; and the output voltage of VVT Sensor is less than 0.3 volt for 4 seconds **Possible Causes:** • Open or short in VVT Sensor circuit for intake camshaft • VVT Sensor for intake camshaft • Camshaft timing gear for intake camshaft • Jumped tooth of timing chain for intake camshaft • ECM
DTC: P0343 **2T CCM, MIL: Yes** **Years:** 2007, 2008 **Models:** All **Engines:** All **Transmissions:** All	**Camshaft Position Sensor 'A' Circuit High Input Conditions:** Engine cranking; and the output voltage of VVT Sensor is more than 4.7 volts for 4 seconds. **Possible Causes:** • Open or short in VVT Sensor circuit for intake camshaft • VVT Sensor for intake camshaft • Camshaft timing gear for intake camshaft • Jumped tooth of timing chain for intake camshaft • ECM
DTC: P0345 **2T CCM, MIL: Yes** **Years:** 2007, 2008 **Models:** All **Engines:** All **Transmissions:** All	**Camshaft Position Sensor 'A' Circuit (Bank 2) Conditions:** Engine cranking; and the PCM detected twelve (12) or more CMP Sensor 'A' (Bank 2) signals during the test. The Left Hand VVT Camshaft Position Sensor consists of a magnet, and a circuit board in which a Magnetic Resistive (MR) device is mounted. The VVT signal plate includes three (3) protrusions on its outer surface. **Possible Causes:** • VVT Sensor signal circuit open, shorted to ground or power • VVT Sensor pulley is damaged/damaged timing chain or gear • VVT Sensor has failed, or the PCM failure
DTC: P0347 **2T CCM, MIL: Yes** **Years:** 2007, 2008 **Models:** All **Engines:** All **Transmissions:** All	**Camshaft Position Sensor 'A' Circuit Low Input (Bank 2) Conditions:** Engine cranking; and the output voltage of VVT Sensor is less than 0.3 volt for 4 seconds **Possible Causes:** • Open or short in VVT Sensor circuit for intake camshaft • VVT Sensor for intake camshaft • Camshaft timing gear for intake camshaft • Jumped tooth of timing chain for intake camshaft • ECM
DTC: P0348 **2T CCM, MIL: Yes** **Years:** 2007, 2008 **Models:** All **Engines:** All **Transmissions:** All	**Camshaft Position Sensor 'A' Circuit High Input (Bank 2) Conditions:** Engine cranking; and the output voltage of VVT Sensor is more than 4.7 volts for 4 seconds. **Possible Causes:** • Open or short in VVT Sensor circuit for intake camshaft • VVT Sensor for intake camshaft • Camshaft timing gear for intake camshaft • Jumped tooth of timing chain for intake camshaft • ECM

DTC	Trouble Code Title, Conditions & Possible Causes
DTC: P0351 **1T CCM, MIL: Yes** **Years:** 2007, 2008 **Models:** All **Engines:** All **Transmissions:** All	**Ignition Coil No. 1 Primary/Secondary Circuit Malfunction Conditions:** Engine started, and the PCM did not detect a change in the IGF signal on the Ignition Coil No. 1 IGF circuit. This engine uses a Direct Ignition (DI) system where one coil is used to fire one cylinder. The coil high-energy secondary wire is connected to one spark plug. If P0351 to P0356 are all set, check for an open/shorted IGF circuit. **Possible Causes:** • IGT1 circuit open or shorted to ground • Ignition Coil No. 1 is damaged or it has failed • Problem present in the Ignition System • PCM failure
DTC: P0351 **1T CCM, MIL: Yes** **Years:** 2007, 2008 **Models:** All **Engines:** All **Transmissions:** All	**Ignition Coil No. 'A' Primary/Secondary Circuit Malfunction Conditions:** Engine started, and the PCM did not detect a change in the IGF signal on the Ignition Coil No. 1 IGF circuit. This engine uses a Direct Ignition (DI) system where one coil is used to fire one cylinder. The coil high-energy secondary wire is connected to one spark plug. If P0351 to P0356 are all set, check for an open/shorted IGF circuit. **Possible Causes:** • IGT1 circuit open or shorted to ground • Ignition Coil No. 1 is damaged or it has failed • Problem present in the Ignition System • PCM failure
DTC: P0352 **1T CCM, MIL: Yes** **Years:** 2007, 2008 **Models:** All **Engines:** All **Transmissions:** All	**Ignition Coil No. 2 Primary/Secondary Circuit Malfunction Conditions:** Engine started, and the PCM did not detect a change in the IGF signal on the Ignition Coil No. 2 IGF circuit. This engine uses a Direct Ignition (DI) system where one coil is used to fire one cylinder. The coil high-energy secondary wire is connected to one spark plug. If P0351 to P0356 are all set, check for an open/shorted IGF circuit. **Possible Causes:** • IGT2 circuit open or shorted to ground • Ignition Coil No. 2 is damaged or it has failed • Problem present in the Ignition System • PCM failure
DTC: P0352 **1T CCM, MIL: Yes** **Years:** 2007, 2008 **Models:** All **Engines:** All **Transmissions:** All	**Ignition Coil No. 'B' Primary/Secondary Circuit Malfunction Conditions:** Engine started, and the PCM did not detect a change in the IGF signal on the Ignition Coil No. 2 IGF circuit. This engine uses a Direct Ignition (DI) system where one coil is used to fire one cylinder. The coil high-energy secondary wire is connected to one spark plug. If P0351 to P0356 are all set, check for an open/shorted IGF circuit. **Possible Causes:** • IGT2 circuit open or shorted to ground • Ignition Coil No. 2 is damaged or it has failed • Problem present in the Ignition System • PCM failure
DTC: P0353 **1T CCM, MIL: Yes** **Years:** 2007, 2008 **Models:** All **Engines:** All **Transmissions:** All	**Ignition Coil No. 3 Primary/Secondary Circuit Malfunction Conditions:** Engine started, and the PCM did not detect a change in the IGF signal on the Ignition Coil No. 3 IGF circuit. This engine uses a Direct Ignition (DI) system where one coil is used to fire one cylinder. The coil high-energy secondary wire is connected to one spark plug. If P0351 to P0356 are all set, check for an open/shorted IGF circuit. **Possible Causes:** • IGT3 circuit open or shorted to ground • Ignition Coil No. 3 is damaged or it has failed • Problem present in the Ignition System • PCM failure
DTC: P0353 **1T CCM, MIL: Yes** **Years:** 2007, 2008 **Models:** All **Engines:** All **Transmissions:** All	**Ignition Coil No. 'C' Primary/Secondary Circuit Malfunction Conditions:** Engine started, and the PCM did not detect a change in the IGF signal on the Ignition Coil No. 3 IGF circuit. This engine uses a Direct Ignition (DI) system where one coil is used to fire one cylinder. The coil high-energy secondary wire is connected to one spark plug. If P0351 to P0356 are all set, check for an open/shorted IGF circuit. **Possible Causes:** • IGT3 circuit open or shorted to ground • Ignition Coil No. 3 is damaged or it has failed • Problem present in the Ignition System • PCM failure

DTC	Trouble Code Title, Conditions & Possible Causes
DTC: P0354 **1T CCM, MIL: Yes** **Years:** 2007, 2008 **Models:** All **Engines:** All **Transmissions:** All	**Ignition Coil No. 4 Primary/Secondary Circuit Malfunction Conditions:** Engine started, and the PCM did not detect a change in the IGF signal on the Ignition Coil No. 4 IGF circuit. This engine uses a Direct Ignition (DI) system where one coil is used to fire one cylinder. The coil high-energy secondary wire is connected to one spark plug. If P0351 to P0356 are all set, check for an open/shorted IGF circuit. **Possible Causes:** • IGT4 circuit open or shorted to ground • Ignition Coil No. 4 is damaged or it has failed • Problem present in the Ignition System • PCM failure
DTC: P0354 **1T CCM, MIL: Yes** **Years:** 2007, 2008 **Models:** All **Engines:** All **Transmissions:** All	**Ignition Coil No. 'D' Primary/Secondary Circuit Malfunction Conditions:** Engine started, and the PCM did not detect a change in the IGF signal on the Ignition Coil No. 4 IGF circuit. This engine uses a Direct Ignition (DI) system where one coil is used to fire one cylinder. The coil high-energy secondary wire is connected to one spark plug. If P0351 to P0356 are all set, check for an open/shorted IGF circuit. **Possible Causes:** • IGT4 circuit open or shorted to ground • Ignition Coil No. 4 is damaged or it has failed • Problem present in the Ignition System • PCM failure
DTC: P0355 **1T CCM, MIL: Yes** **Years:** 2007, 2008 **Models:** All **Engines:** All **Transmissions:** All	**Ignition Coil No. 5 Primary/Secondary Circuit Malfunction Conditions:** Engine started, and the PCM did not detect a change in the IGF signal on the Ignition Coil No. 5 IGF circuit. This engine uses a Direct Ignition (DI) system where one coil is used to fire one cylinder. The coil high-energy secondary wire is connected to one spark plug. If P0351 to P0356 are all set, check for an open/shorted IGF circuit. **Possible Causes:** • IGT5 circuit open or shorted to ground • Ignition Coil No. 5 is damaged or it has failed • Problem present in the Ignition System • PCM failure
DTC: P0355 **1T CCM, MIL: Yes** **Years:** 2007, 2008 **Models:** All **Engines:** All **Transmissions:** All	**Ignition Coil No. 'E' Primary/Secondary Circuit Malfunction Conditions:** Engine started, and the PCM did not detect a change in the IGF signal on the Ignition Coil No. 5 IGF circuit. This engine uses a Direct Ignition (DI) system where one coil is used to fire one cylinder. The coil high-energy secondary wire is connected to one spark plug. If P0351 to P0356 are all set, check for an open/shorted IGF circuit. **Possible Causes:** • IGT5 circuit open or shorted to ground • Ignition Coil No. 5 is damaged or it has failed • Problem present in the Ignition System • PCM failure
DTC: P0356 **1T CCM, MIL: Yes** **Years:** 2007, 2008 **Models:** All **Engines:** All **Transmissions:** All	**Ignition Coil No. 6 Primary/Secondary Circuit Malfunction Conditions:** Engine started, and the PCM did not detect a change in the IGF signal on the Ignition Coil No. 6 IGF circuit. This engine uses a Direct Ignition (DI) system where one coil is used to fire one cylinder. The coil high-energy secondary wire is connected to one spark plug. If P0351 to P0356 are all set, check for an open/shorted IGF circuit. **Possible Causes:** • IGT6 circuit open or shorted to ground • Ignition Coil No. 6 is damaged or it has failed • Problem present in the Ignition System • PCM failure
DTC: P0356 **1T CCM, MIL: Yes** **Years:** 2007, 2008 **Models:** All **Engines:** All **Transmissions:** All	**Ignition Coil No. 'F' Primary/Secondary Circuit Malfunction Conditions:** Engine started, and the PCM did not detect a change in the IGF signal on the Ignition Coil No. 6 IGF circuit. This engine uses a Direct Ignition (DI) system where one coil is used to fire one cylinder. The coil high-energy secondary wire is connected to one spark plug. If P0351 to P0356 are all set, check for an open/shorted IGF circuit. **Possible Causes:** • IGT6 circuit open or shorted to ground • Ignition Coil No. 6 is damaged or it has failed • Problem present in the Ignition System • PCM failure

DTC	Trouble Code Title, Conditions & Possible Causes
DTC: P0357 **1T CCM, MIL: Yes** **Years:** 2007, 2008 **Models:** All **Engines:** All **Transmissions:** All	**Ignition Coil No. 7 Primary/Secondary Circuit Malfunction Conditions:** Engine started, and the PCM did not detect a change in the IGF signal on the Ignition Coil No. 7 IGF circuit. This engine uses a Direct Ignition (DI) system where one coil is used to fire one cylinder. The coil high-energy secondary wire is connected to one spark plug. If P0351 to P0358 are all set, check for an open/shorted IGF circuit. **Possible Causes:** • IGT7 circuit open or shorted to ground • Ignition Coil No. 7 is damaged or it has failed • Problem present in the Ignition System • PCM failure
DTC: P0357 **1T CCM, MIL: Yes** **Years:** 2007, 2008 **Models:** All **Engines:** All **Transmissions:** All	**Ignition Coil No. 'G' Primary/Secondary Circuit Malfunction Conditions:** Engine started, and the PCM did not detect a change in the IGF signal on the Ignition Coil No. 7 IGF circuit. This engine uses a Direct Ignition (DI) system where one coil is used to fire one cylinder. The coil high-energy secondary wire is connected to one spark plug. If P0351 to P0358 are all set, check for an open/shorted IGF circuit. **Possible Causes:** • IGT7 circuit open or shorted to ground • Ignition Coil No. 7 is damaged or it has failed • Problem present in the Ignition System • PCM failure
DTC: P0358 **1T CCM, MIL: Yes** **Years:** 2007, 2008 **Models:** All **Engines:** All **Transmissions:** All	**Ignition Coil No. 8 Primary/Secondary Circuit Malfunction Conditions:** Engine started, and the PCM did not detect a change in the IGF signal on the Ignition Coil No. 8 IGF circuit. This engine uses a Direct Ignition (DI) system where one coil is used to fire one cylinder. The coil high-energy secondary wire is connected to one spark plug. If P0351 to P0358 are all set, check for an open/shorted IGF circuit. **Possible Causes:** • IGT8 circuit open or shorted to ground • Ignition Coil No. 8 is damaged or it has failed • Problem present in the Ignition System • PCM failure
DTC: P0358 **1T CCM, MIL: Yes** **Years:** 2007, 2008 **Models:** All **Engines:** All **Transmissions:** All	**Ignition Coil No. 'H' Primary/Secondary Circuit Malfunction Conditions:** Engine started, and the PCM did not detect a change in the IGF signal on the Ignition Coil No. 8 IGF circuit. This engine uses a Direct Ignition (DI) system where one coil is used to fire one cylinder. The coil high-energy secondary wire is connected to one spark plug. If P0351 to P0358 are all set, check for an open/shorted IGF circuit. **Possible Causes:** • IGT8 circuit open or shorted to ground • Ignition Coil No. 8 is damaged or it has failed • Problem present in the Ignition System • PCM failure
DTC: P0365 **1T CCM, MIL: Yes** **Years:** 2007, 2008 **Models:** All **Engines:** All **Transmissions:** All	**Camshaft Position Sensor 'B' Circuit (Bank 1) Conditions:** Engine started, and the input voltage to ECM remains at less than 0.3 V, or more than 4.7 V for more than 4 seconds when 2 or more seconds have elapsed after turning ignition switch to ON. No VVT Sensor signal to ECM during cranking **Possible Causes:** • Open or short in VVT Sensor for exhaust camshaft circuit • VVT Sensor for exhaust camshaft • Exhaust camshaft • Jumped tooth or timing chain • ECM
DTC: P0367 **1T CCM, MIL: Yes** **Years:** 2007, 2008 **Models:** All **Engines:** All **Transmissions:** All	**Camshaft Position Sensor 'B' Circuit Low Input (Bank 1) Conditions:** Engine started, and the output voltage of VVT Sensor less than 0.3 V 4 seconds. **Possible Causes:** • Open or short in VVT Sensor for exhaust camshaft circuit • VVT Sensor for exhaust camshaft • Exhaust camshaft • Jumped tooth or timing chain • ECM

DTC	Trouble Code Title, Conditions & Possible Causes
DTC: P0368 **1T CCM, MIL: Yes** **Years:** 2007, 2008 **Models:** All **Engines:** All **Transmissions:** All	**Camshaft Position Sensor 'B' Circuit High Input (Bank 1) Conditions:** Engine started, and the Output voltage of VVT Sensor more than 4.7 V 4 seconds. **Possible Causes:** • Open or short in VVT Sensor for exhaust camshaft circuit • VVT Sensor for exhaust camshaft • Exhaust camshaft • Jumped tooth or timing chain • ECM
DTC: P0390 **1T CCM, MIL: Yes** **Years:** 2007, 2008 **Models:** All **Engines:** All **Transmissions:** All	**Camshaft Position Sensor 'B' Circuit (Bank 2) Conditions:** Engine started, and the input voltage to ECM remains at less than 0.3 V, or more than 4.7 V for more than 4 seconds when 2 or more seconds have elapsed after turning ignition switch to ON. No VVT Sensor signal to ECM during cranking **Possible Causes:** • Open or short in VVT Sensor for exhaust camshaft circuit • VVT Sensor for exhaust camshaft • Exhaust camshaft • Jumped tooth or timing chain • ECM
DTC: P0392 **1T CCM, MIL: Yes** **Years:** 2007, 2008 **Models:** All **Engines:** All **Transmissions:** All	**Camshaft Position Sensor 'B' Circuit Low Input (Bank 2) Conditions:** Engine started, and the output voltage of VVT Sensor less than 0.3 V 4 seconds. **Possible Causes:** • Open or short in VVT Sensor for exhaust camshaft circuit • VVT Sensor for exhaust camshaft • Exhaust camshaft • Jumped tooth or timing chain • ECM
DTC: P0393 **1T CCM, MIL: Yes** **Years:** 2007, 2008 **Models:** All **Engines:** All **Transmissions:** All	**Camshaft Position Sensor 'B' Circuit High Input (Bank 2) Conditions:** Engine started, and the Output voltage of VVT Sensor more than 4.7 V 4 seconds. **Possible Causes:** • Open or short in VVT Sensor for exhaust camshaft circuit • VVT Sensor for exhaust camshaft • Exhaust camshaft • Jumped tooth or timing chain • ECM
DTC: P0412 **1T CCM, MIL: Yes** **Years:** 2007, 2008 **Models:** All **Engines:** All **Transmissions:** All	**Secondary Air Injection System Switching Valve 'A' Circuit Conditions:** Engine started, and the Air Injection Control Driver (AID) detects open and short circuits according to the voltages of the air pump terminal (VP) and the Air Switching Valve (ASV) terminal (VV), and transmits diagnostic information as a signal to the ECM. For a short time after cold engine starts, the ECM transmits command signals to the AID to drive the air pump and ASV. **Possible Causes:** • Open in ASV drive circuit • Short between ASV drive circuit and body ground • AID • ASV • ECM
DTC: P0415 **1T CCM, MIL: Yes** **Years:** 2007, 2008 **Models:** All **Engines:** All **Transmissions:** All	**Secondary Air Injection System Switching Valve 'B' Circuit Conditions:** Engine started, and the Air Injection Control Driver (AID) detects open and short circuits according to the voltages of the air pump terminal (VP) and the Air Switching Valve (ASV) terminal (VV), and transmits diagnostic information as a signal to the ECM. For a short time after cold engine starts, the ECM transmits command signals to the AID to drive the air pump and ASV. **Possible Causes:** • Open in ASV drive circuit • Short between ASV drive circuit and body ground • AID • ASV • ECM

DTC	Trouble Code Title, Conditions & Possible Causes
DTC: P0418 **1T CCM, MIL: Yes** **Years:** 2007, 2008 **Models:** All **Engines:** All **Transmissions:** All	**Secondary Air Injection System Control 'A' Circuit Conditions:** Engine started, and The Air Injection Control Driver (AID) detects open and short circuits according to the voltages of the air pump terminal (VP) and the Air Switching Valve (ASV) terminal (VV), and transmits diagnostic information as a signal to the ECM. For a short time after cold engine starts, the ECM transmits command signals to the AID to drive the air pump and ASV. **Possible Causes:** • Open in air pump drive circuit • Short between ASV drive circuit and body ground • Air pump • AID • ECM
DTC: P0419 **1T CCM, MIL: Yes** **Years:** 2007, 2008 **Models:** All **Engines:** All **Transmissions:** All	**Secondary Air Injection System Control 'B' Circuit Conditions:** Engine started, and The Air Injection Control Driver (AID) detects open and short circuits according to the voltages of the air pump terminal (VP) and the Air Switching Valve (ASV) terminal (VV), and transmits diagnostic information as a signal to the ECM. For a short time after cold engine starts, the ECM transmits command signals to the AID to drive the air pump and ASV. **Possible Causes:** • Open in air pump drive circuit • Short between ASV drive circuit and body ground • Air pump • AID • ECM
DTC: P0420 **2T CAT, MIL: Yes** **Years:** 2007, 2008 **Models:** All **Engines:** All **Transmissions:** All	**Catalyst Efficiency Below Normal (Bank 1) Conditions:** DTC P0100, P0101, P0102, P0103, P0110, P0112, P0113, P0115, P0116, P0117, P0118, P0120, P0121, P0122, P0123, P0335, P0340 and P0500 not set, engine started, vehicle driven to a speed of 45-60 MPH at 2500-3000 RPM in closed loop for 3-5 minutes, and the PCM detected too much variation in the voltage amplitudes of the Heated Oxygen Sensor signal (Bank 1). **Possible Causes:** • Catalytic converter is damaged, contaminated or has failed • Front A/FS or rear Heated Oxygen Sensor is contaminated with fuel or moisture • Front A/FS or the rear Heated Oxygen Sensor is loose in its mounting hole • Front A/FS is older (aged) than the rear Heated Oxygen Sensor (Heated Oxygen Sensor is lazy) • Gas leaks at the exhaust manifold or in the exhaust pipes
DTC: P0430 **2T CAT, MIL: Yes** **Years:** 2007, 2008 **Models:** All **Engines:** All **Transmissions:** All	**Catalyst Efficiency Below Normal (Bank 2) Conditions:** DTC P0100, P0101, P0102, P0103, P0110, P0112, P0113, P0115, P0116, P0117, P0118, P0120, P0121, P0122, P0123, P0335, P0340 and P0500 not set, engine started, vehicle driven to a speed of 45-60 MPH at 2500-3000 RPM in closed loop for 3-5 minutes, and the PCM detected too much variation in the voltage amplitudes of the Rear Heated Oxygen Sensor for Bank 2 during the Catalyst Monitor test. **Possible Causes:** • Catalytic converter is damaged, contaminated or has failed • Front A/FS or rear Heated Oxygen Sensor is contaminated with fuel or moisture • Front A/FS or the rear Heated Oxygen Sensor is loose in its mounting hole • Front A/FS is older (aged) than the rear Heated Oxygen Sensor (Heated Oxygen Sensor is lazy) • Gas leaks at the exhaust manifold or in the exhaust pipes
DTC: P043E **2T EVAP, MIL: Yes** **Years:** 2007, 2008 **Models:** All **Engines:** All **Transmissions:** All	**Evaporative Emission System Reference Orifice Clog UP Conditions:** Five hours after the ignition switch is turned off, the leak detection pump creates negative pressure (vacuum) in the EVAP system. The ECM monitors for leaks and actuator malfunctions based on the EVAP pressure. **Possible Causes:** • Canister pump module (Reference orifice, leak detection pump, vent valve) • Connector/wire harness (Canister pump module-ECM) • EVAP system hose (Pipe from air inlet port to canister pump module, canister filter, fuel tank vent hose) • ECM
DTC: P043F **2T EVAP, MIL: Yes** **Years:** 2007, 2008 **Models:** All **Engines:** All **Transmissions:** All	**Evaporative Emission System Reference Orifice High Flow Conditions:** Five hours after the ignition switch is turned off, the leak detection pump creates negative pressure (vacuum) in the EVAP system. The ECM monitors for leaks and actuator malfunctions based on the EVAP pressure. **Possible Causes:** • Canister pump module (Reference orifice, leak detection pump, vent valve) • Connector/wire harness (Canister pump module-ECM) • EVAP system hose (Pipe from air inlet port to canister pump module, canister filter, fuel tank vent hose) • ECM

DTC	Trouble Code Title, Conditions & Possible Causes
DTC: P0441 **2T EVAP, MIL: Yes** **Years:** 2007, 2008 **Models:** All **Engines:** All **Transmissions:** All	**Evaporative Emission Control System Incorrect Purge Flow** The two monitors, Key-Off and Purge Flow, are used to detect malfunctions relating to DTC P0441. The Key-Off monitor is initiated by the ECM internal timer, known as the soak timer, 5 hours after the ignition switch is turned off. The purge flow monitor runs while the engine is running. **Possible Causes:** • Purge VSV • Connector/wire harness (purge VSV-ECM) • ECM • Canister pump module • Leak from EVAP system • Leak from EVAP line (Purge VSV-Intake manifold)
DTC: P0450 **1T CCM, MIL: Yes** **Years:** 2007, 2008 **Models:** All **Engines:** All **Transmissions:** All	**Evaporative Emission Control System Pressure Sensor/Switch Conditions:** The canister pressure Sensor output rapidly fluctuates between less than 315.9 mmHg-a (42.1 kPa-a) and more than 928.4 mmHg-a (123.8 kPa-a), the ECM interprets this as an open or short circuit malfunction in the canister pressure Sensor or its circuit, and stops the EVAP (Evaporative Emission) system monitor. The ECM then illuminates the MIL and sets the DTC. **Possible Causes:** • Canister pump module • EVAP system hose (pipe from air inlet port to canister pump module, canister filter, fuel tank vent hose) • ECM
DTC: P0451 **2T CCM, MIL: Yes** **Years:** 2007, 2008 **Models:** All **Engines:** All **Transmissions:** All	**Evaporative Emission Control System Pressure Sensor Range/Performance Conditions:** The canister pressure Sensor voltage output fluctuates rapidly for 10 seconds; the ECM stops the EVAP system monitor. The ECM interprets this as noise from the canister pressure Sensor, and stops the EVAP system monitor. The ECM then illuminates the MIL and stores the DTC. Alternatively, if the Sensor voltage output does not change for 10 seconds, the ECM interprets this as the Sensor being fixed/flat, and stops the monitor. The ECM then illuminates the MIL and stores the DTC. **Possible Causes:** • Canister pump module • Connector/wire harness (Canister pump module-ECM) • EVAP system hose (pipe from air inlet port to canister pump module, canister filter, fuel tank vent hose) • ECM
DTC: P0452 **2T CCM, MIL: Yes** **Years:** 2007, 2008 **Models:** All **Engines:** All **Transmissions:** All	**Evaporative Emission Control System Pressure Sensor/Switch Low Input Conditions:** Engine started; VSV for vapor pressure switching valve "off"; and the PCM detected an unexpected low voltage condition on the vapor pressure Sensor circuit during the CCM test. **Possible Causes:** • Canister pump module • Connector/wire harness (Canister pump module-ECM) • EVAP system hose (pipe from air inlet port to canister pump module, canister filter, fuel tan vent hose) • ECM
DTC: P0453 **2T CCM, MIL: Yes** **Years:** 2007, 2008 **Models:** All **Engines:** All **Transmissions:** All	**Evaporative Emission Control System Pressure Sensor/Switch High Input Conditions:** Engine started; VSV for vapor pressure switching valve "off"; and the PCM detected an unexpected low voltage condition on the vapor pressure Sensor circuit during the CCM test. **Possible Causes:** • Canister pump module • Connector/wire harness (Canister pump module-ECM) • EVAP system hose (pipe from air inlet port to canister pump module, canister filter, fuel tan vent hose) • ECM
DTC: P0455 **2T CCM, MIL: Yes** **Years:** 2007, 2008 **Models:** All **Engines:** All **Transmissions:** All	**Evaporative Emission Control System Leak Detected (Gross Leak) Conditions:** Leak detection pump creates negative pressure (vacuum) in EVAP system and EVAP system pressure measured. Reference pressure measured at start and at end of leak check. If stabilized pressure higher than [second reference pressure × 0.2], ECM determines that EVAP system has large leak. **Possible Causes:** • Fuel tank cap loose • Leak from EVAP line (Canister – fuel tank) • Leak from EVAP line (Purge VSV – Canister) • Canister pump module • Leak from fuel tank • Leak from canister

DTC	Trouble Code Title, Conditions & Possible Causes
DTC: P0456 **2T EVAP, MIL: Yes** **Years:** 2007, 2008 **Models:** All **Engines:** All **Transmissions:** All	**EVAP System Very Small Leak (0.020") Detected Conditions:** Engine started; IAT Sensor signal from 39-86°F, fuel tank level from 25-75% for 10 seconds, and the PCM detected the EVAP system was unable to hold a specified vacuum level for a set period of time. After the system is purged, the PCM shuts off the VSV for the purge valve to seal the vacuum in the system, and then monitors the increase in pressure in the system. The pressure should increase slowly. If it increases at too fast a rate, this code is set. **Possible Causes:** • Canister Purge valve is damaged, leaking or has failed • Charcoal canister is loaded with fuel or moisture • Fuel filler cap loose, cross-threaded, incorrect part or damaged • Fuel tank is cracked (leaking), or a leak exists in the 'O' ring • Fuel tank pressure Sensor is damaged or has failed • Fuel tank overfill check valve is cracked or is damaged • Fuel vapor line(s), fuel pipes or hoses damaged or leaking • PCM failure
DTC: P0500 **1T CCM, MIL: Yes** **Years:** 2007, 2008 **Models:** All **Engines:** All **Transmissions:** All	**Vehicle Speed Sensor 'A' Conditions:** No TP Sensor codes set, engine runtime 2 seconds with the ECT Sensor more than 132°F and IAT Sensor more than 50°F, P/N switch indicating 'P' or 'N', TP angle less than 13° with the engine speed less than 2350 RPM; or TP angle less than 21° with the engine speed less than 2680 RPM; or TP angle less than 30° with the engine speed less than 2835 RPM; or TP angle less than 30° with the engine speed less than 3250 RPM; and the PCM detected the engine speed was equal or more than the VSS signal speed for 500 ms during testing. **Possible Causes:** • Open or short in speed signal circuit • Wheel speed Sensor • Combination meter • ECM • Skid control ECU
DTC: P0504 **2T CCM, MIL: Yes** **Years:** 2007, 2008 **Models:** All **Engines:** All **Transmissions:** All	**Brake Switch 'A' To 'B' Correlation Malfunction Conditions:** Key on or engine running; brake pedal released, and the PCM detected the STP signal indicated "off" while the ST1 signal also indicated "off". The stoplight switch signal is used to prevent the engine from stalling when the brakes are applied suddenly. The stoplight switch uses a duplex system (STP and ST1 signals). **Possible Causes:** • Stoplight switch signal circuit is shorted to power • Stoplight switch assembly is damaged or shorted • PCM failure
DTC: P0505 **2T CCM, MIL: Yes** **Years:** 2007, 2008 **Models:** All **Engines:** All **Transmissions:** All	**Idle Control System Malfunction Conditions:** Engine started, engine running at idle speed n closed loop, and the PCM detected the Actual Idle Speed was more than 100-200 RPM above or below the Target Idle Speed. A Rotary solenoid type of ISC valve is located in front of the air intake chamber and intake air bypassing the throttle valve is directed to the Intake Air Control (IAC) valve via a passage. The PCM controls the idle speed by regulating the amount of intake air volume that bypasses the throttle valve. **Possible Causes:** • Air Induction system leaks (check for intake manifold leaks) • Air leaks in the PCV system (at the valve or its related hoses) • Throttle body assembly is damaged or has failed • PCM failure
DTC: P050A **2T CCM, MIL: Yes** **Years:** 2007, 2008 **Models:** All **Engines:** All **Transmissions:** All	**Cold Start Idle Air Control System Performance Conditions:** Engine started, engine coolant temperature at −10° to 50°C (14 to 122°F). The DTC will be set after the engine idles for 13 seconds. This code is set when there is insufficient mass air flow at cold start. **Possible Causes:** • Throttle body • Mass air flow meter • PCV hose • Air induction system • VVT system • ECM

DTC	Trouble Code Title, Conditions & Possible Causes
DTC: P050B **2T CCM, MIL: Yes** **Years:** 2007, 2008 **Models:** All **Engines:** All **Transmissions:** All	**Cold Start Ignition Timing Performance Conditions:** Engine started, engine coolant temperature at −10° to 50°C (14 to 122°F). The DTC will be set after the engine idles for 13 seconds. This code is set when there is insufficient ignition timing retard at cold start. **Possible Causes:** • Throttle body • Mass air flow meter • PCV hose • Air induction system • VVT system • ECM
DTC: P0550 **1T CCM, MIL: Yes** **Years:** 2007, 2008 **Models:** All **Engines:** All **Transmissions:** All	**Power Steering Pressure Sensor Circuit Malfunction Conditions:** Key on or engine running; and the power steering oil pressure Sensor voltage less than 0.28 V, or more than 4.9 V for 0.5 seconds. **Possible Causes:** • Open or short in power steering oil pressure Sensor circuit • Power steering oil pressure Sensor • ECM
DTC: P0552 **1T CCM, MIL: Yes** **Years:** 2007, 2008 **Models:** All **Engines:** All **Transmissions:** All	**Power Steering Pressure Sensor Circuit Low Input Conditions:** Key on or engine running; and the PCM detected the Power Steering Pressure (PSP) Sensor was less than 0.26v during the CCM test. **Possible Causes:** • PSP Sensor signal circuit open or shorted to ground • PSP Sensor ground circuit open • PSP Sensor is damaged or has failed • PCM failure
DTC: P0553 **1T CCM, MIL: Yes** **Years:** 2007, 2008 **Models:** All **Engines:** All **Transmissions:** All	**Power Steering Pressure Sensor Circuit High Input Conditions:** Key on or engine running; and the PCM detected the Power Steering Pressure (PSP) Sensor was more than 4.90v during the CCM test. **Possible Causes:** • PSP Sensor signal circuit is shorted to power • PSP Sensor ground circuit open • PSP Sensor is damaged or has failed • PCM failure
DTC: P0560 **2T CCM, MIL: Yes** **Years:** 2007, 2008 **Models:** All **Engines:** All **Transmissions:** All	**System Voltage (Backup Power Circuit) Malfunction Conditions:** Key on or engine running; and the PCM detected an unexpected low voltage condition on the Backup Power Circuit during the test. **Possible Causes:** • Battery backup circuit open between battery and the PCM • PCM failure
DTC: P0604 **1T PCM, MIL: Yes** **Years:** 2007, 2008 **Models:** All **Engines:** All **Transmissions:** All	**PCM Internal Control Module Random Access Memory Processing Error Conditions:** Key on, and the PCM detected a processing error in the Internal Control Module Random Access Memory (RAM) function. **Possible Causes:** • Clear the codes and retest for this code. If the same code resets, substitute a known good control module and retest. If the trouble code is gone, the original PCM failure.
DTC: P0606 **1T PCM, MIL: Yes** **Years:** 2007, 2008 **Models:** All **Engines:** All **Transmissions:** All	**ECM/PCM Processing Error Conditions:** Key on, and the PCM detected a processing error occurred. **Possible Causes:** • Clear the codes and retest for this code. If the same code resets, substitute a known good control module and retest. If the trouble code is gone, the original PCM failure. • Exhaust gas leak • ECM
DTC: P0607 **1T PCM, MIL: Yes** **Years:** 2007, 2008 **Models:** All **Engines:** All **Transmissions:** All	**Control Module Performance Conditions:** Key on, and the PCM detected a performance problem occurred. **Possible Causes:** • Clear the codes and retest for this code. If the same code resets, substitute a known good control module and retest. If the trouble code is gone, the original PCM failure.

DTC	Trouble Code Title, Conditions & Possible Causes
DTC: P060A **1T PCM, MIL: Yes** **Years:** 2007, 2008 **Models:** All **Engines:** All **Transmissions:** All	**Internal Control Module Monitoring Processor Performance Conditions:** Key on, and the PCM detected a performance problem occurred. **Possible Causes:** • ECM
DTC: P060B **1T PCM, MIL: Yes** **Years:** 2007, 2008 **Models:** All **Engines:** All **Transmissions:** All	**Internal Control Module A/D Processing Performance Conditions:** Key on, and the PCM detected a performance problem occurred. **Possible Causes:** • ECM
DTC: P060D **1T PCM, MIL: Yes** **Years:** 2007, 2008 **Models:** All **Engines:** All **Transmissions:** All	**Internal Control Module Accelerator Pedal Position Performance Conditions:** Key on, and the PCM detected a performance problem occurred. **Possible Causes:** • ECM
DTC: P060E **1T PCM, MIL: Yes** **Years:** 2007, 2008 **Models:** All **Engines:** All **Transmissions:** All	**Internal Control Module Throttle Position Performance Conditions:** Key on, and the PCM detected a performance problem occurred. **Possible Causes:** • ECM
DTC: P0617 **1T CCM, MIL: Yes** **Years:** 2007, 2008 **Models:** All **Engines:** All **Transmissions:** All	**Starter Relay Circuit High Input Conditions:** Engine started, engine speed over 1000 RPM, system voltage over 10.5v, and the PCM detected the Starter Motor signal indicated high. **Possible Causes:** • Park/Neutral switch assembly is damaged or it has failed • Ignition switch is damaged or has failed • PCM failure
DTC: P0630 **1T CCM, MIL: Yes** **Years:** 2007, 2008 **Models:** All **Engines:** All **Transmissions:** All	**VIN Not Programmed or Mismatch-ECM/PCM Conditions:** Engine started, engine speed over 1000 RPM, system voltage over 10.5v, and the PCM detected the Starter Motor signal indicated high. **Possible Causes:** • ECM
DTC: P0657 **1T CCM, MIL: Yes** **Years:** 2007, 2008 **Models:** All **Engines:** All **Transmissions:** All	**Actuator Supply Voltage Circuit/Open Conditions:** Key on or engine running; and the PCM detected an unexpected voltage condition on the Actuator Supply Voltage circuit. **Possible Causes:** • ECM
DTC: P0705 **2T CCM, MIL: Yes** **Years:** 2007, 2008 **Models:** All **Engines:** All **Transmissions:** All	**All Range Sensor Circuit (PRNDL) Malfunction Conditions:** Key on or engine running; and the PCM detected simultaneous "on" signals (N, 2, L or R) from the Transmission Range Sensor circuit. The P/N switch indicates "on" whenever the shift lever is in the 'N' or 'P' position. When it is "on", the NSW circuit to the PCM is grounded to chassis ground through the starter motor relay, and reads 0.00v. When the shift lever is in 'R', 'D' or 'L' position, the switch is "off" and the NSW circuit reads 12.0v. When the shift lever is moved from the 'N' to the 'D' position, the PCM uses this signal to air/fuel ratio correction and idle speed control (estimated control) functions. **Possible Causes:** • Park/Neutral switch assembly is shorted • Park/Neutral switch assembly is damaged or has failed • PCM failure.
DTC: P0710 **1T CCM, MIL: Yes** **Years:** 2007, 2008 **Models:** All **Engines:** All **Transmissions:** All	**Transmission Fluid Temperature Sensor Circuit Malfunction Conditions:** Engine started, and the PCM detected the TFT Sensor indicated less than −40°F, or with the engine runtime over 15 minutes, it detected the TFT Sensor indicated more than 300°F. **Possible Causes:** • TFT Sensor signal circuit open, shorted to ground or shorted to system power • TFT Sensor is damaged or has failed • PCM failure

DTC	Trouble Code Title, Conditions & Possible Causes
DTC: P0710 **1T CCM, MIL: Yes** **Years:** 2007, 2008 **Models:** All **Engines:** All **Transmissions:** All	**Transmission Fluid Temperature Sensor 'A' Circuit Malfunction Conditions:** Engine started, and the PCM detected the TFT Sensor indicated less than −40°F, or with the engine runtime over 15 minutes, it detected the TFT Sensor indicated more than 300°F. **Possible Causes:** • Open or short in No 1 ATF temperature Sensor circuit • TFT Sensor is damaged or has failed • PCM failure
DTC: P0711 **1T CCM, MIL: Yes** **Years:** 2007, 2008 **Models:** All **Engines:** All **Transmissions:** All	**Transmission Fluid Temperature Sensor Performance Conditions:** ECT and IAT Sensor signals more than 14°F, engine runtime over 12 seconds, then after the vehicle was driven for more than 6.2 miles and the engine runtime exceeded 20 minutes, the PCM detected the TFT Sensor remained at a value less than 14°F during the CCM test. **Possible Causes:** • TFT Sensor signal circuit has high resistance • TFT Sensor is damaged or has failed (it may be contaminated) • PCM failure
DTC: P0712 **1T CCM, MIL: Yes** **Years:** 2007, 2008 **Models:** All **Engines:** All **Transmissions:** All	**Transmission Fluid Temperature Sensor Circuit Low Input Conditions:** Engine started, engine running, and the PCM detected the Transmission Fluid Temperature (TFT) Sensor indicated a value of more than 284°F for 500 ms in the test. **Possible Causes:** • TFT Sensor signal circuit is shorted • TFT Sensor is damaged or has failed • PCM failure
DTC: P0712 **1T CCM, MIL: Yes** **Years:** 2007, 2008 **Models:** All **Engines:** All **Transmissions:** All	**Transmission Fluid Temperature Sensor 'A' Circuit Low Input Conditions:** Engine started, engine running, and the PCM detected the Transmission Fluid Temperature (TFT) Sensor indicated a value of more than 284°F for 500 ms in the test. **Possible Causes:** • TFT Sensor signal circuit is shorted • TFT Sensor is damaged or has failed • PCM failure
DTC: P0713 **1T CCM, MIL: Yes** **Years:** 2007, 2008 **Models:** All **Engines:** All **Transmissions:** All	**Transmission Fluid Temperature Sensor Circuit High Input Conditions:** Engine started, engine running, and the PCM detected the TFT Sensor indicated a value of less than −40°F for 500 ms in t he test. **Possible Causes:** • TFT Sensor signal circuit open • TFT Sensor is damaged or has failed • PCM failure
DTC: P0713 **1T CCM, MIL: Yes** **Years:** 2007, 2008 **Models:** All **Engines:** All **Transmissions:** All	**Transmission Fluid Temperature Sensor 'A' Circuit High Input Conditions:** Engine started, engine running, and the PCM detected the TFT Sensor indicated a value of less than −40°F for 500 ms in t he test. **Possible Causes:** • TFT Sensor signal circuit open • TFT Sensor is damaged or has failed • PCM failure
DTC: P0717 **1T CCM, MIL: Yes** **Years:** 2007, 2008 **Models:** All **Engines:** All **Transmissions:** All	**All Turbine Shaft Speed Sensor Circuit No Signal Conditions:** No Shift Solenoid or P/N codes set, engine started, P/N switch indicating "off", gear position indicating 2nd, 3rd gear, or in O/D, no gear change occurring, and the PCM detected the Turbine Shaft Speed (ISS) Sensor indicated less than 300 RPM, or more than 1000 RPM for 4 seconds. The PCM detects the rotation speed of the input turbine, and compares the signals from the input turbine speed (NT) Sensor to the counter gear speed Sensor (NC). The PCM uses this signal to detect the shift time so that it can control the engine torque and hydraulic pressure in response to various driving conditions. **Possible Causes:** • Input shaft speed Sensor is damaged or loose • Input shaft speed Sensor signal (NT) circuit open or shorted • Input shaft speed Sensor is damaged or has failed • PCM failure

DTC	Trouble Code Title, Conditions & Possible Causes
DTC: P0722 **1T CCM, MIL: Yes** **Years:** 2007, 2008 **Models:** All **Engines:** All **Transmissions:** All	**Output Speed Sensor Circuit No Signal Conditions:** The speed Sensor SP2 detects the rotation speed of the transmission output shaft and sends signals to the ECM. The ECM determines the vehicle speed based on these signals. An AC voltage is generated in the speed Sensor SP2 coil as the parking gear mounted on the rear planetary gear assembly rotates, and this voltage is sent to the ECM. The parking gear on the rear planetary gear is used as the timing rotor for this Sensor. The gear shift point and lock-up timing are controlled by the ECM based on the signals from this vehicle speed Sensor and the throttle position Sensor signal. **Possible Causes:** • Open or short in speed Sensor SP2 circuit • Speed Sensor SP2 • ECM • Automatic transmission (clutch, brake or gear, etc)
DTC: P0724 **2T CCM, MIL: Yes** **Years:** 2007, 2008 **Models:** All **Engines:** All **Transmissions:** All	**Brake Switch 'B' Circuit High Input Conditions:** Engine started, vehicle driven to cruise speed and then back to idle speed at least 30 times, and the PCM did not detect any change in the Brake Switch 'A' circuit status. The STP 'B' switch signal is used to determine when the brakes have been applied, and to determine the Fuel Cutoff engine speed during periods with the brakes applied. **Possible Causes:** • Stoplight switch signal circuit is shorted to power • Stoplight switch assembly is damaged or shorted • PCM failure
DTC: P0741 **2T CCM, MIL: Yes** **Years:** 2007, 2008 **Models:** All **Engines:** All **Transmissions:** All	**Torque Converter Clutch Solenoid Performance (Shift Solenoid Valve DSL) Conditions:** The ECM uses the signals from the throttle position Sensor, air flow meter, turbine (input) speed Sensor, intermediate (counter) shaft speed Sensor and crankshaft position Sensor to monitor the engagement condition of the lock-up clutch. **Possible Causes:** • Shift solenoid valve DSL remains open or closed • Valve body is blocked • Torque converter clutch • Automatic transaxle (clutch, brake or gear etc.) • Line pressure is too low
DTC: P0746 **2T CCM, MIL: Yes** **Years:** 2007, 2008 **Models:** All **Engines:** All **Transmissions:** All	**Pressure Control Solenoid 'A' Performance (Shift Solenoid Valve SL1) Conditions:** The ECM uses signals from the output shaft speed Sensor and input speed Sensor to detect the actual gear position (1st, 2nd, 3rd, 4th or 5th gear). Then the ECM compares the actual gear with the shift schedule in the ECM memory to detect mechanical problems of the shift solenoid valves, valve body or automatic transmission (clutch, brake or gear, etc.). **Possible Causes:** • Shift solenoid SL1 remains closed • Shift solenoid SR remains open or closed • Valve body is blocked • Automatic transmission (clutch, brake or gear, etc.)
DTC: P0748 **2T CCM, MIL: Yes** **Years:** 2007, 2008 **Models:** All **Engines:** All **Transmissions:** All	**Pressure Control Solenoid 'A' Electrical (Shift Solenoid Valve SL1) Conditions:** This DTC indicates an open or short in the shift solenoid valve SL1 circuit. The ECM commands gear shifts by turning the shift solenoid valves "ON/OFF". When there is an open or short circuit in any shift solenoid valve circuit, the ECM detects the problem, illuminates the MIL and stores the DTC. The ECM performs the fail-safe function and turns the other normal shift solenoid valves "ON/OFF". (In case of an open or short circuit, the ECM stops sending current to the circuit.). **Possible Causes:** • Open or short in shift solenoid valve SL1 Circuit • Shift solenoid valve SL1 • ECM
DTC: P0751 **8T CCM, MIL: Yes** **Years:** 2007, 2008 **Models:** All **Engines:** All **Transmissions:** All	**Shift Solenoid 'A' Signal Range/Performance Conditions:** Engine started, vehicle driven to a speed over 50 MPH, and the PCM detected the Actual gear position did not match the Desired gear position during the CCM test period. The PCM uses inputs from the VSS and Direct Clutch speed Sensor to determine the actual gear position (i.e., 1st, 2nd, 3rd or O/D gear). **Possible Causes:** • SSA control circuit open or shorted to ground • SSA control circuit is shorted to system power (B+) • SSA is damaged or has failed (an electrical fault) • PCM failure

DTC	Trouble Code Title, Conditions & Possible Causes
DTC: P0751 **8T CCM, MIL: Yes** **Years:** 2007, 2008 **Models:** All **Engines:** All **Transmissions:** All	**Shift Solenoid 'A' Signal Range/Performance (Shift Solenoid Valve S1) Conditions:** Engine started, vehicle driven to a speed over 50 MPH, and the PCM detected the Actual gear position did not match the Desired gear position during the CCM test period. The PCM uses inputs from the VSS and Direct Clutch speed Sensor to determine the actual gear position (i.e., 1st, 2nd, 3rd or O/D gear). **Possible Causes:** • Shift solenoid valve S1 remains open or closed • Shift solenoid valve SLT remains open or closed • Body valve is blocked • Automatic transmission (clutch, brake or gear, etc)
DTC: P0756 **8T CCM, MIL: Yes** **Years:** 2007, 2008 **Models:** All **Engines:** All **Transmissions:** All	**Shift Solenoid 'B' Signal Range/Performance Conditions:** Vehicle driven to a speed over 50 MPH, and the PCM detected the Actual gear position did not match the Desired gear position. The PCM uses inputs from the VSS and Direct Clutch speed Sensor to determine the actual gear position (i.e., 1st, 2nd, 3rd or O/D gear). **Possible Causes:** • Shift solenoid valve S2 remains open or closed • Shift solenoid valve SLT remains open or closed • Valve body is blocked • Automatic transmission (clutch, brake or gear, etc)
DTC: P0766 **8T CCM, MIL: Yes** **Years:** 2007, 2008 **Models:** All **Engines:** All **Transmissions:** All	**Shift Solenoid 'D' Performance (Shift Solenoid Valve S4) Conditions:** The ECM uses signals from the vehicle speed Sensor to detect the actual gear position (1st, 2nd, 3rd, 4th or 5th gear). Then the ECM compares the actual gear with the shift schedule in the ECM memory to detect mechanical problems of the shift solenoid valves, valve body or automatic transaxle (clutch, brake or gear etc.). **Possible Causes:** • Shift solenoid valve S4 remains open or closed • Valve body is blocked • Automatic transaxle (clutch, brake or gear etc.)
DTC: P0771 **2T CCM, MIL: Yes** **Years:** 2007, 2008 **Models:** All **Engines:** All **Transmissions:** All	**Shift Solenoid 'E' Performance (Shift Solenoid Valve SR) Conditions:** The ECM uses signals from the output shaft speed Sensor and input speed Sensor to detect the actual gear position (1st, 2nd, 3rd, 4th or 5th gear). Then the ECM compares the actual gear with the shift schedule in the ECM memory to detect mechanical problems of the shift solenoid valves, valve body or automatic transmission (clutch, brake or gear, etc.). **Possible Causes:** • Shift solenoid SR remains open or closed • Shift solenoid SL1 remains closed • Valve body is blocked • Automatic transmission (clutch, brake or gear, etc.)
DTC: P0776 **2T CCM, MIL: Yes** **Years:** 2007, 2008 **Models:** All **Engines:** All **Transmissions:** All	**Pressure Control Solenoid 'B' Performance (Shift Solenoid Valve SL2) Conditions:** The ECM commands gear shifts by turning the shift solenoid valves ON/OFF. When the gear position commanded by the ECM and the actual gear position are not the same, the ECM illuminates the MIL and stores the DTC. **Possible Causes:** • Shift solenoid valve SL2 remains open • Shift solenoid valve SLT remains open or closed • Valve body is blocked • Automatic transmission (clutch, brake or gear, etc.)
DTC: P0778 **2T CCM, MIL: Yes** **Years:** 2007, 2008 **Models:** All **Engines:** All **Transmissions:** All	**Pressure Control Solenoid 'B' Electrical (Shift Solenoid Valve SL2) Conditions:** This DTC indicates an open or short in the shift solenoid valve SL2 circuit. The ECM commands gear shifts by turning the shift solenoid valves "ON/OFF". When there is an open or short circuit in any shift solenoid valve circuit, the ECM detects the problem and illuminates the MIL and stores the DTC. The ECM performs the fail-safe function and turns the other shift solenoid valves in good condition "ON/OFF". (In case of an open or short circuit, the ECM stops sending current to the circuit.). While driving and shifting gears, if the ECM detects an open or short in the shift solenoid valve SL2 circuit, the ECM determines there is a malfunction. **Possible Causes:** • Open or short in shift solenoid valve SL2 circuit • Shift solenoid SL2 • ECM
DTC: P0781 **2T CCM, MIL: Yes** **Years:** 2007, 2008 **Models:** All **Engines:** All **Transmissions:** All	**1-2 Shift Condition:** The gear required by the ECM does not match the actual gear when driving. **Possible Causes:** • Valve body is blocked up or stuck (1-2 shift valve) • Shift solenoid valve SLT remains open or closed • Automatic transmission (clutch, brake or gear, etc.)

DTC	Trouble Code Title, Conditions & Possible Causes
DTC: P0791 **1T CCM, MIL: Yes** **Years:** 2007, 2008 **Models:** All **Engines:** All **Transmissions:** All	**Intermediate Shaft Speed Sensor 'A' Circuit Conditions:** Engine started, engine running during normal driving conditions, the vehicle speed 31 mph or more, the Park/Neutral position switch (STAR) is OFF or the speed Sensor (NC) is less than 300 rpm. **Possible Causes:** • ST solenoid connector is damaged or loose • SL solenoid control circuit open or shorted to system power • SL solenoid is damaged or has failed • PCM failure
DTC: P0793 **1T CCM, MIL: Yes** **Years:** 2007, 2008 **Models:** All **Engines:** All **Transmissions:** All	**Intermediate Shaft Speed Sensor 'A' Conditions:** Engine started, engine running during normal driving conditions, the vehicle speed 31 mph or more, the Park/Neutral position switch (STAR) is OFF or the speed Sensor (NC) is less than 300 rpm. **Possible Causes:** • ST solenoid connector is damaged or loose • SL solenoid control circuit open or shorted to system power • SL solenoid is damaged or has failed • PCM failure
DTC: P0796 **2T CCM, MIL: Yes** **Years:** 2007, 2008 **Models:** All **Engines:** All **Transmissions:** All	**Pressure Control Solenoid 'C' Performance (Shift Solenoid Valve SL3) Conditions:** Engine started, engine running, the gear required by the ECM does not match the actual gear when driving. **Possible Causes:** • Shift solenoid valve SL3 remains open or closed • Valve body is blocked • Automatic transaxle (clutch, brake or gear etc.)
DTC: P0798 **2T CCM, MIL: Yes** **Years:** 2007, 2008 **Models:** All **Engines:** All **Transmissions:** All	**Pressure Control Solenoid 'C' Electrical (Shift Solenoid Valve SL3) Conditions:** Engine started, engine running, the gear required by the ECM does not match the actual gear when driving. **Possible Causes:** • Open or short in shift solenoid valve SL3 circuit • Shift solenoid valve SL3 • ECM
DTC: P0872 **2T CCM, MIL: Yes** **Years:** 2007, 2008 **Models:** All **Engines:** All **Transmissions:** All	**Transmission Fluid Pressure Sensor/Switch 'C' Circuit Low Conditions:** The ATF pressure switch NO 1 OFF is detected twice when any of the gears from 1st to 4th engages normally **Possible Causes:** • ATF temperature Sensor assembly (ATF pressure switch No 1) • Transmission wire • TCM
DTC: P0873 **2T CCM, MIL: Yes** **Years:** 2007, 2008 **Models:** All **Engines:** All **Transmissions:** All	**Transmission Fluid Pressure Sensor/Switch 'C' Circuit High Conditions:** The ATF pressure switch NO 1 ON is detected twice when any of the gears from 1st to 4th engages normally **Possible Causes:** • ATF temperature Sensor assembly (ATF pressure switch No 1) • Transmission wire • TCM
DTC: P0877 **2T CCM, MIL: Yes** **Years:** 2007, 2008 **Models:** All **Engines:** All **Transmissions:** All	**Transmission Fluid Pressure Sensor/Switch 'D' Circuit Low Conditions:** The ATF pressure switch NO 1 OFF is detected twice when any of the gears from 1st to 4th engages normally **Possible Causes:** • ATF temperature Sensor assembly (ATF pressure switch No 1) • Transmission wire • TCM
DTC: P0878 **2T CCM, MIL: Yes** **Years:** 2007, 2008 **Models:** All **Engines:** All **Transmissions:** All	**Transmission Fluid Pressure Sensor/Switch 'D' Circuit High Conditions:** The ATF pressure switch NO 1 ON is detected twice when any of the gears from 1st to 4th engages normally **Possible Causes:** • ATF temperature Sensor assembly (ATF pressure switch No 1) • Transmission wire • TCM

DTC	Trouble Code Title, Conditions & Possible Causes
DTC: P0894 **2T CCM, MIL: Yes** **Years:** 2007, 2008 **Models:** All **Engines:** All **Transmissions:** All	**Transmission Component Slipping Conditions:** The ECM detects a malfunction on SLT, S1, S2, SR, SL1, SL2 or 1-2 shift solenoid valve according to the revolution difference of the turbine and the output shaft, and also by the oil pressure. **Possible Causes:** • Shift solenoid valve SLT remains open or closed • Shift solenoid valve S1, S2, SR, SL1 or SL2 remains open or closed • 1-2 shift valve is stuck • Valve body is blocked • Automatic transmission (clutch, brake or gear, etc)
DTC: P0973 **1T CCM, MIL: Yes** **Years:** 2007, 2008 **Models:** All **Engines:** All **Transmissions:** All	**Shift Solenoid 'A' Circuit Low Input Conditions:** Engine started, engine running during normal driving conditions, and the PCM detected an unexpected low voltage condition on the Shift Solenoid 1 control circuit at least (4) times. **Possible Causes:** • S1 (solenoid) connector is damaged or shorted • S1 (solenoid) control circuit is shorted to ground • S1 (solenoid) is damaged or has failed • PCM failure
DTC: P0974 **1T CCM, MIL: Yes** **Years:** 2007, 2008 **Models:** All **Engines:** All **Transmissions:** All	**Shift Solenoid 'A' Circuit High Input (Shift Solenoid Valve S1) Conditions:** Engine started, engine running during normal driving conditions, and the PCM detected an unexpected high voltage condition on the Shift Solenoid 1 control circuit at least (4) times. **Possible Causes:** • S1 (solenoid) connector is damaged or open • S1 (solenoid) control circuit open or shorted to power (B+) • S1 (solenoid) is damaged or has failed • PCM failure
DTC: P0976 **1T CCM, MIL: Yes** **Years:** 2007, 2008 **Models:** All **Engines:** All **Transmissions:** All	**Shift Solenoid 'B' Circuit Low Input (Shift Solenoid Valve S2) Conditions:** Engine started, engine running during normal driving conditions, and the PCM detected an unexpected low voltage condition on the Shift Solenoid 2 control circuit at least (4) times during testing. **Possible Causes:** • S2 (solenoid) connector is damaged or shorted • S2 (solenoid) control circuit is shorted to ground • S2 (solenoid) is damaged or has failed • PCM failure
DTC: P0977 **1T CCM, MIL: Yes** **Years:** 2007, 2008 **Models:** All **Engines:** All **Transmissions:** All	**Shift Solenoid 'B' Circuit High Input (Shift Solenoid Valve S2) Conditions:** Engine started, engine running during normal driving conditions, and the PCM detected an unexpected high voltage condition on the Shift Solenoid 2 control circuit at least (4) times during testing. **Possible Causes:** • S2 (solenoid) connector is damaged or open • S2 (solenoid) control circuit open or shorted to power (B+) • S2 (solenoid) is damaged or has failed • PCM failure
DTC: P0982 **1T CCM, MIL: Yes** **Years:** 2007, 2008 **Models:** All **Engines:** All **Transmissions:** All	**Shift Solenoid 'D' Control Circuit Low (Shift Solenoid Valve S4) Conditions:** Engine started, engine running ECM detects short in solenoid valve S4 circuit 2 times when solenoid valve S4 is operated. **Possible Causes:** • Short in shift solenoid valve S4 circuit • Shift solenoid valve S4 • ECM
DTC: P0983 **1T CCM, MIL: Yes** **Years:** 2007, 2008 **Models:** All **Engines:** All **Transmissions:** All	**Shift Solenoid 'D' Control Circuit High (Shift Solenoid Valve S4) Conditions:** Engine started, engine running ECM detects open in solenoid valve S4 circuit 2 times when solenoid valve S4 is operated. **Possible Causes:** • Open in shift solenoid valve S4 circuit • Shift solenoid valve S4 • ECM
DTC: P0985 **1T CCM, MIL: Yes** **Years:** 2007, 2008 **Models:** All **Engines:** All **Transmissions:** All	**Shift Solenoid 'E' Circuit Low Input (Shift Solenoid Valve SR) Conditions:** Engine started, engine running ECM detects short in solenoid valve SR circuit 2 times when solenoid valve SR is operated. **Possible Causes:** • SR (solenoid) connector is damaged or open • SR (solenoid) is damaged or has failed • PCM failure

DTC	Trouble Code Title, Conditions & Possible Causes
DTC: P0986 **1T CCM, MIL: Yes** **Years:** 2007, 2008 **Models:** All **Engines:** All **Transmissions:** All	**Shift Solenoid 'E' Circuit High Input (Shift Solenoid Valve SR) Conditions:** Engine started, engine running ECM detects short in solenoid valve SR circuit 2 times when solenoid valve SR is not operated. **Possible Causes:** • SR (solenoid) connector is damaged or open • SR (solenoid) is damaged or has failed • PCM failure
DTC: P0989 **2T CCM, MIL: Yes** **Years:** 2007, 2008 **Models:** All **Engines:** All **Transmissions:** All	**Transmission Fluid Pressure Switch No 3 Circuit Low Conditions:** Transmission fluid pressure switch No 3 is OFF when lock up occurs in response to a lock up request. **Possible Causes:** • ATF temperature Sensor assembly (ATF pressure switch No 3) • Transmission wire • TCM
DTC: P0990 **2T CCM, MIL: Yes** **Years:** 2007, 2008 **Models:** All **Engines:** All **Transmissions:** All	**Transmission Fluid Pressure Switch No 3 Circuit Low Conditions:** Transmission fluid pressure switch No 3 is ON when lock up does not occur. Lock up does not occur when shift solenoid valve (SLU) is requested to turn off in the lock up range. **Possible Causes:** • ATF temperature Sensor assembly (ATF pressure switch No 3) • Transmission wire • TCM

OBD II Trouble Code List (P1xxx Codes)

DTC	Trouble Code Title, Conditions & Possible Causes
DTC: P1340 **1T CCM, MIL: Yes** **Years:** 2007, 2008 **Models:** All **Engines:** All **Transmissions:** All	**Camshaft Position Sensor 'A' (Bank 1 Sensor 2) Conditions:** Engine started; and no camshaft position Sensor signal to ECM during cranking. There is no camshaft position Sensor signal to ECM with the engine speed 600 rpm or more. **Note: This vehicle uses a Coil-On-Plug design Ignition System.** **Possible Causes:** • Open or short in camshaft position Sensor circuit • IGT or IGF signal circuit is shorted to ground • Igniter is damaged or has failed • PCM failure
DTC: P1342 **1T CCM, MIL: Yes** **Years:** 2007, 2008 **Models:** All **Engines:** All **Transmissions:** All	**Camshaft Position Sensor 'A' Low Input (MRE) Conditions:** Engine started; and output voltage of the camshaft position Sensor is less than 0.3v for 4 seconds. **Note: This vehicle uses a Coil-On-Plug design Ignition System.** **Possible Causes:** • Open or short in camshaft position Sensor circuit • IGT or IGF signal circuit is shorted to ground • Igniter is damaged or has failed • PCM failure
DTC: P1343 **1T CCM, MIL: Yes** **Years:** 2007, 2008 **Models:** All **Engines:** All **Transmissions:** All	**Camshaft Position Sensor 'A' Low Input (MRE) Conditions:** Engine started; and output voltage of the camshaft position Sensor is more than 4.7v for 4 seconds. **Note: This vehicle uses a Coil-On-Plug design Ignition System.** **Possible Causes:** • Open or short in camshaft position Sensor circuit • IGT or IGF signal circuit is shorted to ground • Igniter is damaged or has failed • PCM failure
DTC: P1607 **1T CCM, MIL: Yes** **Years:** 2007, 2008 **Models:** All **Engines:** All **Transmissions:** All	**Cruise Control Input Processor Conditions:** The ECM continuously monitors its main and sub CPUs. This self-check ensures that the ECM is functioning properly. If outputs from the CPUs are different and deviate from the standard, the ECM will illuminate the MIL and store a DTC immediately. **Possible Causes:** • ECM

DTC	Trouble Code Title, Conditions & Possible Causes
DTC: P1613 **1T CCM, MIL: Yes** **Years:** 2007, 2008 **Models:** All **Engines:** All **Transmissions:** All	**Secondary Air Injection Driver Malfunction Conditions:** For a short time after cold engine starts, the ECM transmits command signals to the Air Injection Control Driver (AID) to drive the air pump and the Air Switching Valve (ASV). The AID detects open and short circuits according to the voltages at the AID terminals to the air pump and ASV, and the circuit voltage of the AID power source, and transmits diagnostic information as a signal to the ECM. If the Secondary Air Injection (AIR) system circuit or the AID itself malfunctions, the AID sends a malfunction signal (duty signal) as diagnostic information to the ECM (when the system is normal, a system normal signal is sent). **Possible Causes:** • Short in diagnostic information signal circuit (AID-ECM) • Open in diagnostic information signal circuit (AID-ECM) • Open or short in air pump and sir switching valve command signal circuit (AID-ECM) • Open or short in AID power source circuit • Open in AID ground circuit • AID • ECM
DTC: P1614 **1T CCM, MIL: Yes** **Years:** 2007, 2008 **Models:** All **Engines:** All **Transmissions:** All	**Secondary Air Injection Driver Bank 2 Conditions:** For a short time after cold engine starts, the ECM transmits command signals to the Air Injection Control Driver (AID) to drive the air pump and the Air Switching Valve (ASV). The AID detects open and short circuits according to the voltages at the AID terminals to the air pump and ASV, and the circuit voltage of the AID power source, and transmits diagnostic information as a signal to the ECM. If the Secondary Air Injection (AIR) system circuit or the AID itself malfunctions, the AID sends a malfunction signal (duty signal) as diagnostic information to the ECM (when the system is normal, a system normal signal is sent). **Possible Causes:** • Short in diagnostic information signal circuit (AID-ECM) • Open in diagnostic information signal circuit (AID-ECM) • Open or short in air pump and sir switching valve command signal circuit (AID-ECM) • Open or short in AID power source circuit • Open in AID ground circuit • AID • ECM

OBD II Trouble Code List (P2xxx Codes)

DTC	Trouble Code Title, Conditions & Possible Causes
DTC: P2A00 **2T CCM, MIL: Yes** **Years:** 2007, 2008 **Models:** All **Engines:** All **Transmissions:** All	**Air Fuel Sensor (Bank 1 Sensor 1) Signal Slow Response Conditions:** Vehicle driven at cruise speed at over 1400 RPM in closed loop at 60 MPH, and the PCM detected an unexpected voltage condition on the Bank 1 Air Fuel Sensor 1 (AFS1) circuit. **Possible Causes:** • A/F Sensor connector is damaged or loose • A/F Sensor circuit open or shorted, or the Sensor has failed • A/F Sensor heater is damaged or has failed • A/F Sensor heater relay circuit open or the relay has failed • Fuel delivery component has failed (fuel pressure regulator, one or more fuel injectors is leaking or severely restricted) • Induction system problems (air leaks or restricted air filter) • PCM failure
DTC: P2A03 **2T CCM, MIL: Yes** **Years:** 2007, 2008 **Models:** All **Engines:** All **Transmissions:** All	**Air Fuel Sensor (Bank 2 Sensor 1) Signal Slow Response Conditions:** Vehicle driven at cruise speed at over 1400 RPM in closed loop at 60 MPH, and the PCM detected an unexpected voltage condition on the Bank 2 Air Fuel Sensor 1 (AFS1) circuit. **Possible Causes:** • A/F Sensor connector is damaged or loose • A/F Sensor circuit open or shorted, or the Sensor has failed • A/F Sensor heater is damaged or has failed • A/F Sensor heater relay circuit open or the relay has failed • Fuel delivery component has failed (fuel pressure regulator, one or more fuel injectors is leaking or severely restricted) • Induction system problems (air leaks or restricted air filter) • PCM failure

DTC	Trouble Code Title, Conditions & Possible Causes
DTC: P2004 **2T CCM, MIL: Yes** **Years:** 2007, 2008 **Models:** All **Engines:** All **Transmissions:** All	**Intake Manifold Runner Control Stuck Open (Bank 1) Conditions:** When the IMRV opening angle is more than 10° after requesting the IMRV full close. **Possible Causes:** • IMRV motor circuit • IMRV motor • IMRV position Sensor • IMRV • ECM
DTC: P2006 **2T CCM, MIL: Yes** **Years:** 2007, 2008 **Models:** All **Engines:** All **Transmissions:** All	**Intake Manifold Runner Control Stuck Closed (Bank 1) Conditions:** When the IMRV opening angle is less than 35° after requesting the IMRV full open. **Possible Causes:** • IMRV motor circuit • IMRV motor • IMRV position Sensor • IMRV • ECM
DTC: P2009 **2T CCM, MIL: Yes** **Years:** 2007, 2008 **Models:** All **Engines:** All **Transmissions:** All	**Intake Manifold Runner Control Circuit Low (Bank 1) Conditions:** The Intake Manifold Runner Valve (IMRV) motor is driven by the ECM. The IMRV motor opens and closes the IMRV. When the current of the IMRV motor deviates from the standard range, the ECM determines that there is a malfunction. **Possible Causes:** • Open in IMRV motor circuit • IMRV motor • ECM
DTC: P2010 **2T CCM, MIL: Yes** **Years:** 2007, 2008 **Models:** All **Engines:** All **Transmissions:** All	**Intake Manifold Runner Control Circuit High (Bank 1) Conditions:** The Intake Manifold Runner Valve (IMRV) motor is driven by the ECM. The IMRV motor opens and closes the IMRV. When the current of the IMRV motor deviates from the standard range, the ECM determines that there is a malfunction. **Possible Causes:** • Short in IMRV motor circuit • IMRV motor • ECM
DTC: P2014 **2T CCM, MIL: Yes** **Years:** 2007, 2008 **Models:** All **Engines:** All **Transmissions:** All	**Intake Manifold Runner Position Sensor/Switch Circuit (Bank 1) Conditions:** The Intake Manifold Runner Valve (IMRV) motor is driven by the ECM. The IMRV motor opens and closes the IMRV. When the current of the IMRV motor deviates from the standard range, the ECM determines that there is a malfunction. **Possible Causes:** • Open or short in IMRV position Sensor circuit • IMRV position Sensor • ECM
DTC: P2016 **2T CCM, MIL: Yes** **Years:** 2007, 2008 **Models:** All **Engines:** All **Transmissions:** All	**Intake Manifold Runner Position Sensor/Switch Circuit Low (Bank 1) Conditions:** The Intake Manifold Runner Valve (IMRV) motor is driven by the ECM. The IMRV motor opens and closes the IMRV. When the current of the IMRV motor deviates from the standard range, the ECM determines that there is a malfunction. **Possible Causes:** • Open or short in IMRV position Sensor circuit • IMRV position Sensor • ECM
DTC: P2017 **2T CCM, MIL: Yes** **Years:** 2007, 2008 **Models:** All **Engines:** All **Transmissions:** All	**Intake Manifold Runner Position Sensor/Switch Circuit High (Bank 1) Conditions:** The Intake Manifold Runner Valve (IMRV) motor is driven by the ECM. The IMRV motor opens and closes the IMRV. When the current of the IMRV motor deviates from the standard range, the ECM determines that there is a malfunction. **Possible Causes:** • Open or short in IMRV position Sensor circuit • IMRV position Sensor • ECM

DTC	Trouble Code Title, Conditions & Possible Causes
DTC: P2102 **1T CCM, MIL: Yes** **Years:** 2007, 2008 **Models:** All **Engines:** All **Transmissions:** All	**Throttle Actuator Control Motor Circuit Low Input Conditions:** Engine started, throttle control motor output duty cycle at 80% or higher, and the PCM detected the throttle motor current was less than 0.5 amps for 2 seconds. The PCM controls the motor position in order to open and close the throttle valve. The opening angle of the throttle valve is sensed by the TP Sensor mounted on the throttle body. The PCM uses the TP Sensor signal to control the Throttle Valve opening angle (throttle motor) to respond to driving conditions. **Possible Causes:** • Throttle motor connector is damaged or open • Throttle control motor circuit open • Throttle control motor is damaged or has failed • PCM failure
DTC: P2103 **1T CCM, MIL: Yes** **Years:** 2007, 2008 **Models:** All **Engines:** All **Transmissions:** All	**Throttle Actuator Control Motor Circuit High Input Conditions:** Engine started, throttle control motor output duty cycle at 80% or higher, and the PCM detected the throttle motor current was more than 10.0 amps for 600 ms. The PCM controls the motor position in order to open and close the throttle valve. The opening angle of the throttle valve is sensed by the TP Sensor mounted on the throttle body. The PCM uses the TP Sensor signal to control the Throttle Valve opening angle (throttle motor) to respond to driving conditions. **Possible Causes:** • Throttle motor connector is damaged or shorted • Throttle control motor circuit is shorted • Throttle control motor is damaged or has failed • PCM failure
DTC: P2111 **1T CCM, MIL: Yes** **Years:** 2007, 2008 **Models:** All **Engines:** All **Transmissions:** All	**Throttle Actuator Control System Stuck Open Conditions:** Key on or engine started, and the PCM detected the throttle control motor position is stuck open. The PCM controls the motor position in order to open and close the throttle valve. The opening angle of the throttle valve is sensed by the TP Sensor mounted on the throttle body. The PCM uses the TP Sensor signal to control the Throttle Valve opening angle (throttle motor) to respond to driving conditions. **Possible Causes:** • Throttle control motor circuit open • Throttle control motor is damaged or has failed • Throttle body or throttle valve is damaged or has failed
DTC: P2112 **1T CCM, MIL: Yes** **Years:** 2007, 2008 **Models:** All **Engines:** All **Transmissions:** All	**Throttle Actuator Control System Stuck Closed Conditions:** Key on or engine started, and the PCM detected the throttle control motor position is stuck closed. The PCM controls the motor position in order to open and close the throttle valve. The opening angle of the throttle valve is sensed by the TP Sensor mounted on the throttle body. The PCM uses the TP Sensor signal to control the Throttle Valve opening angle (throttle motor) to respond to driving conditions. **Possible Causes:** • Throttle control motor circuit is shorted • Throttle control motor is damaged or has failed • Throttle body or throttle valve is damaged or has failed
DTC: P2118 **1T CCM, MIL: Yes** **Years:** 2007, 2008 **Models:** All **Engines:** All **Transmissions:** All	**Throttle Actuator Control Motor Current Performance Conditions:** Key on or engine started, and the PCM detected an unexpected low voltage condition (open circuit) on the ETCS power source circuit. Battery positive voltage is applied to the +BM circuit of the PCM under both Key on and Key off conditions. **Possible Causes:** • ETCS power source circuit open • PCM failure
DTC: P2119 **1T CCM, MIL: Yes** **Years:** 2007, 2008 **Models:** All **Engines:** All **Transmissions:** All	**Throttle Actuator Control Throttle Body Performance Conditions:** Engine started, and the PCM detected Actual throttle opening angle continued to vary greatly from the Target opening angel. The idle speed on this vehicle is controlled by the Electronic Throttle Control system (ETCS). This system includes a throttle control motor to operate the throttle valve, a throttle position Sensor to detect the accelerator pedal position, and the PCM to control the ETCS and one-valve design of throttle body. The PCM controls this motor in order to control the throttle valve opening to achieve its target speed. **Possible Causes:** • ETCS throttle control system • PCM failure

DTC	Trouble Code Title, Conditions & Possible Causes
DTC: P2120 **1T CCM, MIL: Yes** **Years:** 2007, 2008 **Models:** All **Engines:** All **Transmissions:** All	**Throttle Pedal Position Sensor/Switch 'D' Circuit Malfunction Conditions:** Engine started, and the PCM detected VPA1 signal indicated less than 0.20v while the VPA2 signal indicated over 0.97 degrees, or the VPA1 signal indicated more than 4.80v for 500ms. This system (ETCS) does not use a throttle cable. The Accelerator Pedal Position (APP) Sensor is mounted on the accelerator pedal bracket. It includes two Sensors to detect the accelerator position, and to detect any faults in the APP Sensor or its related circuits. **Possible Causes:** • APP Sensor signal circuit open or shorted to ground • APP Sensor is damaged or has failed • PCM failure
DTC: P2121 **1T CCM, MIL: Yes** **Years:** 2007, 2008 **Models:** All **Engines:** All **Transmissions:** All	**Accelerator Pedal Position Sensor Signal Performance Conditions:** Engine started, IDL signal "off", and the PCM detected the difference between the VPA and VPA2 signal was out-of-range for 2 seconds. This system (ETCS) does not use a throttle cable. The Accelerator Pedal Position (APP) Sensor is mounted on the accelerator pedal bracket. It includes two Sensors to detect the accelerator position, and to detect any faults in the APP Sensor or its related circuits. **Possible Causes:** • APP Sensor is damaged or has failed • PCM failure
DTC: P2122 **1T CCM, MIL: Yes** **Years:** 2007, 2008 **Models:** All **Engines:** All **Transmissions:** All	**Throttle Pedal Position Sensor/Switch 'D' Circuit Low Input Conditions:** Engine started, and the PCM detected VPA1 signal was less than 0.20v while the VPA2 signal indicated over 0.97°for 500ms. This system (ETCS) does not use a throttle cable. The Accelerator Pedal Position (APP) Sensor is mounted on the accelerator pedal bracket. It includes two Sensors to detect the accelerator position, and to detect any faults in the APP Sensor or its related circuits. **Possible Causes:** • APP Sensor signal circuit is shorted to ground • APP Sensor is damaged or has failed • PCM failure
DTC: P2123 **1T CCM, MIL: Yes** **Years:** 2007, 2008 **Models:** All **Engines:** All **Transmissions:** All	**Throttle Pedal Position Sensor/Switch 'D' Circuit High Input Conditions:** Engine started, and the PCM detected VPA1 signal indicated over 4.80v for 2 seconds. This system (ETCS) does not use a throttle cable. The Accelerator Pedal Position (APP) Sensor is mounted on the accelerator pedal bracket. It includes two Sensors to detect the accelerator position, and to detect any faults in the APP Sensor. **Possible Causes:** • APP Sensor signal circuit open • APP Sensor is damaged or has failed • PCM failure
DTC: P2125 **1T CCM, MIL: Yes** **Years:** 2007, 2008 **Models:** All **Engines:** All **Transmissions:** All	**Throttle Pedal Position Sensor/Switch 'E' Circuit Malfunction Conditions:** Engine started, and the PCM detected VPA2 signal indicated less than 0.50v while the VPA1 signal indicated over 0.97 degrees, or the VPA1 signal was more than 4.80v or less than 0.20v for 500ms. This system (ETCS) does not use a throttle cable. The Accelerator Pedal Position (APP) Sensor is mounted on the accelerator pedal bracket. It includes two Sensors to detect the accelerator position, and to detect any faults in the APP Sensor or its circuits. **Possible Causes:** • APP Sensor signal circuit open or shorted to ground • APP Sensor is damaged or has failed • PCM failure
DTC: P2127 **1T CCM, MIL: Yes** **Years:** 2007, 2008 **Models:** All **Engines:** All **Transmissions:** All	**Throttle Pedal Position Sensor/Switch 'E' Circuit Low Input Conditions:** Engine started, and the PCM detected VPA2 signal was less than 0.20v while the VPA1 signal indicated over 0.97°for 500ms. The ETCS does not use a throttle cable. The Accelerator Pedal Position Sensor is mounted on the accelerator pedal bracket. It includes two Sensors to detect the accelerator position or any faults in the APP Sensor or its circuits. **Possible Causes:** • APP Sensor signal circuit is shorted to ground • APP Sensor has failed, or the PCM failure
DTC: P2128 **1T CCM, MIL: Yes** **Years:** 2007, 2008 **Models:** All **Engines:** All **Transmissions:** All	**Throttle Pedal Position Sensor/Switch 'E' Circuit High Input Conditions:** Engine started, and the PCM detected VPA1 signal was over 4.80v or under 0.20v for 2 seconds. This system (ETCS) does not use a throttle cable. The Accelerator Pedal Position (APP) Sensor is mounted on the accelerator pedal bracket. It includes two Sensors to detect the accelerator position, and any faults in the APP Sensor. **Possible Causes:** • APP Sensor signal circuit open • APP Sensor is damaged or has failed • PCM failure

DTC	Trouble Code Title, Conditions & Possible Causes
DTC: P2135 **1T CCM, MIL: Yes** **Years:** 2007, 2008 **Models:** All **Engines:** All **Transmissions:** All	**Throttle Pedal Position Sensor/Switch 'A'/'B' Voltage Correlation Conditions:** Engine started, and the PCM detected the value of the VPA1 signal less the VPA2 was less than 0.02v, or the VPA1 signal was less than 0.20v with the VPA2 signal less than 0.50v for 400-500 ms. This system (ETCS) does not use a throttle cable. The Accelerator Pedal Position (APP) Sensor is mounted on the accelerator pedal bracket. It includes two Sensors to detect the accelerator position, and to detect any faults present in the APP Sensor. **Possible Causes:** • APP Sensor signal circuit open • APP Sensor is damaged or has failed • PCM failure
DTC: P2138 **1T CCM, MIL: Yes** **Years:** 2007, 2008 **Models:** All **Engines:** All **Transmissions:** All	**Throttle Pedal Position Sensor/Switch 'D'/'E' Voltage Correlation Conditions:** Engine started, and the PCM detected the value of the VPA1 signal less the VPA2 was less than 0.02v, or the VPA1 signal was less than 0.20v with the VPA2 signal less than 0.50v for 2 seconds. This system (ETCS) does not use a throttle cable. The Accelerator Pedal Position (APP) Sensor is mounted on the accelerator pedal bracket. It includes two Sensors to detect the accelerator position, and to detect any faults present in the APP Sensor. **Possible Causes:** • APP Sensor signal circuit open • APP Sensor is damaged or has failed • PCM failure
DTC: P2195 **1T CCM, MIL: Yes** **Years:** 2007, 2008 **Models:** All **Engines:** All **Transmissions:** All	**Air Fuel Sensor 1 (Bank 1 Sensor 1) Signal Stuck "Lean" Conditions:** Vehicle speed from 25-87 MPH at over 1500 RPM with the throttle valve open, and the PCM detected the Air Fuel Sensor signal indicated more than 3.80v for 10 seconds. **Possible Causes:** • A/FS1 signal circuit open or shorted to ground • A/FS1 is damaged, contaminated or it has failed • Air induction system is severely restricted • Fuel Control component problems (e.g., low fuel pressure, or one or more severely restricted fuel injectors) • PCM failure
DTC: P2196 **1T CCM, MIL: Yes** **Years:** 2007, 2008 **Models:** All **Engines:** All **Transmissions:** All	**Air Fuel Sensor 1 (Bank 1 Sensor 1) Signal Stuck "Rich" Conditions:** Vehicle speed from 25-87 MPH at over 1500 RPM with the throttle valve open, and the PCM detected the Air Fuel Sensor signal indicated more than 2.80v for 10 seconds. **Possible Causes:** • A/FS1 signal circuit open or shorted to ground • A/FS1 is damaged, contaminated or it has failed • Air induction system is leaking (check for PCV system leaks) • Fuel component problem (high fuel pressure, leaking regulator or a leaking injector) • PCM failure
DTC: P2197 **1T CCM, MIL: Yes** **Years:** 2007, 2008 **Models:** All **Engines:** All **Transmissions:** All	**Air Fuel Sensor 1 (Bank 2 Sensor 1) Signal Stuck "Lean" Conditions:** Vehicle speed from 25-87 MPH at over 1500 RPM with the throttle valve open, and the PCM detected the Air Fuel Sensor signal indicated more than 3.80v for 10 seconds. **Possible Causes:** • A/FS1 signal circuit open or shorted to ground • A/FS1 is damaged, contaminated or it has failed • Air induction system is severely restricted • Fuel component problem (e.g., low fuel pressure, or a severely restricted fuel injector) • PCM failure
DTC: P2198 **1T CCM, MIL: Yes** **Years:** 2007, 2008 **Models:** All **Engines:** All **Transmissions:** All	**Air Fuel Sensor 1 (Bank 2 Sensor 1) Signal Stuck "Rich" Conditions:** Vehicle driven at a steady speed of 25-87 MPH at over 1500 RPM with the throttle valve open, and the PCM detected the Air Fuel Sensor signal indicated more than 2.80v for 10 seconds. **Possible Causes:** • A/FS1 signal circuit open or shorted to ground • A/FS1 is damaged, contaminated or it has failed • Air induction system is leaking (check for PCV system leaks) • Fuel Control component problems (e.g., high fuel pressure, a leaking pressure regulator, one or more leaking fuel injectors) • PCM failure

DTC	Trouble Code Title, Conditions & Possible Causes
DTC: P2237 **1T CCM, MIL: Yes** **Years:** 2007, 2008 **Models:** All **Engines:** All **Transmissions:** All	**Air Fuel Sensor 1 (Bank 1 Sensor 1) Pumping Current Signal Open Conditions:** Vehicle speed from 25-87 MPH at over 1500 RPM with the throttle valve open, and the PCM detected the A/F Sensor AF+ signal was less than 0.50v or more than 4.80v for 5 seconds. **Possible Causes:** • A/FS1 signal circuit open or shorted to ground • A/FS1 is damaged, contaminated or it has failed • A/FS1 heater assembly is damaged or its circuit has failed • A/FS1 heater relay is damaged or has failed • PCM failure
DTC: P2238 **1T CCM, MIL: Yes** **Years:** 2007, 2008 **Models:** All **Engines:** All **Transmissions:** All	**Air Fuel Sensor 1 (Bank 1 Sensor 1) Pumping Current Circuit Low Conditions:** Vehicle speed from 25-87 MPH at over 1500 RPM with the throttle valve open, and the PCM detected the A/F Sensor AF+ signal was less than 0.50v or more than 4.80v for 5 seconds. **Possible Causes:** • A/FS1 signal circuit open or shorted to ground • A/FS1 is damaged, contaminated or it has failed • A/FS1 heater assembly is damaged or its circuit has failed • A/FS1 heater relay is damaged or has failed • PCM failure
DTC: P2239 **1T CCM, MIL: Yes** **Years:** 2007, 2008 **Models:** All **Engines:** All **Transmissions:** All	**Air Fuel Sensor 1 (Bank 1 Sensor 1) Pumping Current Circuit High Conditions:** Vehicle speed from 25-87 MPH at over 1500 RPM with the throttle valve open, and the PCM detected the A/F Sensor AF+ signal was less than 0.50v or more than 4.80v for 5 seconds. **Possible Causes:** • A/FS1 signal circuit open or shorted to ground • A/FS1 is damaged, contaminated or it has failed • A/FS1 heater assembly is damaged or its circuit has failed • A/FS1 heater relay is damaged or has failed • PCM failure
DTC: P2240 **1T CCM, MIL: Yes** **Years:** 2007, 2008 **Models:** All **Engines:** All **Transmissions:** All	**Air Fuel Sensor 1 (Bank 2 Sensor 1) Pumping Current Signal Open Conditions:** Engine started, vehicle driven at a steady speed of 25-87 MPH at over 1500 RPM with the throttle valve open, and the PCM detected the Air Fuel Sensor AF+ signal indicated less than 0.50v or indicated more than 4.80v for 5 seconds. **Possible Causes:** • A/FS1 signal circuit open or shorted to ground • A/FS1 is damaged, contaminated or it has failed • A/FS1 heater assembly is damaged or its circuit has failed • A/FS1 heater relay is damaged or has failed • PCM failure
DTC: P2241 **1T CCM, MIL: Yes** **Years:** 2007, 2008 **Models:** All **Engines:** All **Transmissions:** All	**Air Fuel Sensor 1 (Bank 2 Sensor 1) Pumping Current Circuit Low Conditions:** Engine started, vehicle driven at a steady speed of 25-87 MPH at over 1500 RPM with the throttle valve open, and the PCM detected the Air Fuel Sensor AF+ signal indicated less than 0.50v or indicated more than 4.80v for 5 seconds. **Possible Causes:** • A/FS1 signal circuit open or shorted to ground • A/FS1 is damaged, contaminated or it has failed • A/FS1 heater assembly is damaged or its circuit has failed • A/FS1 heater relay is damaged or has failed • PCM failure
DTC: P2242 **1T CCM, MIL: Yes** **Years:** 2007, 2008 **Models:** All **Engines:** All **Transmissions:** All	**Air Fuel Sensor 1 (Bank 2 Sensor 1) Pumping Current Circuit High Conditions:** Engine started, vehicle driven at a steady speed of 25-87 MPH at over 1500 RPM with the throttle valve open, and the PCM detected the Air Fuel Sensor AF+ signal indicated less than 0.50v or indicated more than 4.80v for 5 seconds. **Possible Causes:** • A/FS1 signal circuit open or shorted to ground • A/FS1 is damaged, contaminated or it has failed • A/FS1 heater assembly is damaged or its circuit has failed • A/FS1 heater relay is damaged or has failed • PCM failure

DTC	Trouble Code Title, Conditions & Possible Causes
DTC: P2252 **1T CCM, MIL: Yes** **Years:** 2007, 2008 **Models:** All **Engines:** All **Transmissions:** All	**Air Fuel Sensor 1 (Bank 1 Sensor 1) Reference Ground Circuit Low Conditions:** Engine started, vehicle driven at a steady speed of 25-87 MPH at over 1500 RPM with the throttle valve open, and the PCM detected the Air Fuel Sensor AF+ signal indicated less than 0.50v or indicated more than 4.80v for 5 seconds. **Possible Causes:** • A/FS1 signal circuit open or shorted to ground • A/FS1 is damaged, contaminated or it has failed • A/FS1 heater assembly is damaged or its circuit has failed • A/FS1 heater relay is damaged or has failed • PCM failure
DTC: P2253 **1T CCM, MIL: Yes** **Years:** 2007, 2008 **Models:** All **Engines:** All **Transmissions:** All	**Air Fuel Sensor 1 (Bank 1 Sensor 1) Reference Ground Circuit High Conditions:** Engine started, vehicle driven at a steady speed of 25-87 MPH at over 1500 RPM with the throttle valve open, and the PCM detected the Air Fuel Sensor AF+ signal indicated less than 0.50v or indicated more than 4.80v for 5 seconds. **Possible Causes:** • A/FS1 signal circuit open or shorted to ground • A/FS1 is damaged, contaminated or it has failed • A/FS1 heater assembly is damaged or its circuit has failed • A/FS1 heater relay is damaged or has failed • PCM failure
DTC: P2255 **1T CCM, MIL: Yes** **Years:** 2007, 2008 **Models:** All **Engines:** All **Transmissions:** All	**Air Fuel Sensor 1 (Bank 2 Sensor 1) Reference Ground Circuit Low Conditions:** Engine started, vehicle driven at a steady speed of 25-87 MPH at over 1500 RPM with the throttle valve open, and the PCM detected the Air Fuel Sensor AF+ signal indicated less than 0.50v or indicated more than 4.80v for 5 seconds. **Possible Causes:** • A/FS1 signal circuit open or shorted to ground • A/FS1 is damaged, contaminated or it has failed • A/FS1 heater assembly is damaged or its circuit has failed • A/FS1 heater relay is damaged or has failed • PCM failure
DTC: P2256 **1T CCM, MIL: Yes** **Years:** 2007, 2008 **Models:** All **Engines:** All **Transmissions:** All	**Air Fuel Sensor (Bank 2 Sensor 1) Reference Ground Circuit High Conditions:** Engine started, vehicle driven at a steady speed of 25-87 MPH at over 1500 RPM with the throttle valve open, and the PCM detected the Air Fuel Sensor AF+ signal indicated less than 0.50v or indicated more than 4.80v for 5 seconds. **Possible Causes:** • A/FS1 signal circuit open or shorted to ground • A/FS1 is damaged, contaminated or it has failed • A/FS1 heater assembly is damaged or its circuit has failed • A/FS1 heater relay is damaged or has failed • PCM failure
DTC: P2401 **1T CCM, MIL: Yes** **Years:** 2007, 2008 **Models:** All **Engines:** All **Transmissions:** All	**Evaporative Emission Leak Detection Pump Stuck OFF Conditions:** The leak detection pump creates negative pressure through the reference orifice. When the system is normal, the EVAP pressure is between 724 to 752 mmHg* and saturated within a minute. If not, the ECM interprets this as a malfunction. The ECM will illuminate the MIL and store a DTC if this malfunction is detected in consecutive drive cycles. **Possible Causes:** • Reference orifice clogged • Reference orifice high flow • Leak detection pump OFF malfunction • Leak detection pump ON malfunction • Vent valve ON (close) malfunction
DTC: P2401 **1T CCM, MIL: Yes** **Years:** 2007, 2008 **Models:** All **Engines:** All **Transmissions:** All	**Evaporative Emission System Leak Detection Pump Control Circuit Low Conditions:** The leak detection pump creates negative pressure through the reference orifice. When the system is normal, the EVAP pressure is between 724 to 752 mmHg* and saturated within a minute. If not, the ECM interprets this as a malfunction. The ECM will illuminate the MIL and store a DTC if this malfunction is detected in consecutive drive cycles. **Possible Causes:** • Reference orifice clogged • Reference orifice high flow • Leak detection pump OFF malfunction • Leak detection pump ON malfunction • Vent valve ON (close) malfunction

DTC	Trouble Code Title, Conditions & Possible Causes
DTC: P2402 **1T CCM, MIL: Yes** **Years:** 2007, 2008 **Models:** All **Engines:** All **Transmissions:** All	**Evaporative Emission Leak Detection Pump Stuck ON Conditions:** The leak detection pump creates negative pressure through the reference orifice. When the system is normal, the EVAP pressure is between 724 to 752 mmHg* and saturated within a minute. If not, the ECM interprets this as a malfunction. The ECM will illuminate the MIL and store a DTC if this malfunction is detected in consecutive drive cycles. **Possible Causes:** • Reference orifice clogged • Reference orifice high flow • Leak detection pump OFF malfunction • Leak detection pump ON malfunction • Vent valve ON (close) malfunction
DTC: P2402 **1T CCM, MIL: Yes** **Years:** 2007, 2008 **Models:** All **Engines:** All **Transmissions:** All	**Evaporative Emission System Leak Detection Pump Control Circuit High Conditions:** The leak detection pump creates negative pressure through the reference orifice. When the system is normal, the EVAP pressure is between 724 to 752 mmHg* and saturated within a minute. If not, the ECM interprets this as a malfunction. The ECM will illuminate the MIL and store a DTC if this malfunction is detected in consecutive drive cycles. **Possible Causes:** • Reference orifice clogged • Reference orifice high flow • Leak detection pump OFF malfunction • Leak detection pump ON malfunction • Vent valve ON (close) malfunction
DTC: P2419 **1T CCM, MIL: Yes** **Years:** 2007, 2008 **Models:** All **Engines:** All **Transmissions:** All	**Evaporative Emission System Switching Valve Control Circuit Low Conditions:** The leak detection pump creates negative pressure through the reference orifice. When the system is normal, the EVAP pressure is between 724 to 752 mmHg* and saturated within a minute. If not, the ECM interprets this as a malfunction. The ECM will illuminate the MIL and store a DTC if this malfunction is detected in consecutive drive cycles. **Possible Causes:** • Canister pump module • Connector/wire harness (Canister pump module-ECM) • EVAP system hose (pipe from air inlet port to canister pump module, canister filter, fuel tank vent hose) • ECM
DTC: P2420 **1T CCM, MIL: Yes** **Years:** 2007, 2008 **Models:** All **Engines:** All **Transmissions:** All	**Evaporative Emission System Switching Valve Control Circuit High Conditions:** The leak detection pump creates negative pressure through the reference orifice. When the system is normal, the EVAP pressure is between 724 to 752 mmHg* and saturated within a minute. If not, the ECM interprets this as a malfunction. The ECM will illuminate the MIL and store a DTC if this malfunction is detected in consecutive drive cycles. **Possible Causes:** • Canister pump module • Connector/wire harness (Canister pump module-ECM) • ECM
DTC: P2431 **1T CCM, MIL: Yes** **Years:** 2007, 2008 **Models:** All **Engines:** All **Transmissions:** All	**Secondary Air Injection System Air Flow/Pressure Sensor Circuit Range/Performance Bank 1 Conditions:** Pressure Sensor indicates less than 45.6 kPa (342 mmHg), or more than 135 kPa (1013 mmHg). **Possible Causes:** • Pressure Sensor • Open or short in pressure Sensor circuit • ECM
DTC: P2432 **1T CCM, MIL: Yes** **Years:** 2007, 2008 **Models:** All **Engines:** All **Transmissions:** All	**Secondary Air Injection System Air Flow/Pressure Sensor Circuit Low Bank 1 Conditions:** While engine running, voltage output of pressure Sensor remains below 0.5 V. **Possible Causes:** • Pressure Sensor • Open or short in pressure Sensor circuit • ECM
DTC: P2433 **1T CCM, MIL: Yes** **Years:** 2007, 2008 **Models:** All **Engines:** All **Transmissions:** All	**Secondary Air Injection System Air Flow/Pressure Sensor Circuit High Bank 1 Conditions:** While engine running, voltage output of pressure Sensor remains above 4.5 V. **Possible Causes:** • Pressure Sensor • Open or short in pressure Sensor circuit • ECM

DTC	Trouble Code Title, Conditions & Possible Causes
DTC: P2436 **1T CCM, MIL: Yes** **Years:** 2007, 2008 **Models:** All **Engines:** All **Transmissions:** All	**Secondary Air Injection System Air Flow/Pressure Sensor Circuit Range/Performance Bank 2 Conditions:** Pressure Sensor indicates less than 45.6 kPa (342 mmHg), or more than 135 kPa (1013 mmHg). **Possible Causes:** • Pressure Sensor • Open or short in pressure Sensor circuit • ECM
DTC: P2437 **1T CCM, MIL: Yes** **Years:** 2007, 2008 **Models:** All **Engines:** All **Transmissions:** All	**Secondary Air Injection System Air Flow/Pressure Sensor Circuit Low Bank 2 Conditions:** While engine running, voltage output of pressure Sensor remains below 0.5 V. **Possible Causes:** • Pressure Sensor • Open or short in pressure Sensor circuit • ECM
DTC: P2438 **1T CCM, MIL: Yes** **Years:** 2007, 2008 **Models:** All **Engines:** All **Transmissions:** All	**Secondary Air Injection System Air Flow/Pressure Sensor Circuit High Bank 2 Conditions:** While engine running, voltage output of pressure Sensor remains above 4.5 V. **Possible Causes:** • Pressure Sensor • Open or short in pressure Sensor circuit • ECM
DTC: P2440 **1T CCM, MIL: Yes** **Years:** 2007, 2008 **Models:** All **Engines:** All **Transmissions:** All	**Secondary Air Injection System Switching Valve Stuck Close Bank 1 Conditions:** Pressure Sensor detects pulsation of exhaust gas despite ECM commanding Air Switching Valve (ASV) to close, while engine running. **Possible Causes:** • ASV • Open or short in ASV circuit • Pressure Sensor • Pressure Sensor circuit • Air Injection Control Driver (AID) • ECM
DTC: P2441 **1T CCM, MIL: Yes** **Years:** 2007, 2008 **Models:** All **Engines:** All **Transmissions:** All	**Secondary Air Injection System Switching Valve Stuck Open Bank 1 Conditions:** Pressure Sensor detects no pulsation of exhaust gas despite ECM commanding Air Switching Valve (ASV) to open, while engine running. **Possible Causes:** • ASV • Open or short in ASV circuit • Vacuum hose (ASV-pressure Sensor) • Air injection hose • Pressure Sensor circuit • Air Injection Control Driver (AID) • ECM
DTC: P2442 **1T CCM, MIL: Yes** **Years:** 2007, 2008 **Models:** All **Engines:** All **Transmissions:** All	**Secondary Air Injection System Switching Valve Stuck Close Bank 2 Conditions:** Pressure Sensor detects pulsation of exhaust gas despite ECM commanding Air Switching Valve (ASV) to close, while engine running. **Possible Causes:** • ASV • Open or short in ASV circuit • Pressure Sensor • Pressure Sensor circuit • Air Injection Control Driver (AID) • ECM
DTC: P2443 **1T CCM, MIL: Yes** **Years:** 2007, 2008 **Models:** All **Engines:** All **Transmissions:** All	**Secondary Air Injection System Switching Valve Stuck Open Bank 2 Conditions:** Pressure Sensor detects no pulsation of exhaust gas despite ECM commanding Air Switching Valve (ASV) to open, while engine running. **Possible Causes:** • ASV • Open or short in ASV circuit • Vacuum hose (ASV-pressure Sensor) • Air injection hose • Pressure Sensor circuit • Air Injection Control Driver (AID) • ECM

DTC	Trouble Code Title, Conditions & Possible Causes
DTC: P2444 **1T CCM, MIL: Yes** **Years:** 2007, 2008 **Models:** All **Engines:** All **Transmissions:** All	**Secondary Air Injection System Pump Stuck On Bank 1 Conditions:** Secondary air pressure more than 2.4 kPa (18 mmHg) despite ECM commanding air pump to turn off. **Possible Causes:** • Short in air pump circuit • Open or short in pressure Sensor circuit • Pressure Sensor • Air Injection Control Driver (AID) • ECM
DTC: P2445 **1T CCM, MIL: Yes** **Years:** 2007, 2008 **Models:** All **Engines:** All **Transmissions:** All	**Secondary Air Injection System Pump Stuck Off Bank 1 Conditions:** Secondary air pressure more than 2.4 kPa (18 mmHg) despite ECM commanding air pump to turn off. **Possible Causes:** • Air pump • Open in air pump circuit • Air injection system piping • Vacuum hose (pressure Sensor-air switching valve) • Pressure Sensor • Open or short in pressure Sensor circuit • Air Injection Control Driver (AID) • ECM
DTC: P2446 **1T CCM, MIL: Yes** **Years:** 2007, 2008 **Models:** All **Engines:** All **Transmissions:** All	**Secondary Air Injection System Pump Stuck On Bank 2 Conditions:** Secondary air pressure more than 2.4 kPa (18 mmHg) despite ECM commanding air pump to turn off. **Possible Causes:** • Short in air pump circuit • Open or short in pressure Sensor circuit • Pressure Sensor • Air Injection Control Driver (AID) • ECM
DTC: P2447 **1T CCM, MIL: Yes** **Years:** 2007, 2008 **Models:** All **Engines:** All **Transmissions:** All	**Secondary Alr Injection System Pump Stuck Off Bank 2 Conditions:** Secondary air pressure more than 2.4 kPa (18 mmHg) despite ECM commanding air pump to turn off. **Possible Causes:** • Air pump • Open in air pump circuit • Air injection system piping • Vacuum hose (pressure Sensor-air switching valve) • Pressure Sensor • Open or short in pressure Sensor circuit • Air Injection Control Driver (AID) • ECM
DTC: P2610 **2T CCM, MIL: Yes** **Years:** 2007, 2008 **Models:** All **Engines:** All **Transmissions:** All	**ECM/PCM Internal Engine Off Timer Performance Conditions:** To ensure the accuracy of the EVAP (Evaporative Emission) monitor values, the soak timer, which is built into the ECM, measures 5 hours (+/−15 minutes) from when the ignition switch is turned off, before the monitor is run. This allows the fuel to cool down, which stabilizes the EVAP pressure. When 5 hours have elapsed, the ECM turns on. **Possible Causes:** • ECM
DTC: P2714 **2T CCM, MIL: Yes** **Years:** 2007, 2008 **Models:** All **Engines:** All **Transmissions:** All	**Pressure Control Solenoid 'D' Performance (Shift Solenoid Valve SLT)Conditions:** The ECM detects a malfunction on SLT (ON side) according to the revolution difference of the turbine and the output shaft, and also by the oil pressure. **Possible Causes:** • Shift solenoid valve SLT remains open or closed • Shift solenoid valve S1, S2, SR, SL1 or SL2 remains open or closed • 1-2 shift valve is stuck • Valve body is blocked • Automatic transmission (clutch, brake or gear, etc)
DTC: P2716 **2T CCM, MIL: Yes** **Years:** 2007, 2008 **Models:** All **Engines:** All **Transmissions:** All	**Pressure Control Solenoid 'D' Electrical (Shift Solenoid Valve SLT) Conditions:** The ECM detects a malfunction on SLT (ON side) according to the revolution difference of the turbine and the output shaft, and also by the oil pressure. **Possible Causes:** • Shift solenoid valve SLT remains open or closed • Shift solenoid valve SSLT • ECM

DTC	Trouble Code Title, Conditions & Possible Causes
DTC: P2725 **1T CCM, MIL: Yes** **Years:** 2007, 2008 **Models:** All **Engines:** All **Transmissions:** All	**All Pressure Control Solenoid 'E' Circuit Malfunction Conditions:** Engine started, engine warmup period completed, gearshift selector in 'P' or 'N', engine speed 500 RPM or more, and the PCM detected that current flowed to the Shift Solenoid (SLN) control circuit for over 1 second. The Shift Solenoid SLN controls the hydraulic pressure acting on the accumulator control valve when gears are shifted in order to provide smooth gear shifting. **Possible Causes:** • Shift solenoid (SLN) connector is damaged or loose • Shift solenoid (SLN) control circuit open or shorted to ground • Shift solenoid (SLN) is damaged or has failed (electrical fault) • PCM failure
DTC: P2740 **1T CCM, MIL: Yes** **Years:** 2007, 2008 **Models:** All **Engines:** All **Transmissions:** All	**Transmission Fluid Temperature Sensor 'B' Circuit Conditions:** The Automatic Transmission Fluid (ATF) temperature Sensor is on the transmission, just in front of the oil cooler inlet pipeline. If the ECM detects an abnormality high ATF temperature near this Sensor, it illuminates the warning light. **Possible Causes:** • Open or short in No 2 ATF temperature Sensor circuit • Transmission wire (No 2 ATF temperature Sensor) • ECM
DTC: P2742 **1T CCM, MIL: Yes** **Years:** 2007, 2008 **Models:** All **Engines:** All **Transmissions:** All	**Transmission Fluid Temperature Sensor 'B' Circuit Low Input Conditions:** Number 2 ATF temperature Sensor resistance is less than 0.046v (25 ohm) for 0.5 seconds or more. **Possible Causes:** • Open or short in No 2 ATF temperature Sensor circuit • Transmission wire (No 2 ATF temperature Sensor) • ECM
DTC: P2743 **1T CCM, MIL: Yes** **Years:** 2007, 2008 **Models:** All **Engines:** All **Transmissions:** All	**Transmission Fluid Temperature Sensor 'B' Circuit High Input Conditions:** When 15 minutes or more have elapsed after engine is started, No 2 ATF temperature Sensor resistance is more than 4.915v (156 k ohms) for 0.5 seconds or more. **Possible Causes:** • Open or short in No 2 ATF temperature Sensor circuit • Transmission wire (No 2 ATF temperature Sensor) • ECM
DTC: P2757 **2T CCM, MIL: Yes** **Years:** 2007, 2008 **Models:** All **Engines:** All **Transmissions:** All	**Torque Converter Clutch Pressure Control Solenoid Performance (Shift Solenoid Valve SLU) Conditions:** Lock up does not occur when driving in the lock up range (normal driving at 50 mph), or lock up remains ON in the lock up OFF range. **Possible Causes:** • Shift solenoid valve SLU remains open or closed • Valve body is blocked • Torque converter clutch • Automatic transmission (clutch, brake or gear, etc) • Line pressure is too low
DTC: P2759 **2T CCM, MIL: Yes** **Years:** 2007, 2008 **Models:** All **Engines:** All **Transmissions:** All	**Torque Converter Clutch Pressure Control Solenoid Control Circuit Electrical (Shift Solenoid Valve SLU) Conditions:** Open or short is detected in shift solenoid valve SLU circuit for 1 second or more while driving. **Possible Causes:** • Open or short in shift solenoid valve SLU circuit • Shift solenoid valve SLU • ECM
DTC: P2769 **2T CCM, MIL: Yes** **Years:** 2007, 2008 **Models:** All **Engines:** All **Transmissions:** All	**Torque Converter Clutch Solenoid Circuit Low (Shift Solenoid Valve DSL) Conditions:** ECM detects short in solenoid valve DSL circuit when solenoid valve DSL is operated. **Possible Causes:** • Short in shift solenoid valve DSL circuit • Shift solenoid valve DSL • ECM
DTC: P2770 **2T CCM, MIL: Yes** **Years:** 2007, 2008 **Models:** All **Engines:** All **Transmissions:** All	**Torque Converter Clutch Solenoid Circuit High (Shift Solenoid Valve DSL) Conditions:** ECM detects short in solenoid valve DSL circuit when solenoid valve DSL is operated. **Possible Causes:** • Open in shift solenoid valve DSL circuit • Shift solenoid valve DSL • ECM

DTC	Trouble Code Title, Conditions & Possible Causes
DTC: P2808 **2T CCM, MIL: Yes** **Years:** 2007, 2008 **Models:** All **Engines:** All **Transmissions:** All	**Pressure Control Solenoid 'G' Performance or Stuck OFF (Shift Solenoid Valve SL4) Conditions:** ECM detects short in solenoid valve DSL circuit when solenoid valve DSL is operated. **Possible Causes:** • Shift solenoid valve SL4 remains open or closed • Valve body is blocked • Automatic transaxle (clutch, brake or gear etc.)
DTC: P2810 **2T CCM, MIL: Yes** **Years:** 2007, 2008 **Models:** All **Engines:** All **Transmissions:** All	**Pressure Control Solenoid 'G' Electrical (Shift Solenoid Valve SL4) Conditions:** ECM detects short in solenoid valve DSL circuit when solenoid valve DSL is operated. **Possible Causes:** • Open in shift solenoid valve DSL circuit • Shift solenoid valve SL4 • TCM
DTC: U0100 **2T CCM, MIL: Yes** **Years:** 2007, 2008 **Models:** All **Engines:** All **Transmissions:** All	**Lost Communication with ECM/PCM 'A' Conditions:** No communication with ECM continues. **Possible Causes:** • ECM • TCM
DTC: U0101 **2T CCM, MIL: Yes** **Years:** 2007, 2008 **Models:** All **Engines:** All **Transmissions:** All	**Lost Communication With TCM Conditions:** The Transmission Control Module (TCM) and ECM perform 2-way communications with each other via the Controller Area Network (CAN). The TCM sends signals to the ECM concerning required engine rpm, required engine torques, warning indicators in the combination meter, DTCs and other data. The ECM sends signals to the TCM concerning engine rpm, opening angles of the throttle valve, temperature of intake air, temperature of engine coolant, engine torques and other data. If the TCM cannot communicate with the ECM, the TCM will conclude that there is a malfunction in the CAN system, illuminate the MIL and store a DTC **Possible Causes:** • Open or short in TCN and ECM circuit • TCM • ECM

SCION

Diagnostic Trouble Codes

25

DIAGNOSTIC TROUBLE CODES ... 25-1

DIAGNOSTIC TROUBLE CODES

OBD II VEHICLE APPLICATIONS

SCION

tC
2007–2008
- 2.4L I4 . . . . 2AZ-FE

xB
2008
- 2.4L I4 . . . . 2AZ-FE

xD
2008
- 1.8L I4 . . . . 2ZR-FE

OBD II Trouble Code List (P0xxx Codes)

DTC	Trouble Code Title, Conditions & Possible Causes
DTC: P0010 **2T CCM, MIL: Yes** **Years:** 2007, 2008 **Models:** xB, xD, tC **Engines:** All **Transmissions:** All	**Camshaft Position "A" Actuator Circuit (Bank 1) Conditions:** Key on or engine running; and the PCM detected an unexpected voltage condition on the VVT Oil Control Valve Bank 1 circuit. The VVT system controls the intake camshaft in order to provide optimal valve timing during all conditions based signals from the ECT, IAT and TP sensor. The VVT regulates the intake camshaft angle using oil pressure through the Oil Control Valve. This results in the relative position between the camshaft and crankshaft to become optimal. The result is higher torque, better fuel economy and low emissions. **Possible Causes:** • OCV assembly connector is damaged or loose • OCV assembly control circuit is open or shorted to ground • OCV assembly is damaged or has failed • PCM has failed
DTC: P0011 **2T CCM, MIL: Yes** **Years:** 2007, 2008 **Models:** xB, xD, tC **Engines:** All **Transmissions:** All	**Camshaft Position 'A' Over-Advanced Or System Performance (Bank 1) Conditions:** Engine started, ECT sensor more than 158°F, vehicle driven at an engine speed of 400-4000 RPM, and the PCM detected the valve timing did not change from the "current" valve timing, or the valve timing remain fixed during testing. The VVT system controls the intake camshaft in order to provide optimal valve timing during all conditions based signals from the ECT, IAT and TP sensor. The VVT regulates the intake camshaft angle using oil pressure through the Oil Control Valve. This results in the relative position between the camshaft and crankshaft to become optimal. The result is better engine torque, fuel economy and lower emissions. **Possible Causes:** • Engine valve timing malfunction • Camshaft timing oil control valve unit is damaged or has failed • PCM or VVT ECM has failed
DTC: P0012 **2T CCM, MIL: Yes** **Years:** 2007, 2008 **Models:** xB, xD, tC **Engines:** All **Transmissions:** All	**Camshaft Position 'A' Over-Retarded (Bank 1) Conditions:** Engine started, ECT sensor more than 158°F, vehicle driven at an engine speed of 400-4000 RPM, and the PCM detected the valve timing did not change from the "current" valve timing, or that the valve timing remain fixed during the test period. The VVT system controls the intake camshaft in order to provide optimal valve timing during all conditions. The VVT regulates the intake camshaft angle using oil pressure through the OCV. This causes the relative position between the camshaft and crankshaft to become optimal. The result is improved engine torque, better fuel economy and low emissions. **Possible Causes:** • Engine valve timing malfunction • Camshaft timing oil control valve unit is damaged or has failed • PCM or VVT ECM has failed
DTC: P0013 **2T CCM, MIL: Yes** **Years:** 2007, 2008 **Models:** xB, xD, tC **Engines:** All **Transmissions:** All	**Camshaft Position "B" Actuator Circuit / Open (Bank 1) Conditions:** Key on or engine running; and the PCM detected an unexpected voltage condition on the VVT Oil Control Valve Bank 1 circuit. The VVT system controls the intake camshaft in order to provide optimal valve timing during all conditions based signals from the ECT, IAT and TP sensor. The VVT regulates the intake camshaft angle using oil pressure through the Oil Control Valve. This results in the relative position between the camshaft and crankshaft to become optimal. The result is higher torque, better fuel economy and low emissions. **Possible Causes:** • Open or short in OCV (for exhaust camshaft) circuit • OCV (for exhaust camshaft) • ECM
DTC: P0014 **2T CCM, MIL: Yes** **Years:** 2007, 2008 **Models:** xB, xD, tC **Engines:** All **Transmissions:** All	**Camshaft Position "B" – Timing Over-Advanced or System Performance (Bank 1) Conditions:** Key on or engine running; valve timing is not adjusted in exhaust valve timing advance range. **Possible Causes:** • Valve timing • Camshaft Timing Oil Control Valve (OCV) (for exhaust camshaft) • OCV filter • Camshaft timing gear assembly (for exhaust camshaft) • ECM

DTC	Trouble Code Title, Conditions & Possible Causes
DTC: P0015 **1T CCM, MIL: Yes** **Years:** 2007, 2008 **Models:** xB, xD, tC **Engines:** All **Transmissions:** All	**Camshaft Position "B" – Retarded (Bank 1) Conditions:** Key on or engine running; valve timing is not adjusted in exhaust valve timing retard range. **Possible Causes:** • Valve timing • Camshaft Timing Oil Control Valve (OCV) (for exhaust camshaft) • OCV filter • Camshaft timing gear assembly (for exhaust camshaft) • ECM
DTC: P0016 **2T CCM, MIL: Yes** **Years:** 2007, 2008 **Models:** xB, xD, tC **Engines:** All **Transmissions:** All	**Camshaft Position-Crankshaft Position (Bank 1 Sensor A) Conditions:** Engine started, engine running, and the PCM detected a deviation between the crankshaft position sensor signal and the VVT Sensor 1 signal during the test period. The crankshaft position (NE) sensor consists of a magnet, iron core and pickup coil. The NE sensor signal plate, installed on the crankshaft-timing pulley, has 34 teeth. This sensor generates 34 signals for each engine revolution. The PCM detects the crankshaft angle and engine speed based on the NE signal. It detects the correct cylinder based on signals from the VVT 1 sensor along signals from the crankshaft position sensor. **Possible Causes:** • Engine valve timing problem • Engine timing chain mechanical problem (skipped teeth or belt) • PCM has failed
DTC: P0017 **2T CCM, MIL: Yes** **Years:** 2007, 2008 **Models:** xB, xD, tC **Engines:** All **Transmissions:** All	**Camshaft Position-Crankshaft Position (Bank 1 Sensor B) Conditions:** Engine started, engine running, and the PCM detected a deviation between the crankshaft position sensor signal and the VVT Sensor 1 signal during the test period. The crankshaft position (NE) sensor consists of a magnet, iron core and pickup coil. The NE sensor signal plate, installed on the crankshaft-timing pulley, has 34 teeth. This sensor generates 34 signals for each engine revolution. The PCM detects the crankshaft angle and engine speed based on the NE signal. It detects the correct cylinder based on signals from the VVT 1 sensor along signals from the crankshaft position sensor. **Possible Causes:** • Engine valve timing problem • Engine timing chain mechanical problem (skipped teeth or belt) • PCM has failed
DTC: P0031 **1T CCM, MIL: Yes** **Years:** 2007, 2008 **Models:** xB, xD, tC **Engines:** All **Transmissions:** All	**Oxygen Sensor (Bank 1 Sensor 1) Heater Circuit Low Code Conditions:** Engine started, and the A/F sensor heater current is less than 0.8A. **Possible Causes:** • Open in A/F sensor heater circuit • A/F sensor heater • EFI relay • ECM
DTC: P0032 **1T CCM, MIL: Yes** **Years:** 2007, 2008 **Models:** xB, xD, tC **Engines:** All **Transmissions:** All	**Oxygen Sensor (Bank 1 Sensor 1) Heater Circuit High Input Conditions:** Engine started, and the A/F sensor heater current is more than 10A. **Possible Causes:** • Short in A/F sensor heater circuit • A/F sensor heater • EFI relay • ECM
DTC: P0037 **1T CCM, MIL: Yes** **Years:** 2007, 2008 **Models:** xB, xD, tC **Engines:** All **Transmissions:** All	**Heated Oxygen Sensor (Bank 1 Sensor 2) Heater Circuit Low Input Conditions:** Engine started, and the HO2 sensor heater current is less than 0.3A. **Possible Causes:** • Open in HO2 sensor heater circuit • HO2 sensor heater • EFI relay • ECM
DTC: P0038 **1T CCM, MIL: Yes** **Years:** 2007, 2008 **Models:** xB, xD, tC **Engines:** All **Transmissions:** All	**Heated Oxygen Sensor (Bank 1 Sensor 2) Heater Circuit High Input Conditions:** Engine started, and the PCM detected the HO2S-12 heater control circuit indicated more than 2.0 amps during the CCM test period. **Possible Causes:** • HO2S-12 heater control circuit is shorted to ground • HO2S-12 heater assembly is damaged or has failed • ECM has failed

DTC	Trouble Code Title, Conditions & Possible Causes
DTC: P0100 **1T CCM, MIL: Yes** **Years:** 2007, 2008 **Models:** xB, xD, tC **Engines:** All **Transmissions:** All	**Mass Or Volume Airflow Sensor Circuit Malfunction Conditions:** Engine started, engine running at under 4000 RPM, and the PCM detected an unexpected low or high voltage condition on the Mass Airflow (MAF) sensor circuit for over 3 seconds during the CCM test. The MAF sensor on this engine includes a hot wire assembly with an air temperature sensor, platinum hot wire and control unit mounted in a plastic housing. This airflow meter works on the principle that the hot wire and temperature sensor located in the intake air bypass of the housing detect any changes in the (incoming) air temperature. **Possible Causes:** • MAF sensor signal circuit is open, shorted to ground or power • MAF sensor ground circuit is open between sensor and ground • MAF sensor power circuit is open (check the power to the relay) • MAF sensor has failed, or the PCM has failed
DTC: P0101 **2T CCM, MIL: Yes** **Years:** 2007, 2008 **Models:** xB, xD, tC **Engines:** All **Transmissions:** All	**Mass Or Volume Airflow Sensor Signal Range/Performance Conditions:** DTC P0100 not set, engine speed under 900 RPM, throttle valve closed, ECT sensor over 158°F, and the PCM detected the MAF was above 2.20v; or with the engine speed over 1500 RPM, the throttle valve closed and the TP sensor over 0.63v, the MAF sensor was less than 1.06v. This airflow meter works on the principle that the hot wire and temperature sensor in the intake air bypass of the housing detect any changes in the (incoming) air temperature **Possible Causes:** • MAF sensor signal circuit is open, shorted to ground or power • MAF sensor is contaminated, damaged or it has failed • PCM has failed
DTC: P0102 **2T CCM, MIL: Yes** **Years:** 2007, 2008 **Models:** xB, xD, tC **Engines:** All **Transmissions:** All	**Mass Or Volume Airflow Sensor Circuit Low Input Conditions:** DTC P0100 not set, engine started, and the PCM detected an unexpected low voltage condition on the MAF sensor circuit during the CCM test period. This airflow meter works on the principle that the hot wire and temperature sensor located in the intake air bypass of the housing detect any changes in the (incoming) air temperature **Possible Causes:** • MAF sensor signal circuit is open or shorted to ground • MAF sensor is contaminated, damaged or it has failed • PCM has failed
DTC: P0103 **2T CCM, MIL: Yes** **Years:** 2007, 2008 **Models:** xB, xD, tC **Engines:** All **Transmissions:** All	**Mass Or Volume Airflow Sensor Circuit High Input Conditions:** DTC P0100 not set, engine started, and the PCM detected an unexpected high voltage condition on the MAF sensor circuit during the CCM test period. This airflow meter works on the principle that the hot wire and temperature sensor located in the intake air bypass of the housing detect any changes in the (incoming) air temperature **Possible Causes:** • MAF sensor signal circuit is shorted to power • MAF sensor is contaminated, damaged or it has failed • PCM has failed
DTC: P0110 **1T CCM, MIL: Yes** **Years:** 2007, 2008 **Models:** xB, xD, tC **Engines:** All **Transmissions:** All	**Intake Air Temperature Sensor Circuit Malfunction Conditions:** Engine started, and the PCM detected an unexpected low or high voltage on the IAT sensor circuit (Scan Tool reads −40°F or 284°F). This sensor is located inside the MAF sensor. **Possible Causes:** • IAT sensor signal circuit is open, shorted to ground or VREF • IAT sensor ground circuit is open between sensor and ground • IAT sensor is contaminated, damaged or has failed • PCM has failed
DTC: P0111 **2T CCM, MIL: Yes** **Years:** 2007, 2008 **Models:** xB, xD, tC **Engines:** All **Transmissions:** All	**Intake Air Temperature Sensor Gradient Too High Conditions:** In duration between engine warmed up and next engine starts, change in Intake Air Temperature (IAT) sensor output below threshold. During engine warming up after cold engine starts, change in IAT sensor output below threshold. **Possible Causes:** • Mass Air Flow Meter
DTC: P0112 **1T CCM, MIL: Yes** **Years:** 2007, 2008 **Models:** xB, xD, tC **Engines:** All **Transmissions:** All	**Intake Air Temperature Sensor Circuit Low Input Conditions:** Key on or engine running; and the PCM detected an unexpected low voltage condition on the IAT sensor circuit for over 500 ms. **Possible Causes:** • IAT sensor connector is damaged (it may be shorted internally) • IAT sensor ground circuit is shorted to ground • IAT sensor is damaged or has failed • PCM has failed

DTC	Trouble Code Title, Conditions & Possible Causes
DTC: P0113 **1T CCM, MIL: Yes** **Years:** 2007, 2008 **Models:** xB, xD, tC **Engines:** All **Transmissions:** All	**Intake Air Temperature Sensor Circuit High Input Conditions:** Key on or engine running; and the PCM detected an unexpected high voltage condition on the IAT sensor circuit for over 500 ms. **Possible Causes:** • IAT sensor connector is damaged (it may be open internally) • IAT sensor ground circuit is open • IAT sensor is damaged or has failed • PCM has failed
DTC: P0115 **2T CCM, MIL: Yes** **Years:** 2007, 2008 **Models:** xB, xD, tC **Engines:** All **Transmissions:** All	**Engine Coolant Temperature Sensor Circuit Malfunction Conditions:** Key on or engine running; and the PCM detected an unexpected "low" or "high" voltage on the ECT sensor circuit (Scan Tool reads less than −40°F or low voltage or more than 284°F for high voltage). **Possible Causes:** • IAT sensor signal circuit is open, shorted to ground or VREF • ECT sensor ground circuit is open between sensor and ground • ECT sensor is contaminated, damaged or has failed • PCM has failed
DTC: P0116 **2T CCM, MIL: Yes** **Years:** 2007, 2008 **Models:** xB, xD, tC **Engines:** All **Transmissions:** All	**Engine Coolant Temperature Sensor Range/Performance Conditions:** Engine started, ECT sensor from 95°F to 140°F, IAT sensor more than 19.9°F, vehicle driven with several changes in the VSS signals, and the PCM detected the ECT sensor signal did not increase more than 37.4°F after engine was started and the test period completed. **Possible Causes:** • Check for problems in the cooling system (i.e., coolant, the fan) • ECT sensor signal circuit or ground circuit has high resistance • ECT sensor is contaminated, damaged or has failed • PCM has failed
DTC: P0117 **2T CCM, MIL: Yes** **Years:** 2007, 2008 **Models:** xB, xD, tC **Engines:** All **Transmissions:** All	**Engine Coolant Temperature Sensor Circuit Low Input Conditions:** Key on or engine running; and the PCM detected an unexpected low voltage condition on the ECT sensor (Scan Tool reads below −40°F). **Possible Causes:** • ECT sensor signal circuit is shorted to ground • ECT sensor is damaged or has failed • PCM has failed
DTC: P0118 **2T CCM, MIL: Yes** **Years:** 2007, 2008 **Models:** xB, xD, tC **Engines:** All **Transmissions:** All	**Engine Coolant Temperature Sensor Circuit High Input Conditions:** Key on or engine running; and the PCM detected an unexpected high voltage condition on the ECT sensor (Scan Tool reads over 284°F). **Possible Causes:** • ECT sensor signal circuit is open • ECT sensor ground circuit is open • ECT sensor is damaged or has failed • PCM has failed
DTC: P011B **2T CCM, MIL: Yes** **Years:** 2007, 2008 **Models:** xB, xD, tC **Engines:** All **Transmissions:** All	**Engine Coolant Temperature/Intake Air Temperature Correlation Conditions:** The monitor runs when the engine started cold after 7 hours or more has lapsed since the engine was stopped (ignition switch turned to OFF) on the previous trip. If the difference between the ECT and the IAT on a cold start exceeds 36°F (20°C), the ECM interprets this as a malfunction in the ECT sensor circuit and IAT sensor circuit, and sets the DTC. **Possible Causes:** • IAT sensor • ECT sensor ground circuit is open • ECT sensor • ECM
DTC: P0120 **2T CCM, MIL: Yes** **Years:** 2007, 2008 **Models:** xB, xD, tC **Engines:** All **Transmissions:** All	**TP Sensor or Switch 'A' Circuit Malfunction Conditions:** Engine started, and the PCM detected the TP sensor signal (VTA) was under 0.10v or over 4.90v. The TP sensor, mounted on the throttle body, detects the Throttle Valve opening angle (0.30v-0.70v closed). The PCM uses this signal for A/F ratio correction and power increase changes during **All** modes of engine operation. **Possible Causes:** • VTA sensor signal circuit is open or shorted to ground • VTA sensor ground circuit is open • VTA power circuit (VREF) is open • TP sensor is damaged or has failed • ECM has failed

DTC	Trouble Code Title, Conditions & Possible Causes
DTC: P0121 **2T CCM, MIL: Yes** **Years:** 2007, 2008 **Models:** xB, xD, tC **Engines:** All **Transmissions:** All	**TP Sensor or TP Switch 'A' Signal Range/Performance Conditions:** Vehicle speed more than 19 MPH at least once; and the PCM detected the TP sensor input was out of the applicable range with the VSS input reading between 30 MPH and 0 MPH. **Possible Causes:** • TP sensor signal circuit open or shorted to ground (intermittent) • TP sensor is loose at it mounting or the throttle is binding • TP sensor is damaged or has failed (perform a sweep test) • PCM has failed
DTC: P0122 **2T CCM, MIL: Yes** **Years:** 2007, 2008 **Models:** xB, xD, tC **Engines:** All **Transmissions:** All	**TP Sensor or TP Switch 'A' Circuit Low Input Conditions:** Vehicle speed more than 19 MPH at least once; and the PCM detected the TP sensor input was out of the applicable range with the VSS input reading between 30 MPH and 0 MPH. **Possible Causes:** • TP sensor signal circuit open or shorted to ground (intermittent) • TP sensor is loose at it mounting or the throttle is binding • TP sensor is damaged or has failed (perform a sweep test) • PCM has failed
DTC: P0123 **1T CCM, MIL: Yes** **Years:** 2007, 2008 **Models:** xB, xD, tC **Engines:** All **Transmissions:** All	**TP Sensor or TP Switch 'A' Circuit High Input Conditions:** Output voltage of VTA1 is 4.535v or more for 2 seconds or more when accelerator pedal is depressed (10 seconds or more when accelerator pedal released). **Possible Causes:** • Throttle position sensor (built into throttle body) • Open in VTA1 circuit • Open in E2 circuit • Short between VC and VTA1 circuit • ECM
DTC: P0125 **1T CCM, MIL: Yes** **Years:** 2007, 2008 **Models:** xB, xD, tC **Engines:** All **Transmissions:** All	**Insufficient Coolant Temperature For Closed Loop Conditions:** ECT sensor less than 19.4°F, engine started, engine runtime over 20 minutes, and the PCM detected the ECT sensor indicated 68°F or less; or with the ECT sensor from 19.4°F to 50.0°F at startup, engine runtime over 5 minutes, the PCM detected the ECT sensor indicated 68°F or less; or with the ECT sensor more than 19.4°F at startup, and the engine runtime over 5 minutes, the PCM detected the ECT sensor signal was 68°F or less; or with the ECT sensor over 50°F at startup, engine runtime over 2 minutes, the PCM detected the ECT sensor signal did not reach 86°F during the CCM test period. **Possible Causes:** • Check the operation of the thermostat (it may be stuck open) • ECT sensor signal circuit has high resistance • ECT sensor has failed • Inspect for low coolant level for an incorrect coolant mixture
DTC: P0128 **2T CCM, MIL: Yes** **Years:** 2007, 2008 **Models:** xB, xD, tC **Engines:** All **Transmissions:** All	**Thermostat System Malfunction Conditions:** Engine started, ECT sensor signal below 140°F at startup, and the PCM detected the ECT sensor did not reach 167°F after the warm-up period expired (engine runtime 5-10 minutes). **Possible Causes:** • Check the operation of the thermostat (it may be stuck open) • ECT sensor is out-of-calibration or skewed • Inspect for low coolant level or for an incorrect coolant mixture
DTC: P0136 **2T CCM, MIL: Yes** **Years:** 2007, 2008 **Models:** xB, xD, tC **Engines:** All **Transmissions:** All	**HO2S-12 (Bank 1 Sensor 2) Circuit Malfunction Conditions:** Vehicle driven to a speed of over 25 MPH while in closed loop, and the PCM detected the HO2S-12 signal was fixed at more than 400 mv, or that it was fixed at less than 600 mv. **Possible Causes:** • HO2S signal circuit is open or shorted to ground • HO2S signal circuit shorted to VREF or system power (B+) • HO2S is damaged, contaminated or it has failed
DTC: P0137 **2T CCM, MIL: Yes** **Years:** 2007, 2008 **Models:** xB, xD, tC **Engines:** All **Transmissions:** All	**HO2S-12 (Bank 1 Sensor 2) Circuit Low Voltage Conditions:** Vehicle driven to a speed of over 25 MPH while in closed loop, and the PCM detected the HO2S-12 signal was fixed at more than 400 mv, or that it was fixed at less than 600 mv. **Possible Causes:** • HO2S signal circuit is open or shorted to ground • HO2S signal circuit shorted to VREF or system power (B+) • HO2S is damaged, contaminated or it has failed

DTC	Trouble Code Title, Conditions & Possible Causes
DTC: P0138 **2T CCM, MIL: Yes** **Years:** 2007, 2008 **Models:** xB, xD, tC **Engines:** All **Transmissions:** All	**HO2S-12 (Bank 1 Sensor 2) Circuit High Voltage Conditions:** High voltage (short), during active air-fuel ratio control, Heated Oxygen (HO2) sensor voltage output 0.59v or more and the target air-fuel ratio is lean. Extremely high voltage (short), HO2 sensor voltage output exceeds 1.2v for more than 10 seconds. **Possible Causes:** • Short in HO2 sensor (sensor 2) circuit • HO2 sensor (sensor 2) • ECM internal circuit malfunction • Air-Fuel Ratio (A/F) sensor (sensor 1)
DTC: P0139 **2T CCM, MIL: Yes** **Years:** 2007, 2008 **Models:** xB, xD, tC **Engines:** All **Transmissions:** All	**HO2S-12 (Bank 1 Sensor 2) Circuit Slow Response Conditions:** Heated oxygen sensor (sensor 2) voltage does not drop below 0.2v immediately after fuel cut starts. **Possible Causes:** • Short in HO2 sensor (sensor 2) circuit • HO2 sensor (sensor 2) • ECM internal circuit malfunction • Air-Fuel Ratio (A/F) sensor (sensor 1)
DTC: P0141 **2T CCM, MIL: Yes** **Years:** 2007, 2008 **Models:** xB, xD, tC **Engines:** All **Transmissions:** All	**Oxygen Sensor Heater Circuit Malfunction (Bank 1 Sensor 2) Conditions:** Engine started, engine running, and the cumulative heater resistance correction value exceeds the threshold. **Possible Causes:** • Open or short in HO2 sensor heater circuit • HO2S sensor heater (sensor 2) • EFI relay • ECM
DTC: P0171 **2T FUEL, MIL: Yes** **Years:** 2007, 2008 **Models:** xB, xD, tC **Engines:** All **Transmissions:** All	**Fuel System Too Lean (Bank 1) Conditions:** DTC P0100, P0101, P0105, P0110, P0115, P0120, P0121, P0130, P0133, P0136, P0135, P0136, P0141, P0151, P0156, P0161, P0300, P0301-P0306, P0440, P0500 and P0505 not set, ECT sensor more than 158°F, vehicle driven at a constant speed of less than 62 MPH with the engine speed over 1500 RPM, and the PCM detected the lean fuel trim correction value was over the limit. **Possible Causes:** • A/FS or HO2S is contaminated, deteriorated or it has failed • Air leaks after the MAF sensor, or in the EGR or PCV system • Base engine "mechanical" fault affecting one or more cylinders • Exhaust leaks located in front of the A/FS or HO2S location • Fuel control sensor is out of calibration (i.e., ECT, IAT or MAP) • Fuel delivery system supplying too little fuel during cruise or idle periods (e.g., faulty fuel pump or dirty, restricted fuel filter) • Fuel injector (one or more) dirty or pressure regulator has failed • Vehicle driven low on fuel or until it ran out of fuel
DTC: P0172 **2T FUEL, MIL: Yes** **Years:** 2007, 2008 **Models:** xB, xD, tC **Engines:** All **Transmissions:** All	**Fuel System Too Rich (Bank 1) Conditions:** DTC P0100, P0101, P0105, P0110, P0115, P0120, P0121, P0130, P0133, P0136, P0135, P0136, P0141, P0151, P0156, P0161, P0300, P0301-P0306, P0440, P0500 and P0505 not set, ECT sensor more than 158°F, vehicle driven at a constant speed of less than 62 MPH with the engine speed over 1500 RPM, and the PCM detected the rich fuel trim correction value was over the limit. **Possible Causes:** • A/FS or HO2S is contaminated, deteriorated or it has failed • Base engine "mechanical" fault affecting one or more cylinders • EVAP system component has failed or canister fuel saturated • Exhaust leaks located in front of the A/FS or HO2S location • Fuel control sensor is out of calibration (i.e., ECT, IAT or MAF) • Fuel delivery system supplying too much fuel during cruise or idle periods (e.g., faulty fuel pump, or faulty pressure regulator) • Fuel injector(s) is leaking or stuck partially open (one or more)
DTC: P0220 **1T CCM, MIL: Yes** **Years:** 2007, 2008 **Models:** xB, xD, tC **Engines:** All **Transmissions:** All	**Throttle/Pedal Position Sensor/Switch "B" Circuit Conditions:** Output voltage of VTA2 quickly fluctuates beyond lower and upper malfunction thresholds for 2 seconds or more when accelerator pedal is depressed (10 seconds or more when accelerator pedal is released). **Possible Causes:** • Throttle position sensor (built into throttle body) • ECM

DTC	Trouble Code Title, Conditions & Possible Causes
DTC: P0222 **1T CCM, MIL: Yes** **Years:** 2007, 2008 **Models:** xB, xD, tC **Engines:** All **Transmissions:** All	**Throttle/Pedal Position Sensor/Switch "B" Circuit Low Input Conditions:** Output voltage of VTA2 is 1.75v or less for 2 seconds or more when accelerator pedal is depressed (10 seconds or more when accelerator pedal is released). **Possible Causes:** • Throttle position sensor (built into throttle body) • Short in VTA2 circuit • Open in VC circuit • ECM
DTC: P0223 **1T CCM, MIL: Yes** **Years:** 2007, 2008 **Models:** xB, xD, tC **Engines:** All **Transmissions:** All	**Throttle/Pedal Position Sensor/Switch "B" Circuit High Input Conditions:** Output voltage of VTA2 is 4.8v or more, and VTA1 is between 0.2v and 2.02v, for seconds or more when accelerator pedal is depressed (10 seconds or more when accelerator pedal is released). **Possible Causes:** • Throttle position sensor (built into throttle body) • Short in VTA2 circuit • Open in VC circuit • Short between VC and VTA2 circuits • ECM
DTC: P0300 **2T MISFIRE, MIL: Yes** **Years:** 2007, 2008 **Models:** xB, xD, tC **Engines:** All **Transmissions:** All	**Multiple Cylinder Misfire Detected Conditions:** DTC P0100, P0101, P0102, P0103, P0105, P0110, P0112, P0113, P0115, P0117, P0118, P0120, P0121, P0122, P0123, P0125, P0335, P0340, P0500, P0505 and P0510 not set, engine started, vehicle driven to a speed of over 3 MPH for 1 minute, and the PCM detected a misfire rate of 1-2% (High Emissions 2T), or a misfire rate of 6-30% (Catalyst Damaging 1T) in two or more cylinders. **Note: If the misfire is severe, the MIL will flash on/off on the 1st trip. Look at the misfire ratio for All of the cylinders on the Scan Tool. The cylinder with the highest misfire ratio should be checked first.** **Possible Causes:** • Air leak in the intake manifold, or in the EGR or PCV system • Base engine mechanical fault that affects two or more cylinders • EGR valve is stuck open or the PCV system has a vacuum leak • Fuel delivery component fault that affects two or more cylinders (e.g., contaminated, dirty or sticking fuel injectors) • Ignition system fault (coil or plug) that affects several cylinders • Mass airflow meter is contaminated, or its signal is out of range
DTC: P0301 **2T MISFIRE, MIL: Yes** **Years:** 2007, 2008 **Models:** xB, xD, tC **Engines:** All **Transmissions:** All	**Cylinder 1 Misfire Detected Conditions:** DTC P0100, P0101, P0102, P0103, P0105, P0110, P0112, P0113, P0115, P0117, P0118, P0120, P0121, P0122, P0123, P0125, P0335, P0340, P0500, P0505 and P0510 not set, engine started, vehicle driven to a speed of over 3 MPH for 1 minute, and the PCM detected a misfire rate of 1-2% (High Emissions 2T), or a misfire rate of 6-30% (Catalyst Damaging 1T) in Cylinder 1. **Note: If the misfire is severe, the MIL will flash on/off on the 1st trip** **Possible Causes:** • Base engine mechanical fault that affects only Cylinder 1 • EGR valve is stuck open or the PCV system has a vacuum leak • Fuel component fault that affects only Cylinder 1 (a contaminated or sticking injector) • Ignition system problem (coil or plug) that affects Cylinder 1
DTC: P0302 **2T MISFIRE, MIL: Yes** **Years:** 2007, 2008 **Models:** xB, xD, tC **Engines:** All **Transmissions:** All	**Cylinder 2 Misfire Detected Conditions:** DTC P0100, P0101, P0102, P0103, P0105, P0110, P0112, P0113, P0115, P0117, P0118, P0120, P0121, P0122, P0123, P0125, P0335, P0340, P0500, P0505 and P0510 not set, engine started, vehicle driven to a speed of over 3 MPH for 1 minute, and the PCM detected a misfire rate of 1-2% (High Emissions 2T), or a misfire rate of 6-30% (Catalyst Damaging 1T) in Cylinder 2. **Note: If the misfire is severe, the MIL will flash on/off on the 1st trip.** **Possible Causes:** • Base engine mechanical fault that affects only Cylinder 2 • EGR valve is stuck open or the PCV system has a vacuum leak • Fuel component fault that affects only Cylinder 2 (a contaminated or sticking injector) • Ignition system problem (coil or plug) that affects Cylinder 2

DTC	Trouble Code Title, Conditions & Possible Causes
DTC: P0303 **2T MISFIRE, MIL: Yes** **Years:** 2007, 2008 **Models:** xB, xD, tC **Engines:** All **Transmissions:** All	**Cylinder 3 Misfire Detected Conditions:** DTC P0100, P0101, P0102, P0103, P0105, P0110, P0112, P0113, P0115, P0117, P0118, P0120, P0121, P0122, P0123, P0125, P0335, P0340, P0500, P0505 and P0510 not set, engine started, vehicle driven to a speed of over 3 MPH for 1 minute, and the PCM detected a misfire rate of 1-2% (High Emissions 2T), or a misfire rate of 6-30% (Catalyst Damaging 1T) in Cylinder 3. **Note: If the misfire is severe, the MIL will flash on/off on the 1st trip.** **Possible Causes:** • Base engine mechanical fault that affects only Cylinder 3 • EGR valve is stuck open or the PCV system has a vacuum leak • Fuel component fault that affects only Cylinder 3 (a contaminated or sticking injector) • Ignition system problem (coil or plug) that affects Cylinder 3
DTC: P0304 **2T MISFIRE, MIL: Yes** **Years:** 2007, 2008 **Models:** xB, xD, tC **Engines:** All **Transmissions:** All	**Cylinder 4 Misfire Detected Conditions:** DTC P0100, P0101, P0102, P0103, P0105, P0110, P0112, P0113, P0115, P0117, P0118, P0120, P0121, P0122, P0123, P0125, P0335, P0340, P0500, P0505 and P0510 not set, engine started, vehicle driven to a speed of over 3 MPH for 1 minute, and the PCM detected a misfire rate of 1-2% (High Emissions 2T), or a misfire rate of 6-30% (Catalyst Damaging 1T) in Cylinder 4. **Note: If the misfire is severe, the MIL will flash on/off on the 1st trip.** **Possible Causes:** • Base engine mechanical fault that affects only Cylinder 4 • EGR valve is stuck open or the PCV system has a vacuum leak • Fuel component fault that affects only Cylinder 4 (a contaminated or sticking injector) • Ignition system problem (coil or plug) that affects Cylinder 4
DTC: P0327 **1T CCM, MIL: Yes** **Years:** 2007, 2008 **Models:** xB, xD, tC **Engines:** All **Transmissions:** All	**Knock Sensor 1 Circuit Low Input (Bank 1 Or Single Sensor) Conditions:** Engine started, vehicle driven with the engine speed over 1200 RPM, and the PCM detected an unexpected voltage condition on the Knock Sensor 1 (KS1) circuit during the CCM test. **Possible Causes:** • Verify that the Knock Sensor (KS) is tightened to specification • Knock sensor signal circuit is open or shorted to ground • Knock sensor signal circuit is shorted to VREF or system power • Knock sensor is damaged or has failed • PCM has failed
DTC: P0328 **1T CCM, MIL: Yes** **Years:** 2007, 2008 **Models:** xB, xD, tC **Engines:** All **Transmissions:** All	**Knock Sensor 1 Circuit High Input (Bank 1 Or Single Sensor) Conditions:** Engine started, vehicle driven with the engine speed over 1200 RPM, and the PCM detected an unexpected voltage condition on the Knock Sensor 1 (KS1) circuit during the CCM test. **Possible Causes:** • Verify that the Knock Sensor (KS) is tightened to specification • Knock sensor signal circuit is open or shorted to ground • Knock sensor signal circuit is shorted to VREF or system power • Knock sensor is damaged or has failed • PCM has failed
DTC: P0335 **1T CCM, MIL: Yes** **Years:** 2007, 2008 **Models:** xB, xD, tC **Engines:** All **Transmissions:** All	**Crankshaft Position Sensor 'A' Circuit Malfunction Conditions:** Engine cranking; and the PCM did not detect any CKP Sensor 'A' signals, or with the engine speed over 600 RPM, it did not receive any CKP sensor signals, or the CKP signal was lost. **Possible Causes:** • CKP Sensor 'A' signal circuit is open, shorted to ground or shorted to system power • CKP Sensor 'A' signal ground circuit is open • CKP Sensor 'A' is damaged or has failed
DTC: P0339 **1T CCM, MIL: Yes** **Years:** 2007, 2008 **Models:** xB, xD, tC **Engines:** All **Transmissions:** All	**Crankshaft Position Sensor 'A' Circuit Intermittent Conditions:** Engine cranking; and the PCM did not detect any CKP Sensor 'A' signals, or with the engine speed over 600 RPM, it did not receive any CKP sensor signals, or the CKP signal was lost. **Possible Causes:** • CKP Sensor 'A' signal circuit is open, shorted to ground or shorted to system power • CKP Sensor 'A' signal ground circuit is open • CKP Sensor 'A' is damaged or has failed

DTC	Trouble Code Title, Conditions & Possible Causes
DTC: P0340 **2T CCM, MIL: Yes** **Years:** 2007, 2008 **Models:** xB, xD, tC **Engines:** All **Transmissions:** All	**Camshaft Position Sensor "A" Circuit (Bank 1 Or Single Sensor) Malfunction Conditions:** Engine cranking; and the PCM did not detect any CMP sensor signals, or with the engine speed over 600 RPM, it did not detect any CMP signals, or the CMP signal was interrupted. **Possible Causes:** • CMP sensor signal circuit is open, shorted to ground or power • CMP sensor signal ground circuit is open • CMP sensor has failed, or the PCM has failed
DTC: P0342 **1T CCM, MIL: Yes** **Years:** 2007, 2008 **Models:** xB, xD, tC **Engines:** All **Transmissions:** All	**Camshaft Position Sensor "A" Circuit Low Input (Bank 1 Or Single Sensor) Conditions:** Engine running, output voltage of CMP sensor less than 0.3v for 4 seconds. **Possible Causes:** • Open or short in CMP sensor circuit for intake camshaft • CMP sensor for intake camshaft • Camshaft timing gear for intake camshaft • Jumped tooth of timing chain for intake camshaft • ECM
DTC: P0343 **1T CCM, MIL: Yes** **Years:** 2007, 2008 **Models:** xB, xD, tC **Engines:** All **Transmissions:** All	**Camshaft Position Sensor "A" Circuit High Input (Bank 1 Or Single Sensor) Conditions:** Engine running, output voltage of 4.7v for 4 seconds. **Possible Causes:** • Open or short in CMP sensor circuit for intake camshaft • CMP sensor for intake camshaft • Camshaft timing gear for intake camshaft • Jumped tooth of timing chain for intake camshaft • ECM
DTC: P0351 **1T CCM, MIL: Yes** **Years:** 2007, 2008 **Models:** xB, xD, tC **Engines:** All **Transmissions:** All	**Ignition Coil "A" Primary/Secondary Circuit Malfunction Conditions:** Engine started, and the PCM did not detect a change in the IGF signal on the Ignition Coil No. 1 IGF circuit. This engine uses a Direct Ignition (DI) system where one coil is used to fire one cylinder. The coil high-energy secondary wire is connected to one spark plug. If P0351 to P0356 are all set, check for an open/shorted IGF circuit. **Possible Causes:** • IGT1 circuit is open or shorted to ground • Ignition Coil No. 1 is damaged or it has failed • Problem present in the Ignition System • PCM has failed
DTC: P0352 **1T CCM, MIL: Yes** **Years:** 2007, 2008 **Models:** xB, xD, tC **Engines:** All **Transmissions:** All	**Ignition Coil "B" Primary/Secondary Circuit Malfunction Conditions:** Engine started, and the PCM did not detect a change in the IGF signal on the Ignition Coil No. 2 IGF circuit. This engine uses a Direct Ignition (DI) system where one coil is used to fire one cylinder. The coil high-energy secondary wire is connected to one spark plug. If P0351 to P0356 are All set, check for an open/shorted IGF circuit. **Possible Causes:** • IGT2 circuit is open or shorted to ground • Ignition Coil No. 2 is damaged or it has failed • Problem present in the Ignition System • PCM has failed
DTC: P0353 **1T CCM, MIL: Yes** **Years:** 2007, 2008 **Models:** xB, xD, tC **Engines:** All **Transmissions:** All	**Ignition Coil "C" Primary/Secondary Circuit Malfunction Conditions:** Engine started, and the PCM did not detect a change in the IGF signal on the Ignition Coil No. 3 IGF circuit. This engine uses a Direct Ignition (DI) system where one coil is used to fire one cylinder. The coil high-energy secondary wire is connected to one spark plug. If P0351 to P0356 are all set, check for an open/shorted IGF circuit. **Possible Causes:** • IGT3 circuit is open or shorted to ground • Ignition Coil No. 3 is damaged or it has failed • Problem present in the Ignition System • PCM has failed

DTC	Trouble Code Title, Conditions & Possible Causes
DTC: P0354 **1T CCM, MIL: Yes** **Years:** 2007, 2008 **Models:** xB, xD, tC **Engines:** All **Transmissions:** All	**Ignition Coil "D" Primary/Secondary Circuit Malfunction Conditions:** Engine started, and the PCM did not detect a change in the IGF signal on the Ignition Coil No. 4 IGF circuit. This engine uses a Direct Ignition (DI) system where one coil is used to fire one cylinder. The coil high-energy secondary wire is connected to one spark plug. If P0351 to P0356 are all set, check for an open/shorted IGF circuit. **Possible Causes:** • IGT4 circuit is open or shorted to ground • Ignition Coil No. 4 is damaged or it has failed • Problem present in the Ignition System • PCM has failed
DTC: P0365 **1T CCM, MIL: Yes** **Years:** 2007, 2008 **Models:** xB, xD, tC **Engines:** All **Transmissions:** All	**Camshaft Position Sensor "B" Circuit (Bank 1) Conditions:** Engine started, no exhaust camshaft position sensor signal to ECM while cranking. Missing exhaust camshaft position sensor signal for 5 seconds at engine speed of 600 rpm or more. Input voltage to ECM remains less than 0.3v, or more than 4.7v for 4 seconds when 2 or more seconds have elapsed after turning ignition switch ON. **Possible Causes:** • Open or short in exhaust camshaft position sensor circuit • Exhaust camshaft position sensor • Camshaft timing gear for exhaust camshaft • Jumped tooth of timing chain for exhaust camshaft • ECM
DTC: P0367 **1T CCM, MIL: Yes** **Years:** 2007, 2008 **Models:** xB, xD, tC **Engines:** All **Transmissions:** All	**Camshaft Position Sensor "B" Circuit Low Input (Bank 1) Conditions:** Engine started, output voltage of exhaust camshaft position sensor less than 0.3v for 4 seconds. **Possible Causes:** • Open or short in exhaust camshaft position sensor circuit • Exhaust camshaft position sensor • Camshaft timing gear for exhaust camshaft • Jumped tooth of timing chain for exhaust camshaft • ECM
DTC: P0368 **1T CCM, MIL: Yes** **Years:** 2007, 2008 **Models:** xB, xD, tC **Engines:** All **Transmissions:** All	**Camshaft Position Sensor "B" Circuit High Input (Bank 1) Conditions:** Engine started, output voltage of exhaust camshaft position sensor 4.7v for 4 seconds. **Possible Causes:** • Open or short in exhaust camshaft position sensor circuit • Exhaust camshaft position sensor • Camshaft timing gear for exhaust camshaft • Jumped tooth of timing chain for exhaust camshaft • ECM
DTC: P0420 **2T CAT, MIL: Yes** **Years:** 2007, 2008 **Models:** xB, xD, tC **Engines:** All **Transmissions:** All	**Catalyst Efficiency Below Normal (Bank 1) Conditions:** DTC P0100, P0101, P0102, P0103, P0110, P0112, P0113, P0115, P0116, P0117, P0118, P0120, P0121, P0122, P0123, P0335, P0340 and P0500 not set, engine started, vehicle driven to a speed of 45-60 MPH at 2500-3000 RPM in closed loop for 3-5 minutes, and the PCM detected too much variation in the voltage amplitudes of the HO2S-12 signal (Bank 1). **Possible Causes:** • Catalytic converter is damaged, contaminated or has failed • Front A/FS or rear HO2S is contaminated with fuel or moisture • Front A/FS or the rear HO2S is loose in its mounting hole • Front A/FS is older (aged) than the rear HO2S (HO2S is lazy) • Gas leaks at the exhaust manifold or in the exhaust pipes
DTC: P043E **2T EVAP, MIL: Yes** **Years:** 2007, 2008 **Models:** xB, xD, tC **Engines:** All **Transmissions:** All	**EVAP Emission System Reference Orifice Clog Up Conditions:** Engine off, key off, a predetermined condition (closed loop, etc) is met, the purge VSV is opened by the ECM and fuel vapors stored in the canister are purged into the intake manifold. The ECM changes the duty cycle ratio of the purge VSV to control the purge flow volume. The purge volume is also determined by the intake manifold pressure. Atmospheric pressure is allowed into the canister through the vent valve to ensure that the purge flow is maintained when negative pressure (vacuum) is applied to the canister. **Possible Causes:** • Reference orifice clogged up • Reference orifice high flow • Leak detection pump OFF malfunction • Leak detection pump ON malfunction • Vent valve ON (close) position

DTC	Trouble Code Title, Conditions & Possible Causes
DTC: P043F **2T EVAP, MIL: Yes** **Years:** 2007, 2008 **Models:** xB, xD, tC **Engines:** All **Transmissions:** All	**EVAP Emission System Reference Orifice High Flow Conditions:** Engine off, key off, a predetermined condition (closed loop, etc) is met, the purge VSV is opened by the ECM and fuel vapors stored in the canister are purged into the intake manifold. The ECM changes the duty cycle ratio of the purge VSV to control the purge flow volume. The purge volume is also determined by the intake manifold pressure. Atmospheric pressure is allowed into the canister through the vent valve to ensure that the purge flow is maintained when negative pressure (vacuum) is applied to the canister. **Possible Causes:** • Reference orifice clogged up • Reference orifice high flow • Leak detection pump OFF malfunction • Leak detection pump ON malfunction • Vent valve ON (close) position
DTC: P0441 **2T EVAP, MIL: Yes** **Years:** 2007, 2008 **Models:** xB, xD, tC **Engines:** All **Transmissions:** All	**EVAP System Incorrect Purge Flow Detected Conditions:** ECT sensor less than 86°F at startup, vehicle driven at 55-60 MPH for 2-3 minutes, and the PCM detected the EVAP canister pressure did not decrease during purge conditions, or it remained too low during purge cutoff conditions. The vapor pressure sensor, VSV for the canister closed valve (CCV) and the VSV for the vapor-switching valve are used to detect EVAP system faults. The PCM closes the CCV and opens the VSV for vapor switching valve to cause an increase in vacuum in the EVAP system. Once the vacuum reaches a certain point, the PCM closes the VSV to test system operation. **Possible Causes:** • Charcoal canister is clogged, loaded with fuel or with moisture • Fuel tank over-fill check valve cracked or damaged • Fuel tank seal leaking, fuel tank cracked or damaged/leaking • Fuel vapor hoses/tubes blocked or restricted, or fuel vapor control valve tube or fuel vapor vent valve assembly blocked • Vacuum hose or tubing cracked, damaged or disconnected • Vapor pressure sensor is damaged or has failed • VSV circuit for the canister purge, VSV for the CCV or the VSV for the pressure switching valve is open or shorted to ground • VSV for the vapor pressure sensor circuit is open or shorted to ground, or the vapor pressure sensor is damaged or has failed
DTC: P0450 **2T CCM, MIL: Yes** **Years:** 2007, 2008 **Models:** xB, xD, tC **Engines:** All **Transmissions:** All	**EVAP Emission Control System Pressure Sensor/Switch Conditions:** Engine started, engine runtime less than 10 seconds since startup, and the PCM detected the Vapor Pressure sensor value was less than −3.5 kPa (−1.0 in. Hg), or the Vapor Pressure sensor was more than or equal to 1.5 kPa (0.4 in. Hg) during testing. The PCM uses the Vapor Pressure Sensor, VSV for the Canister Closed valve and VSV for the Pressure Switching valve to find faults in this system. **Possible Causes:** • Vapor pressure sensor signal circuit open or shorted to ground • Vapor pressure sensor ground circuit is open • Vapor pressure sensor power circuit is open • Vapor pressure sensor is damaged or has failed • PCM has failed
DTC: P0451 **2T CCM, MIL: Yes** **Years:** 2007, 2008 **Models:** xB, xD, tC **Engines:** All **Transmissions:** All	**EVAP Vapor Pressure Sensor Range/Performance Conditions:** Engine started, engine at idle speed with the VSS indicating 0 MPH, VSV for Vapor Switching valve "off", and the PCM detected too much change in the pressure sensor value, or the pressure sensor value equaled the opening value of the charcoal canister. The PCM uses the Vapor Pressure Sensor, VSV for the Canister Closed valve and VSV for Pressure Switching valve to find faults in the system. **Possible Causes:** • Vapor pressure sensor vacuum hoses loose or damaged • Vapor pressure sensor is damaged or has failed • PCM has failed
DTC: P0452 **2T CCM, MIL: Yes** **Years:** 2007, 2008 **Models:** xB, xD, tC **Engines:** All **Transmissions:** All	**EVAP Vapor Pressure Sensor Circuit Low Input Conditions:** Engine started; VSV for vapor pressure switching valve "off"; and the PCM detected an unexpected low voltage condition on the vapor pressure sensor circuit during the CCM test. **Possible Causes:** • Vapor pressure sensor connector is damaged or open • Vapor pressure sensor circuit is open • Vapor pressure sensor is damaged or has failed • PCM has failed

DTC	Trouble Code Title, Conditions & Possible Causes
DTC: P0453 **2T CCM, MIL: Yes** **Years:** 2007, 2008 **Models:** xB, xD, tC **Engines:** All **Transmissions:** All	**EVAP Vapor Pressure Sensor Circuit High Input Conditions:** Engine started; VSV for vapor pressure switching valve "off"; and the PCM detected an unexpected high voltage condition on the vapor pressure sensor circuit during the CCM test. **Possible Causes:** • Vapor pressure sensor connector is damaged or shorted • Vapor pressure sensor circuit is shorted to VREF • Vapor pressure sensor is damaged or has failed • PCM has failed
DTC: P0455 **2T EVAP, MIL: Yes** **Years:** 2007, 2008 **Models:** xB, xD, tC **Engines:** All **Transmissions:** All	**EVAP System Leak Detected (Gross Leak) Detected Conditions:** Engine started; IAT sensor signal from 39-86°F, fuel tank level from 25-75% for 10 seconds, and the PCM detected the EVAP system was unable to hold a specified vacuum level for a set period of time. After the system is purged, the PCM shuts off the VSV for the purge valve to seal the vacuum in the system, and then monitors the increase in pressure in the system. The pressure should increase slowly. If it increases at too fast a rate, this code is set. **Possible Causes:** • Canister Purge valve is damaged, leaking or has failed • Charcoal canister is loaded with fuel or moisture • Fuel filler cap loose, cross-threaded, incorrect part or damaged • Fuel tank is cracked (leaking), or a leak exists in the 'O' ring • Fuel tank pressure sensor is damaged or has failed • Fuel tank overfill check valve is cracked or is damaged • Fuel vapor line(s), fuel pipes or hoses damaged or leaking • PCM has failed
DTC: P0456 **2T EVAP, MIL: Yes** **Years:** 2007, 2008 **Models:** xB, xD, tC **Engines:** All **Transmissions:** All	**EVAP System Leak Detected (Very Small Leak) Detected Conditions:** Engine started; IAT sensor signal from 39-86°F, fuel tank level from 25-75% for 10 seconds, and the PCM detected the EVAP system was unable to hold a specified vacuum level for a set period of time. After the system is purged, the PCM shuts off the VSV for the purge valve to seal the vacuum in the system, and then monitors the increase in pressure in the system. The pressure should increase slowly. If it increases at too fast a rate, this code is set. **Possible Causes:** • Canister Purge valve is damaged, leaking or has failed • Charcoal canister is loaded with fuel or moisture • Fuel filler cap loose, cross-threaded, incorrect part or damaged • Fuel tank is cracked (leaking), or a leak exists in the 'O' ring • Fuel tank pressure sensor is damaged or has failed • Fuel tank overfill check valve is cracked or is damaged • Fuel vapor line(s), fuel pipes or hoses damaged or leaking • PCM has failed
DTC: P0500 **2T CCM, MIL: Yes** **Years:** 2007, 2008 **Models:** xB, xD, tC **Engines:** All **Transmissions:** All	**Vehicle Speed Sensor "A" Circuit Malfunction Conditions:** Engine started, vehicle driven with the engine speed from 1500-5500 RPM and back to idle several times, and the PCM did not receive any VSS signals during the test. The VSS (No.1) assembly outputs a 4-pulse signal for every revolution of the rotor shaft, which is generated by the transmission output shaft via the driven gear. **Possible Causes:** • VSS signal circuit is open between the meter and the PCM • VSS signal circuit shorted to ground between meter and PCM • VSS No. 1 is damaged or has failed • Combination Meter is damaged or has failed • PCM has failed
DTC: P0504 **2T CCM, MIL: Yes** **Years:** 2007, 2008 **Models:** xB, xD, tC **Engines:** All **Transmissions:** All	**Vehicle Speed Sensor "A"/"B" Circuit Malfunction Conditions:** Engine started, the stop light switch is a duplex system that transmits 2 signals: STP and ST1-. These 2 signals are used by the ECM to monitor whether or not the brake system is working properly. If the signals, which indicate the brake pedal is being depressed and released, are detected simultaneously, the ECM interprets this as a malfunction in the stop light switch and sets the DTC. **Possible Causes:** • Short in the stop light switch signal circuit • STOP fuse • Stop light switch • ECM

DTC	Trouble Code Title, Conditions & Possible Causes
DTC: P0505 **2T CCM, MIL: Yes** **Years:** 2007, 2008 **Models:** xB, xD, tC **Engines:** All **Transmissions:** All	**Idle Control System Malfunction Conditions:** Engine started, engine running at idle speed in closed loop, and the PCM detected the Actual Idle Speed was more than 100-200 RPM above or below the Target Idle Speed. A Rotary solenoid type of ISC valve is located in front of the air intake chamber and intake air bypassing the throttle valve is directed to the Intake Air Control (IAC) valve via a passage. The PCM controls the idle speed by regulating the amount of intake air volume that bypasses the throttle valve. **Possible Causes:** • Air Induction system leaks (check for intake manifold leaks) • Air leaks in the PCV system (at the valve or its related hoses) • Throttle body assembly is damaged or has failed • PCM has failed
DTC: P050A **2T CCM, MIL: Yes** **Years:** 2007, 2008 **Models:** xB, xD, tC **Engines:** All **Transmissions:** All	**Idle Control System Malfunction Conditions:** Engine started, engine running at idle speed in closed loop, and the PCM detected the Actual Idle Speed was more than 100-200 RPM above or below the Target Idle Speed. A Rotary solenoid type of ISC valve is located in front of the air intake chamber and intake air bypassing the throttle valve is directed to the Intake Air Control (IAC) valve via a passage. The PCM controls the idle speed by regulating the amount of intake air volume that bypasses the throttle valve. **Possible Causes:** • Throttle body assembly • MAF meter • Air induction system • PCV hose connections • VVT system • Air cleaner filter element • ECM
DTC: P050B **2T CCM, MIL: Yes** **Years:** 2007, 2008 **Models:** xB, xD, tC **Engines:** All **Transmissions:** All	**Cold Start Ignition Timing Performance Conditions:** Engine started insufficient ignition timing retard at cold start. **Possible Causes:** • Throttle body assembly • MAF meter • PCV system • Air cleaner filter element • Air induction system • VVT system • ECM
DTC: P0560 **2T CCM, MIL: Yes** **Years:** 2007, 2008 **Models:** xB, xD, tC **Engines:** All **Transmissions:** All	**System Voltage (Backup Power Circuit) Malfunction Conditions:** Key on or engine running; and the PCM detected an unexpected low voltage condition on the Backup Power Circuit during the test. **Possible Causes:** • Battery backup circuit is open between battery and the PCM • PCM has failed
DTC: P0604 **1T PCM, MIL: Yes** **Years:** 2007, 2008 **Models:** xB, xD, tC **Engines:** All **Transmissions:** All	**Internal Control Module Random Access Memory (RAM) Error Conditions:** Key on, the ECM continuously monitors its internal memory status. This self-check ensures that the ECM is functioning properly. It is diagnosed by internal "mirroring" of the main CPU and sub CPU to detect the Random Access Memory (RAM) errors. If outputs from these CPUs are different and deviate from the standards, the ECM will illuminate the MIL and set a DTC immediately. **Possible Causes:** • ECM
DTC: P0606 **1T PCM, MIL: Yes** **Years:** 2007, 2008 **Models:** xB, xD, tC **Engines:** All **Transmissions:** All	**ECM/PCM Processing Error Conditions:** Key on, and the PCM detected a processing error occurred. **Possible Causes:** • ECM
DTC: P0607 **1T PCM, MIL: Yes** **Years:** 2007, 2008 **Models:** xB, xD, tC **Engines:** All **Transmissions:** All	**Control Module Performance Conditions:** Key on, and the PCM detected a performance problem occurred. **Possible Causes:** • ECM

DTC	Trouble Code Title, Conditions & Possible Causes
DTC: P060A **1T PCM, MIL: Yes** **Years:** 2007, 2008 **Models:** xB, xD, tC **Engines:** All **Transmissions:** All	**Internal Control Module Monitoring Processor Conditions:** Key on, ECM sub CPU error **Possible Causes:** • ECM
DTC: P060D **1T PCM, MIL: Yes** **Years:** 2007, 2008 **Models:** xB, xD, tC **Engines:** All **Transmissions:** All	**Internal Control Module Monitoring Processor Conditions:** Key on, ECM sub CPU error. **Possible Causes:** • ECM
DTC: P060E **1T PCM, MIL: Yes** **Years:** 2007, 2008 **Models:** xB, xD, tC **Engines:** All **Transmissions:** All	**Internal Control Module Throttle Position Performance Conditions:** Key on, ECM main CPU error. **Possible Causes:** • ECM
DTC: P0617 **1T CCM, MIL: Yes** **Years:** 2007, 2008 **Models:** xB, xD, tC **Engines:** All **Transmissions:** All	**Starter Relay Circuit High Input Conditions:** Engine started, engine speed over 1000 RPM, system voltage over 10.5v, and the PCM detected the Starter Motor signal indicated high. **Possible Causes:** • Park/Neutral switch assembly is damaged or it has failed • Ignition switch is damaged or has failed • PCM has failed
DTC: P0630 **1T CCM, MIL: Yes** **Years:** 2007, 2008 **Models:** xB, xD, tC **Engines:** All **Transmissions:** All	**VIN Not Programmed or Mismatch ECM/PCM Conditions:** Key on, engine off, the VIN is not stored in the ECM or the input VIN is not accurate. Input the VIN with the intelligent tester. **Possible Causes:** • ECM
DTC: P0657 **1T CCM, MIL: Yes** **Years:** 2007, 2008 **Models:** xB, xD, tC **Engines:** All **Transmissions:** All	**Actuator Supply Voltage Circuit/Open Conditions:** Key on, engine off, the ECM monitors the output voltage to the throttle actuator. This self check ensures that the ECM is functioning properly. The output voltage is usually 0v when the ignition switch is turned off. If the output voltage is higher than 7v when the ignition switch is turned off, the ECM will illuminate the MIL and set a DTC when the ignition switch is turned on. **Possible Causes:** • ECM
DTC: P0705 **2T CCM, MIL: Yes** **Years:** 2007, 2008 **Models:** xB, xD, tC **Engines:** All **Transmissions:** All	**A/T Range Sensor Circuit (PRNDL) Malfunction Conditions:** Key on or engine running; and the PCM detected simultaneous "on" signals (N, 2, L or R) from the Transmission Range sensor circuit. The P/N switch indicates "on" whenever the shift lever is in the 'N' or 'P' position. When it is "on", the NSW circuit to the PCM is grounded to chassis ground through the starter motor relay, and reads 0.00v. When the shift lever is in 'R', 'D' or 'L' position, the switch is "off" and the NSW circuit reads 12.0v. When the shift lever is moved from the 'N' to the 'D' position, the PCM uses this signal to air/fuel ratio correction and idle speed control (estimated control) functions. **Possible Causes:** • Park/Neutral switch assembly is shorted • Park/Neutral switch assembly is damaged or has failed • PCM has failed.
DTC: P0710 **1T CCM, MIL: Yes** **Years:** 2007, 2008 **Models:** xB, xD, tC **Engines:** All **Transmissions:** All	**Transmission Fluid Temperature Sensor "A" Circuit Malfunction Conditions:** Engine started, and the PCM detected the TFT sensor indicated less than −40°F, or with the engine runtime over 15 minutes, it detected the TFT sensor indicated more than 300°F. **Possible Causes:** • TFT sensor signal circuit is open, shorted to ground or shorted to system power • TFT sensor is damaged or has failed • PCM has failed

DTC	Trouble Code Title, Conditions & Possible Causes
DTC: P0711 **1T CCM, MIL: Yes** **Years:** 2007, 2008 **Models:** xB, xD, tC **Engines:** All **Transmissions:** All	**Transmission Fluid Temperature Sensor "A" Performance Conditions:** ECT and IAT sensor signals more than 14°F, engine runtime over 12 seconds, then after the vehicle was driven for more than 6.2 miles and the engine runtime exceeded 20 minutes, the PCM detected the TFT sensor remained at a value less than 14°F during the CCM test. **Possible Causes:** • TFT sensor signal circuit has high resistance • TFT sensor is damaged or has failed (it may be contaminated) • PCM has failed
DTC: P0712 **1T CCM, MIL: Yes** **Years:** 2007, 2008 **Models:** xB, xD, tC **Engines:** All **Transmissions:** All	**Transmission Fluid Temperature Sensor "A" Circuit Low Input Conditions:** Engine started, engine running, and the PCM detected the Transmission Fluid Temperature (TFT) sensor indicated a value of more than 284°F for 500 ms in the test. **Possible Causes:** • TFT sensor signal circuit is shorted • TFT sensor is damaged or has failed • PCM has failed
DTC: P0713 **1T CCM, MIL: Yes** **Years:** 2007, 2008 **Models:** xB, xD, tC **Engines:** All **Transmissions:** All	**Transmission Fluid Temperature Sensor "A" Circuit High Input Conditions:** Engine started, engine running, and the PCM detected the TFT sensor indicated a value of less than −40°F for 500 ms in t he test. **Possible Causes:** • TFT sensor signal circuit is open • TFT sensor is damaged or has failed • PCM has failed
DTC: P0717 **1T CCM, MIL: Yes** **Years:** 2007, 2008 **Models:** xB, xD, tC **Engines:** All **Transmissions:** All	**A/T Turbine Shaft Speed Sensor Circuit No Signal Conditions:** No Shift Solenoid or P/N codes set, engine started, P/N switch indicating "off", gear position indicating 2nd, 3rd gear, or in O/D, no gear change occurring, and the PCM detected the Turbine Shaft Speed (ISS) sensor indicated less than 300 RPM, or more than 1000 RPM for 4 seconds. The PCM detects the rotation speed of the input turbine, and compares the signals from the input turbine speed (NT) sensor to the counter gear speed sensor (NC). The PCM uses this signal to detect the shift time so that it can control the engine torque and hydraulic pressure in response to various driving conditions. **Possible Causes:** • Input shaft speed sensor is damaged or loose • Input shaft speed sensor signal (NT) circuit is open or shorted • Input shaft speed sensor is damaged or has failed • PCM has failed
DTC: P0724 **1T CCM, MIL: Yes** **Years:** 2007, 2008 **Models:** xB, xD, tC **Engines:** All **Transmissions:** All	**Brake Switch 'B' Circuit High Input Conditions:** Engine started, vehicle driven to cruise speed and then back to idle speed at least 30 times, and the PCM did not detect any change in the Brake Switch 'A' circuit status. The STP 'B' switch signal is used to determine when the brakes have been applied, and to determine the Fuel Cutoff engine speed during periods with the brakes applied. **Possible Causes:** • Stoplight switch signal circuit is shorted to power • Stoplight switch assembly is damaged or shorted • PCM has failed
DTC: P0741 **2T CCM, MIL: Yes** **Years:** 2007, 2008 **Models:** xB, xD, tC **Engines:** All **Transmissions:** All	**Torque Converter Clutch Solenoid Performance (Shift Solenoid Valve DSL) Conditions:** Engine started, lock-up does not occur when driving in lock-up range. Lock-up remains ON in lock-up OFF range. **Possible Causes:** • Shift solenoid valve DSL remains open or closed • Valve body is blocked • Torque converter clutch • Automatic transaxle (clutch, brake or gear etc.) • Line pressure is too low

DTC	Trouble Code Title, Conditions & Possible Causes
DTC: P0741 **2T CCM, MIL: Yes** **Years:** 2007, 2008 **Models:** xB, xD, tC **Engines:** All **Transmissions:** All	**Torque Converter Clutch Solenoid Performance (Shift Solenoid Valve SL) Conditions:** Engine started, lock-up does not occur when driving in lock-up range. Lock-up remains ON in lock-up OFF range. **Possible Causes:** • Shift solenoid valve DSL remains open or closed • Valve body is blocked • Torque converter clutch • Automatic transaxle (clutch, brake or gear etc.) • Line pressure is too low
DTC: P0746 **2T CCM, MIL: Yes** **Years:** 2007, 2008 **Models:** xB, xD, tC **Engines:** All **Transmissions:** All	**Pressure Control Solenoid "A" Performance (Shift Solenoid Valve SL1) Conditions:** Engine started, the gear required by the ECM does not match the actual gear when driving. **Possible Causes:** • Shift solenoid valve SL1 remains open or closed • Valve body is blocked • Automatic transaxle (clutch, brake or gear etc.)
DTC: P0748 **2T CCM, MIL: Yes** **Years:** 2007, 2008 **Models:** xB, xD, tC **Engines:** All **Transmissions:** All	**Pressure Control Solenoid "A" Electrical (Shift Solenoid Valve SL1) Conditions:** Engine started, the ECM checks for an open or short in the shift solenoid valve SL1 circuit while driving and shift gears. The outside signal duty equals to 100 percent. **Possible Causes:** • Open or short in shift solenoid valve SL1 circuit • Shift solenoid valve SL1 • ECM
DTC: P0751 **2T CCM, MIL: Yes** **Years:** 2007, 2008 **Models:** xB, xD, tC **Engines:** All **Transmissions:** All	**A/T Shift Solenoid 'A' Signal Range/Performance Conditions:** Engine started, vehicle driven to a speed over 50 MPH, and the PCM detected the Actual gear position did not match the desired gear position during the CCM test period. The PCM uses inputs from the VSS and Direct Clutch speed sensor to determine the actual gear position (i.e., 1st, 2nd, 3rd or O/D gear). **Possible Causes:** • Shift solenoid valve S1 remains • Valve body is blocked • Shift solenoid valve S1 • Automatic transaxle (clutch, brake or gear etc.)
DTC: P0756 **2T CCM, MIL: Yes** **Years:** 2007, 2008 **Models:** xB, xD, tC **Engines:** All **Transmissions:** All	**A/T Shift Solenoid 'B' Signal Range/Performance (Shift Solenoid Valve S2) Conditions:** Vehicle driven to a speed over 50 MPH, and the PCM detected the Actual gear position did not match the Desired gear position. The PCM uses inputs from the VSS and Direct Clutch speed sensor to determine the actual gear position (i.e., 1st, 2nd, 3rd or O/D gear). **Possible Causes:** • Shift solenoid valve S2 remains open or closed • Valve body is blocked • Shift solenoid valve S2 • Automatic transaxle (clutch, brake or gear etc.)
DTC: P0766 **2T CCM, MIL: Yes** **Years:** 2007, 2008 **Models:** xB, xD, tC **Engines:** All **Transmissions:** All	**Shift Solenoid "D" Performance (Shift Solenoid Valve S4) Conditions:** Engine started, the gear required by the ECM does not match the actual gear when driving. **Possible Causes:** • Shift solenoid valve S4 remains open or closed • Valve body is blocked • Automatic transaxle (clutch, brake or gear etc.)
DTC: P0776 **2T CCM, MIL: Yes** **Years:** 2007, 2008 **Models:** xB, xD, tC **Engines:** All **Transmissions:** All	**Shift Solenoid "B" Performance (Shift Solenoid Valve SL2) Conditions:** Engine started, the gear required by the ECM does not match the actual gear when driving. **Possible Causes:** • Shift solenoid valve SL2 remains open or closed • Valve body is blocked • Automatic transaxle (clutch, brake or gear etc.)
DTC: P0778 **2T CCM, MIL: Yes** **Years:** 2007, 2008 **Models:** xB, xD, tC **Engines:** All **Transmissions:** All	**Shift Solenoid "B" Electrical (Shift Solenoid Valve SL2) Conditions:** Engine started, the ECM checks for an open or short circuit in shift circuit in shift solenoid valve SL2. Hybrid IC for solenoid indicates fail. **Possible Causes:** • Open or short in shift solenoid valve SL2 circuit • Shift solenoid valve SL2 • ECM

DTC	Trouble Code Title, Conditions & Possible Causes
DTC: P0787 **1T CCM, MIL: Yes** **Years:** 2007, 2008 **Models:** xB, xD, tC **Engines:** All **Transmissions:** All	**A/T Shift Timing Solenoid (ST) Circuit Low Input Conditions:** Engine started, engine running during normal driving conditions, and the PCM detected an unexpected low voltage condition on the Shift Timing Solenoid (SL) control circuit at least (4) times during testing. **Possible Causes:** • ST solenoid connector is damaged or shorted • SL solenoid control circuit is shorted to ground • SL solenoid is damaged or has failed • PCM has failed
DTC: P0788 **1T CCM, MIL: Yes** **Years:** 2007, 2008 **Models:** xB, xD, tC **Engines:** All **Transmissions:** All	**A/T Shift Timing Solenoid (ST) Circuit High Input Conditions:** Engine started, engine running during normal driving conditions, and the PCM detected an unexpected high voltage condition on the Shift Timing Solenoid (SL) control circuit at least (4) times during testing. **Possible Causes:** • ST solenoid connector is damaged or loose • SL solenoid control circuit is open or shorted to system power • SL solenoid is damaged or has failed • PCM has failed
DTC: P0793 **1T CCM, MIL: Yes** **Years:** 2007, 2008 **Models:** xB, xD, tC **Engines:** All **Transmissions:** All	**Intermediate Shaft Speed Sensor "A" Conditions:** Engine started, the ECM detects the vehicle speed is 31 mph, Park/Neutral position switch (NSW) is off, and the Speed Sensor (NC) is less than 300 rpm for 5 seconds or more. **Possible Causes:** • Open or short in transmission revolution sensor NC (speed sensor NC) circuit • Transmission revolution sensor NC (speed sensor NC) • ECM
DTC: P0973 **1T CCM, MIL: Yes** **Years:** 2007, 2008 **Models:** xB, xD, tC **Engines:** All **Transmissions:** All	**A/T Shift Solenoid "A" Circuit Low Input Conditions:** Engine started, engine running during normal driving conditions, and the PCM detected an unexpected low voltage condition on the Shift Solenoid 1 control circuit at least (4) times. **Possible Causes:** • SS1 (solenoid) connector is damaged or shorted • SS1 (solenoid) control circuit is shorted to ground • SS1 (solenoid) is damaged or has failed • PCM has failed
DTC: P0974 **1T CCM, MIL: Yes** **Years:** 2007, 2008 **Models:** xB, xD, tC **Engines:** All **Transmissions:** All	**A/T Shift Solenoid "A" Circuit High Input Conditions:** Engine started, engine running during normal driving conditions, and the PCM detected an unexpected high voltage condition on the Shift Solenoid 1 control circuit at least (4) times. **Possible Causes:** • SS1 (solenoid) connector is damaged or open • SS1 (solenoid) control circuit is open or shorted to power (B+) • SS1 (solenoid) is damaged or has failed • PCM has failed
DTC: P0976 **1T CCM, MIL: Yes** **Years:** 2007, 2008 **Models:** xB, xD, tC **Engines:** All **Transmissions:** All	**A/T Shift Solenoid "B" Circuit Low Input Conditions:** Engine started, engine running during normal driving conditions, and the PCM detected an unexpected low voltage condition on the Shift Solenoid 2 control circuit at least (4) times during testing. **Possible Causes:** • SS2 (solenoid) connector is damaged or shorted • SS2 (solenoid) control circuit is shorted to ground • SS2 (solenoid) is damaged or has failed • PCM has failed
DTC: P0977 **1T CCM, MIL: Yes** **Years:** 2007, 2008 **Models:** xB, xD, tC **Engines:** All **Transmissions:** All	**A/T Shift Solenoid "B" Circuit High Input Conditions:** Engine started, engine running during normal driving conditions, and the PCM detected an unexpected high voltage condition on the Shift Solenoid 2 control circuit at least (4) times during testing. **Possible Causes:** • SS2 (solenoid) connector is damaged or open • SS2 (solenoid) control circuit is open or shorted to power (B+) • SS2 (solenoid) is damaged or has failed • PCM has failed

DTC	Trouble Code Title, Conditions & Possible Causes
DTC: P0982 **1T CCM, MIL: Yes** **Years:** 2007, 2008 **Models:** xB, xD, tC **Engines:** All **Transmissions:** All	**Shift Solenoid "D" Control Circuit Low (Shift Solenoid Valve S4) Conditions:** Engine started, the ECM detects short in solenoid valve S4 circuit 2 times when the solenoid valve S4 is operated. **Possible Causes:** • Short in shift solenoid valve S4 circuit • Shift solenoid valve S4 • ECM
DTC: P0983 **1T CCM, MIL: Yes** **Years:** 2007, 2008 **Models:** xB, xD, tC **Engines:** All **Transmissions:** All	**Shift Solenoid "D" Control Circuit High (Shift Solenoid Valve S4) Conditions:** Engine started, the ECM detects short in solenoid valve S4 circuit 2 times when the solenoid valve S4 is not operated. **Possible Causes:** • Short in shift solenoid valve S4 circuit • Shift solenoid valve S4 • ECM

OBD II Trouble Code List (P1xxx Codes)

DTC	Trouble Code Title, Conditions & Possible Causes
DTC: P1607 **1T CCM, MIL: Yes** **Years:** 2007, 2008 **Models:** xB, xD, tC **Engines:** All **Transmissions:** All	**Cruise Control Input Processor Conditions:** Key on, ECM internal error. **Note: The PCM will not store any other codes with this code set.** **Possible Causes:** • ECM

OBD II Trouble Code List (P2xxx Codes)

DTC	Trouble Code Title, Conditions & Possible Causes
DTC: P2102 **1T CCM, MIL: Yes** **Years:** 2007, 2008 **Models:** xB, xD, tC **Engines:** All **Transmissions:** All	**Throttle Actuator Control Motor Circuit Low Conditions:** The throttle actuator duty ratio is 80 percent or more for 2.0 seconds. The throttle actuator current is 0.5A or less for 2.0 seconds. **Possible Causes:** • Open in throttle actuator circuit • Throttle actuator • ECM
DTC: P2103 **1T CCM, MIL: Yes** **Years:** 2007, 2008 **Models:** xB, xD, tC **Engines:** All **Transmissions:** All	**Throttle Actuator Control Motor Circuit High Conditions:** The hybrid IC diagnosis fails, or the hybrid IC current limiter port fails. **Possible Causes:** • Short in throttle actuator circuit • Throttle actuator • Throttle valve • Throttle body assembly • ECM
DTC: P2111 **1T CCM, MIL: Yes** **Years:** 2007, 2008 **Models:** xB, xD, tC **Engines:** All **Transmissions:** All	**Throttle Actuator Control System – Stuck Open Conditions:** Throttle actuator does not close when signaled by ECM. **Possible Causes:** • Throttle actuator • Throttle body assembly • ECM
DTC: P2112 **1T CCM, MIL: Yes** **Years:** 2007, 2008 **Models:** xB, xD, tC **Engines:** All **Transmissions:** All	**Throttle Actuator Control System – Stuck Closed Conditions:** Throttle actuator does not open when signaled by ECM. **Possible Causes:** • Throttle actuator • Throttle body assembly • ECM

DTC	Trouble Code Title, Conditions & Possible Causes
DTC: P2118 **1T CCM, MIL: Yes** **Years:** 2007, 2008 **Models:** xB, xD, tC **Engines:** All **Transmissions:** All	**Throttle Actuator Control Motor Current Range/Performance Conditions:** Engine running, the ECM monitors the battery supply voltage applied to the throttle actuator. When the power supply voltage (+BM) drops below 4v for 0.8 seconds or more, the ECM interprets this as an open in the power supply circuit (+BM). The ECM illuminates the MIL and sets the DTC. **Possible Causes:** • Open in ETCS power source circuit • ETCS fuse • ECM
DTC: P2119 **1T CCM, MIL: Yes** **Years:** 2007, 2008 **Models:** xB, xD, tC **Engines:** All **Transmissions:** All	**Throttle Actuator Control Body Range/Performance Conditions:** Engine running, the throttle opening angle continues to vary greatly from the target opening angle. **Possible Causes:** • ETCS • ECM
DTC: P2120 **1T CCM, MIL: Yes** **Years:** 2007, 2008 **Models:** xB, xD, tC **Engines:** All **Transmissions:** All	**Throttle/Pedal Position Sensor/Switch "D" Circuit Conditions:** Engine running, the VPA1 fluctuates rapidly beyond upper and lower malfunction thresholds for 0.5 seconds or more. **Possible Causes:** • Accelerator Pedal Position (APP) sensor • ECM
DTC: P2121 **1T CCM, MIL: Yes** **Years:** 2007, 2008 **Models:** xB, xD, tC **Engines:** All **Transmissions:** All	**Throttle/Pedal Position Sensor/Switch "D" Circuit Range/Performance Conditions:** Engine running, the VPA1 fluctuates rapidly beyond upper and lower malfunction thresholds for 0.5 seconds or more. **Possible Causes:** • Accelerator Pedal Position (APP) sensor • ECM
DTC: P2122 **1T CCM, MIL: Yes** **Years:** 2007, 2008 **Models:** xB, xD, tC **Engines:** All **Transmissions:** All	**Throttle/Pedal Position Sensor/Switch "D" Circuit Low Input Conditions:** Engine running, the VPA1 0.4v or less for 0.5 seconds or more when accelerator pedal fully released. **Possible Causes:** • Accelerator Pedal Position (APP) sensor • Open in VCP1 Circuit • Open or ground short in VPA1 circuit • ECM
DTC: P2123 **1T CCM, MIL: Yes** **Years:** 2007, 2008 **Models:** xB, xD, tC **Engines:** All **Transmissions:** All	**Throttle/Pedal Position Sensor/Switch "D" Circuit High Input Conditions:** Engine running, the VPA1 0.4v or less for 0.5 seconds or more when accelerator pedal fully released. **Possible Causes:** • Accelerator Pedal Position (APP) sensor • Open in EPA Circuit • ECM
DTC: P2125 **1T CCM, MIL: Yes** **Years:** 2007, 2008 **Models:** xB, xD, tC **Engines:** All **Transmissions:** All	**Throttle/Pedal Position Sensor/Switch "E" Circuit Conditions:** Engine running, the VPA2 fluctuates rapidly beyond upper and lower malfunction thresholds for 0.5 seconds or more. **Possible Causes:** • Accelerator Pedal Position (APP) sensor • ECM
DTC: P2127 **1T CCM, MIL: Yes** **Years:** 2007, 2008 **Models:** xB, xD, tC **Engines:** All **Transmissions:** All	**Throttle/Pedal Position Sensor/Switch "E" Circuit Low Input Conditions:** Engine running, the VPA2 1.2v or less for 0.5 seconds or more when accelerator pedal fully released. **Possible Causes:** • Accelerator Pedal Position (APP) sensor • Open in VCP Circuit • Open or ground short in VPA2 circuit • ECM

DTC	Trouble Code Title, Conditions & Possible Causes
DTC: P2128 **1T CCM, MIL: Yes** **Years:** 2007, 2008 **Models:** xB, xD, tC **Engines:** All **Transmissions:** All	**Throttle/Pedal Position Sensor/Switch "E" Circuit High Input Conditions:** Engine running, the VPA2 is 4.8v or more for 2.0 seconds or more. The VPA1 is between 0.4v and 3.45v for 2.0 seconds or more. **Possible Causes:** • Accelerator Pedal Position (APP) sensor • Open in VPA2 circuit • ECM
DTC: P2135 **1T CCM, MIL: Yes** **Years:** 2007, 2008 **Models:** xB, xD, tC **Engines:** All **Transmissions:** All	**Throttle/Pedal Position Sensor/Switch "A"/"B" Voltage Correlation Conditions:** Engine running, the difference between VTA1 and VTA2 is 0.02v or less for 0.5 seconds or more. Or the VTA1 is 0.2v or less and VTA2 is 1.75v or less for 0.4 seconds or more. **Possible Causes:** • Short between VPA1 and VPA2 circuits • Accelerator Pedal Position (APP) sensor • ECM
DTC: P2138 **1T CCM, MIL: Yes** **Years:** 2007, 2008 **Models:** xB, xD, tC **Engines:** All **Transmissions:** All	**Throttle/Pedal Position Sensor/Switch "D"/"E" Voltage Correlation Conditions:** Engine running, the difference between VPA1 and VPA2 is 0.02v or less for 2.0 seconds or more. Or the VPA1 is 0.4v or less and VPA2 is 1.2 v or less for 2.0 seconds or more. **Possible Causes:** • Short between VPA1 and VPA2 circuits • Accelerator Pedal Position (APP) sensor • ECM
DTC: P2195 **1T CCM, MIL: Yes** **Years:** 2007, 2008 **Models:** xB, xD, tC **Engines:** All **Transmissions:** All	**Air Fuel Sensor 1 (Bank 1 Sensor 1) Signal Stuck "Lean" Conditions:** Vehicle speed from 25-87 MPH at over 1500 RPM with the throttle valve open, and the PCM detected the Air Fuel sensor signal indicated more than 3.80v for 10 seconds. **Possible Causes:** • A/FS1 signal circuit is open or shorted to ground • A/FS1 is damaged, contaminated or it has failed • Air induction system is severely restricted • Fuel Control component problems (e.g., low fuel pressure, or one or more severely restricted fuel injectors) • PCM has failed
DTC: P2196 **1T CCM, MIL: Yes** **Years:** 2007, 2008 **Models:** xB, xD, tC **Engines:** All **Transmissions:** All	**Air Fuel Sensor 1 (Bank 1 Sensor 1) Signal Stuck "Rich" Conditions:** Vehicle speed from 25-87 MPH at over 1500 RPM with the throttle valve open, and the PCM detected the Air Fuel sensor signal indicated more than 2.80v for 10 seconds. **Possible Causes:** • A/FS1 signal circuit is open or shorted to ground • A/FS1 is damaged, contaminated or it has failed • Air induction system is leaking (check for PCV system leaks) • Fuel component problem (high fuel pressure, leaking regulator or a leaking injector) • PCM has failed
DTC: P2237 **1T CCM, MIL: Yes** **Years:** 2007, 2008 **Models:** xB, xD, tC **Engines:** All **Transmissions:** All	**Oxygen (A/F) Sensor Pumping Current Circuit/Open (Bank 1 Sensor 1) Conditions:** Engine running, open in the circuit between terminals AF+ and AF- or the sensor. **Possible Causes:** • Open or short in A/F sensor (Sensor 1) circuit • A/F sensor (sensor 1) • ECM
DTC: P2238 **1T CCM, MIL: Yes** **Years:** 2007, 2008 **Models:** xB, xD, tC **Engines:** All **Transmissions:** All	**Oxygen (A/F) Sensor Pumping Current Circuit Low (Bank 1 Sensor 1) Conditions:** Engine running, AF sensor output drops while engine is running. The voltage at terminal AF+ is 0.5v or less. The voltage difference between terminals AF+ and AF- is 0.1v or less. **Possible Causes:** • Open or short in A/F sensor (Sensor 1) circuit • A/F sensor (sensor 1) • ECM
DTC: P2239 **2T CCM, MIL: Yes** **Years:** 2007, 2008 **Models:** xB, xD, tC **Engines:** All **Transmissions:** All	**Oxygen (A/F) Sensor Pumping Current Circuit High (Bank 1 Sensor 1) Conditions:** Engine running, AF+ voltage is more than 4.5v. **Possible Causes:** • Open or short in A/F sensor (Sensor 1) circuit • A/F sensor (sensor 1) • ECM

DTC	Trouble Code Title, Conditions & Possible Causes
DTC: P2252 **2T CCM, MIL: Yes** **Years:** 2007, 2008 **Models:** xB, xD, tC **Engines:** All **Transmissions:** All	**Oxygen (A/F) Sensor Reference Ground Circuit Low (Bank 1 Sensor 1) Conditions:** Engine running, AF- voltage is 0.5v or less. **Possible Causes:** • Open or short in A/F sensor (Sensor 1) circuit • A/F sensor (sensor 1) • ECM
DTC: P2253 **2T CCM, MIL: Yes** **Years:** 2007, 2008 **Models:** xB, xD, tC **Engines:** All **Transmissions:** All	**Oxygen (A/F) Sensor Reference Ground Circuit High (Bank 1 Sensor 1) Conditions:** Engine running, AF- voltage is more than 4.5v. **Possible Causes:** • Open or short in A/F sensor (Sensor 1) circuit • A/F sensor (sensor 1) • ECM
DTC: P2401 **2T EVAP, MIL: Yes** **Years:** 2007, 2008 **Models:** xB, xD, tC **Engines:** All **Transmissions:** All	**EVAP Emission System Leak Detection Pump Control Circuit Low Conditions:** Engine off, key off, a predetermined condition (closed loop, etc) is met, the purge VSV is opened by the ECM and fuel vapors stored in the canister are purged into the intake manifold. The ECM changes the duty cycle ratio of the purge VSV to control the purge flow volume. The purge volume is also determined by the intake manifold pressure. Atmospheric pressure is allowed into the canister through the vent valve to ensure that the purge flow is maintained when negative pressure (vacuum) is applied to the canister. **Possible Causes:** • Reference orifice clogged up • Reference orifice high flow • Leak detection pump OFF malfunction • Leak detection pump ON malfunction • Vent valve ON (close) position
DTC: P2402 **2T EVAP, MIL: Yes** **Years:** 2007, 2008 **Models:** xB, xD, tC **Engines:** All **Transmissions:** All	**EVAP Emission System Leak Detection Pump Control Circuit High Conditions:** Engine off, key off, a predetermined condition (closed loop, etc) is met, the purge VSV is opened by the ECM and fuel vapors stored in the canister are purged into the intake manifold. The ECM changes the duty cycle ratio of the purge VSV to control the purge flow volume. The purge volume is also determined by the intake manifold pressure. Atmospheric pressure is allowed into the canister through the vent valve to ensure that the purge flow is maintained when negative pressure (vacuum) is applied to the canister. **Possible Causes:** • Reference orifice clogged up • Reference orifice high flow • Leak detection pump OFF malfunction • Leak detection pump ON malfunction • Vent valve ON (close) position
DTC: P2419 **2T EVAP, MIL: Yes** **Years:** 2007, 2008 **Models:** xB, xD, tC **Engines:** All **Transmissions:** All	**EVAP Emission System Switching Valve Control Circuit Low Conditions:** Engine off, key off, a predetermined condition (closed loop, etc) is met, the purge VSV is opened by the ECM and fuel vapors stored in the canister are purged into the intake manifold. The ECM changes the duty cycle ratio of the purge VSV to control the purge flow volume. The purge volume is also determined by the intake manifold pressure. Atmospheric pressure is allowed into the canister through the vent valve to ensure that the purge flow is maintained when negative pressure (vacuum) is applied to the canister. **Possible Causes:** • Reference orifice clogged up • Reference orifice high flow • Leak detection pump OFF malfunction • Leak detection pump ON malfunction • Vent valve ON (close) position
DTC: P2420 **2T EVAP, MIL: Yes** **Years:** 2007, 2008 **Models:** xB, xD, tC **Engines:** All **Transmissions:** All	**EVAP Emission System Switching Valve Control Circuit High Conditions:** Engine off, key off, EVAP pressure change when vent valve is closed (ON) is less than 2.3mmHg. **Possible Causes:** • Pump module (0.02 inch orifice, vacuum pump, vent valve) • Connector/wire harness (Pump module-ECM) • ECM

DTC	Trouble Code Title, Conditions & Possible Causes
DTC: P2610 **2T EVAP, MIL: Yes** **Years:** 2007, 2008 **Models:** xB, xD, tC **Engines:** All **Transmissions:** All	**ECM/PCM Internal Engine Off Timer Performance Conditions:** Engine on, key off, EVAP internal malfunction. **Possible Causes:** • ECM
DTC: P2714 **2T CCM, MIL: Yes** **Years:** 2007, 2008 **Models:** xB, xD, tC **Engines:** All **Transmissions:** A/T	**Pressure Control Solenoid "D" Performance (Shift Solenoid Valve SLT) Conditions:** Engine running, ECM detects malfunction on SLT (ON side) according to difference in revolutions of turbine (input) and output shaft. **Possible Causes:** • Shift solenoid valve SLT remains closed • Valve body is blocked • Torque converter clutch • Automatic transaxle (clutch, brake or gear, etc.)
DTC: P2716 **1T CCM, MIL: Yes** **Years:** 2007, 2008 **Models:** xB, xD, tC **Engines:** All **Transmissions:** A/T	**Pressure Control Solenoid "D" Electrical (Shift Solenoid Valve SLT) Conditions:** Engine running, open or short is detected in shift solenoid valve SLT circuit for 1 second or more while driving. **Possible Causes:** • Open or short in shift solenoid valve SLT circuit • Shift solenoid valve SLT • ECM
DTC: P2769 **2T CCM, MIL: Yes** **Years:** 2007, 2008 **Models:** xB, xD, tC **Engines:** All **Transmissions:** A/T	**Torque Converter Clutch Solenoid Circuit Low (Shift Solenoid Valve DSL) Conditions:** Engine running, ECM detects short in solenoid valve DSL circuit (0.1 second) when solenoid valve DSL is operated. **Possible Causes:** • Short in shift solenoid valve DSL circuit • Shift solenoid valve DSL • ECM
DTC: P2769 **2T CCM, MIL: Yes** **Years:** 2007, 2008 **Models:** xB, xD, tC **Engines:** All **Transmissions:** A/T	**Torque Converter Clutch Solenoid Circuit Low (Shift Solenoid Valve SL) Conditions:** Engine running, ECM detects short in solenoid valve SL circuit (0.1 second) when solenoid valve SL is operated. **Possible Causes:** • Short in shift solenoid valve SL circuit • Shift solenoid valve SL • ECM
DTC: P2770 **2T CCM, MIL: Yes** **Years:** 2007, 2008 **Models:** xB, xD, tC **Engines:** All **Transmissions:** A/T	**Torque Converter Clutch Solenoid Circuit High (Shift Solenoid Valve DSL) Conditions:** Engine running, ECM detects open in solenoid valve DSL circuit (0.1 second) when solenoid valve DSL is not operated. **Possible Causes:** • Short in shift solenoid valve DSL circuit • Shift solenoid valve DSL • ECM
DTC: P2770 **2T CCM, MIL: Yes** **Years:** 2007, 2008 **Models:** xB, xD, tC **Engines:** All **Transmissions:** A/T	**Torque Converter Clutch Solenoid Circuit High (Shift Solenoid Valve SL) Conditions:** Engine running, ECM detects open in solenoid valve SL circuit (0.1 second) when solenoid valve SL is not operated. **Possible Causes:** • Short in shift solenoid valve SL circuit • Shift solenoid valve SL • ECM

DTC	Trouble Code Title, Conditions & Possible Causes
DTC: P2A00 **2T CCM, MIL: Yes** **Years: 2007, 2008** **Models:** xB, xD, tC **Engines:** All **Transmissions:** All	**Air Fuel Sensor (Bank 1 Sensor 1) Signal Slow Response Conditions:** Vehicle driven at cruise speed at over 1400 RPM in closed loop at 60 MPH, and the PCM detected an unexpected voltage condition on the Bank 1 Air Fuel Sensor 1 (AFS1) circuit. **Possible Causes:** • A/F sensor connector is damaged or loose • A/F sensor circuit is open or shorted, or the sensor has failed • A/F sensor heater is damaged or has failed • A/F sensor heater relay circuit is open or the relay has failed • Fuel delivery component has failed (fuel pressure regulator, one or more fuel injectors is leaking or severely restricted) • Induction system problems (air leaks or restricted air filter) • PCM has failed